11th Asia-Pacific Conference on Combustion 2017

Sydney, Australia
10 - 14 December 2017

Volume 1 of 2

ISBN: 978-1-5108-5646-2

Printed from e-media with permission by:

Curran Associates, Inc.
57 Morehouse Lane
Red Hook, NY 12571

proceedings.com

Some format issues inherent in the e-media version may also appear in this print version.

Printed by Curran Associates, Inc. (2018)

Combustion Institute China Section
c/o Section Chair: Fei Qi, Shanghai Jiao Tong University

fqi@sjtu.edu.cn

Additional copies of this publication are available from:

Curran Associates, Inc.
57 Morehouse Lane
Red Hook, NY 12571 USA
Phone: 845-758-0400
Fax: 845-758-2633
Email: curran@proceedings.com
Web: www.proceedings.com

TABLE OF CONTENTS

VOLUME 1

VOLUME 2

11th Asia-Pacific Conference on Combustion,
The University of Sydney, NSW Australia
10th -14th December 2017.

Effects of mixing and reaction timescales on LES/PDF simulations of a turbulent lean premixed jet flame

Hua Zhou[1], Zhuyin Ren[1,2*], David H. Rowinski[3]

[1]Center for Combustion Energy, Tsinghua University,

Beijing, 100084, China

[2]School of Aerospace Engineering, Tsinghua University,

Beijing, 100084, China

[3]Convergent Science Inc., Madison, WI 53719, USA

Abstract

The effects of mixing and reaction timescales on the prediction of a turbulent lean premixed flame exhibiting extinction-reignition feature are investigated through a large eddy simulation (LES)/probability density function (PDF) study of the flame PM1-150 from the Sydney Piloted Premixed Jet Burner (PPJB). It is found that the simulation with the mixing timescale model constant $C_M = 5.0$ significantly overpredicts the reaction progress, while increasing C_M (namely decreasing mixing timescale) or decreasing the overall reaction rate (namely increasing reaction timescale) yields better predictions in terms of species radial profiles. The particle-level statistics further demonstrate that decreasing mixing timescale or increasing reaction timescale changes the turbulence chemistry interaction in a similar manner for the flame PM1-150, indicating Damköhler number (*Da*) is a controlling parameter for the flame characteristics of PM1-150. The findings are consistent with previous sensitivity analysis on C_M and overall reaction rate for the PPJB flames. Further improvement for the transported PDF (TPDF) simulations of this flame should focus on improving mixing timescale model and quantifying any factors that may affect chemical reaction rates.

1 Introduction

Turbulent lean premixed combustion has received increasingly attention in modern combustion devices driven by ever increasingly stringent emission regulations [1]. However, the coupling of highly non-linear chemical reaction and turbulent mixing, involving potentially hundreds of species over a broad range of scales makes the accurate prediction of turbulent premixed combustion a particular challenge.

The Sydney Piloted Premixed Jet Burner (PPJB) was developed to investigate turbulence chemistry interactions at lean and highly turbulent conditions [2-4]. Detailed measurements of both velocity and scalar statistics about PPJB flames have been reported to facilitate model validation for turbulent premixed combustion. The flame PM1-150 of PPJB features a central jet of lean premixed methane-air, surrounded by a hot pilot of stoichiometric methane-air products, which is then surrounded by a hot vitiated coflow of lean hydrogen-air products. The bulk velocity of the central jet is 150 m/s. The PM1-150 exhibits reduced luminosity between the streamwise locations x/D of 15 and 45, and the luminosity increases after x/D of 45. This extinction-reignition feature is believed to result from the intensive turbulence chemistry interactions.

Several computational studies have been carried out to explore the extinction-reignition feature with various turbulent combustion models [4-7]. However, almost all these combustion models overpredict the reaction progress regardless of whether being applied in a RANS or LES context. More recently, transported PDF (TPDF) methods have shown some promise towards better predictions for the flame PM1-150 [8, 9]. A LES/PDF simulation shows quantitatively good agreement for the species radial profiles by increasing the mixing timescale model constant quantifying the mechanical-to-scalar timescale ratio to over 20.0 [9]. A recent particle-level sensitivity analysis [10] reveals that for PM1-150, the overall reaction progress shows negative sensitivity to mixing rate (corresponding to the mixing timescale model constant) and positive sensitivity to overall reaction rate, indicating that a decrease of mixing timescale or an increase of reaction timescale suppresses the reaction progress. In this work, a parametric study has been performed to investigate the effects of these two timescales on the LES/PDF simulations of the flame PM1-150. Preliminary results on species radial profiles and scatter plots are reported.

2 LES/PDF Simulations

The LES/PDF approach employed is based on a two-way coupled hybrid method [11-13] and the subgrid turbulence-chemistry interaction is resolved through the joint composition PDF. Specifically, the LES solver solves the transport equations for mass and momentum with a second-order accuracy method [14], and continuity is enforced through solving the pressure Poisson equation. An additional equation for specific volume is solved to obtain mean density. In the particle solver, each computational particle carries the properties, e.g. composition $\boldsymbol{\phi}$, mass m and position $\boldsymbol{x}$, and the evolution of the particles in physical and composition space is governed by the following stochastic differential equations:

$$d\boldsymbol{x}^* = \left[\widetilde{\boldsymbol{U}} + \nabla(\tilde{\Gamma}_t\bar{\rho})/\bar{\rho}\right]^* dt + \left(2\tilde{\Gamma}_t^*\right)^{\frac{1}{2}} d\boldsymbol{W}^*, \quad (1)$$

$$d\boldsymbol{\phi}^*(t) = \Omega_M^*\left(\widetilde{\boldsymbol{\phi}^*} - \boldsymbol{\phi}^*\right)dt + \left(\nabla\cdot\left(\bar{\rho}\tilde{\Gamma}\nabla\widetilde{\boldsymbol{\phi}}\right)/\bar{\rho}\right)^* \quad (2)$$

Corresponding author. Fax: +00-00-000-000
E-mail address: zhuyinren@tsinghua.edu.cn

$$+S(\phi^*)dt,$$

where $\widetilde{U}$ and $\bar{\rho}$ are filtered velocity and density respectively, $\tilde{\Gamma}$ and $\tilde{\Gamma}_t$ are filtered molecular and turbulent diffusivity respectively, dW^* is an independent Wiener increment, $S(\phi^*)$ is reaction rate and Ω_M^* is mixing rate. The superscript $*$ in (1~2) denote either an individual particle property or a value of the filtered LES field evaluated at the particle's location. According to (2), the molecular transport in physical space is handled by the gradient transport term (the second term on the right hand side), while the molecular mixing in composition space is handled by the Interaction-by-Exchange-with-the-Mean (IEM) model (the first term on the right hand side). This treatment of molecular diffusion allows differential diffusion effects to be accounted and produces no spurious scalar variance [15]. The mixing rate is computed as

$$\Omega_M = C_M(\tilde{\Gamma} + \tilde{\Gamma}_t)/\Delta^2, \quad (3)$$

where Δ is the filter length, and C_M is the mixing timescale model constant representing the mechanical-to-scalar timescale ratio.

An algebraic subgrid model [16] is used to compute turbulent viscosity, and turbulent diffusivity is computed from the turbulent viscosity using a turbulent Schmidt number of 0.4. Molecular diffusivity is computed using the unity Lewis number assumption, therefore, differential diffusion effects are not considered explicitly in this work. A 16-species reduced mechanism, ARM1 for methane [17] is used, along with *In-Situ* Adaptive Tabulation (ISAT) [18, 19] for computational efficiency.

The cylindrical computational domain extends $72D$ in the axial direction and $24D$ in the radial direction, where D is the central jet diameter of 4 mm. The simulations employ a grid of 192×96×48, an ISAT tolerance of 5×10^{-6}, 20 particles per cell, and the statistics are time averaged for ten flow-through times. The numerical settings have been extensively tested in [9] to ensure numerical accuracy.

3 Results and Discussions

3.1 Effects of the mixing timescale

Figure 1 shows the radial profiles of the mean mass fractions of the fuel CH_4 and intermediate species CO. It can be observed that the LES/PDF results with C_M of 5.0 and 20.0 are similar at upstream, and the deviation becomes significant at downstream location of $x/D = 30$. The simulation with $C_M = 5.0$ significantly underpredicts CH_4 at $x/D = 30$, indicating a significant overprediction of the reaction progress. By increasing C_M to 20.0, namely decreasing mixing timescale by a factor of 4, the simulation yields better predictions for CH_4 and CO at further downstream. Note that varying the mixing timescale does not directly affect the predicted means, as the mixing process does not change the mean quantities. The predicted mean quantities is changed indirectly through the variation of the subgrid particle compositions via mixing and subsequently chemical reaction. The increase of CH_4 at downstream locations by increasing C_M indicates that the reaction progress gets suppressed with stronger mixing. This is consistent with the findings in previous particle-level sensitivity analysis on high-speed PPJB flames [10], in which the reaction progress was found to show negative sensitivity to C_M, namely an enhancement of mixing results in the suppression of reaction progress due to the fact that the flame is near the blow-off limit.

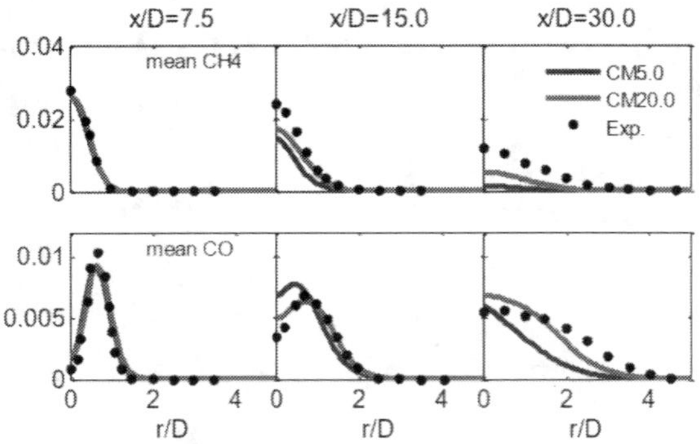

Figure 1: Radial profiles of the mean mass fraction of CH_4 and CO at axial locations of x/D=7.5, 15 and 30, respectively. Blue line: C_M=5.0; Red line: C_M=20.0; Dots: experiment.

Figure 2 shows the radial profiles of the rms mass fractions of CH_4 and CO, with both the total and resolved rms. The total rms of quantity f is approximated by $\langle f^2 \rangle - \langle f \rangle^2 \approx \langle \widetilde{f^2} \rangle - \langle \tilde{f} \rangle^2$, where $\tilde{f}$ represents a mass-weighted spatially averaged quantity, and $\langle f \rangle$ represents an ensemble averaged quantity. The resolved rms of quantity f is computed as $\langle \tilde{f}^2 \rangle - \langle \tilde{f} \rangle^2$. As shown, the difference between the total rms and the resolved rms is obvious when C_M=5.0, indicating a significant portion of the subgrid rms. By increasing C_M to 20.0, the difference between the total rms and the resolved rms becomes negligible as a larger value of C_M results in a faster decay of subgrid variance. In addition, the simulation with C_M =5.0 overpredicts the peak rms of CH_4 and CO at upstream and fail to predict the shape of rms radial profiles at downstream. By increasing C_M to 20.0, better predictions for the peak and the shape of the rms radial profiles can be obtained.

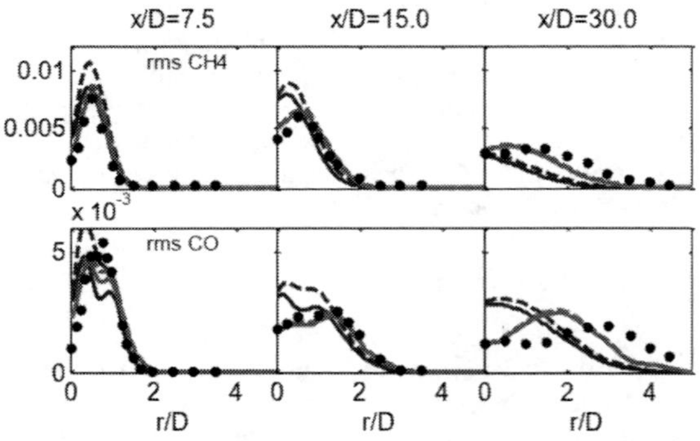

Figure 2: Radial profiles of the rms mass fraction of CH_4 and CO at axial locations x/D=7.5, 15 and 30, respectively. Dashed blue line: total rms with C_M=5.0; Solid blue line: resolved rms with C_M=5.0; Dashed red line: total rms with C_M=20.0; Solid red line: resolved rms with C_M=20.0.

3.2 Effects of the reaction timescale

The effects of reaction timescale are investigated by introducing an attenuation factor R to the reaction source term $S(\phi)$, as the work done by Rowinski et al [8]. Here, the mixing timescale model constant C_M remains to be 5.0, while reducing the reaction rate by a factor of 5 through setting R=0.2 to keep the characteristic Damköhler number (Da) roughly the same as the case with C_M=20.0 and R=1.0.

Figure 3 shows the radial profiles of the mean mass fraction of the CH_4 and CO. As shown, the simulation with C_M=5.0 and R=0.2 yields similar predictions as the one with C_M=20.0 and R=1.0 at x/D of 7.5 and 15, while the simulation with C_M=5.0 and R =0.2 yields slightly better predictions at further downstream. The increase of CH_4 at downstream by attenuating the reaction rate is trivial as the combustion process gets slower with a larger overall reaction timescale. The better agreement for the intermediate species CO profiles at downstream indicates a better prediction for the turbulence chemistry interactions with the attenuated reaction rate.

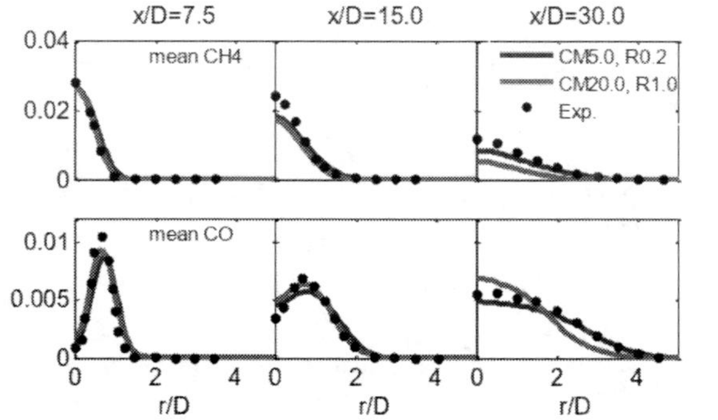

Figure 3: Radial profiles of the mean mass fractions of CH_4 and CO at axial locations x/D= 7.5, 15 and 30, respectively. Blue line: C_M=5.0 and R=0.2; Red line: C_M=20.0 and R=1.0; Dots: experiment.

Figure 4 shows the radial profiles of the rms mass fractions of CH_4 and CO, both the total and resolved rms are shown. The difference between the total and the resolved rms is the subgrid rms. By comparing with Fig. 2, it is found that the case with C_M=5.0 and R=0.2 exhibits much smaller subgrid rms than the case with C_M=5.0 and R=1.0. This can be easily explained as a smaller overall reaction rate results in a smaller contribution from chemical reaction to the source term in scalar variance equation. Similar to the findings from Fig. 3 that the case with C_M=5.0 and R=0.2 yields the similar predictions as the case with C_M=20.0 and R=1.0 at x/D of 7.5 and 15, while the case with C_M =5.0 and R =0.2 yields better predictions at further downstream.

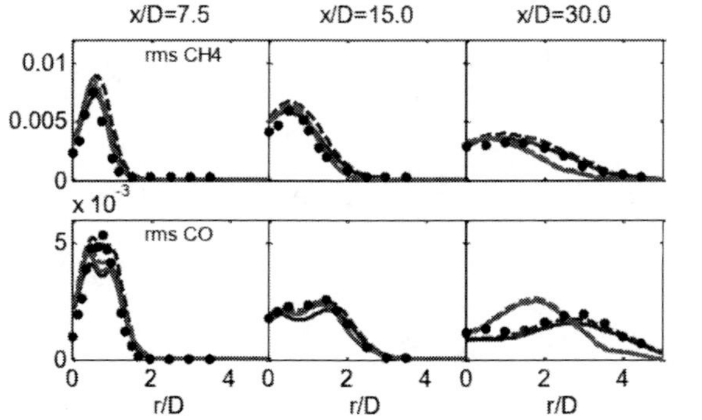

Figure 4: Radial profiles of the rms mass fraction of CH_4 and CO at axial location x/D of 7.5, 15 and 30, respectively. Dashed blue line: total rms with C_M=5.0 and R=0.2; Solid blue line: resolved rms with C_M=5.0 and R=0.2; Dashed red line: total rms with C_M=20.0 and R=1.0; Solid red line: resolved rms with C_M=20.0 and R=1.0.

To further investigate the difference between the configuration of C_M=5.0 and R=1.0, C_M=20.0 and R=1.0, and C_M=5.0 and R=0.2 in terms of turbulence chemistry interactions, the CO mass fractions of computational particles are sampled and conditionally averaged on temperature. Figure 5 shows the scatter plots of $T - CO$, where the sampled particles are at x/D of 15 and whose radial location is between the centerline and the half-width based on the mean jet mixture fraction. It can be observed that the scatter of CO predicted by the case with C_M=5.0 and R=1.0 is significantly larger than the other two cases, indicating stronger turbulence chemistry interactions. By either decreasing the mixing timescale (the case with C_M=20.0 and R=1.0) or increasing the reaction timescale (the case with C_M=5.0 and R=0.2) the scatters of CO are suppressed to a similar level. This may be explained by Damköhler number. Taking the characteristic *Da* of the case with C_M=5.0 and R=1.0 to be 1.0, then the *Da* of the case with C_M=20.0 and R=1.0 is 0.05, and the *Da* of the case with C_M=5.0 and R=0.2 is 0.04. The similarity of *Da* results in the similarity of the CO scatters, indicating that *Da* is a controlling parameter for the flame characteristics of the flame PM1-150.

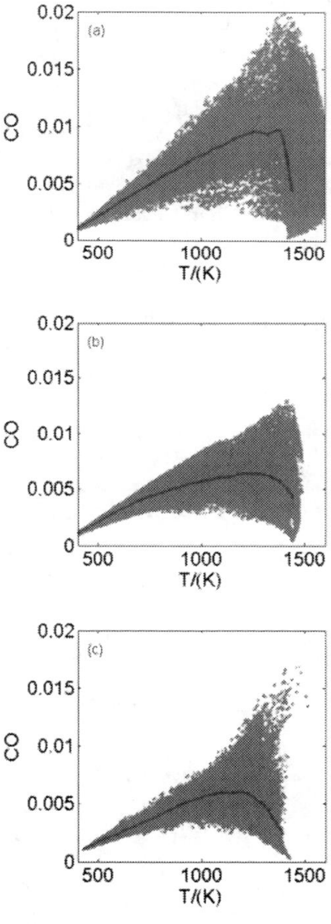

Figure 5: Scatter plots of mass fraction of CO versus temperature. The computational particle are sampled at x/D of 15. Solid red line represents the conditional mean. (a) C_M=5.0 and R=1.0; (b) C_M=20.0 and R=1.0; (c) C_M=5.0 and R=0.2.

3.3 Discussions

The parametric study on mixing timescale model constant C_M shows that C_M has to be increased to over 20.0 to obtain reasonable predictions of the species profiles, indicating that the scalar mixing rate for the flame PM1-150 is governed by the scales other than those in (3). It is known that the mixing timescale of a reactive scalar is governed by both turbulent strain and chemical reaction. However, the effect of chemical reaction is completely ignored in (3). Recently, a mixing timescale model for reactive scalars was proposed [20], which is an algebraic model that linearly blends the mixing timescale formulations in the passive mixing limit and laminar flamelet limit. Therefore, the model accounts for both the effects of turbulent strain and chemical reaction. For the future work, the model performance will be investigated in PM1-150.

4 Conclusions

A parametric study has been carried out to investigate the effects of mixing and reaction timescales on LES/PDF simulations of the flame PM1-150. It is found that the simulation with C_M= 5.0 overpredicts the reaction progress. By increasing C_M to 20.0, namely decreasing mixing timescale by a factor of 4, the predicted combustion process gets suppressed and yields better agreement with experimental data. This is consistent with the findings in previous particle-level sensitivity analysis on high-speed PPJB flames, in which the reaction progress was found to show negative sensitivity to C_M. The effects of reaction timescale are investigated by introducing an attenuation factor R to reaction rate. By attenuating the reaction rate by a factor of 5, LES/PDF simulations yield good predictions for the species radial profiles. The scatter plots of particle-level $T - CO$ reveal that decreasing mixing timescale by a factor of 4 or increasing reaction timescale by a factor of 5 suppresses the CO scatters to a similar level, which can be attributed to the similarity of the Damköhler number, indicating that *Da* is a controlling parameter for the flame characteristics of the flame PM1-150. The further improvement for the TPDF simulations of this flame should focus on improving mixing timescale model and quantifying any factors that may affect chemical reaction rates.

5 Acknowledgment

The work at Tsinghua is supported by National Natural Science Foundation of China 91441202 and 51476087. Simulations are performed with the computational resources of the Tsinghua National Laboratory for Information Science and Technology, and Tianhe 2.

References

[1] T. Bradley and J. Fadok. Advanced hydrogen turbine development update. in ASME Turbo Expo 2009: Power for Land, Sea, and Air. 2009. American Society of Mechanical Engineers.

[2] M.J. Dunn, A.R. Masri, and R.W. Bilger, Combust. Flame., 151 (2007) 46-60.

[3] M.J. Dunn, A.R. Masri, R.W. Bilger, R.S. Barlow, and G.H. Wang, Proc. Combust. Inst., 32 (2009) 1779-1786.

[4] M.J. Dunn, A.R. Masri, R.W. Bilger, and R.S. Barlow, Flow. Turbul. Combust., 85 (2010) 621-648.

[5] C. Duwig, K.J. Nogenmyr, C.K. Chan, and M.J. Dunn, Combust. Theor. Model., 15 (2011) 537-568.

[6] V.N. Prasad, K.H. Luo, and W.P. Jones. LES-PDF simulation of a highly sheared turbulent piloted premixed flame. in 7th Mediterranean Combustion Symposium. 2011.

[7] Y. Chen and M. Ihme. Flame characterization of a piloted premixed jet burner. in Proceedings of the Spring Technical Meeting of the Central States Section of the Combustion Institute. 2012.

[8] D.H. Rowinski and S.B. Pope, Combust. Theor. Model., 15 (2011) 245-266.

[9] D.H. Rowinski and S.B. Pope, Combust. Theor. Model., 17 (2013) 610-656.

[10] H. Zhou, S. Li, Z. Ren, and D.H. Rowinski, Combust. Flame., 181 (2017) 136-148.

[11] H. Wang and S.B. Pope, Proc. Combust. Inst., 33 (2011) 1319-1330.

[12] Y. Yang, H. Wang, S.B. Pope, and J.H. Chen, Proc. Combust. Inst., 34 (2013) 1241-1249.

[13] P.P. Popov, H. Wang, and S.B. Pope, J. Comput. Phys., 294 (2015) 110-126.

[14] C.D. Pierce and P. Moin, J. Fluid. Mech., 504 (2004) 73-97.

[15] R. McDermott and S.B. Pope, J. Comput. Phys., 226 (2007) 947-993.

[16] A.W. Vreman, Phys. Fluids., 16 (2004) 3670-3681.

[17] C.J. Sung, C.K. Law, and J.Y. Chen. An augmented reduced mechanism for methane oxidation with comprehensive global parametric validation. in Symposium (International) on Combustion. 1998. Elsevier.

[18] S.B. Pope, Combust. Theor. Model., 1 (1997) 41-63.

[19] L. Lu and S.B. Pope, J. Comput. Phys., 228 (2009) 361-386.

[20] M. Kuron, Z. Ren, E.R. Hawkes, H. Zhou, H. Kolla, J.H. Chen, and T. Lu, Combust. Flame., 177 (2017) 171-183.

Reviewer #1:

The paper examines the mixing operator in the concept of the LES/PDF. This is an interesting analysis. I am just wondering if the authors can show this analysis for low-speed species such as NO. It is not clear how the total RMS (resolved + subgrid) was computed. For the future work, I also suggest the authors comment on the effects of Ka number if it is possible. Perhaps, it would be interesting if authors can somehow investigate effects of molecular and turbulent diffusivities, separately.

Reply:

It is a good suggestion to show the results of NO. However, the ARM1 mechanism applied in this work does not contain NO mechanism. Analysis on NO can be carried out using mechanisms containing NO as future work.

The total RMS of quantity f is approximated by $\langle f^2 \rangle - \langle f \rangle^2 \approx \langle \widetilde{f^2} \rangle - \langle \tilde{f} \rangle^2$, where $\tilde{f}$ represents a mass-weighted spatially averaged quantity, and $\langle f \rangle$ represents an ensemble averaged quantity. The resolved RMS of quantity f is computed as $\langle \tilde{f}^2 \rangle - \langle \tilde{f} \rangle^2$. This has been added to the revised manuscript.

To investigate the effects of molecular and turbulent diffusivities, the comparison between the molecular diffusivity and turbulent diffusivity will be carried out as the first step in future work.

Reviewer #2:

This paper studies about the relative importance of mixing and chemical timescales and is a significant contribution to the field of PDF methods. Below are few suggestions/questions:

1) Sec 3.2, line 2: Please explain the attenuation factor and how it affects the change in species mass fractions due to chemical reactions. Provide references where R has been used.

2) Pg.3, right column, line 3: The excess scatter in Cm=5 & R=1 case could be due to the non-local IEM model used. One might obtain less scatter with Cm=5 & R=1 with a local model like EMST or MMC. Please discuss about the validity of using such artificial attenuation factor.

Reply:

1) The attenuation factor R attenuates the overall reaction rate, i.e. attenuating the change of ϕ_i due to chemical reaction by a factor of R: $d\phi_i/dt = RS_i(\boldsymbol{\phi})$. We followed the work of Rowinski et al. [1], who once investigated the effect of R on RANS/PDF simulations of the flame PM1-200. The reference has been added to the revised manuscript.

[1] D.H. Rowinski and S.B. Pope, Combust. Theor. Model., 15 (2011) 245-266.

2) The authors agree that mixing models featuring more localness may suppress the scattering. Here IEM models are applied in all computational cases to provide the same basis for comparing the scatters with different *Da* number, and to investigate the effect of *Da* number qualitatively. As suggested by the reviewer, it is worth to investigate the scatter predicted by mixing models featuring more localness, and to compare with experiment as the future work.

Experimental and Numerical Investigation of Spray Characteristics of Butanol-Diesel blends

Sattar Jabbar Murad Algayyim [a, b, *], **Andrew P. Wandel** [a], **Talal Yusaf** [a]

[a] School of Mechanical and Electrical Engineering, University of Southern Queensland,

West Street, Toowoomba, QLD, 4350, Australia

[b] Department of Mechanical, College of Engineering, University of Al-Qadisiyah

Al-Diwaniyah, Iraq

Abstract

Spray characteristics are among the most important factors that affect compression ignition (CI) engines' performance and emission levels. Flow visualisation and optical diagnostics have been widely employed in previous and current research as methods for controlling the combustion processes. This paper investigates the spray visualisation of butanol-diesel blends to determine spray characteristics such as spray penetration (S) and Average Sauter Mean Diameter (ASMD) using Ansys Forte under different ambient pressures and temperatures. The spray results showed that the spray penetration length is decreased as a result of the increased ambient pressure, while it is increased as a result of increased injection pressure of all test fuels. An increase in ambient temperature caused pure diesel penetration to become longer and wider, while butanol-diesel blends penetration becomes shorter. The ASMD of the butanol-diesel blend is higher than that of pure diesel at all operating conditions.

1 Introduction

The growing number of alternative fuels such as alcohol has led to an increased interest in studying the spray and combustion characteristics of these fuels. These alternative fuels can be used alone or mixed with diesel fuel in different ratios. Flow visualisation and optical diagnostics should be employed to help understand the combustion processes. Complete combustion is assisted by maximising the contact the injected fuel has with the available air. Since design and fabrication of an engine with an optical window is a costly and complex option, visualisation techniques have been applied in modified engines with optical access [1, 2] and in a constant volume chamber (CVV) at similar conditions to a real engines [3]. In addition, these visualisation techniques can be used to investigate a wide range of operating conditions and different alternative fuels used later. Some studies have also used a CVV at ambient conditions [4, 5] and made use of software (Forte or KIVA) to simulate the results for a wider range of operating conditions. Investigating, operating conditions can be expensive if carried out experimentally, so effective software can contribute to a saving in both cost and effort. The software can be set for different injection systems and optimal operating conditions.

The key parameters for the visualisation technique can be classified into classes of parameter: (1) macroscopic parameters such as spray penetration can be determined through direct visualisation methods. A charge-coupled device (CCD) is commonly used in research labs for taking spray images [6]. (2) The microscopic parameters such as droplet size and ASMD can be measured using Particle Image Velocimetry (PIV).

Moreover, there are many studies that numerically investigated spray behaviour. These studies used different software packages such Ansys Fluent and Ansys Forte to predict the spray characteristics. Agudelo et al. [7] studied spray behaviour by using the Kelvin-Helmholtz Rayleigh-Taylor (KH-RT) model in Ansys Forte. The simulation parameters that were used are three different injection pressures (400 bar, 500 bar, and 600) bar and three ambient gas pressures (1 bar, 10 bar and 20 bar). The model outputs were spray tip penetration, drop mean diameter and evaporation rate. These results have been validated with experimental data and good agreement obtained. Vijayraghavan and Rutland [8] investigated the effect of physical properties on spray models using Ansys Forte and for non-reacting spray simulations using the KH-RT model. Verma et al. [4] also studied the spray behaviour of gasoline fuel in DI engine using Ansys Forte at atmospheric conditions at an injection pressure of 100 bar and durations of injection of 0.88 ms and 2.4 ms. The result of the simulation had very good agreement with experiments at the same conditions.

This paper investigates the spray characteristics of a butanol-diesel blend experimentally and numerically in Ansys Forte to measure spray penetration length (S) and ASMD under various ambient pressures and temperatures.

2 Experimental Apparatus

2.1 Spray Test Setup

The spray test was carried out on a CVV at atmospheric pressure and room temperature (30 °C). An air-driven high-

Corresponding author.
E-mail address: Sattarjabbarmurad.algayyim@usq.edu.au

pressure fuel pump was used in the fuel injection system, where the fuel was pressurised in a common-rail system and injected using a solenoid Delphi-type injector with six holes (each 0.198 mm in diameter).

A Photron (CCD) camera was used to capture the spray blend images. The camera has a resolution of 1024×1024 pixels and the shutter speed and frame rate were fixed at 1/5,000 s. The camera was synchronised with the injector by using same triggering signal. A Nikon AF Micro-Nikkor lens with a focal length of 60 mm and a maximum aperture of f/2.8D with filter size 62 mm was connected to the camera. An LED light was used for illuminating the fuel spray on each window to ensure constant background light for the camera. The spray characteristics of neat diesel (D) and 20% butanol 80% diesel (B20D80) fuel blends were investigated. Two injection pressures were used 300 bar and 500 bar. Fig. 1 shows the schematic of the CVV with the fuel injection setup.

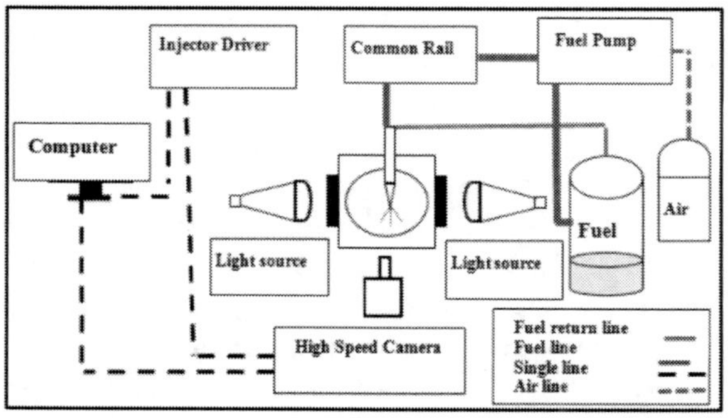

Figure 1. Schematic of CVV with fuel injection system setup with CCD.

2.2 Simulation of Spray Visualisation

2.2.1 Spray Model

Ansys Forte was used to simulate fuel injection with the RNG k-ε turbulence model and an independent spray breakup model proposed by Alaina [9] called the gas jet model. The multi-component vaporisation model solves unsteady vaporisations of single and multi-component fuel droplets with consideration of both normal and flash-boiling vaporisation conditions. The spray was modelled by using an Eulerian-Lagrangian approach and also incorporating multicomponent fuels. The initial spray conditions at the nozzle exit were determined through the specification of a discharge coefficient and the KH/RT breakup model was used for droplet breakup.

Distribution of droplet size at the nozzle exit was specified using a Rosin-Rammler distribution. Velocities of primary parcels (blobs) were calculated as a function of the measured injected-mass flow rate profile. The gas jet model uses a correlation from classical gas-jet model theory to model the drops' relative axial velocity [10].

The gas-jet model was applied with an advanced KH-RT hybrid break-up model. Droplet collisions were modelled using the radius of influence (ROI) collision model.

2.2.2 Spray Simulations

The simulation cases were carried out in a 45-degree sector of the diesel engine (Fig. 2). Non-reactive, low pressure conditions were used since they allow calibration of the model for predicting correct spray dynamics. The chamber was initialized with air at ambient pressure of 1 bar and a temperature of 30°C. A single-hole injector with hole diameter of 0.198 mm was used for the simulation at 300 bar injection pressure. The default values for the model constants were used. Moreover, the spray model parameters for KH size and the time constant were set to 1 and 40 respectively; RT size time and distance constants were 0.15, 1 and 1.9 respectively. Gas entrainment constant for the unsteady gas-jet model was 0.5. The nozzles have a coefficient of discharge of 0.7.

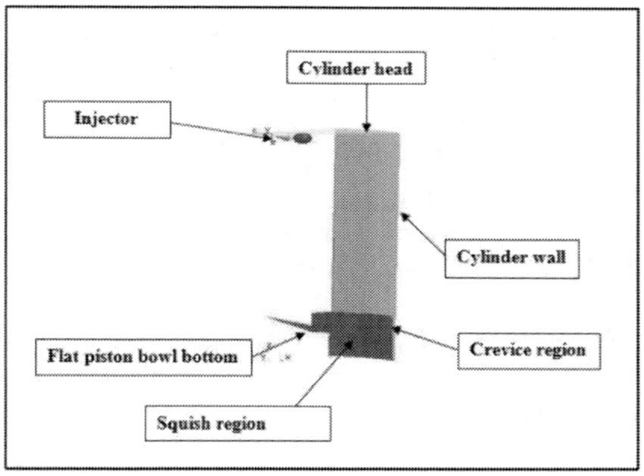

Figure 2. 45 degree sector of engine with injector

3. Result and Discussion

3.1 Spray Characteristics

Figure 3 compares the spray images of D and B20D80 for different injection pressures and times after start of injection (ASOI). It is clearly seen that the spray penetration is increased as a result of increased injection pressure of all test fuels. Adding butanol to diesel resulted in increased spray penetration length due to the low viscosity and high surface tension of butanol. Fig. 4 shows the spray images of neat diesel fuel at 300 bar injection pressure and ambient pressure and temperature (1 bar and 30 °C). Surrogate composition was used for diesel (nC_7H_{16}) species' (n-Tetradcane) physical properties.

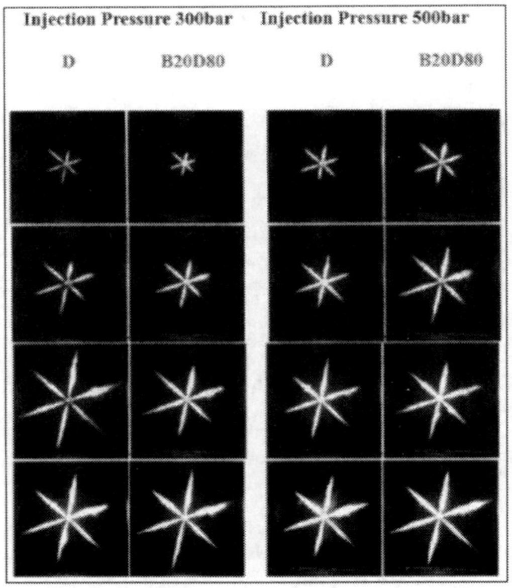

Figure 3. Spray images of test fuels. Rows are (top to bottom) ASOI: 0.5 ms, 0.75 ms, 1 ms, and 1.5 ms.

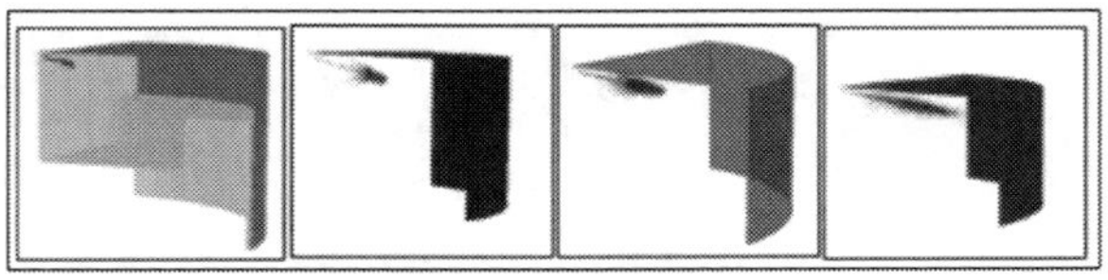

Figure 4. Spray images of diesel at injection pressure 300 bar, ambient conditions using Ansys Forte. Images (left to right) ASOI: 0.5 ms, 0.75 ms, 1 ms, and 1.5 ms.

The spray patterns for the simulations are qualitatively similar to the experiments. For a quantitative measure, a good match in the liquid penetration was found as seen in Fig. 5.

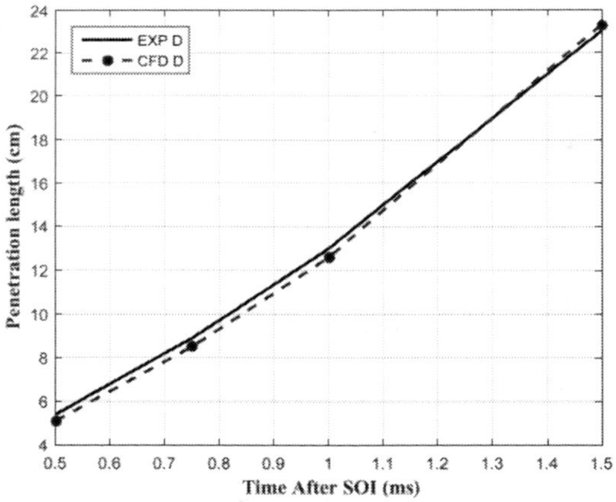

Figure 5. Spray penetration of diesel at ambient conditions.

Figure 6 presents the impact of the ambient pressure on the spray penetration length of diesel (D), 10% butanol-90% diesel (B10D90) and 20% butanol 80% diesel (B20D80). The spray penetration of all test fuels is decreased with increasing ambient pressure (the pressure inside the chamber) because of insufficient radial momentum to overcome penetration resistance and the effect of the pressure difference across the sheet. As a consequence, the spray shoulders become strongly curved and the spray collapses into a form that can ultimately become narrow. The region of spray tip penetration of neat diesel becomes longer and wider as a result of ambient temperature increases, due to the lower in-cylinder gas density at higher temperatures (Fig. 7). However, for butanol-diesel blends, the results were reversed due to the high heat of vaporisation of butanol being more than double that of diesel (Table1) so penetration length will be shorter and plumes will be narrower.

Properties	D	Butanol
Density (kg/L)	0.82-0.85	0.810
Viscosity (mm^2/s) at 40 °C	1.9-4.1	2.22
Surface tension (mN /m)	23.8	24.2
Latent heat (MJ/kg) at 25 °C	270	582

Table 1. Fuel properties.

Figure 8 and Figure 9 present the ASMD of the test fuels under different ambient pressures and ambient temperatures predicted via Ansys Forte. It is clear from Fig. 8 that the ASMD decreases with an increase in the ambient pressure for all test fuels. The ASMD difference among the test fuels is mainly due to the differences in their viscosity and surface tension. A higher viscosity leads to a lower fuel jet velocity, leading to larger droplet size. A lower surface tension makes the spray easier to break up into small droplets. In contrast, no significant difference in the ASMD was found when ambient temperature increased.

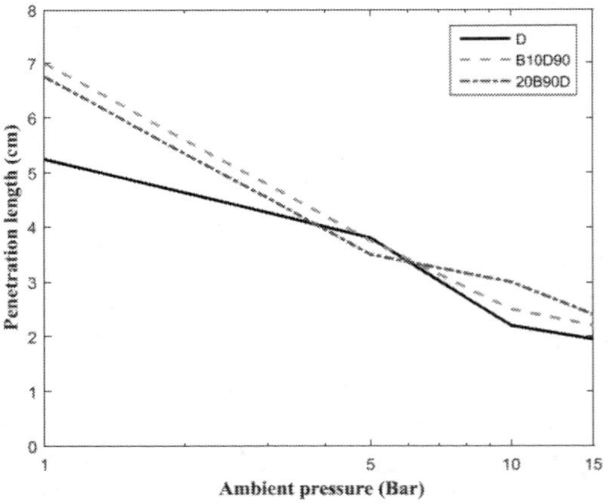

Figure 6. Penetration length under different ambient pressures.

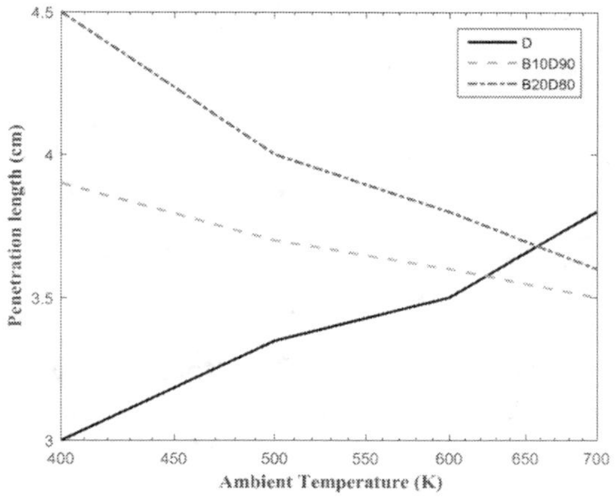

Figure 7. Penetration length under different ambient temperatures.

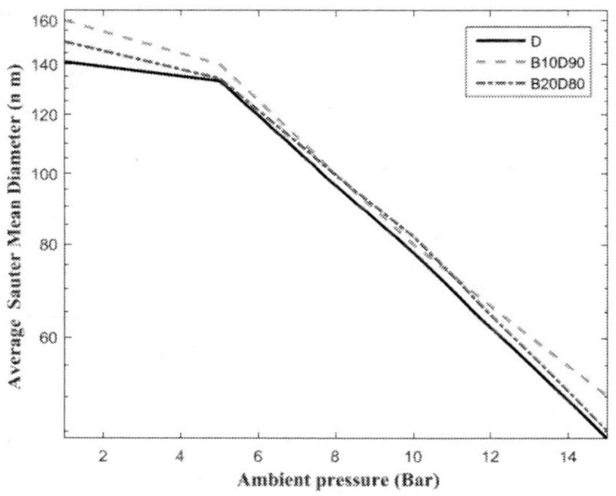

Figure 8. ASMD under different ambient pressures.

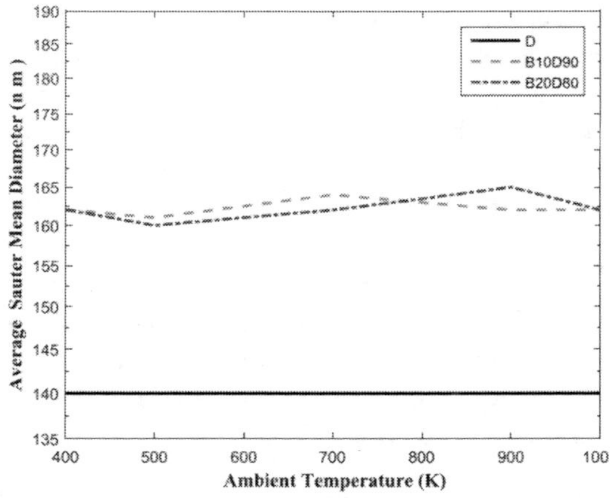

Figure 9. ASMD under different ambient temperatures.

4 Conclusions

- The spray penetration (S) length is decreased as a result of increased ambient pressures of all test fuels.
- The spray penetration (S) length is increased as a result of increased injection pressures of all test fuels.
- Increased ambient temperature causes the spray penetration of pure diesel to become longer and wider. In contrast, the penetration of butanol-diesel becomes shorter.
- The ASMD of the butanol-diesel blend is higher than pure diesel.

5 Acknowledgment

The first author thanks the Iraqi Government Ministry of Higher Education and Scientific Research. The authors are also grateful to (USQ staff) for assisting with experiments.

References

[1] Soid, S. N. and Zainal, Z. A., ENERGY 36 (2) (2011) 724-741.

[2] Merola, S. S., Marchitto, L., Corcione, F., Valentino, G., and Tornatore, C, SAE International Journal, 4 (2) (2011) 2543-2558.

[3] Liu, H., Lee, C., Huo, M., and Yao, M., ENERGY & FUELS 25 (4) (2011) 1837-1846.

[4] Verma, I., Bish, E., Kuntz, M., Meeks, E., Puduppakkam, K., Naik, C., and Liang, L., SAE International Journal, 2016.

[5] Gupta, J. G. and Agarwal, A. K., SAE International Journal, 2016.

[6] Chen, P.-C., Wang, W.-C., Roberts, W. L., and Fang, T., FUEL 103 (2013) 850-861.

[7] Agudelo, J., Agudelo, A., and Benjumea, P., Fac. Ing. Univ. Antioquia 49 (2009) 61-69.

[8] Vijayraghavan Iyengar, S. and Rutland, C., SAE International Journal, 2013.

[9] Abani, N., Kokjohn, S., Park, S., Bergin, M., Munnannur, A., Ning, W., Sun, Y., and Reitz, R. D., SAE Technical Paper, 2008.

[10] Beale, J.C. and R.D. Reitz, ATOMIZATION SPRAY. 9 (6) (1999) 623-650.

11th Asia-Pacific Conference on Combustion,
The University of Sydney, NSW Australia
10th -14th December 2017.

The effect of Butanol-Acetone Mixture-Cottonseed Biodiesel Blend on Spray Characteristics, Engine Performance and Emissions in Diesel Engine

Sattar Jabbar Murad Algayyim [a, b, *], **Andrew P. Wandel** [a], **Talal Yusaf** [a]

[a] School of Mechanical and Electrical Engineering, University of Southern Queensland,
West Street, Toowoomba, QLD, 4350, Australia

[b] Department of Mechanical, College of Engineering, University of Al-Qadisiyah, Iraq

Abstract

Increasing energy demands and more stringent legislation relating to pollutants such as nitrogen oxide (NO_x) and particulate matter (PM) from mineral fuels used in diesel engines have encouraged the use of biodiesel. Biodiesel fuels produced from non-edible oils have properties comparable to diesel fuel, which make them promising alternative fuels. However, there are some drawbacks associated with biodiesel as fuel for compression-ignition (CI) engines such as high viscosity and higher NO_x emissions. Using an alcohol butanol-acetone (BA) or acetone-butanol-ethanol (ABE) mixture is one solution to improve blend efficiency and also to lower NO_x emissions. The aim of this paper is to investigate the impact of a BA or ABE mixture blended with cottonseed biodiesel on spray characteristics, engine performance (in-cylinder pressure, brake power (BP) and specific fuel consumption (SFC)) and emission levels (NO_x and carbon monoxide (CO)). The results demonstrated that BA and ABE decreased biodiesel viscosity and resulted in improved spray characteristics. BP was reduced while SFC was increased. The peak in-cylinder pressure was comparable at a lower engine speed while being slightly lower at 2000 rpm. The maximum reduction in NO_x and CO was shown to be from 10BA90Bd by 13.84% and 41.5% respectively at 2000 rpm.

1 Introduction

Significant energy demands from population growth together with environmental concerns have encouraged researchers to look for renewable energy resources. There are a number of substitute fuels produced from different edible and non-edible resources. Biodiesel is methyl or ethyl ester or fatty acid made from vegetable oils and animal fat. Cottonseed is one non-edible resource of biodiesel production. There are molecular similarities between biodiesel and conventional diesel fuel. Therefore biodiesel can be used directly or as an additive for conventional fuels, increasing the chance of replacing fossil fuels in the near future. However, current limitations of using biodiesel as an alternative fuel for CI engines include: (1) high density and viscosity due to large molecular weight and complex chemical structure [1], which causes obstacles in completing fuel atomisation and combustion [2]; (2) lower heating value due to the oxygen content which produces less engine power [3] and (3) production of higher NO_x emissions (which increases health problems) due to high combustion temperature [4]. There are many ways used to solve these issues including: (1) micro-emulsion (biodiesel washing with water and blending with surfactants) (however, some experimental studies have found that micro-emulsion can lead to negative effects in injection systems due to low lubricity and wear corrosion) [5, 6]; and (2) use of an exhaust gas recirculation (EGR) system. Some experiments have demonstrated an increase in PM and these techniques require engine modification at extra cost. Therefore, using alcohol as an additive to improve biodiesel properties and combustion efficiency is a good option. Bio-alcohols derived through biochemical processes from different biomass resources have a high oxygen content, which helps to complete the combustion [7].

In particular, ethanol and methanol with biodiesel have been extensively investigated [8,9]. However, some studies revealed some disadvantages of ethanol and methanol used as blends for use in diesel engines. Therefore, researchers have suggested the use of butanol as a suitable blend for diesel engines [10]. Butanol presents comparable fuel properties to conventional diesel and these benefits can be utilised to improve the fuel efficiency and can contribute to providing power comparable to regular diesel fuel while producing less emissions [11]. However, the cost of butanol production is the main issue of using it as fuel in internal combustion (IC) engines. Because of the high recovery and purification costs, other fermentation products such as ABE and BA mixtures (the intermediate outcome during the production of butanol) have been proposed as an additive blend. This provides the potential to reduce NO_x and CO emissions in CI engines. ABE is produced in the typical ratio of ABE (3:6:1) [12].

A study by Li et al. [13] found that it was possible to produce a mixture of butanol and acetone (2.9:1) with no ethanol from the fermentation process when cassava was used as substrate in the fermentation. To our knowledge, BA as an additive to biodiesel has not been investigated. The aim of this paper is to investigate the impact of a BA or ABE mixture on cottonseed

Corresponding author.
E-mail address: Sattarjabbarmurad.algayyim@usq.edu.au

biodiesel characteristics and compare them to conventional diesel as a baseline fuel regarding spray characteristics, engine performance and emission levels.

2 Methodology

This Normal butanol (nB) and acetone (A) were used at 99.8% analytical grade. Ethanol (E) was used at 100% analytical grade. A neat cottonseed biodiesel (Bd) was obtained from Queensland University of Technology (QUT). The cottonseed methyl ester fuel was prepared from cottonseed oil via transesterification. Diesel was obtained from a local petrol station in Toowoomba, Australia as a baseline. A BA mixture was prepared with a ratio of 2.9:1 (B:A) by volume and an ABE mixture with a ratio of 3:6:1, which was used to simulate the intermediate fermentation production. Once this was completed, different volumetric blends with neat cottonseed biodiesel were formed: 10% BA or 10% ABE blended with 90% cottonseed biodiesel, referred to as 10BA90Bd and 10ABE90Bd respectively. Table 1shows properties of the fuel blends.

Properties	A	E	nB	Bd	D
Density (kg/L)	0.971	0.795	0.810	0.85	0.85
Viscosity (mm^2/s)	0.35	1.08	2.22	4.4	1.9-4.1
Calorific value (MJ/kg)	29.6	26.8	33.1	36.8	42.8
Surface tension (mN/m)	22.6		24.2	32.4	23.8

Table 1. Fuel properties.

The spray experimental test was carried out on a constant volume vessel (CVV) at atmospheric pressure. An air-driven high-pressure fuel pump was used in the fuel injection system using a solenoid Bosch-type injector with six holes and two different injection pressures (300 bar and 500 bar). A Photron Charge-Coupled Device (CCD) camera was used to capture the spray blend images. The camera has a resolution of 1024×1024 pixels. An LED light was used for illuminating the fuel spray on each window to ensure constant background light for the camera.

The engine test was conducted using a single-cylinder, four-stroke, water-cooled, direct injection (DI) diesel engine. An electrical dynamometer connected to the engine was used to control the load. The crank angles were measured using a crank angle encoder set up on the shaft of the engine. A Kistler 6052C pressure transducer (CT400.17) and charge amplifier connected to a data acquisition system with software (CT 400.09) were used to record cylinder pressure values at one crank angle for 50 cycles each test. The exhaust gas emission was analysed using a Coda gas analyser to measure NO_x, and CO. The test was carried out at a compression ratio of 19:1 with three engine speeds (1400, 2000 and 2600 rpm) at full load. The test began at least 20 minutes before recording commenced. The experiments were carried out in triplicate to reduce the experimental error.

3. Result and Discussion

3.1 Spray Characteristics

Before testing biodiesel blends in an engine, spray visualisation should be understood because fuel spray characteristics result in an air-fuel mixing rate which has a direct impact on the engine performance and emission levels. Spray images of neat diesel (D), neat biodiesel (Bd), 10BA90Bd and 10ABE90Bd blends are illustrated at two injection pressures 300 bar and 500 bar in Fig. 1 and Fig. 2. The main drawback of biodiesel is its higher viscosity and higher surface tension (Table 1) which resulted in a reduction of the Weber number and injection velocity respectively [14, 15]. It also led to an increase in droplet size compared to neat diesel [15]. In addition, the higher viscosity of biodiesel required a high injection pressure to reduce the friction force contact between the nozzle wall and the fuels. The high jet penetration and poor atomisation resulted in an insufficient reaction rate which leads to reduced engine power and increased fuel consumption. Adding ABE or BA mixtures can improve spray penetration due to the benefits of its properties to decrease viscosity and surface tension which results in improving air-fuel mixing.

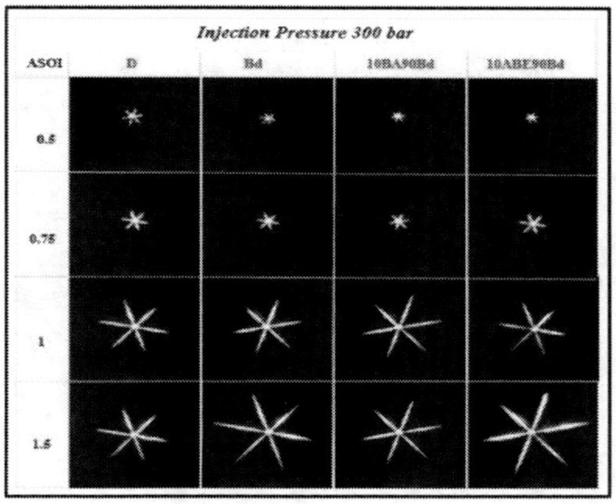

Figure 1. Spray images of different test.

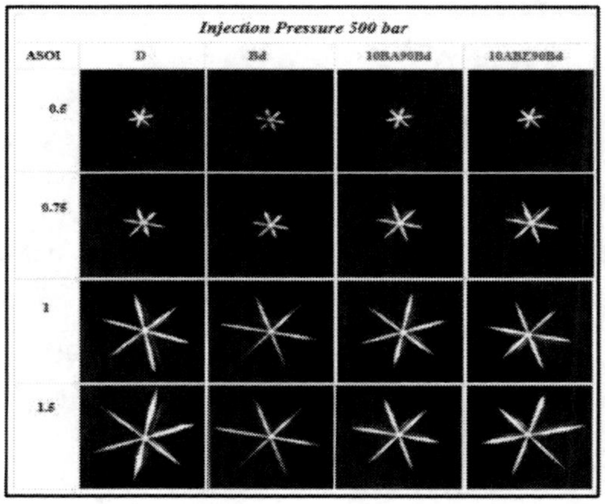

Figure 2. Spray images of different test.

3.2 Engine Performance

3.2.1 In-Cylinder Pressure

Figure 3 presents the relationship between the peak in-cylinder pressure trace and the crank angle of the test fuels at 1400 and 2000 rpm. It can be seen in Fig. 3 that the peak in-cylinder pressure of the BA/ABE-biodiesel blends is comparable with neat diesel at 1400 rpm. In contrast, the peak in-cylinder pressure of 10BA90Bd and 10ABE90Bd was slightly lower compared to that of neat diesel at 2000 rpm. This result was in an agreement with Nabi et al [16] who noticed that the peak in-cylinder pressure of biodiesel was 12% less than diesel.

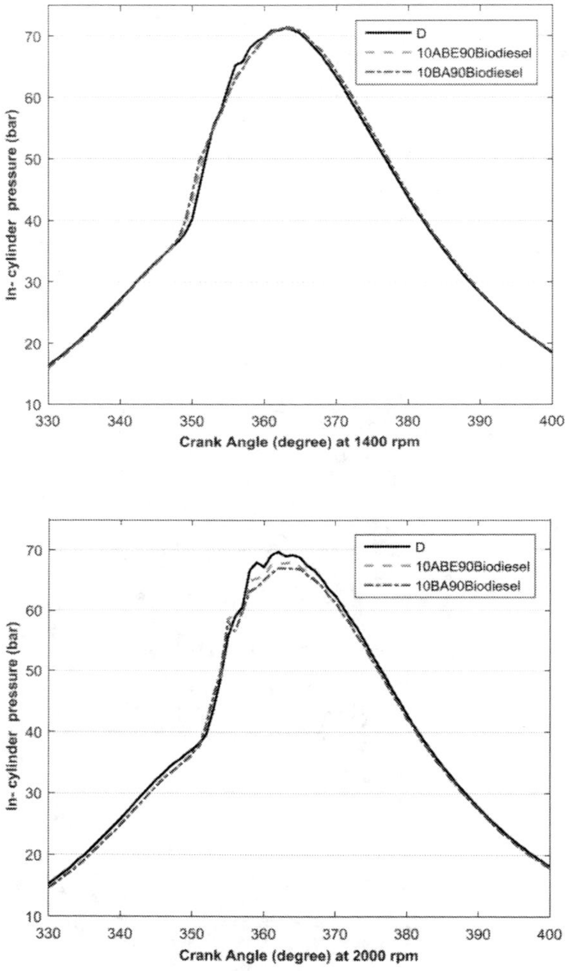

Figure 3. In-cylinder pressure of test fuels (a) 1400 rpm and (b) 2000 rpm.

3.2.2 Brake Power (BP) and Specific Fuel Consumption (SFC)

Figure 4 shows the variation of BP and SFC with the engine speed of the test fuels. It is observed that BP was reduced, while SFC was increased with both fuel blends compared to that of neat diesel due to the low calorific value of the blends (Table 1).

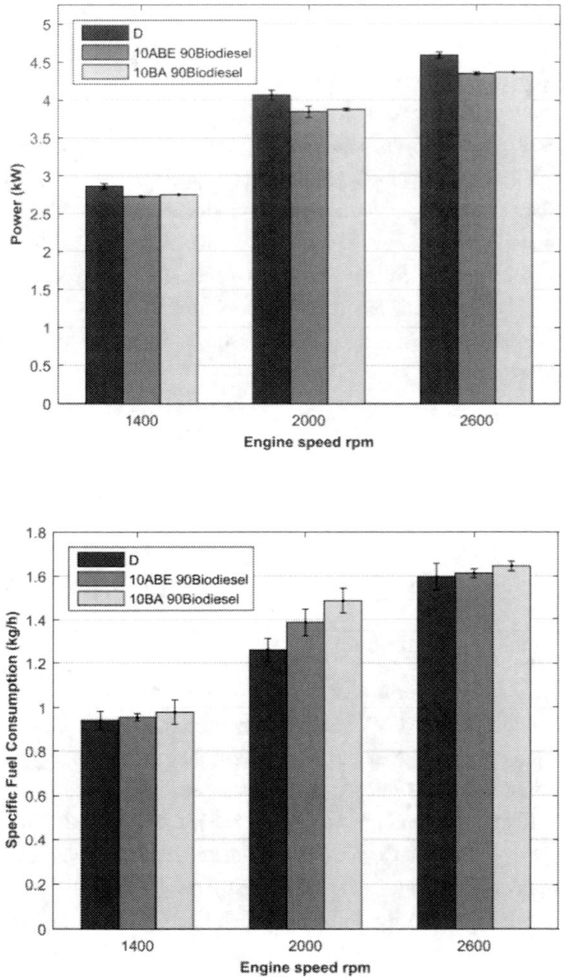

Figure 4. BP and SFC of test fuels.

3.2.3 NO_x and CO Emissions

Figure 5 presents the NO_x and CO emissions of the test fuels at various engine speeds. All BA/ABE-biodiesel blends showed a decrease in NO_x and CO emissions compared to that of neat diesel. The maximum reduction in NO_x and CO was shown for 10BA90Bd by 13.84% and 41.5% respectively at 2000 rpm engine speed. This reduction is due to the BA or ABE mixture having a high oxygen content, which results in decreased exhaust gas temperature (EGT). This reduction in EGT results in reduced NO_x and CO emissions.

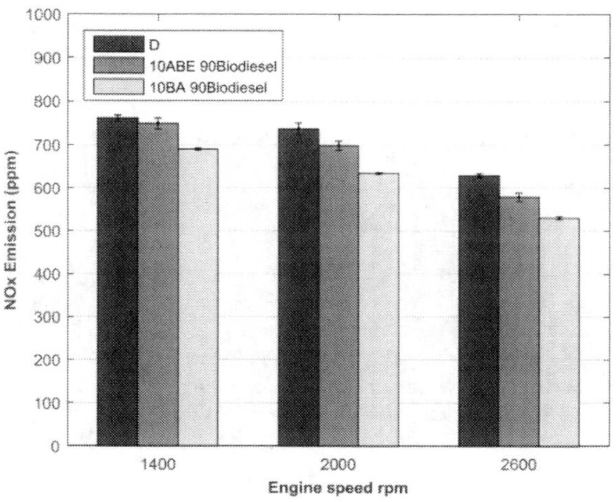

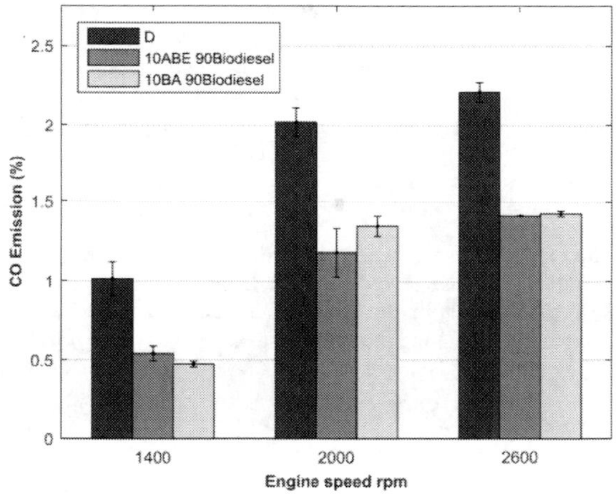

Figure 5. NO_x and CO emissions of test fuels

4 Conclusions

The experimental work has concluded some significant results for the test blends. The results are as follows:

- BA or ABE can enhance the spray characteristics of biodiesel, which results in improved air-fuel mixing.
- The peak in-cylinder pressure of 10BA90Bd was comparable to neat diesel at a lower engine speed.
- All ABE/BA-biodiesel blends showed a decrease in NO_x and CO emissions at all engine speeds. The maximum reduction in NO_x and CO emissions was for 10BA90Bd by 13.84% and 41.5% respectively at 2000 rpm compared to neat diesel.

5 Acknowledgment

The first author thanks the Iraqi Government Ministry of Higher Education and Scientific Research. The authors are also grateful to Dr Saddam Al-Wayzy, Dr Rishabh Choudhury and Mr Robert Kimber (USQ staff) for assisting with experiments.

References

[1] Subbaiah, G. V., Gopal, K. R., Hussain, S. A., Prasad, B. D., and Reddy, K. T., IJORARIAS 3(3) (2010) 334-342.

[2] Koçak M. S., Ileri, E., and Utlu, Z., ENERG FUEL 21 (6) (2007) 3622-3626.

[3] Agarwal, A.K., PROG ENERG COMBUST. 33 (3) (2007) 233-271.

[4] Wu, H.-W. & Z.-Y. Wu, APPL ENERG 104. (2013) 362-370.

[5] Huo, M., Lin, S., Liu, H., and Lee, C.-f. F, FUEL 123 (2014). 218-229.

[6] Nabi, M. N., Rahman, M. M., and Akhter, M. S., APPL THERM ENG 29 (11) (2009) 2265-2270.

[7] Campos-Fernández, J., Arnal, J. M., Gómez, J., and Dorado, M. P., APPL ENERG 95. (2012) 267-275.

[8] Tse, H., Leung, C. W., and Cheung, C. S., ENERGY 83 (2015) 343-350.

[9] Sayin, C., FUEL 89 (11) (2010) 3410-3415.

[10] Laza, T. and Bereczky, Á, FUEL 90(2) (2011) 803-810.

[11] Doğan, O., FUEL 90 (7) (2011) 2467-2472.

[12] Qureshi, N., Saha, B. C., Hector, R. E., Dien, B., Hughes, S., Liu, S., Iten, L., Bowman, M. J., Sarath, G., and Cotta, M. A.

BIOMASS BIOENERG 34 (4) (2010) 566-571.

[13] Li, X., Li, Z.-G., and Shi, Z.-P., Bioresources and Bioprocessing 1 (1) (2014) 1-13.

[14] Gupta, J. G. and Agarwal, A. K., SAE International (2016).

[15] Wang, X., Huang, Z., Kuti, O. A., Zhang, W., and Nishida, K, International Journal of Heat and Fluid Flow. 31(4) (2010) 659-666.

[16] Nabi, M. N., Rahman, M. M., and Akhter, M. S., APPL THERM ENG 29 (11) (2009) 2265-2270.

11th Asia-Pacific Conference on Combustion,
The University of Sydney, NSW Australia
10th -14th December 2017.

Assessment of 3D printing technology for potential application towards manufacturing composite propellants

Anirudha Ambekar[1] and Jai-ick Yoh[1]

[1]Department of Mechanical and Aerospace Engineering, Seoul National University

Seoul, 151-742, Korea

Abstract

Ammonium perchlorate based solid composite propellants are conventionally manufactured by a casting process. The casting of propellants is highly resource intensive and demands a shaped mold for each grain shape to be casted. This limits the number of grain shapes that can be tested with various propellant compositions. As the grain shape directly affects the performance of a propellant charge, innovative grain geometries could provide significant boost in performance. 3D printing techniques are a subset of additive manufacturing or rapid prototyping and provide the capability of quick and economical manufacture. Thus, application of these techniques for composite propellant manufacture has the potential to allow the study of parametric variation of propellant shapes with ease. This article presents an exploratory study aimed at application of 3D printing techniques for manufacturing composite propellants.

1 Introduction

Solid composite propellants are prominent amongst the range of available energetic materials due to their high energy density, simplicity, and relative safety. The conventional method of solid rocket motor manufacturing involves casting of a slurry containing oxidizer, fuel, binder, and curing agents into a mold. This technique relies on molds and shaped patterns to create the grain shape, which subsequently determines the thrust profile for a given rocket. Furthermore, longitudinally varying grain geometry has been proposed [1, 2, 3] for improved grain performance. However, the process of creating these molds and patterns is costly and time consuming. Thus, production and testing novel grain patterns is considerably resource intensive. This is a significant limitation in parametric design, optimization, and testing of solid rocket propellants.

In contrast to conventional casting, additive manufacturing (AM) or 3D printing is inherently capable of a rapid design and manufacturing cycle irrespective of the geometric complexity of objects to be manufactured. In typical 3D printing process, three-dimensional objects are created by successive deposition of layers of a given material, which is typically a thermoplastic or a UV hardening resin. Fused deposition modeling (FDM) and selective laser sintering (SLS) are two AM techniques, which have found widespread usage in many fields. The application of AM technology to solid rocket propellants would provide greater freedom for testing multiple grain patterns, novel compositions, and optimization of solid rocket motors.

Previously, the FDM technology has been used for fabrication of hybrid rocket grain by Fuller et al. [4], Whitmore et al. [5, 6], and Derrick et al. [7]. While, sugar-KNO_3 based composite propellants were manufactured using the SLS technique by Brown et al. [8]. Furthermore, Cattani et al. [9] have recently reported a preliminary study regarding printing energetic composite filaments to be used for 3D printing.

Corresponding author. Fax: +82-2-882-1507
E-mail address: jjyoh@snu.ac.kr

1.1 Basics of 3D printing

The term 3D printing is most commonly used to refer to fused deposition modelling. In this technique, a thermoplastic polymer such as acrylonitrile butadiene styrene (ABS) and polylactic acid (PLA) in the form of a filament is heated to its melting point and pushed through a heated nozzle. The movement of the nozzle is controlled through a computer and molten material exiting the nozzle is laid on a plate to create desired shape. The thermoplastic laid on the plate cools rapidly, below the glass transition temperature of the polymer, and acquires a solid form.

The repeated extrusion and controlled nozzle movement is used to build successive layers of plastic to create desired object. The key requirement for the correct function of a FDM based 3D printer lies in keeping the thermoplastic "*liquid in the nozzle and solid on the build plate*".

1.2 Objective

The objective of the current study may be stated as assessing simple 3D printing technology for manufacturing ammonium perchlorate composite propellants (APCP). The techniques envisaged here seek to replace the conventional casting method and may provide several advantages such as relatively quick design and fabrication of any grain shape, rapid testing with novel compositions, reduced cost due to due reduced tooling, scalability, and accuracy of geometrical patterns, and possibility of creating functionally graded propellants. The 3D printing techniques can potentially also create grains with longitudinally varying geometry.

In addition to being the most common material used for 3D printing, ABS has been considered as a fuel for hybrid [10] as well as composite propellants [11]. Current study conducts a theoretical assessment of the performance of an ABS based propellant with conventional APCP and proposes a fabrication technique for such a propellant.

2 Potential techniques of 3D printing APCP

The origin of this study was merely based on the notion that 3D printers may be used for composite propellant manufacture. The study in this initial stage has been largely exploratory with significant literature review and various trials to ascertain viability of future scope.

A commercial FDM printer (Ecubmaker fantasy II) shown in Fig. 1 was utilized for this study in conjunction with slicing software Cura 15.04 and 3D modelling software freeCAD version 0.16.

Figure 1: Ecubmaker fantasy II FDM 3D printer with assortment of components.

The potential techniques, which may be used to manufacture APCP with a modified 3D printer, include direct printing with AP-HTPB slurry, 3D printing of molds, dual extrusion technique, and solvent extrusion technique. Following paragraphs discuss each of these methods in detail.

2.1 Direct printing with AP-HTPB slurry

Typical composite propellants consist of the oxidizer ammonium perchlorate (AP), aluminium particles as fuel, the binder hydroxyl-terminated polybutadiene (HTPB), and certain curing agents. The ingredients are typically mixed in batch process and the resulting slurry is cast into moulds for curing. The initial viscosity of the slurry [12] is similar to that of molten thermoplastic [13].

An attempt was made to use the 3D printing technology for directly printing the grain shape with the propellant slurry. The 3D printer was modified to interface with a paste extruder. The paste extruder was fabricated in-house using ABS thermoplastic and the 3D printer. Figure 2 shows different views of the paste extruder assembly spate and mounted on the 3D printer. The original design of the paste extruder was tweaked to accept standard 5 ml syringes which would hold the propellant slurry.

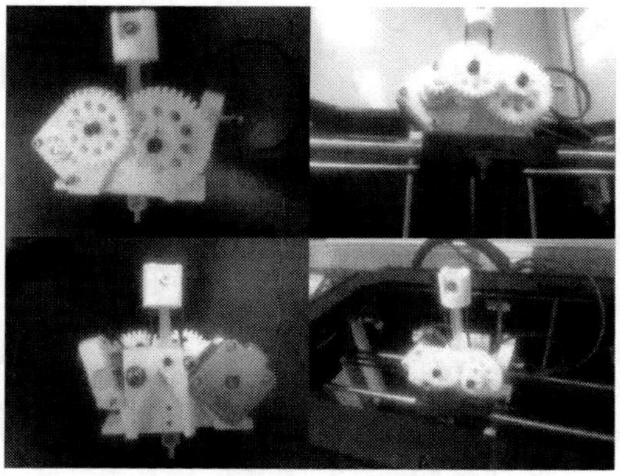

Figure 2: Paste extruder for APCP slurry printing.

However, as the curing time for HTPB bonded propellants is significantly longer [14], the printed shape of the grain was not maintained. Figure 3 shows an attempt of printing the wagon wheel grain with AP-HTPB propellants mixed in 75:25 ratio. However, the shape of the profile could not be maintained due to low viscosity of the propellant composition at this stage of curing.

Figure 3: A trial print of wagon-wheel grain geometry with 75:25 AP-HTPB slurry showing issues due to low viscosity and high curing time.

Therefore, due to the long curing times AP-HTPB based composite propellants cannot satisfy the "*liquid in the nozzle and solid on the build plate*" requirements and use of molds was deemed necessary.

2.2 3D printing of molds

This method provides a simple approach wherein the mold of the desired shape is pre-printed using a standard 3D printing technique. The 3D printing process of the mold provides the opportunity of rapid manufacture of intricate core shapes without the need of complex tooling. The typical 3D printing materials such as ABS and PLA may be utilized for this purpose. Figure 4 shows the 2D and 3D model as well as samples fabricated from ABS plastic for a simple star grain designated 'A' and a helical star grain 'B'.

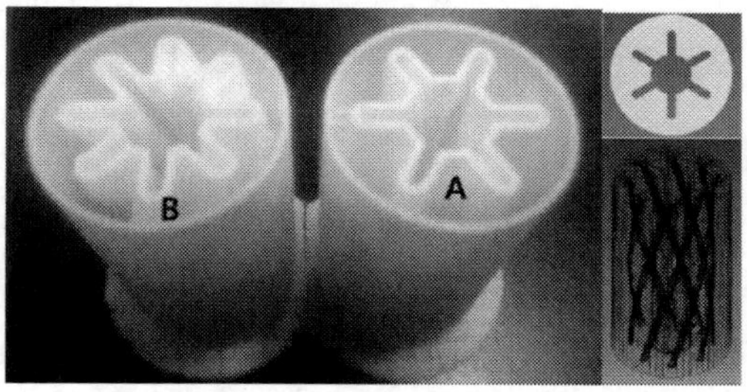

Figure 4: 2D cross-section of a star grain, 3D model a helical star grain, and samples fabricated from ABS plastic.

The fabrication of complex helical geometry was rendered easily with the 3D printing method. Thus, establishing the utility of this method for creating complex grain geometry. With appropriate design process determining the various dimensions of the mold, sufficient mechanical strength can be ensured.

The central core, forming the internal geometry of the propellant grain, may be designed to be removable through techniques such as melting and solvent dissolution or it could be designed to be burnt in place with the AP-HTPB propellant. The outer layer of the mold may be integrated into or function as a part of the motor casing similar to recently 3D printed rocket motor [15].

The technique of 3D printed molds was implemented with two simplified geometries viz. wagon wheel and plane cylinder. The outer diameter of the wagon wheel grain was 30 mm with minimum web thickness of approximately 4 mm and a thickness of 5 mm. the end burning plane cylindrical grain had an outer diameter of approximately 18 mm and height of 8 mm. A hand-mixed composition with AP-HTPB-IPDI in the proportion of 77%:20.5%:2.5% was manually filled in these molds. The filled molds were cured at 60 ° for 8 days and the cured propellant was separated from the mold by cutting. The resulting sample grains have been shown in Fig. 5.

(a) (b)

Figure 5: (a) Cylindrical grain (b) wagon wheel grain.

Although, the grain geometry was successfully created, further development to ensure the elimination of voids and cracks in the propellant matrix is necessary.

2.3 Dual extrusion technique

In this technique, the first extruder would be a conventional heated 3D printer head used for polymer filaments while the second extruder would be an unheated paste extruder. In order to implement this technique, a custom multi-extruder 3D printer will have to be designed and constructed. This technique will allow printing any grain shape accurately without needing intricate molds and patterns of a fixed shape. The 3D printed mold created by the first extruder can be designed to be of any shape and appropriate strength. The quasi-simultaneous construction of the mold and pouring of the propellant slurry would eliminate the possibility of voids or trapped air bubbles typically associated with the filling process of full sized molds.

Furthermore, the mold could be made of a low melting point material to reduce the exposure of the propellant to high temperatures. Candidate materials for this purpose include PLA, paraffin wax, and other commercially available low temperature 3D printing materials.

2.4 Solvent extrusion technique

This technique is particularly applicable to the ABS-based propellants where the slow curing HTPB is replaced with ABS, which can solidify quickly. The similarity between the energetic nature of HTPB and ABS has been established [16]. The theoretical performance of a rocket utilizing ABS-based propellant was estimated using NASA CEA code [17] instead of conventional APCP. The chemical formulae and the heat of formation for both ABS and HTPB were obtained from the literature [16]. The chamber pressure for this hypothetical rocket was assumed 20 bar and complete expansion of the combustion products was assumed to occur in an infinite area combustor. The vacuum specific impulse for a rocket operating on a stoichiometric mixture of AP-HTPB was found to be 210.2 s, while the ABS-based propellant with a AP:ABS molar ratio of 4.05:1 yielded a vacuum specific impulse of 188.85 s, which is comparable to the conventional propellant case.

In addition to being a thermoplastic, ABS is also soluble in acetone, and butanone. This method may be implemented through fewer modifications to the existing printers without a heated extruder. Solvent extrusion technique attempts to create the "*liquid in the nozzle and solid on the build plate*" condition with AP based compositions. The combination of AP, ABS, and a suitable solvent may be optimized such that the resulting slurry will remain in liquid phase within the extrusion nozzle while exposure to ambient air after extrusion will allow evaporation of the solvent and resulting AP-ABS structure will form a composite propellant grain.

Implementing this technique requires investigation into the rheological and evaporative properties of AP-ABS-solvent combinations. The ratio of mixing will have to be optimized for obtaining the correct properties such as viscosity of the semi-liquid mixture and short enough solidification time. As this approach adds the novelty of using a new binder system for AP based composite, a study into reaction of kinetics of the new combination will also need to be carried out.

Preliminary study was conducted with AP-ABS-acetone mixtures, which form a slurry. The slurry contained 1.151 gram ABS, 6 gram AP, and 5 ml Acetone. AP and ABS were mixed in stoichiometric ratio and acetone acted as the solvent. This slurry was subsequently extruded through the paste extruder to create a layer of AP-ABS propellant in the shape of a propellant grain. However, due to the low boiling point of acetone, significant amount of slurry was solidifying at the nozzle of the syringe leading to blockage. This prevented the printing of a proper propellant grain. Future studies may use butanone instead of acetone in order to alleviate this problem.

3 Conclusions and future scope of work

So far, the proposed study was focused on establishing the feasibility of 3D printing technique for composite propellant manufacture. The work done so far has given valuable insights into the problem at hand. The short term objective of the study viz. designing, assembling, and testing of simple hardware related to 3D printer capable of printing conventional AP-HTPB based composite propellants was accomplished. Theoretical calculations comparing the performance of ABS based composite propellants with conventional APCP were also carried out.

Although, direct printing of APCP using 3D printing technique was found to be impractical at this stage, the advantages of 3D printing technique may be utilized through other techniques. The utility of 3D printing technique for rapidly creating intricate molds for complex grain shapes was established by 3D printing simple grain molds and successfully casting small rocket grains. The solvent extrusion technique could not be implemented due to rapid evaporation of acetone solvent.

In conclusion, the study shows that 3D printing techniques demonstrate encouraging prospects for advantageous and successful application towards manufacture of composite propellants.

Future scope of work includes further development with the pre-printed molds. This area demands an investigation into proper method of filling the mold, propellant curing, and separation of the mold and the propellant. The development of a dual extruder printer is the primary requirement for implementing the quasi-simultaneous printing of mold and the propellant while the solvent extrusion technique requires a study into rheology and chemical kinetics of various solvent-based compositions.

The combustion characterization of these propellants through burning rate measurement, calorimetry, and thrust measurements is also a part of the future scope of work.

5 Acknowledgment

This work was financially supported by BK21 PLUS program at the Department of Mechanical and Aerospace Engineering, Seoul National University. Additional support came from

Advanced Research Center Program (NRF-2013R1A5A1073861) contracted through Advanced Space Propulsion Research Center at Seoul National University.

References

[1] F. Dong-Hui, H.F. Yang, Z. Wei-Hua, Proc. Inst. Mech. Eng. Part G J. Aerosp. Eng. 228(7) (2014) 1156–1170.

[2] H.L. Archer Jr, U.S. Patent No. 6,431,072. 13 Aug. 2002.

[3] M. Golafshani, M. Farshchi, H. Ghassemi, J. Propuls. Power, 18(1) (2002) 123–130.

[4] J. Fuller, D. Ehrlich, P. Lu, R. Jansen, J. Hoffman, Advantages of Rapid Prototyping for Hybrid Rocket Motor Fuel Grain Fabrication, 47th AIAA/ASME/SAE/ASEE Joint Propulsion Conference, 2011, pp. 1–10.

[5] S.A. Whitmore, S.D. Walker, D.P. Merkley, M. Sobbi, J. Propuls. Power. 31 (2015) 1727–1738.

[6] S.A. Whitmore, S.L. Merkley, L. Tonc, S.D. Mathias, J. Propuls. Power. (2016) 1–11.

[7] A. Derrick, E. Boyer, B. R. McKnight, J. D. DeSain, J. K. Fuller, K. K. Kuo, B. B. Brady, and T. J. Curtiss, Int. J. Ener. Mat. Chem. Prop., 13(4) (2014) 287-307.

[8] C.B. Brown, E. Chewakin, M. Feldman, A. Lima, N. Lindholm, C. Lipscomb, R. Niedzinski, J. Sobol, Solid Propellant Additive Manufacturing (SPAM), Project Report, University of Colorado Boulder.

[9] P.A. Cattani, T.J. Fleck, J.F. Rhoads, S.F. Son, I.E. Gunduz, Applications of Additive Manufacturing Techniques in Making Energetic Materials The Summer Undergraduate Research Fellowship (SURF) Symposium, 4 August 2016, Purdue University, West Lafayette, Indiana, USA.

[10] T.S. Elliott, B. Jenkins, R. Zeineldin, J. Johnson, M. Simons, J. Godfrey, Additive Manufacturing of Small Scale Rocket Grain Cartridges with Uniformly Distributed Aluminum Particles" 52nd AIAA/SAE/ASEE Joint Propulsion Conference, 2016, pp. 1–7.

[11] B. Clark, Z. Zhang, G. Christopher, and M. L. Pantoya, J. Mater. Sci., 52(2) (2017) 993–1004.

[12] R. Muthiah, V.N. Krishnamurthy, B.R. Gupta, Propellants Explos. Pyrotech. 21 (1996) 186–192.

[13] J. Pasquale, M.G.M. Marascio, J.A. Månson, D. Pioletti, P.E. Bourban, Eur. Cells Mater. 32 (2016) 29.

[14] W. M. Adel, L. Guo-zhu, Developing a Viscoelastic Relaxation Model for AP-HTPB Composite Solid Propellant Based on Experimental Data, 21st AIAA International Space Planes and Hypersonics Technologies Conference. 2017.

[15] Charlie Garcia available at, <http://rocketry.mit.edu/2017/04/100-3d-printed-solid-rocket-motor/>

[16] S. A. Whitmore, Z. W. Peterson, and S. D. Eilers, Analytical and Experimental Comparisons of HTPB and ABS as Hybrid Rocket Fuels, 47th AIAA/ASME/SAE/ASEE Joint Propulsion Conference, 2011 1–48.

[17] Gordon, S. and B.J. McBride, Computer program for calculation of complex chemical equilibrium compositions and applications. 1996.

Kinetics of Hydrogen Abstraction and Addition Reactions of 3-Hexene by ȮH Radical

Feiyu Yang, Wuchuan Sun, Youshun Pan, Fuquan Deng, Xue Jiang, Yingjia Zhang*, Zuohua Huang

State Key Laboratory of Multiphase Flow in Power Engineering, Xi'an Jiaotong University

Xianning West Road, Xi'an, Shaanxi, 710049, People's Republic of China

Abstract

Rate coefficients of H-atom abstraction and addition reactions of 3-hexene by the hydroxyl radical (ȮH) were determined using canonical variational transition-state theory, with potential energy surface evaluated at the CCSD/CBS//BHandHLYP/6-311G (d, p) level and quantum mechanical effect corrected by the multi-dimensional small-curvature tunneling method. Results reveal that accounting for approximate 70% of the overall H-atom abstractions occur in the allylic site *via* both direct and indirect channels. The indirect channel containing two Van der Waals pre-reactive complexes exhibits two times higher rate coefficient relative to the direct one. The ȮH-addition reaction also contains two Van der Waals complexes, and its submerged barrier results in a negative temperature coefficient behavior at low temperatures. In contrast, the ȮH-addition pathway dominates only at the temperatures below 590 K whereas the H-atom abstraction reactions dominate overwhelmingly at the temperature over 590 K. All of the rate coefficients calculated with an uncertainty of a factor of 5 were estimated. Analyses on the potential energy surface and minimum reaction path were also performed in *this study*.

1 Introduction

With the efforts of worldwide researchers, the combustion of alkanes has been thoroughly investigated. However, well-understanding of oxidation and combustion for alkenes [1-3] is still under way. It is well known that the fate of alkenes is of great importance in alkane combustion, particularly in richer flame where alkenes play an important role as intermediates in the fuel oxidation and combustion and as a critical precursor of soot formation. ȮH radical, one of the top vital radicals involved in hydrocarbon combustion [4], dominates the chain propagation, chain branching and even chain termination processes at different temperature regions. The reactions of alkene with ȮH radicals are therefore major reaction pathways under either atmospheric or typical combustion conditions.

Limited research is available with respect to alkene + ȮH reaction systems at present. Tully [5-6] measured the global rate coefficients of ethene/propene/1-butene + ȮH using a laser photolysis/laser-induced fluorescence technique. Together, they also explored the impact of deuteration on their measured results. However, the rate coefficients of individual elementary reactions were unavailable due to the limitation of research conditions. Recently, Huynh et al. [7] studied the kinetics of enol formation from the reactions of propene + ȮH, and pointed out that H-atom abstraction was more important than ȮH-addition in the propene consumption due to the higher branching ratio of the former (more than 90%). Zádor et al. [8] predicted the detailed potential energy surface of terminal addition, central addition and H-atom abstraction of propene + ȮH, concluding that the channel resulting in allyl radicals predominates overwhelmingly. Szori et al. [9] provided a detailed description of allylic H-atom abstraction for propene + ȮH system employing as many as 9 methods and they found that those reactions proceed through two channels, direct and indirect H-atom abstractions. With the occurrence of the reactant-complex, the indirect channel exhibits lower barrier and thus facilitates the reactions. Besides, within the 9 methods adopted, the BHandHLYP method behaves more accurate and more efficient. Sun and Law [10] explored the kinetics of H-atom abstraction reactions of butane isomers by ȮH radicals. Similar to propene, both the direct and indirect allylic H-atom abstraction channels were identified and the latter appeared to be favorable. Among all of the H-atom abstractions, the allylic H-atom abstraction dominates overwhelmingly while vinyl H-atom abstraction is least favored. Greenwald et al. examined the addition of ȮH to ethylene [11] and isoprene [12] at high-pressure limit and as a function of pressure *via* transition stated theory and master equation simulations. A two transition state model was proposed with simple reduction, the computed rate coefficient exhibited reasonable agreement with experiments.

To the best knowledge of the authors, current research only focuses on small alkene + ȮH reactions. However, the detail description of potential energy surface and detailed reaction paths for larger alkene + ȮH reactions are limited. With this in mind, the aim of this study is to examine the 3-hexene + ȮH system *via* the high-level quantum mechanical calculations and sophisticated variational transition state theory [13], and to provide accurate rate constants of the title reactions for kinetic mechanism development of large alkenes, which can be eventually used to descript quantificationally the evolution of alkenes in alkane combustion.

2 Theory

2.1 Electronic structures

The BHandHLYP method, having been proved to be suitable for unsaturated hydrocarbons calculation, uses Becke's half and half nonlocal exchange with the Lee-Yang-Parr (LYP) [14] correlation functional. IRC scan was performed for each transition state at BHandHLYP/6-311G(d, p) level to identify whether the transition state belongs to the target reaction. Additionally, to increase the accuracy, the CCSD method with the correlation-consistent, polarized-valence, double-ζ (cc-pVDZ) and triple-ζ (cc-pVTZ) basis set of Dunning was employed to compute the single point energies which were

Corresponding author. Fax: +86-29-8266-8789
E-mail address: yjzhang_xjtu@xjtu.edu.cn

extrapolated to complete basis set limit (CBS).

2.2 Calculation of rate coefficients

When geometries, frequencies and energies have been obtained accurately, it is also pivotal in selecting appropriate reaction theory to obtain more reasonable rate coefficients. The conventional transition state theory (TST) exhibits the drawback that its transition state only locates at the saddle point and maintains the same when temperature varies. This shortcoming is readily compensated in canonical variational transition state theory (CVT), whose transition state is located at the peak of maximum activation Gibbs free energy change (ΔG).

In *this work*, the partition function of low frequency torsional modes were obtained by employing the segmented reference Pitzer-Gwinn (SRPG) method, which decomposes partition function into the contributions from different wells. Besides, the multidimensional small-curvature tunneling (SCT) method is utilized. Taking the corner-cutting effect into account, SCT correction has been proved to be more sophisticated.

3 Results and discussion

3.1 Potential energy surface

Figure 1 illustrates the PES of H-atom abstraction and $\dot{O}H$-addition reactions predicted at CCSD/CBS level. Note that, the H-atom abstraction reaction can take place at three sites of 3-hexene molecule: methyl site (Rxn1), allyl site (Rxn2) and vinyl site (Rxn3). Rxn1 releases 15.47 kcal/mol heat after surmounting a barrier of 4.52 kcal/mol. However, Rxn3 releases 8.65 kcal/mol heat after surmounting a barrier of 4.76 kcal/mol. Rxn3 is thus not favored neither kinetically nor thermodynamically. On the contrary, Rxn2 is the most favored reaction with the lowest barrier and largest exothermicity. Note that there are two sub-channels of Rxn2, the direct one and the indirect one. Although the indirect sub-channel exhibits higher barrier, as reported by other researchers [10], sometimes it is more favored.

The $\dot{O}H$ radical can also add to 3-hexene *via* a Van der Waals well. Due to the large depth of the well, the height of the barrier becomes lower than entrance energy, which might lead to negative temperature coefficient (NTC) of this reaction.

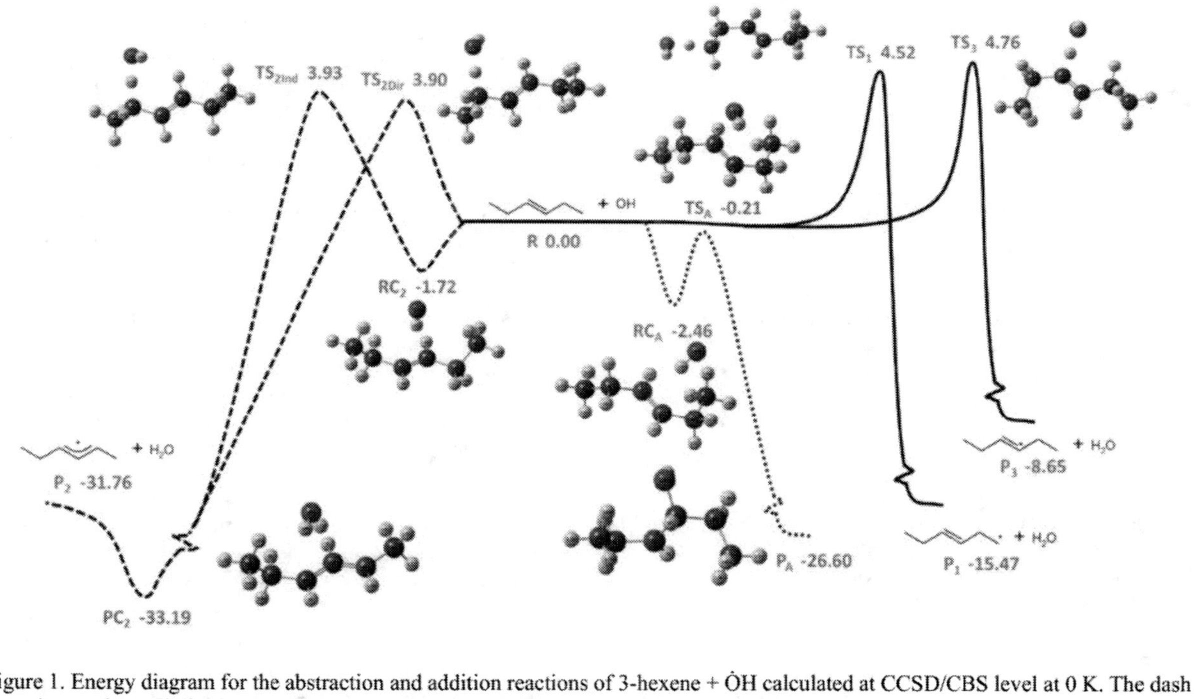

Figure 1. Energy diagram for the abstraction and addition reactions of 3-hexene + $\dot{O}H$ calculated at CCSD/CBS level at 0 K. The dash lines denote the s-allylic H-atom abstraction channels and the dot line denotes the addition reaction.

3.2 MEP analysis

Transition state theory suggests that bond cleaving and forming take place along the minimum energy path (MEP) due to the lowest energy expense. Another important energy named adiabatic ground-state potential energy (V_a^G) equals the sum of VMEP and the zero point energy (ZPE). Since multi-atom molecules oscillate even at 0 K, V_a^G , which takes vibrational information into account, is generally regard as the PES on which reaction, tunneling and non-classical reflection proceed.

Figure 2 compares the MEP and V_a^G between Rxn1, Rxn2DIR, Rxn2IND and Rxn3. All the four reactions share the feature that their MEP and V_a^G are flat at reactant side and steep at product side. There are wells or well-like curves in the V_a^G curves. For vinyl site H-atom abstraction there are two apparent wells, for methyl site H-atom abstraction the wells become weaker. When it comes to allyl site H-atom abstraction, the well is quite weak and many well-like curves can be observed. This might cause from the bond dissociate energy (BDE) of different reactions.

Due to the appearance of wells or well-like curves, the peak of V_a^G curve can shift from the saddle point.

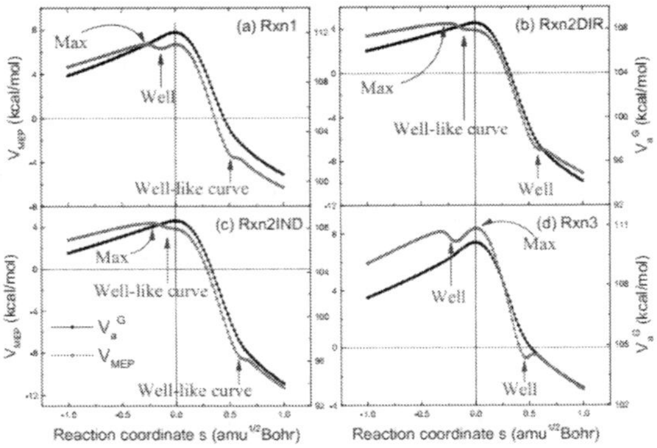

Figure 2. MEP potential energy and the adiabatic ground-state potential energy evaluated at BHandHLYP/6-311G (d, p) level for H-atom abstraction of 3-hexene + $\dot{O}H$.

Figure 3 illustrates the MEP and V_a^G of RxnAdd. The IRC scan was performed for a large of reaction coordinate and two wells were observed at the saddle point at reactant side. Note that the wells are not the Van der Waals wells presented in Fig. 1. The former is herein associated to a chemical process and the later is to a physical one.

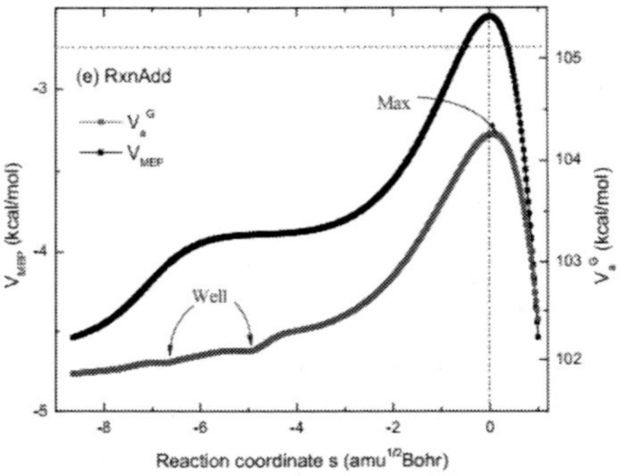

Figure 3. MEP potential energy and the adiabatic ground-state potential energy evaluated at BHANDHLYP/6-311G (d, p) level for $\dot{O}H$ addition reactions of 3-hexene + $\dot{O}H$.

To figure it out which vibration modes contribute to the obvious decrease of ZPE and subsequently result in the reactant-well, the vibrational frequencies of the four geometries with negative slope at reactant-side (s=-0.254, s=-0.265, s=-0.275 and s=-0.286 amu1/2•Bohr) are computed and compared. Figure 4 illustrates the frequencies of the selected geometries. Note that all of the frequencies varies undetectable except the 13th vibration mode, where red shift occurs from 2100 to 2800 cm^{-1} when reaction coordinate increases from -0.254 to -0.286 amu1/2•Bohr. There is no doubt that this red shift is responsible for the sharp ZPE decrease. Likewise, at product-side well, the blue shift of the 13th mode is responsible for the sharp ZPE increase after surmounting the barrier.

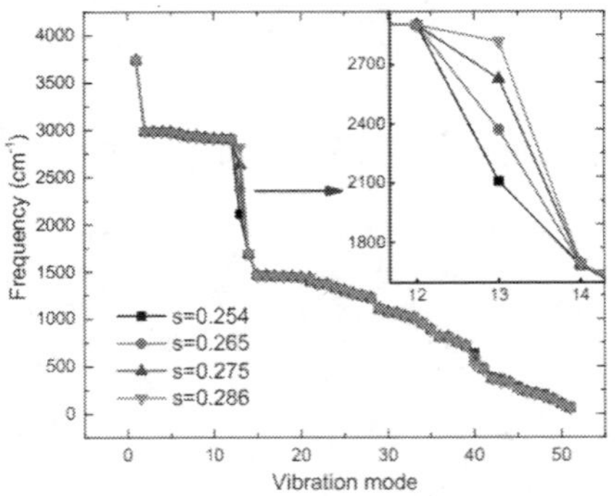

Figure 4. Frequencies of the selected geometries located at the downslope side of reactant-well.

3.3 Rate coefficients

Generally different reaction categories exhibit different competitive capability at specific temperature and the dominated-reaction varies with temperature while typically 3-Hexene + $\dot{O}H$ system is not an exception. To explore the competition among the H-atom abstraction channels, as well as between the H-atom abstraction and the $\dot{O}H$ addition channels, the calculated CVT rate coefficients with SCT correction (kCVT/SCT) are illustrated in Fig. 5. Moreover, the branching ratios are illustrated in Fig. 6.

The $\dot{O}H$ addition reaction dominates the system overwhelmingly at temperatures lower than 590 K while, the H-atom abstraction reactions are predominant at higher temperature region, especially for those proceeding at allylic sites. The branching ratio of $\dot{O}H$ addition reaction reaches 80% at 450 K while the branching ratio of H-atom abstraction reaction reaches 80% at 1000 K. Within 450 – 1000 K the competitiveness of both reaction categories are comparable. Although abstracting hydrogen from methyl carbon exhibits higher barrier than from vinyl carbon, the rate coefficients of those two reactions are comparable. As presented in Fig. 1, the products of H-atom abstraction from methyl site are more stable than that from the vinyl site. Rxn1 is therefore kinetically and thermodynamically more preferential than Rxn3, which is responsible for the larger rate coefficient of Rxn1. The allylic H-atom abstraction reactions exhibit the rate coefficients around ten times higher than non-allylic H-atom abstraction reactions, due to the obviously lower barrier. It is noteworthy that the indirect channel proceeding at allyic site exhibits the rate coefficient around two times higher than the direct channel. Although the barrier of the indirect channel is even slightly higher than the direct channel, the reactant-complex can readily facilitate the H-atom abstraction and increases the rate coefficient. Accounting for a portion of around 70%, the allylic channels overwhelmingly dominate the H-atom abstraction reactions.

Figure 13 compares the rate coefficients of 3-hexene with those of 1-butene and ethylene. As can be noted that most reactions of the same category almost stay in the same magnitude which means our computation are physically valid. For non-allylic H-abstraction reactions, the rate coefficients are comparable, while the rate coefficients of allylic H-abstraction and OH addition reactions are close at high temperature and the difference increases with the decrease in temperature, which could result from the more significant tunneling effect of small mass molecules. It could also be concluded that the appearance of saturated carbon-chain reduced the rate of not only allylic H-atom abstraction but also the $\dot{O}H$ addition reaction.

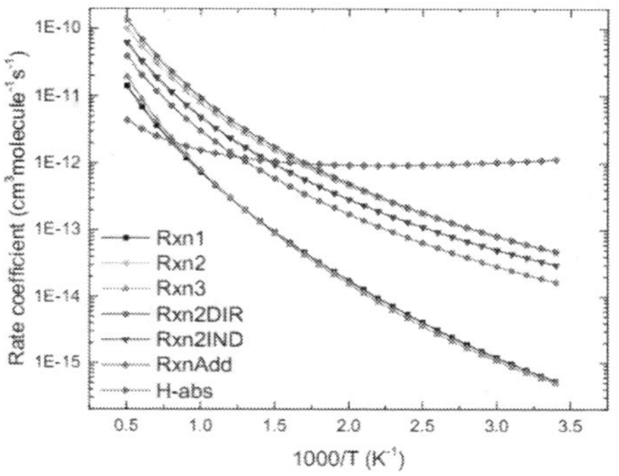

Figure 5. CVT/SCT rate coefficients of H-atom abstraction and $\dot{O}H$ addition reactions for 3-hexene +$\dot{O}H$.

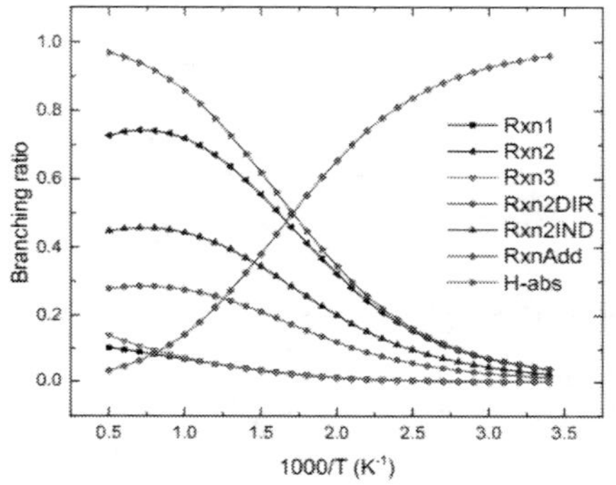

Figure 6. Branching ratios of H-atom abstraction and $\dot{O}H$ addition reactions for 3-hexene +$\dot{O}H$.

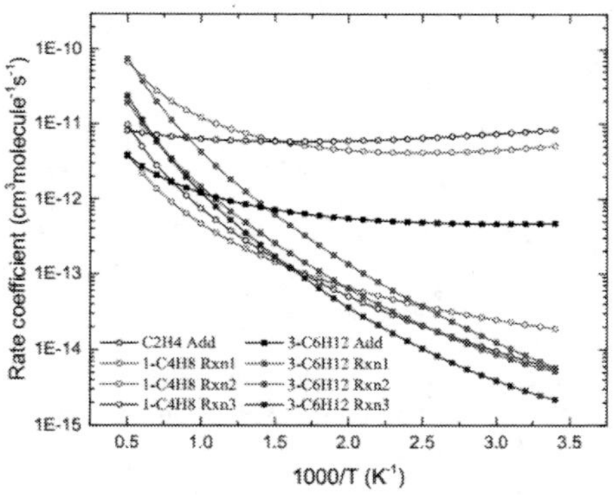

Figure 7. Comparison of rate coefficients of 3-hexene, 1-butene and ethylene + $\dot{O}H$ system.

4 Conclusions

The H-atom abstraction and addition reactions of 3-hexene by $\dot{O}H$ radical was initially investigated using sophisticated quantum mechanical method and variational transition state theory. The stationary points and transition states were optimized at BHandHLYP/6-311G (d, p) level and more accurate energies were evaluated at CCSD/CBS level. The evolution of minimum reaction paths and adiabatic ground-state energy surface were described in details. Rate coefficient calculations were performed using canonical transition state theory with small curvature correction. Due to the similar method of Sun et al [10], the rate coefficients were estimated with an uncertainty factor of 5.

The 3-hexene + $\dot{O}H$ system is dominated by addition reaction when temperature is below 590 K and dominated by H-atom abstraction reactions at over 590 K. Allylic H-atom abstraction overwhelmingly dominates the overall H-atom abstraction with a branching ratio larger than 70%. Due to the reactant-complex, the indirect allylic H-atom abstraction channel exhibits the rate coefficient around two times larger compared to the direct one in spite of its slightly higher barrier. Similarly, the $\dot{O}H$ addition reaction also possesses a reactant Van der Waals well and its submerged barrier makes contribution to the obvious negative temperature dependence at low temperature region.

5 Acknowledgment

The authors acknowledge the financial support of the National Natural Science Foundation of China under Grant No. 91541115 and 91441203.

References

[1] F. Yang, F. Deng, Y. Pan, Z. Tian, Y. Zhang, Z. Huang, Chem. Phys. Lett. 663 (2016) 66-73.

[2] F. Yang, F. Deng, P. Zhang, E. Hu, Y. Cheng, Z. Huang, Energy & Fuels 30 (2016) 5130-5137.

[3] F. Yang, F. Deng, P. Zhang, Z. Tian, C. Tang, Z. Huang, Energy & Fuels 30 (2016) 706-716.

[4] Z. Hong, D.F. Davidson, E.A. Barbour, R.K. Hanson, Proc. Combust. Inst. 33 (2011) 309-316.

[5] F.P. Tully, J.E.M. Goldsmith, Chem. Phys. Lett. 116 (1985) 345-352.

[6] F.P. Tully, Chem. Phys. Lett. 143 (1988) 510-514.

[7] L.K. Huynh, H.R. Zhang, S. Zhang, E. Eddings, A. Sarofim, M.E. Law, P.R. Westmoreland, T.N. Truong, The J. Phys. Chem. A 113 (2009) 3177-3185.

[8] J. Zádor, A.W. Jasper, J.A. Miller, PCCP 11 (2009) 11040-11053.

[9] M. Szori, C. Fittschen, I.G. Csizmadia, B. Viskolcz, J. Chem. Theory Comput. 2 (2006) 1575-1586.

[10] H. Sun, C.K. Law, J. Phys. Chem. A 114 (2010) 12088-12098.

[11] E.E. Greenwald, S.W. North, Y. Georgievskii, S.J. Klippenstein, J. Phys. Chem. A 109 (2005) 6031-6044.

[12] E.E. Greenwald, S.W. North, Y. Georgievskii, S.J. Klippenstein, J. Phys. Chem. A 111 (2007) 5582-5592.

[13] D.G. Truhlar, A.D. Isaacson, B.C. Garrett, Generalized Transition State Theory, CRC, 1985, p. 65-137.

[14] C. Lee, W. Yang, R.G. Parr, Phys. Review B 37 (1988) 785-789.

11th Asia-Pacific Conference on Combustion,
The University of Sydney, NSW Australia
10th -14th December 2017.

A Numerical Study of Flame Instability and Cell Dynamics of Opposed Nonpremixed Tubular Flames near Radiation-Induced Extinction Limit

Hyun Su Bak[1], Chun Sang Yoo[1, 2, *]

[1]Department of Mechanical Engineering, Ulsan National Institute of Science and Technology

50 UNIST-gil Banyeon-ri Eonyang-eup Ulju-gun, Ulsan, 44919, Republic of Korea

[2]School of Mechanical, Aerospace and Nuclear Engineering, Ulsan National Institute of Science and Technology

50 UNIST-gil Banyeon-ri Eonyang-eup Ulju-gun, Ulsan, 44919, Republic of Korea

Abstract

The flame instability and cell dynamics of opposed nonpremixed tubular flames near radiation-induced extinction limit are investigated using high fidelity numerical simulations with various initial conditions (IC). When the radiation effect is considered, two different extinction Damköhler numbers (i.e. large radiation-induced extinction Damköhler number, $Da_{E,R}$, and high stretch-induced extinction Damköhler number, $Da_{E,S}$) are obtained by 1-D analysis with different radiative intensity, I. Two critical Das, $Da_{C,C}$ and $Da_{C,O}$, near $Da_{E,R}$ are identified by the linear stability analysis. In 2-D simulations using the perturbed IC, it is found that near $Da_{E,R}$, the flame instability is categorized into four different regimes with different instability characteristics. When the radiation effect is extremely large, decaying or growing oscillatory instability appears although the stretch rate is high enough, and the cellular instability occurs in the whole Da at which the flame can survive.

1 Introduction

In general, flame instability should be avoided because it is harmful to combustion systems. The cellular instability, however, has been extensively studied due to the needs of its comprehensive understanding and attractive features. The cellular instability is often observed near extinction limit when the effective Lewis number of premixed flames is less than unity.

The extinction phenomena of nonpremixed flames have been investigated by experiments and numerical simulations with various configurations [1-7]. Recently, when the Lewis number is less than unity, the diffusive-thermal (D-T) instability of opposed nonpremixed tubular flames near extinction was investigated using high fidelity numerical simulation with the linear stability analysis [6]. The experiments were compared with 2-D numerical simulations including multicomponent transport and detailed chemical kinetics, and these showed good agreement [7]. In addition, there have been several studies to identify the instability of nonadiabatic flames with radiative heat loss. In a stagnant mixing layer with unity Lewis number, three types of flame evolution, the decaying, diverging and stable limit-cycle solution, were investigated [8]. The structure and dynamic response of nonadiabatic counterflow flames with nonunity Lewis number were identified by 1-D/2-D simulations employing different initial conditions [9].

In the present study, therefore, high fidelity numerical simulations using various initial conditions based on the linear stability analysis are performed to obtain a comprehensive understanding of cellular and oscillatory instabilities of opposed nonpremixed tubular flames near radiation-induced extinction limit.

2 Problem Formulation

To obtain extinction information, 1-D analysis is conducted. As in refs [5, 6, 8, 9], the governing equation including radiative heat loss is formulated as follows:

$$u_r \frac{d}{dr}\begin{pmatrix} T \\ Y_F \\ Y_O \end{pmatrix} = \frac{1}{r}\frac{d}{dr}\left(r\frac{d}{dr}\right)\begin{pmatrix} T \\ Y_F/Le_F \\ Y_O/Le_O \end{pmatrix} + Da Y_F Y_O e^{-T_a/T}\begin{pmatrix} q \\ -\alpha_F \tilde{Y}_{O,2} \\ -\alpha_O \tilde{Y}_{F,1} \end{pmatrix} - I \times Da \begin{pmatrix} T^4 - T_\infty{}^4 \\ 0 \\ 0 \end{pmatrix}, \quad (1)$$

where T, Y_F, and Y_o are the non-dimensional temperature, fuel and oxidizer mass fractions, respectively. To investigate the effects of radiative heat loss, the radiative heat loss term, $I \times Da(T^4 - T_\infty^4)$, is considered in the energy equation of Eq. (1). The I denotes the intensity of radiation, and it is an important nondimensional parameter to show the ratio the radiative heat of transfer thermal energy production. In the present study, the radiation intensity was controlled by using different I values. For the details of parameters and boundary conditions, readers are referred to [6].

The maximum temperature, T_{max}, as functions of Da with four different I, is shown in Fig. 1. These isolas are obtained by solving Eq. (1) using the Newton-Raphson method with a simple continuation algorithm. The upper branch of isolas is well-burning state and the lower branch represents unstable and physically unrealistic. Figure 1 shows that two extinction Damköhler numbers, $Da_{E,R}$ and $Da_{E,S}$, are obtained. The flammable region decreases as I increases. It is noted that for relatively-large $I = 3.76\times10^{-7}$, $Da_{E,S}$ comes close to $Da_{E,R}$, which implies that for this case, the radiation effect cannot be ignored even near high stretch-induced extinction. In the present study, $I = 10^{-8}$ for which the effects of radiation can be ignored near $Da_{E,S}$ is adopted to see the ordinary phenomena. In addition to that, the results of extremely high radiative intensity, $I = 3.76\times10^{-7}$, is only covered in Section 3.3.

* Corresponding author. Fax: +82-52-217-2449
E-mail address: csyoo@unist.ac.kr

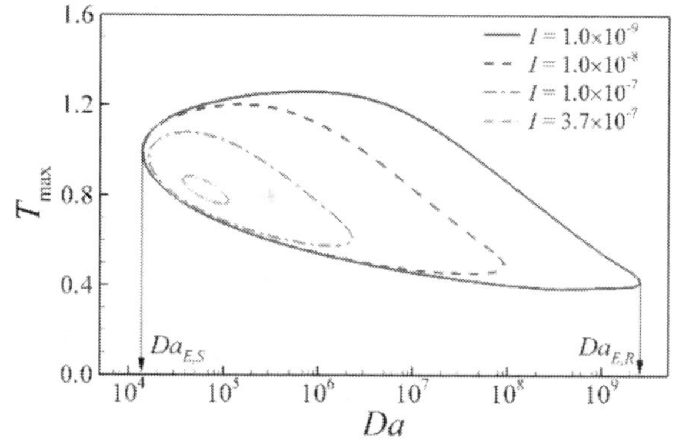

Figure 1: The maximum temperature, T_{max}, as functions of Damköhler number, Da, with different radiative intensity, I.

3 Results and Discussion

3.1 Linear stability analysis

The linear stability analysis helps to predict flame instability characteristics with good agreement [6]. However, the stability analysis near $Da_{E,R}$ is uncommon. In the analysis, the solution variables, T, Y_F, and Y_O, are defined by the sum of the steady solution of the 1-D tubular flame, denoted by an overbar, and small harmonic perturbations:

$$T = \bar{T}(r) + \epsilon T'(r)e^{ik\theta+\lambda t}, Y_F = \bar{Y}_F(r) + \epsilon Y_F'(r)e^{ik\theta+\lambda t},$$
$$Y_O = \bar{Y}_O(r) + \epsilon Y_O'(r)e^{ik\theta+\lambda t} \quad (2)$$

where λ is a complex number whose the real part, λ_R, represents the growth rate, k is the wavenumber, and ϵ is the small perturbation amplitude. Substituting Eq. (2) into Eq. (1) and linearizing the equation, an eigenvalue problem of $(\mathbf{A}-\lambda\mathbf{B})\mathbf{x}' = 0$ is then obtained by discretizing, where $\mathbf{x}'$ is the discretized solution vector including T', Y_F', and Y_O'. The eigenvalues, $\lambda = \lambda_I i + \lambda_R$, of perturbed governing equations play an important role to identify flame instability, because if λ_R is large enough to have an effect to the solution, then the instability can occur. If λ_I is zero, the non- or cellular instability depends on the sign of λ_R. Similarly, if λ_I is nonzero, the growing or decaying oscillatory instability rests on whether λ_R is positive or negative. In addition, the cell number is related by the wavenumber, k_{max}, at which attains its maximum value for a given Da. Therefore, prior to conducting 2-D simulations, the linear stability analysis is conducted to find the critical Damköhler numbers which can distinguish the cellular and oscillatory instabilities.

The largest growth rate, λ_R, as functions of wavenumber, k, and Damköhler number, Da, is as shown in Fig. 2. The graph gradually moves upward with increasing Da, and finally λ_R at $k = 0$ has positive value (Fig. 2c). Two critical Damköhler numbers are obtained: thresholds of cellular instability, $Da_{C,C} = 5.76\times10^7$ (Fig. 2a), and growing oscillatory instability, $Da_{C,O} = 6.17\times10^7$ (Fig. 2b). Therefore, it may be conjectured that the cellular instability occurs if Da is larger than $Da_{C,C}$, and the growing oscillatory instability first appears at $Da_{C,O}$.

The wavenumber, k_{max}, decreases gradually from 28.8 at $Da_{C,C}$ with increasing Da, and λ_R also increases at k_{max}. As mentioned above, k_{max} is related to the cell number. Therefore, it can be expected that the cell number decreases as Da increases. The impact of oscillation also reduces the time for the occurrence of the D-T instability, because the region of positive λ_R increases.

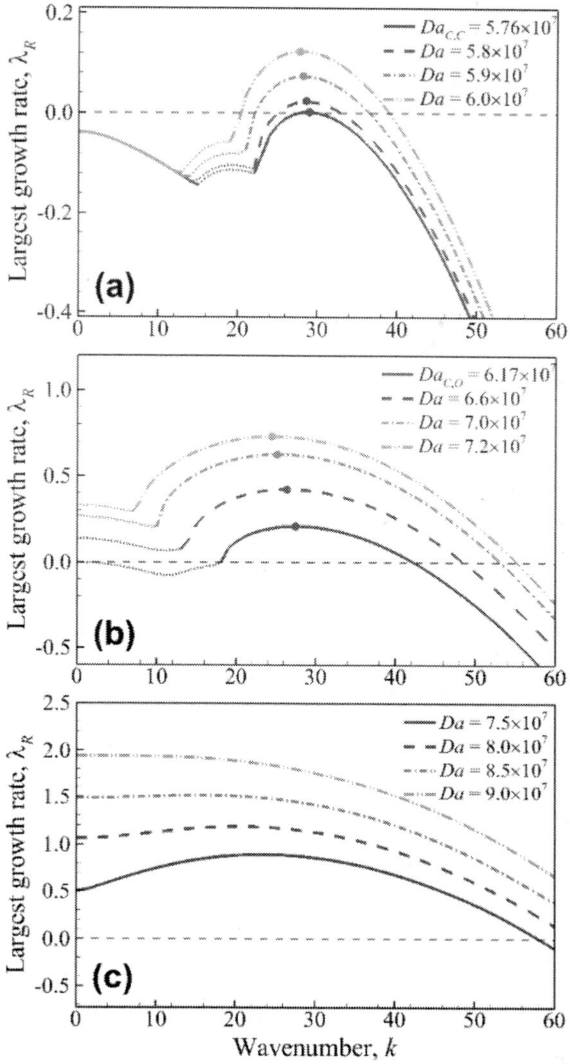

Figure 2: The largest growth rate, λ_R, as functions of wavenumber, k, for different Damköhler numbers, Das. Dotted segments mean the imaginary part of $\lambda \neq 0$.

3.2 Effects of small amplitude disturbance

The fundamental characteristics of nonpremixed tubular flames are obtained using the perturbed initial condition (IC) with a fixed Da. The perturbed IC is initialized by a well-burning tubular flame having small ($\sim O(\varepsilon)$) amplitude disturbance from 1-D solutions. The perturbation amplitude, ε, of 10^{-5} is selected to make a minimized condition.

The variation of the maximum temperature with numerical time is as shown in Fig. 3. The cellular instability can be identified by a sudden rise of temperature. Once the cellular instability occurs, the maximum temperature does not change anymore in time. Note that the oscillatory instability, therefore, can be observed only during the early stage of simulations. The cellular instability can be distinguished by the $Da_{C,C}$ from the linear stability analysis in Fig. 3a. Likewise, it is verified in Fig. 3b that $Da_{C,O}$ divides into decaying and growing oscillatory instability. The oscillatory characteristics such as the amplitude and duration of the oscillation become prominent as the positive region of λ_R increases with increasing Da. As a result, the oscillation promotes the occurrence of the cellular instability. In Fig. 3c, the

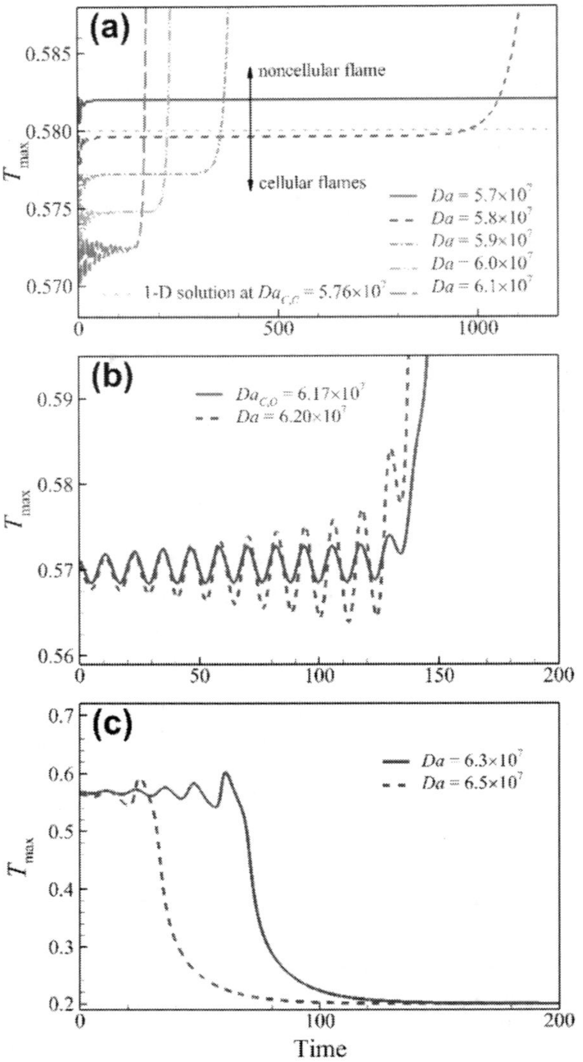

Figure 3: The variation of the maximum temperature, T_{max}, with time: (a) the decaying oscillation, (b) the growing oscillation with cellular instability, and (c) the extinction.

flames of which Da is greater than $Da_{C,O}$ cannot sustain from the growing oscillation, and finally are extinguished although the corresponding 1-D solution exists.

The summary of numerical simulations with linear stability analysis near large radiation-induced extinction Damköhler number, $Da_{E,R}$, is shown in Fig. 4. Depending on the instability characteristics, the vicinity of $Da_{E,R}$ is divided into four regimes by two critical Das: $Da_{C,C}$ and $Da_{C,O}$, and $Da_{E,P}$, which denotes the 2-D extinction Da using the perturbed IC with a fixed Da (=6.28×10^7), as shown in Fig. 4: Regime I. ND ($Da < Da_{C,C}$), Regime II. CD ($Da_{C,C} \leq Da < Da_{C,O}$), Regime III. CG ($Da_{C,O} \leq Da < Da_{E,P}$), and Regime IV. EG ($Da \geq Da_{E,P}$). In first column of table, N, C, and E denote noncellular, cellular, and extinction, respectively. In second column, D and G denote decaying and growing oscillation, respectively. Note that it is reasonable to predict the 2-D simulation outcome wherever the vicinity of $Da_{E,S}$ and $Da_{E,R}$ through a 1-D linear stability analysis with good agreement.

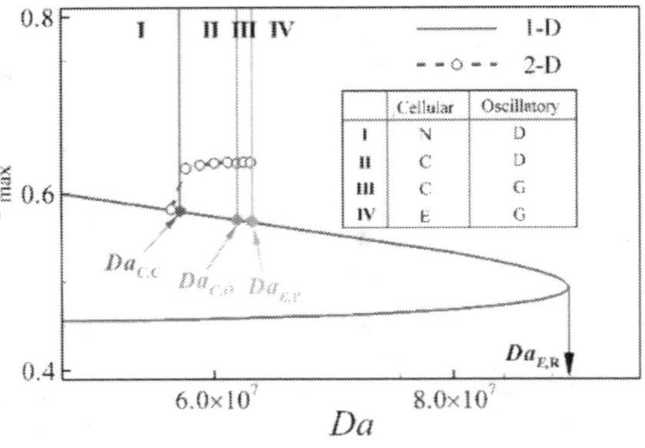

Figure 4: The summary of numerical simulations with linear stability analysis near large radiation-induced extinction Damköhler number, $Da_{E,R}$: **N** (noncellular flame), **C** (Cellular flame), **E** (Extinction), **D** (Decaying), and **G** (Growing)

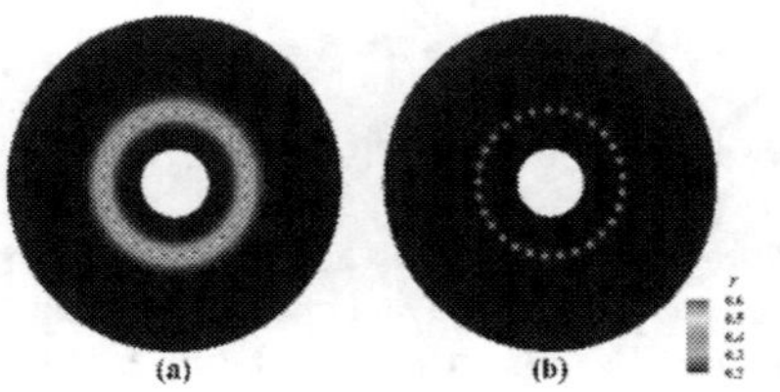

Figure 5: Temperature isocontours using the perturbed IC at (a) $Da = 5.8\times10^7$ and (b) $Da = 1.74\times10^9$.

It is possible to make the cellular flame survive at $Da > Da_{E,R}$, if the Da increases successively with previous solution. To find a global extinction Da, 2-D simulations are conducted using the perturbed IC with increasing Da from $Da_{C,C}$ gradually. Figure 5 shows the temperature isocontours of initial flame at Da 5.80×10^7 and (b) the cellular flame closed to global extinction $Da = 1.74\times10^9$, respectively. Decreasing each cell size as shown in Fig. 5, the maximum temperature does not change until $Da = 10^9$ due to the balance between radiative heat loss and reaction by high Da. However, the flame temperature decreases as Da increases till the global extinction Da, 1.74×10^9.

3.3 Effects of extremely high radiative intensity

Da[a]	k_{max}[b]	N_{cell}[c]	Sign[d]	Oscillation[e]
38000	10.0	E	+	G
38732	10.1	10	+	G
50000	10.8	11	-	D
60000	11.2	11	-	D
70000	11.6	12	-	D
80000	11.8	12	-	D
88600	11.9	12	+	G
90000	11.9	11	+	G
100000	11.8	E	+	G

a. Damköhler number, two extinction Das are 37363 and 112500, respectively.

b. Wavenumber of the maximum largest growth rate.

c. Cell number, E: extinction.

d. Sign of maximum largest growth rate when the imaginary part, λ_I, is not equal to 0.

e. Oscillation characteristics, D: decaying; G: growing.

Table 1: Flame instability characteristics for extremely high radiative intensity, $I = 3.7\times10^{-7}$, in overall Da region.

Generally, the D-T instability only occurs near an extinction limit when the Lewis number is less than unity, because the flame cell formation by fuel focusing effect appears to help that the flame sustains from the harsh environment. However, if the intensity of radiation is extremely high, $I = 3.76\times10^{-7}$ in the present study, the possibility which induce the D-T instability is shown in all domain. The flame instability characteristics for extremely large radiative intensity, $I = 3.76\times10^{-7}$, in the entire range of Da are written in Table 1. Two extinction Das obtained from 1-D simulations are $Da_{E,S} = 37363$ and $Da_{E,R} = 112500$, respectively. The results near $Da_{E,R}$, do not seem to be special. However, near $Da_{E,S}$, the decaying and growing oscillatory instability are observed, and the tendency is similar to the vicinity of a large radiation-induced extinction. Two $Da_{C,O}$ are observed, no $Da_{C,C}$. It is also noticeable that the nonpremixed tubular flames under extremely high radiation always exist with the cellular instability.

4 Conclusions

The flame instability and cell dynamics of nonpremixed tubular flames with radiative heat loss are investigated using 2-D simulations with the linear stability analysis based on 1-D analysis. Two different extinction Das, high stretched-induced and radiation-induced extinction Da, are obtained as various intensities of radiation, I. The perturbed initial condition is used to elucidate the fundamental instability characteristics near $Da_{E,R}$. In conclusion, the vicinity of $Da_{E,R}$ is divided into four regimes as shown in Fig. 4, Regime I. ND ($Da < Da_{C,C}$), Regime II. CD ($Da_{C,C} \leq Da < Da_{C,O}$), Regime III. CG ($Da_{C,O} \leq Da < Da_{E,P}$), and Regime IV. EG ($Da \geq Da_{E,P}$), and these can be predicted by the linear stability analysis with good agreement. The global extinction Da is extended by the cellular instability. If radiative heat loss is relatively too large, $I = 3.76\times10^{-7}$, the radiation effects cannot be ignored near high stretched nonpremixed tubular flames, since the oscillatory instability is observed. In the specific condition, the cellular instability appears in overall Da.

5 Acknowledgment

This work was supported by the Space Core Technology Development Program (No.2015M1A3A3A02027319) and Basic Science Research Program (No.2015R1A2A2A01007378) through the National Research Foundation of Korea grant funded by the Ministry of Science, ICT and Future Planning.

References

[1] M. Short, J. Buckmaster, S. Kochevets, Combust. Flame 125 (2001) 893-905.

[2] S. Hu, P. Wang, R. W. Pitz, Proc. Combust. Inst. 31 (2007) 989-996.

[3] S. Hu, R. W. Pitz, Combust. Flame 156 (2009) 51-61.

[4] S. Hu, R. W. Pitz, Y. Wang, Combust. Flame 156 (2009) 90-98.

[5] S. W. Shopoff, P. Wang, R. W. Pitz, Combust. Flame 158 (2011) 876-884.

[6] H. S. Bak, S. R. Lee, J. H. Chen, C. S. Yoo, Combust. Flame 162 (2015) 4612-4621.

[7] C. A. Hall, R. W. Pitz, Combust. Flame 36 (2017) 1595-1602.

[8] C. H. Sohn, J. S. Kim, S. H. Chung, K. Maruta, Combust. Flame 123 (2000) 95-106.

[9] J. R. Nanduri, C. J. Sung, J. S. T'len, Combust. Theory Model. 9 (2005) 515-548.

11th Asia-Pacific Conference on Combustion,
The University of Sydney, NSW Australia
10th -14th December 2017.

Optimization of Urea Selective Catalytic Reduction Reactor Based on RSM

Wu Yonge[1], Liang Xingyu[1], Wang Yuesen[1], Yu Hanzhengnan[2], Liu Qingling[3], Liu Xikai[1], Cao Xinyi[1]

[1]Affiliation

State Key Laboratory of Engines, Tianjin University, Tianjin, 300072, China

[2]Affiliation

China Automotive Technology and Research Center, Tianjin, 300300, China

[3]Affiliation

Tianjin Key Laboratory of Indoor Air Environmental Quality Control, School of Environmental Science and Engineering, Tianjin University, Tianjin, 300072, China

Abstract

Selective catalytic reduction (SCR) has been the main after-treatment systems to control diesel engine NOx emission in light and heavy duty vehicles. In practical applications, the volume of the SCR system is larger, the initial installation costs are higher, which hinder its widespread use in the real ship. In this paper, a new method of design SCR reactor structure is put forward. In the design process, the performance of the SCR reactor under high, medium and low load is taken into account. In this study, the structural parameters of SCR reactor under high, medium and low loads are respectively optimized by using the Response Surface Methodology (RSM). In this process, the coupling relationship among these structural parameters is considered. The optimal structural parameters of SCR reactor are obtained under different loads. The corresponding optimal structural parameters are verified. The result shows that the errors are within acceptable scope. Then, the optimal structural parameters under each load are brought into other loads to observe the change trend of SCR performance. Through the comparative analysis of the results obtained, the optimum structural parameters of the SCR reactor suitable for the diesel engine are obtained.

1 Introduction

SCR system could convert the nitrogen oxides (NO_x) into nitrogen and water through injecting the urea-water solution (UWS) into the exhaust gas in a certain temperature and the presence of catalyst conditions [1]. In the current research and development of NOx after-treatment methods, SCR is the most mature one, it can effectively reduce NOx emission in reaction environment that flow, temperature and composition of diesel engine exhaust change [2].

The SCR reactor, the substrate of which is a columnar honeycomb porous channel carrier and the catalyst is uniformly coated on the carrier, is the skeleton part of the SCR after-treatment system [3]. There is a contradictory relationship between NOx conversion rate and pressure loss of SCR system. The design requirements of SCR catalyst substrate of diesel engine are: the gas flow resistance is as low as possible in the case where the NOx reduction rate in the exhaust gas is satisfied [4].

Usually, the parameters of catalyst substrate are mainly determined by using a combination of simulation analysis and experiment. Tronconi et al. compared the simulation results with the experimental results of the catalyst substrate and obtained the effect of carrier length and wall thickness on NOx conversion rate [5]. Dhanushkodi et al. carried out simulation research of the catalyst model and analyzed the effect of carrier volume on NOx conversion rate [6]. Wang et al. analyzed the relationship between wall thickness and pressure loss under different coating thickness [7]. For SCR system of marine diesel engines, because of its large volume, doing a large number of experiments to study the impact of its structural parameters on the performance of it to determine these parameters needs huge cost, therefore, numerical simulation plays an increasingly important role.

In addition, design of experiments (DOE) technique has been widely used in internal combustion engine industry. Azael et al. used a full factorial design of experiment and optimization to study multiphase injector modelling for automotive SCR systems. They found that the method applied in this work successfully predicted the performance of SCR system [8]. Chen and Tan carried an optimization study based on Taguchi methods. The use of DOE techniques coupled with CFD tools can help reducing computational costs during stage of an SCR system [9].

In the current application, the larger size of the SCR system is one of the main problems [10]. SCR reactor is one of the main components of the SCR system and it occupies a large proportion of the total volume. Therefore, it is necessary to design an SCR reactor with a compact structure and reliable NOx catalytic performance to reduce the volume of the SCR system and reduce costs. In the previous study, when designing the structure of the SCR reactor, only the effects of the structural parameters of the SCR reactor on the catalyst performance at high load conditions were studied, and the performance of the SCR reactor at medium and low load was not taken into account. Previous studies did not take into account the coupling relationship of the various parameters of SCR reactor, therefore, the structural parameters of the SCR reactor lack in system optimization. In this paper, considering the coupling relationship of each structural parameter, the structure of SCR reactor is systematically optimized to reduce its volume by using RSM under low, medium and high load, which can meanwhile maintain the good performance of the SCR system.

Corresponding author. Fax: +86-022-2789-1285
E-mail address: lxy@tju.edu.cn

2 Methodology

2.1 SCR model

Structural parameters and exhaust parameters are consistent with those in Zhang's work [11].

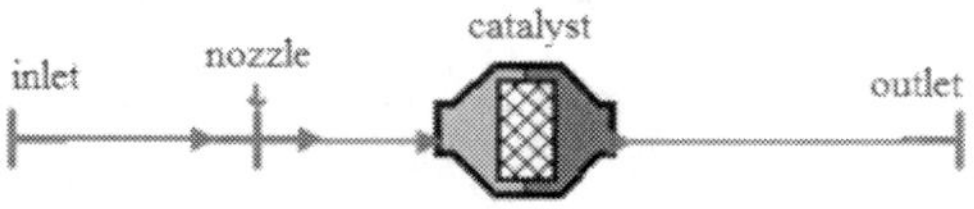

Figure 1: SCR 1D simulation model.

According to the structure parameter, the SCR 1D simulation model is set up using AVL BOOST, as shown Fig. 1. The inlet boundary condition is mainly the mass flow rate. Outlet boundary condition refers to the static pressure. The catalyst structure parameters are set to 1:1. In this study, V_2O_5/TiO_2 catalyst is used and AdBlue is sprayed into exhaust gas.

2.2 Selective catalytic reduction model

SCR reaction mechanism is very complex, the following five total package reactions are used to describe the NH_3 selective catalytic reduction:

$$HCNO+H_2O=NH_3+CO_2 \quad (1)$$

$$4NH_3+4NO+O_2=4N_2+6H_2O \quad (2)$$

$$4NH_3+2NO+2NO_2=4N_2+6H_2O \quad (3)$$

$$8NH_3+6NO_2=7N_2+12H_2O \quad (4)$$

$$4NH_3+3O_2=2N_2+6H_2O \quad (5)$$

The expression of the reaction rate equation of the package reaction can be found in the literature [12]. Kinetic parameters of the SCR reactions are calibrated before optimization.

2.3 Response surface optimization

Before optimizing structure parameters of the SCR reactor, the effects of single structural parameter on NOx conversion rate and pressure drop are studied. The range of structural parameters is determined by this process. Based on this work, the time that is needed during the optimization process is decreased and the accuracy of the optimization results is improved. The parameters scopes of the catalyst are determined as:

A=24617~123088 mm^2; B=100~450 mm; C=140~500 $1/in^2$;

D=0.01~0.15 mm; E=0.10~0.30 mm.

A-Cross-sectional area; B-Length; C-CPSI; D-Coating thickness; E-Substrate wall thickness;

Then, the structural parameters of the SCR reactor under high, medium and low load are optimized respectively by using RSM. Verification of optimum structural parameters of SCR reactor is carried out. The optimum structural parameters of each load are brought into other load, and the change trend of SCR performance is observed.

The optimum structure parameters of SCR reactor suitable for the current diesel engine are obtained by comparing and analyzing the NOx conversion rate and pressure drop of each structure parameter group under each load.

3 Results and Discussion

3.1 Response surface analysis

The RSM is an optimization method based on experimental conditions. It is suitable to solve the problems of nonlinear data. In this paper, it will be used to optimize the structure of SCR reactor. In this progress, emission parameters of 37%, 51%, 71% load of diesel engine are used to represent low, medium and high load. The optimal structural parameters of each load can be obtained by using the RSM. The medium load will be used as an example to illustrate the specific process of optimization.

The experimental design is carried out by using the Box Behnken Design (BBD) experimental design method of 5 factors and 3 levels. The cross-sectional area, length, CPSI, coating thickness and substrate wall thickness are design variables, volume of catalyst, NOx conversion rate and pressure drop are target values. The values of each factor level are respectively listed in Table 1.

Name	Factor	Minimum	Maximum
Cross-sectional area	A	24617	123088
Length	B	100	450
CPSI	C	140	500
Coating thickness	D	0.01	0.15
Substrate wall thickness	E	0.10	0.30

Table 1: level value of each factor of BBD experimental design

Test of each factor and level is carried out by using CFD software. The initial sample number of the five structural parameters is determined. In this study, the initial sample number is 46(46 tests is required.). The NOx conversion rate and pressure loss corresponding to the each initial sample point are obtained and catalyst volume of the corresponding point is calculated. The data for test results are subjected to multiple regression fitting. In this process, the items with little effect on the pressure drop are removed. The regression equation of the pressure drop will be described as an example, it can be described as follow:

Y represents the pressure drop.

$$Sqrt(Y)=6.89302-1.21949E\text{-}004*A+0.059360*B+0.017*C-49.87044*D-24.22172*E-4.10814E\text{-}007*A*B-6.44195E\text{-}007*A*C-7.66752E\text{-}004*A*D-3.82710E\text{-}004*A*E+1.20445E\text{-}004*B*C+0.14486*B*D+0.072294*B*E+0.42563*C*D+0.21201*C*E+212.78952*D*E+2.13468E\text{-}009*A^2-6.62149E\text{-}005B^2$$

1. The size of the P value indicates the significant level of the model and the various factors. P values below 0.05 indicate these factors is significant, namely, these factors has a greater impact on the response value. P values no more than 0.01 suggest that these factors have a highly significant effect on response value. As shown in table 2, p-value of model is $p<0.0001$, which indicates that the model is highly significant. In the case, A, B, C, D, E, AB, AC, AD, AE, BC, CD, CE, A^2, B^2 are significant model terms.

2. The lack of fit indicates the degree of deviation between the fitted regression equation and the test data. As can be seen from Table 2, the lack of fit is not significant, which indicates that the fitted quadratic polynomial model is reasonable.

Variance Source	Sum of squares	df	Mean square	F value	p-value	
Model	6857.75	17	403.40	104.153	<0.0001	signi-ficant
A	2057.13	1	2057.1	531.133	<0.0001	
B	1226.67	1	1226.6	316.715	<0.0001	
C	1616.34	1	1616.3	417.326	<0.0001	
D	616.63	1	616.63	159.209	<0.0001	
E	572.55	1	572.55	147.828	<0.0001	
AB	58.46	1	58.456	15.093	0.0006	
AC	73.34	1	73.336	18.935	0.0002	
AD	27.93	1	27.933	7.212	0.0120	
AE	25.88	1	25.884	6.683	0.0152	
BC	37.78	1	37.777	9.754	0.0041	
BD	14.69	1	14.692	3.793	0.0615	
BE	13.67	1	13.61	3.514	0.0713	
CD	64.71	1	64.714	16.709	0.0003	
CE	59.72	1	59.718	15.419	0.0005	
DE	16.17	1	16.174	4.176	0.0505	
A^2	273.14	1	273.13	70.521	<0.0001	
B^2	57.06	1	57.064	14.733	0.0006	
Residual	108.45	28	3.873			
Lack of Fit	49.31	23	2.144	0.181	0.9982	not signi-ficant
Pure Error	59.14	5	11.828			
Cor Total	6966.19	45				

Table 2: Box Behnken Design - Analysis of variance of pressure drop

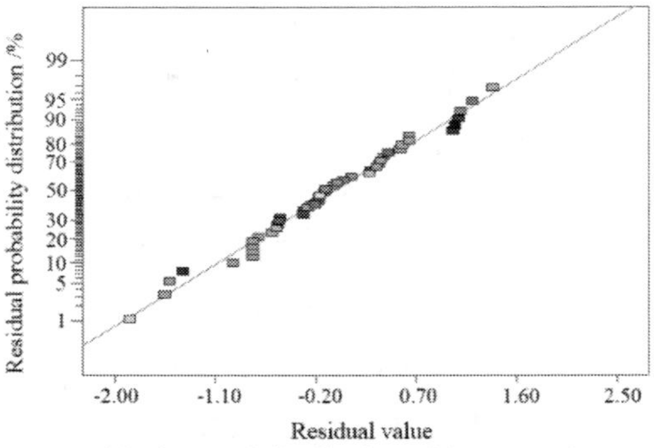

Figure 2: Residual normal distribution of the regression model of pressure drop

Residual is the difference between the simulated test value and the predicted value of the regression model. In the absence of experimental anomalies, the residuals should be consistent with the normal distribution. The points in the Fig. 2 distribute around a straight line or on the line, which indicating that the residuals of this test are normal distribution and the model of the response surface is accurate enough.

Figure 3 shows the comparison between the predicted value of the response surface model and the simulated test value for the pressure drop. The more points that fall on a straight line or around a straight line, the more reliable the regression model is. As can be seen from Fig. 3, this test model can be used to predict the response value.

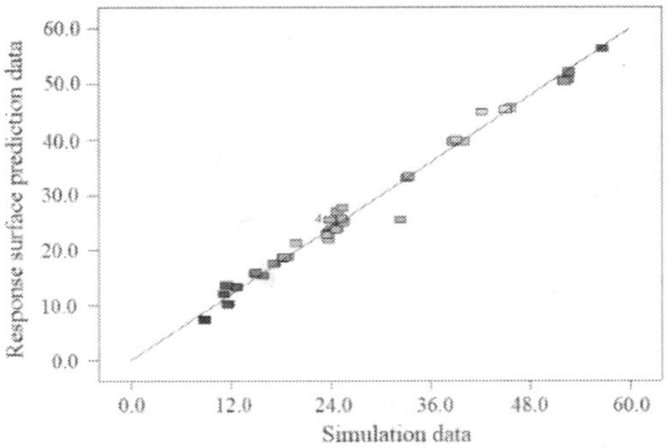

Figure 3: Comparison between the predicted value of the response surface model and the simulated value of the pressure drop

$$\begin{cases} \min V\,(A,B,C,D,E) \\ \min\ \Delta p\,(A,B,C,D,E) \\ \max \eta_{NOx}\,(A,B,C,D,E) \\ \text{S.t.} \\ 24617<A<123088 \\ 100<B<450 \\ 140<C<500 \\ 0.01<D<0.15 \\ 0.10<E<0.30 \end{cases} \quad (6)$$

η_{NOx} -NOx conversion rate. ΔP-Pressure drop of catalyst.

Load/%	37		51		71	
Number	L1	L2	M1	M2	H1	H2
A/dm^2	8.21	6.65	12.31	12.07	12.31	11.54
B/mm	173.3	203.04	111.7	111.1	110.0	110.0
$C/1/in^2$	500	500	200	228	200	200
D/mm	0.02	0.01	0.12	0.12	0.12	0.11
E/mm	0.17	0.10	0.18	0.10	0.18	0.17
V_p/L	14.07	13.18	13.58	13.34	13.40	12.80
$\eta_{NOx,p}/\%$	33.07	32.71	97.28	96.93	94.98	94.86
$\triangle P_p/Pa$	224.00	223.71	107.37	108.79	165.52	151.08
V_s/L	14.22	13.51	13.75	13.42	13.54	12.70
$\eta_{NOx,s}/\%$	31.13	31.81	94.05	94.21	94.39	94.39
$\triangle P_s/Pa$	244.26	243.74	104.92	103.10	162.81	159.25
error of V/%	1.09	2.43	1.29	0.59	1.051	-0.82
error of $\eta_{NOx}/\%$	-6.23	-2.82	-3.44	-2.893	-0.63	-0.50
error of $\triangle P/\%$	8.30	8.22	-2.34	-5.52	-1.66	5.13
Change rate of V/%	14.42	18.72	17.25	19.28	18.54	23.60

Subscript P represents the predicted value. Subscript S represents the simulated value. Error indicates the deviation between the simulated value and the predicted value. Change rate of V represents percentage of volume reduction compared to before optimization. - represents that the predicted value is higher than the simulated value.

Table 3: The optimum structural parameters of the SCR reactor under each load

The objective of catalyst structure optimization is: higher NOx conversion rate, lower pressure loss, smaller volume. The multi-objective optimization problem, the catalyst volume, pressure drop and NOx conversion rate as the target value and the structural parameter as constraint, is established, it can be expressed as (6). The optimum structural parameters of the catalyst are obtained by solving the polynomial under each load.

After solving, the optimal solution of each load is displayed in Table 3.

The CFD software is used to validate the optimum structural parameters and predicted response values. The comparison between the simulated value and predicted value and calculation of the relative error between the two values are carried out. The results are shown in Table 3. The biggest error is 8.30%, but still within the acceptable range. In summary, the results obtained are more accurate. The optimization of the catalyst structure parameters is effective.

3.2 Comparison of optimal structure parameters groups

Load/%	37		51		71	
Number	L1	L2	L1	L2	L1	L2
η_{NOx}/%	31.13	31.81	94.40	94.37	94.29	94.32
△P/Pa	244.26	243.74	476.8	475.8	748.9	747.39
Change rate of NOx/%	-21.35	-24.00	-0.03	-0.02	0.07	0.04
Change rate of △P/%	24.60	24.76	25.63	25.78	25.66	25.81

- indicates that the SCR performance parameters are higher than those before optimization when using the optimal structural parameters.

Table 4: Using the optimum structural parameters of the low load, the SCR performance under high and medium load

Load	37		51		71	
Number	M1	M2	M1	M2	M1	M2
η_{NOx}/%	23.04	24.08	94.05	94.21	93.39	94.39
△P/Pa	53.82	52.90	104.92	103.10	165.3	162.52
Change rate of NOx/%	10.18	6.14	0.32	0.15	-0.04	-0.04
Change rate of △P/%	83.39	83.67	83.64	83.92	83.59	83.87

Table 5: Using the optimum structural parameters of the medium load, the SCR performance under high and low load

Load/%	37		51		71	
Number	H1	H2	H1	H2	H1	H2
η_{NOx}/%	22.86	22.37	94.00	93.83	94.39	94.39
△P/Pa	52.99	51.83	103.29	101.0	162.81	159.25
Change rate of η_{NOx} /%	10.90	12.83	0.37	0.55	-0.03	-0.04
Change rate of △P/%	83.68	84.0	83.89	84.25	83.84	84.19

Table 6: Using the optimum structural parameters of high load, the SCR performance under medium and low load

By analyzing Table 4, 5, 6, it can be seen that the through using the optimum structural parameters of low load, pressure drop can be reduced under each load, at the same time, NOx conversion rate can be improved under low load. When using the optimum structural parameters of medium and high load, only can the NOx conversion rate of high load be improved lightly, while the pressure loss under each load can be significantly reduced.

From the above, the combinations of SCR structure parameters can significantly reduce the volume of SCR reactor. In addition, it can be seen that M2 is the best combination of structural parameters, because it has the least impact on NOx conversion rate under low loads, but also pressure loss caused by the catalyst can be reduced under each load.

4 Conclusions

1. The kinetic parameters of SCR catalyst are calibrated and the NOx conversion rate after calibration is closer to the experimental value. Single parameter study is carried out to shrink the scope of each structure parameter.
2. RSM is used to optimize the structure of SCR reactor under high, medium and low load. The optimized structural parameters of each load are obtained by solving polynomials. At the same time, these parameters are validated by using one-dimensional model, the maximum deviation is 8.30%, but still within acceptable limits.
3. These combinations of the best structural parameters can reduce the size of the SCR reactor. As can be seen from the simulation results, the optimized structural parameters are as follows: cross-sectional area is 120700 mm^2; length is 111.1 mm; CPSI is 228; coating thickness is 0.12 mm; substrate wall thickness is 0.10 mm.

5 Acknowledgment

This work was supported by National Natural Science Foundation of China (No. 51376136 and No. 91641111) and National Sci-Tech Support of China (2016YFC0205304). The author also wants to appreciate the support of high-tech Ship Research Program of MIIT.

References

[1] C.B. Zhu, B.S. Jin, F. Li, et al, Proceedings of the CSEE 27 (29) (2007) 45-50.

[2] J.K. Gong, F.J. Zhang, J.Q. E, et al, Chinese Journal of environment Engineering 6 (12) (2012) 4573-4578.

[3] J.S. Howitt, T.C. Sekella, SAE International, 1974.

[4] G.Y. Zheng, K. Adam, G. Michael, et al, SAE International, 2012.

[5] T. Enrico, B. Alessandra, Catalysis Today 52 (23) (1999) 249-258.

[6] S.R. Dhanushkodi, N. Mahinpey, M. Wilson, Process Safety and Enviroment Protection 86 (4) (2008) 303-309.

[7] Q. Wang, J. Luo, J. Du, Tractor & Farm Transporter 37 (3) (2010) 100-106.

[8] C. Azael, I. Fernando, Computers and Mathematics with Applications 74 (2017) 188-200.

[9] Chyi-Tsong Chen, W.L. Tan, Journal of the Taiwan Insitute of Chemical Engineers 43(2012)409-419.

[10] G.Y. Zheng, G. Timothy, K. Adam, et al, SAE International, 2011.

[11] WJ Zhang, SJ Shuai, HY Dong, JX Wang, Transactions of Csice 25 (5) (2007) 433-438.

[12] JN Chi, HFM Dacosta, Sae Transactons114(4)(2005)449-464.

Numerical Study of Supersonic Flame Characteristics with Elevated Fuel Temperature Conditions at Fixed Mass Flow Rate Condition

Won-Sub Hwang, Seung-Min Jeong, Jeong-Yeol Choi

Pusan National University

Busan, 46241, Republic of Korea

Abstract

A diagnostic investigation regarding the turbulent combustion is analyzed for supersonic co-axial flame. The Mathematical formulation is based on the hybrid RANS/LES formulation for compressible reacting flows. A validation case study is performed for an experimental case with three levels of different grid system with three different high resolution schemes namely 3rd-order MUSCL scheme, a 5th-order WENO scheme and multi-dimensional 5th-order oMLP scheme. The time-averaged results of the experimental case shows the importance of fine grid resolution and high-order accurate numerical scheme for reliable prediction, capturing the fine scale instabilities of the supersonic turbulent combustion. The oMLP scheme exhibits a remarkable performance without a big expense. An analysis of the flow field is carried out to investigate the supersonic turbulent flame structure. The comparison of the combustion parameters including OH mass fraction, scalar dissipation rate and flame index reveals that the supersonic combustion of the validation case has a characteristics of turbulent lifted flame where the combustion is held mostly at premixed mode following turbulence mixing at the shear layer.

1 Introduction

Liquid rocket and scramjet engines are working at high load conditions where the flow becomes turbulent inevitably by the high Reynolds number flow caused by the high mass flow rate and high speed flows. In the meanwhile, the fuel is delivered at high temperature in the regenerative cooling engines or at the prebunred condition in the staged combustion rocket engines. Almost all the liquid rocket engines and the scramjet engines of X-51A employ the regenerative cooling, while many high pressure liquid rocket engines use staged combustion cycle and HyFly demonstration program uses the dual combustion ramjet (DCR) engine. Previously, the possibility of supersonic combustion and the scramjet engine was suspected, but the realization during last decade was greatly helped by the application of high temperature fuel.

However, the effect of the fuel temperature on the supersonic combustion was not studied in detail. It is readily understood that the flame speed is increasing and the combustion stability is enhanced since the reaction rate is the exponential function of temperature. However, a little number of studies are found for the quantitative investigation on the effect of fuel temperature on the turbulent flame structures and stability. In the present study, the effect of fuel temperature on the structure of turbulent supersonic combustion was investigated by LES simulation with detailed chemistry based on the previous study[1] on the experimental case by Evans and Schexnayder[2], where the supersonic combustion of hydrogen with co-flowing vitiated air.

Corresponding author. Fax: +82-51-513-3760
E-mail address: aerochoi@pusan.ac.kr

2 Physical Modeling and Numerical Approaches

The supersonic combustion has the traditional characteristics of turbulent lifted flame where the fuel is burned in the mixing layer with co-flowing supersonic air. The coupled form of species, momentum, energy conservation equations and turbulent transports equations are used as governing equations. Menter's DES (detached eddy simulation) formulation is used to capture the eddy motion in the separated combustion field. Since the air temperature in scramjet is quite high, a high temperature hydrogen combustion model is used, which consists of six reacting species (O, O2, H, H2, OH, H2O) and nitrogen (N2) assumed as inert. The model is subset of the combustion model developed by Singh and Jachimowski[3]. The governing equations were solved in fully coupled manner using fully implicit formulation. The computational code has been used for a supersonic combustor studies and further developed with the oMLP scheme and the modified AUSMPW+ scheme for convective fluxes. Viscous fluxes are formulated by 4th order central difference scheme. The second order implicit time integration is used with sub-iterations for time accurate computation. The code is parallelized by OpenMP for the optimum performance in multi-core SMP (shared memory processors) machines. Further details on the numerical procedure and validation is include in the previous study [1].them.

3 Experimental conditions and Computational Domain

Figure 1 is the schematics of reference experimental condition studied in reference [2]. A comprehensive grid independency study has been studied with 4 levels of different grid for the experimental grid with different high resolution scheme. Fig2. is the smallest grid constructed for the computational domain for the understanding of grid schemes. The grid is clustered to the solid walls and shear layer region. No slip adiabatic condition was assumed for the solid boundaries, while outer boundary is assumes as slip wall for the computational efficiency. The grid independency study exhibits that the combination of AUSMPW+ flux splitting and oMLP scheme has the best resolution, while the time averaged solution is

converging to the experimental value with very fine grid of 1,921×401 grid of main combustor. The combustion efficiency is shown to be a good indicator of grid convergence which converging with the finest grid. Further details are included in the previous paper.[1]

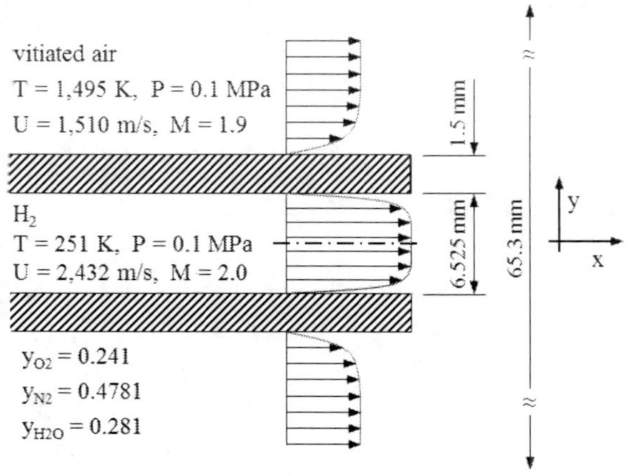

Figure 1: Geometry of the supersonic nozzle and H_2 injector

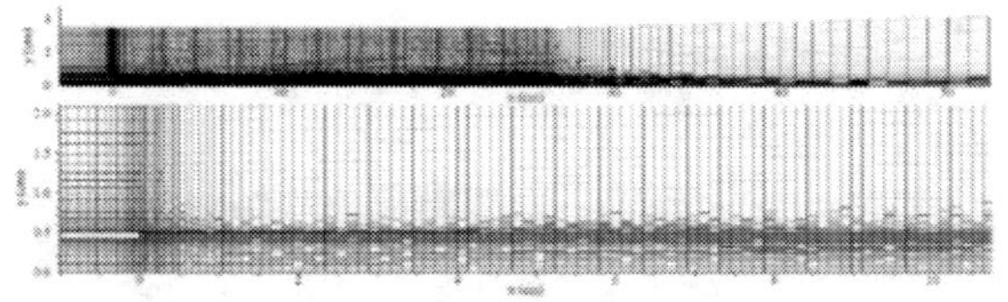

Figure 2: Smallest grid for the grid convergence study. Entire computational domain (upper) and magnified plot around the injector (lower) [1]

4 Conditions of Numerical Experiments

The primary purpose of present study is the investigation of the effect of fuel temperature on the supersonic combustion characteristics, but other flow variables are affected by the nature of compressible flow when the temperature is changed. Therefore, additional constraints should be enforced for the fair comparison. In the present study fuel temperature is changed from 300 to 1,500 K and, while the mass flow rate and flow speed was fixed to the experimental condition. As the fuel temperature, fuel pressure increases and Mach number decays from the reference conditions, as listed in Table 1 for selected cases.

5 Conditions of Numerical Experiments

Instantaneous and time-averaged temperature distributions are plotted in Fig. 3 and 4. As the fuel temperature increase, it is found that the regime changes to under-expansion condition since incoming pressure increases while the incoming Mach number decrease. Combustion is held upon the unsteady supersonic turbulent mixing layer and the enhancement of combustion efficiency is easily noticed with short flame length at high temperature condition. As the fuel temperature increases while the flow speed kept constant, the local Reynolds number is getting reduced resulting larger eddy structures. Also, the shock strength is getting higher by the pressure difference between the fuel and air sides is getting higher. As a result, shock/flame interaction is getting higher and closer to the injector, and the flame is stabilized further downstream after the interaction region.

Table 1: Fuel supply conditions

T (K)	P (MPa)	Mach No.
300	0.119522	1.82939
600	0.239044	1.29357
900	0.358566	1.05620
1,200	0.478088	0.91469
1,500	0.597610	0.81813

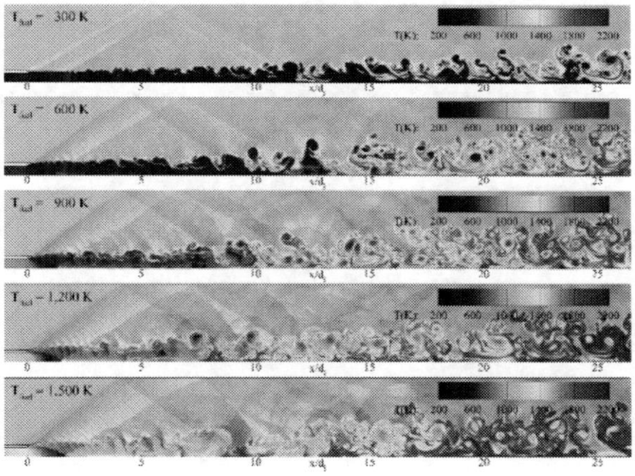

Figure 3: Instantaneous temperature contours

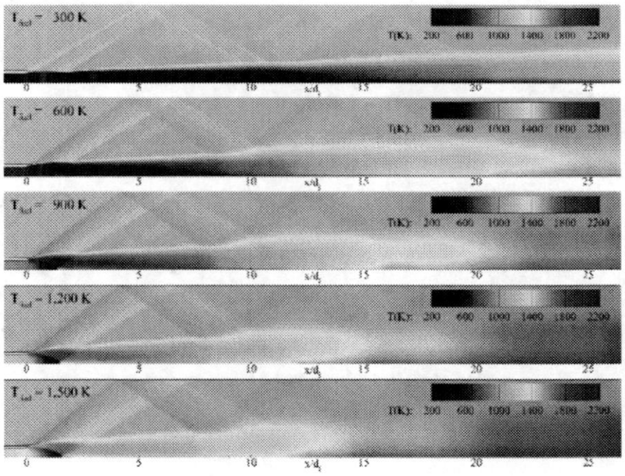

Figure 4: Time-averaged temperature contours

Time averaged distributions of scalar dissipation rate and OH mass fraction are plotted in Fig. 5 and 6. Flame has the characteristics of lifted flame around the 10 cm downstream of nozzle exit where the mixing layer is interacting with shock wave. As the fuel temperature increase, the scalar dissipation rate is maintained at low level and wide distribution those enhancing the combustion efficiency.

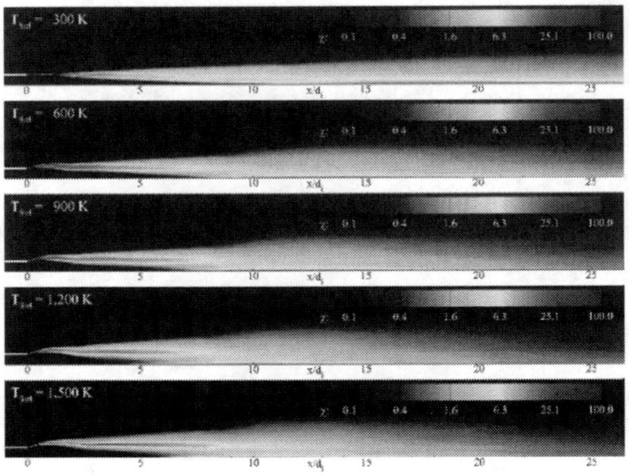

Figure 5: Time-averaged scalar dissipation rate contours

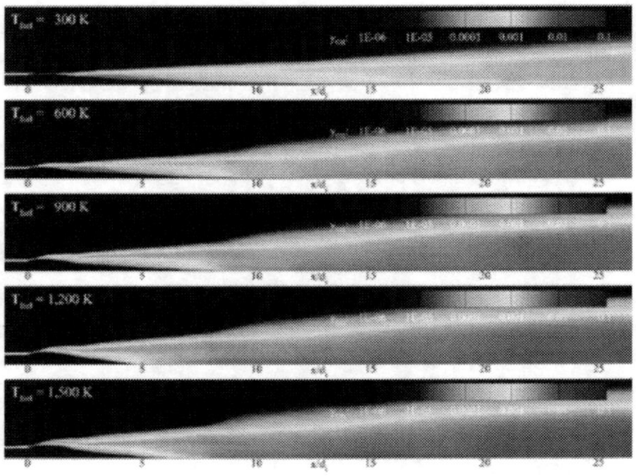

Figure 6: Time-averaged OH mass fraction contours

Fig. 7 and 8 are the time averaged mixture fraction and H2O mass fraction distributions. It is clearly found that there is a transition region between fuel injection and flame stabilization which is getting reduced and closer to the injector lip, but is kept almost constant length around 4 cm and location 10 downsteam of injector lip above 1,000 K. It is considered that combustion efficiency is getting higher as the fuel temeprature is getting higher.

This results implies that the flame characteristics are changed to mixing dominated flame while the supersonic flame shown to have a combined natures of premixed and non-premixed flame at low fuel temperature conditions in the previous study [1]. It is readily understood from the fact that the reaction rate and flame speed is getting higher as fuel temperature increases.

It is expected that the further analyses with flame index and scattered plots of the scalar dissipation rate could show the regime changes from the partially premixed flame to completely mixing dominated flame.

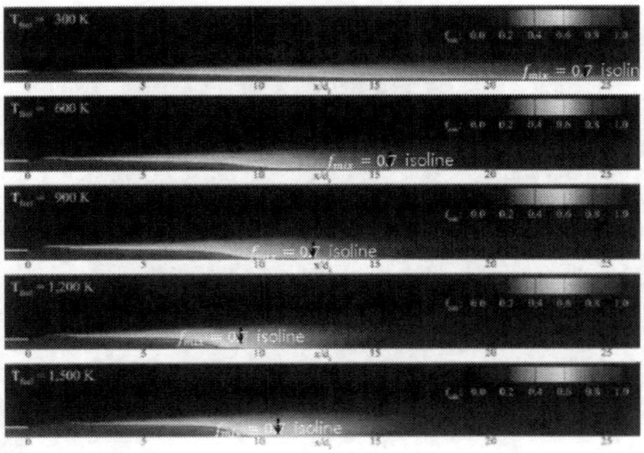

Figure 7: Time-averaged mixture fraction contours

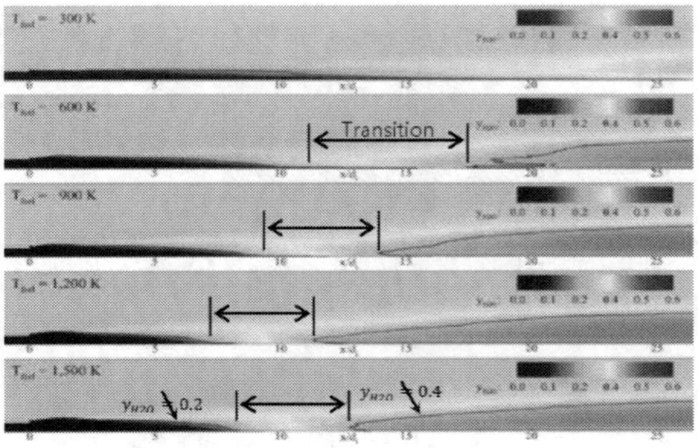

Figure 8: Time-averaged H_2O mass fraction contours

6 Conclusions

A numerical experiment has been carried out to investigate the effect of fuel temperature on the characteristics of supersonic turbulent flame. Additionally, constant mass flow rate of fuel is assumed for fair comparison for the compressible flows, since the temperature effect cannot be isolated, but the other flow properties could be influenced by the change of temperature. As a result, the increase in fuel pressure and the decrease in fuel Mach number is introduced to the flow those further enhances the shock/flame interactions those has a positive effect on the flame stabilization.

Overall effect of fuel temperature could be summarized as regime changes from partially premixed flame to mixing dominated flame by the enhancement of reaction rate and flame speed. That is, fuel and air is burnt as they mixes at high temperature conditions, while the partially premixed nature is preserved in case the fuel temperature is low. Further analysis of the flame index could fortify this conclusion.

7 Acknowledgment

This research work is supported by the Space Core Technology Development Program (2017M1A3A3A03015993 and 2013M1A3A3A02042430) and Space Propulsion Research Center(SPRC) of Advanced Research Center Program (NRF-2013R1A5A1073861) of National Research Foundation(NRF) of Korea, funded by Ministry of Science, ICT and Future

Planning (MSIP) of the Republic of Korea Government.

References

[1] Choi, J.-Y., Han, S.-H, Kim, K.H., Yang, V. "High Resolution Numerical Study on the Coaxial Supersonic Turbulent Flame Structures," AIAA 2014-3745, 50th AIAA/ASME/SAE/ASEE Joint Propulsion Conference, July 28-30, 2014, Cleveland, OH, U.S.A.

[2] Evans, J. S., and Schexnayder, C. J., "Influence of Chemical Kinetics and Unmixedness on Burning in Supersonic Hydrogen Flames," AIAA Journal, Vol. 18, Feb. 1980, pp. 188-193.

[3] Singh, D.J. and Jachimowski, C.J., "Quasigolbal Reaction Model for Ethylene Combustion", AIAA Journal, Vol. 32, No.1, pp213-216.

[4] Huh, H. and Driscoll, J.F., "Shock-wave-enhancement of the mixing and the stability limits of supersonic hydrogen-air jet flames." Proceedings of the Combustion Institute. Vol. 26. No. 2. Elsevier, 1996, pp.2933-2939 .

11th Asia-Pacific Conference on Combustion,
The University of Sydney, NSW Australia
10th -14th December 2017.

Direct Estimation of Edge Flame Speeds from Liftoff Heights of n-Butane and DME

Min-Kyu Jeon and Nam Il Kim

Department of Mechanical Engineering, KAIST,

291 Daehak-ro, Yuseong-gu, Daejeon, 34141, Republic of Korea

Abstract

It is well known that the edge flame speed (EFS) plays important roles on the tribrachial flame stabilization in a mixing layer. There has been a lot of effort to clarify the relation between EFS and the fuel concentration gradient (FCG). Recently, a simple method that could estimate EFS/FCG relationship was developed. Using an ordinary laminar jet flow, lifted flames could be stabilized when Schmidt number was larger than unity. Then, the experimental results of liftoff heights and flow rates could be converted to the EFS/FCG relationship. In this study, the same method was applied to investigate EFS/FCG relationships of n-butane and DME flames. In the process, effective Schmidt number was obtained considering its variation with liftoff height and jet velocity. Eventually, with physical properties of each fuels, EFS/FCG of n-butane and DME flame were directly estimated and compared with that of propane flame. From this study, the direct estimation method for EFS/FCG for n-butane and DME was confirmed.

1 Introduction

Stabilization mechanism of lifted flames has been explained with two similarity solutions of laminar jet flow of a cylindrical round jet [1]. It is known that a lifted flame can be stabilized when Schmidt number is larger than unity ($Sc > 1$) and the edge flame speed (EFS, S^{*}_{edge}) is balanced with the local flow velocity at the flame position. In addition, EFS is known to be varied mainly by the fuel concentration gradient (FCG).

In order to measure EFS, several methods have been suggested. For example, EFS/FCG relationship of methane ($Sc < 1$) was obtained by Ko and Chung [2]. With a high-speed camera, unsteady flame propagation to the tube tip was recorded after ignition at the downstream with a high power laser. The relative velocity of flame to the flow was defined as EFS, and FCG was estimated from the flame scale. After then, the same or similar methods were used to measure EFS/FCG relationship for propane (C_3H_8) flame [3].

Recently, Jeon and Kim [4] developed a simple method to estimate EFS from stable lifted flames of propane just using liftoff heights and jet velocity. During this process, an effective Schmidt number was suggested. In this study, direct method to measure EFS will be adapted to estimate different fuels of normal butane (n-butane, n-C_4H_{10}) and DME with diverse tube diameters. Conclusively, the theory of a lifted flame in a laminar jet will be improved.

2 Experiments

2.1 Experimental Method

Experimental apparatus is shown in Fig. 1. Flow rates of propane, n-butane, and DME were calibrated with a soap-bubble flow meter (Sigma-Aldrich), and they were regulated with mass flow controllers (MFCs, Linetech). The fuel tube was made of stainless steel, and its inner diameter was 0.317 mm. Dehumidified air (< 5% relative humidity) was flowed around the fuel jet. The amount of oxidant from the co-flow air was sufficient while the average velocity was sufficiently small (less than 0.12 m/s) compared to the fuel jet velocity (larger than 4 m/s). An acryl tube (inner diameter = 163 mm and length =1.0 m) was installed at around the co-flow to suppress the disturbance by the surround air. Flame liftoff height was measured with a height gauge equipped with a CCD camera.

2.2 Effective Schmidt number

Lee and Chung [5] suggested an important analytic equation related to the lifted flame. The relationship between bulk average flow velocity, $\overline{u}_e$, and liftoff height, x^* , could be presented as Eq. (1a). To investigate a constant Schmidt number, Sc^0 , the second term in the right hand side of C^* of Eq. (1a) was regarded as constant in previous [5], but this research treated its original form as variables dependent on other terms of Eq. (1b).

$$\ln(x^*) = \frac{2Sc-1}{Sc-1}\ln(\overline{u}_e) + \ln C^* \tag{1a}$$

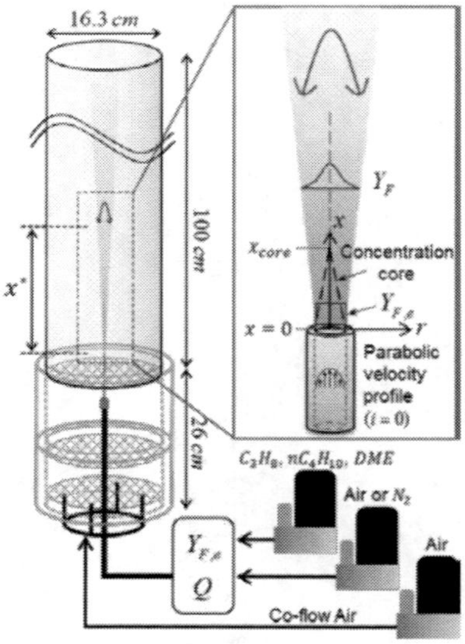

Fig. 1. Experimental apparatus

Corresponding author. Fax: +82-42-350-3210
E-mail address: nikim@kaist.ac.kr

$$\ln x^* = \frac{2Sc^*_{eff}-1}{Sc^*_{eff}-1}\ln \bar{u}_e + \ln\left(\frac{(2Sc^*_{eff}+1)R^2 Y_{F,e}}{8Sc^*_{eff}\mathcal{D}_{FA}Y_{F,st}}\left[\frac{12}{(3+i)(2Sc^*_{eff}+1)S^*_{edge}}\frac{Y_{F,st}}{Y_{F,e}}\right]^{\frac{Sc^*_{eff}}{Sc^*_{eff}-1}}\right) \quad (1b)$$

Recently, Jeon and Kim [4] asserted that Eq. (1a) could not be applied on the EFS/FCG relationship. It is because the EFS will become constant if the second term is not considered. Conclusively, a direct estimation of edge flame speed could be possible. For that, an effective Schmidt number, Sc^*_{eff}, was employed with the assumption that the binary diffusivity between fuel and air was almost constant. By substituting $x^+ \equiv \ln x^*$ and $u^+ \equiv \ln \bar{u}_e$, Eq. (1a) could be replaced as

$$x^+ = f(Sc^*)u^+ + c^+(Sc^*, u^+, x^+)\,. \quad (2)$$

Then, the Schmidt number Sc^* was evaluated though an iteration process. Thus, Sc^* was decomposed into previous term, $\widetilde{Sc}$, and small perturbation term, Sc', like as $Sc^* = \widetilde{Sc} + Sc'$. Then, Eq. (2) could be simplified as

$$\frac{\partial x^+}{\partial u^+} \approx f\left(\widetilde{Sc}\right) + \frac{df\left(\widetilde{Sc}\right)}{d\widetilde{Sc}}Sc' + \frac{\partial c^+\left(\widetilde{Sc}; u^+, x^+\right)}{\partial u^+}\,. \quad (3)$$

$\partial x^+ / \partial u^+$ could be obtained from the best-fit function of experimental results for x^+ and u^+. A polynomial function of u^+ in terms of x^+ was selected. The initial estimation of $\widetilde{Sc}_0$ was obtained as $\widetilde{Sc}_0 = \left(1-[du^+/dx^+]_{fit}\right)/\left(1-2[du^+/dx^+]_{fit}\right)$. With $\widetilde{Sc}_0$, the Starting from the initial value of $\widetilde{Sc}_0$, the first perturbation term Sc'_1 was calculated using Eq. (4)

$$Sc'_n = \left(\widetilde{Sc}_{n-1}-1\right)^2\left[f\left(\widetilde{Sc}_{n-1}\right) - f\left(\widetilde{Sc}_0\right) + \frac{\partial h\left(\widetilde{Sc}_{n-1}, u^+\right)}{\partial u^+}\right] \quad (4)$$

Then, a final solution was derived ($Sc^*_{eff} = \lim_{n\to\infty}\widetilde{Sc}_n$) when $\lim_{n\to\infty} Sc'_n = 0$. Detailed process is given in [4].

3 Result and Discussion

3.1 Experimental Results

Through experiments, the relations between liftoff height, x^*, and the jet velocity, $\bar{u}_e$, could be obtained for stable lifted flames. The results lifted flames of n-butane and DME were measured with the fuel tube ($d_i = 0.317$) as shown in Figs. 2a and 2b, respectively. Volume fraction of fuel, X_F, in the fuel stream from the tube was varied with air premixing. The stable lifted flames could be obtained when X_F were in the range of 0.7 to 1.0 for n-butane, and the flow velocity were about from 6 m/s to 13 m/s. This range of jet velocity was similar to those of propane lifted flames, but the slope of x^* to the variation of $\bar{u}_e$ was steeper than that of propane.

On the other hand, Fig. 2b shows the results of DME lifted flames. The range of jet velocities for stable lifted flames was narrower than those of propane and n-butane. One interesting result is that a distinctive variation in the liftoff height was

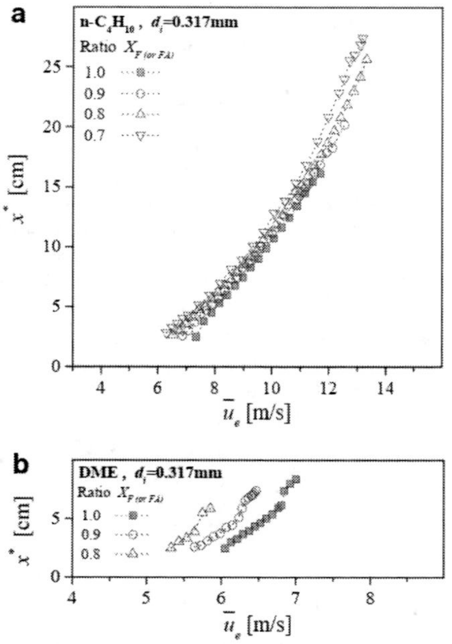

Fig. 2. Liftoff height with jet velocity; (a) n-butane and (b) DME at the tube (d_i= 0.317 mm)

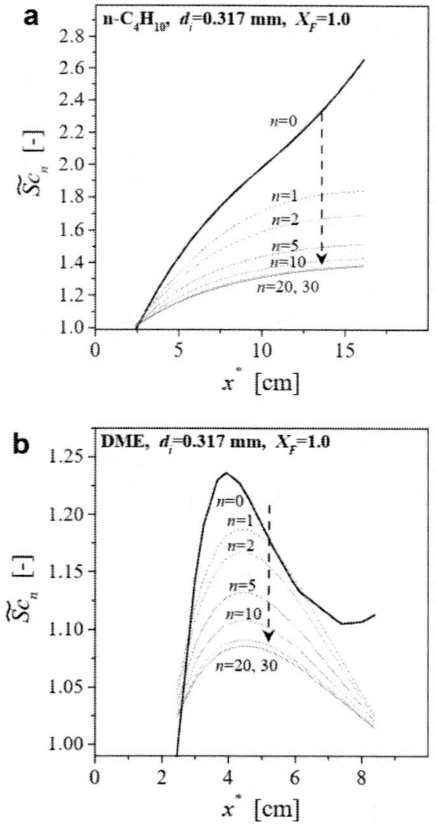

Fig. 3. Effective Schmidt numbers of (a) n-butane and (b) DME at $d_i = 0.317$ mm, X_F=1.0

observed. That is to say, the liftoff height of DME flame suddenly increased at certain jet velocities, and that critical velocity slowly moved downstream (or higher) with the increase of the flow rate. Three effects are suspected. One is the buoyancy effect, another is vortex shedding, and the other is flame merging effect. Related to this issue, a continuous study will be conducted.

3.2 Effective Schmidt number

With an assumption of linear relationship, average Schmidt numbers (Sc^o) regarding n-butane and DME could be obtained. The average values of 1.59 for n-butane and 1.17 for DME in this study were very close to 1.61 for n-butane by Chung and Lee [1] and 1.15 for DME by Xue and Ju [6], respectively. Therefore, the experimental results in this study can be said to be similar to the previous results.

However, the difference of this study is that an effective Schmidt number, Sc^*_{eff} , is used instead of constant one. The effective Schmidt number was first developed in the previous study by Jeon and Kim [4]. Through an iterative process, an eventual Sc^*_{eff} that can satisfy the original solution of Eq. (1a) could be obtained. For that, an initial estimation was conducted using the experimental results of $\ln x^*$ and $\ln \overline{u}_e$ by evaluating 4th-order polynomial best-fit functions. After that, using a perturbation method, Sc^*_{eff} was obtained. The variation of Schmidt numbers through the iteration process are shown in Figs. 3a and 3b for pure n-butane and DME, respectively.

3.3 Edge Flame Speed

To estimate EFS/FCG relationship [4], some physical properties are necessary. They are stoichiometric mass fraction, $Y_{F,st}$, molar weight, MW , binary diffusivity, $\mathcal{D}_{FA}$, and kinematic viscosity of fuel, ν_F . They are summarized in Table 1. Here, ν_F were obtained using some references [7, 8], and $\mathcal{D}_{FA}$ between fuel and air at a given condition was calculated using Chapman-Enskog equation [7, 9]. The Schmidt number of fuel, Sc_F , was compared to estimate the growth of the concentration core. Using these values, the relationship of EFS/FCG were estimated as shown in Figs. 5a and 5b for n-butane and DME, respectively.

Table 1. Properties of propane, n-butane, and DME

At 300 K	$Y_{F,st}$	MW [g/mol]	$\mathcal{D}_{FA}$ [cm²/s]	$\nu_F \times 10^{-6}$ [m²/s]	Sc_F $(\nu_F / \mathcal{D}_{FA})$
C_3H_8	0.0601	44.096	0.1136	4.609	0.4057
$n\text{-}C_4H_{10}$	0.0609	58.123	0.1028	3.208	0.3106
DME	0.1002	46.069	0.1289	4.299	0.3334

With the similarity solution, EFS and FCG could be defined as

$$S^*_{edge} = \frac{3}{2(3+i)}\left(\frac{8}{2Sc+1}\frac{Y_{F,st}}{Y_{F,e}}\right)^{\frac{1}{Sc}}\left(\frac{R^2}{Sc\mathcal{D}_{FA}x^*}\right)^{1-\frac{1}{Sc}}\left(\overline{u}_e\right)^{2-\frac{1}{Sc}} \quad (5)$$

$$\left|\frac{\partial Y_F}{\partial r}\right|^* = \left|-\frac{\partial Y_F}{\partial \xi}\frac{\partial \xi}{\partial r}\right|^* = \frac{3Y_{F,e}Sc^*_{eff}\left(2Sc^*_{eff}+1\right)}{32(3+i)}\left(\frac{\overline{u}_e R}{Sc^*_{eff}\mathcal{D}_{FA}}\right)^3\left(\frac{Rr^*}{x^{*3}}\right)\left(1+\frac{\xi^{*2}}{4}\right)^{-1-2Sc^*_{eff}} . \quad (6)$$

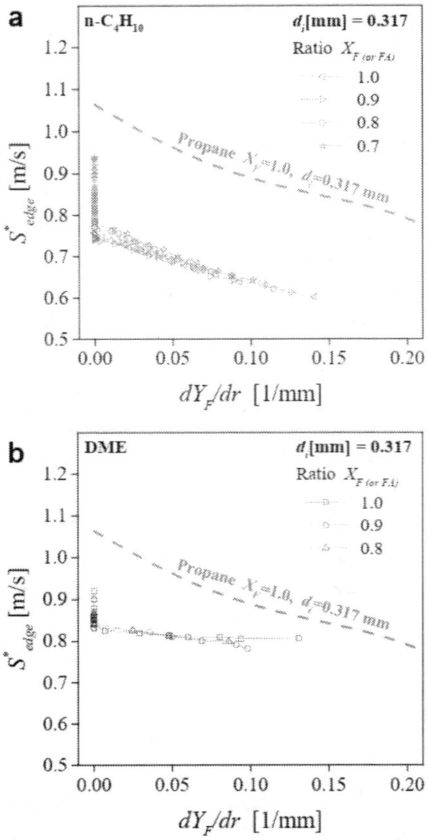

Fig. 5. EFS with FCG for (a) n-butane and (b) DME

First, in Fig. 5a for n-butane, the EFS/FCG relation was compared with that of propane. The value of EFS was slightly lower than that of propane. This result can be explained with the fact that the laminar burning velocity of n-butane is smaller than that of propane. The interception value of EFS at the FCG of zero was about 0.74 m/s. Nevertheless, the slopes of n-butane and propane were very similar. Although $Y_{F,st}$ for propane and n-butane are very similar, Schmidt numbers of fuel were quite different. It may cause difference in the development of the concentration core near the tube exit. It was not considered sufficiently in the previous study.

In addition, in Fig. 5b, the estimated EFS of DME flames were almost constant, and its interception value was about 0.84 m/s at the FCG of zero. Considering the larger laminar burning velocity of DME compared to that of propane, smaller interception point of DME is questionable. One possible reason is concerned with the much larger stoichiometric mass fraction of DME compared to the other fuels due to the DME chemical structure which have an oxygen atom between two methyl groups, C–O-C. Thus, DME have different lean and rich premixed limits compared with long chain carbon such as propane and n-butane. Then the stoichiometric line will be located near the centerline. Therefore, the lifted flame can be easily merged, and can become a wrinkled premixed flame. Then the variation of EFS with FCG can be significantly reduced. It is expected that a better relation can be obtained if a larger tube is used.

4 Conclusions

In this study, laminar lifted flames of n-butane and DME have been investigated. Experimental results of the flow rate and the liftoff height were measured. Based on these experimental results, a method that can estimate EFS/FCG relation was applied for n-Butane and DME flames.

The effective Schmidt number, Sc^{*}_{eff}, suggested by Jeon and Kim [4] was used. Conclusively, the EFS/FCG relation of n-butane and DME could be obtained and they were compared with the results of propane. From this study, the applicability of direct estimation method for EFS/FCG for n-butane and DME was confirmed. Although the previous method using the effective Schmidt number is reasonable in many aspects, it needs to be improved to reflect the effects of concentration core especially when the liftoff height is relatively small. It would also be useful to relate the values of EFS to the steady laminar flame speeds to put these into explaining the stabilization of turbulent lifted flames. It will be investigated in the future.

5 Acknowledgment

This work was supported by a Research Grant from KAIST under the title "Study of flame structures under conditions of high temperature and MILD combustion", and was partly supported by the BK21-plus program.

References

[1] S.H. Chung, B.J. Lee, On the characteristics of laminar lifted flames in a nonpremixed jet, Combust. Flame 86 (1991) 62-72.

[2] Y.S. Ko, S.H. Chung, Propagation of unsteady tribrachial flames in laminar non-premixed jets, Combust. Flame 118 (1999) 151-163.

[3] J. Lee, S.H. Won, S.H. Jin, S.H. Chung, O. Fujita, K. Ito, Propagation speed of tribrachial (triple) flame of propane in laminar jets under normal and micro gravity conditions, Combust. Flame 134 (2003) 411-420.

[4] M.K. Jeon, N.I. Kim, Direct Estimation of Edge Flame Speeds of Lifted Laminar Jet Flames and a Modified Stabilization Mechanism, Combust. Flame 186 (2017) 140-149.

[5] B.J. Lee and S.H. Chung, Determination of Schmidt number of mixed fuels by the characteristics of laminar lifted jet flames, Fuel 85 (2006) 68-74.

[6] Y. Xue, Y. Ju, Studies on the liftoff properties of dimethyl ether jet diffusion flames, Combust. Sci. Technol. 178 (2006) 2219-2247.

[7] S.R. Turns, An Itroduction to Combustion: Concepts and Applications, 3rd ed., McGraw-Hill, New York, 2012. pp. 701-709.

[8] J.R. Andrews, O. Biblarz, Temperature dependence of gas properties in polynomial form. Monterey, California. Naval Postgraduate School, 1981.

[9] M.E. Van Leeuwen, Derivation of Stockmayer potential parameters for polar fluids, Fluid Phase Equilibria, 99 (1994) 1-18.

11th Asia-Pacific Conference on Combustion,
The University of Sydney, NSW Australia
10th -14th December 2017.

A Study on Measurement of Laminar Burning Velocity and Cellular Instability of SNG Fuel with 3% Hydrogen in Spherical Propagating Flames

Dong Chan Kim[1], Jun Ho Song[1], Kee Man Lee[2]

[1]Department of Aerospace Engineering, Sunchon National University

255, Jungang-ro, Suncheon-si, Jeollanam-do, 57922, Republic of Korea

[2]School of Mechanical and Aerospace Engineering, Sunchon National University

255, Jungang-ro, Suncheon-si, Jeollanam-do, 57922, Republic of Korea

Abstract

This article describes laminar burning velocity and cellular instability of synthetic natural gas (SNG) with 3% hydrogen. In this study, experimental apparatus was employed cylindrical combustor, and was carried out with high speed camera and Schlieren systems for visualization of flame shape. Experimental laminar burning velocity was compared with numerical calculation by CHEMKIN with GRI 3.0, USC-II and UC Sandiego mechanism. From results of this work, the laminar burning velocity with increasing pressure was decreased regardless of the change of equivalence ratio. And the cellular instability appeared as the pressure increased except for the rich mixture flames.

1 Introduction

Natural gas and oil, whose demand and price continue to rise, are facing a major problem of depletion of fuel. Conversely, coal with abundant reserves in the USA, Russia and Asia is considered to be one of future energy sources. However, these coals are known to be the main reason of smog. When coal is used, carbon dioxide and impurities such as nitrogen oxides and sulfur oxides are generated, which causes various environmental pollution. To eliminate these problems, there is a method of gasifying coal. The syngas produced by coal gasification at high temperature and pressure. And then, methane and hydrogen were manufactured through the gas purification and synthesis process. The produced gas by this process is called synthetic natural gas (SNG). Through this process, it was possible to remove sulfur, and the effect of reducing carbon dioxide can be expected. And SNG can substitute for natural gas. So, it can solve the above environmental problems. From this point of view, extensive research of SNG is actively conducted [1].

In this study, our group was conducted to know the combustion characteristics of SNG. In order to know the combustion characteristic, there is a method of measuring the laminar burning velocity. The laminar burning velocity was included the basic information on diffusion and reactivity of combustible mixture and it was most important parameter to control the actual combustion phenomena such as blow-off and flashback in operating condition setting and combustor design.

Corresponding author. Fax: +82-61-750-3820
E-mail address: kmlee@scnu.ac.kr

Another combustion characteristic is cellular instability. It have been associated three kinds of instabilities with diffusive-thermal effects that depend on Lewis number, hydrodynamic effects that depends on density and equivalence ratio. These combustion characteristics were experimentally measured, and also compared with numerical calculated data using PREMIX [2] code. It were used to CHEMKIN with GRI 3.0 [3], USC-II [4] and UC Sandiego [5] mechanism.

2 Experimental Set-up and Methodologies

In this study, experimental combustion characteristics were measured using cylindrical combustor. Fig. 1 shows the schematic diagram of experimental systems. The dimensions of inside of the chamber was designed 200 × 220 mm^2, and quartz windows were fitted to both sides of the cylindrical chamber for visualization of the flame. The window size was 150 mm in diameter and 45 mm in thickness. The pressure gauge (Bit02B) was mounted on the combustor. It used to check and control the pressure with micro metering valve. And ignition systems were installed in connection with two tungsten electrodes. The electrodes diameter were 0.5 mm, and spark ignited at the center of combustor. The high speed camera (FASTCAM – 1024 PCI) was used to acquire for spherical flame images and was operated at 6000 frame per second and resolution of 384 × 368 $pixel^2$. Schlieren device was used to show the shape of propagating flame using shadow and was consist of two optical mirrors, 100 W halogen light source, a slit and knife edge.

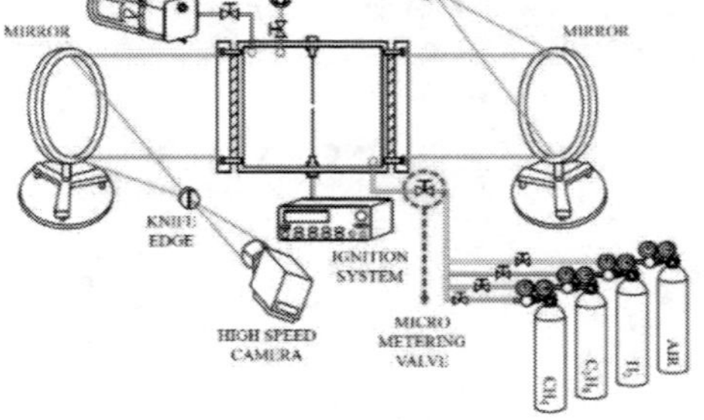

Figure 1: Schematic diagram of experimental systems

The laminar burning velocity measurement using spherical

flame was known as outwardly propagating spherical flame method. Stretched laminar burning velocity at burned mixture, S_b, is defined as dr_f/dt, and as the spherical radius increases, stretch effect increases, and stretch rate is defined as $K = (2/R_f)S_b$. In previous experiments, the suitable flame radius (10 mm < R_f < 25 mm) and theoretical model (NM I [6, 7] : $S_b = S_b^0 - 2S_b^0 L_b/R_f$) for the combustor of our research group were obtained by using the stretched laminar burning velocity in the experiment [8].

In addition, experiments were conducted to increase the initial pressure (0.1, 0.3, 0.5 and 0.7 MPa) to confirm the hydrodynamic instability. This is because the pressure increases, the flame thickness decreases and the density gradient increases with the decrease of flame thickness, which can improve the development of hydrodynamic instability. The flame thickness was calculated as $l_f = \lambda / C_p \rho_u S_u^0$, proposed by C. K. Law [9], and the thermal expansion ratio was expressed as $\sigma = \rho_u/\rho_b$.

Also, the diffusive-thermal instability was checked by varying the equivalence ratio (Φ = 0.8, 1.0, 1.2). The above description and composition ratio of SNG were shown in table 1.

Description	Specification
Fuel (Purity)	CH_4 (99.999%), C_3H_8 (99.5%), H_2 (99.999%)
Oxidizer	High grade purity of Air
Equivalence ratio (Φ)	0.8, 1.0, 1.2
Initial Pressure (MPa)	0.1, 0.3, 0.5, and 0.7
Composition of SNG (%)	$CH_4 : C_3H_8 : H_2 = 91 : 6 : 3$

Table 1: Experimental conditions

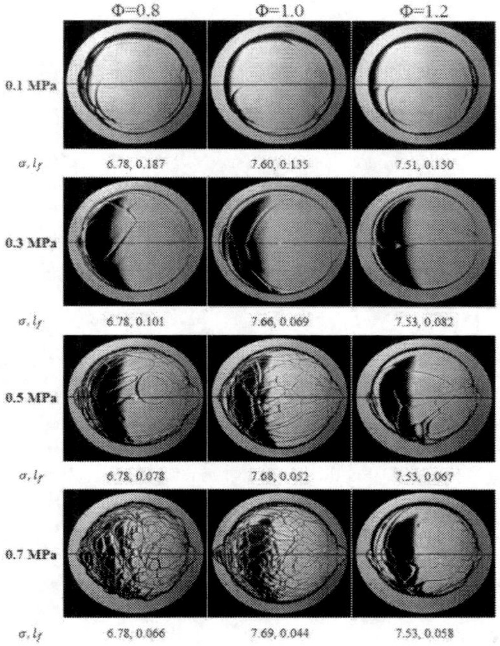

Figure 2: Schlieren images of Φ = 0.8, 1.0, 1.2 at 0.1, 0.3, 0.5, and 0.7 MPa

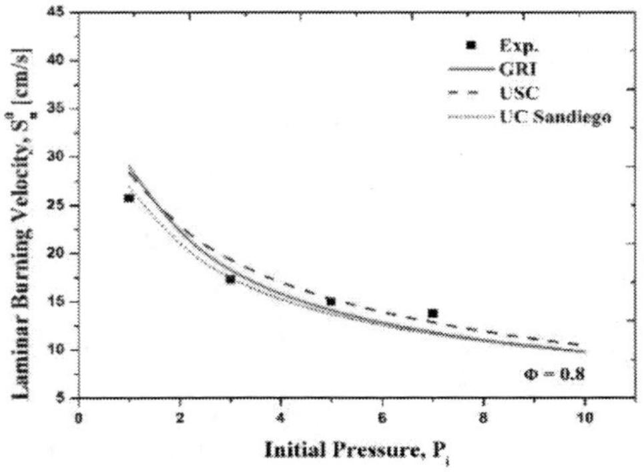

Figure 3: Calculated (solid line: GRI 3.0, dash line: USC, short dot line: UC Sandiego) and experimental (closed symbol) laminar burning velocity as function of initial pressure for equivalence ratio of 0.8

3 Results and Discussion

3.1 Flame shape with increasing pressure and Φ

The spherical propagating flame images of Φ = 0.8, 1.0, 1.2 at 0.1, 0.3, 0.5 and 0.7 MPa are shown in Fig. 2 and express calculated σ and l_f at the bottom of the figure.

A smoothed flames appeared at 0.1 and 0.3 MPa without wrinkles, and wrinkle flames appeared with pressure increases. As shown in the Fig. 2, it can be seen that the hydrodynamic instability promotes, which means that the thermal expansion ratio increases and the flame thickness thinner when the pressure increases.

In terms of equivalence ratio, in cases of the lean and stoichiometry conditions, wrinkle flames appeared as pressure increased, but it was not in rich cases. The reason for the cellular instability in the lean mixture can be explained by effective Lewis number. For example, when Φ = 0.8 of 0.7 MPa, the effective Lewis number is 1.074 and Φ = 1.2 is 1.098. According to the diffusive-thermal instability, it was judged that the wrinkle flames appeared due to the Lewis number of lean condition is smaller than the rich condition.

Previous studies [9] have found that flames are unstable in rich propane mixtures and flames are stable in rich hydrogen mixtures. According to previous studies, the reason why the wrinkle flame does not appear in a rich mixture indicates that the effect of methane and hydrogen on flame stabilization is more dominant than propane.

3.3 Laminar burning velocity of SNG

The calculated and experimental laminar burning velocity of SNG at Φ = 0.8 are shown in Fig. 3. Also, Φ = 1.0 and Φ = 1.2 are shown in Fig. 4 and Fig. 5, respectively. The calculated laminar burning velocity decreased with increasing pressure regardless of the equivalence ratio, and the experimental values were similar to the calculated value. From the results of the laminar burning velocity of calculations and experiments in Fig. 3, 4 and 5, it can be seen that the laminar burning velocity is

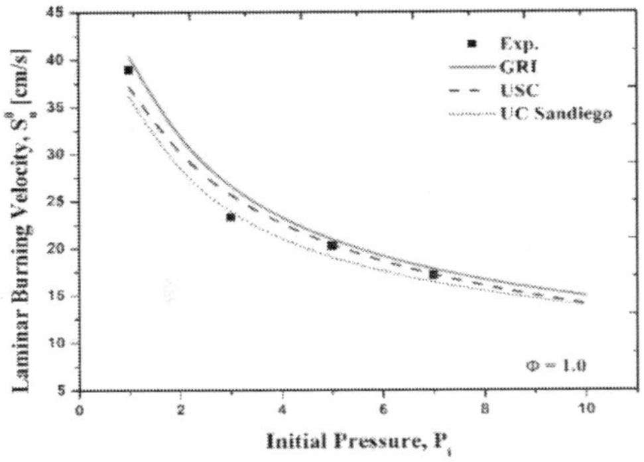

Figure 4: Calculated (solid line: GRI 3.0, dash line: USC, short dot line: UC Sandiego) and experimental (closed symbol) laminar burning velocity as function of initial pressure for equivalence ratio of 1.0

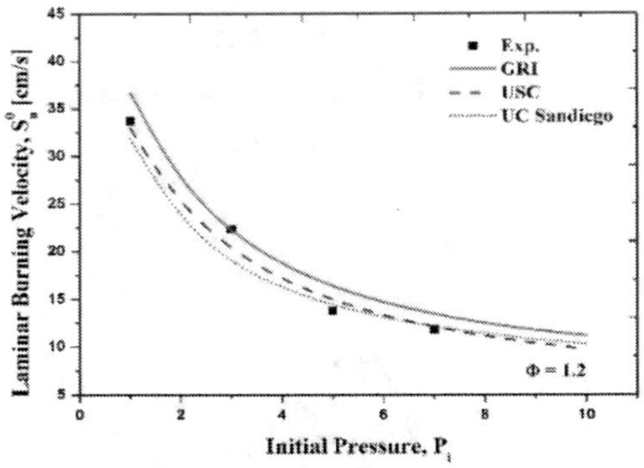

Figure 5: Calculated (solid line: GRI 3.0, dash line: USC, short dot line: UC Sandiego) and experimental (closed symbol) laminar burning velocity as function of initial pressure for equivalence ratio of 1.2

inversely proportional to the initial pressure, because the increase of initial pressure cause the density of the combustion chamber to increase, which reduces the laminar burning velocity.

4 Conclusions

The purpose of this paper is to describe the laminar burning velocity and cellular instability of SNG. The research was carried out using spherical flames with various initial pressures. The conclusions are as follow:

1. As the initial pressure was increased, the laminar burning velocity has decreased, and this tendency was the same even as the equivalence ratio changed.
2. In cases of lean and stoichiometry, cellular instability appeared when the pressure increased, which is expressed as the hydrodynamic instability. The hydrodynamic instability was affected by thermal expansion ratio and flame thickness.
3. Even though increasing pressure, in case of lean condition, cellular instability was observed. The reason for this phenomenon is that the effective Lewis number of lean condition is smaller than the rich condition.
4. In case of rich condition, cellular instability was not observed. This indicates that the effect of methane and hydrogen on flame stabilization is more dominant than propane.

5 Acknowledgment

This work was supported by the Energy Efficiency & Resources Core Technology Program of the Korea Institute of Energy Technology Evaluation and Planning (KETEP), granted financial resource from the Ministry of Trade Industry & Energy, Republic of Korea (No. 2015-2010-103050)

References

[1] M. Chandel, E. Williams, Climate Change Policy Partnership, 2009.

[2] R. J. Kee, J. F. Grcar, M. D. Smooke, J. A. Miller, E. Meeks, Sandia National Laboratories Report SAND85-8240, 1985.

[3] G.P. Smith, D. M. Golden, M. Frenklach, et al., http://www.me.berkeley.edu/gri_mech/.

[4] H. Wang, X. You, A. V. Josh, S. G. Davis, A. Laskin, F. N. Egolfopoulos, C. K. Law, USC Mech Version II, http://ignis.usc.edi/USC_Mech_II.htm.

[5] UC Sandiego mechanism webpage, available at http://web.eng.ucsd.edu/mae/groups/combustion/mechanism.html.

[6] T.Tahtouh, F. Halter and C. Mounaim-Rousselle, Combust. Flame 156 (2009) 1735-1743.

[7] F. Halter, T. Tahtouh and C. Mounaim-Rousselle, Combust. Flame 157 (2010) 1825-1832

[8] D. C. Kim, K. M. Lee, Trans. of the Korean Hydrogen and New Energy Society 27 (6) (2016) 737-746.

[9] C. K. Law, G. Jomaas, J. K. Bechtold, Proc. Combust. Inst. 30 (2005) 159-167.

11th Asia-Pacific Conference on Combustion,
The University of Sydney, NSW Australia
10th -14th December 2017.

Throttling Effects on Dual-fuel Premixed Charge Compression Ignition under Low Load Condition in a Heavy Duty Diesel Engine

Euijoon Shim[1], Hyunwook Park[1], Choongsik Bae[1]

[1]Korea Advanced Institute of Science and Technology (KAIST)

291 Daehak-ro, Yuseong-gu, Deajeon, 34141, Republic of Korea

Abstract

For the last five decades, many engine combustion researchers have been researching about advanced combustion technologies such as homogeneous charge compression ignition (HCCI) and dual-fuel premixed charge compression ignition (Dual-Fuel PCCI) for reducing particulate matter (PM) and nitrogen oxides (NOx) emissions simultaneously in diesel engines. However, the advanced combustion technologies have drawbacks such as high amount of unburned hydrocarbon (UHC) and carbon monoxide (CO) emissions due to the incomplete combustion at all operating ranges especially under the low load conditions. In this study, throttling was adopted to reduce the UHC and CO emissions. Controlling the throttle could improve the exhaust emission and the combustion stability performance by increasing the global equivalence ratio under low load conditions. The results suggest that controlling the charge air quantity is an effective way to improve combustion efficiency (η_c), and to extend low load operating ranges of the DF-PCCI combustion.

1 Introduction

The invention of the internal combustion engine, which is based on fossil fuels, has contributed to the development of human civilization and industrialization in its usefulness. However, because of the harmful emission of exhaust gases and greenhouse gas (GHG) emissions like carbon dioxide (CO_2), each country's government has been fortifying exhaust emissions and CO_2 regulations. [1] In particular, a diesel engine has higher combustion efficiency than a gasoline engine since diesel engines use higher compression ratio. Despite these advantages, diesel engines produce large quantities of particulate matter (PM) resulting from the diffusion flame. Moreover, diesel engines also produce large amounts of nitrogen oxides (NOx) due to the high combustion temperature during the combustion process. In the case of PM, it is known to cause respiratory, cardiovascular disease, and nerve denaturation. Furthermore, NOx can produce photochemical smog, which is known to cause lung diseases and acid rain, resulting in adverse effects on the environment. [2]

As a part of the policy for reducing harmful emissions and GHG, internal combustion engines have been researched in numerous ways for improving emissions and improve efficiency. To achieve high efficiency and low emissions in conventional diesel combustion (CDC), numerous advanced engine technologies such as ultra-high-pressure fuel injection equipment (FIE), exhaust gas recirculation (EGR), and high-efficiency boosting systems have been adopted. Moreover, the after-treatment systems such as diesel oxidation catalyst (DOC), diesel particulate filter (DPF), and selective catalytic reduction (SCR) are actively utilized for fulfilling the emission regulations. However, the after-treatment systems can aggravate thermal efficiency since the exhaust gas pressure is increased. Therefore, many researchers have continued to conduct various research on advanced combustion technologies to minimize harmful exhaust gases production (like PM and NOx emissions) during the combustion process and to improve thermal efficiency without after-treatment systems. [3-5]

Homogeneous charge compression ignition (HCCI), premixed charge compression ignition (PCCI), and low temperature diesel combustion (LTC) have been studied actively to reduce NOx and PM emissions simultaneously during the combustion process. [6-8] However, these advanced combustion technologies have several obstacles such as combustion phase control, combustion stability, narrow operating ranges, and high level of carbon monoxide (CO) and unburned hydrocarbon (UHC) emissions to be commercialized.

The dual-fuel premixed charge compression ignition (DF- PCCI) combustion concept such as reactivity controlled compression ignition (RCCI) [9] and the pilot-DF [10] were invented to overcome several problems in the advanced combustion technology. However, the CO and UHC emission problem is still recognized as the hurdle of dual-fuel PCCI combustion technology. Methane was defined as a severe GHG material by the intergovernmental panel on climate change (IPCC) report, and it was assigned a global warming potential (GWP) factor of 25. [11]

Natural gas (NG) is usually utilized as a low-reactivity fuel in dual-fuel combustion concept due to its high octane number (130). Therefore, NG is effective in resisting auto-ignition during the compression process. The fumigated NG is supplied early to the engine during the intake process for the homogeneous charge. However, as the operating condition move toward the low load condition, the global equivalence ratio becomes too lean for complete combustion in dual-fuel mode. The excessive amount of CO and UHC can be produced under low load conditions.[12]

In this study, the charge air quantity was controlled by adjusting the throttle valve position. It was possible to reduce the incomplete combustion phenomena, and thus CO and UHC emissions can be well oxidized even under low load operating conditions. The intake charge air quantity control strategy will contribute to overcome the commercialization hurdles of dual-fuel PCCI advanced combustion technologies.

2 Methodology

2.1 Experimental setup

The experimental engine system is depicted in the Figure 1. The experiment was conducted in the single cylinder engine which is modified from a 6-cylinder heavy duty diesel engine. In order to adopt dual-fuel system, NG supply system was installed on the intake port. The diesel injection system equipped with a common-rail system was also applied to the single cylinder engine. The NG and diesel were used as respectively the low reactivity fuel and the high reactivity fuel on dual-fuel operating mode. Table 2 shows the chemical characteristics of each fuel. The equation (1) manifest the substitution ratio (SR) which is the fuel quantity ratio between NG and diesel. The equation (2) manifest the combustion efficiency. A mass flow meter (Blonkhorst F-206BI) was used to control the supplied NG mass. The diesel injection quantity, timing, and pressure were controlled by a programmable injector controller (Zenobalti Co., ZB_9012P). The exhaust gas emissions such as CO, CO2, HC, NOx were measured by a gas analyzer (HORIBA MEXA 1500D). The smoke meter (AVL, 415S) was utilized for the soot measurement. The AVL INDI-micro was utilized to investigate in-cylinder combustion phenomenon. The in-cylinder pressure was measured by a piezo type pressure transducer (Kistler 6056A). All of the experimental data were acquired over 100 engine cycles with the 0.2 crank angle degree (CAD) resolution.

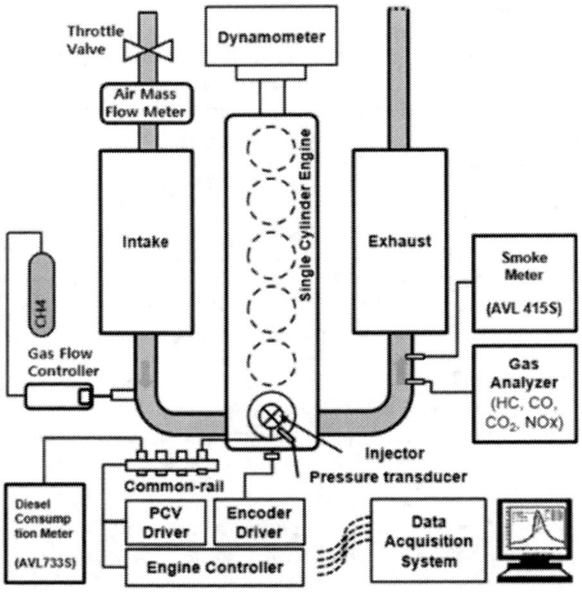

Figure 1: Schematic diagram of experimental engine system

$$SR = \frac{\dot{m}_{CNG}LHV_{CNG}}{(\dot{m}_{CNG}LHV_{CNG})+(\dot{m}_{Diesel}LHV_{Diesel})} \quad (1)$$

$$\eta_C = 1 - \frac{(ISCO \cdot LHV_{CO})+(ISHC \cdot LHV_{DF})}{(\dot{m}_{CNG}LHV_{CNG})+(\dot{m}_{Diesel}LHV_{Diesel})} P_i \quad (2)$$

Corresponding author. Fax: +82-42-350-5023
E-mail address: csbae@kaist.ac.kr

Table 1: Specifications of single cylinder engine

Specification	Value
Engine type	Single cylinder compression ignition type
Displacement [cc]	987
Bore × Stroke [mm]	100 × 125
Con-rod [mm]	202
Compression ratio [-]	17.4
Number of Valves [EA]	4
Piston type	Re-entrant bowl
Diesel Injector	180 MPa 7 Holes - 150 Deg
NG supplier	Bronkhost F-206BI(NG)

Table 2: Fuel properties [13-14]

Specification	Diesel	Natural Gas
Density @ 293 K	0.84 – 0.88	0.79
Cetane Number	52.1	-
RON	-	130
Auto-ignition Temperature[℃]	180 – 220	650
Carbon content [%]	87	75
Lower Heating Value [MJ/kg]	42.5	45.3
Stoichiometry A/F [kg/kg]	14.3	17.2

2.2. Experimental procedure

Previous researches have shown that dual-fuel combustion, such as pilot-DF and RCCI, produces extremely high amount of UHC and CO emissions during the combustion process. [12] Especially under low-load conditions, it is quite difficult to maintain proper combustible mixture due to excessively lean air-fuel equivalence ratio. Throttle is able to reduce the charge air quantity for preventing bulk quenching effect under low-load operating conditions. In this experiment, the charge air quantity was reduced step by step up to 60% of base air quantity by using the throttle valve.

Table 3 summarizes the ranges of main experimental parameters. The base engine, which was a heavy-duty diesel engine, has a maximum torque at 1400 rpm and maximum power at 1800 rpm. All experiments were conducted on the low load operating conditions, indicated mean effective pressure (IMEP) of 0.3 MPa, representing the poor combustion stability and the high CO and UHC emissions for dual-fuel combustion.

Table 3: Test parameter range

Specification	Range
Engine speed (rpm)	1400
Engine Load [IMEP] (MPa)	3
Diesel Injection Timing (CAD BTDC)	30
Diesel Injection Pressure (MPa)	100
Substitution Ratio [SR] (%)	60
AFR [Throttling] (-)	35 ~ 60

3 Results

3.1 Emission and combustion characteristics of baseline test

Figure 2 shows the emission characteristics of the conventional diesel combustion (CDC), and the CNG-diesel dual-fuel premixed charge compression ignition (DF-PCCI). In the case of CDC, the NOx level was recorded quite high such as 6.5 g/kWh due to the high in-cylinder combustion temperature over 2000K. However, CO and THC levels showed negligible results. Also the combustion efficiency was maintained higher than 99% as shown in Figure 5. These results are typical emission characteristics of the CDC. When implementing the DF-PCCI, the NOx level was decreased down to 0.2 g/kWh. This result satisfies the EU-VI emission registration without after-treatment systems. In the DF mode, the natural gas (NG) was supplied as a low reactivity fuel during the intake process. The very early injected NG results in the homogeneous charge within the cylinder. It was possible to reduce the NOx and PM emissions simultaneously by lowering the peak combustion temperature and reducing the local equivalence ratio (Φ_{local}). However, the CO and THC were increased significantly due to the fact that the overly lean mixture results in the bulk quenching effect, which deteriorates complete combustion process and combustion stability. Compared to the CDC, the CO and THC emissions showed about 3.8 and 48 times higher level, respectively. These results strongly agree with previous studies that showed the high level of the CO and THC emissions problem under the low load conditions, which is the hurdle for DF-PCCI commercialization.

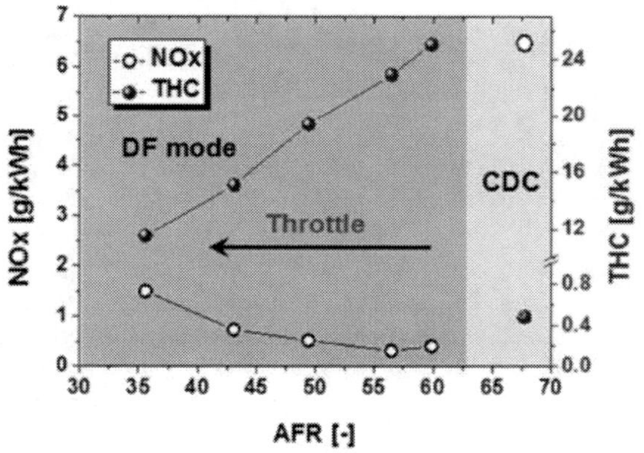

Figure 2: The NOx and THC emission levels of CDC and DF-PCCI combustion mode while varying the AFR

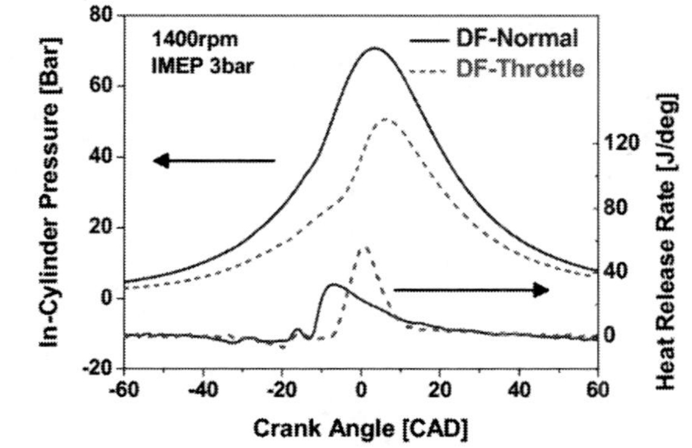

Figure 3: combustion pressure profile and HRR of DF-PCCI normal and throttle mode

3.2 Throttle effects of DF-PCCI combustion characteristics

Figure 2 shows the NOx and THC emission characteristics according to the charge air quantity using the throttle in DF-PCCI combustion. The THC gradually decreased down to 52% compared to CDC by reducing the charge air quantity. The throttling is able to change the global equivalence ratio to rich condition which helps in improving the flammability. On the other hands, as the charge air quantity is reduced, the NOx level was gradually increased. This can be explained by the in-cylinder pressure and heat release rate (HRR) analysis. Figure 3 shows the in-cylinder pressure profile and the HRR of DF-normal and DF-throttle mode. The start of combustion (SOC) timing is almost same for both DF-normal and DF-throttle mode. However, the DF-throttle mode resulted in retarding the timing of high temperature oxidation (HTO), reducing the burn duration, and increasing the peak level of HRR. These factors increase the peak combustion temperature, which is closely related to the NOx formation.

Figure 4 shows the CO and THC emission characteristics of various combustion strategies such as CDC, DF-normal, and DF-throttle. When the DF-PCCI combustion was implemented, the CO and THC emission levels extremely increased about 4 and 50 times higher than the CDC under low load conditions, respectively. The overly lean mixture condition is the main cause of the high CO and THC emissions. For the DF-PCCI combustion, the NG is supplied into the cylinder very early during the intake process so that the NG is well mixed with air.

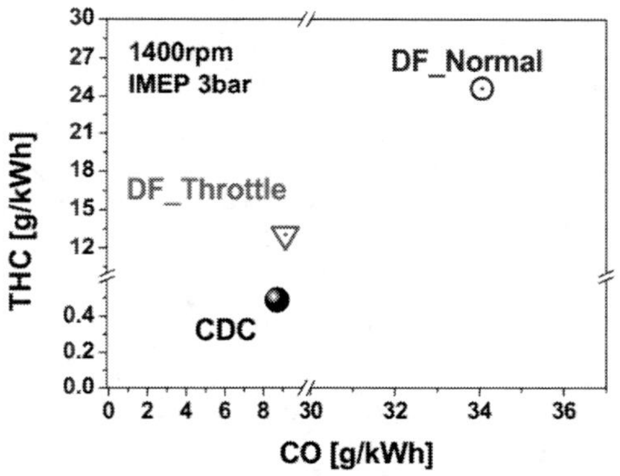

Figure 4: The CO and THC emission levels of various combustion modes

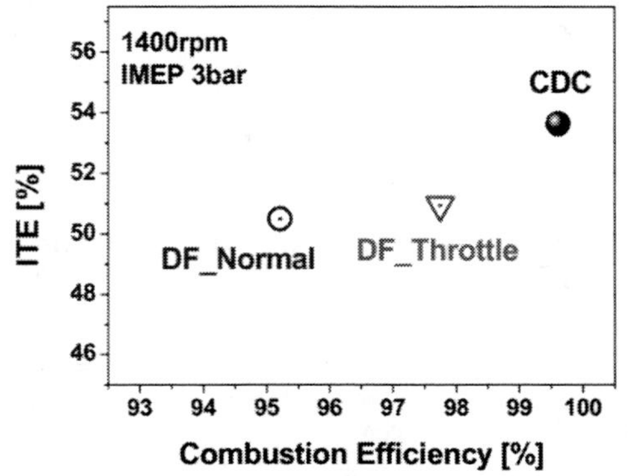

Figure 5: The combustion efficiency and ITE levels of various combustion modes

However, under the low load conditions, the air-fuel equivalence ratio is too lean. Therefore, the lean mixture charge could not ignite or combust completely. Thus it can cause to deteriorate the combustion stability and increase the CO and THC emissions. In this study, the throttle was adopted to control the charge air quantity. The CO and THC emissions in the DF-PCCI combustion dramatically decreased by the reducing charge air quantity. The throttling was able to change the global equivalence ratio (Φglobal) to rich conditions. The level of the CO emission in the DF-throttle mode almost decreased down to the level of the CDC, and the level of the THC emission decreased approximately 50% compared to the DF-normal mode.

Figure 5 shows the indicated thermal efficiency (ITE) and the combustion efficiency (η_c) characteristics of various combustion strategies. The combustion efficiency of DF-PCCI decreased by 95.4% due to the bulk quenching effect. However, adopting the throttle showed that the combustion efficiency was recovered up to 2.5%p by mitigating the incomplete combustion, and the ITE was also slightly improved. As the charge air quantity was reduced, the combustion phase moved toward the effective work region, and the main burn duration shortened as it is shown in figure 3.

4 Conclusions

In this study, the throttling strategy was implemented to reduce the CO and THC emissions under a low load operating condition in DF-PCCI combustion.

1) Implementing the DF-PCCI combustion under the low load condition produces excessive amount of the CO and THC emissions compared to CDC due to the bulk quenching effect.

2) Reducing the charge air quantity through the usage of the throttle can reduce the CO and THC emissions on DF-PCCI combustion by making the air-fuel ratio rich..

3) When the throttle strategy was adopted, the indicated thermal efficiency (ITE) slightly improved due to moving the combustion phase toward the effective work region. The combustion efficiency also recovered by 2.6% by reducing the incomplete combustion.

The method of controlling the charge air quantity is effective in reducing CO and UHC emissions in DF-PCCI combustion. All the experiments were conducted under low load conditions of 3 bar IMEP. In future research, the operating range will be extended to middle load conditions.

5 Acknowledgment

The authors would like to appreciate the Global-Top Project (Development of Advanced Combustion Technology for Global Top Low Emission Vehicle (2016002070001)) of the Ministry of Environment (MOE) of Korea for financial support by the CEFV (Center for Environmentally Friendly Vehicle).

References

[1] International Energy Agency. Office of Energy Technology, R&D., and Group of Eight (Organization). Energy Technology Perspectives, International Energy Agency, 2006.

[2] Wei, L., & Geng, P., A review on natural gas/diesel dual fuel combustion, emissions and performance. Fuel Processing Technology, 142, (2016) 264-278.

[3] Kalghatgi, Gautam T., Per Risberg, and Hans-Erik Ångström. Advantages of fuels with high resistance to auto-ignition in late-injection, low-temperature, compression ignition combustion. No. 2006-01-3385. SAE Technical Paper, 2006.

[4] Kalghatgi, Gautam T., Per Risberg, and Hans-Erik Ångström. Partially pre-mixed auto-ignition of gasoline to attain low smoke and low NOx at high load in a compression ignition engine and comparison with a diesel fuel. No. 2007-01-0006. SAE Technical Paper, 2007.

[5] Inagaki, K., Fuyuto, T., Nishikawa, K., Nakakita, K., & Sakata, I., Dual-fuel PCI combustion controlled by in-cylinder stratification of ignitability. No. 2006-01-0028. SAE Technical Paper, 2006.

[6] Park, S. H., & Yoon, S. H., Effect of dual-fuel combustion strategies on combustion and emission characteristics in reactivity controlled compression ignition (RCCI) engine, Fuel 181 (2016): 310-318.

[7] Akihama, K., Takatori, Y., Inagaki, K., Sasaki, S., & Dean, A. M., Mechanism of the smokeless rich diesel combustion by reducing temperature. No. 2001-01-0655. SAE technical paper, 2001.

[8] Manente, V., Tunestal, P., Johansson, B., & Cannella, W. J. Effects of ethanol and different type of gasoline fuels on partially premixed combustion from low to high load. No. 2010-01-0871. SAE Technical Paper, 2010.

[9] Reitz, R. D., & Duraisamy, G., Review of high efficiency and clean reactivity controlled compression ignition (RCCI) combustion in internal combustion engines, Progress in Energy and Combustion Science 46 (2015): 12-71.

[10] Gebert, K., Beck, N. J., Barkhimer, R. L., & Wong, H. C., Strategies to improve combustion and emission characteristics of dual-fuel pilot ignited natural gas engines. No. 971712. SAE Technical Paper, 1997.

[11] Solomon, Susan (ed.). Climate change 2007-the physical science basis: Working group I contribution to the fourth assessment report of the IPCC. Cambridge University Press, (2007), p. 104.

[12] Florea, R., Neely, G. D., Abidin, Z., & Miwa, J., Efficiency and Emissions Characteristics of Partially Premixed Dual-Fuel Combustion by Co-Direct Injection of NG and Diesel Fuel (DI 2). No. 2016-01-0779. SAE Technical Paper, 2016.

[13] Namazian, Mehdi, and John B. Heywood., Flow in the piston-cylinder-ring crevices of a spark-ignition engine: effect on hydrocarbon emissions, efficiency and power., No. 820088. SAE Technical Paper, 1982.

[14] Yang, B., Wei, X., Xi, C., Liu, Y., Zeng, K., & Lai, M. C., Experimental study of the effects of natural gas injection timing on the combustion performance and emissions of a turbocharged common rail dual-fuel engine, Energy Conversion and Management 87 (2014): 297-304.

11th Asia-Pacific Conference on Combustion,
The University of Sydney, NSW Australia
10th -14th December 2017.

Dynamical Analysis of Turbulent Premixed Hydrogen/Air Flames in the Thin Reaction Zone Regime

Efstathios-Al. Tingas[1], Roman Kashtanov[1,3], Francisco E. Hernández Pérez [1], Hong G. Im[1], Pietro Paolo Ciottoli[2], Riccardo Malpica Galassi[2], Mauro Valorani[2]

[1]King Abdullah University of Science and Technology (KAUST)

Thuwal-Jeddah, 23955-6900, Saudi Arabia

[2]Mechanical and Aerospace Engineering Department, Sapienza University of Rome

Via Eudossiana, 18, 00184 Rome, Italy

[3]Mechanical Department of Aerospace Engineering, Universität der Bundeswehr München

Neubiberg, 85577 Munich, Germany

Abstract

In the current study, tools generated from the computational singular perturbation (CSP) are used to analyze the dynamics that develops in a three-dimensional (3D) turbulent H_2/air flame. On top of these tools, the recently proposed tangential stretch rate (TSR) approach is used to obtain physical understanding of the turbulent flame. The analysis is first conducted on a laminar premixed H_2/air flame followed by the analysis of the 3D turbulent flame at three distinct time steps. Using algorithmic tools, the flame front was identified along with the dominant processes reaction and transport (diffusive/convective) processes. Furthermore, the overall role of kinetics/transport was determined at each part of the flame.

1 Introduction

The investigation of turbulent flames has always been a challenging task. The spectacular increase of the computational power over the last 15 years has enabled the analysis of turbulent flames in the context of direct numerical simulations (DNS). DNS is a very powerful tool in fundamental research of turbulence since it can provide insights that are very difficult or sometimes even impossible to obtain in the lab. Despite the significant advances in computational power, the implementation of detailed chemistry in DNS remains an extremely demanding task that very few researchers can tackle. The great advantage of the use of a detailed chemical kinetics model instead of a reduced or simplified one is that simplified/reduced models aim at very particular features of the system and, therefore, their applicability and validity are by definition limited and maybe questionable. However, obtaining meaningful physical understanding is rather difficult when detailed kinetics models are used. To address this difficulty, various techniques have been proposed.

Computational singular perturbation (CSP) is a method proposed in the late 1980s by Lam and Goussis. Its basis lies on the Geometric Singular Perturbation, proposed in the mid 1970s, exploiting the decomposition of the physical space into fast and slow subdomains, therefore, enabling the construction of reduced models. The success of CSP in constructing reduced models and obtaining physical understanding of reacting flows has been widely tested, from zero-dimensional (0D) autoignition problems to 3D turbulent flames.

In the present work, CSP tools are used to analyze and gain physical understanding on a turbulent 3D H_2/air flame, which was simulated using DNS and detailed chemistry.

2 The Algorithmic Tools

The species and energy equations that govern the evolution of a reactive-transport system can be cast in the general form of (1):

$$\frac{\partial \boldsymbol{z}}{\partial t} = \boldsymbol{L}(\boldsymbol{z}) + \boldsymbol{g}(\boldsymbol{z}) \quad (1)$$

where **g**(**z**) is the chemical source term of the system, **L**(**z**) is a spatial differential operator (convection and/or diffusion), **z** is the (N+1)-dimensional state column vector including the N species' mass fractions and temperature. Based on the CSP methodology [1], the system of (1) can be decomposed into the following form by using the CSP basis vectors $\boldsymbol{a}_i$ (i=1, N+1):

$$\frac{\partial \boldsymbol{z}}{\partial t} = \sum_{i=1}^{N+1} \boldsymbol{a}_i(\boldsymbol{z}) h^i(\boldsymbol{z}) \quad (2)$$

where $h^i = \boldsymbol{b}^i \cdot (\boldsymbol{L}(\boldsymbol{z}) + \boldsymbol{g}(\boldsymbol{z}))$ is the amplitude of the i-th mode and $\mathbf{b}^i$ denotes the dual basis vector $\boldsymbol{b}^i \cdot \boldsymbol{a}_j = \delta_j^i$ [1]. The CSP basis vectors $\mathbf{a}_i$ and covectors $\mathbf{b}^i$ (i=1, N+1) can be approximated to leading order, by the right and left, respectively, eigenvectors of the Jacobian $\mathbf{J}_g$ of **g**(**z**) [1]. Based on the TSR approach, which was recently introduced by Valorani et al. [2,3], the local stretching rate of the dynamics of a system in the direction tangential to the vector field **g**(**z**)+**L**(**z**), can be given by (3):

$$\omega_{\tau,pde} = \sum_{i=1}^{N+1} W_{i,pde} \lambda_i \quad (3)$$

with λ_i being the eigenvalue of the i-th mode, $W_{i,pde}$ the weight of each CSP mode defined by $W_{i,pde} = \frac{h^i}{L+g} \sum_{k=1}^{N+1} \frac{h^k}{L+g} (\boldsymbol{a}_k . \boldsymbol{a}_i)$, and L+g is the norm of the vector field. Thus, it can be concluded that $\omega_{\tau,pde}$ is a linear combination of the timescales of the chemical system and it represents the stretching rate along the unit vector aligned with the state evolution direction due to the reactive-transport system. In case transport is ignored ($\boldsymbol{L}(\boldsymbol{z}) = \boldsymbol{0}$), TSR ($\omega_\tau$) is the stretching rate along the unit vector oriented in the direction of the chemical source term exclusively.

The contribution of each of reaction to the time scale ($\tau_n = \frac{1}{|\lambda_n|}$) of the n-th mode can be assessed through the Time scale Participation Index (TPI) [4]. Additionally, the Amplitude Participation Index (API) measures the relative contribution of the k-th process (reaction or transport term) to the amplitude h^n of the n-th mode [5,6]. On the basis of the API, Valorani et al. introduced the TSR Participation Index ($H_k^{\omega_{\tau,pde}}$), which is a sum of the API of the k-th reaction to all modes, each one weighted by the contribution of that mode to the TSR [2,3]. This tool provides a more generalized view for the contribution of each process to the local evolution of the system. Therefore, the summation of all $H_k^{\omega_{\tau,pde}}$ that relate to reactions can provide a clear understanding about the role of chemical kinetics in the system. Likewise, the summation of the contributions of all the convective/diffusive terms ($H_{Conv}^{\omega_{\tau,pde}}/H_{Diff}^{\omega_{\tau,pde}}$) can clarify the role of convection/diffusion into the system and their sum ($H_{Transport}^{\omega_{\tau,pde}}$) can be used to determine the regions where transport prevails over kinetics. A similar approach has been proposed in the past in the context of CSP [5,7], where the summation of the individual terms was performed over the determined dominant mode.

Due to the large orders of magnitude of the $\omega_{\tau,pde}$ and ω_τ, the logarithmic quantities of these variables will be used as follows: $\Omega = Sign(\omega_\tau)\cdot Log_{10}|\omega_\tau|$, $\Omega_{pde} = Sign(\omega_{\tau,pde})\cdot Log_{10}|\omega_{\tau,pde}|$, $\Delta = Sign(\delta(\omega))\cdot Log_{10}|\delta(\omega)|$.

3 Laminar Flame

A premixed H2/air laminar flame was generated using Chemkin-Pro [8], at p(0)=1 atm, T(0)=300 K, φ=0.7. The detailed mechanism of Burke et al. [9], including 9 species and 23 reversible reactions, was used for all simulations, both laminar and turbulent. The reactions that were found to be the most significant ones and will be further referred to in the current study are summarized in Table 1.

1f	$H+O_2 \rightarrow O+OH$	11f	$H+O_2\,(+M)\rightarrow HO_2\,(+M)$
4f	$H_2+OH\rightarrow H_2O+H$	13f	$HO_2+H\rightarrow 2OH$
4b	$H_2+OH\leftarrow H_2O+H$		

Table 1: The reactions providing significant contribution to the dynamics of the system.

Figure 1 on the left displays the evlution of the species mass fractions agaisnt a temperature based progress variable $c_T = \frac{T-T_r}{T_p-T_r}$, where T_r is the temperature of the fresh reactants and T_p the temperature of the combustion products. It is shown that HO_2 increases earlier than H_2O_2 and OH and OH increases earlier than H radical.

Moreover, Fig. 1 on the right, displays the evolution of Ω, Ω_{pde} and the eigenvalues of the system against c_T. Since the right hand side (rhs) of (1) is zero in a steady system like the considered laminar flame, Ω_{pde} was calculated using diffusion and excluding convection. The region where Ω_{pde} is positive (i.e., $0.012 < c_T < 0.586$) is identified as the flame front, thus establishing the upstream zone for $0.0 < c_T < 0.012$ and the downstream region for $0.586 < c_T < 1.0$. The evolution of Ω indicates that the dynamics of the system when accounting for kinetics and excluding any transport phenomena becomes explosive in a much thinner region compared to the full reactive-transport system. This suggests that the role of diffusion is to enhance the explosive dynamics in the flame front region. In addition, in the first half of the flame front ($0.012 < c_T < 0.285$), Ω_{pde} tracks the fast explosive time scale of the system, later gradually decelerates ($0.285 < c_T < 0.409$), tending to track the slow explosive timescale of the system as a result of the large increase of the amplitude of the slow explosive mode. For instance, when c_T=0.408, the amplitude of the fast explosive mode is $f^{e,f}$=0,79875 while the respective amplitude of the slow explosive mode is $f^{e,s}$=254.76. Afterwards ($0.409 < c_T < 0.586$), Ω_{pde} accelerates until it abruptly turns negative suggesting the end of the flame front. It has to be noted though, that in the whole flame front region the largest contributor to Ω_{pde} is always the fast explosive mode. The difference between Ω and Ω_{pde} can be used to identify regions where the dynamics is kinetically driven, thus offering a ready characterization of the flame topology; when the difference of the two observables is negligible it suggests that the dominant dynamics of the reaction-transport system resembles a lot the dynamics of the reactive system, therefore kinetics is dominant. This is the case for the downstream region, where as it will be also confirmed later, the system is essentially kinetically driven.

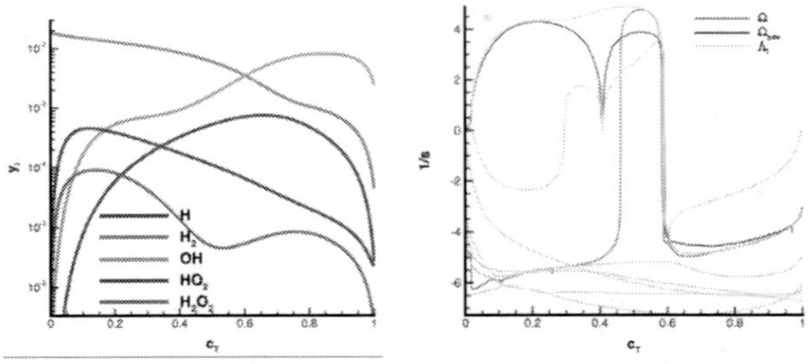

Figure 1: For the laminar premixed H_2/air flame (p(0)=1 atm, T(0)=300 K, φ=0.7). Left: The evolution of species mass fractions against c_T. Right: The evolution of Ω, Ω_{pde} and the eigenvalues Λ_i of the system against c_T.

For the assessment of the role of kinetics/diffusion in each part of the laminar flame, the values of $H_{Chem}^{\omega_{\tau,pde}}/H_{Diff}^{\omega_{\tau,pde}}$ were calculated throughout the computational domain and the results are displayed in Fig. 2. These results indicate that the first and largest part of the flame front ($0.012 < c_T < 0.3274$) is dominated by diffusion; the next and second largest part of the flame front is dominated by kinetics ($0.3274 < c_T < 0.56$) while at the last part, which is confined to a very thin region ($0.56 < c_T < 0.586$), diffusion prevails over kinetics.

In order to gain deeper physical understanding the largest TSR PI are displayed on the right of Fig. 2.

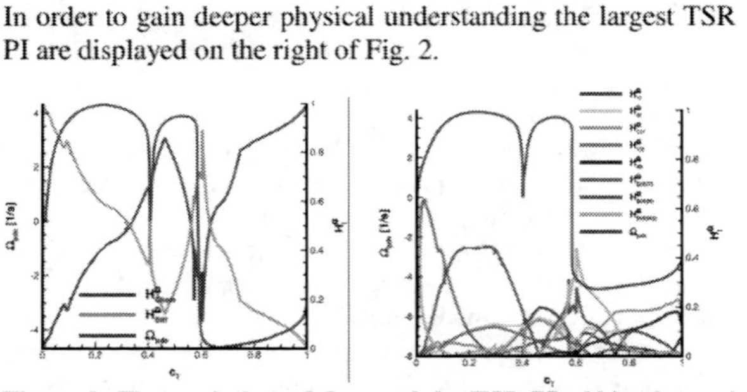

Figure 2: The evolution of Ω_{pde} and the TSR PI of kinetics and diffusion (left) along with the largest individual contributors (right), for the laminar premixed H_2/air flame (p(0)=1 atm, T(0)=300 K, φ=0.7).

In the first part of the flame front, which was found to be dominated by diffusion, the heat diffusion (initially) and the diffusion of H have the largest contributions to the dynamics of the system. In terms of kinetics, the most significant reactions are 4f, 13f and 11f, the first two being chain carrying and branching and the last one being a chain terminating. All these reactions (especially 4f and 11f) are the major exothermic reactions that contribute to the temperature increase in the flame front region. Moreover 11f and 13f involve the consumption of H radicals. Later, in the region where kinetics becomes more important than diffusion to the dynamics of the system, the chain branching 1f becomes the major process, followed by reactions 4f and 13f. Reaction 1f is a highly endothermic reaction, therefore its action at this part of the domain, where temperature has increased sufficiently ($\sim 1200\ K$), can be justified. At the last thin region of the flame front that is prevailed by diffusion, the contributions of the diffusion of H_2 and H gradually increase and later decrease in the downstream region. In the downstream region where the contribution of kinetics is larger than that of diffusion, the chain terminating reaction 11f plays a major role, which is expected considering that the system is driven to equilibrium. Moreover, reactions 4f and 4b gradually equilibrate each other.

4 Turbulent Flame

The DNS solution of a premixed H2/air flame representative of the thin reaction zone regime, previously obtained by Im et al. [10,11], was used for the current analysis. In particular, numerical solutions at three different times were selected and analyzed. For the selection of the time steps, the results displayed in Fig. 3 [10,11] were used. Figure 3 shows the temporal evolution of the volumetric heat release rate normalized by the reference laminar flame value, further scaled by the nondimensional flame surface area. Also, the time axis is normalized by the integral eddy turnover time for each turbulent Passot-Pouquet profile, which is 731 μs. An investigation of the burning rates for a temperature based progress variable value of c=0.1 and c=0.5 reveals that the normalized heat release rate approximates an asymptotic value of 0.88 and 0.95 when c=0.1 and c=0.5, respectively. The three selected times aim to be representative of the aforementioned trend of the burning rate. The first time step was selected to be $t_1 \approx t_l = 731$μs, the second was selected to be $t_2 \approx 2 * t_l = 1462$μs and the third one $t_3 \approx 4 * t_l = 2920$μs.

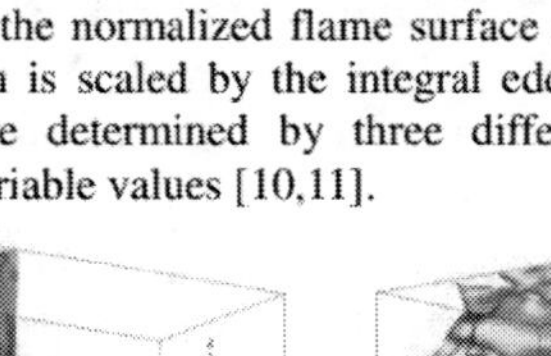

Figure 3: Normalized volume-integrated heat release rate divided by the normalized flame surface area as a function of time, which is scaled by the integral eddy time scale. Flame surfaces are determined by three different choices of the progress variable values [10,11].

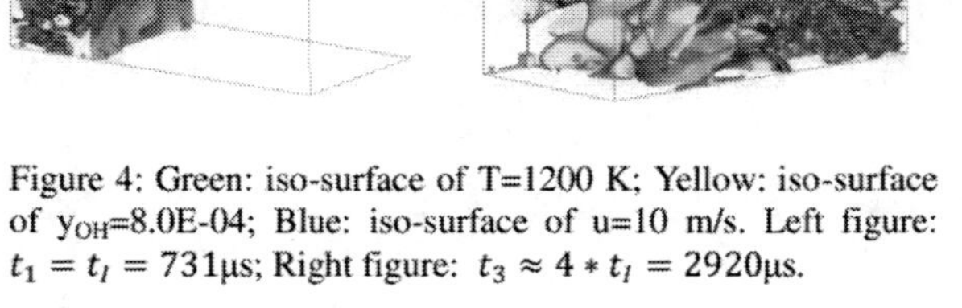

Figure 4: Green: iso-surface of T=1200 K; Yellow: iso-surface of y_{OH}=8.0E-04; Blue: iso-surface of u=10 m/s. Left figure: $t_1 = t_l = 731$μs; Right figure: $t_3 \approx 4 * t_l = 2920$μs.

Figure 4 highlights the effect of the increase of turbulence from t_2 to t_3 on the iso-surfaces of T=1200 K and y_{OH}=8.0E-04, which become highly wrinkled. It is also shown that the iso-surface of y_{OH} detaches from the iso-surface of temperature as a result of the increased flame/turbulence interaction.

For the three time steps under investigation, the conditional averages of Ω and Ω_{pde} were calculated for the whole 3D domain, against c_T. Their results are displayed in Fig. 5, while on Fig. 6, the standard deviation of these observables are illustrated.

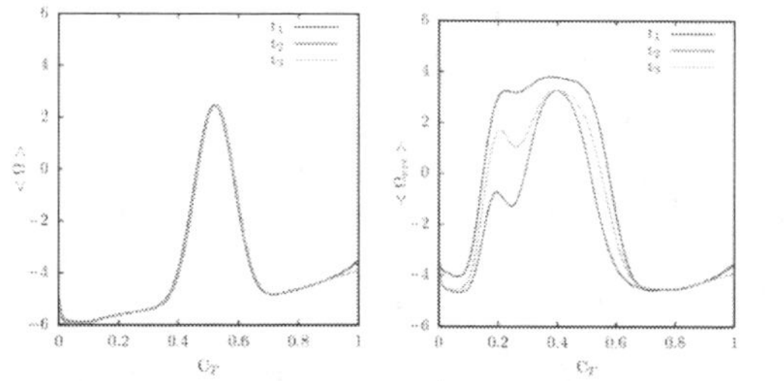

Figure 5: The conditional averages of Ω (left) and Ω_{pde} (right) for the three turbulent time steps under investigation.

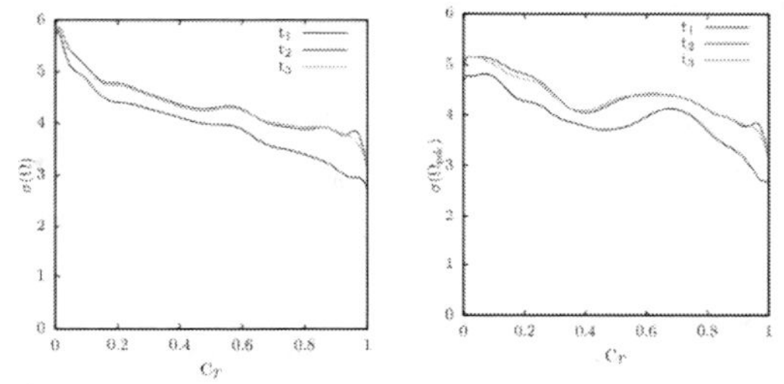

Figure 6: The standard deviation of Ω and Ω_{pde} for the three turbulent time steps under investigation, expressed as $\log_{10}(\sigma(\Omega))$.

Figure 5 proves that Ω, Ω_{pde} are well defined in all time steps, regardless the effect of turbulence. The effect of turbulence on the average of Ω is insignificant, as opposed to Ω_{pde}, which is highly affected, primarily in the flame front region, as this is indicated by the positive values of Ω_{pde}. Additionally, a distinguishable trend of the evolution of Ω_{pde}, is not observed as time increases, since from t_1 to t_2 the region where $\Omega_{pde}>0$ decreases, but from t_2 to t_3 the respective region slightly increases. The difference between Ω and Ω_{pde} in the downstream region enables a characterization of the flame topology which resembles the laminar one.

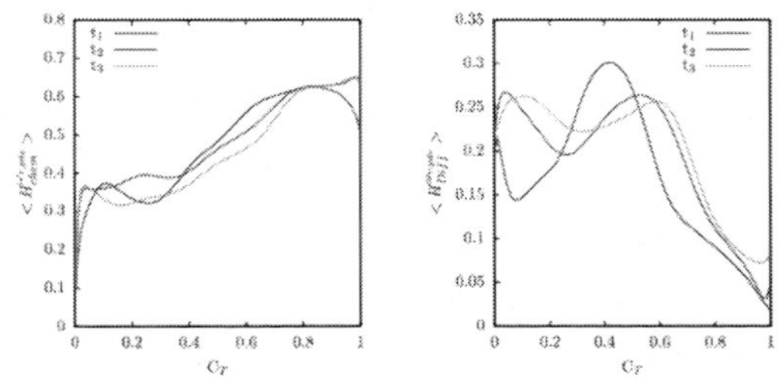

Figure 7: The evolution of TSR PI of kinetics ($H_{Chem}^{\omega_{\tau,pde}}$) (left) and diffusion ($H_{Diff}^{\omega_{\tau,pde}}$) (right) for the three time steps under investigation.

The examination of the standard deviation of Ω and Ω_{pde} reveals that it increases as time increases, although this effect seems to be more pronounced in the case of Ω_{pde}. This is reasonable, considering that Ω_{pde} accounts for the full system while Ω is based on the chemical source term of the system. However, interestingly enough, the dispersion seems to be increased in low temperatures, for both variables.

In order to assess the role of kinetics and diffusion/convection in the system, the conditional averages of $H_{Chem}^{\omega_{\tau,pde}}/H_{Diff}^{\omega_{\tau,pde}}$ against c_T were calculated for all three time steps. Figure 7 shows that, in overall, the effect of kinetics on the system drops as time increases. On the other hand, there is no clear trend of $H_{Diff}^{\omega_{\tau,pde}}$ as time increases, except for the downstream region where it seems that the effect of diffusion decreases.

Additionally, it is shown that in general, the effect of kinetics is more pronounced than diffusion. Although not shown here, it has been found that convection is the dominant process in the preheat and upstream zone . In the flame front region, although kinetics has the largest contribution, the combined effect of diffusion and convection outweighs that of kinetics. It is interesting, though, that the effect of convection is always larger than that of diffusion, even in the flame front region. Finally, in the downstream region, kinetics was found to be the major contributor to the dynamics of the system. All these results are consistent with the findings that were previously reported for the laminar flame.

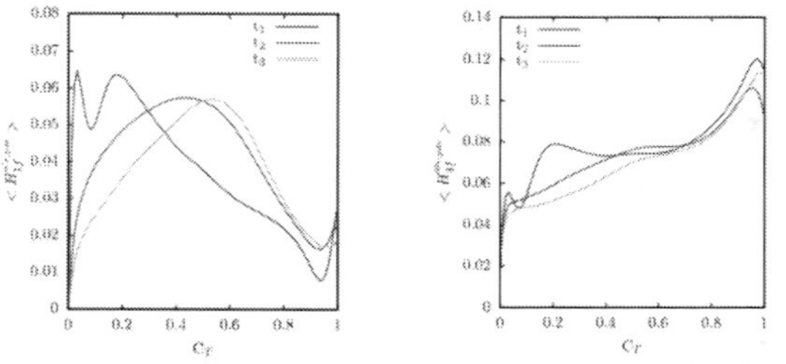

Figure 8: The evolution of TSR PI of reaction 1f ($H_{1f}^{\omega_{\tau,pde}}$) (left) and 4f ($H_{4f}^{\omega_{\tau,pde}}$) (right) for the three turbulent time steps under investigation.

To obtain a more detailed understanding about the individual role of each process, the conditional averages of the largest TSR PI were calculated for the three time steps. Due to the limited space, Fig. 8 includes the TSR PI of reactions 1f and 4f while Fig. 9 includes the TSR PI of thermal convection and diffusion of H.

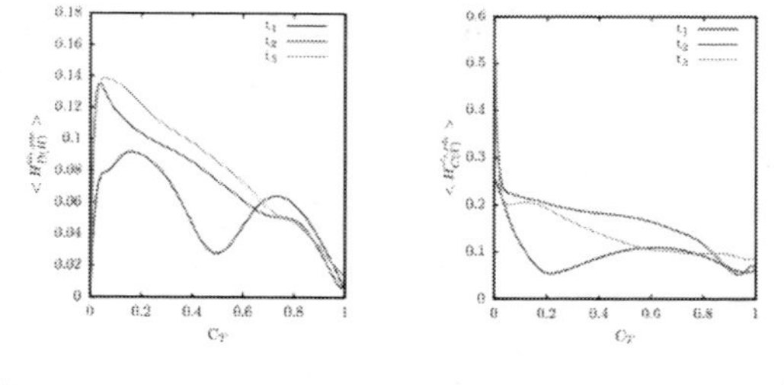

Figure 9: The evolution of TSR PI of diffusion of H ($H_{D(H)}^{\omega_{\tau,pde}}$) (left) and the thermal convection ($H_{C(T)}^{\omega_{\tau,pde}}$) (right) for the three turbulent time steps under investigation.

Overall, the contribution of 4f is larger than 1f, as opposed to the laminar case, where it was found that the contribution of 1f is larger than 4f in the flame front region. Moreover, the contribution of 4f gradually increases from $c_T \approx 0.05$ up to $c_T \approx 0.9$, for all time steps (with a small exception for t_1, where it exhibits a decreasing behavior in the region of $0.2 < c_T < 0.4$). On the other hand, the contribution of 1f initially increases, reaches a maximum in the flame front region and afterwards decreases (with a small exception for t_1 in the region of $0.02 < c_T < 0.2$).

The conditional average of $H_{D(H)}^{\omega_{\tau,pde}}$ reveals that the effect of the diffusion of H decreases as temperature increases. Additionally, a distinguishable trend is not observed as time increases. Thermal convection seems to be the dominant convective process at all time steps, especially in the upstream/preheat zone and its contribution gradually decreases, reaching approximately 10% in the downstream zone.

4 Conclusions

The use of CSP tools and the TSR approach provided a detailed physical understanding on the processes that contribute to the dynamics of a 3D turbulent flame, modeled with detailed hydrogen chemistry. At all the time steps investigated, TSR successfully identified the flame front, along with the upstream and downstream zones. The roles of kinetics and transport to the different parts of the domain were quantified and discussed.

5 Acknowledgment

This work was sponsored by competitive research funding from King Abdullah University of Science and Technology (KAUST). The work of PPC, RMC and MV was also supported by the Italian Ministry of University and Research (MIUR). The simulations presented in this work made use of KAUST Supercomputing Laboratory

References

[1] S. H. Lam, D. A. Goussis, Proc. Combust. Inst. 22 (1989) 931-941.

[2] M. Valorani, S. Paolucci, E. Martelli, T. Grenga, P. P. Ciottoli, Combust. Flame 162 (8) (2015) 2963-2990.

[3] M. Valorani, P. P. Ciottoli, R. M. Galassi, Proc. Combust. Inst. 36 (1) (2017) 1357-1367.

[4] D. Goussis, G. Skevis, in: K. J. Bathe (Ed.) Computational fluid and solid mechanics, Elsevier, Amsterdam, 2005, p. 650-653.

[5] J. Prager, H. Najm, M. Valorani, D. Goussis, Combust. Flame 158 (11) (2011) 2128-2144.

[6] D. Diamantis, E. Mastorakos, D. Goussis, Combust. Theor. Model. 19 (3) (2015) 382-433.

[7] M. Hadjinicolaou, D. Goussis, SIAM J. Sci. Comput. 20 (3) (1998) 781-810.

[8] CHEMKIN-PRO, R. (2011) 15112, Reaction Design. Inc., San Diego, CA.

[9] M. Burke, M. Chaos, Y. Ju, F. L. Dryer, S. J. Klippenstein, Int. J. Chem. Kinet. 44 (7) (2012) 44-474.

[10] H. G. Im, P. G. Arias, S. Chaudhuri, H. A. Uranakara, Combust. Sci. Technol. 188 (8) (2016) 1182-1198.

[11] P. Arias, H. Uranakar, S. Chaudhuri, H. Im, APS Meeting Abstracts (2015, November)

11th Asia-Pacific Conference on Combustion,
The University of Sydney, NSW Australia
10th -14th December 2017.

Preliminary investigation by experiment on the premixed MILD combustion of C_3H_8 in a cylindrical furnace

Kin-Pang Cheong[1, 2], Guochang Wang[1], Bo Wang[1], Jianchun Mi[1]

[1]College of Engineering, Peking University

Beijing, 100871, China

[2]Department of Mechanical and Automation Engineering, The Chinese University of Hong Kong

New Territories, Hong Kong

Abstract

This paper reports the experimental results of the premixed MILD combustion of propane (C_3H_8) in a cylindrical furnace with a single jet burner. Cases under different inlet diameters, thermal inputs and equivalence ratios are investigated. Results show that premixed MILD combustion can be established over wide ranges of inlet velocity and equivalence ratio. However, it becomes unstable when the inlet velocity is less than 20 m/s. In-furnace measurements indicate that temperature and species distribute closely uniformly under premixed MILD combustion. Relatively low exhaust NO emissions (< 50 ppm) are achieved. Under current experimental conditions, the relationship between the NO emission and mean temperature of the exhaust gas may be considered approximately linear. This case is quite similar to the combustion in a well-stirred reactor, whose NO emission also has a nearly linear relationship with the exhaust temperature over the range of 1100~1500K.

1 Introduction

Premixed combustion is of great importance for the fundamental research of combustion [1]. In practical applications, premixed flame burns more sufficiently and produces less NOx [2] and soot [3] versus diffusion flame. On the other hand, MILD combustion as a novel phenomenon for low NOx combustion has been extensively investigated by experiments and numerical simulations [4,5]. However, most of these investigations have focused on non-premixed combustion; consequently, premixed MILD combustion research is limited. This is mainly due to the fact that MILD combustion is discovered and developed from high temperature exhaust gas recycling combustion [5], which cannot be operated with premixed reactants. Although the researchers later discovered that MILD combustion can be established by controlling the inlet momentum rate and internal entrainment without preheating [6–8], the premixed MILD combustion still lacks sufficient attention. Only very few literatures for premixed MILD combustion are available comparing to those of the non-premixed case.

Mi et al. experimentally and numerically investigates the effects of initial condition on the establishment of fully premixed and partially premixed MILD combustion [9–12]. They found that once the inlet momentum is higher than the critical value, the MILD combustion occurs. Increasing the inlet momentum does not significantly alter the combustion and emission characteristics in the furnace, indicating that MILD combustion is a stable combustion mode [9]. Then they further develop a critical Reynolds number standard from the critical momentum criterion, so that the conclusion can be extended to different furnaces [10]. By hydrogen addition, Arghode and Gupta reduced the CO emission from premixed MILD under gas turbine conditions, and achieved higher intensity and better stability [13]. Minamoto et al. [14] used DNS to analyze the micro-scale morphology of the premixed MILD combustion reaction zone and found that there are various structures interacting and predominated by the pancake-like structures. In contrast, the traditional premixed combustion mainly consists of sheet-like structures.

Though these investigations provided some insight into premixed MILD combustion, there are still problems remains unsolved. For example, the mechanism of 'flameless', the details within the MILD combustion furnace, etc. To fill the gap, a laboratory scale platform of MILD combustion is built to further explore the fundamental and practical characteristics of MILD combustion. This paper preliminarily reports some detailed in-furnace temperature and species measurements and NO emissions of premixed MILD combustion of propane.

2 Experimental Setup

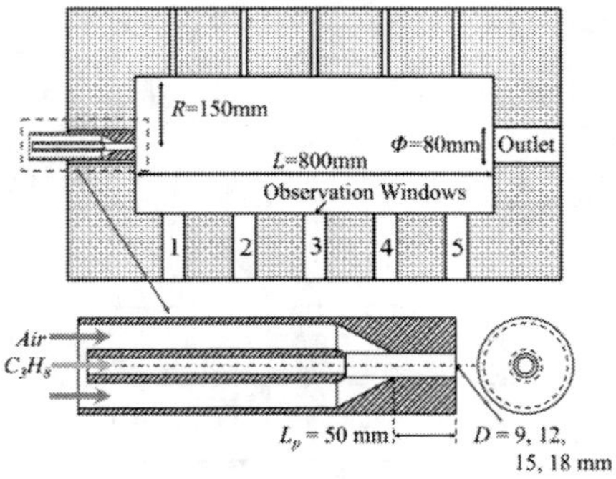

Figure 1: Geometry of the furnace and the burner.

The combustion chamber of the cylindrical furnace is 800 mm in length (x) and 150 mm in radius (R), see Fig. 1. On the furnace, there are five equally spaced pluggable observation windows with quartz glass. On the opposite side of the windows, there are five measuring ports for either temperature or species

Kin-Pang Cheong. E-mail: malcolmcheong@pku.edu.cn
Jianchun Mi. E-mail: jmi@pku.edu.cn

measurements. The diameter of the furnace outlet is 80 mm. During combustion, the furnace is slightly pressurized to prevent ambient air leakage. Figure 1 also presents the details of the single jet burner. Initially, the fuel and the oxidant are separately transported prior to a premixing tube with a constant length of L_p = 50 mm and varying nozzle diameters of D = 9, 12, 15, 18mm. The fuel and air then meet and mix in the tube before entering the chamber. Consequently, the MILD combustion within the furnace is established by the single turbulent jet of premixed reactants. Note that the perfectly mixing at the molecular level between fuel and oxidant might not achieve. This is because the premixing length is limited. Yet, the premixing might still complete since the mixing process continues when the reactant enters the chamber before ignition. So, all the cases of investigation are considered as well premixed combustion. Table 1 summarizes the inlet conditions of the single turbulent jet for the tested cases.

Case	D (mm)	Thermal Input P (kW)	Φ	U (m/s)
1	9	15	1.0~0.6	63.30~103.82
2	12	10, 15, 20	1.0~0.6	35.61~58.40
3	15	15	1.0~0.6	22.79~37.38
4	18	15	0.9	11.89

Table 1: Summary of the inlet conditions of the tested cases.

For the fuel supply, high-pressured liquefied propane (purity 98%) is used after evaporating into gaseous phase and balancing temperature with ambient. The flow rate of the fuel is adjusted by a mass flow controller. The air is supplied by a Roots type blower. The uncertainties in the flow rates of fuel and air are ±0.2% and ±2%, respectively.

Before establishing MILD combustion, the temperature of the furnace has to be warmed up to 1000□ for auto-ignition using conventional combustion, which usually takes about 30 minutes. After switched to MILD combustion, 2.5 hours is required for the furnace to reach the steady state and thermal balance with the environment. During the combustion process, the radial profiles of time-averaged temperature are measured at the five measuring ports by traversing the B-type thermocouples (Pt-30%Rh vs. Pt-6%Rh). In order to characterize the temperature distribution in the furnace in detail, 15 uniformly distributed radial measuring points are made at each measuring port. The mean temperature are determined by averaging the data acquired in 5 minutes after a 5 minutes period for steady-state operation. The in-furnace and exhaust gas concentrations are measured by a TESTO 350 portable flue gas analyzer with a self-designed sampling probe, and all the data are dry-based. Table 2 lists the nominal accuracies and estimated uncertainties of temperature and gas composition measurements.

Data	T (K)	O_2 (vol%)	CO_2 (vol%)	CO (ppmv)	NO (ppmv)
Accuracy	±0.25%	±0.2	±1%	±5%	±1ppm
Uncertainty	±4%	±2.5%	±3.5%	±7.5%	2.5%

Table 2: Nominal accuracies and uncertainties of temperature and species measurements.

3 Results and Discussion

3.1 Observations

Premixed MILD combustion can be established easily for Cases 1~3 over different thermal input rate P (10~20kW) and global equivalence ratio Φ (1.0~0.6). However, for Case 4 (D = 18mm), combustion cannot sustain due to the flashback in the furnace, resulting termination of examining Case 4. Figure 2 presents typical images of the combustion state within the furnace under different inlet diameters (thus different inlet velocities) with thermal input P =15 kW and global equivalence ratio Φ = 0.9. All the photos are taken from window 1 with preset camera settings. As shown in Fig. 2, uniform luminous intensity and no flame front can be observed under MILD combustion. From these cases, we can also see that the minimum inlet velocity required by premixed MILD combustion of propane is about 20m/s, which is quite low comparing to those reported for MILD combustion of methane in the literature [9,15]. This is due to the fact that the density of propane is larger than methane, resulting in a lower velocity for the same momentum rate required for establishing MILD combustion. The minimum momentum rate is about 0.1 kg/s^2.

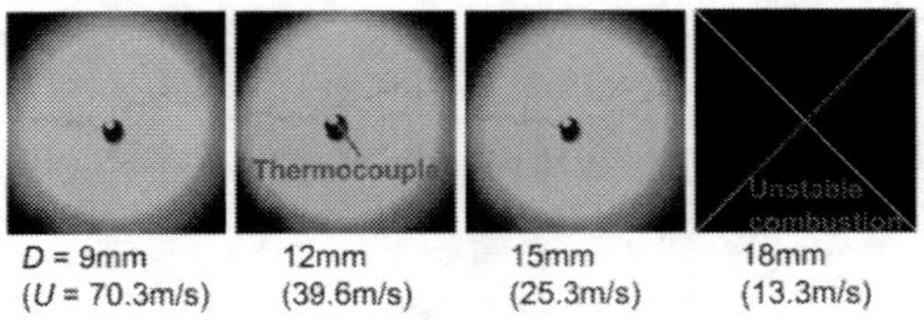

Figure 2: Observations of MILD combustion under different inlet diameters from Window 1 (P = 15kW, Φ = 0.9).

3.2 In-furnace Measurements

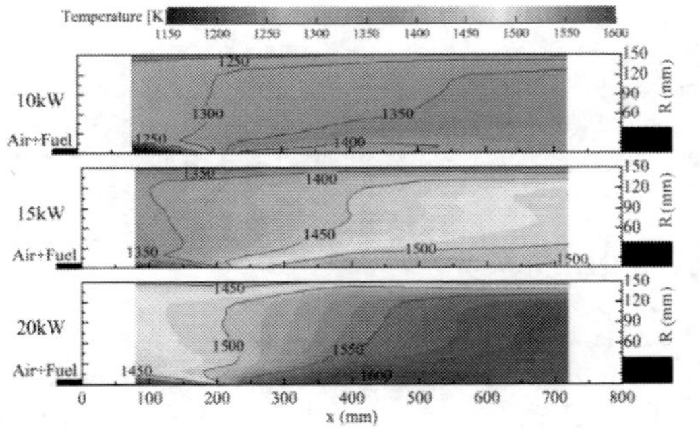

Figure 3: Measured temperature distributions in the furnace under different thermal inputs (D = 12mm, Φ = 0.9).

Figure 3 shows the temperature distributions within the furnace under different thermal inputs with D = 12mm and Φ = 0.9. Similar distributions can be observed as P increases from 10kW to 20kW, resulting in the maximum temperatures of 1400K, 1500K and 1600K, respectively. The temperature differences between the main reaction zone and the wall are less than 150K, indicating the uniformity of the temperature field and the efficient heat transfer for the current MILD combustion established by a single jet burner.

Figure 4 presents the detailed in-furnace concentration distributions of O_2, CO_2, CO and NO with P = 15 kW, D = 12mm and Φ = 0.9. All the species distributed quite uniformly within the furnace, even the high CO concentration region only appears near the nozzle. The species distributions further indicates that the oxidation of propane is quite homogenous under MILD combustion. The NO concentration is lowered in the corresponding region due to the reducibility of CO. Moreover, the small region of high CO concentration also demonstrates the advantage of the premixed MILD combustion comparing to non-premixed combustion: the oxidation mainly proceeds under fuel lean condition once the initial air is excess and consequently soot hardly forms.

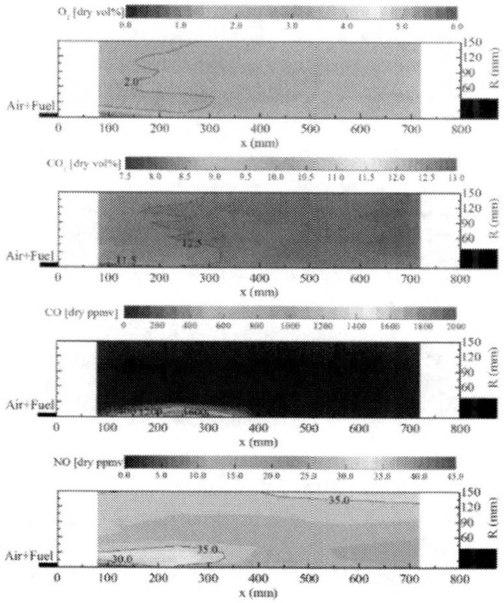

Figure 4: Species field under MILD combustion. From top to bottom: O_2, CO_2, CO, NO. (P = 15kW, D = 12mm, $\Phi = 0.9$)

3.3 Exhaust Emissions

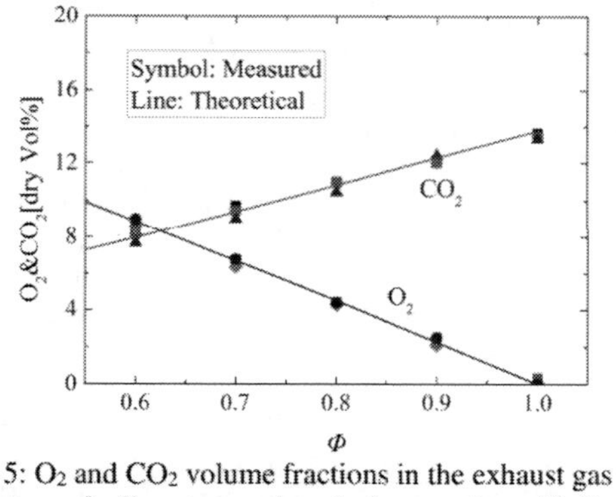

Figure 5: O_2 and CO_2 volume fractions in the exhaust gas under different nozzle diameters and equivalence ratios with P = 15 kW.

Figure 5 presents the effect of equivalence ratio on O_2 and CO_2 mole fractions in the exhaust gas for different cases under P = 15 kW. Measured data are plotted as symbols and the lines represent the theoretical values of complete combustion of propane. As Φ varies from 1.0 to 0.6, O_2 is over supplied and its concentration changes in the exhaust from near 0 to around 9%. Since the amount of CO_2 is the same after complete combustion, the concentration of CO_2 reduces as Φ decreases. It is determined that complete combustion is achieved in the cases of Φ = 0.9~0.6 since the measured CO emissions in the exhaust gas for these cases are less than 10ppm (not shown). The relative error between the measured and theoretical values are within 7%, implying that the current concentration measurements of O_2 and CO_2 are reliable. The error may be caused by the uncertainties in gas supplies, the sealing of the furnace and the errors during the data acquisition.

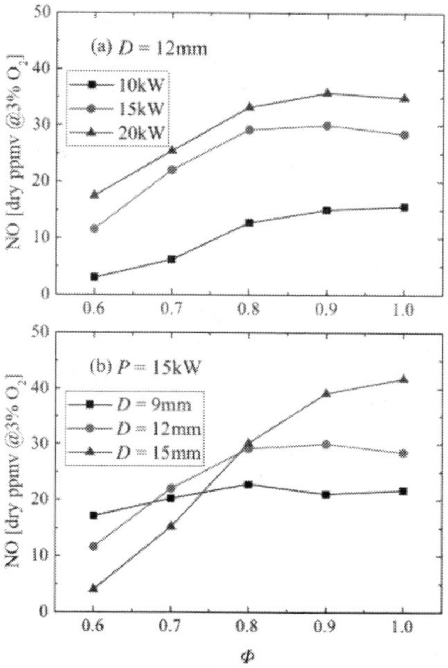

Figure 6: Variations of NO emissions under MILD combustion with different (a) thermal input P and (b) nozzle diameter D.

Figure 6 shows the relationships between Φ and NO emissions from premixed MILD combustion of C_3H_8 under different thermal inputs and inlet diameters (velocities). Evidently, the NO emission is lowered as Φ decreases. Figure 6(a) also indicates that increasing the thermal input (thus furnace temperature) will increase the final NO emission. However, due to the suppression of thermal NO under MILD combustion, the overall NO emission is within 50ppmv for the thermal input up to 20kW. Moreover, Figure 6(b) demonstrates that the NO emission from the case of D = 9mm is much lower than those of D = 12 mm and 15mm in the range of Φ =1.0~0.8; it remains stable over Φ =1.0~0.6. This implies that a high inlet velocity helps to reduce the NO emission from the premixed MILD combustion and stabilizes the combustion under different equivalence ratios. On the other hand, the emissions from the case of D = 15mm reduces rapidly from 40ppmv to a few ppmv as Φ varies from 1.0 to 0.6. The main reason is that the NO emission positively relates with the mean temperature of the exhaust gas, and the latter significantly alters with equivalence ratio and thermal input. To further investigate the relationships, the NO emission versus T plot is displayed in Figure 7.

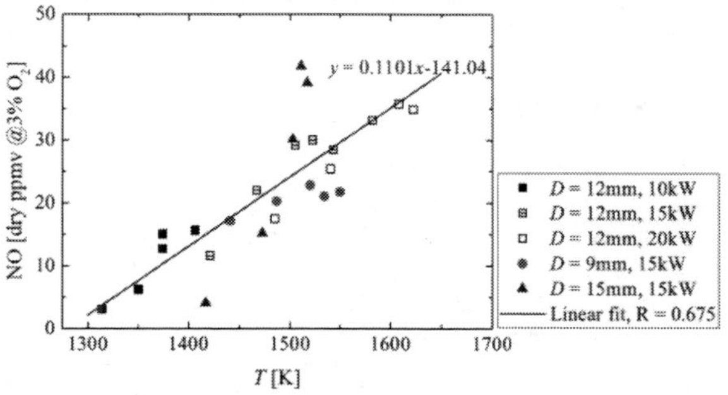

Figure 7: Relationship between mean exhaust gas temperature and NO emission under MILD combustion.

The temperature of the measuring point (x = 720mm, r = 0mm) that is closest to the outlet is chosen as the mean temperature of the exhaust gas $T_{exhaust}$. By plotting the measured NO emissions in Fig. 7, a linear relationship between NO emission and $T_{exhaust}$ can be observed. This finding is similar to that of non-premixed MILD combustion reported by Szegö et al.[7] Figure 7 also shows that the high inlet velocity enhances the linearity of the NO-$T_{exhaust}$ relationship. To better understand this phenomenon, chemical calculations of WSR are performed to simulate the local combustion within the furnace, as shown in Fig. 8.

The species concentrations for WSR are calculated according to the inlet conditions in the experiments. For example, under Φ = 0.9 and 90% dilution, the species concentrations for the inlet of WSR are 3.84%C_3H_8, 28.95%O_2, 9.97%CO_2, 13.30%H_2O and 73.44%N_2. The ratio between the flowrates of reactant and the volume of the furnace is applied as the residence time. The reactor temperature taken for the local temperature varies from 1000K~1300K. The USCMech II [16] together with the NOx chemistry from the GRI-Mech 2.11 [17] is applied as the chemistry mechanism. Since the higher the inlet velocity, the better the mixing within the furnace will achieve. The local dilution ratio in the furnace will tend to be the portion limit of the flue gas under complete mixing with the reactant. The maximum of this limit is about 0.92 for the present test cases. Thus, presuming that MILD combustion occurs after the mixing of reactant and flue gas, the WSR MILD combustion under different dilution ratios are calculated to obtain the NO-$T_{exhaust}$ relationship under the ideal condition, as shown in Fig. 8. Clearly, we can see that the NO emissions under different dilution ratios are all exponentially related to the mean temperature. As the dilution ratio increases, the NO emissions become less sensitive to T. Nevertheless, for the temperature ranging from 1100K~1500K, the NO-T relationship even looks closely linear. This explains the results plotted in Fig. 7 and the better linearity by higher inlet velocity for the NO-$T_{exhaust}$ relationship. Moreover, it also implies that the premixed MILD combustion is similar to the combustion in a well-stirred reactor.

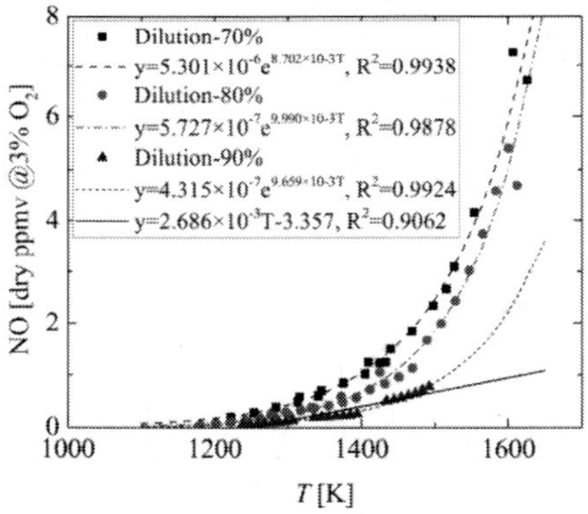

Figure 8: Relationship between mean exhaust gas temperature and NO emission in a WSR under MILD combustion.

4 Conclusions

This paper reports the preliminary investigation of the premixed MILD combustion of propane established by a single jet burner in a laboratory-scale cylindrical furnace. Observations and systematic measurements of in-furnace profiles and exhaust emission of NO are provided under different global inlet conditions. The main conclusions may be drawn below:

(1) Premixed MILD combustion of C_3H_8 can be established over wide ranges of inlet velocity and equivalence ratio. However, combustion becomes unstable when the inlet velocity is less than 20 m/s.

(2) Within the furnace, the temperature and species distributions are quite uniform. The temperature difference in the reaction zone is less than 150 K.

(3) NO emissions for all the tested cases are quite low (< 50 ppm). For the present measurements, the NO emission and mean temperature of the exhaust gas exhibit nearly linear relationship, although they are exponentially related under ideal condition. According to chemical calculations, the premixed MILD combustion is similar to the combustion in a well-stirred reactor, whose NO emission also has an approximately linear relationship with the exhaust temperature over the range of 1100~1500K.

5 Acknowledgment

The authors gratefully acknowledge the support of Natural Science Foundation of China (No. 51276002). Also, KP recognizes the technical support from Beijing Peking University Pioneer Technology Co., Ltd.

References

[1] C.K. Law, C.J. Sung, Prog. Energy Combust. Sci. 26 (2000) 459–505.

[2] M.P.B. Musculus, P.C. Miles, L.M. Pickett, Prog. Energy Combust. Sci. 39 (2013) 246–283.

[3] D.R. Tree, K.I. Svensson, Prog. Energy Combust. Sci. 33 (2007) 272–309.

[4] A. Cavaliere, M. de Joannon, Prog. Energy Combust. Sci. 30 (2004) 329–366.

[5] J.A. Wünning, J.G. Wünning, Prog. Energy Combust. Sci. 23 (1997) 81–94.

[6] S. Kumar, P.J. Paul, H.S. Mukunda, Proc. Combust. Inst. 29 (2002) 1131–1137.

[7] G.G. Szegö, B.B. Dally, G.J. Nathan, Combust. Flame 154 (2008) 281–295.

[8] G.G. Szegö, B.B. Dally, G.J. Nathan, Combust. Flame 156 (2009) 429–438.

[9] J. Mi, P. Li, B.B. Dally, R.A. Craig, Energy Fuels 23 (2009) 5349–5356.

[10] P. Li, J. Mi, B.B. Dally, R.A. Craig, F. Wang, Energy and Fuels 25 (2011) 2782–2793.

[11] P. Li, J. Mi, Flow, Turbul. Combust. 87 (2011) 617–638.

[12] P. Li, F. Wang, J. Mi, B.B. Dally, Z. Mei, Energy Fuels 28 (2014) 2211–2226.

[13] V.K. Arghode, A.K. Gupta, Int. J. Hydrogen Energy 36 (2011) 6292–6302.

[14] Y. Minamoto, N. Swaminathan, S.R. Cant, T. Leung, Combust. Flame 161 (2014) 2801–2814.

[15] A.S. Veríssimo, A.M.A. Rocha, M. Costa, Energy Fuels 25 (2011) 2469–2480.

[16] H. Wang, X. You, A. V Joshi, S.G. Davis, A. Laskin, F. Egolfopoulos, C.K. Law, available at http://www.me.berkeley.edu/gri_mech/.

[17] C.T. Bowman, R.K. Hanson, D.F. Davidson, W.C. Gardiner Jr, V. Lissianski, G.P. Smith, D.M. Golden, M. Frenklach, M. Goldenberg, available at http://ignis.usc.edu/ USC_Mech_II.htm.

11th Asia-Pacific Conference on Combustion,
The University of Sydney, NSW Australia
10th -14th December 2017.

Time-resolved investigation of droplet size and velocity inside diesel fuel sprays

Zehao Feng, Mingzhi Zhang, Jiapei Yang, Chenglong Tang*, Zuohua Huang

State Key Laboratory of Multiphase Flow in Power Engineering,

Xi'an Jiaotong University, Xi'an, 710049, China

Abstract

To better understand the microscopic spray characteristics, measurements of temporal changes in the size and velocity of droplets inside diesel fuel sprays were conducted at a fixed measurement position of 40mm downstream from the injector tip. Particle/droplet image analysis (PDIA) system was applied to obtain the microscopic droplet images in a long time scale (0-10 ms) which consists of continuous injection stage (0-2.1 ms) and the spray free development stage (2.1-10 ms). The results show that the droplet size distribution has a significant variation at the beginning, but basically independent of time after 1.6 ms. The Sauter mean diameter (SMD) increases significantly at the continuous injection stage, but slightly decreases at the free development stage. In addition, it is observed that most droplets' velocity are 2-3 m/s while the maximum value is 25 m/s. An interesting similar bimodal shape of velocity distribution probability curves was found at different time after start of injection. What's more, the scalar velocities gradually decrease as time goes on, but the direction of the velocities is messy due to the air entrainment.

1 Introduction

In essence, combustion process is affected by the spatial and temporal distributions of spray within the combustion chamber [1]. The mixing process of fuel and air in the combustion chamber is one of the key processes governing the combustion and the pollutant formation in diesel engines[2]. A better understanding of the transient characteristics of diesel sprays is essential for further improvement of the engine performance to copy with the requirement of higher thermal efficiency and the increasingly stringent emission regulations.

Research works related to the macroscopic spray development such as the characteristics of the liquid phase and vapor penetration have been conducted by many researchers. These characteristics are now quite well known and have enable the development of very useful 1D spray models [3]. So, modern CFD computations are usable and reliable in the prediction of the spray development, which is very useful in the design of the shape of combustion chamber. However, relatively few studies have focused on the microscopic spray characteristics especially the size and velocity of the droplets, which are also of great importance to evaluate combustion and emission performance and for model validation [4]. Because spray in the chamber is a high-speed and three-dimensional unsteady two-phase flow process, it is a tough challenge to understand the droplets behavior accurately in a complete spray process. At present, phase Doppler anemometry is generally considered the most accurate method for spray characterization. However the measurement accuracy is dependent on a number of parameters including laser power, optical configuration, droplet homogeneity and concentration and requires skilled set-up and operation to obtain accurate results [5]. Another fundamental limitation of PDA is the inability to accurately measure non-spherical droplets as shown by Damaschke et al [6]. As the improved resolution, stability the image-based droplet sizing techniques has intensified significantly in recent years [7]. Particle/droplet image analysis (PDIA) as one of the techniques which uses an automated processing algorithm for analyzing digital images of two-phase flow can be more accurate to determine the properties of individual droplets or particles such as velocity, size, shape and particle concentration over a finite region of interest in the spray compared to PDA.

In the present study, The PDIA method was applied to measure the temporal changes in the velocity and size of droplets in the diesel sprays at a fixed measurement position of 40mm downstream from the injector tip. The measurement was conducted in a long time scale (0-10 ms) which consists of continuous injection stage (0-2.1 ms) and the spray free development stage (2.6-10 ms) and was conducted under constant high injection and ambient pressure which aimed at providing boundary conditions and validation data for spray computer models.

2 Experimental setup

Figure 1 shows the sketch of the test rig. It includes primarily the high pressure fuel supply assembly, synchronization circuit, constant volume chamber and optical path. A commercial high pressure common rail system, including the high pressure pipeline, high pressure pump, fuel filter, common rail and injector with the hole diameter of 0.2 mm.

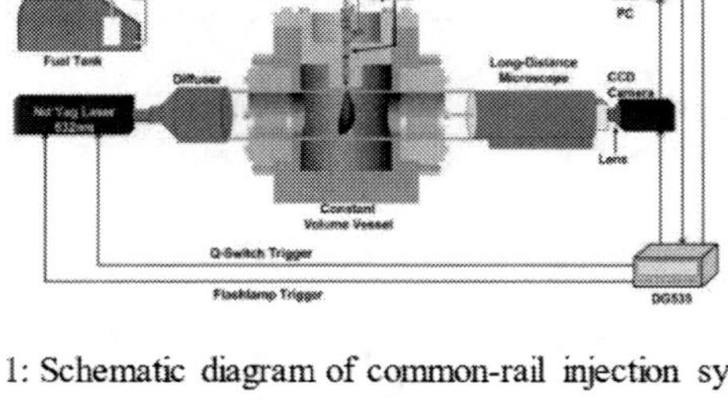

Figure 1: Schematic diagram of common-rail injection system

A double pulsed Nd:YAG laser with the wavelength of 532 nm is used as the light source. A diffusor is attached to the laser for

Chenglong Tang, Associate Professor. Fax: 86-29-8266-5075
E-mail address: chenglongtang@mail.xjtu.edu.cn

the homogeneous illumination. The diameter of the lens at the head of the diffusor is 120 mm, which is larger than the diameter of the chamber window, so that the whole spray flow field is illuminated. A CCD camera (ImagerProSX 5M), attached with a long focus microscope (Queststar QM1) is used to capture the local droplet distribution in spray. The camera and the long focus microscope are mounted on an electric positioner MC600. The precise displacement control (1 μm) of the electric positioner ensures the accurate determination of the scanning target position.

We fixed the measurement window at the location of $Z = 40$ mm below the nozzle tip and $R = 6$ mm from the spray axis. After scaling with a scaling plate (minimum scale of 25 μm and total length of 5 mm), the size of the measuring window is 1.8 × 1.5 mm as shown in fig. 2. The injection pulse width is set as 1.5ms in this study, however the actual injection duration is 2.1 ms more accurate measurement. Fig .3 shows pulse sequences of different trigger signals for microscopic spray characteristics investigation. By synchronization of the laser delay time and the injection width, the picture with droplets at a fixed delay time will be recorded by the long focus microscope and CCD camera at high resolution. In this study, injection and ambient pressure are fixed at 1000 bar and 20 bar separately. To capture the images of the droplets at different phases after the start of the injection (aSOI), the laser delay time (T, see Fig. 3) is gradually tuned with an accuracy of 1 μs. For each constant T, 100 distinct clear pictures after filter were recorded and no less than 6000 droplets were obtained for droplet statistics. A typical image of the local droplets after processed is shown in fig. 2, in which the diameter and velocity of each droplet are shown.

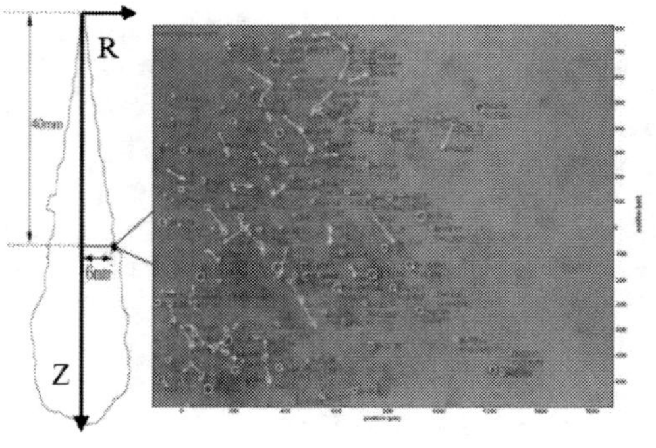

Figure 2: Measurement position and typical image of microscopic characteristics

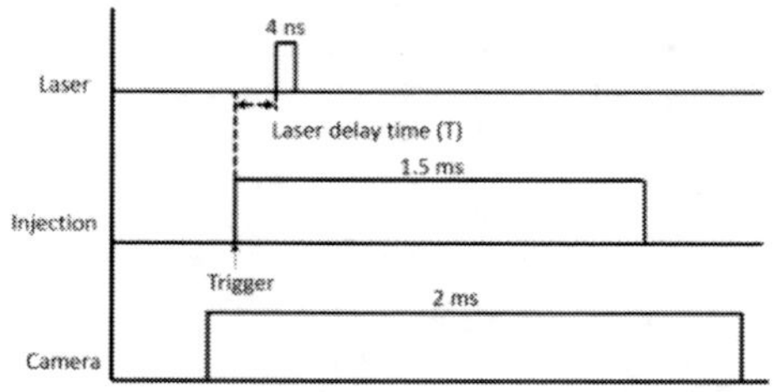

Figure 3: Time sequence control of microscopic characteristics measurement

3 Results and discussion

3.1 Droplet size statistics

Figure 4 shows the temporal droplet size distribution probability curves of different time after start of injection, in which the probability is highest for the intermediate droplet size (10-20 μm), and similar unimodal curves for different time are observed. It is worth noting that the first time when the droplets were captured was at 0.6 ms after injection. In the next 1 ms, the curve is shifted to the larger droplet side obviously. For the small diameter droplets, the probability of $t = 0.6$ ms is the highest and for the big diameter droplets, opposite result was found. In addition, after 1.6 ms, little difference was found between the curves. The variation of maximum probability diameter and its corresponding peak probability with time was shown in Fig. 5. It was seen that the maximum probability diameter increases before 1.6 ms as time goes on, despite of the peak probability decreases. In addition, the two parameters remain basically stable after 1.6 ms.

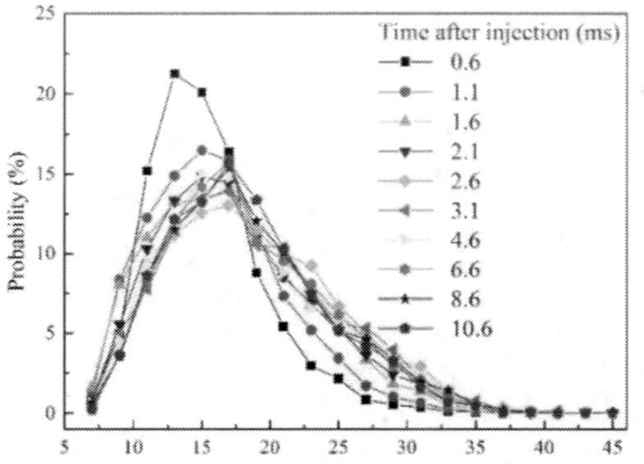

Figure 4: Droplet size distribution of different time after start of injection

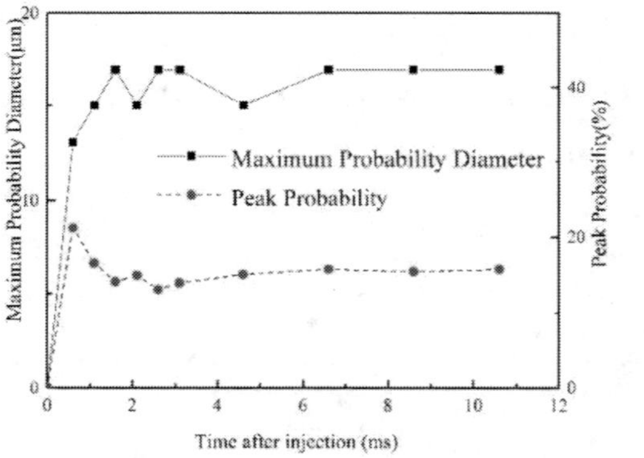

Figure 5: Variation of maximum probability diameter and its probability with the time after start of injection

Figure 6 shows the temporal cumulative volume distribution curves of different time after start of injection. It is known that the cumulative volume distribution is significantly influenced by the large diameter droplets. It can be seen from this figure that the distribution curves are very similar to each other at different moment but there is a little difference between the growth rate. Specially, as time goes on, the curve is shifted to the larger droplet side obviously before 1.6 ms, however, this increase seems to be weakened after that which is consistent with the number fraction shown in fig. 4.

The results of the temporal variation of the Sauter Mean Diameter (SMD) and the local droplet number density which represents the average droplet number of every injection are presented in Fig. 7. The two curves are divided into three stages according to their characteristics. In the first stage (0.6-1.6 ms), it is observed that the SMD are initially low as small droplets enter the measurement volume before the large droplets. This is because at the front of the spray, the relative velocity of gas-liquid two-phase is larger and the interaction is stronger which result in more violent breakup process according to the spray breakup theory. So, more and smaller droplets appear at the most front of spray head. With the development of the spray, the head passes over the measuring window and the first half but not the head appear on the measuring window, the interaction of the gas and liquid phases is weakened, leading to a gradually increase of SMD and the droplets number is gradually reduced. In the second stage (1.6-2.6 ms), it is observed that the SMD continually increase, and the droplet number increase sharply at the beginning and then decrease. In this period, with the development, the second half of the spray body appear on the measuring window, the interaction of the gas and liquid phases is further weakened, but the movement of the droplets is more messy leading to the whole window be covered with droplets, so the droplet number is increased significantly for at the first stage droplets only account for the left half of the window. After 2.6 ms, the spray is in the stage of free development, the body leaves off the nozzle and the downstream droplets is no longer affected by the upstream. It is observed that, the SMD and the droplet number are decreased slowly. This is because the droplets are scattered, the chance of collision between droplets becomes smaller, most of the droplets will remain in their original state. But may be due to the slightly evaporation, the droplet size and number is decreased slightly.

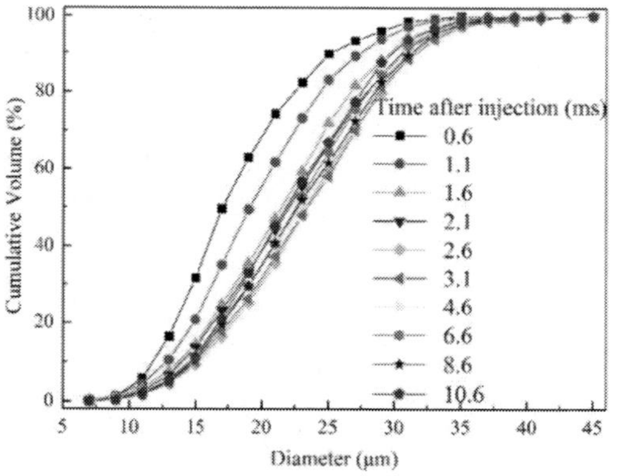

Figure 6: Cumulative volume curves of different time after start of injection

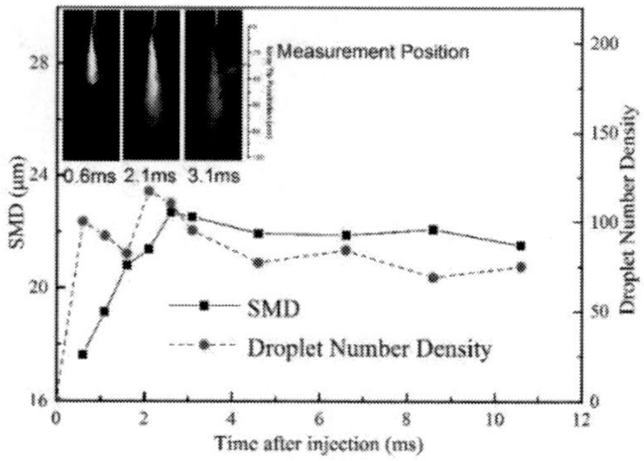

Figure 7: Variation of SMD and droplet number density with time after start of injection.

3.2 Droplet velocity statistics

Figure 8 shows the temporal droplet velocity distribution probability curves of different time after start of injection. Similar bimodal shape before 8.6 ms are observed and after that the first peak is disappeared. It is observed that most droplets' velocity are 2-3 m/s while the maximum value is 25 m/s at the fixed measurement position of 40mm downstream from the injector tip. In addition, the first peak probability is increased as time goes on which indicates that the number of low velocity droplet is increased and at later times after 8.6 ms, most droplets have the velocities of around 1 m/s. It is interesting to note that in the range of 10-15 m/s, a second peak is appeared for all the curves, however the peak value is much lower than the first one and is independent of the different time. This distribution may be attributed to variability in the whole spray development with groups of droplets of different velocities reaching the measurement position.

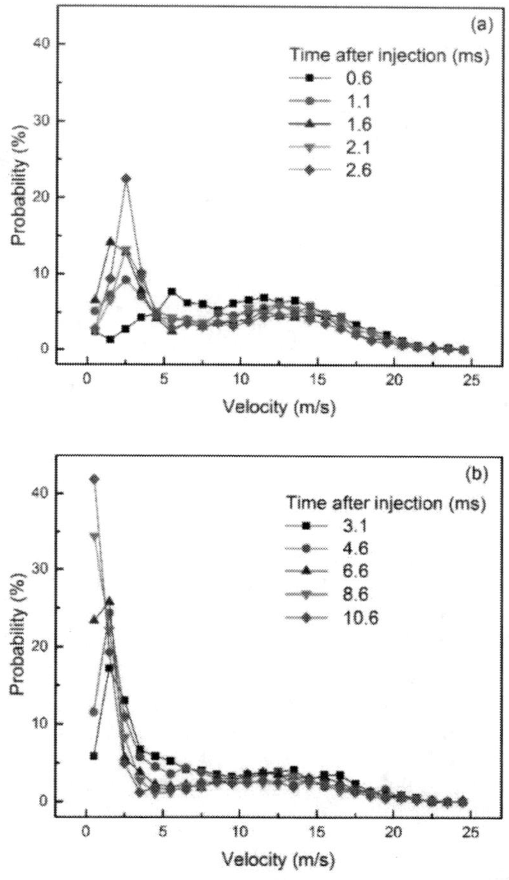

Figure 8: Droplet velocity distributions of different time after start of injection

The results of the variation of the average velocities with time after start of injection are presented in Fig. 9. In this figure, each data point is obtained by averaging all the droplets under each experimental condition. V_R and V_Z are respectively the radial and axial component of velocity on the focal plane, the direction is shown in fig. 2. V is the square root of V_R and V_Z, that is, the droplet synthetic velocity on the focal plane. V_1 and V_2 are respectively the absolute value of V_R and V_Z. When the droplets first appear on the measuring window, the average velocity is about 11 m/s, as time goes on, due to the aerodynamic drag force, the curve presents a monotonically decreasing trend, and the decreasing rate gradually decreases.

Because of the sheer stress between the high speed spray and the quiescent air, many vortexes are formed at the edge of the spray due to the gas entrainment. So, the droplets movement is disorganized at the beginning time just as shown of the V_R and V_Z. At later time, the velocities become relatively stable, for V_R is about 0.2 m/s, V_Z is about 1.5 m/s. This is associated with the decelerated droplets reaching an equilibrium with the entrained air which is in contrast to the early injected droplets whose instantaneous velocities are determined by their initial conditions. In addition, to compare the scalar radial and axial velocities, the droplet velocity is averaged without considering the direction and little difference was found between V_1 and V_2 as shown in fig. 9. It indicates that droplets have a very similar velocity component between axial and radial direction which is caused by the air entrainment.

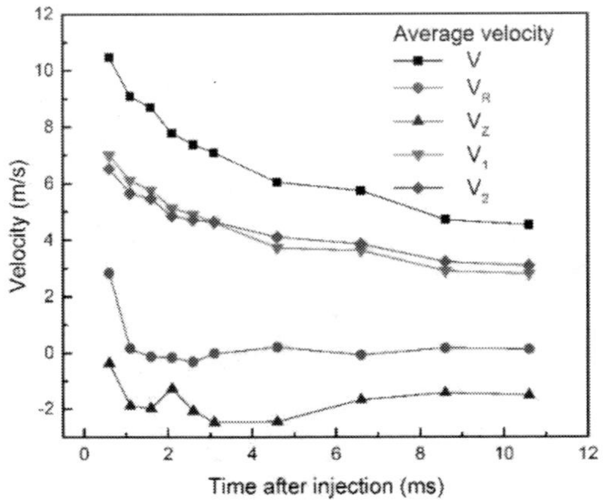

Figure 9: Variation of average velocities with time after start of injection.

4 Conclusions

Investigation of the temporal evolution of droplets size and velocity inside diesel fuel sprays were conducted by PDIA system at an injection pressure of 1000 bar and ambient pressure of 20 bar. Mainly conclusions include:

(1) The droplet size distribution curve is shifted to the larger droplet side obviously at the early time but independent of time after 1.6 ms, and same results were found of cumulative volume distribution. The Sauter mean diameter (SMD) increases significantly at the continuous injection stage , but slightly decreases at the free development stage.

(2) Most droplets' velocity are 2-3 m/s while the maximum value is 25 m/s at the fixed measurement position of 40mm downstream from the injector tip. An interesting similar bimodal shape was found of velocity distribution probability curves at different time after start of injection. What's more, the scalar value of velocities gradually decrease as time goes on, but the direction is messy due to the air entrainment.

5 Acknowledgment

This work is supported by the National Natural Science Foundation of China (91541107, 11572258 and 51406158), China Postdoctoral Science Foundation (2014M562408) the Science Challenge Project (No. TZ2016001), and the Fundamental Research Funds for the Central Universities.

References

[1] A.A. Amer, M.-C. Lai. Time-resolved measurement in transient port injector sprays, SAE, 950509.

[2] N. Kawaharada, D. Sakaguchi, H. Ueki, M. Ishida, Effect of injection pressure on droplet vehavior inside diesel fuel sprays, SAE, 2015-01-1841.

[3] R. Payri, J.P. Viera, H. Wang, L.-M. Malbec, Velocity field analysis of the high density, high pressure diesel spray, International Journal of Multiphase Flow, 80 (2016) 69-78.

[4] A. Sinha, R. Surya Prakash, A. Madan Mohan, R.V. Ravikrishna, Experimental studies on evaporation of fuel droplets under forced convection using spray in crossflow methodology, Fuel, 164 (2016) 374-385.

[5] J.T. Kashdan, J.S. Shrimpton, A. Whybrew, Two-Phase Flow Characterization by Automated Digital Image Analysis. Part 1: Fundamental Principles and Calibration of the Technique, Particle & Particle Systems Characterization, 20 (2003) 387-397.

[6] C. Tropea, G. Gouesbet, G. Gréhan, H. Mignon, N. Damaschke, Response of Phase Doppler Anemometer systems to nonspherical droplets, Applied Optics, 37 (1998) 1752-1761.

[7] J.T. Kashdan, J.S. Shrimpton, A. Whybrew, Two-Phase Flow Characterization by Automated Digital Image Analysis. Part 2: Application of PDIA for Sizing Sprays, Particle & Particle Systems Characterization, 21 (2004) 15-23.

Power generation performance of hydrogen-fueled micro thermophotovoltaic reactor

Chien-Chun KAO[1], Yueh-Heng Li[1, 2*], Jing-Ru Hong[1]

1 Department of Aeronautics and Astronautics, National Cheng Kung University, Tainan, 701, Taiwan, R.O.C.
2 Research Center for Energy Technology and Strategy, National Cheng Kung University, Tainan, 701, Taiwan, R.O.C.

ABSTRACT: A tubular platinum reactor with a perforated annular array enables fuel/air mixtures to exchange sides, thus sustaining flames and preventing heat loss. Consequently, the combustion efficiency and operational range can be enhanced. A hydrogen/air mixture was introduced into inner and outer chambers at different equivalence ratios and flow velocities to chemically and physically investigate the interplay between the chambers. The benefits of hydrogen include a high heating value, flame speed, and diffusion capacity and short chemical reaction time. The coexistence of heterogeneous (surface) and homogeneous (gas) reactions in the micro TPV reactor was examined and elucidated in terms of aerodynamics, mass and heat transfer, and chemical reactivity. Furthermore, TPV reactor with TPV cell arrays was assembled, and the corresponding radiant efficient of the emitter and the overall efficiency of the proposed micro TPV system were determined in this study.

Keywords: Catalytic combustion; platinum; small-scale reactor; thermophotovoltaic; heterogeneous and homogeneous reaction.

1. Introduction

The proliferation of microelectromechanical systems and their demand for miniature power sources, ranging from 10 to 1000 W/kg [1], have accelerated the development of miniature power. In general, power systems employing hydrogen or hydrocarbon fuels provide much higher energy density according to specific mass compared with traditional systems [2, 3]. A series of combustion-based micro power systems have been successfully prototyped, including the micro gas turbine [4], micro-free-piston engine [5], micro thermoelectric device [6] and micro thermophotovoltaic (TPV) system [7-9]. Unlike solar photovoltaics, TPV cells are illuminated by combustion-driven radiation sources. Accordingly, understanding the fundamental characteristics of combustion in the micro scale is the key to improving the system efficiency and optimizing the design.

Some research groups have applied combustion-enhancing methods such as external heating to overcome the heat loss problem, whereas others have experimented with catalyzed combustion to minimize the radical quenching effect. Reducing wall thickness and step height gives rise to a higher wall temperature. For instance, Pan et al. [12] discussed the effects of major micro-combustor parameters on the radiation intensity in micro-TPV system, fuel to oxygen ratio, nozzle to combustor diameter ratio, and wall thickness to combustor diameter ratio. Furthermore, porous media combustion occurs in a three-dimensional solid porous matrix having interconnected pores. This enhances the heat transfer from the burned hot gas to the unburnt mixture through conduction, convection, and radiation. In addition, catalytic micro combustors exhibit wider stability than that of homogeneous micro combustors [15, 16]. Using numerical simulation, Li et al. [17, 18] investigated heterogeneous and homogeneous mechanisms of hydrocarbon fuels combustion enhancement through catalyst segmentation with cavities in a micro reactor.

Regrading to the improvement of TPV power output, the use of spectral-control technique for tailoring the radiation spectrum emitted onto the TPV system is prevailing. A selective emitter capitalizes on suppressing the emission of sub-bandgap photons. According to the Stefan–Boltzmann law, a higher emitter surface temperature leads to higher power output. Therefore, a means for increasing the temperature of the reactor is required. Yang et al. [19] tested different emitting materials for enhancing radiation efficiency. In addition, quartz enables to provide optical transmission from near-ultraviolet to mid-infrared wavelength, and it is effective filter to block low-energy photons [20].

In this study, the interaction between the inner and outer chambers was experimentally and numerically investigated with regards to the effect of fuel concentration. The coexistence of heterogeneous and homogeneous reactions in the micro TPV reactor was examined in terms of aerodynamics, mass and heat transfer, and chemical reactivity. Finally, a TPV reactor with PV cell arrays was assembled, and the corresponding radiant efficient of the emitter and the overall efficiency of the proposed micro TPV system are determined.

2. Experimental Apparatus

Figure 1 shows a schematic diagram of the micro TPV reactor, which consists of a platinum tube with dimensions of 5.3 mm (ID) × 6 mm (OD) × 30 mm (L) and a quartz tube with dimensions of 8 mm (ID) × 10 mm (OD) × 160 mm (L). The quartz tube is mounted outside the platinum tube. Eight perforation holes (1 mm in diameter) are equidistantly placed around the platinum tube at a distance of 5 mm from the bottom. The experiment involved introducing fuel at different equivalence ratios and compositions into the inner and outer chambers to investigate the flame behavior in the micro combustor. The flame behavior was recorded using the Nikon D80 with the aperture and exposure time fixed at F5.6 and 1/100, respectively, for all photographs.

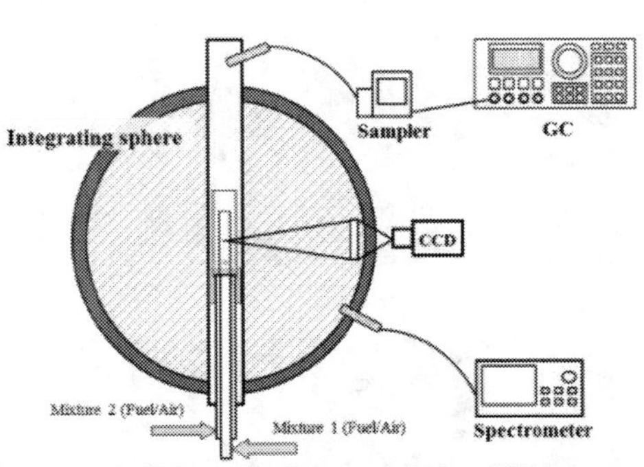

Figure 1. The schematic diagram of micro TPV reactor.

An integrating sphere connected to a spectrometer (USB2000+XR, Ocean Optic Inc.) was employed to measure the irradiation of the emitter, as shown in Fig. 2. The integrating sphere was 15 cm in diameter, and magnesium oxide was deposited on the inner surface. The spectrometer covers wavelengths from 200 to 1050 nm.

In addition, a monochromator (DK240, Spectral Products Company) with an IR photoresistor (AD131-USB, Spectral Products Company) was employed to measure the spectral distribution of the emitter. The IR photoresistor was a convenient computer-controlled photoresistor connected to the monochromator.

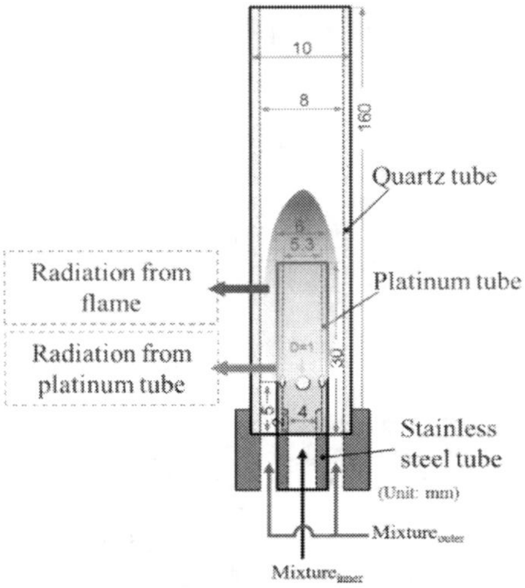

Figure 2. The schematic of the integrating sphere assembling with the spectrometer.

3. Combustion Characteristics of Micro TPV Reactor

3.1 Effect of fuel concentration

In a previous study [18], numerical results demonstrated that the existence of a gap in a segmented platinum tubular reactor can facilitate the induction of a catalytically stabilized thermal (CST) gas reaction in a confined tube. The illumination behaviors of a micro TPV reactor under various fuel deployment conditions at a fixed flow velocity of 6 m/s in the inner and outer chambers were shown in Fig. 3.

The equivalence ratio of one chamber was fixed at 0.3, whereas the other chamber had various equivalence ratios ranging from 0.1 to 0.5.

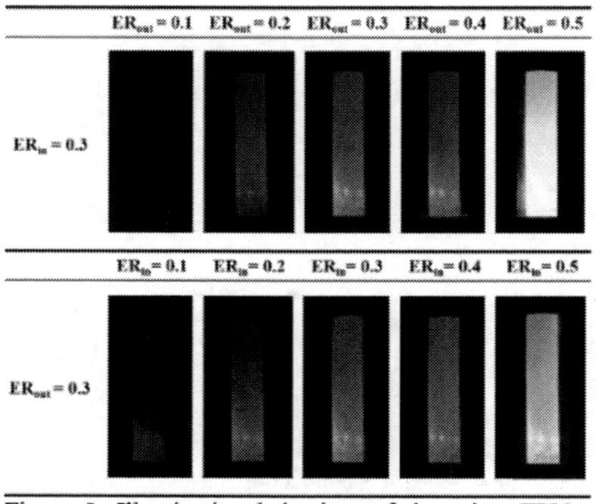

Figure 3. Illumination behaviors of the micro-TPV reactor in various fuel deployments and fixed flow velocity of 6 m/sec.

A commercial code, CFD-ACE+, was modified to incorporate detailed gas-phase and surface reaction mechanisms in CHEMKIN formats for the simulation. H2–air mixtures were individually introduced into the inner and outer chambers at different equivalence ratios. The inlet temperature was 300 K, and a uniform velocity profile (6 m/s) was specified at the inlet. At the exit, a constant ambient pressure of 101 kPa was specified and an extrapolation scheme was used for determining species and temperature. Chemical reaction mechanisms were used in the gas phase as well as on the catalyst surface.

According to Fig. 4, the velocity vectors in the vicinity of the gap flow from the outer chamber to the inner chamber. Characteristically, a mixture with a relatively high equivalence ratio in the outer chamber can induce the CST gas reaction along the outer surface of the upstream platinum tube, and unburnt fuels and radicals are then driven into the inner chamber through the gap to facilitate the gas reaction in the inner chamber. The tendency of unburnt fuels to be drawn to the inner chamber lowers the concentration of remaining fuel in the outer chamber. Therefore, insufficient heat release and the remaining fuel cannot sustain the consequent chemical reaction in the downstream segment of the outer chamber.

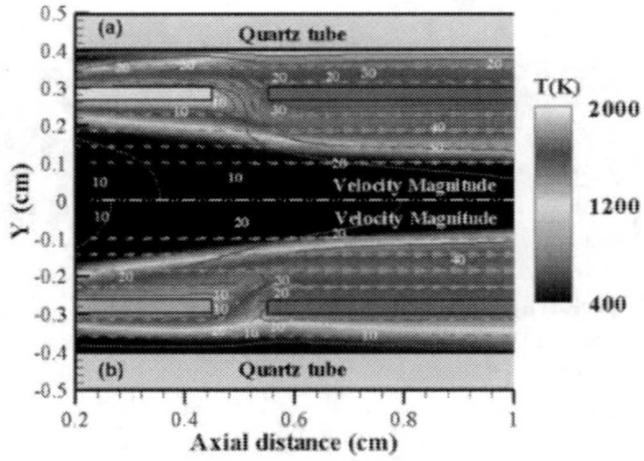

Figure 4. The distributions of temperature with overlaid velocity vectors and velocity magnitude level lines for the cases of (a) ERin = 0.3/ERout = 0.5, and (b) ERin = 0.5/ERout = 0.3, respectively, in the fixed inlet velocity of 6 m/sec in first 1 cm distance away from the inlet.

3.2 Effect of flow velocity

Figure 5 shows the combustion behaviors at the flow velocity ratios of 2 and 0.5 with the inner and outer equivalence ratios of 0.3.

Figure 5. Combustion behavior under the ratios of inner velocity to outer velocity of (a) 2 and (b) 0.5 with the inner and outer equivalence ratio of 0.3.

Figure 6 illustrates the distributions of temperature with overlaid velocity vectors and velocity magnitude level lines for flow velocity ratios. The results elucidate that the flow motion in the vicinity of the gap is predominantly driven not by thermal diffusion and mass diffusion, but by pressure-driven flow dynamics. At the flow velocity ratio of 2, the flame seems to tail toward only the inner surface of the downstream platinum tube. The flow tendency in the gap lowers the fuel concentration in the outer chamber. A low LER along the outer surface of the downstream platinum tube leads to a fuel concentration that is insufficient to induce the CST gas reaction, but is sufficient to induce the surface reaction. By contrast, at the flow velocity ratio of 0.5, the flame seems to tail toward the inner and outer surfaces of the downstream platinum tube. The outer velocity is relatively higher than the inner velocity, indicating that the hydrogen amount is greater in the outer chamber than in the inner chamber. The remaining fuel in the outer chamber is sufficient to induce the CST gas reaction in the downstream segment.

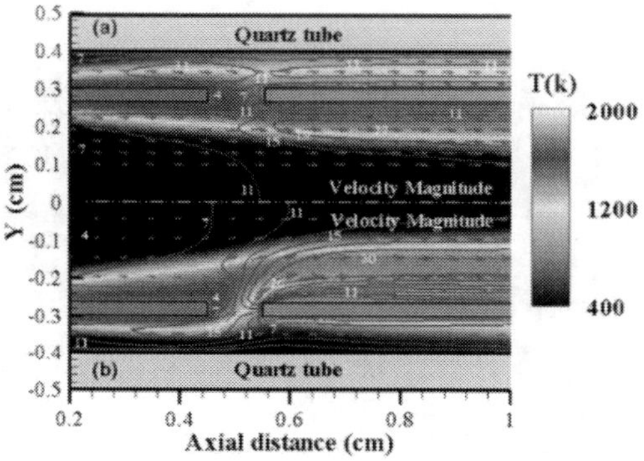

Figure 6. The distributions of temperature with overlaid velocity vectors and velocity magnitude level lines behavior under the ratios of inner velocity to outer velocity of (a) 2 and (b) 0.5 with the inner and outer equivalence ratio of 0.3 in first 1 cm distance away from the inlet.

4. Demonstration of Micro TPV Power System

4.1 Irradiance of micro TPV power system

The overall efficiency of a TPV power system is related to the radiation efficiency of the emitter and quantum efficiency of the TPV cell. Perturbations of combustion and fluid dynamics may lead to non-uniformity of the TPV radiation. Therefore, the proposed micro TPV reactor was placed in the center of the integrating sphere. As the radiation of the combustor scattered uniformly inside the integrating sphere, an optical fiber was used to collect the radiation and sent the signal to the spectrometer to measure the irradiance. Table 1 shows the irradiance and the corresponding radiance efficiency at different equivalence ratios of the fuel/air mixture with the fixed velocity of 6 m/s. Two scenarios were considered: the first was a fixed fuel concentration in the inner chamber, and the second was a fixed fuel concentration in the outer chamber. Regardless of the scenario, the irradiance of the TPV reactor evidently increased as the hydrogen concentration increased. In addition, the maximal irradiance was 6123 W/m^2 and the corresponding radiance efficiency was 1.51% for an ERin of 0.3 and ERout of 0.5.

Table 1. Irradiance and the corresponding radiance efficiency at different equivalence ratios of the mixture with a fixed velocity of 6 m/s.

Fuel/Air Mixtures	Irradiance (W/m2)	Radiation efficiency (%)
H2 (ERin=0.3, ERout=0.3)	1993	0.79
H2 (ERin=0.4, ERout=0.3)	4453	1.42
H2 (ERin=0.5, ERout=0.3)	5162	1.46
H2 (ERin=0.3, ERout=0.4)	4666	1.43
H2 (ERin=0.3, ERout=0.5)	6123	1.51

4.2 Electrical output of micro TPV power system

To evaluate the benefit of improving the proposed micro TPV reactor, photovoltaic cells were employed to collect luminosity and convert it into electricity. It is essential to detect the spectrum distribution of radiant from the emitter and match up the quantum efficiency of photovoltaic cell. In the study, the monochromator (DK240, Spectral Products Company) with IR photoresistor (AD131-USB, Spectral Products Company) was used to measure the spectral distribution of the emitter, ranging from 400 – 1500 nm.

Accordingly, the prototype TPV system comprised the proposed micro TPV reactor with the necessary assemblies surrounded by a GaSb PV cell array, fabricated by JX Crystal Company, as shown in Fig. 7. The TPV circuit consists of 24 gallium antimonide PV cells. Each cell array had an area of 18 cm2 and contained two strings of six series-connected cells in tandem. The open circuit voltage (Voc) and short circuit current density (Isc) of the TPV panel are 23.023V and 2.9875A/cm2, respectively, and the fill factor is 0.733.

Figure 7. The micro-TPV reactor with GaSb PV cell array.

5. Conclusion

The flame stabilization mechanism of a micro TPV reactor was investigated through a simplified numerical case of two platinum segments in tandem. The function of the gap between the two platinum tubes in tandem was addressed by analyzing the fuel concentration and flow velocity effects. In addition, a platinum tube with a ring of holes was fabricated and tested in a series of experiments. The radiation efficiency and effective power efficiency of the micro TPV power system were measured under various experimental conditions. The most crucial results are listed as follows:

It is apparently difference of luminous intensity and temperature distribution of emitter surface in different hydrogen/air deployments. The speculation is that the effect of fuel concentration in two sides of chambers impacts the flame stabilization. Numerical results demonstrated the velocity vectors in the vicinity of the gap flow from the outer chamber to the inner chamber due to the localized pressure imbalance between the two sides of the gap. The tendency of unburnt fuels to be drawn to the inner chamber lowers the concentration of remaining fuel in the outer chamber. Therefore, insufficient heat release and the remaining fuel cannot sustain the consequent chemical reaction in the downstream segment of the outer chamber.

The occurrence of the CST gas reaction in the inner chamber is associated with the chemical reactivity in the outer chamber. Consequently, reactors with high Vin/Vout ratios require relatively rich fuel/air mixtures to sustain the CST gas reaction in the inner chamber in order to overcome the fuel concentration reduction and the heat loss in the outer chamber.

An integrating sphere connected to a spectrometer was employed to measure the irradiation of the emitter. The emitter was placed in the center of the integrating sphere, and the spectrometer was connected to the integrating sphere to obtain the radiation flux. The irradiance of the TPV reactor evidently increased as the hydrogen concentration increased. In addition, the maximal irradiance was 6123 W/m2 and the corresponding radiance efficiency was 1.51% for an ERin of 0.3 and ERout of 0.5.

For the overall efficiency of the TPV system, the maximal power was 0.16 W with the corresponding ηe of 1.13% as ERin was 0.3 and ERout was 0.5. However, when a reverse tube is used to redirect hot flue gas and further heat the platinum tube, the I-V performance for an ERin of 0.3 and ERout of 0.4 was substantially improved, and the maximal power output was 0.19 W with the corresponding ηe of 1.56%.

Acknowledgements

This research was partially supported by the Ministry of Science and Technology under Grant Nos. MOST 104-2228-E-006 -012 and MOST 105-2628-E-006-005-MY3. Computer time and numerical packages provided by the National Center for High-Performance Computing, Taiwan, are gratefully acknowledged

References

[1] D. Dunn-Rankin, E.M. Leal, D.C. Walther, Personal power systems, *Progress in Energy and Combustion Science*, 31 (2005) 422-465.

[2] D.C. Kyritsis, S. Roychoudhury, C.S. McEnally, L.D. Pfefferle, A. Gomez, Mesoscale combustion: a first step towards liquid fueled batteries, *Experimental Thermal and Fluid Science*, 28 (2004) 763-770.

[3] A.C. Fernandez-Pello, Micropower generation using combustion: Issues and approaches, *Proceedings of the Combustion Institute*, 29 (2002) 883-899.

[4] A.H. Epstein, Millimeter-Scale, Micro-Electro-Mechanical Systems Gas Turbine Engines, *Journal of Engineering for Gas Turbines and Power*, 126 (2004) 205-226.

[5] H.T. Aichlmayr, D.B. Kittelson, M.R. Zachariah, Micro-HCCI combustion: experimental characterization and development of a detailed chemical kinetic model with coupled piston motion, *Combustion and Flame*, 135 (2003) 227-248.

[6] S. Yadav, P. Yamasani, S. Kumar, Experimental studies on a micro power generator using thermo-electric modules mounted on a micro-combustor, *Energy Conversion and Management*, 99 (2015) 1-7.

[7] S.K. Chou, W.M. Yang, K.J. Chua, J. Li, K.L. Zhang, Development of micro power generators – A review, *Applied Energy*, 88 (2011) 1-16.

[8] Y.H. Li, Y.S. Lien, Y.C. Chao, D. Dunn-Rankin, Performance of a Mesoscale Liquid Fuel-film Combustion-driven TPV Power System, *Progress in Photovoltaics*, 17 (2009) 327-336.

[9] Y.H. Li, H.Y. Li, D. Dunn-Rankin, Y.C. Chao, Enhancing Thermal, Electrical Efficiencies of a Miniature Combustion-Driven Thermophotovoltaic System, *Progress in Photovoltaics*, 17 (2009) 502-512.

[12] J.F. Pan, J. Huang, D.T. Li, W.M. Yang, W.X. Tang, H. Xue, Effects of major parameters on micro-combustion for thermophotovoltaic energy conversion, *Applied Thermal Engineering*, 27 (2007) 1089-1095.

[15] Y.-H. Li, G.-B. Chen, F.-H. Wu, T.-S. Cheng, Y.-C. Chao, Effects of catalyst segmentation with cavities on combustion enhancement of blended fuels in a micro channel, *Combustion and Flame*, 159 (2012) 1644-1651.

[16] Y.H. Li, G.B. Chen, H.W. Hsu, Y.C. Chao, Enhancement of methane combustion in microchannels: Effects of catalyst segmentation and cavities, *Chemical Engineering Journal*, 160 (2010) 715-722.

[17] W.M. Yang, S.K. Chou, C. Shu, Z.W. Li, H. Xue, Study of catalytic combustion and its effect on microthermophotovoltaic power generators, *Journal of Physics D: Applied Physics*, 38 (2005) 4252.

[18] Y.-T. Wu, Y.-H. Li, Combustion characteristics of a micro segment platinum tubular reactor with a gap, *Chemical Engineering Journal*, 304 (2016) 485-492..

[19] W.M. Yang, S.K. Chou, C. Shu, H. Xue, Z.W. Li, Research on micro-thermophotovoltaic power generators with different emitting materials, *Journal of Micromechanics and Microengineering*, 15 (2005) S239-S242.

[20] E.C. Beder, C.D. Bass, Shacklef.Wl, Transmissivity and Absorption of Fused Quartz Between 0.22 mu and 3.5 mu from Room Temperature to 1500 degrees C, *Applied Optics*, 10 (1971) 2263-8.

11th Asia-Pacific Conference on Combustion.
The University of Sydney, NSW Australia
10th -14th December 2017.

Blowoff Dynamics of Lean Premixed Flames Stabilized on a Meso-scale Bluff-body

Yu Jeong Kim[1], Bok Jik Lee[2], Hong G. Im[1]

[1]Clean Combustion Research Center, King Abdullah University of Science and Technology,
Thuwal 23955-6900, Saudi Arabia

[2]School of Mechanical Engineering, Gwangju Institute of Science and Technology,
Gwangju 61005, Republic of Korea

Abstract

Bluff-bodies have been used as effective flame holders for practical applications of premixed flames. While many studies about bluff-body stabilization have been done both experimentally and numerically, there is still the lack of understanding of blowoff mechanism. Direct numerical simulations are conducted to investigate dynamics of lean premixed hydrogen-air and syngas-air flames anchored on a meso-scale bluff-body flame holder. A two-dimensional square channel of 10 mm-by-10 mm with a square bluff-body of 0.5 mm is considered as a computational domain. The height of domain is chosen as an unconfined condition to minimize the effect of blockage ratio. Density ratio of the flame decreases by preheating to investigate hydrodynamic effects on flame dynamics in both hydrogen-air and syngas-air mixtures as ramping up the mean inflow velocity until blowoff occurs. Phenomenological descriptions of flame dynamics are presented, including the process of blowoff events. The objectives of this study are to understand blowoff mechanism and to investigate the hydrodynamic effects on the flame dynamics.

1 Introduction

Bluff-body flame holders have been served as one of flame stabilization schemes to investigate the mechanism of flame stability in high-speed, premixed combustion systems such as gas turbine engines and propulsion devices. For the bluff-body-stabilized flame, while a bluff-body has an advantage of improving stability from blowoff by creating a recirculation zone behind the body, the bluff-body itself causes the instability into the flame. In small scale combustion devices, flame stability becomes more challenging, and the flame is only stabilized in a critical range of conditions [1, 2]. Therefore, investigating the condition where the flame is stable or extinguishing and the unstable flame dynamics behind the bluff-body is necessary to understand the mechanism of flame stability.

As a well-defined model problem, recent studies [3, 4] on near-blowoff dynamics of bluff-body-stabilized both hydrogen-air and syngas-air flames in a meso-scale channel present complicated sequences of transient events are observed as approaching blowoff limit. In the near-blowoff dynamics, the onset of the first local extinction was correlated with time scale analysis and the local extinction leads to the total extinction as a precursor. They reported that the observed periodic oscillation of local extinction and re-ignition was controlled by ignition process. Although simulations were conducted in two-dimension in a miniature scale domain, their results showed strong similarities with large scale experimental observation. However, the narrow confinement seems to affect the flame instability and wake flows. To reduce the confinement effect, simulations were conducted in an extended channel with the same size of a meso-scale bluff-body for both hydrogen-air and syngas-air mixtures [5]. Several distinct symmetrical and asymmetrical flame behaviors were observed, and intermittent transition between symmetric and asymmetric instabilities and a sequence of local extinction and recovery temporally showed in a near-blowoff condition. In an extended channel, local extinction did not immediately lead to the total extinction. One of the key features in an extended channel is lateral flapping, which is the attached bulk flame moving around in a lateral direction of the domain. To investigate the causes of ultimate extinction, they suggested lateral flapping may play some roles in the ultimate extinction, and time scale analysis was conducted. However, the investigated condition is not sufficient to identify the roles of lateral flapping and to find the causes of ultimate blowoff.

The present study is one of the ongoing works in a reduced confinement channel, investigating flame dynamics in a lower density ratio by preheating. Dynamics of lean premixed flames in both hydrogen-air and syngas-air mixtures are observed by approaching the mean inflow velocity until blowoff limit. The objectives of this study are to investigate the hydrodynamic effect on flame dynamics and to understand blowoff mechanism in a reduced confinement channel by investigating its dynamics.

2 Numerical Method

The present simulations solve fully compressible multi-species reacting Navier-Stokes equations with a finite difference method using the 8th order in space and 4th order in time [6]. Chemical kinetics, thermodynamics, and transport properties are computed using CHEMKIN libraries [7, 8], and the mixture-averaged diffusion model is used. For the mixture, hydrogen-air and syngas-air ($CO:H_2$ = 1:1) with an equivalence ratio of 0.5 are investigated. To solve chemical reaction, detailed mechanisms of hydrogen-air [9] and syngas-air [10] are used.

Figure 1 shows a schematic of a two-dimensional computation domain of a square channel with a meso-scale bluff-body. A bluff-body flame holder is represented by a square cylinder of 0.5 mm (D), whose leading edge is located at 2 mm (4D) downstream of the inflow boundary. Both height and length of a

Hong G. Im. Office: +966-12-808-4726
E-mail address: hong.im@kaust.edu.sa

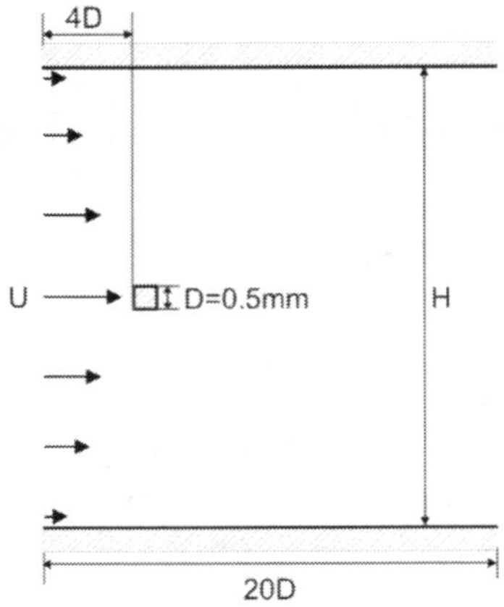

Figure 1: Configuration with dimensions of a two-dimensional channel with a square bluff-body flame holder.

domain are 10 mm (20D), and The height was chosen as an unconfined condition by a convergence test of the Strouhal number. Non-reflecting characteristic boundary conditions [11] are applied to the inflow and outflow boundaries. No-slip and adiabatic conditions are employed to the channel walls and bluff-body surfaces. A fully developed velocity profile with a mean value of inflow velocity, U, is imposed with the preheated inflow temperature of 500 K and atmospheric pressure. A Cartesian grid system with a uniform spacing of Δx = 10 μm is adopted. A typical run with a Cray XC40 cluster (Shaheen II) for time integration up to 10 ms takes approximately 8800 and 16000 CPU hours for hydrogen-air and syngas-air mixtures, respectively. Every simulation was conducted more than 20 ms, which is long enough to settle the transient characteristics and to reach a limit cycle behavior.

3 Results and Discussion

Reacting flow simulations are conducted to investigate flame dynamics of lean premixed both hydrogen-air and syngas-air flames stabilized on a meso-scale bluff-body. Several distinct flame behaviors are observed from a stable state, symmetrical and asymmetrical instabilities as pushing the inflow velocity (U) until blowoff limit in the range of Reynolds numbers based on the size of bluff-body (Re_D) 220 to 810 and 240 to 601 for hydrogen-air and syngas-air, respectively.

As a reminder of the previous non-preheating study of T_{in} = 300 K [5], an intermittent transition of flame behaviors between symmetric and asymmetric shedding with local extinction was observed in the near-blowoff condition. Characteristics time scales were adopted to determine critical oscillation time scale, which was a precursor of blowoff in narrow channel studies. Due to the intermittency of transition, determining critical oscillation time scales was hard. For an alternative proposal, we attempted decreasing density ratio of the flame by preheating based on the background of density ratio effects [12]. Emerson et al. [12] reported a series of experimental works of density ratio effects on bluff-body flow field characteristics in the premixed methane-air mixture as varying density ratio by adjusting flow rate and equivalence ratio, including preheat. They reported that density ratio strongly affects the wake characteristics of bluff-body-stabilized flames. As for decreasing density ratio, flames become the asymmetric shape, and the manner of the flame temporally exhibits limit-cycling. In the present study, density ratios of flame are decreasing from 4.91 to 3.26 for hydrogen-air mixture and 5.15 to 3.38 for the syngas-air mixture.

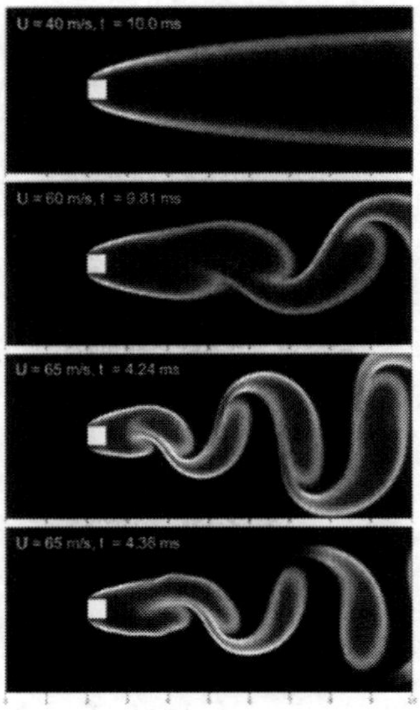

Figure 2: Instantaneous snapshots of iso-contours of hheat release rate for hydrogen-air flames.

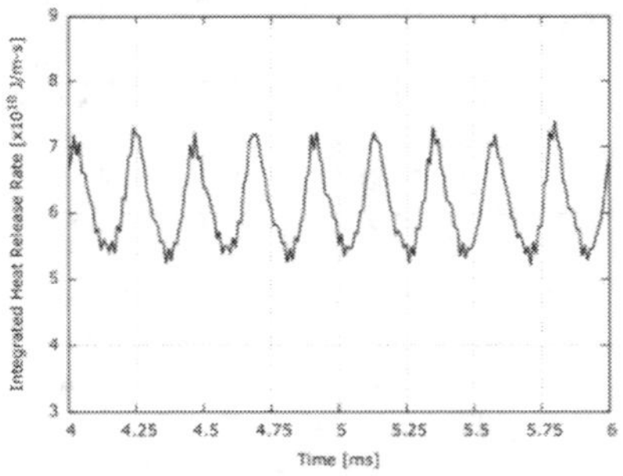

Figure 3: Temporal evolution of domain-integrated heat release rate for hydrogen-air flames at U = 65 m/s.

3.1 Hydrogen-air

For the hydrogen-air mixture, the iso-contours of the subsequent figures represent the magnitude of heat release rate, and the maximum level of that is set at 10^{10} J/m^3-s. Figure 2 represents instantaneous snapshots of heat release rate in the far from the blowoff limit at U = 40, 60, and 65 m/s. The flame is steadily stabilized on the adjacent leading edge of the bluff-body. as before non-preheating case until U = 40 m/s. When the U increases at 45 m/s, symmetric mild fluctuation occurs on the flame sheet at the beginning, and then asymmetric shedding has developed from downstream. As shown in the second snapshot of Fig. 2, the flame behavior is settled in asymmetrical vortex shedding from U = 45 m/s to 60 m/s without any transition and local extinction. During the interval of U, the length of recirculation zone decreases by enhancing asymmetric shedding toward upstream. Ramping up U at 65 m/s, the length of recirculation zone becomes shorter, and the onset of local extinction occurs at the entrainment point in asymmetric shedding. The length of recirculation zone behind the bluff-body periodically fluctuates in a longitudinal direction, showing limit-cycling behavior in a frequency of about 4.5 kHz. This limit-cycling behavior represents a temporal history of domain integrated heat release rate, as shown in Fig. 3. The magnitude of integrated heat release rate depends on the area of the flame sheet in the domain so that the maximum represents the flame with the shortest recirculation zone. When the U increases further at 70 m/s, shown in Fig. 4, the flame behavior

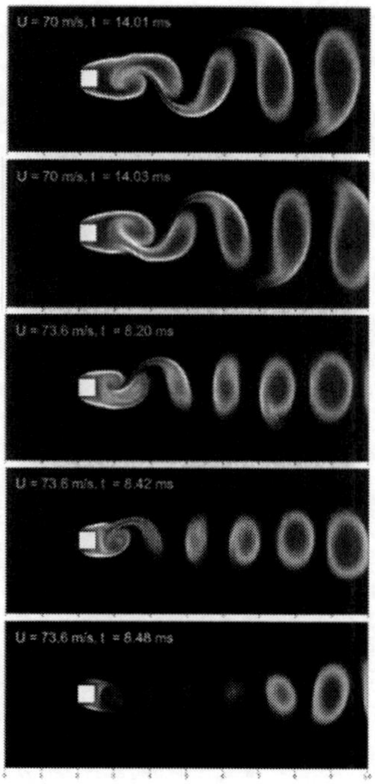

Figure 4: Instantaneous snapshots of iso-contours of heat release rate for hydrogen-air flames at the near-blowoff of U = 70 m/s and blowoff limit of U = 73.6 m/s.

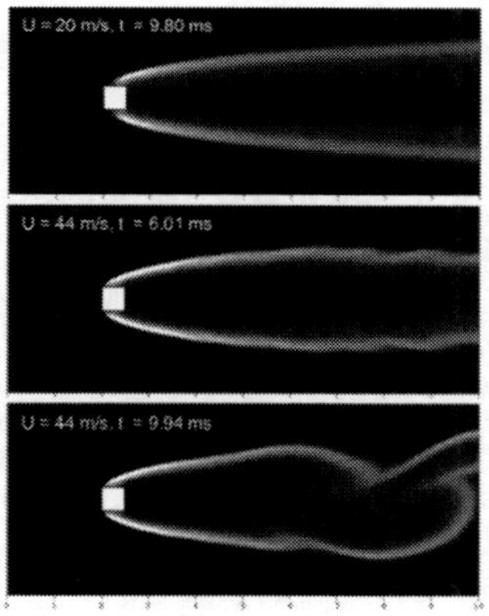

Figure 5: Instantaneous snapshots of iso-contours of heat release rate for syngas-air flames.

is the same with U = 65 m/s, but the length of recirculation zone behind the bluff-body is shorter than that of U = 65 m/s, but there is no variation of the length of recirculation zone. This limit-cycling behavior of vortex street continues showing until U = 73.5 m/s without any transition to other flame behaviors. In other words, there is no intermittent transition between symmetric and asymmetric shedding after the onset of local extinction. The three snapshots of U = 73.6 m/s in Fig. 4 shows a sequence of blowoff event after a few cycles of the vortex street behavior as before. From t = 8.3 ms, the attached flame starts weakening and shrinking, and then the ultimate blowoff happens after t = 8.5 ms.

3.2 Syngas-air

For the syngas mixture, general flame behaviors are similar with hydrogen-air flames as increasing the inflow velocity until blowoff occurs. Figure 5 shows the instantaneous snapshots of the heat release rate at U = 20 m/s and 44 m/s. The maximum level of the magnitude of heat release rate is set at 5^{10} J/m^3-s,

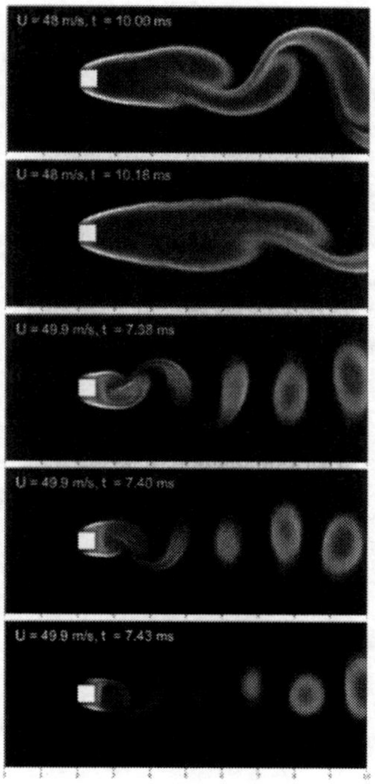

Figure 6: Instantaneous snapshots of iso-contours of heat release rate for syngas-air flames at the near-blowoff of U = 48 m/s and blowoff limit of U = 49.9 m/s

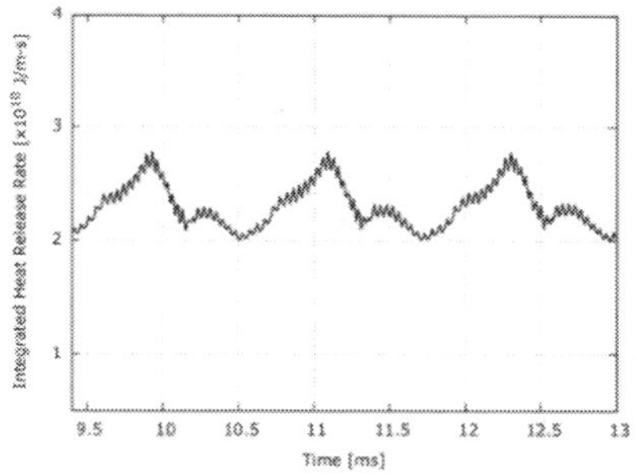

Figure 7: Temporal evolution of domain-integrated heat release rate for syngas-air flames at U = 48 m/s.

which is the half of the hydrogen-air mixture. At the low-velocity range, flames are stabilized on the near leading edge of the bluff-body until U = 20 m/s. From U = 30 m/s to 44 m/s, the mild fluctuation starts developing on the shear layer. At U = 44 m/s, shown in Fig. 5, the asymmetric shedding has developed from downstream without local extinction. When the inflow velocity is pushed at U = 48 m/s, the onset of the local extinction exhibits, followed by vortex street behavior, and the longitudinal fluctuation of recirculation zone is periodically observed as before hydrogen-air flames. Figure 6 shows sequences of local extinction and recovery manner at U = 48 m/s and blowoff events at U = 49.9 m/s. Following the local extinction, the anchored flame tends to recover in a symmetrical shape, but the asymmetric shedding comes back toward the upstream. This periodic event of local extinction and recovery is showing at a frequency of 0.83 kHz, as shown in Fig. 7. At the higher velocity of U = 49.9 m/s, the flame undergoes a few cycles of the same periodic event, the attached flame becomes shrinking, and then the total extinction proceeds.

4 Conclusions

Two-dimensional direct numerical simulations were conducted to investigate the flame dynamics in lean premixed hydrogen-air and syngas (H_2/CO ratio of 1) -air mixtures under an unconfined channel with a meso-scale bluff body. Flame dynamics were observed as increasing the inflow velocity until blowoff limit. The simulations were carried out to understand dynamics of bluff-body-stabilized flames under the strong asymmetrical flame instability and the characteristics of blowoff dynamics in preheating condition. For both hydrogen-air and syngas-air mixtures, the flame behaviors were monotonically enhanced from symmetrical instability to asymmetrical instability as approaching the blowoff limit. From the onset of local extinction, flames temporally underwent limit-cycling behavior until the near-blowoff condition. The intermittent transition of flame behaviors in the previous study was suppressed in this lower density ratio condition. The further quantitative analysis is in progress to explain the detailed observations. The present study was used mixture-averaged diffusion model. Due to the fuel components, the flame structure and stabilization would be affected by differential diffusion. Differential diffusion may affect mixing behind the bluff-body so that it would also affect flame stabilization mechanism. Investigating the effect of differential diffusion on the bluff-body flame stabilization would be one of the further studies.

5 Acknowledgments

This work was sponsored by King Abdullah University of Science and Technology (KAUST), and made use of the computing cluster provided by KAUST Supercomputing Laboratory (KSL).

References

[1] K. Maruta, Proc. Combust. Inst. 43 (2011) 125-150.

[2] S.J. Shanbhogue, S. Husain, T. Lieuwen, Prog. Energy Combust. Sci. 35 (1) (2009) 98-120.

[3] B.J. Lee, C.S. Yoo, H.G. Im, Combust. Flame 162 (2015) 2602-2609.

[4] B.J. Lee, H.G. Im, Proc. Combust. Inst. 36 (2017) 1569-1576.

[5] Y.J. Kim, B.J. Lee, H.G. Im, 55th AIAA Aerospace Sciences Meeting, 2017, AIAA 2017-1103.

[6] J.H. Chen, A. Choudhary, B. de Supinski, M. DeVries, E.R. Hawkes, S. Klasky, W.K. Liao, K.L. Ma, J. Mellor-Crummey, N. Podhorski, R. Sankaran, S. Shende, C.S. Yoo, Comput. Sci. Discov. 2 (2009) 015001.

[7] R.J. Kee, G. Dixon-Lewis, J. Warnatz, M.E. Coltrin, J.A. Miller, A Fortran Computer Code Package for the Evaluation of Gas-phase, Multicomponent Transport Properties, Report No.SAND86-8246, Sandia National Laboratories, 1986.

[8] R.J. Kee, F.M. Rupley, J.A. Miller, Chemkin-II: A Fortran Chemical Kinetics Package for the Analysis of Gas-phase Chemical Kinetics, Report No.SAND89-8009, Sandia National Laboratories, 1989.

[9] J. Li, Z. Zhao, A. Kazakov, F.L. Dryer, Int. J. Chem. Kinet. 36 (2004) 566-575.

[10] J. Li, Z. Zhao, A. Kazakov, M. Chaos, F.L. Dryer, Int. J. Chem. Kinet. 39 (2007) 109-136.

[11] C.S. Yoo, H.G. Im, Combust. Theor. Model. 11 (2) (2007) 259-286.

[12] B. Emerson, J. O'connor, M. Juniper, T. Lieuwen, J. Fluid Mech. 706 (2012) 219-250.

11th Asia-Pacific Conference on Combustion,
The University of Sydney, NSW Australia
10th -14th December 2017.

A Study on Extinction Behavior of Stretched Premixed SNG(H_2-11%) Flames

Seulgi Kim[1], Keunseon Sim[1], Kwangjin Kim[1], Keeman Lee[2*]

[1]Department of Aerospace Engineering, Sunchon National University

255, Jungang-ro, Suncheon-si, Jeollanam-do, 57922, Republic of Korea

[2]School of Mechanical and Aerospace Engineering, Sunchon National University

255, Jungang-ro, Suncheon-si, Jeollanam-do, 57922, Republic of Korea

Abstract

Experimental and numerical studies are conducted to investigate the downstream interaction between premixed flames on SNG(H_2-11%) fuel with variation strain rate. The selected fuel mixture components are CH_4 80%, C_3H_8 9% and H_2 11%. Flame stability diagrams mapping lower and upper limit fuel concentrations for extinction as a function of strain rate are explored. The Lean Extinction Boundaries (after, LEB) are slanted shape, thereby indicating strong interactions, and the Rich Extinction Boundaries (after, REB) present square shape, thereby indicating weak interactions. The detailed explanations on those flame extinction characteristics in the stability diagrams are made through analysis of the flame structures.

1 Introduction

Currently, generated energy was almost dominant for fossil fuel, and environmental problems that are Global Warming and Fine Dust can be caused by generated carbon oxide and impurities such as nitrogen oxides on use of fossil fuel [1]. Thus, reductions of carbon oxide and impurities emissions are becoming a huge global issue. Thus, these problems should be found the solution for emission reduction. However, it is difficult to limit the use of fossil fuels due to the recent increase in energy demand. Thus, research on alternative fuel development or emission reduction is active. In this regard, Synthetic Natural Gas (after, SNG) is becoming a source of alternative gas that reduces emissions and reduces burden for the depletion of fossil fuels [2]. The SNG is produced by this process. The syngas produced by coal gasification at high temperature and pressure. And then, methane and hydrogen were manufactured through the gas purification and synthesis process.

Depending on the characteristic of the combustion of the gas turbine burner, the study of between flames can be an important element in identifying and understanding the basic flame characteristics for developing the gas turbine combustion. The downstream interaction between flames was conducted with hydrocarbon such as C_3H_8, CH_4 [3]. The downstream interaction between flames was well described by Lewis Number (Le). Namely, Le is larger than unity, it means weak interaction. Le is less than unity, it means strong interaction such as CH_4, H_2. The many studies conducted the interaction in rich (lean) – rich (lean) single hydrocarbon fuel [4]. However, studies on characteristics of blending fuel especially H_2 contents have been seldom found yet in the literature.

Thus, in current study, experimental and numerical are conducted to downstream interaction between premixed SNG (H_2-11%) flames. The fuel composition rate is CH_4 80%, C_3H_8 9%, and H_2 11%. Flame stability diagram mappings are lower and upper limit fuel concentration at extinction limit.

2 Experimental and Numerical methods

2.1 Experimental methods

The experiments are carried out in a counter flow burner in premixed flame configuration. Only a brief description is presented in this paper, and more detailed information of the burner may be found in K.S. Sim et al [5]. The separation distance (after, L) between the lower - upper nozzle exit is adjustable but is kept at 15mm. Each nozzle is surround by nitrogen (N_2) curtain flow to eliminate the reactive stream from outer perturbations. The fuel was used SNG (CH_4 80% C_3H_8 9% and H_2 11%) and purities CH_4, C_3H_8 and H_2 are 99.99% and purity of air is 99.95%. The global strain rate (after, a_g) was defined as following (1).

$$a_g = (V_U + V_L)/2L \qquad (1)$$

The V denotes the nozzle axial velocity of mixture and the subscripts, U and L, represent the upper and lower nozzle, respectively. The LEB and REB are attained varying the mixture concentration in SNG/Air (H_2 -11%) with a_g. Reactant volume fraction, Ω_i, was defined by (2).

$$\Omega_{CH4} = \frac{X_{CH4}}{\sum_{i=1}^{5} X_i} \qquad i = CH_4, C_3H_8, H_2, N_2,\ O_2 \qquad (2)$$

The X_i was the mole fraction of species i.

2.2 Numerical methods

This approach has been followed in the development of the OPPDIF code of the CHEMKIN package. In this code, the mathematical description of the combustion field along a stagnation stream line is one-dimension and the model adopted in the study is that developed by Kee et al [6] and extended by Lutz et al [7]. The detailed flame structures of counter-flow flames were numerically investigated by GRI v-3.0[8], USC v-II [9], and UC San Diego[10] for SNG fuel. Deviation of numerical with experimental results are 3.25 % for GRI v-3.0, 2.22 % for USC v-II, and 1.9 % for the

Corresponding author. Fax: +82-61-750-3820
E-mail address: kmlee@scnu.ac.kr

mechanism of UC San Diego, respectively. Thus, adopting the mechanism of UC San Diego (after, UCSD) could be appropriate in the current study.

3 Results and Discussion

3.1 Flame Stability map and extinction behavior

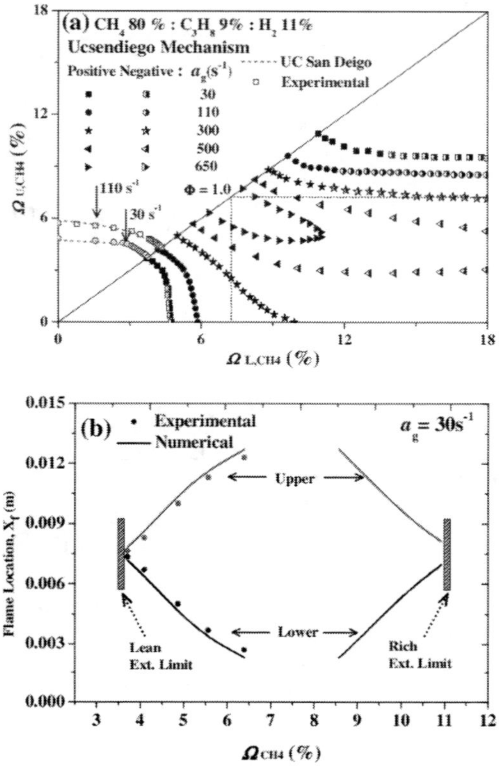

Figure 1: (a) Flame stability diagram mapping the upper and lower fuel concentration for flame extinction in terms of strain rate, (b) Location of flame with methane volume fraction at $a_g = 30s^{-1}$.

Figure 1 (a) shows flame stability maps with lower and upper methane volume fraction at various a_g in interacting premixed SNG/Air (H_2-11%) flames. The open symbols denote experimental results and the dash lines represent numerical results using UCSD mechanism (Reason of use was explained in section 2.2). The solid line is symmetry conditions that can be supplied mixture with the same fuel composition from the upper and lower nozzle. Also, in this study, we plotted the half of data due to the symmetry. The closed symbols is the positive flame propagation velocities while the half closed symbol is negative flame propagation velocities such that weaker flame crosses the stagnation point [11].

The laminar burning velocity in increase of strain rate, generally, is increase for Le less than unity [4]. The methane volume fraction of the lower and upper for REB are reduced with increase of strain rate, whereas LEB are extend at this case. However, in this study, both the REB and LEB are reduced in increase of strain rate. Also, Effective Lewis Number (after, Le_e) is larger than unity for $1.085 < Le_e < 1.198$. The reason was explained in K.S. Sim et al [5]. Fig. 1 (a) shows that the LEB are slanted, it means strong interaction; meanwhile the REB are square shape, it means weak interaction [3]. These are minutely described in Fig. 1 (b).

Figure 1 (b) shows the location of flame with lower and upper symmetry methane volume fraction for fixed $a_g = 30s^{-1}$. The flame location, X_f, was defined as the image processing in Matlab (Experimental) and position of maximum heat release rate (Numerical). The solid line is numerical results, and closed symbol is experimental results. The two flames completely merge at extinction

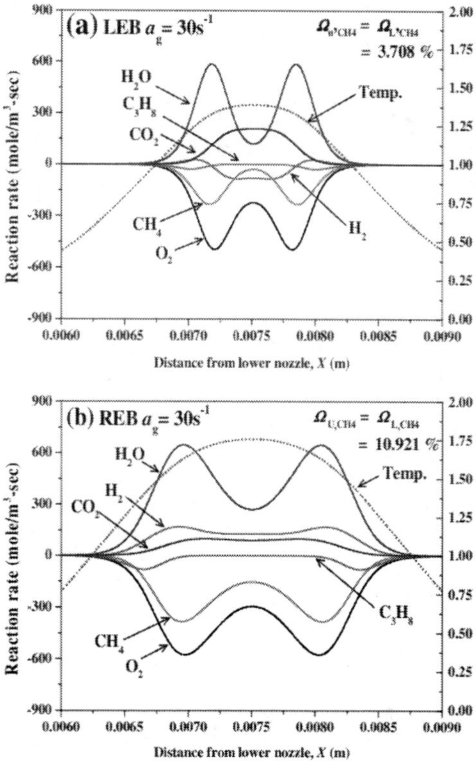

Figure 2: Comparison of flame structure in the symmetry extinction boundaries (a) lean-lean, (b) rich-rich at $a_g = 30s^{-1}$.

for lean mixtures, but are still separated for rich mixtures. These mean strong interaction for lean mixtures, but weak interaction for rich mixtures. The strong interaction can be caused by chemical interaction such as a radical sharing and conductive heat transfer between upper and lower flame. The details will be described for extinction behavior in section 3.2.

3.2 Flame structure

Figure 2 describes the symmetry flame structure (a) lean-lean, (b) rich-rich. The dash line and solid line represents temperature and reaction rate of species, respectively. Fig. 2 (a) shows that the upper and lower flame distance are very close. The maximum temperature presents the downstream region. The CO_2 has flatlands, and H_2O is almost produced flame zone. The CH_4 and C_3H_8 are consumed at near the flame zone, while H_2 is produced at near flame zone, and is consumed at downstream region. These mean that lean-lean condition is strong interaction because chemical interaction such as radical sharing is increase at downstream region [5].

Figure 2 (b) shows that rich-rich symmetry flame structure. The upper- lower flame distances are distant. In the case (b), the flame structure is the similarity to (a) lean-lean case with the exception of H_2. Thus, in (b) the rich-rich case, H_2 is produced at the near the flame zone and without consumed in the downstream region. These mean that rich-rich condition is weakly interaction due to chemical interaction such as radical sharing is weak at downstream region.

Hence, Fig. 2 describes that flame structure with methane volume fraction for $a_g = 30s^{-1}$. In the lean-lean case, it is very strong interaction due to strong radical sharing at the downstream region. In the rich-rich case, weak interaction can be caused by weakly radical sharing like that H_2 is produced without consumed at the downstream region.

4 Conclusions

Experimental and numerical studies are conducted to investigate the downstream interaction between premixed flames on SNG(H_2-11%) fuel with variation strain rate. The composition rate of fuel is CH_4 80%, C_3H_8 9%, and H_2 11%.

The laminar burning velocity is generally increase with strain rate for Le less than unity. The methane volume fraction of the lower and upper for REB are reduced with increase of strain rate, whereas LEB are extended at this case. Also, typically, flame interaction was dominant for Le and chemical reaction such as radical sharing in the downstream region. However, in this study, the LEB were slanted at a_g = 30 s^{-1}(low a_g), implying that strong interaction and in increase of strain rate more the reduction. The REB were square shape at low a_g, it means weakly interaction and was reduced in increase of strain rate. We have assumed that flame interaction was dominant by radical sharing in downstream region instead of Le. In the lean-lean symmetry condition, CH_4 and C_3H_8 were almost consumed at near the flame zone, but H_2 was produced at near the flame zone and consumed at downstream region. However, in the rich-rich symmetry condition, flame structure is the similarity to lean-lean condition with the exception of H_2 that is produced in the downstream region. Thus, LEB were slanted due to strong chemical interaction such as radical sharing in the downstream. The REB were square shape because of weakly chemical interaction in the downstream region.

5 Acknowledgment

This work was supported by the Energy Efficiency & Resources Core Technology Program of the Korea Institute of Energy Technology Evaluation and Planning (KETEP), granted financial resource from the Ministry of Trade, Industry & Energy, Republic Korea.(No.2015-2010-103630)

References

[1] Fotache C.G., Tan T., Sung C.J., C.K. Law., Combust. Flame 120: (2010) 417-426.

[2] Kopy, J., Schil. T.J., Biollaz, S.MA., Fuel. 89: (2010) 1763-1783.

[3] Satoru Ishizuka., C. K. Law., Proc. Combust. Inst. (1982) 327-335.

[4] Junichi Sato., Proc. Combust. Inst. (1982) 1541-1548.

[5] Sim. K.S., Lee. K.M., Korean Soc. Combut. 21(4) (2016) 48-60.

[6] Lutz AE, Kee RJ, Grcar JF, Rupley FM. A fortran program for computing opposed-flow diffusion flames. Sandia National Laboratories Report. SAND 96-8243; 1997.

[7] Kee RJ, Miller JA, Evans GH, Dixon-Lewis G. A computational model of the structure and extinction of strained, opposed flow, premixed methane-air flames. Proc Combustion Inst 1989;22(1):1479-1494.

[8] Smith GP, Golden DM, Frenklach M, Moriarty NW, Eiteneer B, Goldenberg M, et al. GRI-mech. 3.0. http://combustion.berkeley.edu/gri-mech/; 1999

[9] Wang H, You X, Joshi AV, Davis SG, Laskin A, Egolfopoulos F, Law CK. USC mech version II. high-temperature combustion reaction model of H2/CO/C1-C4 compounds. http://ignis.usc.edu/USC_Mech_II.htm/; 2007.

[10] Chemical-Kinetic Mechanisms for Combustion Applications. San Diego Mechanism web page, Mechanical and Aerospace Engineering (Combustion Research), University of California at San Diego. http://combustion.ucsd.edu/; 2014.

[11] S. H. Sohrab., Z. Y. YE., C. K. Law., Proc. Combust. Inst. (1984) 1957-1965.

11th Asia-Pacific Conference on Combustion,
The University of Sydney, NSW Australia
10th -14th December 2017.

Flame Modes and Combustion Characteristics of a Triple Port Burner

Chun-Han Chen[1], Chao-Wei Huang[2], Yueh-Heng Li[1*]

[1]Department of Aeronautics and Astronautics, National Cheng Kung University, Taiwan

[2]Department of Chemical and Material Engineering, National Kaohsiung University of Applied Science, Taiwan

*Corresponding Author, Email: yueheng@mail.ncku.edu.tw

Abstract

Increasing concerns of issues on global warming and climate change have urged stringent expectation on new energy conversion devices for much higher thermal efficiency and carbon dioxide capture and sequestration. Considering the practicability and safety, multi-port burners are widely used in industry to achieve the purpose of safety and emission control. Therefore, in this study co-annular triple port burner was used to investigate the flame phenomenon. The flame phenomenon was observed by varying the velocity of oxidizer stream (mixture of N_2 and O_2) and oxygen concentration ratio in central tube. Flame images showed that the increase in the velocity of oxidizer stream can effectively inhibit the soot formation before the presence of the inverse diffusion flame (IDF). In addition, the hysteresis of flame was also observed by means of adjusting flow velocities in the central tube. The flame modes were classified into six types, namely normal diffusion flame, partially-premixed flame, closed-tip inverse diffusion flame, opened-tip inverse diffusion flame, lifted-off edge flame and plume flame, respectively. Especially, the positions of rich-premixed branch for inner and outer triple flames were reverse because of the layout of the oxidizer stream and fuel stream. The rich-premixed branches for inner flame and outer flame were merged in the downstream and meanwhile trapped soot in flame. It explains the flame with opened-tip and bright flame. The gas analyzer was carried out the measurement of burned gas. As the result of Ω=33%, 40% and 60%, it can be conjectured that the flame temperature increases significantly due to the well combustion reaction when the flame is form of lifted-off flame or IDF, resulting in an increase of the thermal NO_x.

Keywords: Triple port burner; diffusion flame; inverse diffusion flame; oxy-enriched combustion; triple flame.

1 Introduction

Jet flame factures low complexity of the equipment system, high safety and reliability, and low maintenance cost. Accordingly, it prevails to apply in industrial and residential sectors, such as power plant boilers, gas turbine engines, stoves and so on. In addition, its mechanical configuration is straight forward, and often used in the basic research subject for exploring the chemical reaction and combustion behavior. According to the mixing condition of fuel and oxidizer, jet flames could be classified into two groups, one is the premixed flame and the other is the diffusion flame, respectively. In addition, diffusion flames are also divided into two types in accordance with the layout of the oxidizer stream and fuel stream, namely, one is normal diffusion flame (NDF) and the other is inverse diffusion flame (IDF). This study centers on the flame modes and combustion characteristics of inverse diffusion flames in oxygen-rich condition.

In the study of flames, the flame height is an important parameter for comparing the combustion characteristics among different flames. Mikofski et al. [1] validated the applicability of Roper's theory [2~3] to CH_4 / air and C_2H_4 / air IDF and made some modifications to their theory. The OH-PLIF (planar laser induced fluorescence) was employed to gauge flame heights. However, the measured flame heights were compared with the calculated flame height based on modified correlation, and the empirical formula was proposed eventually. Furthermore, the combustion characteristics of NDFs and IDFs were investigated in detail by OH-PLIF, PAH-PLIF, and soot-PLII (planar laser induced incandescence). Mikofski et al. [4] pointed out that the distances among the OH, PAH and soot zones were similar in IDFs and NDFs, but the locations of those zones were inverted in IDFs with regard to NDFs. Liu and Smallwood [5] found that the central air jet has a significant effect on the flame structure and sooting characteristics of the flame in a triple port co-annular burner. Johnson and Sobiesiak [6] studied the hysteresis of methane inverse diffusion flame. Their numerical simulation results suggested that when the local equivalence ratio approaches to unity in the upstream of the flame, the partially premixed flame (PPF) propagates upstream and stabilizes as an IDF.

In order to mitigate the impact of environment, reducing pollutant emission is an important issue. Yamamoto et al. [7] studied the combustion characteristics of the lifted off flame in a triple port co-annular burner. The experimental results showed that the maximum soot concentration is decreased significantly when the inner or outer flame is lifted. On the other hand, nitrogen oxides (NO_x) are decreased only when the outer flame is lifted. Additionally, the particular flame behavior, namely triple flame, can be observed when inner or outer flame is lifted. Yamamoto and Takemoto [8] investigated the behavior of the flame under the same experimental condition in double port co-annular burner and triple port co-annular burner respectively by PAH-LIF and soot-LII technique. The results show that the concentration of PAH and soot is reduced significantly in the triple port co-annular burner. Jung et al. [9] examine changes in oxygen concentration in the oxidizer in inverse diffusion flames, and investigate the soot formation of incipient soot particles. The two-color pyrometry showed that soot temperature along the soot growth path below the flame height was between 2000 K and 2200 K regardless of the oxygen enrichment. C/H ratios of soot particles linearly increased with increasing residence time, and incipient soot particles from inverse diffusion flames were carbonized gradually. Escudero et al. [10] studied the effects of the Oxygen Index (OI) in an ethylene IDF. Both the local and integrated soot quantities increase with OI. The slopes of the soot formation dominating zone increase linearly with the OI. However, the slopes of the oxidation predominant zone do not show a clear behavior and tend to keep approximately constant.

The behavior of leading edge of lifted-off flames was investigated by numerical simulation [11]. The results showed that there is an inward radial velocity component toward the central axis at the leading edge of the outer lifted-off flame, resulting in the acceleration of the flow field in front of the inner lifted off flame. The behavior of inner lifted-off flame reattaching to the nozzle can be observed as the lifted-off height of inner flame increasing. The edge propagation speed decreased outstandingly when the lifted-off height of the inner and outer flames are comparable. However, the effect of the acceleration of the flow field in front of the inner lifted-off flame is also mitigated. Therefore, the edge propagation speed and axial velocity play an important role in the flame reattachment . In this study, the flame modes and combustion characteristics under different condition of oxygen enrichment ratio and velocity ratio were investigated in IDFs. The particular flame modes and their detail flame characteristics in IDFs were studied by a theoretical model which was investigated by Ko et al [12].

2 Research method and apparatus

2.1 Experimental apparatus

The triple port co-annular burner was used to investigate the flame behavior, which is made of stainless steel. The internal oxidizer flow nozzle has an inner diameter of 0.15 cm (rim thickness of 0.03 cm), the fuel flow nozzle has an inner diameter of 0.4 cm (rim thickness of 0.05 cm), the external air flow nozzle has an inner diameter of 5 cm (rim thickness of 0.15 cm), as shown in fig. 1(a). Methane was used for fuel. Figure 1(b) shows a schematic diagram of experimental setup. The gas analyzer (MRU ; Vario plus) was used to measure burned gas.

2.2 Research method and condition

The requirement of gas in this study, including methane, oxygen and nitrogen is supplied by high pressure gas cylinder. Air is supplied by an air compressor and pass through a dehumidifier to eliminate the mist in the air. In this study, the velocity of oxidizer stream and its oxygen concentration were varied and the velocity of fuel stream and air stream were maintained at 25 cm/s and 9.12 cm/s. It is noted that the purpose of air stream is to prevent the disturbance from ambient air. The oxygen enrichment ratio is expressed as:

$$\Omega = \left(\frac{X_{O_2}}{X_{O_2} + X_{N_2}}\right) \cdot 100\%$$

The velocity ratio is expressed as:

$$R = \frac{\text{The velocity of oxidizer stream}}{\text{The velocity of fuel stream}} = \frac{V_{oxidizer}}{V_{fuel}}$$

The behavior of flames were investigated under 7 conditions of oxygen enrichment ratio, which are 21%, 27%, 33%, 40%, 60%, 80% and 100%, the velocity ratio was varied from 0 to 10 and then varied back to 0 under each oxygen concentration conditions.

For the measurement of burned gas, the burned gas, namely O_2 、CO_2 、CO and NO_x were measured by gas analyzer in each velocity ratio (0 to 10). The data was recorded per second by gas analyzer. In addition, the gas analyzer was started to record after wait for 1 minute for making flame stable and record the data for 5 minutes, thus there had 300 data in each velocity ratio and got the average value of these 300 times result.

The theoretical predictions of flame shape are based upon the triaxial burke-schumann methodology which derived from Ko et al. [12]. It is based on the fast chemistry (flame-sheet approximation), with various assumptions such as unity Lewis number, neglect of buoyancy, momentum transfer, shearing force and radiation heat transfer effects. The general solutions are obtained as follows:

$$0 = A'_{0,3} + \sum_{n=1}^{\infty} A'_{n,3} \frac{2}{\Phi_n J_0^2(\Phi_n)} J_0(\Phi_n \xi_f) \exp\left(\frac{Pe - \sqrt{Pe^2 + 4\Phi_n^2}}{2} \eta_f\right)$$

$$A'_{0,3} = B_{0,3} - A_{0,3} = -\tilde{Y}_{3,j} \cdot \left(\frac{u_3}{u_2}\right) - c_1^2 \left[\tilde{Y}_{1,j} \cdot \left(\frac{u_1}{u_2}\right) + \tilde{Y}_{2,j}\right] + c_2^2 \left[\tilde{Y}_{2,j} + \tilde{Y}_{3,j}\left(\frac{u_3}{u_2}\right)\right]$$

$$A'_{n,3} = B_{n,3} - A_{n,3} = c_2 J_1(c_2\Phi_n)\left[\tilde{Y}_{2,j} + \tilde{Y}_{3,j}\left(\frac{u_3}{u_2}\right)\right] - c_1 J_1(c_1\Phi_n)\left[\tilde{Y}_{1,j} \cdot \left(\frac{u_1}{u_2}\right) + \tilde{Y}_{2,j}\right]$$

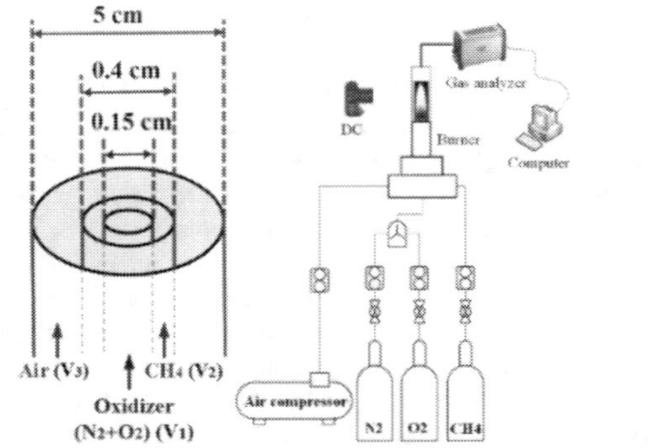

Figure 1. Schematic diagrams of the (a) triple port co-annular burner and (b) experimental setup.

3 Result and discussion

3.1 The flames appearance

The flame appearance change from bright yellow-orange to pale blue gradually with increasing velocity ratio, as shown in fig. 2(a), because the combustion reaction was more complete with increasing oxidizer flow rate so that decreased the soot concentration. Besides, the double flame structure and a bright spot at downstream, as shown in fig. 2(a), also can be observed with increasing velocity ratio. It can be attributed to a fact that the fuel and oxidizer at downstream was partially-premixed so that the local equivalence ratio approaches to unity at the location of bright spot [5, 6]. Figure 3(a) shows the flame appearance under the condition that Ω=33% and R varied from 0 to 10. With increasing the R to 8, 9 and 10, the behavior of inner lifted-off flame occurred. It is supposed that the fuel and oxidizer was partially-premixed at downstream resulted in a flame propagation speed toward upstream. However, the flame is lifted when the flame propagation speed and the velocity of flow field approach to equilibrium. When the oxygen enrichment ratio increased to Ω=60% and the velocity ratio increased to R=4, the IDF can be observed, as shown in fig. 4(a). As mentioned above, when the flow field velocity of partially-premixed flame and flame propagation speed can not approach to equilibrium, the inner flame propagates to upstream and stabilizes at nozzle rim to form an IDF. As the increase of the velocity ratio an opened-tip IDF can be observed, further increased, the inner flame tip spread to both sides. When the velocity ratio varied back to zero (R=10 to 0), as shown in fig. 2(b), 3(b) 4(b) and 5, the hysteresis was occurred except Ω=21%. As this result, it can be conjectured that the hysteresis do not occur when the flame appearance do not have a formation of IDF or lifted-off flame

Corresponding author. Fax: +00-00-000-000
E-mail address: abc@abcde.com

with increasing the velocity ratio (R=0 to 10).

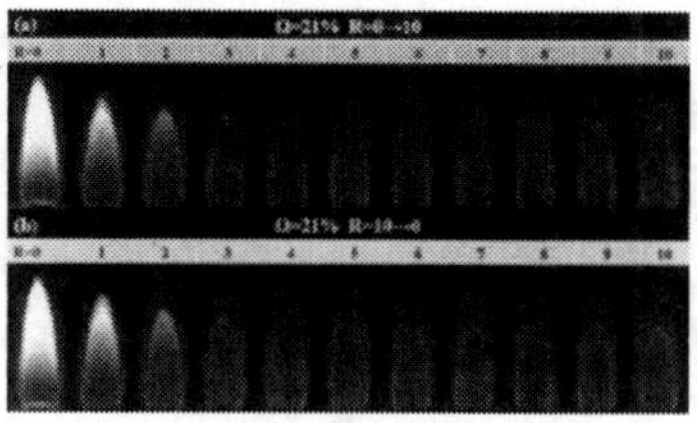

Figure 2. The visible flame appearances for Ω=21% with varying R (a) from 0 to 10 and (b) from 10 back to 0.

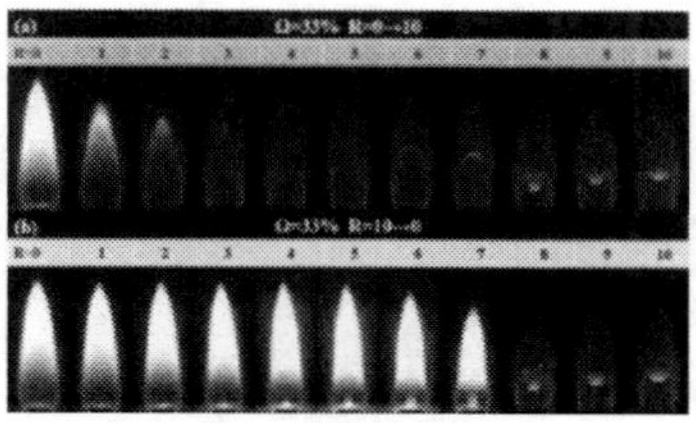

Figure 3. The visible flame appearances for Ω=33% with varying R (a) from 0 to 10 and (b) from 10 back to 0.

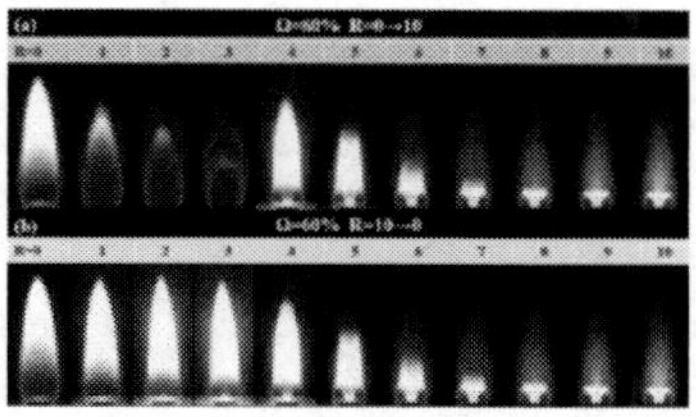

Figure 4. The visible flame appearances for Ω=60% with varying R (a) from 0 to 10 and (b) from 10 back to 0.

Figure 5 shows the different flame modes with various Ω and R. The velocity ratio of formation of IDF decreases as oxygen enrichment ratio increases, as shown in fig. 5(a). As the numerical simulation result that studied by Johnson and Sobiesiak [6], the PPF propagates to upstream and stabilizes as an IDF when the local equivalence ratio approaches to unity. Therefore, the formation of IDF only needs a smaller velocity ratio at a higher oxygen enrichment ratio. Compare fig. 5(a) with (b), the hysteresis can be observed. Additionally, a particular flame appearance, namely plume flame, can be observed, as shown in fig. 5(a), (b) and fig. 4(a) at R=8, 9 and 10.

3.2 Flame modes

The flame modes were classified into six types, namely normal diffusion flame (NDF), partially-premixed flame (PPF), closed-tip inverse diffusion flame (closed-tip IDF), opened-tip inverse diffusion flame (opened-tip IDF), lifted-off edge flame and plume flame, respectively, as shown in fig. 5.

Figure 6(a) shows the comparison diagram between visible image of NDF and the theoretical predictions of flame shape [12]. It is noted that the parameters, ξ and η, were obtained by normalizing the radial position and axial position respectively, thus the location of 0.08 is the location of fuel stream nozzle rim. The flame mode is a normal diffusion flame because the oxidizer was only supplied by air nozzle. The NDF consist of a triple flame structure that consist of three branches, namely rich-premixed branch closed to fuel side, lean-premixed branch closed to air side and diffusion flame tail. In addition, all of them

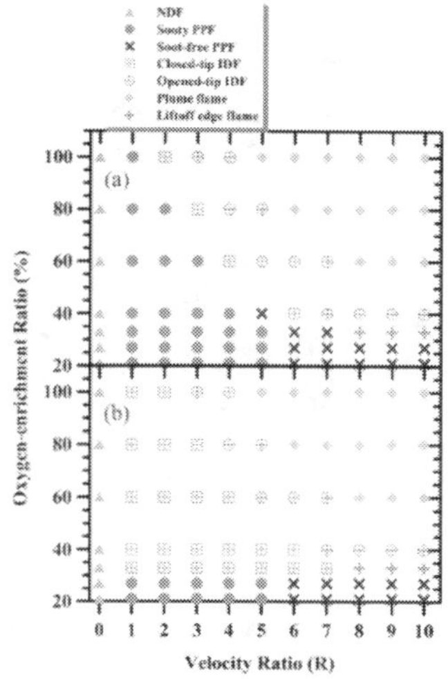

Figure 5. The different flame modes with various Ω and varying R (a) from 0 to 10 and (b) from 10 back to 0.

are spread toward downstream from a location which is the outside of the fuel stream nozzle rim, called this triple point [13]. Compare with visible image of flame, the diffusion flame tail which is in bright blue color and existed at fuel nozzle rim, spreading toward downstream. Furthermore, the pale yellow flame and pale blue flame which is respectively close to fuel side and air side, namely rich-premixed branch and lean-premixed branch. For closed-tip IDF, two flame structures are respectively existed at oxidizer stream nozzle rim and fuel stream nozzle rim and only the outer flame structure is triple flame, as shown in fig. 6(b). It is interesting that the position of inner rich-premixed branch and outer rich-premixed branch are reverse because of the layout of the oxidizer stream and fuel stream. Figure 6(c) shows the comparison diagram of opened-tip IDF. In this flame mode, two triple flame structures are respectively existed at oxidizer stream nozzle rim and fuel stream nozzle rim. Besides, the rich-premixed branches for inner and outer triple flame are merged in the downstream and meanwhile trapped soot in flame. It explains the flame with opened-tip and bright flame. At higher oxygen enrichment ratio and velocity ratio, the plume flame is formed, as shown in fig. 6(d). Both rich-premixed branch and diffusion flame tail are merged. For this reason, the inner flame tip spread to both sides. In addition, the excess of oxidizer react with fuel in the downstream resulted in a bright blue plume flame.

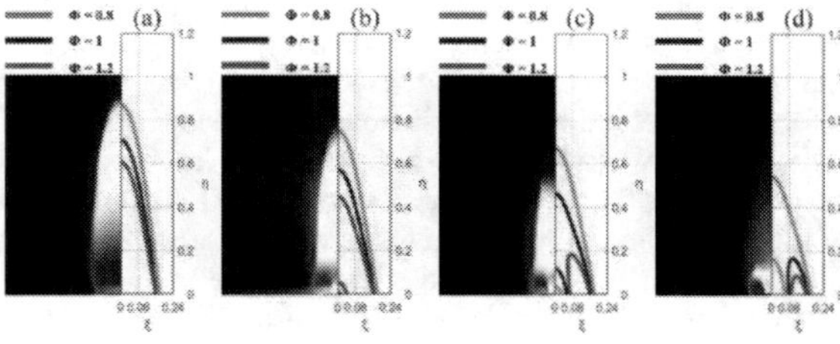

Figure 6. Comparison diagrams between visible images of (a) NDF (b) closed-tip IDF (c) opened-tip IDF (d) plume flame and their theoretical predictions of flame shape.

3.3 Analysis of burned gas

Figure 7(a) shows the average of burned gas concentrations with various velocity ratio. The concentrations of O_2, NO_x and CO_2 do not show a clear behavior and tend to keep approximately constant. However, the CO content decreases as the velocity ratio increase due to more oxidizer injection. For Ω=33%, the CO content is less than Ω=21% at a given velocity ratio and the NO_x content increases slightly after R=7, as shown in fig. 7(b). The CO content decreases to 0 at R=7, as shown in fig. 7(c). It is indicated that the combustion behavior tends to complete reaction. In addition, the NO_x content starts to increase significantly at R=6. Increasing the oxygen enrichment ratio to 60%, the NO_x content also increases strongly when the CO content decreases to 0, as shown in fig. 7(d). Additionally, the concentration of O_2 increases slightly after R=4. For this result, it is confirmed that the excess of oxidizer react with fuel in the downstream resulted in a bright blue plume flame, as shown in fig. 6(d). As the result of Ω=33%, 40% and 60%, it can be conjectured that the flame temperature increases significantly due to the well combustion reaction when the flame is form of lifted-off flame or IDF, resulting in an increase of the thermal NO_x.

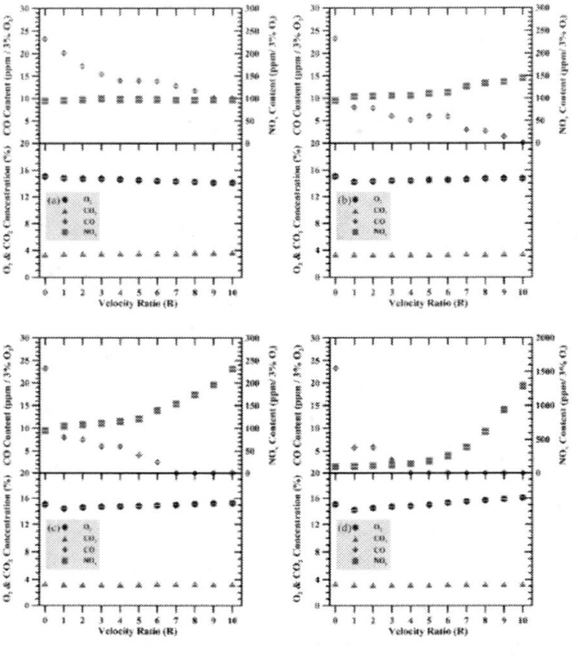

Figure 7. The average of burned gas concentrations with various velocity ratio for (a) Ω=21% (b) Ω=33% (c) Ω=40% (d) Ω=60%.

4 Conclusions

The particular flame behavior can be observed by varying the oxygen enrichment ratio and velocity ratio. Flame images showed that the increase in the velocity of oxidizer stream can effectively inhibit the soot formation before the presence of the IDF. At a given oxygen enrichment ratio, the hysteresis do not occur when the flame appearance do not have a formation of IDF or lifted-off flame with increasing the velocity ratio (R=0 to 10). The flame modes in this study were classified into six types, namely normal diffusion flame (NDF), partially-premixed flame (PPF), closed-tip inverse diffusion flame (closed-tip IDF), opened-tip inverse diffusion flame (opened-tip IDF), lifted-off edge flame and plume flame, respectively. Most of them were in the form of triple flame.

For the measurement of burned gas, the NO_x content increases with increasing the oxygen enrichment ratio and velocity ratio. In contrast, the CO content decreases with increasing the oxygen enrichment ratio and velocity ratio. Increasing the oxygen enrichment ratio to 60%, the concentration of O_2 increases slightly after R=4. For this result, it is confirmed that the excess of oxidizer react with fuel in the downstream resulted in a bright blue plume flame. As the result of Ω=33%, 40% and 60%, it can be conjectured that the flame temperature increases significantly due to the well combustion reaction when the flame is form of lifted-off flame or IDF, resulting in an increase of the thermal NO_x.

References

[1] Mikofski, M.A., Williams, T.C., Shaddix, C.R., Blevins, L.G., "Flame height measurement of laminar inverse diffusion flames." Combustion and Flame (2006), 146: 63-72.

[2] Roper, F.G., "The prediction of laminar jet diffusion flame sizes: part 1. Theoretical Model. " Combustion and Flame (1977), 29:219-226.

[3] Roper, F.G., "The prediction of laminar jet diffusion flame sizes: part 2. Experimental Verification. " Combustion and Flame (1977), 29:219-226.

[4] Mikofski, M.A., Williams, T.C., Shaddix, C.R., Fernandez-Pello, A.C., Blevins, L.G., "Structure of laminar sooting inverse diffusion flames." Combustion and Flame (2007), 149: 463-478.

[5] Liu, F., Smallwood, G.J., "Control of the structure and sooting characteristics of a coflow laminar methane/air diffusion flame using a central air jet: An experimental and numerical study." Proceedings of the Combustion Institute (2011), 33: 1063-1070.

[6] Johnson, M.B., Sobiesiak, A., "Hysteresis of methane inverse diffusion flames with co-flowing air and combustion products." Proceedings of the Combustion Institute (2011), 33: 1079-1085.

[7] Yamamoto, K., Kato, S., Isobe, Y., Hayashi, N., Yamashita, H., "Lifted flame structure of coannular jet flames in a triple port burner." Proceedings of the Combustion Institute (2011), 33: 1195-1201.

[8] Yamamoto, K., Takemoto, M., "Measurement of PAH and soot of diffusion flames in a triple port burner." Fuel Processing Technology Special Issues (2013), 107: 99-106.

[9] Jung, Y., Oh, K.C., Bae, C., Shin, H.D., "The effect of oxygen enrichment on incipient soot particles in inverse diffusion flames" Fuel (2012), 102: 199-207.

[10] Escudero, F., Fuentes, A., Demarco, R., Consalvi, J.L., Liu, F., Elicer-Cortés, J.C., Fernandez-Pello, C., "Effects of oxygen index on soot production and temperature in an ethylene inverse diffusion flame" Experimental Thermal and Fluid Science (2016), 73: 101-108

[11] Yamamoto, K., Isobe, Y., Hayashi, N., Yamashita, H., Suk Ho Chung., "Behaviors of tribrachial edge flames and their interactions in a triple-port burner." Combustion and Flame (2015) 162: 1653–1659

[12] Ko, Y.C., Hou, S.S., Lin, T.H., "Laminar diffusion flames in a mutiport burner." Combustion Science and Technology (2005), 177: 1463-1484.

[13] Chung, S.H., "Stabilization, propagation and instability of tribrachial triple flames." Proceedings of the Combustion Institute (2007), 31: 877-892.Proc. Combust. Inst. 23 (1990) 809-815.

Investigation on Combustion Characteristics of Sewage Sludge Solid Fuels using Pilot-scale Bubbling Fluidized Bed Reactor

Donghee Kim[1, 2], Hyungjun Ahn[1], Kang Y. Huh[2], Youngjae Lee[1]

[1]Thermochemical Energy System, Korea Institute of Industrial Technology

89, Yangdaegiro-gil, Ipjang-myeon, Seobuk-gu, Cheonan-si, Chungcheongnam-do, 31056, Republic of Korea

[2]Department of Mechanical Engineering, Pohang University of Science and Technology

77, Cheongam-ro, Nam-gu, Pohang-si, Gyeongsangbuk-do, 37673, Republic of Korea

Abstract

In this study, combustion characteristics of sewage sludge solid fuel produced by pretreatment based on hydrothermal carbonization process are discussed. In thermogravimetric analysis results, there are two burning sections. This tendency is more prominent in SS compared to WP. As a result of experiment using bubbling fluidized bed reactor, the temperature and pressure of the reactor were kept stable. But CO, NO_x and SO_x emissions of SS were higher than that of WP. Moreover, the ash deposition tendency of SS was predicted highly.

1 Introduction

Nowadays, the consumption of conventional fossil fuel increases continuously due to the increment of global energy demand. The global problem of environment and energy resources has encouraged researchers to investigate the possibility of renewable energy sources such as biomass and waste fuel. But facility and flue gas were badly affected with the use of the renewable energy, there is a lack of research in this area.

A fluidized bed reactor has high heat transfer efficiency compared to a fixed bed reactor. The fuel type is also more flexible in a fluidized bed reactor than a fixed bed reactor. Therefore the pretreated fuels such as biomass and waste fuel are utilized as energy source in fluidized bed reactor[1]. Especially J. Werther and T. Ogada[2] discussed the combustion of sewage sludge in various aspects. Also M. Otero et al.[3, 4] and M. Belen Folgueras et al.[5] discussed co-firing characteristics of sewage sludge and coal using thermogravimetric analysis. And Amand and Leckner[6] studied on the NO and NO_2 formation characteristics during co-firing of sewage sludge, coal and woody biomass. And also they studied on metal component in ashes of sewage sludge, coal and woody biomass including co-firing case[7]. Lastly Tomasz Kupka et al.[8] discussed the ash deposition characteristics of co-firing coal and sewage sludge.

In this study, the combustion characteristics of dried sewage sludge and wood pellet are discussed using experiment result of pilot-scale bubbling fluidized bed reactor. Also through the metal component analysis result of sewage sludge ash, the slagging and fouling tendency is discussed.

Corresponding author. Fax: +82-41-589-8464
E-mail address: leeyj@kitech.re.kr

2 Experimental Method

2.1 Fuel and Bed Material

Generally, swage sludge becomes a high temperature slurry when it with moisture content of 80 to 85% is heated to 230 ~ 250°C. And moisture of sewage sludge vaporizes below the subcritical temperature(374.2°C). The long hydrocarbon chains of cellulose, which is a woody fuel, are decomposed by heat and converted into the low molecular weight carbide. By such a hydrothermal carbonization process, the decarboxylation and dehydration reaction of sewage sludge is induced. Thus oxygen/carbon and hydrogen/carbon ratio is decreased, as the volatile matter decreases, the carbon fixation increases the fixed carbon and increases the calorific value. This study was carried out to investigate the combustion characteristics of sewage sludge solid fuel(SS) produced by pretreatment process based on hydrothermal carbonization. The two SS(SS1, SS2) with different production periods were tested, the WP combustion experiment was also performed for comparative analysis. Table 1 shows results of proximate analysis, ultimate analysis and heating value analysis of tested fuel.

Fuel	SS1	SS2	WP
Ultimate analysis (wt%, daf basis)			
C	69.7	66.4	49.2
H	7.5	4.2	4.3
N	8.9	5.8	1.9
S	0.7	0.8	0.01
O	13.2	22.8	44.2
Proximate analysis (wt%)			
Moisture	17.0	10.4	11.5
Volatile	53.5	53.6	77.0
Fixed carbon	5.6	7.5	11.1
Ash	23.9	28.5	0.4
Heating value (kcal/kg)			
HHV[a]	4550	4610	4350
LHV[b]	4240	4430	4100

a. Higher Heating Value, b. Lower Heating Value.

Table 1: Ultimate, proximate and heating value analysis results of SS1, SS2 and WP.

As a result of ultimate analysis, carbon and nitrogen content of SS were higher than that of WP. Conversely, oxygen content of SS was lower than that of WP, there was no sulfur content in WP. And SS had lower volatile matter and more ash content than WP in proximate analysis.

The mean particle size of the bed material used in this experiment is 460 μm, the particle density is 2800 kg/m^3. Furthermore sphericity of bed material was assumed to be land sand based on the mean particle size and the particle density, corresponds to Geldart B classification.

2.2 Experimental Set Up

Figure 1 shows process schematic diagram of the 100kW$_{th}$ pilot-scale bubbling fluidized bed reactor system. The reactor has a diameter of 400 mm and a height of 4300 mm. On the inner wall of the reactor, 7 K-type thermocouples and 7 pressure transmitters were installed. And also the preheater at the bottom of the reactor was raise the temperature of the reactor. The fuel was crushed to a size of 10 mm or less, and was fed into the reactor at the top of the reactor.

After the reactor, there are two cyclones to capture ashes and scattered bed material. And the reactor pressure was regulated by I.D. fan. Also ashes was captured in 1st cyclone using sintered mesh filter, flue gas sampled from the stack was analyzed using gas analyzer TESTO 350K.

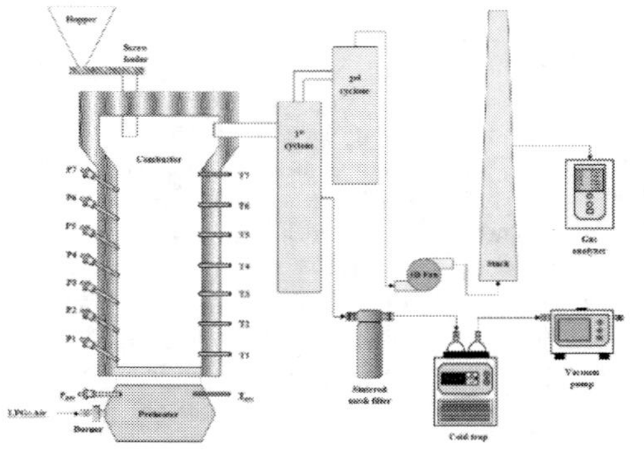

Figure 1: Schematic diagram of pilot-scale bubbling fluidized bed reactor system

2.3 Experimental Conditions

Table 2 shows the experimental conditions of this study. There are 4 cases, 54.5 ~ 96.5 kW$_{th}$ based on the thermal input to the reactor. Equivalence ratio is varied 1.3 ~ 1.6, U$_{mf}$(minimum fluidization velocity) is calculated assuming 800°C air. The fluidization number ranges from 3.8 to 4.5. The burner installed in the preheater was used to raise temperature of the reactor.

	Unit	C1	C2	C3	C4
Fuel type	-	SS2	SS2	SS1	WP
Feeding rate	kg/h	18	13.5	10.3	19
Thermal input	kW$_{th}$	96.5	72	54.5	96.1
Equivalence ratio	-	1.34	1.61	1.6	1.65
Fluidization number[a]	-	4.5	4.05	3.5	3.8
Air flow rate	l/min	2492	2246	1956	2077

a. multiple of minimum fluidization velocity

Table 2: Experimental conditions of bubbling fluidized bed reactor

3 Results

3.1 Thermogravimetric Analysis

Prior to the pilot-scale bubbling fluidized bed reactor experiment, the TGA(thermogravimetric analysis) of SS1, SS2 and WP are performed to investigate combustion profiles of fuel. The samples around 10mg were used, the results of TGA are shown in Fig.2. The analysis were performed in a static atmosphere of air, using a heating rate of 20°C/min from the ambient temperature to 800°C. And the intersection of the burning section's tangent lines was defined as the burning point.

Figure 2 (a) and (b) show TGA results of SS, the 1st burning points are 280°C in both SS1 and SS2. And the 2nd burning points of SS1 and SS2 are 448°C and 398°C respectively. Also the 1st burning point of WP was 325°C, the 2nd burning point was 474°C in Fig.2 (c). The 1st burning section of WP was bigger than the 2nd burning section, but the size of burning sections of SS1 and SS2 are similar compared to WP.

Compared with proximate analysis results of Table 1, the mass loss in the 1st burning is similar to the mass of volatile matter. Therefore combustion of volatile matter is dominant in the 1st burning section. Also the reaction with more complex structure occur in the 2nd burning section[5]. Like this results, there are two burning sections in TGA results. This tendency was more prominent in SS compared to WP.

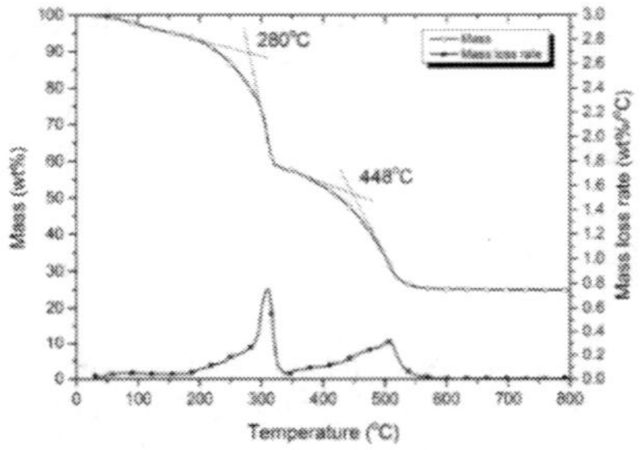

(a) SS1

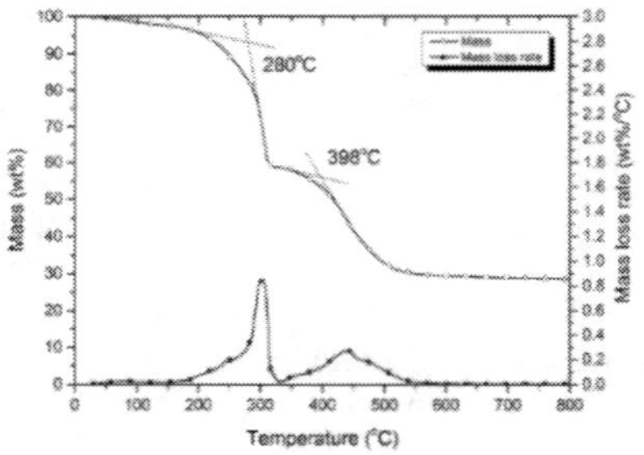

(b) SS2

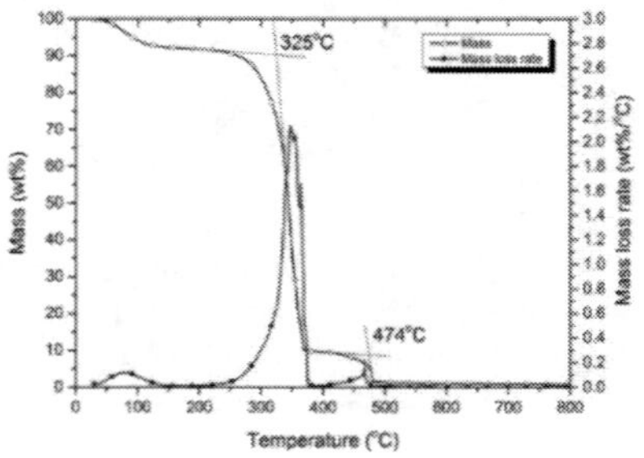

(c) WP

Figure 2: TGA results of SS and WP

3.2 Experimental Result of BFB Reactor

Figure 3 and 4 show the mean temperatures and the mean pressures profile in the reactor according to the height respectively. The mean data to be mentioned in the following measured for 1 hour after reactor stabilization, and stabilization of reactor including the preheating process took about 6 hours. The mean temperatures of the reactor showed a similar trend under all conditions, the highest temperature of the reactor was located at T1 ~ T3 in all cases. This tendency is attributable to the fact that the fluidization by the bed material is the most active at T1 ~ T3. Therefore the almost combustion reaction is formed at T1 ~ T3. As a result of the mean pressure, the qualitatively similar results were obtained as with the mean temperature results.

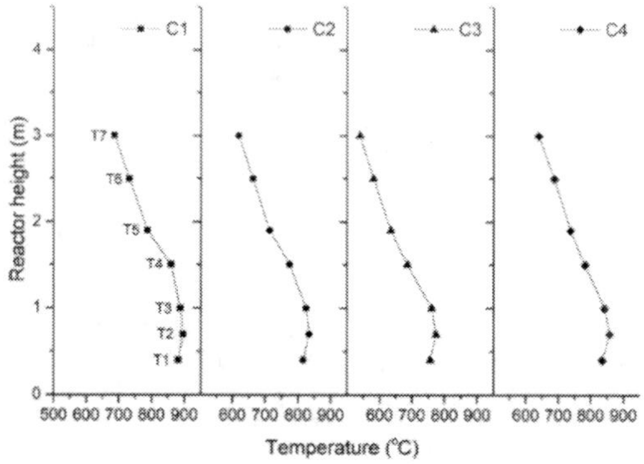

Figure 3: Mean temperature in the reactor according to the reactor height

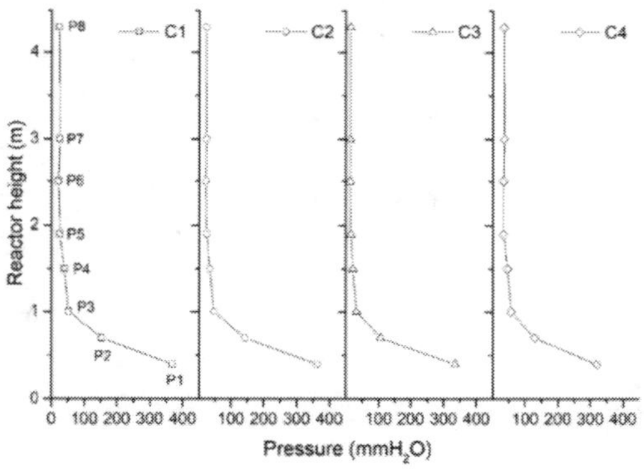

Figure 4: Mean pressure in the reactor according to the reactor height

Table 3 shows that the mean emissions of the flue gas in all cases. The mean emissions of C1 and C2 are qualitatively similar, the NO_x and SO_x emissions are 700 ~ 800 ppm and 300 ~ 400 ppm respectively. The thermal NO_x is generated mainly above 1300°C[9]. Since the inner wall temperature of the reactor did not exceed 900°C, the ratio of thermal NO_x in NO_x is not expected to be high. As a result of the CO emissions to evaluate combustibility, the CO emissions of C3 were 611.1 ppm, it is 4 times higher compared to the average of C1 and C2. It is because of low hardness of SS1 compared to SS2, the pulverized SS1 was fed to the reactor. The pulverized SS1 could not reach the combustion zone of the reactor because of its light weight, and it produced the unburned gas by going straight to the 1st cyclone. Therefore the CO emissions of C3 is much higher than C1 or C2, the size and weight of fuel are important for the top feeding type of bubbling fluidized bed combustion. C4, which uses WP as

Parameter	Unit	C1	C2	C3	C4
O_2	Vol.%	10.0	11.0	11.6	11.5
CO_2		9.6	8.7	8.1	9.1
CO	ppm	135.2	174.1	611.1	88.4
NO_x		789.5	718.4	445.7	64.5
NO		788.5	715.5	439.5	64.4
NO_2		1.0	2.9	6.1	0.1
SO_x		391.4	303.4	209.7	0.1

Table 3: Mean value of flue gas emissions at stack

fuel, shows low CO emissions in contrast to C1 ~ C3. And also the NO_x and SO_x emissions of C1 ~ C3, which use SS as fuel, are higher than C4 using WP as fuel.

3.3 Slagging/Fouling Tendency Analysis

The ash causes the slagging and fouling in the reactor, which greatly affects the life of the combustor and the heat transfer efficiency. Therefore because it is important to evaluate the slagging and fouling tendency, the slagging and fouling tendency indices were calculated using (1) ~ (5) based on the XRF(x-ray fluorescence) analysis results of ashes. The slagging and fouling tendency indices were evaluated based on the evaluation standard of the ash deposition tendency shown in Table 4.

$$B/A = \frac{CaO+MgO+K_2O+Fe_2O_3}{SiO_2+Al_2O_3+TiO_2} \quad (1)$$

$$S/A = \frac{SiO_2}{Al_2O_3} \quad (2)$$

$$I/C = \frac{Fe_2O_3}{CaO} \quad (3)$$

$$TA = Na_2O+K_2O \quad (4)$$

$$BAI = \frac{Fe_2O_3}{K_2O+Na_2O} \quad (5)$$

Index	Tendency		
	Low	Medium	High
Base-to-acid ratio(B/A)	<0.5	0.5<B/A<1	>1
Silica-alumina ratio(S/A)	<0.31 or >3	-	<0.31 or >3
Iron-calcium ratio(I/C)	<0.31 or > 3	-	<0.31 or >3
Total alkali(TA)	<0.3	0.3<TA<0.4	>0.4
Bed agglomeration index(BAI)	-	-	<10

Table 4: Evaluation standard of ash deposition tendency

Prior to the slagging and fouling tendency analysis, the average ash content of SS1 and SS2 is 26.2%, it is about 66 times higher than its WP, which is 0.4%. Therefore when predict the combustion trouble, actual ash deposition tendency and ash content should be considered.

Table 5 shows XRF analysis results of the ashes captured at the 1st cyclone. The results of C1 ~ C3, which use SS as fuel, show a different tendency from C4 using WP as fuel.

Table 6 shows the results of evaluating the slagging and fouling tendency based on the evaluation standard of slagging and fouling tendency of Table 4 and the XRF analysis results of Table 5. All slagging and fouling indices of C1 ~ C3 cases were estimated "high". But B/A ratio and S/A ratio of C4 were estimated "medium" and "low" respectively. This is because the ash content of WP is so low that the portion of the scattered bed material captured at the sintered mesh filter was higher than other cases. Table 5 shows SiO_2 content of C4, which is main component of bed material, was so high than other cases as mentioned earlier. But these slagging and fouling indices are not sufficient to explain the ash deposition tendency, additional experiments are needed.

Composition	Case			
	C1	C2	C3	C4
Na_2O	10.40	9.29	11.48	4.35
MgO	3.90	5.30	5.53	0.87
Al_2O_3	15.61	17.24	17.54	15.24
SiO_2	29.26	27.33	26.32	46.59
K_2O	4.20	4.17	4.04	7.94
CaO	15.86	16.19	15.93	14.48
TiO_2	1.69	1.59	1.49	0.84
MnO	2.70	2.97	2.83	0.94
Fe_2O_3	16.38	15.92	14.84	8.75

Table 5: XRF analysis result of captured ashes

Index	Tendency			
	C1	C2	C3	C4
Base-to-acid ratio(B/A)	1.14 (h)	1.10 (h)	1.14 (h)	0.58 (m)
Silica-alumina ratio(S/A)	1.21 (h)	1.58 (h)	1.50 (h)	3.06 (l)
Iron-calcium ratio(I/C)	0.80 (h)	0.98 (h)	0.93 (h)	0.60 (h)
Total alkali(TA)	21.13 (h)	10.30 (h)	11.90 (h)	11.44 (h)
Bed agglomeration index(BAI)	0.38 (h)	1.18 (h)	0.96 (h)	0.71 (h)

※ Slagging/fouling tendency (h): high, (m): medium, (l): low

Table 6: Evaluation of ash deposition tendency as XRF analysis result of captured ashes

4 Conclusions

In this study, combustion characteristics of sewage sludge solid fuel(SS) and wood pellet(WP) were investigated using pilot-scale bubbling fluidized bed reactor.

As a results of TGA, the 1st burning points of SS were 280°C in both SS1 and SS2. And the 2nd burning points of SS1 and SS2 were 448°C and 398°C respectively. Also the 1st burning point of WP was 325°C, the 2nd burning point was 474°C. And also combustion of volatile matter was dominant in the 1st burning section and the reaction with more complex structure occur in the 2nd burning section. The tendency that exist two burning sections in TGA results was more prominent in SS compared to WP.

According to the inner wall temperatures profile of the reactor, the main combustion reaction took place in T1 ~ T3 section and showed a similar trend qualitatively in all cases. As a measurement result of flue gas, the average NO_x and the CO emissions of SS were 10.1 times and 3.5 times higher than its WP respectively. Especially in WP case, there was hardly SO_x emissions in flue gas. As a result, combustion characteristics of WP was better than SS in aspect of the flue gas emissions.

Lastly as the XRF analysis result of ash captured at the 1st cyclone, the slagging and fouling tendency of SS was evaluated bad. But the ash of WP case had better the slagging and fouling tendency than SS cases, also the amount of the ash was very small.

5 Acknowledgment

This work was supported by the Korea Institute of Energy Technology Evaluation and Planning (KETEP) grant, Funded by the Korea Government Ministry of Trade, Industry, and Energy (Project No.20153010102130)

References

[1] Y.H. Choung, K.C. Cho, D.H. Kang, Y.K. Kim, C.W. Park, K.J. Oh, Korea Society of Environmental Engineers 28 (7) (2006) 746-751.

[2] J. Werther, T. Ogada, Progress in Energy and Combustion Science 25 (1999) 55-116.

[3] M. Otero, C. Diez, L.F. Calvo, A.I. Garcia, A. Moran, Biomass and Bioenergy 22 (2002) 319-329.

[4] M. Otero, L.F. Calvo, M.V. Gil, A.I. Garcia, A. Moran, Bioresource Technology 99 (2008) 6311-6319.

[5] M.B. Folgueras, R.M. Diaz, J. Xiberta, I. Prieto, Fuel 82 (2003) 2051-2055.

[6] B. Leckner, L.-E. Amand, K. Lucke, J. Werther, Fuel 83 (2004) 477-486.

[7] L.-E. Amand, B. Leckner, Fuel 83 (2004) 1803-1821.

[8] T. Kupka, M. Mancini, M. Irmer, R. Weber, Fuel 87 (2008) 2824-2837.

[9] Nitrogen Oxides(NO_x) Why and How They Are Controlled, Report No.EPA-456/F-99-006R, U.S. Environmental Protection Agency, 1999.

11th Asia-Pacific Conference on Combustion,
The University of Sydney, NSW Australia
10th -14th December 2017.

Transition between Swirl and Bluffbody Flame Stabilization in a Novel Ultra-lean Non-premixed Burner

R. Sadanandan[1], A. Chakraborty[2], V.K. Arumugam[2], S.R. Chakravarthy[2]

[1]Department of Aerospace Engineering, Indian Institute of Space Science and Technology
Valiamala P.O., Thiruvananthapuram, Kerala, India – 695 547

[2]Department of Aerospace Engineering and National Center for Combustion Research and Development,
Indian Institute of Technology Madras, Chennai, India – 600 036

Abstract

The flame dynamics and stabilization mechanism in a novel, ultra-lean, non-premixed, model GT burner is experimentally investigated with optical diagnostics. The burner operating with methane as fuel exhibits a convergent-divergent flow field at the burner exit. This results in two flame stabilization zones depending on the global equivalence ratios, Φ_{glob}. At $0.6<\Phi_{glob}<1.0$, the flame stabilizes in the diverging section where the flow field is characterized predominantly by the swirl. The strain rates in these locations are found to be well below the extinctions strain rates for turbulent non-premixed flames. With decreasing Φ_{glob}, the flame is located progressively near the burner lip, where a large recirculation zone of the central bluff body in the burner exists. The upstream marching of the flame despite a decreasing Φ_{glob}, and a corresponding decrease in heat input, is explained in terms of a flow-flame balance along the stoichiometric surface. At still lower Φ_{glob} extending to ultra-lean conditions ($\Phi_{glob} = 0.1$), the flame stabilizes in a region characterized by bluffbody induced toroidal flow structures closer to the burner exit.

1 Introduction

In our efforts to slow down climate change globally there is a growing determination to move away from coal to cleaner options like renewables. However, renewable alone cannot meet the growing demand for energy, which is estimated to rise by as much as 40% in the next 20 years. Here natural gas (NG) is projected as the favourite candidate to bridge the gap with renewable energy sources as it emits half as much CO_2 as coal. So globally there is an urgent need to develop efficient, low emissions burners/combustors that run on NG.

When it comes to reducing pollutants, gas turbine (GT) manufacturers pursue different approaches like fuel flexibility, staged combustion, flameless combustion[1-3] for terrestrial GT engines and twin annular premixing swirler (TAPS), rich-quench-lean and rich dome designs [4,5] in case of aero engines. Here, swirl is the preferred flame stabilizing mechanism. One of the prominent features of a swirl injector is the formation of the central recirculation zone (CRZ) above a certain swirl number S. The flow in this region is associated with high shear stresses, high turbulence intensities, and flame speeds, resulting from the vortex breakdown [6]. In order to achieve low thermal NOx emissions a "quasi-distributed" reaction zone with minimal temperature fluctuations and flat thermal fields is desired. So the optimization of CRZ is critical as it strongly impacts the mixing of the fresh fuel/air with the burned gases.

Considering these parameters, a novel, non-premixed, model GT burner (IIST-GS1) [7] is currently under development at Indian Institute of Space Science and Technology (IIST). Methane is used as the fuel in the present work fuel as NG contains approximately 95 % of methane. The experimental analysis of the flames performed in this work tries to get a deeper insight into the flame dynamics at different operating conditions of the IIST-GS1 swirl burner. The investigation focuses on flow fields, the regions of heat release, and effects of turbulence-chemistry interactions. One specific topic of interest in this study is the mechanism of the transition from swirl to bluff-body stabilization. Despite the enormous progress made in the understanding of turbulent lifted flame stabilization in the last decade a definitive picture is still not in hand. Flame stabilization is investigated by relatively straightforward diagnostics such as direct imaging of the flame's luminosity, OH* chemiluminescence, and snap shot two-component particle image velocimetry (2C-PIV) in the axial plane under isothermal and combustion conditions. These are utilized to comprehend the flow-flame balance that dictates the flame stabilization over the range of global equivalence ratios (Φ_{glob}, based on the inflow mass flowrate of fuel and air) investigated. The results obtained will contribute towards the generation of a comprehensive database that can be used for the verification and improvement of combustion simulation codes in the future. In addition, the developed burner will also be beneficial as a research burner for studying turbulent non-premixed flame stabilization mechanism.

2 Experimental Details

2.1 Burner geometry and operating conditions

The investigated model GT burner (IIST-GS1) is shown schematically in Fig. 1. The swirling motion is imposed on the main air-flow by means of a 60 degree radial swirler consisting of 12 channels. Air at room temperature and pressure, fed from

Corresponding author. Fax: +91-471-2568-406
E-mail address: rajeshsadanandan@iist.ac.in

a plenum of inner diameter 78 mm and a height of 42 mm, passes through the radial swirler and enters into the flame zone through an annular nozzle. The annular space at the burner exit has outer and inner diameters of 42 and 36 mm respectively, resulting in a hydraulic diameter of 6 mm at the nozzle exit.

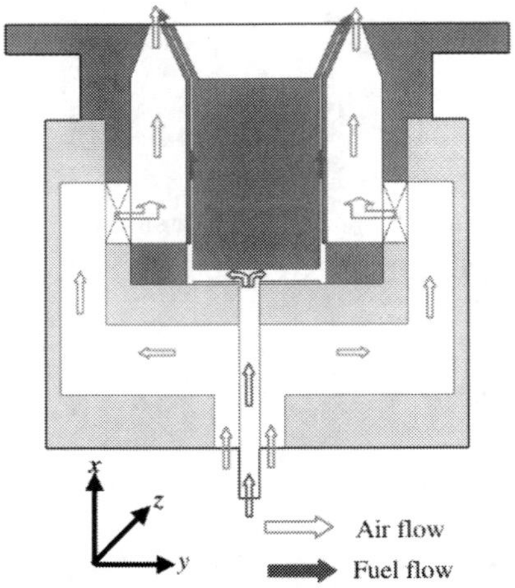

Figure 1: Schematic of the IIST-GS1 swirl burner

The fuel is fed through a non-swirling annual gap of 0.5 mm between the nozzle body and the central bluff body. The central region of the bluff body is recessed in the shape of an inverted truncated cone; the recess height is approximately 11 mm. The swirl number, calculated following the expression derived by Beer and Chigier [5], is 1.59. The nozzle Reynolds number (*Re*) based on the cold inflow bulk velocity and hydraulic diameter is 1649. The mass flows of the gases are controlled by Alicat mass flow controllers. The operating conditions for the investigated flames are summarized in Table 1.

Type	Q_{air}	Q_{fuel}	ϕ_{glob}	T_{ad}	P_{th}
FL-0	100	0	-	-	-
FL-1	100	10.5	1	2227	6.18
FL-2	100	6.3	0.6	1666	3.71
FL-3	100	3.1	0.3	1066	1.85
FL-4	100	1.1	0.1	578	0.62

Q_{air}– air flow rate (slpm); Q_{fuel}– fuel flow rate (slpm); T_{ad} – adiabatic flame temperature (K); P_{th}– thermal power (kW)

Table 1: Investigated flame conditions.

2.2 Optical and laser diagnostic measurements

A Nikon camera (model—D 3200) is used for direct flame imaging at an exposure of 1/60s. The PIV system adopted here consists of a frequency doubled, dual cavity, Nd:YAG laser (Litron make, Bernoulli model, 532 nm wavelength, 75 mJ/pulse energy, 80 μs between pulses), double shutter CCD camera (PCO make, Sensicam model, 1280×1024 pixels resolution), and a sequencer (PIVTEC make, PIV Seq model) to synchronize the laser pulses with the camera exposure timings. The vertical laser sheets, of height ~65 mm and thickness ~1 mm, are aligned parallel to the flow in the *x-y* plane at $z = 0$, and the base of the measurement plane is located 7 mm from the burner exit to avoid laser light interference from the burner geometry. Light scattered from the Al_2O_3 seeded (particle size ~1-10 μm) flow is collected by the camera fitted with a narrow bandpass filter (LaVision, 532±5 nm transmissivity) to reduce the influence of flame luminosity. Approximately 300 image pairs are acquired for each flow condition.

Commercial PIV software (Lavision Davis 8.3) is used for the vector realizations by spatial cross-correlation of the particle images. An adaptive multi-pass vector evaluation technique is employed with interrogation window sizes decreasing from 64×64 to 16×16pixels with 75-50% overlap. This results in a spatial resolution and vector spacing of approximately 2.08 mm and 1.04 mm respectively, and an instantaneous velocity uncertainty of 0.16 m/s.

The chemiluminescence diagnostics consist of an intensified CCD camera (Lavision make, Nanostar model) equipped with a UV lens (Nikon 105). An interference filter band (with a peak λ~310 nm, Dantec) is used to capture the strong OH emissions from the A-X (0, 0) and also to suppress the background radiation. The image intensifier is set to an exposure time of 100 μs and for each test condition, 300 images are taken.

3 Results and Discussion

3.1 Flame luminosity and OH* chemiluminescence

Figure 2 shows the flame luminosity images with varying Φ_{glob}. In all cases, the flame is seen lifted and the location of the flame stabilization depends on Φ_{glob}. Being a lifted flame, a certain degree of fuel-air premixing upstream of the flame base is expected. In addition to the changes in the size and shape of the flame there is clear upstream movement of the flame with decreasing Φ_{glob}. For Φ_{glob}~1, the flame is stable with a shape characteristic of swirling flames. For Φ_{glob} ~0.6-0.4, the flames exhibit bi-stability with low-frequency oscillations. With a further decrease in Φ_{glob}, especially in the region between Φ_{glob} ~0.4-0.1, the flame shape resembles a torroidal structure. The width of the flame base shrinks to a minimum as the Φ_{glob} is lowered from 1.0 to ~0.6, and expands back to nearly the bluff-body diameter with further decrease in Φ_{glob} down to 0.1.

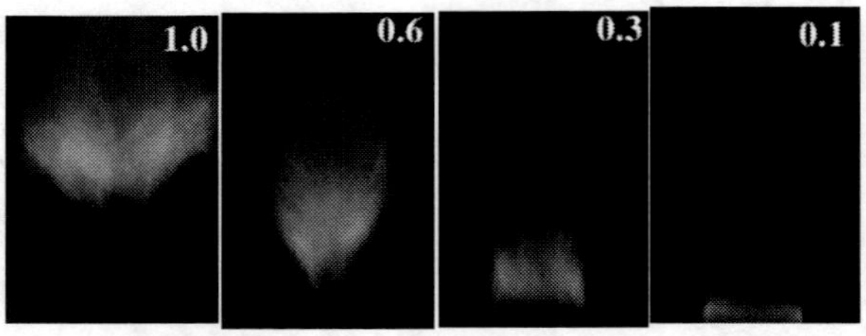

Figure 2. Flame photographs at different Φ_{glob} (values at top right). The field of view is 60 mm x 100 mm.

The very low Φ_{glob} (0.4-0.1) cases are noteworthy. The ignition and stabilization of the flame at such low values is a clear indication of the effect of the burned or exhaust gas recirculation (EGR). The burner geometry is designed to optimize this recirculation by promoting the mixing of the fresh reactants with the burned gases due to the high recirculation rates, thus producing gas temperatures higher than the auto-ignition temperature of the mixture (see further discussion below). This results in the high stability of the flames in this burner even at extremely lean conditions. This is particularly interesting as ultra-low NO_x GT combustors usually require a certain degree of air preheating for sustaining combustion under such lean conditions, which is not necessary in the present burner.

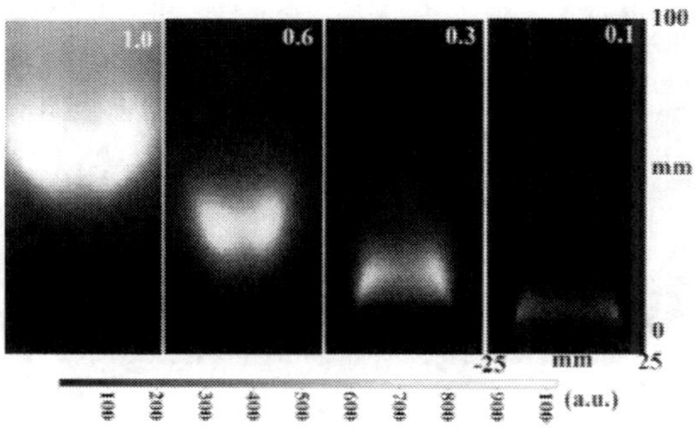

Fig. 3.OH*chemiluminescence distribution at different Φ_{glob} (values at top right).

Figure 3 shows the time-averaged OH* variations with Φ_{glob}. The near-UV spontaneous emission from OH* is visible only in the heat release zone and not in the region of burned gases. OH*chemiluminescence can also be related to the reaction rate or heat release rate [8]. At Φ_{glob}= 1, the heat release zone is lifted to ~50 mm from the burner exit. Comparing the luminosity and the OH* images, it can be seen that the shape of the heat release zones are similar to those found in swirl-stabilized burners down to Φ_{glob}= 0.6. The twin peaks noticed particularly in the Φ_{glob} range of 1.0-0.6 indicate flame stabilization along the stoichiometric surface in the fuel-air mixing region, as in partially premixed flames. Notice that the OH* peaks come together as the flame moves down with the decrease in Φ_{glob} from 1.0 to 0.6.

3.2 Velocity field

Figure 4 shows the time-averaged axial velocity component (V_z) overlapped with the Abel deconvoluted OH* contours (in greyscale) to show the relative position of the heat release zone and the flow field. The flow converges between x = 0 and x = 30 mm due to the geometry of the narrow annular nozzle for the air-flow at the burner exit, separated by the central bluff-body with sizeable recess. The flow subsequently diverges eventually due to effect of the swirl, exhibiting characteristics of a mildly recirculating swirl flow. The central flow recirculation (bubble type vortex breakdown) is visible at the beginning of the diverging mean axial flow field. Such a convergent-divergent flow field influences the fuel-air mixing and the orientation of the stoichiometric surface in the mixing region. The vortical structures at the shear layer produce intense turbulent fluctuations, which can promote mixing between the incoming high-velocity fresh reactants and the recirculating burned gases.

Comparing the Φ_{glob} = 1 case with the isothermal case, there is striking similarity of the convergent-divergent flow pattern up to the region of the OH* contours in Fig. 4. Beyond this, there is noticeable local flow acceleration (in both the axial and radial directions) in the flame zone and downstream due to the thermal expansion of the medium in the reacting flow field. As the flow converges between x = 0 and x = 30 mm, it axially accelerates in the middle, so the flame has to stabilize downstream where the flow diverges subsequently. The flame is anchored at the peak OH* locations over the stoichiometric surface along the inner edge of the shear layer in this region for Φ_{glob} = 1.

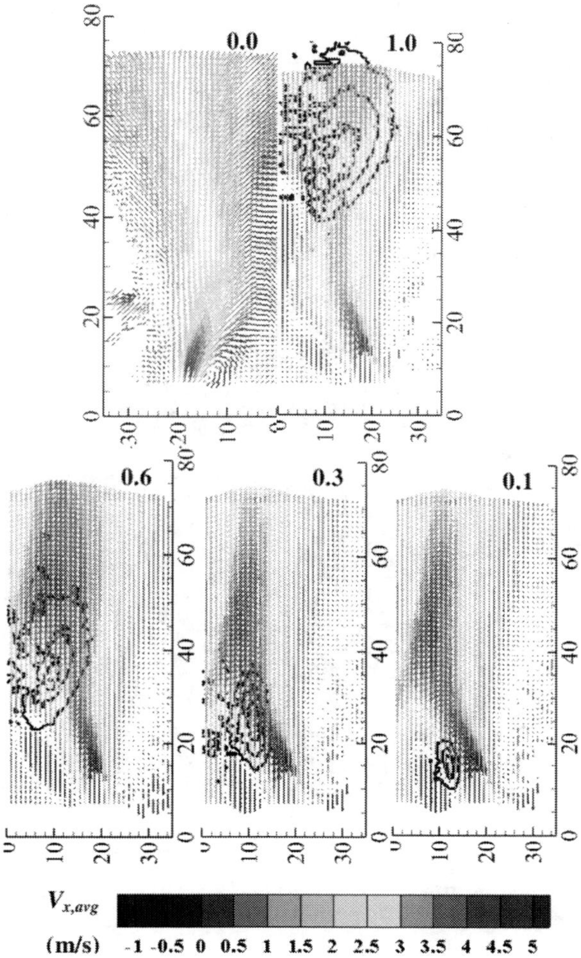

Figure 4. Time-averaged with axial velocity component overlapped with the deconvoluted OH*

To further investigate the flame stabilization, the two dimensional fluid dynamic extensive strain rate field, defined as $S_{xy} = (\partial u/\partial y + \partial v/\partial x)/2$, is shown overlapped with the heat release contours in Fig. 5. The flame is clearly seen to stabilize at regions of low S_{xy}. In comparison with the isothermal strain field, it is evident that this region of low S_{xy} is flame induced. For CH_4/air flames, previous research works on turbulent, non-premixed flows have shown that localized flame extinction occurs when the strain rates are close to 400 s^{-1} [9]. The swirl stabilized flames ($\Phi_{glob} \geq 0.6$) are also seen to stabilize at regions where $S_{xy} < 400\ s^{-1}$.

As the Φ_{glob} is decreased to 0.6, the fuel flow velocity reduces, shifting the stoichiometric surface further inward and away from the high velocity of the shear layer. As such, the fuel-air mixing is less than at the farther location for the higher Φ_{glob}, which further localizes the flame stabilization region (peak OH* in Fig. 3). In addition, the turbulent mixing of fuel, air and burned gases create a large variety of gas compositions and temperatures locally. Mixture fraction measurements using laser Raman scattering by Meier et al. in a partially premixed, swirl stabilized CH_4/air burner has shown local mixture fraction values larger than Φ_{glob} in the flame zone [10]. This, according to the authors, imply a preferential transport of the fuel to the flame axis, leading to an increase in the local flame temperatures above $T_{glob,ad}$ for the Φ_{glob}. Similar turbulence-chemistry interactions and the subsequent increase in flame speed and reactivity in the present case may permit the flame to propagate and anchor upstream, which is close to the most convergent point in the flow. The multi-dimensional effect of this is to decelerate the flow substantially upstream of the flame stabilization point, as seen in Fig. 5.

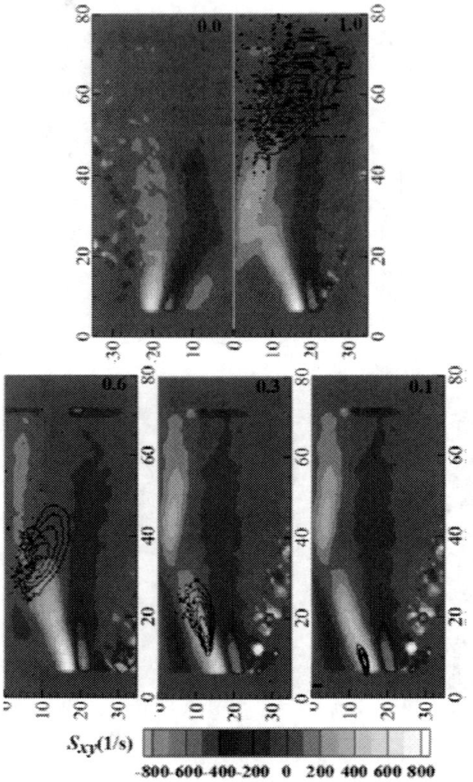

Figure. 5. Time-averaged 2D-principal extensive strain rate field overlapped with the deconvoluted OH* contours.

As the Φ_{glob} is reduced further (0.6-0.4), the situation is progressively conducive for the flame to stabilize much closer to the burner than before. This is because the stoichiometric surface shifts further inward, where the axial velocity progressively decreases as one travels upstream along the centerline, because the flow is convergent here. As the flame moves inward at Φ_{glob} = 0.3 and 0.1, it substantially alters the flow field in its upstream and downstream vicinity. The heat release zone is engulfed by the bluff-body induced recirculating flow field aided by the sizeable recess in the burner geometry. The flame is upstream of the most convergent point, so the swirl effect diverging the burned gas flow further downstream is intact, but does not affect the flame. Thus, the flame is no longer stabilized by the swirl, but more by the bluff-body mechanism rather. Notice that the flame manages to stabilize at regions of high strain rate. The fast mixing of fuel, air, and the burned gas due to the recirculation in this region can lead to a high temperature reactant mixture here, which is kinetically more reactive than the mixtures downstream for $\Phi_{glob} \geq 0.6$. Thus, the flame stabilization at such extreme globally lean conditions is clearly aided by the presence of large-scale vortex structures.

4 Conclusions

Optical/laser diagnostics methods are employed to understand the flame stabilization mechanisms in a novel, ultra-lean, non-premixed, model GT burner at atmospheric conditions. The mean flow-field derived from the 2D-PIV measurements show two competing flow zones, first one characterized by toroidal structures at the burner exit and a second swirling flow with a weak central. The burner exhibits two preferred flame stabilization modes for the investigated operating conditions depending on Φ_{glob}. At Φ_{glob} ~1, the flame exhibits a shape similar to a swirling flame and is stabilized at a region far downstream from the burner exit, where the strain rates are well below the extinctions strain rates for turbulent non-premixed flames. At very lean conditions, the flame stabilizes at fluid dynamically high strain rate regions closer to the burner exit, clearly pointing to flame ignition aided by the reactant mixing with the recirculating burned gases. For Φ_{glob} between 0.4-0.6, bi-stability of the flame is observed with low frequency oscillation between the two stabilization locations.

The stabilization of the flame progressively upstream with decreasing Φ_{glob} is explained by means of a flow-flame balance along the stoichiometric surface in the fuel-air mixing region in the inner parts of the high-velocity shear layer. The unique combination of the annular air swirl nozzle and large bluff-body with sizeable recess causes the convergent-divergent flow field, which allows the flame to shift past the most convergent region from swirl-stabilized to bluff-body stabilized regime as the Φ_{glob} is reduced.

5 Acknowledgment

The National Centre for Combustion R&D is supported by Department of Science and Technology

References

1. J.M. Fleck, P. Griebel, A.M. Steinberg, M. Stöhr, M. Aigner, Experimental investigation of a generic, fuel flexible reheat combustor at gas turbine relevant operating conditions, Proceedings of the ASME, 2010, Paper No. GT2010-22722.

2. F. Guethe, J. Hellat, P. Flohr, J. Eng. Gas Turbines Power 131(2009) 021503 1-7.

3. R. Sadanandan, R. Lückerath, W. Meier, C. Wahl, J. Propul. Power 27(5) (2011) 970-979.

4. H.C. Mongia, TAPS – A 4th Generation Propulsion Combustor Technology for Low Emissions," Proceedings of the AIAA International Air and Space Symposium, 2003, Paper 2003-2567.

5. A.H. Lefebvre, Gas Turbine Combustion, 2nd edition, Taylor and Francis, 1998.

6. Y. Huang, V. Yang, Prog. Energy Combust. Sci. 35 (2009) 293–364.

7. R. Sadanandan, inventor; Indian Institute of Space Science and Technology, applicant. Ultra lean non-premixed gaseous fuel burner. Indian Patent Application No: 6632/CHE/2015

8. H.N. Najm, P.H. Paul, C.J. Mueller, P.S. Wyckoff, Combust. Flame 113 (1998) 312-332.

9. N. Peters, Proc. Combust. Inst. 21(1) (1986)1231-1250.

10. W. Meier, X.R. Duan, P. Weigand, Combust. Flame 144 (2006), pp 225-236.

Flame Behaviors within transient regime in Helium-diluted Laminar-lifted Coflow-jet Flames

S. H. Oh[1], K. H. Van[1], J. Park[1], O. B. Kwon[1], S. H. Chung[2], J. H. Yoon[3], and S. I. Keel[3]

[1]Department of Mechanical Engineering, Pukyong Nat'l University

365, Sinseon-ro, Nam-Gu, Busan, KS012, Republic of Korea

[2]Clean Combustion Research Center, King Abdullah University of Science and Technology

Thuwal, 23955, Saudi Arabia

[3]Environment & Energy Research Division, Korea Institute of Machinery and Materials

156, Gajeongbuk-ro, Yuseong-gu, Daejeon, KS015, Republic of Korea

Abstract

An experiment has been conducted to investigate flame behaviors in helium-diluted laminar lifted coflow jets within transient regime. There were two distinct self-excitation: the one was a buoyancy-driven self-excitation (hereafter, called BDSE) and the other was a capsized flame near flame extinction (hereafter, called CFFE) which shows a strange flame behavior that the lifted flame should be capsized to propagate to upstream like premixed flame. Numerical simulations for steady-state and reacting flows were used to find out phenomenon of BDSE and CFFE. The differences between the two self-excitations were shown and discussed.

1. Introduction

Many studies on laminar non-premixed lifted flame have been widely conducted, because the fundamental characteristics are useful for designing an industrial burner in extended laminar stretched flamelet modeling. Because of the first spot of laminar tribrachial flame in lifted state, the flame behavior of free- or coflow-jet flames is well described by the cold jet theory. Such a lifted flame configuration propagates along a stoichiometric contour line with tribrachial flame structure such that the flame speed is balanced by local flow one. The tribrachial flame propagation speed depends on the mixture strength, Lewis number, heat loss, fuel concentration gradient, and buoyancy [1-4]. Because of such a stabilization mechanism of tribrachial flame, the lifted flame could be self-excited by some factors of them that modify the edge flame speed, buoyancy by a flame flicker, the repetitive interaction of burning rate and buoyancy-driven convection, and conductive heat loss from premixed wings to trailing diffusion flame. The self-excitation with frequency of the order $O(1)$ Hz in laminar lifted coflow-jet flames was defined by buoyancy effect [6-7]. Meanwhile, Füri et al also investigated the self-excitation with same order identified by unbalanced between thermal- and diffusional effect just prior to flame extinction [5]. Won et al extended their work through microgravity experiments as well as numerical simulation caused by the repetitive interaction of burning rate and buoyancy-induced convection [7]. But, the strange flame behavior like a capsized configuration with coupled buoyancy-induced effect has not yet been reported specifically in the previous papers and was shown near the flame extinction. Unlike the previous research that were focused on cause of flame oscillation, this research has proceeded for identification of flame oscillation. To identify this strange flame behavior, the further experiment may be required to find the mechanism and characteristics near the flammability limit in laminar lifted coflow-jet flames.

Corresponding author. Fax: +82-51-629-6150
E-mail address: jeongpark@pknu.ac.kr

2. Experimental facility

The experimental facility consisted of a coflow-jet burner, mass flow controllers, a digital camera system, and a visualization system. As shown in Fig. 1, a nitrogen-diluted propane coflow-jet was injected into the helium-diluted air through fuel tube nozzles with D = 10.25 mm. Propane and helium with a purity of 99.99 % and nitrogen with a purity of 99.95 % were used. A series of glass beads were installed in the lower half of the compartment to suppress external disturbances and to obtain uniform outer jet flows. A cylindrical acrylic compartment with 10 cm diameter and 40 cm length was used to reduce external derangement as well. The fuel nozzle tube length was more than 100 times the nozzle diameter to attain fully developed velocity profiles. The flow rates were controlled precisely by using mass flow controllers and a Flow Manager software (version 3.2). Their nominal accuracies for the full scale flow rate were within 1.0 %. Experiments were conducted by adding helium with a high thermal conductivity to the outer coflow-air to control heat losses from the flames to ambience. The lifted off height was measured with a digital camera (SAMSUNG-HMX-S15BS) with 300 fps attached to a 2-D transfer device. A Matlab-based code was used to analyze flame images. Flame length and lift-height were defined as the brightest points in converted gray-level images.

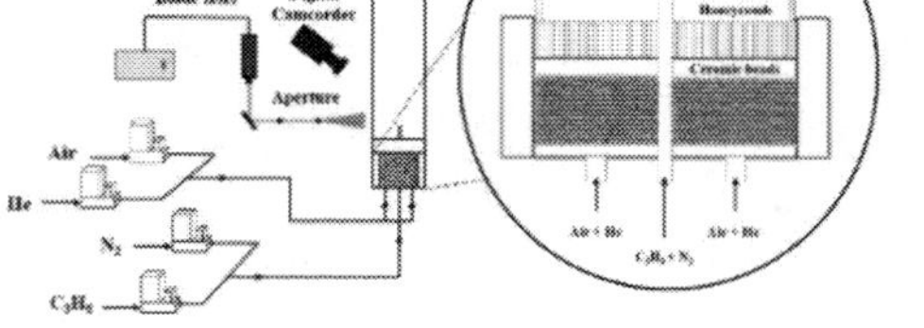

Figure 1: The schematic of experimental setup.

3. Results and discussion

A stability map as a function of fuel mole fraction, $X_{F,O}$, with nozzle exit velocity, U_O at fixed coflow velocity with pure coflow-air is shown in Fig. 2. The BDSE in Fig. 2 is dominantly represented like previous study [6-8]. To control conductive heat loss from the lifted flame to the coflow-air [9-10], the helium volume fraction, Ω_{He} defined as the ratio of the volume flow rate of helium, Q_{He} to the summation of the volume flow rate of helium, Q_{He} and air, Q_{Air} was controlled with a high thermal conductivity. The results of experiments repeated with varying Ω_{He} and U_O at fixed $X_{F,O}$ show that two types of flame behaviors were observed: BDSE(regime I), and CFFE(regime II). The CFFE was observed at $X_{F,O} = 0.14$, such that flame behavior appeared just prior to flame extinction when helium volume fraction was much larger than that of BDSE. The stability maps with varying $X_{F,O}$ were shown in Fig. 3. Note that the fuel mole fractions, in which CFFE is observed, are $X_{F,O} = 0.14$, and 0.15 and nozzle exit velocities are $7 < U_O < 11$. To compare the flame behavior observed here with that reported in Ref. [11], the sequence of BDSE and CFFE was shown in Fig. 4(a) and (b), respectively, during one cycle of oscillation. Unlike two types of BDSE in Fig. 4(a), CFFE has a configuration of flat flame at the downstream and then propagate to upstream like a capsized configuration. To clarify the difference between BDSE and CFFE, variation of flame dimensions with time was investigated as shown in Figure 5. The flame tip height and the flame length in the case of BDSE in Fig. 5(a) are shown to be in phase, whereas the flame tip and base heights are out of phase. Such characteristics are very similar to those for the BDSE observed in the previous studies [6-8]. However, for the CFFE in Fig. 5(b), the flame tip and base were matched to be in phase and flame tip was opened when propagate to the upstream near the fuel nozzle.

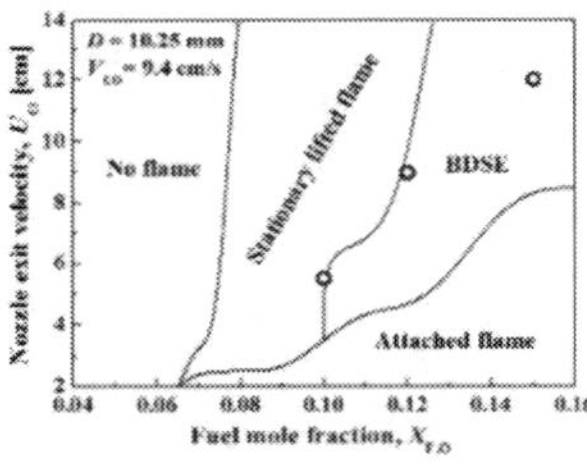

Figure 2: A stability map as function of fuel mole fraction, $X_{F,O}$ with nozzle exit velocity, U_O at fixed coflow-air velocity.

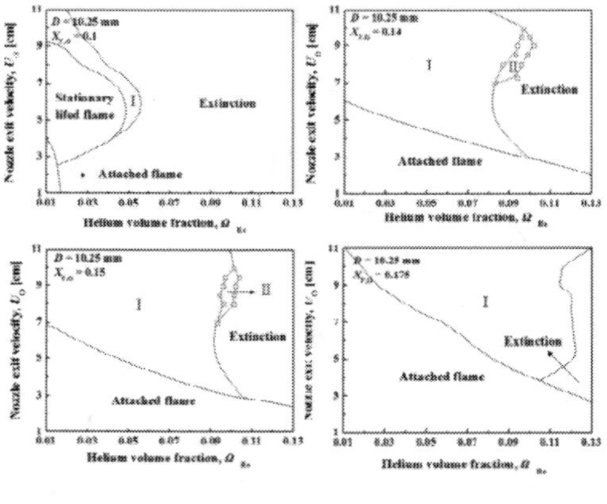

Figure 3: The stability map as a function $X_{F,O}$ with U_O in helium concentration.

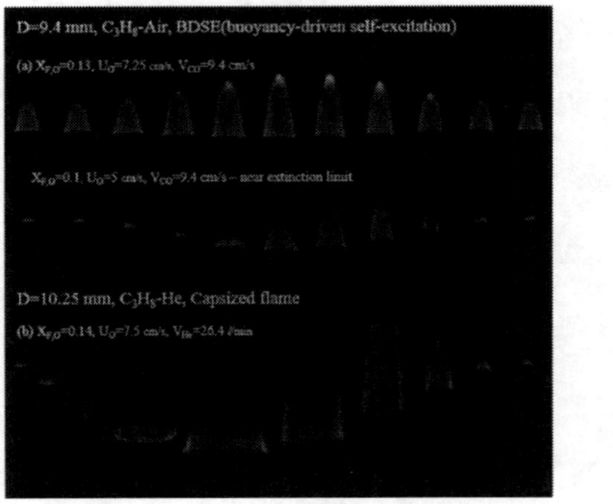

Figure 4: The sequence of (a) BDSE without helium addition reported in Ref [11] and (b) CFFE.

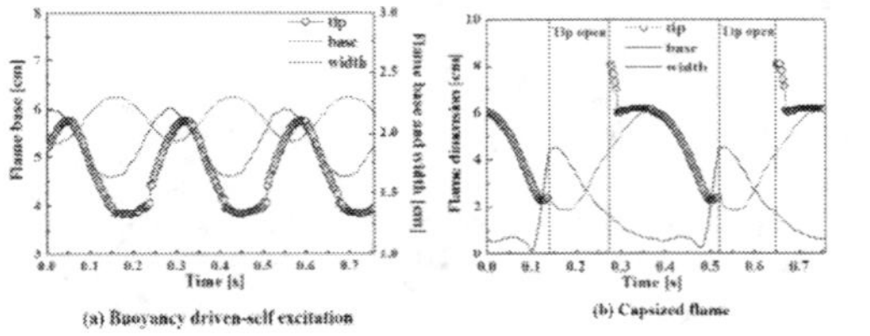

Figure 5: Variation of flame dimensions with time for (a) BDSE and (b) CFFE.

To identify the difference between BDSE and CFFE, two dimensional computations in spherical coordinates with global reaction chemistry were conducted. Two dimensional axisymmetric computations in steady-state and cylindrical coordinates with cold and reacting flow for examining the multidimensional flame near the flame extinction were calculated by using FLUENT 6.3.26. The residual convergence criterion for velocity (in m/s) and species mass fraction was set to 10^{-6} and 10^{-7} for cold flows (10^{-4} and 10^{-3} for reaction flows), respectively. Typically, 10^{+4} iterations. Figure 6 shows a numerical domain in the present computations. As shown in Fig. 6, the coflow burner used in the experiment, the boundary conditions and size of domain were modeled. A mixture in the fuel nozzle with poiseuille flow velocity was injected from the fuel nozzle exit, whereas in coflow-air side with uniform flow velocity was issued. Pressure-outlet condition was adopted at the exit of flow. The size of mesh near the reaction zone was less than 0.1 mm and the discrete ordinate method was adopted to solve the radiative transfer equation and the WSGGM (Weighted-Sum of Gray-Gas Model) domain based was used for computation of the radiative absorption coefficient. The three step overall reaction was used. Figure 7 shows the stoichiometric line with variation of Ω_{He} at normal gravity condition in cold flow. As shown in Fig. 7(a), stoichiometric line was approaching to the fuel nozzle with increasing a helium volume fraction. It should be supposed to be a significant differences of density variation between fuel and ambient air side, in that the molecular weight of helium is significantly low. However, the stoichiometric line is similar with the variation of helium volume fraction in Fig. 7(b) at zero gravity.

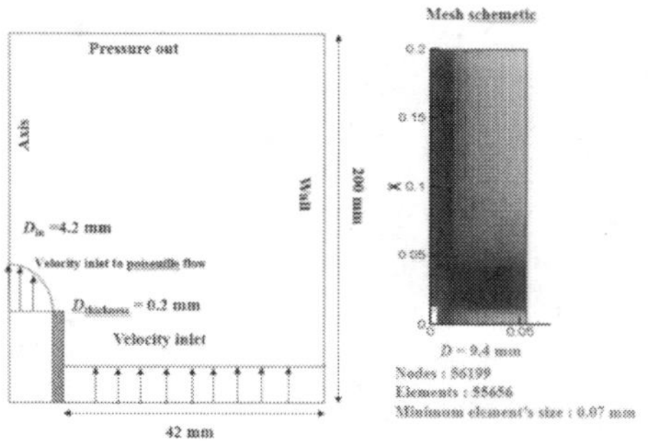

Figure 6: The schematic diagram and boundary conditions of coflow burner.

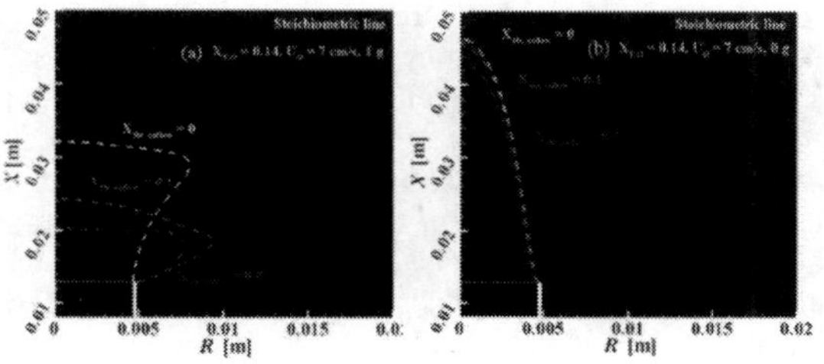

Figure 7: The stoichiometric line with variation of helium volume fraction (a) at normal gravity and (b) at zero gravity in cold flow.

Note that the stoichiometric line was significantly manipulated with the helium volume fractions in only normal gravity. It may be required to clarify the heat release rate which is good indicator on flame position and the stoichiometric line in flame existed with the variation of helium volume fraction. Figure 8 represented heat release contour line and axial velocity with the stream line at $X_{F,O}$ = 0.1, U_O = 7 cm, and V_{CO} = 9.4 cm/s. As shown in Fig. 8(a), in case of pure air passing coflow, the lifted flame configuration maintained the tribrachial flame structure, and the stoichiometric line is along the trailing diffusion flame. It means that the lifted flame is at the position where flame propagation speed was balanced by local flow one. However, in case of Fig. 8(b), the configuration of lifted flame had a bibrachial flame structure, and the stoichiometric line is out of phase with the contour line of heat release rate, in that axial velocity below the lifted flame was zero. Consequently, helium addition to coflow-air makes a lean premixed flame on the side of axial velocity because flame deviate from a stoichiometric line. Further confirmation will be a future work through unsteady state numerical simulation to specifically define alternation of stoichiometric line with each time step.

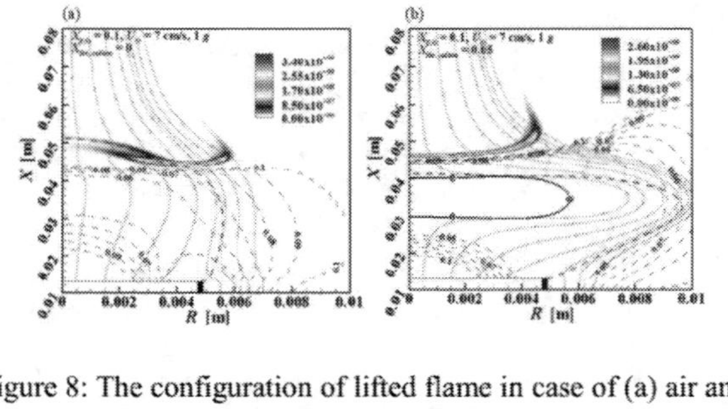

Figure 8: The configuration of lifted flame in case of (a) air and (b) helium with the stream line at $X_{F,O}$ 0.1, U_O = 7 cm, and V_{CO} = 9.4 cm/s.

4. Conclusions

Experiments in laminar lifted coflow-jet propane flames diluted highly with nitrogen were performed to investigate discernible features between BDSE and CFFE. It was shown that phase diagrams of flame dimensions for BDSE and CFFE are quite different, and recognized that the CFFE was mainly observed near the flame extinction. In many aspects, the CFFE observed in this experiment propagates to upstream like lean premixed flame and in those of numerical results, too. However, further confirmation will be a future work through unsteady numerical simulation.

5. Acknowledgment

This work was supported by Near Zero Emission of the Korea Institute of Energy Technology and Planning (KETEP), granted financial resource from Ministry of Trade, Industry & Energy, Republic of Korea. (No.20152010103050)

References

[1] S. H. Chung, B. J. Lee, Combust. Flame 86 (1991) 62-72.

[2] B. J. Lee, S. H. Chung, Combust. Flame 109 (1997) 163-182.

[3] B. J. Lee, M. S. Cha, S. H. Chung, Combust Sci. Technol 127 (1997) 55-70.

[4] B. J. Lee, J. S. Kim, S. H. Chung, Combust. Inst. 24 (1994) 1175-1181.

[5] Marc Furi, Paul Papas, Peter A. Monkewitz, Combust. Inst. 28 (2000) 831-838.

[6] S. H. Won, S. H. Chung, M. S. Cha, B. J. Lee, Combust. Inst. 28 (2000) 2093-2099.

[7] S. H. Won, J. Kim, M. K. Shin, S. H. Chung, O. Fujita, T. Mori, J. H. Choi, K. Ito, Proc. Combust. Inst. 29 (2002) 37-44.

[8] W. J. Lee, J. Park, O. B. Kwon, J. H. Yun, S. L. Keel, Fuel 112 (2013) 366-374.

[9] S. H. Yoon, J. K. Han, J. Park, J. H. Yun, S. I. Keel, Combust. Flame 158 (2011) 1960-1969.

[10] S. H. Yoon, J. Park, O. B. Kwon, J. H. Yoon, S. I. Keel, J. S. Kim, D. S. Bae, Fuel 91 (2012) 51-60.

[11] K. H. Van, W. J. Lee, J. Park, T. H. Kim, J. H. Park, J. Korean Soc. Combust 20 (2015) 43-51.

Tomographic background-oriented schlieren techniques for three-dimensional density field reconstruction in shock-containing flows

R. Kirby[1], D. J. Tan[1], C. Atkinson[1], D. Edgington-Mitchell[1]

[1]Laboratory for Turbulence Research in Aerospace and Combustion,

Department of Mechanical and Aerospace Engineering

Monash University, Victoria, 3800, Australia

Abstract

Two measurement techniques for acquiring pixel displacement data for tomographic background oriented schlieren are compared; the first analyses image pairs of an optimised random-dot pattern using a conventional cross-correlation algorithm. The second evaluates optical flow displacements for image pairs of a fractal texture background composed of multi-scale wavelet noise. Quasi-3D tomographic reconstruction is performed under the parallel-ray assumption using an algebraic method, initialised with large-gradient regions from a filtered back-projection gradient inversion, for improved clarity and noise reduction. Reconstruction of the near-field density of a highly underexpanded jet from an elliptical nozzle of aspect ratio 2:1 is performed using both optical flow and cross-correlation displacement fields. Diffraction effects at the Mach disk are seen to produce strong artefacts in reconstructed gradient fields. Differences in gradient components from optical flow and cross-correlation are effectively smoothed by Poisson integration, yielding similar final density distributions, stressing that selection of a suitably accurate method for integration of the Poisson equation is required to preserve increased fidelity of gradient fields. Good qualitative agreement in shock structure is found with respect to existing studies.

1 Introduction

The inherent complexity of compressible, shock-containing, flows demands three-dimensional visualisation methods, while requiring non-intrusive imaging. The ability to quantify the effect of density gradients on the path of light through an inhomogeneous flow field, more traditionally visualised using schlieren methods at the cost of line-of-sight ambiguity, has encouraged the adoption of the background-oriented schlieren (BOS) method. BOS provides quantifiable path-integrated light ray deflections $\boldsymbol{\epsilon}$, for gradient components local to the ray trajectory $\boldsymbol{p}$,

$$\boldsymbol{\epsilon} = \frac{G}{n_0}\int_p \nabla\rho(\boldsymbol{\xi})\ d\xi \qquad (1)$$

where the above relies on the simple relationship between density and refractive index n for a gas [1], given by the linear Gladstone-Dale relation, under the assumption of constant wavelength illumination

$$\rho(x,t)G(\lambda) = n(\boldsymbol{x},t) - 1 \qquad (2)$$

Corresponding author.
E-mail address: daniel.m.mitchell@gmail.com

for G the Gladstone-Dale constant. Recovery of the spatial density field $\rho(\boldsymbol{x})$ requires inversion of (1) based on knowledge of numerous deflections $\boldsymbol{\epsilon}$. Ray deflections are inferable through background displacements $\Delta\boldsymbol{x}_p$, quantifying the optical distortion between reference and refracted image pairs taken in absence of, or through, the flow object, in turn. Displacement vectors may be evaluated using cross-correlation algorithms typically used in PIV applications [2], or by computing the optical flow between image pairs [3]. Net deflection angles may be related to pixel displacements between reference and displaced image pairs, assuming paraxial rays, and under the small-angle approximation, using setup geometry and image magnification [1].

Experimentally derived BOS deflections may be passed through conventional Fourier-based algorithms such as filtered back-projection (FBP) under the refractionless ray assumption, for rapid reconstruction and resolution of steep gradients. Recovered gradient fields may then be numerically differentiated to form the source term for the Poisson equation, the solution to which returns the spatial refractive index field. Streak artefacts are manifest in FBP, particularly when limited to few viewing angles. At the cost of clarity and ease of implementation, artefacts may largely be avoided by using an algebraic reconstruction technique (ART), in place of FBP, to compute the gradient components. More advanced methods (e.g. [4]) combine an algebraic approach with a ray tracing procedure, which allows the straight ray assumption to be relaxed if refractive effects are considered non-negligible, or where setup dimensions invalidate the parallel-ray simplification. The error resulting from the application of conventional inversion algorithms, under the assumption of non-refracting rays, may be significant for strongly refracting fields.

Early applications of BOS have considered two-dimensional (2D) planar flows, thereby removing the transverse (line-of-sight) density gradient [5]. Studies of three-dimensional (3D) problems predominantly consider time-averaged axisymmetric mean fields, enabling direct tomographic reconstruction using average displacement fields from a single viewpoint, via well-known inverse Radon methods under radial symmetry [6]. Algebraic reconstruction techniques (ART) have been applied successfully in tandem with ray tracing procedures, for non-axisymmetric and time resolved flows, utilising large camera arrays [4] or under assumptions of rotational symmetry [7].

While basic algebraic methods are known to behave robustly when provided with limited data, challenges include poor resolution of steep gradients, and the subsequent loss of clarity in the final refractive index field. Solutions are also sensitive to initial conditions. Hartmann & Seume [8] and Atkinson *et al* [9] report notable improvement through initialisation of ART with a subset of the gradient field extracted from an initial FBP reconstruction.

In this study, the sensitivity of the reconstructed density field to the input displacement field is tested experimentally. Two approaches to extracting the ray displacements are compared, as detailed in §2. A shock-containing supersonic elliptical jet is used as a case study. Elliptical jets are of interest due to their mixing [10] and aeroacoustic properties [11], as is well summarized in Gutmark [12]. Supersonic elliptical jets operating in the underexpanded regime exhibit complex shock structures in the potential core [13], that remain poorly understood [14].

2 Experimental methodology

Inversion of (1) is based on two-dimensional projection data, acquired in practice from a limited number of viewing angles. BOS measurements are performed in the LTRAC supersonic jet facility. The reader is referred to Tan *et al.* [6] for details of the BOS optical layout and lighting. A highly underexpanded jet issuing from an elliptical nozzle of aspect ratio 2:1 serves as the base flow for comparison of measurement techniques. Description of the facility and nozzle geometry may be found in Edgington-Mitchell *et al.* [11]. Experimental parameters are summarised in Table 1, with Reynolds number Re based on the ideally expanded Mach number at a nozzle pressure ratio (NPR) of 4.2, and the equivalent nozzle diameter $D = 2\sqrt{ab}$ for semi-major and semi-minor axes $a = 10$mm, $b = 5$mm respectively. Images are acquired using a PCO4000 camera, with CCD array of 4008x2672 pixels, of size 9μm/px, and a 105mm AF Micro Nikkor lens. An f-stop of 22 is used for maximum depth-of-field. Despite the characteristic timescales of the flow typically being far shorter, a relatively long exposure of 95μs is used in order to maximize signal-to-noise ratio in the image. Since mean displacement fields are used to compute the gradient fields, the inability to resolve fine-scale turbulent features is considered inconsequential. Reference background images are acquired prior to displaced images for each viewing angle. Two hundred displaced samples are acquired during a single run for time-averaging. Each mean field is mirrored across the major and minor axes under the assumption of a quarter-symmetric mean-field, providing 20 effective viewing angles over a half-circle, or 40 views over a full circle, invoking the symmetry of line-integrated projection data. Setup constraints imposed a minimum angular spacing of 10°.

2.1 BOS displacements

Two data sets of the same base flow, each at 10 viewing angles, are obtained; the first using a random-dot BOS background pattern designed according to optimal PIV seeding parameters unique to the optical layout [15]. The seeding density is estimated to be approximately 4.85 particles per 16x16 pixel window. The full background contains 220000 particles of size 70μm. A 21mm sample is enlarged in Fig. 1a. Multigrid cross-correlation is performed using PIVview 2C in Speckle Displacement mode for evaluation of pixel displacement vectors between every displaced image and a single reference image at each view, using a final interrogation window of size 16x16 with an overlap of 50%. A 3x3 pixel median filter is applied across each snapshot, prior to time-averaging at each viewing angle.

Data acquisition is repeated using a fractal texture background composed of multi-scale wavelet noise, for the purposes of optical flow evaluation. A 21mm sample is shown in Fig. 1b. Generation of wavelet noise functions is described in detail by Cook & DeRose [16]. Practical implementation in BOS is discussed by Atcheson, Heidrich & Ihrke [17], who find that wavelet noise backgrounds are able to provide greater accuracy, particularly when combined with multi-scale gradient-based optical flow algorithms [18]. A major advantage of optical flow is the potential to provide dense fields at one vector per image pixel. In this case fields are downsampled to match those obtained via PIVview for comparison. No filtering is applied, which results in slightly noisier gradient fields – an example is provided in Fig. 3. Two- and eight-step downsampled field reconstructions are also compared with the full non-downsampled case.

Experimental Parameters	
Lens-to-background distance Z_B (mm)	562
Background-to-object distance Z_D (mm)	206
Lens focal length f (mm)	105
Spatial Resolution (μm/px)	25.9
Equivalent nozzle diameter D (mm)	14.1
Ambient density ρ_0 (kg/m^3)	1.22
Re	8.74x10^5
NPR	4.2

Table 1: Experimental BOS layout and flow parameters.

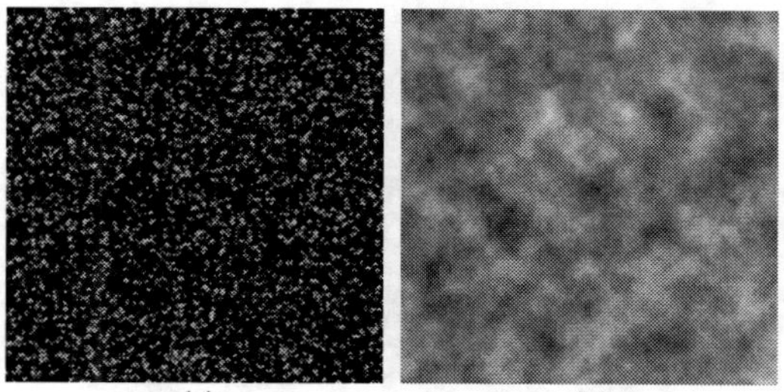

(a) (b)

Figure 1: 21mm background samples for (a) random-dot and (b) wavelet noise.

2.2 Tomographic reconstruction

Similar to the approach of Venkatakrishnan & Meier [5], FBP is performed on deflection data, using Shepp-Logan frequency domain filtering for attenuation of high-frequency noise. A subset of the gradient field obtained using FBP is then passed to a simultaneous ART (SART) scheme for resolution of steep gradients and reduction of artefacts, following improvements seen by Hartmann & Seume [8] and Atkinson *et al* [9].

A masking procedure is used to extract high-gradient regions, assumed to be confined within a variable ellipse whose semi-major and semi-minor axes are defined by the extent of the shear layer along the respective axes. Sobel edge detection is used to identify the outer edge of the shear layer. A threshold is placed over the gradient magnitude to remove low gradients within the flow domain. The relative success of masking is found to be highly sensitive to the structure of the gradient field, and is hence difficult to optimise a priori without knowledge of artefact-free characteristics. For instance, flow features of interest may be removed by placing a threshold on gradient magnitude, which may preserve high-intensity artefacts. Over-thresholding may mitigate any improvement over a homogeneous initial condition, presenting a difficulty in automation further to the uniqueness of each image in scale and intensity.

Iterative SART is performed using AIR Tools MATLAB Package [19], under the assumption of parallel beam tomography. Each axial plane is reconstructed independently. Discrete second-order approximations to the derivative terms are combined to estimate the source term for the 2D Poisson equation. A second-order discretisation of the Poisson equation is solved using an iterative SOR scheme. The convergence tolerance is set to a maximum difference of 10^{-10} in successive values of n at any point. Each plane is solved independently, enforcing ambient Dirichlet boundary conditions over a square domain. The computational cost of Poisson integration may be relieved by reduction of the domain size, as performed in 3D by Atcheson *et al.* [4] through the definition of a visual hull. Noise levels in the resultant solution are also reduced by the consequent elimination of far-field artefacts in recovered gradients, prior to computation of the Poisson source field. In this case, the domain is cropped prior to gradient reconstruction.

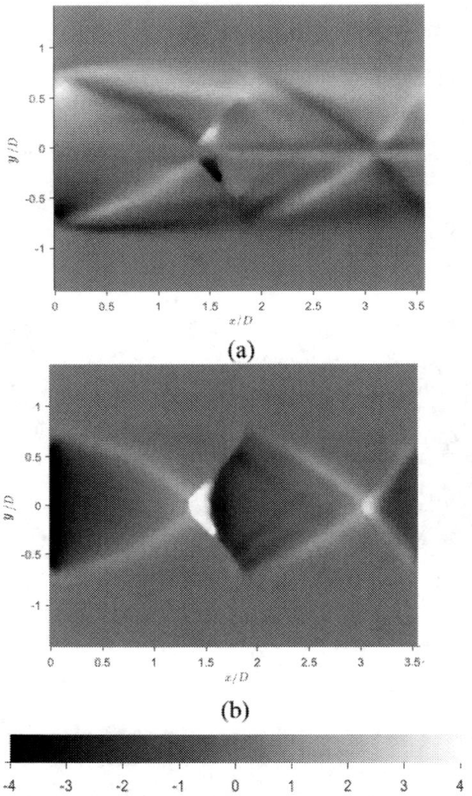

Figure 2: Example time-averaged pixel displacements viewed along the minor axis, obtained using optical flow algorithm (a) Δy (b) Δx, in pixels.

3 Results

3.1 Recovery of spatial gradient components

Gradients are recovered using transverse displacement components, assuming independent parallel-ray projections at each downstream plane. An example case of reconstructed components at a single plane, immediately upstream of the Mach disk, are compared in Fig. 3. This plane is chosen to highlight the dominance of artefacts surrounding the Mach disk and reflected oblique shock, particularly evident using the downsampled optical flow data (Fig. 3a-b). Spatial filtering performed on cross-correlation data is observed to provide smoother fields without significant loss of clarity (Fig. 3c-d).

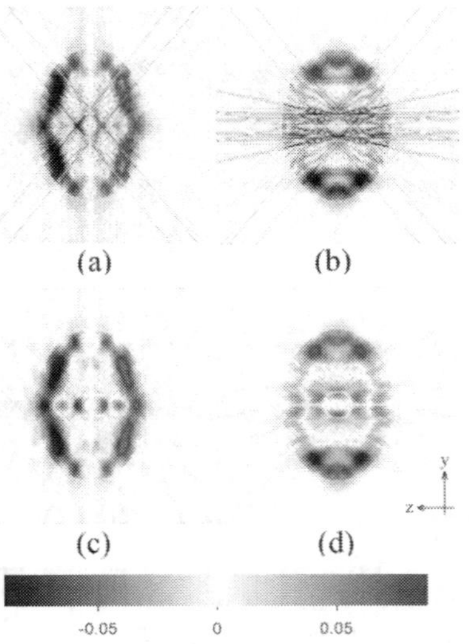

Figure 3: Reconstructed gradient components (viewed from downstream) for a plane just before the Mach disk (x/D=1.42) using optical flow displacements (a) $\partial n/\partial z$ (b) $\partial n/\partial y$ and cross-correlation data (c) $\partial n/\partial z$ (d) $\partial n/\partial y$.

3.2 Quantitative near-field density

Normalised density distributions for the major- and minor-axis planes are shown in Fig. 4. Results are visually indistinguishable for optical flow and cross-correlation methods at equivalent resolution, so only the former is shown. Figure 5 exemplifies differences across the most challenging region for reconstruction, comparing axial distributions just before and after the Mach disk, and illustrating the variable agreement in cross-correlated and optical flow fields. Fine-scale features appear to be better resolved using optical flow in some locations. Improvement from increased resolution is limited; no significant change in reconstructed density is seen between 2-step downsampled and full optical flow fields, which together agree well those from cross-correlation. This likely reflects a resolution limit imposed by finite aperture BOS. Good agreement is seen between noisier optical flow and filtered cross-correlation, and no difference is observed in reconstructions using filtered (not shown) and unfiltered 8-step downsampled data, suggesting that significant smoothing is inherent in the integration method, with consequence to flow features of similar strength.

Good qualitative agreement is observed in relation to PIV velocity distributions at the same flow conditions [10]. A low density shear layer originating behind the Mach disk is observed to persist past the second shock cell, as observed in corresponding high-resolution schlieren images [11]. In comparison with numerical shock structures found at lower NPR by Menon & Skews [13], it is noted that the first axis-switching event is not complete by the second shock reflection. The first shock cell appears to share similarities with numerical results of Li *et al.* [14] for both elliptical and rectangular nozzle geometries at higher NPR, suggesting that slight asymmetries present in real flows may result in significant changes in shock structure.

Artefacts in the reconstructed density field are evident at the axial location of the Mach disk. This is most likely due to diffraction effects at the Mach disk itself. A strong shock oriented at a grazing angle to the direction of light propagation has been observed to diffract incident light in both PIV [20] and PLIF [21] applications. This typically influences measurements once the light propagates past the shock, however the path-integrated nature of BOS results in artefacts crossing the radial extent of the density field in the vicinity of the Mach disk. The effect is intensified by the present symmetry assumption, but remains confined locally under plane-wise reconstruction.

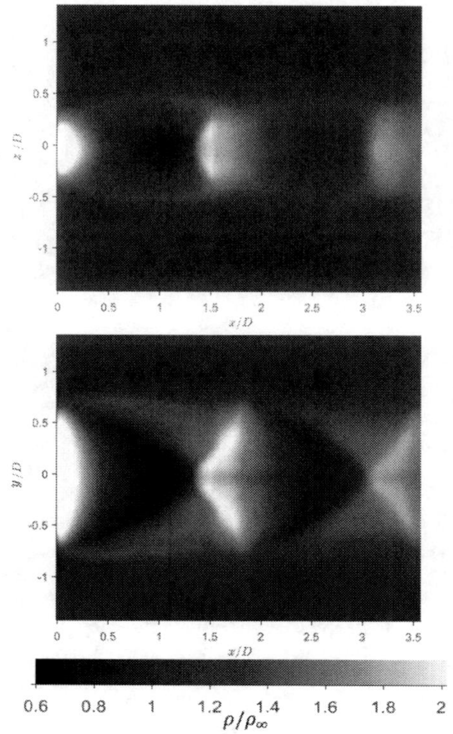

Figure 4: Normalised reconstructed density obtained using optical flow displacements, as viewed along major & minor axes.

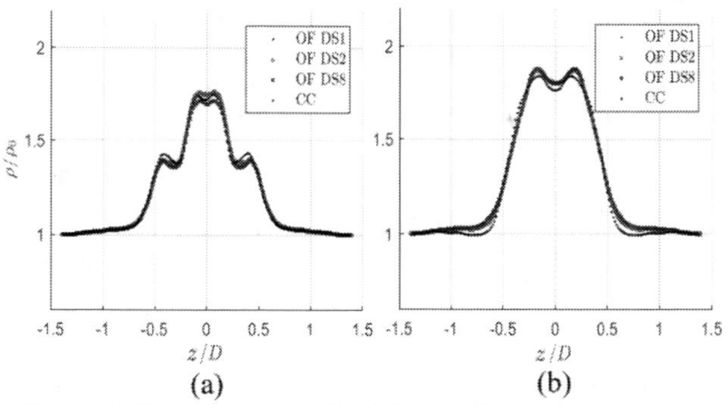

Figure 5: Transverse normalised density distributions along the minor axis for planes (a) x/D=1.42 (b) x/D=1.76, at and just after the Mach disk. OF DS1,-2,-8: full, 2-step & 8-step downsampled optical flow, CC: cross-correlated displacement fields.

4 Conclusions

Several approaches to tomographic background-oriented schlieren have been compared through their ability to resolve the density field of a highly underexpanded elliptical supersonic jet. The optical flow based method proved more susceptible to diffraction effects at the edge of strong shocks. Despite increased noise levels and stronger streak artefacts using unfiltered optical flow data, reconstructed density fields differed only slightly in comparison to those computed using filtered displacements from cross-correlation. Agreement regardless of effective resolution suggest that the benefits of optical flow may be mitigated by lower spatial resolution. Substantial smoothing during the Poisson integration suggests that for fine-scale flow features, the choice or order of integration scheme may be as influential as the technique for displacement evaluation.

5 Acknowledgment

The authors wish to thank T. Knast for his assistance in improving the robustness of the imaging setup. The support of the Australian Research Council is gratefully acknowledged.

References

[1] H. Richard, M. Raffel, Meas. Sci. Technol. 12 (9) (2001) 1576-1585.

[2] J. Soria, Exp. Therm. Fluid Sci. 12 (2) (1996) 221–233.

[3] G. Settles, M.J. Hargather, Meas. Sci. Technol 28 (4) (2017) 042001.

[4] B. Atcheson, I. Ihrke, W. Heidrich, A. Tevs, D. Bradley, M. Magnor, H. Seidel, ACM Trans. Graph. 27 (5) (2008) 132.

[5] L. Venkatakrishnan, G.E.A. Meier, Exp. Fluids. 37 (2) (2004) 237-347.

[6] D.J. Tan, D. Edgington-Mitchell, D. Honnery, Exp. Fluids 56 (11) (2015) 204.

[7] H.M. Lang, K Oberleithner, C.O. Paschereit, M Seiber, Exp. Fluids 58 (7) 88.

[8] U. Hartmann, J.R. Seume, Meas. Sci. Technol. 27 (9) (2016) 097001.

[9] C. Atkinson, S. Amjad, J. Soria, A Tomographic Background-Oriented Schieren Method for 3D Density Field Measurements in Heated Jets. In 11th Asia-Pacific Conference on Combustion, 2017.

[10] D.M. Mitchell, D.R. Honnery, J. Soria, Exp. Fluids 54 (7) (2013) 1578.

[11] D. Edgington-Mitchell, D. Honnery, J. Soria, International Journal of Aeroacoustics, 14 (7), (2015) 1005-1024.

[12] E.J. Gutmark, F.F. Grinstein, Annu. Rev. Fluid Mech. 31 (1) (1999) 239-272.

[13] N. Menon, B.W. Skews, Shock Waves 20 (2010) 175-190.

[14] X. Li, R. Zhou, W. Yao, X. Fan, Appl. Therm. Eng. 125 (2017) 240-253.

[15] M. Raffel, C. Willert, J. Kompenhans, Particle image velocimetry: a practical guide, Springer, Berlin, 1998

[16] R.L. Cook, T. DeRose, ACM Trans Graph 24 (3) (2005) 803–811.

[17] B. Atcheson, W. Heidrich, I. Ihrke, Exp. Fluids 46 (3) (2009) 467-476.

[18] C. Liu, *Beyond Pixels: Exploring New Representations and Applications for Motion Analysis*, PhD thesis, Massachusetts Institute of Technology, Massachusetts, USA, 2009.

[19] P.C. Hansen, M. Saxild-Hansen, J. Comput. Appl Math. 236 (8) (2012) 2167-2178.

[20] G. Bell, J. Soria, D. Honnery, D. Edgington-Mitchell. Particle Image Velocimetry Analysis of the Twin Supersonic Jet Structure and Standing-Wave. In 23rd AIAA/CEAS Aeroacoustics Conference, 2017, 3517.

[21] J. Yoo, D. Mitchell, D.F. Davidson, R.K. Hanson, Exp. Fluids. 49 (4) (2010), 751-759.

11th Asia-Pacific Conference on Combustion,
The University of Sydney, NSW Australia
10th -14th December 2017.

A solar-assisted metal oxide redox system for peak-load shifting in coal-fired power plants

Sike Wu, Cheng ZHOU, Elham DOROODCHI, Behdad MOGHTADERI*

Priority Research Centre for Frontier Energy Technologies & Utilization

Discipline of Chemical Engineering, School of Engineering,

Faculty of Engineering and Built Environment

University Drive, Newcastle, New South Wales, 2308, Australia

Abstract

This study is concerned with a solar-assisted metal oxide redox system integrated with a 500 MWe coal-fired power plant for peak-load shifting applications. The integrated system is a thermochemical energy storage system which stores energy from both the excess electricity and solar irradiance during off-peak hours and recover them during peak hours. The analysis was conducted using the simulation packages Aspen Plus v8.8 and with Co_3O_4/CoO as the metal oxides. The results demonstrate that such energy storage system can achieve an energy storage density of 821 kJ/kg and an optimum round-trip efficiency of 32-46% depending on whether waste heat recovery is implemented. In addition, a dynamic-state case study was conducted for the integrated system operating under the demand profile of a typical Australian summer day. Its preliminary design parameters and operating performances were determined. It was confirmed that the peak electricity demand can be fully met by the energy stored from the off-peak surplus electricity and solar heat, which accounts for 92% and 8% of the total energy input, respectively. The load shifting achieved by the integrated system offers the coal-fired power plant a great flexibility in meeting varied electricity demands.

1 Introduction

A great difficulty faced by the electricity grid is the mismatch between demands and supplies. The electricity demand varies substantially during a day due to human activities [1]. Specifically, the electricity demand usually reaches the highest at about 8-9 pm and the lowest at mid-night. A consequence of this is that the grid has to actively adjust the power supply via the use of smaller standby power plants such as gas turbine and renewable energy systems. Although the newly built coal-fired power plants can achieve minor load variations, the adjustment of the power output of most existing coal-fired power plants is not favored since it will significantly increase operating/maintenance costs primarily on the turbine units.

Energy storage technologies are deemed as the key solution to address the mismatch between electricity demand and supply [2]. The surplus electricity generated during the off-peak time can be stored and then be released to the power grid in peak hours. This is now realized by mainly pumped hydro storage, compressed air storage and batteries. However, these technologies have their limitations such as specific site topology and high capital costs. In contrast, thermochemical energy storage using metal oxide redox system offers another possibility of storing energy at an extremely high energy density and without geographic restrictions. It has a better scalability and cost than those of battery systems. The energy storage density of metal oxide redox systems can be up to 5 and 15 times greater than those of latent and sensible energy storage technologies [3]. It operates at significantly high temperature (> 600°C) compared to other thermal storage technologies enabling a high theoretical conversion efficiency. Moreover, the energy can be stored in long-term at ambient conditions and with a possible production of a high-value added by-product - oxygen [4].

This study proposed a solar-assisted metal oxide redox system for the purpose of peak-load shifting in coal-fired power plants. The metal oxides redox couple used in this study is Co_3O_4/CoO considering its high energy storage density and excellent reaction reversibility [5]. The redox reactions involved in the system can be described as (1) and (2). In the reduction process, Co_3O_4 are reduced at high temperature with the assistance of external thermal energy (e.g. off-peak electricity and/or solar energy). In the following oxidation process, the reduced CoO is oxidized to the initial state and the stored energy is released.

Reduction: $Co_3O_4 \rightarrow 3CoO + 1/2O_2$ $\Delta H=196.2$ kJ/mol (1)

Oxidation: $3CoO + 1/2O_2 \rightarrow Co_3O_4$ $\Delta H=-196.2$kJ/mol (2)

The major drawback of the above cycle when applied in electricity recovery is that only a portion of the stored heat can be converted back into electricity due to the inefficiencies. Hence, we proposed the use of additional solar heat to compensate for such losses with the objective to fully meet the peak demand.

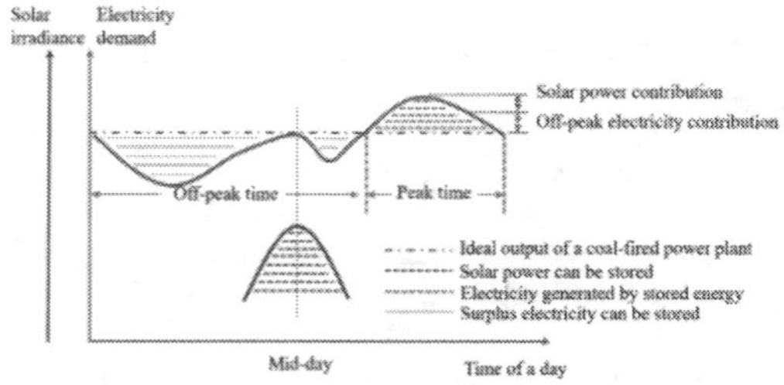

Figure 1: The operation principles of the integrated solar-assisted metal oxide energy storage system

The operating principles of the system can be seen in Figure 1,

Corresponding author. Fax: +61-2-4921-6893
E-mail address: Behdad.Moghtaderi@newcastle.edu.au

which indicates that both of the surplus off-peak electricity and solar heat is stored in the metal oxide redox system in the form of chemical energy. During the peak demand period, the stored chemical energy is then converted back into heat and electricity. By properly sizing the solar-assisted metal oxide redox system, the integrated coal-fired power plant can meet the variation of electricity demand at a constant power output.

Figure 2 illustrates the schematic diagram of the integrated system. As shown in Figure 2, surplus electricity generated by a coal-fired power plant and heat collected from a concentrated solar plant are used for reducing the metal oxides in the reduction reactor. During this process the byproduct - oxygen is released, while electrical and thermal energy are converted and stored in the form of chemical energy. For achieving a better conversion, a vacuum pump is applied to reduce the pressure of the reactor. During the peak time, the reduced material is recycled to a high-pressure oxidation reactor and re-oxidized in the presence of air. The reduced air (i.e. oxygen-deficient) then takes away the oxidation reaction heat and is heated to a high temperature. After that, the high-temperature air drives an air turbine and a generator to generate electricity. In this way, the stored chemical energy is converted back to electricity during peak hours.

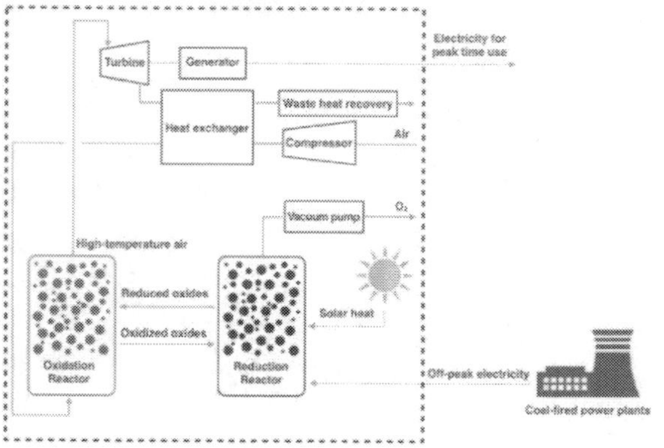

Figure 2: Schematic diagram of the solar-assisted metal oxides energy storage system of coal-fired power plants

2 Methodology

2.1 Process simulation

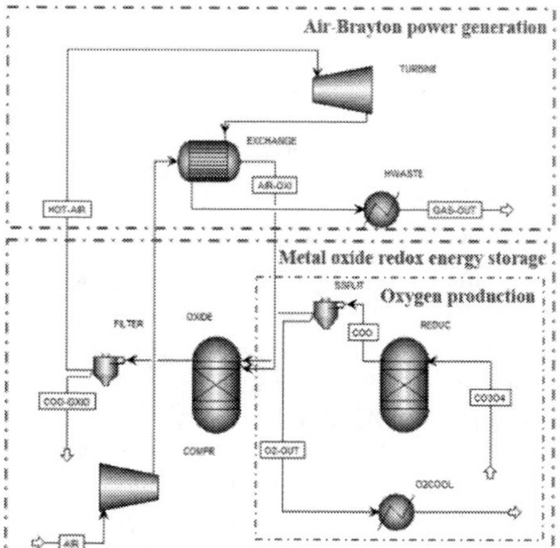

Figure 3: Aspen process model of the proposed energy storage system with heat exchanger.

Aspen Plus v8.8 is employed to simulate the process and conduct thermodynamic studies. Figure 3 shows the Aspen simulation flowsheet of the integrated system, in which the process focuses on the simulation of the metal oxides redox system connected with an air Brayton cycle. Both the systems with or without a heat exchanger were investigated.

In the whole process, air, metal oxides, heat generated by off-peak electricity and solar heat are the main energy and material inputs. For simplicity, Joule heating efficiency (η_{eh}) and the mechanical efficiency of the generator (η_{we}) are assumed to be 100%. The flowrate of the inlet Co_3O_4 is fixed at 1 kmol/hr. The inlet air has a temperature of 20°C and a pressure of 101.325 kPa. The pressure and temperature of the reduction reactor is kept at 1/10 of the ambient pressure and 900°C, respectively. The isentropic efficiency of both the compressor and turbine units is assumed to be 80%.

2.2 Performance indicators

Net power production, power cycle efficiency and specific power production are introduced to evaluate the performance of this energy storage system.

The net power production (P_{net}), as given in (3), is defined as the gross power being generated ($W_{turbine}$) for 1 kmol/hr of Co_3O_4 subtracted by the power consumption of the compressor ($W_{compressor}$) and the vacuum pump (W_{vacuum}). Of which, the power consumption of the vacuum pump (W_{vacuum}) can be calculated according to (4) [6]. In (4), nO_2 refers to the mole quantity of oxygen, T_0 is the reduction reactor temperature, and ΔS is the entropy difference of oxygen under reduction pressure and ambient pressure at T_0.

$$P_{net}=(W_{turbine}-W_{compressor}-W_{vacuum}) \times \eta_{we} \quad (3)$$

$$W_{vacuum}=-nO_2 T_0 \Delta S \quad (4)$$

The energy input of the system is the heat consumed in the reduction reactor ($Q_{reduction}$), and consists of the heat generated by the surplus electricity ($E_{surplus}$) and the received solar heat (Q_{solar}). The power cycle efficiency (η) is calculated by the net power produced of the system divided by the system energy input (5).

$$\eta=\frac{P_{net}}{Q_{reduction}}=\frac{P_{net}}{E_{surplus} \times \eta_{eh}+Q_{solar}} \quad (5)$$

The specific power output ($P_{specific}$) refers to the net power produced per kg/hr of the inlet air flow (mAir) as demonstrated in (6). The specific power output is important for determining the sizes of the turbine and compressor units.

$$P_{specific}=\frac{P_{net}}{mAir} \quad (6)$$

3 Results

3.1 Steady-state analysis

3.1.1 Effects of air flowrate and pressure on power cycle efficiency

Figure 4 gives the relationship between power cycle efficiency and air flowrate for five different compressor outlet pressures (i.e. oxidation reactor pressure). It can be seen that the power cycle efficiency of the systems with or without a heat exchanger show a similar trend, namely the efficiency increases gradually to a peak then declines as air flowrate increases. However, the optimum air flowrate for achieving the peak efficiency is different for the two systems. The reason behind the increasing efficiency at the low air flowrate region is that the air flowrate is not sufficient for a complete oxidation and thus a greater air flowrate increases the heat available for power generation leading to a greater efficiency. On the other hand, when the metal oxides become fully oxidized, an overly supplied air flow can result in a lower turbine inlet temperature which in turns reduces the power cycle efficiency. When it comes to the effect of pressures on the cycle efficiency, Figure 4 shows that a high pressure is beneficial for achieving a greater cycle efficiency if no heat exchanger is involved in the process. However, for the redox energy storage system with a heat exchanger there is an optimum pressure that leads to the maximum efficiency. On one hand, a greater reaction pressure is beneficial for enhancing the oxidation rate according to the Le Chatelier's principle, leading to more heat being released in the reaction. On the other hand, a greater pressure results in a greater compression ratio over the compressor and turbine units which helps boost the power generation. However, when a heat exchanger is involved, an overly high pressure was found to be ineffective mainly due to the inefficiency of the heat exchanger caused by the high temperature outlet of the compressor.

Moreover, according to Fig. 4, the optimum power efficiency of the model with a heat exchanger is generally higher than that of the model without a heat exchanger. This can be explained by the increased heat recovery and hence a reduced amount of waste heat due to the introduction of the heat exchanger. The highest achievable power cycle efficiency (η_{max}) was found to be 32% for the model with a heat exchanger which operates at a compressor outlet pressure of 225 kPa and an air flowrate of 25 kmol/hr. The corresponding net power production was calculated to be 16.97 kWe.

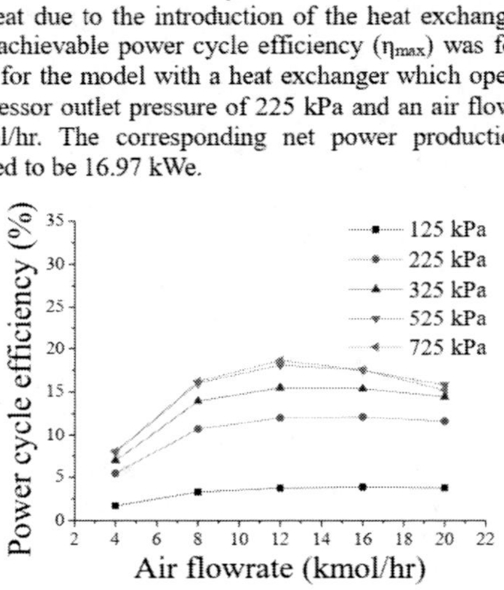

a) Power cycle efficiency (without heat exchanger)

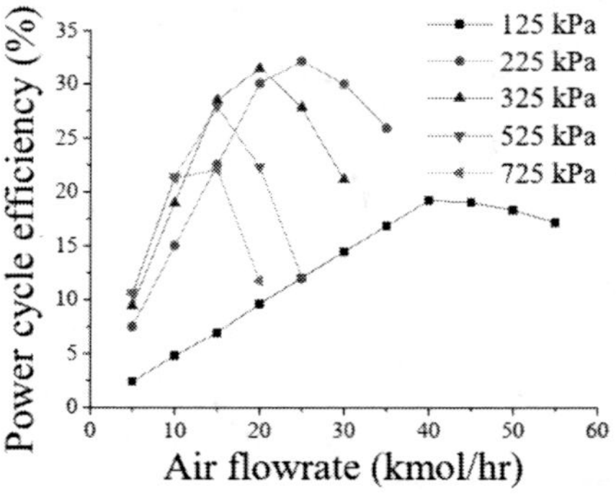

b) Power cycle efficiency (with heat exchanger)

Figure 4: Power cycle efficiency of the integrated systems with or without a heat exchanger under different air flowrates

3.1.2 Effects of turbine inlet temperature on power cycle efficiency

Figure 5 demonstrates the power cycle efficiency as a function of the turbine inlet temperature for the system with a heat exchanger. It was found that the power cycle efficiency increased from 11% to 32% when the turbine inlet temperature increased from 550°C to 900°C. This is expected since a greater turbine inlet temperature leads to a greater efficiency, which then enables the turbine to generate more work under the same pressure.

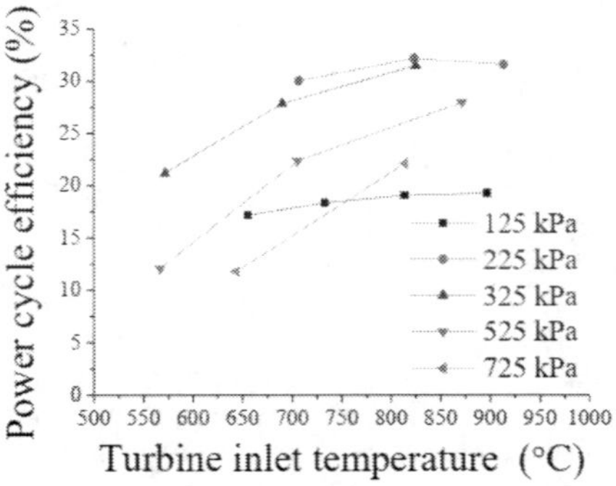

Figure 5: Power cycle efficiency of the integrated system with a heat exchanger under different turbine inlet temperature

3.1.3 Effects of air flowrate and pressure on specific power production

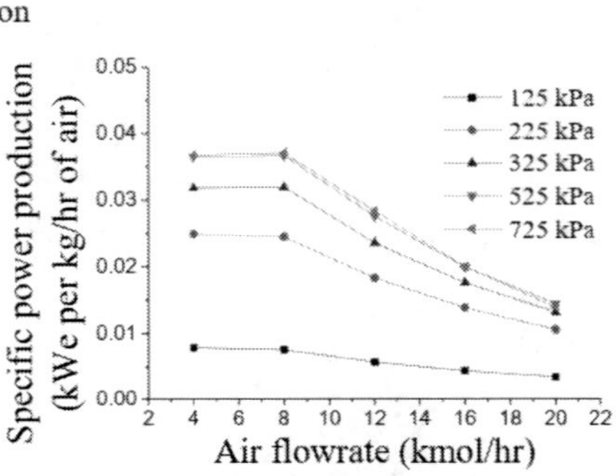

a) Specific power production (without heat exchanger)

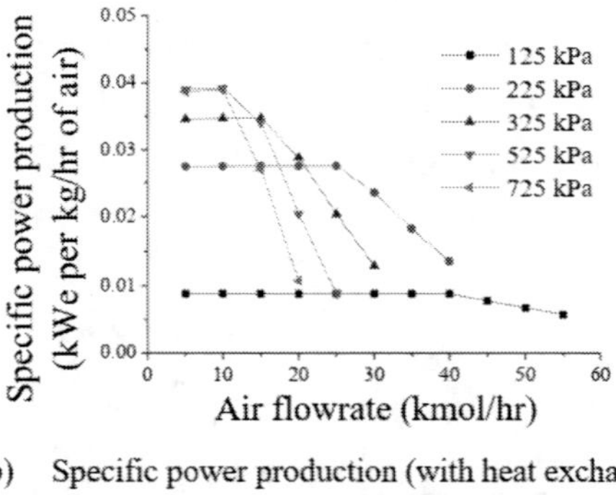

b) Specific power production (with heat exchanger)

Figure 6: Specific power production of the integrated systems with or without a heat exchanger as a function of air flowrate.

Figure 6 shows the specific power production as a function of air flowrate for various oxidation reaction pressures. It was found that the specific power production is directly proportional to the reaction pressure at all air flowrates for the system without a heat exchanger. Similar to efficiency (Figure 4), for the system with a heat exchanger there is an optimum reaction pressure that leads to the maximum specific power production. On the other hand, an increasing air flowrate was found to generally reduce the specific power production. This suggests that, despite a low power cycle efficiency, a system operating under a low air flowrate can lead to a favorable specific power production and thus may result in a smaller turbine/compressor unit as well as a smaller plant footprint. This is a preferred option when specific power production and plant size, rather than efficiency, are the key performance indicators. Our results shows that the achievable specific power production for the redox energy storage system is about 0.01 - 0.04 kWe per kg/hr.

3.2 Dynamic-state analysis

A dynamic-state analysis on the redox energy storage system integrated with a 500 MWe coal-fired power plant was conducted for a typical summer day. Figure 7 illustrates the electricity demand and supply curves during the day. The power output of a coal-fired power plant was kept at a constant value of 500 MWe during the 24 hours, as would be expected in most conventional coal-fired power plants. According to Fig. 7, the power supply during the off-peak hours (0:00 – 16:00 and 23:00 – 24:00) exceeds the electricity demand of the corresponding periods and thus there are surplus electricity that can be stored. During the peak hours (16:00 - 23:00), the electricity generated by the coal-fired power plant cannot meet the peak demand and thus requires additional power generation capacity from peaking plants/energy storage units.

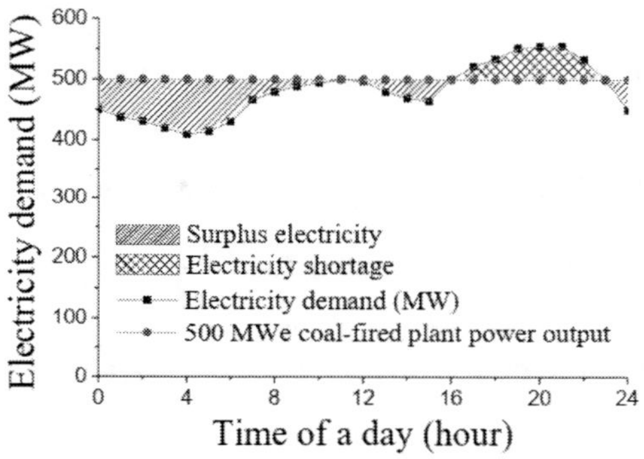

Figure 7: The electricity demand carve of a typical summer day in an area of Australia

The amount of surplus electricity and electricity shortage ($E_{shortage}$) as presented in Figure 7 were calculated by using curve fitting and integral methods. The results were then used to size the redox energy storage system and the additional energy requirement from solar. Table 1 summarizes the preliminary design parameters and operating performances of the redox energy storage system.

Energy balance	
Energy shortage ($E_{shortage}$)	260 MWh
Power cycle efficiency (η_{max})	32%
Surplus electricity ($E_{surplus}$)	748 MWh
Solar heat input (Q_{solar})	64.5 MWh
Reduction reactor net energy consumption	812.5 MWh
Oxidation reactor net heat generation	0[a] MWh
Compressor power consumption	267 MWh
Turbine power production	527 MWh
Waste heat	547 MWh
Operation details	
Reduction reactor temperature	900°C
Reduction reactor pressure	10 kPa
Oxidation reactor temperature	823°C
Oxidation reactor pressure	225 kPa
Energy storage density	821 kJ/kg
Total mass of Co_3O_4	3692 tonnes
Total volume of Co_3O_4	600 m^3
Container size of Co_3O_4	10 m×10 m×6 m
Total production of oxygen production	245 tonnes/day
Total consumption of air	1109 tonnes /day

a. Reaction heat was completely absorbed by air.

Table 1: Energy balance and operation details of the energy storage system

According to Table 1, a total of 812.5 MWh energy are required to store in order to generate 260 MWh electricity during the peak time. Among the input energy, the surplus electricity accounts for nearly 92% while the other 8% energy is supplied by the solar heat. Considering the large amount of the waste heat (547 MWh) in the current design, a heat recovery system is suggested to be employed to further increase the cycle efficiency. The upper limit was estimated to be 46%.

4 Conclusions

The current research studied the integration and performance characteristics of a new solar-assisted metal oxides energy storage system applied to a 500 MWe coal-fired power plant. Using this system, the balance between electricity supply and demand can be better maintained. Our simulation results show that the highest power cycle efficiency of the integrated system is about 32% under a compressor outlet pressure of 225 kPa. Adding a heat exchanger was found to be further improve the power cycle efficiency. The dynamic-state analysis of the integrated system for a typical summer day indicates that about 92% of electricity shortage of the peak hours can be provided by the surplus electricity stored during the off-peak hours. While the rest 8% can be provided by solar heat. In addition to a better match between supply and demand, the redox energy storage system can also produce ~245 tonnes/day of oxygen. Further studies would be to improve power cycle efficiency and reduce system costs especially the cost of metal oxides.

5 Acknowledgment

The authors wish to acknowledge the financial support they received from the University of Newcastle, Australia.

References

[1]. H. Ibrahim, A. Ilinca, J. Perron, Renewable & Sustainable Energy Reviews 12 (5) (2008) 1221-1250.

[2]. A.F., Regin, S.C. Solanki and J.S. Saini, Renewable & Sustainable Energy Reviews 12 (9) (2008) 2438-2458.

[3]. A.H. Abedin, M.A. Rosen, Open Renewable Energy Journal 4 (4) (2011) 42-46.

[4]. B.Wong, Thermochemical heat storage for concentrated solar heat, Report No.GA-C27137, General Atomics, 2011.

[5]. L. Andre, S. Abanades, G. Flamant, Renewable & Sustainable Energy Reviews 64 (2016) 703-715.

[6]. A.J. Schrader, A.P. Muroyama, P.G. Loutzenhiser, Solar Energy 118 (2015) 485-495.

11th Asia-Pacific Conference on Combustion,
The University of Sydney, NSW Australia
10th -14th December 2017.

A numerical study of coal gasification flame with one-dimensional turbulence model

Qi Zhang, Yuxin Wu*, Lv Junfu, Hai Zhang

Key Laboratory for Thermal Science and Power Engineering of Ministry of Education,

Department of Thermal Engineering, Tsinghua University

Tsinghua Yuan 1, Haidian District, Beijing, 100084, PR China

Abstract

In the near nozzle region in an entrained flow coal gasifier, strong interactions exist between coal gasification and turbulent fluctuations. These interactions have essential effects on coal gasification process and shall be considered. Based on proposed particle-eddy interaction model and spatially developing One Dimensional Turbulence (ODT) method, instantaneous coal gasification flames are simulated with a finer mesh resolution up to Kolmogorov scales. Characteristics of coal gasification flames are studied based on instantaneous realization and average results. It is observed that central fuel rich region is formed firstly due to rapid devolatilization process and slower particle dispersion rate in the near nozzle region. Then stronger heterogeneous coal char reactions occur at the edge of gasification flame. As the oxygen diffusion develops, the high temperature region merges at the center line in the developed region. Size-dependent coal gasification process and particle-eddy interactions are also studied in this paper. Concept of particle response time is shown in a straight way by comparing instantaneous scalars of coal particles and local gas phase.

1 Introduction

Coal gasification is an important advanced clean coal utilization technology. However, it is hard to get detailed information near the coal gasification flame through experiments. Thus intensive numerical simulations on entrained flow coal gasifiers have been carried out to study the fundamental physical/chemical processes in coal gasification. It has been noticed that turbulence fluctuations strongly affect motions and reactions of coal particles[1]. Therefore, traditional Reynolds-Averaging-Navier-Stokes equation (RANS) method cannot provide enough turbulence information so that turbulence-reaction coupling effects cannot be considered directly and additional turbulence models have to be involved. To better consider the detailed turbulence effects, Large Eddy Simulation (LES) and Direct Numerical Simulation (DNS) have been used in coal jet flame studies[2, 3]. Due to limitation of computing cost, applications of LES and DNS on coal combustion flames are still limited. In order to decrease computing expense, One Dimensional Turbulence (ODT) is proposed while the main characteristics of turbulence are still retained[4]. ODT is an unsteady simulation of instant flow fields on a one-dimensional domain representing a notional line of sight through a turbulent flow where Kolmogorov scales can be resolved directly. Due to its low dimension limitation, ODT model is only suitable for the flows in which gradients in one direction are much larger than those in other directions. Since ODT captures individual eddy motions, it is appealing to study coal particle combustion or gasification process in turbulent jet flame. Up to now, ODT simulation has been applied to coal combustion[5] successfully, but not to coal gasification yet.

In this work, a particle tracking method in Lagrangian frame is introduced to couple with a spatially developing ODT model. With this model, 2-Dimensional instantaneous coal gasification flames are generated with a full knowledge of turbulent information in all scales. Chemical/physical changes of Coal particles and particle tracks are analyzed to study the effects of turbulent fluctuations and acquire particle dispersion characteristics of different size.

2 ODT and coal gasification process models

2.1 ODT model implementation

The spatially developing turbulent ODT model generates realizations parameterized by (y, x) coordinates, where x is the streamwise coordinate and y is the transverse direction in a jet flame. ODT line is parallel with y coordinate so that turbulence is considered in y direction. The field defined on the one-dimensional domain is evolved by two separate mechanisms. In the first step, a Monte Carlo stochastic procedure is used to identify the distribution of turbulent eddies which represent turbulent mixing and fluctuations in an instantaneous flow field. Then a set of transport equations in one dimension are solved to represent molecular evolution process[6].

For the gas phase, the incompressible transport equations on the ODT line are

$$\frac{\partial(\rho u_1)}{\partial x} + \frac{\partial(\rho u_2)}{\partial y} = S_{\mathrm{m}} \tag{1}$$

$$\frac{\partial(\rho u_1 u_1)}{\partial x} + \frac{\partial(\rho u_1 u_2)}{\partial y} = \frac{\partial}{\partial y}\left(\mu \frac{\partial u_1}{\partial y}\right) + S_{u_1,\mathrm{m}} + B_Y \tag{2}$$

$$\frac{\partial(\rho u_1 Y_k)}{\partial x} = \frac{\partial}{\partial y}\left(\rho D_k \frac{\partial Y_k}{\partial y}\right) + S_{Y_k} + \dot{\omega}_k \tag{3}$$

$$\rho u_1 C_{\mathrm{p}} \frac{\partial T}{\partial x} = \frac{\partial}{\partial y}\left(\lambda \frac{\partial T}{\partial y}\right) + \left(\rho \sum_{k=1}^{n} C_{p,k} D_k \frac{\partial Y_k}{\partial y}\right)\frac{\partial T}{\partial y} + \dot{\omega}' + S_{\mathrm{rad}} + S_T \tag{4}$$

where u_1 and u_2 are streamwise velocity and transverse velocity, S_{m}, $S_{u1,\mathrm{mom}}$, and S_T are particle source terms of mass, momentum in streamwise, the kth species and temperature. S_{rad} is the radiation source term. The buoyancy force in Eq. (2), B_Y, is calculated by $(\rho_b-\rho)gu_1$, where ρ_b is the background gas density. $\dot{\omega}_k$ and $\dot{\omega}'$ are the source terms due to gas phase reactions. Equilibrium assumptions are used to instead solution of gas phase reactions, thus $\dot{\omega}_k$ and $\dot{\omega}'$ are set to 0 in Eq. (3) and Eq.

Corresponding author. Fax: +0086-10-6278-1559(O)
E-mail address: wuyx09@mail.tsinghua.edu.cn

(4). Here, a separate equilibrium solution of 19 species system is adopted.

In order to analyze the role of turbulent mixing in the coal reaction flow, an additional set of transport equations of the mixture fractions of coal are solved,

$$\frac{\partial(\rho u_1 f_k)}{\partial x} = \frac{\partial}{\partial y}\left(\rho D_{fk}\frac{\partial f_k}{\partial y}\right) + S_{f_k} \tag{5}$$

where f_k is the kth mixture fraction, D_{fk} is the diffusion coefficient of the kth mixture fraction. Five mixture fractions are considered, which are the mass fraction of the primary oxidizer, the secondary oxidizer, the moisture, volatile, and char yielding out of the coal particles.

Punctuating this diffusion process, the random events representing advection may be interpreted as the model analogue of individual turbulent eddy events. One eddy event can be expressed symbolically by

$$\begin{aligned} u_i(y) &\rightarrow u_i(f(y)) + c_i K(y) + b_i J(y) \\ \theta_i(y) &\rightarrow \theta_i(f(y)) \end{aligned} \tag{6}$$

where y is the new position of a gas parcel on ODT line whose original position is at $f(y)$ on ODT line, and c_i and b_i are the transforming coefficients of the triplet kernel $K(y)$ and $J(y)$, where $K(y) = y - f(y)$, $J(y) = |K(y)|$. The solution of coefficients c_i and b_i can be found in detail in [7].

The relationship between $f(y)$ and y is expressed by

$$f(y) = y_0 + \begin{cases} 3(y - y_0) & y_0 \le y \le y_0 + \frac{1}{3}l_e \\ 2l_e - 3(y - y_0) & y_0 + \frac{1}{3}l_e \le y \le y_0 + \frac{2}{3}l_e \\ 3(y - y_0) - 2l_e & y_0 + \frac{2}{3}l_e \le y \le y_0 + \frac{2}{3}l_e \\ y - y_0 & \text{otherwise} \end{cases} \tag{7}$$

where y_0 is the eddy starting point and l_e is the eddy length. The effect of triplet map is to shrink the original profile to a third of its original length, then to place three copies on the original domain with the middle copy being reversed[4]. After the triplet map, local gradients are higher while all the scalars are conserved.

The eddy events are sampled from an event rate distribution that reflects the physics governing eddy turnovers. This is implemented using a stochastic eddy selection procedure, and can be found in detail elsewhere[8].

2.2 Particle tracking and particle-eddy interaction model

Particles are tracked in a Lagrangian frame. Each particle is regarded as a point although at some places the ODT cell size will not exceed the particle size. The momentum equation of a particle is given by

$$\frac{du_{p,i}}{dt} = \frac{f_d}{\tau_p}\left(u_{g,i} - u_{p,i}\right) + \frac{g_i(\rho_p - \rho_g)}{\rho_p} \tag{8}$$

where t is the integration time, $u_{p,i}$ is the particle velocity in the ith direction, τ_p is the particle aerodynamic response time, $\tau_p = \rho_p d_p^2/18\mu_g$, and g_i is the gravity force in the ith direction. The drag coefficient f_d is a function of particle Reynolds number.

Eddy effects on the particle dispersion in temporally developing ODT has been discussed[9, 10]. Here, a related particle-eddy interaction model is briefly explained from a spatial perspective and then its incorporation into spatially developing ODT is proposed. The particle-eddy interaction model is based on several assumptions. Firstly, each occurring eddy is supposed to stay at a fixed position at initial height x_0 for a period of time βt_e, where β is a model parameter and t_e is eddy life time. In spatially developing ODT, the eddy surviving time βt_e is represented by an eddy height $\beta u_e t_e$ where u_e is the eddy velocity. As the ODT line advances past the height $x_0+\beta u_e t_e$, this eddy will vanish so that it won't have any further effects on particles. Secondly, although the effect of an eddy on a particle is instantaneous in ODT, this effect has to be calculated by tracking particle motion in an eddy region. The particle integration time is defined as the period when a particle and an eddy occupy the same spatial region while the eddy exists. The eddy spatial region is assumed to be a rectangular region of width l_e and height γl_e, where γ is another model parameter that defines the shape of the eddy. Based on these assumptions, the particle-eddy interaction region is described in Fig. 1.

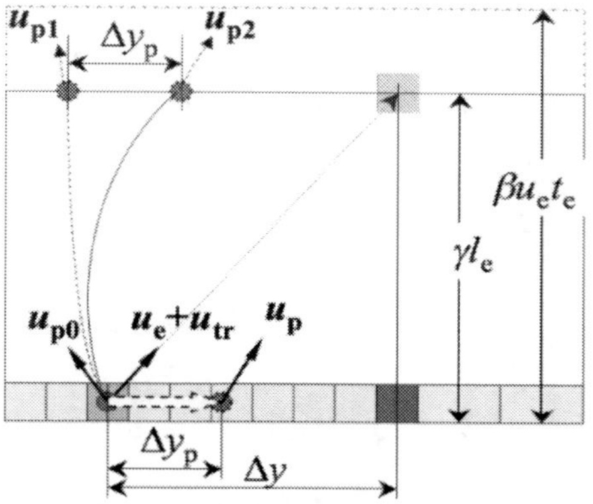

Fig. 1. Particle tracking procedure in the particle-eddy interaction model.

During the particle-eddy interaction period, the particle motion is tracked based on its drag law,

$$\frac{du_{p,i}}{dt} = \frac{f_d}{\tau_p}\left(u_{e,i} + u_{tr,i} - u_{p,i}\right) + \frac{g_i(\rho_p - \rho_g)}{\rho_p} \tag{9}$$

where $u_{e,i}$ is the eddy velocity of the ith component, $u_{tr,i}$ is an additional velocity associated with the triplet map, $u_{tr,1} = 0$ and $u_{tr,2} = \Delta y u_{e,1}/(\gamma l_e)$ where Δy is the displacement of the gas parcel containing the particle by the triplet map effect. In order to avoid the double counting effect[10], a shadow particle tracking calculation is adopted. The track of the shadow particle is shown as the dot-curved line. The final particle position and velocity at any integral time step is then calculated by

$$\begin{aligned} y_p &= y_{p0} + y_{p2} - y_{p1} \\ u_{p,i} &= u_{p0,i} + u_{p2,i} - u_{p1,i} \end{aligned} \tag{10}$$

where y_p is the final particle position at any integration time step, y_{p0} is the initial particle position, y_{p1} is the shadow particle tracking position, y_{p2} is the particle tracking position, $u_{p,i}$ is the final particle velocity in the ith component, $u_{p0,i}$, $u_{p1,i}$ and $u_{p2,i}$ are initial particle velocity, shadow particle tracking velocity and particle tracking velocity respectively. The final eddy effects are then appeared by an instant horizontal displacement of y_p, and instant particle velocity changes.

2.3 Coal gasification process

The coal gasification process is divided by five sub-processes which are heating up of inert particles, evaporation and boiling of water in coal slurry particles, devolatilization and heterogeneous reaction process. The evaporation and boiling of water are predicted by the droplet evaporation and boiling model[11]. The devolatilization process is calculated by the one-step model, and is expressed by

$$\frac{dm_p}{dt} = -A_v \exp\left(-\frac{E_v}{RT_p}\right)(m_p - m_{pc}) \tag{11}$$

where m_p and m_{pc} are the particle mass and the mass of char and ash in the particle[12]. Four heterogeneous reactions are considered to represent the char gasification process,

$$C(s) + O_2 \rightarrow CO_2 \tag{R-1}$$

$$C(s) + CO_2 \rightarrow 2CO \tag{R-2}$$

$$C(s) + H_2O \rightarrow CO + H_2 \tag{R-3}$$

$$C(s) + 2H_2 \rightarrow CH_4 \tag{R-4}$$

For each reaction, the final reaction rate is determined by the diffusion process and the intrinsic chemical reaction rate[13]. The bulk diffusion rate is calculated by

$$R_{i,d} = C_i \frac{[(T_p+T_g)/2]^{0.75}}{d_p} p_i \tag{12}$$

where C_i is the diffusion coefficient of ith reactant, T_p is the particle temperature, T_g is the temperature of the bulk gas phase, d_p is the particle diameter, p_i is the partial pressure of the ith reactant. The apparent chemical reaction rate is calculated by the nth empirical reaction model,

$$R_{i,k} = A_i \exp\left(-\frac{E_i}{RT_p}\right)\left(\frac{p_i}{10^5}\right)^n \tag{13}$$

where A_i and E_i are the pre-exponential and activation energy of the ith reaction, respectively.

3 Simulation results of coal gasification

3.1 Description of Soelberg gasification testbed

The simulated gasifier test rig is built in Utah[14]. The inside diameter of the reactor was 20 cm and the height was 124 cm. The primary stream of premixed coal, O_2 and Ar was preheated to 366 K while the secondary stream of steam to 430 K. The designed coal feed rate was 0.0068 kg/s and the coal-O_2-Ar-steam mass ratio was 1.0 : 0.91 : 0.14 : 0.27. A Utah high-volatile bituminous coal was used. The proximate analysis was 0.1 wt% moisture, 46 wt% volatile, 46.55 wt% fixed carbon and 7.35 wt% ash. The ultimate analysis (db) was 73.1 wt% C, 5.45 wt% H, 12.05 wt% O, 1.4 wt% N and 0.55 wt% S. Detailed description of the experimental work can be found in [15].

The ODT simulation domain is a rectangle with width of 0.424 m and height of 1.2 m. The dimension of dual channel nozzle and the mass flow rate of coal and each reactant are identified based on principle that the momentum flux ratio of inner jet and outside jet is the same between the simulation case and experimental work. In addition, the velocity of the primary jet in simulation is the same with the velocity in experimental work. The velocity of the co-flow is 2 m/s. 18000 particle tracks were used to represent the coal particles. The particle size distribution was 22 wt% small particles with diameter of 24 μm, 67 wt% medium particles with diameter of 42 μm, and 11 wt% large particles with diameter of 78 μm. In this simulation, the eddy factor γ equals to 0.6, and the eddy lifetime factor β equals to 0.03. The average ODT cell size is 40 μm, the average ODT diffusion step is 20 μm. One instantaneous coal flame simulation spends 7 - 9 hours. The average values are acquired based on 240 instantaneous realizations.

3.2 Average coal gasification flames

Figure 2 shows the average scalar profiles of the coal gasification flame. For the coal gasification flame, there is a low-temperature region on the center line between the two high flame temperature regions where char-O_2 reaction undergoes rapidly and O_2 exists. The characteristic of the predicted average temperature profile and the main species profile near the burner region is similar with the experimental measurements[14]. It is interesting that most volatile yields out on the center line at about 0.1 m below the nozzle where the low-temperature region also exits. The reason can be explained from Fig. 2d that as volatile yields out, local O_2 is consumed rapidly and since the surrounding O_2 cannot diffuse inward, a fuel rich region is formed on the center line. Figure 2c shows the profile of the average char source term. The char source appears on the center line at first, however, the maximum char source region appears symmetrically along the center line well below the region where the maximum volatile source yields out.

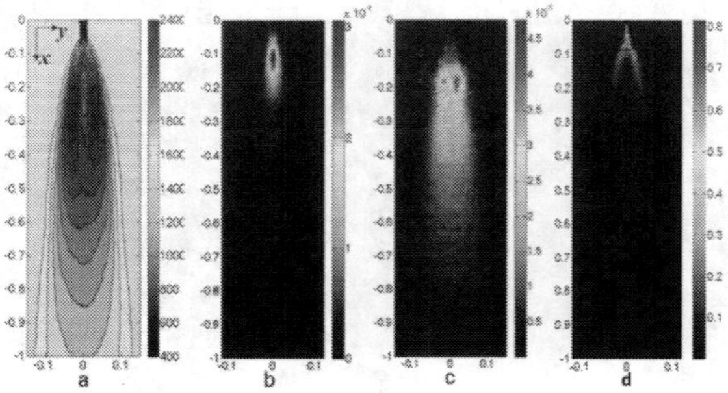

Fig. 2. Average scalar profiles of coal gasification flame. (a) Temperature, (b) volatile source term, (c) char source term, (d) mixture fraction of O_2.

3.3 Particle dispersion and turbulent fluctuations

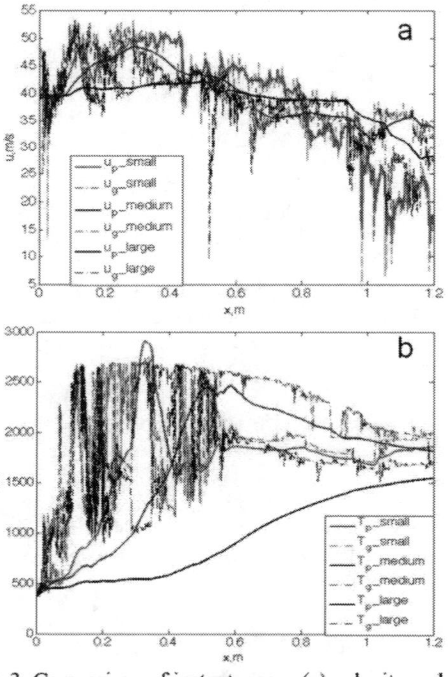

Fig. 3. Comparison of instantaneous (a) velocity and (b) temperature between particles of different size and local gas.

Turbulent fluctuations have important effect on particle motion and gasification process. Figure 3 shows the comparison of instantaneous stremawise velocity/temperature between particles of different size and local gas. For gas phase, the velocity fluctuations are always intensive throughout the furnace while temperature fluctuations in the region of $x > 0.6$ m are much smaller. This interesting phenomenon can be explained from Fig. 4 that most coal reactions have finished before particles reach the height of 0.6 m, thus most particle weights do not decrease further and the strong fluctuations for temperature happen mainly because of the frequent occurance of coal reactions. For particle phase, the responding time of temperature is much longer than that of velocity. As shown in Fig. 3a, the small particles always follow the local velocity fluctuations closely while larger particles miss most of turbulent fluctuations due to a longer responding time. Figure 3b shows that small particles are heated more rapidly and have stronger reactions before they exit the flame region, but for large particles which are heated slowly, they have passed the flame region when heterogeneous reactions occur, thus their char burning rate is suppressed due to the lack of oxygen.

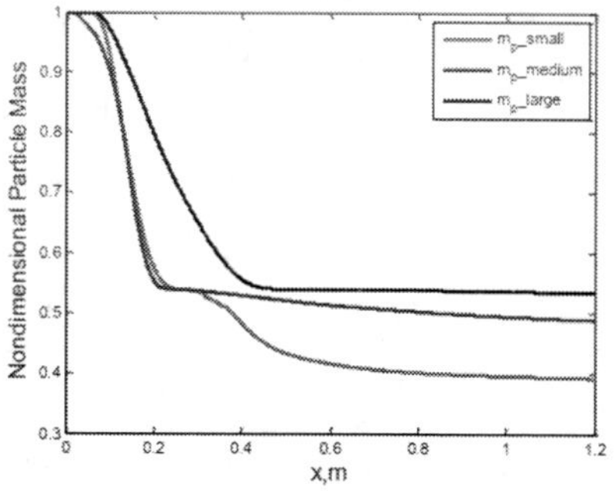

Fig. 4. Particle mass loss of different sizes.

Figure 5 shows the particle number density (PND) profile of different particle size. For all particles, dispersion is limited when $x < 0.4$ m. Small particles spread out at very early stage, but their PND profile at the down flow region is smoother comparing with larger particles.

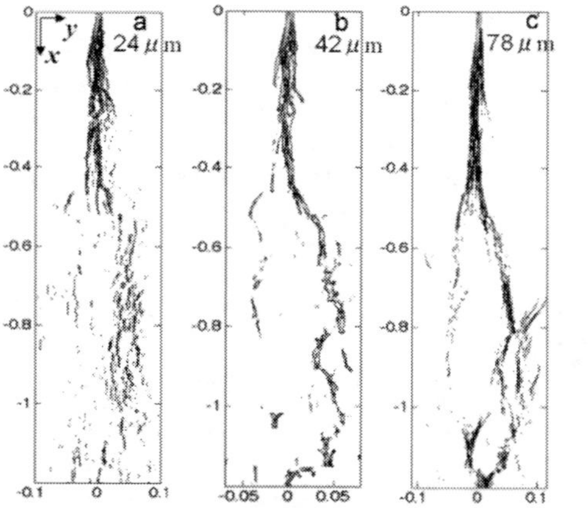

Fig. 5. Instantaneous PND profile of different particle size. (a) 24μm, (b) 42μm, (c) 78μm.

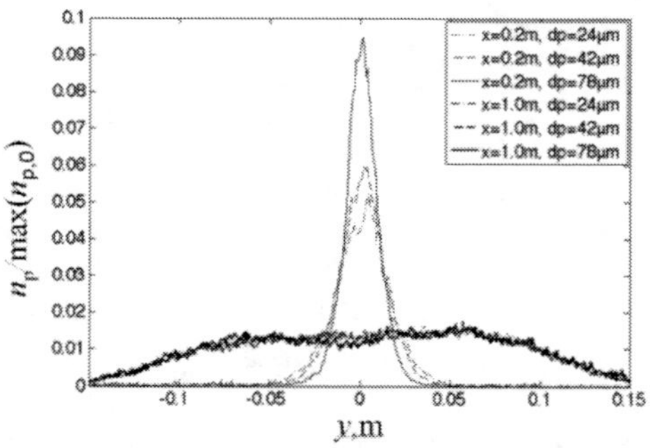

Fig. 6. Non-dimensional average PND of different particle size at different height

Figure 6 shows the average non-dimensional PND profile of different particles at different height which is defined as PND = $n_p(x)/\max(n_{p,0})$ where $n_{p,0}$ is the initial particle number density value on the ODT line. At the near nozzle region($x = 0.2$ m), small particles disperse outward faster than large particles since they response to all eddies immediately. As the jet flow develops, more large eddies occur in the jet region and dominate the macro flow field as well as particle dispersion so that all particles have almost the same PND profile at the developed region($x = 1.0$ m).

4 Conclusions

Based on spatially developing ODT method and proposed particle-eddy interaction model, coal gasification flames are simulated with a finer mesh resolution up to Kolmogorov scales. Central fuel rich region is formed due to rapid devolatilization process and slow particle dispersion rate at the near nozzle region where temperature is lower comparing with the high temperature surrounding regions. Then stronger heterogeneous coal char reactions occur at the edge of gasification flame where oxygen exists. As the oxygen diffusion develops, the high temperature region merges at the center line in the developed region. Particle temperature response time is much longer than particle motion response time. The motion of the small particles response to eddies more rapidly, thus they disperse outward faster than large particles at the near nozzle region. However, particle size has little effect on particle dispersion at the developed region where large eddies dominate both macro flow field and particle dispersion.

5 Acknowledgment

This material is based upon work supported by the Projects of International Cooperation and Exchanges NSFC (Grant NO. 51761125011)

References

[1] Y. Wu, P.J. Smith, J. Zhang, et al., Energ. Fuel 24(2) (2010) 1170-1175.

[2] R. Kurose, H. Makino, Combust. Flame 135(1) (2003) 1-16.

[3] T. Hara, M. Muto, T. Kitano, et al., Combust. Flame 162(12) (2015) 4391-4407.

[4] A.R. Kerstein, J. Fluid Mech. 392 (1999) 277-334.

[5] B. Goshayeshi, J.C. Sutherland, Proc. Combust. Inst. 35(3) (2015) 2829-2837.

[6] T. Echekki, A.R. Kerstein, T.D. Dreeben, et al., Combust. Flame 125(3) (2001) 1083-1105.

[7] W.T. Ashurst, A.R. Kerstein, Phys. Fluids 17(11) (2005) 025107.

[8] D. O. Lignell, D. S. Rappleye, Combust. Flame 159(9) (2012) 2930-2943.

[9] J.R. Schmidt, J.O.L. Wendt, A.R. Kerstein, J. Stat. Phys. 137(2) (2009) 233-257.

[10] G. Sun, D.O. Lignell, J.C. Hewson, et al., Phys. Fluids. 26(10) (2014) 103301.

[11] L.X. Zhou, Theory and Numerical Modeling of Turbulent Gas-Particle Flows and Combustion, Science Press, Beijing, China, 1994, p. 88.

[12] S. Badzioch, P.G.W. Hawksley, Ind. Eng. Chem. Process Des. Dev. 9(4) (1970) 521-530.

[13] I.W. Smith, Proc. Combust. Inst. 19(1) (1982) 1045-1065.

[14] N.R. Soelberg, L.D. Smoot, P.O. Hedman, Fuel 64(6) (1985) 776-781.

[15] L.D. Smoot, B.W. Brown, Fuel 66(9) (1987) 1249-1256.

A sub-grid scale combustion model based on thickened-flame method and REDIM chemistry table

Liang XU[1], Ping WANG[2,*], Qian YU[1], Tian-zeng HOU[1]

[1]School of Energy and Power Engineering, Jiangsu University
Zhenjiang, Jiangsu 212013, China;
[2]Institute for Energy Research, Jiangsu University
Zhenjiang, Jiangsu 212013, China

Abstract

The reaction-diffusion manifold (REDIM) technique is utilized in combination with the dynamically thickened-flame (DTF) model, which forms a new turbulent premixed combustion model, REDIM-DTF. The new model was then used to calculate two turbulent premixed swirling flames in PRECCINSTA combustor via LES, with equivalence ratio equals to 0.75 and 0.83, respectively. The radial profiles of velocity, temperature and CO_2 mass fraction as well as the flame surface area were analyzed and compared with the results from original DTF model and the experimental data. The results are in good overall agreement with corresponding experiment. Since less species transport equations are solved in REDIM-DTF model, the computational efficiency of it is about 15% higher than the original DTF model. In theory, the new model can also be applied to simulate partially premixed flames, as long as an appropriate REDIM chemistry look-up table is used, e.g. two-dimensional table.

1. Introduction

Large eddy simulation (LES) has become increasingly important as a combustion simulation tool in the past few years due to its ability to predict instability, extinction, re-ignition and other instabilities through a suitable sub-grid combustion model [1]. The well-known dynamically thickened-flame (DTF) model can increase flame thickness artificially by modifying the pre-exponential constant of chemical reaction rate and the diffusion coefficient, so that the flame surface can be resolved by the LES grid. However, DTF model also has some shortcomings: the reaction source terms in the species transport equations are calculated by the Arrhenius law. In order to control the computational effort, the DTF model usually considers reduced reaction mechanisms composed of only a few steps.

Based on the Intrinsic Low Dimensional Manifold (ILDM) method [2], Bykov and Maas proposed a new reduction scheme called Reaction-Diffusion Manifold (REDIM) [3,4]. This method takes full account of the molecule diffusion process while reducing the mechanism, which solves the coupling problem between the chemical reaction and the molecular transport process. It is hence also suitable for slow combustion reaction where the diffusion process governs the reaction rate. A distinct number of reduced coordinates can be generated while reducing detailed chemistry mechanism with REDIM. Other variables can be determined through these reduced coordinates and the generated look-up table.

In this paper, a new SGS combustion model, REDIM-DTF, is proposed to combine the REDIM chemistry table and the artificial thickening concept. The new model is first introduced in Sec. 2. Its performance is then validated by simulating two lean premixed swirling flames in the well-known PRECCINSTA combustor [5], with equivalence ratios 0.75 and 0.83, respectively. The calculation results are studied through the comparison with the results of the original DTF model and the experimental data.

2. Sub-grid scale combustion models

2.1. Dynamically thickened-flame model

The thickened-flame model can artificially increase the flame thickness through modifying the pre-exponential constant and diffusion coefficient with a factor F. However, the thickening factor F will make Damköhler number decrease, small-scale eddies can no longer wrinkle the flame front. Colin et al. [6] introduced an efficiency function E to recover an accurate turbulent flame state, whereby the species transport equation needed to be solved in the thickened-flame model reads

$$\frac{\partial \rho Y_k}{\partial t}+\frac{\partial \rho Y_k u_j}{\partial x_j}=\frac{\partial}{\partial x_j}(\rho D_k EF\frac{\partial Y_k}{\partial x_j})+\frac{\boldsymbol{E}}{\boldsymbol{F}}\dot{\omega}_k(Y_i,T) \tag{1}$$

$$F(n)=n\Delta x/\delta_L \tag{2}$$

where Y, $\dot{\omega}$, D and T are the species concentration, reaction rate, diffusion coefficient and temperature, respectively. The value of F can be determined by Eq. (2), where Δx and δ_L are the local grid size and the laminar flame thickness, respectively. By adjusting the factor F, the thickened flame front can be resolved by n grids. The value of n is usually in the range 5-10, here n = 5 was chosen.

A process of dynamically adjusting the F value was introduced in [7], where F = 1 in the place away from the flame front which doesn't need to be thickened. While near the flame front, F will increase accordingly, so that realistic transport coefficients can be acquired. This dynamic process is based on the local reaction rate Q_a of the reaction (a) shown in Eq. (9) in reference [1] and reads

$$F_{dyn}=1+\left(F(n)-1\right)\tanh\left(10Q_a/Q_{a,\max}\right) \tag{3}$$

where $Q_{a,max}$ is the maximum value of Q_a which can be

Corresponding author. Fax: +86 511 88799500
E-mail address: pingwang@ujs.edu.cn (P. Wang)

obtained from a corresponding one-dimensional laminar flame simulation [8]. If the constant 10 in Eq. (3) is adjusted accordingly, the dynamic process can also be calculated by the local reaction rate Q_b of the reaction (b) shown in Eq. (10) in reference [1].

In previous work, the DTF model was utilized in combination with 2sCM2 two-step scheme [9] to calculate the same PRECCINSTA flames, and the results were in good overall agreement with the experimental data [1].

However, this method needs to resolve the multiple species transport equations, and the computational effort is very large. The REDIM-DTF model proposed in this paper only needs to solve one transport equation for CO_2 mass fraction. The other variables in the equation, such as $\dot{\omega}$, ρ, T, and the mass fraction of other species, can be obtained through interpolation in a one-dimensional REDIM table, which effectively reduces the computational effort.

2.2. Reaction-diffusion manifold technique

The REDIM method abandons hypothesis of ILDM method. It takes full account of the coupling process of the molecular transport process and the chemical reaction. The REDIM method is based on solving an evolution equation for a low-dimensional manifold in the thermo-kinetic space [3,4]:

$$\frac{\partial \psi}{\partial t} = \left(1 - \psi_\theta \psi_\theta^{+}\right)\left\{ F(\psi) + \frac{D}{\rho} \chi \circ \psi_{\theta\theta} \circ \chi \right\} \quad (4)$$

where, the thermo-kinetic space is described by a vector Ψ. This vector is a function of one or multiple reduced coordinates, given by the vector θ. The variable ρ is the density, χ is the vector of spatial gradient estimates for θ, $F(\Psi)$ is the vector of the chemical source terms, Ψ_θ is the matrix of partial derivatives of Ψ with respect to θ and Ψ^{+}_θ its Moore–Penrose pseudo-inverse [4], $\Psi_{\theta\theta}$ is the Hessian matrix, D is the molecular diffusion coefficient matrix. The symbol ' ∘ ' is an abbreviation for the multiplication of two vectors with a tensor of third order.

REDIM method can generate a reduced coordinate system of any dimension in theory. The flame calculated in this paper is a fully premixed methane/air flame, the equivalence ratio is constant and there is no local extinction phenomenon, so the use of one-dimensional reduced coordinate system is sufficient, that is, using a thermodynamic look-up table represented only by one parametric variable. The variable is the CO_2 mass fraction, and the mass fraction of other species such as CH_4, CO, O_2, H_2O, and N_2, as well as the density and production rate of CO_2 can be determined by interpolating the thermodynamic look-up table. Figure 1 shows the density and the production rate of CO_2 associated with CO_2 mass fraction given in employed one-dimensional REDIM look-up table. Note: in theory, the table can also be coordinated with other major species, like H_2O.

In LES of turbulent flames, this one-dimensional look-up table cannot be used directly. Because the low-dimensional manifold generated by the REDIM method is a flame structure represented by the reduced method in the thermo-kinetic space, it is necessary to use a suitable method to couple the low-dimensional manifold with the turbulent field. In this paper, REDIM method was combined with the DTF model. The REDIM method was used to reduce the detailed reaction mechanism of methane/air GRI-Mech 3.0 to obtain the one-dimensional reduced manifold parameterized by CO_2 mass fraction. Then the DTF model was employed to describe the distribution of the CO_2 mass fraction in the control body, so that other variables can be determined through interpolation in the one-dimensional REDIM look-up table.

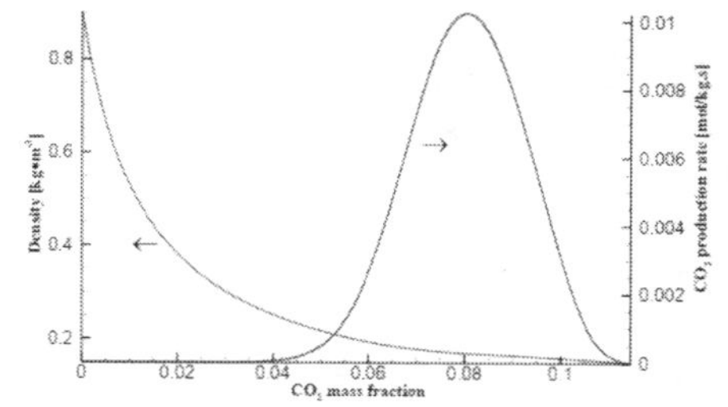

Figure 1: The density and CO_2 production rate associated with the CO_2 mass fraction given in the one-dimensional REDIM look-up table.

3. Burner structure and computational set-up

In order to validate the new REDIM-DTF model, swirling premixed flames in PRECCINSTA burner [5] were simulated by this model via LES. Meier has carried out a detailed experimental study of the burner [10]. The geometry of PRECCINSTA burner is shown in Fig. 2a. It consists of a plenum, a radial swirl injector with 12 passages, a square combustion chamber and a cylindrical exhaust pipe. The reference length used throughout here is the diameter of the vane outlet D = 27.85mm. Methane enters the swirl injector through small holes and mixes fast and well with the incoming dry air before it enters the combustion chamber so that combustion indeed is premixed.

The PRECCINSTA flames considered in this work are two lean premixed swirling flames with global equivalence ratios ϕ = 0.75 and ϕ = 0.83. The P_{in} = 1bar, U_{in} = 11m/s, Re = 30522 for both cases. While T_{in} = 320K for ϕ = 0.75, and T_{in} = 325K for ϕ = 0.83. Detailed experimental data analysis shows that these two kinds of flames (ϕ = 0.75 and 0.83) calculated in this paper do not appear thermo-acoustic instability [10]. Hence, the Low-Mach-number version of the Navier–Stokes equations can be employed. The plenum as well as the atmosphere outside the exhaust pipe are not included in the calculations because sonic waves are eliminated from the equations.

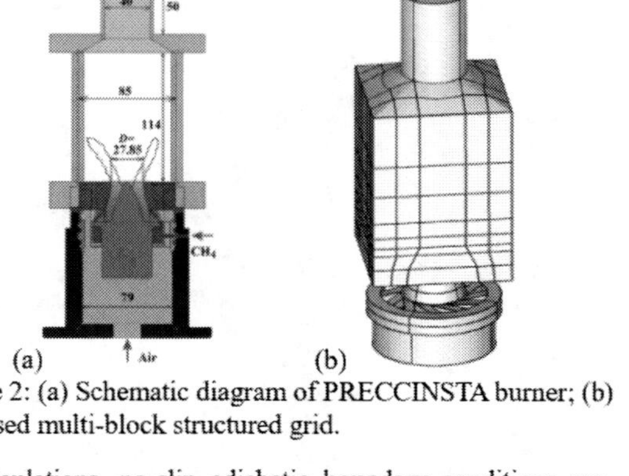

Figure 2: (a) Schematic diagram of PRECCINSTA burner; (b) The used multi-block structured grid.

In calculations, no-slip adiabatic boundary conditions are used for the walls. At the inlet of the vanes, a uniform velocity profile without fluctuation is imposed matching the experimental flow rate. Turbulence develops spontaneously in the vanes and attains the correct magnitudes illustrated by the profiles at x = 1.5mm shown below. The outlet of the computational domain is located at the end of the exhaust pipe, with convective condition imposed there.

The mesh used is a multi-block body-fitted structured grid that covers the inlet vanes and the combustion chamber, with a total of 3.8 million cells in 423 blocks which are shown in Fig. 2b. The meshing is verified to be reasonable in this calculation through detailed grid resolution studies in [11]. In this paper, all simulations were implemented in the in-house Finite-Volume code LESOCC2C [12].

4. Results and analysis

In LES of turbulent flames, the information of turbulent flame surface area is an important issue due to its ability to model the turbulent flame speed. A procedure was introduced in [1] to calculate the surface area of the instantaneous flame front in detail. Figure 3 shows the time history of the summary of the flame surface area. As seen, they vary a lot across the time. It is seen that the total surface area of flame front is slightly larger in the DTF computation for case ϕ = 0.75. The reason for this is not clear. While in case ϕ = 0.83, the results from the REDIM-DTF model and the original DTF model are very close to each other as time goes on.

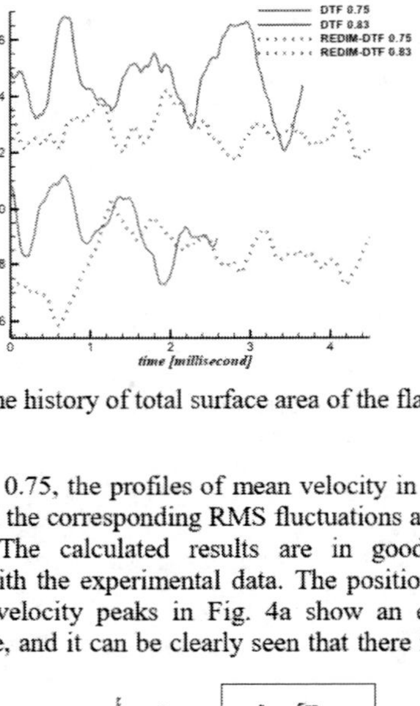

Figure 3: Time history of total surface area of the flame. D^2 = 775.6mm^2.

For case ϕ = 0.75, the profiles of mean velocity in the axial direction and the corresponding RMS fluctuations are shown in Fig. 4. The calculated results are in good overall agreement with the experimental data. The positions of the axial mean velocity peaks in Fig. 4a show an expanded flame surface, and it can be clearly seen that there is a large zone where the velocity belows zero in the middle of the flow field, which is the central recirculation zone (CRZ). There are two shear layers at x = 1.5 mm, one between the annular jet and the surrounding fluid and the other between the jet and the CRZ. Further downstream, the strength of the shear layer becomes weaker, while the fresh reactants are constantly being consumed. The RMS fluctuations of the axial velocity shown in Fig. 4b are also in good agreement with the experimental results, the position of the shear layers is generally predicted to be accurate. At the point near the outlet of cones, the peaks are lower than the experimental results due to the presence of large velocity gradients, so a finer mesh is needed for resloving at that point. Similar observations have been made with the radial and tangential velocity plots, not shown here for conciseness.

Figure 5a and b show the profiles of mean and RMS fluctuations of temperature for case ϕ = 0.83, respectively. As can be seen from the figure, the LES results are in good overall agreement with the experimental data. Further downstream, more fresh reactants are consumed, and both sides of flame front are getting closer. Accordingly, the minimum of mean temperature increases with x. For x > 30mm, the combustion is nearly stopped, so the radial profiles of mean temperature have become smoother. From the peak shift at positions of x = 15mm and 20mm, it can be seen that the experimentally measured flame cone is wider. In the outer region of $x \leq$ 15mm, the temperature predicted by LES is about 400K higher than the experimental result. This is because, during simulation, it is assumed the walls are adiabatic, ignoring the heat loss to the external environment. So this is not the cause of combustion model.

The radial profiles of the RMS fluctuations of temperature shown in Fig. 5b show that in upstream region, the peaks are narrower and the predicted size is slightly higher than the experimental value. Moving downstream, however, the RMS fluctuations of temperature predicted by LES are lower than the experimental measurement, such as the region $x \geq$ 40mm. This is because the predicted flame height is slightly lower than experimental result, so the region x > 40mm is usually in the post flame region, and no flame is found here in LES.

Figure 6a shows the radial profiles of mean mass fraction of CO_2 which is the major species in the post flame regions. It can be seen that the results of LES are in good agreement with the experimental results. The same is true for the profiles of the RMS fluctuations of the CO_2 mass fraction as shown in Fig. 6b, considering the substantial difficulty to capture this quantity correctly.

(a) (b)

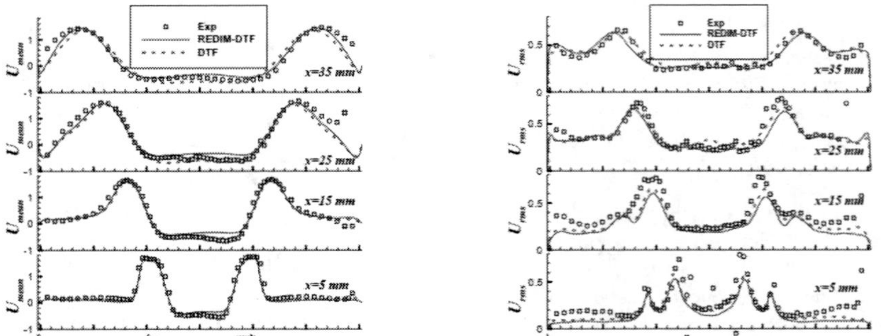

Figure 4: Radial profiles of mean and RMS fluctuations of axial velocity for case ϕ = 0.75. Velocity is normalized by 20m/s.

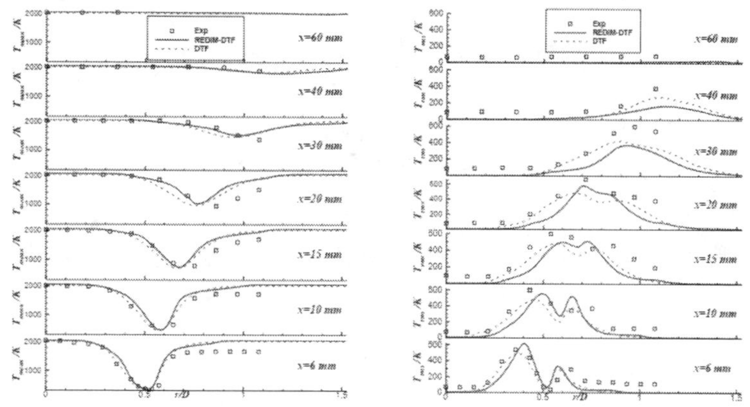

(a) (b)

Figure 5: Radial profiles of mean and RMS fluctuations of temperature for case $\phi = 0.83$.

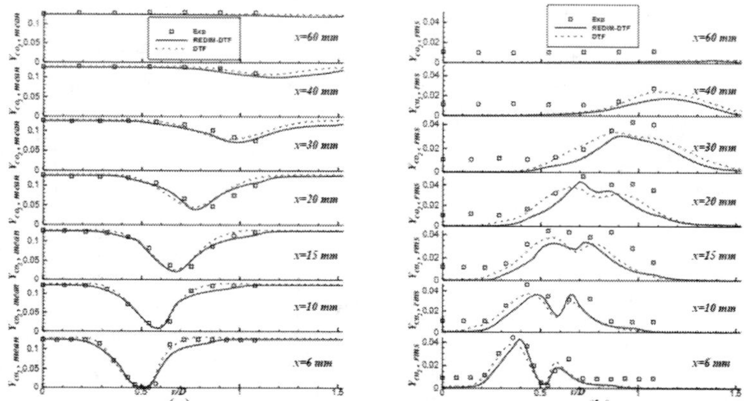

(a) (b)

Figure 6: Radial profiles of mean and RMS fluctuations of the CO_2 mass fraction for case $\phi = 0.83$.

5. Conclusions

In this paper, a new SGS combustion model is proposed by combing the DTF concept and REDIM chemistry table. The REDIM-DTF model is then validated by simulating two lean premixed swirling flames (ϕ = 0.75 and 0.83) in PRECCINSTA combustor. The LES results are checked by comparing with the results obtained with original DTF model and the experimental data. It is shown that the velocity, temperature and CO_2 mass fraction profiles agree well with the experimental results. Additionally, the LES results from the REDIM-DTF model and the original DTF model are very close to each other. However, the computational efficiency of REDIM-DTF is improved, because less species transport equations are solved. In theory, this new model can be applied to simulate more complex flame states, such as partially premixed flames, supposing suitable multi-dimensional REDIM tables are used.

6 Acknowledgement

The support of National Natural Science Foundation of China through project 51576092, the support of Natural Science Foundation of Jiangsu Province through project BK20151344, and the support of PAPD project funded by Jiangsu Higher Education Institution are gratefully acknowledged. All computation is performed on the High Performance Computing Platform of Jiangsu University.

References

[1] P. Wang, J. Fröhlich, U. Maas, Z.X. He, C.J. Wang, Combust. Flame 164 (2016) 329-345.

[2] U. Maas, S.B. Pope, Combust. Flame 88 (1992) 239-264.

[3] V. Bykov, U. Maas, Combust. Theory Model. 11 (6) (2007) 839-862.

[4] V. Bykov, U. Maas, Proc. Combust. Inst. 32 (2009) 561-568.

[5] G. Kelsall, C. Troger, Appl. Therm. Eng. 24 (2004) 1571-1582.

[6] O. Colin, F. Ducros, D. Veynante, T. Poinsot, Phys. Fluids 12 (7) (2000) 1843-1863.

[7] J.P. Légier, T. Poinsot, D Veynante, Proc. of summer program, Center for Turbulent Research, Stanford University. 2000 157-168.

[8] P. Wang, J. Fröhlich, U. Maas, J. Turbul. 11 (11) (2010) 1-19.

[9] L. Selle, G. Lartigue, T. Poinsot, R. Koch, K.-U. Schildmacher, W. Krebs, B. Prade, P. Kaufmann, D. Veynante, Combust. Flame 137 (2004) 489-505.

[10] W. Meier, P. Weigand, R. Giezendanner-Thoben, X.R. Duan, Combust. Flame 150 (2007) 2-26.

[11] P. Wang, J. Fröhlich, U. Maas, In: Proc. of the 5th European Combustion Meeting, Paper 356 Cardiff UK 2011.

[12] P. Wang, N.A. Platova, J. Fröhlich, U. Maas, Int. J. Heat Mass Transf. 70 (2014) 486-495

A Spectral Radius Scaling Semi-Implicit Temporal Scheme for Compressible Reacting Flow with Detailed Chemistry

Qing Xie[1], Zhuyin Ren[1,2*]

[1]School of Aerospace Engineering, Tsinghua University, Beijing, 100084, China

[2]Center for Combustion Energy, Tsinghua University, Beijing, 100084, China

Abstract

A semi-implicit temporal scheme using spectral radius scaling of the chemical Jacobian is proposed for the simulations of compressible unsteady reacting flows with detailed chemistry. The stability and accuracy of the proposed scheme are tested with hydrogen/air combustion using the homogeneous auto-ignition and one-dimensional non-premixed flame and premixed propagating flame. The formal second order of accuracy is demonstrated when the sub-iteration has fully converged. The scheme proves to be stable at a large time step size of 10μs for all the tested cases.

1 Introduction

Time accurate simulation of reacting flow with detailed chemistry is challenging due to the large number of chemical species and the wide range of time scales involved in detailed chemical kinetics [1]. Consequently, many recent efforts have been focused on the development of robust, accurate, and efficient temporal schemes [3]. Among them the semi-implicit scheme has gained significant interest since it is free of the limitation of the minimum reaction time scales compared to the explicit schemes and is superior to the fully implicit schemes because the costly full matrix inversion of the exact chemical Jacobian is avoided. A number of studies focus on the approximation of chemical Jacobian, the triangular approximation has been studied in [5], but the diagonal approximation [6] proves more promising in efficiency. The performance of the diagonal method has been studied using incompressible low Mach flow solvers [8], but it is also essential to assess the accuracy and stability of the diagonal method using the compressible flow solver in high Mach flows.

Here, the performance of a modified diagonal method with enhanced stability is studied using homogeneous auto-ignition cases and 1D laminar premixed and non-premixed flames. And a 1D detonation flow is further simulated to demonstrate the capability of the compressible reacting flow solver.

2 Methodologies

The governing equations for a multi-species compressible reactive system [10] can be written as

$$\frac{\partial \boldsymbol{F}}{\partial t} + \sum_{j=1}^{3} \frac{\partial \boldsymbol{G}_c^j(\boldsymbol{F})}{\partial x^j} = \sum_{j=1}^{3} \frac{\partial \boldsymbol{G}_v^j(\boldsymbol{F})}{\partial x^j} + \boldsymbol{G}_s(\boldsymbol{F}) \quad (1)$$

*Corresponding author: Zhuyin Ren.
E-mail address: zhuyinren@tsinghua.edu.cn

where x^j (j=1,2,3), are the Cartesian coordinates, $\boldsymbol{F} = [\rho, \rho\boldsymbol{u}^T, \rho e_t, \rho Y_k]^T$ are the conserved variables in the form of column vector, ρ is mixture density, $\boldsymbol{u}$ is the velocity vector, $e_t = \sum_{k=1}^{N_s} Y_k(h_k - R_k T) + \frac{1}{2}\boldsymbol{u}^2$ is the specific total energy per unit mass, and Y_k is the mass fraction of species k, h_k is the enthalpy of species k per unit mass, $R_k = \frac{R_0}{W_k}$ is the specific gas constant, R_0 is universal gas constant and W_k is the molecular weight. The convective fluxes $\boldsymbol{G}_c^j(\boldsymbol{F})$ are given by

$$\boldsymbol{G}_c^j(\boldsymbol{F}) = \rho u^j \begin{bmatrix} 1 \\ \boldsymbol{u} \\ h_t \\ Y_k \end{bmatrix} + \begin{bmatrix} 0 \\ p\boldsymbol{e}^j \\ 0 \\ 0 \end{bmatrix} \quad (2)$$

where h_t is the specific total enthalpy of the mixture per unit mass, p is the thermodynamic pressure, and $\boldsymbol{e}^j$ is the unit base vector of the orthogonal Cartesian coordinate system. The viscous and diffusive fluxes $\boldsymbol{G}_v^j(\boldsymbol{F})$ are given by

$$\boldsymbol{G}_v^j(\boldsymbol{F}) = \begin{bmatrix} 0 \\ \boldsymbol{\tau}^j \\ \boldsymbol{u}^T \cdot \boldsymbol{\tau}^j + \boldsymbol{q}^T \cdot \boldsymbol{e}^j \\ -\rho Y_k \boldsymbol{V}_k^T \cdot \boldsymbol{e}^T \end{bmatrix} \quad (3)$$

where $\boldsymbol{\tau}^j = \boldsymbol{\tau} \cdot \boldsymbol{e}^j$, $\boldsymbol{\tau}$ is the viscous stress tensor, $\boldsymbol{q}$ is the heat flux, $\boldsymbol{V}_k$ is the diffusion velocity of species k. The reactive source terms $\boldsymbol{G}_s(\boldsymbol{F})$ are

$$\boldsymbol{G}_s(\boldsymbol{F}) = \begin{bmatrix} 0 \\ 0 \\ 0 \\ \omega_k \end{bmatrix} \quad (4)$$

where ω_k is the mass production rate per unit volume of the kth species due to chemical reactions. The constitutive relations for $\boldsymbol{\tau}$, $\boldsymbol{V}_k$ and $\boldsymbol{q}$ can be found in [10]. Note that Soret and Dufour effects, body force and radiative heat transfer are ignored for simplicity. The species thermodynamic properties, and detailed chemical kinetics for species reaction rates are evaluated through CHEMKIN package [11]. The mixture-averaged approach is adopted for transport coefficients.

The spatial discretization of convective terms is achieved through the HLLC Riemann solver [12] and the third order MUSCL scheme [13] is used to provide a higher spatial resolution. To avoid spurious oscillations, the reconstruction is carried out through characteristic variables [14]. The viscous and diffusive terms are discretized by the second order central difference scheme.

The temporal discretization of the spatially discretized equations is achieved through the Crank-Nicolson scheme. The linear iterative scheme is based on the widely used LUSGS [15] scheme, whose Jacobian coefficient matrix is mainly contributed by the convective terms

$$\Delta\left(\tilde{\boldsymbol{G}}_c^{i+\frac{1}{2}}\right)^{k+1} \approx \left(\boldsymbol{M}_{i+1}^k\right)^- \Delta\boldsymbol{F}_{i+1}^{k+1} + \left(\boldsymbol{M}_i^k\right)^+ \Delta\boldsymbol{F}_i^{k+1} \quad (5)$$

where $\boldsymbol{M} = \frac{\partial G_c}{\partial F}$, $\boldsymbol{M}^{\pm} = \frac{\boldsymbol{M} \pm \sigma \boldsymbol{I}}{2}$. And σ is the spectral radius of $\boldsymbol{M}$, which is vital for the stability of LUSGS.

In current work, inspired by Coakley [16] and Savard et al. [8], the spectral radius scaling is applied to the Jacobians of viscous/diffusive terms and reactive source terms. Since spectral radius scaling results in spherical matrix, all the influence of viscous/diffusive and reactive terms can be lumped together with the convective terms into σ.

The contribution of convective terms to σ is

$$\sigma_C = |\boldsymbol{u}| + a \quad (6)$$

where a is the local speed of sound.

The contribution of viscous/diffusive terms can be evaluated based on the numerical time scales

$$\tau_\mu = \frac{\rho \Delta x^2}{\mu},\ \tau_\lambda = \frac{\rho c_p \Delta x^2}{\lambda},\ \tau_D = \frac{\Delta x^2}{MAX(D_{k,m})} \quad (7)$$

$$\sigma_{VD} = \frac{2\Delta x}{MIN(\tau_\mu, \tau_\lambda, \tau_D)} \quad (8)$$

where μ is shear viscosity, c_p is heat capacity, λ is thermal conductivity and $D_{k,m}$ is the diffusion coefficient of species k with respect to the mixture.

The contribution of reactive source terms to σ is

$$\sigma_R = MAX(\tau_k^{-1})\Delta x,\ \tau_k = \frac{c_k}{\omega_k^D} \quad (9)$$

where ω_k^D, c_k and τ_k are the mole destruction rate, mole concentration and lifetime of species k respectively as in [8]. Then the spectral radius σ accounting for convection diffusion and chemical reactions is expressed as

$$\sigma = \sigma_C + \sigma_{VD} + \sigma_R \quad (10)$$

3 Numerical Tests

3.1 Test Cases

Simulations of homogeneous auto-ignition, 1D laminar diffusion flame, 1D laminar freely propagating flame and 1D detonation flow are performed using the hydrogen mechanism with 9 species and 21 reactions [17].

The constant-pressure auto-ignition simulations are performed with stoichiometric equivalence ratio ϕ=1.0, an initial pressure and temperature of P_0=2atm and T_0=1000K respectively.

The 1D laminar diffusion flame is initialized with uniform pressure and temperature of P_0=2atm and T_0=1000K, respectively at quiescent condition. The initial spatial distributions of the mixtures consist of two different uniform regions separated by a discontinuity, the left side of the discontinuity consists of air with the mole fraction ratio of $X_{O_2}:X_{N_2}$=0.21:0.79, and the right side consists of H_2 and N_2 with the mole fraction ratio of $X_{H_2}:X_{N_2}$=0.42:0.79. The grid spatial resolution is Δx=3.1μm and the overall simulation time is t_s=0.7ms. Both sides of the domain are set to be outflow boundary condition.

The 1D laminar premixed propagating flame is initialized from the incompressible CHEMKIN solution with a stoichiometric hydrogen/air of unburnt temperature 300K and pressure 1atm. The left side of the domain (unburnt) is set to be outflow boundary, and the right side is set to be adiabatic wall boundary. The grid spatial resolution is Δx=3.75μm.

The 1D detonation flow is initialized with the inflowing mixture with temperature, velocity and pressure of 1200K, 1.6×10^3m/s and 1atm respectively. The mixture consists of H_2, O_2 and N_2 with $X_{H_2}:X_{O_2}:X_{N_2}$ =0.3:0.15:0.55. The right boundary is set to be outflow. The grid resolution is Δx=12μm.

3.2 Numerical Stability

The stability analysis of the proposed scheme is performed using the linear method as in [8], where only the chemical Jacobian is considered. The stability matrix can be written as

$$\widehat{\boldsymbol{A}} = \boldsymbol{I} - (\hat{\boldsymbol{J}})^{-1}\left(\frac{1}{\Delta t}\boldsymbol{I} - \frac{1}{2}\boldsymbol{J}_e\right) \quad (1)$$

where $\hat{\boldsymbol{J}} = \frac{1}{\Delta t}\boldsymbol{I} - \frac{1}{2}\boldsymbol{\Lambda}$, $\boldsymbol{J}_e = \frac{1}{\Delta t}\boldsymbol{I} - \frac{1}{2}\boldsymbol{J}_{chem}$, and $\boldsymbol{\Lambda}$ is the diagonal approximation of the exact chemical Jacobian $\boldsymbol{J}_{chem}$. For the original approximation in [8], each entry of $\boldsymbol{\Lambda}$ is calculated using the corresponding species lifetime τ_k. In contrast, for the spectral radius scaling method, all the entries are calculated using the minimum species lifetime $MIN(\tau_k)$.

The maximum spectral radius of $\widehat{\boldsymbol{A}}$, i.e., σ_D from the original approximation in [8] and σ_S using the spectral radius scaling method are shown in Fig. 1. It is shown when Δt is small, there is no difference between σ_S and σ_D, indicating almost identical convergence performance of the two methods. More interestingly, when Δt is large, σ_D yields large excess of unity while σ_S remains bounded, indicating improved stability of the scaling method. It is also noted that the maximum spectral radius is close to unity when Δt is approximately two orders of magnitude larger than the minimum species lifetime (10^{-10}s for auto-ignition test, and 10^{-9}s for diffusion and propagating flame tests), and this feature can be used to estimate the performance limit of the proposed method.

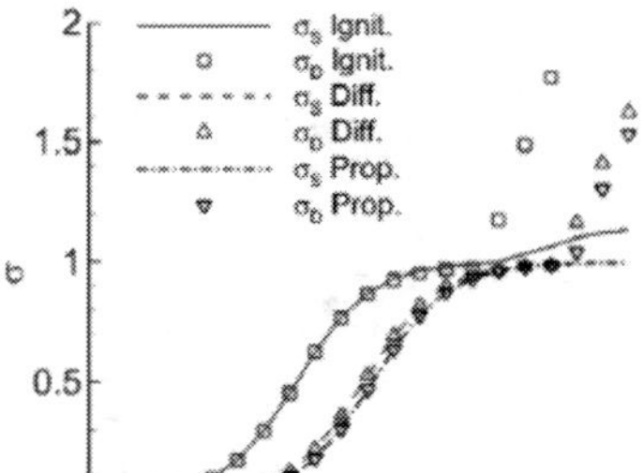

Figure 1: The spectral radius of $\widehat{\boldsymbol{A}}$ against time step size Δt in auto-ignition (Ignit.), 1D diffusion flame (Diff.) and 1D propagating flame (Prop.) testing cases.

The actual stability limits of the semi-implicit scheme are numerically tested in auto-ignition, diffusion flame, and propagating flame cases. The stability limit is defined as the maximum numerical time step size for which the solution remains bounded. For auto-ignition cases the solution refers to the whole ignition process. For diffusion flame and propagating flame cases the solution refers to one step simulation that is initialized from the previously converged solution. The one time step advancement is regarded to be converged if the sub-iterations can decrease the maximum residual of total energy ρe_t by three orders of magnitude.

The theoretical and the numerically determined stability limits of the semi-implicit scheme are listed in Table 1. Note that the theoretical stability limit is determined by σ_D using the original diagonal method, while the numerical limit is determined using the scaling method. As shown, the proposed semi-implicit scheme using scaling method proves to be stable at a large time step size e.g., Δt=1.0×10^{-5}s in all the test cases. Actually the stability limit of the proposed scheme should be larger than 1.0×10^{-5}s. It is worth mentioning that although stable with this large time step size, all the cases need thousands of sub-iterations N_{iter} for the maximum residual of ρe_t being reduced by three orders of magnitude since σ_S is very close to unity and it is computationally inefficient.

Table 1: The theoretical and numerical stability limits of the semi-implicit scheme.

	Numerical	Theoretical
Auto-Ignition	>1.0×10^{-5}s	2.9×10^{-7}s
Diffusion Flame	>1.0×10^{-5}s	2.2×10^{-6}s
Propagating Flame	>1.0×10^{-5}s	2.4×10^{-6}s

3.3 Numerical Accuracy

The impact of the magnitude of sub-iteration residual on the accuracy of temperature profile is first studied using auto-ignition cases. Then, the order of accuracy of the proposed semi-implicit scheme is demonstrated in the 1D diffusion flame and propagating flame.

Figure 2 shows the temperature profiles from auto-ignition obtained under the constraints that the sub-iteration residual of temperature decreases by one, two and three order of magnitude e.g., TOL=10^{-1},10^{-2},10^{-3}. Results from both the scaling method and the original diagonal method are displayed and compared to the exact ones obtained with DVODE [18].

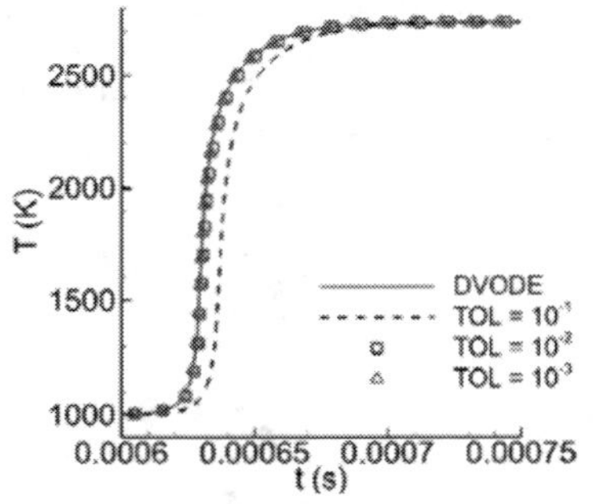

(a) Using the scaling method

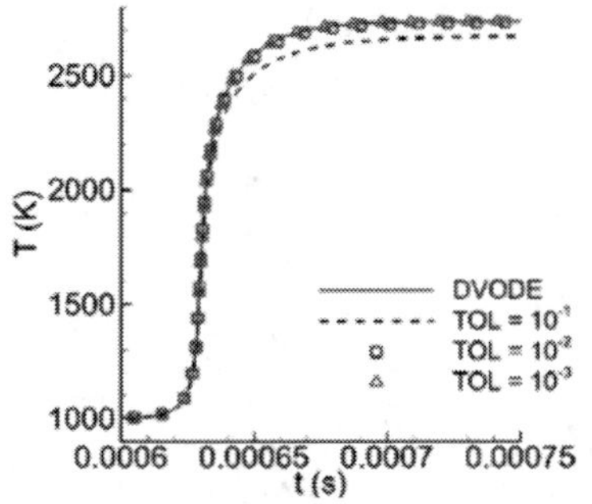

(b) Using the original diagonal method

Figure 2: The temperature profiles from auto-ignition obtained under different constraints on temperature residual, Δt=10^{-7}s.

As shown, both the scaling method and the original diagonal method obtain the accurate solutions when the residual decreases by two or more order of magnitude. But the solutions from the original diagonal method are unable to converge to the correct equilibrium states when the convergence of sub-iteration is not sufficient, see Fig. 2b.

To study the temporal order of accuracy, the solutions $f_{\Delta t}^{\tau}$ are marched to the same time τ=4×10^{-6}s using different time steps Δt, and the pseudo-error is defined as

$$\varepsilon_{\Delta t} = \left| f_{\Delta t}^{\tau} - f_{\Delta t/2}^{\tau} \right| \tag{11}$$

When $\Delta t \to 0$, the plot of $\log(\Delta t) \sim \log(\varepsilon_{\Delta t})$ tends to be a straight line and the slope will be the formal order of the temporal scheme.

For the premixed propagating flame, f is chosen as the spatial position of the flame x_f with T=1700K.

For the diffusion flame, the pseudo-error is defined as the spatial L$_1$ error of f

$$\varepsilon_{\Delta t} = \frac{1}{p} \sum_{i=1}^{p} \left| f_{\Delta t}^{\tau} - f_{\Delta t/2}^{\tau} \right|_i \tag{12}$$

where i denotes the index of spatial locations. And f can be chosen as density, temperature, and mass fractions.

Figure 3 shows the numerical errors against the time step size for the semi-implicit scheme in the flame tests. As shown, the formal second order of accuracy of the proposed scheme is demonstrated.

Note that the convergence of sub-iteration is ensured for these accuracy tests by decreasing the maximum residual of ρe_t by eight orders of magnitude.

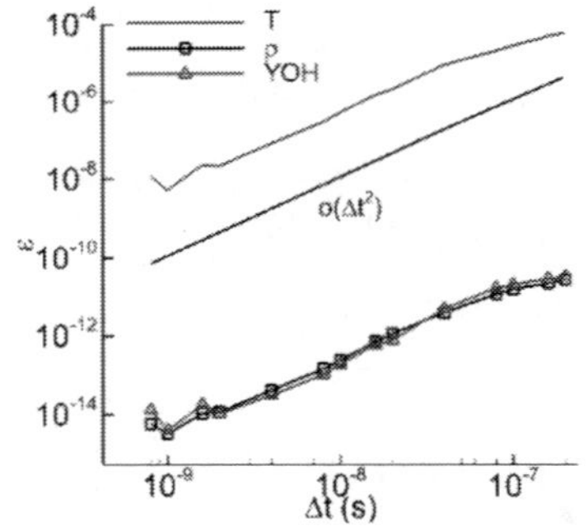

(a) Non-premixed diffusion flame

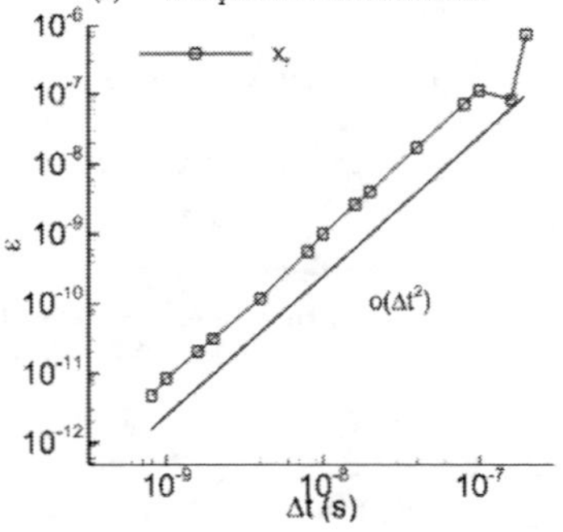

(b) Premixed propagating flame

Figure 3: The numerical errors against the time step size of the proposed scheme in 1D flame tests.

3.4 Detonation

The proposed semi-implicit scheme has been applied to

simulate the fully developed detonation wave as shown in Fig. 4. The detonation is induced by the auto-ignition process downstream of the flow. The propagating speed of the fully developed detonation wave relative to the inflow is 3297.7 m/s.

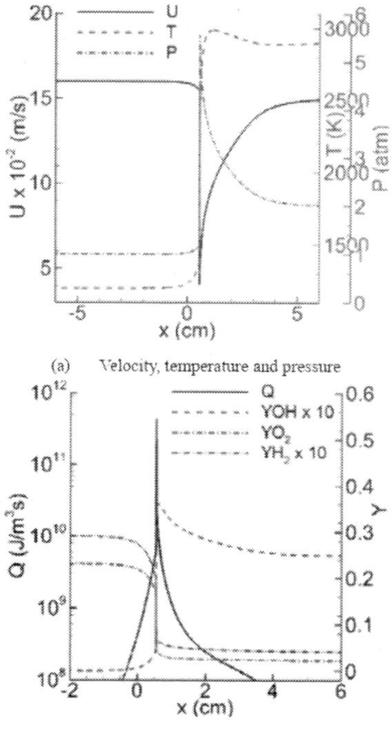

(a) Velocity, temperature and pressure

(b) Heat release rate and mass fractions

Figure 4: Solution profiles of the fully developed detonation flow with the semi-implicit scheme with Δt=10^{-8}s.

4 Conclusions

In this work, an improved diagonal approximation of the chemical Jacobian is developed for unsteady compressible reacting flow simulation. The proposed semi-implicit scheme proves to be stable at large time step size as 10μs. And the formal second order of accuracy of the scheme is demonstrated. The capability of the compressible flow solver using the proposed temporal scheme has been demonstrated in detonation flow simulation.

5 Acknowledgments

The work conducted in Tsinghua University is supported by National Natural Science Foundation of China 91441202 and 51476087. The simulations are performed with the computational resources from the Tsinghua National Laboratory for Information Science and Technology.

References

References

[1]. Law C K. Combustion at a crossroads: Status and prospects[J]. Proceedings of the Combustion Institute, 2007, 31(1): 1-29.

[2]. Lu T, Law C K, Yoo C S, et al. Dynamic stiffness removal for direct numerical simulations[J]. Combustion and Flame, 2009, 156(8): 1542-1551.

[3]. Yoh J J, Zhong X. New hybrid Runge-Kutta methods for unsteady reactive flow simulation[J]. AIAA journal, 2004, 42(8): 1593-1600.

[4]. Yungster S, Radhakrishnan K. A fully implicit time accurate method for hypersonic combustion: application to shock-induced combustion instability[J]. Shock Waves, 1996, 5(5): 293-303.

[5]. Kim S L, Choi J Y, Jeung I S, et al. Application of approximate chemical Jacobians for constant volume reaction and shock-induced combustion[J]. Applied numerical mathematics, 2001, 39(1): 87-104.

[6]. Eberhardt S, Imlay S. Diagonal implicit scheme for computing flows with finite rate chemistry[J]. Journal of Thermophysics and Heat Transfer, 1992, 6(2): 208-216.

[7]. Ju Y. Lower-upper scheme for chemically reacting flow with finite rate chemistry[J]. AIAA journal, 1995, 33(8): 1418-1425.

[8]. Savard B, Xuan Y, Bobbitt B, et al. A computationally-efficient, semi-implicit, iterative method for the time-integration of reacting flows with stiff chemistry[J]. Journal of Computational Physics, 2015, 295: 740-769.

[9]. MacArt J F, Mueller M E. Semi-implicit iterative methods for low Mach number turbulent reacting flows: Operator splitting versus approximate factorization[J]. Journal of Computational Physics, 2016, 326: 569-595.

[10]. Law C K. Combustion physics[M]. Cambridge university press, 2010. p 157.

[11]. Kee R J, Rupley F M, Miller J A. Chemkin-II: A Fortran chemical kinetics package for the analysis of gas-phase chemical kinetics[R]. Sandia National Labs., Livermore, CA (USA), 1989.

[12]. Toro E F, Spruce M, Speares W. Restoration of the contact surface in the HLL-Riemann solver[J]. Shock waves, 1994, 4(1): 25-34.

[13]. Van Leer B. Towards the ultimate conservative difference scheme. V. A second-order sequel to Godunov's method[J]. Journal of computational Physics, 1979, 32(1): 101-136.

[14]. Gerolymos G A, Sénéchal D, Vallet I. Very-high-order WENO schemes[J]. Journal of Computational Physics, 2009, 228(23): 8481-8524.

[15]. Yoon S, Jameson A. Lower-upper symmetric-Gauss-Seidel method for the Euler and Navier-Stokes equations[J]. AIAA journal, 1988, 26(9): 1025-1026.

[16]. Coakley T J. Implicit upwind methods for the compressible Navier-Stokes equations[J]. AIAA journal, 1985, 23(3): 374-380.

[17]. Li J, Zhao Z, Kazakov A, et al. An updated comprehensive kinetic model of hydrogen combustion[J]. International journal of chemical kinetics, 2004, 36(10): 566-575.

[18]. Brown P N, Byrne G D, Hindmarsh A C. VODE: A variable-coefficient ODE solver[J]. SIAM journal on scientific and statistical computing, 1989, 10(5): 1038-1051.

Replies to Comments

Reviwer #1: A clearly written and concise paper.

Reply: Thanks for the patience reviewing this paper.

Reviwer #2: Nice advancement to currently available semi-implicit temporal schemes for some challenging (ch emically) compressible flow dynamics. The following paper should also be cited amongst those related to the robust semi-implicit time integrators developed for chemically reactive flow systems:

J. J. Yoh, X. Zhong, "New Hybrid Runge-Kutta Methods for Unsteady Reactive Flow Simulation," AIAA Journal, 2004, Vol.42: 1593-1600

Reply: Thanks for reminding, this paper has been cited as the third reference paper, which has extended the authors' knowledge of time integration schemes.

11th Asia-Pacific Conference on Combustion,
The University of Sydney, NSW Australia
10th -14th December 2017.

Evolution of Surface Elements of Premixed Flames in Turbulence

Abinesh Mohan[1], Himanshu L.Dave[1], Swetaprovo Chaudhuri[1*]

[1]Dept. of Aerospace Engineering, Indian Institute of Science, Bangalore, Karnataka, 560012, India

Abstract

Direct numerical simulations are performed to generate two different cases of statistically planar turbulent premixed flames. Flame particles, uniformly distributed on a selected isotherm of the flame, are tracked backward in time using Backward Flame Particle Tracking (BFPT) algorithm. Using BFPT, we identify specific regions of an isotherm which would disperse to span the complete surface at a later time. To study the generation process, finite sized triangular elements are created using these flame particles as their vertices and tracked forward in time. A kinematic evolution equation, used in non-reacting turbulent flows, is modified to describe the time evolution of these triangles. Evolution of the triangles is nearly equivalent to that of the portions of the flame surface they represent. This modified equation is used to derive a stretch rate equation for flame surface elements with edge vectors whose mutual orthogonality is not necessarily required. Size and shape of the triangles are measured to quantify flame surface generation process. Based on these observations, the current state of the flame is found to depend on both the instantaneous quantities and flow-flame parameters at an earlier time. Generality of observations is verified by similarity of results from both the DNS cases.

1 Introduction

Lagrangian methods have been used to study material line element and material surface element stretching [1,2], alignment characteristics of eigenvectors of strain-rate tensor [3], and particle-pair dispersion. In these methods, fluid particles, described as mass-less particles that always move with the local fluid velocity vector, are tracked. Similarly, Lagrangian Flame Particle Tracking (FPT) of flame surfaces have been shown to be quite insightful in explicating flow-flame parameters responsible for flame surface annihilation [4], and the non-linear behavior of displacement flame speed (S_d) with respect to curvature [5]. A flame is composed of a collation of isoscalar surfaces. In the present context, the corresponding scalar variable is an active scalar whose behavior is dependent upon the chemical properties of the reactant mixture and the multi-scale velocity field of the turbulent flow. Here, an isotherm is chosen to study the kinematics of premixed turbulent flames with the aid of flame particles. Flame particles are mathematical points defined on an isotherm, and are constrained to remain on their resident isotherm while co-moving with it. As the flame particles move with the isotherm, Eulerian properties such as strain-rate, curvature, flame surface normal, flow velocities, and S_d can be interpolated onto them. Such a procedure provides not only the instantaneous snapshots of these properties, but also their time histories. The current study is focused on a statistically stationary turbulent premixed flame ($\phi = 0.81, Le = 0.84$) whose time averaged surface area remains constant, in a statistical sense. To maintain a statistically constant area, both surface generation and annihilation processes should balance each other. The destruction processes have been explained in earlier works [4,5]. In this article, we investigate the generation process of a flame surface by studying an isotherm very close to the reactant mixture with the help of BFPT and evolution of finite sized triangular elements. The evolution of these triangular elements is described by a first order ordinary differential equation, which is then used to derive a modified form of stretch rate equation for flame surface elements. An expression for mass consumption rate of reactants using the modified stretch rate equation shows the significance of leading regions in determining the characteristics of a turbulent premixed flame. Recent investigation [6] supports this outcome through time correlation between leading edge speed and turbulent consumption speed.

2 Methodology and Techniques

2.1 Backward Flame Particle Tracking (BFPT)

The motion of a flame particle forward in time is given by the following formula [7],

$$\frac{d\vec{X}^F}{dt} = \vec{V}^F = \vec{U}^F + \hat{n}^F S_d \tag{1}$$

where, $\vec{X}^F$ is the particle position, $\vec{U}^F$ is the flow velocity, and $\hat{n}^F$ is the flame surface normal. All the parameters in the rhs are evaluated at time t to compute the particle position at $t + \Delta t$. The presence of S_d differentiates propagating surfaces from material surfaces where $S_d = 0$. With a uniform distribution of flame particles on the isotherm, (1) is numerically integrated forward in time. Unlike fluid particles, flame particles possess a finite life-span, i.e., they cease to exist once their resident flame surface gets annihilated. So, forward tracking of flame particles can be used to investigate flame surface annihilation processes. Generation process of a flame surface cannot be understood through forward tracking of a reasonably finite number of flame particles, hence paving the way for backward tracking. To illustrate the advantages of BFPT over FPT, consider a scenario where the time history of a small portion of a flame surface from time t_0 to t, such that $t_0 < t$, is to be analyzed. In FPT, since the exact locations of the region of interest are not known a priori, particles must be arbitrarily introduced on the surface. Due to flame surface annihilation processes, a large number of particles do not reach that region. In case of BFPT, the exact portion of the flame surface under study is tagged with flame particles, and these particles are tracked backwards to time t_0. Thus, the Lagrangian time history of that portion of the flame surface is

*Corresponding author; E-mail address:
swetaprovo@gmail.com; schaudhuri@aero.iisc.ernet.in

obtained using BFPT, in addition to reduction in computational time in terms of the number of particles tracked.

The BFPT algorithm for the present work has been developed for tracking flame particles backward in time. The method has been adapted from [8] and modified to constrain the particles to always move with their resident isotherm. Flame particles are uniformly but randomly distributed on the chosen isotherm at time t_f. The particle trajectories are then integrated backward in time using (1), but with sign of the time interval reversed resulting in

$$\vec{X}^F_{n-1} = \vec{X}^F_n + \vec{V}^F(\vec{X}^F_{n-1}, t_{n-1})(-\Delta t) \quad (2)$$

where, n corresponds to the current time and Δt is the time interval. Due to its dependence on unknown velocity at $\vec{X}^F_{n-1}$, (2) cannot be explicitly solved. In order to solve (2), two stages of particle position approximation are introduced: (i) estimation, and (ii) correlation. In the estimation stage, the particle is displaced using its instantaneous fluid velocity and time interval Δt. From this intermediate location, a pair of rays are shot in the $\pm\hat{n}^F$ directions. A ray-face intersection algorithm [9] is used to find intersection(s) of these rays with the isotherm, but at an earlier time $t_{n-1}(= t - \Delta t)$ as shown in Fig.1a. Thus, an estimated location $\vec{X}^{F*}_{n-1}$ is found using the velocity $\vec{V}^F(\vec{X}^F_n, t_n)$.

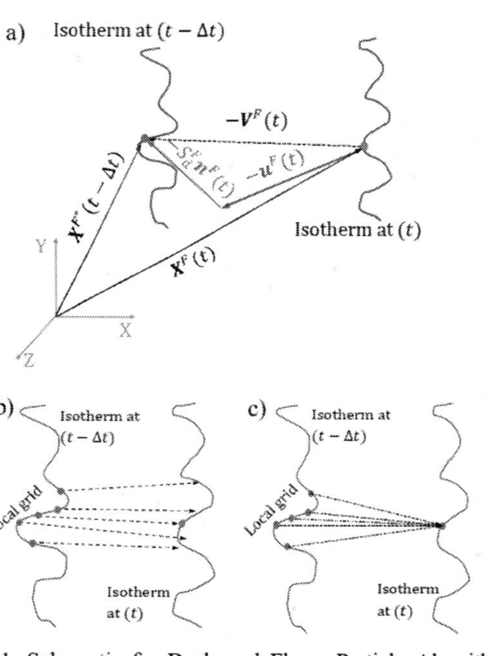

Figure 1. Schematic for Backward Flame Particle Algorithm (BFPT) on a cross-section of the flame surface. a) Estimation step: The tentative flame particle location at an earlier time is estimated; b) Local grid is created around the tentative particle position and projected forward to create array **A**; c) Artificial velocity vectors created at the local grid to generate array **B**. $(t - \Delta t)$ corresponds to $(n - 1)$ and t corresponds to n in (2).

In the correlation stage, a local grid on the isotherm at $(t - \Delta t)$ is generated. These local grid points represent the most probable candidates for identifying $\vec{X}^F_{n-1}$. $\vec{V}^{F*}$ is evaluated at the local grid by interpolation and the array **A** is formed using these velocity vectors. A second array, denoted as **B**, is formed with vectors joining these grid points with the flame particle $\vec{X}^F_n$. These artificial velocity vectors would take the grid points to $\vec{X}^F_n$. A correlation vector **Q** is computed to evaluate the similarity between the two matrices. It is defined as

$$Q^j = \frac{A^j . B^j}{\max\{|A^j|, |B^j|\}^2} \quad (3)$$

where, j is the grid point index.

A correlation coefficient of value unity corresponds to the actual particle under question $(\vec{X}^F_{n-1})$. Due to finite resolution of the local grid considered, and large local velocity gradients, the correlation value will not generally be equal to 1. A grid point that produces a value closest to 1 among others in its set will be chosen as $\vec{X}^F_n$. To quantify the uncertainties introduced, the BFPT tracked particles are used as initial condition for FPT. On comparison, the magnitude of uncertainty between BFPT and FPT results normalized by the laminar flame thickness is found to be significantly less than unity, thereby validating the backward tracking algorithm.

3 Results and Discussion

Only DNS case-I $(Re_\lambda = 85)$ results are presented, while case-II $(Re_\lambda = 98.8)$ results are qualitatively similar. An isotherm close to the reactants, T_0=350K, is selected for BFPT analysis. Flame particles are uniformly but randomly distributed on this isotherm at time t_f, and tracked backward in time to t_i; $(t_i < t_f)$. To investigate the evolution of flame surface elements, these particles are used to form triangles using Delaunay triangulation; Delaunay triangulation is used because it minimizes the generation of highly skewed triangles. These triangular elements represent portions of the flame surface between their vertices. Displacement of the flame particles causes the geometry of the triangles to change, and hence represent corresponding changes to the flame surface. All analyses henceforth are forward in time, i.e., from time t_i to t_f. Figure 2 shows the evolution of flame particles and triangles on the selected isotherm at initial and final tracking times. It can be observed that the flame particles form spatially distinct clusters at time t_i. With progress in time, the triangles along with the isotherm, evolve due to flame-turbulence interactions. Periodic boundary conditions in YZ-plane are implemented in backward tracking algorithm. In this method, additional grid points are added to the computational domain in the transverse directions. Once the flame particles leave the domain from one of the transverse directions, they are forced to reappear from the corresponding opposite face of the domain. A similar operation for triangles will cause its size to rampantly vary. Figure 2b shows that some portions of the flame surface are apparently not spanned by the triangles. This is a visual artifact caused by the periodic boundary condition.

If $\vec{x}_i$ represents position vector of each of the vertices, a triangle can be defined by reduced vectors $(\vec{\rho}^1, \vec{\rho}^2)$ as given in [10],

$$\vec{\rho}^1 = \frac{\vec{x}_2 - \vec{x}_1}{\sqrt{2}}, \ \vec{\rho}^2 = \frac{2\vec{x}_3 - \vec{x}_2 - \vec{x}_1}{\sqrt{6}} \quad (4)$$

$$R = \sqrt{(\vec{\rho}^1)^2 + (\vec{\rho}^2)^2}, \ A = \frac{\sqrt{3}}{2} | \vec{\rho}^1 \times \vec{\rho}^2 | \quad (5)$$

In (5), R is the radius of gyration and A is the area of a triangle. The evolution of the triangles can be described with the aid of reduced vectors and evolution equation [11],

$$\frac{d\rho^\alpha_i}{dt} = \rho^\alpha_j M_{ij} + u^\alpha_i \quad (6)$$

The idea behind (6) is that the velocity field influencing the triangles can be decomposed [11] into three parts: scales smaller than R, scales comparable to R, and scales larger than R. The larger scales simply advect the triangles and do not significantly contribute to their geometric changes.

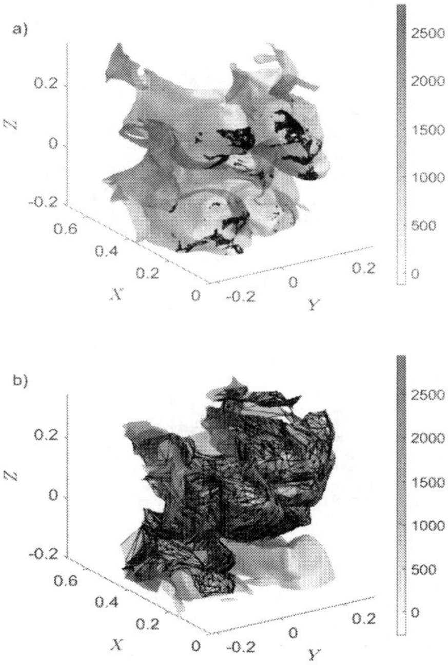

Figure 2: Evolution of Delaunay algorithm generated triangles at times (a) t_i, (b) t_f on T_0=350K isotherm. The axes units are in cm. Colorbar represent x-component of fluid velocity ($cm\ s^{-1}$).

Gradient of the velocity given in (1) is evaluated at the vertices of the triangles and averaged over each of the three vertices of a triangle to form the coarse grained strain-rate tensor **M**. As the size of triangle diminishes, **M** tends to the true velocity gradient tensor $\vec{\nabla}\vec{V}^F$. It is found that the triangles are predominantly below the inertial range of scales through the entire tracking duration, which in turn is smaller than the integral time-scale of the flow. In this case, only the first term on rhs of (6) causing large shape distortions will be dominant, as discussed in [11]. As coherent straining of the triangles is marked by an increase in distortion of the shapes [12,13], the shape of the triangles are also tracked. Triangle shapes are expressed by eigenvalues of a moment-of-inertia like tensor **G**. Its eigenvalues (I_1, I_2; $I_1 \geq I_2$) are normalized by R^2; normalization assures $I_1 + I_2 = 1$. I_2 measures the lateral extent of a triangle with respect to its radius of gyration. A small value of I_2 corresponds to large distortion of the triangle. On the other hand, an additive noise, represented by the second term in (6), opposes the distortion generating multiplicative action of the coarse grained velocity gradient tensor, thereby leading to larger values of I_2. We examined the mean of I_2 conditioned on the initial size of the triangles. The triangles are segregated into three separate bins to study the shape statistics: $\langle R_i \rangle \leq \eta, \eta \leq \langle R_i \rangle \leq 2\eta, 2\eta \leq \langle R_i \rangle \leq 3\eta$, where R_i is the radius of gyration at time t_i. Time evolution of $\langle I_2 | R_i \rangle$ exhibited a decrease for all the triangles, signifying the dominance of the first term in (6). Thus, the second term $\vec{u}$ is neglected on the basis that small scale white noise is not significant during the initial stages of pair separation between two particles; this yields an approximate evolution equation with respect to (6). The large distortions of smaller triangles can be related to the dynamics of the local velocity gradients.

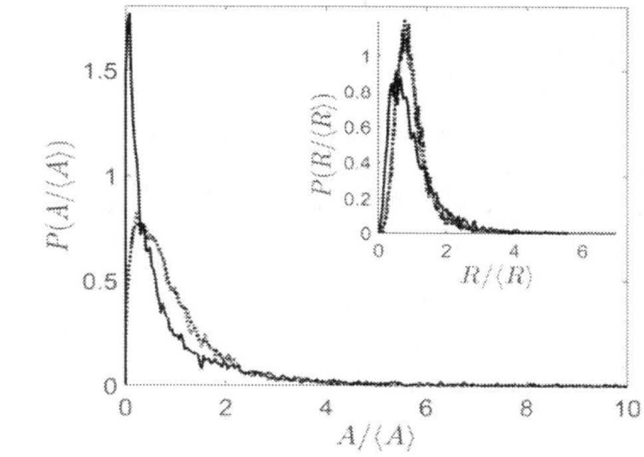

Figure 3: Probability density function (pdf) of A normalized by its mean value. Inset shows pdf of R normalized by its mean. Red line represents the parameters found using (6); Black line represents the parameters evaluated using (5). Solid lines correspond to time t_i, and dotted line correspond to t_f

Availability of the time series of **M** allows the computation of the reduced vectors by integrating (6), after neglecting the second term, forward in time using an Runge-Kutta 5(4) method [14]. Initial condition for $\vec{\rho}$ is evaluated using the triangles at t_i, and the time-step is maintained same as that for backward tracking. It turns out that the kinematics of the triangles, which represent portions of a flame surface, can be described by a simple differential equation, provided the time history of components of **M** are known. Results of the simplified (with $u_i^\alpha = 0$) evolution equation (6) is found to satisfactorily match with the size statistics of the actual triangles evolved by particle tracking. Comparison of R and A evaluated using (5) and that using the simplified form of (6) are shown in Fig.3. From Fig.3, the tail portions of pdf of normalized radius of gyration and area computed from (5) are not captured effectively by the model. However, the magnitude of pdf tails are found to be rather insignificant. In addition, the scale decomposition of the velocity field is not as straight-forward as in non-reacting flows due the presence of the propagation term in the velocity $\vec{V}^F$.

A joint pdf of R and A (not shown) exhibited a positive correlation between these quantities at both t_i and t_f, indicating that either of the quantities can be used to describe the triangle size. Here, evolution of A is considered. As the simplified form of (6) is found to closely represent the evolution of the triangles, it can be used to compute their fractional rate of increase in area, i.e., the stretch rate. The stretch rate equation is given as,

$$K = \frac{1}{A}\frac{dA}{dt} = \left(\delta_{ij} - n_i n_j\right)\frac{\partial u_i}{\partial x_j} + 2S_d\kappa_m \quad (7)$$

where, A is the area of an infinitesimal flame surface element, n is the unit surface normal, κ_m is the curvature of the flame surface at the location of interest. Equation (7) is derived [15,16] by considering mutually perpendicular edges of an infinitesimal flame surface element. With evolution of the surface elements in space and time, the orthogonality constraint will not be guaranteed. In order to evaluate the time history of stretch rate, the following equation is derived,

$$\frac{1}{A}\frac{dA}{dt} = \frac{1}{\sin^2\theta}\{\pi_1 + \pi_2 + \pi_3 + \pi_4\} \quad (8)$$

where, $\pi_1 = \hat{\rho}_i^1 \hat{\rho}_j^1 M_{ij}$; $\pi_2 = -(\hat{\rho}_i^2 \hat{\rho}_j^1 M_{ij})\hat{\rho}_l^1 \hat{\rho}_l^2$

$\pi_3 = \hat{\rho}_i^2 \hat{\rho}_j^2 M_{ij}$; $\pi_2 = -(\hat{\rho}_i^1 \hat{\rho}_j^2 M_{ij})\hat{\rho}_l^1 \hat{\rho}_l^2$

where, θ is the angle between the unit reduced vectors $\hat{\rho}^1$ and $\hat{\rho}^2$. Inclusion of θ relaxes the orthogonality constraint. The $\hat{\rho}_i^\alpha \hat{\rho}_j^\beta M_{ij}$ term represents projection of the coarse grained strain rate tensor on the plane spanned by the two unit vectors $\hat{\rho}^\alpha, \hat{\rho}^\beta$. The equation has two limiting cases: i) if $\theta = 0$ or $\theta = \pm\pi$, the degenerate case where the two reduced unit vectors are parallel to each other; ii) if $\theta = \pi/2$, the equation reduces to (7). Equation (8) gives the evolution of the fractional rate of change of area of the triangles, and hence that of the portion of flame surface they represent. Solution to (8) is given by the following analytical form expressing the time dependent area of each triangle,

$$A = A_0 e^{\int f(t)dt} \tag{9}$$

where, $f(t)$ is the rhs of (8), A_0 is the area of the corresponding triangle at t_i, and the integration in the exponential is performed from time t_i to t_f. Implicit in the above equation is the Lagrangian time history of the triangles and the components of $\mathbf{M}$. Trapezoidal rule is used to evaluate the integral in the exponential. Comparison of area of the triangles evaluated using (5) and that using (9) exhibited a very close match, thereby validating (8).

Finally, the area of flame surface can be used to evaluate the mass consumption rate of the reactant mixture. The mass consumption rate is given by $\dot{m} = \rho_u A' S_L$, where ρ_u is the density of the unburnt reactants, $A' S_L$ is the volume flow rate of the reactant mixture through the entire surface. As the triangles span the entirety of the surface at time t_f, the mass consumption rate expression can be modified using (9) as $\dot{m} = \rho_u \sum A S_d' = \rho_u \sum A_0 e^{\int f(t)dt} S_d'$, where $A S_d'$ is the volume flow rate of reactants through each of the triangles, and is summed over all the triangles. The second equality shows that the consumption rate is influenced by area of the finite sized elements at the leading portions of the isotherm at time t_i. Thus, the current state of the flame surface not only depends on the instantaneous quantities [17], but also on the quantities at an earlier time [18,19]. This observation signifies the importance of such leading portions of the flame surface which would later determine the propagation characteristics of that surface. In addition, this dependence of consumption rate on time history of flame surface elements could be the source of the observed time-delayed interaction between leading edge speed and turbulent consumption speed in [6].

4 Conclusions

In this work, we have investigated the generation process of a flame surface using BFPT and evolution of triangles. The instantaneous flame surface is found to be generated from spatially discrete clusters of flame particles at an earlier time. It is an encouraging observation that the kinematic evolution equation of triangles used in non-reacting flows is applicable to the current scenario. A modified form of stretch rate is derived and validated. Using this equation, the mass consumption rate of the reactant mixture can be expressed as a function of time history of surface-element area and strain-rate, and the instantaneous displacement flame speed. Thus, we provide supporting observation that the leading regions of the flame surface indeed influence the consumption rate, but with a finite time-lag. The current work also prompts a future investigation of evolution of triangles with well-defined initial shapes to understand the geometric structure of the background flow field.

5 Acknowledgment

We thank SERC, IISc for providing computer time. We also thank Dr. Uddipan Mukherjee for helping us with triangular mesh generation.

References

[1] G.K. Batchelor, Proc. R. Soc. London A Math. Phys. Eng. Sci. 213 (1952) 349–366.

[2] S.S. Girimaji, S.B. Pope, J. Fluid Mech. 220 (1990) 427.

[3] R. Ni, N.T. Ouellette, G.A. Voth, J. Fluid Mech. 743 (2014) R3.

[4] S. Chaudhuri, Proc. Combust. Inst. 35 (2015) 1305–1312.

[5] H.A. Uranakara, S. Chaudhuri, H.L. Dave, P.G. Arias, H.G. Im, Combust. Flame 163 (2016) 220–240.

[6] S.H. Kim, Proc. Combust. Inst. 0 (2016) 1–8.

[7] S.B. Pope, Int. J. Eng. Sci. 26 (1988) 445–469.

[8] A. Nahum, A. Seifert, Phys. Rev. E 74 (2006) 16701.

[9] T. Moller, B. Trumbore, J. Graph. Tools 2 (1997) 21–28.

[10] B.I. Shraiman, E.D. Siggia, Phys. Rev. E 57 (1998) 2965–2977.

[11] A. Pumir, B.I. Shraiman, M. Chertkov, Phys. Rev. Lett. 85 (2000) 5324–7.

[12] J.F. Hackl, P.K. Yeung, B.L. Sawford, Phys. Fluids 23 (2011).

[13] P. Castiglione, a Pumir, Phys. Rev. E. Stat. Nonlin. Soft Matter Phys. 64 (2001) 56303.

[14] J.R. Dormand, P.J. Prince, J. Comput. Appl. Math. 15 (1986) 203–211.

[15] S.M. Candel, T.J. Poinsot, Combust. Sci. Technol. 70 (1990) 1–15.

[16] C.K. Law, Combustion Physics, 2002.

[17] R.W. Bilger, S.B. Pope, K.N.C. Bray, J.F. Driscoll, Proc. Combust. Inst. 30 (2005) 21–42.

[18] V.M. Tikhomirov, in:, V.M. Tikhomirov (Ed.), Sel. Work. A. N. Kolmogorov Vol. I Math. Mech., Springer Netherlands, Dordrecht, 1991, pp. 242–270.

[19] A. Amato, M. Day, R.K. Cheng, J. Bell, T. Lieuwen, Proc. Combust. Inst. 35 (2015) 1313–1320.

11th Asia-Pacific Conference on Combustion,
The University of Sydney, NSW Australia
10th -14th December 2017.

Local Three-point Heat Flux Measurement on Sub-millimeter Scale in an Internal Combustion Engine

Kazuhito Dejima[1], Osamu Nakabeppu[2], Yuto Nakamura[1], Tomohiro Tsuchiya[1]

[1]Graduate School of Science and Technology, Meiji University

[2]Department of Mechanical Engineering, School of Science and Technology, Meiji University

1-1-1, Higashimita, Tama-ku, Kawasaki, Kanagawa, 214-8571, Japan

Abstract

To investigate the heat transfer mechanism, a sensor with high spatial and temporal resolution and high durability is necessary. Our research group has been developed a thin film thermal sensor by using the MEMS technologies. The sensor has three resistance temperature detectors with the size of 315 μm on an Al alloy substrate. In this work, heat flux measurement with the MEMS sensor was conducted in the laboratory engine under three equivalence conditions. The heat fluxes were measured for 200 cycles, and were discussed in terms of that on 200-cycle average and in each cycle. As a result, the wall heat flux decreases with the equivalence ratio because the gas temperature becomes low at the lean condition. In addition, the cyclic variation of the heat flux is significant in terms of the magnitude, rise and attenuation, and it was clarified that the instantaneous heat transfer in each cycle is stronger phenomena than that estimated from the averaged data. Furthermore, the instantaneous heat fluxes have different phase in spite of the close range of the measurement points. This result indicates that the combustion flame has turbulence on sub-millimeter scale, namely the distance between the measurement points of 780 μm.

1 Introduction

Investigation of heat transfer characteristics on engine inner wall is important for various areas of engine research. However, the environment in the combustion chamber of the engine is quite harsh, that is high pressure, high temperature and high heat flux load. Furthermore, the combustion is fast and complex phenomena. Therefore, detailed measurement in the combustion chamber is difficult, and the heat transfer mechanism has not been clarified yet. To investigate the heat transfer mechanism, a sensor with high spatial and temporal resolution and high durability is necessary.

The authors have developed an original thin film resistance type heat flux sensor by using MEMS (Micro-Electro-Mechanical Systems) technologies in a national project, SIP [1]. This sensor has fast response speed and three measurement points on sub-millimeter scale. In this paper, the result of wall heat flux measurement under continuous cycle operating conditions in a laboratory internal combustion engine will be reported.

Corresponding author. Fax: +81-44-934-7907
E-mail address: kdejima@meiji.ac.jp

2 Experimental Apparatus and Conditions

2.1 MEMS Three-Point Heat Flux Sensor

Figure 1 shows the thin film resistance type MEMS heat flux sensor on an Al alloy (AC8A) substrate with the diameter of 6 mm. This sensor has three RTDs (Resistance Temperature Detector) of a square 315 μm on a side. These RTDs called Top, Left and Right are on 900 μm diameter circle in rotational symmetry, so the distance between the center of each RTD is approximately 780 μm. The RTD type sensor has an advantage in heat flux calibration because the resistance can generate known heat by self-heating [2]. These RTDs measure the surface temperatures, and instantaneous heat fluxes are calculated through three dimensional heat conduction analysis with boundary conditions given by the surface temperature measurements [3].

This MEMS sensor adopted a feedthrough wiring in order to be installed in the combustion chamber flush with its inner wall surface, and the sensor body is made to be interchangeable with a commercially available heat flux sensor, Medtherm [4]. The response speed of the sensor itself is on the order of 100 kHz, and the sensor is driven under a constant current mode with the cut-off frequency of 10 kHz.

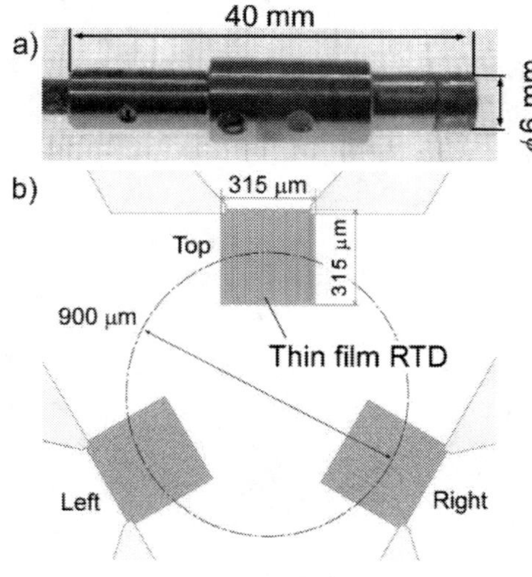

Figure 1: MEMS three-point heat flux sensor. a) Sensor body compatible with a conventional heat flux sensor, b) RTD pattern on the sensor top surface

The sensor is constructed of the AC8A substrate, an Al_2O_3 insulation layer, a Pt RTD layer and an Al_2O_3 passivation layer. Since AC8A is widely used as the engine material, the sensor can measure the temperature variation similar to that of actual engine wall surface. In order to enhance the durability, Al_2O_3 is adopted as the material of the insulation and passivation layer because Al_2O_3 has a thermal expansion coefficient relatively close to that of AC8A.

2.2 Laboratory Engine & Experimental Conditions

The laboratory engine used for this study is a spark ignition type gasoline engine and has four valves and single cylinder. Bore × stroke is 75 mm × 112.5 mm, and compression ratio was 13:1. Figure 2 shows the bottom view of the cylinder head. The MEMS sensor and a pressure transducer were installed in the cylinder head through an adapter, and the Top RTD faced the spark plug.

Table 1 shows engine operation conditions. In all experiments, the indicated mean effective pressure (IMEP) and the engine speed were set to 400 kPa and 2000 rpm, respectively. The equivalence ratio ϕ was varied from 1.0 to 0.6, and the ignition timing was at MBT, that is -20°, -24° and -40° for ϕ = 1.0, 0.8 and 0.6, respectively. A plate for tumble flow enhancement was attached on the intake port in order to prevent a misfire at the lean condition. The temperature data measured with the MEMS sensor were obtained for 200 cycles in every 0.1° crank angle, which was equal to 120 kHz, and the data were averaged into 24 kHz afterward.

3 Local Wall Heat Fluxes Measured with the MEMS Sensor

3.1 Heat Fluxes on 200-cycle Average

Figure 3 shows in-cylinder gas behaviors on 200-cycle average. The compression process starts at -180° after TDC, and the intake valve is closed at -123°. Then, the pressure and temperature gradually rise, and the fuel is ignited with the spark plug. After the ignition, the pressure and temperature rapidly increase and take the peak at around 15°. The maximum pressures are approximately 3.0 – 3.5 MPa in the all conditions. This is caused by the high intake pressure at the lean condition in order to make the IMEP equal. On the other hand, the maximum gas temperature decreases with thinning down the fuel because of the small amount of the fuel and large heat capacity of the intake gas.

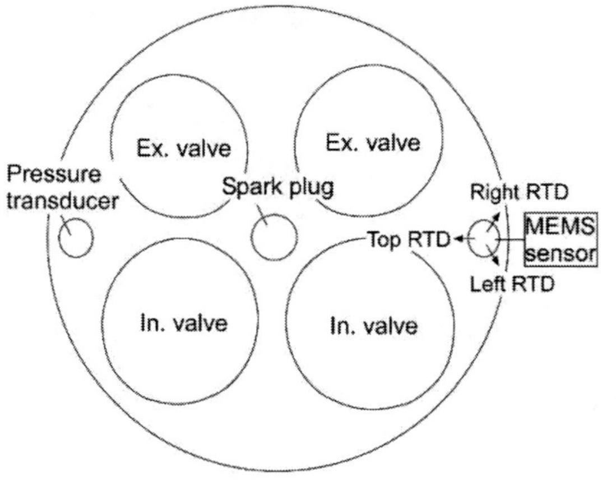

Figure 2: Bottom view of the cylinder head.

The heat fluxes and the wall temperature variations from mean wall temperatures on 200-cycle average measured with the Top RTD of the MEMS sensor are shown in Figure 4. The heat flux and temperature data have sharp drops caused by the startup of a spark circuit The maximum heat fluxes are 1.2 MW/m^2, 1.0 MW/m^2 and 0.7 MW/m^2 at ϕ = 1.0, 0.8 and 0.6, respectively. The heat flux decreases with the equivalence ratio because the gas temperature becomes low in the lean condition. The wall temperature variation is also small in the lean condition due to the small gas temperature swing. In addition, the wall temperature rises only by 7 K or so in spite of the large gas temperature swing of more than 1000 K. This indicates that the heat transfer on the chamber wall is dominated by the gas phase.

IMEP	400 kPa
Engine speed	2000 rpm
Equivalence ratio	1.0, 0.8, 0.6
Ignition timing	MBT (-20°,-24°,-40° after TDC)
Intake valve open/close	-362° / -123° after TDC
Exhaust valve open/close	147° / 349° after TDC
Coolant temperature	80 °C

Table 1: Engine operating conditions

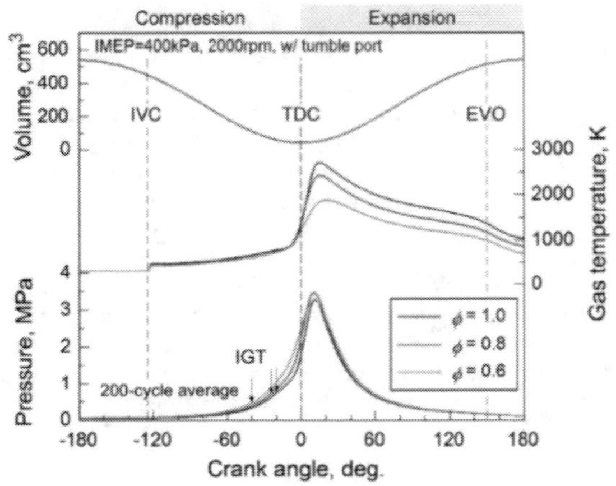

Figure 3: In-cylinder gas pressure, temperature and volume on 200-cycle average at ϕ = 1.0, 0.8, 0.6.

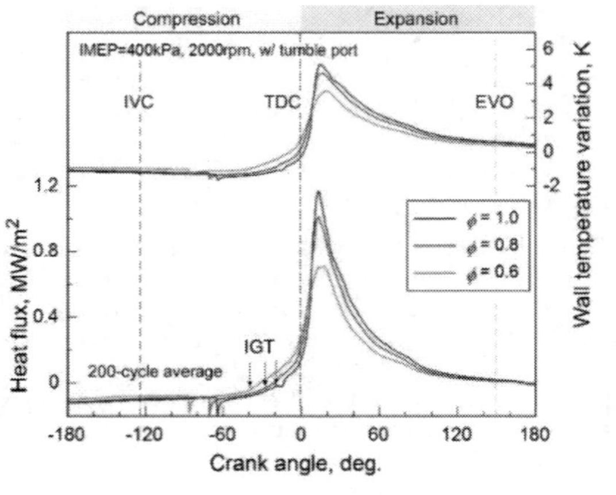

Figure 4: Heat fluxes and wall temperature variations on 200-cycle average at ϕ = 1.0, 0.8, 0.6.

3.2 Instantaneous Heat Fluxes in Each Cycle

In most conventional studies, only wall heat fluxes on average for several dozen or a few hundred cycles were discussed in investigations of wall heat transfer in engines. This is based on the claim that the averaged heat flux indicates typical heat transfer characteristics of the engine. However, the combustion gas has small turbulence and its state changes momentarily. Therefore, not only the mean heat flux but also instantaneous one at each cycle should be analyzed to elucidate the turbulence heat transfer mechanism in the engine. Since the MEMS sensor developed in this study could obtain the heat flux data with high resolution and a low noise level, the instantaneous heat fluxes are discussed in this section.

Figure 5 shows the instantaneous heat fluxes measured with the Top RTD at $\phi = 0.8$ for continuous 5 cycles. The instantaneous heat fluxes show different trend from that on 200-cycle average and the cyclic variation is significant. In addition, the instantaneous heat fluxes have sharper and higher peaks than 200-cycle averaged one. The reason of that is the averaged heat flux becomes smooth due to the shifts of the timing of the heat flux rise caused by the variation of the flame arrival timing. This indicates that the instantaneous heat transfer in each cycle is stronger phenomena occurring in shorter period than that estimated from the averaged data. Furthermore, the instantaneous heat fluxes oscillate around the heat flux peak. These oscillations are caused by the instantaneous local turbulence in the combustion gas. The frequency of the oscillation is 1 – 5 kHz. When the gas flow velocity is assumed to be comparable with the mean piston speed tentatively, the scale of the turbulence in the combustion gas can be estimated at 3.75 mm – 750 µm from the oscillation frequency. If the actual flow speed adjacent to the measurement position, the turbulence scale could be estimated more accurately.

Figure 6 shows spatial variation of the instantaneous heat fluxes at $\phi = 0.8$. The blue, red and green lines mean the heat fluxes simultaneously measured with the Top, Left and Right RTDs respectively, and the black line indicates the 200-cycle averaged one. The peak value of the averaged ones at Top, Left, Right RTDs differ only by several dozens of kW/m^2. On the other hand, the instantaneous heat fluxes have different phase and different magnitude between each other in spite of the close range of the measurement points. It is inferred from this result that the turbulence of the combustion gas has a spatial scale of less than or equal to the distance between the RTDs, namely 780 µm.

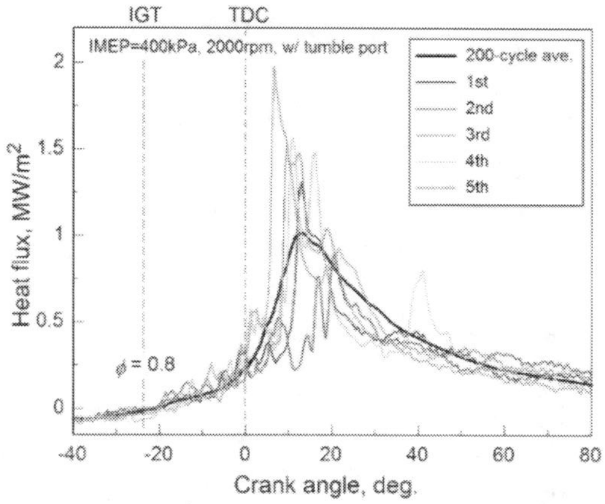

Figure 5: Cyclic variation of the heat flux for 5 cycles at $\phi =$ 0.8 measured with the Top RTD.

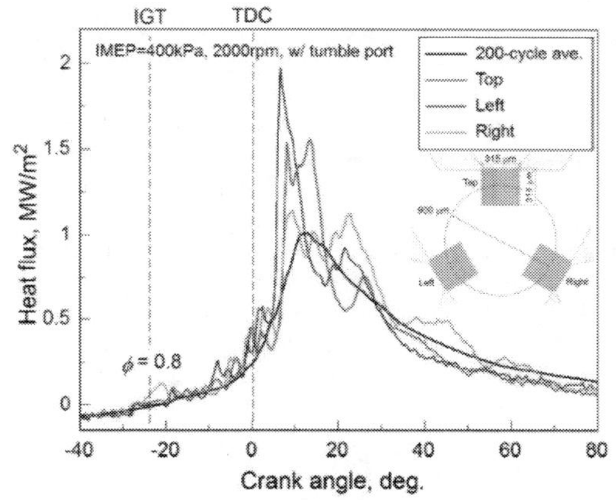

Figure 6: Spatial variation of the heat flux at $\phi = 0.8$.

4 Conclusions

The three-point thermal sensor with the temporal resolution of 10 kHz and the spatial resolution of 315-780 µm was developed by using the MEMS technologies. The heat flux measurements with the MEMS sensor were conducted in the laboratory engine. As a result, the wall heat flux decreases with the equivalence ratio because the gas temperature becomes low at the lean condition. Furthermore, the heat flux variation between each cycle is significant in terms of the magnitude, rise and attenuation. The instantaneous heat fluxes have oscillation with the frequency of 1 – 5 kHz. This means the turbulent scale of the combustion gas is 3.75 mm – 750 µm if the gas flow velocity is equal to the mean piston speed. In addition, the instantaneous heat fluxes measured simultaneously with the three RTDs, and they have different phase in each other in spite of the close range of the measurement points. This result also indicates that the turbulent scale of the combustion gas is on sub-millimeter, namely less than or equal to 780 µm. In future work, the cyclic and spatial variation of the heat fluxes will be analyzed in detail.

5 Acknowledgment

This work was supported by Council for Science, Technology and Innovation (CSTI), Cross-ministerial Strategic Innovation Promotion Program (SIP), "Innovative Combustion Technology" (Funding agency: JST).

References

• Cabinet Office, Government of Japan, Cross-ministerial Strategic Innovation Promotion Program available at <http://www8.cao.go.jp/cstp/gaiyo/sip/>.

• O. Nakabeppu, T. Tsuchiya, Y. Nakamura and K. Nagasaka, Transactions of the JSME, 82, 840 (2016), DOI:10.1299/transjsme.16-00083.

• K. Dejima, O. Nakabeppu, K. Nagasaka, Y. Nakamura and T. Tsuchiya, Proc. of the Ninth International Conference on Modeling and Diagnostics for Advanced Engine Systems (COMODIA 2017), A202 (2017) (accepted).

• Medtherm Co. available at <http://medtherm.com/>.

11th Asia-Pacific Conference on Combustion,
The University of Sydney, NSW Australia
10th -14th December 2017.

Enhanced oxidation of methane (CH_4) by addition of carbon disulfide (CS_2)

Zhe Zeng[1], Bogdan Z. Dlugogorski*[1], Mohammednoor Altarawneh[1]

[1]School of Engineering and Information Technology, Murdoch University

90 South Street, Murdoch, WA, 6150, Australia

Abstract

This contribution reports a promotion effect of carbon disulfide (CS_2) on methane (CH_4) oxidation with experiments conducted in a jet-stirred reactor (JSR). The experiments include neat oxidation of CS_2/O_2 and CH_4/O_2, as compared with co-oxidation of $CS_2/CH_4/O_2$ under stoichiometry condition. The addition of CS_2, in trace amount, leads to a lower ignition temperature of CH_4 at 1210 K, as compared with neat oxidation at 1290 K under equivalent condition. On the contrary, the presence of CH_4 significantly inhibits the oxidation of CS_2 at low temperature range (below 1100 K), implying a latent interaction between CH_4 and sulfur radicals. Kinetic modelling has been conducted for the neat oxidation of CH_4/O_2 and CS_2/O_2, respectively. The accumulation of O radical from CS_2 conversion process initiates the oxidation of CH_4 at lower temperature. We also propose competing channels between consumption and formation of CS_2 through interaction between sulfur radicals and hydrocarbon under limited supply of O_2, accounting for the inhibition effect of CH_4 on CS_2 oxidation. The detailed interaction reactions between hydrocarbon and sulfur radicals requires further investigations to illustrate the role of sulfur impurities in natural gas.

1. Introduction

Reduced sulfur species (H_2S, CS_2 and COS) exist as critical impurities and by-products in a range of industrial processes. The recently extracted and exploited shale gas, coal seam gas and syngas exhibit high sulfur fraction up to several percent [1], making of the so-called sour gas. Although Claus process represents the most significant desulfurising method [2, 3] for raw natural gas, pipeline quality gas usually contains sulfur levels of up to 2,000 ppm [4]. Reduced sulfur species also arise as unwanted products in gasification of coal [5] and biomass fuels [6]. Due to increasing world-wide consumption of natural gas and syngas, understanding the role of sulfur impurities in the commercial use (i.e., burning) of these fuels remain paramount.

Several exploratory studies reported the investigations on H_2S/fuel interactions in combustion processes. Hydrogen sulfide (H_2S) exhibits an inhibition effect on the pyrolysis of n-octane in experiments under high pressure at 70 MPa [7, 8]. Theoretical calculations demonstrate that, H_2S displays distinct inhibition and promotion effects on pyrolysis of C_1 – C_4 hydrocarbons, depending on the strength of S-H and C-H bonds [9]. By introducing O_2, Selim et al. reported the oxidation competition between H_2S and CH_4 in CH_4/O_2 flames doped with H_2S [10, 11]. Both CS_2 and COS form in the co-oxidation of $H_2S/CH_4/O_2$ under fuel rich conditions, suggesting the interaction between SH/SO radicals and aliphatic hydrocarbons. Similarly, by adding SO_2 into an ethylene/O_2 mixture, Abián et al. observed decreased soot emission due to the production of CS_2 and COS [12]. However, the reaction channels leading to the formation of CS_2 and COS remain unknown and open to speculation.

A recent study by Gerson et al. has demonstrated the reduced ignition temperature of CH_4 in the presence of H_2S in experiments involving a high-pressure plug-flow reactor [13]. The oxidation of 1 % (by volume) H_2S commences at 600 K, while a temperature of 775 K affords the consumption of 10 % (by volume) CH_4. However, the conversion of neat CH_4/O_2 mixture does not proceed below 850 K under the same experimental condition. The production of HO_2/OH radical, due to the conversion of H_2S at lower temperature, prompts the enhanced oxidation of CH_4. Similarly, the presence of a trace amount of NO_x reduces the ignition temperature of morpholine by 150 K [14]. Thus, typically, the presence of impurities enhances the oxidation of hydrocarbon fuels.

Recognised as an extremely flammable chemical, CS_2 poses significant fire hazard in industrial processes. Experiments with tubular-flow reactor revealed the low ignition temperature of CS_2 [15]. A subsequent investigation, combining experiments and theoretical calculation elucidated the extreme flammability of CS_2 by proposing intersystem crossing, which reduces the activation barrier of key elementary reactions [16]. The production of CS_2 as by-product in Claus process affects the oxidation of CH_4. However, details of the interaction channel between CS_2 on CH_4 oxidation remain poorly understood.

This study aims to investigate the co-oxidation of $CS_2/CH_4/O_2$, probing the influence of sulfur impurities on the oxidation of natural gas. We collect experimental results for: 1) CH_4/O_2 oxidation, 2) CS_2/O_2 oxidation and 3) $CS_2/CH_4/O_2$ co-oxidation in a jet-stirred reactor (JSR) under stoichiometric conditions from 800 – 1400 K. The kinetic modelling of CH_4/O_2 oxidation [17] and CS_2/O_2 oxidation mechanism [16] in JSR achieves a good agreement with experimental measurements of neat oxidation of CH_4 and CS_2, respectively. The promotion channels for enhanced ignition of CH_4 by CS_2 is proposed to be the build-up of radical pool in CS_2 oxidation. In the meanwhile, we also highlight the inhibited oxidation of CS_2 with presence of CH_4. The interaction between hydrocarbon and sulfur radicals deserves further investigation with experimental and theoretical studies

2. Methodology

2.1 Experimental set-up

Figure 1 depicts the schematic diagram of the experimental apparatus. Three mass flow controllers (MFC) adjust the flow rate of CH_4 diluted in N_2, synthetic air and N_2 (BOC Australia). A programmable syringe pump introduces CS_2 into the flowing stream to obtain a concentration of 50 ± 5 ppm. A heating tape

Corresponding author: Bogdan Z. Dlugogorski
Email address: B.Dlugogorski@murdoch.edu.au

(set at 328 K) controls the gas temperature before the injection point of CS_2 to ensure its complete vaporisation.

The jet-stirred reactor (JSR, built by Monash Scientific, Melbourne Australia) deployed in this work is based on the design developed by Herbinet and Guillaume [18] (Laboratoire Réactions et Génie des Procédés, Nancy, France) that originates from the work of Matras and Villermaux [19]. The JSR sphere (59 mm ID and 107.5 cm^3 volume) and access lines comprise ultrapure (99.999 %) quartz to minimise the catalytic wall reactions. The injector part of the reactor houses four nozzles of 0.3 mm ID each, for the delivery of the inlet gas in turbulent flow condition. This arrangement eliminates temperature and concentration gradients in the reactor resulting in the conditions representing the so-called continuously stirred-tank reactor (CSTR, also denoted as PSR – perfectly stirred reactor) [18]. The JSR rests along the centreline of an electrically heated single-zone furnace (T_1), functioning from 800 – 1400 K with an error of ± 2.5 K. An external thermocouple monitors the reaction temperature T_a in the centre of JSR sphere. The equal temperature reading of T_1 (single zone furnace) and T_a (centre position of the JSR sphere) indicates the homogeneous temperature distribution in the reactor. A constant residence time of 1.0 s is applied to all experiments, by controlling the flow rate of inlet gas.

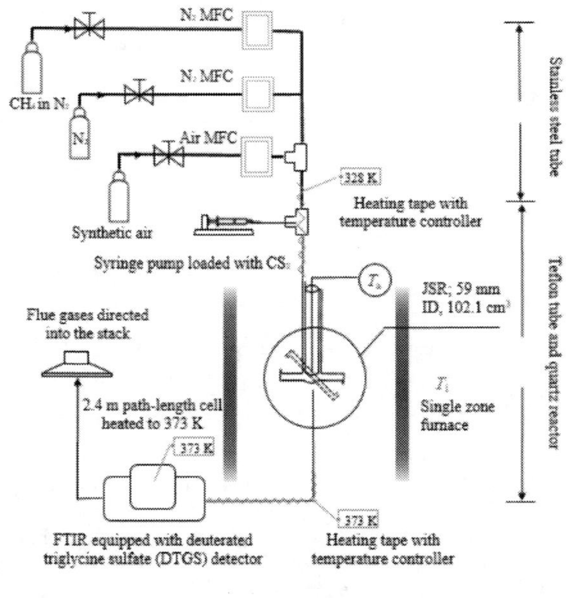

Figure 1. Schematic diagram of experimental setup.

A Fourier transform infrared (FTIR) spectroscope facilitates online monitoring of product species exiting the JSR. The setup incorporated an electrically heated transfer line linking the reactor to a small volume (100 mL) 2.4 m path-length online gas sampling cell (Pike) mounted inside the FTIR compartment. The spectrometer averages 8 accumulated scans per spectrum at 1 cm^{-1} resolution. Both the sample line and the sampling cell operate at 373 K, to avoid the condensation of volatile products prior and during the analysis. The 2.4 m multi path-length gas cell affords the detection and quantitation of species involved in experiments, with a limit of detection of [CS_2] = 0.6 ppm, [SO_2] = 0.5 ppm, [CO] = 2.5 ppm, [COS] = 0.4 ppm, [CH_4] = 8 ppm and [CO_2] = 1.4 ppm based on the background noise of ± 0.005 abs in IR spectra. The experimental uncertainty originates from the accuracy of mass flow controllers (2 %) and syringe pump for the liquid CS_2 (1 %). Other contributing factors include the uncertainty in the temperature of the reaction zone (± 2.5 K) and the online FTIR measurements of species (± 2 %), because of background noise.

Comparative experiments include neat oxidation of CS_2/O_2 and CH_4/O_2, and co-oxidation of $CS_2/CH_4/O_2$. We maintained the concentration of CH_4 at 500 ppm, while a small amount of CS_2 at 50 ppm enters the co-oxidation experiments, and enforced the stoichiometric condition. Extremely small concentration of CH_4 and CS_2 minimises the temperature rise in the reactor due to the exothermic nature of the oxidation reactions. Table 1 lists the composition of reactants in the experiments. The stoichiometric condition of CS_2 oxidation is defined as $CS_2 + 2.5\ O_2 \rightarrow CO + 2\ SO_2$ with no CO_2 produced in experiments under dry condition [16]. The lack of H/OH radical clearly shuts down the reaction channel $CO + OH \rightarrow CO_2 + H$. In term of CH_4 oxidation, the final experimental products include both CO and CO_2 with presence of OH radical.

Table 1. Composition of reactants applied in the JSR experiments

No.	Experiment	Reactants concentration (ppm)
1	Neat oxidation	[CS_2]:[O_2] = 50:125
2	Neat oxidation	[CH_4]:[O_2] = 500:1000
3	Co-oxidation	[CS_2]:[CH_4]:[O_2] = 50:500:1125

2.2 Kinetic modelling

In this study, kinetic modelling of neat CH_4/O_2 oxidation and CS_2/O_2 oxidation has been conducted to validate against our experimental results. The optimised GRI-Mech 3.0 provides the subset of CH_4/O_2 reactions [17], which has been validated against a series of high-accuracy experiments [20,21]. With respect to the CS_2/O_2 subset, we adopt the mechanism proposed in our work, the robustness of which has been proven in TFR and JSR experiments [15, 16]. The literature lacks a CS_2/CH_4 subset, preventing us to obtain a comprehensive co-oxidation mechanism of $CH_4/CS_2/O_2$ oxidation. Chemkin-Pro [22] enables the chemical kinetic modelling, facilitating the application of a perfectly stirred reactor (PSR) model to describe the flow condition operating in our JSR.

3. Results and discussions:

3.1 Neat oxidation of CH_4/O_2 and CS_2/O_2

Figure 2 presents the experimental measurement and kinetic modelling of neat oxidation of CH_4/O_2 and CS_2/O_2, respectively, as function of temperature. Experiments on neat oxidation of methane ([CH_4]:[O_2] = 500 ppm:1000 ppm) in JSR reveal the ignition temperature of 1290 K for the residence time of 1.0 s. At 1310 K, CH_4 converts completely into CO, CO_2 and H_2O (not quantitated). The GRI-Mech 3.0 mechanism [17] as adopted for the kinetic modelling of CH_4 oxidation, yields the onset temperature of 1320 K. The kinetic modelling also captures the accumulation of CO and its further conversion to CO_2, consistent with the experimental measurements. A minor shift of 30 K indicates a good agreement between experiments and kinetic modelling.

The ignition of CS_2/O_2 oxidation (50 ppm:125 ppm) occurs at 850 K, a temperature much lower than that of CH_4 under equivalent conditions. The accumulation and consumption of

COS distinguish it as an important intermediate. We observed a relatively slow conversion process of CS_2 and COS from 850 K – 1100 K. Both CO and SO_2 arise as final products, consistent with the results of a previous study on dry oxidation of CS_2 [16]. The oxidation mechanism of CS_2 developed in our previous work affords the kinetic modelling of CS_2 conversion, predicting the onset temperature of CS_2 oxidation at 900 K with residence time at 1.0 s in JSR model, slightly higher than that of experiments by 50 K. Considering the over-prediction of the ignition temperature of methane of 30 K, we suspect a possible error in the experiments which results in a lower ignition temperature by around 30 K. The uncertainty may originate from the accuracy of mass flow controllers (2 %) and syringe pump for the liquid CS_2 (1 %) which adjust the ratio between CH_4/CS_2 and O_2.

The mechanism also underestimates the production of SO_2 with significant accumulation of $CS/SO/S_2$ radical. Due to the occurrence of intersystem crossing (ISC) between reduced sulfur species and oxygen [16], it is reasonable for us to expect faster conversion from $CS/SO/S_2$ radical to SO_2 with higher reaction rate parameters modified with ISC calculation in future work.

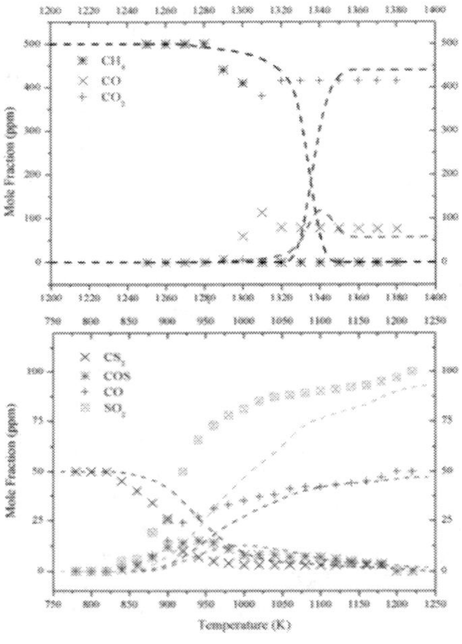

Figure 2. Experimental results (symbols) and kinetic modelling (dashed lines) of neat oxidation of $[CH_4]:[O_2]$ = 500 ppm:1000 ppm (top) and $[CS_2]:[O_2]$ = 50 ppm:125 ppm (bottom), respectively. This figure reports the conversion processes of CH_4 and CS_2 as function of temperature. All experiments are carried out in JSR at a fixed residence time of 1.0 s.

3.2. Co-oxidation of $CS_2/CH_4/O_2$

Our co-oxidation experiments comprise 500 ppm of CH_4 in the presence of 50 ppm CS_2, and 1125 ppm O_2 under stoichiometric condition. The initial concentration of $[CS_2]:[CH_4]:[O_2]$ =50 ppm:500 ppm:1125 ppm), represents the combination of both systems of CS_2/O_2 (50 ppm: 125 ppm) and CH_4/O_2 (500 ppm: 1000 ppm), respectively. Figure 3 demonstrates the IR spectra taken in the co-oxidation experiments of $CS_2/CH_4/O_2$ by an online FTIR spectrometer. The wide absorption range of H_2O in IR spectra clearly hindered the identification of other products at 1230 K. Carbon dioxide arises as the major product due to the presence of H_2O.

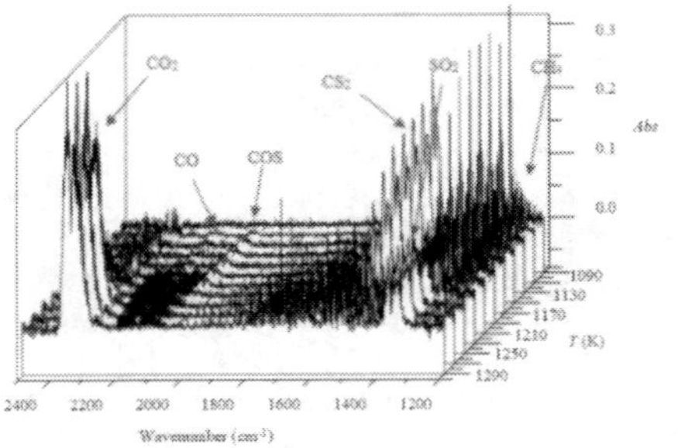

Figure 3. Online FTIR spectra taken for co-oxidation of $CS_2/CH_4/O_2$ (50 ppm: 500 ppm: 1125 ppm) in JSR experiments.

Figure 4 illustrates the precise profile quantitated from the IR spectra as function of temperature. Carbonyl sulfide (COS) appears as an intermediate during the consumption of CS_2. While CO, CO_2, SO_2 and H_2O arise as the reaction products. From the experimental results, we can observe that, the oxidation of CS_2 and CH_4 proceeds from 1150 K to 1250 K. The oxidation of CS_2 commences prior to that of CH_4 at 1150 K, while the oxidation of CH_4 initiates at 1210 K with 40 % CS_2 reacted. The formation of CO and SO_2 accompanies the consumption of CS_2 at 1150 K. Moreover, as H_2O evolves (from oxidation of CH_4), the production of CO_2 prevails over CO at 1250 K. Carbonyl suflide arises as an intermediate during the oxidation of CS_2.

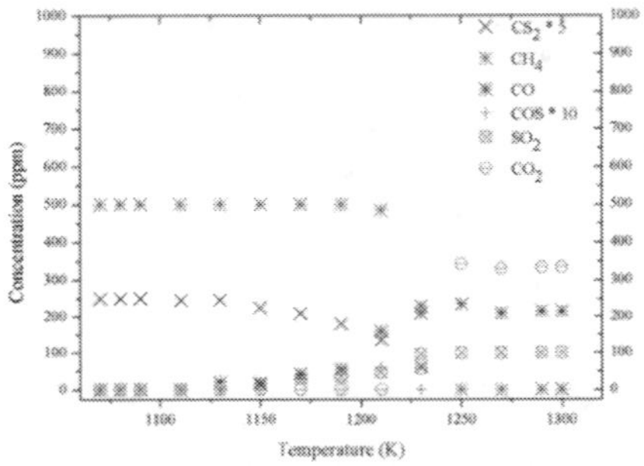

Figure 4. Species profile quantitated from the IR spectra as function of temperature in co-oxidation experiments of $CH_4/CS_2/O_2$ (= 500 ppm:50 ppm:1125 ppm).

The addition of a small amount of CS_2 reduces the ignition temperature of methane from 1290 K to 1210 K. However, the onset temperature for CS_2 increases dramatically from 850 K to 1150 K in co-oxidation with CH_4, suggesting the inhibition effect of CH_4 on the oxidation of CS_2. In Figure 5, we contrast the ignition temperature for CS_2 and CH_4 in both neat oxidation and co-oxidation experiments.

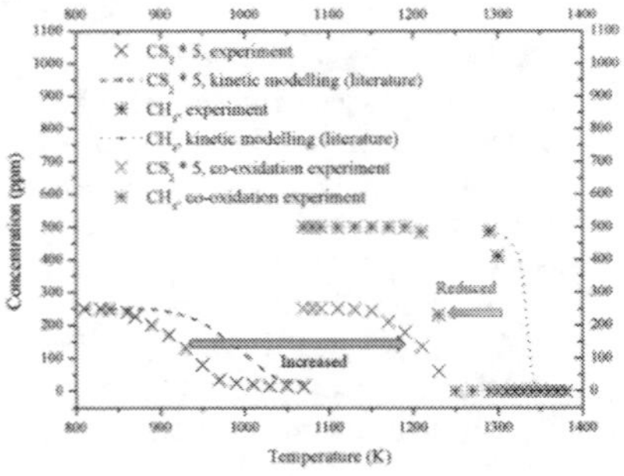

Figure 5. Comparison of the ignition temperature of CS_2 and CH_4 in neat oxidation and co-oxidation experiments.

Here, we highlight three remarks from experiments: 1) The addition of CS_2 reduces the ignition temperature of CH_4; 2) The presence of CH_4 inhibits the oxidation of CS_2 at low temperature; and, 3) The oxidation of CS_2 and CH_4 proceeds synergistically rather than separately, implying the interaction between CS_2 and CH_4. In the next section, we analyse the possible interactive channels explaining the promotion and inhibition effect between $CH_4/CS_2/O_2$.

3.3. Interaction between CH_4 and CS_2 oxidation

As we have observed in experiments, the oxidation of CS_2 arises prior to that of CH_4. The consumption of 40 % CS_2 at 1210 K induces the onset of CH_4 oxidation. This fact prompts us to consider the potential promotion effect of radicals produced in CS_2 conversion on the oxidation process of CH_4. A careful review on CS_2 oxidation mechanism identifies the key chain branching steps in CS_2 oxidation, represented by the reaction R1 and R2 [16]. Atomic O constitutes the most important radical in chain branching steps of CS_2 oxidation, and assists the chain initiation of CH_4 oxidation through R3. The activation energy of R3 fitted at 420 – 1520 K amounts to 32 kJ/mol [23], which is significantly lower than that of original chain initiation step through R4 at 238 kJ/mol [24].

$$S + O_2 \rightarrow SO + O \quad \text{R1}$$

$$SO + O_2 \rightarrow SO_2 + O \quad \text{R2}$$

$$CH_4 + O \rightarrow CH_3 + OH \quad \text{R3}$$

$$CH_4 + O_2 \rightarrow CH_3 + HO_2 \quad \text{R4}$$

Althoug we already have CS_2/O_2 subset and CH_4/O_2 subset at hand, the interactions between CS_2/CH_4 is still missing in the mechanism. We notice that CS_2 and COS arise in combustion systems involving H_2S/SO_2 and hydrocarbons as intermediate [11, 25]. The unburned hydrocarbons react with sulfur radicals, e.g. S, SO and SO_2, to produce CS_2 and COS, as shown below. This latent process may arise and compete with the consumption of CS2, leading to an inhibition effect as observed in experiments.

$$SO/SO_2 + C_xH_y \rightarrow CS + \ldots \rightarrow COS/CS_2 + \ldots$$

Figure 6 displays the schematic diagram summarising the comprehensive oxidation mechanism of $CH_4/CS_2/O_2$ oxidation. The shared radical pool in CS_2/O_2 and CH_4/O_2 subset, as well as the interaction between CS_2 and CH_4 are yet to be incorporated in our deployed mechanism.

Figure 6. Schematic diagram summarising the co-oxidation mechanism of $CS_2/CH_4/O_2$.

4. Conclusions

This study reports the enhanced ignition of CH_4 in presence of trace amount of CS_2 in combustion process. We highlight three remarks from experiments: 1) Trace amount of CS_2 reduces the ignition temperature of CH_4; 2) The presence of CH_4 inhibits the oxidation of CS_2 at low temperatures; and, 3) The oxidation of CS_2 and CH_4 proceeds synergistically rather than separately, suggesting the direct interaction between CS_2 and CH_4. Good agreement has been achieved between kinetic modelling and experimental results for neat oxidation of CH_4 and CS_2. The promotion effect of O radical as produced in the CS_2 oxidation process has been proposed. However, the interaction between CH_4 and sulfur radicals remains missing, preventing us to explain the increased initial temperature of CS_2 oxidation in the presence of CH_4.

5. Acknowledgment

This study has been supported by grants of computing time from the National Computational Infrastructure (NCI) Australia as well as funds from the Australian Research Council (ARC).

References

[1] H. H. Schobert, Chapter 4 - Natural Gas, in The Chemistry of Hydrocarbon Fuels, Butterworth-Heinemann1990, pp. 75-84.

[2] H. Fischer, Production of Sulfur From Claus Process Waste Gas, Google Patents, 1978.

[3] E. Jüngst, W. Nehb, Hydrogen Sulfide to Sulfur (Claus Process), Handbook of Heterogeneous Catalysis, Wiley 2008.

[4] P. Jaramillo, W.M. Griffin, H.S. Matthews, Environ. Sci. Technol. 41 (2007) 6290-6296.

[5] W. Xie, L. Chang, D. Wang, K. Xie, T. Wall, J. Yu, Fuel 89 (2010) 868-873.

[6] X. Meng, W. De Jong, R. Pal, A.H. Verkooijen, Fuel Process. Technol. 91 (2010) 964-981.

[7] V.P. Nguyen, Burklé-Vitzthum, V., Marquaire, P. M., Michels, R., J. Anal. Appl. Pyrolysis 103 (2013) 307-319.

[8] V.P. Nguyen, Burklé-Vitzthum, V., Marquaire, P. M., Michels, R., J. Anal. Appl. Pyrolysis 113 (2015) 46-56.

[9] Z. Zeng, M. Altarawneh, I. Oluwoye, P. Glarborg, B.Z. Dlugogorski, J. Phys. Chem. A 120 (2016) 8941-8948.

[10] H. Selim, A. Al Shoaibi, A.K. Gupta, Ap. En. 92 (2012) 57-64.

[11] H. Selim, A. Al Shoaibi, A.K. Gupta, Ap. En. 88 (2011) 2593-2600.

[12] S.A. Lawton, Combust. Flame 75 (1989) 175-181.

[13] S. Gersen, et al., Energy Fuels 31 (2017) 2175-2182.

[14] I. Oluwoye, B.Z. Dlugogorski, J. Gore, P.R. Westmoreland, M. Altarawneh, Fire Saf. J., in press 2017.

[15] M. Abián, M. Cebrián, Á. Millera, R. Bilbao, M.U. Alzueta, Combust. Flame 162 (2015) 2119-2127.

[16] Z. Zeng, B.Z. Dlugogorski, M. Altarawneh, Fire Saf. J., in press 2017.

[17] G.P. Smith, D.M. Golden, M. Frenklach, N.W. Moriarty, B. Eiteneer, M. Goldenberg, C.T. Bowman, R.K. Hanson, S. Song, W.J. Gardiner, GRI 3.0, Gas Research Institute, Chicago, IL, http://www. me. berkeley. edu/gri_mech, (2000).

[18] O. Herbinet, G. Dayma, Jet-stirred Reactors, Cleaner Combustion, Springer2013, pp. 183-210.

[19] D. Matras, J. Villermaux, Chem. Eng. Sci. 28 (1973) 129-137.

[20] K. Bosschaart, L. De Goey, Combust. Flame 136 (2004) 261-269.

[21] P.A. Berg, D.A. Hill, A.R. Noble, G.P. Smith, J.B. Jeffries, D.R. Crosley, Combust. Flame 121 (2000) 223-235.

[22] Reaction Design, San Diego, CHEMKIN-PRO, 2013.

[23] Miyoshi A, Ohmori K, Tsuchiya K, et al. Chemical physics letters, 1993, 204(3-4): 241-247

[24] Baulch D L, Cobos C J, Cox R A, et al., J. Phys. Chem. Ref. Data, 1992, 21(3): 411-734.

[25] M. Abián, Á. Millera, R. Bilbao, M.U. Alzueta, Fuel 159 (2015) 550-558.

Engine tail pipe particulate emissions and fault detection using an ion current sensor

Rahul Rao and Damon Honnery

Laboratory for Turbulence Research in Aerospace and Combustion

Department of Mechanical and Aerospace Engineering

Monash University, 3800, Australia

Abstract

This paper explores the possibility of the use of an ion current sensor to detect faults in engine combustion processes related to particulate emissions. The paper begins by showing a causal relationship between the signal measured by an ion current sensor and particulate formation. Because of the electrical conductivity of the particulate, a deposition mechanism is proposed as being the cause. The change in ion current signal is shown to be sufficiently sensitive for it to be used to detect faults in fuel injectors and exhaust treatment systems, which give rise to enhanced particulate production.

1 Introduction

There has been considerable research into the use of ion current sensors to monitor engine operation. The universality of the signal for this purpose is evidenced by its suggested use to indicate air to fuel ratio [1], peak cylinder pressure [2], ignition timing [3], torque [4], and NOx [5]. The ability to monitor such a wide range of performance indicators from the one signal stems directly from the mechanism which leads to the formation of ions during combustion [6-7]. While mechanisms developed to model chemi-ion formation can contain up to 400 reactions and almost 60 species [8], for an engine, primary ion sources are the reactions:

$$CH + O \leftrightarrow HCO^{+} + e^{-} \quad (R1)$$

$$HCO^{+} + H2O \leftrightarrow H3O^{+} + CO \quad (R2)$$

$$O2 + e^{-} + M \leftrightarrow O2^{-} + M \quad (R3)$$

In these reactions, rapid conversion of HCO^{+} to $H3O^{+}$ gives rise to $H3O^{+}$ being the major species in the positive ion pool (80%) [8]. Species such as $C2H3O^{+}$, $C3H3^{+}$ and $CH3^{+}$ generally make up most of the ions remaining [8].

The detection of ions via the ion current is typically done using an ion current sensor. Figure 1 shows a sensor used in a diesel engine, inserted via the glow plug mount. The signal produced depends on the voltage applied to the positive electrode (around 120 Volts), electron mobility and electron concentration [7], which depend on fuel type, mixture, temperature and pressure. Current measured also depends on what is known as the sensor geometric factor [7], which depends on the electrode gap size and overall design. For most sensors, the relatively smaller size of the positive electrode used (centre electrode in Fig. 1) usually results in the concentration of the negatively charged species acting as a limiting factor in generation of the current.

Corresponding author.
E-mail address: damon.honnery@monash.edu

In an engine, ion formation via reactions R1 to R3 peaks near the flame front, so the ion current can be used to indicate quantities related to the rate of heat release, Fig. 2. Since heat release is related to time varying quantities such as pressure and temperature the ion current signal can be used to indicate bulk quantities such as fuel to air ratio and torque. The complex chemistry involved also suggests the possibility to detect such things as NOx and soot (particulate matter PM). While the link between the ion current signal and NOx formation has received considerable study [eg 5], the link to soot formation has received less [eg 9].

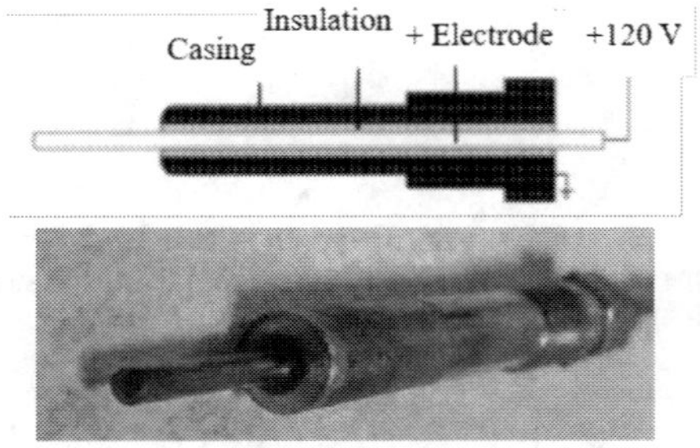

Figure 1: Top: a sketch of the ion current sensor. Ceramic material (insulation) is used to separate the two electrodes. Diameter of the positive electrode tip is approximately 4mm. Bottom: actual sensor showing evidence of soot deposition both on the positive electrode and the insulating ceramic.

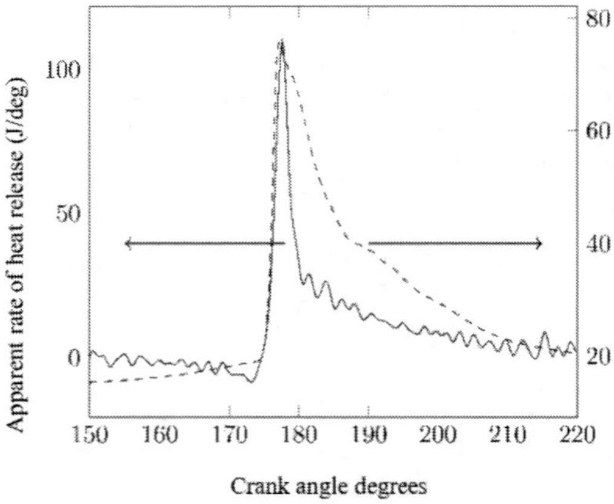

Figure 2: Comparison of heat release rate and ion current at engine speed 1600RPM and 140 Nm of torque.

Ions have been linked to soot formation in a number of studies [10-11]. These studies posit that a charged species is the soot precursor and reactions between this charged species and neutral species such as acetylene produce larger ions. Further reactions with neutral species occur to produce even larger ions, resulting finally in the formation of soot particles. $C3H3^+$ is generally thought to be this precursor species.

Calcote & Keil [11] found that in a pre-mixed acetylene-oxygen flame, soot appearance coincided with a sharp decay in ion concentration. They proposed an ionic mechanism for soot formation beginning with the reaction for the formation of CHO^+, from which $C3H3^+$ can be formed via several mechanisms

With the addition of low molecular weight neutral species, very heavy ions can be produced in this manner. In another study, Calcote & Keil [12] observed ions up to a mass of 557. Michaud et al. [13] observed ions up to a mass of 265 in low pressure acetylene-oxygen flames.

The ionic mechanism for soot formation has not, however, been universally accepted. Hall-Roberts et al. [14] studied the effect of introducing caesium into a pre-mixed acetylene-oxygen flame. It is believed that caesium removes $C3H3^+$ ions from the flame by undergoing ionisation itself to Ce^+. It was found, however, that the addition of caesium had little effect on the amount of soot formed early in the flame. The role of ions in soot formation is still the subject of some debate.

A different way in which soot formation may affect the ion current observed is via the deposition of soot on the ion sensor. Soot is electrically conductive and its deposition may therefore change the geometry of the current path between the two electrodes. Depending on the sensor geometry, soot may entirely bridge the electrode gap, leading to the observation of a non-zero ion current even in the absence of ions. It is possible that this phenomenon could be used to estimate soot mass emissions. Sensors utilising this principle have previously been developed and placed in the exhaust systems of diesel engines [15].

The two mechanisms described above and the possibility of using the in-cylinder ion sensor to indicate soot emissions are explored here.

2 Ion current and the role of soot

To investigate the effect of soot on the ion sensor, the ion current was measured at several engine loads in the Monash Diesel Engine Test Facility [16]. This facility consists of a fully instrumented direct injection, naturally aspirated, 4 litre, 4 cylinder, 4 stroke direct injection diesel engine. The engine can be monitored for steady state performance, including emissions, and in-cylinder analysis. The ion current sensor shown in Fig. 1 was installed in the glow plug mount located in the top of the cylinder head. This position is also used for pressure transducer mounting to record cylinder pressure for heat release analysis. Ion current measured by the sensor is reported as micro-amps for the applied 120 volts. Undiluted tail pipe particulate matter (PM) was measured by a DustTrak DRX8533, calibrated against gravimetric filter measurements [16].

With the sensor in place, the engine was set to a speed of 1200 RPM. The load was then gradually increased from idle to 125 Nm, and then to 200 Nm. The ion current signal for these three conditions is shown in Fig. 4, averaged over 50 cycles. Since the ion current signal is related to rate of heat release via the reaction chemistry, it closely follows the typical heat release rate profile for the three torque settings. A torque dependent offset in the signal is clearly evident during the non-heat release period (ie 140-160 degrees crank) for all three conditions. As shown in Figs 4a and 4b, this offset remains should the engine be quickly switched off during operation, and an offset remains during cold cranking in the absence of both fuel and residual gases. Given these conditions will not give rise to ion formation, any current observed cannot be due to the ions, but must be caused by another electrically conductive material in the cylinder.

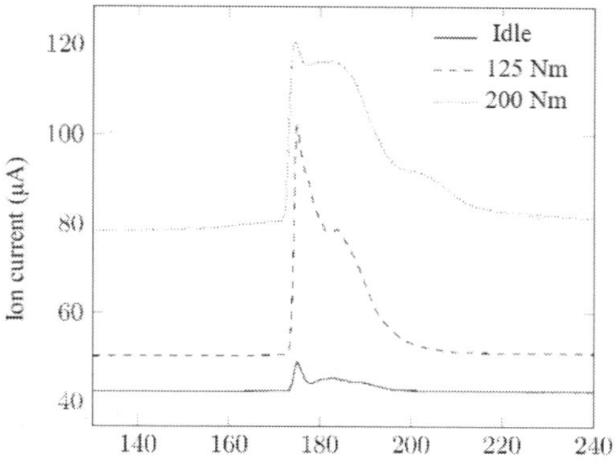

Figure 3: Ion current signal at three torque settings for an engine speed of 1200RPM.

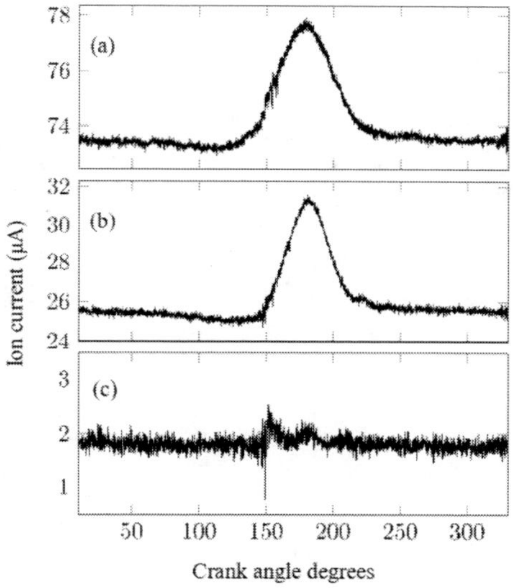

Figure 4: Ion current signal under engine crank for: (a) just after shut down; (b) after engine has cooled; and (c) after the sensor has been cleaned.

On removal of the sensor from the engine, a deposit of soot was observed on the ceramic insulation located between the casing and the centre electrode, Fig. 1. Once cleaned of this deposit, the offset was seen to fall to a near zero value and to remain at this value under repeated cold cranking, Fig. 4c. The variation in ion current shown in Figs 4a and 4b during compression and expansion is likely to result from sensitivity to a combination of temperature and pressure for the soot coated sensor.

When the ion current offset is graphed against tailpipe soot concentration for varying engine-out torque, Fig. 5, a strong correlation ($r^2 = 0.96$) is observed; suggesting the ion current could be used to indicate engine tail pipe PM emissions, this has also been observed by [9]. Despite this result, there is little, if any, understanding of the soot deposition process on the ion current sensor. An additional consideration arises from the fixed position of the sensor; it cannot provide any indication of spatial variation in either ion concentration or PM. This is being the case, the result is likely to depend to a great extent on the sensor and combustion chamber geometry, and may also depend to some extent on the engine speed. In the present case, similar correlations have been shown to occur at different engine speeds, however a full analysis of the deposition process is yet to be undertaken.

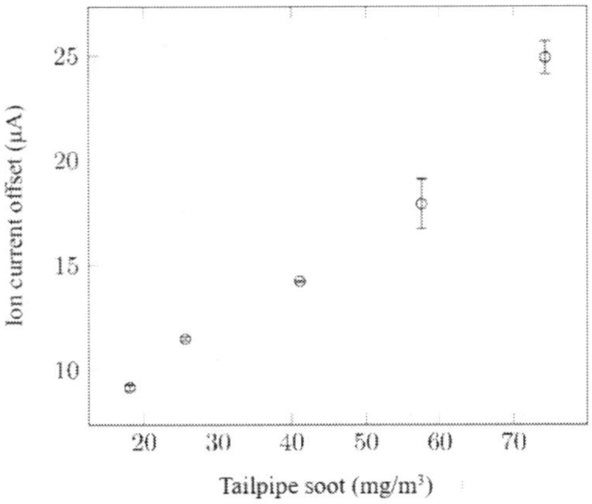

Figure 5: Tailpipe PM (soot) concentration against ion current offset for different loads at an engine speed of 1200RPM.

3 Fault detection

Excessive PM formation in engines can be indicative of poor health of engine components. Two conditions found in engines are conducive to PM formation: low temperature and high equivalence ratio. Pockets of high equivalence ratio mixtures can be found in the combustion chamber when air-fuel mixing is poor. A common reason for poor air-fuel mixing is diesel injector wear. Worn injectors display spray patterns that result in inadequate atomisation of the fuel and reduced air entrainment into the spray, causing incomplete combustion of the fuel.

Given the relationship between the ion current offset and PM emissions from the engine, it is suggested that the ion sensor could be used to diagnose worn injectors. To assess this possibility, the ion current was measured at 1200 RPM for loads of 80 Nm and 200 Nm for injectors in different states of wear. Old injectors were tested on a Bosch EPS200 injector bench tester and found to fail seat leakage and spray pattern tests. These were later reconditioned. The measured PM emissions for these test conditions are shown in Table 1.

Table 1. Tailpipe PM emissions for worn and reconditioned injectors at 1200RPM

Injector	Tailpipe PM (mg/m^3) at 80 Nm	Tailpipe PM (mg/m^3) at 200 Nm
Worn	86.1	295*
Reconditioned	15.4	51.8

*Estimate

The PM emissions for the worn injectors saturated the measurement system at 200 Nm. A rough estimate for PM emissions for this condition from a line of best fit from a correlation between the ion current cycle-minimum value and tailpipe soot at 1600 RPM gave around 295 mg/m^3.

A comparison of the ion current measured when each set of injectors was installed in the engine is shown in Fig. 6. The difference between the signal offset with old injectors and that with reconditioned injectors is clear.

Another cause of high DPM emissions is decreased volumetric efficiency caused by conditions such as clogged air filters or damaged compressor wheels in turbochargers. The lower effective compression ratio leads to a higher equivalence ratio for the same engine out torque, thereby increasing PM production. Restrictions in the exhaust system such as damaged turbines in turbochargers lead to increased back-pressure on the cylinder during the exhaust stroke. This increases the residual gas fraction and decreases the oxygen fraction of the mixture before start of combustion, leading to high equivalence ratios, lower maximum combustion temperatures and the concomitant increase in PM.

Modern diesel-powered vehicles use diesel particulate filters (DPFs) in the exhaust system to trap DPM upstream of the exit to atmosphere. The Hino engine used in this study can be fitted with an underground mine-spec diesel exhaust system, consisting of a scrubber and a filter [17]. Exhaust gas is first bubbled through water in the scrubber and then passed through the filter before exiting to the atmosphere. Both the scrubber and the filter contribute to increased back-pressure on the engine compared to the standard exhaust system. Since higher back pressure increases PM formation, the ion sensor offset should increase.

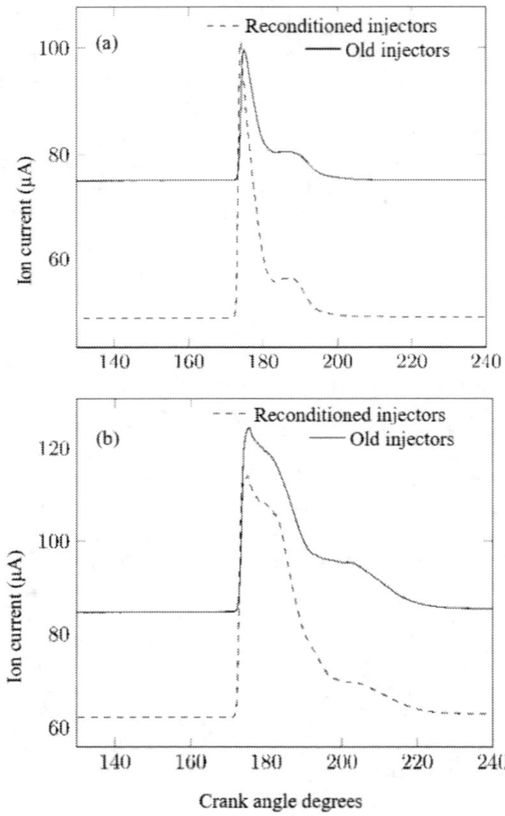

Figure 6: Ion current signal for old (worn) and reconditioned fuel injector: (a) 80 Nm 1200 RPM; (b) 200 NM 1200 RPM.

Figure 7 shows the measured ion current for the same exhaust system with and without a DPF. The higher offset for the exhaust system with the filter is as expected, indicating that the sensor is able to detect changes in the back-pressure experienced by the engine due to changes in the exhaust system. Note also that the additional back pressure for the wet scrubber system itself also adds to the overall soot loading, as evidenced by the increase in signal offset shown in Fig. 7 relative to Fig. 6. An important difference between injector fault detection and back pressure (or inlet pressure) faults is that an injector fault is more likely to indicate an increase in ion current offset for a single cylinder, while a back pressure fault will occur across all cylinders.

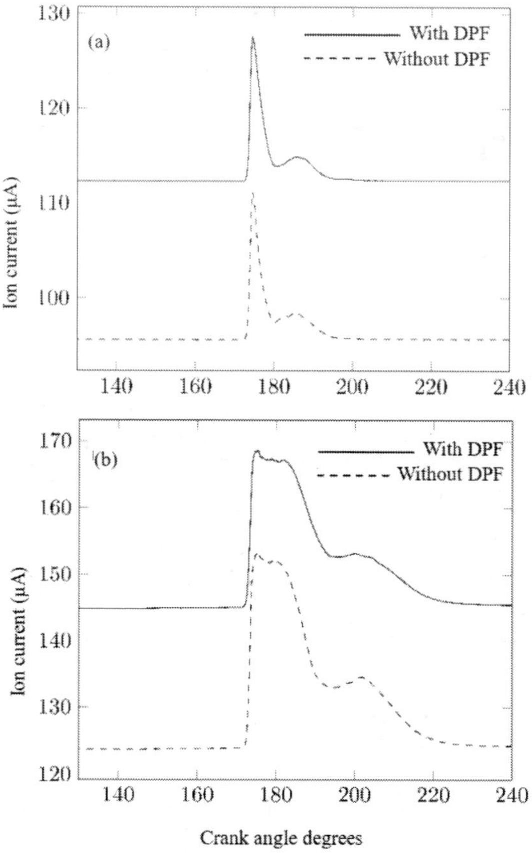

Figure 7: Ion current signal for the test engine operating a wet scrubber fitted with and without the diesel particulate filter: (a) 80 Nm 1200 RPM; (b) 200 NM 1200 RPM.

4 Conclusions

The deposition of electrically conductive soot on the ion sensor has been shown to be the likely cause for the observed correlation between the ion current signal and particulate matter concentration. The sensitivity of the generated signal to particulate matter is sufficient to suggest that it can be used to indicate faults in engine systems, which give rise to enhanced particulate matter. Two such engine operating conditions, worn injectors and an included diesel particulate filter in the exhaust system, were tested. The ion sensor was found able to detect the soot increase for both these conditions. When combined with the ability to detect a range of combustion system parameters, the ion sensor has the potential to act as a universal sensor within an engine control system.

5 Acknowledgment

The first author was a recipient of an Australian Postgraduate Award, and the support of ACARP with the DPF wet scrubber experiments is duly acknowledged.

References

[1] R. Reinmann, A. Saitzkoff, F. Mauss. "Local Air-Fuel Ratio Measurements Using the Spark Plug as an Ionization Sensor", SAE 970856, 1997.

[2] N. Wickstrom, M. Taveniku, A. Linde, M. Larsson, B. Svensson. "Estimating pressure peak position and air-fuel ratio using the ionization current and artificial neural networks" IEEE Conference on Intelligent Transportation System, ITSC'97, 1997, IEEE, pp. 972–977.

[3] M. Glavmo, P. Spadafora, R. Bosch. "Closed Loop Start of Combustion Control Utilizing Ionization Sensing in a Diesel Engine", SAE 1999-01-0549, 1999.

[4] R. Rao, D. Honnery. "The Prediction of Torque in a Diesel Engine Using Ion Currents and Artificial Neural Networks", Int J. Engine Res. 15 (2014) 370-80

[5] R. Rao, D. Honnery. "A study of the relationship between NOx and the ion current in a direct-injection diesel engine", Combust. Flame 176 (2017) 309-17.

[6] H. Calcote. "Ion and electron profiles in flames" Proc. Comb. Inst. 9 (1962) 622–637.

[7] S. Aithal, "Analysis of the Current Signature in a Constant-Volume Combustion Chamber", Comb. Sci. Technol. 185 (2013) 336–349.

[8] R. Rao, D. Honnery. "A simplified ionic mechanism for methane oxidation", Comb Flame 162 (2015) 2928-36.

[9] T. Badawy, F. Estefanous, N. Henein. "Cycle-by-cycle soot estimation in diesel engines". SAE 2013-01-0545.

[10] H. Calcote, D. Olson, and D. Keil. "Are ions important in soot formation?", Energy & Fuels 2 (1988) 494–504.

[11] H. Calcote, D. Keil, et al. "The role of ions in soot formation", Pure Appl. Chem 62 (1990) 815–824

[12] S. Aithal. "Prediction of Voltage Signature in a Homogeneous Charge Compression Ignition (HCCI) Engine Fueled with Propane and Acetylene", Combust. Sci. Tech. 185 (2013) 1184–1201.

[13] P. Michaud, J. Delfau, and A. Barassin. "The positive ion chemistry in the postcombustion zone of sooting premixed acetylene low pressure flat flames", Proc. Comb. Inst. 18 (1981) 443–451.

[14] V. Hall-Roberts, A. Hayhurst, D. Knight, S. Taylor. "The origin of soot in flames: Is the nucleus an ion?", Comb. Flame 120 (2000) 578–584.

[15] A. Warey and M. J. Hall. "Performance characteristics of a new on-board engine exhaust particulate matter sensor', SAE Technical Paper 2005-01-3792, 2005.

[16] V. Rao, D. Honnery. "Application of a multi-step soot model in a thermodynamic diesel engine model", Fuel 135 (2014) 69-78.

[17] V. Rao, D. Dayawansa, D. Honnery. "Reducing diesel particulate matter in underground mines by optimising design and operation of diesel exhaust systems" ACARP project report C23013 MTI/DS/631 December 2015, 117 pages.

Experimental investigation of engine performance, emission and combustion characteristics of tea tree oil blend in a CI engine

S.M. Ashrafur Rahman[1], T.C. Van[1], F.M. Hossain[1], A. Dowell[2], M.A. Islam[1], M.N. Nabi[3], A.J. Marchese[4], J. Tryner[4], K. Suara[5], T. Bodisco[6], T. Rainey[1], Z.D. Ristovski[1, 7], R.J. Brown[1]

[1]Biofuel Engine Research Facility, Queensland University of Technology (QUT)

[2]Southern Cross Plant Science, Southern Cross University, PO Box 157, Lismore, NSW 2480, Australia

[3]School of Engineering and Technology, Central Queensland University, Perth, WA 6000, Australia

[4]Department of Mechanical Engineering, Colorado State University, Fort Collins, Colorado 80523, USA

[5]Environmental Fluid Mechanics Group, Queensland University of Technology (QUT)

[6]School of Engineering, Deakin University, Waurn Ponds, Victoria 3217, Australia

[7]International Laboratory for Air Quality and Health, Queensland University of Technology (QUT)

Abstract

Essential oils are mainly used in the aromatherapy and natural medicine sector. However, as high quality oils are needed in this sector, there is a significant essential oil waste-stream which can be used in transport and agricultural sector specially in diesel engines. The aim of this paper is to evaluate a multi-cylinder diesel engine using tea tree oil (TTO) essential oil blended with diesel. The blend contained 90% diesel and 10% TTO (by volume). The diesel engine used was a 160kw modern high speed common rail turbocharged diesel engine. The engine was operated at 1500 rpm and 4 loads. Property-wise, compared to base diesel, TTO-diesel blend has lower viscosity, higher density significantly lower heating value and cetane number. There was minimal variation of brake thermal efficiency and brake specific fuel consumption between base diesel and TTO-diesel blend. However, in-cylinder pressure for TTO-diesel blend was significantly lower compared to base diesel. In most loads, TTO-diesel blend emitted lower NO_X but significantly higher CO and PM emission.

1 Introduction

Essential oils are natural volatile oils, which are accountable for many of the fragrances produced by plants. These oils can be easily obtained using a variety of different extraction techniques such as distillation or solvent extraction [1]. The main difference between vegetable oils and essential oils is that, vegetable oils are usually derived from the fatty parts of a plant, such as the seed, kernels or nuts, while, essential oils are generally derived from non-fatty areas such as roots, bark, leaves, stems and the aromatic portions of the plant. The market of essential oil is growing rapidly; in 2014 and 2015, the global demand was 165 and 179-kilo tonnes respectively [2, 3]. It is forecasted that this demand will surge at a compound annual growth rate of 8.4% from 2016 to 2024 [3]. This rapid growth can be attributed to the growth in natural therapies industry. However, only pure, finest quality essential oils are used to produce medicine in the health industry. This leaves a huge amount of essential oils, which are considered as not suitable for use, for example, if color is not right or the smell is too harsh. Many small oil producers who fail to maintain this "therapeutic grade" quality oil is thus suffering loss. If these oils can be used to operate a diesel engine, it will immensely help these small producers. The use of essential oils as a diesel fuel substitute is a concept that has not been comprehensively exploited in the past especially- this study aims to evaluate this research gap and find out the performance of essential oils as diesel engine fuel. The essential oil that has been selected for this study is TTO; it was blended with neat diesel and used in a six-cylinder engine to evaluate performance, emission parameters and combustion characteristics of the engine.

Corresponding author. H/P: +61410193428
E-mail address: s2.rahman@qut.edu.au

2 Methodology

The engine experiments were conducted at The QUT Biofuel Engine Research Facility (BERF). The engine specification is given in Table 2 and a schematic diagram of the setup is given in Fig. 1The engine was operated at a consistent speed of 1500 rpm (maximum torque speed), and at three different loads – 25, 50% and 100%. The maximum load of the engine at any specific engine speed is governed by type of fuel used. For this reason, at the commencement of each fuel, the maximum load was determined when the engine was at full throttle running at 1500 rpm (considered as 100% load), and it is from here other load % were determined for that speed.

Model	ISBe220 31
Cylinders	6 in-line
Capacity	5.9L
Bore x Stroke	102x120 (mm)
Max Torque	820Nm @ 1500 rpm
Compression ratio	17.3
Fuel Injection	High pressure common rail
Aspiration	Turbocharged
Dynamometer	Electrically controlled Water Brake
Emission Standard	Euro III

Table 1: Engine Specification

Particle number size distribution was measured with a DMS-500 (Cambustion Ltd) with the heated sample line connected. PM mass was calculated from DMS 500 data by using a re-inversion tool in the DMS data analysis suite (version UIv 7.11), as suggested by Jonathan et al. [4] In this case, a density factor of 2.2E-15 and a power coefficient of 2.65 and 5.2E-16 and 3 were applied to accumulation mode particles and to nucleation mode particles, respectively. A CAI 600 series CO_2 analyzer was used to measure the CO_2, and CO concentration directly from the raw exhaust. Testo 350 was used to measure CO and NO_X emission. Exhaust emissions were sampled from the exhaust manifold via a 0.5 m long stainless steel tube- which was sent to the gas analyzers and DMS500 (through heated sampling line) via a copper tube fitted with a HEPA filter. The engine performance data including in-cylinder pressure, diesel injection timing and degrees of crank angle rotation information were recorded. Further detail of the engine performance measurements can be found in Bodisco and Brown [5].

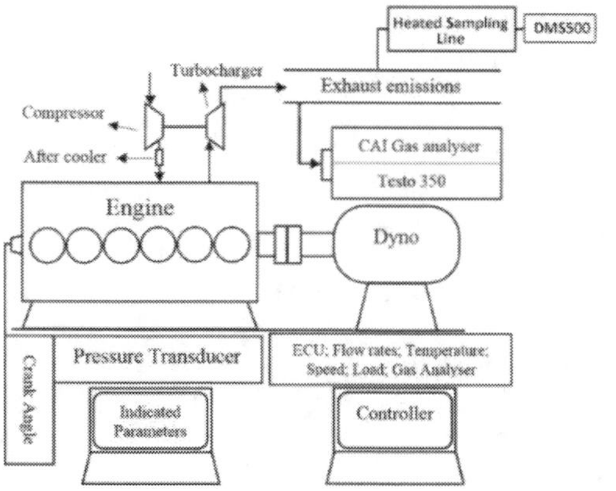

Figure 2: Schematic diagram of engine campaign setup

Properties including density, viscosity, heating value and surface tension were measured at QUT. For measuring properties - a KSV Sigma 702 tensiometer is used to measure the density and surface tension, a cone and plate version of the Brookfield DV-III Rheometer is used to measure the viscosity, and an oxygen bomb calorimeter 6200 model manufactured by Parr Instrument Company is used to measure the heat of combustion or higher heating value (HHV) of fuel samples. Properties measured at Colorado State University- The Advanced Biofuels Combustion and Characterization (ABC2) Laboratory, the Engines and Energy Conversion Laboratory (EECL) at Colorado State University measured Cetane Number with their Qaukesha FIT ignition quality tester. Cetane number is a measurement of diesel fuel ignition and combustion quality thus is an important parameter for an engine test.

3 Result and Discussion

Some of the key properties of base diesel and TTO-diesel blend is listed in Table 2. TTO-diesel blend has lower viscosity, lower heating value and lower cetane number and higher density and higher surface tension. From gas chromatography conducted by Southern Cross Plant Science, Southern Cross University, neat TTO contains around 5.4% oxygen (chemical composition obtained from GC-MS is shown in Table 3).

Property	Unit	100D	10T90D
Density @25°C	Kg/l	0.837	0.841
Kinematic viscosity @40°C	mm^2/s	3.12	2.98
Higher heating value	MJ/kg	45.92	44.98
Lower heating value[a]	MJ/kg	43.95	42.26
Surface tension @25°C	mN/m	26.09	26.32
Cetane Number measured @CSU		43.2	39.1
[a] calculated			

Table 2: Fuel Properties

Compound Name	%
Pinene-A	2.30
Sabinene	0.06
Mycrene	0.69
Phellandrene-A	0.86
Pinene-B	0.61
Terpinene-A	9.69
Cymene-P	2.51
Cineole-1,8	4.41
Terpinene-G	21.50
Isoterpinolene	3.12
Terpinene-4-ol	41.66
Terpineol	4.22
Ledene	1.45
Cadinene-D	1.13

Table 3: Chemical composition of TTO derived from GC-MS

Figure 3 shows the brake thermal efficiency (BTE) and brake specific fuel consumption (BSFC). BTE can be defined as of brake power of a heat engine as a function of the thermal input from the fuel. BTE is used to define the engine how mechanically efficient to convert chemical power of the fuel to useful mechanical power [6]. It is observed that BTE reached a maximum (~37%) at around 50% load for both fuels. This is consistent with other published results [5, 7, 8]. A slight increase in BTE with the TTO-diesel blend was observed. The increase in BTE can be attributed to increasing oxygen content in the blend.

BSFC is typically used for comparing the mass of fuel consumed by engines with output power. It is the ratio between rate of fuel consumption and the brake power. However, from Fig 3, an insignificant variation in BTE was observed with the tested fuels. Variation of BSFC for reference diesel and their blends is 219 – 223.7 g/kWh comparing all loads. Zare et al. [9] reported similar results. Increase in BSFC can be attributed to lower heating value of TTO-diesel blend [10].

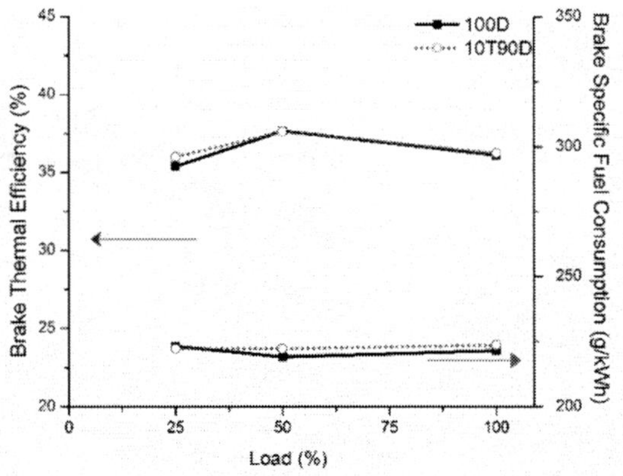

Figure 3: BTE and BSFC for 100D and 10T90D at 1500 rpm and 25, 50 and 100% load

From Figure 4, the in-cylinder pressure with 100D was found to be near 11000 kPa, much higher than 10T90D. The lower in-cylinder pressure with TTO-diesel blend can be attributed to lower energy content (HHV), lower cetane number of the fuel.

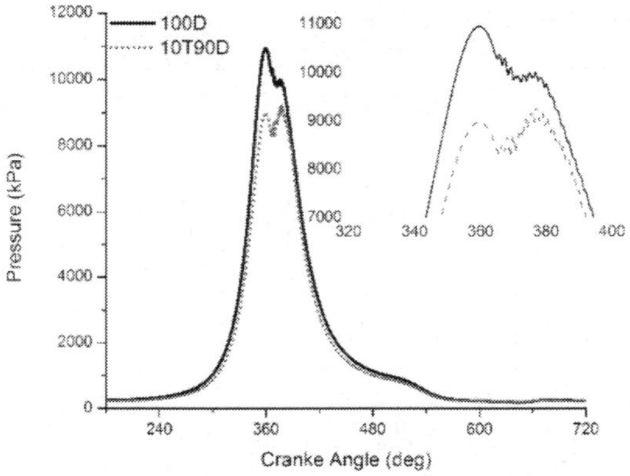

Figure 4: In-cylinder pressure variation for 100D and 10T90D at 1500 rpm and 100% load

From Fig 5 the ignition delay for TTO-diesel blend was slightly longer compared to base diesel due to lower cetane number. Due to being more volatile, TTO evaporates more readily than diesel increasing the quantity of premixed fuel and air in the combustion chamber. The latent heat required to do this corresponds to lower pressure (Fig 4) and higher negative heat release rate (Fig 5), which means the mixture is cooler, not releasing as much heat as base diesel.

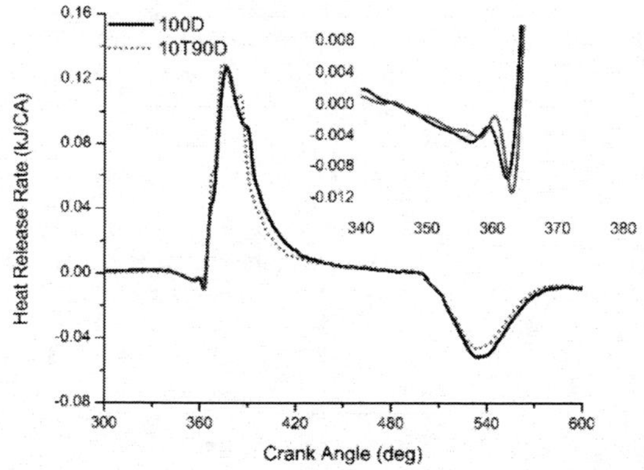

Figure 5: Heat release rate for 100D and 10T90D at 1500 rpm and 100% load

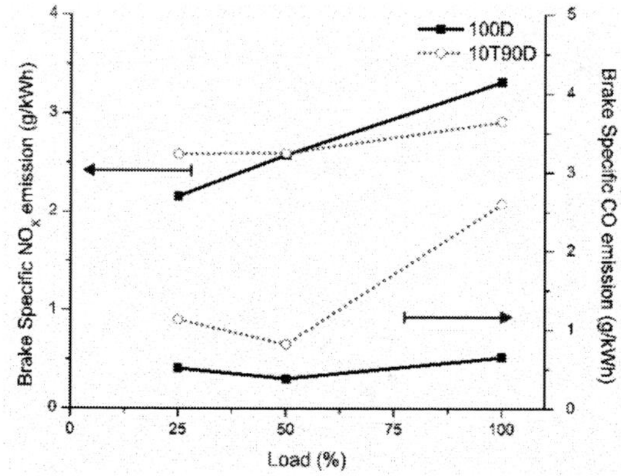

Figure 6: Brake specific NOx and CO emission for 100D and 10T90D at 1500 rpm and 25, 50 and 100% load

Figure 6 shows brake specific NOx and CO emission for 100D and 10T90D at 25, 50 and 100% load. From the figure, it is evident that, compared to base diesel, TTO-diesel blend deteriorates combustion thus CO emission increases although having higher oxygen content. Furthermore, having lower cetane number and due to incomplete combustion, NOx emission decreases compared to base diesel.

Lower cetane number means the fuel will have shorter residence time (chemical) at elevated temperature thus will have lower NOx emission. Fuel properties such as lower viscosity, density and the sound velocity of TTO-diesel blend can also be attributed to lower NOx emission, which is suggested by several researchers [11-13].

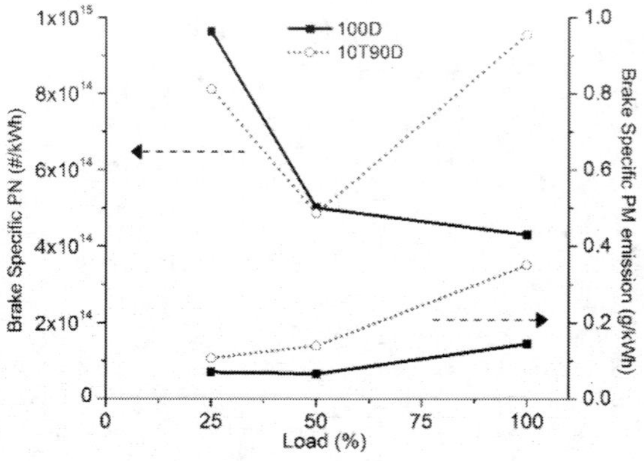

Figure 7: Brake specific PM and PN emission for 100D and 10T90D at 1500 rpm and 25, 50 and 100% load

TTO-diesel blends have lower viscosity and cetane number, which results in higher PM emission at all loads. Less viscous fuels have lower cetane number which results in longer ignition delay, ignition moves more towards TDC. Thus, combustion products, have less resident time at elevated temperature, which results in less oxidation of soot precursors [14]. Consequently, PM emission increases. Furthermore, from Fig 7 brake specific PN for TTO blends were lower at low loads but was significantly higher at higher load compared to neat diesel fuel.

4 Conclusions

The following results are obtained: in the current study, engine performance and exhaust emissions were investigated with TTO-diesel blend in a six-cylinder fully instrumented turbocharged diesel engine. The engine was operated at 1500 rpm and 25, 50 and 100% load. The results of this study maybe summarised as follows:

- TTO-diesel blend has lower viscosity, significantly lower heating value and cetane number, higher density.
- Minimal variation of brake thermal efficiency and brake specific fuel consumption between base diesel and TTO-diesel blend.
- In-cylinder pressure for TTO-diesel blend was significantly lower compared to base diesel.
- Due to having lower cetane, TTO has higher ignition delay compared to base diesel
- In most loads, TTO-diesel blend emitted lower NOx but significantly higher CO and PM emission.
- At higher loads PN significantly increased for TTO blend

5 Acknowledgment

The authors would like to thank the QUT for providing PhD scholarship. The authors also thank Mr. Noel Hartnett for maintaining the engine for research, Michael Bongard, Lauren Butler and Nathaniel Raup for their valuable help in measuring various fuel properties. Lastly, the authors would like to show gratitude to Australian Tea Tree Oil exchange for supplying essential oils for testing.

References

[1] El-Mougy NS. Effect of some essential oils for limiting early blight (Alternaria solani) development in potato field. Journal of Plant Protection Research. 2009;49(1):57-62.

[2] Grandview-Research. Essential Oil Market Analysis By Product (Orange, Corn Mint, Eucalyptus, Citronella, Pepper Mint, Lemon, Clove Leaf, Lime, Spearmint), By Application (Medical, Food & Beverage, Spa & Relaxation, Cleaning & Home) And Segment Forecasts To 2022. 2015.

[3] Grandview-Research. Essential Oil Market Analysis By Product (Orange, Corn Mint, Eucalyptus, Citronella, Peppermint, Lemon, Clove Leaf, Lime, Spearmint), By Application (Medical, Food & Beverage, Spa & Relaxation, Cleaning & Home) And Segment Forecasts To 2024. 2016.

[4] Symonds JP, Reavell KSJ, Olfert JS, Campbell BW, Swift SJ. Diesel soot mass calculation in real-time with a differential mobility spectrometer. Journal of Aerosol Science. 2007;38:52-68.

[5] Bodisco T, Brown RJ. Inter-cycle variability of in-cylinder pressure parameters in an ethanol fumigated common rail diesel engine. Energy. 2013;52:55-65.

[6] Nabi MN, Rahman MM, Islam MA, Hossain FM, Brooks P, Rowlands WN, et al. Fuel characterisation, engine performance, combustion and exhaust emissions with a new renewable Licella biofuel. Energy Conversion and Management. 2015;96:588-98.

[7] Nabi MN, Zare A, Hossain FM, Rahman MM, Bodisco TA, Ristovski ZD, et al. Influence of fuel-borne oxygen on European Stationary Cycle: Diesel engine performance and emissions with a special emphasis on particulate and NO emissions. Energy Conversion and Management. 2016;127:187-98.

[8] Nabi MN, Zare A, Hossain FM, Rahman MM, Stuart D, Ristovski ZD, et al. Formulation of new oxygenated fuels and their influence on engine performance and exhaust emissions. Proceedings of the 2015 Australian Combustion Symposium: The Combustion Institute Australia and New Zealand Section; 2015. p. 64-7.

[9] Zare A, Nabi MN, Bodisco TA, Hossain FM, Rahman M, Ristovski ZD, et al. The effect of triacetin as a fuel additive to waste cooking biodiesel on engine performance and exhaust emissions. Fuel. 2016;182:640-9.

[10] Islam MA, Rahman MM, Heiman K, Nabi MN, Ristovski ZD, Dowell A, Thomas G, Feng B, Alvensleben NV, Brown RJ. Combustion analysis of microalgae methyl ester in a common rail direct injection diesel engine. Fuel, 2015; 143: 351-360.

[11] Nettles-Anderson SL, Olsen DB. Survey of straight vegetable oil composition impact on combustion properties. SAE Technical Paper; 2009; 2009-01-0487, 2009.

[12] Yuan W, Hansen A. Computational investigation of the effect of biodiesel fuel properties on diesel engine NOx emissions. International Journal of Agricultural and Biological Engineering. 2009;2:41-8.

[13] Tat ME, Van Gerpen JH. Speed of sound and isentropic bulk modulus of alkyl monoesters at elevated temperatures and pressures. Journal of the American Oil Chemists' Society. 2003;80:1249-56.

[14] Vallinayagam R, Vedharaj S, Yang W, Roberts WL, Dibble R. Feasibility of using less viscous and lower cetane (LVLC) fuels in a diesel engine: A review. Renewable and Sustainable Energy Reviews. 2015;51:1166-90.

11th Asia-Pacific Conference on Combustion,
The University of Sydney, NSW Australia
10th -14th December 2017.

Analysis of Vortex Dynamics behind Bluff Body Burners using the Discrete Vortex Method

Zetian Ren[1], Haojie Tang[2], Chih-Yu Kuo[3], Min Zhu[1]

[1]Department of Thermal Engineering, Tsinghua University, Beijing, 100084, China

[2]Central Research Academy of DEC, Chengdu, 611731, China

[3]Division of Mechanics, Research Center for Applied Sciences, Academia Sinica, Taipei 115, Taiwan

Abstract

The two-dimensional discrete vortex method is combined with conformal mapping to investigate the vortex dynamics behind bluff body burners. The method is computationally inexpensive and can be easily applied to complicated geometries. The vortex dynamics of three bluff body burners with different geometries are simulated. The vortex shedding frequency is proportional to the inflow velocity. The existence of high temperature flame accelerates the vorticity dissipation thus restrain the vortex instabilities. More large eddies and wrinkles are present under higher harmonic of velocity forcing. When an additional central flow channel exists, the vortex interactions are intense when the forcing of side and central channels are in phase. This work provides a better understanding of the vortex dynamics behind bluff body burners.

1 Introduction

Bluff bodies are widely used to stabilize flames in combustors since they have a recirculation zone behind the burners enhances the entrainment and mixing[1]. The typical flow configuration behind a bluff body includes the shear layer, the recirculation zone, and the wake flow behind. Both the Kelvin-Helmholtz instability in the shear layers and the Karman Vortex Street in the wake flow can influence the hydrodynamic behavior and hence the combustion instabilities. A comprehensive understanding of the vortex dynamics after bluff body burners is desirable to guide the design of stable combustors.

Studies have focused on the vortex dynamics in existence of flames, which are quite different from those in non-reacting flows. Flame lead to higher mixture temperatures and higher viscosities, which significantly increases the vortex decay rate. Also, both the gas expansion and the baroclinic effect near the flame can inhibit the vortex band development, and reduce the flow instabilities. Erickson[2] compared the vorticity fields of non-reacting and reacting conditions, which showed less large scale vortex structures in reacting flows. Shanbhogue[3] measured the convective instabilities in reacting flows in the presence of disturbances and found that the vortex sheds from the edge of the bluff body with the disturbance frequency and transports downstream. The vortex becomes larger when the shear layer area is extended. Ayache[4] simulated the flow field behind an enclosed axisymmetric bluff body using Large-Eddy Simulations to explore the influences of acoustic forces at different frequencies on the flow. Sundaram[5] did three-dimensional, unsteady numerical simulations of turbulent, bluff body stabilized lean premixed flames using the shear stress transport $k-\omega$ model, a detailed combustion mechanism, and the eddy dissipation concept. These numerical studies provided new understanding of the vortex dynamics in bluff body flows.

Corresponding author. Phone: +86-13810086132
E-mail address: zhumin@mail.tsinghua.edu.cn

The vortex dynamics around industrial bluff body burners with complicated geometries, especially with harmonic velocity disturbances, needs further understanding. However, three-dimensional simulations using the direct numerical simulations or LES method are always time consuming and expensive. This study uses a more convenient, computationally inexpensive two-dimensional discrete vortex method[6] to investigate the vortex dynamics after bluff body burners for various conditions. Conformal mapping[7] is also applied to further facilitate the calculations, especially for complicated geometries.

2 Method

2.1 Conformal Mapping

According to Riemann's theorem, a complicated physical domain can be projected into a simple standard plane by conformal mapping. Figure. 1(a) shows the physical z field regarded as a polygon area with 7 vertices and Fig. 1(b) is a standard semi-infinite λ plane. The Schwartz-Christoffel conformal mapping formula for a semi-infinite plane is[7]

$$z = f(\lambda) = K_m \int_0^{\lambda} \prod_{j=1}^{6} (\xi - \lambda_j)^{\alpha_j - 1} d\xi + C_m \quad (1)$$

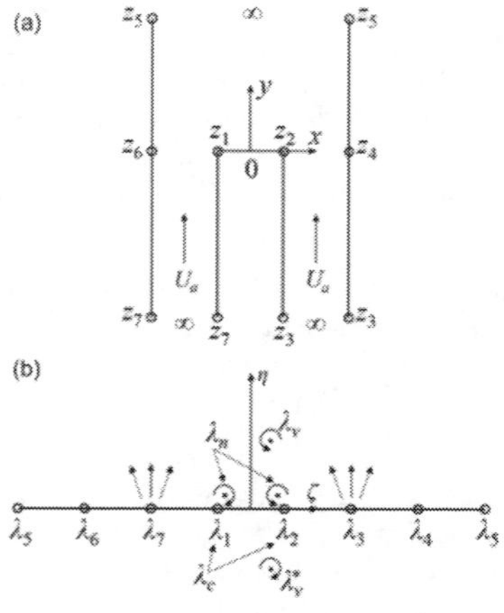

Figure 1. Schematic of the conformal mapping planes. (a) physical plane, (b) standard mapping plane.

where K_m and C_m are coefficients, and α_j are the angles of the polygon vertices in rad. The mapping from the physical plane to the standard plane can also be achieved by using the inverse transformation. The conformal mapping method can be easily applied to complex structures with several air/fuel channels, with different bluff body shapes, or with area expansions.

2.2 Discrete Vortex method

The discrete vortex method mainly describes the mechanisms of vortex element generation, vortex motion and vorticity dissipation. The fundamental assumption is that the viscosity only influences the flow near the flow separation point, while the flow is ideal and incompressible in other areas. The model assumes that a vortex element is shed from the bluff body corner at each calculation time step. The development of the whole flow can then be solved by tracing the motions of all the vortex elements using the Lagrangian method. The vortex elements and the inflow from the two ducts are treated as line vortices and sources. Then there exists a complex potential in both the physical and mapping plane:

$$F(z) = \phi + i\psi = W(\lambda)$$
$$= \sum_{k=1}^{K} \frac{\dot{V}_{s,k}}{\pi} \ln(\lambda - \lambda_{s,k}) + \sum_{k=1}^{M} \frac{\Gamma_{v,k}}{2\pi i} \ln(\frac{\lambda - \lambda_{v,k}}{\lambda - \lambda^*_{v,k}}) \quad (2)$$

where $\Gamma_{v,k}$ and $\lambda_{v,k}$ are the circulation intensities and positions of the M existing vortex elements. $\lambda^*_{v,k}$ is the complex conjugate of vortex element $\lambda_{v,k}$, as sketched in Fig. 1(b). The image system of the vortices is included to the mapping plane, helping to full fill the boundary condition of zero normal velocity on the walls in physical plane. The flow field is driven by two line sources located at the infinity in the two ducts, and $\dot{V}_{s,k} = U_a(b-a)$ denotes the strength of the concentrated line source, where U_a is the inflow velocity and $b-a$ is the width of the two ducts. The velocity at physical point z, can then be obtained from the complex potential,

$$U^*_z = u - iv = \frac{dF}{dz} = \frac{dW/d\lambda}{dz/d\lambda} \quad (3)$$

The circulation Γ_n of the newly generated vortex element is obtained to satisfy the Kutta condition[8],

$$\Gamma_n = \frac{2\pi i(\lambda_c - \lambda_n)(\lambda_c - \lambda^*_n)}{\lambda_n - \lambda^*_n} W'_{\lambda_n,\lambda^*_n}(\lambda_c) \quad (4)$$

where λ_c denotes the corner points, and λ_n denotes the position of the newly generated vortex element after one time step, as sketched in Fig. 1(b). $W'_{\lambda_n,\lambda^*_n}(\lambda_c)$ is the complex potential of the existing vortex elements and line sources.

After their generation from the bluff body, the vortex elements will decay continuously. The existence of high-temperature flames will increase the viscosity and accelerate the vorticity dissipation. In this study, the vorticity dissipation was simulated by a convenient decay model developed by Kiya[9],

$$\frac{\Gamma(t)}{\Gamma(0)} = 1 - \exp\left[-\frac{KD^2}{\nu t}\right] \quad (5)$$

The gas viscosity ν is a function of temperature, so the vortex motion with flame existence is simulated by increasing the viscosity from 1.54×10^{-5} m^2/s for 298 K cold flow to 3.86×10^{-4} m^2/s for 2000 K reacting flow. Although this method is an approximation, it still exhibits the main characteristics of the vortex motion.

The discrete vortex method was validated through the experiment data of a triangular bluff body flow[10]. Table 1 shows that the simulated recirculation zone length L_r/D, the vortex shedding frequency f and the St number agree well with the experiment.

Case	Re	L_r/D	f/Hz	St
Experiment	45000	1.33	105	0.25
Simulation	45000	1.34	106.3	0.256

Table 1: Comparison between experiment[10] and simulation.

3 Results

3.1 Self-induced instability

The flow and vortex dynamics were analyzed for three different bluff body burner structures. For a simple backward step bluff body in an enclosed chamber, Fig. 2 shows the distributions of the point vortices at different phases in one Karman Vortex Street cycle. This is for a cold case with Re = 4000. Starting from phase 0°, the vortex elements shed from the left corner gradually form into a vortex sheet behind the bluff body and then form a large vortex. The circulation intensity of the large vortex increases with time and the large vortex on the left pushes the one on the right to move downstream. The large vortex on the right finally sheds from its vortex sheet at phase 180°. Then, the vortex sheet on the right influences the motion of the large vortex on the left and results in the shedding of the left large vortex. The whole process is repeated to form a stable Karman Vortex Street. By analyzing the velocity fluctuations at the monitor points, the frequency of the unstable vortex street f_{BVK} can be obtained as shown in Fig. 3. The main frequency of 76 Hz and the corresponding high order harmonics are presented. Cases with different inflow velocities are calculated, and Fig. 4 shows that the vortex shedding frequency is

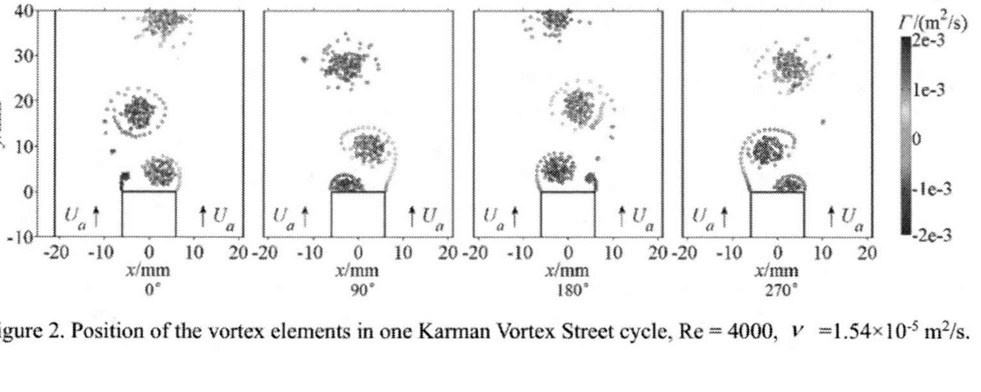

Figure 2. Position of the vortex elements in one Karman Vortex Street cycle, Re = 4000, ν =1.54×10^{-5} m^2/s.

proportional to the velocity of incoming flow U, and the Strouhal number defined by St = $f_{BVK}D/U$ is 0.185, which is a reasonable value compared with literature[11].

We then considered the influence of flame existence on vortex dynamics. The viscosity coefficient is set to 3.86×10^{-4} m²/s to simulate the temperature of 2000 K. The vorticity dissipation is accelerated and the vortex elements disappear from the flow field at about 15mm behind the bluff body, as shown in Fig. 5. The vortex elements cannot be accumulated into large eddies behind the bluff body so the Kaman Vortex Street does not turn up. The distribution of the free vortex elements is stable, thus the velocity field and the circulation of the new vortex elements stay unchanged. The flame front is attached at the bluff body corners and the combustion is quite stable.

3.2 Vortex dynamics with harmonic forcing

In this section, the vortex shedding process and the unsteady movements of the vortex structures with harmonic forcings are studied. The harmonic velocity forcings are introduced to the mean inflow velocity, $U_a(t)=\bar{U}_a+U_a'\sin(2\pi ft)$. Figure. 6 shows the position of the vortex elements under forcing frequencies for a bluff body with a gradually growing width. The viscosity is 3.84×10^{-4} m²/s to simulate the case with flame existence, Re = 12000 and the external forcing amplitude is $U_a'/\bar{U}_a=0.4$. For the case with 60 Hz forcing, the circulations of the new vortices change periodically with the largest at 90° and the smallest at 270°. The vortex elements in the vortex sheets accumulate into two large vortexes before they are separated from the vortex sheets. In premixed combustion, the flame fronts are located between the two vortex sheets, so the coupling between the large vortexes and the reaction zone will change the flame structures and the flame instabilities will be convected downstream by the flow. The result shown in Fig. 6(b) is for a higher forcing frequency of 180 Hz. There are four large vortexes in each vortex sheet, which mean that the vortex element suffers more forcing cycles under high frequency during its lifetime. More large vortexes then mean more wrinkles in the flame fronts. However, more large vortexes also mean that the circulation intensity of each large vortex will decrease and their influences on the velocity field will be

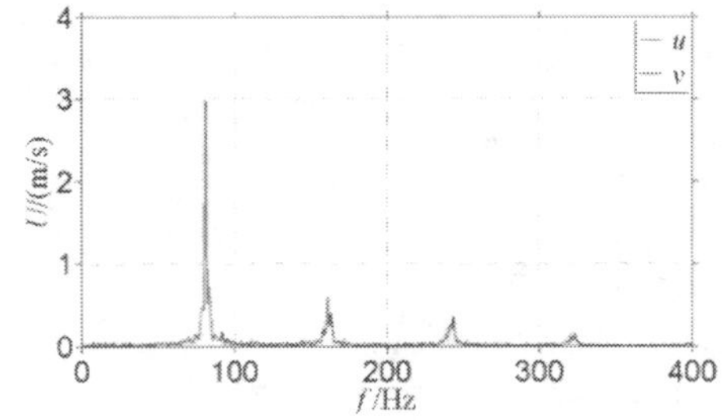

Figure 3. Frequency analysis of velocity components at x = 12 mm, y = 12 mm. Re = 4000, ν = 1.54×10^{-5} m²/s.

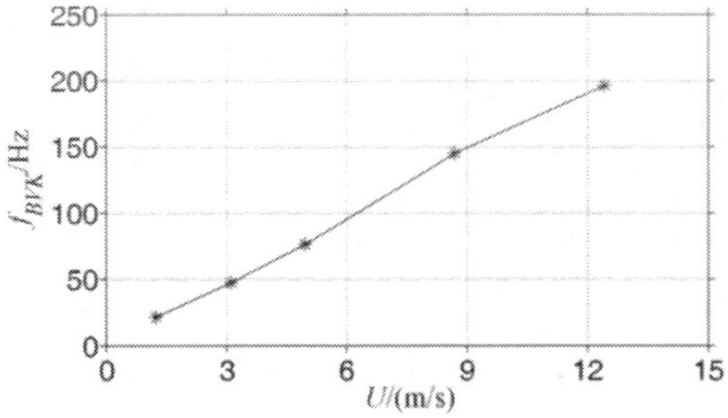

Figure 4. Vortex shedding frequency at different flow velocities.

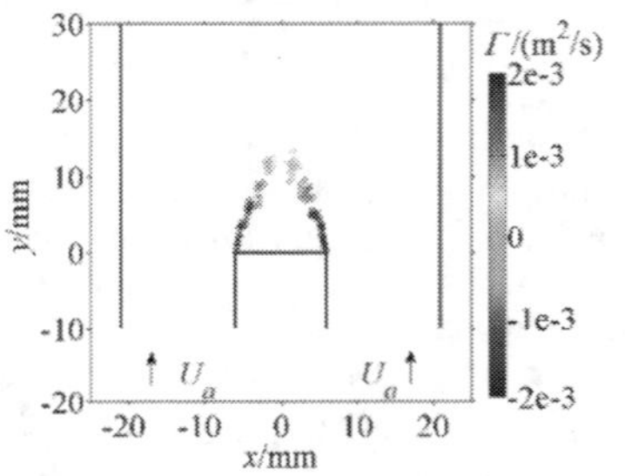

Figure 5. Positions of the vortex elements for high temperature case, Re = 4000, ν = 3.84×10^{-4} m²/s.

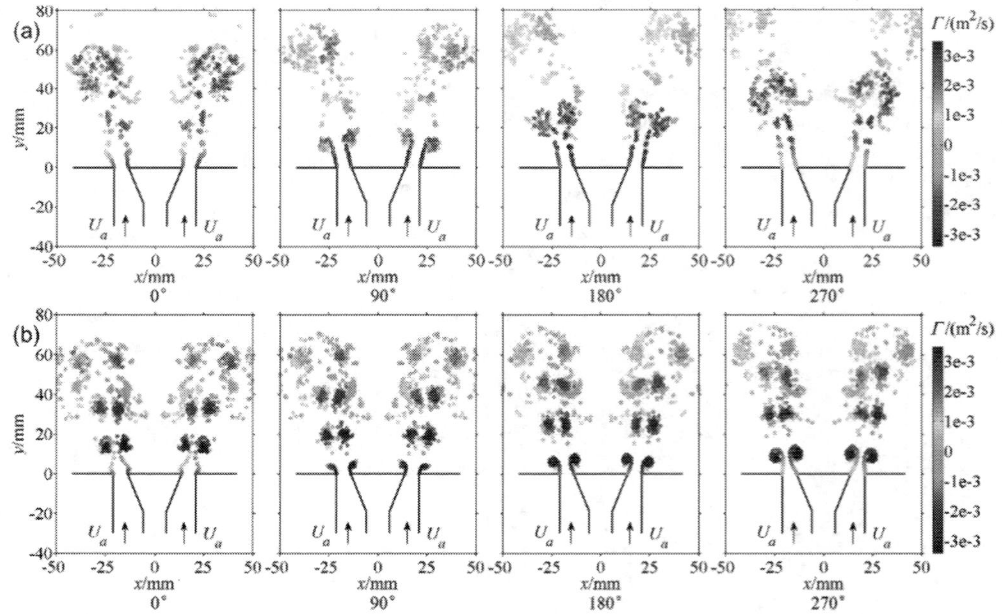

Figure 6. Positions of the vortex elements in one harmonic forcing cycle, Re = 12000, ν = 3.84×10^{-4} m²/s, $U_a'/\bar{U}_a=0.4$, (a) f = 60 Hz, (b) f = 180 Hz.

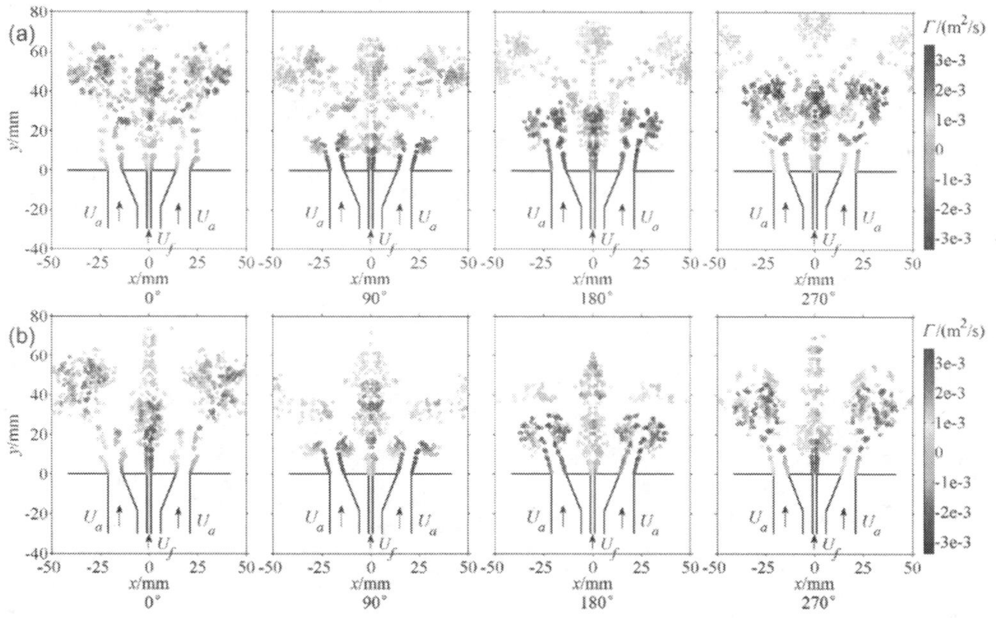

Figure 7. Positions of the vortex elements in one harmonic forcing cycle, Re = 12000, ν = 3.84×10^{-4} m^2/s, f = 60 Hz, $U_f'/\bar{U}_f = U_a'/\bar{U}_a = 0.4$, with U_f' and U_a' are (a) in phase (b) opposed in phase.

smaller. This is one reason for the low-pass property of the disturbed flames as described in previous studies[12].

For the burner with an additional channel in the center of the bluff body, the air flows through the two side channels of the bluff body while the fuel flows through the central channel. The forcing frequencies are 60 Hz and the forcing amplitudes are 0.4 times of the mean velocity $\bar{U}_f = \bar{U}_a = 6.6$m/s . The distributions of the vortex elements in one harmonic forcing cycle are shown in Fig. 7. When the air and fuel channel forcings are in phase, the vortex shedding and motion from the two streams also have the same phase. At 90°, the circulations of the new vortices at the sides and along the center line are both maximized and the large eddies are close to each other; thus, the interactions are strongest at this moment. The intense interactions between the air and fuel eddies will result in better mixing. Figure. 7(b) shows the results when the two channels have opposed forcing phases. When the side flow velocity forcing phase is 90°, the circulations of the large eddies from the two side channels are very large and are close to the vortex sheets from the central channel. This promotes the shedding of the vortex sheets from the central channel, forming an isolated large eddy. At 270°, the situation has reversed which leads to the vortex structures shedding from the two side channels. In this case, the side eddies have only weak interaction with the central eddies, resulting in little mixing effect.

4 Conclusions

The paper provided a convenient, computationally inexpensive two-dimensional discrete vortex method to simulate the vortex dynamics of the bluff body flows. The results revealed that the bluff body burner geometry, Reynolds number, high-temperature flame, and harmonic velocity forcing all have influences on the vortex dynamics, which is important for the fuel-air mixing and flame dynamics of the bluff body stabilized combustion system.

5 Acknowledgment

The authors are grateful for the support from the National Natural Science Foundation of China (Grant No.51376107).

References

[1] S.J. Shanbhogue, S. Husain, T.C. Lieuwen, Progress in Energy and Combustion Science 35 (1) (2009) 98-120.

[2] R.R. Erickson, M.C. Soteriou, P.G. Mehta, The Influence of Temperature Ratio on the Dynamics of Bluff Body Stabilized Flames, Proceedings of the 44th AIAA Aerospace Sciences Meeting and Exhibit, Nevada, US, 2006-753.

[3] S.J. Shanbhogue, *Dynamics of perturbed exothermic bluff-body flow-fields*, Master thesis, Georgia Institute of Technology, Atlanta, US, 2008.

[4] S. Ayache, J. Dawson, A. Triantafyllidis, R. Balachandran, E. Mastorakos, International Journal of Heat and Fluid Flow 31 (5) (2010) 754-766.

[5] S.S. Sundaram, V. Babu, C. Obulesu, R. Sivakumar, Combustion Science and Technology 184 (3) (2012) 351-373.

[6] S.G. Voutsinas, International Journal of Computational Fluid Dynamics 20 (1) (2006) 3-18.

[7] T.A. Driscoll, L.N. Trefethen, Schwarz-Christoffel Mapping, Cambridge University Press, Cambridge, U.K., 2002, p. 10.

[8] D.G. Crighton, Annual Review of Fluid Mechanics 17 (1) (1985) 411-445.

[9] M. Kiya, K. Sasaki, M. Arie, Journal of Fluid Mechanics 120 (1982) 219-244.

[10] A Sjunnesson, C Nelsson, E Max, Laser Anemometry 3 (1991) 83-90.

[11] B. Cantwell, D. Coles, Journal of Fluid Mechanics 136(1) (1983) 321-374.

[12] D. Durox, T. Schuller, S. Candel, Proc. Combust. Inst. 30 (2) (2005) 1717-1724.

The effect of wall heat loss on the flame behavior in a sudden-expansion channel

Ching-Wei Wu[1], Zheng-Yen Lin [1], Sheng-Yen Hsu [1]

[1] Department of Mechanical & Electro-Mechanical Engineering, National Sun Yat-sen University, Kaohsiung

No.70, Lianhai Rd, Gushan Dist, Kaohsiung City 804, Taiwan (R.O.C.)

Abstract

The flame behaviors in a mesoscale one-stepped sudden-expansion channel are numerically investigated by a two-dimensional model without symmetric assumption at the centerline. The channel expansion ratio β is set equal to 2. The complete Navier-Stokes equations and one-step chemical reaction are used. The wall heat loss adjusted by varying the external heat convection coefficient ($h \geq 50$ W/m$^2\cdot$K).

Three types of flame behaviors are found: steady slant-shaped flame, unsteady vibrating flame and periodically oscillating flame. The steady slant flame exists if the heat loss is low (50 W/m^2 K). As the wall heat loss increases, the flame becomes unsteady and vibrating (h = 60 ~ 93 W/m$^2\cdot$K). In addition, if the wall heat loss further increases, the periodically oscillating flame is observed (h > 96 W/m$^2\cdot$K).

1 Introduction

Nowadays, due to the successful development of micro-electro-mechanical systems, the further studies on the micro-and/or meso-scale combustion features have drawn many researches' attention. Some of important flame characteristics and interesting flame behaviors were clarified by the experiments and numerical models [1].

Norton and Vlachos [2] used a two-dimensional unsteady model to investigate the combustion characteristics and flame stability in a micro-scale channel. The wall conductivity and thickness are extremely important to determinate the upstream heat transfer, which affects the flame ignition and stability. Two modes of flame extinction, wall thermal quenching and blow-off, were simulated and identified. In addition, they also found the repeating ignition/extinction flame behavior due to the thermal effect of channel wall. Pizza et al. [3,4] numerically investigated the premixed flames with a wall temperature gradient in two-dimensional channels of different widths. They also found the repeating ignition /extinction flame behavior. Moreover, in their study, they found that different types of flame behavior can exist at a certain inlet flow velocity according to the initial condition.

Ju and Xu [5, 6] studied the flame propagation in a mesoscale tube with the emphasis on the effects of heat loss by wall and the wall-flame interaction via heat recirculation. In their studies, two flame speed regimes were experimentally observed and in good agreement with the theory and numerical simulation. They also found mesoscale flames can propagate faster in less adiabatic tubes at appropriate conditions of flow velocities and wall thermal properties. Wall thermal effect is one of the significant factors on mesoscale flame characteristics. Daou and Matalon [7] studied the premixed flames in small channels with the heat loss on wall. They focused on the heat loss intensity, the channel width, and the mean flow velocity. According to channel width, they discuss two possible modes of extinction: total flame extinction and partial flame depends on heat loss. In Tsai's study [8], symmetric mushroom-shaped flame is found in a small duct and only an asymmetrical slant-shaped flame can exist in a larger one (mesoscale) owing to the Landau-Darrieus instability. The hydrodynamic flame instability still plays a significant role on the flame behavior in mesoscale. Liu and Hsu [9] used a two-dimensional model without symmetry assumption at centerline to investigate flame behaviors in a meso-scale one-stepped sudden-expansion. Four modes of flame behaviors were found: the steady mushroom-shaped flame, the steady tulip-shaped flame, the simple periodically oscillating flame and the complex periodic oscillating flame. The different flames exists by virtue of the mean velocity of inlet, flame instability, wall- quenching effect and symmetry of the flow field.

From the previous studies, the flame dynamic behavior is attributed to the diffusive-thermal instability due to the wall-flame interaction. However, some scholars indicated that the flame instability (Landau-Darrieus instability) also plays an important role in a meso-scale combustion. In this study, the flame behavior in a mesoscale sudden-expansion channel is investigated by changing the magnitude of wall heat loss.

2 Numerical model

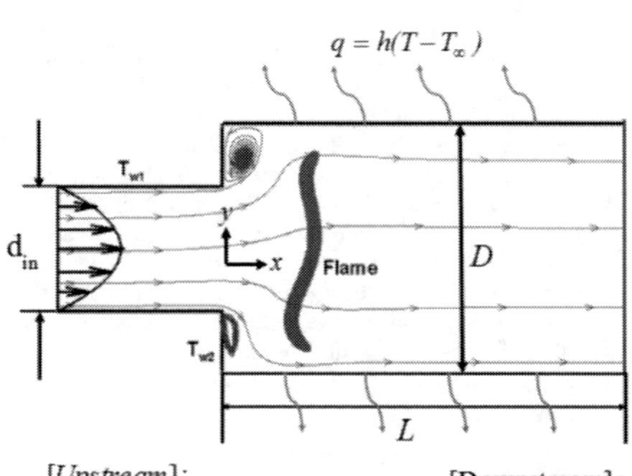

[Upstream]:

$$u = 0.65 \times 1.5[1-(\frac{y}{d_{in}})^2],$$

$$v = 0, Y_i = Y_{i,\infty}$$

$$T_{w1} = T_{w2} = T_\infty = 300K$$

[Downstream]:

$$\frac{\partial u}{\partial x} = 0, \frac{\partial v}{\partial x} = 0,$$

$$\frac{\partial Y_i}{\partial x} = 0, \frac{\partial T}{\partial x} = 0$$

Figure 1. Two-dimensional model configuration and boundary condition.

Corresponding author. Fax: +886-7-525-4299
E-mail address: syhsu@mail.nsysu.edu.tw

The configuration of the two-dimensional model used in this study is shown in Figure 1. The premixed methane-air mixture enters the combustor by a prescribed Poiseuille flow. The wall thickness is neglected. In addition, the upstream narrow duct and the sudden-expansion steps are fixed at a low temperature. The downstream wider duct have wall heat loss by convection. In this study, the flame is expected to exist after the sudden-expansion steps.

The physical model is based on the following assumptions: (1) all species have constant and equal specific heat, (2) the mass diffusion coefficients are equal and the Lewis number is of unity, (3) the radiative heat transfer is not considered, (4) the work due to the pressure drop is ignored, (5) the Soret and Dufour effects are neglected, and (6) no gravity effect. With the above assumptions, the governing equations can be simplified as the following:

Continuity equation:

$$\frac{\partial \rho}{\partial t}+\frac{\partial \rho u}{\partial x}+\frac{\partial \rho v}{\partial y}=0 \tag{1}$$

X-direction momentum equation:

$$\frac{\partial \rho u}{\partial t}+u\frac{\partial \rho u}{\partial x}+v\frac{\partial \rho u}{\partial y}=-\frac{\partial P}{\partial x}+\frac{\partial}{\partial x}(\frac{4}{3}\mu\frac{\partial u}{\partial x}) -\frac{\partial}{\partial x}(\frac{2}{3}\mu\frac{\partial v}{\partial y})+\frac{\partial}{\partial y}(\mu\frac{\partial u}{\partial y})+\frac{\partial}{\partial y}(\mu\frac{\partial v}{\partial x}) \tag{2}$$

Y-direction momentum equation:

$$\frac{\partial \rho v}{\partial t}+u\frac{\partial \rho v}{\partial x}+v\frac{\partial \rho v}{\partial y}=-\frac{\partial P}{\partial y}+\frac{\partial}{\partial y}(\frac{4}{3}\mu\frac{\partial v}{\partial y}) -\frac{\partial}{\partial y}(\frac{2}{3}\mu\frac{\partial u}{\partial x})+\frac{\partial}{\partial x}(\mu\frac{\partial v}{\partial x})+\frac{\partial}{\partial x}(\mu\frac{\partial u}{\partial y}) \tag{3}$$

Energy equations:

$$C_p(\frac{\partial \rho T}{\partial t}+u\frac{\partial T}{\partial x}+v\frac{\partial T}{\partial y})= k(\frac{\partial^2 T}{\partial x^2}+\frac{\partial^2 T}{\partial y^2})+Q\omega_{CH_4} \tag{4}$$

Species equation:

$$\frac{\partial \rho Y_{CH4}}{\partial t}+\frac{\partial \rho u Y_{CH4}}{\partial x}+\frac{\partial \rho v Y_{CH4}}{\partial y}= \frac{\partial}{\partial x}(\rho D\frac{\partial Y_{CH4}}{\partial x})+\frac{\partial}{\partial y}(\rho D\frac{\partial Y_{CH4}}{\partial y})+\dot{\omega}_{CH4} \tag{5}$$

$$\frac{\partial \rho Y_{O_2}}{\partial t}+\frac{\partial \rho u Y_{O_2}}{\partial x}+\frac{\partial \rho v Y_{O_2}}{\partial y}= \frac{\partial}{\partial x}(\rho D\frac{\partial Y_{O_2}}{\partial x})+\frac{\partial}{\partial y}(\rho D\frac{\partial Y_{O_2}}{\partial y})+\dot{\omega}_{O_2} \tag{6}$$

Ideal gas equation:

$$P=\rho RT \tag{7}$$

In this study, the methane-air combustion are modeled by a simple one-step forward reaction and nitrogen is assumed to be an inert gas:

$$CH_4+2(O_2+3.76N_2)\rightarrow CO_2+2H_2O+7.52N_2 \tag{8}$$

The volumetric mass consumption rate of methane is represented by the Arrhenius form of second order:

$$\dot{\omega}_{CH_4}=-Ae^{-E/R_uT}\rho^2 Y_{CH_4}Y_{O_2} \tag{9}$$

Where E indicates the activation energy and A is the pre-exponential factor. The kinetic parameters are from [6] and according to Equation (8), the volumetric mass consumption rate of O_2 is:

$$\dot{\omega}_{O_2}=-4\times\dot{\omega}_{CH_4} \tag{10}$$

In order to correctly simulate the flame behaviors, a fully two-dimensional channel without symmetric assumption at the centerline is applied. It is assumed that the upstream narrow walls (include the steps), the methane-air unburned mixture and ambient temperature are isothermal (T_∞ = 300K). As illustrated in Fig. 1, the inlet velocity at the upstream boundary is prescribed as a Poiseuille flow with an average velocity (u_{ave}), 0.65 m/s. The downstream outlet is set as the gradient free conditions. In this study, the diameter d of the narrow channel and the diameter D of the wide channel are respectively 2.5 mm and 5 mm, so the channel expansion ratio β (D/d) is 2. The length of the downstream duct (L) is 52mm. The simulation results of flame behaviors were carried out without the influence of the downstream boundary condition.

The commercial software FLUENT [10] is used to help with the establishment of the model and the numerical calculation of the governing equations, which are discretized by using the control volume method and the finite-difference technique with the fully implicit and power-law schemes. The SIMPLE algorithm [11] is applied to solve the coupling pressure and velocity fields. The computations are performed by using the time step equal to 0.0001 sec and the finest mesh size (0.0463mm) is clustered in the flame region to provide good numerical results.

3 Result and Discussions

In this study, the heat convection coefficient (h) at the walls downstream wider channel is varied to investigate the flame behavior. In order to discuss the numerical results, the reaction rate higher than 0.95 kg · mol/m^3 · s is defined as the flame zone. In addition, the point closest to the upstream is called the flame front, and the position is measured from the sudden-expansion steps in x axis.

Fig. 2 shows the positions of flame front for different h. It is found that as h becomes larger, the flame becomes unsteady and turns to be a dynamic flame behavior. In the numerical investigation, h starts from a small value of 50 W/m^2 K and a steady flame is observed. As h increases, the wall temperatures become lower. Therefore, the flame becomes unstable and then vibrates (60 < h < 93 W/m^2 K). As h>96 W/m^2 K, the flame turns to oscillate periodically. Three types of flame behaviors are classified in this study and will be discussed in the following.

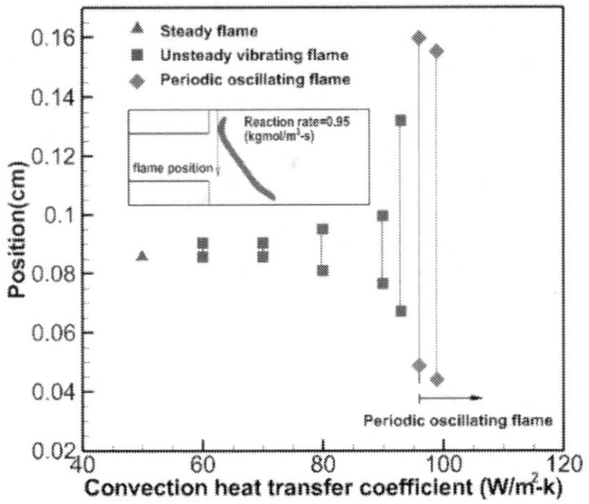

Figure 2: The position of flame front as a function of convection heat transfer coefficient. (For the dynamic flames, the position of flame front varies in a certain range.)

3.1 Steady flame

For the inlet flow velocity (u_{ave} = 65 cm/s), a dynamic flame is found in [9] as the channel walls are isothermal (300 K). However, in this study, a steady slant flame is found when h = 50 W/m^2 · K at the downstream channel, as shown in Fig. 3. It is observed that the flame front does not touch the bottom wall, while the flame tip anchors on the top wall. For such heat loss condition at wall, the wall temperature cannot reach higher enough to ignite the unburnt mixture, but sufficiently causes the flame tip close the wall. Therefore, the boundary layer helps the slant flame stabilize in the sudden-expansion channel.

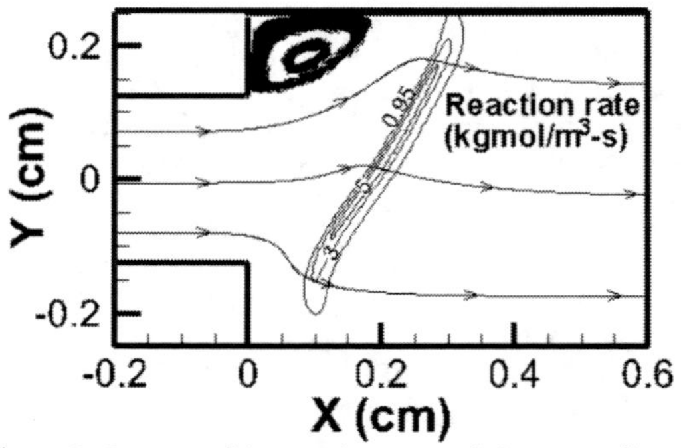

Figure 3. Contour of the reaction rate and the stream lines. (convection heat transfer coefficient =50 W/m^2 · K)

3.2 Unsteady vibrating flame

When h is between 60 and 93 W/m^2 · K (Fig. 2), the vibrating flame is observed. Fig. 4 shows the vibrating flame at h = 90 W/m^2 K. It is found that the flame tip moves away from the wall due to the higher heat loss as compared to Fig. 3. It results in that the flame tip cannot be stabilized in the flow condition in the sudden-expansion channel (The flow tends to form two identical recirculation zones behind the steps.), and therefore, it becomes a vibrating slant flame. The vibrating magnitude increases with h, as shown in Figure 2.

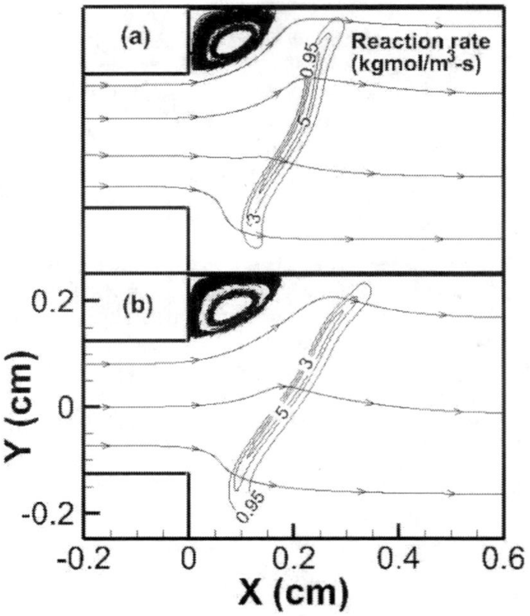

Figure 4. Contours of the reaction rate and the stream lines. (convection heat transfer coefficient =90 W/m^2 · K)

3.3 Periodic oscillating flame

When $h \geq$ 96 W/m^2 · K, the flow pattern of symmetric recirculation zones in the sudden-expansion channel significantly affects the flame behavior. Fig. 5 represents the time evolution of the flame-front position for h=96 W/m^2 · K. One can see that the flame front moves periodically. The flame structures marked in Fig. 5 are showed in Fig. 6.

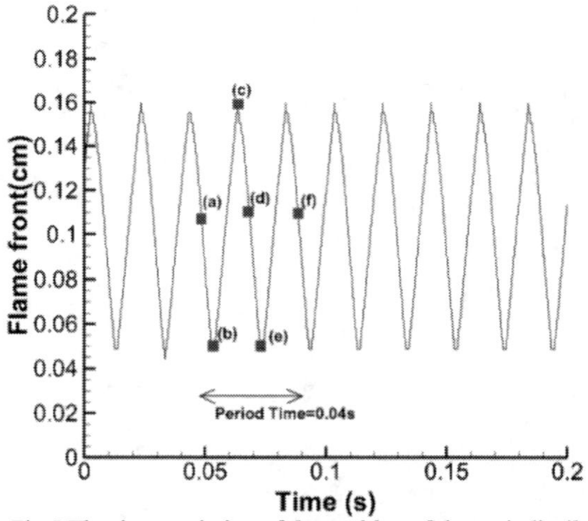

Fig 5.The time evolution of the position of the periodically oscillating flame front. (u_{ave} = 0.65m/s.)

From Fig. 6 (a) to (b), it shows the development of upper asymmetric slant-shaped flame. The flame front moves close to the upper step and the recirculation zone behind the lower step becomes large. After reaching the longest upper slant-shaped flame, the upper flame front and lower flame tip moves respectively downward and upward to form a transient tulip-shaped flame, as shown in Fig 6(c). The position of flame front changes from the upper to the lower. Continuously, the flame front moves to the lower step and the recirculation zone behind the upper step becomes large, as shown in Fig. 6(d). Fig. 6(e) shows the longest lower asymmetric slant-shaped flame. After that, the flame becomes the shape in Fig. 6(f), which is identical to that in Fig. 6(a). So, a period of oscillating flame finishes and the next period begins. The periodically oscillating flame behavior is similar to that found in [9], which is attributed to the interaction of the Landau-Darrieus flame instability and the hydrodynamic flow pattern in the sudden-expansion channel. From Fig. 5, it is found that the period of the oscillating flame is about 0.04 sec (u_{ave} = 0.65 m/s), which the frequency is 25 Hz. The moving position of flame front is from 0.048 cm to 0.16 cm (measured from the steps), and the moving distance is about 0.112 cm. If the wall heat loss further raises to nearly isothermal condition (300K), the oscillating-flame frequency will be close to 33 Hz.

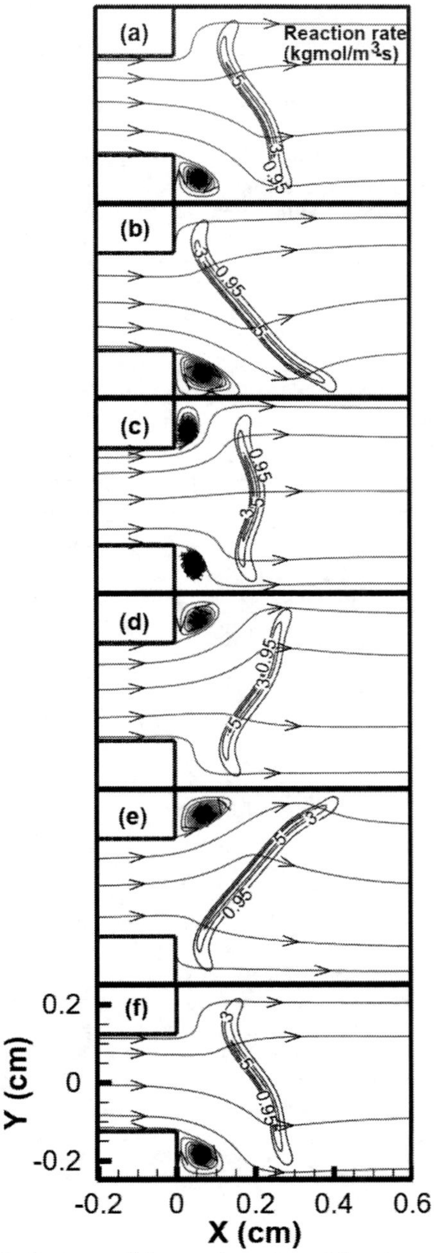

Figure 6. Contours of the reaction rate and the stream lines in various time. (convection heat transfer coefficient=96 W/m^2 · K)

4 Conclusions

In this study, the thermal effect of wall on the flame behaviors in a meso-scale one-stepped sudden-expansion channel (the channel expansion ratio is equal to 2) are numerically investigated by using a two-dimensional model without symmetry assumption at centerline. The wall thickness is neglected and the heat convection coefficient (50-96 W/m^2 K) at the downstream wider channel is varied to study the wall-flame interaction. Three types of flame behavior are found: the steady slant-shaped flame, unsteady vibrating flame and the periodically oscillating flame. The steady slant-shaped flame is formed because the flame tip is anchored on the wall and stabilized by the boundary layer. As the wall heat loss increases, the flame tip moves away from wall. So, the flame becomes unstable and turns to be a vibrating flame (60 W/m^2 · K < h <93 W/m^2 · K). If the wall heat loss further increases (h > 96 W/m^2 · K), the flame becomes a periodically oscillating flame due to the interaction of the Landau-Darrieus flame instability and hydrodynamic flow pattern in the sudden-expansion channel.

5 Acknowledgment

This research was supported by the Ministry of Science and Technology under Project No.: MOST 105-2221-E-110-008-MY2.

References

[1] Y. Yu, K. Maruta, Technology development and fundamental research, Progress in Energy and Combustion Science 37, 2011, pp.669-715

[2] D. G. Norton, D. G. Vlachos, Combustion characteristics and flame stability at the microscale: a CFD study of premixed methane/air mixtures, Chemical Engineering Science 58, 2003, pp.4871 – 4882

[3]G. Pizza, C. E. Frouzakis, J. Mantzaras, A. G. Tomboulides, K. Boulouchos, Dynamics of premixed hydrogen/air flames in mesoscale channels, Combustion and Flame 155, pp. 2-20

[4] G. Pizza, C. E. Frouzakis, J. Mantzaras, A. G. Tomboulides, K. Boulouchos, Dynamics of premixed hydrogen/air flames in microchannels, Combustion and Flame 152, pp. 433–450

[5] Y. Ju, B. Xu, Effects of channel width and Lewis number on the multiple flame regimes and propagation limits in mesoscale, Combustion Science and Technology, Vol. 178, 2006, pp.1723-1753

[6]Y. Ju, B. Xu, Theoretical and experimental studies on mesoscale flame propagation and extinction, Proceedings of the Combustion Institute, Vol. 29,2002, pp.949-956

[7] J. Daou, M. Matalon, Influence of conductive heat-losses on the propagation of premixed flames in channels, Combustion and Flame, Vol.126, 2002, pp.321-339

[8] C.-H. Tsai, The asymmetric behavior of steady laminar flame propagation in ducts, Combustion Science and Technology, Vol.180, 2008, pp.533-545

[9] J.-Y. Liu and S.-Y. Hsu, "The Premixed-Flame Behaviors in Mesoscale Stepped Channel" The Asian Symposium on Computational Heat Transfer and Fluid Flow, Busan, Korea, Nov., 2015..

[10] ANSYS FLUENT 16.Theory Guide.

[11] S.V. Patankar, Numerical heat transfer and fluid flow, Hemisphere, 1980.

11th Asia-Pacific Conference on Combustion,
The University of Sydney, NSW Australia
10th -14th December 2017.

Propane Spray Structure in an Optically Accessible Direct Injection, Spark Ignition Engine: A Post-Processing Algorithm for Planar Laser Mie-Scattering

H.B. Aditiya[1], J.S. Lacey[1], M.J. Brear[1], R.L. Gordon[1], C. Lakey[2], S. Ryan[2], B. Butcher[2]

[1]Department of Mechanical Engineering
University of Melbourne, Victoria 3010, Australia
[2]Ford Motor Company of Australia

Abstract

Liquefied petroleum gas (LPG) coupled with direct-injection (DI) can increase internal combustion (IC) engine efficiency and reduce CO_2 emissions. However, successful implementation of any DI fuelling system requires a thorough understanding of the fuel's spray properties. To this end propane, as a surrogate for LPG, is investigated with planar laser Mie-scattering to study the spray structure in an optically accessible version of a modern, production DI engine. This paper presents the processing algorithm developed to extract quantitative information from the Mie-scattering images.

1 Introduction

The two principal driving factors for IC engine research are the reduction of tailpipe emissions and the increase of engine efficiency. Two promising technologies to achieve these goals are direct injection (DI) and the use of alternative fuels. Direct injection offers several notable benefits. The wall wetting issues found in port fuel injection (PFI) vehicles are largely avoided because the fuel is injected directly in-cylinder. DI has a charge cooling effect that enhances knock resistance, allowing for higher compression ratios and higher thermal efficiency [1]. DI also is suitable for stratified operation, where additional efficiency benefits may be realised due to globally lean operation [2].

LPG is a potential alternative fuel for DISI engines. It has less energy specific CO_2 emissions than gasoline, and a 3-20% higher octane rating than standard RON 91 gasoline, which provides additional protection against knock [3]. Also, LPG is lower in cost per GJ than gasoline.

LPG spray characterisation is necessary to realise a successful DI LPG engine concept because most DI applications are designed for conventional fuels. Fuel spray characterisation is commonly accomplished utilising optical diagnostics techniques in engines with optical access or other spray imaging facilities.

Propane, the primary component of LPG, is heavily flash-boiling at cylinder conditions corresponding to DISI operation [4, 5], and therefore the spray rapidly vaporises. The variability induced by flash-boiling may create image processing challenges for a code written specifically for liquid-phase fuel sprays. Therefore, this study is aimed at providing a post-processing algorithm that can accommodate any phase variability due to highly-flashing fuel sprays. Compared with standard liquid spray image post-processing [6], the proposed method includes a series of corrections due to the optical setup and is suited not only for liquid fuel (e.g. diesel, gasoline) but also fuel that is prone to flash-boiling phenomena (e.g. LPG).

2 Experimental setup

2.1. Optical EcoBoost engine and operating conditions

In this study, DI propane (a surrogate for LPG) spray is investigated in a motoring 2.0L Ford EcoBoost DI engine, which has been modified to have optical access. Cylinder 1 was replaced with a fused silica liner, and the pistons were extended to preserve the factory geometry in the combustion chamber. The design specifications of the engine are listed in Table 1. Propane was injected at 290 bTDC with image timings at 4, 5.6 and 7.2 CAD ASOI. The experimental conditions are summarised in Table 2. 200 images were captured at every engine condition tested to ensure that each data set had statistical significance.

Bore	87.5 mm
Stroke	83.1 mm
Compression ratio	9.3:1
Optical liner material	Fused silica
Injector	7-hole Bosch HDEV 5.1
Fuel injector position	Side mounted
Injector angle orientation	27.5° from piston crown

Table 1. Design specifications of the optical 2.0L Ford EcoBoost engine.

2.2 Mie-scattering setup

The illumination for the planar laser Mie-scattering experiments are provided by a Quantel Brilliant B Nd:YAG laser using a frequency doubler for a 532 nm output. The duration of the laser pulse is 5 ns. The beam was formed into a sheet with a height of approximately 65 mm and a thickness of 1 mm at the focus of the laser sheet. A LaVision Flowmaster CCD camera was positioned orthogonal to the laser sheet, which provided a side view of the propane spray. A 105 mm Nikon lens at f/8 with a 532 nm bandpass filter was focused on the plane that contained the laser sheet. A simplified schematic of the experimental setup is depicted in Fig. 1.

Corresponding author e-mail address:
aharjon@student.unimelb.edu.au

Engine speed (RPM)	600
P_{rail} (bar)	45, 70, 100
T_{fuel} (°C)	30, 50, 85
Intake manifold pressure (bar)	0.6
Start of injection (CAD bTDC)	290
Image timing (CAD ASOI)	4, 5.4, 7.2
Injector pulse width (ms)	2.0

Table 2. Experimental conditions of the DI propane optical investigation

3 Image and camera calibration

3.1 Image distortion correction

Image distortion is introduced not only from the camera lens, but also from the curved, optical cylinder liner, and this effect is strongest at the edges of the cylinder bore. To correct distortion due to lensing effects, the MATLAB Camera Calibrator application was employed using a standard checkerboard pattern placed inside the optical liner. After recording several images of the pattern, the calibration process yields a transformation matrix to correct the image distortion, which was applied to the raw spray image. The algorithm details are omitted here for brevity and readers are referred to section 3 of reference [7] for details of the distortion calibration calculation. It was found that the transformation matrix and subsequent distortion correction were similar when using different sets of calibration images. Also, the primary area of interest is in the middle of the optical liner where the lensing impact of the curved liner is at a minimum.

3.2 Darkfield, lightfield and pixel-to-mm calibration

Darkfield and lightfield calibrations are used to account for the pixel-to-pixel differences of the sensor [8]. The darkfield calibration determines the 'dark' signal count of every pixel. The lightfield calibration quantifies the pixel-to-pixel sensitivity variation and any intensity variation due to vignetting. The darkfield reference image was the average of 100 images captured with the lens cap on. The lightfield reference image was set by averaging the same amount of images with a uniform light source generated by an LED array and a diffuser.

The engine bore dimension was used to scale the image. The edges corresponding to the extended cylinder head bolts at the edges of the bore image were used as a reference scale, as they are a known distance apart. The resulting scale is approximately 75 µm per pixel.

4 Image post-processing

4.1 Darkfield, lightfield calibration and phase determination

Darkfield and lightfield calibrations are first performed by using Eq. (1).

$$I_{cal(n)}(i,j) = [I_{raw(n)}(i,j) - DF(i,j)] \times LFCM(i,j) \quad (1)$$

Here the raw image, I_{raw}, is subtracted with darkfield, DF, which levels the dark count from raw image to zero. The operation is performed pixel-by-pixel, which is indicated by notation i,j, and notation n refers to the image number being processed within the data set. The pixel-to-pixel light

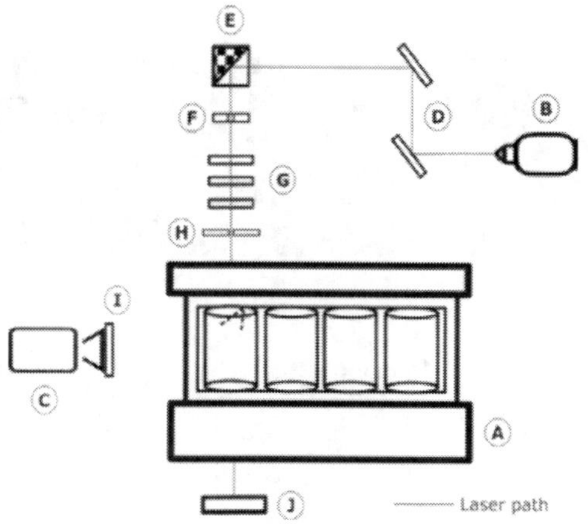

Figure 1. Schematic diagram of the Mie-scattering optical setup. (A) Optical 2.0L Ford EcoBoost engine; (B) Quantel Brilliant B Nd:YAG laser; (C) LaVision Flowmaster CCD camera; (D) periscope; (E) beam splitter (80/20); (F) beam iris; (G) sheet forming optics; (H) laser sheet slit; (I) 532 nm bandpass filter; (J) beam dump.

sensitivity calibration is performed by applying lightfield calibration matrix, $LFCM$, which is prepared by Eq. (2):

$$LFCM(i,j) = \frac{\max(LF)}{LF(i,j)} \quad (2)$$

This matrix contains the ratio of the maximum intensity count of the lightfield, $\max(LF)$, to individual pixel of the lightfield matrix, LF.

At some engine conditions and times after the start of injection (ASOI), liquid propane is not present in the images. As the injector is mounted in the pent-roof of the engine cylinder head and this area is obscured from the experimental field of view, the first 10 mm of spray penetration are not visualised. At some engine conditions, this propagation length provides a sufficient amount of time for the spray to completely vaporise before entering the viewing area. The Mie scattering signal from liquid is significantly stronger than Rayleigh scattering present from in-cylinder vapour, therefore distinguishing between images that contain some liquid or only vapour can be based on signal peaks. With liquid, the maximum intensity counts in an image will be at least 10^2 above the darkfield. In contrast, vapour produces lower maximum intensities, on the order of 10^1 counts above the darkfield. Any frame that has a maximum intensity above the liquid count threshold is deemed to contain liquid. The code extracts spray profiles only from images that are determined to have liquid phase present.

4.2 Liquid phase image processing

4.2.1 Scalar laser energy fluctuation correction

The noise from the laser system, the optics that direct the beam and form the laser sheet and the vibration of the engine apparatus all introduce uncertainty into the illumination that reaches the optical liner from cycle-to-cycle. As Mie scattering is sensitive to the illumination source and experimental setup, this study adopts a simple approach to address these sources of uncertainty in the optical system. Here, information from the reflections of the laser sheet in the optical cylinder liner is utilised. To begin, a set of 200 images is captured with no fuel injection while the engine is motoring at the same speed as the experiments. The reflection intensity variations in these images

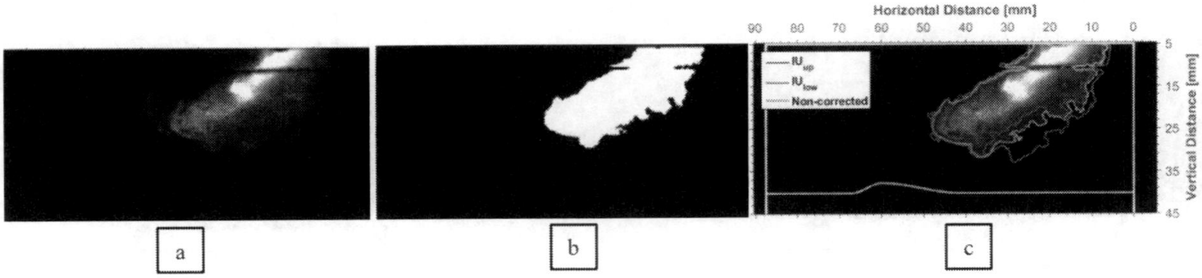

Figure 2. Results from processing a liquid-phase spray at 600 RPM, T_{fuel} 50°C, P_{rail} 45 bar, and 4 CAD ASOI. (a) Raw image; (b) binarised image with threshold level of 3% of non-corrected maximum spray intensity; (c) final processed liquid spray superimposed with liquid-phase extent by IU_{up}, IU_{low} and non-corrected image.

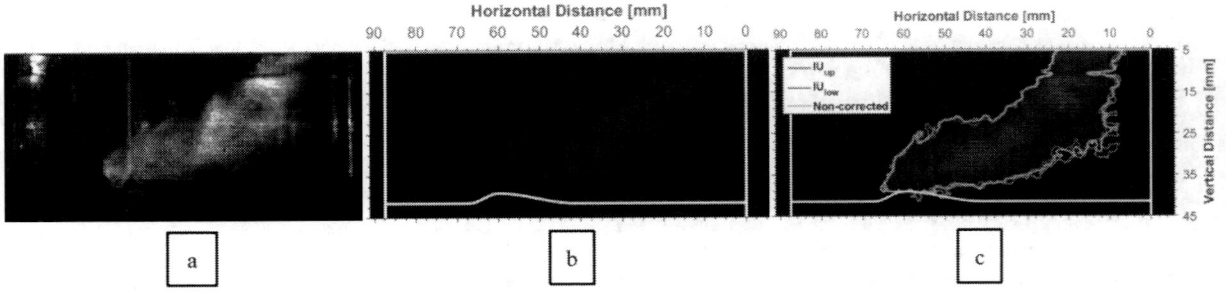

Figure 3. Individual propane images processed at 600 RPM, T_{fuel} = 50°C, P_{rail} = 70 bar and 5.6 CAD ASOI. (a) Contrast-amplified raw image of a vapour plume. This image is brightened by an order of magnitude over the images in (b) and (c); (b) processed from image (a); (c) processed liquid spray image in the same data set.

are representative of the fluctuation of light that reaches the cylinder liner, independent of any fuel spray. Then, a 7x25 pixel window is set at a fixed position in the middle of the reflection region. Pixel counts within this window are summed per frame, and an average intensity integral is determined over the image set ($\bar{W}$). The standard deviation (σ) of the mean is calculated and is used to quantify the uncertainty that the optical system introduces into the illumination of the glass liner. The illumination uncertainty (*IU*) correction factor applied during post-processing is shown in Eq. (3).

$$IU_{up/low} = 1 \pm + \frac{\sigma}{\bar{W}} \qquad (3)$$

Correction factors IU_{up} and IU_{low} characterise the upper and lower limit of the light fluctuation and are used to quantify uncertainty in the extent of the liquid spray.

4.2.2 Liquid-phase boundary detection and spray profiling

The algorithm in this study employs a common method in defining the extent of the liquid-phase proposed by Siebers [6]. A threshold of 3% from the maximum intensity count is used to determine the boundary of the liquid spray, and the upper and lower bounds of this region are determined using the uncertainty factors detailed in the previous section. Binarised images are produced based on these thresholds, then the binary outlines are traced and superimposed on the spray image. An example of the results of this procedure is shown in Fig. 2(c). Horizontal and vertical penetration lengths and spray area are extracted from these defined boundaries. The sizes of the error bars in Fig. 5 are determined by the *IU* correction factors. It was observed that doubling and halving the liquid-phase threshold of 3% of maximum intensity had little impact on the quantitative characteristics of the spray, therefore the results are relatively insensitive to the threshold.

4.3 Vapour phase image processing

Images classified as vapour-only are processed differently than images containing liquid. This algorithm focuses only on reflection exclusion and isolating the vapour-phase spray region. All vapour-only images intensities are scaled by a factor of 10 to ease direct comparison to the images containing liquid. The processed image with aberrations removed is also shown with the same contrast as that of a liquid spray image in Fig.3(b) in order to highlight the substantial difference in brightness between images containing only vapour or any liquid.

4.4 Ensemble averaging

Each processed image set is ensemble averaged. Quantitative values, such as mean penetration length and mean spray area are calculated by taking the average of these quantities in the individual images. The ensemble averaged image with the individual liquid traces superimposed onto it is used as an indication of spray formation variability occurring cycle-to-cycle. The phase of the visualised propane spray is not necessarily uniform throughout all the images in a data set at one condition. Figure 3 shows an instance of an experimental condition where the phase across all the frames is inconsistent. In general, cycle-to-cycle spray formation variability is dependent on in-cylinder flow turbulence and image timing.

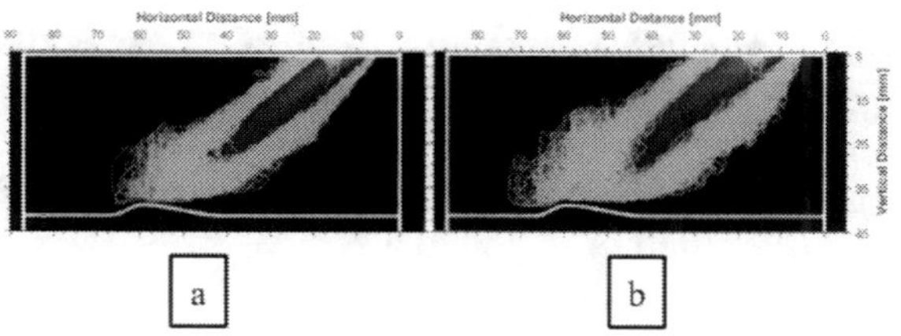

Figure 4. Ensemble averaged of processed images with overlayed individual liquid-phase boundaries at 600 RPM, T_{fuel} 85 °C and 5.4 CAD ASOI. (a) P_{rail} = 45 bar and (b) P_{rail} = 70 bar.

5 Processed DI propane spray characteristics

The post-processing algorithm can provide an indication of cycle-to-cycle spray variation. These cyclic changes are expected due to turbulence present in-cylinder from the piston motion and induction of the air charge. Figure 4 depicts this spray variability by showing individual liquid-phase traces superimposed on the ensemble averaged image. The thickness of the cumulative overlayed traces indicates how the liquid extent varies over the 200 engine cycles imaged, as shown by Fig.4(a) which has smaller cycle-to-cycle variability than Fig.4(b). An increase in rail pressure would normally result in increased penetration length with conventional liquid fuels; however, the ensemble-averaged maximum penetration lengths for DI propane at both 45 and 70 bar are similar. At these two conditions, it seems that the most significant impact of the increasing rail pressure is to increase the variation of the liquid spray area.

The variation of DI propane spray area is non-monotonic with respect to fuel temperature (Fig.5). A monotonic relationship is a common characteristic of conventional fuels where the increase in fuel temperature produces a faster rate of vapourisation and a smaller liquid spray area. With DI propane, as observed from this study, this behaviour does not occur. At T_{fuel} = 30°C and P_{rail} = 70 bar, the spray area is almost zero, indicating most images had no detected liquid. At higher T_{fuel}, the spray area is high because there was always liquid present in-cylinder. This seemingly counterintuitive behaviour may be

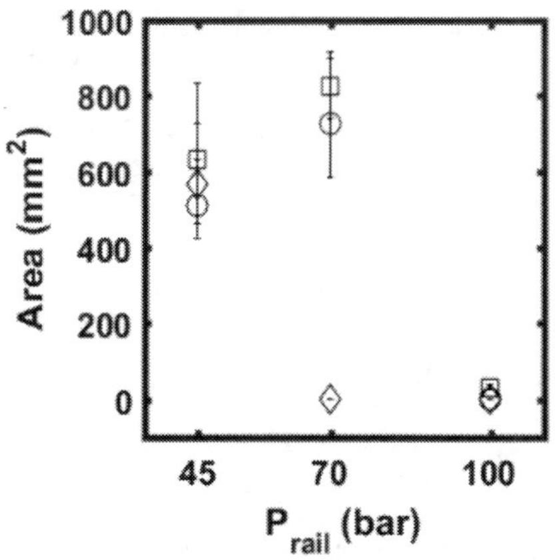

Figure 5. Averaged liquid propane spray area from 200 processed images of engine condition at various rail pressures, 5.6 CAD ASOI, and varying T_{fuel} of 30°C (diamonds), 50°C (squares) and 85°C (circles). Error bars are resulted from IU_{up} and IU_{low} corrections.

a consequence of DI propane operating close to its thermodynamic critical point in the fuel rail (T_c = 96°C and P_c = 42.5 bar). This is not the case with conventional fuels, as they reside in the rail at conditions well below their critical temperatures.

6 Conclusions

A robust in-house algorithm was developed to process highly-flashing DI propane spray in a modified, production engine. It enables the determination of qualitative and quantitative spray characteristics. The image post-processing routine determines the spray phase, accounts for signal fluctuation due to uncertainties introduced by the optical system, outlines the region defining the liquid-phase and determines if only vapour is present in the image. The algorithm can be used to provide quantitative information such as liquid spray area and penetration length, and qualitative comparisons of cycle-to-cycle spray variability.

Acknowledgments

This research was supported by the Advanced Centre for Automotive Research and Testing (ACART, http://www.acart.com.au) and the Australian Research Council. The authors also gratefully acknowledge the Ford Motor Company for their support.

References

[1] Y. Huang, G. Hong, and R. Huang, "Investigation to charge cooling effect and combustion characteristics of ethanol direct injection in a gasoline port injection engine," *Applied Energy,* vol. 160, pp. 244-254, 2015.

[2] F. Zhao, M. C. Lai, and D. L. Harrington, "Automotive spark-ignited direct-injection gasoline engines," *Progress in Energy and Combustion Science,* vol. 25, pp. 437-562, 1999/10/01/ 1999.

[3] K. J. Morganti, T. M. Foong, M. J. Brear, G. da Silva, Y. Yang, and F. L. Dryer, "The Research and Motor octane numbers of Liquefied Petroleum Gas (LPG)," *Fuel,* vol. 108, pp. 797-811, 2013.

[4] J. Lacey, F. Poursadegh, M. Brear, P. Petersen, C. Lakey, S. Ryan, and B. Butcher, "Optical characterization of propane at representative spark ignition, gasoline direct injection conditions," *SAE Technical Paper,* 2016.

[5] H. B. Aditiya, J. S. Lacey, F. Poursadegh, M. J. Brear, R. L. Gordon, C. Lakey, S. Ryan, and B. Butcher, "Preliminary Investigation of Propane Spray Structure in an Optically Accessible Direct-Injection, Spark-Ignition Engine," presented at the 20th Australasian Fluid Mechanics Conference, Perth, Australia, 2016.

[6] D. L. Siebers, "Liquid-phase fuel penetration in diesel sprays," *SAE technical paper,* 1998.

[7] Z. Zhang, "A flexible new technique for camera calibration," *IEEE Transactions on pattern analysis and machine intelligence,* vol. 22, pp. 1330-1334, 2000.

[8] A. P. Hammersley, S. O. Svensson, A. Thompson, H. Graafsma, Å. Kvick, and J. P. Moy, "Calibration and correction of distortions in two dimensional detector systems," *Review of Scientific Instruments,* vol. 66, pp. 2729-2733, 1995.

Application of N-doped Fly Ash Materials with Enhanced Working Capacity for Post-combustion CO_2 Capture

Peng Wang, Chuanwen Zhao*, Yafei Guo, Junjie Yan, Ping Lu

School of Energy and Mechanical Engineering, Nanjing Normal University, Nanjing, 210042, China

Abstract

Large amounts of fly ashes produced from fossil flue-fired power plants potentially threaten the ecological environment. Recycling of the fly ashes as solid sorbents for pollutants control is an economical and environmentally friendly route. In this work, fly ashes (FAs) were modified with tetraethylenepentamine (TEPA), monoethanolamine (MEA), diethanolamine (DEA), polyethyleneimine (PEI), and diethylenetriamine (DETA) to synthesize solid sorbents with enhanced CO_2 capture performances. The synthesized sorbents were characterized by N_2 adsoption/despertion and elemental analyzer. CO_2 capture behaviors of the synthesized sorbents were tested under 60°C, 20%CO_2+15%H_2O using a fixed-bed reactor, and sorbent regeneration behaviors were evaluated under 120°C, 10°C/min and 100%N_2. The sorbent synthesized of ash collected from biomass power plants modified with TEPA (TEPA-FA-4) is identified as the optimum candidate with superior capability for CO_2 capture and sorbent regeneration. The desired sorbent is demonstrated to be stable during 10 cyclic operations. The results of this work can be used as basic date for the technology of CO_2 capture from flue gas in fossil flue-fired power plants using solid sorbents.

1 Introduction

Global warming, believed to be associated with the emission of greenhouse gases, has attracted worldwide attention in recent years [1]. It is imperative to capture CO_2 all over the world. CO_2 capture and storage (CCS) is gaining momentum as a means for combating climate change [2]. Amongst the available CCS technology options, post-combustion CO_2 capture technologies could provide an efficient path for such purpose, since it could be readily realizable for large-scale application in new-built and exist power plants without upgrading and reconstruction [3]. To satisfy the requirement of large quantities flue gas treatment for a typical fossil fuel-fired power plant, the development of economical CCS processes utilizing sorbents with both low-cost and facile regeneration is highly necessary [4]. Therefore, there has been considerable academic and industrial interest in developing promising solid sorbents for effective and retrofittable CO_2 emission reduction application [5-7].

To reduce the overall cost for the CO_2 capture process, calcium, magnesium, potassium, sodium and lithium-rich deprecated alkaline resources such as municipal solid waste incinerator (MSWI) bottom ash [8], steelmaking slags [9], paper mill waste [10-13], biomass waste [14, 15] and mining and industrial waste [16-21] have been utilized as raw materials to synthesize sorbents for CO_2 capture. Particularly, these waste materials have recently been demonstrated to show potential for CO_2 capture [22-25]. This indicates that waste materials could be recycled and utilized as promising CO_2 sorbents, which can provide a low-cost route to reduce CO_2 emissions. Furthermore, the adverse environmental impacts of these solid wastes could also be eliminated.

Fly ash, thought to be an environment pollutant for several decades, is a typical kind of waste material produced in the combustion from thermal power plants. Disposal of fly ash is a serious threat to the environment and hence is a widespread concern for conversion of these wastes into useful products. With their low cost, substantial alkali earth metals, easily modified pore structure and high surface area, some of them were used for CO_2 capture and storage, which was considered to be a promising technology. However, the carbonation capacity and the carbonation conversion of fly ash are generally low and popularization of this technology is limited [26-29].

Recently, several researchers have made progress in amine-tethered sorbents [30-32]. The results indicated that amine-tethered sorbent is promising for large-scale CO_2 capture from flue gas due to its high capture capacity. In order to obtain a sorbent with both high capture capacity and low cost, fly ashes (FAs) were modified with polyamines to synthesize solid sorbents in this work. However, it was unknown whether the CO_2 sorption capacity reaches a high value in this way. It is also unclear which kind of solid amine is the optimal polyamine and which kind of fly ash is optimal support.

To solve the problems above, the CO_2 capture and the sorbent regeneration performances of these sorbents in a simulated flus gas atmosphere were investigated in detail with a modified fixed-bed reactor system. In addition, the cyclic operating performance of the desired sorbent was also investigated. The results of this work can be used as basic date for the technology of CO_2 capture from flue gas in fossil flue-fired power plants using solid sorbents.

2 Experimental Section

2.1 Sorbent Preparation

Four fly ash (FA) samples were obtained from the combustion by-product of fossil flue-fired power plants. For convenient description, these samples were labeled as FA-1, FA-2, FA-3 and FA-4, respectively. FA-1 was collected from MSW power plants. The main composition of FA-2 and FA-3 were coal gangue ash and bituminous coal ash, respectively. FA-4 were obtained from biomass ash power plants. To improve their CO_2 capture capacities, the samples were further modified with

Corresponding author. Fax: +86 25 85481123
E-mail address: cwzhao@njnu.edu.cn

different types of polyamines as tetraethylenepentamine (TEPA), monoethanolamine (MEA), diethanolamine (DEA), polyethyleneimine (PEI) and diethylenetriamine (DETA).

The amine-modified fly ash sorbents were synthesized via the wet impregnation method. A certain gram of amine was dissolved in 50 mL of deionized water under moderate stirring to achieve uniform mixing. Then, 10 grams of the fly ash sample was added to the solution under stirring at ambient temperature for 10 h. The mixture was dried in an oven at 105°C for 6 h and then calcined at 130°C for 4 h in a muffle furnace to obtain the desired sorbent. Raw materials of FAs were derived from farmers or biomass-fired plants. TEPA (technical grade), MEA (AR, 99.0%), DEA (GC, 99.5%), PEI (M.W.1800, 99.0%) and DETA (GC, 99.0%) were purchased from the Aladdin Industrial Corporation.

FA-4 was modified with different amines to synthesize sorbents of TEPA-FA-4, MEA-FA-4, DEA-FA-4, PEI-FA-4 and DETA-FA-4. Their CO_2 capture and sorbent regeneration performances were compared, to enable the optimal amine being identified. Similarly, in order to seek the optimal FA support, different FAs were promoted with the optimal amine to synthesize sorbents of TEPA-FA-1, TEPA-FA-2, TEPA-FA-3 and TEPA-FA-4.

2.2 Characterization

N_2 adsorption-desorption isotherms of the synthesized sorbents were collected using a Tristar II 3020 M instrument (Micromeritics, USA) to calculate their texture parameters including the specific surface area, pore volume and average pore size. The actual amine loading amounts were calculated by measuring nitrogen contents on a Vario EL cube elemental analyzer (Elementar, Germany).

2.3 Apparatus and Procedure

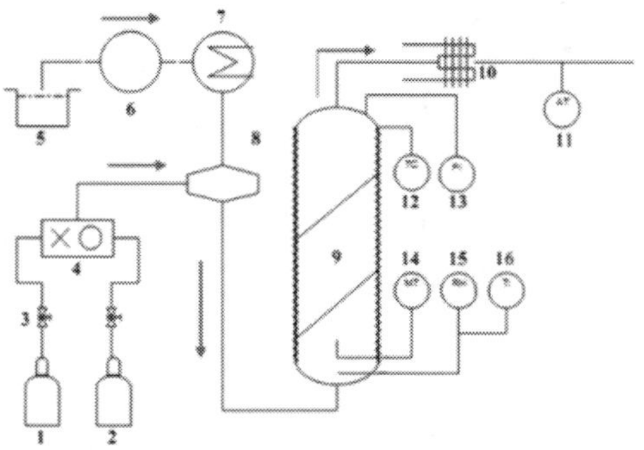

Figure 1: Schematic for the experimental setup. (1) N_2; (2) CO_2; (3) pressure regulating valve; (4) mass flow controller; (5) H_2O; (6) high-precision syringe pump; (7) evaporator; (8) mixture tank; (9) fixed-bed reactor; (10) condenser; (11)gas analyzer; (12) temperature controller; (13) pressure indicator; (14) mass transmitter; (15) relative humidity monitor; (16) temperature indicator.

The CO_2 capture performances of the synthesized sorbents were assessed using a modified fixed bed reactor system coupled with a gas analyzer. The experimental setup is shown in Fig. 1. The flue gas was simulated by a constant flow of CO_2, H_2O and balanced N_2. CO_2 and N_2 were supplied by high-purity cylinders and delivered to the experimental apparatus with mass flow controllers. H_2O was injected using a high-precision syringe pump and then heated to ensure complete vaporization of water. The fixed-bed reactor with an inner diameter of 0.02 m was made of stainless steel and was set in an electric heating element. The electric heating was used to heat the reactor to the given temperature. A gravimetric sensor was used to detect the change of the sample's weight online. A gas analyzer was used to measure the CO_2 concentration of the outlet gas continuously. The relative humidity (RH) and temperature of the simulated atmosphere were also monitored online by the relative humidity monitor and temperature sensor.

2.4 Theoretical analysis

Outlet CO_2 concentration during the CO_2 capture and sorbent regeneration tests was recorded by the gas analyzer, and the CO_2 capture capacity and sorbent regeneration capability could be determined as:

$$q_a = \frac{1}{w} \times \left[\int_0^t Q \times \frac{C_{in} - C_{out}}{1 - C_{out}} \times dt \right] \times \frac{T_0}{T} \times \frac{1}{V_m} \quad (1)$$

$$q_d = \frac{1}{w} \times \left[\int_0^t Q \times C_{out} \times dt \right] \times \frac{T_0}{T} \times \frac{1}{V_m} \quad (2)$$

Where and q_a are the CO_2 capture capacity and sorbent regeneration capability (mmol CO_2/g), respectively. C_{in} and C_{out} are the inlet and out CO_2 concentrations (vol.%), respectively. w denotes the weight of the sample (g), Q is the total gas flow rate (L/min), t is the reaction time (min), T is the reaction temperature (K), T_0 is 273 K and V_m is 22.4 mL/min.

Based on the CO_2 capture capacity and sorbent regeneration capability, the regeneration conversion, η (%) could be presented as:

$$\eta = \frac{q_d}{q_a} \times 100\% = \frac{\int_0^t C_{out} dt}{\int_0^t \frac{C_{in} - C_{out}}{1 - C_{out}} dt} \times 100\% \quad (3)$$

3 Results and Discussion

3.1 CO_2 Capture and Sorbent Regeneration Performances of Amines-modified FA-4

In order to determine the optimal polyamine for synthesizing desired sorbent, FA-4 was modified with different polyamines (TEPA, MEA, DEA, PEI and DETA). The CO_2 capture and sorbent regeneration performances of the amines-modified FA-4 sorbents were investigated. The results are presented in Fig. 2.

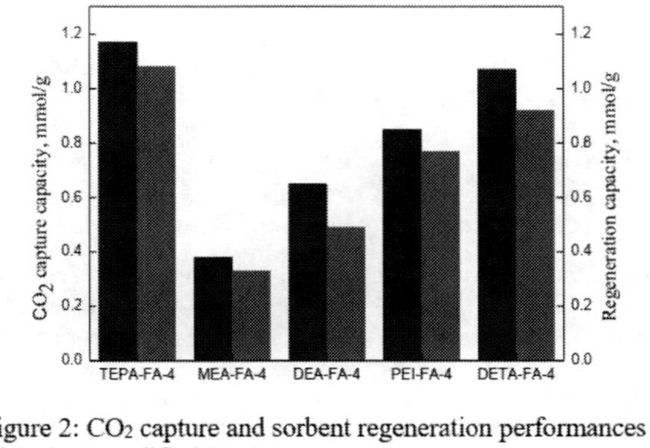

Figure 2: CO_2 capture and sorbent regeneration performances of Amines-modified FA-4

Fig. 2 displays the CO_2 capture capacities and regeneration capabilities of FA-4 modified with different polyamines tested under 60°C, 20%CO_2+15%H_2O for CO_2 capture, and sorbent regeneration behaviors were evaluated under 120°C, 10°C/min and 100%N_2. The CO_2 capture capacities for TEPA-FA-4, MEA-FA-4, DEA-FA-4, PEI-FA-4 and DETA-FA-4 are calculated as 1.17, 0.38, 0.65, 0.85 and 1.07 mmol CO_2/g, respectively. The regeneration capabilities of the five sorbents are 1.08, 0.33, 0.49, 0.77 and 0.92 mmol CO_2/g, the corresponding regeneration conversions are calculated as 92.3%, 86.8%, 75.4%, 90.6% and 86.0%, respectively. It is found that TEPA-FA-4 shows both the greatest CO_2 capture and sorbent regeneration performances amongst these sorbents.

Table 1 lists the actual amine loading amounts and physical properties of FA-4 promoted with different polyamines. The BET surface area and BJH pore volume of FA-4 are 24.04 m^2/g and 25.06 $\times 10^{-3}$ cm^3/g. Both of them for FA-4 show obvious decline as the sample is further modified with different polyamines. The theoretical amine loadings for all the samples are 15 wt.%. As indicated, the actual amine loadings for TEPA-FA-4 could be rather high as 14.40 wt.%. BET surface area and BJH pore volume of MEA-FA-4 could be 2.35 m^2/g and 5.70 $\times 10^{-3}$ cm^3/g, which are the highest of all these samples. However, the use of MEA to modify solid CO_2 sorbent could meet several bottleneck problems. On the one hand, MEA shows strong chemical binding force towards CO_2 molecules and it would consume significant energy load for sorbent regeneration [33, 34]. On the other hand, the boiling point of MEA is rather low, which means that the sorbent is sensitive to temperature and are more likely to suffer amine loss during thermally cyclic operations. The loading of DEA, PEI and DETA on FA-4 has caused obvious decline in BET surface area and BJH pore volume. By contrast, TEPA-FA-4 keeps better physical properties. It is concluded that TEPA-FA-4 presents the excellent CO_2 capture and sorbent regeneration performances amongst these amine-modified sorbents. As a result, TEPA is identified as the optimal polyamine for further investigation.

	Actual amine loading, wt.%	BET surface area, m^2/g	BJH pore volume, $10^{-3}cm^3$/g
TEPA-FA-4	14.40	2.08	4.54
MEA-FA-4	3.70	2.35	5.70
DEA-FA-4	12.65	1.27	1.66
PEI-FA-4	12.23	1.35	1.44
DETA-FA-4	9.82	1.90	2.44

Table 1: Actual amine loadings and physical properties of FA-4 promoted with different polyamines

3.2 CO_2 Capture and Sorbent Regeneration Performances of TEPA-modified FAs

To screen the optimal fly ash sample for sorbent preparation, four FAs were modified with TEPA. Their CO_2 capture and sorbent regeneration performances are demonstrated in Fig. 3.

Fig. 3 shows that the CO_2 capture capacities for TEPA-FA-1, TEPA-FA-2, TEPA-FA-3 and TEPA-FA-4 are calculated as 0.83, 0.35, 0.65 and 1.17 mmol CO_2/g. The regeneration capabilities of the TEPA-FAs sorbents are 0.65, 0.32, 0.55 and 1.08 mmol CO_2/g, their regeneration conversions are then determined as78.2%, 91.4%, 84.6% and 92.3%. This might be attributed to the diverse coupling behaviors between TEPA and the fly ash materials due to their different structure and surface features. It can be thus concluded that TEPA-FA-4 shows the maximum capture capacity and the best sorbent regeneration performance.

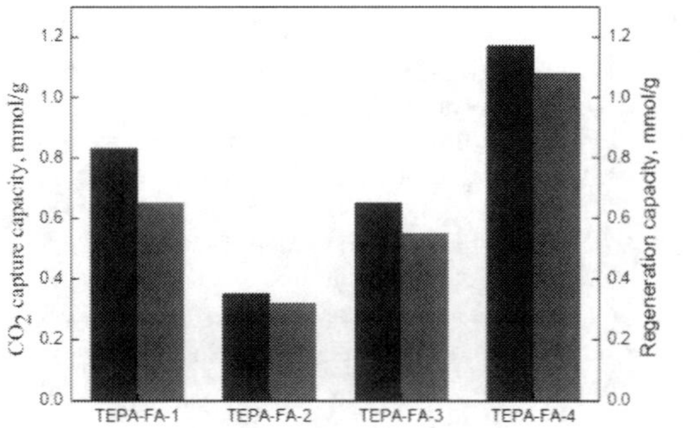

Figure 3: CO_2 capture and sorbent regeneration performances of TEPA-modified FAs

The actual amine loadings for the different TEPA-WAs are 14.35 wt.%, 14.39 wt.%, 10.43 wt.% and 14.40 wt.%, respectively, as listed in Table 2. BET surface area and BJH pore volume of TEPA-FA-4 could be 2.08 m^2/g and 4.54 $\times 10^{-3}$ cm^3/g, which are the highest of all these samples. Except for that of TEPA-FA-4, both BET surface areas and BJH pore volumes of the other three samples show no significant difference. Compared to the raw FAs, amine loading has resulted in obvious reduction in their texture parameters. It is deduced that TEPA presents better coupling characteristics towards FA-4. In other words, the amine groups show more uniform dispersion on FA-4, which provide more active sites for CO_2 sorption. This might be responsible for the superior CO_2 capture and sorbent regeneration performances for TEPA-FA-4. Considering its high CO_2 capture capacity and sorbent regeneration capability, FA-4 is identified as the optimal support. Amongst the five sorbents, TEPA-FA-4 showed superior CO_2 capture capacity and sorbent regeneration capability. Therefore, TEPA-FA-4 is demonstrated as the potential candidate for further investigation.

	Actual amine loading, wt.%	BET surface area, m^2/g	BJH pore volume, $10^{-3}cm^3$/g
TEPA-FA-1	14.35	1.00	0.83
TEPA-FA-2	14.39	0.94	1.36
TEPA-FA-3	10.43	1.01	0.95
TEPA-FA-4	14.40	2.08	4.54

Table 2: Actual amine loadings and physical properties of FAs promoted with TEPA

3.3 Multiple Cycle Behaviors

In order to evaluate the long-term stability of TEPA-FA-4 in cyclic operations, 10 carbonation/regeneration cycles were carried out. The CO_2 capture process of TEPA-FA-4 is carried out under 60°C, 20%CO_2+15%H_2O, and sorbent regeneration behaviors were evaluated under 120°C, 10°C/min and 100%N_2. The CO_2 capture capacities of TEPA-FA-4 changing against the cycle numbers are shown in Fig.4.

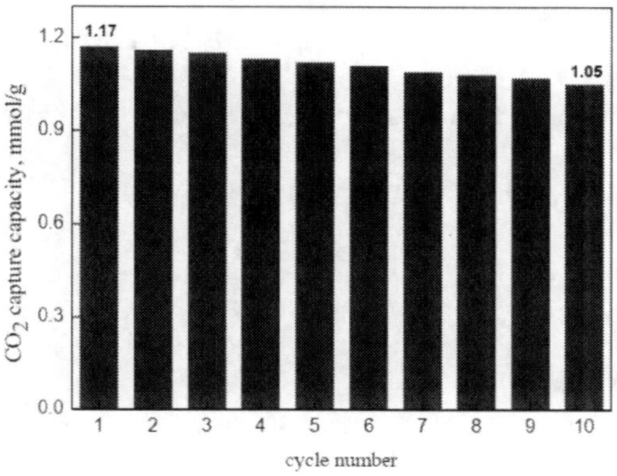

Figure 4: Multiple cycle behaviors of TEPA-FA-4

Fig.4 shows that the CO_2 capture capacity of TEPA-FA-4 slightly decreases from 1.17 to 1.05 mmol CO_2/g after 10 cycles. The working capacity of TEPA-FA-4 during the 10 carbonation/regeneration cycles kept constant, indicating that the sorbent retains stable CO_2 capture performance for long-term cyclic operation.

4 Conclusionsn

Fly ash materials were tried as potential solid sorbents for capturing CO_2 from post-combustion flue gas. To improve their CO_2 capture capacities, the sorbents were further modified with different polyamines. The CO_2 capture and the sorbent regeneration performances of the sorbents were investigated systematically with a modified fixed bed reactor system. Amongst the various amine-modified sorbents, TEPA-FA-4 is determined as the optimal candidate with preferable CO_2 capture capacity and excellent regeneration conversion. Besides, the working capacities of TEPA-FA-4 during the cyclic CO_2 separation are kept constant. Considering the worldwide abundance and recyclability of FA, the as-synthesized amine-modified fly ash samples could be potentially used as low-cost solid CO_2 sorbents for large-scale application in post-combustion flue gas treatment. Meanwhile, this could be a green route to disposing the solid wastes from fossil flue-fired power plants to avoid their potential hazards to environment.

5 Acknowledgment

Financial support from the National Natural Science Foundation of China (U1510129, 51206155) and Qing Lan Project are sincerely acknowledged.

References

[1] B. Li, Y. Duan, D. Luebke, B. Morreale, Appl Energy 102 (2013) 1439-1447.

[2] B. Jiang , X. Wang , M. L. Gray, Y. Duan, D. Luebke, B. Li, Appl Energy, 109 (2013) 112-118.

[3] A.S. Bhown, B.C. Freeman, Environ. Sci. Technol. 45 (2011) 8624-8632.

[4] A. Samanta, A. Zhao, G.K. Shimizu, P. Sarkar, R. Gupta, Ind. Eng. Chem. Res. 51 (2011) 1438-1463.

[5] E.R. Bobicki, Q. Liu, Z. Xu, H. Zeng, Prog. Energy Combust. Sci. 38(2) (2012) 302-320.

[6] A. Sanna, M. Dri, M.R. Hall, M. Maroto-Valer, Appl Energy 99 (2012) 545-554.

[7] A. Kaithwas, M. Prasad, A. Kulshreshtha, S. Verma, Chem. Eng. Res. Des. 90(10) (2012) 1632-1641.

[8] S. Arickx, T. Van Gerven, C. Vandecasteele, J. Hazard. Mater. 137(1) (2006) 235-243.

[9] W.J.J. Huijgen, G.J. Witkamp, R.N.J. Comans, Environ. Sci. Technol. 39 (2005) 9676-9682.

[10] R. Sun, Y. Li, C. Liu, X. Xie, C. Lu, Chem. Eng. J. 221 (2013) 124-132.

[11] Y. Li, C. Liu, R. Sun, H. Liu, S. Wu, C. Lu, Ind. Eng. Chem. Res. 51(49) (2012) 16042-16048.

[12] R. Pérez-López, J. Castillo, D. Quispe, J.M. Nieto, J. Hazard. Mater. 177 (2010) 762-772.

[13] Y. Li, R. Sun, Energy Convers. Manage. 82 (2014) 46-53.

[14] M.G. Plaza, C. Pevida, B. Arias, J. Fermoso, M.D. Casal, C.F. Martín, F. Rubiera, J.J. Pis, Fuel 88 (2009) 2442-2447.

[15] K. Wang, X. Guo, P. Zhao, F. Wang, C. Zheng, J. Hazard. Mater. 189 (2011) 301-307.

[16] S. Teir, S. Eloneva, C.J. Fogelholm, R. Zevenhoven, Appl Energy 86 (2009) 214-218.

[17] D.N. Huntzinger, J.S. Gierke, L.L. Sutter, S.K. Kawatra, T.C. Eisele, J. Hazard. Mater. 168(1) (2009) 31-37.

[18] P.J. Gunning, C.D. Hills, P.J. Carey, Waste Manage. 30(6) (2010) 1081-1090.

[19] G.D. Senior, S.A. Thomas, Int. J. Miner. Process. 78 (2005) 49-61

[20] Y. Li, M. Su, X. Xie, S. Wu, C. Liu, Appl Energy 145 (2015) 60-68.

[21] W. Zhang, Y. Li, L. Duan, X. Ma, Z. Wang, C. Lu, Chem. Eng. Res. Des. 109 (2016) 806-815.

[22] M. Bhagiyalakshmi, L.J. Yun, R. Anuradha, H.T. Jang, J. Hazard. Mater. 175(1) 2010 928-938.

[23] L.Y. Lin, H. Bai, Chem. Eng. J. 197 (2012) 215-222.

[24] L.Y. Lin, H. Bai, Microporous Mesoporous Mater. 170 (2013) 266-273.

[25] H.T. Jang, Y.K. Park, Y.S. Ko, J.Y. Lee, B. Margandan, Int. J. Greenhouse Gas Control 3(5) (2009) 545-549.

[26] W.H. Chen, S.M. Chen, C.I. Hung, Appl Energy, 111(4) (2013) 731-741.

[27] J.G. Lu, C.T. Lu, Y. Chen, L. Gao, X. Zhao, H. Zhang, Z.W. Xu, Appl Energy 115(4)(2014) 573-581.

[28] J.S. Lee, J.W. Lee, Y.T. Kang, Appl Energy, 143 (2015) 119-129.

[29] Goto K. Appl Energy, 111(11) (2013) 710-720.

[30] R.S. Franchi, P.E. Harlick, A. Sayari, Ind Eng Chem Res, 44(2005) 8007-8013.

[31] C. Chen, W. Ahn, Chem. Eng. J, 166 (2011) 646-651.

[32] W. Zeng, H. Bai, Chem. Eng. J, 251(2014) 1-9.

[33] T. Witoon, Mater. Chem. Phys. 137(1) (2012) 235-245.

[34] A.B. Rao, E.S. Rubin, Environ. Sci. Technol. 36(20) (2002) 4467-4475.

11th Asia-Pacific Conference on Combustion,
The University of Sydney, NSW Australia
10th -14th December 2017.

Experimental Study of Spray Characteristics, Engine Performance and Emission Levels of Acetone-Butanol-Ethanol Mixture-Diesel Blends in a Diesel Engine.

Sattar Jabbar Algayyim [a, b, *], **Andrew P. Wandel** [a], **Talal Yusaf** [a], **Ihsan Hamawand** [a], **Saddam Al-Wayzy** [a]

[a] School of Mechanical and Electrical Engineering, University of Southern Queensland, Toowoomba, QLD 4350, Australia.

[b] Department of Mechanical, College of Engineering, University of Al-Qadisiyah
Al-Diwaniyah, Iraq

Abstract

This paper investigates spray and engine performance of an acetone-butanol-ethanol (ABE) mixture blended with diesel fuel in a single-cylinder direct injection (DI) diesel engine. Spray images were evaluated using a high-speed camera under 300 bar injection pressure. Engine performance such as brake power (BP), brake-specific fuel consumption (BSFC) and in-cylinder pressure were measured. Exhaust gas emissions such as oxides of nitrogen (NO_x), carbon monoxide (CO) and unburned hydrocarbon (UHC) were also assessed. The test was carried out at three engine speeds (1400, 2000 and 2600 rpm) at full load. The experiment results showed that: liquid penetration of ABE-diesel is longer than that of diesel. BP of ABE-diesel blends was comparable with pure diesel at 2600 rpm, while the peak in-cylinder pressure was higher compared to diesel at 2000 rpm. UHC and CO emissions were significantly reduced as a result of the addition of ABE to the neat diesel, while NO_x emissions were slightly increased.

1 Introduction

With the high demand for environmental security, more attention is being paid to utilising renewable alternative fuels in diesel engines. Alcohol blends have the potential to reduce exhaust emissions as well as improve fuel efficiency due to their high oxygen content. Using alcohols as additives could also reduce dependence on fossil fuel because the alcohols are derived or produced from renewable materials such as agricultural waste. Ethanol and methanol are being widely researched in diesel engines, but some difficulties have been reported such as storage safety and low cetane number [1].

Currently, the ABE mixture has the potential to be an alternative biofuel due to its manner of production and the advantages of its properties. The volumetric ratio of ABE was 3:6:1 after fermentation processes [2-5]. Several researchers have experimentally investigated ABE blends in constant-volume chamber and diesel engines [6, 7]. These studies demonstrated that: (1) engine efficiency was improved; (2) NO_x and soot emissions were decreased [8].

The aim of this paper is to assess the impact of ABE-diesel blends on spray characteristics, engine performance and emission levels in a DI diesel engine.

* Corresponding author.
E-mail address: sattarjabbarmurad.algayyim@usq.edu.au

2 Methodology

2.1 Fuel Preparation and Properties

Normal butanol (B) and acetone (A) were used at 99.8% analytical grade and obtained from Chem-Supply Australia. Ethanol (E) was used at 100% analytical grade. Diesel was obtained from a local petrol station in Toowoomba, Australia as a baseline. The ABE mixture was prepared with a ratio of 3:6:1 by volume, which was used to simulate the intermediate fermentation production. Then 10% and 20% ABE was blended with diesel, referred to as 10ABE90D and 20ABE80D respectively. Miscibility and stability of the ABE-diesel blends were monitored over a one-month period before the tests were carried out on the engine. The samples were stored in glass bottles and visually observed every week, with all blends maintaining a good homogeneous mixture. Table 1 shows the properties of the separate fuel and blends.

Properties	A	E	B	D	ABE	10ABE	20ABE
Viscosity (mm^2/s) @ 40 °C	0.35	1.08	2.22	1.9-4.1	1	2	1.8
Calorific value (MJ/kg)	29.6	26.8	33.1	42.8	31.4	41.6	40.5
Cetane number	-	8	17-25	48	-	-	-
Surface tension (mN /m)	22.6	-	24.2	23.8	-	-	-
Flash point °C	17.8	8	35	74	-	-	-
Latent heat (MJ/kg) @ 25 °C	518	904	582	270	595	300.4	331.2

Table 1. Fuel properties [8].

2.2 Experimental Apparatus

2.2.1 Spray Test Setup

The spray experimental test was carried out on a constant volume vessel (CVV) at atmospheric pressure. An air-driven high-pressure fuel pump was used in the fuel injection system using a solenoid Bosch-type injector with six holes (each 0.18 mm in diameter) and an injection pressure of 300 bar. A Photron Charge-Coupled Device (CCD) camera was used to capture the spray blends images. The camera has a resolution of 1024×1024 pixels. An LED light was used for illuminating the fuel spray to ensure constant background light for the camera. For each fuel test, the fuel tank and fuel system line were cleaned and emptied and the fuel filter was replaced with a new one. After ensuring all the injection systems were cleaned and emptied, the spray testing started with some preliminary injection tests for at least five minutes before recording the new images.

2.2.2 Engine Test Setup

The engine test was conducted using a single-cylinder, four stroke, water-cooled, DI diesel engine. An electrical dynamometer connected to the engine was used to control the load. The crank angles were measured using a crank angle encoder set up on the shaft of the engine. A Kittler 6052C pressure transducer (CT400.17) and charge amplifier connected to a data acquisition system with software (CT 400.09) were used to record cylinder pressure values at one crank angle revolution for 50 cycles each test. The exhaust gas emissions were analysed using a Coda gas analyser to measure NO_x, CO and UHC. The test was conducted at a compression ratio of 19:1 with three engine speeds (1400, 2000 and 2600 rpm) under full load. The test began at least 20 minutes before recording commenced. The experiments were carried out in triplicate to reduce the experimental error. Table 2 contains the engine specifications. Fig.1 shows operating setting of engine.

Engine model	G.U.N.T. Hamburg
Combustion type	Direct Injection Engine
Number of cylinders	1
Compression ratio	5:1-19:1
Maximum power (kW)	Approx. 6kW
Speed range (rpm)	900-3000
Bore	90mm
Stroke	74mm
Capacity	470cm^3
Maximum compression pressure	60-80 bar
Nozzle injection pressure	300 bar
Injection type	Direct Injection

Table 2. Engine specifications

Figure 1. Operating setting of G.U.N.T engine.

3 Results and Discussion

3.1 Spray Characteristics

The macroscopic characteristics of ABE-diesel blends were obtained using a high speed- camera under various after start of injection (ASOI). Fig. 2 illustrates the spray images analysis from a Bosch type injector. Because the engine used in the experimental test was only equipped with mechanical injectors, the injection pressure used was 300 bar. Liquid spray penetration of ABE-diesel is longer than that of diesel. Fuel properties of blends have a significant impact on liquid penetration especially; under evaporating or burning conditions. According to Table 1 ABE features a much higher latent heat and lower viscosity than pure diesel, which leads to enhanced vaporisation and atomisation. Therefore the penetration length will be shorter and the plume narrower at high ambient pressure and temperature inside the diesel engine cylinder. Because almost all the physical properties change with increased ambient temperature, there is a decrease in viscosity and surface tension while there is an increase in vapour pressure. These changes significantly accelerate the atomisation and evaporation of the liquid spray.

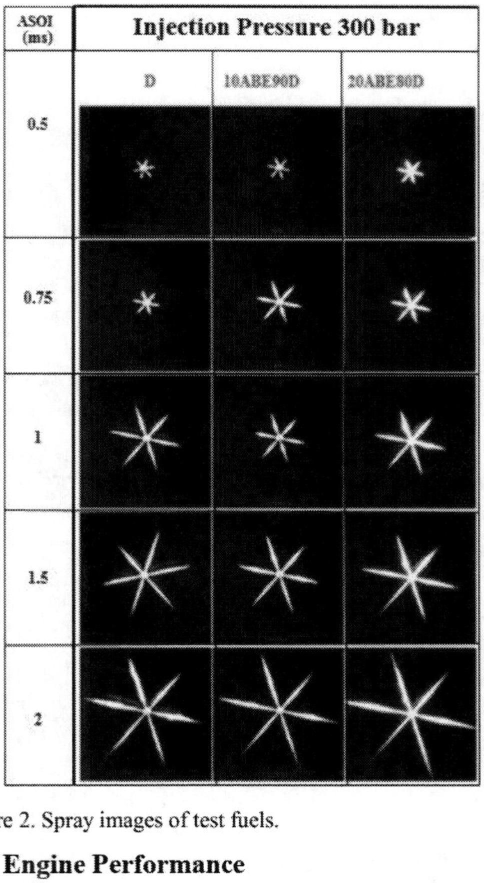

Figure 2. Spray images of test fuels.

3.2 Engine Performance

3.2.1 In-Cylinder Pressure

Figure 3 shows the relationship between the peak in-cylinder pressure trace and the crank angle of the test fuels at 1400 and 2000 rpm. 20ABE80D blend gives a maximum peak in-cylinder pressure compared to neat diesel due to the low cetane number

(CN) of the ABE blend. This results in increased ignition time and rapid in-cylinder pressure increase.

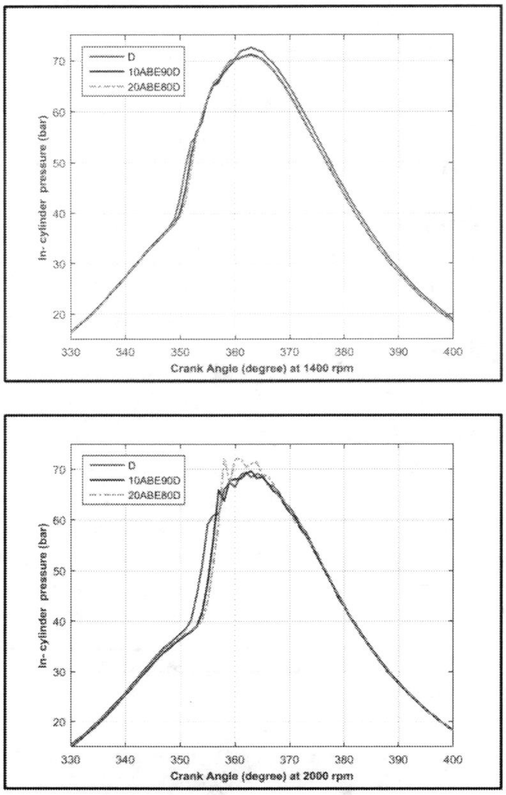

Figure 3. In-cylinder pressure at 1400 rpm and 2000 rpm.

3.2.2 Heat Release Rate (HRR)

Figure 4 presents the heat release rate of the test blends at two engine speeds. It can be seen that the diesel blend showed the highest peak HRR at the low engine speed. In contrast, the maximum HRR of 20ABE-80diesel blend occurred at 2000 rpm engine speed. The peak cylinder pressure (Fig. 3) generally corresponds to the highest HRR.

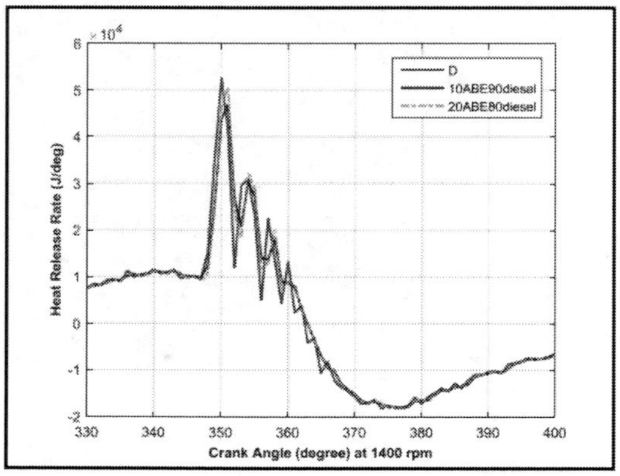

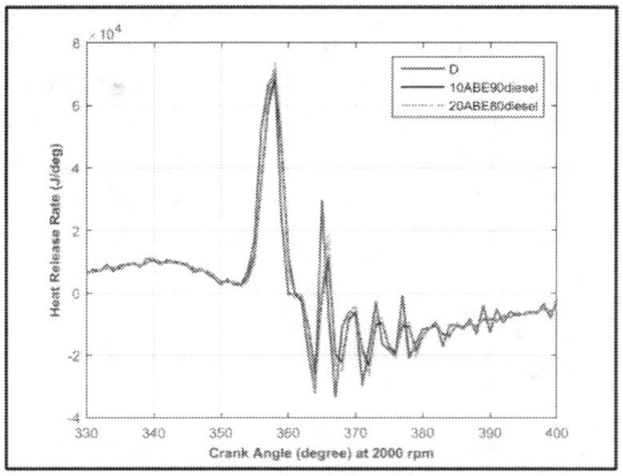

Figure 4. HRR at engine speed 1400 rpm and 2000 rpm.

3.2.3 Brake Power and Brake Specific Fuel Consumption

Figure 5 shows the variation of BP and BSFC with the engine speed of the test fuels. The BP of the ABE-diesel blend showed comparable value with diesel at the high engine speed due to its high oxygen content. BSFC was increased with both fuel blends compared to that of pure diesel due to the low calorific value of the blends (Table 1).

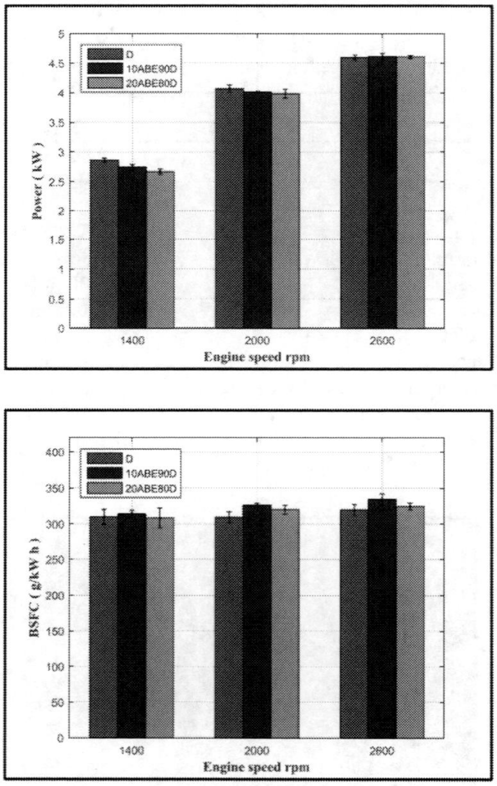

Figure 5. BP and BSFC of test fuels at three engine speeds.

3.2.4 NO_x and CO Emissions

Figure 6 presents the NO_x and CO emissions of the test fuels at various engine speeds. All ABE-diesel blends showed a slight increase in NO_x emissions at all engine speeds. CO emissions were reduced at all engine speeds. This trend could relate to: the high oxygen content and the lower cetane number of the ABE-diesel blends. These complications led to delays in ignition time and resulted in an increase in the premixed zone. This process can increase the local temperature and result in increased NO_x emissions.

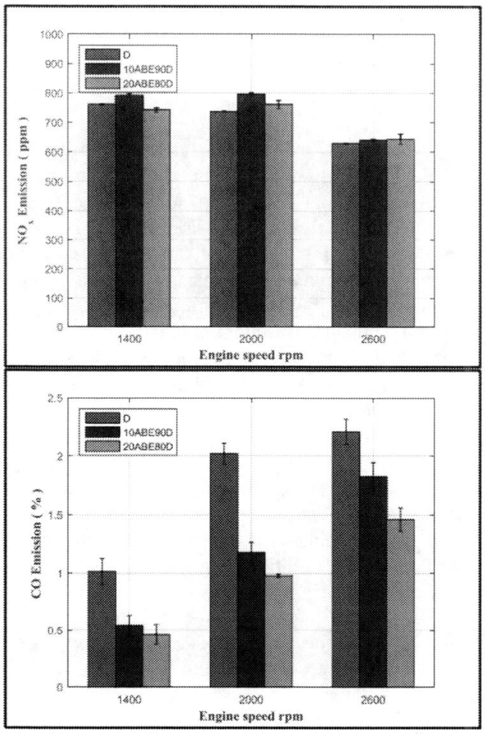

Figure 6. NO_x and CO emissions of test fuels at three engine.

3.2.5 UHC Emissions

The use of ABE-diesel blends decreased the UHC emissions compared to neat diesel at medium and high engine speeds (Fig. 7). This reduction occurred because ABE blends is a type of multi-component fuel with different volatilities, which might produce micro-explosions and thus promote combustion performance. Also, the difference in droplet lifetime between ABE (3.25 s/mm^2) and neat diesel (3.75 s/mm^2) at 823 K affects the reaction time of ABE blends, which results in increased mixing time and leads to complete reaction resulting in decreased UHC emissions [9].

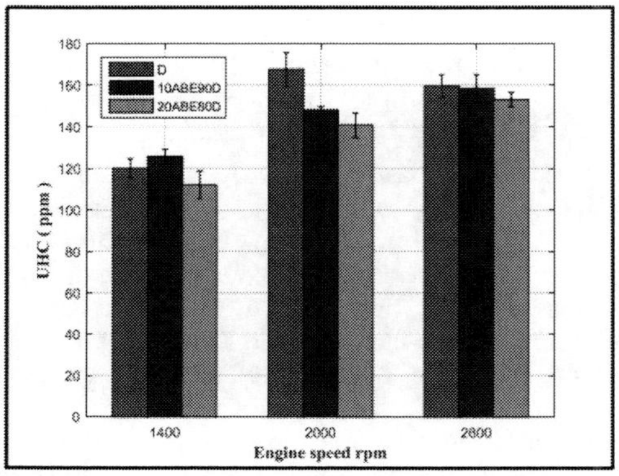

Figure 7. UHC emission of test fuels at three engine speeds.

4 Conclusions

The experimental work has concluded some significant results for the test fuels. The results are as follows:

- Liquid spray penetration of ABE-diesel blend is longer than that of diesel at ambient conditions.
- The BP of the ABE-diesel blends was comparable with neat diesel at the high engine speed, while the peak in-cylinder pressure and HRR were higher compared to diesel at the medium engine speed.
- UHC and CO emissions were significantly reduced as a result of the addition of ABE to diesel blends, while NO_x emissions were slightly increased.

5 Acknowledgments

The first author thanks the Iraqi Government Ministry of Higher Education and Scientific Research. The authors are also grateful to (USQ staff) for assisting with experiments.

References

[1] Shahir, S. A., Masjuki, H. H., Kalam, M. A., Imran, A., and Ashraful, A. M., RENEW SUST ENERG REV 48 (2015) 62-78.

[2] Abd-Alla, M. H., Zohri, A. N., El-Enany, A. W., and Ali, S. M., ANAEROBE 32 (2015) 77-86.

[3] Bellido, C., Infante, C., Coca, M., González-Benito, G., Lucas, S., and García-Cubero, M. T., BIORESOURCE TECHNOL 190 (2015) 332-338.

[4] Cai, D., Zhang, T., Zheng, J., Chang, Z., Wang, Z., Qin, P.-y, and Tan, T.-w., BIORESOURCE TECHNOL 145 (2013) 97-102.

[5] Li, H.-g., Ofosu, F. K., Li, K.-t., Gu, Q.-y., Wang, Q., and Yu, X.-b., BIORESOURCE TECHNOL 172 (2014) 276-282.

[6] Wu, H., Nithyanandan, K., Lee, T. H., Lee, C.-f. F., and Zhang, C., ENERG FUEL 28 (10) (2014) 6380-6391.

[7] Wu, H., Nithyanandan, K., Zhang, J., Lin, Y., Lee, T. H., Lee, C., and Zhang, C., APPL ENERG 149 (2015) 367-378.

[8] Wu, H., Nithyanandan, K., Zhou, N., Lee, T. H., Lee, C.-f. F., and Zhang, C., FUEL 142 (2015) 109-116.

[9] Ma, X., Zhang, F., Han, K., Yang, B., and Song, G., FUEL 160 (2015) 43-49.

A Computational Study of Pre-ignition to Detonation Transition in a One-Dimensional Chamber

Aliou Sow, Mohammed Jaasim, Francisco E. Hernández Pérez, Hong G. Im

Clean Combustion Research Center, King Abdullah University of Science and Technology, 23955-6900, Saudi Arabia

Abstract

Regulations in CO_2 emissions have led to the development of downsized, highly boosted and direct injection technologies that enhance power density and fuel consumption in Spark Ignition (SI) engines. Despite these enhancements, SI engines are prone to exhibit uncontrolled pre-ignition, whose occurrence, specially at low speed and high load, can trigger detonation development which is associated to violent peak pressures and pressure oscillations that can damage the engine. The uncontrolled combustion is a major challenge to overcome for the new generation of SI engines to be widely used. Here, we studied detonation development using high resolution direct numerical simulations of a closed one-dimensional combustion chamber. A highly reactive hydrogen-oxygen mixture with a detailed chemical reaction mechanism is used in this study. Unlike previous studies where bulk mixture properties were arbitrarily defined, we used results from engine simulations obtained with the CONVERGE software to set the initial bulk mixture properties. Depending on the initial conditions two different paths to detonation development were identified: 1) auto-ignition near the wall followed by detonation combustion and 2) transition of the initial propagating flame to detonation combustion mode through flame acceleration.

1 Introduction

SI engines operating under high load and low speed are prone to host an uncontrolled pre-ignition event that can trigger detonation combustion development. Detonation was linked to super-knock [1], a major challenge to overcome in modern SI engines due to its associated high peak and potentially destructive pressures. Super-knock has been the subject of widespread interest over the last decades [2,3,4,5,6,7]. Despite the considerable efforts on the subject, pre-ignition and detonation formation are still not well understood. Experimental evidence has shown that pre-ignition and detonation are closely correlated in SI engines. Although pre-ignition is the precursor of detonation formation, it's not a sufficient condition for detonation development [1]. Furthermore, the underlying mechanism of pre-ignition is poorly understood in SI engines. Nowadays, in the context of SI engines there is no detailed experimental characterization of super-knock formation. Most of the studies used maximum pressure rise or peak in-cylinder pressure to identify super-knock. In [1,2] rapid compression machines were used to study super-knock. The results indicated that, depending on the initial conditions, pre-ignition first triggered hot spots near the wall. In turn, subsequent auto-ignition from the created hot spots generated detonation combustion near the wall. From a numerical viewpoint, Chen et al. [8] investigated end-gas auto-ignition in a closed chamber using a hydrogen/air mixture, pointing out the possibility of auto-ignition followed by detonation development similar to the experimental observations in [1,2]. Although interesting and insightful, their simulation results are of limited applicability based on the fact that mixture inhomogeneity and temperature distribution that are present in engines were neglected in their model. In addition, the domain dimension was not representative of an engine. In order to fill this gap and account for realistic engine conditions, we utilize engine simulation results obtained with CONVERGE [9] to provide the initial conditions (i.e., temperature and equivalent ratio distributions) for high-fidelity direct numerical simulations (DNS). Hence, we start our DNS computations with conditions for end-gas that are likely to be encountered in modern SI engines.

In what follows, the numerical methodology is described, numerical results are presented and discussed, and concluding remarks are provided.

2 Mathematical and numerical model

The one-dimensional (1D) governing equations for a reactive mixture of gases were used as mathematical model. The hydrogen/air kinetics mechanism developed by Burke et al. [10], involving 9 species and 23 reactions, was employed. The computational domain length was set to 0.07353 m and reflecting boundary conditions were specified at the walls. The governing equations were discretized using a 5th order shock-capturing weighted essentially non-oscillatory (WENO) scheme for convective terms [11] and an 8th order finite difference scheme for the viscous ones. Time stepping was performed through a total variation diminishing Runge-Kutta (TVDRK3) scheme. To accurately resolve the flame front and pressure waves, in all simulations performed here a constant mesh of 4 μm and a time step of 8.10^{-10} s was used.

3 Results and discussion

Our interest is on the subsequent evolution of end-gas after a pre-ignition front has been successfully established in the combustion chamber. Occurrence of pre-ignition is out the scope of the current investigation. Therefore, in all numerical experiments conducted in this study, we ignite the mixture by imposing near the right wall a freely propagating flame representing the pre-ignition front. This front propagates from right to left (see Fig. 1 and Fig. 2). Figures 1 and 2 show the initial profiles of temperature and H_2 mass fraction, respectively. Four tests cases (a), (b), (c) and (d) have been studied. The temperature within the bulk mixture is increased

Corresponding author.
E-mail address: aliou.sow@kaust.edu.sa

from case (a) to case (d) but the mass fractions of H_2, O_2, and N_2 have the same distribution.

Moreover, the initial pressure is kept at 3 MPa in all the cases presented here, which represents the top dead center (TDC) pressure in the engine conditions adopted for the study.

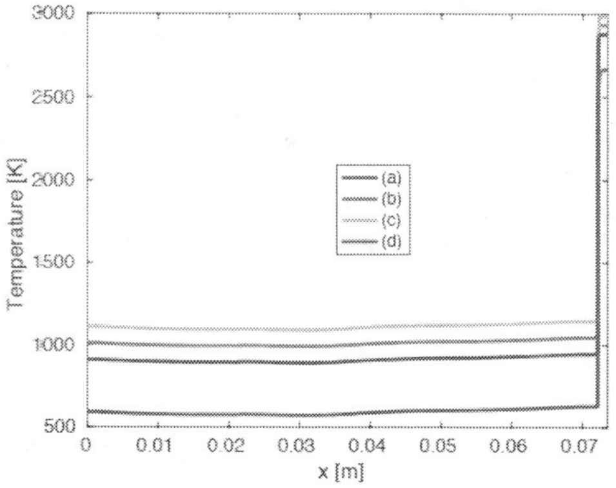

Figure 1: Initial temperature distributions for cases (a)-(d).

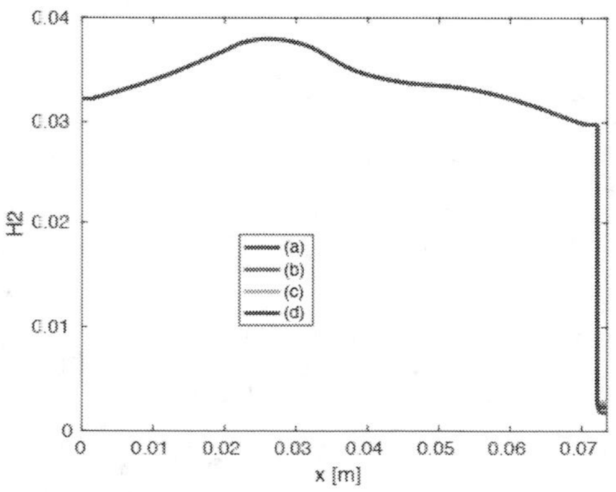

Figure 2: Initial mass fraction distributions of H_2 for cases (a)-(d).

Figure 3 shows the temperature distributions for case (a). The initial bulk mixture temperature is relatively low. This configuration corresponds to a normal combustion mode. The fresh mixture is completely consumed by the propagating flame. Neither endwall auto-ignition nor detonation is observed.

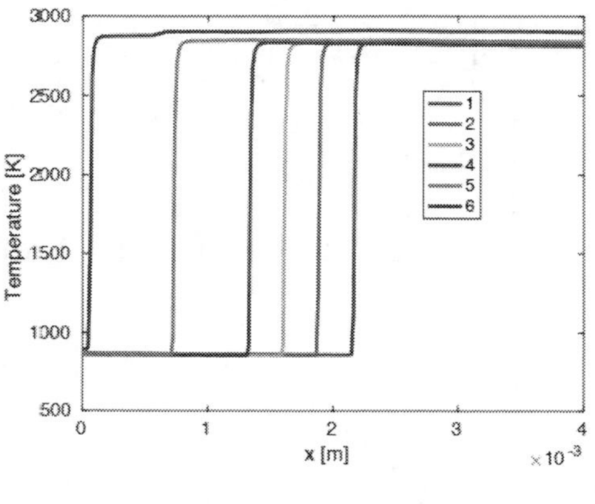

Figure 3: Temperature distribution for case (a). Lines (ms): 1 (2.37), 2 (2.38), 3 (2.39), 4 (2.4), 5 (2.42) and 6 (2.44).

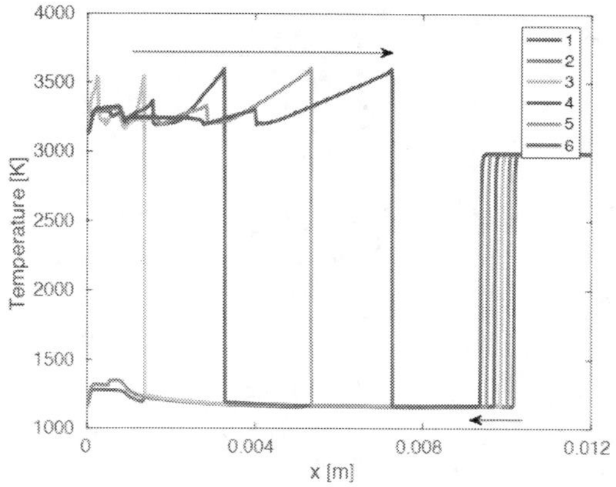

Figure 4: Temperature distribution at different times for case (b). The upper arrow indicates the propagation direction of the detonation front while the lower arrow indicates the propagation direction of the flame front. Lines (ms): 1 (0.749), 2 (0.750), 3 (0.751), 4 (0.752), 5 (0.753) and 6 (0.754).

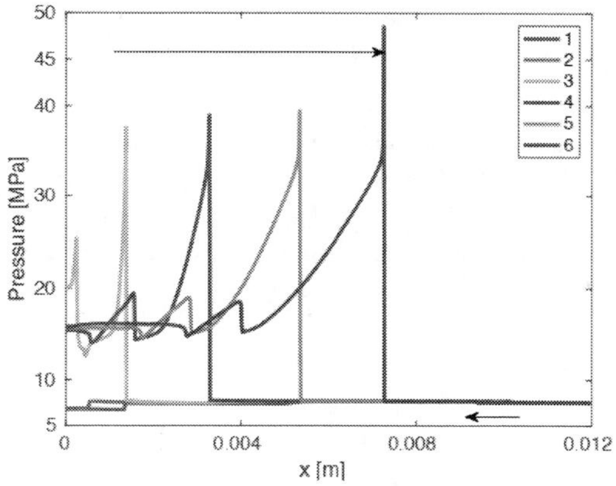

Figure 5: Pressure distribution at different times for case (b). The upper arrow indicates the propagation direction of the detonation front while the lower arrow indicates the propagation direction of the flame front. Lines (ms): 1 (0.749), 2 (0.750), 3 (0.751), 4 (0.752), 5 (0.753) and 6 (0.754).

Results for case (b) are presented in Fig. 4 and Fig. 5. When the flame propagates to around 10 mm a hot spot is generated near the wall. This hot spot quickly initiates an auto-ignition front which in turn evolves into a detonation combustion mode in a very short time. The generated detonation front travels at an average speed of 2000 m/s. During the process, the temperature increases to around 3500 K and a maximum pressure peak (see Fig. 3) of 48 MPa is attained. Such level of pressure peak has been associated to super-knock in previous studies.

Results for case (c) are summarized in Fig. 6 and in Fig. 7. The initial flame front propagates up to approximately 0.055 m before it accelerates and turns into a detonation. The transition to detonation is the result of a shock formation due to the conducive conditions created by the initial flame front acceleration.

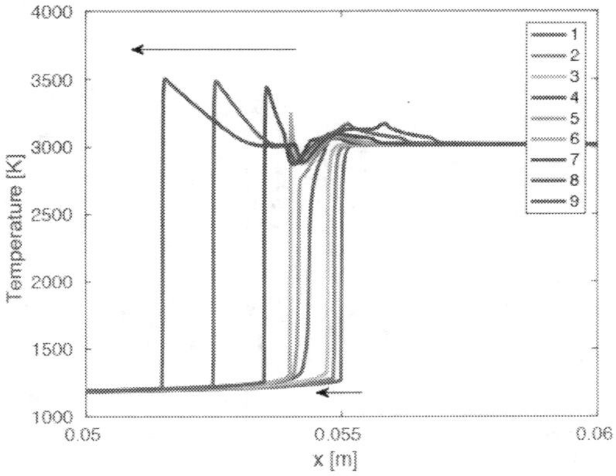

Figure 6: Temperature distribution at different times for case (c). The upper arrow indicates the propagation direction of the detonation front while the lower arrow indicates the propagation direction of the flame front. Lines (ms): 1 (0.132), 2 (0.133), 3 (0.134), 4 (0.135), 5 (0.1352), 6 (0.1353), 7 (0.1355), 8 (0.136) and 9 (0.1365).

The well pronounced detonation wave can be clearly observed from the pressure distribution in Fig. 7. At 0.054 m a peak pressure is generated due to the increased amount of unburned gases consumed by the propagating reaction front. Due to the positive feedback that might start just before the transition process, the pressure pulse generated by the propagating flame and the chemical heat released by the reaction get coupled. The pressure growth rate increases sharply during the detonation development. Pressure peaks of the order of 22.5 MPa are reached.

Initial bulk mixture temperature distributions higher than those in case (c) have been tested in case (d). We find that the run-up distance to detonation shortens compared to case (c) due to the increased reactivity of the bulk mixture.

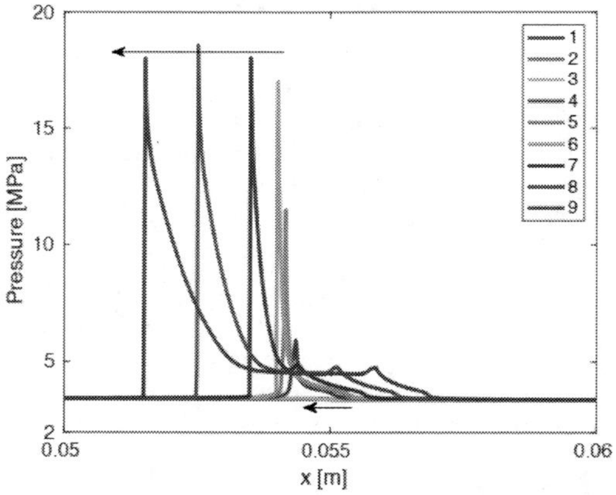

Figure 7: Pressure distribution at different times for case (c). The upper arrow indicates the propagation direction of the detonation front while the lower arrow indicates the propagation direction of the flame front. 1 (0.132), 2 (0.133), 3 (0.134), 4 (0.135), 5 (0.1352), 6 (0.1353), 7 (0.1355), 8 (0.136) and 9 (0.1365).

4 Conclusions

This study investigated detonation development in a closed chamber representative of SI engines. Unlike most numerical studies imposing an initial gradient of temperature/concentration or uniform bulk mixture properties, we considered engine initial conditions that were obtained from auxiliary computations using CONVERGE. Two transition paths to detonation were identified depending on the initial conditions: auto-ignition near the wall followed by detonation development and transition from the initial propagating deflagration to detonation. This study is our first attempt to gain insights into detonation development in SI engines under realistic engine conditions. This research work will be extended to include cooling effects at walls and different fuels such as iso-octane will be also considered.

5 Acknowledgment

The work presented in this paper was sponsored by KAUST and Saudi Aramco under the FUELCOM II Project. The computational work utilized resources from the KAUST Supercomputing Laboratory.

References

[1] Y. Qi, Z. Wang, J. Wang, X. He, Combust. Flame 162 (2015) 4119-4128.

[2] Z. Wang, Y. Qi, X. He, J. Wang, S. Shuai, C. K. Law, Fuel 144 (2015) 222-227.

[3] Z. Wang, Y. Qi, H. Liu, Y. Long, J.-X. Wang, SAE Technical paper 2015-01-0752, 2015, doi:10.4271/2015-01-0752.

[4] Z. Wang, H. Liu, T. Song, Y. Qi, X. He, S. Shuai, J-X. Wang, Int. J. Engine Res. 16 2 (2015).

[5] T. Inoue, Y. Inoue, M. Ishikawa, SAE Technical paper 2012-01-1141 (2012).

[6] X. J. Gu, D. R. Emerson, D. Bradley, Combust. Flame 133 (2003) 63-74.

[7] G. T. Kalghatgi, D. Bradley, Int. J. Engine Res. 13 (4) (2012) 399-414.

[8] H. Yu, Z. Chen, Combust. Flame 162 (2015) 4102-4111.

[9] M. Mubarak Ali, F. Hernandez Perez, S. Vedharaj, R. Vallinayagam, H. G. Im, R. W. Dibble, SAE Technical Paper 2017-01-0686, 2017, doi:10.4271/2017-01-0686.

[10] M.P Burke, M. Chaos, Y. Ju, F. L. Dryer, S.J. Kippenstein, Int.J. Chem. Kinet. (2011).

[11] CW. Shu, Advanced numerical approximation of nonlinear hyperbolic equations, (1988), 325-432.

11th Asia-Pacific Conference on Combustion,
The University of Sydney, NSW Australia
10th -14th December 2017.

Thermometry in Turbulent Dilute Spray Flames using 5kHz CPP-fs-CARS

A. Lowe[1], L. Thomas[2], A. Satija[2], G. Singh[1], R.P. Lucht[2], A.R. Masri[1]

[1]School of Aerospace Mechanical and Mechatronic Engineering, University of Sydney, Sydney, NSW, 2006, Australia

[2]School of Mechanical Engineering, Purdue University, West Lafayette, Indiana, 47906, USA

Abstract

Temperature fields in turbulent dilute spray flames are investigated – for the first time - using chirped-probe-pulse (CPP) femtosecond (fs) coherent anti-Stokes Raman spectroscopy (CARS) at a repetition rate of 5kHz. This ultrafast technique is applied to the Sydney Needle Spray Burner (SYNSBURN™) featuring air-blast atomization of a liquid jet issued from a needle that can be translated within a coflowing air stream. This burner allows direct control of the spray inlet condition for a fixed mass loading, thus regulating spray inhomogeneity between the two extremes of dilute and dense. Various dilute spray flames of acetone for a fixed recess length of 80mm are investigated at different global Weber number or bulk jet Reynolds number. Comparison of mean temperature profiles with thermocouple measurements show very good agreement. In addition, the high data rate of 5kHz provides temporally resolved information about the dynamics of fluid structures in the main jet/pilot and pilot/coflow shear layers. These high bandwidth temperature measurements will aid in turbulent combustion model validation and provide insight to spatio-temporal instabilities in practical combustion devices such as gas turbine combustors and augmentors.

1 Introduction

Reducing the emission of pollutants in modern liquid-fueled engines without compromising efficiency requires, amongst other things, detailed measurements of temperature. Relevant examples range from diesel engines which employ biodiesel blends to direct injection, spark ignited (DISI) engines where liquid fuel spray is guided to achieve mixture stratification with lean overall combustion [1]. Many laser diagnostic methods that would otherwise work in gaseous flames [2] do not work as well in sprays due to the presence of droplets. However, alternative laser diagnostic methods capable of probing denser parts of spray jets [3-4] are becoming available along with numerical codes that can simulate the evolution of sprays [5]. Coherent anti-Stokes Raman spectroscopy (CARS) is one technique capable of providing reliable measurements of temperature in particle-laden flames. An advanced variant of this approach employing a femto-second (fs) laser and a chirped probe pulse (CPP) is employed here. Thermometry using CPP-fs-CARS is already well-established as an accurate thermometry measurement in gas phase reacting flames [6–9]. Variations of CARS schemes have been investigated where multi-pass cavity arrangements and special beam formatting optics were used to obtain discrete 1D probe volumes [10-11]. Numerous CARS approaches have been explored and applied for gas phase diagnostics by using nanosecond (ns) and picosecond (ps) duration laser pulses. Despite these efforts, quantitative femtosecond (fs) measurements in multi-phase flows are lacking in the literature. It is acknowledged that CPP-fs-CARS can provide high bandwidth (1-10kHz), nearly collision-free measurements of temperature and/or species concentrations in reacting spray flows. Therefore, this research aims to employ the CPP-fs-CARS technique to spray combustion environments to investigate new physical phenomena and provide corresponding temperature datasets for future experimental and modeling research efforts.

The Sydney Needle Spray Burner (SYNSBURN™) designed to study turbulent spray flames where the liquid morphology can range between the two extremes of dense and dilute is employed here. This burner has already been extensively characterized in terms of velocity flow-field, size and spatial distribution of droplets as well as complex liquid fragment shapes inherent to denser spray flames [12]. Only dilute spray flames are studied in this paper to illustrate the capability of the technique in providing reliable, time- and space-resolved measurements of temperature in particle-laden flames.

The first section of this paper introduces the design and features of the Sydney Needle Spray Burner developed at the University of Sydney [12]. Several test cases are specified relative to flame stability limits for parametric studies. This is followed by a brief overview of the CPP-fs-CARS system developed at Purdue University [13]. The results section consists of the following: (i) mean temperature fields in various dilute spray flames are discussed in relation to far-field thermocouple measurements (ii) time series and fast-Fourier transform (FFT) analysis of temperature are discussed.

2 Experimental Methodology

2.1 Sydney University Needle Spray Burner

The burner, shown schematically in Figure 1a, uses air-blast to atomize liquid issuing from a needle with a diameter D_l=686μm, centred in a fast coflowing stream of air in a jet diameter, D=10mm. A key feature of the burner is that the needle can be recessed by up to 80mm upstream of the jet exit plane. This is similar, in principle, to the gaseous piloted burner which has been used earlier to investigate the effects of inlet compositional inhomogeneity on flame structure [14]. The air-blasting jet is surrounded by a 3-gas pilot with an outer annulus diameter, D_p=25mm, where premixed hydrogen H_2, acetylene C_2H_2 and air issue under stoichiometric conditions with an unburnt velocity U_{bu}=1.5m/s. The pilot composition is such that its C/H ratio is the same as that of the main acetone fuel (C_3H_6O) and provides a respective heat release of 1.49kW albeit with a peak adiabatic flame temperature of 2512K. This is higher than that of acetone which peaks at 2281K (φ=1.05).

Corresponding author: Albyn Lowe
E-mail address: alow2350@uni.sydney.edu.au

The entire burner assembly is mounted in a wind tunnel providing a coflowing air stream with a mean axial velocity of 1.5m/s.

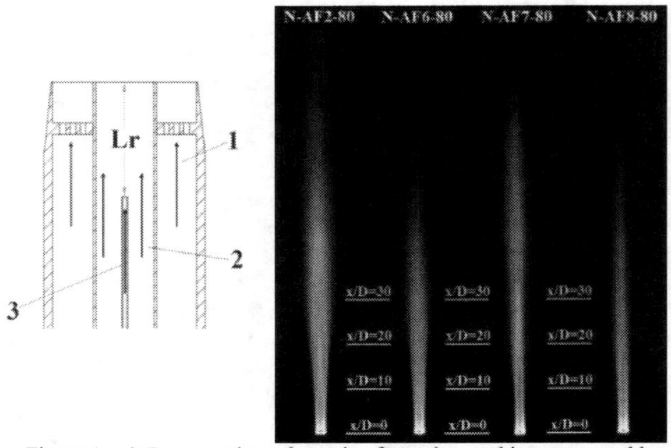

Figure 1 - a) Cross-section schematic of atomizer and burner assembly showing the pilot stream (1), main air jet stream (2) and liquid fuel stream (3). b) Sample long exposure (1s) images of selected acetone flame cases for recess length of 80mm, details in Table 1.

Cases	N-AF2	N-AF6	N-AF7
U_j (m/s)	36	36	60
Re_g	21858	21858	36430
Q (g/min)	75	45	75
We	45	45	125
F/A (by mass)	0.36	0.22	0.22
U_j/U_{bo}	0.5	0.57	0.83

Table 1 - Initial conditions for reacting acetone spray cases studied here.

While the flame studies can range from dense to dilute by simply changing the recess distance of the needle, only dilute flames are studied in this paper with Lr=80mm. This is intentional to demonstrate the capability of the CPP-fs-CARS technique. Photographs of four such flames are displayed in Figure 1b to illustrate the effects of liquid loading and air-blast velocity on global features such as flame appearance and visible length. Selected properties for these cases are specified in Table 1 showing that the bulk air blast velocity or liquid mass flowrate are systematically varied for parametric sensitivity testing. The naming methodology is explained with reference to case N-AF2-80: N = Needle, A = Acetone, F refers to reacting flame. Cases 2 and 6 have the same jet velocity but different liquid loading while the reverse is true for cases 2 and 7. The exit Weber number, of relevance in this burner, is defined as $We = \rho_g U_g^2 D_l/\sigma$ where ρ_g is the gas density, U_g is the bulk gas velocity and σ is the surface tension. The fuel to air ratio (F/A) is defined as the mass fraction of injected liquid over that of the blast air; Re_g refer to the Reynolds numbers of the carrier air. The quantity U_j/U_{bo} is the ratio of bulk jet velocity to blow off velocity; the blow off condition is where the flame is visibly intermittent and detached from the pilot for a specific mass loading.

These flame stability limits for acetone flames are shown in Figure 2 versus liquid mass loading for a recess length of Lr= 80mm. For a fixed recess length of 80mm, the blow-off velocity U_{bo} increases as the fuel mass loading increases. This recess length is sufficient for atomization to complete, such that the impact of evaporation and interphase mixing become more prominent and the flame's behaviour gradually approaches that of a conventional dilute spray.

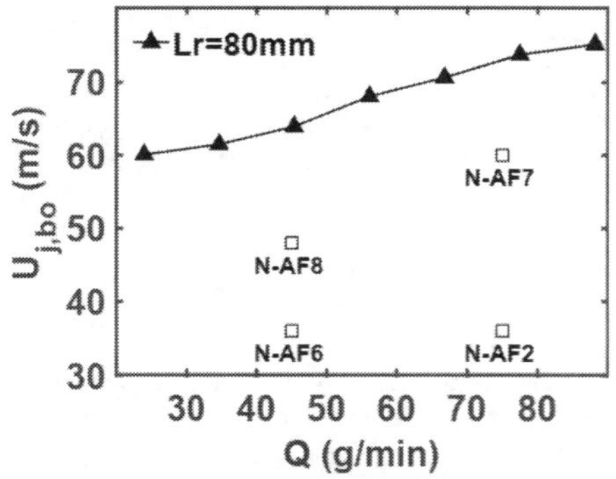

Figure 2 - Blow-off air-blast jet velocity (U_{bo}) plotted vs. liquid acetone mass loading (Q). Boxes specify selected cases defined in Table 1.

2.2 Chirped-Probe-Pulse Femtosecond Coherent Anti- Stokes Raman Scattering

In CPP-fs-CARS, a probe beam is frequency-chirped for single laser shot imaging of the time-dependent frequency spread dephasing of the Raman coherence; generated by a high-power 800nm beam used as the pump and a tunable 983nm beam used as the Stokes for CARS signal generation from the N_2 Q-branch transitions at ~2330 cm^{-1}. Integration times of less than 4 picoseconds (ps) and spatial resolution of 790 μm along beam propagation and 59 μm in the transverse dimension were achieved. The resulting CARS signal at ~ 675nm is spatially and spectrally resolved. The system consists of an amplified titanium-sapphire (Ti:S) fs laser, a frequency shifting optical parametric amplifier (OPA), various routing optics, a calibration flame, a spectrometer and an electron-multiplying charge-coupled device camera (EMCCD). The Ti:S laser (Legend Elite Duo, Coherent Inc) uses chirped-pulse amplification via a pulse stretcher, a regenerative amplifier, a single-pass amplifier, and a compressor to generate mode-locked 60-fs pulses centred on 800nm at approximately 2mJ per pulse at a 5kHz pulse repetition frequency. A beam-splitter routes 80% of the energy to pump the OPA (OPerA Solo, Coherent Inc.) to create 60-fs Stokes pulses centred on 982nm at 50μJ per pulse. The remaining 20% of the energy is used to generate the pump (100μJ) and probe pulses at a 30-70 energy split respectively. The probe pulse is sent through a chirping rod (30cm SF-10 glass) to impart a frequency dependant delay and stretch the 200μJ pulse from 60-fs to ca. 3.5 ps. Delay stages are used to optimize timing overlap of the pump and Stokes pulses and control the probe time delay. All three beams are focused using a broadband-AR-coated 250mm focal length plano-convex (PCX) lens and are positioned such that their relative geometry satisfies energy and momentum conservation for efficient CARS generation at the probe volume using the "folded-BOXCAR" phase matching protocol. The CARS beam is collected and horizontally dispersed by a 500mm Czerny-Turner-type spectrometer and focused onto a back-illuminated EMCCD camera (iXon888, Andor Technologies). The CARS signal is vertically binned on the EMCCD and the 5kHz laser pulse repetition rate is matched to achieve a 5kHz data acquisition rate. An extended explanation of the CPP-fs-CARS system, calibration and generative spectral fitting algorithm are found in [6, 13] and are not the focus here.

3 Turbulent Dilute Spray Flame Results

3.1 Mean Temperature Fields

Mean temperature profiles at x/D=0.4-20 for selected flames in Table 1 are presented in Figure 3 and 4. For acetone, the pilot adiabatic temperature is 2512K and this is detected in all flames at x/D=0.4 to within 5% accuracy. The main jet r/D=-0.5 to 0.5 is close to room temperature at the exit plane and this persists down to x/D=5. From x/D=10-20 the entire radial profiles trend with F/A ratio with N-AF2 exhibiting the hottest flame and N-AF7 the coldest.

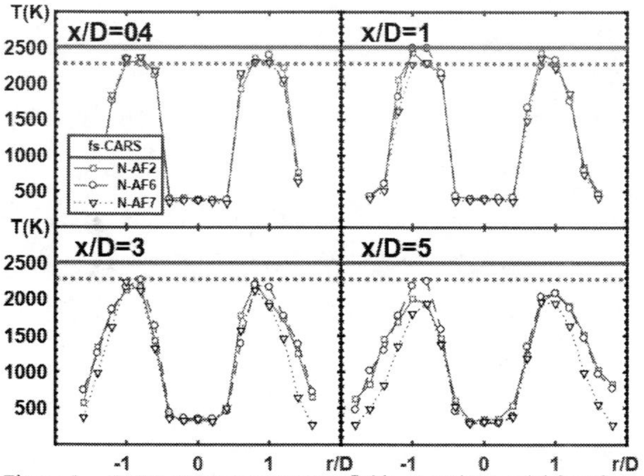

Figure 3 - CARS mean temperature fields at various axial locations (empty markers). The adiabatic temperature of acetone is 2281K (horizontal dashed) and the 3-gas pilot is 2512K (horizontal solid).

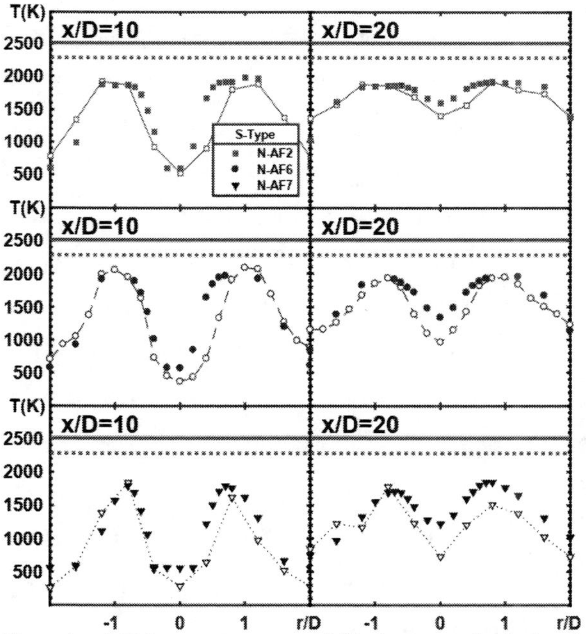

Figure 4 - CARS mean temperature fields (empty markers) plotted against thermocouple temperature fields at x/D=10-20 (solid markers). The adiabatic temperature of acetone is 2281K (horizontal dashed line) and the 3-gas pilot is 2512K (horizontal solid line).

S-type thermocouple measurements sampled at 100Hz (PCI-6035E NI-DAQ) were taken at x/D=10 and 20 (the measurements are limited to regions downstream of the pilot past x/D=5, S-Types have a maximum operating temperature near 1755K). The thermocouple measurements are almost identical to fs-CARS at x/D=10 for all flames. At x/D=20 the thermocouple measurements overshoot the fs-CARS temperatures in central regions r/D=-1 to 1. This may be due to spatial averaging error (SAE) or liquid collecting on the thermocouple bead and is discussed further in section 3.2. The probe volume size becomes especially important in the flame brush region of the flame as the reaction zone moves across the flame front. The fs-CARS probe measures 790μm in length and 59μm in diameter as oppose to S-type bead diameters of approximately 0.3mm. The turbulent length scales in the central flame zone at x/D=20 i.e. the flame brush, may be smaller than the spatial resolution of the thermocouple data at this acquisition rate; we may then observe consistent overshooting of 350K at r/D=0 because the thermal dissipation rate of the bead due to its size causes localized spatial averaging error.

3.2 Statistical Analysis of CARS

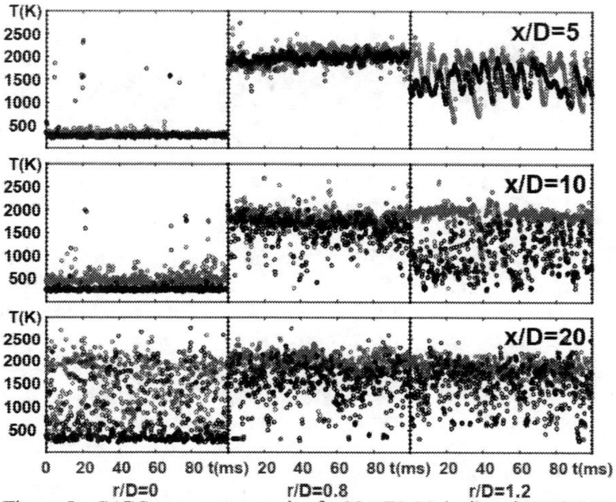

Figure 5 - CARS temperature series for N-AF2-80 (red) and N-AF7-80 (black) at centerline (r/D=0) and in the reaction zone (r/D=0.8-1.2)

The pulse repetition frequency of the fs-CARS system is estimated to be on the same order as that of the largest turbulence timescales in these flames. Large eddies exist on the kHz time scale and the 5kHz data rate supports Fourier analysis up to 2500Hz by the Nyquist criterion. These timescales are on similar orders of magnitude which would allow clear detection of smoothly varying oscillatory structures in some of the temperature time series. In Figure 5, these smoothly varying structures are captured at heights below x/D=20 in the flame/coflow interface past r/D=0.8; however, this does not persist in regions directly over the pilot or in main jet. High scatter temperature series with no detectable oscillatory structures and exhibiting broadband FFTs in Figure 6 are a good indicator of the turbulent induction zones i.e. the flame brush at the edge of the potential core; such areas exist at r/D=0-1.2 for x/D=20 N-AF2-80 and N-AF7-80 and in the main jet /pilot shear layer at r/D=0.6 for x/D=5. Temperature kernel density estimates (KDE) in Figure 7 at these regions exhibit a flat broadband profile stretching the whole temperature range (500-2500K) with little skew. Steady cold regions of the flame are exhibited at centerline up to x/D=10; FFT analysis shows significantly weaker frequencies up to 1kHz at these locations. However, dominant frequencies centered around 130-150Hz exist at r/D=1 and 1.4 for x/D=5. Centerline KDEs of temperature skew on the cold side and approach zero for higher temperatures. Moving downstream at r/D=0.8, indicates a transition from the steady premixed pilot flame of near adiabatic temperature and low intensity broadband frequencies into a turbulent reaction zone for both spray flames. A comparison between N-AF2-80 and N-AF7-80 indicates that the flame spreading rate, centerline temperatures and intensity of oscillating structures driven in the slow coflow/main jet interface (possibly Kelvin-Helmholtz) trend with blast air velocity for a fixed fuel loading.

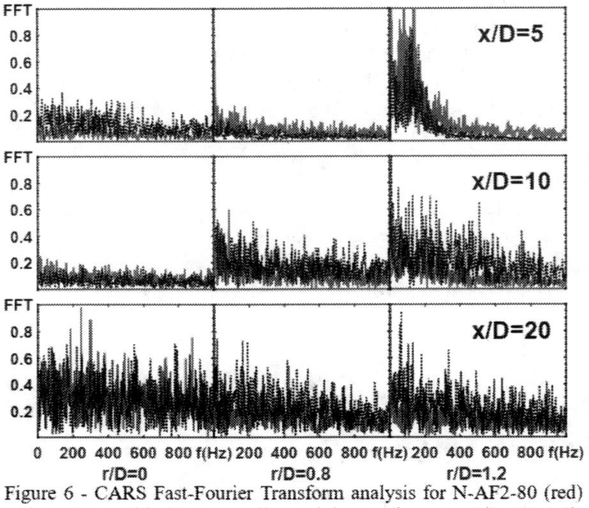

Figure 6 - CARS Fast-Fourier Transform analysis for N-AF2-80 (red) and N-AF7-80 (black) at centerline and the reaction zone (r/D=0.8-1.2)

When comparing to S-type thermocouple measurements and keeping in mind the flame brush; time series analysis indicates that the mean CARS measurements were consistently lower for all flame cases at x/D=20 from r/D=0-0.8 and identical for the same locations in x/D=10 due to spatial averaging error. No emissivity correction was applied to thermocouple measurements. However, a high spread temperature distribution with high intensity broadband frequencies would still result in significant spatially averaging error in the larger thermocouple bead as discussed earlier. This is significant downstream at x/D=20, which is closer to the tip of the potential core at x/D=25 for N-AF2-80 and x/D=35 for N-AF7-80. This is not as prominent at x/D=10 for the inner radial locations where the time series are now concentrated at 500K and 300K for N-AF2-80 and N-AF7-80 respectively; FFTs are of low intensity here as well. This is because the measurement probe is now deemed below this highly turbulent reaction zone.

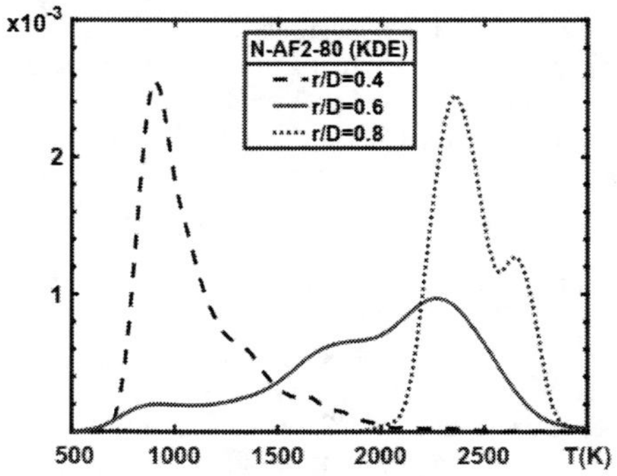

Figure 7 - Kernel Density Estimate (KDE) of CARS at x/D=5

4 Conclusions

Point-temperature measurements integrating over a probe volume 790μm in length and 59μm in diameter are reported for the first time using CPP-fs-CARS at a 5kHz acquisition rate in turbulent dilute spray flames. Spatial and temporal averaging effects are detected and are sensitive to probe volume size. Single-shot temperature measurements effectively resolve strong, non-zero frequency components in the coflow/pilot interface up to 1kHz.

5 Acknowledgment

The Sydney group is supported by the Australian Research Council. The Purdue group is supported by the U.S. Department of Energy, Division of Chemical Sciences, Geosciences and Biosciences under Grant No. DE-FG02-03ER15391 and by the King Abdullah University of Science and Technology under CCF sub- award No. 1975-01.

References

[1] J. Trost, L. Zigan, and A. Leipertz, *Proc. Combust. Inst.*, vol. 34, no. 2, pp. 3645–3652, Jan. 2013.

[2] A. C. Eckbreth, "CARS thermometry in practical combustors," *Combust. Flame*, vol. 39, no. 2, pp. 133–147, Oct. 1980.

[3] M. Linne, *Prog. Energy Combust. Sci.*, vol. 39, no. 5, pp. 403–440, 2013.

[4] C. Schulz and V. Sick, *Prog. Energy Combust. Sci.*, vol. 31, no. 1, pp. 75–121, Jan. 2005.

[5] F. Salehi, M. J. Cleary, and A. R. Masri, SAE Technical Paper, Warrendale, PA, SAE Technical Paper 2016-1–859, Apr. 2016.

[6] C. N. Dennis, C. D. Slabaugh, I. G. Boxx, W. Meier, and R. P. Lucht, *Proc. Combust. Inst.*, vol. 35, no. 3, pp. 3731–3738, 2015.

[7] D. R. Richardson, R. P. Lucht, S. Roy, W. D. Kulatilaka, and J. R. Gord, *Proc. Combust. Inst.*, vol. 33, no. 1, pp. 839–845, 2011.

[8] C. N. Dennis, C. D. Slabaugh, I. G. Boxx, W. Meier, and R. P. Lucht, *Combust. Flame*, vol. 173, pp. 441–453, Nov. 2016.

[9] C. D. Slabaugh, C. N. Dennis, I. Boxx, W. Meier, and R. P. Lucht, *Combust. Flame*, vol. 173, pp. 454–467, Nov. 2016.

[10] J. Bood, C. Brackmann, P.-E. Bengtsson, and M. Aldén, *Opt. Lett.*, vol. 25, no. 20, pp. 1535–1537, Oct. 2000.

[11] M. Afzelius, P.-E. Bengtsson, J. Bood, C. Brackmann, and A. Kurtz, *Appl. Opt.*, vol. 45, no. 6, pp. 1177–1186, Feb. 2006.

[12] A. Lowe, A. Kourmatzis, and A. R. Masri, *Combust. Flame*, vol. 176, pp. 511–520, Feb. 2017.

[13] L. Thomas, A. Lowe, A. Satija, R. P. Lucht, and A. Masri, in *55th AIAA Aerospace Sciences Meeting*, American Institute of Aeronautics and Astronautics.

[14] S. Meares and A. R. Masri, *Combust. Flame*, vol. 161, no. 2, pp. 484–495, Feb. 2014.

11th Asia-Pacific Conference on Combustion,
The University of Sydney, NSW Australia
10th -14th December 2017.

An experimental study of evaporation characteristics of pure and blended biodiesel droplets at high temperatures

Kai Han[1], Zhaojing Ni[1], Hao Chen[1], Bo Pang[1]

[1] School of Mechanical Engineering, Beijing Institute of Technology

5 South Zhongguancun Street, Beijing, 100081, China

Abstract

The evaporation characteristics of pure and blended biodiesel droplets have been investigated experimentally at different ambient temperatures under normal gravity and atmospheric pressure. The droplet suspended on the welding point of a thin K-type thermocouple method coupled with photographic technique is used to characterize simultaneously the histories of droplet temperature and size. The results indicate that pure and blended biodiesel droplet show the same two phase evaporation process as diesel. With the increase of biodiesel content, the maximum increment of squared droplet diameter and droplet temperature at the end of evaporation process increases. In addition, the transient heating phase t_h, equilibrium evaporation phase t_e, droplet lifetime t_t increases with the increase of biodiesel content. With the increase of ambient temperature, the evaporation rate k of blends droplet increase, parameters of droplet temperature and time of evaporation phases of blended biodiesel droplet vary non-linearly with the ambient temperature. In addition, the decelerating effect of adding biodiesel to diesel on evaporation obviously weaken with the increase of ambient temperature.

1 Introduction

Concerns of fossil fuel depletion and environmental degradation have promoted the development of alternative renewable energy. Biodiesel has been suggested as a most promising alternative to the conventional diesel fuel because of their mitigation of greenhouse gases, wide sources of production, being used in compression ignition engine without modification in the engine hardware, and the ability to reduce the pollutant emissions from the engine[1-2].

However, biodiesel is produced from many of vegetable oils or animal fats through a transesterification process with alcohols, which leads to the different physical properties from diesel fuels [3]. The viscosity, density, surface tension, flash point and distillation temperature of biodiesel are distinctly higher than diesel and biodiesel fuels are less volatile than diesel fuels [4-6]. Differences in physical properties affect the evaporation process and consequently the combustion and emission characteristics of the fuel spray combustion in the diesel engines. General results found that the application of biodiesel as a supplementary diesel engine fuel could promote combustion performance and thus leads to low exhaust emissions [7-9].

Corresponding author. Fax: +86-10-68912295
E-mail address: autosim@bit.edu.cn

The evaporation characteristics are very fundamental and important because they affect combustion quality and then engine performance. Hallett and Legault[10] established a model for the evaporation of droplets of biodiesel fuels based on continuous thermodynamics and compared it with the results of experiments on suspended single droplets. Saha et al.[11] established multicomponent evaporation model for pure and blended biodiesel droplets to investigate their evaporation characteristics in high temperature convective environment. Qubeissi et al.[12] established a multi-component model to simulate biodiesel fuel droplet heating and evaporation, which takes into account temperature gradient, recirculation, and species diffusion within droplet. With regard to the investigation of the evaporation characteristics of biodiesel fuel, however, little attention has been paid to experimentally investigate the evaporation characteristics of pure and blended biodiesel.

The droplet suspension technique has been widely used to investigate the evaporation characteristics of fuel droplet, though the fiber conduction would enhance evaporation [13]. A thermocouple is often used as the suspension wire to record the droplet temperature because the history of droplet temperature is an important data to analyze the fuel droplet evaporation process [14-15]. Mura et al. [16] analyzed the emulsion droplet micro-explosion through two experimental approaches of Leidenfrost technique and thermocouple-suspended droplet technique. Watanabe et al. [17] investigated the breakup characteristics of secondary atomization of emulsified fuel droplet suspended on an R-type thermocouple. Therefore it is reasonable to choose the thermocouple as suspension wire.

In this study, the evaporation characteristics of pure and blended biodiesel droplets are investigated by thermocouple-suspended droplet technique at high temperatures. A photographic technique and image processing is used for droplet size determination. In addition, the effects of different blending ratios of biodiesel in blends and ambient temperature are also analyzed.

2 Experimental methods

2.1 Experimental setup and data analysis

The experimental apparatus used in this study has already been detailed described in Ref.[18], hence will be only briefly discussed here. As shown in Fig. 1, the temperature inside the chamber is measured using a 0.5 mm diameter K-type thermocouple and controlled by a PID feedback temperature controller. The maximum attainable temperature of the chamber is 1000 K, and the temperature control accuracy is ±5 K. In order to reduce the effect of the chamber's high temperature on the droplet before the droplet entering the high temperature chamber, the through hole is surrounded by a cooling jacket. The droplets are suspended on the welding point of a thin K-type thermocouple, whose diameter is 127 μm. To control the size of a suspended droplet, the micro syringe with range of 2 μL and accuracy of 0.04 μL is chosen. In this study, the initial diameter of the droplet is generally at the range of 0.91 mm to 0.97 mm. In order to ensure the accuracy of experimental results, each experiment is repeated at least three times. After the suspension of the droplet, the droplet is rapidly moved to the position where the high-speed camera was focused through the electric push device which push speed is set at 300 mm·s^{-1}, and the time of droplet reaching the focus position is less than 0.3 s. The time histories of evaporating droplets are obtained by a high-speed video camera (Photron brand Fastcam SA4) coupled with a micro lens (Nikon Micro-ED 200 mm f/4). The camera and droplet temperature acquisition system are controlled by the synchronous trigger system. A Matlab code is developed to determine droplet size history, from which droplet evaporation rate could be derived. The detailed image-processing can be seen at Ref[19].The error of droplet measurement with this method is estimated within ±3.8%.

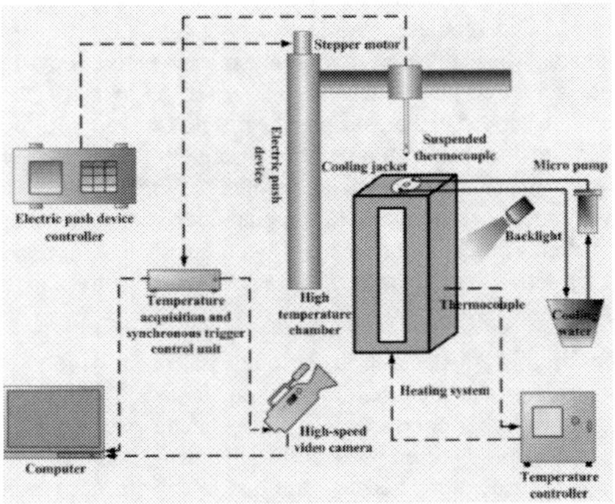

Figure 1:Schematic diagram of the droplet evaporation experimental apparatus

2.2 Fuel properties

In this study, diesel, biodiesel and its blend have been examined. The biodiesel considered in this paper is obtained from soybean. The temperature range of the boiling points for the components constituting soybean methyl ester is narrow and high [6]. According to Chinese National Standard GB/T 2541-1981, GB/T 265-1988, GB/T 386-2010 and GB/T 384-1988, some fundamental physical properties of the commercial diesel and biodiesel employed in this present paper are compared in Table 1.

Properties	Diesel	Biodiesel
Density at 293K (kg·m^{-3})	835.5	878.1
Kinematic viscosity at 293K (mm^2·s^{-1})	4.367	7.292
Kinematic viscosity at 313K (mm^2·s^{-1})	2.293	3.625
Cetane number	53.1	54.6
Low heat value (MJ·kg^{-1})	42.77	37.26
Flash point (°C)	64	160
Boiling point (K)	453-613	601-633

Table 1: Fuel properties of diesel and biodiesel.

3 Results and discussion

3.1 Evaporation characteristics of diesel-biodiesel blends

Figure 2 shows the series of photographs for pure diesel, B30 (30% mass fraction of biodiesel in blends), B50 (50% mass fraction of biodiesel in blends), and pure biodiesel at 650 K. As shown in this figure, these fuel droplets show the same two phase evaporation during the whole evaporation process. The droplet lifetime of diesel, B30, B50 and biodiesel are 4.346 s·mm^{-2}, 4.970 s·mm^{-2}, 5.322 s·mm^{-2} and 6.209 s·mm^{-2}, respectively.

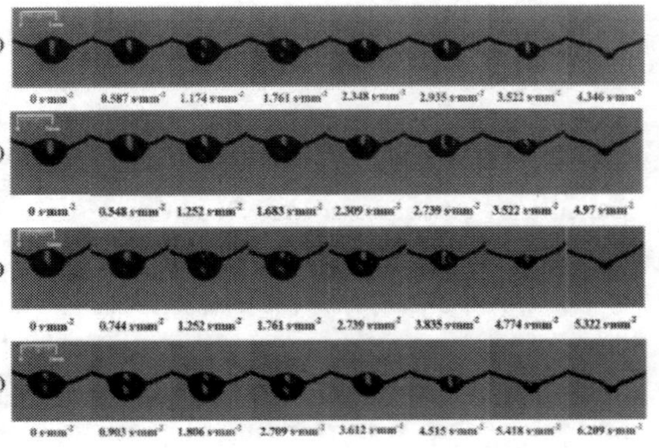

Figure 2: Series of photographs of pure and blended biodiesel droplets at 650 K;(a)diesel, (b)B30, (c)B50, (d)biodiesel.

Figure 3 shows the curves of normalized squared droplet diameter and droplet temperature versus the normalized time under the ambient temperature of 650 K for pure diesel, B30, B50 and pure biodiesel. Those curves can be divided into two phases: transient heating phase (t_h) and equilibrium evaporation phase (t_e)[20]. The transient heating phase (t_h) is defined as the evaporation process before the point of droplet diameter becoming equal to the initial droplet diameter, and equilibrium evaporation phase (t_e) represents the residual evaporation process. During the phase t_h, the heat transferred from ambient is utilized mainly to increase the droplet temperature, and the density of the droplet decreases with the increasing droplet temperature, which leads to the droplet thermal expansion. After that, a balance between thermal expansion and evaporation determines the diameter of droplet. The maximum increment of squared diesel droplet diameter is about 3.5% in relation to initial squared droplet diameter while the maximum increment of squared B30, B50 and biodiesel droplet diameter are about 5.8%, 8.5% and 10.2%, respectively. With the increase of biodiesel content, the maximum increment of squared droplet diameter increases, which is caused by the increase of transient heating phase and the larger decrease of density. As the droplet surface temperature increases and approaches its boiling temperature, evaporation accelerates. In the phase t_e, the temperature inside the droplet reaches a quasi-steady state, the droplet diameter decreases monotonically.

The droplet temperature increases quickly during t_h. After the heating period, diesel droplet firstly enters the equilibrium evaporation phase t_e, which is due to the components of diesel with lower boiling points. In the equilibrium evaporation phase t_e, owing to the multi-component properties of diesel, B30, B50 and biodiesel, their wet bulb temperatures are not a constant value. At the end of evaporation process, the diesel droplet temperature is 589 K, while the B30, B50 and biodiesel droplet temperature are 592 K, 603 K and 615 K, respectively. The maximum boiling point in biodiesel is higher than diesel, therefore, with the increase of biodiesel content, there will be more component with higher boiling point which lead to the increase of droplet temperature at the end of evaporation process.

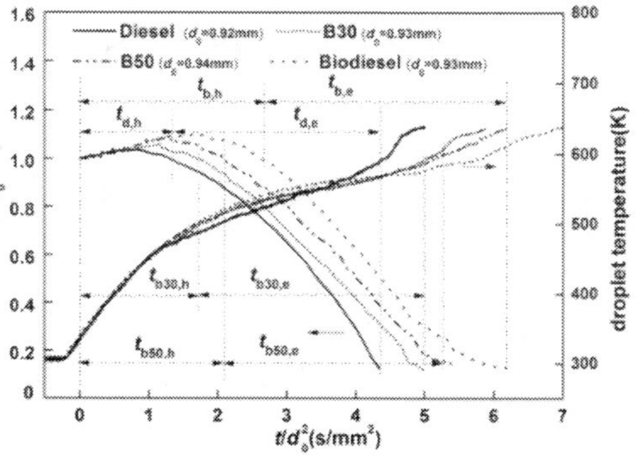

Figure 3: Variations of normalized squared droplet diameter and droplet temperature with the normalized time for pure and blended biodiesel droplets at 650 K.

From Fig. 3, it is obviously noted that with the increase of mass fraction of biodiesel, droplet lifetime increases. Diesel droplet lifetime is the shortest, B30 and B50 droplets lifetime are longer than diesel, but shorter than biodiesel. Since B50 contains more biodiesel fuel, it evaporates at a slower rate than B30. As shown in Fig. 4, with the increase of mass fraction of biodiesel, t_h, t_e, and droplet lifetime t_t increase, which is due to huge difference in the volatility of diesel and biodiesel and the addition of biodiesel reducing the volatility of blended fuels. Compared to diesel fuel, the droplet lifetime of B30 and B50 are increased by 14.9% and 24.1%, respectively.

In order to further analyze the evaporation characteristics of diesel-biodiesel blends, according to the d^2-law, the equivalent evaporation rate k can be extracted from the phase t_e of the evaporation history curve. The slope of this line, which is passing through the phase t_e, is the minus of evaporation rate. As shown in Fig. 5. With the increase of mass fraction of biodiesel, the fuel equivalent evaporation rate k decreases. Compared to diesel fuel, the equivalent evaporation rate k of B30 and B50 are reduced by 6.3% and 10.1%, respectively.

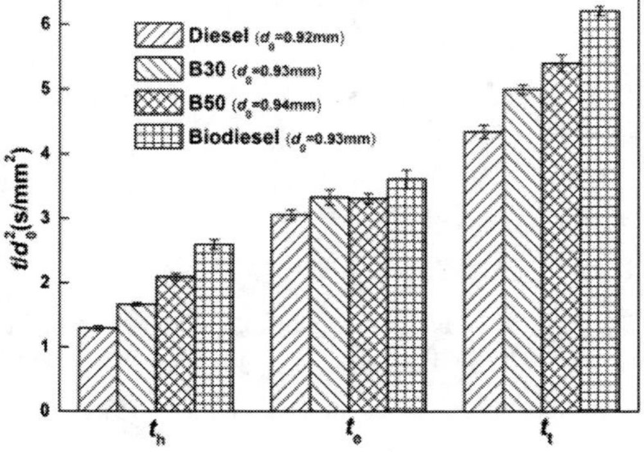

Figure 4: Comparison of characteristic times of pure and blended biodiesel droplets at 650 K.

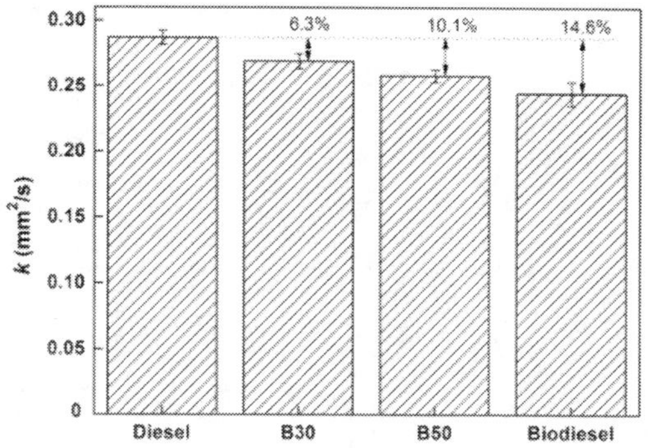

Figure 5: Comparison of equivalent evaporation rate k of pure and blended biodiesel droplets at 650K.

3.2 Effect of ambient temperature on the evaporation of B30 droplet

The influence of ambient temperature on B30 droplet evaporation is investigated at different ambient temperatures (453 K-703 K). Figure 6 shows the variations of normalized squared droplet diameter and droplet temperature versus the normalized time for B30 at 453 K, 553 K, and 650 K. High ambient temperature indicates higher heating rate, and consequently faster evaporation rate. Therefore, with the increase of ambient temperature, the droplet evaporation rate increases, and the droplet lifetime decreases.

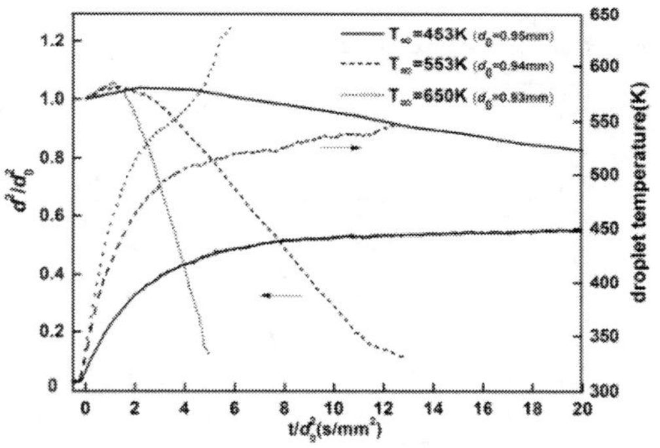

Figure 6: Variations of normalized squared droplet diameter and droplet temperature with the normalized time at 453 K, 553 K, and 650 K.

Figure 7 depicts the equivalent evaporation rate k of biodiesel, diesel and B30 at different ambient temperatures. The equivalent evaporation rate of B30 is between that of diesel and biodiesel. With the ambient temperature increasing, the evaporation rate k of biodiesel, diesel and B30 increase, which can be theoretically analyzed by the Spalding quasi-steady evaporation theory [21]. In addition, with the increase of ambient temperature, the difference in k of biodiesel, diesel and B30 decreases, which because the effect of fuel physical properties on droplet evaporation characteristics decreases as ambient temperature increasing.

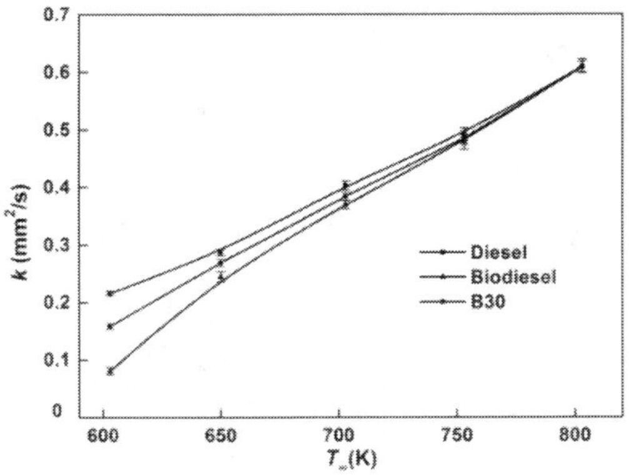

Figure 7: Variations of the equivalent evaporation rate k of biodiesel, diesel and B30 at different ambient temperatures.

The variations of t_h, t_e, t_t, droplet temperature at the end of transient heating phase T_{t1} and equilibrium evaporation phase T_{t2} at different ambient temperatures between 453 K and 703 K for B30 are presented in Table 2. With the increase of ambient temperature, t_h, t_e, and t_t decrease, T_{t1} and T_{t2} increase. However, these parameters do not change linearly with ambient temperature. The reason for the non-linear change of these parameters contains two aspects: on the one hand, the key parameters affecting droplet evaporation, such as the vapor pressure, diffusion coefficient, and specific heat capacity, are changed non-linearly with the ambient temperature; on the other hand, with the increase of ambient temperature, higher evaporation rate causes higher amount vapor annulus to surround the droplet, which limits the increase of droplet temperature and evaporation rate[11].

T_∞ (K)	t_h ($s{\cdot}mm^{-2}$)	t_e ($s{\cdot}mm^{-2}$)	t_t ($s{\cdot}mm^{-2}$)	T_{t1} (K)	T_{t2} (K)
453	6.64	210.26	216.90	429	453
503	4.05	32.23	34.28	456	503
553	2.68	10.48	13.16	476	548
603	2.11	5.62	7.73	489	580
650	1.71	3.29	5.00	499	593
703	1.69	2.43	4.12	514	605

Table 2: Parameters in droplet evaporation process at different temperatures for B30.

4 Conclusions

The evaporation characteristics of pure and blended soybean biodiesel droplets have been investigated by thermocouple-suspended droplet technique at different ambient temperatures. The major conclusions are drawn as follows:

1) Pure and blended biodiesel droplet show the same two phase evaporation process as diesel. With the increase of biodiesel component, the maximum increment of squared droplet diameter and droplet temperature at the end of evaporation process increases.

2) With the increase of biodiesel content, the transient heating phase t_h, equilibrium evaporation phase t_e and droplet lifetime t_t increases. In addition, equivalent evaporation rate k decreases with the increase of biodiesel content.

3) With the increase of ambient temperature, the evaporation rate k of blends droplet increase. In addition, the restrain effect of adding biodiesel to diesel is obviously at lower temperature, but apparently weaken at higher temperature.

4) Parameters of droplet temperature, droplet lifetime, droplet evaporation rate, and time of evaporation phases of blended biodiesel droplet vary non-linearly with the ambient temperature.

5 Acknowledgment

This work is supported by the Nation Natural Science Foundation of China (Grant No. 51306021).

References

• M.M. Gui, K.T. Lee, S. Bhatia, Energy 33(11)(2008) 1646-1653.

• A.K. Agarwal, Prog. Energy. Combust. Sci. 33(3)(2007) 233-271.

• A. Karmakar, S. Karmakar, S. Mukherjee, Bioresource. Technol. 101(19)(2010) 7201-7210.

• P. Benjumea, J. Agudelo, A. Agudelo, Fuel 87(11-12)(2008) 2069-2075.

• E. Alptekin, and M. Canakci, Fuel 88(1)(2009)75-80.

• W. Yuan, A.C. Hansen, Q. Zhang, Fuel 84(7-8)(2005) 943-950.

• J.M. Desantes, R. Payri, A. García, J. Manin, Energy Fuels, 23(6)(2009)3227-3235.

• T.G. Fang, C.F. Lee, Proc. Combust. Inst. 32(2) (2 009) 2785-2792.

• Y.C. Lin, K.H. Hsu, C.B. Chen, Energy 36(1)(2011) 241-248.

• W.L.H. Hallett, N.V. Legault, Fuel, 90(3)(2011) 1221-1228.

• K. Saha, E. Abu-Ramadan, X.G. Li, Appl. Energy 93(5)(2012) 71-79.

• M. A. Qubeissi, S.S. Sazhin, C. Crua, Fuel 154(1) (2015) 308-318.

• K. Han, G.Q. Song, X.K. Ma, B. Yang, Appl. Therm. Eng. 101(2016) 568-575.

• S. Dirbude, V. Eswaran, A. Kushari, Atomization Sprays 21(9)(2011) 787-798.

• I.N.G. Wardana, Fuel 89(3)(2010) 659-664.

• E. Mura, R. Calabria, V. Califano, P. Massoli, Exp. Therm Fluid Sci. 56(4)(2014) 69-74.

• H. Watanabe, T. Harada, Y. Matsushita, H Aoki, T. Miura, Int. J. Heat Mass Transf, 52(15-16)(2009) 3676-3684.

• K. Han, C.L. Zhao, G. Fu, F.J. Zhang, S.P. Pang, and Y.C. Li, Fuel, 157 (2015)270-278.

• X Ma, F Zhang, K Han, Fuel, 160(2015)43-49.

• C. Morin, C. Chauveau, I. Gökalp, Exp. Therm. Fluid Sci. 21(1-3)(2000) 41-50.

• D. B. Spalding, Symposium on Combust, 4(1)(1953) 847-864.

11th Asia-Pacific Conference on Combustion,
The University of Sydney, NSW Australia
10th -14th December 2017.

Characteristics of an Acoustically Forced Non-premixed Jet Flame

K.K. Foo[1], Z.W. Sun[1], P.R. Medwell[1], Z.T. Alwahabi[2], G.J. Nathan[1], B.B. Dally[1]

Schools of [1]Mechanical Engineering and [2]Chemical Engineering,

The University of Adelaide, Adelaide, South Australia, 5005, Australia

Abstract

This study presents a comprehensive characterisation of the local dynamic response of a flame to an acoustic forcing. In the experiments, a laminar non-premixed jet flame was subjected to fuel flow perturbation driven by a loudspeaker. Acoustic forcing of the fuel flow rate is used to phase-lock the periodic flame flicker close to the natural flicker frequency. The temperature and the soot field of the forced flame were measured using planar two-line atomic fluorescence (TLAF) and planar laser-induced incandescence (LII), respectively. Time-resolved LII was employed to measure the diameter of primary soot particles in the flame. The position of the reaction zone of the flame was measured using planar laser induced fluorescence (LIF) of the hydroxyl radical (OH). The resulting images were phase-resolved and averaged to yield a complete cycle of flame flickering with fine temporal and spatial resolution. The results shed light on the interdependency of the measured parameters and show that the soot production is greater for the forced condition than for the steady counterpart.

1 Introduction

Comprehensive and systematic experimental data sets in unsteady flames (flickering, forced and turbulent) are indispensable for achieving a deeper understanding of soot evolution. Additionally, experimental data sets that contain multiple key parameters are necessary for the development, and validation of predictive numerical models. However, such data are quite scarce in the literature owing to the difficulty of measuring multiple parameters in an environment dominated by turbulence, especially those laden with particles. In any turbulent flame, each instantaneous realisation is unique, and none are identical to the average [1]. Thus, it is important that the measurements are both temporally and spatially resolved. The difficulty of integrating such experimental techniques increases significantly with the increase in the number of the parameters of interest. Therefore, the overall objective of the current study is to investigate flame/vortex interactions in unsteady laminar flames with a more comprehensive diagnostic system than is realistic with turbulent flames.

Flows and flames that are acoustically forced at frequencies close to the natural oscillation frequency of the flame can produce large-scale periodic instabilities associated with vortex shedding [2]. Since the unsteadiness is cyclic, measurements can be phase-averaged, which increases the signal-to-noise ratio of the measurements without sacrificing the local characteristics. The periodic, time-varying laminar flame is an intermediate between steady laminar flames and truly turbulent conditions. One regime of turbulent combustion is that of the wrinkled laminar flames [3]. The elementary flame/vortex interactions therefore become an important step in advancing the understanding of turbulent flames and the soot evolution.

Forced laminar flames are highly reproducible, which allows multiple parameters to be measured independently at different times and even in different laboratories, but still combined at the same phase to progressively add to the completeness of the database. The precondition is that the initial and boundary conditions of the flames be well defined [1]. Over the past decades, the National Institute of Standards and Technology (NIST) has invested considerable effort into the investigation of acoustically forced flames [4-6]. Even though the experimental studies conducted on flames stabilised on the NIST burner are extensive, a major limitation for the data set is that the initial flow conditions was not quantified. The only way to reproduce the flickering flames without the initial flow conditions is to match the flame shapes [7], which will always result in inevitable ambiguity.

Connelly [8] has designed a forced flame burner (Yale Burner) aiming to initiate a series of collaborative studies [8-10]. While the experimental study of time-varying flames on the Yale Burner is a significant development, a few key parameters were missing from the data set including the flame temperature, arguably one of the most important parameters controlling soot formation. Thus, the current study aims to provide a more comprehensive data set of an acoustically forced laminar flame and investigate the response of the flame to a time-varying flow field.

2 Experimental details

2.1 Burner

The burner used in the current study was a co-annular axisymmetric burner with a 5.6-mm-diameter fuel tube surrounded by a 74-mm-diameter air co-flow, based on the design of Connelly [8]. A uniform co-flow was generated with a layer of 2-mm-diameter glass beads sandwiched between two layers of hexagonal honeycomb. A loudspeaker was placed inside the plenum of the fuel stream and used to impose a periodic perturbation on the fuel exit profile. The loudspeaker was powered with a signal generator and phase-locked with all the optical diagnostic systems.

The fuel stream had 42% ethylene and 58% nitrogen, by volume, with a total volumetric flow rate of 0.31 standard litres per minute (SLPM), conditioned to standard temperature 20°C and 1 atm pressure. The Reynolds number for the fuel jet is 96, based on the burner diameter and the cold fuel properties. Nitrogen was added to the fuel to limit the maximum soot volume fraction in the forced flame to avoid significant interference from soot to the laser diagnostic measurements. The air co-flow was maintained at a volumetric flow rate of 60 SLPM.

Corresponding author. Mobile +60 450 300 929
E-mail address: kae.foo@adelaide.edu.au

2.2 Photodiode

A high-speed silicon photo detector (DET110, Thorlabs) was used to register the luminosity flickering of the flame. The photodiode was placed at a distance of 300 mm from the axis of the burner and at a height of 80 mm from the burner nozzle so that it captured the whole dynamic range of the flickering flame. The soot luminosity signal was recorded on a computer via a data acquisition device. A total of 23000 data points were recorded at a sampling rate of 1 kHz.

2.3 Constant temperature anemometer

A constant temperature anemometer (CTA) was used to measure the centreline axial velocity of the non-reacting fuel jet. The probe was placed at a height of 2 mm above the burner (HAB). The sensor is calibrated in compressed air, but since the thermal properties of air, ethylene and nitrogen are similar, the error is negligible. The amplitude of the forcing is defined as the ratio of the maximum velocity change to the mean velocity at the centreline, presented as $\alpha = (v_{c,max} - \bar{v}_c)/\bar{v}_c$.

2.4 Temperature measurement

Planar measurement of temperature was conducted using non-linear two-line atomic fluorescence (NTLAF). Atomic indium was chosen as a tracer because of its strong laser-induced fluorescence signal and its sensitivity to flame temperatures between 800 and 2800 K.

The NTLAF technique is based on the optical excitation of two electronic transitional states and the sequential detection of the spectrally shifted emission, denoted as Stokes and anti-Stokes transitions. Two laser wavelengths were chosen to be 410.18 nm and 451.13 nm to induce the Stokes and anti-Stokes processes, and the corresponding fluorescence signals were collected at the opposite wavelengths, 451.13 nm and 410.18 nm, respectively. Two customised narrow band-pass filters were employed for interference suppression, centred at 451.4 nm (full-width at half-maximum, FWHM = 1.32 nm) and 410.4 nm (FWHM = 1.08 nm) following our earlier work [11].

Calibration of the NTLAF measurements were conducted in a non-sooting premixed flat-flame with an equivalence ratio of 2. A 75 µm R-type Pt/Pt-Rh 13% thermocouple was used to measure the flame temperature in the central region of the flat-flame where the temperature is uniform. Radiation correction to the thermocouple reading was applied following a method proposed by Friedman [12]. A correction of 32 K was added to the thermocouple measurements to compensate for the radiation heat loss. The details of the data processing have been reported previously [13].

2.5 Time-resolved Laser-Induced Incandescence (TiRe-LII)

The time-resolved laser-induced incandescence measurements employed the fundamental output (1064 nm) from a Nd:YAG laser. Four LII images were subsequently recorded with an ultra-high-speed camera comprising four intensified CCD camera modules, each viewing through a common lens via an image splitter. A filter centred at 435 nm with a FWHM of 48.4 nm was mounted in front of the lens to minimise the interferences from other optical measurements. The four cameras were delayed at 0, 40, 80 and 120 ns relative to the prompt LII signals and their integration times were set to 40 ns. The diameter of primary particles, d_p, was determined with a look-up table of theoretical intensity ratios of S_2/S_1, S_3/S_1, S_4/S_1, S_3/S_2, S_4/S_2 and S_4/S_3 (where S_i denotes the signal from the ith camera) at the four chosen delay times spanning d_p from 5 to 100 nm in increments of 5 nm. By comparing the ratios of measured LII signal with the theoretically calculated decay curve, each ratio provides an independent measurement of d_p. Averaging all six measurements gives the mean value of d_p in the measured region. The details of this method have been reported previously [14].

The prompt LII signal was used to evaluate the soot volume fraction, f_v. This was calibrated by performing laser beam extinction measurements in a Santoro-type flame.

2.6 Planar Laser Induced Fluorescence of Hydroxyl (OH-PLIF)

The fluorescence of the hydroxyl radical (OH) was excited with a single ultra-violet (UV) laser beam tuned to the $Q_1(8)$ transition of the OH radical (283.5 nm). This transition was chosen because of its temperature insensitivity. The OH fluorescence is reflected with a dichroic mirror and imaged onto an ICCD camera equipped with a 105 mm f/1.2 UV lens after transmission through a UV filter centred at 310 nm with a FWHM = 10 nm.

3 Results and discussion

3.1 Natural flickering flame

The formation of a stratified layer between the hot gases of combustion and the cold quiescent air initiates the flickering phenomenon, which further leads to the Kelvin-Helmholtz instability. Steady flames transition to intermittent and then consistent flickering behaviour with an increase in the exit velocity. The flickering frequency ranges from 10 to 20 Hz. Since the radiation flickering is well correlated with the flame oscillation, the flickering frequency can be derived from the soot luminosity signal. A power spectral density (PSD) measurement was performed to identify the dominant flicker frequency. Figure 1 shows that the flickering flame has a dominant frequency of ~10.8 Hz, with three higher harmonic components. To enhance the reproducibility of the oscillation, the current flame was acoustically forced by a 10 Hz sinusoidal perturbation.

Figure 2 shows that the soot luminosity intensity is greatest at the phase with the lowest centreline exit velocity of the cycle, at phase $\varphi = 270°$. The amplitude of the velocity fluctuation is about 50% at the centreline. At this phase, the exit velocity slows down, while the flame continues to propagate downstream due to buoyancy, stretching the flame, which allows the in-flame soot to grow to its maximum.

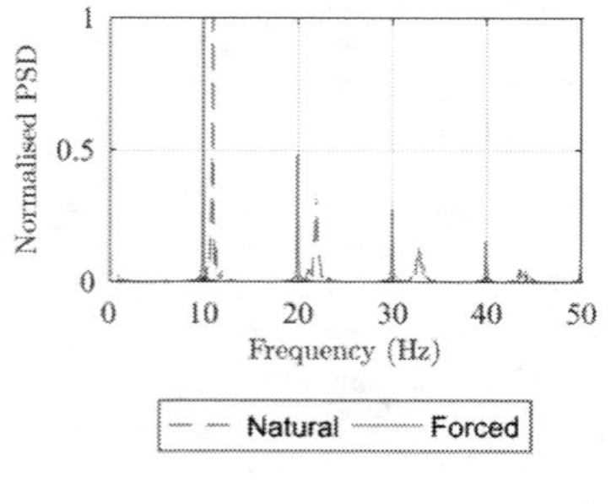

Figure 1 PSD estimates of the natural flickering flame and the acoustically forced flame.

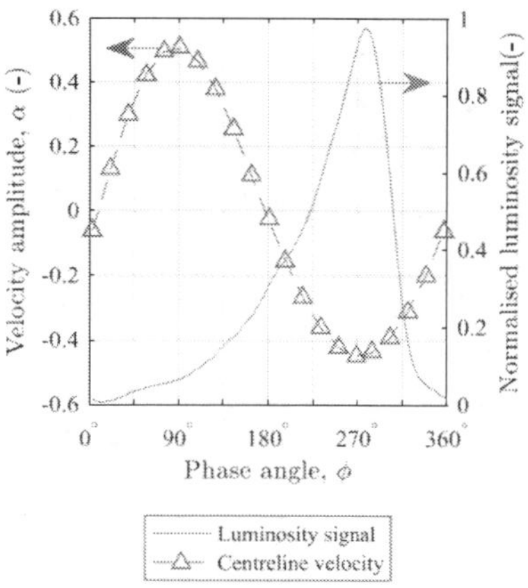

Figure 2 Dependence of the centreline axial velocity 2 mm above the exit plane on phase angle, as measured for the non-reacting flow (ethylene/nitrogen), superimposed on the intensity of the flame luminosity as measured with a photodiode.

3.2 Steady flame

Figure 3 shows the average from 50 single-shot images of flame temperature (T_g), soot volume fraction (f_v), diameter of primary particles (d_p) and OH laser-induced fluorescence for the steady ethylene/nitrogen flame. The flame temperatures presented range from 800 to 2300 K because T_g has to exceed 800 K to generate sufficient anti-Stokes signal to yield meaningful result. Moreover, the measurement on the lean edge of the flame drops rapidly is due to the low population of indium nanoparticles because of oxidation. The soot field has a peak of $f_{v,max} = 1.43\ ppm$ located at the centreline. While $f_{v,max}$ peaks on the centreline, the image of d_p shows that largest soot particles are not only limited to the centreline, but are evenly distributed into the wings. The largest soot particles measured are $d_{p,max} = 78.6\ nm$. The image of OH-PLIF identifies the reaction zone, which closely follows the position of maximum of T_g. The OH-PLIF signal is not quantitative so is presented in arbitrary units.

3.3 Forced flame

Figure 4(a) presents the images of T_g for the forced flame. The temporal evolution of the time-varying flame was collected with a time interval of 12.5 ms (equivalent to a phase angle of 45°). The region of relatively cold gas (T < 800 K) immediately above the burner for φ = 45° to 225° are corresponds to the potential core of the jet. This shows the cold jet penetrates further into the flame zone as the exit velocity is increased. At the phases φ = 180° and 225°, the potential core collapses sharply because of the velocity drop and the flame propagates back upstream. The buoyancy force accelerates the flow downstream particularly in the detached flamelet at the tip.

Figure 4(b) presents the corresponding images of f_v derived from the LII signal. By comparing the profile of f_v and T_g, it is apparent that the soot regions are confined in the regions where T_g ranges from 1600-1800 K. The sudden growth of f_v starts at phase φ = 225° associated with the reduction of exit velocity. The volume-integrated soot volume fraction, $f_{v,int}$ reaches its maximum $3.14 \times 10^{-8}\ m^3$ in the detached flamelet at φ = 315°. In contrast, the total $f_{v,int}$ for the steady flame is only $0.83 \times 10^{-8}\ m^3$. However, taking the average of $f_{v,int}$ across the whole cycle, it is only $1.85 \times 10^{-8}\ m^3$. In this analysis, volume-integration of f_v was first performed for each phase, area-integrated revolving the axis and then integrated over the axial height above the burner. The time-averaged, volume-integrated soot volume fraction was the average of the eight phase measurements.

Figure 4(c) shows the distribution of d_p in the forced flame. At the phases from φ = 45° to 270°, it is apparent that a band of large soot particles are located between the fuel rich zone and oxidation zone. This indicates that the soot is produced at the centreline of the flame and continues to grow while being transported outward radially into the oxidation zone. Beginning of the soot oxidation can be easily identified from the distribution of d_p. The largest diameter of soot particles, $d_{p,max}$ is 86.8 nm, located at the wings of the flame at phase φ = 225°.

Figure 4(d) shows the corresponding images of the OH-PLIF. The comparison of the OH-PLIF and the image of T_g shows good agreement with respect to the outline of the reaction zone. Importantly, the flame is established above the burner throughout the whole cycle, so that no reaction zone is extended into the burner pipe. At phase φ = 315°, the absence of OH between the detached and attached flames shows that the local extinction has occurred between them.

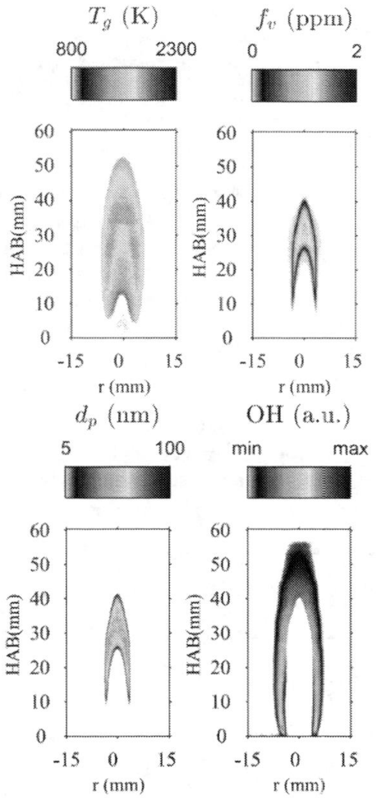

Figure 3 Pseudo-colour images for the steady laminar non-premixed jet flame of (top left) flame temperature, (top right) soot volume fraction, (bottom left) diameter of primary soot particles and (bottom right) qualitative intensity of OH radical. These measurements were collected simultaneously.

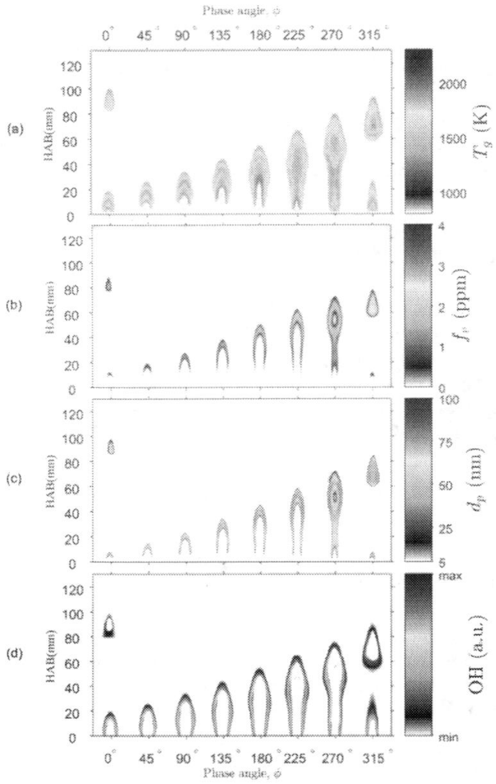

Figure 4 Pseudo-colour images of the acoustically forced flame driven at 10 Hz, showing (a) gas temperature; (b) soot volume fraction; (c) diameter of primary particles and (d) OH-PLIF.

Table 1 compares the peak soot volume fraction measured in the steady and forced flames. Also shown are the volume-integrated soot volume fraction and the maximum diameter of primary particles measured in both flames. The peak soot volume fraction increased by a factor of 2.3. The maximum flame height for the forced flame determined from the OH-PLIF is 99.2 mm, also about two times taller than the steady flame height of 52.1 mm. This indicates that the soot enhancement is caused by the increase in the flame length, and in the in-flame residence time. The unsteadiness therefore decreases the overall mixing rate of the flame.

Table 1 Summary of peak soot volume fraction, volume-integrated soot volume fraction and maximum diameter of primary particles measured in steady and forced flames.

	$f_{v,max}$ (ppm)	$f_{v,int}$ $\times 10^8$ (m^3)	$d_{p,max}$ (nm)
Steady	1.43	0.83	78.6
Forced	3.22	1.85	86.8

4 Conclusions

The current study has presented simultaneous measurements of flame temperature, soot volume fraction, diameter of primary particles and OH-PLIF in the central plane of steady and acoustically forced flames. The peak soot volume fraction and the volume-integrated soot volume fraction in the forced flame increased by a factor of 2.3 relative to the steady flame values. This increase can be attributed to the enhanced air entrainment caused by the vortex rings that also induces the necking. In addition, the residence time in the pinch-off flame is significantly increase due to buoyancy effect, which also helps increasing the overall soot volume fraction. An increase in the flame length implies an increase in the residence time. The role of buoyancy seems to be important and will be the subject of future investigations.

5 Acknowledgment

The authors gratefully acknowledge the Australian Research Council for their financial support.

References

[1] R.S. Barlow, *Proc. Comb. Inst.* **31** (2007) 49-75.

[2] A.W. Strawa and B.J. Cantwell, *J. Fluid Mech.* **200** (1989) 309-336.

[3] P.-H. Renard, D. Thévenin, J.C. Rolon and S. Candel, *Prog. Energy Combust. Sci.* **26** (2000) 225-282.

[4] C.R. Kaplan, C.R. Shaddix and K.C. Smyth, *Combust. Flame* **106** (1996) 89-115.

[5] C.R. Shaddix and K.C. Smyth, *Comb. Flame* **107** (1996) 418-452.

[6] C.R. Shaddix, J.E. Harrington and K.C. Smyth, *Comb. Flame* **99** (1994) 723-732.

[7] G. Papadopoulos, R.A. Bryant and W.M. Pitts, *Exp. Fluids* **33** (2002) 472-481.

[8] B.C. Connelly, *Quantitative Characterization of Steady and Time-Varying, Sooting, Laminar Diffusion Flames using Optical Techniques*, Yale University, 2009 (Ph.D. thesis).

[9] S.B. Dworkin, J.A. Cooke, B.A.V. Bennett, B.C. Connelly, M.B. Long, M.D. Smooke, R.J. Hall and M.B. Colket, *Combust. Theor. Model* **13** (2009) 795-822.

[10] S.B. Dworkin, B.C. Connelly, A.M. Schaffer, B.A.V. Bennett, M.B. Long, M.D. Smooke, M.P. Puccio, B. McAndrews and J.H. Miller, *Proc. Combust. Inst.* **31** (2007) 971-978.

[11] D.H. Gu, Z.W. Sun, G.J. Nathan, P.R. Medwell and Z.T. Alwahabi, *Combust. Flame* **167** (2016) 481-493.

[12] R. Friedman, *Proc. Combust. Inst.* **4** (1953) 259-263.

[13] K.K. Foo, Z.W. Sun, P.R. Medwell, Z.T. Alwahabi, B.B. Dally and G.J. Nathan, *Combust. Flame* **181** (2017) 270-282.

[14] Z.W. Sun, D.H. Gu, G.J. Nathan, Z.T. Alwahabi and B.B. Dally, *Proc. Combust. Inst.* **35** (2015) 3673-3680.

11th Asia-Pacific Conference on Combustion,
The University of Sydney, NSW Australia
10th -14th December 2017.

Investigations on Ignition of Atomized Fuel-Air Mixtures and Liquid Fuel Column-air Combinations by Low Energy Laser Pulses

Awanish Pratap Singh[1], Upasana P. Padhi [1], Harikrishna Tummalapalli[2], Ratan Joarder[1]

[1] Aerospace Engineering Department, Indian Institute of Technology Kharagpur

Hijli, Kharagpur, West Bengal, 721302, India

[2]Altair India Private Limited

Prestige Tech Park, Bengaluru, Karnataka, 560037, India

Abstract

The laser induced technique is used for ignition of the atomized fuel-air mixture and liquid fuel column-air combinations. Low energy laser pulses were used to avoid severe "overkill" of the breakdown zone. However, unfortunately, no display of ignition was observed after performing the set of experiments. It was found that fuel-air mixture cannot be ignited by this technique until and unless the rate of ionization dominates over the recombination rate. A detailed analysis of the experimentally realized cases is provided. To calculate the available energy for ignition accurately, the radiative heat loss from the breakdown zone is planned to be calculated by the finite volume method along with the k-distribution model to calculate the absorption coefficient.

1 Introduction

The invention of the laser has been a turning point in the history of science, which enables the possibility of laser-induced ignition of combustible mixtures. The major advantages of laser ignition are, (i) absence of electrodes which disturb the combustor geometry, (ii) choice of ignition location within the combustor, (iii) precise ignition timing, (iv) controllability of the input energy and its duration, (v) more stable combustion, (vi) higher reliability and durability of the ignition system, (vii) possibility of ignition of leaner mixtures, resulting in lower peaks in the combustor temperatures, (viii) ignition of high-pressure mixtures, etc. [1,2]. Due to such advantages, the laser ignition is under active research.

Higher thermal efficiency and stringent emissions norms are the major forcing factors for the advance research and development of new concept or technique in the field of gas turbine engines. To protect the environment, governmental agencies across all the countries have stringent emissions norms for reducing NO_x and particulate matter (PM) emissions from combustion. In this respect, the laser ignition gained significant pace after the initial study performed by Ramsden and Savic in 1964, just four years after the invention of the laser [3]. The laser ignition can help to reduce the emission of NO_x by burning lean mixtures, which leads to the reduction in overall flame temperatures, and that of PM [2, 4, 5].

In the present study, various atomized fuel-air mixtures and liquid column-air combinations are attempted to be ignited by low energy laser pulses. The atomized fuel-air mixture could be pulsed at a rate up to 0.167 kHz. The primary goals of the study are to understand the transition from a plasma-dominated breakdown to a flame kernel, and then subsequent growth and propagation of the flame from the kernel location to the entire combustion zone.

2 Experimental Setup

A photograph of the experimental setup is shown in Fig. 1. The fuel-air mixture was injected downward from the injector/atomizer mounted at the top of a vertical-flow channel with a hexagonal cross section. All the six sides of the flow channel are covered with glass/quartz windows. The injector along with the metallic pipes are movable in the vertical direction, which is useful for the locating the most reliable sparking location. A continuous, top to bottom air co-flow is established with the help of a honeycomb like structure, and this air allows to control the equivalence ratio of the fuel-air mixture. Also, there is a provision to supply nitrogen into the flow channel to flush out any unburnt fuel vapor and fumes after the experiment. A commercial (PulsaJet: AA10000JJAU-VI, Spray set up: SU1A-SS) full cone spray nozzle is used as the fuel injector which is an electrically actuated air atomizing nozzle and can provide the fastest cycling up to 10,000 cycles per minute. The injector can produce a weakly swirling full-cone spray with a cone angle of approximately 18 degrees. The droplet diameters are in the range of 10-100 μm and flow rates are up to 0.61 l/min at 7 bar. The air and fuel flow rates are controlled by digital mass flow controllers (Air- AALBORG GFC-37; Fuel- Bronkhorst M13 mini CORI-FLOW). Laser induced air breakdown for the ignition of the atomized fuel-air mixture is achieved by focusing the light emitted by a laser (diode-pumped, Q-switched frequency-doubled Nd: YAG from IB Photonics, Spark-S500-SHG). The laser is capable of producing 270 μJ per pulse at 500Hz repetition rate and at a wavelength of 532 nm and pulse width of 370 ps. A photodetector along with an oscilloscope is used before and after the laser pulse to determine the actual output power/energy at ignition location. A high-speed ICCD camera (4 Quick E, Stanford computer optics) is arranged for flow field visualization. A digital delay generator (BNC-745) of femtosecond accuracy is used to trigger the laser, camera and, control the other electronic devices.

Corresponding author. Fax: +91-3222-282242
E-mail address: awanishsingh009@iitkgp.ac.in

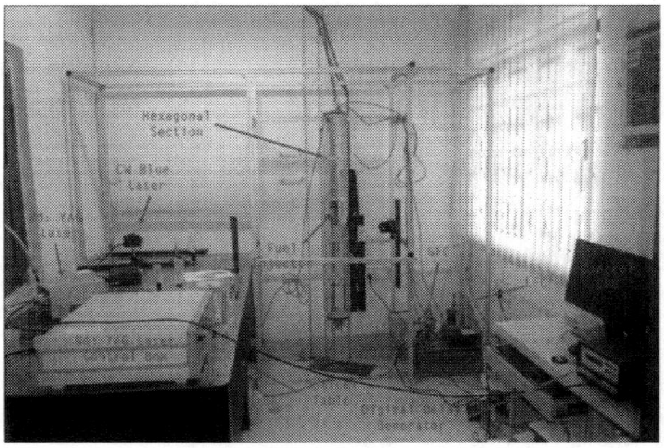

Figure 1: A photograph of the experimental set-up.

3 Methods, Results, and Discussions

The following data are used in the discussions presented later. Threshold intensity for breakdown in air = 10^{12}- 10^{13} W/cm^2, diameter of the focal point (d_f) = $4f\lambda/\pi D$ = 5 μm, peak power (P_{peak}) = pulse energy/pulse width = 729730 W, pulse length (L_p) = speed of light × pulse width = 110 mm, theoretical peak intensity = P_{peak}/focal area (A_f) = 3.60×10^{12} W/cm^2.

3.1 Experiment 1: Ignition of Atomized Ethanol-Air Mixture

In this experiment, near stoichiometric and also with various equivalence ratios, the ethanol-air mixture is injected into the flow channel and subsequently, the laser is focused at various distances downstream of injector exit along the axis of symmetry of the spray using a doublet lens of 150 mm focal length as shown in Fig. 2.

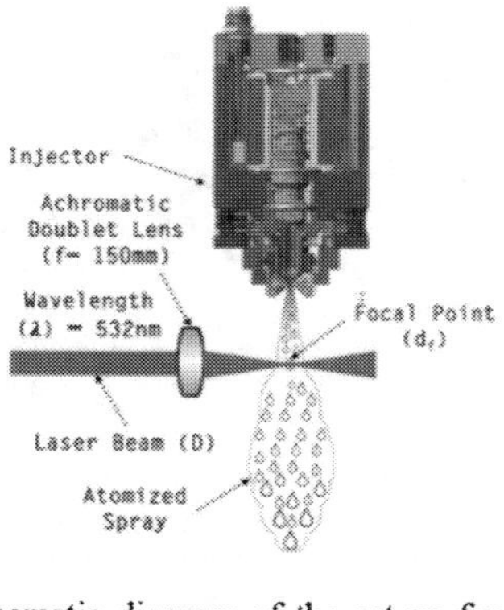

Figure 2: Schematic diagram of the set-up for the ignition of the atomized ethanol-air mixture.

In this set-up, Intermittent reddish light spots appeared in the fuel-air mixture during the experiments with no audible sound due to breakdown. Probably, the actual laser intensity at the focal point was smaller than the threshold intensity needed to create the breakdown in the air as well as an atomized ethanol-air mixture. From the literature [6], it was found that the presence of materials with low ionization potential in the fuel-air mixture can help in reducing the breakdown threshold. That is tried in the next experiment.

3.2 Experiment 2: Ignition of Atomized Ethanol-Air Mixture seeded with Aluminium Particles

In this experiment, near stoichiometric and also with various equivalence ratios, the ethanol-air mixture seeded with aluminium particles (externally) is injected into the flow channel, and subsequently, the laser was focused along the axis of symmetry of the spray at various distances downstream of the injector exit.

In this case, air breakdown was not observed, but the less intense intermittent light was observed during the experiments. However, the created high energy spots were unable to sustain for a long time and ignite the mixture due to insufficient energy. There are many ways to explain why aluminium nanoparticle did not help to ignite the mixture, but the major factor is the recombination rate, which is faster than the ionization rate. The aluminium particle only enhances the ionization rate of the atoms due to more absorption of the photon at the focal point but probably not helpful in reducing the recombination rate. The enhancement in ionization rate with the aluminium particle can be understood by the appearance of less intense light spots at/near the focal point. From the literature [7], it is found that by using two lasers simultaneously, one can reduce the energy required for breakdown and enhance the efficiency of absorption if the lasers are fired within hundreds of nanoseconds interval.

3.3 Experiment 3: Ignition of Atomized Ethanol-Air Mixture with Two Lasers

In this experiment, near stoichiometric and also with various equivalence ratios, the ethanol-air mixture with and without aluminium particle seeding is injected into the flow channel and subsequently two lasers (pulsed green laser of wavelength 532 nm with energy 270 μJ per pulse and CW blue laser of wavelength 405 nm with an average power of 254 mW). were focused along the axis of symmetry of the spray at various distances downstream of the injector exit. This double-laser energy deposition experiment was designed for three different cases with two distinct conditions. A schematic of the set-up is shown in Fig. 3.

In this case also no breakdown was observed, and the continuous laser did not show any visible contribution for the cause. From the literature [8], it is found that 2nd laser is effective only when it is fired within ±200 ns time interval of first laser energy deposition. Therefore, the overall contribution by both lasers in the effective time interval of 400 ns is as follows. Maximum possible energy from green laser = 270 μJ per pulse. Effective maximum possible energy from blue laser = average power × effective time = 254×10^{-3} J/s × 400×10^{-9} s. Maximum possible effective energy from both lasers = 270 μJ + 101.6 nJ = 270.1 μJ. Thus, the energy deposition by the blue laser was not even one μJ in the period of 400 ns. That is why this case behaved like the previous experiments.

3.4 Experiment 4: Ignition of Ethanol in Container

In this experiment, ethanol was filled in a transparent container and afterward the laser energy was deposited in two different ways by focusing the laser beam near the center of the container. First, only the green laser and then both lasers were focused in liquid fuel column as shown in Fig. 4(a). Later, a mesh made of copper wire (diameter less than 1 mm) was placed on the surface of the liquid fuel as shown in Fig. 4(b). The reason to use the mesh near/below the breakdown zone was to transfer the heat energy from the breakdown zone to the neighboring area.

In this case, first, only with the green laser, the breakdown was observed (at the pulsing rate of the laser) along with intense sound and bright spark, which indicates the increase in the probability of creating breakdown due to increase in the absorption of photons. We used three different configurations of focusing the laser, (a) below the liquid surface (Fig. 4(c)), (b) at the liquid surface (Fig. 4(d)), (c) above the liquid surface (Fig. 4(e)). The breakdown was observed only for case (b). Despite highly intense breakdown created on the surface by the laser, the breakdown zone was unable to sustain for a long time, and therefore ignition of the fuel was not observed. The phenomenon was approximately the same when we used both the lasers simultaneously. Later, when the mesh made of copper wire was used in liquid fuel column, the evaporation rate of fuel was found to increase, but not the ionization rate.

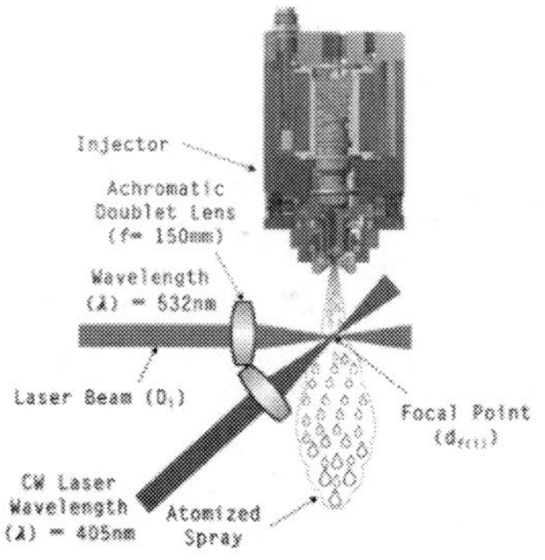

(a)

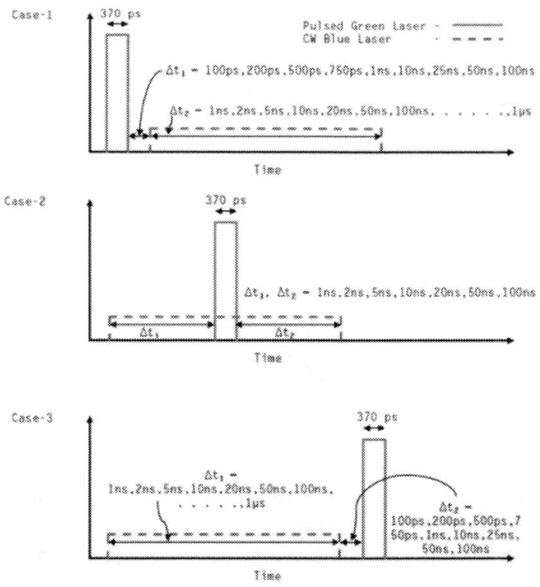

(b)

Figure 3: (a) Schematic of the set-up for ignition of atomized ethanol-air mixture by two lasers, (b) schematic representations of sequence of laser pulses, (i) **case-1:** an initial short duration high peak power pulse (370 ps) is followed by a subsequent laser pulse (Δt_2) of longer duration having significantly lower peak power after a temporal delay (Δt_1), (ii) **case-2**: overlap of long-pulse and short-pulse, (iii) **case-3**: long-pulse (Δt_1) followed by a short pulse (370 ps) after the temporal delay (Δt_2).

The breakdown was observed in these cases due to the increase in the molecular density per unit differential cross-sectional area. In other words, the number of atoms/molecules is increased in the focal volume; it is because atom or molecules inside the liquids are densely packed compared to air/aerosol. The breakdown was only effective near/at the liquid-air interface as shown in the Fig. 4(c), because above the liquid surface, the medium is air with lower molecular density, and below the surface, the part of the conical zone, from the liquid surface to focal point, is absorbing some energy. Consequently, no breakdown occurred due to the less available photon flux.

The breakdown process was mainly consisting of multiphoton ionization, whereas, the inverse bremsstrahlung process was not playing a significant role for further ionization. It is because the higher-energy free electrons do not have enough potential to start the electron avalanche and initiate the ignition. After the energy deposition, the plasma formed at the focal zone was unable to sustain for a long time due to insufficient energy left within the core. Thus, the means of energy transfer from core zone to surrounding was mainly by the molecular collision or atom-atom interaction. The other reason for not observing the inverse bremsstrahlung process was the pulse-width of the laser pulse. Due to short pulse width, there was very less probability of finding the free electron to absorb the photon, because the time scale of laser pulse was approximately the same as the time scale of excitation of the electron.

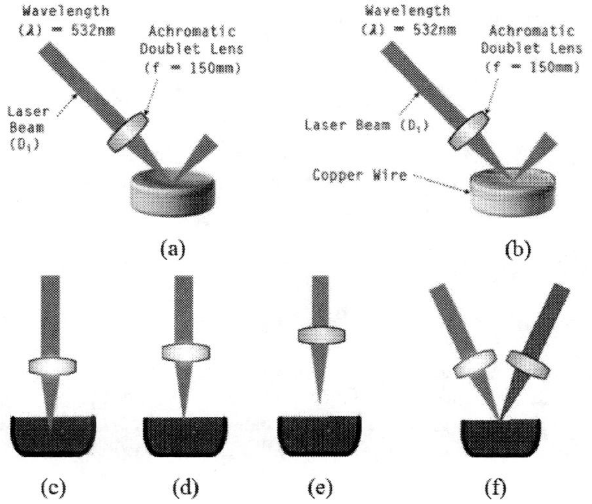

Figure 4: Schematic of the set-up for ignition of liquid fuel in a container (a) with a single laser, (b) single laser with copper wire, (c-e) various focus positions, (f) two lasers focused at the fuel-air interface.

4 Calculation of Radiation Loss

In the high-temperature environments, like the one considered here, radiative heat transfer is the dominating mode of heat transfer. In the current study, radiative heat flux will be estimated numerically using finite volume method [9–12], which gives a right balance between accuracy and computational efficiency [13]. Coelho [14] applied higher order methods in computational fluid dynamics to radiative transfer equation and found that superbee scheme is the most accurate among TVD and NVD methods. The most general form of radiative transfer equation is given by

$$\frac{\partial I_\eta}{\partial s} = k_\eta I_{b\eta} - \beta_\eta I_\eta + \frac{\sigma_{s\eta}}{4\pi}\int_0^{4\pi} I_\eta(\hat{s}_i)\phi_\eta(\hat{s}_i,\hat{s})d\Omega_i$$

which can be solved by the method given in Ref. [9]. The absorption coefficient of gases, unlike solids, varies erratically with wavelength and considering gases to be gray is a now here close reality. Solving an RTE at every line of the absorption spectrum is called the line by line method which most accurate and computationally inefficient. Narrow band and wide band models are great alternatives to the line by line method. Earlier work in narrow band k-distribution models was done in metrology by Lacis *et al.* [15] and Fu *et al.* [16]. In the non-homogeneous domain, RTE is solved using correlated k-distribution [15, 16] or scaled k-distribution. Though correlated k-distribution works very well in meteorology application where the temperature gradients are not so high, however in combustion environments, high-temperatures give rise to 'hotlines' as shown by Soufiani *et al.* [17]. Modest [18] showed that correlated k-distribution assumption could be applied very well in the environments of high-temperature gradients. Calculating absorption coefficient for a mixture of gases is based on observation by Taine *et al.* [19], the transmissivity of mixture of gases is the product of transmissivities of individual species. In a narrow band, adsorption spectrum attains same value many times thus calculating absorption coefficient at all the wavenumbers increases computational time. Following [20], this can be mathematically written as

$$\bar{\tau}_\eta = \frac{1}{\Delta\eta}\int \tau_\eta d\eta = \int_{k\_min}^{k\_max} \tau_k f(k)dk$$

where f(k) is k-distribution given by

$$f(k) = \frac{1}{\Delta\eta}\sum\left|\frac{d\eta}{dk}\right| = \frac{1}{\Delta\eta}\sum_{\eta=\eta_1}^{\eta_2}\delta(k-k_\eta)\left|\frac{d\eta}{dk_\eta}\right|$$

and $\delta(k\text{-}k_\eta)$ is the Dirac delta function. Even this k-distribution oscillates, thus we take a new variable g which is called cumulative k-distribution

$$\bar{\kappa}_\eta = \int \kappa_\eta d\eta = \int kf(k)dk = \int k(g)dg$$

Moreover, $g = \int_0^k f(k')dk'$

Here we follow the correlated k-distribution method by Modest [18], which gives considerably accurate results in high-temperature gradients. Narrow band k-distribution in the case of a mixture of gases is found following the work by Modest *et al.* [21].

$$\tau_{mix} = \prod_{i=1}^{k} \tau_k$$

We know that

$$\bar{\tau}_i = \int_{k\_min}^{k\_max} e^{-kl} f(k)dk = L(f(k))$$

From above two equations and taking the inverse Laplace transform of product gives

$$g_{mix}(k_{mix}) = \int_{g_1=0}^{1}\int_{g_2=0}^{1} H[k_{mix}-(k_1+k_2)]dg_2dg_1$$

$$= \int_{g_1=0}^{1} g_2(K_{mix}-k_1)dg_1$$

5 Conclusions

The present work investigates the ignition of atomized fuel-air mixtures and liquid fuel column-air combinations by low energy laser pulses. The breakdown of the fuel-air mixture was observed in some of the cases, but the breakdown zone was unable to grow and convert itself to a flame due to the following reasons. (i) The recombination rate of ions and electron was dominating over the ionization rate of atom due to frequent interaction of ions and electrons, within the small focal volume, (ii) the ions and electrons generated due to photoionization were unable to efficiently ionize the other nearby atoms by inverse-bremsstrahlung process due to insufficient energy, (iii) the shock wave created due to breakdown was carrying out most of the deposited energy. In order to achieve the primary objective, the future study will be performed with slightly higher energy laser pulses. The radiation loss from the breakdown zone will be calculated as per the method given in Ref [22] with the inclusion of the k-distribution model to accurately predict the energy available for ignition.

References

[1] S.A. O'Briant, S.B. Gupta, S.S. Vasu, Propuls. Power Res. 5 (2016) 1–21.

[2] R.K. Prasad, S. Jain, G. Verma, A.K. Agarwal, Fuel 190 (2017) 318–327.

[3] S.A. RAMSDEN, P. SAVIC, Nature 203 (1964) 1217–1219.

[4] R. Oldenborg, J. Early, C. Lester, Advanced Ignition and Propulsion Technology Program, Los Alamos, NM, 1998.

[5] D. Bradley, C.G.W. Sheppard, I.M. Suardjaja, R. Woolley, Combust. Flame 138 (2004) 55–77.

[6] T.X. Phuoc, Opt. Lasers Eng. 44 (2006) 351–397.

[7] J.W. Early, Multiple Laser Pulse Ignition Method and Apparatus, US5756924 A, n.d.

[8] M.S. Bak, L. Wermer, S. Im, J. Phys. D. Appl. Phys. 48 (2015) 485203.

[9] J.C. Chai, H.S. Lee, S. V. Patankar, J. Thermophys. Heat Transf. 8 (1994) 419–425.

[10] G.D. Raithby, E.H. Chui, J. Heat Transfer 112 (1990) 415.

[11] W. Tian, W.K.S. Chiu, Numer. Heat Transf. Part B Fundam. 47 (2005) 199–211.

[12] J.P. Moder, J.C. Chai, G. Parthasarathy, H.S. Lee, S. V. Patankar, Numer. Heat Transf. Part B Fundam. 30 (1996) 437–452.

[13] S.C. Mishra, M.Y. Kim, S. Maruyama, Int. J. Heat Mass Transf. 55 (2012) 5819–5835.

[14] P.J. Coelho, J. Quant. Spectrosc. Radiat. Transf. 109 (2008) 189–200.

[15] A.A. Lacis, V. Oinas, J. Geophys. Res. 96 (1991) 9027.

[16] Q. Fu, K.N. Liou, J. Atmos. Sci. 49 (1992) 2139–2156.

[17] A. Soufiani, F. André, J. Taine, J. Quant. Spectrosc. Radiat. Transf. 73 (2002) 339–347.

[18] M.F. Modest, J. Quant. Spectrosc. Radiat. Transf. 76 (2003) 69–83.

[19] J. Taine, A. Soufiani, in:, 1999, pp. 295–414.

[20] M. Modest, Radiative Heat Transfer, 2nd ed., Academic Press, 2003.

[21] M.F. Modest, R.J. Riazzi, J. Quant. Spectrosc. Radiat. Transf. 90 (2005) 169–189.

[22] R. Joarder, G.C. Gebel, T. Mosbach, Int. J. Heat Mass Transf. 63 (2013) 284–300.

11th Asia-Pacific Conference on Combustion,
The University of Sydney, NSW Australia
10th -14th December 2017.

Theoretical investigation of the ethylbenzene with NH_2 radical: H abstraction and addition reactions pathways

Kamal Siddique[1], Mohammednoor Altarawneh[1], Anam Saeed[1], Jeff Gore[2] and Bogdan Z. Dlugogorski[1]

[1]School of Engineering and Information Technology, Murdoch University
90 South Street, Murdoch, WA, 6150, Australia
[2]Dyno Nobel Asia Pacific Pty Ltd
Mt Thorley, NSW, 2330, Australia

Abstract

Thermal processes like pyrolysis of biomass and coal account for the interaction of alkylbenzenes with amidogen (NH_2) radical. The NH_2 radical evolves from the oxidation of ammonia and hydrogen cyanide in processes encompassing the transformation of nitrogenous fuels, with significant propositions to atmospheric and combustion systems. In this study, for the first time, we investigate the gas-phase mechanisms governing the reactions of ethylbenzene, as modelled alkylbenzene species, with NH_2 radical. More specifically, we determine H abstractions and addition reactions of ethylbenzene with NH_2. We compute standard reaction ($\Delta_r H^{\circ}_{298}$) and activation ($\Delta^{\ddagger} H^{\circ}_{298}$) enthalpies for H removal from the alkyl side chains (primary H and benzylic H) in ethylbenzene, as well as the addition of NH_2 at the four possible sites (*ipso*, *ortho*, *meta* and *para*) of the ethylbenzene phenyl ring. The kinetic analysis of the interaction of ethylbenzene with NH_2 shows that abstraction of benzylic H atom represents the exclusive plausible corridor at all temperatures. The computed mechanistic and kinetic parameters (fitted in the temperature range of 300-2000 K) entail high accuracy level and corroborate with the available literature measurements.

1 Introduction

Research on the fate of the nitrogenous compounds emissions from pyrolysis and combustion of coal and biomass as well as the detonation of explosives has gained momentum since the implementation of stringent environmental regulations. Understanding of the chemical events contributing to the emission of nitrogen oxides (NO_x) from nitrogen rich fuels covers a big area of combustion and explosive detonation [1]. Collectively, most of the combustion systems pose the concern of NO_x formation that is mainly composed of nitrogen oxide (NO) with the relatively lower amount of nitrogen dioxide (NO_2) The generation and mitigation of nitrogen oxides (NO_x) from solid based fuels and their chemical and physical transformations in the combustion environment are challenging. Numerous studies reported the main and important intermediate during the conversion of fuel-N are HCN, NH, NH_2, and NH_3. Decomposition of hydrogen cyanide, thermal degradation of alkyl amines and oxidation of ammonia are the precursors of the active amidogen intermediate (NH_2 radicals) in the course of chemical transformations of nitrogenous compounds. NH_2 radicals play a vital role in the combustion chemistry underlying co-gasification of coal with biomass, oxidation of volatile nitrogen-linking compounds in biomass, and oxidation and pyrolysis of ammonia [2-5].

Aromatic species constitute a major portion of commercial gasoline, diesel and jet fuels on average of 25%, 33%, and 16%, respectively [6]. In particular, C_8H_{10} aromatics (ethylbenzene and xylene) account for a substantial fraction of petroleum-based fuels. In addition, ethylbenzene has been proposed as an aromatic component of surrogate fuels for kerosene. Ethylbenzene acquired more reactivity due shorter ignition delay times when compared to simple toluene. As the simplest alkylbenzene with the crucial benzylic C–C bond structure, ethylbenzene can be regarded as prototype fuel for *n*-propylbenzene and *n*-butylbenzene which are used to represent aromatic components in kerosene and diesel oil [7]. Also, the inclusion of aromatic based hydrocarbons in gasoline is trending after the omission of lead-based knock suppressors. Generation of ammonia in commercial engines corroborates the reactions that entail the NH_2 radical [8]. We anticipate the coexistence of aromatic alkylbenzenes [9, 10] and NH_2 radical [11] combustion systems [12, 13].

Many researchers have reported the presence of alkylbenzenes and ammonia in several combustion systems. For instance, in a small-scale drop-tube pilot plant, Hernandez et al. [14] study the production of the tar from gasification of biomass. They elucidate the presence of benzene, toluene, and ethylbenzene. Nowakowska et al. [15] reported the formation of alkylbenzene (benzene, ethylbenzene and *n*-propylbenzene) from the thermal decomposition and oxidation of anisole in a jet stirred reactor in a temperature range of 673 to 1173 K. Furthermore, Wang et al. [16] analysed the release of volatile organic compounds (VOCs) from biofuel combustion to identify benzene, toluene, ethylbenzene, propylbenzene and other aromatic compounds. Thermal degradation of fuel-N exist in biomass generates NH_3 and HCN [17, 18]. Decomposition of alkylamines present in biofuels represents a direct corridor for the formation of NH_2 radical [19]. Thus, the above-mentioned studies give the nod to inquire into the interactions of NH_2 radical with alkylbenzenes and the imperative thermo-kinetics of the interactions, for these species come off together in combustion systems and in fuel-processing operations.

Herein, we present the modelling of reactions of NH_2 radical with ethylbenzene in gas phase systems. Explicitly, we probe abstraction (primary and benzylic) and addition (*ipso*, *ortho*, *meta* and *para*) reactions for the ethylbenzene as a modelled alkylbenzene molecules. To the best of our knowledge, no kinetic information exists for reactions of alkylbenzene + NH_2. So, the current investigation amplifies our on-going initiatives to establish an inclusive kinetic database for the reactions of NH_2 with RH species at CBS-QB3 computational method [20-22]. We estimated (i) the kinetic parameters for H abstractions reactions by NH_2 from the side chain in ethylbenzene; (ii) the role of the addition channels with the commonly viable abstraction route from the methyl and ethyl chain; (iii) the influence of the aromaticity on the kinetics and thermodynamics of the reactions between alkanes and NH_2 radicals, and, (iv) the comparison of different radicals related to primary and benzylic H abstraction reactions from ethylbenzene. The presented results in this research work will lay the foundation for the new advancement of robust kinetic models for the pyrolysis of nitrogen-rich fuels.

2 Reaction systems and computational methods

This study elucidates H abstraction reactions entailing the removal of H from the alkyl side chain of the ethylbenzene (primary and secondary C–H sites) by NH_2. Addition pathways include the attachment of NH_2 at *ipso*, *ortho*, *meta* and *para* sites of the aromatic ring in ethylbenzene. Prefixes *p* and *b* are used to denote removal of a hydrogen atom from primary and benzylic C–H sites, respectively, of the ethylbenzene side chain.

$C_6H_5C_2H_5 + NH_2 \rightarrow p\text{-}C_8H_9 + NH_3$ (R1)

$C_6H_5C_2H_5 + NH_2 \rightarrow b\text{-}C_8H_9 + NH_3$ (R2)

$C_6H_5C_2H_5 + NH_2 \rightarrow ipso\text{-}(C_8H_{10})\,NH_2$ (R3)

$C_6H_5C_2H_5 + NH_2 \rightarrow ortho\text{-}(C_8H_{10})\,NH_2$ (R4)

$C_6H_5C_2H_5 + NH_2 \rightarrow meta\text{-}(C_8H_{10})\,NH_2$ (R5)

$C_6H_5C_2H_5 + NH_2 \rightarrow para\text{-}(C_8H_{10})\,NH_2$ (R6)

The Gaussian09 [23] suite of the program was employed to optimise structures of reactants, transition states, and products at the complete basis set CBS-QB3 level of computation. CBS-QB3 composite method evaluates geometries and frequencies at the B3LYP/6-311G(2d,d,p) basis set with the inclusion of 2d and d-polarisation functions on the second and first row elements. A series of successive single point energies is then performed at MP4(SDQ)/6-311G(2d(f),p) and CCSD(T)/6-31+G(d') levels of theory. The CCSD(T) is deemed to be more precise than QCISD(T) for spin-contaminated radicals. Numerous work in the literature has exhibited an acceptable performance of the CBS-QB3 method in determining thermo-kinetic parameters for H abstraction/transfer reactions [24, 25]. We quote all calculated standard activation ($\Delta^{\ddagger}H^{o}_{298}$) and standard reaction ($\Delta_rH^{o}_{298}$) enthalpies at 298.15 K.

We estimated reaction kinetic parameters using the KiSThelP [26] code based on the conventional transition state theory (TST) [27]. We considered one dimensional asymmetrical Eckart tunnelling that accounts for the quantum tunnelling effect on the reaction rate constants [28]. For the total reaction degeneracy, we used the approach of Fernández-Ramos et al. [29]. We conducted hindered-rotor measurements to gauge for the internal rotations in reactants and transition structures. The unique coupling of the internal rotation of the methyl and the ethyl groups in alkylbenzene have been composed in detail in our earlier studies [30, 31]. In this capacity, we execute partial optimisations along the analogous angles for methyl and ethyl group rotation. The two internal rotations of the methyl and the ethyl groups in ethylbenzene conform to vibrational frequencies of 220.8 cm^{-1} (barrier of 14.1 kJ/mol) and 50.9 cm^{-1} (barrier of 4.9 kJ mol-1), respectively.

Literature has reported satisfactory results of the CBS-QB3 method for the H transfer/abstraction reactions. For example, in our previous work of ethylbenzene and *n*-propylbenzene with HO_2 radical, the computed standard enthalpy of formation of ethylbenzene, *n*-propylbenzene and their derived radicals reside within 0.2–2.6 kJ/mol from the analogous experimental values [30, 31]. Furthermore, the mean absolute deviation of bond dissociation enthalpy (BDH) of C–H sites for the titled reactions is 4.5 kJ/mol. We compare the experimental standard reaction enthalpies ($\Delta_rH^{o}_{298}$) with the analogous CBS-QB3 values for Reactions R1 and R2, with the mean unsigned error estimated as 10.0 kJ/mol. The experimental $\Delta_rH^{o}_{298}$ values (literature values) are derived from the experimentally measured standard heat of formation of individual species in each reaction. Our accuracy of the current study is based on the above mentioned benchmarking.

3 Results and discussion

3.1 H abstraction reactions

Ethylbenzene contains three distinct types of C–H bonds, namely, the aromatic C–H bonds on the phenyl ring, benzylic ethyl bonds ($C_6H_5\mathbf{CH_2}CH_3$) and primary methyl bonds ($C_6H_5CH_2\mathbf{CH_3}$). The aromatic C–H bonds are significantly stronger than the alkyl side chain C–H bonds (primary and secondary). The BDH values of the aromatic C–H bonds and side chain C–H sites are approximately 460.0 kJ/mol and 360.0–422.0 kJ/mol, respectively. The contribution of benzene ring hydrogens is negligible towards the overall H abstraction reaction kinetics in the ethylbenzene + HO_2 gas phase system.[30, 31]. So, we only considered H abstraction from the side chain of the ethylbenzene by NH_2. Figure 1 illustrates the transition geometries of the abstraction and addition channels for ethylbenzene + NH_2, while labels of the transition structures reflect reaction numbers.

Table 1 itemises $\Delta^{\ddagger}H^{o}_{298}$ and $\Delta_rH^{o}_{298}$ for Reaction R1-R6. For Reaction R1, $\Delta^{\ddagger}H^{o}_{298}$ for H abstraction from primary C–H bond by NH_2 amounts to 32.4 kJ/mol with the corresponding $\Delta_rH^{o}_{298}$ of -20.7 kJ/mol. The standard enthalpy of activation of H abstraction form benzylic C–H site in Reaction R2 attains a value of 23.9 kJ/mol with an analogous $\Delta_rH^{o}_{298}$ value of -79.6 kJ/mol. Thus, $\Delta_rH^{o}_{298}$ and $\Delta^{\ddagger}H^{o}_{298}$ for ethylbenzene demonstrate a proportional correlation. This trend coincides with the decreasing strength of the C–H bond in ethylbenzene i.e., primary H (421.3 kJ/mol) > benzylic H (359.9 kJ/mol). For example, NH_2 prefers to attack the weakest C–H site (benzylic) that shows the most auspicious reaction pathway with $\Delta^{\ddagger}H^{o}_{298}$ of 23.9 kJ/mol and $\Delta_rH^{o}_{298}$ of -77.5 kJ/mol. We previously noticed a similar propensity of decreasing $\Delta^{\ddagger}H^{o}_{298}$ for abstraction of primary H to benzylic H for the reaction of ethylbenzene + HO_2. For example, H abstraction from benzylic C–H site in ethylbenzene by HO_2 is associated with the most accessible $\Delta^{\ddagger}H^{o}_{298}$ which amounts to 46.8 kJ/mol. Abstraction of a primary H atom incurs $\Delta^{\ddagger}H^{o}_{298}$ of 75.7 kJ/mol [30]. Since literature provides no thermodynamic values for the considered reactions herein, we confer the accuracy of $\Delta^{\ddagger}H^{o}_{298}$ by juxtaposing them with those of other hydrocarbons with similar C–H BDHs. For example, $\Delta^{\ddagger}H^{o}_{298}$ for abstraction of primary H

Corresponding author. Mohammednoor Altarawneh
Phone: +61 8 93607507
E-mail address: M.Altarawneh@Murdoch.edu.au

in ethylbenzene + NH_2 (32.4 kJ/mol) concurs well with our previous work [21] for primary H abstraction in *n*-propane + NH_2 (34.9 kJ/mol). This indicates a nominal influence of the electron delocalised aromatic system on the kinetics of H abstraction by NH_2 and concurs with our previous study (ethylbenzene + HO_2) [30]. The BDH value of primary C–H bond in ethylbenzene (421.3 kJ/mol) [30] resembles well with the corresponding value in propane (428.4 kJ/mol), [21] which justifies a minor disparity in $\Delta^{\ddagger}H^{\circ}_{298}$ for H abstraction from the primary C–H bond among the two molecules.

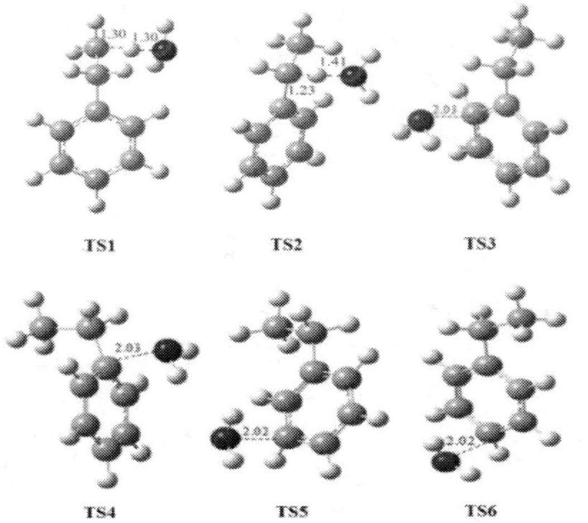

Figure 1: Optimised geometries of the transition states of ethylbenzene + NH_2. Distances are in Å.

3.2 Addition reactions

The standard activation and reaction enthalpies for the addition of NH_2 at the *ipso*, *ortho*, *meta* and *para* sites of ethylbenzene are listed in table 1. Addition of NH_2 at an *ortho* site (R4) of the aromatic ring in ethylbenzene reveals the lowest $\Delta^{\ddagger}H^{\circ}_{298}$ of 24.5 kJ/mol with the highest reaction exothermicity ($\Delta_rH^{\circ}_{298}$ of -43.6 kJ/mol). The formation of the *meta* moiety and the *para* adduct in Reactions R5 and R6 require $\Delta^{\ddagger}H^{\circ}_{298}$ of 33.8 and 34.1 kJ/mol with the corresponding $\Delta_rH^{\circ}_{298}$ of -38.3 and -37.7 kJ/mol, respectively. The standard activation enthalpy for the addition of NH_2 at an *ipso* site in Reaction R3 requires a modest activation enthalpy of 25.2 kJ/mol with $\Delta_rH^{\circ}_{298}$ of -48.9 kJ/mol. In the ethylbenzene + NH_2 system, abstraction of the benzylic H requires the least $\Delta^{\ddagger}H^{\circ}_{298}$ (23.9 kJ/mol) followed by attachment of NH_2 at an *ortho* site (24.5 kJ/mol) and at an *ipso* position (25.2 kJ/mol).

Table 1: Standard activation ($\Delta^{\ddagger}H^{\circ}_{298}$) and standard reaction enthalpies ($\Delta^{\ddagger}H^{\circ}_{298}$) of abstraction and addition channels for ethylbenzene + NH_2. All values are in kJ/mol.

Reactions	Activation enthalpy $\Delta^{\ddagger}H^{\circ}_{298}$	Reaction enthalpy $\Delta_rH^{\circ}_{298}$
R1	32.4	-20.7
R2	23.9	-79.6
R3	25.2	-48.9
R4	24.5	-43.6
R5	33.8	-38.3
R6	34.1	-37.7

The abstraction channel via the primary site exhibits an intermediate barrier of 32.4 kJ/mol as compared to the addition of NH_2 at the *meta* (33.8 kJ/mol) and *para* (34.1 kJ/mol) sites of the aromatic ring of ethylbenzene. Our previous study demonstrated that the abstraction of H from benzene by NH_2 dominates the addition of NH_2 to the benzene ring [32].

3.3 Reaction kinetics

Herein, table 2 presents the kinetic quantities of the H abstraction from primary and benzylic C–H site of ethylbenzene by NH_2 as well as for the addition of NH_2 at four plausible sites on the aromatic ring of ethylbenzene. We didn't address the H abstraction from the aromatic site due to trivial reaction pathway compared to the side chain H atom removal.

The H abstraction from the benzylic site (R2) remains the leading reaction corridor at all temperature range (300–2000 K). The scission of primary H (R1) becomes the second dominant channel. At low temperatures (400-1200 K), the rate of H removal in Reaction R2 is four times the rate of primary H abstraction of R1. At higher temperature (1300-2000 K), the contribution of Reaction R2 is approximately 2 folds of the Reaction R1. The formation of the NH_2-phenyl ring adduct for ethylbenzene shows negligible importance. The overall contribution from the addition of NH_2 at *ipso*, *ortho*, *meta* and *para* sites of the aromatic ring in ethylbenzene combines to a ceiling of 9.0%. The dominance characteristic benzylic H abstraction channel in the case of NH_2 matches the analogous behavior that we found for reactions of HO_2 with ethylbenzene [30]. Lower *A* factors for the addition channels pursue from the lower entropy changes between transition states and separated reactants, in comparison to the abstraction channels.

Table 2: Kinetic parameters for ethylbenzene + NH_2. Reaction rate constants are fitted in the temperature range of 300-2000 K. Values of the Arrhenius parameters (*A* and *Ea*) are in units of cm^3 s^{-1} $molecule^{-1}$ and kJ/mol, respectively.

Reactions	A	E_a
R1	4.53×10^{-11}	41.0
R2	8.65×10^{-11}	36.2
R3	5.26×10^{-13}	52.9
R4	1.20×10^{-12}	52.5
R5	1.31×10^{-12}	61.7
R6	2.30×10^{-12}	61.9

Due to non-availability of the literature data, we compare values of the ethylbenzene + NH_2 with the results of our previous computations for propane + NH_2 [21]. Such comparison is peculiarly notable in the prospect of the very comparable BDHs of the targeted removable H atoms; our computed (CBS-QB3) BDH values of primary H in propane [21] vs ethylbenzene [30] correspond to 428.4 kJ/mol versus 421.3 kJ/mol. The activation energy for H abstraction from the primary position in ethylbenzene (42.5 kJ/mol) concurs with the related value from propane (42.0 kJ/mol) (per one abstractable H atom). Similar outcomes reported for the activation energies of removal of primary hydrogen atoms from propane and butane versus ethylbenzene and *n*-propylbenzene by HO_2 radical [31]. Figure 2 shows the relative importance of primary and benzylic H abstraction from ethylbenzene by different active radicals (NH_2, CH_3 [33], H [34] and HO_2 [35]) present in the combustion systems. We have noticed the second highest reaction rate for benzylic H removal from ethylbenzene by NH_2 radical after the dominant reaction of ethylbenzene + H radical (benzylic H removal).

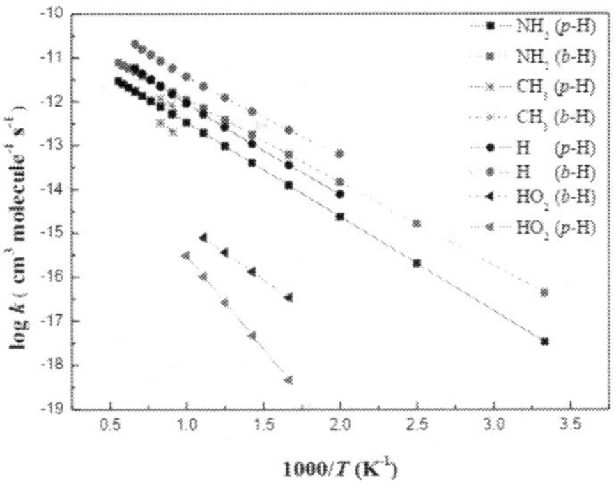

Figure 2. Comparison of primary and benzylic H abstraction reactions from ethylbenzene by NH_2 (current), CH_3 [33], H [34] and HO_2 [35] radicals.

4 Conclusions

This study reports H abstraction and addition reactions channels for the gas phase system of ethylbenzene with NH_2. Abstraction of H from primary and secondary sites incurs lower standard activation and reaction enthalpies contrasting with the addition of NH_2 to the phenyl ring of ethylbenzene. The removal of H atom from the primary site of ethylbenzene and propane acquired similar standard activation and reaction enthalpies based on the similar C–H BDHs. The reaction kinetic shows that the contribution of primary H removal increase with temperature but benzylic H abstraction is the sole dominant channel at all temperature. The sum of contribution from additional reaction pathways amounts to 9.0% for ethylbenzene + NH_2. We found a good comparison of the current kinetic data of H abstraction reaction of ethylbenzene with our previous work of H removal from propane by NH_2. Lastly, the reaction rates for abstraction channels of ethylbenzene + NH_2 hold as significance position as other active radicals in the combustion processes. The thermo-kinetic data presented in this study will facilitate in developing kinetic models for combustion and pyrolysis of nitrogen-rich fuels.

5 Acknowledgment

This study was funded by the Australian Research Council and supported by grants of computing time from the National Computational Infrastructure (NCI), Canberra, Australia and the Pawsey Supercomputing Centre, Perth, Australia. K.S. thanks Murdoch University for a postgraduate scholarship.

References

[1] I. Oluwoye, B.Z. Dlugogorski, J. Gore, S. Vyazovkin, O. Boyron, M. Altarawneh, Environ. Sci. Technol. 51 (2017) 7714-7722.

[2] J.A. Miller, C.T. Bowman, Prog. Energy Combust. Sci. 15 (1989) 287-338.

[3] A. Lucassen, K. Zhang, J. Warkentin, K. Moshammer, P. Glarborg, P. Marshall, K. Kohse-Höinghaus, Combust. Flame 159 (2012) 2254-2279.

[4] T. Mendiara, P. Glarborg, Combust. Flame 156 (2009) 1937-1949.

[5] M.T. Javed, N. Irfan, B. Gibbs, J. Environ. Manage. 83 (2007) 251-289.

[6] H.-P.S. Shen, M.A. Oehlschlaeger, Combust. Flame 156 (2009) 1053-1062.

[7] W. Yuan, Y. Li, G. Pengloan, C. Togbé, P. Dagaut, F. Qi, Combust. Flame 166 (2016) 255-265.

[8] M.A. Sutton, U. Dragosits, Y.S. Tang, D. Fowler, Atmos. Environ. 34 (2000) 855-869.

[9] B. Yuan, Y. Liu, M. Shao, S. Lu, D.G. Streets, Environ. Sci. Technol. 44 (2010) 4577-4582.

[10] Y. Ohmukai, I. Hasegawa, K. Mae, Fuel 87 (2008) 3105-3111.

[11] M. Becidan, Ø. Skreiberg, J.E. Hustad, Energy Fuels 21 (2007) 1173-1180.

[12] C. Li, K. Suzuki, Renew. Sustainable Energy Rev. 13 (2009) 594-604.

[13] K.-M. Hansson, J. Samuelsson, C. Tullin, L.-E. Åmand, Combust. Flame 137 (2004) 265-277.

[14] J.J. Hernández, R. Ballesteros, G. Aranda, Energy 50 (2013) 333-342.

[15] M. Nowakowska, O. Herbinet, A. Dufour, P.-A. Glaude, Combust. Flame 161 (2014) 1474-1488.

[16] S. Wang, W. Wei, L. Du, G. Li, J. Hao, Atmos. Environ. 43 (2009) 4148-4154.

[17] J.A. Miller, R.J. Kee, C.K. Westbrook, Annu. Rev. Phys. Chem. 41 (1990) 345-387.

[18] F.-J. Tian, J. Yu, L.J. McKenzie, J.-i. Hayashi, C.-Z. Li, Energy Fuels 21 (2007) 517-521.

[19] M. Altarawneh, M.H. Almatarneh, A. Marashdeh, B.Z. Dlugogorski, Combust. Flame 163 (2016) 532-539.

[20] M. Batiha, M. Altarawneh, A. Alsofi, M. Al-Harahsheh, I. Altarawneh, S. Alrawadieh, Theor. Chem. Acc. 129 (2011) 823-832.

[21] K. Siddique, M. Altarawneh, J. Gore, P.R. Westmoreland, B.Z. Dlugogorski, J. Phys. Chem. A. 121 (2017) 2221-2231.

[22] K. Siddique, M. Altarawneh, J. Gore, B.Z. Dlugogorski, Proceedings of the Australian Combustion Syposium, December 7-9, University of Melbourne, Vic, (2015) 272-275.

[23] M.J. Frisch, G.W. Trucks, H.B. Schlegel, G.E. Scuseria, M.A. Robb, J.R. Cheeseman, G. Scalmani, V. Barone, B. Mennucci, G.A. Petersson,; et al. Gaussian 09, revision A.1., Gaussian, Inc., Wallingford, CT, 2009.

[24] J.A. Montgomery Jr, M.J. Frisch, J.W. Ochterski, G.A. Petersson, J. Chem. Phys. 110 (1999) 2822-2827.

[25] M. Saeys, M.-F. Reyniers, G.B. Marin, V. Van Speybroeck, M. Waroquier, J. Phys. Chem. A. 107 (2003) 9147-9159.

[26] S. Canneaux, F. Bohr, E. Henon, J. Comput. Chem. 35 (2014) 82-93.

[27] D.G. Truhlar, B.C. Garrett, S.J. Klippenstein, J. Phys. Chem. 100 (1996) 12771-12800.

[28] C. Eckart, Phys. Rev. 35 (1930) 1303.

[29] A. Fernández-Ramos, B.A. Ellingson, R. Meana-Pañeda, J.M. Marques, D.G. Truhlar, Theor. Chem. Acc. 118 (2007) 813-826.

[30] M.K. Altarawneh, B.Z. Dlugogorski, E.M. Kennedy, J.C. Mackie, Combust. Flame 160 (2013) 9-16.

[31] M. Altarawneh, B.Z. Dlugogorski, Combust. Flame 162 (2015) 1406-1416.

[32] M. Altarawneh, A.A.H. Al-Muhtaseb, B.Z. Dlugogorski, E.M. Kennedy, J.C. Mackie, J. Comput. Chem. 32 (2011) 1725-1733.

[33] A. Lifshitz, A. Suslensky, C. Tamburu, Proc. Combust. Inst., 28 (2000) 1733-1739.

[34] C. Ellis, M.S. Scott, R.W. Walker, Combust. Flame, 132 (2003) 291-304.

[35] D. Baulch, C. Cobos, R. Cox, P. Frank, G. Hayman, T. Just, J. Kerr, T. Murrells, M. Pilling, J. Troe, J. Phys. Chem. Ref. Data, 23 (1994) 847-848.

11th Asia-Pacific Conference on Combustion,
The University of Sydney, NSW Australia
10th -14th December 2017.

A numerical study of the auto-ignited laminar lifted methane/hydrogen mixtures in heated co-flow air

Ki Sung Jung[1], Seung Ook Kim[1], Bok Jik Lee[2], Suk Ho Chung[3], Chun Sang Yoo[1*]

[1]Department of Mechanical Engineering, Ulsan National Institute of Science and Technology
Ulsan 44919, Republic of Korea

[2]School of Mechanical Engineering, Gwangju National Institute of Science and Technology
Gwangju 61005, Republic of Korea

[3]Clean Combustion Research Center, King Abdullah University of Science and Technology
Thuwal, Saudi Arabia

Abstract

In the present study, the liftoff characteristics of a laminar HCNG flame are investigated using 2-D numerical simulations by varying the fuel jet velocity. It is observed that the liftoff height decreases with increasing fuel jet velocity. Gradual transition from the tribrachial flame to the MILD combustion is also observed. From both *Da* analysis and transport budget analysis, it is verified that auto-ignition is the main stabilization mechanism of the lifted flame. Additional simulations are carried out with modified hydrogen diffusivity, showing that the decrease of liftoff height is mainly attributed to the high diffusivity of hydrogen molecule. Preferential diffusion effect of hydrogen is clarified by revising the ignition delay time term in the liftoff height relation. It is also shown that the revised ignition delay time increases as the fuel jet velocity decreases since hydrogen ratio R_H decreases at the upstream of flame front with decreasing fuel jet velocity. The high diffusive nature of hydrogen molecule increases the ignition delay time at lower fuel jet velocity, which, in turn, leads to the increase of the liftoff height.

1 Introduction

With increasing demand of reducing pollutant emissions, compressed natural gas (CNG) is regarded as a suitable fuel to replace other fossil fuels because natural gas is abundant in the world and produces less pollutant emissions. However, it is limited to apply CNG to practical engine systems due to its low ignitability. In this regard, hydrogen addition to the natural gas, known as HCNG or H_2CNG, is highlighted to cope with the limitation of methane combustion based on the hydrogen's high flame speed and wide flammability range [1]. Especially, auto-ignition characteristics of methane/hydrogen mixture plays an important role for the ignition processes in diesel engines. In this study, therefore, the characteristics of auto-ignited laminar lifted flames of HCNG in heated co-flow jets are numerically investigated.

According to the previous experimental study, auto-ignited laminar fuel is stabilized either lifted flame or nozzle attached flame shape depending on the initial conditions, and lifted laminar jet flame is categorized into tribrachial flame and MILD combustion regimes [2,3]. Adiabatic ignition delay time is identified as one of the key factors for the stabilization of auto-ignited lifted flame, and the liftoff height relation $H_L \sim U_0 t_{ig,ad}^2$ is derived and correlated well with the various fuels for the tribrachial flames, where H_L is the liftoff height, U_0 is the fuel jet velocity, and $t_{ig,ad}$ is the adiabatic ignition delay time at the stoichiometry condition. Therefore, the liftoff height H_L generally increases with increasing U_0 since $t_{ig,ad}$ has nothing to do with U_0.

In the auto-ignition of HCNG, however, an unusual phenomenon was observed from previous experiment [4] that liftoff height H_L decreases with increasing jet velocity U_0 at relatively low temperature. This behavior was not observed previously, and the reason was not fully understood yet. Note that when initial concentration of hydrogen is relatively low, the methane/hydrogen mixture exhibits the conventional lifted jet flame characteristics such that liftoff height increases with increasing fuel jet velocity. We can observe the unusual decreasing behavior above only when the amount of hydrogen in fuel jet is relatively high. The exact hydrogen ratio level leading to the decreasing liftoff height behavior highly depends on the initial co-flow temperature.

Another unique characteristic of the auto-ignition of HCNG in laminar co-flow jet is that transition from tribrachial flame to MILD combustion occurs when U_0 becomes low enough even though fuel mole fraction in fuel jet is the same.

The purpose of the present study, therefore, is divided into three parts: (1) understanding the stabilization mechanism of HCNG auto-ignited lifted flame, (2) clarifying the decreasing behavior of H_L with increasing U_0, and (3) elucidating the MILD combustion characteristics of HCNG lifted flame.

2 Numerical methods

Two-dimensional numerical simulations are performed using laminarSMOKE [5,6], which is an OpenFOAM based laminar reacting flow solver. The laminarSMOKE code is a numerical framework for the simulation of laminar flames in multi-dimensional geometries with detailed kinetics. The conservation equations of mass, momentum, species, and energy are solved.

In the present study, a skeletal methane-air kinetic mechanism of 30 species and 184 reactions is used [7]. Following the

* Corresponding author. Fax: +82-52-217-2409
E-mail address: csyoo@unist.ac.kr

previous experimental setup [2~4], an axisymmetric domain of 6.65 cm × 30 cm in the r- and z-directions is adopted. In the r-direction, a 100 μm mesh is uniformly distributed within 1 cm, and stretched mesh is applied to the remaining area. In the z-direction, a uniform grid of 100 μm is used. Inlet temperature, T_0, for both fuel jet and coflow is 950 K and jet velocity, U_0, varies from 3 to 25 m/s. Fuel mole fraction at fuel jet, $X_{F,0}$, is 0.2, and the hydrogen ratio in the fuel jet, R_H, is 0.3.

A grid convergence test is carried out for the case with U_0 = 15 m/s. The results show that the present grid size of 100 μm is fine enough to capture the flame structure and predict the liftoff height well compared to those with a finer grid size. Hence, all of the simulations in the present study use the grid system of 100 μm.

3 Results

3.1 General characteristics of lifted flame

Figure 1 shows temperature isocontours for the auto-ignited HCNG lifted flames with various jet velocities. As shown in the figure, the present simulation results can capture the unique characteristic of HCNG combustion that the liftoff height decreases with increasing jet velocity. Here, the flamebase is defined as the lowest z position of λ_{exp}= 0 iso-contour. Note that λ_{exp} is an eigenvalue of the Jacobian, J_w, of the chemical source term ω in a reacting governing equation. For the details of λ_{exp}, readers are referred to [8~]. The point where λ_{exp} crosses zero is defined as a 'crossover point' between non-explosive and explosive regions and it can be used to detect the flamebase, especially for lean mixtures.

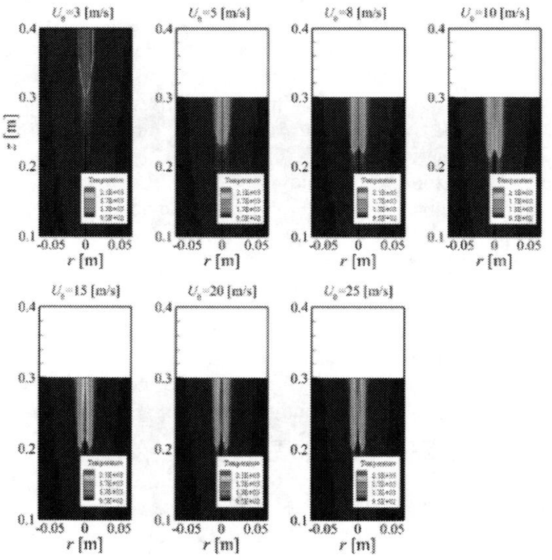

Figure 1: Temperature isocontours with various fuel jet velocities

For cases with relatively low U_0 (equal to or less than 5 m/s), the lifted flames do not feature the typical tribrachial lifted flame shape. This is because they can be categorized into the moderate or intense-low oxygen dilution (MILD) combustion regime, where the conditions of $(T_{max} - T_0)/T_{ig} < 1$ and of $T_{max} >$ 1 are both satisfied.

Figure 2 shows the liftoff height as a function of fuel jet velocity and the corresponding temperature index to determine MILD combustion condition. Two points are noted from the figure. First, as reported in a previous experimental study [4], the liftoff height decreases with increasing U_0, which is an unusual feature of lifted flame behavior. Second, the liftoff height increases significantly under the MILD combustion regime, which will be discussed in the following section.

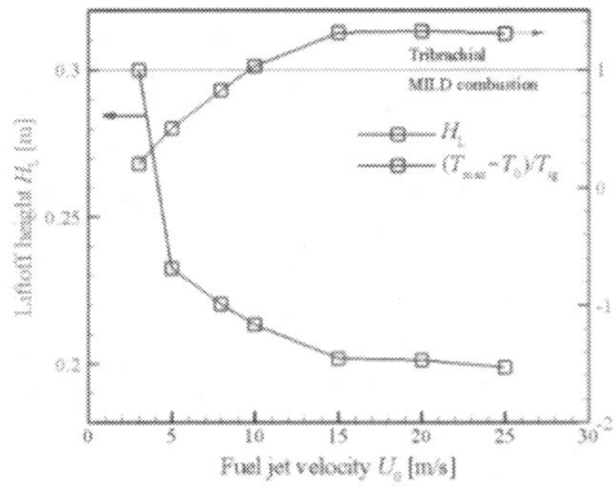

Figure 2: Temperature and temperature index with various fuel jet velocities.

3.2 Flame stabilization mechanism

To understand the unique phenomena of auto-ignited HCNG lifted flames, their flame stabilization mechanism is first elucidated. In the present study, both Damköhler number and transport budget analyses are carried out for all cases to demonstrate how auto-ignition or flame propagation affects the stabilization of the flamebase.

Damköhler number, $\boldsymbol{Da}$, is defined as the ratio of the characteristic flow time to the characteristic ignition delay time, $\boldsymbol{Da = t_{flow}/t_{ig}}$. It is evaluated at every single point in a streamline which passes through the flamebase. The characteristic flow time, $\boldsymbol{t_{flow}}$, is defined as the required time for a fluid particle to arrive at the flamebase starting from each point in the streamline. The characteristic ignition time, $\boldsymbol{t_{ig}}$ is the 0-D ignition delay time at each point. Therefore, when $\boldsymbol{Da > 1}$ region exists at the upstream of flame base, $\boldsymbol{t_{ig}}$ is shorter than $\boldsymbol{t_{flow}}$, implying that auto-ignition can occur at the upstream and the flamebase may be stabilized by auto-ignition. On the other hands, if $\boldsymbol{Da < 1}$ along the streamline from the flamebase, $\boldsymbol{t_{ig}}$ is always larger than $\boldsymbol{t_{flow}}$, indicating that auto-ignition may not be dominant upstream of the flame base. Therefore, the flamebase is stabilized by flame propagation rather than auto-ignition.

Figure 3 shows Da along the streamline passing the flamebase for cases with U_0 = 8 and 25 m/s. From the figure, it is readily observed that Da becomes greater than unity at the far upstream of the flamebase, which implies that the lifted flame is stabilized by auto-ignition rather than flame propagation. Similar trend is observed for the other cases (not shown here).

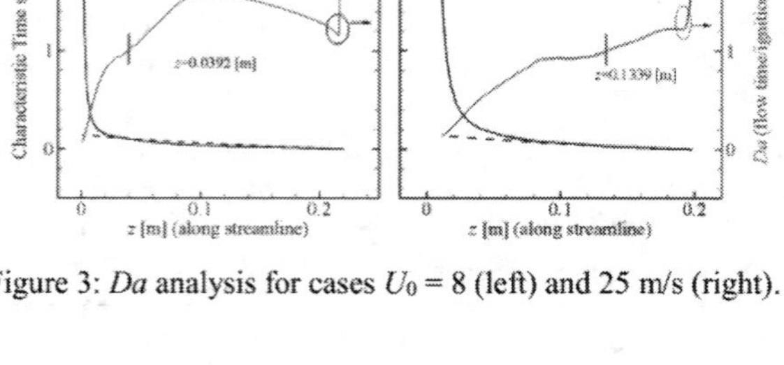

Figure 3: Da analysis for cases U_0 = 8 (left) and 25 m/s (right).

To further identify the stabilization mechanism of the lifted flames, the transport budget analysis is also performed along the streamline passing through the flamebase. Species transport equation for the given species i is given by:

$$\frac{\partial(\rho Y_i)}{\partial t} = -\frac{\partial(\rho Y_i u_j)}{\partial x_j} - \frac{\partial(\rho Y_i V_{j,i})}{\partial x_j} + \omega_i, \quad (1)$$

where ρ is density, Y_i is species mass fraction, $V_{j,i}$ is the diffusion velocity, and ω_i is the species mass reaction rates. The left hand side of Eq. (1) is zero since all of simulations reaches the steady state. The first term on the right hand side indicates the convection, the second term represents diffusion, and the last term is reaction. Each term for OH species is evaluated along the streamline. As in the *Da* analysis, cases with U_0 = 8 and 25 m/s are analyzed and the other cases show similar results. As shown in Fig. 4, the convection term balances reactions term and the diffusion term is relatively small compared to the other terms, verifying that the flamebase is stabilized by auto-ignition rather than flame propagation.

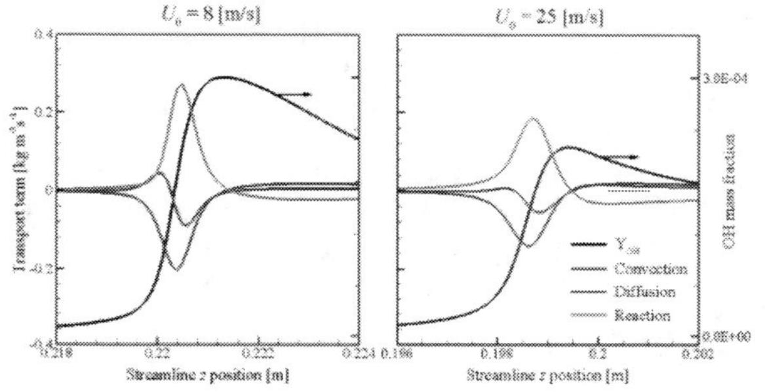

Figure 4: Budget analysis of OH species transport equation and OH mass fraction along the streamline.

3.3 Decreasing behavior of H_L

In the previous experiment [4], it was conjectured that the differential diffusion of the duel fuel, especially the high mass diffusivity of hydrogen, would be primarily attributed to the decrease of the liftoff height with increasing U_0. To verify the effect of hydrogen molecule's high diffusivity on the lifteoff height, additional numerical simulations are performed by setting the mass diffusivity of hydrogen molecule to that of methane. Figure 5 shows the liftoff height and temperature index $(T_{max} - T_0)/T_{ig}$ for cases with real and modified diffusivity of H_2. Except for the MILD combustion regime, modified diffusivity cases show that the liftoff height increases with increasing jet velocity, which follows the conventional liftoff height tendency. This result proves that the high diffusive nature of hydrogen molecule is attributed to the decrease of H_L with increasing U_0. To incorporate the effect of large mass diffusivity of H_2 into the liftoff height model, $H_L \sim U_0 t_{ig,ad}^2$, we come up with a modified ignition delay time. The adiabatic ignition delay time, $t_{ig,ad}$, corresponds to the 0-D ignition delay time of the stoichiometric mixture of fuel jet and coflow air, and hence, the differential diffusion effect of the dual fuel cannot be accounted for in the liftoff height relation. For auto-ignition of single fuel jets such as methane or propane, the conventional liftoff height relation is well correlated with both experimental and numerical results. For dual fuel jets such as HCNG, however, the ignition delay time cannot be determined solely from the fuel jet composition because the hydrogen content varies significantly depending on the location of the flamebase. In this study, the modified ignition delay time, t_{ig}, is obtained by tracing the

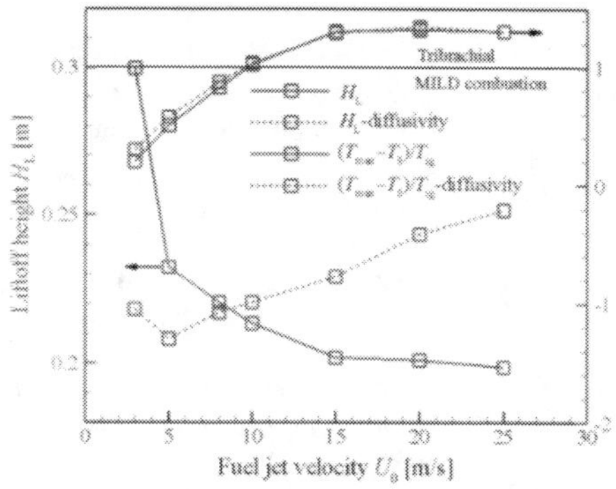

Figure 5: Temperature and temperature index with various fuel jet velocities for both real and modified diffusivity of H_2. Solid line is real, dashed line is modified H_2 diffusivity cases.

initial mixture composition of the flamebase. Based on the mixture contents at the flame base, we can find the mole fraction of initial mixture composition: CH_4, H_2, O_2, and N_2 (referred to hereinafter as 'evaluated initial mixture composition'). Our assumption is that 0-D ignition delay time of evaluated initial mixture composition would represent the characteristic ignition delay time in 2-D domain since it includes the information on the different diffusion effect led by different convection time scale.

Figure 6 shows the modified liftoff height relation $H_L \sim U_0 t_{ig}^2$ and actual liftoff height H_L. It is readily observed from the figure that the modified liftoff height relation is well correlated with both real and modified H_2 diffusivity cases. Note that to find H_L correlation, we divide it into two regimes: tribrachial flame regime and MILD combustion regime.

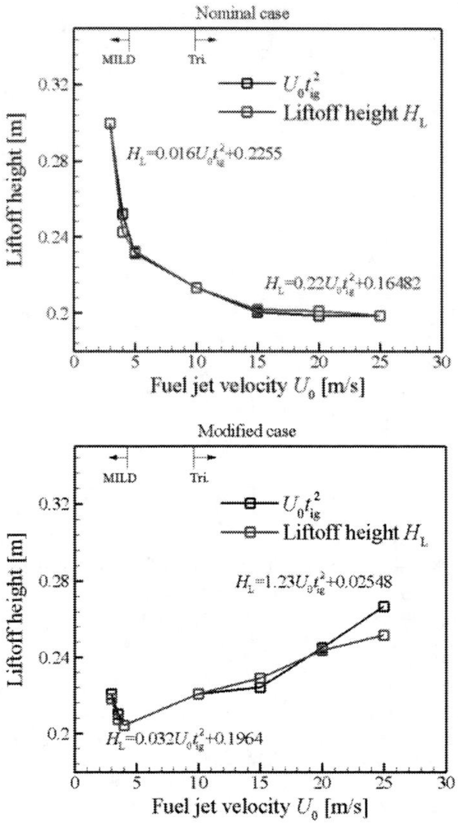

Figure 6: The relationship between $H_L \sim U_0 t_{ig,ad}^2$ and actual liftoff height H_L for real (above) and modified H_2 diffusivity cases (below).

Figure 7 shows the modified ignition delay time, t_{ig}, and hydrogen ratio, R_H, with various jet velocities. R_H is obtained from the evaluated initial mixture composition. It is readily observed from the figure that t_{ig} decreases as U_0 increases. It is of importance to note that t_{ig} significantly increases under the MILD combustion conditions compared to the tribrachial conditions. This is because under the MILD combustion conditions, the flamebase is located at the centerline and the corresponding hydrogen content vanishes there. As shown in Fig. 8, the more hydrogen molecules are included in the mixture or the higher R_H is, the shorter the ignition delay time is.

In summary, under the tribrachial flame conditions, hydrogen molecules can diffuse out more radially with increasing U_0, and subsequently, the R_H at the flamebase increases, leading to the shorter H_L. Under the MILD combustion conditions, the R_H vanishes at the flamebase and hence, the ignition delay time is increased, resulting in the longer H_L. Under either conditions, the large mass diffusivity of hydrogen molecule together with the flamebase location is primarily attributed to the decrease of H_L with increasing U_0.

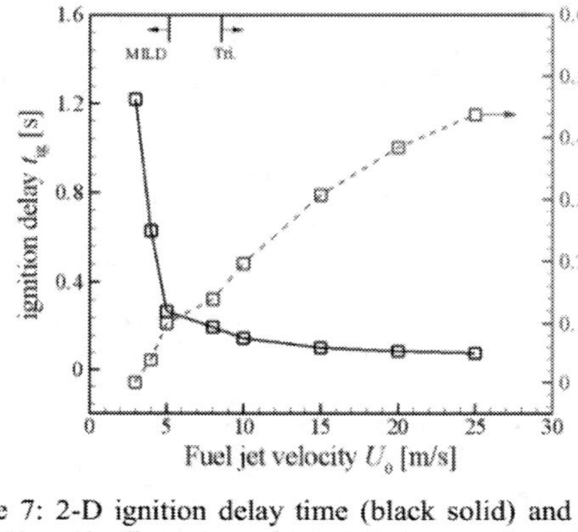

Figure 7: 2-D ignition delay time (black solid) and hydrogen ratio R_H at the flame base (blue dotted) with the different fuel jet velocities.

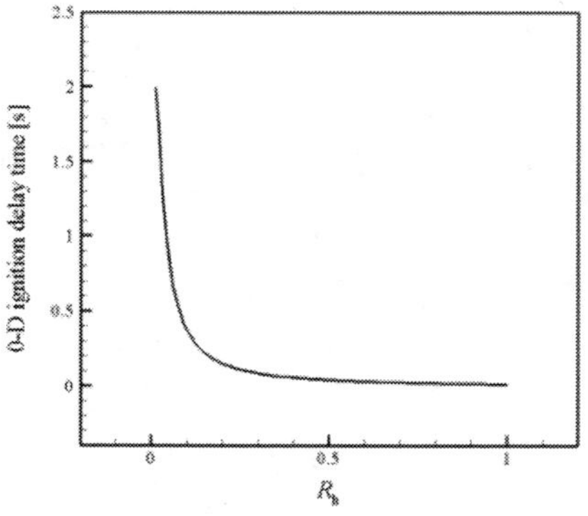

Figure 8: 0-D ignition delay time as a function of the hydrogen ratio R_H. Initial temperature for the 0-D calculation is 950 K, and pressure is 1 atm. Fuel mole fraction of the fuel jet $X_{F,0}$ is 0.2.

Although it is not shown in the present paper, additional numerical simulations were carried out with the initial condition of $T_0 = 1000$ K and $R_H = 0.08$ in fuel jet. The result revealed that the liftoff height is proportional to the fuel jet velocity. It is consistent with the previous experimental results, which showed that liftoff height change trend in HCNG is sensitive to the both initial temperature and hydrogen concentration level in fuel jet.

4 Conclusions

The conclusions of the present study are as follows.

(1) From the *Da* analysis and transport budget analysis, it is found that the lifted flames are stabilized by auto-ignition rather than flame propagation. Therefore, ignition delay time, $t_{ig,ad}$, plays a key role in determining the liftoff height.

(2) The large diffusivity of hydrogen molecule is the main factor to induce the unusual tendency of the liftoff height; H_L decreases as U_0 increases. Depending on U_0, the lifted flames belong to either tribrachial flame or MILD combustion. As such, we could obtain two different H_L correlations corresponding to the tribrachial and MILD combustion regimes. This is because the ignition delay time changes significantly under the MILD combustion conditions (lower R_H), whereas it marginally varies under the tribrachial flame conditions (higher R_H).

(3) The MILD combustion in HCNG occurs when U_0 is relatively low. As the jet velocity becomes low, the hydrogen molecules in the fuel jet are more apt to diffuse out radially and finally the content of hydrogen vanishes at the flamebase. As a result, the ignition of the fuel jet is more delayed and the fuel mole fraction at the flamebase decreases significantly, leading to the combustion mode change from the tribrachical flame to the MILD combustion.

References

[1] C.M. White, R.R. Steeper, A.E. Lutz, Int. J. Hydro. Energy 31 (2006) 1292-1305.

[2] B.C. Choi, K.N. Kim, S.H. Chung, Combust. Flame 156 (2009) 396-404.

[3] B.C. Choi, S.H. Chung, Combust. Flame 157 (2010) 2348-2356.

[4] B.C. Choi, S.H. Chung, Combust. Flame 159 (2012) 1481-1488.

[5] Cuoci A., Frassoldati A., Faravelli T., Ranzi E., Combust. Flame 160 (2013) 870-886.

[6] Cuoci A., Frassoldati A., Faravelli T., Ranzi E., Energy. Fuels 27 (2013) 7730-7753.

[7] Lu, T.F., Law, C.K., Combust. Flame 154 (2008) 761–774.

[8] T.F. Lu, C.S. Yoo, J.H. Chen, C.K. Law, J. Fluid Mech. 652 (2010) 45-64.

[9] Z. Luo, C.S. Yoo, E.S. Richardson, J.H. Chen, C.K. Law, T.F. Lu, Combust. Flame 159 (2012) 265-274.

[10] C.S. Yoo, T.F. Lu, J.H. Chen, C.K. Law, Combust. Flame 158 (2011) 1727–1741.

11th Asia-Pacific Conference on Combustion,
The University of Sydney, NSW Australia
10th–14th December 2017.

How transverse acoustic velocity affects flame response to axial acoustic perturbations

Aditya Saurabh, Christian Oliver Paschereit

Technische Universität Berlin

Müller-Breslau Str. 8, Berlin, 10623, Germany

Abstract

Flame response to transverse acoustic fields has recently emerged as an important aspect concerning the phenomenon of thermoacoustic instability in annular gas turbine combustors. Only few studies exist on the subject and our understanding is consequently limited. In this contribution we provide results from an experimental study of the response of a premixed, swirl-stabilized flame in the presence of transverse acoustic fluctuations. The study reveals that transverse acoustic velocity fluctuations can have a non-trivial impact on flame response. Furthermore, the impact is dependent on the magnitude of transverse acoustic velocity, the phase of transverse fluctuations with respect to axial fluctuations, operating conditions, and even the asymmetry of the flame. The study provides observations on flame response to transverse acoustics that must be investigated in more detail in future studies.

1 Introduction

Thermoacoustic instability in gas turbine (GT) engines, where the acoustic field of the combustor and perturbations in the flame(s) get coupled in a constructive feedback interaction and consequently result in large amplitude pressure oscillations, is one of the most concerning technical issues affecting the development of GT combustors. A key aspect to the coupling is the response of flames to acoustic perturbations. Because this aspect involves intricate interactions among the flame, the hydrodynamic field in the vicinity of the flame, and acoustic perturbations incident on the burner [1]; much of which is yet to be elucidated, the task of obtaining a detailed understanding of the response of technical flame configurations is quite challenging. Furthermore, for low-order [2] models aimed at the prediction, the flame response characteristics is a crucial input that must accurately reflect the dynamics of the flame(s) employed in the system for which the model has been set-up.

An unresolved aspect of flame response, which is also of significant interest to contemporary and future GT combustors, is the response of flames to a multidimensional acoustic field. Previous studies on flame response have largely been focused on planar acoustic waves, with the axis of propagation aligned with the burner axis. The corresponding flame response and coupling mechanisms may be used for combustors if the acoustic mode associated with self-excited thermoacoustic instability is also planar and aligned with the burner axis. However, in modern combustors that employ an annular combustor configuration, thermoacoustic instability is associated with the azimuthal acoustic mode, such that acoustic fluctuations transverse to the flame are responsible for generating flame response—possibly in addition to axial fluctuations generated due the indirect effect of azimuthal mode of the combustor or the plenum.

Corresponding author: aditya.saurabh@tu-berlin.de

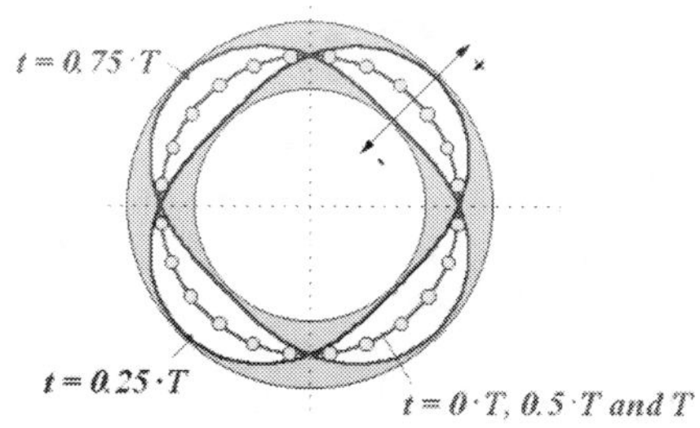

Figure 1: Acoustic pressure profile of an azimuthal acoustic mode at various instants with respect to the time period of the acoustic oscillation, T, and relative to the generic circumferential arrangement of burners in an annular combustors (depicted as yellow circles). From Ref. [3].

Table 1: Transverse acoustic velocities (u') induced during simultaneous axial-transverse(anti-symmetric) forcing. 'low', 'med', and 'high' correspond to the markers used in Fig. 4. Mean values from multiple experiments. Parenthesized values correspond to the range (i.e. difference between maximum and minimum values) of the velocities obtained in experiments with different phase relation between axial and transverse forcing. The air mass flow rate, $\dot{m}_a$ is in $\mathrm{kg\,h^{-1}}$.

$\dot{m}_a$ / swirl	Transverse acoustic velocity, u' ($\mathrm{m\,s^{-1}}$)		
	low	med	high
61 / 0.7	0.81(0.05)	1.98(0.07)	3.05(0.29)
61 / 0.9	0.86(0.04)	2.01(0.10)	3.00(0.26)
92 / 0.7	0.83(0.07)	2.00(0.10)	3.08(0.36)
92 / 0.9	0.86(0.06)	2.05(0.10)	3.21(0.31)

Figure 1 is a schematic from Ref. [3] showing the azimuthal acoustic mode (first harmonic) in an annular chamber with circumferentially-arranged flames. The pressure fluctuations are shown. Velocity fluctuations are in quadrature with the pressure fluctuations. The schematic shows that different flames experience different combinations transverse acoustic velocity and pressure, depending on their location with respect to the spatial profile of the acoustic mode. It is also possible that the mode maybe slowly rotating or spinning [5], in which case individual flames will experience variation in the amplitude and composition of the acoustic mode.

The difference between axially-aligned, planar acoustic field and the scenario in annular combustors, where the flame is under the influence of transverse—and possibly axial—forcing is illustrated in Fig. 2. The planar axial and transverse acoustic fields are denoted by upstream and downstream travelling wave components,

f and g.

In previous studies, it was pointed out that transverse standing acoustic field with the velocity antinode at the burner produces an anti-symmetric flame response [6, 7], such that the modulations in the flame surface do not result in a global heat release rate fluctuations. Based on this result, it was implied that the transverse acoustic velocity is not a relevant factor with respect to flame response, and therefore with respect to self-excited thermoacoustic coupling in annular combustors—unless the flame itself if not symmetric, in which case mutual cancellation of heat release rate fluctuations from different parts of the flame would not occur [8].

Contrary to this implied result, other studies which have investigated the combined effect of transverse and axial acoustic velocity suggest that the transverse velocity has an important role [9, 10]—or in the case of Ref. [10], to reflect the conclusions of the authors more accurately, acoustic pressure is not the only factor governing flame response when transverse acoustics are present.

2 Methods

An atmospheric rig, shown in Fig. 3, was employed for the tests conducted under the framework of this project. The rig consists of a rectangular cuboid enclosure, which acts as the combustion chamber. One face of this combustor is equipped with a quartz glass window for optical access to the flame. On the left and right sides of the combustion chamber are cylindrical ducts with closed ends, each equipped with a loudspeaker assembly. Downstream of the combustion chamber is a short exhaust duct.

Natural gas and air mixture enters the combustor through a radial swirler with adjustable swirler vanes and an annular exit to the combustor (inner, outer diameters being 27.5 mm and 55 mm respectively). This swirler receives the fuel-air mixture from a long inlet duct terminated by an orifice plate. Fuel and air are introduced in a small mixing space further upstream of the orifice plate. With this setup acoustic fluctuations do not affect the fuel air mixing, consequently removing the possibility of equivalence ratio fluctuations entering the inlet duct. The inlet duct is equipped with another loudspeaker assembly. For results presented in this paper, the rig is operated at geometric swirl intensities, **S** = 0.7, 0.9, inlet air flow rates of 61, 92 kg h^{-1}, and at a lean condition with an equivalence ratio, $\phi = 0.8$.

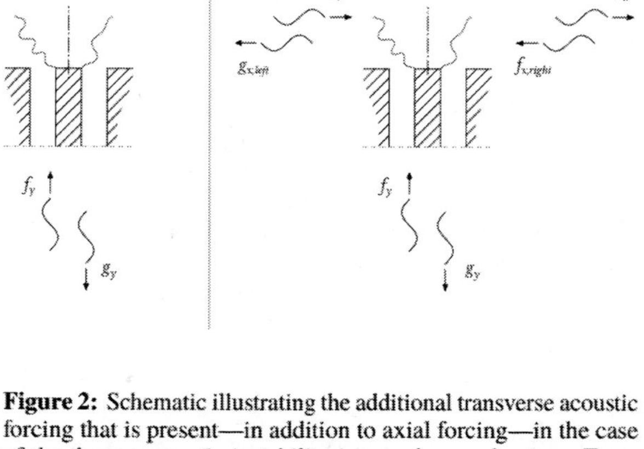

Figure 2: Schematic illustrating the additional transverse acoustic forcing that is present—in addition to axial forcing—in the case of the thermoacoustic instability in annular combustors. From Ref. [4]

2.1 Experiments and diagnostics

The premixed swirl flame that is established in the combustor is acoustically forced using speakers mounted on the inlet plenum (axial forcing), and on the transverse ends (transverse forcing). The amplitude of axial acoustic forcing is fixed such that the velocity fluctuations are close to $v' \sim 5\% v_0$, where v_0 is the bulk flow velocity through the annular burner exit.

Transverse forcing is established such that a velocity antinode exists at the burner axis. To achieve this objective, the speaker transfer function and the reflection coefficient of the transverse end were considered in the generation of voltage input to the speakers, such that acoustic fluctuations on either sides of the burner axis is identical in amplitude and out-of-phase by 180°.

Complex acoustic velocity at the reference planes, as indicated in Fig. 3, was obtained through the Multi-microphone method where the acoustic field in the upstream plenum is assumed planar and composed of forward and backward travelling waves. Multi-microphone method allows for the decomposition of the acoustic field and identification of the mutually-opposite travelling wave components. Linear combinations of the components gives the acoustic pressure and velocity in the respective test rig sections (more details may be found in Ref. [11]).

In experiments concerning simultaneous axial and transverse forcing, which is the focus of the present study, the amplitude of transverse acoustic velocity at the reference location (see Fig. 3) was gradually increased and the relative phase between axial and transverse forcing was varied between 0° and 180° in steps of 45°. Investigations reported here were carried out at 112 Hz. Other forcing frequencies lead to identical inferences [12]. Table 1 provide a summary of simultaneous forcing experiments reported here.

The response of the flame to acoustic fluctuations generated by the speakers is the most important physical measure for the study. For premixed flames, the flame response—or the heat-release rate ($\dot{q}(t)$) generated in response—to acoustic forcing is conventionally obtained through flame chemiluminescence measurements ($I(t)$) as the assumption, $\dot{q}(t) \propto I(t)$ is valid for premixed flames. The same approach is followed here and the flame response is characterized by the flame transfer function in the frequency domain,

$$\mathbf{F}(\omega) = \frac{\hat{\dot{q}}/\dot{q}_0}{\hat{v}/v_0}$$

The flame chemiluminescence is captured by a photo-diode equipped with a narrow-band OH filter and is additionally imaged using a camera, also equipped with an OH filter.

Several repetitions were performed for each experiment in order to obtain confidence intervals for the flame transfer function **F**.

3 Results

The response of the flame in terms of the absolute value and the phase of the flame transfer function, **F**, as defined earlier; to varying combinations of simultaneous axial and transverse forcing is presented in Fig. 4. The operating conditions are indicated within the plots.

The phase has been converted to time-delay in ms. For each frequency-operating condition pair, three transverse velocity amplitudes (indicated in the legend as 'low', 'med', and 'high', with exact values in Tab. 1) were considered. The forcing amplitude was controlled only by feeding a certain input voltage to the speaker amplifier. The mean values with 95% confidence intervals are plotted for each experiment configuration. For comparison, the mean and confidence interval for the corresponding case of pure axial forcing is indicated as a grey band for each condition.

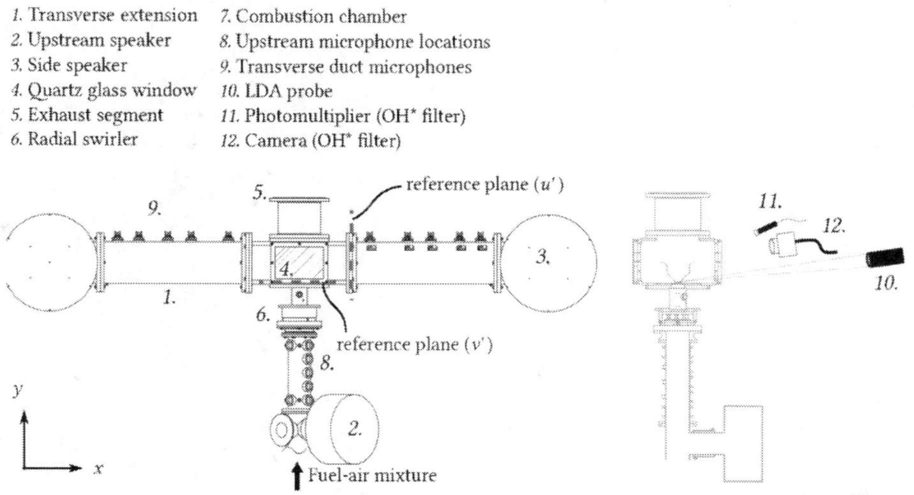

Figure 3: Details of the combustion rig and instrumentation employed in this study.

The y-axis limits for the plots have been kept identical to aid comparison.

It can clearly be identified that the introduction of a transverse acoustic velocity influences the flame response to axial acoustic forcing. In order words, the general case of a flame under the influence of simultaneous axial and transverse velocity forcing in an annular combustor may not be substituted by flame response to axial forcing, as is concluded in several previous investigations.

We can infer from Fig. 4 that the influence of transverse acoustic velocity depends on (a) the operating condition, (b) the amplitude of transverse acoustic velocity, and (c) the relative phase between axial and transverse forcing. The effect of transverse acoustic velocity is prominently to reduce the flame response, suggesting that the antisymmetric nature of flame modulation associated with transverse acoustic velocity might be acting against the axisymmetric modulation due to axial forcing.

The fact that the relative phase between transverse and axial forcing is an important parameter indicates the existence of an interaction between modulations generated by axial fluctuations and those generated by transverse fluctuations: the interaction between the two flame modulation components depends on the timing of fluctuations generated due to transverse forcing with respect to those generated by axial forcing.

Another aspect that can be noted is that the deviation in flame response due to transverse acoustic velocity is not identical at 0° and 180°. A two-fold symmetry in the system (the flame and the forcing) should lead to identical response at 0° and 180°. Because the forcing has been calibrated to result in a symmetric acoustic field on either sides of the combustor, it can be inferred that non-identical deviation results from asymmetries in the flame. This is further elaborated through the subsequent analysis of the coherent fluctuations in the flame surface.

Figure 5 presents the dominant mode pairs obtained from Partial Orthogonal Decomposition (POD) of the chemiluminescence image sequence at 0° and 180° phase difference between axial and transverse forcing. Comparing the two cases, it is immediately clear that the flame surface modulations are asymmetric and that the mode shape is flipped about the burner axis as the relative phase difference changes by 180°. One of the sides of the flame contains large modulations compared to the other side. It can be inferred that if the two sides of the flame are not identical, the spatial integration of flame surface fluctuations, which gives the global flame response, will also be different. This is the reason for the observed difference in the flame response between 0° and 180° relative phase difference.

4 Conclusion

In a practical scenario, when an instability develops in an annular combustor, the location of individual flames relative to the azimuthal acoustic mode can be such that while some flames experience a pressure antinode, others might experience either a velocity antinode or an intermediate acoustic forcing that is composed of simultaneous transverse pressure and velocity oscillations. The azimuthal mode could be standing or rotating/spinning. In addition to the azimuthal acoustic mode of the annular combustor, thermoacoustic instability could involve plenum modes, which would lead to additional axial acoustic oscillations upstream of the flame, not necessarily in-phase with azimuthal acoustic oscillations. Thus, pure transverse acoustic pressure or velocity explores the possible flame-acoustic interaction regimes only partially. A more likely case is the simultaneous transverse and axial acoustic forcing of flames.

In this study we show that the flame response is critically dependent on the transverse acoustic velocity. Specifically, the effect of transverse acoustic velocity on flame response depends on several factors that were identified here: the amplitude of transverse acoustic velocity, its phase with respect to axial acoustic forcing, the operating conditions, and the symmetry of the flame.

It may be concluded from the study that, as the effect of transverse acoustic velocity is amplitude-dependent, and because appears only for finite amplitudes of transverse velocity; it is safe to neglect transverse acoustic velocity for linear assessment of combustor stability. For nonlinear prediction, however, it is important to incorporate the effect of transverse velocity as well.

References

[1] Sébastien Ducruix, Thierry Schuller, Daniel Durox, and Sébastien Candel. Combustion dynamics and instabilities: Elementary coupling and driving mechanisms. *J. Propul. Power*, 19(5):722–734, sep 2003.

[2] B. Schuermans, V. Bellucci, and C. O. Paschereit. Thermoacoustic modeling and control of multi burner combustion

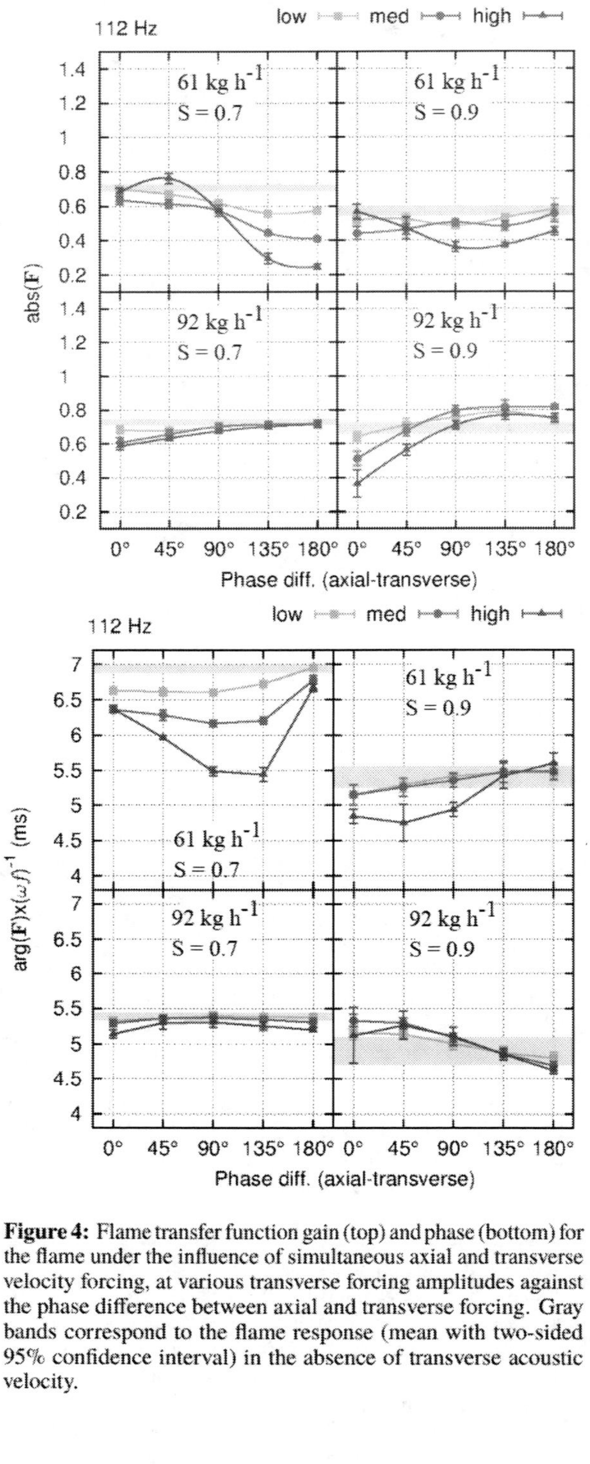

Figure 4: Flame transfer function gain (top) and phase (bottom) for the flame under the influence of simultaneous axial and transverse velocity forcing, at various transverse forcing amplitudes against the phase difference between axial and transverse forcing. Gray bands correspond to the flame response (mean with two-sided 95% confidence interval) in the absence of transverse acoustic velocity.

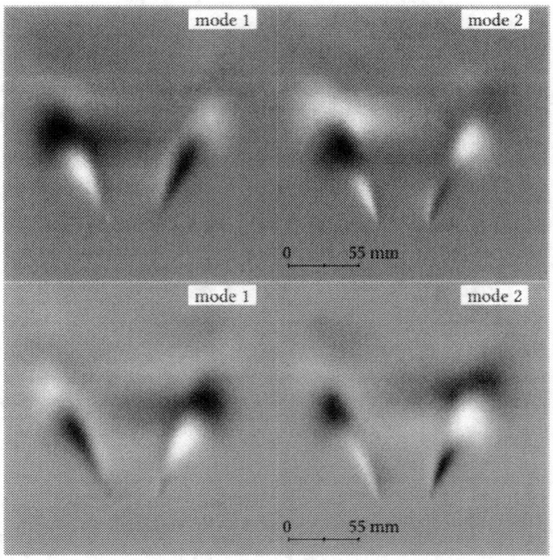

Figure 5: Dominant POD mode pairs for the flame under the influence of simultaneous transverse and axial acoustic forcing with a phase difference of 0° (top) and 180° (bottom).

systems. In *Proc. ASME Turbo Expo, June 16-19, Atlanta, Georgia, USA*, number GT2003-38688, 2003.

[3] P. Berenbrink and S. Hoffmann. Suppression of dynamic combustion instabilities by passive and active means. In *Proc. ASME Turbo Expo, May 8-11, München, Germany*, number 2000-GT-0079, 2000.

[4] Aditya Saurabh and Christian Oliver Paschereit. Dynamics of premixed swirl flames under the influence of transverse acoustic fluctuations. *Combust. Flame*, 182:298–312, aug 2017.

[5] Nicolas Noiray and Bruno Schuermans. Deterministic quantities characterizing noise driven hopf bifurcations in gas turbine combustors. *Int. J. Non Linear Mech.*, 50:152–163, apr 2013.

[6] Aditya Saurabh and Christian Oliver Paschereit. Swirl flow response to transverse and axial acoustic forcing. *JFST*, 9(3):JFST0059–JFST0059, 2014.

[7] Jacqueline O'Connor, Tim Lieuwen, and Michael Kolb. Visualization of shear layer dynamics in a transversely excited, annular premixing nozzle. In *49th AIAA Aerospace Sciences Meeting including the New Horizons Forum and Aerospace Exposition*. American Institute of Aeronautics and Astronautics (AIAA), jan 2011.

[8] V. Acharya, D.-H. Shin, and T. Lieuwen. Premixed flames excited by helical disturbances: Flame wrinkling and heat release oscillations. *J. Propul. Power*, AIAA Early Edition:1–10, 2013.

[9] Martin Hauser, Michael Wagner, and Thomas Sattelmayer. Transformation of transverse acoustic velocity of the burner approach flow into flame dynamics. In *Volume 2: Combustion, Fuels and Emissions, Parts A and B*. ASME International, jun 2012.

[10] Florian Lespinasse, Françoise Baillot, and Toufik Boushaki. Responses of v-flames placed in an HF transverse acoustic field from a velocity to pressure antinode. *Comptes Rendus Mécanique*, 341(1-2):110–120, jan 2013.

[11] C. O. Paschereit, B. Schuermans, W. Polifke, and O. Mattson. Measurement of transfer matrices and source terms of premixed flames. *J. Eng. Gas Turbines Power*, 124(2):239–247, 2002.

[12] Aditya Saurabh, Jonas P. Moeck, and Christian Oliver Paschereit. Swirl flame response to simultaneous axial and transverse velocity fluctuations. *J. Eng. Gas Turbines Power*, 139(6):061502, jan 2017.

11th Asia-Pacific Conference on Combustion,
The University of Sydney, NSW Australia
10th–14th December 2017.

Noise-induced dynamics in a stable thermoacoustic system: Numerical evidence of coherence resonance

Vikrant Gupta[1], Aditya Saurabh[2], Christian Oliver Paschereit[2], Lipika Kabiraj[3]

[1]Southern University of Science and Technology, Shenzhen 518055, Guandong, China

[2]Technische Universität Berlin, Müller-Breslau Str. 8, Berlin, 10623, Germany

[3]Indian Institute of Technology Ropar, Punjab 140001, India

Abstract

Thermoacoustic instability is a feedback coupling phenomenon concerning combustors such as those employed in gas turbine engines. Noise is inherent to such applications and understanding the influence of noise on thermoacoustic coupling is, therefore, of practical importance. In this paper, we focus on noise-induced dynamics in the stable regime—prior to the bistable regime of thermoacoustic instability. We show that even in this regime noise induces a non-trivial dynamics where the system response to noise is most coherent at a certain, intermediate noise intensity. This phenomenon, known as coherence resonance, has been recently demonstrated in experiments. Here, we employ a representative model for thermoacoustic coupling in order to elucidate noise-induced behaviour, in particular coherence resonance.

1 Introduction

In recent experiments [1], it was found that in the subthreshold region of instability—prior to the bistable zone of the corresponding subcritical Hopf bifurcation—noise-induced effects are governed by coherence resonance, where an optimum level of noise-induced coherence appears at an intermediate level of noise intensity. This article investigates this phenomenon through a simple representative thermoacoustic system model. On the basis of stochastic numerical simulations of the model it is shown here that the phenomenon of coherence resonance in combustion systems is a result of thermoacoustic coupling.

Thermoacoustic instability, where a constructive feedback coupling between unsteady heat release rate processes and acoustic oscillations leads to high amplitude self-excited oscillations, is a concerning issue in combustion systems. The fact that noise is inherent to practical combustion systems due to factors such as flow turbulence and combustion noise, makes it important to investigate the role of stochastic effects in addition to the study of the deterministic dynamics of thermoacoustic instability.

The onset of self-excited thermoacoustic oscillations is characterized by either a subcritical or a supercritical Hopf bifurcation. The coupling mechanisms leading to the instability are nonlinear. Therefore, the oscillatory state resulting from the bifurcation could be a limit cycle behaviour, or a more complex state—period-2 oscillations, quasiperiodicity, chaos—resulting from subsequent bifurcations of the limit cycle [2, 3, 4, 5].

In previous investigations, the effects of noise at the onset of the instability and on the characteristics of the self-excited oscillations have been studied [6, 7]. It has been shown that for the case of subcritical Hopf bifurcation, noise could result in transitions between the oscillatory and the steady state within the subcritical zone [8]. Recently, it has also been shown that due to this effect, high amplitude noise levels can cause the subcritical bifurcation to appear as a supercritical scenario [9]. In a few investigations, dealing mostly with liquid propellant combustors and rocket motors involving high frequency instabilities, attention was also given to noise-driven system dynamics before the onset of the instability [10, 11]. The variation that noise brings to system behaviour has also been proposed as a medium to extract deterministic quantities (linear characteristics such as growth/decay rate) from measurements of stochastic fluctuations [11, 12].

In contrast to most of the previous work on noise effects in thermoacoustic systems, the focus in this work is on the effects of noise prior to the bistable zone for the case of a model thermoacoustic system undergoing subcritical Hopf bifurcation. We refer to this region as the *subthreshold* region, a term borrowed from Ref. [13], where stochastic dynamics of model nonlinear oscillators was considered. A representative model of thermoacoustic coupling, which has been extensively investigated in the past for studying deterministic as well as stochastic features of the instability has been employed in this study. The main motivation of the study comes from our recent experiments [1] on a laboratory combustor employing a premixed laminar flame. Here, it was shown that noise leads to a significant system response in the form of induced coherent oscillations. The effect of noise was found to critically depend on the noise intensity and the proximity to the approaching (subcritical) Hopf bifurcation. *Coherence resonance* [14], which refers to the phenomenon where the system response to noise intensity bears a relation similar to the resonance curve (in other words, an optimum response occurs at a certain intermediate level of noise) was identified in the experiments. The experimental work was motivated by observations of similar non-trivial noise effects and coherence resonance in other nonlinear systems [15, 16, 17]. The model employed here is not a replication of the experimental configuration and a direct comparison cannot be expected. Instead, the focus here is on investigating whether the said noise effects are a feature of thermoacoustic coupling.

This article is divided further into two main sections—the description of the model system, including the introduction of the stochastic term, and the corresponding results on the effects of noise at varying parameter values and for varying noise intensities. The main conclusions of the work follow subsequently.

2 Model Description

The model adopted in this article is from Balasubramanian & Sujith [18]. We investigate the scenario of subcritical Hopf bifurcation in this paper. More details, including the case of subcritical Hopf bifurcation, have been presented in a recent publication [19]. This

Corresponding author: lipika.kabiraj@iitrpr.ac.in

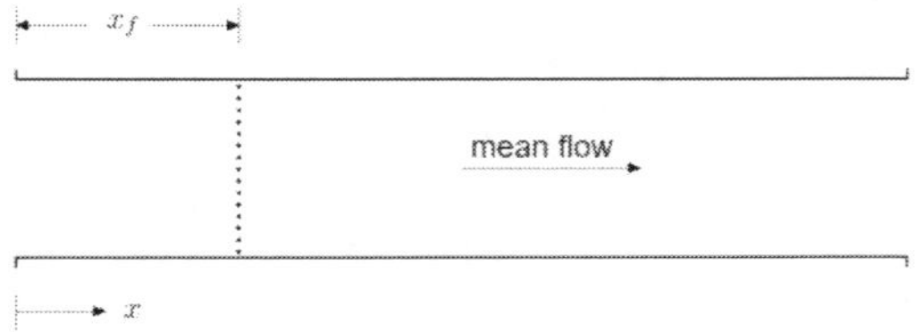

Figure 1: Sketch of the Rijke tube model used in this article. It is a horizontal tube with acoustic open-open boundary conditions.

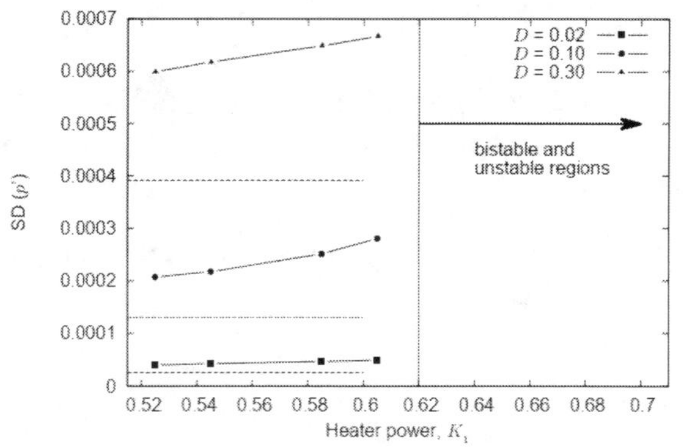

Figure 2: Variation in the standard deviation (SD) of pressure fluctuations with the heater power. The noise response in the absence of the heater is indicated via the dashed horizontal lines.

model simulates a horizontal Rijke tube system with an electric heat source element, and a schematic is shown in Fig. 1.

A constant area duct with heater located at a particular location is considered. The boundary conditions assumed here for acoustic is open-open. Because the tube is horizontal, the natural convection is absent. The mean flow in such a system is maintained by a blower, and the mean heat is controlled by an independent heat source. It is assumed in this model that the mean flow has a rectangular profile and is small as compared to the speed of sound in the tube, the effect of mean temperature gradient in the tube is neglected. The acoustics considered is linear one-dimensional, which is justified in low Mach number cases. The heat release rate ($\dot{Q}$) in the system oscillates because of the oscillating acoustic velocity, and is modelled using a modified King's law [20]. For such a model, the non-dimensional acoustic momentum and energy equations are give as below.

$$\frac{\partial u}{\partial t} = -\frac{1}{\gamma M_a}\frac{\partial p}{\partial x} \tag{1}$$

$$\frac{\partial p}{\partial t} = -\gamma M_a \frac{\partial u}{\partial x} - \zeta p + \dot{Q}\delta\left(x - x_f\right) \tag{2}$$

where u and p are acoustic velocity and pressure oscillations, M_a is the Mach number which is small, γ is the ratio of specific heats of the air at constant pressure and constant volume, respectively. The damping term is represented by the ζp term. The oscillating heat release rate is $\dot{Q} = K\left[\sqrt{\left|\frac{1}{3} + u_f(t-\tau)\right|} - \frac{1}{\sqrt{3}}\right]$, where u_f is the acoustic velocity at the heater location, x_f is the heater location, and τ is the time-delay of heat release rate response to the oscillating acoustic velocity. The non-dimensional heater power K is given as:

$$K = \frac{2(\gamma-1)L_W}{\gamma M_a c_o \bar{p} S\sqrt{3}}\left(T_W - \bar{T}\right)\sqrt{\pi\lambda C_v u_o \bar{\rho}\frac{d_W}{2}} \tag{3}$$

where L_W and d_W are the heater wire length and diameter, respectively, c_o is the speed of sound in the tube, u_o, $\bar{p}$, $\bar{T}$ and $\bar{\rho}$ are the mean velocity, pressure, temperature and density, respectively, in the tube, S is the cross-sectional area of the tube, T_W is the wire temperature, and λ and C_v are the heat conductivity and specific heat constant at constant volume, respectively.

In order to solve the partial differential equations, the spatial dependence of acoustic velocity and pressure are projected on Galerkin base modes. These Galerkin base modes satisfy the acoustic boundary condition at the open-ends of the horizontal Rijke tubes. The decomposition is given as below.

$$u = \sum_{n=1}^{N} a_n \cos(n\pi x), \quad p = \sum_{n=1}^{N} \gamma M_a \frac{b_n}{n\pi}\sin(n\pi x) \tag{4}$$

On substituting this form of acoustic velocity and pressure in equations 1 and 2, they are converted to the following ODE as below.

$$\dot{a}_n = -b_n \tag{5}$$

$$\dot{b}_n = n^2\pi^2 a_n - \zeta_n b_n + \frac{2Kn\pi}{\gamma M_a}\left[\sqrt{\left|\frac{1}{3} + u_f(t-\tau)\right|} - \frac{1}{\sqrt{3}}\right]\sin\left(n\pi x_f\right) \tag{6}$$

where the damping term is given as $\zeta_n = c_1 n + c_2\sqrt{n}$. These equations are solved by using a fourth-order Runge-Kutta method in Matlab. The number of modes used is $N = 10$, which are found sufficient for convergence in Ref. [9]. The parameter values required in the system are set as: $c1 = 0.1$, $c2 = 0.06$, $\tau = 0.16$, $x_f = 0.25$, and the heater power parameter is varied. The heater power parameter is clubbed as $K_1 = \frac{2K}{\gamma M_a}$. For the values of the parameters used in this article, the Hopf bifurcation point is found to be near $K_1 = 0.742$ and the fold bifurcation point is found to be near $K_1 = 0.62$ The noise model used in this article is from Gopal & Sujith [9]. It is an additive Gaussian white noise, and because it is additive it does not affect the system bifurcation. This noise is added to each of the b_n mode, i.e. to the pressure modes in the system.

$$\dot{b}_n = n^2\pi^2 a_n - \zeta_n b_n + \frac{2Kn\pi}{\gamma M_a}\left[\sqrt{\left|\frac{1}{3} + u_f(t-\tau)\right|} - \frac{1}{\sqrt{3}}\right]\sin\left(n\pi x_f\right) + D\xi_n \tag{7}$$

where D is a measure of noise intensity, and ξ_n is of the size N-by-1. Each of the N components of ξ_n is Gaussian white noise with zero mean and a standard deviation proportional to the square root of the numerical time-step Δt. The noise response is solved using a stochastic Runge-Kutta method as explained in Appendix-A of Ref [9].

3 Results

In the results that follow, noise-induced features for the mentioned parameter values in the stable regime, for different levels of noise amplitude have been explored. For these results, statistical and spectral features of pressure fluctuations, p', at the heater location have been presented. The focus is on the qualitative variations in the system response with noise intensity at the chosen parameter values with different levels of noise.

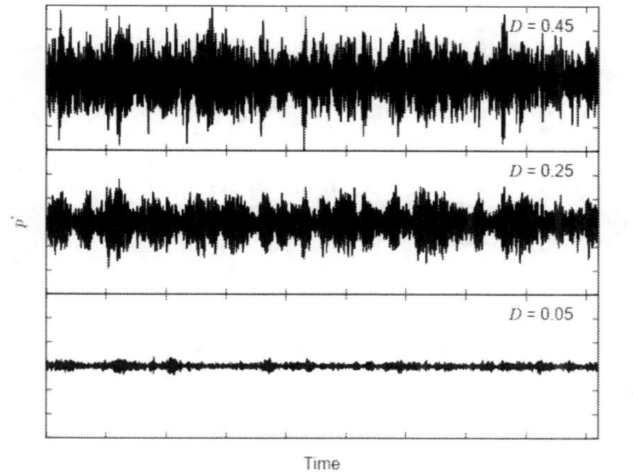

Figure 3: Pressure time trace for varying noise intensities, for $K_1 = 0.585$.

The pressure fluctuations are first characterized in terms of the standard deviation as shown in Fig. 2. The variation of this quantity with the proximity of the system to the bistable region (with change in the heater power parameter K_1) and the variation with noise level is shown. The corresponding reference noise levels obtained in the absence of the heater are simultaneously indicated (dashed lines). First, as expected, a higher noise level leads to a higher amplitude of pressure fluctuations. But the difference between the reference noise level and the corresponding response with the heater is observed to be larger for higher noise levels. Furthermore, this difference gradually increases as the system is brought closer to the bifurcation. For $D = 0.02$, this difference is not visible, but is clearly observed for $D = 0.10$ and $D = 0.30$.

In the time series, this response is observed as intermittent bursts of oscillatory behaviour. In Fig. 3, the pressure time trace for three levels of noise intensity, for $K_1 = 0.585$ are presented. The vertical axis range is kept identical for all the cases to aid comparison. Increasing noise level appears to increase the amplitude of induced fluctuations during individual bursts, and possibly increases the frequency of occurrence of such bursts.

The noise-induced coherent oscillations that appear as bursts contribute to the spectral content of the fluctuations as seen in Fig. 4. Again, the spectral content in fluctuations at three noise intensities has been plotted. The peak frequency in the noise-induced response is close to the frequency of the deterministic limit cycle oscillations (0.517) that appear at the Hopf bifurcation. It is clearly observable that noise affects the peak height. It also affects the peak width, as can be identified by comparing the two cases with similar peak heights; and causes slight changes in the peak frequency. The curve in red corresponds to the noise intensity that leads to an optimum induced coherence. This is discussed next on the basis of a spectral quality measure, β.

The system response to noise in the stable region is a competition between two mechanisms: the noise itself, and the noise-induced coherent oscillations. Both, the spectral width and the height are influenced by the two. The spectral quality or alternatively, the coherence factor [15, 17, 21], $\beta = H \times (\Delta f / f_p)^{-1}$, where, H, Δf and f_p are the peak height, peak width at half-maximum and peak frequency, respectively, is used to track the relative variation in the two at different noise levels and parameter values. The peak width and height were extracted from a Lorentzian fit to the spectral peak [16].

Figure. 5 presents the coherence factor for the investigated parameter values for different levels of noise. The curves indicate

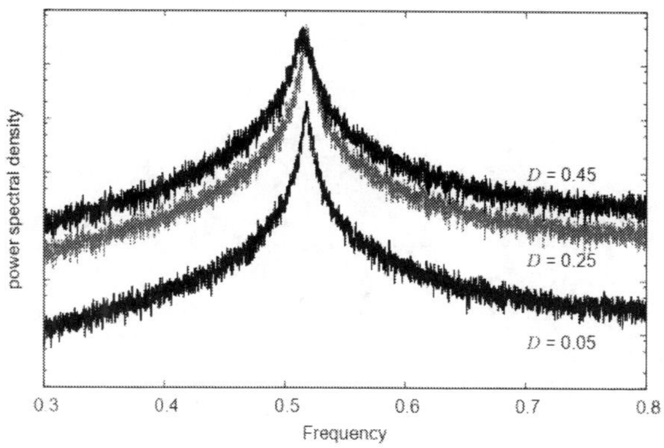

Figure 4: Spectra of pressure oscillations. The red curve corresponds to noise intensity close to the optimum level.

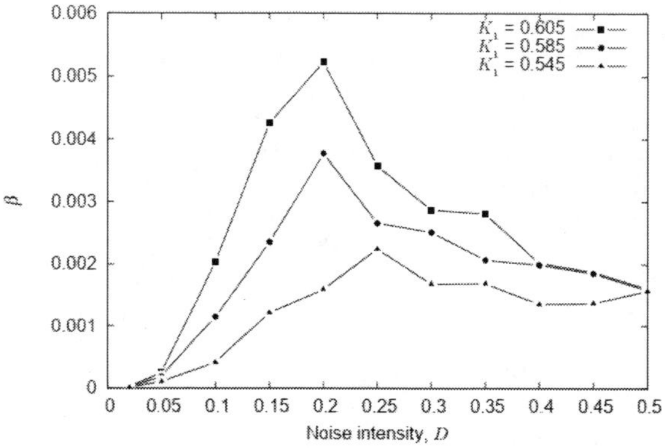

Figure 5: Coherence factor calculated for different noise levels and heater powers.

that the induced coherence first increases with increasing noise level. At a certain noise intensity, an optimum level of coherence is induced. Noise becomes more prominent than the induced coherence at higher intensities. Hence, there is an intermediate level of noise for each case, which corresponds to an optimum in the induced coherence in the system. Furthermore, for parameter values closer to the bifurcation, the induced coherence is higher. A shift in the intermediate level of noise which leads to the optimum response to higher noise intensities as the distance from the bifurcation increases, is also apparent in the plot.

These trends that have been observed in the model are identical to the trend previously reported in experiments on a laminar premixed flame combustor [1]. The fact that qualitative and quantitative characteristics of noise-induced response depend on the proximity of the system to the Hopf bifurcation, and that it is marked by oscillatory bursts—in experiments and in the numerical model—indicate that the response is associated with thermoacoustic coupling.

The spectra is related to the autocorrelation of the fluctuations, and similar trend as mentioned above is observed in the autocorrelation curves (Fig. 6). From discussions so far, it is expected that noise-induced oscillations result in a slower decay of the autocorrelation. However, as seen in the spectral characteristics, beyond a certain noise intensity the induced coherence decreases. Correspondingly, in the autocorrelation, the slowest decay is obtained at an intermediate level of noise.

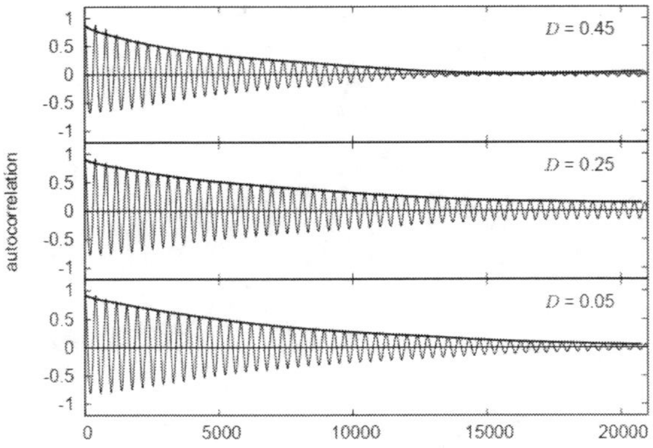

Figure 6: Autocorrelation from pressure time trace for different levels of noise. $K = 0.545$

Conclusions

The effects of noise on a thermoacoustic system in the subthreshold regime associated with subcritical Hopf bifurcation were investigated in a representative model. The addition of a stochastic term results in induced coherent oscillations that appear as intermittent bursts in the system. This effect is a function of the amplitude of noise level, and the proximity of the system to the approaching bifurcation: The induced coherence is higher for parameter values closer to the saddle-node point. The trend in system response with respect to the noise amplitude is interesting to note: An optimum coherence is induced at an intermediate level of noise. This result indicates that noise-induced dynamics in the thermoacoustic system corresponds to the phenomenon known as coherence resonance, which has been shown to govern stochastic dynamics of other nonlinear systems in the subthreshold regime in literature. The study was motivated by recent experimental findings and the results on the model reported here correspond qualitatively to the observations of the experiments.

The study confirms that the noise-induced phenomenon, coherence resonance is inherent to thermoacoustic coupling. A direct practical implication of coherence resonance in combustors is that for a constant level of noise, as the system approaches Hopf bifurcation, noise will induce coherent dynamics that will be observed as intermittent oscillatory behaviour. Such stochastic behaviour could be interpreted as a precursor to the instability.

References

[1] L. Kabiraj, R. Steinert, A. Saurabh, and C. O. Paschereit. Coherence resonance in a thermoacoustic system. *Phys. Rev. E*, 92(4), 2015.

[2] H. Gotoda, H.i Nikimoto, T. Miyano, and S. Tachibana. Dynamic properties of combustion instability in a lean premixed gas-turbine combustor. *Chaos*, 21(1):013124, 2011.

[3] L. Kabiraj, A. Saurabh, P. Wahi, and R. I. Sujith. Route to chaos for combustion instability in ducted laminar premixed flames. *Chaos*, 22:023129, 2012.

[4] L. Kabiraj and R. I. Sujith. Nonlinear self-excited thermoacoustic oscillations: intermittency and flame blowout. *J. Fluid Mech.*, 713:376–397, 2012.

[5] K. Kashinath, I. C. Waugh, and M. P. Juniper. Nonlinear self-excited thermoacoustic oscillations of a ducted premixed flame: bifurcations and routes to chaos. *J. Fluid Mech.*, 761:399–430, 2014.

[6] V. S. Burnley and F. E. C. Culick. Some dynamics of acoustic oscillations with nonlinear combustion and noise. *Int. J. Energetic Materials and Chemical Propulsion*, pages 998–1013, 1997.

[7] T. C. Lieuwen. Statistical characteristics of pressure oscillations in a premixed combustor. *J. Sound Vib.*, 260(1):3–17, 2003.

[8] I. Waugh and M. P. Juniper. Triggering in a thermoacoustic system with stochastic noise. *Intl. J. Spray Combus. Dyn.*, 3(3):225–242, 2011.

[9] E. A. Gopalakrishnan and R. I. Sujith. Effect of external noise on the hysteresis characteristics of a thermoacoustic system. *J. Fluid Mech.*, 776:334–353, 2015.

[10] F. E. C. Culick, L. Paparizos, J. Sterling, and V. Burnley. Combustion noise and combustion instabilities in propulsion systems. In *Proc. AGARD Conf. Combat Aircraft Noise*, Neuilly sur Seine, France, 1992.

[11] C. Seywert. *Combustion instabilities : issues in modeling and control*. PhD thesis, California Institute of Technology, 2 2001.

[12] N. Noiray and B. Schuermans. Deterministic quantities characterizing noise driven hopf bifurcations in gas turbine combustors. *Intl. J. Non-Linear Mech.*, 50:152–163, 2013.

[13] A. Zakharova, T. Vadivasova, V. Anishchenko, A. Koseska, and J. Kurths. Stochastic bifurcations and coherence-like resonance in a self-sustained bistable noisy oscillator. *Phys. Rev. E*, 81(1), 2010.

[14] A. S. Pikovsky and J. Kurths. Coherence resonance in a noise-driven excitable system. *Phys. Rev. Lett.*, 78(5):775–778, 1997.

[15] I. Z. Kiss, John L. Hudson, G. J. E. Santos, and P. Parmananda. Experiments on coherence resonance: Noisy precursors to hopf bifurcations. *Phys. Rev. E*, 67(3), 2003.

[16] O. V. Ushakov, H.-J. Wünsche, F. Henneberger, I. A. Khovanov, L. Schimansky-Geier, and M. A. Zaks. Coherence resonance near a hopf bifurcation. *Phys. Rev. Lett.*, 95(12), 2005.

[17] A. Zakharova, A. Feoktistov, T. Vadivasova, and E. Schöll. Coherence resonance and stochastic synchronization in a nonlinear circuit near a subcritical hopf bifurcation. *Eur. Phys. J. Spec. Top.*, 222(10):2481–2495, 2013.

[18] K. Balasubramanian and R. I. Sujith. Thermoacoustic instability in a rijke tube: Non-normality and nonlinearity. *Phys. Fluids*, 20(4):044103, 2008.

[19] Vikrant Gupta, Aditya Saurabh, Christian Oliver Paschereit, and Lipika Kabiraj. Numerical results on noise-induced dynamics in the subthreshold regime for thermoacoustic systems. *Journal of Sound and Vibration*, 390:55–66, 2017.

[20] M. A. Heckl. Non-linear acoustic effects in the rijke tube. *Acta Acustica united with Acustica*, 72(1), 1990.

[21] A. Neiman, P. I. Saparin, and L. Stone. Coherence resonance as noisy precursors of bifurcations in nonlinear dynamical systems. *Phys. Rev. E*, 56(1):270–273, 1997.

11th Asia-Pacific Conference on Combustion,
The University of Sydney, NSW Australia
10th-14th December 2017.

Flame Instability of Synthetic Liquefied Petroleum Gas and Natural Gas on Ceramic Porous Burner

Amornrat Kaewpradap[1] and Sumrerng Jugjai[1]

[1]Combustion and Engine Research Laboratory (CERL), Department of Mechanical Engineering, Faculty of Engineering, King Mongkut's University of Technology Thonburi

Pracha Uthit Road, Bang Mod, Thung Khru, Bangkok, 10140, Thailand

Abstract

This study focused on the flame instability between synthetic natural gas (NG) and liquefied petroleum gas (LPG) on ceramic porous burner (CPB). In order to investigate natural gas and LPG combustion flames used significantly for industrial part in Thailand, gas composition of CH_4/CO_2 (NG1: 76.0/24.0, NG2: 84.1/15.9) and C_3H_8/C_4H_{10} (LPG1: 70/30, LPG2: 60/40 and LPG3:40/60) were varied as synthetic gas. The cell size, blown-off flame limit, power spectral density and chaotic behavior were resulted as flame instability. When equivalence ratios increased cell sizes were smaller for both NG and LPG flames. At Φ=0.75, greater cell size, higher power spectral density and more complicated attractor of NG1 and NG2 flames compared to LPG1, LPG2 and LPG3 were obtained owing to decrease of instability intensity. In addition, flammability limit of NG2 and LPG3 flames on CPB were wider than of other NG and LPG flames. At Φ>0.75, the radiation effects also played important role for LPG1, LPG2 and LPG3 because higher heat of combustion enhanced burner temperature. It was concluded that LPG3 flame had more stable range compared to LPG1 and LPG2 flames on CPB. For more applications with NG combustion on CPB, NG2 flames could be applied on lean combustion with wider blown-off limit at Φ<0.70 due to increase of burning velocity.

1 Introduction

In the past, combustion was always derived from fossil fuel which emitted the greenhouse gases (GHG). In order to reduce emission gases, liquefied petroleum gas (LPG) and natural gas (NG) are taken place of heavy carbon fuel. Generally, LPG is applied widely in residential part whereas NG is employed significantly for industrial part [1]. Following the energy development plan form Ministry of Energy, Thailand between 2015-2036 [2]-[3], natural gas was mainly used in power plants, gas plant, transportation and others as 59%, 21%, 14% and 6%, respectively. There are many sources of natural gas in Thailand which they are grouped by location as eastern and western sources. NG is mainly composed of methane (CH_4) and carbon dioxide (CO_2) that have two important sources on western and western of Thailand categorized by CH_4/CO_2 (NG1: 76.0/24.0, NG2: 84.1/15.9) [4]. For LPG, the main compositions are propane (C_3H_8) and butane (C_4H_{10}) which have three compositions of C_3H_8/C_4H_{10} (LPG1: 70/30, LPG2: 60/40 and LPG3:40/60) in Thailand [5]. To decrease more emission such greenhouse gases, the lean combustion system was interesting to investigate for this study. However, the lean combustion is the burning of fuel under excess air ratio which brings about the unstable flame such as cellular flame due to diffusive-thermal instability [6]. Experiment of flame instability was studied by Markstein [7]. There are three basic phenomena of intrinsic instability as diffusive-thermal instability, hydrodynamic instability and body-force instability [8]. Cellular flame is induced by diffusive-thermal instability where the mass diffusion of deficient reactant is in excess of the heat diffusion owing to stabilizing and destabilizing influence at Lewis number Le>1.0 and Le<1.0 [9]. In this present, cordierite ceramic porous burner with low is applied widely for LPG combustion to increase thermal efficiency due to low thermal expansion and excellent thermal shock resistance. In the past, investigation of porosity effects in porous burners was studied and the average porosity leads to reduction of the emission [10]. Kadowaki S. studied time series analysis on the emission of light from methane-air lean premixed flames [11]. The previous study, flame instability on porous burner such as McKenna Flat burner was investigated by oxyhydrogen/air with carbon dioxide dilution [12]. The flame instability of CH_4/H_2/air non-adiabatic flat flames was also investigated. Consequently, it is important to study flame instability of varied compositions for synthetic LPG and NG with lean combustion for application on ceramic porous board.

2 Methodology

Flame instability of NG and LPG combustion on ceramic burner was followed by experimental setup as shown in Fig.1 and Fig.2. For synthetic NG, two main compositions as NG1 and NG2 used in this study, were illustrated in Table 1. This study was also experimented with three compositions of LPG as shown in Table 2. Following the experimental setup, cordierite ceramic burner with 60 mm in diameter was applied. NG1, NG2, LPG1, LPG2 and LPG3 mixtures flowed from CH_4, CO_2, C_3H_8, C_4H_{10} and air compressor through digital mass flow controller to burner at 45 L/min of total gas flow rate.

Corresponding author. Fax: +66-02-470-9111
E-mail address: amornrat.kae@kmutt.ac.th

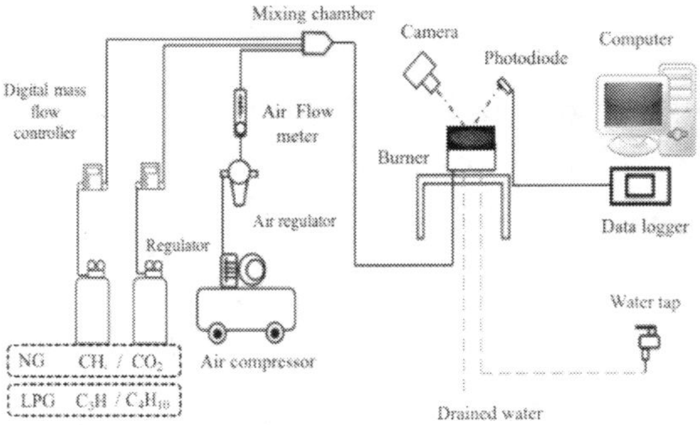

Figure 1: Experimental setup.

Figure 2: Cordierite ceramic porous burner.

Natural Gas Source	CH_4 [%]	CO_2 [%]
NG1	76.0	24.0
NG2	84.1	15.9

Table 1: The compositions of natural gas for NG1 and NG2 sources.

Liquefied Petroleum Gas Source	C_3H_8 [%]	C_4O_{10} [%]
LPG1	70.0	30.0
LPG2	60.0	40.0
LPG3	40.0	60.0

Table 2: The compositions of liquefied petroleum gas for LPG1, LPG2 and LPG3 sources.

2.1 Cell size analysis

The cell size of cellular flame is the distance between top of cellular flame front as shown in Fig. 3. The flame instability intensity was investigated such a cell size of cellular flame which can be described by Eq. (1).

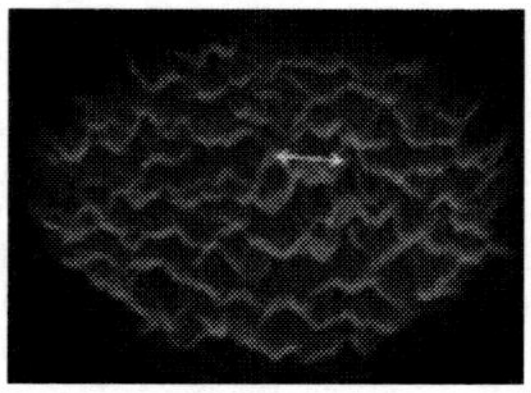

Figure 3: Cell size of cellular flame.

$$X_a = X_p(D_a/D_p) \quad (1)$$

X_a = Actual cell size

X_p = Measured cell size

D_a = Actual burner diameter

D_p = Measured burner diameter

2.2 Power spectral density analysis

The light emission detected from photo-diode was analyzed by Fast Fourier Transform (FFT) to obtain the shaft peak frequency (f_1). Moreover, the shaft peak frequency had the relationship with time delay (τ) as shown in Eq. (2).

$$\tau = \frac{1}{4f_1} \quad (2)$$

Use the SI system of units.

2.3 Chaotic behavior analysis

The chaotic behavior was analyzed by Takens' embedding theorem[13] with time delay obtained from power spectral density. The embedding, m=3 was set in vector $\vec{V}(t)$ followed by Eq. (3) to create attractor.

$$\vec{V}(t) = [y(t), y(t+\tau), y(t+2\tau), \cdots, y(t+(m-1)\tau)] \quad (3)$$

3 Results and discussions

3.1 Cell size

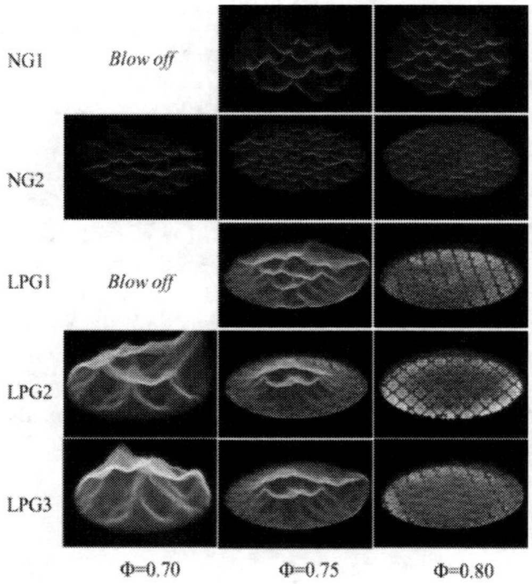

Figure 4: Photos of NG1, NG2, LG1, LPG2 and LPG3 flames at Φ=0.70, Φ=0.75 and Φ=0.80.

Figure 4 shows photos of NG1, NG2, LG1, LPG2 and LPG3 flames at Φ=0.70, Φ=0.75 and Φ=0.80. When the equivalence ratio increased from Φ=0.70-0.80, cell size of cellular flames were smaller due to weak of diffusive-thermal instability. For NG combustion, NG2 flames were existed in lean combustion between Φ=0.70-0.80 whereas NG1 flames were blown off at Φ<0.75. At Φ=0.75 cell size of NG2 flames were smaller compared to NG1 flames because higher CH_4 composition decreased strengthen of flame instability. For LPG combustion, smaller cell size and wider range of blown-off limit at Φ<0.7 were observed for LPG 3 flames due to higher burning velocity.

The results were also found that the radiation effects played important role at Φ>0.75 for LPG1, LPG2 and LPG3 because higher heat of combustion reduced flame instability due to increase of burner temperature.

Φ	Cell size [mm]				
[-]	NG1	NG2	LPG1	LPG2	LPG3
0.70	Blow off	10.60	Blow off	Blow off	16.62
0.75	11.94	8.15	7.65	7.57	4.64
0.80	7.74	7.10	Radiation	Radiation	Radiation

Table 3: The cell sizes of NG1, NG2, LG1, LPG2 and LPG3 flames for Φ=0.70, Φ=0.75 and Φ=0.80.

3.2 Power spectral density

For NG combustion, Fig. 5 shows higher power spectral density of NG2 flame when the equivalence ratio increases from Φ=0.70-0.80. At Φ=0.75, the power spectral density of LPG3 flame was higher compared to LPG1 and LPG2 flames as shown in Fig 6. It is because higher sharp peak frequency induced by increase of CH_4 and C_4H_{10} compositions for NG and LPG flames, respectively.

3.3 Chaotic behavior

As the relationship of sharp peak frequency, time delay was calculated from Eq. (2) and reconstructed attractor was created by Eq. (3). For NG flame, smaller reconstructed attractor was observed when the equivalence ratio increased from Φ=0.70-0.80 as illustrated in Fig.7. At Φ=0.75, the attractor of LPG3 flames were more complicated than of LPG1 and LPG2 flames as shown in Fig 8. It was resulted that lower time delay brought about smaller and smoother reconstructed attractor owing to decrease of instability intensity.

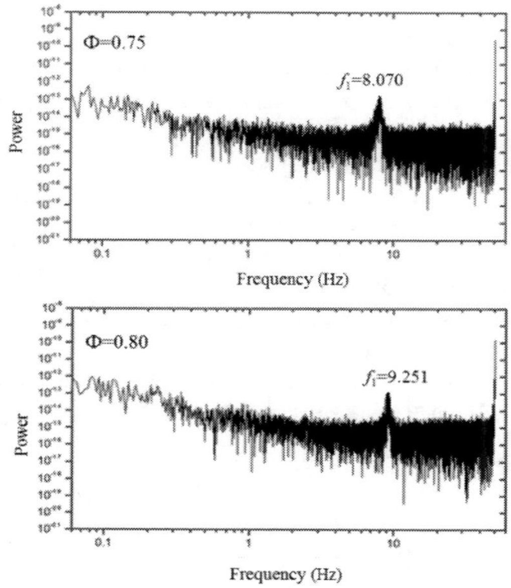

Figure 5: Power spectral density of NG1 flames at Φ=0.75 and Φ=0.80.

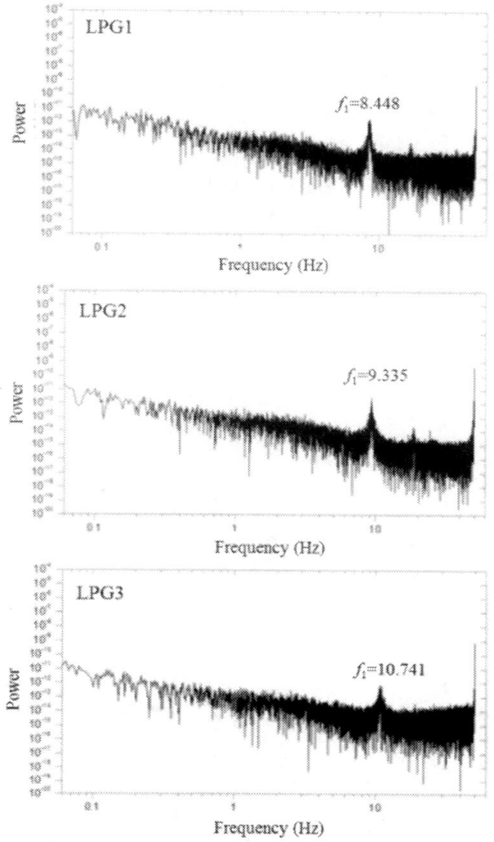

Figure 6: Power spectral density of LPG1, LPG2 and LPG3 flames at Φ=0.75.

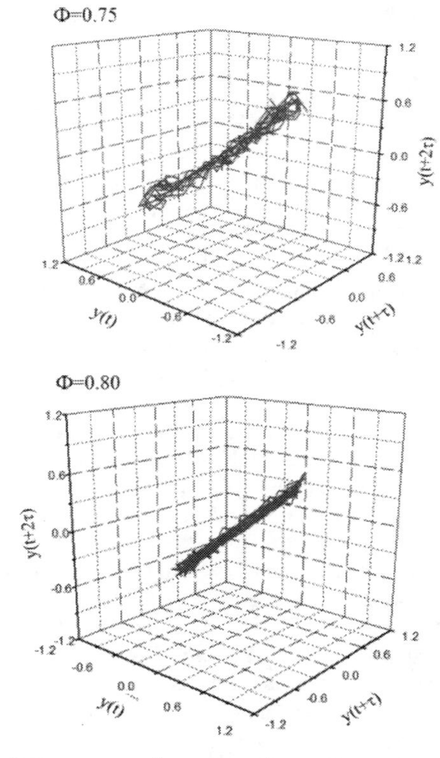

Figure 7: Reconstructed attractor of NG1 flames at Φ=0.75 and Φ=0.80.

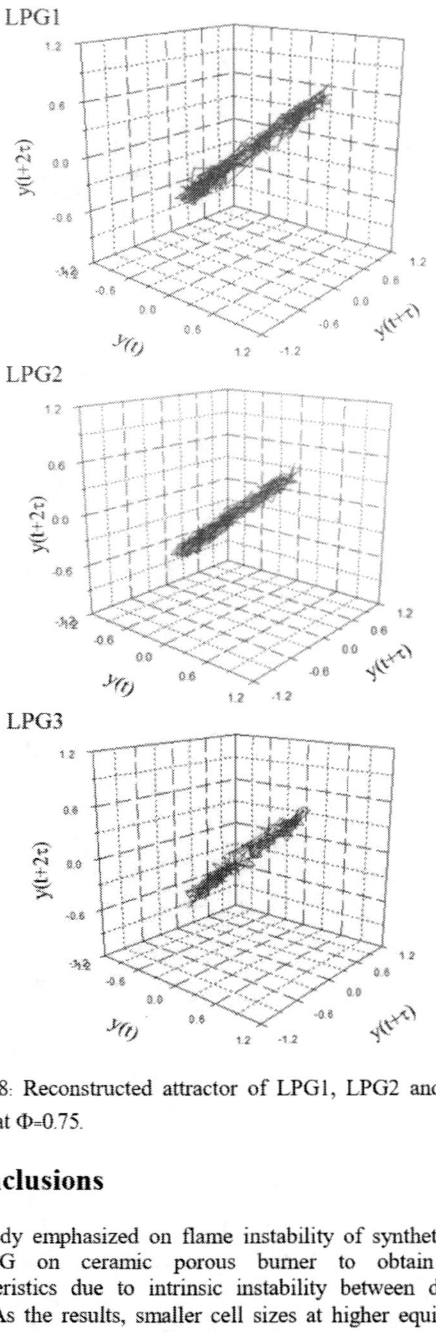

Figure 8: Reconstructed attractor of LPG1, LPG2 and LPG3 flames at Φ=0.75.

4 Conclusions

The study emphasized on flame instability of synthetic LPG and NG on ceramic porous burner to obtain flame characteristics due to intrinsic instability between different gases. As the results, smaller cell sizes at higher equivalence ratios were observed for both NG and LPG gas flames due to diffusive-thermal instability. At the same equivalence ratio, greater cell size, higher power spectral density and more complicated attractor from NG flames were compared to LPG flames due to decrease of instability intensity. As the flammability limit, blown-off flame range of NG2 and LPG3 were wider for NG and LPG flames, respectively. It was also found that the radiation effects played important role for merely LPG flames because higher heat of combustion enhanced burner temperature. Consequently, it was concluded that LPG3 flame had more stable range compared to LPG1 and LPG2 flames on ceramic porous burner. For wider application on NG lean combustion, NG2 flames were be applicable on CPB, with wider blown-off limit at Φ<0.70 due to enhanced burning velocity.

5 Acknowledgment

The authors would like to thank to Office of the Higher Education Commission, Thailand Research Fund (TRF: IRG5780005), for financial support and Department of Mechanical Engineering, Faculty of Engineering, King Mongkut's University of Technology Thonburi (KMUTT) for facility supports.

References

[1] Department of organization activity, (2012), *Trend of global energy and overview 2040*, ESSO (Thailand) Public Company, pp. 15-51.

[2] Ministry of Energy, (2013). *Production and import of natural gas in 2013*, Thailand's Energy Situation in 2013.

[3] Energy Policy and Planning Office, (2013), *Energy Statistics of Thailand.*

[4] A. Kaewpradap and S. Jugjai, (2016), Effects of CH_4 and CO_2 on intrinsic instability of synthetic Thai natural gas flame, *The 7th TSME International Conference on Mechanical Engineering, Thailand.*

[5]World LPG Association, Exceptional energy, [Online], Available:\http://www.wlpga.org/about-lpg/why-use-lpg.

[6] A.A. Konnov, (2005), Model of Cellular Instability of Flames of Ternary Mixtures, *Combustion, Explosion and Shock Waves*, vol.41, No.5, pp.496-503.

[7] G.H. Markstein, (1952), Instability phenomena in combustion waves, *Proceedings of the Combustion Institute*, vol. 4, pp.44-59.

[8] Sivashinsky GI, (1983), Instabilities, pattern formation, and turbulence in flames, *Annual Review of Fluid Mechanics*, vol.15, pp.179-199.

[9] Sivashinsky G., (1977), Diffusive-thermal theory of cellular flames, *Combustion and Science Technology*, vol. 15, pp.137-145.

[10]Fatemeh Bahadori, Khadijeh Mirza, Alireza Behroozsarand and Sima Rezvantalab, (2015), Investigation of porosity effects in the emission of pollutants in porous burners, *Journal of Chemical Technology and Metallurgy*, vol. 50, pp.171-175.

[11] Kadowaki S. and Ohashi H., (2012), Shape and fluctuation of cellular premixed flames: lean combustion system of CH4/O2/CO2 mixtures, *Journal of Visualization*, vol.16, Issue 4, pp.351-356.

[12]A. Kaewpradap and S. Kadowaki, 2016, "Instability Influenced by CO2 and Equivalence Ratio in Oxyhydrogen Flames on Flat Burner", Combustion Science and Technology, vol. 189, Issue 3, pp.438-452.

[13]F.Takens, A 1981, Detecting strange attractors in turbulence, In D.A.Rand and L.-S. Young, Dynamical Systems and Turbulence, Lecture Notes in Mathematics, vol. 898, Springer-Verlag, pp. 366-381.

11th Asia-Pacific Conference on Combustion,
The University of Sydney, NSW Australia
10th -14th December 2017.

Autoignition behavior of Fuel Rich Natural Gas/ Air Combustion Product Jet Discharged into Quiescent Air

Saeedreza Zadsirjan[1], Sadegh Tabejamaat[1], Masoud E. Attarzadeh[1]

[1]Amirkabir University of Technology,424 Hafez Ave, Tehran, Iran, 15875-4413. +98 (21) 64540

Abstract

A turbulent non-premixed autoignition phenomena for a jet of high temperature fuel rich combustion product of natural gas/air discharging to the air has been investigated. The Fuel rich mixture burns in a pre-combustor and discharged through a tube to the quiescent air. The autoignition behavior of the jet is observed and categorized into 1-No ignition 2- Anchored diffusion flame 3- Lifted Diffusion Flame 4- Unstable Diffusion Flame 5- Random autoignition kernel. The autoignition kernel formation locus is investigated in different jet Reynolds number and temperature.

1 Introduction

The ignition triggered by a discharging jet is remarkable subject in industry regarding aspects of safety, emission reduction, or performances. The jet and its environment temperature, chemical composition, velocity as well as jet diameter are governing parameters alongside the turbulent transient or stationary flow field which make it much complicated case for scientific interpretation [1-11].

There is vast research on autoignition phenomenon as result of transient jet or sudden injection as well as stationary jet ignition. The effect of a jet temperature and diameter on the ignition of vaporized hydrocarbon fuels such as n-octane, n-decane, jp6, which were mixed with air, were studied by Cato and Kutcha [12]. This study showed that increase in a jet diameter could decrease a minimum required jet temperature to ignite a mixture. Regarding fireproof enclosure, Phillip[13] studied a minimum diameter of a hot inert-gas jet which could ignite a fuel-air mixture.

Sadanandan el. al. [14] investigated a jet of hot exhaust gas impinging through a narrow nozzle into a quiescent hydrogen/air mixture. High-speed laser-induced fluorescence (LIF) image sequences of the hydroxyl radical (OH) and laser Schlieren methods are used to measure the spatial and temporal evolution of the ignition process. This study indicated that a probability of mixture ignition increases with lower jet velocity a well as higher jet diameter. Also, the turbulent and chemistry interaction is the dominant phenomenon in autoignition and kernel ignition distance for mixture ignited by hot jet.

The autoignition of acetylene, released from a finite-sized circular nozzle into a turbulent coflow of hot air confined in a pipe, was the subject of Markides and Mastorakos research [15]. They reported the turbulent intensity reduction and enhancement in characteristic length scale, decreases the autoignition length. But, higher jet velocity results in higher autoignition length. The autoignition length is a distance of kernel formation from the jet nozzle.

Markides and Mastorakos [16] studied non-premixed autoignition phenomenon, by a jet of hydrogen/methane mixture injected into a hot air turbulent co-flow. Effects of temperature, velocity and composition of both coflow and jet were investigated. Four different regimes were reported in this study: no ignition, random spots, flashback, and lifted flame.

In this paper, an autoignition behavior of a jet of high temperature fuel rich combustion product which is discharged through a nozzle into the quiescent air is investigated. A pre-combustor is used to produce the combustible high-temperature stationary jet, by burning the fuel rich mixture of natural gas and air. The discharging jet may re-ignite and forms a reactive zone downstream of the nozzle exit. This observation is classified to five category based on the extinction (no flame), ignition or flame formation. The fifth category, Random autoignition kernel, is the next subject of this research. The autoignition behavior is investigated with respect to the jet Reynolds number, pre-combustor equivalence ratio, and jet exit temperature using three different jet.

2 Experimental apparatus and operation

The experimental setup includes air electric compressor, natural gas feed line, volumetric flow measurement tools, pre-combustor, K-type thermocouple, and Nikon V1 camera.

Figure 1 illustrates details of pre-combustor chamber. An electrical air compressor and natural gas feed line supply the mixture for pre-combustor (1) with room temperature air and natural gas. The air compressor stores compressed air up to 10 bar in 300L storage and delivers up to 250 standard liter per minute (SLM) air and the natural gas feed line delivers up to 40 SLM of natural gas. Four rotameters are used to measure volumetric flow rates of air and natural gas feed lines. Because of the pressure difference at rotameters inlet, and room pressure, the correction factor is applied to the flow rate measurement. The rotameters of the pre-combustor feed line measure the range of 0-400 with an accuracy of 5 SLM and the range of 0-50 with the accuracy of 1 SLM for air and natural gas, respectively. The rotameters of the torch igniter feed line measure the range of 0-10 with an accuracy of 0.2 SLM. A manifold feeds the air and fuel stream (2,3) into a swirl burner (17) in pre-combustor to burn fuel rich mixture in the pre-combustor. The pre-combustor is made of the 47 cm height and 20.32 cm (8 inch) diameter steel tube. The internal wall is covered by two layer of dense ceramic fiber as a thermal insulator (8) so that the internal diameter of the pre-combustor is 15 cm. an igniter (6) is mounted to chamber body to ignite the mixture in pre-combustion chamber. The

Corresponding author. Fax: +98 (21) 64540
E-mail address: s_zadsirjan@aut.ac.ir

igniter is made up of air and natural gas inlets, spark generator, spark igniter, igniter chamber and mounting. The high temperature combustion product generated by igniter, discharge to pre-combustion chamber helps to ignite the burner (17). A ignition is verified by visor (7) and measuring the temperature of pre-combustion discharging jet by k-type thermocouple (11) located at the center of nozzle exit plane. The pre-combustion product is discharged though nozzle with 15 cm length.

Due to a large thermal inertia of the chamber, it takes about an hour to reach the quasi/steady state jet exit temperature. The measurement takes place when the discharging jet is in the quasi/steady state, e.g., constant temperature, mass flow rate and Reynolds number. The operational points were repeated three times to assess the repeatability of test setup and results. The pre-combustor high thermal inertia helps to achieve a range of jet exit temperature at quasi steady state condition.

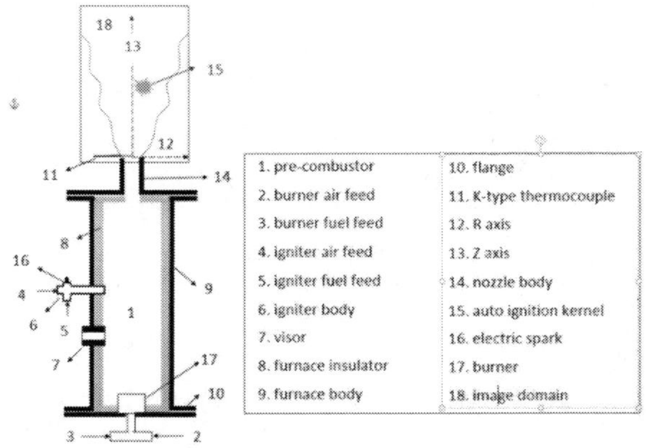

Figure 1: schematic the pre-combustor used to produce hot fuel rich jet

As the inlet condition of pre-combustion chamber is known as well as exit jet temperature, therefore the state of high temperature discharging jet is determined by thermochemical calculation. The residence time of the chamber is in the O(-1) second and the combustion chemical time scale is in the O(-3) second, thus it is reasonable to assume that the jet mixture is in thermochemical equilibrium. NASA CEA online version was used for thermochemical calculation.

The jet Reynolds number was defined as Re=(ρVD)/(μ), where ρ is density, V is mean jet discharging velocity, D is jet diameter equal to nozzle internal diameter and μ is dynamic viscosity. NASA CEA provides density and dynamic viscosity of mixture. The mean jet discharging velocity is calculated based on the conservation of mass as the inlet mass flow rates are known.

3 Result and discussion

The hot combustible jet is produced by the pre-combustor burns fuel rich mixture of the natural gas and air. The jet Reynolds number varies from 2200 to 7700 that is discharged to the quiescent room temperature air. The pre-combustor equivalence ratio varies from less than unity up to 2.2. Afterward for the sake of simplicity, equivalence ratio is used instead of the pre-combustor equivalence ratio.

3.1 Flame behavior

The methane/ air mixture burns within the pre-combustor at total equivalence ratio of 0.97 to 1.77 and discharge through nozzle to quiescent air with the jet Reynolds number varies between 5500 to 6000. The photography of the downstream of nozzle in gray scale visual range is used to record the behavior of the jet. This observation can be qualitatively categorized into five types of behavior

1-"No ignition" no flame or ignition kernel is observed in this type, Figure 2-A. It occurs in two conditions. First, when the equivalence ratio is unity or less, Second, when the combustion product is highly fuel rich which leads to low temperature jet exit temperature.

2-"Anchored diffusion flame" is nearly attached flame which is appeared when the equivalence ratio is slightly greater than unity, Figure 2-B,C,D. the lift-off height is approximately less than a jet diameter.

3-"Lifted diffusion flame" is lifted diffusion flame. The flame is stationary, no flame extinction or blow out occurs and the lift-off height is larger than jet diameter, Figure 2-E.

4-"Unstable diffusion flame" the extinction and re-ignition happen repetitively and lift-off height is obviously larger than Lifted Diffusion Flame, Figure 2-F.

5-"Random autoignition kernel" is autoignition kernel which randomly takes place, Figure 2-G. The autoignition kernel is convecting reactive zone which is not form the turbulent flame. This kernel is convected downstream and extinguished, finally. Formation of the autoignition accompanies with a loud noise as reported in [16].

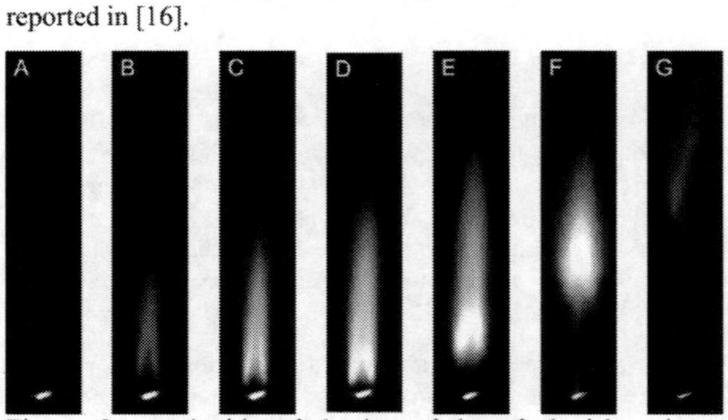

Figure 2: autoignition behavior of hot fuel rich exhaust combustion product jet to quiescent air for five different fuel to air ratio in the pre-combustor, Re=5500-6000. φ_A=0.97 , φ_B=1.02,φ_C=1.13,φ_D=1.34,φ_E=1.61,φ_F=1.72, φ_G=1.77.

3.2 Autoignition kernel

In this section, the report focuses on the fifth category using three different jet to study the effect of jet Reynolds number, equivalence ratio of pre-combustor and jet exit temperature on "Random autoignition kernel". Specification of jets are presented Table 1.

Test setup	Re	φ	nozzle dia. (mm)	min. Temp. K	max. Temp. K
C2	5300	1.35	13	1080	1141
C3	6400	1.46	13	1150	1323
C4	10400	1.66	13	1280	1344

Table 1 Test set up for random autoignition kernel

The high-speed luminous film is used to capture the random

autoignition kernel in the jet downstream flow field. Each film includes five seconds of the jet flow at constant jet Reynolds number and jet exit temperature. The in-house code is developed to post process the films to capture auto-ignition kernels. A filter is implemented to omit the radiation of soot in the jet flow. Outputs of the film post processing are height, lifetime and number of autoignition kernels. Figure 4 is illustrate one of the captured autoignition kernels which is convected downstream of nozzle and quenched.

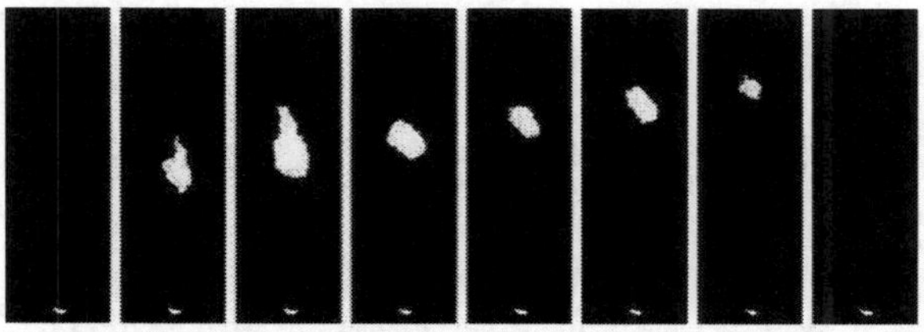

Figure 4: Random autoignition kernel formation, evolution, and extinction captured on high-speed film with 2.5 ms time resolution. The bright zone in the bottom is thermocouple located at the nozzle exit as a reference point of jet exit.

Mastorakos [17] introduce the most reactive mixture fraction to describe the most suitable ignition location in diffusion reactive configuration. If fuel temperature is more than oxidizer temperature, the ignition start from the rich side [18]. Thus, the local temperature and mixture fraction are playing the major role to determine the most reactive mixture fraction. In the present case, the turbulent jet of fuel rich combustion product is mixed with air, so the jet Reynolds number, temperature, and composition are key parameters for autoignition kernel locus

3.3 Autoignition locus

•Jet Reynolds number effects

The reputation of autoignition kernel formation in axial and radial projection can be used as criteria to determine the proper zone for autoignition kernel formation. To find this zone, the probability distribution (P.D.) of autoignition is defined. The mathematical definition of P.D. in axial projection is

$$P.D.(h_0) = \frac{\int_{h_0}^{h_0+\Delta h} H(h)dh}{\int_0^{\infty} H(h)dh} \tag{1}$$

Function *H(h)* represents the number of autoignition kernels that takes place in distance *h* from the jet exit. Δh is equal to one exit jet diameter D. So the probability distribution of random autoignition kernel is defined as the number of autoignition kernels that take place in radial or axial projection to the total number of autoignition. The maximum of P.D. may interpret as the most suitable zone for autoignition in axial or radial direction. Figure 5 illustrates the probability distribution (P.D.) of random autoignition kernel in axial (left) and radial (right) projection. Comparison of maximum P.D. in different jet Reynolds number, indicate that higher Re shifts the proper height further from jet exit. Also, higher jet Re number decreases the maximum P.D. of axial projection, corresponding to the longer proper axial distance that ignition could take place. For instance, in C3 with Re=6400, the ignition approximate height start from Z/D = 5 up to Z/D = 10, but in C4 with Re=10400 the ignition approximate height start from Z/D = 6 up to Z/D = 14. Besides, decrease in maximum P.D. of radial projection in higher Re number is equivalent to more off-axis ignition kernel formation (radial projection).

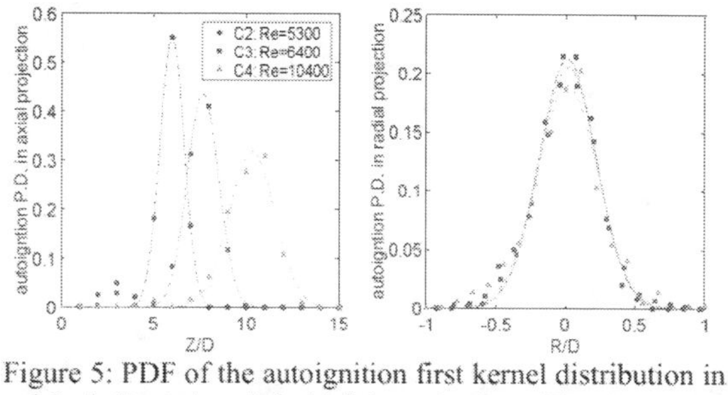

Figure 5: PDF of the autoignition first kernel distribution in the axial (left) and radial (right) projection. The lines are the Gaussian curve fit of probability distribution data.

•Jet temperature effect

In the vast study of the autoignition, the temperature is introduced as the key parameter to define the autoignition. Based on the Arrhenius equation, temperature directly affects the reaction rate of a combustible mixture. Sequentially, temperature growth, border the flammability limit of mixture in both lean and rich sides [19].

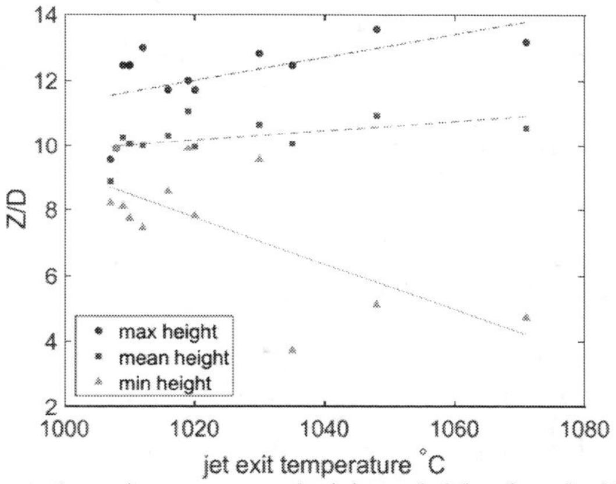

Figure 6: maximum, mean and minimum height of autoignition kernel formation with jet temperature variation in case C4.

Therefore by increasing the jet exit temperature, the rich flammability limit develops to upstream (richer zone) and the lean flammability limit develops to the downstream (leaner limit). This behavior of flammability limit with respect to the jet exit temperature is indicated in Figure 6 for case C4. The autoignition kernel takes place between minimum (rich) and maximum (lean) height of kernel formation. As it is shown, rising the jet exit temperature, broadening the flammability limit by increasing maximum and decreasing minimum location of ignition kernel formation. But the mean height of the autoignition kernel formation is almost constant. The mean height of ignition kernel in three different jets Re number 5300, 6400 and 10400 and temperature ranges is depicted Figure 7. Each point indicates the mean height of all ignition kernel takes place at the constant temperature. This figure shows the mean autoignition height is mostly effected by the jet Re number rather than jet temperature. Higher jet Re number leads to higher ignition kernel formation height, but the variation of jet exit temperature in each case almost makes no noticeable change in ignition kernel height.

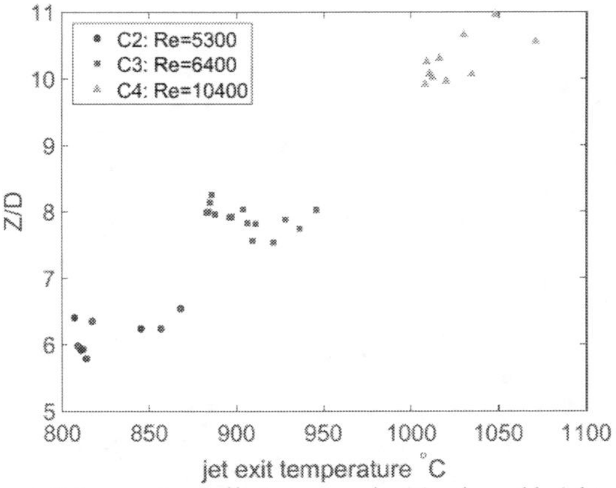

Figure 7: temperature effect on mean ignition kernel height.

4 Conclusions

The fuel rich combustion product of the natural gas and air burns in a pre-combustor and discharged to the quiescent air. The autoignition behavior of the combustible jet is qualitatively categorized into five types based on the jet Reynolds number, temperature, and pre-combustor equivalence ratio, 1-No ignition 2- Anchored diffusion flame 3- Lifted Diffusion Flame 4- Unstable Diffusion Flame 5- Random autoignition kernel. The Random autoignition kernel is the main intention of this report. Brief results of the random autoignition kernel are:

•The higher jet Re number increases the ignition kernel height and broads the axial zone that ignition may occur. Also, it causes more off-axis ignition kernel formation.

•The jet temperature rise up, develops the possible zone for autoignition kernel formation to the downstream (richer zone) as well as upstream (leaner zone). But, the mean height of ignition kernel formation is strongly effected by jet Reynolds number rather than jet exit temperature.

References

[1] M. de Joannon, P. Sabia, G. Sorrentino, and A. Cavaliere, "Numerical study of mild combustion in hot diluted diffusion ignition (HDDI) regime," Proc. Combust. Inst., vol. 32, no. 2, pp. 3147–3154, 2009.

[2] P. Sabia, M. de Joannon, A. Picarelli, and R. Ragucci, "Methane auto-ignition delay times and oxidation regimes in MILD combustion at atmospheric pressure," Combust. Flame, vol. 160, no. 1, pp. 47–55, 2013.

[3] G. Sorrentino, D. Scarpa, and A. Cavaliere, "Transient inception of MILD combustion in hot diluted diffusion ignition (HDDI) regime: A numerical study," Proc. Combust. Inst., vol. 34, no. 2, pp. 3239–3247, 2013.

[4]G. Gentz, B. Thelen, M. Gholamisheeri, P. Litke, A. Brown, J. Hoke, and E. Toulson, "A study of the in fl uence of ori fi ce diameter on a turbulent jet ignition system through combustion visualization and performance characterization in a rapid compression machine," Appl. Therm. Eng., vol. 81, pp. 399–411, 2015.

[5]G. Gentz, M. Gholamisheeri, and E. Toulson, "A study of a turbulent jet ignition system fueled with iso-octane : Pressure trace analysis and combustion visualization," Appl. Energy, vol. 189, pp. 385–394, 2017.

[6] Maxson, J., Hensinger, D., Hom, K., and Oppenheim, "Performance of multiple stream pulsed jet combustion systems," SAE Technical Paper, 1991.

[7]J. A. Maxson and A. K. Oppenheim, "pulsed jet combustion-key to a refinement of the stratified charge concept," pp. 1041–1046, 1990.

[8] and A. K. O. Hensinger, D. M., J. A. Maxson, K. Hom, "Jet plume injection and combustion," SAE Technical Paper, 1992.

[9]H. K. Murase E, "Control of the Start of HCCI Combustion by Pulsed Flame Jet," SAE Technical Paper 2002-01-2867; 2002.

[10]H. K. Murase E, "Enhancement of Combustion by Injection of Radicals.," SAE Technical Paper 2001-01-0194; 2000.

[11]O. A. Murase E, Ono S, Hanada K, "Pulsed Combustion Jet Ignition in Lean Mixtures," SAE Technical Paper 942048; 1994.

[12]J. M. K. Ralph J. Cato, "HOT GAS IGNITION TEMPERATURES OF HYDROCARBON FUEL VAPOR-AIR MIXTURES," 1965.

[13]H. Phillips, "Ignition in a transient turbulent jet of hot inert gas," Combust. Flame, vol. 19, no. 2, pp. 187–195, 1972.

[14]R. Sadanandan, D. Markus, R. Schießl, U. Maas, J. Olofsson, H. Seyfried, M. Richter, and M. Alden, "Detailed investigation of ignition by hot gas jets," Proc. Combust. Inst., vol. 31 I, pp. 719–726, 2007.

[15]C. N. Markides and E. Mastorakos, "Experimental investigation of the effects of turbulence and mixing on autoignition chemistry," Flow, Turbul. Combust., vol. 86, no. 3–4, pp. 585–608, 2011.

[16]C. N. Markides and E. Mastorakos, "An experimental study of hydrogen autoignition in a turbulent co-flow of heated air," Proc. Combust. Inst., vol. 30, no. 1, pp. 883–890, 2005.

[17]E. Mastorakos, "Ignition of turbulent non-premixed flames," Prog. Energy Combust. Sci., vol. 35, no. 1, pp. 57–97, 2009.

[18]M. de Joannon, G. Sorrentino, and A. Cavaliere, "MILD combustion in diffusion-controlled regimes of Hot Diluted Fuel," Combust. Flame, vol. 159, no. 5, pp. 1832–1839, 2012.

[19]I. Wierzba and V. Kilchyk, "Flammability limits of hydrogen-carbon monoxide mixtures at moderately elevated temperatures," Int. J. Hydrogen Energy, vol. 26, no. 6, pp. 639–643, 2001.

11th Asia-Pacific Conference on Combustion,
The University of Sydney, NSW Australia
10th-14th December 2017.

Detonation in Ammonia-based Mixtures

R. Mével[1,2,*], J. Melguizo-Gavilanes[3] and N. Chaumeix[4]
[1]Center for Combustion Energy, [2]Department of Automotive Engineering, Tsinghua University
30 Shuang Qing road, 100084 Beijing, China
[3]California Institute of Technology, Pasadena, USA
1200 E. California Blvd, MC 205-45, Pasadena, California 91125, USA
[4]ICARE-CNRS
1C avenue de la recherche scientifique 45071 Orléans Cedex 2, France

Abstract
Despite numerous potential industrial application of ammonia (NH_3), the risk of detonation in NH_3-based mixtures has not been characterized in details. In the present study, the velocity and cell size of detonation propagating in NH_3-O_2 and NH_3-N_2O mixtures have been measured. Mixtures containing nitrous oxide as an oxidant demonstrated higher sensitivity to detonation than mixtures containing oxygen. The semi-empirical correlation from Ng was employed to predict the detonation cell size. It provided reasonable predictions with maximum deviation below a factor of 2. The detonation chemical structure was characterized using steady one-dimensional ZND simulations. For mixtures containing nitrous oxide, the production of OH radical is controlled by the reaction $N_2O+H=N_2+OH$ whereas for mixtures containing O_2, the controlling reaction is $H+O_2=O+OH$. Consistent with the experimental soot foils, the calculated stability parameters enable classifying the studied ammonia-based mixtures as highly unstable.

1 Introduction

Ammonia is a very important industrial chemical used in its pure form (e.g. as fertilizer and alkaline cleansers), and as a feedstock for a wide variety of chemicals (e.g. ammonium sulfate/nitrate/hydrogen phosphate and urea). Also, its thermodynamic properties, make ammonia attractive in the framework of industrial refrigeration [1]. The manufacture of fertilizers is by far the most important use of ammonia at present [2]. However, the possibility of using it as a fuel has recently sparked new interest since its combustion does not produce CO_2 or unburnt hydrocarbons. The combustion characteristics of ammonia based mixtures have been studied using a number of experimental techniques. Available data include ignition delay-times and species profiles obtained in shock-tubes [3, 4], burning speeds and flame structures [5, 6], among others. Although some detonation cell size measurements have been performed [7, 8], the number of available data is very limited compared to more common fuels such as hydrogen and hydrocarbons. The purpose of the present study is to obtain detonation cell sizes for undiluted NH_3-O_2 and NH_3-N_2O mixtures over a wide range of equivalence ratios and initial pressures.

*Corresponding author: mevel@mail.tsinghua.edu.cn

2 Materials and Methods

2.1 Materials

Mixtures were prepared in a 50 L stainless steel cylinder from NH_3, O_2 and N_2O cylinders using the partial pressure method. Once the appropriate composition was obtained, the gases were allowed to mix through diffusion for at least four hours. The detonation tube was a 4.6 m long stainless steel tube with 78 mm inner diameter. Prior to each experiment, the tube was vacuumed to a pressure below 15 Pa. A one meter long Schelkin spiral with a blockage ratio around 0.5 was attached at one end of the tube. A flame was spark-ignited in the mixture and the spiral enables the transition to detonation of the flame. In order to measure the detonation velocity, the other end of the tube was equipped with 7 shock detectors located 15 or 30 cm from each other. The uncertainty on the detonation velocity was on the order of 1 %. For each experiment, a soot foil was placed at the opposite end of the tube to register the cellular pattern.

2.2 Methods

The steady one-dimensional ZND model as described by Shepherd [9] was used to study the structure of detonation propagation in ammonia-based mixtures. The detailed reaction model of Mével et al. [10, 11] was employed to describe the chemical kinetics of the mixtures. While this reaction model was mostly developed for hydrogen-oxygen and hydrogen-nitrous oxide combustion, some data for mixtures containing NH_3 were also used for its validation.

3 Results and Discussion

3.1 Experimental results

As shown in Figure 1 and Figure 2, detonation velocity was measured at T_1=295 K, for NH_3-O_2 and NH_3-N_2O mixtures in the ranges: Φ=0.3-1.5 and P_1=40-100 kPa.

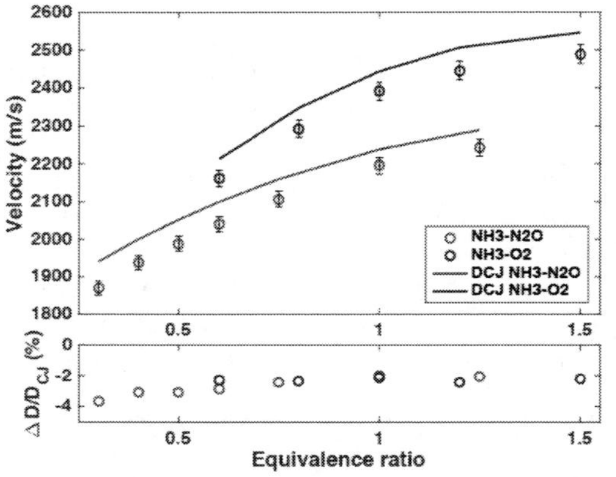

Figure 1: Experimental and CJ detonation velocity and velocity deficit as a function of Φ for NH_3-O_2 and NH_3-N_2O mixtures. Φ=0.3-1.5; P_1=71 kPa; T_1=295 K.

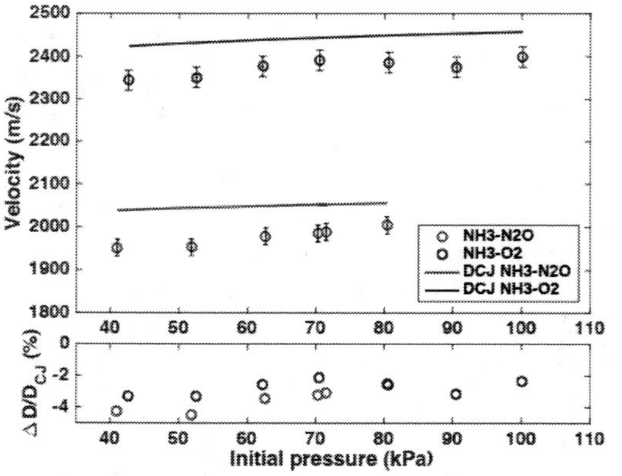

Figure 2: Experimental and CJ detonation velocity and velocity deficit as a function of P_1 for NH_3-O_2 and NH_3-N_2O mixtures. Φ=0.5 and 1 for mixtures with O_2 and N_2O, respectively; P_1=40-100 kPa; T_1=295 K.

Detonation propagating in NH_3-O_2 mixtures demonstrate much higher velocity than detonation propagating in NH_3-N_2O mixtures. The experimental values agree with the Chapman-Jouguet velocities within a few percent. The mean velocity deficit observed was around 3%. For leaner mixtures and at lower initial pressures, either accelerated deflagration or overdriven detonations were observed indicating either that the detonation tube used was too short or that the mixtures were not sensitive enough to obtain self-sustained detonations.

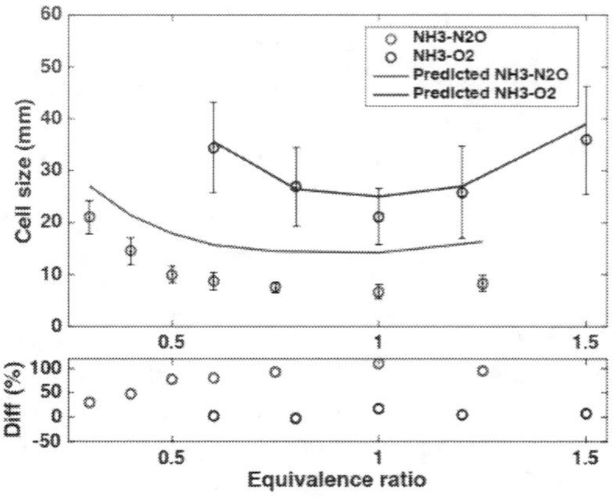

Figure 3: Experimental and predicted detonation cell size as a function of Φ for NH_3-O_2 and NH_3-N_2O mixtures. Φ=0.3-1.5; P_1=71 kPa; T_1=295 K.

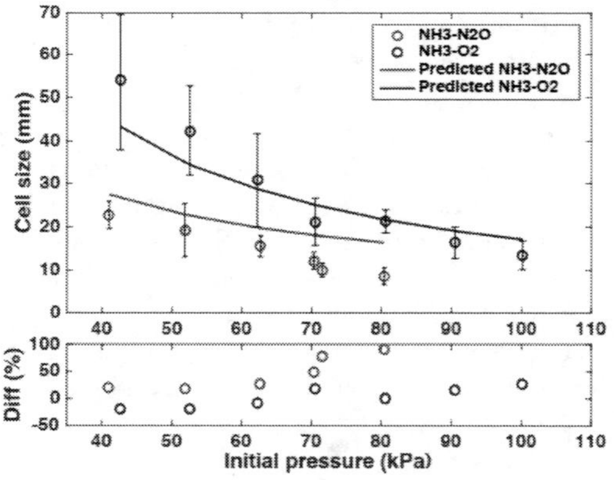

Figure 4: Experimental and predicted detonation cell size as a function of P_1 for NH_3-O_2 and NH_3-N_2O mixtures. Φ=0.5 and 1 for mixtures with O_2 and N_2O, respectively; P_1=40-100 kPa; T_1=295 K.

Detonation cell size (λ) measurements were performed in the same range of conditions, see Figure 3 and Figure 4. For both mixtures, the classical evolution of the cell size was observed: (i) smallest cell size at Φ=1; (ii) U-shape for the curve λ=f(Φ); and (iii) quasi-exponential decrease of λ with initial pressure increase. Mixtures with N_2O are much more sensitive to detonation than mixtures with O_2 (i.e. λ is almost three times smaller at Φ=1 and P_1=70 kPa). The predictions from the semi-empirical model of Ng [12] are also shown in these figures. Maximum deviation between the predicted and experimental cell width was close to 100%.

3.2 ZND structure

Figure 5 shows the ZND species profiles obtained for both NH_3-N_2O and NH_3-O_2 mixtures. Consistent with the cell size measurements and predictions, mixtures containing nitrous oxide have shorter induction length. These mixtures also exhibit lower maximum OH mole fractions and oxidant final mole fraction.

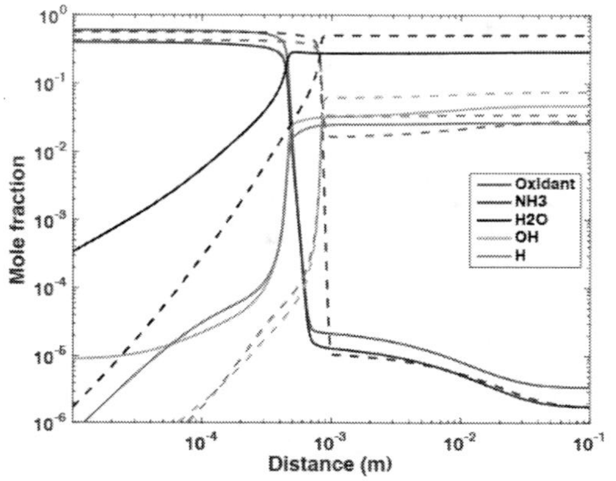

Figure 5: ZND structure of detonation propagating in NH_3-N_2O (solid lines) and NH_3-O_2 (dashed lines) mixtures. Φ=1; P_1=70 kPa; T_1=295 K.

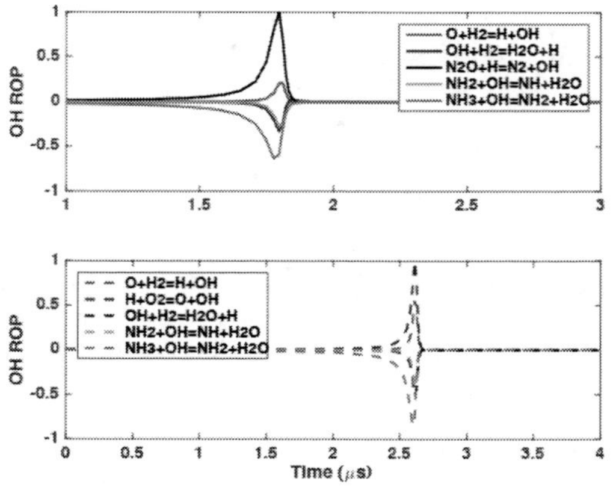

Figure 6: OH rate of production in ZND detonation propagating in NH_3-N_2O (top) and NH_3-O_2 (bottom) mixtures. Φ=1; P_1=70 kPa; T_1=295 K.

Figure 6 shows the rate of production of OH radical. When nitrous oxide is present in the mixture, the reaction N_2O+H=N_2+OH is the main production pathway. When O_2 is used as the oxidant, the chain branching processes, H+O_2=OH+O and O+H_2=OH+H, dominate the production of OH.

3.3 Detonation stability

For all initial conditions studied, the experimental soot foils appeared very irregular as seen in Figure 7.

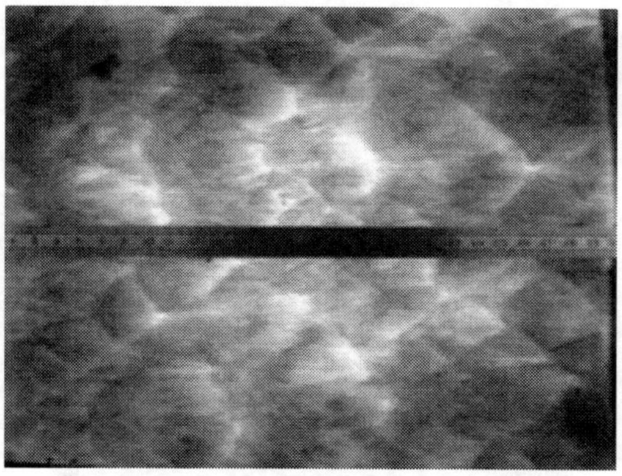

(a) Φ=1; T_1=295 K; P_1=51.8 kPa

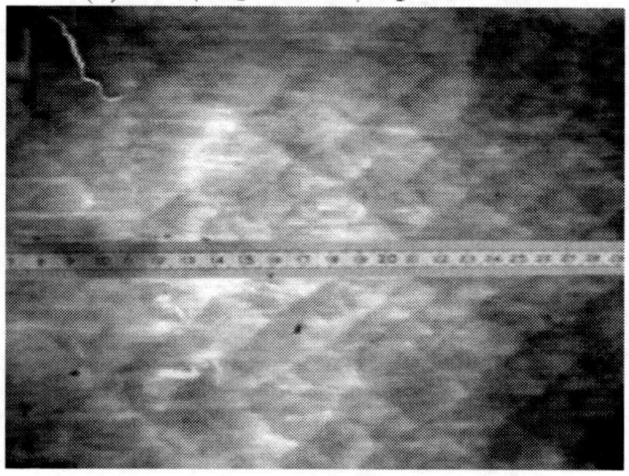

(b) Φ=1; T_1=295 K; P_1=41 kPa

Figure 7: Soot foil obtained for NH_3-O_2 (a) and NH_3-N_2O (b) mixtures. Propagation from left to right.

The reduced activation energy, Θ, and Ng stability parameter, χ, enable a quantitative assessment of the detonation stability. Θ and χ are respectively given by

$$\Theta = \frac{1}{T_{VN}} \cdot \left(\frac{ln\tau_{-1} - ln\tau_{+1}}{1/T_{-1} - 1/T_{+1}} \right), \quad (1)$$

and

$$\chi = \Theta \cdot \Delta_i \cdot \frac{\dot{\sigma}_{max}}{u'_{CJ}}, \quad (2)$$

with

$$\dot{\sigma} = \sum_{i=1}^{N} \left(\frac{\bar{W}}{W_i} - \frac{h_i}{C_P T} \right) \frac{\omega_i}{\rho} \quad (3)$$

T_{VN} is the von Neumann temperature; τ is the auto-ignition delay-time; the subscript ±1 indicate that the parameters have been calculated considering variation of 1% of the detonation speed with respect to the CJ velocity; Δ_i is the induction length; $\dot{\sigma}_{max}$ is the maximum thermicity; u'_{CJ} is the flow velocity in the CJ plan; N is the number of species; $\bar{W}$ is the mean molar mass; W_i is the molar mass; h_i is the enthalpy; C_P is the heat capacity at constant pressure; T is the temperature; ω_i is the source term; ρ is the density; and the subscript i represents the ith species.

Figure 8 and Figure 9 show the calculated Θ and χ for the ammonia-based mixtures along with the stability curves and values for some mixtures commonly investigated in detonation studies. These results allow to classify the studied ammonia based mixtures as highly unstable, which is consistent with the irregular soot foils shown in Figure 7.

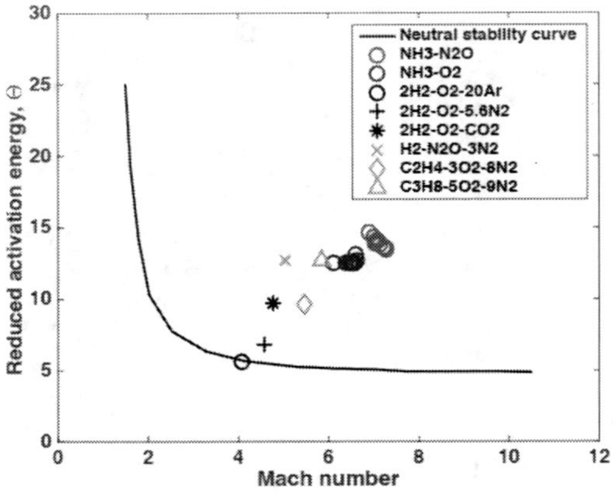

Figure 8: Stability diagram based on reduced activation energy for a variety of mixtures including NH_3-O_2 and NH_3-N_2O. For NH_3-based mixtures: Φ=0.3-1.5; P_1=40-100 kPa; T_1=295 K.

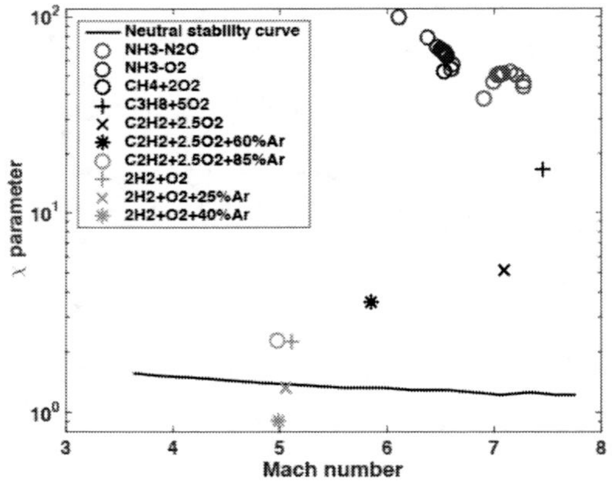

Figure 9: Stability diagram based on χ parameter for a variety of mixtures including NH_3-O_2 and NH_3-N_2O. For NH_3-based mixtures: Φ=0.3-1.5; P_1=40-100 kPa; T_1=295 K.

4 Conclusions

The characteristics of detonation propagating in NH_3-O_2 and NH_3-N_2O mixtures were obtained experimentally and through steady one-dimensional ZND simulations. For both mixtures, detonations appear very unstable with the NH_3-N_2O mixtures being the most sensitive to detonation. Future work will focus in developing reduced reaction models to study detonation in NH_3 based mixtures through unsteady two-dimensional numerical simulation.

Acknowledgements

This work was partially supported by the French "Ministère de l'Éducation Nationale, de l'Enseignement Supérieur et de la Recherche" (ATER funding) and by Centre National de la Recherche Scientifique (CNRS Orléans).

References

[1] Industrial uses of ammonia, `http://www.easychem.com.au/monitoring\discretionary{-}{}{}and-management/maximising-production/industrial-uses-of-ammonia`.

[2] The essential chemical industry online - ammonia, `http://www.essentialchemicalindustry.org/chemicals/ammonia.html`.

[3] Fujii N., Miyama H., Koshi M. and Asaba T., Proceedings of the Combustion Institute, 18 (1981) 873–883.

[4] Davidson D., Kohse-Hoinghaus K., Chang A. and Hanson R., International Journal of Chemical Kinetics, 22 (1990) 513–535.

[5] Cohen L., Fuel, 34 (1955) 123–127.

[6] Zakaznov V. and Kursheva L., Zhurnal Prikladnoi Khimii, 53 (1980) 1865–1867.

[7] Kaneshige M. and Shepherd J.E., Detonation database, Tech. Rep. FM97-8, GALCIT (1997).

[8] Akbar R., Kaneshige M., Schultz E. and Shepherd J., Detonations in h$_2$-n$_2$o-ch$_4$-nh$_3$-o$_2$-n$_2$ mixtures, Tech. Rep. FM-97-3, Explosion Dynamics Laboratory, California Institute of Technology (1997).

[9] Shepherd J.E., Progress in Astronautics and Aeronautics, 106 (1986) 263–293.

[10] Mével R., Javoy S., Lafosse F., Chaumeix N., Dupré G. and Paillard C.E., Proceedings of The Combustion Institute, 32 (2009) 359–366.

[11] Mével R., Ph.D. thesis, Université d'Orléans (2009).

[12] Ng H.D., Ju Y. and Lee J.H.S., International Journal of Hydrogen Energy, 32 (2007) 93–99.

11th Asia-Pacific Conference on Combustion,
The University of Sydney, NSW Australia
10th-14th December 2017.

Reduction Methodology for Detailed Kinetic Mechanisms: Application to n-Hexane-Air Hot Surface Ignition

S. Coronel[1,*], J. Melguizo-Gavilanes[1], D. Davidenko[2], R. Mével[3,4], J. E. Shepherd[1]
[1]California Institute of Technology, Pasadena, USA
1200 E. California Blvd, MC 205-45, Pasadena, California 91125, USA
[2]The French Aerospace Lab ONERA, Palaiseau, France
[3]Center for Combustion Energy, [4]Department of Automotive Engineering, Tsinghua University
30 Shuang Qing road, 100084 Beijing, China

Abstract

A reduction method for detailed kinetic mechanisms is presented. The reduction procedure was performed by using the ignition delay time as the target parameter. Overall, good qualitative and quantitative comparison was observed between the reduced mechanism results and experimental parameters such as ignition delay time, laminar burning speed, and concentrations in a jet-stirred reactor. The resulting reduced mechanism was used in multi-dimensional numerical simulations to predict the ignition behavior and thresholds of reactive gas adjacent to hot surfaces. Errors of less than 10% were observed between the numerical and experimental ignition thresholds indicating the adequacy of the reduction procedure.

1 Introduction

Accidental hot surface ignition is a problem of interest in the manufacturing, nuclear and mining sectors, and in aviation [1]. Significant progress has been made to experimentally understand hot surface ignition, in particular as it is applied to aviation safety [2, 3]. Typically, n-hexane has been used as a single component surrogate of kerosene since it exhibits a relatively high vapor pressure which facilitates testing at ambient conditions. Collecting test data is time consuming and costly since a large number of hot surface geometries and mixture conditions mus be tested. While performing multi-dimensional numerical simulations is an appealing alternative to experiments; predictive simulations require the use of detailed chemical mechanisms to predict the ignition and flame propagation events. Detailed reaction models for hydrocarbon fuels typically consist of hundreds of species and thousands of reactions. It is not feasible with existing computational resources to use these mechanisms in direct numerical simulations to model engineering tests or applications. An alternative is to use a reduced chemical mechanism that will still predict the appropriate target parameters such as ignition delay time of a full detailed mechanism. The current study describes the reduction for a n-hexane chemical mechanism. Comparisons are made between, 0D and 1D models that use the reduced mechanism and experimental results, and finally the resulting mechanism is applied to several multi-dimensional numerical simulations of thermal ignition.

*Corresponding author: coronel@caltech.edu

2 Materials and Methods

A modified version of the reduction procedure presented in [4] was employed. The reduction is controlled by the accuracy of one target phenomenon, the auto-ignition process. More specifically, the target parameters are: time to peak thermicity, maximum thermicity, equilibrium temperature, equilibrium mean molar mass, and temporal profiles of, thermicity, temperature, and mean molar mass. The reduction is performed as follows:

1. Calculate the error in the target parameters introduced by deactivating one species and all associated reactions; the error score at each deactivation step is given by
$$s_q = \epsilon_q/\epsilon_q^*, \quad (1)$$
where,
$$\epsilon_q = |q - q^{\mathrm{ref}}|/q^{\mathrm{ref}}. \quad (2)$$
q^{ref} is the target parameter calculated by using the complete detailed mechanism, q is the target parameter with the reduced mechanism, and ϵ_q^* is a user defined error limit. The error score is calculated across various conditions of initial temperature, T_0, initial pressure, P_0, and equivalence ratio, Φ. The sum S corresponds to the $\sum s_q$ across all the conditions tested.

2. Reactive species and all associated reactions and return to step 1 to deactivate the next species, repeat until S has been computed for each deactivated species.

3. Eliminate a single species with the lowest S value and all associated reactions and return to step 1. This sequence is applied until S reaches a user defined value of S^*.

The procedure described above was applied to the detailed chemical mechanism of Mével [5] which consists of 531 species and 2628 reactions. The mechanism was reduced for the following range of conditions: $T_0 = 800 - 1600$ K, $P_0 = 1$ atm, and $\Phi = 0.9$. The final reduced mechanism consists of 62 species and 223 reactions.

3 Results and Discussion

Several experimental parameters were compared against calculations performed in Cantera [6] using the reduced mechanism. Shock tube ignition delay time data were compared against numerical values obtained in a constant volume adiabatic reactor, experimental flame speeds were compared against 1D freely propagating flame calculations and jet-stirred reactor (JSR) experimental mole fractions were compared against perfectly-stirred reactor (PSR) model calculations.

3.1 Ignition Delay Time

n-Hexane shock tube ignition delay time data were taken from [7, 8]. Figure 1 shows comparisons between the model and experimental results at $\Phi = 0.5$ and 1.0. The experimental values lie significantly above the model results at $\Phi = 2.0$. Overall, at $P = 350$ kPa, the root-mean-square error (RMSE), is 29 μs, 131 μs, and 132 μs at $\Phi = 0.5$, 1.0, and 2.0, respectively.

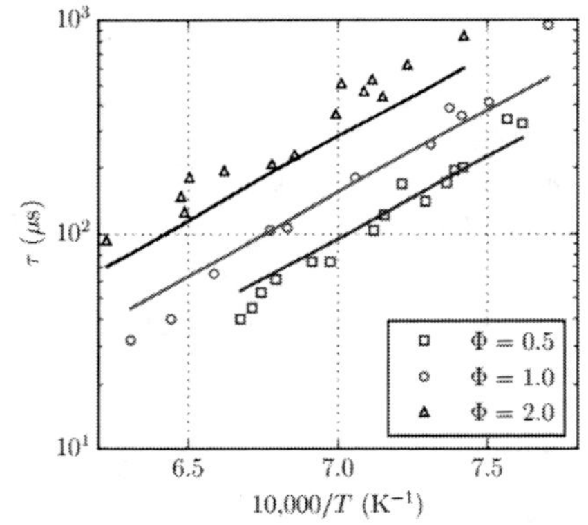

Figure 1: Experimental [7] (symbols) and model (solid lines) ignition delay time of *n*-hexane-O_2-Ar at $P = 350$ kPa. Argon mass fraction is 0.96.

Figure 2 shows experimental and simulated ignition delay time results at $P = 1.27$ and 6.13 MPa at $\Phi = 0.5$. The experimental values lie very close to the model results at $P = 1.27$ MPa. At $P = 6.13$ MPa, the model reproduces the ignition delay times at high temperatures and deviates slightly from the experimental values at lower temperatures. At $P = 1.27$ and 6.13 MPa, the RMSE is equal to 47 μs and 107 μs, respectively.

3.2 Laminar Burning Speed

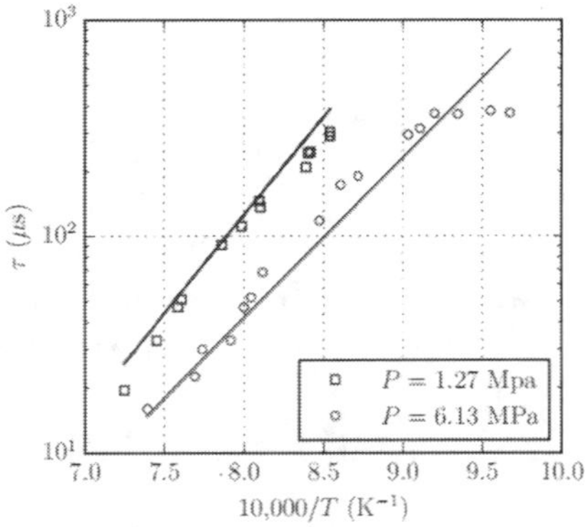

Figure 2: Experimental [8] (symbols) and model (solid lines) ignition delay time of *n*-hexane-air at $\Phi = 0.5$.

After the chemical mechanism was reduced with the ignition delay time as the target parameter, the reduced mechanism was applied to 1D freely propagating flame simulations and the results are compared against *n*-hexane-air experimental laminar burning speeds obtained from [9, 10]. Figure 3 shows that the model over-estimates the experimental values for lean conditions and lies within the uncertainty of the experimental values for rich conditions. The RMSE is 3.6 cm/s and 2.5 cm/s when comparing the model against the results from [9] and [10], respectively. Although the mechanism was reduced by using the auto-ignition process as the target phenomenon, reasonable agreement is observed between the experimental laminar burning speeds and the 1D freely propagating flame simulations. This suggests that if the laminar burning speed is chosen as the target phenomenon, then reasonable agreements between the model and the ignition delay time would also be obtained.

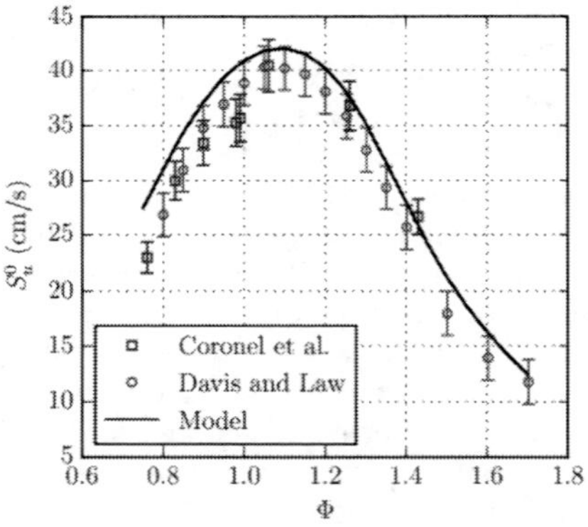

Figure 3: Experimental [9, 10] (symbols) and calculated (solid line) laminar burning speeds of *n*-hexane-air at $T =$ 300 K and $P = 100$ kPa.

3.3 Jet-Stirred Reactor (JSR)

A perfectly-stirred reactor model was used with the reduced chemical mechanism to compare against jet-stirred reactor experimental results from [11]. The experimental values were obtained for the oxidation of

n-hexane at $P = 10$ atm and $\Phi = 1.0$. The normalized root-mean-square error (NRMSE) shown in Table 1 is calculated for species mole fractions shown in Figs. 4 and 5.

Species	NRMSE	Species	NRMSE
n-C_6H_{14}	0.35	H_2	0.5
O_2	0.15	CH_4	1.4
H_2O	0.48	CH_2O	0.4
CO_2	0.7	C_3H_6	1.2
CO	1.2	CH_3CHO	0.6
C_2H_4	10.6	C_4H_8-1	0.4

Table 1: NRMSE between jet-stirred reactor experimental values and model results.

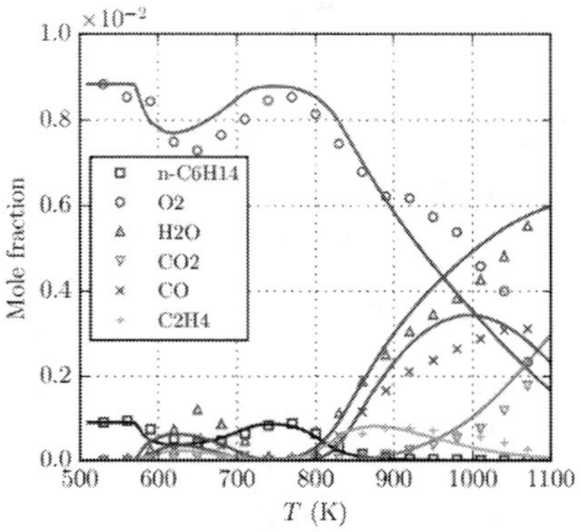

Figure 4: Experimental [11] (symbols) and modeling (solid lines) results for n-hexane oxidation in a jet-stirred reactor at $P = 10$ atm and $\Phi = 1.0$; only major species are shown.

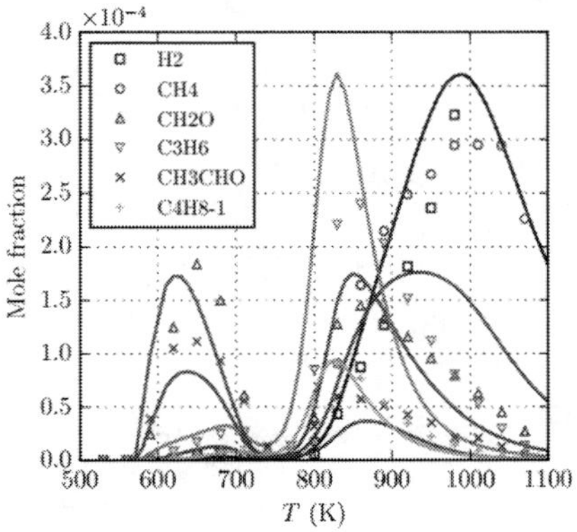

Figure 5: Experimental [11] (symbols) and modeling (solid lines) results for n-hexane oxidation in a jet-stirred reactor at $P = 10$ atm and $\Phi = 1.0$; only minor species are shown.

Based on the values of NRMSE, the model increases in accuracy for the major species in the following order: $C_2H_4 \rightarrow CO \rightarrow CO_2 \rightarrow H_2O \rightarrow n\text{-}C_6H_{14} \rightarrow O_2$. The model increases in accuracy for the minor species in the following order: $CH_4 \rightarrow C_3H_6 \rightarrow CH_3CHO \rightarrow H_2 \rightarrow CH_2O \rightarrow C_4H_8\text{-}1$

3.4 Multi-Dimensional Numerical Calculations

Finally, the reduced mechanism was applied to two-dimensional numerical simulations of thermal ignition using the OpenFOAM framework [12]. The motion, transport and chemical reaction in the gas were modeled using the variable-density, reactive Navier-Stokes equations with temperature-dependent transport properties. Differential diffusion was included using a constant non-unity Lewis number for each species. A detailed description of the model can be found in [13]. The problem simulated is ignition of n-hexane-air at $\Phi = 0.9$, $T_0 = 300$ K, and $P_0 = 100$ kPa, by heated surfaces of $6 - 10$ mm in size. Several configurations, shown in Fig. 6, were tested. Filled contours of CO_2 mass fraction are shown in Fig. 6; the images correspond to ignition events taking place when the surface temperature is at the ignition threshold.

A commercially available glow plug (9.3 mm in height and 5.1 mm in diameter) was simulated in Fig. 6 (a). A temperature ramp of 220 K/s was imposed at the surface; the heat from the surface diffuses into the gas and is convected upwards forming a hot plume. Ignition occurred in a stagnation point located above the top surface of the glow plug. The concentration of CO_2 was used to pinpoint and visualize the ignition location. A moving hot sphere (6 mm in diameter) was simulated in Fig. 6 (b). The sphere had a surface temperature of 1300 K and was moving at 2.4 m/s. A thermal boundary layer developed around the moving hot sphere. At the temperature and velocity given, the flow is laminar and axisymmetric with a toroidal vortex in the wake. The flow also has a separation region (marked by the high concentration of CO_2) due to adverse pressure gradients. It is in this region of flow separation were ignition takes place. A vertical cylinder with a heated area of 10 mm in height was simulated in Fig. 6 (c). A temperature ramp of 220 K/s was imposed at the surface. Thermal and hydrodynamic boundary layers grow upwards along the cylinder surface. Ignition takes place in the gas at a height corresponding with the top boundary of the heated surface. Finally, a horizontal cylinder (10 mm in diameter) was simulated in Fig. 6 (d). A temperature ramp of 220 K/s was imposed at the surface. There is growth of the thermal and momentum boundary layers from the front stagnation point to the rear stagnation point. Ignition takes place in the rear stagnation point where the temperature gradients are smallest.

Geometry	Exp. (K)	Calc. (K)	$\Delta T/T_{\text{exp}}$
Hor. Cyl. [14]	1180	1093	0.07
Vert. Cyl. [14]	1270	1191	0.06
Sphere [3]	1224	1300	0.06
Glow Plug [15]	1275	1162	0.09

Table 2: Comparison of experimental and numerical results for all hot surface geometries tested.

Experimental and predicted ignition thresholds for all the geometries considered are shown in Table 2. Normalized differences, $\Delta T/T_{\text{exp}}$, were calculated between

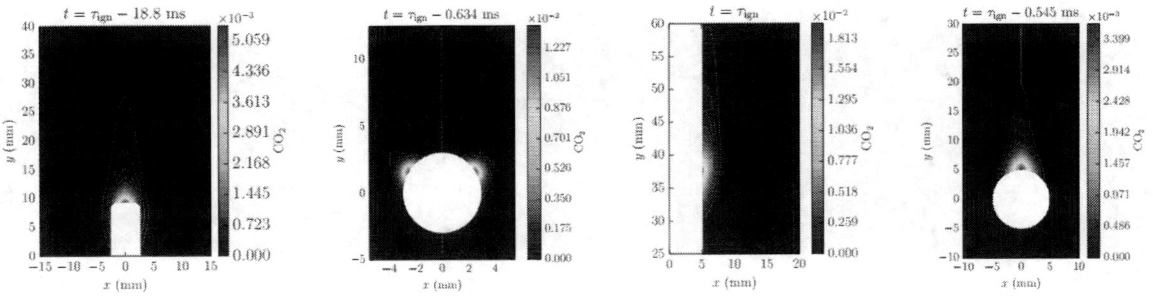

(a) Glow plug (b) Moving sphere (c) Vertical cylinder (d) Horizontal cylinder

Figure 6: CO_2 mass fraction fields at the ignition event using a (a) glow plug, (b) moving sphere, (c) vertical cylinder, and (d) horizontal cylinder.

the ignition thresholds obtained numerically, T_{calc}, and experimentally, T_{exp}; ΔT is equal to $|T_{\text{exp}} - T_{\text{calc}}|$. Regardless of the geometric configuration of the hot surface, the difference between the numerical and experimental results was less than 10%. The multi-dimensional simulations indicate that ignition can be predicted using the reduced n-hexane mechanism presented in this study.

4 Conclusions

A reduction methodology for detailed kinetic mechanisms was described and applied to obtain a reduced mechanism for n-hexane. The reduced mechanism predictions using a constant volume adiabatic reactor showed good quantitative comparison with experimental ignition delay time measurements. Although not reduced for flame speed calculations, the reduced mechanism performs well when applied to a 1D freely propagating flames. Jet-stirred reactor results indicated good quantitative comparison for the major and minor species presented. Finally, the reduced mechanism was applied to a two-dimensional simulation of ignition by heated surfaces and resulted in differences of less than 10% when compared against experimental ignition measurements.

5 Acknowledgements

This work was carried out in the Explosion Dynamics Laboratory of the California Institute of Technology, and was supported by The Boeing Company through a Strategic Research and Development Relationship Agreement CT-BA-GTA-1.

References

[1] Stack B., Sepeda A. and Moosemiller M., Process Safety Progress, 33 (2014) 19–25.

[2] Boettcher P.A., Ph.D. thesis, California Institute of Technology (2012).

[3] Coronel S., Ph.D. thesis, California Institute of Technology (2016).

[4] Davidenko D., Mével R. and Dupré G., in Proceedings of the European Combustion Meeting 2009.

[5] Mével R., Chatelain K., Boettcher P.A. and Shepherd J.E., Fuel, 126 (2014) 282–293.

[6] Goodwin D.G., Moffat H.K. and Speth R.L., Cantera: An object-oriented software toolkit for chemical kinetics, thermodynamics, and transport processes, `http://www.cantera.org` (2017), version 2.3.0.

[7] Mével R., Niedzielska U., Melguizo-Gavilanes J., Coronel S. and Shepherd J.E., Combustion Science and Technology, 188 (2016) 2267–2283.

[8] Zhukov V.P., Sechenov V.A. and Starikovskii A.Y., Combustion and Flame, 136 (2004) 257–259.

[9] Coronel S., Lapointe S., Mével R. and Shepherd J., in Preparation for Fuel.

[10] Davis S.G. and Law C., Combustion Science and Technology, 140 (1998) 427–449.

[11] Zhang K., Banyon C., Togbé C., Dagaut P., Bugler J. and Curran H.J., Combustion and Flame, 162 (2015) 4194 – 4207.

[12] Weller H.G., Tabor G., Jasak H. and Fureby C., Comput. Phys., 12 (1998) 620–631.

[13] Melguizo-Gavilanes J., Coronel S., Mével R. and Shepherd J., International Journal of Hydrogen Energy, 42 (2017) 7380–7392.

[14] Boeck L., Meijers M., Kink A., Mével R. and Shepherd J.E., Combustion and Flame, (2017), in press.

[15] Melguizo-Gavilanes J., Nové-Josserand A., Coronel S., Mével R. and Shepherd J.E., Combustion Science and Technology, 188 (2016) 2060–2076.

11th Asia-Pacific Conference on Combustion,
The University of Sydney, NSW Australia
10th -14th December 2017.

The evolution of nascent soot formation in methane-ethylene, ethane-ethylene and propane-ethylene laminar premixed flames

Can Shao[1,2], He Lin[2], Baiyang Lin[2], Mani Sarathy[1], Zhen Huang[2]

[1]King Abdullah University of Science and Technology, Clean Combustion Research Centre,
Thuwal, 23955-6900, Saudi Arabia

[2]Shanghai Jiao Tong University, Key Lab. for Power Machinery and Engineering of M.O.E,
800 Dongchuan RD. Minghang District, Shanghai, 200240, China

Abstract

The evolution of primary soot particles (particle size distribution, number density and volume fraction) in a laminar premixed flame was investigated in this paper. It was found that all the blended flames have bimodal particle size distribution. The blending of propane greatly enhanced the nucleation of soot in the flame, and the strong continuous nucleation was also observed in the downstream of the flame. The mobility volume fraction and the number density of propane-ethylene flame are the biggest comparing to the other two flames. Calculations show that more PAHs are formed in propane-ethylene flame, resulting in the promotion of soot particles formation.

1 Introduction

The vast majority of PM2.5 pollution in the cities comes from the burning of fossil fuels in the engines [1, 2]. It is of great significance to study the mechanism of soot particles formation in the combustion process of engines to alleviate urban air pollution. Though significant progress has been demonstrated in soot formation mechanism [3-14] over last two decades, critical gaps remain in many areas [13]. The mechanism of soot nucleation and soot growth, that reflects the coagulation of the soot particles as well as the particle surface growth, has to be studied thoroughly with both theoretical and experimental methodologies [15]. Nowadays, the technique of probe sampling coupled with mobility sizing is applied to obtain the evolution of particle size distribution functions (PSDFs) of nascent soot from particle nucleation to mass growth which can give the quantitative understanding of soot formation processes.

Present work mostly focuses on the effect of equivalence ratio and temperature on soot particle evolution, while fuel blending has been seldom studied. The impact of fuel blending on soot formation is certainly worthy of attention. The chemical compositions of actual fuels (such as gasoline and diesel) are very complicated, and it is very difficult to study the soot formation mechanism directly. Meanwhile, in the beginning of combustion progress, these will decompos into methane, ethane, propane, ethylene and other small molecules of hydrocarbons. Therefore, in this study, the evolution of soot particle in alkane-ethylene premixed Burner Stabilized Stagnation (BBS) flames was studied using probe sampling technique coupled with mobility sizing. Cross comparisons with respect to the PSDFs of soot particles were approached between methane-ethylene flame, ethane-ethylene flame and propane flame. The purpose of the study is to give an insight into the impact of different alkanes' doping on nucleation and growth of soot particle. To get a deep understanding of evolution on the PSDFs, the mole fraction of some important PAHs were calculated with KAG mechanism [16] in this paper.

2 Experimental and Computation Method

As it has been described in detail in previous papers [15, 17, 18], the experimental system was shown in Fig.1. Briefly, the mobility measurement of PSDF of nascent soot particles in ethylene-alkane flames were carried out using the BSS flames configuration, in which the sampling tube is embedded in a stagnation plate above the flame. Then the nascent soot particles were inhaled into the sampling orifice which was 130 um in diameter and immediately diluted by high speed of cold nitrogen (30L/min, STP) for minimizing the coagulation losses of particles in the sampling channels. A fraction of the diluted sample (1.5L/min, STP) was analyzed by TSI SMPS system consisting of a 3085 nano-Differential Mobility Analyzer (DMA), a Kr-85 Neutralizer and a 3776 ultrafine Condensation Particle Counter (PCP). As the particle size smaller than 10 nm was overestimated by Cunningham slip correction [19-21], correction of mobility diameter based on the nano-particle transport theory was needed.

In the current study, the flame temperature was precisely measured by an S-type fine wire thermocouple, which was coated with the Y/Be/O mixture to prevent high-temperature oxidation reaction on thermocouple surface. The emissivity of Y/Be/O coating ranges from 0.3 to 0.6, therefore, the Shaddix procedure [22] was used for temperature correction. The radiation-corrected temperature was obtained by averaging the temperatures corrected with the upper and lower limits of emissivity [17, 23-25]. On the other hand, the temperature profiles were calculated using the pseudo one-dimensional model in OPPDIF code coupled with USC Mech II [26] consisting 111 species and 784 reactions. As it has been reported that flame temperature can influence the evolution of nascent soot greatly, it is quite essential to keep the flame temperature profiles accordant during the PSDF measurement to eliminate

Corresponding author: He Lin. Fax: +86-021-34207774
E-mail address: linhe@sjtu.edu.cn

the effect of temperature on PSDFs. The peak temperatures of the three flames are shown in Table 1.

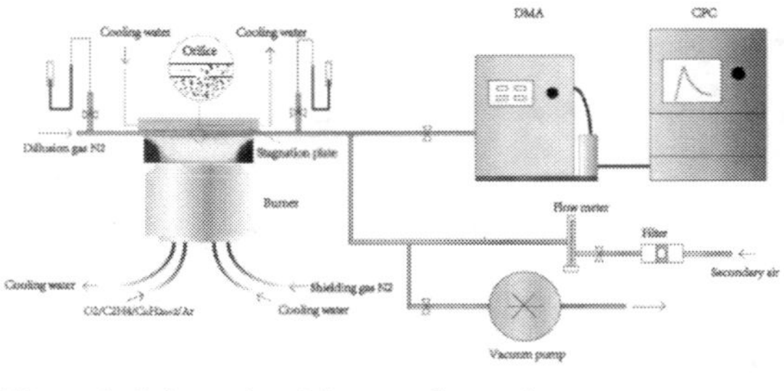

Figure 1: Schematic of the experimental setup.

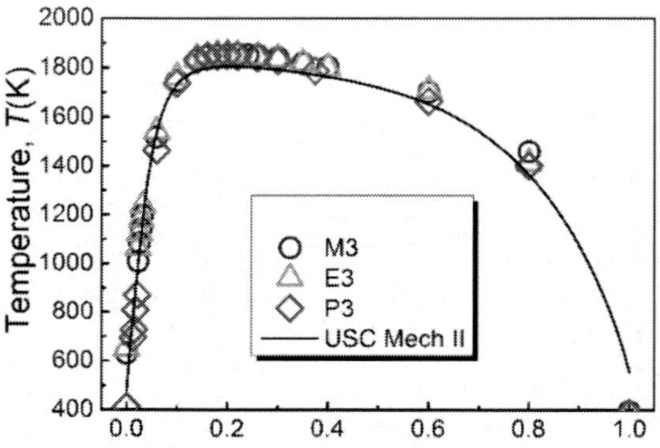

Figure 2: The comparison of measured temperature profiles.

The concentrations of PAHs in blending flames were calculated by CHEMKIN-PRO software package coupled with KAP mechanism [16]. The simulation of PAHs concentration assists in understanding the changes of PSDFs from a different perspective. The KAP mechanism contains 572 species and 3379 reactions, which can predict PAHs from benzene (A1) to coronene (A7). In this simulation, the temperature profiles measured with thermocouple were used as the input temperature of the Fit Temperature Model in CHEMKIN-PRO. In this study, the flames with a same equivalent ratio of 2.07 were investigated. The blending ethylene flames with 20% (mole percent) of alkanes respectively were named M3, E3 and P3.

Table 1: Summary of the flames studied.

Flame	Mole Fraction				Cold Gas	$T_{f,max}$(K)
	Alkanes	C_2H_4	O_2	Ar	Velocity	
M3(methane)	0.0339	0.1358	0.2303	0.6	8	1849±82
E3(ethane)	0.0320	0.1279	0.2401	0.6	8	1853.5±82.5
P3(propane)	0.0302	0.1209	0.2489	0.6	6.5	1851±2

As shown in Table 1, the velocities of cold gas of the flames were 8.0 cm/s, 8.0cm/s and 6.5cm/s, respectively. The peak temperatures of those flames are close to 1850K. Fig. 2 shows that the temperature profiles of the blending flames are almost identical and are highly consistent with the simulated temperature profile. By unifying the equivalence ratio and the temperature profile of the blending flames, fuel type is the only factor that can potentially affect the PSDFs.

3 Results and discussion

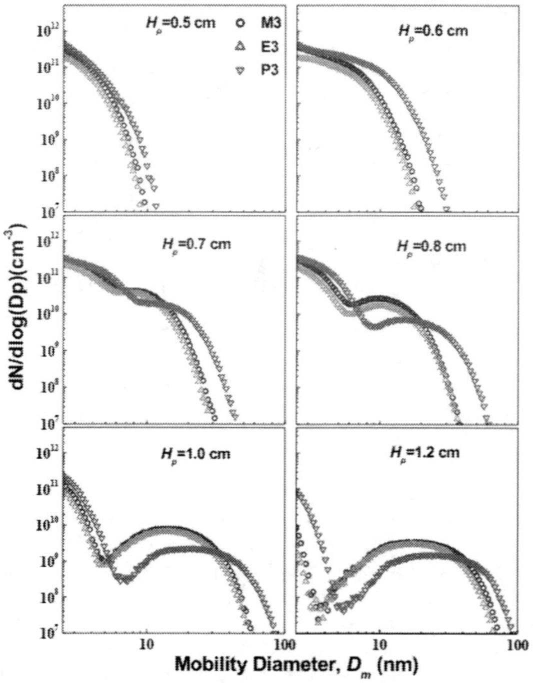

Figure 3: The PSDFs measured for the flames studied.

The PSDFs of the flames are shown in Fig. 3. The nucleation tendency in M3 and E3 flames decreases with the height (H_p) increasing. In the flame P3, the nucleation tendency is high even at large height. The bimodal distribution characterized by a nucleation tail for the small particles and a lognormal part for the large particles is caused by soot nucleation and soot size growth. The bimodality is mostly postponed in the P3 flame as it has larger nucleation zone and mass growth zone. The nucleation process is greatly promoted by the propane doping. The number density in nucleation region in P3 flame is near one order of magnitude larger than M3 and E3 flame. As shown in Fig. 4, the PSDFs troughs move to large particle size side in the order: P3>M3>E3.

The mobility volume fractions of the soot particle was calculated by integrating the PSDFs overall particle size measured (>2.5nm) and shown Fig. 4 as a function of H_p. It indicates that the mobility volume fraction of P3 flame is generally larger than those of M3 and E3 flame. That is due to its high number density of nucleation sized particles and wider size distribution, as shown in Fig. 3. While the mobility volume fraction of M3 flame is slightly larger than E3 flame.

The total number densities of the flames at each H_p above the burner are shown in Fig. 5, which computed based on the PSDFs data. The total number density of M3 flame is slightly larger than the E3 flame. Similarly with the tendency of mobility volume fraction, the total number density of P3 flame is markedly larger than those of M3 and E3 flame. Also, the total number density of M3 flame is slightly larger than the E3 flame. The nucleation sized particles contribute a lot to number density, as they are much more than mass growth sized particles.

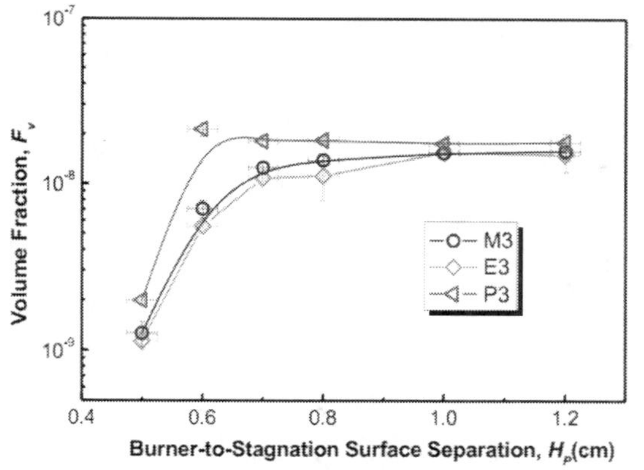

Figure 4: The comparison of the mobility volume fraction of flames.

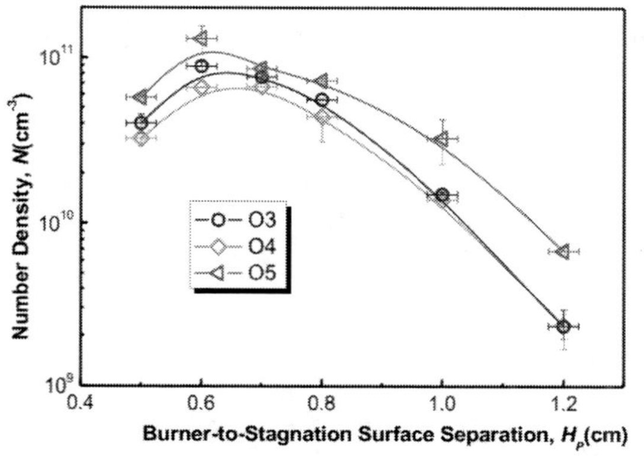

Figure 5: The comparison of mobility volume fraction of flames.

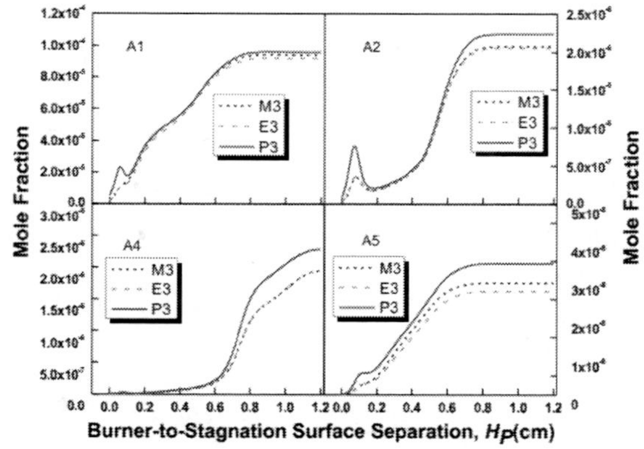

Figure 6: The comparison of the computed benzene (A1), naphthalene (A2), pyrene (A4) and benzopyrene (A5) mole fraction profile of the flames.

The quantitative investigation on the formation of polycyclic aromatic hydrocarbons (PAHs) will contribute to better understanding of the soot particle evolution in flames. The computed mole fractions of A1, A2, A4 and A5 in alkane-ethylene flames were shown in Fig. 6. The quantity of A1 in the different flame is closed to each other. However, the mole fraction from A2-A5 is quite different, and is ranked as the in the following order: P3>M3>E3, which is consistent with the ranking of the mobility volume fraction shown in Fig. 4. Pyrene (A4) is usually regarded as the key PAH in the initial stage of soot nucleation, and the dimerization reaction of A4 is considered as the first step of soot nucleation as noted in many celebrated soot models. It means that the strength of soot nucleation process in flame is in proportion to concentration of A4 [13]. As reported before [15,23], alkane doping on ethylene can promote the CH3's and C3H3's contribution to A4 formation, therefore leads to more soot formation.

4 Conclusions

The evolution of the mobility PSDFs of nascent soot in alkane-ethylene premixed laminar flames was studied through the BSS flame configuration. The following conclusions can be made:

1. The PSDFs of soot particles in alkane-blending flames behave bimodality. The propane-ethylene flame has the largest nucleation zone and mass growth zone.
2. The mobility volume fraction of the propane-ethylene is the biggest one among the three flames studied, and the methane's is just slightly bigger than ethane – ethylene flame.
3. The number density of propane-ethylene flame is much higher than methane-ethylene flame and ethane-ethylene flame.
4. Computed results of the PAHs indicate that in propane-ethylene flame the PAHs formation is promoted compared to the other two flames. This results in more soot particle formation (both in volume fraction and number density).

5 Acknowledgment

This work was supported by the National Natural Science Foundation of China (51210100 and 9441129), the National Basic Research Program of China (2013CB228502) and the Program of Introducing Talents of Discipline to Universities ('111 Project'-B13018). We also acknowledge Competitive Research Funding from the King Abdullah University of Science and Technology (KAUST) and the Clean Combustion Research Center.

References

[1] F. Karagulian, C.A. Belis, C.F.C. Dora, A.M. Prüss-Ustün, S. Bonjour, H. Adair-Rohani, M. Amann, Atmospheric Environment, 120 (2015) 475-483.

[2] P. Pant, R.M. Harrison, Atmospheric Environment, 77 (2013) 78-97.

[3] I.M. Kennedy, Progress in Energy and Combustion Science, 23 (2) (1997) 95-132.

[4] K.H. Homann, Angewandte Chemie International Edition, 37 (18) (1998) 2434-2451.

[5] D.B. Kittelson, Journal of aerosol science, 29 (5) (1998) 575-588.
[6] H. Richter, J.B. Howard, Progress in Energy and Combustion science, 26 (4) (2000) 565-608.
[7] M. Frenklach, Physical Chemistry Chemical Physics, 4 (11) (2002) 2028-2037.
[8] C.K. Law, C.J. Sung, H. Wang, T. Lu, AIAA journal, 41 (9) (2003) 1629-1646.
[9] Z. Mansurov, Combustion, Explosion and Shock Waves, 41 (6) (2005) 727-744.
[10] C.S. McEnally, L.D. Pfefferle, B. Atakan, K. Kohse-Höinghaus, Progress in Energy and Combustion Science, 32 (3) (2006) 247-294.
[11] A. D'Alessio, A. D'Anna, P. Minutolo, L. Sgro, H. Bockhorn, A. D'Anna, A. Sarofim, H. Wang, Combustion Generated Fine Carbonaceous Particles, in, KIT Scientific Publishing, Karlsruhe, Germany, 2009.
[12] H. Wang, A. Abid, H. Bockhorn, A. D'Anna, A. Sarofim, H. Wang, Combustion generated fine carbonaceous particles, in: Proceedings of an International Workshop held in Villa Orlandi, Anacapri (May 13–16, 2007), Karlsruhe University Press, Karlsruhe, 2009, pp. 367-384.
[13] H. Wang, Proceedings of the Combustion Institute, 33 (1) (2011) 41-67.
[14] H. Bockhorn, Soot formation in combustion: mechanisms and models, Springer Science & Business Media, 2013.
[15] C. Shao, B. Guan, B. Lin, H. Gu, C. Gu, Z. Li, H. Lin, Z. Huang, Fuel, 186 (2016) 422-429.
[16] S. Park, Y. Wang, S.H. Chung, S.M. Sarathy, Combustion and Flame, 178 (2017) 46-60.
[17] C. Gu, H. Lin, J. Camacho, B. Lin, C. Shao, R. Li, H. Gu, B. Guan, Z. Huang, H. Wang, Combustion and Flame, (2016).
[18] H. Lin, C. Gu, J. Camacho, B. Lin, C. Shao, R. Li, H. Gu, B. Guan, H. Wang, Z. Huang, Combustion and Flame, 172 (2016) 365-373.
[19] Z. Li, H. Wang, Physical Review E, 68 (6) (2003) 061206.
[20] Z. Li, H. Wang, Physical Review E, 68 (6) (2003) 061207.
[21] Z. Li, H. Wang, Journal of aerosol science, 37 (1) (2006) 111-114.
[22] C.R. Shaddix, Correcting thermocouple measurements for radiation loss: a critical review, in, Sandia National Labs., Livermore, CA (US), 1999.
[23] S. Yoon, S. Lee, S. Chung, Proceedings of the Combustion Institute, 30 (1) (2005) 1417-1424.
[24] A. Abid, E. Tolmachoff, D. Phares, H. Wang, Y. Liu, A. Laskin, Proceedings of the Combustion Institute, 32 (1) (2009) 681-688.
[25] J. Camacho, C. Liu, C. Gu, H. Lin, Z. Huang, Q. Tang, X. You, C. Saggese, Y. Li, H. Jung, Combustion and Flame, 162 (10) (2015) 3810-3822.
[26] H. Wang, X. You, A.V. Joshi, S.G. Davis, A. Laskin, F. Egolfopoulos, C.K. Law, USC Mech Version II. High-Temperature Combustion Reaction Model of H2/CO/C1-C4 Compounds, in, 2007.

11th Asia-Pacific Conference on Combustion,
The University of Sydney, NSW Australia
10th -14th December 2017.

Effect of oxidation deposition by supercritical kerosene on heat transfer in a coiled tube

Xinyan Pei[1], William L. Roberts[1], Lingyun Hou[2]

[1] King Abdullah University of Science and Technology, Clean Combustion Research Center

Thuwal, 23955-6900, Saudi Arabic

[2]Tsinghua University, School of Aerospace Engineering

Haidian District, Beijing, 100084, China

Abstract

The deposition in fuel cooling systems remains a challenge to the development of active cooling technologies for air-breathing engines. In this study, we investigated the influence of the secondary flow and heat transfer of a supercritical kerosene in a coiled tube on oxidation deposition both in experimental and numerical methods. The coiled heated tube reactor under the constant heat flux and the flow rate was applied to simulate the heat-exchanger in an aero-engine cooling system. The distributions of dynamic pressure, temperature, velocity and turbulence kinetic energy were studied in a circular cross-section located in the middle of the tube. The microstructure morphology, elements of the deposition, and residual fuel compositions have been measured by SEM, XPS and GC-MS. Secondary flows induced in the coiled tube enhance the heat transfer greatly and slightly increase the deposition rate, resulting in a linear wall temperature profiles and a uniform coking distribution along the tube. The modified heat transfer correlation of the supercritical RP-3 in the coiled tube was fitted at different flow rates and heat fluxes.

1 Introduction

When the aviation fuel acts as the coolant in the engine cooling system, it reacts with dissolved oxygen and oxygenated compounds in the temperatures ranging from 423K to 723K to form coke deposition on the inner surface of the tube, which can deteriorate the heat transfer performance and even block fuel lines [1]. Thus, the thermal stability and heat transfer characteristics studies are significant for the application of active fuel cooling technology. The previous studies on the oxidation deposition properties of liquid fuel were mainly conducted in straight tubes [2-5]. The geometry of the tube defines the flow regime, which has much impact on the deposition process and heat transfer characteristics. The helically coiled tube is widely used as a heat exchanger in the various cooling systems. However, many researchers only focused on the heat transfer investigation of supercritical fluids such as water and CO_2 [6-8]. Therefore, the deposition process and heat transfer characteristics of supercritical kerosene in the cooled helically coiled tube are still deficient. In this paper, the experimental and numerical studies are carried out on the oxidation deposition and heat transfer of supercritical aviation kerosene of RP−3 in the coiled tube.

2 Experimental and Computation Method

Corresponding author: Xinyan Pei. Fax: +86-010-62772157
E-mail address: xinyan.pei@kaust.edu.sa

2.1 Experimental set up

A standard Chinese aviation kerosene (RP-3) was used as the coolant fuel in this study [9]. The properties of RP-3 are given in Table 1. Details of the experimental setup and measurement systems are provided in Figure 1. The amount of deposition was evaluated by weighing method [10]. A coiled heated tube reactor under the constant heat flux and the flow rate was applied to simulate the heat-exchanger in an aero-engine cooling system. The detailed experimental parameters and boundary conditions are listed in Table 2. The helically coiled tube is made of stainless steel with 3000mm long, 23 cycles, an inner diameter of 2mm, an outer diameter of 3mm, a coil pitch P of 22mm and a coil radius R of 50mm. One cycle geometry of the coiled tube is given in Fig. 2. Heat transfer coefficient of the experiment can be calculated as $h = q/(T_w - T_b)$, where q is the heat flux, which depends on the heating power and excludes the heat loss. The Nusselt number (Nu) is defined as $(h*d)/\kappa$. The results of the error analysis on the measured parameters and transfer formula are presented in Table 3.

Properties	Values
Critical Pressure (MPa)	2.39
Critical Temperature (K)	645.5
Density (g/cm^3)	0.7913
Flash point (K)	323
Distillation range (K)	436-485
Relative Molecular weight (g/mol)	148.33

Table 1: Properties of RP-3 aviation kerosene.

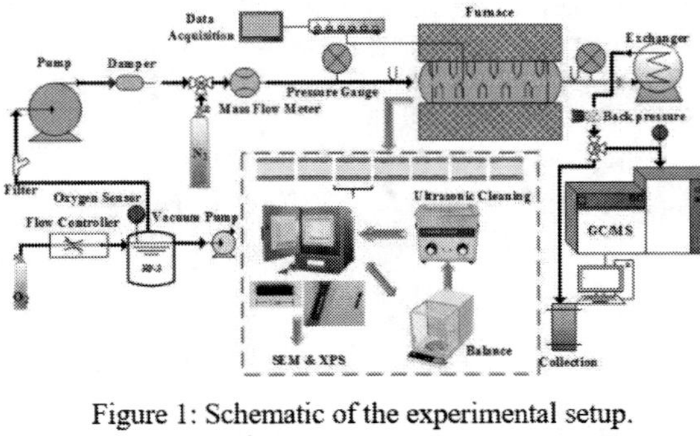

Figure 1: Schematic of the experimental setup.

Parameters	Values
Heat flux (kW/m^2)	40
Heating time (h)	1.75
Inlet mass flow rated (g/s)	1
Pressure (MPa)	3
Inlet Temperature (K)	300
Maximum fuel temperature at outlet (K)	678

Table 2: Experimental parameters.

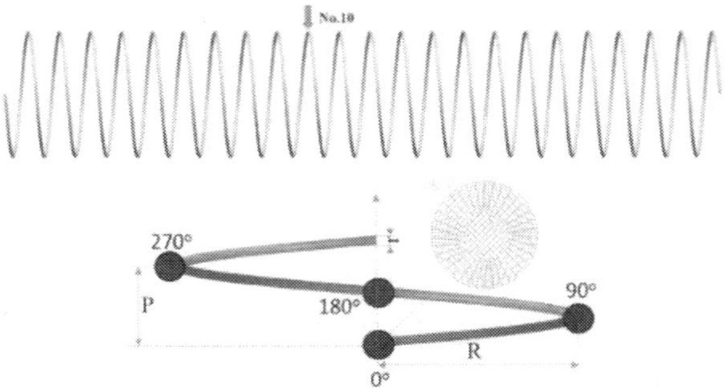

Figure 2: Schematic geometry of the coiled and cross section grid.

Items	Error (%)
Heat flux	5.055
Temperature difference	7.1
Heat transfer coefficient	10
Mass flow rate	±0.2
Pressure	±1%

Table 3: Error analysis.

2.2 Numerical model

Commercial software, Ansys Fluent [11], was utilized to solve the steady and three-dimensional flow as well as heat transfer in the coiled tube model. The turbulent model of standard k- ε was applied by considering the transition from laminar flow to turbulent flow. The governing equations in finite volume form were solved iteratively by using a second-order upwind scheme. In addition, the SIMPLEC pressure-velocity coupling scheme was employed in the solution of the momentum equations. When the normalized error residuals for all of the calculated variables reduced to six orders of magnitude below their maxima, the solution was deemed to be converged. The boundary conditions of the simulation are coincidence to the experiment. The mesh of the cross section of the coiled tube is the o-grid shown in Fig. 2 and independency study on mesh has been investigated on three different densities of mesh. As depicted in Fig. 3, the thermal physical properties of the aviation kerosene are temperature-dependent and defined by using a three-component surrogate model in the simulation. When the flow state is close to the critical point, small changes in pressure or temperature result in large changes in the thermal properties and the fuel becomes a supercritical fluid with gas-like diffusivity, density and viscosity. The processes of the deposition and the chemical reactions were neglected in the simulation.

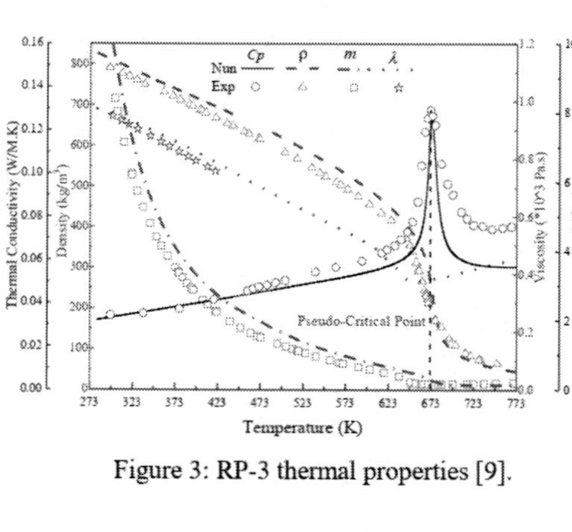

Figure 3: RP-3 thermal properties [9].

3 Results and discussion

Numerical wall temperature distribution along the coiled tube is gradually increased in Fig. 4. The dynamic pressure, temperature distribution, secondary flow streamlines together with velocity and turbulence kinetic energy, are reported in Fig. 5 at the cross sections (0°) of No.10 cycle of the coiled tube, shown in Fig. 2. The dynamic pressure contour in Fig. 5a illustrates the degree of the secondary flows, which are defined as the difference between the total pressure and the static pressure divided by the maximum velocity head ($\frac{1}{2}\rho v_{max}^2$) of the cross section [12]. Secondary flows displace the position of the maximum total pressure towards the outside of the tube. The flow is fully developed along the coiled and secondary flows dominate the flow field. In Fig. 5b, the temperature contour of the cross section is slightly changed in the core but changes fast near the wall and the maximum temperature takes place near the inner side. Secondary flows lead to a pair of symmetrical vortices in the cross-section and the fluid trajectory is in the form of a double coil in Fig. 5c. The unbalanced centrifugal forces of the main flow result in the shift of the maximum velocity fields to the outer wall. The changes of turbulence kinetic energy are shown in Fig. 5d. The turbulence kinetic energy near the wall retains a relatively high level. The high turbulence kinetic energy region is located at the outer side of the sections due to the great gradient of velocity.

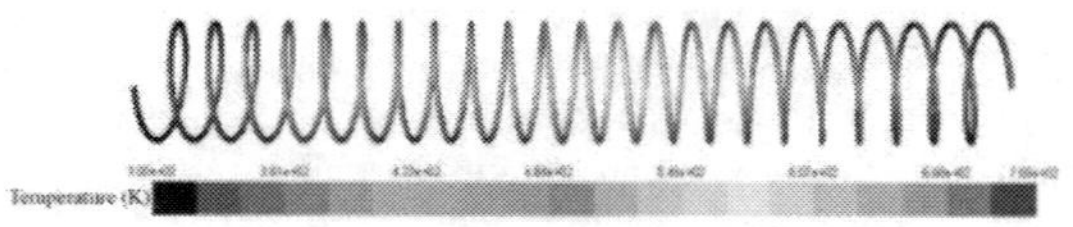

Figure 4: The outer wall temperature contour.

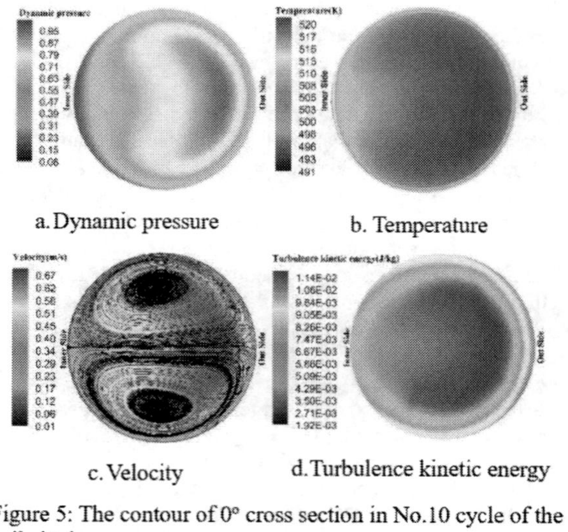

Figure 5: The contour of 0° cross section in No.10 cycle of the coiled tube.

Figure 6 shows the wall temperature and deposition profiles along the tube as the increase of heating time. The local wall temperature linearly increases along the tube. The extent of wall temperature profile increases as heating progressed. There is small fluctuation in the distribution of mass of deposition along the tube. In Fig. 7, the increase rate of *Re* along the axial of the tube is enhanced with the fuel temperature increasing because the viscosity of the supercritical fluid decrease inducing an increase in the velocity after the critical temperature. In our previous study [13], we conducted the deposition experiment in a straight tube reactor with the same heating time, flow rate and outlet fuel temperature to the coiled tube experiment. In comparison with the situation of the coiled tube, the wall temperature distribution along the straight tube has much fluctuation and its values have nearly 100K higher than that of the coiled tube because secondary flows enhance the heat transfer of the flow in the coiled tube. In addition, the inlet effect is not very obvious on the deposition near the inlet because of high turbulent of the flow in the coiled tube comparing to the straight tube. The average deposition rate of the coiled tube is 0.2034mg/cm^2, which is higher than that of 0.1813mg/cm^2 in the straight tube. Because under the turbulent condition of relatively high Re number, the heat and mass exchanges are enhanced and there is a thinner boundary layer, which benefits the deposition process [10].

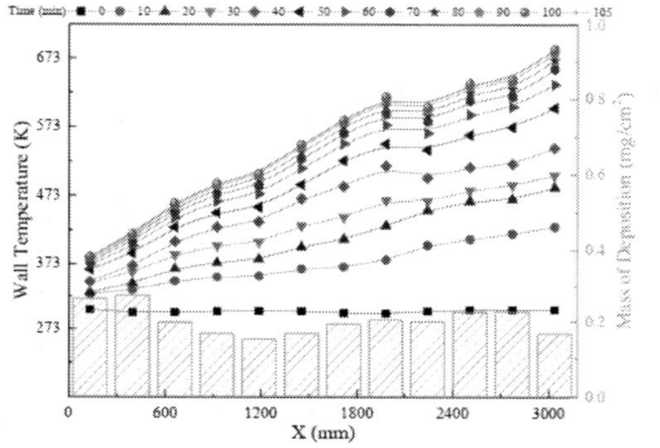

Figure 6: Outer wall temperature and deposition profiles at different locations.

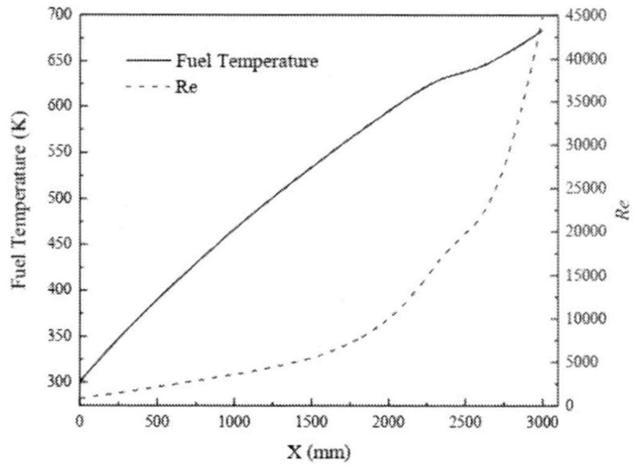

Figure 7: The fuel temperature distribution and Re number along the axial of the reaction section.

The composition comparison of original and residual fuel after heating analyzed by GC/MS is presented in Fig. 8. Original RP-3 is primarily composed of straight-chain alkanes (54.4 wt%), cycloalkanes (21.3 wt%) and aromatic hydrocarbons (21.5 wt%). It is noted that the amount of straight chain alkanes decreases almost 50% after oxidation deposition reactions and there is a significant increase in oxygen-containing compounds compared with the original kerosene.

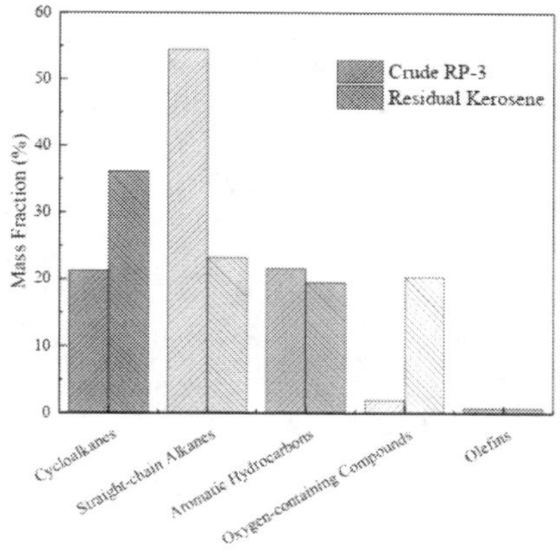

Figure 8 : Comparison of original and residual fuel compositions.

To facilitate data comparisons, the inner surface of a reactor tube before a coking trial was analyzed by SEM and a typical image is presented in Fig. 9. The original surface exhibits random ravines and embossments, which will tend to promote the deposition of insoluble substances due to an enlarged surface area. In Fig. 8b, the deposit is composed of amorphous and flocculent particles. XPS analysis of the deposition sample obtained from fuel containing a saturated dissolved oxygen is shown in Table 4. These data show relatively high C and O levels of 65.49% and 26.26%, respectively. Based on fuel compositions and deposition elements analysis, it is proved that the oxygen-containing compounds have a critical impact on the formation of oxygen deposition in the oxidation deposition mechanism.

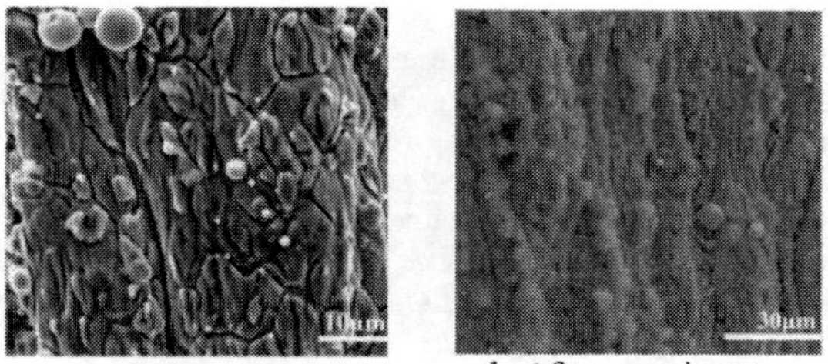

a. Before experiment b. After experiment

Figure 9: SEM micrographs of the inner wall of the coiled tube.

Elements	Si	S	C	N	O	Cr	Fe
Atomic	0.82	0.78	65.49	1.95	26.26	0.45	4.25

Table 4: Elemental composition of the deposition

The heat transfer has been investigated at different flow rates from 0.5 g/s to 3 g/s and heat fluxes from 25 kW/m^2 to 65 kW/m^2. The flow regime in the coiled is dominated by the secondary flow, which effect can be taken into account by the Dean number, defined as $De = Re\sqrt{\frac{r}{R}}$ [14]. The correlation of the coiled tube heat transfer based on a general correlation for forced convection heat transfer that is the Dittus-Boelter correlation as $Nu_o = 0.0243 Re_b^{0.8} Pr_b^{0.4}$ [15]. Supercritical properties of the fuel are taken into account by the function of the boundary layer. The Nusselt number correlation of supercritical fuel in the coiled tube is modified as below:

$$\frac{Nu}{Nu_o}=0.0741\left(\frac{\rho_f}{\rho_b}\right)^{0.2942}\left[1+21.4\left(\frac{d}{x}\right)\right]\left(1+De^{-0.7993}\right)\left[1+11.75\left(\frac{Nu}{Nu_o}\right)^{1.102}\right]$$

In which, the exit effect is considered and corrected by $\left[1+21.4\left(\frac{d}{x}\right)\right]$, in which d is the inner diameter of the tube and x is the position of the thermocouple. In the subscript, f represents the considering supercritical properties by the boundary layer effect, such as $\rho_f=\frac{\int_{T_b}^{T_w}\rho(T)dT}{T_w-T_b}$ and b is a calculated value; w is the measured wall temperature. The comparison of the results calculated by the modified correlation equation and the experimental data is plotted in Fig. 10. The error between the experimental and calculated values is within ±10%.

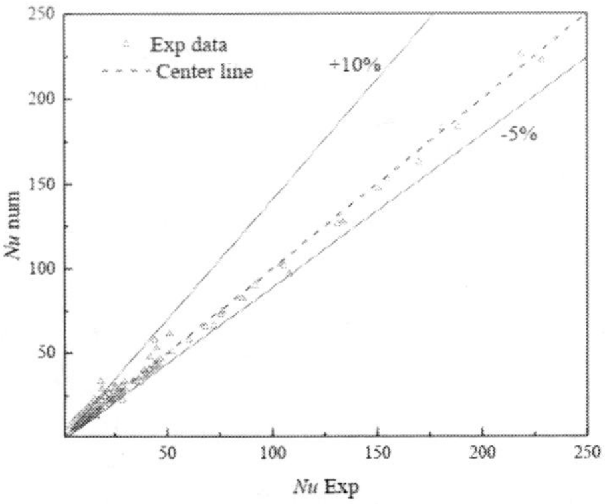

Figure 10: Comparison of the result calculated by the modified correlation equation and experimental data in the coiled tube.

4 Conclusions

A coiled tube with 3000mm long and 23 cycles has been employed as the heat-exchanger to investigate the deposition and heat transfer characteristics of supercritical aviation fuel RP-3 both in experimental and numerical methods. The flowing experiments show that the relative high Re number and turbulent flow in the coiled tube enhance considerably in the heat transfer and slightly in the deposition rates. According to fuel compositions and deposition elements analysis, it can be concluded that the oxygen-containing compounds have a critical impact on the formation of oxygen deposition in the oxidation deposition mechanism. The effects of secondary flows in the coiled tube were illustrated numerically in the distribution of dynamic pressure, temperature, velocity and turbulence kinetic energy in the circular cross-section located in the middle of the tube. Finally, considering the effects of supercritical properties of the fuel, boundary layer, secondary flow and exit effect, the correlation of heat transfer in the coiled tube is modified based on the experimental data of the aviation fuel of RP-3.

Acknowledgments

This work was supported by the National Natural Science Foundation of China (91641114). We also acknowledge Competitive Research Funding provided by the Clean Combustion Research Center, King Abdullah University of Science and Technology (KAUST).

Nomenclature

d	Inner diameter	mm
De	Dean number	
h	Heat transfer coefficient	W/(m^2.K)
Nu	Nusselt number	
P	Coil pitch	mm
Pr	Prandtl number	
q	Wall heat flux	kW/m^2
r	Inner radius	mm
R	Coil radius	mm
Re	Reynolds number	
T	Temperature	K
v	Velocity	m/s
x	Axial position	mm
ρ	Density	kg/m^3
κ	Thermal conductivity	W/m.K
Subscripts		
f	Supercritical effect	
b	Bulk fuel	
w	Wall	

References

[1] T. Edwards, J. Propul. Power, 19 (6) (2003) 1089-1107.
[2] X.Y. Pei, L.Y. Hou, Z.Y. Ren, Energy & Fuels, (2017).
[3] Z. Tao, Y.C. Fu, G.Q. Xu, H.W. Deng, Z.X. Jia, Energy & Fuels, 28 (9) (2014) 6098-6106.
[4] H. Jiang, J. Ervin, Z. West, S. Zabarnick, J. Thermophys. Heat Transf., 27 (4) (2013) 668-678.
[5] N.J. Kuprowicz, S. Zabarnick, Z.J. West, J.S. Ervin, Energy & Fuels, 21 (2) (2007) 530-544.
[6] K.Z. Wang, X.X. Xu, C. Liu, W.-J. Bai, C.-b. Dang, International J. Heat and Mass Transfer, 108, Part B (2017) 1645-1655.
[7] W. Zhang, S. Wang, C. Li, J. Xu, Appl. Therm. Eng., 88 (2015) 61-70.
[8] I.L. Pioro, R.B. Duffey, Nuclear Engineering and Design, 235 (22) (2005) 2407-2430.
[9] X.Y. Pei, L.Y. Hou, C.K. Mo, N. Dong, J. Aerospace Power, 30 (9) (2015) 2122-2128.
[10] X.Y. Pei, L.Y. Hou, Z.Y. Ren, Energy & Fuels, 29 (9) (2015) 6088-6094.
[11] A.F. Ansys, ANSYS inc, (2011) 390-391.
[12] M. Rowe, J. Fluid Mechanics, 43 (4) (2006) 771-783.
[13] X.Y. Pei, L.Y. Hou, Fuel Processing Technology, 142 (2016) 86-91.
[14] X.Y. Pei, L.Y. Hou, Fuel, 167 (2016) 68-74.
[15] R.H.S. Winterton, International J. Heat and Mass Transfer, 41 (4–5) (1998) 809-810.

11th Asia-Pacific Conference on Combustion,
The University of Sydney, NSW Australia
10th -14th December 2017.

Turbulence Model Effects on Multiple-Swirl Flame Aerodynamics

Sandeep Jella[1,2], Wing Yin Kwong[3], Jeffrey Bergthorson[1], Gilles Bourque[2], Adam Steinberg[3]

[1]McGill University
845 Sherbrooke St W, Montreal, Quebec, H3A 0G4, Canada

[2]Siemens Canada Limited
9545 Cote-de-Liesse, Dorval, Quebec, H9P 1A5, Canada

[3]University of Toronto Institute for Aerospace Studies
4925 Dufferin St, North York, Ontario, M3H 5T6, Canada

Abstract

Reynolds-Averaged Navier-Stokes (RANS) turbulence models are commonly used for design of gas turbine combustion systems which feature strongly anisotropic flow dynamics. Advanced turbulence models provide closures for the individual Reynolds stresses or by relying on scale resolution. In this paper, a typical multiple-injector configuration is studied using two RANS and two Large Eddy Simulation (LES) turbulence models in Ansys Fluent, a commercial CFD code. Model validation data consists of 10 kHz Stereo-Particle Image Velocimetry (S-PIV) and planar Laser-Induced Fluorescence of OH (OH-PLIF). All turbulence models predict the mean velocity field reasonably well, but the fluctuating velocity field and flame brush is predicted best by LES with explicit transport of sub-grid kinetic energy.

1 Introduction

A major requirement of injector design in lean, technically-premixed gas turbines, is the accurate prediction of aerodynamic features (shear layers, wakes) characteristic of the flame-stabilizer. Typically, a high speed jet on the order of 50-100 m/s issues from the premixer into the combustion chamber and is decelerated rapidly to prevent an excessively lifted flame leading to incomplete combustion. Since chamber residence times are very short (e.g. to minimize thermal NO), the flame anchoring aerodynamics need to be effective (to minimize CO). A controlled separation is often sought to minimize pressure losses. Swirl and sudden expansion are commonly used but this results in geometry-induced flow anisotropy which poses a challenge to computational models used for optimizing these features. Standard industrial (i.e. RANS) models often assume isotropy of Reynolds stresses and include empirical constants to compensate. For widely used combustion models – such as those based on presuming a pdf shape and progress variables, c, – the turbulence model governs the mean reaction rate through the variance, c''^2, of the progress variable. The focus of this paper is the net effect of the turbulence model on predictions of the flow dynamics and consequently, the mean reaction rate. Four models of increasing complexity (and computational cost) were chosen to model an industrially-relevant flame configuration.

Corresponding author
E-Mail: sandeep.jella@siemens.com

2 Experiment and Model Description

The Linear Array Swirler system (Fig. 1 can incorporate real gas-turbine complexities (pressure, preheat, partial-mixing etc.) in a systematic manner. In the premixed version, a methane-air mixture is supplied through a common plenum to five gas-turbine swirler injectors set at a pitch of 1.34 times the diameter of each swirler. The confinement was sized to provide realistic chamber residence times at full flow rates.

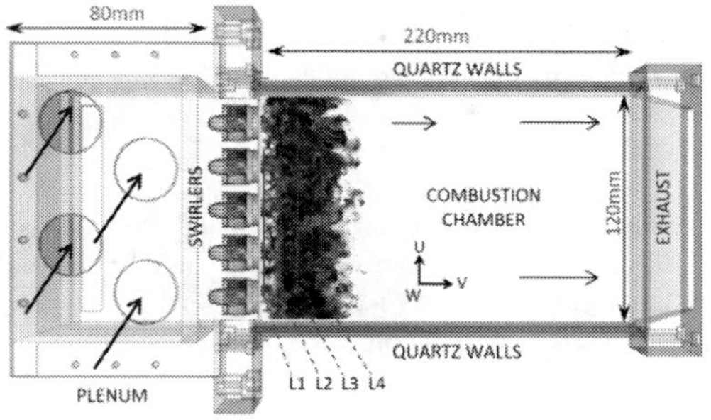

Figure 1: Linear Swirler Array Rig Schematic - Flame depicted in blue as volumetric rending of $c = 0.5 - 0.9$ from an LES. U,V,W are the measured velocity components

Simultaneous 10kHz S-PIV and OH-PLIF was carried out to measure the three component velocity field (U,V,W) and determine the heat release zones for the three central nozzles corresponding to area of 62×41mm with a spatial resolution of 1.7mm. The maximum uncertainty in each velocity vector from the PIV was on the order of 0.027m/s with the average being less than 0.005m/s. The four red lines labeled L1 (swirler exit) to L4 (neighboring jets merging) in Fig. 1 represent the axial locations where measurements were used to validate the CFD predictions. The relevant flow rates and premixed flame properties are summarized in Table 1.

Table 1: Boundary Conditions

T_{preheat}	$\dot{m}$	δ_{f}	T_{flame}	Φ	Da/Ka
500	11.1	0.71	1732	0.55	0.14/121

UNITS: mm, bar, g/s, Kelvin

A polyhedral-hexahedral computational mesh of approximately 9 million cells was found sufficient for LES to show negligible variation in the second order (rms) statistics. An LES index of quality (LESIQν) as defined in [1]) of more than 85% of the turbulence kinetic energy was achieved by this mesh, though the effects of numerical dissipation were not included. Perfectly cubic cells of characteristic length 0.5mm were used in the flame zone, 0.25 mm polyhedral cells were used to mesh the swirler vane passages. Favre-filtered/averaged conservation equations were solved using a second order upwinding scheme for all scalars. In the case of LES, a bounded central differencing scheme was used for the convective terms in the momentum equation. The LES flow-time step was determined by a maximum CFL number of 0.667 (3e-6s) while in RANS a time-step of 3e-5s was chosen. Flow statistics were collected until the Reynolds stresses converged.

3 Physical Models

Turbulence

All turbulence models have to address closures for flow (Reynolds) stress terms ($\overline{u_i u_j}$) that appear when the exact Navier-Stokes equations are averaged (RANS) or filtered (LES). These velocity correlations are representative of the intermittent motions of coherent structures with kinetic energy taken from the bulk flow and strongly augment or attenuate reaction rates which are exponentially dependent on the associated thermo-chemical fluctuations. Four turbulence models were investigated – two in RANS and two in LES. For brevity we focus on the closures for $\overline{u_i u_j}$, described below in increasing level of fidelity to the real physics. The 'turbulent' viscosity, μ_t, is how the turbulence model impacts the momentum equations.

k-ω SST

Menter's RANS-SST model [2] is exemplary of two-equation models assuming a linear relation between the Reynolds stresses and the mean strain-rate field, $\overline{S}_{ij}$, using an isotropic eddy viscosity (Boussinesq hypothesis).

$$
\begin{aligned}
\overline{u_i u_j} &= \frac{2}{3}k\delta_{ij} - 2\nu_{\mathrm{T}}\overline{S}_{ij} \\
\nu_{\mathrm{T}} &= \frac{k}{\omega}\mathrm{MIN}\left(\alpha^*(\mathrm{Re}_t), \frac{a_1\omega}{||S_{ij}||F(k,\omega,y,\mu)}\right)
\end{aligned} \tag{1}
$$

The functions F and α^* in Eq. 1 allow the SST model to transition in a manner consistent with the change of ratio of production-to-dissipation from low (i.e. near wall) to high (free shear) Reynolds numbers.

k-ω Explicit Algebraic Reynolds Stress Model (EARSM)

The RANS k-ω Explicit Algebraic Reynolds Stress Model (EARSM) model is based on Wallin and Johansson's work [3, 4]. Explicit expressions for the Reynolds stresses tensor are obtained assuming Rodi's weak equilibrium hypothesis between the Reynolds stresses and the mean velocity gradient (and a non-linear stress-strain relationship) [5]:

$$
\begin{aligned}
\overline{u_i u_j} &= \frac{2}{3}k\delta_{ij} + \frac{a_{ij}}{2k} \\
\nu_{\mathrm{T}} &= \frac{k}{\omega}
\end{aligned} \tag{2}
$$

where $a_{ij}/2k$ is the normalized anisotropy tensor calculated explicitly as a root of a tensor polynomial decomposition [4]. The resulting algebraic expressions provide the individual Reynolds stresses, which are used to close the production term for the k equation in Menter's k-ω BSL (baseline) model. This is used to obtain the turbulent viscosity in the usual manner.

Localized Dynamic Sub-grid Kinetic Energy Model (LDKM)

In LES a transported equation for sub-grid k is used and can include a production term instead of being purely dissipative (e.g. Smagorinksy) but uses the Boussinesq assumption at the SGS scale to calculate τ_{ij} the sub-grid stress:

$$
\begin{aligned}
\tau_{ij} \equiv \overline{\rho}\left(\overline{u_i u_j} - \overline{u}_i\overline{u}_j\right) &= \frac{2}{3}\rho k\delta_{ij} - 2C_k\rho k^{1/2}\Delta\overline{S}_{ij} \\
\nu_{\mathrm{T}} &= C_k k^{1/2}\Delta
\end{aligned} \tag{3}
$$

The constant C_k is calculated dynamically using the procedure discussed by Kim and Menon [6]. The model has the advantage of accounting for history and local effects. It avoids numerical issues due to negative turbulence viscosity when backscatter occurs. A conventional turbulence viscosity is used at the SGS level but this effect is at a much smaller scale than in the RANS context in the preceding paragraphs. Due to direct resolution of the Reynolds stresses (up to the mesh resolution), closure of SGS terms for k scale can be much simpler compared to RANS and has the advantage of being functional on resolved scale quantities. The model degenerates to zero turbulence viscosity (laminar flow) when all velocity gradients are resolved.

Wall-Adaptive Large Eddy Simulation (WALE)

Solving transport equations and/or dynamic evaluations to obtain model constants at the sub-grid scale level is often said to be computationally expensive for LES methods in addition to DNS mesh requirements at the wall. Nicoud *et al.* [7] proposed the algebraic WALE model to reproduce correct behavior of flow decay near the wall without an exceedingly fine mesh and obtain the turbulence viscosity from resolved velocity gradients:

$$
\nu_{\mathrm{T}} = (C_w\Delta)^2\frac{(S^d_{ij}S^d_{ij})^{3/2}}{(\overline{S}_{ij}\overline{S}_{ij})^{5/2} + (S^d_{ij}S^d_{ij})^{5/4}} \tag{4}
$$

where S^d_{ij} is the traceless, symmetric part of the *square* of the velocity gradient tensor and is calculated as a function of the resolved strain and vorticity fields. Like the LDKM model, it degenerates to a laminar flow when necessary. Sub-grid stress modeling is not required and a transport equation is not solved.

Combustion

A flamelet-progress variable method was selected where a reaction-progress variable, $c = Y_{\mathrm{CO}} + Y_{\mathrm{CO_2}}$ and its variance, c''^2 characterizes the flame evolution. The mean reaction rate is modeled as $\overline{\omega}_c = \int_0^1 \dot{\omega}_c(\zeta)\, P(\zeta)\, d\zeta$, where $\dot{\omega}_c(\zeta)$ is obtained in composition space (c-space) using *flamelet generated manifolds* [8, 9]. The thermo-chemistry is tabulated after convolving with a beta-pdf shape for c. Of the terms in the variance transport, the scalar dissipation rate (SDR) of c''^2 is critical and has been the subject of recent research [10]. In the present work, a linear relaxation model (i.e. a passive scalar closure) has been retained for all turbulence models. Whilst this can be a issue for variance predictions in RANS in the flamelet regime, it is of interest to see whether LES is as sensitive. In LES, the reactive SDR is

– to an extent – resolved, and takes into account the effect on large eddy dynamics by heat release. From this perspective, a comparison to a RANS model using a passive closure for the SDR is not entirely appropriate.

4 Results

Flow

Figure 2 shows the comparison between models and experiment for the time-averages of the three velocity components for the central three swirlers. All turbulence models predict the swirler exit flow ($U_0 \sim 40$ m/s) and its development reasonably well but exhibit inaccuracies at the point where the reverse flow transitions to positive, gaining a 'net' swirling motion at the same time (Location L4). This is more evident when inspecting the U (transverse) and W (out-of-plane, swirl-related) profiles. Both RANS models under-predict the strength of recirculation (Location L1 and L2 in the V profile). In general, the flow retains 'memory' of its exit – a V-shaped profile for each swirler (artifact of the 2D slice) and wakes in between the swirlers. These small wakes are only seen in the L1 profiles as they dissipate rapidly, but the the swirler exit profile remains well-defined until a distance of approximately 3 times the swirler diameter – at which point the flow is essentially a large, slow mass of swirling post-flame material.

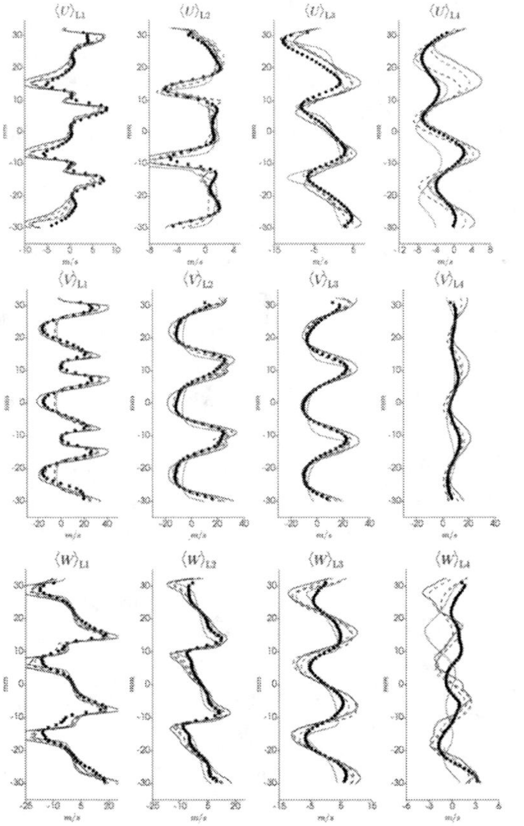

Figure 2: Velocity (top: U, middle: V (Axial), bottom: W) RANS (- - -SST, ——EARSM), LES (- - -LDKM, ——WALE), • Experiment. Flow is from left to right (i.e. L1 to L4).

Figure 3 compares the predictions of local turbulence intensity where u_0' includes all three fluctuating components of velocity. The peak values appear to be on the order of 20% of the inlet bulk velocity (U_0) close to the swirler exit. After additional generation of turbulence due to jet-jet interaction and flame-induced flow acceleration, the flow settles down to a spatially uniform value (L4) that then decays through the rest of the chamber. The SST model consistently under-predicts the intensity while the WALE model consistently over-predicts it. The EARSM and LDKM models perform best but the anisotropic part of the Reynolds stress tensor differentiates them further (Fig. 4).

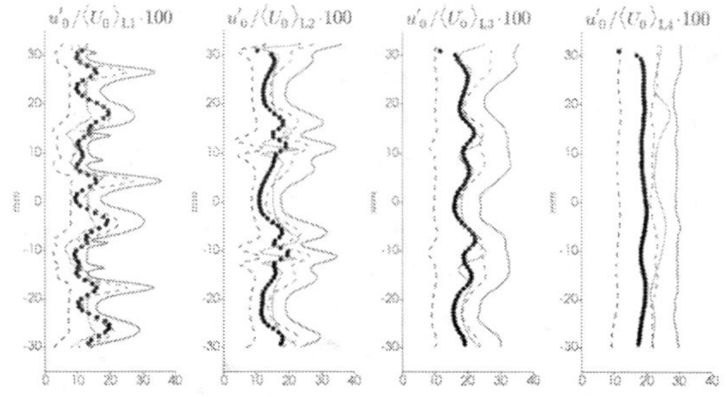

Figure 3: Turbulence Intensity

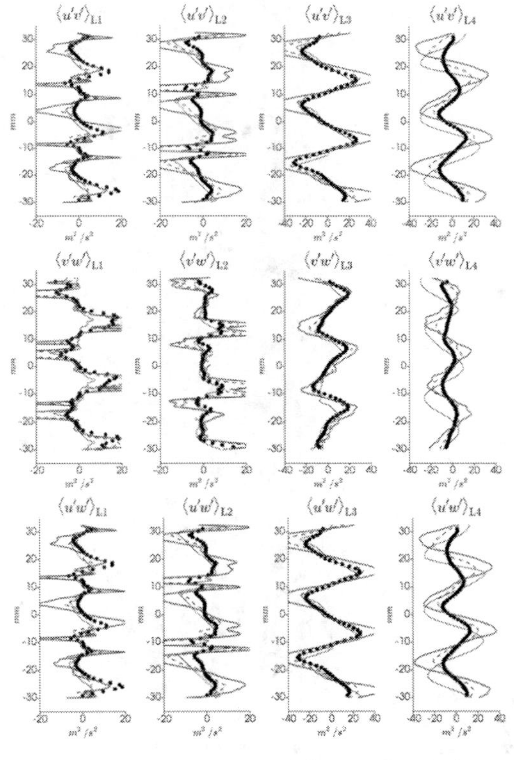

Figure 4: Shear Stresses (top: $u'v'$, middle: $u'w'$, bottom: $v'w'$). Note that scales are chosen differently for L1 and L2 for clarity. Flow is from left to right.

The nozzle-to-nozzle interaction (jets merging) reaches a a maximum at L3 and corresponds to the maximum level of activity amongst the shear stresses, even though little change is seen in the level of normal stresses (c.f. u_0' in Fig. 3 at L3). This implies that energy redistribution effects are dominant over kinetic energy production at L3. It is plausible that the cause for this is the simultaneous effects of jet-jet interaction and heat release. The shear stresses also show a directional bias at L4 as higher magnitudes in the y-axis range $-30 < y < 0$. This is in the direction of a global swirl imposed by the co-swirling jets and the fact that it is mostly evident in the $\overline{u'w'}, \overline{u'v'}$ components indicates heat and mass redistribution in the transverse direction is favored. While the LDKM model provides the best description of these complex trends, the EARSM provides a surprisingly good description (up to L3) if a little over-emphatic about the global swirl. WALE over-predicts the stresses, a consequence of pulling too little kinetic energy from the bulk-flow and does not properly predict the directional bias of the anisotropy at L4.

Combustion

Fig. 5 compares the experimentally derived flame location (using a thin flame assumption on the OH gradient) to the models. For a preliminary, qualitative check, the distribution of the chemical source of c is used as an indicator of flame location in the CFD. All models predict similar flame lengths and location of maximum interaction ($\sim$ 3 nozzle diameters downstream) where the flow anisotropy reaches a peak.

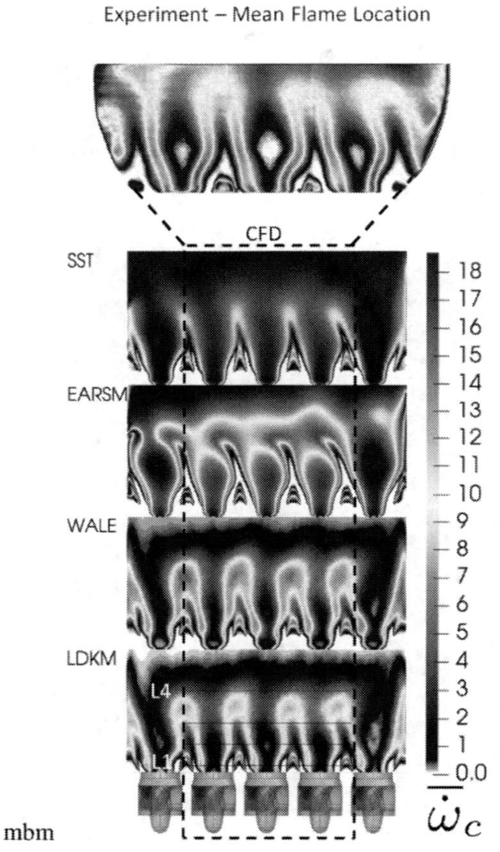

Figure 5: Qualitative estimate of Flame location. Black dashed line is imaged section. Flow is from bottom to top.

Turbulence is predominantly controls flame stabilization given the low Da and Ka (Table. 1). Both RANS models show a very thin flame brush at the base of the flame prior to where the jets merge (L3-L4). The SST consistently predicts a 'flamelet'-like brush as it under-predicts the flow fluctuations as shown earlier. In the L3-L4 region, the EARSM predicts a stronger net-swirl of the bulk flow than the LES or the experiments and artificially increases the reaction-rate. This is evident from the model's mean U and W profiles in Fig. 2 at L3 and the over-predicted anisotropy of the $\overline{u'w'}$ and $\overline{v'w'}$ stresses at the same location in Fig. 4, which emphasizes the role of swirl components (W and w'). The errors in the WALE model appear to lead to an increased reaction rate compared to the LDKM model. The LDKM model better satisfies causality of flame-flow interaction as it matches the turbulence kinetic energy as well as the normalized shear stress values. Flame lengths correspond to the experiment but other issues such as presuming the pdf and chemical sources of variance will have first-order effects and these must be investigated next.

5 Conclusions

Turbulence model effects were investigated for an industrially-relevant gas turbine flame configuration featuring multiple, interacting flames. Regions of increased flow anisotropy directly corresponded to the flame-flame interaction and implies that RANS models must account for the flow anisotropy in addition to reactive scalar effects in the variance. To this end, LES modeling with transported sub-grid kinetic energy performs best overall but algebraic modeling of the Reynolds stresses appears to be a reasonable compromise.

Acknowledgments

The experimental work in this paper was funded by the Natural Sciences and Engineering Research Council of Canada and Siemens Canada under the CRD program. HPC resources were provided by a Compute Canada Grant. Computations were performed on the GPC supercomputer at the SciNet HPC Consortium at the University of Toronto.

References

[1] J. Meyers and B. J. Geurts. *Quality and reliability of large-eddy simulations*. Springer, 2008.

[2] F. Menter, M Kuntz, and R Langtry. *Turbulence, heat and mass transfer* 4 (2003).

[3] S. Wallin and A. V. Johansson. *Journal of Fluid Mechanics* 403 (2000).

[4] F. Menter, A. Garbaruk, and Y Egorov. *Progress in Flight Physics* 3 (2012).

[5] S. B. Pope. *Turbulent flows*. Cambridge University Press, 2000.

[6] S Menon and W.-W. Kim. *AIAA, 35th Aerospace Sciences Meeting and Exhibit - AIAA Paper 97-0210* (1997).

[7] F. Nicoud and F. Ducros. *Flow, turbulence and Combustion* 62 (1999).

[8] J. van Oijen, F. Lammers, and L. de Goey. *Combustion and Flame* 127 (2001).

[9] P.-D. Nguyen, L. Vervisch, V. Subramanian, and P. Domingo. *Combustion and Flame* 157 (2010).

[10] N. Swaminathan and K. N. C. Bray. *Turbulent Premixed Flames*. Cambridge University Press, 2011.

11th Asia-Pacific Conference on Combustion,
The University of Sydney, NSW Australia
10th -14th December 2017.

Flame Chemiluminescence Measurements of a Laminar Forced Flame Interacting with a Cold Wall

Jacob E. Rivera, Robert L. Gordon, Mohsen Talei

Department of Mechanical Engineering
The University of Melbourne, Parkville, Victoria, 3010, Australia

Abstract

This study presents flame chemiluminescence measurements of a flame undergoing transient flame-wall interaction. The atmospheric, fully premixed, forced laminar burner is designed to investigate the flame quenching behaviour as the flame oscillates. Near the wall, there is low contrast, increasing the uncertainty in edge detection. The image processing algorithm is designed to detect the flame edge given this uncertainty. An Abel inversion was also applied to account for the limitations of line-of-sight imaging, resulting in more consistent edge-detection. The definition of the flame edge location is also a source of uncertainty in the measurement of quenching distance. Two different definitions were compared. The flame-wall distance at the flame tip is more invariant across the whole cycle, than the minimum flame edge to wall distance. The resulting Peclet numbers are comparable to those reported in the literature. The quenching distance was observed to change with axial velocity modulations, suggesting a link between these two parameters.

1 Introduction

Compliance with regulatory emissions standards is a key consideration of modern combustor design, especially for power generation gas turbines. Strict NO_x regulations have forced these lean premixed gas turbines to operate with lower combustion temperatures, increasing their CO emissions. Hence, it is important to understand the fundamental mechanisms that produce these emissions, such as inefficient fuel atomisation and reactant mixing, and flame-wall interaction (FWI). Early work on FWI linked this effect to pollutant species associated with combustion inefficiency [1], however the difficulty in experimentally measuring near-wall quantities have confined most of these early studies to numerical simulation [2]. These studies established the quenching distance (δ_Q) as an important parameter that affects both the dynamics and the emissions of the flame near the wall.

The quenching distance is the distance of the flame from the wall at the point of quenching [3]. This parameter's most obvious consequence is the wall heat transfer, which is inversely proportional to quenching distance [4]. Apart from the potential damage to the combustor liner, a small quenching distance quenches the flame near the wall via a loss of enthalpy [5]. This loss of enthalpy activates radical recombination reactions near the wall. This depletes the flame of active radicals and leaves un-oxidised intermediates [6], linking the quenching distance with emissions. Turbulent vortices complicate this by quenching multiple regions of the flame simultaneously [7].

Previous studies of quenching distances have also used flame chemiluminescence measurements (CL) [8, 9]. Their work used a constant volume chamber to examine the propagation of a spark ignited flame towards a cold obstacle. This means that the flames interacted with the wall at a single point in time. In this case, the flame interacts with the wall continuously. This increases the uncertainty of experimentally determining quenching distances due to the low contrast of the near-wall region. This low contrast is a product of reduced reaction rate and flame thickening as a result of FWI [10].

This study presents the image processing and quenching distance quantification of flame CL images within one cycle of a lean, forced laminar flame. Several stages of edge detection were carried out to account for low contrast near the wall, and image artefacts. An Abel inversion was also applied to account for the limitations of line-of-sight imaging. These techniques were applied to one whole cycle, and the quenching distances are compared to values in the literature.

2 Experimental Methods

2.1 Burner Configuration

Figure 1a shows the burner, which is a modified version of the research group's forced laminar burner [11-13]. The modified version includes a centrally mounted water-cooled tube which represents the wall that induces the FWI in the test section. The test section is confined by a quartz tube allowing optical access. The laminar flame stabilises on a 25 mm circular port at the end of a tapered nozzle. The flame exhibits a variety of shapes, largely a function of equivalence ratio (ϕ) and flow velocity (u). The flame geometry used in this study is shown in Fig. 1b. The inlet conditions are $\phi \approx 0.81$ and $u \approx 1.24\, m/s$. While the burner configuration is axisymmetric,

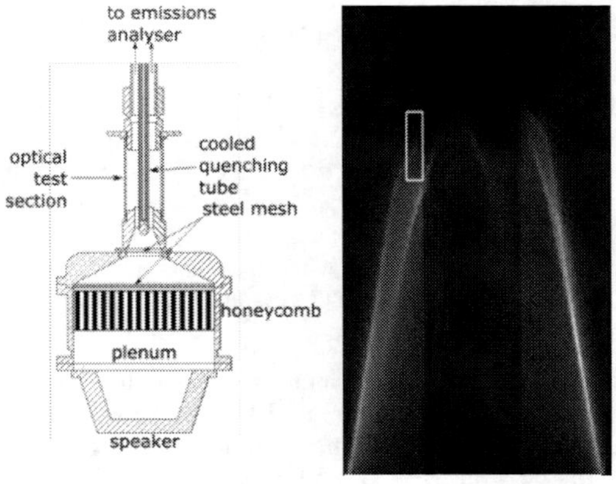

(a) (b)

Figure 1: Burner schematic (a) and the flame shape used (b) showing the region of interest (ROI).

Corresponding author. Phone: +61-03-8344-8876
E-mail address: jrivera@student.unimelb.edu.au

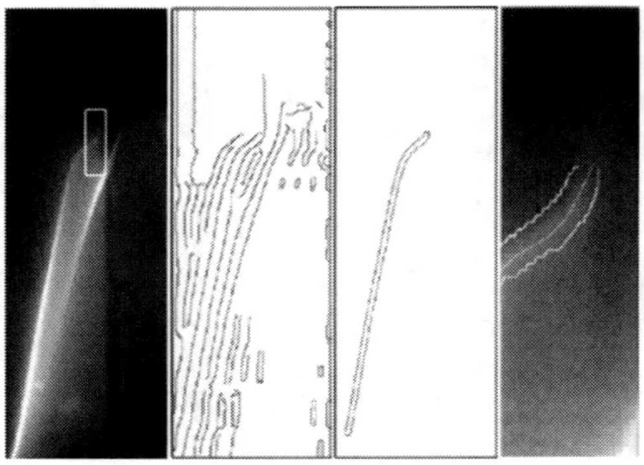

(a) (b) (c) (d)

Figure 2: An example of a cropped image with the ROI (a), single stage Canny edge detection result (b), closed boundary result (c) and the final detected edges in the ROI (d). Figures (a) and (d) are shown with artificially enhanced contrast.

imperfections in the current version of the water-cooled tube creates "hot-spots", which leads to flame wrinkling.

The fuel (liquid petroleum gas (LPG)) and air are metered by two MKS mass flow controllers (model 100B and 1599A respectively), and are mixed approximately 3m upstream of the plenum to ensure full premixed-ness. Once in the plenum, the reactants pass through honeycomb and a series of steel meshes to remove flow unsteadiness. A speaker (WOOFER L8S800 8 in) is used to force the flame. The speaker is driven by a sinusoidal signal, and produces sinusoidal axial velocity modulations. The driving signal is used as a reference for the camera trigger system. Full details of the burner configuration, as well as general flame behaviour, can be found in Ref. [13].

2.2 Flame Imaging

The images taken are single shot images at eight points in the 10 Hz cycle, with a 500 μs exposure time. These images were captured using a 30 Hz CCD camera (Point Grey GS3-U3-14S5M-C), with a 25 mm f/1.4 lens (TECHSPEC Ci Series). The camera was placed approximately 200 mm from the central plane of the burner, with a 35 mm field of view.

The camera was triggered using an in-house timing system, which phase-locks the trigger signal to the forcing signal of the speaker. This allows the camera to capture the flame at different points in its forcing cycle, for any arbitrary forcing frequency. This system is based on the our previous work [12].

2.3 Image Correction and Calibration

To account for dark noise and vignetting respectively, dark and white field corrections were conducted on all images. The pixel distance is $0.077\ mm/pix$. For the purpose of this discussion, the resolution error used is based on the half pixel distance ($\pm 0.04\ mm$). However, improved characterisation of the resolution uncertainty, via determination of the Modulation Transfer Function with the USAF-1951 target [14], will be performed for future work.

The images were cropped in the horizontal direction at the left nozzle edge and the centre of the wall. The region of interest (ROI) is the near wall region highlighted in Fig. 1b. The ROI is 65×20 pixels, or in millimetres, $5 \times 1.54\ mm$. The left edge of the wall is the right edge of the ROI.

2.4 Flame Surface Detection

The Canny edge detector from the $MATLAB$ image processing toolbox was used to initially detect the edges of the cropped image (Fig. 2a). As seen in Fig. 2b, the longest lines represent the edge of the flame, but many other spurious edges are also detected by this process. Some of these edges are due to the flame wrinkling forming edges out of the plane of interest. Others are from reflections and other image artefacts. Ideally, a threshold would be applied based on the length of the lines which would remove any line less than a certain length. However, due to the low contrast of the near wall region, there are many small disconnected edges detected in the ROI that are part of the flame surface. Hence, the image processing algorithm needs to remove the spurious edges, while retaining the small disconnected edges in the ROI that are part of the flame surface.

The algorithm uses the near-nozzle region as an anchor to determine the start of the flame surface, since this region has high contrast. The algorithm then finds the two left-most binary signals that are within one flame thickness of the stabilisation point. The flame thickness is estimated by the user and is an input into the algorithm. The coordinates of these two binary signals are stored as part of the flame edge. The algorithm then loops through the rows and takes the two binary signals that are within one flame thickness of the previous row's flame edge coordinates. This is repeated until the flame height is reached. Once the flame height is reached, all pixels outside of the flame edge coordinates are set to zero. The flame edges are then connected by linear interpolation, forming the closed boundary shown in Fig. 2c. This closed boundary is then filled, and used as a mask applied to the cropped image.

Once the mask has been applied, the peak signal can be found, rejecting out of plane flame edges and image artefacts. The closed boundary shown in Fig. 2c represents the edge of the flame where the peak signal gradient exists. Both the peak signal and the peak signal gradient have been used to define the flame front in previous flame CL studies [8], and the results for these are shown in Fig. 2d.

2.5 Effect of Abel Inversion

The line-of-sight nature of flame CL introduces error since the image is a planar projection of a 3D object. The field of view

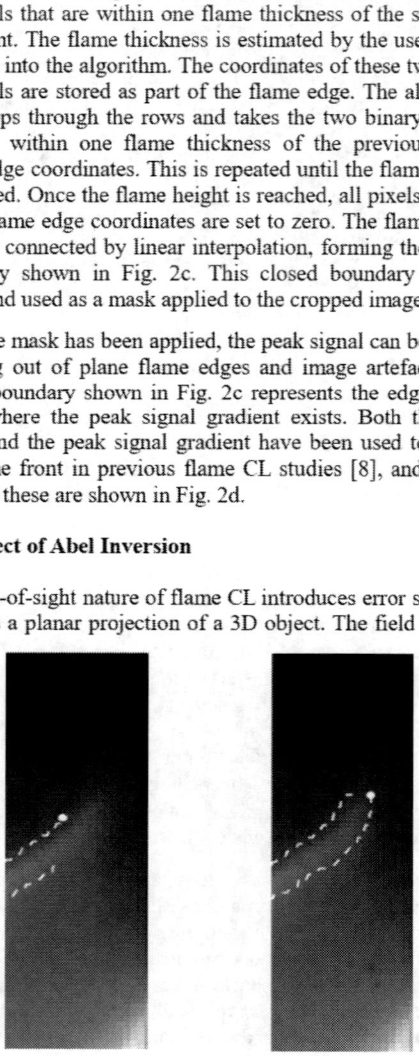

(a) (b)

Figure 3: Flame edge in the ROI for the image without Abel inversion (a) and with Abel inversion (b). Measured quenching distances are highlighted. Images are shown with enhanced contrast.

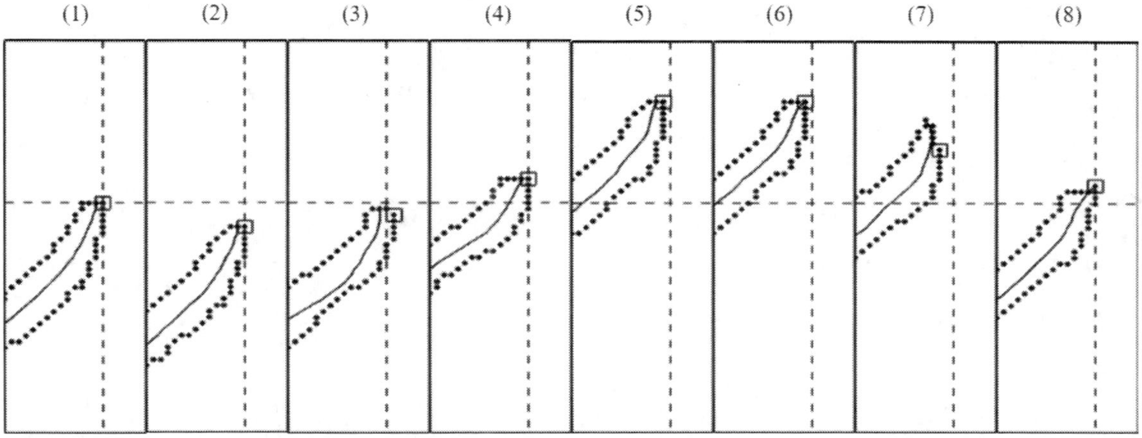

Figure 4: The ROI of all eight frames, with the flame edge results (dots) and the flame tip results (solid lines). The squares represent the δ_Q of each frame. The dashed lines reference the δ_Q of the first frame.

is larger than the nozzle diameter. This can introduce uncertainty when attempting to detect the flame edge.

To reduce this effect, an Abel inversion was performed before the edge detection. Using the method from Refs. [15, 16], an example of the results of the Abel inversion are shown in Fig. 3. It shows that the Abel inverted image (Fig. 3b) allows the flame surface to be more consistently captured, than the raw image (Fig. 3a). The inverted image is also more robust to low contrast. It tracks the flame front well for all images tested, whereas the raw image detects a false edge for two images. The following sections only present the results of the Abel inverted images.

3 Results

3.1 Quenching Distance

Figure 4 shows the change in both flame height and flame-wall distance for each of the eight frames of the cycle. These are summarised in Fig 5. The flame height is based on the coordinates of the flame tip. The two quenching distance measurements are based on the following definitions of the flame-wall distance:

1) The minimum distance of the flame edge from the wall (flame edge).
2) The minimum distance of the peak signal from the wall (flame tip).

The flame height shows a sinusoidal shape in response to the axial velocity modulations introduced by the speaker. Given the low frequency forcing signal ($10\,Hz$), the flame exhibits large macroscopic motion. This agrees with previous studies of the response of laminar flames to low frequency forcing [11, 12, 17, 18].

The quenching distances measured show little difference between the two definitions. The flame tip results are generally further from the wall than the flame edge results, while the flame edge results show larger variation across the cycle. Although for either case, the variations are much smaller compared to the variation of the flame height. The mean quenching distance across the cycle of the flame tip and flame edge results are $0.56\,mm$ and $0.49\,mm$ respectively.

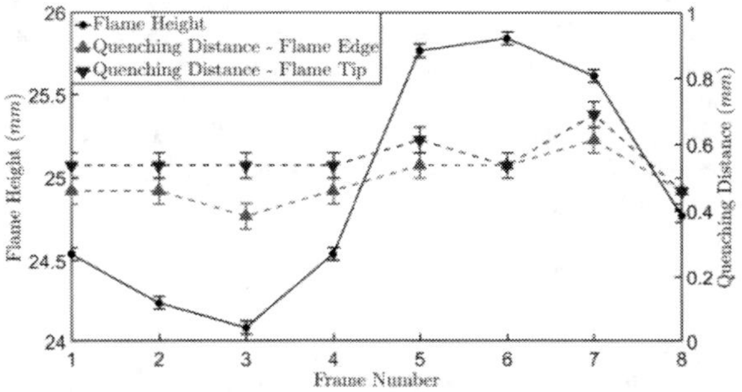

Figure 5: Flame height measurement (-) and quenching distance measurements (--), for each of the flame-wall distance definitions, across each of the eight frames.

The measured quenching distances (Fig. 5) show some resemblance to the shape of the flame height. This suggests that the quenching distance of a forced flame is affected by axial velocity modulations. However, further work is required to confirm this conclusion.

3.2 Peclet Number Comparison

Table 1 compares the flame edge quenching distances from Fig. 5, with side-wall quenching configuration values from the literature [8, 19]. The quenching distance was normalised with the Peclet number (Pe). This accounts for the differences in both the fuel (methane v. LPG) and equivalence ratio (1.0 v. 0.8) used [20]. The Peclet number used in the normalisation is defined as [3]:

$$Pe_Q = \frac{\delta_Q}{\delta_l^0} \tag{1}$$

$$\delta_l^0 = \frac{\alpha}{s_l^0} \tag{2}$$

The quenching Peclet number (Pe_Q) is the ratio of the quenching distance (δ_Q) and the laminar, unstretched, adiabatic diffusive flame thickness (δ_l^0). This diffusive flame thickness, defined as the ratio of the thermal diffusivity (α) and the laminar, unstretched, adiabatic flame speed (s_l^0), was calculated using the same method as [3]. The calculation was performed using CHEMKIN [21] and the LLNL C1-C4 mechanism [22], with propane as the surrogate for LPG.

Study	Fuel – ϕ	$\delta_Q\ (mm)$	Pe_Q
Flame Edge Result	LPG – 0.80	0.49*	8.15
Flame CL – CVC [8]	Methane – 1.00	0.32*	8.5
PIV-OHPLIF – V-Flame [19]		0.34	6.2
Theoretical [3]	–	–	7

Table 1: Quenching distance and Peclet number comparison, (*) indicates mean values. Values used are for side-wall quenching configurations.

The values used to compare with the results are from a flame chemiluminescence (CL) study [8], and a PIV-OHPLIF study [19], as well as the accepted theoretical value [3]. The flame CL study used a constant volume chamber (CVC) to examine the propagation of a spark-ignited flame along a cold wall. Hence, the flame interacts with the wall at only a single point. In contrast, the PIV-OHPLIF study used a V-Flame with one side continuously impinging on a cold wall. Hence, the latter configuration is closer to this study's configuration, but uses the former's measurement technique.

As seen in Table 1, the flame edge result is 16% larger than the theoretical value. The flame CL result is 21% larger, and the PIV-OHPLIF is 11% smaller than the theoretical value. This suggests that the measurements and the image processing algorithm give reasonable results for quenching distance. However, further work is required to confirm these results, as well as to explore the relationship of the forcing frequency on the quenching distance.

4 Conclusions

Image processing and flame chemiluminescence measurements have been conducted on an atmospheric, laminar, lean, premixed flame undergoing transient flame-wall interaction. Using a water-cooled tube, centrally mounted in an optically accessible test section, the flame was forced using a speaker in the plenum. Flame chemiluminescence images were obtained at eight points in one low frequency cycle, and processed to calculate the quenching distance.

Compared to previous flame chemiluminescence measurements of quenching distance in the literature, this burner presented several new challenges which the image processing algorithm addressed. The low contrast in the region of interest, the presence of "hot spots" and the limitations of line-of-sight imaging increased the uncertainty of measuring the quenching distance.

Two different methods of defining flame-wall distance at the point of quenching were tested. It was found that using the flame tip definition, produced less varying quenching distances across the cycle compared to using the flame edge definition. The resulting Peclet numbers are comparable to flame chemiluminescence and advanced laser diagnostic measurements from the literature, relative to theoretical values. Finally, the quenching distance was observed to change with the flame height. This may suggest a link between the axial velocity modulation and the flame behaviour near the wall.

5 Acknowledgment

Jacob E. Rivera acknowledges the contribution of Jan P. Felger and Samuel M. Wiseman for building the timing system. Robert L. Gordon acknowledges the support of the Newton International Fellowship.

References

[1] C. K. Westbrook, A. A. Adamczyk, and G. A. Lavoie, *Combust. Flame* 99 (40) (1981) 81-99.

[2] A. Dreizler and B. Böhm, *Proc. Combust. Inst.* 35 (1) (2015) 37-64.

[3] T. Poinsot, D. Haworth, and G. Bruneaux, *Combust. Flame* 95 (1993) 118-132.

[4] B. Boust, J. Sotton, S. A. Labuda, and M. Bellenoue, *Combust. Flame* 149 (3) (2007) 286-294.

[5] I. S. Wichman and G. Bruneaux, *Combust. Flame* 103 (4) (1995) 296-310.

[6] P. Popp and M. Baum, *Combust. Flame* 108 (3) (1997) 327-348.

[7] G. Bruneaux, T. Poinsot, K. Akselvoll, and J. H. Ferziger, *Combust. Flame* 107 (1-2) (1996) 27-36.

[8] M. Bellenoue, T. Kageyama, S. A. Labuda, and J. Sotton, *Exp. Therm. Fluid Sci.* 27 (3) (2003) 323-331.

[9] J. Sotton, B. Boust, S. A. Labuda, and M. Bellenoue, *Combust. Sci. Technol.* 177 (7) (2005) 1305-1322.

[10] A. Gruber, R. Sankaran, E. R. Hawkes, and J. H. Chen, *J. Fluid Mech.* 658 (2010) 5-32.

[11] N. Karimi, M. J. Brear, S. H. Jin, and J. P. Monty, *Combust. Flame* 156 (11) (2009) 2201-2212.

[12] S. M. Wiseman, M. J. Brear, R. L. Gordon, and I. Marusic, *Combust. Flame* 183 (2017) 1-14.

[13] J. E. Rivera, R. L. Gordon, M. Talei, and M. J. Brear, *Proc. 20th AFMC*, AFMS, Perth, Australia, 2016.

[14] G. D. Boreman and S. Yang, *Applied Optics* 34 (34) (1995) 8050-8052.

[15] G. Pretzler, *Z. Naturforsch. A.* 46 (7) (1991) 639-641.

[16] G. Pretzler, H. Jäger, T. Neger, H. Philipp, and J. Woisetschläger, *Z. Naturforsch. A.* 47 (9) (1992) 955-970.

[17] T. Schuller, D. Durox, and S. M. Candel, *Combust. Flame* 134 (1-2) (2003) 21-34.

[18] M. Talei, M. J. Brear, and E. R. Hawkes, *Combust. Flame* 159 (2) (2012) 757-769.

[19] C. Jainski, M. Rißmann, B. Böhm, J. Janicka, and A. Dreizler, *Combust. Flame* 183 (2017) 271-282.

[20] J. Andrae, P. Björnbom, L. Edsberg, and L.-E. Eriksson, *Proc. Combust. Inst.* 29 (1) (2002) 789-795.

[21] CHEMKIN-PRO 15131. Reaction Design: San Diego, 2013.

[22] N. M. Marinov, W. J. Pitz, C. K. Westbrook, A. M. Vincitore, M. J. Castaldi, S. M. Senkan, and C. F. Melius, *Combust. Flame* 114 (1-2) (1998) 192-213.

11th Asia-Pacific Conference on Combustion,
The University of Sydney, NSW Australia
10th -14th December 2017.

Kinetic Modelling and Experimental Study of Small Esters: Methyl Acetate and Ethyl Acetate

Ahfaz Ahmed[1], Marco Mehl[2], Nitin Lokachari[3], Elna J.K. Nilsson[4], Alexander A. Konnov[4], Scott W. Wagnon[2], William J. Pitz[2], Henry J. Curran[3], William L. Roberts[1], S. Mani Sarathy[1]

[1]King Abdullah University of Science and Technology, Clean Combustion Research Center
Thuwal-23955, Kingdom of Saudi Arabia
[2]Lawrence Livermore National Laboratory, Livermore, California-94550, United States of America
[3]Combustion Chemistry Center, National University of Ireland, Galway, Ireland
[4]Combustion Physics, Lund University, Lund, Sweden

Abstract

A detailed chemical kinetic mechanism comprising methyl acetate and ethyl acetate has been developed based on the previous work by Westbrook *et al.* [1]. The newly developed kinetic mechanism has been updated with new reaction rates from recent theoretical studies. To validate this model, shock tube experiments measuring ignition delay time have been conducted at 15 & 30 bar and equivalence ratio 0.5, 1.0 and 2.0. Another set of experiments measuring laminar burning velocity was also performed on a heat flux burner at atmospheric pressure over wide range of equivalence ratios [~0.7-1.4]. The new mechanism shows significant improvement in prediction of experimental data over earlier model across the range of experiments.

1 Introduction

Global energy trends show continuous increase in consumption of bio-based fuels for transportation sector [2]. This trend supports growing efforts to mitigate climate change by offsetting the consumption of fossil fuels. At present, a wide variety of bio-based fuels are proposed as alternatives to fossil fuels. Esters represent one such an important class of bio-derived fuels with potential to supplement gasoline as well as diesel fuel supply. While longer chain esters such as methyl oleate, methyl linoleate etc. are found suitable for diesel engine applications [3], smaller esters with higher RON values are found suited to gasoline range applications [4]. In principle, esters can be synthesized using various feasible pathways including vegetable oils, animal fats [5], algae [6] etc. Esters as a bio-based fuel makes an attractive alternative to fossil fuels due to its suited combustion properties and with proven synthesis pathways. However in the literature, there has been more focus on larger esters [3] and studies on combustion characteristics of smaller esters are limited. In past a couple of flame studies have been conducted with these fuels by Gasnot *et al.* [7], Dayma *et al.*[8] and Badawy *et al.*[9] in which various parameters like flame speeds, low pressure oxidation were investigated, Gasnot *et al.* [7] also proposed a kinetic model. There have been a few studies investigating ignition behavior of these fuels, Zhang *et al.* [10] studied high temperature ignition of methyl acetate in a shock tube and developed a kinetic mechanism. Kumegh *et al.* [11] also studied the high temperature ignition of methyl acetate and ethyl acetate in a shock tube, and proposed a kinetic model entailing these fuels. There is another work by Yang *et al.* [12] studying pyrolysis and oxidation of methyl acetate in low pressure flames, a mechanism was also proposed. There have been some other attempts to study these fuels blends in engines [13, 14] as well. Additionally, there have been some theoretical studies discussing the rates of H atom abstraction from fuel [15], decomposition of fuel [16] and fuel radicals [17].

In this study, a detailed chemical kinetic model for methyl and ethyl acetate (Fig. 1) has been developed. This model is advanced from the mechanism proposed for laminar premixed flames by Westbrook and coworkers in 2009 [1]. Acetates studied in this work are both high RON fuels with suitable physical and chemical properties [Table 1] to be considered as potential fuels in advanced gasoline engines [4]. Shock tube experiments measuring ignition delay time have been conducted at 15 & 30 bar and equivalence ratio 0.5, 1.0 and 2.0. Another set of experiments measuring laminar burning velocity have also been performed on a heat flux burner at atmospheric pressure over wide range of equivalence ratios. The model developed in this work shows good agreement with ignition data and laminar burning velocity data across the temperature and equivalence ratio range respectively.

Fuel	RON	MON	Density [Kg/m3]	LHV [KJ/Kg]
Methyl acetate	120	120	927.4	17.9
Ethyl acetate	118	120	894.6	21.34

Table 1: Properties of Methyl acetate and Ethyl acetate[4]

Methylacetate **Ethyl acetate**

Figure 1: Structure of Methyl acetate and Ethyl acetate

2 Experiments

In this study methyl and ethyl acetate have been studied in a shock tube facility and heat flux burner for ignition delay time and laminar flame velocity measurements, respectively. The following section discusses the experimental details and measurement procedure for the two facilities.

2.1 High Pressure Shock Tube [HPST]

The high temperature (1000-1400 K) ignition delay data in HPST has been measured behind reflected shock wave at 15 and 30 bar and equivalence ratio 0.5, 1.0, 2.0 as descripted in Table 2. This shock tube is made of 9 m long section of steel with uniform cross-section of 63.5 mm inner diameter. The length of shock tube is divided into three sections; driver section [3 m], driven section [5.7 m] and a section with double diaphragm. This double diaphragm section [30 mm] separates the driver and driven section and enables improved control of the generated shock wave. In this study helium or mixture of helium and nitrogen is used as driver gas in the shock tube during the experiments.

P/ bar	Phi	MA/mol %	O2/mol %	N2/mol %
15, 30	0.5	2.91	20.388	76.699
	1.0	5.65	19.807	74.514
	2.0	10.714	18.75	70.535
	Phi	EA/mol %	O2/mol %	N2/mol %
	0.5	2.05	20.568	77.375
	1.0	4.03	20.15	75.815
	2.0	7.749	19.373	72.88

Table 2: Details of shock tube experiments of methyl acetate [MA] and ethyl acetate [EA].

During experiments, pressure-time histories are measured using the Kistler 603B pressure transducer mounted on the end wall. The ignition delay in the HPST is defined as the time interval between two sharp pressure rises; one of them due to the shock wave reaching the end-wall and other due to initiation of ignition. The acceptable error for the measured pressures behind the reflected shock wave was ±0.5 bar. The NUI Galway HPST has been further discussed in detail in Nakamura *et al.*[18]

2.2 Heat Flux Burner

Laminar burning velocities of methyl acetate/air and ethyl acetate/air were determined using the heat flux method at Lund University. Measurements were performed at atmospheric pressure, initial gas mixture temperatures 298 and 338 K, and equivalence ratios in the range 0.7-1.4 for methyl acetate and 0.7-1.3 for ethyl acetate. The experimental setup and measurement method have been described in detail by Alekseev et al. [19], and hence it is briefly discussed in this work.

The heat flux method for determination of laminar burning velocities was first presented by de Goey and co-workers [20] in 1993. The method was originally introduced for gaseous fuels and was later extended to liquid fuels. The heat flux method is based on the principle that; at adiabatic burning velocity conditions there is no net heat transfer between the flame and the burner plate and hence temperature of the burner plate remains uniform. One of the advantages this method offers is the measurement of laminar burning velocity in a stretch free flame under adiabatic conditions, hence no corrections for flame stretch are required.

The composition of fuel/oxidizer mixture is metered using Mass Flow Controllers (MFCs) for metering the air, a Cori-Flow liquid mass flow controller for the liquid fuel and a Controlled Evaporator Mixer (CEM) for evaporation of the fuel. Uncertainties in gas mixture composition as well as in the laminar burning velocity is assessed as described by Alekseev *et al.* [19]. However no uncertainty quantification analysis is done in this work.

3 Kinetic Modelling

Detailed chemical kinetic mechanism has been developed for methyl and ethyl acetate based on previous work by Westbrook et al [1]. The former mechanism has been updated with additional reaction rates and reaction classes based on theoretical calculations and structurally similar molecules. In present work Aramcomech 2.0 [21] has been used as the base chemistry.

3.1 Methyl acetate

Kinetic reaction scheme of methyl acetate comprises of H abstraction reactions followed by decomposition of fuel and fuel radical. Rates of H abstraction reaction by OH and HO_2 are calculated in this work by Dr. Carlo Cavallotti [politecnico di Milano] at CCSD(T) energy levels. However H abstractions by H, O, CH_3 are taken from Tan *et al.* [15]. The fuel decomposition rates are taken from Peukert *et al.* [16] and the fuel radical decomposition reaction rates are from another study by Tan *et al.* [17]. In the reaction scheme the decomposition of fuel and the fuel radicals lead to formation of formaldehyde and some unstable radicals eventually producing CH_3, CO and CO_2 as shown in Fig.2. Further consumption of formaldehyde and other products has been dealt in the base chemistry reactions.

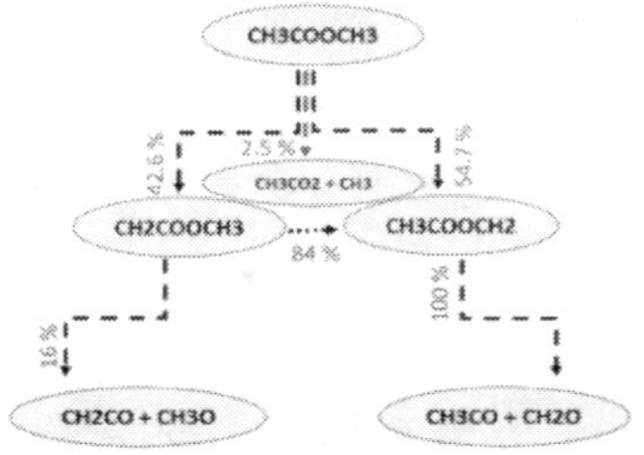

Figure 2: Reaction path analysis of methyl acetate at10% of fuel conversion time. P=15 bar, 1200 K and phi =1.

3.2 Ethyl acetate

The kinetic mechanism of ethyl acetate involves additional reaction pathways as compared to methyl acetate, owing to its relatively larger size. Its oxidation chemistry comprises of unimolecular fuel decomposition and H abstraction reactions by various radicals from fuel. The unimolecular decomposition of ethyl acetate has been studied by Swihart *et al.*[22] using pulsed laser powered homogeneous pyrolysis. They utilized this technique for studying rates of gas phase decomposition reactions. The measurements revealed that the major products of breakdown are acetic acid and ethene. The chemistry of acetic acid involves H atom abstraction followed by fuel radical decomposition. The ethene chemistry is described in detail within the base chemistry and is not discussed here. Some other minor products were also observed as a result of abstraction of primary, secondary H atoms and breakage of methyl, ethyl groups from the fuel molecule. The fuel radicals produced after H atom abstractions along with products of decomposition of C-H bond further undergoes radical decomposition to produce smaller radicals. Figure 3 describes major reaction pathways in

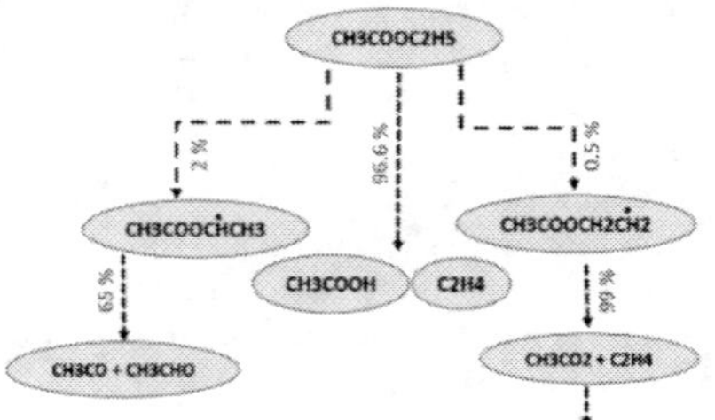

Figure 3: Reaction path analysis of ethyl acetate at 10% of fuel conversion time. P=15 bar, 1200 K and phi =1.

E-mail address: ahfaz.ahmed@kaust.edu.sa

the oxidation chemistry of ethyl acetate. The kinetic mechanism developed in this work only deals with high temperature chemistry as both the experiments discussed in this work are conducted in the high temperature region. The thermodynamic properties are retained from the previous work [1] The transport parameters are recalculated in this work using correlations from Wang *et al.* [23] and Dooley *et al.* [24].

4 Results

The proposed kinetic mechanism has 1958 species and 8098 reactions. Aramcomech 2.0 [21] has been used as base mechanism and all the species with 6 carbon and less have been retained to allow complete account of reactivity. As explained in the experimental sections, ignition and burning velocity data have been acquired at NUI Galway and Lund University, respectively. In this section comparison of developed mechanism is presented with experimental data. The simulations have been performed using Chemkin Pro 18.1. Shock tube simulations were performed with closed homogenous batch reactor model. The ignition delay simulations for methyl acetate and measured data is shown in Fig. 4 and 5. It is evident that the newly proposed model captures the ignition data quite well across all equivalence ratios, pressure and temperature ranges. While the model detailed in [1] under predicts the ignition across all conditions. Comparisons of ignition data and simulations for ethyl acetate is presented in Fig. 6 and 7. Here the old model by Westbrook et al. under predicts the ignition delay at high temperature and over predicts ignition delay at temperatures lower than 1150 K.

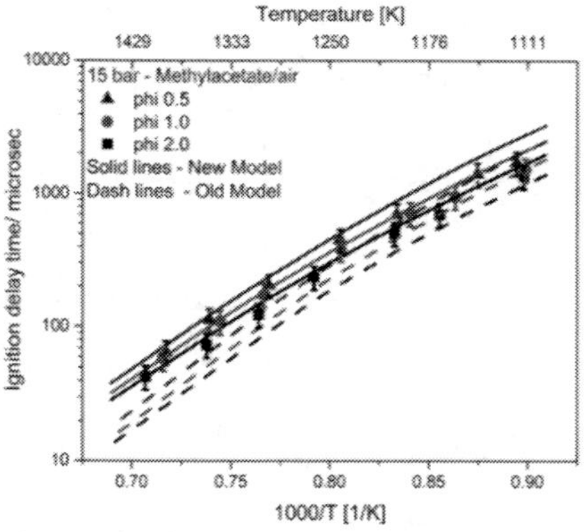

Figure 4: Comparison of methyl acetate ignition delay data with proposed model [solid lines] and model by Westbrook et al. [dash lines] at 15 bar, equivalence ratio = 0.5, 1.0, 2.0.

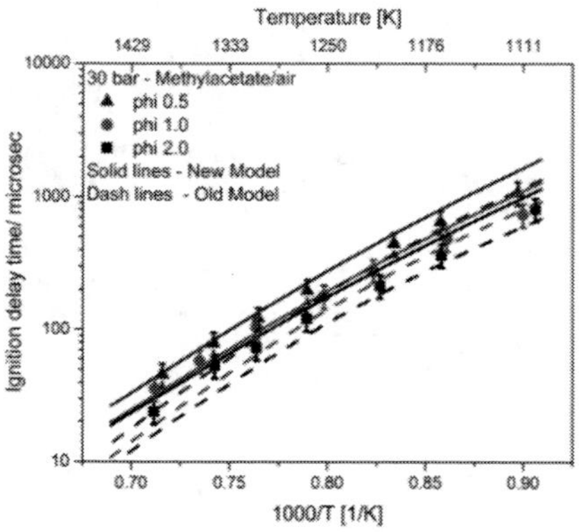

Figure 5: Comparison of methyl acetate ignition delay data with proposed model [solid lines] and model by Westbrook et al. [dash lines] at 30 bar, equivalence ratio = 0.5, 1.0, 2.0.

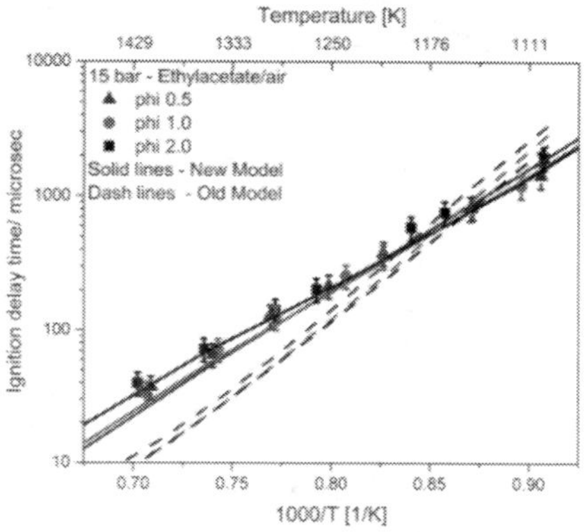

Figure 6: Comparison of ethyl acetate ignition delay data with proposed model [solid lines] and model by Westbrook et al. [dash lines] at 15 bar, equivalence ratio = 0.5, 1.0, 2.0.

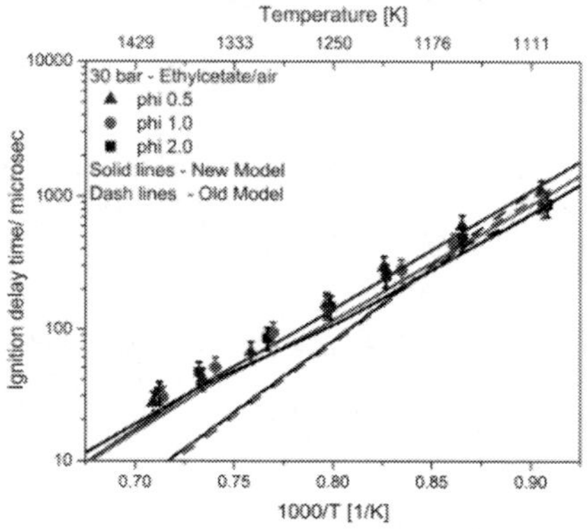

Figure 7: Comparison of ethyl acetate ignition delay data with proposed model [solid lines] and model by Westbrook et al. [dash lines] at 30 bar, equivalence ratio = 0.5, 1.0, 2.0.

Burning velocity simulations have been conducted at $T_{unburned}$ = 298 K and 338 K and $P_{unburned}$ = 1 atm with premixed laminar flamespeed model in Chemkin Pro 18.1. The comparisons between experimental data and simulations are presented in Figure 8 and 19 for methyl acetate and ethyl acetate, respectively. The Westbrook model over predicts the burning velocity for both the fuels over the entire equivalence ratio range. However the model developed in this study predicts the burning velocity within 10% of measured data at all conditions.

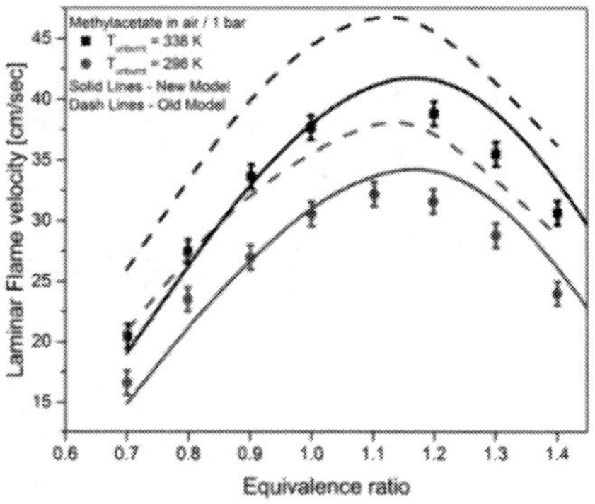

Figure 8: Comparison of methyl acetate burning velocity data with proposed model [solid lines] and model by Westbrook et al. [dash lines] at 1 bar, 298 K, 338 K and equivalence ratio = 0.7 -

1.4.

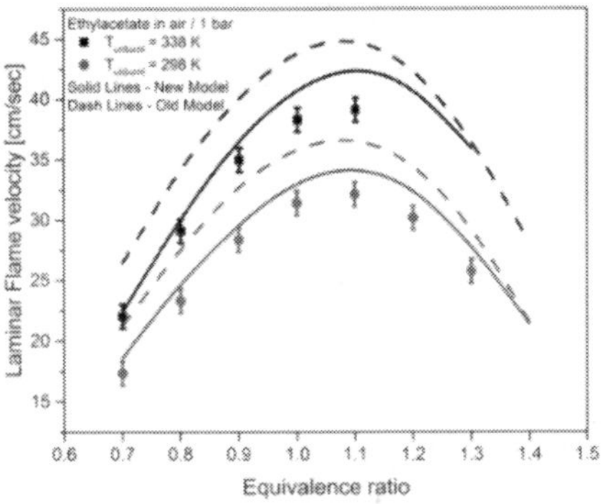

Figure 9: Comparison of ethyl acetate burning velocity data with proposed model [solid lines] and model by Westbrook et al. [dash lines] at 1 bar, 298 K, 338 K and equivalence ratio = 0.7 - 1.3.

5 Conclusions

A detailed mechanism for small esters, methyl acetate and ethyl acetate has been developed. Experiments were conducted to measure ignition delay time at NUI Galway and laminar burning velocity at Lund University. Comparisons between the model and experimental data is conducted and model was found to agree well with the experimental data over the entire experimental range.

The modifications made in the mechanism could not be presented here due to limitation of space. The corresponding author could be contacted for obtaining the mechanism and thermochemistry data files.

6 Acknowledgment

- The authors at KAUST acknowledge funding support from Future Fuels Program.
- The authors at NUI Galway recognize funding support from Science Foundation Ireland via their Principal Investigator Program through project number 15/IA/3177.
- The work by authors at LLNL was performed under the auspices of the U.S. Department of Energy under Contract DE-AC52-07NA27344.
- The authors at Lund University acknowledge financial support from the Centre for Combustion Science and Technology (CECOST), and Swedish Research Council (VR) via project 2015-04042.

References

[1] C.K. Westbrook, W.J. Pitz, P.R. Westmoreland, F.L. Dryer, M. Chaos, P. Osswald, K. Kohse-Höinghaus, T.A. Cool, J. Wang, B. Yang, N. Hansen, T. Kasper, A detailed chemical kinetic reaction mechanism for oxidation of four small alkyl esters in laminar premixed flames, Proceedings of the Combustion Institute 32 (2009) 221-228.
[2] BP Energy Outlook 2017 edition, British Petroleum, 2017.
[3] J.Y. Lai, K.C. Lin, A. Violi, Biodiesel combustion: advances in chemical kinetic modeling, Progress in Energy and Combustion Science 37 (2011) 1-14.
[4] R.L. McCormick, G.M. Fioroni, L. Fouts, J. Yanowitz, E. Polikarpov, K. Albrecht, D.J. Gaspar, J. Gladden, A. George, Selection Criteria and Screening of Potential Biomass-Derived Streams as Fuel Blendstocks for Advanced Spark-Ignition Engines, SAE International Journal of Fuels and Lubricants 10 (2017).
[5] G. Knothe, J. Krahl, J. Van Gerpen, The biodiesel handbook, Elsevier2015.
[6] T.M. Mata, A.A. Martins, N.S. Caetano, Microalgae for biodiesel production and other applications: a review, Renewable and sustainable energy reviews 14 (2010) 217-232.
[7] L. Gasnot, V. Decottignies, J. Pauwels, Kinetics modelling of ethyl acetate oxidation in flame conditions, Fuel 84 (2005) 505-518.
[8] G. Dayma, F. Halter, F. Foucher, C. Mounaim-Rousselle, P. Dagaut, Laminar Burning Velocities of C4–C7 Ethyl Esters in a Spherical Combustion Chamber: Experimental and Detailed Kinetic Modeling, Energy & Fuels 26 (2012) 6669-6677.
[9] T. Badawy, J. Williamson, H. Xu, Laminar burning characteristics of ethyl propionate, ethyl butyrate, ethyl acetate, gasoline and ethanol fuels, Fuel 183 (2016) 627-640.
[10] Z. Zhang, E. Hu, C. Peng, X. Meng, Y. Chen, Z. Huang, Shock Tube Measurements and Kinetic Study of Methyl Acetate Ignition, Energy & Fuels 29 (2015) 2719-2728.
[11] B. Akih-Kumgeh, J.M. Bergthorson, Experimental and Modeling Study of Trends in the High-Temperature Ignition of Methyl and Ethyl Esters, Energy & Fuels 25 (2011) 4345-4356.
[12] X. Yang, D. Felsmann, N. Kurimoto, J. Krüger, T. Wada, T. Tan, E.A. Carter, K. Kohse-Höinghaus, Y. Ju, Kinetic studies of methyl acetate pyrolysis and oxidation in a flow reactor and a low-pressure flat flame using molecular-beam mass spectrometry, Proceedings of the Combustion Institute 35 (2015) 491-498.
[13] F. Contino, F. Foucher, C. Mounaïm-Rousselle, H. Jeanmart, Experimental Characterization of Ethyl Acetate, Ethyl Propionate, and Ethyl Butanoate in a Homogeneous Charge Compression Ignition Engine, Energy & Fuels 25 (2011) 998-1003.
[14] R. Chandra, R. Kumar, Fuel Properties of Some Stable Alcohol–Diesel Microemulsions for Their Use in Compression Ignition Engines, Energy & Fuels 21 (2007) 3410-3414.
[15] T. Tan, X. Yang, C.M. Krauter, Y. Ju, E.A. Carter, Ab Initio Kinetics of Hydrogen Abstraction from Methyl Acetate by Hydrogen, Methyl, Oxygen, Hydroxyl, and Hydroperoxy Radicals, The Journal of Physical Chemistry A 119 (2015) 6377-6390.
[16] S.L. Peukert, R. Sivaramakrishnan, M.-C. Su, J.V. Michael, Experiment and theory on methylformate and methylacetate kinetics at high temperatures: Rate constants for H-atom abstraction and thermal decomposition, Combustion and Flame 159 (2012) 2312-2323.
[17] T. Tan, X. Yang, Y. Ju, E.A. Carter, Ab Initio Unimolecular Reaction Kinetics of CH2C(=O)OCH3 and CH3C(=O)OCH2 Radicals, The Journal of Physical Chemistry A 119 (2015) 10553-10562.
[18] H. Nakamura, D. Darcy, M. Mehl, C.J. Tobin, W.K. Metcalfe, W.J. Pitz, C.K. Westbrook, H.J. Curran, An experimental and modeling study of shock tube and rapid compression machine ignition of n-butylbenzene/air mixtures, Combustion and Flame 161 (2014) 49-64.
[19] V.A. Alekseev, J.D. Naucler, M. Christensen, E.J. Nilsson, E.N. Volkov, L.P.H. de Goey, A.A. Konnov, Experimental uncertainties of the heat flux method for measuring burning velocities, Combustion Science and Technology 188 (2016) 853-894.
[20] L. De Goey, A. Van Maaren, R. Quax, Stabilization of adiabatic premixed laminar flames on a flat flame burner, Combustion science and technology 92 (1993) 201-207.
[21] Y. Li, C.-W. Zhou, K.P. Somers, K. Zhang, H.J. Curran, The oxidation of 2-butene: A high pressure ignition delay, kinetic modeling study and reactivity comparison with isobutene and 1-butene, Proceedings of the Combustion Institute 36 (2017) 403-411.
[22] M.T. Swihart, R.W. Carr, Pulsed laser powered homogenous pyrolysis for reaction kinetics studies: Probe laser measurement of reaction time and temperature, International journal of chemical kinetics 28 (1996) 817-828.
[23] H. Wang, M. Frenklach, Transport properties of polycyclic aromatic hydrocarbons for flame modeling, Combustion and flame 96 (1994) 163-170.
[24] S. Dooley, M. Uddi, S.H. Won, F.L. Dryer, Y. Ju, Methyl butanoate inhibition of n-heptane diffusion flames through an evaluation of transport and chemical kinetics, Combustion and Flame 159 (2012) 1371-1384.

11th Asia-Pacific Conference on Combustion.
The University of Sydney, NSW Australia
10th -14th December 2017.

Autoignition of Low to High Octane Gasolines

A. Farooq[1,*], B.R. Giri[1], M. AlAbbad[1], S.M. Sarathy[1], H. Curran[2]

[1]Clean Combustion Research Center, Division of Physical Sciences and Engineering, King Abdullah University of Science and Technology (KAUST), Thuwal, 23955-6900, Saudi Arabia

[2] Combustion Chemistry Centre, National University of Ireland Galway, Galway, Ireland

Abstract

Ignition delay times of two high-octane gasolines (RON: 91 and 97.5) and two low-octane gasolines (RON: 70.3 and 71.8) were studied in a high-pressure shock tube and in a rapid compression machine. The high-octane gasolines were oxygenated and contained 5 – 10 % ethanol. Experiments were carried out over a wide range of temperatures (700 – 1300 K), at two pressures (20 and 40 bar) and at two equivalence ratios (0.5 and 1.0). All fuels exhibited very similar reactivity at high temperatures. The RON-dependence was strongest in the NTC region while the sensitivity (RON – MON) effects were seen both in the intermediate- and low-temperature regions. Binary (PRF), tertiary (TPRF) and multi-component surrogates were formulated to simulate the reactivity of these gasolines. It was observed that PRF surrogates captured the reactivity of low sensitivity (and low octane) gasolines adequately while multi-component surrogates were needed to fully describe the autoignition behavior of high sensitivity (and high octane) gasolines. Experimental and simulation results are compared across wide variety of gasoline ignition delay data. Fuel composition effects on autoignition were elaborated with the help of detailed chemical kinetic simulations.

1 Introduction

Many advanced combustion engine (ACE) technologies such as homogeneous charge compression ignition (HCCI) and reactivity controlled compression ignition (RCCI) have been proposed [1-3], which offer the possibility of achieving high efficiency and low NO_x and soot emissions. Moreover, in any development of ACE technologies, a detailed understanding of the influence of fuel chemistry is needed. Furthermore, because the new combustion modes used for ACEs are based on the autoignition of a reactive mixture and it is difficult to control combustion phasing in ACEs, a fundamental understanding of the chemical kinetics of fuel combustion is essential in the development and optimization of these technologies [4]. However, to develop a detailed reaction mechanism comprising all of the components present in a real fuel (e.g., gasoline) in order to predict its ignition behaviour is an intractable problem. Owing to this complexity, a popular approach to describe the combustion behaviour of gasoline is to use a surrogate fuel composed of several reference components.

The purpose of this work is to study the merits of surrogate formulation strategies in homogenous environments for widely varying gasolines. For this, ignition delay times of two high-octane gasolines (RON: 91 and 97.5) and two low-octane gasolines (RON: 70.3 and 71.8) were studied in a high-pressure shock tube and in a rapid compression machine. Binary (PRF) and multi-component surrogates were formulated to simulate the reactivity of these gasolines. It was observed that PRF surrogates captured the reactivity of low sensitivity (and low octane) gasolines adequately while multi-component surrogates were needed to fully describe the autoignition behavior of high sensitivity (and high octane) gasolines.

1.1 *High Octane gasolines*

The two certification oxygenated high octane gasolines used in this work were supplied by Coryton Advanced Fuels and Haltermann Solutions. It can be seen from Table 1 that both Haltermann (HG) and Coryton (CG) are high octane (RON > 91) gasolines. Moreover, the two gasolines have high sensitivities as well (7.6 and 10.9 respectively). It can also be seen from the table that the two gasolines have some ethanol (oxygenated) content present in them as well.

Several surrogates are tested for these gasolines are also listed in Table 1. The TPRF (toluene/PRF) surrogate blends, referred to as HG-TPRF and CG-TPRF, were formulated utilizing the methodology developed by Kalghatgi et al. [5] to match the RON and MON of the target gasolines. The Quaternary (ethanol/TPRF) surrogates, referred to as HG-NUIG and CG-NUIG, were formulated using the methodology proposed by Anderson et al. [6] for estimating the octane number of alcohol-gasoline blends. The multicomponent surrogates, referred to as HG-KAUST and CG-KAUST, were developed using the surrogate formulation strategy proposed by Ahmed et al. [7] is based on optimization of surrogate palette species to match the target properties of the real fuel. The multi-component surrogate composition are also listed separately in Table 2.

*Corresponding author:
E-mail address: Aamir.Farooq@kaust.edu.sa

Table 1. Comparison of properties of Haltermann and Coryton gasolines and their respective surrogates. Compositions are given in mole %. (a) calculated value.

	Haltermann (HG)	HG-TPRF	HG-NUIG	HG-KAUST
RON	91	91[a]	91[a]	91.5[a]
MON	83.4	83.4[a]	85.1[a]	82.8[a]
sensitivity	7.6	7.6	5.9	8.7
density (kg/m^3)	755	742	756	739
H/C	1.97	1.96	1.97	1.94
avg. mol. wt	88.8	105.4	93.9	85.3
n-alkanes	12.2	17.0	14.4	16.3
iso-alkanes	26.1	54.0	40.3	22.5
aromatics	22.9	29.0	27.3	24.2
oxygenates	16.8		18.0	14.7
alkenes	6.4			6.9
cycloalkanes	15.6			15.4
	Coryton (CG)	**CG-TPRF**	**CG-NUIG**	**CG-KAUST**
RON	97.5	97.5[a]	97.5[a]	99.2[a]
MON	86.6	86.6[a]	89.8[a]	89.9[a]
sensitivity	10.9	10.9	7.7	9.3
density (kg/m^3)	749	828	770	743
H/C	1.776	1.41	1.83	1.83
avg. mol. wt	90.6	95.1	98	86.9
n-alkanes	10.1	14.3	10.3	10.4
iso-alkanes	31.9	8.1	41	29.2
aromatics	33.6	77.6	39.9	34.3
oxygenates	8.2		8.8	9.3
alkenes	11.2			12.5
cycloalkanes	5			4.3

Table 2. Composition of the multi-component surrogate mixtures in mole fractions.

Palette compound	HG-KAUST	CG-KAUST
n-Butane	0.0715	0.0727
2-Methyl butane	0.1382	0.0865
Ethanol	0.1469	0.0934
Cyclopentane	0.1540	0.0000
1,2,4-Trimethylbenzene	0.2418	0.0000
1-Hexene	0.0691	0.1246
n-Heptane	0.0912	0.0311
2,2,4-Trimethylpentane	0.0873	0.2061
Toluene	0.0000	0.3426
cyclohexane	0.0000	0.0430

1.2 *Low octane gasolines*

The two low octane FACE gasolines I and J, were acquired from Conoco Philipps Chemical Company and Fuel compositions and other relevant properties are listed in Table 3. These gasolines were compared with the multicomponent surrogate (composition listed in Table 4) simulations formulated based on Ahmed et al. [7] and with the experiments and simulations of a simple PRF surrogate.

Table 3. Properties of FACE gasoline I and J, multi-component and PRF surrogates. Hydrocarbon types are given in mol%.

	FACE I	FGI-KAUST	FACE J	FGJ-KAUST	PRF 70
RON	70.3	70.7	71.8	70.6	70
MON	69.6	68.4	68.8	66.5	70
Sensitivity	0.7	2.3	3	4.1	0
AKI	69.9	69.6	70.3	68.6	70
Avg. mol.	95.5	98.9	100.2	103.2	109.7
n-alkanes	14	12	31.5	35	33
iso-alkanes	70	72	32.4	35	67
Cycloalkanes	4	6	2.4	-	0
Aromatics	5	4	30.6	30	0
Olefins	7	6	0.6	-	0
Unidentified	-	-	2.5	-	-

Table 4. Composition (mole%) of the multi-component and PRF surrogates.

	FGI-KAUST	FGJ-KAUST	PRF 70
n-butane		10	-
2-methylbutane	11	-	-
2-methylhexane	27	23	-
1-hexene	6	-	-
n-heptane	12	25	33
2,2,4-trimethylpentane	34	12	67
cyclopentane	6	-	-
1,2,4 trimethylbenzene	4	30	-

2 Experimental Facilities and Methods

All experiments were performed in the high pressure shock tube and rapid compression machine facilities at the King Abdullah University of Science and Technology (KAUST) and National University of Ireland Galway (NUIG). Details of these facilities and experimental methods are discussed in [8]. For all experiments, gasoline/air mixtures were prepared manometrically in a magnetically-stirred mixing vessel. The mixing vessel and shock tube (or RCM) were pre-heated to 120 ^{0}C. The fuel was injected directly in the heated mixing vessel and allowed to vaporize completely before adding O_2 and N_2 gases. Some experiments were carried out at lower pre-heating temperatures (~ 100 ^{0}C) which produced very consistent results with 120 ^{0}C heating. These experiments ensured that the chosen pre-heating temperature is sufficient for adequate vaporization of the fuel and there were no condensation issues.

3 Results and Discussion

The ignition delay experiments and simulations were undertaken at a wide range of test conditions, however, only a subset of the data will be presented here for brevity. Throughout this section, solid symbols and lines represents shock tube data and simulations and open symbols and lines represents those of RCM respectively. Experimentally, ignition delay times were inferred using pressure or OH* measurements for shock tube and pressure measurements for RCM. For the kinetic simulations, ignition delay times were inferred using the temperature or OH rise.

3.1 *High octane gasolines*

As described previously, three surrogates were formulated for both Haltermann and Coryton gasolines. The performance of various surrogates is analysed in detail with the help of chemical kinetic simulations. Additionally, the Haltermann surrogates were compared experimentally by measuring ignition delay times in the RCM at nominal pressure of 20 bar and near-stoichiometric; the results are plotted in Fig. 1. The experimental scatter points (Fig. 1(b)) show that all surrogates display similar ignition delay times and do a reasonable job of capturing the gasoline ignition data between 700 and 900 K. At temperatures less than 700 K, the 4-component surrogate (HG-NUIG, blue triangle) matches the data more closely compared to the other two surrogates. At temperatures higher than 850 K, however, the HG-NUIG surrogate appears to be a little slower (longer ignition delay times) than the gasoline. At low temperatures (< 750 K), HG-TPRF surrogate is the most reactive followed by HG-NUIG surrogate and then HG-KAUST surrogate. High paraffinic content in the HG-TPRK mixtures leads to faster low-temperature reactions, while a large fractions of aromatics (1,2,4-TMB), in addition to cycloalkane (cyclopentane) and olefins, in the HG-KAUST surrogate leads to relatively slow reactivity at low temperatures. The HG-TPRF simulations predict the HG-TPRF experimental data quite well, which indicates that the n-heptane, iso-octane and toluene chemistries are adequately modelled in the KAUST gasoline surrogate model [9], at least at these specific conditions. The HG-NUIG simulations predict the HG-NUIG experimental data reasonably well at high and low temperatures but over-predict the data in the intermediate temperatures of 700 – 800 K. This discrepancy may perhaps be attributed to the deficiency in the ethanol chemistry or the cross-reactivity of ethanol with n-heptane, iso-octane or toluene. Similar intermediate temperature over-prediction is observed by comparing the HG-KAUST simulations and HG-KAUST experimental ignition delays. Fig. 2 shows a similar comparison for Coryton gasoline and its surrogates. It can be seen from the figure that a multi-component surrogates better reproduces the ignition requirements of Coryton gasoline and a simple TPRF surrogate underpredicts its ignition requirements at low and NTC temperatures.

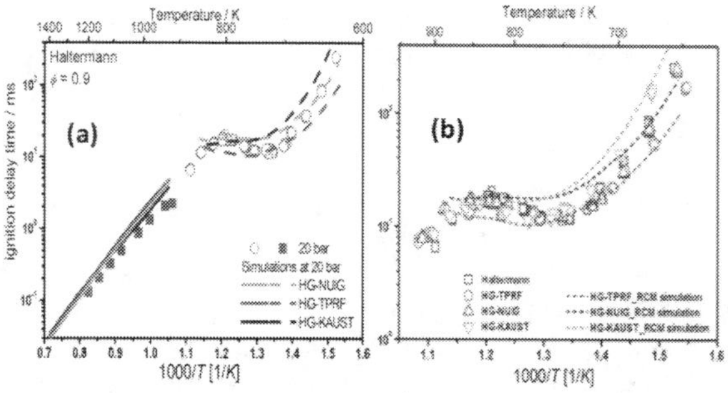

Figure 1. Comparison of the experimental ignition delay results for Haltermann gasoline and its surrogates at 20 bar. (a) Equivalence ratio is φ = 0.9 for Haltermann gasoline and (b) φ = 1.0 for the surrogates. For the details of the surrogates, refer to Table 3. The dashed lines are chemical kinetic simulations using the gasoline surrogate model [9].

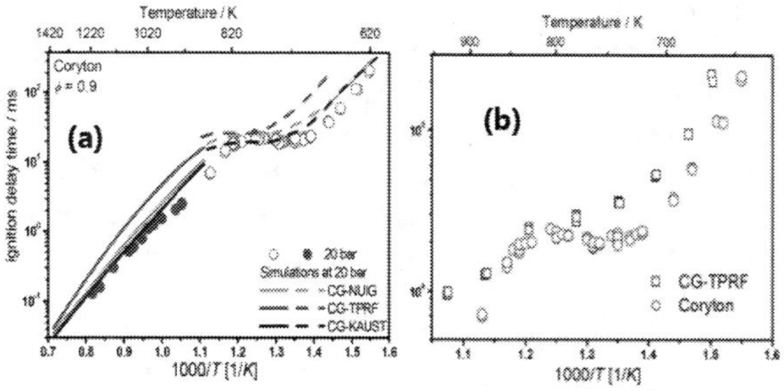

Figure 2. Comparison of the experimental ignition delay results for Coryton gasoline and its surrogates at 20 bar. (a) Equivalence ratio is φ = 0.9 for Coryton gasoline and (b) φ = 1.0 for the TPRF surrogate. For the details of the surrogates, refer to Table 3. The dashed lines are chemical kinetic simulations using the gasoline surrogate model [9].

3.2 *Low octane gasolines*

Fig. 3 shows the ignition delay times of the two low octane gasolines and the PRF surrogate. It can be seen from Fig. 3(a) the two gasoline with varying compositions but similar octane ratings yield similar ignition delay times and also that the PRF surrogate adequately captures the ignition requirements of these low octane gasolines. It can be seen from Fig. 3(b) the multicomponent surrogate and PRF simulations agree well with each other and with the experimental data in the high and low temperatures, but there is some discrepancy in the NTC region which we think will require mechanism refinement.

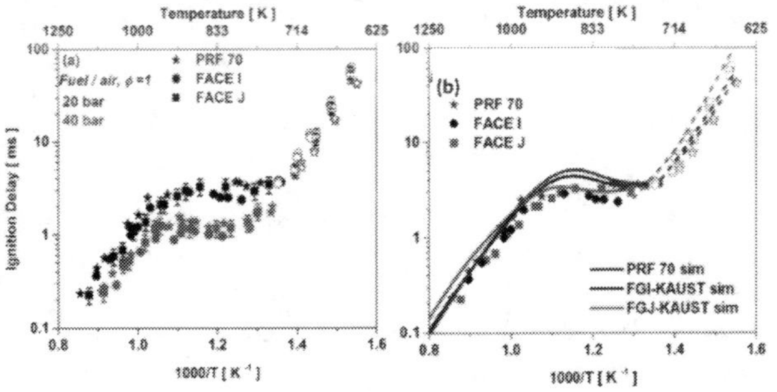

Figure 3. Comparison of ignition delay measurements of PRF 70, FACE I and FACE J (a) only experiments (scatter) and (b) simulations (lines) and experiments for φ = 1.0 at (a) 20 bar.

4 Conclusions

Utilizing the ignition delay times of two high and two low octane gasolines, we have tried to evaluate the surrogate complexity requirements of these gasolines in homogenous environments. We have shown that for medium sensitivity high octane Haltermann gasoline (S = 7.6, RON = 91), a TPRF surrogate does a reasonable job to capture the ignition requirements and other complex surrogates further improve the agreement with the gasoline data. We have also shown that for high sensitivity high octane Coryton gasoline (S = 10.9, RON = 97.5), TPRF surrogate fails to capture the ignition requirements and more complex surrogates better emulates the reactivity. For low octane low sensitivity gasolines, FACE I and FACE J, a simple binary PRF surrogate satisfactorily captures the ignition requirements. It should also be noted that for each case discussed here, multicomponent surrogates performs better or as good as simple surrogates and may enable better kinetic understanding; looked in a different way, depending on the gasoline complexity and fidelity required in simulation, one can get away with using a simple surrogate in simulations.

5 Acknowledgment

Research reported in this work was funded by King Abdullah University of Science and Technology (KAUST).

6 References

[1] G.T. Kalghatgi, Proc. Combust. Inst. 35 (2015) 101.

[2] R.D. Reitz, G. Duraisamy, Prog. Energy Combust. Sci. 46 (2015) 12.

[3] S. Saxena, I.D. Bedoya, Prog. Energy Combust. Sci. 39 (2013) 457.

[4] G.T. Kalghatgi, R.A. Head, SAE paper 2004-01-1969 (2004).

[5] G. Kalghatgi, H. Babiker, J. Badra, SAE International Journal of Engines 8 (2015) 505.

[6] J.E. Anderson, U. Kramer, S.A. Mueller, T.J. Wallington, Energy Fuels 24 (2010) 6576.

[7] A. Ahmed, G. Goteng, V.S. Shankar, K. Al-Qurashi, W.L. Roberts, S.M. Sarathy, Fuel 143 (2015) 290.

[8] T. Javed, C. Lee, M. AlAbbad, K. Djebbi, M. Beshir, J. Badra, H. Curran, A. Farooq, Combustion and Flame 171 (2016) 223.

[9] S.M. Sarathy, G. Kukkadapu, M. Mehl, T. Javed, A. Ahmed, N. Naser, A. Tekawade, G. Kosiba, M. AlAbbad, E. Singh, S. Park, M.A. Rashidi, S.H. Chung, W.L. Roberts, M.A. Oehlschlaeger, C.-J. Sung, A. Farooq, Combust. Flame 169 (2016) 171.

11th Asia-Pacific Conference on Combustion,
The University of Sydney, NSW Australia
10th -14th December 2017.

Stabilization of Turbulent Auto-Igniting Hydrogen Jets Issuing in a Hot Vitiated Coflow

A.R.W. Macfarlane[1], M.J. Dunn[1], M. Juddoo[1], A.R. Masri[1]

[1]School of Aerospace, Mechanical and Mechatronic Engineering,

The University of Sydney, NSW 2006, Australia

Abstract

This paper presents an experimental study of a hydrogen fuelled jet issuing into a large coaxial vitiated hot coflow of lean premixed H_2/air combustion products. Hydrogen flames have been well studied by numerical and experimental groups and have provided a platform for understanding the process and characteristics of auto-ignition. This paper provides a new high speed study of hydrogen flames to place ignition events (kernels) and flame base dynamics into temporal context. This is done by identifying the interaction, location and consumption of ignition kernels relative to the flame base.

The hydrogen jet is diluted with three parts nitrogen by volume and flame lift-off heights are measured for a range of coflow temperatures. To gain insight into auto ignition dynamics OH PLIF was imaged at 10 kHz to determine a formation map, formation rates and kernel growth characteristics. Ignition location and rates are compared to mean OH fields to identify the correlation to stability between a relatively high and stable coflow temperature (T_c=1070 K) and relatively low and unstable coflow temperature (T_c =1045 K). Mean OH structures do not vary greatly between high and low coflow temperatures. However a reduction in coflow temperature increases the axial range for auto-ignition events to occur.

1 Introduction

The use of hydrogen in combustion based applications is of great interest and importance owing to no emission of carbon based products. These applications include exhaust gas recirculation (EGR) that can be used to control the combustion duration, NO_X emissions [1] and increase power output [2] from SI hydrogen engines. Furthermore the use of hydrogen in HCCI engines is being explored to reduce NO_X emissions [3]. The additional benefit of EGR is to produce stabilised flames within highly turbulent flows. This stabilisation mechanism of EGR is a simple attribute that can be explored within the coflow burner.

The use of hydrogen in a hot coflow burner enables conditions that replicate the recycling of exhaust gases in varying combustion applications. A major focus of this burner and of this paper is the examination of the stabilising mechanism; focusing on the interaction of auto-igniton and flame base stability and further the temporal evolution of auto-ignition. The concept of auto-igntion stabilisation differs from commonly accepted flame propagation, which occurs along contours of mean stoichiometric mixture at the position where the axial mean velocity balances the turbulent burning velocity. A summary of stabilising mechanisms has been given by Lyon et al. [4].

By using the asymmetric coflow burner design simplified boundary conditions and 2-D numerical simulations can be explored. The nature of auto-igniting flames for the coflow burner additionally enables the study of low temperature chemistry whilst decoupling the kinetics from complex swirling flow such as those found in industrial applications.

Multiple groups have used hydrogen fuel in the coflow burner, which was initially explored by Cabra et al [5] and other groups [6-8] to analyse auto-ignition and flame base dynamics. The studies have used OH chemiluminesce and OH PLIF as is done in this study to diagnose these flames. In more recent times methane jets have been used to analyse the interaction of kernels and the flame base [9, 10]. Previous work has identified the interaction and importance of kernels on the flame base for Dimethyl ether flames at various coflow temperatures [11].

The stabilisation mechanism of these flames has been examined extensively by numerical groups [12-14]. They have identified for hydrogen flames through HO_2 autoigntion indicators that auto-ignition is the main stabilising mechanism for high and low coflow temperatures. The formation of ignition events and inherently the stability of the flames have shown to initiate in areas of low scalar dissipation and for lean mixture fractions (most reactive) [15, 16]. The formation and location of ignition kernels in low scalar dissipation regions has been verified experimentally [17]. A review on the formation of kernels and the flames stabilisation has been done by Mastorakos [18] for hot coflow flames.

This paper addresses the interest in the temporal correlation of flame base stability/location and the formation of auto ignition kernels using high speed chemiluminescence and OH-PLIF diagnostics.

2 Setup

The burner employed is very similar to that used by Cabra and Dibble [4]. The wide coflow diameter of 197 mm has ~1800x1.6 mm diameter holes which stabilize lean H_2/air flames. The coflow products shroud the fuel jet at a constant burnt velocity of 3.5 m/s for all cases. The temperature of the coflow is controlled by varying the equivalence ratio, with the range covered in the present study between T_c=1040 K to T_c=1070 K. The entire burner assembly is placed in a wind tunnel providing a coflowing stream of air at a fixed velocity of 0.8 m/s. The N_2 diluent is kept constant within the central fuel jet (D_j=4.45) while maintaining a constant 107 m/s bulk velocity. The operating conditions are summarised in Table 1 for the fuel jet and a single T_c=1070 K coflow case ($\O_c$=0.26) where the flow of air/H_2 is varied to alter

A.R.W Macfarlane
Email Address: amac7548@uni.sydney.edu.au

the flames equivalence ratio and therefore coflow temperature. The cases of further study for the OH-PLIF setup is for a relatively stable flame at T_c=1070 K and unstable flame T_c=1045 K.

Table 1 Jet and coflow conditions

	Fuel Jet	Coflow
V(m/s)	107	3.5
T (K)	293	1040
QH2 (SLM)	25	169
QN2 (SLM)	75	-
Qair (SLM)	-	1664

3 Experimental method

Flame Lift off heights (L_H) were determined using a digital camera with a 10 second exposure. The long exposure time is successful to image relative lift off heights. However to capture kernel ignition and growth a significantly faster speed of 10 kHz is required. Therefore a Lavision CMOS camera at 10 kHz was used to image Chemiluminescence and OH-PLIF (not simultaneous).

The OH radical is given as the marker for kernel intiation and growth, it was fluoresced using an Edgewave Nd-YAG laser (Edgewave IS811-E) that pumps a tuneable dye laser (Sirah Credo). This system produced a wavelength of 283nm that overlapped onto the strong $Q_1(6)$ transition for $A^2\Sigma^+ \leftarrow X^2\Pi(1,0)$, excitation, where the energy was measured at 0.14 mJ/pulse. The OH PLIF signal was collected with a 96 mm $f_{\#}$1.65 UV Nikkor lens resulting in a mapped pixel resolution of 56 µm/pixel.

Owing to the sensitivity of hydrogen flames to the coflow temperature a thermocouple is used to verify repeatability between sets of results. Through the use of ALICAT mass flow controllers a small flow rate fluctuation and in turn coflow temperature oscillation is observed of approximately ±6 K.

4 Lift off heights

Figure 1a show chemiluminescent images from which lift off heights were obtained. It can be seen that a decrease in temperature increases the lift off height indicated by the horizontal line. It is noted that lift off is defined by the location of the steepest axial gradient that lies on a vertical line connecting the furthest upstream location of signal above the background.

It is seen in Figure 1a that the flames are relatively short (~40 *x/D* or 180mm) and the flame brush tip downstream doesn't vary greatly with lift off height, that is the flame gets shorter proportionally to lift off height. Figure 1b shows a strong dependence for the lift off heights versus coflow temperature. Furthermore it is very sensitive to temperature around T_c=1050 K seen by the small temperature range of 40 K. That is a 'seated' flame is found at T_c=1070 K and a slight decrease in coflow temperature produces a highly lifted and unstable flame at T_c=1040 K. Figure 1b additionally shows a comparison against the work by Gordon et al. [6] and Cabra et al. [5]. There is a small variation in the coflow temperature, however owing to the sensitivity of the temperature range this is expected.

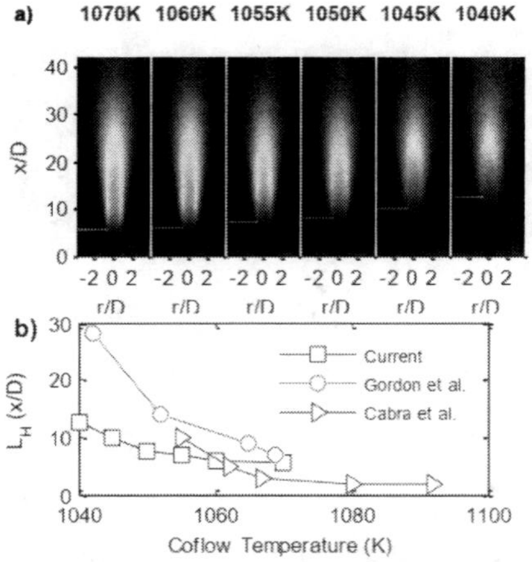

Figure 1a) Digital images for lift off height variations with coflow temperatures, horizontal lines indicate the lift off height location. b) Comparative lift off heights for the current study and those by Gordon et al. and Cabra et al.

5 Temporal lift off heights

Imaging chemiluminescence at 10 kHz for these hydrogen flames has shown to be an excellent way to identify kernel and flame base interaction. Figure 2 shows the temporal profile for the flame base and kernel interaction over 80ms for two coflow temperatures: Tc=1040 and Tc=1070.

Having achieved good resolution from the chemiluminescence camera the identification and tracking of ignition kernels is possible. Initiation is the first identification of a new kernel event, where its axial location is based on the furthest upstream point of the kernel. Due to the steady downstream drift of a somewhat symmetrically growing kernel, the centroid is able to identify and track a growing kernel shot to shot. Consumption of the kernel occurs as the new or growing kernel connects to the flame base.

Auto-ignition is defined as the stabilisation mechanism here as the merging of kernels 'propagates' the flame base upstream. This process of kernels forming upstream of the advecting flame base and merging with it repeats and forms a sawtooth profile for both flames seen in Figure 2. The flame base advects downstream at a somewhat constant rate of 8 m/s and 5.6 m/s for Tc=1045 K and Tc=1070 K respectively seen in Figure 2. The upstream formation and merging of ignition kernels is what cause the flame base to 'jump' upstream. The reduced flame base range and slower downstream advection of the Tc=1070 K flame base, despite sitting in an upstream region with higher localised stream wise velocities, indicates increased stability for this hotter coflow flame.

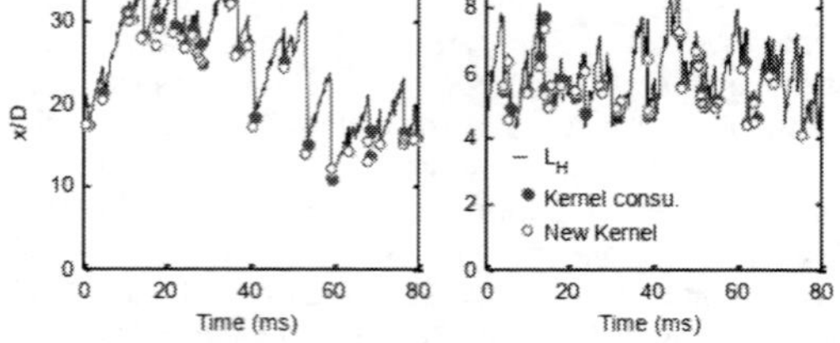

Figure 2 Temporal flame base and kernel interaction (kernel initiation/consumption) T_c=1045 K (Left) T_c=1070 K (right)

6 OH-PLIF

The mean and instantaneous structure of flames of selected flames is reported here with a particular focus on the upstream region close to the stabilisation point. Full flame mean images are given for two coflow temperatures: T_c=1045 K and T_c=1070 K.

6.1 Mean OH structure

Mean images are given below for the stable and unstable flames: T_c=1045 K and T_c=1070 K respectively. The structure of OH signal between the high and low coflow is similar, while intensity and downstream formation location differ.

The relative intensity is lower for the reduced coflow temperature and is related to the 'flickering' on and off of the flame (higher RMS) vastly reducing the mean intensity by ~35%. Furthermore the lower coflow temperature has a higher lift off height of ~10 *x/D* opposed to ~5 *x/D* indicated by the horizontal dashed line comparable to the digital lift off heights in Figure 1.

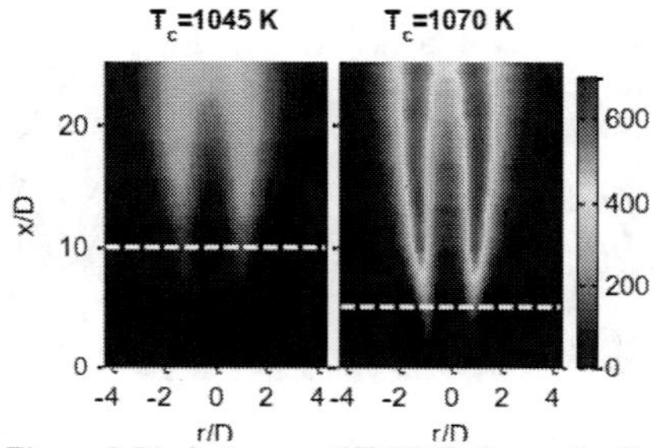

Figure 3 Stacked mean OH-PLIF flames for T_c=1045 K (left) and T_c=1070 K (right). The dashed horizontal line indicates the mean OH lift off height

6.2 Instantaneous 10 kHz OH-PLIF images

Four consecutive images are presented in Figure 4 for T_c=1070 K and T_c=1045 K. The top of the image shows an OH structure identified as the flame base (touching the top of the small ROI PLIF image) and the detached upstream OH objects are defined as ignition OH-kernels (indicated with an arrow). The ignition kernels form at 100 μs in both rows and grow in the consecutive images.

There is seemingly little variation in size and intensity between ignition kernels between high and low coflow temperatures further identified by the similar OH structure from the mean fields in Figure 3.

It is emphasised here that a high coflow temperature, T_c =1070 K, has a decrease in fluctuating flame base and kernel structures. However the ignition kernels have similar intensities between high and low coflow temperatures with the center of the kernel being most intense at approximately 600 counts.

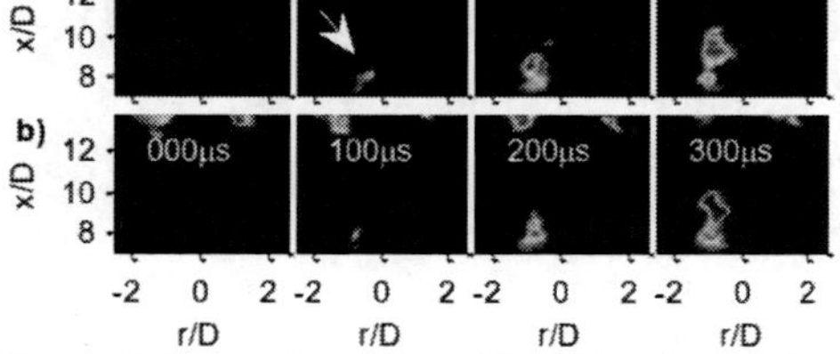

Figure 4 Consecutive images identifying kernel formation and growth for: T_c =1045 a) T_c =1070 b)

6.3 OH-Kernel formation map

By tracking the centroid of kernels it is possible to define the first stage of ignition. The centroids location of the ignition events were binarised in an ignition map then binned at a 1.8mm^2 bin size seen in Figure 5. This map denotes the location and the density formation of ignition events i.e. ignition/second.mm^2.

The vertical dashed centreline indicates that kernels form outside the jet center towards the coflow shear layer with minimal events occurring along the centerline. With reference to the dotted horizontal line indicating mean lift of heights from Figure 3 it can be seen as expected the highest formation density is around the mean flame base and slightly upstream.

The reduction in coflow temperature (T_c =1045 K) produces a more unstable/fluctuating flame and therefore increases the axial length for formation of kernels. The higher T_c =1070 K coflow is more stable and never advects far downstream and therefore ignition events are highly focused further upstream.

The relative location of the flame base to ignition events identifies that both of these flames are stabilised by auto-ignition. It is hypothesised that a reduction in coflow temperature has increased dependence to auto ignition and a high enough coflow temperature is theorized to be stabilized by free flame propagation.

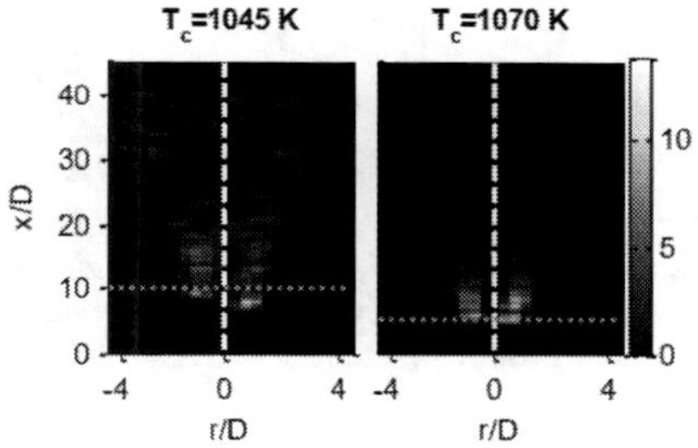

Figure 5 Formation density map. The colour bar denotes the formation rate per mm^2 (ign/s.mm^2), vertical dashed line markes the centreline, the horizontal dotted line showcases the mean OH lift off height

6.4 OH-Kernel formation rates

Axial formation rates seen in Figure 6 are formed by collapsing the radial dimension from Figure 5 and placed in 3 x/D bins. They show as seen in Figure 5 that the peak formation rates occur upstream at ~4 *x/D* for T_c=1070 K and ~10 *x/D* for T_c=1045 K. However the higher coflow temperature T_c=1070 K has a more defined peak. The gradual downstream transition (flatter profile) of auto ignition events for the T_c=1045 K again emphasises the transient nature of the flame and the ability for ignition events to form along almost the entire length of the flame.

The total planar ignition events are 3294 ign/s and 1534 ign/s for T_c=1045 K and T_c=1070 K respectively. It is noted that these are planar measurements and a radial correction can be applied however it won't account for growth into plane. The movement of kernels into or out of plane are acknowledged however have been identified not to be significant. However into planar growth from kernels that initiate out of plane may cause significant biasing and artificially increase formation rates. Further volumetric experiments are required to obtain a true flame kernel ignition rate.

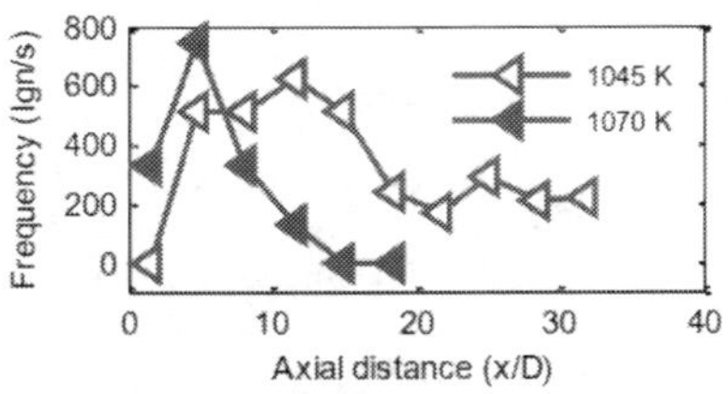

Figure 6 Axial Formation rates for a T_c=1045 K and T_c=1070 K coflow temperature for 3x/D axial bins.

6.5 OH-Kernel growth rate

Having temporally resolved the formation and growth of ignition kernels in these flames it is possible to identify how they evolve. It is noted here that 10 kHz is only able to capture up to three stages (200 μs) of a kernels lifetime for a 107 m/s jet to remain statistically resolved.

Figure 7a show the kernel size with respect to kernel lifespan. It is interesting that both T_c=1045 K and T_c=1070 K initiate with similar sizes, approximately 2mm^2, and grow at similar rates. Whilst the growth rates are similar between low and high temperature cases the higher coflow temperature produces kernels that initiate and grow with an increased aspect ratio. The kernel aspect ratio is defined by the axial length divided by the radial width; this identifies that the low temperature produces more spherical kernels (using the assumption that kernels are axisymmetric along the PLIF principle access) with an aspect ratio closer to one. Since the higher coflow temperature T_c=1070 K has a similar growth rate however the aspect ratio is greater, the kernels must be evolving at a faster rate axially observed in Figure 7b.

As above with the formation rates, growth rates may be biased by in or out of plane movement. It however is verified that within the PLIF plane the kernel centroid motion follows axial streamlines with minimal deviation. Additionally, growth rate is a mean value and hence are statically resolved by both in and out of plane motion. Volumetric chemiluminescence indicates a similar linear growth of kernels with time (not shown here).

The products from both coflow temperatures are similar with ~15 % O_2 and the jets have the same composition, so their stoichiometric mixture fractions are very similar. However the higher coflow temperature produce kernels with larger aspect ratios seen in Figure 7b and therefore it can be assumed that kernels are initiating further upstream in regions of steeper radial mixture fraction gradients. Furthermore, they grow with an increasing aspect ratio (increasing axial length) which can be assumed to be preferentially along similar mixture fracture iso-contours. The more spherical nature of the T_c =1045 K kernels and symmetrical growth (constant aspect ratio) identifies that the kernels grow and initiate in more gradual mixture fraction gradient regions.

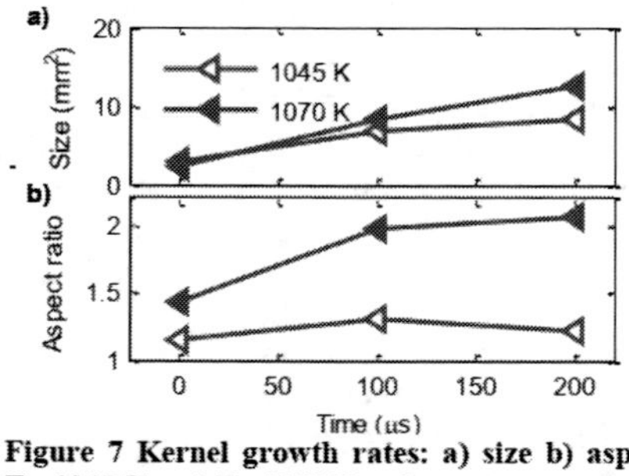

Figure 7 Kernel growth rates: a) size b) aspect ratio for a T_c=1045 K and T_c=1070 K coflow

7 Conclusion

This paper presented temporally resolved experimental results at 10 kHz for auto igniting hydrogen flames. Lift off heights were used to characterise hydrogen flames within the hot coflow burner for a sensitive temperature range of T_c=1040 K to T_c=1070 K. The T_c=1070 K coflow produced a more stable 'seated' flame while T_c=1040 K was more lifted and highly unstable.

High speed 10 kHz chemiluminescence has proved to be very insightful to identify the necessity of kernels to 'stabilise' these auto-igniting flames. The 10 kHz OH-PLIF setup was successful in identifying relative kernel formation locations and rates for high and low coflow temperatures. Mean flames indicated that the OH structure is similar between the high and low coflow temperatures.

Formation rates and the location of kernels relative to the flame base indicate that both flames are stabilised by ignition events. Reduced coflow temperatures lead to an increased axial formation rate at almost the entire length of the flame, leading to the conclusion that these flames have increased dependence on auto-ignition. Finally the growth characteristics of kernels lead to the recognition that increased coflow temperature forms kernels in regions of steeper mixture fraction gradients.

8 References

[1] J.W. Heffel, Int. J. Hydrogen Energy, 28 (2003) 1285-1292.
[2] C.M. White, R.R. Steeper, A.E. Lutz, Int. J. Hydrogen Energy, 31 (2006) 1292-1305.
[3] P.G. Aleiferis, M.F. Rosati, Combust. Flame, 159 (2012) 2500-2515.
[4] K.M. Lyons, Prog. Energy Combust. Sci., 33 (2007) 211-231.
[5] R. Cabra, T. Myhrvold, J.Y. Chen, R.W. Dibble, A.N. Karpetis, R.S. Barlow, Proc. Combust. Inst., 29 (2002) 1881-1888.
[6] S.H.S. R. L. Gordon, A. R. Masri and R. W. Bilger Proceedings of the 5th Asia-Pacific Conference on Combustion, University of Adelaide, 333-336 (2005).
[7] B. Johannessen, A. North, R. Dibble, T. Løvås, Combust. Flame, 162 (2015) 3210-3219.
[8] C.N. Markides, E. Mastorakos, Proc. Combust. Inst., 30 (2005) 883-891.
[9] C.M. Arndt, J.D. Gounder, W. Meier, M. Aigner, Appl. Phys. B, 108 (2012) 407-417.
[10] E. Oldenhof, M.J. Tummers, E.H. van Veen, D.J.E.M. Roekaerts, Combust. Flame, 160 (2013) 1428-1440.
[11] A.R.W. Macfarlane, M.J. Dunn, M. Juddoo, A.R. Masri, Proc. Combust. Inst.
[12] A.R. Masri, R. Cao, S.B. Pope, G.M. Goldin, Combustion Theory and Modelling, 8 (2004) 1-22.
[13] R.R. Cao, S.B. Pope, A.R. Masri, Combust. Flame, 142 (2005) 438-453.
[14] R.L. Gordon, A.R. Masri, S.B. Pope, G.M. Goldin, Combustion Theory and Modelling, 11 (2007) 351-376.
[15] R. Hilbert, D. Thévenin, Combust. Flame, 128 (2002) 22-37.
[16] E. Mastorakos, T.A. Baritaud, T.J. Poinsot, Combust. Flame, 109 (1997) 198-223.
[17] C.M. Arndt, M.J. Papageorge, F. Fuest, J.A. Sutton, W. Meier, M. Aigner, Combust. Flame, 167 (2016) 60-71.
[18] E. Mastorakos, Prog. Energy Combust. Sci., 35 (2009) 57-97.

11th Asia-Pacific Conference on Combustion,
The University of Sydney, NSW Australia
10th -14th December 2017.

Study of Single Oil Droplet Induced Auto-Ignition of Hexadecane and Engine Oil

Sumit Maharjan[1], Long Jiang[1], Yasser Al Qahtani[2], William L. Roberts[1]

[1]King Abdullah University of Science and Technology, Clean Combustion Research Center

Thuwal, 23955-6900, Saudi Arabia

[2]Saudi Aramco, Research and Development Center

Dhahran, 31311-8974, Saudi Arabia

Abstract

Pre-ignition (PI) has been considered as a bottleneck to further improvement in a Gasoline Direct Injection (GDI) engines. Researchers have shown lubricant oil droplet as one of the probable source of PI occurrence. A sophistically engineered constant volume combustion chamber (CVCC) has been built which is capable of operating at elevated temperature and pressure and is equipped with a single oil droplet generator to study the characteristics of the droplet-induced PI in SI engine. In this evolving study, in an attempt to understand the occurrence of PI, different experiments are being carried out considering the importance of droplet size, equivalence ratio, temperature, pressure and fuel composition. In the current study, the effect of pressure on the ignition of two different oil droplet has been highlighted.

1. Introduction

A major proportion of all the passenger cars are powered by spark ignited engines, and there were more than 1.2 billion cars by 2015. Automotive manufactures are continually aspiring towards improving the performance and efficiency of internal combustion engines to reduce fuel consumption and emissions. Downsized boosted Gasoline Direct Injection (GDI) engines allows similar power from smaller displacement engines, reducing throttling losses over the load-speed diagram. However, such technology remains constrained by several abnormal combustion phenomena, such as knock and pre-ignition (PI) [1, 2].

PI is an event where the air-fuel mixture ignites prior to firing the spark plug. It could be initiated by an external ignition source such as hot spots in the chamber, carbon particle deposits from the previous cycle and so on. The phenomenon is highly stochastic, but is more frequent at lower engine speeds and higher loads. With extreme downsizing and downspeeding of spark-ignited engines purported as the next step to efficiency improvement, engine operation at these conditions will be more frequent. Issues of PI become a bottleneck to further improvement in such case, more because a single PI cycle may be catastrophic to engine health, resulting in molten exhaust valves, holes in piston and broken connecting rod etc. As opposed to the problem of knock, PI occurrence cannot be controlled by changing the spark timing or reducing compression ratios.

The impact of engine lubricants on the occurrence frequency of the PI is receiving considerable attention by several research groups. These lubricants exhibit high cetane numbers (CN) in the range between 78 and 89. Under certain conditions, the lubricant droplets released from the cylinder liner and create a localized ignitable region that is susceptible to ignition prior to the defined spark timing. The impact of lubricant reactivity, however, is yet to be fully investigated [3]. Engine parameters also have some influence on PI frequency. For example, increasing the A/F ratio, boost pressure, or the energy flux tends to promote pre-ignition [4].

Previously, Dahnz et al. [5] investigated PI occurrences in highly supercharged SI engines both experimentally and numerically and reported that the most probable explanation for the occurrence of PI could be the release of lubricant oil droplets from the cylinder liner. According to the authors, PI was engine dependent and fuel injection on cylinder liners also showed the increase in PI occurrence. Also, researchers at Toyota Motors showed that there is minimal possibility for an oil droplet released from the crevice area to ignite at the same cycle in SI engines due to the low temperature of the droplet. However, it might be heated during combustion and carried on to the next cycle and become an ignition source [6]. Amann et al. [1], based on their findings, also suggested the lubricant or fuel based hydrocarbon deposit on the piston crevices to be the leading cause of PI.

In this study, hexadecane and engine oil droplets are used, to simulate PI precursors. The effect of pressure at an elevated temperature is investigated for both sources. The overall objective of this study is to investigate the tendency of lubricant oils' ignitability and its probability in leading to a fully developed flame. It should be taken as an initial study, showcasing the capabilities of indigenously developed PI vessel.

2. Experimental Setup

A pre-ignition constant volume (4L) chamber made of stainless steel 316L (Fig. 1) that will allow introduction of a single droplet of lubricating oil has been built. The length and outer and inner diameter of the vessel are 320 mm, 200 mm and 140 mm respectively. The vessel is pressure tested to 70 bar and can is capable of operating at 500 °C. It is equipped with three orthogonal pairs of quartz windows that provide optical access to allow photographic (high-speed Schlieren imaging and chemiluminescence) and laser (OH-PLIF) diagnostics.

Corresponding author.
E-mail address: long.jiang@kaust.edu.sa

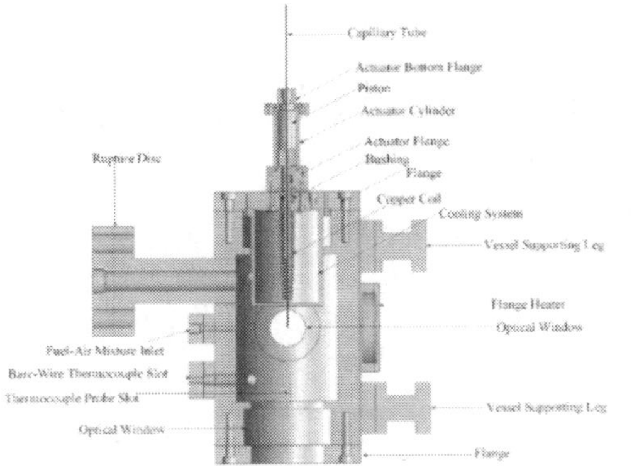

Figure 1. Schematic of PI vessel (cross section view).

In order to simulate the droplet-induced PI event, a droplet injection system is incorporated in the modification of the combustor. The droplet generator has been designed in such a way that it insulates oil from the elevated reactant temperature. This design includes the ability to water cool the hypodermic needle for the high initial reactant temperatures cases. The oil delivery system is driven by a syringe pump which provides precise control of the mass of lubricating oil. The oil droplet is suspended on the junction of a thermocouple wire where the instantaneous droplet can be measured throughout the oil droplet lifetime. The images of different droplet size for engine oil and hexadecane are shown in fig. 2 respectively.

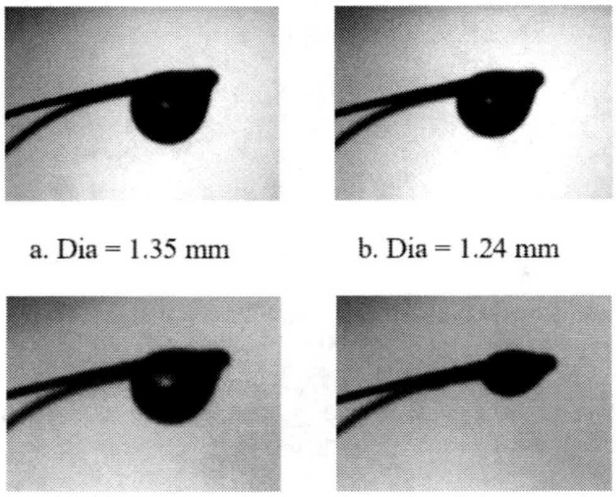

a. Dia = 1.35 mm b. Dia = 1.24 mm

a. Dia = 1.20 mm b. Dia = 0.82 mm

Figure 2. Droplet suspension with different diameter (lubricant oil: top, hexadecane: bottom)

The vessel is equipped with dynamic and static pressure sensors, temperature sensors, safety valve, and rupture disk. Multiple heating sources such as in-line heater with power rating of 3 kW and 4 heating tapes with power rating of 0.7 kW each has been employed for heating up the vessel. The piston which extends and retracts the needle inside the vessel is actuated by pneumatically controlled solenoid valves.

Two different oil sources were used in this study. First, lubricant oil surrogate investigated in this study was Hexadecane ($C_{16}H_{34}$) which has auto-ignition temperature of 215 °C. Second, an engine oil with SAE grade 15W-40. The working vessel available at KAUST is shown in Fig. 3.

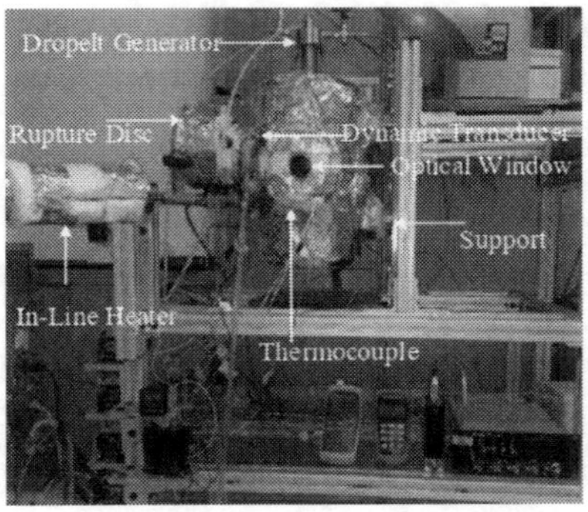

Figure 3. PI Vessel at KAUST.

3. Preliminary Result

3.1 Effect of pressure on lubricant surrogate ($C_{16}H_{34}$) ignition in ambient air.

The effect of pressure on ignition of single droplet of hexadecane in ambient air was investigated. The experiments were carried out at 2 bar, 3 bar, 4 bar and 8 bar gas pressure. The vessel was heated to 400 °C before introducing the droplet. Two sets of experiment, each comprised of 20 consecutive runs were carried out for each set pressure. The average volume injected to get an ignition for each set pressure is given in table 1.

Table 1: Average droplet size required to get an ignition for hexadecane.

Pressure (bar)	2	3	4	8
Volume (μl)	3.56	3.20	2.89	2.31
Diameter (mm)	1.89	1.80	1.77	1.64

Ignition was observed for all tested pressure conditions. Fig. 4 (a) shows the initiation of ignition of the lubricant oil surrogate. By calculating the average volume for each set pressure, it was noted that highest volume of oil surrogate required to get an ignition was at 2 bar and as the pressure increased to 8 bar, the volume requirement decreased. At 8 bar pressure, the volume of oil droplet required for ignition was significantly low compared to the former ones. The flame was short and yellow and the amount of carbonaceous particles was sparse (Fig. 6).

3.2 Effect of pressure on engine oil ignition in ambient air.

The effect of pressure on ignition of single droplet of engine oil was investigated. The experiments were carried out at 2 bar, 3 bar and 4 bar gas pressure. The ambient gas temperature was at 400 °C prior to introducing the oil droplet. The average volume injected to get an ignition for each set pressure is given in table 2.

Table 2: Average droplet size required to get an ignition for engine oil.

Pressure (bar)	2	3	4
Volume (μl)	1.50	0.80	0.30
Diameter (mm)	1.42	1.15	0.83

It was observed that the minimum volume required to get an ignition for the engine oil decreased with the increasing pressure. The flame was yellow and sooty and the amount of carbonaceous particles was dense, as compared to hexadecane droplet ignition (Fig. 4).

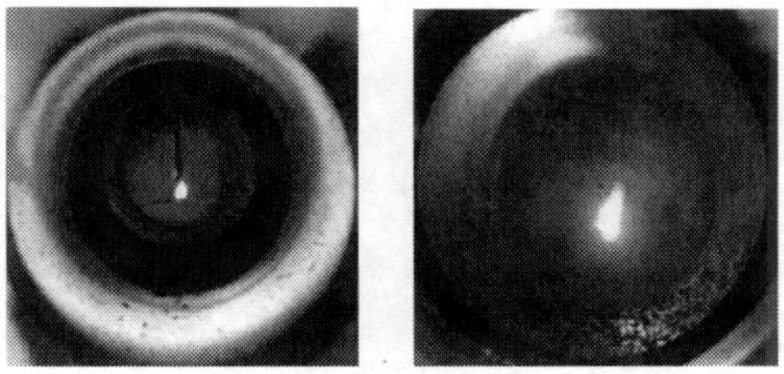

Figure. 4. Ignition of (a) hexadecane droplet (left) (b) engine oil droplet (right).

4. Conclusion

The experiments were conducted at different pressures and 400 °C ambient temperature. Hexadecane and engine oil (following SAE 15W40) were investigated in this study. In all the cases, ignition of hexadecane and engine oil were observed. In both of the cases, the volume required to get an ignition decreased with the increasing pressure. The flame size was much bigger for engine oil compared to the hexadecane, although the volume required at same pressure condition, were lower. Also, the flame was sooty for an engine oil compared to hexadecane.

This is an on-going study and further research will be thoroughly carried out to investigate the tendency of lubricant oils' ignitability and its probability in leading to a fully developed flame. Effect of various parameters like the temperature (T) and the pressure (P) of the mixture, the equivalence ratio of the mixture (Φ) of the bath gas (fuel-air surrounding the droplet), and the temperature (T_d) and the size (D_d) of the droplet with different fuels will be investigated as well.

5. Acknowledgment

This project is supported by Saudi Aramco under FUELCOM II project and Clean Combustion Research Center, King Abdullah University of Science and Technology.

References

[1] M. Amann, D. Mehta, T. Alger, SAE Int. J. Engines (4) (2011) 274-285.

[2] B. Lewis, G. Von Elbe, Combustion, Flames and Explosions of Gases, Third ed., Academic Press, Inc., Orlando, FL.1987.

[3] M. Amann, T. Alger, SAE Int. J. Fuels Lubr. (5) (2012) 760-771.

[4] M. Amann, T. Alger, D. Mehta, SAE Int. J. Engines (4) (2011) 235-245.

[5] C. Dahnz, K.-M. Han, U. Spicher, M. Magar, R. Schiessl, U. Maas, SAE Int. J. Engines (3) (2010) 214-224.

[6] M. Ohtomo, H. Miyagawa, M. Koike, N. Yokoo, K. Nakata, SAE Int. J. Fuels Lubr. (7) (2014) 673-682.

11th Asia-Pacific Conference on Combustion,
The University of Sydney, NSW Australia
10th -14th December 2017.

Large Eddy Simulation of a Dual Swirl Gas Turbine Model Combustor with Self-excited Thermo-acoustic Instability

Zhi X. Chen[1] **and N. Swaminathan**[1]

[1] Department of Engineering, Cambridge University

Trumpington Street, Cambridge, CB2 1PZ, UK

Abstract

Large eddy simulation is performed for stable and unstable swirl flames established in a partially premixed gas turbine model combustor. A flamelet closure is used for the sub-grid combustion process including both premixed and non-premixed mode contributions. Cold flow study is conducted to investigate mesh sensitivity on the flow field in the combustion chamber. Two reacting cases, one stable flame and another unstable, are computed to assess the model abilities of predicting flame structure and thermo-acoustic instability. The computed results of velocity, temperature and OH concentration agree well with the experimental data. The self-excited instability is also predicted in the simulation of the unstable flame, however, the oscillating frequency is significantly under-predicted.

1 Introduction

The rapidly increasing demand for high-energy density power generation but with more stringent regulations on industrial pollution inevitably drive the gas turbine (GT) design towards high fuel efficiency and low pollutant emissions. Recent advances in computing power allow Large Eddy Simulation (LES) to be used for practical combustors with complex geometries [1]. Compared to the Reynolds-averaged Navier-Stokes (RANS) approach, LES is expected to be more accurate and reliable especially for flows with large-scale unsteadiness such as swirling flows in GT combustors. The typical cell size in LES grid is larger than the reaction zone thickness and thus a sub-grid scale (SGS) combustion model is required to account for the SGS turbulence/chemistry interaction. As GT combustion often involves complex geometry leading to heavy computations, simple but robust model with low computational cost is of particular interest in practice. Flamelet approach is a suitable candidate for this because of the decoupling of chemistry and turbulent flow calculations [2].

Lean premixed combustion has the promising features of high efficiency with low emissions [2]. A well-known lean premixed GTMC is the *PRECCINSTA* burner [3], which was scoped to tackle the thermo-acoustic dynamics in lean swirl flames and also to present a good database for model validations. A number of studies were conducted using different combustion models [4, 5], assuming a homogeneous mixture inside the combustion chamber. However, it was pointed out in [3] that the imperfect mixing exists in this burner resulting in a particular type of thermo-acoustic instability under certain operational conditions. This is further confirmed in a recent numerical study [6] showing that the self-excited instability phenomenon occurs only when the inhomogeneous mixture is considered.

To further investigate the partially premixing effect, Meier *et al.* [7] modified the flow intake part of the *PRECCINSTA* burner to a dual swirl configuration under partially premixed conditions. Three flames, designated as flame A , B and C, were investigated with flame A being stable, flame B showing thermo-acoustic instability and flame C approaching blow-off. An earlier study by See and Ihme [8] assessed the Flamelet/Progress Variable (FPV) and Filtered Tabulated Chemistry (F-TACLES) models for flame A and showed fairly good comparisons for the velocity, temperature and major species. However, notable differences were found near the inner swirler where strong partial premixing exists at the base of the lifted flame. Donini et al. [9] also computed Flame A using a 5-D Flamelet Generated Manifold (FGM) approach including heat loss effects. However, considerable deviations from the measured data were observed for velocity and temperature suggesting that heat loss effect is not of main importance for this combustor. Further modelling efforts are needed to gain further insight for this flame. Moreover, the unstable flame B has not yet been studied numerically so far and self-excited instability is challenging for modelling.

Thus, the first objective of this paper is to assess the performance of our flamelet approach under complex geometry and flow conditions in a dual swirl GTMC experimentally investigated by Meier *et al.* [7]. This approach views partially premixed flames as a collection of unstrained laminar premixed flamelets with varying equivalence ratios and turbulent fluctuation is accounted for through a presumed joint probability density function (PDF). The non-premixed effect is also included through an algebraic closure involving the scalar dissipation rate of mixture fraction. Secondly, it is of interest to evaluate the model ability to capture self-excited thermo-acoustic instability in this confined combustor.

2 Modelling Methodology

The LES modelling framework is briefly described here as elaborate details are in [10, 11].

2.1 LES formulation

The Favre-filtered conservation equations of mass, momentum and thermo-chemical enthalpy are solved. The SGS stresses are modelled using the standard Smagorinsky model [12]. In LES, chemical reactions are sub-grid phenomena and thus SGS combustion model is required. The flamelet model used for this study maps all thermo-chemical quantities into a low-dimensional manifold characterised by mixture fraction, Z, and a reaction progress variable, c, and utilises a presumed joint sub-grid PDF for Z and c. The mixture fraction is defined using Bilger's formula. The normalised form of progress variable, $c = \psi / \psi^{\mathrm{Eq}}$, where $\psi = Y_{\mathrm{CO}} + Y_{\mathrm{CO_2}}$ and ψ^{Eq} is its equilibrium value for the local mixture [10, 11] is used. The first two moments of Z and c, filtered mean and SGS variance, are solved using their transport equations [11]. The unclosed terms in these equations are the SGS scalar dissipation terms, $\widetilde{\chi}_{Z,\mathrm{sgs}}$ and $\widetilde{\chi}_{c,\mathrm{sgs}}$, for mixture fraction and progress variable respectively, and the reaction source terms, $\overline{\dot{\omega}_c^*}$ and $\overline{c\,\dot{\omega}_c^*}$ in the progress variable equations. These terms require

Corresponding author.
E-mail address: zc252@cam.ac.uk

modelling and their closure models are described next.

2.2 SGS combustion model

The SGS scalar dissipation rate of mixture fraction is modelled using a linear relaxation approximation: $\overline{\rho}\widetilde{\chi}_{Z,\mathrm{sgs}} \equiv \overline{\rho D_Z (\nabla Z \cdot \nabla Z)} - \overline{\rho}\widetilde{D}_Z \left|\nabla \widetilde{Z}\right|^2 = C_Z(\mu_t/\Delta^2)\widetilde{Z''^2}$, where D_Z is the molecular diffusivity of Z and the model coefficient is taken to be $C_Z = 2.0$ [13]. For progress variable, its dissipation rate model needs to include the combustion effects [2] and thus the model proposed by Dunstan et al. [14] is used:

$$\widetilde{\chi}_{c,\mathrm{sgs}} = \mathscr{F}_0\left[2K_c^*(S_L^0/\delta_L^0) + (C_3 - \tau C_4 \mathrm{Da}_\Delta)(2u'_\Delta/3\Delta)\right]\widetilde{c''^2}/\beta'_c, \quad (1)$$

where the SGS velocity $u'_\Delta = |\widehat{\widetilde{u}} - \widetilde{u}|$ is obtained using a test filter width of $\widehat{\Delta} = 2\Delta$. The function $\mathscr{F}_0 = \left(1 - e^{-0.75\Delta/\delta_L^0}\right)$ ensures that $\widetilde{\chi}_{c,\mathrm{sgs}}$ approaches 0 when (Δ/δ_L^0) is infinitely small. The parameters, C_3, C_4, K_c^*, τ, S_L^0 and δ_L^0 are untunable and depend on local thermo-chemical and fluid-mechanical quantities. The model constant of $\beta'_c = 7.5$ is used following the previous study [11].

As the combustion is partially premixed in this study, a sub-grid model accounting for different combustion mode contributions is used. The filtered reaction rate model is written as [11]

$$\overline{\dot{\omega}_c^*} = \overline{\dot{\omega}}_c + \overline{\dot{\omega}}_{\mathrm{np}} = \overline{\dot{\omega}}_c + \overline{\rho D_Z (\nabla Z \cdot \nabla Z) \frac{c}{\psi^{\mathrm{Eq}}} \frac{d^2\psi^{\mathrm{Eq}}}{dZ^2}}, \quad (2)$$

where $\overline{\dot{\omega}}_c$ and $\overline{\dot{\omega}}_{\mathrm{np}}$ signify premixed and non-premixed combustion mode contributions respectively. The premixed contribution term is modelled as

$$\overline{\dot{\omega}}_c = \overline{\rho} \int_0^1 \int_0^1 \frac{\dot{\omega}_c(\xi,\zeta)}{\rho(\xi,\zeta)} P_\Delta(\xi,\zeta)\; d\xi\, d\zeta\,, \quad (3)$$

where ξ and ζ are the sample space variables for Z and c respectively. The flamelet reaction rate, $\dot{\omega}_c$, and density, ρ, are computed using unstrained 1D laminar flame calculations for different mixture fractions covering the flammable range. To account for the SGS fluctuations of both Z and c, which we assume to be independent for this study, the SGS joint PDF, $P_\Delta(\xi,\zeta)$, is modelled as $P_\Delta(\xi,\zeta) = P_\Delta(\xi;\widetilde{Z},\widetilde{Z''^2})P_\Delta(\zeta;\widetilde{c},\widetilde{c''^2})$ using a presumed Beta distribution for both PDFs. A simple algebraic model is used for the non-premixed term in Eq. (2) written as

$$\overline{\dot{\omega}}_{\mathrm{np}} \simeq \overline{\rho}\widetilde{c}\widetilde{\chi}_Z \int_0^1 \frac{1}{\psi^{\mathrm{Eq}}(\xi)} \frac{d^2\psi^{\mathrm{Eq}}(\xi)}{d\xi^2} \widetilde{P}_\beta(\xi)\, d\xi, \quad (4)$$

where $\widetilde{\chi}_Z$ is the total scalar dissipation rate including both the resolved and SGS parts: $\widetilde{\chi}_Z = \widetilde{D}_Z(\nabla\widetilde{Z} \cdot \nabla\widetilde{Z}) + \widetilde{\chi}_{Z,\mathrm{sgs}}$.

The reaction term, $\overline{c\,\dot{\omega}_c^*}$, in the SGS variance equation of c and other themo-chemical quantities, such as species mass fractions, are calculated similarly as Eq. (3) and the filtered mixture density is computed using the ideal gas state equation. The flamelet look-up table is constructed using four control parameters, $\widetilde{Z}$, $\widetilde{Z''^2}$, $\widetilde{c}$ and $\widetilde{c''^2}$ discretised over $50 \times 15 \times 75 \times 40$ points.

3 Experimental Configuration and Numerical Setup

Figure 1a shows a schematic of the dual swirl burner setup of the experiments [7]. The injector comprises two coaxial swirling air nozzles and an annular ring of fuel jet nozzles. The exit plane of the inner swirler and fuel jets is 4.5 mm below the outer swirler exit, which is defined as $h = 0$ in the axial direction. Dry ambient air was supplied through the inlet located at the bottom of the plenum, and the inner and outer air nozzles diameters are 15 and 25 mm, respectively. Non-swirling methane was supplied through 72 square channels (0.5×0.5 mm) and these small squares are retained in the numerical grid to capture the correct mixing. Other geometrical details of the combustion chamber are marked in Fig. 1a. In this study, three test cases are simulated including one non-reacting and two reacting flow cases as detailed in Table 1. Figure 1b shows the filtered reaction rate iso-surface of $\overline{\dot{\omega}_c^*} = 200$ kg/m^3/s, coloured by temperature at a typical instant for flame A and the axial locations for reacting flow measurements [7] are also highlighted.

Table 1: Summary of the computed test cases [7].

Case	Φ_{glob}	$\dot{m}_{\mathrm{plenum}}$ (g/s)	$\dot{m}_{\mathrm{jet}}$ (g/s)	Reynolds number	Swirl number
Non-reacting	—	19.74	1.256	≈ 58,000	≈ 0.9
Flame A	0.65	18.25	0.697	≈ 58,000	≈ 0.9
Flame B	0.75	4.68	0.205	≈ 15,000	≈ 0.55

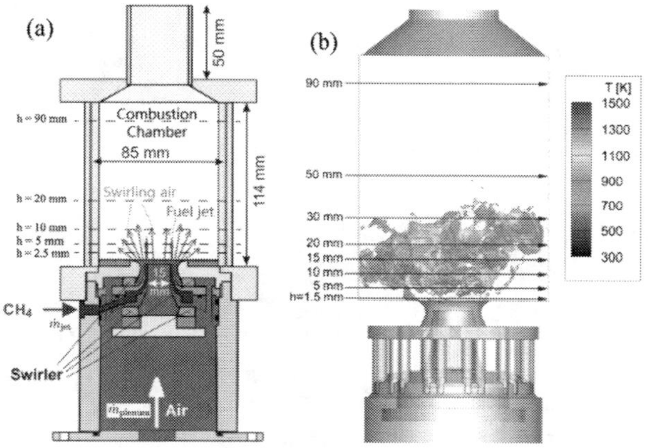

Figure 1: Schematic of (a) the DLR dual swirl burner setup [7] and (b) iso-surface of the filtered reaction rate, $\overline{\dot{\omega}_c^*} = 200$ kg/m^3/s, coloured by temperature.

The numerical boundary conditions at the flow inlets are given using the data listed in Table 1. Non-slip wall condition is employed for the combustor confinement walls. The computational domain is extended to the far field downstream of the outlet duct to avoid numerical reflection of the acoustic waves. Thus, an unstructured numerical grid with tetrahedral mesh cells is used and more detailed are given later while discussing the mesh sensitivity. The CFD toolbox OpenFOAM 2.3.0 with second-order accuracy in time and space is used. The LES statistics are collected over a physical time period of 0.1 s, which ensures convergence of statistics in the entire combustion chamber. The simulations are performed on the ARCHER UK National Supercomputer using 1080 cores and 1 ms of physical time for the reacting case requires about 2 hours wall-clock time. The reacting flow simulation is initiated on a fully developed mixing solution.

4 Results and Discussion

4.1 Non-reacting flow and mesh sensitivity

The non-reacting flow velocity measurements [15] were taken at five axial locations as marked in Fig. 1a. Both measured values of mean and r.m.s. axial velocities are compared with the LES results in Fig. 2. Three sets of LES results shown are for three numerical grids, G1, G2 and G3, having 12, 15 and 20 million mesh cells respectively. A strong grid sensitivity is observed in Fig. 2 when the grid size increases from G1 to G2. The comparison between the experimental data [15] and LES results is very good for G2, significantly improved from G1. However, no evident difference is found when further increasing the refinement level from G2 to G3, both agreeing well with the measurements.

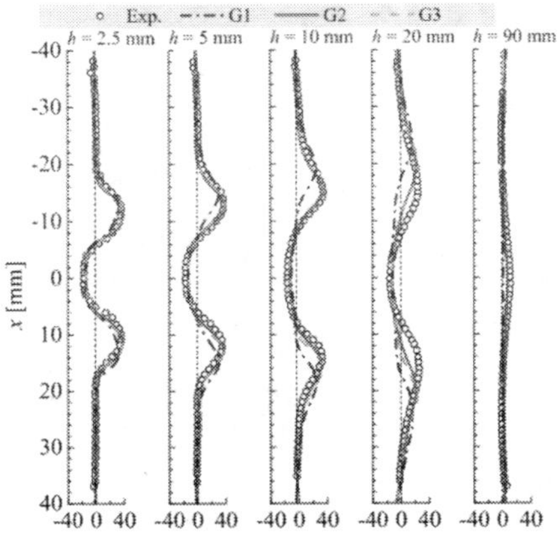

(a) Averaged axial velocity [m/s]

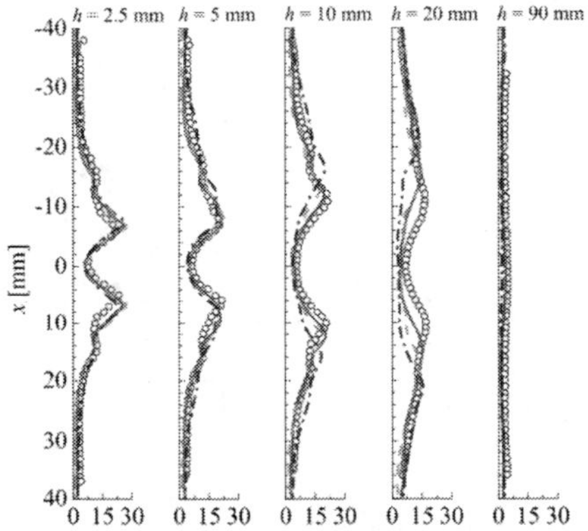

(b) r.m.s. axial velocity [m/s]

Figure 2: Cold flow comparison of axial velocity profiles: (a) mean values (b) r.m.s fluctuations, obtained from the experimental measurements [15] and the LES using grid G1, G2 and G3.

The measurements taken at h = 2.5 mm allows for examining the LES prediction at the combustor inlet. Figure 2a shows that the measured mean axial velocity peak is about 40 m/s. The LES profile agrees well with the experimental data at this near-field location for grids G2 and G3 suggesting that the flow motions inside the swirlers, upstream of the combustion chamber, are well captured. However, for coarse grid G1 it under-predicts the peak value with the peak located further away from the centre axis, i.e. x = 0 mm. This becomes more evident in the downstream showing an approximately 80% over-prediction for the inner recirculation zone (IRZ) width at h = 20 mm for G1. In contrast, this IRZ width is predicted reasonably well for grids G2 and G3. Thus, grid G2 with 15 million tetrahedral cells is chosen for the subsequent reacting flow simulations which are discussed in next subsection.

4.2 Reacting flow cases

The mid-plane contour of filtered reaction rate, $\overline{\dot{\omega}_c^*}$, is shown in Fig. 3a for flame A along with three mixture fraction iso-lines corresponding to stoichiometry (black), lean (red) and rich (black) flammability limits. The reaction zone is mainly located in the shear layer between the IRZ and main inflow stream within the flammability limits, showing a V shape with a small tip pointing towards the outer recirculation zone (ORZ) near the chamber wall. This tip is formed because of the flammable mixture pockets shed from the main swirling flow, which first hit the side wall and then recirculate down into the ORZ creating a small reaction zone. This outer reaction zone is more visible in Fig. 3b comparing the averaged experimental OH-PLIF image [7] with the computed mean OH signal, $\langle \widetilde{S}_{OH} \rangle$. This OH signal is calculated using the local OH mole fraction and temperature obtained from the LES using [16] $\widetilde{S}_{OH} \propto [OH]\,\widetilde{T}^{1-\beta}$, $-2.0 \leqslant \beta \leqslant 1.0$, where the parameter is $\beta = 0$ for this study. The overall V shape of the OH distribution with high concentration in the IRZ is captured reasonably well in the LES. Considerable discrepancy is found in the downstream region where the OH-PLIF signal is weaker than that for the flame base, whereas the computed $\widetilde{S}_{OH}$ value remains high suggesting an over-estimated equilibrium value of OH in the flamelet calculation. The flame lift-off height, L_f, defined as the distance between the fuel jet exit and the most upstream point of the central reaction zone, was measured to be about 10 mm in [7] as marked in Fig 3b. The computed value of L_f is about 11.5 mm, in good agreement with the measurements.

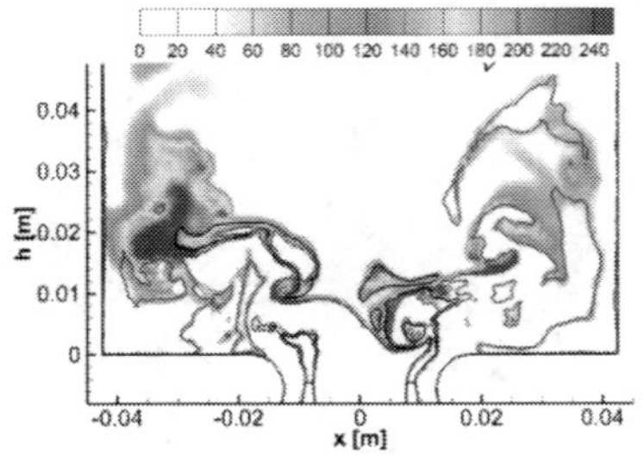

(a) $\overline{\dot{\omega}_c^*}$ (kg/m^3/s)

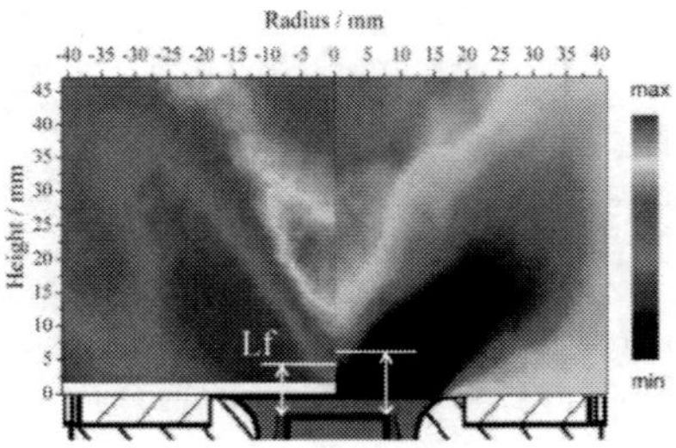

(b) OH (A.U.)

Figure 3: Mid-plane contours for flame A: (a) filtered reaction rate and mixture fraction (b) comparison of time-averaged OH concentration obtained from experiments (left) [7] and LES (right).

Figure 4 compares the radial variation of computed temperature with the measured values [7] for flame A. The averaged temperature, $\langle \widetilde{T} \rangle$, is predicted quite well for all the streamwise locations shown. The LES results are about 30% higher than the measured value for the radial positions, $10 < r < 20$ mm at h = 20 mm and for $15 < r < 30$ mm at h = 30 mm. The average burner outlet temperature was about 1900 K in the experiments, which is predicted very well in the simulations. The trend of temperature r.m.s. evolution in the radial and axial directions is also captured.

The prediction of the IRZ length is of main importance for swirl flames and it is quite challenging for unstable flames experiencing longitudinal instabilities such as flame B. Figure 5 compares the computed and measured mean centreline axial velocity for both

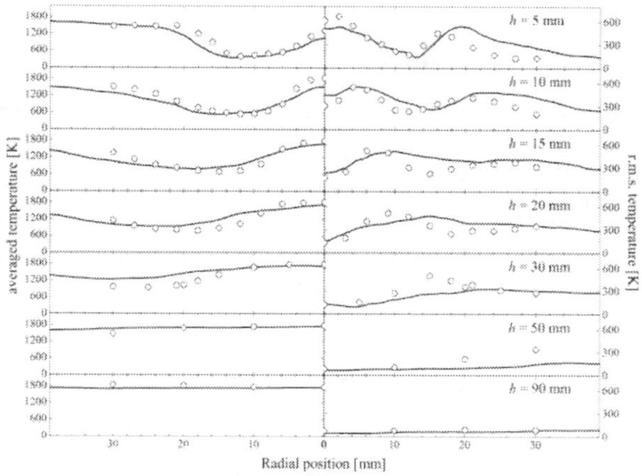

Figure 4: Radial profile of temperature: mean (left) and r.m.s. (right). Symbol: experiment [7]. Line: LES.

flames A and B. It can be seen that the computed IRZ length is in excellent agreement with the measured value for both flames. The centreline variation in flame A is predicted very well despite that the computed value decreases slightly faster at the beginning near the nozzle. This is improved in flame B and the overall agreement is very good for this unstable flame suggesting that the unsteady moment of the flow field caused by thermo-acoustic instability is captured in the simulation.

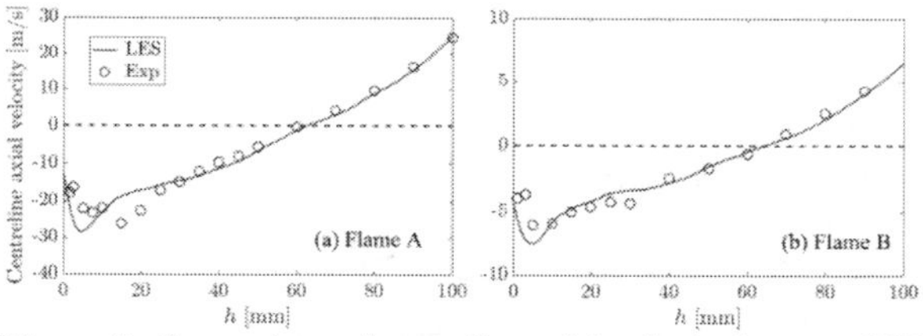

Figure 5: Comparison of centreline axial velocity between LES and experiments [7].

In the experiments [7], the pressure fluctuation was measured using a microphone placed in the plenum. A probe is used at the same location in the LES and the time series of the probed pressure fluctuations are presented in Fig. 6 for both flames A and B. The spectra analysis of the probed data using the Fast Fourier Transform (FFT) technique is also shown in this figure. Similar to the experimental observations [7], no clear periodic signal is found for flame A as the flame is stably burning and thus no peak can be identified in the spectrum. In contrast, for the unstable flame B, there is a pronounced periodic variation of the pressure signal resulting in a peak frequency of about 180 Hz. This suggests that the combustion process undergoes self-excited instability which is captured by the simulation with no inlet perturbation imposed. However, the computed frequency, 180 Hz, at which the instability occurs, is quite different from the measured value about 290 Hz [7]. The reason for this difference still remains unclear and needs further investigation. Nevertheless, the modelling framework used in this study performs quite well for predicting the flow and flame structures inside this complex gas turbine model combustor and also shows potential for self-excited thermo-acoustic instability problems.

5 Conclusions

Partially premixed methane-air swirl flames in a gas turbine model combustor have been simulated using LES in this study. Two flames, one stable and the other with thermo-acoustic instability,

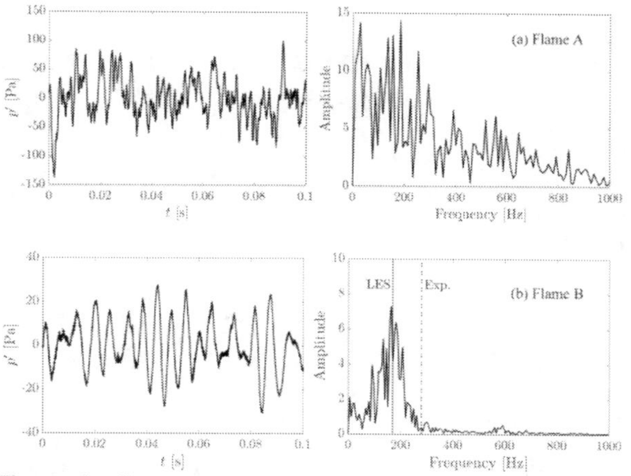

Figure 6: Computed pressure fluctuations in the plenum for flames A and B: time series (left) and spectra (right).

are investigated and the simulation results agree well with the experimental data for statistics of velocity, temperature and OH radical concentration. The self-excited longitudinal instability is also captured in the LES, however, the oscillating frequency is under-predicted being about 180 Hz, compared to 290 Hz observed in the experiments. A possible reason for this under-prediction is the inconsistency of acoustic boundary conditions between the numerical simulation and experiments and this will be further investigated in a future study.

Acknowledgement

ZC and NS acknowledge support of Mitsubishi Heavy Industries. NS also acknowledges the support from EPSRC.

Selected References

[1] L.Y.M. Gicquel, G. Staffelbach, T. Poinsot, Prog. Energy Combust. Sci. 38 (2012) 782–817.

[2] N. Swaminathan, K.N.C. Bray (Eds.), Turbulent Premixed Flames, Cambridge University Press, 2011.

[3] W. Meier, P. Weigand, X.R. Duan, R. Giezendanner-Thoben, Combust. Flame 150 (2007) 2–26.

[4] B. Fiorina, R. Vicquelin, P. Auzillon, N. Darabiha, O. Gicquel, D. Veynante, Combust. Flame 157 (2010) 465–475.

[5] V. Moureau, P. Domingo, L. Vervisch, Combust. Flame 158 (2011) 1340–1357.

[6] B. Franzelli, E. Riber, L.Y.M. Gicquel, T. Poinsot, Combust. Flame 159 (2012) 621–637.

[7] P. Weigand, W. Meier, X.R. Duan, W. Stricker, M. Aigner, Combust. Flame 144 (2006) 205–224.

[8] Y.C. See, M. Ihme, Proc. Combust. Inst. 35 (2015) 1225–1234.

[9] A. Donini, R.J.M. Bastiaans, J.A. van Oijen, L.P.H. de Goey, Flow Turbul. Combust. 98 (2017) 887–922.

[10] Z. Chen, S. Ruan, N. Swaminathan, Combust. Flame 162 (2015) 703–716.

[11] Z. Chen, S. Ruan, N. Swaminathan, Proc. Combust. Inst. 36 (2017) 1645–1652.

[12] J. Smagorinsky, Mon. Weather Rev. 91 (1963) 99–164.

[13] H. Pitsch, H. Steiner, Phys. Fluids 12 (2000) 2541.

[14] T.D. Dunstan, Y. Minamoto, N. Chakraborty, N. Swaminathan, Proc. Combust. Inst. 34 (2013) 1193–1201.

[15] A. Widenhorn, B. Noll, M. Aigner, AIAA Paper 647 (2009).

[16] Y. Minamoto, N. Swaminathan, Combust. Flame 161 (2014) 1063–1075.

11th Asia-Pacific Conference on Combustion,
The University of Sydney, NSW Australia
10th -14th December 2017.

Influence of Vortical Structures on Ignition in DNS of a Turbulent Mixing Layer with Non-Premixed H_2/air Combustion

Tatsuya Yonemura[1], Takashi Ohta[1], Yasuyuki Sakai[1]

[1]University of Fukui

Bunkyo 3-9-1, Fukui, 910-8507, Japan

Abstract

To investigate the dynamic behavior of vortical structures along with heat release, we performed a direct numerical simulation of a time-developing three-dimensional turbulent mixing layer together with non-premixed H_2/air combustion using a detailed chemical reaction scheme. We focus on the relationship between the developing vortical structures and high-heat-release-rate regions before ignition. In a turbulent mixing layer, ignition spots are not located inside roller vortices but around rib vortices, where the ignition kernels are intensified. According to the history of the distribution of the heat-release rate, the ignition kernels concentrate around ribs as a result of flow around developing and pairing rollers, which result in the localization of the ignition kernel.

1 Introduction

For turbulent non-premixed combustion, turbulence structures strongly influence ignition spots and combustion intensity since turbulence controls the degree of the mixing of fuel and the oxidizer. As a consequence, the variation of physical parameters (e.g., viscosity, specific heat capacity, and diffusion coefficient) resulting from the chemical reaction alters, in turn, the development of turbulence. This sequence of events strengthens the relationship between turbulence structures and chemical reactions compared with the case of premixed combustion. Consequently, it is difficult to predict the ignition and quenching spots of non-premixed combustion in turbulence.

Direct numerical simulation (DNS) has previously been performed to investigate the relationship between turbulent structures and chemical reactions in detail. For example, Im et al. [1] carried out the DNS of two-dimensional homogeneous turbulence on non-premixed H_2/air combustion. By adopting a detailed combustion reaction scheme, they showed that stronger turbulence slightly delays the ignition time compared with weak and moderate turbulence. They also observed that the ignition kernel tends to be located in the region where the mixing layer is convex toward the fuel side, because of the high diffusivity of the hydrogen molecules. Hilbert et al. [2] performed similar simulations to Im et al. using a detailed combustion reaction scheme, and revealed that the first ignition spot always occurs for the lowest value of scalar dissipation rate along a well-defined mixture fraction. They also showed that the influence of the turbulent Reynolds number on the ignition time is very small, whereas a large turbulence intensity leads to a large increase in the peak heat release rate. Doom et al. [3] performed a DNS of three-dimensional isotropic turbulence with non-premixed H_2/air combustion, and also with a detailed combustion reaction scheme. They showed that ignition occurs in fuel lean regions, with low vorticity and high temperature and a low scalar dissipation rate around the "most reactive" mixture fraction. Therefore, DNS is an effective technique to investigate the relationship between turbulence and combustion. However, to realize the prediction and control of combustion in turbulent flow, we focus on vortical structures peculiar to each turbulent flow.

We performed a DNS of a time-developing three-dimensional turbulent mixing layer along with non-premixed H_2/air combustion, and observed the distribution of high-heat-release-rate regions as ignition spots. In particular, in focusing on the developing vortical structures, we investigate the dynamic characteristics of the ignition kernels prior to ignition.

2 Simulation method

2.1 Governing equations of fluid flow

We solved the compressible total-mass conservation equation, momentum equations, energy equation for temperature, equation of states for the ideal gases and mass conservation equation for species k as the governing equations for compressible flow. In the simulation, the variations of mass fraction and heat release rate from the chemical reaction are considered in the mass conservation equation for species k and the energy equation at each time step, respectively. The thermodynamic parameters in the governing equations are dynamically estimated depending on the distributions of temperature and pressure. Moreover, because the thermo-diffusion effect is not negligible in light species, such as H and H_2, we take this effect into account in the mass conservation equation for species.

2.2 Numerical method for unsteady flow

We have developed a new method to simulate compressible multi-phase flow with combustion by extending a method for compressible flow at low Mach number [4]. The time integral is calculated based on a pressure equation in the fractional-step method. The convection and viscosity terms of the momentum equation are integrated explicitly using the second-order Adams-Bashforth method. The pressure-gradient term and the mass conservation equation are integrated implicitly using the backward-Euler method. The spatial derivatives are approximated with the second-order central-finite-difference method.

Corresponding author. E-mail address: t-ohta@u-fukui.ac.jp

No.	reaction	No.	reaction
R1	$O_2 + H = OH + O$	R12	$OH + HO_2 = H_2O + O_2$
R2	$H_2 + O = OH + H$	R13	$HO_2 + HO_2 = H_2O_2 + O_2$
R3	$OH + OH = O + H2O$	R14	$HO_2 + HO_2 = H_2O_2 + O_2$
R4	$H_2O + H = H_2 + OH$	R15	$OH + OH\ (+M) = H_2O_2\ (+M)$
R5	$H_2O_2 + OH = H_2O + HO_2$	R16	$H_2O_2 + H = OH + H_2O$
R6	$H + OH + M = H_2O + M$	R17	$H_2O_2 + H = HO_2 + H_2$
R7	$O_2 + H + M = HO_2 + M$	R18	$H + O + M = OH + M$
R8	$O + HO_2 = O_2 + OH$	R19	$O + O + M = O_2 + M$
R9	$H + HO_2 = OH + OH$	R20	$H + H + M = H_2 + M$
R10	$H + HO_2 = H_2 + O_2$	R21	$H_2O_2 + O = OH + HO_2$
R11	$H + HO_2 = H_2O + O$		

Table 1: Detailed chemical reaction scheme for hydrogen.

2.3 Detailed combustion reaction scheme

We extracted reactions related to hydrogen from a detailed combustion reaction scheme for iso-octane [5] as shown in Table 1, and used these reactions as a model to predict H_2/air combustion. The extracted combustion reaction scheme involves nine species (H_2, O_2, H_2O, OH, H, O, HO_2, H_2O_2, N_2) and 21 reactions, and takes into account the forward and backward reactions. The reaction rate of species k, $\dot{\omega}_k$, is estimated with the modified Arrhenius law at all grid points at each time step during the simulation.

3 Simulation conditions

In the simulation, we target the time-developing turbulent mixing layer shown in Fig. 1, where the spatial coordinates x, y and z are defined with the corresponding velocity components of u, v, and w, respectively. Two fluids are initially in separate regions and contact with each other at the plane $y = 0$. The computational domain size is L_x =10.0 mm, L_y = 6.015625 mm, and L_z =5.0 mm. We use staggered grids, with the number of computational grids set to 256×308×256. We apply periodic boundary conditions in the x and z directions, a convective outflow condition related to the pressure as the non-reflecting condition, and the free-slip boundary condition at the top and bottom boundaries. We impose the temperature T =300 K, and mass fractions of species, Y_{H_2} = 0.023, Y_{O_2} = 0, and Y_{N_2} = 0.977 on the fuel side denoted as the blue region, and T = 1100 K, Y_{H_2} = 0, Y_{O_2} = 0.233, and Y_{N_2} = 0.767, on the air side as initial uniform profiles. The initial pressure in the region is 1 atm and a hyperbolic tangent function is used to generate the initial velocity, temperature and mass-fraction profiles.

The heated air and cold hydrogen diluted with nitrogen flow in the opposite direction to each other along the plane $y = 0$. The Reynolds number based on the initial momentum thickness is defined as $Re_m^0 = \Delta U \delta_m^0 / \nu$, where ΔU, δ_m^0 and ν are the velocity difference across the two fluids, the initial momentum thickness, and mixture kinetic viscosity at $y = 0$, respectively. Here, we set ΔU = 74.0446 m/s with Re_m^0 = 100, and superimpose a disturbance of sine waves and weak white noise on the initial velocity profile to force the transition to turbulence. In addition, we performed a simulation of fluid at rest with the same configuration for comparison.

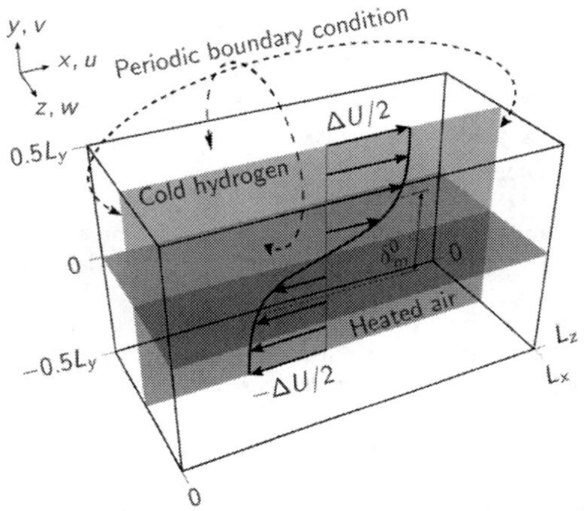

Figure 1: Computational domain and boundary conditions.

4 Results and Discussion

4.1 Variations of maximum heat release rate

Figure 2 shows temporal variations of the maximum heat release rate in the developing turbulent mixing layer and in fluid at rest. The increase tendency and the maximum value change depending on the condition of the flow fluid. We can expect the relationship between the vortical structures and the reaction promotion. In the following, we investigate the developing vortices and heat release rate in the flow filed.

4.2 Ignition around rib vortices

Figure 3 shows vortical structures and high-heat-release-rate regions at t/t_{ign} = 1.01. Here, t_{ign} is an ignition time defined by the maximum heat release rate q_{max} in each computational region as

$$\left.\frac{\partial^2 q_{max}}{\partial t^2}\right|_{t=t_{ign}} = 0. \quad (1)$$

Green isosurfaces indicate a constant z component of vorticity ω_z, while blue and yellow isosurfaces indicate constant x components of vorticity ω_x. We observe two roller vortices in

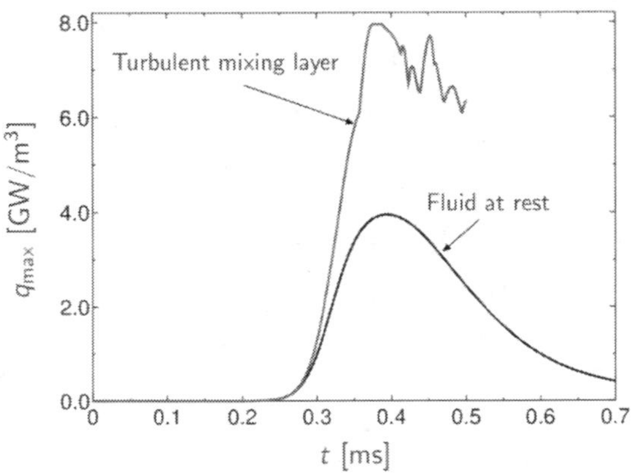

Figure 2: Temporal variations of maximum heat release rate.

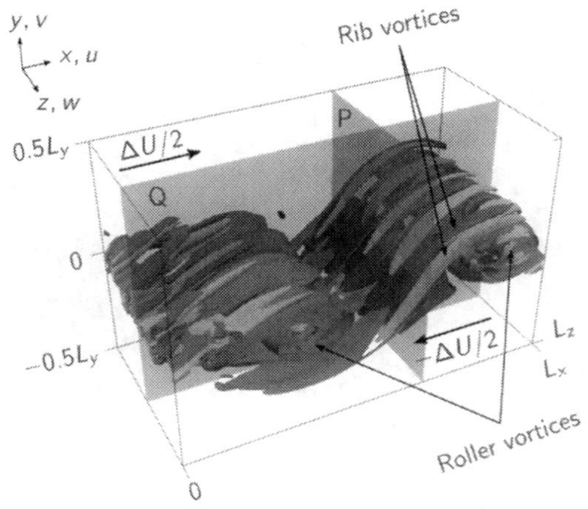

Figure 3: Vortical structures and high-heat-release-rate regions in a turbulent mixing layer (yellow: $\omega_x \delta_m^0 / \Delta U = 0.06$, blue: $\omega_x \delta_m^0 / \Delta U = -0.06$, green: $\omega_z \delta_m^0 / \Delta U = -0.1$, red: $q = 1.0\ \mathrm{GW/m^3}$) at $t/t_{ign} = 1.01$.

the mixing layer and rib vortices between the rollers in the z direction, which rotate alternately in opposite directions, and detect the rollers stretching the ribs, which are known to be the most typical vortical structures in turbulent mixing layers. We detect no high-heat-release regions inside the rollers, but only around the ribs, with regions shaped similarly to the ribs, and a three-dimensional appearance in our flow.

4.3 Tracking of ignition kernels

Figure 4 shows the distribution of the heat release rate on the plane P ($x/L_x = 0.65$) illustrated in Fig. 3. The heat release rate changes in the z direction. We focus on the plane Q at $z/L_z = 0.5$ shown in Fig. 3 to investigate the dynamics of the ignition kernels up to the ignition.

Figure 5 shows the distributions of the heat release rate and contours of the z component of vorticity ω_z at $t/t_{ign} = 0.32$, 0.48, 0.64, 0.8 and 1.01 on the plane Q (see Fig. 3). We define high-heat-release-rate regions as the "ignition kernel", in which obvious ignition is expected to occur, and focus on the temporal development of the ignition kernels. At first, as shown in Fig. 5(a), there are some kernels at the bottom of rollers on the fuel-lean side. At $t/t_{ign} = 0.48$, kernels, which are marked as A and B, are formed around ribs as shown in Fig. 5(b) due to flow around developing and pairing rollers, and then move to the bottom of the pairing rollers at $t/t_{ign} = 0.64$ (Fig. 5(c)). As shown in Fig. 6(a), at $t/t_{ign} = 0.48$, in both kernels A and B, reactions R1, R4 and R7 among all reactions relatively contribute to total heat release rate. In particular, the heat release rate of R7 is approximately 80 percent of total heat release rate. These facts are the same in case of fluid at rest. Next, kernel A is stretched gradually along with the rollers, so that the gradient of the scalar field becomes larger. The kernel disappears, while kernel B continues to intensify as it moves along the bottom of a roller as shown in Fig. 5(d). Finally, the intense kernel is located around the ribs between the rollers, and becomes an ignition spot in this simulation. At $t/t_{ign} = 1.01$, as shown in Fig. 6(b), the reactions in kernel B are similar to those of fluid at rest. In this process, rather than the difference of reactions in kernels, the developing vortical structures in the turbulent flow concentrate the distribution of ignition spots. The chemical reaction in kernel A is retarded by developing vortical structures.

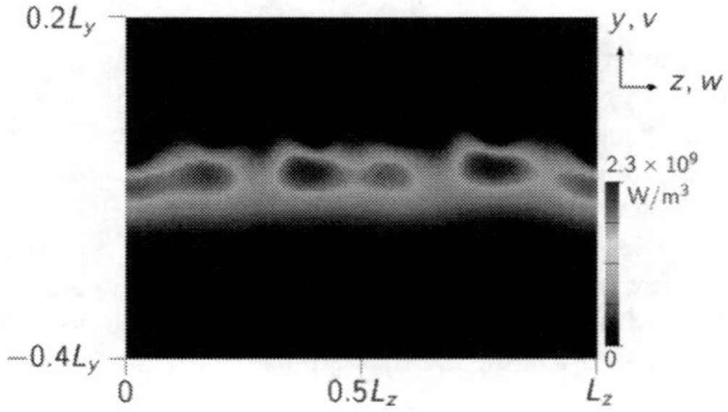

Figure 4: Distribution of high-heat-release-rate regions on the plane P ($x/L_x = 0.65$) at $t/t_{ign} = 1.01$.

5 Conclusion

We performed a DNS of a time-developing three-dimensional turbulent mixing layer with non-premixed H_2/air combustion using a detailed combustion reaction scheme. We then investigated the temporal development of vortical structures and ignition kernels with a high-heat-release rate in the flow up to ignition, and found that ignition spots were strongly related to developing rib vortices rather than roller vortices. The localization of the ignition was caused by flow around developing and pairing rollers.

References

1. H.G. Im, J.H. Chen, C.K. Law, Proc. Combust. Inst. 27 (1998) 1047-1056.
2. R. Hilbert, D. Thévenin, Combust. Flame 128 (2002) 22-37.
3. J. Doom, K. Mahesh, 47th AIAA Aerospace Science Meeting including The New Horizons Forum and Aerospace Exposition, American Institute of Aeronautics and Astronautics, Orlando, Florida, 2009, 240.
4. T. Ohta, T. Kajishima, K. Mizobata, K. Nakamura, Flow, Turbulence and combust. 89(3) (2012) 435-448.
5. H.J. Curran, P. Gaffuri, W.J. Pitz, C.K. Westbrook, Combust. Flame 129(3) (2002), 253-280.

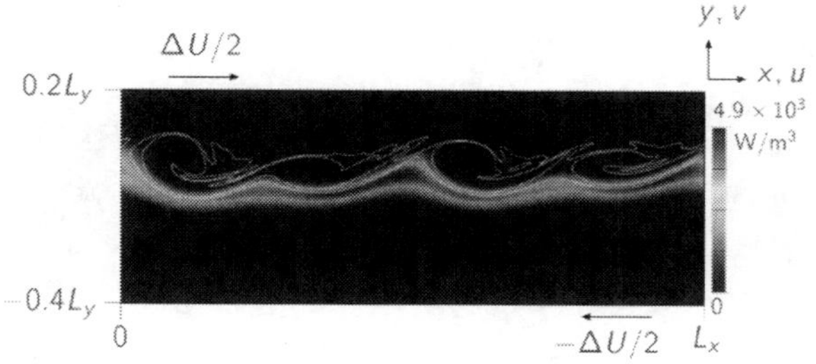

(a) $t/t_{\mathrm{ign}} = 0.32$

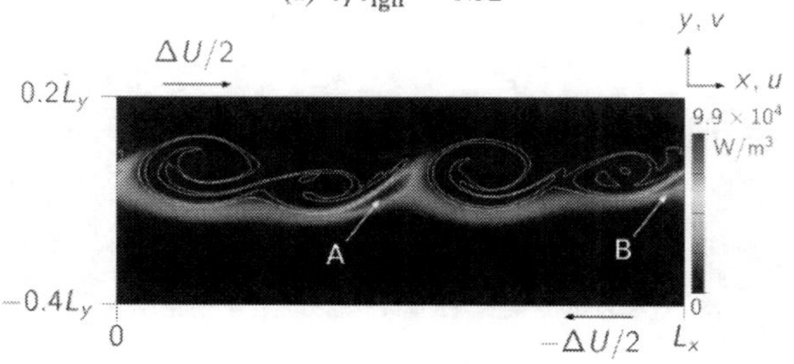

(b) $t/t_{\mathrm{ign}} = 0.48$

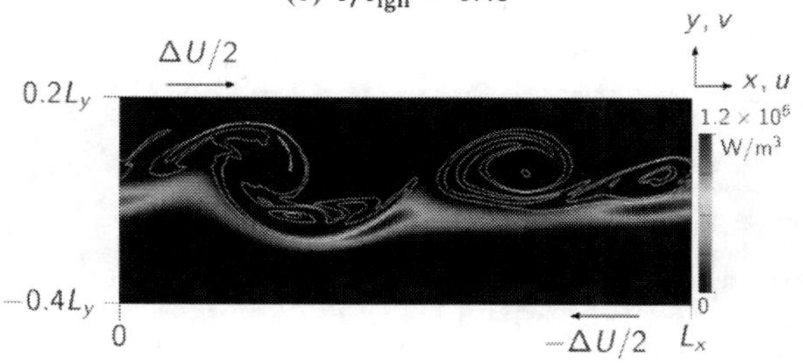

(c) $t/t_{\mathrm{ign}} = 0.64$

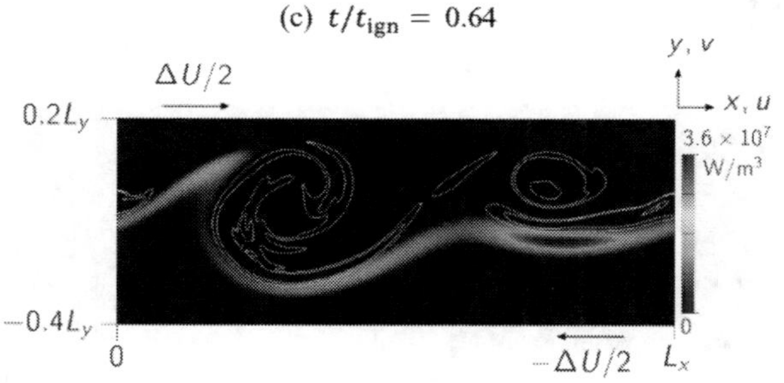

(d) $t/t_{\mathrm{ign}} = 0.8$

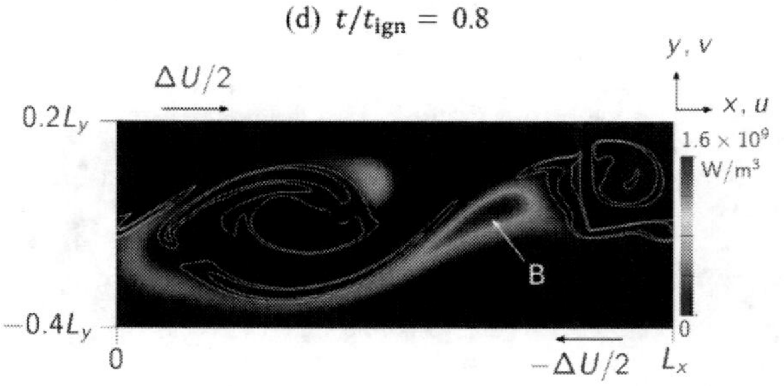

(e) $t/t_{\mathrm{ign}} = 1.01$

Figure 5: Instantaneous distributions of high-heat-release-rate regions and roller vortices ($\omega_z \delta_m^0/\Delta U = -0.05$ and -0.1) at $t/t_{\mathrm{ign}} = 0.32, 0.48, 0.64, 0.8$ and 1.01 on the plane Q ($z/L_z = 0.5$).

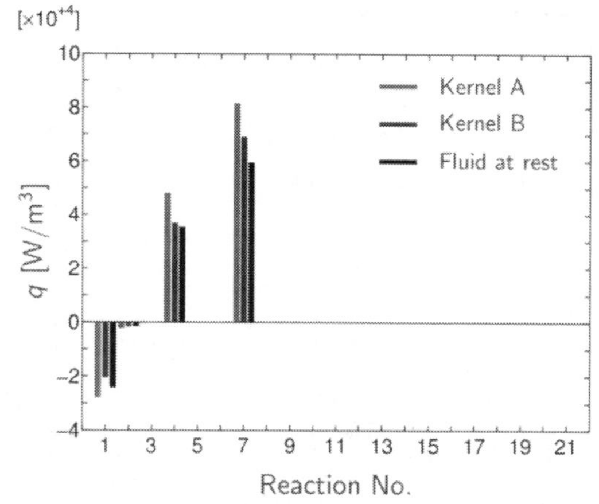

(a) $t/t_{\mathrm{ign}} = 0.48$.

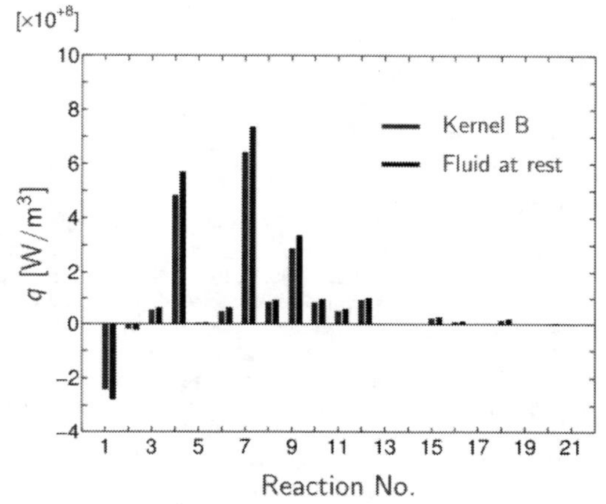

(b) $t/t_{\mathrm{ign}} = 1.01$.

Figure 6: Comparison of heat release rates due to each reaction.

11th Asia-Pacific Conference on Combustion,
The University of Sydney, NSW Australia
10th -14th December 2017.

A Kinetic Study of Pyrolysis of Pine Wood using a Thermogravimetric Analyser and a Combined Model-free and $y(\alpha)$Master Plot Method

Zhezi Zhang, Mingming Zhu*, Pengfei Liu, Hendrix Y Setyawan and Dongke Zhang

Centre for Energy (M473), The University of Western Australia, 35 Stirling Highway, Crawley, WA 6009, Australia

Abstract

Pyrolysis of pine wood was investigated using a non-isothermal TGA technique at heating rates ranging from 1 to 10K min^{-1}. A kinetic model was also developed to reveal the pyrolysis mechanism. The pyrolysis process was modelled as a single step reaction, in which the activation energy and pre-exponential factor were obtained using a combined model-free and y(α) master plot method based on the experimental data. The activation energy was almost constant at 178.5 kJ mol^{-1} for pine wood in the pyrolysis conversion range of 0.05 to 0.8, validating the single-step model in the major pyrolysis region of this material. The kinetic modelling results using the method in the present study showed good agreement with the experimental data, suggesting that the present method can serve as an accurate yet simple tool to perform the kinetic analysis and to predict the pyrolysis behaviour of the studied material.

1 Introduction

Pyrolysis, a series of complex thermal decomposition reactions under oxygen-depleted conditions, is one of the appealing approaches to utilising biomass, such as agricultural residues and forestry wastes, for power and heat generations in off-grid rural areas [1, 2]. Simple and adequate kinetic models to predict the thermal decomposition behaviour of a given biomass feedstock are necessary for reactor design, plant operation and optimisation and, therefore, they have been subjected to numerous studies in the literature [3-12]. The single-step reaction model (SSRM), which assumes the pyrolysis as a single step reaction, has emerged as one of the promising models due to its simplicity and acceptable accuracy. The global reaction governing the pyrolysis process is usually viewed as the reaction in which the raw biomass thermally decomposes into the volatile matters and biochar. The rate equation of the single step biomass pyrolysis is given in (1).

$$\frac{d\alpha}{dT} = \frac{1}{\beta}k(T)f(\alpha) = \frac{1}{\beta}Aexp(\frac{-E_a}{RT})f(\alpha) \qquad (1)$$

where α is the degree of conversion of the pyrolysis process, T the particle temperature (K), β the linear heating rate (K min^{-1}), $k(T)$ the Arrhenius rate constant, $f(\alpha)$ the reaction model (The commonly used ones in the literatures were summarised in Ref [13]), A the pre-exponential factor (min^{-1}), E_a the activation energy (kJ mol^{-1}) and R is the universal gas constant (8.314J K^{-1} mol^{-1}), respectively.

Nevertheless, accurate measurement of the kinetic triplets, E_a A and $f(\alpha)$, of biomass pyrolysis processes in the SSRM represents a major challenge.

The model-free methods, which do not require any assumption of the reaction model $f(\alpha)$, have gained widespread currency and even become a routine journey in the pyrolysis kinetic studies for determining the activation energy [4, 6, 14, 3, 15-18]. The commonly used methods to calculate the model free activation energy are Friedman, FWO, KAS and modified KAS methods [14, 15, 19, 20]. However, it is common that many publications stop short here without pursuing the pre-exponential factor and reaction model and leave a one thirdly told story to the readers. Some researchers attempted to use the model-fitting method, as expressed in (2), to determine the kinetic parameters by choosing one of the reaction models from literature to process the data obtained from a single TGA experimental run.

$$\ln\left(\frac{\beta}{f(\alpha)}\frac{d\alpha}{dT}\right) = \ln(A) - \frac{E_a}{R}\frac{1}{T} \qquad (2)$$

However, such approach not only requires great computational effort, but also suffers from the challenge in dissociating the effects of temperature dependence $k(T)$ and the reaction model $f(\alpha)$ as both temperature and degree of conversion vary as a function of time simultaneously in a single TGA experiment. Thus, almost any $f(\alpha)$ can satisfactorily fit experimental data at the cost of drastic variations in the activation energy and pre-exponential factor [17]. It is interesting to note that although the pre-exponential factor and activation energy vary widely with $f(\alpha)$, they all demonstrated a strong linear correlation known as the compensation effect [21, 22]. It was suggested that the pre-exponential factor can be determined also in a model-free manner once the model-free activation energy and the linear correlation are known. Subsequently, the reaction model can be found as the best matching $f(\alpha)$ by substituting the obtained A_0 and $E_{a,0}$ value into (3) [16, 23, 24]. However, others criticised that without a better explanation from the fundamental level, such phenomena were caused by undetected thermal lag or simply due to the incorrect computational procedures and/or assumptions [25-27].

$$f(\alpha) = \beta\left(\frac{d\alpha}{dT}\right)_\alpha [A_0exp(-\frac{E_{a,0}}{RT_\alpha})]^{-1} \qquad (3)$$

$$y(\alpha) = \beta\left(\frac{d\alpha}{dT}\right)_\alpha exp\left(\frac{E_{a,0}}{RT_\alpha}\right) = Af(\alpha) \qquad (4)$$

The $y(\alpha)$ master plot method allows the reaction model to be determined once the model-free activation energy was obtained. As expressed in (4), the experimental curves on the left hand side will be compared against the theoretical $f(\alpha)$ plots on the right hand side to find the best match model [28]. However, such approaches were rarely adopted in the pyrolysis kinetics studies.

The present contribution describes a combined model-free and y(α) master plots method for determining the complete set of kinetic triplets of biomass pyrolysis in a thermogravimetric analyser. The kinetic parameters obtained were validated by comparing the model predictions against the experimental results.

2 Experimental

In order to obtain the true kinetic parameters, great attentions and efforts need to be paid to eliminate or at least minimise the heat and mass diffusion limits prior to the experimentation. As illustrated in Fig. 1, the pyrolysis process in the TGA apparatus can be described as 4 distinct steps; 1) Chemical reaction step: at molecular level, volatile matters are first produced from the bond cleavages of the cellulose, hemicellulose and lignin

Corresponding author. Mingming Zhu Fax: +61-8-6488 7622
E-mail address: mingming.zhu@uwa.edu.au

chemical structure as the particle temperature increases; 2) Intra-particle diffusion step: the produced volatile matters diffuse through the micro-pores/tracheids/vessels and left the individual biomass particles; 3) Inter-particle diffusion step: the volatile matters travel through the bed of biomass samples; 4) Bulk diffusion step: volatile matters escape from the surface of the sample bed to the surrounding. Suuberg et al. found that the apparent activation energy of cellulose pyrolysis at high heating rate is almost the same to the latent heat of tar vaporisation and suggested that high heating rate enhances the interparticle diffusion limitations of cellulose pyrolysis [29]. Moreover, due to the poor thermal conductivity of the biomass material, the temperature gradient within the particle and the difference between the particle and the recorded furnace temperature increase with the increasing furnace heating rate. It has been reported that the thermal decomposition of biomass materials with particle sizes up to 0.2 mm may be kinetically evaluated up to ca. 800K without accounting for internal heat transport restrictions [10]. Thus, a good TGA run should consist of 1) a purging gas to sweep the volatiles away from the sample bed; 2) a thin layer of the evenly distributed sample at the bottom of the crucible; 3) using small particle size sample; 4) slow heating rate.

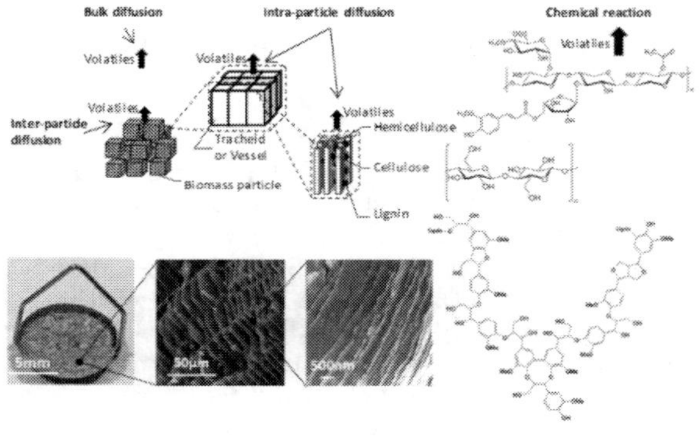

Figure 1: A multiscale schematic illustration of biomass pyrolysis

2.1 Materials

The pine wood sample in the size fraction of 50 - 200μm obtained locally was used for experimentation. The proximate and ultimate analyses of the materials are shown in Table 1. The as-received samples were dried in an oven at 378K overnight and then stored in a desiccator prior to use.

Characteristics \ Sample	Composition (wt%)
Volatile matter	85.5
Fixed carbon	14.0
Ash	0.5
Carbon	46.33
Hydrogen	8.16
Nitrogen	0.09
Sulphur	0.03
Oxygen[a]	45.39

a. calculated by differences

Table 1: Proximate and ultimate analysis of pine wood

2.2 Experimental setup

Pyrolysis experiments were performed using a TGA (TA Instrument Q5000IR), operating at five different heating rates, namely 1, 2.5, 5, 7.5 and 10K min^{-1}, respectively. A thin layer of the sample (around 5mg) was distributed evenly at the bottom of the sample crucible and placed in the TGA. The sample was first heated at 20K min^{-1} from room temperature to 378K and held at that temperature for 10 minutes to remove any remaining surface moisture from the sample. The sample was then heated at the desired heating rate to the final pyrolysis temperature of 823K. During the heating, the mass and temperature of the sample were continuously recorded. Ultra-high purity nitrogen was used as a purge gas at a constant flow rate of 150ml min^{-1}.

In the actual biomass pyrolysis process, moisture, ash, and fixed carbon may be considered not to participate in chemical reactions. The moisture and ash contents of the same material may vary from time to time. Therefore, the TG curves were normalised to subtract the effect of the moisture and ash. The mass loss during the heating period from ambient to 378K was regarded as the moisture content in the sample, and the ash content was derived from the proximate analysis. Thus, the TG curve was transformed to the degree of conversion (α) vs. temperature plot, with $\alpha = (m_d - m_i)/(m_d - m_{ash})$, where m_d is the mass of the dry sample, m_i the instantaneous mass remaining at time, t, and m_{ash} is the mass of the ash.

3 Kinetic analysis

The SSRM was used, and the activation energy was determined using the model-free method while the pre-exponential factor and the reaction model were determined using the $y(\alpha)$ master plot method. The model-free activation energy $E_{a,0}$ derived from each method (Friedman, FWO, KAS and modified KAS method) were compared and evaluated for their dependency on the degree of conversion.

$$\ln[y(\alpha)] = \ln\left[\beta_i \left(\frac{d\alpha}{dT}\right)_{\alpha,i} exp\left(\frac{E_{a,0}}{RT_{\alpha,i}}\right)\right] = \ln(A_i) + \ln[f(\alpha)] \quad (5)$$

Once the model free activation energy was obtained, by substituting the $E_{a,0}$ value into (1) and rearranging the equation, the $y(\alpha)$ master plot function can be formulated in (4). Equation 5 can be arrived by taking the natural logarithm on both side of (4). Thus, the left hand side of (5) can easily be determined from the model-free activation energy and experimental values of $(d\alpha/dT)_{\alpha,i}$ and $T_{\alpha,i}$ at any given heating rate of β_i. The resulting $\ln\left[\beta_i \left(\frac{d\alpha}{dT}\right)_{\alpha,i} exp\left(\frac{E_{a,0}}{RT_{\alpha,i}}\right)\right]$ vs conversion curves must be checked for any significant variation among themselves with heating rate, in order to confirm the single dependency of the $\ln[y(\alpha)]$ on α. The shape of the right hand side of the (5) vs conversion curve is entirely determined by the shape of the theoretical reaction model, as shown in Fig. 2 (a) to (f), while the pre-exponential factor A_i, as a constant, shifts the curve up or down. Therefore the pre-exponential factor and the true reaction model can be determined and identified simultaneously as the best fitting parameters of $\ln\left[\beta_i \left(\frac{d\alpha}{dT}\right)_{\alpha,i} exp\left(\frac{E_{a,0}}{RT_{\alpha,i}}\right)\right]$ vs α curves by using the "Fitting Function Builder" in OriginPro 8.5 software.

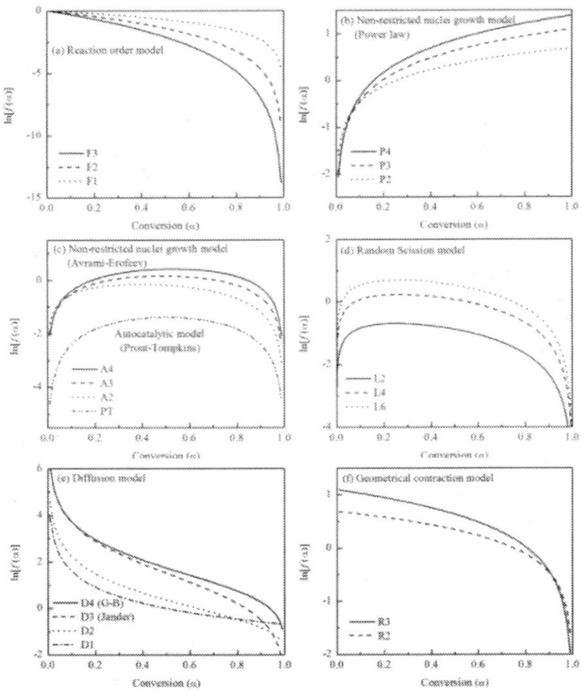

Figure 2: Theoretical $\ln[f(\alpha)]$ plots of the reaction models summarized in Ref.

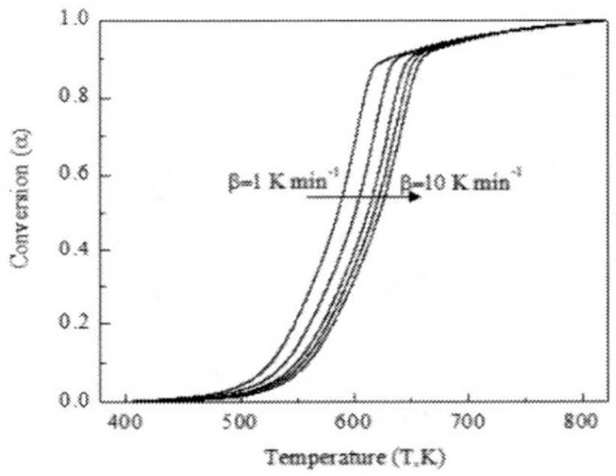

Figure 3: α vs. temperature curves for the pyrolysis of all samples at different heating rates

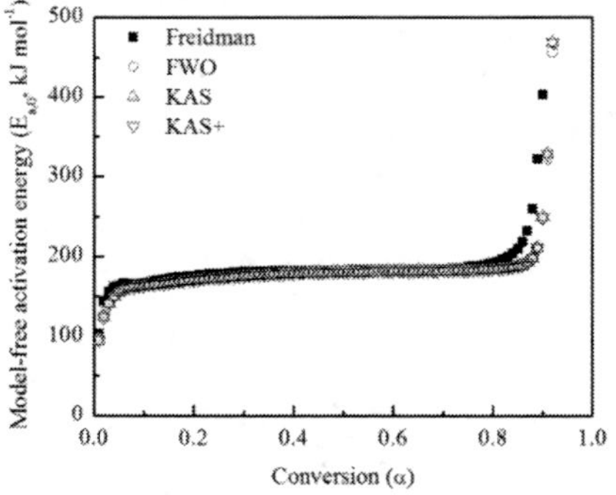

Figure 4: Plots of model-free activation energy as a function of conversion for pine wood pyrolysis

4 Results and Discussion

The degree of conversions (α) as a function of temperature for all samples is shown in Fig. 3 which is then analysed according to the aforementioned model-free methods to determine the model-free activation energy (E_{a0}). Figure 4 shows the relationship of the model-free activation energies of pine wood as a function of the degree of conversion, determined from the four model-free methods mentioned above. In the conversion range of $0.05 \leq \alpha \leq 0.8$, little dependence of E_{a0} on α indicates that there exists a high probability for the presence of a single-step reaction model. Thus supporting the single step pyrolysis reaction assumption used in the SSRM and more importantly, the E_{a0} values obtained were independent of the method used and thus can be used for further kinetic analysis. In the present study, the model-free activation energy for pine wood is 178.5 kJ mol^{-1}.

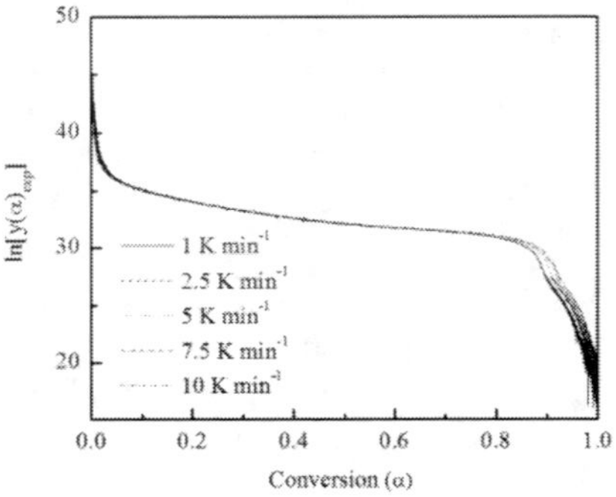

Figure 5: $\ln[y(\alpha)_{exp}]$ vs conversion plots for pine wood at different heating rates

As shown in Fig. 5, the experimental $\ln[y(\alpha)]$ vs α plots of the pine wood samples at all five heating rates are obtained following the steps described in the kinetic analysis section. First, it is clear that no significant variations of the $\ln[y(\alpha)]$ plots can be observed at the chosen heating rates and therefore confirms the single dependency of the $\ln[y(\alpha)]$ on the degree of conversion. By comparing these experimental $\ln[y(\alpha)]$ plots with the theoretical the $\ln[f(\alpha)]$ plots of different reaction models (Fig. 2), it can be easily found that the shape of these experimental $\ln[y(\alpha)]$ plots is similar to that of theoretical $\ln[f(\alpha)]$ plots of the diffusion models. Thus, only the diffusion controlled models (D1 to D4) need to be considered when constructing the "Fitting Functions" in OriginPro 8.5 software, as expressed in (6) below.

$$\ln[y(\alpha)_{exp}] = \ln(A) + \ln[f(\alpha)_{Dn}] \quad (6)$$

where $f(\alpha)_{Dn}$ are the diffusion models described in Ref [13].

Heating rate (K min^{-1})	Pre-exponential factor (min^{-1})	R^2
1	3.03×10^{13}	0.95
2.5	2.87×10^{13}	0.95
5	2.79×10^{13}	0.96
7.5	2.95×10^{13}	0.96
10	2.90×10^{13}	0.96

Table 2: Pre-exponential factors of pyrolysis of pine wood at different heating rates

Under all heating rate conditions used in this study, the D3 diffusion mechanism is the best model to describe the pine wood pyrolysis process. The pre-exponential factors of pyrolysis of pine wood at different heating rates were summarised in Table 2. It is clearly evidenced that the pre-exponential factor is almost the same under different heating rate conditions. Therefore, the average value of 2.908×10^{13} min^{-1} was used for model validation.

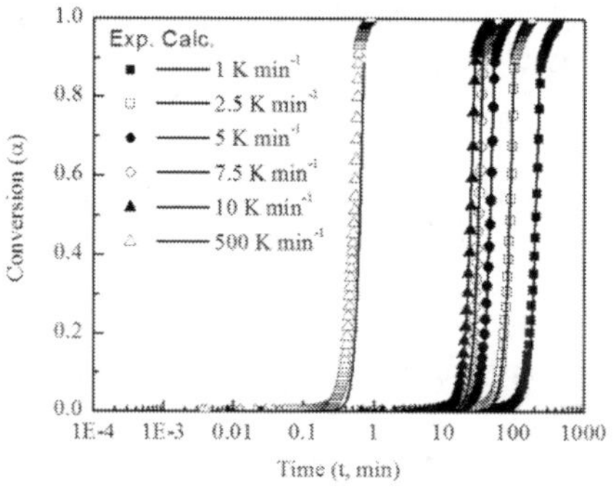

Figure 6: The α vs temperature curves of the pyrolysis of pine sawdust at different heating rates (Symbols: experimental data; lines: calculated data)

The last step was to substitute the above obtained model-free activation energy ($E_{a,0}$), pre-exponential factor (A) and the D3 diffusion reaction model into the SSRM model as expressed in (7).

$$\frac{d\alpha}{dT} = Aexp\left(\frac{-E_{a,0}}{RT}\right)\{1.5(1-\alpha)^{2/3}[1-(1-\alpha)^{1/3}]^{-1}\} \quad (7)$$

Equation (7) was solved using the Runge-Kutta method (ODE45 function) in MATLAB. Figure 6 compares the calculated conversion curves against the experimental results of the pyrolysis of pine wood samples at various heating rates and, clearly, good agreement is evidentover the range of TGA heating rates examined, indicating that the kinetic model used in the current study is capable of predicting the overall thermal decomposition behaviour of the pine wood samples used. Moreover, the results in author's recent study revealed that during the pyrolysis process, lignin first underwent the process of softening/melting and migration, and finally deposition on the surface of cell walls of biomass particles. The melted lignin createda viscous liquid, serving as a barrier delaying the releases of volatiles from the solid substrate. The removal of lignin diminished such an effect and enhanced the transportation of the pyrolysis products to escape from the solid matrix [30].

5 Conclusions

Kinetics of pyrolysis of pine wood samples was determined using the combined model-free and y(α)master plot methods based on the single step reaction model. The kinetic parameters thus obtained were then used to predict the pyrolysis behaviour of the pine wood samples. The activation energy obtained in the present study was independent of the complex reaction models used in the literature. The pre-exponential factor and reaction model (D3 diffusion model) were determined simultaneously without making any assumptions. The current model was shown to predict the pyrolysis behaviour of the biomass samples at the pyrolysis conditions tested with good agreement with the experimental data. The combined model-free and y(α) master plot method deployed in this study is not only simple but also very accurate way to determine the kinetic parameters.

6 Acknowledgment

Financial and other supports have been received from the Australian Research Council under the ARC Linkage Projects Scheme (ARC Linkage Project LP100200135), BHP Billiton Iron Ore Pty Ltd, Ansac Pty Ltd and ENN Group.

References

1. L. Shen and D. Zhang, Fuel 82 (4) (2003) 465-472.
2. L. Shen and D. Zhang, Fuel 84 (7–8) (2005) 809-815.
3. K. Slopiecka, P. Bartocci, and F. Fantozzi, Applied Energy 97 (0) (2012) 491-497.
4. J.E. White, W.J. Catallo, and B.L. Legendre, Journal of Analytical and Applied Pyrolysis 91 (1) (2011) 1-33.
5. G. Varhegyi, M.J. Antal, T. Szekely, and P. Szabo, Energy & Fuels 3 (3) (1989) 329-335.
6. D. Chen, Y. Zheng, and X. Zhu, Bioresource Technology 131 (0) (2013) 40-46.
7. T. Sonobe and N. Worasuwannarak, Fuel 87 (3) (2008) 414-421.
8. H. Zhou, Y. Long, A. Meng, Q. Li, and Y. Zhang, Thermochimica Acta 566 (0) (2013) 36-43.
9. K. Miura, Energy & Fuels 9 (2) (1995) 302-307.
10. G.M. Simmons and M. Gentry, Journal of Analytical and Applied Pyrolysis 10 (2) (1986) 117-127.
11. D.P. Ross, C.A. Heidenreich, and D.K. Zhang, Fuel 79 (8) (2000) 873-883.
12. C.A. Heidenreich, H.M. Yan, and D.K. Zhang, Fuel 78 (5) (1999) 557-566.
13. A. Khawam, and Douglas R. Flanagan, The journal of physical chemistry B 110 (35) (2006) 17315-17328.
14. T. Ozawa, Thermochimica Acta 203 (0) (1992) 159-165.
15. M.J. Starink, J Mater Sci 42 (2) (2007) 483-489.
16. S. Vyazovkin, A.K. Burnham, J.M. Criado, L.A. Pérez-Maqueda, C. Popescu, and N. Sbirrazzuoli, Thermochimica Acta 520 (1–2) (2011) 1-19.
17. S. Vyazovkin and C.A. Wight, Thermochimica Acta 340–341 (0) (1999) 53-68.
18. F. Yao, Q. Wu, Y. Lei, W. Guo, and Y. Xu, Polymer Degradation and Stability 93 (1) (2008) 90-98.
19. H.L. Friedman, Journal of Polymer Science Part C: Polymer Symposia 6 (1) (1964) 183-195.
20. H.E. Kissinger, Analytical Chemistry 29 (11) (1957) 1702-1706.
21. A.I. Lesnikovich and S.V. Levchik, Journal of Thermal Analysis 27 (1) (1983) 89-93.
22. T. Bligaard, K. Honkala, A. Logadottir, J.K. Nørskov, S. Dahl, and C.J.H. Jacobsen, The Journal of Physical Chemistry B 107 (35) (2003) 9325-9331.
23. B. Janković and B. Adnađević, International Journal of Chemical Kinetics 39 (8) (2007) 462-471.
24. P. Budrugeac, J.M. Criado, F.J. Gotor, J. Malek, L.A. Pérez-Maqueda, and E. Segal, International Journal of Chemical Kinetics 36 (6) (2004) 309-315.
25. M.E. Brown and A.K. Galwey, Thermochimica Acta 387 (2) (2002) 173-183.
26. R. Narayan and M.J. Antal, Industrial & Engineering Chemistry Research 35 (5) (1996) 1711-1721.
27. R. Agrawal, Journal of Thermal Analysis 31 (1) (1986) 73-86.
28. J. Málek, Thermochimica Acta 138 (2) (1989) 337-346.
29. E.M. Suuberg, I. Milosavljevic, and V. Oja, Symposium (International) on Combustion 26 (1) (1996) 1515-1521.
30. Z. Zhang, M. Zhu, P. Hobson, W. Doherty and D. Zhang, Contrasting the pyrolysis behaviour of selected biomass and the effect of lignin, in 42nd International Technical Conference on Clean Energy 2017: Clearwater, Florida USA.

11th Asia-Pacific Conference on Combustion,
The University of Sydney, NSW Australia
10th -14th December 2017.

Inference and combination of missing data sets for investigation of H_2O_2 thermal decomposition rate uncertainty

Tiernan A. Casey, Habib Najm

Combustion Research Facility, Sandia National Laboratories
7011 East Ave., Livermore CA 94551, USA

1. Abstract

Prescribing uncertainty measures to rate expressions is crucial for performing useful predictive combustion computations. Raw experimental measurement data, and its associated noise and uncertainty, is typically unavailable for most reported investigations of elementary reaction rates, making the direct derivation of the desired joint uncertainty structure of the parameters in rate expressions difficult. To approximate this uncertainty structure we construct an inference procedure, relying on maximum entropy and approximate Bayesian computation methods, and using a two-level nested Markov Chain Monte Carlo algorithm, to arrive at a joint density on rate parameters and missing data. This method employs the reported context of a specific experiment to construct a set of hypothetical experimental data profiles consistent with the reported statistics of the data, in the form of error bars on rate constants at the experimental temperatures. Bayesian inference can then be performed using these consistent data sets as evidence to determine the joint posterior density on the rate parameters for any choice of chemical model. The method is also used to demonstrate the combination of missing data from different experiments for the generation of consensus rate expressions using these multiple sources of experimental evidence.

2. Introduction

Chemical kinetic mechanisms consist of uncertain parameter specifications from a variety of sources, such as elementary rate measurements, ab initio quantum chemistry calculations, and rate-rule constructions. The effort to understand these data sources and their consequence on combustion modeling from an uncertainty quantification (UQ) perspective has gained increased recent attention [1]. From the point of view of UQ, an appropriate treatment of uncertainty would involve analysis of this underlying uncertain data which are typically unavailable or unreported in investigations of elementary reactions, with only the fitting parameters of the data and perhaps some statistics on these parameters given as a measure of the uncertainty of the investigation. A method for inferring missing data subject to constraints was proposed by Berry et al. [2], where a data space is explored in a Bayesian framework with a likelihood constructed to enforce statistics on the data, resulting in a maximum entropy-based (MaxEnt [3]) solution for the posterior distribution of the data. This method has been demonstrated for model problems [4], applied to single-step reaction rate parameters [5], and used to infer real data given experimental statistics [6]. In this paper we apply the method to infer missing data from a shock tube experiment measuring the rate of H_2O_2 decomposition. The thermal decomposition of H_2O_2 is an important process in hydrocarbon combustion playing a particularly crucial role in providing a source of radicals at high pressure where it controls the 3rd explosion limit in the H_2-O_2 system, and also as a branching reaction in intermediate-temperature hydrocarbon oxidation. We use the inferred data to reveal the structure of the posterior PDF of the low-pressure limit Arhennius parameters of the Lindemann reaction model that fit this data (treating the high pressure limit parameters as certain) and utilize samples from this PDF as input for a forward model to demonstrate the necessity of knowledge of the parameter correlation structure for performing UQ investigations of combustion processes of interest.

3. Experimental data

To investigate the thermal decomposition reaction:

$$H_2O_2 + M \longrightarrow 2\,OH + M$$

we consider the reported rate data from a recent shock tube study [7]. This set of experiments employed laser absorption to measure the time decay of H_2O_2 concentration resulting from the thermal decomposition occurring in the high temperature region behind reflected shock waves in an Ar diluent in the temperature range 930-1235 K. The raw experimental data are unreported but statistics of the data are presented in the form of nominal values and symmetric error bars (see Fig. 1) on k(T) values for 11 temperatures at low pressure conditions ($\approx$1–2 atm) and 6 temperatures at higher pressure conditions ($\approx$10 atm). We will initially concentrate solely on the reported results at low pressure conditions and later treat the higher pressure results as a separate experiment.

Since no explicit assumption is made as to how the k(T) are distributed, we choose to interpret these reported statistics as constraints on the moments of k(T) resulting from parameter inferences on hypothetical data sets. The reported nominal values are interpreted as means, while the error bars are treated as either 1 or 2 standard deviation measures on an inferred k(T) ensemble. Interpretation of the error bars as 1 or 2 standard deviations is chosen as an explicit definition of the error bars is not provided in the target experiment. The

Corresponding author: Tiernan A. Casey
E-mail address: tcasey@sandia.gov

data fitting procedure performed by the experimentalists to construct the rates was carried out using a calibration chemical mechanism [8] of their choosing in conjunction with a specified thermochemistry database [9]. For the purposes of performing data fitting through Bayesian inference we assume an experimental data noise model in the form of standard normally distributed additive noise, ε, with a constant nominal strength σ:

$$[H_2O_{2\text{data}}]_{i,j} = [H_2O_{2\text{model}}]_{i,j} + \sigma\varepsilon_{i,j} \tag{1}$$

where the indices {i, j} span all the data points of the experiment in temperature and temporal space respectively, and ε is independently and identically distributed across the indices {i, j}.

4. Bayesian inference

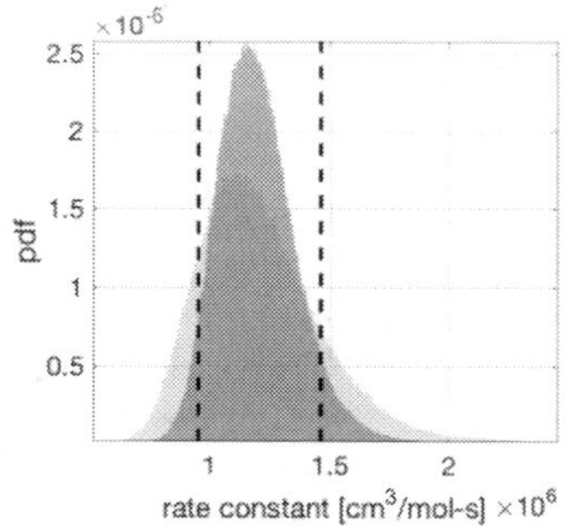

Figure 1: Comparison of a reported constraint from the target experiment in the form of symmetric error bars on the value of a rate constant (dashed lines) to the inferred values (PDFs). The PDFs correspond to inferrences performed with interpretations of the constraint as 1 (light grey) and 2 (dark grey) standard deviations of the uncertain parameter.

The inference method proceeds as follows. An inference on the experimental data is constructed in a Bayesian framework (eq. 2) which proposes data, **D** (see Figs. 2a, 2b), and measures its consistency with the N reported experiment statistics, S (see Fig. 1),

$$p(\mathbf{D}|S) = \frac{p(S|\mathbf{D})\pi(\mathbf{D})}{p(S)} \tag{2}$$

in the context of approximate Bayesian computation (ABC)[10]. An ABC likelihood, $p(S|\mathbf{D})$, is constructed using Gaussian error kernels (eq. 3) which penalize proposed data whose statistics ($S_\mathbf{D}$) do not match those of the experiment ($S_{\text{experiment}}$).

$$p(S|\mathbf{D}) = \prod_{i=1}^{N} \exp\left[-\delta_i\left(\frac{S_{i,\mathbf{D}} - S_{i,\text{experiment}}}{S_{i,\text{experiment}}}\right)\right] \tag{3}$$

The δ_i values in the ABC likelihood expression represent the degree to which consistency is demanded for each statistic. The ABC likelihood requires the determination of statistics on the Arrhenius parameters of the proposed data, which are produced by performing a second inner Bayesian inference (eq. 4)

$$p(\lambda|\mathbf{D}) = \frac{p(\mathbf{D}|\lambda)\pi(\lambda)}{p(\mathbf{D})} \tag{4}$$

on the Arrhenius parameters using the assumed error model described in eq. 1. λ in eq. 4 is the vector of inferred parameters consisting of the Arrhenius parameters of the rate constant and the data error strength parameter (i.e. error standard deviation), σ. Equation 4 shows the parameter likelihood construction:

$$p(\mathbf{D}|\lambda) = \frac{1}{\sqrt{2\pi}\sigma}\exp\left(-\frac{\left([H_2O_{2\text{data}}]_{i,j} - [H_2O_{2\text{model}}]_{i,j}\right)^2}{2\sigma^2}\right) \tag{5}$$

The nested inferences are performed using Markov Chain Monte Carlo (MCMC) sampling, with single-site Metropolis sampling employed for the outer data chain and adaptive Metropolis sampling used for the inner parameter chain. For a data vector **D** (representing the vector of all experimental data points across all temperatures), a set of constraints S, is applied to the Arrhenius parameters associated with the data. These constraints are the means with which the experimental error is imposed explictly on the data inference. In this case S is the mean and standard deviation of the k(T) values computed using the inferred Arrhenius parameters that fit the proposed data. $\pi(\mathbf{D})$ is a prior PDF on the data vector.

As each inner parameter chain must be run long enough to generate sufficient samples to calculate converged statistics on those parameters, and the evaluation of the parameter likelihood at each point on the inner MCMC chain involves the solution of an ODE system, considerable computational effort is required. To accelerate the evaluation of the ODE system, a surrogate model is constructed using a polynomial chaos expansion (PCE) to represent the model output (i.e. H_2O_2 concentration as a function of time, see Figs. 2a, 2b) with a stochastic dimension reflecting the uncertain rate constant of interest, using a separate PCE to construct a surrogate for each experimental temperature considered.

On completion of the outer MCMC data chain, we arrive at a set of hypothetical experimental data sets consistent with the experimentally reported statistics. This set of data is the least constrained representation of possible data that could have been measured by the experimentalists given their reported statistics. Performing Bayesian inference to estimate Arrhenius parameters using the chemical model employed by the experimentalists and these data sets delivers the joint uncertainty structure of the parameters and enables estimation of k(T) uncertainty at arbitrary temperatures (see Fig. 2c). However, since we now have access to a representation of raw experimental data, we are free to choose any chemical model (i.e. chemical mechanism) to use for fitting this data to determine the parameters of the reaction of interest. Rather than using the mechanism employed by experimentalists to perform the required fitting, we choose to use the model of Li et al. [11] which is used

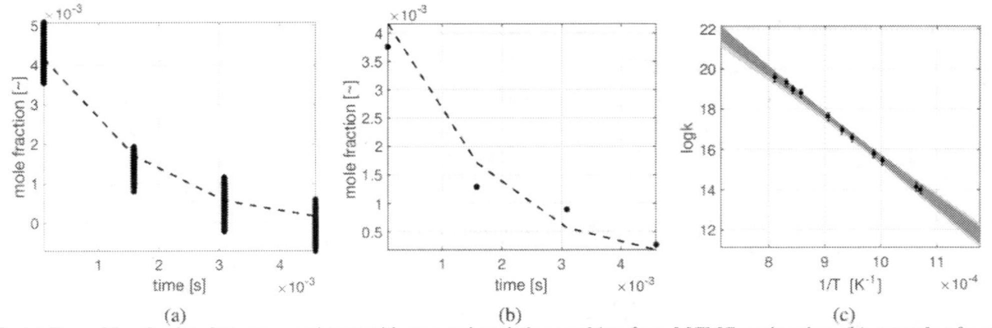

Figure 2: (a) Ensemble of noisy data sets consistent with reported statistics resulting from MCMC exploration, (b) example of a single noisy data set, (c) comparison of reported nominal k(T) values and their uncertainties (symmetric error bars) compared to confidence intervals of k(T) computed using the inferred Arrhenius parameters using 1 (light grey) and 2 (dark grey) standard deviations as the uncertainty measure in the data inferences. The dashed lines in (a) and (b) are interpolants of nominal noiseless data for comparison, generated using the ODE model and employing the nominal Arrhenius parameters reported by the experimentalists.

extensively in combustion modeling of H_2-O_2 systems. In this way the reported results of the experiment are reinterpreted in a way that ensures consistency with a choice of modeling context. Each consistent data set is now refit using this new chemical model using Bayesian inference to arrive at a posterior PDF on the parameters. Treating each of these posterior PDFs as a possible representation of the parameter uncertainty, we combine the PDFs using linear pooling to arrive at a consensus pooled parameter posterior PDF. Figures 3a, 3b show the results of this procedure for different interpretations of the constraints, indicating a strong degree of correlation between the Arrhenius parameters (presented as log(A) and E_a) in the joint posterior density.

5. Forward UQ in combustion modeling

To demonstrate the importance of having knowledge of the joint parameter distribution we perform an uncertainty study of a system relevant to this reaction, namely the 3rd explosion limit of H_2-O_2 mixtures. To generate the characteristic Z-shaped explosion limit boundary, homogeneous ignition calculations are performed over a range of pressures to classify the temperature boundary separating explosion from non-explosion. Figure 4a shows confidence intervals of the explosion limit curve drawing samples from a joint PDF determined using the data inference procedure applied to low pressure results from the target experiment. Comparing this result to evaluations of the explosion limit model using input samples taken from parameter PDFs that are independently distributed, but with identical means and marginal variances (i.e. an uncorrelated joint PDF), demonstrates the drastic over-prediction of the uncertainty in the location of the explosion limit boundary in temperature space when the correlation of the parameter uncertainty is ignored (Fig. 4b). This over-prediction of the uncertainty arises from the over and under-prediction of the value of the rate of the H_2O_2 decomposition reaction when Arrhenius parameter value pairs are chosen in directions transverse to the direction of correlation, which correspond to the largest departures from the nominal rate value (Fig. 3c).

To further extend this methodology to multiple sources of experimental evidence, the hypothetical data sets discussed previously, generated using low pressure results from the target experiments, are supplemented with hypothetical data from a second separate inference conducted using high pressure experimental points with their reported statistics as constraints. Bayesian inference is then performed on combined instances of these two data sets (low and high pressure), with the resulting parameter PDFs linearly pooled to deliver a consensus parameter PDF across all the hypothetical data. The results for the explosion limit model using this parameter PDF determined from combined learning is shown in Fig. 4c. The resulting uncertainty is quite similar to the case computed using information from the low pressure data only, indicating that constraints at multiple pressures inform the parameter uncertainty in similar ways, and also suggests the increased influence of the 11 low pressure constraints over the 6 high pressure constraints in the combined learning.

6. Conclusion

We present the application of a data inference procedure to generate missing experimental data from an experiment consistent with the reported statistics in the form of nominal values and error bounds of rate constants at the experimental temperatures. The resulting pooled joint posterior PDF of the Arrhenius parameters reveals strong correlation between log(A) and E_a. Samples from this correlated joint PDF are propagated through a forward model for predicting the explosion limit boundary in the H_2-O_2 system, and compared to the result of sampling from independently distributed normal PDFs, highlighting the necessity of the correlated density for forward UQ. Besides delivering the joint PDF on the parameters of a reaction from a single experimental investigation, the natural extension of combining data from multiple experiments is presented. This approach presents an objective framework for the assimilation of evidence from different sources to constrain the uncertain representation of a reaction of inter-

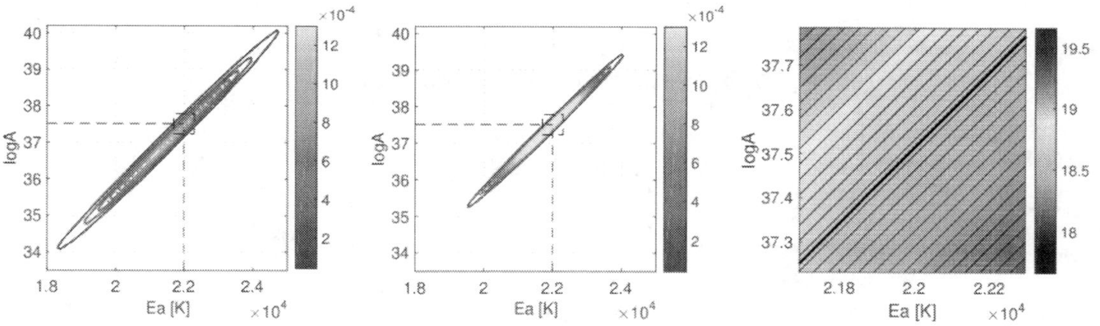

(a) (b) (c)

Figure 3: (a) Correlated posterior Arrhenius parameter PDF computed using low pressure shock tube data and reported error bars interpreted as a 1 standard deviation constraint. Dashed lines indicate best-fit parameters determined by the experimentalists using least-squares fitting. The dashed rectangle is the uncertainty of the least-squares procedure (i.e. fitting nominal values only). (b) Posterior PDF with error bars interpreted as a 2 standard deviation constraint. (c) Direction of correlation implied by the Arrhenius model. The solid black line corresponds to a low pressure experiment temperature of 1166.6 K.

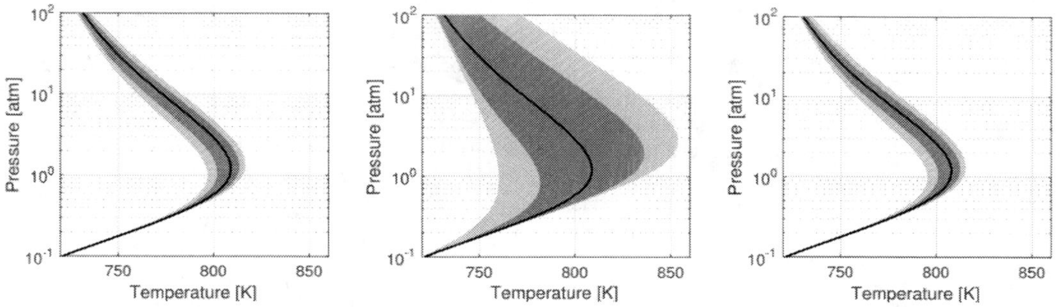

(a) (b) (c)

Figure 4: (a) Explosion limit Z-curve near the third limit for stoichiometric H_2-O_2 mixtures. The solid black curve uses nominal rate values and the shaded areas are 95% confidence intervals using samples drawn from pooled parameter posteriors using 1 (light grey) and 2 (dark grey) standard deviation interpretations of the constraints. (b) Explosion limit curve calculated using samples from independently distributed normals with means and marginal variances identical to the pooled posteriors. (c) Explosion limit curve calculated using the samples from the pooled PDF computed using combined learning from the two experiments (low and high pressure).

est in the context of any choice of model by combining evidence at the data level.

7. Acknowledgements

This work was supported by the US Department of Energy (DOE), Office of Basic Energy Sciences (BES) Division of Chemical Sciences, Geosciences, and Biosciences. Sandia National Laboratories is a multi-mission laboratory managed and operated by National Technology and Engineering Solutions of Sandia, LLC., a wholly owned subsidiary of Honeywell International, Inc., for the U.S. Department of Energy's National Nuclear Security Administration under contract DE-NA-0003525.

References

[1] H. Wang, D. A. Sheen, Progress in Energy and Combustion Science 47 (2015) 1–31.

[2] R. Berry, H. Najm, B. Debusschere, H. Adalsteinsson, Y. Marzouk, Journal of Computational Physics 231 (2012) 2180–2198.

[3] E. Jaynes, Probability Theory: The Logic of Science, G.L. Bretthorst, Ed, Cambridge University Press, Cambridge, UK, 2003.

[4] K. Chowdhary, H. N. Najm, Statistics and Computing 26 (2016) 149–169.

[5] H. N. Najm, R. D. Berry, C. Safta, K. Sargsyan, B. J. Debusschere, Int. J. Uncertainty Quantification 4 (2014) 111–132.

[6] M. Khalil, K. Chowdhary, C. Safta, K. Sargsyan, H. Najm, Proceedings of the Combustion Institute 36 (2017) 699 – 708.

[7] M. B. Sajid, E. Es-sebbar, T. Javed, C. Fittschen, A. Farooq, International Journal of Chemical Kinetics 46 (2014) 275–284.

[8] Z. Hong, D. F. Davidson, R. K. Hanson, Combustion and Flame 158 (2011) 633–644.

[9] R. Kee, F. Rupley, J. Miller, UC-4, Sandia National Laboratories (1987).

[10] M. A. Beaumont, W. Zhang, D. J. Balding, Genetics 162 (2002) 2025–2035.

[11] J. Li, Z. Zhao, A. Kazakov, F. L. Dryer, International journal of chemical kinetics 36 (2004) 566–575.

11th Asia-Pacific Conference on Combustion,
The University of Sydney, NSW Australia
10th -14th December 2017.

Numerical Simulation of Burned Gas-Wall Interaction on Stagnation Methane-Air Premixed Flame

T. Ishida[1], N. Hayashi[1]

[1]Dept. Mech. Eng., Kanagawa Institute of Technology

1030 Shimo-Ogino, Atsugi, Kanagawa, 243-0292, JAPAN

Abstract

In this paper, interaction between flames and wall, especially in burned gas region is discussed using numerical simulation with surface reaction model. Flames in combustors are often surrounded by walls, and thus it is important to evaluate the influence of wall on the flame. The influence becomes significant when the small scale combustor because of the large surface-to-volume ratio. OH, H and O radicals are important for the surface reaction, which causes radical quenching. OH radicals exist from reaction zone to burned gas region. Thus, a stagnation flow premixed flame was used for the investigation of the flame-wall interaction phenomenon in burned gas region. The conditions of stagnation wall temperature are changed from 700 K to 1000 K, and equivalence ratio and inlet velocity is fixed. For the investigation of the influence of surface reaction, the active wall (with surface reaction) and the inert wall (without surface reaction) for stagnation plate are compared. As a result, the wall surface temperature is not influenced from surface reaction. On the wall surface chemical species are absorbed by the surface reaction and the concentration difference of OH by wall temperature is the largest.

1 Introduction

Flames in combustors are often surrounded by walls. Therefore, the flame-wall interaction is one of the key issue in design of combustors. The influence of the wall is important especially for small scale combustors due to their large surface-to-volume ratio. The influence includes heat loss and radical quenching by surface reaction. When the radicals are absorbed to the wall and react, they become lower activation species, and the flame is weakened or local extinction may occur. The phenomenon is called radical quenching. These effects cause local extinction and increase exhaust gas. Moreover, the radical quenching mechanism have not been clear yet. Maruta et al. [1] have researched oscillation flame in micro tubes with an axial wall temperature gradient. Bai et al. [2] estimated the influence of Lewis number and the wall temperature to flame speed with heat loss and radical quenching. Rebecca et al. [3] have investigated the influence of the wall on n-heptane premixed flame with the objective of wall temperature and surface reaction for H, O and OH radicals. Saiki et al. [4] have measured the adsorption behaviors of H radicals on the wall surface using the information of molecular beams of radical and square wave molecular beam modulation under ultrahigh vacuum. Therefore, understanding of combustion phenomenon near wall is very important for practical use of microscale combustors.

In the above researches, OH and H radicals are noted to be especially important in the surface reaction. Here, OH radicals mainly exist from flame front to burned gas region. Thus, we make attention behaviors of radicals in burned gas region. In this study, the numerical simulation with or without surface reaction of stagnation flow methane-air premixed flame was conducted, and the influence of the surface reaction and the temperature of stagnation wall is discussed.

2 Analytical model and numerical method

The stagnation flow is modelled (Fig. 1). A stagnation plate with isothermal condition is placed at 10 mm from the burner. The coordinate axes are the x in the vertical direction and the y axis in the horizontal direction of the stagnation plate, with the stagnation plate as the origin. Methane-air mixture injected from the nozzle, and the inlet velocity is fixed at 1.4 m/s. The pressure is atmosphere. The calculation area is 10 mm in the x direction, and the calculation lattice is 500. The lattice spacing is 20 μm at equal intervals in the x direction.

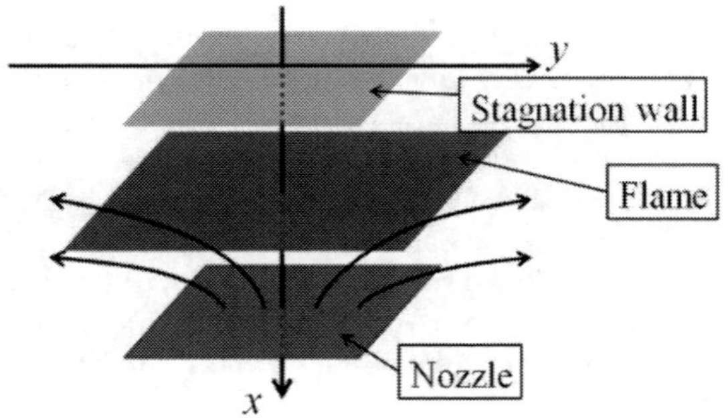

Figure 1: Analytical model of stagnation premixed flame

Surface reaction is considered on the stagnation plate. The Raimondeau's surface reaction mechanism is used [5]. In this reaction mechanism, O, H, OH and CH_3 radicals are considered as absorption reactions and stable molecules such as H_2, H_2O, O_2, CH_4 and C_2H_6 are formed. The inert wall condition (without surface reaction) is also conducted. The wall temperature, T_w, is 700, 800, 900, and 1000 K. In this paper, the results for T_w = 700 and 1000 K are mainly discussed.

Corresponding author. Fax: +81-46-291-3314
E-mail address: s1681006@cce.kanagawa-it.ac.jp

The thermodynamic properties for the species are obtained from the CHEMKIN database [6][7]. This assumes the thermodynamic properties as temperature dependent function under constant pressure. And it is given as a multinomial of temperature given the least square method in the appropriate temperature range for the value on JANAF table [8]. GRI-Mech 3.0 [9], from which the reactions icluding N atom was removed, was used for elementary reaction model. It consists of 36 chemical species and 219 elementary reactions. Simplified Transport Model approached by Smooke et al. [10] is applied.

3 Results and Discussion

First, Figs. 2 and 3 show the temperature distributions. Here, the flame front exists around x = 1.9 mm, and the position is almost the same for all conditions. As shown in Fig. 3, the temperature near the wall increases with the increase in wall temperature, and there is no difference between active or inert wall conditions. Therefore, the surface reaction is not cause the difference of temperature distributions. Therefore, we can separate the influence of surface reaction from influence of wall temperature.

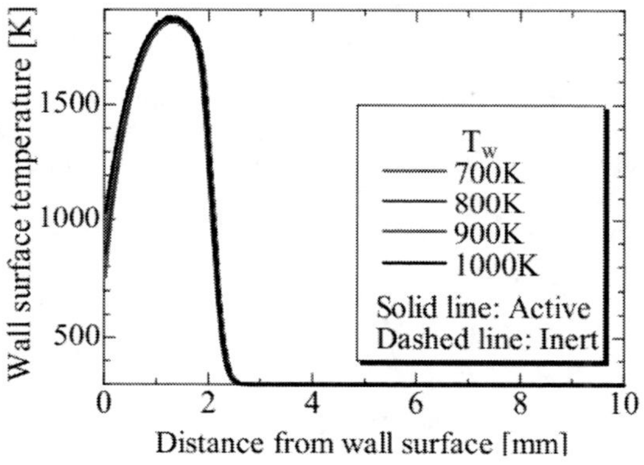

Figure 2: Distribution of temperature with or without surface reaction

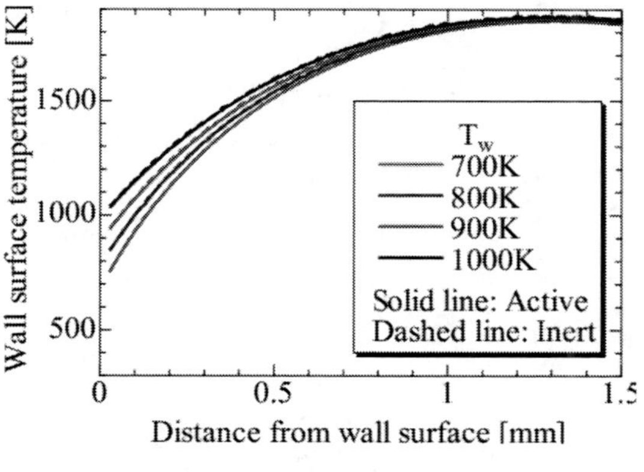

Figure 3: Distribution of temperature near the stagnation wall with or without surface reaction

Figure 4 is the molar concentration distribution of CH_3. CH_3 almost exists in the reaction zone. Thus, the concentration is zero near the wall, and the difference between active or inert wall hardly occurs.

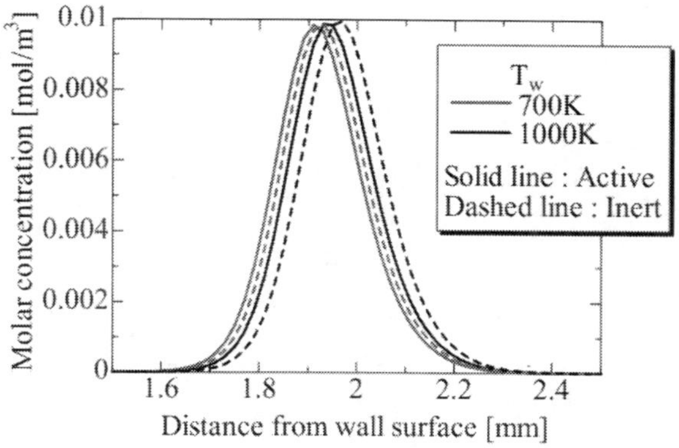

Figure 4: Distribution of molar concentration of CH_3

The molar concentration distributions of H, O and OH for inert wall conditions are shown in Fig. 5. From the wall surface to the reaction zone, the concentration increases with the increase in distance from the stagnation wall. Since the molar concentration of OH is larger than that of O and H, and the concentration gradient is large. Figure 6 shows the radical distributions of each species near the wall. Near the wall surface, the concentrations for all species is also large when the wall temperature is large. Particularly for H, the difference of concentration is significant near the wall surface. The concentration of O is always smaller than OH and H. OH is a large concentration gradient at either wall temperature and gently converges as compared to H and O. Differense in the concentration of H for wall temperature is the largest in all species. And it occurs at x = 0.6 mm. Differense in the concentration of O and OH occurs x = 0.9 and 1.4 mm, respectively. Terefore, we should attend behavior of H against temperature near the wall surface.

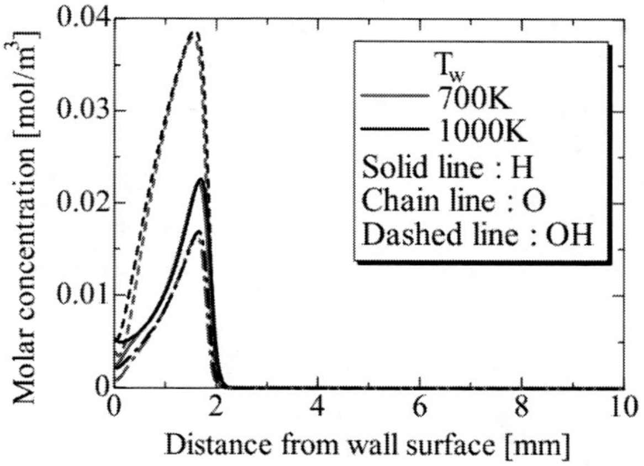

Figure 5: Distribution of molar concentration without surface reaction

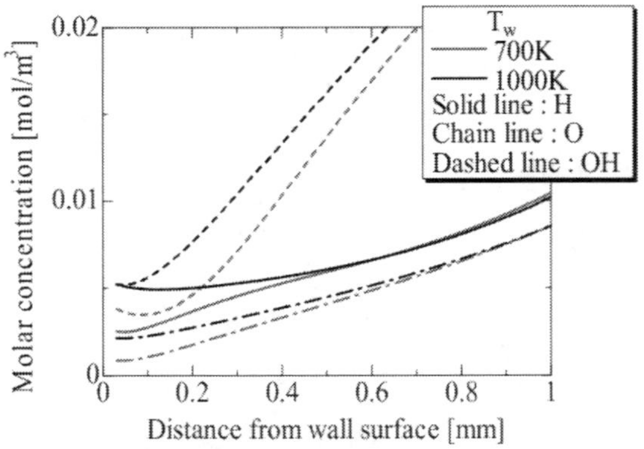

Figure 6: Distribution of molar concentration near the stagnation wall without surface reaction

Figure 7 shows the molar concentration distributions of H, O, OH for active wall conditions. The concentration increases with the increase in distance from the wall surface in the burned gas region, The molar concentration of OH is larger than O and H, and the concentration gradient is also large. Figure 8 shows distributions of the concentration of each species near the wall. Near the wall surface, the molar concentration for all species increases with the increase in wall surface temperature. On the wall surface, the concentrations of these radicals aproach to zero for all temperature coditions. Also the concentration gradient of OH is the largest, and the gradient of T_w = 700 K is smaller than that of T_w = 1000 K until x = 0.2 mm, because the OH chemical reaction in gas phase is more active for T_w = 1000 K. There is hardly the difference in molar concentration of H by the wall surface temperature. The difference in molar concentration of O occurs by the wall surface temperature near the wall surface, but concentration difference is smaller than that of OH, and converges at a position closer to the wall surface. This means that the reaction related to O which contributes to the wall surface temperature consists in a small range near the wall surface as compared with OH.

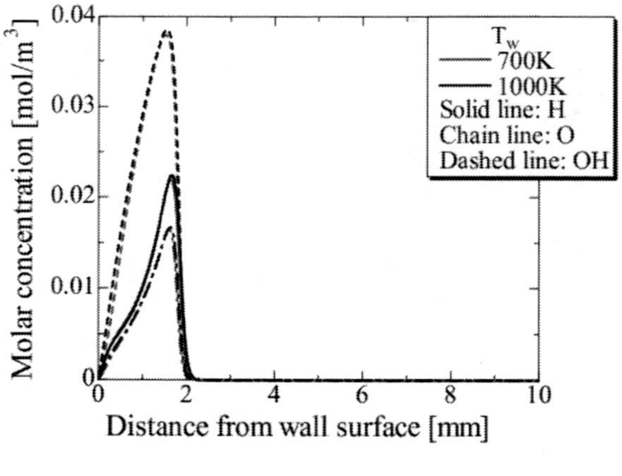

Figure 7: Distribution of molar concentration with surface reaction

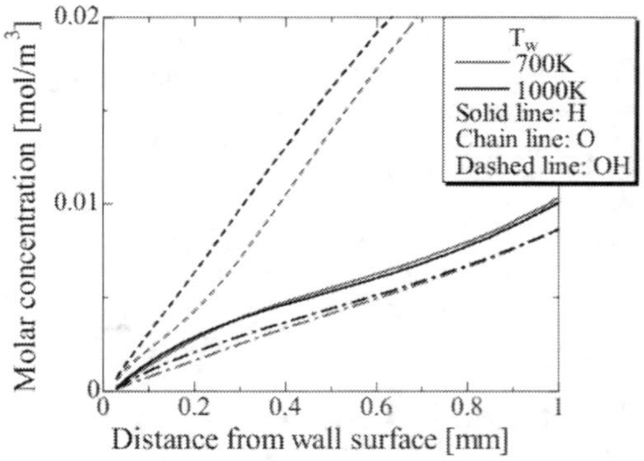

Figure 8: Distribution of molar concentration near the stagnation wall with surface reaction

Figure 9 shows the molar concentration distribution of OH with or without surface reaction conditions. In both cases, the concentration increases as the wall temperature increases. The concentration near the wall aproaches to zero for the active wall condition, because OH is absorbed by the wall surface. However, the difference of concentation occurs as the wall surface temperature without sarface reaction near the wall. The concentration gradient for active wall is larger than that for inert wall. When the distance from the wall surface increases, the concentration difference with or without the surface reaction decreases and eventually fit. Therefore, the influence of the surface reaction appears from the wall surface to x = 0.3 ~ 0.4 mm.

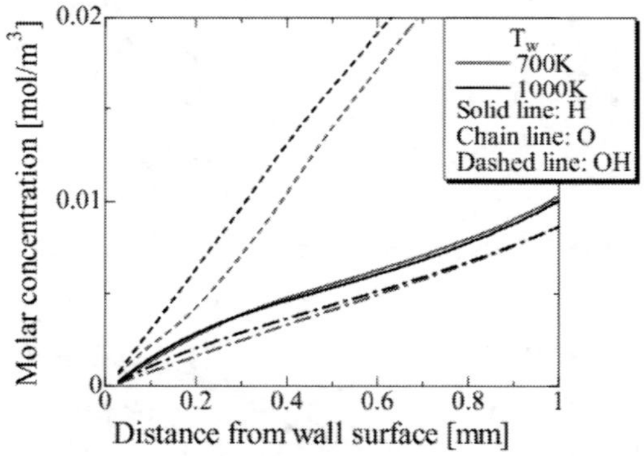

Figure 8: Distribution of molar concentration near the stagnation wall with surface reaction

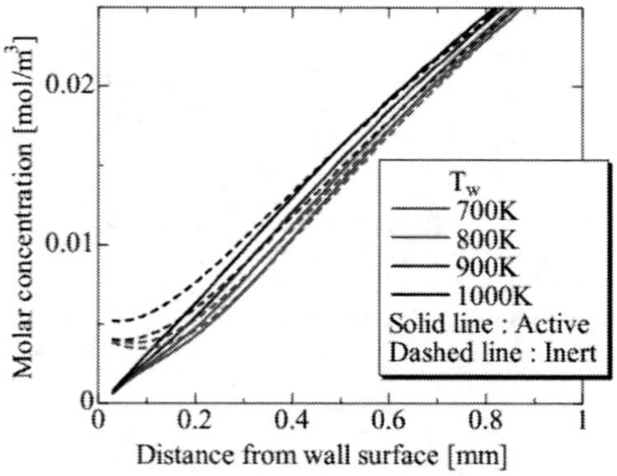

Figure 9: Distribution of molar concentration of OH near the stagnation wall with or without surface reaction

4 Conclusion

Numerical analysis was carried out using a stagnation flow model of methane-air premixed flame due to clarify the influence of the wall on the flame. We investigated H, O, and OH radicals which are important for surface reaction, depending on the difference in wall temperature and with or without surface reaction of the wall which causes radical quenching.

1. For the inert wall conditions, the molar concentration of OH is larger than that of O and H, and the concentration gradient is also large. The concentration near the wall also increases as the wall surface temperature is high for all species, especially the difference in concentration of H is large. The concentration gradient of OH is large at either wall surface temperature and gently converges as compared with H and O.
2. For active wall conditions, the concentration converges to zero regardless of the wall surface temperature on the wall surface. The molar concentration near the wall also increases with the wall surface temperature increase. The concentration gradient of OH is the largest in each species, and the concentration gradient of T_w = 700 K near the wall is smaller than T_w = 1000 K. It is considered that there is a reaction of OH which contributes to the wall surface temperature in that range.
3. For active wall conditions, the molar concentration of OH approaches to zero on the wall. On the other hand, for inert wall conditions, the difference occurs for wall temperature. As the distance from the wall increases, the concentration difference with or without surface reaction decreases and eventually converges.

5 Acknowledgment

This work was supported by JSPS KAKENHI Grant Number 26820062.

References

[1] K. Maruta, T. Kataoka, N.I. Kim, S. Minaev, R. Fursenko, Proceedings of the Combustion Institute 30 (2005) 2429-2436.
[2] B. Bai, Z. Chen, H. Zhang, and S. Chen, Combustion and Flame 160 (2013) 2810-2819.
[3] R. Owston, V. Magi, J. Abraham, Combustion and Flame 148 (2007) 127-147.
[4] Y. Saiki, National Heat Transfer Symposium of Japan (In Japanese) 54 (2017).
[5] S. Raimondeau, D. Norton, D.G. Vlachos, R.I. Masel, Proceeding of the Combustion Institute 29 (2002) 901-907.
[6] R.J. Kee et al., Sandia Report, SAND 86-8246 (1986).
[7] R.J. Kee et al., Sandia Report, SAND 89-8009 (1989).
[8] NIST, JANAF thermochemical tables, Third Edition (1990).
[9] G.P. Smith, D.M. Golden, M. Frenklach, N.W. Moriarty, M. Eiteneer, C.T. Goldenberg, R.K. Bowman, S. Hanson, W. song, V.V. Lissianski and Z. Qin, "GRI-Mech 3.0," [Online]. Available:http://combustion.berkeley.edu/Combustion_Laboratory/gri-mech/.
[10] M.D. Smooke et al., Reduced Kinetic Mechanisms and Asymptotic Approximations for Methane-Air Flames, Springer-Verlag, (1991) 1-28.

11th Asia-Pacific Conference on Combustion,

The University of Sydney, NSW Australia
10th -14th December 2017.

Impact of nitric oxide on the low temperature oxidation of iso-octane in a pressurized flow reactor

Zhongyuan Chen, Zhewen Lu, Hao Yuan, Yi Yang*, Michael Brear

Department of Mechanical Engineering, University of Melbourne

Parkville, Victoria, 3010, Australia

Abstract

Nitric oxide (NO) is present in the residual gases inside reciprocating engines, and has previously been reported to have a complex impact on fuel autoignition. This paper therefore investigates the impact of NO on the low temperature oxidation of iso-octane in a pressurized flow reactor (PFR). Experiments were conducted at 550-900K, a fixed reactor pressure of 10 bar and equivalence ratio of 0.058 with NO added at 0, 20, and 200 ppm in the mixture. Results show that NO promotes isooctane oxidation reactivity at temperatures higher than 700K, as indicated by the higher CO formation. However, an overall inhibiting effect is observed at lower temperatures, whereas for some temperatures, opposing NO impacts are observed for the 20 and 200 ppm cases. Kinetic simulations are then conducted using two existing mechanisms from the literature. Both mechanisms capture the promoting impact of NO at the higher temperatures but fail to reproduce the more complex impact at lower temperatures.

1 Introduction

Autoignition is a critical process in conventional and emerging engines. Autoignition of unburnt fuel/air mixture can induce knock in spark ignition (SI) engines, limiting the engine compression ratio and therefore thermal efficiency. In compressed ignition (CI) engines, such as diesel, homogeneous-charge compression-ignition (HCCI) engines and reactivity control compression-ignition (RCCI) engines, autoignition is the primary process controlling the start of combustion and subsequent heat release rates [1-3], which significantly affects the engine performance, efficiency, and emissions.

Autoignition is driven by chemical kinetics of fuel oxidation, which is highly dependent on species composition and the thermodynamic conditions in the reacting system. Nitric oxide (NO) typically exists in engine cylinder as one of leftover products from previous combustion cycles, either as the residual gas or from using exhaust gas recirculation. Although generally at a concentrations of no more than hundreds of ppm, this residual NO has been reported to have significant and complicated impact on autoignition in SI engines [4-7] and HCCI engines [8-10].

Investigations of NO/hydrocarbon chemistry have been conducted in kinetic experiments, including flow reactors [11-15] and stirred reactors [16, 17]. Most of these studies have focused on small hydrocarbons, whereas larger hydrocarbons of interest to liquid fuels have been rarely studied [18, 19].

This work aims to explore the impact of NO on the low temperature oxidation of a gasoline primary reference fuel, iso-octane. Experiments are conducted in a pressurized flow reactor (PFR) which are modelled with CHEMKIN simulations using existing NO/hydrocarbon mechanisms. The impact of NO on iso-octane oxidation has been only reported in a jet stirred reactor [18].

2 Experimental

The experiment study is conducted in a turbulent pressurized flow reactor recently established in the University of Melbourne. Details of this facility can be found in [20], with major features summarized below.

2.1 PFR setup

Layout of the PFR facility is shown in Fig. 1. Air is supplied from an oil-free compressor to the reactor after heated to a specified temperature. High pressure nitrogen from a nitrogen bottle is used to pressurize liquid fuel in the fuel tank, and to vaporize the liquid fuel. The vaporized fuel and hot air mix rapidly in a novelly designed mixer at the inlet of reactor.

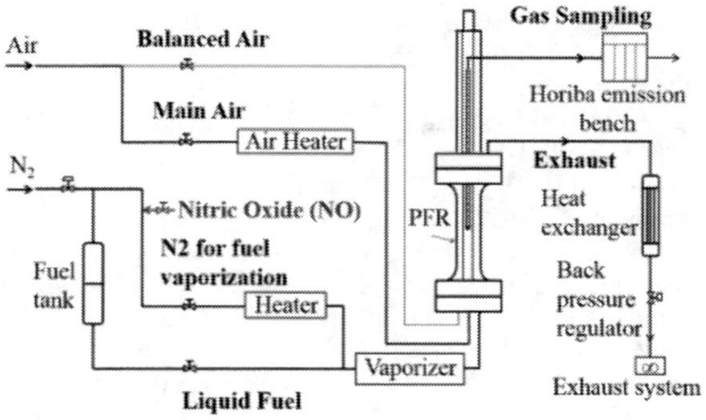

Fig. 1 Schematic of the pressurized Flow Reactor system.

Fast mixing is confirmed by injecting inert CO_2 at 900 K and 10 bar which indicated that complete mixing is achieved at ~ 50 mm from the mixture [20]. The reactor is a 25 mm i.d., 1000 mm long, quartz tube housed in a pressure vessel with a design pressure of 50 bar at 1000 K. The reactor pressure is controlled by an automatic back pressure valve on the exhaust line. A "hot

Corresponding author: Yi Yang, phone: (+61) 3 83448104

E-mail address: yi.yang@unimelb.edu.au

water" cooled sampling probe is installed on a linear actuator which allows gas collection along the centerline of the reactor. The sampling probe is also equipped with three different sized K-type thermocouples installed axisymmetrically for gas temperature measurement accounting for thermocouples radiative losses. The reaction gas temperatures are derived from the three measured temperatures and the junction size ratios without assuming the wall emissivities and convection heat transfer coefficients. Details of the temperature correction can be found in [20], where the uncertainty of corrected temperature is estimated to be ±5 K at 900 K (smaller at lower temperatures). The sampled gas is sent to an engine emission bench for CO, CO2 and NO analyses, as well as to a gas chromatograph for more detailed species analysis (not reported in this work). The CO reported here is measured by a non-dispersive infrared analyzer at a resolution of 20 ppm. NO is introduced to the reactor from a gas bottle via the cold N2 line (see Fig. 1). NO addition is metered with a needle valve and monitored via a chemiluminescence detector in an emission bench.

2.2 Experimental Conditions

The PFR experimental conditions used in this work are listed in Table 1. With these conditions, developing turbulent flow is generated in the reactor, establishing essentially uniform (time-averaged) radial profiles for temperature, species and velocity, a major feature of turbulent plug flow.

Typically a 2~3 hours warming period is required to reach these conditions. During the warming up, hot air (~1100K, 10bar) flows through the PFR, with the wall heaters surrounding the quartz tube also turned on. The thermal equilibrium is considered to be achieved when the temperature inside the PFR reaches a constant value (drift less than 2°C in 30 minutes). In this work, the temperature measured at the inlet of the reactor (0mm position) is used as the nominal temperature for reporting the reaction temperature. The local gas temperature is obtained from the three K-type thermocouples on the sampling probe, which moves along the reactor to obtain the reaction temperature profile. Details of the temperature correction and the reactor validation can be found in [20].

Table 1: PFR experiment conditions for isooctane oxidation

Reactor pressure (bar)	10
Reactor nominal temperature (K)	550 - 900
Air flow rate (g/s)	6.00
Nitrogen flow rate (g/s)	0.33
Reynolds number	11000 - 8000
Fuel/nitrogen pressure (bar)	20 -21
Equivalence ratio	0.058
NO addition in PFR (ppm)	0, 20, 200

3 Results and Discussion

3.1 PFR Results

The PFR experiment results in term of CO formation are shown in Fig. 2. CO is reported as the common indicator of fuel's low temperature oxidation reactivity. The results are obtained at a residence time of 0.21 s, which is collated from individual experiment results at each reactor temperature. It is obvious that neat iso-octane without NO addition exhibits a typical negative temperature coefficient (NTC) behavior from 650K to 850K. The CO peak at ~650K indicates the peak of iso-octane's low temperature reactivity which then decreases as temperature increases and vanishes at 750K. At temperature higher than 850K, CO start to form again indicating the oxidation chemistry enters the intermediate temperature regime where the Arrhenius's law resumes the ruling of the global reaction rate.

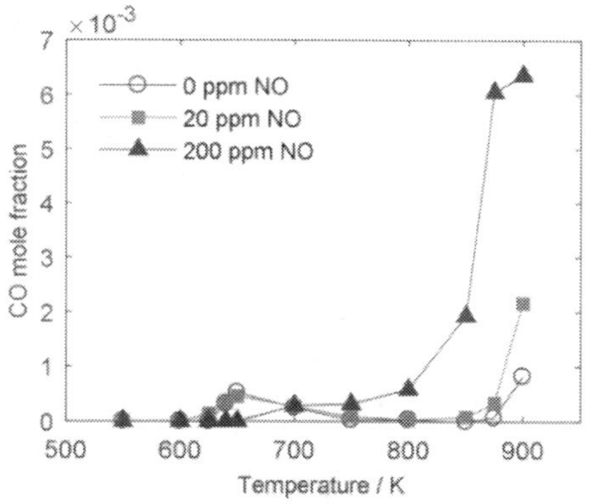

Fig. 2 Measured CO mole fraction for iso-octane in PFR with different NO additions from 550K to 900K at 10bar, $\phi = 0.058$ and $\tau = 0.21$s.

With 20 ppm NO addition, the CO formation remains similar to the baseline case (0 ppm NO) at temperatures lower than 800K, but is promoted at higher temperatures (>800K). The effect is more complex at 200 ppm NO addition. As shown in Fig. 2, at temperatures higher than 700K, 200 ppm NO substantially promotes the CO formation and this promotion effect increases monotonically as temperature increases. However, at temperatures lower than 700K, 200 ppm NO generally retards the CO formation resulting in nearly zero CO formation. This temperature dependent effect of NO counteracts the NTC behavior of iso-octane and the NTC behavior is essentially removed with 200 ppm NO addition (Fig. 2). Similar NO impact on the NTC behavior of n-heptane is recently observed in a rapid compression machine experiment [21].

Detailed illustrations of the NO impact is shown in Fig. 3 (a) and (b) at two selected temperatures, 900K and 625K, respectively. At 900K, without NO addition, CO only starts to form after ~500mm in a relatively slow rate. The addition of 20 ppm modestly promotes the CO formation, whilst 200 ppm NO significantly advanced the formation of CO which reaches a plateau at ~500mm, indicating the conversion to CO2 (not shown here). At 625K, a more interesting impact of NO is observed, in that 20 ppm NO promotes the CO formation but 200 ppm NO completely shuts down all the reactivity of iso-octane and produces no measurable CO.

Moreac et al. studied the impact of NO on iso-octane oxidation in a jet stirred reactor (JSR) [18] and found similar results. Experiments were conducted at 650-1000K, and a fixed reactor pressure of 10 bar, equivalence ratio of 1 and residence time of 1s with NO added at 0 and 50 ppm in the mixture. They found that with 50 ppm NO added the onset of reaction started ~20 K lower relative to the 0 ppm case, and the promoting effect became stronger as temperature increased, which is consistent with the promoting effect of 20 ppm NO at 625K and higher temperatures in this work. Consistent results were also observed in a recent engine study by the authors, which showed that the impact of NO varies with both temperatures and the amounts of NO [6]. At an engine intake temperature of 52°C, NO advanced knock-onset timing at addition levels < 200 ppm, but retarded it at higher levels. As the intake temperature increased, the promoting effect of NO became stronger.

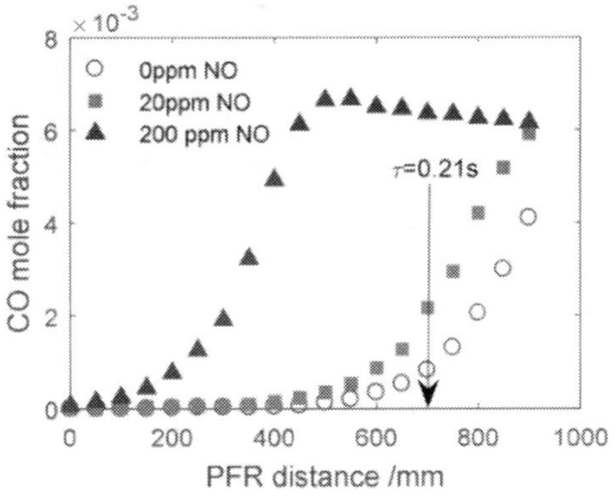

(a) 900K

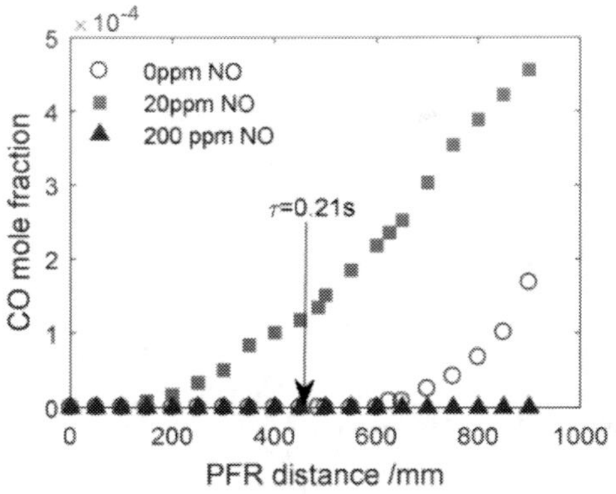

(b) 625K

Fig. 3 Measured CO mole fraction for iso-octane in PFR with different NO additions at (a) 900K and (b) 625K, at 10bar, $\phi = 0.058$ and $\tau = 0.21$s.

3.2 Simulation

Chemical mechanisms for NO/hydrocarbon interactions have been proposed by several groups [9, 19, 22-24]. Two mechanisms are used for the kinetic modelling for the PFR experiments. The mechanism from Contino et al. [9] is for NO-PRF (mixture of iso-octane and n-heptane) and was validated primarily with HCCI experiments. The mechanism from Anderlohr et al. [23] is for NO-TRF (mixture of toluene and PRF) and was validated with a JSR and HCCI experiments. Simulations are conducted in the plug flow reactor model in CHEMKIN PRO [25] using the temperature profiles corrected with the three-thermocouple method as the input. Start of the reaction is set at the mixer plane. Simulation results using these two mechanisms are shown in Fig. 4 (a) and (b) respectively.

It is observed that Contino's mechanism well reproduced the trend of the NO impact at temperatures higher than 700K, but failed to reproduce the inhibiting impact of NO at lower temperatures. This is potentially because it incorrectly reproduced the baseline isooctane oxidation. As shown in Fig. 4 (a) no CO formation was predicted by the model at temperature lower than 650K, therefore the inhibition effect of NO, if any, was not able to be exhibited. Anderlohr's mechanism also reasonably predicted the promoting effect at high temperatures (>800K) but over-estimated the inhibition effect at 700K and 750K where no CO formation is produced at 200 ppm NO addition.

Overall, the two mechanisms captured the promoting effect of NO on CO formation at high temperatures, but need improvement at low temperatures. More importantly, improvement is needed for the baseline hydrocarbon mechanisms which should be correctly reproduced as the basis for the NO/fuel interactions.

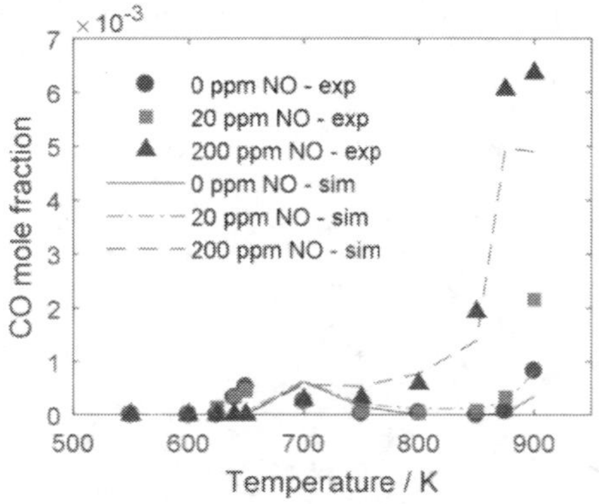

(a) Simulation using mechanism from Contino.

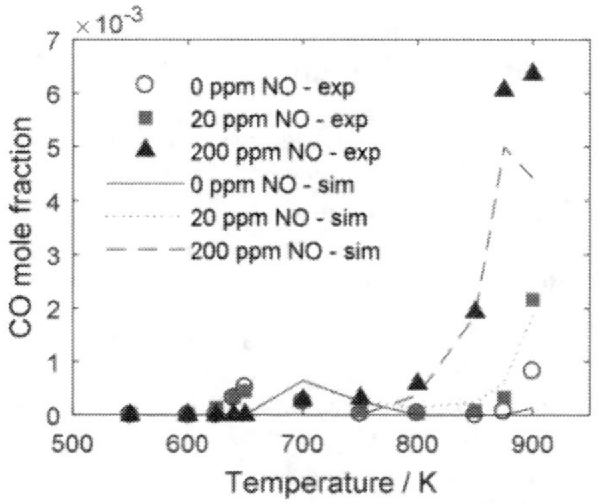

(b) Simulation using mechanism from Anderlohr.

Fig. 4 Simulated CO mole fraction for iso-octane in PFR with different NO additions from 550K to 900K at 10bar, $\phi = 0.058$ and $\tau = 0.21$s. (a) Contino's mechanism [9], (b) Anderlohr's mechanism.

3.3 Discussion

Previous studies [9, 19, 22-24] suggest that NO affects the hydrocarbon's low temperature oxidation mainly through its reaction with small radicals and some fuel-derived intermediates. As illustrated in Fig. 5, hydrocarbon oxidation without NO starts with H-abstraction from the fuel (RH), followed by the ROO and QOOH steps, leading to degenerate chain branching. Adding NO then introduces new reactions. For example, NO reacting with HO2 (R1) produces a reactive OH radical which accelerates oxidation. On the other hand, NO reacting with OH (R2) produces a less reactive HONO radical which slows down oxidation. Another important reaction is NO with RO2 radicals (R3), which in some cases promotes ignition, due to RO2 being converted to a more reactive RO radical, but in other cases inhibits ignition, due to consuming the ROO radical which can otherwise react in the low temperature branching channel and accelerate autoignition.

$$NO + HO2 \leftrightarrow NO2 + OH \quad (R1)$$

$$NO + OH + M \leftrightarrow HONO + M \quad (R2)$$

$$NO + RO2 \leftrightarrow NO2 + RO \quad (R3)$$

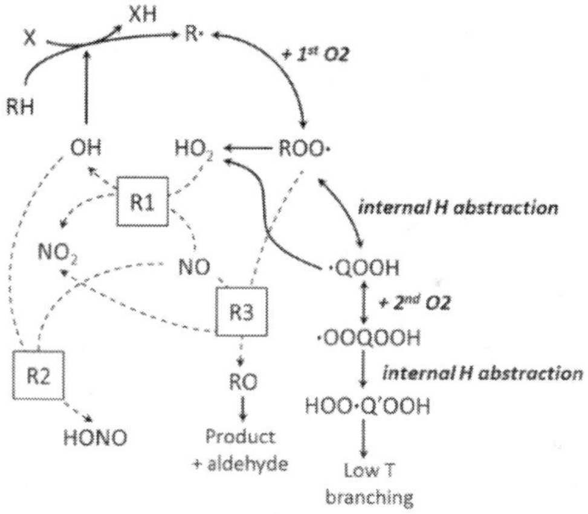

Fig. 5 Schematic of low temperature hydrocarbon oxidation without NO (full line) and with NO (dash line).

At low temperature of 625K in the PFR, the addition of the first 20 ppm could favor R1 involving HO_2, and produce the promoting effect. However, at higher NO addition levels, NO could start to react through reactions R2 and R3. The reaction R2 is apparently inhibiting for the overall reactivity as mentioned above. The reaction R3 could be on the inhibition side as isooctane can react along low temperature oxidation channels shown in Fig. 5. The conversion of $C_8H_{17}O_2$ to $C_8H_{17}O$ therefore could reduce low temperature degenerate branching flux and consequently slow down the autoignition process, particularly at high NO addition levels, as observed in the 200 ppm case in the PFR experiment. It is noted that although all three reactions (R1-R3) could occur with any NO additions, simulation in a closed homogeneous reactor showed that the reaction rate ratios of R1 over R2 decreased as NO concentrations increased at temperature of 650 K [6] and more inhibition was produced at higher NO concentrations. As temperature increases, less RO2 is formed because reaction R + O2 = RO2 shifts to the reverse direction at higher temperatures. Consequently, more HO2 is formed from the decomposition of QOOH and also ROO as shown in Fig. 5 [4, 6]. With abundant HO2 around, R1 could remain important even at high NO addition levels and produces stronger promotion impact. Moreover, higher temperatures tend to reverse the combination reaction R2 and reduce its inhibition effect on oxidation [26]. R3 in this case become less relevant due to the decomposition of RO2. These arguments are in line with the observations in our CFR engine experiments [6, 27] and the strong promoting effect of NO on iso-octane oxidation observed at high temperatures in this work.

4 Conclusions

This paper investigated the impact of nitric oxide (NO) on low temperature oxidation of iso-octane in a pressurized flow reactor (PFR). The CO mole fraction was measured as an indicator of the oxidation reactivity, in air/N2/iso-octane mixtures with 0, 20 and 200 ppm NO additions over 550-900K, a pressure of 10bar and an equivalence ratio of 0.058. Experimental results showed that NO promoted the CO formation at higher temperatures but generally inhibited CO formation at lower temperatures. This temperature-dependent impact of NO significantly weakened the NTC behavior of iso-octane oxidation.

Kinetic simulations using literature mechanisms from Contino et al. [9] and Anderlohr et al. [23] showed that they generally captured the promoting impact of NO at higher temperatures. However, they did not reproduce the more complex impact of NO at lower temperatures. Current understanding of NO/hydrocarbon interactions is used to qualitatively interpret these observed trends.

5 Acknowledgement

This research was supported by the Advanced Centre for Automotive Research and Testing (ACART, www.acart.com.au), the Australian Research Council. The authors also thank Mr. Alistair Knox for his technical support in building the facility.

References

[1] Dec JE. Proc Combust Inst. 2009;32:2727-42.
[2] Kokjohn S, Hanson R, Splitter D, Reitz R. International Journal of Engine Research. 2011;12:209-26.
[3] Reitz RD, Duraisamy G. Progress in Energy and Combustion Science. 2015;46:12-71.
[4] Roberts PJ, Sheppard CGW. SAE Int J Engines. 2013;6:2028-43.
[5] Morganti KJ, Brear MJ, da Silva G, Yang Y, Dryer FL. Proc Combust Inst. 2015;35:2933-40.
[6] Chen Z, Yuan H, Yang Y, Brear M. 10th Asia Pacific Conference on Combustion (2015) paper 240.
[7] Burluka AA, Liu K, Sheppard CGW, Smallbone AJ, Woolley R. SAE transactions. 2004;113:1873-89.
[8] Masurier J-B, Foucher F, Dayma G, Dagaut P. Proc Combust Inst. 2015;35:3125-32.
[9] Contino F, Foucher F, Dagaut P, Lucchini T, D'Errico G, Mounaïm-Rousselle C. Combust Flame. 2013;160:1476-83.
[10] Risberg P, Johansson D, Andrae J, Kalghatgi G, Björnbom P, Ångström H-E. Warrendale, PA: SAE International; 2006.
[11] Ano TA, Dryer FL. Proc Combust Inst. 1998;27:397-404.
[12] Bromly JH, Barnes FJ, Mandyczewsky R, Edwards TJ, Haynes BS. Proc Combust Inst. 1992;24:899-907.
[13] Nelson PF, Haynes BS. Symposium (International) on Combustion. 1994;25:1003-10.
[14] Prabhu SK, Bhat RK, Miller DL, Cernansky NP. Combust Flame. 1996;104:377-90.
[15] Hori M, Koshiishi Y, Matsunaga N, Glaude P, Marinov N. Proc Combust Inst. 2002;29:2219-26.
[16] Dagaut P, Nicolle A. Combust Flame. 2005;140:161-71.
[17] Dagaut P, Lecomte F, Chevailler S, Cathonnet M. Combust Sci Technol. 1999;148:27-57.
[18] Moréac G, Dagaut P, Roesler JF, Cathonnet M. Combust Flame. 2006;145:512-20.
[19] Dubreuil A, Foucher F, Mounaı¨m-Rousselle C, Dayma G, Dagaut P. Proc Combust Inst. 2007;31:2879-86.
[20] Lu Z, Cochet J, Leplat N, Yang Y, Brear M. Measurement Science and Technology, 2017; 28: 105902.
[21] Chen Z, Zhang P, Yang Y, Brear MJ, He X, Wang Z. Combust Flame. 2017;186:94-104.
[22] Glaude P-A, Marinov N, Koshiishi Y, Matsunaga N, Hori M. Energy & fuels. 2005;19:1839-49.
[23] Anderlohr JM, Bounaceur R, Pires Da Cruz A, Battin-Leclerc F. Combust Flame. 2009;156:505-21.
[24] Faravelli T, Frassoldati A, Ranzi E. Combust Flame. 2003;132:188-207.
[25] Reaction Design CHEMKIN-RPO release 15112 (30-Jan-2012)
[26] Andrae JCG. Energy & Fuels. 2013;27:7098-107.
[27] Chen Z, Foong TM, Yuan H, Yang Y, Brear M. 19th Australasian Fluid Mechanics Conference. Melbourne, Australia: Australasian Fluids Mechanics Society; 2014.

A Study of Low Temperature Oxidation of N-heptane and Iso-octane in a Pressurised Flow Reactor

Zhewen Lu, Zhongyuan Chen, Yi Yang*, Michael Brear

Department of Mechanical Engineering

The University of Melbourne, Parkville VIC 3010 Australia

Abstract

A turbulent pressurised flow reactor (PFR) has recently been established at the University of Melbourne for investigating hydrocarbon oxidation chemistry. This paper reports a study in this reactor on n-heptane and iso-octane at temperatures of 550-900 K, fixed equivalence ratio of 0.058 and a reactor pressure of 10 bar. Carbon monoxide is measured along the reactor as an indicator of the low temperature oxidation reactivity. The CO concentrations measured at a fixed resident time (0.21 s) show negative temperature coefficient (NTC) behaviour for both n-heptane and isooctane. Gas temperatures are also measured along the reactor and corrected for thermocouple radiative losses, which is then input into the Chemkin PFR model for kinetic simulation. Detailed combustion mechanisms from the literature are used in the modelling, reproducing the NTC behaviour and the CO formation at most temperatures tested. These results further validate the latest mechanisms of n-heptane and iso-octane.

1 Introduction

With the advances in engine technologies and increasing uses of alternative fuels, developing combustion chemistry models that can be used for the new engines and fuels becomes imminent. Combustion chemistry is inherently challenging to study, particularly due to the large number of species and reactions involved. Achieving this task requires acquisition of quality experimental data. In this regard, a turbulent pressurised flow reactor (PFR) has been established at the University of Melbourne for investigating oxidation chemistry at up to 50 bar and 1000 K.

PFR is a continuous flow device in which well-mixed fuel and oxidiser flow and react at the same time. The reacting mixtures are highly diluted, so that the kinetics are slow and the reaction zone is long [1]. Each location along the flow centreline thus represents a unique reaction time. This one-dimensional steady, spatial problem can be converted to a zero-dimensional temporal problem given knowledge of the gas velocity. Compared to compression-ignition experiments, e.g. shock tubes and rapid compression machines (RCMs), which are useful for obtaining ignition delay data, PFR experiments are more adapted for species concentration measurements in steady flow environment.

The oxidation chemistry of n-heptane and iso-octane are studied in this work, which are primary reference fuels (PRF) for octane rating in internal combustion engines. Many kinetic studies have been performed to these fuels, including shock tubes [2,3], RCMs [4,5], jet-stir reactors [6,7], laminar flames [8,9], etc. These experiments were often conducted at high temperatures for iso-octane due to its relatively low reactivity in low temperature oxidation. Studies in flow reactors are also reported for n-heptane and iso-octane oxidation [10,11], but the results at elevated pressures with lean mixtures are relatively limited. Notably, validation against PFR experiments was missing in the latest mechanisms of n-heptane [12] and isooctane [13].

In this work, oxidation of the two compounds was studied in a PFR at temperatures from 550-900 K, 10 bar, and equivalence ratio of 0.058. Kinetic modellings were performed using some of the best validated reaction mechanisms in the literature, including the latest development.

2 Experimental and modelling methods

2.1 Experimental setup and reaction conditions

The experiments in this study were conducted in the PFR at the University of Melbourne, which has been reported in detail in [14]. The layout of this PFR facility is shown in Fig. 1. A pre-heating method is adopted in the system, by which the reactants are heated separately to the desired temperature, and then mixed to the desired composition. Air as a dilution gas is supplied from an oil-free compressor and is heated to the reaction temperature before entering the reactor. Liquid fuels are pressurized by high pressure bottled nitrogen, and vaporized by hot nitrogen in a cross flow. The mixed fuel vapour and nitrogen are maintained at a temperature that avoids fuel condensation and pyrolysis before entering the reactor. Air is supplied at a large excess relative to the fuel in order to minimise the temperature rise resulted from reaction heat release. The air flow is measured by a flow meter and controlled by a needle valve. The fuel flow is derived from temporal measurement of the fuel tank weight. The reactor pressure is controlled by a backpressure regulator on the exhaust line downstream of a heat exchanger.

The PFR is a quartz tube with 25 mm i.d. and 1000 mm length, housed in a pressure vessel with a design pressure of 50 bar at 1000 K. Surrounding the quartz tube are cylindrical heaters and insulations to compensate the heat loss. The quartz tube inlet is connected to a novel designed mixer for rapid mixing of fuel and air streams. Fast mixing is confirmed by injecting inert CO_2 at

* Corresponding author. Phone: +61 3 8344 8104
E-mail address: yi.yang@unimelb.edu.au

900 K and 10 bar which indicated that complete mixing is achieved at ~ 50 mm from the mixer [14].

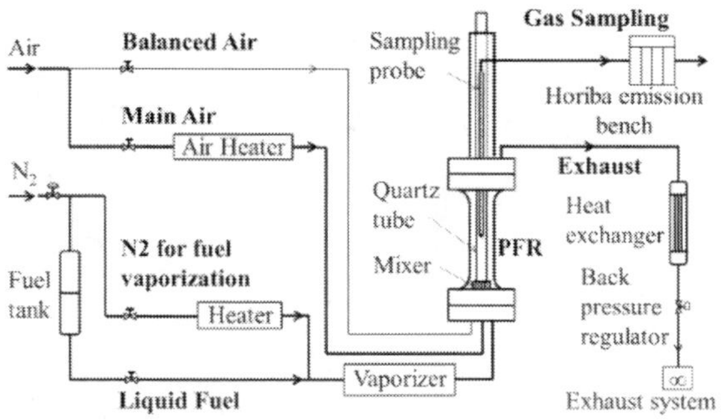

Figure 1 Schematic of the PFR system layout at the University of Melbourne

A water-cooled sampling probe is travelling along the centreline of the quartz tube, continuously collecting the gas samples and measuring the reacting temperatures. The sampled gas is transferred via a heated line (at 473K) to a Horiba emission bench where carbon monoxide is measured by a non-dispersive infrared (NDIR) analyser at a resolution of 20 ppm. A novel three-thermocouple method is used for gas temperature measurement, which uses three K-type thermocouples of different junction sizes installed axisymmetrically on the sample probe. The reaction gas temperatures are derived from the three measured temperatures and the junction size ratios without assuming the wall emissivities and convection heat transfer coefficients. Using this method, the gas temperature profile along the reactor centreline is obtained at each test condition. Details of the temperature correction can be found in [14], where the uncertainty of corrected temperature is estimated to be ±5 K at 900 K (smaller at lower temperatures).

During each PFR experiment, the flow rates of air, nitrogen, and fuels, and the reactor pressure, are held constant. Before introducing fuel, the quartz tube reaction zone is maintained at a temperature slightly higher than the PFR nominal temperature (defined by the temperature at the quartz tube inlet). This provides an almost adiabatic non-reacting environment. After the fuel is injected and a steady state is reached, the gas temperatures and CO readings are recorded at various locations along the reactor centreline using the sampling probe. Table 1 shows the experimental conditions used in this work.

Table 1 Experimental conditions

PFR pressure (bar)	10
PFR nominal temperature (K)	550 - 900
Entire resident time (sec)	0.44 - 0.27
Air mass flow rate (g/s)	6
N_2 mass flow rate (g/s)	0.32
Reynolds number in reaction zone	11000-8000
Fuel/N_2 pressure (bar)	21
Fuel/N_2 vaporisation temperature (K)	500
Equivalence ratio	0.058

In this study, two types of results were reported, namely the "reactivity result" and the "species evolution result". For the "reactivity result", the reacting gas was sampled at a fixed reaction residence time of 0.21 s at different PFR nominal temperatures. As the nominal temperature changes, actual sampling location varies in order to maintain the same residence time. The required locations corresponding to this residence time were calculated using the non-reacting flow properties corrected for the PFR nominal temperatures. Although the local temperature can increase due to reaction heat release, which increase the actual axial velocity, this impact is secondary and the variation of actual residence times is less than ±2.5% for all nominal temperatures tested. For the "species evolution result", reacting gas was sampled within the reaction zone from 5 to 905 mm downstream the mixer at each PFR nominal temperature. This type of experiment can give detailed temporal evolutions of reaction products. In both types of experiments, gas temperatures were measured along the reactor at an interval of 50 mm. A complete temperature profile was thus made available for kinetic modelling at each test condition.

2.2 Kinetic modelling method

Simulations in this study were carried out using the Plug Flow Reactor model in Chemkin Pro [15]. The temperature profiles corrected with the three-thermocouple method were input to the model to solve the energy equation. The following detailed mechanisms of n-heptane and isooctane were used.

Table 2 Selected mechanisms of h-heptane and iso-octane

	Mechanism	Origin
n-heptane	Mehl et al. [16]	LLNL
	Zhang et al. [12]	NUI-Galway
iso-octane	Mehl et al. [17]	LLNL
	Atef et al. [13]	KAUST

A note to these mechanisms is that they share a common origin from the mechanisms by Curran et al. for n-heptane [18] and iso-octane [19], which are commonly regarded as the most validated mechanisms for the two compounds. The LLNL mechanisms here used [16,17] are further updated from the original version, whereas the mechanisms from NUI [12] and KAUST [13] are more recent development and both have been further validated with new experimental data from shock tubes, jet stir reactors and laminar flames.

3 Results and discussions

3.1 N-heptane

This section presents the n-heptane oxidation results in terms of CO formation. Figure 2 presents the reactivity results of n-heptane at residence time of 0.21 s as a function of PFR nominal temperatures. Evident negative temperature coefficient (NTC) behaviour is demonstrated by n-heptane at this condition. At lower temperatures (550-600 K), no CO formation is detected, indicating that n-heptane is not reactive. As temperature increases, oxidation starts and the reactivity rapidly increases until the CO production reaches ~ 2500 ppm at 650 K, which corresponds to the peak of the low temperature degenerate branching. Nevertheless, as temperature further increases, the CO formation decreases during 650 – 800 K, showing that the n-heptane oxidation is going through the NTC regime where the global reaction rate is supressed by the increase of temperature. At even higher temperatures (above 800 K), CO mole fraction starts to increase again and eventually reaches the second peak at 900 K (approximately 5800 ppm). Figure 2 also shows the modelling results with the LLNL and NUI mechanisms. Overall, both mechanisms reproduce the CO results very well, in particular in capturing the NTC behaviour. The NUI mechanism produces an excellent agreement at lower temperature (550-650 K) but somewhat underpredicts CO at higher temperatures. The LLNL mechanism predicts a slightly earlier start in the low temperature oxidation and underpredicts the CO during the NTC regime, but produces a closer agreement at temperature > 800 K. Both mechanisms are short in reproducing the high CO formation at temperature > 850 K. The LLNL mechanism also shows a decrease of CO mole fraction at 900K due to conversion to CO2.

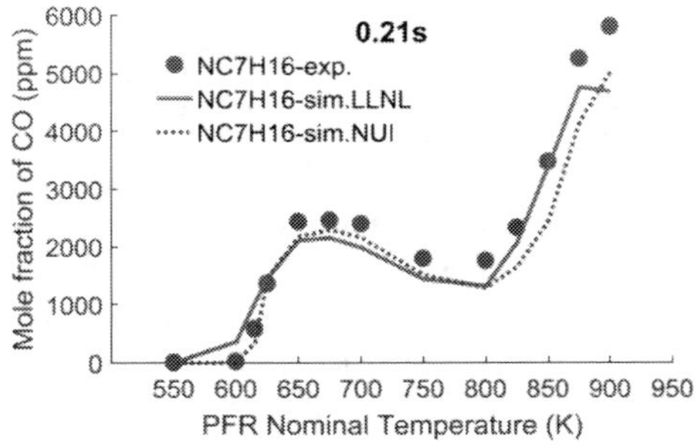

Figure 2 Reactivity mapping for n-heptane in the PFR 10 bar, ϕ = 0.058, τ = 0.21 s.

The CO evolution histories are reported at selected temperatures as shown in Fig. 3. Figure 3a presents the CO concentration profile along the reactor at 625 K. At this temperature, the n-heptane reactivity is dominated by the low temperature chemistry, where CO starts to form after 200mm and gradually increase along the reactor. Simulation with the NUI mechanism reproduces the experiment excellently, whereas the LLNL mechanism predicts a quicker start of CO formation. At 750 K (Fig. 3b) where the oxidation is within the NTC regime, the CO formation shows a rapid rise right after the mixer and then a nearly flat profile along the reactor. Both mechanisms capture this trend very well. At 875 K where the oxidation enters the intermediate temperature regime, Fig. 3c shows the CO formation increases steadily along the reactor until reaching a plateau near the end of the reactor, where CO starts to form CO2. A maximum CO of ~ 5800 ppm is produced, corresponding to 80% of carbon in the supplied fuel. The simulation results of LLNL mechanism match the experiment well during the CO rise but underpredict the CO by roughly 1000 ppm at the plateau. The NUI mechanism, on the other hand, is less reactive compared to the experiment and also underpredicts the maximum CO formation.

3.2 Iso-octane

In this section, the isooctane oxidation results are presented. Figure 4 shows the reactivity of iso-octane at a constant residence time (0.21 s) in terms of CO formation. Similar to n-heptane, iso-octane also demonstrates low temperature reactivity as well as the NTC behaviour, although the overall reactivity is much lower. The low temperature reactivity starts at > 625 K and reaches the peak at ~ 675 K. The oxidation then enters the NTC regime, from

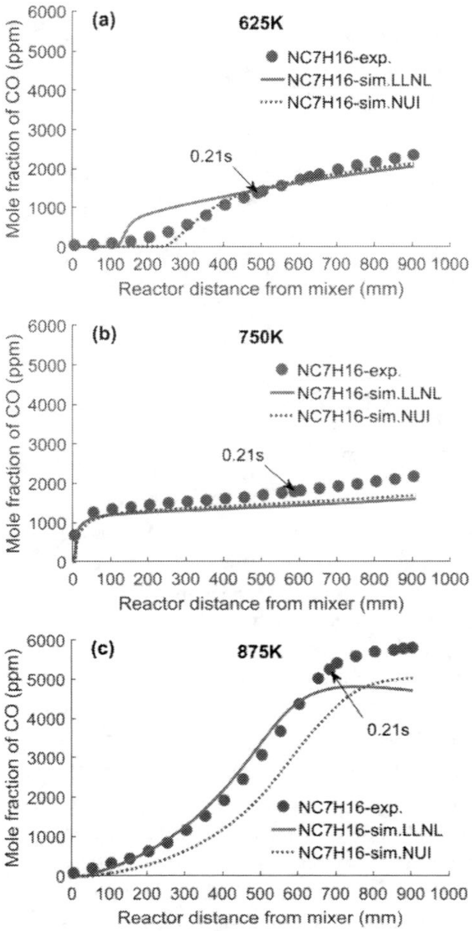

Figure 3 CO evolution profile for n-heptane oxidation in the PFR at 10 bar, ϕ = 0.058, 625 K (a), 700 K (b), and 875 K (c).

675-750 K, where the CO formation declines to zero and remains dormant until 850K. At temperatures above 875K, the iso-octane becomes much more reactive and CO formation takes off rapidly. Kinetic modelling results using LLNL and KAUST mechanisms both capture the low temperature reactivity and the NTC behaviour. In particular, the LLNL mechanism gives a better agreement with experiment throughout the temperature range, where the KAUST mechanism performs well at higher temperatures, but underpredicts the reactivity in the NTC regime.

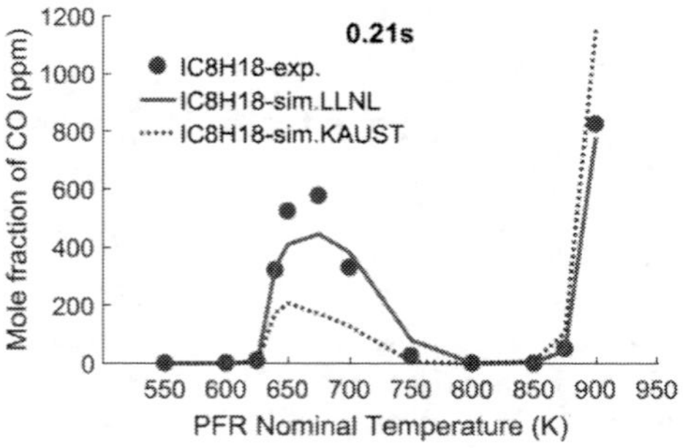

Figure 4 Reactivity mapping for iso-octane in the PFR 10 bar, ϕ = 0.058, τ = 0.21 s.

Detailed CO evolution histories of isooctane oxidation are presented for temperatures of 640, 700 and 900 K in Figs. 5a, 5b, and 5c, respectively. At 640 K (Fig. 5a), a gradually increase of CO mole fraction is observed from experiment. The LLNL mechanism captures the experiment well despite minor discrepancies at the start and end of the reactor. On the other hand, the KAUST mechanism consistently underestimates the CO formations along the reactor. Figure 5b shows the CO evolution profile at 700 K which is in the NTC regime. The CO mole fraction starts to rise right after the mixer until a third of reactor length, and then remains at ~ 350 ppm till the end of the reactor. LLNL mechanism captures this behaviour very well whereas the KAUST reproduces less than half of the CO formation throughout the reactor. At 900 K, where the oxidation enters the intermediate temperature regime, the CO formation stays at zero until 450 mm from the mixer, followed by a steady increase to ~ 4000 ppm at the end of the reactor. Both LLNL and KAUST mechanisms successfully reproduce the results whilst LLNL mechanism performs even better.

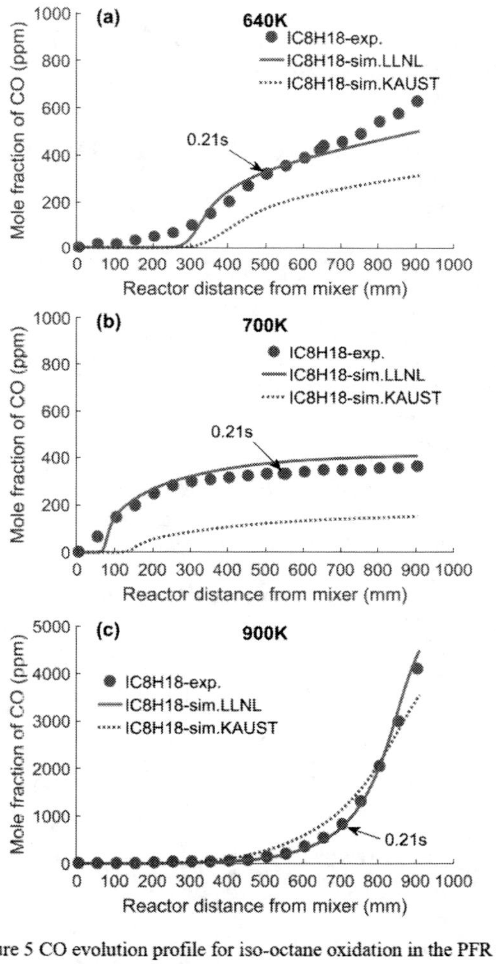

Figure 5 CO evolution profile for iso-octane oxidation in the PFR at 10 bar, ϕ = 0.058, 640 K (a), 700 K (b), and 900 K (c).

4 Conclusions

Low temperature oxidation of n-heptane and iso-octane was studied in a pressurised flow reactor over 550–900 K, 10 bar, and an equivalence ratio of 0.058. Despite the major difference in the oxidation reactivity, both n-heptane and iso-octane demonstrate evident low temperature reactivity and NTC behaviour over a similar temperature range at the resident time of 0.21 s. Consistent behaviours are also observed at 0.27 s which is not included in this work. Along the reactor, the CO evolution exhibits distinctively different behaviours as the oxidation enters different temperature regimes. Modelling of n-heptane oxidation using two published mechanisms shows that both mechanisms produce good agreement with the experiment. However, the NUI mechanism performs slightly better at lower temperatures whilst the LLNL mechanism gives closer agreement at higher temperatures. For iso-octane oxidation, the LLNL mechanism also performs better than the latest KAUST mechanism, and agrees excellently with experiment over the temperatures studied. This work fills a gap in the literature in validating these latest mechanisms for n-heptane and isooctane using PFR experiments.

Acknowledgments

This project is supported by the University of Melbourne, the Australian Research Council (DP140100846) and the Advanced Centre for Automotive Research and Testing (ACART, www.acart.com.au).

References

[1] F.L. Dryer, K. Brezinsky, Combust. Sci. Technol. 45 (1986) 199–212.
[2] J. Herzler, L. Jerig, P. Roth, Proc. Combust. Inst. 30 (2005) 1147–1153.
[3] D.F. Davidson, B.M. Gauthier, R.K. Hanson, Proc. Combust. Inst. 30 (2005) 1175–1182.
[4] R. Minetti, M. Carlier, M. Ribaucour, E. Therssen, L.R. Sochet, Combust. Flame 102 (1995) 298–309.
[5] S.M. Walton, X. He, B.T. Zigler, M.S. Wooldridge, A. Atreya, Combust. Flame 150 (2007) 246–262.
[6] O. Herbinet, B. Husson, Z. Serinyel, M. Cord, V. Warth, R. Fournet, P.-A. Glaude, B. Sirjean, F. Battin-Leclerc, Z. Wang, M. Xie, Z. Cheng, F. Qi, Combust. Flame 159 (2012) 3455–3471.
[7] P. Dagaut, M. Reuillon, M. Cathonnet, Combust. Sci. Technol. 103 (1994) 315–336.
[8] A.J. Smallbone, W. Liu, C.K. Law, X.Q. You, H. Wang, Proc. Combust. Inst. 32 (2009) 1245–1252.
[9] S.G. Davis, C.K. Law, Symp. Int. Combust. 27 (1998) 521–527.
[10] C.V. Callahan, T.J. Held, F.L. Dryer, R. Minetti, M. Ribaucour, L.R. Sochet, T. Faravelli, P. Gaffuri, E. Rani, Symp. Int. Combust. 26 (1996) 739–746.
[11] J.-S. Chen, T.A. Litzinger, H.J. Curran, Combust. Sci. Technol. 156 (2000) 49–79.
[12] K. Zhang, C. Banyon, J. Bugler, H.J. Curran, A. Rodriguez, O. Herbinet, F. Battin-Leclerc, C. B'Chir, K.A. Heufer, Combust. Flame 172 (2016) 116–135.
[13] N. Atef, G. Kukkadapu, S.Y. Mohamed, M.A. Rashidi, C. Banyon, M. Mehl, K.A. Heufer, E.F. Nasir, A. Alfazazi, A.K. Das, C.K. Westbrook, W.J. Pitz, T. Lu, A. Farooq, C.-J. Sung, H.J. Curran, S.M. Sarathy, Combust. Flame 178 (2017) 111–134.
[14] Z. Lu, J. Cochet, N. Leplat, Y. Yang, M.J. Brear, Meas. Sci. Technol, (2017), 28:105902.
[15] CHEMKIN-PRO, San Diego, 2010.
[16] M. Mehl, W.J. Pitz, C.K. Westbrook, H.J. Curran, Proc. Combust. Inst. 33 (2011) 193–200.
[17] M. Mehl, H.J. Curran, W.J. Pitz, C.K. Westbrook, in:, Eur. Combust. Meet., Vienna, Austra, 2009.
[18] H.J. Curran, P. Gaffuri, W.J. Pitz, C.K. Westbrook, Combust. Flame 114 (1998) 149–177.
[19] H.J. Curran, P. Gaffuri, W.J. Pitz, C.K. Westbrook, Combust. Flame 129 (2002) 253–280.

11th Asia-Pacific Conference on Combustion,
The University of Sydney, NSW Australia
10th -14th December 2017.

Oxidations of Ethanol and Gasoline Reference Fuels in a Pressurized Flow Reactor

Hao Yuan, Zhewen Lu, Zhongyuan Chen, Yi Yang, Michael J. Brear

Department of Mechanical Engineering, University of Melbourne, Parkville, Victoria, 3010, Australia

Abstract

Ethanol has been commonly added into gasoline for its excellent anti-knock performance in spark ignition engines. However, the chemical interactions between ethanol and hydrocarbon compounds are not fully understood. The paper reports preliminary results on the oxidation chemistry of ethanol and gasoline reference fuels in a recently commissioned turbulent pressurized flow reactor (PFR) at the University of Melbourne. The measured species profiles from the PFR are then compared to equivalent simulations with several kinetic mechanisms from the literature. Good agreement between simulation and experiment is observed for neat isooctane and ethanol, but not for toluene. Furthermore, agreement with experiment becomes increasingly poor as the reference fuel contains more toluene. This indicates that further investigation into toluene combustion and its potential interactions with other compounds is required.

1 Introduction

Ethanol is commonly used as a gasoline blending component around the world. In addition to being a renewable fuel, ethanol is attractive for its low autoignition reactivity that yields a high octane number and superior anti-knock performance in spark ignition engine combustion. Octane number measurements have been conducted for mixtures of ethanol with commercial gasoline [1, 2], primary reference fuels (PRFs, mixture of isooctane and n-heptane) [3-6], and toluene reference fuels (TRFs, mixture of toluene and PRFs) [3-6]. Significant non-linear blending behaviors have been reported, and both penalty and bonus in octane rating can be resulted depending on the base gasoline's hydrocarbon composition. Chemical interactions are therefore suggested between ethanol and hydrocarbons during autoignition processes. However, few studies have been conducted to clarify the underlying chemistry.

This paper reports preliminary results on autoignition chemistries of ethanol and hydrocarbon mixtures, using a recently built turbulent pressurized flow reactor (PFR) at the University of Melbourne. Combustion chemistry of ethanol, PRFs and TRFs, are studied via detailed species concentration measurements in the PFR. Existing mechanisms for neat compounds and simple mixtures are used to kinetically model the PFR results.

Combustion chemistry of representative hydrocarbons for gasoline, such as isooctane, n-heptane and toluene, has been studied by many groups in various fundamental combustion apparatuses. The most cited reaction mechanisms for isooctane and n-heptane are from Curran et al. [7, 8], which have been updated recently by Atef et al. [9] and Zhang et al. [11], respectively, with significant amount of new experimental data. However, the PFR experiment results were missing in validating the latest mechanism. Combustion mechanism for toluene has also been extensively studied, although a consensus similar to that with PRFs has not been reached. The latest developments include those by Metcalfe et al. [12] and Yuan et al. [13]. Apart from hydrocarbons, comprehensive mechanisms of ethanol oxidation has been developed by Marinov [14], and recently updated by Mittal et al. [15]. In contrast to the progress in understanding neat compounds, kinetic models for fuel mixtures are relatively sparse, among which the gasoline surrogate mechanisms proposed by Mehl et al. [10], and Andrae [16] are the most widely used.

2 Experimental and Numerical Methods

2.1 Experimental Method

A schematic of the PFR facility used in this work is shown in Fig.1. The air is supplied by an oil-free compressor and enters the flow reactor after being heated to a specified temperature. Fuel is supplied from a tank pressurized by high pressure nitrogen and is vaporized by hot nitrogen in a cross flow. Air and fuel flow rates are controlled to obtain a fixed equivalence ratio, ~0.058, for all fuels tested. The vaporized fuel and hot air mix rapidly in a novelly designed mixer at the inlet of reactor. Fast mixing is confirmed by injecting carbon dioxide at 900 K and 10 bar which indicated that complete mixing is achieved at approximately 50 mm from the mixer [18]. The reactor is a 25 mm i.d., 1000 mm long, quartz tube housed in a pressure vessel with a design pressure of 50 bar at 1000 K. The reactor pressure is controlled by an automatic back pressure valve on the exhaust line.

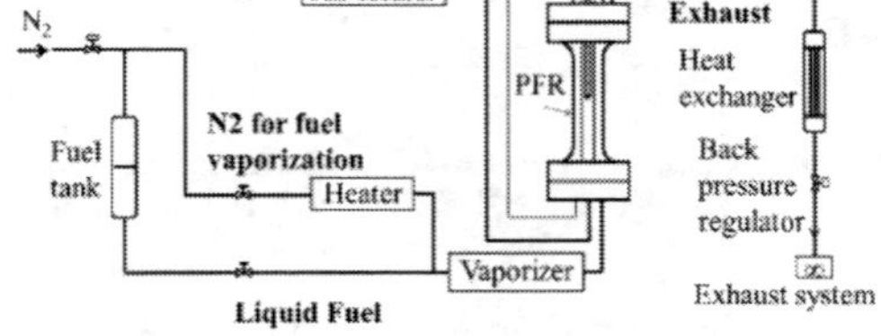

Figure 1: Schematic of the PFR system

A "hot water" cooled sampling probe is installed on a linear actuator which allows gas collection along the centerline of the reactor. The sampling probe is also equipped with three different sized K-type thermocouples having uncertainties of ±0.25% for gas temperature measurement accounting for thermocouples radiative losses. The sampled gas is transferred

Corresponding author. Tel: +61-03-83448104
E-mail address: yi.yang@unimelb.edu.au

via a heated line (at 473K) to a Horiba emission bench where CO is measured by a non-dispersive infrared (NDIR) analyzer with a resolution of 20 ppm. A novel three-thermocouple method is used for gas temperature measurement, which uses three K-type thermocouples of different junction sizes (d_1, d_2, and d_3) installed axisymmetrically on the sample probe. The reaction gas temperatures are derived from the three measured temperatures (T_1, T_2, and T_3) and the junction size ratios without assuming the wall emissivities and convection heat transfer coefficients, and the equation is given by [18]:

$$T_{gas} = \frac{T_1 - \left(\frac{d_1}{d_2}\right)^{2/5} T_2 - \left[T_1 - \left(\frac{d_1}{d_3}\right)^{2/5} T_3\right]\left(\frac{T_1^4 - T_2^4}{T_1^4 - T_3^4}\right)}{1 - \left(\frac{d_1}{d_2}\right)^{2/5} - \left[1 - \left(\frac{d_1}{d_3}\right)^{2/5}\right]\left(\frac{T_1^4 - T_2^4}{T_1^4 - T_3^4}\right)} \quad (1)$$

Using this method, the gas temperature profile along the reactor centerline is obtained at each test condition. Details of the temperature correction can be found in [18], where the uncertainty of the corrected temperature is estimated to be ±5 K at 900 K (smaller at lower temperatures).

The PFR operating conditions in this work are listed in Table 1. The test fuels and reaction mechanisms for kinetic modeling are shown in Table 2. All fuels are tested at 900K except for neat toluene which was tested at 930K due to its low reactivity.

Table 1: Experimental conditions for the PFR study

Reactor pressure (bar)	10
Reactor nominal temperature	900-930
Equivalence ratio	0.058-0.060
Air flow rate (g/s)	6
Nitrogen flow rate (g/s)	0.32
Reynolds number in PFR	8000
Fuel/nitrogen pressure (bar)	20-21
Fuel/nitrogen temperature (K)	500

Table 2: Test fuels and reaction mechanisms for modeling

Fuel	Mechanism	Composition
Isooctane	Atef et al. [9]	isooctane
	Mehl et al. [10]	gasoline surrogate
	Andrae [16]	gasoline surrogate
Ethanol	Marinov [14]	ethanol
	Mittal et al. [15]	ethanol
	Mehl et al. [10]	gasoline surrogate
	Andrae [16]	gasoline surrogate
Toluene	Metcalfe et al. [12]	toluene
	Yuan et al. [13]	toluene
	Mehl et al. [10]	gasoline surrogate
	Andrae [16]	gasoline surrogate
PRF91 TRF91-15	Mehl et al. [10]	gasoline surrogate
TRF91-30 TRF91-45	Andrae [16]	gasoline surrogate

PRF91: 91% isooctane, 9% n-heptane
TRF91-15: 72.6% isooctane, 12.4% n-heptane and 15.0% toluene
TRF91-30: 53.2% isooctane, 17.0% n-heptane and 29.8% toluene
TRF91-45: 34.7% isooctane, 20.3% n-heptane and 45.0% toluene
All composition by volume [3].

2.2 Kinetic Modeling Method

Simulations in this study are carried out using the Plug Flow Reactor model in Chemkin Pro [17]. The temperature profiles corrected with the three-thermocouple method were inputs to the model to solve the energy equation, and the start of reaction was set at the mixer plane. The measured CO and the fuel mole fractions are modelled with the state of the art chemical mechanisms as listed in Table 2. Except for the mechanisms by Mehl et al. [10] and Andrae [16] which were developed for multi-component gasoline surrogates, all mechanisms are for neat compounds.

3 Results and Discussions

3.1 Neat Fuels

As one of the primary reference fuels for gasoline, isooctane is first tested in this study. The measured CO and isooctane profiles are plotted in Fig. 2 as a function of the flow distance. The CO formation remains at zero until 400 mm whereas isooctane starts to react at ~ 200 mm. These results are modelled using the latest isooctane mechanism from Atef et al. [9] and two gasoline surrogate mechanims from Mehl et al. [10] and Andrae [16], the results of which are also shown in Fig. 2. Among these mechanisms, the one from Mehl et al. [10] best reproduces both the CO and fuel profiles, where Andrae mechanisms [16] slightly underpredicts, and Atef et al.'s mechanism [9] overpredicts the isooctane oxidation reacitivity at the given PFR condition.

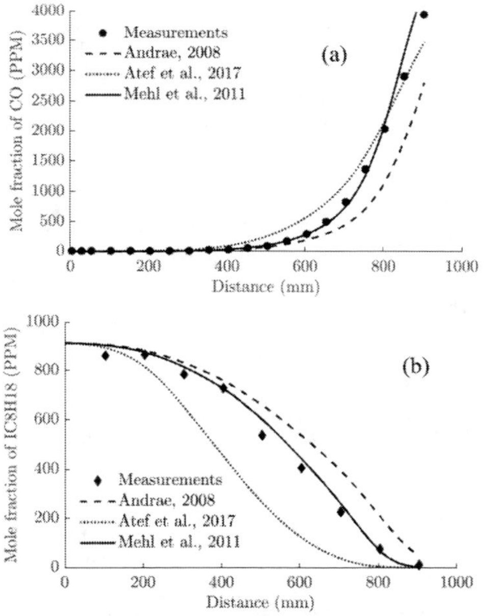

Figure 2: The measurements and modelling results of (a) CO and (b) isooctane from the neat isooctane oxidation experiment at 900K and 10bar

The CO and fuel profiles for ethanol are shown in Fig. 3, together with the modelling results with three mechanisms. Marinov mechanism [14] is one of the earliest comprehensive mechanisms for ethanol, which under-predicts the ethanol's oxidation reactivity at this condition. Three more recent mechanisms from Mehl et al. [10], Andrae [16], and Mittal et al. [15] perform similarly, which produce excellent agreement for CO and ethanol except for CO near the end of reactor.

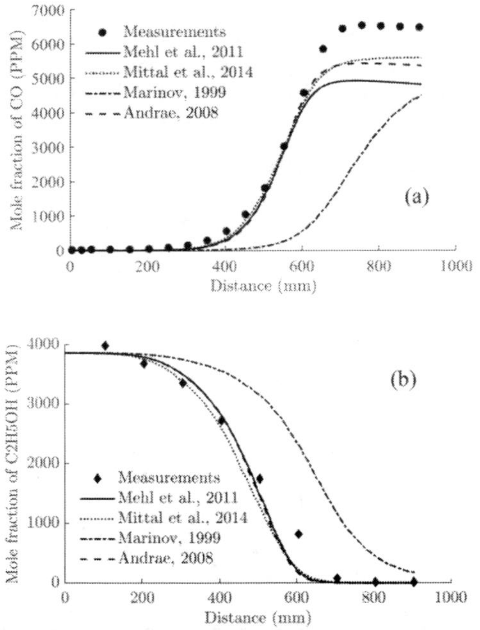

Figure 3: The measurements and modelling results of (a) CO and (b) ethanol from the neat ethanol oxidation at 900K and 10bar

The agreement suggests that the models accurately capture the chemistry where the fuel is converted to CO. The discrepancy in the CO plateau value is significant and the reason causing it is unclear. Similar discrepancies have been observed with other fuels when CO reaches the plateau and CO2 starts to rise (not shown here). One speculation is that the kinetics of CO to CO2 conversion may need to be reevaluated for the ultra-lean oxidation conditions used here.

Comparing with isooctane and ethanol, toluene has much lower reactivity, and its experiment was conducted at a higher temperature, 930K. Even at this temperature, the measured CO concentrations are significantly lower than those of isooctane and ethanol, as shown in Fig.4.

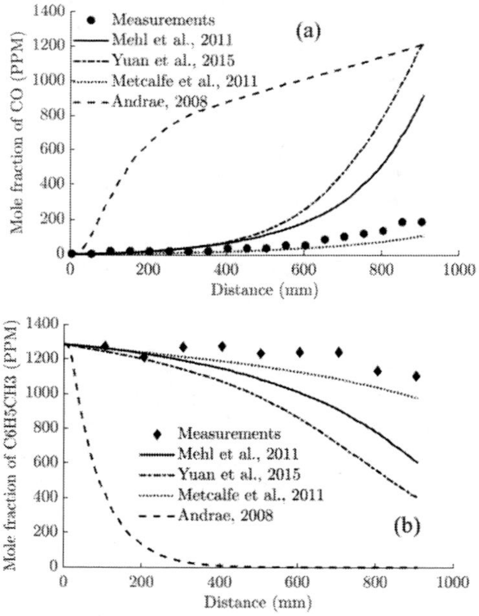

Figure 4: The measurements and modelling results of (a) CO and (b) toluene from toluene oxidation at 930K and 10bar.

3.2 Fuel Mixtures

Practical fuels are all mixtures. Nevertheless, the mechanism of a mixture is not the simple addition of individual mechanisms, because cross reactions can take place between fuel components [19]. Four gasoline reference fuels were formulated here, including PRF91, and TRF91 with 15%, 30%, and 45% toluene by volume, to examine the existing gasoline surrogate mechanisms. The RONs of these gasoline surrogates are all ~ 91 [3], similar to the regular grade gasoline in Australian market.

The CO profile from PRF91 experiment and its modelling results with two gasoline surrogate mechanisms are compared in Fig. 5. It is clear that both mechanisms predict the measured CO profile pretty well, and the mechanism from Mehl et al. [10] is more reactive than Andrae's [16], which is consistent with the modeling results with neat isooctane in Fig. 2.

Apart from PRF91, the TRF-based fuels were formulated with the different amount of toluene content. With 15% toluene added, the simulation from the two gasoline surrogate mechanisms become worse compared to the results of PRF91, as shown in Fig. 6, which is probably due to their problematic toluene sub-mechanisms. As the toluene concentration increases to 30%, both mechanisms over-predict the measurements even further, in particular the Andrae's mechanism [16], as shown in Fig. 7. This over-prediction should not be surprising given that the Andrae's mechanism [16] performs the worst for neat toluene oxidation (Fig. 4). With the further increased toluene content, the issues with the toluene sub-mechanisms are amplified, represented by the largest differences between the measured CO profile and the model prediction, as shown in Fig. 8. The increased discrepancies shown from Fig.5 to 8 indicated that the two existing gasoline surrogate mechanisms have difficulties to predict the reactivity of toluene containing fuel mixtures.

Although the mechanism from Metcalfe et al. [12] performs significantly better than the two gasoline surrogate mechanisms in terms of modeling the reactivity for neat toluene, it has not been incorporated in a gasoline surrogate mechanism. It is subject to further study to determine how this neat toluene mechanism will perform when used for modeling the reactivity of mixtures.

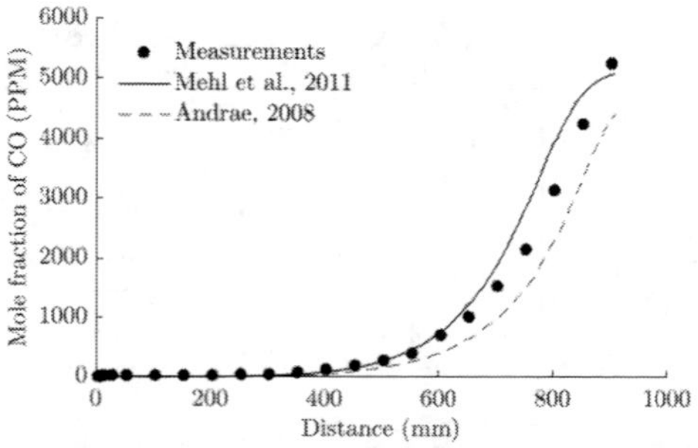

Figure 5: The measurements and modelling results of CO from the PRF91 oxidation at 900K and 10bar

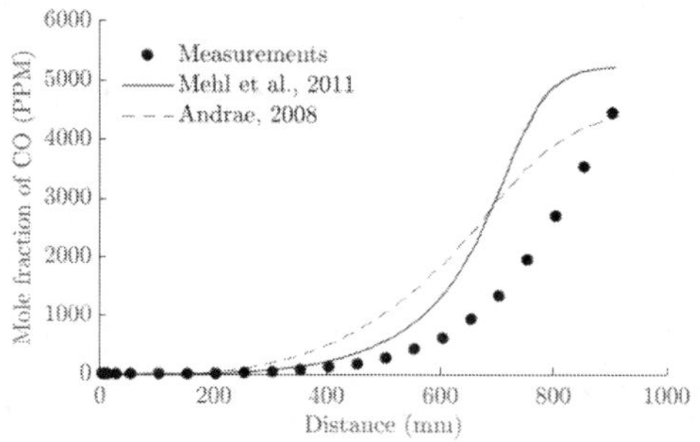

Figure 6: The measurements and modelling results of CO from the TRF91-15 oxidation at 900K and 10bar

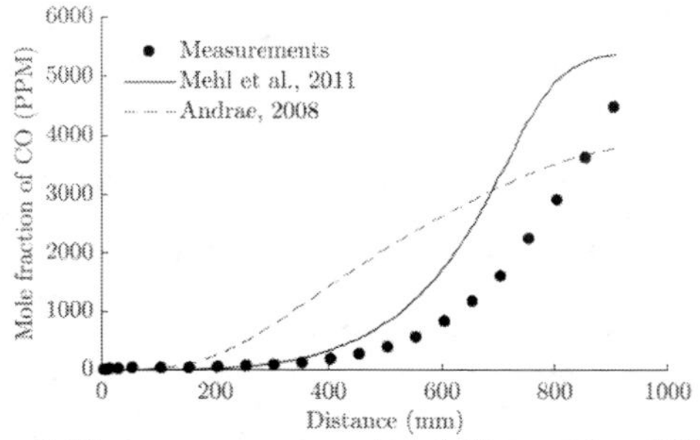

Figure 7: The measurements and modelling results of CO from the TRF91-30 oxidation at 900K and 10bar

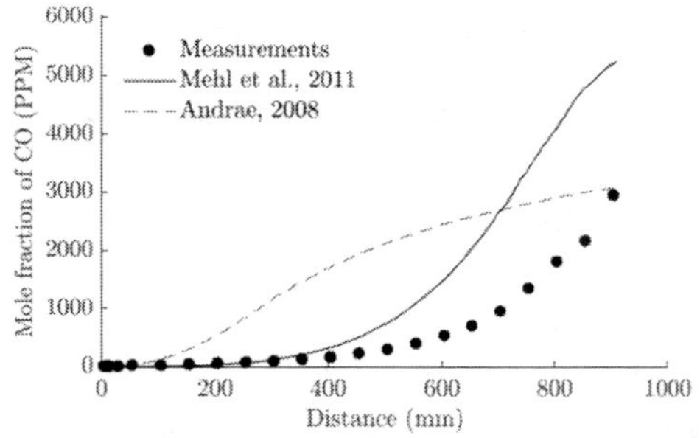

Figure 8: The measurements and modelling results of CO from the TRF91-45 oxidation at 900K and 10bar

Since the toluene mechanism appears problematic based on the experimental and modelling results presented in this study, simulation of ethanol and TRFs mixtures also resulted in poor agreement with the experiment, particularly for mixtures of high toluene concentrations (not reported here). Global sensitivity analysis and reaction pathway analysis found that in the better performed Mehl et al.'s model [10], the rate constants of several high-sensitivity reactions containing toluene or toluene-like radicals are considerably different from the latest experimental and computational values, which likely resulted in the discrepancies reported in this work. The future work is to develop a new gasoline surrogate mechanism which contains a validated toluene sub-mechanism and is able to model not only the oxidation of hydrocabron mixtures, but also that of ethanol containing fuels.

4 Conclusions

This study examined the oxidation of ethanol and gasoline surrogates in a turbulent pressurized flow reactor. The CO and fuel molar fractions are measured along the reactor length. Most of the experimental results are reproduced by the existing kinetic models, particularly those containing mostly isooctane and ethanol. Significant discrepancies between experiment and modelling is however observed for neat toluene and mixtures containing toluene. This indicates that further investigation is required on toluene oxidation and its potential interaction with other compounds.

5 Acknowledgments

This research was generously supported by the Advanced Centre for Automotive Research and Testing (acart.com.au) and the Australian Research Council.

References

[1] J. E. Anderson, T. G. Leone, M. H. Shelby et al., SAE Technical Paper, 2012-01-1274, 2012.

[2] J. E. Anderson, U. Kramer, S. A. Mueller et al., Energy & Fuels, 24(12):6576–6585.

[3] T. Foong, K. J. Morganti, M. J. Brear et al., Fuel, 115:727–739, 2014.

[4] H. Solaka, M. Tuner, B. Johansson, SAE Technical Paper, 2013-01-0903, 2013.

[5] I. Truedsson, M. Tuner, B. Johansson et al., SAE Technical Paper, 2013-01-1669, 2013.

[6] S. Aronsson, M. Tuner, and B. Johansson, SAE Technical Paper, 2014-01-1303, 2014.

[7] H. J. Curran, P. Gaffuri, W. J. Pitz et al., Combust. Flame, 129(3):253–280, 2002.

[8] H. J. Curran, P. Gaffuri, W. J. Pitz et al., Combust. Flame, 114(1):149–177, 1998.

[9] N. Atef, G. Kukkadapu, S. Y. Mohamed et al., Combust. Flame, 178:111–134, April 2017.

[10] M. Mehl, W. J. Pitz, C. K. Westbrook et al., Proc. Combust. Inst. 33(1):193–200, 2011.

[11] K. Zhang, C. Banyon, J. Bugler et al., Combust. Flame, 172:116–135, 2016.

[12] W. K. Metcalfe, S. Dooley, and F. L. Dryer, Energy & Fuels, 25(11):4915–4936, 2011.

[13] W. Yuan, Y. Li, P. Dagaut et al., Combust. Flame, 162(1):3–21, 2015.

[14] N. M. Marinov, Int. J. Chem. Kinet., 31(3):183–220, 1999.

[15] G. Mittal, S. M. Burke, V. A. Davies et al., Combust. Flame, 161(5):1164–1171, 2014.

[16] J. C. G. Andrae, Fuel, 87(10–11):2013–2022, 2008.

[17] CHEMKIN-PRO 15131, Reaction Design: San Diego, 2013

[18] Z. Lu, J. Cochet, N. Leplat et al., Meas. Sci. Technol. 28: 105902, 2017

[19] J.C.G. Andrae, P. Björnbom, R. Cracknell et al., Combust. Flame, 149(1-2):2-24, 2007

11th Asia-Pacific Conference on Combustion,
The University of Sydney, NSW Australia
10th -14th December 2017.

Kinetic Modeling of Engine Combustion: an Uncertainty Analysis

Song Cheng, Yi Yang, Michael J. Brear

Department of Mechanical Engineering, the University of Melbourne,
Parkville, Victoria, 3010, Australia

Abstract

Prediction of engine combustion to some specified degree of accuracy requires kinetic models with sufficiently certain parameters. It is an open question as to whether a kinetic model that has been extensively validated by high-quality kinetic experiments is capable of reproducing engine combustion phenomenon to within typical levels of measurement uncertainty. This paper therefore introduces a method that can be used to evaluate the uncertainty in simulating engine combustion processes using the uncertainties that are embedded in the chemical mechanism. The method is based on a deterministic framework of uncertainty quantification, and is termed *bound-to-bound data collaboration*. This method can be used to quantify the engine modeling uncertainty propagated from the uncertainties in the kinetic mechanisms, and also allows examination of the impact of engine operating conditions on engine modeling uncertainty.

1 Introduction

Numerical modeling of practical combustion processes requires detailed kinetic mechanisms as input. The development of such a detailed kinetic mechanism often involves aggregating a large number of chemical species and elementary reactions. The rate constants of the individual elementary reactions, as determined from fundamental experiments, theoretical calculations, or estimations from analogous reactions, are often of significant uncertainties. The overall uncertainty of the mechanism accumulated from that of individual reactions can be notable. Take the following high temperature chain branch reaction as an example,

$H + O_2 \rightarrow O + OH \ (R1)$

an uncertainty of 10-15% in this rate constant, k_1, will result in an overall uncertainty of 4-6% in the laminar flame speed simulation for ethylene [1].

When many elementary reactions are put together to form a combustion mechanism, the uncertainty of the mechanism in modeling physical processes may not be linearly propagated from that of individual elementary reactions. This considers that reaction kinetics is highly non-linear and for some cases could even be non-monotonic. Quantification of such uncertainty propagation is therefore essential for application of these detailed mechanisms [2].

Further, detailed combustion mechanisms are generally validated with kinetic experiment data, such as flame speeds, ignition delay times (IDT) and species profiles, all the experiments have their own uncertainty in measurements, therefore, it is unclear whether the overall uncertainty of a validated mechanism contents the degree of confidence and reliability required for practical applications such as engine combustions [3-5]. Very recently Cheng et al. [6] modelled the knock onset timing of pentane isomers in a spark-ignition (SI) engine using some well-validated mechanisms, and further implied that the deficiency in the accuracy of Rapid Compression Machines (RCM) experiments may not allow the mechanisms to model the knock onset timing within the engine operating tolerance.

Following this thought, this paper introduces a method that investigates the uncertainty propagation from the rate constants of elementary reactions to kinetic modeling of engine combustion. The estimation includes the uncertainty in the kinetic experiment measurements as well as in determining the engine in-cylinder conditions that affect the combustion kinetics. First, the method will be demonstrated with a simplified problem. Then, application of the method to simulating the knock onset timing in an engine is then discussed, which includes more practical considerations than the simple example to handle the detailed reaction mechanism and the impacts of engine in-cylinder conditions.

2 Method for Uncertainty Quantification

The method used in this work is the so called bound-to-bound (B2B) data collaboration method [3, 4, 7, 8], which is one of the most commonly used approaches for deterministic uncertainty quantification (UQ) of combustion mechanisms. An outline on how this method works is shown in Fig. 1. Each step is detailed in the following. A virtual mechanism containing

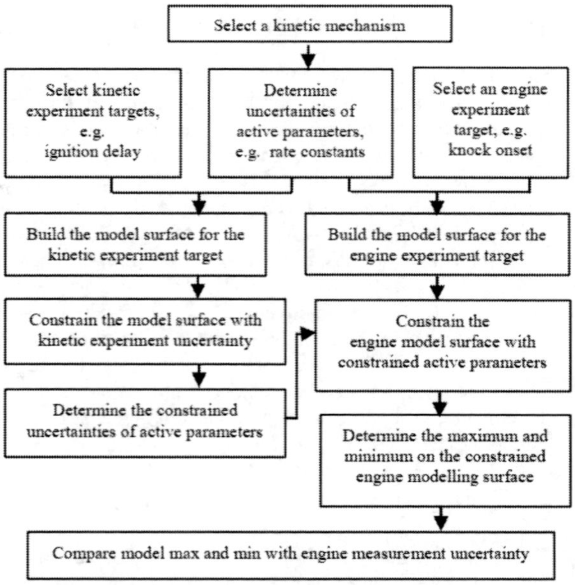

Figure 1. The flow diagram of the proposed method.

Corresponding author. Phone: (+61) 3 8344 8104
E-mail address: yi.yang@unimelb.edu.au

two elementary reactions is used as an example.

2.1 Select a mechanism and determine the uncertainty

A reaction mechanism is selected as the starting point of the uncertainty analysis for kinetic modeling. This mechanism should be the best validated mechanism for the studied fuel. The purpose is to allow one to determine if the best existing mechanism can give satisfactory performance in engine combustion modeling. Also, doing so ensures that the uncertainties of the individual elementary reaction rate constants can be reliably rendered from literature.

Here for the example mechanism consisting of two elementary reactions, we consider that the model uncertainty originates from the rate constants of the two reactions, k_1, k_2, and their corresponding uncertainties, $[k_{1,min}, k_{1,max}]$, $[k_{2,min}, k_{2,max}]$, which form an 2-dimensional "hypercube" denoted as:

$$H = \{x \in R^2 : k_{j,min} \leq k_j \leq k_{j,max}, j = 1,2\}$$

Figure 2 illustrates the 2-dimensional hypercube H_1. Each dimension of the hypercube corresponds to the uncertainty of one elementary reaction rate constant. Assuming that this mechanism is well validated and the uncertainty of each reaction rate constant is accurately determined. Although the recommended mechanism might be located at the intersection of the two recommended k value lines, i.e. the black dot in Fig. 2, any point inside this hypercube still represents a physical allowable mechanism within the uncertainty of this mechanism.

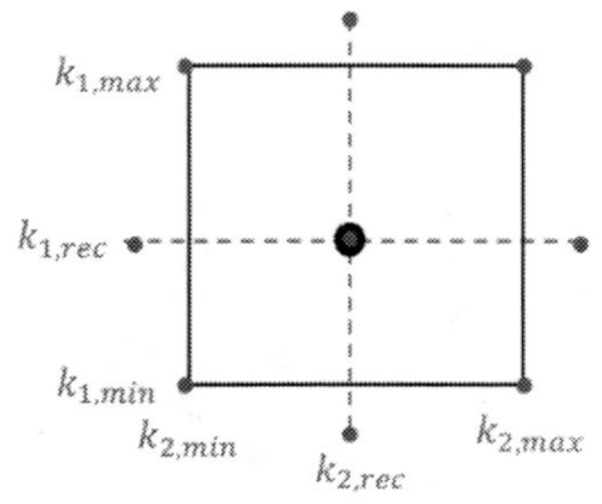

Figure 2. The 2-D "hypercube" H_1 formed by the two rate constants of an example mechanism consisting of two elementary reactions. $k_{1,rec}$ and $k_{2,rec}$ are the recommended rate constants from literature, $[k_{1,min}, k_{1,max}]$ and $[k_{2,min}, k_{2,max}]$ are their corresponding uncertainties.

Therefore, if one intends to simulate a kinetic process, using any allowable mechanism in the hypercube H_1 is physically legitimate. The largest difference in the modeling results using these allowable mechanisms is thus considered as the uncertainty of the model.

2.2 Select a kinetic experiment target and build the model surface

The next step is to determine the uncertainty of the model which is propagated from uncertainties in elementary reactions. To do this, kinetic experiment targets, such as ignition delay time (IDT), need to be selected for modeling. For the sample problem, an IDT from a high-quality RCM experiment, y_1, is selected from literature. The reported measurement uncertainty, $[y_{1,lower}, y_{1,upper}]$, is also obtained. As aforementioned, if the simulated IDTs using all the allowable mechanisms in H_1 are obtained, the uncertainty of the IDT modeling can be determined by evaluating the maximum and minimum among the simulated IDTs. These simulated IDTs are obtained by running a number of RCM simulations, termed computer experiments, which can be realized using software such as CHEMKIN [9]. Given that it is impossible to perform such simulations on every possible point inside the hypercube, discrete points are selected to run the simulation to cover the targeted model surface. Techniques such as Central Composite Design [10, 11], shown as the red dots in Fig. 2, are used to assist the selection.

Figure 3 depicts the model surface created from this method. The plane formed by k_1 and k_2 is the 2-D hypercube illustrated in Fig. 2. The blue dots above the k_1k_2 plane represent the CHEMKIN simulated IDTs using the 2-reaction mechanism at the selected (k_1 ,k_2) values in Fig. 2 (red dots). To create a continuous IDT surface, interpolation between the discrete design points is conducted using fitting techniques. This fitted surface is called the model surface, $\varphi_1(x)$, as shown in Fig. 3, for the selected RCM experiment on H_1.

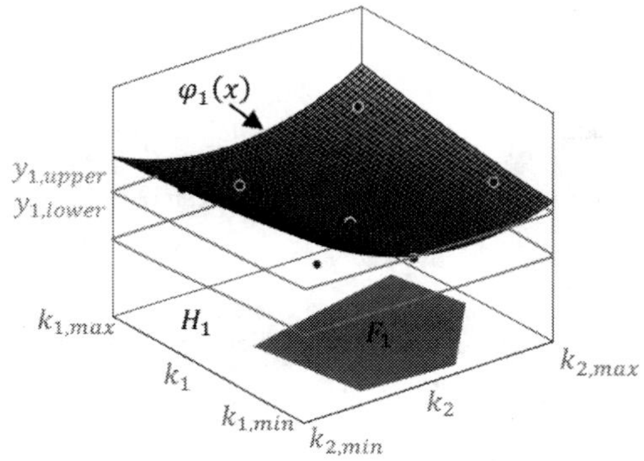

Figure 3. The model surface $\varphi_1(x)$ of the selected RCM experiment on the hypercube H_1 illustrated in Fig. 2.

2.3 Constrain the model uncertainty with kinetic experiment's uncertainty.

After building the fitted response surfaces, the uncertainty of the model can be reduced by constraining the model surface with experimental uncertainty. This is achieved by satisfying an inequality constraint $y_{1,lower} \leq \varphi_1(x) \leq y_{1,upper}$, where $x \in H_1$. Doing so, the model surface $\varphi_1(x)$ is trimmed by the planes of $y = y_{1,lower}$ and $y = y_{1,upper}$, as shown in Fig. 3. The projected area of the trimmed model surface on H_1 is called the feasible set F_1, denoted in red in Fig. 3. Obviously, F_1 is smaller than the original hypercube H_1, indicating that the prior uncertainties of k_1 and k_2 are curtailed by the constraint from experimental uncertainties.

2.4 Propagating the model uncertainty to engine modeling uncertainty

To propagate the uncertainty of a kinetic mechanism to the uncertainty of engine modeling, engine experiment targets need to be selected for modeling, e.g. knock onset timing. For the example problem, the model surface $\varphi_2(x)$ for an engine's knock onset is developed, as shown in Fig. 4.

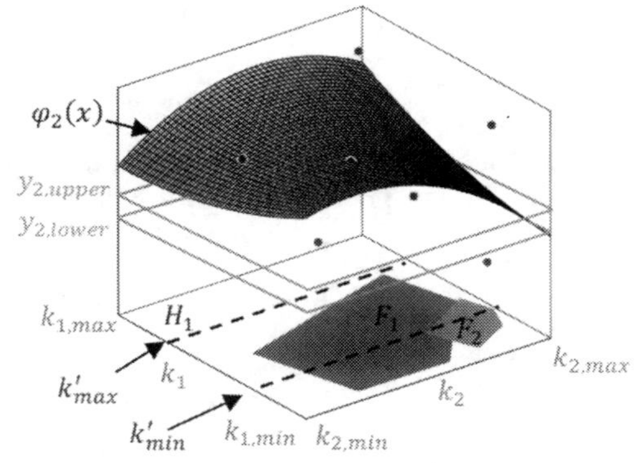

Figure 4. The model surface $\varphi_2(x)$ of an engine's knock onset on the hypercube H_1 illustrated in Fig. 2. F_2, denoted in green, is the feasible set obtained by constraining the model surface $\varphi_2(x)$ with the engine knock onset measurement uncertainty $[y_{2,lower}, y_{2,upper}]$.

Similar to $\varphi_1(x)$, $\varphi_2(x)$ is fitted from the blue dots which are obtained by running the engine simulations at the designed points on the hypercube H_1(red dots in Fig. 2).

The uncertainty of the engine knock onset modeling can then be determined with different constraints. At the base level, the engine modeling uncertainty, σ_2 , is propagated from the uncertainty in the rate constants, which can be determined by:

$$\sigma_2 = [\min(\varphi_2(x)), \max(\varphi_2(x))], x \in H_1$$

Further, the engine modeling uncertainty is constrained by the RCM measurement uncertainty, which can be determined by:

$$\sigma_2' = [\min(\varphi_2(x)), \max(\varphi_2(x))], x \in F_1$$

σ_2' can be subsequently compared with the engine operating uncertainty, e.g. cycle to cycle variations, to determine the degree of reliability of the kinetic mechanism for engine knock onset modeling.

3 Application to Practical Cases

3.1 Handling large kinetic mechanisms

A detailed combustion mechanism of practical fuels often involves thousands of elementary reactions. Application of the above method that considers the uncertainty of each elementary reaction is general impractical and unnecessary. The number of model parameters should be reduced without ill-conditioning the model responses so that the computational cost can be reduced. Fortunately, it has been proved that the model response often does not depend on all the rate parameters at a given condition, but only on a small group of parameters, called active parameters. These active parameters can be obtained through sensitivity analysis of different kinetic experiment targets, e.g. IDT, species formation, flame speeds, etc.

Once the active parameters are decided, model surfaces can then be developed on the high dimensional (> 2-D) hypercube generated with the active parameters, for multiple experiment targets. Again, construction of the model surfaces is based on the simulation at selected design points and surface fitting of these simulated results. Despite the fact that these model surfaces are highly distorted and difficult to visualize, overfitting can be avoided by increasing design points and validated by cross validation.

When a large kinetic mechanism is used, computational tools is required to implement the computer experiments. For kinetic experiment targets, CHEMKIN PRO [9] has been used to run the computer experiments for various flame speeds, ignition delay times, and CO formation in a Jet Stirred Reactor. For engine experiment, multi-dimensional engine models are not used for this problem for the prohibitive computational cost with detailed chemistry. Instead, a quasi-dimensional engine combustion model from Foong et al. [6, 12, 13] is used for the computer experiments to model the knock onset timing. This model incorporates the generic two-zone model concept for SI engine combustion [14] and considers detailed fuel combustion chemistry. The program is implemented in MATLAB with Cantera [15] embedded as kinetics solver. The actual engine experiment is conducted in a cooperative fuel research (CFR) engine, which is a standard engine for gasoline octane number measurements and has been carefully calibrated for the in-cylinder conditions [16].

After the computer experiments completed at the selected design points, the fitting for the model surface is also more complicated when more model parameters are taken into account. Higher order fitting typically gives smaller fitting error but also increases computational intensity on post-processing as well as the chance of overfitting. Frenklach et al. [3, 4] has proposed that quadratic fitting of the response surfaces, as below, is sufficient for fitting fidelity.

$$\varphi_i(x) = c_0 + \sum_i c_i x_i + \sum_{i \le j} c_{i,j} x_i x_j$$

Therefore, quadratic equations are used in this work to fit all the model surfaces.

3.2 Including engine in-cylinder conditions as model parameter

Engine combustion has more complicated boundary and initial conditions than the kinetic experiments. These conditions affect fuel autoignition kinetics and thus need to be considered in the knock onset modeling. In this work, three in-cylinder conditions are considered as active parameters in constructing the model surfaces of knock onset timing, including in-cylinder heat transfer coefficient multiplier, temperature at the intake valve closure (T_{IVC}), and in-cylinder concentrations of nitric oxide (NO) from residual gas. Each of the three conditions is investigated over a range of values that correspond to the prescribed uncertainties in their determination.

3.3 Further reducing engine modeling uncertainty

The uncertainty of the engine modeling can be further reduced with more constrained active parameter uncertainties. These further constraints can be obtained by using more experimental data (either kinetic experiments or engine experiments), or reducing the uncertainties of active parameters and/or experimental measurements.

Take the Fig. 3 and 4 as an example. If more experiment targets are used, the active parameters will be more constrained in their uncertainty, resulting in smaller feasible set F_1. In fact, when the engine experiment is included, the feasible set becomes $F_1 \cap F_2$, which is clearly more constrained than the original F_1. Similarly, if the uncertainty of active parameters is reduced, for example, k_1 reduced to $[k'_{min}, k'_{max}]$ (Fig. 4) with more accurate theoretical calculations, the feasible set will be further constrained and thus model engine knock with reduced uncertainties.

4 Conclusion

This paper introduced a method that evaluated the uncertainty in simulating engine combustion processes using the uncertainties that are embedded in the chemical mechanism of the fuel combustion. The method was based on a deterministic framework of uncertainty quantification, and was termed bound-to-bound data collaboration. This method quantifies the engine modeling uncertainty propagated from the uncertainties in the kinetic mechanisms, and also allows examination of the impact of engine operating conditions on engine modeling uncertainty. Results using this method to examine engine knock in a CFR engine will be published in the future.

5 Acknowledgement

This research was supported by the Advanced Centre for Automotive Research and Testing (ACART, www.acart.com.au) and the Australian Research Council.

References

[1] D. A. Sheen, H. Wang, Combust. Flame 158 (12) (2011) 2358-2374.

[2] M. Frenklach, A. Packard, G. Garcia-Donato, R. Paulo, J. Sacks, SIAM/ASA J. Uncertain. Quantif. 4 (1) (2016) 875-901.

[3] M. Frenklach, H. Wang, M. J. Rabinowitz, Prog. Energ. Combust. 18 (1) (1992) 47-73.

[4] X. You, T. Russi, A. Packard, M. Frenklach, Proc. Combust. Inst. 33 (1) (2011) 509-516.

[5] D. A. Sheen, X. You, H. Wang, T. Løvås, Proc. Combust. Inst. 32 (1) (2009) 535-542.

[6] S. Cheng, Y. Yang, M. J. Brear, D. Kang, S. Bohac, A. L. Boehman, Proc. Combust. Inst. 36 (3) (2017) 3499-3506.

[7] R. Feeley, M. Frenklach, M. Onsum, T. Russi, A. Arkin, A. Packard, J. Phys. Chem. A 110 (21) (2006) 6803-6813.

[8] R. Feeley, P. Seiler, A. Packard, M. Frenklach, J. Phys. Chem. A 108 (44) (2004) 9573-9583.

[9] CHEMKIN-PRO 15131, Reaction Design: San Diego, 2013.

[10] G. E. Box, N. R. Draper, Empirical model-building and response surfaces, Wiley New York, 1987.

[11] G. E. Box, W. G. Hunter, J. S. Hunter, Statistics for experimenters: an introduction to design, data analysis, and model building, JSTOR, 1978.

[12] H. Yuan, T. M. Foong, Z. Chen, Y. Yang, M. Brear, T. Leone, J. E. Anderson, Modeling of Trace Knock in a Modern SI Engine Fuelled by Ethanol/Gasoline Blends, SAE Technical Paper 2015-04-14, 2015.

[13] T. M. Foong, M. J. Brear, K. J. Morganti, G. da Silva, Y. Yang, F. L. Dryer, Energy & Fuels 31 (3) (2017) 2378-2389.

[14] J. B. Heywood, Internal combustion engine fundamentals, Mcgraw-hill New York, 1988.

[15] D. Goodwin, An open-source, extensible software suite for CVD process simulation, Division of Engineering and Applied Science, California Institute of Technology Report No. 40, 2003.

[16] T. M. Foong. On the autoignition of ethanol/gasoline blends in spark-ignition engines. PhD Thesis, The University of Melbourne, Melbourne, Australia, 2013.

11th Asia-Pacific Conference on Combustion,
The University of Sydney, NSW Australia
10th -14th December 2017.

A Comparison of Complex Chemistry Mechanisms for Hydrogen Methane Blends Based on the Sandia / Sydney Bluff-Body Flame HM1

H. H.-W. Funke[1], N. Beckmann[1], S. Abanteriba[2]

[1]Aachen University of Applied Sciences

Hohenstaufenallee 6, Aachen, 52064, Germany

[2]Royal Melbourne Institute of Technology

124 La Trobe Street, Melbourne, Victoria, 3000, Australia

Abstract

The paper presents numerical simulations of the Sandia / Sydney bluff-body stabilized flame HM1 [1] fueled with a mixture of 50 vol.% hydrogen and 50 vol.% methane. A variety of 7 detailed reaction mechanisms ranging from a skeletal mechanism with 16 species to a comprehensive mechanism with 118 species are tested. The reactive flow regime is solved by an unsteady RANS approach. Turbulence chemistry interactions are treated by the Eddy Dissipation Concept. A comparison of the different reaction mechanisms is carried out based on the axial temperature distribution, radial species distributions, the integral exhaust gas species mass flows and the required computation time. The results are compared to the available experimental data. Aside from the 2 basic reaction mechanisms there is generally good agreement between the experimental and the numerical results. In addition, differences among the more elaborated reaction mechanisms are small. For the chosen approach, a 30 species mechanism could be identified that achieves nearly the same results as the well-established GRI 3.0 at 40% less computational effort. A 24 species mechanism shows only small deviations at 80% less computational effort, which makes it highly attractive for large parametric studies.

1 Introduction

The research on renewable energy sources is focused in particular on the reduction of the greenhouse gases carbon-dioxide (CO_2) and nitrogen oxides (NO_x). Hydrogen-based gas turbine systems can play an important role for non-CO_2 and low-NO_x-emission combustion. As bridging technology towards a hydrogen based energy supply, fuel mixtures of CH_4 and H_2 and the fuel flexible operation of gas turbines are of importance.

During the design process of gas turbine combustors, CFD tools are applied for the numerical prediction of the combustion performance, exhaust emissions, thermal loads, effectiveness of cooling methods, combustion instabilities, thermoacoustic effects etc. For the analysis of single design cases or academic combustion problems, sophisticated modeling approaches like Large Eddy or Direct Numerical Simulations are suitable. These provide the most realistic reproduction of the flow and combustion process. Due to the excessive computational expanse that is necessary to resolve turbulence with little or no modelling, these methods are not feasible for large parametric studies even if simplified combustion models are applied.

For this application, steady or unsteady RANS simulations are more suitable, due to the lower computational demand. Here turbulence models are needed since turbulence effects are not resolved by the numerical grid. Flow solvers can be coupled with a wide variety of combustion models ranging from tabulated chemistry approaches (Flamelet Generated Manifold, Steady Laminar Flamelet etc.) to complex chemistry models applying detailed reaction mechanisms to solve the reactive flow regime. The aim of this paper is to combine an unsteady RANS modeling approach with detailed chemistry and to identify suitable combustion mechanisms for simulating the reactive flow regime of a combustor fueled with hydrogen methane mixtures.

Corresponding author. Fax: ++49.241.6009 52680

E-mail address: funke@fh-aachen.de

2 Source of Validation

As source of validation the bluff-body stabilized flame HM1, investigated in detail by the University of Sydney, is used [1]. For this study, it offers a desirable similarity to genuine gas turbine combustors used in industry (recirculation areas that act as flame holders, fuel injection etc.) and a broad database of experimental results as well as well-defined boundary conditions. The geometry features a bluff-body diameter of 50mm and a jet diameter of 3.6mm. The combustor is fueled with a mixture of 50 vol.% CH_4 and 50 vol.% H_2.

3 Numerical Set-Up

The present study focuses on the impact of different detailed reaction mechanisms with variable complexity on the numerical solution of the presented experimental test case. The reactive flow regime is solved by the commercial CFD-code Star-CCM+ with application of a three-dimensional, unsteady, pressure based RANS solver using the standard k-ε-turbulence model with all y+ wall treatment. For better prediction of the turbulence dissipation within the recirculation areas of the bluff-body flame, the $C_{1\varepsilon}$ constant has been changed to a value of 1.6 based on [2]. The chemical reactions of the combustion process are solved by a complex chemistry model described in detail in [3]. The 7 reaction mechanisms applied by the complex chemistry model for this study range in complexity between 16 and 118 species and between 41 and 777 reactions respectively (cf. Table 1).

The mechanisms chosen are the GRI 3.0 (5), derivatives of the GRI 3.0 and GRI 1.2 ranging down from 30 to 17 species (2)-(4), a skeletal reaction mechanism with 16 species (1) and two comprehensive mechanisms by Dagaut (6), reduced by necessity analysis targeting Sulphur containing species, and the National University of Ireland Galway (7). The GRI 3.0 (5) and the mechanism by Dagaut (6) provide detailed mechanisms for nitrogen oxide formation. For the remaining mechanisms, a simplified thermal NO model based on the extended Zeldovich is applied. For modelling of turbulence-chemistry interactions, the Eddy Dissipation Concept (EDC) is used.

No.	Name	Number of species	Number of reactions	Color (Figure 2, 3 & 5)	References
1	Star-CCM+ internal CH_4 mechanism	16	41		[4]
2	Star-CCM+ internal CH_4 mechanism	17	73		[5]
3	DRM22	24	104		[6]
4	Star-CCM+ internal CH_4 mechanism	30	184		[7]
5	GRI 3.0	53	325		[8]
6	Dagaut 2003	110	777		[9]
7	National University of Ireland Galway	118	665		[10]

Table 1: Summary of the investigated detailed reaction mechanisms, number of species and reactions with color code

The computational domain is composed of a fluid region with a cut-out, that represents the bluff-body (Figure 1). Due to the fully symmetrical nature of the bluff-body flame a 2-D axisymmetric mesh is generated. The mesh density is varied between the fuel pipe, flame region and co-flow to achieve good convergence and grid independence with a total cell count of approx. 48,000 cells, including a prism layer mesh to resolve the boundary layer.
The boundary conditions used in the simulations have been specified based on the experimental data [1].

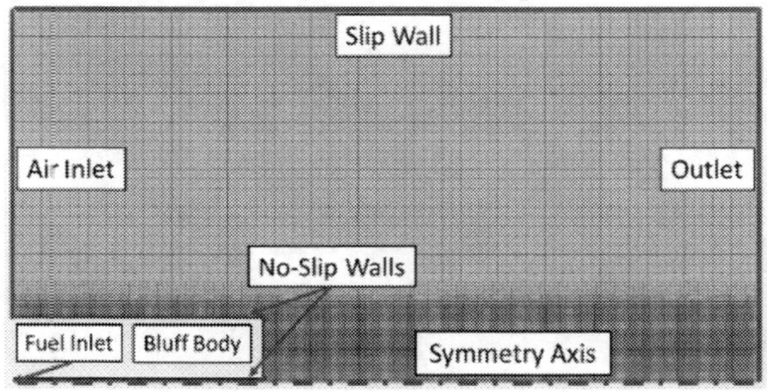

Figure 1: General view of the 2D axisymmetric computational domain and the simulation grid

4 Results

In Figure 2, the simulated temperature distribution of the HM1 flame is displayed along with the simulated and measured axial temperature profile for the domain spreading between 0.1m upstream and 0.2m downstream of the bluff-body face. All investigated reaction mechanisms predict nearly the same axial temperature profile that is shifted slightly downstream in axial direction when compared to the experimental results.

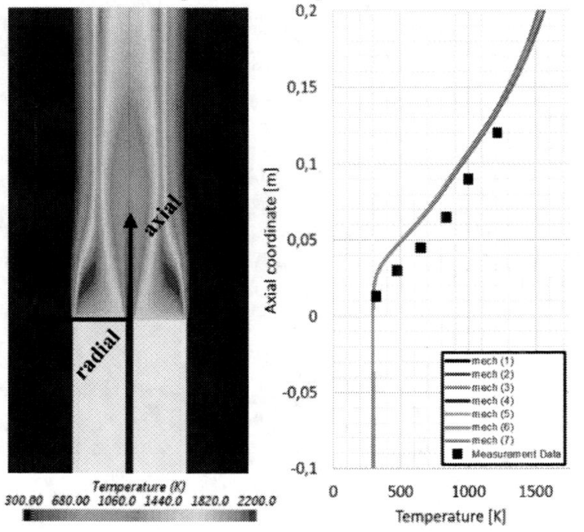

Figure 2: Simulated temperature distribution (left) and simulated and measured axial temperature profile (right)

This shift might be a result of the chosen simplified turbulence modelling approach, which is constant between all investigated combustion models.
The following evaluation of the radial species profiles obtained with the 7 different reaction mechanisms is performed at the axial location x=0,013m. Here the impact of the turbulence model constant $C_{1\varepsilon}$ on the velocity and species profiles is small compared to farther downstream locations.
Figure 3 a) and b) show the radial species profiles of O_2 and H_2O. The simulated major exhaust emissions are in very good agreement to each other and to the measurement data.
The species profiles of H_2 and CH_4 are displayed in Figure 3 c) and d). There is a slight underestimation of the concentration level in front of the bluff-body face. In the wake region of the bluff-body mixing of fuel and co-flowing air happens by transport processes e.g. convection and diffusion. Apart from present measurement errors, the chosen turbulence modelling approach can reduce the mixing or spreading effect of fuel components in the vortex region, leading to deviations. Additionally the peak concentration of H_2 inside the fuel jet is slightly overestimated by all applied combustion mechanisms. The peak concentration of CH_4 is well modeled.
As Figure 3 e) shows, CO_2 emissions are in good agreement between simulation and experimental results. Significant differences between the individual reaction mechanisms can be found for the CO profile, displayed in Figure 3 f). The differences are due to the varying complexity of the CO chemistry integrated in the investigated reaction mechanisms. The number of reactions involving CO range from 8 for mechanism (1) and 92 for mechanism (6).
The highest deviations between the experimental and simulation results are present for the peak OH concentrations in the shear layer between the bluff-body wake and the co-flowing air (cf. Figure 3 g)). The experiment shows a small peak at this location which is significantly overestimated by the coinciding simulation results. The results suggest that especially OH radicals are highly sensitive towards the chosen turbulence modelling approach independent of the significantly varying combustion mechanisms.
The radial distribution of NO is displayed in Figure 3 h). All models applying the thermal NO approach (1,2,3,4,7) yield comparable results that overestimate the experimental results slightly. The more elaborate mechanisms GRI 3.0 (5) and the mechanism proposed by Dagaut (6) incorporate additional formation pathways (thermal NO, NH, NNH, N_2O and prompt NO_x), that overestimate the NO emissions drastically when implemented with the chosen modelling approach. As discussed earlier, the temperature profiles are dependent of the turbulence modelling approach, since it influences the mixing process between the vortex structures and the central jet. In comparison to the experimental results, the chosen approach leads to higher vortex temperatures whereas the central fuel jet temperature is reduced. Especially the formation of thermal NO is promoted under elevated temperatures, which explains the overestimation of the experimental results by all numerical models.

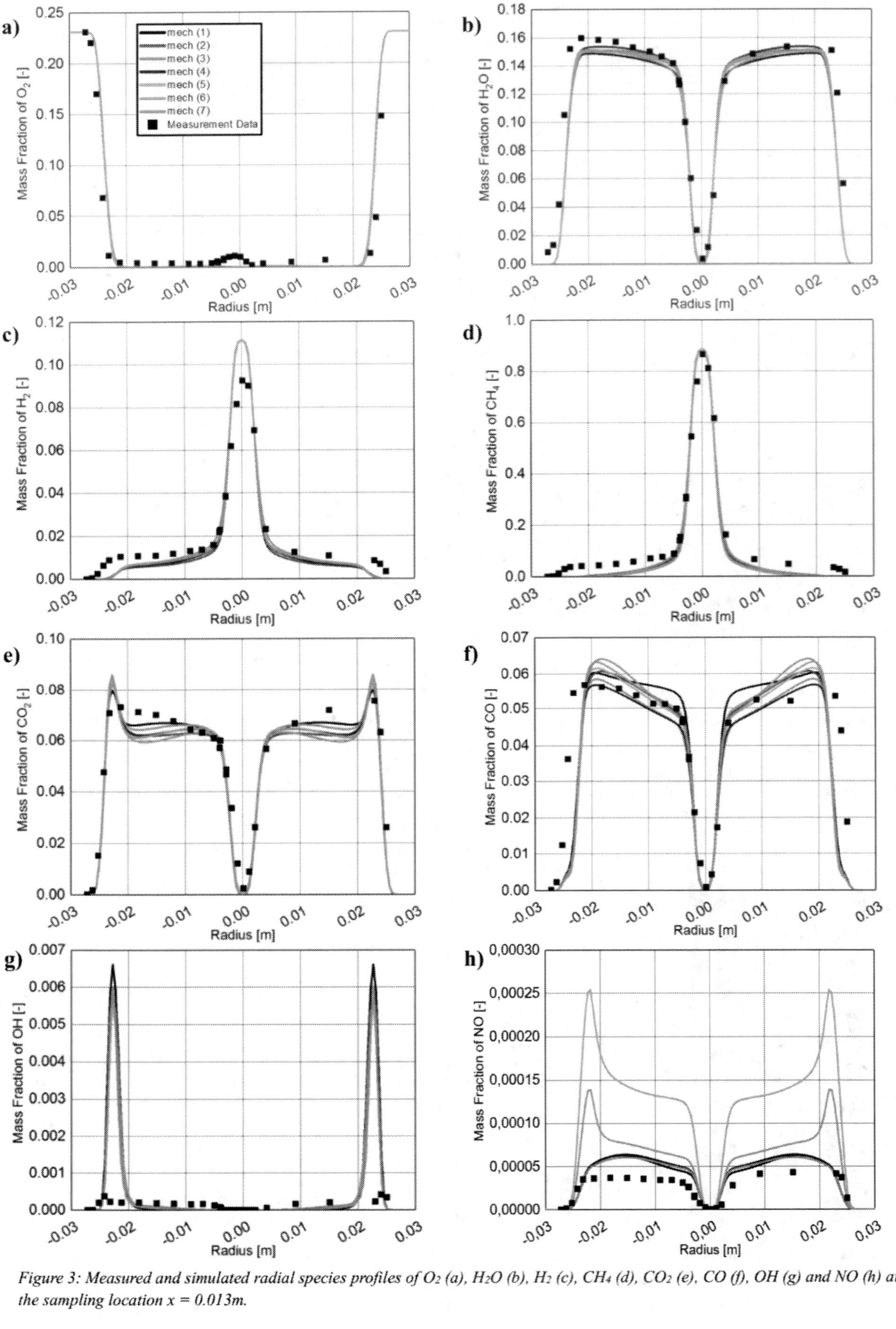

Figure 3: Measured and simulated radial species profiles of O_2 (a), H_2O (b), H_2 (c), CH_4 (d), CO_2 (e), CO (f), OH (g) and NO (h) at the sampling location x = 0.013m.

For evaluating the differences in the overall combustion progress within the simulated domain, an analysis is carried out based on the integral exhaust mass flows per species. This parameter isolates the total production or consumption of a single species independent of the overall exhaust gas composition. It is calculated from the mass flow averaged mass fraction of the individual species at the domain outlet and the total exhaust gas mass flow.

Figure 4 shows the relative deviation of the investigated mechanisms for the different species exhaust mass flows with respect to the simulation result of the well-established GRI 3.0 (5) mechanism.

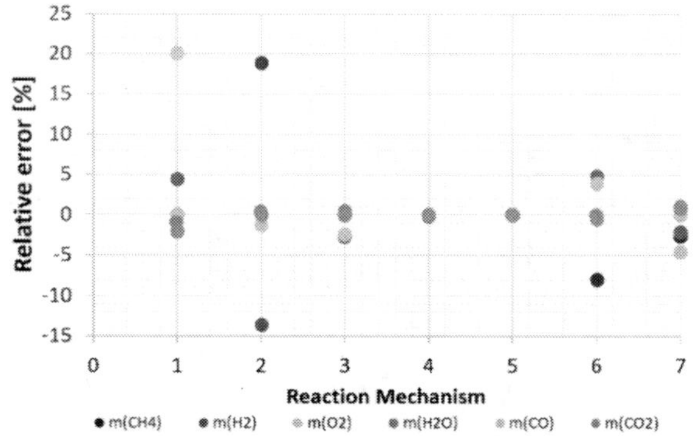

Figure 4: Relative error of integral species mass flows m(i) with reference to GRI 3.0 (5) at the domain outlet

Concerning the main exhaust emissions O_2, H_2O and CO_2 all mechanisms except for (1) show deviations less than 1% between each other. The deviations of the GRI 1.2 and GRI 3.0 derivatives (2-4) with respect to the GRI 3.0 results (5) decrease with increasing species count. The deviations produced by mechanism 3 do not exceed 2.7% whereas mechanism 4 shows nearly identical results as the GRI 3.0 (rel. error < 0.11%).

Since the more elaborate mechanisms 5 – 7 have been developed independently with varying validation goals, the obvious deviations are comprehensible. The results of these 3 mechanisms fall in an error margin of less than 8%.

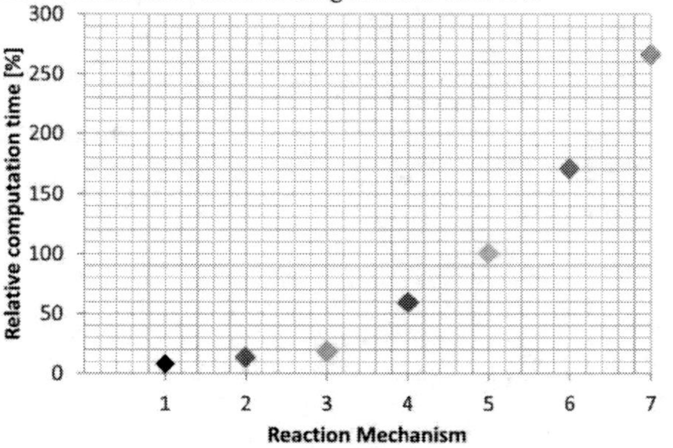

Figure 5: Relative computation time per time step for different reaction mechanisms with GRI 3.0 (5) as reference

An important aspect of the investigations carried out in this study is the identification of a reaction mechanism that delivers the best compromise between accuracy and computational time. Figure 5 shows a comparison of the relative computation times with reference to GRI 3.0 (5). It is evident that mechanism (4), which showed good representation of the investigated test case, reduces the computational time by approx. 40% in comparison to the GRI 3.0. However, it is also worth mentioning, that mechanism (3), which showed promising results too, reduces the computational time by approx. 80%. With the minor deviations displayed in Figures 2 – 4 it is highly attractive for further analysis based on an application oriented test case.

5 Conclusion

For analyzing the performance of different detailed reaction mechanisms, numerical simulations of the Sandia / Sydney bluff-body stabilized flame HM1 fueled with a mixture of 50 vol. % hydrogen and 50 vol. % methane have been carried out. A variety of 7 detailed reaction mechanisms ranging from a skeletal mechanism with 16 species to a comprehensive mechanism with 118 species have been evaluated. All reaction mechanisms represent the measured radial species profiles at the chosen location of x=0.013 well, with the largest deviations occurring for CO, OH and NO. Concerning the integral exhaust mass flows per species, generated by the individual mechanisms, the solution of the GRI 3.0 and GRI 1.2 derivatives approach the solution of the unmodified GRI 3.0 mechanism.

As computation time is a limiting factor, mechanism (3) and (4) are most attractive for further evaluation since they reduce the computation time by 40-80% compared to the GRI 3.0 mechanism while achieving comparable accuracy.

6 Acknowledgment

The numerical combustion and flow simulations presented in this paper have been carried out with the STAR-CCM+ software of Siemens PLM. Their support is gratefully acknowledged.

References

[1] A. R. Masri, "Sydney Bluff-body flame HM1 test data" available at http://sydney.edu.au/engineering/aeromech/thermofluids/bluff.htm

[2] B. Dally, D. Fletcher and A. Masri, "Flow and mixing fields of turbulent bluff-body jets and flames," Combustion Theory and Modelling, no. 2, pp. 193-219, 1998.

[3] H. H.-W. Funke, N. Beckmann, J. Keinz, S. Abanteriba Comparison of Numerical Combustion Models for Hydrogen and Hydrogen-Rich Syngas Applied for Dry-Low-NOx Micromix Combustion, ASME Turbo Expo 2016, GT2016-56430, Seoul, South Korea, June 2016.

[4] B. Yang and S. B. Pope, "An investigation of the Accuracy of Manifold Methods and Splitting Schemes in the Computational Implementation of Combustion Chemistry," Combustion and Flame, pp. 16-32, 1998.

[5] R. Sankaran, "Structure of a spatially developing turbulent lean methane-air Bunsen Flame," Proceedings of the Combustion Institute, no. 31, pp. 1291-1298, 2007.

[6] University of California, Berkeley, available at http://combustion.berkeley.edu/drm/

[7] T. Lu and C. Law, "A criterion based computational singular perputation for the identification of quasi steady state species: A reduced mechanism for methane oxidation with NO chemistry," Combustion and Flame, no. 154, pp. 761-774, 2008.

[8] University of California, Berkeley, available at http://combustion.berkeley.edu/gri-mech/version30/text30.html

[9] Dagaut, P., Lecomte, F., Mieritz, J. and Glarborg, P. (2003), Experimental and kinetic modeling study of the effect of NO and SO2 on the oxidation of CO H2 mixtures. Int. J. Chem. Kinet., 35: 564–575. doi:10.1002/kin.10154

[10] E.L. Petersen, D.M. Kalitan, S. Simmons, G. Bourque, H.J. Curran, J.M. Simmie, "Methane/Propane Oxidation at High Pressures: Experimental and Detailed Chemical Kinetic Modelling," Proceedings of the Combustion Institute, 31: 447-454, 2007.

Preliminary comparison of chemical heat storage systems for saving exhaust gas energy in gasoline and diesel engines

Duc Luong Cao[1], Guang Hong[1], Tuan Le Anh[2]

[1]University of Technology Sydney

15 Broadway, Ultimo, NSW 2007, Australia

[2]Hanoi University of Science and Technology

1 Dai Co Viet Road, Hanoi, Vietnam

Abstract

Thermal energy storage has become more important to utilizing the wasted energy and improving the overall efficiency of energy systems. This study was aimed to develop a chemical heat storage system using magnesium hydroxide ($Mg(OH)_2$) and its dehydration and hydration reactions to recover the thermal energy wasted by the exhaust gas in internal combustion (IC) engines. Experiments and simulations were conducted on a gasoline engine (Toyota Aurion 2GR-FE 3.5L) and a diesel engine (D1146TI) to compare the efficiency of chemical heat storage (CHS) technology in two difference kinds of IC engines. With the same CHS system and working time, the amount of energy stored by the reactor in the heat storage process in the gasoline engine is higher than that in a diesel engine because the average temperature of the exhaust gas in the gasoline engine is higher than that in the diesel engine. Alternatively the time for storing the same amount of heat energy in the gasoline engine will be shorter than that in the diesel engine.

1 Introduction

In today's modern life, IC engines are still widely used in many fields, such as transportation, construction or agricultural sectors. However, a significant amount of fuel energy is lost as wasted heat through exhaust gas and cooling systems in IC engine. If the heat loss in the exhaust gas can be stored and used, the overall efficiency of an IC engine will be increased. When a certain amount of exhaust gas heat could be recovered, the fuel could be saved and the environment pollution could be reduced.

This research is focused on applying CHS to store energy wasted by the exhaust gas in IC engines. The chemical material adopted in the CHS device in this research is Magnesium hydroxide ($Mg(OH)_2$) based on its reversible reaction as follow:

$$Mg(OH)_2(s) \leftrightarrow MgO(s) + H_2O(g) \quad (1)$$

In the heat storage process, Magnesium hydroxide absorbs the wasted heat of the exhaust gas and converts to magnesium oxide and water vapor in the dehydration reaction in the reaction chamber. MgO is retained inside the reactor chamber, and the water vapor produced from the chemical reaction is moved into a condenser and condensates to the liquid state.

In the heat output process, the water vapor flows into the reactor chamber. The reaction takes place between the MgO and the water vapor to become $Mg(OH)_2$ and heat energy is released in this reaction.

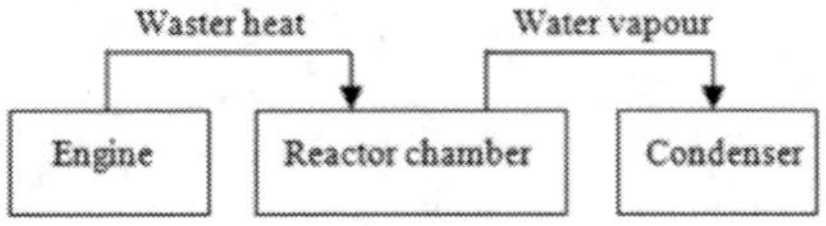

Figure 1: The heat storage process [1]

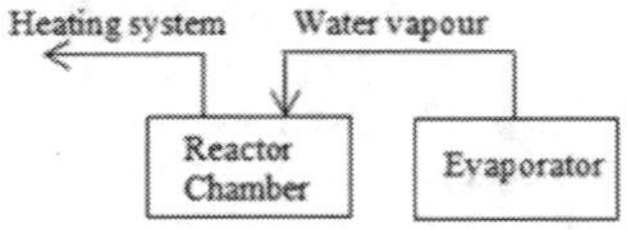

Figure 2: The heat output process [1]

The thermal conductivity of the packed bed of $Mg(OH)_2$ pellets is too low within 0.15 – 0.16 W/m.K [2], which affects the heat absorption capacity of chemical material and thereby decreases the efficiency of the reactor. To increase the heat transfer efficiency of the chemical material, a new $Mg(OH)_2$ compound suggested by Massimiliano was the combination of $Mg(OH)_2$ and expanded graphite (EG) with the mass mixing ratio 8:1 and in the block state (EM8 block). As reported in [2], the advantages of EM8 block, compared with pure $Mg(OH)_2$, include:

- Higher thermal conductivity: The thermal conductivity of EM8 block is about ten times that of the pure $Mg(OH)_2$ pellets.
- Higher density: The density of EM8 block is 1.6 times that of $Mg(OH)_2$ pellets.
- Reduced void fraction of the bed: it will enhance the contact between the packed material and the inner surface and consequently improve the thermal conductivity of the reactor

Parameters	Unit	Value
Mass mixing ratio (r_{mix})		8:9
Density of bed	g/cm^3	1.002
Working temperature	K	297-800
Reaction enthalpy	kJ/mol	81
Thermal conductivity	W/m.K	1.5 – 1.7
Heat storage capacity in 1 hour	MJ/m^3	700

Table 1: The main properties of EM8 block [2].

E-mail address: ducluong.cao@student.uts.edu.au

2 Potential applications of CHS technology to IC engines

2.1 Applications of CHS to vehicle engines

In IC engine vehicles, the cold start process is the major source for pollutant emissions, including uncombusted hydrocarbons (HC), carbon monoxide (CO), and particulate matter (PM). Besides, The fuel consumption of the vehicle engine is therefore increased during the cold start. Kunze [3] presented results that showed the difference in the fuel consumption between a cold start and a warm start of a gasoline engine. When the temperature increased from 25^0C to 90^0C, the fuel consumption decreased by 10% in the engine starting process.

The other potential application of CHS is for the catalyst converters. The lower limit of optimal temperature (for purification performance) for a catalyst converter is 150^0C [4]. The temperature of exhaust gas immediately after starting the engine is around 100^0C, which is not enough for the operation of the catalyst. Because of this the efficiency of the catalyst at the low temperature is very low as shown in figure 3.

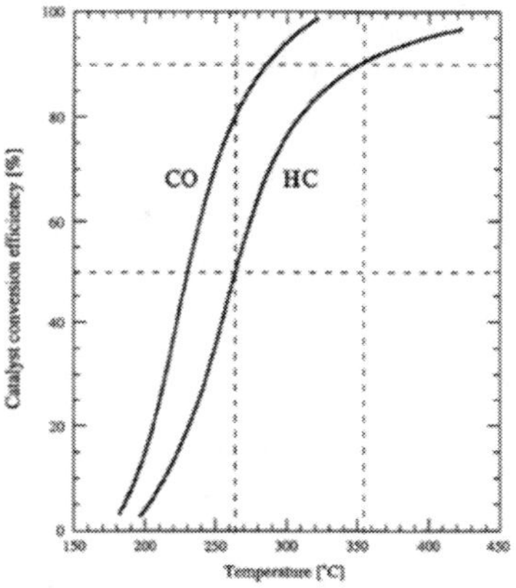

Figure 3: Dependence of catalytic convertor efficiency for HC and CO on temperature [5].

Beside the catalyst, the low temperature in the cold-start process affects the performance of the lubricant. In the low temperature, the viscosity of the lubricant increases, it makes the friction coefficient to decrease and reduction of the engine efficiency in cold start conditions.

In summary, for IC engine vehicles, The CHS technology can be applied to:

- heating the catalyst in the cold-start process.
- heating the lubricant in the cold-start process in the low temperature condition.
- defogging .

2.1 Applications of CHS in ybrid vehicles

Similar applications of CHS can be extended to hybrid vehicles. In the cold weather, a hybrid vehicle needsenergy for vehicle's climate control system. The energy of the exhaust gas of the IC engine can be utilized directly to increase the temperature in the cabin. However, the hybrid vehicle may need more energy stored for heating purpose in the cold weather.

Furthermore, the battery in a hybrid vehicle can be highly affected by low temperature. In the cold temperature, the change of the viscosity inside the battery (in the electrolyte) may lead to sluggish ion transport. The mobility of the ions decreases in cold temperature, causing a rise in internal resistance. The increase in the inside resistance of the battery leads to the lower battery performance.

In summary, in the cold weather, with the hybrid vehicles, CHS technology can be applied to:

- assisting vehicle's climate control system.
- heating the electrical battery.
- heating the IC engine or keeping it warm.

3 Experiments

Experiments were conducted on two engines: a gasoline engine (6-cylinder spark ignition Toyota Aurion engine at UTS engine laboratory, and a diesel engine (D1146TI engine equipped on buses in Vietnam) at Hanoi University of Science & Technology.

(a) (b)

Figure 4: Photos of the tested engines (a) Toyota Aurion, (b) D1146TI diesel engine.

Parameters	Unit	Gasoline engine	Diesel engine
Number of cylinders		6	6
Number of strokes		4	4
Bore	mm	94	111
Stroke	mm	83.10	139
Displacement volume	cc	3456	8071
Compression ration		10.8:1	16.7:1
Maximum power	kW	200@6200 rpm	151@2200 rpm
Maximum Torque	N.m	336@4700 rpm	735@1400 rpm

Table 2: The major specifications of tested engines.

The engine power is calculated based on the engine speed and torque. From the exhaust gas temperature and its components (were acquired by Horiba MEXA-584L gas analyser in the gasoline engine and the AVL test-bed in the diesel engine), heat loss is calculated by the summing the energy of components.

Properties	Unit	Gasoline engine	Diesel engine
Temperature	K	1028	704
CO	ppm	1000	31
CO_2	ppm	155000	50379
HC	ppm	9.9	183
NO_x	ppm	2335.7	1330
O_2	ppm	2000	98151
N_2	ppm	707714	806164
H_2O	ppm	130912	43762

Table 3: Data of the exhaust gas measurements.

The engine thermal efficiency and the heat lost in the exhaust gas were calculated with the experimental data. The exhaust gas energy was calculated based on the temperature and components of the exhaust gas. Heat energy of exhaust gases are the sum of the energy of components and are shown in figure 5.

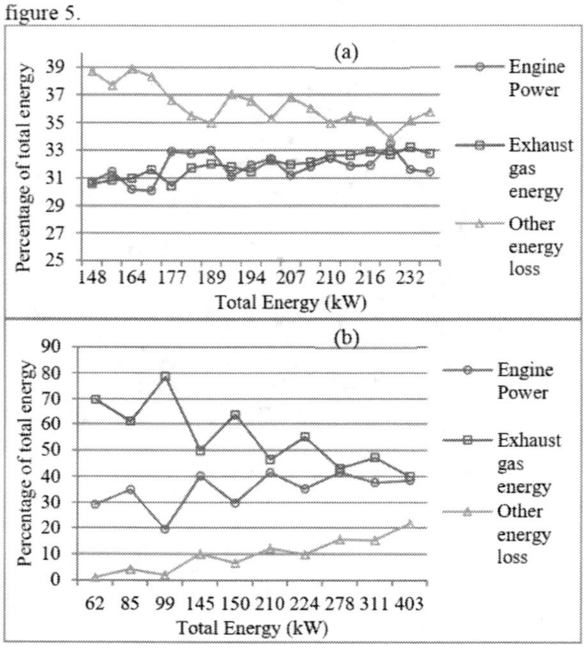

Figure 5: The energy consumption of gasoline engine (a) and diesel engine (b).

As shown in Figure 5, the energy loss due to the exhaust gas in the gasoline engine is smaller than that in the diesel engine. However, in almost operation conditions it is higher than 30%. In the diesel engine, in some condition the energy loss in the exhaust gas can increase to 70%.

4 Reactor design

Figure 6 shows the structure of the reactor designed based on the experimental results. The reactor consists of two tubes: inner and outer tubes. EM8 block is stored inside the inner tube, and the exhaust gas flows in the space between the two tubes.

Parameters	Unit	Value	Parameters	Unit	Value
Inner tube			Inlet, outlet and vapour ports		
Diameter	mm	100	Diameter	mm	88.9
Thickness	mm	3.05	Thickness	mm	3.05
Outer tube			Wings		
Diameter	mm	160	Thickness	mm	3.05
Thickness	mm	3.05			

Table 4: The dimensions of the reactor

In the heat storage process, the exhaust gas flows in the space between the two tubes from the inlet port to the outlet port while heat is transferred from the exhaust gas to EM8. With the heat energy provided by the exhaust gas, the dehydration reaction of $Mg(OH)_2$ takes place inside the inner tube. Water vapor from the reaction flows out of the reactor from the vapor port placed at the top of the reactor. In the space between two tubes, two wings are designed to make the temperature inside the reactor becomes even, and to increase the moving time of the exhaust gas flow in the reactor thereby to increase the heat transfer efficiency of the reactor.

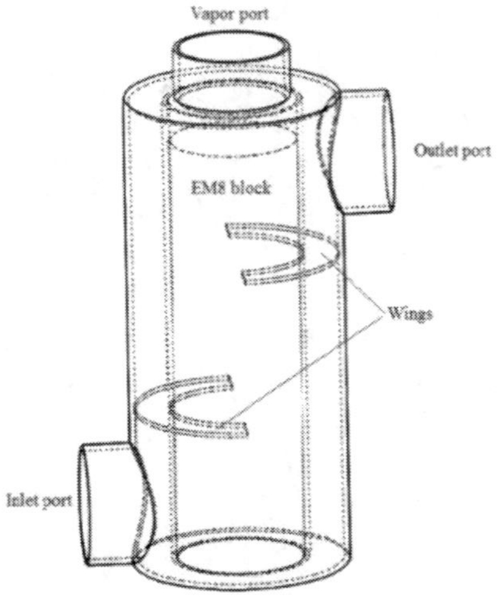

Figure 6: The reactor design

In the heat output process, the water vapor from the evaporator (water liquid is reheated by an electrical resister and evaporates at the evaporation pressure) flows to the reactor through vapor port. The hydration reaction is taken place inside the reactor and heat transfer from thermochemical material to fresh air through the inner tube wall. Low-temperature fresh air comes to the reactor at the inlet port, receives heat from thermochemical material and higher temperature fresh air moves out of the reactor at the outlet port to transfer to the heating system.

5 Simulation results of the heat storage process.

Simulation was performed to estimate the performance of the CHS and determine the dimensions. In simulations it was assumed that the reactor be placed right after the engine and in the heat storage process and that the temperature and the components of the exhaust gases at the inlet of the reactor be the same as that at the engine exhaust port. The measured volume ratios of the exhaust gases are listed in Table 3.

ANSYS Fluent was used to simulate the gas flows and heat transfer in the reactor. As the temperature of the exhaust gas at the engine exhaust port is high, the mixed thermal condition (combination of heat convection and radiation) model was chosen to simulate the heat transfer between the exhaust gas and the reactor walls. EM8 block is assumed as a solid material, so the heat transfer process inside the EM8 block is assumed as the heat conduction process with the properties given in table 1.

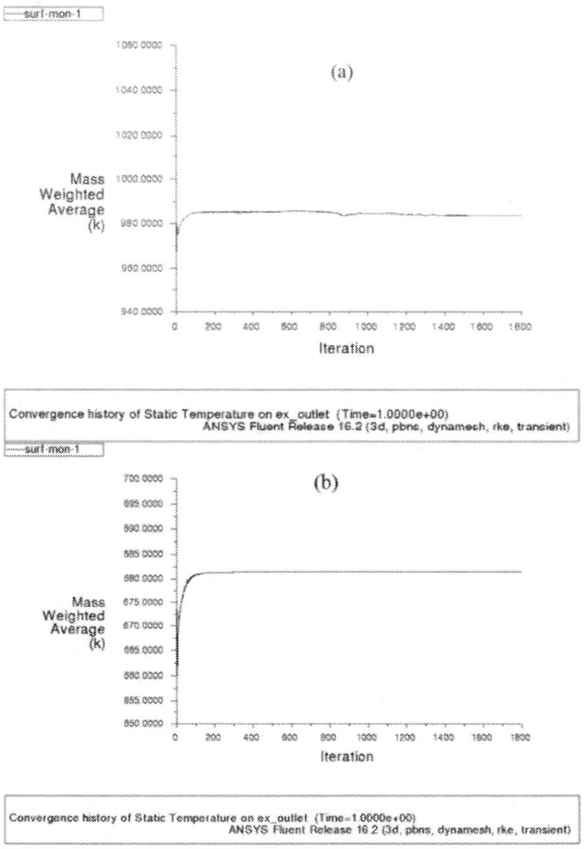

Figure 7: The temperatures of exhaust gas at the reactor outlet, (a) gasoline engine, (b) diesel engine.

As shown in figure 7, in the first 200s when the exhaust gas moves into the reactor, the heat energy of the exhaust gas is transferred to the reactor to increase the temperature of the reactor, the chemical material inside the reactor to a level required for chemical reaction. After that when the temperature of the reactor and chemical material becomes stable, the heat energy is 100% used for chemical reaction. The heat storage capacity of the chemical material is assumed as constant and so is the heat storage capacity of the reactor. This results in constant temperature of the exhaust gas at the outlet as shown in figure 7.

Based on the temperature of the exhaust gas, the heat energy of the exhaust gas at the outlet is calculated. The reactor is assumed as an insulated device so the energy stored inside the reactor is equal to the energy change of the exhaust gas between inlet and outlet of the reactor.

The efficiency of the reactor is defined as the percentage of heat energy of exhaust gas stored in the heat storage process. As shown in table 5, the efficiency of the reactor, 5.88%, in the gasoline engine is higher than that, 3.42%, in the diesel engine. The main reason is the average temperature of the exhaust gas of the gasoline engine is higher than that of the diesel engine. The reaction rate of the chemical material highly depends on its temperature. It increases with the increase of the temperature of the reaction material.

Properties	Unit	Gasoline engine	Diesel engine
Heat energy at the reactor inlet	kW	61.9	97.48
Temperature at the reactor outlet	K	985	681.5
Heat energy at the reactor outlet	kW	58.3	94.14
Energy stored in the reactor	kW	3.6	3.34
The reactor efficiency	%	5.88	3.42
Volume of EM8 block	dm^3	2.51	2.51
Mass of EM8 block	kg	2.51	2.51

Table 5: Energy storage in the reactor in the heat storage process.

The reaction of $Mg(OH)_2$ is only efficient when the temperature of the material is higher than 297^0C [6] (reaction temperature). With the diesel engine, at some low load conditions, the temperature of the exhaust gas is smaller than 297^0C and using magnesium hydroxide CHS system will be not efficiency in these cases. With gasoline engine, in almost operation conditions, the exhaust gas temperature is higher than 300^0C so using CHS system is efficient.

6 Conclusions

Using CHS system to store heat energy of exhaust gas is feasible with the efficiency of the gasoline and diesel engines are 5.88 % and 3.42 %, respectively with the same reactor and operation condition. The efficiency of the reactor in the gasoline engine is higher because its exhaust gas temperature is higher.

In some operation conditions, the temperature of exhaust gas of the diesel engine was lower than 297°C which is the optimal temperature of $Mg(OH)_2$. The overall efficiency of the reactor in the diesel engine is low due to the low exhaust gas temperature.

References

1. Duc.L.Cao, G.Hong, J.Wang, Preliminary investigation to the feasibility of chemical heat storage for saving the exhaust gas energy in a spark ignition engine, in 20th Australasian Fluid Mechanics Conference. December 2016: Perth, Australia.
2. Massimiliano Zamengo, Y.K., A study on heat tranfer-enhanced composites for a Magnesium Oxide/Water Chemical Heat Pump. 2014.
3. Klaas Kunze, S.W., Irina Lade and Johann Tonhauser, A systematic analysis of CO2-reduction by an optimized heat supply during vehicle warm-up. SAE technical paper series, 2006. 2006-01-1450.
4. Satoshi Hariu, T.Y., Yukihiro Noguchi, Hiroyasu Kawouchi, Hideaki Suzuki, Chemical Heat Storage Device. 2016: United States.
5. Roberts, A., R. Brooks, and P. Shipway, Internal combustion engine cold-start efficiency: A review of the problem, causes and potential solutions. Energy Conversion and Management, 2014. 82: p. 327-350.
6. W.Wongsuan, S.K., P.Neveu, F.Meunier, A review of chemical heat pump technology and application. 2001

11th Asia-Pacific Conference on Combustion,
The University of Sydney, NSW Australia
10th -14th December 2017.

Lift-off Characteristics of n-Heptane Dilute Spray Flames in Vitiated Co-flows

Huifeng Gong, Qiong Yang, Zongjie Hu, Liguang Li

School of Automotive Studies, Tongji University

4800 Cao'an Road, Shanghai, 201804, China

Abstract

This paper presents experimental investigations about lift-off characteristics of n-heptane dilute spray jet flames in vitiated co-flows. Averaged lift-off heights are determined with flame luminous images under a co-flow temperature range of about 1200 ~ 1600 K. The influences of jet velocity, global equivalence ratio and initial droplet size on flame lift-off heights are examined separately. Results show that the relationship between lift-off height and co-flow temperature could be well expressed with an Arrhenius-like expression. The effect of spray jet velocity on lift-off height is co-flow temperature dependent, and a critical temperature exists. Increasing jet velocity causes an increase in lift-off height when co-flow temperatures are lower than the critical temperature, while for higher co-flow temperature conditions, the trend reverses. With fixed jet velocity, increasing the global equivalence ratio of spray could lower the flame lift-off height. Compared with large droplet size, sprays of small droplet size tend to have lower lift-off heights, but the differences are reduced as the co-flow temperature or global equivalence ratio increases.

1 Introduction

Spray combustion is widely used and of great importance in many practical applications, such as diesel engines, GDI engine, gas turbine engines, ramjets. Lift-off flames generally characterize the spray combustion, where auto-ignition and flame stabilization are key problems. Extensive investigations, both experimental and numerical, have been conducted for gaseous fuels like hydrogen and methane [1-4], and many theories have been proposed to reveal fundamental understanding of lift-off flames [5-7]. Unlike the gaseous fuel jet, spray combustion is more complex due to its two-phase nature. The evaporation of dispersed droplets and the mixing of fuel vapor and surrounding air strongly affect the initiation of combustion and stabilization of flame, dominating the combustion regimes in spray flames [8-11]. The droplet dynamics with varied sizes, gas-liquid interactions and the coupling with chemistry still need detailed investigations to further understand spray combustion.

However, in practical applications, the liquid breakup and atomization processes are usually inherent and essential in spray formation [12, 13], which make a dense spray region and involve great sophistication that challenges both the current experimental and numerical investigations. More fundamental researches on dilute sprays are performed with simple and well-defined boundary conditions and initial conditions which facilitate both diagnostics and modeling. Extensive studies have been performed in the University of Sydney with a combined configuration of vitiated co-flow and piloted spray burner, showing that spray flames have unique features because of the existence of droplets [14-16].

This paper presents investigations on lift-off flames of n-heptane dilute sprays in vitiated co-flows. Flame lift-off heights are measured with luminous flame images. Several key parameters, like jet velocity, spray global equivalence ratio and spray initial droplet size, are varied to examine their effects on the evolution of flame lift-off height as a function of co-flow temperature.

2 Experimental Method

A general layout of the experimental setup is shown in Fig. 1. The system basically consists of two parts: a burner and a dilute spray generator. The burner is very similar to the Cabra and Dibble burner [17], which provides a hot vitiated co-flow of combustion products from burning premixed hydrogen/air mixture. The spray generator uses ultrasonic atomization to pre-atomize the liquid fuel to micron size droplets, which are carried by air at room temperature through a nozzle tube. The carrier air and the liquid fuel are provided separately and quantitatively, which gives the flexibility of varying the jet velocity and global equivalence ratio at the nozzle exit. Two ultrasonic nebulizers are used to generate sprays of different droplet size. One is an ultrasonic nozzle at 96 kHz, which produces relatively large droplets. The other is a set of four ultrasonic discs at 1.7 MHz, which produces much smaller droplets. The inner diameter of nozzle tube is 5 mm in, and the nozzle exit is 70 mm higher than the burner perforated plate. To prevent the hot co-flow heating effects on the spray, a thin cooling water jacket covers the nozzle tube. The co-flow temperature is measured using a thermocouple (Pt-30%Rh/Pt-6%Rh) with a bead diameter of about 0.25 mm, and the temperature value is corrected for radiation loss with emissivity determined to be 0.2 by infrared thermography.

Flame lift-off heights are determined with luminous images of the n-heptane spray lift-off flames. 50 images are taken for each case, and the average lift-off height is calculated from the images. Key parameters including jet velocity, global equivalence ratio and initial spray droplet size are investigated to determine their effects on flame lift-off height as a function of co-flow temperature. Boundary conditions of the flames are shown in Tab. 1. Each parameter is examined separately and independently for its effect on flame lift-off heights under a sequence of co-flow temperatures. Jet velocities are examined with three values: 20 m/s, 40 m/s, 80 m/s. The spray global equivalence ratios at the nozzle exit are determined by the combination of flow rates of carrier air and fuel, which are 1.0, 2.0, 4.0 and 8.0. Spray droplet size is varied by using ultrasonic nozzle and ultrasonic disc, and the initial mean droplet size d_{32}

Corresponding author: Zongjie Hu. Fax: +86-21-6958-9978
E-mail address: zongjie-hu@tongji.edu.cn

at the nozzle exit are about 30 μm and 10μm, respectively. The co-flow temperature is varied from about 1200 K to 1600 K with temperature interval of about 25 K.

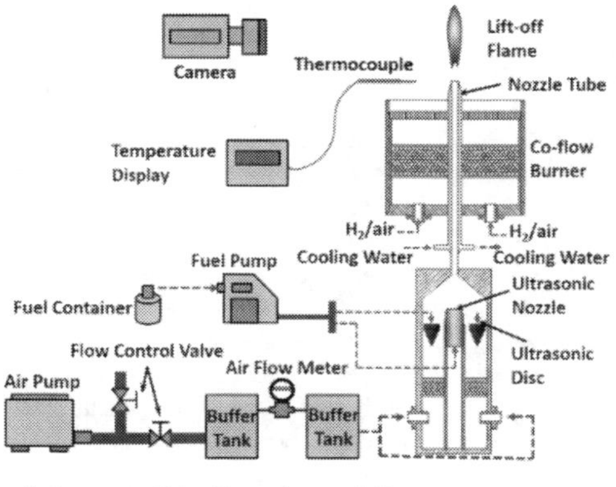

Figure 1: Layout of the Experimental Setup

Jet velocity U_0 [m/s]	20/40/80
Global equivalence ratio ϕ [-]	1.0/2.0/4.0/8.0
Initial mean droplet size d_{32} [μm]	~30, ~10
Co-flow temperature T_c [K]	~1200-~1600

Table 1: Boundary conditions for n-heptane spray flames

3 Results and Discussions

3.1 Dependence of Lift-off Height on Temperature

Fig. 2 shows typical flame images under a sequence of co-flow temperatures for the case: U_0 = 40 m/s, ϕ = 2.0, d_{32} = 30 μm. Generally blue lift-off flames are presented, and as co-flow temperature increases, the flames get brighter and lift-off heights decrease steeply. The curve of normalized lift-off height by nozzle tube inner diameter L_h/D versus co-flow temperature T_c is shown in Fig. 3, indicating that flame lift-off height is very sensitive to the co-flow temperature.

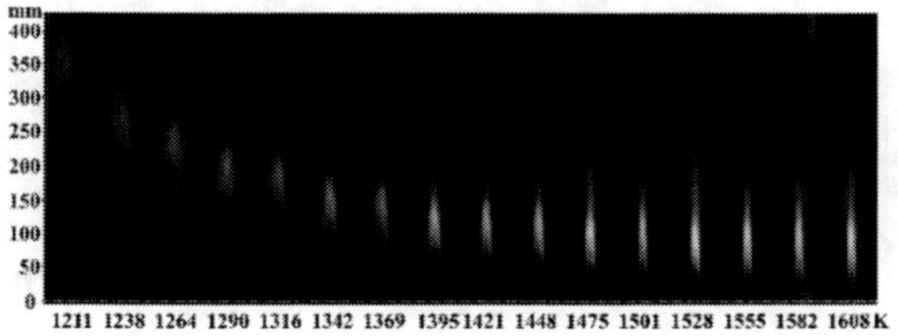

Figure 2: A typical sequence of flame images with varied co-flow temperatures (U_0 = 40 m/s, ϕ = 2.0, d_{32} = 30 μm)

Flame lift-off height is the distance that spray jet travels within the duration of ignition delay, which is commonly considered to follow an Arrhenius scaling expression, and consequently the lift-off height is also expected to have similar scaling expression in some literatures. In this paper, a similar Arrhenius-like expression is used to fit the flame lift-off height as a function of co-flow temperature, as shown below:

$$L_h/D = a \cdot e^{\frac{b}{T_c}} \cdot T_c^n \quad (1)$$

Conducting logarithmic transformation,

$$\ln(L_h/D) = A + B \cdot (1000/T_c) + C \cdot \ln(1000/T_c) \quad (2)$$

where $\ln(L_h/D)$ is the normalized lift-off height in logarithmic format, $1000/T_c$ is the inverse co-flow temperature, A, B and C are constants.

Fitting data in Fig. 3 with Eq. (2) to get the fitting expression:

$$\ln(L_h/D) = -47.7 + 55.8(1000/T_c) - 30.4\ln(1000/T_c) \quad (3)$$

where the goodness-of-fit R^2 is 0.9998. The fitting line is shown in Fig. 4.

Similar fittings are also conducted for other cases, and the R^2 are generally larger than 0.99, showing that this Arrhenius-like expression represents the relationship between the lift-off height and co-flow temperature quite well.

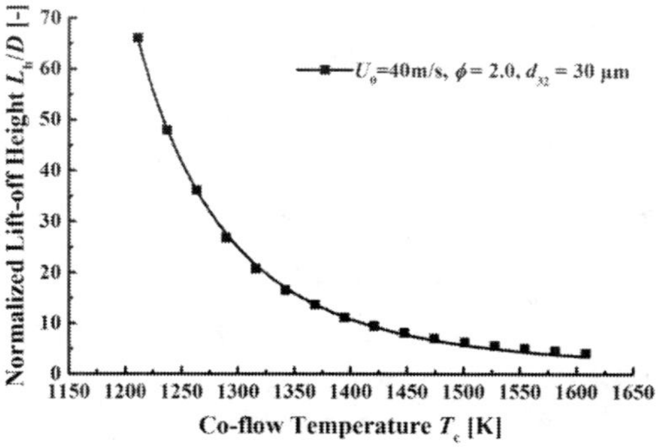

Figure 3: Flame lift-off height decreases steeply with increased co-flow temperature (U_0 = 40 m/s, ϕ = 2.0, d_{32} = 30 μm)

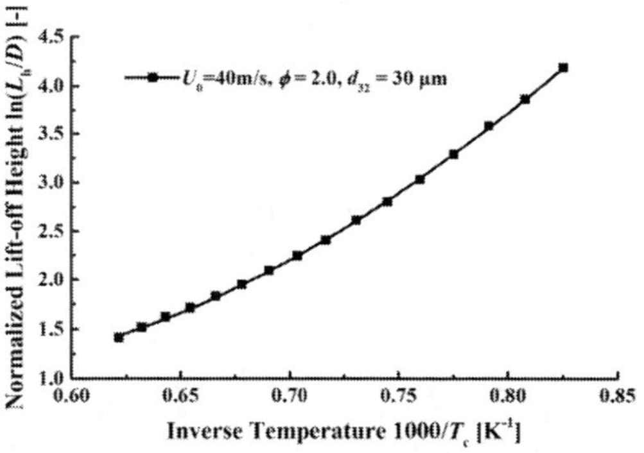

Figure 4: Arrhenius fitting plot of normalized lift-off height versus co-flow temperature (U_0 = 40 m/s, ϕ = 2.0, d_{32} = 30 μm)

3.2 Effects of Jet Velocity on Lift-off Height

Jet velocity determines the spray jet turbulence intensity, which strongly influences the spray evaporation, mixing and combustion processes. Fig. 5 shows the effects of jet velocity on the evolution of flame lift-off height as a function of co-flow temperature. The curves indicate that the effect of jet velocity on flame lift-off height is not consistent within the whole range of experimental co-flow temperature conditions. Instead, a critical temperature exists, and under conditions where co-flow temperatures are lower than the critical temperature, the flame lift-off height increases with jet velocity, however, when co-flow temperatures are higher, the effect of jet velocity on flame lift-off height is completely opposite. This phenomenon is quite different from that of gaseous fuel jets and diesel sprays [18, 19], where the flame lift-off height is considered to proportionally increase with the jet velocity.

The auto-ignition of a spray basically combines physical process and chemical process. Generally, the physical process is the liquid fuel evaporation/mixing process, which prepares the ignitable mixture, while the chemical process covers the initiation of chemical reactions of ignitable mixture before combustion. The duration of these two processes contribute the ignition delay, and the spray jet travels downstream during the ignition delay time to form the lift-off height.

High jet velocity, on one hand, makes the spray jet travels downstream much faster, leading to a larger lift-off height, but on the other hand, high jet velocity enhances shear interactions between the jet and the surrounding hot co-flow, which intensifies the evaporation and mixing processes, and reduces the physical delay time to shorten the lift-off height. A trade-off is expected, and the critical temperature is the balance point. Under conditions of relatively lower co-flow temperature, the chemical delay time is long and is the main contributor of the total ignition delay compared with the physical part, as a consequence, the reduction of physical delay time due to high jet velocity cannot sufficiently shorten the total ignition delay, so the lift-off height grows with the increased jet velocity. However, under high co-flow temperature circumstances, the chemical delay time decreases sharply, and the physical delay is the main contributor of the total ignition delay, so the reduction of physical delay by high jet velocity can shorten the total ignition delay enough, and the lift-off height is reduced as the jet velocity increases.

The critical temperature varies under different spray global equivalence ratios. In Fig.5 (a) where ϕ is 1.0, the critical temperature could be calculated to be about 1520 K. While in Fig. 5 (b) with ϕ = 2.0, the critical temperature decreases to be about 1270 K. In Fig. 5 (c) with the highest ϕ of 4.0, the critical temperature is about 1340 K, a little higher than that in ϕ = 2.0, but the difference between two jet velocities, as seen in Fig. 5 (c), is not distinct within low co-flow temperature range, this critical temperature could not be a solid value . Generally, the critical temperature is decreased at high global equivalence ratios, indicating that higher global equivalence ratio helps reduce the lift-off height. Independent examination of the effect of spray concentration is conducted in the next section.

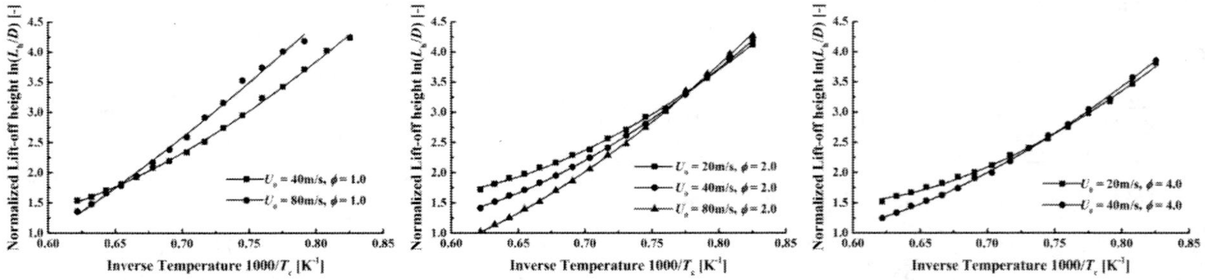

(a) Global equivalence ratio ϕ = 1.0 (b) Global equivalence ratio ϕ = 2.0 (c) Global equivalence ratio ϕ = 4.0

Figure 5: Effects of jet velocity on flame lift-off height as a function of co-flow temperature under constant global equivalence ratio (Arrhenius-like expression fitting, d_{32} = 30 μm) : (a) ϕ = 1.0; (b) ϕ = 2.0; (c) ϕ = 4.0

3.3 Effects of Equivalence Ratio on Lift-off Height

The global equivalence ratio of spray is another key parameter that influences the spray auto-ignition characteristics. Fig. 6 presents the effects of global equivalence ratio on the evolution of flame lift-off height as a function of co-flow temperature for large droplet size of n-heptane sprays under fixed jet velocities. The curves show that the rise of spray global equivalence ratio leads to a decline in the lift-off height as could be expected. This trend is consistent for three fixed jet velocities within the whole tested co-flow temperature range.

Sprays of large global equivalence ratio provide more fuel droplets, which vaporize more when traveling through the nozzle tube and issuing into the hot co-flow. Vaporization of droplets is believed not to be obviously restrained by high vapor concentration because of the dilute nature of the spray. The spray vaporizes much faster with more droplets in it from the perspective of the total amount of vapor generated within unit time. Faster vaporization of spray provides the most ignitable mixtures for auto-ignition earlier as the jet travels downstream, so the lift-off height is expected to be lower.

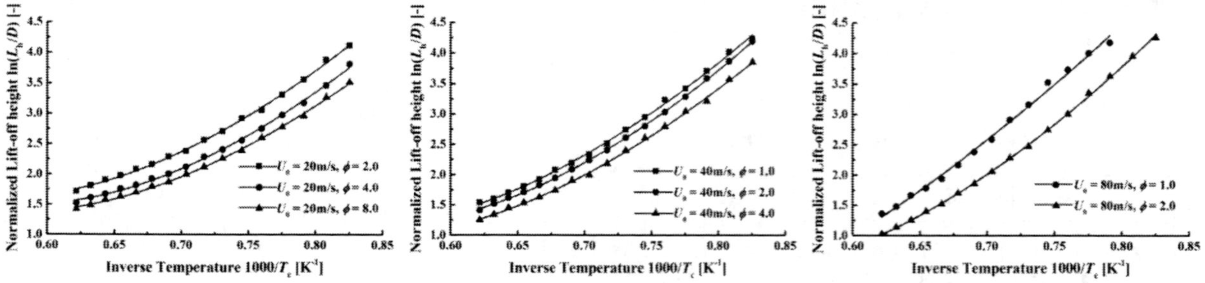

(a) Jet velocity U_0 = 20 m/s (b) Jet velocity U_0 = 40 m/s (c) Jet velocity U_0 = 80 m/s

Figure 6: Effects of global equivalence ratio on flame lift-off height as a function of co-flow temperature under constant jet velocity (Arrhenius-like expression fitting, d_{32} = 30 μm): (a) U_0 = 20 m/s; (b) U_0 = 40 m/s; (c) U_0 = 80 m/s

3.4 Effects of Initial Droplet Size on Lift-off Height

Fig. 7 shows the effects of initial droplet size on the lift-off height with three cases at fixed jet velocity and equivalence ratio. The data show that the spray of small droplet size has shorter lift-off height, which could be expected because of the faster vaporization of small droplets. However, the difference from droplet size is reduced as the co-flow temperature increases. Similar trends could be seen in all three cases as shown in Fig. 7 (a), (b) and (c).

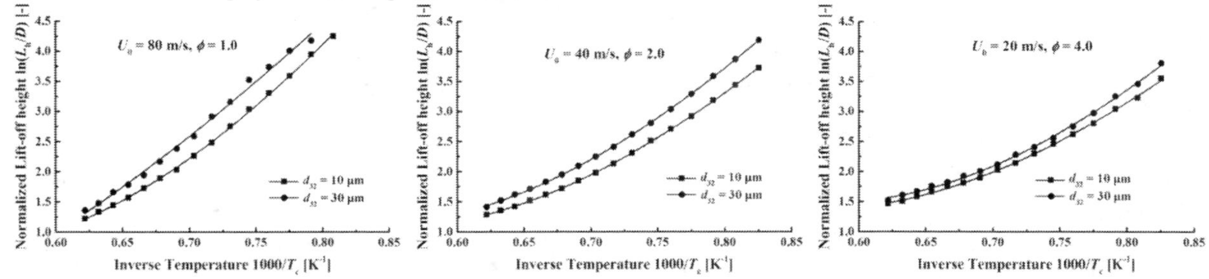

(a) Global equivalence ratio ϕ = 1.0; Jet velocity U_0 = 80 m/s

(b) Global equivalence ratio ϕ = 2.0; Jet velocity U_0 = 40 m/s

(c) Global equivalence ratio ϕ = 4.0; Jet velocity U_0 = 20 m/s

Figure 7: Effects of initial droplet size on flame lift-off height as a function of co-flow temperature under constant jet velocity and global equivalence ratio(Arrhenius-like expression fitting): (a) U_0 = 80 m/s, ϕ =1.0; (b) U_0 = 40 m/s, ϕ =2.0; (c) U_0 = 20 m/s, ϕ=4.0

4 Conclusions

Lift-off heights of n-heptane dilute spray flames in vitiated co-flows were investigated within a co-flow temperature range of 1200~1600K. Results indicate that the flame lift-off height is very sensitive to co-flow temperature, which could be expressed well with an Arrhenius-like expression. The influence of spray jet velocity on lift-off height is temperature-dependent. There exists a critical temperature, which decreases with spray global equivalence ratio. When the co-flow temperature is lower than the critical temperature, the lift-off height increases with increasing jet velocity, while the trend reverses when the co-flow temperature is higher than the critical value. The increase of spray global equivalence ratio leads to a reduced lift-off height. Flames of the sprays with smaller droplet size tends to have lower lift-off heights, while the reduction decreases as the co-flow temperature rises.

5 Acknowledgment

This work is supported by the National Natural Science Foundation of China (NO. 51106113).

References

[1] R.R. Cao, S.B. Pope, A.R. Masri, Combust. Flame 142 (4) (2005) 438-453.

[2] C.N. Markides, E. Mastorakos, Proc. Combust. Inst. 30 (1) (2005) 883-891.

[3] R.L. Gordon, A.R. Masri, E. Mastorakos, Combust. Flame 155 (1) (2008) 181-195.

[4] P. Domingo, L. Vervisch, D. Veynante, Combust. Flame 152 (3) (2008) 415-432.

[5] C.J. Lawn, Prog. Energy Combust. Sci. 35 (1) (2009) 1-30.

[6] K.M. Lyons, Prog. Energy Combust. Sci. 33 (2) (2007) 211-231

[7] E. Mastorakos, Prog. Energy Combust. Sci. 35 (1) (2009) 57-97.

[8] A. Sadiki, W. Ahmadi, M. Chrigui, et al, in: B. Merci, D. Roekaerts, A. Sadiki, Experiments and Numerical Simulations of Diluted Spray Turbulent Combustion, Springer, Dordrecht, 2011, p. 69.

[9] A. Sadiki, W. Ahmadi, M. Chrigui, in: B. Merci, D. Roekaerts, A. Sadiki, Experiments and Numerical Simulations of Diluted Spray Turbulent Combustion, Springer, Dordrecht, 2011, p. 111.

[10] H. Olguin, E. Gutheil, Combust. Flame 161 (4) (2014) 987-996.

[11] H. Olguin, E. Gutheil, in: B. Merci, E. Gutheil, Experiments and Numerical Simulations of Turbulent Combustion of Diluted Sprays, Springer, Cham, 2014, p. 79.

[12] C. Dumouchel, Exp. Fluids 45 (3) (2008) 371-422.

[13] D.R. Guildenbecher, C. López-Rivera, P.E. Sojka, Exp. Fluids 46 (3) (2009): 371-402.

[14] W. O'Loughlin, A.R. Masri, Combust. Flame 158 (8) (2011) 1577-1590.

[15] W. O'Loughlin, A.R. Masri, Flow Turbulence Combust 89 (1) (2012) 13-35.

[16] A.R. Masri, A. Kourmatzis, W. O'Loughlin, et al, in: B. Merci, E. Gutheil, Experiments and Numerical Simulations of Turbulent Combustion of Diluted Sprays, Springer, Cham, 2014, p. 1.

[17] R. Cabra, J.-Y. Chen, R.W. Dibble, et al, Combust. Flame 143 (4) (2005) 491-506.

[18] Y. Wu, Y. Lu, I.S. Al-Rahbi, et al, Int. J. Hydrogen Energ. 34 (14) (2009) 5940-5945.

[19] L.M. Pickett, D.L. Siebers, C.A. Idicheria, SAE Technical Paper No. 2005-01-3843.

11th Asia-Pacific Conference on Combustion,
The University of Sydney, NSW Australia
10th -14th December 2017.

An experimental investigation of the heat transfer performance in a hybrid solar receiver combustor operating with the solar-only and combustion-only modes

Alfonso Chinnici[1], Zhao F. Tian[1], Graham J. Nathan[1], Bassam B. Dally[1]

[1]School of Mechanical Engineering, Centre for Energy Technology, The University of Adelaide
The University of Adelaide, South Australia, 5005, Australia

Abstract

This paper reports an experimental investigation of the heat transfer characteristics in a laboratory-scale hybrid solar receiver combustor (HSRC) when operated under either combustion or solar-only mode. It aims to investigate the performance of the HSRC under these two modes of operation. A conventional swirl-stabilised natural gas flame and a 5-kW$_{el}$ xenon-arc solar lamp simulator were used as energy source for the combustion and solar modes, respectively. The influence of the mode of operation and heat transfer fluid (HTF) flow rate on the thermal efficiency, heat losses, heat flux distribution within the cavity were investigated. Despite the different contributions from the two dominant modes of heat transfer, namely radiation and heat convection, under the different regimes of operation, it was found that the device can achieve similar thermal efficiency in both modes (considering heat recovery from the exhausts). Also, the heat flux distributions on the HTF coils were found to be significantly different under the two modes of operation.

1 Introduction

The concept of hybridising two thermal energy technologies namely; concentrated solar thermal, CST, and combustion is receiving growing attention due to the complementary nature of these two energy sources [1-3]. The use of concentrated solar radiation, CSR, in power generation and industrial process heat applications offers low net greenhouse gas emissions but presently suffers from relatively high cost and the challenge of managing the intermittent and variable nature of the resource. In contrast, the combustion of fossil fuels offers broad availability and low fuel cost, but at the expense of high CO_2 and pollutant emissions [4]. Alternative fuels, such as biomass and solar fuels, offer lower net CO_2 emission although it suffers from seasonal fluctuations and relatively higher cost. An opportunity to reduce the cost and increase the penetration of CST technology is therefore to combine it with established technologies employing fossil, or renewable fuels. Such hybrid systems offer the potential to lower the cost of solar thermal power, due to the reduced infrastructure and potential thermodynamic synergies [5]. However, further research is required to develop hybrid systems that are both energetically efficient and technologically robust.

Recently, direct integration of a solar receiver with a combustor into a single device, which is used to harness both energy sources (solar and combustion), has been proposed as a potential means to lower the levelised cost of energy by overcoming limitations of current hybrid solar-combustion systems requiring the connection of two different devices [6-7]. This novel concept of hybrid receiver-combustor, termed the Hybrid Solar Receiver Combustor, HSRC, allows a single device to operate in three modes: solar-only (when solar radiation is abundant), combustion-only (in the absence of solar energy) and a mixed-mode (a combination of both solar and combustion, to manage short and long-term variability of the solar source). Previous techno-economic studies [8-11] have shown that the HSRC offers advantages over existing hybrids of shared infrastructure and reduced start-up and shut down losses. Also, a recent numerical study [12-13] provided a detailed heat transfer analysis and estimate of the thermal performance for both the solar-only and combustion-only modes of operation of the device. It also identified the key geometrical and operative parameters influencing the thermal performance of the device. However, no experimental data on the thermal performance of the HSRC operating under the different modes of operation have been reported in literature, so that a comprehensive understanding of its heat transfer characteristics is not presently available. Furthermore, experimental data are required to develop accurate modeling and scaling/optimisation tools of the device in order to progress its development.

To meet the aforementioned needs, the current study reports a systematic experimental assessment of the influence of the mode of operation and operative conditions with the aim of understanding their influence on the multiple performance criteria of thermal efficiency, heat losses and flux distribution within a laboratory-scale HSRC.

2 Methodology

2.1 Experimental arrangement and operative conditions

A schematic diagram of the laboratory-scale HSRC is presented in Figure 1. It features an insulated cavity with an aperture to let in concentrated solar radiation. The cavity body is a 300-mm diameter (*D*), 700-mm length (*L*), cylinder/cone made of alumina, with a 60-mm diameter aperture (d_{ap}). Conduction heat losses are minimized by lining the cavity with 60% SiO_2 – 30% Al_2O_3 – 10% ZrO_2 ceramic thermal insulation. The outer shell is made of 316L stainless steel. The heat is transmitted to the heat transfer fluid (HTF) through four helical-coils (made of 253MA stainless steel) positioned along the x-axis of receiver (260 mm inner coil diameter, d_{coil}, 18 and 20 mm pipe diameter and thickness, respectively). The outlet temperature of the HTF (air) from each coil was monitored by N-type thermocouples. The aperture is closed by a conventional swirl burner (geometrical Swirl number, S_g = 0.5) when the device is operated under combustion-only mode. The inner cavity temperature (alumina layer) was monitored employing N-type thermocouples,

Corresponding author. Dr. Alfonso Chinnici
Tel: +61883132307
E-mail address: alfonso.chinnici@adelaide.edu.au

positioned along the x-axis of the receiver. The device can also operate under MILD combustion regime employing six auxiliary fuel and air injectors. The outlet section of the receiver consists of a cone made of high-temperature stoneware clay (the shape was optimised for MILD combustion operations). Electronic mass flow controllers (ALICAT MCR-series) were used to control the flow rates of fuel, combustion air and HTF. For the combustion-only mode, natural gas (LHV = 45.8 MJ/Kg, composition: CH_4 = 92%, C_2H_6 = 4.5%, inerts = 3.5% by mole) was employed as energy source. The thermal capacity and the equivalence ratio were set to 8 kW and 0.9, respectively. No air combustion preheating and/or heat recovering from the exhaust gas were considered here. For the solar-only mode, a 5-kW$_{el}$ short arc xenon lamp (Osram XBO 5000W/HBM OFR unit) close-coupled with an ellipsoidal reflector [14] was used as energy source. In the present study, the focal point was positioned 90 mm far from the aperture plane (diverging beam). With this arrangement, the estimated energy introduced into the cavity through the aperture, $\dot{Q}_{solar,in}$, was estimated to be 0.5 kW, based on the heat flux profiles reported by [15]. The outlet section of the solar receiver was not closed, resulting in additional radiative and convective heat losses. For both modes of operation, the total HTF flow rate was varied in the range 200-600 standard liter per minute (slpm) and the tilt angle was fixed to 0° (horizontal position). Time-averaged temperatures were measured and recorded under steady-state conditions. In the present study, steady-state conditions were assumed when the values of all the measured temperatures did not vary of up to 10°C in 0.5 hr. The warm-up time of the receiver to reach steady-state operation from a cold state was of about 3 hr.

2.2 Heat Transfer analyses

For the combustion-only mode, the overall energy balance of the system is calculated as follows:

$$P_{in} = \dot{Q}_{abs} + \dot{Q}_{ex} + \dot{Q}_{cond} \quad (1)$$

where P_{in} is the total thermal input (comprising the fuel, $\dot{Q}_{fuel,in}$, and air) and $\dot{Q}_{abs}$ is the rate of energy absorbed by the HTF. The terms $\dot{Q}_{ex}$ and $\dot{Q}_{cond}$ are the sensible heat in the exhausts and the conductive heat losses, respectively. The overall thermal efficiency of the system, η_{th}, was calculated as follow:

$$\eta_{th} = \frac{\dot{Q}_{abs}}{\dot{Q}_{fuel,in}} \quad (2)$$

The heat losses terms were estimated by monitoring the exhausts gas temperature (employing a N-type thermocouple) and the outer shell temperature (employing an infrared thermometer).

For the solar-only mode, the overall energy balance of the system is calculated as follows:

$$\dot{Q}_{solar,in} = \dot{Q}_{abs} + \dot{Q}_{rad} + \dot{Q}_{conv} + \dot{Q}_{cond} \quad (3)$$

where $\dot{Q}_{rad}$ and $\dot{Q}_{conv}$ are the radiative and convective heat losses, respectively. The term $\dot{Q}_{rad}$ was estimated with the following equation:

$$\dot{Q}_{rad} = \varepsilon_{eff}\ \pi(r_{ap}^2 + r_{out}^2)\sigma T_c^4 \quad (4)$$

where T_c is the nominal cavity temperature, ε_{eff} is the emittance of the cavity receiver and σ is the Stefan-Boltzmann constant. The convective losses were obtained as difference from the overall energy balance. The thermal efficiency of the solar cavity receiver was obtained as follows:

$$\eta_{th} = \frac{\dot{Q}_{abs}}{\dot{Q}_{solar,in}} \quad (5)$$

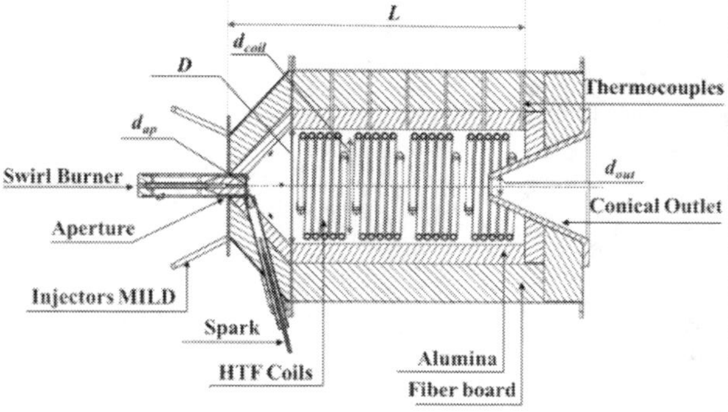

Figure 1: Schematic diagram of the laboratory-scale HSRC (side-view).

3 Results and discussions

Figure 2 presents the measured inner cavity temperature (alumina layer) along the length of the receiver as a function of the HTF flow rate and mode of operation. It can be seen that the maximum temperature occurs close to the aperture plane (x/L ≈ 0.1) for the combustion-only mode while it is located further downstream (x/L ≈ 1) for the solar-only mode. It is worth noting that this trend is different to that generally generated by a heliostat field (for which the maximum peak of absorption is close to the aperture of the receiver). This difference is attributed to the particular shape of the beam of the solar simulator used in this study. Noteworthy, that a 20 kW$_{el}$ metal halide lamp system (consisting of 3 lamps) is being constructed and will be used for further experiments in the future. Also, it was found that for both modes of operation, the maximum and mean cavity temperatures decrease with an increase of HTF flow rate due to an increase of the total heat absorbed through the coils.

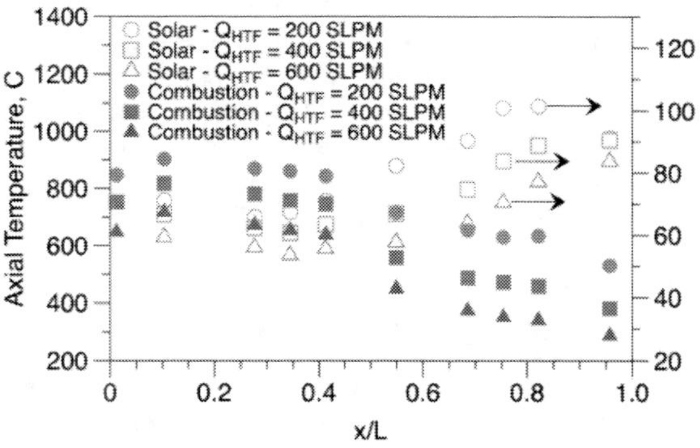

Figure 2: Inner cavity temperature (alumina) along the length of the receiver as a function of the HTF flow rate and mode of operation.

Figure 3 presents the measured thermal efficiency and the maximum outlet temperature of the HTF from the coils, $T_{max,HTF}$, as a function of the HTF flow rate, for all the experimental cases investigated. For the combustion-only mode, the values of η_{th} calculated by assuming a partial heat recovery of the sensible heat in the exhausts through an external heat exchanger (effectiveness = 0.9) are also reported. It can be seen that solar-only mode operations are characterized by a higher η_{th} compare to that of combustion mode without heat recovery of the exhaust gases (while similar performance can be potentially achieved by recovering a portion of the sensible heat). Also, η_{th} increases with an increase of the HTF flow rate (up to ≈70% with $T_{max,HTF}$ ≈670°C, and up to ≈90% with $T_{max,HTF}$ ≈100°C, for the combustion-only and solar-only mode, respectively.

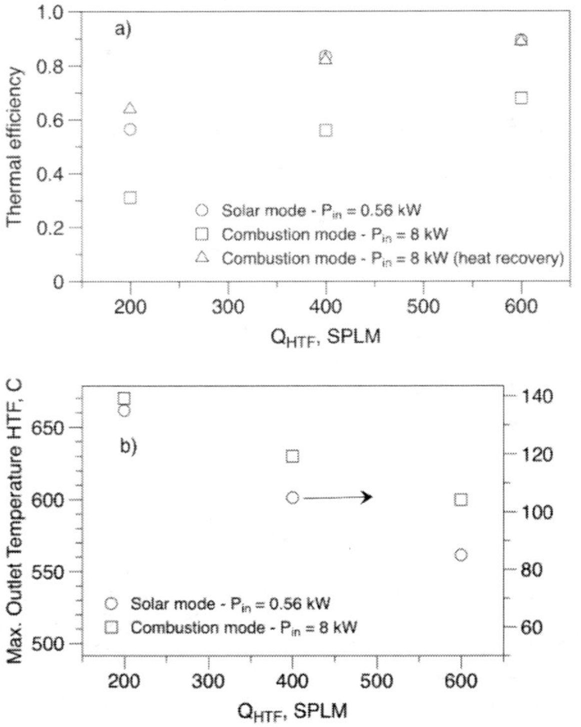

Figure 3: a) Thermal efficiency and b) maximum outlet HTF temperature as a function of the HTF flow rate for the combustion and solar modes of operation.

Figure 4 presents the distribution of the absorbed power and heat losses inside the cavity as a function of the HTF flow rate and mode of operation. For the combustion mode of operation, it can be seen that $\dot{Q}_{ex}$ represents the main heat loss for all the conditions analysed (up to 50% of the total energy input) while $\dot{Q}_{cond}$ is significant only for low HTF flow rates (up to 15% of the total energy input). For the solar mode of the operation, it was found that $\dot{Q}_{rad}$ is significantly lower than $\dot{Q}_{conv}$ and $\dot{Q}_{cond}$ for all the cases investigated. In particular, $\dot{Q}_{rad}$ was only ≈1% of the total energy input (due to a low value of T_c) while $\dot{Q}_{conv}$ was the main heat losses through the cavity, accounting for up to 25% of the total energy input. For both mode of operation, it can be seen that an increase in HTF flow rate leads to an increase in $\dot{Q}_{abs}$. In particular, the calculated average heat transfer coefficient was found to increase with HTF flow rate (from ≈12W/m^{2}K for Q_{HTF} = 200 slpm up to ≈20W/m^{2}K for Q_{HTF} = 600 slpm), indicating a higher heat transfer rate for higher HTF flow rates.

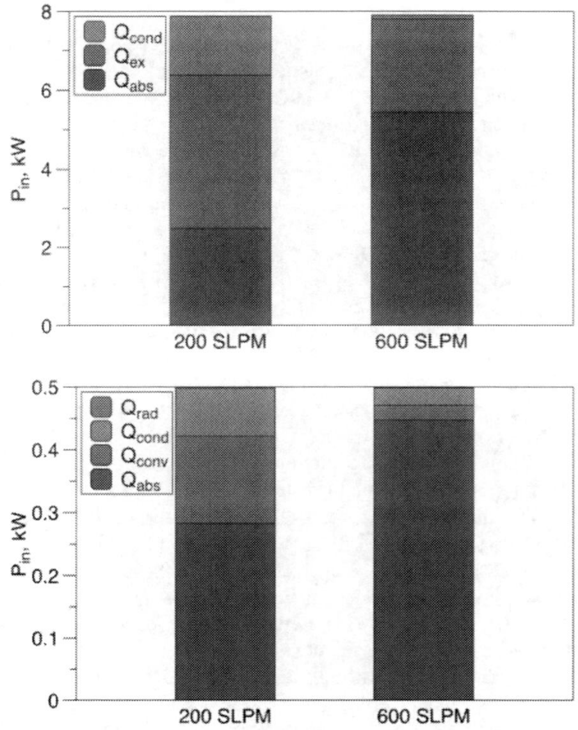

Figure 4: Heat absorbed and heat losses as a function of the HTF flow rate for the combustion and solar modes of operation.

Figure 5 presents the average axial distribution of heat flux to the HTF coils normalised by the peak flux, q/q_{max}, for all the cases investigated. It can be seen that the heat flux distributions on the HTF coils were found to be significantly different under the two modes of operation. In particular, for the combustion mode, the peak flux occurs close to the aperture plane, while it occurs further downstream for the solar-only mode. This demonstrates the importance of considering the difference in heat flux distribution in the material selection and design strategies.

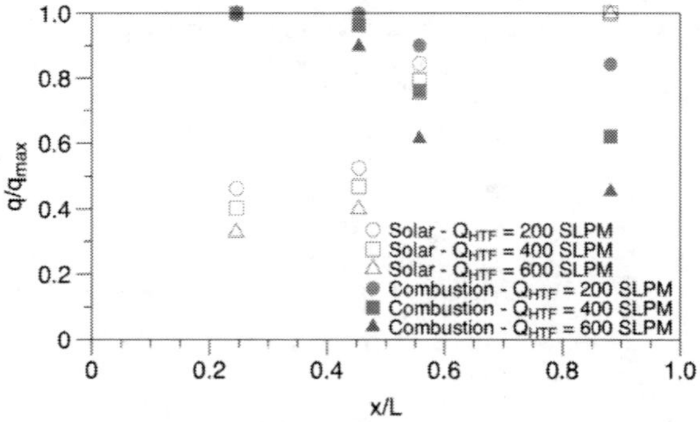

Figure 5: Normalised heat flux distribution along the length of the receiver as a function of the HTF flow rate and mode of operation.

4 Conclusions

The key outcomes from the experimental investigation on the thermal performance of a laboratory-scale HSRC operating with the solar-only and combustion-only modes are as follows:

- The thermal efficiency, η_{th}, is calculated to be lower for the combustion-only mode of operation than for the solar-only mode;

- A recover of the sensible heat from the exhaust gas is required to potentially provide similar heat transfer performance under the two modes of operation;
- The combustion-only (employing conventional combustion) and solar-only modes of operation exhibit a different trend in the axial distribution of heat flux on the HTF coils. In addition, the magnitude of the normalised peak is lower for the combustion mode than for the solar mode.

5 Acknowledgment

The support of the Australian Research Council and of FCT Combustion Pty. Ltd. and of Vast Solar Pty. Ltd. through the ARC Linkage scheme LP110200060 is gratefully acknowledged.

References

• Ordorica-Garcia, G. et al., 2011. Novel integration options of concentrating solar thermal technology with fossil-fueled and CO_2 capture processes. Energy Proc. 4, 809–816.

• Moore, J., Apt, J., 2013. Can hybrid solar–fossil power plants mitigate CO2 at lower cost than PV or CSP? Environ. Sci. Technol. 47 (6), 2487–2493.

• Nathan, G.J. et al., 2017. Research challenges in combustion and gasification arising from emerging technologies employing directly irradiated concentrating solar thermal radiation. Proc. Combust. Inst. 36 (2), 2055–2074.

• Kreith, F., Goswami, D.Y., 2007. Handbook of Energy Management and End Use Efficiency. CRC Press.

• Peterseim, J.H. et al., 2013. Concentrated solar power hybrid plants, which technologies are best suited for hybridisation? Renewable Energy 57, 520–532.

• Nathan, G.J. et al., 2013. A hybrid receiver-combustor. US Patent 2015/0054284. A.R. I.P. Ltd.

• Nathan, G.J. et al., 2014. Economic evaluation of a novel fuel-saver hybrid combining a solar receiver with a combustor for a solar power tower. Appl. Energy 113, 1235–1243.

• Lim, J.H. et al., 2016a. Analytical assessment of a novel hybrid solar tubular receiver and combustor. Appl. Energy 162, 298–307.

• Lim, J.H. et al., 2016b. Impact of start-up and shut-down losses on the economic benefit of an integrated hybrid solar cavity receiver and combustor. Appl. Energy 164, 10–20.

• Lim, J.H. et al., 2016c. Techno-economic assessment of a hybrid solar receiver and combustor. AIP Conf. Proc. 1734, 070020.

• Lim, J.H. et al., 2016d. Assessment of the potential benefits and constraints of a hybrid solar receiver and combustor operated in the MILD combustion regime. Energy 116 (Part 1), 735–745.

• Chinnici, A. et al., 2017. Comparison of system performance in a hybrid solar receiver combustor operating with MILD and conventional combustion. Part II: Effect of the combustion mode. Sol. Energy 147:479-488.

• Chinnici, A. et al., 2017. Comparison of system performance in a hybrid solar receiver combustor operating with MILD and conventional combustion. Part I: Solar-only and combustion-only employing conventional combustion. Sol. Energy 147: 489-503.

• Dong, X., et al. 2015. Concentric multilayer model of the arc in high intensity discharge lamps for solar simulators with experimental validation. Sol. Energy 122: 293-306.

• Craig, R.A., 2010. Investigating the Use of Concentrated Solar Energy to Thermally Decompose Limestone. Ph.D. thesis, The University of Adelaide, Adelaide, SA.

11th Asia-Pacific Conference on Combustion,
The University of Sydney, NSW Australia
10th -14th December 2017.

Experimental Study on Velocity Deficits of Near-Limit Detonation Waves Initiated in Narrow Channels

Yuefei Xiong[1], Hung-Ping Chan[2], Ming-Hsun Wu[2*], Jiang Qin[1], Chien-Fu Chen[3]

[1]Institute of Advanced Power, Harbin Institute of Technology

Hexing, Harbin, Heilongjiang, 150001, China

[2]Department of Mechanical Engineering, National Cheng Kung University

1 University Rd, Tainan, 70101, Taiwan

[3]Institute of Applied Mechanics, National Taiwan University

No. 1, Sec. 4, Roosevelt Rd, Taipei, 10617, Taiwan

Abstract

The effect of channel size and equivalence ratio of ethylene/oxygen mixtures on detonation velocity deficits were experimentally investigated in miniature square channels in this paper. The premix mixtures were spark ignited under ambient pressure and temperature, and detonation waves were formed through deflagration-to-detonation transition (DDT) process in the smooth and straight channel in which no cross-sectional area change existed. The reaction wave in the small channels went through flame acceleration, detonation transition, and overdriven detonation before stably propagated along the chanel at a value lower than the Chapman-Jouguet velocity. As the channel size decreased from 1.2 mm to 0.6 mm, both DDT time and DDT distance reduced while the detonation velocity deficit increased. In the 0.8 mm channel, the propagation speed of detonation wave dropped after a certain distance after detonation transition, and then fluctuated around a new value. Low-speed detonation mode appeared when the channel size was 0.6 mm. Moreover, the detonation velocity deficit gradually increased with the decrease of equivalence ratio within the test range. Non-dimensional detonation velocities V/V_{CJ} becomes invariant as non-dimensional number D_H/λ is larger than 1.4. But the deficit is not only a function of D_H/λ for near-limit conditions when the channel dimension approaches or becomes smaller than the cell size.

1 Introduction

The phenomenon that experimentally measured detonation wave propagation velocities are less than the corresponding Chapman-Jouguet (CJ) values is widely observed. The deficit becomes more significant as the near-limit conditions for detonation propagation are approached [1-3]. The discrepancy is likely results of frictional drag on the wall and heat transfer to the wall, which are neglected in CJ velocity (V_{CJ}) calculations using classical theory proposed by Zel'dovich [4]. Many researchers have studied this phenomenon and several theories have been proposed to calculate the deficit value. A more realistic model in which the effects of boundary layer had been considered was proposed by Fay [5]. In Fay's theory, the divergence of streamlines due to the existence of boundary layer causes fluid movement from the main stream to the boundary layer, resulting in a negative displacement thickness. Other models can be found in Refs. [6] and [7].

Successful validation of the models mentioned above were reported in some works, but there were experiments showing contradicting results. Kitano et al. [8] studied detonation velocity deficits in small diameter round tubes. By reducing the initial pressure and changing the equivalence ratio of mixtures, the near-limit conditions were approaching. Their results showed that velocity deficits obtained in experiments were well predicted by the modified ZND model with full detailed chemical kinetics. They also found that the cell size increased as the initial pressure reduced. Gao et al. [9] carried out an experimental investigation of detonation limits in hydrogen-oxygen-argon mixtures. Two mixtures representing 'stable' and 'unstable' mixture, respectively. Mixtures in circular tubes with three different diameter were tested. Their results revealed the difference between 'stable' and 'unstable' mixtures. The detonation velocity deficits predicted by Fay's theory agreed well with the experimental results for 'stable' mixture, but deviated from the experimental results for 'unstable' mixture. This indicated that the inherent instabilities of detonations are also important for the propagation of detonation wave. Identical findings were obtained by Camargo et al [10]. Since detonation velocity deficits predicted by these theories do not coincident with experimental data precisely for all cases, some more realistic models need to be developed.

Another issue is to find an appropriate criteria to define the detonation limits. Some researchers considered the onset of spinning detonation as the detonation limit because of the instability of the spinning detonations and their frequent appearance in experiments [11]. However, spinning detonation may be very stable under certain conditions [12]. Different from this phenomenological criteria, a criteria based on the non-dimensional number, d/λ (the ratio of tube diameter, d, to detonation cell size, λ, which represented the number of cells across the tube diameter) seems to be more objective and practical. Lee et al. [13] used d/λ to characterize the mixture by investigating the near-limit behavior of the detonation velocity. As the limit was approached, rapid decrease of the detonation velocity occurred within a narrow range of values of d/λ. They also found this narrow range corresponded to the onset of spinning detonation. Gao et al. [9] suggested that hydraulic diameter was a more appropriate value for the calculation of d/λ when non-circular channels were used. Camargo et al. [10]

*Corresponding author. Fax: +886-6-235-2973
E-mail address: minghwu@mail.ncku.edu.tw

found that for stable mixtures, Fay's theory could predict the detonation limits accurately. But for unstable mixtures, the relationship, $D_H/\lambda \approx 1/\pi$, seemed to be a more appropriate criteria. Ishii et al. [14] found that a combination of the calculated velocity deficit and the number of cells benefited the prediction of propagation limits of detonations. Their work offered another way of thinking on the prediction of detonation limits.

Near-limit propagation characteristics were also investigated in relevant works. Wu & Wang [15] experimentally studied the reaction propagation scenarios of ethylene/oxygen mixtures in transparent capillary tubes. Several terminal reaction propagation modes, including near C-J, low-speed, galloping, and quasi-detonation, were observed. Zhang et al. [16] investigated the detonation velocity fluctuation phenomenon for stoichiometric methane-hydrogen-oxygen mixture under near-limit conditions in small diameter round tubes. Six different detonation modes were recognized in their experiments. Their results indicated that the influence of boundary layer on detonation velocity deficits was more significant for small diameter tubes than big ones. They also studied the near-limit characteristics of methane–oxygen detonation [17]. Smoked foil was used to record the cellular structure of detonation wave under different conditions. By this method, they obtained clear pictures of the multi-, double- and single-headed structure. The influence of wall roughness on detonation limits was also investigated in hydrogen-oxygen mixture [18]. By inserting a Schelkin spiral with different wire diameter and pitch, rough walls with different roughness were generated. Experimental results indicated that wall roughness had a contradictory effect on detonation limits under different conditions. The roughness could benefit the detonation and extend the limits only when it was small than a certain value. With the wall becoming rougher and rougher, the inhibition effect of wall on detonation limits was more and more significant and obvious. Other experimental research can be found in paper [19-21].

The purpose of the present study is to investigate the detonation velocity deficits in small square channels. Different from previous experiments in which detonation wave is first initiated in a larger channel before transmiting into the smaller tube, channel, or annular gap for deficit measurements, detonation were initiated through DDT process in the same narrow channel in the present study. This detonation initiation method eliminate potential interference of detonation wave propagation during detonation transmission. Instead of changing d/λ through reducing the initial pressure as most works in the literature, the equivalence ratio of the ethylene/oxygen mixture initially at ambient temperature and pressure was altered to vary λ in the present work while initial pressure was kept constant. Influence of initial pressure on heat transfer characteristics can therefore be minimized with the present approach.

2 Experimental Setup

The present small channel assembly is shown in Fig. 1. The detonation channel sandwiched by two glass windows is fabricated on a stainless steel plate using electro-discharge machining (EDM). The smooth channel is square in cross-section with an overall length of 650 mm. Four channel sizes, 0.6, 0.8, 1.0, and 1.2, were investigated. A pair of holders fix and seal the channel and glass windows. Two ignition electrodes are epoxied in the small holes drilled at the middle of each glass window. On one of the glass windows, two holes are drilled at the ends for mixture feeding and purging.

Figure 2 shows our experimental apparatus. Ethylene and oxygen are chose as fuel and oxidant respectively. Their flow rates are regulated by electronic mass flow controllers (1179A, MKS Instruments) resulting in equivalence ratio uncertainty smaller than ±0.01. A spark generator is used to generate sparks whose energy is enough to ignite the mixture but far less than the value needed for direct detonation initiation. Identical setups were utilized in Refs. [22].

Figure 1: Detonation channel assembly

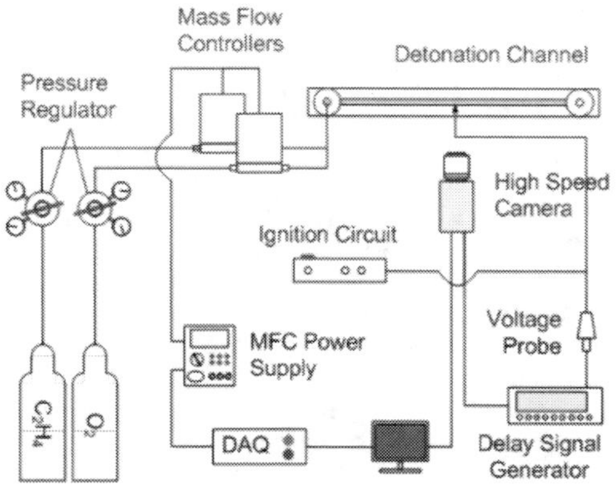

Figure 2: Experimental apparatus

Reaction propagation processes are recorded by a high-speed camera (Phantom v12.1, Vision Research). The highest frame rate for the recordings is 270,000 fps. The high-speed camera is triggered at the instant when voltage breakdown across the electrodes occurs. Fig. 3 shows a collection of pictures captured by the high-speed camera. A Matlab program was developed to recognize the flame front and record its position. The reaction front velocity was calculated by the second order central difference algorithm. Since the error of position of flame front recognized by the program is one pixel, the corresponding uncertainty of reaction front velocity is 38 m/s.

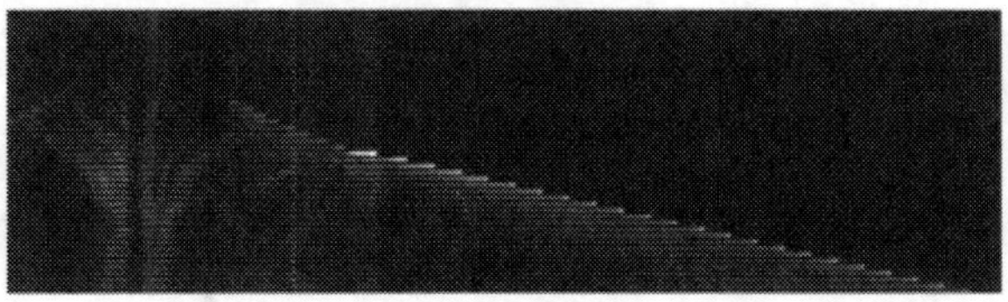

Figure 3: Pictures of detonation flame captured at different instants

In our experiments, the initial pressure and temperature are the ambient pressure and temperature. So the near-limit condition can be approached only by reducing the detonation channel size and changing the equivalence ratio. In fact, these methods are adopted and their influence on the detonation velocity deficits is investigated. The detonation channel sizes vary from 1.2 mm to 0.6 mm which are smaller than values found in previous published papers. The range of equivalence ratio is from 0.5 to 1.7.

3 Results and Discussions

3.1 Effects of Channel Size

Reaction front velocity evolutions corresponding to different channel sizes are shown in Fig. 4. The equivalence ratio of mixtures was kept at stoichiometric for all tests. When channel sizes are 1.2 mm and 1.0 mm, the stable detonation wave propagation velocities are nearly the same, indicating that the detonation propagations are still far from the limiting state.

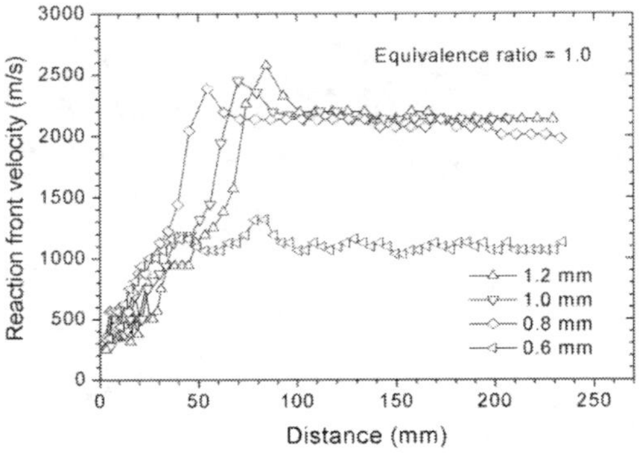

Figure 4: Velocity evolutions of the reaction fronts in channels with different sizes

In the 0.8 mm channel, there exists a drop on detonation wave propagating velocity after steady propagation over a certain distance, and velocity fluctuation occurs thereafter, as shown in Fig. 5. The behavior of velocity fluctuation seems like the stuttering mode except that the amplitude of velocity fluctuation in the present experiments is smaller than those reported in Ref. [16]. This experimental observation reveals that the detonation waves under this near-limit condition is unstable.

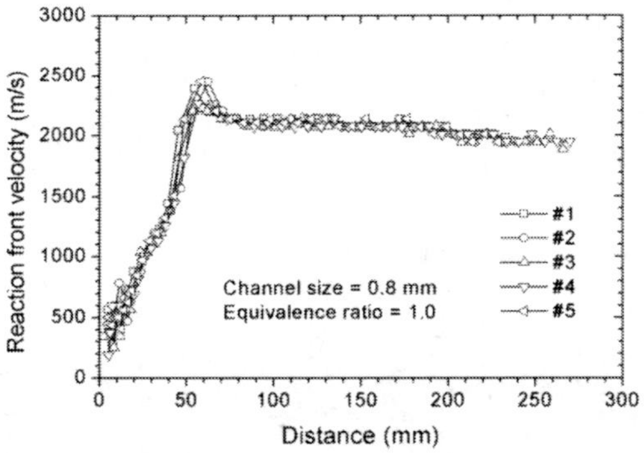

Figure 5: Velocity evolutions of the reaction fronts in 0.8 mm channel

Low-speed mode is observed for detonation waves propagating in the 0.6 mm channel. The reaction front velocity increases rapidly after ignition, then fluctuates at around 0.46 V_{CJ}. The amplitude of velocity fluctuation gradually decreases with the detonation propagating. A longer channel will be needed in the future to verify if the oscillation will completely diminish, and the velocity eventually becomes fully steady.

Figure 6 shows the dependence of detonation velocity deficit on channel size. Since the detonation wave velocity suddenly reduces after steady propagation over a certain distance when 0.8 mm width channel is used, the mean value of velocities after the reduction is chose to calculate the velocity deficit. The result shows detonation velocity deficit increases as the channel size reduces.

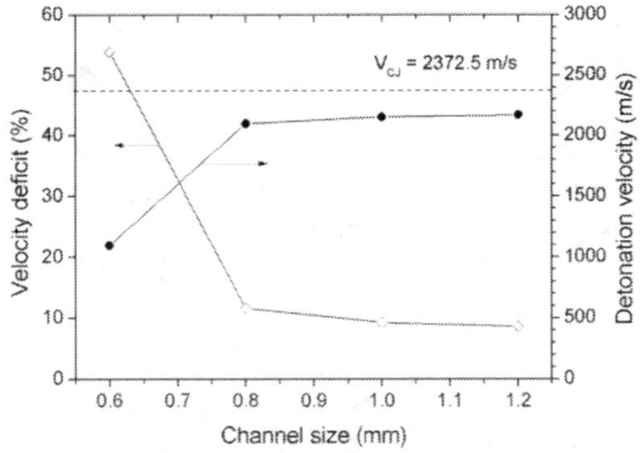

Figure 6: The effect of channel size on the detonation velocity deficit

3.2 Effects of Equivalence Ratio

Typical reaction front velocity evolutions for mixtures of various equivalence ratios from fuel lean to fuel rich are plotted in Fig. 7. As depicted in the figure, the detonation wave propagating velocity increases with equivalence ratio of the mixture, while DDT distance becomes shorter.

Figure 8 shows the dependence of detonation velocity deficit on equivalence ratio. The experimentally measured detonation velocities and ideal CJ velocities are also shown in the figure for comparison. The deficit is around 10 % except for mixtures close to the lean limit. The measured and ideal detonation velocity both increase with the equivalence ratio, but the deficit gradually reduces as the mixture becomes richer.

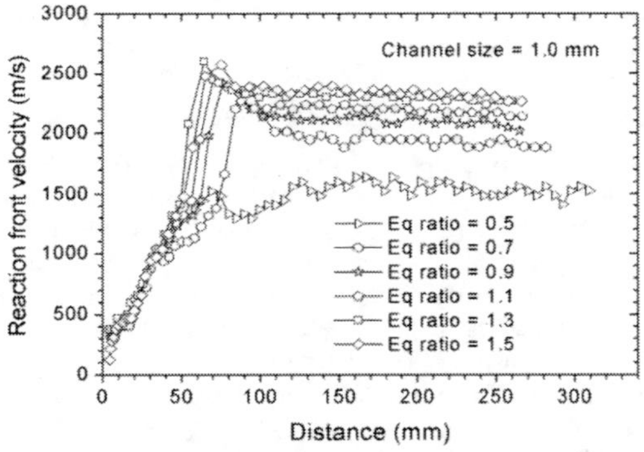

Figure 7: Velocity evolutions of the reaction fronts under different equivalence ratios

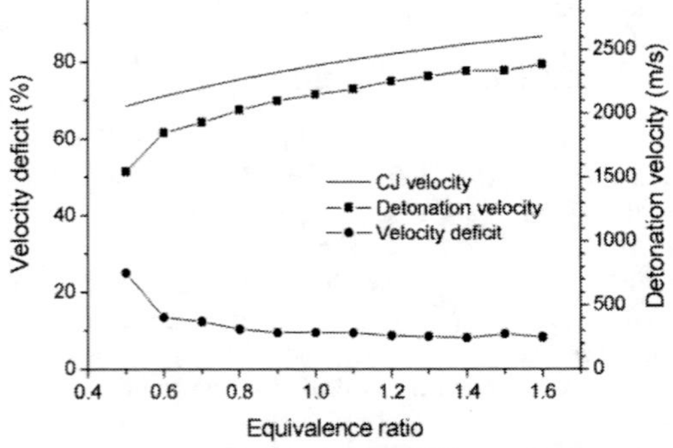

Figure 8: The effect of equivalence ratio on the detonation velocity deficit

In order to verify the dependence of V/V_{CJ} on D_H/λ, relationship between cell size and equivalence ratio of ethylene/oxygen mixture under ambient pressure and temperature reported in Refs [23-25] are compiled in Fig. 9, and a second order polynomial fitting curve is obtained for calculating D_H/λ. The resulting V/V_{CJ} versus D_H/λ plot is shown in Fig. 10. Results reported in Refs [1] are also plotted in the figure for comparison. The figure shows that V/V_{CJ} is approximately 0.9, and almost invariant if D_H/λ is larger than 1.4. The deficit is also not affected by the way D_H/λ is varied when the hydraulic diameter is 1.4 times larger than the cell size. However, significant discrepancies on V/V_{CJ} exist in the region where the hydraulic diameter approaches the cell size. D_H/λ alone is not sufficient for determining detonation deficit.

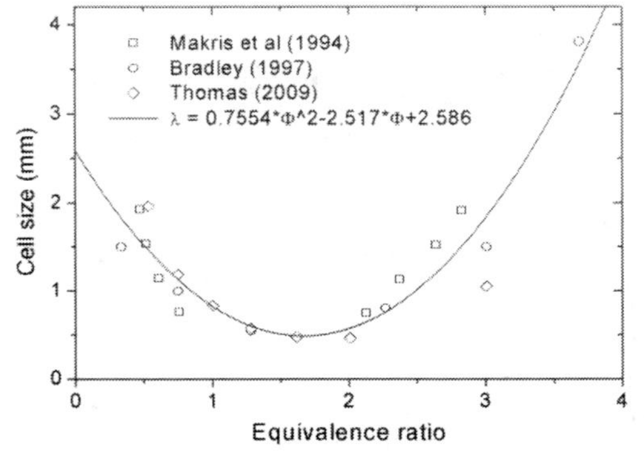

Figure 9: Variation of detonation cell size with equivalence ratio for ethylene/oxygen mixtures under ambient pressure and temperature

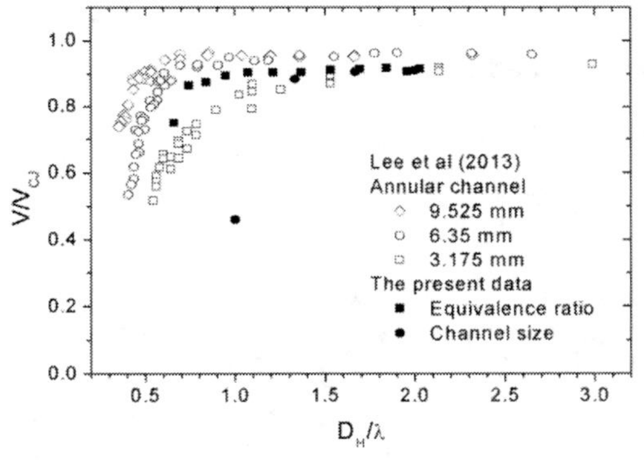

Figure 10: The dependence of non-dimensional detonation velocity V/V_{CJ} on non-dimensional number D_H/λ

4 Conclusions

Effects of channel size and equivalence ratio on the detonation velocity deficit were experimentally investigated by applying high-speed cinematography on recording reaction wave propagating in transparent channels. As the channel size reduces, detonation velocity deficit gradually increases. With 0.8 mm channel, unstable detonation wave propagation is observed. Rapid increase of velocity deficit occurs when channel size is smaller than 0.8 mm. Low speed detonation mode appears in the 0.6 mm channel. The detonation velocity increases as the equivalence ratio of the mixture increase from fuel lean to fuel rich within the test range between 0.5 and 1.7. However, the velocity deficit is smaller in richer mixtures. Dependence of non-dimensional detonation velocity V/V_{CJ} on non-dimensional number D_H/λ is analyzed and compared with data reported in the literature. It is found that the non-dimensional detonation velocity approximately equals to 0.9 for channels with hydraulic diameter larger than 1.4 times of the cell size. As the ratio between hydraulic diameter and cell size becomes smaller than 1.4, universal correlation between V/V_{CJ} and D_H/λ does not exist.

5 Acknowledgment

Financial supported from Ministry of Science and Technology, Taiwan under grant MOST 106-3113-E-006-012 was gratefully acknowledged.

References

[1] J.H.S. Lee, A. Jesuthasan, H.D. Ng, Proc. Combust. Inst. 34 (2013) 1957-1963.
[2] K. Ishii, K. Itoh, T. Tsuboi, Proc. Combust. Inst. 29 (2002) 2789-2794.
[3] B. Zhang, X. Shen, L. Pang, Int. J. Hydrogen Energ. 40 (2015) 15078-15087.
[4] Y.B. Zel'dovich, NACA Technol. Mem. 1261 (1950) 542-568.
[5] J.A. Fay, Phys. Fluids 2 (3) (1959) 283–289.
[6] J. Dove, T.D. Tribbeck, Astronaut. Acta 15 (1970) 387–397.
[7] J. Yao, D.S. Stewart, Combust. Flame 100 (4) (1995) 519–528.
[8] S. Kitano, M. Fukao, A. Susa, et al, Proc. Combust. Inst. 23 (2) (2009) 2355-2362.
[9] Y. Gao, H.D. Ng, J.H.S. Lee, Shock Waves 27(2) (2017) 199-207
[10] A. Camargo, H.D. Ng, J. Chao, et al, Shock Waves 20 (2010) 499-508.
[11] I.O. Moen, M. Donato, R. Knystautas, et al, Proc. Combust. Inst. 18 (1981) 1615–1622.
[12] O.V. Achasov, O.G. Penyazkov, Shock Waves 11 (2002) 297–308.
[13] J.H.S. Lee, A. Jesuthasana, H.D. Ng, Proc. Combust. Inst. 23 (2) (2009) 2355-2362.
[14] K. Ishii, M. Monwar, Proc. Combust. Inst. 33 (2011) 2359-2366.
[15] M.H. Wu, C.Y. Wang, Proc. Combust. Inst. 33 (2011) 2287-2293.
[16] B. Zhang, C. Wang, X. Shen, et al, Int. J. Hydrogen Energ. 41 (39) (2016) 17750-17759.
[17] B. Zhang, X. Shen, L. Pang, et al, Fuel 177 (2016) 1-7.
[18] B. Zhang, Combust. Flame 169 (2016) 333-339.
[19] J. Chao, H.D. Ng, J.H.S. Lee, et al, Proc. Combust. Inst. 32 (2009) 2349-2354.
[20] K. Ishii, H. Gr¨onig, Shock Waves, 8 (1998) 55-61.
[21] S. Jackson, B.J. Lee, J.E. Shepherd, Combust. Flame 167 (2016) 24-38.
[22] M.H. Wu, W.C. Kuo, Combust. Flame 159 (2012) 3414-3422.
[23] A. Maris, T.J. John, H.S. Lee, et al, Proc. Combust. Inst. 25(1) (1994) 65-71.
[24] G.O. Thomas, J. Hazard. Mater. 163 (2009) 783-794.
[25] C.M. Bradley, A kinetic approach to detonation in gaseous hydrogen- and hydrocarbon-oxygen systems, PhD Thesis, Imperial College of Science and Technology, London, UK, 1997.

11th Asia-Pacific Conference on Combustion,
The University of Sydney, NSW Australia
10th -14th December 2017.

Measurements of Laminar Burning Velocity of Gasoline Surrogate Fuel /Air/EGR Gas Mixtures

Shota Doi[1], Hirokazu Uesaka[1], Ryosuke Matsui[1], Masamichi Matsuura[1], Ryunosuke Okazaki[1], Hidefumi Kataoka[1], Daisuke Segawa[1]

[1]Osaka Prefecture University

1-1 Gakuen-cho Naka-ku, Sakai, Osaka, 599-851, Japan

Abstract

In this study, the effects of EGR (Exhaust Gas Recirculation) ratio on laminar burning velocity were studied for gasoline surrogate fuel S5R /air mixtures. S5R is liquid fuel which composed with iso-octane, n-heptane, methylcyclohexane, toluene and diisobutylene. The test mixtures were obtained by using controlled evaporation and mixing system. The stretched laminar burning velocity of mixtures was determined from the measured pressure histories in a spherical vessel. The effect of stretch rate on the laminar burning velocity was analyzed using nonlinear method. The spherical vessel was made of 7075 aluminium alloy, with the inner diameter of 70 mm. The experiments of flame propagation were conducted in a short-drop free-fall laboratory facility that provides 0.6 s of microgravity condition. The laminar burning velocities decreased with EGR ratio increases. The laminar burning velocity ratio decreased with the adiabatic flame temperature ratio decrease.

1 Introduction

EGR (Exhaust Gas Recirculation) is used as an effective method for reducing NOx emissions of an internal combustion engine. Improvement of thermal efficiency is expected by using lean burn with EGR [1]. However, EGR worsens the ignitability and decreases the flame propagation speed. Therefore, technological development for enhanced ignitability and combustion promotion is an important task. For the academic support of such technological development, it is important to clarify the effects of EGR on the laminar burning velocity.

In general, laminar burning velocity is a combustion characteristic value determined by the composition, temperature, and pressure of a mixture, and is also used to verify the chemical reaction model. The effects of dilution gas [2]-[8] on the laminar burning velocity have been studied. The several equations have been proposed for the effect of EGR on the laminar burning velocity [2],[7]-[8].

One way to measure the laminar burning velocity is the constant volume spherical vessel method to determine the laminar burning velocity from the analysis of the behavior of the spherical propagation flame in a spherical vessel. In our previous study, measuring the laminar burning velocities and validation of estimation equation for lean and diluted propane/air mixtures that many results are reported, have been conducted using spherical vessel in microgravity[9].

For simulation of gasoline engine, it is necessary to construction of combustion model and verification using a gasoline. However, since gasolines are complex mixtures of hundreds of hydrocarbon, it is difficult to simulate using real gasoline. On the other hand, it is generally agreed that their combustion properties can be reproduced by simpler fuel surrogates containing few components. In our previous study, the effects of equivalence ratio and pressure on the laminar burning velocity of S5R [10] consisting of five components and designed to match regular Japanese gasoline with octane number (RON, MON) was studied. In this study, to clarify the effects of EGR gas on laminar burning velocity of S5R/air mixtures, we conducted the experimental study using spherical combustion vessel under microgravity.

2 Experimental Apparatus and Procedure

Figure 1 shows a schematic diagram of the experimental apparatus. The spherical vessel was made of 7075 aluminium alloy, with the inner diameter of 70 mm. The pressure transducer (Kistler, 6052C) and the ignitor have been installed in the spherical vessel. The pressure transducer was connected by wire to the data logger (Keyence, NR-2000), through the charge amplifier (Kistler, 5018) on the ground. Two stainless steel electrodes (diameter 1.2 mm) with a 5 mm gap were used for ignition at the center of the vessel. The spark energy was determined by integration of the measured voltage multiplied by the measured current during the spark discharge. The voltage and current were measured using the high voltage probe (Tektronix, P6015A) and the current probe (Pearson Electronics, 110) with digital oscilloscope (Yokogawa Meters & Instruments, DL7200). An electric heater located around the spherical vessel was used to heat test mixtures to initial temperature.

In this study S5R was used for test fuel. S5R is liquid fuel which composed with iso-octane, n-heptane, methylcyclohexane, toluene and diisobutylene. S5R has same research octane number (RON=90.8) and motor octane number (MON=82.9) with Japanese regular gasoline. Table 1 shows the component of S5R. EGR gas is simulated with 15 % carbon dioxide and 85 % nitrogen.

The test mixtures were obtained by using controlled evaporation and mixing system. Mass flow rate of S5R and synthetic air (79 vol. % N_2 and 21 vol. % O_2) were measured

Corresponding author. Fax: +81-72-254-9370
E-mail address: kataoka@me.osakafu-u.ac.jp

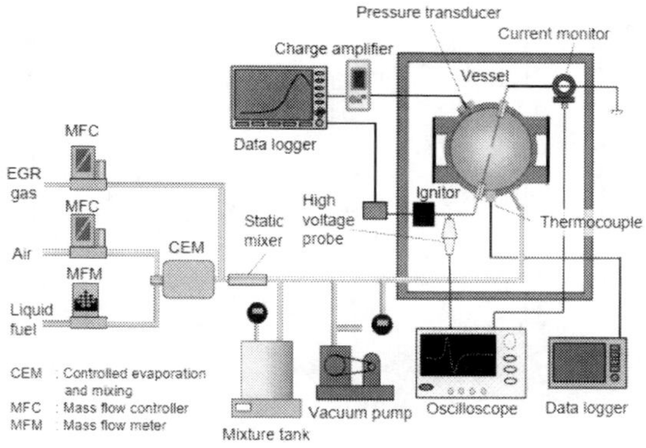

Figure 1: A schematic diagram of the experimental apparatus.

Table 1: Component of S5R.

Component	Vol. %
iso-Octane (2,2,4-Trimethylpentane)	29.0
n-Heptane	21.5
Methylcyclohexane	5.0
Toluene	30.5
Diisobutylene (TMP-1 and TMP-2)	14.0

by a coriolis flow meter (Bronkhorst mini CORI-FLOW 5 g/h) and a thermal mass flow controller (Bronkhorst EL-FLOW 0.5 ln/min) to convey the injected fluids to CEM. S5R was vaporized in a controlled evaporator mixer (Bronkhorst) and mixed with synthetic air. Mixtures were introduced into stainless steel mixture tank through heated stainless tube and ware stirred constantly.

Equivalence ratio was set to 0.8 and 1.0. EGR ratio was varied from 0 to 0.33. The unburned gas temperature was 360 K and the pressure was 0.1, 0.3, 0.5 MPa. The EGR rate is the mass fraction of EGR gas used for dilution in the total mixture.

Microgravity experiments were carried out at about 0.6 s drop tower with the free-fall distance of 1.8 m in Osaka Prefecture University. A falling package with a drag shield is released from the top of the drop tower, and experiments are performed during the fall.

Flame radius, r, and stretched laminar burning velocity, S_u, were calculated from Equation (1) and (2) [11].

$$\frac{r}{r_v} = \left[1 - \left(\frac{p_i}{p}\right)^{\frac{1}{\kappa}} \frac{p_e - p}{p_e - p_i}\right]^{\frac{1}{3}} \quad (1)$$

$$S_u = \frac{r_v}{3(p_e - p_i)}\left(\frac{p_i}{p}\right)^{\frac{1}{\kappa}}\left[1 - \left(\frac{p_i}{p}\right)^{\frac{1}{\kappa}} \frac{p_e - p}{p_e - p_i}\right]^{-\frac{2}{3}} \frac{dp}{dt} \quad (2)$$

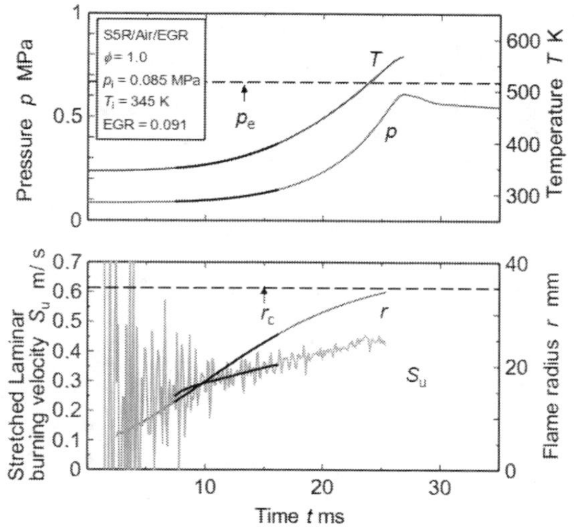

Figure 2: Histories of pressure, p, temperauue, T, flame radius, r, and stretched laminar burning velocity, S_u.

where, p, p_i, p_e, κ, r_v, and t are pressure in the spherical vessel, initial pressure, adiabatic maximum pressure, specific heat ratio, equivalent radius of the spherical vessel, and time after signal for ignition respectively. In this study, $2r_v$ is 70.15 mm. Adiabatic maximum pressure was calculated using CHEMKIN-PRO [12]. On the supposition of adiabatic compression, temperature of unburnt gas was calculated from the pressure history. The specific heat ratio was calculated from specific heat at constant pressure as a multinomial expression of the unburnt gas temperature. Pressure history of 90~150% for estimated pressure were approximated for quartic equation to reduce noise influence. Figure 2 shows the histories of pressure, p, unburnt gas temperature, T, flame radius, r, and stretched laminar burning velocity, S_u, in under the normal gravity. Initial pressure was 0.085 MPa, initial temperature was 345 K and equivalence ratio was 1.0. The dashed line denotes adiabatic maximum pressure, p_e, and equivalent radius of the spherical chamber, r_v, in Fig. 2. The maximum pressure was smaller than the adiabatic pressure due to heat loss.

Effects of stretch cannot be negligible because spherically flame has curvature and propagates with its effects. To estimate laminar burning velocity without effects of stretch, relationship between stretch due to curvature and laminar burning velocity for each instant of time was investigated. Linear function which Clavin[13] et al. calculated from asymptotic analysis and nonlinear function by Kelley and Law et al.[14] Equation (3) is known for relational expression between laminar burning velocity and stretch rate. We can apply nonlinear extrapolation when effects of stretch rate on laminar burning velocity are small. Tanoue et al. investigated propagating flame velocity of ethanol, n-heptane and isooctane, and they reported that there was about 20 % difference of propagating flame velocity between velocities linear extrapolated and one nonlinear extrapolated [15]. In this study, we calculated laminar burning velocities which have no effects of stretch by nonlinear extrapolation. The results are shown in Fig. 3. We measured stretch rate at each flame radiuses using equation (4).

$$\left(\frac{S_u}{S_L}\right)^2 \cdot \ln\left(\frac{S_u}{S_L}\right)^2 = -2 \cdot \frac{L_u \cdot \alpha}{S_L} \quad (3)$$

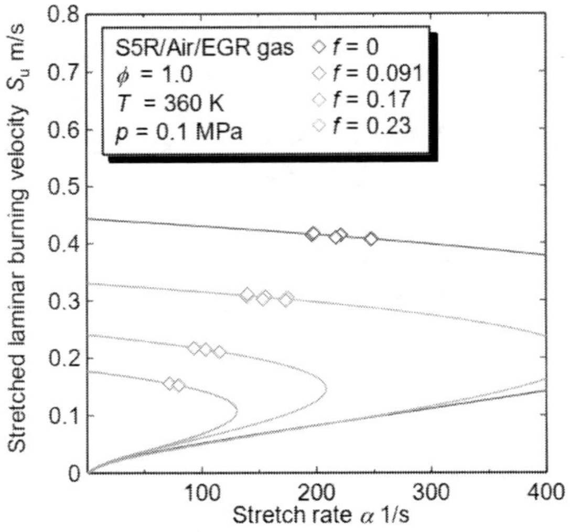

Figure 3: Effects of stretch and EGR ratio on laminar burning velocities at 0.1 MPa and 360 K.

$$\alpha = \frac{2}{r} \cdot \frac{dr}{dt} \tag{4}$$

where α is stretch rate at each flame radiuses, S_L is laminar barning velocity with no stretch, proportionality constant, L_u, is Markstein length which was deduced by nonlinear extrapolation from relationship between the flame velocity and flame stretch rate.

In the authors' past work of Ref. 16, we conducted measurements of laminar burning velocity of propane/air mixtures using two different inner diameter vessels. The differences between using one vessel and using two vessels were small. Measurements of laminar burning velocity of propane/air mixtures using two vessels will form the subject of a future study.

In figure 3, there were no experimental results deviating from the extrapolated line. Therefore, it was assumed that there was no transition to cellular flame in this study.

Laminar burning velocities were calculated using CHEMKIN-PRO. The reaction mechanism was the detailed reaction mechanism Ver. 1β of S5R by Miyoshi. All the laminar burning velocities were calculated with the same conditions of gradient and curvature leading to a similar number of grid points for all conditions of around 200.

3 Results and Discussion

Figure 4 shows the relation between laminar burning velocity and dilution by EGR gas for equivalence ratio is 1.0. Solid lines show the calculation results in fig.4. At 0.1 MPa, flame propagation was not observed for the EGR ratio was 0.29 or more. In this study, laminar burning velocities were found to decrease with increase EGR ratio. When the equivalent ratio was 0.8, the same tendency was shown.

Figure 5 shows the effects of EGR ratio on the ratio of the laminar burning velocity with and without dilution. The pressure dependence and the equivalence ratio dependence were observed. Therefore, it was suggested that the effects of EGR on the laminar burning velocity can't be estimated using the EGR rate.

Figure 6 shows a comparison between the results and the

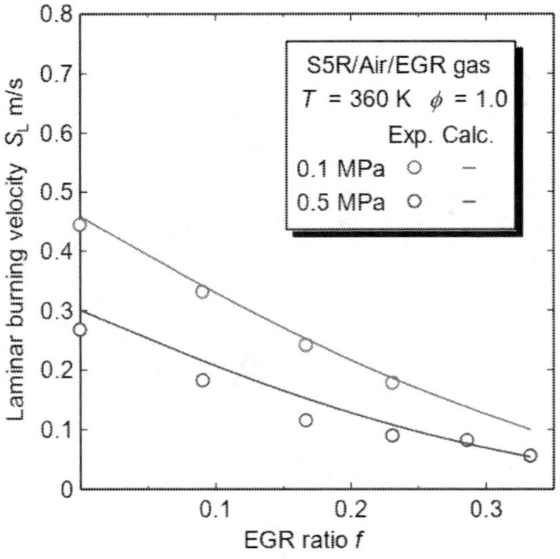

Figure 4: Measured laminar burning velocities at unburned mixtures temperature of 360 K and at pressures of 0.1 and 0.5 MPa.

estimation equation for equivalence ratio is 1.0.

Metghalch and Keck[2] proposed the equation as shown in (5) for iso-octane/air/EGR gas,

$$S_L/S_{L,f=0} = 1 - 2.1f \tag{5}$$

where, S_L, $S_{L,f=0}$, f are laminar burning velocity, laminar burning velocity(EGR mass rate = 0) and EGR gas mass rate. EGR gas component is 15 % carbon dioxide and 85 % nitrogen.

Rhodes and Keck[7] proposed the equation as shown in (6) for RMFD303/air/EGR gas

$$S_L/S_{L,f'=0} = 1 - 2.06f'^{0.733} \tag{6}$$

where, $S_{L,f'=0}$, f' are laminar burning velocity(EGR volume rate = 0) and EGR gas volume rate. EGR gas component is 20 % carbon dioxide and 80 % nitrogen.

Kian Eisazadeh-Far, Metghalchi and Keck [8] proposed the equation as shown in (7) for JP-8/air/EGR gas,

$$S_L/S_{L,f'=0} = (1 - f')^{3.52} \tag{7}$$

EGR gas component is 14 % carbon dioxide and 86 % nitrogen.

For EGR ratio was less than 0.33, the experimental results and calculated results were almost between Metghalch and Keck's Equation and Rhodes and Keck's Equation.

Figure 7 shows the effects of an adiabatic flame temperature ratio on the laminar burning velocity ratio. The adiabatic flame temperature, $T_{adiabatic}$, was calculated using CHEMKIN-PRO. The laminar burning velocity ratio decreased with the adiabatic flame temperature ratio decrease. It was suggested that the effects of EGR ratio on the laminar burning velocity can be arranged by using adiabatic flame temperature ratio. A detail analysis between the laminar burning velocity ratio and adiabatic flame temperature ratio will form the subject of a future study.

4 Conclusions

In this study, the effects of EGR ratio on laminar burning velocity were studied for S5R/air mixtures. The laminar burning velocities have been calculated by analyzing the

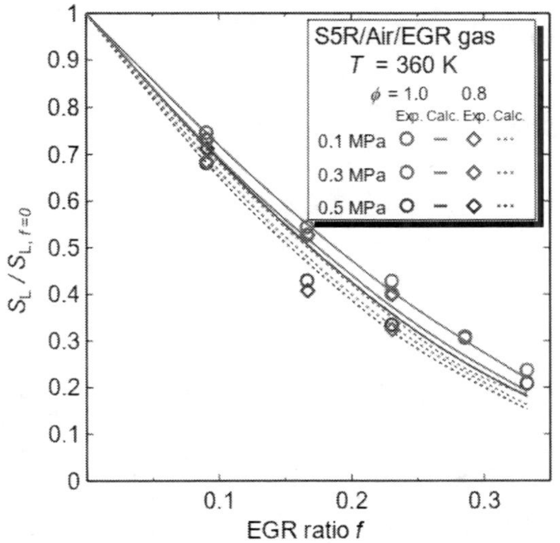

Figure5: Effects of the EGR ratio on the laminar burnin g velocity at 360 K.

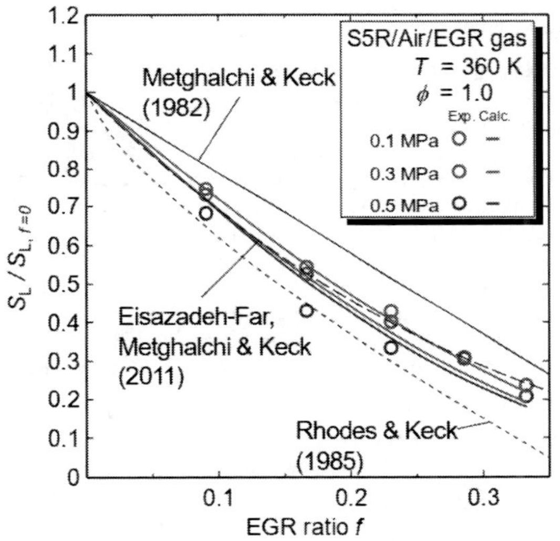

Figure 6: Comparison between the results and the estim ation equation.

pressure histories in the constant volume combustion vessel. In this study, equivalence ratio of S5R/air mixtures set to 0.8 and 1.0. EGR ratio was 0 to 0.33. The experiments were conducted in a short-drop free-fall laboratory facility that provides 0.6 s of microgravity condition. The following conclusions are drawn from the present study:

1. The laminar burning velocities decreased with EGR ratio increases in the lean conditions.
2. The laminar burning velocity ratio decreased with the adiabatic flame temperature ratio decrease.

5 Acknowledgment

A part of this work was supported by Council for Science, Technology and Innovation(CSTI), Cross-ministerial Strategic Innovation Promotion Program (SIP), "Innovative Combustion Technology" (Funding agency: JST).

References

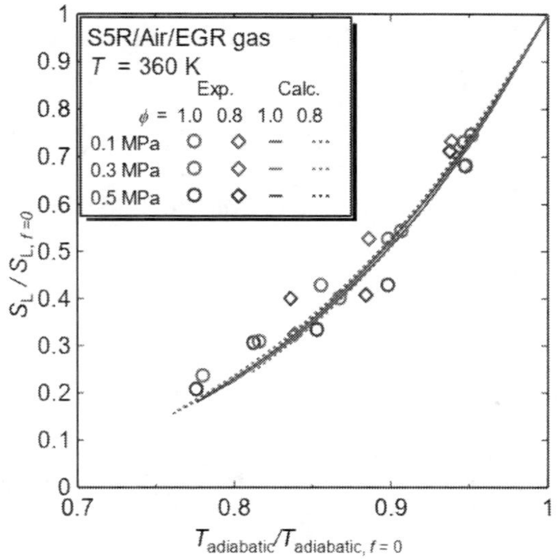

Figure 7: Effects of adiabatic flame ratio on the laminar burning velocity.

[1] Y. Daisho, Journal of Society of Automotive Engineers of Japan 70 (9) (2016) 4-11.

[2] M. Metghalchi, and J. C. Keck, Combust. Flame 48 (1982) 191-210.

[3] F. Rahim, K. Eisazadeh-Far, F. Parsinejad, R. J. Andrews, and M. Metghalchi, Int. J. of Thermodynamics 11 (2008) 151-160.

[4] J. X. Zhou, M. Cordier, C. Mounaïm-Rousselle, and F. Foucher, Combust. Flame 158 (2011) 2375-2383.

[5] T. Tahtouh, F. Halter, E. Samson, and C. Mounaïm-Roussellea, International Journal of Hydrogen Energy 34 (2009) 8329-8338.

[6] L. Qiao, Y. Gu, W. J. A. Dahm, E. S. Oran, and G. M. Faeth, Combust. Flame 151 (2007) 196-208.

[7] D. B. Rohodes and J. C. Keck, SAE Technical paper (1985) 850047.

[8] K. Eisazadeh-Far, A. Moghaddas, H. Metghalchi, and J. C. Keck, Fuel 190 (2011) 1476-1486.

[9] S. Shibata, H. Uesaka, R. Matsui, H. Kataoka, and D. Segawa, Twenty-sixth Internal Combustion Engine Symposium (in Japanese). (2015) No.20158044.

[10] H. Uesaka, R. Matsui, S. Doi, M. Matsuura, H. Kataoka, and D. Segawa, Proceedings of The Ninth International Conference on Modeling and Diagnostics for Advanced Engine Systems, (2017) B-312.

[11] B. Lewis and G. Elbe, Combustion, Flames and Explosions of Gases (2nd edition), 1961, p. 374.

[12] CHMKIN-PRO, Reaction Design: San Diego

[13] P. Clavin, Progress in Energy and Combustion Science 11 (1985) 1-59.

[14] A. P. Kelly, and C. K. Law, Combust. Flame 156 (2009) 1844-1851.

[15] K. Tanoue, Y. Kato, R. Shimada, K. Iwashimizu, K. Miyawaki, F. Shimada, and J. Hashimoto, Transactions of the Japan Society of Mechanical Engineers Series B 78 (2012) 1432-1440.

[16] S. Doi, M. Matsuura, H. Uesaka, R. Matsui, H. Kataoka, and D. Segawa, Proceedings of the fifty-fourth symposium (Japanese) on combustion, (2016) No. 117.

11th Asia-Pacific Conference on Combustion,
The University of Sydney, NSW Australia
10th -14th December 2017.

Effects of Non-uniform Ignition on Ignition Delay Measurement in Shock Tube

Chengyang Huang, Chengken Qi, Zheng Chen

SKLTCS, Department of Mechanics and Engineering Science, College of Engineering, Peking University, Beijing, 100871, China

Abstract

Shock tube experiments are widely used to measure the ignition delay time, which can be used to validate and develop chemical mechanisms for different fuels. In this study, one-dimensional simulations are performed to study the non-uniform ignition of hydrogen/oxygen/argon and n-heptane/oxygen/nitrogen behind a reflected shock wave. The objective is to examine the influence of non-uniform ignition and detonation development on the determination of ignition delay time. It is observed that there is detonation development behind the reflected shock. The mechanism for detonation development is interpreted and it is mainly due to the coherent interaction between the pressure wave and chemical reaction. Moreover, ignition delay times measured at different distances from the end wall are compared to that in a homogeneous system. It is found that the deviation decreases with dilution and it changes linearly with the distance from the reflecting wall. A correlation is obtained from theoretical analysis and it can accurately predict the deviation between the measured ignition delay time and that in a corresponding homogeneous system.

1 Introduction

The ignition delay measured in shock tube experiments is popularly used to develop and validate chemical kinetic mechanisms [1]. Accurate measurement of ignition delay is extremely important since it is difficult to constrain the uncertainty of chemical models using experimental data with large uncertainty. In ideal shock tube experiments, the test-gas mixture behind the reflected shock should be stationary and uniform if conditions are ideal. However, in practical shock tube experiments, there are non-ideal behaviors due to boundary layer growth, boundary layer/reflected shock interaction, and inhomogeneous ignition [2, 3].

In the literature, there are many studies on the ignition behavior in mixtures behind a reflected shock. For examples, Strehlow and coworkers [4, 5] identified three types of ignition behavior in a shock tube. Voevodsky and Soloukhin [6] found weak and strong ignition regimes based on the second explosion limit of hydrogen/oxygen mixtures. Meyer and Oppenheim [7] showed that the flame kernels started at a distance away from the reflecting wall in the weak ignition regime. Oran et al. [8, 9] conducted simulations for cases in weak and strong regimes of hydrogen/oxygen/argon mixtures; and they found that the ignition delay time at a distance away from the reflecting wall is much smaller than the calculated chemical induction time. Consequently, the constant volume or constant pressure assumption is not appropriate when dealing with the behavior of mixtures behind reflected shock. Yamashita et al. [10] found that there were ignition positions around central axis of the shock tube and triple point trajectory, and that ignition can be caused by bifurcated shock waves behind the reflected shock. Melguizo-Gavilanes and Bauwens [2] showed that the propagation of reflected shock itself induces gradient of ignition delay. Grogan and Ihme [3] found that multi-dimensional effects can cause non-uniform ignition. Javed and Badra [11] observed that pre-ignition occurs and it greatly affects the accuracy of ignition delay times.

Most of studies mentioned above focused on boundary layer and reflected shock interaction, localized ignition kernels and the multi-dimensional effects that cause non-uniform ignition. Besides, the test-gas mixture is consecutively compressed by the reflected shock. The ignition delay is spatially non-uniform since the mixture closer to the end wall is earlier compressed by the reflected shock. According to Zel'dovich's reactivity gradient mechanism [12] and Lee's SWACER mechanism [13] detonation may develop at certain value of the ignition delay gradient. Consequently, the ignition delay time measured in shock tube experiments is different from that in a homogeneous system, either at constant-volume or at constant-pressure [2].

The objective of this work is to examine the influence of non-uniform ignition and detonation development on the determination of ignition delay time in a shock tube. First, the mechanism for detonation development is interpreted. Then, ignition delay times measured at different distances from the end wall are obtained and compared to that in a homogeneous system. The deviation is analyzed. Finally, a correlation is derived from theoretical analysis to accurately predict the deviation between the measured ignition delay time and that in a corresponding homogeneous system.

2 Numerical model and specifications

One-dimensional simulation is conducted to study the transient ignition and detonation development process after a reflected shock. The length of the computational domain is 60 cm. The shock reflects on the adiabatic wall located at the left boundary. The right side is nonreflecting boundary condition. The 1D direct detonation initiation process is simulated using the in-house code A-SURF [14]. The conservation equations for unsteady, 1D, compressible, multi-component, reactive flow in a cylindrical coordinate system are solved in A-SURF using the finite volume method. A-SURF has been successfully used in previous studies on flame propagation, end-gas autoignition and detonation development. The details on governing equations, numerical methods and code validations can be found in Refs [14]. Detailed hydrogen mechanism [15] and skeletal n-heptane mechanism [16] are used in simulation. In order to efficiently

Corresponding author. Fax: +86-10-6276-6232
E-mail address: hacy1578@pku.edu.cn

resolve the reaction front, shock wave and detonation, dynamically adaptive mesh refinement strategy is adopted. The finest mesh size is 4 μm.

case	Ar (vol. %)	T_S (K)	P_S (atm)	t_{ex} (μs)	ξ
1	0	1034.7	1.49	0.8	1.64
2	40	1033.2	1.38	2.2	1.53
3	70	1033.3	1.29	8.4	1.44
4	90	1032.2	1.23	63.2	1.37

Table 1: Four mixtures and their states and ignition properties after the reflected shock.

We consider hydrogen/oxygen mixture with different amounts of argon dilution and the n-heptane/oxygen mixture with nitrogen dilution. The hydrogen/oxygen/argon mixtures shown in table 1. The subscript s represent the state behind the reflected shock. t_{ex} is the excitation time of mixture. ξ is the ratio of local acoustic velocity to the front velocity.

3 Results and discussion

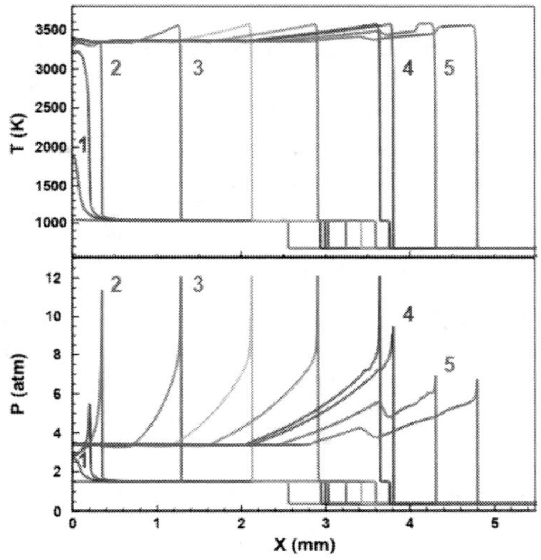

Figure 1: Temporal evolution of temperature and pressure distributions for case 1.

We first consider the results of hydrogen/oxygen mixture with different amounts of argon dilution. For case 1 without argon dilution, the temperature and pressure's development processes as shown in Fig. 1. The mixture's temperature and pressure behind the reflected shock is 1034.7 K and 1.49 atm. When the shock reflecting at the end wall, timing starts. At t=49.59 μs (line #1), ignition starts near the reflecting wall, and the front propagates to the right. With the coupling and interacting of pressure wave and the reaction, detonation wave structure first develops (line #2) at x=0.33 cm. Then the detonation wave is well formed (line #3), which propagation speed is near CJ detonation speed. When the reaction front brushes x=3.79 cm, the detonation wave catches up with the reflected shock (line #4), and forms a new detonation wave.

Similar results are obtained for case 2 and case 3. With the increase of argon dilution, detonation development becomes slower due to lower heat release rate. When argon dilution reaches 90% as in case 4, Fig. 2 shows that a fully developed detonation does not occur since the ignition front propagation velocity is much lower than the CJ detonation velocity.

It is mentioned earlier that ξ is the ratio of local acoustic velocity to the front velocity. According to the theory of Zel'dovich [12], the front could interact with the pressure wave, when the front velocity approaching local acoustic velocity, easily lead to the detonation initiation. From the calculated results, ξ of every cases is between 1 and 2, indicating that the front velocity is closer local acoustic velocity. The studies of Bradley et al. [17, 18] pointed out that another dimensionless time scale ε should be introduced, which is inversely proportional to the excitation time of the mixture. When the rate of argon dilution increases, the excitation time increases and the value of ε decreases. When ξε is below critical value, the detonation wave can't form. Therefore, when the argon dilution is high enough, developed detonation structure does not occur.

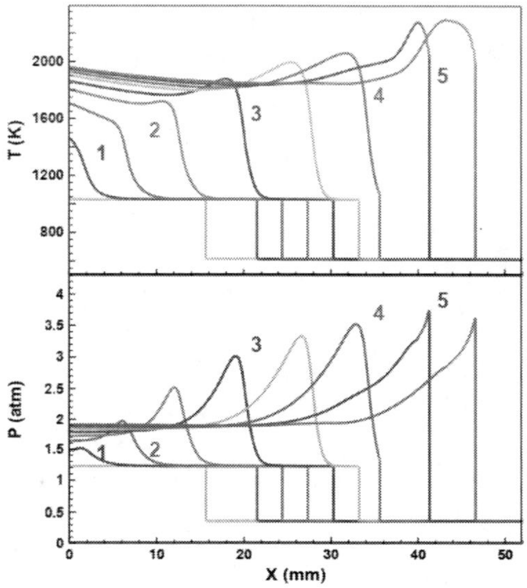

Figure 2: Temporal evolution of temperature and pressure distributions for case 4.

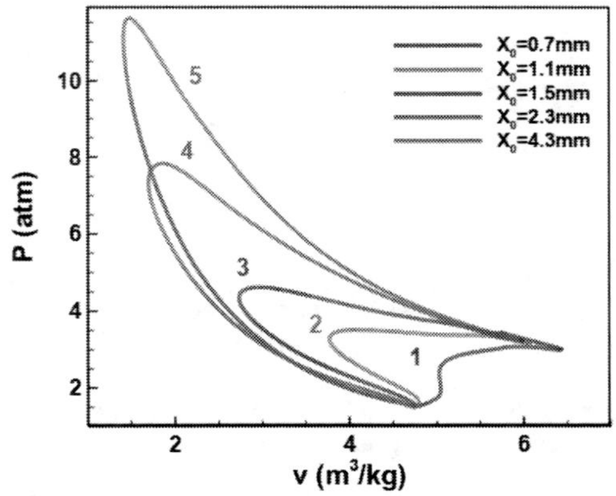

Figure 3: P-v diagram for particles at different initial positions for case 1.

In order to further explain ignition and detonation initiation, the mixture particles were tracked and the P-v diagram is obtained. The results for case 1 are plotted in Fig. 3. It is observed that particles initially at X_0=0.7 mm go through four stages: compression, constant volume combustion, expansion and compression again. The mixture is compressed by reflected shock and the constant-volume ignition starts, resulting in temperature and pressure rise. Then, the temperature and pressure decrease due to thermal expansion. Finally, the mixture is compressed again due to ignition in neighboring mixture.

However, the particles initially at X_0=1.1 mm go through three stages: combustion, expansion and compression. The combustion is not in constant volume. For particles initially at X_0=1.1~4.3 mm, the compressible effects on mixture are intensified, until the detonation is formed. The particles at different initially position following different processes. It is clear that constant volume or constant pressure combustion assumption is invalid for all particles. Compared to case 1, case 2 and case 3 have similar processes. As for case 4, it is found that the compressible effects on mixture is intensified with the increasing initial positions of particles, but the detonation is not formed in this case.

As mentioned above, the particles at a different distance away from the reflecting wall following different process for every cases. The combustion behaviors behind the reflected shock is non-uniform, but is followed by compression and expansion. The mechanism of reaction behind the reflected shock is quite complex, the uniform assumption is not appropriate.

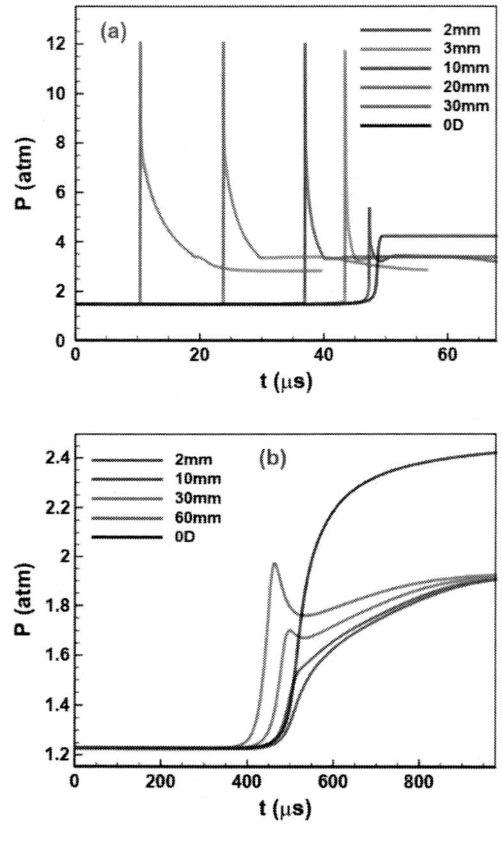

Figure 4: Temporal evolution of pressure at different positions and that for 0D homogeneous ignition: (a) case 1, (b) case 4.

To further reveal the difference between ignition delay time recorded in a shock tube and that in a homogeneous system, the pressure history recorded at different positions is compared to that in a 0D homogeneous system. The results in shown in Fig. 4 for cases 1 and 4. It is noted that timing starts when the shock reflects on the end wall and that the ignition delay is defined as the time when maximum pressure rise rate occurs. Figure 4(a) shows that the ignition delay decreases as the distance away from the reflecting wall increases, and that they are all much earlier than that for the 0D homogeneous ignition system. Besides, due to detonation development, the peak pressure at different positions is much large than the equilibrium pressure for the 0D homogeneous ignition system. The pressure evolution law of case 2 and case 3 are similar to it in the case 1. When detonation is not formed for case 4, Fig. 4(b) indicates the ignition delay measured at different positions is close to that for 0D homogeneous ignition system. In a short summary, the ignition delay at a distance away from the reflecting wall is smaller than the induction time in the 0D homogeneous system; and the difference decreases as the argon dilution increases.

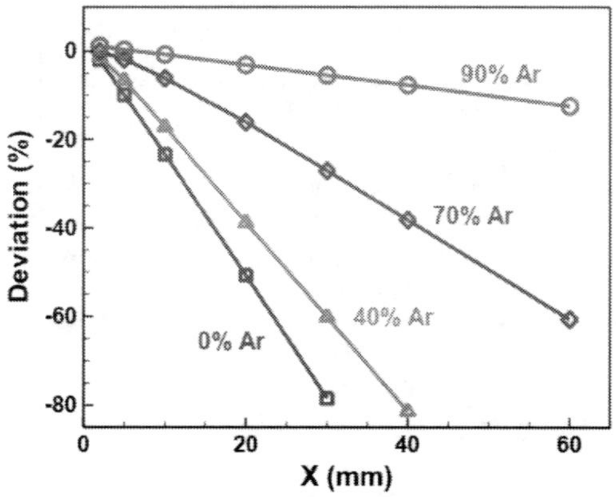

Figure 5: The deviation between ignition delay at different positions and that in the homogeneous ignition system for hydrogen/oxygen/argon mixture.

Figure 5 summarizes the influence of pressure recording position and argon dilution on the deviation between the measured ignition delay and that in a homogeneous system. The derivation is defined as $\tau_{ig}(x)/\tau^0_{ig}$ -1, where $\tau_{ig}(x)$ is the ignition delay measured at different positions in the shock tube and τ^0_{ig} is the ignition delay in 0D homogeneous system. Figure 5 shows that the deviation is negative, indicating that the ignition delay measured at different positions in the shock tube is smaller than that in the 0D homogeneous system. It is observed that deviation changes linearly with the distance away from the reflecting wall. Besides, the absolute value of deviation decreases with dilution.

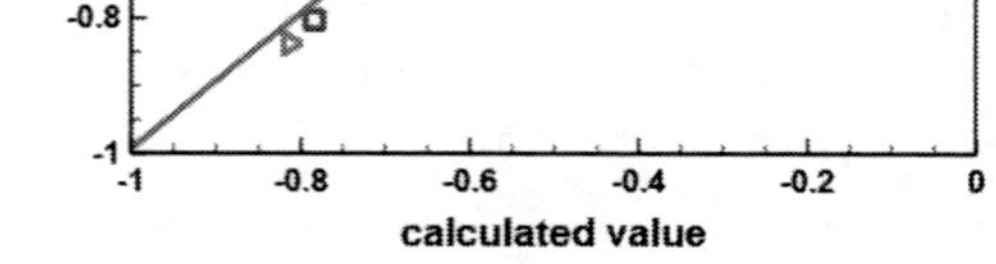

Figure 6: Comparison between theoretical prediction and calculated deviation for hydrogen/oxygen/argon mixture.

To explain such observation, we conducted theoretical analysis for the non-uniform ignition process. We first consider the ignition delay at one position is the time difference between reflected shock and reaction front reaches this position. For simplicity, we assume that the velocity of reflecting shock (u_{shock}) and reaction front (u_{front}) are constant. Then we can obtain the following expression for the deviation between the ignition delay the ignition delay measured at different positions, $\tau_{ig}(x)$, and that 0D homogeneous system, τ^0_{ig},

$$\frac{1}{\tau^0_{ig}}[\tau_{ig}(x)-\tau^0_{ig}]=\frac{1}{\tau^0_{ig}}(\frac{x}{u_{front}}-\frac{x}{u_{shock}}+\Delta t) \quad (1)$$

in which Δt is the difference between τ^0_{ig} and the ignition delay time at the reflecting wall. Figure 6 shows that the correlation can accurately predict the deviation between the measured ignition delay time and that in a corresponding homogeneous system.

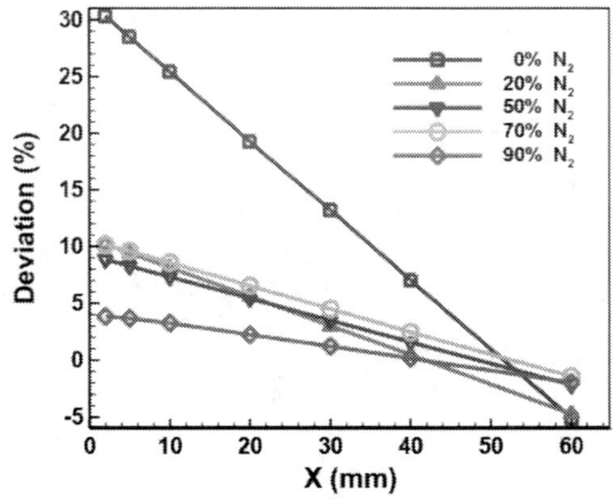

Figure 7: The deviation between ignition delay at different positions and that in the homogeneous ignition system for n-heptane/oxygen/ nitrogen mixture.

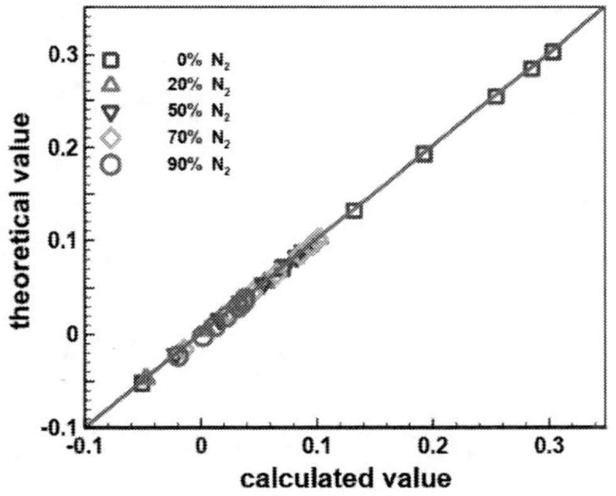

Figure 8: Comparison between theoretical prediction and calculated deviation for n-heptane/oxygen/ nitrogen mixture.

The results for n-heptane/oxygen/nitrogen mixtures are shown in Figs. 7 and 8. Positive deviation is observed and this needs to be interpreted. Nevertheless, the deviation still changes linearly with the distance from the reflecting wall and it can be accurately predicted by the theoretical correlation in (1).

4 Conclusions

Numerical simulation for non-uniform ignition in 1D shock tube of $H_2/O_2/Ar$ and $nC_7H_{16}/O_2/N_2$ mixtures are performed considering detailed chemistry. It is observed that detonation is formed behind the reflected shock at low dilution due to spatially non-uniform ignition after the reflected shock. Ignition delay times measured at different distances from the end wall are compared to that in a homogeneous system. The derivation is shown to decrease with dilution and to change linearly with the distance from the reflecting wall. A correlation is obtained from theoretical analysis, indicating that the velocity of reaction front directly affects the ignition time. Comparison between theoretical prediction and calculated deviation for $H_2/O_2/Ar$ and $nC_7H_{16}/O_2/N_2$ mixtures demonstrates that the correlation can accurately predict the deviation between the measured ignition delay time and that in a corresponding homogeneous system.

5 Acknowledgment

This work was supported by National Natural Science Foundation of China (No. 91541204).

References

[1] R.K. Hanson and D.F. Davidson, Progress in Energy and Combustion Science. 44 (2014) 103-114.

[2] J. Melguizo-Gavilanes and L. Bauwens, Shock Waves. 23 (2013) 221-231.

[3] K.P. Grogan and M. Ihme, Proceedings of the Combustion Institute. 35 (2014) 2181-2189.

[4] R.A. Strehlow and A. Cohen, 5 (1962) 97-101.

[5] R.B. Gilbert and R.A. Strehlow, Aiaa Journal. 4 (1966) 1777-1783.

[6] V.V. Voevodsky and R.I. Soloukhin, Symposium on Combustion. 10 (1965) 279-283.

[7] J.W. Meyer and A.K. Oppenheim, Symposium on Combustion. 13 (1971) 1153-1164.

[8] E.S. Oran, et al., Combustion & Flame. 48 (1982) 135-148.

[9] E.S. Oran and J.P. Boris, Combustion and Flame. 48 (1982) 149-161.

[10] H. Yamashita, et al., Combustion and Flame. 159 (2012) 2954-2966.

[11] T. Javed, et al., Combustion Science and Technology. (2016)

[12] Y.B. Zeldovich, Combustion and Flame. 39 (1980) 211-214.

[13] J.H. Lee, R. Knystautas, and N. Yoshikawa, Acta Astronautica. 5 (1978) 971-982.

[14] P. Dai and Z. Chen, Combustion and Flame. 162 (2015) 4183-4193.

[15] J. Li, et al., International Journal of Chemical Kinetics. 36 (2004) 566–575.

[16] S. Liu, et al., Combustion and Flame. 137 (2004) 320-339.

[17] D. Bradley, et al., Sae Technical Papers. 2002-01-2868 (2002)

[18] X.J. Gu, D.R. Emerson, and D. Bradley, Combustion and Flame. 143 (2005) 227-245.

11th Asia-Pacific Conference on Combustion,
The University of Sydney, NSW Australia
10th -14th December 2017.

Numerical Study and HCHO-PLIF Measurement of Wall-Stabilized Cool Flames

Minhyeok Lee[1], Yong Fan[1], Christopher B. Reuter[2], Yiguang Ju[2], Yuji Suzuki[1]

[1]The University of Tokyo

Hongo 7-3-1, Bunkyo-ku, Tokyo 113-8656, Japan

[2]Princeton University

Princeton, NJ 08544, USA

Abstract

Wall-stabilized cool flames were studied by both numerical and experimental methods. One- and two-dimensional numerical simulations in the axisymmetric condition were performed to estimate the characteristics of the cool flame, such as flammability and temperature/species distributions. A nozzle burner and heating plate were designed and fabricated for the experimental investigation of the cool flame. In the experiment, a wall-stabilized cool flame was observed for the first time through formaldehyde Planar Laser-Induced Fluorescence (HCHO-PLIF) measurements. The spontaneous initiation of lean cool flames at high wall temperatures, which was expected from *S*-curve calculations of the cool flame branch, was also successfully confirmed.

1 Introduction

The cool flame is a flame formed at relatively lower temperatures than conventional hot flames, as its maximum flame temperature is generally not higher than 1000 K [1]. It is a phenomenon characterized by the slight emission of faint bluish chemiluminescence, and it is governed by low-temperature oxidation chemistry via oxygen addition to fuel radicals (RH) and subsequent isomerization and radical branching. The cool flame and its low-temperature chemistry (LTC) are responsible for knocking phenomena in spark ignition engines and auto-ignition in the compression ignition engines [1]. It has been revealed that cool flames are also related to the two-stage heat lease process in homogeneous charge compression ignition (HCCI) engines [2]. Therefore, understanding the cool flame and LTC are crucial in terms of improving the thermal efficiency of these engines and reducing pollutants.

Existence of the cool flame has been first reported more than a century ago [3]. However, most of the observed cool flames are transient phenomena just before igniting into the hot flame or periodic emergence continuing ignition and extinction. One of the main difficulties to achieve self-stabilizing cool flames is in its initiation process owing to its long induction time period. Plasma- and ozone-assisted combustion have been tried to prompt LTC [4,5], but their complex coupling with radicals makes it difficult to understand the fundamental chemistry.

In our previous study [6], the ozone was used only for the ignition of the cool flames. Once the cool flame started to stabilize, the injection of the ozone was interrupted. The ozone-less cool flames remained self-sustaining in a relatively low strain rate range compared with the ozone-assisted cool flames. The low-temperature reactivity of various n-alkane fuels was evaluated by measuring the extinction strain rate and compared with the numerical simulation results based on various mechanisms.

In the previous cool flame studies, flame-wall interactions have been intentionally avoided to clearly define the boundary conditions, for instance, by introducing the counterflow burner. However, from the practical point of view, considering the existence of the wall in the vicinity of flames can be crucial. Furthermore, flame-wall interactions (FWI) involving cool flames have not been considered seriously, though recent studies show some noticeable radical quenching effects in the micro flow channel [7]. We have previously reported the basic characteristics of the wall-stabilized cool flames based on the one-dimensional simulation [8]. In this paper, the numerical simulations are expanded to two dimensions in order to capture spatial distributions along the wall. We also focus on the experimental formation and observation of wall-stabilized cool flames, including the design of the experimental setup.

2 Numerical Method and Experimental Setup

We consider the impinging of a fuel and oxygen mixture onto a heated wall as shown in Fig. 1. Dimethyl ether (DME) is used as the fuel due to its simple chemical structure but relatively strong low-temperature reactivity. For the one-dimensional numerical simulations, both a modified PREMIX code with an improved arc-length continuation method [9] and CHEMKIN-Pro are used. A two-dimensional simulation using Fluent is also performed to observe the species distributions along the wall. In both simulations, the premixed DME/O_2 flow field propagating toward the stagnation point/plane is assumed to be axisymmetric. Constant temperature and zero species flux at the chemically inert wall are assumed for the boundary conditions.

Even though DME is the smallest carbon compound possessing strong LTC, there still exists a large gap in its estimated low-temperature reactivity. Figure 2 shows the calculated ignition delay time of a DME and O_2 mixture (equivalence ratio: 0.5) under the isobaric condition. Six recent models [10-15] were tested. As shown in the graph, there are discrepancies up to one order of magnitude in the mechanisms, especially in the

Corresponding author. Fax: +81-3-5841-6419
E-mail address: ysuzuki@mesl.t.u-tokyo.ac.jp

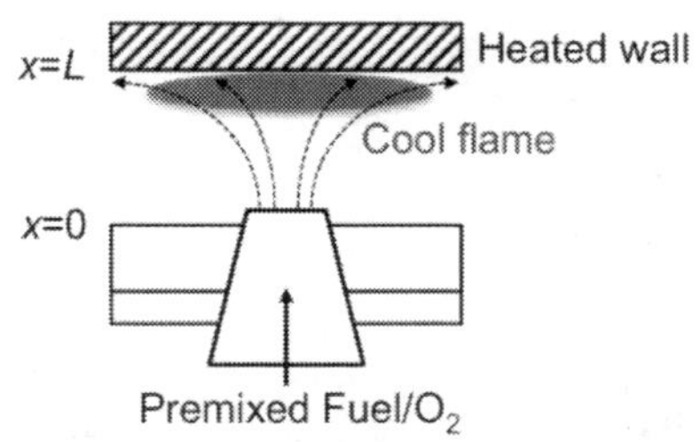

Figure 1: Schematic of the burner configuration and the stabilization of the cool flame near the wall.

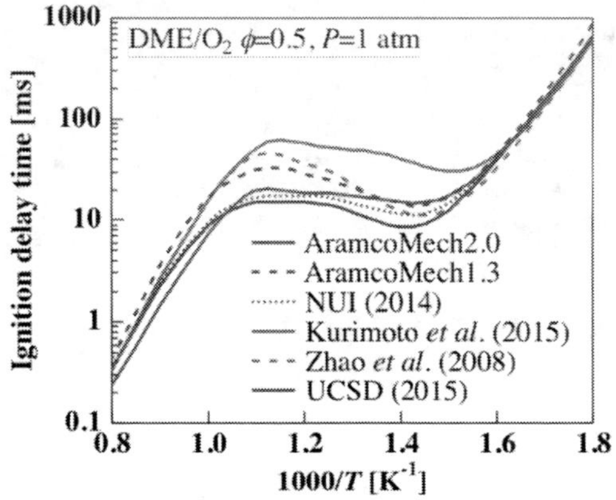

Figure 2: Ignition delay time calculation with various DME oxidation models [10-15].

negative temperature coefficient (NTC) region that is important to the formation of the cool flame. Furthermore, the low-temperature reactivity cannot be defined with only the ignition delay time [6]. For example, the AramcoMech2.0 shows the shortest homogeneous ignition delay times, but it estimates quite low flammability in terms of the cool flame extinction.

Therefore, we have conservatively chosen two models that show relatively weak low-temperature reactivity as defined not by the ignition delay time but by the cool flame flammability. One is a reduced version of the AramcoMech2.0 [10], whose species and reactions involving molecules with 3 or more carbon atoms are eliminated to save the computational cost. The reduced model shows the exact same profile as the original model in terms of both the ignition delay time and the cool flame extinction. The other one is the updated model from Zhao *et al.* [11] by Kurimoto *et al.* [12], of which the low-temperature reactivity is weakened by modifying QOOH branching ratio. Details of two models are shown in Table 1.

For the experiment, a stainless steel nozzle with a contraction ratio of 12 was designed. The main flow is introduced from 4 cross-directions and then passes through a honeycomb and 2 screen meshes. An outer nozzle for the co-flow is also prepared to minimize the disturbance from the ambient flow. The outlet diameters of the inner and the outer nozzle are 10 and 20 mm, respectively.

A heating plate mainly consists of two parts: a brass circular plate with two concentric grooves to bury heaters, and a stainless cover to fix the heaters. A disc-shape plate (ϕ70×3 mm^2) was also designed to make it separable from the center of the whole plate (ϕ120 mm). There exists a 10 mm distance between the surface of the plate and the main heater to create a

Model Number	AramcoMech2.0		Kurimoto *et al.* [12]
	Original [10]	Reduced	
Carbon No.	C_0–C_5	C_0–C_2	C_0–C_2
Species	493	86	55
Reactions	2716	490	290

Table 1: Details of two DME oxidation mechanisms.

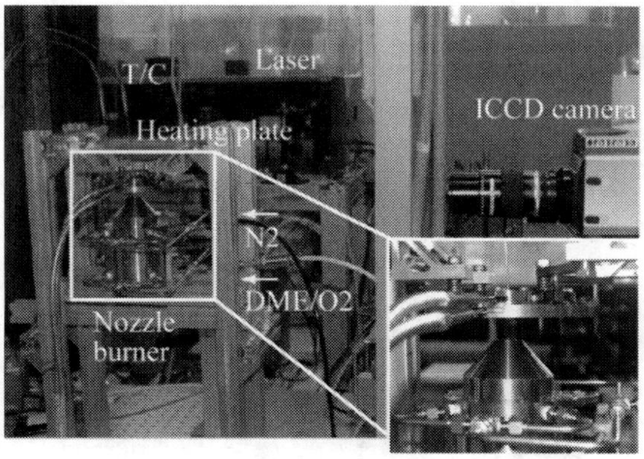

Figure 3: Experimental setup for stabilizing and observing the wall-stabilized cool flame.

uniform temperature distribution on the wall through sufficient thermal conduction. Two spirally wound wire heaters (main: SQ60/400 W, sub: SQ70/450 W, Sakaguchi E.H Voc Corp.) are mounted in the plate, and their input powers are individually controlled by using a manual voltage slider (YS265, Yamabishi-denki). The wall temperature is monitored by two K-type thermocouples, one at the center and one 36 mm away from the center. The thermocouples reside in a pair of holes drilled 1 mm away from the wall surface.

The set of nozzles and heating plate are fixed on an optical base, which maintains a 15 mm distance from the nozzle tip to the wall. The flow rates of DME, O_2 and N_2 are controlled with mass flow controllers (5100/3660 model, Kofloc). DME and O_2 are premixed before flowing into the nozzle and pass through a filter with 40 µm pores. The entire setup is installed under a suction hood and surrounded by noncombustible dark curtains.

For the excitation of HCHO in the HCHO-PLIF measurement, the 3rd harmonic of a pulsed Nd:YAG laser (LPY 400, Lambda Physik) at 355 nm wavelength is used. The laser beam with an energy intensity of 20 mJ/pulse is converted into a laser sheet using a cylindrical convex lens. The laser sheet is adjusted to pass through the flame in the gap between the wall and the nozzle. The fluorescence from the excited HCHO is captured with an ICCD camera (iStar DH340T, Andor) through a 430 nm band-pass filter (FWHM: 40 nm). A photograph of the entire experimental setup is shown in Fig 3.

3 Results

Figure 4 shows the calculated *S*-curve of wall-stabilized cool flames when the wall temperature is 700 K. In the moderately lean condition at an equivalence ratio of 0.5, it shows the typical *S*-curve with several bifurcation points. However, in the much leaner case at an equivalence ratio of 0.2, the cool flame branch is separated from the hot flame branch, and the high temperature ignition (HTI) point is no longer visible.

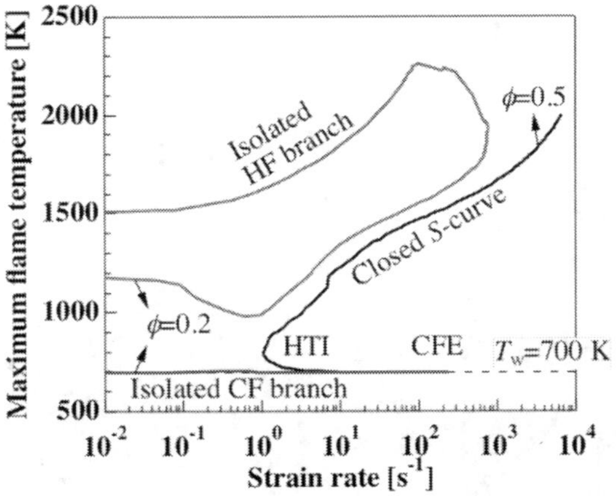

Figure 4: *S*-curve of wall-stabilized DME/O_2 cool flames calculated with Kurimoto *et al.* model.

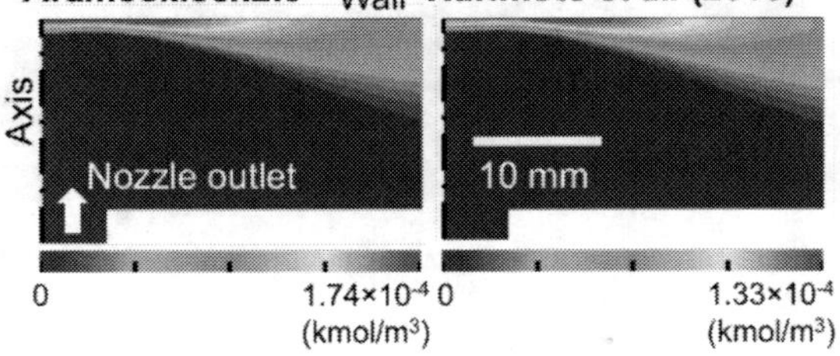

Figure 5: Molar concentration of HCHO calculated with AramcoMech2.0 (left) and Kurimoto *et al.* model. (right)

As reported in the previous studies [8], at high wall temperatures above that of the flame temperature, the cool flame branch does not show bifurcation but stretches without showing a clear low temperature ignition (LTI) point. In contrast with the numerical results calculated with the AramcoMech1.3 or Zhao *et al.* models that tend to over-estimate the DME low-temperature reactivity, the maximum flame temperature computed in the Kurimoto *et al.* model is generally lower than the wall temperature even in the low strain rate region. Because of this, it is more difficult than usual to define the maximum flame temperature. Therefore, the extinction point cannot be judged only by the temperature profile, but also the concentration of major product such as HCHO has to be checked. Disappearance of the LTI limit implies the possibility of the cool flame initiation without ignition process, so that the addition of the ozone is no more required. Two different DME cool flame initiation phenomena, an ignition at lower temperatures (LTI) and an aforementioned direct transition at higher temperatures (sans LTI), have been predicted in the previous numerical simulations [8,16]. However, the Kurimoto *et al.* model predicts that cool flames hardly exist when the wall temperature is lower than 700 K, and accordingly only the latter case can happen.

The two-dimensional calculation also reveals the stabilization of the cool flame by the heated wall. The axisymmetric model mimicking the experimental setup in Fig. 3 is created. The nozzle contraction part is also included to reproduce the outlet velocity profile. The simulation domain between the nozzle and the wall that the actual reaction happens is 15 mm×30 mm. Figure 5 shows the calculated concentration distribution of HCHO. The wall temperature, the outlet velocity and the equivalence ratio are 700 K, 50 cm/s, and 0.2, respectively. Even though the absolute concentration values are somewhat

	DME	O_2	N_2
Flow rate [L/min]	0.147	2.21	1.0
Outlet velocity [cm/s]	50		~ 10

Table 2: Initial condition for DME, O_2, and N_2 flows to initiate the cool flame.

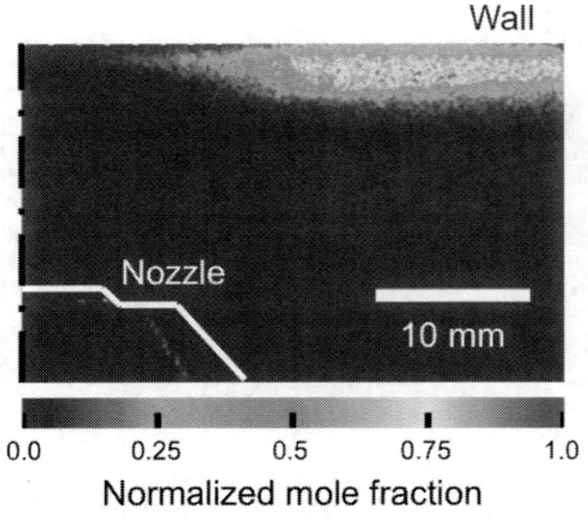

Figure 6: Concentration distribution of HCHO measured by PLIF.

different, both models show similar distribution along the wall, with the maximum HCHO concentration located approximately 10 mm away from the centerline. The molar concentration of HCHO on the order of 10^{-4} kmol/m^3 is high enough to detect through the PLIF measurement [17].

In the experiment, we increase the wall temperature to 700 K by controlling two power input units. Temperature distribution on the wall surface was checked with a thermography (FSV-1200 L16, Apiste), and remained within ±3 K at 700 K. After stabilizing the wall temperature, the O_2 main flow and N_2 co-flow are introduced. Due to the cooling effect by the unheated room temperature fluids impinging onto the wall surface, the wall temperature slightly starts to decrease. However, due to high enough thermal capacity of the designed heating plate, the decline remains less than 2 K. The flow rate of DME is gradually increased until it reaches to the target equivalence ratio. The flow rates used to initiate the cool flame are represented in Table 1. We could observe a gradual wall temperature increase at the rate of ca. 1.2 K/min, which begins right after introducing DME into the O_2 flow. It is interpreted that there is a small amount of heat production from the cool flame. Furthermore, this result explains the aforementioned stretch of the cool flame branch in the *S*-curve. The ignition limits of the cool flame disappear, resulting in the direct transition from premixed DME/O_2 mixtures into the cool flame. In the highly lean condition, the chemiluminescence from the cool flame in a visible ray range is extremely weak.

Figure 6 shows the HCHO-PLIF measurement results under the experimental condition of Table 2. The equivalence ratio is 0.2, the same condition as the numerical calculation case. The position of the wall and the nozzle were determined through image processing captured without laser irradiation. Even though the fluorescence was not captured in the vicinity of the centerline due to imperfect alignment, a concentration distribution along the wall was successfully observed. The location that HCHO shows the maximum concentration is also slightly far away from the wall position if compared to the simulated HCHO distribution. Outside the centerline region, the overall spatial distribution shows some resemblance to the simulation results shown in Fig 5. Further investigation such as the influence of the equivalence ratio and the strain rate (outlet velocity) will follow through the quantitative analysis of the PLIF measurement.

4 Conclusions

A cool flame was stabilized by impinging coaxial jets of a DME/O_2 mixture and N_2 onto a heated wall. In the numerical calculations, two DME oxidization mechanisms were used. The simulation results showed the formation of a cool flame that is initiated and subsequently stabilized by the hot wall. To reproduce these results, an experimental setup including a burner with double coaxial nozzle and a heating plate mounted with two spirally wound wire heaters was designed and fabricated. The initial condition to initiate the cool flame was set to an equivalence ratio of 0.2, an outlet velocity of 50 cm/s, and a wall temperature of 700 K. It was observed that the wall temperature started to increase after introducing the premixed DME/O_2 mixture, implying some slight heat release from the cool flame. The formation of HCHO along the wall observed in the PLIF measurements showed modest agreement with the numerical estimations. Our future works will focus on the investigation of not only dynamics and chemistry of the cool flames, but also the cool flame-wall interactions under the head-on quenching configuration.

Acknowledgment

This work is supported through JSPS KAKENHI Grant Number 15H03927 and 15H05508. ML is supported through the Leading Graduate Schools Program, "Global Leader Program for Social Design and Management," by MEXT Japan. CBR is supported by the DoD National Defense Science and Engineering Graduate (NDSEG) Fellowship program.

References

[1] P.G. Lignola, E. Reverchon, Prog. Energy Combust. Sci. 13 (1) (1987) 75-96.

[2] F.L. Dryer, Proc. Combust. Inst. 35 (2015) 117-144.

[3] H. Davy, Philos. Trans. R. Soc. London 107 (5) (1817) 77-85.

[4] W. Sun, S.H. Won, Y. Ju, Combust. Flame 161 (8) (2014) 2054-2063.

[5] S.H. Won, B. Jiang, P. Diévart, C.H. Sohn, Y. Ju, Proc. Combust. Inst. 35 (2015) 881-888.

[6] C.B. Reuter, M. Lee, S.H. Won, Y. Ju, Combust. Flame 179 (2017) 23-32.

[7] S. Wan, Y. Fan, K. Maruta, Y. Suzuki, 36th Int. Symp. Combust., Seoul, Korea, 2016, Work-in-Progress Poster 2P138.

[8] M. Lee, C.B. Reuter, Y. Suzuki, Y. Ju, 1st Asian Conf. Therm. Sci., Jeju, Korea, 2017, ACTS-P00731.

[9] Y. Ju, H. Guo, K. Maruta, F. Liu, J. Fluid Mech. 342 (1997) 315-334.

[10] U. Burke, K.P. Somers, P. O'Toole, C.M. Zinner, N. Marquet, G. Bourque, E.L. Petersen, W.K. Metcalfe, Z. Serinyel, H.J. Curran Combust. Flame 162 (2) (2015) 315-330.

[11] Z. Zhao, M. Chaos, A. Kazakov, F.L. Dryer, Int. J. Chem. Kinet. 40 (1) (2008) 1-18.

[12] N. Kurimoto, B. Brumfield, X. Yang, T. Wada, P. Diévart, G. Wysocki, Y. Ju, Proc. Combust. Inst. 35 (2015) 457-464.

[13] W.K. Metcalfe, S.M. Burke, S.S. Ahmed, and H.J. Curran, Int. J. Chem. Kinet. 45 (10) (2013) 638-675.

[14] U. Burke, K.P. Somers, P. O'Toole, C.M. Zinner, N. Marquet, G. Bourque, E.L. Petersen, W.K. Metcalfe, Z. Serinyel, H.J. Curran, Combust. Flame 162 (2) (2015) 315-330.

[15] J.C. Prince, F.A. Williams, Combust. Flame 162 (10) (2015) 3589-3595.

[16] Y. Ju, C.B. Reuter, S.H. Won, Combust. Flame 162 (10) (2015) 3580-3588.

[17] T. Shimizu, H. Nakamura, T. Tezuka, S. Hasegawa, K. Maruta, Energy & Fuels 31 (3) (2017) 2298-2307.

11th Asia-Pacific Conference on Combustion,
The University of Sydney, NSW Australia
10th -14th December 2017.

Numerical Simulation of LPG – Hydrogen Jet Diffusion Flames

[1]Muthu Kumaran S*, [1]Vamsi Krishna Ch. and [2]Vasudevan Raghavan

[1]Indian Institute of Technology Madras

Chennai, Tamil Nadu, 600036, India.

[1,2]Indian Institute of Technology Madras

Chennai, Tamil Nadu, 600036, India.

Abstract

LPG (Liquefied Petroleum Gas) is a high potential fuel gaining popularity in domestic household and automobile sector in recent times in India. Research works in the field of flame structure and its stability are scarce. Therefore, study of its flame structure and stability characteristics is important to understand and design its burners. Since it is a multi-component fuel, proper kinetics mechanism, at least a single step global mechanism, for simulating LPG flames should be proposed. In this paper, numerical simulations of LPG jet diffusion flames are attempted with formulated single step chemistry. Experiments are done in the lab with a burner of diameter 0.4 mm and photographs of flames are taken for different fuel jet velocity in order to study the flame heights for each fuel velocity. Numerical simulations are done by modeling LPG as a single component fluid (70% butane and 30% propane). Numerical results are compared with corresponding experimental data and are validated. The mechanism is tested for its ability to predict the flame length obtained from experiments using a criterion of LPG mass fraction. Numerical simulations are then carried out to study LPG- hydrogen jet diffusion flames.

1. Introduction

Liquefied Petroleum Gas (LPG) is widely used as cooking fuel in several countries including India. LPG is a multicomponent fuel; the major constituents being propane, butane and trace amounts of saturated and unsaturated hydrocarbons. The composition of LPG varies from one country to other. Understanding LPG flame characteristics and transient instability phenomena such as liftoff and blow off are essential for good burner design.

Considerable amount of literature is available on LPG flame stabilization studies, most of them are experimental. Kiran et al. [1] studied liftoff height, flame length and blow off velocity of LPG jet diffusion flames. Kiran and Mishra [2] investigated the stability of LPG-air flames in premixed and triple flame configurations. Khan et al. [3] analyzed the variation of flame length and liftoff heights in LPG-H_2 jet flames and reported a reduction in flame length and liftoff height for higher percentage of hydrogen in the mixture. Mahesh et al. [4] conducted experiments to determine the flame stability of LPG-air flames in inverse diffusion configuration in a back-step burner and reported an increase in the blow off velocity with increase in jet velocity. Kalghati [5] investigated jet diffusion flames fuelled by hydrogen, propane, methane and ethylene, for a large range of nozzle exit velocities and nozzle diameters. He also generated classical experimental data on liftoff heights. He observed that the liftoff height varies as a linear function of the jet exit velocity and nearly independent of the nozzle diameter. It was concluded that the flame liftoff height is dependent on the laminar burning velocity. Maike-Lye [6] investigated the liftoff heights for methane, ethylene and natural gas jet flames which were partially premixed with air and reported a linear dependence of flame length on jet exit velocity.

It is clear from the above studies that although immense experimental data is available, numerical simulation of LPG jet diffusion flames has not been attempted. This is due to complexity of the fuel and its kinetics. In this study, a single step global chemistry is formulated and used in simulations. The lab scale experimental flame length data are compared with the numerical results for validating the single step mechanism.

2. Numerical Methodology

Commercial CFD software ANSYS – FLUENT 16.1, which employs finite volume method, is used to study LPG diffusion flames. Appropriate models available in ANSYS – FLUENT along with User Defined Functions (UDF) and kinetics sub-model are used. Temperature and concentration dependent thermo-physical properties are used. Multi-component mass diffusion and optically thin approximation based radiation sub-model are employed. Steady-state governing equations are solved. A second order upwind scheme for convective terms and SIMPLE algorithm for pressure-velocity coupling are employed for discretizing the governing equations.

Laminar species transport model with volumetric reactions is used. Full multicomponent diffusion equation along with a diffusion energy source and thermal diffusion (Soret effect) are solved. The net rate of formation of several species in the reaction is solved using stiff chemistry solver. Single step global chemistry is used for estimating the reaction rate. Based on the single step kinetics for propane and butane, reported by Westbrook and Dryer [7], where the activation energy is the same for both fuels, single step chemistry parameters are estimated and used.

Thermal radiation absorption by species such as CO_2 and H_2O has been modeled using an optically thin approximation based model following Barlow et al. [8]. Here, the radiation between the flame and the cold surroundings is estimated using Planck's

*Corresponding author.
E-mail address: kumaranms92@gmail.com

absorption coefficients of the participating species. The heat loss to the surrounding due to thermal radiation is expressed in the form

$$q_R'' = 4\sigma(T^4 - T_b^4)\sum p_i a_{p,i}, \quad (1)$$

where σ is the Stefan-Boltzmann constant (5.67×10^{-8} W/m^2-K^4), T is the temperature in K, T_b is the ambient temperature in K, p_i is the partial pressure of species i in atmospheres, a_{pi} is the Planck mean absorption coefficient (in atm^{-1}m^{-1}) for species i, which is calculated as a function of mole fractions [8]. This model is implemented using an UDF in FLUENT. Multi-block structured grid with non-uniform spacing in both axial and radial directions is used. Mesh is refined near the walls and the burner exit to capture the steep gradients in the flame. The number of cells in the domain is 170000. The computation domain along with boundary conditions is shown in Figure 1.

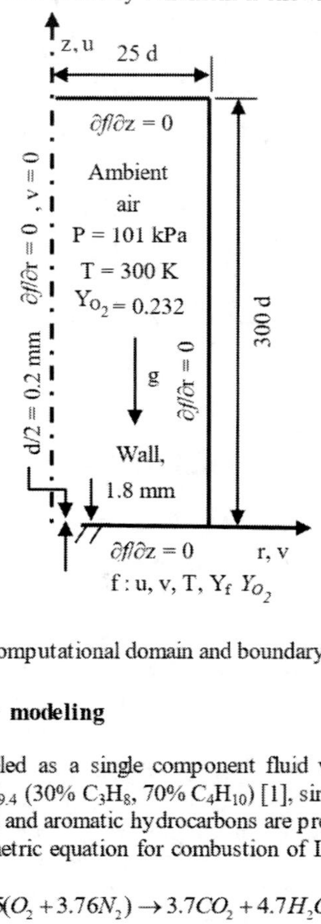

Figure 1: Computational domain and boundary conditions.

2.1 Property modeling

LPG is modeled as a single component fluid with molecular formula $C_{3.7}H_{9.4}$ (30% C_3H_8, 70% C_4H_{10}) [1], since, only traces of unsaturated and aromatic hydrocarbons are present.
The stoichiometric equation for combustion of LPG is as given below

$$C_{3.7}H_{9.4} + 6.05(O_2 + 3.76N_2) \rightarrow 3.7CO_2 + 4.7H_2O + 22.752N_2$$

The thermodynamic properties are calculated as given below [9]. Critical temperature is determined using Kay's rule.

$$T_{c,LPG} = \sum Y_i T_{c,i} \quad (\mathrm{i} = C_3H_8, C_4H_{10}), \quad (2)$$

$$0.5 < \frac{T_{c,i}}{T_{c,j}} < 2 \text{ and } 0.5 < \frac{p_{c,i}}{p_{c,j}} < 2,$$

where Y_i is the mass fraction of ith species, T_c is the critical temperature of ith species (K), p_{ci} is the critical pressure of ith species (N/m^2) and T_c is the critical temperature of LPG. Critical pressure of LPG, $p_{c\,LPG}$ is calculated by modified Prausnitz and Gunn combination method [9].

$$p_{c,LPG} = \frac{R_u T_{c,LPG} \sum Y_i T_{c,i}}{\sum Y_j V_{c,j}}, \quad (3)$$

where V_c is the critical volume of LPG (m^3). The enthalpy H (kJ/kg) and entropy S (kJ/kg-K) of LPG is calculated from its components, H_i and S_i as,

$$H = \sum Y_i H_i \quad (4)$$

$$S = \sum Y_i S_i \quad (5)$$

The acentric factor (ω) of the mixture is evaluated as,

$$\omega_{LPG} = \sum Y_i \omega_i . \quad (6)$$

After evaluating the acentric factor, the Lennard-Jones parameters have been calculated using the relations,

$$\sigma\left(\frac{p_c}{T_c}\right)^{1/3} = 2.3551 - 0.087\omega \quad (7)$$

$$\frac{\varepsilon}{kT_c} = 0.7915 + 0.1693\omega. \quad (8)$$

Here, σ is the Lennard-Jones critical distance (m) and ε/k is the Lennard-Jones critical temperature (K). The Lennard-Jones parameters are employed in the calculation of kinematic viscosity (μ), and binary diffusion coefficient (D_{ij}), from kinetic theory [9]. The specific heat of LPG, $c_{p,LPG}$ (kJ/kg), is calculated from the specific heat of its components, $c_{p,i}$.

$$c_{p,LPG} = \sum Y_i c_{p,i} . \quad (9)$$

The kinetic parameters are estimated using Kay's rule:

$$K_{LPG} = \sum X_i K_i , \quad (10)$$

where X_i is the mole fraction of ith species, K_i is the rate coefficient of species i and K_{LPG} is the rate coefficient of LPG, at a given temperature. For individual species,

$$K_i = A_i \exp\left(\frac{-E_{a,i}}{R_u T}\right), \quad (11)$$

where A_i is the pre-exponential factor (m^3/kmol)$^{1-a-b}$(1/s), E_a is the activation energy (kJ/kg), R_u is the universal gas constant (8.314 kJ/kmol-K), T is the temperature of mixture (K), a and b are the order of fuel and oxidizer respectively. The values of a and b are assumed to be the same as that of butane-air single step reaction [7]. It should be noted that E_a is the same for butane and propane and A_{LPG} has been effectively calculated by this approach.

Taking natural logarithm of Eqn. 11, eqn. (12) is obtained.

$$\ln K_{LPG} = \ln A_{LPG} - \frac{E_{a,LPG}}{R_u T} \quad (12)$$

Equation (12) is of the form y = mx + c. A graph is plotted between ln (K_{LPG}) and (1/T). The slope is (E_a/R_u) and intercept is (ln A) of the LPG mixture, which are obtained from graph. A value of 1.256×10^8 for (E_a/R_u) and a value of 4.335×10^9 for pre-exponential factor (A) is used in the present simulation.

3. Results and discussion

3.1. Flame length

Lab-scale experiments are conducted using a simple orifice burner having an exit diameter of 0.4 mm. For varying LPG flow rates, flame photographs are captured. Figure 2 shows the flame photographs for different LPG jet velocities. It is seen that the flames length increases as the jet velocity is increased. The flame is anchored close to the burner wall for all velocities as shown in Figs. 2(a), 2(b), 2(c) and 2(d). Table 1 gives the experimental value of flame length, for cases simulated in the present study.

Figure 2: Photographs of LPG jet diffusion flames for various jet velocities (a) 6 m/s (b) 7.3 m/s (c) 8.5 m/s (d) 9.7 m/s.

Velocity (m/s)	Flame length (mm)
6	43
7.3	52
8.5	59
9.7	67
11	74

Table 1: Experimental data for LPG jet diffusion flames.

Figure 3 presents flame lengths obtained from simulations compared against the experimental results. Flame length from numerical simulation is obtained by following the contour of fuel mass fraction of value 0.03. This criterion is used for delineating the flame zone. A similar criterion has been reported by Wade and Gore [10] in their experiments on acetylene-air jet diffusion flames. It is observed that flame length varies linearly with jet velocity. Moreover, the numerical trend matches quite well with the experimental trend.

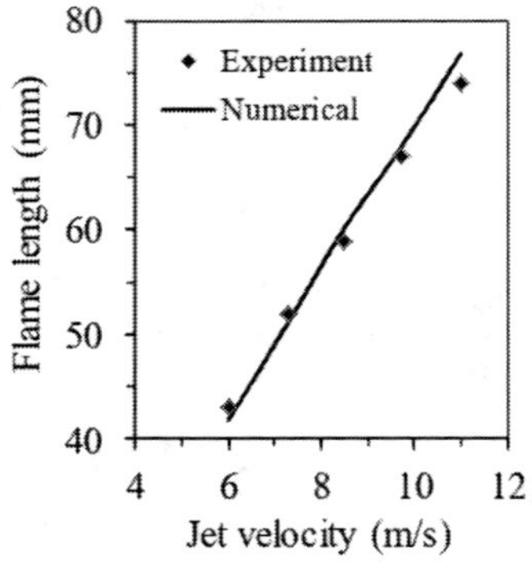

Figure 3: Flame length for LPG-air jet diffusion flames.

Figure 4 presents the isotherms for different fuel velocities. It is observed that as the velocity is increased, the peak temperature, indicated by the contour level 5, increases axially for laminar flames as shown in Figs. 4(a), 4(b) and 4(c).

The flame length from simulations is shown using black contour line of LPG mass fraction with a value of 0.03, superimposed over the temperature contours in Fig. 4. It is observed that the radial extent of the flame slightly decreases as the fuel velocity is increased. This is because as the jet momentum is increased the entrainment of ambient air increases and hence the flame width decreases.

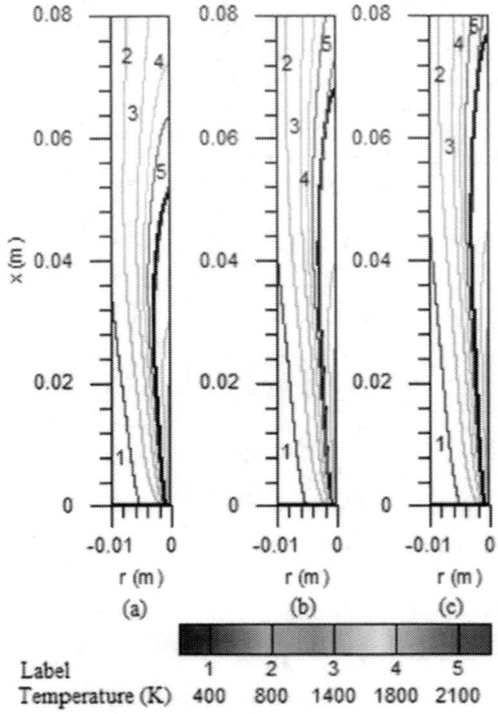

Figure 4: Contours of static temperature for LPG diffusion flames (a) 7.3 m/s (b) 9.7 m/s and (c) 11 m/s. Black lines: contour lines of LPG mass fraction 0.03.

In order to carry out simulations of LPG-H_2 jet diffusion flames, one more species, H_2 and a global mechanism has been added, which is given as,

$$H_2 + 0.5(O_2 + 3.76\,N_2) \rightarrow H_2O + 1.88\,N_2.$$

The rate of this global reaction [7] with thermodynamic properties of additional species is available in FLUENT. A case with mixture velocity of 6 m/s has been chosen to simulate LPG - H_2 jet diffusion flames. Hydrogen is introduced into the mixture in volumetric proportions of 15%, 30%, 45% and 60%. The total volumetric flow rate of the mixture is kept constant.

Figure 5 presents the isotherms for the cases having different hydrogen fractions. The same criterion of LPG mass fraction of 0.03 has been used to delineate the flame length for these cases also, as shown by the black contour lines in Fig. 5. It is clear that the flame zone comes closer to the burner exit as the percentage of hydrogen in the fuel mixture is increased. The flame length decreases notably as shown in Fig. 5. Higher reactivity of hydrogen causes this trend. This has been experimentally observed by Khan et al. [3] using a different burner. Since laminar flame length is directly proportional to the volumetric flow rate of fuel, a large difference in the flame length is observed as compared to the results of Khan et al. [3]. However, the numerical trend matches with experimental trend quite well.

The diffusivity of hydrogen is higher when compared to hydrocarbon fuels. Therefore, the rate at which oxygen diffuses from the ambient in the presence of hydrogen is also higher. Further, the stoichiometric oxidizer fuel ratio decreases. Both these factors tend to reduce the flame length as given by the correlation proposed by Ropar, which is given as,

$$L_f \approx \frac{Q_F}{D_F \ln(1+1/S)}, \quad (13)$$

where Q_F is the volumetric flow rate of mixture in m^3/s, D_F is the diffusivity of the fuel mixture into air (m^2/s) and S is the molar stoichiometric oxidizer fuel ratio.

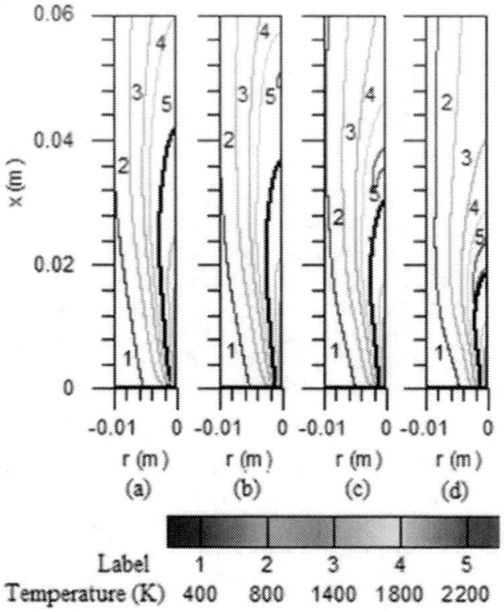

Figure 5: Contours of static temperature (a) 0% (b) 15% (c) 30% and (b) 60% H_2 by volume. Black lines: contour lines of LPG mass fraction 0.03.

It is clear that the proposed single step kinetics is able to predict the flame heights reasonably well. The prediction can be further improved using detailed kinetics

3.2. Flame structure

Figure 6 presents the radial profiles of mass fractions of reactants and products in a pure LPG flame. These are extracted at an axial location of 10 mm from the nozzle exit. It is seen that the fuel mass fraction decreases as it reaches the reaction zone. Finite amount of oxygen leaks through the reaction zone. It is evident that both CO_2 and H_2O attain their maximum around the reaction zone. Finite amounts of CO_2 and H_2O diffuse through the reaction zone and reach the reactant side. Figure 7 shows the radial profiles of mass fractions of reactants and products in LPG-H_2 flame, where 30% H_2 by volume is present in the mixture. These are extracted at an axial location of 0.5 mm from the nozzle exit, since in the presence of H_2, flame zone comes closer to nozzle exit. It is visible that due to the addition of hydrogen, the oxygen diffusion has increased significantly. Hydrogen is consumed near to the axis and LPG is consumed farther away. Product mass fractions reach the maximum values close to the consumption zone of hydrogen.

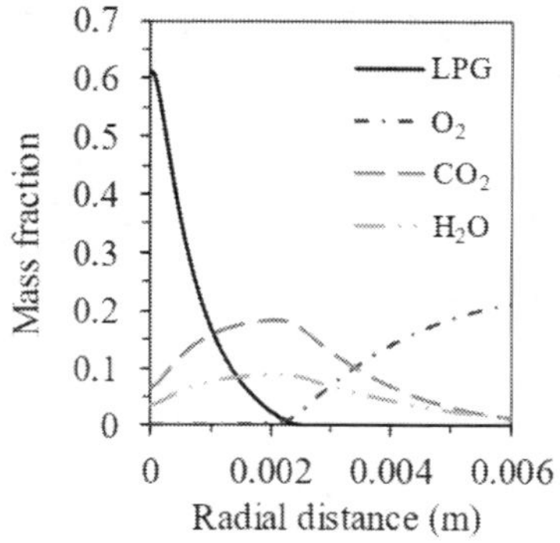

Figure 6: Profiles of LPG, O_2, CO_2, and H_2O mass fractions in pure LPG jet diffusion flame.

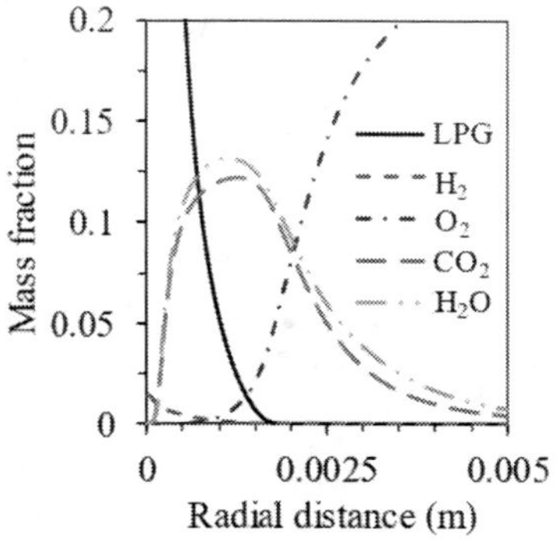

Figure 7: Profiles of LPG, H_2, O_2, CO_2, and H_2O mass fractions in LPG-H_2 jet diffusion flame with 30% H_2 in the fuel mixture.

4. Conclusions

LPG-air diffusion flames are simulated by modeling LPG, which is a multi-component fuel, as a single component fuel. The thermodynamics and kinetic parameters are derived from its major constituents - propane and butane. Single-step global chemistry is derived and used in simulations. Numerical results are validated against experimental data. The numerical model using the derived single step chemistry is able to capture the flame length of LPG diffusion flames quite reasonably. Model also predicts the decreasing flame length trend when hydrogen is added to LPG. The kinetics parameters can be further fine-tuned and improved using opposed flow flame experiments.

References

[1] D.Y. Kiran, D.P. Mishra, Fuel 86 (10-11) (2007) 1545-1551.

[2] D.P. Mishra, Int. J. Chem. Reactor Eng. 7 (2009) 64.

[3] N. Khan, K.R.V. Manikantachari, R. Dutta Roy, V. Raghavan, Asian J. Sci. Res. 6(2) (2013).

[4] S. Mahesh, D.P. Mishra, Fuel 87 (2008) 2614–2619.

[5] G.T. Kalghatgi, Combust. Sci. Technol. 26 (1981) 233–239.

[6] R.C. Maike-Lye, J.A. Hammer, Proc. Combust. Inst. 22 (1988) 817–824.

[7] C.K. Westbrook, F.L. Dryer, Prog. Energy Combust. Sci. 10 (1984) 1-57;

[8] R.S. Barlow, A.N. Karpetis, J.H. Frank, J.Y. Chen, Combust. Flame 127 (2001) 2102-2018.

[9] Robert C. Reid, John M. Prausnitz, Bruce E. Poling, The Properties of Gases and Liquids, Tata McGraw Hill, 1987.

[10] R.W. Wade, J.P. Gore, Nat. inst. Stand. Technol. 5904 (1996) 41-42

11th Asia-Pacific Conference on Combustion,
The University of Sydney, NSW Australia
10th -14th December 2017.

Effects of Discharge Frequency on Ignition Behaviors of DBD for Lean Methane/Air Mixtures

Shinji Nakaya[1], Taichi Kobayashi[1], Shingo Iseki[1] ,

Mitsuhiro Tsue[1], Masaaki Kono[2], Kanehito Nakamura[2]

[1]The University of Tokyo

7-3-1 Hongo, Bunkyo-ku, Tokyo 113-8656, JAPAN

[2] Nippon SOKEN Inc.

14 Iwaya, Shimohasumi-cho, Nishio-shi, Aichi 445-0012, JAPAN

Abstract

Ignition process of lean mixture by dielectric-barrier-discharge has been investigated for methane/air mixtures. The frequency of the applied voltage was varied, and the effects of the frequency on the ignition performance and discharge characteristics were studied at various pressure based on optical measurements. Emission spectra from dielectric barrier discharge were also measured with an echelle grating spectrometer. Then, ignition probability was investigated, at different pressure. Ignition performance was examined under the same period of the discharge or the same number of the discharge. Results indicated that emission time of plasma increased with increasing the ambient pressure, indicating the formation of thermal plasma. When the ignition period was fixed at 1ms, the ignition probability became higher with an increase in the frequency at pressure of 50 and 100 kPa. The ignition probability was also investigated with a fixed discharge count at various frequencies. Even for the same discharge count, an increase in the frequency resulted in the improvement to the ignition performance.

1 Introduction

In order to increase the flame propagation speed, turbulence with swirl and tumble flows is introduced into the combustor of spark ignition engines for lean combustion. In such environments, it is difficult to ignite the mixture. Therefore, an improvement to the ignition performance for lean and diluted mixtures is required to increase the thermal efficiency of engines. Although electric spark ignition has been used widely for conventional automobile engines, high spark energy is required to enhance the ignition performance of the electrical spark ignitions for lean mixtures, resulting in damages to the electrode and large heat loss. Thus, novel ignition methods as with laser breakdown spark ignition [1], repetitive short pulse ignition [2], microwave plasma assisted ignition [3], low temperature plasma ignitions [4,5] have been tested to establish a high-performance ignition method. In the present study, we focus on the low temperature plasma ignition using a dielectric barrier discharge (DBD). Plasma assisted combustion [7,8] has been tested to enhance the reactivity of the mixture and increase the flame speed. For this purpose, low temperature plasma discharge has been used to form radicals, excited molecules and electrons, resulting in the increase in the reaction rates. The enhancement of chemical reaction rates using plasma-assisted combustion are achieved by the combination of the thermal effects and chemical effects, which are strongly related to the plasma conditions. Most combustion studies on plasma-assisted combustion have been conducted at low pressure less than the atmospheric pressure, where low temperature plasma can be formed easily. However, the mixture is ignited at high pressure in engine condition, and the understanding on the low temperature plasma formations under such environment is rather limited. In the present study, DBD at high pressure was studied in order to obtain fundamental knowledge on formation characteristics of the low and the high temperature plasmas, and pressure condition where the low temperature plasma is formed was identified. Then, the effects of discharge frequency on Ignition characteristics were also investigated experimentally in the condition where the low temperature plasma was formed.

2 Experimental

A schematic of the experimental setup is shown in Figure 1. Ignition experiments were performed in a combustion chamber with a constant volume. The combustion chamber has several flanges with quartz glasses. Through an optical window, discharge processes were observed using a high-speed video camera (Vision Research, Phantom V2511) with the frame speed of 50000 fps. Methane/air mixtures were prepared in a mixture tank with a stirrer. Then, the mixture was introduced to the combustion chamber. The equivalence ratio tested in the present study was less than 0.6. Three pressure transducers (KEYENCE, AP-33, AP-30 and AP-14S) were equipped with the tank and the concentration of the mixture was determined from the partial pressure of the components. The same mixture was introduced to the chamber and ignition tests were conducted under the same equivalence ratio. A spark plug was mounted on a flange of the side wall, and the central electrode was insulated by a ceramic, which regulated the amount of the current and enabled to form the DBD. Streamer was formed between the central electrode and the outer electrode. The spark plug was connected to a high-voltage and high-frequency AC power supply. The primary AC

Corresponding author. Fax: +81-3-5841-6633
E-mail address: tsnakaya@mail.ecc.u-tokyo.ac.jp

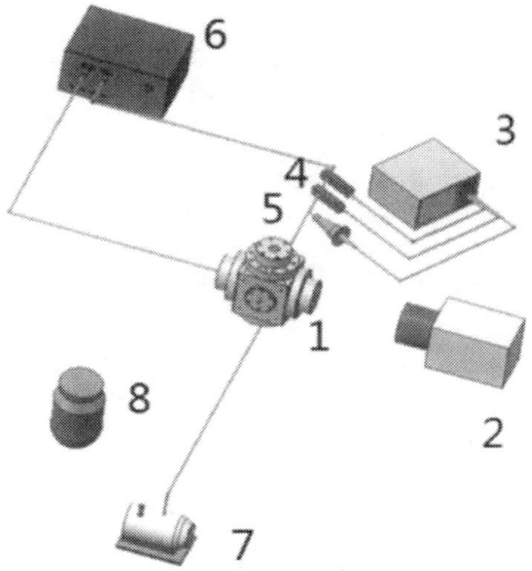

Figure 1: Schematic of experimental setup. 1.Combustion chamber, 2. High-speed video camera, 3. Oscilloscope, 4. Current probes, 5. High voltage probe, 6. Power supply, 7. Vacuum pump, 8. Mixture tank

current was applied to the circuit with a coil and amplified at the resonant frequency. The maximum output voltage of the power supply was ±40kVp-p. In the present study, frequencies of the current were 600, 800 and 1030kHz, and two input signals to control the discharge period and frequency were produced by a pulse generator (BNC, 555). Using this system, the number of the current pulses applied to the plug was controlled. Through the optical window, plasma emissions were also measured with an echelle grating spectrometer (Bunkokeiki, EMDP-100). Emissions from plasma was directed into the inlet of the optical fiber with a reducing optical system consisting of lens with the focal lengths of f=45mm and f=20mm. In the present study, emissions ranging from 400nm to 900nm were measured.

Characteristics of DBD were investigated in air at various pressures from 10 to 3000kPa at room temperature. In this case, the primary voltage of the alternate current was fixed at 120Vp-p or 150Vp-p. The streamer formations were observed with the high-speed video camera, and the lifetime of plasma emissions was evaluated. Thereafter, ignition experiments were performed at 50kPa and 100kPa for methane/air mixtures to investigate the performance of the low temperature plasma. Also, the effects of the discharge frequency on the lean limit were investigated.

3 Results and discussion

3.1. DBD characteristics

Firstly, the effects of pressure on DBD were focused. Figure 2 shows the instantaneous images of DBD measured by a high-speed video camera at various pressures. In this figure, the exposure time of the high-speed video camera ranged from 4.5 to 19.5μs, depending on the intensity of the emission. In this case, the primary voltage of the circuit was 120Vp-p, and the discharge frequency was 640kHz. Thus, multiple discharges for many cycles were recorded on one image because high-frequency AC was applied. It seemed that one discharge was achieved for nanoseconds. The total discharge period of AC was 1.0ms. It was noted that no streamer formations were observed for the primary voltage of 120Vp-p at 1000kPa. At less than

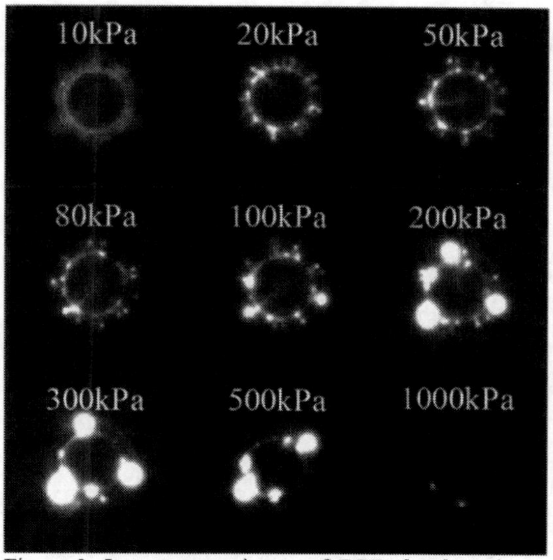

Figure 2: Instantaneous image of DBD for the primary voltage of 120Vp-p at various pressure from 10kPa to 1000kPa.

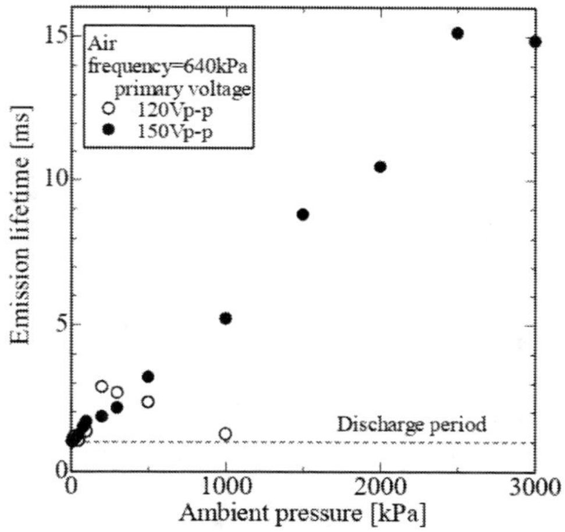

Figure 3: Relationship between plasma emission lifetime and ambient pressure.

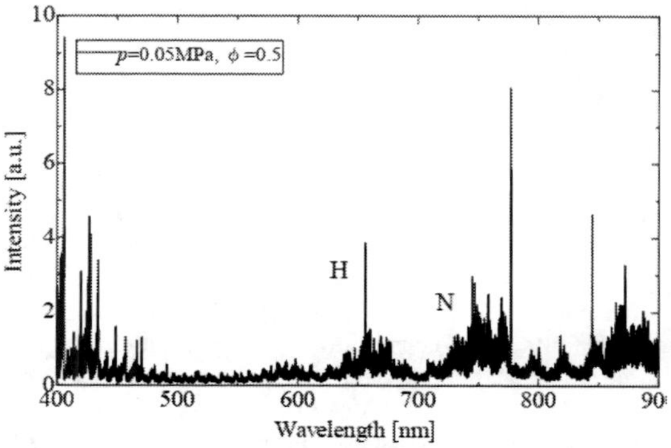

Figure 4: Emission spectra from DBD at 50kPa for methane/air mixture of φ=0.5.

1000kPa, it can be seen that the number of streamers increases with a decrease in the pressure. Above 200kPa, the strong emission of the plasma kernel was observed locally between the

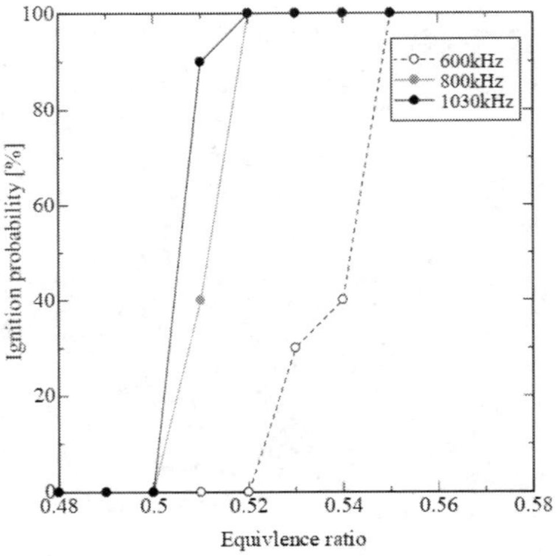

Figure 5: Ignition probability for methane/air mixtures at 100kPa.

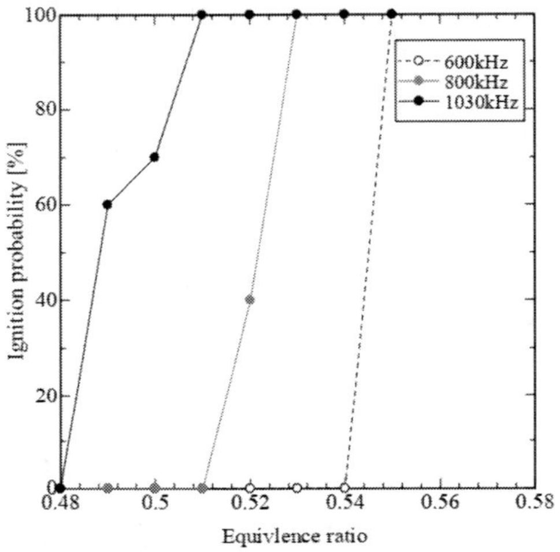

Figure 6: Ignition probability for methane/air mixtures at 50kPa.

electrodes. At high pressure over 100kPa, low temperature plasma of the streamer discharge could change to thermal equilibrium plasma. Starikovskaia reported that streamer discharges are observed for the applied voltage of 10 to 100kVp-p at 0.01 and 0.1 MPa [4]. The same tendency was confirmed by the results in Figure 2 at the voltage of 25 to 35kVp-p and higher pressure.

Figure 3 shows emission lifetime of plasma for the primary voltage of 120Vp-p and 150Vp-p at various pressures. In the case of 150Vp-p, discharge was observed even at 3000kPa although no discharge was observed in the case of 120Vp-p at the pressure. It can be seen that emission periods were approximately 1ms at the pressure of 100kPa or less, which is equal to the discharge duration of current. When the pressure was over 100kPa, the plasma lifetime was greater than the discharge duration, indicating that the thermal equilibrium plasma was formed. In the case of 150Vp-p, the applied voltage ranged from 35 to 40kVp-p. An increase in the applied voltage resulted in the extension of the upper limit of pressure for discharge. However, it resulted in an increase in the period of plasma emission with ambient pressure. Active thermal plasma seemed to be formed for 150Vp-p because the plasma lifetime became longer than the total discharge time with an increase in the pressure. Even at 3000kPa, although the discharge time was only 1ms, plasma emission was observed for 15ms. Even after the discharge, plasma seemed to exist and high temperature was sustained.

At less than or equal to 100kPa, the discharge period and the plasma emission period was almost the same. Thus, it was elucidated that non-thermal plasma was formed during the discharge time. In order to verify the plasma condition, emission spectra of DBD was measured at 50kPa for methane/air mixture with the equivalence ratio of φ=0.5. In this measurement 200-time accumulation of the emission was recorded on ICMOS because emission of DBD was weak. The primary voltage was 200Vp-p, and the discharge period was 2ms. It can be seen that there were strong emissions of atomic line as with H (e.g. 434nm, 656nm), O (e.g. 750nm, 777nm and 840nm) and N (e.g. 747nm). Also, it is noted that no continuum emissions of black body were observed indicating that the plasma was non-thermal.

3.2. Ignition performance of low temperature plasma

From the experimental results above, low temperature plasma could be formed at 100kPa or less. As described above, the purpose of the present study was to investigate the ignition performance of the low temperature plasma excluding the effects of the thermal plasma. The radical formations in low temperature plasma was influenced by the discharge frequency. In order to investigate the effects of the DBD frequency on ignition performance of the low temperature plasma, ignition experiments were conducted for lean methane/air mixtures at 100kPa and 50kPa. The applied voltage to the electrode was fixed to 35±1kVp-p and the total discharge period was also fixed at 1ms. The frequency of the alternate current was varied to 600, 800 and 1030kHz. Totally, 10 runs of discharge were conducted at the same equivalence ratio. Figures 5 and 6 show relationships between the ignition probability and the equivalence ratio of the methane/air mixture at 50kPa and 100kPa. It can be seen that the lean limit was extended with an increase in the discharge frequency. Since the discharge period was fixed at 1ms in this case, higher frequency resulted in more discharge counts. Therefore, applied energy to the mixture was higher, resulting in the extension of the lean limit. At 50kPa, successful ignitions were observed at even less than the equivalence ratio of 0.5. The lean limit of the equivalence ratio at 50kPa was less than that at 100kPa. This could be due to the induced flow by the buoyancy. Buoyancy becomes stronger with an increase in the pressure, and

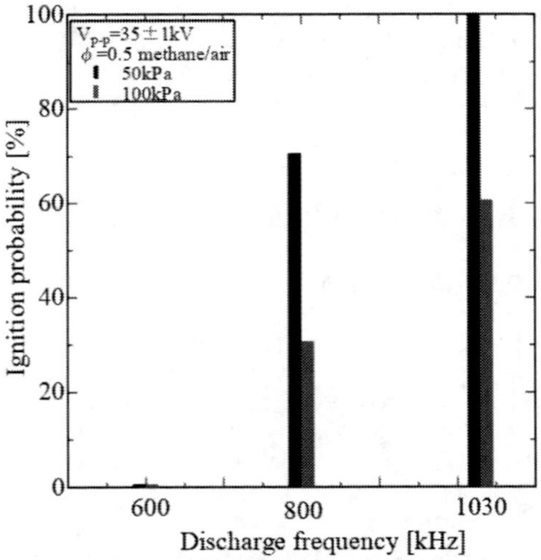

Figure 7: Ignition probability for methane/air mixtures at 50kPa at the same discharge count.

this slight flow fields seemed to inhibit the ignition [10,11]. Ignition experiments were also conducted for lean methane/air mixtures under the same count of the discharge pulse. The discharge durations for 600kHz, 800kHz and 1030kHz are 1.7167, 1.2875 and 1.0000 ms, respectively. Figure 7 shows ignition probability at 50kPa and 100kPa for methane/air mixtures with the equivalence ratio of 0.5. Ignition characteristics near the lean limit were investigated. Even under the same condition of the discharge counts, the ignition probability became higher with an increase in the discharge frequency. For 600 kHz, no ignition was observed as shown in Fig. 7 although leaner mixtures were ignited in the case of 1030 kHz as can be seen in Fig. 6. In the case of 800 kHz, no ignition was observed for the discharge duration of 1ms. However, an increase in the discharge duration to 1.2875ms resulted in successful ignition of the mixture. In the case of 1030kHz, the ignition performance was the best at the same discharge counts in the present study. It indicated that an increase in the discharge frequency resulted in the increase in the ignition performance. Under this condition, streamer was formed during the short period of nanosecond when the electrons were formed by low temperature plasma reactions. During the period of no discharge, electrons were absorbed by ions. When the frequency became higher, electrons might be still alive when the next discharge came. In this case, electron productions seemed to be accelerated, resulting in the further production of active radicals. Therefore, high-frequency discharges seem to show the better ignition performance. This paper indicated that high frequency discharge has advantage on the production of active radicals.

4 Conclusions

A discharge and an ignition processes of the DBD were investigated in a combustion chamber with a constant volume. As a result of this study, conclusions below were obtained.

1. Low temperature plasma was formed at 100kPa or less using the ignition system of this study.

2. Plasma lifetime was longer than the discharge period with an increase in the ambient pressure especially over 1000kPa for the same primary voltage.

3. At 50kPa, continuum emission was not observed, and it indicated that non-thermal plasma was obtained. At 100kPa, the low temperature plasma seemed to be formed because the plasma life time was same as the total discharge period of AC.

4. For the same discharge period of 1ms, higher discharge frequency resulted in better ignition performance at 50kPa and 100kPa.

5.Even under the same discharge counts, higher discharge frequency resulted in better ignition performance at 50kPa and 100kPa.

5 Acknowledgment

We would like to thank Yoichi Sugita, Teiji Iwata, Takeo Okunuki and Masafumi Utsumi for their valuable support in designing and fabricating the experimental setup. We would also thank Xiao Jing Gu for their support to the experiments.

References

[1] Morsy, M.H., Review and recent developments of laser ignition for internal combustion engines applications, Renew. Sust. Energ. Rev. 16 (2012) 4849-4875.

[2] Tanoue, K., Kuboyama, T., Moriyoshi, Y., Hotta, E., Imanishi, Y., Shimizu, N., Iida, K., Development of a Novel Ignition System Using Repetitive Pulse Discharges: Application to a SI Engine., SAE Int. J. Eng. 2(1) (2009) 298-306.

[3] Wolk, B., DeFilippo, A., Chen, J-Y., Dibble, R., Nishiyama, A., Ikeda, Y., Enhancement of flame development by microwave-assisted spark ignition in constant volume combustion, Combust. Flame, 160 (7) (2013)1225-1234.

[4] Starikovskaia, S.M., Topical review: Plasma assisted ignition and combustion. J. Phys.D: Appl. Phys., 39 (2006) R265–R299.

[5] Starikovskii, A.Y., Plasma supported combustion, Proc. Combust. Inst., 30 (2) (2005) 2405-2417.

[6] Starikovskiy, A., Aleksandrov, N., Plasma-assisted ignition and combustion, Prog. Ener. Combust. Sci. 39 (2013) 61-110.

[7] Ombrello, T., Won, S.H., Ju, Y., Williams, S., Flame propagation enhancement by plasma excitation of oxygen. Part II: Effects of O2(a1Dg), Combust. Flame, 157 (2019 1906-1915.

[8] Sun, W., Uddi, M., Ombrello, T., Won, S.H., Carter, C., Ju, Y., Effects of non-equilibrium plasma discharge on counterflow diffusion flame extinction, Proc. Combust. Inst., 33 (2) (2011) 3211-3218.

[9] Ju, Y., Sun, W., Plasma assisted combustion: Progress, challenges, and opportunities, Combust. Flame, 162 (2015) 529-532.

[10] Nakaya, S., Kobayashi, Y., Tsue, M., Ignition behaviors of pyrolyzed component and air near the lean limit under microgravity condition available from parabolic flights, Int. J. Microgravity Sci. Appl., 32 (2015) p320404.

[11] Nakaya, S. Iseki, S. Gu, XJ., Kobayashi, Y., Tsue, M., Flame kernel formation behaviors in close dual-point laser breakdown spark ignition for lean methane/air mixtures, Proc. Combust. Inst. 36 (3) (2017) 3441-3449.

11th Asia-Pacific Conference on Combustion,
The University of Sydney, NSW Australia
10th -14th December 2017.

On the Use of Sparse-Lagrangian MMC-LES for Simulation of a Piloted Coal Jet

L. F. Zhao[1*], M. J. Cleary[1], O.T. Stein[2], A. Kronenburg[2]

[1] The University of Sydney, School of Aerospace, Mechanical and Mechatronic Engineering

[2]Institut für Technische Verbrennung, Universität Stuttgart

Abstract

An advanced, multiphase turbulent combustion model is implemented to simulate the piloted, dilute coal combustion (PCC) which was experimentally investigated by Hwang et al. in 2005. To simulate the carrier phase reaction and its corresponding turbulent flow, a hybrid combination of sparse-Lagrangian multiple mapping conditioning (MMC) and large eddy simulation (LES) is adopted. The main development in this paper is the utilization of a dispersed Lagrangian model in capturing the evolution of coal particles, incorporating a mixture-fraction-based model for sub-grid heat and mass transfer between the dispersed and carrier phases. In this first attempt to model a coal flame by MMC-LES the pyrolysis gases are assumed to be pure methane which has the advantage of being the same as the pilot fuel thus permitting the use of a single mixture fraction to characterize the combustion and condition the MMC mixing model. Additionally, char reactions are neglected in common with other attempts to model this flame reported in the literature. Comparisons are made between the multiphase MMC-LES predictions and experimental data for the flame temperature and some reactive species profiles. Generally, good qualitative agreement is observed while there are quantitative differences.

1 Introduction

The World Energy Outlook projects that the consumption of coal will increase continuously throughout the first half of the 21st century, and most of this increased consumption will take place in utility power stations in developing Asian countries [1]. The pulverized coal combustor is preferred over other types, such as the fluidized bed combustor, as it provides higher oxidation intensity and stronger heat transfer [2]. Therefore, enhancing our understanding and predictive capability of pulverized coal combustion remains an important research topic. With the advancements in state-of-art numerical modelling techniques as well as the rapid increase in computing power offered by modern CPUs, computational fluid dynamics (CFD) has emerged as a useful tool for analysing combustion systems. In particular, large eddy simulation (LES) has become an increasingly adopted method for its potential of finding the balance between computational cost and simulation accuracy.

Pulverised coal particles undergo a series of processes before turning into non-flammable residuals. These include convective and radiative heating of coal, coal devolatilisation, burning of volatile matter and char conversion [3]. Accompanying these is coal particle dispersion. Each process is affected by turbulence, and in the context of LES, sub-grid modelling is required.

The experimental flame case selected for this study is the pulverized coal jet of Hwang et al. from 2005 and 2006 [4, 5]. Prior to those experiments, the same research group conducted the first combustion LES for a solid fuel dispersion, in which a flamesheet model based on the conserved scalar approach was adopted [6]. In subsequent work, Kurose and colleagues updated their investigations of turbulent pulverized coal combustion both experimentally and numerically including a comparison between RANS and LES [7]. Various other research groups have since performed LES of Hwang's flame [8-12]. A simplified three-step global reaction mechanism of a postulate volatile species obtained from chemical percolation pyrolysis (CPD) modelling is combined with the eddy break-up model in [8, 9]. A more realistic way of estimating the unsteady and inhomogeneous coal devolatilisation rate is achieved by Rieth et al. who applied CPD modelling to every coal particle on-the-fly [10]. A flamelet approach for modelling the sub-grid gas phase reactions is adopted in [10, 11]. In the most recent study, a flamelet-progress variable approach is adopted with the progress variable describing the extent of coal combustion [12].

The strategy for capturing the turbulence-chemistry interaction is a key aspect of combustion modelling. In multiphase combustion there is also the added necessity of modelling the interactions between turbulence and interphase heat and mass transfer. The mixture fraction, as a conserved scalar, is the foundation of many turbulent, non-premixed combustion models [13]. In the present work, we use a hybrid numerical model involving Eulerian LES and stochastic Lagrangian multiple mapping conditioning (MMC) [14] for turbulent gas phase chemical reaction. It is a transported probability density function (PDF) approach [15] combined with conserved scalar concepts adopted from conditional moment closure (CMC) [16]. The novelty of this study is to apply MMC-LES to a reacting coal flame for the first time. A particular focus is placed on the formulation of the sub-grid heat and mass transfer between the solid fuel particles and the carrier gas. We use a conserved scalar form of the Spalding transfer model. Validation of the overall model is carried out by comparing simulation results against the experimental data for temperature and species.

2 Model Development

2.1 Eulerian LES Formulation

The Favre filtered transport equations for mass, momentum and reference mixture fraction are

$$\frac{\partial\bar{\rho}}{\partial t}+\frac{\partial\bar{\rho}\tilde{u}_i}{\partial x_i}=\tilde{\bar{S}}_\rho, \quad (1)$$

$$\frac{\partial\bar{\rho}\tilde{u}_j}{\partial t}+\frac{\partial\bar{\rho}\tilde{u}_i\tilde{u}_j}{\partial x_i}=-\frac{\partial\bar{P}}{\partial x_j}+\frac{\partial}{\partial x_i}\left(\tilde{\tau}_{ij}-\tau_{ij}^{sgs}\right)+\tilde{\bar{S}}_u, \quad (2)$$

$$\frac{\partial\bar{\rho}\tilde{f}}{\partial t}+\frac{\partial\bar{\rho}\tilde{u}_i\tilde{f}}{\partial x_i}=-\frac{\partial}{\partial x_i}\left(\tilde{J}_{f,i}-J_{f,i}^{sgs}\right)+\tilde{\bar{S}}_f. \quad (3)$$

* Corresponding author. Phone: (+61) 2 9351 5360

E-mail address: lzha7583@uni.sydney.edu.au

The resolved stress and diffusive flux are denoted as $\tilde{\tau}_{ij}$ and $\bar{J}_{f,i}$ in the above equations, respectively. A standard eddy viscosity model is applied here to emulate the sub-grid scale stress, τ_{ij}^{sgs}, with a classical Smagorinsky closure, which has a constant model coefficient of 0.167. The corresponding sub-grid scale diffusive mass flux is closed by a similar eddy diffusivity model with $J_{f,j}^{sgs} = -\bar{\rho}\mathcal{D}_t \frac{\partial \tilde{f}}{\partial x_j}$, where the turbulent diffusivity is given by the ratio of turbulent viscosity and the turbulent Schmidt number, which has been set to 0.4.

The influence of solid fuel particles on the mass, momentum and reference mixture fraction of the LES solved carrier phase are represented by the filtered source terms $\bar{\tilde{S}}_\rho$, $\bar{\tilde{S}}_u$ and $\bar{\tilde{S}}_f$ in the above equations. These are formulated in Section 2.4.

2.2 Stochastic MMC Formulation

A sparse-Lagrangian version of the filtered density function (FDF) with an MMC mixing closure is employed for the sub-grid reacting scalar fluctuations. The FDF is defined jointly for the list of species, $\boldsymbol{Y}$, and enthalpy, h. We let $\boldsymbol{\phi} = (\boldsymbol{Y}, h)$ and give it the sample space, $\boldsymbol{\psi}$. The single-phase FDF equations are derived in [17], while here two additional interphase transfer terms are added on the RHS:

$$\frac{\partial F}{\partial t} + \frac{\partial}{\partial x_i}\left(\tilde{u}_i F - \bar{\rho}\,\mathcal{D}_e \frac{\partial}{\partial x_i}\left(\frac{F}{\bar{\rho}}\right)\right) + \frac{\partial}{\partial \Psi_n}(W_n(\boldsymbol{\Psi})F) + \frac{\partial^2}{\partial \Psi_n \Psi_m}\left(\overline{\mathcal{D}\frac{\partial \phi_n}{\partial x_i}\frac{\partial \phi_m}{\partial x_i}\bigg|\boldsymbol{\Psi}} F\right) = \frac{\partial}{\partial \Psi_n}(S_n \Pi(\boldsymbol{\Psi})F) + \Pi(\boldsymbol{\Psi})F. \quad (4)$$

S_n represents the difference of the nth reactive scalar between the solid fuel particle and the surrounding gas at the particle position and $\Pi(\boldsymbol{\Psi})$ is the conditional mass transfer rate per unit gaseous volume. The closure of these two terms is detailed in the following section. The second term on the LHS of (4) represents resolved-scale convection and effective diffusion. The third term on LHS is the reaction term which appears in closed form, and the fourth term is the sub-grid conditional scalar dissipation that is closed with the MMC mixing model.

The FDF equation has a high degree-of-freedom (determined by the number of reacting scalars). To avoid the intractable cost of solving (4) via an Eulerian finite volume approach, it is replaced by the following equivalent set of stochastic differential equations solved for an ensemble of Lagrangian notional particles (denoted by superscript *sp*):

$$dx_i^{sp} = \left(\tilde{u}_i + \frac{1}{\bar{\rho}}\frac{\partial}{\partial x_i}(\bar{\rho}\mathcal{D}_e)\right)^{sp} dt + \left(\delta_{ij}\sqrt{2\mathcal{D}_e}\right)^{sp} d\omega_i, \quad (5)$$

$$d\phi_n^{sp} = \left[W_n^{sp} + M_{n,G}^{sp} + S_n^{sp}\Pi^{sp}\right]dt, \quad (6)$$

$$dm^{sp} = \frac{m^{sp}\Pi^{sp}dt}{1 - \Pi^{sp}dt}. \quad (7)$$

Equation (5) provides the movement of notional particles in 3-dimensional space with $d\omega_i$ being a random walk. The variation in composition and standardized enthalpy due to chemical reactions, radiation, mixing (to account for the sub-grid conditional scalar dissipation) and interphase transfer are encapsulated in (6). Mixing is modelled by MMC whereby particle pairs are formed to be local in the reference mixture fraction space. Full details can be found in [17]. Equation (7) accounts for variation in mass and emulates the last term on the RHS of (4).

2.3 Lagrangian Fuel Particle (LFP) Model

2.3.1 Particle dispersion model

The model for solid fuel particle dispersion follows the approach introduced by Dukowicz [18]. It has been widely used for modelling the dispersion of both liquid sprays [19] and solid particles [8]. Fuel particles are transported by a set of Lagrangian equations utilising a relaxation method to compute the particle acceleration [20].

2.3.2 Interphase heat and mass transfer model

A model for sub-grid heat and mass transfer between the dispersed and carrier phases is developed following Spalding's approach [21]. By expressing the Spalding transfer number, B, in terms of conserved scalars, namely the mixture fraction for mass transfer and standardised enthalpy for heat transfer, we arrive at two equivalent expressions for the mass flux across the diffusive layer surrounding a coal particle [22]:

$$\dot{m}_D'' = Sh\frac{\rho\mathcal{D}}{d}\ln(1 + B_M), \quad (8)$$

$$\dot{m}_D'' = Nu\frac{\rho\alpha}{d}\ln(1 + B_H), \quad (9)$$

where Sh and Nu are the Sherwood and Nusselt numbers and d is the particle diameter. The Spalding transfer numbers from heat and mass perspectives are derived from references [3] and [19], using the mixture fraction, ξ, and standardised enthalpy, h, which both vary monotonically across the diffusive layer regardless of the presence of a flame:

$$B_M = \frac{\xi_G - \xi_S}{\xi_S - 1}, \quad (10)$$

$$B_H = \frac{h_G - h_S}{h_S - h_D + q_D - q_R}. \quad (11)$$

The subscripts G, S and D represent the quantities in the gas phase at the outer edge of the diffusive layer around the solid fuel particle, the gas phase at the particle surface, and within the fuel particle itself. The two intensive terms, q_D and q_R, describe the internal heating of a fuel particle and heat loss or gain due to radiation.

By equating equations (8) and (9) and incorporating equations (10) and (11), the fuel particle internal heating can be obtained as,

$$q_D = \frac{h_G - h_S}{B_M} - (h_S - h_D) + q_R. \quad (12)$$

The unity Lewis number assumption has been made during the derivation. Then the unsteady temperature and mass of a fuel particle is given by

$$m_D C_{p,D}\frac{dT_D}{dt} = \dot{m}'' A q_D, \quad (13)$$

$$\frac{dm_D}{dt} = -\dot{m}'' A. \quad (14)$$

Where A is the surface area. A simple temperature driven Arrhenius model is adopted to predict the rate of pyrolysis,

$$\dot{m}_D'' = \Lambda \exp\left(-\frac{E_a}{R_u T_D}\right)\frac{m_v}{A}, \quad (15)$$

where R_u is the universal gas constant, m_v represents the remaining volatile content within a fuel particle and T_D and A are the current fuel particle temperature and surface area. The pre-exponential factor Λ and activation energy of pyrolysis E_a are empirical inputs of this model. The pre-exponential factor and activation energy are set to $4.474 \times 10^3 \mathrm{sec}^{-1}$ and $1.9188 \times 10^7 \mathrm{J/(kmolK)}$, respectively, according to the chemical percolation devolatilisation (CDP) model calibration made for the Newland bituminous coal [9].

Significant amount of carbon was discovered at the end of the flow field indicating that the heterogeneous surface reaction is very limited in the piloted jet configuration [5]. Therefore the corresponding model is not included in this work.

The main attraction of such a model utilizing conserved scalars for diffusive layer heat and mass transfer is its independence of chemical reactions. The above model formulation will hold regardless of the presence of envelope flame, which may form around a fuel particle leading to an increase of fuel particle temperature and change in chemical composition within the diffusive layer.

2.4 Coupling Models

2.4.1 LFP to LES coupling

The mass and momentum exchange along the trajectory of a fuel particle generates source terms for filtered mass, momentum and reference mixture fraction for the finite volume field. Particle-Source-In Cell is employed to formulate these source terms [23],

$$\tilde{\bar{S}}_p = \tilde{\bar{S}}_f = \frac{1}{V_\Delta} \sum_{fp=1}^{N_{fp}} \dot{m}_D^{fp}, \quad (16)$$

$$\tilde{\bar{S}}_{u,i} = \frac{1}{V_\Delta} \sum_{fp=1}^{N_{fp}} \left(-m_D^{fp} a_{D,i}^{fp} + \dot{m}_D^{fp} u_{D,i}^{fp}\right). \quad (17)$$

Here V_Δ is the filter volume. In (17), the first term on the RHS represents the momentum exchange caused by drag force (with $a_{D,i}^{fp}$ being the acceleration of fuel particle), while the second denotes the momentum changes as a result of interphase mass transfer.

2.4.2 LFP to MMC coupling

The effect of solid fuel particles evolution is accounted for by adding source terms in the stochastic differential equations for reactive scalars (6) and total mass (7).

$$S_h^{sp} = (h_D - q_D + q_R) - h_G, \quad S_y^{sp} = Y_{D,\alpha} - Y_{G,\alpha}, \quad (18)$$

$$\Pi^{sp} = \frac{dm^{sp}}{(dm^{sp} + m^{sp})dt}. \quad (19)$$

3 Experimental and Numerical Details

Fine grained coal particles (Newland bituminous type) entrained by air at room temperature are injected through a 6mm (D) diameter nozzle which is surrounded by a 0.5mm thick annular methane pilot [4, 5]. Free entrainment of fresh air is allowed through the co-flow area. The detailed flow rate information is summarised in Table 1.

Parameter	Value
Coal mass flow rate	1.49×10^{-4} kg/s
Air flow rate	1.80×10^{-4} m^3/s
Methane flow rate (pilot)	2.33×10^{-5} m^3/s
Jet Reynolds number	2544

Table 1: Experiment inflow conditions [4].

The multiphase MMC-LES model is implemented numerically based on an open-source CFD platform called OpenFOAM (version 2.4.x). A structured hexahedral mesh is used with maximum resolution of 0.04D around the jet shear layer and grid stretching in radial direction only. Detailed chemistry is considered via a reduced mechanism called DRM-22, involving 24 species (including N_2 and Ar) and 104 reactions [24].

4 Results

4.1 Solid fuel particle dispersion

The predicted mean velocity profile shown in Fig. 1 agrees reasonably well with the experimental results. Slight jet acceleration can be seen from the mean velocity profile in between $z/D = 5$ and 10, which is believed to be caused by the density change when hot volatile matter starts to form at the surface of coal particles. The experimental velocity fluctuations also increase at around the same location supporting this view. Although small in magnitude, there is an initial RMS mismatch between the simulation and experimental data which may be attributed to the inflow turbulent structure. Since there is no experimental measurements available at the nozzle exit plane, modellers have to obtain their own settings for the turbulent inflow model they use. The inflow velocity variance ($0.1\ m^2/s^2$) is determined according to the suggestion from the 2nd Workshop on Measurements and Simulation of Coal and Biomass Conversion [25]. The stokes number of this flow is enclosed by the range from 0.1 to 10.

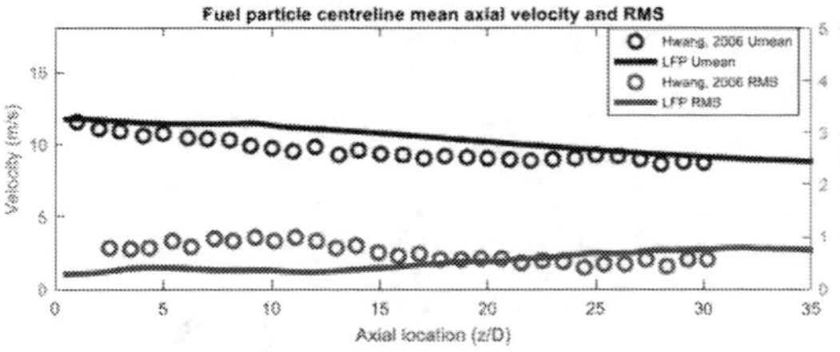

Figure 1: Centreline mean and RMS of coal particle velocity.

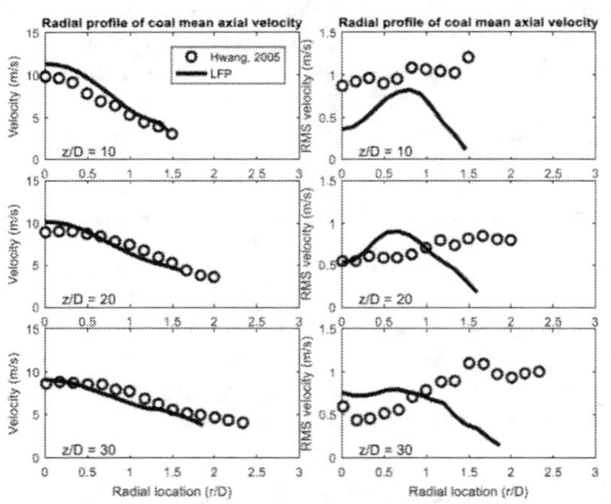

Figure 2: Radial profile of coal particle mean and RMS velocity.

The radial profiles of fuel particle mean axial velocity and RMS (Fig. 2) show reasonable agreement with the experimental data. Note that the slight over prediction of centreline velocity shown in Fig. 1 is also visible here for axial locations of $z/D = 10$ and 20. The spreading of solid fuel particles in the radial direction at $z/D = 20$ and 30 locations is slightly under predicted. Due to the lack of detailed information regarding the co-flow velocity, a 0.5m/s axial co-flow velocity is added to stabilise the flow and may lead to a reduction of strain and jet spreading in the shear layer. Another factor that may contribute to the under-prediction of jet spreading is the choice of the inflow turbulent integral length scale. The effects of inflow length scale (from 0.3D to 0.5D) in multiphase reacting flow has been discussed in [9], indicating a positive correlation with downstream axial momentum loss along the centreline. A lower integral length scale (0.3D) is selected in this simulation to minimise the parameter differences with the other modelling groups [9].

Comparing against the experiment, Fig. 3 shows a similar coal particle temperature evolution along the domain centreline. The initial apparent temperature undershoot in the simulations is likely to be due to uncertainty in the experimental data which relied on line of sight pyrometry and was affected by coal and soot particle radiation as well as the surrounding hot methane flame [4, 9], therefore, the upstream experimental temperature will only be used as a reference. The general trends of O_2 and CO_2 are captured by the simulation although the simulated consumption of O_2 and formation of CO_2 occur too rapidly. Due to the detailed reaction mechanism employed in this simulation, the relatively low centreline gas temperature prevents the oxidizer from being completely consumed downstream. The discrepancy in O_2 content at around $z/D = 10$ is probably caused by the predicted rapid jet break-up, which can be observed in Fig. 1 resulting in intense mixing between the hot flue gas from the pilot flame and the coal particles and forming a fuel rich environment. Subsequent rapid oxidation occurs between the volatile released at the hot coal surface (at around 1000K) and the fresh air. However, the excessive volatile suffocates the flame at the centre, preventing the reactions from reaching equilibrium, and therefore there is a slight under-prediction of the downstream CO_2 content.

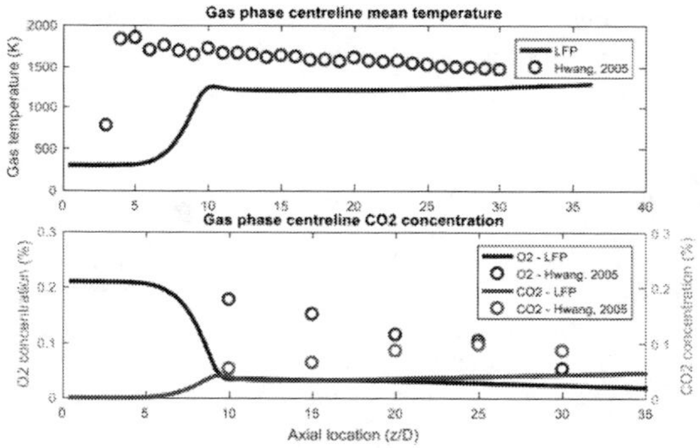

Figure 3: Mean centreline coal particle temperature (top) and gas phase mole fractions of O_2 and CO_2.

Figure 3 demonstrates that the model is capable of predicting the correct trend for major reactive species in the selected pulverised coal flame. With an appropriate model for envelope flames, which may be formed within the coal particle diffusive layer, more energy is expected to be released in the form of heat and therefore boosts the production of CO_2 in the centre. This will be investigated in future work.

5 Conclusions

A multiphase turbulent combustion model based on the conserved scalar approach, incorporating a state-of-art combustion model (MMC) is implemented in a LES context and tested against a dilute, piloted pulverised coal flame. Satisfactory agreement with the experimental results is observed, however there is still room for further improvement. A more realistic devolatilisation product will be considered, as well as the effect of surface reaction.

6 Acknowledgment

We acknowledge the support from ARC under grant number DP130100763, and the use of the HPC system (Artemis) at the University of Sydney.

References

1. IEA, *World Energy Outlook 2014*. International Energy Agency: 2014.
2. A. Williams, M. Pourkashanian, and J. Jones, Prog. Energ. Combust. Sci., 2001. **27**(6): p. 587-610.
3. S.R. Turns, *An introduction to combustion*. McGraw-hill Singapore: 2012.
4. S.M. Hwang, R. Kurose, F. Akamatsu, H. Tsuji, H. Makino, and M. Katsuki, Energ. Fuels, 2005. **19**(2): p. 382-392.
5. S.M. Hwang, R. Kurose, F. Akamatsu, H. Tsuji, H. Makino, and M. Katsuki, JSME Int. J., 2006. **49**(4): p. 1316-1327.
6. R. Kurose and H. Makino, Combust. Flame, 2003. **135**(1): p. 1-16.
7. R. Kurose, H. Watanabe, and H. Makino, KONA Powder and Particle Journal, 2009. **27**: p. 144-156.
8. B. Franchetti, F.C. Marincola, S. Navarro-Martinez, and A. Kempf, Proc. Combust. Inst. , 2013. **34**(2): p. 2419-2426.
9. O. Stein, G. Olenik, A. Kronenburg, F.C. Marincola, B. Franchetti, A. Kempf, M. Ghiani, M. Vascellari, and C. Hasse, Flow, Turbul. Combust., 2013. **90**(4): p. 859-884.
10. M. Rieth, A. Clements, M. Rabaçal, F. Proch, O. Stein, and A. Kempf, Proc. Combust. Inst. , 2017. **36**(2): p. 2181-2189.
11. J. Watanabe, T. Okazaki, K. Yamamoto, K. Kuramashi, and A. Baba, Proc. Combust. Inst. , 2017. **36**(2): p. 2155-2163.
12. X. Wen, K. Luo, H. Jin, and J. Fan, Combustion Theory and Modelling, 2017: p. 1-29.
13. H. Pitsch, Annu. Rev. Fluid Mech., 2006. **38**: p. 453-482.
14. A.Y. Klimenko and S. Pope, Phys. Fluids, 2003. **15**(7): p. 1907-1925.
15. S.B. Pope, Prog. Energ. Combust. Sci., 1985. **11**(2): p. 119-192.
16. A.Y. Klimenko and R.W. Bilger, Prog. Energ. Combust. Sci., 1999. **25**(6): p. 595-687.
17. M. Cleary and A. Klimenko, Phys. Fluids, 2011. **23**(11): p. 115102.
18. J.K. Dukowicz, J. Computational Phys., 1980. **35**(2): p. 229-253.
19. S. De, K. Lakshmisha, and R.W. Bilger, Combust. Flame, 2011. **158**(10): p. 1992-2008.
20. M.N. Khan, L.F. Zhao, M.J. Cleary, R.W. Bilger, O.T. Stein, and A. Kronenburg, Proc. 19th Australiasian Fuild Mech. Conf., 2014: p. No. 101.
21. D.B. Spalding, *Some fundamentals of combustion*. Vol. 2. Academic Press: 1955.
22. M.J. Cleary, Proc. Australian Combust. Symposium, 2015: p. 36-47.
23. M. Sharma and D. Stock, J. Fluids Eng., 1977. **325**.
24. A.Kazakov and M.Frenklach. Available from: http://www.me.berkeley.edu/drm/.
25. C.Hasse, A.Kempf, A.Kronenburg, P.Pepiot, C.Shaddix, O.Stein, and H.Watanabe. *2nd Workshop on Measurement and Simulation of Coal and Biomass Conversion*. 2017 [cited 2017; Available from: http://www.cbc.uni-due.de/?file=workshop.

11th Asia-Pacific Conference on Combustion,
The University of Sydney, NSW Australia
10th -14th December 2017.

Effects of Karlovitz number on Localised Forced Ignition of Stratified Combustible Mixtures: A Numerical Investigation

Dipal Patel[1], Jiawei Lai[2]

[1]University of Ontario Institute of Technology

2000 Simcoe Street North, Oshawa, ON, L1H 74K, Canada

[2]cmcl innovations

Sheraton House, Castle Park, Cambridge, CB3 0AX, UK

Abstract

Localised forced ignition of turbulent stratified mixtures (with $\langle\phi\rangle = 1.0$, $\phi' = 0.3$, $l_\phi/l_f = 2.6$) has been analysed based on direct numerical simulations (DNS) for different values of Karlovitz number (Ka) corresponding to thin reaction zone in the turbulent regime diagram. The initial values of turbulent fluctuations (i.e. $u'/S_{b(\phi=1)}$) and the integral length scale of turbulence (i.e. L_{11}/l_f) have been modified to bring about the change in Ka. The localized ignition is accounted for by a source term in the energy transport equation which deposits energy over a specified time interval. The energy supplied by the ignitor that leads to successful ignition and self-sustained combustion have been identified for both homogeneous and stratified mixtures based on DNS. It has been found that combustion takes place predominantly under premixed mode of combustion and percentage of heat release due to premixed mode of combustion increases with increasing Ka. An increase in Ka has been shown to have adverse effects on burned gas mass. Furthermore, a stratified combustion mixture has been found to be more favorable choice over homogeneous mixtures for a given turbulent flow condition.

1 Introduction

Localised forced ignition (e.g. spark or laser ignition) of inhomogeneous mixtures has a number of important applications ranging from Gasoline Direct Injection (GDI) engines to high-altitude relight in aero-gas turbines. A comprehensive review can be obtained from [1]. A number of previous analyses concentrated on localized forced ignition of inhomogeneous mixtures where the mixture inhomogeneity is characrterised by a gradient of mean equivalence ratio. By contrast, locased forced ignition of stratified mixtures characterized by a constant global mean value of equivalence ratio (i.e. $\langle\phi\rangle = 1.0$) with a non-zero value of equivalence ratio fluctuation (i.e. $\phi' = 0.3$) taking full advantage of flammability limits is rarely analysed.

The mechanisms of localized forced ignition and the minimum ignition energy (MIE) for laminar and weakly turbulent flows has been experimentally analysed in details by Lefebvre and his co-workers [2] but relatively limited effort has been directed to localized forced ignition of both homogeneous and stratified mixtures under highly turbulent environments where the internal structure of the flame is severely modified by energetic turbulent eddies and flame kernel originated from localized forced ignition is subjected to a high rate of turbulent heat transfer. Furthermore, the mixture inhomogeneity can potentially increase the extent of flame wrinkling and may significantly affect the cycle-to-cycle variation of heat release [3,4]. The flame-turbulence interaction in turbulent premixed combustion is often characterized in terms of a non-dimensional number known as Karlovitz number $Ka = [u'/S_{b(\phi=1)}]^{3/2}/\sqrt{L_{11}/l_f}$ (where L_{11} and l_f is the longitudinal integral length scale and Zel'dovich flame thickness respectively. Alternatively, Ka can be interpretend as the ratio of flame thickness (l_f) to tke Kolmogorov length scale (η) (i.e. $Ka \sim l_f^2/\eta^2$).

On the regime diagram for premixed turbulent combustion [5] the condition $100 > Ka > 1$ corresponds to thin zone regime (present study), where turbulent eddies penetrate into the flame and significantly affect the preheat zone as the reaction later thickness remains smaller than the Kolmogorov length scale. In the stratified mixtures, the length scale of mixture inhomogeneity is defined using Taylor micro-scale as $l_\phi = \sqrt{6\langle[\phi - \langle\phi\rangle]^2\rangle/\langle\nabla[\phi - \langle\phi\rangle]\cdot\nabla[\phi - \langle\phi\rangle]\rangle}$, where angle bracket indicates the global mean evaluated over the whole computational domain [6]. The equivalence ratio variation is initialized using a random distribution of ϕ following a bi-modal distribution (for $\langle\phi\rangle = 1.0$, $\phi' = 0.3$, $l_\phi/l_f = 2.6$). Many previous analyses [1-6] focused on various aspects of localized forced ignition of both homogeneous and inhomogeneous mixtures, but the effects of Ka on localized forced ignition of stratified combustible mixture are yet to be analysed using DNS data. This gap has been addressed here by analyzing the influences of Ka on localized forced ignition of both homogeneous and stratified combustible mixtures using three-dimensional DNS data.

Homogeneous ($\phi = 1.0$) and Stratified mixtures with ($\langle\phi\rangle = 1.0$, $\phi' = 0.3$, $l_\phi/l_f = 2.6$)			
Ka	$u'/S_{b(\phi=1)}$	L_{11}/l_f	Case
7.2	8	10	TR_1
20.3	16	10	TR_2
37.2	24	10	TR_3

Table 1: Simulation parameters.

Corresponding author. Tel: +1-905-721-8668 ext. 5713
E-mail address: dipal.patel@uoit.ca

2 Mathematical Background and Numerical Implementation

A modified single step chemical mechanism ($\text{Fuel} + s \cdot \text{Oxidiser} \rightarrow (1+s) \cdot \text{Products}$) has been considered for the purpose of computational economy, where s is the mass of oxidizer consumed per unit mass of fuel consumption under the stoichiometric condition [7]. The activation energy and heat of combustion are taken to be function of ϕ [7] so the realistic ϕ dependence of unstrained laminar burning velocity $S_{b(\phi)}$ can be obtained. The equivalence ratio ϕ can be expressed as $\phi = \text{AFR}/\text{AFR}_{st}$ (where AFR and AFR_{st} are the actual and stoichiometric air-fuel ratio and) and also in terms of mixture fraction $\xi = [Y_F - Y_O/s + Y_{O\infty}/s]/[Y_{F\infty} + Y_{O\infty}/s]$ and stoichiometric mixture fraction $\xi_{st} = Y_{O\infty}/[sY_{F\infty} + Y_{O\infty}]$ as $\phi = [1-\xi_{st}]\xi/[1-\xi]\xi_{st}$ where $Y_{F\infty}$ and $Y_{O\infty}$ are the fuel and oxidizer mass fractions in pure fuel and air stream respectively, and Y_F and Y_O are the local fuel and oxidizer mass fractions respectively. In the present analysis $\xi_{st} = 0.055$ ($\text{AFR}_{st} = 17.16$) represents methane-air binary mixture. The extent of the completion of the chemical reaction is quantified using a reaction progress variable $c = (Y_{Fu} - Y_F)/(Y_{Fu} - Y_{Fb})$ (or $c = [\xi Y_{F\infty} - Y_F]/[\xi Y_{F\infty} - \max[0, (\xi - \xi_{st})]Y_{F\infty}]$) [6,8] where Y_{Fu} and Y_{Fb} are the fuel mass fractions in the unburned and burned gases respectively. According to this c rises from 0 in the fully unburned reactants to 1.0 in the fully burned products.

The localized ignition is modelled by using a source term q''' in the energy transport equation, which deposits energy in the radial direction from the center of the ignitor in the following manner [6,8,9]:

$$q'''(r) = A_q \exp\left[-\frac{1}{2}\frac{r^2}{R^2}\right] \tag{1}$$

where r is the radial direction from the center of the ignitor and R is the characteristic width of energy deposition (i.e. ignitor dimension), which is taken to be normalized with $l_f = \alpha_{T0}/S_{b(\phi=1)}$ (where α_{T0} is the thermal diffusivity of the unburned mixture). The constant A_q can be evaluated by the volume integration: $\dot{Q} = \int_V q'''\, dV$, where

$$\dot{Q} = a_{sp}\rho_0 C_P \tau T_0 \left[\frac{4}{3}\pi l_f^3\right]\left(\frac{H(t) - H(t - t_{sp})}{t_{sp}}\right) \tag{2}$$

is the ignition power with $\tau = 3.0$ is the heat release parameter [6,8,9]. The parameter a_{sp} determines the total energy deposited by the ignitor and the Heaviside function $H(t)$ and $H(t - t_{sp})$ ensure the ignition source term remains operational for the energy deposition time duration t_{sp}. The parameter t_{sp} can be expressed as $t_{sp} = b_{sp}t_f$, where b_{sp} is the energy deposition duration parameter and $t_f = l_f/S_{b(\phi=1)}$ is the characteristic chemical time scale. For the present analysis $a_{sp} = 38.6$, $b_{sp} = 0.2$, $R = 2.45l_f$ are taken (following [6,8] to ensure self-sustained combustion) and Lewis number of all species is taken to be unity. The ignitor is placed at the center of the computational domain and the ignition model used here only addresses the thermal aspect of the localized forced ignition and the details of spark formation (momentum modification contribution, plasms formation and shock wave) are kept beyond the scope of the present analysis to keep it computationally feasible. The pseudo-spectral method proposed by Eswaran and Pope [10] was used for generating initial ϕ distribution following bi-modal distribution.

A compressible three-dimensional DNS code SENGA [11] was used to carry out the simulations under decaying turbulence in a domain of size $51l_f \times 51l_f \times 51l_f$, which further descretised by a Cartesian grid of size $310 \times 310 \times 310$ with uniform grid spacing ensuring at least 10 grid points within $\delta_{th(\phi=1)} = [T_{ad(\phi=1)} - T_0]/[max|\nabla \hat{T}|_L]$ where $\hat{T}$ is the instantaneous temperature. The boundaries in the x_1- direction are taken to be partially non-reflecting (using Navier-Stokes Characteristic Boundary Conditions), and all other boundaries are periodic [12]. A standard pseudo-spectral method [13] is used to initialize the turbulent velocity fluctuations. A high order finite difference scheme is used for spatial differentiation and third-order low storage Runge-Kutta scheme is used for explicit time-advancement [14].

The simulation parameters are listed in Table 1. The parameter cases listed in Table 1 corresponds to thin reaction zone regime according to the regime diagram by Peters [5]. All simulations have been carried out for $t = 8.90t_{sp} = 1.8t_f$, and this simulation time has been found sufficient to determine self-sustained combustion.

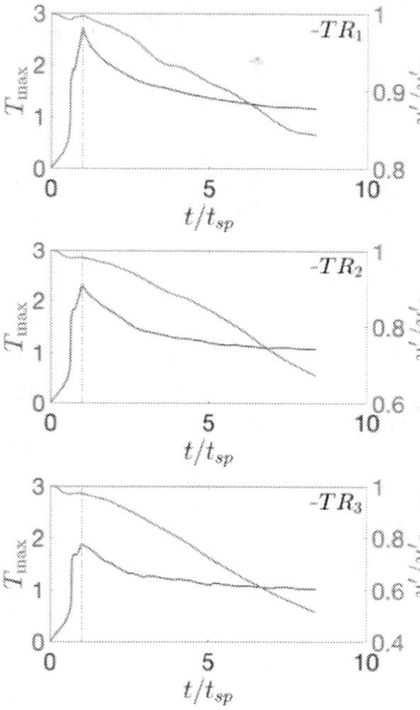

Figure 1: Temporal evolution of T_{max} and corresponding turbulence decay profile for all stratified mixture cases from Table 1 with cyan dashed-broken line shows $t = t_{sp}$.

3 Results and Discussion

3.1 Temporal evolution of maximum temperature

The temporal evolutions of the maximum values of non-dimensional temperature (i.e. $T_{max} = [\hat{T}_{max} - T_0]/[T_{ad(\phi=1)} - T_0]$) and corresponding turbulence decay profile for all stratified mixture cases from Table 1 are shown in Fig. 1. Figure 1 shows that T_{max} rises with time until $t = 1.0t_{sp}$ due to energy deposition during $0 < t < t_{sp}$. Additionally, Fig. 1 shows that at $t = t_{sp}$, the T_{max} attains its peak value after the ignitor has been switched off. The ignition event can be determined from Fig. 1 by the sharp change in slope in the temporal evolution of T_{max} at $t = t_{sp}$, which is accomplished by a thermal runaway leading to a value of $T_{\max}$ around 1.0 at $t \gg t_{sp}$, indicating self-sustained combustion following successful ignition. Additionally Fig. 1 shows that with increasing Ka, the peak value of $T_{\max}$ at $t = t_{sp}$ reduces for a given amount of energy deposition (i.e. $a_{sp} =$ 38.6). This suggests that high value of temperature during ignition can be avoided for higher values of Ka, which ultimately reduces thermal NO_x emission and still achieve self-sustained combustion [15,16].

The surface area to volume ratio of the hot gas kernel decreases with the expansion of the flame kernel. In this configuration u' decays with time and the eddy thermal diffusivity $D_t \sim u' L_{11}$ decreases with time. This along with decreasing surface to volume ratio with time leads to a reduction in the magnitude of heat transfer rate from the hot gas kernel as time progresses, which leads to more favorable condition for thermal runaway in the cases of high values of $T_{\max}$. Therefore the duration over which the peak value of $T_{\max}$ is obtained after $t = t_{sp}$ depends on the background fluid turbulence.

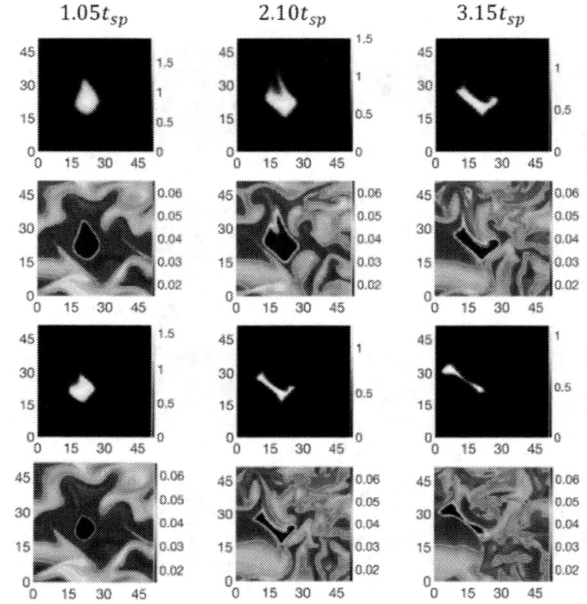

Figure 2: Distribution of non-dimensional temperature T and Y_F on the central $x_1 - x_2$ plane showing initial kernel growth for the cases TR_2 and TR_3 from Table 1 (cyan line shows $\xi = \xi_{st}$, additionally 1st & 3rd row shows T field for Homogenous mixture case TR_2 whereas 2nd & 4th row shows Y_F field for Stratified mixture case TR_3).

3.2 Spatial distribution of temperature and fuel mass fraction

The distribution of T and Y_F at initial kernel development times (i.e. $t = 1.05t_{sp}, 2.10t_{sp}, 3.15t_{sp}$) for the cases TR_2 and TR_3 are shown in Fig. 2. It can be seen from Fig. 2 that the contours of T remain approximately spherical (this effect is stronger in TR_3 case with stratified mixture) during the period of energy deposition but they become increasingly wrinkled as time progresses. Before the ignition phase, the evolution of T is principally determined by the diffusion of deposited energy, whereas after ignition, the evolution of isotherms depends on the magnitude of the reaction rate at the local mixture composition and flame stretch induced by the background fluid motion.

Figure 2 shows that Y_F is depleted at the regions associated with high values of T due to consumption of fuel as a result of chemical reaction. It can be seen from Fig. 2 that the level of non-uniformity in the ϕ distribution decreases as time progresses for stratified mixture cases. Additionally Fig. 2 shows that at high temperature region where chemical reaction takes place has been fragmented in the case TR_3 at $t > t_{sp}$ due to penetration of energetic eddies into the flame. This fragmented high temperature region in case TR_3 is indicative of disturbance of the reaction zone by turbulence under high values of Ka.

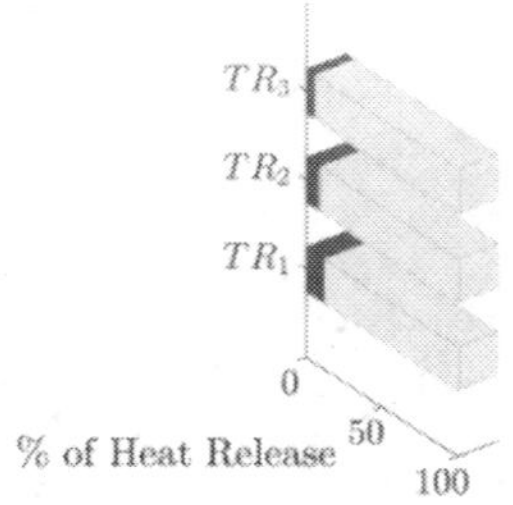

Figure 3: Percentage of overall heat release arising from premixed and non-premixed mode of combustion at $t =$ $8.40t_{sp}$ for all stratified mixture cases from Table 1.

3.3 Percentage of heat release

The percentage of overall heat release arising from premixed (i.e. $I_c > 0$) and non-premixed (i.e. $I_c < 0$) modes of combustion for all stratified mixture cases from Table 1 at $t =$ $8.90t_{sp}$ are shown in Fig. 3 where $I_c =$ $[\nabla Y_F \cdot \nabla Y_O]/[|Y_F||Y_O|]$ is the flame index, which assumes positive and negative values for premixed and non-premixed mode of combustion respectively. It is evident from Fig. 3 that the chemical reaction takes place predominantly in premixed mode but some pockets of non-premixed combustion can also be found. However, the probability of finding $I_c < 0$ decreases with increasing time due to mixing process (not shown here). Figure 3 shows the percentage of heat release from the premixed mode of combustion is greater in the TR_3 case in comparison to cases TR_1 and TR_2. The effects of mixing are stronger for higher values of u' and $Ka =$ $[u'/S_{b(\phi=1)}]^{3/2}/\sqrt{L_{11}/l_f}$, suggesting $I_c < 0$ increases with decreasing value of Ka. Furthermore, the decay rate of ϕ' increases with increasing u' as turbulent straining acts to generate scalar gradient, which in turn increases the rate of micro-mixing [6].

3.4 Temporal evolution of burned gas mass

Although Fig. 1 provides the signature of successful ignition, it is necessary to assess the temporal evolution of burned gas mass increasing with time in order to ascertain the possibility of obtaining self-sustained combustion following successful ignition. The extent of burning can be characterized by the burned mass m_b with $c \geq 0.9$ [6,8,9]. The temporal evolution of burned gas mass normalized with the mass of an unburned gas sphere with a radius equal to l_f (i.e. $M_b =$ $[m\_b(c \geq 0.9)]/[(4/3)\pi\rho_0 l_f^3]$) for all cases listed in Table 1 is shown in Fig. 4. A comparison between Figs. 1 and 4 reveals that M_b increases with time in the cases where $T_{\max}$ settles to a value close to 1.0 at $t \gg t_{sp}$ indicating self-sustained combustion without any external energy addition following successful ignition, except TR_3, where M_b starting to decrease at $t \geq 3\,t_{sp}$.

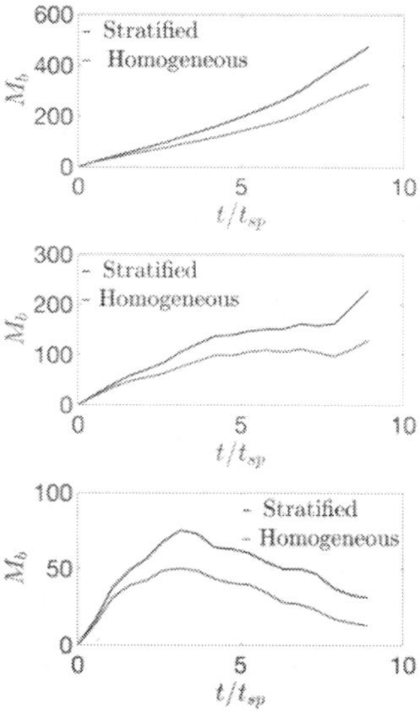

Figure 4: Temporal evolution of M_b for all the cases from Table 1.

It can be seen from Fig. 4 that M_b for TR_3 case does not monotonically increase with time similar to other cases, the heat transfer from hot gas kernel overcomes the chemical heat release at early stages of flame evolution, which give rise to a reduction of M_b at $t \geq 3\,t_{sp}$. Moreover the TR_2 case shows non-monotonic behavior of M_b at $t \geq 7t_{sp}$, where the rate of turbulent heat transfer reduces with time due to the reduction in the magnitude of eddy thermal diffusivity $D_t \sim u'L_{11}$ as a result of decaying turbulence level (see Fig. 1), and the chemical heat release starts to exceed the heat transfer rate from hot gas kernel at $t > 8\,t_{sp}$ for case TR_2, which gives rise to an increase in M_b again. The probability of finding high value of c decreases with increasing due to $u'/S_{b(\phi=1)} \sim Ka^{1/2}Re_t^{1/4}$ enhanced heat transfer rate from hot gas kernel. The reduction in burning rate with high value of u' in stratified mixture is found to be consistent with previous experimental [17] and computational [6] findings. Importantly, Fig. 4 shows that it is possible to achieve higher value of M_b in the stratified mixture compare to the corresponding homogeneous mixture case for a given realization of ϕ distribution. Moreover, the stratified mixture with the choice of $\phi' = 0.3$ takes full advantage of the flammability limit and yet shows higher extent of burning for a given realization of ϕ distribution. It is well known that combustion succeeds only for some realisations even when the turbulent flow statistics remain unaltered [18]. This aspect is particularly prevalent in the cylinder of IC engines due to cycle-to-cycle variations, therefore analysis for more realisations of initial ϕ distribution will be necessary, which will form the basis of future investigation.

4 Conclusions

The effects of Karlovitz number Ka, turbulent velocity fluctuations on the localized forced ignition of both homogeneous and stratified mixtures has been investigated using three-dimensional DNS. The ignition process is modelled by a Gaussian power source in the energy conservation equation, which deposits energy over a stipulated time interval. The flame resulting from localized forced ignition shows a predominantly premixed mode of combustion, although some pockets of non-premixed mode of combustion have also been observed. It has been found that high values of temperature during ignition can be avoided for higher values of Ka for both homogeneous and stratified mixtures. Moreover, an increase in turbulent velocity fluctuation u' decreases the extent of burning and may lead to misfire for large value of u'. The detrimental effects of high values of u' are more critical for high values of Ka. It is possible to obtain a higher value for the extent of burning in the stratified mixture compare to corresponding homogeneous mixture even for a higher value of Ka. Although the qualitative nature of flame propagation following localized forced ignition is not likely to change in the presence of detailed chemical kinetics, three-dimensional DNS with detailed chemistry and transport will be required to achieve more comprehensive physical understanding and accurate qualitative predictions.

5 Acknowledgment

The authors are grateful to SHARCNET for providing computational support.

References

1. E. Mastorakos, Prog. Energy Combust. Sci., 35 (2009) 57-97.
2. A.H. Lefebvre, Gas Turbine Combustion, 2nd Edition, Taylor & Francis, Philadelphia, 1998.
3. N. Swaminathan, R. Grout, E. Mastorakos, Direct simulation of forced ignition in stratified turbulent combustion, ,3rd European Combustion Meeting, Greece, 2007.
4. C. Pera, S. Chevillard, J. Reveillon, Combust. Flame 160 (2013) 1020-1032.
5. N. Peters, Turbulent Combustion, Cambridge University Press, U.K. 2000.
6. D. Patel, N. Chakraborty, Combust. Theo. Mod., 18(6) (2014) 627-651.
7. E. Tarrazo, A. Sanchez, A. Liñán, FA. Williams, Combust. Flame, 147 (2006) 32-38.
8. D. Patel, N. Chakraborty, Inter. Jour. Spray Combust. Dynamics, 7(2) (2015) 151-174.
9. N. Chakraborty, E. Mastorakos, R.S. Cant, Combust. Sci. Technol., 179 (2007) 293-317.
10. V. Eswaran, S.B. Pope, Phys. Fluids, 31 (1988) 506-520.
11. K.W. Jenkins, R.S. Cant, DNS of Turbulent Flame Kernels, Proceedings of the Second AFOSR Conference at Rutgers – The State University of New Jersey, New Brunswick, Springer Netherlands (1999), 191-202.
12. T. Poinsot, S. Lele, Jour. Comput. Physics, 101 (1992) 104-129.
13. R.S. Rogallo, NASA Technical Memorandum 81315, NASA Ames Research Centre, 1981.
14. A.A. Wray, Monimal Storage time Advancement schemes for spectral methods, NASA Ames Research Centre, 1990.
15. J.E. Dec, Proc. Combust. Inst. 32 (2009) 2727-2742.
16. A.A.A. Abuelnour, M.A. Wahid, H.A. Mohammed, A. Saat, Inter. Jour. Energy Research, 38 (2014) 827-846.
17. B. Renou, E. Samson, A.M. Boukhalfa, Combust. Sci. Technol., 176 (2004), 1867-1891.
18. S.F. Ahmed, R. Balachandran, E. Mastorakos, Proc. Combust. Inst., 31(1) (2007) 1507-1513.

11th Asia-Pacific Conference on Combustion,
The University of Sydney, NSW Australia
10th -14th December 2017.

Impact of Torrefaction on Flash Pyrolysis Products of Biomass and Comparison to Fossil Fuels

Stefan Pielsticker[1], Benjamin Gövert[1], Georg Möller[1], Oliver Hatzfeld[1], Reinhold Kneer[1]

[1]Institute of Heat and Mass Transfer (WSA), RWTH Aachen University

Augustinerbach 6, Aachen, North Rhine-Westphalia, 52056, Germany

Abstract

A substitution of fossil fuels with alternative biofuels combined with carbon capture and storage technology can be used for the reduction of CO_2 emissions from pulverized fuel furnaces. Pretreating raw biomass via torrefaction improves its properties concerning energy density, grindability and storage. Furthermore, the torrefaction process influences the composition and amount of released devolatilization products under flame conditions.

In this study, the effect of torrefaction of beech wood on flash pyrolysis products is investigated. Rhenish lignite and Colombian bituminous coal are chosen as reference fossil fuels. Flame equivalent conditions are approximated with a small scale fluidized reactor located inside an electrically heated furnace. Temperature is varied from 873 K to 1473 K while atmosphere is kept constant at 100 vol.-% N_2.

Experiments show, that torrefaction has a significant impact on pyrolysis products and which behave under reaction conditions quite similar to Rhenish lignite. Furthermore, a relation between yields of hydrocarbons, temperature and fuel is derived.

1 Introduction

Carbon caption and storage (CCS) is a possibility to decrease CO_2 emissions from pulverized fuel fired furnaces. A substitution of fossil fuels by biomass-derived fuels can lead to a further reduction of carbon net emissions and enables a negative carbon balance [1]. Burning of raw biomass instead of coal comes with several challenges, as biomass has a lower energy density and worse grindability and storage properties. Torrefaction, which means pretreatment of raw biomass at temperatures between 473 and 573 K and residence times up to 60 min in inert atmosphere, helps to improve these properties and helps to use torrefied biomass as a substitute fuel for pulverized fuel boilers [2]. To optimize or redesign (torrefied) biomass fired furnaces by means of numerical simulations (e.g. [3]), knowledge about pyrolysis and char combustion kinetics and products is required. This concerns changes in ignition behavior, burnout times and pollutant formation. For high temperatures (> 973 K) literature data, especially on detailed gas analysis for high particle heating rates is scarce. This brings the risk of large discrepancies between experimental data and models used [4].

Reported data on pyrolysis products from torrefied biomass is either determined from experiments with low heating rates in thermogravimetric analyzers [4, 5] or with high heating rates in drop tube reactors, fixed bed [6] reactors or heated foil reactors [7]. Most authors obtained a higher char yield for the torrefied biomass compared to the untreated one. Simultaneously, lower gas and tar yields are reported [6, 7, 8]. Concerning gas composition, different observations were made. For the release of CO, Wafiq et al. [8] reported a lower yield for torrefied biomass, while Li et al. [9] observed higher CO yields, especially for samples torrefied at temperatures above 523 K. Furthermore, Tsalidis et al. [7] stated that the release of CO from raw and torrefied biomass strongly depends on pyrolysis temperature and can show different trends. For CO_2, Li et al. [9] observed a decreased yield for torrefied biomass, which was explained with decarboxylation reaction during the torrefaction process. The effect of torrefaction on hydrocarbons has not been studied very intense. From the results of Wafiq et al. [8] and Li et al. [9] a decreased release for torrefied samples is expected, but the effect of pyrolysis temperature cannot be derived.

The objective of this study is to identify the effect of torrefaction on flash pyrolysis products compared to raw biomass and conventional fossil fuels. As reference, Colombian bituminous coal and Rhenish lignite are chosen as fossil fuels and beech wood is selected as an exemplary biomass. The torrefaction process is undertaken with a laboratory screw reactor [10], while flash pyrolysis products are determined with a small scale fluidized bed reactor (FBR) equipped with ex-situ FTIR gas analysis [11].

2 Experimental Investigation

Flash pyrolysis products of four different fuel samples are investigated with a small scale fluidized bed reactor coupled with ex-situ FTIR gas analysis. The sample of torrefied biomass was produced in a continuously working screw reactor before. All fuels used were characterized using proximate, ultimate and particle size analysis.

2.1 Fuel Pretreatment and Characterization

In this study four different fuels are investigated. Rhenish lignite (RL) and Colombian bituminous coal (CC) are chosen as representatives for conventional fossil fuels. As alternative biofuel beech wood is selected. Besides raw biomass (B), also self-made torrefied samples (TB) are investigated to quantify the pretreatment effect. For the torrefaction process a continuously working screw reactor is used. A detailed description can be found in [10]. In the reactor, wood chips are pretreated at 573 K with a residence time of 30 min in pure nitrogen atmosphere.

Corresponding author. Fax: +49-241-80-92143
E-mail address: pielsticker@wsa.rwth-aachen.de

All fuels are ground and sieved. For pyrolysis experiments a mesh size fraction of 90–125 µm is chosen to ensure homogeneous conditions at particle level. Results of proximate and ultimate analysis of all fuels are summarized in Tab. 1.

	B	TB	RL	CC
C[a]	48.86	62.28	68.87	78.72
H[a]	5.64	4.81	4.81	5.26
N[a]	0.23	0.35	1.28	2.11
S[a]	0.02	0.01	0.26	1.09
O[a,b]	45.25	32.55	24.78	12.82
Ash[c]	0.36	0.96	5.78	7.01
Moisture[d]	7.28	1.64	11.26	3.63
Volatiles[a]	85.34	58.58	52.12	40.52
Fixed carbon[a,b]	14.66	41.42	47.88	59.48

a. dry ash free; b. from difference; b. c. dry; d. as analyzed

Table 1: Proximate and ultimate analysis of raw (B) and torrefied (TB) biomass from beech wood as well as Rhenish lignite (RL) and Colombian bituminous coal (CC) (weight %)

2.2 Fluidized Bed Reactor for Flash Pyrolysis

For the determination of pyrolysis products, a small scale fluidized bed reactor inside an electric heated furnace, both shown in Fig. 1, is used. A detailed description of construction, operation, data acquisition, evaluation procedure and error analysis can be found elsewhere [11].

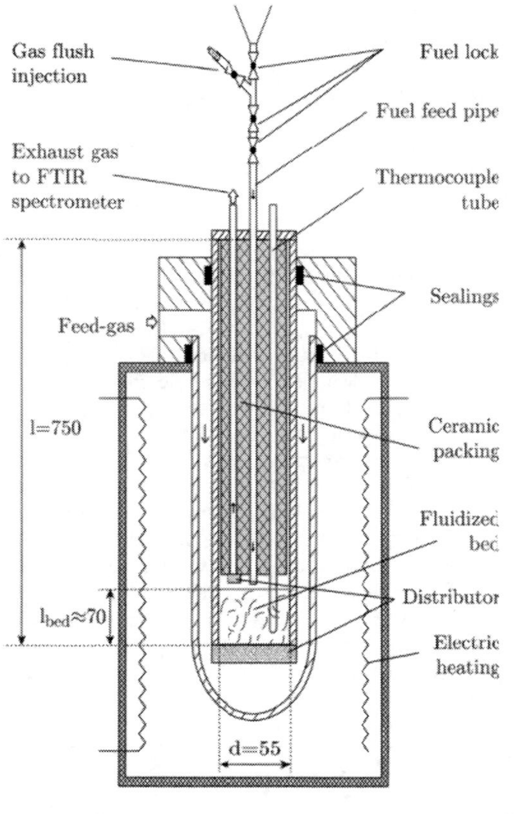

Figure 1: Scheme of the fluidized bed reactor used for flash pyrolysis experiments (dimensions in mm)

The main part of the reactor consists of two axial Al_2O_3 pipes, mounted in a stainless steel reactor head. The annular gap between the pipes serves as heat up zone for the fluidizing gas flow (100 l/h N_2, standard conditions), while the inner part encloses the fluidized bed consisting of Al_2O_3 particles (d_{50}=116 µm). Fuel samples (15 mg) are supplied to the reactor in batches. Particle transport from the fuel lock to the bed is forced with a gas flush from (10 ml N_2) to achieve fast heating rates of approximately 10^4 K/s [11]. Product gases are captured just above the bed and then transported to an FTIR spectrometer. Sampling line is heated to 453 K to avoid condensation of H_2O and tar components. The ceramic packing reduces dead volume, while the distributor in front of the gas sampling line allows for higher volume flows without particle entrainment. Both increase the gas exchange rate and thus reduce secondary pyrolysis reactions. Time dependent gas concentrations are measured with a Gasmet DX-2000 FTIR spectrometer, which analyzes in the wavenumber range from 600 to 4200 cm^{-1}, a spectral resolution of 8 cm^{-1} and a sampling frequency of 0.5 Hz.

Based on the time dependent concentration profiles $c_i(t)$, mass specific pyrolysis yields y_i are calculated according to (1)

$$y_i = \frac{\rho_{N_2} \cdot M_i}{m_{fuel} \cdot M_{N_2}} \cdot \int_{t_{start}}^{t_{end}} \frac{\dot{V}_{N_2}(t)}{1-\sum_{j=1}^{n} c_j(t)} \cdot c_i(t)\, dt \quad (1)$$

with injected fuel mass m_{fuel}, molar mass M and volume flow $\dot{V}_{N_2}$ and density ρ_{N_2} of N_2. Yield of the remaining char is derived by adding O_2 to the fluidizing gas stream and measuring the burning products CO_2, CO, H_2O, SO_2 and NO_x. Under the assumption, that oxygen content in the char can be neglected, char yield can be calculated by the sum of C, H, S and N. Due to particle sticking to the feeding pipe walls, particle entrainment, condensation of heavy tars and undetected H_2, total mass recovery fluctuates around 90%. Results therefore are normalized to the recovered mass.

3 Results

The following section presents mass specific yields calculated from (1). Yields are recorded for all four fuels in the temperature range from 873 to 1473 K. Points in the plot represent the mean of 10 single batches and error bars indicate the corresponding standard deviation, respectively. Section 3.1 deals with cumulative yields, while section 3.2 focuses on exemplary gas species.

3.1 Cumulative Products

Figure 2 shows the cumulative yields divided in the three main components char, tar and gas. In contrast to most other studies, all these three components are determined via ex-situ gas analysis and not by gas flow condensation (tar) or solid probe sampling (char). Yield of char is measured by burning the residual solids remaining in the FBR in a N_2/O_2 mixture. Yield of tar is estimated summing up all measured gas species with a boiling point above 293 K, except water vapor.

For all fuels, **char** yield shows a strictly decreasing trend with increasing pyrolysis temperature. Thereby raw biomass (B) at about 1173K and Colombian bituminous coal (CC) at 1273 K reach a state of stable char yield, while those of torrefied biomass (TB) and Rhenish lignite (RL) still show a decreasing trend up to the maximum investigated temperature of 1473 K. Comparing absolute values, an increase of char with fixed carbon content is seen, but for all fuels the char yields are lower than those from proximate analysis, which is related to the higher heating rate [9]. Absolute values of torrefied biomass char yield are always a little bit higher than reported in literature which is most likely explained by the lower torrefaction temperatures (523K [6, 8], 563 K [7]) compared to 573 K in this study. Furthermore, an increase of char yield can be observed for all temperatures comparing TB with B, as several authors reported before for selected temperatures [6, 7, 8].

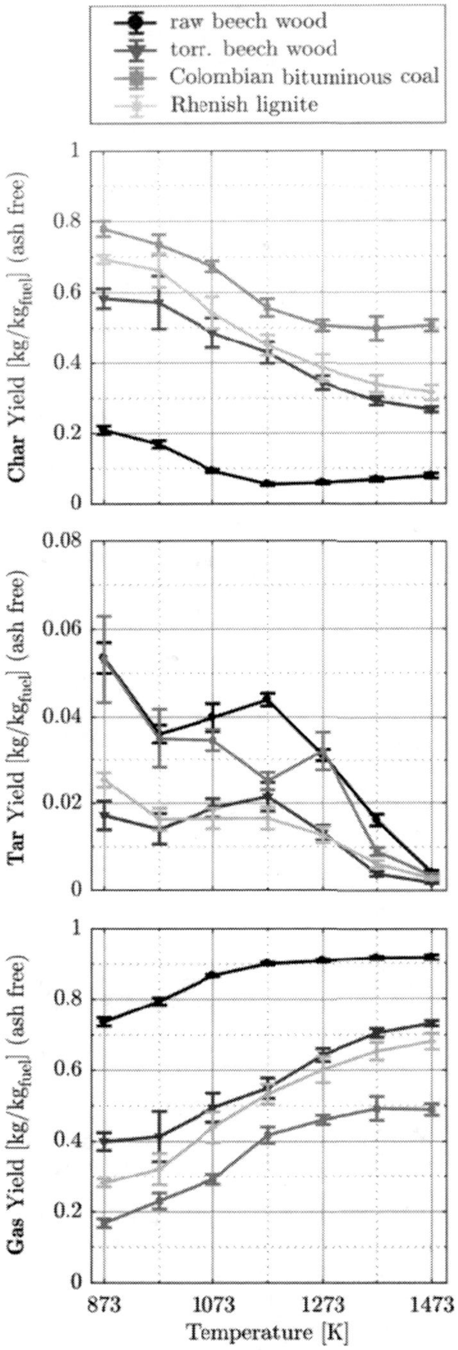

Figure 2: Experimentally derived yields of char (obtained from burnout), tar (sum of measured components with boiling temperature > 293 K, except H_2O) and gas (rest of measured components) for the four investigated fuels. Error bars indicate standard deviation of 10 samples.

Similar to the char yield, the overall trend of detected **tar** decreases with higher pyrolysis temperature. But for B and TB a local minimum of tar yield around 973 K and a local maximum at 1173 K are observed similar to the study from Tsalidis et al. [7]. In contrast to char, no clear trend with regard to fuel rank can be observed, as B and CC show comparatively high yields whereas TB and RL show low yields. Tar yields in this study are between those reported in literature. Deviation of tar yields is mainly caused by different gas residence times under hot conditions and the relating secondary pyrolysis and tar cracking reactions.

Detected **gas** yields show a strictly increasing trend with increasing temperature and level out for B and CC, as those profiles represent approximately the inverse of char yield for high temperatures. Compared to the study of Wafiq et al. [8], measured yield of the torrefied sample (0.51 @ 1073 K [8]) corresponds well with TB, but yield of raw biomass (0.62 @ 1073 K [8]) is much lower than those obtained for B. A much better agreement can be found with Raut et al. [6] who reported yields for raw biomass of 0.88–0.91 at 1073 K. Also the slightly higher yields of 0.57–0.62 for torrefied biomass are plausible to data of TB when the lower torrefaction temperature in [6] is taken into account.

3.2 Gas Species

After discussion of cumulative yields in the previous section, this paragraph focuses on the release of single gas species. In total 30 different gas species were extracted from the measured IR spectra. The exact number of selected of gases depends on the specific fuel used and reaches typically values of 15 to 20. Different gas selections are required to adapt the fuel specific releases. For example additional sulphur compounds (COS, CS_2) are added for the fossil fuels CC and RL. Yields of the three main species CO, CO_2 and CH_4 as well as the yields of hydrocarbons C_2H_4, C_3H_6 and C_6H_6 are shown in Fig. 3.

Carbon monoxide is identified as the most present pyrolysis gas for all fuels at almost all temperatures. Maximum yields are detected for B at highest investigated temperature, where CO counts for more than 80% of released pyrolysis gases. For low temperatures, CO_2 also provides a significant contribution to the gas yield. Thereby the CO_2 release increases with fuel rank from CC (0.045) over RL (0.09) and TB (0.13) to B (0.22). A comparison of the CO_2 yield released from raw and torrefied biomass reveals a decrease (TB<B), which can be attributed to decarboxylation reactions of acid groups during the torrefaction process [9]. The finding corresponds to a strongly decreasing yield of acetic acid ($C_2H_4O_2$) in the flash pyrolysis products. Above 1173 K the amount of released CO_2 drops for all fuels monotonically, which can be related to secondary Boudouard gasification reactions of CO_2 with the remaining char particles. Furthermore, local maxima can be found only for the fossil fuels located at approximately 1073–1173 K. Hydrocarbon yields have in common, that either a maximum is obtained at a certain species specific temperature or a monotonic decrease can be observed. In this case the species specific maximum might be outside the investigated temperature range. Regarding results, fuel affects only peak height but not the maximum related temperature, which depends only on the gas species itself. Thereby components with high bonding energies e.g. triple bonds (maximum of C_2H_2 (not shown) at 1250 K) or aromatic ring structures (maximum of C_6H_6 at 1173) have the highest related temperatures. This trend can be continued by going from alkenes (e.g. C_2H_4 at 1073 K) to alkanes (C_2H_6 (not shown) at 973 K).

In comparison with data from Wafiq et al. [8], only the CO_2 yield shows good agreement with experimental derived yields

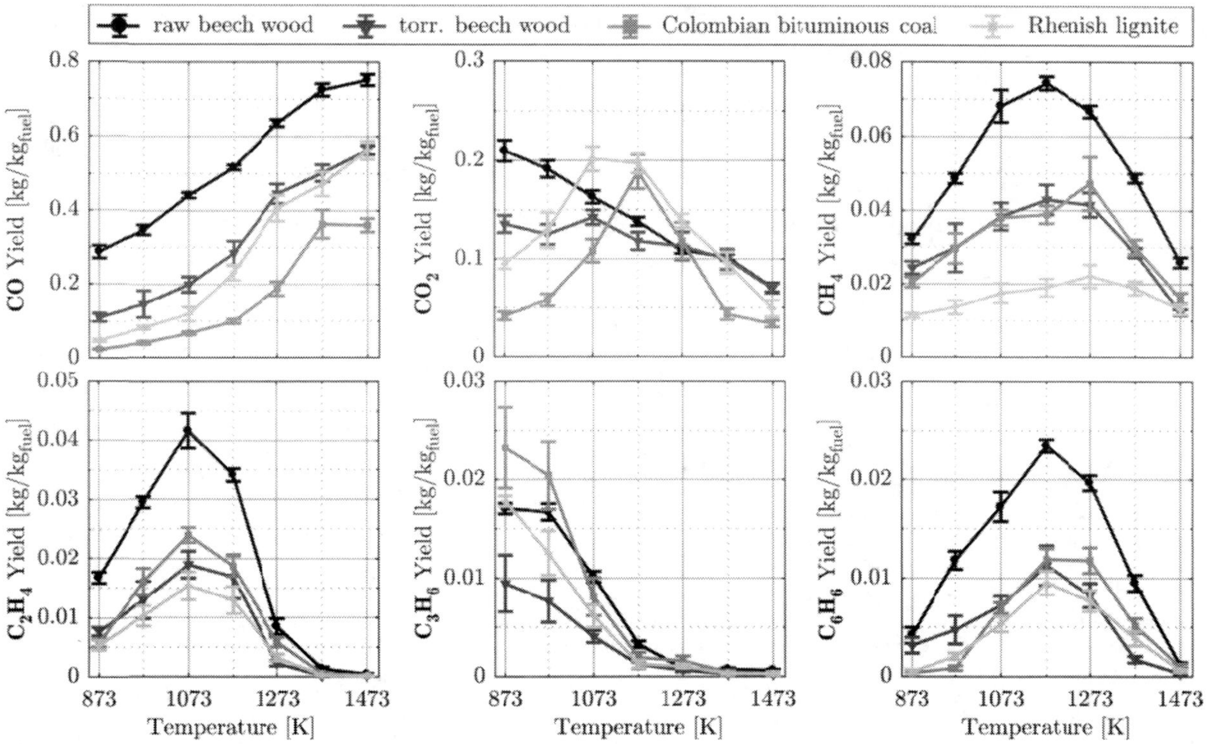

Figure 3: Experimentally derived species yields (exemplary chosen for CO, CO_2, CH_4, C_2H_4, C_3H_6 and C_6H_6) for all investigated fuels. Error bars indicate standard deviation of 10 samples.

of this study. For CO, CH_4 and C_2H_4 similar trends concerning the torrefaction impact are observed. Nevertheless absolute values differ up to the factor of two. The predicted temperature dependency for CO from Tsalidis et al. [7] cannot be found in the present results as CO yields of TB are lower than those of B for all temperatures. An agreement with Tsalidis et al. [7] was found for the CO_2 yield, where no dependency of torrefaction above approximately 1073K was found. Comparing the derived trends with the Neves model used in [4], high deviations are observed, especially for CO and CH_4.

4 Conclusions

In this study, pyrolysis yields of raw and torrefied biomass as well as two representative fossil fuels were determined in the temperature range from 873 to 1473 K. To approximate boundary conditions of actual pulverized fuel burners, a small scale fluidized bed reactor was used. Based on an ex-situ FTIR gas analysis, a mass balance around the reactor was carried out to derive mass specific yields and cumulative yields of char, tar and gas.

The results have shown, that the torrefaction process has a significant influence of the released pyrolysis products and the remaining char. The torrefied sample in this study (573 K) has shown quite similar behavior to investigated Rhenish lignite, especially for yields of char, gas and CO. Known literature trends concerning char, tar and gas yields have been confirmed. Furthermore, answers to open questions regarding single species temperature trends have been given. Above all, it was shown, that all hydrocarbons have a yield maximum at a certain temperature. This temperature only depends on bonding energies in the species, while yield magnitude is a function of fuel.

5 Acknowledgment

The authors would like to thank the German Research Foundation (DFG) for funding Stefan Pielsticker and Benjamin Gövert within the SFB/Transregio 129 "Oxyflame".

References

[1] J. S. Tumuluru, R. D. Boardman, C. T. Wright, J. R. Hess, Energies 5 (10) (2012) 3928-3947.

[2] P. C. A. Bergman, A. R. Boersma, R. W. R. Zwart, J. H. A. Kiel, Torrefaction for biomass co-firing in existing coal-fired power stations, ECNC05013, Energy research Centre of the Netherlands, 2005.

[3] J. Li, A. Brzdekiewicz, W. Yang, W. Blasiak, Applied Energy 99 (2012) 344–354.

[4] F. Barontini, E. Biagini, L. Tognotti, Characterization of the Devolatilization Products of Selected Second Generation Biofuels, 18th IFRF Members Conference, Freising, Germany, 2015, Paper 28.

[5] J. Wannapeera, B. Fungtammasan, N. Worasuwannarak, Journal of Analytical and Applied Pyrolysis 92 (1) (2011) 99-105.

[6] M. K. Raut, P. Basu, B. Acharya, International Journal of Renewable Energy & Biofuels, Vol. 2016 (2016) 1-13.

[7] G.-A. Tsalidis, K. Voulgaris, K. Anastasakis, W. De Jong, J. H. A. Kiel, Energy & Fuels 29 (9)(2015) 5825-5834.

[8] A.Wafiq, D. Reichel, M. Hanafy, Fuel 179 (2016) 156-167.

[9] J. Li, G. Bonvicini, L. Tognotti, W. Yang, W. Blasiak, Fuel 122 (2014) 261-269.

[10] A. Ohliger, M. Förster, R. Kneer, Fuel 104 (2013) 607-613.

[11] S. Pielsticker, B. Gövert, T. Kreitzberg, M. Habermehl, O. Hatzfeld, R. Kneer, Fuel 190 (2017) 420-434.

11th Asia-Pacific Conference on Combustion,
The University of Sydney, NSW Australia
10th -14th December 2017.

On MILD Combustion in a Perfectly Stirred Reactor with Exhaust Gas Recirculation

Yang Zhang*, Yuxin Wu, Hai Zhang, Qing Liu, Junfu Lv, Guangxi Yue

Key Laboratory for Thermal Science and Power Engineering of Ministry of Education

Department of Thermal Engineering, Tsinghua University, Beijing 100084, China

Abstract

The characteristics and criteria of the moderate and intensive low oxygen dilution (MILD) combustion in a perfectly stirred reactor (PSR) were studied with special attentions paid to the effect of exhaust gas recirculation (EGR). The well-known PSR S-curve theory was expended to the cases with EGR and the mathematical criteria for MILD combustion in a PSR with EGR was addressed. Results demonstrated that EGR helped establish a MILD combustion by first decreasing the self-ignition temperature of the mixture and second reducing the temperature increase with respect to the temperature of initial fresh mixture. Based on the PSR MILD combustion criteria, the combustion regime map diagram was exhibited, classifying 5 different combustion regimes: 1) no combustion; 2) unstable combustion; 3) feedback combustion; 4) MILD combustion and 5) high-temperature combustion. Including EGR in the process considerably enlarged the area the MILD combustion regime in the combustion regime map.

1 Introduction

The MILD (moderate and intensive low oxygen dilution) combustion, also known formerly as HiTAC (high-temperature air combustion) or FLOX (flameless oxidation), is one of the most promising clean combustion modes. The MILD combustion process is originated by the de-NO_x motivation. In early practice in 1990s [1], low NO_x combustion was achieved by increasing not only the jet velocity for a better mixing of the reactant and hot product but also air temperature for a stable combustion. Consequently, MILD combustion is featured with a highly-preheated reactant, low oxygen mole fraction in the reaction zone, relatively uniform heat release and non-visible flame surfaces [2, 3]. If well organized, a combustion device using MILD combustion is able to increase the thermal efficiency by 30% [4], and effectively reduce NO_x emission [4, 5]. Although recent studies [6] showed that the highly-preheated reactant and the extensive low oxygen dilution are indeed not essential for the establishment of MILD combustion, the terminology "MILD" is kept to indicate this unique combustion regime.

There are two widely accepted definitions for MILD combustion. Definition #1 is from the practice and defines that a combustion process is named MILD combustion when the fresh reactant temperature (T_u) is higher than the mixture self-ignition temperature (T_{si}) whereas the maximum allowable temperature increase (ΔT) with respect to the fresh reactant temperature during the combustion process is lower than mixture self-ignition temperature, i.e., $T_u>T_{si}$ and $\Delta T< T_{si}$ [7]. Definition #2 is from the perfectly stirred reactor (PSR) study. In PSR, MILD combustion is defined as a combustion process in which the ignition-extinction S-curve is stretched to monotonic one and the ignition state and the extinction state no longer exist [8]. Theoretical analysis showed that relation (1) holds for Definition #2.

$$\frac{E_a}{R_u} \leq 4T_u\left(1+\frac{c_pT_u}{q_cY_u}\right) \tag{1}$$

where E_a denotes the activation energy, R_u the universal gas constant, c_p the heat capacity, q_c the heat value of the fuel by mass and Y_u the fuel mass fraction in the fresh reactant.

Based on these two definitions, a number of studies in terms of the characteristics (e.g., [5, 9, 10]) and establishment conditions (e.g., [6, 11, 12]) of MILD combustion were conducted. Some researchers have drawn diagrams (e.g., [1, 6, 7, 12]) to classify the MILD combustion and conventional combustion regimes, seeking to obtain insights into MILD combustion process.

Exhaust gas recirculation (EGR) is a broadly used technology to decrease NO_x emission of a combustion process [13, 14]. In practice, 10% – 40% of EGR is found to be very effective in the suppression of NO_x formation [14]. When EGR is applied, the characteristics and criteria of MILD combustion is still unclear. Thus, the aim of this paper is to theoretically study the effect of EGR on the characteristics and criteria of MILD combustion. The investigation and analysis is based on a PSR model. One should note that the EGR concept in this study is actually different from the one in [1]. EGR in this study indicates introducing the tail exhaust gas (after heat loss or done the work, normally with a low temperature) to the fresh reactant, which is similar to the FGR (Flue gas recirculation) process of a boiler or the EGR process of an engine. EGR in [1] means the recirculation of the hot combustion product by the flow entrainment, which happens in almost every combustor.

2 A PSR model with EGR

The combustion process can be abstracted to a simple one as demonstrated in Fig. 1. Assuming that the heat exchange between the fresh reactant and recirculated exhaust gas are extremely fast such that they will have the same temperature T_u before they enter the reactor, the energy equation and species equation can be thus written as (2) and (3).

*Corresponding author. Fax: +86-10-6278-1743
E-mail address: yang-zhang@tsinghua.edu.cn

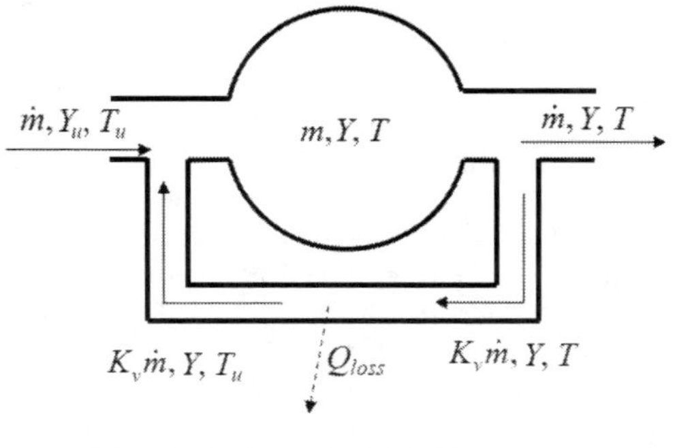

Figure 1: PSR physical model with EGR

Energy equation:

$$mc_p\frac{dT}{dt}=\dot{m}c_pT_u+K_v\dot{m}c_pT_u-(1+K_v)\dot{m}c_pT+Vq_cM\omega \tag{2}$$

Species equation:

$$\begin{aligned} m\frac{dY}{dt}&=\dot{m}Y_u+K_v\dot{m}Y-(1+K_v)\dot{m}Y-VM\omega \\ &=\dot{m}Y_u-\dot{m}Y-VM\omega \end{aligned} \tag{3}$$

where m, T, Y are respectively the mixture mass, temperature and fuel mass fraction inside the PSR, $\dot{m}$ the mass flux, c_P the heat capacity, V the volume of the reactor, t the time, M the molecular weight of the fuel, q_c the heat of combustion, ω the reaction rate, and K_v the EGR ratio.

Further assuming that the reaction is a one-step, one-order reaction, thus the reaction rate can be written as (4).

$$\omega=\frac{\rho Y}{M}B\exp(-\frac{E_a}{R_uT}) \tag{4}$$

where ρ is the mixture density and B the pre-exponential factor.

Equations (2) and (3) can be normalized as (5) and (6).

$$\frac{d\tilde{T}}{d\tilde{t}}=\tilde{T}_u-\tilde{T}+Da_c\tilde{Y}\exp(-\frac{\tilde{T}_a}{\tilde{T}}) \tag{5}$$

$$\frac{d\tilde{Y}}{d\tilde{t}}=\frac{1}{1+K_v}(1-\tilde{Y})-Da_c\tilde{Y}\exp(-\frac{\tilde{T}_a}{\tilde{T}}) \tag{6}$$

where $\tilde{T}=c_pT/(q_cY_u)$, $\tilde{Y}=Y/Y_u$, $t_r=m/[(1+K_v)\dot{m}]$, $\tilde{t}=t/t_V$, $Da_c=Bt_V$ and $T_a=E_a/R_u$. Note that the $\tilde{T}_a$ value of normal fuels (H_2, CH_4) is in the order of $10^{0.5}\sim10^1$, and the $\tilde{T}_a$ involved in this study was chosen within this range.

Using the coupling function approach and considering the steady state solution, (5) and (6) lead to (7) and (8).

$$\tilde{T}_u-\tilde{T}+\frac{1}{1+K_v}(1-\tilde{Y})=0 \tag{7}$$

$$\tilde{Y}=1+(1+K_v)(\tilde{T}_u-\tilde{T}) \tag{8}$$

Substituting (8) into (5), we will have the general equation controlling the temperature of the process, as (9).

$$\tilde{T}_u-\tilde{T}_S+Da_c[1+(1+K_v)(\tilde{T}_u-\tilde{T}_S)]\exp(-\frac{\tilde{T}_a}{\tilde{T}_S})=0 \tag{9}$$

where $\tilde{T}_S$ denotes the steady solution of (9).

3 Effect of EGR on MILD combustion process

Computed based on (9), Fig. 2 demonstrates $\tilde{T}_S$ as a function of $\tilde{T}_u$ when Da_c and $\tilde{T}_a$ are respective constants. It can be seen that when $K_v = 0$, the $\tilde{T}_S \sim \tilde{T}_u$ curve exhibits the so-called S-curve featured by an ignition turning point and an extinction turning point. According to Definition #2, this process is a conventional combustion process rather than a MILD one. As K_v increases to $K_v = 0.2$, the ignition turning point and the extinction turning point merge to only one point (critical point) and the S-curve is stretched to a monotonic one. $\tilde{T}_u$ of the critical point is thus defined as the self-ignition temperature $\tilde{T}_{si}$ of the PSR. In this case of $K_v = 0.2$, $\tilde{T}_{si}$ approximates 0.5. Along the upper branch ($\tilde{T}_u>\tilde{T}_{si}$) of the curve of $K_v = 0.2$, although ΔT of the curve of $K_v = 0.2$ is lower than that of $K_v = 0.0$, ΔT of the curve $K_v = 0.2$ (for instance $\Delta T\sim0.7$ at $\tilde{T}_u = 0.8$) is still greater than $\tilde{T}_{si}$, indicating that the case of $K_v = 0.2$ is not MILD combustion as per Definition #1. Further increasing K_v results in a further decrease of ΔT, and consequently some points along the curve of $K_v = 0.4$ falls into the MILD combustion regime.

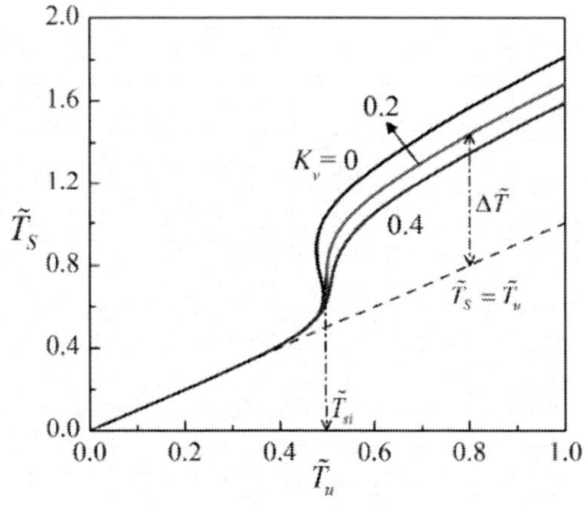

Figure 2: $\tilde{T}_S$ as a function of $\tilde{T}_u$ ($\tilde{T}_a$= 3.2, Da_c = 18.2)

The self-ignition point in Fig. 2 physically implies that a tiny temperature increase of the fresh mixture beyond this critical temperature will result in a significant temperature increase of the mixture. That is the self-sustained combustion.

In order to further study the effect of EGR on $\tilde{T}_{si}$, $\tilde{T}_S$ plots as a function of Da_c at $\tilde{T}_a$= 3.2 and $\tilde{T}_u$ = 0.5 is given in Fig. 3. When $K_v = 0.2$, the $\tilde{T}_S \sim Da_c$ curve exhibits the critical monotonic feature, implying that the ignition will eventually occur as the residence time is sufficiently long, and the transition from unburned state to burned state is gradual. If K_v is lower, say $K_v = 0$, the two singular turning points emerge and there are sudden "jump" and "drop" when the transition between the unburned and burned states occurs. If K_v is greater, say $K_v = 0.4$, $\tilde{T}_S$ is further decreased and the monotonic curve is easier to achieve.

Note that $\tilde{T}_{si}$ is physically equal to the $\tilde{T}_u$ value at which the critical monotonic $\tilde{T}_S \sim Da_c$ curve is achieved in Fig. 3. Thus, $\tilde{T}_{si}$ can be mathematically solved by setting

$$\frac{d(\ln Da_c)}{d\tilde{T}_S}=0 \tag{10}$$

Thus, (11) is obtained.

$$\left(1/\tilde{T}_a + 1 + K_v\right)\tilde{T}_S^2 - \left[1 + 2\left(1 + K_v\right)\tilde{T}_u\right]\tilde{T}_S + \tilde{T}_u\left[1 + \left(1 + K_v\right)\tilde{T}_u\right] = 0 \quad (11)$$

The critical monotonic $\tilde{T}_S \sim Da_c$ curve equals to that the quadratic equation only has one root, resulting in

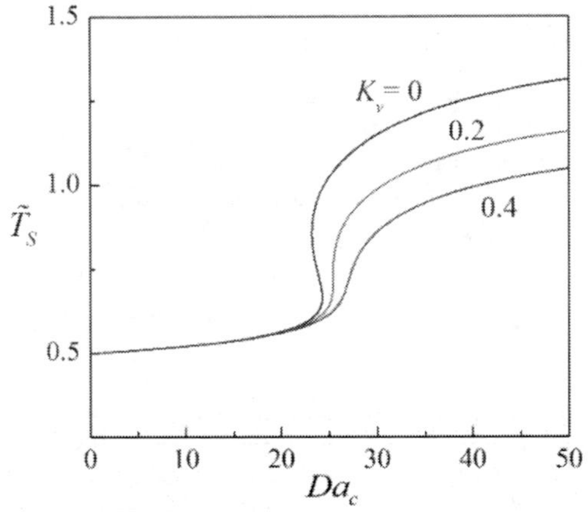

Figure 3: $\tilde{T}_S$ as a function of Da_c ($\tilde{T}_a = 3.2$, $\tilde{T}_u = 0.5$)

$$\tilde{T}_{si} = \frac{-1 + \sqrt{1 + (1 + K_v)\tilde{T}_a}}{2(1 + K_v)} \quad (12)$$

and the criterion for MILD combustion of Definition #2 with EGR is (13)

$$\tilde{T}_a \le 4\tilde{T}_u\left[1 + \left(1 + K_v\right)\tilde{T}_u\right] \quad (13)$$

Letting $K_v = 0$, (13) becomes (14), which is exactly the same with (1) in Definition #2.

$$\tilde{T}_a \le 4\tilde{T}_u\left(1 + \tilde{T}_u\right) \quad (14)$$

Figure 4 demonstrates the $\tilde{T}_{si}$ variation as a function of K_v according to (12). As K_v increases, $\tilde{T}_{si}$ slightly decreases, suggesting that $\tilde{T}_{si}$ is easier to achieve if K_v is greater.

Equation (13) indicates that the right hand side becomes greater as K_v increases, implying that MILD combustion is easier to establish if EGR is applied, being consistent with the results in Figs. 2, 3 and 4.

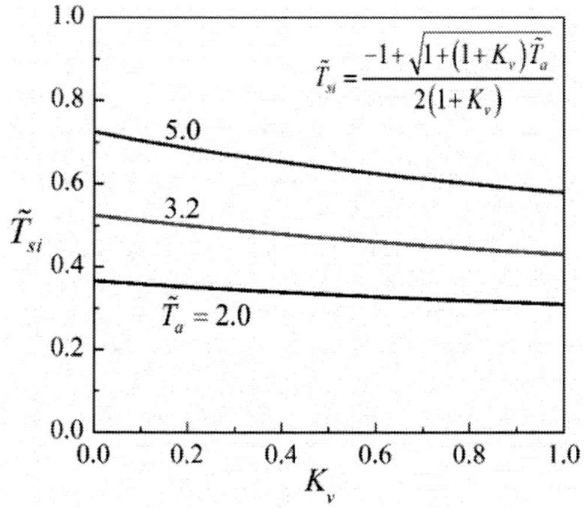

Figure 4: $\tilde{T}_{si}$ variation as a function of K_v

In common combustion process, $\tilde{T}_a$ is normally depended on the fuel type, and $\tilde{T}_u$, Da_c and K_v are operation parameters. Substituting (12) into (9), the $\tilde{T}_u \sim Da_c$ relation that satisfy Definition #1 of MILD combustion, $\Delta T < T_{si}$ can be numerically solved. Thus, incorporating both Definitions #1 & #2, a combustion regime map diagram, in terms of the operation parameters $\tilde{T}_u$ and Da_c can be obtained, as shown in Figs. 5 and 6. Five combustion regimes are specified. They are 1) no combustion; 2) unstable combustion; 3) feedback combustion; 4) MILD combustion and 5) high-temperature combustion. The mathematical criteria and characteristics of each combustion regime are listed as follows:

1) **No combustion:** $\tilde{T}_u < \tilde{T}_{si}$ and $Da_c < Da_{c,E}$, basically only exists slight oxidation, but no intensive combustion.
2) **Unstable combustion:** $\tilde{T}_u < \tilde{T}_{si}$ and $Da_{c,E} < Da_c < Da_{c,I}$, small temperature perturbation will results in sudden ignition or extinction.
3) **Feedback combustion:** $\tilde{T}_u < \tilde{T}_{si}$ and $Da_c > Da_{c,I}$, although the temperature of the unburned fresh mixture is low, the residence time is sufficient high to induce a stable combustion, producing high-temperature products.
4) **MILD combustion:** $\tilde{T}_u > \tilde{T}_{si}$ and $Da_c < Da_c|_{\Delta\tilde{T}/\tilde{T}_{si}=1}$, the temperature of the unburned fresh mixture is relatively high but the combustion is rather moderate.
5) **High-temperature combustion:** $\tilde{T}_u > \tilde{T}_{si}$ and $Da_c > Da_c|_{\Delta\tilde{T}/\tilde{T}_{si}=1}$, high-temperature preheated intensive combustion, producing very hot products.

Comparing Figs. 5 and 6, one can tell that the area of MILD combustion regime at $K_v = 0.3$ is significantly larger than that at $K_v = 0.3$, indicating that EGR benefits the establishment of MILD combustion. This result is consistent with the findings in Figs. 2, 3 and 4. This finding need to be further verified by experimentation and detailed numerical modelling with full description of the chemical kinetics and molecular transport, which are the future efforts.

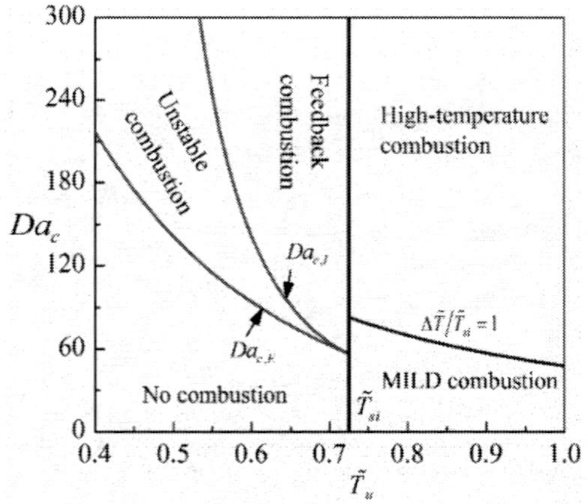

Figure 5: Combustion regime diagram at $K_v = 0$, $\tilde{T}_a = 3.2$

($Da_{c,I}$: ignition Damköhler number; $Da_{c,E}$: extinction Damköhler number)

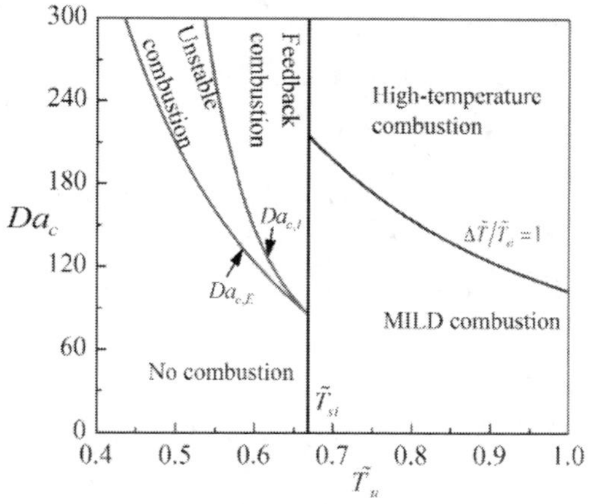

Figure 6: Combustion regime diagram at $K_v = 0.3$, $\tilde{T}_a = 3.2$

($Da_{c,I}$: ignition Damköhler number; $Da_{c,E}$: ignition Damköhler number)

4 Conclusions

MILD combustion in a PSR with EGR was studied. In particular, the S-curve theory was extended to the PSR combustion with EGR. Theoretical analysis showed that EGR helped establish a MILD combustion by decreasing the self-ignition temperature of the mixture and reducing the temperature increase before and after burning. Based on the PSR MILD combustion criteria, the combustion regime map diagram was exhibited, classifying 5 different combustion regimes: 1) no combustion; 2) unstable combustion; 3) feedback combustion; 4) MILD combustion and 5) high-temperature combustion. Including EGR in the process considerably enlarged the area the MILD combustion regime in the combustion regime map, confirming that EGR benefits the establishment of MILD combustion.

5 Acknowledgment

This work is financially supported by the National Natural Science Foundation of China (NSFC No. 51476088).

References

[1] J.A. Wünning, J.G. Wünning, Prog. Energy Combust. Sci. 23 (1997) 81-94.

[2] B. Danon, A. Swiderski, W. de Jong, W. Yang, D.J.E.M. Roekaerts, Combust. Sci. Technol. 183 (2011) 686-703.

[3] P. Li, B.B. Dally, J. Mi, F. Wang, Combust. Flame 160 (2013) 933-946.

[4] H. Tsuji, A. Gupta, T. Hasegawa, Crc Press, (2003).

[5] H. Zhang, G. Yue, J. Lu, Z. Jia, J. Mao, T. Fujimori, T. Suko, T. Kiga, Proc. Combust. Inst. 31 (2007) 2779-2785.

[6] F. Wang, J. Mi, P. Li, Energy & Fuels 27 (2013) 3488-3498.

[7] A. Cavaliere, M. de Joannon, Prog. Energy Combust. Sci. 30 (2004) 329-366.

[8] M. Oberlack, R. Arlitt, N. Peters, Combustion Theory and Modelling 4 (2000) 495-509.

[9] M.M. Noor, A.P. Wandel, T. Yusaf, International Journal of Automotive and Mechanical Engineering 6 (2012) 730-754.

[10] F. Wang, P. Li, Z. Mei, J. Zhang, J. Mi, Energy 72 (2014) 242-253.

[11] P. Li, J. Mi, B. Dally, F. Wang, L. Wang, Z. Liu, S. Chen, C. Zheng, Science China Technological Sciences 54 (2011) 255-269.

[12] M.J. Evans, P.R. Medwell, H. Wu, A. Stagni, M. Ihme, Proc. Combust. Inst. 36 (2017) 4297-4304.

[13] E. Hu, Z. Huang, B. Liu, J. Zheng, X. Gu, Int. J. Hydrogen Energy 34 (2009) 1035-1044.

[14] E. Hu, Z. Huang, B. Liu, J. Zheng, X. Gu, B. Huang, Int. J. Hydrogen Energy 34 (2009) 528-539.

11th Asia-Pacific Conference on Combustion,
The University of Sydney, NSW Australia
10th -14th December 2017.

Observation of Different Self-ignition Features of High Pressurized Hydrogen Released through a Tube

Hyoung Jin Lee[1], Sang Yoon Lee[2], In-Seuck Jeung[2*]

[1]Department of Aerospace Engineering, Inha University

Inha-ro, Nam-gu, Incheon, 22212, Korea

[2]Department of Aerospace Engineering, Seoul National University

Gwanak-ro, Gwanak-gu, Seoul, 08826, Korea

Abstract

Utility value of high pressure hydrogen is considerable, but there still exists a safety issue due to its possibility of self-ignition. And this are related to the burst conditions strongly because it depends on the flow development after rupture. To understand this relationship, the experimental data including visualization images for the whole region of downstream tube were obtained from immediately after rupture. In this study, three different ignition features were observed from the visualization images and their effects on the self-ignition were investigated in detail.

1 Introduction

Hydrogen is receiving much attention as a next-generation energy carrier, but it needs to be compressed or liquefied due to its low volumetric density. At present, most systems prefer the compressed form over the liquefied form because the latter needs additional energy to maintain it in a tank at low temperatures. However, high-pressure hydrogen presents potential hazards because of its strong flammability, e.g., the wide range of flammability. In fact, according to past researches, over 90% of hydrogen related accidents have been due to ignition, of which 60% of the ignition sources have yet to be identified [1]. Furthermore, it is reported that 90% of hydrogen related accidents have been initiated from gas handling systems such as pipes, flanges and valves[2].Like this, utility value of high pressure hydrogen is considerable, but there still exists a safety issue due to its possibility of self-ignition. Thus, many studies have been conducted in order to reveal the ignition mechanism and to prevent the phenomenon. In present, there seems to be a high probability of diffusion ignition through previous studies [3-13]. Extensive experimental data have been reported in recent years and they are increasing the possibility of diffusion ignition for the self-ignition of high pressure hydrogen. Especially, the visualization images around the tube brought deeper understanding of the mechanism [5-7]. These results showed a general tendency that the possibility of self-ignition can increase as a burst pressure is high, a diameter and length of the tube is small and long. Besides, several numerical simulations have also been conducted to aid limitative experimental data [8-12]. The numerical visualization brought an intuitive grasp on the phenomenon [9-10].

Corresponding author. Fax: +82-2-880-7387
E-mail address: enjis@snu.ac.kr

From previous results, the ignition mechanism and essentials to initiate the ignition are quite clear as followings. First, a high burst pressure is necessary so that an air can be heated enough. Second, a sufficient tube length that a mixing region induced by flow interactions can be developed is needed. Despite of these observations, understanding of the phenomenon has still remained imperfect because a flow development with failure conditions of a rupture disk is not known. In fact, the ignition characteristics are not always the same, even though main burst conditions such as a burst pressure and extension tube are the same. This is because an initial flow can be different as an open condition is varied, which can result in a different ignition feature. Regarding to this, numerical study showed different ignition feature with burst conditions [14], but it has not been reported through experimental studies yet.

In this study, the authors want to reveal the mechanism of different ignition features and their effect on the self-ignition through theoretical analysis and visualization images. For this purpose, the authors have investigated an initial flow immediately after failure of a rupture disk and its effect on the self-ignition as observing visualization images for the whole downstream tube. Also, the authors want to investigate the different ignition patterns with various burst conditions. To do this, burst conditions were changed by varying open area ratio. Different open area ratio can simulate the different border of a disk after a failure in practical usages. Experimental study was conducted to achieve these goals using a rupture disk and a rectangular extension tube.

2 Experimental Setup

Experimental apparatus consists of a hydrogen booster, a storage tank, a rupture disk and an extension tube. Mylar polyester film of various thicknesses was used to control the burst pressure as a rupture disk. A pressure was measured on the storage tank wall to measure a burst pressure. The extension tube has 10 mm by 10 mm square shaped cross-section and its length is 200 mm. Window glass substitute for both side walls to obtain visualization images. Open area of a disk is basically same as the cross-sectional area of the extension tube, but it can be varied by placing a plate with a smaller hole behind of a disk. The several plates that the area ratio of the hole to the tube cross-section is 0.125~1.0 were applied to change open area. The detailed view on the test model including the plates is shown in Fig. 1. Pressure on the lower wall surface and light on the upper wall surface were measured and direct and shadow images as flow visualizations were obtained by experimental data. All measurement systems were synchronized by a triggering signal. Six piezoelectric pressure transducers and photodiodes were installed to measure the pressure and ignition position. First pressure transducer located 21 mm from a rupture disk was used as a triggering source to excite other devices. Next transducers are installed every 34.8 mm. Shadowgraph and direct images were simultaneously obtained using two high-speed cameras.

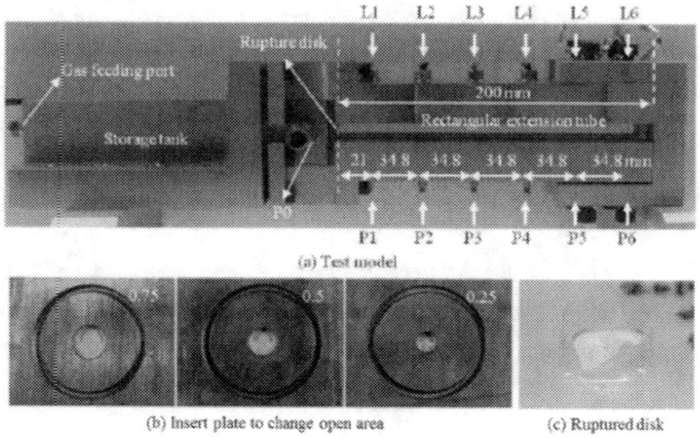

(a) Test model

(b) Insert plate to change open area (c) Ruptured disk

Figure 1, Test Model

3 Result and Discussion

3.1 Visualization images of self-ignition

Figure 2 shows the typical direct and shadowgraph images for the self-ignition at the burst pressure of 9.1 MPa.and open area ratio of 1.0. The overall process of self-ignition and flame propagation, which is main ignition in Fig. 2, is as follows. An initial flat disk is spherically expanded gradually as a pressure of storage tank increases, and then ruptured. After that, hydrogen gas discharges abruptly and very small spherical shock is generated through partially opened area. The generated shock is propagated to the wall as forming a thin mixing region that a density gradient exists. The shock arrived at the wall is reflected to the center again and spherical shock becomes flatter and closer to plane wave. Incident shock is propagated downstream gradually and ignition near the boundary layer is initiated. After that the flame is developed to center region gradually. The ignition is initiated near the boundary layer first and it is expanded to the center of tube.

Additionally, another two ignition feautres were observed as follows; First, initial ignition and flame was observed. The ignition was initated near the diaphragm immediatedly after rupture, but the flame disappred soon. This initail ignition and flame was generated more strongly so that the burst disk can be burned. However, although the initial ignition and flame near the disk is generated strongly and fast, this does not affect next self-ignition mechanism. Figure 3 shows the initial ignition images at the higher burst pressure. As a second igntion and flame, local ignition and flame was observed. Local ignition is also initiated from a strong mixing region, which is induced by complex flow interactions at a local point. After that, it is propagated shortly, then disappeared soon or merged to the ignition region initiated near the boundary layer. Figure 4 shows the another visualization images at the burst pressure of 9.0 MPa and open area ratio of 0.75. Simlaraly to the Fig. 2, initail, local and main ignition and flame propagations are observed. Unlike to the initial ignition, this local ignition was observed very randomly.

3.2 Theoretical analysis for initial ignition

To understand the initiation mechanism of initial ignition, simple theoretical analysis was conducted. Temperature and pressure values calculated from the thermodynamic conditions of driver and driven gases using the shock tube theory. Then, the temperature of a mixture was computed by considering isochoric mixing of hydrogen and air. This series of calculations was iterated for several equivalence ratios which are assumed for mixtures, and finally temperature values of mixtures were prepared. The induction time of the mixture could be calculated from the temperature and pressure values of mixtures obtained in the previous step and the selected equivalence ratios. Figure 5 shows the result of this analysis. The values indicated at each line tell selected equivalence ratios of mixtures which are formed immediately after the bursting. Each line means the mixtures of which equivalence ratios are identical with different bursting

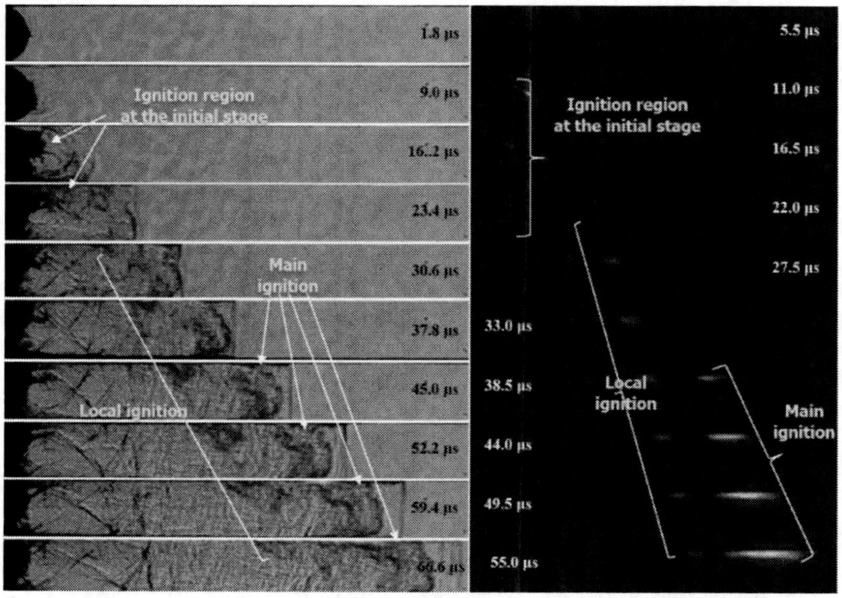

Figure 2, Visualization images for self-ignition at burst pressure of 9.1 MPa and open area ratio of 1.0

(a) Burst pressure of 8.9 MPa (no initial ignition)

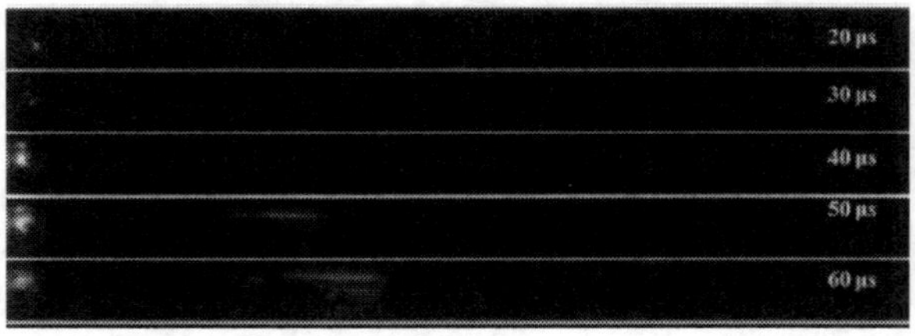

(b) Burst pressure of 10.9 MPa (Strong initial ignition)

Figure 3, Initial flame images with burst pressure

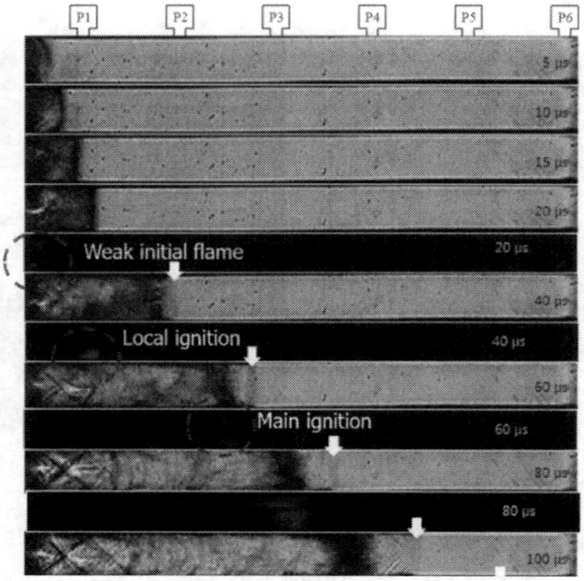

Figure 4, Visualization images for self-ignition at burst pressure of 9.0 MPa and open area ratio of 0.75

pressures. The induction length corresponding to a specific induction time was acquired from the Mach number and the sound speed of a related mixture. If equivalence ratio is kept constant, the mixture is easier to be ignited with higher bursting pressure because the temperature of mixture increases. Therefore, the induction length is inversely proportional to the bursting pressure. In the result, there are two groups, dotted and solid. The first is lower equivalence ratios than 0.05. Because mixtures in this group has comparatively small amount of hydrogen, equivalence ratios is dominant to induction time. Thus, this causes the shorter induction length with the larger equivalence ratio. Secondly, for the larger equivalence rations than 0.1, opposite tendency are observed. In this group, the decrease of temperature is more critical. As a result, the line moves to the right with larger equivalence ratio whereas the dotted line moves to the left. Another feature of this group is that the left parts of lines are almost attached to the y-axis. Because there is enough fuel, a mixture can be ignited easily if a certain condition of tem-perature and pressure is met and there is no big difference in ignition lengths between mixtures over this point. This second region is related to the initial ignition and flame. Initial ignition can be initiated when burst pressure is above 80 bar and equivalence ratio is above 0.05. For example, induction distance can be very short, 9mm, because the temperature of mixture is sufficient high, 1337K, if burst pressure is 80 bar and equivalence ratio is about 0.1.

3.3 Analysis for local ignition

Figure 6 shows the self-ignition limit for the case of extension tube length of 200 mm. The possibility of self-ignition becomes lower as open area ratio decreases. Therefore, it can be thought

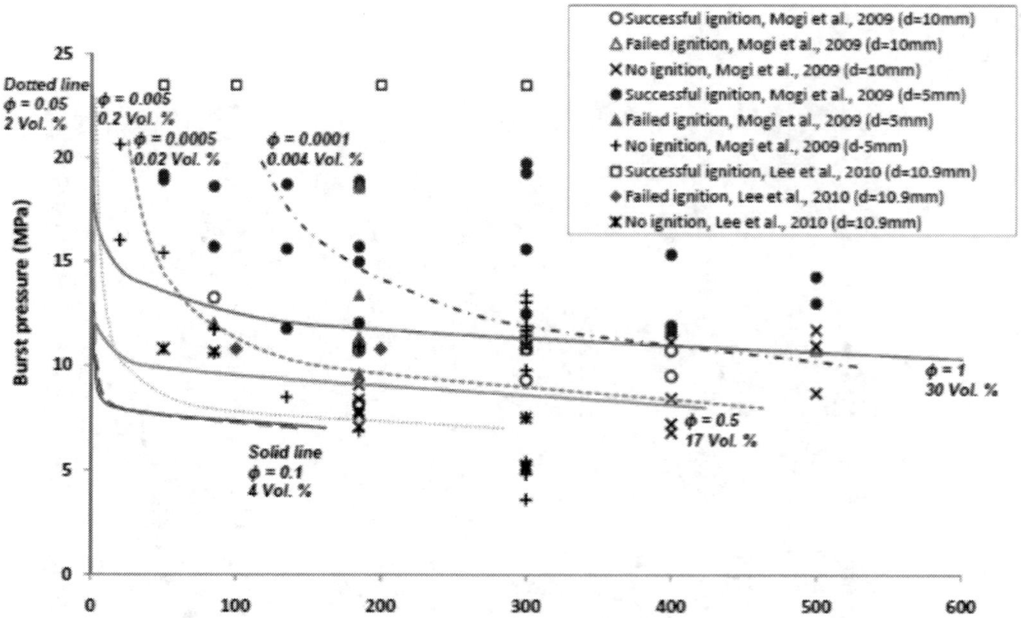

Figure 5, Induction length vs. bursting pressure with experimental data

that small open area ratio can be helpful to prevent unintended ignition. However, the self-ignition can be initiated when the burst pressure is high sufficiently.

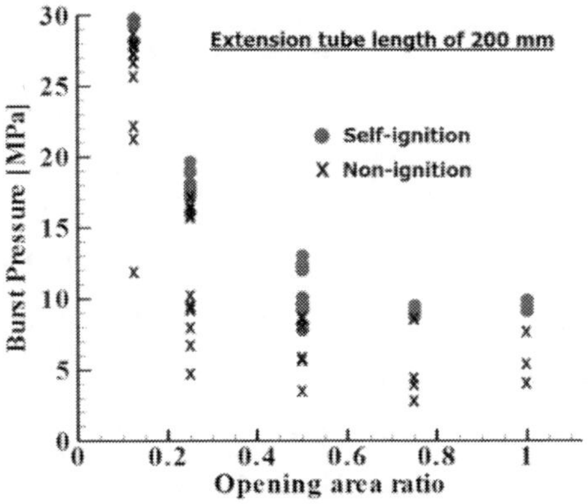

Figure 6, self-ignition limit with open area ratio

Figure 7 shows the local maximum pressure value with open area ratio and burst pressure. Red and solid symbolic line represents the case of self-ignition, and dashed and open symbolic line corresponds to the non-ignition. Difference of the pressure value is quite large with the self and non-ignition conditions at the open area ratio of 0.5 and 0.25. And pressure increase is shown between position 2 and 3 and shock speed is also increase. This is because the planar shock is formed completely and this was confirmed from the visualization images. Therefore, downstream of position 2 has the mixing and heating conditions that self-ignition can be initiated. However, this tendency is not preserved at the open area ratio of 0.125. As shown in Fig. 7(c), the difference of pressure value is not large between the case of self and non-ignition. The reason is not clear because of the lack of visualization images, but it is possible that the local ignition as shown in Fig. 5 might be propagated.

4 Conclusions

Self-ignition characteristics were investigated when high pressure hydrogen was released suddenly through a tube into the air by the failure of a rupture disk. The characteristics were analyzed using the visualization images for the whole tube from the rupturing instant of the disk and the pressure data and light signals on the wall surface. The visualization images showed that the initial ignition region that has not been observed in the previous results exists near the disk. However, this initial ignition cannot affect the whole process of the self-ignition as it is not propagated downstream and disappeared when the burst pressure is about 10 MPa. After that, as already known ignition mechanism, another ignition region near the boundary layer is formed and then expands to the center. The experimental images showed that the ignition is always started from the boundary layer and this is the most important thing for the self-ignition. It is possible that the self-ignition can be initiated when the burst pressure is high enough at small open area. When the area ratio is higher than 0.25 it seems that the mechanism of self-ignition initiated from the heating and mixing induced by the shock and flow interactions is preserved from the pressure data inside the tube. However, at open area ratio of 0.125, it is difficult to classify the limit of burst pressure that the self-ignition can be initiated and it seems that the mechanism might be different. The reason is unclear, but it is plausible that the self-ignition can be induced from the local ignition.

References

[1] B.P. Xu, J.X. Wen, V.H.Y. Tam, Int. J. Hydrogen Energy 36 (3) (2011) 2637-2644.
[2] Y. Suwa, Int. Ass. for hydrogen energy, 2006.
[3] Wolanski P, Wojcicki S. Proc. Combust. Inst. 14: (1973) 1217-1223.
[4] Dryer FL, Chaos M, Zhao Z, Stein JN, Alpert JY, Homer CJ. Combust. Sci. Tech. 179 (4) (2007) 663-694
[5] Mogi T, Wada Y, Ogata Y, Hayashi AK. Int. J. Hydrogen Energy 34 (2009) 5810-16
[6] H.J. Lee, Y.R. Kim, S.H. Kim, I.S. Jeung, Proc. Combust. Inst. 33 (2011) 2351-58.
[7] Kim YR, Lee HJ, Kim SH, Jeung IS. Proc. Combust. Inst. 34 (2013) 2057-64
[8] Xu BP, El H, Wen JX, Dembele S, Tam VHY, Donchev T. J. Loss Prevent Proc. 21 (2008) 205-13
[9] Wen JX, Xu BP, Tam VHY. Combust. Flame 156 (2009) 2173-89
[10] Lee BJ, Jeung IS. Int. J. Hydrogen Energy. 34 (2009) 8763-69
[11] Bragin MV, Molkov VV. Int. J. Hydrogen Energy 36 (2011) 2589-96
[12] Yamada E, Kitabayashi N, Hayashi AK, Tsuboi N. Int. J. Hydrogen Energy 36 (2011) 2560-66
[13] Golovastov S, Bocharnikov V. Int. J. Hydrogen Energy 37 (2012) 10956-62
[14] Lee HJ, Park JH, Kim SD, Kim SH, Jeung IS. Proc. Combust. Inst. 35 (2015) 2173.

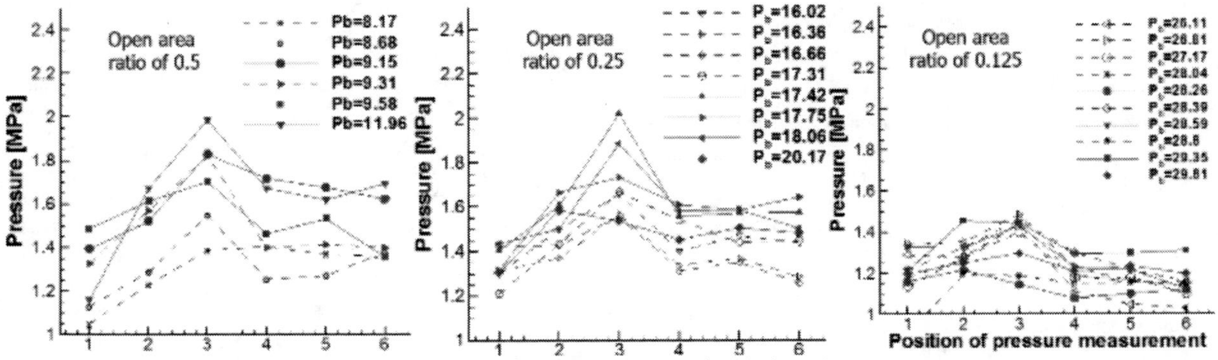

Figure 7 Local Maximum Pressure inside the Tube

Dynamical Analysis of Methanol-Dimethyl Ether Autoignition at Engine-Relevant Conditions

Wonsik Song, Efsthatios-Al. Tingas, Hong G. Im

Clean Combustion Research Center (CCRC), King Abdullah University of Science and Technology (KAUST)
Thuwal 23955-6900, Kingdom of Saudi Arabia

Abstract

We have computationally investigated the auto-ignition of a methanol-air mixture with the addition of dimethyl ether at a high pressure and low-temperature environment aiming at its utilization to HCCI engines. It is found that the addition of dimethyl ether substantially reduced the ignition delay of the methanol-air mixture. This is caused by the inclusion of a specific chemical species and reaction of the dimethyl ether combustion. Comprehensive understanding of chemical pathways, as well as the important species and reactions, have been identified with the help of tools derived from the computational singular perturbation method.

1 Introduction

Homogeneous charge compression ignition (HCCI) engines have been considered the next-generation vehicle engines by adopting the merits from the spark-ignition (SI) and compression-ignition (CI) engines. However, one of the major drawbacks is that there is no external ignition source such as a spark in SI engines or injectors in CI engines thereby the ignition relies only on the chemical process of the combustible mixture. This makes the operation of HCCI challenging especially adjusting the ignition timing or combustion phase. A conventional approach to these problems in HCCI engines is to add diluents such that the additives change the reaction rates of the mixture. Recently, a newly proposed approach is to use multi-fuels that have entirely different ignition properties.

Methanol (CH3OH, MeOH) has been widely known as a clean fuel in that it has no carbon-carbon bonds and contains an extra oxygen atom, which allows smokeless combustion. However, the high octane number (~109), high volatility, and high latent heat of vaporization of MeOH have been paid more attention to its use in spark ignition engines [1,2]. A remedy to this situation can be to use blends of methanol with other fuels. A highly recommended fuel is dimethyl ether (CH3OCH3, DME) due to its close characteristics to diesel fuel such as a high cetane number (55-60) and a low boiling temperature. This provides a short ignition delay time, fast mixture formation, and also excellent cold-start characteristics, implying that DME can be a perfect ignition improver.

There has been a large number of experimental research investigating the combustion behaviors with the use of methanol and dimethyl ether blends [1-3]. However, the computational studies are rarely reported especially for the comprehensive understanding how the MeOH-fueled mixture ignites with the addition of DME. In this work, we present the effects of DME addition to MeOH-air mixtures in HCCI engine environments. Analyses with tools generated from the computational singular perturbation (CSP) allow the in-depth understanding of the combustion process of MeOH-air with the addition of DME. The analysis is performed by homogeneous adiabatic constant volume autoignition so that the chemical kinetics is not affected by any transport effect.

2 Governing Equations and CSP Tools

The homogeneous adiabatic constant volume auto-ignition of a mixture is governed by the species and energy equations, which can be expressed by the combined form as follows:

$$\frac{d\mathbf{z}}{dt} = \mathbf{g}(\mathbf{z}) = \sum_{k=1}^{2K} \hat{\mathbf{S}}_k R^k \tag{1}$$

where $\mathbf{z} = [\mathbf{y}, T]^T$ is the (N+1)-dimensional state column vector that includes species mass fractions ($\mathbf{y}$) and temperature (T), $\hat{\mathbf{S}}_k$ is the generalized stoichiometric coefficients vector of the k-th reaction and R^k is the related reaction rate. Here, we consider the forward and backward reactions separately such that (1) is integrated over time for $2K$ terms with the K being the number of unidirectional reactions.

In CSP, the governing equation can be cast in the following form by introducing a new set of basis vector:

$$\frac{d\mathbf{z}}{dt} = \mathbf{g}(\mathbf{z}) = \sum_{n=1}^{N-E+1} \mathbf{a}_n f^n \tag{2}$$

where $\mathbf{a}_n$ is the (N+1)-dimensional CSP column basis vector of n-th mode and f^n is corresponding amplitude defined as:

$$f^n = \mathbf{b}^n \cdot \mathbf{g}(\mathbf{z}) = \sum_{k=1}^{2K} \left(\mathbf{b}^n \cdot \hat{\mathbf{S}}_k\right) R^k \tag{3}$$

In (2) and (3), N, E, and K indicate the number of species, elements, and reactions of the chemical reaction mechanism, respectively, which corresponds to 253, 6, and 1542 for the present mechanism, which has been validated against a wide range of initial conditions and experiments apparatus [4]. $\mathbf{b}^n$ is the dual basis vector of $\mathbf{a}_n$ satisfying $\mathbf{b}^i \cdot \mathbf{a}_j = \delta^i_j$ [5,6]. The CSP basis vectors $\mathbf{a}_n$ and $\mathbf{b}^n$ can be approximated by the right and left eigenvectors of the Jacobian of $\mathbf{g}(\mathbf{z})$.

The timescale of each mode is expressed by the relation $\tau_n = |\lambda_n|^{-1}$ (n = 1, ..., N-E+1), where λ_n is the n-th non-zero eigenvalue of the Jacobian $\mathbf{J}$ of $\mathbf{g}(\mathbf{z})$. The eigenvalue can be directly calculated by the Jacobian decomposition, $\lambda_n = \beta^n \cdot \mathbf{J} \cdot \alpha_n$ where α_n and β^n are the n-th right (column) and left (row) eigenvector of the Jacobian $\mathbf{J}$ of the system, respectively.

When system (2) exhibits M timescales of dissipative

Corresponding author. Office: +966-12-808-4726
E-mail address: hong.im@kaust.edu.sa

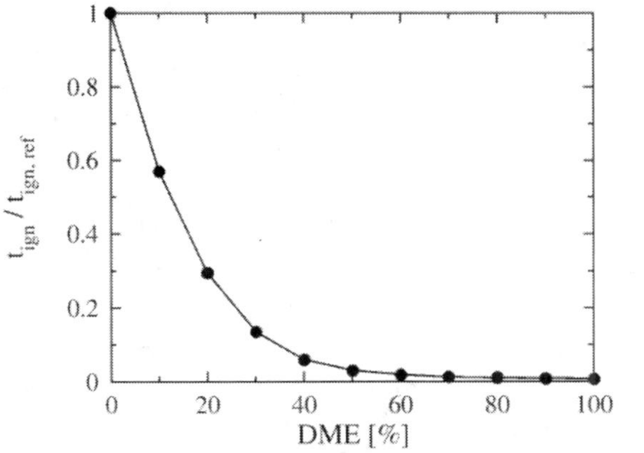

Figure 1: Normalized ignition delay time of DME addition to the MeOH-Air mixture by the ignition delay of the pure-MeOH, at P(0) = 30 atm, T(0) = 800 K, and ϕ = 0.6.

nature and much faster than the rest, then the reduced system is obtained in the form of (4):

$$f^j \approx 0 \ (j=1,M) \qquad \frac{d\mathbf{z}}{dt} = \sum_{i=M+1}^{N+1} \mathbf{a}_i f^i \tag{4}$$

Of all modes that develop in a system, of particular interest are the explosive ones, i.e., modes that tend to drive the system away from equilibrium and are directly related to ignition phenomena.

Considering the Jacobian $\mathbf{J} = grad\left(\hat{\mathbf{S}}_1 R^1 + \cdots + \hat{\mathbf{S}}_{2K} R^{2K}\right)$, the n-th eigenvalue of the system can be expressed by (5):

$$\lambda_n = \beta^n \cdot \sum_{k=1}^{2K} grad\left(\hat{\mathbf{S}}_k R^k\right) \cdot \alpha_n = c_1^n + \cdots + c_{2K}^n \tag{5}$$

where c_k^n measures the contribution of k-th reaction to the n-th eigenvalue. A positive (negative) c_k^n implies that k-th reaction favors (opposes) the explosive (dissipative) character of the n-th timescale τ_n.

Equation (5) suggests the introduction to the timescale participation index (TPI):

$$J_k^n = \frac{c_k^n}{\left|c_1^n\right| + \cdots + \left|c_{2K}^n\right|}, \ (n=1,\ldots,N-E+1, \ k=1,\ldots,2K) \tag{6}$$

and by definition, $\sum_{k=1}^{2K}\left|J_k^n\right| = 1$. J_k^n measures the relative contribution of the k-th reaction to the n-th eigenvalue λ_n, and therefore, τ_n.

3 Results and Discussions

Figure 1 shows the normalized ignition delay of the MeOH-air mixture by the ignition delay of the pure MeOH (DME0) at P(0) = 30 atm, T(0) = 800 K, and ϕ = 0.6 with the addition of DME by volume. It was found that as DME was added to the MeOH-air mixture, the ignition delay time was significantly reduced, and 40% of DME addition (DME40) resulted in about 95% shortened ignition delay compared to the pure-MeOH-air mixture. A noticeable feature of DME0's ignition was that it exhibited a single stage ignition only, while the DME100 showed a two-stage ignition. This feature affected the ignition of MeOH-air mixture so that from DME30, it started having a two-stage ignition.

For the practical use of MeOH-DME-air mixture in a six cylindered HCCI engine operating at 1200 rpm, the required time to ensure combustion during a particular crank angle degree (CAD) can be approximated by the following formula:

$$t = \frac{CAD}{6 \times 1200} \tag{7}$$

and with CAD 20, it requires about 3.3 ms, which corresponds to DME50. Thus, in the following, we present how DME related chemistry, influences the MeOH ignition by investigating the ignition of DME0, DME30, and DME50. For the analysis, tools derived from the computational singular perturbation are employed, so that the important reactions that play a crucial role in the MeOH-DME-air ignition can be identified.

3.1 Dynamics of MeOH-DME-Air Ignition

The analysis of all modes revealed that throughout the whole process in each case, the dominant mode of the system was found to be the fast explosive one. Therefore, the fast explosive timescale was determined to be the characteristic time of each system.

Figure 2 displays the temporal evolution of the explosive timescale (τ_e) and temperature for selected cases: (a) DME0, (b) DME30, and (c) DME50 at P(0) = 30 atm, T(0) = 800 K, and ϕ = 0.6. Points P1 – P4 on each figure indicate the points on which the TPI values were measured, and they are considered to be representative of the changes in the dominant dynamics of the system. Black and blue lines represent the explosive timescales and temperature, respectively. It is shown that by increasing the initial concentration of DME, the initial value of the fast explosive time scale becomes faster and later it disappears and briefly afterward re-appears gradually accelerating. The disappearance and reappearance of the fast explosive timescale as a set with a slow one, suggest the existence of a two-stage ignition, which becomes evident

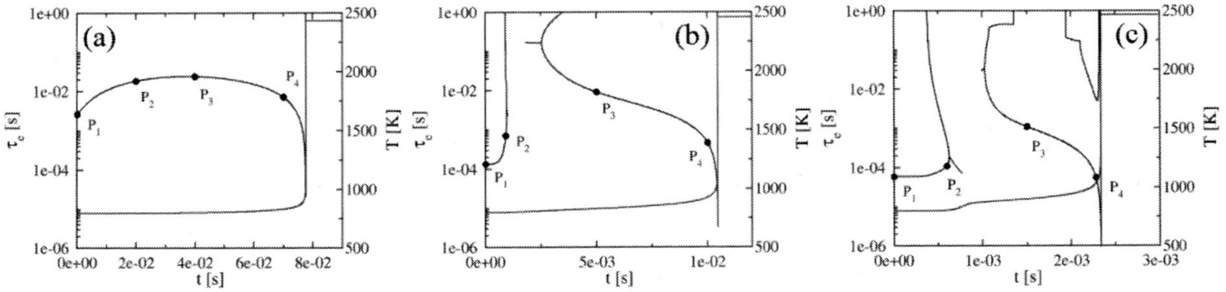

Figure 2: Temporal evolution of the explosive timescales and temperature of (a) pure-MeOH, (b) 0.7MeOH+0.3DME, and (c) 0.5MeOH+0.5DME at P(0) = 30 atm, T(0) = 800 K, and ϕ = 0.6.

No.	Reaction	No.	Reaction
16/17	HO2+HO2<=>H2O2+O2	432	CH3OCH3+OH<=>CH3OCH2+H2O
19	H2O2(+M)<=>OH+OH(+M)	435	CH3OCH3+HO2<=>CH3OCH2+H2O2
72	CH2O+OH<=>HCO+H2O	438	CH3OCH3+O2<=>CH3OCH2+HO2
76	CH2O+HO2<=>HCO+H2O2	443	CH3OCH2+CH3O<=>CH3OCH3+CH2O
120	CH3OH+OH<=>CH2OH+H2O	455	CH3OCH2O2<=>CH2OCH2O2H
122	CH3OH+O2<=>CH2OH+HO2	456	CH2OCH2O2H=>OH+CH2O+CH2O
123	CH3OH+HO2<=>CH2OH+H2O2	457	CH2OCH2O2H+O2<=>O2CH2OCH2O2H
149	CH3+O2(+M)<=>CH3O2(+M)		

Table 1: Reactions providing significant contribution to the generation of the fast explosive timescales of MeOH-DME-air auto-ignition at P = 30 atm, T = 800 K, and ϕ = 0.6.

especially when the initial concentration of DME is larger than 50%. The faster the fast explosive timescale in both stages, the more visible the two-stage ignition is.

In particular, in the DME0 case, an explosive timescale ($\tau_{e,f}$) exists from the very beginning of the process. Initially, it gradually increases and after roughly P3, it gradually decreases. After P4 this explosive timescale rapidly accelerates and afterward rapidly decelerates before it meets another (slower) explosive timescale and both disappear. The two explosive timescales disappear when the temperature is about to reach its maximum gradient, i.e., the ignition delay time of the process.

In the DME30 case, a drastically shortened ignition delay time could be seen. As it will be shown next, the decreased value of the ignition delay time is directly related to the profile of the fast explosive timescale. In particular, a fast explosive timescale exists from the beginning of the process, but it has a mush lower value compared to the DME0 case. Also, unlike to the DME0 case, the fast explosive timescale, after a very brief period of time that it remains roughly constant, it rapidly decelerates and meets a slow explosive timescale before they both disappear. The disappearance of both explosive timescales coincides with a temperature increase that it is small enough so that the process cannot be evidently identified as a two-stage ignition process. Yet, as it will be shown next, the increase of the initial concentration of DME can magnify this phenomenon, so that the two-stage process becomes clearly evident.

Indeed, as Fig. 2 shows, when the initial concentration of DME becomes 50%, the temperature profile exhibits clearly a two-stage ignition process. The profiles of the fast explosive timescales look very much alike to those observed in the DME30 case. In particular, each stage is characterized by a distinct set of explosive timescales. However, the fast explosive timescale is always faster when compared to the respective timescale of the DME30 case. This explains the shorter ignition delay time that the DME50 case attains. In the following chapter, we present how DME chemistry affects the MeOH-air ignition using the explosive timescale and corresponding CSP diagnostics.

3.2 CSP Diagnostics

For a better understanding of the reactions that dominate the

	P1		P2		P3		P4	
time [s]	0.000E-03		20.000E-03		40.000E-03		70.000E-03	
τ [s]	2.653E-03		18.467E-03		24.185E-03		7.034E-03	
T [K]	800.0		800.8		804.2		833.1	
TPI	123f	0.5036	19f	0.3289	19f	0.3358	19f	0.4077
	19f	0.4946	123f	0.3225	123f	0.2839	123f	0.1648
			16f/17f	-0.3182	16f/17f	-0.2837		

Table 2: Timescale participation indices (TPI) of MeOH-air mixture on the points shown in Fig. 2(a).

	P1		P2		P3		P4	
time [s]	0.000E-03		0.930E-03		5.000E-03		10.000E-03	
τ [s]	0.134E-03		0.712E-03		9.060E-03		46.800E-05	
T [K]	800		800.38		828.3		923.1	
TPI	432f	0.2114	432f	0.2308	19f	0.1658	19f	0.28
	120f	-0.2009	120f	-0.2179	456f	-0.1301	432f	0.1083
	435f	0.1548	456f	-0.1094	432f	0.1061	16f/17f	-0.07127
	457f	0.1227	435f	0.1077	120f	-0.0893	120f	-0.06237
	456f	-0.117	457f	0.1073	435f	0.08454		
	455f	0.07247	16f/17f	-0.0944	16f/17f	-0.0705		

Table 3: Timescale participation indices (TPI) of 30% DME-added MeOH-air mixture on the points shown in Fig. 2(b).

dynamics of the systems under investigation, the reactions with the largest TPI values are summarized in Table 1. In the following, the "*f* " and "*b*" attached to the reaction number stand for forward and backward, respectively.

Table 2 includes the CSP diagnostics for the DME0 case. It is shown that initially reaction 123f is the largest participating reaction to the explosive eigenvalue of the MeOH-air mixture, followed by 19f, both of them favoring the explosive character of the explosive timescale, thus favoring ignition. HO2 produced by the chain initiation of MeOH reacts with the reactant CH3OH, forming CH2OH and H2O2 through 123f. Then, H2O2 third-body reaction forms two highly reactive OH radicals through 19f, largely contributing to the fast explosive timescale. The chain branching reaction by 19f is observed from the beginning of the ignition process, and it becomes the largest contributor after the temperature starts to rise significantly. At P2 and P3, reactions 16f/17f are competing with reaction 123f, both consuming HO2 radicals, thus, favoring the dissipative character of the explosive timescale, therefore, opposing ignition. The increased contribution of reactions 16f/17f and the decreased contributions of 123f and 19f, explain the deceleration of the explosive timescale at this part of the process. At the end of the process, near point P4, when temperature has increased sufficiently, the contribution of reaction 19f (which has a high activation energy) becomes the dominant one, largely promoting ignition.

Considering that DME is in general more ignitable than MeOH, the DME-related reaction pathways, when compete with the MeOH-related ones, were found to favor ignition and have positive TPI values as opposed to those of MeOH which were found to have negative TPI values. This is indicated in the results included in Table 3 for the DME30 case. In particular, at P1, the largest contributors to the fast explosive timescale are reactions 432f, 435f and 120f by 21.14%, 15.48% and -20.09%, respectively. Reactions 432f and 435f include the decomposition of DME with OH and HO2, respectively, while reaction 120f competes with 432f since it includes the decomposition of MeOH through OH, therefore explaining its negative TPI value. Reaction 455f, which was found to be a major contributor to the generation of the fast explosive timescale favoring its explosive character, is a very important

	P1		P2		P3		P4	
time [s]	0.000E-03		0.600E-03		1.500E-03		2.280E-03	
τ [s]	0.059E-03		0.111E-03		1.100E-03		0.056E-03	
T [K]	800		803.8823272		899.0		1032.3	
TPI	432f	0.1806	432f	0.2173	19f	0.1953	19f	0.3089
	120f	-0.1716	120f	-0.1948	432f	0.1077	432f	0.06878
	455f	0.16	456f	-0.123	456f	-0.0904	18f	0.06175
	457f	0.1233	457f	0.1168	16f/17f	-0.0629	76f	0.05018
	456f	-0.1108	455f	0.0902	435f	0.06114		
	435f	0.08001			120f	-0.0598		
					72f	-0.052		

Table 4: Timescale participation indices (TPI) of 50% DME-added MeOH-air mixture on the points shown in Fig. 2(c).

reaction in the low temperature oxidation of DME since it isomerizes the CH3OCH2O2 radical to form CH2OCH2O2H. The last one can either react with molecular oxygen through reaction 457f and produce O2CH2OCH2O2H or decompose into two molecules of CH2O and one OH radical through reaction 456f. Both 456f and 457f were found to have considerable contributions to the explosive timescale. However, since they compete, they have opposite signs of TPI values. Reaction 457f leads to a more reactive pathway than 456f, therefore the former one favors and the latter one opposes ignition, as this is indicated by their negative TPI values. The deceleration of the fast explosive timescale at the end of the first stage is mainly due to the increasing influence of reaction 16f/17f and the decreasing effect of 455f. Reactions 16f/17f were found to produce the deceleration of the explosive timescale in the DME0 case, while the decreasing effect of 455f has already been reported to be associated with the two-stage ignition in the low temperature oxidation of DME [7]. The reappearance of the fast explosive timescale, gradually accelerating, is accompanied by the large contribution of reaction 19f, which produces OH radicals, thus promoting ignition. This is the largest difference compared to the first stage. The positive contributions of reactions 432f and 435f, drop significantly while the competition between 456f and 457f does not exist anymore, since the contribution of 456f is the second most significant at point P3, while the contribution of 457f becomes trivial. Reaction 120f continues to compete with 432f, but its negative contribution is significantly decreased. Finally, the contribution of 16f/17f slightly decreases and at point P3 reaches -7.05%. The further acceleration of the explosive timescale at the last part of the process is mainly due to the effect of reaction 18f/19f, as this is indicated by the results displayed at the point P4, where it can be seen that the contribution of 18f/19f reaches 31.37% and becomes the dominant contributor to the explosive timescale. Additional contribution is provided by reaction 432f, favoring the explosive character of the fast explosive timescale. The major opposition is provided by reaction 16f/17f and 120f, with their contributions being slightly decreased compared to the previous point. At this part of the process, the contributions of 456f and 435f becomes insignificant, the first one because it relies on the formation of the CH2OCH2O2H intermediate which is inherently unstable at high temperatures and the second one because the decomposition of DME is favored through reaction 435f that involves OH, a radical in more abundance and more reactive than HO2, at this part of the process.

Table 4 includes the CSP diagnostics for the DME50 case. Comparing the results with the respective results of case DME30, it can be seen that at point P1, the largest difference relates to the increased influence of reaction 455f, which contributes 16.00% to the fast explosive timescale. This explains the faster value of the explosive timescale at the initiation of the process. The larger contribution of 455f is consistent with previously reported study on the low temperature ignition of DME, where it was found the contribution of this reaction was dominant at the initial part of the process. The competition between 432f and 120f, as well as between 456f and 457 still hold with roughly the same values compared to the DME30 case, but the contribution of 435f decreases significantly to 8.01% compared to the value of 15.38% that it obtained at P1 in the DME30 case. The CSP diagnostics between DME30 and DME50 at point 2 indicate the dominant pathways remain the same with the exception of reaction 455f, which provides larger contribution to the explosive timescale in the DME50 case. Also, the effect of 16f/17f is insignificant in the DME50 case, as opposed to the DME30 case, where it was found to be -9.44%. This explains why the DME50 reaches first-stage ignition earlier than the DME30 case does. The resemblance of the CSP diagnostics between DME30 and DME50 applies in the second stage as well. In particular, the investigation of P3 and P4 of DME50 reveals that the effect of 18f/19f increases, the contributions of 456f, 120f and 16f/17f (all of them opposing the explosive character of the fast explosive timescale) decrease. Apparently, the effect of hydrogen-related chemistry becomes more pronounced in the second ignition stage, as the initial concentration of DME increases, which is consistent with previous research where it was found that the second stage in the low temperature oxidation of DME is dominated largely by reaction 18f/19f and at the end of the process the dominant reactions are exclusively hydrogen and CO-to-CO2 related [7].

4 Conclusions

In the present research, we investigated the effect of dimethyl ether addition on the ignition delay time of the methanol-air mixture at high pressure and low temperature conditions. It was found that as dimethyl ether was added to the methanol-air mixture, the ignition delay time was substantially decreased. The significant decrease in the ignition delay time of the mixture was found to relate to the emergence of a two-stage ignition. Since DME is more ignitable than MeOH, the DME pathways that are activated by the DME addition are found in general to favor the explosive character of the fast explosive timescale and compete with the respective pathways of MeOH. During the first and second stages, reactions 455f and 18f/19f are found to play key roles, favoring the explosive character of the explosive timescale, therefore promoting ignition, with their effect increased as the initial concentration of DME increased.

5 Acknowledgment

Research reported in this publication was supported by the King Abdullah University of Science and Technology (KAUST)

References

[1] M. Yao, Z. Chen, Z. Zheng, B. Zhang, and Y. Xing, Fuel 85 (14-15) (2006) 2046–2056.

[2] Z. Zheng, M. Yao, Z. Chen, and B. Zhang, SAE International, 400 Commonwealth Drive, Warrendale, PA, United States, 2004, pp. 2004–01–2993.

[3] T. Shudo, Int. J. Veh. Des., 41 (1-4) (2006) 206–226.

[4] W.K. Metcalfe, S.M. Burke, S.S. Ahmed, H.J. Curran, Int. J. Chem. Kinet. 45 (10) (2013) 638–675.

[5] S.H. Lam and D.A. Goussis, Proc. Combust. Inst. 22 (1988) 931–941.

[6] S.H. Lam and D.A. Goussis, Int. J. Chem. Kinetics 26 (4) (1994) 461–486.

[7] E.-A. Tingas, D.C. Kyritsis, D.A. Goussis, Combust. Theory Model. 21 (1) (2016) 1–28.

11th Asia-Pacific Conference on Combustion,
The University of Sydney, NSW Australia
10th-14th December 2017.

Direct Numerical Simulation of the Transition from a Laminar Cool n-Heptane/Air Ignition Front to a Distributed Premixed Turbulent Cool Flame

B. Savard[1,*], **H. Wang**[1], **E. R. Hawkes**[1,2]

[1]School of Mechanical and Manufacturing Engineering, University of New South Wales

Sydney, NSW, 2052, Australia

[2]School of Photovoltaic and Renewable Energy Engineering, University of New South Wales

Sydney, NSW, 2052, Australia

Abstract

The first direct numerical simulation of a 3D n-heptane/air premixed turbulent cool flame in turbulence intensity typical of the thin reaction zones regime is presented. Chemistry is described by a 129-species, 1234-reaction mechanism (reduced from CaltechMech). Under the current conditions (1 atm, 650 K, and equivalence ratio of 0.7) the initial laminar cool flame is strongly affected by auto-ignition, which is expected to occur under engine conditions, and has an ignition front (or wave) structure. As the turbulent flame develops, turbulent diffusion becomes sufficiently large to initiate self-propagation of the cool flame. The flame is observed to propagate upstream steadily until it reaches the inlet. The steady-state turbulent flame is found to be highly distorted and have a highly distributed reaction zone. As a result, the chemical source terms and heat release rate conditioned on temperature are strongly reduced (more than 50%) compared to the reference laminar cool flame. However, this strong turbulence does not affect the global chemical pathways, but only their magnitude and locations in temperature space.

1 Introduction

Turbulent premixed flames play a major role in modern internal combustion engines and gas-turbine combustors. These flames often involve large hydrocarbon fuels (*e.g.* n-heptane, n-dodecane, etc.) and typically fall in the thin reaction zones (TRZ) regime [1, 2]. A particularity of these fuels is their low-temperature chemistry (LTC) behavior including a two-stage ignition in the negative temperature coefficient (NTC) region [3]. Under such thermochemical conditions, cool flames can be observed. With their chemical kinetics only involving LTC, these flames are in general associated with low heat release rate and, hence, a small increase in temperature across the flame. However, their products have increased reactivity (in terms of high-temperature chemistry) compared to the reactants. As such, in gasoline engines, cool flames are suspected to largely affect the occurrence of knock [4], while in diesel engines (locally partially premixed), they are argued to strongly affect the stabilization of spray flames [5]. Understanding premixed cool flames in turbulence conditions associated with the TRZ regime can therefore lead to a significant impact on engine design.

Laminar premixed cool flames have been studied widely theoretically [6], numerically [7–10], and experimentally [8, 11]. In particular, unstrained self-propagating (ignition independent) premixed cool flames have been identified numerically in configurations where the flow residence time is sufficiently smaller than the first-stage ignition delay time [7, 8]. However, as temperature increases to values relevant to engine conditions (but remains sufficiently low to be in the NTC region), the first-stage ignition delay time decreases and, as a result, auto-ignition can strongly affect the propagating character of cool flames, leading to an ignition front rather than a self-propagating flame behavior [7, 8].

Partially premixed turbulent cool flames have been investigated in the context of jet and mixing layers under diesel engine relevant conditions [12, 13]. While results were reported mainly on the time evolution of the cool flame in mixture fraction space, it was found that its propagation character (ignition front *vs.* self-propagating flame) can have a strong impact on the time of occurrence and location of second-stage ignition kernels [13]. However, further investigation of purely premixed turbulent cool flame structures is required.

Premixed turbulent flames in the TRZ regime have been studied both experimentally (*e.g.* [14, 15]) and numerically (*e.g.* [16–18]) in the past few years. In this regime, thickened preheat zones with thin reaction zones have been clearly observed (*e.g.* [15, 18]). While the reaction zone remains thin, it was shown to be strongly affected by turbulence, especially with large hydrocarbon fuels [18–22]. The effect of LTC and first-stage ignition of the reactants (without cool flames) on this behavior has been investigated recently in [23] and it was found to be negligible for the conditions considered. Nevertheless, to the best of the authors' knowledge, the effect of turbulence on purely premixed cool flames has never been investigated.

In light of the above discussion, the main objective of this work is to perform the first DNS of a n-heptane/air premixed turbulent cool flame in turbulence intensity typical of the TRZ regime. The thermodynamic conditions are the same as those used in [23] and are chosen such that the first-stage ignition delay time is short (such that auto-ignition can substantially affect the cool flame) and significantly shorter than the second-stage ignition delay time. With this premixed turbulent cool flame DNS, the sub-objectives are to, first, assess the effect of turbulence on the propagation character of the cool flame, and, second, asses the flame structure in both physical and temperature space as well as the effect of turbulence on chemical pathways.

2 Numerical Approach

The numerical approach, briefly reviewed here, follows that of [18–20, 22].

2.1 Flow configuration

Figure 1 presents a schematic diagram of the flow configuration. A statistically-planar, freely-propagating flame was chosen in order to isolate the effects of turbulence on the flame from mean flow shear and curvature effects.

*Corresponding author.
E-mail address: b.savard@unsw.edu.au.

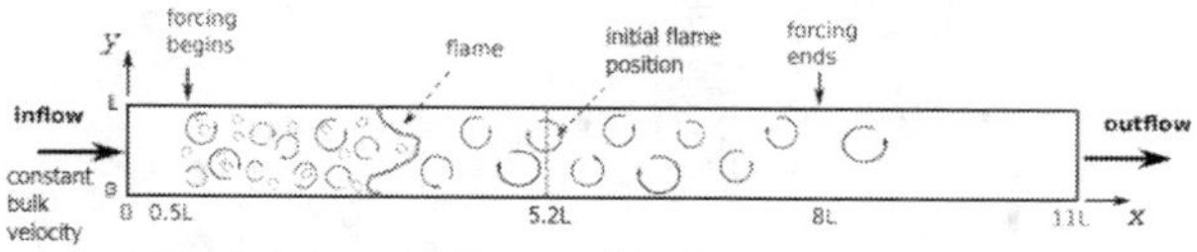

Fig. 1: Schematic diagram of the flow configuration.

Periodic boundary conditions in both the y- and z-directions and inflow/outflow in the x-direction are used. The unburnt gas is injected at constant bulk velocity, which corresponds to that of the reference laminar ignition front, with a low turbulent kinetic energy (TKE), such that there are no negative inlet velocities. Away from the inflow and outflow regions, velocity field forcing maintains a constant TKE across the flame (see Fig. 1).

A planar laminar LTC ignition front (more details in the next section) was initially superimposed to a turbulent flow field and the simulation was run until the flame reached the inlet. As will be described in Section 3.1, the flame reached a statistically steady state after 15 eddy turnover times (15τ). As a result, for statistics in the steady state, the first 15 eddy turnover times were ignored and the data were collected over the next 30 eddy turnover times.

2.2 Governing equations and chemistry model

The low Mach number Navier-Stokes equations are solved numerically using the energy conservative, finite difference code NGA [24]. A semi-implicit Crank-Nicolson time integration is used [25, 26]. Mixture-averaged diffusivities are used in the present simulations (more details on the implementation in [27]).

The chemical kinetics mechanism is CaltechMech [28] from which aromatic species and associated reactions have been removed (this is justified by the lean mixture considered in this work). This reduced chemical model contains 129 species and 1234 elementary reactions (forward and backward rates counted separately).

2.3 Simulation parameters

2.3.1 Reference laminar ignition front

The thermodynamic conditions are defined by a pressure $P = 1$ atm, an unburnt temperature $T_u = 650$K, and an n-heptane/air mixture with an equivalence ratio $\phi = 0.7$. At these conditions (which correspond to the NTC region), there is a clear separation between the first-stage ignition delay time and the second-stage ignition delay time (a factor of 10).

Under these conditions, a stable freely propagating cool flame solution could not be obtained. Since the cool flame characteristic time is of the order of the first-stage ignition delay time, the solution is strongly dependent on the simulation domain length. As such, a stable flame can only be attached to the inlet (with heat loss to the inlet) or lifted. In the latter case, as the inflow velocity increases, the flame loses its propagating character (attributed to heat and species diffusion), and becomes an ignition front, approaching the 0D ignition solution in residence time space. A thorough analysis of this transition is beyond the scope of the present work and the reader is referred to Ref. [29] for a detailed analysis on the topic.

The present laminar cool flame (or ignition front) is chosen such that the ratio of the front thickness to the lift-off length is suitable for the turbulent flame simulation. An inflow velocity of 0.5 m/s was identified as to provide such a suitable flame. The flame/ignition front is simulated in a 1D domain using NGA until a steady state is reached. The domain length is chosen to be sufficiently long so that the reacting front remains in the domain and short enough so that high-temperature ignition does not occur inside this domain (achievable due to the large separation between first- and second-stage ignition delay times). The flame/ignition front solution is presented in residence time space and compared to that of the corresponding 0D ignition in Fig. 2. These profiles indicate that the reaction zone behaves closely like an ignition front. (This statement will be further supported in the next section.) Nevertheless, the products of this ignition front show non-negligible differences with the products

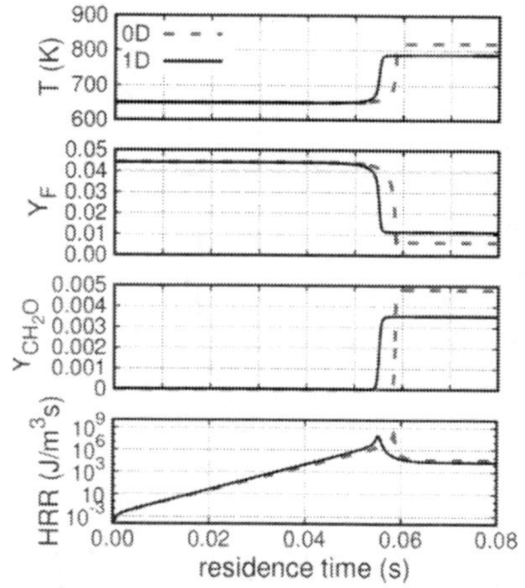

Fig. 2: Reference laminar LTC ignition front structure in residence time space compared to that of the corresponding 0D first-stage ignition.

of the 0D first-stage ignition. This is expected to be due to diffusion, which remains non-negligible in the present ignition front (but not as important as in a self-propagating flame). A detailed quantitative description of this phenomenon would require further analysis, which should be performed in future work.

2.3.2 Turbulent flame

The turbulence quantities, listed in Table 1, are chosen such that 1) the ratio of integral length scale l to the reference laminar LTC ignition front thickness l_{ref} (taken here as the thickness of the fuel consumption layer since it is larger than the thermal thickness based on maximum temperature gradient) is unity (as previously done in [18, 20, 22] for hot flames) and 2) the ratio of RMS velocity fluctuation to the reference laminar ignition front velocity is such that the turbulent flame falls high in the TRZ regime. Note that while the regime diagram proposed by Peters [30] considers only self-propagating flames rather than ignition fronts, the values for the Karlovitz number in the unburnt and at the reaction zone, Ka_u^* and Ka_δ^*, are provided as a reference.

u_{ref} (m/s)	0.5	Ka_u^*	41
l_{ref} (mm)	1.1	$Ka_{\delta,u}^*$	830
u'/u_{ref}	5.6	Ka_δ^*	270
l/l_{ref}	1.0	$Re_{t,u}$	53

Table 1: Simulation parameters. $Re_{t,u} = u'l/\nu_u$, $Ka_u^* = t_{\mathrm{ref}}/t_{\eta_u}$, with the flame time $t_{\mathrm{ref}} = l_{\mathrm{ref}}/u_{\mathrm{ref}}$ and the Kolmogorov characteristic time evaluated as $t_{\eta_u} = (\nu_u l/u'^3)^{1/2}$, u_{ref} is the laminar reference ignition front velocity, u' the rms of the velocity fluctuations, l the integral length scale, and ν_u the kinematic viscosity in the unburnt gas. $Ka_\delta^* = \delta_{\mathrm{ref}}^2/\eta_\delta^2$, where δ_{ref} is the reference reaction zone thickness (here $\delta_{\mathrm{ref}} = l_{\mathrm{ref}}$) and η_δ is the kinematic viscosity evaluated at the reaction zone. $Ka_{\delta,u}^* = \delta_{\mathrm{ref}}^2/\eta_u^2$. In all cases, $L = 5.26l$, and a uniform $128^3 \times 11$ grid is used. The definitions of Ka_u^* and Ka_δ^* are chosen to be consistent with those of Ref. [18, 20, 22, 31].

3 Results and Discussion

3.1 Propagating turbulent flame

Figure 3 presents the time evolution of the flame from its initial state until the end of the simulation. Figure 3(a) shows a volume rendering of the fuel burning rate at four instants corresponding to $t/\tau = 0, 15, 30, 45$. It is clear that the reaction zone, which is initially thin and planar, becomes highly distributed within the first 15 eddy turnover times and moves upstream towards the inlet. The distributedness of the reaction zone will be further discussed in the next section. Figure 3(b) presents the time evolution of the flame position. It can be readily observed that after a short transient period ($t/\tau < 15$) the flame propagates steadily upstream until it reaches the inlet (around $t/\tau = 45$). Figure 3(c) further highlights that the instantaneous propagation speed (based on fuel consumption) fluctuates around a constant value over the period $15 < t/\tau < 45$. Its mean value, normalized by the reference laminar speed is provided in Table 2. These results indicate that the initial reaction front, which is ignition controlled, transitions to a self-propagating flame with a turbulent flame speed independent from the mixture residence time. To the best of the author's knowledge, such a transition for either cool or hot flames has never been reported in the literature. Occurrence of this transition may have significant impact on engine knock in gasoline engines and spray stabilization in diesel engine, as discussed in the Introduction.

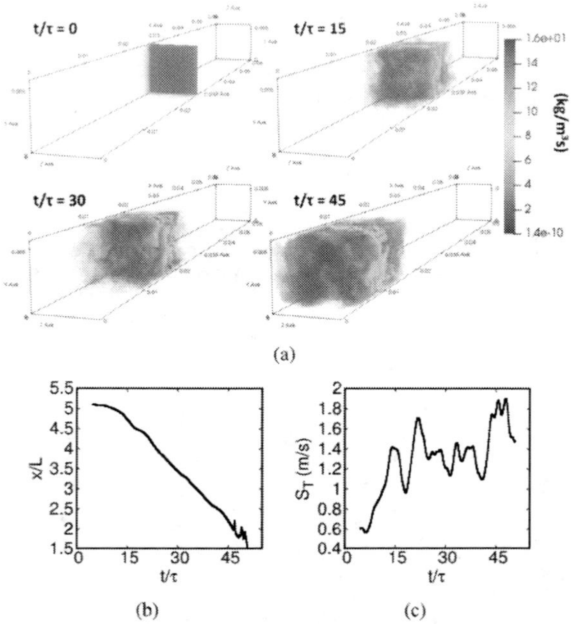

Fig. 3: Time evolution of the turbulent cool flame. (a) Volume rendering of the fuel burning rate at four simulation times. τ is the integral eddy turnover time. (b) Position of the flame front (based on integrated progress variable) *vs.* time. (c) Turbulent flame speed (obtained by integrating the fuel consumption rate over the entire domain) *vs.* time.

$S_T/u_{\rm ref}$	A_T/A	I_0	$\delta_T/\delta_{\rm ref}$
2.7	4.3	0.63	7.2

Table 2: Global statistics of the steady-state turbulent cool flame.

This transition to a self-propagating turbulent flame is expected to be caused by turbulent mixing, which effectively increases the diffusivity of the thickened flame front. As this turbulent diffusion becomes sufficiently large to transport heat and radicals upstream of the reaction zone, the flame can propagate independently of the auto-ignitive character of the reactants. Quantifying this turbulent diffusion in the current flame is however beyond the scope of this study, as the choice of length and time scales to appropriately define turbulent diffusion in premixed flames has been discussed theoretically [30, 32] but not formally identified yet.

3.2 Flame characteristics and chemical pathways

Figure 4 presents contours of formaldehyde mass fraction (a product of the cool flame) and fuel burning rate on a 2D slice. As mentioned in the previous section, it is clear that the flame is strongly distorted and has a highly distributed reaction zone.

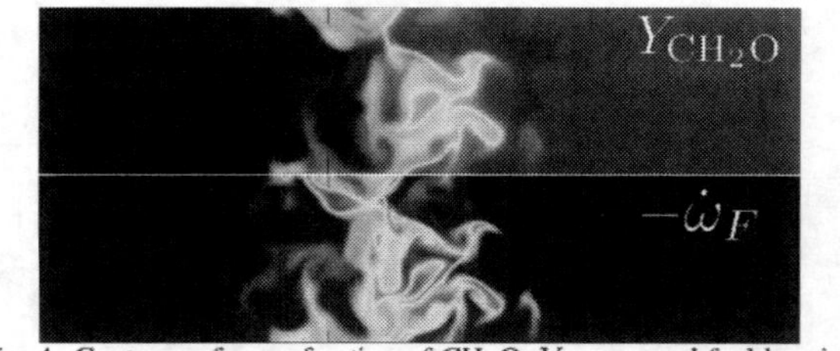

Fig. 4: Contours of mass fraction of CH_2O, Y_{CH_2O}, and fuel burning rate, $\dot{\omega}_F$, on a two-dimensional slice of the domain (only a part of the domain length is shown) for the cool turbulent flame. The range for Y_{CH_2O} are $[0, 3.5 \times 10^{-3}]$. The range for $\dot{\omega}_F$ is [-13.5, 0] kg/m^{3}s.

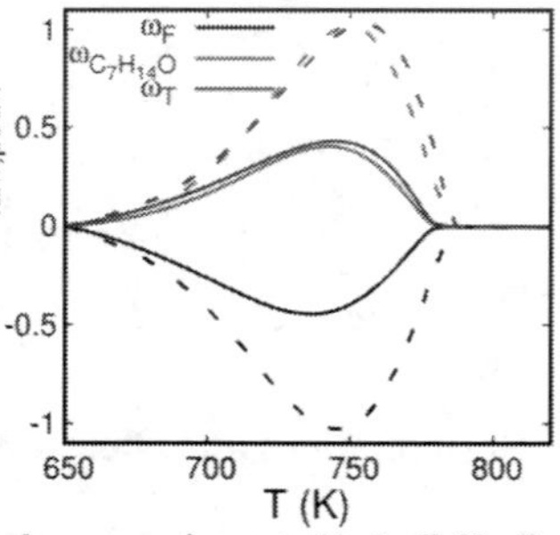

Fig. 5: Mean fuel consumption rate ($\dot{\omega}_F$), $C_7H_{14}O$ production rate ($\dot{\omega}_{C_7H_{14}O}$), and heat release rate ($\dot{\omega}_T$) conditioned on temperature. Solid lines: turbulent. Dashed lines: reference laminar ignition front.

The distributedness of the reaction zone is quantified by computing the ratio of the turbulent reaction zone thickness $\delta_T = \mathrm{Vol}\left(\dot{\omega}_F > 0.05\dot{\omega}_{F,\rm lam,peak}\right)/L^2$ to that of the reference laminar ignition front $\delta_{\rm ref}$. (Note that this definition, used in [20], is only formally valid in Damköhler's small-scale limit [33], which is approached by the current flame.) Such a large ratio, as listed in Table 2, is characteristic of a highly distributed flame, and can be attributed to the present large Ka_δ [20]. The particularity of the present flame is that it is highly distributed for $l/l_{\rm ref}$ and $u'/u_{\rm ref}$ ratios corresponding to the TRZ regime (previously reported distributed flames [20, 34, 35] had significantly larger turbulence intensity, placing them far beyond the TRZ regime in Peters' diagram [30]). This is due to the weak increase in temperature (and hence kinematic viscosity) across the flame [20] and the large reaction zone thickness compared to the thermal thickness. This result suggests that turbulent cool flames are more prone than hot flames to be distributed under engine turbulence conditions.

The distortion of the flame front is quantified by the turbulent surface area (based on the isothermal surface corresponding to the peak burning rate) normalized by the domain cross-section which is provided in Table 2. The large ratio is consistent with previous hot flame results obtained at comparable Ka_δ and $l/l_{\rm ref}$ ratios [20].

The burning efficiency factor computed as $I_0 = S_T/u_{\rm ref}\left(A_T/A\right)^{-1}$ is found to be particularly low (0.63). While it can be partly attributed to differential diffusion [19, 31], it is suspected that the transition from an ignition front to a propagating flame also contributes to this reduction. This suggests a strong reduction in mean burning rate compared to the laminar reference. This is clearly shown in Fig. 5, which presents the mean fuel burning rate, source term of $C_7H_{14}O$ (a product of LTC), and heat release rate conditioned on temperature. The magnitudes are consistently reduced by more than 50%. Another clear observation is the shift of the respective peaks to lower temperatures in the turbulent flame.

Finally, while turbulence significantly affects the magnitude of the chemical source terms, the global chemical pathways remain virtually unaffected, as evidenced in Fig. 6(a). More specifically, this carbon flow, which is integrated over the entire domain and averaged in time, indicates that the fuel reaches

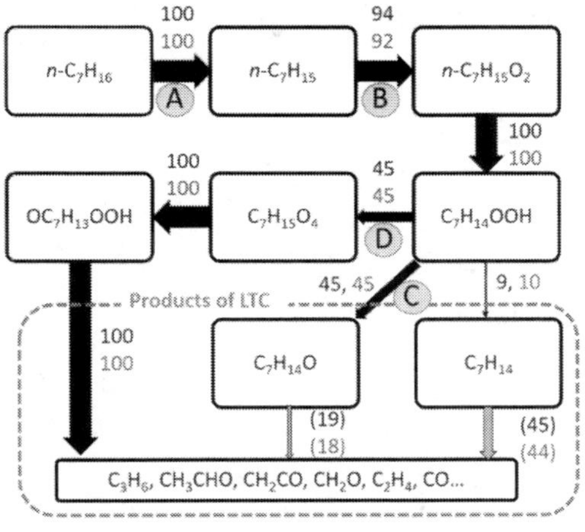

(a)

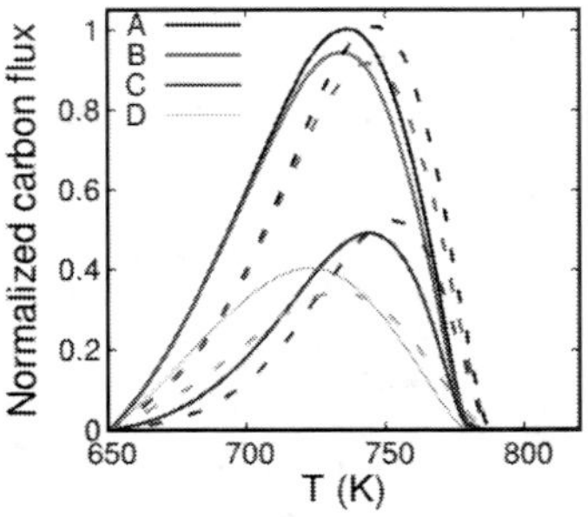

(b)

Fig. 6: Chemical pathways. (a) Carbon flow from the fuel to the products of LTC integrated over the whole domain and averaged in time. Black: turbulent. Red: laminar. Numbers without parenthesis indicate percentage of consumption in the direction indicated by the arrow. Numbers in parenthesis indicate percentage of consumption in the direction of the arrow (only possible direction) over production of the species preceding the arrow. (b) Molar consumption rate corresponding to the labels (A-D) in subfigure (a) conditioned on temperature and normalized by the peak consumption rate of the fuel. Solid lines: turbulent. Dashed lines: laminar.

globally the products of LTC through the same pathways in the same proportions in both the laminar and the turbulent cases. In contrast, Fig. 6(b) presents the mean molar consumption rates of key reactions labeled in Fig. 6(a) normalized by the peak fuel consumption rate and conditioned on temperature. First, consistent with Fig. 5, a clear shift towards lower temperatures is observed in the turbulent flame for all reactions. Second, the relative magnitudes are slightly affected in the turbulent flame. Most noticeable is the increase of pathway D (see Fig. 6(a)) at lower temperatures, pathway which leads to the formation of small LTC products only.

4 Conclusions

The first DNS of a 3D n-heptane/air premixed turbulent cool flame in turbulence intensity typical of the TRZ regime showed that turbulent diffusion can cause a laminar LTC ignition front to transition to a self-propagating front. In addition, even under such moderate turbulence intensity, the premixed cool flame was found to be highly distributed. This was attributed to the small temperature ratio across the cool flame, which fails to re-laminarize the flow, as typically seen with turbulent hot flames. The distribued reaction zone was associated with a significant reduction (more than 50%) in mean chemical source terms and heat release rate (conditioned on temperature) compared to the reference laminar LTC ignition front. However, this strong turbulence-chemistry interaction did not affect the global chemical pathways, but only their magnitude and locations in temperature space.

5 Acknowledgements

This work was supported by the Australian Research Council. BS gratefully acknowledges funding from the Fonds de recherche du Québec - Nature et technologies (Postdoctoral fellowship B3) and the authors are thankful to the Academic Computer Center in Gdańsk, where the simulations were performed.

References

[1] Linse D., Hasse C. and Durst B., Combust. Theory Model., 13 (2009) 167–188.
[2] Mounaïm-Rousselle C., Landry L., Halter F. and Foucher F., Proc. Comb. Inst., 34 (2013) 2941–2949.
[3] Curran H.J. Gaffuri P., Pitz W. and Westbrook C., Combust. Flame, 114 (1998) 149–177.
[4] Sun W., Won S.H., Gou X. and Ju Y., Proc. Comb. Inst., (2015).
[5] Musculus M.P.B., Miles P.C. and Pickett L.M., Progress in Energy and Combustion Science, 39 (2013) 246–283.
[6] Griffiths J.F. and Scott S.K., 13 (1987) 161–197.
[7] Ju Y., Reuter C.B. and Won S.H., Combust. Flame, 162 (2015) 3580–3588.
[8] Zhao P., Liang W., Deng S. and Law C.K., Fuel, 166 (2016) 477–487.
[9] Pan J., Wei H., Shu G., Chen Z. and Zhao P., Combust. Flame, 174 (2016) 179–193.
[10] Ju Y., Combust. Flame, 178 (2017) 61–69.
[11] Reuter C.B., Won S.H. and Ju Y., Combust. Flame, 166 (2016) 125–132.
[12] Krisman A., Hawkes E.R. and Chen J.H., J. Fluid. Mech., 824 (2017) 5–41.
[13] Krisman A., Hawkes E.R., Talei M., Bhagatwala A. and Chen J.H., 36 (2017) 3567–3575.
[14] Osborne J., Ramji S., Carter C. and Steinberg A., Proc. Comb. Inst., 36 (2017) 1835–1841.
[15] Wabel T., Skiba A., Temme J. and Driscoll J., Proc. Comb. Inst., 36 (2017) 1809–1816.
[16] Hawkes E., Chatakonda O., Kolla H., Kerstein A. and Chen J., Combust. Flame, 159 (2012) 2690–2703.
[17] Aspden A., Day M. and Bell J., Proc. Comb. Inst., 35 (2015) 1321–1329.
[18] Savard B., Bobbitt B. and Blanquart G., Proc. Comb. Inst., 35 (2015) 1377–1384.
[19] Savard B. and Blanquart G., Combust. Flame, 162 (2015) 2020–2033.
[20] Lapointe S., Savard B. and Blanquart G., Combust. Flame, 162 (2015) 3341–3355.
[21] Aspden A., Bell J., Day M. and Egolfopoulos F., Proc. Comb. Inst., 36 (2017) 2005–2016.
[22] Savard B., Lapointe S. and Teodorczyk A., Proc. Comb. Inst., 36 (2017) 3543–3549.
[23] Savard B. and Teodorczyk A., Proceedings of 8th European Combustion Meeting, Dubrovnik, Croatia (2017).
[24] Desjardins O., Blanquart G., Balarac G. and Pitsch H., J. Comput. Phys., 227 (2008) 7125–7159.
[25] Pierce C.D., Ph.D. thesis, Stanford University (2001).
[26] Savard B., Xuan Y., Bobbitt B. and Blanquart G., J. Comput. Phys., 295 (2015) 740–769.
[27] Burali N., Lapointe S., Bobbitt B., Blanquart G. and Xuan Y., Combust. Theory Model., 20 (2016) 632–657.
[28] Blanquart G., CaltechMech v2.3. Available at http://theforce.caltech.edu/resources/.
[29] Krisman A., Hawkes E. and Chen J., Combust. Flame., (2017) under review.
[30] Peters N., Turbulent Combustion, Cambridge: Cambridge University Press (2000).
[31] Lapointe S. and Blanquart G., Combust. Flame, 167 (2016) 294–307.
[32] Kerstein A., Physical Review. E,, 64 (2001) 066306.
[33] Damkhöler G., Z. Elektrochem., 46 (1940) 601–652.
[34] Aspden A., Day M. and Bell J., J. Fluid Mech., 680 (2011) 287–320.
[35] Wang H., Hawkes E.R., Zhou B., Chen J.H., Li Z. and Aldén M., Proc. Comb. Inst., 36 (2017) 2045–2053.

11th Asia-Pacific Conference on Combustion,
The University of Sydney, NSW Australia
10th -14th December 2017.

Effect of Diameter on Flame Spread Over Electrical Wires with Applied AC Electric Fields

S. H. Park[1], S. J. Lim[2], J. Park[1], O. Fujita[3], S. H. Chung[4]

[1]Department of Mechanical Engineering, Pukyong National University, Busan, Korea

[2]Advanced Combustion Laboratory, Korea Institute of Energy Research, Daejeon, Korea

[3]Division of Mechanical & Space Engineering, Hokkaido University, Sapporo, Japan

[4]King Abdullah University of Science and Technology (KAUST), Clean Combustion Research Center (CCRC), Thuwal, Saudi Arabia

Abstract

An experimental study on spreading flame over electrical wire, which was insulated by Polyethylene(PE) and had different diameters, was conducted with applied AC electric field. The result showed that the flame spread rate decreased in increase of the diameter of insulator at a fixed electric field. The flame spread rate exhibited increase or decrease tendency with applied AC electric field, having three distinct regimes depending on applied voltage and frequency. In each regime, the flame spread rate was characterized by physical parameters of applied electric fields and wire dimensions, and the behaviors could be explained by a thermal balance mechanism.

1 Introduction

Space technologies have been developed in various areas such as satellite, space hotel, international space station and space food. Electrical wire fire in space can result in extensive damages and fatalities, e.g., the Apollo 1 and the Mir Space Station. In this sense, developing fire safety code is an essential requirement for space development. Flame spread over electrical wire can be influenced by several factors, such as material properties of the inner core and outer insulator, gravity level, ambient flow, and pressure [1-6]. Wire fire can be caused by unexpected electrical short and/or overheating of wire. Thus, it can be actually under influence of electric fields. Even if an electrical shortage disconnects an electrical circuit during the initial and propagation stages of a wire fire, a wire can still be under the influence of an electric field. This type of electrode configuration behaves like an open circuit when voltage is applied. In such a case, the charged particles generated through chemi-ionization and subsequent ion chemistry in the reaction zone of a spreading flame can be influenced by the electric field through the Lorentz force. However, studies on the effect of applied electric fields on flame spread over electrical wire has been limited to two papers [7, 8]. Here, we report the effect of insulator diameter in flame spread over electrical wire with applied electric fields.

2 Experimental facility

Figure 1 shows a schematic of experimental apparatus, which consists of an electrical wire, a wire holder, an AC power supply, and a video camera. Polyethylene-insulated electrical wires of 400 mm in length housed 0.5 mm diameter (D_C) NiCr core and were coated with insulators of 0.8, 1.1 and 1.5 mm in diameter (D_{OUT}). The wire was installed on a wire holder made of nonconductive acetal resin. One end of a wire was fixed to the wire holder and the other end of wire was connected to a spring in order to prevent bending of the wire due to thermal expansion during flame spread. The wire and wire holder were surrounded by acetal mesh to prevent outside disturbances. One end of the electrical wire was connected to high voltage terminal of the power supply and the other terminal was linked to building ground, resulting in one electrode configuration. The voltage (V_{AC}) and frequency (f_{AC}) applied to the wire were in range of 1 ~ 5 kV and 10 ~ 1000 Hz. The flame was ignited by an igniter which was placed on an air cylinder. In order to minimize the interaction between the ignition system and applied AC electric fields, igniter was removed away from the wire after ignition. The wire length was 247 mm, excluding the portion connected to the wire holder. Initial 70 mm from the one end of wire and final 10 mm were excluded due to ignition transition and interaction of the wire holder, respectively, leaving 167 mm of wire length available for flame spread. A video camera was triggered to capture spreading flame. Images were analyzed using a Matlab-based code.

3 Results and Discussion

3.1 Overall feature of spreading flame

Figure 2 shows representative flame images for several diameters with various applied f_{AC} at V_{AC}=4 kV. For f_{AC} = 10 Hz case, the size of spreading flame increased in increase of wire diameter due to growth of amount of insulator. In case of f_{AC} = 30 Hz, the spreading flame were tilted to bare wire compared with f_{AC} = 10 Hz. Also, the sizes of flames became smaller compared with the f_{AC} = 10 Hz. At f_{AC}=200 Hz, the size of flame became bigger compared with f_{AC} = 30 Hz and it was decreased again with an increased applied f_{AC} up to 1000 Hz. At f_{AC}= 600 (1000) Hz, the spreading flame was extinguished at D_{OUT} = 0.8 (1.1) mm. The results show that spreading flame does not extinguish at all frequencies and voltages in range of the current electric fields for D_{OUT} = 1.5 mm but extinguish at high frequencies and voltages for D_{OUT} = 0.8 and 1.1 mm. Figure 3 shows temporal positions of flame front X with time at several diameters for V_{AC} = 4kV and f_{AC} = 10 Hz (a) and at

Corresponding author. Fax: +82-51-629-6126
E-mail address: jeongpark@pknu.ac.kr

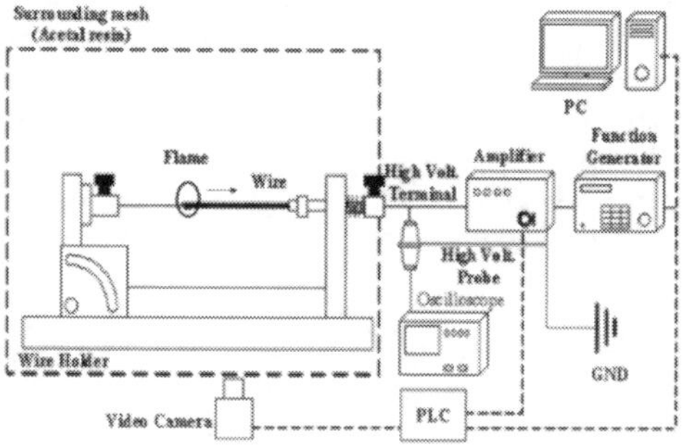

Figure 1: Schematic illustration of experimental setup

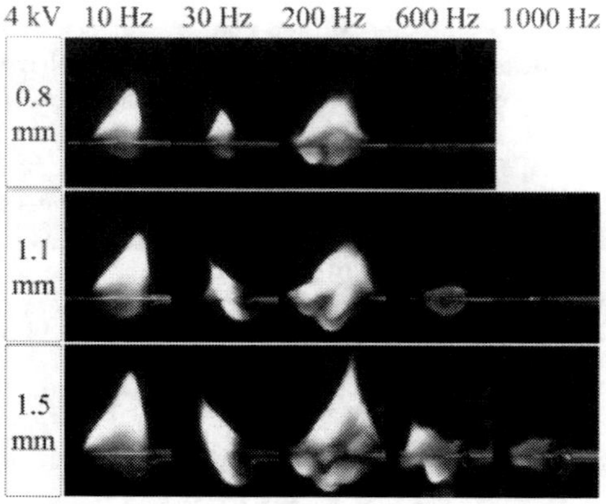

Figure 2: Instantaneous flame images with applied frequencies for several wire diameters at V_{AC}=4kV

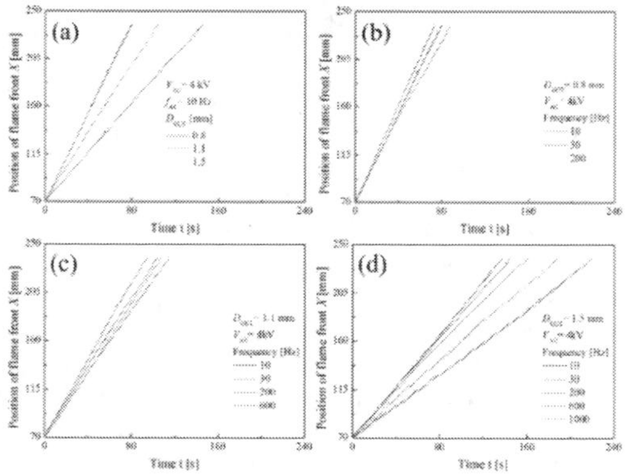

Figure 3: Temporal position of flame front for (a) several diameters of wire at V_{AC}=4 kV, f_{AC}=10 Hz and various frequencies for (b) 0.8, (c) 1.1, (d) 1.5 mm in diameter

various frequencies for (b) D_{OUT} = 0.8, (c) 1.1, (d) 1.5 mm. The position of flame front was identified as the flame front in contact with the electrical wire. In figure 3(a), the flame position varies almost linearly during flame spread. Thus, flame spread rate (FSR), $S_W = \Delta X/\Delta t$, was defined reasonably as the overall rate of change in the position of flame front with time. The FSR increased in increase of wire diameter. The results show that flame front spreads sensitively to applied frequency and voltage in addition to wire diameter. Thus, further detailed analysis can be required, and presented in later section.

3.2 Flame spread rate

Figure 4 shows variations in FSR against f_{AC} at several voltages for (a) D_{OUT} = 0.8, (b) 1.1, and (c) 1.5 mm. The FSR, for the baseline case ($S_{W,0}$) with no applied electric field, was

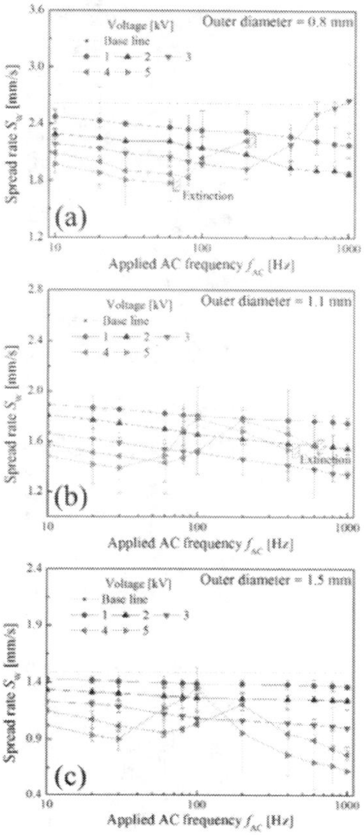

Figure 4: Flame spread rate against applied frequency at several voltages for (a) D_{OUT}= 0.8, (b) 1.1, and (c) 1.5 mm

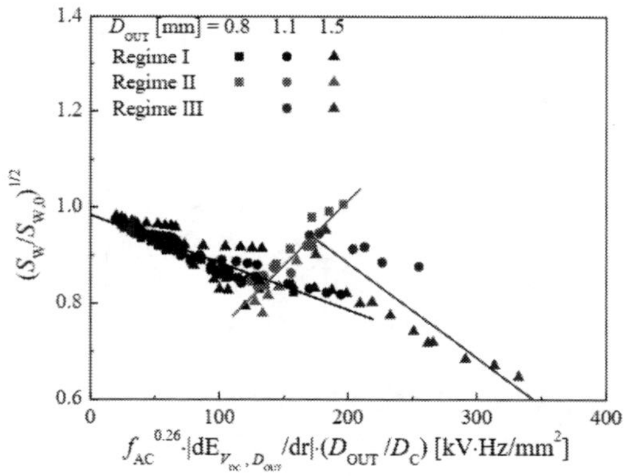

Figure 5: Characterization in flame spread rate for each regime

indicated with a black dot line. For D_{OUT} = 0.8 mm in Fig. 4(a), the FSR decreased in increase of f_{AC} at V_{AC} = 1 and 2 as well as at $f_{AC} \leq$ 200 Hz for V_{AC} = 3 kV and at $f_{AC} \leq$ 80 Hz for V_{AC} = 4 kV while the flame extinguished at f_{AC}> 60 Hz for V_{AC} = 5 kV. While, it increased again at $f_{AC} \geq$ 200 Hz for V_{AC} = 3 kV, at $f_{AC} \geq$80 Hz for V_{AC}=4 kV. Note that all the FSRs with applied electric fields were smaller than that for the baseline case, except at V_{AC}=3 kV and f_{AC}=1000 Hz. For D_{OUT} = 1.1 and 1.5 mm in Figs. 4(b) and (c), the FSRs with applied electric fields decreased monotonously in increase of f_{AC} up to V_{AC} = 3 kV. However, when V_{AC} increased further for D_{OUT} = 1.1 and 1.5 mm, those decreased at $f_{AC} \leq$ 60 (30) Hz and 4 (5) kV, increased at 60 (30) $\leq f_{AC} \leq$ 200 (100) Hz for V_{AC} = 4 (5) kV,

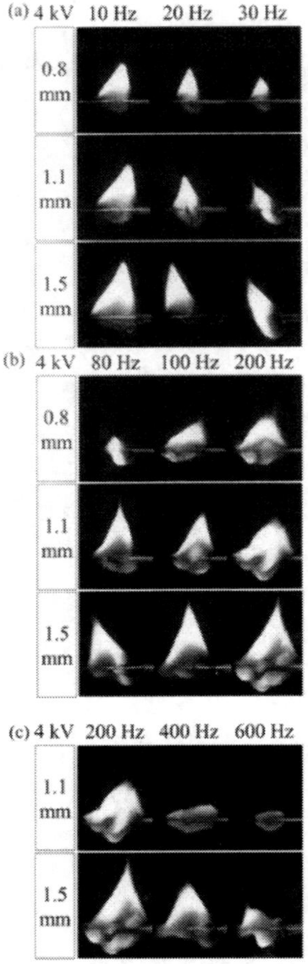

Figure 6: Representative flame images at (a) Regime I, (b) Regime II and (c) Regime III with applied electric fields and wire diameter

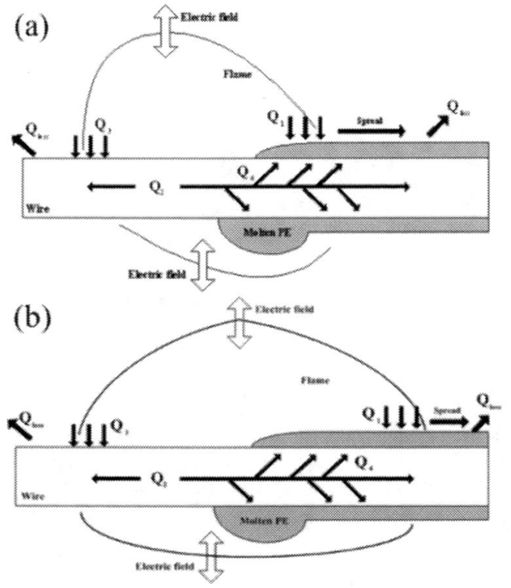

Figure 7: Schematic illustration of thermal balance mechanism in spreading flame

and then decreased again at $f_{AC} \geq 200$ (100) Hz for $V_{AC} = 4$ (5) kV, respectively. For D_{OUT}=1.1 (0.8) mm, the flame extinguished at $f_{AC} > 600$ (200) Hz for $V_{AC} = 4$ kV and $f_{AC} >$ 400 (60) Hz for V_{AC}=5 kV. The flames did not extinguish in range of the current electric fields for D_{OUT} = 1.5 mm. Extinction limit was extended to higher frequency when the wire diameter increased. Considering the behaviors of the FSR with applied electric fields, it could be grouped into three distinct regimes depending on wire diameter as well as V_{AC} and f_{AC}: the FSR with V_{AC} and f_{AC} decreased monotonously in regime I, increased in regime II and then decreases again in regime III.

3.3 Characterization of flame spread rate

As mentioned above, the FSRs were influenced by wire diameter, applied V_{AC} and f_{AC}. Considering the relationship between the wire diameter and effect of electric field, variation in electric field intensity inside PE is important. The applied V_{AC} has a relationship with the rate of reduction of intensity electric field in PE depending on the wire diameter. Slope of intensity electric field, $dE_{V_{DC}D_{OUT}}/dr$, was defined as the overall rate of change in the intensity electric field inside PE with radial direction. In characterization, V_{AC} could be replaced with an absolute values of $\left|dE_{V_{DC}D_{OUT}}/dr\right|$. Intensities of electric field were calculated by using FEMM-4.2v for 0.8, 1.1 and 1.5 mm in diameter [9]. The FSRs could be characterized well by applied f_{AC} and absolute values of slope of intensity electric field as well as wire dimensionless in each regime.

$$C^* = f_{AC}^{0.26} \times \left|\frac{dE_{V_{DC}D_{OUT}}}{dr}\right| \times \left(\frac{D_{OUT}}{D_C}\right) \left[\frac{\mathrm{kV\times Hz}}{mm^2}\right] \quad (1)$$

$$S^* = \left(\frac{S_W}{S_{W,0}}\right)^{\frac{1}{2}} \quad (2)$$

Here, the absolute values of slope of intensity electric field inside PE was calculated by using FEMM-4.2v for 0.8, 1.1, and 1.5 mm in diameter [9]. Figure 5 shows that the behaviors of FSR are well characterized by C* and S*, i.e., S* = -0.00033C* +0.98 (Correlation coefficient R=0.89) in regime I, S* = -0.00089C*+0.48 (R=0.90) for regime II, and S* = -0.00066 C*+1.28 (R=0.90) for regime III, respectively.

3.4 Thermal balance mechanism

Figure 6 exhibits the representative flame images with applied electric fields in regime I (a), regime II (b) and regime III (c). In regime I, the size of spreading flames became smaller in increase of V_{AC} and the flames were tilted to the bare wire in increase of V_{AC} and several diameters at a fixed f_{AC} as shown in Fig. 6(a). In regime II, the sizes of spreading flames were larger in increase of f_{AC} at a fixed V_{AC} as shown in Fig. 6(b). While, in regime III, the sizes were smaller again compared to those in regime II as shown in Fig. 6(c). Figure 7 illustrates the schematic representations of thermal balance mechanism for (a) regime I, III and (b) regime II. For the cases of regime I and III, the spreading flame The heat transfer from the flame to unburned PE (Q1) is reduced in the situation of the leaning of flame to the bare wire. Moreover, as mentioned before, the flame size decreases appreciably with increasing V_{AC} and f_{AC}, this decreasing the heat transfer from the flame to the burnt wire (Q3). Subsequently, the decrease in Q3 causes reduction in the heat transfer to the molten PE through the wire (Q2) and thereby to unburned wire (Q4). Therefore, there will be reduction in the total heat transfer from the flame to PE, leading to the decrease of production and evaporation rate of molten PE. Consequently, the FSR becomes slower [7]. In previous studies, convective heat transfer was calculated [10], and Temperature profile for the wire in currents was calculated [11]. In the case of regime II case, The FSRs increases due to increasing increase in total heat transfer from the flame to PE with Q1 and Q3.

4 Conclusions

An experimental study on flame spread over electrical wires with various insulator diameters under applied AC electric field was investigated in range of V_{AC}=1 ~ 5 kV and f_{AC}=10 ~ 1000 Hz. The results show that extinction limit (FSR) was extended (increased) in increase of wire diameter. The spreading flame was influenced significantly by wire diameter, applied V_{AC} and f_{AC} via changing the shape and size of flame. The FSRs having distinct three regimes, could be characterized by V_{AC}, f_{AC} and wire dimensions as well as absolute values of slope of intensity electric field inside PE. In each regime, the changing FSR could be explained by thermal balance mechanism.

5 Acknowledgment

This research was supported by SGER Program through the National Research Foundation of Korea (NRF) funded by the Ministry of Education.

References

[1] Fujita O, Nishizawa K, Ito K. (2002). Effect of low External Flow on Flame Spread over Polyethylene-insulated Wire in Microgravity. Proc. Combust. Inst. 29:2545
[2] Nakamura Y, Yoshimura N, Ito H, Azumaya K, Fujita O. (2009). Flame Spread over Electric Wire in Sub-atmospheric Pressure. Proc. Combust. Inst. 32:2559
[3] Umemura A, Uchida M, Hirata T, Sato. (2002). Physical Model Analysis of Flame Spreading along an Electrical Wire in Microgravity. Proc. Combust. Inst. 29:2535
[4] Onishi Y, Fujita O, Agata K, Takeuchi H, Nakamura Y, Ito H, Kikuchi M. (2010) Observation of Flame Spreading over Electric Wire under Reduced Gravity Condition Given by Parabolic Flight and Drop Tower Experiments. Transactions of the Japan society for Aeronautical and Space Sciences, Aerospace Technology Japan. Vol.8:19
[5] Nakamura Y, Yoshimura N, Matsumura T, Ito H, Fujita O. (2008). Opposed-wind Effect on Flame Spread of Electric Wire in Sub-atmospheric Pressure. Journal of Thermal Science and Technology, Vol.3:430
[6] Kikuchi M, Fujita O, Ito K, Sato A, Sakuraya T. (1998). Experimental study on flame spread over wire insulation in microgravity. Proc. 27th In. Comb. Symp.:2507
[7] Kim MK, Chung SH, Fujita O. (2011). Effect of AC electric fields on flame spread over electrical wire. Proc. Combust. Inst. 33:1145
[8] Lim SJ, Kim MK, Park J, Fujita O, Chung SH. (2104). Flame spread over electrical wire with AC electric fields: Internal circulation, fuel vapor-jet, spread rate acceleration, and molten insulator dripping. Combustion and Flame.
[9] Finite Element Method Magnetics v-4.2.
[10] S.W. Churchill, H.H.s. Chu. (1975) Correlating equations for laminar and turbulent free convection from a horizontal cylinder, Heat Mass Transfer 18(9), 049-1053.
[11] H. He, Q. Zhang, R. Tu, L. Zhao, J. Liu, Y. Zhang (2016) Molten thermoplastic dripping behavior induced by flame spread over wire insulation under overload currents. Journal of Hazardous Material 320, 628-634.

11th Asia-Pacific Conference on Combustion,
The University of Sydney, NSW Australia
10th -14th December 2017.

Numerical study on non-uniform detonation initiation in $H_2/O_2/N_2$ mixture

Yuan Wang [1], Wang Han [1], Ralf Deiterding [2], Zheng Chen [1]

[1]SKLTCS, Department of Mechanics and Engineering Science, College of Engineering, Peking University, Beijing 100871, China

[2]Aerodynamics and Flight Mechanics Research Group, University of Southampton, Highfield Campus, Southampton SO171BJ, United Kingdom

Abstract

In this study, two-dimensional numerical simulations considering detailed chemistry and transport are conducted to investigate the non-uniform detonation initiation. The high-pressure and high-temperature detonation initial regime is imposed by a sinusoidal disturbance. It is shown that the disturbance promotes the onset of detonation. For disturbance with smaller wave length, the distance of detonation initiation becomes shorter due to stronger shock wave interaction. Moreover, the influence of wall boundary layer on detonation initiation is examined. At the boundary layer, the kinetic energy transforms into the thermal energy. Hence, autoignition happens near the wall, which is accompanied by pressure wave generation. The pressure wave collision induces local high temperature and local explosion, which eventually results in detonation. Therefore, both disturbance and wall boundary layer promote detonation initiation.

1 Introduction

Detonation initiation is a fundamental combustion problem, which is related to the development of high-performance detonation engines and the control of accidental explosion. The disturbance, instability development and boundary layer were considered as the key mechanism of detonation initiation [1]. Therefore, many studies investigated the influence of these factors on detonation initiation.

Chue et al. [2] used one-dimensional numerical simulation to examine the influence of periodic perturbation in density on detonation development. They found that the frequency of applied perturbation plays an important role in direct detonation initiation. Mazaheri [3] and Ng and Lee [4] found that local density disturbances can promote the onset of detonation due to the amplification of local disturbance or instability. Radulescu et al. [5] examined the influence of cellular instability on the direct initiation of 2D weakly unstable detonations. Their results indicate that multi-dimensional perturbations and cellular instabilities inhibit the initiation of inviscid detonations. Ng et al. [1] noticed that a large wave length of perturbation was used by Radulescu et al. [5] and thereby they examined the effects of perturbation wavelength and channel size on direct initiation of 2D weakly unstable detonations. They found that high frequency, small amplitude perturbations can generate fine scale instabilities which can accelerate the heat-release rate and thereby promote the onset of detonation. Li et al. [6] examined the effects of spatial heterogeneity on near-limit propagation of a pressure-dependent detonation. They found that the detonation wave speed depends on the wave length of heterogeneity. Recently, Qi and Chen [7] assessed the effects of local temperature perturbation on 1D direct detonation initiation. They found that at small amplitude of temperature perturbation, i.e., a cold/hot spot, can promote/inhibit direct detonation initiation. Such unexpected observation was shown to be caused by the opposite effects of temperature perturbation: local low temperature reduces the chemical reaction rate while it also increases the local volumetric energy density when the pressure remains unchanged [7].

One-step chemistry was used in most of studies mentioned above. However, as demonstrated by Mazaheri [3] and emphasized by Lee and Higgins [8], successful detonation initiation can always be achieved due to the absence of the cross-over temperature in one-step chemistry model. Therefore, one-step chemistry cannot predict a distinct value for the critical initiation energy below which no detonation occurs and it has limitation in the study of detonation initiation. Therefore, in the present study, a detailed chemical mechanism for hydrogen oxidation is considered in the simulation and it consists of 13 species and 25 elementary reactions [9]. Two-dimensional numerical simulations are conducted. The objective is to investigate the influence of disturbance and wall boundary layer on detonation initiation.

2 Numerical model and methods

The numerical setup for the initiation of a 2D detonation in $H_2/O_2/N_2$ mixture with mixing zone non-uniformity is sketched in Fig. 1. The parallel mesh refinement framework AMROC [10] is adopted to simulate the multi-component compressible Navier-Stokes equations. This efficient and adaptive solver has been extensively validated for supersonic combustion and detonation problems [11, 12]. The length of the computational domain is x=60 cm; and the height is y=5 cm. adiabatic, reflective boundary conditions (ARBC) are used for left and right sides. Up and down sides use periodic boundary condition (PBC) or ARBC to investigate the effect of wall boundary layer on detonation initiation. Initially, a static stoichiometric H_2/O_2 mixture with 70% nitrogen dilution (i.e., H_2:O_2:N_2=2:1:7) is uniformly distributed in the whole computation domain. To initiate detonation in the 2D channel, a thin region of high temperature (T_L=800 K) and high pressure (P_L=50 atm) is placed near the left wall. The mixture on the right side is at the temperature of T_R=298 K and pressure of P_R=0.2 atm. A

Corresponding author. Fax: +86-10-6276-6232
E-mail address: wang_yuan@pku.edu.cn

disturbance is introduced on the surface of high energy slab at x=1 cm and a sinusoidal perturbation with the amplitude of A and wavelength of λ is imposed in the y-direction, A=2 mm was fixed in the whole simulation.

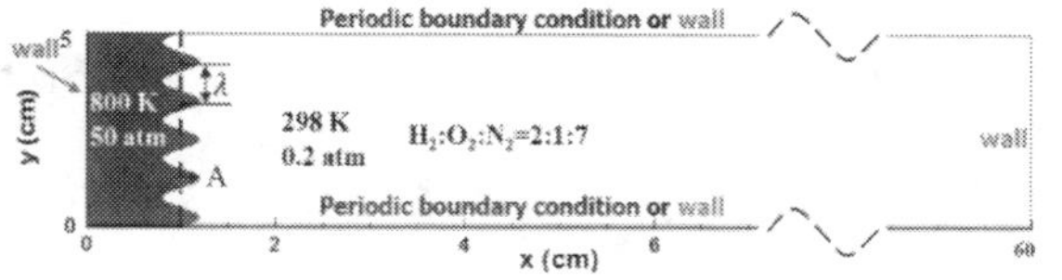

Figure 1. Schematic of mixing zone non-uniformity for detonation initiation.

To accurately and efficiently resolve detonation initiation and propagation, adaptive mesh refinement is used. The spatial mesh size before refinement is 0.5 mm × 0.5 mm; and the finest mesh size is 15.6 μm × 15.6 μm. The initial mesh distribution is shown in Fig. 2. The gradients of temperature, density and pressure are used to control the dynamic refinement and coarseness. Grid convergence is ensured through comparison among results from different finest mesh size of 31.2, 15.6 and 7.8 μm.

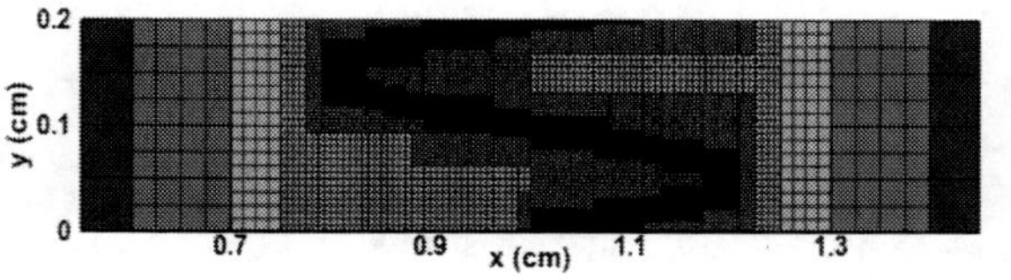

Figure 2. Distribution of initial mesh with six level mesh refinement for A=2 mm and λ=2 mm.

3 Results and discussion

3.1 Effects of disturbance (without wall)

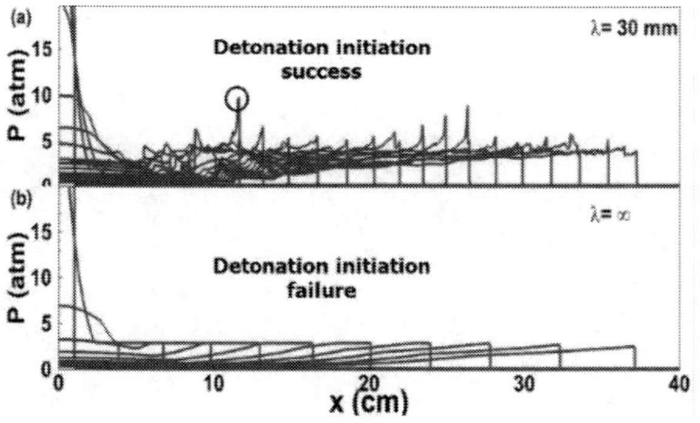

Figure 3. Temporal evolution of pressure distributions with (a) λ=30 mm and (b) λ=∞ (i.e., without disturbance) using PBC.

First we consider detonation initiation using PBC (without wall) at up and down sides. The effects of disturbance on detonation initiation are assessed. Figure 3 shows the temporal evolution of pressure distributions along the propagation direction at the channel center. For λ=∞ (i.e., without disturbance), the unburned gas behind the leading shock is compressed and its temperature and pressure reach to 956 K and 3.3 atm, respectively, at which it will take a long time for appearance of autoignition. As a result, the reaction front of autoignition cannot couple with leading shock, and the pressure at leading shock front continually attenuates. Therefore, detonation initiation fails. For λ=30 mm, the pressure attenuates initially and leading shock propagates to right with a constant value of pressure. This corresponds to the so-called quasi-steady state. When the leading shock is at x=11.9 cm, the peak value of pressure at leading shock front increases to 11 atm. Hence, detonation initiation succeeds. Therefore, disturbance with λ=30 mm promotes the onset of detonation.

Figures 4-7 further show the process of detonation initiation with λ=30 mm. The evolution of temperature distribution is shown in Fig. 4. The curved leading shock results from initiation disturbance propagates and interacts with each other, leading to local autoignition around t=40 μs. The reaction front and shock wave propagates to right as a same velocity. This corresponds to the so-called quasi-steady state with time less than 67 μs as shown in Fig. 5. For t>67 μs, more heat release from autoignition results in stronger pressure wave. The pressure wave compresses the unburned gas to high temperature and thereby accelerates heat release. Consequently, such coherent coupling between heat release and pressure wave induces sequential autoignition and eventually successful detonation initiation around t=74 μs.

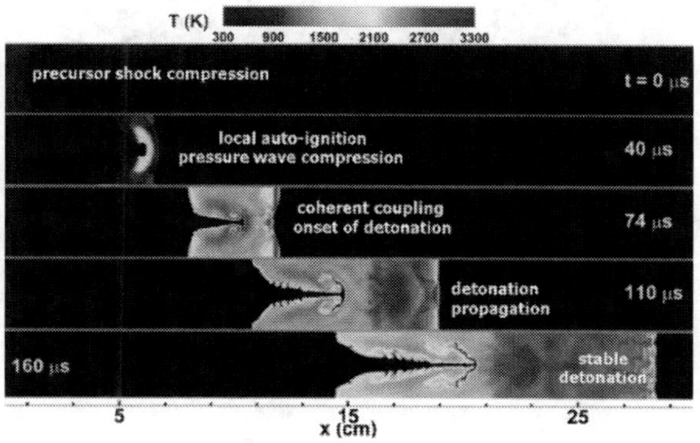

Figure 4. Evolution of temperature contour with λ=30 mm using PBC.

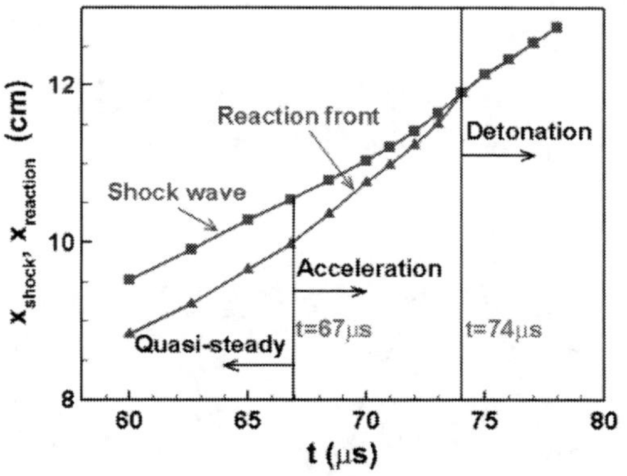

Figure 5. Shock wave and reaction front trajectories with λ=30 mm using PBC.

To further interpret the autoignition and detonation initiation, we plot the density gradient in Fig. 6. Initially, the leading shock is generated due to pressure discontinuity. The velocity and intensity of leading shock along x-direction at the channel center are larger than that near the up and down sides due to the initial disturbance (first frame in Fig. 4 and Fig. 6). Thereby, the interaction of leading shock generates transverse waves and slip lines (second frame in Fig. 6). Then the transverse waves propagate upward and downward and compress the unburned gas. When the transverse wave compresses unburned gas at the second time, autoignition happens (third frame in Fig. 6) which generates strong pressure wave (fourth frame in Fig. 6). The pressure wave further compresses the unburned gas between leading shock and reaction front. The autoignition zone increases and reaction front finally couples with the leading shock, indicating the formation of detonation (third frame in Fig. 4 and fifth frame in Fig. 6). Finally, the stable detonation propagates to right side (last frame in Fig. 4 and 6). The triple points are generated through the interaction among incident shock, Mach stem and transverse wave. Regular movement of triple points forms cellular structure of detonation as shown in Fig. 7. At the initial stage of detonation formation, the cellular structure is unsteady and the cellular size is much larger than that in steady stage. The steady cellular size is about 1.1 cm.

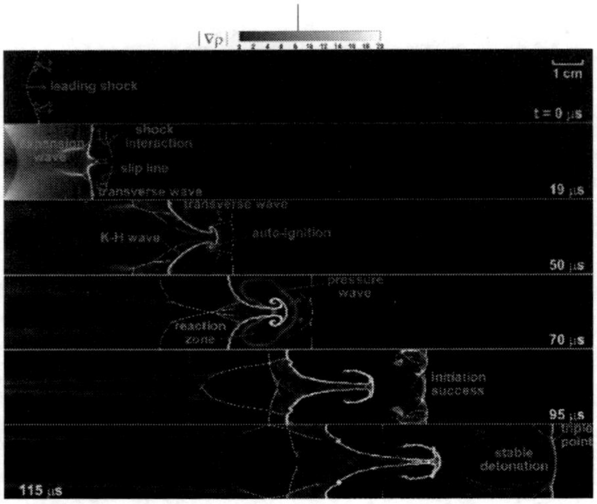

Figure 6. Evolution of density gradient contour with λ=30 mm using PBC.

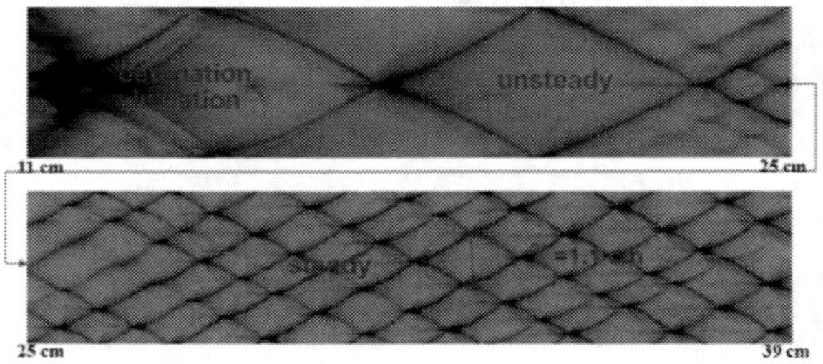

Figure 7. History of shock pressure with λ=30 mm using PBC.

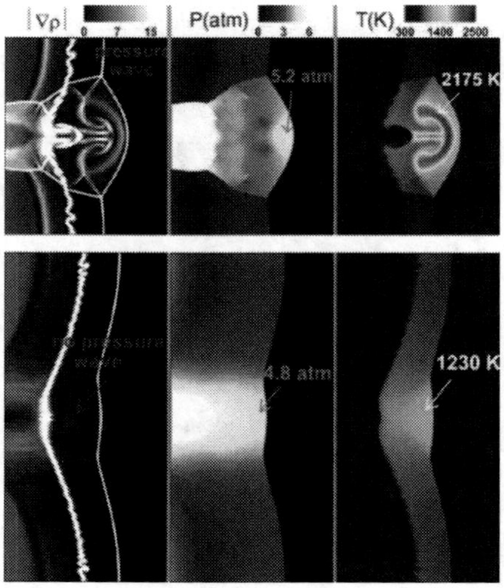

Figure 8. Comparison of density gradient, pressure and temperature contours with different wave lengths at the same time; up: λ=15 mm and down: λ=30 mm.

We study detonation initiation process with different wave lengths of disturbance. Figure 8 shows the density gradient, pressure and temperature contours with two wave lengths of λ=15 mm and λ=30 mm. For smaller wave of λ=15 mm, autoignition happens behind the leading shock wave and both temperature and pressure are higher than those of λ=30 mm. Consequently, the distance of detonation initiation becomes shorter for smaller wave length of disturbance, as shown in Fig. 9.

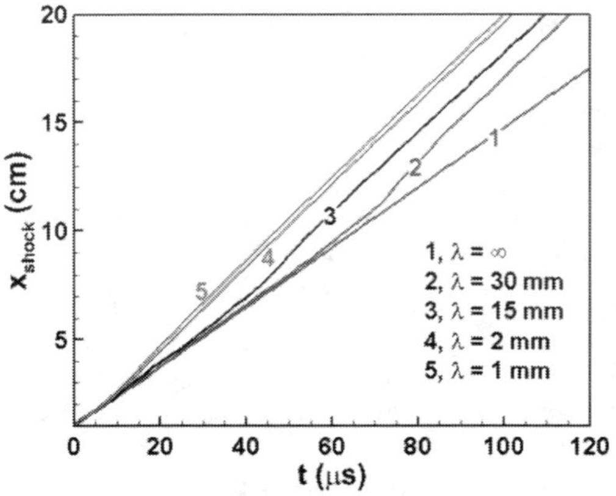

Figure 9. Shock wave trajectories for different wave length.

Figure 9 shows that for λ=∞, shock wave propagates as a constant speed initially and no autoignition happens. Thereby, detonation initiation fails. For λ=30 mm, reaction front couples with leading shock wave at t=74 μs, the propagation speed of shock wave increases suddenly at this moment and successful detonation initiation is achieved. As the wave length of disturbance decreases, the distance of detonation initiation decreases. Therefore, the distance and duration of detonation initiation can be shorten through decreasing wave length of disturbance.

3.2 Effects of wall boundary layer

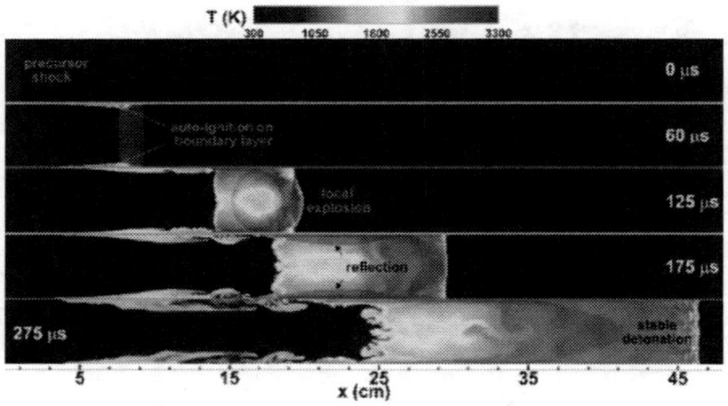

Figure 10. Evolution of temperature contour without disturbance using ARBC.

In this subsection, detonation initiation using ARBC (with wall) is considered. Figure 10 shows that at t=60 μs, autoignition happens near the walls, and the local temperature is above 2000 K. There are two reasons for the occurrence of autoignition near the walls. First, the leading shock compresses the unburned gas and increases its temperature. Second, the kinetic energy transforms into thermal energy due to the wall boundary layer, which further increases the temperature. The autoignition generates pressure wave propagating outwardly, which induces sequential autoignition events. Meanwhile, the pressure wave accumulates and propagates to right side. At t=125 μs, local explosion occurs at the channel center (third frame in Fig. 10).

Successful detonation initiation is eventually achieved around t=175 μs.

The detailed explosion process is shown in Fig. 11. At t=100 μs, the maximal pressure is above 7 atm. Then the pressure waves close and collide with each other which makes local explosion occur at t=115 μs. An explosion ball is formed and similar phenomenon was observed by Han et al. [13]. The blast wave generated by explosion propagates outwardly and reflects on the wall. Interaction of reflective blast wave and leading shock wave eventually forms detonation.

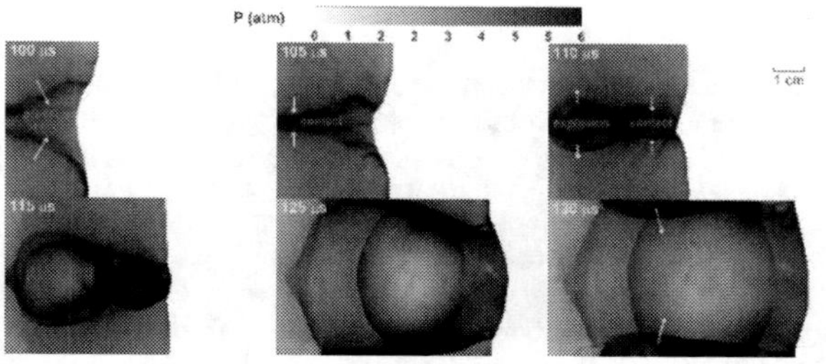

Figure 11. Amplification of pressure contour without disturbance using ARBC.

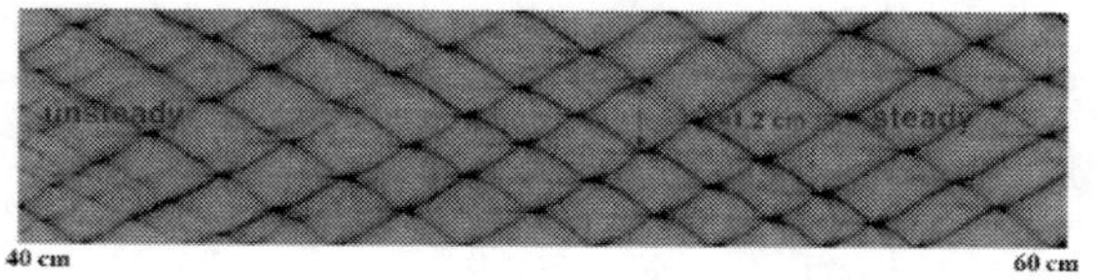

Figure 12. History of shock pressure without disturbance using ARBC.

At the initial stage of detonation initiation, cellular structure is not in steady state and both cellular size and peak pressure change as shown in Fig. 12. For x<47 cm, the cellular structure is irregular and the peak pressure is low. After the transition period, steady detonation forms and the cellular size and peak pressure become constant.

4 Conclusions

Two-dimensional numerical simulation for detonation initiation in $H_2/O_2/N_2$ mixture is conducted. Detailed chemistry is considered and block-structured adaptive mesh refinement is used to efficiently resolve the detonation development. A disturbance is introduced on the surface of high pressure and high temperature initiation region. Periodic and adiabatic reflective boundary conditions are used near the up and down sides. The influence of disturbance and wall on detonation initiation is examined. It is observed that the disturbance induces complex shock wave interaction, which promotes autoignition and onset of detonation. Disturbances with different wave lengths are considered. It is found that the wave length of the disturbance has great impact on direct detonation initiation. The smaller the perturbation wavelength, the stronger the shock wave interaction and thereby the shorter the detonation development distance and time. Furthermore, in the wall boundary layer, the kinetic energy transforms into thermal energy. As a result, autoignition happens in the vicinity of walls which generates pressure wave and local explosion, leading to the detonation formation. Therefore, both disturbance and wall promote the onset of detonation.

Acknowledgment

This work is supported by National Natural Science Foundation of China (Nos. 91541204, 51322602). Zheng Chen thanks Professors Matei I. Radulescu and Hoi Dick Ng for helpful discussions. Yuan Wang thanks Dr. Wang Han, Mr. Jiaping You and Mr. Mingyi Luan for helpful discussions.

References

[1] H.D. Ng, C.B. Kiyanda, G.H. Morgan, N. Nikiforakis. The influence of highfrequency instabilities on the direct initiation of two dimensional gaseous detonations. 25th International Colloquium on the Dynamics of Explosions and Reactive Systems, Leeds, UK; 2015.

[2] R. Chue, J. Lee, F. Zhang, Transition from fast deflagration to detonation under the influence of periodic longitudinal perturbations, Shock Waves 5 (1995) 159-167.

[3] K.M. Body, Mechanism of the onset of detonation in blast initiation, (1997).

[4] H. Ng, J. Lee. The influence of local disturbances on the direct initiation of detonations. 18th International Colloquium on the Dynamics of Explosions and Reactive Systems, Seattle, USA; 2001.

[5] M.I. Radulescu, G.J. Sharpe, C. Law. Effect of cellular instabilities on the blast initiation of weakly unstable detonations. 21st International Colloquium on the Dynamics of Explosions and Reactive Systems. Poitiers, France; 2007.

[6] J. Li, X. Mi, A.J. Higgins, Effect of spatial heterogeneity on near-limit propagation of a pressure-dependent detonation, Proc. Combust. Inst. 35 (2015) 2025-2032.

[7] C. Qi, Z. Chen, Effects of temperature perturbation on direct detonation initiation, Proc. Combust. Inst. 36 (2017) 2743-2751.

[8] J. Lee, A. Higgins, Comments on criteria for direct initiation of detonation, Philosophical Transactions of the Royal Society of London A: Mathematical, Physical and Engineering Sciences 357 (1999) 3503-3521.

[9] J. Li, Z. Zhao, A. Kazakov, F. Dryer, An Updated Comprehensive Kinetics Model of H2 Combustion, Chemical and Physical Processes in Combustion, (2003) 169-172.

[10] R. Deiterding. Block-structured adaptive mesh refinement-theory, implementation and application. Esaim: Proceedings, 34 (2011): 97-150.

[11] R. Deiterding, A parallel adaptive method for simulating shock-induced combustion with detailed chemical kinetics in complex domains, Computers & Structures 87 (2009) 769-783.

[12] R. Deiterding, High-Resolution Numerical Simulation and Analysis of Mach Reflection Structures in Detonation Waves in Low-Pressure H2–O2–Ar Mixtures: A Summary of Results Obtained with the Adaptive Mesh Refinement Framework AMROC, Journal of combustion 2011 (2011).

[13] W. Han, Y. Gao, C.K. Law, Flame acceleration and deflagration-to-detonation transition in micro-and macro-channels: An integrated mechanistic study, Combust. Flame 176 (2017) 285-298.

11th Asia-Pacific Conference on Combustion,
The University of Sydney, NSW Australia
10th -14th December 2017.

Propagation behaviors of twin premixed methane flame in a counterflow annular slot-burner under DC electric fields

Sung Hwan Yoon, Min Suk Cha

King Abdullah University of Science and Technology (KAUST), Clean Combustion Research Center (CCRC) and Physical Science and Engineering Division (PSE), Thuwal 23955-6900, Saudi Arabia

Abstract

Under an electric field, hydrocarbon flames generate bidirectional ionic wind due to the movement of positive and negative ions to the cathode and the anode, respectively. Such bulk motion significantly influences combustion stability in positive or negative sense. However, a detail mechanism of ionic wind on premixed flames is still not clear. Here, we described the effects of direct current (DC) electric fields on propagating twin premixed methane flames in a counterflow, annular slot burner with particular emphasis on change of a flame displacement speed. Applying DC to a lower part of the burner with a grounded upper part, electric field lines were perpendicular to the moving direction of premixed-edge-flames propagating along the annular slots. To provide the propagation speeds of premixed edge-flames with respect to the unburned flow, a PIV (Particle Image Velocimetry) was adopted. Based on measured flame displacement speeds and unburned flow velocities in electric fields, we will produce corrected propagation speed of the premixed-edge-flames.

1 Introduction

A flame speed measurement is of fundamental importance for understanding of combustion process since premixed flame is that a wave of chemical reaction inherently propagates to the unburned gas mixture. It is therefore significant to estimate an accurate the flame speed measurement experimentally to improve performance of industrial combustor engines consisting of fuel consumption rate, torque and generating power.

A two-dimensional slot burner had been considered as the optimized experimental set-up due to having zero flow velocity of unburned gas mixture at the stagnation plane (i.e., displacement flame velocity could be considered as propagation speed of an edge-flame) [1]. Recently, Cha and Tran [2] designed a new type of slot burner, a counterflow annular slot burner, to allow us to measure more accurate flame speed eliminated both lengthwise convection and strain. They suggested that a flow velocity measurement be certainly required for flame speed measurement since the unburned flow field is significantly modified by a gas expansion from the burned side despite zero flow velocity at a cold flow such as the stagnation plane in the two-dimensional slot burner.

Meanwhile, to develop a prompt control method of flame speed is required as the demand of high-performance and high-efficiency engines (e.g., hybrid car) significantly increases. Nowadays, an electrically assisted combustion is getting spotlight as new combustion technology [3-7]. It is possible not only to make quick combustion response without any changes of fuel composition and combustor structure [3, 4] but also to enhance combustion stability [7].

Basically, positive (e.g., CHO^+ and H_3O^+) and negative (e.g., electrons, O_2^-, and OH^-) charge carriers generated in the reaction zone of a hydrocarbon flame via chemi-ionization migrate to lower and higher electrical potential side by the Lorentz force, respectively [8]. Thereby, a bulk motion generates due to a momentum transfer from charge carriers to neutral molecules called as ionic wind. Tran and Cha [3] found that the edge-flame velocity was unaffected in electric field, on the other hand, only displacement edge-flame velocity was considerably varied by the effect of ionic wind originating from the movement of positive and negative ions. This result implies that flow modification due to ionic wind is the key parameter and the other effects such as thermal and chemical problems could be negligible in electric field. However, their study was limited in non-premixed flames, specific global strain rate, σ, and applied direction of electric field with the same orientation as flame propagation so that detail physicochemical mechanism of ionic wind is still open.

Here, following our previous studies [2-4], we have presented effects of DC electric field applied perpendicular to the direction of edge-flame propagation on displacement flame velocity of twin premixed flames. Furthermore, the extent of ignitable limit at high strain rate in electric fields will be briefly discussed.

2 Experimental Setup

The experimental apparatus is schematically outlined in Fig. 1. The axisymmetric counterflow, annular slot burner was composed of three co-axial annular slots in both upper and lower burners. A radius and a circumference of the central annular slots are 96.5 and 606 mm, respectively. The inner and outer slots from the central one provided N_2 sheath flows to prevent external disturbance. Each slot was 2-mm wide separated by 2-mm walls, as shown in Fig. 1. The gap distances, d, between the upper and lower burners was 9 mm. Flow strain was easily expected in a radial direction under cold flow with practically zero azimuthal velocity due to the nature of axisymmetry. The premixed gas composed of methane, oxygen and nitrogen was supplied via the central annular slots of upper and lower burners to generate twin premixed flames. To permit unidirectional flame propagation, the central annular slot was divided by 4 Sections as shown in Fig. 1; each Section (i.e., one quarter of the entire annular) was connected with different mass flow controllers. In addition, layers of packed steel beads inside the burner facilitated a uniform flow distribution at slot exits.

To apply an electric field, a DC power supply (Trek, 40/15H-

Corresponding author.
E-mail address: min.cha@kaust.edu.sa

CH_4	O_2	N_2	Φ	Q	S_L [cm/s]	d [cm]	V_{DC} [kV]	σ [1/s]
0.07	0.14	0.79	1	11.33	12.68	0.9	± 1, ± 0.5, 0	till non-ignitable

Table 1. Experimental condition

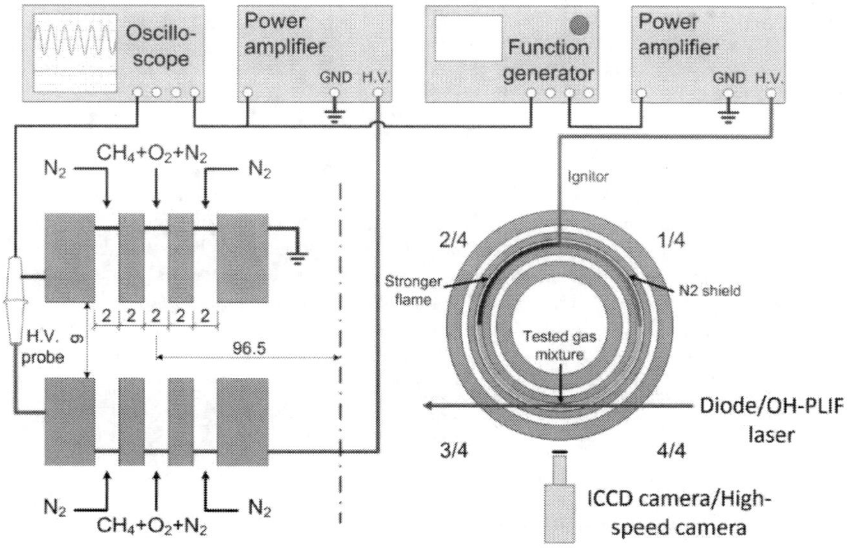

Figure 1: Schematic of experimental setup.

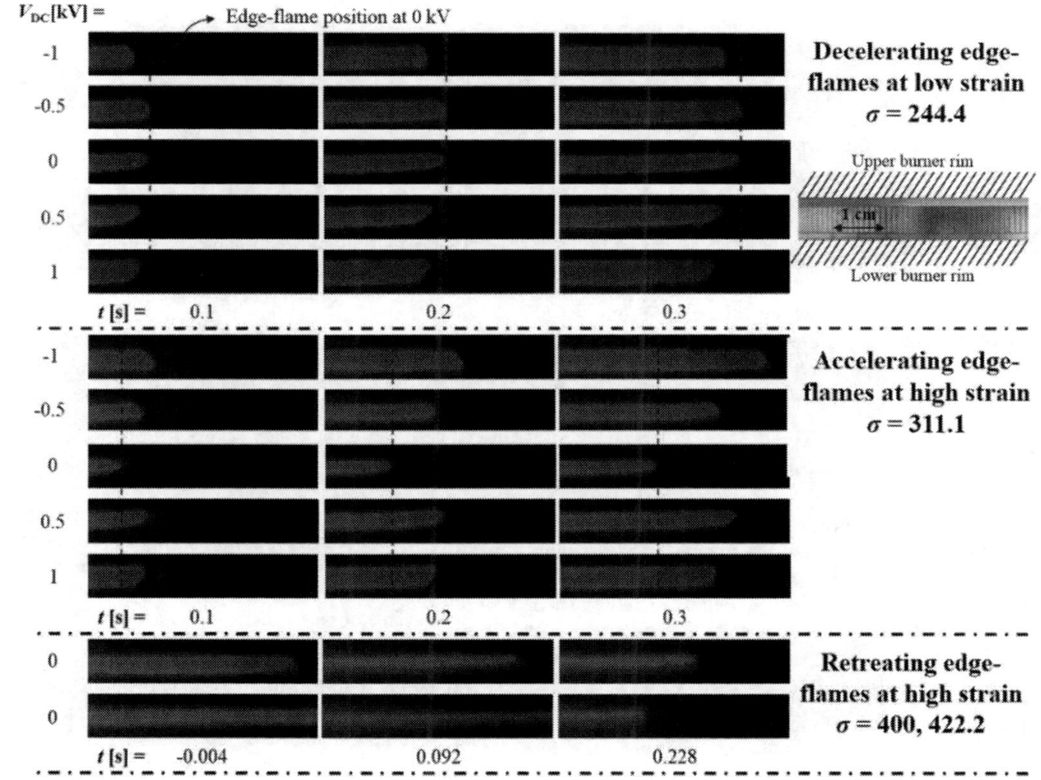

Figure 2: Sequential images of twin premixed edge-flames showing representative characteristic behaviors for different global strain rates (σ). Applied DC voltage (V_{DC}) shown at left side of each image, decelerating and accelerating edge-flames propagate from left to right, and retreating edge-flames propagate from right to left. Each image is the same scale with scaled ruler image on right side.

CE) was used. A high-voltage was supplied to the lower part of the burner and the upper part was grounded, creating electric field lines perpendicular to the direction of edge-flame propagation. The amount of applied voltage (V_{DC}) was controlled using a function generator (NF, WF1974). The applied voltage was obtained by an oscilloscope (Tektronix, MSO2024) using a 1000:1 high-voltage probe (Tektronix, P6015A).

In this study, we tested single gas mixture, 7% (CH_4), 14% (O_2) and 79% (N_2), with positive and negative DC voltages according to various strain rate, σ (= $2U_{jet}/d$, where U_{jet} is slot jet velocity) as shown in Table 1. In Section 1/4, N_2 shield played role as preventing bidirectional flame propagation. A flammable gas mixture, 9.5% (CH_4), 19% (O_2) and 71.5% (N_2), was placed in Section 2/4, for ignition of tested gas mixture. In Section 3/4 and 4/4, the tested gas mixture was provided. When the flame was achieved by ignitor located between Section 1/4 and 2/4, the strain rate of stronger flame and tested flame was matched each other. Here, we defined ignitable condition as the ignited flame of Section 2/4 kept propagating through tested Sections. Otherwise, the tested gas mixture was designated as non-ignitable condition. The time-dependent behaviors of the propagating edge-flames were recorded by a high speed camera (Photron, FASTCAM-ultima-APX).

To visualize the distribution of OH radicals, a planar laser-induced fluorescence (PLIF) technique was also adopted. This setup consisted of a Nd:YAG laser (Continuum, Powerlite DLS 9010), a tunable dye laser (Continuum, ND6000) with rhodamine 590 (Exciton), and a frequency doubling unit (Continuum, UVT-3). The wavelength of the UV output was 282.93 nm tuned at the Q 1 (6) transition. Instantaneous images of flames were captured using an intensified CCD (ICCD) camera (Princeton Instruments, PI-MAX3) with a UV-Micro-APO 105 mm f/4 lens at a 500-µs gate width to obtain a chemi-luminescence of edge-flame.

To obtain the speed of edge-flame propagation, particle image velocimetry (PIV) was adopted to the unburned flow. For Mie scattering images, a continuous wave Diode laser was used. Oil-based seeding droplets were used to prevent clogging the bead layers inside the burner.

3 Results and Discussion

Figure 2 shows the representative characteristic behaviors of twin premixed edge-flames with selected applied voltages for different global strain rates. Three distinct types of behaviors were identified; (1) decelerating edge-flames at low strain (2) accelerating edge-flames at high strain with applied voltages; and (3) retreating edge-flames at high strain. Decelerating and accelerating edge-flames propagated from left to right, and retreating edge-flames propagated from right to left. The numerical values at the bottom of each case indicate time. When the advancing edge-flames reached the view area of the high speed camera (i.e., left side of each picture), t = 0.0 s. For the retreating edge-flames without electric field, t = 0.0 s when switched off electric field after advancing edge-flame with electric field.

As shown the Fig. 2, the flame speed with the electric field decreased comparing with baseline flame with no electric field at lower strain rate (σ = 244.4 s^{-1}). At higher strain rate (σ = 311.1 s^{-1}), however, the flame speed with the field advanced. Note that the trailing flame body thicknesses with electric field had thicker than that with no electric field regardless of strain rate. To clarify this result, the displacement edge-flame velocity (U_d) are plotted

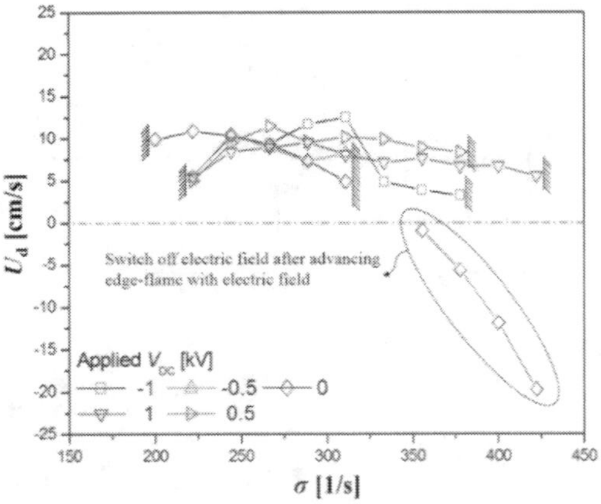

Figure 3: Effect of electric field with various DC voltages on displacement edge-flame velocity (U_d) according to various global strain rate (σ). Hatched lines indicate ignitable limits with each applied V_{DC}. The symbols in dotted circle area correspond to negative displacement edge-flame velocity formed in 0 kV after advancing edge-flame with electric field for a given strain rate.

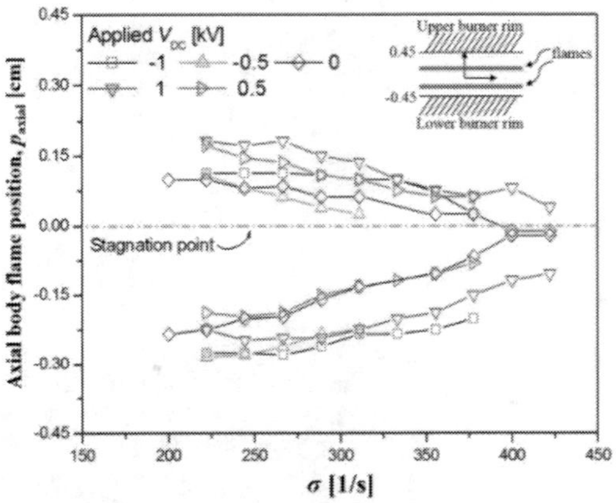

Fig. 4. Effect of electric field with various DC voltages on axial body flame position (p_{axial}) according to various global strain rate (σ). Dashed dotted line indicates stagnation point and inserted drawing describes an axis of coordinate.

in Fig. 3 as a function of global strain rate, σ, with various applied V_{DC}. Hatched lines indicate both low-σ and high-σ ignitable limits. Low-σ ignitable limits with applied voltages shrank comparing with that of no electric field. Here, the tested gas mixture with all of V_{DC} was to be non-ignitable at σ = 222.2 s^{-1}. However, high-σ ignitable limits with applied DC expanded comparing with that of no electric field.

Interestingly, we found another phenomenon, negative displacement edge-flame velocity at 0 kV with switch off electric field, after advancing edge-flame with electric field for a given strain rate indicated the symbols in dotted circle area (also, see Fig. 2). This observation was only acquired at high-σ, could not be obtained at out of low-σ ignitable limit of 0 kV.

Axial body flame position (p_{axial}) with a function of global strain rate is described in Figure 4. Dashed dotted line indicates stagnation point and inset describes an axis of coordinate. Note that the gap between upper and lower axial trailing body flame positions indicates the trailing body thickness of twin flames. On overall trend regardless of applied V_{DC}, most of trailing body thickness was squeezed towards high strain region. Electric field causes to broaden body thickness of twin flames in all of strain

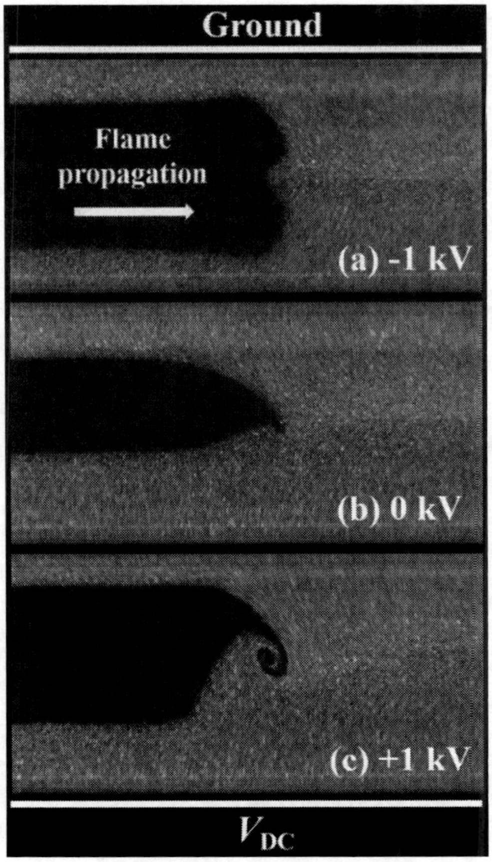

Figure 5: Mie scattering images showing dynamic behaviors of flow field by ionic wind effects for selected applied voltages at $\sigma = 311.1\ s^{-1}$: (a) -1 kV (b) 0 kV (c) +1 kV. The height of each image corresponds to the gap distance of the burner.

rates and polarity. Therefore, we may conclude that increased heat losses to the burner rim due to the expanded body thickness in electric field cause contracted low-σ ignitable limits. For high-σ ignitable limits in electric field, the expanded body thickness may support to reduce suffering from high strain rate.

The Mie scattering images describing the modification of flow field ahead of edge-flames for selected applied voltages are shown in Fig. 5. Considerably modified flow field with the propagating edge-flames at $V_{DC} = \pm 1$ kV as compared to that at $V_{DC} = 0$ kV. Here, the solid black area indicates the flame surface area overlaid with preheating zone because the seeded oil droplets evaporated at around 230 °C, which shows up iso-temperature contours. According to our previous study [2], significant modifications to unburned flow fields caused by thermal expansion from the propagating nonpremixed edge-flames were reported despite no electric field. The propagating premixed twin-edge-flame also can be observable with the modification of flow field by thermal expansion as shown in Fig. 5 (b). The stagnation line temporally varied, not aligned toward flame propagation, by a leading front likely a hook-shaped edge. At V_{DC} = - 1 kV, the leading edge shows a concave structure toward unburned region and propagates while keeping a suction with fresh gas mixture. At V_{DC} = + 1 kV, a spiral wave appears in front of the leading edge and extremely influences flow field of unburned side in spite of far region from the edge-flame.

In near future, we will discuss detailed physicochemical mechanisms of electric field with PIV and OH PLIF results.

4 Acknowledgment

Research reported in this publication was supported by Competitive Research Funding from King Abdullah University of Science and Technology (KAUST).

References

[1] M.S. Cha, P.D. Ronney, Combust. Flame 146 (1-2) (2006) 312-328.

[2] M.-V. Tran, M.S. Cha, Combust. Flame 162 (12) (2015) 4671-4672.

[3] M.-V. Tran, M.S. Cha, Combust. Flame 173 (2016) 114-122.

[4] M.-V. Tran, M.S. Cha, Proc. Combust. Inst. 36 (2017) 1421-1430.

[5] D.G. Park, B.C. Choi, M.S. Cha, S.H. Chung, Combust. Sci. Technol. 186 (4-5) (2014) 644-656.

[6] M.S. Cha, Y. Lee, Ieee Trans. Plasma Sci. 40 (12) (2012) 3131-3138.

[7] M.K. Kim, S.H. Chung, H.H. Kim, Combust. Flame 159 (3) (2012) 1151-1159.

[8] A.B. Fialkov, Prog. Energy Combust. Sci. 23 (5-6) (1997) 399-528.

11th Asia-Pacific Conference on Combustion,
The University of Sydney, NSW Australia
10th -14th December 2017.

Investigation of ignition characteristics for hydrocarbon fuels by using RCER

Koji Fujino, Naoto Izumi, Takahiro Kondo, Tatsuo Oguchi*

Toyohashi University of Technology

1-1 Hibarigaoka, Tempaku-cho, Toyohashi, Aichi, 441-8580, Japan

Abstract

The ignition characteristics of toluene reference fuel (TRF, mixing fuel of n-heptane and toluene) were studied using repetitive compression expansion reactor (RCER). New vaporized fuel supply system using a spray nozzle was developed in this study. Binary mixed sample gas was vaporized from the premixed liquid. The vaporizer makes it possible to supply fuels which have a lower vapor pressure as homogeneously mixed gas so that the experiments were performed in homogeneous charge compression ignition (HCCI) condition. Ignition timings were measured as the ignition characteristics of TRF mixing ratio. Zero-dimensional simulation in HCCI condition was also performed with detailed chemical reaction models. Comparing the results with experimentally obtained ignition timings, it was found that the dependence of ignition timing on the mole fraction of toluene in TRF was similar. Inflection points were also found in the simulation, however, it could not obtain clearly in the experimental results because of the difficulty for the setting. Differences between three models using in the simulation were also discussed.

1 Introduction

Hydrocarbon fuels such as gasoline have been used in internal combustion engine. For improvement of engines, the reaction characteristics of the hydrocarbon in combustion have been investigated. One of them, self-ignition characteristics, have been studied for the reduction of knocking and abnormal combustion [1]. A better understanding of reaction characteristics of hydrocarbon in fuels leads to improve the efficiency of the internal combustion engine and reduces the environmental pollution.

Homogeneous charge compression ignition (HCCI) combustion has been attracting attention due to high combustion efficiency and low emissions [2]. However, HCCI process still has many problems in practical application. HCCI is also one of the important phenomena in combustion chemistry as it is the basic process for understanding the chemical reaction mechanisms of ignition and combustion processes [3]. Therefore, the reaction mechanism of hydrocarbons in HCCI combustion has been investigated.

Several devices such as a Rapid Compression Machine (RCM) or a Shock Tube (ST) are used for measuring ignition delay which is one of the combustion characteristics of hydrocarbons in homogeneous charged compression. Chemical reaction models in HCCI condition have also been verified using these devices. However, RCM and ST experiments are limited to single shot. Furthermore, the mixing ratio of fuel gasses that can be introduced into the devices is also limited by the low vapor pressure of the fuels. Most of the hydrocarbon fuels have low vapor pressure. Thus another measurement system is required to extend measurable hydrocarbon to lower vapor pressure species.

Repetitive compression-expansion reactor (RCER) as a new measurement system has been developed by the authors in recent years [4]. Because of that mechanism of RCER similar to motoring engine, compression for the reaction is repeated. Thus the combustion characteristics can be obtained as the average of repetitive data. The reactor condition, the intake gas flow rate, and the fuel consumption rate are attentively controlled. Fuel and air are premixed by gas supply system. Therefore, the combustion occurs in HCCI condition same as RCM.

In the previous study, the liquid fuel was vaporized by bubbling system. However, it was difficult to vaporize the low vapor pressure fuels by the bubbling system. Recently, the spraying vaporizer was developed as a gas supply system. The vaporizer sprays and vaporizes the liquid fuel. This makes it possible to continuously supply fuels with lower vapor pressure to RCER.

In this study, ignition characteristics of a toluene reference fuel (TRF) were measured using RCER. For comparison of the experimental results to simulation, chemical kinetic simulations as HCCI condition were conducted using detailed chemical reaction mechanism. The comparison was performed to study the tendency of ignition characteristics.

2 Experiment

2.1 Experimental methods

RCER was used in this study to measure auto-ignition timing on the compression. RCER was developed by modifying commercially available single cylinder diesel engine (YANMAR, TF70V-E). The experimental apparatus is shown in Fig. 1. The crank shaft was forcibly rotated at a speed of 600 rpm by an electric motor. The displacement volume of a cylinder is 382 cm^2 and the compression ratio is 27.5. The original fuel injection system was completely removed from the diesel engine system because all fuels and gases were supplied as premixed gases from an intake port in this study. The intake gas was heated to 360 K in the reservoir tank. The initial temperature of the gas in the cylinder is defined as the intake gas temperature. The intake pressure was measured at just before the intake valve. The measured intake pressure corresponds to the pressure in the reactor when the intake valve is open. Thus the initial pressure of the reactor is defined as intake pressure. The initial pressure was set to 25 kPa, 30 kPa, and 35 kPa. The exhaust port was evacuated by a rotary pump to remove exhaust gas completely. The cylinder wall temperature was controlled by a water cooling system set at 330 K.

Corresponding author. E-mail address: oguchi@tut.jp

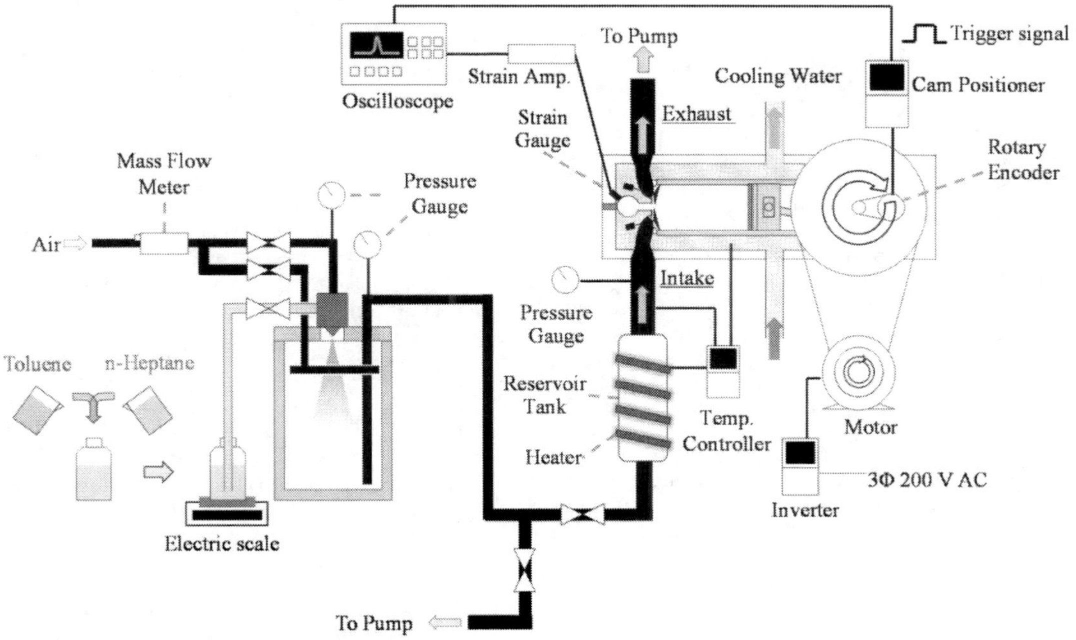

Figure 1: Experimental apparatus of RCER with sample gas supply system.

The liquid fuel was continuously sprayed at 0.5 MPa in the vaporizer tank (3.4 L). The air was added to the tank through the air blowing tube. The fuel gas stays in the tank for a few seconds and it is premixed with air. The total air flow rate was monitored using the mass flow meter. A decrease of weight of liquid fuel is recorded by the electronic scale. The typical recorded data of decrease of weight is shown in Fig. 2. Liquid fuel is constantly reduced. The consumption rate of the fuel was calculated from the mass reduction rate. The equivalent ratio was determined from air flow rate and fuel consumption rate. All experiments were carried out in a stoichiometric ratio.

The pressure profile in the reactor is observed using the oscilloscope via the strain gauge as a pressure transducer attached to the sub combustion chamber. The profile was averaged 64 times. In all experiments, the time axis is defined as the crank angle. The cam positioner measures the crank angle and sends a trigger signal to the oscilloscope. The zero-degree of the crank angle is defined as the compression top dead center. The ignition timing is defined as a crank angle when the pressure in the reactor begins sharply increasing. The ignition timings as ignition characteristics are compared with simulation results. In this study, TRF was used as fuel. TRF is a blended fuel of n-heptane (C_7H_{16}) and methyl-benzene (C_7H_8, following toluene). These components were mixed well in the liquid phase before spraying. In the following section, the notation of the molar fraction in percent ratio is defined as TRFn. The value of "n" is the mole fraction of toluene.

2.2 Computational methods

The chemical kinetic simulation was carried out to compare with experimental results. The simulation was conducted using CHEMKIN-PRO software.[4] Homogeneous compression and expansion reactor (HCCI) model without heat loss was used as an adiabatic zero-dimensional reactor model. Thus, the temperatures and pressures as well as the concentration of chemical species were homogeneously changed during the compression and expansion process. The initial condition was set to the same condition of experiments.

Three of the chemical reaction models were used in this study. The first model is the Gasoline surrogate details reaction mechanism version 1.0 (SIP-Gd1.0-S03, following SIP model) developed by A. Miyoshi. [6]. SIP model corresponds to the three-component common surrogate fuel including n-heptane, toluene, and 2,2,4-trimethylpentane (iso-octane). The second model is the model obtained from the literature by Mehl et al. [7] (following LLNL model). The third model is the model of n-heptane from the literature by Curran with co-workers [8] (Following Curran model). The Curran model is modified based on LLNL model. The ignition crank angle is defined a crank angle when the temperature in reactor reached to 1500 K. The temperature of ignition clank angle is defined as a representative temperature of the simulated hot flame. The simulation was carried out in range of 0 to 100% of TRFn.

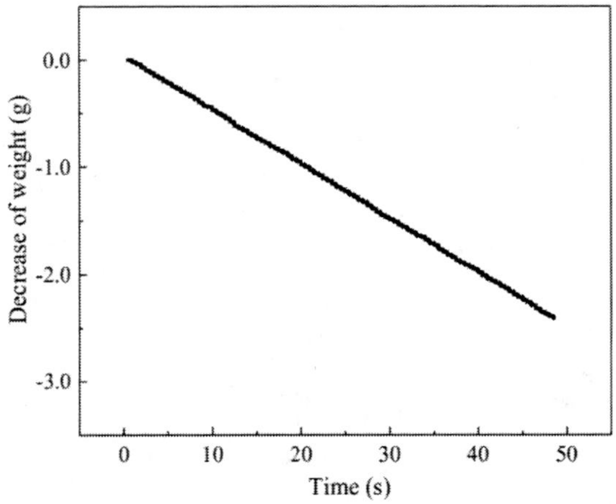

Figure 2: A typical decrease of mixed fuel (n-heptane 50 mol% and toluene 50 mol%) supplied to the vaporizer.

3 Results and discussion

The pressure profiles obtained by the experimental measurement of the ignition characteristics are shown below. Figure 3 shows the experimental results of TRF in the case of 25 kPa as the initial pressure. In the graph, the horizontal axis gives the crank angle and vertical axis represents the pressure in the reactor. The ignition occurred at the crank angle when the pressure sharply increases. The ignition timing was retarded when the mixing ratio of toluene increased in the result.

Simulations were carried out by each three chemical reaction models of SIP, LLNL, and Curran. The simulation using Curran model was performed for only n-heptane (TRF0) because this model is revised version of LLNL model for n-heptane. Figure 4 shows a comparison of ignition crank angle between experiments and simulations at each initial pressure in the reactor. The vertical axis represents the ignition crank angle and the horizontal axis shows the value of TRFn.

The tendency of experimental ignition timing against mole fraction of toluene is approximately linear above 10 mol% of toluene and below 5 degrees of crank angle. Significantly retarded ignition timings over 5 deg. of crank angle may not be discussed because of the strong effect of heat loss from the wall.

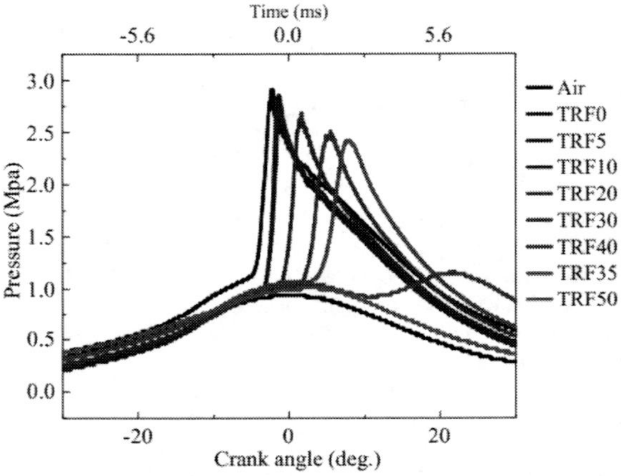

Figure 3: Pressure profile measured using RCER for TRF with the initial pressure of 25 kPa.

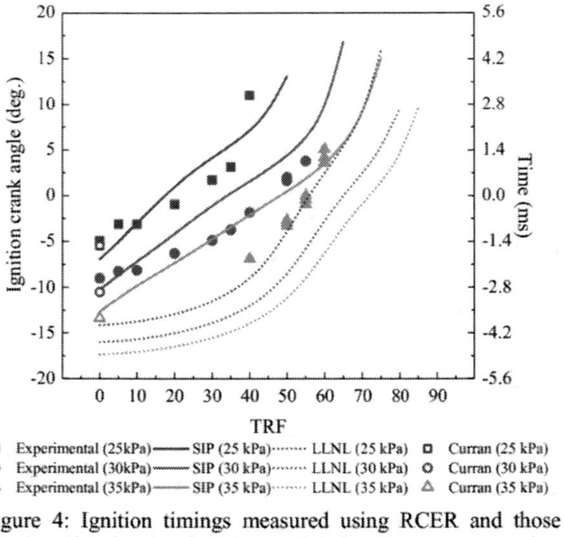

Figure 4: Ignition timings measured using RCER and those simulated by chemical reaction models for TRF at 25, 30, and 35 kPa of initial pressure.

In a comparison of experimental results with simulation, the general ignition crank angle of experiments are close to SIP model in the case of 25 kPa and 30 kPa as initial pressure. These linear tendencies against mixing ratio of toluene are also similar. Although LLNL model is significantly faster than experiments, the dependence of ignition timing on the mole fraction of toluene is considerably similar. The ignition crank angle of TRF0 in the case of 25 kPa corresponds to Curran model very well. The ignition timing of low mixing ratio of TRF was not measured because the pressure in the reactor in the case of 35 kPa reaches unacceptable pressure when ignition occurs. The inflection points are also found in SIP model. However, this tendency could not be compared because inflection points in experiments were not clearly obtained as mentioned above.

It is suggested that SIP model and Curran model are enough to predict the ignition timings of this experimental work. In contrast to these models, LLNL model predicts faster ignition timing of n-heptane. Both of SIP and LLNL models roughly predict the trend of our experiments. However, the two models predict different tendencies of ignition timing against initial pressure, respectively. The inflection points seems to depend on the crank angle, which is around 5 deg., for the SIP model prediction. In contrast, LLNL model predicts similar inflection points as TRF35 in all cases. Unfortunately, the experimental results in this work show different trends between the two cases of 25 and 30 kPa and the case of 35 kPa, so that the validity of the models are not clear. More detail analysis for this problem is needed in future work.

4 Conclusions

The validation study on chemical reaction models for TRF was performed. The ignition timing of TRF was measured by using RCER and the chemical kinetic simulation was also carried out using three models. SIP model as well as Curran model for the n-heptane combustion reproduce well for the observed ignition timing in the experiments. In addition, SIP and LLNL models roughly predict the dependence of ignition timing against TRFn. However, it was found that the effect of the initial pressure is different between SIP model and LLNL model. The effect could not be validated in this study because the trends of the ignition timing of the experimental results were not similar between the different initial pressures.

References

[1] C. K. Westbrook, Proc. Combust. Inst. 28 (2000) 1563-1577.

[2] T. Tsurushima, Proc. Combust. Inst. 32 (2009) 2835-2841.

[3] A. Paykani, A. H. Kakaee, P. Rahnama, R. D. Reitz, Int. J. Engine Res. 17 (2016) 481-524.

[4] a) N. Izumi, Y. Wada, M. Oshima, Y. Sakai, T. Oguchi, Proc. Asia-Pacific Conference on Combustion, 10 (2015) 238. b) N. Izumi, Y. Wada, K. Fujino, T. Oguchi, Int. Symp. Comb., WIP 2P006 (2016).

[5] Chemkin-Pro 18.1, ANSYS Inc.: San Diego, 2017.

[6] A. Miyoshi, Y. Sakai, Proceedings of the 2017 JSAE Annual Congress (Spring), (2017) 311.

[7] M. Mehl, W. J. Pits, C. K. Westbrook, H. J. Curran, Proc. Combust. Inst. 33 (2011) 193-200.

[8] K. Zhang, C. Banyon, J. Bugler, H. J. Curran, A. Rodriguez, O. Herbinet, F. Battin-Leclerc, C. B'Chir, K. A. Heufer, Combust. Flame 172 (2016) 116-135.

11th Asia-Pacific Conference on Combustion,
The University of Sydney, NSW Australia
10th -14th December 2017.

Measurement of solid particulate matter deposited on the bottom of fuel pool during pool diffusion combustion

Yoshihiro Kobayashi[1*], Masahiro Taguchi[2], Masataka Arai[1]

[1]Tokyo Denki University

Senju-Asahi-cho 5, Adachi-ku, Tokyo, 120-8551 Japan

[2]Isuzu advanced engineering center, Ltd

Tsuchidana 8, Fujisawa-shi, Kanagawa, 252-0881 Japan

Abstract

Particulate matter (PM) deposited on the piston surface of internal combustion engine is undesirable due to it being linked to increase knocking propensity, hydrocarbon (HC) emissions and PM emissions. In this study, mass of solid PM which was deposited on the bottom of fuel pool during laminar diffusion combustion of liquid fuel was investigated. Test fuels were n-Heptane (C_7H_{16}), n-Octane (C_8H_{18}), iso-octane (C_8H_{18}), benzene (C_6H_6) and toluene (C_7H_8). PM deposit formation rates from aromatic hydrocarbons fuels such as benzene and toluene were higher than those of n-octane and iso-octane. Deposit formation rate in benzene flame that was heated by a coil heater was also measured and compared with that of the normal flame combustion. It was confirmed that PM deposit formation rate in benzene flame with heating was less than the normal flame. PM deposition rate increased with decreasing of fuel consumption rate at both of the normal and heated benzene flames. Its rate had clear relationship with temperature gradient near fuel surface. It was estimated that PM formed near fuel surface penetrated into the liquid pool and deposited on the pool bottom.

1 Introduction

In recent years, it has been known that nanometer size particulate matter (nano-PM) is formed in combustion field such as diesel and gasoline engines[1]–[5]. The weight ratio of nano-PM in engine emission was very low. However, it seemed that considerable number of nano-PM was formed in engine. Since nano-PM is harmful to human health, new technologies should be established to reduce exhaust emission of the nano-PM. Further, some of PM was deposited on the piston and cylinder surface, and it caused another troubles such as knocking and HC emission increase.

As for a direct injection engine, fuel film combustion coupling with PM formation sometimes occurred on the cylinder and piston walls, and the resulting PM deposits might be formed on the wall surface[6]. The deposits have significant effect on surface heat transfer alteration[7],[8] and also PM emissions[9],[10].

In order to investigate formation process of PM and its characteristics, a small pool combustion equipment in which stable diffusion flame was formed by liquid fuel, was developed[11],[12]. Characteristics and compositions of PM exhausted from a laminar diffusion flame were investigated using this equipment system. Moreover, PAH and soot inside the flame were investigated using LIF, LII and gas analysis methods[12].

Stable pool flame could be formed using the pool combustion equipment in more than an hour of combustion period, however fuel pool was sometimes contaminated by a long time combustion operation (over 2 hours). Figure 1 shows the example of fuel pool contamination after diffusion combustion for hexane and benzene fuels. Pool flame of benzene caused obvious contamination comparing to hexane flame. Color of contamination substance was brown or black, and depends on fuel. The fuel pool contamination might be caused by penetration of solid particle from the flame into the fuel pool.

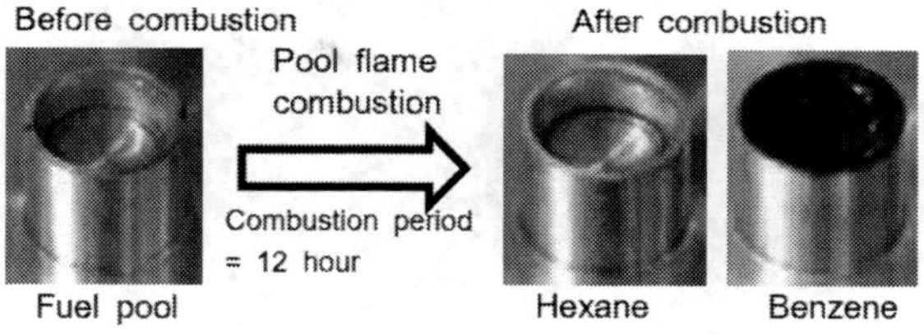

Figure 1: Fuel pool contamination

Since the similar phenomena of fuel contamination may occur during liquid film combustion in internal combustion engine, it is thought that fuel pool contamination research gives us available information to obtain knowledge about PM deposit formation relating fuel film combustion. Research of PM in liquid fuel pool and PM deposit on the fuel bottom at lower temperature condition was especially important for cold start deposits in internal combustion engine.

In this study, PM deposits on the bottom of fuel pool during pool diffusion combustion with liquid fuels were investigated. Moreover, the influence of flame hearing on deposit formation in the fuel pool was clarified to understand the driving force of PM penetration into the fuel pool.

2 Experimental apparatus and method

2.1 Pool combustion equipment

A pool combustion equipment was used to form a laminar

Yoshihiro Kobayashi. Fax: +81-3-5284-5466
E-mail address: ykoba@mail.dendai.ac.jp

diffusion flame of liquid fuel. Figure 2 shows the scheme of pool combustion equipment. This equipment consisted of a fuel tank, a fuel supplying system, a fuel pool and an electric microbalance. The fuel was supplied from the fuel tank through a float chamber. Inner diameter of the fuel pool was 16 mm. A dish (inner diameter was 14 mm and 1.5 mm in depth) as solid particle sampling was set on the pool bottom. The float chamber could keep the fuel surface at a constant level. The fuel consumption rate m_f (and flame length L_f) was controlled by an adjustment of the fuel surface level. The weight of the equipment was measured by the electric microbalance and the fuel consumption rate was derived from the weight loss of it. Here, location of pool rim top level and pool center was adopted as the coordinate origin (z = 0mm and r = 0mm). Therefore, the practical fuel surface located lower than pool rim top, and it was indicated by negative position of z.

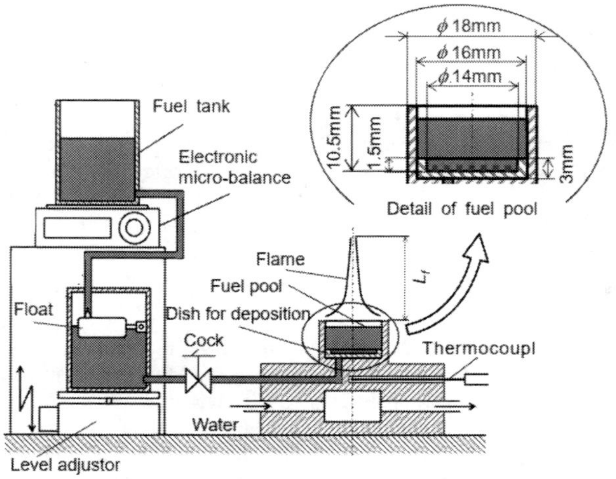

Figure 2: Pool combustion equipment

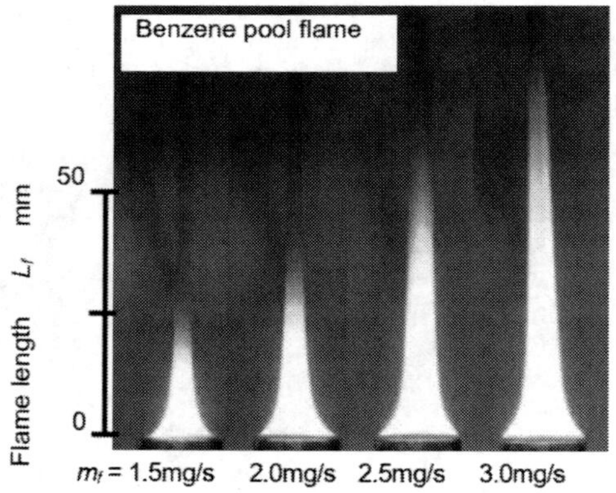

Figure 3: Direct photographs of benzene pool flame

A laminar diffusion flame was stationary formed with this equipment (To form stable flame, the ambient air was rectified by the punching metal). Flame length L_f was defined as a distance from the top of pool rim to the flame tip. Figure 3 shows direct photographs of stable benzene flame formed by the pool combustion equipment. From direct photograph of benzene pool flame, incandescence of soot combustion and visible smoke at downstream of the flame tip was observed. As shown in Fig. 3, L_f increased with increasing of m_f.

2.2 Pool flame heating

Pool flame and surrounding air entrained in the flame were heated by coil heater. The scheme of pool flame heating was shown in Fig. 4. Photographs of normal pool flame and heated flame at m_f = 2.5 mg/s using benzene were also shown in this figure. Height of coil heater was 42 mm, and its inner diameter was 50 mm. Heater surface temperature was set at around 500 °C. Normal benzene flame was typical diffusion flame having open flame tip and PM was exhausted from the tip. When this flame was heated, flame length decreased slightly but no obvious change of flame appearance was observed.

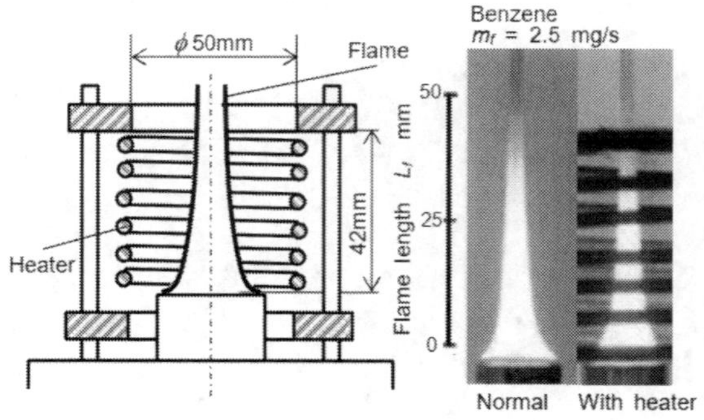

Figure 4: Heating unit and heated flame

Temperature distributions of normal and heated benzene flames of m_f = 2.5 mg/s were measured by R type thermocouple (50 μm). The temperature measurement of sooting flame was slightly difficult because the soot particles adhered around the thermocouple. When the thermocouple was inserted into the flame the thermocouple temperature increased and reached a maximum, after which it decreased because of the adherence of the soot particles. Here, we assumed that the maximum temperature was equivalent to the actual temperature of the flame. The results along the center axis were indicated in Fig. 5. When the benzene pool flame was heated by the coil heater, temperature at upstream part of the flame increased, and flame length decreased slightly. It means that the more active oxidation by heating occurred. Moreover, since the evaporation of fuel in the pool was promoted by heating, fuel surface level shifted to lower position than the normal flame.

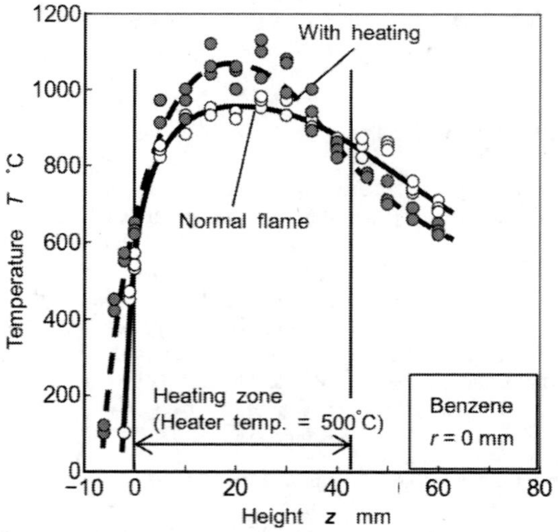

Figure 5: Temperature distribution along the center axis

2.3 Sampling methods of solid PM deposited on the fuel pool

Solid particle measurement method was shown in Fig. 6. Main steps from flame formation to flame extinction were as follows.

First step was a flame adjustment period (t_a). It was a transition period from ignition to steady combustion (t_a was around 20 sec). The second was a steady state combustion period (t). When the combustion operation continued over 2 hours, stable flame could not be maintained by deposition of solids on the pool rim. So, flame is forcibly extinguished after every one hour operation and the pool rim was cleaned up. Total steady combustion period was defined t = 1 hour x repeat number (n).

The final step was an aftertreatment period (t_{at}). Flame was forcibly extinguished by a cover plate, and fuel supply was stopped. After then, fuel in the pool was evaporated and dried up by room temperature condition. After sufficient dry-up period (over 20 hours), mass of the residue on dish in the fuel pool was measured by the electronic micro-balance. There were two kinds of residue substances. One is the submerged and settled on the bottom dish and the other was suspended in the liquid fuel. The residue obtained by this treatment was the total mass of above two kinds of substances. Here the total of these substances was regarded as the deposits that was penetrated from the flame into liquid pool and finally settled on the bottom.

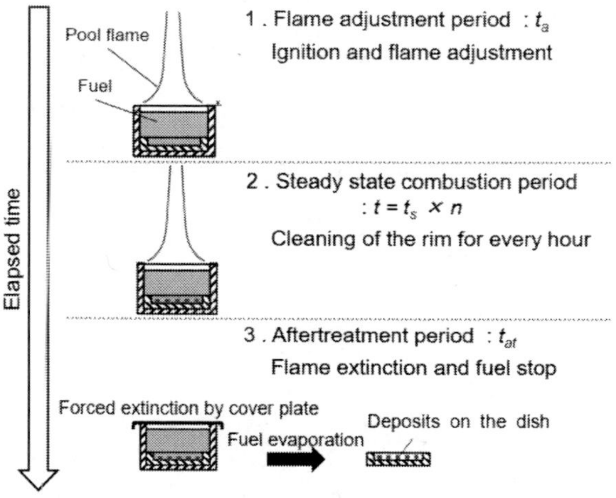

Figure 6: Solid particle measurement method in the fuel pool

3 Results and discussion

3.1 Solid PM deposited on the fuel pool bottom

Relations between total steady combustion period and mass of deposit on the fuel pool bottom were investigated. Deposits obtained for various fuels such as paraffin group (n-heptane and n-octane), iso-alkane (iso-octane) and aromatic group (benzene and toluene) are shown in Fig. 7. The color of solid deposit substance was dark brown and looked like PM deposits.

As for the paraffin fuels such as n-heptane (t = 8 hours) and n-octane, fuel pool was not contaminated after pool combustion of long time operation, and deposit mass could not be detected. As for iso-octane and aromatic fuels, fuel pool was contaminated by pool combustion flame. Deposit mass increased with an increase of steady combustion period.

Deposit mass from iso-octane was lower than those of aromatic fuels. Deposit mass of benzene pool flame was the highest among paraffin, iso-alkane and aromatic groups. Furthermore, deposit mass of benzene (n_c = 6) pool flame was higher than that of toluene (n_c = 7). Fuel depending tendency of deposit mass was similar to result of PM exhaust mass from the tip of pool flame reported by our previous study[12]. However, deposit rate was considerably lower than the PM exhaust rate. For example, deposit rate from benzene flame was around 0.0001 mg/s but PM exhaust rate for the same flame was around 0.25 mg/s.

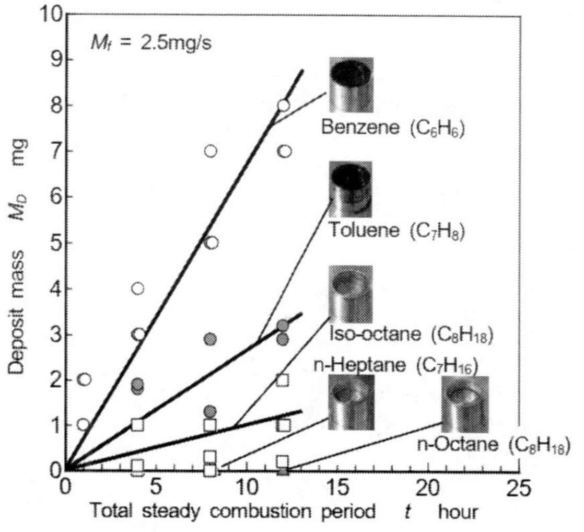

Figure 7: Relation between deposit mass and chemical structure

3.2 The influence of flame heating on deposit

Figure 8 shows relationship between fuel consumption rate and deposit rate on the pool bottom at aromatic fuels such as benzene and toluene. Result of benzene flame with heating condition was also shown in this figure.

As for the normal benzene flame, deposit rate at m_f = 1.0 mg/s was around 3×10^{-4} mg/s. Deposit rate at toluene flame was lower than that of benzene flame. Both deposit rates of benzene and toluene flames decreased with an increase of m_f. From our previous study, PM emission rate from the tip of benzene flame increased with an increase of m_f. Decreasing tendency of deposit rate relating fuel consumption was not agreed with that of PM exhaust rate. Further in the case of benzene flame under heating condition, deposit rate was lower than that of normal benzene flame.

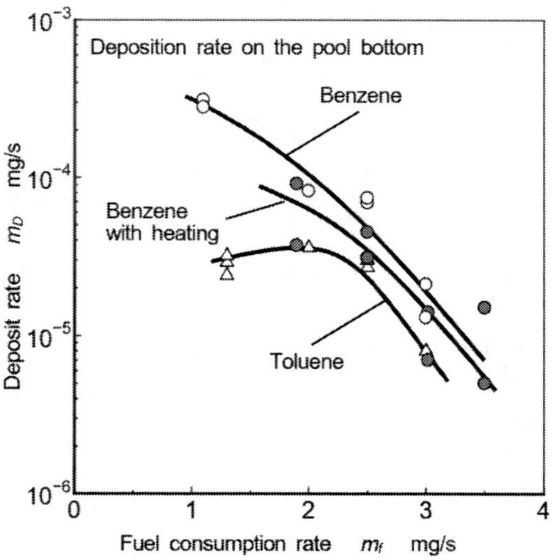

Figure 8: Relation between deposit rate and fuel consumption rate

Figure 9 shows in-flame temperature distributions on the center axis near fuel surface of benzene flames at m_f = 2.0, 2.5 and 3.0 mg/s. As for the m_f = 2.0 mg/s shown in upper part of the figure, temperature at benzene flame with heating condition was higher than that of the normal flame of benzene.

Fuel surface position at heated flame was lower than that of the normal flame. Temperature gradient (dT / dz) at just above the fuel surface under flame heating was lower than that of the normal flame. Generally, nano scale particles were affected by thermophoresis in higher temperature gradient field. It might be considered that PM particles formed in fuel vapor region were moved into the fuel pool by thermophoresis. Lower deposit rate under heating condition might be caused by a decrease of weak thermophoresis effect corresponding to low temperature gradient.

As for m_f = 2.5 mg/s shown in the middle part of the figure, temperature gradient under heating condition was also lower than that of the normal flame. Under normal flame condition, fuel surface position was around z = -2.1 mm, and was higher position than that of m_f = 2.0 mg/s. However, flame length was elongated with m_f increasing, and maximum temperature position shifted to downstream. Therefore, temperature gradient under the normal flame condition at m_f = 2.5 mg/s might be lower than that of 2.0 mg/s. Furthermore, the temperature gradient under the normal flame condition at m_f = 3.0 mg/s shown in the lower part seemed to be lower than that of m_f = 2.5 mg/s.

Deposit rate shown in Fig. 8 had clear relationship with temperature gradient near fuel surface. Then, it was estimated that PM formed near fuel surface penetrated into the liquid pool and deposited on the pool bottom. Driving force of this penetration might be the thermophoresis diffusion of PM near fuel surface but we need more study to confirm the thermophoresis diffusion and behavior of PM that moved and captured by liquid surface.

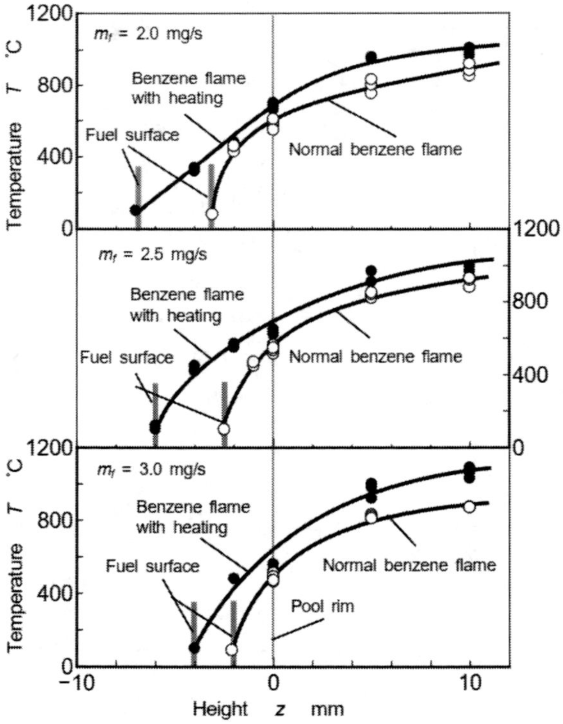

Figure 9: In-flame temperature distribution near fuel surface (r = 0mm)

4 Conclusions

Solid PM deposited on the bottom of fuel pool during a pool diffusion combustion with liquid fuel was measured. PM deposition under flame heating condition was also measured and compared with the normal flame deposit. The main results are summarized as follows.

(1) Solid PM deposit rate of aromatic group was the highest among paraffin, iso-alkane and aromatic groups. Furthermore, PM deposit rate of benzene (C_6H_6) flame was higher than that of toluene (C_7H_8).

(2) PM deposit rate at lower fuel consumption rate was high, and PM deposit rate decreased with increasing of fuel consumption rate.

(3) PM deposited of benzene pool flame in heating condition was lower than that of the normal pool flame.

References

[1] Kittelson,D.B., Journal of Aerosol Science, 29 (5/6) (1998) 575-588

[2] Y. Zhang, R. Zhang, L. Rao, D. Kim and S. Kook, Fuel, 194 (2017) 423-435

[3] V.Seller, E.Boeckmann and P.Eilts, SAE paper No.2008-01-0335 (2008)

[4] R.L.Vander Wall, D.L.Dietrich, Appl. Optics, 34 (1995) 1103-1107.

[5] P.J.Shayler, H.K.Ng, SAE paper, 2004-01-0116 (2004).

[6] Yusmady M. A., M.Arai, FUEL, 89 (2010) 934-942

[7] Yamada Y, Emi M, Ishii H, Suzuki Y, Kimura S, Enomoto Y. JSAE Rev 2002;23:415–21.

[8] Kalam MA, Masjuki HH. Biomass Bioenergy 27 (2004) 289–297

[9] Zerda TW, Yuan X, Moore SM, Leon CA, Leon Y. Carbon 37 (1999) 1999–2009.

[10] Caceres D, Reisel JR, Sklyarov A, Poehlman A. J Eng Gas Turb Power 125 (2003) 358–364.

[11] Y.Kobayashi, T.Furuhata, K.Amagai, M.Arai: Combust. Flame, 154 (2008) 346-355.

[12] Y.Kobayashi, K. Miyanabe, M. Arai, JSAE paper, 43 (6) (2012) 1257-1262

11th Asia-Pacific Conference on Combustion,
The University of Sydney, NSW Australia
10th -14th December 2017.

Direct Numerical Simulation of H2 Effect on Autoignition of DME/Air Mixture under HCCI Engine Conditions

Tai Jin[1,2], Xujiang Wang[2], Kai Hong Luo[2*], Kun Luo[1], Jianren Fan[1]

[1]State Key Laboratory of Clean Energy Utilization, Zhejiang University
Hangzhou, 310027, China

[2]Department of Mechanical Engineering, University College London
London, WC1E 7JE , United Kingdom

Abstract

The effects of hydrogen addition on the multistage autoignition of a lean DME/Air mixture under elevated pressure are investigated by direct numerical simulations (DNSs). The results show that hydrogen addition has a negative effect in the DME low temperature ignition, leading to a declined maximum of the heat release rate (HRR) and a retarded ignition timing. The ignition at higher temperature is significantly delayed by hydrogen addition with increase of the maximum of heat release rate. DNS for thermally stratified turbulent ignitions shows an advanced first stage ignition, yet retarded with addition of hydrogen. The mean HRR in the turbulent cases is more spread out over time while its peak is significantly reduced. The mean HRR during the second and third stage ignition overlap in time, and therefore only a single peak is observed for the thermally stratified cases. And the overall pressure rise rate is smoothed. DNS of the ignition characteristics of uncorrelated and negatively correlated thermally- and compositionally- stratified mixtures is also briefly discussed.

1 Introduction

Homogeneous charge compression ignition (HCCI) engine is a viable new concept for next-generation internal combustion engines, with the potential to improve fuel-consumption economy relative to spark-ignition engines, to reduce particulate emissions relative to diesel engines and to reduce nitrogen oxides (NOx) relative to both spark-ignition and diesel engines[1]. Dimethyl ether (DME) as an alternative fuel for the engines has received extensive attentions due to its favorable chemical characteristics[2]. However, the ignition timing and combustion phase of DME in HCCI combustion process are hard to control because DME fuel is very easy to be self-ignited. An adjustment of the proportion of two kinds of fuels with different reactivities has been proposed as a common method for controlling of the ignition timing in the HCCI combustion[3].

Hydrogen (H_2) has been used in DME HCCI engine to control the ignition timing and expand the operating range of the engines. A thorough understanding of the combustion properties of DEM/H_2 is crucial for developing advanced DME-based combustion engines and corresponding operating strategies. Pan et al.[4] conducted shock-tube experiments to measure the ignition delay of DME/H_2/air mixtures. It has been found that the ignition delay time of DME/H_2 binary fuel blends has complex dependence on the temperature. Yu et al.[5] measured the laminar flame speeds of DME/H_2/air flames at different temperatures, equivalence ratios, and fuel blending ratios. They found that the laminar flame speeds increased with an increase of H_2 fraction and initial temperature. It was also found that the chemical kinetic effects induced by H_2 addition played a dominant role in the increase of the laminar flame speed compared with thermal and diffusive effects. Simulations of the DME/H_2 flames are also conducted with detailed chemical reaction mechanisms. Liu et al.[6] simulated the effects of H_2 addition to the DME base flame. The results indicated that the reduction of CH_3OCH_3 mole fraction in the blend is the dominant factor for the reduction of CH_3OCH_3 and CO mole fractions in the flame. The H, O and OH radicals increased when H_2 was added, and these radicals promoted the combustion process. Hu et al.[7] simulated the autoignition characteristics of DME/H_2/Air blends in homogeneous charge compression ignition combustion engine. They found that the hydrogen addition can retard the autoignition timing of DME in HCCI combustion engine.

A key characteristic of all fuels in the HCCI context is their response of thermal stratification. Thermal stratification always occurs naturally in HCCI engines due to wall-heat transfer and imperfect mixing with residuals. It was found that thermal stratification had the potential to introduce a range of ignition delay times into the mixture, thus preventing the undesirable high pressure-rise rate (PRR) that would result from the simultaneous autoignition of the whole fuel charge[1]. Many computational studies of HCCI combustion using multidimensional direct numerical simulations (DNSs) have been conducted to elucidate the fundamental combustion characteristics of various fuel/air mixtures under HCCI conditions including hydrogen[8-10], dimethyl ether (DME)[11, 12], n-heptane[13], iso-octane[14], primary reference fuel (PRF)[15, 16], ethanol and biodiesel[17]. From these studies, the general characteristics of HCCI combustion have been elucidated: thermal stratification in an HCCI engine cylinder can spread out the PRR under high load conditions by changing the combustion mode of spontaneous auto-ignition into a mixed combustion mode of spontaneous auto-ignition and deflagration. Stratified-charge compression ignition combustion has been devised as another promising remedy for the problems in HCCI combustion. It has been found that fuel inhomogeneities may induce a sequential ignition event, leading to a prolonged combustion process.

Corresponding author. Fax: +44-20-7679-3916
E-mail address: K.Luo@ucl.ac.uk

The objectives of this paper are to investigate the combined effects of thermal stratifications and H_2 addition on autoignition and flame propagation in the NTC region for a lean DME/air mixture at high pressure. The initial turbulence and stratification length scales are chosen to mimic HCCI preignition conditions. The isentropic compression heating resulted from piston motion is also incorporated through inert density source terms in the governing equations.

2 Numerical methods and computational setup

The full compressible Navier-Stokes, species, and energy equations for a reacting gas mixture are solved with an in-house DNS code. A fourth-order explicit Runge-Kutta method for time integration and an eighth-order central differencing scheme for spatial discretization are used with a tenth-order filter for removing spurious high frequency fluctuations in the solution. The mixture is assumed to be a perfect gas and the mass diffusion of species is assumed to obey the Fickian law. CHEMKIN software libraries are linked to the current code to evaluate reaction rate, as well as thermodynamic and transport properties for both the individual species and the mixture-averaged. A reduced chemical mechanism for DME oxidation consisting of 30 chemical species is employed[12]. Detailed description of the DNS code is given in our previous studies[18, 19] .

The computation domain is a constant-volume enclosure. Periodic boundary conditions are employed at all boundaries. To mimic the pressure rise due to piston motion in practical engines, inert source terms are added to the respective mass, species, momentum and energy conservation equations. The density source terms is derived in [12], assuming that the compression process is locally isentropic. The engine parameters used to evaluate the change in volume for all cases are, as follows. The engine speed is 1200 rpm, the compression ratio is 18, and the ratio of connecting rod length to crank-shaft length is 3.2. The initial crank angle is set at 36 CAD before TDC.

Table 1: Simulations parameters for different cases

Case	n_{DME}: n_{H2}	u' (m/s)	T' (K)	ϕ'	L_{11} (mm)
1-a	1:0	0.5	25	0	0.5
3-a	1:1	0.5	25	0	0.5
3-b	1:1	0.5	25	0.05	0.5
3-c	1:1	0.5	25	0.05	0.5
5-a	1:2	0.5	25	0	0.5
5-b	1:2	0.5	15	0	0.5

The initial mean conditions for all the cases studied are set to be $\bar{T} = 677.73\text{K},\ \bar{\phi} = 0.3,\ P = 10.834$ atm. The composition of the DME/H_2/Air mixtures investigated in this study varies with different blending ratio of hydrogen, in order to investigate the effect of hydrogen addition without changing the total equivalence ratio. The mole fraction ratio of H_2 and DME (n_{H2}:n_{DME}) is 0, 0.5,1,1.5,2.0 for Cases 1-5 respectively. Two dimensional simulations are conducted to study the turbulence, temperature stratification effects on the autoignition of the blending fuel. The mean flow is quiescent with isotropic turbulent velocity fluctuations prescribed by the Passot-Pouquet turbulent kinetic energy spectrum function. Random temperature field is superimposed on the mean temperature field, with a similar spectrum of the turbulence kinetic energy. The various turbulence parameters used for different cases are listed in Table 1. At the TDC prior to the main auto-ignition event, different $T-\phi$ correlations may exist due to such factors as fuel delivery strategy, injection timing, amount of EGR, intake charge heating, and wall heat loss. Two most-probable scenarios are examined here as negatively correlated $T-\phi$ fields (Case 3-b) and uncorrelated $T-\phi$ fields (Case 3-c). To understand different magnitude temperature fluctuations effects on the autoignition characteristics, Case 5-b with a smaller T'=15 K is simulated.

3 Results and discussion

3.1 H2 effects on homogeneous autoignition

Prior to examining the turbulent cases, simulations of the HCCI combustion of DME/H_2/Air mixtures are conducted using the single-zone, time-dependent volume, adiabatic, and zero-dimensional reactor in CHEMKIN pro package with SENKIN approach. These results can provide a baseline reference for the 2-D turbulent simulations, and also provide a chemical description of DME/H_2/Air mixtures autoignition without the additional complexities of molecular and turbulent transport.

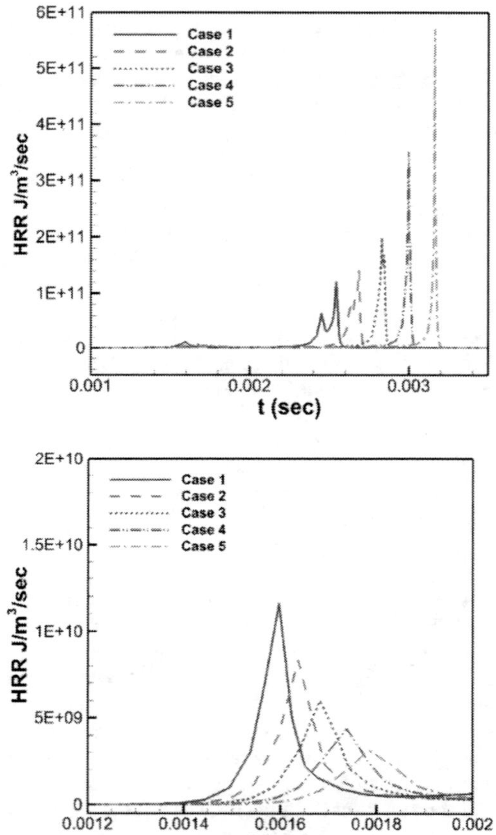

Figure 1: Heat release rate evolution for mixtures with different hydrogen blending ratios

The time evolution of heat release rate during the HCCI combustion process for mixtures with different hydrogen blending ratios are demonstrated in Fig.1. It can be observed that DME and all the blends calculated exhibit a typical multi-stage heat release. For DME only Case 1, it presents two maximum of heat release rate at the high temperature reaction (HTR), which means three-stage ignition. With the increase of the hydrogen blending ratio, the two peaks of heat release rate at HTR would overlap and the "three-stage" heat release phenomenon disappears. It also can be found that both the low temperature reaction (LTR) and the high temperature reaction (HTR) are delayed with the increase of the hydrogen blending ratio. The peak value of the heat release rate of low temperature reaction decreases monotonously with the increase of the hydrogen blending ratio. The declined maximum of the heat release rate and the retarded ignition timing at LTR indicate that the addition of hydrogen has a negative effect in the DME low temperature ignition. This is mainly because that hydrogen acts as competitor with DME during the LTR process through consuming OH radicals ($CH_3OCH_3+OH=CH_3OCH_2+H_2O$, $H_2+OH=H_2O+H$). The ignition at higher temperature are significantly delayed by hydrogen addition, however, the maximum of heat release rate increases a lot for higher hydrogen blending ratio. It indicates that the hydrogen addition promote the HTR by making the burning process more intense. This is due to the oxidation of H_2 which provides OH radicals during the HTR process[7].

3.2 Stratification effects on autoignition

To highlight the stratification effect on autoignition, Figure 2 shows the temporal evolution of volume averaged pressure and mean heat release rate (HRR) for the turbulent cases with temperature inhomogeneities together with the corresponding 0-D homogeneous ignitions simulations. It can be found that the mean HRR in the turbulent cases is more spread out over time while its peak is significantly reduced compared to the corresponding values of 0-D autoignition. And the combustion duration is longer. The ignition occurs sooner for the first stage in the turbulent cases than the 0-D ignition. This is due to the temperature inhomogeneities, leading to higher temperature in some locations and shorter ignition delay. The first stage ignition is found to be retarded with addition of hydrogen. It is also noted that the mean heat release rate during the second and third stage ignition overlap in time for the Cases 1-a and 3-a, and therefore only a single peak is observed, where two maximum of heat release rate can be observed during the HTR for the 0-D ignition. Two significant rising of pressure can be found for the 0-D ignition, which correspond to the LTR and HTR. The pressures after the LTR and HTR are both decreased with the addition of hydrogen. For the turbulent cases, the overall pressure rise rate is smoothed.

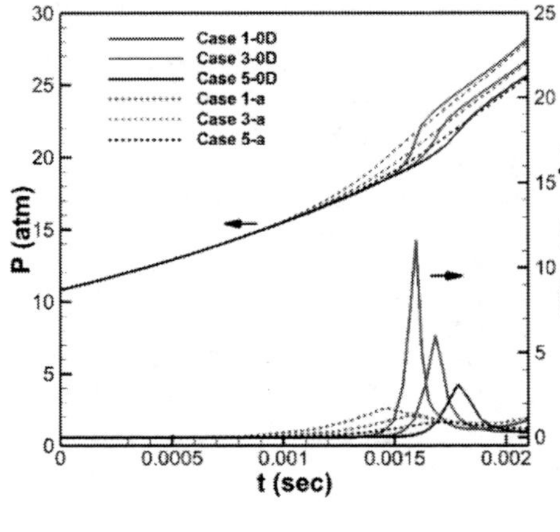

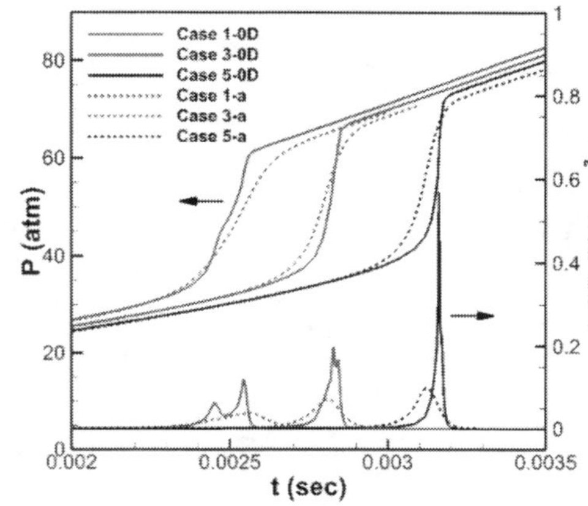

Figure 2: Time evolutions of mean Heat release rate for mixtures with different hydrogen blending ratios

Figure 3 shows the isocontours of heat release field for the three cases at times corresponding to the onset of ignition, the peak first stage ignition and second stage ignition. The onset of ignition correspond to the high temperature locations. It is significantly retarded for Cases 3-a&5-a with hydrogen addition. For the peak first stage ignition, the HRR is much higher for the Case 1-a with DME only than cases with hydrogen blends. It can be observed that the first stage heat release occurs in thin reaction fronts for all three cases. For the peak second stage ignition, the HRR is much higher for the Case 5-a than the other two cases. Heat release occur in thicker reaction fronts for Cases 3-a&5-a than that in Case 1-a.

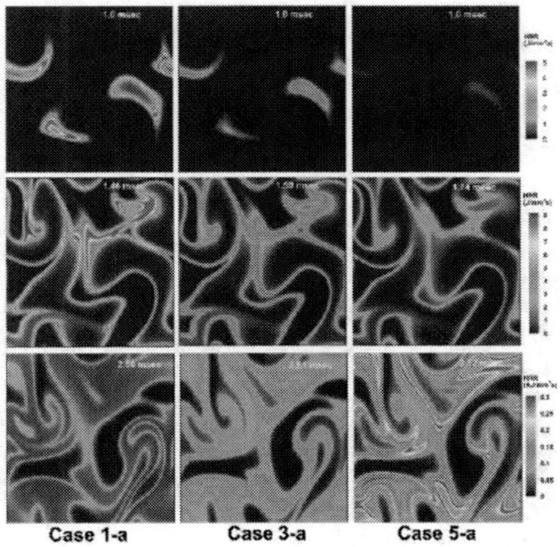

Figure 3: Isocontours of heat release rate at times of the onset of ignition (first row), peak first stage ignition (middle row) and peak second stage ignition (last row)

To understand the effect of two different $T-\phi$ correlations on the ignition of the mixture, the temporal evolution of the mean pressure and HRR for the negatively-correlated (Case 3-b) and uncorrelated (Case 3-c) cases is shown in Fig.4. It can be found that for the three turbulent cases, the heat release rate profile is quite similar, with sooner first stage ignition and reduced maximum of HRR during the low temperature reaction. However, the HRR profiles appear quite different at the high temperature reaction. For Case 3-c with uncorrelated $T-\phi$ fields, the second stage reaction has been advanced significantly, with a reduced maximum HRR. For Case 3-b with negatively correlated $T-\phi$ fields, the volume averaged pressure and heat release rate appear to be consistent with the 0-D ignition, with a high maximum HRR at the high temperature reaction.

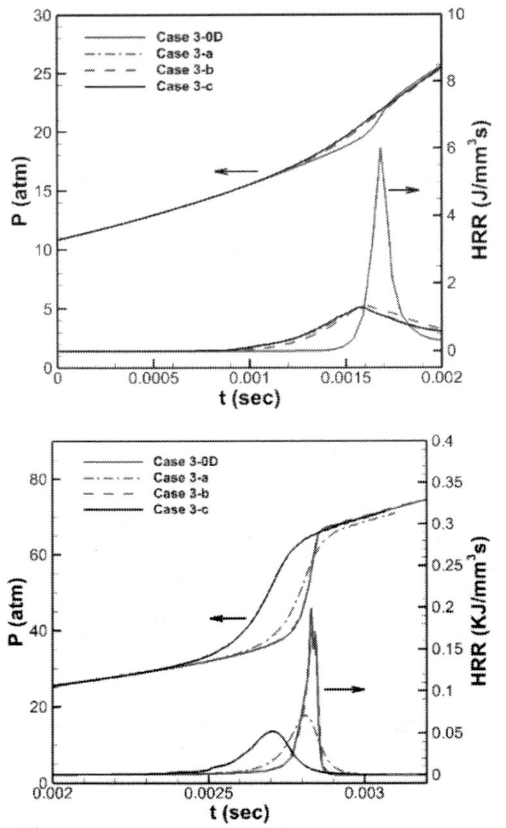

Figure 4: Time evolutions of mean Heat release rate for mixtures with different temperature and composition stratifications

4 Conclusions

This paper reports direct numerical simulations of autoignition of dimethyl ether (DME) and hydrogen blends in high pressure stratified turbulent conditions relevant to HCCI engines. Inert mass source terms were added to the governing equations to mimic isentropic pressure rise that occurs in a practical internal combustion engine. Simulations of 0-D homogeneous autoigniton of the DME/H_2/Air blends are first conducted. It is found that hydrogen addition has a negative effect in the DME low temperature ignition, leading to a declined maximum of the heat release rate and a retarded ignition timing. The ignition at higher temperature is significantly delayed by hydrogen addition, however, the maximum of heat release rate increases a lot for higher hydrogen blending ratio.

The ignition characteristics of thermally stratified mixtures is then investigated. It can be found that the mean HRR in the turbulent cases is more spread out over time while its peak is significantly reduced compared to the corresponding values of 0-D autoignition. The ignition occurs sooner for the first stage in the turbulent cases, however, it is still retarded with addition of hydrogen. The mean heat release rates during the second and third stage ignition overlap in time, and therefore only a single peak is observed for the thermally stratified cases. And the overall pressure rise rate is smoothed.

The ignition characteristics of thermally- and compositionally-stratified mixtures are briefly discussed. The negatively-correlated and uncorrelated cases present a similar advanced first stage ignition and reduced maximum of HRR during the low temperature reaction as the thermally-stratified only case. For the uncorrelated case, the high temperature reaction is significantly advanced, while the volume averaged pressure and heat release rate appear to be consistent with the 0-D ignition for the negatively correlated case. The detailed ignition characteristics would be further studied.

5 Acknowledgment

This work is supported by the National Natural Science Foundation of China (No. 51576176, 91541202) and the Special Program for Applied Research on Super Computation of the NSFC-Guangdong Joint Fund (the second phase). Supercomputing time on ARCHER is provided by the "UK Consortium on Mesoscale Engineering Sciences (UKCOMES)" under the UK Engineering and Physical Sciences Research Council Grant No. EP/L00030X/1.

References

[1] H. Zhang, E.R. Hawkes, J.H. Chen, S. Kook, Proceedings of the Combustion Institute, 34 (2013) 803-812.
[2] Y. Wang, H. Liu, X. Ke, Z. Shen, Appl. Therm. Eng., 119 (2017) 373-386.
[3] A.A. Hairuddin, T. Yusaf, A.P. Wandel, Renewable and Sustainable Energy Reviews, 32 (2014) 739-761.
[4] L. Pan, E. Hu, F. Deng, Z. Zhang, Z. Huang, Int. J. Hydrogen Energy, 39 (2014) 19212-19223.
[5] H. Yu, E. Hu, Y. Cheng, K. Yang, X. Zhang, Z. Huang, Energy & Fuels, 29 (2015) 4567-4575.
[6] J. Liu, H. Wang, M. Ouyang, Int. J. Hydrogen Energy, 36 (2011) 15860-15867.
[7] E. Hu, Y. Chen, Y. Cheng, X. Meng, H. Yu, Z. Huang, Journal of Renewable and Sustainable Energy, 7 (2015) 063121.
[8] E.R. Hawkes, R. Sankaran, P.P. Pébay, J.H. Chen, Combust. Flame, 145 (2006) 145-159.
[9] J.H. Chen, E.R. Hawkes, R. Sankaran, S.D. Mason, H.G. Im, Combust. Flame, 145 (2006) 128-144.
[10] F. Zhang, R. Yu, X.S. Bai, Energy Procedia, 61 (2014) 1536-1539.
[11] H.A. El-Asrag, Y. Ju, Combust. Flame, 161 (2014) 256-269.
[12] G. Bansal, A. Mascarenhas, J.H. Chen, Combust. Flame, 162 (2015) 688-702.
[13] C.S. Yoo, T. Lu, J.H. Chen, C.K. Law, Combust. Flame, 158 (2011) 1727-1741.
[14] C.S. Yoo, Z. Luo, T. Lu, H. Kim, J.H. Chen, Proceedings of the Combustion Institute, 34 (2013) 2985-2993.
[15] F. Zhang, R. Yu, X.S. Bai, Proceedings of the Combustion Institute, 35 (2015) 2975-2982.
[16] S.O. Kim, M.B. Luong, J.H. Chen, C.S. Yoo, Combust. Flame, 162 (2015) 717-726.
[17] M.B. Luong, T. Lu, S.H. Chung, C.S. Yoo, Combust. Flame, 161 (2014) 2878-2889.
[18] T. Jin, K. Luo, S. Lu, J. Fan, Int. J. Hydrogen Energy, 38 (2013) 9886-9896.
[19] K. Luo, H. Wang, F. Yi, J. Fan, Energy & Fuels, 26 (2012) 6118-6127.

11th Asia-Pacific Conference on Combustion,
The University of Sydney, NSW Australia
10th -14th December 2017.

The Effect of Carbon Dioxide Diluted on Combustion Characteristic with a Tubular Flame Burner

Jie Hu[1], Baolu Shi[2], Kazuhiro Hayashida[1], Dasukei Shimokuri[3]

[1]Department of Mechanical Engineering, Kitami Institute of Technology
165 Koen-cho, Kitami, Hokkaido, 090-8507, Japan
[2]School of Aerospace Engineering, Beijing Institute of Technology
No.5 ZhongGuanCun South Street, Haidian, Beijing, 100081, China
[3]Department of Mechanical Science and Engineering, Graduate of School of Engineering, Hiroshima University
1-4-1, Kagamiyama, Higashihiroshima, Hiroshima, 739-8527, Japan

Abstract

In this study, the effect of carbon dioxide diluted (30%~60% in fuel volume) on the combustion characteristics with a tubular flame burner have been systematically investigated, including the flame appearances, the stable flame regions and the axial temperature distributions of the burned gases. For a point of practical view, the exhaust emissions of NO_x and CO at the burner exit have also been examined. The main results obtained are as follow: (1) The tubular flame can be successfully obtained for a wide range in the equivalence ratio and the flow rate of the combustible mixture. However, for $\alpha \geq 0.6$, the rich flammability limits are slightly lower than that of $\alpha = 0.3$. (2) Detailed measurements of the burned gas temperature at the central axis against the various equivalence ratio show that the burned gas temperature is decreased with increasing CO_2 content in the fuel. (3) More detailed measurements on the exhaust emission show that the value of NO_x is reduced approximately to a half value of $\alpha = 0.3$ with increasing CO_2 content to 60%. Thus, the tubular flame burner has the potential for achieving a stable biogas combustion as well as reducing NOx emission.

1 Introduction

Concern with the continuous increasing fossil fuel consumption, the cost of energy and the pollutant emissions has promoted the utilization of renewable energy sources, such as biogas [1-2]. The primary component of biogas is methane (40%~80% by volume depending on the method of biogas production) and carbon dioxide, while minor constitutes may be water vapor (H_2O), nitrogen (N_2) and hydrogen (H_2). Due to these of inert gases, the heating value of biogas is approximately a half value of natural gas of 23.0 MJ/m^3. This means that to achieve the same thermal output a burner must supply 73% more biogas (by volume). This increase in flow rate comparable with same heating output leads to the problem of flame stability related to utilization of biogas as fuels. Many studies have been proposed for improving the flame stabilization, such as lifted diffusion flames, turbulent combustion [3-5]. However, it is a complicated question in these theories as the relevance of each of stabilization mechanisms. To assure utilization of biogas efficiency and stability, the effects of carbon dioxide dilution in methane fuel require a consideration investigation for optimizing biogas burner.

Recently, a new element of flame has attracted attention, that is a tubular flame [6-11]. This tubular flame can be established with a swirling flow type burner and with a counter flow type porous cylinder burner. It has been found that the leanest and richest flammability limits as well as the maximum allowable dilution limits obtained is larger than that of the stander method for the determination of flammability limits by using a propagating flame tube. That can be attributed to two main reasons: (1) the heat loss behind the flame is almost negligible due to its symmetrical temperature distribution. Thus, the flame is almost adiabatic except a radiative heat loss from the hot burned gas; (2) the flame is also aerodynamically stable according to the Rayleigh stability criterion because the flow field consists of an inner burned gas region of low density and an outer unburned gas region of high density.

In this study, in order to achieve a stable combustion with using biogas fuel, a simulated biogas under different CO_2 content by volume in fuel have been introduced into tubular flame burner, then the flame appearances, the stable flame region, the axial temperature distribution of the burned gas and the exhaust emissions have been investigated.

2 Experimental

Figure 1 shows the schematic drawing of the tubular flame burner used in this study. This burner is made of quartz glass, which has two rectangular slits around inner diameter. The inner diameter of burner is 30 mm and its length is 150 mm, while the slit length and width are 70 mm and 2 mm, respectively. To permit viewing and photographing the flames, a round quartz window is installed at the closed end of burner.

Figure 2 shows a schematic of the experimental apparatus. The fuel used is methane/carbon dioxide mixture while the oxidizer used is air. The flow rate of each gas was metered with calibrated float rotameters, well mixed, and then fed to the burner. The unit of flow rate is m^3/h at the STP condition. The flame

Corresponding author. Fax: +81-157-26-9213
E-mail address: jiehu@mail.kitami-it.ac.jp

was photographed using two conventional digital cameras to obtain the front view and the side view through the burner simultaneously. In this study, the volumetric fraction of carbon dioxide in the fuel α is given by the following equation

$$\alpha = \frac{Q_{CO_2}}{Q_{CH_4} + Q_{CO_2}} \tag{1}$$

here Q [m^3/h] is the flow rate of each gas.

Temperatures were determined with a silica-coated Pt/Pt-13%Rh thermocouple (wire diameter: 0.2 mm); but no correction for thermometric errors due to radiation has been made. On the other hand, the combustion gas was sampled with a quartz probe (2 mm in inner diameter) and the concentration of NO_x and other species (CO and CO_2) were determined 100 mm downstream the closed end of burner. NO_x was determined with a chemiluminescence formula NO_x meter (Shimadzu, NOA-7000) and the concentrations of CO and CO_2 were determined with an infrared gas analyzer CO/CO_2 meter (Shimadzu, CGT-7000).

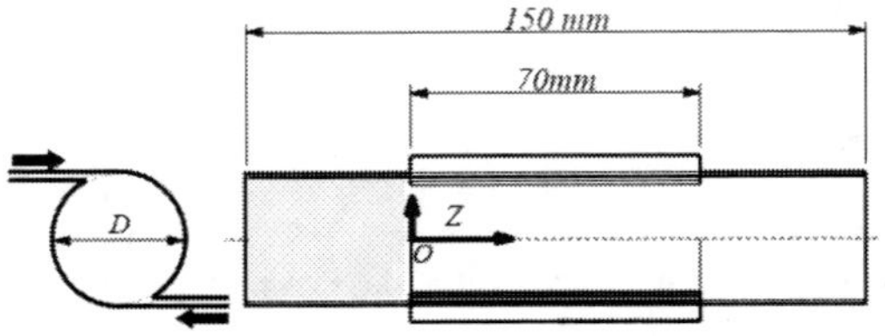

Figure 1. Schematic of the tubular flame burner.

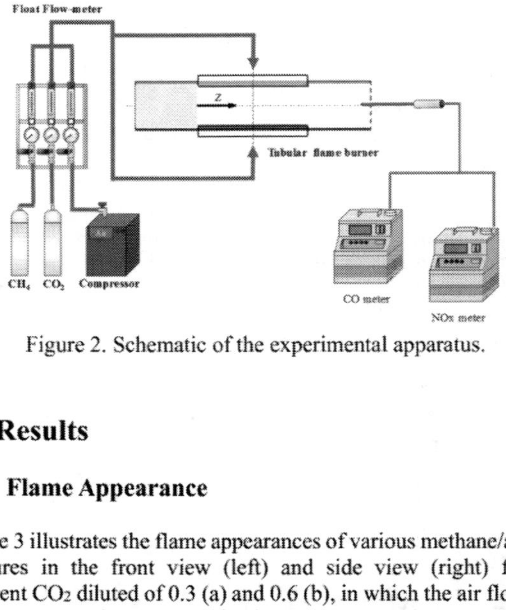

Figure 2. Schematic of the experimental apparatus.

3 Results

3.1 Flame Appearance

Figure 3 illustrates the flame appearances of various methane/air mixtures in the front view (left) and side view (right) for different CO_2 diluted of 0.3 (a) and 0.6 (b), in which the air flow rate for combustion is maintained at a constant value Q_{air} =6.0 m^3/h.

In the case of α = 0.3, it is seen that tubular flames of cylindrical cross section and constant diameter with a blue flame luminosity can be established (Fig. 3a). As seen in the left end pictures, the flame diameter becomes larger as the stoichiometric condition is approached (Fig 3a middle). This is simply because the burning velocity increases as stoichiometry is approached. The flame moves toward the unburned gas side, i.e., the wall. On the other hand, the right end pictures in Fig. 3a, show that a long, constant diameter flame is formed in the combustion tube. Its length is almost remained at constant as

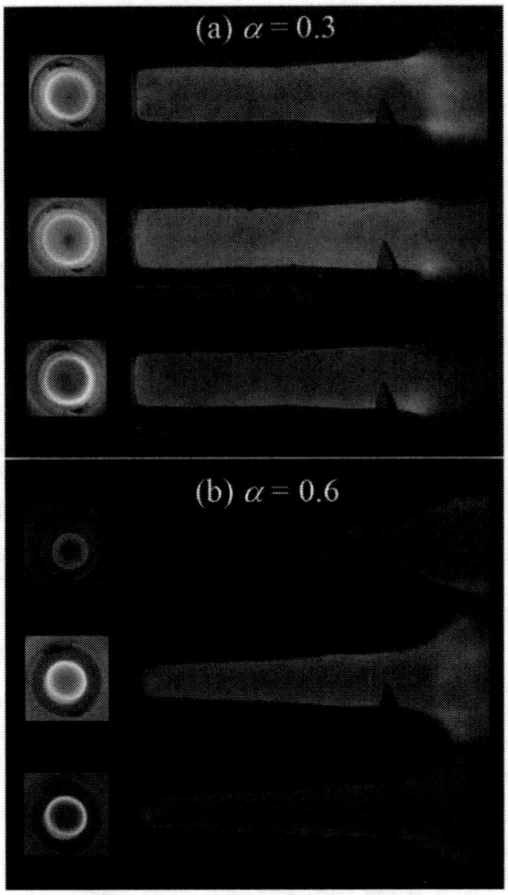

Figure 3. Flame appearances of carbon dioxide diluted in fuel. (a) α = 0.3; (b) α = 0.6; upper, Φ = 1.2; middle, Φ = 1.0; lower, Φ = 0.8; Q_{air} =6.0 m^3/h.

the stoichiometric condition is approached. However, it is noted that the tubular flames are firmly established at the rich and stoichiometric mixtures, whereas the flame luminosity is weakened as the closed end is approached in the lean mixtures.

When α is increased to 0.6, it is seen that tubular flames of cylindrical cross section and constant diameter with a blue flame luminosity can be established (Fig. 3b). However, it is clearly shown in the right end pictures that the length of the stoichiometric flame is greater than that of α = 0.3 (Fig. 3a middle). The luminosity of the lean flame is weaker and the diameter is also smaller than those of α = 0.3, lean flame (Fig 3b lower). The luminosity of the rich flame gets weaker as the tangential slits are approached. The luminous flame disappears around the injection region for rich mixtures, as shown in Fig 3b upper.

3.2 The Stability Limits

Next, the stability limits of tubular flames under different CO_2 content have been determined by varying the air and fuel flow rates. The results are shown in Fig. 4.

In the case of α = 0.3, the equivalence ratios at the lean and rich extinction limits are respectively about 0.5 and 1.4 at the air flow rate of 4.8 m^3/h. With an increase in the air flow rate, the equivalence ratio at the lean and rich limits is almost kept constant at Φ = 0.5 and Φ = 1.4, respectively. With increasing α to 0.4, the equivalence ratios at the lean and rich extinction limits are almost as same values as those of α = 0.3, lean and rich mixtures. With further increase of α to 0.5 and 0.6, the equivalence ratios at the rich extinction limits are significantly shifted to 1.2 and 1.1, respectively. These shifts may be attributed to different diffusivities of reactants. Since the effect of preferential diffusion based on different mass diffusivities and the resulting effect of different Lewis numbers, the flame is prone to be extinguished with getting close to the stagnation [12]. In table 2, Lewis numbers estimated for various α at the rich equivalence ration of $\Phi = 1.2$ are shown. Here, Lewis number is defined as

$$Le = \frac{\alpha_m}{D_m} \quad (2)$$

where α_m is the thermal diffusivity of the unburned mixture and D_m is the diffusion coefficient of the reactant. For example, in the case of $\alpha = 0.3$, the Lewis number is equal to unity, whereas its value is sufficiently less than unity for $\alpha = 0.6$. Hence, the stable tubular flame cannot be obtained under the rich mixtures when $\alpha \geq 0.6$.

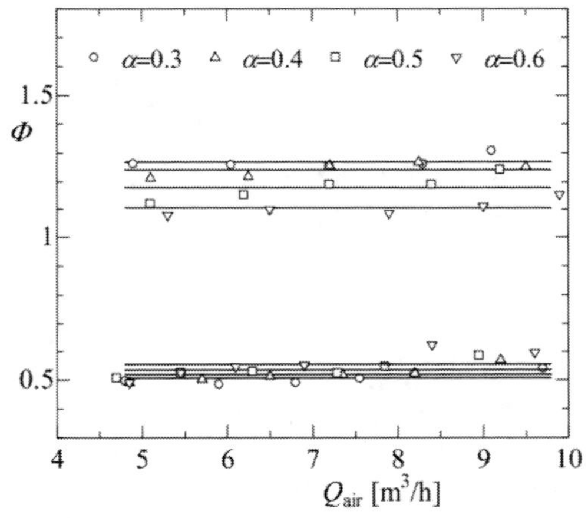

Figure 4. Mapping of stability limits for various carbon dioxide diluted in fuel.

α	0.3	0.4	0.5	0.6
Le	1.000	0.984	0.962	0.933

Table 2
Lewis numbers for various CO_2 content

3.3 Axial Temperature Distribution

Figure 5 shows the variations of temperature distributions along the burner axis for various carbon dioxide volumetric fractions, while the air flow rate is maintained at a constant value Q_{air} =6.0 m^3/h and the equivalence ratio is kept at $\Phi = 0.8$.

In the cases of $\alpha = 0.3$ and 0.4, the burned gas temperature T gradually increases from the tangential slit of burner and reaches its maximum of 1620 °C and 1570 °C, respectively. On the other hand, for the cases of $\alpha = 0.5$ and 0.6, the temperature gradually decreases and reaches its minimum of 1480 °C and 1440 °C, respectively. It seems that the burned gas temperatures are almost uniform and greater than 1400 °C near the exit of the burner. It is, however, interesting to note that the burned gas temperature is decreased with increasing α.

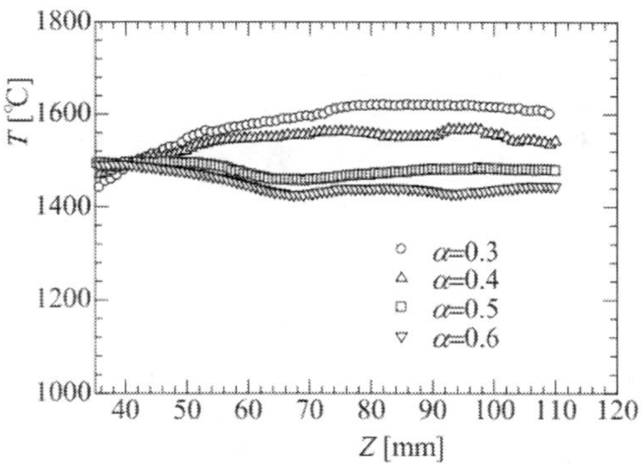

Figure 5. Variations of axial temperature distribution with α (Q_{air} =6.0 m^3/h, $\Phi = 0.8$).

3.4 The Burned Gas Temperature

Figure 6 shows the burned gas temperatures with various equivalence ratios obtained under different of carbon dioxide volumetric fractions at $Z = 35$ mm., in which the air flow rate is kept as same condition as that of Fig. 5.

As seen in Fig. 6, the maximum temperature occurs near the stoichiometric conditions for all the cases. By increasing or decreasing Φ, namely temperature T at lean or rich mixture is gradually decreased. In case of α = 0.3, the burned gas temperature T at stoichiometric condition is reached to 1478 °C and decreased to 1255 °C for lean condition and 1431 °C for rich condition, respectively. When α is increased to 0.4, the maximum temperature is almost kept as the same value as that of $\alpha = 0.3$. T is decreased to 1231 °C at $\Phi = 0.56$ and 1299 °C at $\Phi = 1.27$. When α is further increased to 0.5, T is about 1181 °C, 1466 °C and 1313 °C for lean, stoichiometric and rich condition, respectively. Finally, when α is kept at 0.6, the temperature is 1095 °C, 1433 °C, 1283 °C for $\Phi = 0.56$, 0.95 and 1.1, respectively. Thus, the burned gas temperature is significantly decreased with increasing α.

It is noted that the measured temperature is 200~300 °C lower than adiabatic flame temperature obtained by use of the CEA program [13]. For example, at Φ = 0.8, the temperatures measured is 1451 °C, and the adiabatic flame temperature estimated is 1655 °C for $\alpha = 0.3$. The main reason is that the radial temperature distributions of the tubular flame becomes M-shaped [10-11].

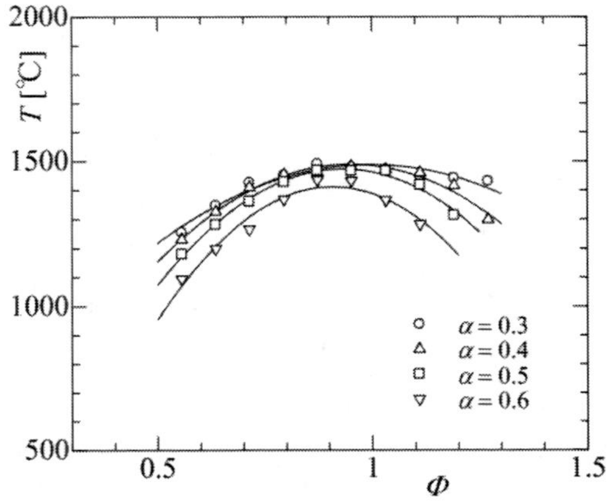

Figure 6. Variations of the burned gas temperature with the equivalence ratio (Q_{air} =6.0 m^3/h).

3.5 NOx and CO

Finally, the emissions have been examined at the exit of burner, in which α is selected as a parameter. The measured NO_x values are converted to the values at 5% O_2 concentration on the basis of O_2 concentration determined in the exhaust gas through the following relation

$$[NO_x] = \frac{0.2095 - 0.05}{0.2095 - [O_2]_m} [NO_x]_m \qquad (3)$$

in which $[O_2]_m$ and $[NO_x]_m$ refers to measured O2 and NO_x volume concentrations in the combustor exit, respectively.

Figure 7 shows the variations of NO_x and CO emissions and the temperature at the exit of burner with the volumetric fraction of carbon dioxide. It is seen that with an increase of the carbon dioxide volumetric fraction, the burned gas temperature is decreased. Accordingly, the NO_x value is gradually decreased with increasing α. For example, the measured NO_x value is 14 ppm for $\alpha = 0.3$, whereas the NO_x value is decreased to 6 ppm for $\alpha = 0.6$. It is important to note that the CO concentration around the burner exit is closed to 0 % for all the cases of α = 0.3, 0.4, 0.5 and 0.6; hence, the combustion seems almost complete. As well known, there are two major routes for NO formation, the thermal route (Zeldovich) and the prompt route (Fenimore). As previously mentioned, in all the cases of $\alpha = 0.3$, 0.4, 0.5 and 0.6, the temperatures exceed 1400 °C near the exit of the burner. Thermal-NO formation rate is significant only at high temperatures greater than 1800 K [14]. Thus, it seems that most of NOx for all the case of $\alpha = 0.3$, 0.4, 0.5 and 0.6 is formed through the thermal route of the Zeldovich mechanism.

As well known, there are two major factors in reducing NO formation rate in the thermal route (Zeldovich mechanism). One is the reduction of temperature and the other is the reduction of oxygen concentration, both of which can be quantitatively evaluated by considering the rate limiting, elementary reaction step $O + N_2 \rightarrow NO + N$. These reductions of oxygen concentration and temperature can describe the drastic reduction of NOx through the thermal route [15-16]. Thus, the present tubular flame burner with using biogas has the potential for reducing NO_x emissions.

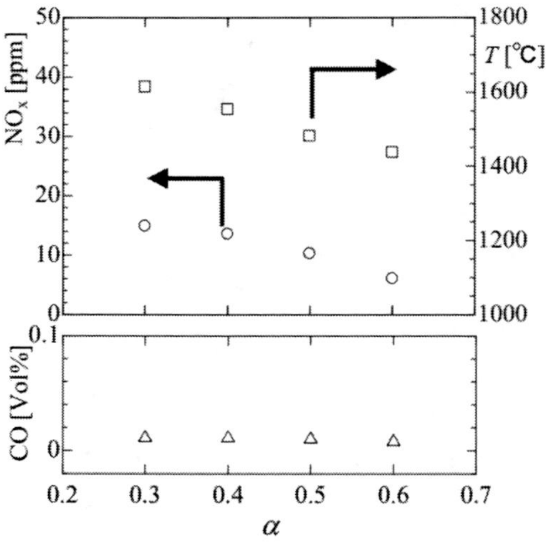

Figure 7. Variations of exhaust emission and temperature with α at the exit of burner (Q_{air} =6.0 m^3/h, Φ = 0.8).

4 Conclusions

In this study, a simulated biogas, which consists of various methane and carbon dioxide contents, is injected to burn in a tubular flame burner. The combustion characteristics under various CO_2 diluted are investigated, including the flame appearances, the stable flame region and the axial temperature distribution of the burned gas. An additional measurement on NO_x emission has also been made. The following are specific conclusions obtained.

(1) When $\alpha < 0.6$, a stable combustion with a simulated biogas can be successfully obtained for a wide range in the equivalence ratio and the flow rate of the combustible mixture. When $\alpha \geq 0.6$, a stable tubular flame cannot be obtained at rich side.

(2) Detailed measurements on the burned gas temperature at the central axis against the various equivalence ratio show that the burned gas temperature is decreased with increasing α.

(3) More detailed measurements on the exhaust emission show that the value of NO_x is approximate a half value of $\alpha = 0.3$ with increasing α from 0.3 to 0.6.

Thus, the tubular flame burner has the potential for achieving a stable biogas combustion as well as reducing NOx emission.

References

[1] J. Chomiak, J.P. Longwell, A.F. Sarofim, Prog. Energy. Combust. Sci. (15) (1989) 109-129.

[2] L. Pizzuti, C.A. Martins, P.T. Lacava, Renew. Sust. Energ. Rev. (62) (2016) 856-865.

[3] L. Vanquickenborne, A.V. Tiggelin, Combust. Flame (10) (1966) 59-69.

[4] H. Eickhoff, B. Lenze, W. Leuckel, Proc. Combust. Inst. 20 (1985) 311-318

[5] S.H. Won, J. Kim, K.J. Hong, M.S. Cha, S.H. Chung, Proc. Combust. Inst. 30 (2005) 339-347.

[6] S. Ishizuka, Proc. Combust. Inst. 20 (1985) 287-294.

[7] M. Nishioka, T. Takeno, S. Ishizuka, Combust. Flame (73) (1988) 287-301.

[8] H. Kobayashi, M. Kitano, Combust. Flame (76) (1989) 89-105.

[9] P.A. Libby, N. Peters, F.A. Williams, Combust. Flame (75) (1989) 265-280.

[10] S. Ishizuka, Prog. Energy Combust. Sci. (19) (1993) 187-226.

[11] K. Yamamoto, S. Ishizuka, T. Hirano, Proc. Combust. Inst. (25) (1994) 1399-1406.

[12] S. Ishizuka, R. Hagiwara, M. Suzuki, A. Nakamura, O. Hamaguchi, Trans. JSME 65 (639) (1999) 3845-3852.

[13] NASA, The NASA Computer Program CEA 1995.

[14] A.F. Sarofim, R.C. Flagan, Prog. Energy Combust. Sci. (2) (1976) 1-25.

[15] S.R. Turns, An Introduction to Combustion, second ed., McGraw Hill, Boston, 2000, p. 130-133, 168-171, 573-584.

[16] J. Warnatz, U. Mass, R.W. Dibble, Combustion, 3rd ed., Springer-Verlag, Berlin, 2001, p. 237-256.

11th Asia-Pacific Conference on Combustion,
The University of Sydney, NSW Australia
10th -14th December 2017.

Computed NO Emission Characteristics of CH_4/NH_3 and H_2/NH_3 Opposed-Jet Diffusion Flames

Kai-Chieh Chia, Hsin-Yi Shih, Yin-Cheng Ko

Department of Mechanical Engineering, Chang Gung University

Guishan District, Taoyuan City 33302, Taiwan

Abstract

The combustion and NO emission of CH_4/NH_3 and H_2/NH_3 blended fuels as alternative energy sources were studied. By using the configuration of opposed-jet diffusion flames, the effects of ammonia concentrations on the flame structures, chemical reaction and NO formations of CH_4 and H_2 were investigated and compared. The numerical model consists of a revised OPPDIF program and narrowband radiation model with detailed chemical kinetics. The analyses of the H_2 diffusion flames showed that the NO formation is dominated by the thermal-NO due to high flame temperature. While for CH_4 diffusion flames, the variations of the NO emission and flame temperature with strain rates are similar only when the strain rate is above 100 s^{-1}. A great amount of NO are generated by prompt-NO through CH_3, CH_2 and CH radicals, and only small amount of NO are from thermal route. Since the H_2 flame has a higher temperature, the NO production rate is higher than the CH_4 flame. For the studies of H_2/NH_3 and CH_4/NH_3 blended fuels, the maximum flame temperatures decrease with the increase of the NH_3 concentrations, but the net production rates of NO increase with NH_3 addition. The NO formation route is mainly due to the fuel-NO which comes from chemical reactions of NH_3.

1 Introduction

In recent years, energy and environmental issues have received global attention. To reduce the use of carbonaceous fuels, the application of hydrogen and hydrogen blended fuels are considered one of the major energy sources in the future. Since 2001, hydrogen-based research and technology development has also opened the so-called "hydrogen economy" era [1]. That is to say hydrogen can replace the previous fossil energy-based lifestyle and other related issues. There are many advantages of hydrogen as a fuel, such as burning speed, low ignition energy, high heating value, wide flammable range and other characteristics. Besides the cost of hydrogen production, hydrogen itself is colorless and odorless, so it is difficult to detect when it leaks, which is a safety concerns in hydrogen storage and transportation. Therefore, adding ammonia in hydrogen has been proposed [2]. The chemical structure of ammonia has no carbon composition, so there are no extra carbon emissions when H_2/NH_3 blended fuels are burn. In addition, due to the pungent smell, ammonia can be easily detected while leaking. The high ignition energy of ammonia can also reduce the risk of mixed gas explosion.

On the other hand, methane (natural gas) is widely used nowadays and its carbon emissions are less compared to other petrochemical fuels, and it can also be obtained from the fermentation of biomass waste. In the past, hydrogen transportation using natural gas pipeline has been proposed and hydrogen addition to methane is also one of the applications to enhance combustion. Therefore, the addition of ammonia to methane will also be considered a possibility for alternative energy source as well.

The use of ammonia as a fuel was successfully applied to the spark ignition engine, diesel engine, and the gas turbine engine during 1960s [3]. Since the hydrogen energy era has been promoted, ammonia as a potential hydrogen carrier has also been proposed [4]. As ammonia is easily to obtain in nature [5], the application of ammonia combustion research has become more popular. In 2012, Praveen Kumar [6] used Hencken burner and swirl-stabilized burner to analyze the combustion and NO emissions of H_2/NH_3 and CH_4/NH_3 laminar and turbulent flames at different ratios of NH_3. Not only the best ratio was obtained, they found that 100% NH_3 combustion is feasible. For the application of internal combustion engines, Lee et al. [7] have shown that the addition of appropriate ammonia can enhance the thermal efficiency of the internal combustion engine. Besides, ammonia itself is carbon-free, with high ignition temperature and low activity, so it can reduce the production of CO_2 while mixed with CH_4 and increase the safety of storage and reduce the laminar flame speed while mixed with H_2 [8]. Reiter and Kong [9] found that up to 95% of NH_3 addition in conventional diesel engines can reduce NO_x production. Um et al. [10] studied the combustion stability and NO emission of the NH_3 addition to the H_2/air diffusion flames. Lee and Kwon [11] studied the effects of NH_3 on NO_x production in synthetic fuels, and results showed that in the high strain rate, the temperature was relatively low, but the production of NO_x has increased. This phenomenon indicated that the NO_x formation in syngas fuels with the addition of NH_3 is dominated by chemical reactions, not temperature alone. Lee et al. [12] then further found out that the emission of NO_x in the fuel rich condition was much less than the fuel lean condition for hydrogen/air premixed flames.

From the literatures, the application H_2/NH_3 or CH_4/NH_3 combustion is feasible and has the possibility of replacing gasoline and diesel in the internal combustion engines. However, the NO_x formation and emission characteristics need to be further understood. Therefore, this study will investigate NO formation characteristics of H_2/NH_3 and CH_4/NH_3 blended fuel combusted with air in the configuration of opposed-jet diffusion flames at different strain rates.

2 Numerical model and method

In this study, the flame structures, chemical reactions and NO formation were calculated with the opposed-jet diffusion flames configuration as shown in figure 1. One side of opposing jets is the fuel either H_2/NH_3 or CH_4/NH_3, and the other side is air, which will produce a diffusion flame near the stagnation plane between two nozzles. The physical model assumes laminar axisymmetric stagnation point region and it

Corresponding author: Prof. Hsin-Yi Shih
E-mail address: hyshih@mail.cgu.edu.tw

can be simplified into one-dimensional governing equations by boundary layer theory, which is essentially a set of simultaneous nonlinear boundary value problems. In general, there are two different formations: one specifies the jet velocities and the distance between two nozzles and the other one is to given a strain rate (velocity gradient) directly. Usually, giving the velocities and spacing of the jets is more appropriate to compare the experimental results. However, the strain rate formulation is more consistent with the theory and convenient to parametric study. Therefore in this study, the model of strain rate approach is employed, and the detailed governing equations and boundary conditions can be found in the literature [13, 14].

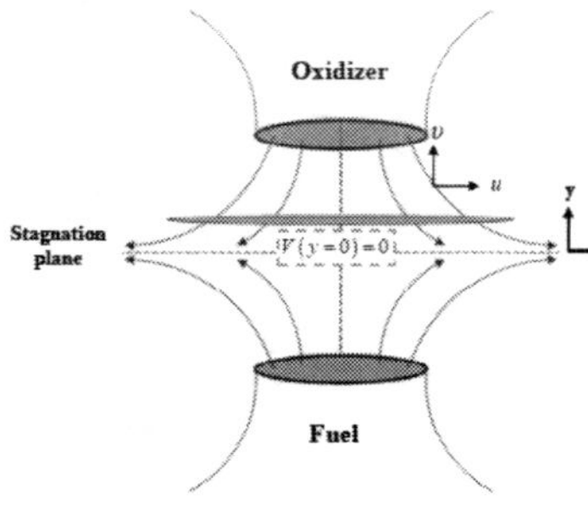

Figure 1: The schematic of opposed-jet diffusion flame.

The computational program OPPDIF [15] developed by Sandia Laboratories was revised in this work with strain rate formulation, which coupled with statistic narrowband radiation model [16]. The flame radiation was calculated for CO, CO_2, and H_2O. The chemical reactions and the thermal transport properties were computed by CHEMKIN [17] and TRANSPORT [18] sub-programs. The detailed chemical mechanism GRI-Mech 3.0 [19] was used, which contains 53 species and 325 reaction steps.

3 Results and discussions

3.1 Strain rate effects on NO formation

First, the flame structures and NO production rates for H_2/air and CH_4/air diffusion flames at different strain rates were analyzed. Figure 2 shows the variations of maximum flame temperatures and net NO production rates with strain rates from 10 s^{-1} to 1000 s^{-1} for H_2/air diffusion flames. The results show that the highest flame temperature is near the strain rate of 50 s^{-1}, and then it decreases gradually toward both higher and lower strain rates. As for net NO production rates, it exhibits similar trend of the flame temperature variation, indicating the characteristics of thermal-NO. However, at high strain rate, the temperature is lower than that at low strain rate, but the NO production rate is higher. Further analysis reveals that although the contribution fom thermal-NO is relatively smaller at high strain rate, NNH-intermediate route becomes more important. In addition, at low strain rate, the reburn route where NO is converted to NH through HNO is intensely active, resulting in a dramatical drop of NO production at low strain rate.

Figure 3 shows the variations of maximum flame temperatures and net NO production rates with strain rates for CH_4/air diffusion flames from 10 s^{-1} to 350 s^{-1}. The results show that the temperature increases as the strain rate decreases, and it reaches the highest near the strain rate of 10 s^{-1}. Although it is not shown in the figure, the temperature is expected to decrease with further decrease of strain rates due to flame radiation loss at lower strain rates. However, the net NO production rates exhibit non-monotonic behavior, even the variation of maximum flame temperature is monotonic in the computed range of strain rates. The results show that the temperature is lower at strain rate of 100 s^{-1} than that at 10 s^{-1}, but NO production rate is higher. Further analysis indicates prompt-NO is the main reason behind this, where NO is formed through CH group. Besides, NO is generated more through NH and HNO at strain rate of 100 s^{-1}.

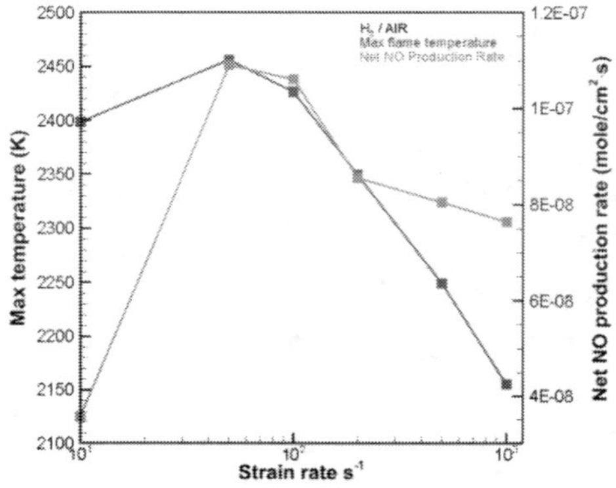

Figure 2: Variations of maximum flame temperature and net NO production rate with strain rate for H_2/air diffusion flames.

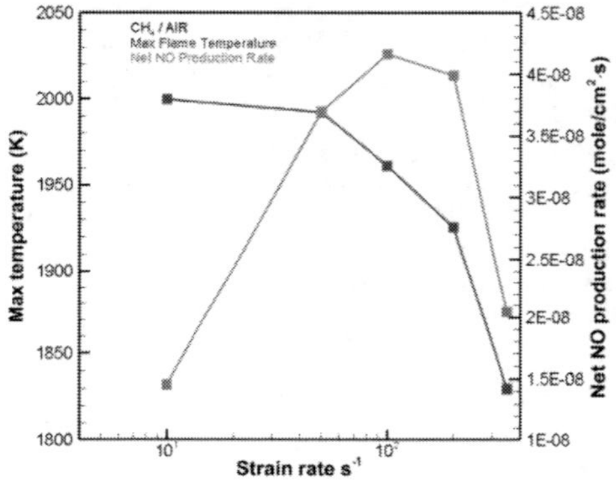

Figure 3: Variations of maximum flame temperature and net NO production rate with strain rate for CH_4/air diffusion flames.

3.2 NH_3 addition effects on NO formation

Here the NH_3 addition on CH_4/air and H_2/air diffusion flames are investigated. Figure 4 shows the changes of maximum flame temperature and net NO production rate with NH_3 percentage for H_2/NH_3 blended fuels at strain rates of 10, 100, 500, and 1000 s^{-1}. Due to lower heating value of NH_3, the maximum flame temperature decrease as NH_3 concentration increases. However, the net NO production rate increases with NH_3 percentage. Recall that for H_2/air diffusion flames, NO formation is dominant by thermal-NO, and the variation of NO production rates with strain rates is non-monotonic. With the addition of NH_3, the flame temperature is the highest at 10 s^{-1} and the NO production rate is just the opposite, higher temperature flame with lower NO production rate among these four strain rates. Therefore, NO emission with NH_3 addition is mainly dominated by chemical reaction, where the fuel-NO reaction pathway becomes more significant.

Figure 5 shows the variations of maximum flame temperature and net NO production rate with NH_3 percentage for CH_4/NH_3 blended fuels at strain rates of 10, 50, and 100 s^{-1}. Like the trend of H_2/NH_3 diffusion flames, NO production increases with NH_3 concentration. However, at certain NH_3 percentage, higher flame temperature with higher NO production rate, which is not attributed to thermal-NO, but to fuel-NO and prompt-NO. The detailed NO reaction pathways will be discussed in the next section. Note that the NO production rate of CH_4/NH_3 diffusion flame is always greater than that of H_2/NH_3 flame under the same strain rate and NH_3 concentration by comparisons of figure 4 and figure 5.

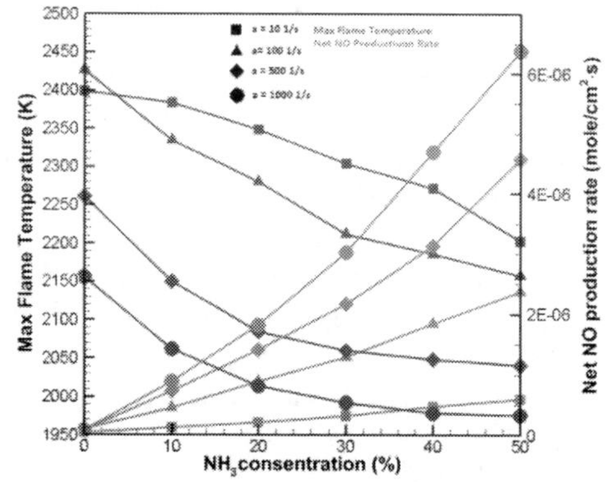

Figure 4: The variations of maximum flame temperature and net NO production rate with NH_3 concentration for H_2 diffusion flames.

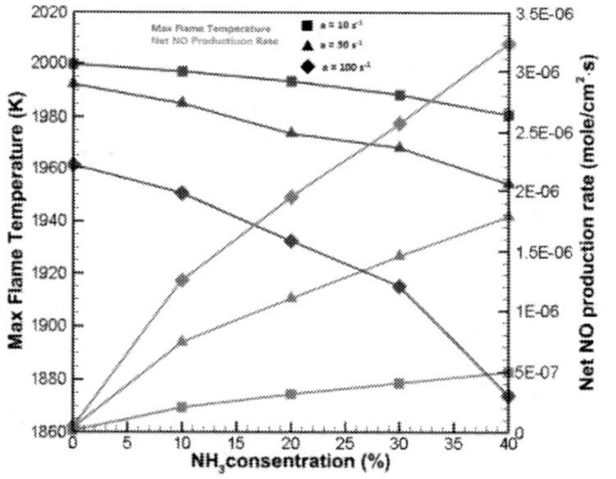

Figure 5: The variations of maximum flame temperature and net NO production rate with NH_3 concentration for CH_4 diffusion flames.

3.3 NO reaction pathways

The NO reaction pathways are analyzed, where the primary, secondary and other paths are identified. The values of the net reaction rates to produce the species are also represented. Figure 6 shows the NO reaction pathways of the H_2/air diffusion flame (without the NH_3 addition) at strain rate of 100 s^{-1}. The analysis shows that the main source of NO formation is from N_2 which is converted to NO by thermal route due to high temperature. The secondary NO formation path is through the NNH-intermediate route, where NNH can be converted to NO directly or changed to NH, and then decomposed into N to produce NO. Note that some of the NO can be consumed back to NH through HNO (reburn route) as shown in the figure. Figure 7 is the NO reaction pathways for H_2/NH_3 diffusion flame at strain rate of 100 s^{-1}, where NH_3 concentration is 40%. Analysis shows that NO is mainly produced by fuel-NO route. NH_3 is transformed into NH_2 and then into NH, which reacts with H_2O to form HNO and then produce NO. Also, NH is decomposed into N then produced NO. NH_3 become the main source of NO production.

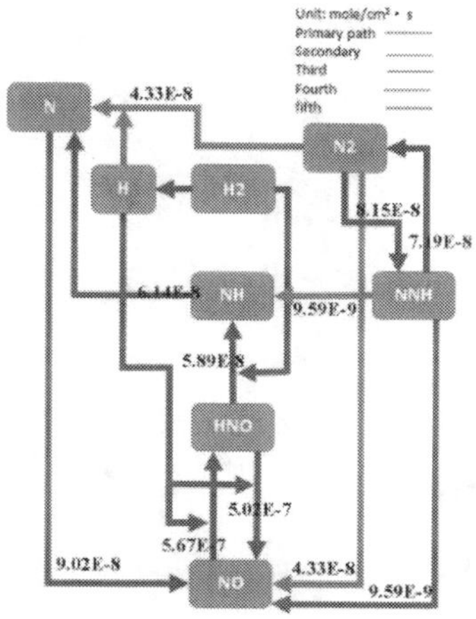

Figure 6: NO reaction pathways of H_2/air diffusion flame (a=100 s^{-1}).

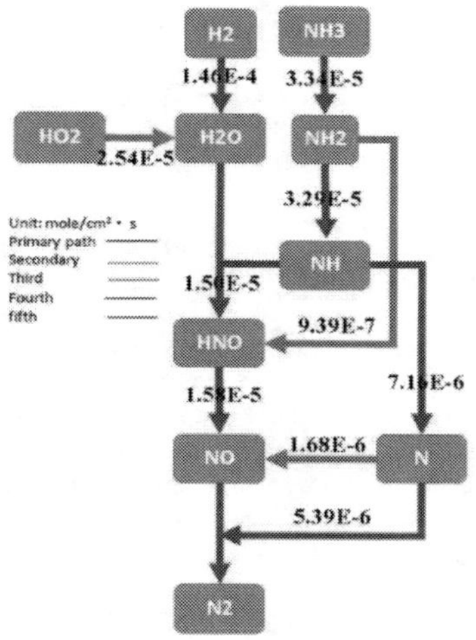

Figure 7: NO reaction pathways of H_2/NH_3 diffusion flame (a=100 s^{-1}, 40% NH_3).

Figure 8 then presents NO reaction pathways of CH_4/air diffusion flame (without addition of NH_3) at strain rate of 100 s^{-1}. In this case, NO is mainly produced through the CH radicals-related reactions, the route of prompt-NO. The sequential reactions to form HCN, NCO, NH and HNO are the primary routes to produce a large amount of NO. Figure 9 then shows the NO reaction pathways of CH_4/NH_3 diffusion flame with 40% NH_3 at strain rate of 100 s^{-1}. Besides the prompt-NO similar to the previous case, where NO is generated from a series of reactions involving CH radicals, HCN, NCO, NH, and HNO. The fuel-NO route becomes very important. NO is mainly produced from NH_3, which is transformed to NH and then HNO. Finally, HNO produces significant amount of NO.

Comparing these reaction pathways, without the addition of NH_3, NO formation in H_2 diffusion flame is mainly by thermal route. For CH_4 diffusion flame, NO is generated first thermally and then produce a greater amount of NO through reactions with CH-radicals. While with the addition of NH_3, NO is mainly generated through the NH_3 conversion, regardless H_2 or CH_4 diffusion flames.

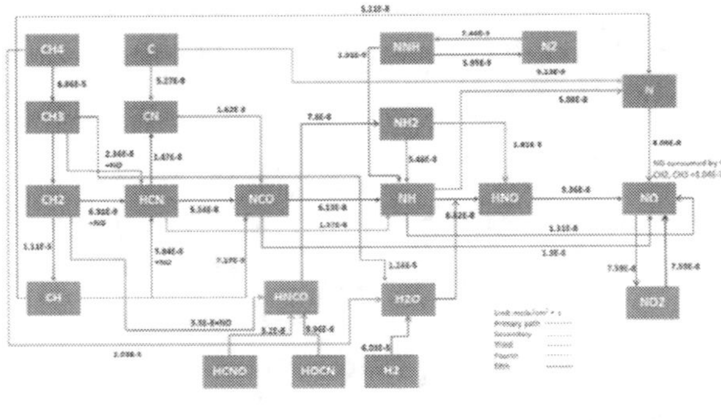

Figure 8: NO reaction pathways of CH_4/air diffusion flame ($a = 100\ s^{-1}$).

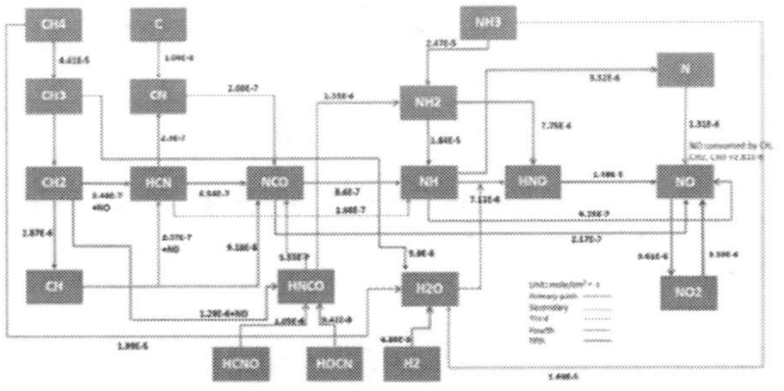

Figure 9: NO reaction pathways of CH_4/NH_3 diffusion flame ($a = 100\ s^{-1}$, 40% NH_3).

4 Conclusions

The NO formation characteristics of H_2/NH_3 and CH_4/NH_3 opposed-jet diffusion flames were studied and analyzed. H_2 diffusion flame has a higher flame temperature near the strain rate of 50 s^{-1}, and NO formation is dominated by thermal route. For CH_4 diffusion flame, temperature near strain rate of 10 s^{-1} is the highest and temperature decreases as strain rate increases, but NO production rates appear non-monotonic with the strain rates due to NO production by prompt-NO and consumption by reburn route. With the addition of NH_3, the maximum flame temperature decreases, but NO production rate increases with the increase of NH_3 concentration. From the NO reaction pathways, NO is generated primary through fuel-NO route, where NH_3 is transformed into NH_2 and then NH, which reacts with H_2O to form HNO and then produce substantial amount of NO.

5 Acknowledgment

This study is supported by the Ministry of Science and Technology, Taiwan under MOST 105-2815-C-182-039-E.

References

[1] National Hydrogen Vision Meeting, U.S. Department of Energy, Washington, DC, Nov. 15-16, 2001.

[2] Zamfirescu C, Dincer I. Using ammonia as a sustainable fuel. J Power Sources 2008; 185:459-65.

[3] Bomelburg HJ. Use of ammonia in energy-related applications, plant/operations progress. Plant/Operations Progress 1982; 1(3):175–180.

[4] Veziroglu T, Sahin S. 21st century's energy: hydrogen energy system. Energy Conversion and Management 2008; 49(7):1820-1831.

[5] Brandhorst H, Baltazar-Lopez M, Tatarchuk B, Cahela DR, Barron T. Ammonia its transformation and effective utilization. In: 6th International energy conversion engineering conference, Cleveland, Ohio; 2008, p. 1-11.

[6] Praveen K. An experimental and numerical study of NOx formation mechanisms in NH3-H2-Air flames (2012). Graduate Theses and Dissertations. Iowa State University, U.S.A.

[7] Lee JH, Lee SI, Kwon OC. Effects of ammonia substitution on hydrogen/air flame propagation and emission. Int J Hydrogen Energy 2010; 35:11332-41.

[8] Law CK, Kwon OC. Effects of hydrocarbon substitution on atmospheric hydrogen–air flame propagation. International Journal of Hydrogen Energy Volume 29, Issue 8, July 2004, Pages 867–879.

[9] Reiter AJ, Kong SC. Combustion and emissions characteristics of compression-ignition engine using dual ammonia-diesel fuel. Fuel 2011; 90:87-97.

[10] Um DH, Joo JM, Lee S, Kwon OC. Combustion stability limits and NOx emissions of nonpremixed ammonia-substituted hydrogen-air flames. Int J Hydrogen Energy 2013:14854–14865.

[11] Lee S, Kwon OC. Effects of ammonia substitution on extinction limits and structure of counterflow-nonpremixed hydrogen/air flames. Int J Hydrogen Energy 2011; 36:10117-28.

[12] Lee JH, Lee SI, Kwon OC. Effects of ammonia substitution on hydrogen/air flame propagation and emission. Int J Hydrogen Energy 2010; 35:11332-41.

[13] Smooke MD, Giovanbibli V. Formulation of the premixed and non-premixed test problems. Lecture Notes in Physics, Ser. 384, Chap. 1, Springer-Verlag, New York, 1991.

[14] Shih HY and Hsu JR. A computational study of combustion and extinction of opposed- jet syngas diffusion flames. Int J of Hydrogen Energy 2011; 36: 15868-15879.

[15] Lutz AE, Kee RJ, Grcar JF, Rulple FM. Oppdif: a fortran program for computing opposed-flow diffusion flames. Sandia National Laboratories Report SAND96-8243, 1996.

[16] Bedir H, Tien, JS, Lee HS. Comparison of different radiation treatments for aonedimensional diffusion flame. Combustion Theory and Modelling 1997; 1:395-404.

[17] Kee RJ, Rulpey FM, Miller JA. Chemkin II: a fortran chemical kinetics package for the analysis of gas-phase chemical kinetics. Sandia National Laboratories Report SAND89- 8009, 1989.

[18] Kee RJ, Dixon-Lewis G, Warnatz JA, Coltrin ME, Miller JA. Fortran computer code package for the evaluation of gas-phase multi-component transport properties. Sandia National Laboratories Report SAND86-8246, 1986.

[19] Gas Research Institute, available from: http://www.me.berkeley.edu/gri_mech/

11th Asia-Pacific Conference on Combustion,
The University of Sydney, NSW Australia
10th -14th December 2017.

Ignition Delay Times of Kerosene Blended with a Bio-Fuel Measured in a Shock Tube

Hee Sun Han[1], Chul Jin Kim[1,2], Chae Hoon Sohn[1*], Jeongsik Han[3], Byung Hun Jeong[3]

[1]Department of Mechanical Engineering, Sejong University
Seoul, 05006, Korea
[2]Howon University
Jeollabuk-do, Gunsan, 54058, Korea
[3]Agency for Defense Development
Daejeon, 305-600, Korea

Abstract

Ignition delay time of kerosene blended with a bio-fuel is measured experimentally by a shock tube for a comparative study with general aviation fuels. The blended fuel of kerosene (Jet A-1) and a bio-fuel shows NTC (negative temperature coefficient) behavior as kerosene does. The blended fuel is similar to Jet A-1 in terms of H/C ratio, density, and heat of combustion. But, it differs from Jet A-1 in ignition characteristics and sooting propensity. The measured ignition delay times of the blended fuel are compared with those of jet fuels over a wide range of temperature at 20 atm and it is found that the blended fuel has significantly short ignition delay time at lower temperature than 1,000 K.

1 Introduction

Ignition delay is one of important parameters considered for fuel selection in various combustors and for their design in engineering aspects. And, academically, the data for ignition delay have been used for kinetic target data to develop, evaluate, and improve detailed reaction models [1-3].

Aviation fuels used for jet and rocket engines are liquid mixtures composed of a large number of hydrocarbon components. To predict propulsion performance of such jet and rocket fuels, their physical and chemical properties must be found. Because there are many parameters relevant to the properties, several critical parameters have been proposed to characterize a specific fuel with ease and they were used to configure a surrogate fuel for the specific real fuel [4, 5]. The critical parameters or properties are called combustion property targets (CPTs) [4] and e.g., they are H/C ratio, molecular weight, derived cetane number (DCN), threshold sooting index (TSI), and so on.

Jet and rocket fuels are from petroleum although they are more complex than a single component fuel with a simple chemical structure. Fortunately, these petroleum-derived fuels have the similar characteristics to each other and various correlations to evaluate their properties have been studied and proposed [6]. The correlations are effective to predict approximately properties although they have relatively simple formulas.

Recently, liquid fuels from various origins or sources are studied for the purpose to replace the existent petroleum-derived fuels. One of them is a bio-fuel from plants or animal [7]. These efforts are desirable in the aspect of fuel flexibility, where energy sources are not confined to fossil fuels.

In this regard, feasibility tests for application of a bio-fuel to jet engines are required. And, in this work, a liquid mixture of kerosene and a bio-fuel is studied experimentally to find out its major properties and ignition delay, and the measured ones are compared with those of kerosene.

2 Experimental Method

The liquid mixture fuel is composed of 50% Jet A-1 and 50% bio-fuel in volume fraction, which is under development for a pilot plant scale test using vegetable plant oil. Here, it is called Bio-Jet Fuel. This bio-blended fuel and 100% Jet A-1 used inside Korea are tested. Regarding major combustion properties, their DCN and smoke point (SP) are measured, and then, ignition delay times are measured by a shock tube.

2.1 Measurement of DCN and SP

DCN is measured by using an ignition quality tester (IQT) according to a standard method, ASTM D6890 [8]. In this experiment, a small amount of a fuel is injected into the chamber with a specified condition of 21.1 atm and 818.15 K and then, it is ignited spontaneously in ignition delay. Depending on the ignition delay, the number or DCN is calculated and it is inversely proportional to ignition delay.

Smoke point is a parameter relevant to sooting propensity, of which measurement method is described in ASTM D1322 [9]. It is expressed by a distance or height of flame where soot generation is initiated. The higher length indicates lower sooting propensity.

2.2 Measurement of Ignition Delay Time

Ignition delay time is measured by a standard shock tube, which is shown in Fig. 1. A test fuel is homogeneously mixed in a driven section and it undergoes two shock waves of an incident and a reflected shock wave, leading to the target pressure and temperature for test behind a reflected shock wave.

*Corresponding author. Fax: +82-2-3408-4333
E-mail address: chsohn@sejong.ac.kr

The shock tube is 1.45 m and 5.85 m long in a driver and a driven section, respectively. Their diameters are 66.9 mm and 64.7 mm. An insert part is inserted between two sections for rupture of a diaphragm.

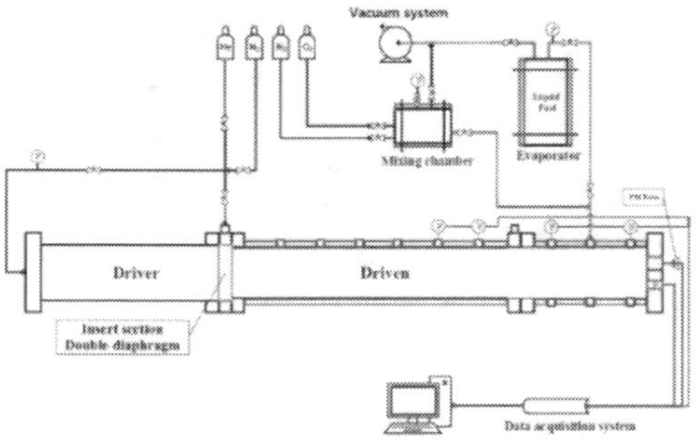

Figure 1: Schematic diagram of the experimental apparatus (shock tube).

For experiments, a pre-vaporized fuel, oxygen, and nitrogen are supplied into the driven section and the tailored mixture of helium and nitrogen is stored in the driver section. The heater is installed to vaporize a fuel, by which temperature can reach 150 °C at the full operation. To prevent the vaporized fuel from being condensed, two sections and feeding lines are thermally heated. A test fuel is Jet A-1 only and it is mixed with air at a stoichiometry, i.e., unity equivalence ratio, $\phi = 1$. For the test of the mixture of Jet A-1 and a bio-fuel, i.e., Jet-Bio, the same oxidizer of air and unity equivalence ratio are adopted for a comparative study with Jet A-1. The target pressure is selected as 20 atm and the target temperature ranges from 695 K to 1,080 K.

Along the tube, four pressure transducers are installed to estimate propagation speed of a shock wave and one pressure sensor is at the end wall of the driven section to measure the pressure behind a reflected shock wave, which must be as close to the target pressure as possible for reproducibility of the data. To monitor ignition process, a photo-multiplier tube measures a signal from the mixture behind a reflected shock wave, which means emission of CH* radical generated in chemical reaction during ignition process. Abrupt increase in the signal indicates that successful ignition is induced. The measurement of ignition delay with the shock tube was validated in our previous work [10].

3 Results and Discussion

3.1 DCN and SP

The measured DCNs and SPs are summarized in table 1. The Jet-Bio has larger DCN, which means shorter ignition delay, by 33 %. And, its smoke point is higher and soot generation would be reduced. Most of bio fuels don't have aromatic content as a component, which reduces soot formation. The measured SP for the Jet-Bio agrees well with this well-known trend.

From comparison of DCN of Jet A-1 with that of Jet-Bio, the current bio-fuel shortens ignition process. But, in the experiments with the IQT, liquid fuel is injected through a nozzle and thus, elementary physical processes or phenomena such as atomization, vaporization, and mixing are involved before chemical reaction. All the processes are relevant to initiation of reaction, i.e., ignition. Accordingly, fundamental ignition delay should be measured by a shock tube. Nevertheless, DCN provides us useful information on general or overall ignition characteristics.

From table 1, the Jet-Bio can be characterized to have reduced sooting propensity and short ignition delay compared with Jet A-1, a typical aviation fuel for jet propulsion engines in Korea.

Properties	Fuel	
	JetA-1	Jet-Bio
DCN (Derived Cetane Number)	46.6	62
SP (Smoke Point)	25 mm	45 mm

Table 1: Measured DCNs and SPs for the test fuels of Jet A-1 and Jet-Bio.

3.2 Ignition Delay Time

First of all, the pressure histories in time are measured for two test fuels and are shown in Figs 2 and 3, respectively. For the two fuels, the target pressure of 20 atm is attained with negligible error and temperatures are similar to each other for comparison. At 911 K, the Jet A-1/air mixture maintains almost constant pressure and then, pressure increases abruptly at a moment. At the same time, the signal for CH* emission shows the same. From the signal for CH* emission, ignition delay is estimated 2.16 ms for this condition. On the other hand, the Jet-Bio shows the similar pressure signal to Jet A-1 and the only difference is in ignition delay time. The latter delay time is shortened significantly by 44 %. It can be predicted from the DCNs for the two fuels reported in the preceding section and ignition delay agrees with the behavior of DCN.

Next, ignition delay times for the two fuels are measured over a wide range of temperature and are shown in Fig. 4 with previous results from Wang et al. [11]. Before we see the data for Jet-Bio, the present data for Jet A-1 are compared with the results from Wang et al. [11], who tested with Jet A not Jet A-1. There is slight difference between Jet A and Jet A-1 in their properties. Accordingly, this verification of the present data may work well. As well known, kerosene or Jet A shows NTC (negative temperature coefficient) behavior around temperature of 1,000 K [12]. The present experiments measured 4 data in the temperature range from 707 K to 1,047 K, which covers the temperature range for the NTC behavior. The present data follows well the Wang et al.'s data over the temperature range and the NTC behavior can be identified clearly.

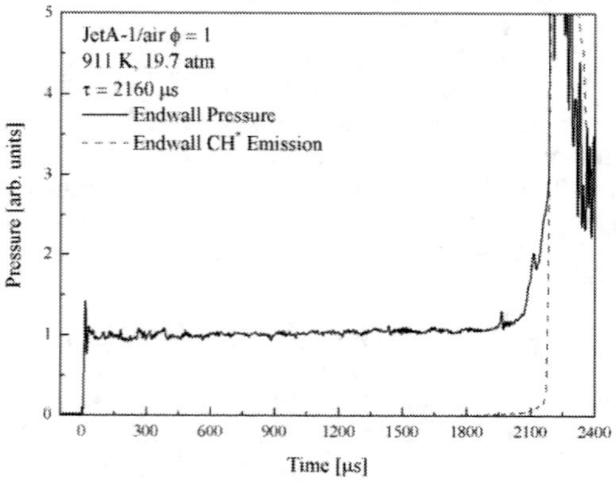

Figure 2: Signals of pressure and CH* emission for JetA-1/air at 911 K, 19.7 atm, and $\phi = 1$.

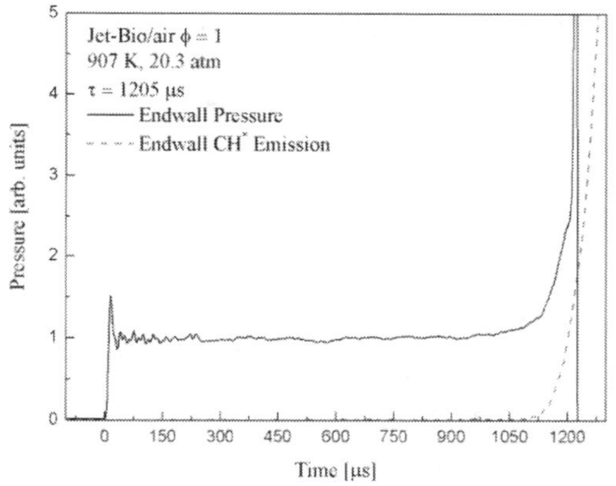

Figure 3: Signals of pressure and CH* emission for Jet-Bio/air at 907 K, 20.3 atm, and $\phi = 1$.

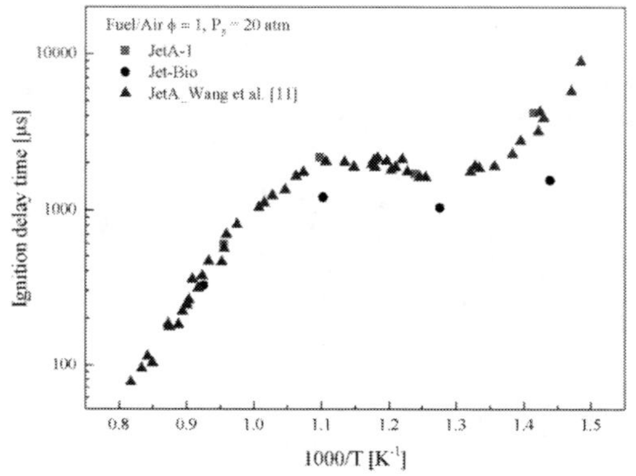

Figure 4: Ignition delay times of three mixtures of stoichiometric fuel/air at 20 atm for JetA-1 and Jet-Bio from the present and for Jet A from Wang et al. [11].

The Jet-Bio is tested in the similar temperature range to that of Jet A-1. Basically, ignition delay of Jet-Bio has the same trend as that of Jet A or Jet A-1. Especially, at a high temperature region over 1,000 K, ignition delays for the fuels are almost the same as each other. But, as temperature falls down below 1,000 K, ignition delay of the Jet-Bio is shortened considerably, nearly by 50 %, and this significant decrease in ignition delay continues to a low temperature region after the NTC region. Accordingly, this quantitative change in ignition delay induced by a bio-fuel should be considered in design of a jet engine combustor fueled by bio-fuels, especially, when it is operated at low temperature.

In the preceding section, significantly increased DCN was measured for the Jet-Bio. The shock-tube data implies that the temperature of 818.15 K in the IQT belongs to a lower temperature region in the test of this Jet-Bio fuel. This implication might depend on the particular fuel and can't be generalized. Accordingly, both measurements of DCN and ignition delay time are required to characterize comprehensive ignition characteristics of liquid fuels.

4 Conclusions

Ignition delay time of kerosene blended with a bio-fuel is measured experimentally by a shock tube for a comparative study with general aviation fuels. The usual aviation fuel, Jet A-1 used in Korea, and a new bio-synthetic fuel of the mixture of Jet A-1 and a bio-fuel have been tested to find ignition characteristics.

The new fuel, Jet-Bio, is being developed for application to a jet engine for fuel flexibility and low emission. From the experimental results, it has lower sooting propensity and higher derived cetane number than Jet A-1, which means shorter ignition delay. From shock-tube experiments, the Jet-Bio follows the NTC (negative temperature coefficient) behavior similarly to Jet fuels. But, at low temperature below 1,000 K, it has significantly short ignition delay, which is reduced by 50 %. It agrees well with DCN results compared with Jet A-1.

The current bio-fuel under development might be used as an alternative aviation fuel for fuel flexibility and contribute to reduction of soot emission when it is mixed with usual jet fuels because its combustion properties follow fundamentally those of jet fuels. But, its different ignition characteristics must be considered for design of a jet engine combustor optimized for the new fuel.

5 Acknowledgment

This work was supported by the National Research Foundation of Korea (NRF) grant funded by the Korean government (MSIP) (Grant No. NRF-2015M1A3A3A02009957). HSH and CHS were partially supported by Agency for Defense Development of Korea under the contract UD150039GD.

References

[1] C.W. Gao, A.G. Vandeputte, N.W. Yee, W.H. Green, R.E. Bonomi, G.R. Magoon, H.W. Wong, O.O. Oluwole, D.K. Lewis, N.M. Vandewiele, K.M.V. Geem, Combust. Flame 162 (8) (2015) 3115-3129.

[2] D.F. Davidson, S.C. Ranganath, K.-Y. Lam, M. Liaw, Z. Hong, R.K. Hanson, J. Propul. Power, 26 (2) (2010) 280-287.

[3] D.F. Davidson, J. Shao, T. Parise, R.K. Hanson, Shock Tube Measurements of Jet and Rocket Fuel Ignition Delay Times, 55th AIAA Aerospace Sciences Meeting, AIAA, 2017, p. 1798-1814.

[4] S. Dooley, S.H. Won, M. Chaos, J. Heyne, Y. Ju, F.L. Dryer, K. Kumar, C.-J Sung, H Wang, M.A. Oehlschlaeger, R. J. Santoro, T. A. Litzinger, Combust. Flame 157 (12) (2010) 2333-2339.

[5] S. Honnet, K. Seshadri, U. Niemann, N. Peters, Proc. Combust. Inst. 32 (2009) 485-492.

[6] A.G. Goossens, Prediction of Molecular Weight of Petroleum Fractions, Ind. Eng. Chem. Res. 35 (1996) 985-988.

[7] H.E. Saleh, Fuel 90 (2011) 421-429.

[8] ASTM D6890. Standard test method for determination of ignition delay and derived cetane number (DCN) of diesel fuel oils by combustion in a constant volume chamber. ASTM International, West Conshohocken, PA; 2014.

[9] ASTM D1322. Standard Test Method for Smoke Point of Kerosine and Aviation Turbine Fuel. ASTM International, West Conshohocken, PA; 2014.

[10] H.S. Han, Y.G. Wang, C.J. Kim, C.H. Sohn, J. Korean Soc. Combust. 22 (1) (2017) 8-13.

[11] H. Wang, M.A. Oehlschlaeger, Fuel 98 (2012) 249-258.

[12] M.P Halstead, L.J. Kirsch, C.P. Quinn, Combust. Flame 30 (1977) 45-60.

11th Asia-Pacific Conference on Combustion,
The University of Sydney, NSW Australia
10th -14th December 2017.

CO_2 participation in chemical reaction in a CH_4 flat flame under $O_2/CO_2/H_2O$ environment

Hirotatsu Watanabe[1], Takashi Komeno[1], Shunsuke Sugai[2]

[1]Tokyo Institute of Technology

2-12-1 Ookayama, Meguro-ku, Tokyo, 152-8500, Japan

Abstract

In this study, CH_4 oxidation in O_2/CO_2, $O_2/CO_2/H_2O$, $O_2/CO_2/Ar$ environments was investigated in a CH_4 flat flame under fuel-rich conditions. In addition to the experiment, calculations were performed using premixed flame model with different kinetics. Both the experiment and the calculation showed that CO_2 was consumed in the CH_4 flat flame and unburnt hydrocarbon such as CH_4 and C_2H_2 significantly decreased in O_2/CO_2 environments. This clearly indicated that CO_2 was a reactive gas. Calculation also showed that OH radicals, which were active in hydrocarbon oxidation, significantly increased in O_2/CO_2 environment due to CO_2 reactivity mainly through $CO_2 + H \rightarrow CO + OH$. Basically, similar trends such as lower unburnt hydrocarbons also appeared in $O_2/CO_2/H_2O$ environment; however, a consumption of CO_2 in $O_2/CO_2/H_2O$ environment was lower than that in O_2/CO_2 environments, suggesting that H_2O reactivity inhibited CO_2 reactivity. Thus, calculated OH concentration in $O_2/CO_2/H_2O$ environment was lower than that in O_2/CO_2 environment, leading to a slightly increasing in unburnt hydrocarbons in $O_2/CO_2/H_2O$ environments, whereas it was still much lower than air environments.

1 Introduction

Carbon capture and storage (CCS) technologies have gained interest as a promising option, which has a potential to reduce CO_2 emission. Oxy-fuel combustion with CCS is a promising approach for CO_2 capture in power generation [1,2]. Under oxy-fuel combustion, instead of using air as oxidizer, a mixture of O_2 and recycled flue gases is used to control the combustion temperature. Flue gas recycling generally occur before or after moisture removal. The former is called as wet recycle and the latter is the dry recycle. O_2/CO_2 combustion, in which the recycled gas is assumed to be only CO_2, is a basic form of oxy-fuel combustion with dry recycle. Recently, the difference between O_2/CO_2 and air combustion has been well studied in coal combustion [3-8] and turbulent premixed flames [9-12]. Several studies showed CO_2 is not inert but participates in chemical reactions primarily through the reaction, $CO_2 + H \rightarrow CO + OH$ (R1) in O_2/CO_2 environments [6,7,13,14]. Our previous studies showed the chemical role of CO_2 was to advance CH_4 oxidation during fuel-rich combustion [6]. A large part of fuel conversion occurs in the gas phase in coal combustion; therefore, understanding of CH_4 oxidation is also important in coal combustion. Meanwhile, the wet recycle mode leads to water vapor levels up to approximately 15-35 % [15-17]. However, impact of H_2O dilution in O_2/CO_2 combustion has not been well discussed. Further studies are necessary to apply the combustion characteristics in O_2/CO_2 environments to those in $O_2/CO_2/H_2O$ environments because H_2O is also reactive gas as well as CO_2.

The aim of this work is to study CH_4 oxidation in O_2/CO_2 and $O_2/CO_2/H_2O$ environment experimentally and numerically by using a flat flame, and find out differences caused by H_2O dilution in O_2/CO_2 environment. CO_2 participation in chemical reaction is also discussed.

2 Experiment

Figure 1 shows a schematic diagram of the experimental setup of the CH_4 flat flame. The flat flame is very useful for one-dimensional analysis. H_2O was continuously supplied from the syringe pump to the evaporator. The vapor was mixed with gas mixture at 200 °C. Temperature profiles was also measured using a thermocouple. Details of experimental procedure were reported previously [6,7].

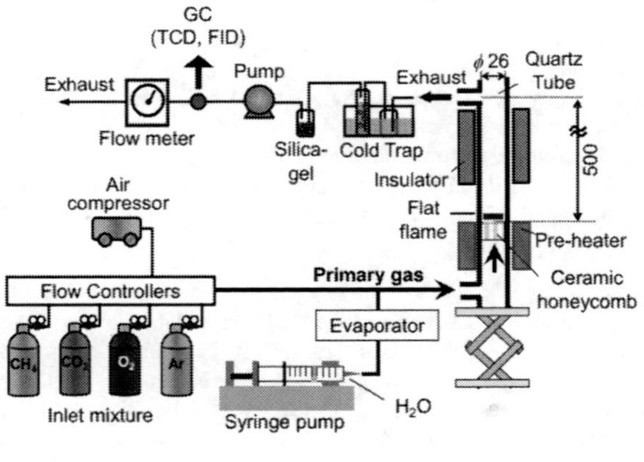

Figure. 1 Schematic diagram of the flat flame reactor

Exhaust gas concentrations were measured by gas chromatography (GC-8A, Shimadzu Corp.). Total flow rate of exhaust gas was measured by wet-gas flowmeter (W-NK-1A, Shinagawa Co. Ltd.). Gas production ratio (PR_{gas}), given as Eq. (1), means carbon conversion ratio of CH_4 to each exhaust gas (CO, CO_2, hydrocarbons) on a carbon molecule basis. Production ratios (PR) of C1 gas and C2 gas were defined as:

Corresponding author. Fax: +81-3-5734-2179
E-mail address: watanabe.h.ak@m.titech.ac.jp

$$PR_{\mathrm{C1gas}} = \frac{\left(Q_{\mathrm{C1,gas}}\right)_{\mathrm{outlet}}}{\left(Q_{\mathrm{CH_4}}\right)_{\mathrm{inlet}}} \times 100 \quad (1)$$

$$PR_{\mathrm{C2gas}} = \frac{\left(2 \times Q_{\mathrm{C2,gas}}\right)_{\mathrm{outlet}}}{\left(Q_{\mathrm{CH_4}}\right)_{\mathrm{inlet}}} \times 100 \quad (2)$$

where Q indicates the molar flow rate of gas.

In O_2/CO_2, $O_2/CO_2/H_2O$, $O_2/CO_2/Ar$ environments, exhaust gas contained inlet CO_2. Thus, production ratio of CO_2, which was defined as PR_{CO2}, was calculated by another formula, as shown below:

$$PR_{\mathrm{CO_2}} = \frac{\left(Q_{\mathrm{CO_2}}\right)_{\mathrm{outlet}} - \left(Q_{\mathrm{CO_2}}\right)_{\mathrm{inlet}}}{\left(Q_{\mathrm{CH_4}}\right)_{\mathrm{inlet}}} \times 100 \quad (3)$$

Table 1 shows primary gas compositions at the excess O_2 ratio (λ) of 0.65. O_2 ratio in O_2/CO_2 environment was set to 0.3 to stabilize the flame. H_2O ratio in $O_2/CO_2/H_2O$ environment was 0.11, which was close to that in the wet recycle mode [15-17].

Table 1 Primary gas composition ($\lambda = 0.65$)

	CH_4 [ml/min]	O_2 [ml/min]	CO_2 [ml/min]	H_2O [ml/min]	Ar [ml/min]	N_2 [ml/min]
O_2/CO_2	250	325	760	0	0	0
$O_2/CO_2/H_2O$	250	325	640	120	0	0
$O_2/CO_2/Ar$	250	325	640	0	120	0
Air	250	325	0	0	0	1225

3 Calculation

The CH_4 oxidation in the flat flame was studied with detailed chemical reaction kinetics. A premixed flame model was used in the calculations of CHEMKIN-PRO. The calculation domain was the same as shown in Fig. 1. It is well-known that GRI-Mech 3.0 which consists of 53 species and 325 elementary reactions is well-optimized for CH_4 oxidation in air environment. In addition to GRI-Mech 3.0, chemical kinetics model proposed by Mendiara and Glarborg's (97 species and 779 elementary reactions) [14] and AramcoMech 1.3 (253 species and 1542 reactions) [18] were also used. The conservation equations for steady quasi one-dimensional diffusion or premixed flames were solved using CHEMKIN-PRO. The energy equation was not solved, and measured temperature distributions were used in the calculation using premixed flame model. Moreover, inert H_2O (physical properties are the same as H_2O, but it is inert) was used in the inlet H_2O to study the chemical role of H_2O. And, amount of CO_2 involved in the reaction was discussed by labelling the inlet CO_2 (like 13C labelled carbon). The equilibrium calculation was also performed to discuss kinetics in various models.

4 Results and discussion

Figure 2 shows the measured temperature profiles in the flat flame reactor. Because the flat flame is formed around $x = 0$ mm, the gas temperature increases rapidly at this point. The difference of gas temperature between O_2/CO_2 and air environments is not significant regardless of higher heat capacity of CO_2 because inlet O_2 concentration was 30 vol% in O_2/CO_2 environments and the electric furnace provides sufficient heat for the experiment. The effect of temperature difference can be removed in this study.

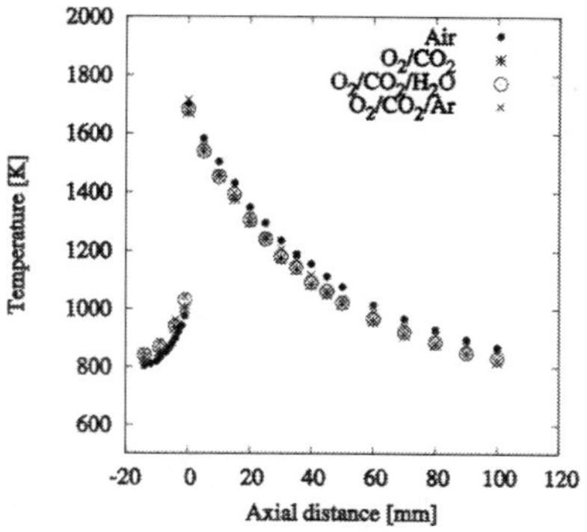

Figure 2: Measured temperature profiles at $\lambda = 0.65$

Figure 3 shows the measured and calculated overall gas production rates in various environment. The measured total production rate, which corresponds to carbon balance, is higher than 95 mol% in all condition, indicating that inflow and outflow are well balanced in the experiment. Overall, experiment and calculation show similar trends. In O_2/CO_2, $O_2/CO_2/H_2O$ and $O_2/CO_2/Ar$ environments, CO_2 production ratio is negative, indicating that inlet CO_2 participates in chemical reactions. Calculations show that CO_2 is consumed mainly through:

$$CO_2 + H \rightarrow CO + OH \quad (R1)$$

as previously reported [6,7]. When inlet CO_2 was labelled in the calculation using GRI-Mech 3.0, it is shown that the ratio of CO produced from inlet CO_2 to all CO was approximately 67 %. This also supports that CO_2 is reactive gas. Compared with O_2/CO_2 environment, CO_2 production rate increases in $O_2/CO_2/H_2O$ environment. Considering that H_2O is reactive and Ar is inert gas, H_2O reactivity inhibits CO_2 participation in chemical reaction. Moreover, when inert H_2O is used in the inlet gas in the calculation, CO_2 production rate increases, also indicating the H_2O inhabitation effect. When the experiments are compared with the calculations, the calculation using Aramco Mech 1.3 well reproduces measured gas production ratios.

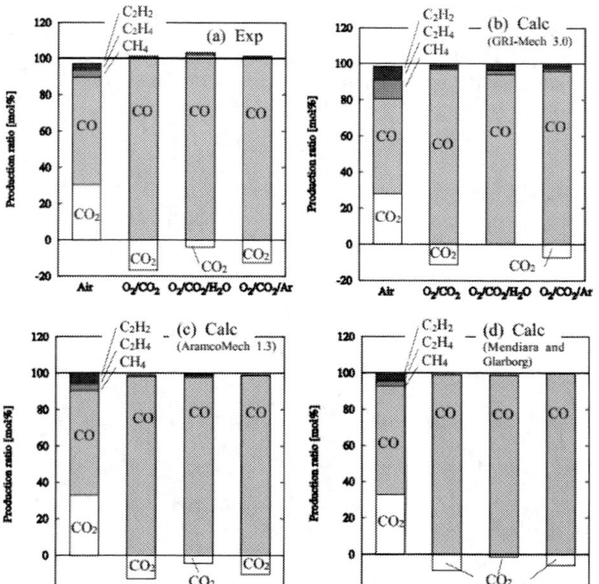

Figure 3: Overall gas production ratios (experiment and calculation using premixed flame model)

Figure 4 shows calculated gas production ratios using equilibrium calculation. In the equilibrium calculation, hydrocarbons are almost completely oxidized, and CO_2 and CO production ratios are similar in each case. Insignificant difference in equilibrium calculations is observed between reaction models of GRI-Mech 3.0, Aramco Mech 1.3 and Mendiara and Glarborg's mechanisms, indicating that differences appeared in Fig. 3 are caused by not thermodynamics parameters but kinetics in each model.

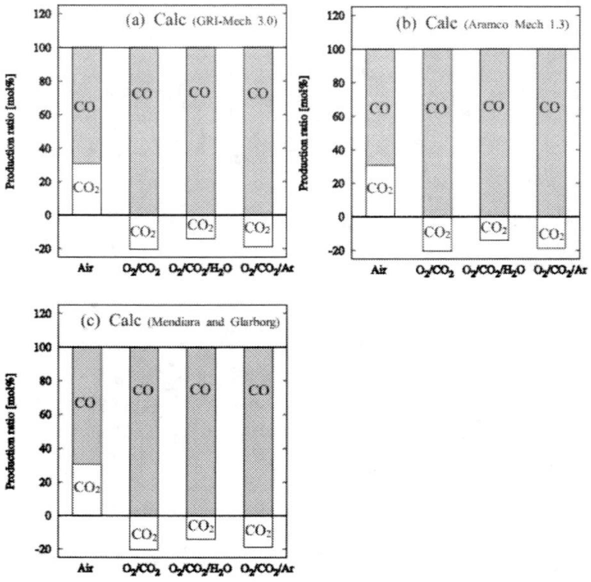

Figure 4: Gas production ratios in equilibrium calculation

Figure 5 shows calculated OH radical profiles in air, O_2/CO_2, $O_2/CO_2/H_2O$ and $O_2/CO_2/Ar$ environments using Aramco Mech 1.3. Here, temperature profiles of O_2/CO_2 combustion are used in all cases to study chemical reaction only. As previously described, an abundant OH radicals are produced in O_2/CO_2 combustion mainly through R1. However, when O_2/CO_2 environments is diluted by Ar, OH radicals are significantly reduced. OH profiles in $O_2/CO_2/H_2O$ environment is lower than those in $O_2/CO_2/Ar$ environment, indicating that H_2O reactivity inhibits OH formation through R1 under this condition.

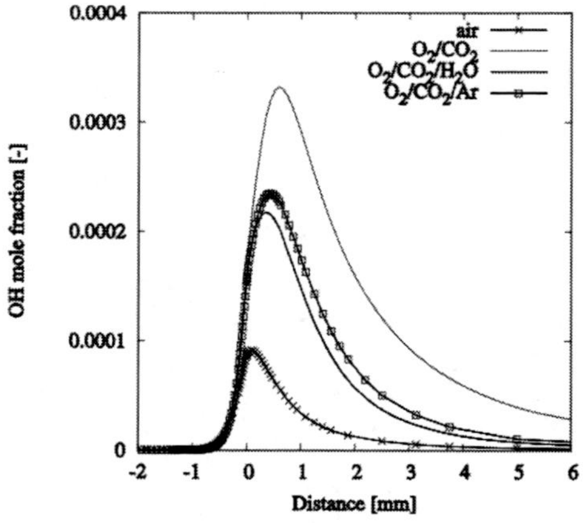

Figure 5: Calculated OH radical profiles (premixed flame model using Aramco Mech 1.3)

Figure 6 shows measured and calculated hydrocarbon production ratios in various environments. Calculations using Aramco Mech 1.3 well agree with experiments. Compared with air environments, unburnt hydrocarbon such as CH_4 and C_2H_2 significantly decreases in O_2/CO_2 environments because of abundant OH radicals. Unburnt hydrocarbons in $O_2/CO_2/H_2O$ are still much lower than those in air environments; however, unburnt hydrocarbons are slightly increased by diluting O_2/CO_2 environment with H_2O. One possible explanation is the decreasing OH radicals in $O_2/CO_2/H_2O$ environments as shown in Fig. 5. In other word, this would be the result of competitive reaction between dry and wet reforming.

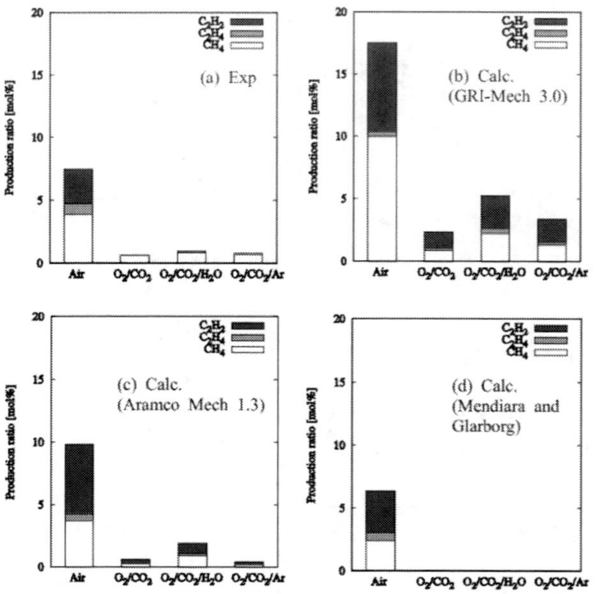

Figure 6: Hydrocarbon production ratios (experiment and calculation using premixed flame model)

4 Conclusions

In this study, CH_4 oxidation in O_2/CO_2, $O_2/CO_2/H_2O$, $O_2/CO_2/Ar$ environments was investigated experimentally and numerically. It was shown that CO_2 was consumed in the CH_4 flat flame and unburnt hydrocarbon such as CH_4 and C_2H_2 significantly decreased in O_2/CO_2 environments. Calculation using the labelled CO_2 showed the ratio of CO_2 produced from inlet CO_2 to all produced CO. These results clearly indicated that CO_2 was a reactive gas. Overall, calculations using GRI-Mech 3.0, AramcoMech 1.3 and Mendiara and Glarborg's mechanism showed similar trends with experiments. Especially, calculation using AramcoMech 1.3 well reproduced measured gas production ratio. Lower unburnt hydrocarbons also appeared in $O_2/CO_2/H_2O$ environment; however, a consumption of CO_2 in $O_2/CO_2/H_2O$ environment was lower than that in O_2/CO_2 environments, suggesting that H_2O reactivity inhibited CO_2 reactivity. Thus, calculated OH concentration in $O_2/CO_2/H_2O$ environment was lower than that in O_2/CO_2 environment, leading to a slightly increasing in unburnt hydrocarbons in $O_2/CO_2/H_2O$ environments, whereas it was still much lower than air environments.

5 Acknowledgment

This study was partly supported by a JSPS Grant-in-Aid for Young Scientists (A) (No. 24686027).

References

[1] T. Wall, Proc. Combust. Inst. 31 (2007) 31-47.

[2] H. Kvamsdal, K. Jordal, O. Bolland, Energy 32 (2007) 10-24.

[3] R. K. Rathnam, L. K. Elliott, T. F. Wall, Y. Liu, B. Moghtaderi, Fuel Process. Tech. 90 (2009) 797-802.

[4] J. Brix, P. A. Jensen, A. D. Jensen, Fuel 89 (2010) 3373-3380.

[5] T. Maffei, R. Khatami, S. Pierucci, T. Faravelli, E. Ranzi, Y. A. Levendis, Combust. Flame 160 (2013) 2559-2572.

[6] H. Watanabe, F. Arai, K. Okazaki, Combust. Flame 160 (2013) 2375-2385.

[7] H. Watanabe, K. Okazaki, Proc. Combust. Inst. 35 (2015) 2363-2371.

[8] J. Hees, D. Zabrodiec, A. Massmeyer, S. Pielsticker, B. Gövert, M. Habermehl, O. Hatzfeld, R. Kneer, Combust. Flame 172 (2016) 289-301.

[9] H. Kobayashi, H. Hagiwara, H. Kaneko, Y. Ogami, Proc. Combust. Inst. 31 (2007) 1451-1458.

[10] A. P. Shroll, S. J. Shanbhogue, A. F. Ghoniem, J. Hees, D. Zabrodiec, A. Massmeyer, S. Pielsticker, B. Gövert, M. Habermehl, O. Hatzfeld, R. Kneer, Combust. Flame 172 (2016) 289-301.

[11] H. Watanabe, S. J. Shanbhogue, S. Taamallah, N. W. Chakroun, A. F. Ghoniem, Combust. Flame 174 (2016) 111-119.

[12] S. Taamallah, N. W. Chakroun, H. Watanabe, S. J. Shanbhogue, A. F. Ghoniem, Proc. Combust. Inst. 36 (2017) 3799-3807.

[13] P. Glarborg, L. L. B. Bentzen, Energy Fuels 22 (2008) 291-296.

[14] T. Mendiara and P. Glarborg, Combust. Flame 156 (2009) 1937-1949.

[15] M. Anheden, U. Burchhardt, H. Ecke, R. Faber, O. Jidinger, R. Giering, H. Kass, S. Lysk, E. Ramström, J. Yan, Overview of operational experience and results from test activities in Vattenfall's 30 MWth Oxyfuel Pilot Plant in Schwarze Pumpe, Energy Procedia 4 (2011) 941-950.

[16] Y. Hu, J. Yan, Characterization of flue gas in oxy-coal combustion processes for CO_2 capture, Applied Energy 90 (2012) 113-121.

[17] B. Yi, L. Zhang, F. Huang, Z. Mao, C. Zheng, Effect of H_2O on the combustion characteristics of pulverized coal in O_2/CO_2 atmosphere, Applied Energy 132 (2014) 349-357.

[18] W. K. Metcalfe, S. M. Burke, S. S. Ahmed, H. J. Curran, Int. J. Chem, Kinetics 45 (2013) 638-675.

The principle of determining the optimized operating parameters based on the adopted fuel property in RCCI engines

Yaopeng Li, Ming Jia *, Yachao Chang, Maozhao Xie

School of Energy and Power Engineering, Dalian University of Technology, Dalian 116024, China

Abstract

Both the gasoline/diesel (G/D) and methanol/diesel (M/D) reactivity controlled compression ignition (RCCI) engines are optimized at low-to-medium load. The results indicated that the impact of fuel property on the determination of the optimal parameters is significant. It is found that homogeneous fuel/air distribution with high premixed ratio and advanced start of injection (SOI) is beneficial for the improvement of engine performance for both G/D and M/D RCCI. Meanwhile, the high initial temperature is indispensable to avoid the excessively late CA50 and the deteriorated combustion efficiency. Fuel reactivity and combustion rate are two important parameters for the fuel in RCCI engines. Due to the lower reactivity and the faster combustion rate of methanol than those of gasoline, lower premixed ratio and more retarded SOI are adopted in M/D RCCI than G/D RCCI to advance the CA50 and alleviate the fast combustion rate. Moreover, higher exhaust gas recirculation (EGR) rate is required for M/D RCCI to reduce its fast combustion rate. Boosting the in-cylinder pressure (P_{IVC}) has two contradictory effects on the combustion temperature and fuel efficiency. Owing to the later CA50 in M/D RCCI, the higher P_{IVC} is favorable for M/D RCCI, while lower P_{IVC} is preferable for G/D RCCI.

1 Introduction

The more and more stringent emission regulation has become a severe challenge for internal combustion engines (ICE), especially the upcoming limit for carbon dioxide (CO_2). Therefore, it is very imperative to improve the in-cylinder combustion process of ICE to reduce the dependence on aftertreatment. Recently, low temperature combustion (LTC) has been intensively investigated due to the simultaneously satisfactory fuel efficiency and exhaust emissions. The most representative regimes of LTC include homogeneous charge compression ignition (HCCI) [1], premixed charge compression ignition (PCCI) [2], and reactivity controlled compression ignition (RCCI) [3].

Compared to HCCI or PCCI/PPC, more effective control over ignition timing and combustion rate can be achieved in RCCI engines by adjusting the ratio of the low-to-high reactivity fuel and the start of injection (SOI) of the high-reactivity fuel [4]. Consequently, higher operating loads and lower cycle-to-cycle variations are realized in RCCI. Moreover, flexible types of fuel, not only the petroleum-based fuels but also various renewable fuels, can be applied in RCCI engines with minor engine modifications. In the study of Gao et al. [5], ethanol and butanol were respectively used as the premixed fuel in a RCCI engine. The results showed that the best strategy to achieve good fuel efficiency and exhaust emissions over the full load range was to adopt butanol at low-to-medium loads and ethanol at high loads. Paykani et al. [6] investigated the effects of the diesel injection strategy on the combustion and emission characteristics in a natural gas/diesel RCCI engine. It is concluded that combustion phasing was dominated by the natural gas fraction, and it was also slightly affected by the second diesel injection timing and the diesel split fraction as the multiple injections were employed. Fueling with different low-reactivity fuels, the different combustion characteristics of RCCI were numerically studied by Zou et al. [7]. It was found that the combustion duration of methanol/diesel RCCI was much shorter than those of gasoline/diesel and n-butanol/diesel RCCI, and all of them were very sensitive to the variation of intake temperature.

Massive studies about RCCI engines fueled with various fuels have been conducted in the previous literatures, it can be found that RCCI engines with different fuels exhibit considerably different combustion and emission characteristics. As a result, when different fuels are employed in RCCI engines, their respectively optimal operating parameters are quite different from each other, even at the same operating condition. Although the optimal operating parameters for different fuels in RCCI have been explored before, the impacts of the adopted fuel property on the determination of the optimal operating parameters have not been well understood yet, which will be investigated in this study by comparing different fuels.

Gasoline has been intensively studied as the low-reactivity fuel in RCCI engines, so it is used in this study as the referential low-reactivity fuel. Meanwhile, methanol becomes an attractive renewable fuel due to its wide sources. Besides, no carbon-carbon bonds and an extra oxygen atom in the methanol molecule enhance the complete oxidation of fuel. In addition, the high latent heat of evaporation of methanol is beneficial to reduce combustion temperature and NO_x emissions. The research octane number (RON), auto-ignition temperature, and lower heating value (LHV) of methanol are quite different from those of gasoline. Therefore, methanol is also chosen as the low-reactivity fuel for comparison with gasoline. due to its significantly different fuel properties from gasoline.

First, the optimized operating parameters are obtained for both gasoline/diesel (G/D) and methanol/diesel (M/D) RCCI. When different fuels are employed in RCCI engines, it is extremely difficult and expensive to search the optimal operating parameters for each fuel through experiment. In contrast, the numerical optimization using genetic algorithm (GA) has become an efficient way for engine design by optimizing multiple objectives simultaneously with low cost and high accuracy. In this study, coupling the non-dominated sorting genetic algorithm II (NSGA-II) [8] with the three-dimensional computational fluid dynamics (CFD) model, the operating parameters of the G/D and M/D RCCI are respectively optimized at low-to-medium load. Second, the optimized parameters for G/D and M/D RCCI are compared to reveal the impacts of fuel property on the determination of the optimal operating parameters in RCCI engines. Based on this study, when a new fuel is adopted in RCCI, the optimal operating parameters can be inferred according to its fuel property.

2 Numerical models

The three-dimensional CFD code, KIVA-3V [9], with the improved sub-models was used to the study a RCCI engine fueled with different fuels, in which the re-normalization group (RNG) k-ε turbulence model [10] and hybrid KH (Kelvin Helmholtz)-RT (Rayleigh Taylor) breakup model [11] are respectively introduced.

The CHEMKIN solver [12] was coupled with the KIVA-3V code to reproduce the ignition and combustion processes of various fuels. n-Heptane and iso-octane (primary reference fuel, PRF) were respectively used as the chemical surrogate of diesel and gasoline. A reduced PRF oxidation mechanism [13] consisting of 41 species and 124 reactions was used in this study. The chemical mechanism was constructed based on the decoupling methodology, in which a detailed methanol mechanism is included, thus the PRF mechanism is capable of reproducing the oxidation characteristics of methanol, n-heptane, iso-octane, and their blends. In terms of the NO_x and soot predictions, the extended Zeldovich NO_x sub-mechanism and an improved phenomenological soot model [14] were integrated into the chemical mechanism.

3 Engine specifications

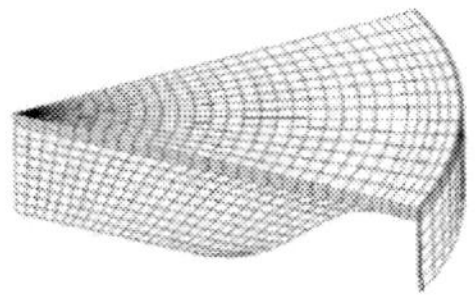

Fig. 1 Computational mesh at top dead center (TDC)

The tested engine was modified from a high-speed direct-injection (HSDI) diesel engine by adding an independent low-pressure injection system to supply the low-reactivity fuel to the intake manifold for RCCI combustion. A high-pressure injector with seven holes was vertically mounted in the axis of the cylinder for the in-cylinder diesel injection. As shown in Fig. 1, a wide and shallow piston bowl was employed, which was designed for the reduction of heat transfer losses and the improvement of fuel efficiency. Compromising the computational cost and accuracy, a 1/7th sector was used in the simulation corresponding to one injection spray. The cell number at intake valve closing (IVC) is 18290, and the average cell size is ~2 mm. Since the early injection timing of diesel is usually employed in RCCI combustion for the realization of the premixed combustion, low injection pressure of diesel was adopted to avoid serious spray/wall impingement. The detailed specifications of the test engine are listed in Table 1.

The experimental data derived from the experiment conducted by Dempsey et al. [15] is used to validate the numerical models. Due to the limited space, the validation results can be found in Ref. [16]. In the simulation, only the working period from IVC to exhaust valve opening (EVO) is calculated without considering the intake and exhaust processes. It is also assumed that the fuel/air mixture in the cylinder at IVC is fully homogeneous.

Table 1. Specifications of the test engine

Displaced volume	0.477 L
Bore × stroke	82×90 mm
Compression ratio	17.3:1
Bowl geometry	Wide/shallow
Injection pressure	500 bar

4 Optimization method

The fast non-dominated sorting genetic algorithm (NSGA) II was coupled with KIVA-3V to optimize the performance of the RCCI engine tested in this study. In NSGA II, each case is assigned a rank according to the number of cases that it dominates [17]. The case who dominates more cases has the priority to be chosen, and it is assigned a lower rank. The cases with rank 1 are called the Pareto Front, which are a group of cases that are non-dominated by any other case. During the optimization process, the rank and crowding distance of each case are firstly determined according to the optimized objectives as described above. Then, the selection for the next generation is conducted, which obeys the following two rules: (1) Cases with lower rank have the priority to be selected; (2) For the cases with the same rank, the one with larger crowding distance prefers to be chosen, which can enrich the diversity of the population. After the selection process, crossover and mutation with a certain probability are employed to these chosen cases to create the next generation. The selection operation restarts to these new generation cases, and the process is repeated until the Pareto Front is found out.

In this study, the equivalent indicated specific fuel consumption (EISFC), RI, and NOx emissions are chosen as the optimization objectives. Five important operating parameters are optimized simultaneously including the premixed ratio, SOI of diesel, EGR rate, initial temperature and initial pressure at intake valve closing (IVC) (named initial temperature (T_{IVC}) and initial pressure (P_{IVC}) below for brevity). The definition of ringing intensity (RI) is calculated as [18]:

$$\mathrm{RI} \approx \frac{1}{2\gamma} \cdot \frac{(0.05 \cdot (dp/dt)_{max})^2}{p_{max}} \cdot \sqrt{\gamma R T_{max}} \quad (1)$$

where $(dp/dt)_{max}$, p_{max}, and T_{max} are the maximum pressure rise rate, the maximum pressure, and the maximum temperature, respectively; γ is the ratio of specific heat, and R is the ideal gas constant.

The definition of EISFC is as follows:

$$\mathrm{EISFC} = \frac{M_d \cdot H_d + M_p \cdot H_p}{H_d} \times \frac{1}{\int_{IVC}^{EVO} p \cdot dV} \quad (2)$$

where M and H are the mass and lower heating value of fuel, the subscript d and p respectively denote diesel and the premixed fuel (i.e., gasoline or methanol used in this study). W_i is the indicated work from IVC to EVO, which is the integral of the in-cylinder pressure and volume. Since the lower heating value of methanol is much lower than those of gasoline and diesel, the premixed energy ratio (R_P) is used in this study, which is calculated as:

$$R_p = \frac{M_p \cdot H_p}{E_{tot}}$$

$$= \frac{M_p \cdot H_p}{M_p \cdot H_p + M_d \cdot H_d} \quad (3)$$

Corresponding author: Ming Jia
Fax: +86-13842870115
E-mail address: jiaming@dlut.edu.cn

where E_{tot} is the total energy of the premixed and direct-injected fuels in one cycle. In both G/D and M/D RCCI combustion tested in this study, E_{tot} is kept constant at 498 J/cyc, and the corresponding IMEP is around 4.7 bar. Four constraints are used in NSGA II to eliminate the unreasonable and unachievable cases in practical engines. The detailed settings for population, generation, constraints, and the ranges of the operating parameters are retained same for G/D and M/D RCCI.

5 Results and discussion

After continuous evolution in the generic algorithm, the population gradually evolves toward the lower EISFC, RI, and NO_x emissions. So the optimized cases with the best engine performance (i.e. the Pareto Front cases) are obtained for G/D and M/D RCCI. For these Pareto Front cases, only the cases who meets EISFC<190 g/kWh, RI<5 MW/m^2, and NO_x<0.4 g/kWh are retained, as shown in Fig. 3. It can be seen that both G/D and M/D RCCI combustion follow the trade-off relationship between EISFC and RI, while the RI of G/D RCCI is much lower than that of M/D RCCI. The NO_x emissions are kept very low level for most of the optimized cases of M/D RCCI, except for Cases A, B, and C, and the reason will be discussed later. Overall, for both G/D and M/D RCCI, the NO_x emissions are maintained well below the Euro VI emission standard by adopting the high R_P (larger than 60%) and the advanced SOI of diesel (earlier than 18 °CA BTDC), as can be seen in Fig. 4.

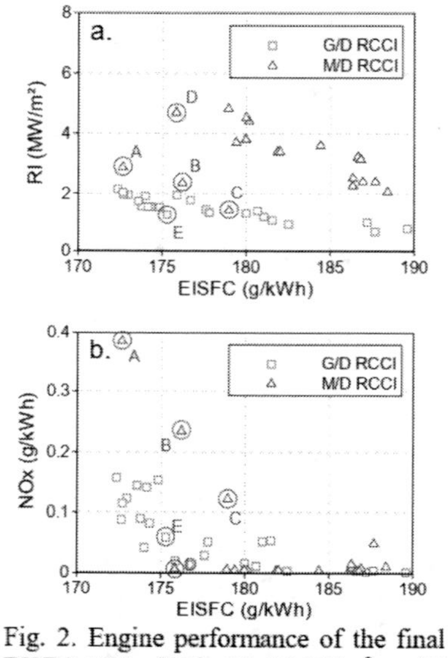

Fig. 2. Engine performance of the final optimized cases with EISFC<190 g/kWh, RI<5 MW/m^2, and NO_x<0.4 g/kWh

Although the ranges of the operating parameters are kept the same during the optimization process, the engine performance of G/D and M/D RCCI is quite different as seen in Fig. 2. This is because different optimized operating parameters are respectively chosen in G/D and M/D RCCI as depicted in Fig. 3, which is related to the different fuel properties. From Fig. 3, it is found that higher R_P and the more advanced SOI are desired in the optimization results of both G/D and M/D RCCI to achieve fully homogeneous fuel distribution. Meanwhile, in order to avoid the excessively late combustion phasing with sufficiently premixed fuel, the high T_{IVC} (close to the upper boundary of T_{IVC}) is chosen.

With respect to P_{IVC}, it is interesting to note that most of the P_{IVC} in G/D RCCI are lower than 1.5 bar, whereas those of M/D RCCI are close to 1.8 bar except for the three cases with the solid symbols shown in Fig. 3. Due to the lower RON of methanol than gasoline, higher P_{IVC} is adopted in M/D to achieve the desirable combustion phasing. However, as seen in Fig. 4, excessively high P_{IVC} reduces the combustion temperature and inhibits the auto-ignition front propagation, which leads to the deteriorated combustion efficiency.

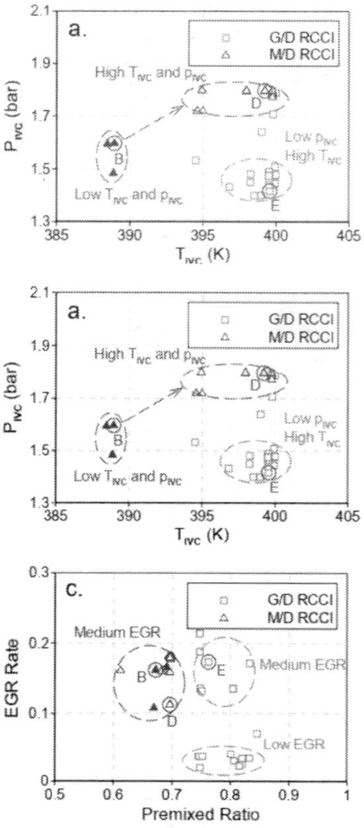

Fig. 3 Distributions of the optimized parameters for the final optimized cases in G/D and M/D RCCI (solid symbols are Cases A, B, and C, Cases B, D, E are marked by red circles)

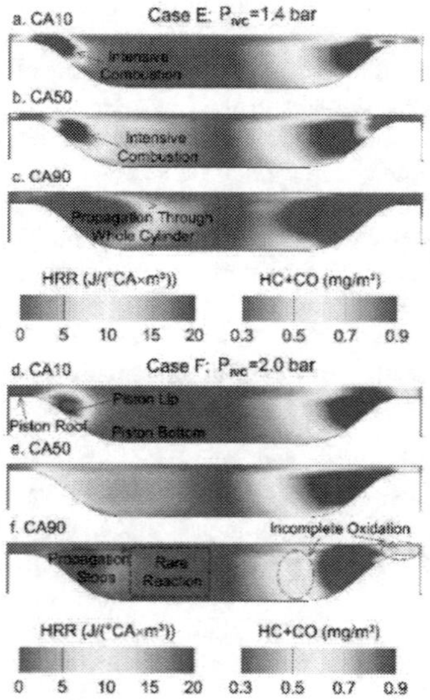

Fig. 4 In-cylinder distributions of HRR (left half cylinder) and unburned mass (right half cylinder) for cases with different P_{IVC}

On one hand, the lower R_p and more retarded SOI are employed in M/D to advance the ignition timing. On the other hand, the lower R_p and more retarded SOI can enhance the in-cylinder fuel and reactivity stratifications, which alleviates the fast combustion rate of M/D RCCI. Similarly, the higher EGR rate in M/D RCCI is also to reduce the combustion rate.

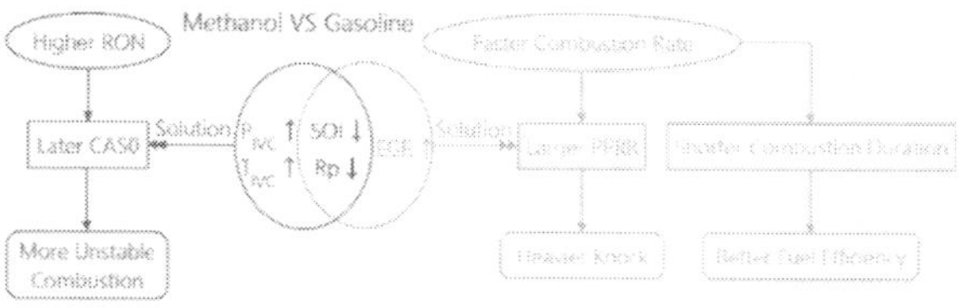

Fig. 5 Impacts of fuel property on the determination of the optimized parameters ("SOI ↓" means retarding SOI)

Overall, the optimal operating parameters for RCCI combustion is considerably affected by the fuel property. Based on above discussion, the impacts of fuel property on the determination of the optimized parameters are summarized in Fig. 5. Owing to the higher RON of methanol than gasoline, higher T_{IVC}, larger P_{IVC}, lower R_P, and more retarded SOI are imperative for M/D RCCI than G/D RCCI to avoid the excessively late CA50. Moreover, compared to G/D RCCI, the stratified combustion with the lower R_P and the more retarded SOI, as well as the higher EGR rate are required in M/D RCCI to alleviate the faster combustion rate and the higher RI.

4 Conclusions

Based on the multi-dimensional simulation and GA, the operating parameters of G/D and M/D RCCI engine are optimized to achieve the satisfactory performance at low-to-mid loads. The impacts of fuel property on the determination of the optimized parameters are concluded as follows:

Without violating the RI limit, more homogeneous fuel distribution with the higher R_P and the more advanced SOI is beneficial for the improvement of the engine performance for both G/D and M/D RCCI. Meanwhile, high T_{IVC} is indispensable to avoid the excessively late CA50 and deteriorated combustion efficiency, especially for M/D RCCI.

The fuel reactivity and combustion rate are two important properties for the fuel in RCCI engines. Because of the lower reactivity and the faster combustion rate of methanol than gasoline, the lower R_P and the more retarded SOI should be introduced in M/D RCCI than those in G/D RCCI to advance the ignition timing and reduce the combustion rate.

Boosting P_{IVC} has two contradictory effects on the combustion temperature and the fuel efficiency. Due to the later CA50 in M/D RCCI, the positive effect dominates in M/D RCCI, and the negative effect outweighs in G/D RCCI. Therefore, the higher P_{IVC} is favorable for M/D RCCI, while the lower P_{IVC} is preferable for G/D RCCI.

5 Acknowledgment

This work was financially supported by the National Natural Science Foundation of China (Grant No. 54176020).

References

• Onishi S, Jo SH, Shoda K, et al. Active thermo-atmosphere combustion (ATAC)-a new combustion process for internal combustion engines. *SAE Technical Paper* 790501, 1979.

• Inagaki K, Fuyuto T, Nishikawa K, et al. Dual-fuel PCI combustion controlled by in-cylinder stratification of ignitability. *SAE Technical Paper* 2006-01-0028, 2006..

• Kokjohn S, Reitz R. Characterization of dual-fuel PCCI combustion in a light-duty engine. *IMEM*, 2010.

• Reitz RD, Duraisamy G. Review of high efficiency and clean reactivity controlled compression ignition (RCCI) combustion in internal combustion engines. *Prog Energy Combust Sci*. 2015,46:12-71.

• Gao T, Reader G, Tjong J, et al. Energy Efficiency Comparison between Butanol and Ethanol Combustion with Diesel Ignition. *SAE Technical Paper* 2015-01-0859, 2015.

• Paykani A, Kakaee A-H, Rahnama P, et al. Effects of diesel injection strategy on natural gas/diesel reactivity controlled compression ignition combustion. *Energy*. 2015,90:814-26.

• Zou X, Wang H, Zheng Z, et al. Numerical Study of the RCCI Combustion Processes Fuelled with Methanol, Ethanol, n-Butanol and Diesel. *SAE Technical Paper* 2016-01-0777; 2016.

• Deb K, Pratap A, Agarwal S, et al. A fast and elitist multiobjective genetic algorithm: NSGA-II. *IEEE Trans Evol Computat*. 2002,6(2):182-97.

• Amsden A. KIVA3V. A Block-Structured KIVA Program for Engines with Vertical or Canted Valves. *Los Alamos National Lab.*, NM (USA); 1997.

• Han Z, Reitz RD. Turbulence modeling of internal combustion engines using RNG k-ε models. Combust Sci Technol 1995;106(4e6):267e95.

• Ricart LM, Reltz RD, Dec JE. Comparisons of diesel spray liquid penetration and vapor fuel distributions with in-cylinder optical measurements. J Eng Gas Turbines Power 2000;122(4):588e95

• Kee RJ, Rupley FM, Miller JA. Chemkin-II: A Fortran chemical kinetics package for the analysis of gas-phase chemical kinetics. *Sandia National Labs.*, CA (USA); 1989.

• Liu Y, Jia M, Xie M, et al. Enhancement on a skeletal kinetic model for primary reference fuel oxidation by using a semidecoupling methodology. *Energy Fuels*. 2012,26(12):7069-83.

• Jia M, Peng Z, Xie M. Numerical investigation of soot reduction potentials with diesel homogeneous charge compression ignition combustion by an improved phenomenological soot model. *Proc Inst Mech Eng D J Auto Eng*. 2009,223(3):395-412.

• Dempsey AB, Walker NR, Gingrich E, et al. Comparison of low temperature combustion strategies for advanced compression ignition engines with a focus on controllability. *Combust Sci Tech*. 2014,186(2):210-41.

• Li Y, Jia M, Chang Y, et al. Towards a comprehensive understanding of the influence of fuel properties on the combustion characteristics of a RCCI (reactivity controlled compression ignition) engine. *Energy*, 2016, 99: 69-82.

• Shi Y, Reitz RD. Optimization study of the effects of bowl geometry, spray targeting, and swirl ratio for a heavy-duty diesel engine operated at low and high load. *Int J Eng Res*. 2008,9(4):325-46.

• Eng J. Characterization of pressure waves in HCCI combustion. *SAE Technical Paper* 2002-01-2859; 2002.

11th Asia-Pacific Conference on Combustion,
The University of Sydney, NSW Australia
10th -14th December 2017.

Laser-Induced Incandescence in Turbulent Non-Premixed Flames at Elevated Pressure

Wesley Boyette[1], Emre Cenker[1], Thibault Guiberti[1], William Roberts[1]

[1]Clean Combustion Research Center, King Abdullah University of Science and Technology
Thuwal 23955-6900, Saudi Arabia

Abstract

There is general interest in understanding sooting behavior in canonical flames at elevated pressures. In this study, a new apparatus is used to evaluate sooting turbulent non-premixed flames at 1 atm and 3 atm, with constant Reynolds number. Laser-induced incandescence is used to assess the distribution and amount of soot in both flames. The results show that increasing the pressure has the effect of creating larger and more intense instantaneous regions of soot, reducing the length of the non-sooting region immediately downstream of the nozzle, and increasing the amount of soot by more than an order of magnitude.

1 Introduction

Soot particles are a common product of modern combustion systems and have been proven to be a serious concern for both the environment [1] and public health [2]. Such concerns are not local, but global, due to the ability of the particles to be transported over long distances by air currents [3]. Furthermore, the effects are not temporary, because these particles contribute to global climate change by adversely affecting the local radiative properties of air [4]. Deposition of ultrafine soot particles in the lungs can results in chronic respiratory illnesses [5]. As a results of these issues and the ubiquitous presence of soot in modern societies, a thorough understanding of soot formation processes and ways to reduce soot emissions are of paramount importance [6].

The International Sooting Flame (ISF) Workshop was established with the aim of developing predictive models of sooting flames by examining well-defined and well-characterized target flames. The workshop is divided into three sections; laminar flames, turbulent flames, and pressurized flames and sprays. The pressurized section includes laminar flames and turbulent premixed flames. The first target flame in the pressurized section is a turbulent premixed swirling flame operated at DLR-Stuttgart at pressure up to 5 atm. Measurements in this flame include techniques such as LII, CARS, OH-LIF, PAH-LIF, and PIV [7, 8, 9]. The second target flame in the pressurized section is a laminar coflow C_2H_4 diffusion flame for which the soot volume fraction and soot particle size have been characterized up to 16 atm [10]. There are a few other laminar flames included in the pressurized flames section of the workshop. The turbulent flames section includes four non-premixed flames, however, there are no turbulent non-premixed flames in the pressurized flames section of the ISF.

There are very few examples of experimental studies of turbulent non-premixed flames at elevated pressures. Perhaps the earliest example was a study of soot formation in small turbulent ethylene-air diffusion flames at pressures up to 8 atm [11]. Laser extinction measurements were used to find that the soot yield at the axial location of peak soot volume fraction increases as $p^{1.4}$. This rig could only accommodate flames with lengths less than 300 mm, so conditions remained relatively moderate with the maximum tube diameter being just over 1 mm, and the maximum jet Reynolds number being 21,100. Other studies of turbulent diffusion flames were produced by the high-pressure rig of J. B. Moss for which flames of kerosene were explored at pressures up to 6.4 atm [12] and later up to 13 atm [13]. The rig was also used for soot studies in methane flames up to 3 atm [14]. This rig had an inner diameter of 155 mm and a vertical movement range of 300 mm. To address the dearth of experimental data in turbulent non-premixed flames at elevated pressures (both sooting and non-sooting), a new rig has recently been commissioned in the high-pressure combustion lab at the Clean Combustion Research Center at KAUST. This apparatus was designed to accommodate very long flames and large burners to avoid some of the limitations of other rigs.

2 Experiment

2.1 High Pressure Combustion Duct

Experiments are performed in the high pressure combustion duct (HPCD). The HPCD is an 8-meter tall stainless steel apparatus designed for investigating large flames at high pressures. The inside diameter is 400 mm, allowing for a wide variety of burner configurations. The design pressure is 40 atm. The test section features four orthogonal arms capped with UV fused silica windows, with a transmission of over 90% from the UV up to 2 μm. The test section also includes two windows oriented at 40 degrees off-axis. A back-pressure regulator in the exhaust section of the duct allows for control of the pressure electronically. A system of compressors supply dry air continuously at a rate of over 0.5 kg/s at 45 atm. Burners may be installed on a translation stage inside the duct, capable of 1500 mm of vertical movement, allowing most flames of interest to be probed from nozzle to tip. Flame ignition is provided by a translating spark igniter for conditions up to 6 atm or it may be induced by the laser used for LII in this study at higher pressures.

Wesley Boyette
E-mail address: wesley.boyette@kaust.edu.sa

2.2 Flame Conditions

The flames investigated in the high pressure combustion duct in this study are part of a family of flames referred to as "KEN" flames, where KEN is an acronym for KAUST Ethylene-Nitrogen. In this family of flames, the composition of the main jet is maintained at 35% C_2H_4 and 65% N_2 by volume. The purpose of the high nitrogen dilution originally was to reduce the soot loading enough to make SMPS measurements of the time-averaged soot particle size distribution possible [15]. Now the N_2 dilution is helpful in keeping the soot loading manageable even as the pressure of the flames increases (enhancing the soot as will be seen in the next section). The flames are piloted with a set of laminar pilot flames composed of premixed C_2H_4 and air at an equivalence ratio of 0.9. The central tube inner diameter is 3.3 mm and the pilot plate has three concentric rings of holes with various diameters and sits 5 mm below the squared-off tip of the central tube. The burner geometry is identical to that described in Zhang et al. 2011 [16], which produces a flame that was recently one of the target turbulent flames in the International Sooting Flame 3 (ISF-3) Workshop. That flame has been characterized by OH-PLIF [16], as well as hybrid fs/ps rotational CARS and laser-induced incandescence (LII) [17]. In the KEN burner, a co-annular flow of air is provided by a 3D printed wind-tunnel with an exit diameter of 250 mm, featuring a honeycomb section and four separate wire mesh flow conditioners as well as a converging exit section. In addition to the conditioned coflow, an airflow of similar velocity is generated in the annular area between the burner and the wall of the HPCD for the purposes of stabilizing the coflow and flame, preventing soot deposition on the windows, and upstream cooling. A schematic of the KEN burner inside the HPCD is given in Fig. 1.

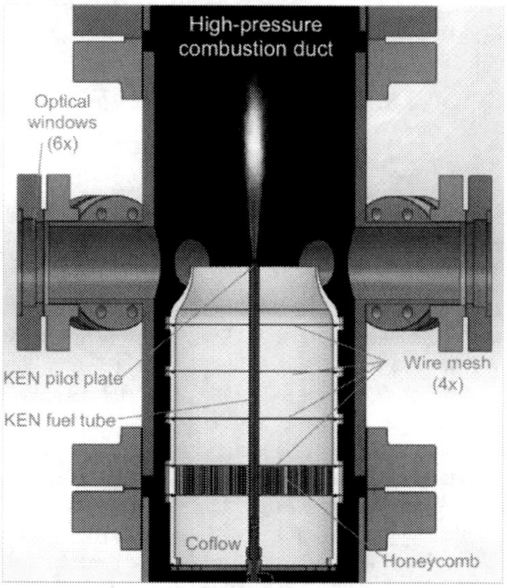

Figure 1: KEN burner and air coflow inside the KAUST high pressure combustion duct.

The flames investigated in this study have an identical composition and Reynolds number (10,000) based on the jet diameter. For the flame with the designation KEN 1A, the pressure inside the duct is atmospheric, whereas for KEN 3A, the pressure inside the duct is 3 atm, thus reducing the bulk jet velocity by a factor of 3 relative to 1A. In both cases, the pilot power is 18% of the heat release of the main jet. The air coflow velocity is maintained at 0.6 m/s for 1A and, to maintain the mass flow rate among conditions, the coflow velocity is a factor of three lower for 3A. The operating conditions are summarized in Table 1. The local turbulence in these flames has not been characterized.

Flame	p (atm)	Re_D	U_j (m/s)	U_{cf} (m/s)
1A	1	10,000	37.9	0.6
3A	3	10,000	12.6	0.2

Table 1: Operating conditions for flames KEN 1A and KEN 3A.

2.3 Laser-Induced Incandescence (LII)

For the LII, the fundamental of a Nd:YAG laser (Continuum Surelite III) at λ = 1064 nm is used for excitation. The laser runs at a repetition rate of 10 Hz and has a pulse width of 5 ns. It is capable of 850 mJ per pulse. The beam is formed into a sheet of roughly 80 mm in height.

Calibration of the LII signal is achieved by simultaneous laser extinction measurements. A HeNe (λ = 632.8 nm) is used for this purpose. A portion of the beam is reflected onto a photodiode as the incident beam before passing through the duct and the remaining signal is collected on the other side of the duct with an integrating sphere and photodiode. Filters are used to reduce interference from other sources of light. The attenuated laser signal is captured by the photodiodes and serves as a marker for the incandescence. The time-averaged voltage recorded immediately before this marker is used for I and I_0. Recordings of I and I_0 without a flame present are used to correct for background intensity. The soot volume fraction, f_v, is related to the ratio of the intensities via equation (1). In this study, a value of K_e = 5.01 is used [18]. Two-dimensional images of the LII signal at the location of the HeNe allows a replacement of f_v by the integral of the intensity along the optical path and L is the addition of lengths of soot pockets in each image.

$$\log\left(\frac{I}{I_0}\right) = -K_e \frac{f_v}{\lambda} L \qquad (1)$$

Images of the flame are captured with a PCO HSFC-PRO camera system, which houses four intensified cameras in a single box. Two channels are used in this study. The first channel records an image immediately before the LII signal, thus capturing the spontaneous emission to be subtracted from the LII image as background. The second channel has a 300 ns delay relative to the first channel and is timed to acquire the peak LII intensity. The camera system is used with a Nikon 105 mm f/4 UV-MICRO-APO lens. Both channels have a Semrock FF01-655/40-50 filter.

1500 images are recorded at four different stages of each flame to map the 2D LII intensity over the entire flame. An analysis of the standard deviation of the intensity at multiple locations shows that 1500 samples are sufficient for statistical convergence. During processing, a flat field image for each camera is used to eliminate variations in pixel sensitivity. The reference camera images are also corrected to account for different camera sensitivities.

3 Results and Discussion

Images of each flame are shown in Fig. 2 for a comparison of their visual characteristics. The images were recorded with a Canon EOS 7D DSLR and cover a length of roughly 40 diameters of the flame. The first characteristic of note is the bluish pre-sooting region that is present in flame 1A. This region is almost non-existent in the image for flame 3A, indicating that soot inception occurs much further upstream, most likely due to the lower bulk jet velocity. The other characteristic of note is that flame 3A is apparently much more luminous (even with different shutter speeds) than 1A, indicating that the increase in pressure has an effect on the propensity to soot.

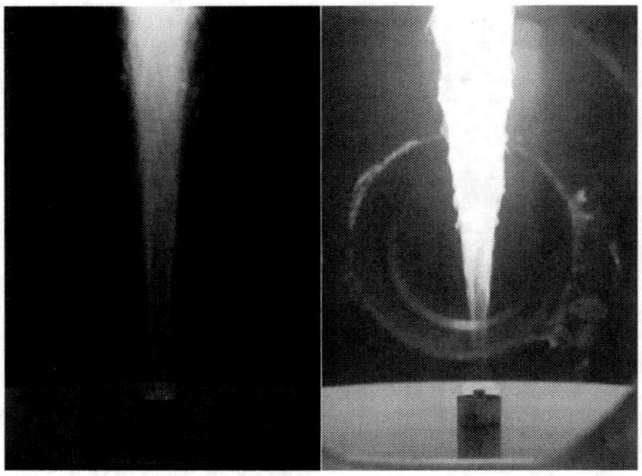

Figure 2: Images of KEN flames 1A (left) and 3A (right).

Processing of the LII images for KEN 1A resulted in 1500 images that were corrected for pixel intensity variations and contributions from spontaneous emissions, and the intensity was scaled with the simultaneous extinction measurements.

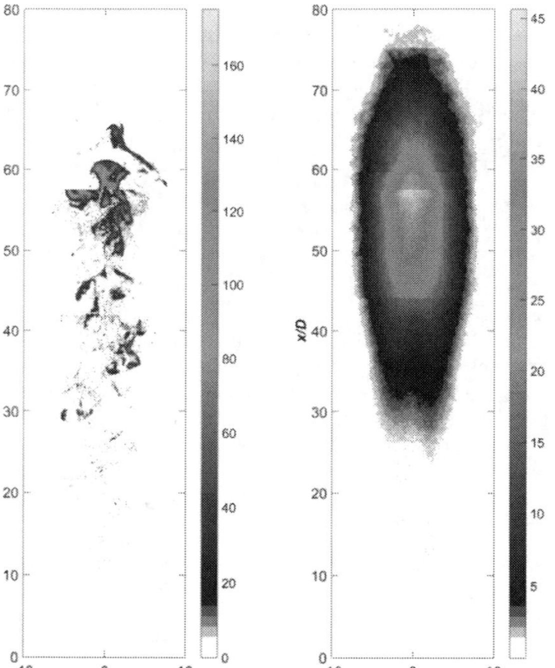

Figure 3: Instantaneous (left) and mean (right) LII intensity in KEN 1A.

Results from randomly selected instantaneous images at each of four sampling locations along the flame are shown on the left in Fig. 3. On the right side of Fig. 3 is the resulting mean LII intensity for KEN 1A.

The instantaneous images of KEN 1A show that soot tends to exist in small isolated pockets throughout most of the flame. This leads to barely detectable extinction measurements below $x/D = 50$. However, from previous measurements in this flame [15], it is known that the soot particles that do exist in this region of the flame tend, on average, to be relatively small. Previous work [19] has shown that LII may not be an effective technique for detecting such particles. The mean LII intensity shows that almost no detectable soot exists below $x/D = 30$, which is easily understood by considering the image in Fig. 2. The peak mean intensity occurs along the flame centerline at a distance from the nozzle of approximately $x/D = 55$.

As with the LII images for KEN 1A, 1500 images were processed at four locations in KEN 3A and randomly selected instantaneous (left) and mean (right) LII intensities are shown in Fig. 4.

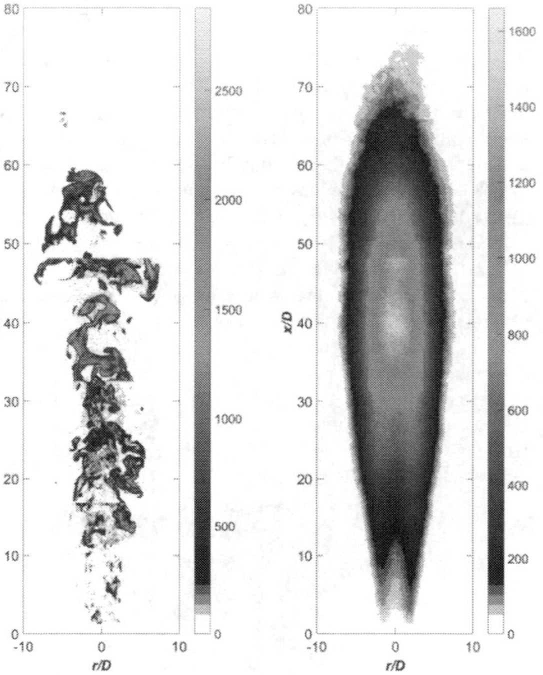

Figure 4: Instantaneous (left) and mean (right) LII intensity in KEN 3A.

Figure 4 has many notable features, particularly in comparison to the atmospheric case from Fig. 3. From the instantaneous images, we see that KEN 3A features much larger, denser, and more connected regions of soot than KEN 1A. As expected from the images in Fig. 2, soot is present almost immediately downstream of the nozzle exit. By comparing the mean LII intensities of the flames, the amount of soot in the flame at 3 atm is over an order of magnitude greater than 1A. The peak mean intensity occurs along the flame centerline at $x/D = 40$. Interestingly, at distances $x/D < 30$, the peak mean intensity occurs slightly off of the centerline.

The increase in the amount of soot with increasing pressure is not unexpected. Extinction measurements from the Flower study found an exponential relationship between soot volume fraction and pressure with an exponent of 1.4 from 1 atm up to 5 atm [11]. In the Brookes and Moss study with methane/air diffusion flames at 1 atm and 3 atm, extinction measurements also indicate an increase in soot volume fraction by an order of magnitude [14]. Studies in laminar diffusion flames reveal an increase in soot volume fraction roughly proportional to the square of the pressure [10]. Thus, the results presented in this study are consistent with other findings.

It is worth considering other effects that may be influencing the formation of soot when comparing KEN 1A to KEN 3A. In particular, while the Reynolds number is kept constant, the Damköhler number, which is directly proportional to the flow time scale, is most certainly higher in the KEN 3A case. A DNS study of soot formation in *n*-heptane/air turbulent non-premixed jet flames investigated Damköhler number effects and found that an increase in Da by a factor of four was associated with an increase in soot volume fraction of more than one order of magnitude [20]. In light of this evidence, the observed increase in soot volume fraction in this study may be due not only to increasing pressure but also increasing Damköhler number.

4 Conclusions

Sooting turbulent non-premixed flames composed of ethylene-nitrogen in an air coflow have been investigated at 1 atm and 3 atm, while maintaining a constant Reynolds number of 10,000 based on the central tube diameter. Laser-induced incandescence was employed with simultaneous extinction measurements to measure the instantaneous 2D soot distribution through the center of each flame. 1500 images were collected at four axial locations along each flame to allow for statistically converged quantification of the entire flame.

The 1 atm case exhibits smaller pockets of soot in the instantaneous images and a soot-free region for almost 30 diameters downstream of the nozzle. The 3 atm flame has much larger regions of soot in the instantaneous images and the sooting region begins almost immediately downstream of the nozzle. In both cases, the peak mean LII intensity occurs along the centerline, with the axial location of the peak slightly closer to the nozzle in the 3 atm case. The peak mean LII intensity is more than an order of magnitude greater in the 3 atm case than the 1 atm case. This is expected, but the increase may be due to a higher Damköhler number, in addition to the higher pressure.

Future work should seek to extend this knowledge to higher pressures. Cases in which the bulk jet velocity, as opposed to the Reynolds number, is preserved as the pressure increases should be explored to reduce the effect of changing Damköhler number. Experiments in which the Damköhler number and the Reynolds number are preserved should be considered as well, although this is admittedly difficult as it would require changing the burner tube diameter with each pressure setting and scaling the pilot flame accordingly.

5 Acknowledgment

The research reported in this publication was supported by funding from King Abdullah University of Science and Technology (KAUST).

References

1. P.R. Buseck, M. Pósfai, Proc. Natl. Acad. Sci. U.S.A. 96 (1999) 3372-3379.
2. T.J. Grahame, R. Klemm, R.B. Schlesinger, J. Air Waste Manage. Assoc. 64 (6) (2014) 620-660.
3. M.-T. Chuang, C.-T. Lee, C.C.-K. Chou, N.-H. Lin, G.-R. Sheu, J.-L. Wang, S.-C. Chang, S.-H. Wang, K.H. Chi, C.-Y. Young, H. Huang, H.-W. Chen, G.-H. Weng, S.-Y. Lai, S.-P. Hsu, Y.-J. Chang, J.-H. Chang, X.-C. Wu, Atmos. Environ. 89 (2014) 507-516.
4. S. Menon, J. Hansen, L. Nazarenko, Y. Luo, Science 297 (2002) 2250-2253.
5. A. Wierzbicka, P.T. Nilsson, J. Rissler, G. Sallsten, Y. Xu, J.H. Pagels, M. Albin, K. Österberg, B. Strandberg, A. Eriksson, M. Bohgard, K. Bergemalm-Rynell, A. Gudmundsson, Atmos. Environ. 86 (2014) 212-219.
6. H. Wang, Proc. Combust. Inst. 33 (2011) 41-67.
7. K.P. Geigle, R. Hadef, W. Meier, J. Eng. Gas Turbines Power 136 (2014) 021,505.
8. K.P. Geigle, M. Köhler, W. O'Loughlin, W. Meier, Proc. Combust. Inst. 35 (2015) 3373-3380.
9. K.P. Geigle, W. O'Loughlin, R. Hadef, W. Meier, Appl. Phys. B Lasers Opt. 119 (2015) 717-730.
10. S.A. Steinmetz, T. Fang, W.L. Roberts, Combust. Flame 169 (2016) 85-93.
11. W.L. Flower, Proc. Combust. Inst. 22 (1988) 425-435.
12. K.J. Young, C.D. Stewart, J.B. Moss, Proc. Combust. Inst. 25 (1994) 609-617.
13. B.A. Fischer, J.B. Moss, Combust. Sci. Technol. 138 (1998) 43-61.
14. S.J. Brookes, J.B. Moss, Combust. Flame 116 (1999) 49-61.
15. S. Chowdhury, W.R. Boyette, W.L. Roberts, J. Aerosol Sci. 106 (2017) 56-67.
16. J. Zhang, C.R. Shaddix, R.W. Schefer, Rev. Sci. Instrum. 82 (2011) 074,101.
17. S.P. Kearney, K.N. Hoffmeister, D.R. Guildenbecher, C. Winters, T.W. Grasser, J.C. Hewison, Hybrid fs/ps Rotational CARS Temperature and Oxygen Measurements and Soot LII Measurements in a Turbulent C2H4-Fueled Jet Flame, AIAA (2016) 2016-0281.
18. J. Zerbs, K.P. Geigle, O. Lammel, J. Hader, R. Stirn, R. Hadef, W. Meier, Appl. Phys. B Lasers Opt. 96 (2009) 683-694.
19. V. Krüger, C. Wahl, R. Hadef, K.P. Geigle, W. Stricker, M. Aigner, Meas. Sci. Technol. 16 (2005) 1477-1486.
20. A. Attili, F. Bisetti, M.E. Mueller, H. Pitsch, Proc. Combust. Inst. 35 (2015) 1215-1223.

Reducing Combustor Length within a Mach 12 Shape-Transitioning Scramjet

Will O. Landsberg[1], Dr Vincent Wheatley[1], Prof. Michael K. Smart[1], Dr Ananthanarayanan Veeraragavan[1]

[1]The Centre for Hypersonics, The University of Queensland

St Lucia, Brisbane, QLD, 4072, Australia

Abstract

With scramjets displaying improving performance within the wider literature, attention is turning towards airframe-integrated systems for access-to-space. The Mach 12 Rectangular-to-Elliptical Shape-Transitioning (M12REST) engine has been designed to operate at upper flight speeds, leading towards operable hybrid launch vehicles. At these speeds, internal components must be tuned to experience minimal losses and heating loads. To this end, the present work, guided by prior simulations, both shortens the M12REST combustion chamber by 57.5%, and removes the legacy backwards facing step. The newly configured combustor was examined experimentally at on-design conditions within the T4 Stalker Tube at The University of Queensland. Heat transfer analysis indicated significant combustion of hydrogen fuel was achieved through the combustor, with substantial pressure rises measured on the thrust nozzle surface.

1 Introduction

In recent years, hybrid rocket-scramjet-rocket launch vehicles have been predicted to offer significant efficiency and operability benefits over current rocket-based systems. For small satellite launches, Smart and Tetlow [1] indicated payload mass fractions of 1.47% were achievable for these hybrid systems. While comparing favourably to the 0.9% offered by standard rocket-based systems, these studies depended on a scramjet engine capable of accelerating to Mach 12 speeds, and few studies have characterised such engines.

Leading towards this aim, the initial REST scramjet was developed by Smart [2]. This engine smoothly integrates to the flight vehicle underside, with the vehicle forebody performing initial shock compression and the rectangular capture area enabling side-by-side mounting. The fixed geometry elliptical combustor provides improved structural efficiency, while highly swept leading edges and a notched cowl permit engine operation over a range of Mach numbers [3].

At flight speeds, however, significant drag and heating loads are incurred by all flow path surfaces. While inlet operability is key to scramjet performance, combustors are often designed conservatively, relying on turbulent mixing to promote sufficient combustion performance. This approach however typically suffers from increased heating and viscous drag loads due to the conservatively long combustors, with previous work examining drag-reducing film injectors to improve system performance [4]. While successful in fuelling the engine, drag and heat load reductions were small [5]. As such, new combustor designs are required to carefully balance combustion performance with internal drag, with the ideal system performance likely sacrificing some combustion potential for a reduction in total drag. Combustor length and geometry are thus the focus of the present work, using an experimental approach to examine supersonic combustion at on-design conditions.

2 Experimental Set-up & Methodology

2.1 Flow Path Geometry & Experimental Model

The present work examines the M12REST flow path. The experimental test model is a 50% scale engine, designed to operate at 50 kPa dynamic pressure at a 6° angle of attack, and was originally studied in the work of Wise (2015) [4]. The engine consists of four primary segments, as shown in Figure 1. The flow path begins with a 500 mm forebody, commencing at a sharp leading edge and fitted with swept ramp boundary layer trips at $x = 235$ mm to ensure turbulent flow enters the inlet. The inlet is 476.2 mm in length when measured from the compression-surface leading edges to the throat. The rectangular inlet smoothly transitions to an elliptical isolator region, with the final shape transition completed by 505.8 mm, with cross-sectional area of 430 mm^2 (compression ratio of 6.61).

The studies of [4, 5] utilised a combustor inclined 6° to the inlet, redirecting flow to the nominal flight direction. The combustor began 530.87 mm downstream of the inlet leading edge, and joined the isolator via a 1.25 mm circumferential backward facing step. This step expanded the area to 530.5mm^2, with the combustor maintaining a 161 mm long constant area section, before diverging at 1.6° for 121 mm. The combustor merged smoothly to the nozzle, which expanded the flow uniformly over 125 mm, with 1:10 area ratio relative the engine throat.

The current work modifies the combustor of the M12REST engine. The combustor backward facing step was originally included to facilitate fuel injection via a ring of porthole injectors with the aim to reduce viscous drag via boundary layer combustion. However, the computational studies of [5] indicated injected fuel was swept away from the combustor walls via native vortex features in the M12REST flow path, minimising any drag reduction. Additionally, the work of [6] indicated that improved mixing performance may be possible where fuel was injected far upstream of this step, inviting its removal. While novel explorative fuel injection methods were explored in [5,7], the expense of model redevelopment ensured that subsequent

Corresponding author.
E-mail address: w.landsberg@uq.edu.au

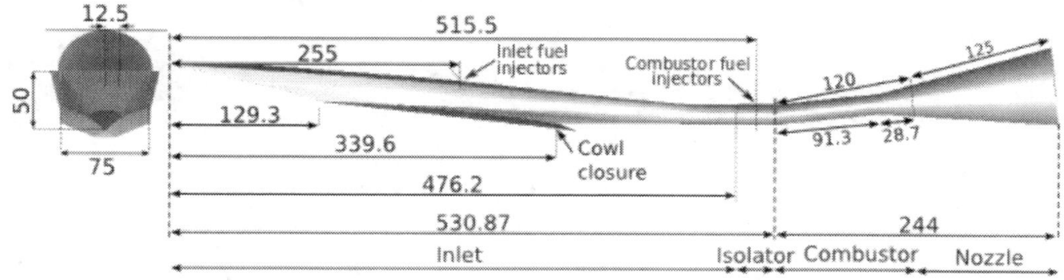

Figure 1: M12REST flow path schematic - Front (left) and side (right) views [dimensions in mm]

experimental studies maintained this backward facing step legacy component. The present work removes this step, maintaining a constant 430 mm^2 combustor area through 91.3 mm, before diverging at the same angle as the nozzle for 28.7 mm. This minimised the length of the diverging portion of the combustor, with the new total combustor length of 120 mm (6.8 combustor heights) set, achieving a reduction in length of 57.5%. The flow path schematic is given in Figure 1 (the forebody is excluded for clarity), with the model as mounted in the test section given in Figure 2.

Figure 2: M12REST engine mounted in T4 test section

The engine is capable of injecting hydrogen fuel at both inlet- and combustor-based locations. These are located 255mm and 515.5mm downstream of the inlet leading edge respectively. The inlet injectors comprise of 3 × 2 mm diameter flush wall portholes on the bodyside surface, located symmetrically about the engine centreline with a jet-to-jet spacing of 12.5 mm. Additional hydrogen fuel is injected at the combustor entrance via 5×0.8 mm flush wall injectors (as examined in [5], see Figure 3), with the combustor-based injectors tailored to leverage favourable interactions with the native flow field. All injectors are inclined 45° to their local wall normal.

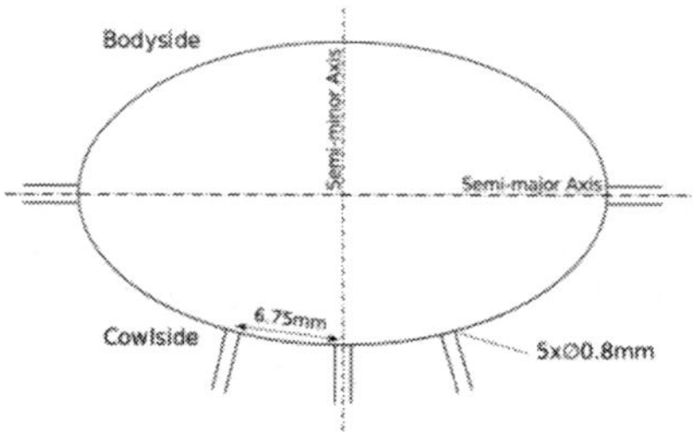

Figure 3: Combustor-based fuel injection configuration [5]

2.2 Facility & Flow Conditions

Flight representative flow conditions are provided by The University of Queensland's (UQ) T4 Stalker Tube facility (T4, see Figure 4). T4 operates using a free piston driver to compress and generate high temperatures and pressures within the driver [8]. This driver subsequently ruptures a 6mm un-scored, bright form steel diaphragm, with the generated shock propagating through the shock tube. This shock reflects at the end of the shock tube, subsequently stagnating the flow and rupturing a 0.1 mm Mylar diaphragm to provide nozzle supply conditions. T4 is capable of generating nozzle supply enthalpies ranging from 3-12 MJ/kg, with stagnation pressures of up to 90 MPa.

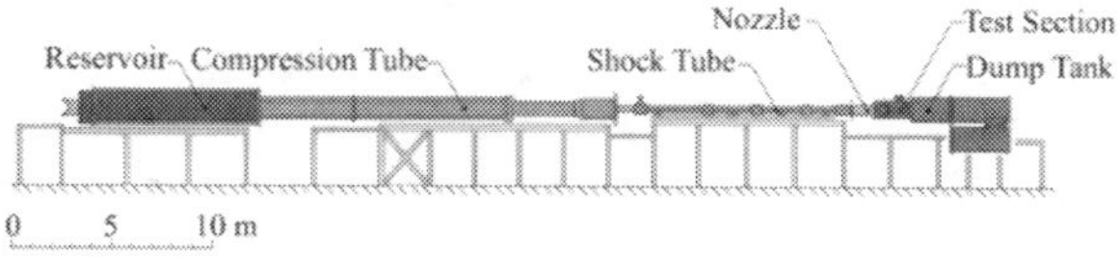

Figure 4: T4 Stalker Tube schematic, reproduced from [9]

Due to the total pressure limitations of T4, the previous studies of [4, 5] examined the M12REST model in semi-free-jet mode. In this mode, conditions are designed to replicate the post 6° forebody shock conditions during Mach 12 flight, with the model forebody inclined 0° within the T4 test section. Nozzle supply conditions are expanded through the T4 Mach 10 contoured nozzle. To quantify conditions at the nozzle exit plane, simulations were performed using UQ's in-house code, NENZFr [10], which calculates nozzle-supply conditions using ESTCj [11] before expanding these conditions through the nozzle geometry via a 2D, axisymmetric RANS simulation using Eilmer3 [12]. Supply pressure was typically 84 MPa, with supply enthalpy of 7 MJ/kg. Nominal nozzle exit conditions are given in Table 1. Air and nitrogen were used as test gases, examining both active and suppressed combustion. As the freestream Mach number determined with NENZFr was higher than would be experience behind a 6° oblique shock at Mach 12, the model was inclined at 1.6° to the freestream. The presently examined conditions are adapted from those of [4, 5].

Variable	Value	Variable	Value
T	376 K	M	9.4
p	1030 Pa	q	63 kPa
ρ	9.54 g/m^3	Re	1.58×10^6
U	3630 m/s	H	6.91 MJ/kg

Table 1: Nominal T4 nozzle exit conditions

2.3 Instrumentation

Data was collected via 16 National Instruments (NI) PXI-6133 data acquisition cards, using a NI PXI-Controller mounted in a NI PXI-1045 chassis. The model was instrumented using 43 Kulite® XTEL-190 (M) piezo-resistive pressure transducers, ranging in design pressures from 25-100 psi. Two Kulite® sensors are located on the forebody, eleven through the inlet, with seven sensors located on each of the combustor body- and cowlside surfaces, and eight on each nozzle surface. A 100 psi and a 500 psi sensor were employed in the inlet- and combustor-based fuel plenums respectively to measure fuel injection rates. Heat transfer was measured via flush mounted, thin-film heat transfer gauges (TF) manufactured in-house at UQ. These fast response sensors measure changes in resistance with temperature versus time of a thin nickel film deposited on a quartz substrate. Details on manufacture and performance are available in [4]. TF's were mounted on the forebody to characterise the boundary layer, with additional TF's also utilised to measure heat transfer through the inlet and combustor.

3 Results

Results are presented for three T4 experiments (shots), as listed in Table 2. Fuel equivalence ratios are provided for each specific injection location, as well as the total.

Shot	Test gas	Φ_{inlet}	Φ_{com}	Φ_{total}
12001	Air	-	-	0.0
11996	N_2	0.35	1.00	1.35
11995	Air	0.34	0.98	1.32

Table 2: Shot data and fuelling conditions

Shot 12001 provides the baseline performance for the unfuelled flow path. Shot 11996 provides a means to assess the measured pressure and heat transfer increases as a direct result of mass addition through both inlet- and combustor-based fuel injection. Nitrogen test gas suppresses combustion, and hence this shot may be compared against fuel-into-air shots to directly assess the extent of combustion. Shot 11995 is fuelled similarly to 11996; however, this shot utilised air as the test gas to examine combustion performance.

3.1 Pressure

Mean pressure data is represented as in (1):

$$\frac{p}{p_1}(t) = \frac{p(t+t_d)}{p_s(t)} \times \frac{p_{s,\mathrm{nom}}}{p_{1,\mathrm{nom}}} \tag{1}$$

Due to the short test time available, a slug tracing technique is utilised to track the passage of the test gas through the engine. Data recorded at each pressure sensor is normalised against the measured supply pressure, with the data for each sensor shifted in time by their individual trigger delay (t_d) from the supply trigger. The trace is then multiplied by the ratio of the mean supply pressure ($p_{s,\mathrm{nom}}$) to the mean forebody pressure ($p_{1,\mathrm{nom}}$, the first forebody pressure sensor - located 162.5mm from the forebody leading edge). The mean of this time-shifted and normalised signal (p/p_1, Eq. (1)) is taken during the selected test time to account for shot-to-shot variance and permit direct comparisons. Figure 5 compares pressure distributions through the engine for each shot, with the flow path schematic visualising each sensor's location. Uncertainty bounds are given as detailed in [4].

Bodyside refers to pressure measured along the flight-vehicle-side of the engine, with the cowlside representing the lower surface. Injection of fuel at the inlet in the combustion suppressed shot induces negligible pressure increase over the compression surface. While incipient combustion does occur here in 11995, more significant combustion-induced pressure rises occur later on through the combustion chamber. While the major shock structures are maintained across all three shots, the combustion process in 11995 gives rise to additional flow features. At approximately 600 mm from the leading edge, a substantial pressure rise is experienced, propagating across both body- and cowlside surfaces. As the flow expands at the combustor exit, this pressure relaxes to near unfuelled levels before maintaining increased pressure across both body- and cowlside nozzle surfaces compared to the non-combusting cases. One final difference is noted at the cowlside nozzle surface, with a spike in pressure measured by the third last pressure sensor.

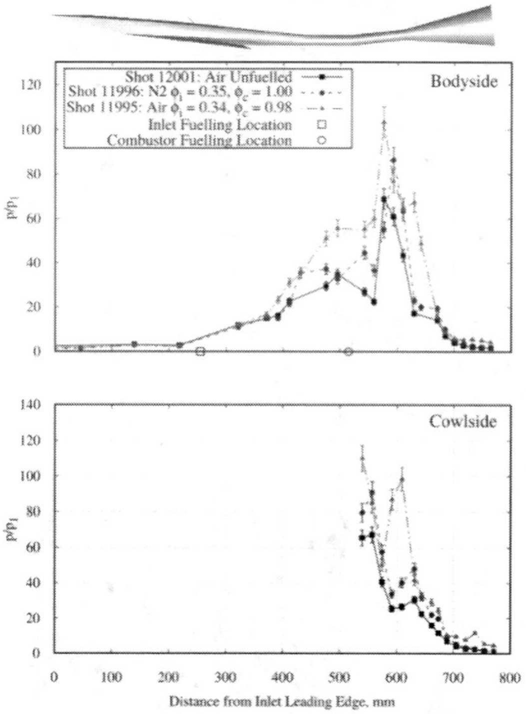

Figure 5: Mean pressure normalised to p_1

The examined cases indicate that a mixing length of only 120 mm within the combustor is adequate to provide a measurable pressure rise on the thrust nozzle. With pressure rises experienced on both the body- and cowlside nozzle surfaces, it is proposed that combustor-injected fuel achieved significant mixing and combustion through its reduced length. Future work will examine performance at differing equivalence ratios, as well as inlet-only versus combustor-only fuelling, with direct comparisons made against the previously installed combustor.

3.2 Heat Transfer

Heat transfer data is presented via the modified Stanton number (St), as in [4]:

$$St = \frac{Q_w}{\rho u(H-h_w)} \tag{2}$$

This method normalises the measured heat flux (Q_w, see (3)) against the freestream density (ρ) and velocity (u), together with the measured supply enthalpy (H) and h_w which refers to enthalpy of air at the model wall temperature (assumed constant at 300 K during the short test time).

$$Q_w = \frac{\sqrt{\rho c k_T}}{\sqrt{\pi}\alpha_R V_0}\sum_{i=1}^{n}\frac{V(t_0)-V(t_{i-1})}{(t_n-t_i)^{1/2}+(t_n-t_{i-1})^{1/2}} \tag{3}$$

To establish the state of the forebody boundary layer, Figure 6 compares measured heat flux to laminar and turbulent Stanton number distributions (as calculated in [4] using the Cebeci and Bradshaw boundary layer code [13]). Uncertainty bounds for St are given as detailed in [4].

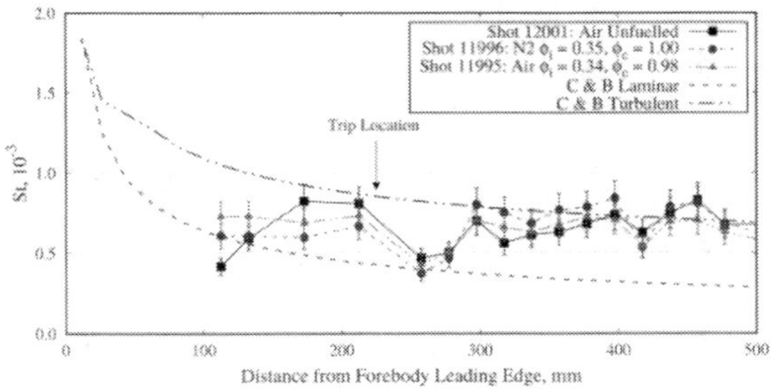

Figure 6: Mean heat transfer along model forebody

It may be seen that upstream of the boundary layer trip, calculated St remains close to laminar levels. Downstream of the trip, heat transfer levels stabilise about the analytical turbulent levels by $x = 350$ mm. As such, it is determined that the boundary layer enters the inlet in a turbulent state. Heat transfer levels in the combustor are presented in Figure 7.

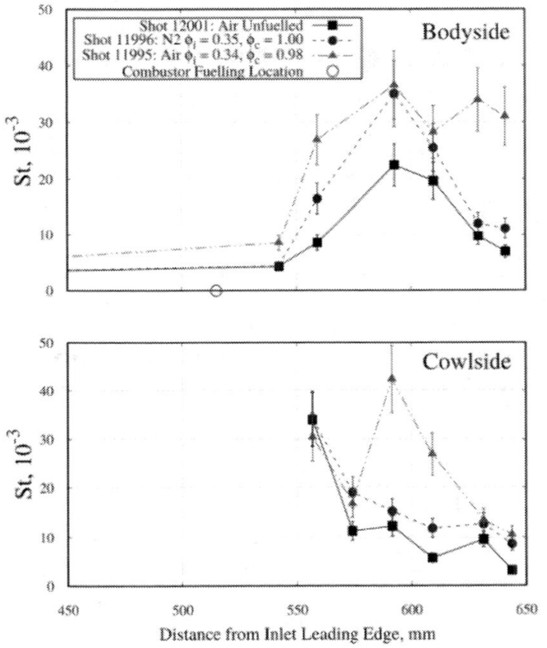

Figure 7: Mean heat transfer in M12REST combustor

Shot 11996 experiences greater heat transfer than the unfuelled shot, indicating that mass addition and fuel-jet bow-shocks induced increased heat loads. Significantly greater heat loads are experienced for the combusting shot (11995) at both the body- and cowlside surfaces, with greater differences experienced in the latter half of the combustor. It is noted that the cowlside surface displays combustion induced heat transfer further downstream than the bodyside surface, but this cowlside-rise does reach a greater margin over 11996 than the bodyside surface.

4 Conclusions

For scramjets to operate as part of access-to-space systems, internal components must be tuned such that losses and heat loads are minimised. Leading towards this, the M12REST combustor length was reduced in length by 57.5%, with the legacy backward facing step also removed. Following experimental analysis, performance metrics indicated significant combustion was achieved throughout the combustor with combined inlet- and combustor-based fuel injection. With this performance, substantial pressure rises were measured over the thrust nozzle.

5 Acknowledgment

This research was supported under the Australian Research Council's Discovery Projects funding scheme (project number DP130102617). This research was undertaken with the assistance of resources provided at the NCI National Facility and the Pawsey Supercomputing Centre through the National Computational Merit Allocation Scheme supported by the Australian Government. W. Landsberg acknowledges support through the Australian Government Research Training Program Scholarship. A. Veeraragavan acknowledges support from the Advance Queensland Research Fellowship for this work.

References

[1] M.K. Smart, M.R. Tetlow, J. Spacecraft & Rockets, 46 (1) (2009) 117-125. doi:10.2514/1.38784

[2] M.K. Smart, J. Propulsion & Power, 15 (3) (1999) 408-416. doi:10.2514/2.5459

[3] Y. Moule, M.K. Smart, J. Propulsion & Power, 29 (1) (2013) 282-285. doi:10.2514/1.B34563

[4] D.J. Wise, *Experimental Investigation of a Three Dimensional Scramjet Engine at Hypervelocity Conditions*, PhD thesis, The University of Queensland, Australia, 2013. doi:10.14264/uql.2015.465

[5] J.E. Barth, D.J. Wise, V. Wheatley, M.K. Smart, 20th AIAA International Space Planes and Hypersonic Systems and Technologies Conference, Scotland, 2015. doi:10.2514/6.2015-3614

[6] W.O. Landsberg, N.N. Gibbons, V. Wheatley, M.K. Smart, A. Veeraragavan, 21th AIAA International Space Planes and Hypersonic Systems and Technologies Conference, China, 2016. doi: 10.2514/6.2017-2389

[7] W.O. Landsberg, V. Wheatley, A. Veeraragavan, AIAA Journal, 54 (12) (2016) 3692-3700. doi: 10.2514/1.J054815

[8] R. J. Stalker, Aeronautical Quarterly, 17 (4) (1966) 351-370. doi:10.1017/S0001925900003966

[9] L. Doherty, *Experimental Investigation of an Airframe Integrated 3-D Scramjet at a Mach 10 Flight Condition*, PhD thesis, The University of Queensland, Australia, 2013. doi:10.14264/uql.2014.382

[10] L. Doherty, F. Zander, P. Jacobs, R. Gollan, W. Chan, R. Kirchhartz, NENZF-r: Non-Equilibrium Nozzle Flow Reloaded. A User Guide, Tech. Rep. 2012/08, The University of Queensland, Australia.

[11] P. Jacobs, R. Gollan, D. Potter, F. Zander, D. Gildfind, P. Blyton, W. Chan, L. Doherty, Estimation of High-Enthalpy Flow Conditions for Simple Shock and Expansion Processes using the ESTCj Program and Library, Tech. Rep. 2014/01, The University of Queensland, Australia.

[12] R. Gollan, P. Jacobs, Int. J. Numerical Methods in Fluids, 73 (1) (2013) 19-57. doi:10.1002/fld.3790

[13] T. Cebeci, P. Bradshaw, Physical and Computational Aspects of Convective Heat Transfer, Springer-Verlag, 1984. doi:10.1007/978-1-4612-3918-5

11th Asia-Pacific Conference on Combustion,
The University of Sydney, NSW Australia
10th -14th December 2017.

Investigation of the Stability and Behavior of Lifted Tri-brachial Flames

Narayan P. Sapkal[1], Kyu Ho Van[1], Jeong Park[1*], Suk Ho Chung[2]

[1]Pukyong National University

Yongdang Campus, (48547) 365, Sinseon-ro, Nam-Gu, Busan, South Korea

[2]King Abdullah University of Science and Technology

Thuwal 23955, Saudi Arabia

Abstract

This paper presents the experimental and numerical study of laminar lifted flames of methane diluted with nitrogen in co-flow jets. Liftoff height characteristics of such flames were observed with the variation of fuel nozzle exit velocity. These lifted flames show the decreasing liftoff height behavior by increase of fuel nozzle exit velocity (at low nozzle exit velocity). Schmidt number for these flames was less than unity. This study is fully focused on lifted flames with decreasing liftoff height behavior only. From numerical simulations it was clarified that under normal gravity the buoyancy convection could be the reason for such decreasing liftoff height behavior of lifted methane jet flames.

1 Introduction

Flame stabilization is one of the important issue in designing industrial burners for gas turbines or boiler. In this regard, laminar lifted non-premixed jet flames have been studied profusely. Detailed studies on such lifted flames shows that, by increase of jet velocity, nozzle attached flame was lifted off from the nozzle and became a lifted flame [1-3]. These lifted flames have tri-brachial flame structure which consists of the rich and lean premixed flame wings and trailing diffusion flame, all extending from a single location. The balance between tri-brachial flame speed and local flow velocity is known as the stabilization mechanism. The correlation of liftoff height, jet velocity and nozzle diameter was derived based on the similarity solutions of velocity and concentration [1,2]. Also, the characteristics of triple flames have been studied in mixing layers by considering the effects of heat generation [4,5], mixture fraction gradient [6], and buoyancy [7]. In addition, lifted flame is found to be stable or unstable when the local flow velocity along the stoichiometric contour decreases or increases with the axial distance in laminar free jets, which corresponds to the cases when Schmidt number, *Sc* of fuel is larger or smaller than unity [2]. Based on the cold jet similarity solutions, experimentally it was shown that propane and n-butane fuels having $Sc > 1$, exhibited stable lifted flames, while no stable lifted flames were observed for methane and ethane fuels having $Sc < 1$, in free jets. However, stationary lifted flames were observed for methane diluted with nitrogen in the near field of co-flow jets under normal gravity due to the influence of buoyancy [8].

By considering those studies on lifted flames, this work is presented to clarify the decreasing liftoff height behavior of lifted methane jet flames. Results demonstrates that buoyancy convection may play a significant role in such lifted flames.

2 Experiment and Numerical Approach

Figure 1, shows the schematic diagram of experimental set up, which consists of a co-flow burner, flow control system and visualization system. The burner had a central fuel nozzle with the diameter of 0.95 mm and its length was 100 times the inner diameter of the nozzle for the flow inside to be fully developed. The co-flow air was supplied to a coaxial nozzle with inner diameter of 92 mm through glass beads and honeycomb for the flow inside to be uniform. A Quartz tube of 40 cm in length was placed on the burner to minimize the outside disturbances. A digital video camera (SONY, HDR-SR12) was placed for visualization. Liftoff heights were measured by the cathetometer. Lifted flames observed through experimental study were also identified by numerical simulations. To simulate the lifted flames in co-flow, steady state axisymmetric cylindrical coordinates were solved for the momentum, species and energy equations. Computations involving reduced kinetic mechanism were performed. Methane-air two step overall reaction mechanism was developed. Detailed schematic diagram of co-flow burner with boundary conditions is as shown in Fig. 1(b).

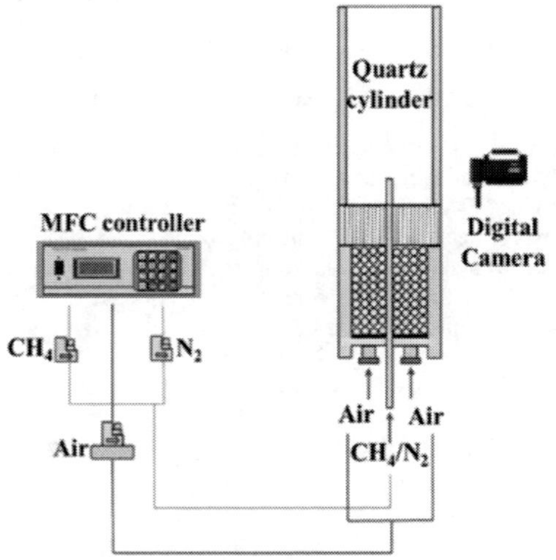

(a)

Corresponding author. Fax: +82-51-629-6126
E-mail address: jeongpark@pknu.ac.kr

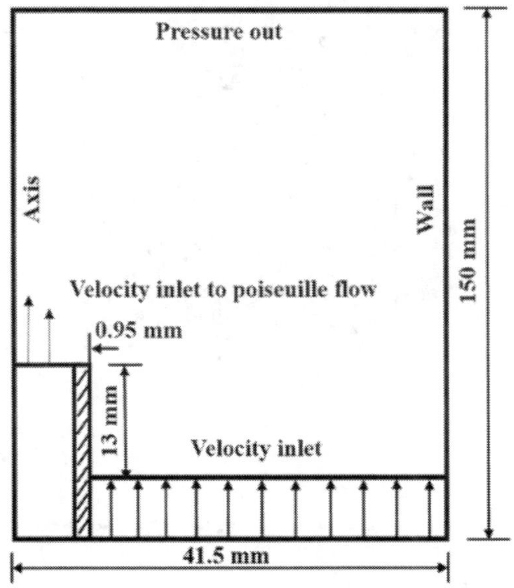

(b)

Figure 1: Schematic diagrams of (a) experimental set up (b) boundary conditions of co-flow burner.

3 Results and Discussions

3.1 Behavior of Lifted Flames

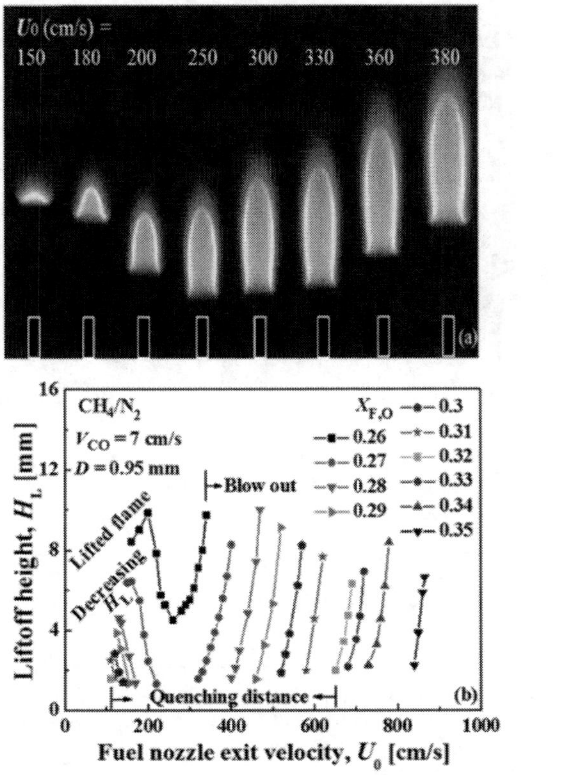

Figure 2: (a) Direct photographs of lifted methane jet flames diluted with nitrogen for $X_{F,O}$ = 0.27 and at various fuel nozzle exit velocities (b) Liftoff height variation of lifted flames with fuel nozzle exit velocity for methane diluted with nitrogen (Sc < 1) at various $X_{F,O}$.

Lifted flames in laminar jets of methane diluted with nitrogen can be found using a nozzle diameter, D = 0.95 mm as shown in the Fig. 2, as a function of liftoff height H_L and fuel nozzle exit velocity, U_0 at various fuel mole fractions, $X_{F,O}$. Co-flow velocity V_{CO} = 7 cm/s was uniform throughout the experiment. Direct photographs of lifted methane jet flames are shown in Fig. 2 (a), for $X_{F,O}$ = 0.27 and with the variation of U_0. Initially flame was lifted at U_0 = 150 cm/s with H_L = 6.38 mm. But H_L was decreased with more increase in U_0 up to 250 cm/s at constant $X_{F,O}$ as shown in the Fig. 2(a). Afterwards, flame was again lifted at U_0 = 300 cm/s with continuous increase in H_L behavior. Note that, present study is mainly focused on the lifted flames with decreasing H_L behavior only. Such lifted flames exhibited the triple flame structure in decreasing H_L region and flame length increases appreciably with increase of U_0 in the range of 150 to 250 cm/s. Figure 2(b) shows the detail plot of both decreasing and increasing regimes of H_L behavior lifted flames at various $X_{F,O}$. The decreasing H_L behavior lifted flames regimes were observed in the range of $X_{F,O}$ = 0.26 to 0.32. The maximum liftoff height in the decreasing region was in the range of several millimeters to 9.86 mm. The Sc no for such lifted flames was in the range of $0.687 < Sc < 0.689$. Therefore, further investigations may require to clarify the reason behind decrease in H_L behavior with increase of U_0 (at low U_0).

3.2 Investigation of Decreasing H_L Behavior

The laminar lifted flames with decreasing H_L are discussed in

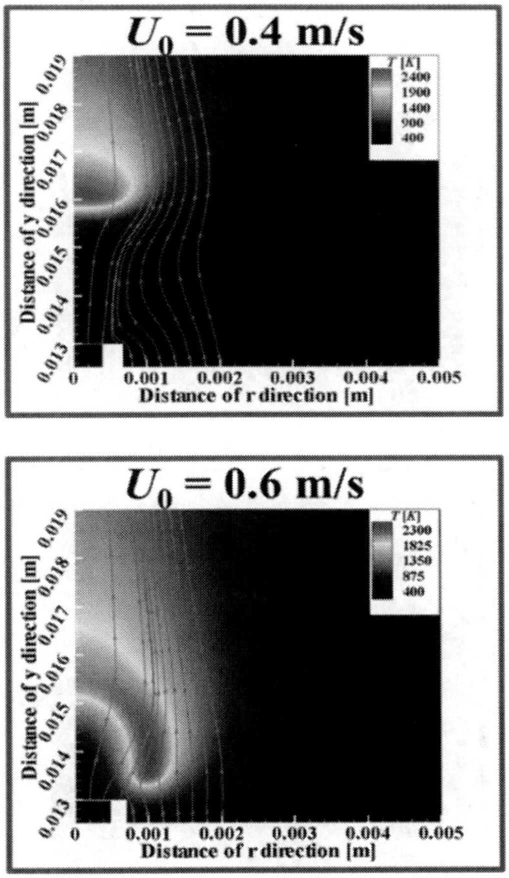

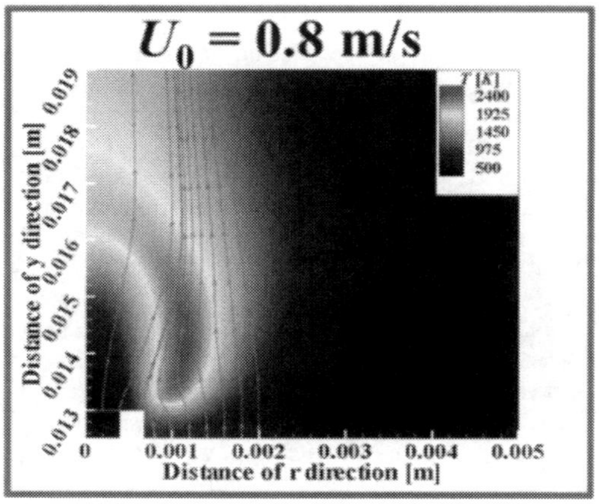

Figure 3: Temperature contour as the function of nozzle exit velocity with velocity contour and stream line for $X_{F,O}$ = 0.24.

previous section. Now it has to be addressed that why such flames were lifted with decreasing H_L behavior in laminar jets. According to stabilization mechanism (balance of edge flame speed to the local flow one), edge flame speed has to increase even at low U_0. To clarify this phenomenon, further study was made with numerical approach. Initially to observe such lifted flames with decreasing H_L behavior, numerical simulations were conducted for $X_{F,O}$ = 0.24 and with change in U_0. Figure 3 shows the temperature contour as the function of nozzle exit velocity, U_0 with velocity contour and stream line at $X_{F,O}$ = 0.24. Similar results were obtained using simulation as that of experimental one. Lifted flames was observed for low U_0 = 0.4 m/s and at $X_{F,O}$

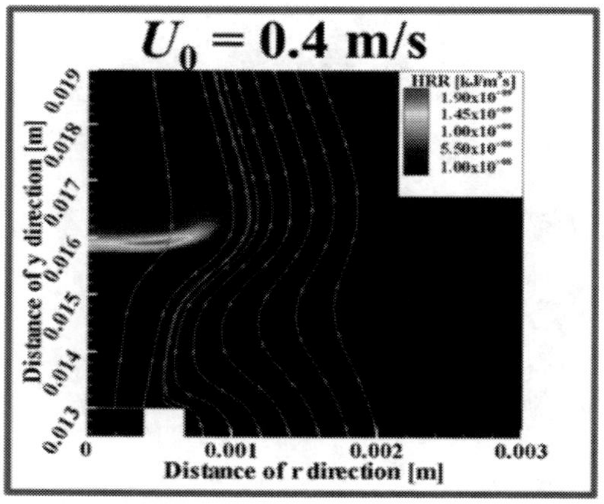

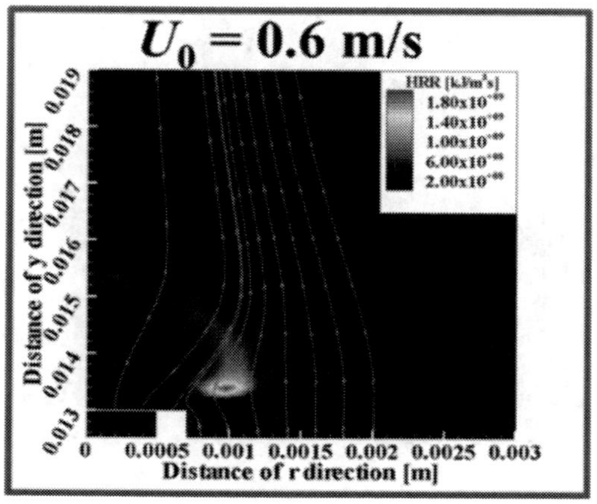

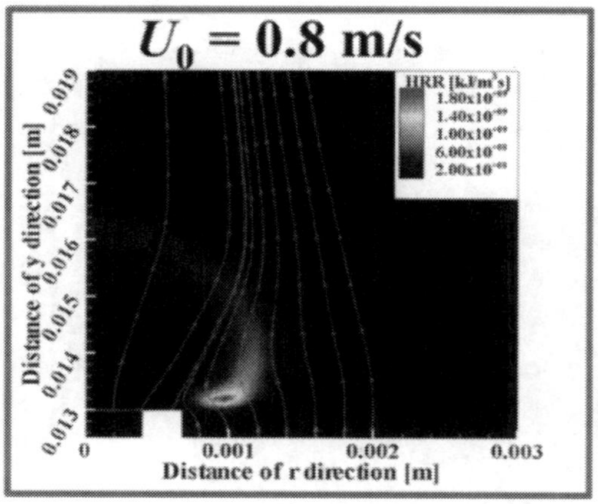

Figure 4: Effect of buoyancy convection as a heat release rate in methane jet flames diluted with nitrogen as a function of U_0 and at constant $X_{F,O}$ = 0.24, in views of stream lines.

= 0.24. But with increase in U_0, lifted flames come towards the nozzle and H_L was decreased for U_0 = 0.6 m/s and 0.8 m/s respectively.

Furthermore, the influence of buoyancy convection was investigated as it is correlated with heat release rate [7]. Figure 4 shows, the effect of buoyancy convection and thereby heat release rate as a function of U_0 and at constant $X_{F,O}$ = 0.24, in views of stream lines. Results demonstrates that, at low U_0 = 0.4 m/s flame was lifted due to buoyancy convection and thereby heat release rate in such a way that, stream lines are deflected towards the centerline causes the increase in the reactant fluxes to the edge flame increasing the reaction rate and hence edge

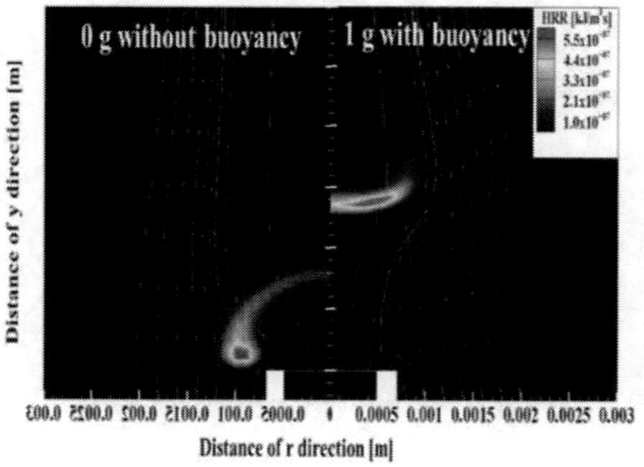

Figure 5: Comparison of H_L with and without buoyancy at $X_{F,O}$ = 0.24.

flame speed. Heat release rate is evaluated as follows:

$$Q = \sum_{\alpha=1}^{N} \omega h_{f\alpha}^{0} \qquad (1)$$

However, in case of large U_0 = 0.6 m/s and 0.8 m/s for constant $X_{F,O}$ = 0.24, flame H_L was decreased respectively in such a way that, streamlines diverge outward and influence of buoyancy convection may diminished. These results from simulations were well matched with experimental one; H_L were decreased with increasing U_0 at constant $X_{F,O}$.

Finally, simulations were performed by considering with and without gravity condition as shown in the Fig. 5. Without gravity condition, streamlines are diverged outward and flame was not lifted. However, under normal gravity condition streamlines are deflected towards the centerline and these flames were lifted due to the influence of buoyancy convection and thereby heat release rate. In this manner, stabilization of lifted methane jet flames with decreasing H_L behavior was explained at low U_0 due to the influence of buoyancy convection under normal gravity.

4 Conclusions

This study demonstrates the lifted flames of methane jet diluted with nitrogen with decreasing H_L behavior by increase of U_0 (at low U_0) and at constant $X_{F,O}$. Such lifted flames were identified both experimentally and numerically for $D = 0.95$ mm. Results shows that, due to the influence of buoyancy convection and thereby heat release rate these flames were lifted at low U_0. Furthermore, H_L was then decreased with increase in U_0 at constant $X_{F,O}$ due to suppress in buoyancy convection effect. In addition, liftoff heights were compared with and without gravity condition.

5 Acknowledgment

This research was supported by the National Research Council of Science and Technology (NST) granted by the Government of Korea (MSIP) (Grant no. CRC-15-07-KIER).

References

[1] S. H. Chung, B.J. Lee, Combust. Flame 86 (1991) 62-72.

[2] B. J. Lee, S.H. Chung, Combust. Flame 109 (1997) 163-172..

[3] Y.S. Ko, S.H. Chung, Combust. Flame 118 (1999) 151-163

[4] G.R. Ruetsch, L. Vervisch, A. Lin˜a´n, Phys. Fluids 7 (1995) 1447-1454.

[5] T. Plessing, P. Terhoeven, N. Peters, M.S. Mansour, Combust. Flame 115 (1998) 335–353.

[6] J.W. Dold, Combust. Flame 76 (1989) 71-88.

[7] J.Y. Chen, Tarek Echekki, Combust. Theory and Modelling, 5:4 (2001) 499-515.

[8] S.H. Won, J. Kim, K.J. Hong, M.S. Cha, S.H. Chung, Proc. Combust. Inst. 30 (2005) 339-347.

Theoretical Analysis for the Exergy Loss Mechanisms of *n*-Heptane/Air Mixtures under LTC Conditions: A Parametric Study

Daojian Liu, Hu Wang, Xumin Zhao, Yuanyuan Zhao, Mingfa Yao*

State Key Laboratory of Engines, Tianjin University

No. 92 Weijin Road, Nankai District, Tianjin 300072, PR China

Abstract

In this study, the detailed exergy loss mechanisms under low temperature combustion (LTC) conditions had been demonstrated. A qualitative exergy loss Φ-T map was created by performing 0-D calculations using n-heptane as the diesel surrogate. The results showed that high exergy loss is generated under the conditions of equivalence ratio under 2.0 and temperature around 1000-2400 K. In the condition of lean gas mixture and temperature above 1600 K the incomplete combustion exergy is relatively low. On the exergy loss Φ-T map, the exergy loss region shows a peninsula shape. Then implications of the Φ-T map to internal combustion (IC) engines were discussed. The combustion path by promptly mixing fuel and air before ignition, accompanied with a certain level of exhaust gas recirculation (EGR), is identified as a high-efficiency and clean combustion mode.

1 Introduction

Low temperature combustion (LTC) modes, e.g. HCCI, PCCI, RCCI, etc. are promising strategies to simultaneously reduce NO*x* and soot emissions [1]. In the meantime, LTC technology can achieve equivalent or even higher thermal efficiency compared to conventional diesel engines [1-3]. However, the combustion efficiency and indicated thermal efficiency of internal combustion (IC) engines are determined by the first law of thermodynamics. Previous studies have demonstrated that about 20-25% of the total input fuel exergy is destroyed due to the inherent thermodynamic irreversibility along the whole cycle [4-6]. Therefore, exergy analysis based on the second law is necessary to further improve the energy utilization efficiency of the IC engines with new technologies.

For typical LTC engines, high level of exhaust gas recirculation (EGR) prolongs mixing time and results in relatively low combustion temperature. Recently, Caton *et al.* [7] explored the effects of EGR on exergy efficiency in a LTC engine. It was concluded that with the heating effect of EGR on the intake charge, which correspondingly led to an increase in combustion temperature, the exergy destruction decreases. As LTC regime shows significantly different combustion characteristics from conventional diesel combustion (CDC), Li *et al.* [8] compared the energy and exergy distributions of CDC, HCCI, and RCCI combustion and found that HCCI achieves the highest energy and exergy efficiency, and CDC performs the worst. At the same time, previous literatures have demonstrated that the combustion process in engines occupies most of the exergy destructions when compared to pumping and fuel/air mixing processes [6, 9]. Hence, two main factors can increase the second law efficiency of engines: (1) minimizing the exergy losses during combustion process; (2) maximizing the work extraction. Yan *et al.* [10] simplified the engine combustion process as adiabatic constant-volume (CV) combustion. By designing the initial conditions, the total exergy loss decreases 15% of fuel chemical exergy. Furthermore, they proposed an RM-HCCI (Reformed molecule HCCI) combustion principle, that achieved the theoretical second law efficiency over 65% [11]. However, the detailed exergy loss mechanisms for these advanced combustion strategies have not been well revealed.

To explore the maximum theoretical exergy efficiency of the combustion process, Lutz *et al.* [12] proposed a conceptual high-temperature reaction process that the reactions are thermodynamically reversible. In this manner, the entropy generation is minimized. Caton [13] conducted a parametric analysis on the exergy destruction of CV combustion, with special emphasis on the results implicated to IC engines. It was concluded that the exergy losses decrease with increasing reactant temperatures and equivalence ratios, while modestly depended on reactant pressure. Although these results are derived from idealized combustion processes, it has significant implications to LTC engines. On one hand, in LTC engines ignition and combustion are mainly controlled by chemical kinetics [1, 14]. Once ignition occurs combustion develops quickly and volumetrically without visible premixed or diffusion flame [1, 15, 16]. Hence, during the combustion process of LTC mode irreversibility is mainly dominated by chemical reactions. On the other hand, according to the conceptualized equivalence ratio-temperature (Φ-T) map [17], when LTC mode is operated, the in-cylinder charge is diluted and the combustion temperature is reduced to suppress NO*x* and soot formations. It is seemly opposite to the purpose of minimizing the exergy losses during combustion process. Therefore, to explore the detailed exergy loss mechanisms under LTC conditions, a quantitative exergy loss Φ-T map is created by performing 0-D calculations using n-heptane as the diesel surrogate. The implications of Φ-T map to IC engines are discussed, which will contribute to identify the ideal combustion region for LTC regimes.

2 Methodology

In this study, the modeling of 0-D homogeneous closed system was conducted at constant pressure and temperature to eliminate effects of the changing temperature during the combustion process on entropy production. The temperature and pressure of reference dead state were taken as T_0=298 K and p_0=1.0 atm, and the molar composition was 20.35% O_2, 75.67% N_2, 3.03% H_2O, 0.3% CO_2, and 0.92% various other substances such as argon. The exergy loss is determined by the Gouy-Stodla equation. The entropy generation inside the

Corresponding author. Fax: +22-27-383-362
E-mail address: y_mingfa@tju.edu.cn

domain of a closed system due to chemical reactions [18] is defined by Eq. (1).

$$\frac{dS}{dt} = -V\frac{\sum_i \mu_i r_i}{T} \quad (1)$$

where μ_i and r_i are the chemical potential and molar reaction rate of the ith species, respectively.

From the perspective of exergy balance, the exergy loss due to combustion is calculated by Eq. (2).

$$Ex_{\text{loss}} = Ex_{\text{init}} - Ex_{\text{fin}} - Ex_{\text{H}} - Ex_{\text{W}} \quad (2)$$

where subscripts 'init' and 'fin' denote the initial and final state of the system, Ex_{H} and Ex_{W} denote the exergy transferred by the heat transfer and work, respectively.

Therefore, the exergy loss fraction (ELF) and incomplete combustion exergy fraction (EICF) can be respectively expressed as Eq. (3) and Eq. (4).

$$\text{ELF} = \frac{Ex_{\text{loss}}}{Ex_{\text{init}}^{\text{ch}}} \times 100\% \quad (3)$$

$$\text{EICF} = \frac{Ex_{\text{fin}}^{\text{ch}}}{Ex_{\text{init}}^{\text{ch}}} \times 100\% \quad (4)$$

where superscripts 'ch' denotes the mixture's chemical exergy.

The initial temperature and equivalence ratio ranges were 800-2800 K and 0.25-4, respectively, which cover the typical IC engine combustion conditions. The detailed n-heptane chemical mechanism from the National University of Ireland (NUI) Galway including 1268 species and 5336 reactions was selected as it is an updated kinetic mechanism and has been well validated [19]. The calculation of the combustion was performed with the SENKIN code of the CHEMKIN package [20]. The local entropy production and exergy balance analysis applied at different instants during combustion process were conducted coupled with a simple procedure, which computed a database for a range of temperature and equivalence ratio generated from the constant temperature and pressure batch reactor solution. Through a time step sensitivity analysis, the non-dimensional time step, $\Delta t/\tau_{\text{res}}$, was set to 0.001, which rendered the relative errors of entropy generation less than 0.5%. Here, τ_{res} is the residence time.

3 Results and discussion

3.1 Exergy loss Φ-T map

Figure 1 shows the exergy loss Φ-T map represented by colours and contours. The ELF is calculated at the conditions of p=100 bar, τ_{res}=1.5 ms and [O_2]=21% (oxygen concentration) which correspond to the in-cylinder fuel combustion pressure and duration (at 1400 rpm) without EGR. On the map with the equivalence ratio under 2.0 and temperature around 1000-2400 K, the exergy loss region shows a peninsula shape. The ELF in this region is about 15-20%, which is consistent with the finding of previous numerical and theoretical studies for IC engines [6, 9, 21]. While in the region of rich gas mixture, it produces much less exergy loss than the lean gas mixture even the temperature is fairly low. To further elucidate the exergy loss characteristics in the map, examples of the exergy loss dependence on temperature and equivalence ratio are depicted in Fig. 2(a) and Fig. 2(b), respectively. Figure 2(a) shows the exergy loss dependence on temperature under the conditions of p=100 bar, Φ=0.55 and [O_2]=21%. With the temperature lower than 1100 K, the exergy loss strongly depends on energy conversion ratio (ECR), which is the ratio of the integrated heat release within the residence time to fuel low heat value (LHV). In the temperature region about 850-1000 K, the exergy loss also shows the negative temperature coefficient (NTC) characteristic, i.e., in the NTC region the fuel ECR decreases as the temperature increases. Further increasing the temperature over 1100 K, the exergy loss is simultaneously controlled by ECR and temperature, leading to the ELF nearly unchanged. When the temperature is higher than 1600 K, since the ECR is independent on temperature, the ELF decreases monotonously with increasing temperature. Figure 2(b) shows the exergy loss dependence on equivalence ratio under the conditions of T=1600 K, p=100 bar, τ_{res}=1.5 ms and [O_2]=21%. In the region of lean gas mixture, the ELF decreases slightly with increasing equivalence ratio. While in the region of rich gas mixture (Φ>1), the exergy loss strongly depends on ECR.

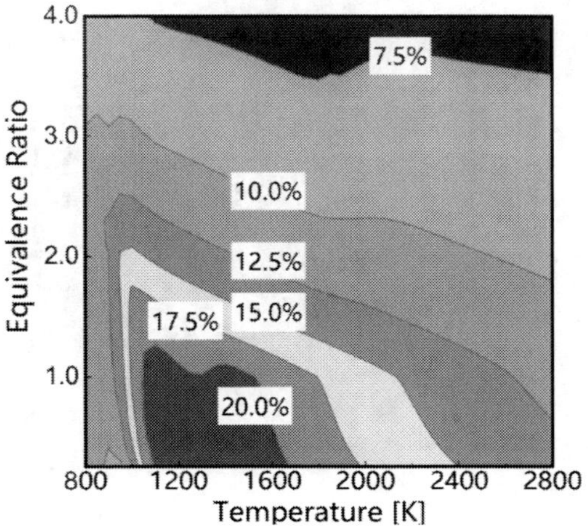

Figure 1: Exergy loss Φ-T map

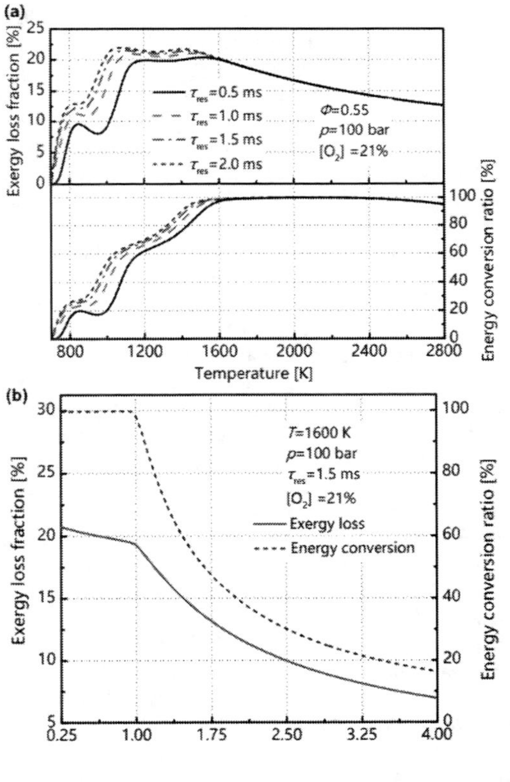

Figure 2: Exergy loss dependence on (a) temperature and (b) equivalence ratio

As depicted in Fig. 2(b), the low exergy loss in the region of rich gas mixture is obtained by paying a price of relatively low fuel ECR. To represent the characteristics of incomplete combustion exergy under various temperature and equivalence ratio conditions, the incomplete combustion exergy Φ-T map represented by colours and contours is created and shown in Fig. 3. Low incomplete combustion exergy region is located in the region of lean gas mixture and temperature above 1600 K. In the low temperature and lean gas mixture region, the incomplete combustion exergy strongly depends on temperature. While in the high temperature and rich gas mixture region, it strongly depends on equivalence ratio.

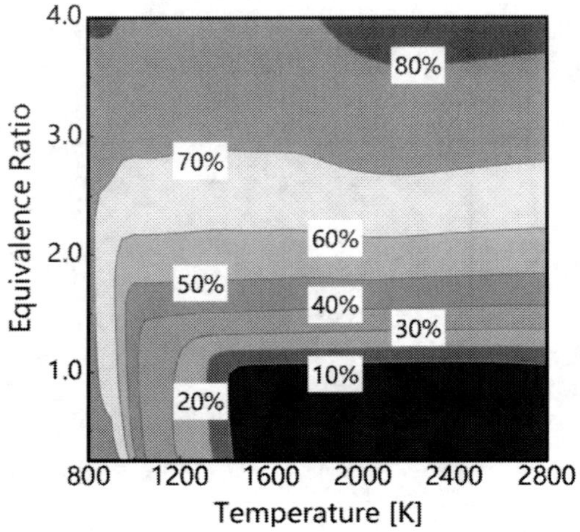

Figure 3: Incomplete combustion exergy Φ-T map

To expand the suitability of the exergy loss Φ-T map to various engine conditions, the 15% ELF limit of the exergy loss peninsula is calculated at different residence time, pressure, and oxygen concentration, which are shown in Fig. 4. Here, the decreased $[O_2]$ represents the increased EGR under the hypothesis of using nitrogen as the diluted gas. Figure 4(a) shows the 15% ELF limit of the exergy loss peninsula at various residence time. The limit is only expanded to the low temperature side slightly with increasing residence time, which remain nearly unchanged with residence time longer than 1.5 ms. Therefore, the residence time of 1.5 ms should be selected, which is also corresponding to the fuel combustion duration (at 1400 rpm). Figure 4(b) shows the 15% ELF limit of the exergy loss peninsula at different pressures. The limit is also only expanded to the low temperature side with increasing pressure. This is mainly because the low temperature reactions are more sensitive to pressure than high temperature reactions [19]. Therefore, the pressure of 100 bar is used, which corresponds to the in-cylinder combustion pressure (at compression ratio=17). Figure 4(c) shows the 15% ELF limit of the exergy loss peninsula at different oxygen concentrations. The limit only changes at the low temperature region, which shifts to high temperature region with decreasing $[O_2]$. This is mainly because the low temperature reactions are more sensitive to oxygen concentration than high temperature reactions [19]. Therefore, the $[O_2]$ of 21% (EGR=0%) is used to avoid the effect of EGR on the exergy loss region limit. From the above discussions, it is concluded that the exergy loss Φ-T map for n-heptane at p=100 bar, τ_{res}=1.5 ms and $[O_2]$=21% shown in Fig. 1 is reasonable and should have a universal application. However, although the Φ-T map gives insight into the exergy loss under static conditions, the isothermal and isobaric conditions employed are severe approximation to a dynamic heterogeneous engine environment characterized by spatial and temporal variation of temperatures and mixing states and significant thermal and concentration stratification.

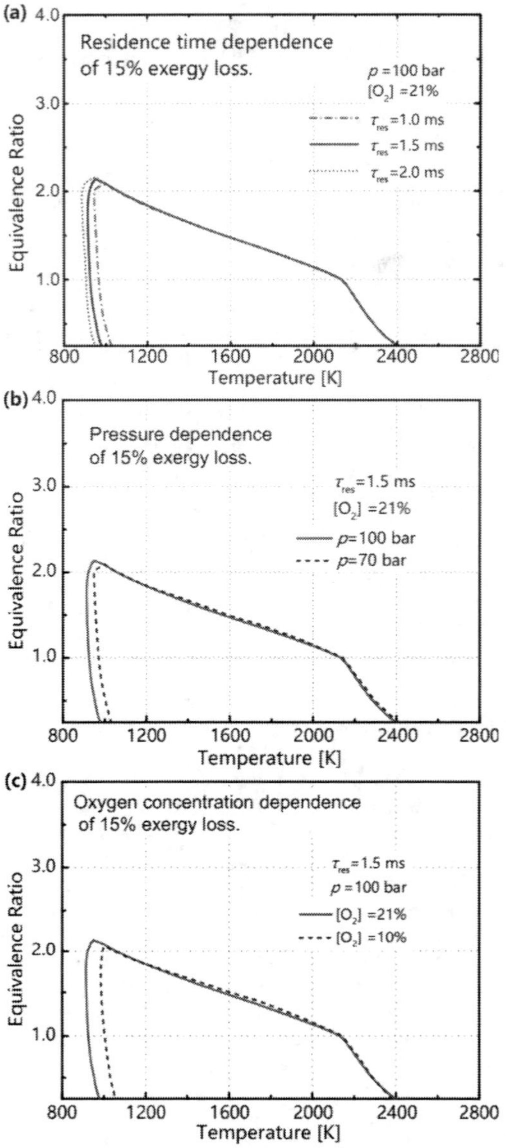

Figure 4: Dependence of 15% ELF limit on (a) residence time, (b) pressure, and (c) oxygen concentration

3.2 Implications to IC engines

The current study is a theoretical analysis and does not involve any experimental measurements; however, the results should have significant implications to real IC engine situations. Here, the 15% ELF limit of the exergy loss peninsula of this study and the 1% soot formation limit and 500 ppm NO*x* formation region from Kitamura *et al.* [17] are integrated into one Φ-T map, as shown in Fig. 5. With the combustion temperature increased, the exergy loss decreases. It is observed that significant amount of NO*x* is produced when the flame temperature is higher than 2200 K. In Fig. 5(a) the flame temperature of CDC path and LTC path with EGR [22] are compared. As expected, the CDC path crosses both the soot and NO*x* formation region. With increasing EGR, the flame temperature decreases and exergy loss increases. However, strong concentration and temperature stratifications exist in CDC mode and LTC mode with low level of EGR, which produce visible diffusion flame [15, 16]. Thus, it produces high proportion of exergy losses due to the mass and heat transfer

[23], which should be further determined. However, the exergy loss due to the mass and heat transfer can be reduced by improving the fuel/air mixing before ignition occurs. As shown in Fig. 5(b), although the HCCI combustion plotted as the shot dash dot line requires completely mixed fuel/air mixture before ignition, it crosses high exergy loss region. Therefore, prompt fuel and air mixing with the equivalence ratio decreased to about 2.0 is required to avoid both the soot formation and the exergy loss peninsulas, and so is a certain level of EGR to avoid the NOx formation. In other words, the postulated combustion path presented as the shot dash line in Fig. 5(b) can be a high-efficiency and clean combustion mode, but the potential advantages of this concept require further works.

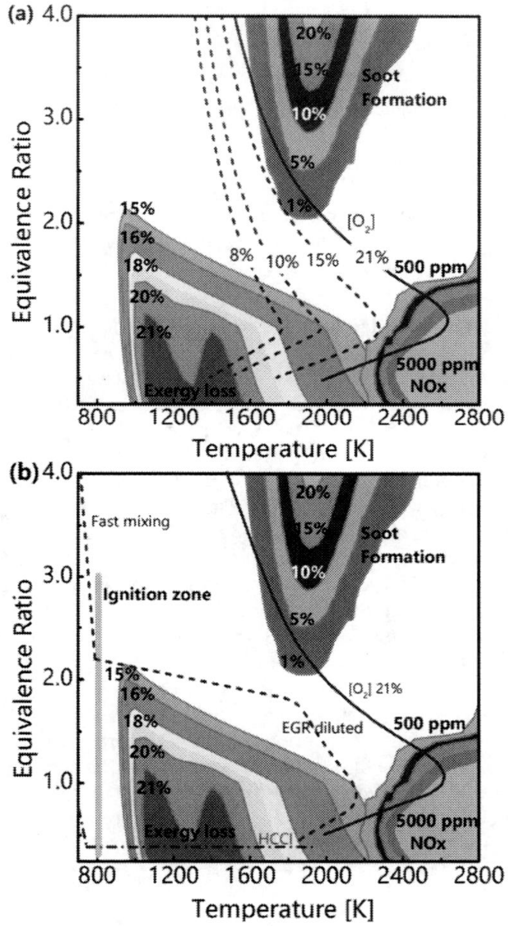

Figure 5: (a) Integrated exergy loss, soot formation and NOx Φ-T map and (b) implications of Φ-T map to IC engines

4 Conclusions

The exergy loss mechanisms for low temperature combustion was studied by performing 0-D calculations under isothermal and isobaric reacting conditions with detailed n-heptane chemical kinetics, based on which a qualitative exergy loss Φ-T map was created. The results showed that high exergy loss is generated under the conditions of equivalence ratio under 2.0 and temperature around 1000-2400 K. With the condition of lean gas mixture and temperature above 1600 K the incomplete combustion exergy is relatively low. On the exergy loss Φ-T map, the exergy loss region shows a peninsula shape. Then the implications of the Φ-T map to IC engines were discussed. The combustion path with prompt fuel and air mixing before ignition, along with a certain level of EGR, was identified as a high-efficiency and clean combustion mode. However, the potential advantages of this concept still require further works.

5 Acknowledgment

This work was supported by the Major Research Plan of the National Natural Science Foundation of China (No. 91541205).

References

[1] A.K. Agarwal, A.P. Singh, R.K. Maurya, Prog. Energy Combust. Sci. 61 (2017) 1-56.

[2] K. Epping, S. Aceves, R. Bechtold, J.E. Dec, SAE Technical Paper, 2002, 2002-01-1923.

[3] R.D. Reitz, G. Duraisamy, Prog. Energy Combust. Sci. 46 (2015) 12-71.

[4] C. Rakopoulos, E. Giakoumis, Prog. Energy Combust. 32 (1) (2006) 2-47.

[5] P. Flynn, K. Hoag, M. Kamel, R. Primus, SAE Technical Paper, 1984, 840032.

[6] C. Rakopoulos, E. Giakoumis, SAE Technical Paper, 2005, 2005-01-1131.

[7] J. Zheng, J.A. Caton, Energy 38 (1) (2012) 78-84.

[8] Y. Li, M. Jia, Y. Chang, S.L. Kokjohn, R.D. Reitz, Appl. Energy 180 (2016) 849-858.

[9] M. Razmara, M. Bidarvatan, M. Shahbakhti, R. Robinett, Appl. Energy 183 (2016) 1389-1403.

[10] F. Yan, W. Su, Energy Fuels 28 (10) (2014) 6635-6643.

[11] F. Yan, W. Su, SAE Technical Paper, 2015, 2015-01-1751.

[12] A.E. Lutz, R.S. Larson, J.O. Keller, Int. J. Hydrogen Energy 27 (10) (2002) 1103-1111.

[13] J.A. Caton, Energy 25 (11) (2000) 1097-1117.

[14] J.E. Dec, Proc. Combust. Inst. 32 (2) (2009) 2727-2742.

[15] S. Singh, R. Reitz, M. Musculus, T. Lachaux, Int. J. Engine Res. 8 (1) (2007) 97-126.

[16] S. Singh, M.P. Musculus, R.D. Reitz, Combust. Flame 156 (10) (2009) 1898-1908.

[17] T. Kitamura, T. Ito, J. Senda, H. Fujimoto, Int. J. Engine Res. 3 (4) (2002) 223-248.

[18] A. Sciacovelli, V. Verda, E. Sciubba, Renew. Sustainable. Energy Rev. 43 (2015) 1167-1181.

[19] K. Zhang, C. Banyon, J. Bugler, H.J. Curran, A. Rodriguez, O. Herbinet, F. Battin-Leclerc, C. B'Chir, K.A. Heufer, Combust. Flame 172 (2016) 116-135.

[20] R. Kee, F. Rupley, J. Miller, CHEMKIN-II: A FORTRAN Package for the Analysis of Gas-Phase Chemical Kinetics, Report No. SAND89-8009B, Sandia National Laboratory, 1989.

[21] P. Chavannavar, J. Caton, Proc. Inst. Mech. Eng. Part A 220 (7) (2006) 655-668.

[22] S. Kook, C. Bae, P.C. Miles, D. Choi, L.M. Pickett, SAE Technical Paper, 2005, 2005-01-3837.

[23] K. Nishida, T. Takagi, S. Kinoshita, Proc. Combust. Inst. 29 (1) (2002) 869-874.

11th Asia-Pacific Conference on Combustion,
The University of Sydney, NSW Australia
10th -14th December 2017.

Effects of Thermal Properties on Low Pressure Deflagration Limit of AP

Kumar Nagendra[1], Naveen Kumar[2], Kumar Ishitha[1] and P.A. Ramakrishna[1]

[1]Department of Aerospace Engineering, Indian Institute of Technology, Madras

Chennai, Tamilnadu, 600036, India

[2]Department of Aerospace Engineering, Madras Institute of Technology

Chennai, Tamilnadu, 600044, India

Abstract

Composite propellant contains Ammonium Perchlorate (AP) as a chief ingredient and its combustion characteristics are mostly controlled by the fraction and particle size of AP present in it. The deflagration behavior of AP is studied to get an insight into complex processes in composite propellant combustion. One specific characteristic of AP is that it does not exhibit self sustained combustion below a certain pressure commonly known as low pressure deflagration limit (LPDL) which is the subject of current study. This paper reports the computational studies that have been carried out to explore the effect of thermal properties on LPDL of AP. Two cases have been explored which differ in the manner in which the heat flux is imposed at the surface. Case I corresponds to constant incident flux which is cut off sharply, while Case II corresponds to constant incident heat flux followed by linear decrease to zero. The computational results obtained are in good qualitative agreement with the experimental results for case I. It has been shown through computations that combustion remains stable near the LPDL region.

Nomenclature

AP	Ammonium Perchlorate
APP	final decomposition products of AP (NH3(gas) + HClO4 (gas))
APD	final decomposition products of AP combustion obtained from chemical equilibrium calculations
A_{DB}	Denison and Baum parameter
C_{PC}	specific heat of condense phase
C_{pg}	specific heat of gas phase
E_S	condense phase activation energy
E_g	gas phase activation energy
f	fraction of AP decomposing at regressing surface to APD
k_c	thermal conductivity of condensed phase
k_g	thermal conductivity of gas phase
n_k	order of reaction
Q_S	heat of pyrolysis
$\dot{q}_s''$	incident surface heat flux
R	universal gas constant
T	temperature
T_S	surface temperature
T_f	adiabatic flame temperature
$T_{-\infty}$	initial temperature
t	time
u_0	velocity in x direction
x	direction perpendicular to regressing surface
α_{DB}	Denison and Baum parameter
ρ_g	gas phase density

Corresponding author.
Kumar Nagendra (Ph.D scholar, IIT Madras)
E-mail address: nag.aero@gmail.com

1 Introduction

Ammonium Perchlorate (AP) being a chief ingredient of composite propellant has been studied for a long time by various researchers. The low pressure deflagration limit (LPDL) of AP is the pressure below which self sustained combustion does not exist, has been investigated over a period of time by various researchers. These are summarized in Fig. 1. Friedman et al. [1] has reported the LPDL of AP to be 45 bar. However, Levy and Friedman [2] later found the LPDL to be 22 bar, when they have used a more efficient ignition technique. Hightower and Price [3] reported the LPDL of AP as 19 bar for a single crystal AP. Boggs [4] in his studies of single AP crystals has reported the LPDL of AP 300 psia (~ 20.5 bar). Thus, the fact that LPDL of AP lies around 20 bar at room temperature is well agreed by all the researchers. Nir [5] studied the effects of various physical properties on the LPDL of pressed AP pellets and concluded that LPDL decreases with increase in the strand density, the burning surface area, the sample initial temperature and the AP particle size.

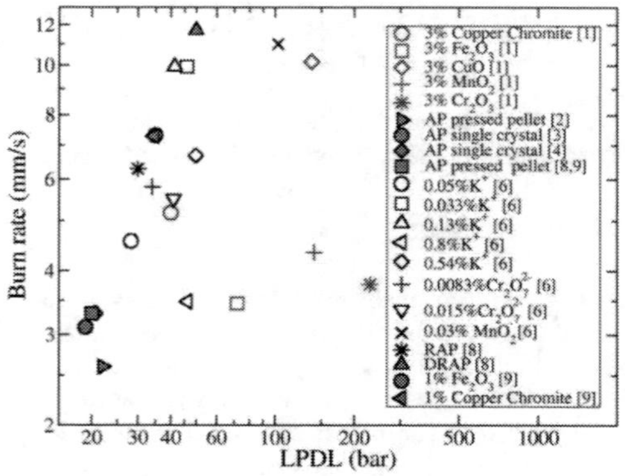

Figure 1: Burn rate at lowest possible pressure at which self sustained combustion occur.

The LPDL has been reported in the literature [1-4,6-9] to vary with the addition of small amount of impurities or additives. The burn rate corresponding to the minimum possible pressure at which AP exhibited self sustained combustion namely LPDL is shown in Fig. 1.

Various researchers have done extensive work to explain the cause of LPDL. Heat loss to the surroundings [10], loss of melt layer at low pressure [3-5] and burning surface to act as a heat sink [11] were the reasons argued in literature for the existence of LPDL of AP. De luca [12] has provided the evidence for importance of condensed phase near LPDL. In a recent work, Ramakrishna et al. [13] have attributed the existence of LPDL to the transient conduction in the condensed phase, heat loss to the surroundings accompanied by the loss of melt layer. All the above explanations appear to provide some insight into the LPDL of pure AP. However, the same explanation, fails to explain the sudden increase in the LPDL of AP when very small quantity of additives were added to AP [1,2,6-8]. Ishitha and Ramakrishna [8,9] have shown that the thermal properties changes drastically with addition of additives. Strahle [14] has reported that the existence deflagration limit is due to the combustion becoming unstable. However, his work as stated by him is more directly applicable to explaining mesa burning of AP.

From all the above, it still is not clear enough as to why LPDL of AP exists and why does it have such a high value of 20 bar. Ramakrishna et al. [13] and De luca [12] have suggested the unsteady heat conduction in the condensed phase as an important factor for the existence of LPDL of AP. This explanation brings the parameters which effects the transient heat conduction in the condensed phase, viz., thermal properties of AP. Also, considering the arguments of Ishitha and Ramakrishna [8] on thermal properties of AP being affected by certain additives, computational studies have been carried out to explore the variation of LPDL with thermal properties such as specific heat and thermal conductivity.

2 Mathematical model

The mathematical model used for the current study was the one developed by Ramakrishna et al. [13]. This combustion model of Ramakrishna et al. [13] for AP has been shown to predict the all steady state features of AP combustion (such as burn rate, pressure index and temperature sensitivity). The AP surface decomposition model used by Ramakrishna et al. [13] was as follows:

$$AP \longrightarrow (1-f)\,APP + (f)APD \qquad (1)$$

Further discussion on decomposition model at regressing surface will be presented in a later section. Gas phase reaction was composed of single step chemistry wherein APP decomposes to APD. The boundary conditions used in the present study were identical to those used by Ramakrishna et al. [13]. The thermodynamic parameters were also identical to one used by Ramakrishna et al. [13] unless otherwise specified.

In an experiment, the AP pellets were ignited with a heat flux source at the surface and self sustenance of the combustion was checked at different pressures. However, most of combustion models including the one used by Ramakrishna et al. [13], the simulations for the steady state solution were generally performed from a given initial surface temperature and a thermal profile in the condensed phase known as initial solution. This was done primarily because if no solution is specified and if an appropriate heat flux is not applied at the surface the condensed phase would act like a heat sink. Second reason being, these initial conditions were close to the final solution for steady state and hence convergence is achieved in less time thereby reducing the computational time. In the present study, the LPDL of AP was suspected to be due to an unsteady process as pointed by Ramakrishna et al. [13] and De luca [12] (transient heat conduction being playing a role in LPDL) and might be due to intrinsic stability as suggested by Strahle [14]. Therefore, the combustion simulations here were performed by specifying a heat flux at the surface which is close to how experiments are conducted. In the present study, the initial temperature in condensed and gas phase was kept as 300 K which is generally a physical scenario. This technique allows the unsteady heat conduction in the condensed phase and stability of the combustion can be explored during transients. Similar technique was used by Ishitha and Ramakrishna [8]. They have shown the LPDL of AP (20 bar) can be obtained in this manner.

3 Results and Discussion

The simulations were started with given incident heat flux added to surface heat balance as given in (2) with the solid and gas phase at the initial temperature and the solution was allowed to evolve in time. Two cases have been explored in the present study. Case I was a constant heat flux incident on the surface for a given time and cut off sharply as shown in Fig. 2. The second case (case II) considered was a constant heat flux incident for a given time and then allowed to decrease linearly with time to zero as shown in Fig. 2. The total time for which heat flux was incident on the surface was 95 ms for both the cases.

$$k_g \frac{\partial T}{\partial x}\bigg|_0^{0+} = k_c \frac{\partial T}{\partial x}\bigg|_0^{0-} + \rho_g u_0 Q_s + \dot{q}_s'' \qquad (2)$$

where, $\dot{q}_s''$ is varied as shown in Fig. 2 for the two cases.

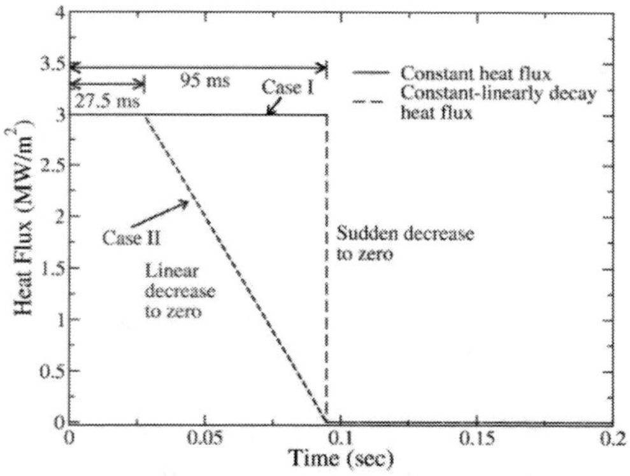

Figure 2: Incident heat flux variation for both the cases

Thermal conductivity of 0.3 W/m-K and specific heat of 1240 J/kg-K was used as thermal properties for AP by Ramakrishna et al. [13] to predict the LPDL of 20 bar and other combustion characteristics of AP. In order to find dependence of LPDL on thermal properties, the thermal conductivity was varied from 0.2 to 0.4 W/m-K, keeping the specific heat of condensed phase constant at 1240 J/kg-K. In another set of computations, the specific heat of condensed phase was varied form 1000 to 1450 J/kg-K, keeping the thermal conductivity constant at 0.3 W/m-K. Figure 3 shows the variation of LPDL with thermal conductivity and specific heat of condensed phase for case I and case II.

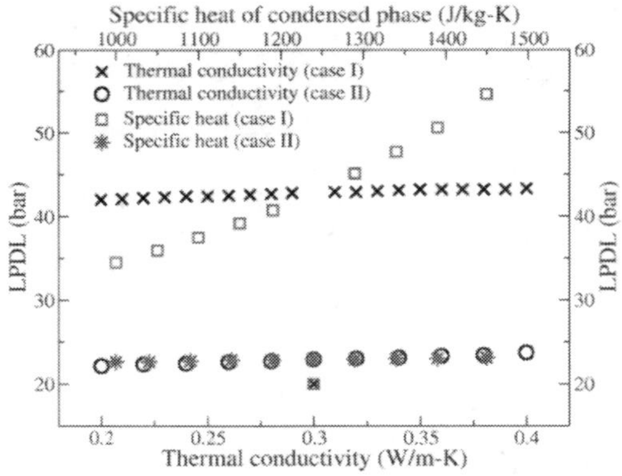

Figure 3: Variation of LPDL with thermal conductivity and specific of condensed phase

It can be seen from Fig. 3 that LPDL of AP changes drastically when the thermal conductivity and specific heat are varied slightly from 0.3 W/m-K and 1240 J/kg-K respectively for case I. However, if one selects a thermal conductivity greater than 0.32 or less that 0.28 W/m-K, the LPDL remains almost constant at around 42 bar. The LPDL is seen to vary significantly with specific heat of condensed phase for case I. The individual values of thermal conductivity and specific heat in Fig. 3 are converted to thermal diffusivity considering the density of condensed phase as 1957 kg/m^3. A comparison between the simulated and experimentally obtained LPDL [8,9] is shown in Fig. 4. The way thermal conductivity and specific heat act on LPDL as seen from Fig. 3, provides two values for given a thermal diffusivity as shown in Fig. 4. These values are limits for LPDL and LPDL from experimental data is expected to lie within these limits. It can be seen from Fig. 4 that the computational results obtained for case I are in good qualitative agreement with literature [8,9].

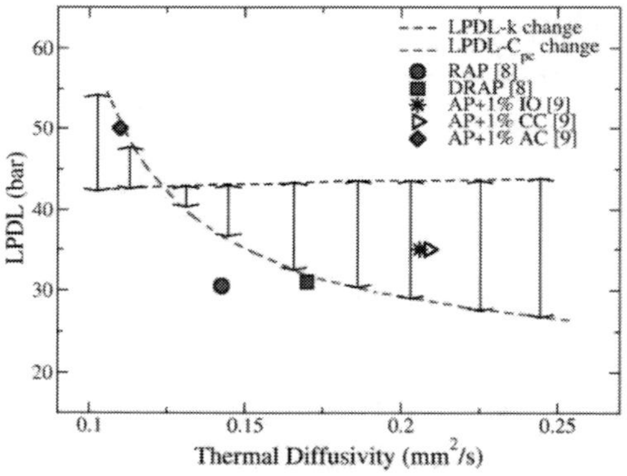

Figure 4: LPDL comparison with experimental data

However, for case II, it is evident from Fig. 2 that the LPDL remains essentially constant irrespective of changes to specific heat and thermal conductivity. LPDL for the thermal properties used by Ramakrishna et al. [13] was found to be 22.9 bar in case II. The case I and case II differ in the manner in which heat flux is imposed on the surface. In case II, the incident heat flux gradually decreases to zero. This points to the fact that the LPDL might be an artifact of how the ignition is done. Some evidence of this is found in literature also. Friedman et al. [1] initially found the LPDL of AP to be 45 bar which was reduced to 22 bar when a better ignition technique was used by Levy and Friedman [2]. Even in the recent studies of Ishitha and Ramakrishna [8], have reported a similar phenomena with K$^+$ doped AP. It seems possible that the ignition energy source used was decaying gradually and thereby reducing the LPDL. These results of case II need further investigation and experiments should be carried out to establish the same.

In order to get further insight on the processes near LPDL, the burn rate and surface temperature variation was studied. Figure 5 shows the burn rate and surface temperature variation near the LPDL limits (LPDL being 47.7 bar for case I), for specific heat of 1340 J/kg-K and thermal conductivity of 0.3 W/m-K while Fig. 6 shows the same for case II. The point E in Fig. 5 represents the end of ignition or switching off the incident heat flux. It can be seen from Fig. 5 that the burn rate and surface temperature decreases drastically once the incident heat flux is switched off. The steady state burn rate and surface temperature is obtained for 48 bar, while extinction is seen for 47 bar after switching off the incident surface heat flux.

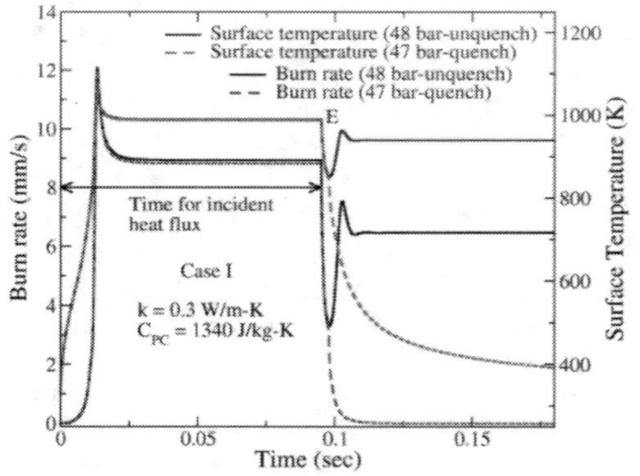

Figure 5: Burn rate variation with time with and without incident heat flux for case I.

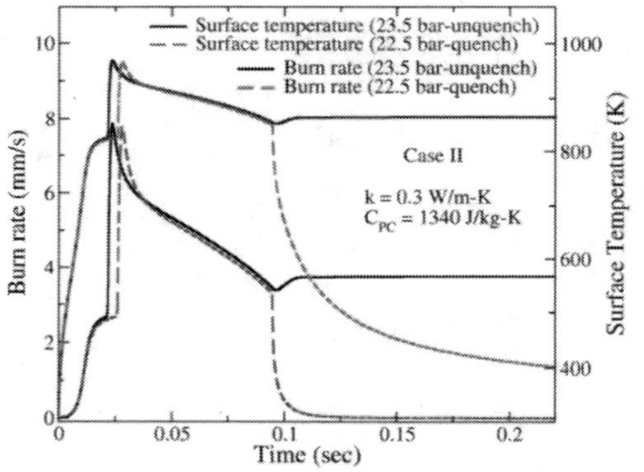

Figure 6: Burn rate variation with time with and without incident heat flux for case II.

It can be seen from Fig. 6 that burn rate and surface temperature decreases gradually as the heat flux for case II is linearly brought to zero. The extinction observed in both the cases is primarily due to the AP surface decomposition model used here (similar to the one used by Ramakrishna et al. [13]). In this model, f the fraction of AP decomposing at surface to the surface heat release. The surface heat release as described by (3-5). The surface heat release varies from -205 kJ/Kg (exothermic) at $f = 0.6$ to 1.85 MJ/Kg (endothermic) at $f = 0$ depending on the surface temperature.

$$f = 0.6 \; for \; T_s > 850K \quad (3)$$

$$f = 0.6 - 0.025(850 - T_s) \qquad for \; 825 < T_s < 850K \quad (4)$$

$$f = 0 \; for \; T_s < 850K \quad (5)$$

Thus, when the surface temperature decreases below 850 K, the AP surface decomposition starts to change from exothermic to endothermic. Once the surface temperature goes below 825 K, f becomes zero which reduces the surface temperature drastically and extinction will follow. This model was shown to predict the LPDL of 20 bar from a given initial solution by Ramakrishna et al. [13]. The same model as described above was used by Ishitha and Ramakrishna [8] to predict the LPDL of RAP and DRAP with constant incident heat flux wherein they had found a good qualitative comparison with experimental results.

Further, following the remarks of Strahle [15] regarding LPDL, stability of combustion was analyzed on a Denison and Baum neutral stability curve [15] near the pressures where AP ceases to burn. The non dimensional parameters of Denison and Baum curve [15] A_{DB} and α_{DB} were evaluated using (6) and (7).

$$A_{DB} = E_s\left(T_s - T_{-\infty}\right)/RT_s^2 \quad (6)$$

$$\frac{1}{\alpha_{DB}} = \frac{C_{pc}(T_s - T_{-\infty})\epsilon}{C_{pg}T_f} \quad where, \; \epsilon = \frac{n_k + 2}{2} + \frac{E_g}{2RT_f} \quad (7)$$

In order to check this hypothesis, Denison and Baum parameters A_{DB} and α_{DB} were evaluated for pressures 47 bar (quench case) and 48 bar (burning case) during the entire burning time including the ignition. These computed parameters are plotted with Denison and Baum neutral stability curve [15] as shown in Fig. 7.

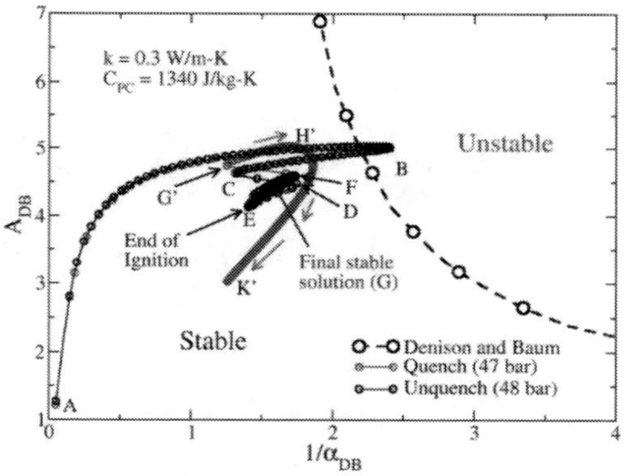

Figure 7: Stability behavior for C_{PC} = 1340 J/kg-K and k = 0.3 W/m-K

The path followed is A-B-C-D-E-F-G for the pressure of 48 bar when the self sustained combustion occur. However, for the case of 47 bar which leads to extinction, the path followed is A-B-C-D-E-F-G'-H'-K'. The point B which lies in the unstable region is merely due to the initial transients during ignition when heat flux is imposed on the surface. The Point E corresponds to the condition when the surface heat flux is reduced to zero or in other words it represents end of ignition. From this point (E) the operating conditions instantly shifts to point F. This corresponds to a sudden and drastic decrease in surface temperature as shown in Fig. 7. The self sustained combustion from point F would lead to self deflagration of AP towards final stable solution (G). However, for quench case, the path followed is F-G'-H'-K' leads to extinction. It should be noted that the path followed in both the cases including point E remains in stable regime as evident from the Fig. 7. It clearly shows that the combustion remains stable near the LPDL. Similar analysis was performed for the case II and results were identical to those depicted in Fig. 7 for the specific heat of 1340 J/kg-K and thermal conductivity of 0.3 W/m-K.

4 Conclusions

The unexplained variation of LPDL of AP with the inclusion of small amounts of additives observed in literature has been explained with the variation of thermal properties of AP. The simulations of LPDL have been carried out by supplying the heat flux at the surface for ignition. This should be considered in LPDL simulations as this technique is much closer to experiments. Two cases for incident heat flux are considered. Case I being constant heat flux with a sharp cut off while the case II is constant heat flux followed by a linear decay. The simulated LPDL is in good qualitatively agreement with the experiments for case I. The stability analysis shows that combustion does not become unstable near the LPDL. Case II provides almost constant LPDL at 23 bar. This points to the fact that the ignition technique used to ignite the AP might have an effect on the LPDL and it needs to be investigated further.

5 Acknowledgment

The authors would like to express their gratitude to Mr. Chaitanya V. for engaging discussions during the study.

References

[1] R. Friedman, R.G. Nugent, K.E. Rumbel, A.C. Shurlock, Proc. Combust. Inst. 6 (1957) 612-618.

[2] J.B. Levy, R. Friedman, Proc. Combust. Inst. 8 (1961) 663-672.

[3] J.D. Hightower, E.W. Price, Proc. Combust. Inst. 11 (1967) 463-472.

[4] T.L. Boggs, AIAA Journal 8 (1970) 867-873.

[5] E.C. Nir, Combust. Flame 20 (1973) 419-435.

[6] T.L. Boggs, E.W. Price, D.E. Zurn, Proc. Combust. Inst. 13 (1) (1971) 955-1008.

[7] T.L. Boggs, K.J. Kraeutle, D.E. Zurn, AIAA Journal 10 (1) (1972) 15-16.

[8] Kumar Ishitha, P.A. Ramakrishna, J. Propul. Power 20 (2) (2014) 277-284.

[9] Kumar Ishitha, P.A. Ramakrishna, Combust. Flame 161 (2014) 2717-2728.

[10] W.E. Johnson, W, Nachbar, Proc. Combust. Inst. 8 (1962) 678-689.

[11] H.Y. Sohn, Combust. Sci. & Techol. 10 (3) (1975) 137-154.

[12] L. De Luca in: L. De Luca, E. Price, M. Summerfield, (Eds.), Nonsteady Burning and Combustion Instability of Solid Propellants. AIAA, Washington DC, 1992, p. 519.

[13] P.A. Ramakrishna, P.J. Paul, H.S. Mukunda, J. Propul. Power 22 (3) (2006) 661-668.

[14] Warren C. Strahle, AIAA J. 9 (4) (1971) 565-569.

[15] M.R. Denison, E. Baum, Am. Rocket Society J. 31 (1961) 1112-1122.

11th Asia-Pacific Conference on Combustion,
The University of Sydney, NSW Australia
10th -14th December 2017.

Investigation of Wall Chemical Effect on Weak Flame with GC and PLIF

Sui WAN[1], Yong FAN[1], Kaoru MARUTA[2, 3], Yuji SUZUKI[1]

[1] The University of Tokyo
Hongo 7-3-1, Bunkyo-ku, Tokyo 113-8656, Japan
[2] Tohoku University
Katahira 2-1-1, Aoba, Sendai 980-8577, Japan
[3] Far Eastern Federal University
Sukhanova St., Vladivostok 690090, Russia

Abstract

Wall chemical effect on DME/air weak flames in a micro channel with a streamwise temperature gradient was investigated. Wall surface materials of quartz and SUS321 were examined. Streamwise distributions of DME, CO and CO_2 in the weak flame were measured with the gas chromatography (GC), while HCHO and OH distributions were measured through planar laser-induced fluorescence (PLIF). It is found that species distributions in both the low- and middle-to-high-temperature oxidation zones of the weak flame are significantly changed for different wall surface materials. A possible surface reaction mechanism of DME synthesis from adsorbed CO, HCHO and H_2 is discussed for the SUS321 surface.

1 Introduction

Understanding flame-wall interaction (FWI) is of fundamental importance in internal combustion systems [1-2]. Previous studies [3-5] on normal flames suggest that the wall chemical effect exists only at high wall temperatures ($\geq$ 1073 K), while the wall thermal effect is dominant at low wall temperatures. Since the wall temperature in most applications is below 1073 K, the wall chemical effect on the high temperature oxidation is considered as negligible except microscale combustion.

However, the above conclusion obtained from only studies of high-temperature normal flames is not necessarily true for flames governed by the low-temperature oxidation. Understanding low-temperature oxidation is important for the control of ignition, knocking and emission in the engine. Maruta et al. [6, 7] proposed weak flame in the micro flow reactor, a stationary flame with spatially separated low-, middle- and high-temperature oxidation zones, as a tool for reaction mechanism studies.

In the present study, wall chemical effects on both low- and middle-to-high-temperature oxidations in the weak flame are investigated, providing measured streamwise distributions of DME, CO, CO_2, HCHO and OH over quartz and SUS321 surfaces through either PLIF or GC. A possible surface reaction mechanism for SUS321 was proposed.

2 Experimental & Computational Methods

Corresponding author: Sui WAN. Fax: +81-03-5841-6419
E-mail address: swan@mesl.t.u-tokyo.ac.jp

Figure 1a shows the quartz micro flow reactor employed in the present study. It has a 100 mm long rectangular inner channel with a cross-section of 1.5 mm × 5 mm. The top and bottom walls are detachable. 150 nm-thick SUS321 film is alternatively deposited onto the top and bottom walls by vacuum arc plasma deposition (ARL-300, ULVAC). By replacing the top and bottom quartz walls with/without the SUS321 coating, the wall chemical boundary is changed, but identical wall thermal boundary is kept. The roughness of both the quartz and SUS321 surfaces are within 2 nm, thus the effect

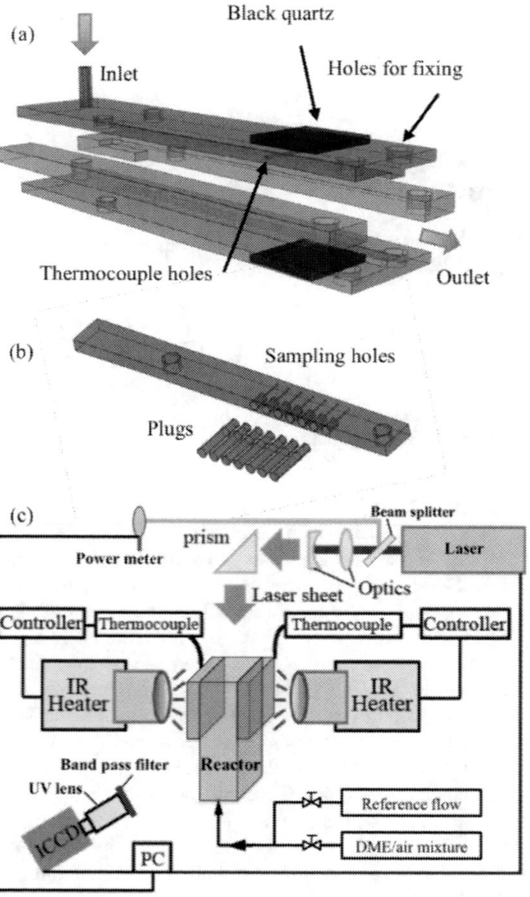

Figure 1: (a) Schematic of the micro flow reactor. (b) Channel sidewall with sampling holes for the GC measurement. (c) Schematic of the experimental setup for the PLIF measurements.

of roughness can be neglected. A black quartz (Nb-doped) layer is fused on the backside of both the top and bottom quartz walls at the downstream part for IR lamp heating. The wall temperature at the heated spot is monitored by two R-type thermocouples in the quartz plate, and kept at 1200 K. A streamwise wall temperature profile with a gradient of approximately 380 K/cm is formed along the channel. The weak flame of DME/air is formed at an extremely low flow velocity, U_{in}, of 1.5 cm/s and $\phi = 0.85$, where low-temperature oxidation zone (cool flame) and middle-to-high-temperature zones (blue and hot flames) are separated in the streamwise direction.

A sidewall with 7 sampling holes is employed for the gas chromatography (GC) measurements as shown in Fig. 1b. The spacing between these sampling holes is 5 mm, and the last one in the downstream is located at the position of maximum wall temperature. The holes are sealed with quartz plugs to prevent leakage when not in use. Gas sampled from the weak flame was analyzed by GC (SHIMADZU GC-2010). Internal standard method with 3.2 % of Kr is employed to suppress the uncertainty due to unwanted air intake during the sampling process. The mean inlet flow velocity and the equivalence ratio are kept the same. A Rtx-642Sil MS column and the mass spectrometer (MS) are used for DME. A Micropacked ST column and the thermal conductivity detector (TCD) are used for CO and CO_2. The sampling speed is less than 0.3% of the main flow, thus no significant influence was found on the combustion in the channel. The concentrations of DME and CO are calibrated with the standard gases. For CO_2, the relative concentration is estimated using the signal peak area.

A schematic of the experimental setup for the PLIF measurement is illustrated in Fig. 1c. A frequency-doubled Nd-YAG laser-pumped pulsed dye laser (SCANmate2C, Lambda Physik) is employed for OH excitation at 283.553 nm, while the HCHO is excited by the frequency-tripled Nd-YAG laser (355 nm). The laser beam is divided into two by the beam splitter. One beam of 2% laser energy is reflected to the power meter for monitoring the laser energy, while the other is shaped into a 0.3 mm thick laser sheet by using a combination of cylindrical lenses. Fluorescence from either OH or HCHO is captured by using an ICCD camera (iStar340T, ANDOR) equipped with an f = 105 mm UV lens and band-pass filters. The fluorescence intensities are proportional to the molar concentration of each species within an uncertainty of 10 % for OH and 15 % for HCHO. The main source of the uncertainty is the temperature dependency of the fluorescence.

A bubbling system is used to generate vaporized reference gas mixture from the liquid tank filled with either acetone or formalin for compensating the non-uniformity of the laser energy in the PLIF measurements. Generated acetone/air gas is used as the reference for OH-PLIF, and $HCHO/H_2O/N_2$ gas for HCHO-PLIF. The concentration of acetone and HCHO in the reference gases is calculated from the saturated vapor pressure. Since the fluorescence intensity, I, is proportional to the molar concentration, C, when the laser energy is constant, the following expression is derived:

$$\frac{I_{tgt}}{C_{tgt}} = a \cdot \frac{I_{ref}}{C_{ref}}, \quad (1)$$

where ref stands for the reference gas, and tgt stands for the target gas. When the target species is the same as the species detected in the reference flow, the constant, a, equals 1 and the absolute concentration of target species is obtained. Absolute HCHO concentration is obtained in this way, but only a relative concentration is obtained for OH.

Two-dimensional numerical simulation of the DME weak flame is conducted in FLUENT 18.0 with detailed gas-phase mechanisms [8] and the surface reaction mechanisms proposed in this study. The computational domain is half of the channel (0.75 mm × 45 mm) and 12 × 300 grids are used. Unequally spaced grids are employed in the wall-normal direction and the

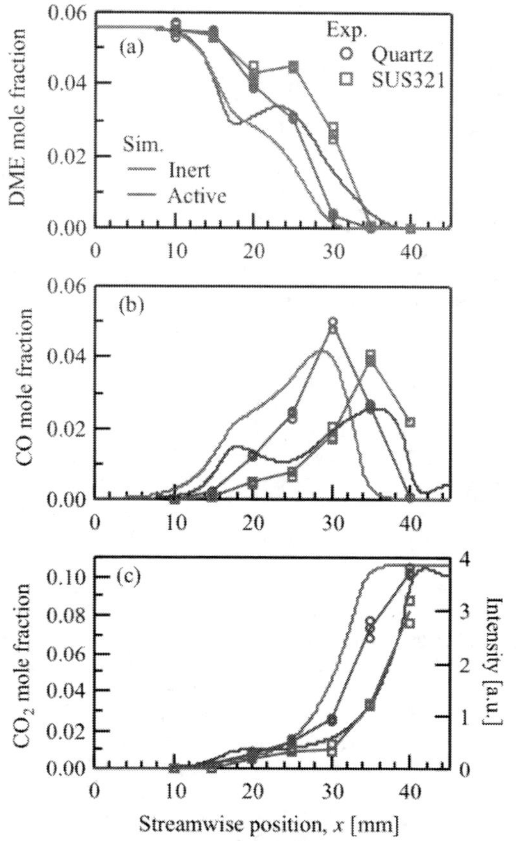

Figure 2: Streamwise species distributions measured in the DME/air weak flame in quartz and SUS321-coated channels in comparison with simulation results for inert and active surfaces. (a) DME, (b) CO, (c) CO_2.

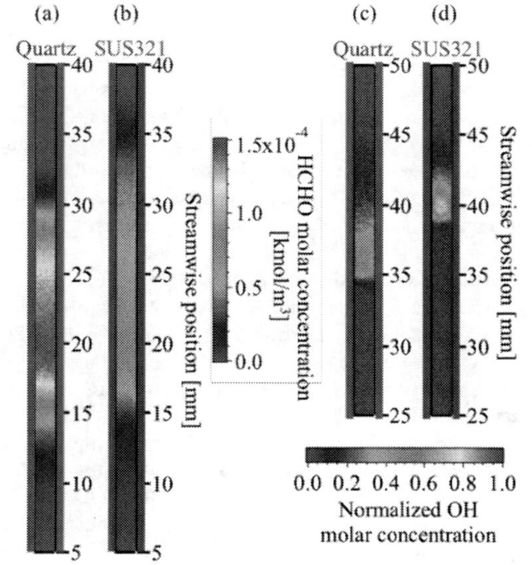

Figure 3: (a) HCHO and (b) OH distributions for the DME/air weak flame (U_{in} = 1.5 cm/s, $\phi = 0.85$) in quartz and SUS321-coated channels.

first grid point in the wall-normal direction is 12 μm away from the wall. An inflow boundary of DME/air mixture is adopted at the inlet. A convective boundary is adopted at the outlet. The wall is a non-slip boundary with the same wall temperature profile as that measured in the experiment.

3 Results and Discussion

DME, CO and CO_2 concentrations measured by GC are shown in Fig. 2. The consumption of DME (Fig. 2a), as indicated by the decrease in its mole fraction, starts at x = 15 mm (625 K) and the decrease trends in both quartz and SUS321-coated channels are similar till x = 20 mm (730 K). After that position, while the DME mole fraction decreases monotonically to nearly 0 at x = 30 mm (1030 K) in the quartz channel, the DME mole fraction in the SUS321-coated channel, on the other hand, increases by 0.002 from x = 20 to 25 mm, and remains a relatively high value of 0.022 even at x = 30 mm. As the DME is consumed, CO is generated. It is shown in Fig. 2b that the CO mole fraction in both the quartz and SUS321-coated channels starts to increase at x = 15 mm (625 K), but the increase is slower in the SUS321-coated channel. As a result, the largest concentration of CO is observed at x = 35 mm (1150 K) in the SUS321-coated channel, which is more downstream than that observed at x = 30 mm (1030 K) in the quartz channel. CO vanishes at x = 40 mm (1170 K) in the quartz channel, while it still has relatively high concentration at x = 40 mm (1170 K) in the SUS321-coated channel. Overall, the CO generation is shifted downstream in the SUS321-coated channel. In consequence, the rising of CO_2 concentration is also delayed in the SUS321-coated channel as shown in Fig. 2c.

2-D distributions of the OH and HCHO molar concentration are shown in Fig. 3, while the streamwise distributions along the centerline of the channel are plotted in Fig. 4. In the quartz channel, HCHO has a maximum concentration of ~1.45×10^{-4} kmol/m^3 at x = 20 mm (730 K), then decreases rapidly, and finally vanishes completely at x = 34 mm (1130 K). On the other hand, in the SUS321-coated channel, the HCHO peak is reduced by 50%, and decreases slowly until x = 38 mm (1170 K). The OH peak position is located at x = 36 mm (1160 K) in the quartz channel, while in the SUS321-coated channel the peak shifts to x = 40 mm (1170 K) with a slightly higher maximum concentration.

It is well known that the oxidation of DME proceeds in the following four steps: (1) H abstraction of DME; (2) HCHO formation through β-scission; (3) partial oxidization of HCHO to produce CO; (4) consumption of CO and CH_4; (5) formation of CO_2. These steps, which are spatially separated in the weak flame, are also identified from measured distributions of DME, HCHO, CO, CO_2 and OH in the present study. As compared with the case in the quartz channel, in the SUS321-coated channel, a higher DME concentration and a lower HCHO concentration are observed in the low-temperature oxidation zone, while the oxidation of HCHO and CO as well as the generation of CO_2 and OH are shifted to the higher temperature region, which indicates that the low-temperature oxidation is changed by the SUS321 coating. As a result, a delay of the hot flame ignition occurs.

There are two possible causes for higher DME concentration observed in SUS321-coated channel: (1) Lower DME consumption rate in gas-phase, (2) DME re-forming (synthesis) on the SUS321 surface. At this temperature region, the only DME consumption in gas-phase is through the H-atom abstraction. A reaction pathway analysis conducted at 820 K shows that 80% of DME is consumed through reactions with OH and H. A simulation using the wall radical quenching model [5] with an initial sticking coefficient of 1.0 for surface reactions of OH, H, O, CH_3 was conducted. However, no obvious difference in the streamwise distribution of species was observed between the inert channel and the channel with wall radical quenching (not shown here). This is due to extremely low OH and H concentrations in the low temperature region. OH and H generated from the gas-phase are immediately consumed before diffusing to the wall surface. Therefore, surface adsorptions of OH and H in the low-temperature region do not affect the DME consumption in gas-phase.

Therefore, the possibility left is the DME synthesis over the SUS321 surface. To date, no surface reaction mechanism which includes DME synthesis is available for stainless in the case of combustion. On the other hand, DME synthesis from CO hydrogenation has been intensively studied in the case of catalytic reactions [9]. This synthesis scheme has two essential steps: (1) CO, HCHO and H_2 are adsorbed onto the metallic site to form CH_3OH; (2) CH_3OH is adsorbed onto the acid site to form DME. Also considering the analogy with DME surface reaction on other metals such as Ni, we propose a possible surface reaction mechanism based on the Ni mechanism and DME synthesis with revisions as described below. For the reaction of H_2, O_2 and CO over SUS321, the reaction rate data from the CH_4 reforming and oxidation over Ni [10] is adopted with only the H_2 adsorption rate reduced by 2 orders. At first, a sticking coefficient, S_0, of 0.01 for H_2, as given in ref. [10] is used, the hot flame is ignited over the surface in the simulation, which conflicts with the experiment observation. Thus, S_0 for H_2 is reduced to 0.0001. The activation energies for the HCO hydrogenation reactions to CH_3OH are taken from DFT study on Ni(111) [11]. The activation energies for the DME synthesis steps from CH_3OH are set to 0, assuming these reactions are not the rate-limiting steps. The pre-exponential factors for all elementary reactions in the HCO hydrogenation to CH_3OH and the DME synthesis from CH_3OH are estimated using the transition state theory. A value of 3.7×10^{21} cm^{2}mol^{-1}s^{-1} is used for the bimolecular reactions.

Sensitivities of species distributions to S_0 for CO, HCHO and CH_3OH are examined by varying each S_0 (data not shown here). Finally, S_0 for CO, HCHO and CH_3OH are obtained as 0.3, 0.0005 and 0.05 by fitting the experimental data. It is found that a S_0 of 0.1 for CO is inadequate to reproduce the DME increment observed in the SUS321-coated channel. Also, the predicted CO concentration in the low-temperature region of active channel is higher than the inert channel case, instead of lower in the experiment. On the other hand, when S_0 for CO is set to 1.0, the shifting of hot flame position in the simulation is twice of that observed in the experiment. Therefore, a proper value of S_0 for CO is between 0.1 and 1.0. The S_0 for HCHO is adjusted in a way that the HCHO maximum concentration is reduced to 50% of that in the inert channel, as in the experiment. When S_0 = 0.0001, the HCHO maximum concentration is still 70% of that in the inert channel. On the other hand, when S_0 = 0.001, the HCHO maximum concentration becomes so low that the CO peak in the hot flame region disappears. Therefore, a proper value of S_0 for HCHO is between 0.0001 and 0.001. When S_0 < 0.01 for CH_3OH, the formation of DME is inadequate. On the other hand, when $S_0 \geq 0.01$, DME, CO and CO_2 concentrations are no longer sensitive to the S_0 for CH_3OH. As such, S_0 for CH_3OH is set to 0.05 in current model. Since the gas-phase reaction mechanism is not perfect yet, especially for the low temperature oxidation, species distributions somehow changes quantitatively when different gas-phase mechanisms are applied, and none of these mechanisms can perfectly predict the measured species distribution in the quartz channel (not shown). It is impractical to obtain optimized S_0 by fitting the experimental data to simulations using an imperfect gas phase mechanism. Thus, the focus of this work is a qualitative prediction of the species distribution in the SUS321-coated channel by our proposed surface reaction mechanism.

Simulated streamwise distributions of DME, CO, CO_2, HCHO and OH in the inert channel and the active channel with our DME surface reaction mechanism (referred as the active channel hereafter) are plotted in Figs. 2 and 4 in comparison with the experimental data. It is shown that simulation results for the inert channel have some discrepancy from the experimental data in the quartz channel, but are in qualitative agreement. It is shown in the simulation results of the active channel that a local high DME concentration and low CO and HCHO concentrations exist in the NTC region, and the CO and HCHO extended to the high temperature oxidation zone leads to a delay of OH and CO_2 generation. Overall, simulation results of the active channel successfully reproduce the qualitative tendency of the experimental data of the SUS321-coated channel, showing possibility of DME synthesis in the low temperature oxidation region.

Figure 5 compares the net reaction rate of DME in the inert and active channels. In order to compare the DME consumption rate in the gas-phase with the generation rate over the surface,

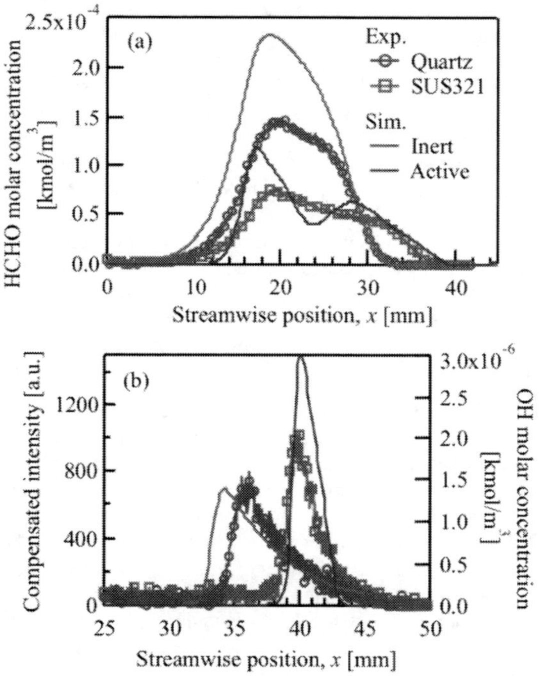

Figure 4: Streamwise species distributions measured in the DME/air weak flame in quartz and SUS321 channels in comparison with simulation results for inert and active surfaces. (a) HCHO, (b) OH.

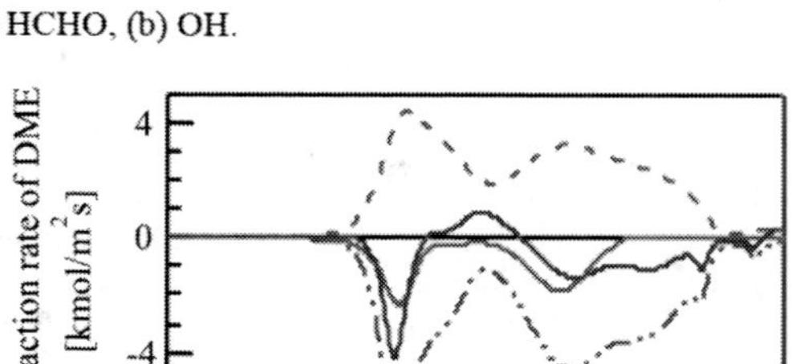

Figure 5: Net reaction rate of DME in the inert and active channels.

the net reaction rate of HCHO for gas-phase reactions is integrated in the wall normal direction to obtain the same unit as the surface reaction. The red solid line is the result for the inert channel while the blue solid line is for the active channel. The two dash lines show the reaction rates of gas-phase and surface, respectively. It is found that surface reactions have two effects on the DME oxidation: one is the promotion of the DME consumption in gas phase since HCHO is removed from the gas-phase; the other is the DME reformation via surface. According to the simulation result, before x = 17.5 mm, the total DME consumption rate in the active channel is larger than that in the inert channel, since less HCHO in the gas-phase reduces the competition to react with OH and H. In x = 19 ~ 26 mm, an increase of the total DME concentration is observed as shown in Fig.2a. It should be noted that x = 19 ~ 26 mm is the NTC zone of gas-phase and a large amount of HCHO, CO and H_2 exist here, which means the gas-phase reaction is weak and the surface reaction is relatively strong. Therefore, it is reasonable that the reformation of DME occurs in such an oxidizing atmosphere.

4 Conclusions

The wall chemical effect of quartz and SUS321 surfaces on the DME weak flame has been investigated via GC and PLIF measurements in a rectangular micro flow reactor. It is found that, comparing with that in the quartz channel, in the SUS321-coated channel, DME concentration is increased in the NTC region (>730 K) by the wall chemical effect, while CO and HCHO are decreased. Peak of OH is shifted downstream. The wall chemical effect may exert strong influence on the low-temperature oxidation, which changes the species pool in the cool flame and leads to a modification of the hot flame region downstream. In preliminary simulation, it is conjectured that the DME synthesis from the HCHO, CO and H_2 occurs over the SUS321 surface.

Acknowledgment

We thank Professor O. Deutschmann and his group members for assistance with the surface reaction mechanism. This work is supported through JSPS KAKENHI Grant Number 15H03927 and 15H05508.

References

• A. Bohlin, M. Mann, B.D. Patterson, A. Dreizler, C.J. Kliewer , Proc. Combust. Inst., 35 (2015) 3723-3730.

• K. Maruta, Proc. Combust. Inst., 33 (2011) 125-150.

• C.M. Miesse, R.I. Masel, C.D. Jensen, M.A. Shannon, M. Short, AIChE J., 50 (2004) 3206-3214.

• K.T. Kim, D.H. Lee, S. Kwon, Combust. Flame, 146 (2006) 19-28.

• Y. Saiki, Y. Fan, Y. Suzuki, Combust. Flame, 162 (2015) 4036-4045.

• K. Maruta, T. Kataoka, N.I. Kim, S. Minaev, R. Fursenko, Proc. Combust. Inst., 30 (2005) 2429-2436.

• H. Oshibe, H. Nakamura, T. Tezuka, S. Hasegawa, K. Maruta, Combust. Flame, 157 (2010) 1572-1580.

• U. Burke, K.P. Somers, P. O'Toole, et al., Combust. Flame, 162(2) (2015) 4036-4045.

• G.R. Moradi, S. Nosrati, F. Yaripor. Catal. Commun., 8(3) (2007) 598-606.

• E.S. Hecht, G.K. Gupta, H. Zhu, et al. Appl. Catal. A-Gen., 295(1) (2005) 40-51.

• I.N. Remediakis, F. Abild-Pedersen, J.K. Nørskov. *J. Phys. Chem. B*, 108(38) (2004) 14535-14540.

On-line product analysis of coal and corn residue co-pyrolysis using photoionization mass spectrometry

Jun-Jie Weng[1], Yue-Xi Liu[1, 2], Ya-Nan Zhu[3], Yang Pan[3], Zhen-Yu Tian[1, 2]

[1]Institute of Engineering Thermophysics, Chinese Academy of Sciences, Beijing 100190, China

[2]University of Chinese Academy of Sciences, Beijing 100049, China

[3]National Synchrotron Radiation Laboratory, University of Science and Technology of China, Hefei, 230029, China

Abstract

The pyrolysis processes of coal, corn, and coal-corn blends have been studied with an online pyrolysis photoionization time-of-flight mass spectrometry (Py−PI−TOFMS). The mass spectra at different organic pyrolysing temperatures (from 400 to 800 °C) as well as time-evolved profiles of several selected species during pyrolysis process were measured. The pyrolysis products such as alkanes, alkenes, dienes, phenols, aromatics, as well as NH_3 and H_2S were detected. As temperature rises, the relative ion intensities of high molecular weight products decrease, while the signals of aromatics, such as benzene and toluene, increase significantly. During the co-pyrolysis, coal can promote the reaction temperature of cellulose in corn. The time-evolved profiles demonstrate that the coal can affect the pyrolysis rate of cellulose, hemicellulose, and lignin of corn in blends. According to this work, the potential for process control to get desired products is possible by monitoring reaction parameters, such as temperature and reaction time.

1 Introduction

Coal combustion leads to the increase of greenhouse gas emissions as well as air pollutants, such as NO_x, SO_2, etc. Compared to coal, biomass captures increasing attracting attention due to its renewability, low NO_x emissions, as well as zero-carbon production [1]. However, the disadvantages of biomass, such as difficulties in collection and transportation, seasonal availability, low energy density and etc., limit its wide application [2]. The most promising solution to figure out these drawbacks is to co-utilize with coal, such as co-combustion, co-gasification and co-pyrolysis [3]. Pyrolysis is the first step in the thermal conversion process, which has an important influence on the subsequent transformation. Thus, a better understanding of the mechanisms underlying co-pyrolysis of coal and biomass can be of great benefit for predicting the process performance in an industrial reactor.

Considerable efforts have been made to study the pyrolysis characteristic of coal and biomass. Thermogravimetric analysis (TGA) is the widely used, simplest and the most effective technique to obtain the pyrolysis profiles of a fuel, such as kinetic parameters, rate constant k and activation energy E_a [3-6]. Normally, pyrolysis behavior is assessed by performing the analysis in an inert (nitrogen or argon) atmosphere. However, TGA cannot reveal the chemical structures of pyrolysis products. Therefore, other analytical methods were applied to characterize the products. Recently, pyrolysis−gas chromatography/mass spectrometry (Py−GC/MS) was used due to its high sensitivity and good performance in separation [7-9]. Ohra-aho et al. studied the catalytic pyrolysis of Kraft lignin and Scots pine wood over three different catalysts, HZSM-5, Y-zeolite and Pd/C [8]. However, Py−GC/MS suffers from the limitation that some polar and/or involatile pyrolysis products may be lost [10]. Besides TGA and Py−GC/MS, Fourier transform infrared spectrometry (FTIR) [11] and ^{13}C nuclear magnetic resonance (^{13}C NMR) [12] were also applied to study the pyrolysis of coal and biomass. Most analytical methods regarding organic compounds involve the pretreatment and isolation processes, which probably disturb the original compositions of gas-phase products and obtain questionable data [10]. To overcome these difficulties, mass spectrometers combined with soft ionization methods, such as vacuum ultraviolet (VUV) photoionization, have been used to analyze the pyrolysis products of coal and biomass [10, 13-15]. Compared to electron ionization (EI), fragment-free mass spectra can be obtained via near-threshold photoionization. Besides, few works focus on rapid online analysis of coal and biomass varied with pyrolysis temperatures and reaction time simultaneously, which can permit a better understanding of the mechanisms underlying co-pyrolysis of coal and biomass.

The aim of this work is to study the pyrolysis products of coal, corn, and coal-corn blends using a pyrolysis photoionization time-of-flight mass spectrometry (Py−PI−TOFMS) with a krypton discharge lamp as ionization source. The mass spectra of coal, corn, and coal-corn blends pyrolysis products at different organic pyrolysing temperatures from 400 to 800 °C and the time-evolved profiles of major species were obtained online, and these information can provide a better understanding of the mechanisms underlying coal and biomass co-pyrolysis, which would offer technical support for the development of efficient clean-combustion techniques of coal and biomass.

2 Experimental Methods

2.1 Samples

The coal samples were bought from Shenmu Energy Developments Ltd. Company and grinded into powder with particle size of 90~100 μm. The corn samples, grows on Shijiazhuang Shilong farm, were supplied by Hebei Zhonggu Food Ltd. Company (Shijiazhuang, China) and grinded into powder with particle size of 90~100 μm. The coal and corn sample (50:50, wt %) were prepared by thorough mixing prior to experiments. An analytical balance (AL 204, Mettler Toledo, Switzerland) was used to weigh 20±0.1 mg of samples for pyrolysis. Thermogravimetric analysis experiments were

Corresponding author. Fax: +86-10-82543184
E-mail address: tianzhenyu@iet.cn

performed in a TG/DTA6300 (PerkinElmer, USA). In all experiments, the temperature was raised from room temperature to 950 °C at a heating rate of 10 °C/min. The reacting atmosphere was nitrogen, which streamed at a constant volume flow rate of 100 mL/min (at STP). The results of proximate and ultimate analysis of the samples are listed in Table 1.

Sample	Proximate analyses (wt.%)				Ultimate analysis (wt.%, daf)						
	M_{ad}	A_{ad}	V_{ad}	FC_{ad}	C_{ad}	H_{ad}	O_{ad}	N_{ad}	S_{ad}	M_{ad}	A_{ad}
Coal	6.53	10.41	32.40	50.66	66.70	4.11	10.79	1.04	0.42	6.53	10.41
Corn	7.01	1.61	84.29	8.30	50.03	6.57	43.02	0.18	0.40	7.02	1.31

Table 1: The proximate and ultimate analysis results of coal and corn samples
M: moisture. A: ash. V: volatile matter. FC: fixed carbon.

2.2 Online Py−PI−TOFMS

The details of the Py−PI−TOFMS setup have been described elsewhere [16], and only a brief description is given here. The experimental setup includes a tubular furnace, a transfer line for introducing pyrolysis products, and a homemade photoionization TOFMS. The pyrolysis temperatures were in the range of 400−800 °C, with an increment of 100 °C. After the temperature reached the set point, the sample was introduced into the middle of the furnace by a quartz sample pole. The transfer line was heated to 250 °C by a heating tape to prevent product condensation. In the photoionization region, pyrolysis products were ionized by the vacuum ultraviolet (VUV) light (10.6 eV) emitted from a krypton discharge lamp.

3 Results and Discussion

3.1 TG analysis

The TG and DTG curves of coal, corn and coal-corn blends are displayed in Fig. 1. In general, there are three major stages during the pyrolysis of the three samples. For the first stage, the release of water and ash by evaporation and devolatilization reactions occurred. The second stage is assigned to the decomposition of samples, including the release of gaseous product, tar, and semi-char formation. The third stage can be attributed to the formation of char and the evolution of H_2, CO_2, and CO [17].

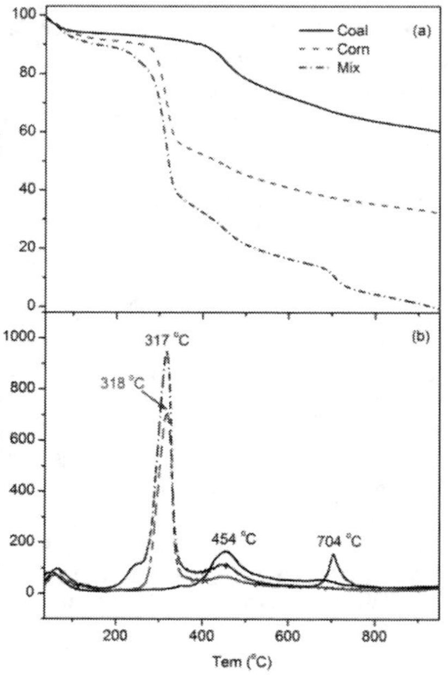

Fig. 1 TG and DTG curves of coal, corn and coal-corn blends.

As shown in Fig. 1a, the pyrolysis reaction rates of corn and coal-corn blends are much faster than that of coal. At the end of pyrolysis, the conversion rates of coal and corn are around 40% and 70%, respectively. However, almost 100% coal-corn blends can be pyrolysed completely, which can be mainly attributed to the different chemical structures and compositions for coal and corn samples. From Fig. 1b, the temperatures of the maximum mass loss rate for corn and coal-corn blends are almost the same (around 318 °C), which are 136 °C lower than that for coal (454 °C). In fact, the interactions occurred from 300 °C to 500 °C, corresponding to the end of corn devolatilization and the beginning of coal decomposition. The volatile oxygen-containing compounds can provide more oxygen for the consumption of coal in blends, which would enhance the conversion rate of coal. Meanwhile, coal can release a lot of heat for corn, and this would also enhance the conversion rate of corn.

3.2 Pyrolysis products

Figure 2 presents the mass spectra of pyrolysis products from coal, corn and coal-corn blends at different temperatures. Each spectrum was recorded once the sample boat was introduced to the middle of the tubular furnace. Compared to electron ionization mass spectra of coal and biomass pyrolysis in previous study [18, 19], fewer fragments are observed in this work. The structural assignments of main pyrolysis products were characterized according to the measured molecular weights and previous literatures [10, 20].

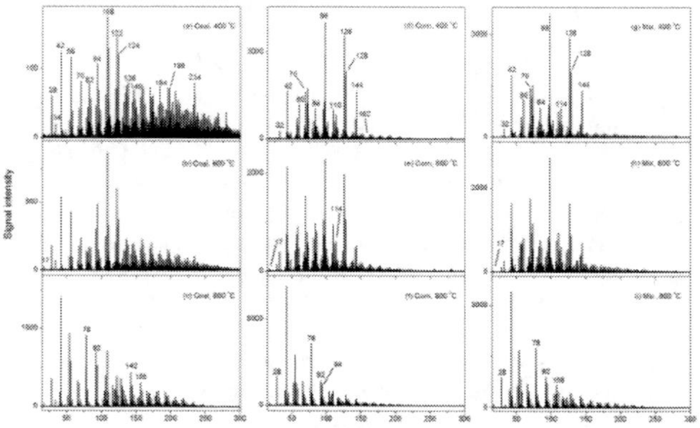

Fig. 2 Mass spectra of coal, corn and coal-corn blends at 400, 600, and 800 °C.

The dominant pyrolysis products of coal are aromatics (*m/z* = 78, 92, 108, 122, 124, etc.), which shows the same result as previous study [21]. The other evolved products includes alkanes (*m/z* = 58 and 72), alkenes (*m/z* = 28, 42, 56, 70, and 84), and dienes (*m/z* = 68 and 82). Besides alkanes and alkenes, the nitrogen- and sulfur-containing compounds, such as NH_3 (m/z = 17) and H_2S (m/z = 34), were also detected. For corn pyrolysis, the relative ion intensities of pyrolysis products with high m/z region (200–300) decreases apparently compared to coal. Similar to other biomass, the main components of corn include cellulose, hemicellulose, and lignin [22]. Cellulose is a polysaccharide consisting of a linear chain of more than ten thousands of levoglucosan ($C_6H_{10}O_5$, m/z = 162) units [23]. During corn pyrolysis, the most abundant peaks (m/z = 98, 126, 128 and 144) are from the dehydration of levoglucosan [23]. The characteristic products from hemicellulose (m/z = 114) and lignin (m/z = 110, 136, 138, etc.) were also observed. During the co-pyrolysis of coal-corn blends, the relative ion intensities are more close to those of corn residue below 600 °C, and more close to those of coal beyond 600 °C, which is due to the lower ignition temperature and activation energy for corn than coal [22].

3.3 Pyrolysis products versus temperature

As seen from Fig. 2, temperature affects the relative abundance of pyrolysis products obviously. The peak areas of several pyrolysis products from coal, corn and coal-corn blends are normalized to the most abundant peak from 400–800 °C, and the results are shown in Fig. 3.

For coal pyrolysis, at relatively low temperatures, the main products are phenols and alkenes. The relative ion intensities of the phenols, such as phenol (m/z = 94) and cresol (m/z = 108), increase from 400 to 700 °C and then decrease slightly, which is due to the rapid increase of cracking reaction rates at high temperatures [24, 25]. In the case of alkenes, such as ethylene (m/z = 28), propene (m/z = 42), and butylene (m/z = 56), their relative ion intensities almost increase all the time as pyrolysis temperature rises, indicating that they were from the decomposition of the coal macromolecule. However, the intensity of butylene decreases at 800 °C due to the cyclization reactions occurring at relatively high temperature. As for aromatics (benzene at m/z = 78, toluene at m/z = 92, naphthalene at m/z = 128, and C2 alkyl naphthalene at m/z = 156), when the temperature increases from 400 to 800 °C, their intensities increase significantly result from the decomposition of coal macromolecule and secondary reactions.

During corn pyrolysis, the products (m/z = 98, 126, 128 and 144) from cellulose reach their maximum at 400 °C due to the thermal decomposition of cellulose. The intensities of above products decrease due to the thermal cracking as temperature rises. The compound from hemicellulose comes to its maximum at 500 °C, indicating the reaction rate of hemicellulose is slower than that of cellulose. The characteristic products from lignin (m/z = 94 and 108) increases to their peak values from 400 to 700 °C, and decrease at higher temperature, showing that the reaction rate of lignin is slower than that of hemicellulose. The intensities of propene (m/z = 42) and butylene (m/z = 56) rise to maximum at 700 °C due to the decomposition of the coal macromolecule, and then decrease because of the cyclization reactions at a relatively high temperature. The trend of aromatics (m/z = 78, 92, and 156) is the same as that of aromatics from coal pyrolysis.

In the case of co-pyrolysis of coal-corn blends, the main products (m/z = 98, 126, 128 and 144) are from cellulose at low temperature. The intensities of above products come to their maximum at 500 °C and then reduce, indicating that coal can increase the reaction temperature of cellulose of corn in blends, compared to pure corn. The trends of hemicellulose (m/z = 114) and lignin (m/z = 94 and 108) in the blends are the same as those of corn with temperature increase. For alkenes, such as propene (m/z = 42), and butylene (m/z = 56), their relative ion intensities are more close to those in coal at a relatively high temperature. Similar to the cases of pure coal and corn, the intensities of aromatics (m/z = 78, 92, and 156) increase significantly as temperature rises. As the reaction proceeds, pyrolysis products contain decreasing oxygen until the formation of oxygen-free aromatics. In fact, both radicals and small molecular weight hydrocarbons contribute to the formation of oxygen-free aromatics through combination, which is similar to the formation of soot during combustion [26]. Thus, the corn can provide more oxygen for coal, leading to the complete pyrolysis of coal in blends. Meanwhile, coal can provide more heat for corn, and coal has the greater influence on the pyrolysis of cellulose than those of hemicellulose and lignin of corn in blends.

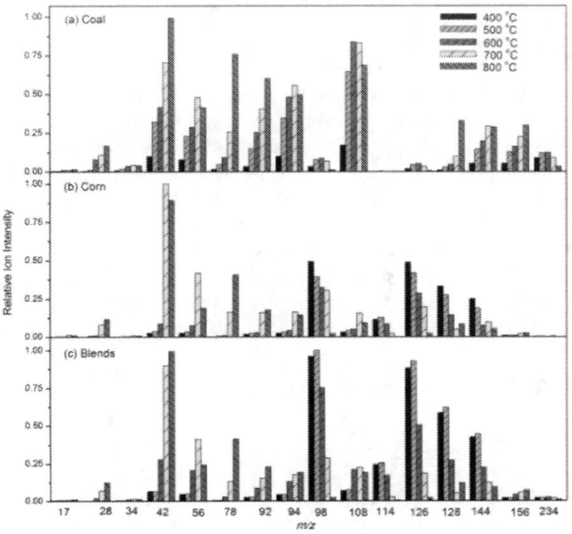

Fig. 3 Effects of the reaction temperature on the relative intensities of major pyrolysis products from coal, corn, and coal-corn blends.

3.4 Pyrolysis products versus reaction times

Figure 4 illustrates the time-evolved profiles of major pyrolysis products of coal, corn and coal-corn blends at the temperature of 500 °C. The spectra were recorded for the time range of 400 s with 5 s intervals, with the background noise subtracted. As shown in Fig. 4a, it takes 80 s for almost all the products of coal pyrolysis to reach the maximum at the same time. In fact, the time-evolved profiles of some selected pyrolysis products of all samples at 400–800 °C were also obtained (not shown), and the results show that the rise time to the maximum is significantly promoted with the increase of temperature due to the higher reaction rate of the rupture of bonds, aromatization, and polymerization [16]. For pyrolysis products of corn illustrated in Fig. 4b, it takes 36 s for the products (m/z = 98, 126, and 114) from cellulose and hemicellulose to reach the maximum. However, the rise-time to peak is about 40 s for the compound (m/z = 94) from lignin, indicating that lignin in corn is slower to pyrolyze than cellulose and hemicellulose. In fact, gasification of lignin particles would take some time to complete [10]. With regard to co-pyrolysis of coal-corn blends, all the products appear and come to the maximum at the same time of around 34 s, which is earlier than the cases of pure coal and corn. The coal and corn in blends both contribute to the formation of products at

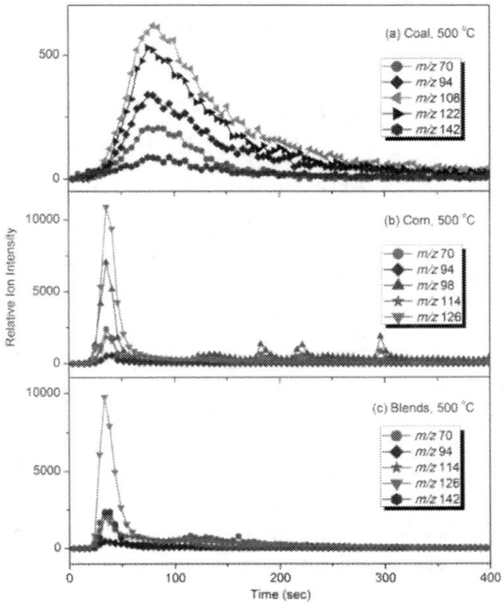

Fig. 4 Time-evolved profiles of major pyrolysis products of coal, corn, and coal-corn blends at 500 °C.

m/z = 70, 94, 126, 142, etc.). The pyrolysis process of coal and corn can be promoted by each other in blends.

4 Conclusions

The real-time and online detection of pyrolysis products of coal, corn, and coal-corn blends have been studied by Py–PI–TOFMS technique. The fragment-free mass spectra of coal, corn, and coal-corn blends pyrolysis products at different pyrolysis temperatures from 400 to 800 °C were measured. The time-evolved profiles of major species were also obtained during the pyrolysis processes. At relatively low temperatures, the main products of coal pyrolysis include phenols and alkenes, while the major products from corn pyrolysis are from the dehydration of levoglucosan. During the co-pyrolysis of coal-corn blends, the relative ion intensities are more close to those of corn below 600 °C, and more close to those of coal beyond 600 °C. As temperature rises, the relative ion intensities of high molecular weight products decrease for coal pyrolysis, while the signals of aromatics, such as benzene and toluene, increase significantly. The same trend can also be obtained for corn pyrolysis. In the case of coal-corn blends, the result shows that coal can promote the reaction temperature of cellulose of corn. The time-evolved profiles reveal that the coal can affect the pyrolysis rate of cellulose, hemicellulose, and lignin from corn in blends. This work shows that Py–PI–TOFMS is a powerful approach to permit a better understanding of the mechanisms underlying the co-pyrolysis of coal and biomass.

5 Acknowledgment

The authors thank the financial support from the Natural Science Foundation of China (No. 51506200), the Ministry of Science and Technology of China (2017YFA0402804) and Recruitment Program of Global Youth Experts.

References

[1] M. Balat, M. Balat, E. Kırtay, H. Balat, Energ. Convers. Manage. 50 (2009) 3158-3168.
[2] A.V. Bridgwater, J. Anal. Appl. Pyrol. 51 (1999) 3-22.
[3] D.K. Seo, S.S. Park, Y.T. Kim, J. Hwang, T.-U. Yu, J. Anal. Appl. Pyrol. 92 (2011) 209-216.
[4] M. Varol, A.T. Atimtay, B. Bay, H. Olgun, Thermochim. Acta 510 (2010) 195-201.
[5] D.K. Park, S.D. Kim, S.H. Lee, J.G. Lee, Bioresource Technol. 101 (2010) 6151-6156.
[6] D. Vamvuka, N. Pasadakis, E. Kastanaki, Energy Fuels 17 (2003) 549-558.
[7] Y. Bai, L. Yan, G. Li, R. Zhao, F. Li, Fuel 134 (2014) 368-374.
[8] T. Ohra-aho, J. Linnekoski, J. Anal. Appl. Pyrol. 113 (2015) 186-192.
[9] J. Kaal, S. Wagner, R. Jaffe, J. Anal. Appl. Pyrol. 118 (2016) 181-191.
[10] J.J. Weng, L.Y. Jia, Y. Wang, S.B. Sun, X.F. Tang, Z.Y. Zhou, K. Kohse-Höinghaus, F. Qi, Proc. Combust. Inst. 34 (2013) 2347-2354.
[11] E.B. Ledesma, M.A. Kalish, P.F. Nelson, M.J. Wornat, J.C. Mackie, Fuel 79 (2000) 1801-1814.
[12] W.J. DeSisto, N. Hill, II, S.H. Beis, S. Mukkamala, J. Joseph, C. Baker, T.-H. Ong, E.A. Stemmler, M.C. Wheeler, B.G. Frederick, A. van Heiningen, Energy Fuels 24 (2010) 2642-2651.
[13] T. Streibel, R. Geissler, M. Saraji-Bozorgzad, M. Sklorz, E. Kaisersberger, T. Denner, R. Zimmermann, J. Therm. Anal. Calorim. 96 (2009) 795-804.
[14] S. Gao, Y. Zhang, Y. Li, J. Meng, H. He, J. Shu, Int. J. Mass Spectrom. 274 (2008) 64-69.
[15] A. Fendt, T. Streibel, M. Sklorz, D. Richter, N. Dahmen, R. Zimmermann, Energy Fuels 26 (2012) 701-711.
[16] Y.N. Zhu, X.M. Chen, Y.Z. Wang, W. Wen, Y. Wang, J.Z. Yang, Z.Y. Zhou, L.D. Zhang, Y. Pan, F. Qi, Energy Fuels 30 (2016) 1598-1604.
[17] D. Wu, G. Liu, R. Sun, Energy Fuels 28 (2014) 3024-3035.
[18] C. Mukarakate, A.M. Scheer, D.J. Robichaud, M.W. Jarvis, D.E. David, G.B. Ellison, M.R. Nimlos, M.F. Davis, Rev. Sci. Instrum. 82 (2011) 033104.
[19] M.-J. Jeon, J.-K. Jeon, D.J. Suh, S.H. Park, Y.J. Sa, S.H. Joo, Y.-K. Park, Catal. Today 204 (2013) 170-178.
[20] J.J. Weng, L.Y. Jia, S.B. Sun, Y. Wang, X.F. Tang, Z.Y. Zhou, F. Qi, Anal. Bioanal. Chem. 405 (2013) 7097-7105.
[21] J.P. Mathews, A.L. Chaffee, Fuel 96 (2012) 1-14.
[22] A.O. Aboyade, J.F. Goergens, M. Carrier, E.L. Meyer, J.H. Knoetze, Fuel Process Technol. 106 (2013) 310-320.
[23] R.W. Jones, T. Reinot, J.F. McClelland, Energy Fuels 24 (2010) 5199-5209.
[24] S.Y. Liu, W.Y. Li, K.C. Xie, Energ. Source 25 (2003) 479-488.
[25] E.B. Ledesma, N.D. Marsh, A.K. Sandrowitz, M.J. Wornat, Proc. Combust. Inst. 29 (2002) 2299-2306.
[26] M.W. Jarvis, T.J. Haas, B.S. Donohoe, J.W. Daily, K.R. Gaston, M.R. Nimlos, Energy Fuels 25 (2011) 324-336.

Iron oxide composite catalyst for the environmental emission control

Patrick Mountapmbeme Kouotou[1,3,*], Achraf El Kasmir[1], Muhammad Waqas[1,2], Zhen-Yu Tian[1,2,*]

[1]Institute of Engineering Thermophysics, Chinese Academy of Sciences, Beijing 100190, China

[2]University of Chinese Academy of Sciences, Beijing 100049, China

[3] Higher Institute of the Sahel, University of Maroua, P.O. Box.: 46 Maroua, Cameroon

Abstract

Iron oxides (Fe_xO_y) have been prepared by chemical vapor deposition and were characterized for physiochemical properties via XRD, XPS, BET and TPR analyses. The catalytic activity towards the oxidation of volatile organic compounds (VOCs: C_2H_2, Ethanol (EtOH) and their mixtures) was investigated. The best results, in terms of the total oxidation of VOCs, were achieved with the composite sample. The redox properties, the higher amount of electrophilic oxygen suggested to play key roles in the reaction sequences.

1 Introduction

Nowadays, one of the most efficient ways of volatile organic compounds (VOCs) removal is catalytic combustion [1]. In this process, the total oxidation of diluted fuel occurs at relatively low temperatures with low emissions of NOx and unburned fuels [2, 3]. This technique offers the possibility of treating waste streams with different VOCs concentrations and effluent flow rates [4]. Thus, catalytic oxidation of VOCs remains an important research topic at present [5, 6].

Typically, catalysts containing both precious metals and transition metal oxides (TMOs) have been found to catalyze many oxidation reactions [7, 8]. Precious metals are successfully used for emission controls. However, they are too expensive and undergo deactivation by chlorinated compounds in the gas stream [9]. As alternatives TMOs have been developed [10]. TMOs catalysts exhibit the advantages of being available, less expensive and more resistant to deactivation by poisoning [11]. Among the TMOs, iron oxides and their composite have been found to be good candidates as cheap and efficient catalysts, especially in environmental catalysis [12, 13]. It has been reported that iron oxides may own promising catalytic activity towards the total oxidation of VOCs [9]. Among iron oxides, Fe_3O_4 and α-Fe_2O_3 have been extensively studied, due to their importance in environmental chemistry as an active agent, in corrosion science and also due to their magnetic/electronic properties [14]. Fe_3O_4 phase has been investigated heavily in the forms of natural single crystals and crystalline thin films grown on metals and oxide substrates [15]. While, α-Fe_2O_3 the most stable phase has been mostly used as catalyst supporter [7]. As an example, the mechanism by which metal oxides supported on α-Fe_2O_3, catalytically enhances the oxidation of alcohol was studied [16]. It was found that, isopropanol oxidation on CuO yielded primarily redox products (dehydrogenation to acetone), whereas that on Fe_2O_3 yielded both redox products and acidic products (dehydration to propylene). Several investigations on the oxidation of EtOH on SnO_2/Fe_2O_3 Au/Fe_2O_3, and CuO/Fe_2O_3 were reported [7, 17, 18]. The mixed metal oxide catalysts were more active than either pure Fe_2O_3 or pure SnO_2 or CuO catalysts, likely due to greater lattice oxygen mobility in the mixed oxide catalysts.

Despite intense research devoted to catalytic application of iron oxides, scarce study on the catalytic oxidation of VOCs by a mixture of both active Fe_3O_4 and stable α-Fe_2O_3 was reported. It is therefore important to explore to possibility to develop iron oxide catalyst composite (Fe_xO_y) by combining both iron oxides phase, expecting an improvement of the activity and the stability of the obtained catalyst composite.

2 Experimental

Synthesis of iron oxides thin films: For the catalyst synthesis, the feedstock composed of iron acetylacetonates [$Fe(acac)_2$ and $Fe(acac)_3$] was dissolved in tetrahydrofuran. The pulsed-spray evaporation (PSE) delivery system was achieved with an opening time of 2 ms and a frequency of 4 Hz. The feedstock was injected as a fine spray into a 30 cm-long evaporation chamber kept at 200 °C. The resulting vapor was transported to the deposition chamber by a flow of N_2 and O_2. The depositions were achieved on stainless steel substrate heated in the temperature range of 350-500 ° C. The total pressure in the reactor was varied from 25 to 50 kPa.

Characterization: The crystalline phase of the grown films was identified using X-ray diffraction (XRD, Brucker D8 Focus with Cu Kα radiation). The chemical composition was determined by XPS (ESCALAB 250 Xi). The redox properties were investigated by temperature-programmed reduction (TPR) using emission FTIR as an in situ monitoring technique. TPR test was performed in the temperature range of 100-500°C using a temperature ramp of 5°C/min in an atmosphere of argon containing 5vol% of hydrogen.

Catalytic test: The catalytic test toward the oxidation of acetylene (C_2H_2), ethanol (EtOH) and their mixture was carried out in a quartz flow reactor connected to the FT-IR apparatus. 20±0.2mg of iron oxide were deposited on grid mesh of stainless steel substrates. For the evaluation of the catalytic activity, the temperature of the reactor was raised with a ramp of 3°C min^{-1} for all experiments using a HT60 controller (Horst). In each experiment, the inlet gas mixture was composed of 1% VOC (C_2H_2 or EtOH or their mixture) and 10 % of oxygen dilute in argon in a total flow of 15 ml min^{-1}. FTIR spectrometer (Brucker VERTEX 70) with spectral range of 400-4000 cm^{-1} with a spectral resolution of 0.04 cm^{-1} was used for exhaust gas analysis. Details of the data treatment were described elsewhere [19].

Corresponding authors. Fax: +86-10-82543184
E-mail address: patrick@iet.cn / tianzhenyu@iet.cn

3 Results and discussion

3.1 Catalysts phases and structures analysis

The XRD patterns and reference are presented in Fig. 1. Three spectra can be clearly identified. From Fig. 1a, the diffraction peaks characteristics of the α-Fe_2O_3 crystalline phase are presented, in good agreement with the reference data for the α-Fe_2O_3 phase (JCPDS card 39-0238). The spectrum with characteristic peaks of the Fe_3O_4 (Magnetite) phase is observed on the top of the XRD patterns (Fig. 1c). The corresponding diffraction peaks match well with the data for the cubic spinel structured Fe_3O_4 (JCPDS card 65-3107). It is worth to mention that, on the above mention XRD spectra (Fig. 1a and c), no additional peaks belonging to other iron oxide phases were observed, indicating the presence of a single crystalline phase with good crystallinity and purity. Conversely, different crystalline phases are observed on the XRD spectrum represented by the Fig. 1b, indicating the presence of mixed iron oxide phases composed of mainly α-Fe_2O_3 and Fe_3O_4. The rather broad XRD peaks that can be observed suggest the occurrence of small crystallites with an average size calculated according to the Scherrer equation are in the range 30-40 nm. Similar BET surface areas were estimated to in the range of 5–9 m^2/g.

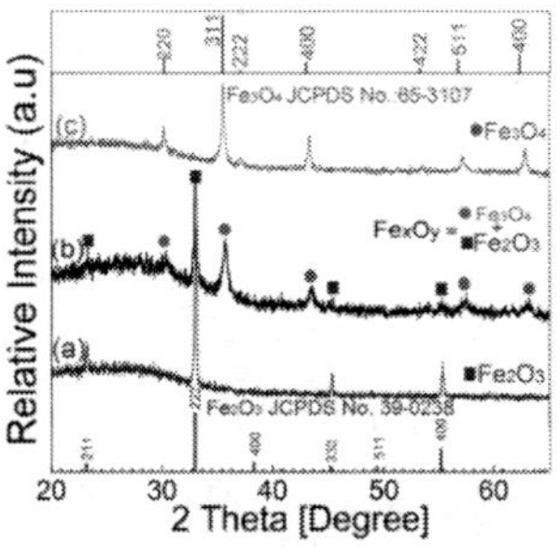

Figure 1: The XRD patterns of the as-prepared (a) α-Fe_2O_3, (b) Fe_xO_y and (c) Fe_3O_4 thin films catalysts.

3.2 Surface chemical composition

Chemical compositions of the catalysts were study by XPS. Figure 2 shows the XPS spectra in the Fe 2p (Fig. 2a) and O 1s (Fig. 2b) binding energy (BE) regions. Fig. 2a displays the Fe 2p XPS spectra of the samples in the 700–740 eV range. The spectrum of α-Fe_2O_3 catalyst shows the Fe $2p_{3/2}$ peak at 709.77 eV and the Fe $2p_{1/2}$ peak at 723.12 eV with respective satellites at 715.61 and 729.67 in agreement with previous reported values [20]. Compared with the Fe 2p spectra characteristic of Fe_xO_y and Fe_3O_4, a slight shift toward higher binding energy is observed. The BE at 711.04 eV (Fe $2p_{3/2}$) and 723.78 eV (Fe $2p_{1/2}$) with weak satellites peaks at 717.86 and 733.38 eV corresponding to the mixed oxide sample (Fe_xO_y) [21], are higher than those of Fe $2p_{3/2}$ (710.01 eV) and Fe $2p_{1/2}$ (723.24) obtained for Fe_3O_4 catalyst. Fe_3O_4 for satellites are respectively observed at 715.23 and 729.37 eV in agreement with the literatures [22]. The highest BE for the Fe_xO_y sample might be due to the co-presence of both α-Fe_2O_3 and Fe_3O_4 phases.

The O 1s spectra shown in Fig. 2b present two main peaks located in the binding energy range of 529-531 and 532-536 eV. Different features, depending on both the chemisorbed oxygen species (O_{Ads}) and the lattice oxygen (O_{Lat}), have been observed after the deconvolution of both prominent peaks indicating the existence of several kinds of surface oxygen species. The peak at 529.97 eV corresponds to the O_{Lat} (O^{2-}), whereas the signal at 531.36–531.66, 532.66-533.81 and 535.12-535.89 eV can be

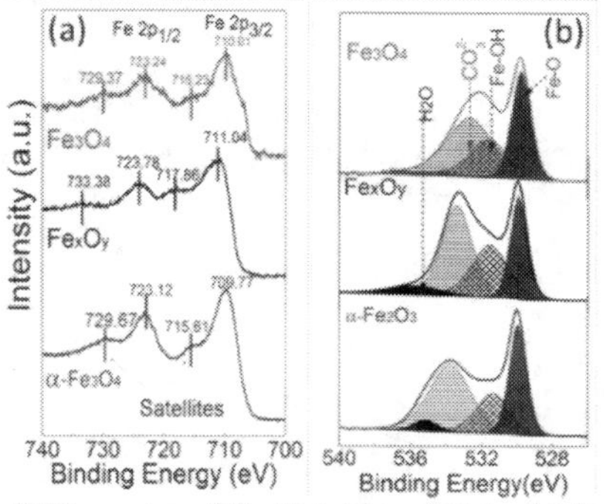

Figure 2: XPS spectra of Fe (2p) (a) and O 1s (b) for α-Fe_2O_3, Fe_3O_4 and Fe_xO_y samples respectively from the bottom to the top.

attributed to surface-adsorbed oxygen species i.e. OH^-, CO_3^{2-} and moisture (H_2O from air) respectively in agreement with literature [23]. It has been reported that, the latter have greater mobility than O_{Lat} and may give rise to beneficial spillover phenomena at the solid surface of TMOs [24]. It is generally accepted that the electrophilic O_{Ads} species play an important role in the deep oxidation of VOCs [25]. The comparison of the O_{Ads}/O_{Lat} indicate that, the highest O_{Ads}/O_{Lat} ratio (1.96) is attained for the Fe_xO_y sample, while those of α-Fe_2O_3 (1.42) and Fe_3O_4 (1.43) were less high, thus confirms the large amount of adsorbed oxygen species at the surface of Fe_xO_y. It is worth noting that O^{2-} ions are nucleophilic reagents and are usually responsible for selective oxidation reactions. Conversely, it has been reported that, electrophilic oxygens (O_2^{-2}, O^- and O^{2-}), which implies the formation of radicals, are highly reactive and give rise to total oxidation [25].

3.3 Surface reducibility

TPR profiles of the catalysts are shown in Fig. 4. The reduction profiles for all samples obtained by integrating the two specific emission bands (at 455 and 530 cm^{-1}) of the Fe_3O_4 sample upon heating under H_2 flow are shown in Fig. 4. Starting from 225 °C the total reduction of pure α-Fe_2O_3 is achieved at ~430 °C. The high temperature at which reduction is obtained for α-Fe_2O_3 catalyst can be due to the subsequent multiple reduction of from α-Fe_2O_3 to Fe_3O_4 and FeO [26]. However a shift of the reduction temperature towards the low values is noticed for the iron oxide mixed catalyst (Fe_xO_y) containing α-Fe_2O_3 and Fe_3O_4, suggesting that the combination of both iron oxide phases decrease the temperature at which reduction occurs, as indicated by the lowest temperatures obtained at ~360 °C. A tentative explanation of this finding can be made in respect to the earlier published work in the literature. In fact, Fe_2O_3 is well known as an excellent catalyst support for noble metals and TMOs catalysts [7, 18], because of their ability to decrease the reduction temperature of the supported active phase. As an example, Minico et al. reported the influence of

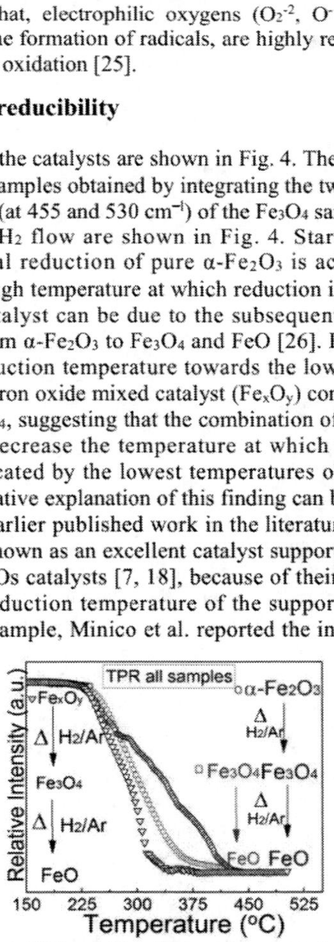

Figure 4: TPR profile of iron oxides samples

Fe_2O_3 in the TPR results of gold (Au)-doped Fe_2O_3 (Au/Fe_2O_3) catalysts, and they attributed the lower reduction temperatures to the effect of Au and Fe_2O_3 interactions on lattice oxygen mobility [7]. Also it has been demonstrated by Litt et al. [18] and Cao et al. [27] that the presence of CuO on Fe_2O_3 decreases the temperature at which reduction occurs. By analogy, it is likely that the presence of both Fe_3O_4 and Fe_2O_3 in the same phase might cause the decreases of the strength of nearby Fe–O bonds, thus increasing the mobility of lattice oxygen.

3.4 Catalytic tests

The light-off curves of the conversion of the selected VOCs and their mixture over iron oxides thin films are reported in Fig. 5. Figure 5a shows the C_2H_2 conversion. All the prepared catalysts display positive C_2H_2 conversion trends with the increases of the reaction temperature. However, different conversion values can be observed: from the Fig. 5a, it is observed that the C_2H_2 conversion over Fe_3O_4 and Fe_xO_y start at ~130 °C, whereas, that of α-Fe_2O_3 begin at 175 °C and the complete conversion is achieved at 205, 215 and 275 °C respectively. The following reactivity order was observed: Fe_xO_y>Fe_3O_4>α-Fe_2O_3. As presented in Fig. 5b, similar catalytic behavior was achieved after three running cycles and no deactivation was observed in terms of total C_2H_2 conversion indicating that, the as-prepared catalysts are stable and reusable. From the catalysis experiments results, it is suggested that, the presence of α-Fe_2O_3 in the same phase with Fe_3O_4 enhances the performance of the resulted catalyst, indicating that the resulted oxide composite (Fe_xO_y) present better performance than pure α-Fe_2O_3 or Fe_3O_4 alone.

To further illustrate the catalytic performance of the as-prepared iron oxides samples, the oxidation of EtOH over the entire samples was performed. Fig. 5c presents the catalytic activity catalytic activity of pure α-Fe_2O_3 and Fe_3O_4 for comparison purpose. Similar catalytic behavior was observed for EtOH total oxidations. The activity of pure α-Fe_2O_3 is the lowest, follow by Fe_3O_4, while the Fe_xO_y catalysts composite present the highest activity and the following performance trend was observed: Fe_xO_y>Fe_3O_4>α-Fe_2O_3. This result further confirms that, there is a synergistic effect between α-Fe_2O_3 and Fe_3O_4, which strongly affect the catalytic performance. Since industrial emissions is generally composed of mixtures of VOCs, it was important for a possible catalyst applications, to test the performance of the catalyst prepared in this study toward catalytic oxidation of acetylene and ethanol as a representative VOCs mixture. The catalytic conversions of the VOC mixture (C_2H_2/EtOH) over the prepared catalysts are reported in Fig. 5d. Again, Fe_xO_y is the most active catalyst, followed by Fe_3O_4 and α-Fe_2O_3.

To understand the difference observed in the performance of the three catalysts and particularly the composite phase, a study of the factors influencing the catalytic oxidation on the surface of TMOs is needed. Based on the characteristic observations, the catalytic performance of iron oxides catalysts could be correlated with O_{Ads} on the surface, the porous-like morphology and attractive redox properties. From a viewpoint of oxygen species participating in the catalysis, O_{Ads} (and/or surface) oxygen and O_{Lat} are supposed to contribute to the suprafacial and intrafacial processes, respectively. XPS results revealed the presence of cations (Fe^{2+} and Fe^{3+}) and anions (O^{2-}, OH^- as well as $CO_3^{2-}$and H_2O) at the surface of the entire samples. Therefore, the activity of pure single iron oxide phases (Fe_3O_4 and α-Fe_2O_3) and their composites (Fe_xO_y) should be dominated by the exposed fraction at their surface. Therefore, we will discuss how either the redox properties or the O_{Ads} might be involved in the VOCs catalytic oxidation process in this work. According to Vedrine et al. [28] the Mars-van-Krevelen (MvK) mechanism is a probable mechanism for many catalytic oxidation reactions of VOCs over mixed metal oxide catalysts. MvK mechanism requires that, O_{Lat} from the bulk catalyst are involved in the reaction. Therefore, high O_{Lat} mobility is favorable. The presence of Fe_3O_4 co-existing with α-Fe_2O_3 in the same phase likely increases the O_{Lat} mobility, as implied in TPR analyses, and so enhances the catalytic activity of α-Fe_2O_3. In the other side, it has been reported that, surface oxygen species play a key role on catalytic activity in oxidation reactions, especially at temperatures below 400 °C [29]. At low temperatures, only the outer catalyst layer may be equilibrated with oxygen in the gas-phase; at temperatures high enough, instead, surface defects diffuse into the bulk and an equilibrium is obtained between the gas phase and the solid oxide [5]. Hence, support the finding of Ruitenbeek et al. [30] who suggested that, oxidation reaction mechanisms on solid catalysts may adapt as a function of temperature. G.I. Golodets [31] as far as he is concerned, proposed an overall classification of oxidation mechanisms as a function of temperature, according to which surface redox-cycles (MvK-type mechanism) occur at moderate temperatures (300–500 °C), and are replaced at higher temperatures by radical processes in the gas-phase. He also identified a "transitional" region between 100-300 °C in which a complex surface oxidation/re-oxidation mechanisms take place. The latter is probably responsible for the formation of the total oxidation products. Similarly, according to Boreskov's classification [32], oxidation reactions occur via an associative (concerted) or redox (step-wise) mechanism: the former is characterized by lower activation energies, whereas the latter involves lattice oxygen and displays higher activation energies. The occurrence of one or other of the two mechanisms depends above all on the oxygen storage capacity (OSC) and lattice oxygen availability of the catalyst. In the present case, over the temperature range of

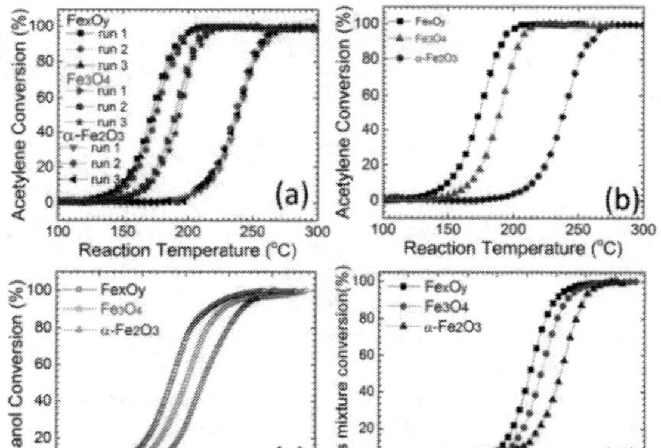

Figure 5: Catalytic results: (a) C_2H_2 conversion, (b) reproducibility for three successive C_2H_2 oxidation runs, (c) EtOH oxidation and (d) VOC mixture oxidation (C_2H_2 and EtOH) a versus the reaction temperature over the iron oxides thin films catalysts.

100–350 °C, the best performances were achieved with Fe_xO_y, which showed the highest amount of adsorbed oxygen species (O^{2-}, OH^- as well as CO_3^{2-} and H_2O), and the lowest reduction temperature (higher O_{Lat} mobility). Both the Fe_3O_4 and α-Fe_2O_3 catalysts with a lower amount of surface oxygen species and higher reduction temperature (lower O_{Lat} mobility), displayed less performance than Fe_xO_y, respectively. By comparing the catalytic performance data and characterization results obtained in our present studies, one can realize that there was a clear correlation of O_{Ads} species concentration, and low-temperature reducibility with the catalytic performance. Therefore, it is suggested that the good catalytic performance of Fe_xO_y composite, was associated with its higher surface oxygen species concentration, and better low-temperature reducibility.

4 Conclusions

Iron oxides catalysts (α-Fe_2O_3, Fe_xO_y and Fe_3O_4) has been prepared by PSE-CVD and systematically characterized. All the samples were tested for the total oxidation of VOCs. The best performance in terms of total VOC total conversion to CO_2 and H_2O, were achieved with the composite catalyst (Fe_xO_y) which possess the highest amount of adsorbed oxygen (OH^-, CO_3^{2-} species) and the lowest reduction temperature, beneficial for the total oxidation of VOCs. Interestingly, the selected sample has exhibited high activity towards the total oxidation of VOCs mixture. More importantly, Fe_xO_y catalyst shows excellent catalytic activity during the combustion of the mixture of VOCs (conversion to CO_2 = 96.8% ± 0.5) and no deactivation occurred during the reaction. The overall findings in this work suggest that, active and stable TMOs based catalysts can be produced via PSE-CVD confirming the technique as promising one for the materials synthesis.

5 Acknowledgment

The authors thank the financial support from the Ministry of Science and Technology of China (2017YFA0402804) and Recruitment Program of Global Youth Experts. PMK is grateful for the support by Chinese Academy of Sciences visiting Professorship for Senior International Scientist (Grant No. 2015PT016).

References

[1] L.F. Liotta, Appl. Catal. B: Environ., 100 (2010) 403-412.
[2] K. Everaert, J. Baeyens, J. Hazard. Mater., 109 (2004) 113-139.
[3] C. Gennequin, S. Siffert, R. Cousin, A. Aboukaïs, Top. Catal., 52 (2009) 482-491.
[4] M. Hakim, Y.Y. Broza, O. Barash, N. Peled, M. Phillips, A. Amann, H. Haick, Chem. Rev. , 112 (2012) 5949-5966.
[5] M. Piumetti, D. Fino, N. Russo, Appl. Catal. B: Environ., 163 (2015) 277-287.
[6] P.M. Kouotou, Z.Y. Tian, H. Vieker, A. Beyer, A. Gölzhäuser, K. Kohse-Höinghaus, J. Mater. Chem. A, 1 (2013) 10495-10504.
[7] S. Minico, S. Scire, C. Crisafulli, R. Maggiore, S. Galvagno, Appl. Catal. B: Environ., 28 (2000) 245-251.
[8] P.M. Kouotou, Z.Y. Tian, Physica Status Solidi (A), 212 (2015) 1508-1513.
[9] H.L. Tidahy, S. Siffert, F. Wyrwalski, J.F. Lamonier, A. Abouka, Catal. Today, 119 (2007) 317-320.
[10] G. Ertl, H. Knözinger, F. Schüth, J. Weitkamp, Handbook of Heterogeneous Catalysis, second ed., Weinheim, Wiley-VCH, 2008, pp. 2394–2411.
[11] M. Piumetti, B. Bonelli, M. Armandi, L. Gaberova, S. Casale, P. Massiani, E. Garrone, Micropor. Mesopor. Mat., 133 (2010) 36-44.
[12] P.M. Kouotou, H. Vieker, Z.Y. Tian, P.H.T. Ngamou, A.E. Kasmi, A. Beyer, A. Gölzhäuser, K. Kohse-Höinghaus, Catal. Sci. Technol., 4 (2014) 3359-3367.
[13] W.-D. Oh, Z. Dong, Z.-T. Hu, T.-T. Lim, J. Mater. Chem. A, 3 (2015) 22208-22217.
[14] V. Idakiev, T. Tabakova, A. Naydenov, Z.Y. Yuan, B.L. Su, Appl. Catal. B: Environ. , 63 (2006) 178-186.
[15] Q. Zeng, J. Bai, J. Li, L. Xia, K. Huang, X. Li, B. Zhou, J. Mater. Chem. A, 3 (2015) 4345-4353.
[16] D. Kulkarni, I.E. Wachs, Appl. Catal. A: Gen., 237 (2002) 121-137.
[17] G. Litt, C. Almquist, Appl. Catal. B: Environ., 90 (2009) 10-17.
[18] V.V. Kovalenko, M.N. Rumyantseva, A.M. Gaskov, E.V. Makshina, V.V. Yushchenko, I.I. Ivanova, A. Ponzoni, G. Faglia, E. Comini, Inorg. Mater. , 42 (2006) 1088-1093.
[19] Z.Y. Tian, P.H.T. Ngamou, V. Vannier, K. Kohse-Höinghaus, N. Bahlawane, Appl. Catal. B: Environ., 117-118 (2012) 125-134.
[20] Z. Cheng, Z. Gao, W. Ma, Q. Sun, B. Wang, X. Wang, CrystEngComm, 15 (2013) 8166-8172.
[21] H. Zhao, Y. Wang, Y. Wang, T.Cao, G. Zhao, Appl. Catal. B: Environ., 125 (2012) 120-127.
[22] R. Dedryvere, M. Maccario, L. Croguennec, F.L. Cras, C. Delmas, D. Gonbeau, Chem. Mater., 20 (2008) 7164-7170.
[23] P.M. Kouotou, Z.Y. Tian, H. Vieker, A. Beyer, A. Gölzhäuser, K. Kohse-Höinghaus, J. Mater. Chem. A, 1 (2013) 10495-10504.
[24] S.C. Kim, W.G. Shim, Appl. Catal. B: Environ., 98 (2010) 180-185.
[25] S. Rousseau, S. Loridant, P. Delichere, A. Boreave, J.P. Deloume, P. Vernoux, Appl. Catal. B: Environ., 88 (2009) 438-447.
[26] M. Khoudiakov, M.C. Gupta, S. Deevi, Appl. Catal. A: Gen., 291 (2005) 151-161.
[27] J.-L. Cao, Y. Wang, X.-L. Yu, S.-R. Wang, S.-H. Wu, Z.-Y. Yuan, Appl. Catal. B: Environ., 79 (2008) 26-34.
[28] J.C. Vedrine, G. Coudurier, J.-M.M. Millet, Catal. Today 33 (1997) 3-13.
[29] Z.-Y. Fei, B. Sun, L. Zhao, W.-J. Ji, C.-T. Au, Chem. Eur. J., 19 (2013) 6480-6487.
[30] M. Ruitenbeek, A.J.v. Dillen, F.M.F.d. Groot, I.E. Wachs, J.W. Geus, D.C.K. ingsberger, Top. Catal. , 10 (2000) 241-254.
[31] G.I. Golodets, New developments in selective oxidation, in: G. Centi, F. Trifirò (Eds.), Studies in Surface Science and Catalysis, vol. 55, Elsevier, Amstedam, 1990, p. 693.
[32] G.K. Boreskow, Catalytic activation of dioxygen, in: J. Anderson, M. Boudart (Eds.), Catalysis Science and Technology, 3, Springer-Verlag, Heidelberg, 1982 (Ch. 3).

11th Asia-Pacific Conference on Combustion,
The University of Sydney, NSW Australia
10th -14th December 2017.

Theoretical Analysis of Fire Development in a closed Compartment

Vasily Novozhilov[1]

[1]Centre for Environmental Safety and Risk Engineering (CESARE)

Victoria University

Melbourne, VIC 8001, Australia

Abstract

Flashover is a major mechanism of fire propagation in compartments. Several models have been proposed to describe this phenomenon using dynamical model approach. The present paper develops a model of this type for a particular case of fire development in a tightly closed compartment. The novelty of the study is in the obtained new approximate analytical solution for the smoke filling problem. This is an essential part of the flashover model. The types of fuels (solid materials) for which the model is particularly suitable are identified. Major outcome of the study is prediction of flashover times as a function of key physical parameters, such as fire heat release rate.

1 Introduction

Fire flashover development via formation of hot smoke and radiation from smoke layer to virgin fuel is a specific form of thermal explosion (i.e. positive feedback loop). This flashover mechanism is well known [1]. Motivation of the present paper is to consider such a scenario in fully closed compartments. Currently, there is a drive towards production and use of solid fuels, alternative to traditional, for clean energy generation. Examples of such solid fuels are biomass fuels, including Refuse Derived Fuels (RDF) [2,3]. The studies of Koseki [2] and Murasawa and Koseki [3] have demonstrated that biomass fuels may develop thermal explosion-type spontaneous ignition if stored in large quantities. One of the modes of transportation of such fuels are closed cargo compartments of ships. In such circumstances consideration of flashover development in a closed compartment becomes a relevant practical problem.

Scenario considered in the present paper is that initial fire starts in a closed compartment (e.g. ship cargo space) by the thermal explosion mechanisms described in [2,3]. Generated hot smoke acts as a source of radiation igniting remaining fuel, i.e. resulting in flashover. The proposed mathematical model of flashover is based on a newly obtained analytical solution for smoke filling in the closed compartment. This solution serves as a basis to predict radiative feedback from the smoke layer to virgin fuel, which eventually allows flashover times to be estimated. The model describes the dependencies of the process on all the key physical par\meters. Of particular interest is reported dependence of fire flashover times on the initial fire heat release rate. Throughout the paper waved variables are dimensional, and unwaved non-dimensional.

Corresponding author. Fax: +61-3-9919-8058
E-mail address: vasily.novozhilov@vu.edu.au

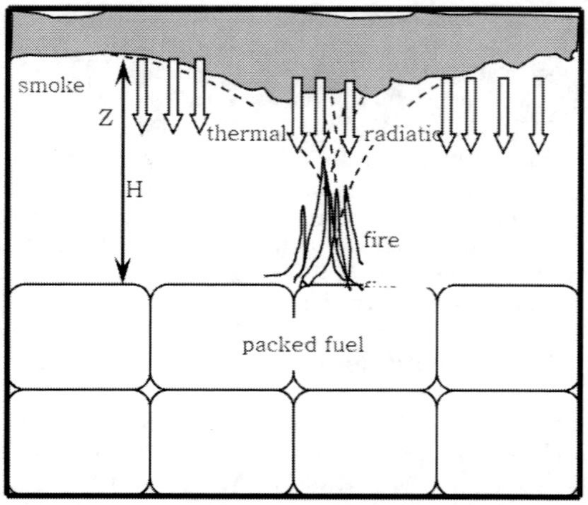

Figure 1: Flashover development scenario in a closed compartment.

2 Mathematical Model

Initial fire, resulting from the autoignition (thermal explosion) of the portion of the fuel is assumed to have constant Heat Release Rate (HRR). This fire may develop in either flaming or smoldering mode of combustion. Scenario that converts this initial fire into catastrophic flashover scenario is considered. We assume the compartment to be fully closed. In practice, some leaks may exist that initially allow oxygen for combustion to diffuse inside the compartment over a long time. However, in the smoke filling problem considered below the effect of such leaks is negligible due to a limited duration of the filling process.

In order to analyse the problem some preliminaries from fire dynamics are needed to obtain a solution for the smoke filling problem in a compartment shown in Figure 1. The problem is considered using the zone approach [4,5], with the commonly adopted correlations for the mass flow rate and excess temperature in fire plume [4]

$$\tilde{\dot{m}}_p(\tilde{z}) = \beta\left(\frac{\tilde{\rho}_0^2\tilde{g}}{\tilde{c}_p\tilde{T}_0}\right)^{1/3}\cdot\tilde{\dot{Q}}^{1/3}\cdot\tilde{z}^{5/3} \qquad (1)$$

$$\tilde{\theta}_p(\tilde{z}) = \beta^{-1}\left(\frac{\tilde{T}_0}{\tilde{g}\tilde{c}_p^2\tilde{\rho}_0^2}\right)^{1/3}\cdot\tilde{\dot{Q}}^{2/3}\cdot\tilde{z}^{-5/3} \qquad (2)$$

Here $\tilde{\dot{Q}}$ is fire heat release rate, $\tilde{z}$ is height above fire source.

Using scaling conventional in the zone fire modeling approach [4,5] and adopting subscripts "1" and "2" for the properties of the bottom and the upper layers, respectively, the following set of non-dimensional equations can be derived.

$$\rho_1 T_1 = P \tag{3}$$

$$\rho_2 T_2 = P \tag{4}$$

$$\frac{d}{d\tau}(\rho_1 y) = -\dot{m}_p(y) \tag{5}$$

$$\frac{d}{d\tau}(\rho_2(1-y)) = \dot{m}_p(y) \tag{6}$$

$$\frac{d}{d\tau}(Py) = -\gamma \dot{m}_p(y) T_1 - (\gamma-1)P\frac{dy}{d\tau} \tag{7}$$

$$\frac{d}{d\tau}(P(1-y)) = \dot{Q} + \gamma \dot{m}_p(y) T_1 + (\gamma-1)P\frac{dy}{d\tau} \tag{8}$$

Here $y(\tau) = \tilde{z}/\tilde{H}$ is non-dimensional height of the smoke interface. The set of equations consists of the equations of state (equations (3) and (4)), conservations of mass (equations (5) and (6)) , and conservations of energy for the lower (cold) and upper (hot) layers, respectively.

Solution of the set of equations (3-8) cannot be derived here in detail due to lack of space. The necessary remarks are that from adding up the equations (7) and (8) it is easily follows that the pressure in fully closed compartment grows linearly with time

$$P = 1 + \dot{Q}\tau \tag{9}$$

Next, either (7) or (8) becomes the equation for $y(\tau)$:

$$\gamma P\frac{dy}{d\tau} + \frac{dP}{d\tau}y + \gamma \dot{m}_p(y) T_1 = 0 \tag{10}$$

This equation can be further simplified by neglecting the ratio $\varepsilon(\tau) = (T_1 - 1)/\theta_p(y)$ in comparison with $y(\tau)$. This is plausible since $\varepsilon(0) = 0$ and as the smoke descends $\theta_p(y)$ grows along with T_1. The result is the integrable equation

$$\gamma P\frac{dy}{d\tau} + \gamma \dot{m}_p(y) + \dot{Q}y = 0 \tag{11}$$

with the solution

$$y(\tau) = \delta^{-3/2}\dot{Q}\left[\left(1+\frac{\dot{Q}^{2/3}}{\delta}\right)(1+\dot{Q}\tau)^{2/(3\gamma)} - 1\right]^{-3/2} \tag{12}$$

where $\delta = \beta\gamma^{2/3}$.

The power dependencies with the exponents $(-5/3)$ and $2/(3\gamma)$ $(\gamma \approx 1.4)$ can be replaced with a very good accuracy by (-2) and $(1/2)$ in $\theta_p(y)$ and $y(\tau)$ dependencies (equations (2) and (12) resepectively), using constant fitting coefficients of the order of unity, With this modification solutions of the equations (3-6) can be obtained in analytical form as well:

$$\int_0^\tau \dot{m}_p(y(\tau))d\tau = a_1\left\{\left(\frac{1}{(1-b)} - \frac{1}{\left(\sqrt{1+\dot{Q}\tau} - b\right)}\right) + \left(\frac{b}{2(1-b)^2} - \frac{b}{2\left(\sqrt{1+\dot{Q}\tau} - b\right)^2}\right)\right\}$$

$$a_1 = \beta\gamma^{-1/3}\delta^{-3}\dot{Q}^{4/3}\left(1+\frac{\dot{Q}^{2/3}}{\delta}\right)^{-3}; \quad b = \left(1+\frac{\dot{Q}^{2/3}}{\delta}\right)^{-1}$$

$$\rho_1 = \frac{\delta^{3/2}\dot{Q}^{-1}\left(1 - \int_0^\tau \dot{m}_p(y(\tau))d\tau\right)}{\left[\left(1+\frac{\dot{Q}^{2/3}}{\delta}\right)(1+\dot{Q}\tau)^{2/(3\gamma)} - 1\right]^{-3/2}} \tag{13}$$

$$T_1 = \frac{(1+\dot{Q}\tau)\left[\left(1+\frac{\dot{Q}^{2/3}}{\delta}\right)(1+\dot{Q}\tau)^{2/(3\gamma)} - 1\right]^{-3/2}}{\delta^{3/2}\dot{Q}^{-1}\left(1 - \int_0^\tau \dot{m}_p(y(\tau))d\tau\right)} \tag{14}$$

$$\rho_2 = \frac{\int_0^\tau \dot{m}_p(y(\tau))d\tau}{1 - \delta^{-3/2}\dot{Q}\left[\left(1+\frac{\dot{Q}^{2/3}}{\delta}\right)(1+\dot{Q}\tau)^{2/(3\gamma)} - 1\right]^{-3/2}} \tag{15}$$

$$T_2 = \frac{(1+\dot{Q}\tau)\left(1 - \delta^{-3/2}\dot{Q}\left[\left(1+\frac{\dot{Q}^{2/3}}{\delta}\right)(1+\dot{Q}\tau)^{2/(3\gamma)} - 1\right]^{-3/2}\right)}{\int_0^\tau \dot{m}_p(y(\tau))d\tau} \tag{16}$$

Solution to the smoke filing problem is illustrated in Figures 2 and 3. Here the positions of the smoke interface and excess smoke temperature are shown, in non-dimensional variables.

Mathematical description of flashover mechanism identified above requires suitable model for radiative flux emitted by the smoke layer. Radiation from the smoke layer is predominantly due to soot.

Soot emission is controlled by the absorption coefficient which can be approximated empirically as [6]

$$\tilde{k}_s = 1264 f_v \tilde{T} \tag{17}$$

where f_v is soot volume fraction (numerical coefficient in the equation (17) is dimensional).

The latter is calculated as follows

$$f_v(\tilde{t}) = \frac{\tilde{s}\tilde{\dot{m}}\tilde{t}}{\tilde{\rho}_s\tilde{S}(\tilde{H}-\tilde{z})} \tag{18}$$

and in non-dimensional form

$$f_v(\tau) = \frac{s\dot{m}\tau}{\rho_s(1-y)} \tag{19}$$

where S is soot conversion factor (soot yield) of the fuel, $\dot{m}$ is non-dimensional mass burning rate of the fuel, ρ_s is non-dimensional soot density.

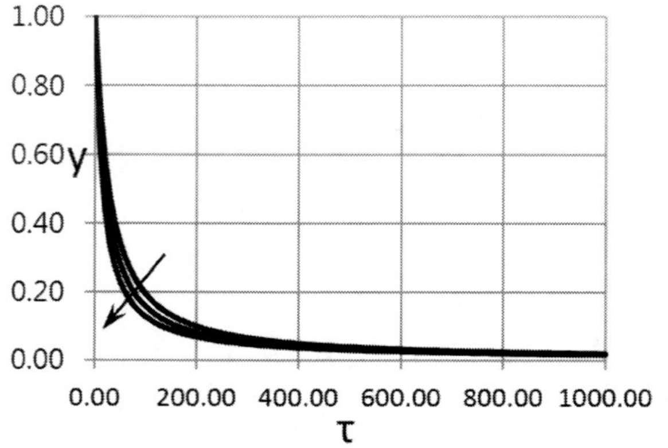

Figure 2: Dynamics of the smoke layer interface. Arrow points towards increase of $\dot{Q}=4.55\cdot10^{-3}$; $\dot{Q}=9.10\cdot10^{-3}$ and $\dot{Q}=1.82\cdot10^{-2}$.

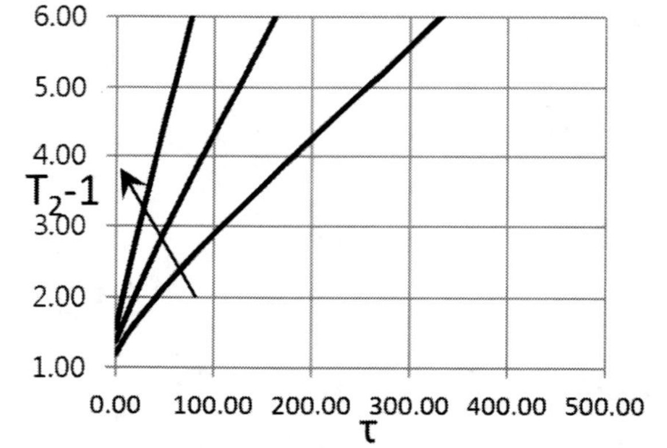

Figure 3: Dynamics of the smoke (upper layer) layer temperature. Arrow points towards increase of $\dot{Q}=4.55\cdot10^{-3}$; $\dot{Q}=9.10\cdot10^{-3}$ and $\dot{Q}=1.82\cdot10^{-2}$.

Emissivity of the smoke (upper) layer can be written, following Drysdale [7] as

$$\varepsilon_1 = 1-\exp\left(-\tilde{k}_s\left(\tilde{H}-\tilde{S}\right)\right) \tag{20}$$

Taking into account relations (17-19) this expression can be written as

$$\varepsilon_1(\tau) = 1-\exp\left(-E\frac{s\dot{m}\tau}{\rho_s}T_2\right) \tag{21}$$

where $E = 1264\tilde{H}\tilde{T}_0$.

View factor for the calculation of radiative heat exchange between the smoke layer and the fuel surface is taken as that for the case of two infinitely long parallel planes [8].

Then, neglecting in the first approximation the effect of the fuel temperature rise on heat exchange, the radiative flux received by the material can be written (in dimensional form) as

$$\tilde{\dot{q}}'' = -\frac{\tilde{\sigma}\left(\tilde{T}_2^4-\tilde{T}_0^4\right)}{\left(\frac{1}{\varepsilon_1}+\frac{1}{\varepsilon_2}-1\right)} \tag{22}$$

where ε_1 and ε_2 are emissivities of the smoke layer and the fuel surface, respectively.

This expression can be used in the well known [9] identity

$$\tilde{T}_s(\tilde{t})-\tilde{T}_0 = -\frac{1}{\sqrt{\tilde{\lambda}\tilde{\rho}\tilde{c}}}\frac{1}{\sqrt{\pi}}\int_0^{\tilde{t}}\frac{\tilde{\dot{q}}''(\tilde{\xi})}{\sqrt{\tilde{t}-\tilde{\xi}}}d\tilde{\xi} \tag{23}$$

relating surface temperature of the material T_s with the heat flux received by the material.

In non-dimensional form (and using straightforward change of integration limits) relation (23) can be written in the following form

$$\theta_s(\tau)-1 = \frac{\sqrt{\tau}}{\sqrt{\pi}}\int_0^1\frac{\frac{\partial\theta}{\partial\xi}(0,\varsigma\tau)}{\sqrt{1-\varsigma}}d\varsigma \tag{24}$$

and further, using (21,22) in the final form

$$\theta_s(\tau)-1 = \frac{Bi_R}{\sqrt{\pi}}\sqrt{\tau}\int_0^1\frac{\left[T_2^4(\varsigma\tau)-1\right]}{\left(\frac{1}{\varepsilon_1(\varsigma\tau)}+\frac{1}{\varepsilon_2}-1\right)}\frac{d\varsigma}{\sqrt{1-\varsigma}} \tag{25}$$

where Bi_R is a parameter similar in nature to conventional Biot number, but based on comparison of conductive and radiative thermal resistances.

Equation (25) represents final form of the mathematical model, from which the time for the surface temperature to reach any given value can be calculated.

Certain "critical" level of the material surface temperature can be considered as an ignition (and therefore in the context of the present paper flashover) criterion, in accordance with the reasonably accurate approach adopted in the fire science [1,7].

In the present study the value of ignition temperature is set at $\theta_s^{ign} = 2.5$ which corresponds to dimensional ignition temperature of about 750 C. This is generally consistent with the reported values for pilot-free ignition of cellulose containing materials .

3 Results and Discussion

With the adopted ignition criterion equation (25) allows the fuel ignition (i.e. flashover) times to be predicted for different sets of controlling parameters. The dependencies on the most important parameters are presented below.

In the non-dimensional formulation, behavior of the system is being essentially controlled by the following parameters: $\dot{m}$, E, Bi_R, s. The Heat Release Rate $\dot{Q}$ is a function of the two former parameters $\dot{Q}=\dot{Q}(\dot{m},E)$ (since E is dependent on the spatial scale $\tilde{H}$). These parameters vary with variations of compartment geometry, fire intensity and fuel properties.

The most important parameters are those characterising fire intensity, that is mass burning rate $\dot{m}$ and HRR of the initial fire $\dot{Q}$.

Non-dimensional ignition time as a function of non-dimensional mass burning rate is presented in Figure 4.

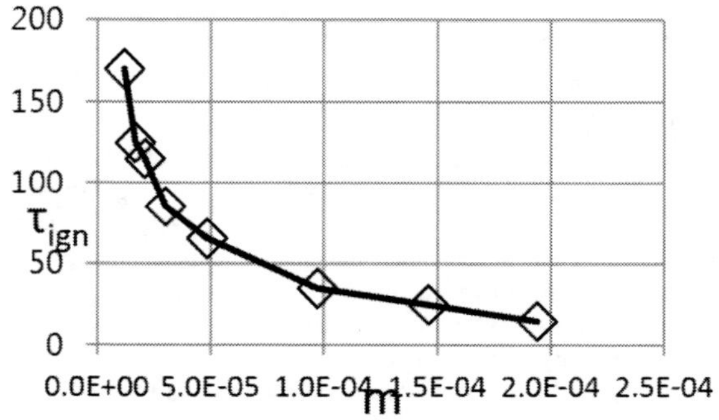

Figure 4: Non-dimensional ignition time as a function of the non-dimensional mass burning rate $\dot{m}$.

As expected, ignition time descreases with the increase in the mass burning rate. However, the rate of decrease drops substantially as the mass burnung rate of the initial fire becomes larger.

Corresponding plot in dimensional variables is shown below as Figure 5.

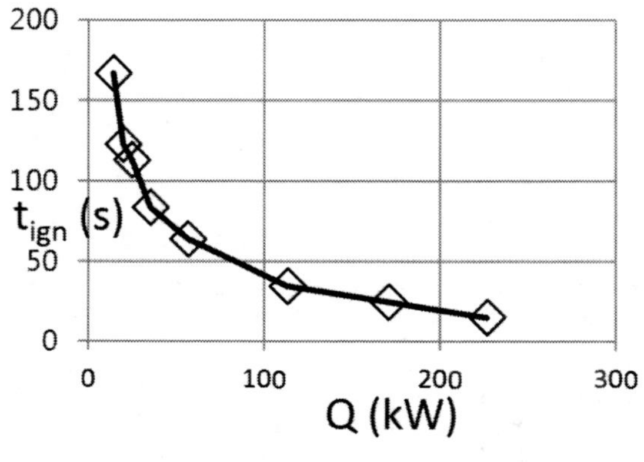

Figure 5: Dimensional ignition time as a function of the initial fire HRR $\dot{Q}$. $\tilde{H}=3m$; Compartment floor area $\tilde{S}=16m^2$

Experimental validation of the model is planned as a follow up on the present research, and will presumably require tightly closed compartment with one transparent wall. Some artificial darkening of the smoke is also probably required to allow reasonably accurate visual observations of smoke layer thickness to be taken. This will avoid the need for more sophisticated smoke detection techniques.

4 Conclusions

The present study considers fire hazards associated with storage or transportation of some modern solid fuels, for example biomass fuels, including RDFs, in closed compartments.

Such hazards originate from the ability of the above fuels to experience spontaneous ignition as a results of microbial fermentation followed by chemical oxidation. This mechanism may lead to fire of limited size, for example in a closed cargo compartment of a ship.

It is shown that such initial fire can lead to catastrophic flashover via the mechanism associated with the radiative feedback from the smoke layer formed by initial fire.

A mathematical model describing smoke filling process along with the positive feedback loop of thermal exchange between the smoke and virgin fuel is developed. The model allows closed form analytical solution for the temperature evolution of virgin fuel to be obtained. Consequently flashover time can be predicted as a function of key physical parameters that control fire dynamics.

In particular dependencies of flashover time on initial fire intensity are presented in both non-dimensional and dimensional forms. Dependencies on other parameters can be equally easily obtained from the developed model.

Practical significance of the model is its ability to directly evaluate fire risks associated with storage and/or transportation of particular types of fuels.

References

[1] V. Novozhilov, J. Eng. Math. 67 (4) (2010) 387-400.

[2] H. Koseki, Energies 4 (2011) 616-627.

[3] N. Murasawa, H. Koseki, Energies 8 (2015) 5143-5158.

[4] B. Carlsson, J.G. Quintiere, Enclosure Fire Dynamics, CRC Press, Boca Raton, U.S., 2000, p. 181.

[5] V. Novozhilov, Fire Saf. J. 47 (2012) 16-17.

[6] V. Novozhilov, B. Moghtaderi, D.F. Fletcher, J.H. Kent, Fire Saf. J. 27 (1) (1996) 69-84.

[7] D. Drysdale, An Introduction to Fire Dynamics, Wiley, Chichester, U.K., 2011, p. 35.

[8] F.P. Incropera, D.P. DeWitt, Fundamentals of Heat and Mass Transfer, Wiley, New York, U.S., 2002, p. 700.

[9] L.D. Landau, E.M. Lifshits, Fluid Mechanics, Butterworth-Heinemann, Oxford, U.K., 1987, p. 288.

11th Asia-Pacific Conference on Combustion,
The University of Sydney, NSW Australia
10th -14th December 2017.

Effect of fuel variation, plate material, and thickness on dynamics of precursors to blow out of shear layer stabilized premixed flame

Arun K Ampi[1] and T M Muruganandam[2]

[1]Research Scholar, NCCRD & Dept. Aerospace Engg, Indian Institute of Technology Madras, Chennai-36, India

[2]Associte Professor, NCCRD &Dept. Aerospace Engg, Indian Institute of Technology Madras, Chennai-36, India

Abstract

High-Speed chemiluminescence is used to study the dynamics of precursors to blow out of a shear layer stabilized flame. The flame stabilized on a splitter plate and a thin metal sheet is studied in detail. Different materials (SS and copper) are used for the flame holders, and also two fuels, LPG and CH_4 are used. Shape of local lifted flame was decided by the Lewis number and differential diffusion effects. The blowout limits were decided by heat loss to the flame holders, and the velocity in the wake. It is observed that the number and duration of precursors vary with fuel, plate material and material thickness. This suggests that there is a coupled effect of these parameters on the blowout dynamics.

1 Introduction

Generally, it is well known that local loss of stabilization is the starting stage for blow-out process for a shear layer stabilized flame[1]. Understanding of blowout process will help design sensing technology for prediction of incipient blowout. Thus it is essential to characterize the blowout process for various possibilities of fuels and flame stabilization geometries. This work focuses on doing this for shear layer stabilized flames on thin plates.

Law [2], Yokomori *et al.*[3], [4], Mallens *et al.*[5] Kawamura *et al.*[6], [7] studied the effect of flame stretch, and flame curvature on extinction. They also studied the effect of preferential diffusion and differential diffusion with respect to Lewis number on local flame strength. Study of flame stabilization by thin wires was done by Lewis and von Elbe [8]. This was one of the pioneering works in flame blow off and they elaborate the Karlovitz flame stretch theory and explain blow off of flames in terms of critical velocity gradient near the plate surface. Stabilization and blow out process in a lean premixed flame downstream of a heat conducting perforated plate was numerically analyzed by Kedia et al.[9]. They observed that the anchoring mechanism of a bluff body stabilized flame was heat loss to the burner plate and in near blowout dynamics, heat loss may not play a significant role.

Several researchers [10]–[15] have studied the effect of flame stretch, curvature and preferential diffusion on flame temperature/strength. The summary of these works are given below. Curved flame is more sensitive to stretch compared to planar flames and flame curvature effect dominates over flow non-uniformity. Preferential and differential diffusion determined the final temperature of the flame.

Blow off dynamics of inverted flame was experimentally analyzed by Edmondson and Heap [16]. They have done experiments on CH_4/air flame stabilized on thin plates of different thicknesses. They found that, flame on thinner plates blows off at lower boundary velocity gradients, implying it is a weaker stabilization. Trevino *et al.* [17] showed that heat transfer to stabilization plate is necessary for inverted flame stabilization. Sung *et al.* [18] did an experimental study on an inverted flame and they showed that heat loss to the wall has negligible influence on flame topology during the blow off process. While there are studies that explained blowout and its cause in detail, there is no work which details the dynamics of precursor events close to blowout, and its dependence on plate and fuel.

This study focuses on understanding the blowout process and the dynamics of precursor events associated with flat plate and thin metal sheet stabilized flames. The focus is on the effect of fuel, plate material and plate thickness on precursor duration and numbers near blowout. This information can help develop sensors to detect incipient blowout.

2 Experimental Setup

Experiments were done in a premixed unconfined burner, in which the premixed jet exhausted into open atmosphere, shown in Figure 1. Air and fuel inlet assembly help the proper mixing inside the mixing chamber. Both air and fuel inlets have perforated plates blocking the large jet, in order to break fuel jet into smaller jets, which criss-cross with similar small air jets, helping in large scale mixing. Proper mixing was assured by providing enough length within the mixing chamber in order to achieve a homogenous mixture at the nozzle exit. Contracting nozzle provides an accelerating flow at the burner exit, with a uniform velocity profile near the stabilization zone for the study, along with reducing flow fluctuations. The burner nozzle exit had an area of 80 mmx20 mm, with the plate or thin sheet protruding above by 15 mm.

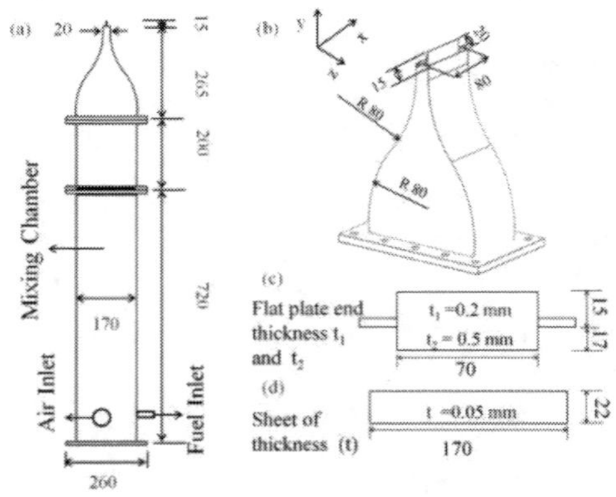

Figure 1: Schematic of (a) rectangular flat plate burner, (b) top section of the burner and (c) flat plate and (d) thin sheet.

Methane and Liquefied petroleum gas (LPG) are the two types of fuels used for the experiments. The nominal composition of LPG is 40% propane and 60% butane. Plates of two different materials, stainless steel (SS) and copper (Cu), were used with two different thicknesses for flame stabilization in the burner.

Corresponding author. +919444297637
E-mail address: aeroarunkampi@gmail.com

Both SS and copper had a thickness of 1 mm and its leading edge and trailing edges are tapered to 0.5 mm and 0.2 mm respectively. The tapering was done to avoid a recirculation zone behind the plate tip. The thin sheets used were 0.05 mm foils which were held in tension to keep them straight. The edge of the thin sheet is sharp and straight like razor edge, and there is no burr along the edge.

To understand the flame behavior during the loss of stabilization, 8-bit resolution high-speed cameras (IDT NX4-S3) was used. IDT camera has a CMOS sensor of size 1024*1024 pixels. The frame rate was 300 fps. A filter at 430 nm ±10 nm (50% transmission) was used to isolate CH* chemiluminescence. However, since the flame was predominantly blue and the flame was very weak, the filter was not used. From previous chemiluminescence literature [19] it is known that very lean gaseous hydrocarbon flames are dominated by CH* with a little bit of CO_2*. For the purpose of this study which uses these images for finding the flame front, CO2* which is present in the product side will not cause any trouble, and CH* which is emitted from the reaction zone will be the main source.

3 Results and Discussion

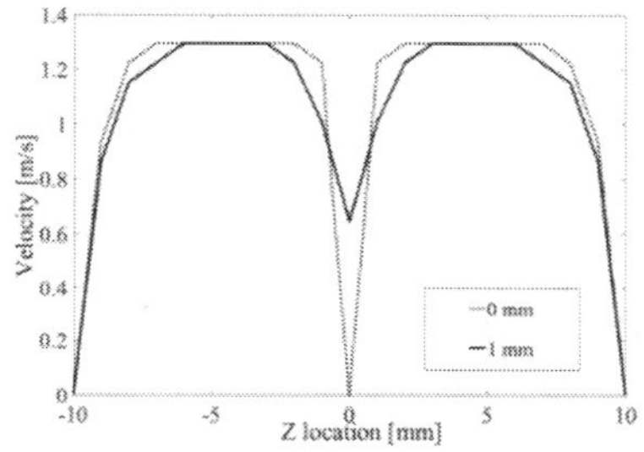

Figure 2: Cold flow velocity distribution downstream of the plate in a laminar stream of air.

Velocity distribution across the plate was measured along z-direction by using a 1 mm diameter pitot probe and a micro-manometer having a precision of ±0.03 m/s. The micro-manometer averages the velocity over 5 s. Figure 2 presents the velocity profile for a cold flow of air at 100 lpm at x=35 mm, and y=0 & 1 mm. It can be observed that the velocity profile is uniform at 1.28 m/s near the middle one-third region. It can be noted that the velocity profile gives almost uniform velocity in some portions and the velocity decreases in the plate wake and the jet shear layers. The inner wall boundary layers have grown to around 2.5 - 3 mm on either side.

Figure 3: (a) Raw image, (b) flame edge, and (c) composite image, of flame stabilized by plate wake (plate edge is shown by a red line).

Figure 3 shows an image of flame. Both images 'a' & 'c' are multiplied by a factor, for better visibility. Since the flame edge is very weak, but important in this study, it is enhanced by edge detection 'b' through image processing, superimposed on the flame image 'a' to outline the flame edge with clarity, as shown in Figure 3(c). Plate edge obtained from background images is also superimposed on every image as a red line. The images shown here onwards are the composite image of enhanced raw image, the edge image and the plate edge image.

3.1 Effect of Fuel

High speed flame luminescence movie images were analyzed qualitatively for identifying changes in structure/shape of flame front during the loss of stabilization. Two fuels viz., LPG and CH_4 are used separately for the experiments. Each experiment had constant flow rates with equivalence ratio (ϕ) maintained constant for over a minute.

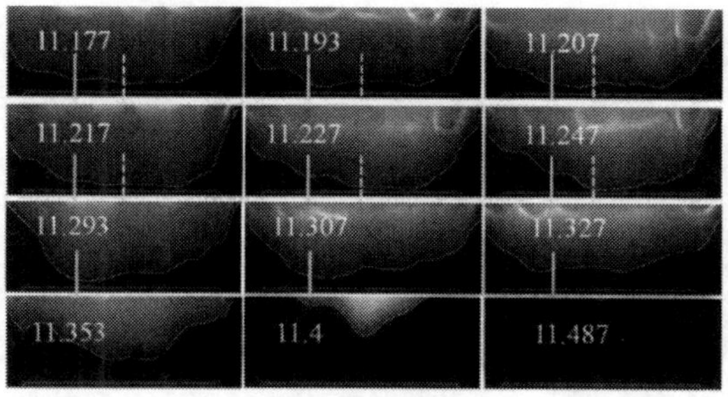

Figure 4: Chemiluminescence image sequence of loss of stabilization at ϕ =0.807 (LPG). Only selected images are shown from a sequence of images recorded at 300fps (Δt=3.33 ms). Frame times [s] are shown on the images.

Figure 4 shows a sequence of chemiluminescence images of a LPG flame during the loss of stabilization and blow out. Camera exposure time was 2000μs. The images were multiplied by a factor of 10 for better visibility. The vertical solid line in the images marks a local lift off event. At time 11.177s, near the line, the flame front moves back towards the plate. During the time 11.217s to 11.293s, the flame lifts off locally from the surface and returns back to plate lip. This type of local loss and regain of stabilization is named as precursor event. While this is happening here, another smaller precursor event may also be observed near the mid length of the plate, marked by dashed line. After 11.293s, the flame lifts off and never returns to the plate. From 11.487s onwards, there is no flame, implying complete blowout of the flame.

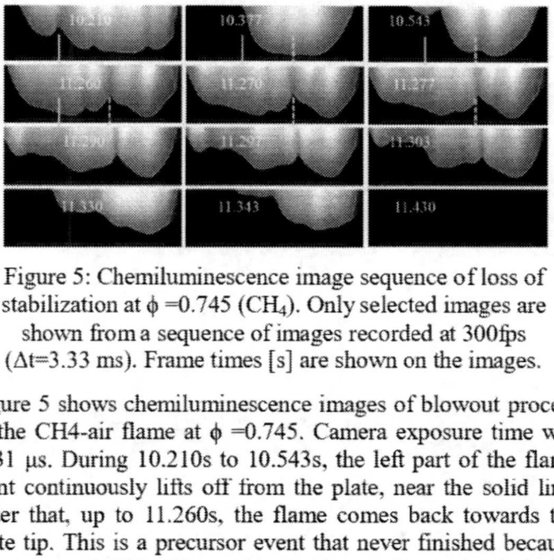

Figure 5: Chemiluminescence image sequence of loss of stabilization at ϕ =0.745 (CH_4). Only selected images are shown from a sequence of images recorded at 300fps (Δt=3.33 ms). Frame times [s] are shown on the images.

Figure 5 shows chemiluminescence images of blowout process of the CH4-air flame at ϕ =0.745. Camera exposure time was 3331 μs. During 10.210s to 10.543s, the left part of the flame front continuously lifts off from the plate, near the solid line. After that, up to 11.260s, the flame comes back towards the plate tip. This is a precursor event that never finished because the next partial lift off started before it ended. Another local lift off can also be seen from 10.543s till 11.303s, near the dashed line. After 11.303s, flame continuously lifts off from the plate and never returns back to the plate. At 11.430s blow out occurred. The whole process of blowout is similar to that for LPG fuel.

As equivalence ratio decreases to very lean value, the flame finds it very difficult to sustain itself against incoming perturbations and starts to respond to disturbances by local lift off (precursors). It was shown in an earlier work [20]by the authors that these events are due to upstream flow disturbances destabilising the flame anchor points. One can notice that the shape of the flame is different between the precursors of the two fuel cases. CH_4

case has more kinks in the lifted part of the flame while the LPG flame has more smooth flame front. Lean LPG and CH_4 have Lewis number of greater (Le > 1) and less (Le< 1) than one, respectively. Differential diffusion factor, (the ratio of diffusivity of deficient reactant, Di, to that of excess reactant, Dj), Di/Dj < 1 for lean LPG and Di/Dj >1 for lean CH4. The flame shape in this burner in stable condition is a 'V' flame. It is evident from the chemiluminescence images that LPG flame has low curvature of inverted V, while the CH4 flame has high curvature. This is because, when there is any disturbance, the flame front curves, and differential and preferential diffusion processes of species favor increase in flame speed for LPG flame, while it decreases flame speed for CH_4 flame.

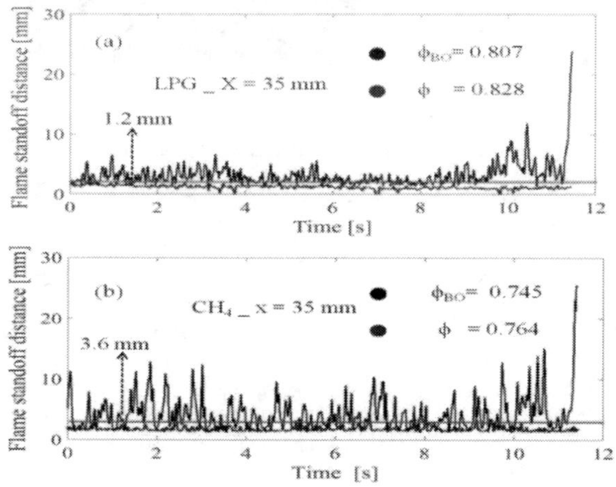

Figure 6: Variation of flame standoff distance at the mid length of plate (x = 35 mm) during blow out, (a) LPG and (b) CH_4. Equivalence ratios are shown on the plots.

Figure 6 shows the temporal variation of flame standoff distance during blowout, and stable flame condition, for both fuels. It can be seen that the flame movement is large near blowout compared to the same at stable condition, for both fuels. However, one can see that the flame motion is much larger for CH_4 compared to LPG fuel. The red horizontal lines in the plot will be used in section 3.4.

3.2 Effect of Plate Material

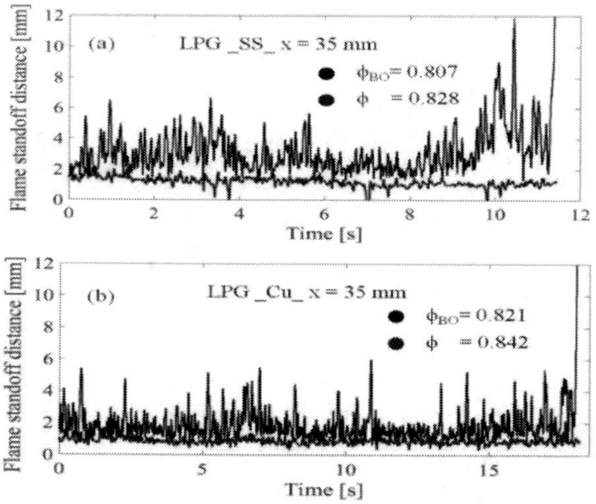

Figure 7: Variation of flame standoff distance at x = 35 mm during blow out, for two different plate materials (a) SS and (b) Cu. Equivalence ratios are shown on the plots.

Two materials, viz., stainless steel (SS) and copper (Cu) were used for stabilization in the burner. The thermal conductivity of stainless steel and copper, are 16 W/(m K) and 401 W/(m K), respectively. Figure 7 shows the variation of flame standoff distance for different plate materials at mid length of the plate. The fuel was LPG for both and the plate thicknesses were the same. Far away from blowout, both flames behave similar to earlier cases for stable conditions. Near blowout, it is evident that the flame stabilized on Cu plate moves less distance during precursor, compared to that on SS plate. It is to be noted that the blowout limit for Cu is higher as Cu absorbs more heat from the flame than SS. Thus Cu plate is hotter than SS plate. This means that the gas that flows along the plate is heated more by Cu than by SS, causing higher flame speed for Cu case. Thus the flame is stronger and does not move farther away from the plate lip.

3.3 Effect of Plate Material Thickness

The previous sections explained the variation of flame fronts with respect to the variations of fuel and plate material. The plates were 1 mm thick and tapered to 0.5 mm at the lip earlier. To study the effect of plate thickness variation, a thin sheet of metal, of thickness 0.05 mm, is used now for stabilization. The blowout equivalence ratios for the plate and sheet stabilised cases for both materials and both fuels are listed in Table 1. It was observed that the thin sheet stabilised flames blowout at higher equivalence ratios than their plate counterparts. This is in line with observations from literature [16] where, as the plate thickness decreased, the flame blows out easily. This was seen for both the materials SS and Cu. The precursor behavior showed that the sheet stabilised cases had longer precursor events compared to the plate stabilised cases. Due to space constraints, this data is not presented in plot form, but is presented in the next section as statistics. This behavior due to plate thickness can be attributed to higher velocities in the sheet case compared to the plate. Thus the flame is more unstable, and moves larger distances.

Table 1: Variation of blow-out equivalence ratio with fuels and materials of different material thickness.

Material Thickness	SS-LPG ϕ_{BO}	Cu-LPG ϕ_{BO}	SS-CH_4 ϕ_{BO}	Cu-CH_4 ϕ_{BO}
Flat plate	0.807	0.821	0.745	0.753
Sheet	0.869	0.899	0.771	0.758

There is one difference shown by the thin sheet compared to the plate case. When material changes from SS to Cu, the ϕ_{BO} increases for LPG, but decreases for CH_4. This is a coupled effect of plate material, thickness and the fuel change and needs further explanations.

3.4 Statistics of precursors

Table 2: Number, total duration and mean duration of precursors at blowout conditions listed in Table 1.

x [mm]	# Prec SS | Cu	τ_{Prec} (ms) SS | Cu	# Prec SS | Cu	τ_{Prec} (ms) SS | Cu
	LPG (plate)		CH_4(plate)	
17.5	16 | 33	708.6 | 341.5	4 |45	2861.8 | 215.2
35	53 | 143	208.1| 69.2	23 | 82	468.4 | 83.66
52.5	33| 49	339.6 | 217.7	7 | 43	1612.9 | 225.9
	LPG (sheet)		CH_4 (sheet)	
17.5	54 | 71	94.93 | 85.21	123 | 41	56.79 |161.95
35	66 | 68	67.92 | 99.61	43 | 67	54.62 | 64.28
52.5	61 | 59	65.19|94.29	62| 53	73.64 | 57.92

To quantify the dynamics of precursor events, one needs to have a threshold to mark the start and end of precursor events. In this study, the thresholds are obtained by comparing the simultaneous schlieren (along x-direction) and chemiluminescence images (along z-direction) during blowout process. Threshold is defined to be twice the flame thickness (from literature), at that equivalence ratio. Threshold values in figure 6 are 1.2 mm for LPG and 3.6 mm for CH_4. Table 2 presents the number of precursors (#Prec) and mean duration (τ_{Prec}) for all the conditions shown in Table 1.

LPG has higher ϕ_{BO} and flame speed methane. It can be seen that CH4 has longer but lesser number of precursors than LPG. This may be due to (preferential/differential) diffusion based decrease of flame speed for methane when lifted off. The cases near edges have longer and more number of precursors due to dilution effect from atmospheric air. For the plate case, Cu has shorter precursors than SS, for both fuels. There is radiation from flame heating the plate lip which passes the heat to plate upstream, and the incoming higher speed flow in the disturbance, picks up the excess heat, and strengthens the flame. This effect is stronger in Cu than SS, causing the lifted flame to reattach quicker in case of Cu.

In the case of sheet, radiation received and the conducted heat will be very small compared to plate. Thus heat feedback is not predominant for this case. The sheet was held such that it does not extend upstream of the jet lip. This offers very little blockage for the perturbation, and also the boundary layer developed is smaller than that for plate. Thus stronger perturbations and thus more precursors are expected for the sheet case. However, they are short lived because the movement away from the sheet is lower due to higher flame speeds for these higher ϕ_{BO} cases. However, the sheet case with CH_4 appears to give a different trend, as observed earlier in Table 1, and this needs further investigations.

Conclusion

Near blowout dynamics of shear layer stabilized flame from a straight, thin, metal plate/ sheet, of material SS/Cu, with CH_4/LPG flames was analyzed. High-speed chemiluminescence imaging was used to monitor the temporary flame lift-off from the lip and reattachment prior to blowout in detail.

When fuel was changed from LPG to CH_4, it was found that Thermo-Diffusive (Lewis Number) dictated the shape of the flame front during the precursor events. CH_4 flame had kinks in the flame front, while LPG flame was smooth. Change in plate material caused changes in ϕ_{BO} based on the heat conduction to plate, causing Cu flame to have higher ϕ_{BO}. When plate thickness decreases, ϕ_{BO} increases, and the number and duration of precursors increases. However, there are differences in trends when comparing the same for CH4 fuel, and thin sheet. There appears to be a coupling between the flame radiation, the plate material and plate thickness, which is not explored yet. The threshold may also have to be decided by other criteria in the future.

References

[1] S. Nair, W. I. Controls, F. Collins, T. Lieuwen, and I. Introduction, "Near-Blowoff Dynamics of a Bluff-Body Stabilized Flame," *J. Propuls. POWER*, vol. 23, no. 2, 2007.

[2] C. K. Law, "Dynamics of stretched flames," *Symp. Combust.*, vol. 22, no. 1, pp. 1381–1402, 1989.

[3] T. Yokomori and M. Mizomoto, "Flame temperatures along a laminar premixed flame with a non-uniform stretch rate," *Combust. Flame*, vol. 135, no. 4, pp. 489–502, 2003.

[4] T. Yokomori and M. Mizomoto, "Interaction of adjacent flame surfaces on the formation of wrinkling laminar premixed flame," *Proc. Combust. Inst.*, vol. 29, pp. 1511–1517, 2002.

[5] R. M. M. Mallens, L. P. H. D. E. Goey, and C. K. Law, "A Numerical Study of the Stabilisation of a Lean Laminar Premixed V-Shaped Flame A Numerical Study of the Stabilisation of a Lean Laminar Premixed V-Shaped Flame," vol. 2202, no. July, 2017.

[6] T. Kawamura, K. Asato, T. Mazaki, T. Hamaguchi, and H. Kayahara, "Explanation of the blowoff of inverted flames by the area-increase concept," *Combust. Flame*, vol. 35, pp. 109–116, 1979.

[7] T. Kawamura, K. Asato, and T. Mazaki, "Reexamination of the blowoff mechanism of premixed flames—inverted flames," *Combust. Flame*, vol. 45, pp. 225–233, 1982.

[8] B. Lewis and G. von Elbe, "Combustion , Flames and Explosions of Gases Third Edition," 1987.

[9] K. S. Kedia and A. F. Ghoniem, "Mechanisms of stabilization and blowoff of a premixed flame downstream of a heat-conducting perforated plate," *Combust. Flame*, vol. 159, no. 3, pp. 1055–1069, 2012.

[10] C. K. Wu and C. K. Law, "On the determination of laminar flame speeds from stretched flames," *Symp. Combust.*, vol. 20, no. 1, pp. 1941–1949, 1985.

[11] M. Mizomoto and H. Yoshida, "Effects of Lewis number on the burning intensity of Bunsen flames," *Combust. Flame*, vol. 70, no. 1, pp. 47–60, 1987.

[12] C. K. Law, P. Cho, M. Mizomoto, and H. Yoshida, "Flame curvature and preferential diffusion in the burning intensity of bunsen flames," *Symp. Combust.*, vol. 21, no. 1, pp. 1803–1809, 1988.

[13] M. Mizomoto, Y. Asaka, S. Ikai, and C. K. Law, "Effects of preferential diffusion on the burning intensity of curved flames," *Symp. Combust.*, vol. 20, no. 1, pp. 1933–1939, 1985.

[14] C. W. Choi and I. K. Puri, "Contribution of curvature to flame-stretch effects on premixed flames," *Combust. Flame*, vol. 126, no. 3, pp. 1640–1654, 2001.

[15] C. W. Choi and I. K. Puri, "Response of flame speed to positively and negatively curved premixed flames Response of flame speed to positively and negatively curved premixed flames," *Combust. Theory Model.*, vol. 7, no. 2, pp. 205–220, 2003.

[16] H. Edmondson and M. P. Heap, "Blowoff of Inverted Flames," *Combust. Flame*, vol. 194, pp. 191–194, 1971.

[17] N. Trevino ,C; donnerhack, S;Peters, "LDA Measurements in the premixed V flame stabilized in the wake of a flat palte boundry layer," *Combust. Flame*, vol. 85, pp. 505–510, 1991.

[18] C. J. Sung, C. K. Law, and A. Umemura, "On adiabatic stabilization of inverted flames," *Symp. Combust.*, vol. 24, no. 1, pp. 205–212, 1992.

[19] T. M. Muruganandam, B. H. Kim, M. R. Morrell, V. Nori, M. Patel, B. W. Romig, and J. M. Seitzman, "Optical equivalence ratio sensors for gas turbine combustors," *Proc. Combust. Inst.*, vol. 30, no. 1, pp. 1601–1608, 2005.

[20] Arun K Ampi and T M Muruganandam, "High speed diagnostics of Blow out precursors in Plate stabilized flame," in *ECM*, 2015.

Uncertainty Quantification of Autoignition Kinetics Using Sparse Polynomial Chaos

Jon Dewitt E. Dalisay*,1, Edwin N. Quiros1

1Department of Mechanical Engineering, University of the Philippines Diliman

Magsaysay Ave., Quezon City, Metro Manila, 1101, Philippines

Abstract

Quantification of prediction uncertainties is requisite for the development of truly predictive combustion models. Parameter uncertainties, such as those associated with reaction rate coefficients, give rise to combustion quantities of interest, such as ignition delay, that are uncertain. In this work, we assessed the applicability of polynomial chaos expansions based on least angle regression (LAR PCE) [1] for the uncertainty propagation (UP) and global sensitivity analysis (SA) of autoignition kinetics. The UP applicability assessment was done on atmospheric, stoichiometric constant volume 0-D autoignition of methane over initial temperatures of 1000-1800 K, and the global SA applicability assessment was done on the same pressure and equivalence ratio conditions but only at an initial temperature of 1000 K. The chemical kinetic mechanism GRI-Mech 3.0 [2] was used for all simulations and direct Monte Carlo results served as references for comparison. We demonstrated the effectiveness of LAR PCE due to its ability to construct accurate sparse PCEs. Further, we carried out UP on a comprehensive range of operating conditions. Given our findings, we anticipate that LAR PCE will be used in the future for the UP and global SA of very large combustion kinetic models involving complex fuels.

1 Introduction

In the assessment of model predictive ability, one cannot neglect the variability in model parameters. Uncertainty quantification (UQ) is a paradigm and set of techniques that addresses such concern. UQ is currently gaining traction in combustion research since academia and industry alike have already acknowledged the information gap brought about by classical validation procedures in model development [3].

In experimental chemical kinetics, measured rate coefficients for elementary reactions are usually characterized by uncertainty factors u that bound the possible values [4]:

$$u = k_{\max}/k_0 = k_0/k_{\min} \quad (1)$$

where k_0 is the nominal, $k_{\max}$ is the maximum, and $k_{\min}$ is the minimum rate coefficient. This definition is hand-in-hand with a symmetric lognormal rate coefficient distribution. These uncertain rate coefficients are taken as model input and their uncertainties are propagated to model predictions. After UP, one can do global SA in order to determine the contributions of input to output variances.

Tomlin [5] has elaborated the utility of UP and global SA in combustion while Wang and Sheen [6] have reviewed the current state of UQ in combustion research. The reader is referred to their works for a better appreciation of the growing subfield of combustion UQ.

1.1 Overview of this Work

This work aims to answer the question: *is full UQ tractable when one is dealing with high-dimensional chemical kinetic models?* We assessed and used a relatively recent UQ approach that was created with the primary goal of tackling high-dimensional models. We expect that demonstrating its utility in the full UQ of a combustion kinetic model will open avenues for previously intractable chemical kinetics problems. To the best of our knowledge, a fully-uncertain rate coefficient space has never been used before for mechanisms with reaction counts on the order of 100; all studies so far involved pre-reduction of the input space through screening methods [4].

1.2 Uncertainty Quantification

Sampling-based methods, widely known as Monte Carlo methods, are the most straightforward and robust way of doing UQ in any context. Model surrogation methods, on the other hand, are an attractive alternative. They involve the generation of computationally inexpensive "metamodels" as approximations. The method of polynomial chaos expansions (PCE) [7] makes use of orthogonal multivariate polynomial bases for surrogation and is therefore a spectral method. Because of its spectral nature, global SA, as discussed in 1.3, can easily be done without doing additional model realizations.

Suppose the model input—rate coefficients of chemical kinetic mechanisms in this work—is expressed as a vector $\mathbf{X} \in \mathbb{R}^M$ of random variables where M is the dimensionality of the input space. Moreover, the model output—ignition delays of 0-D constant volume combustion simulations in this work—is $Y = M(\mathbf{X})$ where M is any map—a functional involving the 0-D constant volume reacting system in this work—between the input and the output. The PCE of $M(\mathbf{X})$ is [8]:

$$Y = M(\mathbf{X}) \approx M^{\mathrm{PC}}(\mathbf{X}) = \sum_{\alpha \in \mathcal{A}} y_\alpha \Psi_\alpha(\mathbf{X}) \quad (2)$$

where $M^{\mathrm{PC}}(\mathbf{X})$ is the order-truncated PCE of model output, $\mathcal{A} \subset \mathbb{N}_0^M$ with $\mathcal{A}^{M,p} = \{\boldsymbol{\alpha} \in \mathbb{N}_0^M : |\boldsymbol{\alpha}| \le p\}$, $|\boldsymbol{\alpha}|$ is the l^1 norm of the multi-index $\boldsymbol{\alpha}$, p is the maximum polynomial order, and $\boldsymbol{\Psi}_\alpha$ is the multivariate polynomial basis built from univariate polynomial bases that are orthogonal with respect to marginal probability density functions (*pdfs*) of the input variables. The total number of PCE coefficients is $P = (M + p)!/(M!\,p!)$ [7].

The primary problem in PCE metamodeling is the determination of the mode strengths. There are intrusive methods which let uncertainties to be directly calculable

*Corresponding author. Mobile: +63 (917) 506-2876
E-mail address: jedalisay@up.edu.ph

although governing equations need to be reformulated, and non-intrusive ones, e.g. sampling-based ordinary least-squares (OLS) minimization [9], that treat the models as black boxes. Once a metamodel has been generated, one can easily estimate valuable statistical and sensitivity information about the output.

1.2 Global Sensitivity Analysis

Global SA can be used for the exploration of the whole input space because input and output variables are treated probabilistically. A popular global SA method is the use of Sobol' indices [10] for quantifying sensitivities. This is achieved through the Sobol' decomposition of output [11]:

$$Y = \mathcal{M}(\mathbf{X}) = \mathcal{M}_0 + \sum_{i=1}^{M} \mathcal{M}_i(X_i) + \sum_{i=1}^{M-1}\sum_{j=i+1}^{M} \mathcal{M}_{ij}(X_i, X_j) + \cdots + \mathcal{M}_{1,\dots,M}(X_1,\dots,X_M) \quad (3)$$

which has associated Sobol' indices defined as:

$$S_{i_1,\dots,i_s} = \mathrm{Var}(\mathcal{M}_{i_1,\dots,i_s})/\mathrm{Var}(Y) \quad (4)$$

where the subscripts belong to a general index set $1 \le i_1 < \cdots < i_s \le M$ and $s \in \mathbb{N}_1 : s \le M$. These indices satisfy the conservation property:

$$1 = \sum_{i=1}^{M} S_i + \sum_{i=1}^{M-1}\sum_{j=i+1}^{M} S_{ij} + \cdots + S_{1,\dots,M} \quad (5)$$

The first-order Sobol' indices S_i represent the fractions of the total variance of Y that are due to input X_i alone. Following this interpretation, the second-order Sobol' indices S_{ij} represent the fractions of the total variance of Y that are due to the combined effects of input X_i and X_j. The logic is same for higher-order Sobol' indices. PCEs, being function decompositions, can always be converted into a Sobol' decomposition. Thus, Sobol' indices can be determined analytically from PCE which makes PCE metamodeling very attractive.

1.3 Recent Applications of UQ to Chemical Kinetics

Recently, Reagan et al. [12] quantified the output uncertainties of supercritical water oxidation, specifically species time evolutions in 0-D isothermal plug flow reactor simulations and species spatial profiles in 1-D premixed flame simulations. A bit later, Le Maitre et al. [13] proposed the use of multi-wavelet PCE bases in the prediction of chemical kinetic quantities of interest in order to address the stability problems brought about by intrusive PCE. Their multi-wavelet approach involves the subdivision of the input hyperspace into "blocks" and the use of low-order PCEs representing uncertain model response over these blocks. They tested their approach on H_2-O_2 combustion and they found out that the multi-wavelet approach entailed a higher computational expense although the numerical instability problem was solved. They also proposed an adaptive block refinement strategy, instead of having a static block pre-decomposition, to reduce the computational cost. Najm [14] extended their work, using the multi-wavelet approach on the non-isothermal ignition of a CH_4-air system while also highlighting the importance of quantifying copulae between the marginal *pdf*s of input parameters. They carried out a computational experiment on uncertain species time evolution data and used an adaptive Markov chain Monte Carlo (MCMC) procedure for the Bayesian inference of the joint *pdf* of uncertain input parameters.

2 Methods

In this work, we propose to assess and use a relatively recent non-intrusive regression-based UQ method that has not been applied to the UQ of chemical kinetics yet, as far as we know. Blatman and Sudret [1] proposed the application of least angle regression (LAR) for the problem of determining PCE coefficients. LAR is an algorithm quite similar to OLS in spirit, but with an additional restriction that the number of predictors must be minimized. In the PCE context, "predictors" are the multivariate polynomials that comprise the polynomial chaos basis. Blatman and Sudret [1] used LAR to develop *sparse* PCEs, i.e. PCEs using only a portion of the full basis thus requiring significantly fewer model realizations.

2.1 Least Angle Regression for Polynomial Chaos Expansions (LAR PCE)

Given an experimental design of size N, the initial conditions of the algorithm are: $\mathbf{y}_\alpha = \mathbf{0}_\alpha$, the *candidate set* is Ψ_α such that $\alpha \in \mathcal{A}$, the *active set* is $\emptyset$, and the residuals vector is $\mathbf{R}_0 = \left[\mathcal{M}(X^{(1)}) \cdots \mathcal{M}(X^{N})\right]^T$. The LAR algorithm is as follows [8]:

1. Find the multivariate polynomial Ψ_{α_1} most correlated with the residuals vector $\mathbf{R}_0$.
2. The coefficient y_{α_1} is increased or decreased depending on the sign of the correlation while the new residuals vector $\mathbf{R}_1$ and its correlations to all multivariate polynomials in the PCE basis are continuously calculated. The "moving" stops when another multivariate polynomial is transferred into the active set.
3. The coefficients y_{α_1} and y_{α_2} are moved while maintaining their equicorrelation.
4. Steps 2 and 3 are repeated until the size of the active set is $\min(P, N-1)$.

After the iterative process is finished, one has a partial PCE basis of size $N-1$ if the experimental design is smaller than the basis size, i.e. $N-1 < P$. This makes it possible for LAR to produce a sparse PCE basis.

2.2 Kullback-Leibler (K-L) Divergence

Since predictions are uncertain, it is more appropriate to use a measure that takes the *pdf* profiles of the "truth" and the "model" into consideration instead of using nominal errors to assess fit. In the current context, the "truth" *pdf* is most appropriately from a direct Monte Carlo sampling of the chemical kinetic model and the "model" is the constructed PCE metamodel for a specific quantity of interest. In this work, we use the K-L divergence [15] for assessing fit:

$$D_{\mathrm{KL}}\left(\mathcal{M}|\mathcal{M}^{\mathrm{PC}}\right) = \int_{-\infty}^{\infty} \mathcal{M}(\mu)\log\left[\frac{\mathcal{M}(\mu)}{\mathcal{M}^{\mathrm{PC}}(\mu)}\right]\mathrm{d}\mu \quad (6)$$

2.3 Case Study: Constant-volume Autoignition of Methane

GRI-Mech 3.0 is used as the chemical kinetic mechanism in this work. Ignition delay predictions of constant volume methane autoignition simulations over a comprehensive range of initial conditions are the quantities of interest considered. Our hypothesis is that the iterative creation of a sparse basis is akin to an automatic global screening of input variables in terms of contribution to the uncertainty of model predictions.

A computational experiment matrix has been prescribed for constant volume autoignition with initial temperatures $T_i = 1000 - 1800$ K in 80 K increments, $P_i = 1, 10, 60, 100$ atm, and $\phi = 0.5, 1, 2$. The rate coefficients of the whole 325-reaction set ($M = 325$) were declared as the uncertain input parameters. All of the rate coefficients were assumed to be lognormally-distributed around their nominal values with an uncertainty factor of 2 corresponding to $\pm 3\sigma$. The multivariate PCE basis is built from a tensor product of univariate Hermite polynomials. Standard normal distributions were acquired from transforming the original lognormal distributions, thus justifying the use of Hermite polynomials. Since we used LAR PCE in this work, we were able to afford high-dimensional UQ with full exploration of the input space. The PCEs were truncated to have a maximum order of $p = 2$ with effect sparsity in mind, as elaborated in [6]. Given the input dimensionality and maximum PCE order, a total of 53,301 PCE coefficients are present in the PCE of the quantity of interest Y.

The UP applicability assessment was done on a subset of the computational experiment matrix involving atmospheric and stoichiometric autoignition. LAR PCE results were compared to reference direct Monte Carlo sampling results taken as truth. 75000 samples from Latin hypercube sampling (LHS) of the input space were used for the reference model predictions.

Following the UP applicability assessment is the global SA applicability assessment. Global SA results of LAR PCE were compared against bootstrapped global SA results using direct Monte Carlo sampling for a single quantity of interest where the operational parameters are T_i = 1000 K, P_i = 1 atm, and $\phi = 1$. The use of a single quantity of interest is due to the fact that using direct Monte Carlo sampling for global SA is very computationally expensive. We used N = 1000 LHS samples for every input variable, compliant with the recommendation of Turanyi and Tomlin [4]. Bootstrap resampling was included in order to provide 95% confidence intervals of the Monte Carlo global SA results. 1000 samples with replacement were used for the bootstrap resampling as suggested by Archer et al. [16].

Closing this work is the application of LAR PCE to the UP of methane autoignition subject to the set of conditions in the declared computational experiment matrix.

One of the primary limitations of this work is the use of constant volume conditions for the simulation of autoignition delay which is only valid for fast autoignition situations. Therefore, large deviations from experimental results at large time scales are expected. Nevertheless, the deviations will not diminish the results of applicability assessments.

2.4 Details of Implementation

All of the code was written in MATLAB R2015b. Existing libraries were used for most of the work: Cantera [17] for chemical kinetics calculations and UQLab [18] for UQ.

3 Results

Fig. 1 shows the comparison between direct LHS histograms and kernel density-fitted Y_{PCE} results for $T_i = 1000$ K. Visually, it is evident that the LAR PCE metamodels were able to reproduce the *pdfs* of the reference model predictions. However, there are some discernible aberrations near the peaks for almost all of the cases which reflected on the K-L divergence. The K-L divergence is minimum at $T_i = 1800$ K, with $D_{KL} = 9.85 \times 10^{-5}$, and maximum at $T_i = 1320$ K with $D_{KL} = 4.60 \times 10^{-4}$.

It should be noted that all of the LAR PCE metamodels for the given sets of operating conditions had a final $N = 1625$ experimental design conforming to a maximum error as dictated by an adaptive algorithm we implemented. Given such a small experimental design—1625 vs. 75000, a ~98% reduction of the required model realizations—it was certainly an important feat even though minute aberrations were present. Moreover, if we had used a full quadrature-based method we would have needed $(p+1)^M = 1.044 \times 10^{156}$ realizations! Sparse grids won't help much; all quadrature-based methods suffer from the curse of dimensionality [1].

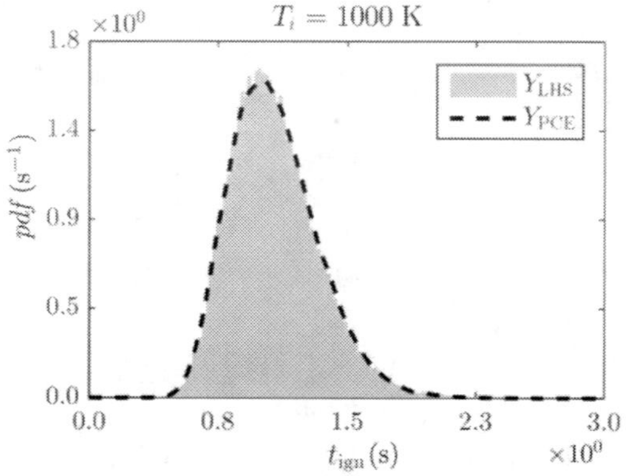

Figure 1: Respective histogram and *pdf* of reference model and LAR PCE metamodel predictions at $T_i = 1000$ K.

We now move forward to global SA applicability assessment. The baseline direct LHS which needed a total of 327000 samples ran unbearably longer compared to LAR PCE which virtually had no additional computational expense.

Fig. 2 shows a very good agreement between direct LHS and LAR PCE results for total sensitivities. In fact, all of the LAR PCE results for the 10 reactions the output is most sensitive to were within the 95% confidence bounds of the direct LHS. For the next 10 most sensitive reactions, there is still good agreement between direct LHS and LAR PCE results. While LAR PCE underpredicted the total Sobol' indices of some reactions belonging to this set, it was able to rank the reactions accurately up until the 16th reaction the output is most sensitive to. Note, however, that even the most sensitive reaction in the second set contributes possibly only up to around 1.8% of the output variance. LAR PCE failed to quantify sensitivity accurately. Not only that, it also failed in ranking reactions correctly for the 21st to 40th reactions the output is most sensitive to. These results show that using LAR PCE is feasible for global SA if the goal is to quantify variance contributions and rankings of the reactions the output is most definitely sensitive to—which is the goal of global SA most of the time.

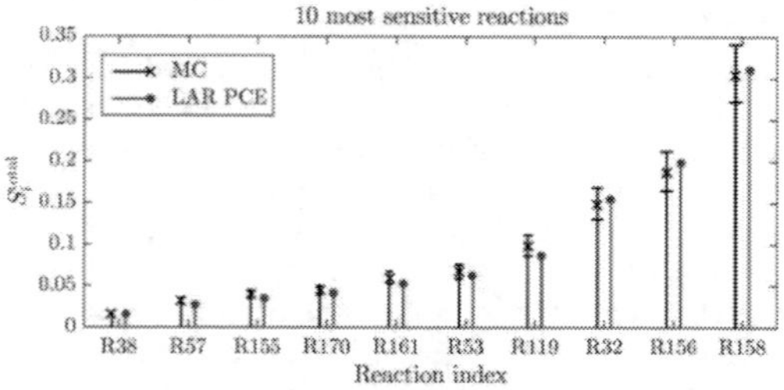

Figure 2: S_i^{total} of the 10 reactions the output is most sensitive to at $T_i = 1000$ K, $P_i = 1$ atm, and $\phi = 1$. Direct LHS results with 95% confidence intervals are displayed.

Fig. 3 shows the UP results for all the condition sets with $\phi = 1$. The plots for other ϕ's look similar because the trends are the same but are not shown here. The most uncertain condition is the lowest pressure $P_i = 1$ atm and temperature $T_i = 1000$ K. Conversely, the least uncertain condition is the highest pressure $P_i = 100$ atm and temperature $T_i = 1800$ K. The "error bars" drawn on Fig. 3 are 99.7% confidence intervals of uncertain output. UP through LAR PCE afforded us the ability to place these intervals on computational results in all of condition sets with minimal cost.

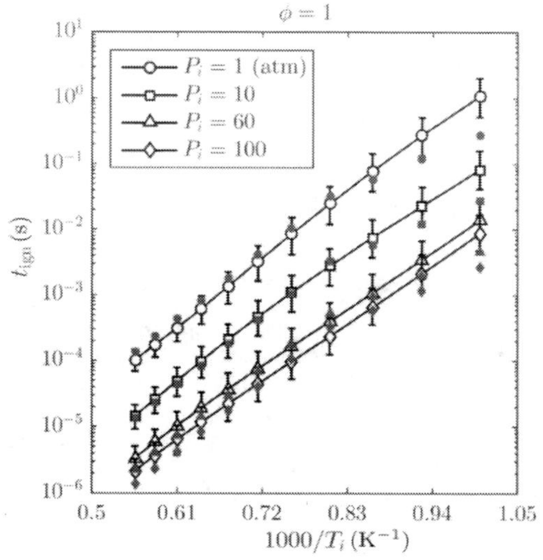

Figure 3: $\phi = 1$ UP results for methane autoignition with superimposed evaluations of an empirical correlation.

Evaluations of an empirical correlation by Li and Williams [19] (filled marks in red) are superimposed on the model predictions. One possible source of deviation is the non-ideal character of shock tube ignitions; they are neither constant volume nor constant pressure phenomena. The deviations at high pressures, and consequently low ignition delay magnitudes, are expected because ignition delay measurements are either based on peak pressure rise or OH radical emission which are both different from the $T_i + 400$ criterion used in this work. The differences between criteria are pronounced at low magnitudes of ignition delay. Further, the deviations at low temperatures are expected since GRI-Mech 3.0 was not optimized against experimental measurements at such conditions. Nevertheless, it should be noted that the evaluations of the empirical correlation by Li and Williams [19] are mostly in good agreement with model predictions in this work within 99.7% confidence intervals.

4 Conclusions

It was demonstrated in this work that LAR PCE is more than capable of carrying out UP and global SA on high-dimensional chemical kinetic models. Even though GRI-Mech 3.0 with 325 reactions is still "low-dimensional" compared to chemical kinetic mechanisms for complex fuels, its dimensionality is quite high in the UQ context. We showed that LAR PCE is, in fact, very effective compared to traditional direct Monte Carlo sampling, needing less than 2.2% as many model realizations as direct Monte Carlo while being able to identify the 16 most sensitive reactions in the case study we presented. Even though LAR PCE could not discriminate among the sensitivities of the rest of the reactions, it was argued that this is a non-issue because of the small output variances contributed by those reactions the response is insensitive to.

In addition to the demonstration of its utility, we used LAR PCE to do UP on a comprehensive operational condition space of autoignition with ignition delays as quantities of interest. Thus, we were able to generate information that was unprecedented in the literature because of the previous stigma on doing UQ with high-dimensional input spaces while considering the full input space as uncertain. UQ was generally thought of as very computationally expensive by the combustion community, which we expect to change upon the publication of this work.

5 Acknowledgments

This work was supported by the Engineering Research and Development for Technology (ERDT) program of the Philippines' Department of Science and Technology (DOST).

References

[1] G. Blatman, B. Sudret, J. Comput. Phys. 230 (6) (2011) 2345-2367.

[2] G.P. Smith, D.M. Golden, M. Frenklach, N.W. Moriarty, B. Eiteneer, M. Goldenberg, C.T. Bowman, R.K. Hanson, S. Song, W.C. Gardiner, Jr., V.V. Lissianski, Z. Qin, available at http://www.me.berkeley.edu/gri_mech/.

[3] C.J. Roy, W.L. Oberkampf, Comput. Methods Appl. Mech. Eng. 200 (25-28) (2011) 2131-2144.

[4] T. Turányi, A.S. Tomlin, Analysis of Kinetic Reaction Mechanisms, Springer-Verlag, Berlin, Germany, 2015, pp. 102, 87.

[5] A.S. Tomlin, Proc. Combust. Inst. 34 (2013) 159-176.

[6] H. Wang, D.A. Sheen, Prog. Energy Combust. Sci., 47 (2015) 1-31.

[7] D. Xiu, G.E. Karniadakis, SIAM J. Sci. Comput., 24 (2) (2002) 619-644.

[8] S. Marelli, B. Sudret, UQLab user manual – Polynomial Chaos Expansions, Report No. UQLab-V0.9-104, ETH Zurich, 2015.

[9] S.S. Isukapalli, *Uncertainty Analysis of Transport-Transformation Models*, PhD thesis, Rutgers, The State University of New Jersey, New Brunswick, New Jersey, 1999.

[10] I.M. Sobol', Math. Model. Comput. Exp. 1 (4) (1993) 407-414.

[11] S. Marelli, C. Lamas, B. Sudret, UQLab user manual – Sensitivity Analysis, Report No. UQLab-V0.9-106, ETH Zurich, 2015.

[12] M.T. Reagan, H.N. Najm, R.G. Ghanem, O.M. Knio, Combust. Flame 132 (3) (2003) 545-555.

[13] O.P. Le Maitre, H.N. Najm, P.P. Pébay, R.G. Ghanem, O.M. Knio, SIAM J. Sci. Comput. 29 (2) (2007) 864-889.

[14] H.N. Najm, B.J. Debusschere, Y.M. Marzouk, S. Widmer, O.P. Le Maitre, Int. J. Numer. Methods Eng. 80 (2009) 789-814.

[15] S. Kullback, R.A. Leibler, Ann. Math. Stat. 22 (1) (1951) 79-86.

[16] G.E.B. Archer, A. Saltelli, I.M. Sobol', J. Stat. Comput. Simul. 58 (2) (1997) 99-120.

[17] D.G. Goodwin, H.K. Moffat, R.L. Speth, available at http://www.cantera.org/.

[18] S. Marelli, B. Sudret, Vulnerability, Uncertainty, and Risk, American Society of Civil Engineers, 2014, pp. 2554-2563.

[19] S.C. Li and F.A. Williams, J. Eng. Gas Turbines Power 124 (3) (2002) 471.

11th Asia-Pacific Conference on Combustion.
The University of Sydney, NSW Australia
10th -14th December 2017.

An Experimental Study on Heavy Fuel Oil Droplet Combustion

Abdulrahman Alkhateeb[1], Paolo Guida[2], Eid Barakati[3], Alaaeldin Dawood[1], Ayman M. Elbaz[1],

William L. Roberts[1]

[1]Clean Combustion Research Centre, King Abdullah University of Science and Technology, 23955-6900 Thuwal, Saudi Arabia.

[2]DCMIC "Giulio Natta", Politecnico di Milano, Piazza Leonardo da Vinci, 32, 20133 Milano, Italy.

[3]Research and Development Centre, Saudi Electricity Company, 23955-6900 Thuwal, Saudi Arabia.

Abstract

This experimental work was carried out to understand the fundamental combustion characteristics of a single Heavy Fuel Oil (HFO) droplet in a thermal environment. In particular, it was explored how different variables such as droplet size, environmental temperature, and air flow rate affect parameters of engineering interest like cenosphere structure, ignition delay time (IDT) and burning time (BT). Both high speed images of the combustion phenomenon and temperature record for liquid and gas phase have been obtained for different droplet sizes. The exposition of the HFO droplet to hot oxidizing environment can lead to two different phenomena, the formation of a globules carbonaceous structure in absence of combustion, or the ignition of the vapor cloud that surround the droplet with consequent growth of a solid shell with a porous structure. Both structures found at the end of the process are called "cenosphere" literally "hollow sphere". Moreover, a physical explanation to the obtained behavior was elaborated based on previous works and observations. Unpredictable behaviors of IDT and BT with respect to the HFO droplet size were recorded. However, the trends seem to be approximately directly proportional to the droplet size.

1 Introduction

The interest of understanding the combustion characteristics of low grade fuel that can be utilized as a source of energy have been increased widely especially in oil production countries. In Saudi Arabia, for example, a significant portion of the energy demand for electricity generation are met by the combustion of Heavy Fuel Oil (HFO). The reason behind burning HFO rather than conventional light oil is due to its lower production cost and higher availability. However, the use of HFO has been declined after reaching the peak in late 1970s due to its adverse environmental effects and equipment technical failure [1].

HFO has poor quality as the fuel is extracted from distillation process of crude oil refining residue. Therefore, it is a complex heterogeneous mixture of saturated, unsaturated and aromatic hydrocarbon molecules with high viscosity, high molecular wieght and high percentage of asphaltene and sulfur contents [2]. Accordingly, HFO combustion results in producing particulate emissions such as smoke, cenosphere and ash residues along with hydrogen sulfide and nitrogen oxide gases.

Two typical types of cenosphere morphologies can be found in earlier research. The traditional morphology, which is widely observed [3], has a carbonaceous and porous shell. A common explanation [4] for how this kind of structure is formed is due to volatile components of HFO at the droplet surface evaporate and form a vapor layer surrounding the droplet. As a result, the more viscous components remaining at the surface begin to form a shell, which prevents the emission of vapor formed by the volatile inner components. With a diffusion flame formed outside the vapor layer, droplet temperature increases, accelerating vaporization and vapor accumulation inside the shell. Eventually, the inside vapor ejects out through the weakest point on the shell, forming the porous structure, Fig 1 left.

The other type of morphology is mentioned as globules, which can reach a diameter, much larger than that of traditional cenosphere [5]. The globules structure is hollow and contain many bubbles. Seen from outside, the cenosphere is a skeletal structure, attached by a membrane among those frames, Fig 1 right. The structure is mostly composed of Silica and Alumina as these are the remaining elements after the combustion of the majority of the carbon atoms.

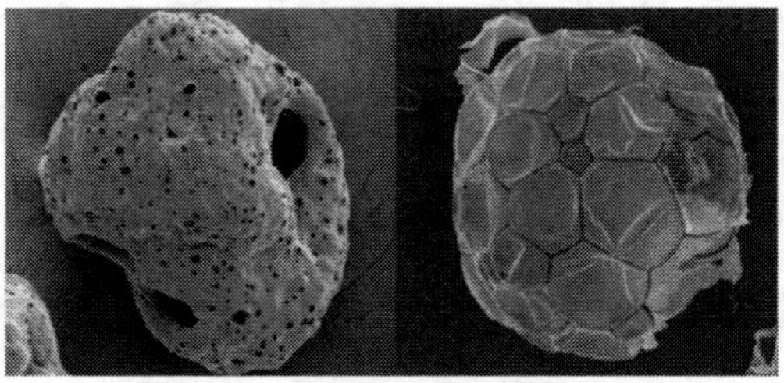

Figure 1: Cenosphere (Left: porous shell structure, Right: skeletal membrane structure).

2 Experimental Setup

HFO sample was collected from Shuaibah power plant in Saudi Arabia. The HFO in Saudi Arabia has an asphaltene content range between 7% to 15% and sulfur content of 3 wt% to 4 wt% and this is a high content range of asphaltene and sulfur comparing to that of in the conventional light oil products [2].

Corresponding author. Fax: +966-53-5253-500
E-mail address: abdulrahman.khateeb@kaust.edu.sa

Physical and chemical properties of the HFO used in this work are listed below in Table 1. More details about the fuel characteristics can be found in the work of Elbaz et al [2] as samples from same HFO source were tested by analytical techniques such as TG/DTG, FT-ICR and NMR spectroscopy.

Table 1: HFO Properties.

Properties	Results
Density at 25 °C (kg/m^3)	970.5
Kinematic viscosity at 40°C	617.740
Sulfur (wt%)	3.292
Asphaltenes (%)	8.2
Vanadium (mg/kg)	18.0
Carbon (wt%)	85.0
Hydrogen (wt%)	10.89
Oxygen (wt%)	0.030
Nitrogen (wt%)	0.239

Figure 2 is a schematic of the experimental setup showing various components. Preheated air co-flow by an inline heater (Sylvania heater, F074719, 8000 Watt) goes upwards through a ceramic honeycomb from the bottom of a hinged tube furnace (Lindberg/blue, HTF55000 series) through quartz tube (134 mm inner diameter and 750 mm length). The co-flow air is assumed to be uniform after that and can be controlled with flow rate up to 130 l/min. Once the ambient air temperature profile inside the electrical furnace reaches required values, an HFO droplet is suspended on a thermocouple junction (Type S, 75 μm diameter, Omega P10R-003).

The thermocouple is inserted inside a two-hole ceramic tube and connected to a computer for data processing and temperature readings. The droplet is positioned into the preheated electrical furnace by a 3D mechanism through an air cooling channel. This channel will keep the droplet temperature low until reaching the burning point inside the furnace. High speed camera (LaVision, Imager HS) is used to record the whole HFO droplet combustion process and a data acquisition card to read the temperature history inside the droplet from the thermocouple simultaneously.

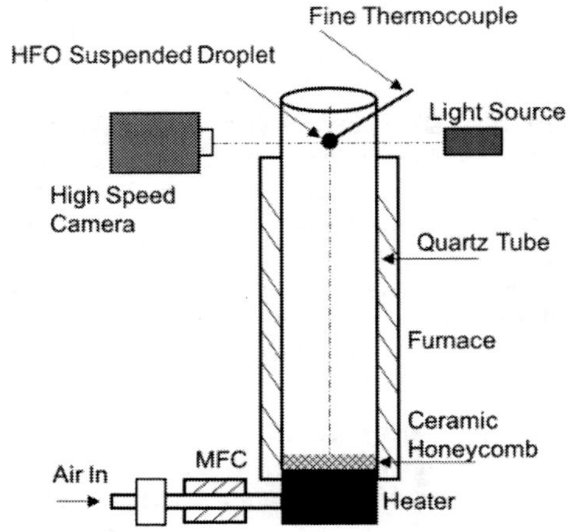

Figure 2: Experimental setup

3 Results and Discussion

Figure 3 shows temperature history of HFO droplets exposed to four different ambient air temperatures starting with 600°C and reaching up to 680°C. This temperature range was carefully chosen after trial and error attempts until meeting a critical ignition condition point which causes the HFO droplet combustion.

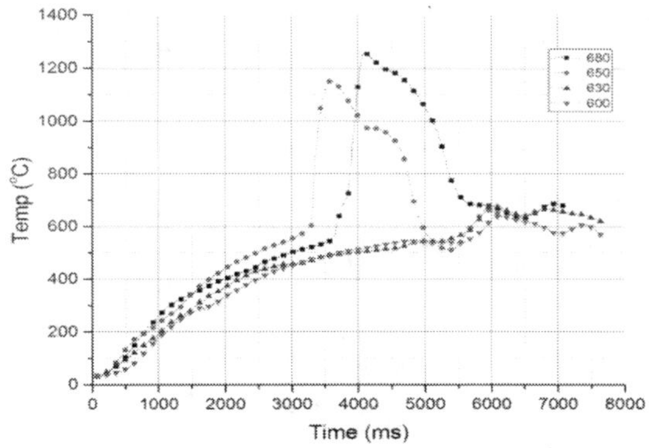

Figure 3: Temperature history profile of HFO droplet with different environmental temperature.

At high ambient air temperatures (i.e. 680°C and 650°C), the temperature profile behaves differently comparing to that of at lower temperatures (i.e. 630°C and 600°C). It was observed that the sudden increase in the temperature profile is a sign of the HFO droplet ignition and ejection of viscous residue which splashes through the flame front due to the formation of vapor bubbles. This can be explained as the droplet started releasing volatile materials once it was located inside the furnace. After a certain time depending on the droplet size and the furnace environmental temperature, the ambient air and vaporized components mixture reached the critical ignition condition which causes the formation of a stable flame as the one in Fig 4. Cenosphere formation with porous shell structure was produced at the end of the droplet combustion at these temperatures.

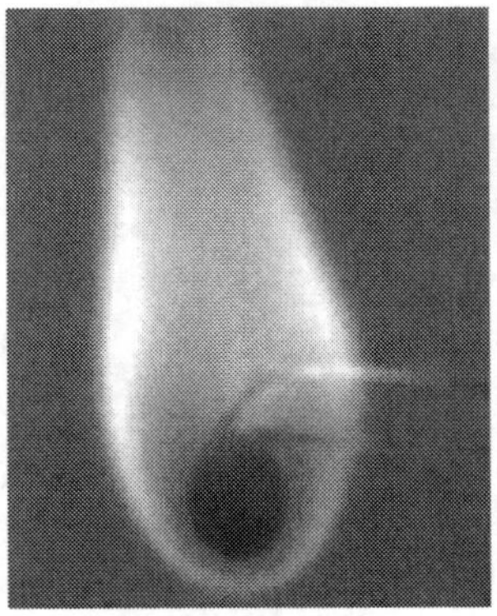

Figure 4: Stable flame of HFO droplet with environmental temperature of 680°C.

On the other hand, the HFO droplet does not ignite at ambient air temperature below 650°C. This can be explained as the ambient air temperature and the vaporized components from the HFO droplet did not reach the critical ignition condition which causes the droplet gases ignite. However, the droplet size doubled due to the emission of the vapor formed by the volatile inner components which could not escape due to the high viscosity of the surface layer leading to the globules cenosphere formation with a skeleton-membrane structure.

Figure 5 shows the thermal behavior of the liquid phase, during the first three seconds the temperature of the droplet increases. It is possible to observe the evaporation of lighter components that form a vapor cloud around the droplet. During this stage the diameter of the droplet remains constant, no splashing nor micro explosion occur at this step, the weight loss due to the evaporation process is balanced from the expansion of the liquid phase at high temperature. The second step accordingly with the ambient air parameters and the initial size of the droplet starts with the ignition of the surrounding gases or with direct cenosphere formation. The ignition coincides with a peak in the first derivative of the temperature with respect to the time. From the high-speed images and the gas phase thermocouple reading the ignition instant is quite clear while from the liquid phase thermocouple reading the ignition is delayed because of the time required to increase the droplet temperature. The high temperature reached inside the droplet can generate either the evaporation of the fuel inside the droplet or the activation of several reactive pathways such as polymerization and pyrolysis.

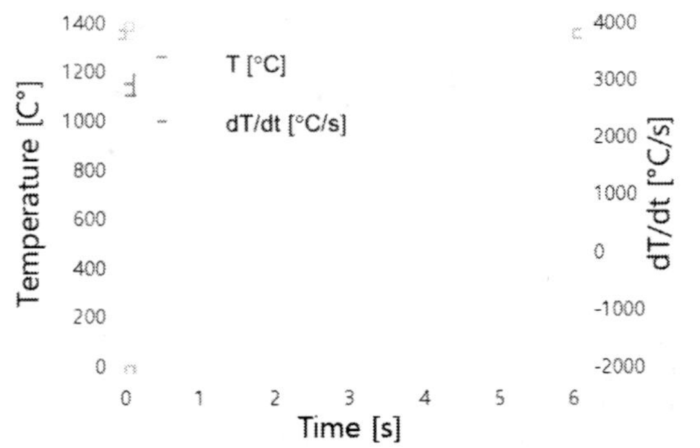

Figure 5: (a) Temperature profile of the HFO droplet "T" [°C] (b) Rate of change of temperature "dT/dt" [°C/s].

The HFO droplet expansion and contractions alternates during the combustion process. The evaporation of the volatile species inside the droplet, in particular is responsible for the huge growth of the droplet size as shown in Fig 6.

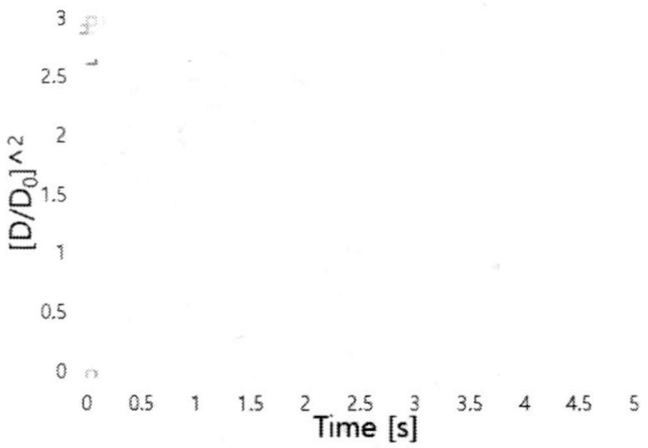

Figure 6: Diameter behavior in time.

The diameter, after the ignition, grows with an exponential behavior up to 3 times the initial diameter, Ikegami et al [5] related this phenomenon to the Lewis number "(1)". Lewis number is defined as the ratio between thermal conductivity and mass diffusivity.

$$Le = \frac{\alpha}{\Gamma} \quad (1)$$

Where α is the thermal conductivity and Γ is the mass diffusivity. For this kind of fuel the heat transfer is much higher than the mass transfer, so $Le >> 1$ and the volatile species are trapped inside the droplet at high temperature, hence these species evaporate and react while the heavier ones such as asphaltenes form a solid shell. Eventually, cenospheres with porous shell structure are always smaller than their initial droplet size while those with skeleton-membrane shell structure are much larger than their initial droplet size.

The Ignition Delay Time (IDT) is here calculated as the time from the instant a droplet is introduced into a hot thermal environment to the instant a flame is detected in the vicinity of the droplet as suggest by Aggarwal et al [6]. Moreover, the burning time is the time occurs from the ignition to the extinction of the flame. These values have been computed from the positive and negative peaks in the derivative of the gas-phase temperature profiles.

Figure 7 illustrates the relation between the HFO droplet size and the ignition delay time in seconds.

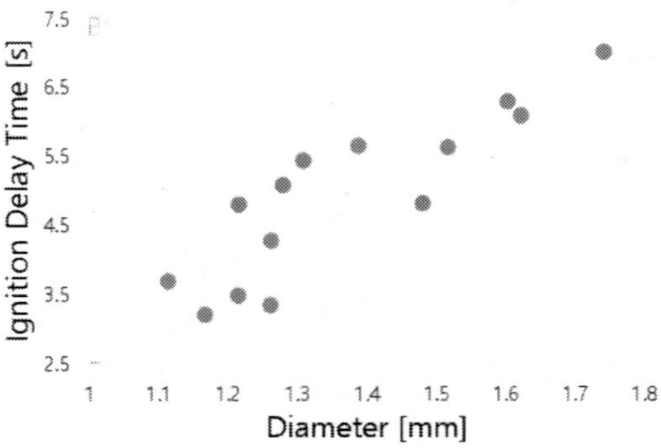

Figure 7: IDT for different initial diameters

The population of sizes tested is limited. However, the trend obtained from the IDT vs diameters graph shows approximately direct proportional relationship. The lower limit is due to the fact that under a certain volume, droplets do not ignite. This behavior can be associated to the effect of the thermocouple that becomes too relevant for smaller diameters. While the upper limit is due to the fact that the weight force is higher than the force generated by the surface tension and therefore, the droplet may fall from the thermocouple before the ignition.

The tendency to reach the high IDT can be justified from fluid dynamic considerations, ignition takes place in the region immediately above the droplet, and this region is a stagnant region. In fact, the velocity magnitude is lower in that point, so the fuel vapors are delivered slower through the bulk region. The bigger is the stagnant region the higher is the probability to have a flammable mixture. The formation of this mixture is directly proportional to the droplet size. Accordingly, more time is needed for ignition and this can be the reason behind the greater ignition delay for bigger droplet over 1.5 mm. From the other side a smaller droplet has a lower inner temperature gradient that reduces secondary reactions promoting pure evaporation.

Burning time present unpredictable behavior as well, however its trend tends to be directly proportional to the initial droplet diameter if average values are taken in consideration as shown in Fig. 8.

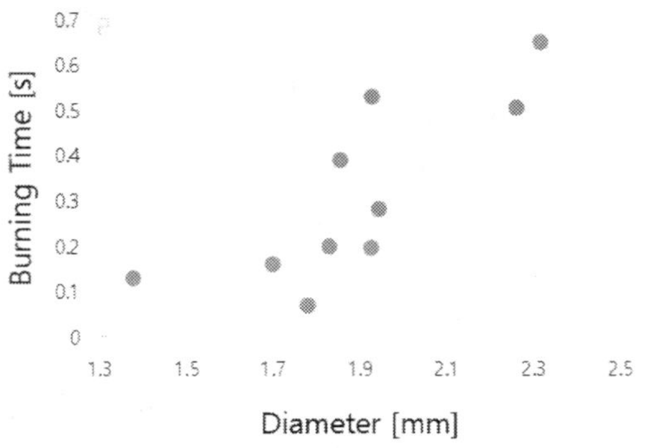

Figure 8: Burning time for different initial HFO droplet diameters.

The average burning time for HFO droplet with a diameter between 1.6mm and 2mm is 300ms. This is the time from the droplet is ignited until the flame is vanished.

4 Conclusions

In this work, the influence of various parameters, in particular droplet size and environmental temperature were investigated to better understand their effect on HFO droplet combustion focusing on cenosphere morphology, ignition delay time (IDT) and burning time (BT).

The results show that for same air co-flow rate and initial droplet size, cenosphere morphology changes from porous shell to skeleton-membrane structure when ambient air temperature becomes lower. From these comparisons, it is shown that relatively high initial droplet temperature is required to be fulfilled to meet the critical ignition condition which helps to generate ignition and as a result a porous shell cenosphere, otherwise it will be possible to find cenospheres with skeleton-membrane structure with no ignition.

It was also observed that the ignition delay time and burning time have unpredictable behavior versus the HFO droplet size. However, the trends seem to be directly proportional when average values are considered. For HFO droplet size between 1.6mm and 2mm, the IDT and BT have values of almost 6s and 300ms respectively.

It should be mentioned that the phenomenon analyzed in this study is particularly complex and the aim of this study is to maximize the knowledge of Heavy Fuel Oil droplet combustion. The complexity is due to the complex chemical reactions both in gas and liquid phase, the complexity of the HFO mixture, and the cenosphere formation.

In the future work, the information collected from this study will be used to develop mathematical models for CFD analysis and to optimize the existing burning processes.

5 Acknowledgment

Research reported in this publication was supported by the Saudi Electricity Company (SEC) in collaboration with the Clean Combustion Research Center (CCRC) at King Abdullah University of Science and Technology (KAUST).

References

[1] N. J. Marrone, I. M. Kennedy, F. L. Dryer, Coke formation in the combustion of isolated heavy oil droplets. Combustion science and technology 36 (3-4) (1984) 149-170.

[2] A. Elbaz, A. Gani, N. Hourani, A. Emwas, S. Sarathy and W. Roberts, TG/DTG, FT-ICR mass spectrometry, and NMR spectroscopy study of heavy fuel oil. Energy & Fuels 29 (12) (2015) 7825-7835.

[3] D. L. Urban, F. L. Dryer, New results on coke formation in the combustion of heavy-fuel droplets. Symposium (International) on Combustion 23 (1) (1991) 1437-1443.

[4] D. L. Urban, S. Huey, F. Dryer, Evaluation of the coke formation potential of residual fuel oils. Symposium (International) on Combustion 24 (1) (1992) 1357-1364.

[5] M. Ikegami, G. Xu, K. Ikeda, S. Honma, H. Nagaishi, D. L. Dietrich, Y. Takeshita, Distinctive combustion stages of single heavy oil droplet under microgravity. Fuel 82 (3) (2003) 293-304.

[6] Aggarwal, K. Suresh, Single droplet ignition: theoretical analyses and experimental findings. Progress in Energy and Combustion Science 45 (2014) 79-107.

11th Asia-Pacific Conference on Combustion,
The University of Sydney, NSW Australia
10th -14th December 2017.

A discussion on the simplified volatile flame model for a solid particle

Jiwoong Kim, Xiaoyang Niu, Sangmin Choi*

Mechanical Engineering Department, KAIST

Daejeon, 34141 Korea

Abstract

Recent progress in experimental observation of volatile flames of coal particles renewed the interest on the modeling description of the single solid fuel particle combustion. A simplified modeling concept of volatile flame is based on the hypothesis of the thin flame sheet which surrounds the particle, thereby forming a volatile cloud. Pulverized fuel particles in a hot gas environment exhibit this complex interaction of the combined heat transfer, mass diffusion, and reactions of devolatilization and oxidation. This paper reviewed the main concepts of the simplified volatile flame model and showed the cases of quantitative evaluation of the parameters used in the physical modeling descriptions. Particle heat-up was modeled from the energy balance, volatiles released from the particle and oxygen diffused from the environment would react stoichiometrically at the flame, which would govern the mass and energy conservation. Although the formulation is complexly interrelated, and highly sensitive in a certain parameter range, mathematical modeling enabled to identify effects of the major parameters. Some of the observable quantities from the experiments were comparatively discussed with the parameters which could be quantified from the modeling.

1 Volatile flame of solid fuel

Single particle combustion studies have been the building block for understanding the solid fuel behavior. Solid fuel in industrial application is typically ground into fine particles, which warrants the idealization of single particle combustion environment, where a single cold particle is introduced into a hot environment and undergoes a sequence of particle heat-up, devolatilization and subsequent combustion of volatiles, and then char combustion. Volatile flame of a solid fuel particle represents the state during the combustion of volatiles, in such a manner that volatiles released form the particle stay adjacent to the particle.

Formation of volatile flame is a result of mass release through devolatilization. The released volatile is spreading out toward the flame zone, where oxygen is available. This concept of diffusion limited flame model has been applied to the single coal particles [1,2,3], suggesting a visual image of progressing volatile flame, for the simplified geometry of particle and flame. Volatile flame is continuously progressing after formation, through growth and decay.

Experimental observation has provided reasonable appearance, demonstrating the presence of the volatile flame [4,5]. Appearance of the volatile flame of solid fuel was characteristically distinctive, with clearly observable luminous flame, as compared to the flame as a result of evaporation of liquid droplets. However, visualized shape of the volatile flame was not close enough to discuss the concentrically spherical geometry of the volatile flame model. Recent progress on the experimental observation enabled to reproduce almost spherically symmetrical shape of the flame [6,7]. Progress of the flame appearance appears in Fig. 1 and is graphically summarized in Fig. 2. This renewed the interest on the modeling description of the solid fuel combustion, based on the simple geometry.

This paper reviewed the conceptual development of the volatile flame model of a single solid particle. Modeling simplifications and hypothetical assumptions to reach a simplified version of the model were discussed. Despite the simplicity of the modeling concept, and also some of the parameter values are highly sensitive, some of the important combustion parameters were identified, and their values were meaningful, as compared to the experimental results.

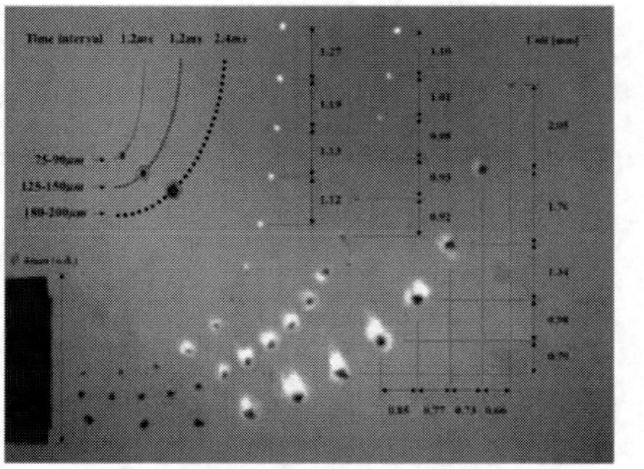

Figure 1: High speed and high magnification visualization of the single coal particle injected into a hot environment. Pulverized coal particles were introduced into a hot gas of 1,200K, 30% O_2.[6]

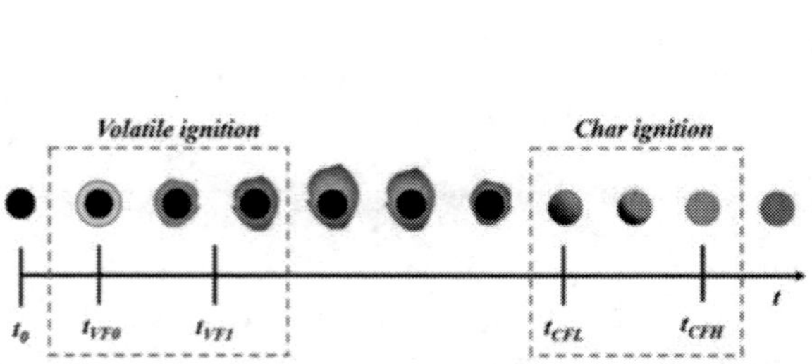

Figure 2: Schematic representation of the time-wise progress of the particle behavior during the particle heat-up and volatile formation phase.

***Corresponding author: smchoi@kaist.ac.kr**

2 Modeling Description

2.1 Overall

Modeling of the volatile flame considers the life-cycle particle behavior from the heat-up, followed by the physico-chemical processes of mass release and combustion, which are mutually interacting. Purpose of the modeling is to construct a network of interaction.

Modeling description is associated with inevitable simplifying assumptions and hypotheses. For the current discussion of the volatile flame, a one dimensional description in radial coordinate is selected, based on the assumption of spherical symmetry for the phenomena. The properties of the particle is homogeneous and uniform inside the particle, i.e., no temperature gradient inside the particle. Components of the volatile are simple mixture of gaseous volatiles and solid (soot-like) volatiles, for both of which would react instantaneously and completely at the flame.

2.2 Particle Heat-up

For a cold particle introduced into a hot gas environment, dynamic progress of the particle temperature can be described as,

$$\frac{d}{dt}(mcT_p) = \dot{Q}_{\substack{net\ to\ particle \\ (vol+flame+gas+wall)}} - \dot{m}_{vol}\Delta H_{vol} \qquad (1)$$

where the total sum of heat transferred from the environment would contribute to increase the particle temperature. The surroundings may include the hot gas, the flame, and the wall, while the presence of the flame would influence the arrangement of heat transfer. From the particle point of view, reaction heat of evaporation and devolatilization may be considered, when necessary.

Without the formation of the flame, the particle would expect a simple configuration of a single particle in an infinite medium of gas. Conductive heat transfer is considered dominant, after checking the contribution of radiation in the given conditions, which depends on the radiative properties including the emissivity and temperatures of the particle and the environment.

When a volatile flame is formed, the heat transfer to the particle is a little bit more complicated, in that total amount of heat transferred to the particle is the sum of heat from the flame, the gas, and the participating medium of soot inside the volatile flame. This configuration of heat transfer requires the treatment of mixed mode of heat transfer of conduction and radiation [2,3]. Also, quantification of heat transfer would require to define the flame temperature, which will be discussed again in the following section.

Particle heat-up time is defined as a measure of quantitative parameter which represents the transient process. As a simple relation for an ideally simplified condition, the heat-up time is given as the ratio of mass and heat capacity divided by the thermal conductivity of the gas and the characteristic length scale.

2.3 Devolatilization

Devolatilization is a highly temperature dependent process; nothing happens until reaching a certain temperature range, once started they progress very fast, and the process is self-exhausting, i.e., the process reaches its completion asymptotically. The process is, however, difficult to generalize, and a variety of modeling relations have been proposed. Here in this work, the basic format with a single kinetic rate coefficient is adopted.

$$\dot{m}_{vol} = -A\exp\left(-\frac{E}{RT_p}\right) m_{vol,\,remaining} \qquad (2)$$

Here, we note that the temperature dependency of the rate equation is given in the activation energy, and the pre-exponential factors. In a typical observation, the volatilization is complete in a few milliseconds of time. We also note here that the current model does not consider individual components of the volatile products.

2.4 Volatile Flame

Formation and progress of the volatile flame is the combined result of the devolatilization and the combustion of volatiles. Volatile material from the particle is emitted and is transported outward until reaching the flame. The volatile flame is hypothesized as of an infinitely thin flame sheet. Mass release rate is simply represented with a single rate coefficient, and the material composition is simplified to be of gaseous volatiles and soot volatiles, with a simple mass ratio of 30% soot and 70% volatiles[3]. Further details of soot number density, size, and emissivity are required for quantification of radiation. Volatiles stay inside the flame, and oxidation is complete at the thin flame sheet. Chemical reactions of both the gaseous volatiles and soot volatiles are fast enough to warrant the oxygen diffusion limited combustion.

Considering the mass and energy conservations, one can derive relations for the flame size and for the flame temperature. Figure 3 shows the single particle environment, (a) for the case of before, and (b) after the formation of volatile flame.

Location of the flame sheet is determined from the mass balance at the flame. Fuel mass which is transported by diffusion of the volatiles released from the particle, and the oxygen mass which is transported by the diffusion of oxygen from the gas environment would match stoichiometrically at the flame sheet. Equation (3) represents one of those relations [3], as written in a dimensionless radius form of flame to particle,

$$\frac{r_f}{r_p} = \frac{\dot{m}_{vol}}{4\pi r_p \rho_g D_{O_2} \ln\left(1+\frac{Y_{O_2}}{\eta}\right)} \qquad (3)$$

Radius of the flame is proportional to the devolatilization rate, and is inversely related with the oxygen concentration, as expected from Y_{O2} .

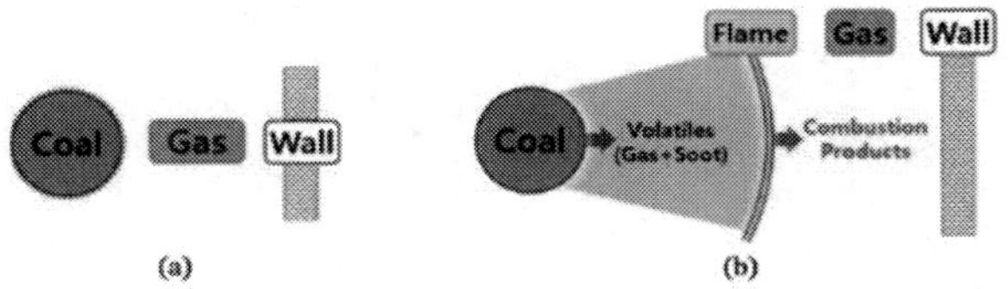

Figure 3: Schematic representation of the particle, flame, and a hot gas environment. (a) before and (b) after the formation of the flame

Flame temperature is defined from the energy balance at the flame. Energy generated by the combustion at the flame must be equal to the net balance of heat transfer across the flame. Net amount of conduction and radiation transfer from the flame to the environment-side, and net heat transfer to the particle-side could be quantified along with the flame temperature. In fact, the term of flame temperature is fictitious, which is used only to quantify the heat transfer rate from the flame to the respective counterpart. An energy balance relation across the flame sheet may be given in the form as in Eq. (4)

$$\dot{m}_{vol}\,\Delta H_{\substack{vol\\comb}} = \dot{Q}_{flame\ to\ env} + \dot{Q}_{flame\ to\ particle} \quad \text{- - - -} \ (4)$$

Here, both the heat transfer from the flame to the particle and that to the surrounding environment are basically related with the flame temperature. Outward heat transfer from the flame would be in the mode of conduction from the flame to the gas and radiation from the flame to the surrounding gas and wall. Inward heat transfer is determined through the volumetrically participating medium of soot and volatile in the concentric mantle of the volatile volume inside the flame. This volume must be what can be observed during the volatile combustion phase. The particle, flame and the volume of volatile would interact in the mode of radiative transfer in the gas environment and in the wall environment. It must be noted that a relatively simplified description of volatile flame would involve mixed mode of heat transfer in complex configuration.

3 Example Case for Quantitative Discussion

3.1 Example Case

In the previous chapter, volatile flame process was described along with modeling concepts and hypotheses. To appreciate the significance of the major parameters, following example case was chosen for the quantitative discussion. The typical cases of volatile flames considered were for the case of pulverized coal (typical diameter of ~80 micron at 1,240K with 21% oxygen). Typical heat-up rates ~300K in ~10 msec.

3.2 Modes of Heat Transfer

When a particle is located in a hot gas environment, the particle receives heat from the gas until equilibrium. All modes of heat transfer from the adjacent environment would participate in heat transfer. When the particle is in motion relative to the gas, one might expect convective heat transfer. However, when the particle size is small, the conduction mode dominates. In the current case of pulverized coal scale, major mode of heat transfer would be conduction and radiation.

The heat transfer environment for a volatile flame is directly affected by the presence of the flame as summarized in. Fig. 3. For the case of volatile flame, one must be aware of the presence of the flame and the radiative participation of the volatile medium inside the flame. Based on the hypothetical definition of the flame as in the previous sections, quantitative evaluation of relative significance has been attempted. For the case of T_p=700K, T_{gas}=1,240K and T_{flame}=1,520K, contribution of inward to outward direction heat flux to the total was checked to be roughly 60 to 40%. Also for each direction, contribution of conduction to the radiative heat flux was checked to be roughly 60 to 40%. Detailed elaboration is considered to be out of scope of the current report.

3.3 Particle Temperature History

Particle temperature during the initial heat-up followed by the period of volatile flame is expressed in Figure 4. Upon introduction into a hot environment, the particle temperature increases steadily, almost at a constant rate, until the appearance of the volatile flame. This state is roughly corresponding to the point of ignition, following the definition of ignition as the initiation point of visible volatile flame. Corresponding particle temperature in this case would be 640K. From Eq.(1) with the corresponding values of physical parameters, time constant τ_1 is evaluated to be ~10 msec., which is considered to be plausible.

When a particle is surrounded by a volatile flame, flame temperature, which is expected to be higher than the gas temperature, would accelerate the heat–up. Heat transfer from the flame to the particle and the environment would match the heat generated by the combustion at the flame. Particle temperature keeps rising at a relatively faster rate than in the case of no-flame. Shown in Fig. 4 is one case selected for discussion. The particle temperature keeps rising when it is surrounded by a flame, whose temperature is higher than the gas temperature. As the flame is re-shrinking, following the pace of the mass release, the flame approaches the surface of the particle, which would push the particle temperature close to that of flame. Quantitative determination of the particle temperature during the volatile release is associated with considerable uncertainties.

Effect of flame is evident, and volumetric luminous radiation from the soot is characteristic. Complex interdependence of conduction and radiation must be evaluated. However, for the fine particle geometry contribution of conduction is still strong.

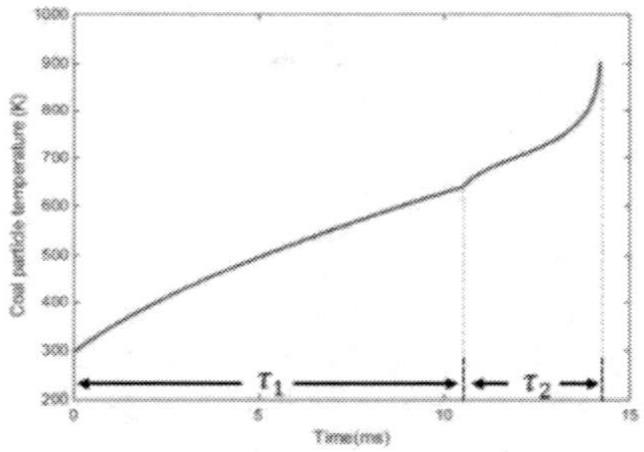

Figure 4: Particle temperature history, Period of particle heat-up before the ignition is τ_1, and the period of volatile flame is τ_2.

3.4 Progress of Volatile Flame

Release rate of volatiles is very sensitive on the particle temperature. Once initiated, the mass release is accelerated to reach its maximum, and then decreases. Temperature dependency is mainly the result of the high activation energy in Eq.(2). In this case, at 83 MJ/kmol, and pre-exponential factor of $10^9 s^{-1}$, time constant as defined as the inverse of kinetic rate value was quantified to be 3 msec, which can be compared with the value in Fig. 4 and 5. Integrating the mass release rate over the time period would represent the total mass release. One must note that the particle temperature is relatively still lower than the gas temperature during the mass release, starting from around 640K.There is no more volatiles left before the particle temperature approaches the gas temperature.

One can easily notice the direct relation between the mass release and the progress of the flame radius, as given in Eq.(3). Formation and progress of the flame size is represented from the model prediction. The typical size of the flame can be checked with the parameter values in Eq.(3). For the case of oxygen concentration of 21%, particle diameter of 80μm, and the relative mass release rate of $100s^{-1}$, calculated value of scaled diameter is 8.

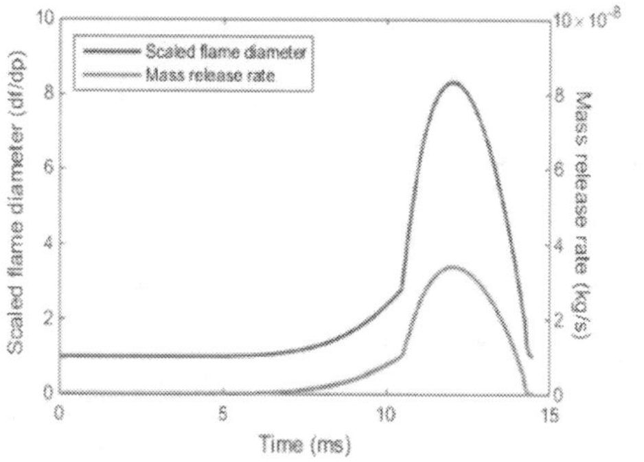

Figure 5: Devolatilization mass release rate as calculated from the particle temperature history and the scaled flame diameter.

3.5 Flame Temperature

Temperature of the flame was defined from the energy balance at the flame, as summarized in Eq.(4). Quantification of energy transfer from the flame is, however, sensitive on the temperature; radiative transfer is proportional to T^4, and conduction heat transfer is related with $\Delta T/\Delta x$ and the characteristic length scale is very short in the particle level. Also, amount and composition of volatiles, which would be related with the particle temperature, are subject to considerable uncertainties. Basic hypothesis of the flame model that all volatiles are instantaneously and completely burn-out at the flame is not fully guaranteed. All these uncertainties are associated in quantification of the flame temperature.

Figure 6 shows one case of calculated flame temperature. The flame temperature is defined from the point of ignition. Note that the particle temperature, at the ignition point, is considerably lower than the gas temperature [8]. However, calculated flame temperature is higher than the gas temperature by some 300K. The flame temperature, which is based on the thermal balance, would maintain the value until the point of extinction, which is related with the disappearance of the volatile flame.

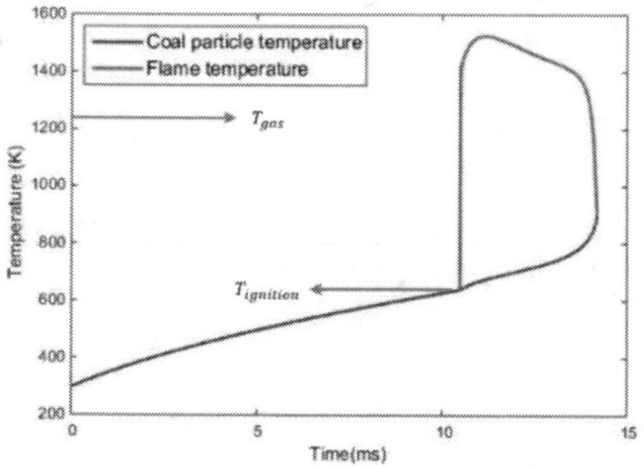

Figure 6: Presenting particle and flame temperature along with the gas temperature. Ignition temperature is defined as the particle temperature, where significant volatile is observed.

It is also noted that temperatures of the particle and the flame predicted from the model should be evaluated with caution, that the modeling hypotheses and selection of physical parameters are not free from uncertainties.

4 Conclusions

Pulverized fuel particles in a hot gas environment exhibit complex interaction of the combined heat transfer, mass diffusion, and mass and chemical reactions. A simplified modeling concept of the volatile flame is based on the hypothesis of the thin flame sheet which surrounds the particle, thereby forming a volatile cloud.

Basic structure of the volatile flame model is a combination of the particle heat-up model and the flame model. Transport of heat and mass between the particle and the flame are governed by the mass and energy conservation. Devolatilization model controls the mass release from the particle. At the typical temperature region, this process is complete within a few milliseconds.

Detailed description of the phenomena is based on the intuitive reasoning on the physico-chemical processes. However, individual steps of modeling are not clearly established, and some of the hypotheses require special attention. Overall, modeling is not unequivocally complete.

Volatile formation behavior, in terms of formation of volatile flame, volumetric radiation encircling the flame, progress of the size of the flame, could be reproduced from the model prediction. Effect of parameters, which included the gas temperature T_{gas}, oxygen concentration x_{O2}, particle diameter d_p, and fuel type, were expected from the model description.

Model predictions of the heat-up time, or time duration of the volatile flame, and the apparent flame size of the volatile could be compared to the measurement data. Considering the level of uncertainties embedded in the modeling and its physical parameters, comparisons were considered encouraging.

5 Acknowledgment

Authors acknowledge the support of BK21.

References

[1] L.D. Timothy, A.F. Sarofim, J.M. Beer, Proc. Combust. Inst. 19 (1982) 1123-1130

[2] V.S. Gururajan, T.F. Wall, J.S. Truelove, Combust. Flame Vol. 72, 1988, pp. 1-12.

[3] C.W. Lau, S. Niksa, Combust. Flame. Vol. 95, 1993, pp. 1-21.

[4] C.R. Shaddix, A.Molina, Proc. Comb. Inst. 32(2009) 2091-2098..

[5] R Khatami. C. Stivers, Y. Levendis, Combust. Flame, Vol. 159, 2012, pp.3554-3568.

[6] H. Lee, S. Choi, Combust. Flame, Vol. 162, 2015, pp. 2610-2620.

[7] C. Mock, H. Lee, S. Choi. V. Manovic, Fuel, 200, 2017, 20-34

[8] A Ponzio, S Senthoorselvan, W Yang, W Blasiak, Fuel, 87, 2008. pp. 974-987.

11th Asia-Pacific Conference on Combustion,
The University of Sydney, NSW Australia
10th -14th December 2017.

Combustion Behavior of Partial Substitution of Solid Fossil Fuel by Biomass

Shuhn-Shyurng Hou[1], Wei-Cheng Huang[1], Ta-Hui Lin[2,3]

[1] Department of Mechanical Engineering, Kun Shan University

Tainan 71070, Taiwan, R.O.C.

[2] Department of Mechanical Engineering, National Cheng Kung University

Tainan 70101, Taiwan, R.O.C.

[3] Research Center for Energy Technology and Strategy, National Cheng Kung University

Tainan 70101, Taiwan, R.O.C.

Abstract

In this experimental study, the combustion characteristics of torrefied biomass co-firing in a pulverized-coal furnace were investigated. Torrefied biomass residues (oil palm empty fruit bunches called OPEFB) were used as a substitute for coal. To examine the feasibility of partial substitution of coal by biomass, coal only and torrefied OPEFB/coal blend with 20% substitution ratio were comparatively investigated. The results showed that torrefied biomass co-firing with 20% substitution ratio in a pulverized-coal furnace can produce stable combustion, similar temperature distributions and lower emission levels of CO_2, as well as comparable emission levels of CO, NO and SO_2 when compared to burning coal only.

1 Introduction

Biomass is currently the most widely used form of renewable energy. The co-firing of biomass with coal is one of the most attractive and easily implemented biomass energy technologies, and has become a feasible alternative to existing pulverized-coal power plants. In recent years, the co-firing of biomass with coal has been widely investigated [1-6], with most studies having focused on evaluating the effects of co-firing on boiler efficiency [7], combustion characteristics [7], and gaseous emissions [8].

Oil palm empty fruit bunches (OPEFB) are inedible biomass and therefore has no competition with food purposes [9]; consequently, it holds great potential to be used as an alternative renewable-energy fuel for combustion processes in industrial furnaces and boilers. Recently, biomass and coal co-firing technologies have received much attention [7,8]. However, to the best of our knowledge, there have been no experimental studies to date that have explored the combustion characteristics, including temperature distributions and gaseous emissions, when co-firing torrefied OPEFB in a pulverized-coal furnace. To fully understand the torrefaction-based co-firing system, further experimental investigations should be conducted in a real furnace or boiler before its commercial application.

Corresponding author. Fax: +886-6-205-0509
E-mail address: sshou@mail.ksu.edu.tw

In this study, the combustion characteristics and emitted pollutants of co-firing torrefied OPEFB and coal were investigated experimentally in a 300 kW_{th} (thermal power) pulverized-coal furnace with no modifications to the combustion system. Our goal was to explore the feasibility of using torrefied OPEFB as a substitute for coal with 20% substitution ratio.

2 Fuel Properties

OPEFB torrefied at 280 °C exhibits almost black color. By comparison, the color of coal is only slightly darker than that of torrefied OPEFB. Proximate and ultimate analysis data as well as energy content (MJ/kg) of coal and torrefied OPEFB are shown in Table 1. The proximate analysis results show that while the ash content of torrefied OPEFB is higher than coal, both the volatile matter and fixed-carbon are lower. The ultimate analysis gives the composition of the biomass in wt% of carbon, hydrogen, oxygen, sulfur and nitrogen. From the ultimate analysis, it was found that in comparison to coal, the weight percent of carbon in the torrefied OPEFB is lower, the hydrogen content is slightly lower, and the oxygen content is similar.

The particle size of coal or biomass is an important factor influencing flame stability and the combustion characteristics. Raw biomass was pretreated by the torrefaction process (torrefied at 280 °C) previously. After pulverizing the coal and torrefied biomass separately, the biomass and coal particles were sieved and selected to ensure a suitable size of 75 μm

Table 1: Chemical compositions and energy contents of coal and torrefied oil palm empty fruit bunches (OPEFB).

Fuel source	Coal	OPEFB
Proximate analysis (dry basis)		
Volatile matter (wt%)	50.11	40.46
Fixed carbon (wt%)	47.01	36.17
Ash (wt%)	2.88	23.37
Ultimate analysis (dry basis)		
Carbon (wt%)	71.51	52.29
Hydrogen (wt%)	5.05	4.04
Oxygen (wt%)	19.40	19.75
Nitrogen (wt%)	1.00	0.75
Sulfur (wt%)	0.17	0.07
Higher heating value (MJ/kg)	22.71	18.53

(200 mesh) for stable and proper combustion performance. The torrefied OPEFB and coal particles were then adequately blended. Finally, the various mixtures were introduced into the combustion chamber by primary air, and combusted with secondary air.

3 Experimental Setup

3.1 Multi-fuel Combustion Test Furnace

The schematic of the combustion test furnace is shown in Fig. 1. It consists of a rotary coal feeder with a maximum feeding rate of 40 kg/h, an adjustable swirl burner on the top of the furnace (as shown in Fig. 2), a vertical cylindrical refractory-lined radiant chamber of 560 mm inner diameter and a height of 3 m, and a flue gas desulfurization (FGD) unit with a maximum $Ca(OH)_2$ feeding rate of 10 kg/h. The bag filter is made of ceramic filters with a dust loading of 2000 mg/m^3, a pressure drop of 150 mmAq, and a maximum temperature resistance of 400 °C. A heat exchanger (HEX) was installed at the downstream of the system to cool down the flue gas and remove the water content in the flue gas.

The measurement configuration of the test furnace is shown in Fig. 2. The vertical combustion chamber is made up of 5 sub-sections, each one equipped with an R-type thermocouple to measure the changes of the wall temperature of the chamber. Each sub-section is equipped with observation windows for visual inspection of the flame. The furnace has a maximum thermal loading of 300 kW$_{th}$ and it can be supplied with fuel oil, emulsion fuel, pulverized coal, or biomass fuel. The facility was designed to maintain a positive pressure at the exit to ensure zero air leakage into the facility [10].

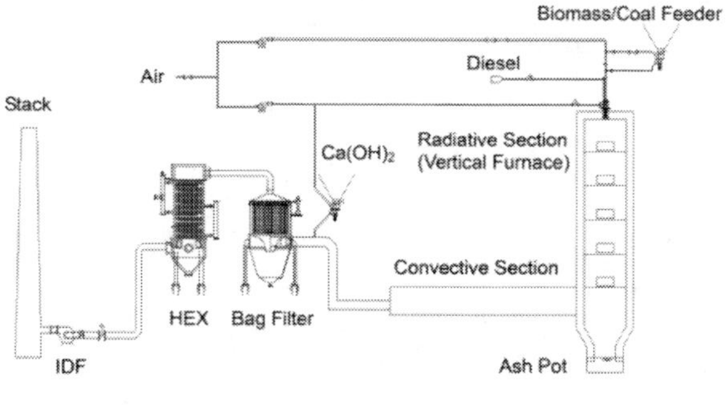

Figure 1: Schematic of 300 kW$_{th}$ multi-fuel test furnace.

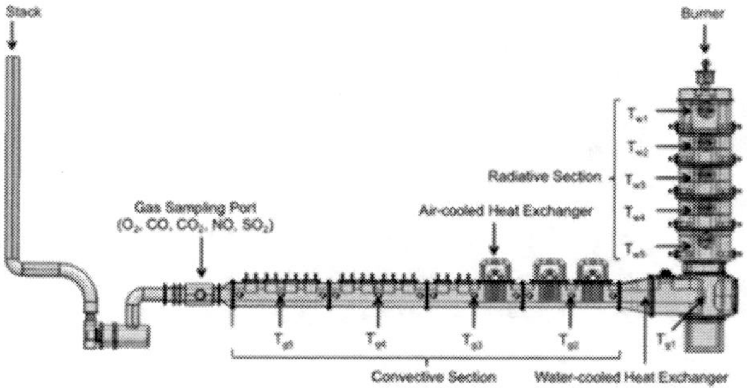

Figure 2: Measurement port configuration of test facility.

The temperature was measured with both R-type and K-type thermocouples. More specifically, five R-type thermocouples (T_{w1} to T_{w5}) were placed from top to bottom in the radiative section to measure the wall temperature of each section in the vertical furnace. Meanwhile, five K-type thermocouples (T_{g1} to T_{g5}) placed from upstream to downstream were employed to measure the gas temperatures at various positions along the convective section. Fuji Electric Analytical Analyzers were employed to analyze the flue-gas emissions, including O_2, CO, CO_2, NO, and SO_2. Gas samples were conditioned by a flue-gas sample conditioning system through cooling, drying and filtering before entering the analyzers. Thus, the emissions are expressed as a volume concentration on a dry basis in this study.

3.2 Experimental Procedures

The combustion experiments were performed using a vertical furnace, as shown schematically in Fig. 1. Three experimental processes were carried out. Firstly, diesel was used as the fuel in the preheating procedure, which took around 22 hours to ensure that every sub-section of the vertical furnace had reached a steady temperature distribution. After the pre-heating process, the diesel fuel was switched to the test solid fuels to perform combustion tuning. Finally, the optimum operating conditions (which met the minimum residual oxygen in the flue gas required for complete combustion) obtained from the combustion-tuning procedures were adopted for the stable combustion experiments.

4 Results and Discussion

4.1 Combustion Tuning

After completing the preheating process, the next step was combustion tuning. The main purpose of combustion tuning is to reduce the air-flow rate to achieve the minimum excess oxygen in the flue gas required for complete combustion of the fuel. The minimum amount of air required for complete combustion of a fuel is called the stoichiometric air or theoretical air; air in excess of the stoichiometric air, is called excess air. The minimum excess oxygen requirement was identified as the critical point at which the CO emission levels in the flue gas were maintained at a minimum. For complete combustion, the excess air must meet or exceed the critical point; otherwise, incomplete combustion occurs, which leads to high levels of CO emissions and the occurrence of smoke. In the experiments, the air-flow rate in each case was tuned to meet the minimum excess oxygen requirement, and at least 30 minutes stable operation was maintained for each case before data collection began.

Figure 3 shows the typical combustion-tuning process used in this study, where 20% substitution ratio was burned with a feed rate of 28.48 kg/h. The minimum oxygen required for complete combustion was determined by gradually adjusting the amount of combustion air. Variations of the emissions with residual oxygen measured in the flue gas are shown in Fig. 3. As can be seen, decreasing the flow rate of the combustion air from case A (381 Nm^3/h) to case G (261 Nm^3/h) resulted in a decrease of the residual oxygen concentration from 9.5% to 4%. When the combustion air exceeded the critical point for complete combustion (case F (282 Nm^3/h), the CO emissions were nearly 0 ppm; however, when the amount of combustion air was below the critical point, large amounts of CO were generated (e.g., case G). On the other hand, formation of SO_2 depended on the sulfur content of the fuel itself; as such, the SO_2 emissions maintained a near-constant value when using the same fuel. When burning 20% torrefied biomass + 80% coal, case F had the optimal amount of combustion air, with an airflow rate of 282 Nm^3/h, at which point the CO emission was low. Therefore, the case-F operating condition was taken as the optimum air-supply rate for the minimum excess-oxygen requirement with low flue-gas heat loss. In the present study, the air-feed rates for the minimum excess-oxygen requirements were 282 and 303 Nm^3/h for 20% substitution ratio and coal alone, respectively. That is, 20% substitution experienced a higher equivalence ratio (0.708) compared to coal (0.667) under the stable complete-combustion conditions. Here the equivalence ratio is defined as the actual fuel-to-air ratio based on the mass divided by the stoichiometric fuel-to-air ratio.

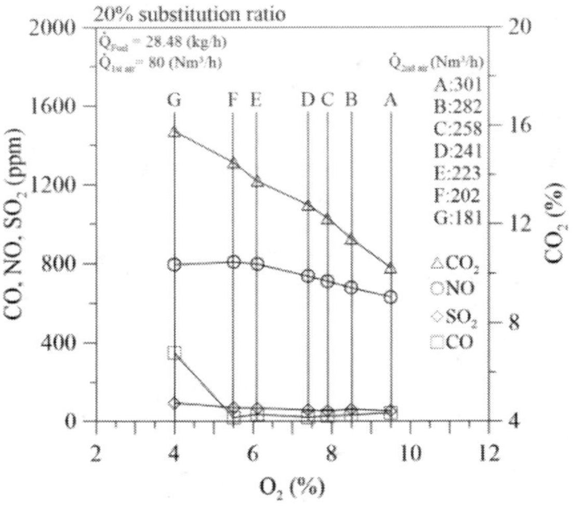

Figure 3: Variations of emissions with residual oxygen in the flue gas during the combustion-tuning process of torrefied biomass.

4.2 Temperature Distributions and Emissions

Temperature versus time at different positions (T_{w1}-T_{w5} and T_{g1}-T_{g5}) under the stable complete-combustion conditions with the minimum residual oxygen in the flue gas required for complete combustion for coal alone and 20% substitution ratio are shown in Figs. 4 and 5, respectively. The corresponding average temperatures at these positions are also illustrated in Table 2. As was shown in Table 1, the torrefied biomass has a lower carbon content and heating value than coal. As could be expected, at the fixed fuel-supply rate of 28.48 kg/h, and with an increasing torrefied biomass proportion from 0% to 20% in the fuel blends, the flame temperature gradually reduced, and consequently, so did the wall temperature. In comparing the results in Figs. 4 and 5, as well as Table 2, it was found that T_{w2} (the wall temperature in the primary combustion zone) reduced only slightly from 1173 to 1149 °C when the torrefied biomass content was increased from 0% to 20%. Similarly, the wall temperatures in the other subsections (namely T_{w1}, T_{w3}, T_{w4}, and T_{w5}) of the radiative section presented the same trend. The gas temperatures at different positions (T_{g1}-T_{g5}) in the convective section also showed the same trend of a slight drop in temperature with increasing torrefied biomass content from 0% to20%.

Figures 6 and 7 present the concentrations of the flue gas composition versus time for burning the coal alone and the torrefied biomass/coal blend (20% substitution ratio) under the stable complete-combustion conditions. The corresponding average emissions are illustrated in Table 3. The values of CO, CO_2, SO_2 and NO reported in this study are all corrected to 6% O_2 [10]. As can be seen, the O_2 contents in the respective flue gases were found to be nearly the same. That is, with increases in substitution ratio from 0% to 20%, the O_2 concentrations in the corresponding flue gases are 5.1% and 5.4%, respectively. The CO_2 concentrations in the flue gases are also maintained at constant value, i.e., 14.49% for coal alone and 14.18% for 20% substitution ratio. Biomass is carbon neutral and renewable. Therefore, net CO_2 emission of burning 20% torrefied biomass + 80% coal is lower than coal alone. Incomplete combustion causes CO emission. Under the stable complete-combustion conditions, all experiments were tuned to the minimum residual oxygen required for complete combustion; therefore, the CO concentrations in both cases were low. It was found that with an increase in the torrefied biomass content from 0% to 20%, CO emissions are 38.0 and 39.2 ppm, respectively, for coal alone and 20% substitution ratio (shown in Table 3), indicating the occurrence of complete combustion for the fuels. Moreover, co-firing biomass/coal blend with 20% biomass emits comparable NO and SO_2 emission levels compared with firing coal alone under the optimum operating condition which meets the minimum residual oxygen in the flue gas required for complete combustion.

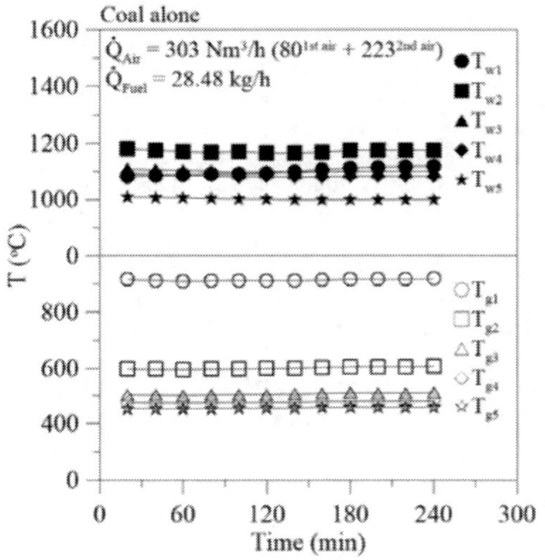

Figure 4: Temperature versus time for coal under stable-combustion conditions.

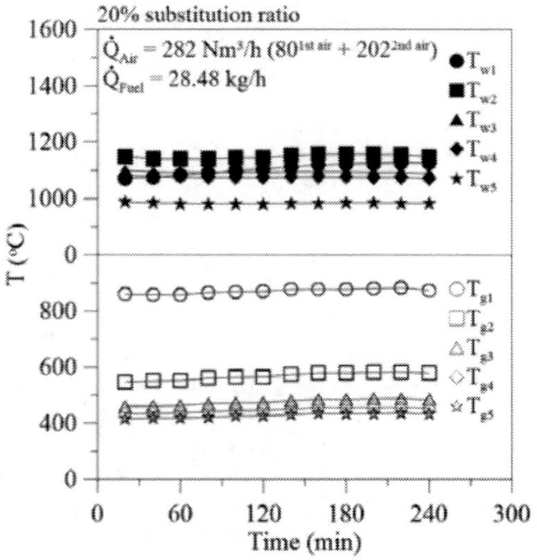

Figure 5: Temperature versus time for 20% biomass + 80% coal under stable-combustion conditions.

Table 2: Variations in average temperature distributions and gas emissions for coal alone and biomass/coal blend under stable-combustion conditions.

Fuel	Coal alone	20% substitution ratio
T_{w1}	1100	1105
T_{w2}	1173	1149
T_{w3}	1101	1094
T_{w4}	1085	1075
T_{w5}	1002	984
T_{g1}	914	871
T_{g2}	601	568
T_{g3}	506	476
T_{g4}	476	448
T_{g5}	455	428
O_2 (%)	5.1	5.4
CO (ppm)	38	39.2
CO_2 (%)	14.49	14.18
NO (ppm)	820	790
SO_2 (ppm)	76	81

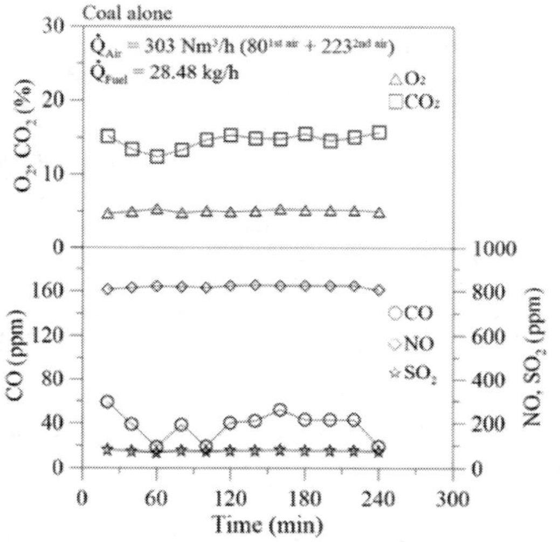

Figure 6: Emisions versus time for coal under stable-combustion conditions.

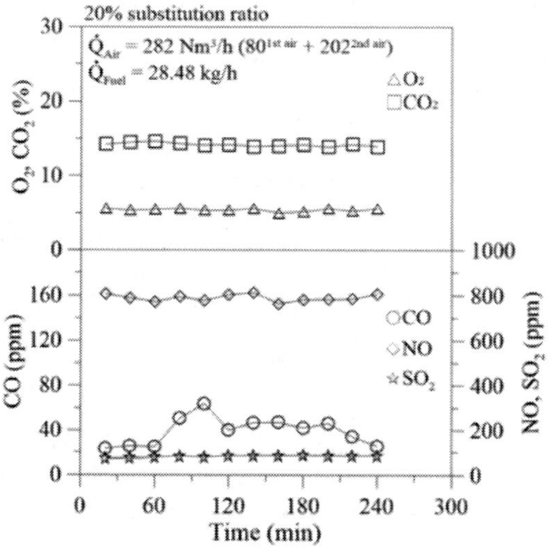

Figure 7: Emissions versus time for 20% biomass + 80% coal under stable-combustion conditions.

4 Conclusions

The combustion characteristics and emitted pollutants of co-firing torrefied biomass and coal were investigated experimentally in a 300 kW_{th} pulverized-coal furnace without any modifications to the combustion system. Coal alone and torrefied biomass/coal blend with 20% substitution ratio were comparatively investigated by focusing on the wall temperatures in the radiative section of the furnace, the gas temperatures in the convective section, and the emission products from combustion. The results of torrefied biomass co-firing with 20% substitution ratio in a pulverized-coal furnace showed that both the wall temperature and gas temperature drop only slightly, which are very close and exhibit similar furnace temperature distribution. Furthermore, co-firing biomass/coal blend with 20% substitution ratio emits comparable CO, NO and SO_2 emission levels compared with firing coal alone but lower net CO_2 emission because biomass is carbon neutral and renewable.

5 Acknowledgment

The authors would like to thank the Ministry of Science and Technology, Taiwan, ROC, for financially supporting this research under grant MOST 105-3113-E-006-004.

References

[1] M. Sami, K. Annamalai, M. Wooldridge, Prog. Energy Combust. Sci. 27(2) (2001) 171–214.

[2] K. Savolainen, Appl. Energy 74(3) (2003) 369–381.

[3] C. Zhou, G. Liu, X. Wang, C. Qi, Bioresour. Technol. 218 (2016) 418–427.

[4] T.Y. Mun, T.Z. Tumsa, U. Lee, W. Yang, Energy 116 (2016) 954–962.

[5] G. Wang, J. Zhang, J. Shao, Z. Liu, G. Zhang, T. Xu, H. Lin, , Energ. Convers. Manage. 124 (2016) 414–426.

[6] J.M. Jones, A.B. Ross, E.J.S. Mitchell, A.R. Lea-Langton, A. Williams, K.D. Bartle, Fuel 195 (2017) 226–231.

[7] J. Li, A. Brzdekiewicz, W. Yang, W. Blasiak, Appl. Energy 99 (2012) 344–354.

[8] J. Li, A. Brzdekiewicz, W. Yang, W. Blasiak, Fuel 102 (2012) 624–633.

[9] D.R. McIlveen-Wright, Y. Huang, S. Rezvani, Y. Wang, Fuel 86 (2007) 2032–2042.

[10] W.C. Huang, S.S. Hou, T.H. Lin, Fuel 208 (2017) 71–81.

11th Asia-Pacific Conference on Combustion,
The University of Sydney, NSW Australia
10th -14th December 2017.

Effects of Adiabatic Temperature and Chemical Composition on Soot Formation in Laminar Diffusion Flames

Awais Ashraf[1], Daniel Bartos[1], Matthew J. Dunn[1], Assaad R. Masri[1]

[1]School of Aerospace, Mechanical and Mechatronic Engineering,

The University of Sydney

Sydney, NSW, 2006, Australia

Abstract

This paper studies the effects of adiabatic temperature and chemical composition on the formation of nanoparticles in a laminar diffusion flame. Time resolved laser induced fluorescence and laser induced incandescence are employed to study concentration, size and growth of particles. Three axisymmetric coflow laminar diffusion flames are investigated by laser diagnostics. The base flame is a jet of 60% ethylene and 40% nitrogen. Second flame is 60% ethylene and 40% carbon dioxide. Third flame is 60% ethylene and 40% argon. Two categories of particles are selectively excited and their responses are analyzed. Three flames have comparable adiabatic temperatures. Nitrogen and argon diluted flames have quite similar response in all cases. As per results, soot concentration in carbon dioxide diluted flame would be significantly less than that of in other two flames. The size of particles in carbon dioxide diluted flame is less than that of in other two flames. The difference in carbon dioxide diluted flames is due to different chemical reactions involved.

1 Introduction

The formation of particles in flames remains an active area of research driven by the necessity to reduce or eliminate soot emissions due to their adverse health effects [1]. Of particular importance is the need to control the number and size of particles formed and this requires a thorough understanding of the mechanisms of formation, growth and oxidation. Extensive work already exists on this topic, and the reader is referred to the several thorough reviews on the topic [2-5].

An issue that has not received much attention is the effects of various diluents on the soot and soot precursor evolution in laminar flames. This is important due to the related impact on the thermochemistry particularly when the C/O ratio is changed. Diluents may be introduced in the coflow or fuel streams and may change the temperature and composition fields depending on whether they are potentially chemically active such as carbon dioxide or inert such as argon.

This paper classifies combustion formed nanoparticles in terms of their spectral properties, temporal emission and size (diameter). Structures with sizes above 10 nm, and ability to absorb in infrared and incandesce are called soot. Structures with size up to 10 nm, and ability to absorb in ultraviolet and fluoresce are termed nanoparticles. Structures and molecules with sizes less than 1 nm, and an ability to absorb in the UV and fluoresce are called polycyclic aromatic hydrocarbons (PAHs). The number density of nanoparticles, compared to soot, is almost an order of magnitude higher but their mass contribution is negligible [6].

Several studies [6-9] have employed in-situ laser diagnostic techniques; laser induced incandescence (LII) and laser induced fluorescence (LIF). LII is a reliable and efficient technique to track mature soot particles. When using LII technique, soot particles absorb the laser beam, heat up and emit black body like radiation which is analyzed to reveal characteristics of the soot particles. LIF is used to study polycyclic aromatic hydrocarbons (PAHs) and nanoparticles. When using LIF, target nanoparticles are irradiated by a UV laser beam. The wavelength of the UV beam is capable of matching the broad transition energy required to elevate electrons of the target particles from one electronic state to another (excitation phase). During de-excitation phase, the excited atoms emit a spectrum of radiation (fluorescence). UV laser induced LIF signals can be attributed to a variety of compounds in flames, for example, H_2O, CO_2, PAHs and nanoparticles. With appropriate spectral filtering, and temporal detection different classes of PAHs and nanoparticles can be selectively excited and analyzed [8].

In a series of papers Bartos et al. and Sirignano et al. have investigated nanoparticle and soot particles using UV based laser diagnostics in Yale and Mckenna burners [6, 7, 9]. These studies suggest that LIF generated upon UV excitation in these flames is due to nanoparticles rather than PAHs, due to the longer time decay constant of the LIF emission signal found matching that of nanoparticle timescale (2-10 ns) rather than the much shorter PAH (50-500 ps). Studies conducted by Bejaoui eta al. [8] and Vander Wal et al. [10] suggest that LIF induced by UV laser is due to PAHs based partially on the spectral emission.

This study explores the effects of chemical composition and adiabatic temperature on the formation, growth and destruction of nanoparticles and soot. Three different axisymmetric laminar diffusion jet coflow flames are studied where the fuel consisting of 60% ethylene is diluted with 40% of either Ar, N_2 or CO_2 Flame length, luminosity, adiabatic temperature of the three flames are compared. Flow rates in fuel and coflow ducts are the same in all cases. In-situ laser diagnostics is applied to study the characteristics of flame particles. Laser induced fluorescence (LIF) and Laser induced incandescence (LII) are employed in this study. Two categories of particles, nanoparticles and soot, are excited and their response is recorded in the temporal domain. Subsequently, the response is processed to determine the relative concentration of nanoparticles and soot in the flames.

Corresponding author.
mash9852@uni.sydney.edu.au

2 Experimental Setup

2.1 Burner and flames

The axisymmetric laminar Yale burner adopted by the International Sooting Flame Workshop (ISF) was used. The burner had a fuel tube and coflow honeycomb. The fuel tube was a 4mm ID 0.038 mm wall thickness, surrounded by 75/4.76 mm OD/ID concentric co-flow honeycomb duct. Three 60% ethylene diluted laminar diffusion flames were studied at atmospheric pressure, fig 1. Nitrogen dilution was used as a base case (reference flame) with argon and carbon dioxide used as well. Fuel/diluent mixture was flowing through the fuel duct. Air was used as an oxidizer. Flow rates of fuel, diluent and air were 0.158, 0.106 and 100 L/s. Nitrogen diluted flame had the maximum length. Carbon dioxide diluted flame had the minimum adiabatic temperature. Argon diluted flame had the maximum luminosity. The luminosity of the flames was based on radially integrated RGB values. The adiabatic temperatures, flame lengths and relative luminosities of the flames are listed in Table 1. The adiabatic temperatures have been simulated using ANSYS Fluent software. The higher luminosity of argon flame, compared to nitrogen flame, can be related to higher adiabatic temperature. However, the reduced luminosity of carbon dioxide flame, compared to nitrogen flame, is mainly due to different chemical reactions as explained in section 4.2.

2.2 Laser sources

Two different laser sources were used to probe particles. A UV range laser source (266 nm and 80 ps pulse) was used to target nanoparticles. An IR range laser source (1064 nm and 8 ns pulse) was used to target soot. UV source was 4th harmonic of a mode-locked Nd:YAG laser (Ekspla PL2251 Series Laser). Its probe diameter was 250 µm resulting in a fluence of 0.57 mJ/cm^2. IR source was fundamental harmonic of Nd:YAG laser (Spectra-Physics Quanta-Ray Pro-350). Its probe diameter was 450 µm resulting in a fluence of 1.15 mJ/cm^2. 1064 nm laser was delayed by 800 ns from 266 nm pulse to allow the same volume to be probed with both lasers.

2.3 Data Acquisition System

The data acquisition system consisted of four PMTs, a spectrometer and an oscilloscope. Excited particles emit response in the form of a radiation spectrum. The response of the excited particles was reflected from a series of parabolic mirrors and then focused on the grating of a spectrometer which resolved the incoming response (radiation spectrum) into the component wavelength bands. PMTs collected the component wavelength bands in the time domain. At a later stage, the temporal response was analyzed to study the evolution and destruction of the excited particles. PMTs collected the response in narrow bandwidths such as PMT1 (266 ± 15 nm), PMT2 (350 ± 15 nm), PMT3 (445 ± 15 nm) and PMT4 (575 ± 15 nm).

Signals from the four PMTs were recorded using a Tektronix oscilloscope (25 Giga-sample/s, 4 GHz bandwidth). At each measurement location, 500 instantaneous acquisitions were captured and stored in wfm file format. The files were subsequently processed in MATLAB (background subtraction, signal smoothing and averaging). Then, a deconvolution process was applied to extract exponential decay function of the material being probed. An integrated signal can represent the overall response from the excited particles. The magnitude of the integrated signal is related to the volume fraction of particles being probed.

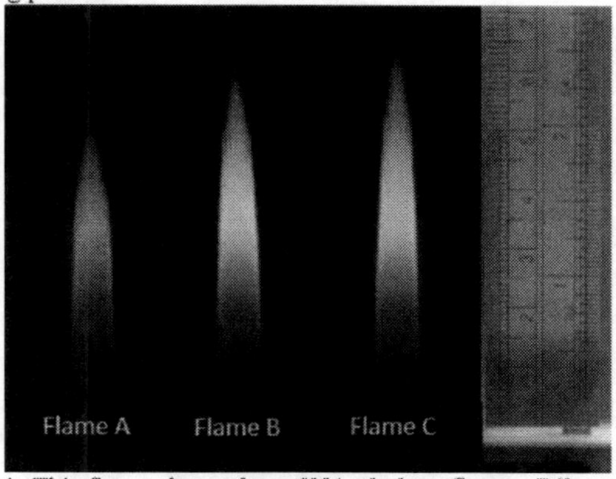

Fig 1: This figure shows three 60% ethylene flames. Diluents in flame A, B and C are carbon dioxide, argon and nitrogen respectively. All flames have regular air in oxidizer stream.

Table 1: comparison of physical properties of three flames

Physical properties	Chemical composition 60% C_2H_4 +		
	40% CO_2	40% N_2	40% Ar
Adiabatic Temp	2300 K	2343 K	2360 K
Flame length	4.7 cm	6.1 cm	5.7 cm
Relative luminosity	0.75	1	1.07

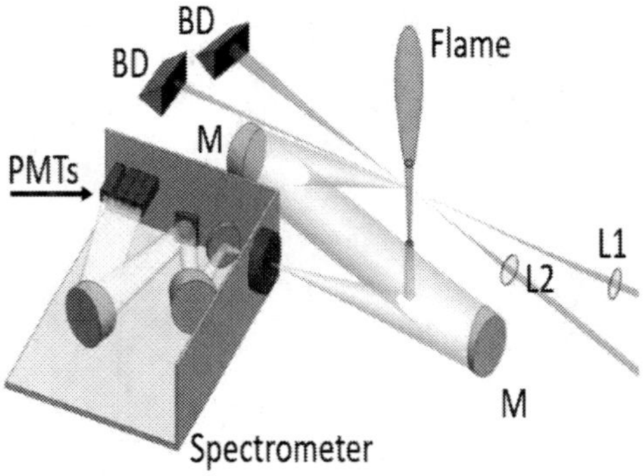

Figure 2: Experimental setup consists of laser sources (L1, L2), lens, parabolic mirrors, half wave plates, spectrometer, photomultipliers (PMTs)

3 UV and IR Excitation Results

Soot particles excited at 1064 nm emit incandescence signal (LII) which is collected at PMTs. The LII signal is proportional to soot volume fraction in the probed area. Curves in fig 3 compare the LII signal of three flames.

The plot area in Fig. 3 can be divided into two regions; a non-sooting region and a sooting region. For nitrogen and argon diluted flames, sooting region starts from 27.5mm HAB onwards. However, for carbon dioxide diluted flame, sooting region starts from 30 mm HAB onwards. In the non-sooting region, negligible LII signal is collected. In the sooting region, LII signal increases with flame height until it reaches maximum sooting region. For nitrogen, argon and carbon dioxide diluted flames, the maximum sooting region lies at HAB equal to 37.5 mm. After this height, oxidation region starts where the LII signal reduces with flame height. The sooting region of carbon dioxide diluted flame is significantly compacted compared to other flames but the maximum sooting region is almost the same 37.5mm HAB.

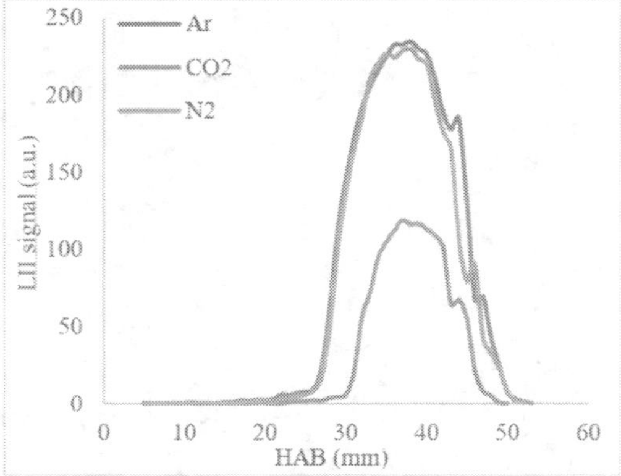

Figure 3: This figure shows integrated LII signal at PMT4 (575 nm). Although these curves relate to soot volume fraction yet they do not represent absolute soot volume fraction.

In N_2 and Ar diluted flames, the LII signal levels are quite similar. However, CO_2 diluted flame shows 48% less LII signal than that of N_2 diluted flame. As the UV laser (266 nm) can excite PAHs, nanoparticles and soot particles the response collected at PMTs may comprise of LIF from PAHs and nanoparticles, LII from soot, and interference from fragmented C_2 and C_3. Curves in Fig. 4 show the signal collected at PMTs upon UV excitation.

In the non-sooting region, the response from CO_2 diluted flame is almost 28% less than that of N_2 diluted flame. In the sooting region, the response from CO_2 diluted flame is almost 67% less than that of N_2 diluted flame.

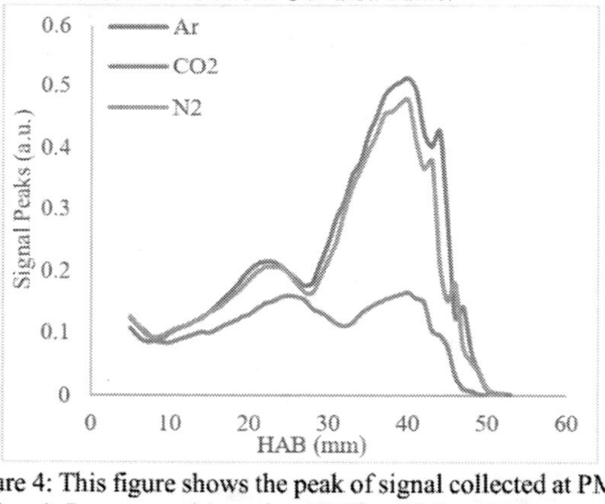

Figure 4: This figure shows the peak of signal collected at PMT2 (350 nm). In non-sooting region (HAB < 27.5 mm), these curves relate to the concentration of nanoparticles but they do not represent the absolute concentration of nanoparticles. In the sooting region (HAB> 27.5 mm), signal comprises of LIF, LII and interference from C_2 and C_3.

4 Discussion

4.1 Effects of the Adiabatic temperature

There is a small difference among the adiabatic temperatures of the three flames. The difference is due to different specific heat values of diluents. The adiabatic temperature of CO_2 diluted flame is only 70 K less than that of N_2 diluted flame but the difference in responses is significant. It is highly unlikely that the 70 K temperature difference would significantly suppress the response of CO_2 diluted flame.

4.2 Effects of the chemical composition

Addition of a diluent gas to a fuel can have three effects; diluent effect (reduced percentage of fuel in reactants), thermal effect (reduced adiabatic temperature of flame), and chemical effect (different products). Ar and N_2 are inert gases which impart the diluent and thermal effects but not the chemical effects. CO_2 is a reactive gas. At higher temperatures, it dissociates. CO_2 reacts with hydrogen radicals to form hydroxyl radicals [11]

$$CO_2 + H \rightarrow OH + CO \qquad (1)$$

The chemical reaction (1) is responsible for the reduction in the concentration of soot in CO_2 diluted flame. The H radical promotes the formation and growth of soot precursors. OH radical destroys (oxidizes) soot precursors. In the CO_2 diluted flame, the H radical is consumed and the concentration of OH radical is enhanced. Therefore, the concentration of soot precursors is reduced which results in reduced incipient soot, reduced nucleation and reduced soot volume fraction. The chemical reaction (1) can explain reduced soot formation in the sooting region. But due to low temperature in the non-sooting region, this reaction is unlikely to be significant.

4.3 Bimodal behavior upon UV excitation

Figure 4 shows two humps in all curves. In case of nitrogen and argon diluted flames, first hump peaks at HAB = 23 mm and second hump peaks at HAB = 40 mm. The first hump can be related to LIF signal due to PAHs and nanoparticles. As moving downstream from the burner, PAHs are converting into nanoparticles which increases the LIF signal until HAB = 23 mm. Moving downstream, nanoparticles begin converting into liquid like incipient soot. The incipient soot is not excited by UV laser, therefore, there is a drop in LIF signal. The region where incipient soot occurs is term "dark region" [10].

As nanoparticles can occur simultaneously with soot particles [6] and UV can excite both nanoparticles and soot, it is possible that signal in the sooting region comprises of LIF from PAHs and nanoparticles, LII from soot, and interferences from fragmented C_2 and C_3. However, it is unclear what would be the relative contributions of LIF, LII and interferences from C_2 and C_3.

In the sooting region, the signal increases with HAB until oxidation region is reached, where signal decreases with height. By comparing Fig. 3 and Fig. 4, it seems that the second hump in Fig. 4 shows a reasonable correlation with 1064 nm LII in Fig. 3. Therefore, it is possible LII may be dominant in the second hump region.

4.4 Size of particles

For particles much smaller than the incident wavelength, emitted scattering signal is directly proportional to 6th power of the particle's diameter. This is true for a single particle. For an aggregate of particles, this proportion (scattering signal α D^6) changes but the overall trend of scattering signal still relates to the size of material under consideration. Discussion in this section assumes that scattering signal is directly proportional to D^6 of particles under study.

Figure 5 shows scattering signal (collected at PMT1, 266 nm) at different heights above burner. The intensity of the scattering signal increases with HAB. It can be inferred that the size of particles is increasing with height above burner. PAHs (diameter < 1 nm) stake up to form nanoparticles (diameter ranging 1 to 10 nm). In the non-sooting region, scattering signal strength of CO_2 diluted flame is lower which may be an indication that nanoparticles in CO_2 diluted flames are smaller compared to nanoparticles in other two flames.

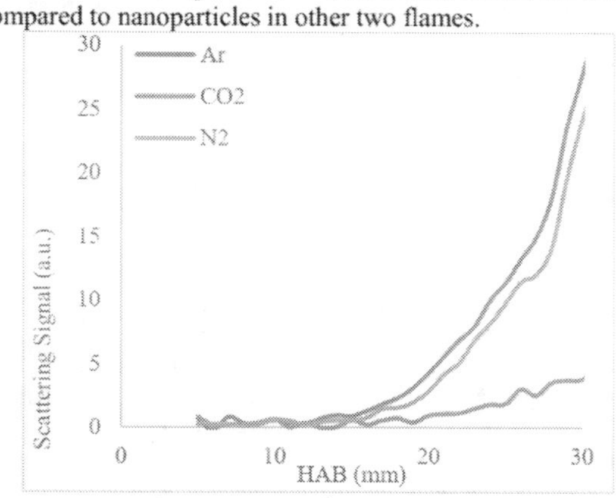

Figure 5: Scattering Signal (α D^6) vs. HAB – in the non-sooting region

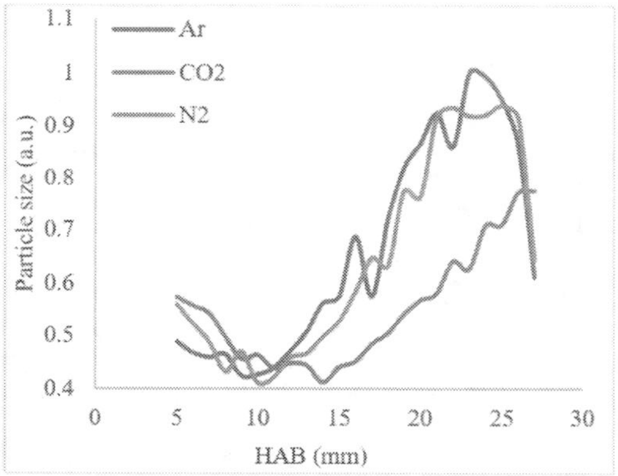

Figure 6: Estimation of particle size vs. HAB. This graph shows a normalized trend of particles' growth in size in the non-sooting region.

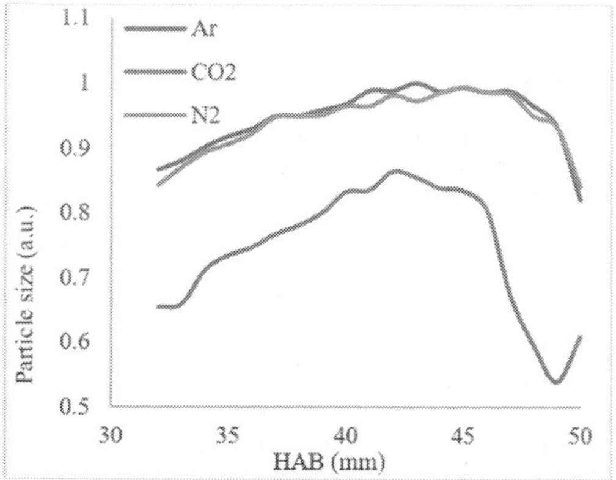

Figure 7: Estimation of particle size vs. HAB. This graph shows a normalized trend of particles' growth in size in the sooting region.

As scattering signal is proportion to D^6 (6th power of diameter) and LII peak is proportional to D^3 (third power of diameter), the ratio of scattering signal to LII peak would relate to volume (D^3) of the particle. Cube root of the ratio is plotted in figure 6 and 7. It is likely that chemical effect of CO_2 suppresses the growth of soot precursors and soot particles.

4 Conclusion

Three different flames were studied to investigate the effects of adiabatic temperature and chemical composition on formation, growth and destruction of nanoparticles and soot. It is found that chemical composition can exert a high impact on concentration and size of soot. Carbon dioxide diluted flame has shown 48% less LII signal as compared to nitrogen diluted flame. LIF signal is primarily due to nanoparticles, not PAHs.

5 Acknowledgment

Research at the University of Sydney was supported by the Australian Research Council

References

1. Chen, R., et al., Beyond PM2.5: The role of ultrafine particles on adverse health effects of air pollution. Biochim Biophys Acta, 2016. 1860(12): p. 2844-55.
2. D'Anna, A., Combustion-formed nanoparticles. Proceedings of the Combustion Institute, 2009. 32(1): p. 593-613.
3. Skeen, S.A., G. Yablonsky, and R.L. Axelbaum, Characteristics of non-premixed oxygen-enhanced combustion: II. Flame structure effects on soot precursor kinetics resulting in soot-free flames. Combustion and Flame, 2010. 157(9): p. 1745-1752.
4. Wang, H., Formation of nascent soot and other condensed-phase materials in flames. Proceedings of the Combustion Institute, 2011. 33(1): p. 41-67.
5. Michelsen, H.A., et al., Laser-induced incandescence: Particulate diagnostics for combustion, atmospheric, and industrial applications. Progress in Energy and Combustion Science, 2015. 51: p. 2-48.
6. Sirignano, M., et al., Detection of nanostructures and soot in laminar premixed flames. Combustion and Flame, 2017. 176: p. 299-308.
7. Bartos, D., et al., Tracking the evolution of soot particles and precursors in turbulent flames using laser-induced emission. Proceedings of the Combustion Institute, 2017. 36(2): p. 1869-1876.
8. Bejaoui, S., et al., Laser induced fluorescence spectroscopy of aromatic species produced in atmospheric sooting flames using UV and visible excitation wavelengths. Combustion and Flame, 2014. 161(10): p. 2479-2491.
9. Sirignano, M., et al., Detection of aromatic hydrocarbons and incipient particles in an opposed-flow flame of ethylene by spectral and time-resolved laser induced emission spectroscopy. Combustion and Flame, 2012. 159(4): p. 1663-1669.
10. Vanderwal, R., Simultaneous Laser-Induced Emission of Soot and Polycyclic Aromatic Hydrocarbons Within a Gas-Jet Diffusion Flame. Combustion and Flame, 1997. 109: p. 399-414.
11. Karataş, A.E. and Ö.L. Gülder, Effects of carbon dioxide and nitrogen addition on soot processes in laminar diffusion flames of ethylene-air at high pressures. Fuel, 2017. 200: p. 76-80.

11th Asia-Pacific Conference on Combustion,
The University of Sydney, NSW Australia
10th -14th December 2017.

Effect of injection strategies on preignition tendency in a turbocharged single cylinder engine

Eshan Singh[1], Adrian Ichim[1], Kai Morganti[2], Robert W. Dibble[1]

[1]King Abdullah University of Science and Technology

Thuwal, Makkah Province, 23955, Saudi Arabia

[2]Saudi Aramco R&DC, Dhahran 31311, Saudi Arabia

Abstract

Preignition is an abnormal combustion that has recently emerged as a bottleneck to downsizing, and efficiency improvement in spark ignited engines. Fuel-oil interaction on the cylinder liner is thought to be a major cause of the phenomenon. In this context, the current work uses varying injection strategies, leading to varying levels of homogeneity and fuel impingement on the cylinder liner, to {understand} establish potential causes of the preignition phenomenon. The injection is varied from one extreme of fuel-liner interaction to another; going from central direct injection to port injection of gaseous fuel and various combinations thereof. It was observed that preignition tendency is heavily dependent on the amount of fuel hitting the cylinder liner. As the phenomenon is stochastic, the frequency of its occurrence can be correlated with the fuel-liner interaction, provided all other test variables are held constant. The results motivated us to test the engine for maximum preignition-limited indicated mean effective pressure (IMEP) using dual fuel mode of gasoline and methane.

1. Introduction

With increasingly stringent global emission standards, one of the approaches used to reduce tailpipe CO_2 emissions is improved engine efficiency [1]. Spark ignited engines form a major proportion of over a billion vehicles on the planet [2]. Moreover, the scope for efficiency improvement is greater when it comes to spark ignited engines. One of the most widely accepted method for efficiency improvement is engine downsizing and downspeeding [3]. The trend is conspicuous with large V8 engines being increasingly replaced by 4 or 6 cylinder engines [4]. However, these smaller displacement engines suffer from reduced peak torque. To compensate for this, downsized engines are usually boosted, which allows similar torque from a smaller displacement engine. However, when running an engine at such high specific outputs, stochastic events of abnormal combustion with very high peak pressure and oscillations called 'super knock' may occur. These knocking events have their origins in another stochastic phenomenon, called 'preignition' [5]. As the name suggests, ignition of the charge before the spark timing is defined as preignition. Although the early combustion is usually similar to a spark ignited flame front, the more premature timing of the preignition event may lead to end-gas autoignition well before top dead center (TDC). This can be catastrophic for engine hardware, often leading to melted exhaust valves and piston, along with connecting rod failure [6].

One of the most prevalent theory on the occurrence of preignition is that fuel spray hits the cylinder liner and interacts with the lubricant oil layer. An fuel-oil mixture having lower surface tension and viscosity would allow a droplet to be pitched into the combustion chamber, owing to inertial effects, when piston is decelerating near top dead center [7]. Other than that the droplet mixture needs to be combustible at the temperature-pressure condition of the charge at the time the droplet is launched [8]. Several speculative theories surround this part of the problem, i.e. initiation of ignition and thereafter propagation of the flame. Several studies suggest that catalytic action of the detergent additives in the lubricant (Calcium and Barium compounds) may be responsible for the initiation phase, while others point to deposits catching fire [9-11].

In this regard, the current study aims to observe the effect of injection strategies, thereby the effect varying amount of fuel-liner interaction, on the preignition tendency of a turbocharged spark ignited engine.

2. Methodology

Experiments were performed with a single cylinder turbocharged engine. The details of the engine are provided in Table 1. The engine features a four valve pent-roof combustion chamber with central and side direct injection. The engine also incorporates two port-fuel injectors, which were used for injecting liquid and gaseous fuels. A port oil injector, a part of the original setup, was replaced with natural gas injector that was operated at 6.2 bar (90 psi) injection pressure.

Combustion system	4-valve pent roof
Cylinder bore	82 mm
Stroke	86 mm
Swept volume	454 cm^3
Engine speed	2000 rpm
Compression ratio	9.7:1
Valve type	4 valve DOHC
IMEP	~20 bar

Table 1: Engine specifications and operating conditions.

The fuel used in this study is Coryton gasoline (CG), which contains 5% ethanol by volume (v/v). The fuel properties that are most relevant to preignition tendency are listed in Table 2. Commercial SAE 5W30 lubricant was used for all tests.

RON	95
MON	85
Specific gravity	0.7485
Aromatics (% v/v)	30.5
Olefins (% v/v)	8.2
Ethanol (% v/v)	5.0

Corresponding author. Phone: +966-544-511-843
E-mail address: eshan.singh@kaust.edu.sa

Table 2: Coryton gasoline fuel properties.

A distillation test based on ASTM D86 [12] was performed for the fuel used. Fig. 1 shows the vapor temperature with volume fraction distilled. The boiling point bracket is typical of gasoline range.

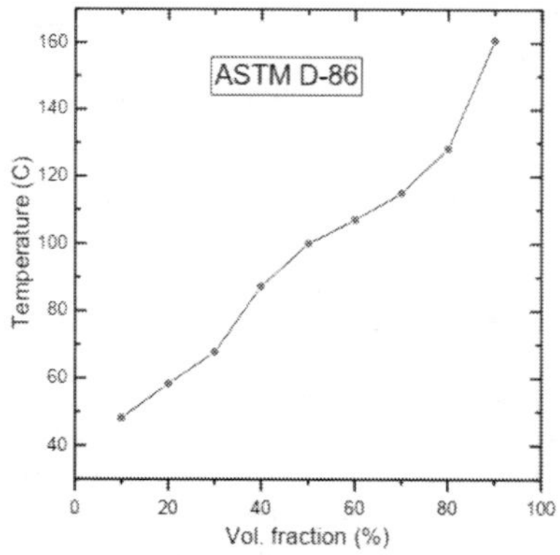

Figure 1: Distillation properties of the Coryton gasoline.

The reduction in ignition delay times with the addition of lubricant oil was quantified using an Ignition Quality Tester (IQT) in accordance with ASTM D6890, [13]. The ignition delay time of the fuel decreases exponentially with addition of lubricant (Fig. 2).

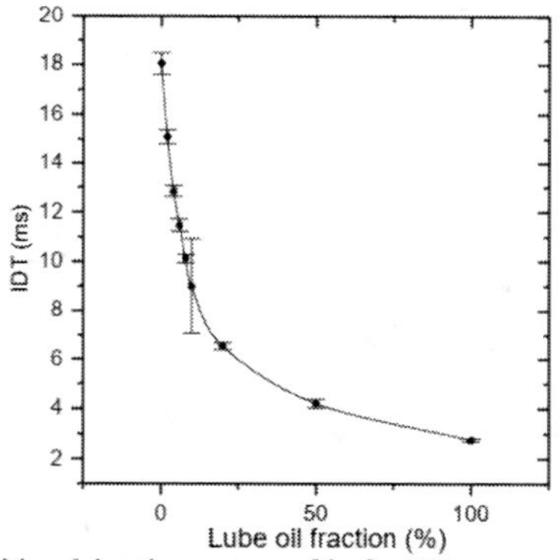

Figure 2: Ignition delay time measured in the IQT.

The gaseous fuel used in this study is methane, which served as a surrogate for natural gas. The list of injection strategies that were tested included:

- Central direct injection,
- Side mounted direct injection,
- Port injection of liquid fuel,
- Port injection of gaseous fuel,
- Central DI, two split injections,
- Central DI, three split injections.

The hypothesis is that injection strategies leading to higher fuel impingement could lead to higher preignition frequency. In this context, side direct injection should be the worst offender, followed by central direct injection, and port injection of liquid and gaseous fuels.

Studies were also done with dual-fuel mode using gasoline and methane. Methane (RON and MON > 130) is used to suppress knock and maintain peak thermal efficiency with increasing load. The engine was operated at full-throttle using gasoline only, and boosting was increased in steps of 0.05 bar. When the engine became knock limited, the fraction of methane was increased, while maintaining lambda=1 (by reducing the amount of gasoline).

3. Results

3.1 Ignition delay of Lubricant-fuel mixture:

Analyzing the ignition delay data, it is plausible that even a small amount of lubricant in fuel drastically affects its ignition characteristics. An IQT might be a better representation of the combustion initiated from droplets, since the injection pressure is low enough to yield droplet-combustion of the injected fuel. Moreover, if it is a fuel-oil mixture that takes part in preignition initiation, the ratio of fuel to oil in the mixture will be highly stochastic. It can, however, be taken with certainty that the droplet cannot be pure fuel, because the fuel RON=95, resists end-gas autoignition. The operating conditions are chosen such that the fuel-air mixture does not knock (let alone autoignite). Due to the physical properties, viz. surface tension and viscosity, it is also difficult for pure oil to be scraped from the liner surface and form a combustible droplet, hence preignition by pure oil is also not likely. A mixture of the two, possessing correct physical properties to allow the droplet formation and further entering the combustion chamber and chemical properties allowing droplet to ignite, and create a propagating flame; is a preignition event. The preignition event may lead to super knock if required conditions are attained.

3.2 Preignition tendency of direct injectors

Initially, tests were performed using direct injection, in order to ascertain whether the central or side injection location performs better with regard to preignition tendency. The tests were run for 2000 consecutive cycles and the number of preignition events were counted. This was repeated three times and the average of three is reported in the Fig. 3 In order to establish the occurrence of a preignition event, CA05 (crank angle at which 5% of the total heat is released) is taken into consideration. Any cycle with CA05 earlier than 5σ (5 x standard deviation) is defined as a preignition event. This limit allowed the inclusion of all preignition events, while accounting for the natural cyclic variability of spark ignited engines.

Fig. 3 shows the difference in preignition events in the two injector locations at intake pressures of 2 and 2.1 bar. As opposed to the hypothesis, the side injection provides a much lower preignition count. When examining the injectors used for the study, it was found that the two injectors have more or less same maximum liquid penetration length (110mm for side versus 100 mm for central). Hence, considering the same amount of fuel injected for maintaining lambda=1, the amount of distance traversed by the side mounted injector is twice that of central mounted injector. For the given geometry, bore of 82mm, and similar maximum spray penetration length, a centrally mounted injector's spray has to travel the radius of the cylinder to reach the liner, while a side mounted injector's spray has to travel the full diameter. This translates to a smaller amount of fuel in liquid state in case of side mounted injector, when and if the spray reaches the wall. Moreover side injector is a 7-hole injector of smaller diameter compared to 6-hole central injector (0.17mm compared to 0.18mm). A 15° bend angle in case of central injector may have also led to fuel hitting the liner directly.

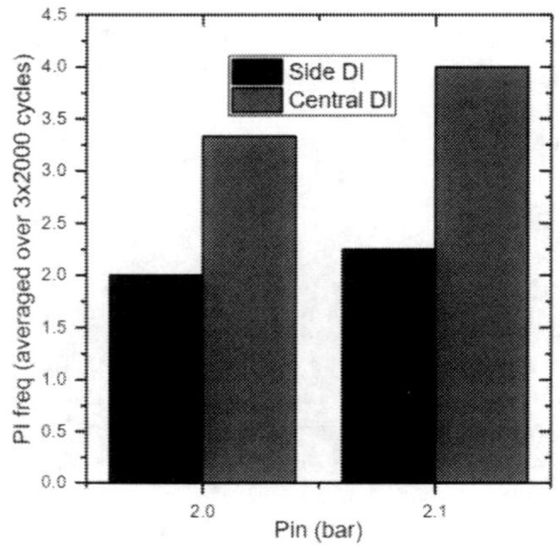

Figure 3: Preignition frequency comparison between side and central injection.

3.3 Preignition tendency with varying injection strategies:

Thereafter, the same fuel was tested in the port injection configuration. This led to a much lower preignition tendency compared with the central DI configuration. This time, the same operating conditions are maintained for 6000 cycles and averaged over two runs. What is surprising is the tendency of port injection to have more preignition events than side mounted direct injection. One needs to consider that the amount of fuel injected doubles for 2 bar intake pressure (maintaining Lambda =1), hence a lot of fuel is port injected while the intake valve is open. Previous researchers have shown that port injection with the intake valve open is more or less similar to direct injection in terms of the charge cooling benefits [14]. Additionally, the low injection pressure (~5 bar) translates into "blobs" of liquid rather than a fine spray (as in high pressure direct injection ~ 150 bar). This results in more liquid fuel entering the combustion chamber at the time of intake valve closing.

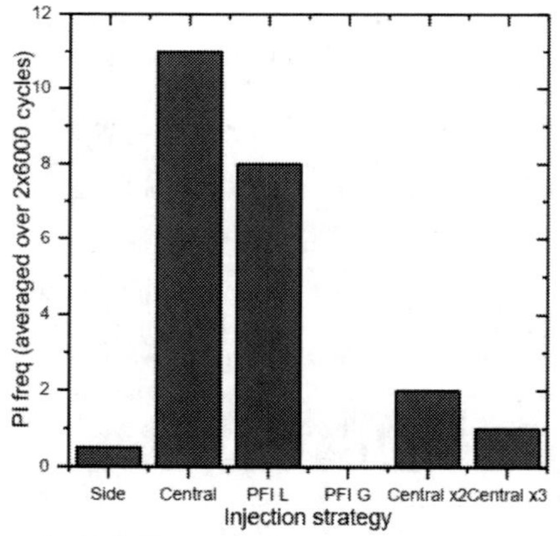

Figure 4: Preignition frequency comparison in various tested injection strategies.

The gaseous fuel methane was used in another port injector and there were no preignition events registered at the equivalent operating conditions. One needs to consider that injecting methane, also changes the fuel chemistry. Methane has a much higher octane number (RON~130). The volumetric efficiency of the engine is also reduced when using gaseous fuel as close to 10% of the cylinder volume is taken up by methane at lambda=1. Further experiments using similar fuel chemistry (port injection of pre-vaporized ethanol) are underway and will be part of a future research topic. The highly effective nature of gas injection does point towards the fact that curtailing fuel impingement over the liner will discourage preignition. However, several researchers have observed preignition events occurring in natural gas injection. The mechanisms behind such events is thought to be different [15].

Splitting the injection into multiple events will essentially supply a lower amount of fuel in each injection pulse, and hence lower the amount of liquid phase fuel that reaches the liner [16]. In this regard, the injection event was split into two pulses of equal width. This led to a considerable reduction in the preignition frequency (Fig. 4). Only the central injection was used for the split injection strategy, to witness some palpable effect of the methodology applied.

Next, the injection event was split into three pulses of equal width. As expected, the preignition frequency was further reduced. One of the disadvantages noticed in splitting the injection event was an increased coefficient of variance (CoV), which can be taken care of by further tweaking the injection amount and in-cylinder flow. Previous studies have also shown that preignition occurrences can be reduced using triple injection [17]. All the values are shown in Fig. 4.

3.4 Use of dual fuel operation at boosted conditions

Having ascertained the benefits of using gaseous injection, the engine was tested to its limit for preignition. A dual-fuel mode operation was employed to extract the maximum benefit of gas injection. The engine was operated at maximum efficiency (and maximum brake torque, MBT) spark timing. Hence, the spark timing was varied to maintain CA50 at 8 CAD aTDC (within 0.5 CAD, as shown in Fig. 5), which corresponded to MBT timing.

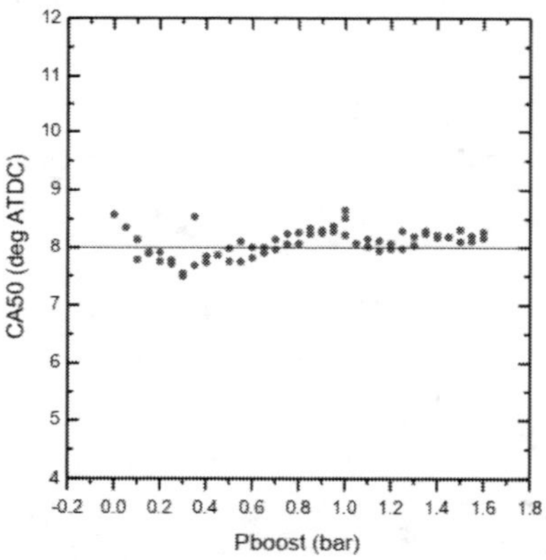

Figure 5: CA50 was maintained close to 8 CAD aTDC.

Fig. 6 shows the fuel injection duration for CG and methane (NG) in dual fuel operation. NG is used to displace a fraction of CG when engine is knock limited (KI>0.5 bar). Thereafter boost is increased (along with increase in CG to maintain Lambda=1). Once knock limited again CG is used to displace CG to get the KI back to 0.5 bar. This results in four data point for each boost level (2 in red, 2 in black).

The corresponding knock intensity is given in Fig. 7. Using this strategy, an impressive indicated mean effective pressure (IMEP) of 32 bar was achieved. Standard preignition tests are performed at 20 bar IMEP (P_{in} 2 bar), while this strategy allows operation at much higher intake pressures (P_{in} 2.6 bar) and specific output, without compromising on performance (maintaining MBT) and total energy (using high calorific CG at all points).

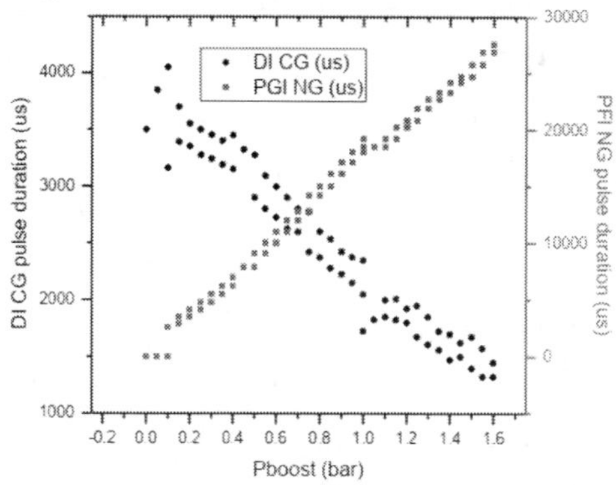

Figure 6: Fuel injection width with varying intake pressure.

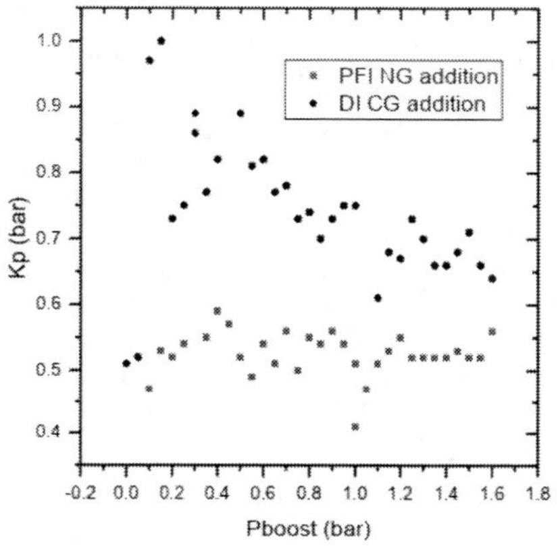

Figure 7: Knock intensity at different boost levels.

The experiments were stopped only due to hardware limitations, i.e. the peak cylinder pressure limit of the engine was 150 bar. Nevertheless, no signs of preignition or knock were observed while operating at CA50 ~ 8 CAD aTDC. This may be due to the reduced liquid fuel impingement on the cylinder liner. As the boost pressure is increased, the liquid mass is reduced (Fig. 6. At maximum boost, the liquid mass injected is half of the initial amount for more than twice the amount of air. This reduces the likelihood of fuel-wall wetting, and hence no preignition was observed.

4. Conclusions

Various injection strategies were assessed with regard to preignition tendency. Initial testing in an IQT revealed that the ignition delay time decreases exponentially with small addition of lubricant oil in the fuel. The correct mixture of fuel-oil is highly stochastic and may not be fixed over different preignition events. Mixture homogeneity is a major factor in establishing the preignition tendency of the engine. Injection strategies leading to more direct impingement of liquid fuel will lead to more preignition tendency. Central direct injection is the worst offender, while gaseous injection through the intake port showed no preignition events. Since the experiments were undertaken with a clean engine, this suggests that the preignition events can be attributed to fuel-oil interaction on the cylinder liner. Using an optimized strategy of liquid and gaseous fuel injection, very high power density can be generated without preignition or knock.

5. Acknowledgment

The authors wish to acknowledge funding from the KAUST Clean Combustion Research Center and Saudi Aramco under the FUELCOM II program. The authors also thank Nimal Naser for enlightening discussions on the project.

References

[1] F. Harvey, "Paris climate change agreement: the world's greatest diplomatic success," The Guardian, vol. 14, p. 15, 2015.

[2] D. Sperling and D. Gordon, Two billion cars: driving toward sustainability: Oxford University Press, 2009.

[3] R. D. Reitz, "Directions in internal combustion engine research," Combustion and Flame, vol. 1, pp. 1-8, 2013.

[4] L. Carley. (2015). Vanishing V8s: Downsizing Engines. Available: http://www.enginebuildermag.com/2015/06/vanishing-v8s-downsizing-engines/

[5] G. T. Kalghatgi and D. Bradley, "Pre-ignition and 'super-knock' in turbo-charged spark-ignition engines," International Journal of Engine Research, vol. 13, pp. 399-414, 2012.

[6] C. Dahnz, K.-M. Han, U. Spicher, M. Magar, R. Schießl, and U. Maas, "Investigations on pre-ignition in highly supercharged SI engines," SAE International Journal of Engines, vol. 3, pp. 214-224, 2010.

[7] C. Dahnz and U. Spicher, "Irregular combustion in supercharged spark ignition engines – pre-ignition and other phenomena," International journal of engine research, 2010.

[8] M. Magar, U. Spicher, S. Palaveev, M. Gohl, G. Müller, C. Lensch-Franzen, et al., "Experimental Studies on the Occurrence of Low-Speed Pre-Ignition in Turbocharged GDI Engines," SAE International Journal of Engines, 8(2015-01-0753), pp.495-504. 2015.

[9] K. Fujimoto, M. Yamashita, S. Hirano, K. Kato, I. Watanabe, and K. Ito, "Engine Oil Development for Preventing Pre-Ignition in Turbocharged Gasoline Engine," SAE International Journal of Fuels and Lubricants, 7(2014-01-2785), pp.869-874. 2014.

[10] T. Kuboyama, Y. Moriyoshi, and K. Morikawa, "Visualization and Analysis of LSPI Mechanism Caused by Oil Droplet, Particle and Deposit in Highly Boosted SI Combustion in Low Speed Range," SAE International Journal of Engines, vol. 8, pp. 529-537, 2015.

[11] Y. Okada, S. Miyashita, Y. Izumi, and Y. Hayakawa, "Study of Low-Speed Pre-Ignition in Boosted Spark Ignition Engine," SAE International Journal of Engines, 7(2014-01-1218), pp.584-594. 2014.

[12] A. International, "ASTM D86, Standard Test Method for Distillation of Petroleum Products at Atmospheric Pressure," ed: ASTM International West Conshohocken, PA, 2004.

[13] N. Naser, S. Y. Yang, G. Kalghatgi, and S. H. Chung, "Relating the octane numbers of fuels to ignition delay times measured in an ignition quality tester (IQT)," Fuel, vol. 187, pp. 117-127, 2017.

[14] S. Cho, N. Kim, J. Chung, and K. Min, "The Effect of Ethanol Injection Strategy on Knock Suppression of the Gasoline/Ethanol Dual Fuel Combustion in a Spark-Ignited Engine," SAE Technical Paper 0148-7191, 2015.

[15] J.-M. Zaccardi and D. Serrano, "A Comparative Low Speed Pre-Ignition (LSPI) Study in Downsized SI Gasoline and CI Diesel-Methane Dual Fuel Engines," SAE International Journal of Engines, vol. 7, pp. 1931-1944, 2014.

[16] M. Lv, Y. Li, X. Hu, Y. Liu, and L. Chang, "Super-knock Suppression Using Split Injection in a Turbo-Charged GDI Engine," in MATEC Web of Conferences, 2016.

[17] A. F. Arnold Kaden, Helmut Gildein, Eberhard Kraus, Uwe Schaupp, "Knocking simulation at Mercedes-Benz – application in series production development," presented at the Knocking in Gasoline Engines, 2013.

11th Asia-Pacific Conference on Combustion,
The University of Sydney, NSW Australia
10th -14th December 2017.

Flame Stabilization and Emission Characteristics of Ammonia Flames in a Micro Gas Turbine Combustor

Ekenechukwu C. Okafor[1], Kazuma Sakai[1], Akihiro Hayakawa[1], Taku Kudo[1], Osamu Kurata[2], Norihiko Iki[2], Hideaki Kobayashi[1]

[1] Institute of Fluid Science, Tohoku University,

2-1-1 Katahira, Aoba-ku, Sendai 980-8577, Japan

[2] Research Institute for Energy Conservation, National Institute of Advanced Industrial Science and Technology (AIST)

Tsukuba, Ibaraki, Japan

Abstract

In order to efficiently utilize ammonia as a gas turbine fuel, it is relevant to understand the combustion characteristics of ammonia in a gas turbine combustor. In this study stabilization and emission characteristics of non-premixed ammonia flames were investigated experimentally using a micro gas turbine swirl burner. The effects of ammonia injection angle, global equivalence ratio of the non-premixed flame, and ambient pressure on flame stabilization and emissions of unburned NH_3 and NO were investigated at gas inlet temperature of 298 K. One dimensional numerical calculations using ANSYS Chemkin PRO were performed to extend understanding of the experimental results. Ammonia flames were stabilized over a range of equivalence ratios. An increase in the ammonia injection angle to the burner axis resulted in increased flame stability and more efficient combustion. Ammonia emission increased with an increase in global equivalence ratio. On the other hand, NO emission decreased with global equivalence ratio owing to the increased reduction of NO by amine radicals. With an increase in pressure, NO emissions decreased due to a decrease in H and OH radicals concentration.

1 Introduction

Earlier applications and studies of ammonia as a fuel reported challenges associated with ammonia combustion which discouraged further interest in NH_3 combustion applications [1, 2]. Ammonia has a small burning velocity that is about five times smaller than that of methane, a narrow flammable range, a high minimum ignition energy and a nitrogen atom in its molecule. These may result to low heat release rate, poor flame stabilization characteristics, low combustion efficiency and high fuel NO emission etc. [2, 3].

However, interest in ammonia as a hydrogen carrier has recently been kindled in a search for safer and cheaper means of hydrogen transportation and storage. This has been accompanied by a renewed interest in NH_3 as a fuel. Recent studies indicate that NH_3 may be a viable carbon-neutral fuel for gas turbine applications. Somarathne et al. [4] and Hayakawa et al. [5] showed in their numerical and experimental studies, respectively, that premixed NH_3-air flames can be stabilized using swirl burners even at high turbulence intensities and elevated pressures. Kurata et al. [6] at the National Institute of Advanced Industrial Science and Technology (AIST), Japan recently demonstrated efficient generation of 44.4 kW from ammonia-fired 50 kW class micro gas turbine plant for the first time ever using a swirl burner. Kurata et al. [6] employed non-premixed combustion, in which the fuel was injected vertically along the burner central axis with a surrounding swirling air flow. Non-premixed combustion is preferred to premixed combustion in gas turbine applications because it mitigates flashback and may ensure better flame stability.

Therefore, recent developments indicate that the use of swirling flows is effective for stabilizing premixed and non-premixed NH_3 flames. Swirl primarily creates an internal recirculation zone which circulates hot products and active radicals from the downstream to the upstream of the burner. Consequently, preheating and continuous ignition of the unburned mixture is enhanced, enabling the stabilization of even flames with small flame speeds such as ammonia. NO emission however remains an important challenge in ammonia combustion applications. Kurata et al. [6] reported values of NO emission with 16% O_2 conversion in the range of 1000 ppm and they recognized that the NO emission was higher than that from natural gas-fired gas turbine burners.

In order to improve the stabilization and emission characteristic of ammonia-fired gas turbines combustors, it is important to clarify the characteristics of swirl stabilized ammonia-air flames.

In the present study, which was conducted in collaboration with AIST, a similar burner as the one used in [6] was employed to investigate the effects of fuel injection angle, θ, global equivalence ratio, Φ and ambient pressure, P on the stabilization and emission characteristics of non-premixed ammonia flames. In addition, numerical simulations were conducted to further the understanding of the experimental results.

2 Experimental Facility and Methods

Figure 1 is a schematic diagram of the micro gas turbine burner which was operated in non-premixed mode in this study. The burner had detachable fuel injectors, surrounded by a swirler for air injection. Fuel injectors with injection angles of 0°, 30°, 45°, and 60° to the burner central axis were investigated. The injectors had twelve 2.4mm diameter injection holes. The swirler

Corresponding author. Fax: +81-22-217-5323
E-mail address:okafor@flame.ifs.tohoku.ac.jp

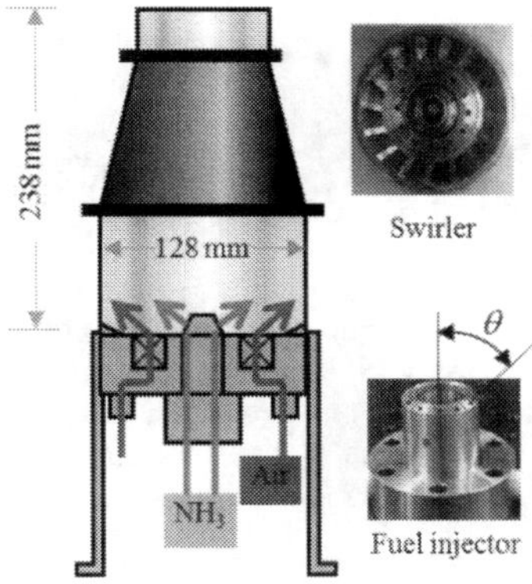

Fig. 1 The micro gas turbine swirl combustor. The angle θ represents the injection angle

had 16 blades with outer and inner diameters, D_o and D_i, of 61.3 and 46 mm, respectively. Swirl number, S, which represents the intensity of the swirling motion was 0.88 as calculated by (1) [7]

$$S = \frac{2}{3}\left[\frac{1-(D_i/D_o)^3}{1-(D_i/D_o)^2}\right] tan\alpha \qquad (1)$$

Here, α is the vane angle of the swirler which was 45°. The actual liner of the burner used by Kurata et al. [6] had holes meant for film cooling of the liner and also for secondary air injection downstream of the burner. However, a model liner was employed in this study for the purpose of simplicity as shown schematically in Fig.1. The model liner had similar shape and dimensions as the actual combustor liner but with no holes. In addition, the top and bottom cylindrical parts of the model liner was made of quartz glass in order to allow for flame visualization.

For the elevated pressure experiments, the burner was placed inside a high pressure combustion rig shown schematically in Fig. 2 where the combustion pressure was controlled up to 0.35 MPa, same as the combustion pressure in [6]. Air at 298 K was supplied to the burner from a 4.5 MPa and 7.5 m^3 air tank with a high-pressure compressor. Ignition was done remotely using a glow plug and hydrogen gas supplied from pressurized hydrogen cylinders. Gaseous ammonia (> 99.9%) was then supplied to the burner from pressurized cylinders while the hydrogen supply is turned off. Air was also supplied to the base of the chamber to facilitate scavenging of the exhaust gases. The flow rates of air and ammonia were measured with mass flow controllers. The chamber pressure was manually controlled using valves mounted on the exhaust pipe. The top section of the chamber and the exhaust pipe were water-cooled.

Emissions were measured using an FTIR (BOB-2000FT) gas analyzer with a scanning frequency of 5 Hz. Exhaust gases were sampled from the burner exit using a 10 mm diameter stainless steel pipe perforated at regular intervals. The line from the sampling probe to the FTIR was maintain at about 191 K to avoid condensation of water vapour along the line and ensure that chemical reactions were quenched. The sampling flow rate was 2 L/min. The measured species concentration in the exhaust gases were allowed to attain an approximately steady value before data logging was done. For each measurement, data was logged for 2 minutes and the average wet value is reported.

One dimensional NH_3-air premixed flames were simulated using ANSYS CHEMKIN-PRO [8] to study the effects of Φ and pressure on NO concentration in the flame. Owing to the effects of turbulence and heat losses, quantitative comparison of the results of the experiment and the simulations may not be possible.

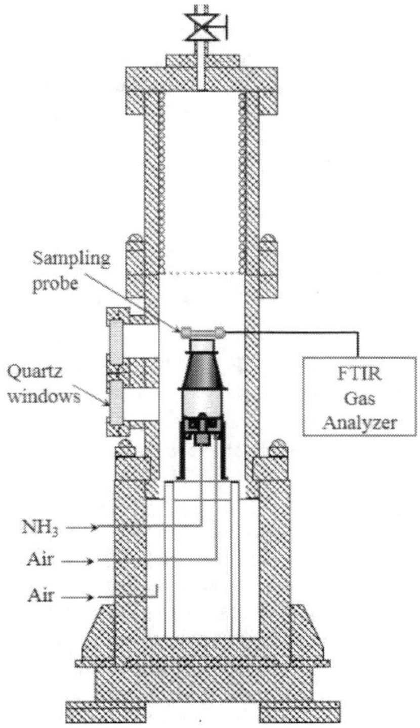

Fig 2. The high pressure chamber

However, the chemical residence time in the burner was considered sufficient to allow an explanation of the trends in the experimental result with the chemistry of one dimensional laminar flames. The reaction mechanism by Tian et al. [9] was employed.

3 Results and Discussion

3.1 Effects of Ammonia Injection Angle on Flame Stabilization and Emissions at 0.10 MPa

It was confirmed in this study that the fuel injection angle is critical to achieving efficient and clean combustion in the combustor. As shown in Fig. 3, vertical fuel injection resulted in lifted flames with heights more than that of the liner. However, slanting fuel injection resulted in more stable flames with heights shorter than the liner. Flame stability increased with an increase in injection angle.

Figure 4 shows the lean and rich blow off limits of the flames. The blow off limits were obtained by increasing or decreasing the air flow rate at a constant ammonia flow rate until the flame blew off. The regions between each pair of curves represent the

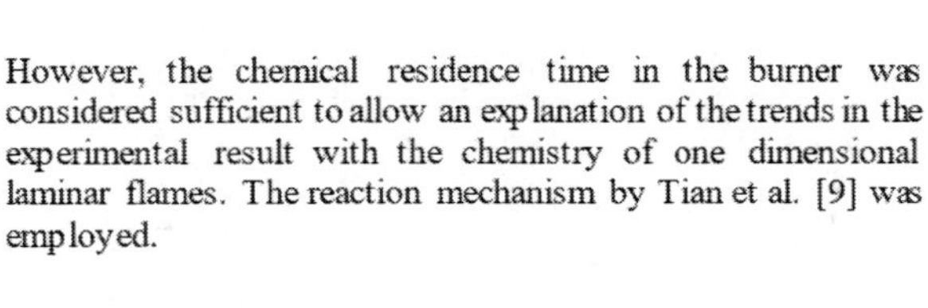

Figure 3 Images of the swirl stabilized ammonia flame at Φ = 1.0 and mean air flow velocity, U_{mean} of 2.62 m/s. Note that the orange chemiluminescence is due to the NH2 α band and the super-heated H_2O vapour spectra.

region of flame stability. The 0° injection angle had the richest blow off limits at low air flow velocities because a large part of the fuel did not mix with air, rather flowed upward to the burner exit without burning. The stability region for the 0° injection angle was the narrowest at high air flow velocities. Vertical fuel injection discouraged proper mixing of the fuel stream and the surrounding swirling air stream, and thus resulted to poor flame stability and large NH_3 emission. For inclined fuel injection, the rich stability limit was extended as injection angle increased. Owing to an increase in injection angle and the consequent increase in the horizontal component of the fuel velocity, turbulent mixing may be enhance at the base of the liner by an increased impingement of the fuel jet on the surrounding swirling air flow.

Figure 5 shows the variations of NH_3 and NO emissions with θ and Φ. The air flow rate was kept constant while the ammonia flow rate was varied in order to vary the global equivalence ratio, Φ. Ammonia emission in the case of vertical fuel injection, θ= 0° was more than the highest NH_3 measuring scale of the FTIR gas analyzer (5000 ppm) at all studied Φ. Therefore, the emissions in that case are not reported in the figure. At a given Φ, NH_3 emission decreased with an increase in θ. This may be due to enhanced mixing at the base of the liner. In other words, an increase in the injection angle may lead to more efficient combustion owing to enhanced mixing. Turbulence and mixing may also be enhanced by an increase in the inlet velocities. An increase in the mean air inlet velocity up to 14 m/s with inclined fuel injection also resulted to a decrease in NH_3 emission at constant Φ. On the other hand, NO emission increased with an increase in θ. The upward trend in NO emission could be due to an increase in NO production with increased NH_3 consumption. Also, with increased NH_3 consumption less NH_3 is available for selective non-catalyzed reduction of NO downstream of the liner thus leading to more NO emission.

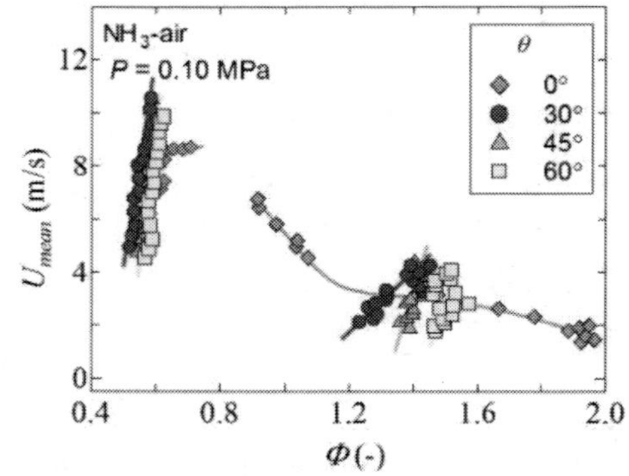

Figure 4 Flame stabilization limit of the non-premixed NH_3-air flames for different injection angles.

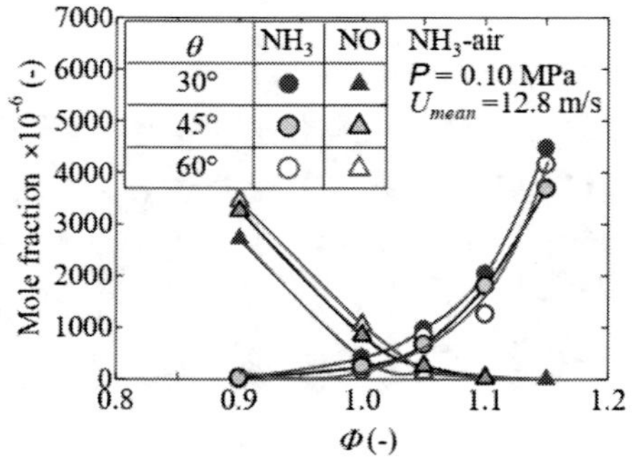

Figure 5 Variations of NH_3 and NO emissions with global equivalence ratio and injection angle.

3.2 Effects of Global Equivalence Ratio on NH_3 And NO Emissions

As shown in Fig. 5, NH_3 emission was low at lean condition but increased steeply at rich conditions with an increase in Φ. On the other hand, high levels of NO emission was recorded at lean condition. NO emission however decreased steeply with an increase in Φ. NH_3 and NO concentrations 4cm from the peak heat release point in one-dimensional simulation of premixed ammonia-air flames showed a similar trend with equivalence ratio. The decrease of NO emission with equivalence ratio could be due to the reduction of NO by the increasing NH_i (i= 2, 1, 0) radicals in the rich flame. NO is predominantly reduced through reaction with the amine radicals [10].

In the micro gas turbine plant operation [6], the global equivalence ratio near the burner exit (primary combustion zone) is rich while that near the liner exit (secondary combustion zone) is lean owing to the injection of secondary air downstream of the liner. It can be understood from Fig. 5 that when the primary zone's Φ is sufficiently rich (>1.1), NO production in the zone will be significantly low. However, the high concentration of unburned NH_3 from the primary zone, when eventually burned in the secondary zone, may result in significant NO production and emission from the combustor. NH_3 and NO emissions were found to be of the same order of magnitude at slightly rich global equivalence ratios. This range of slightly rich equivalence ratios where both NH_3 and NO emissions are low present potentials for simultaneous reduction of NH_3 and NO emission from the combustor.

Hayakawa et al. [5] also observed a similar variation of NH_3 and NO emissions with equivalence ratio in premixed ammonia-air flames stabilized with a smaller swirl burner.

3.3 Effects of Pressure on NH_3 and NO Emissions

Gas turbine combustors are operated at elevated pressures. The pressure ratio employed in the micro gas turbine plant was 3.5 [6]. Therefore, the effects of combustor ambient pressure on NH_3 and NO emissions were investigated in this study at lean and rich conditions. The inlet velocity was kept constant in these experiments to ensure a constant fluid residence time with an increase in pressure. It was confirmed in the present study that the ambient pressure has significant effects on unburned NH_3 and NO emissions from the combustor.

Figure 6 shows the variation of NH_3 emissions with ambient pressure for the different injectors at equivalence ratio of 0.8. NH_3 emission decreased with an increase in pressure. This may be due to increased ammonia consumption resulting from faster burning rate. The mass consumption rate of fuels in turbulent combustion may increase with an increase in pressure owing to

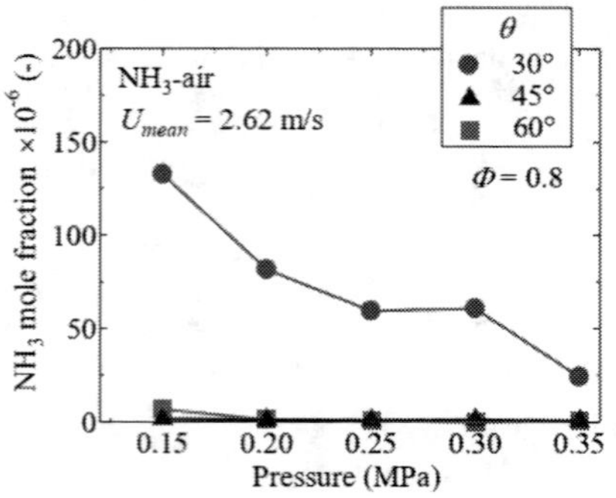

Figure 6 Variation of NH_3 emission with ambient pressure.

increased flame front wrinkling by turbulence, flame instability, and thermo-diffusive effects [11]. Somarathne et al. [4], in their numerical simulation of ammonia-air premixed flames in a gas turbine-like combustor, reported a decrease in unburned ammonia emission with pressure and attributed the result to an increase in the Damköhler number. An increase in the Damköhler number indicates a decrease in chemical reaction time relative to the fluid residence time, suggesting an increased rate of fuel consumption for constant residence time.

On the other hand, measured NO emission decreased with an increase in ambient pressure as shown in Fig. 7 and Fig 8 for θ = 0.8 and 1.2, respectively. The numerical simulations also predicted a decrease in NO concentration with an increase in pressure, thus agreeing with the trend in the experiments. NO concentration has a strong negative sensitivity to the rate constant of the reaction, $H+OH+M=H_2O+M$. In other words, an increase in the forward reaction rate of this reaction which consumes H and OH radicals leads to a decrease in NO concentration because a significant amount of NO production in NH_3 flames proceeds through the reaction of HNO with OH and H. An increase in pressure enhances the forward reaction rate of three-body pressure-sensitive reaction, thereby promoting H and OH consumption and consequently leading to a reduction in NO production and emission from the flame [10].

Observe that the 30° angle of injection resulted to the lowest NO emission at global equivalence ratio of 0.8. This may be due to the relatively high NH_3 emission recorded in the case of this injector. Higher NH_3 emission may contribute to lower NO emission due to NO reduction by the NH_3. At global equivalence ratio of 1.2, NH_3 emission for the three injection angles were more than the measuring range of the gas analyzer.

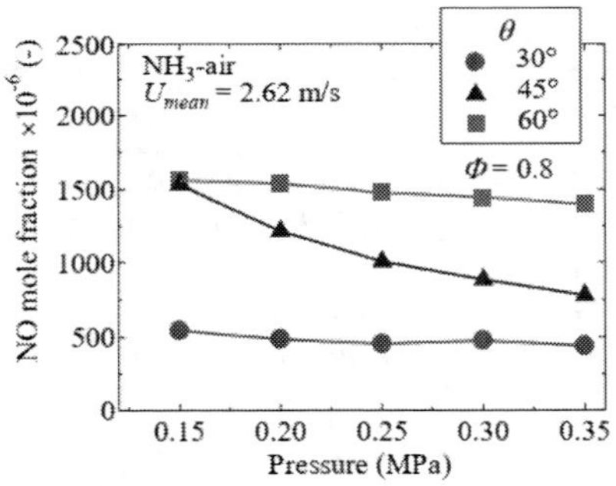

Figure 7 Variation of NO emission with pressure at Φ = 0.8

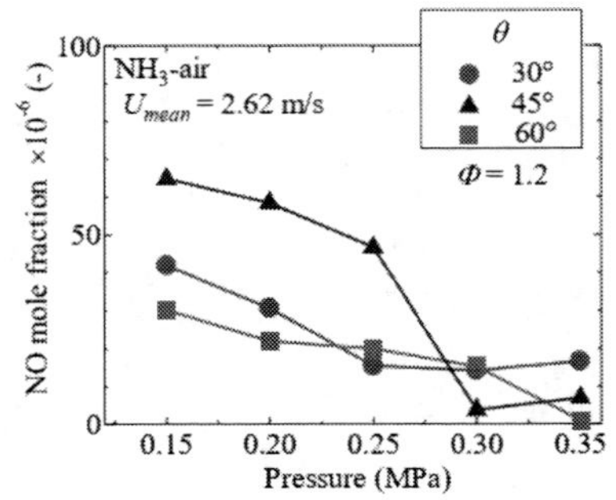

Figure 8 Variation of NO emission with pressure at Φ = 1.2

4 Conclusions

The flame stabilization and emission characteristics of the NH_3 gas turbine swirl burner used at AIST [6] was investigated in this study. The following conclusions can be drawn from the results.

1. It was confirmed that the injection angle is critical to achieving efficient and clean combustion in the combustor. Flame stability may increase with an increase in injection angle owing to increased mixing of the fuel and air. The vertical fuel injection strategy used in [6] resulted in the largest emission of unburned fuel. With an increase in injection angle however, NH_3 emission decreased indicating more efficient combustion. NO emission increased with injection angle due to the increased NH_3 consumption.
2. Unburned ammonia emission increased while NO emission decreased with equivalence ratio. A range of slightly rich equivalence ratios where both NH_3 and NO emissions were low was identified. Operation of the NH_3 gas turbine within this range of slightly rich equivalence ratios may ensure low NO emissions from the combustor.
3. An increase in ambient pressure resulted to a reduction in both NH_3 and NO emissions. NH_3 emission may decrease with pressure due to increased burning rate. On the other hand, NO emission may decrease with pressure as a result of decrease in H and OH radicals which promote HNO and NO production in the flames.

5 Acknowledgment

This research was supported by the Council for Science, Technology and Innovation (CSTI), the Cross-ministerial Strategic Innovation Promotion Program(SIP), "Energy Carriers" (Funding Agency: The Japan Science and Technology Agency (JST)).

References

[1] J.T. Gray, E. Dimitroff, N.T. Meckel, R.D. Quillian Jr., 1966 Automotive Engineering Congress and Exposition, SAE Paper 660156 (1966).

[2] F.J. Verkamp, M.C. Hardin, J.R. Williams, Proc. Combust. Inst. 11 (1967) 985-992.

[3] Hayakawa, T. Goto, R. Mimoto, Y. Arakawa, T. Kudo, H. Kobayashi. Fuel 159 (2015) 98-106.

[4] K.D.K.A.Somarathne, A. Hayakawa, H. Kobayashi, J.Fluid Sci. Tech. 11 (4) (2016) 16-00126.

[5] Hayakawa, Y. Arakawa, R. Mimoto, K.D.K.A. Somarathe, T. Kudo, H. Kobayashi, 21th World Hydrogen Energy Conference, (2016) 626-627.

[6] Kurata, N. Iki, T. Matsunuma, T. Inoue, T. Tsujimura, H. Furutani, H. Kobayashi, A. Hayakawa, Proc. Combust. Inst. 36 (2017) 3351–3359.

[7] A.H. Lefevle. D.R. Ballal. Gas Turbine Combustion (3rd Ed.), CRC Press Taylor & Francis, London, 2010, p. 143.

[8] CHEMKIN-PRO 17.2, ANSYS, Inc.: San Diego, (2016)

[9] Z. Tian, Y. Li, L. Zhang, P. Glarborg, and F. Qi, Combust. Flame 156 (2009) 1413–1426.

[10] Hayawaka, T. Goto, R. Mimoto, T. Kudo, H. Kobayashi, Mech Engineering J 2 (1) (2015), No. 14-00402.

[11] T. Kitagawa, T. Nakahara, K. Maruyama, K. Kado, A. Hayakawa, and S. Kobayashi, Int. J. Hydrogen Energy 33 (2008) 5842–5849.

11th Asia-Pacific Conference on Combustion,
The University of Sydney, NSW Australia
10th -14th December 2017.

Pyrolysis of dieldrin and formation of toxic products

W.N.K. Dharmarathne, J.C. Mackie, E.M. Kennedy, M. Stockenhuber

Process Safety and Environmental Protection Group, Faculty of Engineering

The University of Newcastle, Callaghan NSW 2308

Australia

Abstract

This article investigates the gas phase pyrolysis of dieldrin, one of the most common derivatives of the cyclodiene-based group of pesticides, used extensively in the agricultural sector until its use was banned in most countries (including Australia) in the 1980s. The objective of the present study was to identify its thermal decomposition pathways and to understand the reaction pathways leading to the formation of toxic species including polychlorinated dibenzo-p- dioxins (PCDD), polychlorinated dibenzofuran (PCDF) and their precursors. In order to achieve this aim, experiments were conducted using a tubular silica reactor placed in a furnace, which was fed with a dilute stream of dieldrin in 99.999% He bath gas. The VOC products and gaseous species generated during decomposition were identified and quantified. Thermal decomposition of dieldrin commenced at 673 K, via a retro Diels-Alder elimination reaction and, hexachlorocyclopentadiene (HCCP), benzene and carbon monoxide (CO) were the initial decomposition products identified. As the decomposition temperature increased to approximately 773 K, secondary decomposition products comprising HCl, pentachlorostyrene (PCS), hexachlorostyrene (HCS), polychlorinated benzenes (CBz), and trace amounts of polychlorinated naphthalene (PCN), which are toxic environment pollutants, were identified. Polychlorinated benzenes, which were detected in this pyrolysis study, are acknowledged precursors for PCDD/F formation when O_2 is present in the reacting stream.

1. Introduction

Organochlorine pesticides denote a group of well-known environmental contaminants. Dieldrin and aldrin are two organochlorine insecticides that were introduced to the agriculture sector in the early 1950s having been discovered by Julius Hyman [1]. Dieldrin was used for the control of a wide range of pests and insects in households and to protect field crops [2, 3]. This insecticide is considered to act as a neurotoxin to insects and binds to the membrane of the nervous system of the animal [2]. Aldrin has also been used extensively as a soil-based insecticide because it has a higher vapor pressure than dieldrin. However, aldrin readily converts into dieldrin once it enters the environment or body [1], and therefore, dieldrin is often detected in locations where aldrin was originally applied. However, due to the accumulation of residues of dieldrin in the environment, the use of this insecticide has been strongly regulated. Aldrin was synthesized by the Diels-Alder condensation of hexachlorocyclopentadiene (HCCP) with bicyclo[2.2.1]-2,5-heptadiene and the epoxidation of aldrin was used in the manufacture of dieldrin. Dieldrin is a white crystalline compound.

Corresponding author.
E-mail address: eric.kennedy@newcastle.edu.au

To our knowledge, in the only investigation of the thermal decomposition of dieldrin, Kennedy et al. [4] studied the combustion products of dieldrin at 1173 K undergoing reaction during a 15 minute residence time, and where experimental conditions simulated those in a waste incinerator. CO, CO_2, HCl, Cl_2 and O_2 were the combustion products at 900 °C with 97.6% recovery of CO_2 and no data for products detected at other temperatures were reported. These authors did not attempt to measure the emission of PCDD/F and precursors in their study. Degradation products of the pesticide depended on temperature, residence time and composition of the fed gas stream.

Recognising that the production and usage of dieldrin have been banned for many years, it can still be found in a broad range of environments all over the world [5, 6]. Recently, Wang et al.[5] reported that elevated levels (average 160 pg m^{-3}) of dieldrin were detected in the Australian atmosphere and this value is the highest reported level in the world to date. All organochlorine pesticides, including dieldrin, are widely distributed throughout different parts of the environment, and thermal treatment is one of the methods which could be used to decompose these substances. However, incineration may generate toxic products during treatment, and some decomposition products generated may be more toxic than the original substances. In addition, considerable quantities of these insecticides are known to be stored, and it is vital to understand the thermal behavior of these pesticides in case of an accidental fire that might arise in these storage facilities.

The thermochemical parameters of dieldrin relating to the formation of toxic gaseous, PCDD/F, and their precursors have not yet been studied. Furthermore, a knowledge of the thermal behavior of dieldrin may provide an understanding of the fate of these compound in the environment. From this perspective, the objective of the present study was the identification and quantification of degradation products of dieldrin during thermal decomposition, under non-oxidative conditions.

2. Experimental

2.1 Experimental setup

Technical grade dieldrin ($C_{12}H_8Cl_6O$, 96% purity) was obtained from Sigma-Aldrich and ultrahigh purity He gas (99.999%) used as a bath gas to establish non-oxidative reaction condition. With details described elsewhere [7, 8], the bench scale apparatus consisted of a three-zone furnace, high purity quartz reactor (10 mm i.d) and product collection systems. Dieldrin was loaded into a quartz boat (3 mm in length) and installed in the entry zone of the furnace which was maintained at 533 K to engender a slow and controlled rate of evaporation of dieldrin in the bath gas. The average sublimation rate of reactant is 1.5 mg/min and feed concentration was 2500-3000 ppm. An empty impinger submerged in a glycol bath at 263 K was connected downstream of the reactor to trap VOC products and later extracted with dichloromethane (DCM). Following the trap, the remaining gaseous product species passed through a Fourier Transform Infrared (FTIR) spectrometer equipped with a 0.19 m path length PTFE cell. Each gas sample was scanned 16 times in the spectral range of 500- 4000 cm^{-1} at a resolution of 1 cm^{-1}. We studied the effect of temperatures in the range of 623 -923 K, at a residence time of 5 s and each experiment lasted 1 hour. Connection tubes (PTFE), reactor and rods were cleaned by a solution of DCM and acetone between each experiment to recover any product species and to avoid contamination of products in subsequent runs. For the detection and accurate quantification of benzene, separate experiments were carried out with the exhaust gases bubbled through a chilled solvent trap which contained 5 ml of hexane.

2.2 Product analysis

The analysis of VOC products was performed on an Agilent 7200 quadrupole time-of-flight GC/MS, equipped with a DB-5MS column (30 m length, 0.25 mm i.d., 0.25 μm film thickness). For the VOC analysis, oven temperature program 30°C (held for 5 min), then a rise to 290 °C at 6 °C min^{-1} (held for 1 min). Pyrolysis products were identified by matching the mass spectra with NIST library and reference authentic standards. The identification and quantification of gaseous species employed a QASoft software package which contains reference spectra and region integration routine for quantitative analysis.

3. Results and discussion

3.1 Gas products analysis

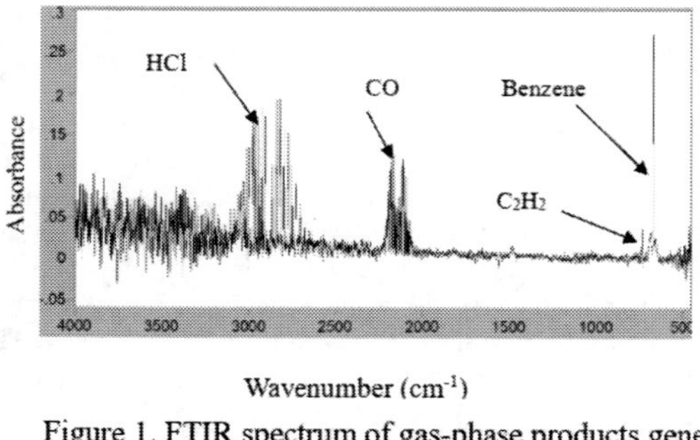

Figure 1. FTIR spectrum of gas-phase products generated during decomposition of dieldrin at 873 K

Figure 1 displays a typical FTIR spectrum with identified gaseous species (HCl, CO, C_2H_2, and benzene) from the experiment performed at 873 K. HCl was observed as a dominant gaseous product at high temperatures and first detected at around 773 K as a secondary product. At a temperature range of 823-873 K, the yield of HCl rapidly increased and reached a maximum yield of 1.39 mol/mol of dieldrin at 923 K (fig 2). CO was detected as a primary combustion product at 673 K, exhibiting a rapid increase in concentration in the temperature region of 773-873 K and a maximum CO yield of 0.98 mol/mol of dieldrin was detected at 923 K. This value is a good indicator of mass balance in our system as dieldrin contains only one oxygen atom, and it converts to CO in pyrolysis. In contrast, the trend of benzene yield shows a more moderate increase after the reaction temperature reached 773 K.

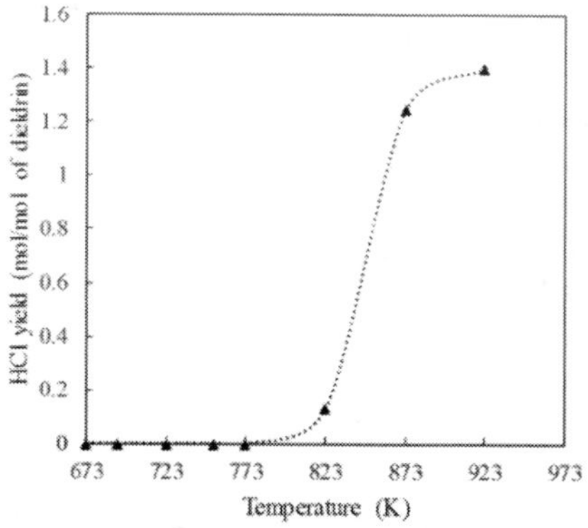

Figure 2. Yield-temperature profile of HCl

3.2 VOC products analysis

Figure 3 presents typical GC-MS chromatograms for experiments performed at 773 K and 923 K with identified species and retention times. All product species were identified by injecting authentic reference standards except for hexachlorostyrene (HCS) and pentachlorostyrene (PCS).

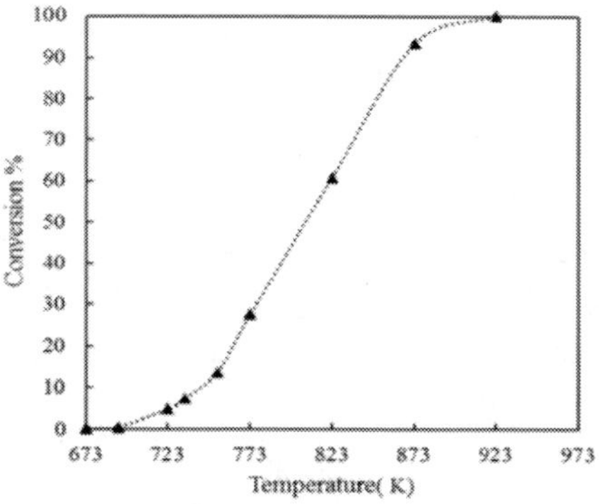

Figure 4. Conversion percentage of dieldrin as a function of temperature

Fig. 4 presents the conversion percentage of dieldrin over the temperature range of 673 to 973 K. Under pyrolytic conditions, the decomposition of dieldrin started at 673 K and hexachlorocyclopentadiene (HCCP), benzene, and carbon monoxide (CO) were the first identified products at 673 K. It is suggested that initial decomposition proceeds via the retro Diels-alder elimination reaction to HCCP, which is the lowest energy decomposition pathway[9]. HCCP is used in the production of several organochlorine pesticides including endosulfan, aldrin, dieldrin, chlordane and heptachlor [10]. The conversion of

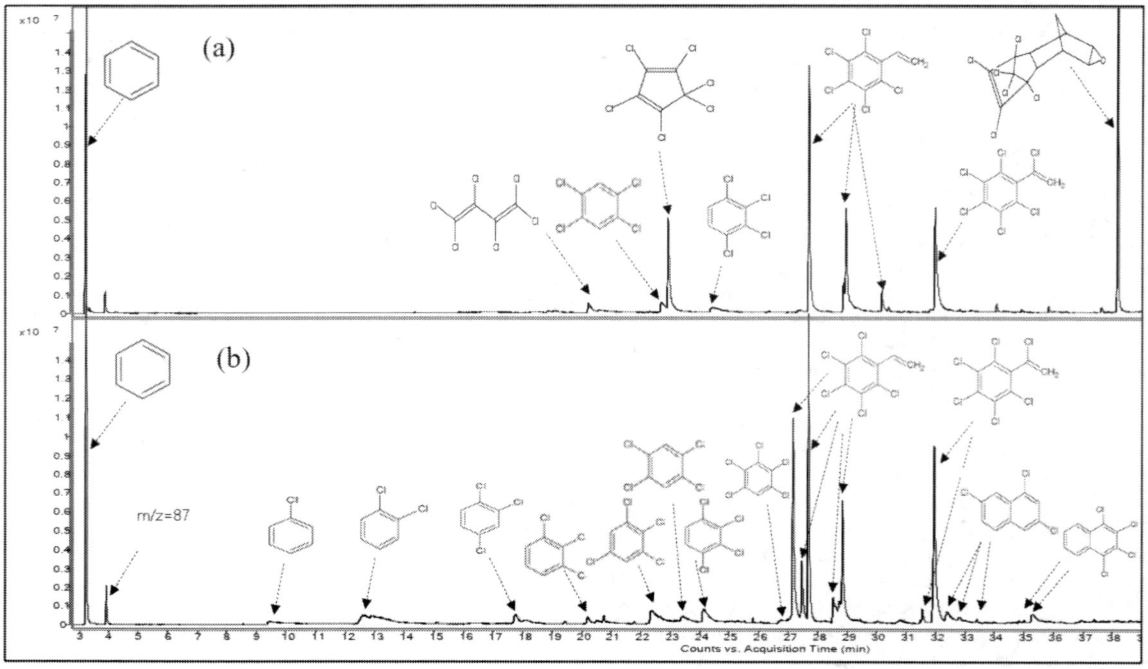

Figure 3. Assignment of peaks in the zoomed chromatogram from the analysis of pyrolysis products of dieldrin (a) at 823 K (b) at 923 K

dieldrin under non-oxidative conditions initially undergoes a slow rate of decomposition and then, above 773 K, a rapid elevation in conversion between 773-873 K and almost complete decomposition of dieldrin is observed at 923 K. In our previous studies [7, 11], we have investigated the pyrolysis of endosulfan which is another cyclodiene pesticide, and decomposition starts at a temperature as low as 573 K, via a retro Diels-Alder elimination to HCCP. It seems that pyrolysis of dieldrin requires a higher activation energy than endosulfan. For example, at 723 K, 86% of endosulfan decomposed, while only 5% of dieldrin decomposed at the same temperature under the same experimental conditions.

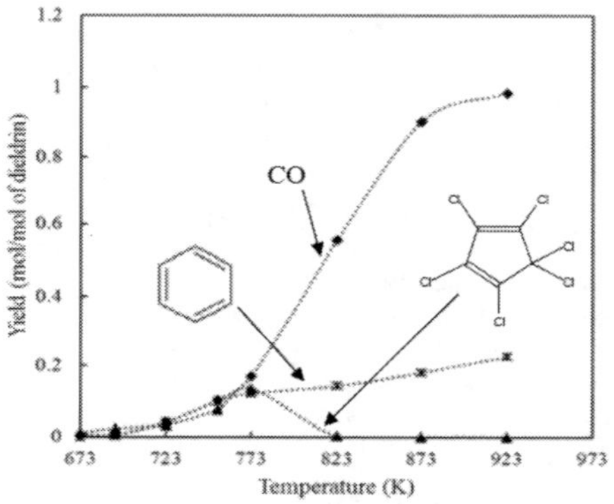

Figure 5. Yield-temperature profiles of selected products species, benzene, HCCP and carbon monoxide

Figure 5 presents the yield temperature profile of CO, HCCP, and benzene as a mole fraction of initial dieldrin loading. The yield of CO, HCCP and benzene are approximately equal at temperatures up to 773 K but this trend deviates at high temperatures. It can be seen that secondary decomposition reactions starts around 773 K with arising secondary products (HCl, PCS and HCS) and elevated levels of CO. According to our theoretical calculations, retro Diels-Alder elimination of dieldrin produces HCCP and norbornadiene-oxide (C_7H_8O) [9]. Norbornadiene-oxide can further decompose to CO and benzene. A maximum yield of HCCP was observed at 773 K (0.13 mol/mol of dieldrin), following which the yield declines and, eventually it is completely absent as a product at 823 K. HCCP was the major primary product from the pyrolysis of endosulfan at 773 K with the yield in excess of 0.55 mol/mol of endosulfan and it too completely disappeared at 823K [7]. This suggests that retro Diels-Alder elimination is a primary decomposition pathway of dieldrin pyrolysis as it was for endosulfan.

Pentachlorostyrene (PCS) and hexachlorostyrene (HCS) are the secondary combustion products, first detected at around 723-773K and their yields increase with rising temperatures, exhibiting a steep rise between 823-923 K (Figure 6). A lack of commercially available reference standards for these compounds prevented us from performing quantitative analysis. The mass spectra of PCS and HCS were not found in the NIST library, and these species were identified by comparison of m/z data and their abundance with literature values [12, 13]. From the pyrolysis of dieldrin, almost five isomers of PCS and two isomers of HCS are detected while PCS is the dominant product in the system. In our previous work [7], pyrolysis of endosulfan produced tetrachlorostyrene (TCS) and PCS which were detected at temperatures greater than 648 K. TCS was the dominant product in the pyrolysis of endosulfan. As far as we know, polychlorinated styrenes were never used as commercial products, and these compounds may be produced as by-products of incineration processes involving chlorinated compounds. Polychlorinated styrenes have the potential for high bioaccumulation, and there have been a number of studies indicating that these polychlorinated styrene compounds exhibit adverse effects on health and environment [14, 15].

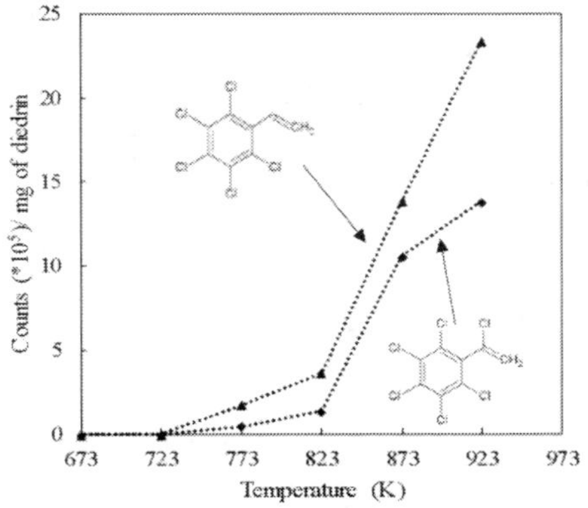

Figure 6. Yield-temperature profiles of PCS and HCS

In addition to the above products, at high temperatures, we have detected trace quantities of several isomers of trichloro-naphthalene and tetrachloro-naphthalene (Fig. 4). Polychloronaphthalenes (PCNs) are environmentally persistent, bioaccumulative and toxic industrial compounds, and researchers have reported that PCNs exhibit toxic equivalents comparable to or even higher than PCDD/F [16].

3.3 Formation of PCDDF and their precursors

During studies on the pyrolysis of dieldrin, all species in the mono-chlorinated to penta-chlorinated benzene series were detected, and the formation of these compounds was first detected at temperatures higher than 773 K, as presented in Fig. 4. The yield of these chlorinated benzenes decreases with decreasing number of chlorine atoms. Chlorinated benzenes are well-recognized precursors for PCDD/F formation when oxygen is present in the reacting system [17, 18]. In a subsequent investigation, we have studied the oxidative decomposition of dieldrin, and significant yields of PCDD/F were detected, comprising all the range of mono to octa-chlorinated dibenzofurans and dibenzo-p-dioxins. In the oxidation of dieldrin [9], we have also discovered a range of chlorinated phenols as PCDD/F precursors.

4. Conclusions

In this contribution we have studied the pyrolysis of dieldrin, an important member of the cyclodiene pesticide group. Unlike the pyrolysis of endosulfan, the non-oxidative decomposition of dieldrin commences at a comparatively higher temperature. We have identified the initial steps of the pyrolysis reaction, which involve a retro Diels-Alder elimination reaction to HCCP and norbornadiene oxide which can further decompose to produce CO and benzene. At temperatures exceeding 723 K, secondary combustion products start to form including PCS, HCS, chlorinated benzenes, and minor quantities of chlorinated naphthalenes. All these compounds have been identified to be bioaccumulative and toxic. Oxidative decomposition of dieldrin favours the formation of a broad range of PCDD/F which are under current investigation. The burning or uncontrolled combustion of dieldrin and dieldrin-contaminated materials should be avoided as it generates toxic compounds upon their combustion.

5. Acknowledgment

The authors would like to thank the Australian Research Council for financial support of this project and WNK thanks the University of Newcastle, Australia for a postgraduate research scholarship.

References

[1] J.L. Jorgenson, Environmental Health Perspectives, 109 (2001) 113-139.
[2] Gladstone C.M , W. J.L, Journal of Agricultural and Food Chemistry 25 (1977) 489-493.
[3] D.E. Stevenson, E.F. Walborg Jr, D. Warner North, R.L. Sielken Jr, C.E. Ross, A.S. Wright, Y. Xu, L.M. Kamendulis, J.E. Klaunig, Toxicology Letters, 109 (1999) 123-186.
[4] M.V. Kennedy, B.J. Stojanovic, F.L. Shuman, Journal of Agricultural and Food Chemistry, 20 (1972) 341-343.
[5] X. Wang, K. Kennedy, J. Powell, M. Keywood, G. R, P. Thai, P. Bridgen, S. Broomhall, C. Paxman, F.K. Waniae, J.F. Muellera, Environ. Sci –Processes & Impacts, 17 (2015) 525-532.
[6] B.T. Hargrave, L.A. Barrie, T.F. Bidleman, H.E. Welch, Environmental Science & Technology, 31 (1997) 3258-3266.
[7] N.K. Dharmarathne, J.C. Mackie, E.M. Kennedy, M. Stockenhuber, Proceedings of the Combustion Institute, 36 (2017) 1119-1127.
[8] N.K. Dharmarathne, J.C. Mackie, E.M. Kennedy, M. Stockenhuber, Environmental Science & Technology, 50 (2016) 10106-10113.
[9] W.N.K. Dharmarathne, J.C. Mackie, E.M. Kennedy, M. Stockenhuber. To be published.
[10] P. Robitschek, C.T. Bean, Industrial & Engineering Chemistry, 46 (1954) 1628-1632.
[11] J.C. Mackie, W.N.K. Dharmarathne, E.M. Kennedy, M. Stockenhuber, Proceedings of the Australian Combustion Symposium, (2015).
[12] D. Carrizo, J.O. Grimalt, Journal of Chromatography A, 1216 (2009) 5723-5729.
[13] M. Coelhan, I. Reil, G. Rimkus, H. Parlar, Environmental Science & Technology, 34 (2000) 4695-4700.
[14] I. Chu, V.E. Secours, D.C. Villeneuve, V.E. Valli, Journal of Toxicology and Environmental Health, 10 (1982) 285-296.
[15] M. Tarkpea, I. Hagen, G.E. Carlberg, P. Kolsaker, H. Storflor, Bulletin of Environmental Contamination and Toxicology, 35 (1985) 525-530.
[16] H. Park, J.-H. Kang, S.-Y. Baek, Y.-S. Chang, Environmental Pollution, 158 (2010) 1420-1427.
[17] R.J. Crawford, S. Lutener, H. Tokunaga, Canadian Journal of Chemistry, 55 (1977) 3951-3954.
[18] W.F. Schmidt, C.J. Hapeman, L.L. McConnell, S. Mookherji, C.P. Rice, J.K. Nguyen, J. Qin, H. Lee, K. Chao, M.S. Kim, J. Agric. Food Chem, 62 (2014) 2013-2030.

11th Asia-Pacific Conference on Combustion,
The University of Sydney, NSW Australia
10th-14th December 2017.

A Comparative Study of a Novel Multi-swirl Lean Direct Injection and Conventional Single Swirl Gas Turbine Burner for Reduced Emissions and Combustion Instability

Deepika, V., Chakravarthy, S.R., Muruganandam, T.M., Raja Bharathi, N

Department of Aerospace Engineering &

National Center for Combustion Research and Development (NCCRD)

Indian Institute of Technology Madras

Chennai, Tamil Nadu, India

Abstract

A gas turbine combustor design must take into consideration performance parameters like combustion efficiency, pressure drop across the combustor, stability limits, exit temperature distribution, emissions, combustion instability, etc. for a good design. This paper discusses the comparison of a novel multi-swirl burner* employing lean direct injection (LDI) with a conventional premixed single swirl gas turbine engine burner in terms of the various performance parameters. Pressure drop is highly reduced and the stability limits are enhanced using the novel multi-swirl LDI burner developed. The LDI burner achieves a uniform mixing of fuel and air, eliminating hot spots in combustion; leading to reduced overall emissions. Results also show that the LDI burner is less susceptible to combustion instability. These advantages of the multi-swirl LDI burner make it a potential contender to be employed in the combustor of gas turbine engines.

1 Introduction

A gas turbine combustor must satisfy a set of broad ranging requirements for a good design; typical considerations are good combustion efficiency, ease of ignition, wide stability limits, low pressure drop across the combustor, exit temperature distribution suited to maximize the lives of turbine blades, low smoke and gaseous pollutants, and reduced combustion instability. With the existing combustor designs, an improvement in one aspect of performance often requires a compromise in some other aspect, for example, emission of NO_x can be reduced by designing for a lean combustion but this impacts flame stabilization which increases other emissions like CO or unburned hydrocarbon (UHC) and also aggravates the problem of combustion instability. Thus new combustor designs have to be developed, resolving conflicts.

Various efforts in this direction have been discussed in the literature. Tacina [1] reviews various low NO_x concepts. Lean premixed pre-vaporized (LPP) burner concept is a widely discussed technique for low NO_x emission in the literature. But the low NO_x potential of LPP is set back by the disadvantages of narrow stability limits and its susceptibility to auto ignition, flashback and combustion instability [1, 2]; leading to focus on other techniques as LDI. Fric *et al.* [3] proposes a combustor that incorporates LDI using fuel and air jets injecting tangentially into the same cross-sectional area of the chamber. Following this, various configurations of LDI involving multiple swirlers were studied [4, 5] with the focus on achieving low NO_x. Most of the work discussed in the literature focuses primarily on NO_x reduction, but performance of the burner in terms of other aspects as pressure losses, stability limits, and combustion instability characteristics has not been discussed much. Addressing combustion instability is paramount for both land based and aviation gas turbine engines considering the catastrophe it can cause. Therefore efforts in the novel burner* developed are towards achieving low overall emissions and curtailing combustion instability, simultaneously addressing other aspects of burner performance.

The novel burner thus developed, named LDI-4C consists of multiple swirlers meshed together very closely throughout the length of the liner. Fuel is injected directly into the swirling air stream at the center of each individual swirlers. This work is an extension of the work done by Preethi *et al.* [6], where the LDI burner consisted of multiple swirlers arranged in a single plane at the head end with fuel injections in each swirler. The multi-swirl mesh burner discussed in the current paper involves swirlers in different planes, all along the liner. A comparison is drawn between the performances of the LDI-4C burner with a conventional single swirler premixed gas turbine burner, as it is prototypical of the LPP concept. Attempt is to compare the burners for all the performance parameters.

2 Experimental Set-up and Procedures

2.1 Experimental Set-up

Figure 1(a) shows the experimental set-up for testing of the LDI-4C burner. It consist of a settling chamber of internal diameter (I.D.) 320 mm (10 mm thick wall) which is fed by a high pressure air supply line. Following that is a sudden contraction to an 86x86 mm square inlet duct, providing a contraction ratio of 10.87. This is done to have an acoustically open boundary condition at the inlet. The 150 mm long inlet duct houses flow straighteners and meshes of 600 and 400 μm size for reducing the turbulent structures in the flow. Downstream of that is an extension duct 70 mm long, preceding the test section that houses the burner. The LDI-4C burner is provided with a fuel supply arrangement supplying fuel at the center of each of the individual swirlers of the multi-swirler array. The liner in the test section is simulating one

Corresponding author E-mail address: deepikav91@gmail.com
***Patent pending**: Application number: 201641024206, July 2016, Chennai
PCT application number: PCT/IN2017/050291

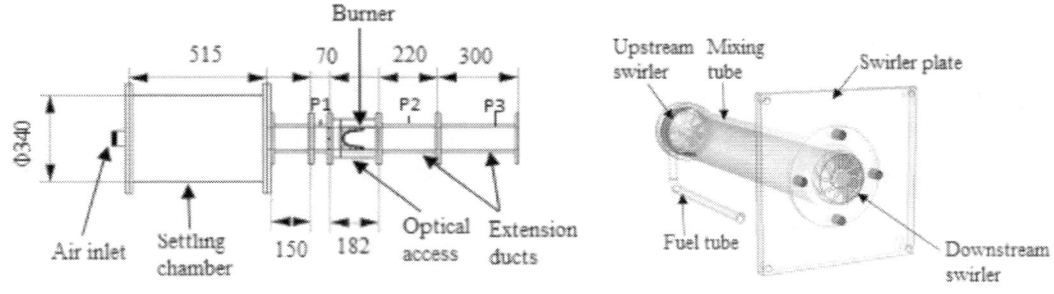

(a) (b)

Figure 1: (a) Experimental set-up for testing of the burners (dimensions in mm), (b) Single swirler burner assembly

sector of an annular combustor with similar swirl-mesh configuration. Downstream of the test-section are added extension ducts. Overall configuration/length of the set-up is retained same for testing of the single swirl conventional gas turbine burner, and the assembly shown in Fig. 1(b) is mounted in the same location as the LDI-4C burner such that plane of the single swirl burner matches the most upstream location of the LDI-4C burner. The complete single swirler burner assembly fits inside the duct same as the LDI-4C set-up, maintaining overall duct configuration the same from acoustics point of view, which is a square duct of side 86 mm and total length 922 mm with acoustically open boundary conditions at both ends. The single swirl burner has a 200 mm long tube of 24 mm I.D., the upstream end of which is a 10-vane swirler with a 45° vane angle. Fuel is injected at the center of this swirler and is mixed with air in the tube before it reaches the second swirler with a vane angle of 45° at the downstream end. The flame is anchored downstream of this swirler. Air and fuel (CH_4) flow rates are controlled through separate Alicat Scientific mass flow controllers of accuracy 1 %.

2.2 Measurement Procedures

Characterization of the burners is done in terms of pressure drop, lean blow-out limit, exit temperature and pattern factor, emissions – CO_2, CO, UHC, NO_x, soot and combustion instability. Pressure drop is measured using a DP Measurement-TT570-S micro manometer connected to the pressure port, P1, of I.D. 2.5 mm. The micro-manometer can measure pressures in the range 1-7.5 kPa with an accuracy of $\pm$ 1 % of reading. An S-type thermocouple with a bead diameter of 0.5 mm is used to measure temperature at the exit plane. Temperature measurement is made at a 9x9 grid spanning the cross-sectional area at the exhaust. Mass flow weighted mean of the temperatures thus measured is the average exit temperature. Gaseous emission measurements are made at the center line of the burner at the exit plane using Indus Scientific, FGA53X gas analyzer. Soot is measured at 390 mm downstream of the settling chamber employing laser extinction method as described in Shaddix *et al.* [7]. Unsteady pressure measurements for studying combustion instability are made at three locations P1, P2 and P3 in Fig. 1(a) along the length of the combustor L (922 mm), corresponding to 0.2, 0.55 and 0.9 L, respectively. However only the results obtained at P2 are discussed in this paper, as the unsteady pressure measurements by all the transducers are qualitatively similar; so the one recording maximum amplitudes is chosen for ease of comparison. The measurements are made using PCB103B02 piezoelectric transducers of sensitivity 217.5 mV/kPa. Unsteady pressure data is acquired at a sampling rate of 10 kHz for 3 seconds. The burners are tested at different equivalence ratios from 0.8 to blow-out limit for flow rates corresponding to velocities of 1.4 to 5.4 m/s in the duct upstream of the burner (600 - 2400 lpm of air at standard conditions). The maximum test velocity is matched to the average bulk velocity of air upstream of the combustor of GE-F404 engine, as calculated from the specifications of the engine obtained from Hill and Peterson [8].

3 Results and Discussions

The individual performance parameters for the two burners are presented in this section, in the order below: pressure drop across the burner, lean blowout limit, gaseous emissions, exit temperature, pattern factor, soot emissions, and combustion instability. Pressure drop across the burner represents the resistance offered to the flow of air across the burner. Higher pressure drop leads to higher specific fuel consumption. The maximum pressure drop (for highest test velocity) was found to be 0.2 % for the LDI-4C burner. The single swirl burner on the other hand, had a pressure drop of 5.4 %, even in the lowest velocity case. Lefebvre [9] states it to be 4-8 % for conventional combustors. Low pressure drop of LDI-4C burner is because of the different swirl mesh arrangement which gives higher flow area for air through the burner.

Lean blow-out (LBO) is the lower limit of overall equivalence ratio (ϕ) at which the burner has a stable flame. For LDI-4C burner, the equivalence ratio where partial blowout of flame starts occurring at individual swirlers is taken as the LBO limit. Experiments show (Fig. 2(a)) the LDI-4C burner to have lower LBO limits compared to single swirl burner. LBO limit for LDI-4C is roughly half of that for single swirl burner for all the velocity cases tested. This is made possible by the distributed flame holding at each individual swirlers, with lower local velocities, and diffusion mode of flame in LDI-4C.

Inlet conditions for comparison of emissions from the two burners is chosen such that there is a fully anchored flame in both the burners in the chosen regime. A fixed ϕ of 0.7 is chosen to compare emissions results. This restricts the maximum inlet velocity to 3.2 m/s for a stable flame in the single swirl burner. It is seen from Fig. 2(b) that the LDI-4C burner achieves a combustion efficiency on par with the single swirl burner as observed from the CO_2 concentration in the exhaust gases, at higher velocities. This is further affirmed by the low CO and UHC concentrations; both formed as a result of improper mixing and incomplete combustion. Less than 0.01 % of CO and 6 ppm of UHC in all the test cases indicate good combustion efficiency for both the burners.LDI-4C is also very effective in reducing NO_x. Emission index of NO_x (g NO_x/kg fuel) is less than 1 for all the test cases (Fig. 2(c)). This is similar to the single swirl burner cases, showing no NOx penalty is incurred by using LDI-4C burner instead of single

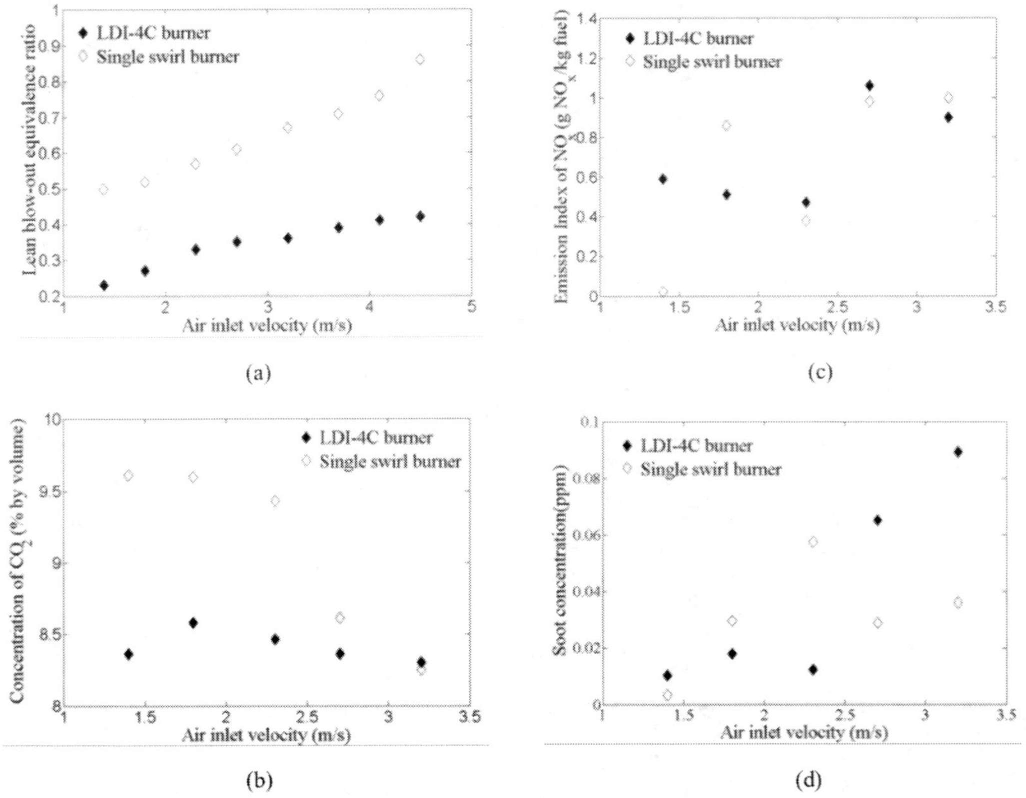

Figure 2: Comparison of the burners in terms of: (a) Lean blow out limit, (b) CO_2 emission, (c) Emission Index of NO_x, (d) Soot

swirl burner. The distributed injection of fuel eliminates the possibility of any zone having fuel rich mixture, which would have the propensity for increased NO_x. It is also substantiated by the uniform temperature distribution at the exit for the LDI-4C burner; a necessity for reducing the differential thermal loading on turbine blades. The extent of uniformity in exit temperature distribution is quantified by 'pattern factor' defined in [9] as:

$$Pattern\ factor = \frac{T_{max} - T_4}{T_4 - T_3} \quad (1)$$

where T_4 is the mass flow weighted mean exit temperature, T_{max} is the maximum exit temperature and T_3 is the temperature of the inlet air. The pattern factor is observed to be less than 0.2 for all the velocity cases for LDI-4C burner (Fig.3), implying a uniform exit temperature distribution.

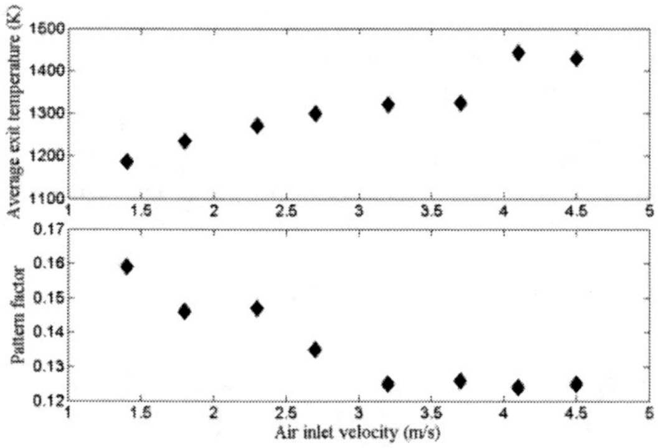

Figure 3: Exit temperature and pattern factor for LDI-4C burner

It is also observed that with increase in air inlet velocity, the average exit temperature obtained is higher indicating a better efficiency at higher velocities. This could be due to increased turbulence assisting mixing of fuel and air. However it can only be confirmed with diagnostic studies on the burner for velocity and mixing fields. It is also seen that the pattern factor values are significantly low at higher velocities indicating a more uniform mixing of fuel and air than that at lower velocities. This inference can be further substantiated by the decreasing difference between the CO_2 and NO_x values for the two burners at higher velocities. Another emission of concern is the particulate matter, predominant composition of which is soot. Soot is formed in significant quantities in a diffusion flame compared to premixed flames. The concentration depends on the extent of fuel-air mixing in diffusion flames and the fuel used. As the LDI-4C burner is a diffusion flame burner; it is intuitive that the flame will be sooty as against the single swirl premixed burner. But on the contrary, it is observed that the flame is non-sooty with negligible quantities of soot recorded and a completely blue flame observed. Fig. 2(d) shows that the LDI-4C burner is comparable to premixed flame in terms of soot; concentration is less than 1ppm in all the cases. This low soot emission may also be due to the methane fuel used. All the measurements in Fig. 2 were repeated four times and uncertainty levels due to test repeatability was found to be less than ±5 % for all the parameters.

Although both the burners compare well in emission characteristics; the single swirl burner is marked by strong instability characteristics, which is a huge set-back for the acceptance of LPP concept as a successful low NO_x technology for gas turbines. It is seen from Fig. 4 that the amplitudes of pressure oscillations for LDI-4C burner is lesser by a

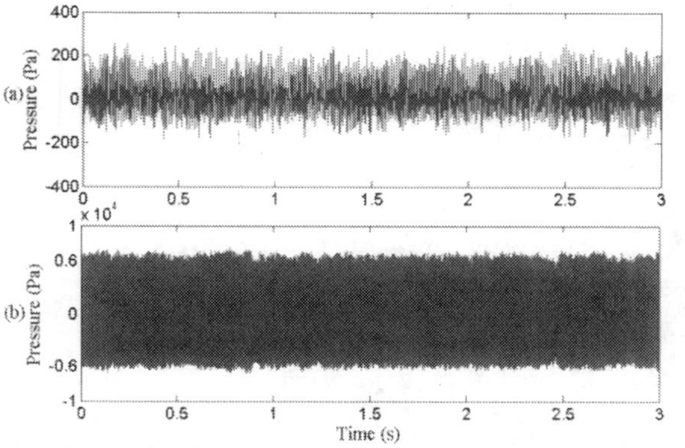

Figure 4: Unsteady pressure data at P2 for $\phi = 0.7$ and 600 lpm air: (a) LDI-4C burner and (b) Single swirl burner

factor of 30 compared to that of single swirl burner. The unsteady pressure data is post processed to obtain the frequency spectra shown in Figs. 5 & 6. It is seen that the LDI-4C burner gives a wide spectrum of frequencies indicating there is no single dominant frequency (Fig. 5). This is against the sharp contrast of single swirl burner where a highly dominant peak is obtained at a particular frequency (Fig. 6). This behavior is the same for different velocities studied. Thus LDI-4C is found to be less susceptible to combustion instability compared to single swirl burner.

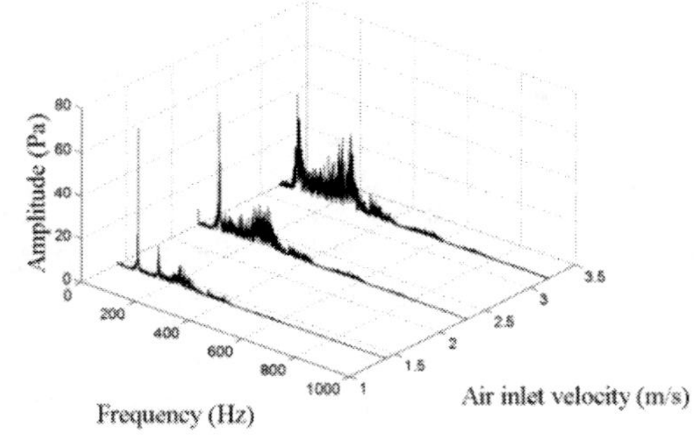

Figure 5: Spectra obtained for LDI-4C burner at $\phi = 0.7$ for different inlet velocities

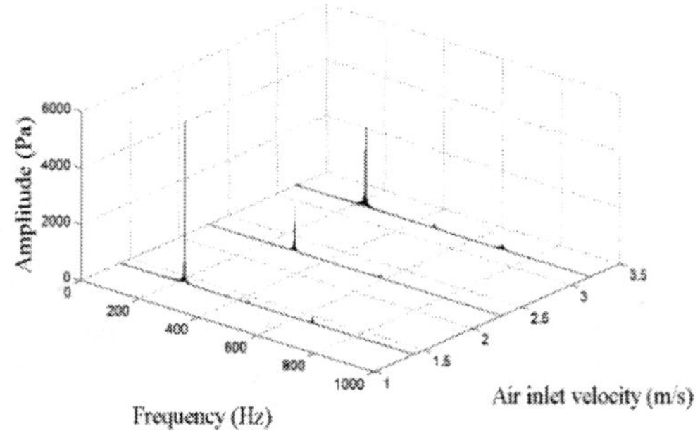

Figure 6: Spectra obtained for single swirl burner at $\phi = 0.7$ for different inlet velocities

4 Conclusions

The novel LDI -4C burner, having multiple swirlers arranged closely, offers very less pressure drop across the burner; thus would reduce specific fuel consumption compared to the conventional single swirl burner. Uniform and rapid mixing of fuel and air achieved by the local fuel injection at each swirler throughout the burner makes possible stable flame over a wide range of equivalence ratios; providing uniform temperature distribution with no hot spots, reduced gaseous emissions and a non-sooty flame. The burner can operate at very lean conditions of fuel compared to the conventional single swirl burner. Stable lean burning makes feasible operation of the combustor at any desired equivalence ratio as against the conventional gas turbine combustor which has to be operated at slightly rich/stoichiometric conditions for stable flame holding; increasing the possibility of NO_x. This also eliminates the need for secondary and tertiary air as required in conventional combustors for completing the combustion and cooling the mixture; this would also reduce the length of the combustor required. In addition, the LDI -4C burner is less susceptible to combustion instability. Thus the LDI-4C burner addresses effectively the different aspects of efficient combustor design for low emissions and reduced combustion instability establishing its supremacy over the single swirl premixed burner.

References

[1] R. Tacina., Low NO_x Potential of Gas Turbine Engines, 28th Aerospace Sciences Meeting, Reno, Nevada, USA,1990, Paper No. AIAA-90-0550.

[2] S.M. Correa, Power Generation and Aero propulsion Gas Turbines: From Combustion Science to Combustion Technology. Proceedings of Combustion Institute 27, (1998) 1793-1807.

[3] T.F. Fric and A. Gulati, Low Emission Combustor Having Tangential Lean Direct Injection, United States Patent No. 5, 479, 781, 1996.

[4] J. Cai., S.M. Jeng, and R. Tacina, Multi-Swirler Aerodynamics: Comparison of Different Configurations, Proceedings of ASME Turbo Expo, Amsterdam, Netherlands, 2002, Paper No. GT-2002-30464.

[5] R. Tacina, C. Mao and C. Wey, Experimental Investigation of a Multiplex Fuel Injector Module with Discrete Jet Swirlers for Low Emission Combustors, 42nd Aerospace Sciences Meeting and Exhibit, Reno, Nevada, USA, 2004, Paper No. AIAA–2004–0185

[6] R.S. Preethi, S.R. Chakravarthy, T.M. Muruganandam, Rajesh Kanna, A Novel Multi-Swirler Lean Direct Injection Combustor for Low Pollutant Emissions, 10th Asia Pacific Conference on Combustion, Beijing, China, 2015.

[7] C.R. Shaddix and K.C. Smyth, Laser Induced Incandescence Measurements of Soot Production in Steady and Flickering Methane, Propane, and Ethylene Diffusion Flames, Combustion and Flame, 107(4) (1996) 418-452.

[8] P.G. Hill, C.R. Peterson, Mechanics and Thermodynamics of Propulsion, Addison-Wesley Publishing Company, California, USA, 1992

[9] A. Lefebvre, D. Ballal and D. Bahr, Gas Turbine Combustion: Alternative Fuels and Emissions, CRC Press, Boca Raton, Florida, USA, 2010.

11th Asia-Pacific Conference on Combustion,
The University of Sydney, NSW Australia
10th -14th December 2017.

Linear instability and DC shift in tactical missile solid rocket motors – a computational study

Vishal Wadhai[1*] and Varunkumar S[1]

[1] Thermodynamics and Combustion Engineering Laboratory,
Department of Mechanical Engineering,
Indian Institute of Technology, Madras, Chennai 600 036, India

Abstract

A CFD framework is developed for simulating the unsteady flow through the port of a side-burning solid rocket motor to capture the 'longitudinal mode acoustic instability'. The flow is modeled as inviscid and axisymmetric, simulated by solving the Euler equations in ANSYS Fluent. Propellant gasification is modeled as mass inlet boundary condition and the mass flux is taken as a function of the static pressure at this boundary. Combustion-acoustic coupling in the linear phase is accounted for by a response function formulation in which the burn rate fluctuations is calculated as a product of frequency response and corresponding pressure perturbation amplitude in the frequency domain. This is shown to lead to an exponential growth of pressure perturbations in the initial phase as observed in actual motor firings. This then transitions into a limit cycle unlike the DC shift observed in experiments. By accounting for the extinction-re-ignition of propellant subject to high amplitude pressure oscillations known to lead to burn rates as high as ten times the mean value for a fraction of the wave time period transition to DC shift is captured – a propellant centric theory for this phenomenon as opposed to earlier ones based on steep fronted waves.

1 Introduction

Solid rocket motors (SRMs) used in tactical missile applications frequently encounter longitudinal mode acoustic instability. Typical chamber pressure-time trace in such unstable motors show a phase of exponential growth of oscillations (indicative of linear instability) followed by a sudden increase in mean chamber pressure (up to 30% of design pressure), termed DC shift, accompanied by large amplitude oscillations. An example can be found in Fig. 2 of [1]. Exponential growth of the pressure wave (linear instability) occurs when amplification rate of pressure perturbations due to combustion-acoustics coupling (α_{prop}) exceeds the decay rate (α_{damp}) due to different damping mechanisms present in the motor. The coupling between propellant combustion and chamber acoustics is quantified by the frequency response (R_p) of the propellant as given by (1).

$$R_p = \frac{r'/\bar{r}}{p'/\bar{p}} \qquad (1)$$

where, r and r' and p and p' are the mean and fluctuation in propellant burn rate and pressure respectively. Obtaining R_p is a problem in itself and has been addressed in the past by many researchers, for instance see [2, 3]. Current work focuses on obtaining pressure-time trace for an actual rocket motor assuming that the propellant frequency response is known. Earlier attempts along this direction include the work reported in [4, 5]. These efforts have focused on the gas-dynamic aspects of the unsteady port flow assuming a passively responding propellant. This later led Flandro and group [6] to conclude that DC shift is a result of propagation of steep-fronted shock like waves through the port. But comparison of pressure-time behavior, including the post DC shift phase, at head and aft end clearly shows sinusoidal oscillations of frequency corresponding to longitudinal chamber standing modes with clear 180° difference in phase [7]. This indicates that the spatial gradients in pressure along the length of the motor still follows the chamber standing modes and hence much less than compared to shock like waves. Also, recent analysis of experimental *p-t* traces by [3] indicates that as the amplitude of pressure oscillations grow, the propellant is subjected to increasing levels of pressurization and de-pressurization, which, especially beyond the point of DC shift, can be sustained only with peak propellant burn rates as high as 100 mm/s (about 10 times the mean) during pressurization and quenched conditions (zero burn rate) during de-pressurization. This clearly indicates the active role of propellant in triggering and sustaining the unsteady flow in the port and not vice-versa. A simple QSHOD based model is developed by [3] to explain this non-linear propellant burn behavior and the results of this are reproduced in Fig. 1, where the burn rate of a typical composite propellant subjected to a high amplitude pressure oscillation is shown to go through cycles of quenching and re-ignition accompanied by sharp burn rate increase up to 10 times the mean for a short duration.

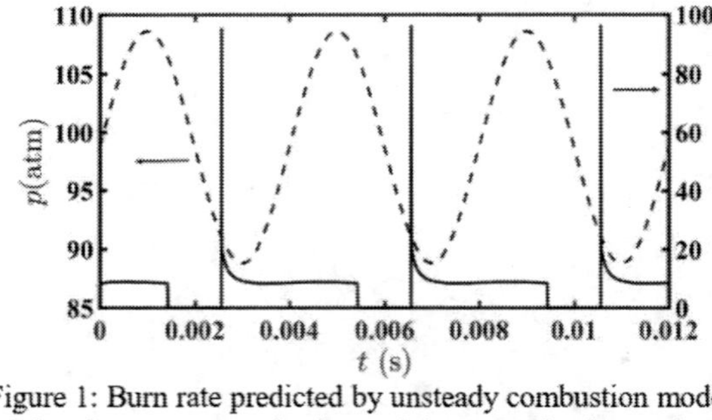

Figure 1: Burn rate predicted by unsteady combustion model in [3] for alternate pressurization and de-pressurization cycle.

By combining the linear behavior with this, exponential growth of pressure perturbation followed by DC shift is captured using a CFD simulation in the current work.

Nozzle damping and flow turning losses are the dominant decay mechanisms in a SRM using non-aluminized propellants. A detail description of damping mechanisms can be found in [2, 5, 8]. Nozzle damping is loss of acoustic energy because of partial reflection of the acoustic wave at the nozzle end. It is most dominant damping mechanisms for the longitudinal mode

*Corresponding author.
E-mail address: vishalwadhai87@gmail.com

of oscillation. Apart from this, flow turning losses are important whenever acoustic velocity is parallel to the burning surface [2]. Flow turning loss is acoustic energy lost in doing work on the combustion product entering radially in the chamber in order to turn them into the axially oscillating flow. Analytical expression of nozzle damping and flow turning losses are reported in [4] as,

$$\alpha_{ND} = \frac{2V_{inj}}{R}; \ \alpha_{FT} = \frac{V_{inj}}{R} \quad (2)$$

where, V_{inj} is the velocity of injection of the gases from the propellant surface and R is the port radius. With increase in motor mean pressure the V_{inj} decreases leading to a decrease in damping and hence the stability margin. Experimental results from [9] are consistent with this expectation – motors are more prone to instability at higher pressures.

Current work deals with numerical simulation of SRMs used in tactical missiles in line with the methodology developed in [3]. The CFD framework developed in [3] to estimate damping rates is extended here to account for coupling between propellant combustion and acoustics with the aim of capturing the linear and non-linear behavior observed in experiments.

2 Numerical Model

The motor geometry used for the numerical simulation is shown in Fig. 2 which can be considered a canonical configuration for studying instability.

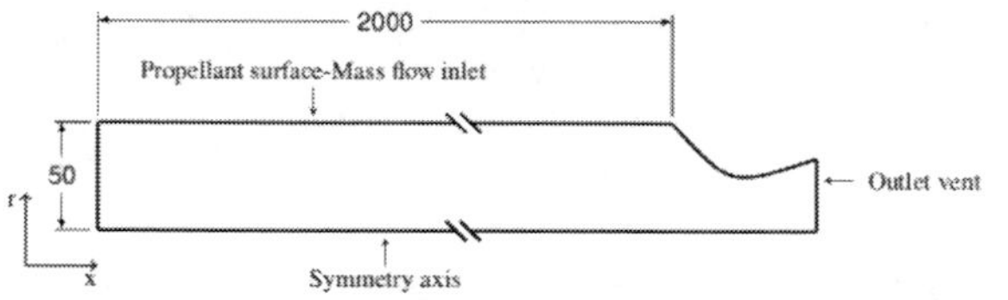

Figure 2: Geometry of SRM used in numerical simulation. All dimensions given are in mm.

The first longitudinal mode ($f = a/2L$) observed in SRMs used in tactical missiles is generally in the range of 200 to 500 Hz corresponding to chamber temperature of 3000 K (chamber temperature) and a motor length in the range of 1-3 m. Based on this a length of 2 m was chosen here (Fig. 2). Simulations were performed for two mean pressures, 70 and 120 atm with a corresponding throat radius of 22.6 and 18.7 mm (obtained from steady state mass-balance).

Propellant gasification processes are confined to a thin layer above the solid surface (< 50 micron) and its effect on the port flow dynamics can be accurately simulated by assuming the combustion zone to be infinitely thin. Hence a mass inlet boundary condition is used at the propellant surface, through which gases at 3000 K enter the port. Mass flux from the propellant surface was calculated using following procedure. The mass flux along inlet boundary has two components – a steady and a fluctuating component as shown in (3).

$$\dot{m}_b(x) = \bar{\dot{m}}_b(x) + \dot{m}_b'(x) \quad (3)$$

Steady component of mass flux is given by (4),

$$\bar{\dot{m}}_b(x) = \rho_p \bar{r}(x); \quad \bar{r}(x) = a\bar{p}(x)^n \quad (4)$$

where, ρ_p (1664 kg/m^3) is the density of propellant, a (1.95) is a burn rate constant and n (0.3) is pressure index. Fluctuating component of mass flux is obtained from fluctuating component of burn rate using following procedure. Fluctuating component of burn rate (r (x, t)) can be calculated from the definition of frequency response (1). Consider a position x on the propellant surface. The static pressure at this point varies with time and is denoted by p (x, t). The mean burn rate is a function of the mean pressure, p(x) and given by (4). The fluctuation in pressure, p'(x, t) is decomposed into respective Fourier modes as follow,

$$p'(x,t) = p_1'(x,t) + p_2'(x,t) + p_3'(x,t) \quad (5)$$

Only first three modes were considered as amplitudes of higher modes were found to be negligible. Using the definition of frequency response, fluctuations in mass flux can be expressed as,

$$\dot{m}_b'(x,t) = \rho_p \frac{\bar{r}(x)}{\bar{p}(x)} \sum_{i=1}^{3} R_{p,i} p_i'(x,t) \quad (6)$$

In general, magnitude and phase of frequency response are different for different modes. Typical magnitude of frequency response known to cause large amplitude oscillations is 3 to 5 for first mode [10]. The propellant has peak frequency response in the range of 200 to 300 Hz, corresponding to the conduction time scale (α/r^2) of a few milliseconds. The response values for higher modes are lower as the trend follows an inverted bell curve. Typical hot gas temperature is 3000 K with a molecular weight of 25 g/mol and ratio of specific heat of 1.2. The corresponding acoustic velocity is 1095 m/s. For the chosen length of 2 m (see Fig. 2) and 1095 m/s sound speed, the first three longitudinal modes will have the following frequencies – 270 Hz (fundamental), 540 Hz (first harmonic) and 810 Hz (second harmonic). Two sets of frequency response, showing magnitude and phase corresponding to first three longitudinal modes of motor considered (mode I – 270 Hz, mode II- 540 Hz, mode III – 810 Hz), used in current work are listed in Table 1.

Case	Magnitude			Phase (°)		
	I (270 Hz)	II (540 Hz)	III (810 Hz)	I (270 Hz)	II (540 Hz)	III (810 Hz)
1	3	1.5	1	0	-15	0
2	5	3.3	3	0	-15	0

Table 1: Values of magnitude and phase of frequency response for I, II and III modes used in the calculations.

The flow of combustion products through the chamber is treated as inviscid, axisymmetric ideal gas flow. Although actual flow is viscous and turbulent, the effect of turbulence on the longitudinal turbulent mode can be ignored as the length scales over which the turbulent flows interact with the mean flow is very small (less than port diameter of motor) compared to the wavelength of the longitudinal mode acoustic waves (few meters). All calculations are performed using ANSYS Fluent® by axisymmetric, transient, density-based solver (time domain simulation). Compatibility of the fluent solver to capture standing waves was tested by [3]. It is reported that at least second order accurate temporal and spatial discretization is required to capture the standing waves without any numerical dissipation. Steady state solutions were obtained for different grid sizes for the motors designed at 70 and 120 atm (cases used in further analysis). Steady state solution from optimum grid size was used as an initial condition to carry out the unsteady analysis by introducing perturbations. Time step independence studies were carried out and a time step of 1×10^{-5} s was chosen such that it accurately captured the growth of fundamental and harmonics.

3 Results

3.1 Decay Rate

Framework used here is capable of capturing decay of the acoustic wave due to nozzle damping and flow-turning losses as these damping mechanisms are inviscid in nature. In order to calculate the decay rate (α_{damp}) of the acoustic waves, the simulation was carried out assuming quasi static propellant response, as in (4). With this, a decaying pressure fluctuations time trace obtained at the head end of the motor at a mean pressure of 120 atm is shown in Fig. 3. The phase difference of 180° was obtained between head end and aft end pressure wave. Table 2 shows a comparison of decay rate obtained from CFD calculations at a mean pressure of 70 and 120 atm with the decay rate calculated using analytical expressions (2). Total decay rate obtained from CFD calculation matches with the sum of decay rate from nozzle damping and flow turning losses. This validates the use of CFD to solve flow through the port of SRM which is considered for current analysis.

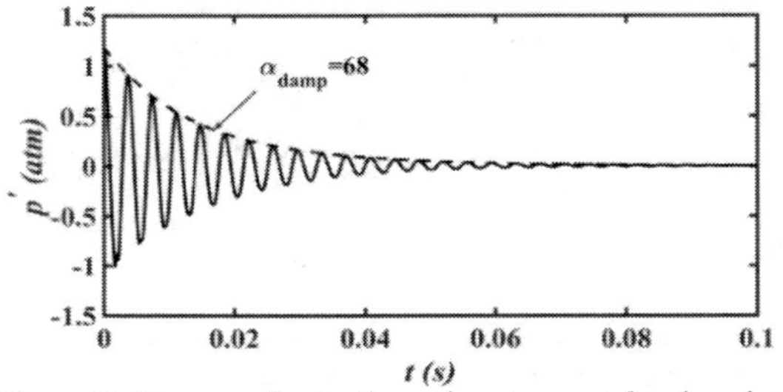

Figure 3: Pressure fluctuations time trace at head end of the motor at mean pressure of 120 atm.

Pressure (atm)	V_{inj} (mm/s)	α_{damp} (s^{-1})	
		$3V_{inj}/R$	CFD
70	1.64	94	91
120	1.13	65	68

Table 2: Comparison of Coefficient of damping obtained from analytical expression and CFD calculations.

3.2 Linear Instability

Numerical simulation was carried out by adding combustion acoustic coupling in the form of fluctuations in mass flux as defined by (6) in the boundary condition to the motor chamber. Pressure time trace was obtained for the motor running at 120 atm. Figures 4 and 5 show an exponential growth of acoustic wave at the head end of the motor for the frequency response values shown in Table 1 for case 1 and case 2, respectively. Net growth rate, $\alpha_{net}=\alpha_{prop}-\alpha_{damp}$, calculated during initial exponential growth for both cases, is shown in Fig. 6. For case 1 (response of 3 at mode I), α_{prop} slightly exceeded α_{damp} causing the net growth of the acoustic wave.

The transition of exponential growth into the limit cycle is obtained for case 2 (response of 5 at I mode) within 0.3 s (Fig. 5). It can be seen in Fig. 6 that net growth rate with the response of 5 is higher than that with response of 3, as expected. Modal analysis of pressure wave shown in Fig. 5 is presented in Fig. 7. All three modes can be seen to grow leading to typical *N*-shaped wave formed at the limit cycle. While growth mechanism is still linear, a limit cycle is obtained. This indicates that beyond a certain amplitude nonlinear gas-dynamic processes begin to play a role. This could be a combination of the following two mechanisms – 1) the damping rates could have become amplitude dependent and 2) coupling of the fundamental mode with higher harmonics leading beyond a certain amplitude leading to transfer of energy across modes (as can be seen from Fig. 7). This is a general feature in the time evolution of self-excited unstable systems (see for instance [11]).

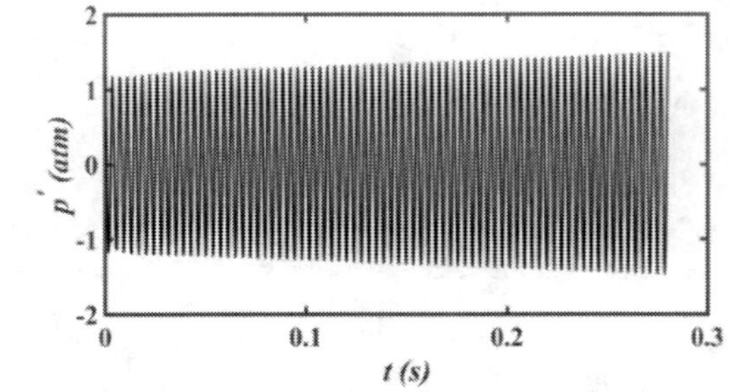

Figure 4: Pressure fluctuations time trace at head end of the motor at mean pressure of 120 atm for case 1.

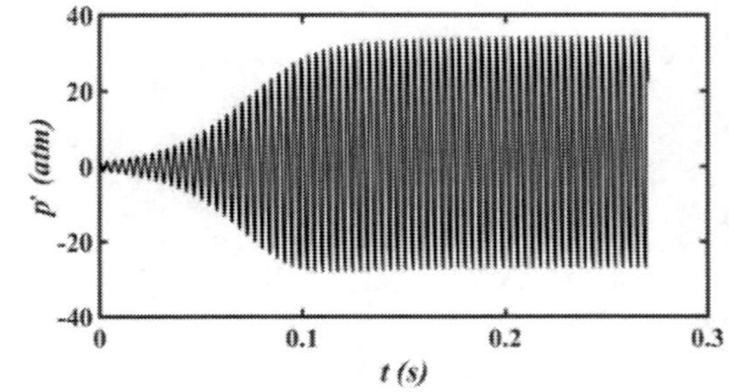

Figure 5: Pressure fluctuations time trace at head end of the motor at mean pressure of 120 atm for case 2.

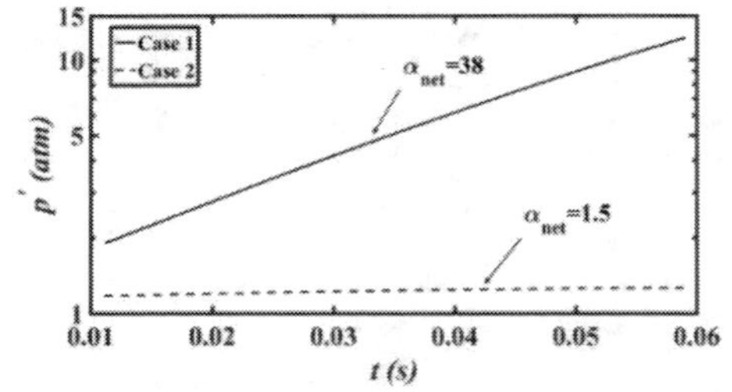

Figure 6: Growth rate of the pressure wave shown in Figs. 4 and 5.

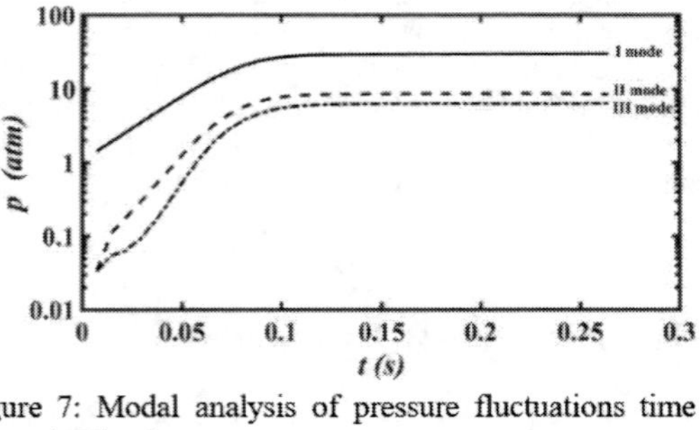

Figure 7: Modal analysis of pressure fluctuations time trace shown in Fig. 5.

4 DC Shift

Pressure time traces obtained in the previous section using the response function approach (Figs. 4, 5 and 7) show exponential growth transitioning into limit cycle without any sign of DC shift. High amplitude de-pressurization induced extinction-re-ignition cycle is a possible cause of DC shift. As shown in [3] this behavior is connected to the extinction of propellant when subjected to a de-pressurization rate exceeding the critical value resulting in burn rate behavior shown in Fig. 1. To test this hypothesis the following modified procedure was used to calculate burn rate fluctuation. Until the point when the de-pressurization rate is greater than critical value, burn rate is calculated using frequency response approach explained earlier. Once the de-pressurization rates decrease below the critical value the burn is assumed to follow the behavior shown in Fig. 1. In the case presented in the previous section (Fig. 5) de-pressurization rate as high as 20 atm/ms is achieved during the limit cycle. Critical de-pressurization rates are usually in the range of 2 to 8 atm/ms [12] which suggest the implementation of the effect of de-pressurization on the propellant burning is essential. Quenching followed by rising of burn rate to 100 mm/s for the critical de-pressurization rate of 8 atm/ms was implemented in the boundary condition. Pressure time trace using this modified procedure with a mean pressure of 120 atm is shown in Fig. 8 and clearly, shows a shift in mean i.e. DC shift. The current results are found to qualitatively agree with the experimental results in [9] – for instance, increase in mean pressure by around 30% due to DC shift is captured. Quantitative comparison requires much more information on the propellant composition etc. and response function values, which is generally not available.

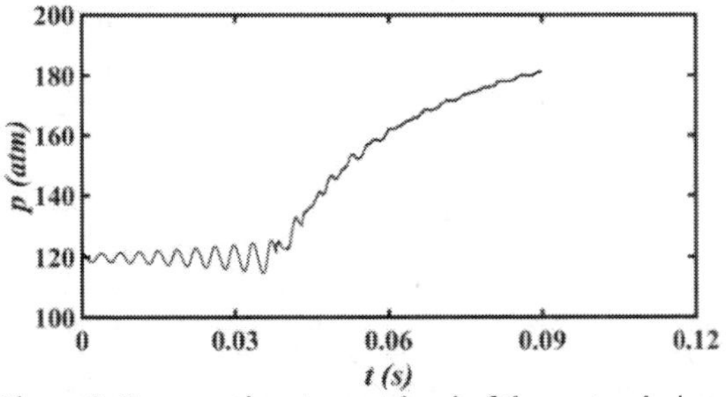

Figure 8: Pressure time trace at head of the motor designed for mean chamber pressure of 120 atm showing DC shift.

As the standing wave forms in the motor chamber with a crest at the head end and trough at the aft end or vice versa (changes with time), one-half of the propellant grain burns under pressurization while another half under de-pressurization. The amplitude of oscillation varies as per spatial variation in standing wave. This creates different de-pressurization rates in different sections of propellant at a particular instant of time. For accumulation of enough mass to cause DC shift a large section of the propellant has to undergo extinction. This is indeed the case with a critical de-pressurization rate of 8 atm/ms. With 15 atm/ms for instance (results not shown here) the magnitude of DC shift is not significant and the *p-t* trace is close to that of a limit cycle. These aspects require further exploration.

5 Conclusions

A CFD framework is developed to analyze the stability of solid rocket motor for a known frequency response of the propellant. The framework is validated by showing that numerically obtained decay rate closely matches with that calculated from analytical expression. It is shown that growth of higher harmonics is responsible for the formation of the *N*-shaped wave. The necessity to account for the effect of high amplitude oscillations on propellant deflagration to capture DC shift is brought out. Implementation of extinction followed by very high burn rate after a critical de-pressurization has shown to cause a shift in mean pressure. Further work is required to reproduce high amplitude oscillations post DC shift.

References

[1] F. S. Blomshield, Historical perspective of combustion instability in motors – Case studies, 37th Joint Propulsion Conference and Exhibit, 2001, p. 3875.

[2] F.E.C. Culick, V. Yang, in: L. De luca, E. W. Price, M. Summerfield (Eds.), Nonsteady Burning and Combustion Stability of Solid Propellants, AIAA Progress in Astronautics and Aeronautics, Washington, DC,1992, p. 719.

[3] Varunkumar, S. and H. S. Mukunda, Progress report on solid rocket combustion instability modelling, Technical Report IITRPR-SMMEE-2013-001, IISc & IIT Ropar, 2013.

[4] J. D. Baum, J. N. Levine, Acta Astronautica, 13(6-7) (1968) 339-348.

[5] J. F. Guéry, Numerical Modeling of Internal Flow Aerodynamics. Part 2: Unsteady Flows, SNPE CENTRE DE RECHERCHES DU BOUCHET VERT-LE-PETIT FRANCE, 2004.

[6] G. A. Flandro, S. R. Fischbach, J. Majdalani, Physics of Fluids, 19(9) (2007) 094101.

[7] A. S. Iyer, V. K. Chakravarthy, S. Saha, D. Chakraborty, The Aeronautical Journal, 120(1231) (2016) 1425-1445.

[8] F. A. Williams, Combustion Theory: the fundamental theory of chemical reacting flow systems, Addssison-Wesely, 1965, p. 3.

[9] F. S. Blomshield, C. J. Bicker, R. A. Stalnaker, High pressure pulsed motor firing combustion instability investigations, 33rd Joint Propulsion Conference and Exhibit, 1997, p. 3253.

[10] Arvind I., S. Varunkumar, H. S. Mukunda, Progress on solid propellant combustion instability studies at IISc, Technical report, Indian Institute of Science, Bangalore, 2013.

[11] Li, Shihuai, and Dan Zhao, Energy conversion and management, 75 (2013) 336-347.

[12] L. Strand, A. Schultz, G. Reedy, Determination of solid-propellant transient regression rates using amicrowave Doppler shift technique, 8th Joint Propulsion Specialist Conference, 1972, p. 1118.

n-Heptane cool flame chemistry: unraveling intermediate species measured in a stirred reactor and motored engine

Zhandong Wang [a*], Bingjie Chen [a], Kai Moshammer [b,i], Denisia M. Popolan-Vaida [c,d], Salim Sioud [e], Vijai Shankar Bhavani Shankar [a], David Vuilleumier [b], Tao Tao [b, h], Lena Ruwe [f], Eike Bräuer [f], Nils Hansen [c], Philippe Dagaut [g], Katharina Kohse-Höinghaus [f], Misjudeen A. Raji [e], S. Mani Sarathy [a]

[a] *King Abdullah University of Science and Technology (KAUST), Clean Combustion Research Center (CCRC), Thuwal 23955-6900, Saudi Arabia*

[b] *Combustion Research Facility, Sandia National Laboratories, Livermore, CA 94551, USA*

[c] *Department of Chemistry, University of California, Berkeley, CA 94720*

[d] *Chemical Sciences Division, Lawrence Berkeley National Laboratory, Berkeley, CA 94720, USA*

[e] *King Abdullah University of Science and Technology, Analytical Core Lab, Thuwal 23955-6900, Saudi Arabia*

[f] *Department of Chemistry, Bielefeld University, D-33615 Bielefeld, Germany*

[g] *Centre National de la Recherche Scientifique (CNRS), INSIS, 1C, 45071, Orléans Cedex 2, France*

[h] *Center for Combustion Energy and Department of Thermal Engineering, Tsinghua University, Beijing 100084, PR China*

[i] *Physikalisch-Technische Bundesanstalt, Bundesallee 100, 38116 Braunschweig, Germany*

Abstract

This work identifies classes of cool flame intermediates from *n*-heptane low-temperature oxidation (e.g., < 750 K) in a jet-stirred reactor (JSR) and a cooperative fuel research (CFR) engine. The sampled species from the JSR were analyzed using a synchrotron vacuum ultraviolet radiation photoionization time-of-flight molecular-beam mass spectrometer and an atmospheric pressure chemical ionization orbitrap mass spectrometer; the latter was also used to analyze the sampled species from the CFR engine. The products can be classified by species with molecular formulas of $C_7H_{14}O_x$ (x=0-5), $C_7H_{12}O_x$ (x=0-4), $C_7H_{10}O_x$ (x=0-4), C_nH_{2n} (n=2-6), C_nH_{2n-2} (n=4-6), $C_nH_{2n+2}O$ (n=1-4, 6), $C_nH_{2n}O$ (n=1-6), $C_nH_{2n-2}O$ (n=2-6), $C_nH_{2n-4}O$ (n=4-6), $C_nH_{2n+2}O_2$ (n=0-4, 7), $C_nH_{2n}O_2$ (n=1-6), $C_nH_{2n-2}O_2$ (n=2-6), $C_nH_{2n-4}O_2$ (n=4-7), and $C_nH_{2n}O_3$ (n=3-6). The identified intermediate species include mainly alkene, dienes, aldehyde/keto compounds, olefinic aldehyde/keto compounds, diones, cyclic ethers, peroxides, acids, and alcohols/ethers. Reaction pathways forming intermediates with the same carbon number as *n*-heptane are proposed and discussed. These experimental results should be helpful in the development of kinetic models for *n*-heptane and longer-chain alkanes.

1 Introduction

An substantial body of literature exists to describe the details of the auto-ignition of the most common fuel surrogate (*i.e.*, *n*-heptane) under conditions relevant to modern and future engine concepts; these studies report fuel conversion, stable products, and a number of intermediates under various conditions in jet-stirred reactors (JSRs) [1-6], flow reactors [7-9], and motored engines [10]. Reviewing this existing *n*-heptane oxidation data reveals that a comprehensive analysis of the pool of reactive intermediates is still scarce, especially under low-temperature and high-pressure conditions; preventing an in-depth understanding of relevant mechanistic pathways and key rate coefficients for reactions controlling the ignition and pollutant formation [11]. Assigning correct structures to such reactive intermediate species is challenging both on experimental and theoretical grounds to identify all relevant molecular conformations [12].

This study attempts to further clarify the intermediate species pool of *n*-heptane low-temperature oxidation in a JSR at atmospheric pressure and in a cooperative fuels research (CFR) engine at higher pressure. The products were analyzed using by synchrotron vacuum ultraviolet radiation photoionization time-of-flight molecular-beam mass spectrometry (SVUV-PI-MBMS), and a experimental setup comprising an atmospheric pressure chemical ionization (APCI) orbitrap mass spectrometer (OTMS). The elemental composition of the intermediates during *n*-heptane JSR oxidation was confirmed by the APCI-OTMS, and the species pool from *n*-heptane low-temperature oxidation in the CFR engine was found to be in agreement with those observed in the JSR. These efforts have made it possible to categorize the intermediate species, deduce their probable structures, and propose their formation mechanisms. The results suggest the combination of SVUV-PI-MBMS and APCI-OTMS to be effective for clarifying the species pool of low-temperature

Corresponding author. Zhandong Wang
E-mail address: zhandong.wang@kaust.edu.sa

oxidation products of long-chain alkanes. The different ionization process and high mass resolution of the APCI-OTMS could prove highly valuable in low-temperature oxidation studies. Furthermore, the combination of an ideal reactor (JSR) and real engine (CFR) shows that complex low-temperature reaction mechanisms exists in both fundamental and practical combustion devices.

2 Experimental method

In this work, a JSR was coupled to the time-of-flight molecular-beam mass spectrometer with synchrotron vacuum ultraviolet radiation as the photoionization source [13]. The mass spectrometer has a sensitivity of 1 ppm, a mass resolving power of ~3500, and a dynamic range of six orders of magnitude. The experiments were performed at Terminal 3 of the Chemical Dynamics Beamline of the Advanced Light Source at the Lawrence Berkeley National Laboratory. Details of the experimental setup can be found in the literature [13]. Stoichiometric *n*-heptane (1%)/O_2 (11%)/Ar (78%) mixtures were investigated at quasi-atmospheric pressures of 0.933 bar and a residence times 2 s.

Another JSR was coupled with a Thermo LTQ Velos Orbitrap mass spectrometer equipped with an IonMax APCI ion source at King Abdullah University of Science & Technology (KAUST), Saudi Arabia. Stoichiometric *n*-heptane (1%)/O_2(11%)/Ar (78%) mixtures were again investigated at atmospheric pressure and a residence times of 2 s. The Orbitrap automated gain control (AGC) target was set to $1x10^6$ charges for full scan and the microscan was set to 500 ms. The mass scan range was set to cover the *m/z* range of 50–300. The calibration of the LTQ-Orbitrap mass analyzer was performed in positive ESI ionization mode, according to the manufacturer's guidelines. The Orbitrap mass spectrometer was operated using XCalibur software. Analytes were detected in the Orbitrap at a mass resolving power of 100.000 and a sensitivity range of 1-5 ppm. The JSR and the sampling method were similar to those used by Dagaut *et al.* [14]. The products were sampled using a quartz tube at the outlet of the reactor and ionized by the APCI; then the ions were sucked into the system using a skimmer with a positive mode for analysis. In addition to the on-line analysis, off-line measurements were also performed. The gas mixture from the outlet of the JSR passed through a U-tube and was condensed by liquid nitrogen for one hour. We note that using liquid nitrogen to condense products in oxidation experiment should be very careful because liquid nitrogen can also condense oxygen into liquid oxygen. The products were then collected by dissolving them in methanol and injecting them into an APCI ionization source with a syringe. The comparison showed that intermediates from both on-line and off-line analysis were in good agreement.

Tunable synchrotron vacuum ultraviolet light was used in a single photon ionization technique. A neutral species is ionized when the photon energy is higher than its ionization energy, resulting in the molecule peak ($M^{+\cdot}$) and/or its fragments. The APCI is a soft-ionization method with low fragmentation risk; it is mainly used with polar and relatively nonpolar compounds, generally producing mono-charged ions. The analyte was in the gaseous form in this work and its ionization was accomplished using an atmospheric pressure corona discharge (5-7 µA). The APCI spectra provide an easily identifiable protonated molecular ion peak $[M+H]^+$, which allows for determination of the molecular mass [15, 16].

Similar to the off-line analysis of the intermediates formed in the JSR, exhaust gas from *n*-heptane oxidation in a motored CFR engine was analyzed. The engine characteristics and operation conditions are presented in Table 1. The CFR engine is a variable compression ratio engine designed and used for measuring the fuel's octane number. The present experiment uses this apparatus as a high-pressure, variable-volume homogenous reactor. The engine was operated in a manner to study products of low-temperature reactivity, and hence ignition was avoided.

Table 1. CFR engine specifications and test conditions.

Engine type	Single Cylinder
Injection system	Port fuel injection
Bore	82.55 mm
Stroke	114.3 mm
Connecting rod length	254 mm
Compression ratio	8.3
Engine speed	600 RPM
Fuel/air Eq. ratio	0.3
Intake pressure	98 kPa
Inlet air temperature	27±1 °C

The intake air to the engine flows through a mass flow meter and is preheated to the set point temperature. *n*-Heptane was injected before the intake valve through a port fuel injector. The fuel was supplied to the injector from a fuel tank pressurized by helium. The mixture formed is assumed to be thoroughly homogeneous at the start of compression. The engine was not connected to an emission analyzer, and the fuel flow through the injector was not measured. Therefore, the equivalence ratio was set using a lambda sensor while operating the engine under normal hot ignition and combustion conditions. The assumption was that complete combustion occurred when the compression ratio was increased to a point where Crank Angle (CA) 50 occurred at top dead center (TDC). This approach is not ideal when operating the engine in HCCI mode due to reduced combustion efficiency. However, the aim of this experiment is to detect species pool during low-temperature oxidation in engine relevant conditions. The compression ratio was then reduced to a point where hot ignition was avoided, and allowed to motor until the coolant temperature reached a steady value before the samples were collected. Low temperature fuel reactivity was determined from first law heat release analysis using the in-cylinder pressure that was recorded using a Kistler pressure transducer at a resolution 0.05° degree of crank angle. We assumed an adiabatic core to perform the heat release analysis.

The exhaust gas mixture from the CFR engine passed through a U-tube immersed in acetone-dry ice mixture for one hour to condense the intermediate species. Our test shows that the coolant of liquid nitrogen or acetone-dry ice mixture has no effect on the result. The samples were dissolved in methanol and analyzed by the APCI-OTMS.

3 Results and discussion

The products can be classified by species with molecular formulas of $C_7H_{14}O_x$ (x=0-5), $C_7H_{12}O_x$ (x=0-4), $C_7H_{10}O_x$ (x=0-4), C_nH_{2n} (n=2-6), C_nH_{2n-2} (n=4-6), $C_nH_{2n+2}O$ (n=1-4, 6), $C_nH_{2n}O$ (n=1-6), $C_nH_{2n-2}O$ (n=2-6), $C_nH_{2n-4}O$ (n=4-6), $C_nH_{2n+2}O_2$ (n=0-4, 7), $C_nH_{2n}O_2$ (n=1-6), $C_nH_{2n-2}O_2$ (n=2-6), $C_nH_{2n-4}O_2$ (n=4-7), and $C_nH_{2n}O_3$ (n=3-6). The intermediate species include mainly alkene, dienes, aldehyde/keto compounds, olefinic aldehyde/keto compounds, diones, cyclic ethers, peroxides, acids, and alcohols/ethers. In general, the species pool detected by the APCI-OTMS from *n*-heptane oxidation in JSR and CFR engine is similar to that detected by the SVUV-PI-MBMS from *n*-heptane oxidation in JSR.

Compared to the earlier work of Herbinet *et al.* [4], a more comprehensive intermediate species pool was detected because of the higher mass resolution of the mass spectrometer and higher fuel concentration. Compared to the high-pressure oxidation data of *n*-heptane by Dagaut *et al.* [3] and Boehman *et al.* [10], the more detailed analysis of the intermediate species pool during the CRF engine sheds new insights into the low-temperature oxidation chemistry of *n*-heptane.

The focus of this work is directed toward intermediates with the molecular formula of $C_7H_{14}O_x$ (x=0-5) and $C_7H_{12}O_x$ (x=0-4), which highlights the first, second, and third O_2 addition reaction mechanism and bimolecular reactions of these intermediates.

1. $C_7H_{14}O_x$ intermediates (x=0-5)

A detailed reaction scheme for the low-temperature oxidation of alkanes was proposed in recent work by Wang *et al.* [17, 18] on 2,5-dimethylhexane and 2-methylhexane. The reaction scheme includes the first, second, and third O_2 addition reactions, which produced intermediates with the same carbon number of the fuels, e.g., $C_nH_{2n}O_x$ (n is carbon number of the fuel, x=0-5). Scheme 1 presents an analogous reaction mechanism for *n*-heptane; important intermediates ($C_7H_{14}O_x$, x=0-5) are highlighted in blue and boxed. They include heptenes (C_7H_{14}, *m/z* 98), C_7 cyclic ethers ($C_7H_{14}O$, *m/z* 114), C_7 olefinic hydroperoxides ($C_7H_{14}O_2$, *m/z* 130), C_7 keto-hydroperoxides and/or hydroperoxy cyclic ethers ($C_7H_{14}O_3$, *m/z* 146), C_7 olefinic dihydroperoxides ($C_7H_{14}O_4$, *m/z* 162), and C_7 keto-dihydroperoxides and/or dihydroperoxy cyclic ethers ($C_7H_{14}O_5$, *m/z* 178). These intermediates, except for $C_7H_{14}O_4$, have been detected in both the JSR and CFR engine experiments. The PI-MBMS and OTMS results are in good agreement and confirm the elemental composition of these intermediates. The formation of these intermediates is dependent on the pool of R, ROO, QOOH, OOQOOH, $P(OOH)_2$, and $OOP(OOH)_2$ radicals, whose reactions, such as O_2 addition, decomposition, and isomerization, are temperature-dependent. The exothermic O_2 addition reactions are thermodynamically unfavorable as temperature increases, while the endothermic decomposition reactions are favored at higher temperatures.

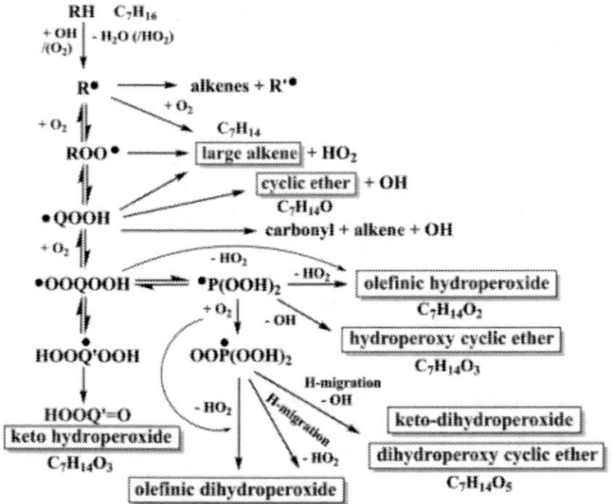

Scheme 1. Reaction scheme for *n*-heptane low-temperature oxidation. Represented intermediates (i.e., $C_7H_{14}O_x$, x=0-5) are boxed and highlighted in blue.

2. $C_7H_{12}O_x$ intermediates (x=0-4)

The intermediates with molecular formula of $C_7H_{12}O_x$ (x=0-4) were also measured by PI-MBMS and OTMS during *n*-heptane JSR and CFR engine oxidation. The signal profiles of the $C_7H_{12}O_4$ (*m/z* 160) intermediates with temperature are similar to those of $C_7H_{14}O_3$ and $C_7H_{14}O_5$. In accordance with previous work by this group [17, 18], it is proposed that this molecule contains one hydroperoxy group and two carbonyl groups (e.g., diketo-hydroperoxides), or one hydroperoxy group, one carbonyl group, and one cyclic ether group (keto-hydroperoxy cyclic ether). These could be produced from OH radical-assisted water elimination of the $C_7H_{14}O_5$ intermediates (*i.e.*, keto-dihydroperoxides and/or dihydroperoxy cyclic ethers); or via H-atom abstraction from $C_7H_{14}O_5$ intermediates at carbon sites not connected to the hydroperoxy group and subsequent cyclization to form $C_7H_{12}O_4$ intermediates. Here, two other pathways are proposed for the production of $C_7H_{12}O_4$ intermediates. The first sequence could be H-atom abstraction from $C_7H_{14}O_3$ intermediates (i.e., C_7 keto-hydroperoxides and hydroperoxy cyclic ethers), O_2 addition to the radical site, intramolecular isomerization, and subsequent decomposition. As a second possibility to produce the $C_7H_{12}O_4$ intermediates, H-abstraction from cyclic ethers, subsequent O_2 addition, isomerization, and decomposition, may also be considered.

$C_7H_{12}O_3$ species (*m/z* 144) could be produced by H-atom abstraction from the $C_7H_{14}O_3$ intermediates and C_7 cyclic ethers. They may also be produced by H-atom abstraction from $C_7H_{14}O_2$ intermediates (e.g., C_7 olefinic hydroperoxide) at the carbon sites not connected to the hydroperoxy group, O_2 addition, and subsequent decomposition. Furthermore, heptenes (C_7H_{12}) are important intermediates during *n*-heptane low-temperature oxidation. Following the low-temperature reactions in Scheme 1, the oxidation of heptenes also produces $C_7H_{12}O_3$ intermediates.

The signal intensity of the $C_7H_{12}O_2$ intermediates (*m/z* 128) is much higher than those of the $C_7H_{12}O_3$ and $C_7H_{12}O_4$ intermediates, indicating that they are important products of *n*-heptane oxidation. These intermediates were also observed in a recent *n*-heptane oxidation study and were reported as C_7 diones [4]. Products with the same carbon skeleton as the fuel, and containing two carbonyl groups, have been reported in various hydrocarbon oxidation studies [4, 19-21]. Recently, Ranzi *et al.* [22] and Pelucchi *et al.* [23] proposed a probable pathway to the formation of diones, e.g., the H-atom abstraction of the keto-hydroperoxide at the α-position to the hydroperoxyl group followed by β-scission. Including these pathways in kinetic models improved predictions of C_4 and C_7 diones in *n*-butane [22] and *n*-heptane [23] low-temperature oxidation. Improvements were also observed in the prediction of C_5 diones in *n*-pentane oxidation [24], but discrepancies still existed.

Another probable pathway for the formation of the $C_7H_{12}O_2$ intermediate is proposed here. First, H-atom abstractions from C_7 keto-hydroperoxides at sites β-, γ-, and/or δ- to the hydroperoxyl group, and further cyclization, produce keto cyclic ethers (e.g., with one carbonyl and one cyclic ether group). Second, the subsequent reactions of C_7 cyclic ethers also form diones and keto cyclic ethers. The production of diones from cyclic ether oxidation was observed by Vanhove *et al.* [25] and Antonov *et al.* [26] during tetrahydrofuran oxidation.

The $C_7H_{12}O$ intermediates (*m/z* 112) could be heptenones, cyclic ethers with one double bond, and/or heptenals, which can be produced from the C_7 keto-hydropoperoxides, C_7 cyclic ethers, and C_7 heptenes. They can also be produced from H-atom abstractions of $C_7H_{14}O_2$ intermediates (e.g., C_7 olefinic hydroperoxide) at the carbon sites not connected to the hydroperoxy group and subsequent cyclization. The *m/z* corresponding to $C_7H_{12}O$ was also observed by Herbinet *et al.* [4]. The C_7H_{12} intermediates (*m/z* 96) could be heptadienes and could be produced by H-atom abstractions from heptenes and subsequent β-scission reactions [27]. The *m/z* corresponding to

C_7H_{12} was not observed by Herbinet *et al.* [4] during the PI-MBMS measurement, but 1,4-heptadiene was detected by the GC analysis.

4 Conclusions

In this work, the low-temperature oxidation of *n*-heptane was studied in a JSR and a CFR engine. The intermediate pool of species was measured using SVUV-PI-MBMS and APCI-OTMS. The results from both techniques are in good agreement, and the combination of these two diagnostics clarified the elemental composition of the oxidation intermediates. Agreement of the distribution of the *n*-heptane low-temperature oxidation intermediates in real engines with those in an ideal reactor is demonstrated for the first time under similar experimental conditions.

The species pool includes intermediates with the same carbon skeleton as the fuel, such as $C_7H_{14}O_x$ (x=0-5), and $C_7H_{12}O_x$ (x=0-4), highlighting the first, second, and third O_2 addition reaction mechanism and bimolecular reactions of these intermediates. Large numbers of C_1-C_6 intermediates were observed, resulting from the decomposition of C_7 intermediates. The intermediates can be classified into alkenes, dienes, aldehyde/keto compounds, olefinic aldehyde/keto compounds, diones, cyclic ethers, peroxides, acids, and alcohols/ethers. The results from this work will contribute to the understanding of the reaction mechanism of *n*-heptane and long-chain alkane oxidation and to further kinetic model development, especially at low-temperature.

5 Acknowledgment

This work was initiated by the Clean Combustion Research Center with funding from King Abdullah University of Science and Technology (KAUST) and Saudi Aramco under the FUELCOM program. Research reported in this publication was also supported by competitive research funding from KAUST. The work of N.H. and K.M. was supported by Division of Chemical Sciences, Geosciences and Biosciences, BES/USDOE. D.M.P.V. was supported by the U.S. Department of Energy (DOE), Office of Science, Office of Basic Energy Sciences, under contract no. DE-AC02-05CH11231, the gas phase chemical physics program through the Chemical Sciences Division of Lawrence Berkeley National Laboratory (LBNL). L.R., E.B., and K.K.H. are grateful for German DFG project under contract KO1363/31-1. P.D. has received funding from European Research Council under FP7/2007-2013/ERC grant 291049-2G-CSafe. Tao Tao is supported by China Scholarship Council (Grant No. 201506210349). Sandia is a multi-mission laboratory operated by Sandia Corporation, a Lockheed Martin Company, for the National Nuclear Security Administration under contract DE-AC04-94-AL85000. The Advanced Light Source is supported by the Director, Office of Basic Energy Sciences, of the U.S. Department of Energy under Contract No. DE-AC02-05CH11231

References

[1] A. Chakir, M. Bellimam, J.C. Boettner, M. Cathonnet, Int. J. Chem. Kinet. 24 (1992) 385-410.
[2] P. Dagaut, M. Reuillon, M. Cathonnet, Combust. Sci. Technol. 95 (1993) 233-260.
[3] P. Dagaut, M. Reuillon, M. Cathonnet, Combust. Flame 101 (1995) 132-140.
[4] O. Herbinet, B. Husson, Z. Serinyel, M. Cord, V. Warth, R. Fournet, P.-A. Glaude, B. Sirjean, F. Battin-Leclerc, Z. Wang, M.F. Xie, Z.J. Cheng, F. Qi, Combust. Flame 159 (2012) 3455-3471.
[5] H.M. Hakka, R.F. Cracknell, A. Pekalski, P.A. Glaude, F. Battin-Leclerc, Fuel 144 (2015) 358-368.
[6] K. Zhang, C. Banyon, J. Bugler, H.J. Curran, A. Rodriguez, O. Herbinet, F. Battin-Leclerc, C. B'Chir, K.A. Heufer, Combust. Flame 172 (2016) 116-135.
[7] C.V. Callahan, T.J. Held, F.L. Dryer, R. Minetti, M. Ribaucour, L.R. Sochet, T. Faravelli, P. Gaffuri, E. Rani, Proc. Combust. Inst. 26 (1996) 739-746.
[8] T.J. Held, A.J. Marchese, F.L. Dryer, Combust. Sci. Technol. 123 (1997) 107-146.
[9] D.B. Lenhert, D.L. Miller, N.P. Cernansky, K.G. Owens, Combust. Flame 156 (2009) 549-564.
[10] J.P. Szybist, A.L. Boehman, D.C. Haworth, H. Koga, Combust. Flame 149 (2007) 112-128.
[11] K. Kohse-Höinghaus, Chem. Eur. J. 22 (2016) 13390-13401.
[12] K. Moshammer, A.W. Jasper, D.M. Popolan-Vaida, A. Lucassen, P. Diévart, H. Selim, A.J. Eskola, C.A. Taatjes, S.R. Leone, S.M. Sarathy, Y. Ju, P. Dagaut, K. Kohse-Höinghaus, N. Hansen, J. Phys. Chem. A 119 (2015) 7361-7374.
[13] K. Moshammer, A.W. Jasper, D.M. Popolan-Vaida, Z. Wang, V.S. Bhavani Shankar, L. Ruwe, C.A. Taatjes, P. Dagaut, N. Hansen, J. Phys. Chem. A 120 (2016) 7890–7901.
[14] P. Dagaut, M. Cathonnet, J.P. Rouan, R. Foulatier, A. Quilgars, J.C. Boettner, F. Gaillard, H. James, J. Phys. E: Sci. Instrum. 19 (1986) 207–209.
[15] A. Raffaelli, A. Saba, Mass Spectrom. Rev. 22 (2003) 318-331.
[16] J.A. Syage, J. Am. Soc. Mass Spectrom. 15 (2004) 1521-1533.
[17] Z. Wang, L. Zhang, K. Moshammer, D.M. Popolan-Vaida, V.S. Bhavani Shankar, A. Lucassen, C. Hemken, C.A. Taatjes, S.R. Leone, K. Kohse-Höinghaus, N. Hansen, P. Dagaut, S.M. Sarathy, Combust. Flame 164 (2016) 386-396.
[18] Z. Wang, S.Y. Mohamed, L. Zhang, K. Moshammer, D.M. Popolan-Vaida, V.S. Bhavani Shankar, A. Lucassen, L. Ruwe, N. Hansen, P. Dagaut, S.M. Sarathy, Proc. Combust. Inst. 36 (2017) 373–382.
[19] O. Herbinet, F. Battin-Leclerc, S. Bax, H.L. Gall, P.-A. Glaude, R. Fournet, Z. Zhou, L. Deng, H. Guo, M. Xie, F. Qi, Phys. Chem. Chem. Phys. 13 (2011) 296-308.
[20] A. Rodriguez, O. Herbinet, Z. Wang, F. Qi, C. Fittschen, P.R. Westmoreland, F. Battin-Leclerc, Proc. Combust. Inst. 36 (2017) 333–342.
[21] A. Rodriguez, O. Herbinet, X. Meng, C. Fittschen, Z. Wang, L. Xing, L. Zhang, F. Battin-Leclerc, J. Phys. Chem. A 121 (2017) 1861–1876.
[22] E. Ranzi, C. Cavallotti, A. Cuoci, A. Frassoldati, M. Pelucchi, T. Faravelli, Combust. Flame 162 (2015) 1679–1691.
[23] M. Pelucchi, M. Bissoli, C. Cavallotti, A. Cuoci, T. Faravelli, A. Frassoldati, E. Ranzi, A. Stagni, Energy Fuels 28 (2014) 7178–7193.
[24] J. Bugler, A. Rodriguez, O. Herbinet, F. Battin-Leclerc, C. Togbé, G. Dayma, P. Dagaut, H.J. Curran, Proc. Combust. Inst. 36 (2017) 441–448.
[25] G. Vanhove, Y. Yu, M.A. Boumehdi, O. Frottier, O. Herbinet, P.-A. Glaude, F. Battin-Leclerc, Energy Fuels 29 (2015) 6118-6125.
[26] I.O. Antonov, J. Zádor, B. Rotavera, E. Papajak, D.L. Osborn, C.A. Taatjes, L. Sheps, J. Phys. Chem. A 120 (2016) 6582-6595.
[27] F. Battin-Leclerc, A. Rodriguez, B. Husson, O. Herbinet, P.-A. Glaude, Z. Wang, Z. Cheng, F. Qi, J. Phys. Chem. A 118 (2014) 673-683.

11th Asia-Pacific Conference on Combustion,
The University of Sydney, NSW Australia
10th -14th December 2017.

Soot Precursor Evolution in Diffusion Flames with Different Sooting Propensities

Daniel Bartos[1], Matthew Dunn[1], Mariano Sirignano[2], Andrea D'Anna[2], Assaad R. Masri[1]

[1] School of Aerospace, Mechanical and Mechatronic Engineering, the University of Sydney, Sydney Australia

[2] Dipartimento di Ingegneria Chimica, dei Materiali e della Produzione Industriale, Universita`degli Studi di Napoli 'Federico II, Napoli, Italy

Abstract

This paper presents a combination of time resolved laser-induced fluorescence (LIF) and laser-induced incandescence (LII), induced respectively by 266 nm and 1064 nm laser sources to investigate soot and soot precursor particles. Measurements were made in two laminar co-flowing diffusion flames; a methane flame deemed non-sooting by LII and an obviously sooting, non-smoking ethylene flame. LIF intensity, decay time and general spectral shape are used to characterize soot precursor particles. The majority of LIF found in both flames is thought to originate from nanostructures rather than gas phase polycyclic aromatic hydrocarbons (PAH). Both flames show similarities in upstream regions in terms of fluorescence spectrum and decay times, indicating similarities in particle structure. However, in the ethylene flame there is a continuous increase in LIF lifetimes and a redshift in the spectrum that highlights a transition towards larger more rigid structures and larger aromatic compounds normally found in these structures. LIF intensities in the ethylene flame indicate that these nanostructures are found in high concentrations in sooting flames prior to the onset of soot.

1 Introduction

Improving current understanding of soot formation and oxidation mechanisms in combustion is increasingly important considering that regulations on the emission of particles are becoming more stringent. Satisfying such regulations necessitates an improved understanding of particle evolution processes in combustion. While extensive research has already been undertaken on this topic, the mechanisms leading to soot inception remains vague and hence forms the subject of the current work [1-5].

In this paper, the term "soot" is used generally in reference to both primary particles and aggregates defined by their ability to absorb in the infrared and incandesce. Other "nanostructures" are conversely defined by their inability to absorb in the infrared and capability to fluoresce under UV excitation. Although the fluorescence from these structures is from aromatic groups, an important distinction is made between gas phase PAH and those bound in condensed phase nanostructures.

Solid particles and aggregates absorbent in the visible and infrared spectral region are known to be well tracked by LII. While the field is still an evolving area of research, LII is generally attributed to soot particles and aggregates in the order of tens to hundreds of nanometers that are sufficiently graphitic to be heated up to temperatures of 3500-4000K [4, 5]. On the other side, there is proof of the existence of a class of compounds, generally identified as soot precursors, in the range of sizes between the largest gas phase PAHs and the smallest solid particles tracked by LII. Compared to soot, the number density of soot precursor particles is more than an order of magnitude higher while their contribution to the total mass concentration is practically negligible [1]. Soot precursor particles are also often referred to as condensed phase, nanostructures, nanoparticles or nano-organic carbon (NOC). LIF and UV extinction techniques have been used to detect such particles [1, 4, 6-9]. However, LIF measurements have also been extensively used to measure gas phase PAH and thus LIF emissions within flames are often attributed to the gas phase species [2, 10-14]. It is likely that LIF measurements, particularly in sooting flames, detect a combination of PAHs and nanostructures. LIF spectrum produced by a 266 nm laser attributed to both gas phase PAH and nanostructures show strong similarities and can be defined as a broadband signal stretching from the UV up to around 550 nm, with a peak often between 350 and 400 nm [10, 14-17].

Time resolved 266 nm LIF measurements are used in the present work to delineate between PAH LIF and that originating from nanostructures. PAH fluorescence decay times decrease rapidly at flame temperatures [18] and thus longer LIF decay times are attributed to nanostructures. When fluorophores are bound within semi-ridged structures they are less able to give roto-vibrational movements and are less able to lose internal energy through non-radiative means resulting in a longer fluorescence lifetime [17]. As structures become more solid, the fluorescence quantum yield decreases as internal conversion of the energy absorbed by the photon increases, inevitably leading to thermal emission (LII). This paper aims to shed some light on this transition from gas phase PAH through nanostructures to soot. Presented here are axial trends along laminar diffusion flames that have previously been chosen as target flames for the International Sooting Flame Workshop (ISF) [19].

2 Experimental Setup

In this study time resolved laser induced emissions (LIE) are employed to probe soot and soot precursor species in co-flowing laminar diffusion flames. The experimental setup used has been described in detail in previous work [20] and thus

Corresponding author.
daniel.bartos@sydney.edu.au

only essential details are given here.

The flames are probed with two different laser sources, namely 266 nm (80 ps pulse) and 1064 nm (8 ns pulse). The first pulse is due to the 4th harmonic (266 nm) of a mode-locked Nd:YAG 10Hz pulsed laser (Ekspla PL2251 Series Laser). The beam focus is measured to have a 250 μm diameter resulting in a fluence of 1.2 J/cm^2. The use of picosecond pulses (compared to earlier measurements using a typical 8 ns pulse [17]) is necessary to avoid distortion of the short LIF lifetimes expected from fluorescing species at high temperatures [18]. The second laser is the fundamental output (1064 nm) of a 10Hz, Nd:YAG pulsed laser (Spectra-Physics Quanta-Ray Pro-350) and was used to exclusively excite LII from soot. The probe diameter was measured to be 450 μm resulting in a fluence of 0.6 J/cm^2.

The collection system consists of four fast photomultiplier tubes (PMTs) (Hamamatsu H10721-210 and H10721-20) situated at the focal plane of a 1/2 m spectrometer. The four PMTs simultaneously collect time-resolved signals in four distinct spectral bands: PMT1 @ 266±15 nm, PMT2 @ 350±15 nm, PMT3 @ 445±15 nm, PMT4 @ 575±15 nm. Data from PMT1 is not presented in this body of work. PMT2 and PMT3 are used measure 266 nm LIF in the UV and the visible spectrum to observe the broadband aromatic fluorescence seen in previous works [14, 16, 17]. Although LII from either 266 nm or 1064 nm laser pulse will produce a quasi-blackbody curve extending down to wavelengths in the UV, PMT4 was dedicated to monitor LII. The placement of the PMTs at these spectral bands also avoids the majority of known carbon cluster interferences [21], however at high soot volume fractions some interference may be present. The signals from the four PMTs were recorded using a high sample-rate (25Giga-samples/s), high bandwidth (4GHz) oscilloscope (Tektronix DSA70404C). The entire system has been spectrally calibrated to allow direct comparison between PMT wavelength bands. Light from the probe volume is focused onto the 20 μm spectrometer slit by a series of two parabolic mirrors collecting at 30 degrees off parallel to the laser path. The resultant collection system is very sensitive and is able to detect soot volume fractions in the order of 10 ppb. 1064 nm LII signals are used only as a marker of soot and hence volume fraction is not presented here but is well documented elsewhere [22, 23].

Two co-flowing diffusion flames are investigated, namely a marginally sooting methane flame and an obviously sooting, non-smoking ethylene flame. Both fuels are diluted with nitrogen with a fuel volume fraction of 65% for methane [24] and 32% for ethylene [22, 23]. The burner consists of a 4 mm ID jet for the fuel mixture surrounded by a 74 mm diameter co-flow of air. The flow rates for both the fuel jet mixture and the co-flow are 35cm/s. Both these flames are target flames for the ISF [19] and the burner design is given in more detail in work by the Yale group [23].

3 Results and Discussion

Presented here are peak signals and decay times measured along the centerline of the two investigated flames. Figure 1 shows an example raw 266 nm LIF signal taken from the methane flame. The temporal peak that occurs only a few nanoseconds after the laser pulse, measured in volts, is used to measure LIF intensity. Voltage signal units reported on the y-axis are arbitrary but are presented here so relative comparisons can be made between the two flames. Decay times are defined as the peak normalized integral. The LIE signals are integrated from 0 ns to 200 ns and then divided by the peak. Decay times calculated this way have a linear relationship with an actual single exponential decay time, provided that the actual decay time is not substantially less than the FWHM of the system response responsible for the convolution of the signal. The measured system response is also shown in Fig. 1 and has a FWHM of 2ns. Consequently, decay times less than 1 ns should be treated with caution, however none are observed in this study. The linearly determined decay times (τ_{eff}) are presented herein. Also shown in Fig. 1 is the single convoluted exponential fit corresponding to the peak normalized integral method just described. This method provides a very good fit to the LIF signal and is acceptable for these signals where no obvious bi-exponential features are present. A maximum error of 18% was found compared to a bi-exponential convoluted fit used in previous work [20]. Error bars will be presented on all decay times reported here. This simplified method was preferred in the absence of a more robust multi-exponential fitting procedure necessary when dealing with signals that are obviously multi-component, particularly in sooting regions of flames.

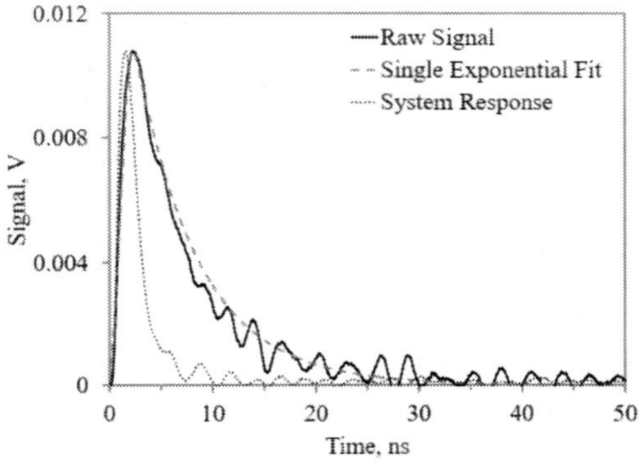

Figure 1: LIF Signal versus time measured by PMT @ 445nm in methane flame (Solid) alongside single exponential decay fit (dashed) corresponding to Integral Over Peak decay time approximation. Measured system response (dotted) is also presented.

LIE measurements along the centerline of the 65% methane flame are presented versus HAB in Fig. 2a while LIF decay times are reported in Fig. 2b. Although this flame has a small, faint luminous tip, LII from 1064 nm was not detected, suggesting that the local soot concentration is below the detection limit as evidenced in Fig. 2a.

LIF signal emerges with reaction marked by the rise in temperature computed by Kempema et al. [24]. LIF intensity continues to increase towards the end of the flame before dropping off in the oxidation zone. This LIF has often been attributed to PAH molecules known to formed in the inception processes of soot [2, 10, 12, 14]. However, this increase in LIF cannot be solely explained by a concentration increase of PAH because of the corresponding increase in decay time seen in Fig. 2b. Also presented here are the effective decay times measured by Ossler et al. [18] of PAHs naphthalene, fluorene, anthracene and pyrene plotted at the corresponding temperatures within the flame. PAH fluorescence decay times decrease rapidly in the presence of flame temperatures and their effective lifetimes are reduced to values less than 1ns at temperatures approaching 1200K. Decay times measured in this methane flame continue to increase beyond 1 ns into regions of significantly higher temperatures for both LIF @ 350 nm and 445 nm. At a height above burner (HAB) = 18 mm it can be seen that both decay times measured are outside the realm of molecular PAH lifetimes and thus must consist of larger structures similar to those measured in premixed flames [20]. In fact, polarization anisotropy measurements [6, 8, 9] have confirmed the presence of fluorescent particles with much larger effective diameters than molecular PAH in non-sooting regions of diffusion and premixed flames. Decay times are a complex function related to many variables including temperature, the concentration of quenching species and particle structure [17, 18, 25] and thus attempts to attribute decay times to an average particle size would be speculative, without further information. However, it would be reasonable to assume that these nanostructures are somewhere between 1 nm and 20 nm diameter as previously observed in both diffusion [6, 9] and premixed flames [8, 20]. The emission spectrum of these particles does not change along the flame as determined by the ratio between LIF @ 445 nm and @ 350 nm. This ratio remains close to 1 indicating that the fluorescent species are primarily comprised of smaller aromatic species around the size of pyrene with a peak in the spectrum expected to be somewhere between 350 nm and 445 nm [10, 15, 26]. This spectrum is more skewed towards PAHs with a small number of rings when considering the redshift of PAH LIF at high temperatures [18].

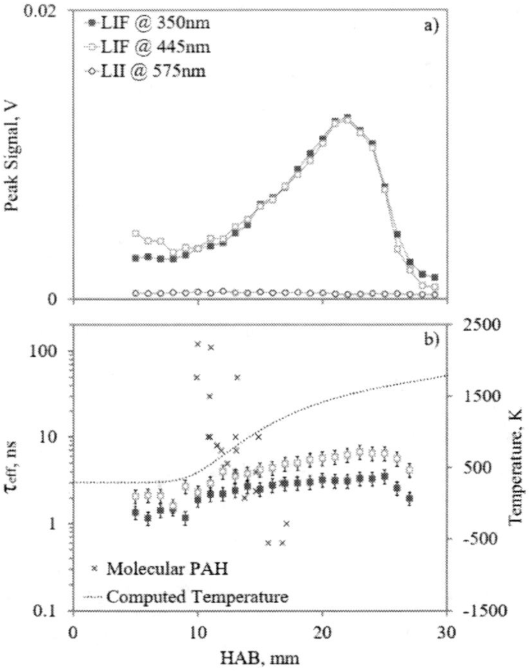

Figure 2: a) Peak LIE centerline measurements against HAB in 65% methane flame. LIF @ 350 nm and 445 nm and 1064 nm LII @ 575 nm. b) Calculated LIF decay times (τ_{eff}) plotted alongside molecular PAH decay times measured by Ossler et al. [18] corresponding to the computed temperature (dotted) of the flame from the work of Kempema et al. [24].

Figure 3 presents the same signals for the 32% ethylene flame. Due to the wavelength and intensity of the 266 nm laser used, photofragmentation of ethylene was observed close to the burner exit as noted by a broadband signal proportional to ethylene concentration recorded by all the PMTs. Assuming that 100% of the LIE signal recorded by all PMTs before the reaction zone is due to fragmentation, the peak signals are corrected based on ethylene concentration calculated using the computed temperature taken from the work of Smooke et al. [23]. It is noted that this correction does not remove any contribution this fragmentation signal has to the temporal decay. However, it was found at HAB > 12 mm that the ethylene concentration is negligible along with any interference of fragmentation on the remaining LIF signals.

The 32% ethylene flame is obviously sooting with LII observed from HAB = 18.5 mm to the tip at 26 mm and shows strong agreement with previous measurements [23]. In the sooting region of this flame the LIE signals from 266 nm become a combination of LIF, LII and potentially interferences from sublimated species such as C_2 and C_3 [21]. This impact of the emerging 266 nm LII signal can be seen in Fig. 3b via the steep increase in decay time measured between HAB = 19 mm to 21 mm. Interpreting the decay time in the sooting region becomes difficult and the current method is no longer appropriate due to obvious multi-exponential components to the signal. Even though the integrated signal and thus the calculated decay time are greatly affected by the LII component of this signal, the peak of the signal is still very much dominated by LIF. Decay times presented in Fig. 3b are only considered with confidence between HAB = 12 mm and 19 mm. Even within this shorter window, it is possible to distinguish a trend similar to that seen in the methane flame: increasing LIF decay time with increasing temperature which is also plotted from Smooke et al. [23]. LIF decay times in this flame are again much larger than that expected of molecular PAH at these temperatures and thus can be attributed to condensed phase nanostructures.

Much like what is observed in the methane flame, LIF intensity increases along the ethylene flame to around HAB = 15 mm. LIF intensity then decreases again, however well before the oxidation zone indicated by the decrease in LII and also predicted oxidation region by Smooke et al. [23]. Additionally LIF @ 445 nm begins to decrease slightly later than LIF @ 350 nm resulting in an increase in the ratio between LIF @ 445 nm and @ 350 nm at HAB > 15 mm. This indicates a shift in the spectrum towards longer wavelengths and consequently an increase in the size of the dominant aromatic island responsible for fluorescence, similar to that observed by Hayashida et al. [12]. The change in the structure of these particles may also contribute to decrease in the quantum yield from 266 nm excitation that would explain the decreasing LIF signal in this region. However, it is not clear if this is a general transition towards larger aromatics before graphitization, or a consumption in the process of primary particle surface growth. Particles formed in lower temperature pyrolytic zones have been attributed to smaller aromatic species expected in loose dimer or trimer structures, whereas red-shifted spectra and longer decay times indicate a more stacked, graphite like structure [17].

The signal valley created between the peaks of UV LIF and LII signals is often referred to as the "dark region" and has been observed in a number of studies [12, 14, 27]. These trends show strong resemblance to UV absorbing nano-organic carbon (NOC) measurements conducted by D'Anna et al. [7] in diffusion flames that have particle size estimated to be between 2 nm and 4 nm. Complementary investigations by the Naples group [9] show that particles in this "dark region" have a higher quantum yield at longer wavelengths that correspond to particles up to 20 nm. In comparison to the marginally sooting methane flame, the precursor particle sizes responsible for fluorescence in the ethylene flame would thus be conceivably larger. Also, considering that fluorescence intensities and lifetimes increase towards much larger values in the ethylene flame. LIF signals peaks at 0.055V compared to 0.012V for the methane showing a dramatically higher concentration of these precursor particles that lead to the formation of soot in the ethylene flame. In the ethylene flame it is possible to see an increase in the τ_{eff} decay times up to 5.8 ns and 9.3 ns from LIF @ 350 nm and 445 nm respectively in the region of confidence before LII. This is significantly larger than that seen in the methane flame and perhaps highlights the transition of these nanostructures from condensed phase precursor particles to solid incipient soot particles.

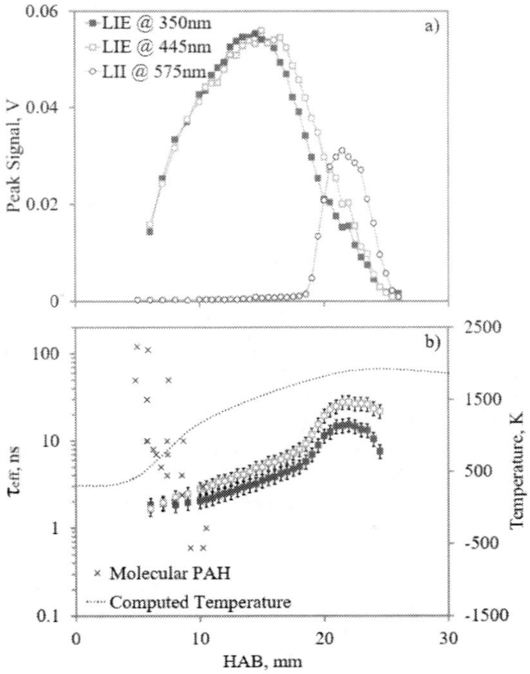

Figure 3: a) Peak LIE centerline measurements against HAB in 32% ethylene flame. b) Calculated LIF decay times (τ_{eff}) plotted alongside molecular PAH decay times measured by Ossler et al. [18] corresponding to the computed temperature (dotted) of the flame from the work of Smooke et al. [23].

4 Conclusions

LIF measurements in combination with LII measurements have proven valuable in capturing the transition of aromatic based nanostructures to soot. Measurements in sooting and non-sooting co-flowing diffusion flames were analyzed. With time resolved measurements, LIF contribution primarily from aromatic nanostructures was determined in both the non-sooting 65% methane flame and sooting 32% ethylene flame. LIF has been attributed primarily to nanostructures due to analysis of measured decay times compared to those expected from PAHs at flame temperatures. Both flames show some similarities in terms of spectrum and decay times in upstream regions. However, just prior to the onset of soot found in the ethylene flame there is a change in the structure of the fluorescent nanostructures that indicate a shift towards PAH constituents with a larger number of rings or more stacked structures. The peak nanostructure size is also expected to be larger in the sooting ethylene flame.

The use of 266 as a probe in the ethylene flame has some challenges with regards to ethylene photofragmentation and LII found in the sooting region. Although this casts some doubt on the LIF decay time in some regions, there is still a large window in the flames that remains undisturbed. The LIF intensities in the ethylene flame are more than 4 times that of the methane which indicates a significantly higher nanostructure concentration that in turn leads to the formation of soot.

5 Acknowledgment

This research is supported by the Australian Research Council. Additionally, acknowledgment is made for the financial support of the MSE-CNR project 2013-2014 "Miglioramento dell'efficienza energetica dei sistemi di conversion locale di energia".

References

[1] A. D'Anna, Proc. Combust. Inst., 32 (2009) 593-613.
[2] P. Desgroux, X. Mercier, K.A. Thomson, Proc. Combust. Inst., 34 (2013) 1713-1738.
[3] H. Wang, Proc. Combust. Inst., 33 (2011) 41-67.
[4] H.A. Michelsen, Proc. Combust. Inst., 36 (2017) 717-735.
[5] H.A. Michelsen, C. Schulz, G.J. Smallwood, S. Will, Prog. Energy Combust. Sci., 51 (2015) 2-48.
[6] M. Commodo, F. Ossler, C. de Lisio, A. D'Anna, P. Minutolo, Combust. Sci. Technol., 184 (2012) 916-928.
[7] A. D'Anna, A. Rolando, C. Allouis, P. Minutolo, A. D'Alessio, Proc. Combust. Inst., 30 (2005) 1449-1456.
[8] A. Bruno, C. de Lisio, P. Minutolo, A. D'Alessio, Combust. Flame, 151 (2007) 472-481.
[9] M. Commodo, L.A. Sgro, X. Wang, C. De Lisio, P. Minutolo, Proc. Combust. Inst., 34 (2013) 1845-1852.
[10] S. Bejaoui, X. Mercier, P. Desgroux, E. Therssen, Combust. Flame, 161 (2014) 2479-2491.
[11] K. Hayashida, T. Mogi, K. Amagai, M. Arai, Fuel, 90 (2011) 493-498.
[12] K. Hayashida, K. Amagai, K. Satoh, M. Arai, Journal of Engineering for Gas Turbines and Power, 128 (2006) 241.
[13] R.L. Vander Wal, Combust. Flame, 112 (1998) 607-616.
[14] R.L. Vander Wal, K.A. Jensen, M.Y. Choi, Combust. Flame, 109 (1997) 399-414.
[15] R. Sun, N. Zobel, Y. Neubauer, C. Cardenas Chavez, F. Behrendt, Optics and Lasers in Engineering, 48 (2010) 1231-1237.
[16] A. D'Anna, M. Commodo, M. Sirignano, P. Minutolo, R. Pagliara, Proc. Combust. Inst., 32 (2009) 793-801.
[17] M. Sirignano, A. Collina, M. Commodo, P. Minutolo, A. D'Anna, Combust. Flame, 159 (2012) 1663-1669.
[18] F. Ossler, T. Metz, M. Aldén, Appl. Phys. B, 72 (2001) 465-478.
[19] G. Nathan, available at <http://www.adelaide.edu.au/cet/isfworkshop/>.
[20] M. Sirignano, D. Bartos, M. Conturso, M. Dunn, A. D'Anna, A.R. Masri, Combust. Flame, 176 (2017) 299-308.
[21] P.-E. Bengtsson, M. Aldén, Combust. Flame, 80 (1990) 322-328.
[22] P.B. Kuhn, B. Ma, B.C. Connelly, M.D. Smooke, M.B. Long, Proc. Combust. Inst., 33 (2011) 743-750.
[23] M.D. Smooke, M.B. Long, B.C. Connelly, M.B. Colket, R.J. Hall, Combust. Flame, 143 (2005) 613-628.
[24] N.J. Kempema, M.B. Long, Appl. Opt., 53 (2014) 6688-6697.
[25] I.B. Berlman, Handbook of fluorescence spectra of aromatic molecules, Elsevier Science, Burlington, 2012.
[26] Z. Chi, B.M. Cullum, D.L. Stokes, J. Mobley, G.H. Miller, M.R. Hajaligol, T. Vo-Dinh, Spectrochimica Acta Part A: Molecular and Biomolecular Spectroscopy, 57 (2001) 1377-1384.
[27] Y. Kobayashi, T. Furuhata, K. Amagai, M. Arai, Combust. Flame, 154 (2008) 346-355.

11th Asia-Pacific Conference on Combustion,
The University of Sydney, NSW Australia
10th -14th December 2017.

Experimental Study on Breakup and Ignition of Water Emulsified n-Dodecane Impinging on a Hot Surface

Gyu Min Jang[1], Nam Il Kim[1]

[1] Korea Advanced Institute of Science and Technology

291 Daehak-ro, Yuseong-gu, Daejeon, 34141, Republic of Korea

Abstract

Water emulsified fuels are used in many combustion systems including internal combustion engines and gas turbines to suppress NOx emission and flame characteristics. When emulsified fuel droplets impinge onto a high temperature surface such as the wall of a combustion chamber, their breakup and ignition phenomena may differ from those of ordinary fuels. In this study, water emulsified n-dodecane was prepared with various mixing ratios, and its single droplet was impinged onto a hot plate surface that consists of a quartz plate channel and a methane-air flame inside. Thus, the surface temperature could be increased up to about 1080K. After impinging of emulsion droplet, detailed phenomena of breakup and ignition were experimentally investigated, and were visualized using a high-speed camera. The effects of the droplet size, velocity, and volume fraction of n-dodecane, X_{dod}, were investigated. When the surface temperature was higher than the boiling temperature, various breakup-patterns could be observed depending different conditions. In addition, diffusion flame could be observed after breakup of the emulsion droplet. The conditions and mechanisms of ignition were investigated.

1 Introduction

Recently, water addition has been used in many combustion systems to suppress NOx emission by reducing the flame temperature. This water addition into combustors was mainly achieved in two ways. One is injecting water into a combustion chamber and the other is using water-oil emulsion. The merit of using water-oil emulsion is that the same water injective device can be used [1]. Therefore, water-oil emulsion is a promising technique that can be used for various combustion systems including internal combustion engines and gas turbines. Usually, the water-oil emulsion is classified into two types; a water-in-oil (W/O) structure and an oil-in-water (O/W) structure. The W/O structure includes a number of micro-scale water drops in a continuous oil liquid pool while the O/W structure includes many micro-scale oil drops within a continuous water liquid pool.

Previous studies suggested that W/O emulsions could be used as suitable emulsified fuels. When O/W emulsion is used, the content of oil is too low to burn, and water will directly contact to a fuel injector and the burner surface directly. That may cause serious wear and damage combustors. Therefore, the W/O emulsions have mostly been used as emulsified fuels of combustion devices [2]. The difference of combustion between a pure droplet and an emulsion droplet is the occurrence of the internal explosion phenomenon, which is called as micro-explosion. Due to the difference in the boiling temperature of water and oil, this micro-explosion phenomenon is induced by preemptive nucleation of dispersed water drops. The drastic expansion of a water vapor induces breakup of the emulsion droplet in the combustion chamber, and it is known as the secondary atomization [3]. However, the combustion mechanism of emulsified fuel is very complex compared to the ordinary droplet combustion. When emulsified fuel droplets impinge onto a high temperature surface such as a wall of combustion chamber, the breakup and ignition phenomena may differ from those of ordinary fuels. It is due to the difference in the thermodynamic properties and the internal structure of emulsion. To understand the combustion phenomena of water-oil emulsion, additional studies on breakup and ignition of a single droplet emulsion have been conducted in this study.

2 Experimental setup

The emulsion was prepared by mixing n-dodecane and distillated water. The X_{dod} is defined as the volume faction of n-dodecane to total mixture. An impeller type mixer was used, and it was operated for 30 minute with 2500rpm. Surfactant (Span® 80, Sigma-Aldrich) was added to the mixture of n-dodecane and water to improve the stability. Volume fraction of surfactant was approximately 1.0%, and the effects of surfactant on the breakup and ignition were neglected. So, the emulsion state could maintain more than 1-hour without being stratified.

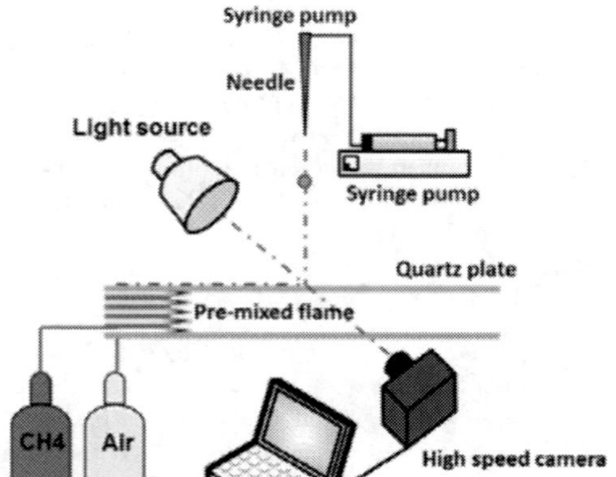

Figure 1: Schematic of experimental setup

Corresponding author. Fax: +82-42-350-3211
E-mail address: nikim@kaist.ac.kr

To get a sufficiently high temperature on the surface, a quartz plate channel equipped with a slot burner of a premixed flame was fabricated as shown in figure 1. A premixed flat flame was formed at one side of the channel, and the hot product gas passed through the channel. The flow rates for the methane-air premixed flame were 2.4 l/min of methane and 26 l/min of air. It was possible to increase the surface temperature up to about 1080K at the near location of the premixed flame. Also the behaviors of the impinging droplet could be observed directly from the bottom side through the transparent quartz plates. To inject the fuel emulsion, a syringe needles (inner diameter of 0.21, 0.26, 0.32 mm) and a syringe pump (flow rate of 2 ml/min) were used. When the smaller diameter tube was used, the smaller droplet jet could be injected.

2 Results and discussion

When the W/O emulsion (X_{dod}=90%) droplet impinge on a hot surface (T_s=1053K), two phenomena could be observed as shown in figure 2 and figure 3. These phenomena were affected by the droplet size, which was varied by selecting different needle diameters. With a needle of a small diameter, small-size droplets were injected consecutively, and each droplet was divided into 2 or 3 small droplets after impinging on the surface as shown in figure 2. On the contrary, significant droplet breakup phenomena occurred when a large-size droplet felled on the hot surface, it was shattered into a lot of tiny secondary droplets. After that, it could be immediately ignited as shown in figure 3 in some experimental conditions. Thus, it can be explained that the surface tension rather than the momentum mainly dominated the mechanism of small droplet impingement. It is advantageous to increase the impact momentum for better atomization performance.

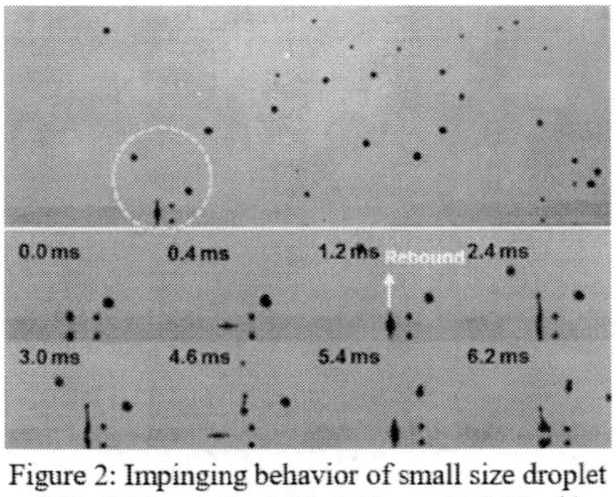

Figure 2: Impinging behavior of small size droplet (I.D.: 0.21mm, D_0: 0.76~1.03mm, V=2.07m/s)

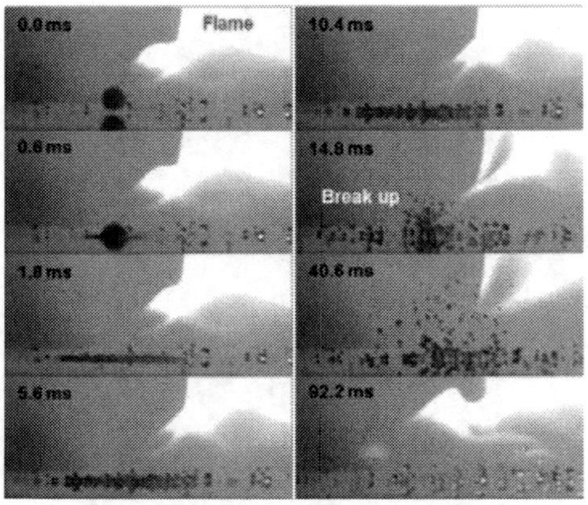

Figure 3: Impinging behavior of large size droplet (I.D.: 0.26mm, D_0: 2.89~2.94mm, V=2.35m/s)

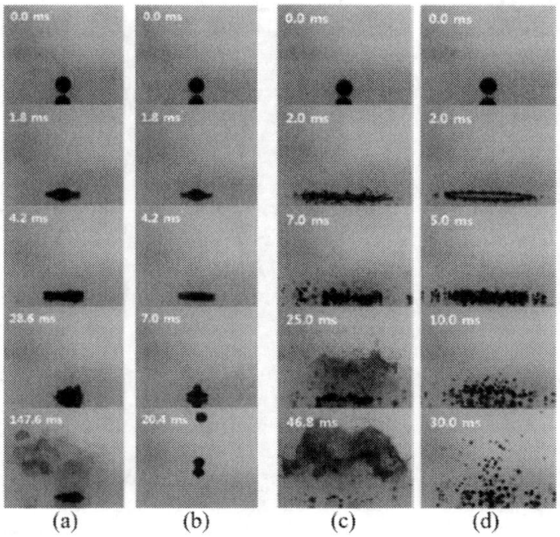

Figure 4: Effect of impinging velocity and surface temperature about impinging behavior
(a): T_s=640K, V=0.62m/s, (b): T_s=1081K, V=0.62m/s,
(c): T_s=640K, V=2.11m/s, (d): T_s=1081K, V=2.11m/s

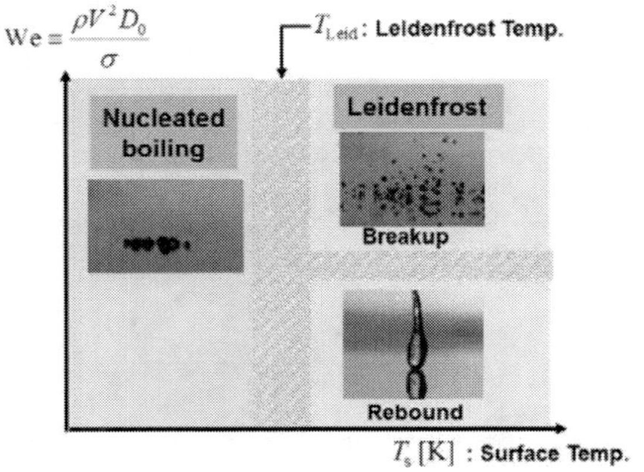

Figure 5: Regime map of post-impingement behavior of droplet

Figure 4 shows the behavior after impinging of emulsion (X_{dod}= 90%) droplets when the diameter of droplet was about 1.8 mm. If the surface temperature was relatively low as like in figure 4(a) and 4(c), nucleate boiling was observed. When the droplet hit the surface, it began to evaporate while staying on the surface. However, when the temperature was relatively high, droplet did not adhere to the surface. At a condition of high temperature surface, if a droplet had lower impinging velocity as shown in figure 4(b), that droplet rebounded and was separated into two or three secondary droplets after impinging. On the other hand, at a relatively high impinging velocity corresponding to figure 4(d), it was clearly observed that numerous secondary droplets appeared after impinging on the hot surface. The rebound of the droplet was thought to be related to the Leidenfrost phenomenon [4]. When the droplet impacts on the hot surface, a vapor layer is generated through a rapid boiling process. Due to this vapor layer, the droplet could hover above the surface rather than getting a direct contact to the surface. When the impact momentum was increased at the same surface temperature, at which the Liedenfrost phenomenon occurred, droplet was separated into numerous secondary droplets.

The post-impinging behaviors of droplet were determined by

the impact momentum and the surface temperature. The rebound and breakup were determined according to the impact momentum when the surface was above the triggering temperature for the Leidenfroast. Beneath that temperature, the nucleate boiling was dominant as shown in figure 5. The triggering temperature for Leidenfroast was also determined by content of water in the emulsion. Figure 6 presents the temperature criterion of the Leidenfroast, and it decreased as the water content increased. It may be due to the vaporization of water drops in the W/O structure because boiling temperature of water (293K) was lower than that of n-dodecane (489K). The emulsion droplet tends to evaporate easily if the emulsion has high water fraction, and Leidenfroast phenomenon will easily occur at relatively lower temperature conditions.

When a droplet impinges onto the hot surface with different water content, the breakup process was visualized in figure 7 for conditions of the same impinging speed (V=0.62m/s) and the same surface temperature (T_s=1081K). As a result, the number and size of the secondary droplets could be distinguished depending on the water composition in the emulsion. When the water content was higher, the lower number of secondary droplets could be produced, and their average size also increased. Furthermore, flame could not be

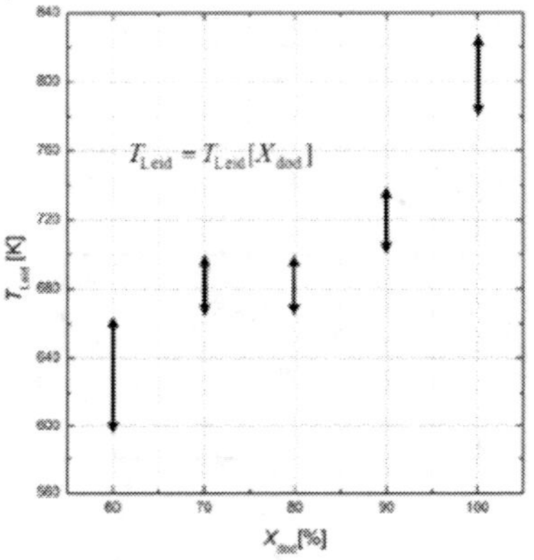

Figure 6: Onset Leidenfroast temperature with content of water

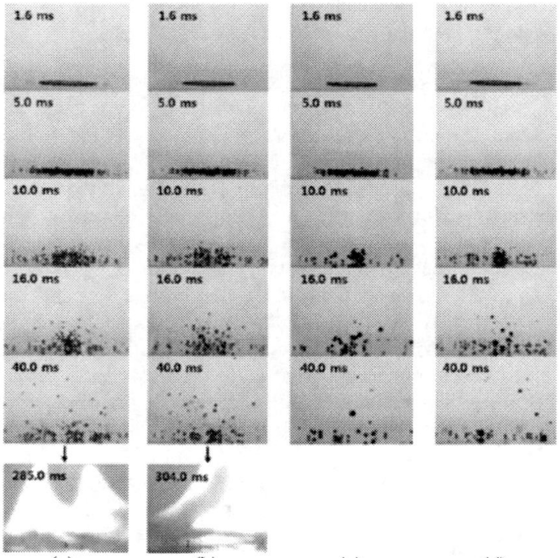

(a) (b) (c) (d)

Figure 7: Breakup and ignition of droplet after impinging hot surface depending on X_{dod}
(a): X_{dod}=95%, (b): X_{dod}=90%, (c): X_{dod}=70%, (d): X_{dod}=50%

observed if the n-dodecane content was less than 80%. These results indicate that if the viscosity of emulsion was high (it corresponds to higher water content), breakup and separation to the secondary droplets were suppressed. Actually, viscosity of emulsion increased as the content of water increased. Also, dispersed water droplets will eventually reduce the concentration of fuel and oxidant, and that will suppress the occurrence of flames. The probability of flame occurrence drastically increased when the n-dodecane content was higher than 80%. Therefore, excessive water content may deteriorate the liquid atomization and also may hinder the flame stability.

3 Conclusions

In this study, impinging of emulsion droplets on the high temperature surface was investigated. An experimental comparison was conducted for the effects of the impact momentum, surface temperature, and water content. For a more efficient atomization performance, intense momentum is required when the droplets impinge against a hot surface. If the water content increased, the Leidenfroast phenomenon occurred at a lower surface temperature due to the micro-explosion or vaporization of dispersed water drops in the emulsion. Intensive momentum and the surface temperature higher than the criterion for the Leidenfroast could break up the original droplet into numerous secondary droplets. However, excessive water content may deteriorate the liquid atomization, and also may hinder the flame stability.

5 Acknowledgment

This work was partly supported by the research grant funded by MHPS.

References

[1] A.M. Ithnin, H. Noge, A.K. Hasannuddin, W. Jazair, J. Energy Inst. 87 (2014) 273-288.

[2] V.N. Antonov, Chemistry and Technology of Fuels and Oils 19 (1983) 606-609.

[3] H. Watanabe, Y. Suzuki, T. Harada, Y. Matsushita, H. Aoki, T. Miura, Energy 35 (2010), 806-813

[4] C. Bai, A.D. Gosman, SAE Technical Paper 950283 (1995).

11th Asia-Pacific Conference on Combustion,
The University of Sydney, NSW Australia
10th-14th December 2017.

Dynamics of Burning of Laminar Liquid Pool Flames

Mayur Anvekar[1], T M Muruganandam[1], V Raghavan[2]

[1]Department of Aerospace Engineering

Indian Institute of Technology Madras, Chennai, 600036, India

[2]Department of Mechanical Engineering

Indian Institute of Technology Madras, Chennai, 600036, India

Abstract

Pool fire is a non-premixed flame established on the surface of a horizontal liquid fuel pool. Analysis of fire scenario and determining the fire safety aspects are mostly done using lab-scale liquid fuel pool fires. Literature reports, a huge amount of research work on pool fires, where the characteristics such as flame length and burning rate have been reported for several scenarios involving different types of liquid fuels. A pool fire can be laminar or turbulent based on the pool diameter and type of fuel. These flames are highly oscillatory and have stable, intermittent and plume regions. Understanding the characteristics of these regions is extremely important. Based on this introduction, a fundamental study involving two alcohol fuels, methanol and ethanol, burnt in circular trays having diameters of 0.05 m and 0.1 m, in a confined environment, has been carried out. Basic burning characteristics such as flame height, mass burning rate, flame width variation and flame puffing frequency have been measured quantitatively. Apart from this, entrainment characteristics have been studied using shadowgraph technique, which has helped in visualizing the flame vortex shedding and measuring the shedding frequency.**Keywords**: Pool flames; Alcohols; Flame puffing; Mass burning rate; Vortex shedding; Entrainment; Shadowgraph

1 Introduction

Pool flame is a buoyant diffusion flame sustained on top of a horizontal evaporating liquid pool. Since the fuel is in liquid state, the rate limiting primary process is the vaporization of the liquid fuel. A pool flame is a type of heterogeneous diffusion flame. Industries and laboratories, which have fuel and solvent handling and storage areas, generally have spillage too. Over such fuel or solvent spillages, an energy source can produce a diffusion flame, when the fuel pool is heated up to its fire point temperature. Initially the flame spreads over the pool and once the entire pool surface is ignited, almost steady burning rate is achieved depending upon the depth of the fuel pool, while the initial spread is transient. However, thissustained diffusion flame can become highly oscillatory depending upon the surrounding convective conditions. Under normal gravity conditions, oscillations in small pool flames are assisted by buoyancy and entrainment processes. Interesting dynamics are observed in such flames, which need to be studied systematically.

The main characteristic of pool flames are their spread burning rates. While the flame spread occur quite rapidly based on convective conditions, the burning rate is a slower process. This depends upon the volatility of the fuel. In general, a pool flame has a jet flame like structure. Flame height is another main feature of the pool flame apart from the burning rate. In an oscillating (or puffing) flame, average flame height is recorded using series of direct flame photographs or using temperature measurements. Usually, the liquid fuel contained in a metallic pan is used in laboratory experiments. The flame characteristics are dependent upon the type of the fuel, size of the pool, thermal conditions of the pool pan and pool level within the pan (called ullage). Several researchers have done experiments to understand the pool flame at different working conditions. Unsteady burning rate of small sized hydrocarbon pool flames have been studied in the past [1, 2, 3]. The burning rates for almost same sized pool and same fuel showed variations as a result of surrounding convective conditions, pan wall conditions and ullage. For instance, as the flame anchors near the pan wall, the pan wall gets heated up and this in turn heats up the liquid fuel. The evaporation process, resulting from surface heating, transitions to boiling process due to heating by pan walls.

At room temperature the liquid pool starts evaporating from its liquid phase due to its low boiling point and starts diffusing into atmosphere, meanwhile air starts diffusing into evaporated fuel forming a zone of combustible mixture. When heat energy is supplied at this location, flame is established. The flame gets anchored at the junction where the air and fuel mixture is stoichiometric. A part of heat energy produced from the flame is transferred to the liquid pool which evaporates the fuel from pool to the flame, whereas the remaining heat energy gets transferred to the surroundings. The pool flame naturally occurs without any metal boundary but during experimentation the pool must be constrained; hence a metallic pan of particular diameter is used.

Preheating of fuel and transition from evaporation to boiling of liquid pool were reported and the effect of uncooled pool pan on fuel evaporation was also highlighted by past researchers. The heat feedback to the pool surface and the combustion rate of medium scale pool firewhich is an unsteady parameter were studied [2, 4, 5].In the study, they characterized the flame physically based on its combustion rate and explained the ventilation effect on flame structure for under-ventilated and well-ventilated conditions. Mass burning rate of large pool fire, unsteady burning rates and flame pulsation are also studied[6, 7, 8]. The fuel burning rate was measured by placing the pool on a load cell and the mass burning rate as a function of heat supplied to pool was discussed. The three burning modes namely convective, radiative-laminar and turbulent were also reported. The characteristics of medium scale pool fire burning and initial fuel temperature effects on pool fire burning were studied [1, 9], explaining the flame height, flame shape and pulsating frequency. The zones in flame height were also

Corresponding author.
E-mail address:murgi@ae.iitm.ac.in

distinguished as persistent, intermittent and plume. Froude model, which defines the dominating velocity in pool fire,was explained and ways to measure the turbulent flame height was also reported.

In this paper, a fundamental study of burning features of pool flame of 0.05m and 0.1m in diameter pool, fuelled by methanol and ethanol, is presented. The flame characteristics namely, the flame height, puffing frequency, burn rate and flame vortex shedding are presented. Experiments are conducted under quasi-steady conditions by maintaining the liquid pool level constant. Phenomenally laminar flames with transient oscillations have been observed. As the pool diameter is less than 0.15m, the Froude number is less than 1 (Fr<1), making it buoyant laminar flame.

2 Experimental setup

Two metal pool pans of diameters 0.05m and 0.1m, and depth of 0.015m have been fabricated. A provision for continuoussupply of the liquid fuel from a relatively bigger reservoir to the pan has been made to maintain the pool level constant, as shown in Fig. 1.To avoid overheating of pan wall and boiling of fuel due to heat transfer from the flame,the pan wall is held at an almost constant temperature of 27°C by circulating water using a cooling jacket around the pool pan and a chiller unit. As shown in Fig.1, the entire setup is placed within a transparent enclosure (0.95x0.95m, height 0.5m, top open to atmosphere).This avoids external disturbance imparted to the flame. The transparent walls help in providing optical access for carrying out diagnostics related tovelocity and species measurements, as a future work. To visualize the flame dynamics for smallduration, a high speed camera is used at high resolution. In order to measure the burn rate, the amount of fuel supplied from reservoir to the pool pan is measured using precise instrumentation.

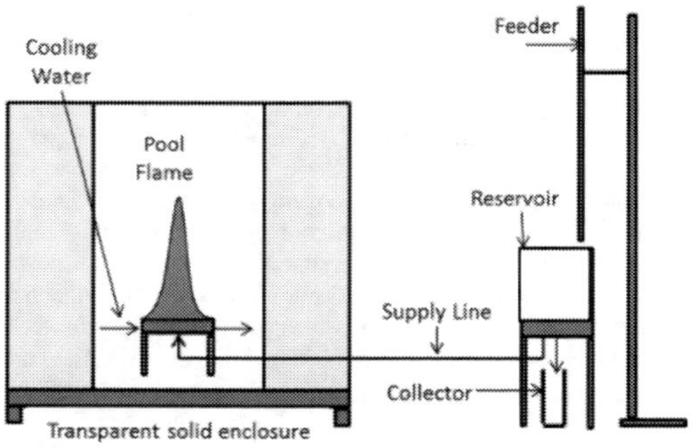

Figure 1: Pool flame experiment setup

A standard scale has been placed behind this convective laminar pool flame and is used as reference image in image processing to measure the pool flame height. Initially, the pool is filled with known amount of fuel up to the brim of the pan, which is kept connected to the reservoir. Upon consumption from the pan, the fuel will be continuously supplied to the pan at the rate equal to its consumption. In fact, the reservoir is designed in such a way that, the fuel supply from feeder will be in excess and only the quantity in the pan that is consumed will be supplied to the pan and the remaining will be drained out into a collector (Fig. 1).

An ignition source is used near the top surface of the pool to initiate a flame. The pilot ignition source is removed once the flame is established. The fuel reaches the flame zone by evaporation and air from ambient reaches the flame zone (through diffusion and entrainment) thus flame is sustained. Based on the amount of fuel supplied from reservoir and the amount of excess fuel collected in the collector,the burn rate is estimated precisely.

The images usedto measure the flame height as a function of timeare captured using a DSLR camera at 60 fps.Shadowgraph, which is one of the flow visualization techniques, is used to visualize the density gradient in a reactive flow. Shadowgraph on pool flame is obtained using 2 concave mirrors (focal length 3m), a light source with slit and a high speed camera. The high speed camera Photron SA1.1 is used at 832x1016 resolution and 1000 fps (frames per second) to acquire the images using a 100mm macro lens of F/2.8. In shadowgraph technique, the flame as well as the plume is visualized. The shadow of pool flame with bright background is captured at high frame rate. The flow field having vortices are also captured using high speed camera and their characteristics are determined.

3 Results and Discussions

Using the high speed images with high pixel resolution in comparison to the scale, the flame height is measured for both pool flames. The flame height of 50 mm diameter pan is found to beshorter when compared to the100mm diameter pan flame, due to the change in amount of fuel evaporation into the flame zone as shown in Figs. 2 and3. The variations in the flame height for a particular pool flame are primarily due to variation in evaporation rate of fuel. This is as a result of fluctuations in air entrainment rate, heat feedback to the fuel surface and fluctuations in the mixing process. The continuous and intermittent zones of flame are visible in both the pans. The uncertainty incurred in the measurement of flame height is found to be1.6%based on the pixel calibration in image processing.

In Figures 4(a) and 4(b), the variation of flame height with time is shown for 100 mm and 50 mm pool cases, respectively. From the high speed imaging of pool flame, it is clear that the flame height is oscillating with time. The sharp peaks in graph are due to flame tip puffing, which arises for short duration.

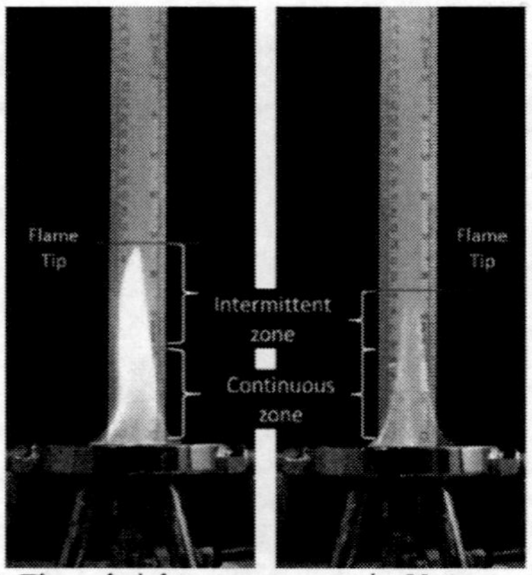

Figure 2: Flame height measurement in 50 mm pool (left - ethanol, right - methanol)

The average value of ratio of flame height to pool diameter (H/D) for methanol small pool flame is 2 and for Ethanol is 1.5, reported by A. Hamins [1], whereas in this paperH/D is 2.08 and 2.34 for 100 mm and 50 mm methanol pool pans, and 1.4, 1.45 for ethanol pool pans. This also shows that the H/D value for a pool flame within laminar zone is almost constant.The magnitude of flame height pulsation is not constant as the flame is unsteady primarily due to buoyancy. The pulsating frequency in the past work was found to be 11-12 Hz for convective laminar pool fire reported by Hayasaka [6] and schalike [7], where as in this paper, itis found to be 10-13 Hz for all the pool flames, based on the displacement of flame tip captured using high speed imaging.

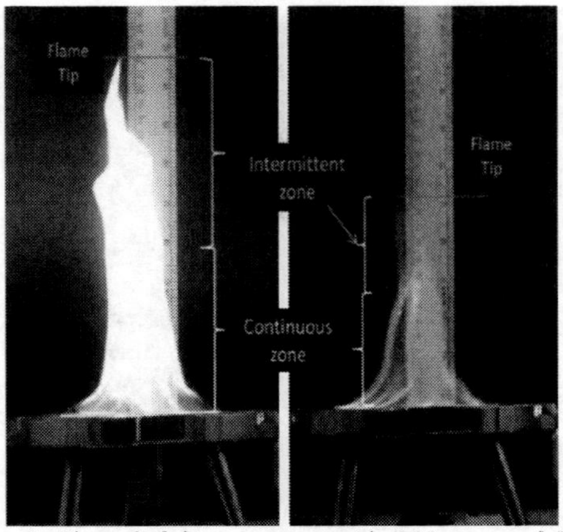

Figure 3: Flame height measurement in 100 mm pool (left - ethanol, right - methanol)

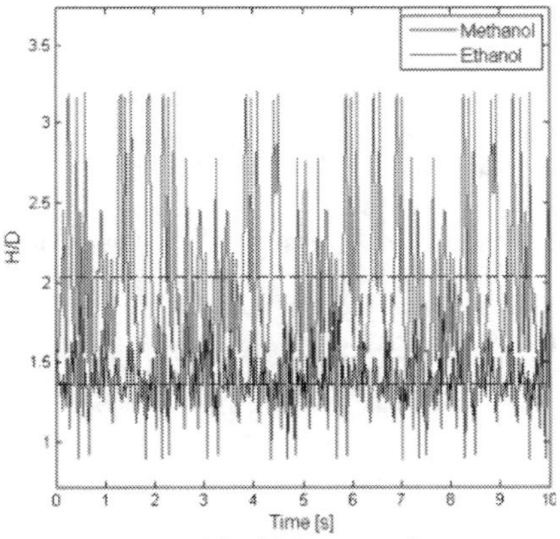

(a) 100 mm pool

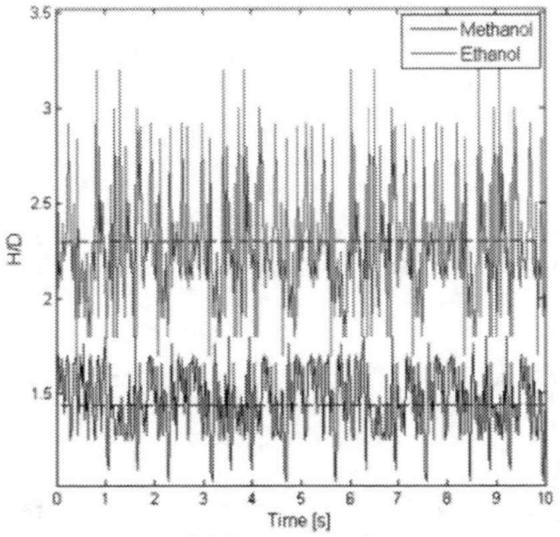

(b) 50 mm pool

Figure 4: Non-dimensional flame height vs. time

To calculate the mass burning rate and its variation, experiment is carried out for 2 minutes, 4 minutes and 6 minutes. As shown in Figs. 5(a) and 5(b), the variation in burn rate with respect to time is very less. As the burn rate is directly proportional to the pool pan size, the 100 mm pool flame has almost 3 times the burn rate of 50 mm pool flame. The uncertainty in the measurement is evaluated and error bars are plotted to indicate it. The major error is from the collector, which is open to atmosphere leads to evaporation of volatile fuel. This was measured by conducting another experiment on the evaporation rate of fuel, which is approximately 0.0007 g/s. The average mass burning rate, when the experiment is conducted at different time durations is found to be the same, as shown by discontinuous line. The burn rate of 50 mm methanol pool flame was reported as 0.016 g/s, by H. Hayasaka [6], and 0.02 g/s by K. Mudan [10], where as in this study it is found to be 0.0198 g/s. However, their cases were uncooled pool pan.

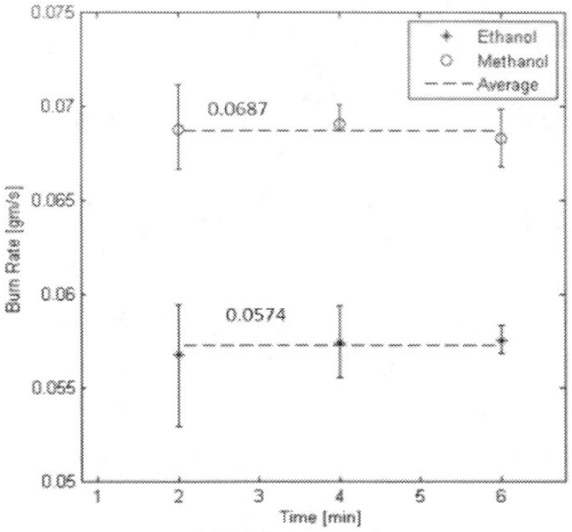

(a) 100 mm pool

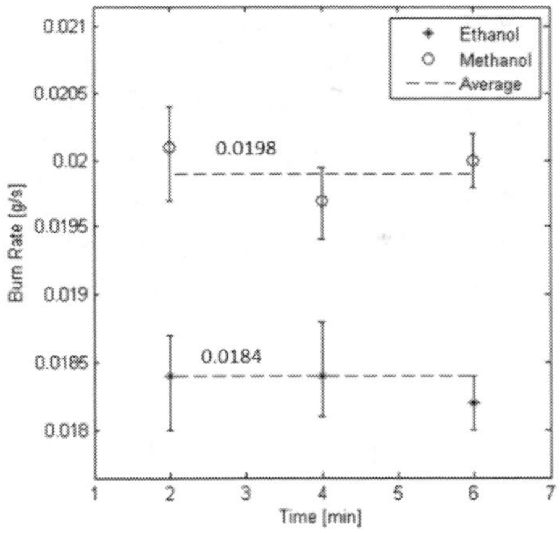

(b) 50 mm pool

Figure 5:Mass burning rate vs. time

Shadowgraph images are shown in Figs.6(a), 6(b),6(c) & 6(d). The flame vortex size near the pan brim is small when compared to a location at a certain height. This is because of the fact that, in the plume region, air entrainment can be stronger due to buoyancy driven by the combustion products. The vortex shedding velocity has been measured by tracking the plume location from two images at time t1 & t2 as shown below.The measurement is restricted to a certain height above pan rim because of the fluctuations present in the intermittent zone. Within certain flame height, the vortex shedding velocityfor 50mm pool flame is found higher than that of 100 mm pool as shown in Fig.7.

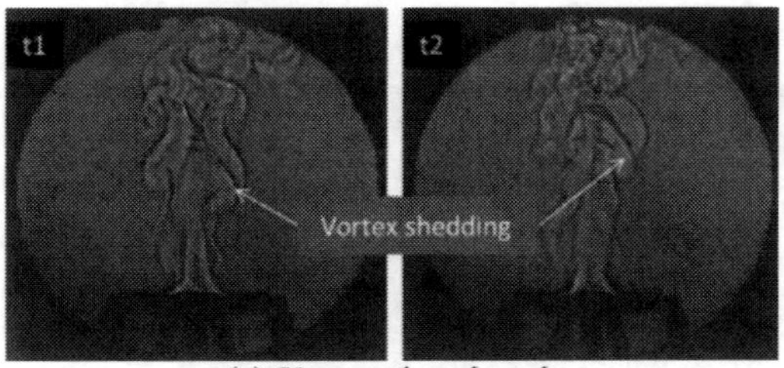

(a) 50 mm ethanol pool

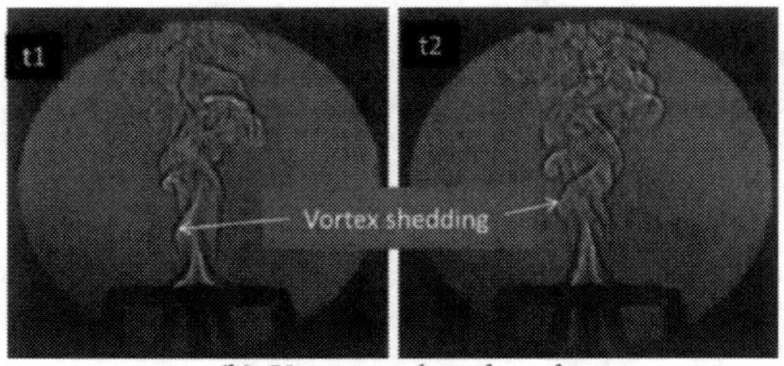

(b) 50 mm methanol pool

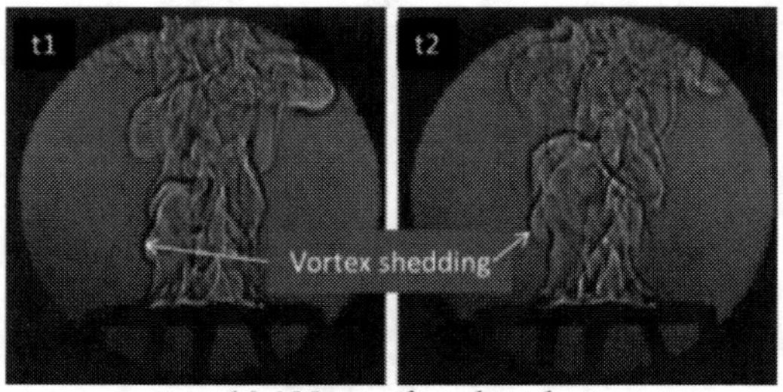

(c) 100mm ethanol pool

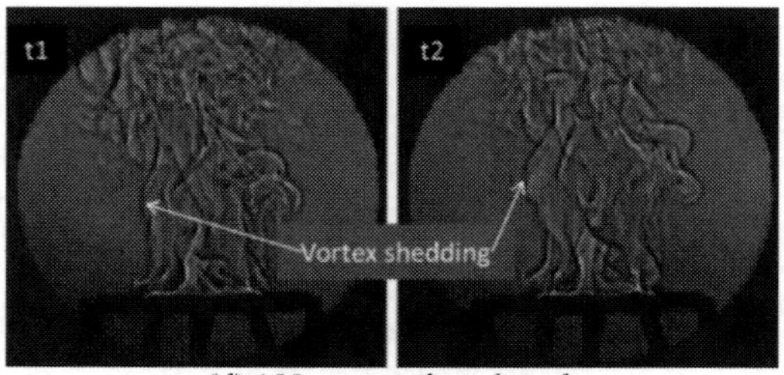

(d) 100 mm methanol pool

Figure 6: Shadowgraph images

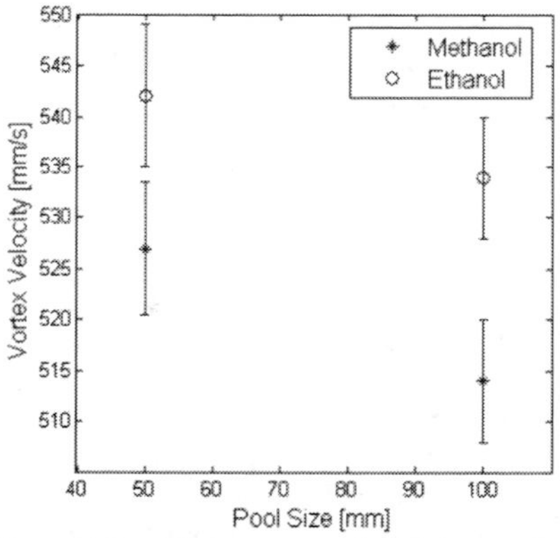

Figure 7: Pool flame vortex shedding velocity

The values of the flame height, the burning rate and vortex shedding velocity are listed in the table below, for the two pool diameters and fuels ethanol and methanol.

	50 mm methanol	50 mm ethanol	100 mm methanol	100 mm ethanol
Burn Rate [g/s]	0.0198	0.0184	0.0687	0.0574
Flame H/D	1.45	2.35	1.4	2.1
Vortex shedding velocity [m/s]	0.527	0.543	0.512	0.534

Table 1: Pool Flame parameters

4 Conclusions

The basic burning characteristics of ethanol and methanol pool flames are reported. The flame heights of 50mm and 100mm pools have been measured as a function of time under well ventilated conditions and by maintaining constant pool level. The pulsating frequency of pool flame is found to be almost the same irrespective of the pool diameter and the type of fuel. The mass burning rate has very less fluctuations in its magnitude over different burning periods.The flame vortex shedding velocities are measured at good pixel resolution considering possible errors. The flame pulsating frequency and vortex shedding frequency can be related in future.

5 Acknowledgments

The Authors are very grateful to NCCRD, IITM for providing research support.

References

1. Hamins, A.,Kashiwagi, T., andBuch, R., 1995,Characteristics of pool fire burning, Fire resistance of Industrial Fluids.
2. Truchot, B., Durussel, T., and Duplantier, S., 2010,Combustion rate of medium scale pool fire, an unsteady parameter.International symposium on Safety Science and Technology, p. 575-585.
3. Babrauskas, V. 1983. Estimating large pool fire burning rates, Fire Technology, vol. 19, p. 251-261.
4. Hamins, A., Klassen, M. E., Gore, J. P., Fischer, S. J., and Kashiwagi, T., 1993, Heat feedback to the fuel surface in pool fire, Combustion science and Technology, vol. 10, p. 1-27.
5. Yumoto, T., 1987, Burning rate and thermal radiation in surface combustion, doctoral thesis, p.57.
6. Hayasaka, H., 1997, Unsteady burning rates of small pool fires. Fire Safety Sciene: Proceedings of the fifth international symposium, p. 499-510.
7. Schalike, S., Mishra, K. B.,Malow, M., Berger, A., Dieter, K.,Wehrstedt and Schonbucher, A., 2013,Mass burning rate of large pool fire- Experimental study and modeling.Chemical Engineering Transactions, p. 853-858.
8. Buckmaster, J. and Peters, N. 1986. Flame pulsation and instability, Twenty first International symposium on Combustion, The Combustion Institute, p.1829.
9. Cheng, B., Lu, S., Li, C., Kang, Q., Lecoustre, V., 2011, Initial Fuel temperature effects on burning rates of pool fires, J. Hazard.Mater., p. 369
10. Mudan, K., 1984, Thermal radiation hazards from hydrocarbon pool fires, Progress in energy and combustion science, p, 59-80.

11th Asia-Pacific Conference on Combustion,
The University of Sydney, NSW Australia
10th -14th December 2017.

ON THE DYNAMICS OF INSTABILITY MITIGATION BY ACTUATING THE SWIRLER IN A COMBUSTOR

Ankit K Dutta, Gopakumar R., Rahul B. V., Jasmeet Singh, Swetaprovo Chaudhuri

Department of Aerospace Engineering,
Indian Institute of Science
C V Raman Avenue, Bangalore, Karnataka, 560012, India

Abstract

In this paper, we present a detailed and novel analysis of the mitigation mechanism of instability in a lean premixed, swirl-stabilized, labscale combustor by actuating the swirler. It has been reported in our previous work that increasing the swirler rotation rate mitigates the self-excited thermo-acoustic instability in a model lab-scale combustor, over a range of conditions. Here, it is found that for a given period of observation, instead of a continuous and gradual decrease in the time localized pressure amplitude from the fully unstable state towards the fully mitigated state, the fraction of the time during which instability is present is reduced. With increasing swirler rotation rates, the instability becomes more bursty and its frequency decreases progressively. Such an intermittent route to instability mitigation could be attributed to the background turbulent flow field and is reminiscent of the intermittent opposite transition (implemented by changing the Reynolds number) from a fully chaotic state to a fully unstable state as recently discovered in Nair et al.[1]. An attempt is made to model the behavior of pressure oscillations using the well-established mean-field Kuramoto model. The variation of the order parameter r, which measures synchronization of the oscillators provides critical insights on the transition from the unstable, intermittent to stable states.

1 Introduction

Modern gas turbine combustors are adopting lean premixed combustion to meet stringent emission regulations on NOx, UHC (unburned hydrocarbon) etc. [2]. With the adoption of this technology, the combustors are getting susceptible to flame blow-off, flashback, and thermo-acoustic instabilities. The first two problems occur when the flame speed deviates from local flow speed and can be avoided by fixing the operating regime. Thermoacoustic Instabilities are high amplitude low-frequency pressure oscillations which originate due to the in-phase coupling of heat-release rate and pressure fluctuations inside the combustor. These instabilities are triggered by flame-acoustic interaction, mixture equivalence ratio fluctuations, large vortex shedding etc. [3–8]. The periodic fluctuation of pressure and temperature, caused by the instabilities results in premature and catastrophic failure of the combustor.

Thermo-acoustic instabilities can be mitigated either by modifying the combustor (passive control) or by using an externally excited auxiliary device (active control). This paper reviews an active control technique, which is realized by actuating the otherwise static swirler to mitigate the instabilities. We observed that the combustor approaches a stable state through an intermittent route from an unstable state [1,9]. This intermittent state is characterized by bursts of high amplitude pressure oscillations, which makes it distinct from combustion noise and instabilities.

To understand and model the intermittency, we used Kuramoto model [10] in this paper. This model is useful for visualizing the behavior of a population of oscillators in macroscopic level. A turbulent flame can be assumed as an ensemble of small oscillating flamelets of different natural frequencies oscillating in the circumferential direction. The synchronization of these flamelets creates a strong instability, which is reflected in the pressure signal. Similarly, lack of synchronization in the flamelets leads to intermittency and then to stable conditions. The frequency spectrum of the flamelets was derived from the experimentally obtained pressure data and used as an input to the model.

2 Experimental method

The combustor used for the present study along with the diagnostics used is detailed in Fig. 1. It houses a 30 degree, axial, 8 blade swirler, whose annular speeds could be varied using a stepper motor, secured using a shaft coupling arrangement. Metered quantities of methane and air are mixed to a molecular level in a premixing chamber and fed into the settling tube (150mm long, 32mm dia.) through the 4 circumferential ports with orifices. The mixture expands into the optically accessible cylindrical combustion chamber (d = 46mm, h = 60mm, made from quartz) through the swirler which is recessed by 2mm. A 2m long aluminum tube is mounted over the quartz tube to generate the self-excited acoustic modes. Tests were performed at a bulk Reynolds number of 7274 and equivalence ratio of ϕ = 0.71 which corresponds to a thermal power of 5.67kW.

Flow rates of methane and air are controlled separately using Alicat MCR series mass flow controllers with an uncertainty of ±0.8%. Acoustic pressure measurements at 1kSa/s were recorded using KISTLER piezo-resistive pressure transducer with an uncertainty of ±0.2% (range of 0-3 bar absolute pressure).

Stereoscopic (2D3C) PIV measurements were performed on x-y plane illuminated by a 527 nm Nd:YLF laser (Photonics

Corresponding author. Phone: 9180-2293-2875
E-mail address: schaudhuri@aero.iisc.ernet.in

Industries DM series) sheet of sub-millimeter thickness. These velocity measurements were done after seeding the flow-field with 1μm alumina particles. A bypass system for blocking the air through the seeder is used to reduce clogging of optical windows by the seeding particles when images were not captured. CMOS cameras were mounted on the same side of the laser sheet, one in forward and the other in backward scatter configuration. Temporal synchronization of acoustic and velocity signatures was done using LaVision Programmable Timing Unit (PTU) and DaVis 8.3.1 software.

Vector fields were computed from the illuminated tracer particles using stereo cross-correlation technique provided with DaVis 8.3.1 software. Multi-pass iterations were performed with interrogation window sizes varying from 96x96 to 64x64 with an overlap of 60%.

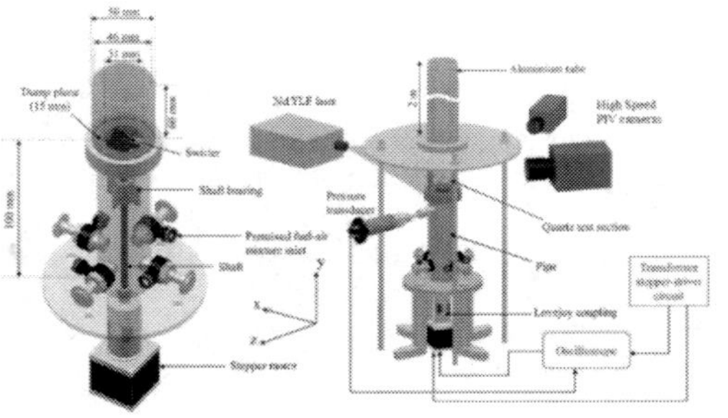

Figure 1: Experimental setup used (on left) with the diagnostic equipment in place (on right).

2.2 Kuramoto model

Kuramoto model [10–12] basically shows the behavior of a population of oscillators at the macroscopic level. This model assumes a phase coupling between the oscillators, oscillating at different frequencies proportional to the sine of their phase difference. Mathematically, it is defined as

$$\dot{\theta}_i = \omega_i + \frac{K}{N}\sum_{j=1}^{N}\sin(\theta_j - \theta_i) \qquad (1)$$

Here, $\theta_i(t)$ is the phase of the oscillators, ω_i is the natural frequency of oscillators, N is the number of oscillators, and K is the coupling or mixing parameter between two oscillators. Equation (1) can be simplified by formulating an order parameter r and average phase Ψ as,

$$re^{i\Psi} = \frac{1}{N}\sum_{j=1}^{N} e^{i\theta_j} \qquad (2)$$

The order parameter r is a complex variable, which reaches 1 when the oscillators are fully synchronized and 0 as they get totally asynchronized. Substituting (2) in (1) changes the equation to,

$$\dot{\theta}_i = \omega_i + Krsin(\Psi - \theta_i) \qquad (3)$$

Equation (3) represents the coupling between the oscillators is proportional to the sine of the difference between average phase and oscillator phase.

3 Results

3.1 Experimental findings

Our lab scale model of a turbulent pre-mixed combustor shows a reduction in first mode amplitude and an increment of induced swirl with increasing swirler rotation rates [13]. Though, the amplitude of pressure oscillations reduces monotonically with swirler rotation, a closer examination of pressure data reveals the presence of intermittent bursts at intermediate rotation rates. Figure 2 depicts temporal evaluation of pressure, obtained at three different swirler rotation rates for 0.8 seconds. At 0 rpm, there is steady pressure oscillation of 5 kPa amplitude, which represents the thermoacoustic instabilities in the combustor. At 2100 rpm, the pressure oscillations become aperiodic and its amplitude got reduced. It is also the stable operating condition we have achieved so far. But at 1800 rpm, we observed aperiodic oscillations followed by a burst of periodic oscillations [1,9,14,15]. The red and black boxes in the Fig. 2, corresponds to periodic and aperiodic oscillations. Surprisingly, the amplitude of periodic and aperiodic oscillations matches with 0 rpm and 2100 rpm counterparts. This periodic burst of oscillations in an aperiodic pressure data is known as intermittency.

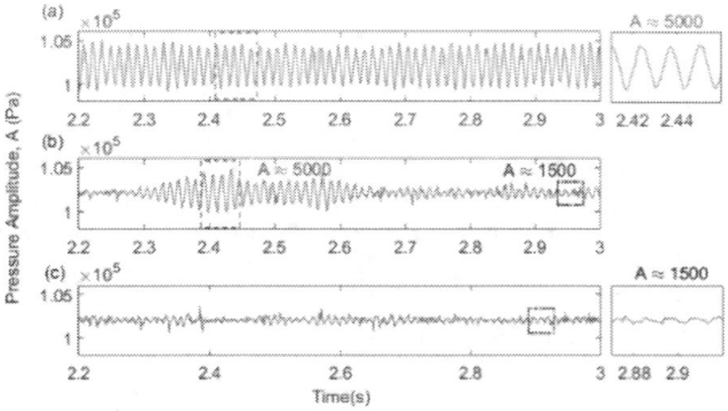

Figure 2: Pressure oscillation signal for different swirler rotation rates at a) 0 rpm, b) 1800 rpm and c) 2100 rpm.

To understand the nature of pressure oscillations more clearly, we used Short Time Fourier Transform (STFT). 4096 FFT points were taken from the pressure data sampled at 1 kHz, with small time window size of 256 points and 250 overlapping points for better resolution. Figure 3, shows the STFT of the signal for three different swirler rotation rates. The unstable condition at 0 rpm, shows a narrow continuous band of the frequency spectrum with time in Fig. 3a. At this stable condition, only weak spurts in frequencies are visible. For 1800 rpm, Fig. 3b shows bursts of first mode frequencies separated by aperiodic oscillations. The second mode frequencies have also started to appear in the spectrum. Figure 3c shows a reduction of overall strength in the first mode frequency, with an increase in second mode frequency for 2100 rpm.

3.2 Modelling of intermittency

Now, to understand intermittency in our combustor using Kuramoto model, we need to specify the oscillators. We assumed our cylindrical swirling flame as an ensemble of a finitely large number of two-dimensional flamelets, uniformly distributed along the circumference of the combustor. Each of the flamelets can be assumed as an oscillator releasing heat in a periodic manner and thereby contributing towards the global pressure variation. As such these oscillators can oscillate at their individual natural frequencies ω. The oscillators under suitable conditions can oscillate in same phase and frequency to create large scale oscillation. They can also oscillate at different phases and frequencies to create combustion noise. Now, the amount of synchronization in these oscillators is modeled using mean-field Kuramoto model.

To calculate the number of oscillators, present in the combustor, we assume their width to be similar with the Kolmogorov length-scales. The Kolmogorov length-scale is estimated from the integral length scales using (4) for all the three operating conditions from high-speed PIV data. The rms value of flow-field calculated from the PIV data, is used to calculate turbulent Reynolds number from (5). The integral length scale is taken as one-sixth of the dump plane diameter.

$$\frac{\eta}{L} = Re^{-0.75} \quad (4)$$

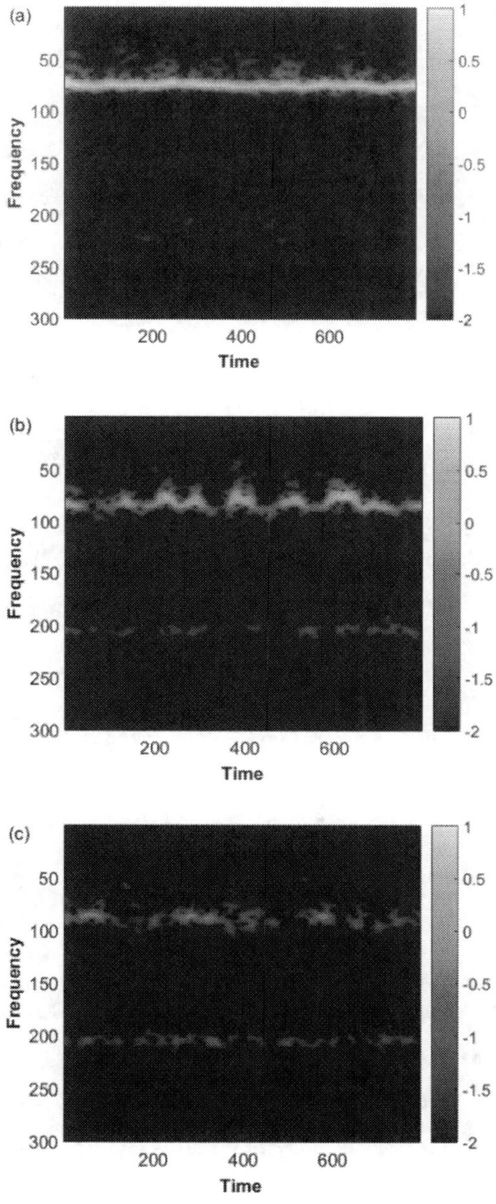

Figure 3: Short Time Fourier Transform of the pressure signal. (a) At 0rpm swirler rotation, narrow band of frequency spectrum shows combustion instabilities. (b) Intermittent operation at 1800 rpm, shows the occurrence of bursts of high amplitude oscillations as high-intensity spots. Low-intensity spots correspond to combustion noise. The second mode of pressure oscillations is more visible. (c) Stable combustor operations at 2100 rpm show low-intensity FFT throughout the time. The second mode becomes more comparable with the first mode.

$$Re = \frac{u_{rms} * L}{\nu} \quad (5)$$

The number of the oscillators are then calculated as the maximum number that can be placed circumferentially in the combustor. The mixing parameter K is a function of swirler rotation rate (Ω) given as,

$$K = \frac{\Omega_0 - \Omega}{150} \quad (6)$$

Where, Ω_0 is 2400 rpm, a theoretical limit where first and second mode amplitude becomes equal, so further instability mitigation is impossible. The constant is chosen arbitrarily.

Figure 4 shows the frequency distribution corresponding to the unstable case at 0 rpm, obtained from the time averaged STFT plot. This frequency PDF is used as our input for the frequency distribution ω. We kept distribution same for all the cases, as our combustor possesses the capability to switch between stable and unstable operation regimes. The oscillator number and mixing parameter values for the three cases are listed in Table 1.

Case	Oscillator number (N)	Mixing parameter (K)
0rpm	3670	16
1800rpm	5100	4
2100rpm	5140	2

Table 1: List of oscillator number and mixing parameter for different swirler rotation rates.

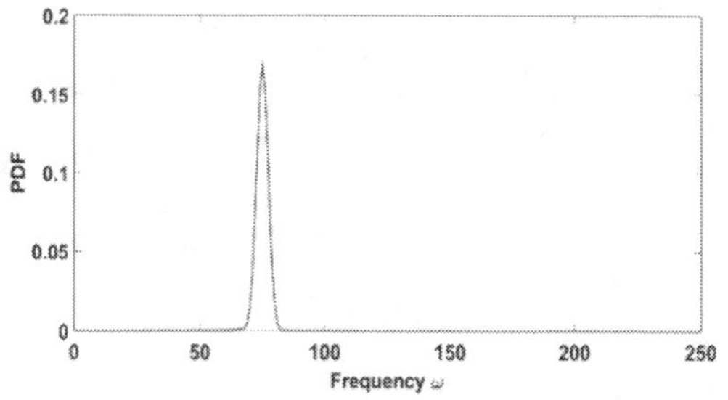

Figure 4: Mean PDF of pressure oscillation data for 0 rpm swirler rotation rate

Using mean-field Kuramoto model, we compute the temporal evolution of order parameter r, for the three cases with variable oscillator number N and mixing parameter K as described earlier. The plot of r with time is given in Fig. 5. The oscillator number keeps increasing with increase in turbulent intensity as the swirler rotation speed increases. For 0 rpm case in Fig. 5a, the order parameter increases to ~0.97 and remained constant after that. The oscillators of different frequencies get coupled due to high mixing parameter and tends to a mean phase angle. Evolution of r for 2100 rpm in Fig. 5c, shows a chaotic behavior around ~0.2, a clear indication of asynchronization. Ideally, order parameter for Kuramoto model should go to zero, for asynchronous oscillator population. This does not happen in our case, as the instabilities can be attenuated even further, which can be inferred from the experimental pressure data showing a weak first mode frequency. Figure 5b, shows the order parameter for 1800 rpm, which keeps oscillating around 0.6. This behavior shows the partial synchronization within the oscillator population. As, they become more and more synchronized their collaboration creates an effect similar to bursts in pressure data. This order parameter behaves non-linearly with time, and also shows a wide range of variations.

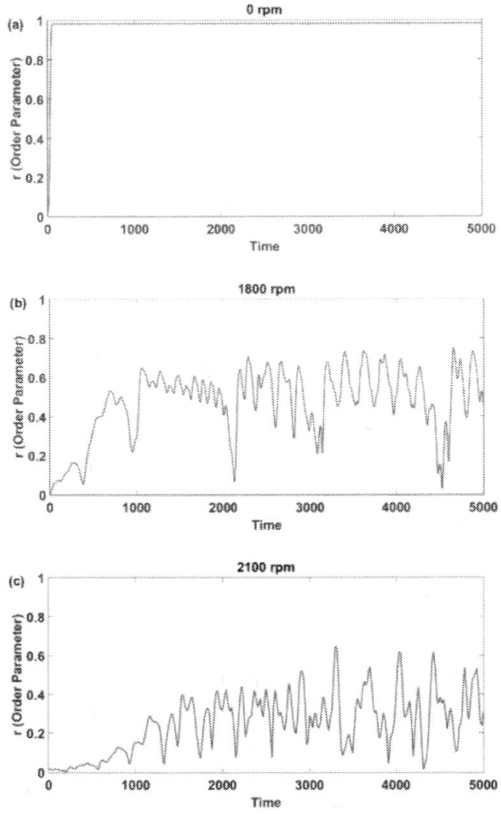

Figure 5: Evolution of order parameter r, the measure of the oscillator synchronization, for the three cases of the swirler rotation. (a)The system is synchronous for 0 rpm unstable combustor operation. (b) At 1800 rpm, some population of oscillators becomes synchronous for some duration before going back to chaos. (c) 2100 rpm, shows similar behavior as 1800 rpm but in smaller magnitude.

4 Conclusions

Here, we have presented dynamics of swirling flame with varying swirler rotation rates. The experimental results for pressure data obtained at three different swirler rotation rate, i.e. 0 rpm, 1800 rpm, 2100 rpm, is presented in the paper. We observe that the amplitude of pressure oscillations does not reduce monotonically with the increase in swirler rotation. Instead, the time duration for which high-amplitude pressure fluctuations are present decreases. At 2100 rpm these fluctuations are attenuated almost completely. Hence, the unstable combustor reaches its stability through intermittency, as the swirler rotation rate increases.

These intermittent dynamics is modeled using mean-field Kuramoto model. The cylindrical swirling flame is considered as a group of two-dimensional flamelet oscillators, which synchronize when they are coupled by strong interactions and become asynchronous as interaction decreases. Order parameter, a measure of synchronization, can help us explain the different type of synchronizations that occurs at unstable, stable and intermittent cases. So, Kuramoto model evolves as a powerful tool to model transition from an unstable state of combustion instabilities to a stable state of combustion noise, through the burst of intermittent oscillations.

5 Acknowledgment

Support of Department of Science and Technology, India Sponsored, National Center for Combustion Research and Development is gratefully acknowledged.

References

[1] V. Nair, G. Thampi, R.I. Sujith, J. Fluid Mech. 756 (2014) 470–487.

[2] A.H. Lefebvre, Gas Turbine Combustion, CRC press, 1998.

[3] H.M. Altay, D.E. Hudgins, R.L. Speth, A.M. Annaswamy, A.F. Ghoniem, Combust. Flame 157 (2010) 686–700.

[4] T. Lieuwen, H. Torres, C. Johnson, B.T. Zinn, Trans. Soc. Mech. Eng. J. Eng. GAS TURBINES POWER 123 (2001) 182–189.

[5] T.J. Poinsot, A.C. Trouve, D.P. Veynante, S.M. Candel, E.J. Esposito, J. Fluid Mech. 177 (1987) 265–292.

[6] S. Sivasegaram, J.H. Whitelaw, Combust. Flame 85 (1991) 195–205.

[7] C.O. Paschereit, E. Gutmark, W. Weisenstein, Phys. Fluids 11 (1999) 2667–2678.

[8] S. Ducruix, T. Schuller, D. Durox, S. Candel, J. Propuls. Power 19 (2003) 722–734.

[9] V. Nair, R.I. Sujith, Chaos An Interdiscip. J. Nonlinear Sci. 23 (2013) 33136.

[10] Y. Kuramoto, I. Nishikawa, J. Stat. Phys. 49 (1987) 569–605.

[11] J.A. Acebrón, L.L. Bonilla, C.J.P. Vicente, F. Ritort, R. Spigler, Rev. Mod. Phys. 77 (2005) 137.

[12] Y. Kuramoto, in:, Int. Symp. Math. Probl. Theor. Phys., 1975, pp. 420–422.

[13] R. Gopakumar, S. Mondal, R. Paul, S. Mahesh, S. Chaudhuri, Combust. Flame 165 (2016) 361–363.

[14] V. Nair, R.I. Sujith, J. Fluid Mech. 747 (2014) 635–655.

[15] V. Nair, G. Thampi, S. Karuppusamy, S. Gopalan, R.I. Sujith, Int. J. Spray Combust. Dyn. 5 (2013) 273–290.

Schlieren CT Measurement of 3D Density Distributions of Flame Kernels of Spark-Ignited Direct-Injection of Free, Cavity-Guided and Plane-Guided Fuel Jets

Ahmad Zaid Nazari[1], Yojiro Ishino[1], Takanori Motohiro[1], Ryoya Yamada[1], Yuta Ishiko[1], Yu Saiki[1]

[1]Nagoya Institute of Technology

Gokiso-cho, Showa-ku, Nagoya, Aichi, 466-8555, Japan

Abstract

Measurement of three-dimensional (3D) density distribution of the initial flame kernels induced by spark-ignition of three type of fuel jets escaping into an ambient space from direct injection nozzle has been conducted. Propane gas is used as a fuel. First, a flame kernel from a spark-ignited free jet is investigated. Next two types of wall-guided (cavity and plane) are used to model fuel jets controlled by a piston bowl shape cavity in a gasoline direct injection(GDI) engine. For 3D observation of flame kernels of fuel jets, non-scanning 3D-CT (three-dimensional computer tomography) technique using a delicate multi-directional quantitative schlieren system with flash light source is employed. By using ML-EM (Maximum Likelihood-Expectation Maximization) an appropriate CT algorithm the 3D density distribution data are successfully reconstructed. The 3D data of initial flame kernel after ignition is of importance to further understanding of the complicated ignition process of the GDI engine.

1 Introduction

Schlieren technique is a common tool for imaging and measuring phenomena in non-uniform density flows in transparent media. Recently, this technique is reviewed in [1] and is shown that schlieren photography is not qualitative visualization method only but has a power of quantitative measurement. In previous works [2,3], by employing and developing a non-scanning 3D-CT technique using a multi-directional quantitative schlieren system with flash light source, measurement of instantaneous density distribution of several type of flame in laminar and turbulent flows have been successfully obtained. In the present study, the mentioned method has been used to investigate the three-type of flame kernels from spark-ignited of free, cavity-guided and plane-guided fuel jets issuing from direct injection nozzle. It is the first time, which the mentioned technique with some improvement for projection images (density thickness images) is used to measure instantaneous density distribution of flame kernels.

2 Experimental Apparatus Set Up and Procedure

Corresponding author. Fax: +81-52-735-5354
E-mail address: ishino@nitech.ac.jp

2.1 Direct Injection Nozzle and Ambient Condition

In the automotive engine, gasoline direct injection(GDI) system is a recent generation of technology to improve engine efficiency [4]. In the present investigation, a direct injection nozzle (injector) is used to measure the density distribution of flame kernels from spark-ignited direct-injection of fuel jets. Figure 1 indicates the schematic diagrams of the injector, fuel jets in three configurations. The fuel injector (Mitsubishi Electric Corporation, EAT-317, installed in Suzuki K6A engine) has 4 holes × 2 rows (8 holes) to generate two fuel jets (in 40° interval) for injecting in two intake paths of the K6A engine. The holes have an exit diameter of 0.15 mm. In this study by blocking 4 holes of the injector, one fuel jet flowing vertically downward is obtained by installing the fuel injector at 20° as shown in Fig. 1. Propane gas is used as a fuel. Figure 1(a) indicates spark ignition on the free fuel jet to obtain a flame kernel. Two types of wall-guided (cavity and plane) are used to model fuel jets controlled by a piston bowl shape cavity in a GDI engine (Fig. 1(b) and (c)). The cavity portion is a hemispherical shape having a diameter of 8mm, a depth of 4mm. The position of the spherical center represents the position of the cavity. In the present investigation, the ambient condition is recorded as ambient temperature 25.6°C and ambient (back) pressure 0.101MPa.

2.2 Multi-Directional Quantitative Schlieren Optical System and Coordinate System

The center of the injector exit has been selected as an origin of *XYZ*-coordinate system for injection system. The origin of the multi-directional optical system (*xyz*-coordinate) coincides with *XYZ*-coordinate system as shown in Fig. 2(a). The *x*-axis is located between camera No.10 (θ = -4.5°) and camera No.11 (θ = +4.5°). Figure 2((b) and (c)) illustrates the concept of 20-directional schlieren camera. In the camera system, target flame kernel/non-uniform density field is observed from 180° direction by numerous schlieren optical system simultaneously. Twenty systems capture the multi-directional views in the 20 angle positions from θ = -85.5° to 85.5° at an interval of 9°. Here angle θ is defined as the horizontal angle from the *x*-axis. The single unit of quantitative schlieren system in the multi-directional schlieren camera is diagramed in Fig. 2(d). This system is composed of two convex achromatic lenses of 50mm in diameter and 300mm in focal length, a light source unit, a vertical knife edge and a digital camera. The light unit is xenon flashtube, full spectrum white with 35μs duration. Between cameras No.13 and No.14 a high-speed camera (HSC) (Photron made FASTCAM-512 PCI 2K) as a single direction schlieren

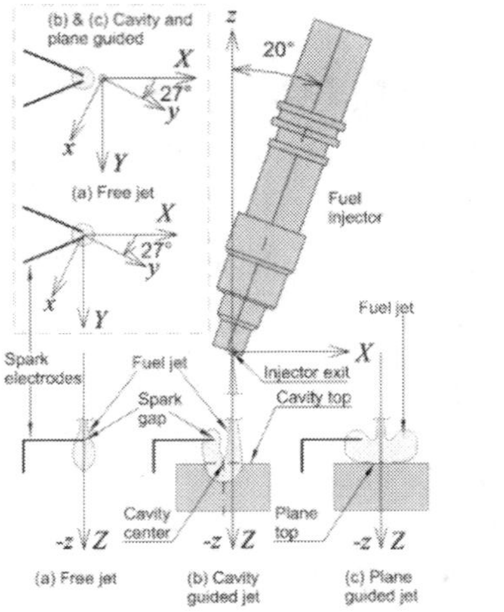

Figure 1: Schematic of fuel jet injection system

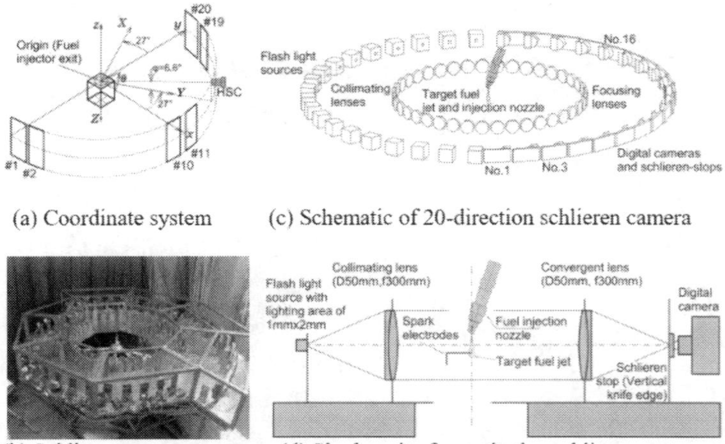

(a) Coordinate system (c) Schematic of 20-direction schlieren camera

(b) Schlieren camera system (d) Single unit of quantitative schlieren system

Figure 2: Multi-direction schlieren and coordinate system

imaging device is installed in the direction of a horizontal angle $\theta = 27°$ and a dip angle $\varphi = 6.6°$ that simultaneously is used with the 20-directional schlieren camera as shown in Fig. 2(a). The ignition electrode (nichrome wire, 0.4mm in diameter and 1mm gap) centerline is in the direction of the horizontal angle $\theta = 27° + 90°$. The previously mentioned fuel injection coordinates (*XYZ*) are set to this angle. The lag timings of the fuel jet injection, the ignition, the flash of the light source and the shutter operation timing of the shooting camera group for simultaneous schlieren photography are set by an electronic circuit. To avoid malfunction due to noise generated by the high voltage spark circuit, visible light laser diode and optical fiber are used for signal transmission to the fuel injector drive circuit, the spark circuit and the flash circuit.

2.3 Image Processing

In the present report, as an improvement point from the previous reports [2,3] for projection images, the derived deviation density is considered from arbitrary reference density (ρ_{ref} =1.3 kg/m^3 in this report). Figure 3 shows conceptual frame work for the formation of the projection image (density thickness) in the proposed schlieren 3D-CT system. Detailed descriptions of the various sections of Fig. 3((a)-(h)), along with the supporting formula, are given in the previous reports [2,3] and many other references cited therein. Here, for brevity, the main points of Fig. 3 with consideration of new improvement are explained as follow. Density, deviation density and density thickness with asterisk sign (*) express the actual values and without this sign represent derived values

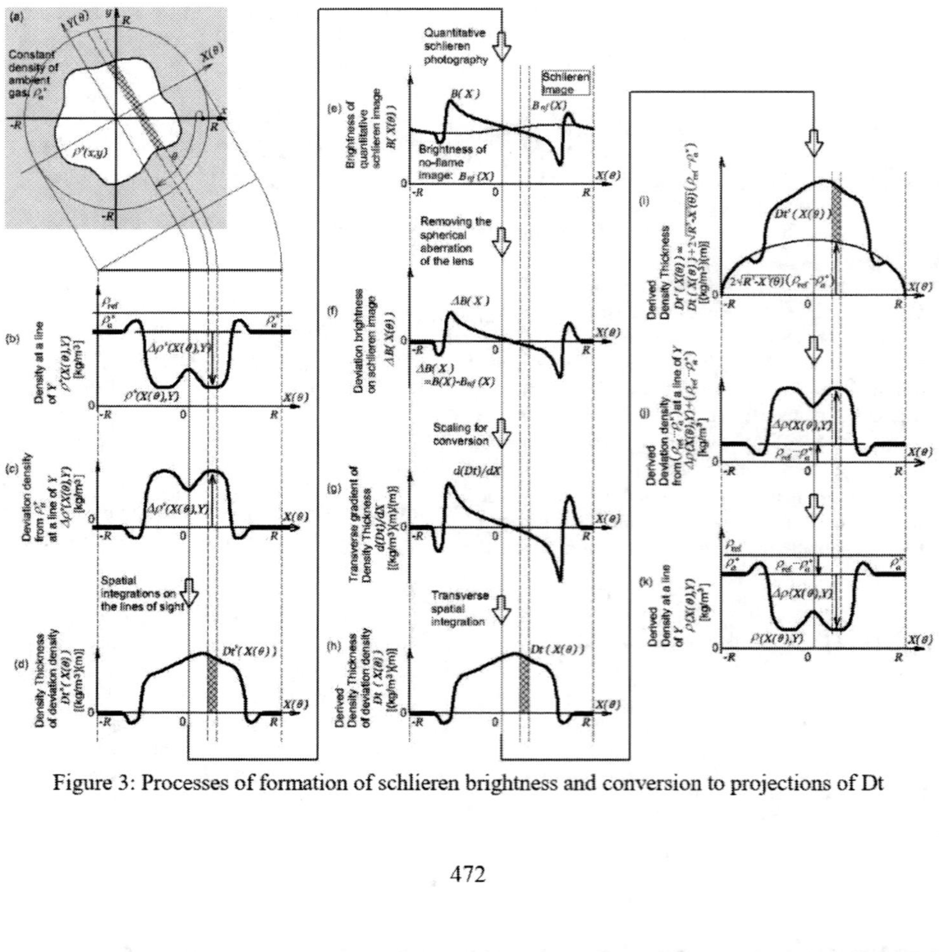

Figure 3: Processes of formation of schlieren brightness and conversion to projections of Dt

from the CT reconstruction process. The first goal is obtaining density thickness of deviation density $Dt^*(X(\theta))$ which shown in Fig. 3(d). The density thickness of deviation density $Dt^*(X(\theta))$ obtains automatically from schlieren observation (Fig. 3(e)-(h)) by spatially-integration of deviation density $\Delta\rho^*(X(\theta),Y)$ along the line of sight. For obtaining density thickness of deviation density $Dt(X(\theta))$, first, two sets of schlieren images "with target" and "without target" are obtained from 20 directions. Next, the brightness of "with target" images shifted and scaled based on "without target" images and the Gladstone-Dale constant of air. $Dt(X(\theta))$ is reproduced by transverse-integration of the final set of schlieren images (Fig. 3(h)). Finally, the "density thickness" $Dt'(X(\theta))$ is obtained by adding the thickness of (ρ_{ref} - ρ_a*) to the deviation density thickness $Dt(X(\theta))$ on peripheral of observed range of radius R (Fig. 3(i)), as is expressed by (1).

$$Dt'(X(\theta)) = Dt(X(\theta)) + 2\sqrt{R^2 - X(\theta)^2}(\rho_{ref} - \rho_a^*) \quad (1)$$

2.4 CT Reconstruction

ML-EM method [2,3], is employed for CT reconstruction. The CT procedure is carried out in each horizontal plane of z-axis for reconstruction of deviation density distribution $\Delta\rho(x,y)$ from linear data set of density thickness $Dt'(X(\theta))$. Finally, 2D density distribution (Fig. 3(k)) is obtained as follow.

$$\rho(x,y) = \rho_{ref} - [\Delta\rho(x,y) + (\rho_{ref} - \rho_a^*)] \quad (2)$$

and

$$\rho(x,y) = \rho_a^* - \Delta\rho(x,y) \quad (3)$$

The 2D distributions $\rho(x,y)$ is accumulated in layers to form 3D-CT distribution $\rho(x,y,z)$. In the present study, the "density thickness" projection images of 500(H)×500(V) pixel (20mm×20mm) are used for CT reconstruction to produce 3D data 500(x)×500(y)×500(z) pixel (20mm×20mm×20mm). The voxel size is 0.04mm in each direction.

3 Results and Discussion

The time series formation of spark-ignited flame kernels for cavity-guided fuel jet is shown in Fig. 4. There is a time lag of 0.5ms between each stage of fuel injection(a), ignition(b),

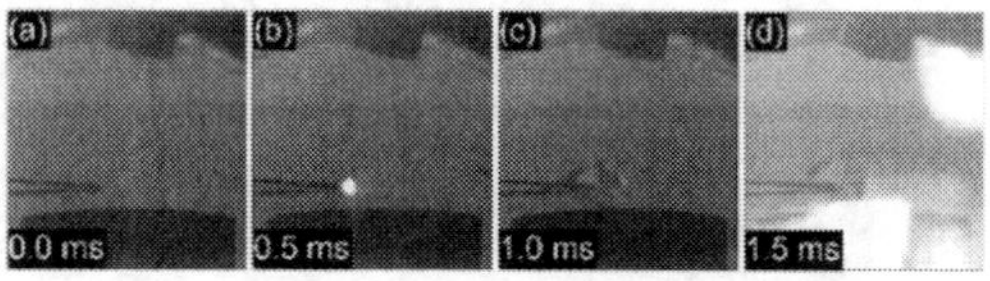

Figure 4: High speed schlieren images of formation of spark ignited flame kernel of cavity guided fuel jet

flame kernel formation(c) and light source flash time(d) for all flame kernels. The position of the ignition electrode for free jet and wall-guided are (X,Y,Z)=(0mm,0mm,12mm) and (X,Y,Z)=(-4mm,0mm,12mm) respectively, and the position of the cavity center and top of plane are (X,Y,Z)=(-2mm,0mm,15mm). In another word, ignition electrode (a black strip in the results (see Fig. 6)) and wall-guided top are located below injector at z = -12mm and z = -15mm respectively.

3.1 Quantitative Schlieren Brightness Images and Density Thickness Images

Here, for brevity, and due to page limitation, the only sample set of images of some directions and some positions are shown in the following figures. Figure 5(left) indicates sample set of quantitative schlieren images of deviation brightness in schlieren images (as shown in Fig. 3(e)) for the spark-ignited flame kernels of free fuel jet (1st row images (a)), cavity guided fuel jet (2nd row images (b)) and plane guided fuel jet (3rd row images (c)). The numbers inserted each picture express the shooting angles θ. The images of Fig. 5(left) are processed by the above-mentioned conversion procedure to projection images of density thickness $Dt'(X(\theta))$, as shown in Fig. 5(right).

3.2 3D-CT Reconstruction of Flame Kernels and Density Distribution Diagrams

Using the density thickness images of Fig. 5(right) as projections for CT reconstruction, 3D instantaneous density distribution of flame kernels of fuel jets has been successfully obtained. Figure 6(a)-(f) gives one sample of vertical and one sample of horizontal distributions of density for different positions for all type of flame kernels. The darker parts are related to lower density area with higher temperature and vice versa. The corresponding density contour diagrams of each

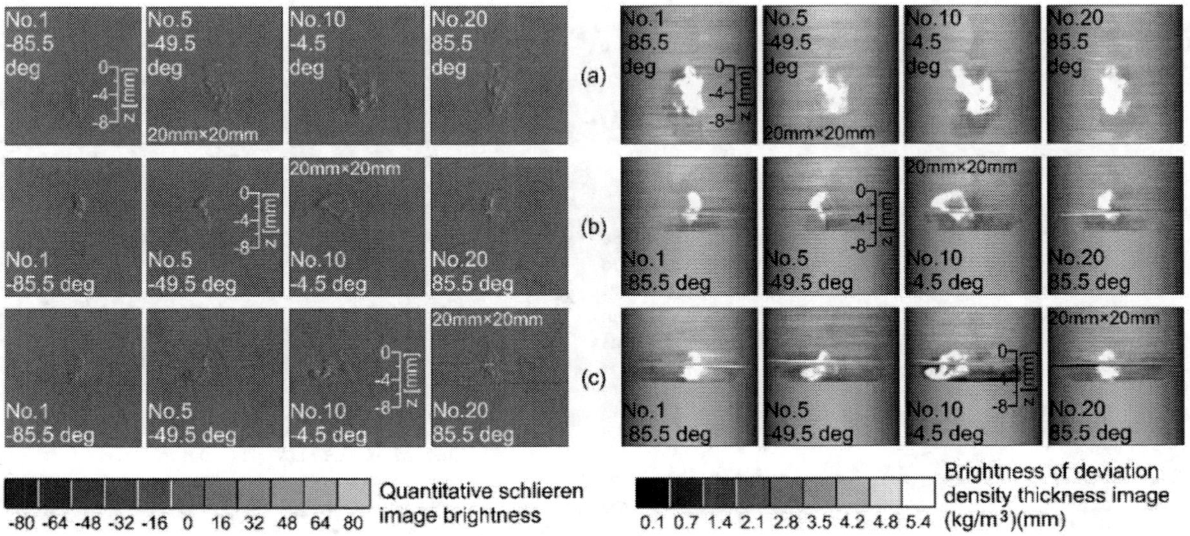

Figure 5: Sample set of quantitative schlieren images (left set) and sample set of density thickness images (right set) of flame kernels for the free fuel jet (1st row images (a)), cavity guided fuel jet (2nd row images (b)) and plane guided fuel jet (3rd row images (c))

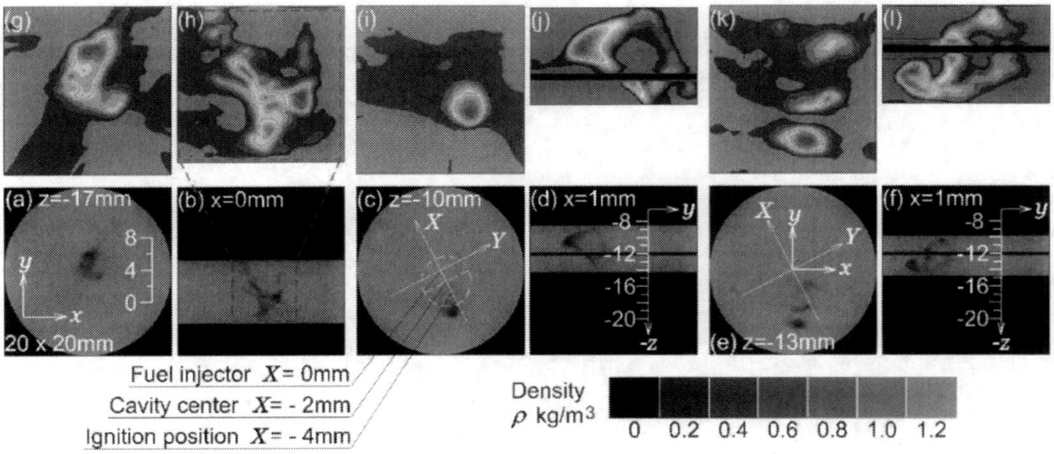

Figure 6: Sample set of horizontal and vertical distributions of CT density distribution (2nd row images) and corresponding density contour diagrams (1st row images) of flame kernels for the free fuel jet((a), (b), (g), (h)), cavity guided fuel jet ((c), (d), (i), (j)) and plane guided fuel jet ((e), (f), (k), (l))

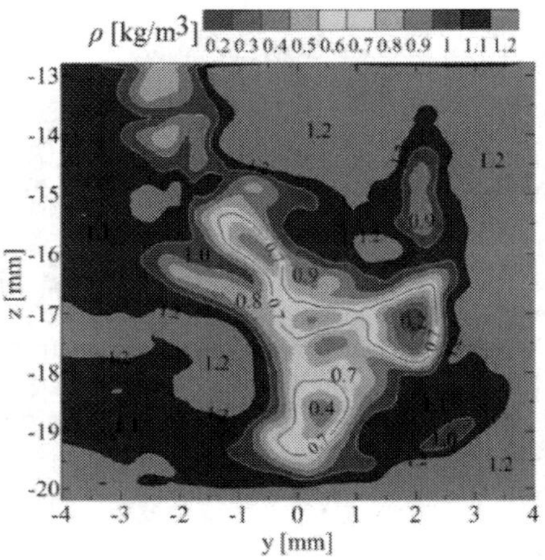

Figure 7: Density contour diagram of flame kernel of the free fuel jet corresponding to Fig. 6 (b) and (h) at x=0mm

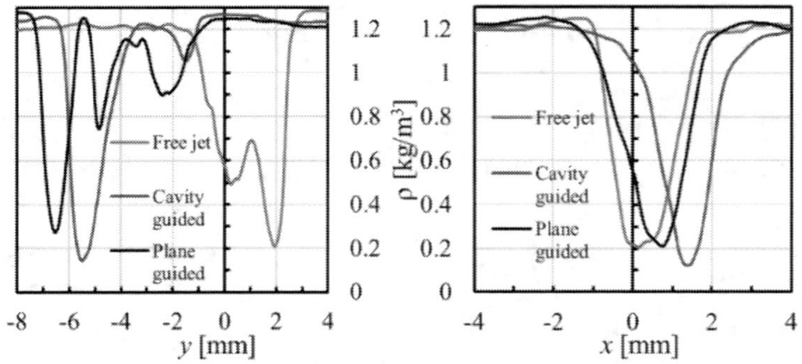

Figure 8: Superposition of axial density distributions of flame kernels
Left(y-axis): free (x = 0mm, z = -17mm), cavity guided (x = 1mm, z = -10mm) and plane guided (x = 1mm, z = -13mm) fuel jets
Right(x-axis): free (y = 2mm, z = -17mm), cavity guided (y = -5mm, z = -10mm) and plane guided (y = -6.5mm, z = -13mm) fuel jets

cross-section are depicted above them (Fig. 6(g)-(l)). For better evaluation of the reconstructed density values, the density contour diagram of Fig. 6(b) in larger scale and superposition of axial density distribution diagrams are illustrated in the Fig. 7 and Fig. 8 respectively. Generally, results clearly demonstrate the minimum density values are about 0.2 kg/m^3 for burnt gas.

4 Conclusions

By employing the non-scanning 3D-CT technique using a 20-directional quantitative schlieren system, CT reconstruction of the instantaneous 3D density distribution of the spark-ignited flame kernels in propane/air fuel jets from direct injection nozzle have been successfully obtained. The 3D structure of initial flame kernel after ignition is of great importance in understanding the ignition phenomena and can be served as validation data for numerical simulations.

5 Acknowledgment

This research work was partly supported by The Nitto Foundation and JSPS Grants-in-Aid for Scientific Research (C)16K06118.

References

[1] G. S. Settles, M. J. Hargather, A Review of Recent Developments in Schlieren and Shadowgraph Techniques, Measurement Science and Technology 28 (2017) 042001.

[2] Y. Ishino, N. Hayashi, I. F. Bt Abd Razak, T. Kato, Y. Kurimoto, Y. Saiki, 3D-CT (Computer Tomography) Measurement of an Instantaneous Density Distribution of Turbulent Flames with a Multi-Directional Quantitative Schlieren Camera (Reconstructions of High-Speed Premixed Burner Flames with Different Flow Velocities), Flow, Turbulence and Combustion 96 (2016) 819–835.

[3] Y. Ishino, N. Hayashi, Y. Ishiko, A. Z. Nazari, K. Nagase, K. Kakimoto, Y. Saiki, Schlieren 3D-CT Reconstruction of Instantaneous Density Distributions of Spark-Ignited Flame Kernels of Fuel-Rich Propane-Air Premixture, Proc. of ASME Heat Transfer Summer Conference, USA, 2016, p. 1-10.

[4] F. Zhao, M. -C. Lai, D.L. Harrington, Automotive Spark-Ignited Direct-Injection Gasoline Engines, Progress in Energy and Combustion Science 25 (1999) 437–562.

11th Asia-Pacific Conference on Combustion,
The University of Sydney, NSW Australia
10th -14th December 2017.

Numerical Investigation of Heating of an Object in a Pool Fire Using Different Liquid Fuels

Alagani Harish[1], Vasudevan Raghavan[1]

[1]Indian Institute of Technology Madras

Chennai, Tamil Nadu, 600036, India

Abstract

Numerical investigation of heating of a mild steel slab in a pool fire using different liquid fuels is reported in the present study. The liquid fuels used are diesel, jatropha, methanol and n-heptane. A pool of 0.5 m × 0.5 m × 0.2 m is considered. A mild steel slab of 0.16 m × 0.16 m × 0.4 m is used to study the heat transfer characteristics of the pool fire. Fire Dynamic Simulator (FDS) is used for solving mass, momentum, energy and species conservation equations. Infinite rate chemistry model is used for modelling the reaction. One quarter of the 3D domain is simulated by assuming two symmetry planes along the length and width. Igniter particles with a temperature of 1600°C are used to onset ignition, except in methanol. The igniter particles are removed at 25 s. Simulation is carried out for 180 seconds after the removal of ignition source to investigate the heating of the object during that period. It is observed that the object is heated uniformly in the pool fires fueled by diesel, jatropha and n-heptane. The mass loss rate of n-heptane and diesel is higher than other fuels. The flame from n-heptane and diesel engulfs the object completely.

1 Introduction

Heating of objects by fires is an important aspect to be analyzed in fire science. The object can be of any size, shape and material, such as a cement structure, iron block, a combustible solid material, or artillery. During fire accidents, the materials in the vicinity of fire heat up. This may initiate a new fire, if the materials in the surroundings are combustible, and transfer the heat to the surroundings materials. Exposure of an explosive container or artillery, accidentally to a fire will result in severe fire accidents causing immense loss of life and property. The external heat flux incident on the artillery container and the time of exposure dictates the start of reactions. A study called cook-off test, in which the artillery is exposed to different heat fluxes for different duration, is done to determine the temperature at which reaction starts. The cook-off decides whether the munition is suitable for production, transport and storage without causing any damage. Two types of cook-off tests, namely fast cook-off and slow cook-off test exist. In slow cook-off test, the artillery was subjected to low heating rates for long periods to investigate the hazard. If the artillery was subjected to high heating rates to access the hazard, such a test is called fast cook-off test [1].

Corresponding author. Fax: +91-44-22574652
E-mail address: harishalagani@gmail.com

A summary of experimental and numerical techniques for fast cook-off test was given by Cocchairo [2]. Effect of contact temperature and time on the initiation of cook-off using 7.62 mm cartridge case is reported by Hameed et al. [3]. Double based propellant was used in the cartridge for cook-off test. They found that the reactions triggered when the gun barrel temperature was in the range of 151.4°C to 153.4°C. The reaction was initiated in less than 300 s of the exposure to heat flux. Behavior of confined HMX energetic material when subjected to heat flux was studied numerically by Gross et al. [1]. A one-dimensional transient model was solved. Semi global and detailed chemistry models, for condensed phase and gas phase, respectively were used. They found that the accuracy of the results depends on the boundary conditions applied to the container. Both experimental and numerical assessment of danger in using AP/HTPB propellant was done by Yang et al. [4]. They reported the ignition delay times with increasing external heat flux and corresponding shell temperatures. With increasing heat flux, the shell temperature also increased and the reaction started in a lesser time. Thus, it is clear that the outer surface temperature plays a crucial role in initiating reactions inside the artillery. Spatial variations of heat flux and temperature in large pool fire were measured by Bainbridge and Keltner [5]. They found that the heat flux and temperature depend on wind conditions. At all conditions, the average heat flux exceeds the critical level for transporting radioactive materials. Thomas and Victor [6] collected and analyzed heat flux data over a large pool diameter of 7.92 m for the purpose of numerical validation. Average heat flux across the pool become steady and the surrounding regions receive high heat fluxes in 100 s after ignition. From the literature review, it is clear that the surface temperature of the object is an important parameter to assess the artillery hazard. This in turn depends on the incident heat fluxes. Pool fires have high energy content. Experiments are generally difficult to conduct, as the optimal location to place the object is not known apriori. They are expensive also, in terms of fuel requirements and instrumentation. Thus, it will be beneficial to investigate the optimal location and the fuel type using numerical simulations. So, in the present study, a detailed numerical investigation of heating of an object in a pool fire using different liquid fuels is reported.

2 Numerical Methodology

Numerical investigation is carried out using Fire Dynamics Simulator (FDS) developed by NIST [7]. FDS solves a type of Navier-Stokes equations simplified for low speed, thermally-driven flows. FDS is focused mainly on smoke and heat transfer from large fires such as building fires. This software uses an explicit predictor-corrector scheme, which is second order accurate in both space and time. Turbulence is modelled using Large Eddy Simulation (LES). A single step, mixing controlled reaction model (infinite rate chemistry) with three species (air, fuel and products) is used for modelling combustion. Constant thermo-physical properties, calculated at a given temperature, are used. The one-dimensional heat conduction equation, normal to the solid surface, is solved in solid objects. Both natural and forced convection correlations are used for heat transfer through convection over solid surfaces. A thin layer approximation is used for radiation heat transfer from or to solids. Figure 1 shows the computational domain with boundary conditions. A pool of 0.5 m × 0.5 m × 0.2 m is considered. An object 0.16 m × 0.16 m × 0.4 m, made up of mild steel, is used. As the attention is given to study the effect of flame height on heating the object, soot formation is not taken into account. It will have a significant effect on mass burning rate of the fuel. The object is placed at a height of 0.2 m from the pool surface. The radiation feedback from the object is considered. Two symmetry planes as shown in Fig.1, at x = 0 m and y = 0 m, are assumed to reduce the computational effort and time. The three dimensional symmetric computational domain has been divided into 28 volumes and each volume is meshed differently. Quadrilateral mesh is used in each volume. A minimum grid size of 2.08 cm × 2.08 cm × 2 cm has been used and a total of 27,732 cells have been used for all the cases.

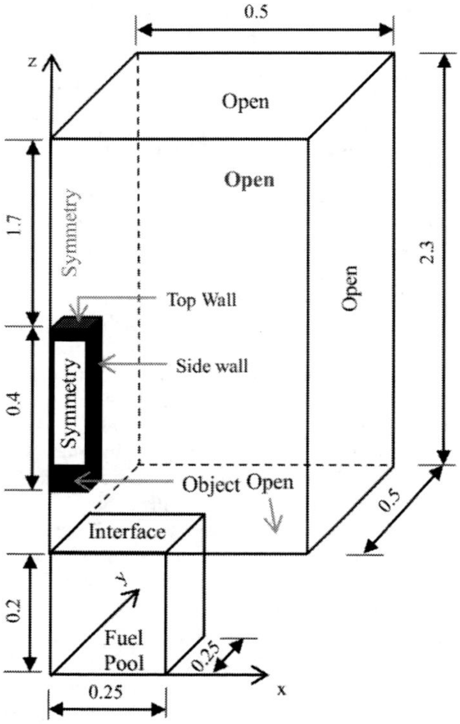

Figure 1: Computational domain with boundary conditions

Open type boundary is exposed to atmosphere. Pressure specified condition is prescribed here. Gases are allowed to flow in and out of an open boundary. When the flow is into the domain, the temperature and mass fractions value corresponds to the specific boundary value. If the flow is going out, the values in the adjacent cell are taken and extrapolated. The sides and bottom of the fuel pool are assumed to be adiabatic. The interface between the liquid and gas is dealt using the following conditions: Clausius-Clapeyron equation is used to find the mole fraction of fuel vapor over the surface.

$$X_{F,l} = \exp\left[-\frac{h_{fg} W_F}{R} \left(\frac{1}{T_s} - \frac{1}{T_b} \right) \right], \qquad (1)$$

where h_{fg} is latent heat of vaporization, W_F is molecular weight of fuel, T_s is the surface temperature and T_b is the boiling point of the fuel. The mass flux during evaporation of fuel is calculated using Stefan diffusion equation [8], given as,

$$\dot{m}'' = h_m \frac{\bar{p}_m W_F}{R T_g} \ln\left[\frac{X_{F,g} - 1}{X_{F,l} - 1} \right]; \quad h_m = \frac{Sh D_{l,g}}{L}, \qquad (2)$$

where $\bar{p}_m$ is pressure, T_g is gas temperature and $X_{F,g}$ is the mole fraction of fuel vapor in the adjacent cell to the pool surface. The liquid is treated as a thermally-thick and heat conduction to the pool is calculated. Using the interface condition given in (1) and (2), the mass burning rate of methanol pool of 15 cm diameter has been validated against an in-house lab-scale experimental data. Igniter particles having a temperature of 1600 °C, is placed near the pool surface to initialize vaporization of fuel. For methanol this is not done because sufficient vapors form at room temperature to sustain a flame. The ignitor particles are removed at 25 s of the simulation. The simulations are carried out for 180 s after the removal of ignitor particles. The results are presented at the time instant of 205 s.

3 Results and Discussions

As mentioned earlier, four liquid fuels, namely diesel, jatropha, methanol and n-heptane, are used. Properties of fuels are taken from literature [9-14]. Table 1 reports the properties of the fuels along with mass loss rates and flame heights. Diesel, n-heptane and jatropha have higher mass loss rates. The associated flame heights are also taller than methanol, because of higher amounts of C and H. The difference in the mass loss rate is attributed to the latent heat of vaporization. Despite having a low boiling point, since methanol has a higher latent heat of vaporization, it has lower mass loss rate. Figure 2 shows the variation of mass loss rate with time. It is seen that the mass loss rate is almost steady over the entire burning duration, after the removal of heater particles, for all the cases.

	Mass loss rate (g/s)	Flame height (m)	Heating value (kJ/kg)	h_{fg} (kJ/kg)	T_{boil} (K)
Diesel	4.37	>2.3	42500	270	533
Jatropha	3.55	2.2	40120	257	632
Methanol	1.05	0.63	19800	1101	337.7
n-Heptane	6.48	>2.3	44924	316.7	371.4

Table 1: Average flame height for different fuels

Figure 3 presents the temperature distribution on the right side surface of the solid at the time instant of 205 s. It is observed that the heating up of the surface by diesel, jatropha and n-heptane flames is more uniform when compared to methanol. In cases of diesel, jatropha and n-heptane flames, it is seen that the higher temperature region on the surface of the object are spread over the entire length, whereas in cases of methanol flame, the maximum temperature region is seen near the bottom of the surface.

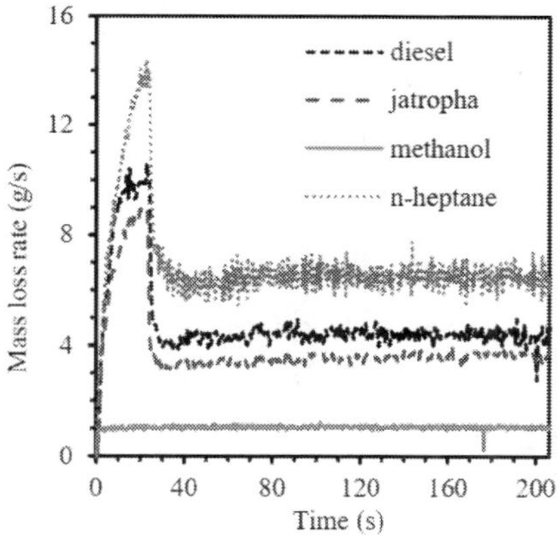

Figure 2: Mass loss rates for different fuels obtained from FDS

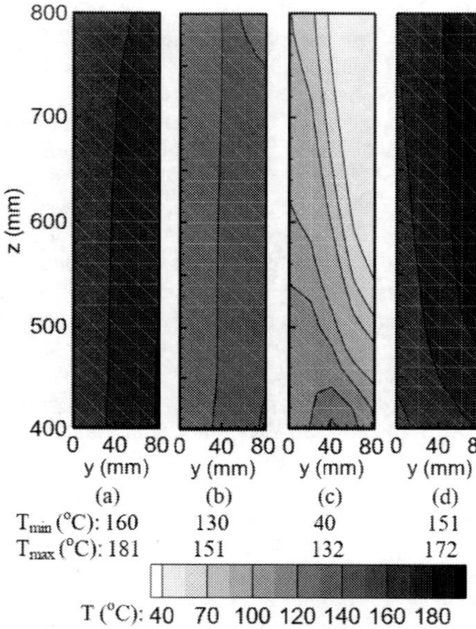

Figure 3: Temperature contours on the right surface of the solid heated by (a) diesel, (b) jatropha, (c) methanol and (d) n-heptane pool flames

Figure 4 shows the net heat flux on the right surface after heating for 180 s. The net heat flux on the surface is higher when the flame height is higher. Lesser heat flux values on the solid surface are observed for the case of methanol; nearly 3 times lower than the minimum heat flux imparted by the flames from other fuels. Similar to temperature distribution (Fig. 3) the heat flux is high along the length for diesel, jatropha and n-heptane flames, whereas the maximum heat flux is distributed only near the bottom surface for the case of methanol, because of its lower flame height. This is the reason why the temperature (Fig. 3) and heat flux (Fig. 4) on top of the surface is low while heating with methanol.

Figure 5 shows the temperature distribution on the bottom surface of the solid being heated. In contrary to the observation on the right side surface, methanol flames have heated the bottom surface to higher temperatures than by the remaining fuels. As the flame height is lower for methanol, the flame impinges on the bottom surface and thereby heating it up at a higher rate.

Figure 6 shows the temperature variation on the top surface of the solid. From Fig 6, fuels burning with higher flame heights are able to heat up the upper surface better. It is evident that diesel and n-heptane, which have the highest flame length heats up the top surface to a temperature higher than that by other fuels. This is because the flame engulfs the object completely and heats up the surface more uniformly.

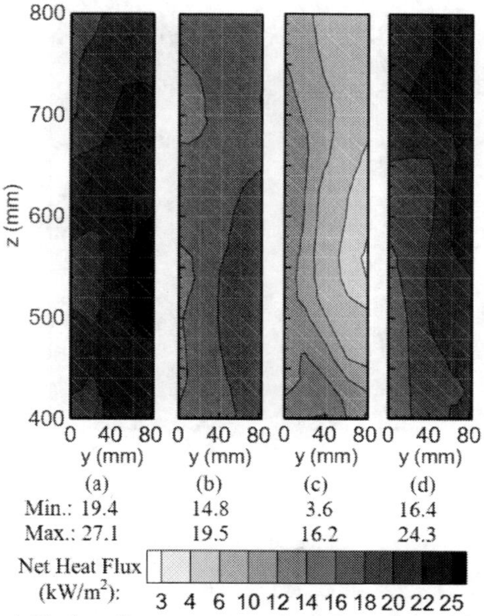

Figure 4: Net heat flux contours on the right surface of the solid heated by a) diesel, (b) jatropha, (c) methanol and (d) n-heptane pool flames

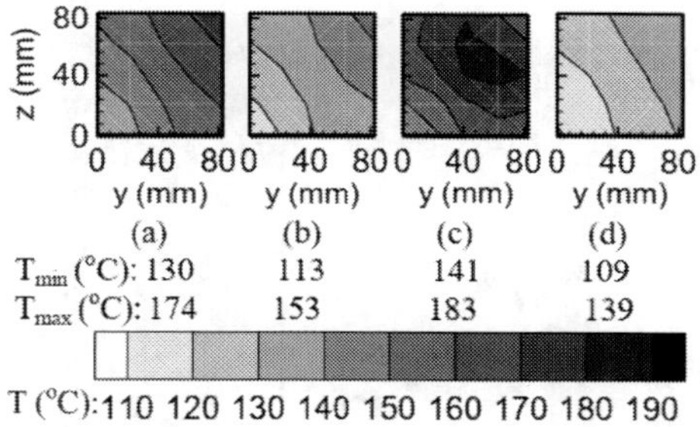

Figure 5: Temperature contour along the bottom surface of the solid heated by a) diesel, (b) jatropha, (c) methanol and (d) n-heptane pool flames

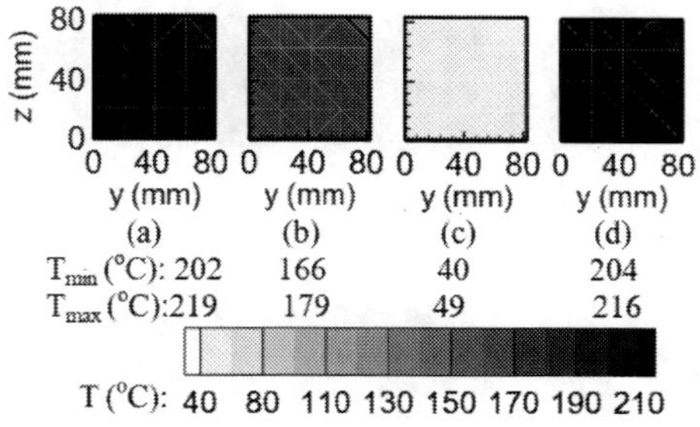

Figure 6: Temperature contour along the top surface of the solid for different fuels of (a) diesel, (b) jatropha, (c) methanol and (d) n-heptane pool flames

Figures 7 and 8 shows the grayscale temperature contours and velocity vectors for different cases. In Figs. 7(a), 7(b) and 8(b) the flames from diesel, n-heptane and jatropha, are shown, respectively. It is clear that flame impinges at the bottom surface of the object, moves sideways and raises up surrounding the object in a recirculation zone. The flame envelopes the object completely. Therefore, the heating of the object occurs in an uniform manner over the solid. On the other hand, the flame from methanol, as shown in Fig. 8(a), ends up heating up the object primarily at its bottom surface. Therefore, the object is not heated uniformly like in the cases of diesel and n-heptane.

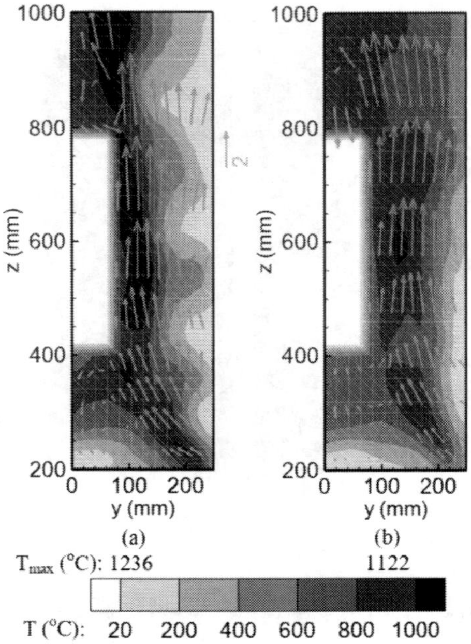

Figure 7: Averaged temperature contours and velocity vectors along the plane at y=0 m for the case of (a) diesel and (b) n-heptane pool flames

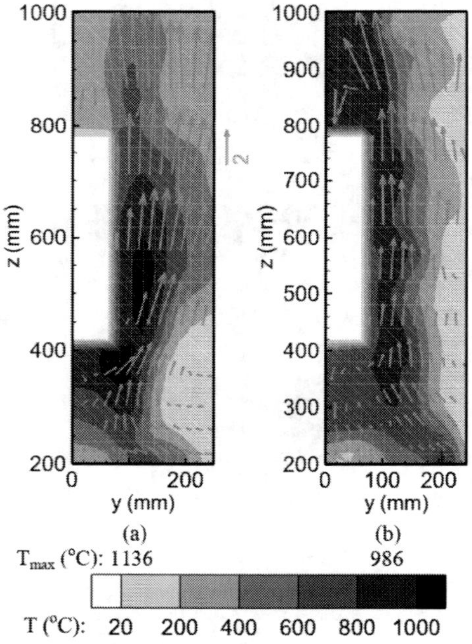

Figure 8: Averaged temperature contours and velocity vectors along the plane y=0 m for (a) methanol and (b) jatropha pool flames

4 Conclusions

The effects of the type of liquid fuel used to heat a solid body are studied numerically using FDS. It is observed that the resultant flame height plays a major role in heating up of the object uniformly. Diesel, jatropha and n-heptane fuels have high flame heights and are able to heat the solid uniformly than methanol. The object receives heat all through its surface from a flame, which is able to engulf it. Smaller flame obtained from methanol heats up the bottom of the sample better than other fuels. However due to lesser flame height, the top and sides of the object are not heated properly. In summary, it can be concluded that for faster and uniform heating, flame from heavy and volatile fuels are required.

5 Acknowledgment

Authors acknowledge fund from Armament Research Board, India.

References

[1] Matthew L. Gross, Karl V. Meredith, Merrill W. Beckstead, Fast Cook-off Modelling of HMX, Combustion and Flame, 162 (2015) 3307-3315.

[2] J.E. Cocchiaro, Subscale Fast cookoff Testing And Modelling for the Hazard Assessment of Large Rocket Motors, Report No. CPTR 72, Chemical Propulsion Information Agency, March 2001.

[3] Amer Hameed, Mathew Azavedo, Philip Pitcher, Experimental investigation of a cook-off temperature in a hot barrel, Defence Technology, 10 (2014) 86-91.

[4] Hou-Wen Yang, Yong-Gang Yu, Rui Ye, Xiao-Chun Xue, Wen-Feng Li, Cook-off Test and Numerical Simulation of AP/HTPB Composite Solid Propellant, Journal of Loss Prevention in the Process Industries, 40 (2016) 1-9.

[5] B. L. Bainbridge and N. R. Keltner, Heat Transfer to Large Objects in Large Pool Fires, Journal of Hazardous Materials, 20 (1988) 21-49.

[6] Thomas Blanchat, Victor Figueroa, Large-Scale Open Pool Experimental Data and Analysis for Fire Model Validation and Development, Fire Safety Science-Proceedings of the Ninth International Symposium, pp. 105-116.

[7] Kevin B. McGrattan, Randall J. McDermott, Craig G. Weinschenk, Glenn P. Forney, 2013,Fire Dynamics Simulator, Technical Reference Guide, Sixth Edition, Special Publication (NIST SP) - 1018.

[8] R. Taylor and R. Krishna. Multicomponent Mass Transfer. Wiley-Interscience, 1993.

[9] K. Annamalai, I. K. Puri, Combusiton Science and Engineering, CRC Series in Computational Mechanics and Applied Analysis, CRC Press, 2007, p.988-989.

[10] Chen Z, Wang X, Pei Y, Zhang C, Xiao M, He J, Experimental investigation of the performance and emission of diesel engines by novel emulsified diesel fuel. Energy conversion Management, 95 (2015) 334-41.

[11] Chen Z, Wang X, Pei Y, Zhang C, Xiao M, He J, Experimental investigation of the performance and emission of diesel engines by novel emulsified diesel fuel. Energy conversion Management, 95 (2015) 334-41.

[12] C. A. Nieto de Castro, S.F. Y. Li, A. Nagashima, R. D. Trengove and W. A. Wakeham. Standard Reference data for the thermal conductivity of Liquids, Journal of Physical chemistry Reference Data, 15 (1986) 1073-86.

[13] Robert C. Reid, John M. Parausnitz, Bruce E. Poling, The properties of gases and liquids, Fourth edition, McGraw-Hill,Inc.

[14] More Manjunath Sahadev, Evaporation and Burning studies of Neat biodieselsand Their Diesel Blends, Master of Science, IIT Madras, Chennai, India, 2015.

Multi-phase Numerical Study of Evaporation of Suspended Biodiesel Droplets

Saroj Ray* **and Vasudevan Raghavan**

Department of Mechanical Engineering, Indian Institute of Technology Madras,

Chennai, Tamil Nadu, 600036, India

Abstract

The paper presents a multi-phase numerical model to simulate evaporation of a transient spherical liquid fuel droplet. Variable thermo-physical properties in both liquid- and vapor-phases, as functions of temperature and species concentrations, are considered. The influence of suspension bead has been considered in the model. The model has been validated against the experimental data available in literature. The validated model is used to study the vaporization characteristics of biodiesel fuel droplets such as jatropha metyl ester (JME), karanja methyl ester (KME) and palm methyl ester (PME) at atmospheric pressure. The effects of convection velocity and the presence of bead on evaporation characteristics are studied systematically. Surface regression rate and droplet lifetime have been predicted and analyzed. Results show that the presence of bead increases the liquid temperature and internal circulation inside the liquid. The maximum change in droplet lifetime due to presence of bead has been found to be less than 5% of total droplet lifetime.

1 Introduction

Evaporation studies of liquid fuel droplets are important in applications such as internal combustion engines, industrial oil furnaces and gas turbine engines. Biodiesel is an important alternative fuel and it is renewable. It also has lower emission features. Biodiesel is a multicomponent fuel and usually contains 6 to 7 fatty acid components. There is lack of study in literature on numerical simulation of evaporation of biodiesel. The suspended droplet technique is the most common method used to study transient droplet evaporation. However the influence of suspender is neglected in many of the theoretical and numerical analysis [1-5]. Raghuram et al. [5] numerically studied evaporation of blended droplet in laminar convective environment. However, the influence of suspension fiber was not included in the model. Shih et al. [6] numerically investigated suspended droplet evaporation in a laminar convective environment. The influence of the suspension fiber on the vaporization characteristics was included in the model. However, their model was not validated against experimental data and the analysis was qualitative. Strotos et al. [7] have studied the evaporation characteristics of binary hydrocarbon droplets using Volume of Fluid (VOF) method to track the interface. Their model included the influence of suspension fiber. A disadvantage of their model is that it is much expensive as their simulations take around 1–3 CPU months.

In this paper, the influence of the suspension bead on the vaporization characteristics of biodiesel such as jatropha metyl ester (JME), karanja methyl ester (KME) and palm methyl ester (PME) are studied. The evaporation characteristics of droplet with bead and without bead are compared against experimental data from literature [8]. The effect of convection velocity and that of bead on evaporation is studied in detail.

2 Numerical model

A two phase numerical model of droplet evaporation in spherical coordinates is developed. All the thermo-physical properties of species in gas- and liquid-phases are evaluated as functions of temperature and composition using correlations from Reid et al. [9]. As biodiesel is a multicomponent fuel, its properties are calculated by applying mixing rule to fatty acid methyl ester properties. The composition of biodiesel is taken from literature [8].

The multi-phase numerical model is based on the following assumptions: (a) laminar and axisymmetric flow, (b) droplet remains spherical throughout its lifetime (low Weber number), (c) negligible thermal radiation, (d) Dufour and Soret effects are negligible in both phases, (e) pressure diffusion is negligible in both phases and (f) viscous dissipation and pressure work are negligible.

Navier-stokes (NS) equations along with energy and species balance equations in both gas- and liquid-phases are discretized using finite volume method (SIMPLEC) [10]. The discretized equations are solved using Alternating Direction Implicit (ADI) method with Tri-Diagonal Matrix Algorithm (TDMA), using an in-house FORTRAN code. The discretization scheme is first order accurate in time and second order accurate in space coordinates. At interface, conditions such as continuity of temperature, no slip, continuity of normal shear stresses, species balance, energy balance and Raoult's law are applied. Figure 1 shows the computational domain along with boundary conditions.

To take in account of mixed convection, velocity induced by natural convection is modeled as per equation (1) and is added to the free stream velocity.

$$U_{NC} = \sqrt{g\beta\Delta T d}\,, \qquad (1)$$

where, d is instantaneous droplet diameter, g is acceleration due to gravity, ΔT is the difference between the ambient (T_∞) and the surface temperature (T_s), and β is the coefficient of thermal expansion [$2/(T_\infty + T_s)$]. The effective free stream velocity is calculated as, $U_\infty = U_{FC} + U_{NC}$, where, U_{FC} is force convection velocity. This effective free stream velocity is used at inlet boundary condition in gas-phase.

*Corresponding author. E-mail address: sarojr19@gmail.com

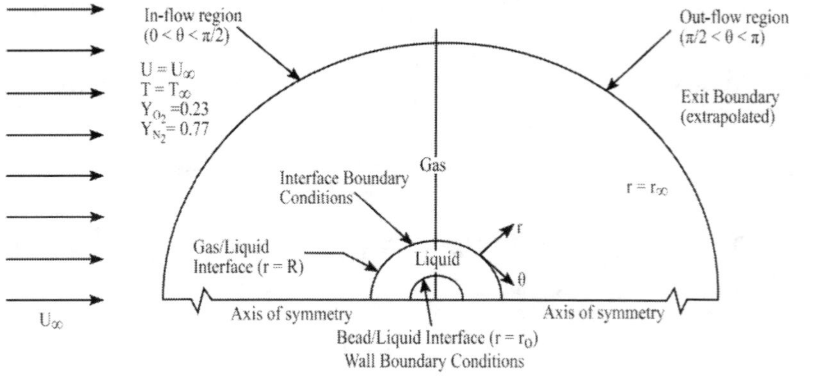

Figure 1: Schematic of the computational domain with boundary conditions.

3 Results and discussion

3.1 Validation

Results from numerical model have been validated against the experimental data reported in Manjunath et al. [8]. This is presented in Fig. 2, where temporal variation of surface regression is shown. In these cases, the initial diameter of droplet is around 2 mm and bead diameter is 1 mm. Manjunath et al. [8] carried out evaporation studies of neat biodiesel and their blends with diesel using suspended droplet technique, at a temperature less than the boiling point. They optically measured the surface regression rate. The agreement between the predicted and measured result is found to be quite good. The validated model is used to study the vaporization characteristic of biodiesel such as JME, KME and PME.

3.2 Parametric studies

For the parametric study, two droplet diameters, 3 mm and 1 mm, are used. The droplet of 3 mm diameter is suspended on a bead of diameter 1.5 mm and 1 mm droplet is suspended on 0.4 mm bead. Three biodiesels, namely, JME, KME and PME, have been studied and their evaporation features are analyzed. The ambient is taken as 300°C and 1 atmospheric pressure. The effects of free stream velocity on the droplet lifetime in the presence and absence of bead are investigated.

3.2.1 Effect of free stream velocity

To study the effect of free stream velocity six forced convection velocity values, 0.05 m/s, 0.5 m/s, 1 m/s,1.5 m/s, 3 m/s and 5 m/s, are considered. The droplet diameter considered is 3 mm, suspended bead diameter as 1.5 mm. Figure 3 shows the surface regression curves of JME, KME and PME droplets evaporating at six different free stream velocity (from 0.05 m/s to 5 m/s) at 300°C and 1 atmospheric pressure. The surface regression rates for the cases with bead are compared against the cases without bead. From Fig. 3, it is clear that droplet lifetime for the cases without bead are always longer than those with bead; the presence of bead reduces the droplet lifetime.

In Figure 4, difference in droplet lifetime in without and with bead is plotted against free stream velocity. Droplet lifetime is the time instant, at which droplet diameter is extrapolated to zero. Droplet lifetime with bead is denoted as t_b and that without bead as t_{nb}. As the free stream velocity increases, the difference in droplet lifetime with bead and without bead decrease; influence of bead on the droplet lifetime decreases at high free stream velocity.

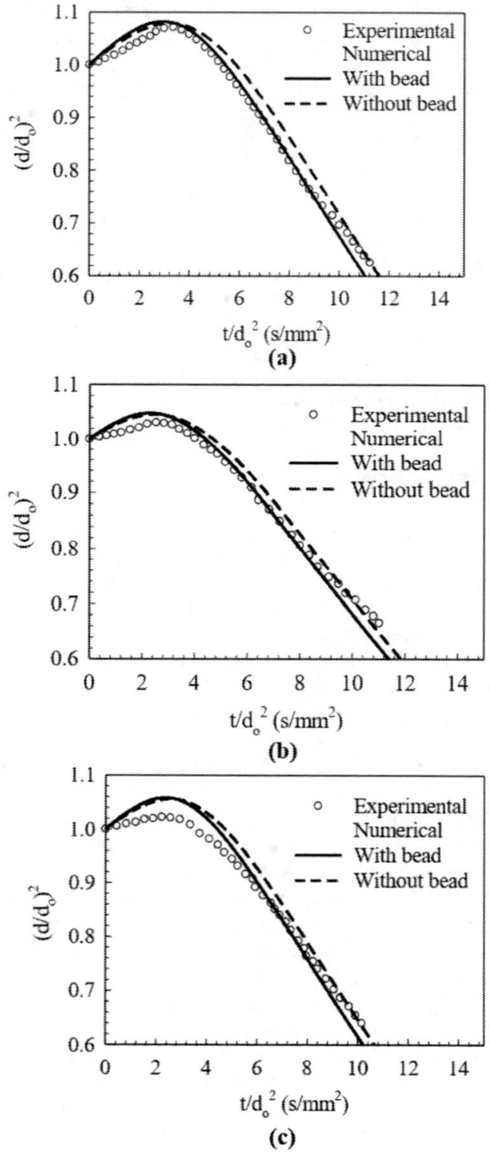

Figure 2: Comparison of surface regression of (a) JME, (b) KME and (c) PME predicted by numerical model against experimental data of Manjunath et al. [8] at 250°C.

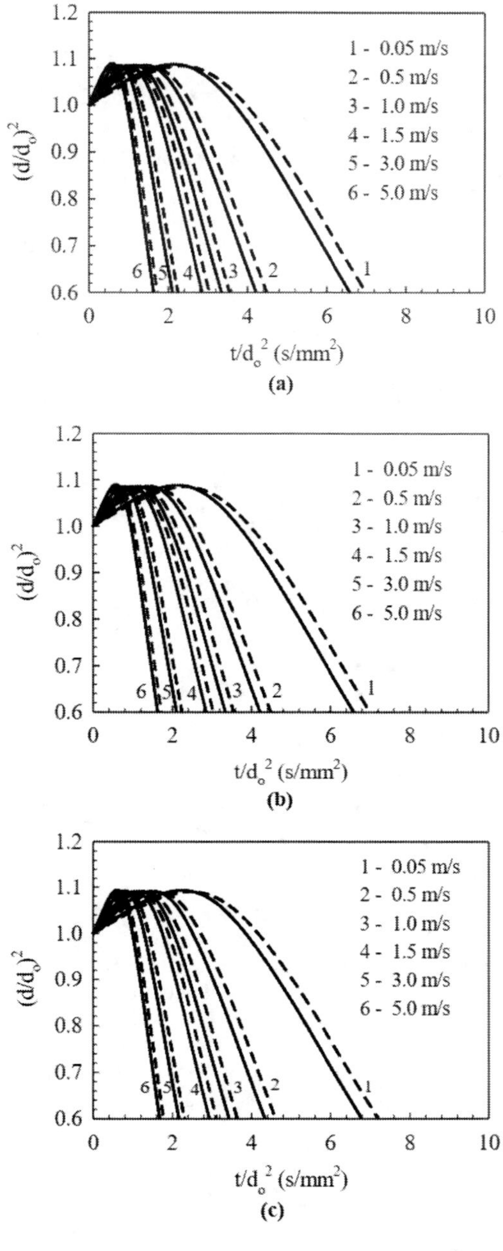

Figure 3: Comparison of surface regression of (a) JME, (b) KME and (c) PME predicted by numerical model at 300°C; d_o = 3 mm, d_b = 1.5 mm (Legends : Solid lines : With bead; Hatched lines : Without bead).

3.2.2 Effect of bead to droplet size

To study the bead to droplet size, the droplet diameter is changed from 3 mm to 1 mm and bead diameter change from 1.5 mm to 0.4 mm. For droplet diameter of 1 mm and bead diameter of 0.4 mm, bead volume to initial droplet volume ratio is 0.064. Whereas for droplet diameter of 3 mm and bead diameter of 1.5 mm have bead volume to initial droplet volume ratio is 0.125.

In Figure 5 percentage difference in droplet lifetime is plotted against Reynolds number (Re) for two bead to droplet sizes. Reynolds number is calculated using effective free stream velocity. It is observed from Fig.5 that as bead to droplet size decreases, the percentage difference between droplet lifetime with bead and without bead decreases, i.e. smaller bead to droplet size causes small change in droplet lifetime. It is seen from Fig. 5 that the maximum change in percentage droplet lifetime between cases without bead and with bead is about 5%.

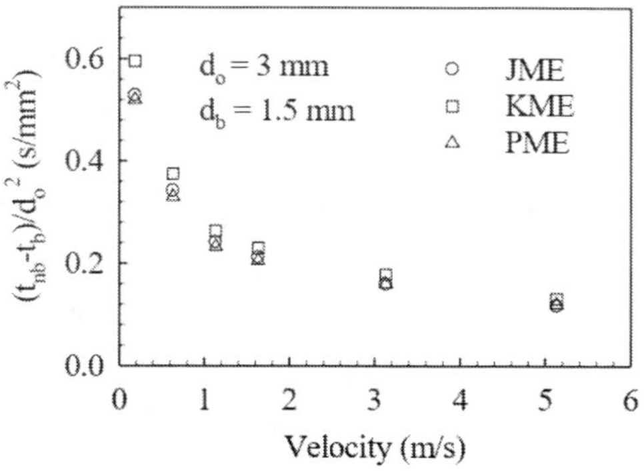

Figure 4: Variation of difference of droplet lifetime without bead and with bead with freestream velocity at T_∞ = 300°C.

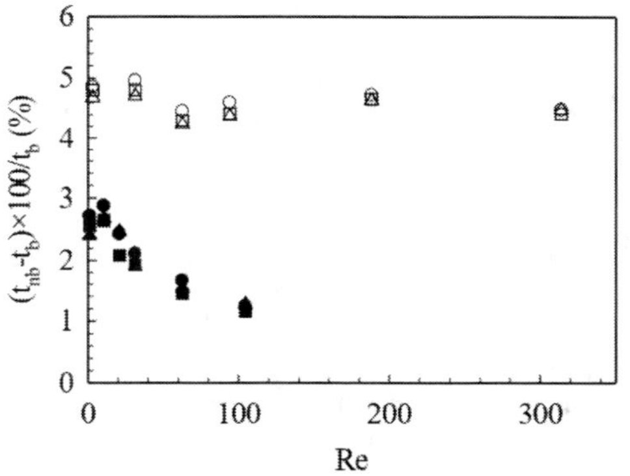

Figure 5: Variation of percentage difference of droplet lifetime without bead and with bead with Reynolds number at T_∞ = 300°C (Legends : O: JME; □: KME; △ : PME; Filled symbols : d_o = 1 mm, d_b = 0.4 mm; Empty symbols : d_o = 3 mm, d_b = 1.5 mm).

Figures 6 and 7 show streamlines and isotherms of KME with and without bead at two time instants. It is observed from Fig.6 that the internal flow of droplet with bead is restricted to a small annular region than that without bead. For same initial droplet diameter, droplet with bead has smaller amount of liquid fuel than that without bead. Since at the beginning the ambient temperature and droplet temperature are the same for both cases, equal amount of heat is transferred. This results in higher heating for droplet with bead during initial period of evaporation as smaller amount of liquid fuel has to be heated. This higher droplet temperature causes higher evaporation rate for droplet with bead. The higher liquid temperature in presence of bead is observed from Fig. 6. Also, at latter instant of time, t = 36 s (half of the lifetime), the liquid is at uniform temperature for the case with bead, whereas still temperature gradient exists for the case without bead. The uniform liquid temperature with bead is attributed to the presence of stronger vortices inside droplet.

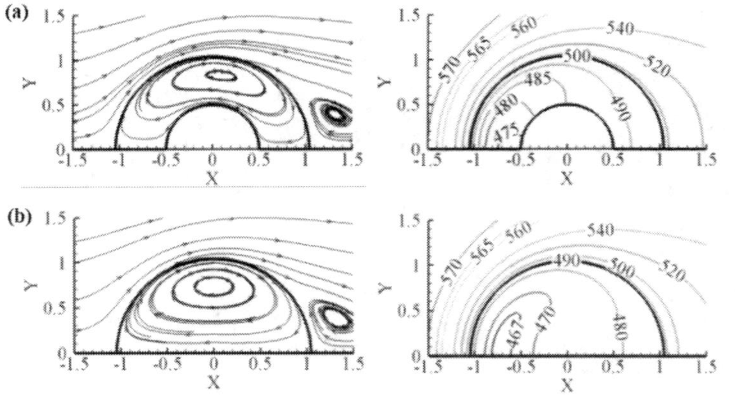

Figure 6: Streamlines (left) and isotherms (right) of KME at one-fourth of the lifetime (a) with bead and (b) without bead; t = 18 s, d_o = 3 mm, U_{FC} = 0.5 m/s, Re_o = 39.8 and T_∞ = 300°C.

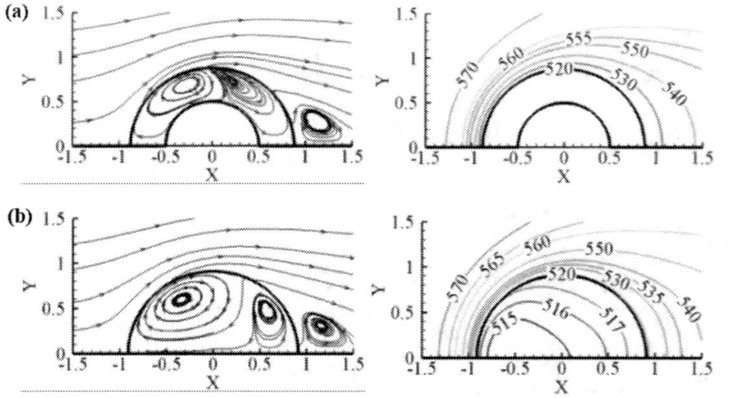

Figure 7: Streamlines (left) and isotherms (right) of KME at half of the lifetime (a) with bead and (b) without bead; t = 36 s, d_o = 3 mm, U_{FC} = 0.5 m/s, Re_o = 39.8 and T_∞ = 300°C.

4 Conclusions

A multi-phase numerical model, which influences suspended bead and variable thermo-physical properties in both liquid- and vapor-phases as functions of temperature and species concentrations, is used to predict vaporization characteristics of suspended biodiesel droplets. The numerical results are compared against the experimental data available in literature and comparison shows good agreement. The model can be used at a temperature below the critical temperature, at low and moderate pressures, provided suitable properties are known. The validated model is used to study the effects of convection velocity and the presence of bead on vaporization characteristics of biodiesel fuel droplets at atmospheric pressure. Simulations using this model take only about 5 to 20 hours of physical time. It has been found that bead significantly affect the lifetime of suspended droplets.

At low freestream velocity, influence of bead on the droplet lifetime is high. Additionally presence of bead in a suspended droplet reduces the droplet lifetime. The internal flow of droplet with bead is found to be restricted to a small annular region than that without bead. For same initial droplet diameter, droplet with bead has smaller amount of liquid fuel than that without bead. Since at the beginning the ambient temperature and droplet temperature are the same for both cases, equal amount of heat is transferred. This results in higher heating for droplet with bead during initial period of evaporation as smaller amount of liquid fuel has to be heated. This higher droplet temperature causes higher evaporation rate for droplet with bead. Thus, smaller bead is found to cause small change in droplet lifetime. Moreover, there is only a marginal change in evaporation characteristics for different fuel namely JME, KME and PME under similar condition, as there is not much difference in their properties such as boiling point (618.9 K, 626.4 K, 616.6 K), molecular weight (289.8, 298.8, 288.5) and latent heat of vaporization (189.5 kJ/kg, 189.8 kJ/kg, 193.7 kJ/kg).

References

[1] A. Daif, M. Bouaziz, X. Chesneau, A. Ali Cherif, Exp. Therm Fluid Sci. 18 (1999) 282–290.

[2] C. H. Chiang, M. S. Raju, W. A. Sirignano, Int. J. Heat Mass Transfer 35 (5) (1992) 1307-1324.

[3] S. S. Sazhin, Prog. Energy Combust. Sci. 32 (2) (2006) 162-214.

[4] S. Raghuram, V. Raghavan, D. N. Pope, G. Gogos, Int. J. Heat Mass Transfer 55 (25) (2012) 7949-7957.

[5] S. Raghuram, V. Raghavan, D. N. Pope, G. Gogos, Int. J. Multiphase Flow 52 (2013) 46-59.

[6] A. T. Shih, C. M. Megaridis, Combust. Flame 102 (3) (1995) 256-270.

[7] G. Strotos, M. Gavaises, A. Theodorakakos, G. Bergeles, Fuel 90 (4) (2011) 1492-1507.

[8] M. Manjunath, V. Raghavan, P. S. Mehta, Int. J. Heat Mass Transfer 88 (2015) 28-41.

[9] R. C. Reid, J. M. Prausnitz, B. E. Poling, The properties of gases and liquids, fourth ed., McGraw-Hill Inc., New York, 1987.

[10] S.V. Patankar, Numerical Heat Transfer and Fluid Flow, Hemisphere Publishing Corp, New York, 1980.

11th Asia-Pacific Conference on Combustion,
The University of Sydney, NSW Australia
10th -14th December 2017.

Burning of Two Arbitrary Sized n-Heptane Droplets Suspended in a Grid Generated Turbulent Jet Flow Field

P Senthil Kumar[1], Gunamani Nath[2], V Raghavan[1], T Sundararajan[1]

[1] Indian Institute of Technology Madras

Chennai, Tamil Nadu, 600036, India.

[2] National Institute of Technology Silchar

Silchar, Assam, 788010, India.

Abstract

Burning of two n-heptane droplets placed side-by-side in a convective flow having significant turbulence fluctuations is studied using suspended droplet technique. The centers of the suspenders are aligned in this configuration. Burning of similar droplets under normal gravity without forced convection is also studied. The turbulence in the jet flow field is introduced using grids. The droplet separation distance has been maintained constant, but the initial droplet sizes are varied. The burning rates have been estimated from the droplet regression curves. The burning rates of the binary droplet have been compared with the burning rates of a single droplet with a similar initial diameter. In this configuration, for the separation distance considered, the burning rate of the larger droplet has been found to be always lesser than that of the small droplet. Direct flame photographs reveal that for side-by-side burning of two droplets in a natural convective environment, the flame always merges. But in the turbulence flow field, merging of flames is intermittent.

1 Introduction

An isolated droplet burning in a turbulent flow field has been investigated by several researchers in the past [1-4]. These literature mainly focus on a single droplet suspended using low thermal conductivity suspenders. In realistic spray combustion devices, droplets are often surrounded by multiple droplets with arbitrary sizes embedded within a turbulence flow field. Such a configuration, in its simplistic form of binary droplets burning together, may be easily studied by using two droplets burning on two suspenders of different sizes placed at various separation distances. Such techniques have been adopted by researchers in the past to investigate the interference effects between burning droplets. The burning rates predicted for binary droplets system are strongly affected by the droplet interference effects and they may significantly vary from that of a single isolated droplet combustion. In such conditions the correlations developed for single droplet studies may not be useful in modeling a realistic spray combustion system. Further, these analyses have been often conducted in a laminar free stream or in a natural convective environment [5-9]. In the present study, burning of two n-heptane droplets placed side-by-side in a convective flow having significant turbulence fluctuations is studied using suspended droplet technique. The centers of the suspenders are aligned in this configuration. Burning of similar droplets under normal gravity without forced convection is also studied. The grid generated turbulent flow field has been well characterized using hot wire anemometer.

Corresponding author. Fax: +91-44-22574650
E-mail address: raghavan@iitm.ac.in

2 Experimental Methodology

The setup consists of a vertical wind tunnel, with a convergent nozzle of diameter, D = 47.5 mm, located at its end as shown in Fig. 1. Grids of two configurations, reported in Table 1, are employed to generate turbulence. The grid opening dimension is represented as 'M' and the blocked width is referred to as 'W'. The solidity ratio (or the blockage ratio) is defined as $\sigma = 1 - [M^2/(2W+M)^2]$. The grid is mounted in flush with the nozzle exit. The mass flow rate of the air is controlled using a digital mass flow meter (ALICAT make). The approaching flow at 6D from the nozzle exit is characterized using two-component hot wire anemometer (mini-CTA, DANTEC). Complete details of the present experimental setup have been presented elsewhere [10]. The calibration and measurement techniques are reported in Senthil et al. [11].

GRIDS	M (mm)	2W (mm)	σ	U (m/s)	T.I (%)
G1	1.5	1	0.64	0.413	24
G2	0.75	0.75	0.75	0.410	33

Table 1: Grid configuration and flow details.

In order to suspend the droplets at 6D from the nozzle exit, two quartz suspenders with bead diameters of 0.68 mm and 2 mm have been used. The suspenders are mounted as shown in Fig.1. The distance between the suspenders (△S) is maintained constant as 0.82 mm. This separation distance is chosen to facilitate auto-ignition of the smaller droplet, once the larger droplet is ignited using a pilot. The ignition lag between the droplets are recorded carefully and only those cases showing similar ignition lag timing have been considered for further analysis.

Three cases have been considered: [a] case with no forced flow condition (NFC), [b] case with grid 1 (G1) mounted at the nozzle exit with a mean convective velocity, U, of 0.41 m/s and a turbulence intensity ($T.I = u_{rms}/U$) of 24% and [c] case with grid 2 (G2) case with U = 0.41 m/s and T.I = 33%. The mean flow velocities in the cases of G1 and G2 are maintained the same and the turbulence intensities have been varied. The maximum uncertainties in the mean flow velocity and u_{rms} components have been estimated as 1.24% and 5.49%, respectively.

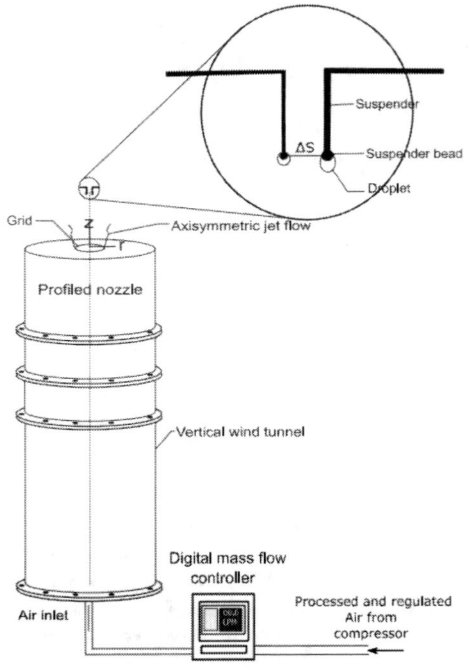

Figure 1: Present experimental setup.

Two n-heptane droplets are manually introduced on to the beads of two suspenders using a glass syringe. The initial droplet diameters are measured using image technique. The larger droplet is ignited using a pilot flame from a small butane torch, which has been momentarily introduced to onset the ignition. The estimation of droplet diameter from a high-definition video has been elaborated in Manjunath et al. [12].

3 Results and Discussion

The burning rates have been estimated from the droplet regression curves. Figure 2 shows the regression curves for two different sized droplets burning adjacent to each other separated by a distance of $\triangle$S. Figure 2 shows the case of no forced flow case, NFC.

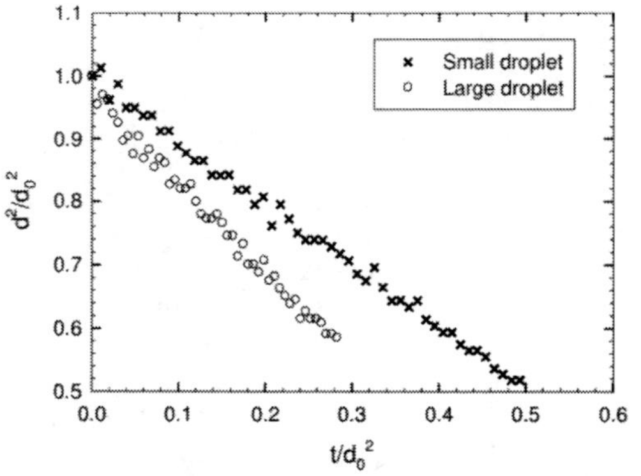

Figure 2: Droplet regression curves for different size droplets burning in a quiescent environment.

In Fig. 2, the square of instantaneous droplet diameter (d^2) and the time (t) have been normalized using the square of the corresponding initial droplet diameter (d_0^2). The burning rates have been estimated by fitting a third order polynomial to the regression curves. The polynomial is differentiated to get the instantaneous slope of $(d/d_0)^2$ vs. t/d_0^2 curve and the instantaneous data has been integrated over the entire time period. This methodology is adopted to account for the non-linearity in the droplet regression curves observed especially in the cases of G1 and G2.

3.1 Single droplet studies

Single droplet studies for varying droplet sizes have been conducted for comparing the burning rates of similar sized droplets burning independently and with a neighboring droplet. Figure 3 shows the variation of single droplet burning rates for the case of G1 for droplets with different initial diameters. Single droplet experiments are conducted for all the flow conditions and their burning rates (K_S) are used as a normalizing parameter. As shown in Fig. 3, the burning rates (K_S) increases with initial droplet diameter. The variation of K_S with the initial droplet diameter is found to be quadratic in nature. Two curve-fits one for lower and another for higher initial droplet diameter ranges are obtained from this data and will be used to estimate K_S when the initial diameter is known. This value of K_S is used to normalize the burning rate of a similar sized droplet burning with a neighboring droplet.

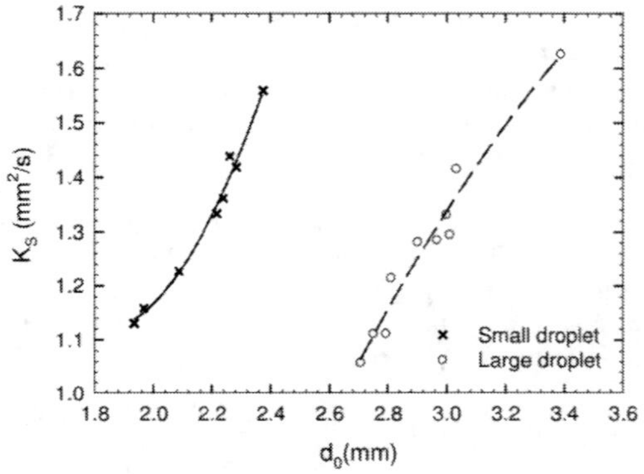

Figure 3: Variation of burning rate as a function of initial droplet diameter for a single droplet burning under G1 condition.

3.2 Binary droplet system

Two droplets with different initial diameters (due to the difference in the bead size) are placed at a fixed separation distance of 0.82 mm. Piloted ignition is onset to the large droplet for all the cases. The heat transfer from the flame automatically ignites the small droplet after a small time lag, which is the additional time taken for by the smaller droplet to get ignited. This varies for each case, especially due to the differences in the flow field between the cases, NFC, G1 and G2. It should be noted that the ratio, $d_r = d_S/d_L$, defined as the ratio of initial diameter of the small droplet to that of the large droplet, varies for each experiment. Thus, d_r also forms a parameter affecting the droplet burning characteristics. As mentioned earlier, the burning rates (K) have been normalized with the corresponding single droplet burning rates (K_S) with almost the same initial diameter.

In Figs. 4(a) - 4(c), the ratio $K/K_S < 1$ indicates that the burning rate is lower than the corresponding burning rate of an isolated droplet burning with almost the same initial diameter. In this experimental configuration, where the centers of the beads of two suspenders have been aligned, the flame from the larger droplet provides hot air flow towards the smaller droplet burning in its side. This, therefore, facilitates faster burning of the smaller droplet and slower burning of the larger droplet, slower than the isolated droplet case in most of the scenario.

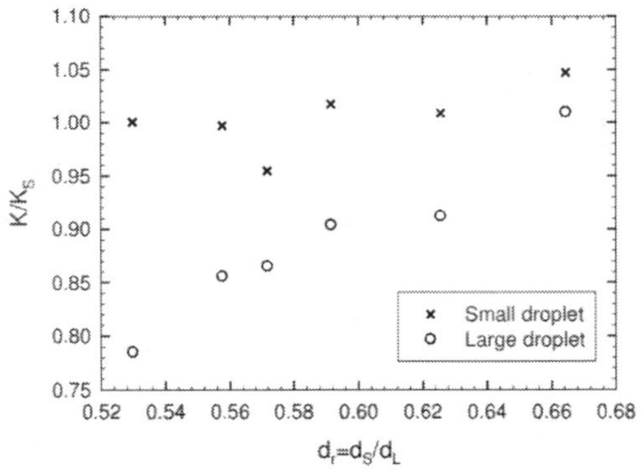

(a) NFC

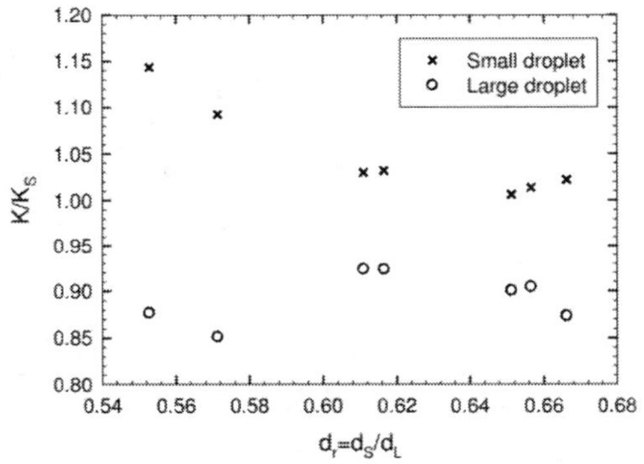

(b) G1, U=0.413 m/s and T.I = 24%

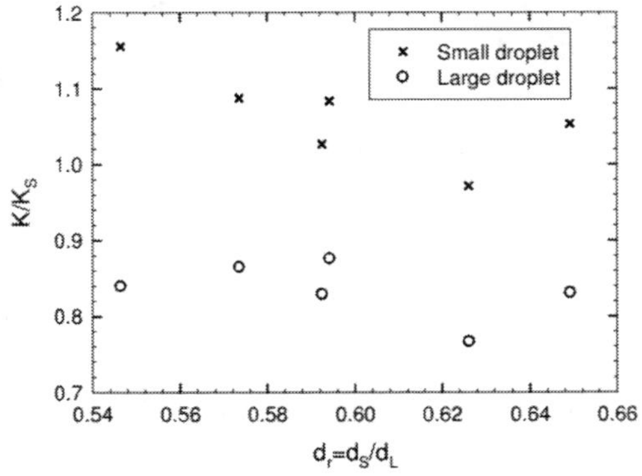

(c) G2, U=0.410 m/s and T.I = 33%

Figure 4(a)-(c): Variation of normalized burning rates with initial droplet diameter ratio for different flow conditions.

Figure 4(a) shows that under natural convective environment, the burning rates of small droplets are almost constant for the given range of droplet diameter ratios and are more or less equal to that of the isolated droplet burning rates. The burning rate of large droplet, on the other hand, increases with increasing diameter ratio, starting from a value that is lesser that its isolated counterpart. It attains the burning rate of smaller droplet (or that of the isolated droplet) at higher diameter ratios. Figures 4(b) and 4(c) show the variation of normalized burning rates with diameter ratio for the cases of G1 and G2, respectively. For smaller d_r, an enhancement in the the burning rates has been observed for both smaller and larger droplets. As d_r is increased, the burning rate of small droplet approaches that of the isolated droplet. The increasing trend in K for larger droplet, as observed in the case of NFC, is not observed in the cases of G1 and G2. Thus, it is clear that turbulent flow field is able to modify the burning characteristics, especially that of the larger droplet. Further the case the grid G2 is able to produce higher burning rates when compared to that of G1. The ignition time lag for the NFC cases is found to be shorter in comparison to the cases, G1 and G2. In the case of NFC, the ignition of the small droplet is almost instantaneous. The flow field is responsible for this and apparently grid generated turbulence plays significant role. This aspect has not been discussed here as more data have to be collected to provide quantitative information.

The direct flame photographs for the scenario of burning of two droplets burning in natural convective environment is shown in Fig. 5. It is apparent that the flames from both droplets merge and continues to burn in the similar fashion throughout the entire droplet life time. The higher burning rate of smaller droplet is due to enhanced heat transfer as well as proper transport of the fuel vapor in to the flame zone due to the resultant flow field. In this configuration, heat transfer for larger droplet is not enhanced due to the presence of smaller droplet, as well as the fuel vapor from larger droplet surface has not been able to be transported efficiently so as to enhance the burning rate.

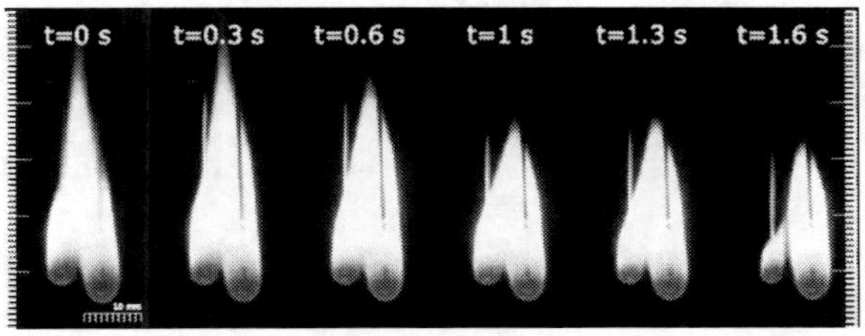

Figure 5: Direct flame photographs of two droplets burning in a quiescent environment.

Figure 6 presents the direct flame photographs for the case, G2, at various time instants. Unlike the case of NFC, the merger of the flames is not observed at all time instants and separate flames are observed at quite a few time instants. That is, the merger of the flames is intermittent. The flames are also quite oscillatory due to the presence of turbulent flow field. As discussed above, turbulence is able to enhance the burning rates for all diameter ratios. The trend that the smaller droplet has higher burning rate than that of the larger droplet, is preserved here. The reasons are very much the same as discussed for the case of NFC.

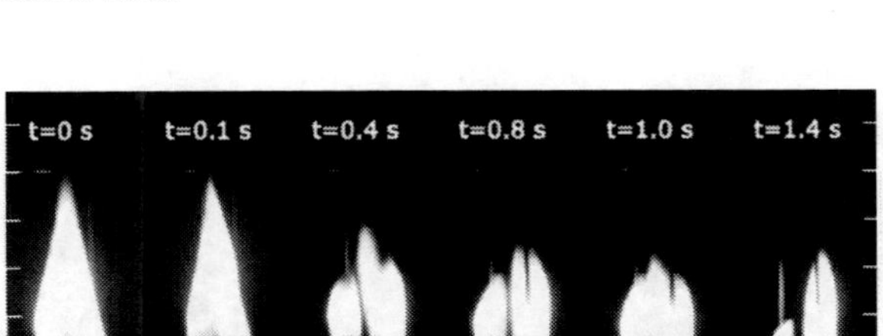

Figure 6: Direct flame photographs of two droplets burning for G2 case.

4 Conclusions

The burning of arbitrary sized droplets adjacent to each other in grid generated turbulent jet flow field is investigated using the suspended droplet technique. Two types of grids have been used to impart well-controlled turbulent field, where the droplets burn. A fixed separation distance between the larger and the smaller droplet has been maintained by suspending them on two quartz suspenders with beads. The centers of the beads have been aligned. Cases of natural convective burning have also been studied. The following are the major conclusions derived from the present study.

(1) In the configurations studied, the smaller droplet gets enhanced heat and oxygen flow due to the presence of the larger droplet burning in its side. This enhances the burning rate of the smaller droplet close to or higher than that of an isolated droplet having a similar initial diameter, based on the ratio of the initial diameters of the smaller and larger droplets.
(2) Larger droplets burn slower than the smaller droplet and also than an isolated droplet of similar initial diameter. Reduction in heat transfer as well as oxygen transport (which affects fuel depletion from the droplet surface) form reasons for this trend.
(3) In natural convective flow field, larger droplet shows increasing trend in its burning rate with diameter ratio. However, for the cases with grid generated turbulence, no such trend is observed.
(4) Flames from two droplets are seen to be always merged in the case of natural convective burning. However, in turbulent flow fields, the flames are oscillatory and the merger is also intermittent.

References

[1] Ohta, Y., Shimoyama, K., and Ohigashi, S., *Bulletin of JSME* 18-115:47-56 (1975).

[2] Birouk, M., Chauveau, C., and Gökalp, I., *Proceedings of the Combustion Institute* 28-1:1015-1021 (2000).

[3] Birouk, M., and Toth, S. L., *Flow, Turbulence and Combustion* 94-4:843-858 (2015).

[4] Senthil Kumar P., Raghavan, V., & Sundararajan, T. Experimental Study of Burning Characteristics of an Isolated Liquid Fuel Droplet in Grid Generated Turbulent Flow Field, ICLASS, Taiwan, 2015.

[5] Brzustowski, T. A., Twardus, E. M., Wojcicki, S., & Sobiesiak, A. (1979). Interaction of two burning fuel droplets of arbitrary size. *AIAA J, 17*(11), 1234-1242.

[6] Rex, J. F., Fuhs, A. E., & Penner, S. S. (1956). Interference effects during burning in air for stationary n-heptane, ethyl alcohol and methyl alcohol droplets. *Jet Propulsion, 26*(3), 179-187.

[7] Kanevsky, J. "Interference during burning in air for nine stationary fuel droplets arranged in a body-centered cubic lattice." *Jet Propulsion* 26.9 (1956): 788-788.

[8] Miyasaka, Kenji, and Chung K. Law. "Combustion of strongly-interacting linear droplet arrays." *Symposium (International) on Combustion*. Vol. 18. No. 1. Elsevier, 1981.

[9] Nuruzzaman, A. S. M., A. B. Hedley, and J. M. Beer. "Combustion of monosized droplet streams in stationary self-supporting flames." *Symposium (International) on Combustion*. Vol. 13. No. 1. Elsevier, 1971.

[10] Senthil Kumar, P., V. Raghavan, and T. Sundararajan. "The effects of grids on the near-field evolution of turbulence in a low-Reynolds number jet." *Physics of Fluids* 27.11 (2015): 114101.

[11] Senthil Kumar P, Effects of grid generated turbulence on the evaporation and burning of isolated droplets in a jet flow field, PhD Thesis, Indian Institute of Technology Madras, Chennai, India, 2017.

[12] Manjunath, M., Raghavan, V., and Mehta, P. S., International Journal of Heat and Mass Transfer 88: 28-41 (2015).

11th Asia-Pacific Conference on Combustion,
The University of Sydney, NSW Australia
10th -14th December 2017.

On the extinction strain rates of counter-flow diffusion flames

S M Ali[1*] and Varunkumar S[2]

[1]Department of Aerospace Engineering, IIT Madras

[2]Department of Mechanical Engineering, IIT Madras

Chennai, 600036, India

Abstract

Extinction strain rate ‘a’, like laminar flame speed, is a fundamental characteristic of fuel-oxidizer combination. Available data for ‘a’ from counter flow diffusion flame experiments from earlier studies using opposed jet (straight and contour nozzles) and Tsuji burners have been compiled; data from a Tsuji burner developed in our laboratory to cross-check earlier results is also included in the analysis. Geometric and associated boundary conditions and strain rate definitions are shown to be responsible for the observed variations in ‘a’ obtained using specific burner for a given fuel-oxidizer combination. Measured variations in ‘a’ with geometric parameters, cylinder diameter in case of Tsuji type and nozzle separation in opposed jet type with straight nozzle, are shown to tend to similar asymptotic values (about 375 s^{-1} for CH_4-Air, for instance), which can be taken to be the extinction strain rate. Numerical results obtained using FLAMEMASTER [1] is shown to follow the same asymptotic trend. The opposite trend, that is, decreasing extinction global strain rate with increasing L/D observed by Sarnacki et al [2] using contour nozzles is due to the use of actual nozzle separation distance instead of the effective separation distance known as *free floating length,* L_{FF} in calculating strain rate. The corrected strain rates obtained by using the effective separation distance is shown to be consistent with the other results and the asymptotic value matches reasonably well with local strain rates reported in Sarnacki et al [2].

1 Introduction

Extinction strain rate of non-premixed flames is a fundamental characteristic of fuel-oxidizer combination, similar to flame speed of premixed fuel-oxidizer mixture. It is used as an optimization parameter for developing chemical kinetic mechanisms (both detailed and short), which in turn are used for generating flamelet libraries for turbulent diffusion flame simulations. Hence it is crucial to accurately determine extinction strain rates of various fuel-oxidizer combinations using experiments. Counterflow configuration is usually used for determining the extinction strain rate. A non-premixed flame established between opposing nozzles or around a porous cylinder/sphere placed in a free-stream (Tsuji type burner) is subject to increasing strain rates by increasing jet/free-stream velocities till the point of extinction. The strain rate in the case of Tsuji type burners is taken to be $2V_{ox}/R$, where, V_{ox} is the free-stream oxidizer velocity and R is the radius of the porous cylinder/sphere [3, 4, 5]. In the case of opposed jet burners the strain rate ‘a’ is usually calculated using (1) derived by assuming large Reynolds Number, inviscid flow self-similarity solution with plug flow boundary conditions where mixing layer is treated as thin layer at the stagnation plane [6].

$$a = \frac{2V_{ox}}{L}\left[1 + \frac{V_F}{V_{ox}}\sqrt{\frac{\rho_F}{\rho_{ox}}}\right] \quad (1)$$

where, L is the separation distance between the nozzles, V_F and V_{ox} is the fuel and oxidizer jet velocity at nozzle exit , ρ_F and ρ_{ox} are the fuel and oxidizer densities respectively. In most experiments it is ensured that the fuel and oxidizer boundary momentum fluxes are equal, that is, $\rho_F V_F^2 = \rho_{ox} V_{ox}^2$, and hence the formula for strain rate reduces to $4V_{ox}/L$.

1.1 Earlier studies

Table 1 contains experimentally and numerically determined values of extinction strain rates taken from various earlier studies for different fuel-Air combinations. Available information on the dimensions along with the type of burner is also included in Table 1. It is clear that the measured extinction strain rates depend on the burner geometry – porous cylinder diameter in the case of Tsuji type burners and separation distance and nozzle diameter in case of opposed jets. The measurements are also dependent on the type of nozzle used (straight or contoured). Kee et al [7] attributed these variations to the implicit assumption of either potential or plug flow like boundary conditions in theoretical formulations from which the expression for strain rate are derived. They have developed a theoretical framework to overcome these limitations by explicitly prescribing realistic boundary conditions at the inlets and used the same to predict extinction strain rates of counterflow premixed flames with reasonable success. Using PIV, Sarnacki et al [2] have shown that irrespective of the burner geometric parameters (L and D), the local strain rate measured close to the stagnation plane is invariant, as can be expected from theoretical considerations. Also, by systematically varying the L/D ratio from 0.88 to 1.52, they have shown that the global strain rate ($4V_{ox}/L$) decreases and asymptotically tends to the local value. This asymptotic value obtained by Sarnacki et al [2] for CH_4-air is 354 ±36 s^{-1} which closely matches with the global strain rate of 375 s^{-1} obtained

*Corresponding author.
E-mail address: smughees.ali@gmail.com

Fuel	Experiments			Numerical	
	Configuration		Strain Rate (s^{-1})	Strain Rate (s^{-1})	
(100% CH_4)-Air	Tsuji Type D = 60, 30, 15 mm		320, 375, 375 [3, 4]	460, 320, 350, 400-405 [8, 9]	
	Opposed Jets			Plug Flow	Potential Flow
	L (mm)	D (mm)			
	25, 10, 10, 11, 11	15, 10, 19, 8, 6	271, 380*,610, 400*, 375*, 370* [10, 8,11,12, 13]	391*, 223* [8, 13]	509*, 494*, 393* [8,12, 13]
	15	Variable	335 [14]	456-470 [14]	
(100% C_3H_8)-Air	Tsuji Type D = 60, 45, 30, 15 mm		365, 420, 486, 650 [5]	-	
	Opposed Jets			Plug Flow	Potential Flow
	L (mm)	D (mm)			
	25, 11,15	15, 6,variable	305, 410 , 385 [10, 13, 14]	-	-
(12% &14% H_2 Rest N_2)- Air	-	-	-	-	78, 139 [15]
(15%, 17%, 18.5% & 20% H_2 Rest N_2)-Air	22	12	292, 515, 801, 1527 [16]	-	-
(100% CH_3OH)- Air	Tsuji Type		-	1200 [17]	
(16% to 30% CH_3OH Rest N_2)- Air	Opposed Jets			Plug Flow	Potential Flow
	L (mm)	D (mm)			
	10	-	65 to 470 [18]	65 to 470 [18]	-

Table 1: Summary of extinction strain rates (s^{-1}) for methane, propane with different geometric configurations.

* Represent local extinction strain rate.

by Tsuji [3, 4] and in the current work. Similar observations apply to C_3H_8-Air flames, barring the anomalous 650 s^{-1} obtained by Tsuji [5].

Another anomaly observed is related to the variation of global strain rate with L/D – experiments of Sarnacki et al [2] indicate that the global strain rate at extinction decreases with increasing L/D and tends to the invariant local strain rate value. On the other hand numerical results obtained using FLAMEMASTER [1] as a part of the current work shows the opposite trend, that is, global extinction strain rates increase with increase in L/D and eventually tends to an asymptotic value. Local strain rates calculated from numerical results at the point of maximum temperature is invariant as expected. This anomaly is later shown to be due to the use of actual nozzle separation distance instead of effective length.

It is pertinent to point out that work on counterflow flames was initiated at our lab with the aim of experimentally determining the extinction strain rates of syngas flames to be used for development and optimization of simplified kinetic mechanisms for CFD simulations. Towards this a Tsuji type burner was designed and developed. Attempts to validate the extinction strain rate results obtained for CH_4-Air flames led to an exploration of earlier results and in turn to the variations and anomalies discussed above. In this paper preliminary experimental results obtained from the laboratory Tsuji burner are presented. A *free floating length* L_{FF} based procedure is used to correct the global strain rate values obtained by Sarnacki et al [2] and is shown to resolve the observed discrepancy in the variation of global extinction strain rate with L/D between experiments and numerical results obtained as a part of the current study.

2 Experiments

A counterflow Tsuji type burner setup [3, 4, 5] as shown in Fig. 1 was designed and developed. It consist of a combustion chamber with dimensions 12 cm × 18 cm × 3 cm (l x h x w) with four different porous cylinders 3 cm long and diameter 1.5 cm, 3 cm, 4.5 cm, 6 cm respectively. For flame visualization, a quartz glass window of dimension 4 cm × 9 cm is fixed on the combustion chamber. Air is supplied from the bottom through a convergent nozzle. The lower section of the nozzle is attached to air settling chamber with dimensions 25 cm × 25 cm× 25 cm. Flow straighteners are fixed inside the settling chamber to reduced local turbulence and maintain a uniform laminar flow at the nozzle exit. Mass flow controllers (MFCs) are used at the upstream of fuel and air supply for metering and controlling supply of air and fuel.

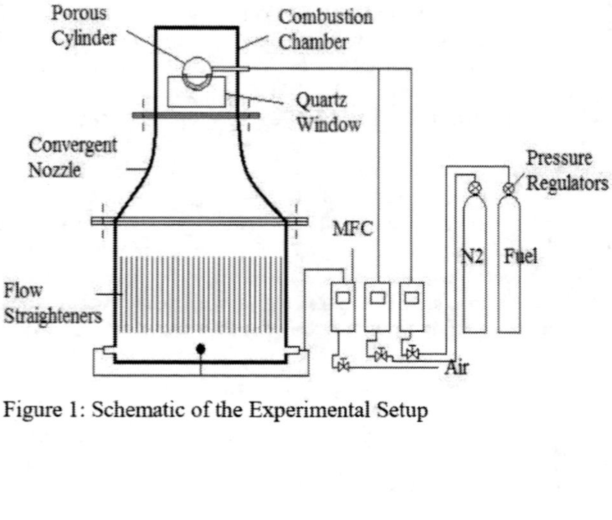

Figure 1: Schematic of the Experimental Setup

Nozzle separation L 'mm'	Global Extinction Strain Rate 'a' (s^{-1})							
	CH_4 – Air		C_3H_8 – Air		$20\ H_2/80N_2$ – Air		CH_3OH – Air	
	San Diego	GRI Mech.3.0	San Diego	GRI Mech.3.0	San Diego	GRI Mech.3.0	San Diego	GRI Mech.3.0
5	546	503	909	1032	1470	1813	2575	1913
10	590	540	957	1081	1564	1881	2675	1991
20	611	566	984	1113	1618	1942	2700	2029
30	619	580	993	1125	1626	1960	2744	2060
40	624	578	997	1132	1638	1962	2738	2059
50	627	580	1005	1134	1637	1959	2753	2072

Table 2: Computations for variation of global Strain Rate 'a' with nozzle separation distance L

Fuel is supplied from pressurized (up to 150 bars) gas cylinders of 47 liters capacity. Two stage pressure regulators fixed with gas cylinders are used and the upstream is maintained at 0.5 bars. Mixture of methane (99.50% purity) diluted with nitrogen (99.99% purity) is used as a fuel. Moisture filter is used to remove moisture from the air supplied to the combustion chamber which maintains less than 1% of moisture by volume in the supplied air. Humidity sensor is used to measure moisture content of the air supplied.

The global strain rate to characterize the Tsuji type flame is defined as '$2V_{ox}/R$'. Fuel flow velocity from the porous cylinder is maintained in such a way that the flame extinction is free from any thermal quenching. The measured extinction strain rate is shown in Fig. 2. The data available from earlier studies for CH_4-Air and C_3H_8-Air from Tsuji's experiments is also included in the figure [3, 4, 5]. It can be seen, as the porous cylinder diameter D reduces the extinction strain rate increases irrespective of the fuel. The data from Tsuji [3, 4] indicate that as D decreases the strain rate tend to an asymptotic value for methane air (375 s^{-1}). The measured value from the current experiments with 1.5 cm diameter burner is also 375 s^{-1}.

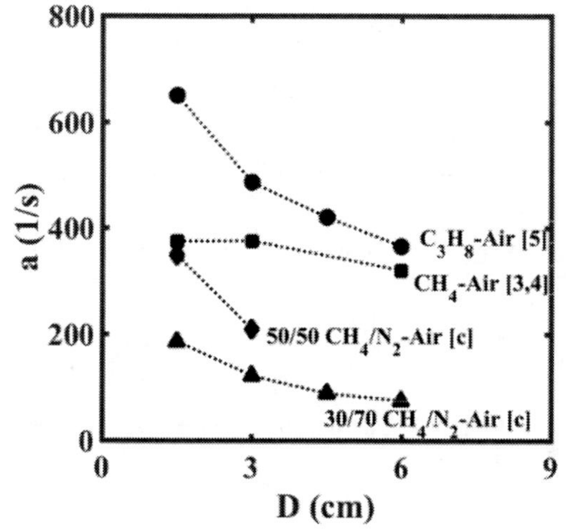

Figure 2: Global Extinction Strain rate variation with burner diameter 'D' for Tsuji type burner; [c] – current work.

To investigate further the geometric dependence of strain rate, numerical simulations were performed with different nozzle separation distances.

3 Numerical Calculations

Two chemical-kinetic mechanisms, GRI-MECH 3.0 [19] and San Diego [20], are used here. San-Diego mechanism consists of 57 species and 235 elementary reaction steps and GRI-MECH 3.0 mechanism consists of 53 species and 325 elementary reaction steps. These mechanisms are simplified by removing elementary reactions steps involving four or more carbon atoms. The fuels used for computation are methane, propane, hydrogen and methanol. Numerical computations are performed on counterflow configuration using computer program FLAMEMASTER [1]. The conservation equations in 1-D for mass, momentum, and energy along with species equations used here are described in detail elsewhere [21]. Thermal diffusion is included with buoyancy neglected. Steady plug flow boundary conditions are employed with mass fluxes and mass fractions of fuel and oxidizer defined at these boundaries. Initial temperature for both fuel and oxidizer is taken to be 300 K.

The distance between the nozzles "L" is varied from 5 mm to 50 mm to show the effect of nozzle separation distance on the extinction conditions. Table 2 shows the computation results of the variation of global strain rate 'a' (calculated using (1)) with the nozzle separation distance for the different fuels. As the nozzle separation distance increases the global strain rate increases and reaches an asymptotic value but the trends observed by Sarnacki [2] are opposite to this. This is discussed further in the next section. The strain rate at L = 50 mm is higher by 16% for CH_4, 10 % for C_3H_8, 11% for $20H_2/80N_2$ and 8% for CH_3OH when compared to L = 5 mm. As expected the scalar dissipation rate $|\nabla Z|^2$, where, Z is the mixture fraction, at the point of maximum temperature is invariant with nozzle separation distance L. This means that the flame structure is invariant of nozzle separation distance even though extinction strain rate is varying with L.

4 Discussions

Global extinction strain rate variation with L/D from Sarnacki et al [2] is shown in Fig. 3. These strain rate values are calculated using $a = 4V_{ox}/L$ since momentum balance has been ensured in these experiments ($\rho_F V_F^2 = \rho_{ox} V_{ox}^2$). This expression for 'a' assumes that the flow is self-similar. But for a given burner diameter there exist a minimum value of L only beyond which this assumption is valid. In case of contour nozzle used by Sarnacki et al [2] this value is experimentally determined as 'L $\simeq$1.8 D' and is termed as free floating length L_{FF}. As can be seen from Fig. 3, range of L/D values investigated in [2] is between 0.88-1.52 and hence below the minimum required (which is 1.8). Therefore the L in the expression for 'a' should be replaced by the effective length, L_{FF}.

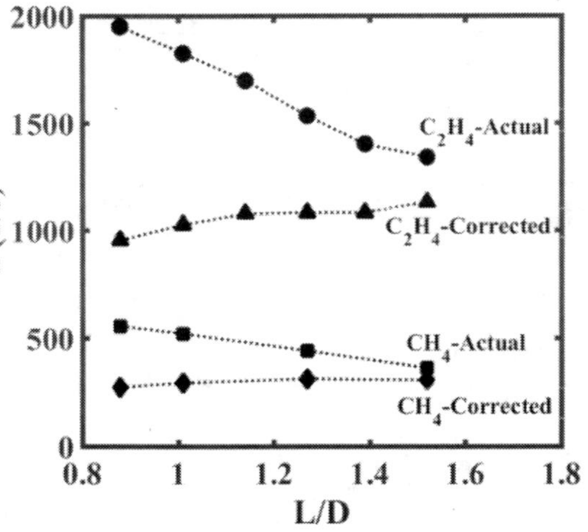

Figure 3: Experimental value of actual and corrected global strain rate at extinction [2]

The corrected strain rate variation calculated using (2) is shown in Fig. 3 and can be seen to follow the same trend observed in experiments.

$$a_{corrected} = a_{actual} \frac{L/D}{1.8} \qquad (2)$$

Similar observations apply to the other hydrocarbon fuels reported in [2]. Given that the correction procedure involving free floating length is based on the structure of the flow, it can be expected to be applicable to hydrogen as well as oxygenated fuels (including preheated cases). Computational results for hydrogen and methanol shown in Table 2 is consistent with results for hydrocarbon fuels - increasing extinction strain rate with L later tending to an asymptotic value. This clearly indicates that the observed differences in experimental results are due to the self-similarity assumption used to extract the critical strain rates and hence independent of the fuel used.

5 Conclusions

Extinction strain rates measured using different types of burners were compiled to bring out the geometric dependence. Certain asymptotic trends discernible from the data were brought out and this asymptotic value is shown to match closely with the local strain rate measurements of Sarnacki [2]. A simple correction procedure is suggested based on *free floating length* and shown to establish consistency between experiments and computations. More experiments are required to further verify these ideas - if found to be valid, then simple Tsuji type burners can be used for accurately determining the extinction strain rates and using this the intrinsic kinetics of fuel-oxidizer combinations with confidence.

5 Acknowledgments

The authors thank Mr. Rohit Khare (IITM) and Dr. Krithika Narayanaswamy (IITM) for helping with FLAMEMASTER simulations and insightful discussions.

References

1. FlameMaster V3.3.10: A c++ program for 0D combustion and 1D laminar flames calculations. Available at http://www.itv.rwth-aachen.de/fileadmin/Downloads/Software/FlameMaster/FlameMasterV3.3.10.zip
2. Sarnacki BG, Esposito G, Krauss RH, Chelliah HK, Combust. Flame 159 (2012) (3) 1026-43
3. Tsuji H, Yamaoka I, In Symposium (International) on Combustion 12(1969)997-1005
4. Tsuji H, Progress in energy and Combust. Science 8(2) (1982) 93-119
5. Tsuji H, Yamaoka I, In symposium (International) on Combustion 11(1967) 979-984
6. Seshadri K, Williams FA, International Journal of Heat and Mass Transfer 21 (1978)(9) 251-3
7. Kee RJ, Miller JA, Evans GH, Dixon-Lewis G, In Symposium (International) on Combustion 22 (1989) (1) 1479-1494
8. Chelliah HK, Law CK, Ueda T, Smoke MD, Williams FA, In Symposium (International) on Combustion 23 (1991) 503-511
9. Dixon-Lewis G, David T, Gaskell PH, Fukutani S, Jinno H, Miller JA, Kee RJ, Smooke MD, Peters N, Effelsberg E, Warnatz J, In Symposium (International) on Combustion 20 (1985) 1893-1904
10. Puri IK, Seshadri K, Combust. Flame 65(2) (1986) 137-50
11. Papas P, Fleming JW, Sheinson RS, In Symposium (International) on Combustion 26 (1996) 1405-1411
12. Du J, Axelbaum RL, In Symposium (International) on Combustion 26 (1996) 1137-1142
13. Chen R, Axelbaum RL, Combust. Flame 142 (2005) (1) 62-71
14. Chen b RH, Lattimer C, Roberts WL, Combustion Science and Technology 160 (2000) 103-18
15. Sohn CH, Chung SH, Combust. Flame. Apr 30;121(1):288-300
16. Trees D, Brown TM, Seshadri K , Smooke MD, Balakrishnan G, Pitz RW, Giovangigli V, Nandula SP, Combust Science and Technology. 104 (1995) (4-6) 427-39
17. Chen JY, Combustion Science and Technology. 78(1991) 127-45
18. Seiser R, Humer S, Seshadri K, Pucher E, Proceedings of the Combustion Institute 31(1)(2007) 1173-80.
19. Gregory P. Smith, David M. Golden, Michael Frenklach, Nigel W. Moriarty, Boris Eiteneer, Mikhail Goldenberg, C. Thomas Bowman, Ronald K. Hanson, Soonho Song, William C. Gardiner, Jr., Vitali V. Lissianski, and Zhiwei Qin http://www.me.berkeley.edu/gri_mech/
20. "Chemical-Kinetic Mechanisms for Combustion Applications", San Diego Mechanism web page, Mechanical and Aerospace Engineering (Combustion Research), University of California at San Diego (http://combustion.ucsd.edu).
21. H. Pitsch, *Entwicklung eines Programmpaketes zur Berechnung eindimensionaler Flammen am Beispiel einer Gegenstrom diffusions flamme* [Diplomarbeit], Rheinisch-Westfalische Technische Hochschule Aachen, Aachen, Germany 1993

11th Asia-Pacific Conference on Combustion,
The University of Sydney, NSW Australia
10th -14th December 2017.

Comparison of chemical mechanisms for *n*-dodecane at engine conditions using an unsteady flamelet model

Armin Wehrfritz[*1], **Bruno Savard**[1], **and Evatt R. Hawkes**[1,2]

[1]School of Mechanical and Manufacturing Engineering, University of New South Wales
Sydney, NSW, 2052, Australia
[2]School of Photovoltaic and Renewable Energy Engineering, University of New South Wales
Sydney, NSW, 2052, Australia

Abstract

In the present study, unsteady flamelet calculations were used to compare chemical mechanisms for *n*-dodecane at thermochemical conditions corresponding to the Engine Combustion Network Spray A reference case. The calculations were carried out for a broad range of scalar dissipation rates and the results were analysed in terms of ignition delay times and location in composition space. The results indicate only a weak dependency of the ignition delay times on the scalar dissipation rate, up to a certain value for which ignition cannot occur any more (χ_{ign}). While this behaviour is consistent for all investigated mechanisms, the value of χ_{ign} varies between them. A more detailed investigation of the effect of mixing showed a significant broadening of the heat release profile in composition space. Consequently, the ignition location shifts to much richer mixtures, even for low scalar dissipation rates. This shift is predicted for all mechanisms, though the extend varies between the mechanisms. The results reveal valuable information with respect to the ignition characteristics of each mechanism, but also indicate that further analysis is needed.

1 Introduction

Long chained hydrocarbons form the major components of fuels used in many combustion engines, such as gasoline, diesel or jet engines. The chemical composition of these fuels is complex and hence their thermophysical and chemical properties are difficult to access. In order to accurately define the fuel properties for experimental and computational combustion research, surrogate fuels, such as the Primary Reference Fuels (PRF) for gasoline [1], are widely used.

Diesel fuel has been represented by *n*-heptane in a wide range of experimental (e.g. [2, 3]) and computational studies (comprehensive lists can be found in [4, 5]) and hence also the first target conditions within the Engine Combustion Network (ECN) [6] were using *n*-heptane. A particular advantage of *n*-heptane is the availability of tractable and widely tested chemical mechanisms. However, the carbon chains of typical diesel fuel components range from 10 to 25 carbon atoms [7] and hence the thermophysical properties, especially with respect to evaporation and boiling, are not well represented by *n*-heptane. *n*-Dodecane was identified as a more suitable surrogate for diesel fuel due to the longer carbon chain and the well defined thermophysical properties [7].

Typical surrogate fuels for jet engine applications include *n*-decane, *n*-dodecane, and trimethylbenzene, among various other components [8, 9]. Thus also the chemical mechanisms for jet fuel often include *n*-dodecane. However, the reaction models for pyrolysis of the large *n*-alkanes in these mechanisms are often strongly simplified [10]. In many cases this is sufficient for steady state combustion devices such as gas turbines and jet engines. For reciprocal combustion engines, however, the combustion takes place in an inherently unsteady process and hence the ignition characteristics and low-temperature chemistry become increasingly important. Another major issue arises due to the significantly higher operating pressures of diesel engines, compared to the conditions for which the jet fuel mechanisms are developed. In recent years, several chemical mechanisms which include low-temperature chemistry pathways for *n*-dodecane have been developed [11–16], partly due to the collaborative effort within the ECN.

Comprehensive comparisons of the available mechanisms with respect to the homogeneous ignition delay times have been presented at ECN workshops [17, 18], as well as in recent studies [19, 20]. The various mechanisms have also been used to simulate the Spray A experiments in various multi-dimensional spray combustion studies (e.g. [21–24]). The multi-dimensional spray combustion simulations involve many complex processes and either various modelling assumptions or an extremely high computational cost in the case of direct numerical simulations. While significant differences between results using different chemical mechanisms are observed in these studies, the complexity of the flows involved makes it difficult to ascribe definitively the root causes of the observed differences. As such, comparing the mechanisms in a simpler environment is desirable, to offer a means to help interpret the more complex spray flame results. One simple approach is to employ a homogeneous ignition configuration; however this does not represent the conditions of an igniting fuel jet well where the transport of heat and radicals plays an important role in the ignition behaviour [25, 26].

The objectives of the present study are to compare the ignition characteristics of the available *n*-dodecane mechanisms (including low-temperature pathways) at the ECN Spray A conditions. The unsteady flamelet model [27] is applied in order to account for the unsteady ignition and mixing processes in a canonical combustion configuration. The considered mechanisms are listed in Tab. 1. In addition, ignition delay times obtained with the de-

	Species	Reactions
Luo [11]	105	678
Narayanaswamy [12]	255	1512
Polimi [13]	77	733
Cai [14]	56	197
Wang [15]	100	553
Yao [16]	54	269

Table 1 – *n*-Dodecane mechanisms

*Corresponding author.
E-mail address: armin.wehrfritz@unsw.edu.au

tailed mechanism (2755 species and 14314 reactions) developed at the Lawrence Livermore National Laboratory (LLNL) [28] are presented as a reference.

2 Methods

In the present study, the unsteady flamelet model developed by Pitsch and Peters [27] is adopted. The governing equations, in the absence of differential diffusion, for this canonical combustion configuration are

$$\frac{dY_i}{dt} = \frac{\chi}{2}\frac{d^2Y_i}{dZ^2} + \omega_i \tag{1}$$

$$\frac{dT}{dt} = \frac{\chi}{2}\frac{d^2T}{dZ^2} + \omega_T \,, \tag{2}$$

where the scalar dissipation rate is prescribed using the following model [25]:

$$\chi(Z) = \chi(Z_{\text{ref}})\frac{f(Z)}{f(Z_{\text{ref}})} \tag{3}$$

$$f(Z) = \exp\left(-2\left[\text{erfc}^{-1}(2Z)\right]^2\right) \,. \tag{4}$$

In order to facilitate a consistent comparison between mechanisms at predefined reference values, the scalar dissipation rate profile is kept constant throughout the ignition process. Differential diffusion processes are neglected in the present study as the conditions for which the chemical mechanisms are utilised, are highly turbulent, and thus the unity Lewis number assumption is well justified [25]. The unsteady flamelets were computed with FlameMaster [29]. The homogeneous ignition delay times were obtained using the Cantera toolbox [30].

The conditions considered for the present study correspond to the ECN Spray A baseline case, as well as a parameter sweep over oxidiser temperature for the homogeneous case (i.e. $\chi = 0\,\text{s}^{-1}$). For the baseline case, the ambient temperature and density are $900\,\text{K}$ and $22.8\,\text{kg/m}^3$, respectively. The oxidiser composition is assumed as $15\,\%$ O_2 and $85\,\%$ N_2, the pressure is obtained invoking the ideal gas law ($\approx 5.96\,\text{MPa}$) and kept constant for all equivalence ratios. The temperature of the liquid n-dodecane fuel is $363\,\text{K}$. Since only the gas phase is considered in present the calculations, the enthalpy of the fuel stream is adjusted to account for the heat of evaporation. Since this leads to an effective temperature of $181.9\,\text{K}$ for the gaseous fuel and thus temperatures beyond the validity range of the NASA polynomials, the fuel stream boundary conditions are set to the values at mixture fraction 0.75, yielding a mixture temperature of $326\,\text{K}$. The calculations are initialised by assuming adiabatic mixing of the fuel and oxidiser streams.

3 Results

The homogeneous ignition delay time τ and corresponding equivalence ratios Φ as a function of oxidiser temperature are plotted in Fig. 1. The ignition delay time is defined by the time of the maximum consumption rate of H_2O_2, an intermediate species produced between first and second stage ignition. In the homogeneous reactor case, this ignition delay time definition is in good agreement with the definitions used within the ECN based on the maximum rate of change in temperature.

As it has been observed in various studies [17–19], the results show significant differences between the mechanisms, where the LLNL, Luo, Narayanaswamy and Wang mechanisms predict

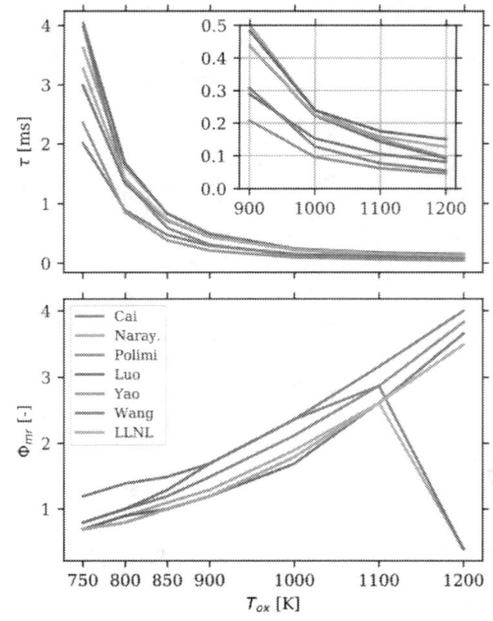

Figure 1 – Homogeneous ignition delay time τ (top) and corresponding equivalence ratio Φ (bottom). The inset shows a detailed view on the ignition delay times at high oxidiser temperatures.

generally a longer ignition delay than the Yao and Cai mechanisms. The Polimi mechanism ranges mostly between these two groups. The differences in ignition delay between the mechanisms range from $2\,\text{ms}$ at $750\,\text{K}$ to approximately $0.1\,\text{ms}$ at $1200\,\text{K}$ oxidiser temperature. With respect to the most reactive equivalence ratio, the Cai, Wang and Polimi mechanisms ignite at richer mixtures compared to the LLNL, Luo, Narayanaswamy and Yao mechanisms. The outliers at $T_{\text{ox}} = 1200\,\text{K}$ for the Wang and Narayanaswamy are due to a shorter ignition delay for the lean single-stage ignition compared to the two-stage ignition at rich conditions.

While the evaluation of the mechanisms in terms of homogeneous ignition characteristics gives a first indication of their performance in a more realistic configuration, it cannot account for effects due to mixing. To investigate the effect of mixing, the results from the unsteady flamelet calculations are presented in the following. Figure 2 shows the first (τ_1) and second (τ) stage ignition delay time as a function of scalar dissipation rate for the 900 K-oxidiser temperature conditions. The detailed LLNL mechanism is not considered for the analysis using the unsteady flamelet model, due to the high computational cost associated with these calculations. Neither is the Wang mechanism for which solutions could only be obtained at a limited number of scalar dissipation rates.

The results for the second-stage ignition are consistent with the homogeneous case, and the ignition delay time shows only a weak dependency on the scalar dissipation rate at lower values. The scalar dissipation rate at which no high temperature flame can be initiated (χ_{ign}) varies significantly between the mechanisms, where the highest value was obtained for the Yao mechanism (40 - $50\,\text{s}^{-1}$). The results for the Luo mechanism indicate a value for χ_{ign} between 10 and $20\,\text{s}^{-1}$, whereas the other mechanisms lie in between. While the results allow for a first estimate of this value a more rigorous estimation will be left to a future study. With respect to the first-stage ignition the results indicate a longer ignition delay for the Luo and Narayanaswamy mechanism, where

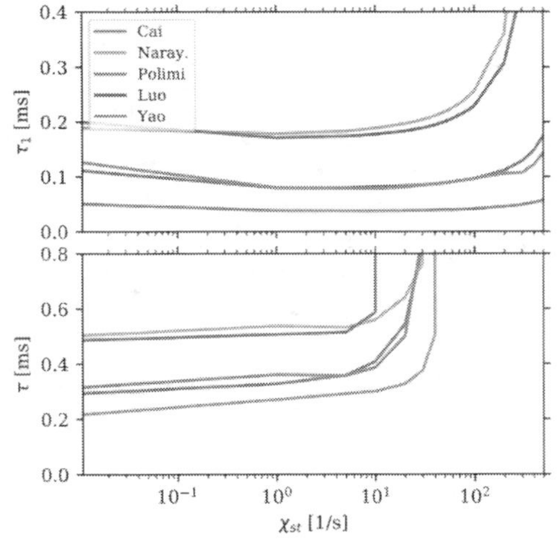

Figure 2 – First (top) and second (bottom) stage ignition delay time as a function of scalar dissipation rate.

the Polimi mechanism predicts the shortest first-stage ignition delay times. Here the first-stage ignition is defined by the time when 20 % of the maximum alkylperoxy radical mass fraction ($Y_{C_{12}H_{25}O_2}$) is reached. It is interesting to note that the first-stage ignition delay times slightly decrease at first with increasing scalar dissipation rate and only increase at very high values, until not even first-stage ignition can occur any more. This trend was not observed for the second-stage ignition delay time, which increased slightly already for low scalar dissipation rates compared to the homogeneous case.

To investigate the effect of mixing in more detail, the heat release rate in Φ–t space is shown in Fig. 3 for three scalar dissipation rates. Only results for the Yao mechanism are shown for brevity. For the homogeneous case ($\chi = 0\,\mathrm{s}^{-1}$) the first and second stage ignition delay times, indicated by the dashed and solid white lines, respectively, are strongly dependent on the equivalence ratio. The first and second stage ignition merge for $\Phi > 2$ and no significant heat release is observed for $\Phi > 3$. However, already at a very low scalar dissipation rate of $1\,\mathrm{s}^{-1}$, the heat release characteristics changes considerably due to transport in composition space. The heat release profile widens and the ignition delay times at richer mixtures reduce significantly. A further increase in scalar dissipation rate yields a flat profile for the second stage ignition delay time with respect to Φ. Also, the low and high temperature ignition remain clearly separated up to $\Phi = 5$. It should be noted that while results for the other mechanisms similarly show a strong effect of increased scalar dissipation rate (i.e. mixing) significant differences are observed in terms of ignition location in Φ-space and response to scalar dissipation rate (as indicated by the results shown in Figs. 1 and 2).

To illustrate the differences between the mechanisms in their response to mixing, the maximum temperature value alongside the equivalence ratio during ignition is shown in Fig. 4 for the homogeneous ($\chi = 0\,\mathrm{s}^{-1}$) and the $\chi_{st} = 10\,\mathrm{s}^{-1}$ cases. The temperature trace for the homogeneous case is in line with the most reactive equivalence ratio shown in Fig. 1. Also the ignition location in Φ-space for both cases using the Yao mechanism are consistent with the results shown in Fig. 3. The temperature evolution in time is different for each mechanism, and thus two reference times are marked in the plots by circles ($t = 0.3\,\mathrm{ms}$) and squares ($t = 0.5\,\mathrm{ms}$). It is interesting to note that the path

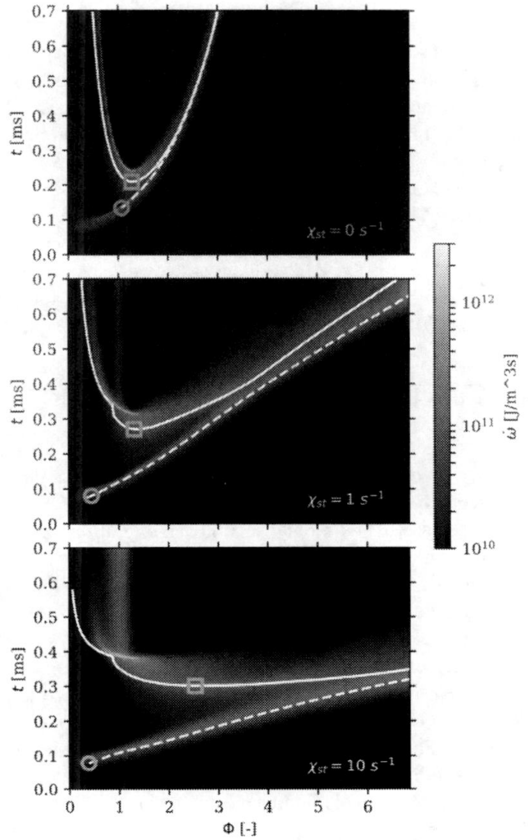

Figure 3 – Heat release rate in Φ–t space for the Yao mechanism and an oxidiser temperature of 900 K. The solid white line marks the maximum consumption rate of H_2O_2 and the dashed white line the maximum $C_{12}H_{25}O_2$ mass fraction. The cyan symbols denote the first (circles) and second (squares) stage ignition.

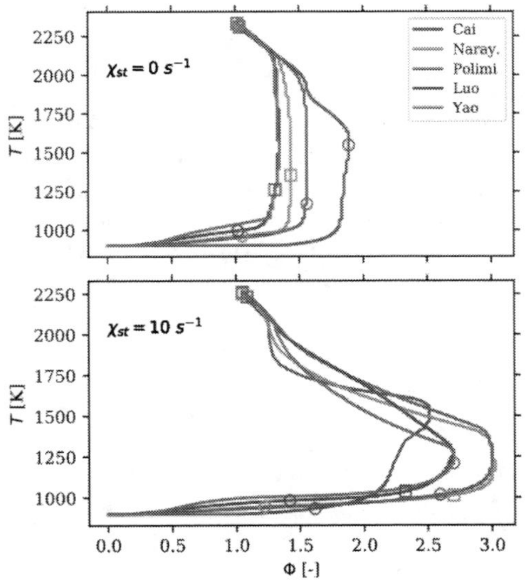

Figure 4 – Maximum temperature value alongside the equivalence ration during ignition for the homogeneous ($\chi = 0\,\mathrm{s}^{-1}$; top) and the $\chi_{st} = 10\,\mathrm{s}^{-1}$ (bottom). The circles and squares denote two reference times, $t = 0.3$ and $0.5\,\mathrm{ms}$, respectively.

of the maximum temperature in Φ-space extends to much higher equivalence ratios for the $\chi_{st} = 10\,s^{-1}$ compared to the homogeneous case for all mechanisms. However, the response to mixing is different for the various mechanisms. Most notably the maximum temperature trace for the Cai mechanism does not extend as far to rich mixtures as the other mechanisms for the $\chi_{st} = 10\,s^{-1}$ case, which is contrary to the homogeneous ignition behaviour where ignition took place in richer conditions compared to the other mechanisms. The differences in the response to mixing might have important implications when real world configurations, such as the ECN Spray A case, are considered, in particular with respect to the flame lift-off height. However, a more detailed investigation and discussion is beyond the scope of the present paper and left for a more comprehensive study in the future.

4 Conclusions

In the present study, results from unsteady flamelet calculations are used to compare chemical mechanisms for *n*-dodecane at thermochemical conditions corresponding to the ECN Spray A reference case. The first and second stage ignition delay times show only a weak dependency on the scalar dissipation rate, up to a certain value for which ignition cannot occur any more (χ_{ign}). While this behaviour is consistent for all mechanisms, the value of χ_{ign} varies considerably between mechanisms.

A more detailed investigation of the effect of mixing showed a significant broadening of the heat release profile in Φ-space. Consequently, the ignition location shifts also to much richer mixtures, already for low scalar dissipation rates. While all mechanisms similarly predict this shift, the response to an increased scalar dissipation rate is significantly different. The investigations show that further analysis and characterisation of the chemical mechanisms is needed in order to interpret the results from multi-dimensional spray combustion and engine simulations using these mechanisms.

5 Acknowledgment

This work was supported by the Australian Research Council (ARC). The research benefited from computational resources provided through the National Computational Merit Allocation Scheme, supported by the Australian Government. The computational facilities supporting this project included the Australian NCI National Facility, the partner share of the NCI facility provided by Intersect Australia Pty Ltd., the Peak Computing Facility of the Victorian Life Sciences Computation Initiative (VLSCI), Pawsey awarded through the National Computational Merit Allocation Scheme with funding from the Australian Government and the Government of Western Australia.

References

1. H. J. Curran, P. Gaffuri, W. J. Pitz, C. K. Westbrook, *Combustion and Flame* **129**, 253–280 (2002).
2. D. Siebers, B. Higgins, "Flame Lift-Off on Direct-Injection Diesel Sprays Under Quiescent Conditions", SAE Technical Paper 2001-01-0530 (SAE International, Warrendale, PA, 2001).
3. L. M. Pickett, D. L. Siebers, *Journal of Engineering for Gas Turbines and Power* **127**, 187–196 (2005).
4. Y. Pei, E. R. Hawkes, S. Kook, *Flow, Turbulence and Combustion* **91**, 249–280 (2013).
5. S. Bhattacharjee, D. C. Haworth, *Combustion and Flame* **160**, 2083–2102 (2013).
6. *Engine Combustion Network* (2017; `http://www.ca.sandia.gov/ecn/`).
7. J. T. Farrell *et al.*, "Development of an Experimental Database and Kinetic Models for Surrogate Diesel Fuels", SAE Technical Paper 2007-01-0201 (SAE International, Warrendale, PA, 2007).
8. S. Honnet, K. Seshadri, U. Niemann, N. Peters, *Proceedings of the Combustion Institute* **32**, 485–492 (2009).
9. S. Dooley *et al.*, *Combustion and Flame* **157**, 2333–2339 (2010).
10. X. You, F. N. Egolfopoulos, H. Wang, *Proceedings of the Combustion Institute* **32**, 403–410 (2009).
11. Z. Luo *et al.*, *Combustion Theory and Modelling* **18**, 187–203 (2014).
12. K. Narayanaswamy, P. Pepiot, H. Pitsch, *Combustion and Flame* **161**, 866–884 (2014).
13. A. Frassoldati *et al.*, *Combustion and Flame* **162**, 3991–4007 (2015).
14. L. Cai, L. C. Kröger, H. Pitsch, presented at the, 15th International Conference on Numerical Combustion.
15. H. Wang, Y. Ra, M. Jia, R. D. Reitz, *Fuel* **136**, 25–36 (2014).
16. T. Yao *et al.*, *Fuel* **191**, 339–349 (2017).
17. E. R. Hawkes, "ECN 4 – Topic 4 Chemistry Effects on Ignition", Kyoto, Japan, 2015.
18. E. R. Hawkes, B. Somers, "ECN 5 Topic 4/5 Ignition and Flame Structure - Model Results & Analysis (in Spray A)", Detroit, MI, USA, 2017.
19. J. M. Desantes, J. J. López, J. M. García-Oliver, D. López-Pintor, *Combustion and Flame* **182**, 76–89 (2017).
20. M. Ghaderi Masouleh, A. Wehrfritz, O. Kaario, H. Kahila, V. Vuorinen, *Fuel* **191**, 62–76 (2017).
21. Y. Pei, E. R. Hawkes, S. Kook, G. M. Goldin, T. Lu, *Combustion and Flame* **162**, 2006–2019 (2015).
22. Y. Pei *et al.*, *Combustion and Flame* **162**, 4442–4455 (2015).
23. A. Wehrfritz, O. Kaario, V. Vuorinen, B. Somers, *Combustion and Flame* **167**, 113–131 (2016).
24. D. K. Dalakoti, E. R. Hawkes, M. S. Day, J. B. Bell, presented at the, 8th European Combustion Meeting.
25. N. Peters, *Turbulent Combustion* (Cambridge University Press, 2000).
26. R. N. Dahms, G. A. Paczko, S. A. Skeen, L. M. Pickett, *Proceedings of the Combustion Institute* **36**, 2615–2623 (2017).
27. H. Pitsch, N. Peters, *Combustion and Flame* **114**, 26–40 (1–2 1998).
28. S. M. Sarathy *et al.*, *Combustion and Flame* **158**, 2338–2357 (2011).
29. H. Pitsch, *FlameMaster, A C++ Computer Program for 0D Combustion and 1D Laminar Flame Calculations*, version 3.3.10, 2017.
30. *Cantera - An Object-Oriented Software Toolkit for Chemical Kinetics, Thermodynamics, and Transport Processes*, version 2.3, 2017.

11th Asia-Pacific Conference on Combustion,
The University of Sydney, NSW Australia
10th -14th December 2017.

The Effect of Dimethyl Ether Addition on Sooting Limits in Non-premixed Counterflow Flames

Zepeng Li[1], Yu Wang[2], Suk Ho Chung[1], William L. Roberts[1],

[1] Clean Combustion Research Center, King Abdullah University of Science and Technology
Thuwal, Saudi Arabia

[2] Hubei Key Laboratory of Advanced Technology for Automotive Components, School of Automotive Engineering, Wuhan University of Technology, Wuhan, China

Abstract

Regarded as one of the most promising alternative fuels, dimethyl ether (DME) plays a positive role in the reduction of combustion emissions, e.g., NO_x, SO_x, and soot particles. The effect of DME on the sooting limits was investigated systematically in counterflow diffusion flames of DME and ethylene mixtures. The primary technique conducted to rate sooting tendency was laser extinction and scattering. The critical fuel mole fraction ($x_{f,cr}$) and critical oxygen mole fraction ($x_{o,cr}$) at the point which soot started to appear in the elastic light scattering signal were recorded to evaluate sooting limits. In this work, the sooting limit map of ethylene/DME mixtures was developed based on the $x_{f,cr}$ and $x_{o,cr}$. The highest sooting tendency appeared at around 6% of DME mixing ratio (β). When β>6%, the addition of DME decreases sooting propensity; to the contrary, when β< 6%, the addition of DME increases sooting propensity. Dilution conditions also affected the impact of DME addition. Additionally, sooting temperature index was introduced to rate sooting, where the limiting flame temperatures (T_f) were computed with Chemkin.

1 Introduction

On the one hand, combustion of fossil fuels and biomass provides energy for industrial or individual usage; on the other hand, the pollutant emissions from this process, e.g., NO_x, Peroxyacetyl Nitrate (PAN) and particles, are hazardous to the environment and human health [1]. As a byproduct of incomplete combustion, soot particles, or black carbon, attract increasing attentions as extensive researches exploring serious health and environmental effects from soot particles. Sound actions to inhibit the formation or enhance the consumption of small particulate are essential to reduce the emissions of polycyclic aromatic hydrocarbons (PAHs) and soot particles.

Soot emission reduction requires a comprehensive understanding of the mechanism soot formation and oxidation. Fuel structure is one of the most important factors in this process. Several quantitative sooting metrics were pronounced to evaluate sooting tendency among a wide range of hydrocarbon fuels, e.g., TSI, YSI. The threshold soot index (TSI) introduced by Calcote and Manos [2] and the yield soot index (YSI) introduced by McEnally and Pfefferle [3] are mainly based on coflow flames or smoke points of pure fuels and air. Current research in combustion field requires more information on the effect of dilution and fuel doping on the sooting tendency in diffusion flames. Furthermore, separation of soot formation and soot oxidation will also facilitate the quantification of the sooting tendency of various fuels. The sooting limit map defined as the critical oxygen mole fraction related to the fuel mole fraction at soot inception point is detailed discussed in [4, 5]. A sooting temperature index and a sooting sensitivity index were introduced to quantify sooting propensity and its dependence on strain rate as well [5]. These indexes point to a better method to evaluate sooting tendency in counterflow flames via light extinction and scattering technique.

For the sake of the reduction of particles and NOx emissions, alternative fuels and fuel additives came into the perspective of the public and were practically utilized during the last several decades. In advantage of the low auto-ignition temperature and low particulate emissions, dimethyl ether (DME, CH_3OCH_3) is considered as one of the most promising fuel additives or alternative fuels [6]. Previous studies have evaluated the synergistic effect of dimethyl ether as a substituent on soot formation in counterflow flames; they found that small amount of DME addition might enhance soot formation in non-premixed flames even though DME is generally regarded as a clean fuel [7, 8]. In this regards, this work looked into the effect of DME addition on sooting limits in non-premixed counterflow flames with a set of DME mixing ratios (0-50%) and nitrogen dilution ratios in the fuel stream (Xf). The synergistic effect of DME was highlighted, and the sooting limit map was drawn to provide complementary information.

2 Methodology

2.1 Experimental apparatus

The experimental setup consisted of a counterflow burner with corresponding flow control system, and a laser scattering/extinction system. The counterflow burner had a pair of diverging and converging nozzles with the exit diameter of 10 mm. The separation distance between these two nozzles was 8 mm. Each nozzle had a concentric slit with the width of 2 mm to provide nitrogen flow as the shield, such that the secondary flame result from ambient air was avoided. The fuel flow coming from the lower nozzle was a mixture of nitrogen, ethylene and dimethyl ether (DME) with various dilution ratios

Corresponding author: Zepeng Li
E-mail address: Zepeng.li@kaust.edu.sa

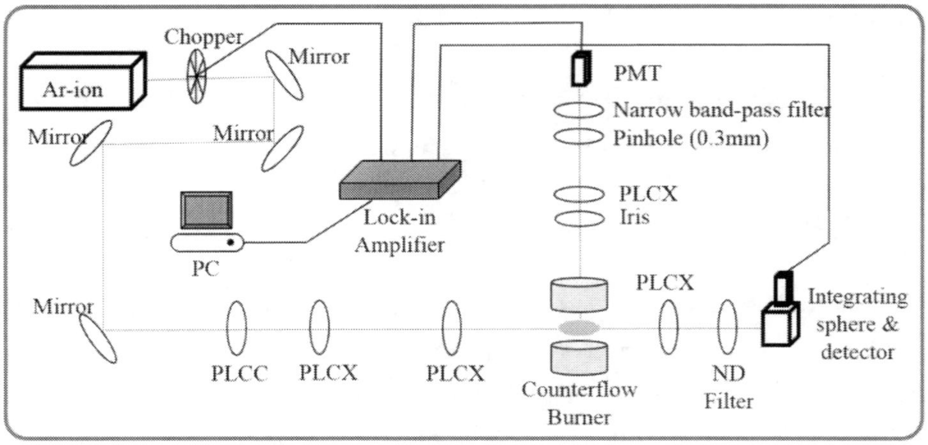

Figure 1: Schematics of the laser scattering and extinction system. PLCC: planoconcave lens; PLCX: plano-convex lens; ND filter: neutral density filter; PMT: photo-multiplier tube

and DME mixing ratios. The dilution ratio was represented by the total mole fraction of fuels, Xf. The DME mixing ratio (β) was defined as the ratio between the volumetric flow rate of DME and volumetric flow rate of fuel mixture (ethylene and DME), that is:

$$\beta \equiv \frac{[CH_3OCH_3]}{[CH_3OCH_3]+[C_2H_4]} \quad (1)$$

The oxidizer flow coming from the upper nozzle was a mixture of nitrogen and oxygen, the mole fraction ratio of oxygen was noted as Xo. To reduce temperature variation from buoyancy, the upper nozzle was cooled by water. Both of the nozzles had an exit velocity of 20 cm/s. The flow rates of the gases were controlled by thermal-based mass flow controllers (MFC, MKS Instruments), which was calibrated by the laboratory primary gas flow calibrator (Definer 220). The whole burner system was supported by an automated two-dimensional translation stage controlled by a Labview program to acquire sufficiently precise burner position. The increments of the data recorded were 0.1 mm.

The schematic of the light extinction and scattering technique is shown in Fig. 1, with more details can be found in [5]. An Argon-ion laser (514.5 nm) operated at the power of 0.4 W as the light source. A lock-in amplifier coupled with a mechanical chopper at the frequency of 1310 Hz was utilized to improve the signal-to-noise ratio. Three kinds of spherical lenses (PLCC, PLCX, PLCX) in the upstream of the flame were used to acquire a tiny laser beam with uniform energy distribution. The diameter of the laser beam at the center of the flame was 0.097 mm. To record the extinction signal, a photodiode (Thorlabs, DET100A) was mounted on an integrating sphere. A photo-multiplier tube (Hamamatsu H7827, PMT) was used to detect the scattering signal at 90° of the incident light. The PMT was calibrated by a set of neutral density filters, and the variance was better than 0.986. Compared with LII measurement, the scattering intensity tends to rise a little earlier [4]. It is because in the LII measurement, we assume the soot is a black body while soot is kind of transparent at the sooting limit cases. Also, soot is very small at its inception and thus it cools quickly, which makes it hard to capture the signal in the LII.

To avoid the spurious laser reflection from other surfaces affect the scattering signal, three gases (nitrogen, propane, and helium) with known scattering cross-sections were employed to the calibrate the scattering signal. Note that helium has very low cross-section ratio relative to other gases (≈0), it was regarded as the background scattering signal. However, the background value taken in this method was over estimated to some extent, it was inevitable to overrate this ratio. After subtracting the scattered intensity of helium, the ratio of the scattered intensities of propane and nitrogen was around 13.66, which was consistent with [9].

Through the comparison of the Rayleigh scattering of gas molecules and the Mie scattering of soot particles, we can determine the sooting limits ($x_{f,cr}$ and $x_{o,cr}$). The method of the determination of critical sooting limits applied in this work is fully described in [4]. Fig. 2 shows the scattering intensity when the fuel mole fraction (Xf) equals to 0.85. Since the increment of mole fraction of oxygen (Xo) is 0.01, the average value between the non-sooting case (Xo=0.17) and sooting case (Xo=0.18) is calculated as the sooting limit, or critical oxygen mole fraction ($x_{o,cr}$). This increment leads to the accuracy of ±0.005.

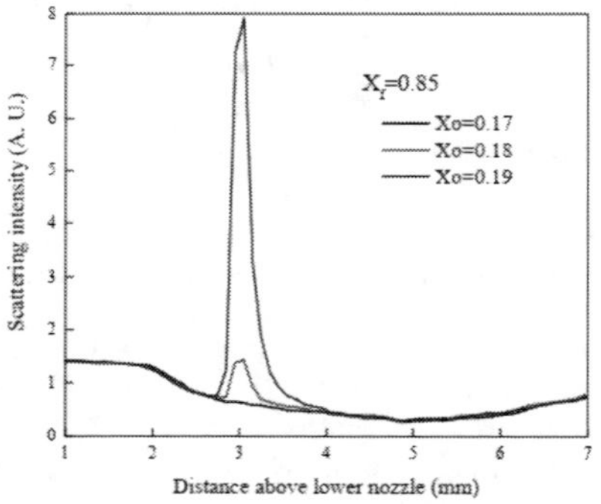

Figure 2: Determination of sooting limit

2.2 Numerical simulation

Considering that the flame temperature profiles are important in sooting behaviors [10], flame temperatures were computed using the Opposed-flow Flame package of Chemkin Pro [11]. The inlet temperature was set at the lab temperature (298K), and the pressure is 1 atm. Aramco Mech 1.3 [12] was employed as the reaction kinetic model. Because the flames studied in this research were close to the sooting limit conditions, a soot model was not included as the effect of soot on the temperature profiles was negligible.

3 Results and Discussions

Figure 3 shows the sooting limit map of ethylene/DME with different DME mixing ratios (β). The whole Xo-Xf plane is divided into two regions by the sooting limit curves. The upper-right region is the sooting region, while the lower-left region is the non-sooting region. The area of the sooting region can represent the sooting tendency, which means that a larger area of a sooting region refers to a higher sooting tendency. Among the cases at β=0, 0.1, 0.2, 0.3, 0.4, 0.5, it is evident that as DME mixing ratio increases, the sooting limit curves moves towards a upper-right direction, which leads to the reduction of the sooting region area. Hence, it infers that the addition of DME into ethylene inhibits soot formation in counterflow diffusion flames. If we look at the details when β≤0.1, the sooting limit behaviors are different. Based on the stoichiometric mixture fraction, the sooting limit curves is separated by Soot Formation (SF) flame and Soot Formation/Oxidation (SFO) flame. In SF part, the curve at β=0.05 gives the maximum area of the sooting region; while in SFO part, three curves at β=0, 0.02, 0.05 are overlapped. A previous study [4] demonstrated that the sooting limits depend critically on fuel structure and subsequently on the flame temperature in SF flames, while the sooting limits only depend on flame temperature in SFO flames [4]. It means that both thermal and chemical effect contributes to the sooting limits in SF flame, while only thermal effect contributes to the sooting limits in SFO flame. Since a slight change of β leads to a negligible change in flame temperature, the variation of thermal effect in this case (β=0, 0.02, 0.05) is considerably small. Therefore, in SFO flames, the sooting limit curves of β=0, 2%, 5% are overlapped. Meanwhile, in SF flames, there is an increasing trend of sooting when comparing the sooting limit curves of β=0, 2%, 5%. Based on this comparison, it is reasonable to infer that the chemical effect plays a key role in enhancing soot formation at soot onset conditions when adding small amounts of DME into ethylene flames.

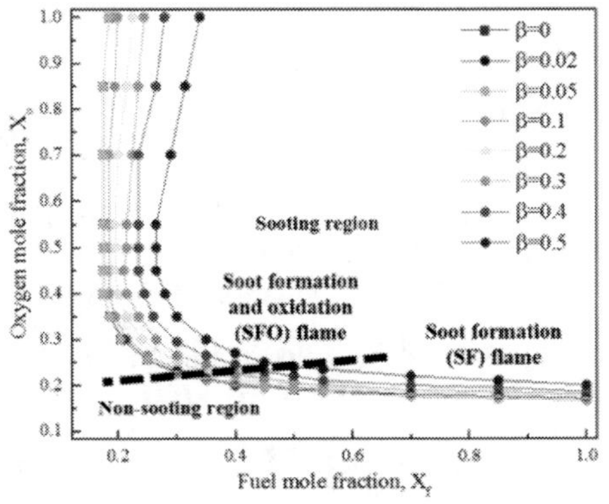

Figure 3: Sooting limit map of ethylene/DME with different DME mixing ratios (β)

Figure 4 adds more evidence to the statement that the maximum sooting propensity occurs at β≈5%. The critical oxygen mole fraction ($x_{o,cr}$) means the minimum amount of oxygen needed to produce soot at a fixed fuel mole fraction (Xf). An increase of critical oxygen mole fraction with larger DME mixing ratio (β) illustrates a lower sooting tendency as β increases. Among the cases at β=0, 0.05, 0.1, 0.2, 0.3, 0.4, 0.5, $x_{o,cr}$ is the minimum at β=0.05. Additionally, the sooting limits are sensitive to dilution, and this propensity magnifies as Xf decreases. When Xf changes from 1 to 0.85, the variation of Xo is much smaller than Xf changes from 0.45 to 0.35.

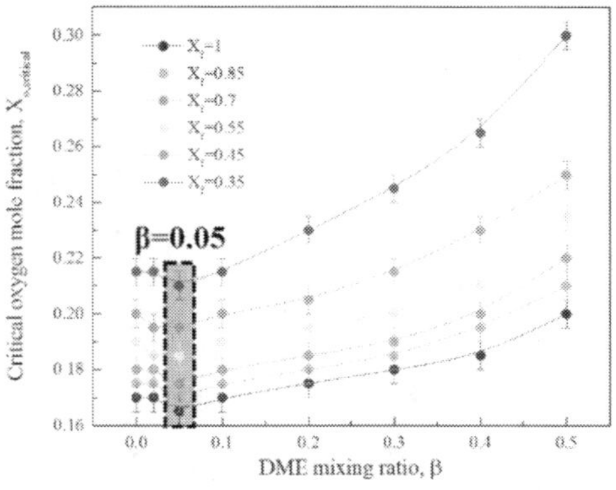

Figure 4: Critical oxygen mole fraction as a function of DME mixing ratios (β)

To acquire the critical β at the highest sooting propensity, the profiles of scattering signal at β=0~0.1 with an increment of 0.01 were recorded. A normalized maximum scattering intensity was selected as a reference to order the sooting tendency. The maximum scattering intensity is the net peak value of the scattering signal curve at which Xo that soot starts to be produced. In Fig. 2, there is no soot formed at Xo=0.17, and soot appears at Xo=0.18. The maximum scattering intensity is the maximum value of the curve at Xo=0.18 subtracting the curve at Xo=0.17. In this method, the normalized maximum scattering intensity adds more details to the sooting limit, and a higher intensity infers a higher sooting trend. Three repeated experiments have the same peak at β=0.06 (Fig. 5). It demonstrates that highest sooting tendency appears at around 6% of DME addition. Adding DME to non-premixed ethylene flames enhances soot formation when β <0.06, while further addition inhibits soot formation. Yoon etc. [7] explained the synergistic effect of mixing DME with ethylene on PAHs and soot formation. An increase of methyl radicals decomposed from DME in the low-temperature region leads to a higher concentration of C_2H_2, which facilitates the formation of PAHs.

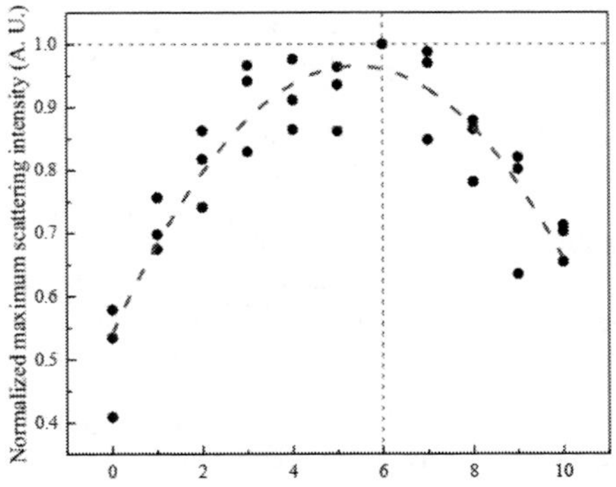

Figure 5: Maximum scattering intensity as a function of DME mixing ratios (β)

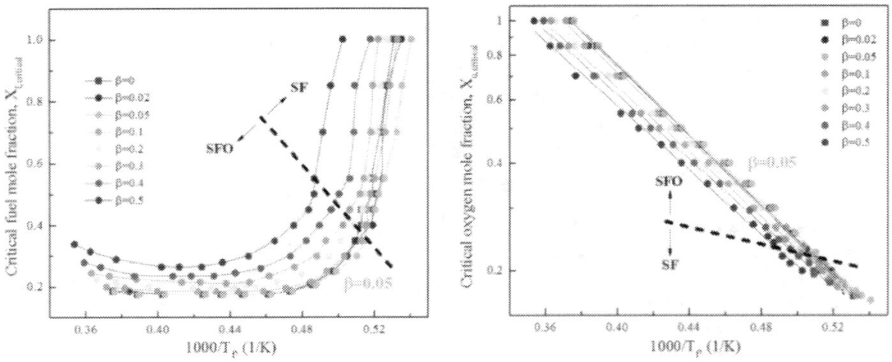

Figure 6: Sooting temperature index. Critical fuel mole fraction and oxygen mole fraction with respect to flame temperature (T_f) at different DME mixing ratios (β)

For a further understanding of the thermal effect on the sooting limit, the critical fuel mole fraction and oxygen mole fraction as a function of maximum flame temperature (T_f) with different DME mixing ratios (β) are shown in Fig. 6. A higher sooting propensity is observed at β=0.05 compared with other experimental cases. At Xo=1, the value of $1000/T_f$ ranges from 0.354 to 0.376 for all βs, the variation is 0.0222; at Xf=1, this value ranges from 0.505 to 0.541, the variation is 0.0382. It indicates that the sensitivity of the sooting conditions to the fuel composition is reduced in SFO flames due to the oxidation process of soot. Furthermore, in SF flames, there exists the threshold temperature (or minimum limiting temperature) at soot onset. If the flame temperature is below this threshold, the flame will be non-sooting. A higher limiting temperature leads to a lower sooting propensity. In Fig. 6, the flame at β=0.5 has the maximum limiting temperature, while the flame at β=00.05 has the minimum limiting temperature. In SFO flames, Xf remains reasonably insensitive to the wide range of flame temperatures.

4 Conclusions

A sooting limit map of ethylene/DME flames with DME mixing ratio ranging from 0-50% was drawn using laser scattering and extinction technique in counterflow diffusion flames. The results demonstrate the synergistic effect of DME addition in non-premixed ethylene flames. The highest sooting tendency appears at around 6% of DME mixing ratio (β). When β is smaller than 6%, the addition of DME increases sooting tendency; when β is larger than 6%, the addition of DME reduces sooting tendency. A stronger impact of DME addition is found in higher N_2 dilution ratios in the fuel nozzle, which means that at lower Xf, more significant variation in the sooting limits (critical oxygen mole fraction) appears as DME added. Therefore, the sooting limits are sensitive to dilution, and this propensity magnifies as Xf decreases. However, the behaviors of the DME effect on sooting limits in SF flame and SFO flame are slightly different. Through comparison of this difference, the chemical effect plays a key role in enhancing soot formation at soot onset conditions when adding small amounts of DME into ethylene flames.

5 Acknowledgment

We acknowledge the support from Clean Combustion Research Center, King Abdullah University of Science and Technology (KAUST), Saudi Arabia.

References

[1] I.M. Kennedy, Proc. Combust. Inst. 31 (2007) 2757–2770.
[2] H.F. Calcote, D.M. Manos, Combust. Flame 49 (1983) 289–304.
[3] C.S. McEnally, L.D. Pfefferle, Combust. Flame 148 (2007) 210–222.
[4] P.H. Joo, Y. Wang, A. Raj, S.H. Chung, Proc. Combust. Inst. 34 (2013) 1803–1809.
[5] Y. Wang, S.H. Chung. Combust. Flame 161.5 (2014): 1224-1234.
[6] S. Maji, S. Ahmed, W. A. Siddiqui, S. Aggarwal, A. Kumar, American Journal of Environmental Protection, 3.2 (2015) 48-52.
[7] Yoon, S.S., D.H. Anh, S.H. Chung, Combust. Flame 154.3 (2008): 368-377.
[8] M. Sirignano, M. Salamanca, A. D'Anna, Fuel 126 (2014): 256-262.
[9] F. Fuest, R.S. Barlow, J.Y. Chen, A. Dreizler, Combust. Flame 159 (2012) 2533–2562.
[10] Ö.L. Gülder, D.R. Snelling, Combust. Flame 92 (1993) 115–124.
[11] Chemkin Pro: a chemical kinetics package for the analysis of gas-phase chemical kinetics, Reaction Desig. Release 15112 (2011).
[12] W. K. Metcalfe, S. M. Burke, S. S. Ahmed, H. J. Curran, International Journal of Chemical Kinetics, 45.10 (2013) 638-675.
[13] D.X. Du, R.L. Axelbaum, C.K. Law, Proc. Combust. Inst. 22 (1989) 387–394.

11th Asia-Pacific Conference on Combustion,
The University of Sydney, NSW Australia
10th -14th December 2017.

Mode-Switching Behaviour of Preheated and Diluted Flames in a Stagnation Burner

Bin Jiang, Robert. L. Gordon, Mohsen. Talei

Department of Mechanical Engineering

University of Melbourne, Parkville, 3010, Australia

Abstract

The combustion mode of preheated and diluted methane/air flames is studied with a one-dimensional (1D) stagnation flame model. The species transport budget is used to determine the combustion mode between premixed flame propagation and autoignition. The autoignition delay time decreases with increasing the dilution level, pressure, oxygen concentration, or adding minor species. The combustion mode can then change from premixed flame propagation to autoignition. This behaviour agrees with previous studies on autoignition. Increasing the inlet velocity has been found to switch the combustion mode from premixed flame propagation to autoignition and back to premixed flame propagation. This double mode switch may be directly related to a competition between autoignition delay time and flame speed of premixed flames as well as features of diluted flames such as the weakened reaction zone.

1 Introduction

Environmental concerns and emissions regulation have motivated researchers to develop novel combustion techniques for achieving ultra-low emissions. Recently, Moderate or Intense Low oxygen Dilution (MILD) combustion has emerged as a promising solution to achieve ultra-low nitrogen oxides (NOx) and carbon monoxide (CO) as well as to maintain high thermal efficiency, fuel flexibility and low noise emission for gas turbines (GTs) [1]. Understanding MILD combustion has become an area of active research, such as the definition of MILD regime [2] and the distinction of combustion regimes in MILD-like combustion [3]. MILD combustion may be possible to be achieved in gas turbines by diluting fresh reactants with hot combustion products. This dilution can be established with internal recirculation or staged combustion. The FLameless OXidation COMbustion (FLOXCOM)[4], LEan Azimuthal Flame (LEAF) combustor [5] and Colorless Distributed Combustion (CDC) [6] use internal recirculation. FLameless OXidation (FLOX) [7] and the Alstom/GE GT24/26 [8] use staged combustion, although [8] does not operate in true MILD condition.

The combustion mode is important for the modelling of the combustion process. Combustion models for premixed flame propagation and autoignition have traditionally been developed for a single combustion mode and may not be suitable when the combustion system can feature different modes.

Mode switching between premixed flame propagation and autoignition can affect flame stabilization, especially in terms of combustor stability and operating flexibility. For example, a premixed propagating flame will move downstream of the injection point, while an autoignition flame moves upstream of the injection point in a secondary stage of a staged GT combustor at elevated pressures [9].

The presence of a wall can introduce effects related to strain rate, heat loss and the boundary layer [10], which may affect the combustion mode. In MILD or staged combustion, the internal recirculation can stabilize the flame close to a wall. This is especially true for the jet in cross flow configuration in a staged combustor, which can induce flame-wall interaction under different momentum ratios of the jet and cross flow [11]. Furthermore, the reaction zone in MILD combustion is generally distributed, such that compared to traditional GT combustors the contact area of the flame and combustor wall may increase. Therefore, it is important to understand wall effects on preheated and diluted flames. The combustion mode of a stagnation flame is studied in this paper under different conditions relevant to a wall-bounded MILD GT combustor, drawing on the dilution method from Sidey et al. [12] and the transport budget from Gordon et al. [13].

2 Numerical Methods

The stagnation-flame reactor of CHEMKIN PRO [14] was used to simulate the effect of a wall on a steady, preheated and diluted flame. The computational domain was 2 cm long from the burner exit to the stagnation plate. The wall was considered to be inert with a constant temperature (300K). In order to ensure an accurate convergence and grid independent solution, the adaptive meshing criteria for both the largest relative gradient and curvature were 0.02.

Methane (CH_4) was used as the fuel with the gas-phase oxidation described by GRI 3.0 [15]. All simulations were performed with mixture-averaged transport properties, and with the Soret effect considered.

Dilution was achieved by mixing stoichiometric CH_4/air at 300K and combustion products at equilibrium temperature. The degree of mixing is defined by the "dilution level" ζ, which is the ratio of hot product mass flow rate to the fresh reactant mixture. The mixture temperature and oxygen (O_2) mole fraction at different dilution levels are shown in Fig. 1. The detailed dilution method can be found in our previous work [12]. To verify the numerical model, simulation results from Sidey et al. [12] were reproduced, showing good agreement [16].

Corresponding author.
E-mail address:jiangb1@student.unimelb.edu.au

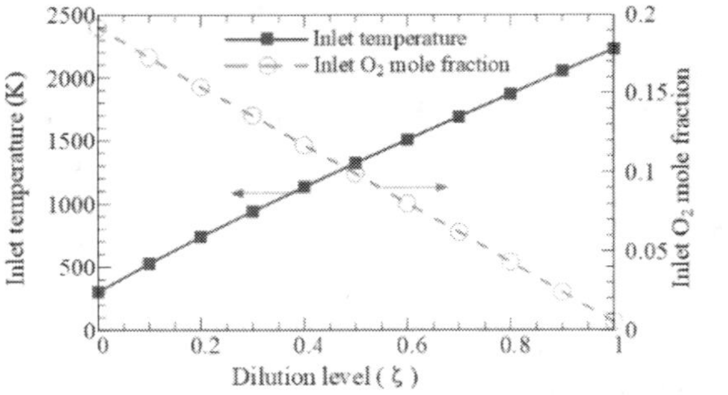

Figure 1: Mixture temperature and O_2 molar fraction profiles under different dilution levels.

3 Indicator for Combustion Modes

The species transport budget is used in this paper to identify the combustion mode [13]. In an autoignition flame, the mass diffusion in the reaction zone is relatively smaller than the mass convection and reaction. However, the amount of mass diffusion in a premixed flame is significant in both the preheat and reaction zones. The transport budget is calculated based on the steady-state species conservation equation given by:

$$-\rho u\frac{dY_k}{dx}-\frac{d}{dx}(\rho Y_k V_k)+W_k\dot{\omega}_k=0 \tag{1}$$

The first term on the left-hand side (LHS) of the equation represents the mass convection (C), which is the summation of mass convection in axial (x) and radial directions. The second term represents the molecular diffusion (D), and the third is the reaction term (R). Here, ρ is the mixture density, u is the axial velocity, Y_k is the mass fraction of a species k, W_k is the molecular weight of a species k, $\dot{\omega}_k$ is the net reaction rate of a species k, and V_k is the diffusion velocity given by:

$$V_k=-\frac{D_{km}}{X_k}\frac{dX_k}{dx}-\frac{D_k^T}{\rho Y_k}\frac{1}{T}\frac{dT}{dx} \tag{2}$$

where X_k is the mole fraction of a species k, D_{km} is the ordinary mixture-averaged diffusion coefficient, D_k^T is the thermal diffusion coefficient, T is the mixture temperature.

Equation (2) shows that there is a double derivative in the diffusion term, which magnifies any discretization and numerical noise. Hence, this term is smoothed by using the Robust LOcally regrESSion (RLOESS) method [17].

Figure 2(a) shows the OH transport budget of a conventional premixed flame at ζ=0. Significant mass diffusion in both the preheat and reaction zones are visible as marked with (1) and (2), respectively. Fig. 2(b) shows the OH transport budget of an autoignition flame at ζ=0.7 with negligible mass diffusion.

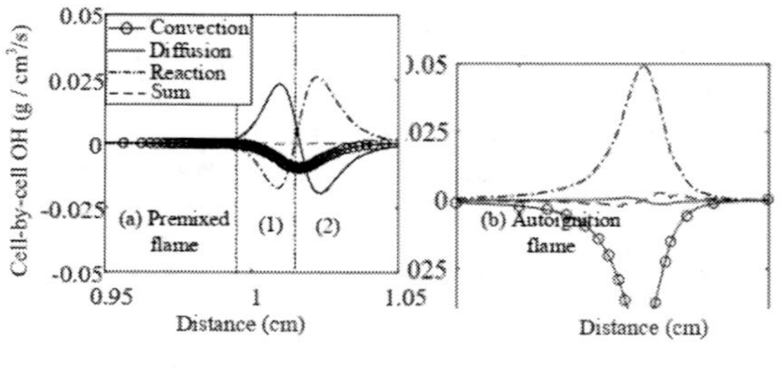

Figure 2: OH transport budget of a premixed flame at ζ=0 (a) and an autoignition flame at ζ=0.7 (b).

4 Results

4.1 Effects of Dilution

OH transport budgets of two dilution levels are presented in Fig. 3. At ζ=0.5, the flame is stabilized by the flame propagation indicated by the balance between mass diffusion and convection in the preheat zone. At ζ= 0, the mass diffusion in the preheat zone is nearly zero, indicating that the flame is stabilized as an autoignition flame. The mixture temperature increases with increasing dilution levels (Fig. 1), thus decreasing the autoignition delay time (Equation 3 [18]). Conversely, the decrease of oxygen concentration at a high dilution level should increase autoignition delay time. However, its effects on autoignition delay time appear to be smaller than the effect of mixture temperature.

$$t=1.77\times10^{-14}\exp(18693/T)[O_2]^{-1.05}[CH_4]^{0.66}[HC]^{-0.39} \tag{3}$$

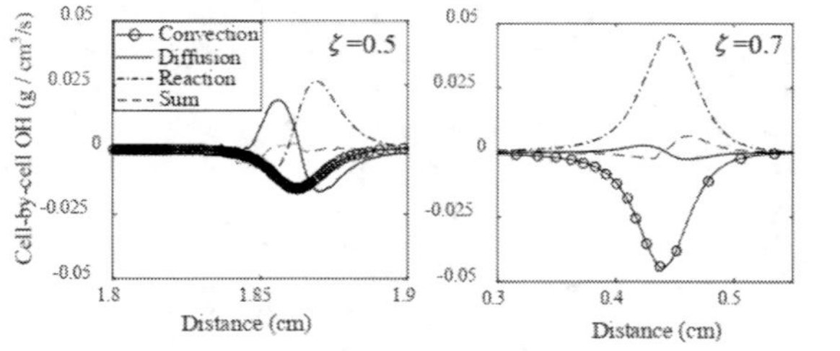

Figure 3: Dilution level (ζ) effects on the OH transport budget at an equivalence ratio (Φ)=1, inlet velocity (V_{in})=50m/s, combustor pressure (P)=1 atm.

With using the combustion products as the diluent, minor species in the products can affect the species transport budget. To study this, a simulation was conducted with only major combustion products as the diluent, i.e. CH_4, O_2, N_2, CO_2, H_2O, CO, H_2 and NO. Figure 4 shows that the combustion with all minor species in the diluent exhibits an autoignition mode, while the combustion with only major species is stabilized by premixed flame propagation. This result reflects that the minor species can reduce the autoignition delay time, which is consistent with the observations in Ref. [12].

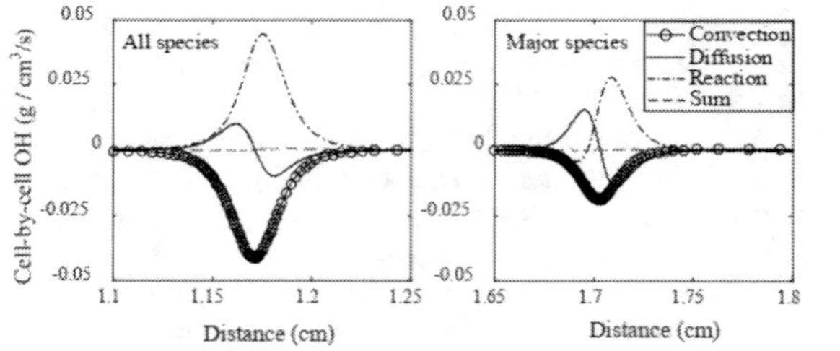

Figure 4: Effects of the minor species on OH transport budget at Φ=1, ζ=0.6, V_{in}=20m/s, P=1atm.

4.2 Effects of Combustor Pressure

Figure 5 shows that the diffusion term decreases with increasing pressure, which promotes the transition from a premixed flame to an autoignition flame. The main reason for the mode switching is that autoignition delay time decreases with increasing pressure as reported from the Ref. [19]. The flame moves upstream at elevated pressures, which agrees with Ref. [9].The peak value of the transport budget at a high pressure is nearly six times larger than at the low pressure, as the unburned flow velocity and density of upstream flame increases with pressure as reported in Ref. [9].

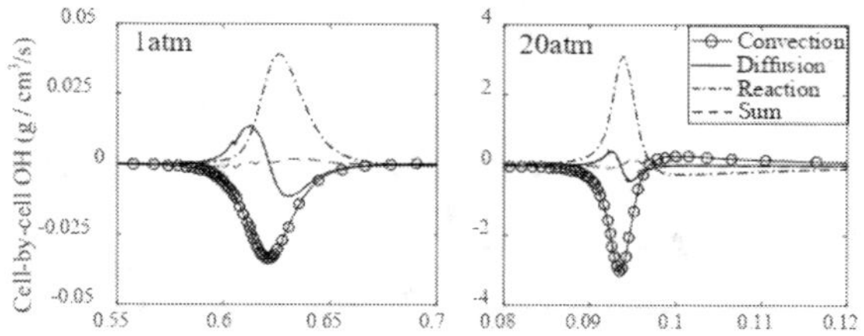

Figure 5: Effects of the pressure on OH transport budget at ζ =0.6, Φ=1, V_{in}=10m/s.

4.3 Effects of Oxygen Concentration

MILD combustion usually has oxygen concentrations from 3% to 12% [4], while in a conventional GT with lean premixed combustion, the O_2 mole fraction is around 15%. To eliminate the effects of other parameters, the diluent was set to be pure N_2, and the CH_4 mole fraction is fixed at 0.03. The compositions of reactants for the two different oxygen mole fraction cases are $CH_4/O_2/N_2$=0.03:0.06:0.91 at Φ=1 and $CH_4/O_2/N_2$= 0.03:0.2:0.77 at Φ =0.3. To clearly show the mode-switching behaviours for two different O_2 mole fractions, the mixture temperature was chosen to be 1400K. In the secondary combustor of a staged GT, the reactant can be preheated above 1000 °C, which is representative of staged combustion condition [8]. As seen in Fig. 6, the combustion mode switches from premixed flame propagation to autoignition with increasing oxygen concentration, since the autoignition delay time decreases with increasing the oxygen concentration, which agrees with the prediction of Equation (3). The peak of the reaction term at 0.06 O_2 mole fraction is two times smaller than the peak value with 0.2 O_2 mole fraction. The reaction zone is weakened by decreasing the O_2 mole fraction, which is consistent with Ref. [20].

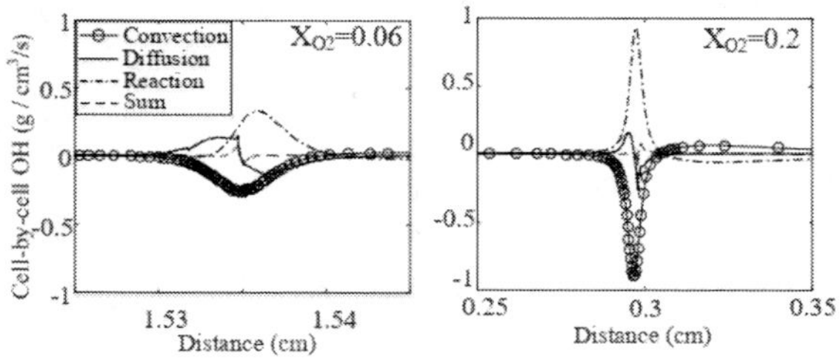

Figure 6: Effects of O_2 mole fraction on OH transport budget at mixture temperature (T_{in})=1400K, V_{in}=10m/s, P=10 atm.

4.4 Effects of Inlet Velocity

The inlet velocity is changed by varying the reactant mass flow rate. Below 20m/s inlet velocity, the mass transport in the preheat zone is balanced by mass diffusion and convection, and reaction is negligible as shown in Fig. 7. Therefore, the combustion is stabilized by premixed flame propagation. As the inlet velocity varies from 20m/s to 30m/s, the OH transport budget is balanced by mass convection and reaction with small contributions from mass diffusion, which indicates an autoignited combustion mode. At a high inlet velocity of 80m/s, the diffusion term again starts to dominate the preheat zone, and thereby the flame is understood to be stabilized by premixed flame propagation. As seen in Fig.7, unlike the other three inlet velocities, the OH transport budget at 80m/s inlet velocity has non-zero components near the wall. As the flame stabilises closer to the wall, a decreasing temperature induced by the wall heat transfer activates radical recombination reactions[10].

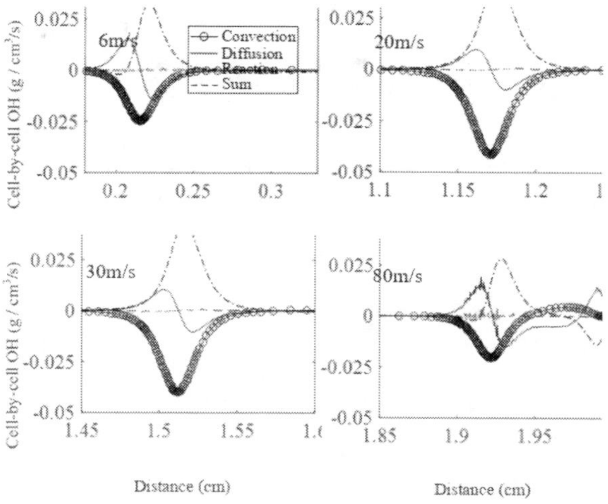

Figure 7: Effects of the inlet velocity on OH transport budget at Φ=1, ζ=0.6, P=1atm.

6 Discussion

With increasing dilution levels, combustor pressure and oxygen concentration, the combustion mode switches from premixed flame propagation to autoignition. This might be attributed to the competition between premixed flame propagation timescale and the autoignition delay time.

At ζ =0.6, increasing the inlet velocity switches the combustion mode from premixed propagation to autoignition and then back to premixed propagation. Figure 8 shows the variation of the burning velocity, defined as the gas velocity at the peak of heat release rate, as a function of inlet velocity [21]. With increasing the inlet velocity, the burning velocity increases almost linearly at low inlet velocities (below 12m/s) and reaches its maximum around 25m/s. After this point, the burning velocity decreases with increasing the inlet velocity.

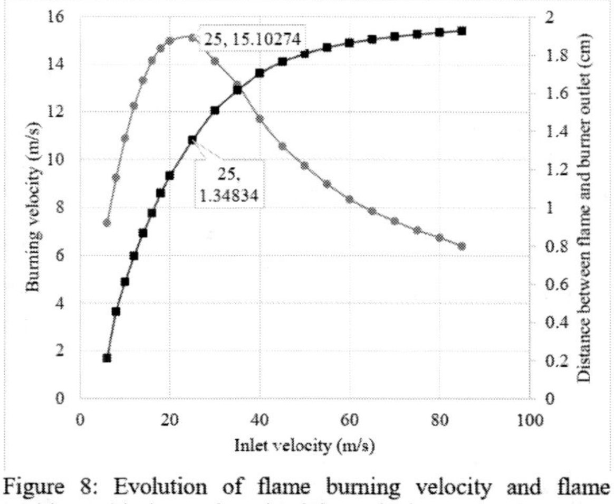

Figure 8: Evolution of flame burning velocity and flame position with increasing the inlet velocity at ζ =0.6, Φ=1, P=1atm.

Figure 8 shows that the flame front (measured from location of the peak heat release rate) is closer to the wall as the inlet velocity increases. At low inlet velocities, the burning velocity is a strong function of the inlet velocity, because the flame is still far from the cold wall. However, at some distances from the wall (around 22 unstretched flame thickness (0.3mm)), the flame appears to be affected by the presence of the wall, and the burning velocity features a decreasing trend.

Two important conclusions can be made from this observation. Firstly, the linear relationship between the burning velocity and the inlet velocity proposed in the previous study [21] does not hold for high values of inlet velocity. The second conclusion is related to the convection term in Equation 1. Figure 7 shows that the diffusion term features insignificant variation for different velocities, while the convection term is a strong function of the burning velocity as expected. Increasing the inlet velocity leads to the convection term dominating the diffusion term, promoting an autoigntion combustion mode. However, closer to the wall, the convection term becomes weaker again and therefore a premixed flame is established before extinction occurs.

Combustion-mode switching is also related to the competition of autoignition delay time and flame speed of premixed flames. Residence time increases with increasing flame-nozzle distance, while decreases with increasing inlet velocity. Figure 8 shows that increasing the inlet velocity increases the flame-nozzle distance but at decreasing rate due to increasing local strain rates. Therefore, at intermediate inlet velocities (20m/s to 30m/s), the residence time can approach the autoignition delay time.

Combustion-mode switching is an important feature of diluted flames that can have significant implications for modelling. Compared with conventional flames, the reaction zone of preheated and diluted flames is generally weaker due to low oxygen concentrations, such that stretch may have a stronger effect on the reaction zone. Therefore, linear models in the stretched flame theory may not capture the behaviour correctly.

7 Conclusion

Combustion modes of preheated and diluted flames have been investigated using the CHEMKIN stagnation burner geometry. The species transport budget is used to identify changes in the combustion mode induced by varying the inlet velocity, dilution level, diluent composition, combustor pressure and oxygen concentration. Increasing dilution level, pressure and oxygen concentration, and adding minor species in the diluent promotes autoignition, which is consistent with previous studies. With increasing the inlet velocity, the combustion mode switches from premixed flame propagation to autoignition, and can change back to premixed flame propagation. This double mode switching correlates with changes in the burning velocity, which can affect the relative magnitudes of diffusion and convection terms in the species transport budget. Changes in the strain rate may also affect the timescale competition. In the future, transient models will be used to study the temporal evolution of combustion mode, as well as the response of the combustion mode to changes in the local properties of diluted flames, such as weakened reaction zones.

8 Acknowledgment

Bin Jiang acknowledges the financial support of the China Scholarship Council. Robert Gordon receives support from the Royal Academy of Engineering as a Newton International Fellowship Alumnos.

References

[1]. A. Cavaliere, M. de Joannon, Prog. Energy Combust. Sci. 30 (4) (2004) 329-366.

[2]. M.J. Evans, P.R. Medwell, H. Wu, A. Stagni, M. Ihme, Proc. Combust. Inst. 36 (3) (2016) 4297-4304.

[3]. O. Schulz, T. Jaravel, T. Poinsot, B. Cuenot, N. Noiray, Proc. Combust. Inst. 36 (2) (2017) 1637-1644.

[4]. G. A. Rao, Y. Levy, 8th HiTACG conference, Poznan (p. 13) (2010).

[5]. J.A.M. Sidey, I.E. Helou, R.L. Gordon, G. Bourque, E. Mastorakos, 11th International ERCOFTAC Symposium on Engineering Turbulence Modelling and Measurements. (2016).

[6]. A.E.E. Khalil, A.K. Gupta, Fuel. 195, 2017, 113-122.

[7]. T. Roediger, O. Lammel, M. Aigner, C. Beck, W. Krebs, J Eng Gas Turbine Power. 135 (3) (2013) 1-12.

[8]. D. Guyot, G. Tea, C. Appel, J Eng Gas Turbine Power. 138 (5) (2016) 051503.

[9]. P. Habisreuther, F.C.C. Galeazzo, C. Prathap, N. Zarzalis, Combust. Flame 160 (12) (2013) 2770-82.

[10]. F.N. Egolfopoulos, H. Zhang, Z. Zhang, Combust. Flame 109 (1) (1997) 237-52.

[11]. R. Sullivan, B. Wilde, D.R. Noble, K. Periagaram, J.M. Seitzman, T.C. Lieuwen, Proc. Combust. Inst. 34 (2) (2013) 3203-10.

[12]. J. Sidey, E. Mastorakos, R.L. Gordon, Combust. Sci. Technol. 186 (4-5) (2014) 453-65.

[13]. R.L. Gordon, A.R. Masri, S.B. Pope, G.M. Goldin, Combust. Flame 151 (3) (2007) 495-511.

[14]. CHEMKIN-PRO 15131. Reaction Design: San Diego, 2013.

[15]. G P. Smith, D M. Golden, M. Frenklach, N W. Moriarty, B. Eiteneer, M. Goldenberg, C T. Bowman, R K. Hanson, S. Song, W C J. Gardiner, V. Lissianski & Z. Qin, Nd GRI 30. http://wwwmeberkeleyedu/gri_mech/.

[16]. B. Jiang, R. Palulli, J. E. Rivera, R. L. Gordon, and M. Talei, 20th Australasian Fluid Mechanics Conference (AFMC), Perth. (2016) 8–11.

[17]. W.S. Cleveland. J Am Stat Assoc. 74 (368) (1979) 829-836.

[18]. L.J. Spadaccini, M.B. Colket, Prog. Energy Combust. Sci. 20 (5) (1994) 431-60.

[19]. L. H. Coweli, A. H. Lefebvre. SAE Technical Paper, No. 860068. (1986).

[20]. P.R. Medwell, P.A.M. Kalt, B.B. Dally, Combust. Flame 148 (1) (2007) 48-61.

[21]. C.K. Law, C.J. Sung, Prog. Energy Combust. Sci. 26 (4) (2000) 459-505.

11th Asia-Pacific Conference on Combustion,
The University of Sydney, NSW Australia
10th -14th December 2017.

Numerical simulation of combustion of energetic components

Bohoon Kim[1] and Jai-ick Yoh[1]

[1]Department of Mechanical and Aerospace Engineering, Seoul National University

Seoul, Korea 151-744

Abstract

A detailed hydrodynamic simulation that requires an accurate reproduction of shock-induced detonation was conducted for design of an energetic component system. A detailed hydrodynamic simulation was used to validate the reactive flow model for predicting the shock propagation in an explosive train configuration and to quantify the shock sensitivity of the energetic materials. The energetic component system is composed of four main components, namely a donor unit (HNS+HMX), a bulkhead (STS), an acceptor explosive (RDX), and a metalized propellant ($BKNO_3$) for gas generation. The pressurized gases generated from the burning propellant were purged into a 10 cc release chamber for study of the inherent oscillatory flow induced by the interferences between shock and rarefaction waves. In this paper, a step-by-step numerical description of detonation of high explosive components, deflagration of propellant component, and deformation of metal component is given in order to facilitate the proper implementation of the outlined formulation into a shock physics code for a detailed hydrodynamic simulation of the energetic component system.

1 Introduction

The multi-material modeling and simulation is a daunting task for any single-phase hydrodynamic solver or Lagrangian structural code. Conventional multi-phase solvers also fall short in handling of such tasks that require precise capturing of dynamic interface between the reacting energetic material and inert structure as well as resolving explosively fast time scale of chemical reaction associated with energetic substances which are typically experimentally calibrated and modeled to reproduce a series of calibration tests according to various military standards. In order to achieve this, a solver must be written in a fully integrated framework that strongly interacts with both fluid and solid. Furthermore, the reaction models must be built for each energetic material with unique composition and calorimetric properties. A high order numerical method must be utilized to track the rapid temporal transition during the chemical reaction which is also stable enough for smoothly linking the strain dynamics from the inert, elastic, and plastic neighboring structures.

One needs to adapt a stable multi-material interface handling algorithm based on a strongly-coupled fluid-structure interaction framework while accurately solving the chemical response of the involved high explosives, propellants, and reactive substances. In addition, the equations of state for each energetic material as well as the inert materials need to be modeled to reproduce the full scale dynamic behavior of the energetic system arranged in a train configuration.

In this paper, we have developed a multi-material hydrodynamic code with a hybrid particle level set method for shock-to-detonation transition of energetic materials in a train configuration. A level set based reactive ghost fluid method that imposes physical boundary conditions at material interfaces is developed to reproduce the experimental data of the closed chamber test. A full-scale simulation is conducted to analyze the performance characteristics of a pyrotechnic system [1] in a closed chamber test, which requires a precise recreation of the shock-induced detonation processes. Test involves detonation of a composite donor consisting of HNS (hexanitrostilbene, $C_{14}H_6N_6O_{12}$) and HMX (cyclotetramethylene-tetranitramine, $C_4H_8N_8O_8$), pressure attenuation in a stainless steel gap domain (STS), detonation of an acceptor RDX (cyclotrimethylene-trinitramine, $C_3H_6N_6O_6$), and ultimately a deflagration of a gas-generating propellant (boron potassium nitrate, $BKNO_3$). The purging gases into a 10 cc dump chamber are subjected to a subsequent monitoring of the flow oscillations due to the inherent shocks and rarefactions. The test results are used to validate the full-scale simulation results presented in this study.

2 Numerical approach

The dynamic response of both energetic and inert components of the given system can be described by the compressible form of the governing equations. The stress tensor that describes the structural response of stainless steel is comprised of a deviatoric stress and a hydrostatic pressure [2]. The Mie-Gruneisen equation of state (EOS) defines the pressure attained by the gap between donor-acceptor pair, while the Jones-Wilkins-Lee (JWL) EOS determines the explosive pressure resulting from the hydro-thermodynamic state. The rate of chemical reaction was based on the ignition and growth kinetics [3]. The interface between any two materials was tracked through a hybrid particle level set method, while material properties in the vicinity of an interface were determined through the ghost fluid method. Additional description of the method can be referred from Ref. 2.

2.1 Governing equations

The compressible Navier-Stokes equations in two-dimensional coordinate system reflect the conservations of mass, momentum and energy as shown in Eq. (1), (2), and (3), respectively.

$$\frac{\partial \vec{U}}{\partial t}+\frac{\partial \vec{E}}{\partial r}+\frac{\partial \vec{F}}{\partial z}=\vec{S}\left(\vec{U}\right) \quad (1)$$

Corresponding author. Fax: +82-2-882-1507
E-mail address: jjyoh@snu.ac.kr

$$\vec{U} = \begin{bmatrix} \rho \\ \rho u_r \\ \rho u_z \\ \rho E \\ \rho \lambda_i \end{bmatrix}, \quad \vec{E} = \begin{bmatrix} \rho u_r \\ \rho u_r^2 + p \\ \rho u_r u_z \\ u_r(\rho E + p) \\ \rho \lambda_i u_r \end{bmatrix}, \quad \vec{F} = \begin{bmatrix} \rho u_z \\ \rho u_z u_r \\ \rho u_z^2 + p \\ u_z(\rho E + p) \\ \rho \lambda_i u_z \end{bmatrix} \tag{2}$$

$$\vec{S} = \begin{bmatrix} -\frac{\rho u_r}{r}\varphi \\ \frac{\eta(s_{rr} - s_{\theta\theta}) - \rho u_r^2}{r}\varphi + \eta\left(\frac{\partial s_{rr}}{\partial r} + \frac{\partial s_{zr}}{\partial z}\right) \\ \frac{\eta s_{zr} - \rho u_r u_z}{r}\varphi + \eta\left(\frac{\partial s_{rz}}{\partial r} + \frac{\partial s_{zz}}{\partial z}\right) \\ \frac{\eta(u_r s_{rr} + u_z s_{rz}) - u_r(\rho E + p)}{r}\varphi + \eta\left(\frac{\partial(u_r s_{rr} + u_z s_{rz})}{\partial r} + \frac{\partial(u_r s_{zr} + u_z s_{zz})}{\partial z}\right) \\ \dot{w}_i \end{bmatrix} \tag{3}$$

Where, $\varphi = 0$ for rectangular and $\varphi = 1$ for cylindrical coordinates. Also, $\eta = 0$ for energetic materials while $\eta = 1$ for inert materials. Here, r, θ, z are the cylindrical coordinate variables and ρ is density, u_z, u_r are velocity components in axial and radial coordinates. $E = e + (u_z^2 + u_r^2)/2$ is the total energy per unit mass with the specific internal energy e, and p is the hydrostatic pressure, respectively. The temperature is derived from the relationship, $e = CvT$. Reaction rate, $\dot{w}_i \equiv \partial\lambda_i / \partial t|_{Chem} = f(T,p)$, is described by the experimentally tuned Arrhenius kinetics for the deflagration to detonation transition also is defined by the pressure based ignition and growth mechanisms via a series of unconfined rate stick tests for the shock to detonation transition. It chosen based on its feasibility to accurately resolve key length and time scales involved in the deflagration or detonation phenomena.

2.2 Constitutive relations

The reactive flow model is described by the rate law that consists of both ignition and growth steps suggested by Kim et al. [3] as shown in Eq. (4).

$$\frac{d\lambda}{dt} = I(1-\lambda)\mu^a + G(1-\lambda)p^b \tag{4}$$

Here, p is pressure, t is time, while ρ_0 and ρ denote the initial and current density respectively. Constants I, a, G, and b are the unknown parameters while λ is the burned mass fraction that represents reaction progress such that $\lambda = 0$ is unreacted and $\lambda = 1$ is reacted state. The compression was defined as $\mu = \rho/\rho_0 - 1$. Four unknown parameters of major significance in view of detonation were determined by a series of standardized unconfined rate stick experiment performed [3].

For the RDX, the constants of ignition I and growth G were set to 5.8×10^7 s^{-1} and 3.8×10^8 s^{-1}Mbar^{-b} respectively. The pressure sensitivity b was 1.1, and the compression sensitivity a was 4.0. The rate law of HMX was modeled by Eq. (5), with the rate parameters of $I = 4.4\times10^7$ s^{-1}, $a = 4.0$, $G = 8.5\times10^8$ s^{-1}Mbar^{-b}, and $b = 2.0$.

$$\frac{d\lambda}{dt} = I(1-\lambda)^{0.222}\mu^a + G(1-\lambda)\lambda^{0.666}p^b \tag{5}$$

$BKNO_3$ is a pyrotechnic propellant used in the solid fuel igniters, thermites, and gas generators, etc. The chemical reaction of $BKNO_3$ is classified as a deflagration, and the reaction rate [4] is calculated using Eq. (6).

$$\frac{d\lambda}{dt} = \rho_0 Z\lambda \exp(-E_a/(\bar{R}T)) \tag{6}$$

Here, the thermal response of $BKNO_3$ is simulated using the Arrhenius law, and related parameters are quantified by the DSC (Differential Scanning Calorimetry) analysis which measures the calorific heat of chemical reaction. The reaction parameters of $BKNO_3$ are $Z = 9.2\times10^9$ s^{-1}, $E_a = 1.8\times10^5$ kJ/kmol.

In order to describe both the unreacted and the product state of an energetic material, the equations of state defining the pressure are necessary. The Jones-Wilkins-Lee (JWL) form of equations [5] are shown in Eq. (7) and (8) for unreacted reactant and gaseous product of high explosives.

$$p_{\text{unreacted}} = \frac{\omega e_0}{(\rho_0/\rho)} + A\left(1 - \frac{\omega}{R_1(\rho_0/\rho)}\right)e^{-R_1(\rho_0/\rho)} + B\left(1 - \frac{\omega}{R_2(\rho_0/\rho)}\right)e^{-R_2(\rho_0/\rho)} \tag{7}$$

$$p_{\text{reacted}} = Ae^{-R_1(\rho_0/\rho)} + Be^{-R_2(\rho_0/\rho)} + \frac{C}{(\rho_0/\rho)^{\omega+1}} \tag{8}$$

Here, ρ_0 and ρ are the initial and current densities respectively. A, B, C, R_1, and R_2 are the material dependent JWL parameters with ω being the Gruneisen coefficient of the explosive. These parameters were obtained by fitting the JWL EOS to the cylinder expansion test results or thermodynamic equilibrium calculation of CHEETAH [6], where $e_0 = \rho_0 C_v T$ refers to the thermodynamic energy in GPa.

As for the gap and bulkhead materials, Mie-Grüneisen EOS is adopted to calculate the pressure of non-reactive materials [7] shown in Eq. (9).

$$p_{\text{non-reactive}} = \Gamma_0 e_0 + \begin{cases} \dfrac{\rho_0 c_0^2 \mu\left[1 + \left(1 - \frac{\Gamma_0}{2}\right)\mu\right]}{\left[1 - (s_0 - 1)\mu\right]^2} & if\ \mu > 0 \\ c_0^2 \rho_0 \mu & if\ \mu < 0 \end{cases} \tag{9}$$

3 Results

The train configuration consisted of a donor charge of HNS+HMX, a bulkhead of STS, an acceptor charge of RDX, and the pyrotechnic propellant $BKNO_3$. In the donor charge, HNS is a primer to a detonating HMX for generating a booster pressure. The ensuing pressure is progressively attenuated through the bulkhead. The bulkhead length is a critical parameter for initiating the acceptor charge and subsequently igniting the gas-generating propellant. As the sympathetic pressure to be transferred to an acceptor charge depends on the thickness of the bulkhead, an accurate analysis is required for prediction of the shock-to-detonation transition process of a composite system arranged in a train configuration. In order to quantify the detonation response of high explosives by a shock impact in such a system, we set out to clarify the relationship between ignition sensitivity and threshold pressures.

Figure 1 shows the pyrotechnic initiator-chamber assembly on the left and the computational domain on the right. The donor-gap-acceptor train configuration is utilized to ignite the propellant for gas generation into a chamber. A pressure sensor measured the central wall pressure fluctuation during the event.

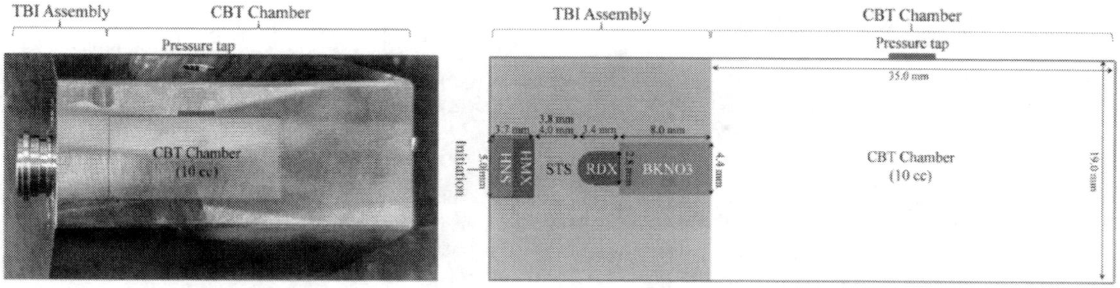

Figure 1: Test specimen (left) and computational domain (right) for closed chamber test.

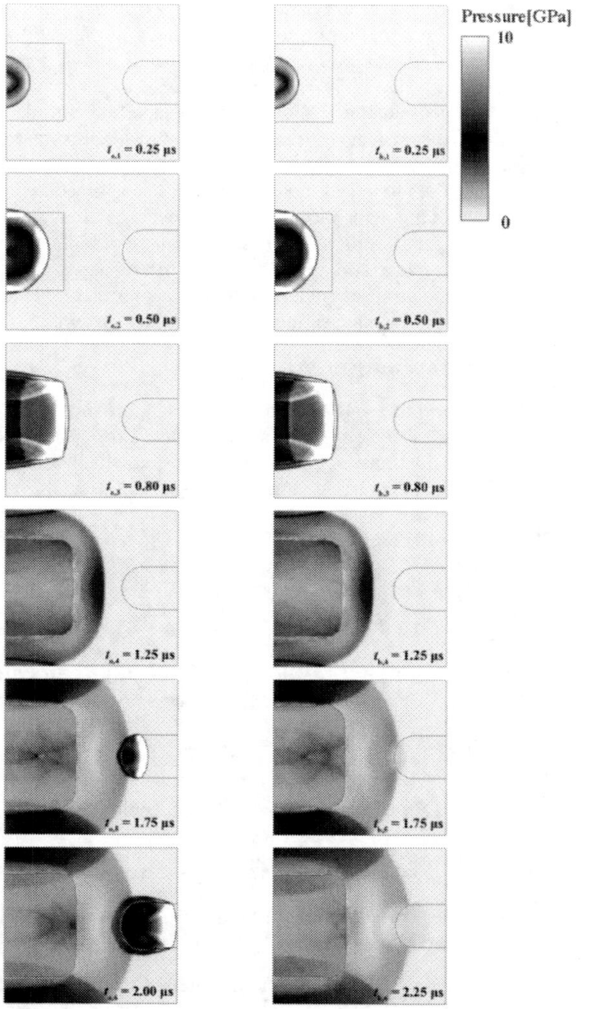

Figure 2: Timed images of calculated pressure contour at 3.8 mm (left), and 4.0 mm (right) bulkhead thicknesses of the explosive train configuration. Precisely tracked energetic and inert boundaries are shown in solid lines.

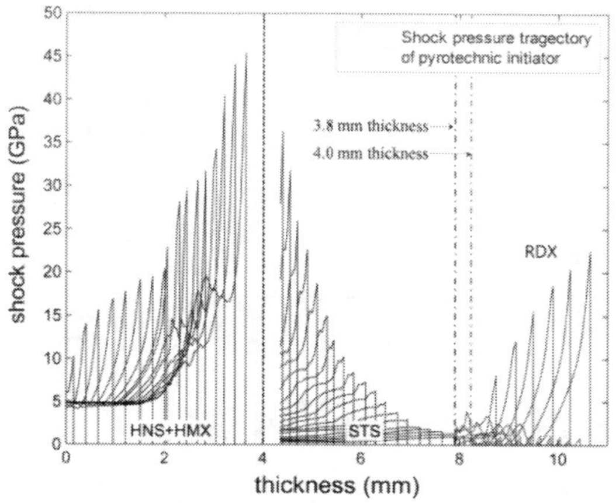

Figure 3: Calculated shock pressure trajectory in donor / bulkhead / acceptor of the explosive train configuration.

Figure 2 shows the calculation results on Go/No-go reactions in the explosive train configuration with two bulkhead thicknesses (3.8 mm and 4.0 mm). This reflects the observation of positive and negative responses at around 3.9 mm in the closed chamber test. The sequential reaction of the explosives to donor (HNS+HMX) / bulkhead (STS) / acceptor (RDX) occurs by the shock wave propagation.

When a high-pressured wave front reaches the acceptor, the chemical reaction is initiated and the thermal energy generated by the reaction supports the propagation of detonation wave. The peak pressure of shock wave is gradually decreased when it passes through the non-reactive STS bulkhead, because the compression consumes energy. The Go reaction can be observed if the shock wave with the minimum pressure to ignite the acceptor is reached. Otherwise, a No-go reaction appears. This result is attributed to the pressure attenuation occurring more as the bulkhead thickness increases. The pressure profile along the centerline of the explosive train configuration is shown in Fig. 3. The ZND pressure of the donor is about 45 GPa, and the pressure gradually decreases as it passes through the STS bulkhead. The acceptor is initiated at around 1 GPa. Therefore, the energy loss caused by the shock wave passing through the STS and the critical initiating pressure of the high explosive are two important factors for determining the reliable operation of the pyrotechnic train.

Figure 4 shows that reactive flow motion generated by the pyrotechnic initiator composed of detonator (HNS+HMX) / bulkhead (STS) / acceptor (RDX) / pyrotechnic propellant ($BKNO_3$), flowed into a 10 cc closed chamber. The bulkhead thickness is set to 3.8 mm. The explosive train consisting of detonator, bulkhead and acceptor completes its reaction in about 2.0 μs. Then, another 40 μs is consumed to fully deflagrate the propellant, $BKNO_3$. The hot product gas fills the closed chamber in approximately 50 μs. In the figure, a shadowgraph is shown in the upper, and pressure is shown in the lower half of each timed image. The release pressure wave into the chamber starts to bounce off of the right-end wall, which is repeating at every 130 μs.

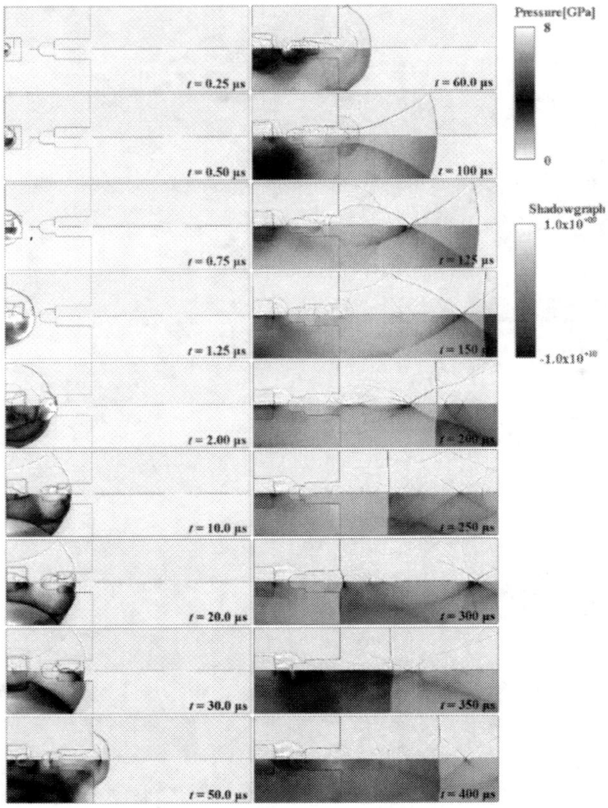

Figure 4: Shown schlieren and pressure evolution in time for entire PMD-chamber assembly that shows a detonator (HNS+HMX), bulkhead (STS), acceptor (RDX), and gas-generating propellant ($BKNO_3$).

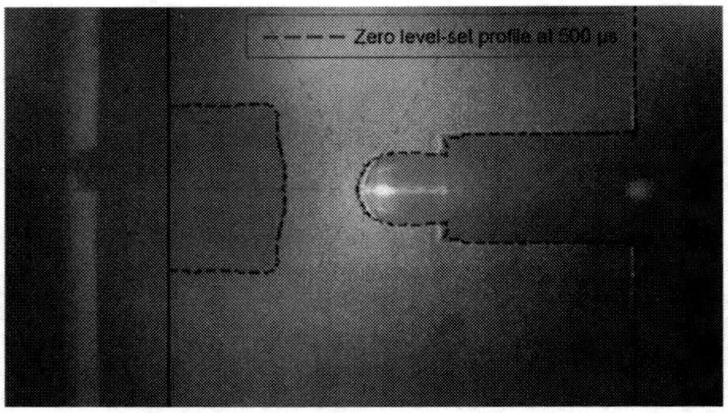

Figure 5: Shape comparison of cavity within PMD between calculation and experiment.

The predicted shape change of the inert-reactant boundaries after the test is quite intriguing. As the pressure is attenuated in the bulkhead as such the most of donor pressure is transmitted into the acceptor, causing a higher compression of the gap during successive acceptor initiation. Donor and acceptor expand in all directions due to the internal pressure buildup. A cross sectional view of an actual sample is shown in Fig. 5. The transmitted shock pressure becomes increasingly high near the central axis of the donor and acceptor due to a highly reactive detonation front in the center. The resulting curvature of STS is therefore concave and it seems to have evolved into a very similar shape between the experiment and calculation. The dynamic shape change of the stainless casing for donor, acceptor, and propellant is also compared between actual photographic image and numerical result from the particle level set tracking. The donor and acceptor cavities are expanded in all directions due to detonation pressure buildup. The resulting cavity shape change is quite precisely predicted from the present simulation as depicted in Fig. 5.

4 Conclusions

A detailed discussion on the hydrodynamic modeling of a pyrotechnic device is presented. The numerical methodology for tracking multi-material boundary interactions as well as severely transient fluid-structure coupling between high explosive charges and metal bulkhead are described. The material modeling techniques for both reactive and inert materials are also explained in regards to their empirical and theoretical bases. The performance of a resulting energetic component system arranged in a train configuration is demonstrated via the pressure measurement of a high pressure burned gases of the $BKNO_3$ propellant, inside a dump chamber. The developed method is shown to be reliable for reproducing the shock-to-detonation transition and material-to-material shock attenuation that govern the performance of the whole system. As the numerical modeling of the energetic component system involving detonation of high explosives, deflagration of propellant, and structural deformation of bulkhead is systematically and thoroughly discussed in this paper, one can implement the outlined contents into a shock physics code for a full scale hydrodynamic simulation of a similar system.

5 Acknowledgment

Authors would like to thank the financial supports from Agency for Defense Development and Hanwha Corp. contracted through the Institute of Advanced Aerospace Technology at Seoul National University. Additional support from the Next Generation Space Propulsion Research Center (NRF-2013R1A5A1073861) at Seoul National University is warmly acknowledged.

References

[1] B. Kim, M. Kim, J.J. Yoh, Proc. Combust. Inst. 36 (2017) 2699-2707.

[2] B. Kim, J. Park, J.J. Yoh, J. of Applied Phys. 119 (14) (2016) 145902.

[3] B. Kim, J. Park, K. Lee, J.J. Yoh, J. of Applied Phys. 116 (2) (2014) 023512.

[4] H.S. Lee, 45th AIAA JPC AIAA2009-5191 (2009) 1-11.

[5] P.C. Souers, S. Anderson, J. Mercer, E. McGuire, P. Vitello, Propel., Explo., Pyrotech., 25 (2) (2000) 54-58.

[6] L.E. Fried, W.M. Howard, P.C. Souers, Cheetah 2.0 User's Manual, Report No. UCRL-MA-117541, Lawrence Livermore National Laboratory, 1998.

[7] D.J. Steinberg, Equation of State and Strength Properties of Selected Materials, Report No. UCRL-MA-106439, Lawrence Livermore National Laboratory, 1996.

11th Asia-Pacific Conference on Combustion,
The University of Sydney, NSW Australia
10th -14th December 2017.

Flame Displacement Speed Characteristics of Turbulent Premixed Flames in a Constant Volume Vessel

Basmil Yenerdag, Yuki Minamoto, Masayasu Shimura, Mamoru Tanahashi

Tokyo Institute of Technology

2-12-1 Ookayama, Meguro, Tokyo 152-8550, Japan

Abstract

Direct numerical simulation is performed for the stoichiometric hydrogen–air mixture with decaying homogeneous isotropic turbulence of which the Reynolds number based on Taylor micro scale is 97.1. Initial distributions of species are assumed to be uniform. A detailed kinetic mechanism is used to represent the hydrogen–air reactions. The flame ignites from the high temperature kernel at the center of the computational domain, and then propagates with the distortion by the turbulent motion. In this study, the propagation characteristics of a hydrogen-air premixed flame in a constant volume vessel are investigated. The contribution of reaction $S_{d,R}$, normal diffusion $S_{d,N}$ and tangential diffusion $S_{d,T}$ to the local flame displacement speed S_d is studied. As the pressure increases in the vessel, the mean S_d also increases. The contribution of $S_{d,T}$ to S_d is significant due to the existence of high negative mean curvature just after the flame ignition. However, this contribution decreases and becomes negligible as the global diameter of the flame increases. The mean S_d as a function of wall normal distance is also investigated and it is found that S_d is almost constant away from the wall, and decreases drastically approximately 1 thermal flame thickness away from the wall.

1 Introduction

Designing high thermal efficiency and low emission combustion devices such as spark ignition (SI) engines is a challenging task as emission standards become more stringent. One of the challenges to meet the standards is to reduce pollutants of hydrocarbon fuels such as CO, CO_2. Hydrogen as the next generation fuel has gained particular interest as it has different characteristics and various advantageous over hydrocarbon fuels and more importantly no CO and CO_2 emissions. These features show that hydrogen may contribute to clean combustion technology in the future.

Direct numerical simulation (DNS), a promising tool for developments and validations of turbulent combustion models, has been used in many previous studies to investigate propagation characteristics and turbulence-flame interactions of premixed flames [1-12]. The flame displacement speed S_d, which is an important quantity for turbulent combustion modeling, has been studied for premixed flames in different flow configurations such as planar flames [7-9], jet flames [10], spherical flames [11-13], by employing either two- or three-dimensional DNS with simple or detailed chemistry. It was shown that in three-dimensional spherical flames, the mean negative curvature significantly affects the flame propagation characteristics [11-13]. These spherical flames studies consider single-step chemistry and constant transport coefficients. Although their results show general good agreement with the studies using detailed chemistry in two-dimensional cases [7-8], employing three-dimensional DNS considering a detailed chemistry with full transport coefficients could provide a deeper understanding of propagation behavior of premixed flames. Moreover, the aforementioned DNS studies consider constant pressure conditions. Whereas in a constant volume vessel such as SI engines, increase in the mean pressure during combustion could influence propagation characteristics of the flame significantly via temporal evolution of thermo-chemical properties. And local flame displacement speed under such conditions, to the best of our knowledge, has not been studied by means of DNS, therefore this should be investigated. In this study, three-dimensional DNS of a stoichiometric hydrogen-air premixed flame considering a detailed kinetic mechanism is conducted to investigate flame propagation characteristics in a constant volume vessel.

2 Direct Numerical Simulation

The details of numerical methods are shown in the previous study [14]. The governing equations are conservation equations of mass, momentum, energy and species concentration. The Soret effect, the Dufour effect, pressure gradient diffusion, bulk viscosity and radiative heat transfer are assumed to be negligible. Spatial discretization is conducted by the fourth order central finite difference scheme. The fourth order compact finite difference filter is used to eliminate unphysical oscillations with higher spatial frequencies than the spatial resolution [15]. The third order Runge–Kutta scheme is used for time integration. Only the reaction source terms in species

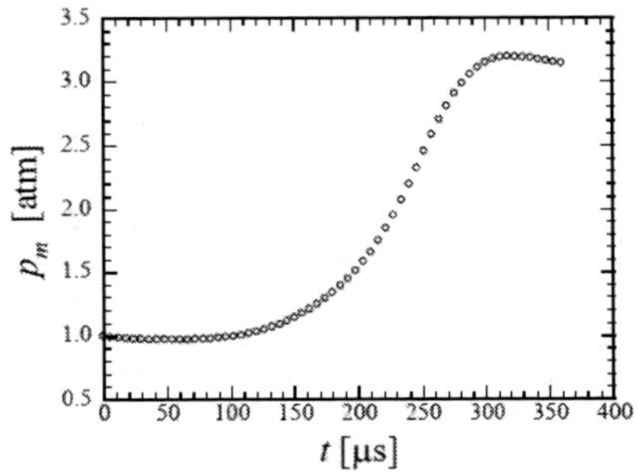

Figure 1: Temporal development of the mean pressure

Corresponding author. Fax: +81-3-5734-2505
E-mail address: ybasmil@navier.mes.titech.ac.jp

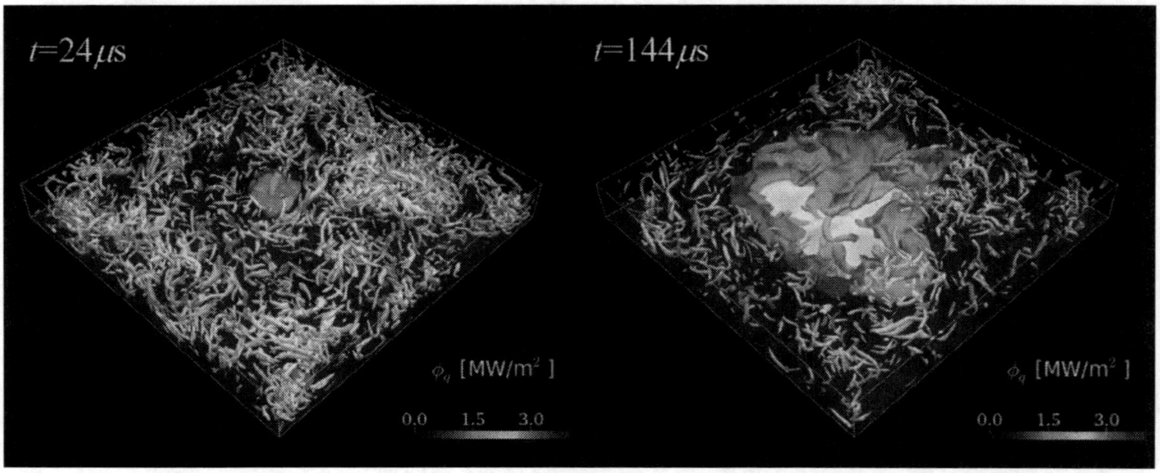

Figure 2: Temporal development of turbulence and flame structures. White and gray iso-surfaces indicate the second invariant of velocity gradient tensor at 1.0×10^{10} and 1.0×10^{9}; note that gray surfaces are drawn only in the burnt side for visibility. Orange iso-surface represents the temperature at 1282 K.

conservation equations are advanced by the implicit method [16]. For the boundary condition, isothermal and no-slip conditions given by the Navier-Stokes characteristic boundary condition (NSCBC) [17, 18] are imposed on the walls and the wall surfaces are assumed to be chemically inert. The initial distributions of species and temperature are uniform. A detailed kinetic mechanism that includes 12 reactive species and 27 elementary reactions is applied for hydrogen-air flame [19]. Temperature dependence of viscosity, thermal conductivity and diffusion coefficients is taken into account [20, 21]. The Reynolds number based on Taylor micro scale (λ) for initial condition is set to 97.1. The grid points in the direction of x, y and z of the constant volume vessel (20mm×5mm×20mm) are 1025×257×1025. The preheating temperature T_{pre}, the wall temperature T_{wall}, the initial pressure P_{ini} and the equivalence ratio are set to 700 K, 450 K, 0.1 MPa and 1.0, respectively. Other important numerical parameters are u_{rms}/S_L=1.80 and l/δ_F = 281, where u_{rms}, S_L, l and δ_F denote the root-mean-square of turbulent velocity, the laminar burning velocity, the integral length scale and the Zel'dovich thickness. The condition is classified into corrugated flamelets regime in turbulent combustion diagram. Since the initial preheated temperature is set to 700 K, the mean pressure P_m increases by a factor of 3, as shown in Fig. 1.

3 Local Flame Displacement Speed

Figure 2 shows the flame and vortical structures indicated by iso-surfaces of temperature and second invariant (Q) of the velocity gradient tensor. The color bar shows the minimum and the maximum values of wall heat flux through the lower wall and the details can be found in our previous study [14]. The temperature iso-surface (orange) indicates T = 1282 K at which the heat release rate peaks in the corresponding laminar flame, hence in the remainder of this paper, this isosurface T = 1282 K is selected for the flame front analysis such as curvature and S_d. The mixture ignites from the high temperature kernel at the center of the computational domain, and then flame front propagates with distortion by the turbulent motion. In the burnt side, for a better visualization, the value of iso-surface of Q is set to one tenth of that in the unburnt side because the vortices are weakened due to expansion and viscosity increase, which is shown by density of vortical structures. As the global radius of the flame increases with flame propagation, the probability of flame front convex towards burnt region increases, and this phenomenon can be seen by the probability density function (pdf) of the mean flame front curvature (k) in Fig. 3a. k is defined as.

$$k = \nabla \cdot \boldsymbol{n};\ \boldsymbol{n} = \frac{\nabla T}{|\nabla T|} \quad (1)$$

The flame front convex towards the unburnt side has negative values. The most expected and the mean k is negative, which is the characteristic of spherical flames. The density weighted displacement speed S_d constitutes of reaction $S_{d,R}$, normal diffusion $S_{d,N}$ and tangential diffusion $S_{d,T}$ terms, and can be written as:

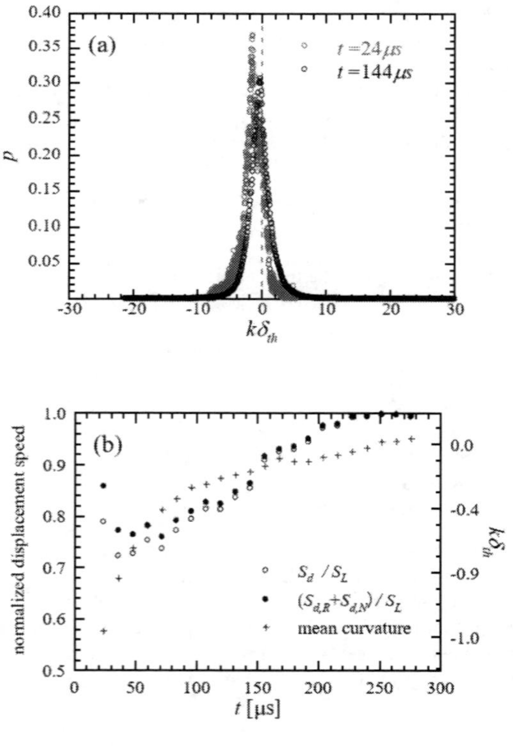

Figure 3: Pdf of the mean curvature (a), displacement speed term and mean curvature as a function of time (b)

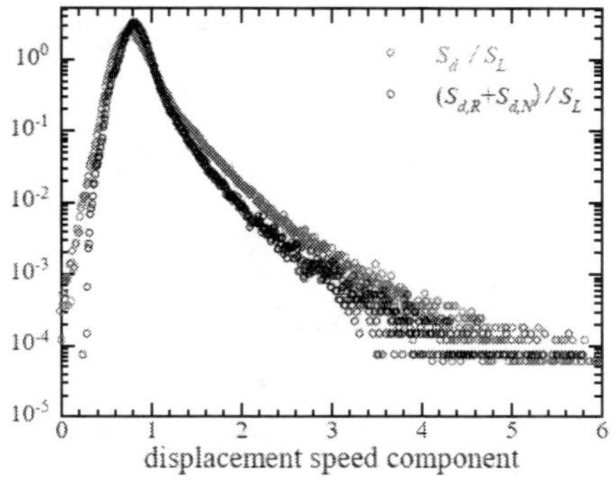

Figure 4: Pdf of displacement speed terms at t=144μs

$$S_{d,R} = \frac{\Delta H}{\rho_u c_p |\nabla T|} \quad (2)$$

$$S_{d,N} = \frac{\boldsymbol{n}\cdot\nabla(\lambda \boldsymbol{n}\cdot\nabla T)}{\rho_u c_p |\nabla T|} \quad (3)$$

$$S_{d,T} = \frac{\lambda}{\rho_u c_p}\nabla\cdot\boldsymbol{n} \quad (4)$$

$S_{d,N}$ is mostly negative, $S_{d,R}$ is the most dominant quantity out of the three terms and has positive values. $S_{d,T}$ is only the function of flame front curvature. Figure 3b shows the displacement speed terms and mean curvature as a function of time. Here, the displacement speed and mean curvature are normalized by S_L and thermal flame thickness δ_{th} in the corresponding laminar flame, respectively. It should be noted that, for corresponding laminar flame, S_L and δ_{th} changes with the pressure increase. Therefore, one-dimensional DNS of laminar flame propagation is conducted by adopting pressure and temperature averaged in the unburnt side of the vessel as the inlet condition, to obtain the corresponding S_L and δ_{th} at each time. Before flame impinging the upper and the lower walls, the flame propagates in a spherical shape on average. This is when t < 80μs and it can be seen that the mean S_d is less than unity, about $0.7S_L$ during this period of time. This characteristic was also observed for spherical flames considering single step chemistry [13] and is different from statistically planar flames [22] in which the density weighted S_d is close to unity, in the range of $0.9S_L$ - $1.0S_L$. The contribution of $S_{d,T}$ to S_d is significant due to the existence of high negative mean curvature just after the flame ignition. As can be seen in Fig. 3b, the difference between S_d and $S_{d,R}$+ $S_{d,N}$ is about $0.1S_L$ as the curvature is relatively large. However, this contribution decreases and becomes negligible as the global diameter of the flame increases. As the flame propagates in a circular disc shape on average and parallel to the upper and the lower walls, the mean curvature also decreases, and the mean S_d increases and reaches about $0.9S_L$ - $1.0S_L$. As shown in Fig. 4, the pdf of S_d and $S_{d,R}$+ $S_{d,N}$ shows good agreement. This indicates that the contribution of tangential diffusion to S_d on average is limited despite the existence of non-zero k. Its effects on S_d locally, however, is relatively higher at highly curved regions. This can be seen in Fig. 5. Figure 5a-b show the joint pdf of S_d and $S_{d,R}$+ $S_{d,N}$ vs. k. S_d shows positive correlations with positive curvature and increases up to 3.5 S_L in the large positive k regions. In the negative k regions, S_d values are around $0.5S_L$ and a relatively weaker negative correlation can be observed. This is mostly attributed to the response of $S_{d,R}$, and therefore $S_{d,R}$+ $S_{d,N}$ to curvature. From the definition, $S_{d,R}$ is a function of heat release rate ΔH, and normal temperature gradient $|\nabla T|$ which is directly related to δ_{th}. Flame thickness increases significantly (low $|\nabla T|$) in the large positive k regions, therefore contributes to very high $S_{d,R}$ and also $S_{d,R}$+ $S_{d,N}$ values. Flame thickness also increases comparably in the large negative k regions, however, ΔH is relatively lower (30-40%) in the negative k regions than that in the positive k regions. This results in relatively lower $S_{d,R}$ and also $S_{d,R}$+ $S_{d,N}$ values. Since $S_{d,T}$ is directly related to flame curvature, $S_{d,T}$ is positive (negative) at positive (negative) curvature, hence its contribution to S_d, is relatively high at large curvature regions. It is clear that at negative (positive) curvature regions, S_d is relatively lower (higher) than the sum of $S_{d,R}$ + $S_{d,N}$.

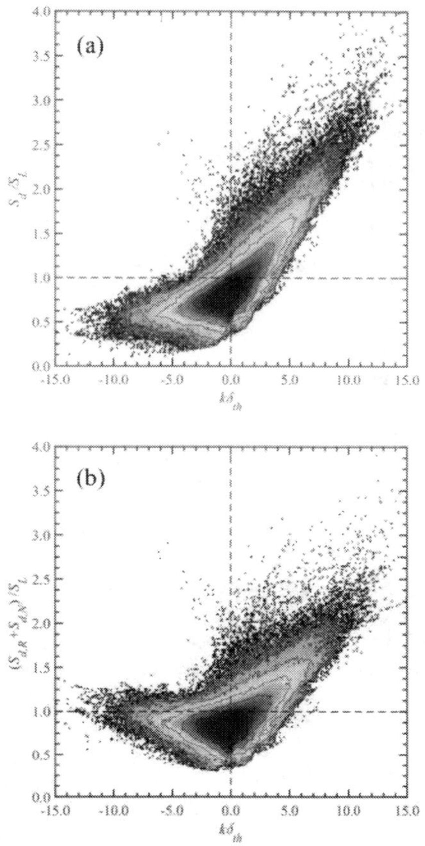

Figure 5: Joint pdf of S_d vs. k (a) and $S_{d,R}$ + $S_{d,N}$ vs. k (b) at t=144μs

As wall surfaces significantly affect the local flame displacement speed, the mean S_d as a function of wall normal distance is also investigated. Here, S_d is averaged in the lower half of the domain and at every 0.2 mm in the y direction. Figure 6 shows the mean S_d as a function of wall normal distance which is normalized by using thermal flame thickness for each time. The mean S_d away from the wall region, where y > δ_{th}, is almost constant. When y < δ_{th}, however, the mean S_d decreases drastically. It is shown that, despite the increase in the mean pressure in the vessel, this distance can be scaled using the thermal flame thickness in the corresponding laminar flame.

4 Conclusions

In this study, three-dimensional DNS of a stoichiometric hydrogen-air premixed flame considering a detailed kinetic

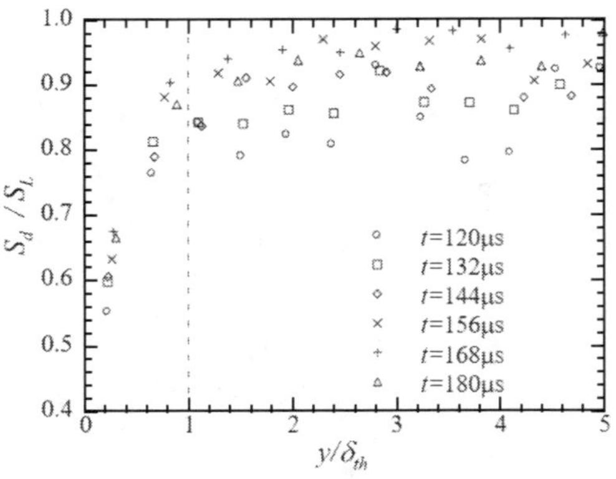

Figure 6: The mean S_d as a function of wall normal distance.

mechanism is conducted to investigate flame propagation characteristics in a constant volume vessel. As the mean pressure increases in the vessel, the mean S_d also increases. The contribution of $S_{d,T}$ to S_d is significant due to the existence of high negative mean curvature just after the flame ignition. However, this contribution decreases and becomes negligible as the global diameter of the flame increases. The mean S_d as a function of wall normal distance is also investigated and it is found that S_d is almost constant away from the wall, and decreases drastically approximately 1 laminar thermal flame thickness away from the wall.

5 Acknowledgment

This work was partly supported by Council for Science, Technology and Innovation (CSTI), Cross-ministerial Strategic Innovation Promotion Program (SIP)"Innovative Combustion Technology" (Funding agency: JST).

References

[1] K. Bray, Proc. Combust. Inst., 17 (1979) 223-233

[2] M. Tanahashi, M. Fujimura, T. Miyauchi, Proc. Combust. Inst., 28 (2000) 529-535

[3] A. Poludnenko, E. Oran, Combust. Flame, 157 (2010) 995-1011

[4] A. Gruber, R. Sankaran, E.R. Hawkes, J.H. Chen, J. Fluid Mech., 658 (2010) 5-32

[5] Y.S. Shim, S. Tanaka, M. Tanahashi, T. Miyauchi, Proc. Combust. Inst., 33 (2011) 1455-1462

[6] T. Jin, K. Luo, S. Lu, J. Fan, Int. J. Hydrogen Energy 38 (2013) 9886-9896

[7] J.H. Chen, H.G. Im, Proc. Combust. Inst., 27 (1998) 819-826.

[8] T. Echekki, J.H. Chen, Combust. Flame, 118 (1999) 308-311

[9] N. Chakraborty, R.S. Cant, Phys. Fluids, 17 (2005) 105105

[10] H. Wang, E.R. Hawkes, J.H. Chen, Combust, Flame 180 (2017) 110-123

[11] N. Chakraborty, M. Klein, R.S. Cant, Proc. Combust. Inst., 31 (2007) 1385-1392

[12] M. Klein, N. Chakraborty, R.S. Cant, Flow Turb. Combust., 81 (2008) 538-607

[13] K.W. Jenkins, M. Klein, N. Chakraborty, R.S. Cant, Combust. Flame, 145 (2006) 415-434

[14] B. Yenerdag, Y. Minamoto, Y. Naka, M. Shimura, M. Tanahashi Int. J. Hydrogen Energy 41 (2016) 9679-9689

[15] S.K. Lele, J. Comput. Phys., 103 (1992) 16-42

[16] P. Brown, G. Byrne, A. Hindmarsh, SIAM J. Sci. Stat. Comput., 10 (1989) 1028-1051

[17] T. Poisont, S. Lele J. Comput. Phys., 101 (1992) 104-129

[18] M. Baum, T. Poisont, D. Thevenin, J. Comput. Phys., 116 (1995) 247-261

[19] E. Gutheil, G. Balakrishnan, F. Williams, in: N. Peters, B. Rogg (Eds.), Reduced Kinetic Mechanisms for Applications in Combustion Systems, Springer, Berlin Heidelberg, 1993, p.177

[20] R.J. Kee, G. Dixon-Lewis, J. Warnatz, M.E. Coltrin, J.A. Miller, A Fortran Computer Code Package for the Evaluation of Gas Phase Multicomponent Transport Properties, Report No. SAND86-8246, Sandia National Laboratories, 1986.

[21] R.J. Kee, G. Dixon-Lewis, J. Warnatz, M.E. Coltrin, J.A. Miller, A Fortran Chemical Kinetics Package for the Analysis of Gas Phase Chemical Kinetics, Report No. SAND89-8009B, Sandia National Laboratories, 1989.

[22] N. Chakraborty, R.S. Cant, Combust. Flame, 137 (2004) 129-147

11th Asia-Pacific Conference on Combustion,
The University of Sydney, NSW Australia
10th-14th December 2017.

Investigation of Combustion and Thermodynamics Processes in a Micro Internal Combustion Swing Engine

Ning Du[1, 2], Xiong Zhou[1, 2], Wenhu Han[1], Wenjun Kong[1, 2*]

[1]Key Laboratory of Light-Duty Gas-Turbine, Institute of Engineering Thermophysics, Chinese Academy of Sciences, No. 11 West Road 4th-Ring North, Beijing, 100190, China

[2]University of Chinese Academy of Sciences

No.19(A) Yuquan Road, Beijing, 100049, China

Abstract

The combustion and thermodynamics processes of micro internal combustion swing engine are studied experimentally and numerically in this work. It has been found that flame in cylinders usually propagates in a narrow channel of millimeter magnitude. Because of heat loss at walls, the combustion temperature is not high enough so that soot can be formed in the process of flame propagation. In thermodynamic cycles, there exists an optimum value of ignition advance coefficient at which efficiency will be the largest. However, huge leakages of mass and heat result in lower thermal efficiency. This provides an important basis for future researches on the improvement of swing engine performance.

1 Introduction

As the development of science and technology, the miniaturization and portability of electronic products has become a development trend. It puts forward higher requirements on the energy density and storage time of energy supply systems. The micro energy power systems based on hydrocarbon fuel combustion can meet these requirements. In 1996, Epstein proposed the concept of micro energy power systems for the first time [1]. He designed a micro turbine engine with a compressor diameter of 8mm and a turbine diameter of 6mm [2, 3]. Although the micro turbine engine has no effective output due to heat dissipation and leakage, the researchers have made efforts on examinations and experiments of individual component, including turbines, compressors and combustion chambers. The Combustion Laboratory of the University of California at Berkeley proposed a micro energy power system using a micro rotor engine as driving power [4]. Their goal is to create a micro rotor engine power system which is able to output power at milliwatts level. In order to provide a reference for this kind of the micro rotary engine, firstly they designed a small rotor engine with a cylinder size of about 1000-1700mm^3. Experimental test showed that output power is 2.7W at speed ~9300RPM with the hydrogen/air mixture. The test showed that heat loss and leakage are key factors determining the successful operation of an engine. The University of Minnesota proposed a micro free piston engine based on homogeneous compression ignition [5, 6]. Their goal is to produce a micro homogeneous compression ignition free piston engine with an output power of 10W. A single shooting experiment was carried out by Aichlmayrl [7] to study the combustion of the gas in the cylinder of the micro free piston engine.

Mijit [8] and Dahm [9] proposed a micro swing engine, consisting of a center swing arm and a cylinder. It's a free swinging piston engine, having the following merits: (1) swing engines, with higher power density, divide the cavity into four small cylinders to make good use of the cavity space; (2) swing engines can directly convert reciprocating motion into electrical work, without other auxiliary structures. This operation is more stable; (3) due to the fact that the micro swing engine is free tilting, the movement process has no dead center, and cold start can be carried out, which greatly simplifies the complexity of the system. However, most of researches on micro swing engines just stay in numerical and theoretical investigations. For example, Gu [10] calculated theoretically the influence of engine size, combustion duration and load on the engine performance. Zhu [11] introduced the control system of swing engine in detail. However, related experimental studies are very few. The goal of this work is to design and manufacture a two stroke micro swing engine with output power of 100W. Studies of engine thermodynamic process with different ignition advance angles and butane-air combustion in cylinder are conducted experimentally and numerically.

2 Method

2.1 Experimental setup

Figure 1 shows a schematic of the micro internal combustion swing engine (MICSE).The swing arm divides the cavity into ABCD four small cylinders. The operating principle of MICSE was described in detail in the reference [10]. Table 1 gives the system parameters for the micro swing engine. In order to provide an experimental reference for the continuous operation of swing engines, a single thermal cycle test is designed in this paper. Figure 2 shows a schematic of the experimental setup. The B and D cylinders were connected with the environment through spark plug holes during the experiment. The A cylinder and C cylinder are respectively used as a combustion cylinder and pneumatic cylinder. The fuel is a premixed gas of butane and air and the driving gas of pneumatic cylinder is compressed air with 4atm pressure. At the beginning of the experiment, the compressed air in the C cylinder drives the swing arm to rotate which compresses the gas in the A cylinder. Then the spark plug ignites. High pressure and temperature gas produced by

Corresponding author. Tel: +86-10-8254-3042
E-mail address: wjkong@iet.cn

the fuel burning drives swing arm rotation and work done. When the swing arm turns on the exhaust port, the exhaust is exhausted. A thermodynamic cycle is completed. The thermal characteristics of the engine are studied by changing the ignition advanced angle. The angular velocity of the swing arm can be measured by the output pulse frequency of the encoder which connected to the engine shaft. The ignition signal is triggered by the single chip microcomputer. The pressure and angle in the experiment are measured by the pressure sensor and encoder respectively. The ignition position is measured by a high voltage passive probe. All sensor output data is collected by oscilloscope. The combustion process is captured by a high-speed camera with the number of frames 5000fps.

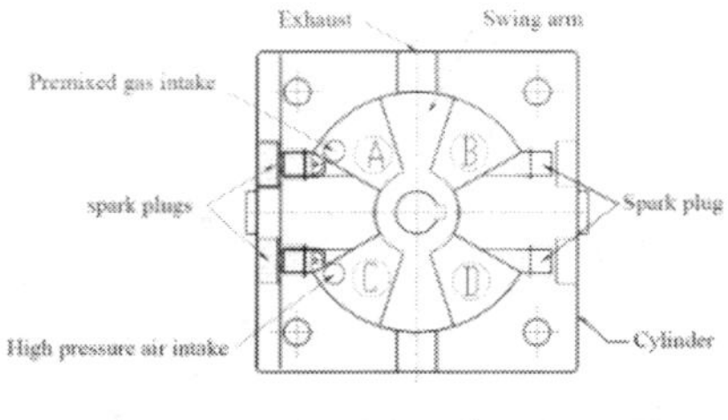

Figure 1: The schematic of the micro internal combustion swing engine

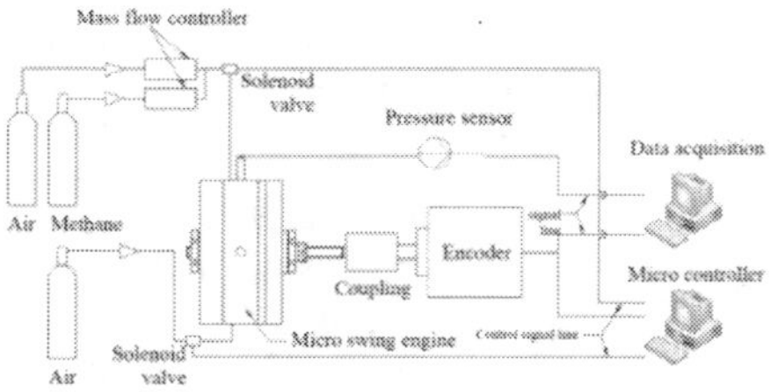

Figure 2 The schematic of the experimental setup.

Table 1: Parameters of MICSE

Parameter/unit	Values
d_i/mm	2
d_c/mm	45.7
h/mm	15.2
θ_e/°	15
θ_a/°	36
θ_c/°	120
Φ	1
T_i/K	300
J/kg·m2	0.000045

Remarks: d_i-diameter of the Intake port or exhaust port; d_c-diameter of the chamber; h-thickness of the chamber; θ_e-angle of the exhaust port; θ_a-angle of the swing arm; θ_c-angle of the chamber; Φ-the equivalence ratio; T_i-inlet temperature; J-external mass load coefficient.

Equation (1) gives the definition of the ignition advance angle Δ in the experiment. The peak pressure refers to the maximum pressure measured by the engine during adiabatic compression. Figure 3 shows the ignition position of this experiment.

$$\Delta = \begin{cases} \dfrac{pressure}{pressure\ peak\ value} - 1 & \text{Compression process} \\ 1 - \dfrac{pressure}{pressure\ peak\ value} & \text{Expansion process} \end{cases} \quad (1)$$

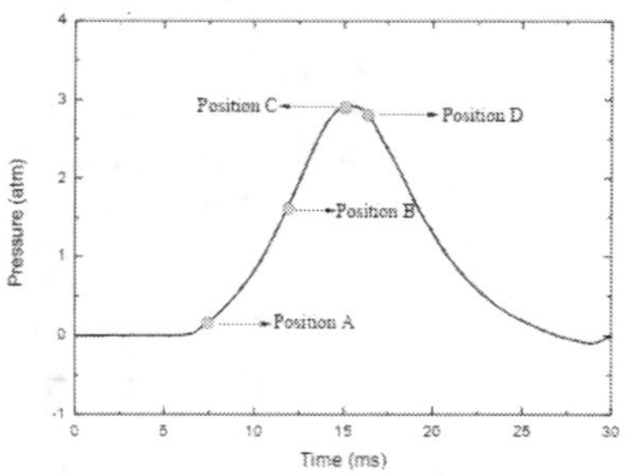

Figure 3: The ignition position of the experiment: Position A/Δ= −0.93; Position B/Δ= −0.58; Position C/Δ= 0; Position D/Δ= 0.06;

2.2Computational method

The governing equations of micro swing engines and the calculation methods of gas properties are given in the literature [10]. In this paper, the numerical simulation of single thermal cycle is carried out by the above calculation method. The energy conservation equation is as follows:

$$\dot{E} = -p\dot{V} - \dot{Q} + \sum_i \dot{m}_i h_i \quad (2)$$

Where $\dot{E}$ is the internal energy in the chamber, p and V are time-varying chamber pressure and volume,$\dot{Q}$ is the heat loss rate to the chamber walls, which can be calculated by the Newton cooling formula. The surface heat transfer coefficient of cylinder inner wall can be computed in detail in the reference [12]. $\dot{m}_i$ and h_i are the mass and specific enthalpy of the i-th inflow stream into chamber or outflow stream from the chamber. The Webb function is used for calculating the fuel combustion in a cylinder: $x_b = 1 - x_{u0}\exp(-a(\frac{t}{\tau_b}))$, x_b is the burned mass fraction;x_{u0} is the initial unburned mass fraction; τ_b is the specified combustion duration. The leakage model was described in detail in the reference [13].The mass flow rates of the intake and exhaust are instead modeled using quasi-one-dimensional isentropic flow theory.

Figure 4 shows a typical P-V diagram, comparison with the experimental results at the ignition advance angle Δ=0. As can be seen from the diagram, the calculated results are in good agreement with the experimental results. The reliability of the zero dimensional computation models is confirmed.

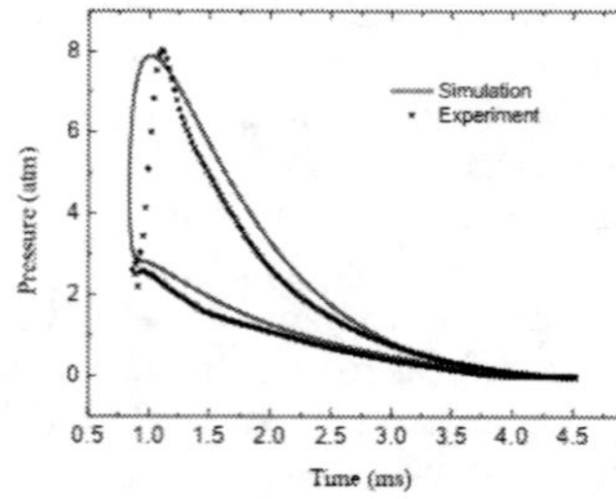

Figure 4: Comparison between the measured and predicted from the model at Δ= 0

3 Results and discussions

3.1 Combustion process

Figure 5 shows combustion process pictures in the engine

cylinder with different ignition advance angles. The time interval between each two pictures is 0.4ms. We take the Δ=0 as a typical example to analyze the flame propagation in the cylinder. When ignition occurs, combustible gas near the spark plug burns and its temperature rises due to the spark plug continuing discharge. As a result, the temperature of soot formation reaches, and hence flame color near the spark plug is yellow, shown in Fig 5 (b). However, the blue flame propagates constantly, due to heat loss in narrow channel (~2-3mm). Due to $Q_{\text{loss}} \propto d^{-2}$[14], d is pipe diameter. The channel width is approximately the order of millimeter magnitude, which is 1% of conventional dimension. Therefore, heat loss at walls is two orders of magnitude on a conventional scale. The heat loss caused the flame temperature lower than the soot formation temperature (1400K). Hence, there is no soot formation during propagation, shown in Fig. 5 (b). Similar processes can be found in Fig. 5 (a-d). Figure 6 shows temperature history for different advance angles in a thermodynamic cycle. We can observe that the temperature is less than 1400K except for the case Δ=-0.93. This also provides evidence for the above explanation. When the Δ is equal to -0.93, the combustion occurs in compression process because of premature ignition. This makes the temperature in the cylinder considerably higher than the other three conditions. From Fig 5 (d), soot formation in the cylinder can be observed. This confirmed that the temperature in the cylinder is larger than 1400K at this working condition.

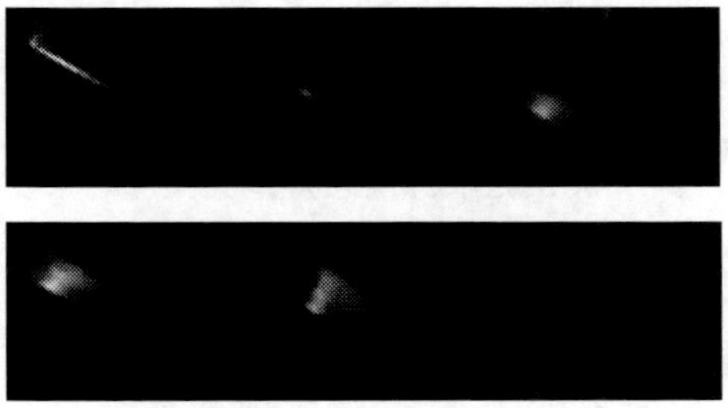

(a)Δ= 0.06

(b)Δ= 0

(c)Δ= −0.58

(d)Δ= −0.93

Figure 5: Combustion process for different ignition advance angles

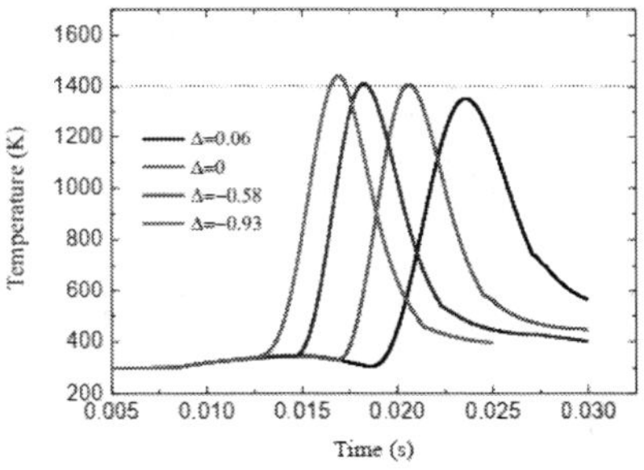

Figure 6: Temperature histories for different advance angles in a thermodynamic cycle.

3.2Thermodynamics process

Figure 7 shows the result of indicated mean effective pressure (IMEP) and indicated thermal efficiency (ITE) for different ignition advance angles in the experiment. The IMEP rises as the ignition advance angle increases; however it starts to drop at Δ>-0.3. Change in thermal efficiency has consistent trend with that of the IMEP. This is reasonable because burning of premixed gas takes a certain time. The spark advance shortens the time during which fuel undergoes from ignition to complete combustion. Hence, it promotes and renders the combustion process closer to ideal constant volume combustion, which is beneficial to improve the running performance of the engine. However, the thermal efficiency of the engine is low. The relationship of heat loss and relative leakage ($\mathrm{M} = \frac{\int_0^t \mathrm{m_l dt}}{\mathrm{m_0}} \times$ 100%, where m_l is the mass leakage rate; m_0 is the initial mass of the gas in the cylinder) along with the ignition advance angle is obtained by calculation of zero-dimensional model, as shown in Figs. 8 and 9. As the ignition advance angle increases, the heat loss decreases and its maximum reaches 53% of total released energy, while the relative leakage increases. At Δ=0.006, the relative leakage reaches finally 55.45% of the initial mass. Therefore, the heat loss and leakage loss are the major reasons resulting in low thermal efficiency.

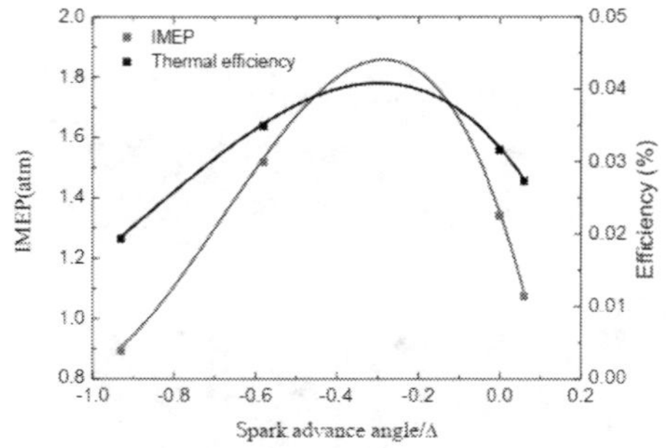

Figure 7: IMEP and ITE at various ignition advance angles

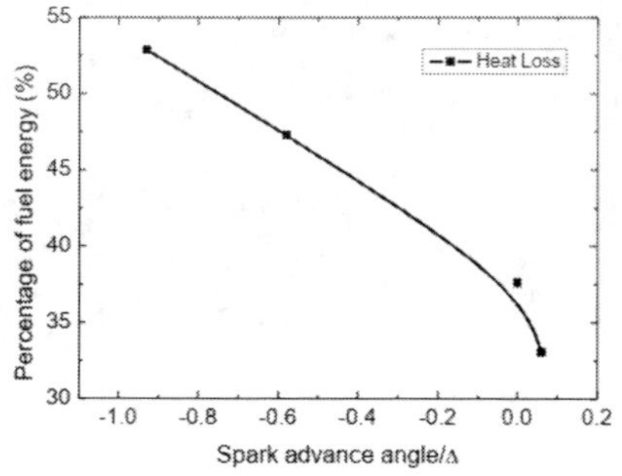

Figure 8: Heat loss at various ignition advance angles

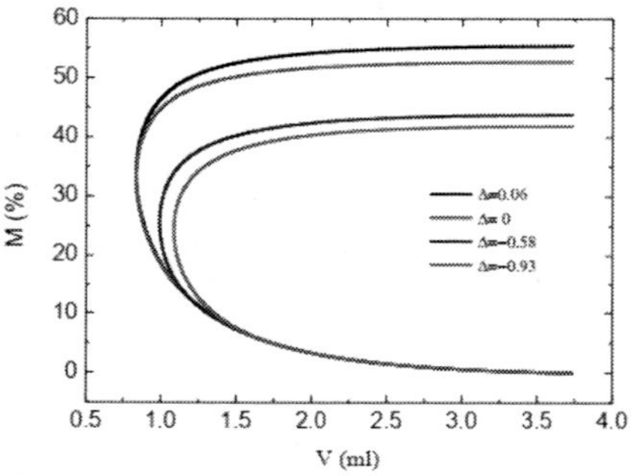

Figure 9: Relative leakage vs.volume

4 Conclusions

Combustion and single thermodynamic cycle of micro swing engine are studied by experiments and simulations. It is found that in the process of flame propagation, heat loss leads to lower flame temperature as compared with that in adiabatic combustion so that soot formation cannot be generated. Moreover, there exists an optimum ignition advance angle that favors a maximum thermal efficiency. Reducing the heat loss and leakage is beneficial to improve engine performance. This provides an important reference for subsequent engine optimization and improvement.

5 Acknowledgment

Financial supports from the National Basic Research Program of China (No: 2014CB239603) is gratefully acknowledged.

References

[1] A.H. Epstein, S.D. Senturia, A. Ayon, et al, Power MEMS and microengines; proceedings of the Solid State Sensors and Actuators, 1997 TRANSDUCER'S97 Chicago, 1997 International Conference on, F, 1997 [C]. IEEE

[2] A.H. Epstein, J. Eng. Gas Turbines Power 126(2004)205–27.

[3] K. Fu, A. Knobloch, B. Cooley, D.C. Walther, D. Liepmann, A.C. Fernandez-Pello, et al, Micro-scale combustion research for applications to MEMS Rotary IC Engines. NHTC2001-11212. Proceedings of 35th ASME 2001 National Heat Transfer Conf. Anaheim, CA; 2001.

[4] K.Fu, J.K. Aaron, C.M. Fabian, C.W. David, F.P. Carlos, et al, Design and experimental results of small-scale rotary engines, In: Proceedings of 2001 ASME international mechanical engineering congress and exposition, New York, USA; 2001.

[5] H.T. Aichlmayr, D.B. Kittelson, M.R. Zachariah, Chem. Eng. SCI. 57(19) (2002) 4161-4171.

[6] H.T. Aichlmayr, D.B. Kittelson, M.R. Zachariah, Chem. Eng. SCI. 57(19) (2002)4173-4186.

[7] H.T. Aichlmayr, D.B. Kittelson, M.R. Zachariah, Combust. Flame, 135(3) (2003) 227-248.

[8] K. Mijit. Design, analysis, and experimentation of a micro internal combustion swing engine, PhD thesis, University of Michigan, Ann Arbor, USA, 2000.

[9] W. Dahm, J. Mijit, R. Mayor, et al, Micro internal combustion swing engine (MICSE) for portable power generation systems[C], 40th AIAA Aerospace Sciences Meeting & Exhibit. 2002: 722.

[10] Y. Gu. Gasdynamic modeling and parametric study of mesoscale internal combustion swing engine/generator systems, PhD thesis, University of Michigan, Ann Arbor, USA, 2006.

[11] H. Zhu. Design, modeling and control of micro internal combustion swing engine (MICSE) systems, PhD thesis, University of Michigan, Ann Arbor, USA, 2006.

[12] Y.Y. Wu, B.C. Chen, F.C. Hsieh, et al, Int. J. Heat Mass Tran, 52(7) (2009) 1875-1886.

[13] X. Zhou, Z. Zhang, W. Kong, et al, Appl. Therm. Eng. 106 (2016) 674-680.

[14] K. Maruta, Proc. Combust. Inst. 33(1) (2011)125-150.

11th Asia-Pacific Conference on Combustion,
The University of Sydney, NSW Australia
10th -14th December 2017.

Application of Multiple Mapping Conditioning Mixing Model to Non-reacting Spray A

A. Varna*1, A. Wehrfritz[1], E. R. Hawkes[1,2], M. J. Cleary[3], T. Lucchini[4], and G. D'Errico[4]

[1]School of Mechanical and Manufacturing Engineering, University of New South Wales
Sydney, NSW, 2052, Australia

[2]School of Photovoltaic and Renewable Energy Engineering, University of New South Wales
Sydney, NSW, 2052, Australia

[3]School of Aerospace, Mechanical and Mechatronic Engineering, University of Sydney
Sydney, NSW, 2006, Australia

[4]Department of Energy, Politecnico di Milano
Milan, 20156, Italy

Abstract

The further development of compression-ignition engines would be facilitated by the availability of sufficiently general combustion models capable of accounting for all relevant in-cylinder processes, which include multi-mode combustion, fast and slow kinetics, and significant interactions of turbulence and chemistry. The transported probability density function approach (TPDF) is one method that, in principle, can account for all these phenomena. The TPDF approach requires a model for molecular mixing to represent transport in composition space. Experience with nonpremixed and premixed flames has shown that the property of localness (i.e. mixing occurring between fluid regions having similar compositions) is beneficial for model performance. On the basis of this observation, a novel multiple mapping conditioning (MMC) based mixing model has been proposed recently in which mixing is conditioned on a reference scalar that is statistically similar, but not identical to mixture fraction. The model has been found to perform well in the context of single-phase nonpremixed combustion due to its localness property. In the present work, the MMC mixing model is validated against the non-reacting Engine Combustion Network (ECN) spray-A case to test its applicability to multi-phase flows. Mass consistency is a major issue in PDF algorithms and a robust coupling between the spray and gas field ensures that the mass conservation law is not violated. The simulation results are validated using liquid length, vapour penetration and mixture-fraction field data. The degree of control over the obtained localness is assessed by comparing statistics of mixture fraction and the reference scalar.

1 Introduction

Increase in the demand for energy and transportation has increased the dependency on combustion systems while stringent emission regulations have lead to the development of efficient and "clean" engines. Often, selective catalytic reduction of the exhaust gas and in-cylinder combustion strategies are combined to meet the strict emission norms. Various combustion strategies have been proposed such as homogeneous-charge compression-ignition (HCCI) and low-temperature combustion (LTC) where the engines are typically operated near the stability limit. Under such operating conditions, which are prone to local flame extinction and re-ignition, turbulence-chemistry interaction (TCI) plays an important role. TCI has been well understood in the context of atmospheric flames and numerous computational fluid dynamic (CFD) studies employing various combustion models [18] have been conducted. However, TCI under diesel engine conditions is an ongoing research topic. The Engine Combustion Network (ECN) is a collaboration of several international research institutes which focuses on the experimental and numerical investigation of diesel engine combustion. Various target conditions for diesel spray combustion have been defined and investigated within the ECN. A wide range of combustion models has been applied to the ECN sprays, including perfectly stirred reactor (PSR) models [10], flamelet models [22], conditional moment closure (CMC) [23] and transported probability density function (TPDF) methods [14]. It is observed that the global quantities like ignition delay and lift-off length are reasonably well predicted by tuning the parameters of relatively simple combustion models, whereas the prediction of emissions [1], flame structure [14] and multiple injection cases [2] are more sensitive to the interactions between turbulence and chemistry. Most of the TPDF calculations of ECN sprays use either the simple interaction by exchange with the mean (IEM) mixing model [3] or the advanced Euclidean minimum spanning tree (EMST) model [19]. A new multiple mapping conditioning (MMC) [8] based mixing model has been developed recently [20] and tested in the context of Sandia flames D-F [21] which exhibit significant TCI. The new MMC mixing model can control the unconditional and conditional variance of the species mass fractions independently using two different model parameters, whereas the IEM and EMST models have just one parameter to control these statistics. The new MMC model, unlike IEM, has the localness property which is necessary to capture TCI and does not underpredict the conditional fluctuations like the EMST model [9]. The objective of this study is to apply the new MMC mixing model in a Reynolds-Averaged Navier-Stokes (RANS) context to simulate the non-reacting ECN spray-A conditions. The fuel spray is modelled using a standard discrete particle method for the liquid phase, whereas the gas phase is modeled using a finite volume (FV) solver coupled with a Lagrangian PDF solver. The statistics of the mixture fraction and reference variable are compared to verify that the first and second order moments are similar as expected by the MMC mixing model. The simulations are validated by comparing the liquid length, vapour penetration and the mixture-fraction field against the experimental data.

*Corresponding author.
E-mail address: a.varna@unsw.edu.au

2 The MMC mixing model

The transport equation for the one-point, one-time joint Eulerian composition PDF is

$$\frac{\partial \rho f_\phi}{\partial t} + \frac{\partial \rho \tilde{u}_i f_\phi}{\partial x_i} + \frac{\partial \omega_k f_\phi}{\partial \psi_k} + \frac{\partial S_k f_\phi}{\partial \psi_k} = -\frac{\partial}{\partial x_i}\left[\langle u_i' | \psi \rangle \rho f_\phi\right] + \frac{\partial}{\partial \psi_k}\left[\left\langle \frac{\partial J_i^k}{\partial x_i} | \psi \right\rangle f_\phi\right], \tag{1}$$

where ρ and $\tilde{u}_i$ are the density and Favre mean velocity, respectively. The symbols ω_k, S_k and J_i^k represent the chemical reaction source term, spray source term and molecular flux of the composition space vector ϕ_k, respectively. f_ϕ is the PDF of ϕ_k which includes chemical species mass fractions and mixture specific enthalpy, and ψ_k is the sample space for ϕ_k. The symbol $\langle | \rangle$ and the single prime represent conditional Reynolds averages and unconditional Favre fluctuations, respectively. The TPDF method [6] has been shown to be well suited for capturing TCI, as the chemical source term does not require a closure model. It is not restricted to any particular combustion regime and can be applied to nonpremixed, premixed and partially-premixed combustion. It can be used to simulate both fast and slow time scale processes. However, the turbulent flux and the transport in composition space represented by the first and second terms on the right hand side (RHS) of Eq. (1), respectively, are unclosed. The gradient model is typically used to calculate the turbulent flux [16] and has been found to be sufficiently accurate [6]. On the other hand, models for the transport in composition space, i.e. mixing models, are the subject of ongoing research in order to overcome the known limitations of the current models, see e.g. [20].

The Lagrangian implementation of the TPDF method is computationally efficient [16] and is widely used in modelling combustion systems. The evolution of the Lagrangian PDF particles is given by:

$$dx_i^* = \left(\tilde{u}_i + \frac{1}{\bar{\rho}}\frac{\partial \Gamma_{eff}}{\partial x_i}\right)dt + \sqrt{\frac{2\Gamma_{eff}}{\bar{\rho}}}dW_i, \tag{2}$$

$$d\phi_k^* = d\phi_{k,\omega}^* + d\phi_{k,s}^* + d\phi_{k,mix}^* \ , \tag{3}$$

where $W_i(t)$ is an independent Wiener process, * denotes particle properties, $d\phi_{k,\omega}^*$ and $d\phi_{k,s}^*$ represent the change in composition vector due to chemical reactions and evaporation of fuel droplets, respectively. The effective diffusivity Γ_{eff} is calculated as:

$$\Gamma_{\text{eff}} = \mu/\sigma_\phi + \mu_T/\sigma_{T\phi}, \tag{4}$$

where μ, μ_T, σ_ϕ and $\sigma_{T\phi}$ denote molecular and turbulent dynamic viscosities, and molecular and turbulent Schmidt numbers, respectively.

The third term on the RHS of Eq. (3) represents the change in composition due to mixing and various mixing models have been tested for the modelling of ECN spray A [14]. In this study, a new MMC mixing model developed recently [20] is applied to ECN spray A. The MMC mixing model equations are:

$$d\xi^* = d\xi_s^* + d\xi_{mix}^* \ , \tag{5}$$

$$d\xi_{mix}^* = -\frac{C_\phi}{\tau}(\xi^* - \tilde{\xi})dt + b_o\sqrt{\frac{2C_\phi\widetilde{\xi'^2}}{\tau}}dW_\xi \ , \tag{6}$$

$$d\phi_{k,mix}^* = -\frac{C_\phi}{2(1-r_t^2)\tau}(\phi_k^* - \widetilde{\phi_k|\xi})dt \ , \tag{7}$$

where ξ is the reference variable whose statistics are similar to that of the mixture fraction, Z, for nonpremixed and partially-premixed combustion ($\tilde{\xi} \approx \widetilde{Z}$, $\widetilde{\xi'^2} \approx \widetilde{Z'^2}$). The terms $d\xi_s^*$ and $d\xi_{mix}^*$ represent the change in reference variable due to evaporation of fuel droplets and mixing, respectively, $\tilde{\xi}$ and $\widetilde{\xi'^2}$ represent Favre mean and Favre variance of the reference variable, and $\widetilde{\phi_k|\xi}$ is the conditional Favre mean value of a scalar ϕ_k conditioned on the reference variable. The constants b_o, C_ϕ and r_t are the model parameters and it is observed that $b_o = \sqrt{0.5}$ is required to ensure $\widetilde{\xi'^2} \approx \widetilde{Z'^2}$ [20]. The scalar dissipation rate constant C_ϕ is found to be in the range [0.6,2.4] [4] for passive scalars like mixture fraction. In this study the commonly used value of $C_\phi = 2$ is used. The parameter r_t is the target correlation coefficient and controls the dissipation rate of conditional fluctuations. It is hence a measure of localness of the mixing model where $r_t = 0$ implies that the MMC model does not posses localness property due to which it behaves like the IEM model. For $r_t \to 1$ the model has extreme localness properties similar to the EMST model. In our previous study on Sandia flames D-F [21], a value of $r_t = 0.935$ was found to work well and the same value will be used in this study.

This work is the first application of the new MMC mixing model to multi-phase flows in which the reference variable has a source term due to evaporating fuel droplets. In the current study, a straight forward approach for the incorporation of a source term to the reference variable in the new MMC model is chosen. In a first step, the mixture fraction source term from the liquid spray is mass-weighted and added to the reference variable ξ of each PDF particle. During the subsequent mixing step, the new ξ value is determined as in the original MMC model (Eq. (6)). Next, all the PDF particles in a given FV cell are sorted in the order of increasing ξ and the scalar values, ϕ_k, are updated using Eq. (7) in which the conditional Favre mean is approximated by [17]:

$$\widetilde{\phi_k|\xi}^n \approx \frac{m^{n+1}\phi_k^{n+1} + m^{n-1}\phi_k^{n-1}}{m^{n+1} + m^{n-1}}, \tag{8}$$

where m^{n-1} and m^{n+1} are the masses of neighbouring particles of the nth particle in ξ-space. The turbulent time scale, τ, is calculated using the turbulent kinetic energy, k, and its dissipation rate, ε, as $\tau = k/\varepsilon$.

3 Numerical setup and experimental data

The calculations are carried out using a two-way coupled RANS-PDF solver implemented in OpenFOAM-2.4.x [13]. The FV solver determines the pressure, mean velocity and turbulence quantities (k, ε, μ_T) which are passed to the Lagrangian PDF solver. The PDF solver calculates the species mass fraction and total enthalpy, which are passed back to the FV solver where the density is updated using the equivalent enthalpy method [12]. Further details of the algorithm can be found in [21]. A first-order discretisation in time and second-order discretisation in space is used for the FV calculations. The liquid fuel spray is modelled using a standard discrete particle method. The Kelvin-Helmholtz Rayleigh-Taylor (KHRT) model is used to account for droplet breakup. Standard models for evaporation and heat transfer are employed; further information regarding the spray submodels can be found in [11]. The spray mass and enthalpy source terms are mass-weighted and added to all the PDF particles in a given cell and this procedure is found to work well in TPDF methods [5]. The standard k-ε [7] with $C_{\varepsilon 1} = 1.57$ is used in the present study.

The computational domain consists of a 2D axisymmetric wedge mesh with a length of 108 mm along axial direction and 60 mm along radial direction. The non-uniform mesh has 7040 cells with refinement near the centerline and nozzle exit.

The non-reacting ECN spray-A baseline case is selected for validation of the computational results in the current study. In the experimental setup, an n-dodecane fuel spray is injected into a constant volume chamber by a single hole injector. The ambient temperature and density of 900 K and 22.8 kg/m^3, respectively, are achieved in a pre-burn combustion event. Further details of the experimental setup and measurement techniques can be found in [15]. The nominal experimental conditions, as listed in Table 1, are used as boundary and initial conditions for the computational setup. The computational results are validated against experimental data for the liquid length and vapour penetration, as well as radial profiles for mixture fraction.

Parameter	Value
Ambient temperature	900 K
Ambient pressure	6 MPa
Ambient density (calculated)	22.8 kg/m^3
Nozzle diameter	0.084 mm
Fuel temperature	363 K
Injection pressure	150 MPa
Injection duration (ramp profile)	6 ms
Injected mass	13.77 mg

Table 1 – Experimental conditions

4 Results and Discussion

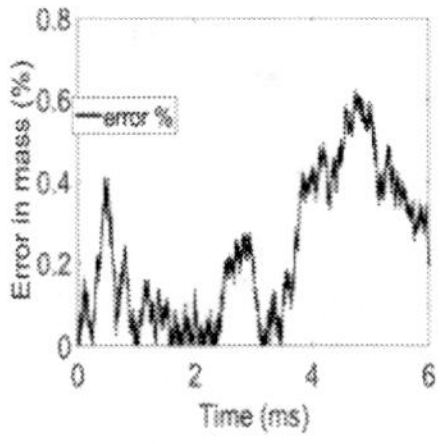

Figure 1 – The time evolution of mass percentage error.

Special care has to be taken to ensure mass, momentum and energy conservation between the (discrete) liquid phase and the (continuous) gas phase in fuel spray simulations. The injected liquid fuel droplets breakup and rapidly decelerate due to drag from the surrounding gas phase. The droplets are further subject to heat transfer and evaporation. This interaction of the liquid and gas phase has to be accounted for via source terms in the transport equations for mass, momentum and enthalpy. In the current setup, gas phase properties are obtained in a coupled RANS-PDF solver where mass and energy conservation has to be ensured in both the FV and PDF solver. Figure 1 shows the temporal evolution of the absolute percentage difference between the gas mass calculated by the FV and PDF solvers and it is seen that there is good mass consistency between two solvers.

Figure 2 shows that there is a good agreement between the liquid length and vapour penetration predictions of CFD and the experiments. In the present study, the liquid length is defined as the maximum distance of 99% liquid mass to the injection location; the vapour penetration is defined as the maximum axial of a FV cell which has a mixture fraction value of 10^{-3}. (Note that the turbulence and spray parameters are adjusted in order

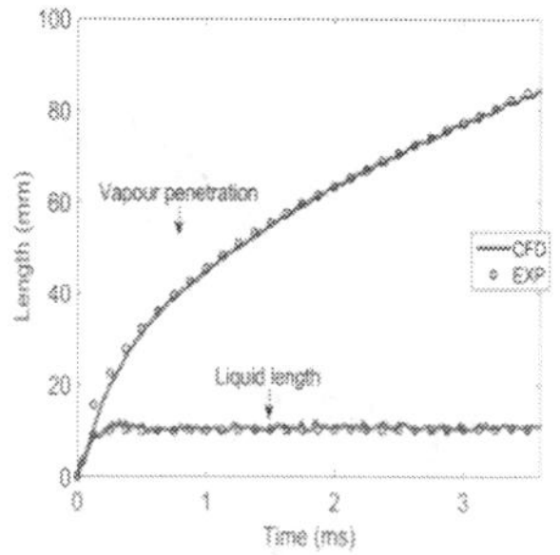

Figure 2 – Comparison of liquid length and vapour penetration between CFD and experiments.

to achieve this good agreement, as is standard practice in spray modelling.)

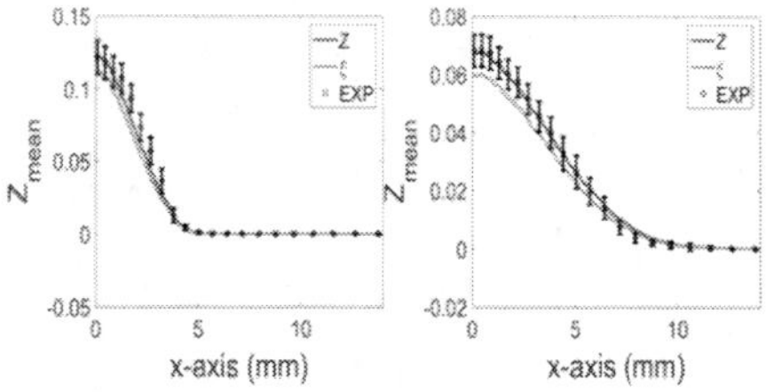

(a) $z = 25$ mm **(b)** $z = 45$ mm

Figure 3 – Validation of radial profiles of $\widetilde{Z}$ and $\widetilde{\xi}$ using experimental data at two axial locations.

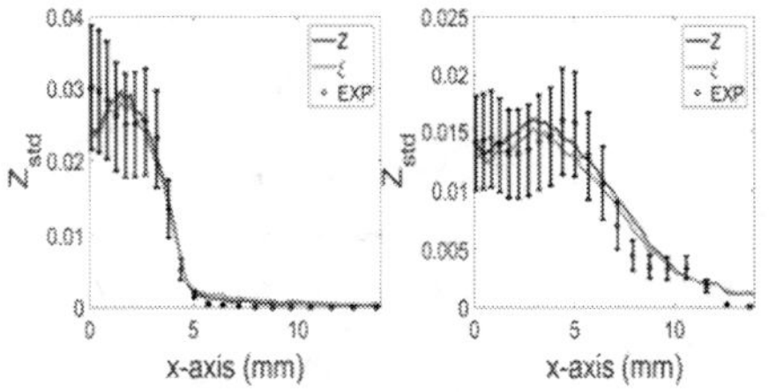

(a) $z = 25$ mm **(b)** $z = 45$ mm

Figure 4 – Validation of radial profiles of $\widetilde{Z'^2}$ and $\widetilde{\xi'^2}$ using experimental data at two axial locations.

The new MMC mixing model is devised such that $\widetilde{\xi} \approx \widetilde{Z}$ and $\widetilde{\xi'^2} \approx \widetilde{Z'^2}$. The MMC model becomes more physically intuitive and it is easier to determine the model constants when the statistics of the reference variable and the mixture fraction are similar. It has been established that $\widetilde{\xi} \approx \widetilde{Z}$ and $\widetilde{\xi'^2} \approx \widetilde{Z'^2}$ in single-phase nonpremixed combustion [21]. However, it remains to be seen whether this holds true in cases where Z and ξ have a source term due to the vapourisation of liquid fuel droplets. Figure 3 shows the radial profiles of the mean mixture fraction, $\widetilde{Z}$, and the mean reference variable, $\widetilde{\xi}$. It is seen that $\widetilde{Z}$ predicted by CFD agrees well with that of the experiments at $z = 25$ mm and $z = 45$ mm. Figure 4 shows that $\widetilde{Z'^2}$ is also predicted accurately. It can also be seen that there is only a minor deviation between $\widetilde{Z}$ and $\widetilde{\xi}$, and $\widetilde{Z'^2}$ and $\widetilde{\xi'^2}$ at $z = 25$ mm and $z = 45$ mm which can be attributed to numerical error. Hence, the MMC mixing model's expectation of $\widetilde{\xi} \approx \widetilde{Z}$ and $\widetilde{\xi'^2} \approx \widetilde{Z'^2}$ is valid in the case of multi-phase flows.

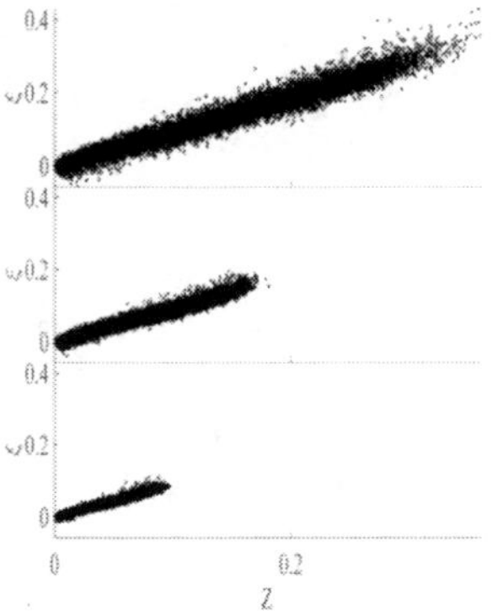

Figure 5 – Scatter plots of ξ v/s Z at z = 15 mm (top), z = 25 mm (centre) and z = 45 mm (bottom).

Next, the localness property of the new MMC model is assessed for multi-phase flows. Figure 5 shows the scatter plots of ξ versus Z at different axial locations where all particles at a given axial location are considered. It is seen that an almost linear mapping between ξ and Z at all the three axial locations is achieved. Hence, localization in ξ-space imply a high degree of localisation in Z-space thereby confirming the localness property of the MMC mixing model for a multi-phase flow. Further work is necessary to evaluate the localness property of the MMC model under reacting conditions.

5 Conclusions

A RANS-PDF algorithm is implemented and applied to simulate multi-phase flows. The computational results are validated against experimental data for the non-reacting ECN spray-A conditions. The gas phase is modelled using a coupled finite volume/Lagrangian TPDF approach and the liquid fuel spray is described by a standard discrete particle method.

The increase in mass due to evaporation of liquid fuel is captured consistently by the finite volume and PDF solvers. There is good agreement between CFD and experiments in terms of vapour penetration and liquid length. The computed radial profiles of the mixture fraction mean and variance are found to be satisfactory for the objectives of the current study.

This is the first study in which a recently developed MMC mixing model [20] is applied to multi-phase flows. It is shown that the new MMC mixing model is suitable for cases in which the reference variable has a source term. It is shown that the conditions $\widetilde{\xi} \approx \widetilde{Z}$ and $\widetilde{\xi'^2} \approx \widetilde{Z'^2}$ apply in multi-phase flows as well. An almost linear mapping between the reference variable and the mixture fraction is observed, implying that the localisation in the reference variable space ensures the localness property of the mixing model. Further research is required to determine the localness property of the new MMC mixing model in reacting flows.

6 Acknowledgements

This work is supported by the Australian Research Council.

References

[1] Bolla, M., Chishty, M. A., Hawkes, E. R., Chan, Q. N. and Kook, S., Proceedings of the Combustion Institute, **36**, 2016, 3551–3558.

[2] Bolla, M., Chishty, M. A., Hawkes, E. R. and Kook, S., International Journal of Engine Research, **18**, 2017, 6–14.

[3] Dopazo, C. and O'Brien, E. E., Acta Astronautic, **1**, 1974, 1239–1266.

[4] Eswaran, V. and Pope, S., Physics of Fluids, **31**, 1988, 506.

[5] Ge, H. W. and Gutheil, E., Combustion and Flame, **153**, 2008, 173–185.

[6] Haworth, D. C., Progress in Energy and Combustion Science, **36**, 2010, 168–259.

[7] Jones, W. and Launder, B., International Journal of Heat and Mass Transfer, **15**, 1972, 301–314.

[8] Klimenko, A. Y. and Pope, S. B., Physics of Fluids, **15**, 2003, 1907–1925.

[9] Krisman, A., Tang, J. C. K., Hawkes, E. R., Lignell, D. O. and Chen, J. H., Combustion and Flame, **161**, 2014, 2085–2106.

[10] Lucchini, T., D'Errico, G., Ettorre, D. and Ferrari, G., SAE Int. J. Fuels Lubr., **2**, 2009, 966–975.

[11] Maghbouli, A., Lucchini, T., D'Errico, G., Onorati, A., Malbec, L.-M., Musculus, M. and Eagle, W., SAE Technical Paper No. 2016-01- 0577.

[12] Muradoglu, M., Pope, S. B. and Caughey, D. A., Journal of Computational Physics, **172**, 2001, 841–878.

[13] OpenFOAM, *www.openfoam.org*, 2017.

[14] Pei, Y., Hawkes, E. R., Bolla, M., Kook, S., Goldin, G. M., Yang, Y., Pope, S. B. and Som, S., Combustion and Flame, **168**, 2016, 420–435.

[15] Pickett, L. M., Manin, J., Genzale, C. L., Siebers, D. L., Musculus, M. P. B. and Idicheria, C. A., SAE International Journal of Engines, **4**, 2011, 764–799.

[16] Pope, S. B., Progress in Energy and Combustion Science, **11**, 1985, 119–192.

[17] Pope, S. B., Physics of Fluids, **25**, 2013, 110803.

[18] Straub, C., De, S., Kronenburg, A. and Vogiatzaki, K., Combustion Theory and Modelling, **20**.

[19] Subramaniam, S. and Pope, S. B., Combustion and Flame, **115**, 1998, 487–514.

[20] Varna, A., Cleary, M. J. and Hawkes, E. R., Combustion and Flame, **181**, 2017, 342–353.

[21] Varna, A., Cleary, M. J. and Hawkes, E. R., Combustion and Flame, **181**, 2017, 354–364.

[22] Wehrfritz, A., Kaario, O., Vuorinen, V. and Somers, B., Combustion and Flame, **167**, 2016, 113–131.

[23] Wright, Y. M., De Paola, G., Boulouchos, K. and Mastorakos, E., Combustion and Flame, **143**, 2005, 402–419.

11th Asia-Pacific Conference on Combustion,
The University of Sydney, NSW Australia
10th -14th December 2017.

Laminar lifted flames in diesel engine conditions

D.K. Dalakoti[1], A. Wehrfritz[1], B. Savard[1], H. Wang[1] and E.R. Hawkes[1,2]

[1] School of Mechanical and Manufacturing Engineering, The University of New South Wales
Sydney, NSW, 2052, Australia

[2]School of Photovoltaic and Renewable Energy Engineering, The University of New South Wales
Sydney, NSW, 2052, Australia

Abstract

We present results from direct numerical simulations (DNS) of laminar lifted two-dimensional (2D) *n*-dodecane flames at thermochemical conditions corresponding to the Engine Combustion Network (ECN) target flame known as Spray A, which canonically represents conditions in a diesel engine. Simulations were performed for three distinct inlet velocities. Results indicate that the flames present a multibrachial structure. The number of branches depends on the inlet velocity and changes from five branches for the 2.5 m/s case to four branches for the 0.6 m/s case. For the 2.5 m/s case the flame consists of a typical triple flame (consisting of a diffusion, a rich premixed, and a lean premixed flames) with additional upstream low temperature chemistry (LTC) and high temperature chemistry (HTC) branches. With decreasing inlet velocity, the triple flame moves closer to the upstream HTC branch, ultimately overtaking it for an inlet velocity of 0.6 m/s. This indicates that the triple flame propagation velocity is O(1 m/s). These results provide an important baseline to compare against a 3D turbulent Spray A flame and investigate how turbulence changes the flame structure and stabilisation mechanism.

1 Introduction

The ever-increasing energy demand and stringent emission norms dictate that we continue to improve our combustion engines. High temperature and pressure conditions present in engines make it difficult to conduct detailed experiments and our understanding of combustion in practical conditions is therefore limited.

Flame properties like lift-off height have been shown to impact the flame structure and the resulting emissions [1]. A long lift-off height gives more time for fuel-air to mix, which is presumed to result in more premixed-mode burning over non-premixed mode, and in burning in relatively leaner mixtures. These factors result in lower emissions of soot and NOx, but may increase unburned hydrocarbon emissions, cycle to cycle variability, combustion noise, and may lower combustion efficiency. Fundamental to predicting the lift-off height and the overall flame structure is a comprehensive understanding of the stabilisation mechanism of the flame and the emissions it generates.

Previous work [2-4] suggests that in lifted flames in non-autoignitive conditions, stabilisation is achieved for low jet velocities by the propagation of partially premixed edge-flames (which correspond to the intersection point of triple flames that consist of a diffusion, a rich premixed, and a lean premixed flames), while at higher jet velocities, flammable regions become broad and stabilisation is achieved by the propagation of essentially premixed turbulent flames. The situation for diesel engines is much more complicated, not only because autoignition occurs, but because it occurs in two distinct stages with a cool flame preceding a hot one.

Experiments offer one alternative to understand combustion in diesel engine conditions. For instance, the Engine Combustion Network (ECN) [5] has been formed with the collaboration of over 30 partners. ECN's reference Spray A flame has generated a lot of attention as a representative case for combustion in partially premixed conditions with moderate exhaust gas recirculation (EGR).

Experimental measurements in high temperature and pressure conditions are difficult, however, and the available data related to the flame structure and stabilisation mechanism is mostly qualitative in nature. OH* chemiluminescence and/or OH planar laser-induced fluorescence (PLIF) and CH_2O PLIF are typically employed to image high-temperature combustion zones and regions undergoing low-temperature chemistry (LTC) respectively. Planar laser-induced incandescence (PLII) is used to image soot-containing regions in planes [6-8], while extinction measurements are used to gain information on total amounts of soot in lines of sight [9]. Experimental studies conducted using these methodologies have shown that the combustion in these conditions occur in two stages. Shortly after injection, LTC occurs which is identified by the presence of CH_2O species. LTC starts at the edge of the jet and subsequently covers the whole jet. Second stage ignition occurs near the tip of the jet. This then develops into a flame which propagates upstream until the quasi-steady lift-off height of the flame. The quasi-steady flame shows distinct regions of LTC and high temperature chemistry (HTC) marked by CH_2O and OH PLIF respectively [6-8]. Conceptual models of combustion at diesel engine relevant conditions by Dec et al [10], Musculus et al. [11] and Idicheria et al. [12] have proposed the quasi-steady flame to consist of a region of LTC, a rich premixed flame in the jet core, a soot region and a diffusion flame shrouding the jet.

Similar structures are also observed in numerical studies using Reynolds-averaged Navier-Stokes (RANS) simulation and large-eddy simulation (LES) frameworks [13-15]. However, there remain many open questions, as it is difficult to draw detailed conclusions from these studies due to the various assumptions involved in RANS and LES models. In light of these limitations, a 3D spatially developing DNS of Spray A was performed to enable a detailed understanding of the combustion physics in diesel engine conditions [16]. Figure 1 presents the contours of mass fraction of OH and CH_2O from the 3D results. It is readily observed that the flame is rather complex. To interpret the results, especially the stabilisation mechanism, we need simplified reference cases, in which the stabilisation mechanism can be readily understood, to compare against the 3D flame. Laminar lifted 2D flames provide such a suitable reference.

Recently a few DNS studies of laminar lifted flames of Dimethyl Ether (DME) in high temperature and pressure conditions have emerged [17-19]. These studies have shown the flame to contain multiple branches, where the number of these branches depend on the thermochemical and flow conditions. These studies serve as a reference for 3D flames [20] and help in understanding turbulence chemistry interaction and the impact of turbulence on the flame structure. Although these studies are very informative they are not directly relevant for diesel engine combustion due to differences in fuel and thermochemical conditions. Given the importance and interest in the Spray A reference case and the absence of studies of such laminar flames at Spray A conditions we present results from simulations of laminar lifted *n*-dodecane (a diesel fuel surrogate) flames in diesel engine conditions.

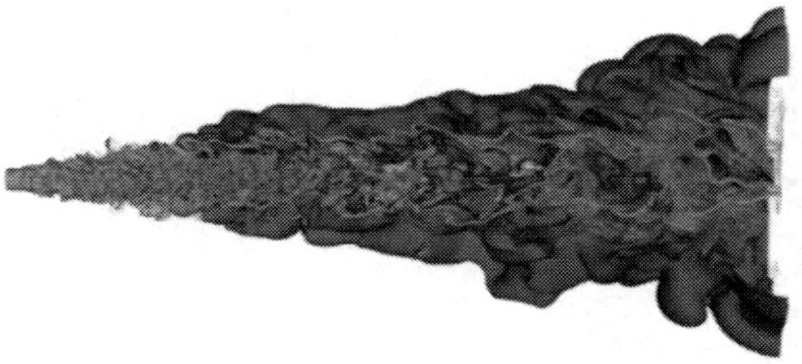

Figure 1: Contours of CH_2O and OH (pink) from 3D turbulent lifted flame in diesel engine conditions.

The main objectives for this study are:

1. to understand the 2D laminar flame structure in Spray A conditions which will eventually serve as a baseline to compare with 3D flames;
2. to understand the impact of inflow velocity on the flame structure; and
3. to assess the effect of the flame structure on the flame stabilisation mechanism.

2 Methodology

The flames were simulated using a low Mach number DNS code called LMC [21]. Using LMC has the advantage that it removes the acoustic signals from the simulation and the simulation time step is governed by advective scales rather than acoustic scales, permitting larger time steps and thus reducing the computational cost for low Mach number problems. LMC also has adaptive mesh refinement (AMR) capabilities. Using AMR, the grid can selectively be refined in certain regions where important features occur. LMC uses a second order projection method to integrate the momentum equation. A spectral deferred correction method is used to integrate the species and energy equations [21].

The simulation domain is a 2D cartesian box of dimensions $10D*20D$ where D is the fuel inlet width. The domain was initialised with fuel on the left side and oxidizer on the right. The left boundary has a symmetry boundary condition and the entire simulation domain is shown in Figure 2. A simple hyperbolic tangent profile was prescribed as the inlet mixing profile. Both fuel and oxidiser have the same initial axial velocity. This was done to remove the shear and keep the flow laminar. Three different inlet velocities were simulated in order to span different combustion regimes. The simulation parameters are presented in Tab. 1. A hot spot was initialised in the domain to quickly ignite the mixture and let the flame develop. The simulation was run until steady state was reached. A 53 species *n*-dodecane chemical mechanism [22] was used. This chemical mechanism has both low- and high-temperature chemical pathways and has been previously used in several studies under similar conditions [14]. The problem was solved on a 3-level dynamic grid giving an effective resolution of 0.6 microns, which sufficiently resolved all length scales.

Property	Value
Fuel inlet width	0.085 mm
Oxidizer Composition	15% O2 85% N2
Pressure	60 atm
Jet composition	*n*-dodecane (45%), O_2 (9.3%), N_2 (45.7%)
Oxidizer Temperature	900K
Jet Temperature	470K
Inlet Velocity	2.5, 1.0 and 0.6 m/s

Table 1: Simulation parameters

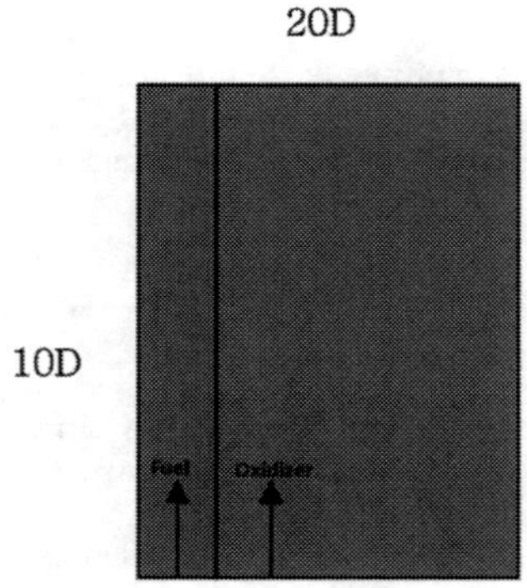

Figure 2: Simulation domain

3 Results

Figure 3 presents heat release rate contours for three inlet velocities (2.5, 1.0 and 0.6 m/s). Solid white lines in this figure represent mixture fraction iso-lines of 0.046 (stoichiometric), 0.064 and 0.1 from right to left respectively. Overall the flame presents a multibrachial structure. For the 2.5 m/s case, the flame presents a five-branch structure (branches numbered in the figure). The flame structure for the lower inlet velocity cases will be discussed later. Figure 4 presents the mass fraction of OH, a high temperature marker, $C_{12}H_{25}O_2$, a low-temperature chemistry marker [11], and H_2O_2, an intermediate species whose consumption is used to detect HTC [23]. These species are presented only for the 2.5 m/s case for

Corresponding author.
E-mail address: d.dalakoti@unsw.edu.au

brevity. Branch 1 in Fig. 3 corresponds to the location of mass fraction of $C_{12}H_{25}O_2$ and is thus associated with LTC. This branch starts at a lean mixture and moves towards richer mixtures with downstream distance. Branch 4 (along the stoichiometric mixture fraction line) corresponds to the location of OH and is thus due to HTC. This branch corresponds to the diffusion flame. H_2O_2 is consumed just upstream of branch 2, 3 and 5 suggesting that these branches are due to HTC. Branch 2 starts at very rich mixtures and divides into two separate segments downstream. One of the segments moves towards richer conditions and merges with the LTC branch, while the other segment moves towards comparatively leaner mixtures downstream and joins the edge flame (defined here as the triple flame structure consisting of branch 3,4 and 5) near the most reactive mixture fraction [23] location (0.064). Visually the flame structure is similar to a triple flame [24] with branches 3 and 5 corresponding to rich premixed and lean premixed HTC flames, respectively, and branch 4 being the diffusion flame. In addition to a typical triple flame, two additional upstream branches are observed in the present case. Such a flame structure was also observed in the DME studies by Deng et al. [17, 18].

Decreasing the inlet velocity changes the flame structure. For the 1.0 m/s case the edge flame has moved closer to the upstream HTC branch. Otherwise the flame structure is similar to that in the 2.5 m/s case. Further decreasing the inlet velocity to 0.6 m/s causes the edge flame to overtake the upstream HTC branch. For this case, the flame presents a classical triple flame structure with an additional upstream LTC branch. This is consistent with the observations by Deng et el. [17, 18] for laminar DME flames.

The observed flame structure has important implications for the flame stabilisation mechanism. For the 2.5 m/s case the presence of an upstream HTC branch indicates that this branch is at least partially responsible for flame stabilisation. The corresponding branch in the DME flame in the studies by Deng et al [17, 18] was attributed to HTC ignition and the flame was said to be stabilised by autoignition. Detailed characterization of this branch is beyond the scope of this paper and will be discussed in future work. With decreasing inlet velocity, the edge flame propagates to an upstream location, and eventually overtakes the HTC branch (branch 2). This indicates that edge-flame propagation starts to dominate with decreasing inlet velocity until it finally becomes the sole flame stabilisation mechanism.

The present cases can be used to estimate the displacement velocity of *n*-dodecane edge flames in the present conditions. The edge flame in these cases is propagating in the products of LTC which are expected to increase the propagation velocity. To quantify the propagation velocity, we extract the displacement velocity [24] of OH at the flame edge, defined here as the point of intersection of iso-lines of stoichiometric mixture fraction and mass fraction of OH at 3.2e-4 (20% of the maximum in the domain). This velocity differs from the inlet velocity due to the LTC heat release which causes flow acceleration and divergence of flow ahead of the flame edge which decreases the velocity along the propagation direction [20]. For the present simulations, these velocities are 1.56, 1.32 and 1.0 m/s for the 2.5, 1.0 and 0.6 m/s cases respectively. This gives an order of magnitude estimate of the propagation velocity of an edge flame in the products of LTC under Spray A conditions.

These results provide an important baseline to compare the 3D flame structure and stabilization mechanism. Comparison of the local velocities near the flame edge in the 3D turbulent case with the velocities obtained in the present cases will provide insights into the possible flame stabilisation mechanism in the

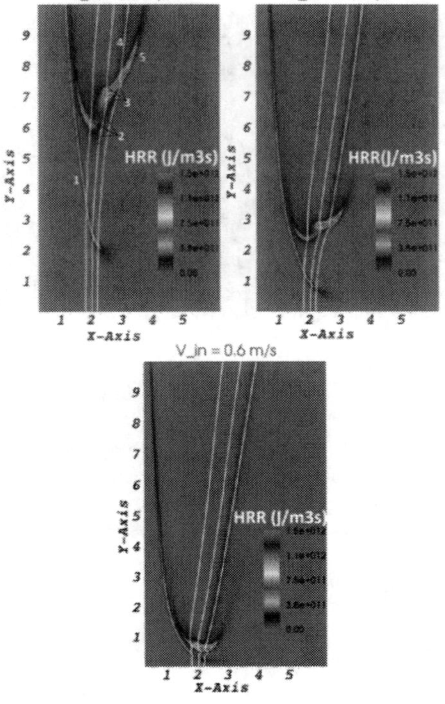

Figure 3: Heat release rate plots. White lines are mixture fraction iso-lines at 0.045, 0.064 and 0.1 (from right to left). The axes are labelled in 100 microns.

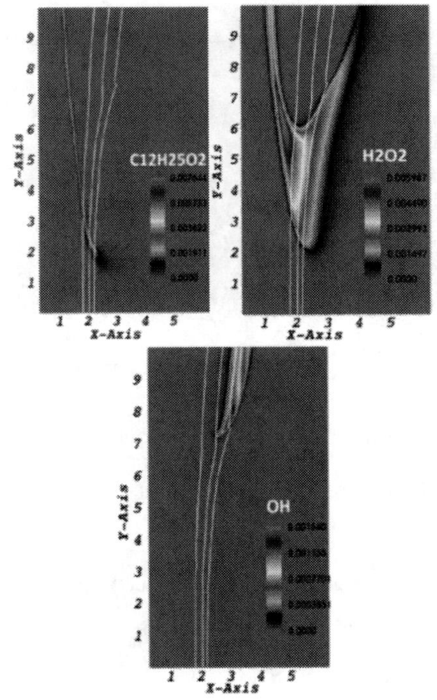

Figure 4: Mass fraction of $C_{12}H_{25}O_2$, OH and H_2O_2 for 2.5 m/s case. The axes are labelled in 100 microns.

3D case. A flame stabilised by edge flame propagation should have a displacement velocity of the same order of magnitude as that obtained here (for 0.6 m/s case) while an autoignition front can in principle sustain any velocity given that the residence times are sufficient to reach the ignition delay.

4 Conclusions

In the present work, 2D DNS of laminar lifted *n*-dodecane flames in Spray A conditions were performed. The main observations are listed below:

- The flames present a multibrachial structure. The number of branches depends upon the inlet velocity and changes from five branches for the 2.5 m/s case to four branches for the 0.6 m/s case.
- The presence of an upstream HTC branch suggests that for the 2.5 m/s case the flame has a hybrid stabilisation mechanism with contributions from both the upstream HTC branch and edge flame propagation.
- For the 0.6 m/s case, the flame is stabilised by an edge flame propagating into the products of LTC.
- The displacement velocity for edge-flame propagation is of the order of 1 m/s. This presents a baseline case to compare the edge flame displacement speed in the turbulent case.

5 Acknowledgment

This work was supported by the Australian Research Council (ARC). The research benefited from computational resources provided through the National Computational Merit Allocation Scheme, supported by the Australian Government. The computational facilities supporting this project included the Australian NCI National Facility, the partner share of the NCI facility provided by Intersect Australia Pty Ltd., the Peak Computing Facility of the Victorian Life Sciences Computation Initiative (VLSCI), Pawsey awarded through the National Computational Merit Allocation Scheme with funding from the Australian Government and the Government of Western Australia.

References

[1] L. M. Pickett et al., SAE Technical Paper (2005-01-3843) (2005).

[2] K. M. Lyons, Progress in Energy and Combustion Science 33 (2) (2007) 211-231.

[3] C. J. Lawn, Progress in Energy and Combustion Science 35 (1) (2009) 1-30.

[4] S. Karami et al., Journal of Fluid Mechanics 777 (2015) 633-689.

[5] https://ecn.sandia.gov/diesel-spray-combustion/target-condition/spray-ab/.

[6] L. M. Pickett et al., SAE International Journal of Engines 3 (2010-01-2106) (2010) 156-181.

[7] S. A. Skeen et al., Proceedings of the Combustion Institute 35 (3) (2015) 3167-3174.

[8] N. Maes et al., Combustion and Flame 174 (2016) 138-151.

[9] S. A. Skeen et al., Extinction-based imaging of soot processes over a range of diesel operating conditions, 8th US National Combustion Meeting, 2013.

[10] J. E. Dec, Proceedings of the Combustion Institute 32 (2) (2009) 2727-2742.

[11] M. P. B. Musculus et al., Progress in Energy and Combustion Science 39 (2-3) (2013) 246-283.

[12] C. A. Idicheria and L. M. Pickett, SAE Technical Paper(2006-01-3434) (2006).

[13] C. Gong et al., Applied Energy 136 (2014) 373-381.

[14] Y. Pei et al., Combustion and Flame 162 (5) (2015) 2006-2019.

[15] Y. Pei et al., Combustion and Flame 162 (12) (2015) 4442-4455.

[16] D. K. Dalakoti et al., Direct numerical simulation of two-stage combustion and flame stabilisation in diesel engine-relevant conditions, International Colloquium on the Dynamics of Explosions and Reactive Systems, Boston USA, 2017.

[17] S. Deng et al., Combustion and Flame 162 (9) (2015) 3437-3445.

[18] S. Deng et al., Combustion and Flame 162 (12) (2015) 4471-4478.

[19] A. Krisman et al., Proceedings of the Combustion Institute 35 (1) (2015) 999-1006.

[20] Y. Minamoto and J. H. Chen, Combustion and Flame 169 (2016) 38-50.

[21] A. Nonaka et al., Combustion Theory and Modelling 16 (6) (2012) 1053-1088.

[22] T. Yao et al., Fuel 191 (2017) 339-349.

[23] E. Mastorakos, Progress in Energy and Combustion Science 35 (1) (2009) 57-97.

[24] T. Echekki and J. H. Chen, Combustion and Flame 114 (1) (1998) 231-245.

11th Asia-Pacific Conference on Combustion,
The University of Sydney, NSW Australia
10th -14th December 2017.

An MMC-LES Simulation of Turbulent Piloted Flames using a Shadow Position Reference Variable

S. Galindo-Lopez[1], F.Salehi[1], M.J. Cleary[1], A.R. Masri[1], A.Y. Klimenko[2]

[1]School of Aerospace, Mechanical and Mechatronic Engineering, The University of Sydney

Sydney, NSW, 2006, Australia

[2]School of Mechanical and Mining Engineering, The University of Queensland

Brisbane, QLD, 4072, Australia

Abstract

This paper presents large eddy simulations (LES) of piloted partially premixed turbulent flames with homogeneous inlets. It is a first attempt to implement Shadow Positions (SP) [1] as reference variable in the MMC-LES modelling framework. The Sydney piloted burner is employed here although, at this stage, without any compositional inhomogeneity at the exit plane. The Shadow Position vector is used as reference variable for conditioning in the mixing model and a flame sheet approximation is made to treat the chemistry. Comparisons between the predictions and the experimental measurements are made. The flame structure is adequately predicted and it serves as a good starting point to expand the use of SP in multiple mapping conditioning (MMC) to compute the structure of premixed flames [2] and hence mixed-mode flames similar to those produced in the Sydney Burner when composition inhomogeneity is introduced.

1 Introduction

The use of computational fluid dynamics for the simulation of turbulent reacting flows is a key tool for the development of cleaner, fuel flexible combustion devices. With the advent of supercomputing and efficient algorithms, problems that were inconceivable to study a few decades ago are now within our reach. Moreover, the increase in computational power and memory capacity has opened the possibility to study the multi scale characteristics of turbulent flows using Direct Numerical Simulations (DNS) or Large Eddy Simulations (LES). However, in turbulent combustion the presence of chemical reactions introduces a more profound scale separation whereby chemical scales usually are smaller than the smallest turbulent scales and the typical mesh resolution used in practical simulations is coarser than what is required; hence, combustion models are essential for the development of combustion technology. Multiple combustion modelling approaches have been proposed ranging from variants of the laminar flamelet concept [3-5], conditional approaches [6-8], and probability density function (PDF) methods [9, 10]. While most of these models were developed in the Reynolds Averaged Simulations (RAS) context, extension and application to LES is the focus of current research. Two advantages for LES are that only the smaller scales are required to be modelled and LES permits the possibility to study unsteady phenomena [11].

Despite the success of many such combustion models, they still face important challenges to address the more complex and computationally demanding flames present in practical combustion devices. For example, flamelet approaches are computationally efficient but make strong *a priori* assumptions about the structure of the flame hampering the universality of the model, while PDF models are very general in their derivation yet a high computational cost is associated with them. A good modelling platform to achieve greater flexibility while keeping computational cost reasonable is multiple mapping conditioning (MMC) [12]. In its stochastic form, MMC combined with LES, is a filtered density function (FDF) approach with a high quality mixing model that allows for the use a sparse set of particles. Additionally, MMC permits us to address some of the difficulties present in other mixing models improving the range of applicability [13].

At the core of the sparse-Lagrangian multiple mapping conditioning model framework is the use of reference variables. These reference variables mimic or simulate quantities of importance to describe the thermochemical state of the mixture and play a role in the mixing operation by enforcing localness in an extended space comprised of reference variables and physical location [13]. Simulations using mixture fraction-like reference variables have been performed for non-premixed flames [13-17] with success and more recently applied to simulate flames with mixed-mode combustion [18] where it was shown that extra modelling efforts are required to capture the premixed structure of the flame. In recent years a new reference variable known as Shadow Position was proposed by Pope [1] for non-premixed flames and modified by Sundaram and Klimenko [2] to be used in premixed flames. Thus it seems logical to use SP as a platform to expand the existing MMC-LES implementation.

The longer term objective of this research is to develop SP in the MMC-LES approach for the prediction of mixed-mode combustion. The Sydney burner with inhomogeneous inlets [19, 20] will be used as a platform for validation. This paper only includes simulations of a partially-premixed diffusion flame with no inlet inhomogeneity as a first step in the development of an MMC model for premixed and mixed mode combustion.

Corresponding author. Phone: (+61) 2 93515092

Email: sgal8842@uni.sydney.edu.

2 MMC model equations

The equations representing an MMC model for the FDF of scalars using Shadow Position as a reference variable can be written as

$$d\mathbf{X}^* = \left(\tilde{\mathbf{U}} + \frac{1}{\bar{\rho}} \nabla \bar{\rho} \Gamma_{eff} \right) dt + \left(2\Gamma_{eff} \right)^{1/2} d\omega \quad (1)$$

$$d\mathbf{Z}^* = \tilde{\mathbf{U}} dt + \frac{\bar{a}}{T_L} \left(\mathbf{Z}^* - \mathbf{X}^* \right) dt + \bar{b} \left(2\Gamma_{eff} \right)^{1/2} d\omega' \quad (2)$$

$$d\phi^* = \left[-\frac{\bar{c}}{\tau} \left(\phi^* - \left\langle \phi^* \mid \mathbf{Z}^*, \mathbf{X}^* \right\rangle \right) + \mathbf{S} \right] dt \quad (3)$$

Here, the asterisk indicates that the quantity is stochastic and an isotropic Wiener process is denoted by $\boldsymbol{\omega}$. Equations (1), (2) and (3) model the physical location **X**, shadow position **Z** and reactive scalars $\boldsymbol{\phi}$ respectively. The evolution of the reactive scalars in (3) is due to chemical reactions **S**; and a mixing operation, which is represented here using *interaction-by-exchange with the mean* (IEM) conditioned on the SP reference variable. The diffusivity and the turbulent time scale are defined in LES as

$$\Gamma_{eff} = \nu + \nu_T \text{ and } \frac{1}{T_L} = \frac{\Gamma_{eff}}{\Delta^2} \quad (4)$$

and $\bar{a}$, $\bar{b}$ and $\bar{c}$ are model coefficients with their specification further explained in [1], τ is modeled as in [21], Δ is the filter size and ν_T is the turbulent viscosity.

3 Test case formulation

3.1 Burner setup

A thorough description of the burner, its stability characteristics and study case is reported in [22] and only a brief account is given here. The complete data set for a range of flames with varying inlet conditions is also posted on the web and may be made available on request [23]. Figure 1 shows an image of the burner that consists of two concentric tubes. The inner pipe with an inside diameter of Di=4mm (wall thickness of .25mm) and an outer pipe with an inside diameter of Do=7.5mm (wall thickness of 0.25mm). The outer tube is shrouded by a five gas pilot (5GP) stream of a stoichiometric mixture of acetylene, hydrogen, oxygen, nitrogen and carbon dioxide. This is designed to have the same C/H ratio as methane while at the same time matching its adiabatic flame temperature. The inner tube can slide within the annulus so that, at the right recess distance, the fuel/air issuing from the exit plane may be compositionally inhomogeneous. Near-homogeneous partially premixed conditions may be obtained at sufficiently large recession distances. The burner assembly is centred in a wind tunnel with a square cross section of 15x15cm providing an ambient temperature uniform air co-flow, which is fixed here at Uc=15 m/s.

The flames issued by the burner have a volumetric air to fuel ratio in the inner tube and the annulus is V_A/V_F=2. The fuel used here is methane. The selected case for study corresponds to the FJ configuration where fuel is injected in the central pipe and air in the annulus and a departure from blow-off such that U_J/U_{BO}=0.7, which corresponds to a bulk jet velocity of 59 m/s at the exit plane of the burner. The recess distance (Lr) is 300mm thus issuing a near-homogeneous partially premixed flame.

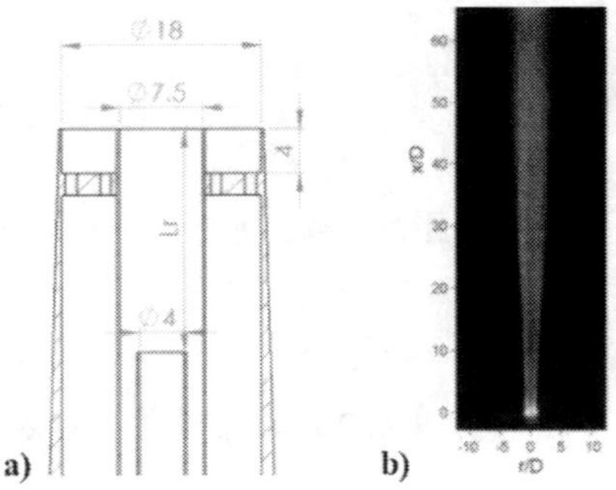

Figure 1: a) Inhomogeneous inlets burner cutout b) Homogenously partially premixed flame [22].

3.2 Numerical setup

The MMC-LES-SP model has been implemented into a new C++ object oriented code called mmcFoam that is compatible with OpenFOAM suite of solvers. The new sparse-Lagrangian MMC classes are coupled with OpenFOAM's existing low-Mach, compressible LES solver with the standard Smagorinsky subgrid model. There is two-way coupling between the LES and MMC so that the fields remain consistent. The filtered velocity, pressure and turbulence parameters computed by LES, are interpolated at particle locations and used to integrate particle movement followed by particle mixing and reaction substeps. Density coupling is achieved through an adaptation of the equivalent enthalpy method [13]. The source term is computed as

$$\bar{\rho} \frac{h_s - \tilde{h}_s^E}{\tau_{\mathrm{rel}}} \quad (5)$$

$\bar{\rho}$ is the filtered density, $\tilde{h}_s^E$ is the Eulerian filtered equivalent enthalpy, τ_{rel} is relaxation time and h_s is a target enthalpy. The relaxation time is determined by numerical conditions and in this work set as 10 times the numerical time step. In contrast to previous MMC-LES simulations, where the target enthalpy values are calculated using flamelet profiles in what could be called a flamelet parametric regression, the target enthalpy is computed as conditional means conditioned on mixture fraction using a kernel estimation that draws ideas from smoothed particle hydrodynamics [24].

For the computation of mixing operation, particles are sorted in the **[X, Z]** space using a k-d tree and the conditional mean in the IEM operator is computed using a two consecutive particle weighted mean. This is the same procedure used in other MMC simulations [13, 16-18] with the difference that SP replaces the mixture fraction for conditioning and is very similar to the near neighbor implementation used in [1].

The computational domain consists of a 3D rectangular mesh that extends 50 jet diameters axially and 10 jet diameters in both transverse directions. The mesh consists of 1.4 million cells and it is refined in the jet and pilot regions resulting in a mesh size of 0.4 mm cubed near the centreline. In the sparse-Lagrangian MMC, 1 particle per 8 Eulerian LES cells (1 L/8E) is used. Particle number control is associated with supercells consisting of 4 × 4 × 8 LES cells with an average of 16 particles in each.

Zero-gradient pressure boundary conditions are applied at the inlets and fixed total pressure conditions at the domain sides and outflows. The composition at the burner exit is set to a constant homogeneous mixture fraction (ξ = 0.22) and synthetic turbulence with intensity of 8% and an integral length scale of 1/8th of the jet diameter for the turbulent velocity inflow. Both values are found to give the correct jet breakup location. For the pilot and coflow streams a constant flat profile is imposed for composition, temperature and velocity. Combustion chemistry is modeled using the idealized flame sheet concept used in [1, 13] where reactive species are piecewise linear functions of the mixture fraction. Reactions which instantaneously convert the reactive species to its burnt state value occur only within a reaction zone described by the mixture fraction ξ and a single progress variable Y, whose equilibrium concentration is a unique function of ξ, $\mathbf{R} \equiv (\{\xi, Y\} \mid 0.03 < \xi \leq 0.132, Y \geq 0.55)$. In this case, a broader reaction region is used to account for the effects of partial premixing.

For the current simulations, we use the suggested value ($a = 1$) for the free model parameter in the Shadow Position model and the parameters $\bar{a}$, $\bar{b}$ and $\bar{c}$ are computed using the equations presented in [1]. Simulations were initialized and ran until a statistically steady state was reached and statistics where collected for 9 representative flow through times. The simulations were performed using 24 cores with run time of 40 hours per simulation.

4 Results

In this section some qualitative analysis is done followed by a quantitative comparison with experimental data.

4.1 Qualitative comparison

Figure 2 shows a side by side plot of a slice of the domain at one time step to compare the topology of the flame in physical space (Fig 2.a) with SP space (Fig 2.b). The particles are colored based on their mixture fraction value; with blue for mixture fraction values to the left of the reactive region, green for particles to right of **R** and red for the chemically active particles. At first glance, little difference is noted. However, there are some pockets in the slice where it can be seen that the reaction region is more localized (clustered) in SP space showing that mixing using SP enforces stronger compositional localness.

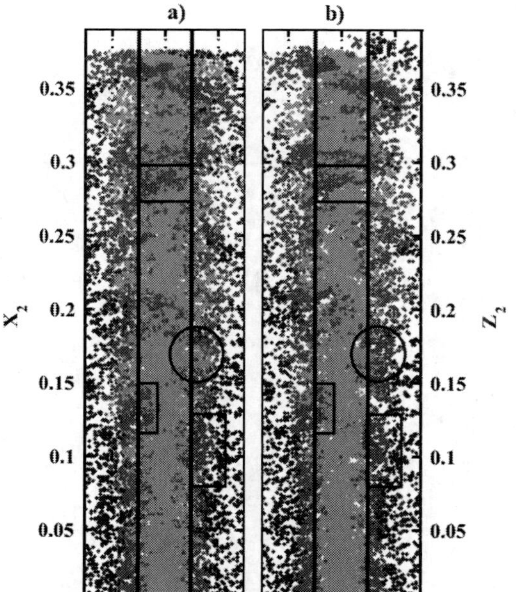

Figure 2: particle distribution of a thin slice of the domain. a) Particle distributions in physical space **X**. b) Particle distribution in shadow position space **Z**. Blue markers represent fuel lean particles outside of the reacting region ($\xi < 0.03$), red markers are particles inside **R** ($0.03 < \xi < 0.132$) and green markers are particle on the rich side of the reacting region ($\xi > 0.132$).

4.2 Radial profiles

First the results are examined in the form of radial profiles for mixture fraction and temperature. In addition to those obtained using SP (MMC-LES-SP) results for the same numerical set up but using the Eulerian Mixture fraction (MMC-LES-*f*) as reference variables are presented for comparison.

Figure 3 presents computed and measured mean and RMS radial profiles of mixture fraction. Agreement of the mean profiles with the experiments is good with some differences near the exit of the burner (x/D = 1) with a profile that is more diffused than the experiments. When comparing with the results using the mixture fraction approach, the difference is very little. However, it appears that near the exit of the burner the mixture fraction approach provides a better localization. Looking at the RMS profiles, good qualitative agreement with experiments is observed but with an overall over prediction at all axial locations.

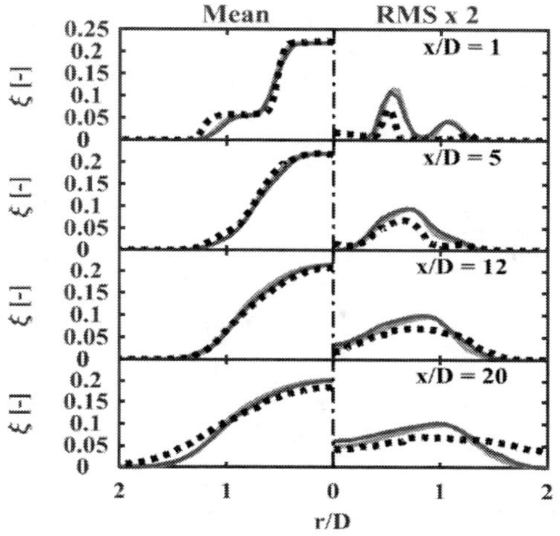

Figure 3: radial profiles of mixture fraction at different axial locations. Squares: experiments; Green line: MMC-LES-f; Red line: MMC-LES-SP.

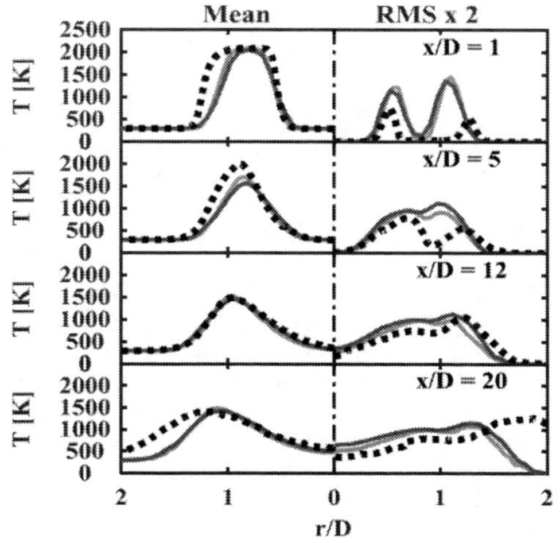

Figure 4: Radial profiles of temperature at different axial locations. Squares: experiments; Green line: MMC-LES-*f*; Red line: MMC-LES-SP.

Figure 4 shows a comparison for the mean and RMS fluctuations of temperature at various axial locations in the flame. Notwithstanding some differences in the peaks, it is evident that the trends are well reproduced and the profiles show similar trends to the mixture fraction profiles, which is expected for this case since the latter is used to describe the scalar structure of the flame. The broader shear layers produced in the simulations result in a thinner pilot at x/D = 1 and lower maximum mean temperatures at x/D =5. The difference between the two simulations is very small with the SP simulations giving slightly better agreement for temperature RMS away from the exit plane.

It is worth noting that the computed results near the exit plane of the burner depend more on the boundary conditions used for the simulations. In our simulations a constant flat profile for the pilot and the coflow was imposed and future work will use a velocity profile in accordance with the profile reported in the experiments.

4.3 Conditional profiles

To further the analysis, comparison of conditional statistics for the SP results with experiments is performed. Figure 5 shows conditional mean and RMS of temperature. Excellent agreement with the experiments is observed for mean profiles at x/D = 5 and x/D = 10. The profiles of temperature RMS follow the correct trends with under prediction of the RMS to the right of the stoichiometric mixture fraction ($\xi = 0.055$). The higher temperature RMS at x/D =5 is a result of the strong mixing between pilot and coflow as observed in the radial profiles above.

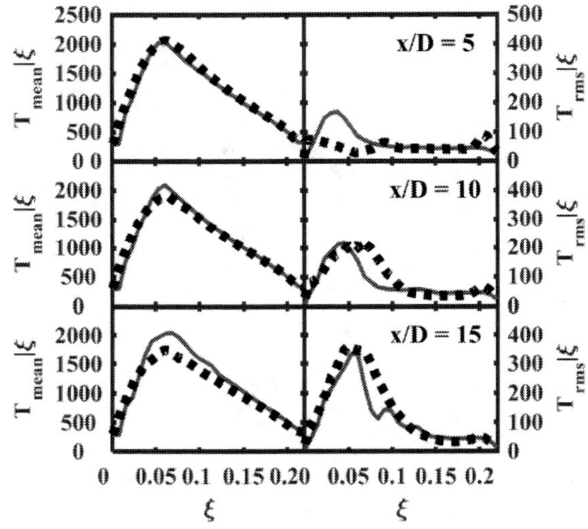

Figure 5: conditional mean and RMS for temperature at different axial locations. Squares: experiments; Red line: MMC-LES-SP.

5 Conclusions

The SP reference variable has been implemented for the first time in the MMC-LES approach and used to compute the structure of a turbulent partially premixed piloted flame with compositionally homogenous inlet. Results are satisfactory within the simplified set up used for the simulations. Comparison with the mixture fraction reference variable approach (MMC-LES-*f*) is also performed to further asses the capacity of SP to impose composition localness. Assessment of the model using finite rate chemistry as well as the influence of the model parameter *a*, is required and future work will address this issues. The usefulness of the shadow position reference variable for conditioning purposes is demonstrated and provides a good avenue for future work towards the computation of the structure of turbulent partially premixed and mixed mode flames using MMC. The next logical step will be the validation against turbulent premixed flames with thin premixed flamelets. The simulations of such flames present a big challenge for PDF/FDF methods with traditional mixing models as the requirement for composition localness is much stronger to avoid unphysical mixing across the flame front. To do so, the alternative SP reference variable proposed by Klimenko [2] for the simulation of thin premixed flamelets will be used.

6 Acknowledgment

This work is supported by the Australian Research Council. Mr Galindo-Lopez is supported by the Colombian Administrative Department of Science, Technology and Innovation – COLCIENCIAS. The authors acknowledge the Sydney Informatics Hub and the University of Sydney's high performance computing cluster Artemis for providing the high performance computing resources that have contributed to the research results reported within this paper.

References

[1] S.B. Pope, Phys. Fluids, 25 (2013) 110803.

[2] B. Sundaram, A.Y. Klimenko, Proc. Combust. Inst., 36 (2017) 1937-1945.

[3] N. Peters, Prog. Energy Combust. Sci., 10 (1984) 319-339.

[4] H. Pitsch, Steiner, Physics of Fluids (1994-present), 12 (2000) 2541.

[5] E. Knudsen, H. Pitsch, Combust. Flame, 159 (2012) 242-264.

[6] A.Y. Klimenko, R.W. Bilger, Prog. Energy Combust. Sci., 25 (1999) 595-687.

[7] S. Navarro-Martinez, A. Kronenburg, F.D. Mare, Flow, Turbul. Combust., 75 (2005) 245-274.

[8] H. Steiner, W.K. Bushe, Phys. Fluids, 13 (2001) 754-769.

[9] P.S. B., Prog. Energy Combust. Sci., 11 (1985) 119.

[10] H.D. C., Prog. Energy Combust. Sci., 36 (2010) 168.

[11] H. Pitsch, Annu. Rev. Fluid Mech., 38 (2006) 453-482.

[12] A.Y. Klimenko, S.B. Pope, Physics of Fluids (1994-present), 15 (2003) 1907-1925.

[13] M.J. Cleary, A.Y. Klimenko, Phys. Fluids, 23 (2011) 115102.

[14] Y. Ge, M.J. Cleary, A.Y. Klimenko, Proc. Combust. Inst., 34 (2013) 1325-1332.

[15] F. Salehi, M.J. Cleary, A.R. Masri, Y. Ge, A.Y. Klimenko, Proc. Combust. Inst., 36 (2017) 3577-3585.

[16] A. Varna, M.J. Cleary, E.R. Hawkes, Combust. Flame, 181 (2017) 342-353.

[17] A. Varna, M.J. Cleary, E.R. Hawkes, Combust. Flame, 181 (2017) 354-364.

[18] S. Galindo, F. Salehi, M.J. Cleary, A.R. Masri, Proc. Combust. Inst., 36 (2017) 1759-1766.

[19] S. Meares, V.N. Prasad, G. Magnotti, R.S. Barlow, A.R. Masri, Proc. Combust. Inst., 35 (2015) 1477-1484.

[20] R.S. Barlow, S. Meares, G. Magnotti, H. Cutcher, A.R. Masri, Combust. Flame, 162 (2015) 3516-3540.

[21] S. Vo, O.T. Stein, A. Kronenburg, M.J. Cleary, Combust. Flame, 179 (2017) 280-299.

[22] S. Meares, A.R. Masri, Combust. Flame, 161 (2014) 484-495.

[23] A.R. Masri, available at <http://web.aeromech.usyd.edu.au/thermofluids/database.php>

[24] J.J. Monaghan, Rep. Prog. Phys., 68 (2005) 1703.

11th Asia-Pacific Conference on Combustion,
The University of Sydney, NSW Australia
10th -14th December 2017.

Properties of in-cylinder fuel reformation and ignition characteristics of $CO/H_2/CH_4$ mixtures

Yuki Murakami[1], Hisashi Nakamura[1], Takuya Tezuka[1], Susumu Hasegawa[1], Go Asai[2], Kaoru Maruta[1,3]

[1] Institute of Fluid Science, Tohoku University,
2-1-1, Katahira, Aoba, Sendai 980-8577, Japan
[2]YANMAR CO., LTD.,
2481, Umegahara, Maibara, Shiga, 521-8511, Japan
[3] International Combustion and Energy Laboratory, Far Eastern Federal University,
Russky Island, Vladivostok, Russia

Abstract

In-cylinder fuel reformation and reactivity of reformate gases were investigated. Compositions of reformate gases derived from in-cylinder fuel reformation were numerically investigated over a wide range of initial temperatures and equivalence ratios. The computational results showed that products of fuel reformation were mainly CO and H_2 at low equivalence ratios. At high equivalence ratios, on the other hand, mole fractions of CH_4 and C_2H_4 became large, H_2 mole fraction became small and CO mole fraction was almost constant. To investigate the effects of CH_4 on reactivity of reformate gas, CO, H_2 and CH_4 were chosen as representative components of reformate gases in this study. Then four types of $CO/H_2/CH_4$ compositions were selected to investigate the effects of H_2-to-CH_4 ratios on reactivity at a constant fraction of CO. For the investigation on the ignition properties, a micro flow reactor with a controlled temperature profile was adopted and weak flames of $CO/H_2/CH_4$/air mixtures were observed. Experimental and computational results showed that the positions of weak flames shifted to the downstream side with the increase of CH_4 fraction, which means reactivity of $CO/H_2/CH_4$/air mixtures was reduced.

1 Introduction

Due to strict regulations regarding environmental pollutions for transport machines, engines with a fuel reformate system are being developed to enhance their thermal efficiencies and to reduce emissions. These engines are composed of two cylinders as shown in Fig. 1. Fuel reformation takes place at very high equivalence ratios (e.g. $\phi > 4$) with liquid fuels in the first cylinder. And then, the reformate gas and secondary air are introduced into the second cylinder where combustion occurs at near stoichiometry to obtain power. There are two potential advantages of this system. One is that combustion with small NO_x and soot emission can be achieved by controlling the condition of reformation and combustion, so that after-treatment is unnecessary or minimized. The other advantage is that the enhancement of thermal efficiencies of the system can be realized by recirculating the thermal energy of exhaust gases to fuel reformation. However, there would be large variations in compositions which include small hydrocarbons in addition to the main components of reformate gases, CO and H_2, due to conditions of fuel reformation. These variations in reformate gas compositions may lead to a considerable deviation in the reactivity of reformate gases. From this perspective, further investigations on the compositions and the reactivity of reformate gases are necessary for the design and the development of engines with in-cylinder fuel reformation.

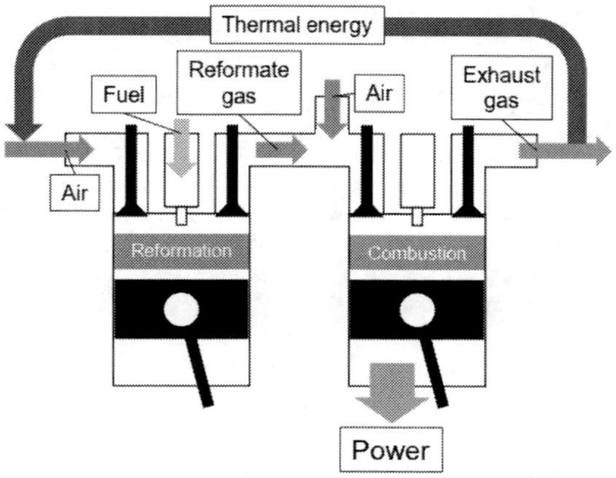

Figure 1: Schematic of a fuel reformate engine.

The reactivity of CO/H_2 mixtures with a wide variety of CO-to-H_2 ratios have been extensively investigated [1–4]. While CO/H_2 mixtures with separate additions of other components have also been studied [4–6], there are very limited studies focusing on realistic compositions of reformate gases. Petersen et al. experimentally and computationally studied ignition delay times of realistic compositions which contains a considerable amount of small hydrocarbons, N_2, CO_2, H_2O [7–8] and other impurities [8] for gas turbine conditions. The results showed considerable deviation in ignition delay times between neat CO/H_2 mixtures and those with other components, especially with small hydrocarbons. However, these studies focused on reformate gas derived from stationary reformation plants with coal and biomass. Compositions of reformate gas derived from in-cylinder reformation, whose residence time is much shorter than those of stationary reformation plants, with liquid hydrocarbons and their reactivity have not been studied.

In the present study, non-dimensional, adiabatic, variable-volume computations were firstly conducted to trace time histories of pressure, temperature and species compositions in fuel reformation processes. For the fuel, *n*-Tridecane was selected as a representative of diesel fuel. The equivalence ratio and initial gas temperature dependence on the reformation of *n*-Tridecane/air mixtures were investigated.

To focus on the effect of small hydrocarbons on the reactivity of CO/H_2 mixtures, CO, H_2, and CH_4 were chosen as representative components of reformate gases based on the computational results of the in-cylinder fuel reformation. Four kinds of simplified compositions were employed as follows: $CO/H_2/CH_4$ = 50/50/00, 50/45/05, 50/25/25 and 50/00/50. The reactivity of $CO/H_2/CH_4$/air mixtures at stoichiometric conditions were experimentally and computationally investigated using a micro flow reactor with a controlled temperature profile (MFR) [9].

2 Computational method for in-cylinder fuel reformation

Computations for fuel reformation of *n*-Tridecane/air mixtures were conducted by the AURORA package of ANSYS CHEMKIN-PRO v17.2 [10]. The specification of the fuel reformation engine is shown in Table 1. Computations were conducted within a range of crank angle from -140 to 125 degrees, and compositions of reformate gases at the final state corresponding to a crank angle of 125 degrees were investigated. Initial pressure was set to 1.115 atm. Equivalence ratios and initial gas temperatures were varied from ϕ = 1 to 10 and from T_i = 320 K to 1000 K, respectively. To investigate fundamentals of in-cylinder reformate gas, the initial temperature was changed up to 1000 K which is much higher than the initial temperature in the practical condition (e.g. 550 K in maximum). LLNL mechanism for C8−C16 *n*-Alkanes [11] was employed for the chemical kinetic model.

3 Experimental and computational method for fuel reactivity investigation

A schematic of the MFR is shown in Fig. 2. A quartz tube with an inner diameter smaller than the quenching diameter is employed as a reactor channel. A temperature gradient is given by an externally provided heat. A premixed fuel/air mixture is introduced into the reactor from the lower temperature side. The gas temperature strongly depends on the given wall temperature and flames of preheated mixtures in the laminar flow field under a constant pressure are observed. Past studies have reported that three types of flames are observed in MFR depending on the inlet mixture flow velocity: normal flames in the high flow velocity regime; flames with repetitive extinction and ignition (FREI) in the intermediate flow velocity regime, and weak flames in the low flow velocity regime [12–13]. A special attention was paid to weak flames in this study because it has been theoretically shown that the weak flame branch corresponds to the ignition branch of the Fendell curve [14]. Weak flames are stabilized in the upstream side at the low temperature region of the reactor when the reactivity of a mixture is high. This feature of weak flames enables to visually show the reactivity of mixtures [15-16]. Therefore, the weak flame methodology is a powerful method for investigating the reactivity of mixtures, particularly, ignition characteristics.

Yuki Murakami. +81-22-217-5296
E-mail address: murakami@edyn.ifs.tohoku.ac.jp

Table 1: Specifications of the fuel reformate engine.

Compression ratio	18.0 [-]
Volume at TDC	89.908 [cm3]
Stroke ratio	3.271 [-]
Speed	1800 [rpm]

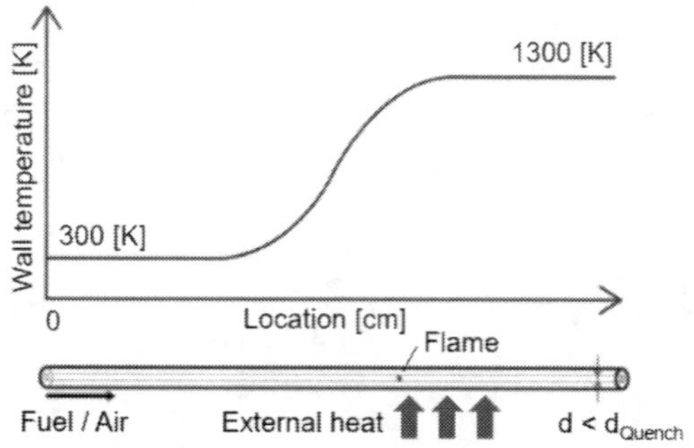

Figure 2: Schematic of experimental setup of MFR and a given wall temperature profile.

In this work, the reactivity of reformate gases were investigated by analyzing weak flames in MFR. A quartz tube with an inner diameter of 1 mm was employed as a reactor channel. The tube was heated by a H_2/air flat-flame burner to form a stationary temperature profile from 300 K to 1300 K along the inner surface of the channel. Premixed $CO/H_2/CH_4$/air mixtures at the stoichiometric condition were supplied to the reactor at an inlet flow velocity of 3 cm/s. A digital still camera with a band-pass filter (λ = 431.2 nm) was used to observe weak flames and the exposure time was set to be 3 minutes. The flame position is defined as the peak of the brightness profile in weak flame images.

For the validation of chemical kinetic models with experimental results, computations for weak flames were also conducted. The PREMIX package in ANSYS CHEMKIN-PRO v17.2 [10] was used for computations and one-dimensional steady computations were conducted. A term of heat transfer between gas and wall was added to the energy equation [9]. The wall temperature profile measured in the experiment was employed in the heat transfer term of the computations. The flame positions in the computations were defined at the peak of heat release rate (HRR) profiles. In this study, three chemical kinetic models were employed: USC Mech Version II of 109 species and 784 reactions [17], San Diego Mechanism of 50 species and 244 reactions [18] and AramcoMech 1.3 of 253 species and 1542 reactions [19].

4 Results and discussion

4.1 Compositions of products by in-cylinder fuel reformation

Figure 3 shows the mole fractions of major components at the final crank angle in the in-cylinder reformation. The mole fractions of CO and H_2 are maximum at $\phi \approx 3$ above $T_i \approx 500$ K, and then decrease with the increase of the equivalence ratio. The CO mole fraction is almost constant above $\phi \approx 5$ and $T_i \approx 500$ K while the H_2 mole fraction further decreases with the increase of the equivalence ratio. CH_4 and C_2H_4 are not produced below $\phi \approx$

4 while their mole fractions largely increase above $\phi \approx 5$ and $T_i \approx 500$ K. C_2H_2 is also formed at high equivalence ratio-high initial gas temperature region. However, the region of C_2H_2 production is much narrower and the amount is much smaller than those of CH_4 and C_2H_4. C_2H_6 is hardly produced in any equivalence ratio-initial temperature condition. These results indicate that CO and H_2 are mainly produced from $\phi \approx 2$ to 5 above $T_i \approx 500$ K, while the productions of small hydrocarbons such as CH_4 and C_2H_4 become larger above $\phi \approx 5$ and $T_i \approx 500$ K. Fig. 4 shows the maximum in-cylinder temperatures with the variation of initial temperatures and equivalence ratios. Fig. 4 indicates that the maximum in-cylinder temperatures are below 1700 K in high equivalence ratio region (approximately higher than 7) at all the initial temperatures. At low initial temperatures around 500 K, the region where the maximum in-cylinder temperatures are below 1700 K is also seen at lower equivalence ratios ($5 < \phi < 7$). According to the ϕ-T map introduced by Kamimoto and Bae [20], the production of soot is extremely small below 1700 K at any equivalence ratio considered in this work. Thus, no soot formation would be observed in such low maximum in-cylinder temperatures. Since the present computations of in-cylinder reformation is adiabatic for

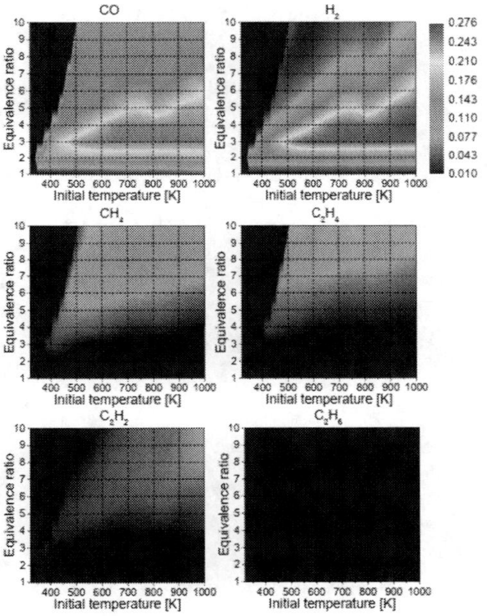

Figure 3: Mole fraction of products by in-cylinder fuel reformation.

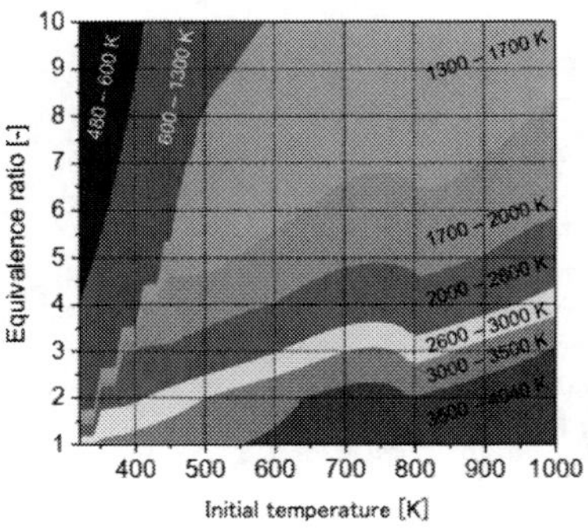

Figure 4: Maximum in-cylinder temperatures with variations of initial temperatures and equivalence ratios.

simplicity, an actual range of equivalence ratio and initial temperature to attain low maximum in-cylinder temperatures is expected to expand compared with the present computations. In fact, engine experiments with n-heptane ($\phi = 2$ to 5 and $T_i = 400$ K to 500K) showed no soot formation [21].

4.2 Representative compositions of reformate gases

Considering the amount of small hydrocarbons in reformate gases, effects of these substances on the reactivity of CO/H_2 mixtures must be investigated. As a first step, this study focused on CH_4 from small hydrocarbons and investigated effects of the CH_4 mole fraction on the fuel reactivity. From a practical perspective, the temperature range below $T_i \approx 550$ K was considered. At the higher initial temperature regime, compositions largely varied with the increase of equivalence ratios. Especially at $T_i \approx 500$ K, the CO mole fraction is almost constant and the CH_4/H_2 mole fraction relatively increases/decreases with the increase of equivalence ratio above $\phi \approx 5$. Fuel compositions of reformate gases were simplified as follows: the CO mole fraction in fuel mixtures was constant (50%) and the ratio of CH_4 to H_2 was varied. In the experiments, four kinds of simplified compositions were employed as follows: $CO/H_2/CH_4$ = 50/50/00, 50/45/05, 50/25/25 and 50/00/50.

4.3 Experimental and computational reactivity analysis by MFR.

Weak flame images of $CO/H_2/CH_4$/air mixtures obtained in the experiment are shown in Fig. 5. Flame positions are shifted to the high temperature region with the increase of the CH_4 mole fraction. Approximately 220 K difference in the wall temperatures at the flame positions is confirmed between 50/50/0 and 50/0/50 mixtures. These trends indicate that the reactivity of $CO/H_2/CH_4$/air mixtures are considerably reduced

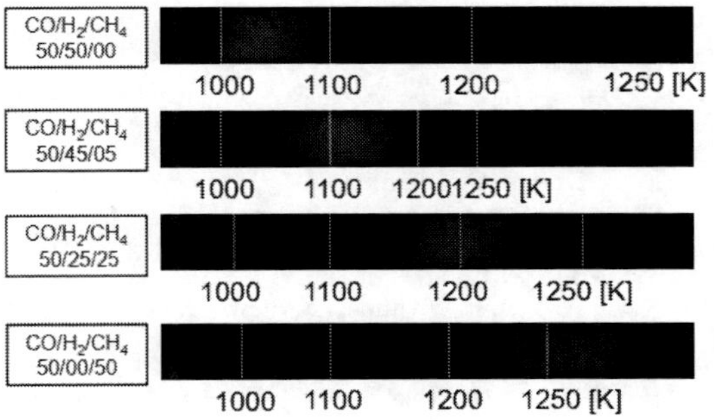

Figure 5: Weak flame images of $CO/H_2/CH_4$/air mixtures with varied compositions.

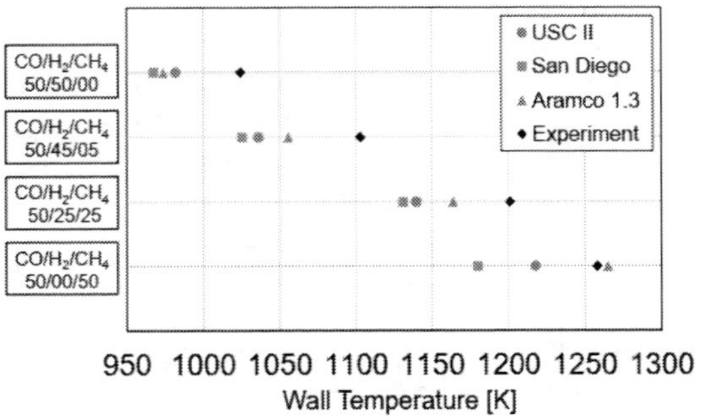

Figure 6: Comparison of experimental and computational results of weak flame positions.

by the existence of CH_4 in the mixtures.

The comparison of experimental and computational results of weak flame position for each mixture composition is shown in Fig. 6. Computational results showed that the flame positions are shifted to the downstream side of the reactor with all the chemical kinetic models. This behavior qualitatively agrees with the experimental results. Computational weak flame positions are mostly located at a more upstream side of the reactor where the wall temperature is lower than the experimental results. This implies that every chemical kinetic model employed in this study may overestimate the reactivity of $CO/H_2/CH_4$/air mixtures. AramcoMech 1.3 showed a relatively good agreement with experimental results compared with the other mechanisms. The computational weak flame positions deviate more among each other when the CH_4 mole fraction is larger.

4 Conclusions

The effects of equivalence ratios and initial gas temperatures on products of the in-cylinder fuel reformation were computationally investigated. Results showed that the mole fractions of major components in reformate gases, namely CO and H_2, are maximum at $\phi \approx 3$ above $T_i \approx 500$ K, and then decrease with the increase of the equivalence ratio. On the other hand, the ratios of small hydrocarbons such as CH_4 and C_2H_4 to H_2 significantly increase at high equivalence ratios and high initial temperatures. For the investigation on the effects of small hydrocarbons on the reactivity, $CO/H_2/CH_4$ mixtures with four kinds of simplified compositions were chosen as follows, $CO/H_2/CH_4$ = 50/50/00, 50/45/05, 50/25/25 and 50/00/50. The reactivity of $CO/H_2/CH_4$/air mixtures were examined with weak flames in a MFR. Experimental and computational results with a MFR showed that the weak flame positions largely shifted to the downstream side of the reactor with the increase of the CH_4 mole fraction. This indicates that the reactivity of $CO/H_2/CH_4$/air mixtures decreases with the increase of the CH_4 mole fraction.

References

[1] H. Sun , S.I. Yang , G. Jomaas , C.K. Law, Proc. Combust. Inst. 31 (2007) 439–446 .

[2] M.C. Krejci, O. Mathieu, A.J. Vissotski, S. Ravi, T.G. Sikes, E.L. Petersen, A. Keromnes, W. Metcalfe, H.J. Curran, J. Eng. Gas Turb. Power 135 (2013) 021503.

[3] X. Li, X. You, F. Wu and C. K. Law, Proc. Combust. Inst., 35, (2015) 617–624.

[4] H. Nakamura, H. Takahashi, T. Tezuka, S. Hasegawa, K. Maruta and K. Abe, Combust. Flame 172 (2016) 94–104 .

[5] J. Natarajan, T. Lieuwen, J. Seitzman, Combust. Flame 151 (2007) 104–119 .

[6] J. Santner , F.L. Dryer, Y. Ju, Proc. Combust. Inst. 34 (2013) 719–726.

[7] O. Mathieu, E.L. Petersen, A. Heufer, N. Donohoe, W. Metcalfe, H.J. Curran, F. Güthe, G. Bourque, J. Eng. Gas Turb. Power 136 (2014) 011502.

[8] O. Mathieu, M.M. Kopp, E.L. Petersen, Proc. Combust. Inst. 34 (2013) 3211–3218.

[9] K. Maruta, T. Kataoka, N. I. Kim, S. Minaev and R. Fursenko, Proc. Combust. Inst. 30 (2005) 2429–2436.

[10] ANSYS CHEMKIN-PRO v17.2, Reaction Design: San Diego, 2016.

[11] LLNL Mechanism, available at <https://combustion.llnl.gov/archived-mechanisms/alkanes/c8c16-nalkanes>.

[12] Y. Tsuboi, T. Yokomori and K. Maruta, Proc. Combust. Inst. 32 (2009) 3075–3081.

[13] H. Oshibe, H. Nakamura, T. Tezuka, S. Hasegawa, K. Maruta, Combust. Flame 157(8) (2010) 1572–1580.

[14] S. Minaev, K. Maruta and R. Fursenko, Combustion Theory and Modelling 11 (2007) 187–203.

[15] M. Hori, A. Yamamoto, H. Nakamura, T. Tezuka, S. Hasegawa and K. Maruta, Combust. Flame 159(3) (2012) 959–967.

[16] T. Kamada, H. Nakamura, T. Tezuka, S. Hasegawa and K. Maruta., Combust. Flame, 161(1) (2014) 37–48.

[17] USC Mech Version II., available at

< http://ignis.usc.edu/USC_Mech_II.html.>

[18] San Diego Mechanism, available at <http://web.eng.ucsd.edu/mae/groups/combustion/mechanism.html.>

[19] AramcoMech 1.3, available at <http://c3.nuigalway.ie/Mechanism_release/download.html>.

[20] T. Kamimoto, M. Bae, High combustion temperature for the reduction of particulate in diesel engines, SAE Paper (1988) 880423.

[21] G. Asai, Y. Watanabe, S. Ishiguro, G. Shibata, H. Ogawa, Y. Kobashi, T. Suzuki, , 2017 JSAE Annual Congress (Spring), JSAE paper (2017) 20175377.

11th Asia-Pacific Conference on Combustion,
The University of Sydney, NSW Australia
10th -14th December 2017.

On the Joint Statistics of Mixture Fraction and Reaction Progress Variable in Mixed Modes of Combustion

H.C. Cutcher[1], A.R. Masri[1], R.S. Barlow[2], G. Magnotti[2]

[1]School of Aerospace, Mechanical and Mechatronic Engineering

The University of Sydney, NSW, 2006, Australia

[2]Combustion Research Facility

Sandia National Laboratories, Livermore, CA 64550, USA

Abstract

This paper presents joint statistics of mixture fraction and an oxygen-based reaction progress variable in turbulent piloted flames of methane with compositionally inhomogeneous inlets. Previous studies of these flames have shown that inhomogeneity at the exit plane increases flame stability which has been attributed to premixed-stratified combustion occurring close to the jet exit. It is confirmed that for these partially premixed cases, through joint statistics and conditional probabilities, flammable mixtures close in the pilot-jet mixing layer are significantly more likely than in an equivalent homogeneous flame. This increased probability of high progress variable samples reinforces previous theories of upstream heat release supporting global flame stability in jets with inhomogeneous inlets. The correlation between mixture fraction and the reaction, which is an important parameter in the modeling of mixed-mode combustion is also discussed. This correlation also change with the introduction of compositional inhomogeneity.

1 Introduction

Compositional inhomogeneity is common in practical turbulent combustion systems due to the inherent difficulty associated with achieving uniform mixing within limited times or lengths. The possible formation of acoustic instabilities and flow pulsations only adds to this problem with their tendency to form concentration gradients [1,2]. As such understanding the mechanisms that drive stability of these flames is an integral part of improving their design. The Sydney inhomogeneous piloted burner, which is the subject of this paper, allows a closer look at mixed mode combustion and the effect it can have on improving flame stability. A comparison of multiple inlet conditions is possible by using two concentric pipes within the pilot stream where the inner pipe which carries fuel, can be recessed within the outer pipe which carries air, hence introducing compositional inhomogeneity at the jet exit plane.

Earlier papers [3–6] have shown that for a number of fuels, the flame stability is significantly improved at an intermediate recess distance of the fuel pipe, and hence at some optimal level of compositional inhomogeneity in the inlet profile. At such conditions, previous measurements of species concentration and temperature have shown that fluid mixtures in the region adjacent to the hot pilot stream are more likely to be within the fuel flammability limits. This is in contrast to the standard piloted jet case, when the mixture at the jet exit plane is homogeneous, such that diffusion and mixing are required before combustion can take place. The assumption that has been made previously is that this increased probability of flammable mixtures upstream in the flame results in greater levels of heat release near the jet exit, enhancing overall flame stability.

The use of a recently developed oxygen based progress variable designed for use with Raman and Rayleigh diagnostic measurements and multi-level conditioning has allowed a closer look at these flames. The use of greater conditioning confirms earlier theories on the importance of inlet inhomogeneity in piloted jet flames, and subsequently analogous systems.

2 Experimental Setup

The burner used in this study is shown in Figure 1 and consists of two concentric tubes surrounded by an 18mm diameter pilot. The 4mm inner diameter (0.25mm wall thickness) central tube can be recessed within the ID 7.5mm (0.25mm wall thickness) outer tube. By controlling the recession distance, *Lr*, the homogeneity at the exit plane can be varied from fully non-premixed at Lr = 0mm to near homogeneous at Lr = 300mm, with intermediate distances exhibiting partially premixed mixtures at the exit plane. For the flames studied in this paper, fuel issued from the central jet and air from the outer jet although the reverse configuration has been examined previously [3–5]. The burner was placed in 300 mm x 300 mm wind tunnel with a co-flow of velocity 15 ms^{-1}.

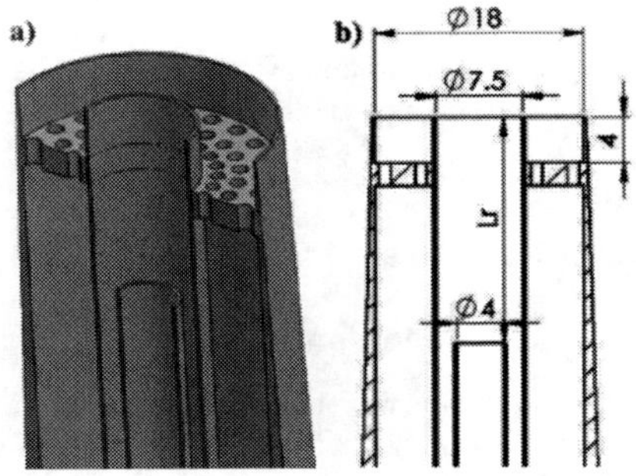

Figure 1 : Isometric (a) and schematic (b) view of the modified Sydney piloted jet burner.

Corresponding author. Hugh Cutcher
E-mail address: hcut6115@uni.sydney.edu.au

Four flames are investigated here, listed in Table 1, three having inhomogeneous mixtures at the inlet but varying departures from blow-off and one with near homogeneous inlet conditions. Note the full case names are listed in the table but for brevity abbreviated versions are used later on with the cases distinguished by their velocity, U_j, and recession distance, *Lr*. All flames have an air/fuel volume ratio of $V_A/V_F = 2$ using methane as fuel. The pilot consists of a mixture of acetylene, hydrogen, air, nitrogen and carbon dioxide, provides a heat release rate of $H_r = 2.2$kW, and is designed to have the same C/H ratio and adiabatic equilibrium temperature as a stoichiometric methane/air mixture. The bulk jet and blow off velocities, U_j and U_{bo}, listed in Table 1 are calculated from the total volumetric flowrate at the exit plane.

Cases	Lr (mm)	U_j (m/s)	U_{bo} (m/s)	U_j/U_{bo} (%)	Re_j
FJ200-5GP-Lr300-59	300	59	84.3	70	27600
FJ200-5GP-Lr75-57	75	57	114.3	50	26800
FJ200-5GP-Lr75-80	75	80	114.3	70	37500
FJ200-5GP-Lr75-103	75	103	114.3	90	48300

Table 1: Table of selected cases. The three inhomogeneous cases have black text and the one homogeneous case has blue.

The multi-scalar measurements presented here were taken at the Turbulent Combustion Laboratory (Sandia National Laboratories) using combined line imaging of Raman and Rayleigh scattering, and laser induced fluorescence of CO [7,8]. This provides single shot profiles of temperature and 8 major species (O_2, N_2, CH_4, H_2O, H_2, CO_2 and CO) across 6 mm long segments. Groups of 300-500 shots were taken at radial locations every 3mm along the laser axis from 6mm on the negative side of the centerline to beyond the flame on the positive side, giving at least 600 shots at each location. Additional shots were also taken at regions of interest but to avoid biasing the probability distribution functions presented later these are not included in the analysis presented in this paper. Measurements were taken a various axial locations from x/D = 1 to 30 although not all are presented here as the focus is on upstream locations.

Both the Raman and Rayleigh line imaging used four frequency doubled Nd:YAG lasers combined to achieve ~1.8J/pulse. To excite CO a 230.1 nm UV laser beam aligned along the same axis as the Nd:YAG laser beams. The signal from the scattering and fluorescence were collect using three CCD cameras after passing through a pair of 150 mm achromats with an optical resolution of 60μm. Measurements were collected at a 20 micron resolution, the oversampling rate chosen to maximise signal to noise ratio and take full advantage of the optical resolution [9].

To resolve the Raman-Rayleigh-LIF measurements at the fine resolution a wavelet denoising algorithm was applied. This signal processing algorithm uses a wavelet adaptive thresholding and reconstruction method and has been shown to give superior performance to other algorithms [9].

3 Joint Statistics

3.1 Variable Definitions

While there are many way to calculate the reaction progress variable from measured reactive scalars, a definition based on oxygen mass fraction was chosen. This was due to ease of calculation over other definitions using temperature or other species measurable in Raman/Rayleigh experiments [10]. The progress variable, c_o, is evaluated using two separate formulae to account for different fully-burnt states for lean and rich mixtures:

$$c_o = \frac{Y_{CO_2}\left(\frac{wO_2}{wCO_2}\right)+Y_{CO}\left(\frac{wO}{wCO}\right)+Y_{H_2O}\left(\frac{wO}{wH_2O}\right)}{Y_{CO_2}\left(\frac{wO_2}{wCO_2}\right)+Y_{CO}\left(\frac{wO}{wCO}\right)+Y_{H_2O}\left(\frac{wO}{wH_2O}\right)+Y_{H_2}\left(\frac{wO}{wH_2}\right)+Y_{CH_4}\left(\frac{2wO_2}{wCH_4}\right)};\ \varphi < 1 \quad (1a)$$

$$c_o = \frac{Y_{CO_2}\left(\frac{wO_2}{wCO_2}\right)+Y_{CO}\left(\frac{wO}{wCO}\right)+Y_{H_2O}\left(\frac{wO}{wH_2O}\right)}{Y_{CO_2}\left(\frac{wO_2}{wCO_2}\right)+Y_{CO}\left(\frac{wO}{wCO}\right)+Y_{H_2O}\left(\frac{wO}{wH_2O}\right)+Y_{O_2}};\ \varphi \geq 1 \quad (1b)$$

Mixture fraction was calculated from the major species' mass fraction using Bilger's formula [11].

3.2 Probability Distributions

When the recession distance is Lr = 75mm the blow-off velocity is about 36% higher than when for near homogeneous inlet conditions at Lr = 300mm. This improved stability has been attributed to the increased probability of flammable samples near the pilot [3,5], an argument that is supported by analysis of alternate fuels [6]. The following analysis of joint and conditional probability distributions aims to confirm that theory. Joint probability distributions of the oxygen based progress variable, c_o, and mixture fraction, ξ, at x/D = 1, 5, 12 and 20 can be seen in Figure 2 from which the following can be noted:

- In the upstream the three inhomogeneous flames are characterised by mixtures near stoichiometric rising sharply between c_O = 0 and 1, while the homogeneous flame shows a more gradual rise in progress variable as mixtures go from the fully mixed mixture fraction to stoichiometric. As the axial distance increases, the differences start to disappear as the inhomogeneous flames begin to look more similar to the FJ200-5GP-59mps-Lr300 case.
- At axial locations x/D = 1, 5, and 12 the homogeneous case has a probability peak near $c_O \sim 0$ and $\xi \sim 0.2$ not found in the three other cases due to their wider mixture fraction range (unburnt mixtures were observed up to $\xi \sim 0.8$).
- The mixing rate can be seen to vary slightly with velocity at x/D = 1, where samples lean of stoichiometric become increasingly common as velocity increases.
- Increasing signs of local extinction can be seen as axial distance increases, visible as partially burnt samples within the flammable limits. Examination at x/D = 20 demonstrates the effect of increasing percentage of blow off velocity, with the two cases at 70% U_j/U_{bo} (FJ-59mps-Lr300 and FJ-80mps-Lr75) showing similar levels of local extinction.

While the differences caused by increased velocity and differing exit plane mixture homogeneity are clearly visible in Figure 2, the cause of increased flame stability for intermediate recession distances, *Lr*, is not so well defined. To help demonstrate that increased probability of flammable samples near the pilot is the cause, Figure 3 shows joint PDFs of c_o and ξ conditioned on radial location at x/D = 1. Samples were grouped into 0.5mm wide radial bands from r=2.75mm to 5.25mm, getting farther away from the centerline of the jet from left to right, as indicated in the bottom row of the figure. This range of radial locations encompasses mixtures purely from the interior of the jet out to fully-burnt mixtures solely from the pilot, ensuring the pilot-jet mixing layer is fully represented. Note the joint PDFs are normalized to the number of samples within each radial band. From this figure a few key observations are:

- The bands closest and farthest from the jet centerline demonstrate the distributions expected of samples from the interior of the jet and the pilot respectively. The only significant difference between the homogeneous and inhomogeneous flames within these two bands being the mean mixture fraction in the jet, with mixtures much closer to stoichiometric for the partially premixed flames.
- As the distance from the centerline increases the distribution moves from $c_o = 0$ to 1 and the mean mixture fraction decreases for all flames although, in the case of the inhomogeneous flames there is a smaller change in mixture fraction distribution.
- The intermediate radial bands (3.25mm<r<4.75mm) are where significant differences can be seen between the inhomogeneous and homogenous flames. The inhomogeneous flames show distributions with a greater percentage of samples within the flammable range and closer to $c_o = 1$, suggesting that the flammable samples are burning in the upstream and providing additional stability to the flame.

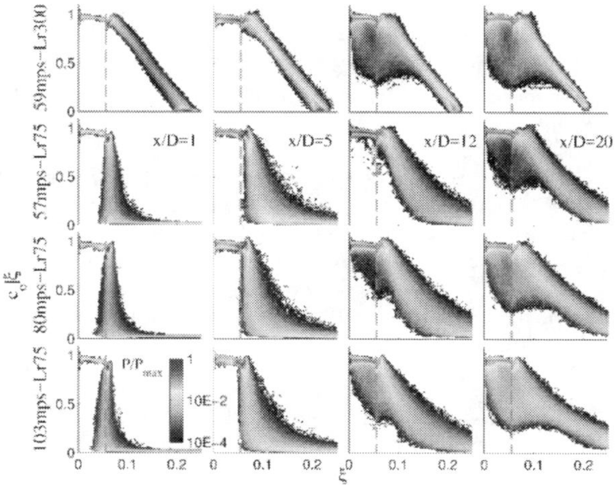

Figure 2: Joint probability distribution of progress variable and mixture fraction at x/D=1, 5, 12 and 20. The vertical dashed line marks the stoichiometric mixture fraction, ξ =0.055.

Results presented in Figs. 2 and 3 also show that both the conditional and unconditional correlations between the reaction progress variable and mixture fraction transition from being strong to weak depending on the inhomogeneity. Homogeneous fluid mixtures (being in cases FJ-59mps-Lr300 or the downstream location of the other three inhomogeneous cases listed in Table 1) show strong c-ξ correlation deteriorates significantly as compositional inhomogeneity increases. This conclusion is highly relevant to the modeling of turbulent flames with mixed –modes.

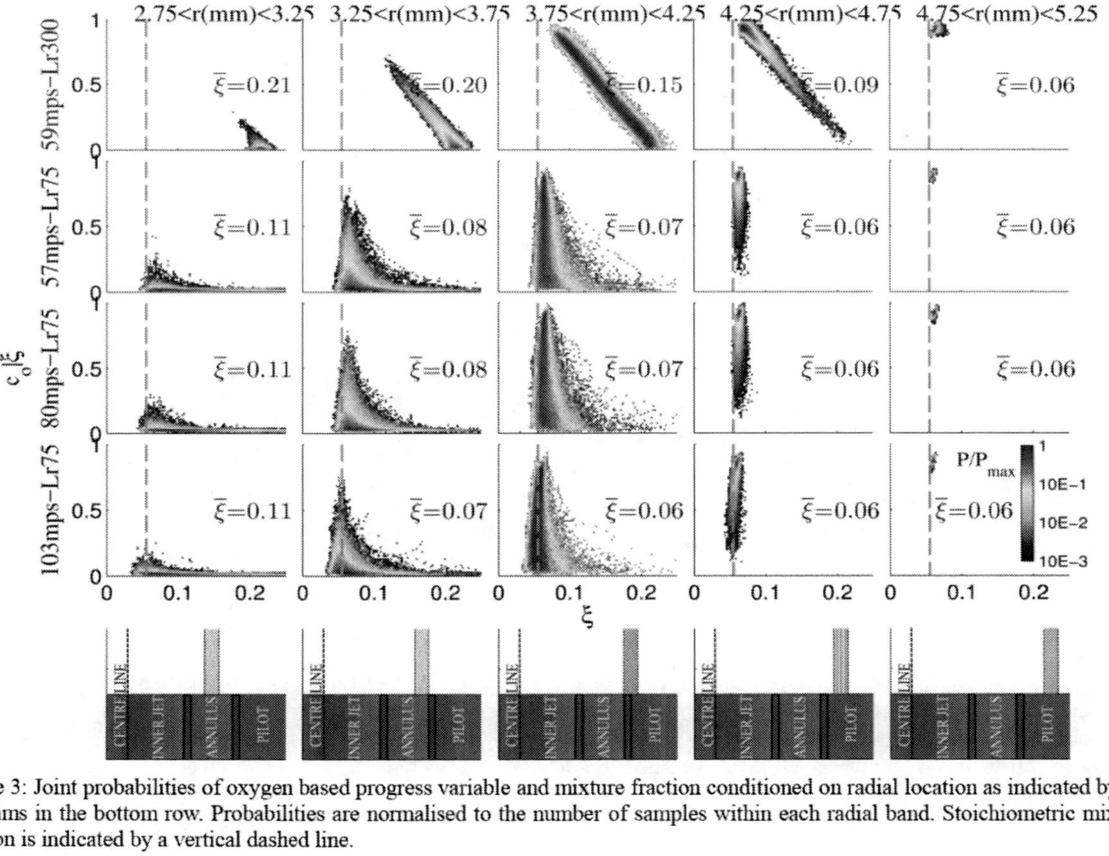

Figure 3: Joint probabilities of oxygen based progress variable and mixture fraction conditioned on radial location as indicated by the diagrams in the bottom row. Probabilities are normalised to the number of samples within each radial band. Stoichiometric mixture fraction is indicated by a vertical dashed line.

To help validate the additional probability of flammable mixtures burning or having been burnt at the pilot-jet interface in the inhomogeneous flames, as seen above, Figure 4 shows conditional probabilities of progress variable for samples conditioned on radial location and flammable mixture limits ($0.035<\xi<0.09$) and normalised to the number of samples with each radial band. Note the y-scale is logarithmic, a necessary feature to observe the relative difference in probability of flammable mixtures between the homogeneous and inhomogeneous flames. There are three key features to take away from this figure:

- As axial distance increases the probability of flammable mixtures existing decreases and the distribution across progress variable range becomes more even. This is due to richer mixtures from the jet become more prevalent and the pilot and jet become more mixed.
- Signs of flammable mixtures in FJ200-5GP-Lr300-59 are not visible until the third left most radial band (3.75<r(mm)<4.25) where mixing with the pilot begins to take place. The probability of flammable mixtures is also far lower than the inhomogeneous flames except in the interior of the pilot.
- There is no clear trend for increasing velocity and FJ200-5GP-80-Lr75 typically shows the largest probability of samples close to fully burnt. This could be due to the differing degrees of mixing caused by velocity having as significant effect as the increased turbulence.

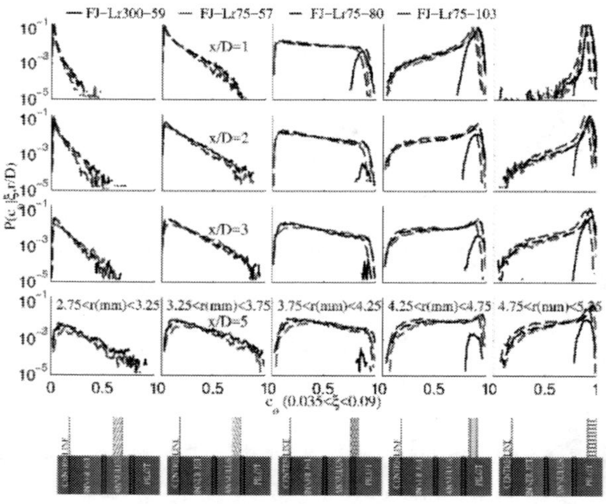

Figure 4: Probabilities of oxygen based progress variable conditioned on radial location and mixture fraction (flammable range). Probabilities are normalised to the number of samples within each radial band.

4 Conclusions

A look at an oxygen based progress variable showed the difference in flame stability for a set of flames with varying degrees of inlet homogeneity and departure from blow-off. The previously held theory of increased flame stability due to flammable mixtures near the pilot in a piloted jet burner is confirmed by conditional statistics. The difference in probabilities of mixtures within flammable limits occurring near the pilot, and their distribution in progress variable space, clearly points to increased upstream heat release as the mechanism for improved flame stability for the flames with compositionally inhomogeneous inlets. Bulk jet velocity has been shown to only have a minor effect on upstream reactivity. The correlation between mixture fraction and reaction progress variable is much stronger in compositionally homogeneous mixtures.

5 Acknowledgment

Work at the University of Sydney was supported by the Australian Research Council. Work at Sandia was supported by the Division of Chemical Sciences, Geosciences and Biosciences, Office of Basic Energy Sciences, US Department of Energy. Sandia National Laboratories is a multiprogram laboratory operated by Sandia Corporation, a Lockheed Martin Company, for the United States Department of Energy under contract DE-AC04-94-AL85000. Contributions by Bob Harmon in support of these experiments are gratefully acknowledged.

References

[1] P. Kutne, B.K. Kapadia, W. Meier, M. Aigner, *Proc. Combust. Inst.*, 33, (2) (2011) 3383–3390.

[2] W. Meier, X.R. Duan, P. Weigand, *Combust. Flame*, 144, (1–2) (2006) 225–236.

[3] R.S. Barlow, S. Meares, G. Magnotti, H. Cutcher, A.R. Masri, *Combust. Flame*, 162, (5) (2015) 3516–3540.

[4] S. Meares, A.R. Masri, *Combust. Flame*, 161, (2) (2014) 484–495.

[5] S. Meares, V.N. Prasad, G. Magnotti, R.S. Barlow, A.R. Masri, *Proc. Combust. Inst.*, 35, (2) (2014) 1477–1484.

[6] T.F. Guiberti, M. Juddoo, D.A. Lacoste, M.J. Dunn, W.L. Roberts, A.R. Masri, *Proc. Combust. Inst.*, 36, (2) (2017) 1777–1784.

[7] M.S. Sweeney, S. Hochgreb, M.J. Dunn, R.S. Barlow, *Combust. Flame*, 159, (9) (2012) 2896–2911.

[8] R.S. Barlow, G.-H. Wang, P. Anselmo-Filho, M.S. Sweeney, S. Hochgreb, *Proc. Combust. Inst.*, 32, (1) (2009) 945–953.

[9] M.S. Sweeney, S. Hochgreb, M.J. Dunn, R.S. Barlow, *Combust. Flame*, 160, (2) (2013) 322–334.

[10] R.S. Barlow, G. Magnotti, H.C. Cutcher, A.R. Masri, *Combust. Flame*, 179, (2017) 117–129.

[11] R.W. Bilger, S.H. Starner, R.J. Kee, *Combust. Flame*, 80, (2) (1990) 135–149.

11th Asia-Pacific Conference on Combustion,
The University of Sydney, NSW Australia
10th-14th December 2017.

Effects of Heat Loss at Walls on Flame Acceleration and Deflagration-to-Detonat ion Transition

Wenhu Han[1], Ning Du[1, 2], Wenjun Kong[1, 2, *], Chung K. Law[3, 4, *]

[1]Key Laboratory of Light-Duty Gas-Turbine, Institute of Engineering Thermophysics, Chinese Academy of Sciences, Beijing 100190, China

[2] University of Chinese Academy of Sciences, Beijing 100049, China

[3] Center for Combustion Energy, Tsinghua University, Beijing 100084, China

[4] Department of Mechanical and Aerospace Engineering, Princeton University, Princeton, NJ 08544, USA

Abstract

This paper studies the effects of heat loss at walls on flame propagation and deflagration-to-detonation transition (DDT) in micro-scale channel by high-resolution simulation. Numerical results show that wall heat loss leads to appearance of oscillating flames. This oscillating flame is attributed to the competition between viscosity and heat loss effects. Moreover, the occurrence of DDT with heat loss depends non-monotonically on the channel width.

1 Introduction

Studies have shown that flame acceleration and DDT in tubes and channels depend strongly on the tube diameter and channel width [1, 2]. In relatively wide macro-scale channels (~10mm), turbulent flame acceleration and reactivity gradient mechanism is the main reason for DDT [3]. Furthermore, numerical simulations have indicated that channel width affect not only the run-up distance and the mechanism of DDT [4, 5], but also the detonation mode [6]. However, as the channel width decreases to the micro scale (~0.1mm), heat conduction and viscosity at wall become important for flame acceleration and DDT [1, 7]. Previous study has discussed the effect of viscosity on flame acceleration and DDT in micro-scale channels [8], and found that viscosity has positive effect on flame acceleration, while it also led to the deficit of detonation velocity. In addition, heat loss at walls is also critical for flame propagation and DDT in a micro-scale channel. Ott et al. [9] found numerically "Jerky" flame in micro-scale channels with isothermal wall. Recently, Wu et al. [10] studied experimentally ethylene-oxygen DDT in capillary channels, and observed oscillating flames. They attributed the behavior to heat loss at the walls. This work simulates the entire process of DDT in micro-scale channels by solving the reactive flow Navier-Stokes equations, and reveals the respective roles of viscosity and wall heat conduction on the flame propagation and DDT.

2 Numerical model and specifications

A combustion wave propagating in micro-scale channel is studied by direct numerical simulation of the two-dimensional Navier-Stokes equations for a compressible reactive flow. The energy release rate is modeled by a one-step Arrhenius kinetics, which is assumed to be of first order with respect to the deficient reactant and of 2nd order with respect to the density. To numerically solve the governing equations, we apply fifth-order local characteristics based on the WENO conservative finite difference scheme [12] to discretize the advection term and the sixth-order central difference to the diffusion term, with third-order TVD Runge-Kutta time discretization.

The channel width W is given as $2f_l$ (f_l, flame thickness), $4f_l$, $8f_l$ and $16f_l$, and the Biot number Bi is 0, 7.5 and 10, respectively. A planar flame initiated by a weak ignition source, with expansion ratio of 10.0, is set at the left. In the unreacted mixture, the initial velocity, temperature and pressure are 0.0m/s, 300K and 1atm, respectively. Other parameters are specified as, specific ratio γ=1.3, heat release $Q=18R_uT_0/M$, activation energy $E_a=20R_uT_0$, Le=1 and Pr=0.75. Both the upper and lower boundaries are non-slip, adiabatic and heat loss walls, with $\partial T/\partial y = 0$ for adiabatic walls and $\partial T/\partial y = -\mathrm{Bi}\,(T - T_0)$ for heat-loss walls, where the Bi number denotes heat loss.

3 Results and discussions

3.1 Comparison of DDTs in opened and closed-end channels with heat loss at walls

Figure 1 shows the entire flame propagation and DDT in an opened-end channel for $W=8f_l$ and Bi=7.5. It is seen that the flame propagation consists of several stages: exponential acceleration, jerky flame, oscillating flame and detonation. Wu et al. [10] observed experimentally the oscillating flame in capillary channels. However, whether the oscillating flame can develop into DDT was no observed. The present results show the the oscillating flame can transit to DDT. Figure 2 shows that, compared with those in the opened-end channel, the appearance of the oscillating flame is delayed, in the closed-end channel, while DDT appears earlier. Consequently, in the following discussions we shall focus mainly on flame propagation and DDT in closed-end channels with heat loss.

3.2 Effect of heat loss on DDT in closed-end channels

Figure 3 shows the flame-tip position as a function of time for different channel widths and Bi numbers. It is seen that for

Corresponding author. Tel: +86-10-8254-3042
E-mail address: wjkong@iet.cn; cklaw@princeton.edu

W=8f_l the flame takes on different behaviors as the Bi number increases. For Bi=0, the flame accelerates and transits directly

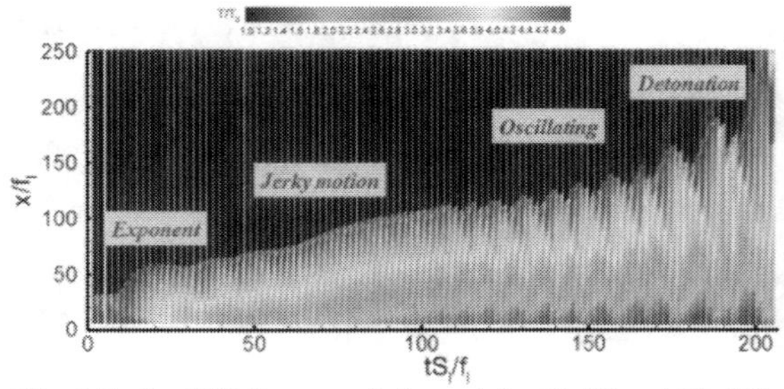

Fig. 1 Entire DDT in opened channel for W=8fl and Bi=7.5.

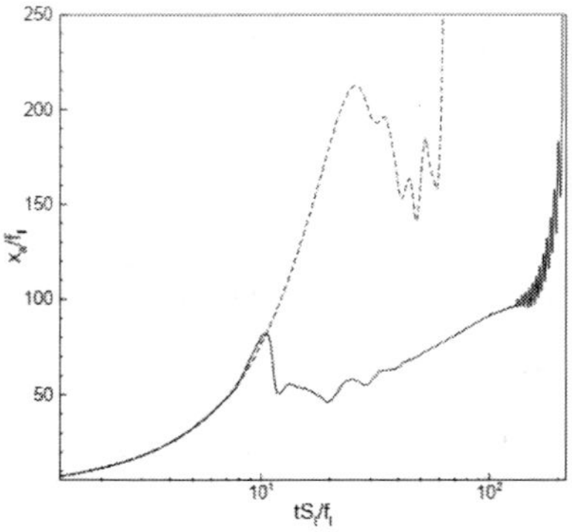

Figure 2: Flame propagation for for W=8f_l and Bi=7.5: solid lines (opened-end); dash lines (closed-end).

to detonation at x ~160. For Bi=7.5, the flame accelerates at x <220, while its acceleration rate is lower than that in the case with adiabatic walls (Bi=0); as a result, the flame begins to retreat at x ~220 because overall heat loss at walls accumulates as the product region becomes longer. In the process of the retreat, an oscillating flame forms and induces pressure pulse ahead at it; while DDT occurs at x ~160, it is late as compared with that with Bi=0. For Bi=10, the flame acceleration rate decreases further and the flame begins to retreat at tS_l/f_l =31.2; subsequently it takes on oscillating propagation and falls globally back; and DDT does not occur. Therefore, increase in Bi number (more heat loss) leads to the reduction of flame acceleration rate and suppresses significantly the occurrence of DDT.

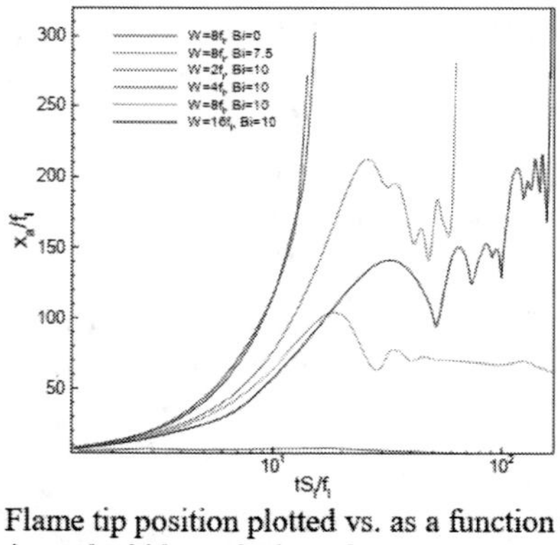

Figure 3: Flame tip position plotted vs. as a function of time for different channel widths and Bi numbers.

For Bi=10, the flame takes on different behaviors and the occurrence of DDT changes considerably as the channel width increases. For W=2f_l, heat loss arrests the flame propagation, causing it to quench eventually. As the width increases to 4f_l, the increase in the flame surface due to elongation caused by viscosity effect makes the flame accelerate and transit directly to detonation, although heat loss also plays a role in the flame propagation. This indicates that the role of viscosity is dominant. For W=8f_l, the flame acceleration rate is reduced; the flame starts to fall back at x ~125 and subsequently retreats after several oscillations and DDT does not occur. This indicates that with increasing channel width, the acceleration caused by viscosity effect is not enough to compensate for the deficit of flame speed caused by heat loss at the walls. Consequently, at x ~30 the flame starts to oscillate and fall back; and DDT is not observed. For W=16f_l, the flame acceleration rate decreases further because viscosity is weaker in the wider channel; at x ~160 the flame begins to retreat and assume oscillatory propagation. After strongly oscillating flame, DDT occurs. Consequently, oscillating flame and DDT is attributed to the competition between viscosity and heat loss effects. The occurrence of DDT depends non-monotonically on the channel width. With increasing channel width, the competition leads to different flame propagation behaviors and DDTs. Wu et al. [10] observed oscillating flame in their experiments of DDT in micro-scale channels. Ott et al. [9] observed numerically "Jerky" flame.

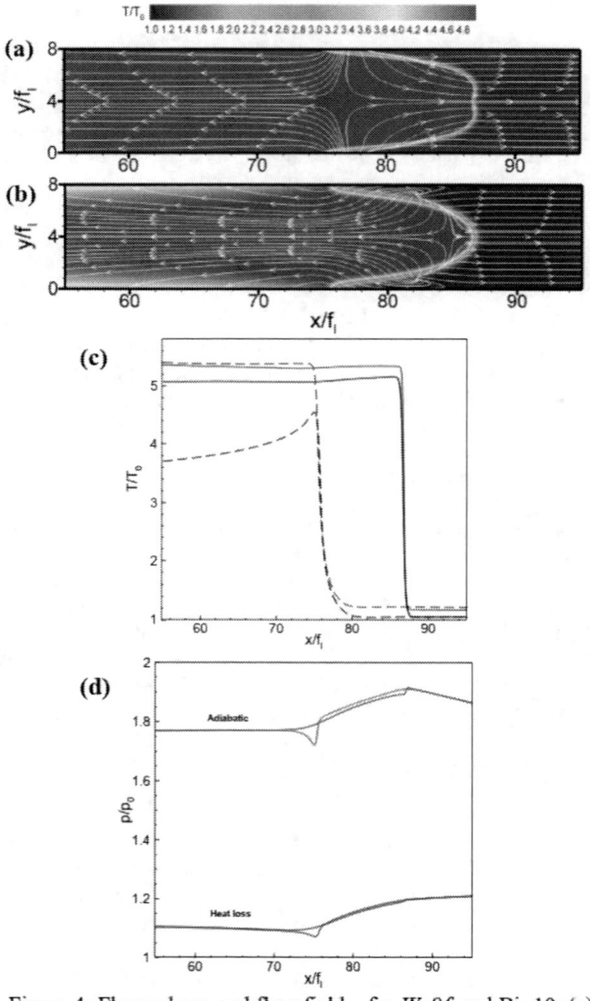

Figure 4: Flame shape and flow field for W=8f_l and Bi=10: (a) adiabatic, (b) heat loss, (c) temperature along wall (dash line) and axis (solid line); (d) pressure along wall (dash line) and axis (solid line).

We take W=8f_l and Bi=10 to compare the adiabatic and

heat-loss cases, as shown in Fig. 4. It is seen that, for the adiabatic case, viscosity effect causes flame elongation and increases the flame surface, leading to flame acceleration. There is a cusp on the front of the flame, which was explained by Bychkov et al. [13, 14]. For heat-loss case the reflow position shifts to the flame tip, and the downstream product has streamlines that are not parallel to the wall, as compared with that in the adiabatic case. As a result, the temperature near the boundaries is lower than that in the center. Moreover, the heat loss also leads to complicated flow between the flame surface and walls, which destroys considerably the adiabatic combustion. Consequently, the flame temperature at the channel center with heat loss is lower than that in the adiabatic case; see Fig. 4(c). This results in a decrease in the flame acceleration rate and the appearance of an oscillating flame. From Fig. 4(d), the pressure in front of the flame has positive gradient in the adiabatic channel, while an inverse pressure gradient appears for with heat-loss. The inverse pressure gradient leads to the reflow of unreacted gas in front of the flame and eventually causes the flame to retreat.

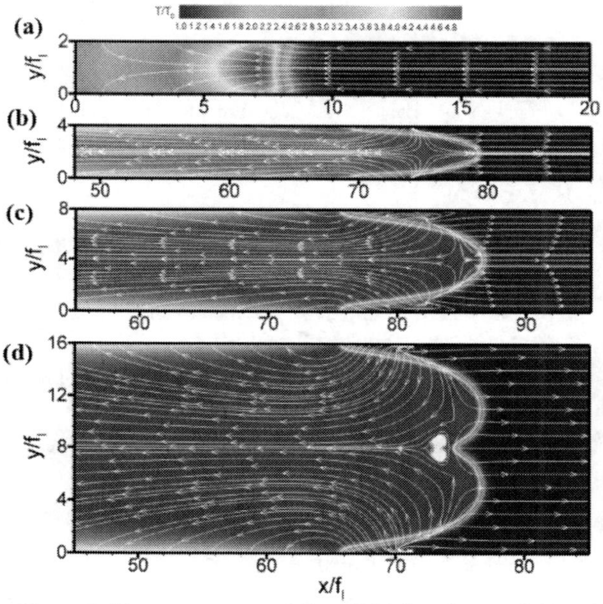

Figure 5: Flame structure and flow field for Bi=10: $2f_l$ (a), $4f_l$ (b), $8f_l$ (c), $16f_l$ (c).

Figure 5 shows the flame structure and the flow field for Bi=10. It is seen that for W=$2f_l$, the unreacted gas in front of the flame reflows into the front due to strong heat loss at the walls and eventually the flame quenches. For W=$4f_l$, the reflow positon shifts to the flame tip and weakens the flame acceleration rate; while the flame can still accelerate because it is elongated considerably due to viscosity effect. For W=$8f_l$, the reflow position shifts further to the flame tip; the reflow in boundary layers between the flame surface and walls reduces significantly the flame acceleration rate; see Fig. 2. For W=$16f_l$, the elongation and hydrodynamic instability lead to a cusp on the flame front. Hence, this front shape is beneficial for the flame acceleration, although the role of viscosity weakens and the substantial reflow is also caused by the heat loss. This indicates that in wider channels hydrodynamic instability influences flame propagation and DDT, in addition to heat loss and viscosity effects.

3.3 DDTs with oscillating flame

For W=$16f_l$ and Bi=10, the entire DDT consists of stages of accelerated flame, oscillating flame and transition-to-detonation. Hence, we chose this case to reveal the formation mechanisms of oscillating flame and DDT by inspecting front structures and flow field distributions, as shown in Fig. 6. It is seen that at tS_l/f_l =27.6, the cumulated heat loss leads to the reflow of all products behind the front; the gas between flame surface and walls takes on an inverse flow; as a results, flame propagation is rather slow. At tS_l/f_l =32.7-34.7, the downstream products and the upstream unreacted gas reflow and global combustion rate is lowered significantly; part of the released heat is lost to the walls. Hence, the flame stagnates and the jerky flame appears, see Fig. 6(c, d). Then the flame rapidly retreats, back to 120 at tS_l/f_l =45.6, as shown in Fig. 6(e). During the retreat, wall friction strengthens the reaction rate in the boundary layer. At tS_l/f_l =49.7 the flame in the middle rapidly falls back, while that near the walls stay up; hence the flame surface area and heat release increase, see Fig. 6(f, g). At tS_l/f_l =52.9, the flame near the walls accelerates and advances forwards again; forming a tulip flame. At tS_l/f_l =55.5, the flame accelerates, shown in Fig. 6(h). In the process of the acceleration, the flame in the middle of channel is fast, while that near walls is relatively slow. The flame surface decreases, see Fig. 6(i, j), such that the released heat is not sufficient to sustain the flame propagation. Hence, the flame retreats again, and eventually forms the oscillating flame shown in Fig. 3.

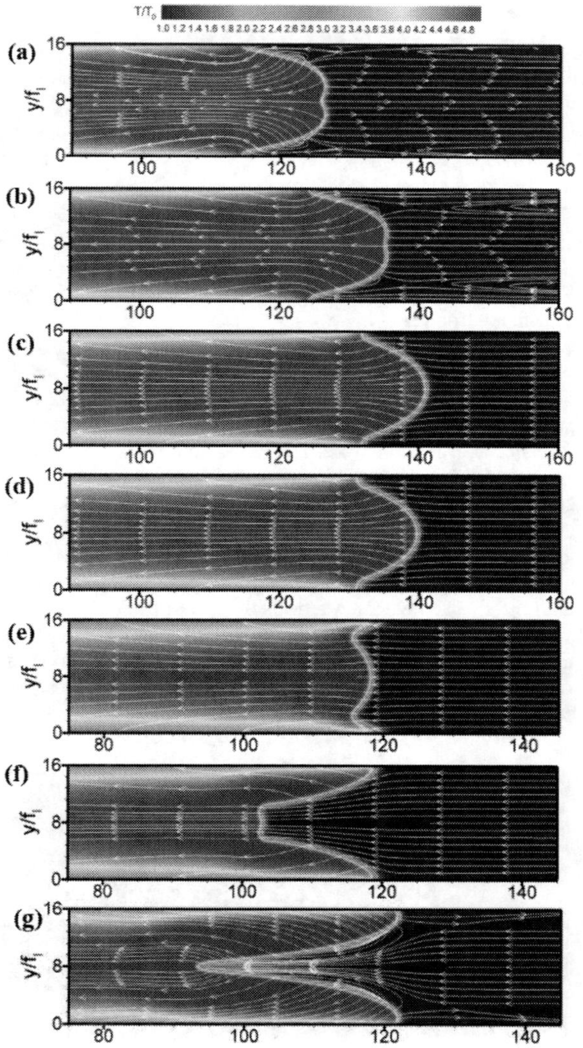

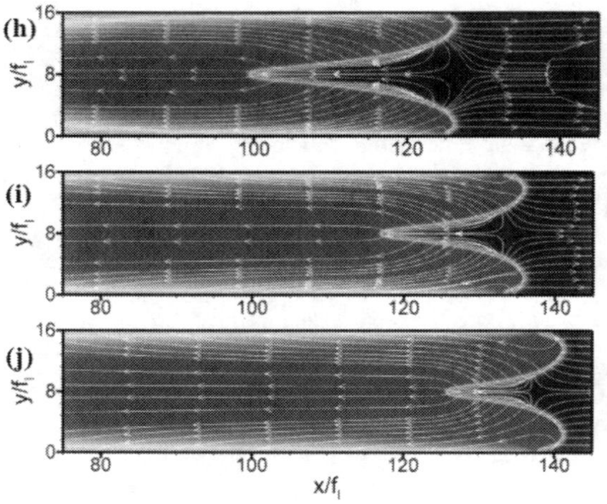

Figure 6: Flame structure and flow field: (a)-(h) tS_l/f_l=22.4, 28.1, 32.7, 34.7, 45.6, 49.6, 50.3, 52.9, 55.5 and 56.8.

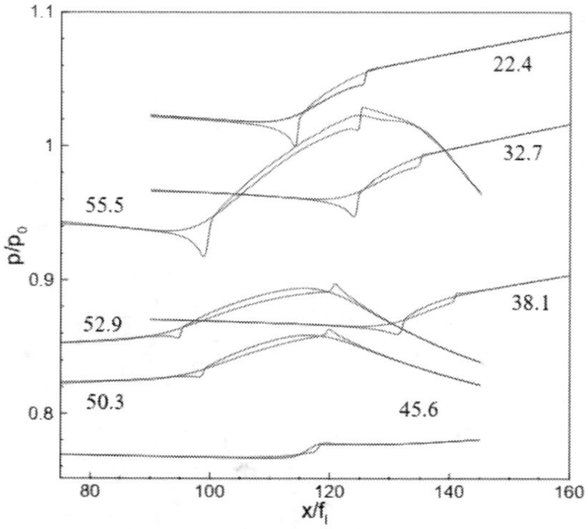

Figure 7: Pressure distributions along axis (blue) and wall (red) near flame fronts:

Figure 7 shows the pressure distribution at the axis and the wall near the flame front. It is seen that at tS_l/f_l =22.4-32.7 the pressure decreases globally; and the inverse pressure gradient forms as the flame retreats, leading to reflow of the products and unreacted gas. At tS_l/f_l = 45.6, the flame rapidly falls back and the inverse gradient weakens. At tS_l/f_l =50.3-52.9, the flame advances and a positive pressure gradient forms. At tS_l/f_l =55.5, the oscillating flame rapidly accelerates and induces a pressure wave in front of the flame. The pressure wave superimposes unto the shock wave that preheats the gas in front of the flame and reduces the induction time. As a result, wall friction leads to local explosion of the unreacted gas between the flame surface and walls; eventually the explosion develops into detonation, as shown in Fig. 8.

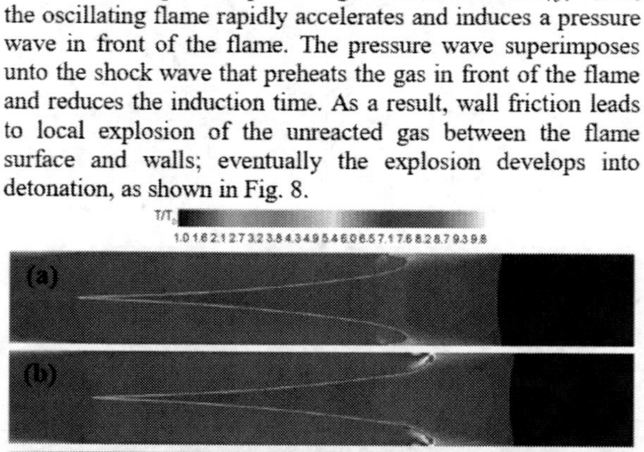

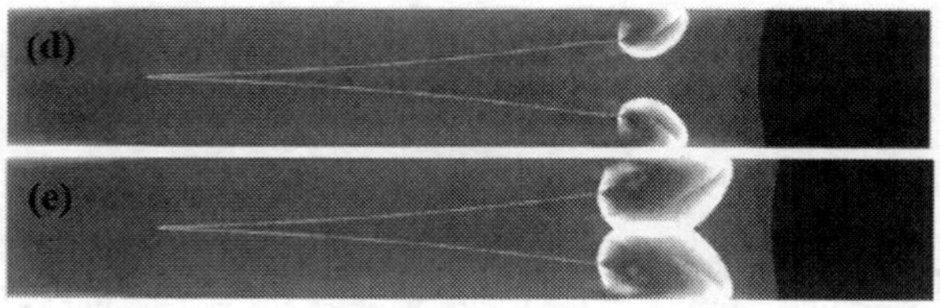

Figure 8: Deflagration to detonation transition: (a)-(e) tS_l/f_l =161.2, 162.1, 162.5, 162.6 and 162.7.

4 Conclusions

This work examines numerically flame propagation and DDT in micro-scale channel, with viscosity and heat loss walls. Results show that wall heat loss leads to the appearance of an oscillating flame. This oscillating flame is attributed to the competition between viscosity and heat loss effects; the occurrence of DDT depends non- monotonically on the channel width.

5 Acknowledgement

This work was supported by the National Natural Science Foundation of China under grant No. 91441131 and the National Key Basic Research Program of China (No. 2014CB239603)

References

[1] M.H. Wu, M.P. Burke, S.F. Son, R.A. Yetter, Proc. Combust. Inst. 31(2007):2429-2436.

[2] D.M. Valiev, V. Bychkov, V. Akkerman, L.E. Eriksson, Physical Review E. 80(2009):036317.

[3]L. Kagan, G. Sivashinsky, Combust. Flame 134(2003) 389-397.

[4] M.A. Liberman, M. Kuznetsov, A. Ivanov, I. Matsukov, Phys. Lett. A. 373(2009): 501-510.

[5] I. Brailovsky, L. Kagan, G. Sivashinsky. Combustion waves in hydraulically resisted systems. Philosophical Transactions the Royal Society A. 370(2012):625-646.

[6] N. Tsuboi, Y. Morii, A.K. Hayashi, Proc. Combust. Inst. 34(2013):1999-2007.

[7] I. Brailovsky, V. Goldshtein, I. Schreiber, G.I. Sivashinsky, Combust. Sci. Tech. 124(1997): 145-65

[8]W.H. Han, Y. Gao, C.K. Law, Combust. Flame 176 (2017):285-298.

[9] J.D. Ott, E.S. Oran, J.D. Anderson Jr, AIAA J 41(2003): 1391-1396.

[10] M.H. Wu, C.Y. Wang, Proc. Combust. Inst. 33(2011): 2287-2293.

[11] G.S. Jiang, C.W. Shu, J. Comput. Phys. 126(1996): 202-236.

[12] C.W. Shu, S. Osher, J. Comput. Phys. 77(1988):439-448.

[13] V. Bychkov, S. Golberg, M. Liberman, L.E. Eriksson, Phys. Rev. E. 54 (1996): 3713.

[14] V. Bychkov, S. Golberg, M. Liberman, A. Kleev, L.E. Eriksson, Combust. Sci. Tech. 129 (1997): 217-242.

Flame Propagation and Development of Cellular Structures of CH4/Air Premixed Flame in a Disk Burner of Variable Mesoscale Gap

Hyejin Jang and Nam Il Kim

Department of mechanical engineering, KAIST

291 Daehakro, Yuseong-gu, Daejeon, 34141, South Korea

Abstract

Unsteady flame propagation has been an important issue in combustion studies, and it is affected by the configuration of the combustion space. With the interests on small scale combustion devices as an energy source of small systems, flame propagations in small scale combustors have been studied. However, the effects of length scale on the unsteady flame propagation have not been clarified yet. In this study, unsteady flame propagation after ignition was investigated within a disk burner. Unlike previous a few studies using a disk of a fixed small scale, the disk diameter of this study was increased to 36 cm, and the disk gap could be varied continuously with the resolution of 0.1 mm. Thus, the scale effects on the unsteady flame propagation could be investigated for a longer period. In each equivalence ratio, the flame propagation velocity became slower as the gap distance was reduced, but it did not have a linear correlation. When the average propagation velocity decreased, flame front transformed to a wrinkled structure. When the number of cells along the flame front increased furthermore, its average propagation velocity could increase. Near the quenching distance, the number of cells increased especially near the stoichiometry. Through this study, the effects of gap scale on the characteristic structural transition of premixed flames could be distinguished, and the developing mechanism of various unsteady flame structures within a mesoscale disk burner could be discussed.

1 Introduction

Propagation velocity of a premixed flame has been an important issue in the combustion studies, and it has been studied with various methods. One representative method was using a constant volume chamber, in which premixed mixture was filled [1,5]. Usually, the flame propagation after spark ignition in a confined volume chamber was monitored with a high speed camera, and it was analyzed to evaluate the laminar burning velocity [2,6]. The other is using a tube or a channel, filled with premixed mixture. Before 1980s, many studies were conducted under the name of 'tulip flames' or 'finger flames' and numerical simulations have also been conducted [3]. One result was that the propagation velocity and flame shape were significantly affected by the geometry or scale of the combustion space. Similar phenomena have been studied related to the development of small scale combustion device or 'micro-combustors' during the last decades [2].

Corresponding author. Fax: +82-42-350-3211
E-mail address: nikim@kaist.ac.kr

Apart from these studies, from the aspect of combustor design, understanding the scale effects on the flame propagation and the flame characteristics is important. It is because designing a combustor can be said as a process of consecutive decision of various length scales related to combustion phenomena. Regarding this issue, one recent study could show overall variation of the flame propagation velocity using an annular stepwise diverging tube (ASDT) [4,7]. Experimental results were described as a concentration-length-velocity diagram (or CLV-diagram). The CLV-diagram of methane-air premixed flames is shown in Fig. 1a [4]. It shows that one among concentration, length scale, and flame propagation velocity (FPV) is affected by two others. FPV with different length scales at each equivalence ratio is orthogonal projection of CLV-diagram and shown in Fig. 1b [4]. Although various discussions could be possible regarding the flame propagation of steady flames, unsteady flame propagation could not be described without using the CLV-diagram.

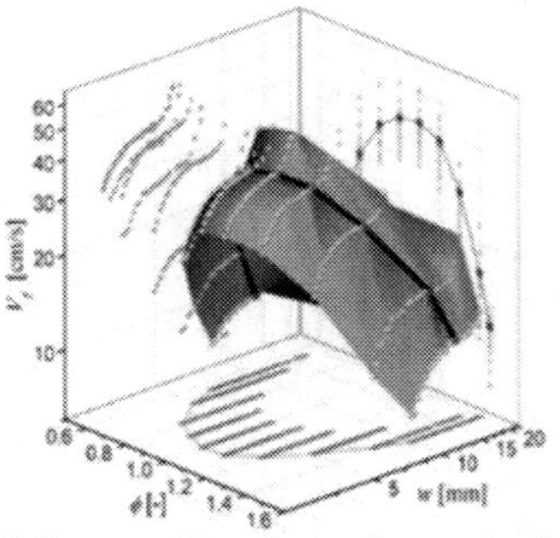

(a) CLV-diagram of methane-air premixed flames

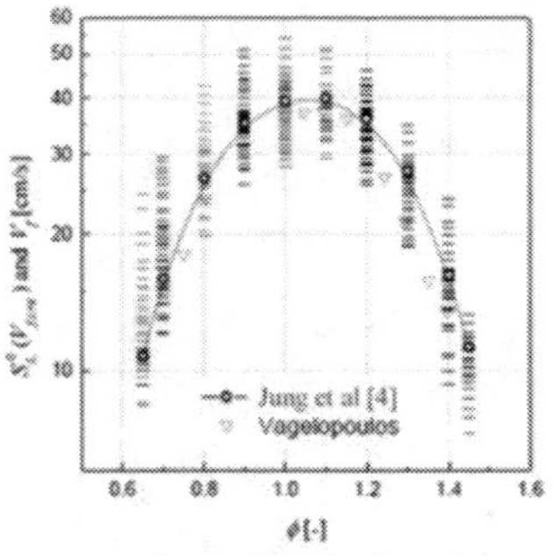

(b) Flame propagation velocity of methane-air flames

Figure 1: (a) Concentration-Length-velocity diagram (CLV-diagram) and (b) flame propagation velocity compared to laminar burning velocity of methane-air premixed flames [4]

Therefore, in this study, unsteady flame propagation after spark ignition was investigated within a disk burner. Although there have been a few studies using small disk burners [2], they were too small to investigate unsteady propagation including structural transition affected by the length scale. Therefore, a disk burner was newly designed. Two novel things of the new burner are that the diameter of the disk is much larger than the previous studies, and the gap between two disk plates could be varied with fine-resolution. Using this burner, the unsteady flame propagation and the growth of the cellular structure will be investigated.

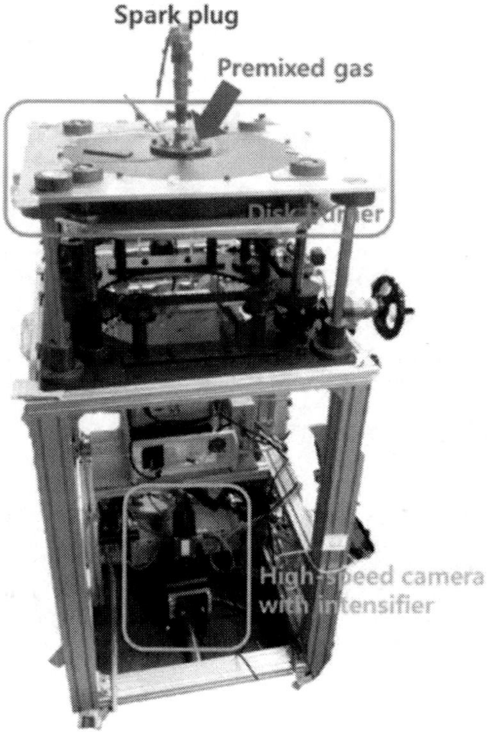

(a) The picture of disk burner

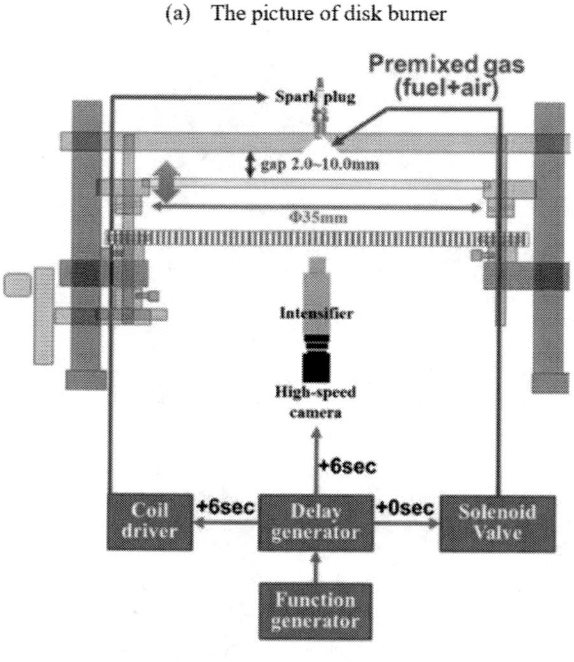

(b) Schematic of (a)disk burner

Figure 2: Disk burner of variable mesoscale gaps

2 Experimental methods

The outlook of the experimental setup is shown in Fig. 2a. A disk burner was placed at the top, and the other experimental devices were placed under the disk burner. This experimental setup was schematically described in Figure 2b. The diameter of the disk was 36 cm, and the disk gap could be varied with the resolution of 0.1 mm, from 2 mm to 8 mm. The disk burner was filled with methane/air mixture at the atmospheric pressure and the room temperature. The equivalence ratio was between 0.7 and 1.3. For the elimination of the flow disturbance of the premixed gas, spark ignition was delayed by 6 seconds after the premixed gas supply through a solenoid valve had been stopped. Flame propagation was monitored with an intensified high-speed camera, triggered with the spark plug at the same time. The images were analyzed to evaluate its propagation velocity and to capture distinctive structural transition.

Figure 3 shows the flammable range of the disk burner with the variation of the equivalence ratio and the disk gap. At the region above the limit line, flames could propagate. The red dotted line represents the flammability limit of this study, and the blue dotted line represents that of Y. Jung et al [4]. The minimum distance was 2.9mm at the stoichiometric condition of this study. It is similar to the quenching distance [4], but the limits of equivalence ratio was mostly affected by the ignition energy of the spark. The flammable range may be widened if the ignition energy is increased.

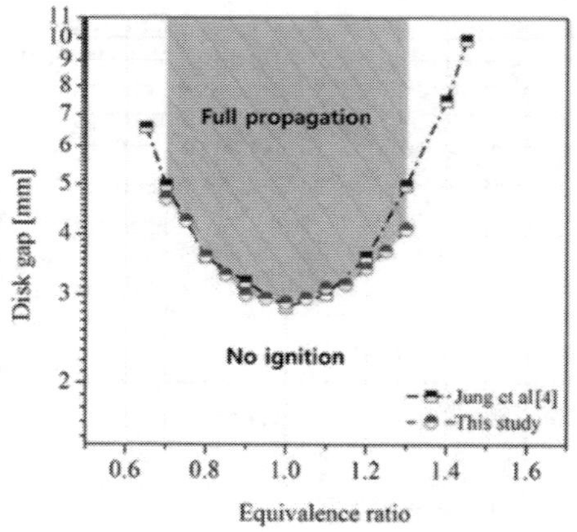

Figure 3: Flammable range of disk burner

3 Flame propagation velocity

3.1 Flame propagation velocity at different disk gaps

Using the unsteady flame propagation images, flame propagation velocities were measured. In the image processing, the flame front was captured and the inner surface area was estimated. From this area, a corresponding radius is calculated, and its time variation was defined as the effective velocity. The effective velocity graph had significant noise until 20ms due to the spark plug ignition.

Each plot of Figure 4 shows flame propagation distance at equivalence ratio, ϕ, of 0.7, 1.0 and 1.3 with different gap scales, L, of 4, 4.5, 5, 5.5, 6, 6.5, 7, 8mm. At each ϕ, flame

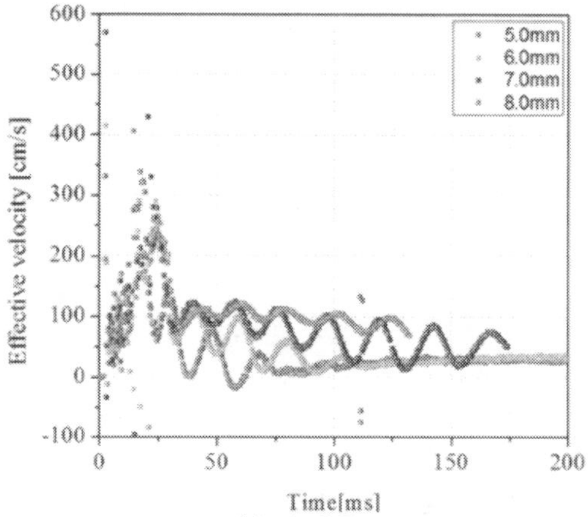

(a) ϕ =0.7

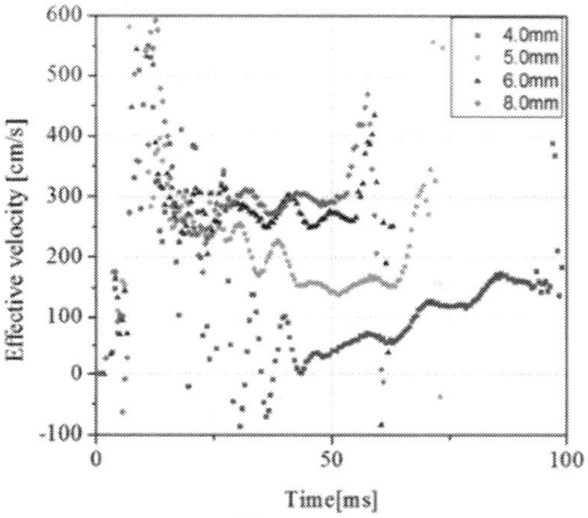

(b) ϕ =1.0

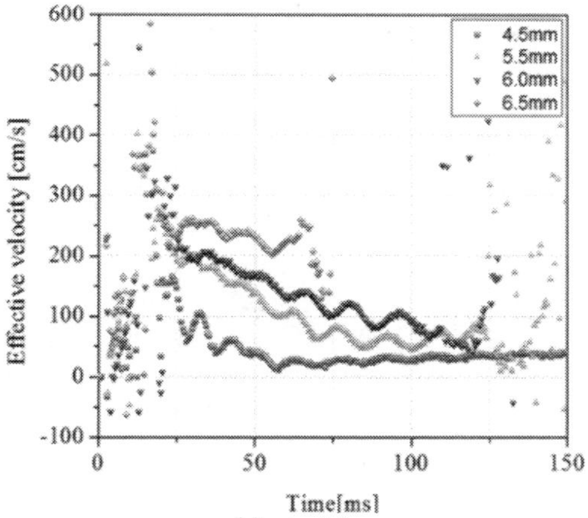

(c) ϕ =1.3

Figure 4: Flame propagation velocity from 4mm to 8mm for each equivalence ratio (a) ϕ =0.7, (b) ϕ =1.0, (c) ϕ =1.3

propagation velocities decreased as the disk gap became smaller. In addition, the flame propagation velocities oscillated after 25ms. The oscillation of propagation velocities at lean and rich conditions(ϕ =0.7, 1.3) became stronger at broader gap sizes and weakened at narrower gap sizes.

Each stabilized velocity at a lean (ϕ =0.7, L=5.0mm) and a rich (ϕ =1.3, L=4.5mm) conditions was about 30cm/s and 38cm/s. To compare them with Fig 1b, they were similar with the maximum flame speed at each equivalence ratio. The maximum flame speed in Fig 1b was flame propagation speed with most negative stretch.

At stoichiometric condition (ϕ =1.0), the propagation velocity vibrated stronger than lean and rich conditions(ϕ =0.7, 1.3). The noisy propagation velocity at the end of the disk was concerned with discontinuous image hidden behind the fringe of the burner.

3.2 Effect of cells on propagation velocity

The experiment was performed under ϕ =1.3 and L=4.5 mm condition. Figure 5a shows the flame shape at the moment of the generation of cellular structure. The number of cells increased as effective radius of flame front increased in Fig 5b with time interval 25ms.

In Figure 6, the blue dots indicate the effective velocity and the red dots indicate the number of cells at flame front. The effective velocity is calculated from the area inside the cellular flame front. The number of the cells was also estimated from the image processing, in which the wrinkling larger than flame thickness and the both curvature convex up were counted. The number of cells graph had significant noise until 20ms due to the spark plug ignition.

Figure 5 and 6 show that flame propagation velocity is slowed instantaneously down when the cellular structure starts to grow, or the cellular structure is formed when the flame propagation speed was decreased. However, when the number of cells started to increase, the flame propagation velocity slightly increased and maintained, and again increased and maintained. This is one example of cell number effect. It is expected that the number of cells affects the flame propagation velocity.

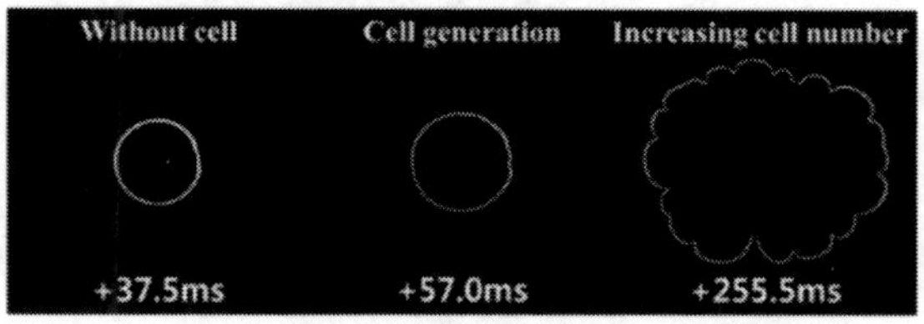

(a) Cell generation

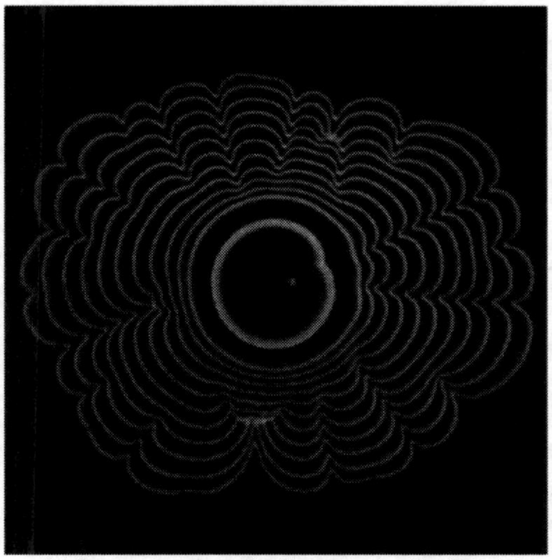

(b) Image merging of cell generation(Δt=25ms)

Figure 5: These are (a)flame propagating moment and (b)overlapped moments at equivalence ratio of 1.3 and gap distance of 4.5mm.

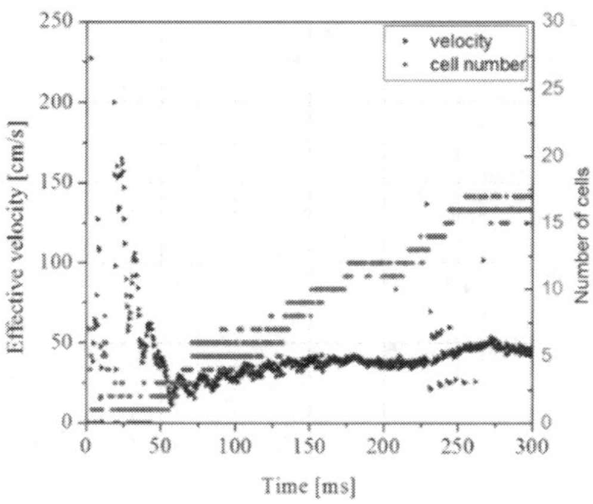

Figure 6: This is flame propagating moment of figure5. Blue dot graph indicates mean propagation velocity along 360 degrees and the red dot graph indicates the number of cells.

4 Conclusions

To investigate the effects of disk gap and cellular structure on flame propagation velocity, a disk burner was designed and fabricated. As the disk gap was getting larger, the flame propagation velocity got faster and it oscillated stronger. The cellular structure of flame front was affected by the first slowdown of the flame propagation. However, the flame propagation could approach to a certain velocity with oscillations at the second generation of the cellular structure.

5 Acknowledgment

This study was partly supported by the BK21-plus program that was granted by Korean Government.

References

[1] Andrews, G. E., and D. Bradley. "The burning velocity of methane-air mixtures." Combustion and Flame 19.2 (1972): 275-288.

[2] Liqiao Jianga, Hang Sub, Haolin Yanga, HaiLiang Caob, Daiqing Zhaoa, "Flame propagation characteristics of n-butane/air mixture in a micro gap constant volume chamber," 10th ASPACC – July 19-22, 2015 – Beijing

[3] Dunn-Rankin, D., P. K. Barr, and R. F. Sawyer. "Numerical and experimental study of "tulip" flame formation in a closed vessel." Symposium (International) on Combustion. Vol. 21. No. 1. Elsevier, 1988.

[4] Jung, Yongjin, Min Jung Lee, and Nam Il Kim. "Propagation and quenching of premixed flames in a concentration-length-velocity diagram." Proceedings of the Combustion Institute 36.3 (2017): 4243-4251.

[5] Metghalchi, M. A. K. J., and J. C. Keck. "Laminar burning velocity of propane-air mixtures at high temperature and pressure." Combustion and flame 38 (1980): 143-154.

[6] Huang, Zuohua, et al. "Measurements of laminar burning velocities for natural gas–hydrogen–air mixtures." Combustion and Flame 146.1 (2006): 302-311.

[7] Liu, Zhao, and Nam Il Kim. "An assembled annular stepwise diverging tube for the measurement of laminar burning velocity and quenching distance." Combustion and Flame 161.6 (2014): 1499-1506.

11th Asia-Pacific Conference on Combustion,
The University of Sydney, NSW Australia
10th -14th December 2017.

Pyrolysis of dieldrin and formation of toxic products. II. Reaction mechanism.

J.C. Mackie*, W.N.K. Dharmarathne, E.M. Kennedy, M. Stockenhuber

Process Safety and Environmental Protection Group, Faculty of Engineering

The University of Newcastle, Callaghan NSW 2308

Australia

Abstract

The mechanism of pyrolysis of dieldrin, a now-banned member of the cyclodiene group of pesticides, has been studied by quantum chemical calculation. The pyrolysis is initiated by retro Diels-Alder dissociation into hexachlorocyclopentadiene and norbornadiene oxide. At low temperatures (623-800 K) intramolecular processes lead to benzene, CO and H_2 as major products. Above ~800 K hexachlorocyclopentadiene undergoes rapid Cl bond fission and Cl atoms abstract H atoms from dieldrin forming HCl and lead to four different types of dieldrin radicals, one of which undergoes a complex series of intramolecular rearrangements and fissions, producing pentachlorostyrene, ketene, acetylene and Cl atom as products. These products accord with results of previous experimental pyrolysis experiments conducted on dieldrin.

1 Introduction

Cyclodienes are a well-known group of pesticides which are produced by the Diels-Alder condensation reaction between hexachlorocyclopentadiene (HCCP) and a substrate [1]. This group includes aldrin, dieldrin and endosulfan. Although the first two members were banned several years ago, and endosulfan is being phased out, large stockpiles, especially of endosulfan, are still in existence as well as large areas of contaminated vegetation and soils [2]. Accidental combustion of cyclodienes through wildfire or fires in stockpiles can produce significant amounts of highly toxic substances. There have been few detailed studies of toxic products of combustion of cyclodienes; one notable exception is the study by Chopra et al [3] in 1978 in which a large number of chlorinated benzenes and other aromatics was identified in the pyrolysis of endosulfan at 1173 K. Nevertheless, until our recent study of oxidative decomposition of endosulfan [4] there had been no reports of the emission of the highly toxic polychlorinated dibenzodioxins and dibenzofurans (PCDD/F) from combustion of cyclodienes. Our present experimental paper on pyrolysis of dieldrin [5] reports on the production of several toxic PCDD/F precursors. A better understanding of the mechanism of formation of toxic products, especially dioxins and their precursors is essential for determination of safe strategies for destruction of stockpiles of cyclodienes.

Our experimental study [5] of pyrolysis of dieldrin revealed the formation of several chlorinated aromatic toxic precursors of PCDD/F. This study was carried out in a silica flow reactor between 623 and 923 K and at a residence time of 5.0 s. Two distinct temperature regions and sets of major products were observed – a low temperature region from 623 K to about 800 K and a high temperature region from about 800 K to 923 K. Major products at low temperatures comprised benzene, CO, HCCP and traces of octachloronaphthalene. At high temperatures, pentachlorostyrene (predominantly), hexachlorostyrene (lesser amounts), acetylene, CO and HCl were detected.

In order to unravel the complex chemistry involved in both low and high temperature regions, we have used quantum chemical calculation to identify major reaction pathways, intermediates and final products. We are aided in our pathways identification by a knowledge of the rates and mechanism of decomposition of HCCP from our previous study [6].

2 Computational Methodology

The Gaussian 09 suite of programs [7] was used for all quantum chemical computations. Optimization of species geometry was performed using the density functional M06-2X and the basis set 6-31+G(d,p). Improved energies were calculated at the M06-2X/GTLarge//M06-2X/6-31+G(d,p) level of theory. Transition states (TS) have been identified for each reaction step and the initial and final states linked by these TSs have been validated by IRC calculation [7].

3 Results and discussion

3.1 Low temperature pyrolysis mechanism

From our earlier studies [5,8], we have found that pyrolysis is initiated by retro Diels-Alder decomposition into HCCP and a substrate. In the case of dieldrin, the substrate is norbornadiene oxide (see Fig. 1a). The activation energy for the decomposition was found to be 228 kJ mol^{-1}, somewhat higher than the 212 kJ mol^{-1} activation energy calculated for endosulfan. Hence dieldrin decomposition is predicted to commence at slightly higher temperatures than endosulfan in agreement with experiment [5,8].

As HCCP does not undergo significant Cl bond fission until a temperature of about 800 K is attained [6], the low temperature region is dominated by intramolecular reactions involving norbornadiene oxide (see Fig. 1b). This oxide first undergoes a concerted rearrangement to form a bicyclic aldehyde. This rearrangement, although not studied in the gas phase has been carried out experimentally in acid-catalysed liquid phase [9]. A facile rearrangement then takes place to form cyclohexa-2,4-diene aldehyde. Next, a hydrogen transfer to the ring leads to the elimination of CO and formation of 1,4-cyclohexadiene. 1,4-elimination of H_2 finally produces benzene.

*Corresponding author.
E-mail address: John.Mackie@newcastle.edu.au

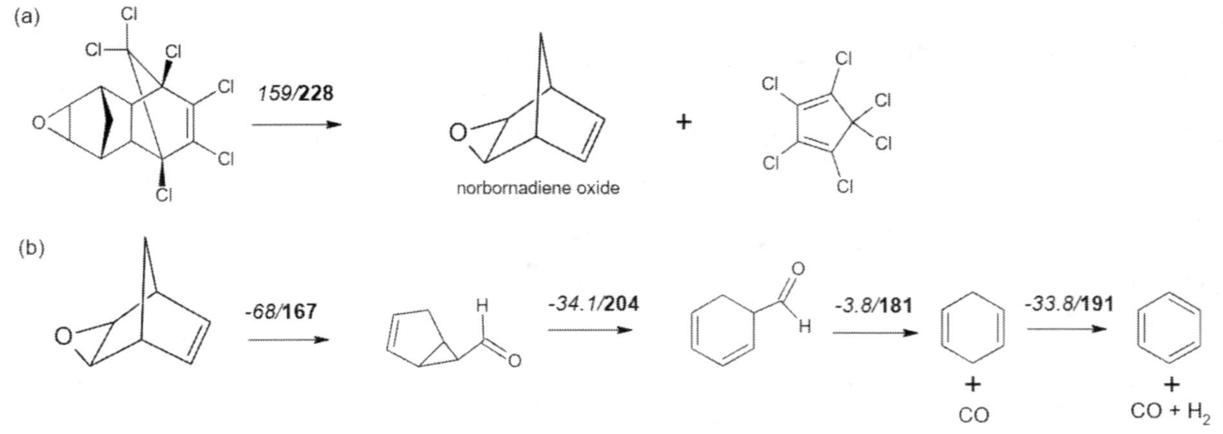

Figure 1 (a): Initial decomposition of dieldrin into norbornadiene oxide and hexachlorocyclopentadiene. (b): Low temperature decomposition mechanism of norbornadiene oxide. Numbers in italics are reaction energies; in bold, barrier heights. Both in kJ mol^{-1} at 0 K.

After the initial barrier to formation of the bicyclic aldehyde is surmounted, the reaction to the final products of benzene, CO and H_2 should be facile and take place via chemical activation.

It should be pointed out that the retro Diels-Alder decomposition is reversible and equilibration will take place, such that the overall decomposition of dieldrin in the low temperature region will be slow. Between 673 and 773 K, the standard Gibbs free energy, $\Delta_r G^o_{c,298}$, ranges from 92.63 to 80.50 kJ mol^{-1}. It is only at temperatures where significant concentrations of Cl atoms become available from HCCP bond fission that dieldrin decomposition becomes accelerated.

3.2 High temperature pyrolysis mechanism

From our previous study [6] the rate constant for Cl fission from HCCP was found to be $k = 1.45\times10^{15}$ exp(-222 ± 9 kJ mol^{-1}/RT) s^{-1}. Thus, in our reactor [5], at 5 s residence time, Cl fission in HCCP becomes very rapid at temperatures above about 800 K. As a consequence, Cl atoms will attack residual dieldrin, abstracting an H atom to form HCl and a dieldrin radical.

There are, in fact, four different dieldrin radicals which may be produced by Cl atom abstraction. These are sketched in Fig. 2.

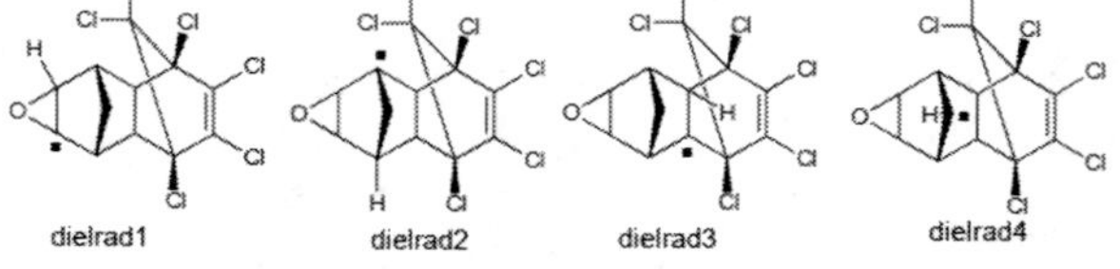

Figure 2: The four dieldrin radicals arising from H abstraction by Cl atom.

Formation of these radicals is facile with abstraction reaction energies between -18 and 22 kJ mol^{-1} (at 0 K) with very small or zero barriers.

The pyrolysis mechanism has been studied by investigating the subsequent decomposition reactions of each of these four radicals. In general, secondary decomposition of a radical takes place by rupture of a so-called β-bond, i.e., a bond adjacent to the α-bond on which the radical site is located. The new radical formed by the fission is also subjected to β-bond fission and the process continued until final products are produced or, in many cases, an intractable intermediate with no readily ruptured bonds is produced. Such a structure is considered to be a "dead end" and hence, an invalid reaction pathway.

It was found that just one of the four radicals underwent decomposition reaction resulting in final products. This was the radical labelled in Fig. 2 as dielrad1. The mechanism derived by quantum chemical calculation therefore commences with dielrad1 and is shown schematically in Fig. 3. In this figure Cl atoms are shown in green, O in red, C in grey and H in white.

Decomposition of dielrad1 commences with β-bond fission which converts the epoxy (-O-) moiety into a carbonyl (C=O) group which leads to facile elimination of the oxygen atom. This is not possible with the other three radicals, and is probably the reason why dieldrad1 undergoes a complete passage to the final observed products. Next an H atom transfer from the bridging methylene (-CH_2-) group to the carbon atom adjacent to the carbonyl group precedes a C-C bond fission which then enables cleavage of the bond releasing ketene (CH_2=C=O). After a further C-C fission, a structure comprising a cyclopentadienyl radical attached to a pentachlorocyclopentadienyl radical results. Ring opening in the latter then takes place resulting in a cyclopentadiene with a chlorinated open chain radical attached. Direct opening of the cyclopentadiene ring has too large a barrier but after an H transfer this ring opens with a barrier of 229 kJ mol^{-1}, the largest barrier in the entire mechanism. A further H transfer to the allyl carbon (=C=) prepares the way for fission of acetylene to take place. Next, a facile rotation of the remaining moiety places the terminal vinyl radical conveniently to undergo six-membered ring closure with a very modest barrier. Finally the extremely weakly bonded Cl atom in the trichlorocyclohexadienyl moiety easily fissions to produce the final product, the major observed species, pentachlorostyrene.

It should be pointed out that ketene, predicted by this mechanism, was not observed experimentally amongst the products. This is because ketene is a highly reactive species and decomposes rapidly on most surfaces. Our silica flow reactor with its relatively long residence time would make detection of ketene unlikely. Its initial products of decomposition are CO and the methylene biradical (:CH_2). The latter rapidly forms ethylene and acetylene, both of which are observed.

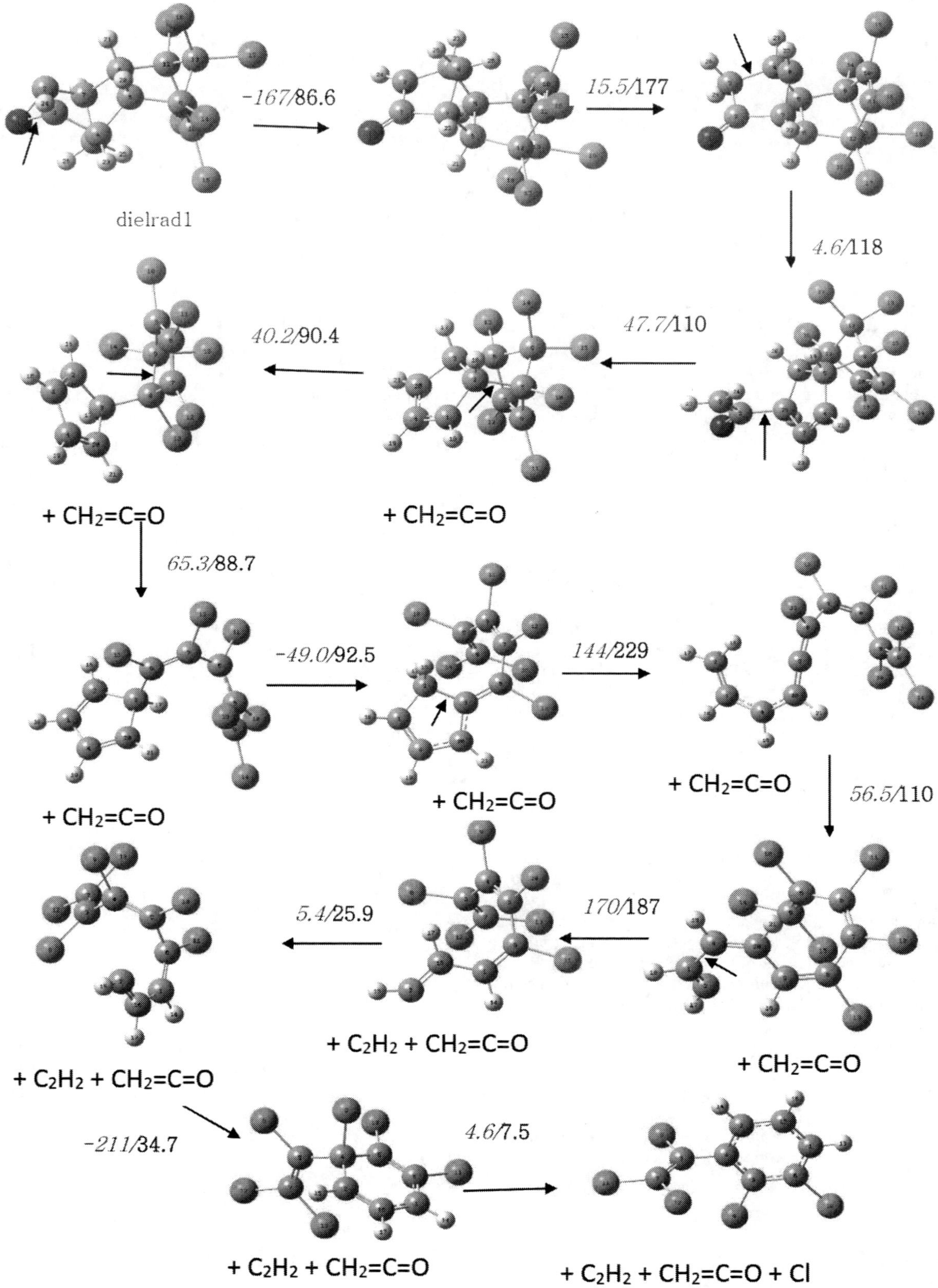

Figure 3: Mechanism of decomposition of dielrad1. Figures in italics: Reaction energies. In bold: Barrier heights both at 0 K. Arrows pointing to bonds represent sites of β-bond fissions.

Hexachlorostyrene is also observed experimentally at high temperatures. This can arise through the propensity of Cl atom to add to ring carbon atoms and displace H atoms. Experimental yields of hexachlorostyrene are greatest at temperatures above which hexachlorocyclopentadiene has completely decomposed. Traces of octachloronaphthalene are also observed in the high temperature region. In our previous study of pyrolysis of HCCP we identified the mechanism of formation of octachloronaphthalene as arising from the recombination of two pentachlorocyclopentadienyl radicals resulting from fission of a Cl atom.

Octachloronaphthalene is an acknowledged persistent toxic industrial contaminant [10]. The polychlorostyrenes, though not industrial products, arise as byproducts of other chlorinated substances, especially from incineration, and are bioaccumulative and potential endocrine disruptors [11].

In addition to its uses in unravelling the complex pathways of thermal decomposition of dieldrin and similar cyclodienes, quantum chemical calculation of the thermodynamic functions of transition states, reactants and products should enable future development of detailed reaction kinetic schemes. Detailed kinetic modelling should be useful in determining combustion conditions which minimize the emissions of toxic products and precursors of PCDD/F to enable safe strategies to be developed for destruction of unwanted pesticides.

4 Conclusions

Thermal decomposition of dieldrin has been shown by quantum chemical computation to be initiated by retro Diels-Alder decomposition into hexachlorocyclopentadiene and norbornadiene oxide. As this is a reversible reaction, equilibration will take place at temperatures below which hexachlorocyclopentadiene can fission a chlorine atom. This leads to a low temperature mechanism in which norbornadiene oxide decomposes by intramolecular rearrangement and loss of CO. The other major products under these conditions are benzene and molecular hydrogen. Once temperatures are attained whereby hexachlorocyclopentadiene rapidly fissions Cl atoms (around 800 K in a flow reactor of 5 s residence time), a high temperature free radical mechanism takes over in which Cl atoms abstract an H atom from dieldrin, forming HCl which becomes a major product. Of four possible dieldrin radicals, we find that just one of these, viz., dielrad1, can undergo a sequence of reactions of modest energy barriers to lead to the major observed high temperature products pentachlorostyrene and acetylene. Ketene was also predicted to be a major product at high temperatures but was not observed experimentally because of its considerable reactivity. Rather, its products of decomposition, CO, acetylene and ethylene, are observed.

5 Acknowledgment

The authors thank the University of Newcastle, Australia for a postgraduate research scholarship to N.K.D. This research was undertaken with the assistance of resources from the National Computational Infrastructure (NCI), which is supported by the Australian Government.

References

[1] G. Matolcsy, M. Nadasy, V. Andriska, Pesticide Chemistry, Elsevier: Amsterdam, The Netherlands, 1988.

[2] X. Wang, K. Kennedy, J. Powell, M. Keywood, P. Thai, P. Bridgen, S. Broomhall, C. Paxman, F.K. Waniae, J.F. Muellera, Environ. Sci –Processes & Impacts, 17 (2015) 525-532.

[3] N. M. Chopra, B. S Campbell, J. C.Hurley, J. Agric. Food Chem. 26 (1978) 255-258.

[4] N.K. Dharmarathne, J.C. Mackie, E.M. Kennedy, M. Stockenhuber, Environmental Science & Technology, 50 (2016) 10106-10113.

[5] N.K. Dharmarathne, J.C. Mackie, E.M. Kennedy, M. Stockenhuber, submitted to 11th Asia-Pacific Conference on Combustion, The University of Sydney, NSW, Australia, 10th-14th December, 2017.

[6] N.K. Dharmarathne, J.C. Mackie, E.M. Kennedy, M. Stockenhuber, J. Phys. Chem. A. 121 (2017) 5871-5883.

[7] M. J. Frisch, G. W. Trucks, H. B. Schlegel et al, Gaussian 09, Revision D.01, Gaussian, Inc., Wallingford CT, 2013.

[8] N.K. Dharmarathne, J.C. Mackie, E.M. Kennedy, M. Stockenhuber, Proceedings of the Combustion Institute, 36 (2017) 1119-1127.

[9] R. Grigg, D. Shelton, Chem. Communications (1971) 1247-1248.

[10] H. Park, J.-H. Kang, S.-Y. Baek, Y.-S. Chang, Environmental Pollution, 158 (2010) 1420-1427.

[11] D. Carrizo, J. O. Grimalt, J. Chromatography A 1216 (30) (2009) 5723-5729.

11th Asia-Pacific Conference on Combustion,
The University of Sydney, NSW Australia
10th -14th December 2017.

Mapping of instantaneous fuel concentration using a bundled LIBS plug

Hyung Min Jun[1], Hyunwoo Kim[1], Jai-ick Yoh[1]

[1]Department of Mechanical and Aerospace Engineering, Seoul National University

1 Gwanakro, Gwanakgu, Seoul 151-742, South Korea

Abstract

Multiple portable LIBS plugs composed of photodiodes and bandpass filters have been designed to measure instantaneous 2-dimensional fuel concentration in hydrocarbon-fueled spray flame field. Plasma emission spectra were analyzed to measure the fuel concentration without using the complex conventional laser-induced breakdown spectroscopy (LIBS) system which consists of spectrometer and ICCD camera. Bandpass filters were used to collect only the spectral lines pertinent to fuel concentration from the whole spectra. Hydrogen and oxygen atomic lines were chosen to measure the said concentration. H/O signal intensity ratio was calculated by dividing H signal intensity ratio with O signal intensity ratio. Calibration curve between H/O signal intensity ratio and equivalence ratio was constructed. Beam splitters and Nd:YAG laser system was designed to make multiple plasmas simultaneously at different locations. The idea of measuring instantaneous 2-dimensional fuel concentration in the hydrocarbon-fueled combustion chamber with multiple LIBS plugs may be applied for instantaneous monitoring of fuel distribution inside the high-speed air breathing combustion engine.

1 Introduction

Evenly distributed concentration of fuel is very important factor for reliable ignition inside combustors of air breathing engines. Especially for high-speed engines such as scramjet engines, fuel injection and mixing are important issues as their residence time is of the order of few milliseconds [1]. In order to promote mixing in scramjet engines, cavity flame holder has been developed in recent [2-3]. Also, to verify distribution of fuel inside scramjet engine combustor, experimental and computational studies [4] have been conducted. However, as these methods are aimed at predicting the flow inside combustor, they cannot diagnose fuel distribution inside operating combustor in real time. Furthermore, in a highly turbulent flow, experimental measurements or computational prediction cannot be exact. This limitation brings out the need for instantaneous measurement inside actual engines.

The present study suggests new method for instantaneous mapping of fuel distribution inside combustor using simplified laser-induced breakdown spectroscopy (LIBS) system. LIBS has been widely used to instantaneously detect various materials in severe environment with simple system configuration. In this method, a concentrated laser dissociates and excites molecules and generates plasma at one point. Excited ions and electrons return to their ground states as the plasma cools down, and emit light with respective atomic frequencies. In conventional LIBS system, plasma emission is collected with spectrometer and ICCD camera and analyzed to give whole spectra of atomic lines. In previous studies, LIBS has been applied to reacting flows for the measurement of local equivalence ratio [5-14] and gas concentrations [15-17]. In these studies, atomic lines of H, C, O and N and molecular bands of CN and C2 were obtained in hydrocarbon fuel-air mixtures. O and N signal intensities represent air concentration while C and H signal intensities correspond to hydrocarbon molecules of fuel.

In order to simplify the system, we decided not to extract the whole spectral signal from plasma. Calibration curves between H/O intensity ratio and equivalence ratio have been established previously [6-11]. With these results, we designed a simple LIBS plug that can measure equivalence ratio from H/O intensity ratio without using spectrometer and ICCD camera. The LIBS plug consisted of a lens, two bandpass filters for hydrogen as well as oxygen, and two photodiodes. Further simplification was obtained when; only one lens was used to concentrate the laser beam and to collect plasma emission simultaneously. Calibration curve between H/O intensity ratio and equivalence ratio was newly established with the LIBS plug. H/O intensity ratio mapping was done with this scheme for an actual spray flame. In order to obtain an instantaneous 2-dimensional fuel concentration map, measurement system that uses beam splitters and other LIBS plugs is planned for future. This new LIBS system could get information from several points at the same time with simple and stable configuration while conventional LIBS system can only obtain information from one point at once with complex and delicate arrangement. The detection of fuel distribution inside a combustor within millisecond will allow feedback control of the combustor. In recent studies, inside scramjet engine combustor, there are diverse spray fuel nozzles in different locations [4]. Fast detection technique and feedback control has possibility to overcome the issue of non-evenly distributed fuel inside scramjet engine combustor.

2 Experimental Setup

This study consists of three different experimental setups. Initially a simplified LIBS plug was designed and manufactured. Setup for H/O intensity ratio and equivalence ratio was constructed and experiments were conducted, with the same filters and photodiodes used in the LIBS plug. Finally, concentration of fuel from spray nozzle was mapped and plan for multiple simplified LIBS plugs were constructed using mirrors and beam splitters.

2.1 Simplified LIBS plug

Figure 1 shows the schematic of the simplified LIBS plug. In order to obtain O and H atomic spectra, two different bandpass

Corresponding author. Fax: +82-2-882-1507
E-mail address: jjyoh@snu.ac.kr

filters were used in the LIBS plug. Bandpass filter with center wavelength (CWL) of 656nm and full width-half-max (FWHM) of 10nm was used to obtain H atomic signal while filter with CWL of 780nm and FWHM of 10nm was used to obtain O atomic signal. Both filters have diameter of 12.5mm. An uncoated UV silica plano-convex lens (LA4078, Thorlabs) with diameter of 50.8mm was used to concentrate laser beam as well as to collect the plasma emission. The lens diameter had to be bigger than the sum of diameters of laser beam and bandpass filters in order to collect plasma beam in parallel direction. Photodiodes used in the LIBS plug are Si photodiodes (FDS010, Thorlabs). Photodiode's rise time was 1 ns and peak responsivity is 0.44 A/W at 730 nm. Each photodiode was connected to an oscilloscope (Teledyne LeCroy, WAVESURFER 64MXS-B) with a BNC cable.

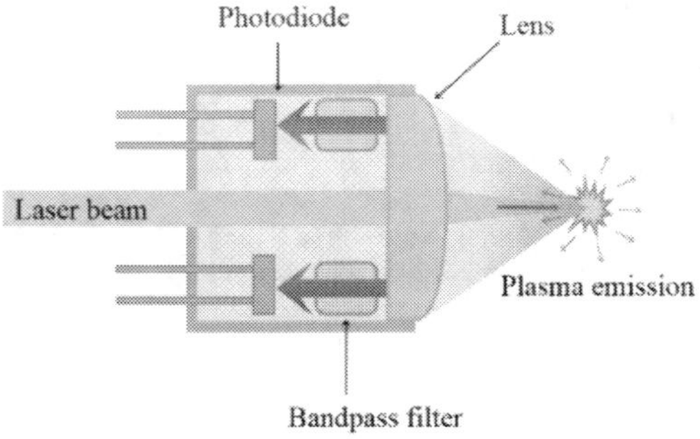

Figure 1: Schematic of the cross plane of simplified LIBS plug.

2.2 Equivalence ratio calibration

Gasoline-air mixture in uniform droplet stream was used to construct calibration curve between H/O ratio and equivalence ratio. Ultrasonic vibrating ceramic plate nebulizer with 1.65 MHz frequency generated uniform droplet stream. Fuel flow rate was fixed at 3ml/min and equivalence ratio of mixture was controlled by varying the air flow rate. The air flow rate was in range from 8 L/min to 28 L/min and the equivalence ratio was ranged from 0.93 to 3.24. Filters and photodiodes utilized in the calibration experiment had the same specification as the simplified LIBS plug. However, in order to make the calibration curve more accurate, diameter of filters was chosen to be 50.8 mm and mounted photodiodes were used instead of unmounted photodiodes. Nd:YAG laser (Continuum, Surelite I) with wavelength of 1064 nm and pulse duration of 5 ns was used to generate plasma for the entire experiment. Plasma was generated within the mixture of certain equivalence ratio and the light emitted from the plasma was captured with two mounted photodiodes. Signal intensity was averaged in time between 250 ns to 1 us. Intensity in first 250 ns was negligible because it was significantly affected by plasma continuum. Signal after 1 us was also negligible because plasma emission drops remarkably after the time period.

2.3 Concentration mapping and scheme of instantaneous mapping

Final tests were carried out to observe the concentration of fuel in different locations in a liquid spray. The liquid fuel was sprayed by a siphon nozzle (Delavan 30609-2) and a map of H/O intensity ratio was drawn using the measurement setup used for the calibration.

Also, instantaneous mapping of equivalence ratio was planned future work. In order to make uniform plasma energy at each point, beam splitters and mirrors will be located in designated position as shown in Fig. 2. On each plasma position, 100 mJ of laser pulse energy will be delivered. A total of six plasmas will be generated with six identical LIBS plugs. Since each LIBS plug needs two BNC cable connection, three oscilloscopes are to be controlled at the same time.

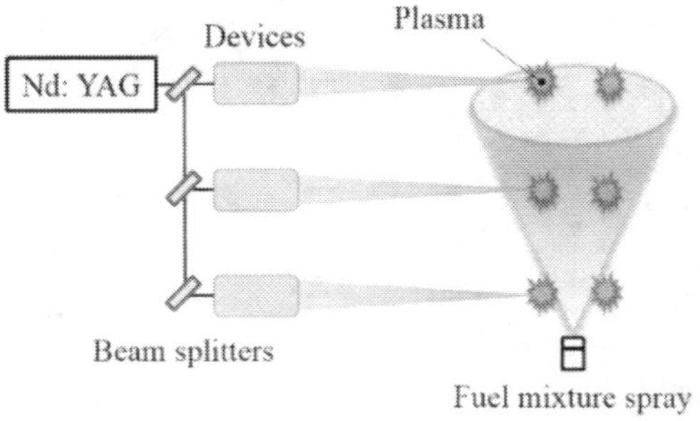

Figure 2: Schematic of multiple LIBS plugs.

3 Results

3.1 Calibration curve

Using LIBS plug, H and O signals are measured with different equivalence ratios. Gasoline-air mixture in uniform droplet stream's equivalence ratio was controlled with gasoline flow rate. Equivalence ratios were selected based on change of gasoline flow rate (2.3, 3.45, 4.6, 5.75, 6.9, 8.05, 9.2 ml/min). Initial gas pressure was 1atm and temperature was 300 K while 160mJ of laser energy was delivered to make proper plasma. For each equivalence ratio, 30 single shots are collected and averaged. As a result, the calibration graph between H/O intensity ratio and equivalence ratio was obtained as shown in Fig. 3. Calibration graph showed linearity with coefficient of determination (R^2) of 0.9498. Error bars shown in the Fig. 3 are based on standard deviation.

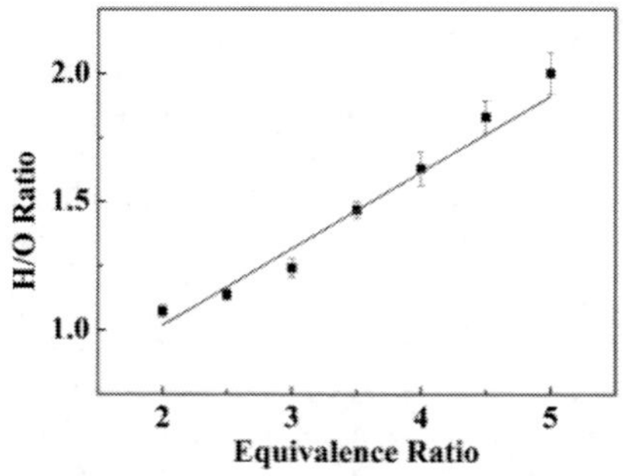

Figure 3: Calibration curve between H/O intensity ratio and equivalence ratio.

As the photodetectors receive signal right after plasma's formation and continue to receive for a long time, the meaningful signals from whole signal needs to be extracted. Immediately after the formation of the plasma, the light intensity was observed to be very strong for every frequency. This was attributed to the plasma background continuum and it persists for about 250 ns. Furthermore, as atomic signals tend to disappear after 1 μs, we collected and averaged signals from 250 ns to 1 μs. The measured signals were averaged and then averaged H signal intensity was divided by averaged O signal intensity. Using the calibration curve in Fig. 3, we continued to the next step.

3.2 Spray fuel concentration mapping results

Using H and O signal intensities, fuel concentration mapping of fuel spray field was possible. With the siphon nozzle spray, liquid gasoline was combined with air in the form of a spray. Flow rate of gasoline was 10 ml/min and flow rate of air was 10 L/min. The mapping was done at atmospheric pressure and temperature. The laser energy delivered to make plasma was 160 mJ. H and O signals were measured from various locations using single LIBS plug. Spray nozzle was moved to measure properties from different locations. Measurements were performed from the height of 0 mm above the siphon nozzle to the 48 mm above the siphon nozzle and vertical spacing between each measurement was 4 mm. Along the x-axis, a distance of 20 mm from the center of nozzle on each side was measured with the interval of 2 mm. The measured H/O intensity ratio is shown in Fig. 4 (a). The LIBS plug signal acquired from two points are shown in Fig. 4 (b) and (c). Fig. 4 (b) is signal measured at 20mm above the siphon nozzle and 0mm distance from the center of nozzle along the x-axis. Fig. 4 (c) is signal measured at the same height with Fig. 4 (b) and 20mm distance along the x-axis. Line colored in black shows the photodiode signal intensity collected through 656 nm filter which means H signal and red line represents signal collected through 780 nm filter which means O signal. To avoid intervention of plasma background emission, signals between 250 ns and 1 μs are selected. From Fig. 4 (b) we can recognize that H signal shows higher intensity than O signal while H signal from Fig. 4 (c) shows lower intensity than O signal. The difference of total signal intensity between two figures is because of the droplet's interference. When laser is focused on gasoline droplet stream, plasma has longer radial length than without droplets [18]. As our LIBS plug detects single point, emission energy disperses when plasma has longer shape.

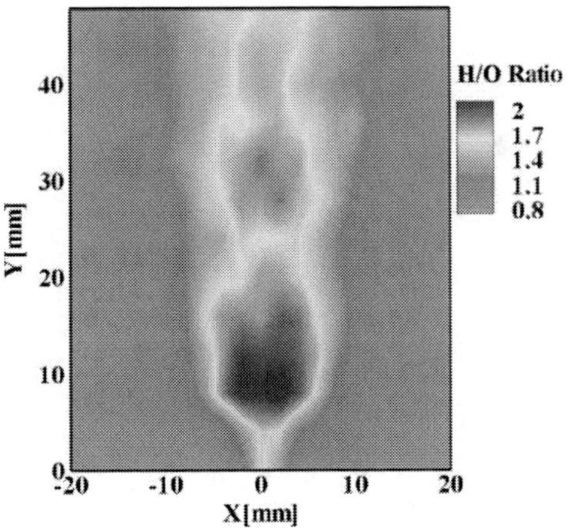

(a)

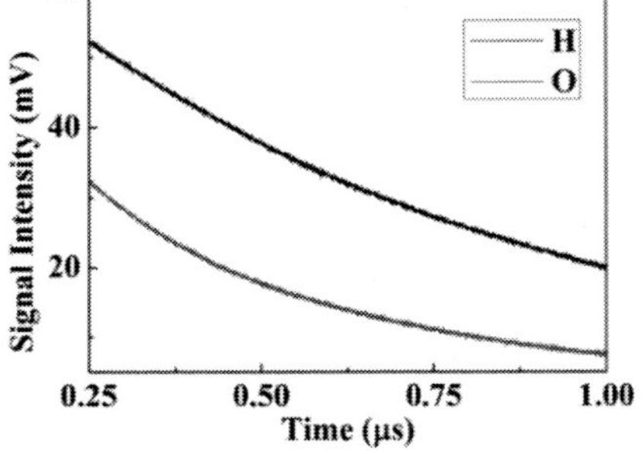

(b)

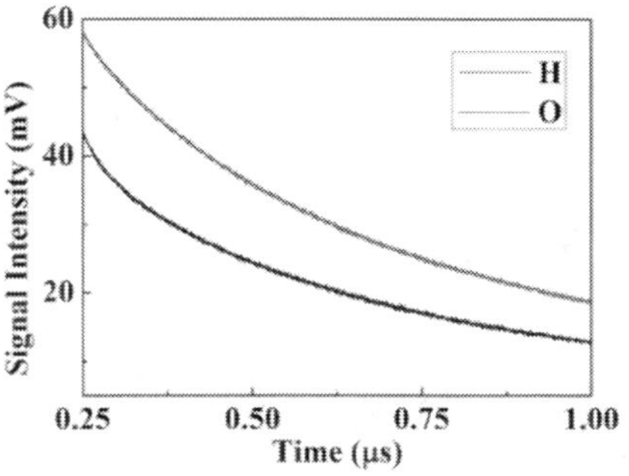

(c)

Figure 4: (a) Mapping of siphon spray H/O intensity ratio using single LIBS plug. (b) LIBS plug signal at 20mm above siphon nozzle and 0mm x-axis distance from the center of the nozzle. (c) LIBS plug signal at 20mm above siphon nozzle and 20mm x-axis distance from the center of the nozzle.

Since our new LIBS plug has small volume, multiple LIBS plugs' plasma can be located in the distribution field of fuel. And as the H/O intensity ratio shows strong linearity with equivalence ratio, it is possible to estimate concentration shape of fuel in the field. A setup using multiple LIBS plugs is currently being developed in order to measure instantaneous 2-dimensional image of fuel concentration. It is expected to measure the distribution of fuel with remarkably short time duration using the proposed system.

4 Conclusions

Simple LIBS plug with bandpass filters and photodiodes was developed, manufactured and tested. The LIBS plug is capable of detecting H and O atomic plasma emission signals at the same time without using complex LIBS configuration such as spectrometer and ICCD camera. Calibration curve between H/O intensity ratio and equivalence ratio was made using the same bandpass filters and photodiodes. With this simple setup, H/O intensity ratio map on siphon spray was made.

Instantaneous 2-dimensional mapping of fuel concentration method was also designed. Spray mapping of the whole region can be performed with the bundled LIBS plug. Once fuel mapping with one LIBS plug is accomplished, validation for multiple LIBS plugs will be possible. Stabilization and miniaturization of the LIBS plug is another issue being tackled currently. Using smaller LIBS plug will allow us to measure properties with short interval of each point. Different filters and more photodiodes can be added in the LIBS plug to measure different properties in the same location.

5 Acknowledgment

This work was supported by Advanced Research Center Program (NRF-2013R1A5A1073861) through the National Research Foundation of Korea (NRF) grant funded by the Korea government (MSIP) contracted through Advanced Space Propulsion Research Center at Seoul National University. And, this research was supported by Basic Science Research Program through the National Research Foundation of Korea (NRF) funded by the Ministry of Education (NRF-2016R1D1A1A02937421).

References

[1] A.B. Yakar, R.K. Hanson, Journal of Propulsion and Power,

17 (4) (2001) 869-877.

[2] M.R. Gruber, J. Donbar, C.D. Carter, K.Y. Hsu, Journal of Propulsion and Power 20 (5) (2004) 769-778.

[3] M.R. Gruber, R.A. Baurle, T. Mathur, K.Y. Hsu, Journal of Propulsion and Power 17 (1) (2001) 146-153.

[4] X.P. Li, W.D. Liu, L.C. Yang B. An, Y. Pan, J.J. Zhu, Experimental Investivgation on Fuel Distribution Using Kerosene-PLIF in a Scramjet Combustor with Dual Cavity, 21st AIAA International Space Planes and Hypersonics Technologies Conference, 2017.

[5] S. Joshi, D.B. Olsen, C. Dumitrescu, P.V. Puzinauskas, A.P. Yalin, Applied spectroscopy 63 (5) (2009) 549-554.

[6] P. Stavropoulos, A. Michalakou, G. Skevis, S. Couris, Chemical physics letters 404 (4) (2005) 309-314.

[7] P. Stavropoulos, A. Michalakou, G. Skevis, S. Couris, Spectrochimica Acta Part B: Atomic Spectroscopy 60 (7) (2005) 1092-1097.

[8] A. Michalakou, P. Stavropoulos, S. Couris, Applied Physics Letters 92 (8) (2008) 081501.

[9] K.E. Eseller, F.Y. Yueh, J.P. Singh, Applied optics 47 (31) (2008) G144-G148.

[10] A.E. Majd, A.S. Arabanian, R. Massudi, M. Nazeri, Applied spectroscopy 65 (1) (2011) 36-42.

[11] T.X. Phuoc, F.P. White, Fuel 81 (13) (2002) 1761-1765.

[12] T.X. Phuoc, Optics and Lasers in Engineering 44 (6) (2006) 520-534.

[13] J. Kiefer, J. Tröger, Z. Li, M. Aldén, Applied Physics B: Lasers and Optics 103 (1) (2011) 229-236.

[14] S. Zhang, X. Yu, F. Li, G. Kang, L. Chen, X. Zhang, Optics and Lasers in Engineering 50 (6) (2012) 877-882.

[15] B. McGann, C.D. Carter, T. Ombrello, H. Do, Combustion and Flame 162 (12) (2015) 4479-4485.

[16] H. Do, C.D. Carter, Q. Liu, T.M. Ombrello, S. Hammack, T. Lee, Proceedings of the Combustion Institute 35 (2) (2015) 2155-2162.

[17] M. Kotzagianni, R. Yuan, E. Mastorakos, S. Couris, Combustion and Flame 167 (2016) 72-85.

[18] S. H. Lee, H. Do, J. J. Yoh, Combustion and Flame 165 (2016) 334-345

11th Asia-Pacific Conference on Combustion,
The University of Sydney, NSW Australia
10th -14th December 2017.

Effects of humidity on the aging of the pyrotechnic compositions ZPP and THPP

Juyoung Oh[1], Anirudha Ambekar[1], Yoocheon Kim[1], Jai-ick Yoh[1]

[1]Department of Mechanical and Aerospace Engineering, Seoul National University

Seoul, 151-742, South Korea

Abstract

This paper studies the effects of humidity and aging period on two pyrotechnic initiators viz. titanium hydride potassium perchlorate (THPP) and zirconium potassium perchlorate (ZPP) subjected accelerated aging. Knowing the combined aging effects for these pyrotechnic initiators are important because they are connected to the performance, lifetime, and safety of the materials. Experiments were carried out using a Differential Scanning Calorimeter (DSC), and the heat flow for each sample was measured over the designated temperature range. In addition, thermal behavior parameters such as the heat of reaction and peak temperature were obtained by using the AKTS software program. These parameters were compared for various aging periods and relative humidity levels. The results clearly show the different effects of humidity levels and aging duration on each sample. Additionally, Laser Induced Breakdown Spectroscopy (LIBS) was used to investigate the mass gain phenomenon of ZPP. Therefore, humidity aging is seen to be an important factor in aging research of common pyrotechnic initiator compositions namely ZPP and THPP.

1 Introduction

Pyrotechnic compositions or propellants are generally composed of three main components: fuel, an oxidizer, and a binder. The safety aspects of these are very important because they are widely used in military or defense industries. However, these are prone to degradation of performance by environment elements such as temperature, humidity, vibration effects [1].

When these materials are aged during storage or transportation, chemical structure difference can occur over small features such as cracks of the particle surface. Although, they may seem to be trivial, but such changes have been known to cause enormous accidents since the past. [2]

Titanium hydride potassium perchlorate (THPP) and zirconium potassium perchlorate (ZPP) are commonly used as pyrotechnic initiators. THPP consists of potassium perchlorate as an oxidizer, titanium hydride as fuel, and a Viton binder. ZPP is composed of zirconium as metal fuel, potassium perchlorate as an oxidizer and a Viton binder. They are generally found in the defense, aerospace, automotive industries. Therefore, it is important to know how their properties and reaction kinetics when accelerated aging or humidity aging was applied.

Several studies on the effects of aging on pyrotechnic materials or propellants have been reported previously. Wang et al. [2] investigated the influences of combinative effect of temperature and humidity on the thermal stability of pyrotechnic mixtures. By analysing the thermal behavior of each compositions and calculating the critical temperature of thermal explosion, it was possible to find the correlation between relative humidity and thermal stability. Also, they found the pyrotechnic composition with maximum thermal stability.

Analysing the thermal behavior is a suitable technique for observing the aging phenomena and other characteristics of the pyrotechnic mixtures. Generally, thermal behavior as well as thermal decomposition characteristics of pyrotechnic compositions consisting of fuel and an oxidizer can be obtained by DSC or thermos gravimetric analysis (TGA) [3, 4]. Although there have been many studies on the aging characteristics of pyrotechnic compositions, there are few studies dealing with humidity aging. In addition, there have been no investigations on humidity and accelerated aging for THPP and ZPP. Therefore, the aim of this study is to analyse the reaction changes of THPP and ZPP affected by humidity and aging period by using DSC.

2 Experimental

2.1 Sample preparation

A total of 10 samples were tested in this study and as listed in Table 1.

Sample type	Humidity and temperature conditions for aging	Duration of aging	n
THPP	75 °C and RH 90 %	20 days	1
	75 °C and RH 0 %	0 day	2
		14 days	3
		42 days	4
ZPP	75 °C and RH 90 %	20 days	5
	75 °C and RH 80 %	20 days	6
		40 days	7
	75 °C and RH 0 %	0 day	8
		14 days	9
		42 days	10

Table 1: All samples used in this study

The THPP and ZPP samples subjected to accelerated aging at 75 °C were sorted according to the relative humidity and finally classified by the aging duration. Henceforth, each sample will be referred by the serial number (n) instead of the name. The THPP mixture was composed of 32% TiH_2, 63% $KClO_4$ and 5 % Viton b. ZPP consists of 52% Zr, 42% $KClO_4$, 5% Viton b, and 1 % graphite.

Corresponding author. Fax: +82-2-882-1507
E-mail address: jjyoh@snu.ac.kr

2.2 Equipment

All samples were tested by using the DSC-3 instrument from Mettler Toledo. This calorimetry device can measure enthalpy changes in samples due to changes in their physical and chemical properties as a function of temperature and time. The net heat generated for each sample and the reaction rate were obtained by using the AKTS software program for the experimental results.

DSC analysis was performed at a heating rate of 2 °C/min in temperature range of 30 °C to 600 °C under nitrogen atmosphere (85 ml/min). The mean mass of the samples used in this experiment has been listed in Table 2. Nearly all samples were tested more than twice. Especially, test with sample 7 was repeated five times. Here, m_b is the net average mass of the sample before the experiment and m_a indicates the average mass of the sample after the experiment. Samples were sealed in 40 μm standard pierced aluminum pan.

n	Sample mass (mg)	Total mass - m_b (mg)	Total mass - m_a (mg)	Mass loss (%)
1	1.00	50.88	50.88	0.00
2	1.31	51.02	50.86	12.21
3	1.25	51.09	50.92	13.60
4	1.21	51.20	51.09	9.09
5	1.36	51.35	51.26	6.62
6	1.07	50.89	50.83	5.18
7	1.29	50.88	51.02	-11.41
8	0.99	50.75	50.61	13.56
9	1.25	51.09	51.89	14.87
10	1.20	50.92	50.74	15.29

Table 2: Mass and mass loss data of all samples

The phenomenon of mass increase of the ZPP composition during the DSC experiment was analyzed by using LIBS technique. A Q-switched Nd:YAG laser was used in the LIBS system (RT-250Ec, Applied Spectra Inc.) that operates at 1064nm, 4-7 ns pulse duration. The pulse energy of the laser range was from 25 mJ to 30 mJ. This spectrometer has 6 channels to improve the resolution and can cover the wavelengths from 190 nm to 1040 nm as detected spectrum range. The resolution of the spectrometer is less than 0.1 nm from ultraviolet to visible range and 0.12 nm from visible to near infrared range. The gate delay was 0.5 μs gate and the gate width was set as 1.05 ms. The pressure in the chamber was maintained at 760 Torr. The ZPP samples were exposed 10 times to the laser beam.

2.3 Reaction rate calculation

As shown in Fig.1, the baseline could be obtained by involving the tangent lines superposition on each side of the exothermic peak of the DSC signal. The reaction rate at the time can be determined by dividing heat flow value which is the difference between the DSC signal of the sample and the base line to the heat generation in the entire process. [5] The values of reaction rate, baseline, and heat of reaction were obtained by using AKTS software program.

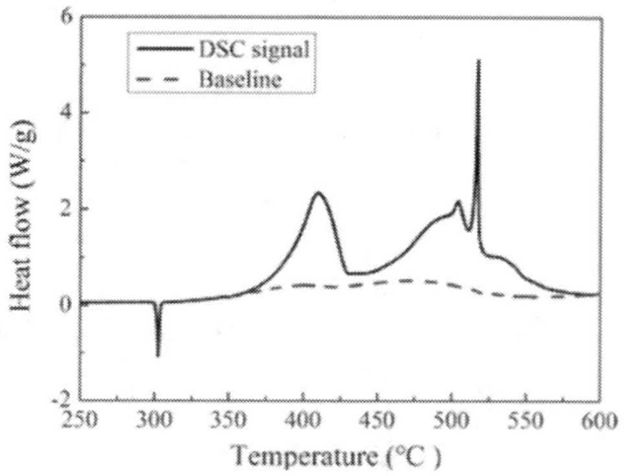

Figure 1: The example of the DSC signal line and base line of the THPP sample aged at 75 °C-RH 90% for 20 days

3 Results and discussion

3.1 Experimental results

Table 3 shows the heat of reaction, peak temperatures, and onset temperature for four THPP samples. Fig. 2 shows the effects of the humidity and aging period on the THPP samples. Here, n is the sample number, ΔH is the heat of reaction, T_p indicates peak temperature, and T_o is onset temperature of reaction. As the aging days increased, the heat of reaction tended to decrease.

Sample type	n	ΔH (J/g)	T_p (°C)	T_o (°C)
THPP	1	3847.60	418.80, 515.11	335.60
	2	5175.16	409.70, 517.10	353.07
	3	4814.49	418.27, 519.24	310.13
	4	4568.55	420.43, 518.43	310.13

Table 3: Heat generation data of THPP samples

Figure 2(a) shows two curves for THPP samples which were subjected to accelerated aging and humidity aging (sample 3, sample 1). The sample 3 as well as sample 1 displayed two exothermic peaks. The peak temperatures for the sample 1 were observed to be similar to that of sample 3. Therefore, it can be seen that humidity aging does not affect the peak temperature. The onset temperatures of sample 1 was higher than that of sample 3. On the other hand, heat of reaction of sample 1 was lower than sample 3. In the case of reaction rate, sample 1 was higher than sample 3 initially and lower for the second exothermic peak. Therefore, the effect on moisture itself could be found in reaction of heat, reaction rate and onset temperature.

Figure 2(b) shows samples with accelerated aging. The curves in this graph represent the non-aged sample (sample 2) and the samples aged at 75 °C for 14 days and 42 days (sample 3, sample 4). In this case, the longer the period of aging days of the sample, the higher the first peak temperature and the reaction rate. Also, heat of reaction decreased with the aging period.

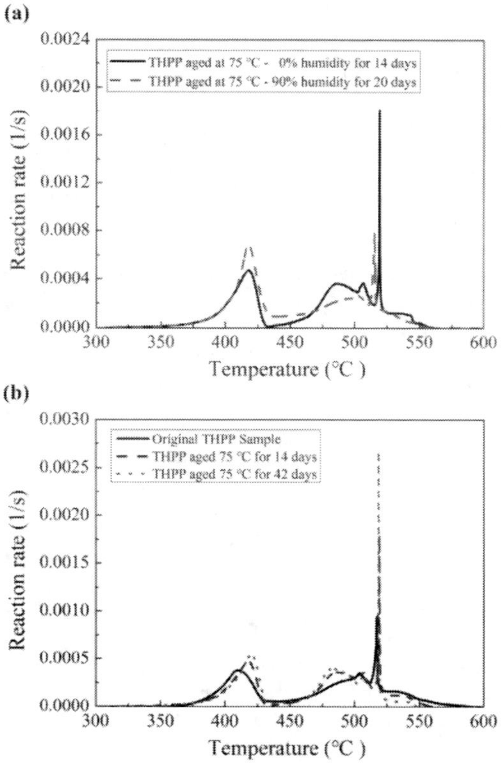

Figure 2: Reaction rate of THPP samples (a) humidity effect, (b) aging period effect except relative humidity

Table 4 shows the heat of reaction, peak temperature, and onset temperature for six ZPP samples. When humidity aging or accelerated aging was applied to the ZPP sample, the heat of reaction tended to be lower than that of the original value. Also, the longer the aging period, the lower the heat of reaction. Under the effect of humidity aging, when the number of days of aging was the same, higher relative humidity resulted in, lower heat release.

Sample type	n	ΔH (J/g)	T_p (°C)	T_o (°C)
ZPP	5	2190.64	415.77	328.20
	6	2449.06	413.83, 483.83	345.00
	7	2010.80	473.33, 511.50	325.73
	8	2629.81	448.10, 504.20	345.27
	9	2529.75	374.43, 444.61, 504.80	312.20
	10	2107.88	374.10, 449.10, 503.83	310.67

Table 4: Heat generation data of ZPP samples

Figure 5(a) shows three curves for the ZPP samples. Two of them are samples with different relative humidity (80%, 90%) which were aged at 75 °C for 20 days (sample 6, sample 5). The remaining sample was aged at 75 °C for 14 days (sample 9).

Each sample had a different number of peaks. Sample 9, 6 and 5 displayed the number of 3, 2 and 1 exothermic peak respectively. As the moisture content increased the number of exothermic peaks decreased. The main exothermic peak temperature and the heat released in the case humidity aged samples were lower than the 0% humidity sample, but main reaction rate was higher.

Meanwhile, amongst the samples with different humidity levels, the first peak did not show a significant difference. However, the special point was found at the second peak. The second peak means the decomposition reaction of Potassium Perchlorate ($KClO_4$). [6] The sample with 90% relative humidity had only first peak. Therefore, it could be assumed that $KClO_4$ was affected by humidity level.

Figure 5(b) shows three curves, two of them are ZPP samples with different aging period (20 days, 40 days) aged at 75 °C and relative humidity 80% (sample 6, sample 7) and the last one is non-aged original ZPP sample (sample 8). All the samples in the graph displayed two exothermic peaks. In the case of heat of reaction, both of the humidity aged samples showed a lower heat value than the original sample. Especially, sample aged for longer period of time had the lower heat value.

A remarkable point was found in the sample 7, which was that the sample mass increased after the experiment. As seen in Table 2, every other sample showed the mass loss. However, the mass of sample 7 was observed to increase by 11.41% compared to the initial sample mass. One of the most likely reasons for the increase in mass was bonding of atmospheric oxygen molecules with the residual sample and increase the sample mass.

In order to confirm the degree of oxidation before and after the DSC experiment, LIBS experiments were performed on Sample 5, 6 and 7. In Fig. 3, the peak indicates the oxygen in ZPP samples. The large difference in oxygen intensity means increased oxidation of zirconium. In Fig. 4, the y-axis shows the difference in oxygen intensity between samples 5, 6, 7 before and after the DSC, and the x-axis shows the mass change of the samples before and after the DSC. The difference in oxygen intensity was noticeable when the number of days of aging increased compared to when humidity was increased. Thus, it could be seen that zirconium oxidation was affected by humidity but is more sensitive to aging period. As a result, due to the oxidative effect after the experiment, 20-day humidity aged samples shows reduced mass loss value compared to accelerated aged samples. Sample humidity aged for 40 days, which was aged longer than other samples, increased in mass.

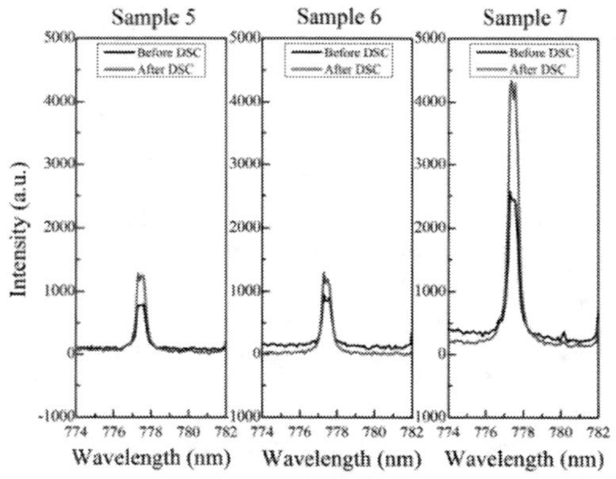

Figure 3: Oxygen intensity peak of sample 5, 6, and 7

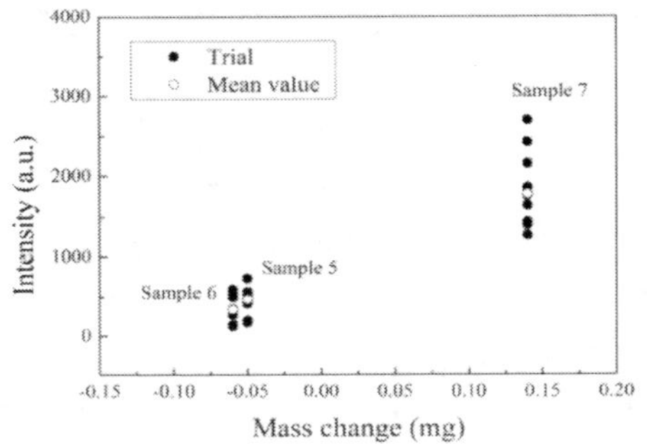

Figure 4: Oxygen intensity difference of sample 5, 6, and 7 with mass change

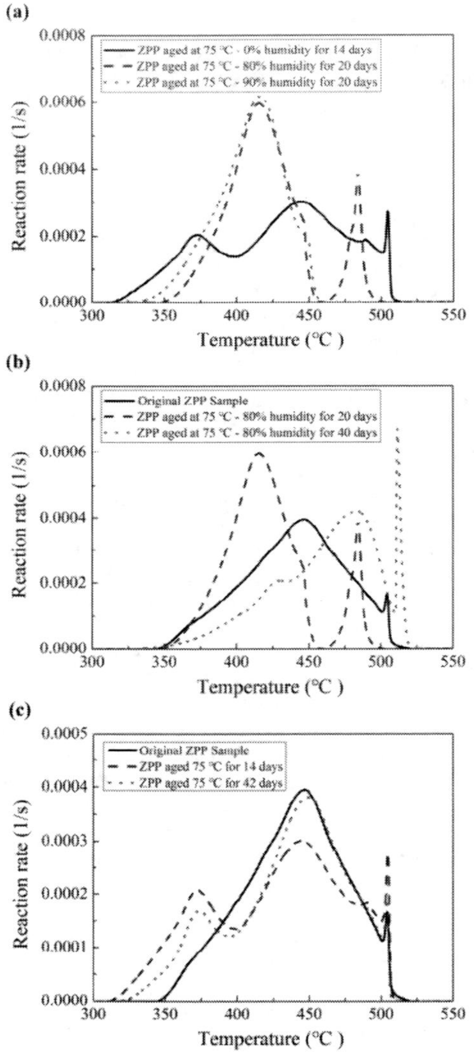

Figure 5: Reaction rate of ZPP samples (a) humidity effect, (b) aging period effect with relative humidity, (c) aging period effect except relative humidity

When accelerated aging was applied to ZPP samples at 75 °C in the case of Fig. 5(c), the first exothermic peak appeared at 374 °C. The second peak and the third peak of the accelerated aged samples were approximately identical when compared to the original sample. Accelerated aged THPP also decreased the released heat with increasing aging days. Finally, the onset temperature tended to decrease when accelerated aging was applied.

4 Conclusions

DSC experiments were conducted using THPP and ZPP samples corresponding to different humidity levels and aging periods. From the DSC data, thermal behavior parameters could be obtained and studied the effects of above conditions to the samples.

In the THPP sample, 90% relative humidity sample had lower heat of reaction and reaction rate. In other words, when aged under high humidity conditions, THPP may cause decreased heat released than the unaged sample. Therefore, humidity aging was deduced to be significantly affect the performances of THPP. In the case of where aging duration was increased, the reaction rate of the samples at the main exothermic peaks increased slightly with the duration of aging. Also, the first exothermic peak tended to shift to higher side, and lower side at the second peak.

The ZPP samples subjected to humidity aging for 20 days displayed a decreased heat of reaction when compared to the sample with 0% humidity. In addition, when the humidity increased, the heat value tended to decrease. Simultaneously, the main exothermic peak temperatures were observed to be decreased. This phenomenon corresponded to significant changes in the sample's performance. In addition, it could be seen that when ZPP is aged at the humid environment, the aging period is more influential on the ZPP oxidation than the humidity.

The humidity aging study was conducted with a limited number of humidity levels, so it was difficult to obtain detailed changes in the chemistry of the initiators due to humidity. Future studies with more samples with larger parametric variation would yield a more accurate picture. However, the current study has established that observable changes in the performance of THPP and ZPP occur due to humidity aging. As humidity aging is an important factor in aging research of THPP and ZPP, further studies will be carried out.

5 Acknowledgment

Authors acknowledge the support from Hanwha Corporation for providing ZPP and THPP samples. Additional support came from Advanced Research Center Program (NRF-2013R1A5A1073861) contracted through Advanced Space Propulsion Research Center at Seoul National University.

References

[1] W. de Klerk, W. Colpa, P. van Ekeren, J. Therm. Anal. Calorim., 85 (1) (2006) 203–7.

[2] L. Wang, X. Shi, W. Wang, J Therm. Anal. Calorim., 117 (2) (2014) 985–992.

[3] E.S. Freeman, V.D. Hogan, D.A. Anderson, Combust. Flame, 9 (1) (1965) 19-23.

[4] S.M. Pourmortazavi, M. Fathollahi, S.S. Hajimirsadeghi, S.G. Hosseini, Thermochim. Acta, 443(1) (2006) 129-131.

[5] Y. Kim, J. J. Yoh, and J. Park, J. Propul. Power, 32 (3) (2016) 777-784.

[6] J. Kim, T. Seo, S. Ko, B. Ryu, J. Kor. Soc. Propul. Engin., 19 (5) (2015) 15-21.

11th Asia-Pacific Conference on Combustion,
The University of Sydney, NSW Australia
10th -14th December 2017.

Downward Flame Spread over Electric Wire under Various Oxygen Concentration

Yusuke Konno[1], Katsuya Tsuzuki[1], Nozomu Hashimoto[1], Osamu Fujita[1]

[1]Division of Mechanical and Space Engineering, Hokkaido University

Kita 13 Nishi 8, Kita-Ku, Sapporo-city, Hokkaido, 060-8628, JAPAN

Abstract

Correlation between flame size and flame spread rate over laboratory electric wire samples (0.8 mm thick) were investigated in various oxygen concentration. Low-density polyethylene (LDPE) insulated NiCr and copper (Cu) wires were used in this work. For NiCr wire, flame spread rate and flame length monotonically increased as the oxygen concentration increased. On the other hand, for Cu wire, flame length showed a peak value at 25% oxygen concentration and flame spread rate decreased in the range from 25% to 31% oxygen concentration. The experimental results showed the heat conduction through the metal core obviously affects the flame length and we found the region where the flame length plays a dominant role rather than the flame temperature on flame spread phenomena. For Cu wire case, presence of three types flame spread modes is proposed in terms of oxygen concentration: (1) gas phase controlling region, (2) metal core controlling region, and (3) overheated region. Finally, mechanism of flame spreading over electric wire is discussed based on the scale model introduced in this study.

1 Introduction

Ignition and subsequent flame spreading over electric wire is a potential cause of fire. To improve fire safety, it is important to understand fire behavior over the electric wire. Over the years researchers have studied flame spread over solid fuel to know the effect of various factors such as ambient flow velocity, flow direction, pressure, oxygen concentration, the size of sample, the type of material, and gravity level [1]–[4].

Fujita et al. [5][6] investigated the flame spread over PE and ethylene tetrafluoroethylene (ETFE) insulated wire (~ 1 mm thick) in microgravity under varying ambient flow velocity and reported that the flame spread rate in microgravity achieved the maximum peak in the extremely low ambient flow velocity condition due to the cylindrical curvature of the sample surface. Nakamura et al. [7] found that flame spread rate over electric wires increases with decreasing ambient pressure and increasing inner metal core conductance. Hu et al. [8] revealed that downward flame spread rate with Cu wire is faster than that with NiCr wire. On the other hand, flame spread rate with NiCr wire is faster than that with Cu wire in a horizontal flame spread case. As it can be seen, a lot of unique feature of flame spread over electric wire were revealed.

In this study, we investigated about the effects of the flame size and the heat conduction through the metal core on flame spread over electric wires under various oxygen concentration to gain the fundamental knowledge of wire combustion. The controlling mechanism of flame spread over electric wires were also discussed based on the scale modeling of the spreading flame.

Corresponding author. Fax: +81-11-706-6385
E-mail address: ofujita@eng.hokudai.ac.jp

2 Experiment

The diagram of the experimental apparatus is presented in Figure 1. This experimental apparatus was designed to conduct the downward flame spread test for a test specimen in accordance with the ISO 4589-2 Oxygen Index Test [9]. It consists cylindrical combustion chimney. The test section itself consists of a 500 mm long PMMA pipe, with an inner diameter of 60 mm. A test specimen is supported vertically in a mixture of oxygen and nitrogen flowing upward. The upper end of the specimen was ignited with an electrically heated kanthal wire coil. A spreading flame was recorded by a digital video camera. The oxygen concentration was set in the range from 15% to 37%. The mixed gas was uniformly supplied to the test section by glass beads and honeycomb at the bottom of the flow duct. The ambient flow velocity was fixed at the speed of 15 cm/s at 1 atm.

The sample wires were manufactured for laboratory experiments. Low-density polyethylene (LDPE), which is a

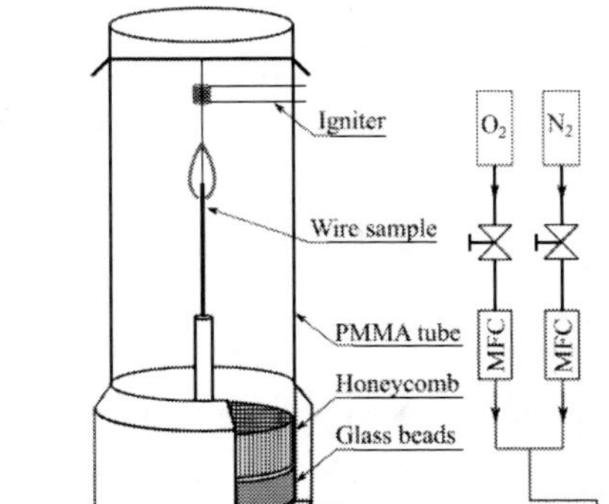

Figure 1: Schematic of experimental apparatus.

	LDPE	Cu	NiCr
Density, ρ [kg/m^3]	940	8880	8670
Specific heat, c [kJ/kg/K]	2.15	0.39	0.44
Thermal cond., λ [W/m/K]	0.30	398	17.4
Melting point, T_m [°C]	105~110	-	-
Pyrolysis temp., T_p [°C]	387	-	-
LOI [%]	18	-	-

Table 1: Physical properties of insulations and metal core.

widely used thermoplastic material for wire insulations, was chosen as insulation material. Two types of the metal core: a high conductivity Cu core and a low conductivity NiCr core, were selected to investigate the effect of the thermal conductivity of a metal core on flame spreading. The physical properties of LDPE, Cu, and NiCr are shown in Table 1. The thermal conductivity of Cu is about 23 times larger than that of NiCr. The wire dimension is 0.5 mm and 0.15 mm for core diameter and insulation thickness, respectively.

3 Results

Flame shape and flame spread rate of two types of wire are compared. 3-5 tests were repeated for each test condition to take the reproducibility. The error bar appears in this study indicates the standard error. The images taken by DV-CAM was always analyzed after the flame reached 5 cm far from the igniter to ignore the effect of initial ignition condition.

Figure 2 shows the typical flame shape observed during the flame spread under various ambient oxygen concentration. The flame consists of blue flame and luminous flame and the length of blue flame becomes short with increasing oxygen concentration. Flame becomes more luminous as oxygen concentration increases. Moreover, it turned out that the length of glowing metal core in the downstream extended as oxygen concentration increased. Particularly NiCr core seems more heated leading to bright red color than Cu wire case. This is an evidence that the low thermal conductivity core cannot conduct heat well into both the upstream and downstream direction. Another feature is when the oxygen concentration exceeds 25% for Cu wire and 29% for NiCr wire, the molten insulation accumulated in the burning zone started boiling and small fire balls popped out from the pyrolysis zone during the spread (Note that this phenomenon is different from bursting bubbles observed during the flame spread over PMMA).

3.1 Flame length and flame radius along the electric wires

Figure 3 (a) and (b) show the flame length and flame radius as function of oxygen concentration along Cu and NiCr wire. Flame size was measured from the visible flame area during the spread. Basically, flame size was always larger with Cu wire than that with NiCr wire. Regarding flame radius, dependency on oxygen concentration is not significant. Further, large fluctuations in relatively high oxygen concentration is caused by boiling of molten insulation. On the other hand, flame length greatly depends on oxygen concentration. For Cu wire case, flame length showed the maximum peak at 25%, and flame length was independent of oxygen concentration in the range from 31 to 37%. For Ni-Cr wire case, although there is a small step around 29%, flame length monotonically increased with oxygen concentration.

3.2 Flame spread rate as function of oxygen concentration

The measured flame spread rate for each condition is plotted in Fig. 3. (c). The trend of flame spread rate with NiCr wire is similar to that of flame length. For Cu wire case, flame spread rate also showed the peak around 25% as well as flame length. However, it started to increase again from 31% oxygen concentration.

Generally, it is expected that high oxygen concentration enhances flame spread since flame temperature increases with oxygen concentration. It is obvious from the universal expression for diffusion flame spread [1] and the previous work for opposed-flow flame spread over thick PMMA and thin paper [2] that oxygen have the positive effect on flame spread. Moreover, as it can be seen in Fig. 2, flame consists almost

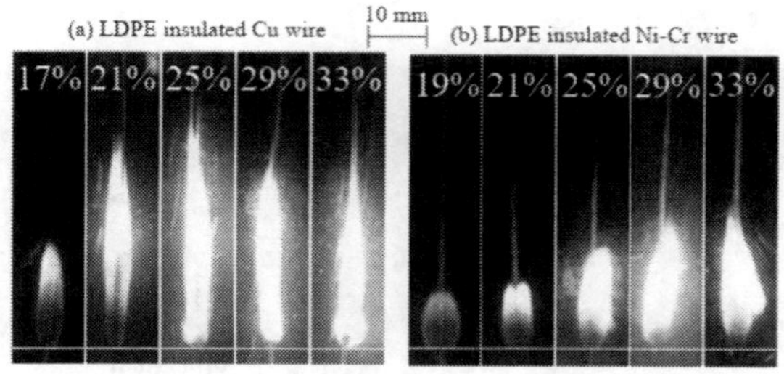

Figure 2: Direct flame images for Cu and Ni-Cr wire sample. The yellow lines reflect the position of flame leading edge and the values written at the top represent ambient oxygen concentration value.

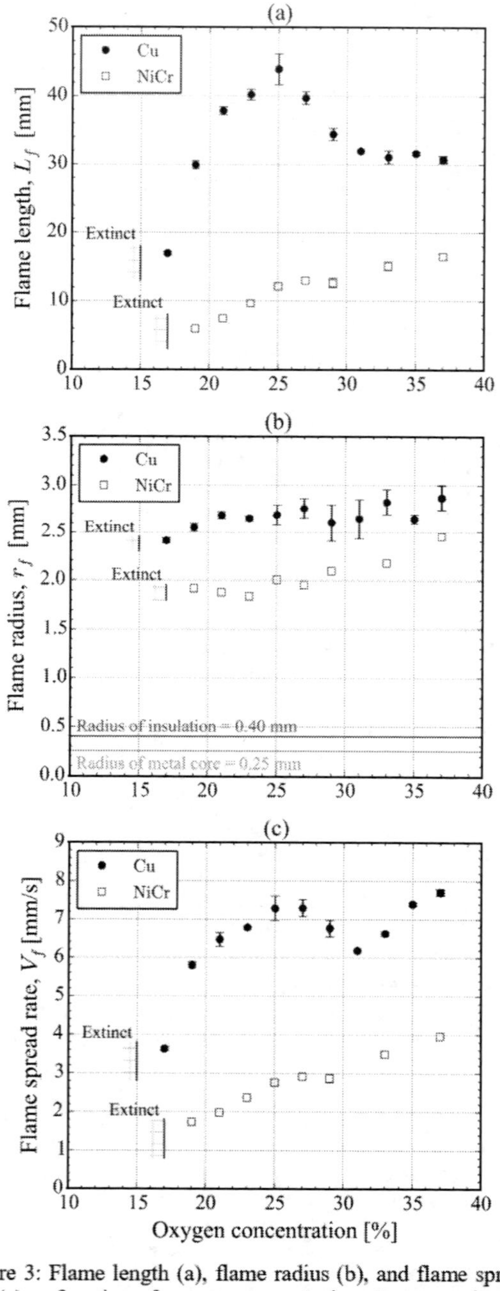

Figure 3: Flame length (a), flame radius (b), and flame spread rate (c) as function of oxygen concentration. Orange and green line in Fig. (b) reflects the radius of metal core and insulation.

only of luminous flame under high ambient oxygen concentration. Thus, radiation from luminous flame to unburned insulation acted as an additional heat source to sustain the flame spread. Therefore, the trend of flame spread obtained with Cu wire is an unexpected finding. This finding provides essential information to understand the controlling mechanism of flame spread over electric wire. In the next section, the effect of oxygen concentration and the conductivity through the metal core on the flame spread over electric wire is discussed to provide insight of flame spreading over Cu core wire.

4 Discussion

Scale model for flame spread over electric wire are constructed. Figure 4 shows the conceptional description of downward spreading flame over electric wire. To simplify the analysis, following assumptions are applied: (1) ignore the dripping of the molten insulation; (2) the wire insulation is thermally thin and consumed uniformly with spreading flame; (3) the combustion is complete, i.e., all the fuel is oxidized in the flame.

4.1 Flame length (L_f) and flame radius (r_f)

The theoretical expression for the flame radius r_f and the flame length L_f of downward spreading flame over electric wires were modelled following Bhattacharjee et al. [10]. In this work, r_f is defined as the distance from the center of the wire to the outer edge of the flame in the radial direction and L_f is defined as the distance between the trailing and leading edges of the flame in the axial direction.

The amount of gasified fuel generated per unit time from the insulation $\dot{m}$ can be calculated by

$$\dot{m} \sim A_s V_f \rho_s \quad (1)$$

where A and ρ are the cross-sectional area and density; V_f is flame spread rate; subscript s shows the insulation, respectively.

The mass conservation equations which derive r_f and L_f can be expressed as follows:

$$\pi r_f^2 V_r \rho_g Y_{O_2,\infty} \sim \nu \dot{m}; \quad \Rightarrow r_f \sim \sqrt{\frac{\nu \dot{m}}{\pi V_r \rho_g Y_{O_2,\infty}}} \propto \sqrt{\frac{\nu V_f}{V_r Y_{O_2,\infty}}} \quad (2)$$

$$2\pi r_f L_f D \rho_g \left.\frac{\mathrm{d}Y_{O_2}}{\mathrm{d}r}\right|_{r=r_f} \sim \nu \dot{m}; \quad \Rightarrow L_f \sim \frac{\nu \dot{m} \ln(r_f/r_s)}{2\pi D \rho_g Y_{O_2,\infty}} \propto \frac{\nu V_f \ln r_f}{Y_{O_2,\infty}} \quad (3)$$

where $Y_{O_2,\infty}$, D and ν are mass fraction of oxygen, diffusion coefficient, and stoichiometric ratio; V_r is the oxidizer gas velocity relative to the flame which is defined as $V_r = V_f + V_g$; subscript g shows the gas phase. The gradient of oxygen concentration in Eq. (3) can be calculated by a one-dimensional mass transfer equation in the cylindrical coordinate with two boundary conditions of $Y_{O_2}(r = r_f) = Y_{O_2,\infty}$ and $Y_{O_2}(r = r_s) = 0$.

Consequently, Eqs. (2) and (3) suggest that increasing oxygen concentration directly makes flame size (r_f and L_f) small. This is a reasonable relation between the oxygen concentration and diffusion flame. However, it should be noted that the numerator of Eqs. (2) and (3) has V_f i.e. the amount of gasified fuel.

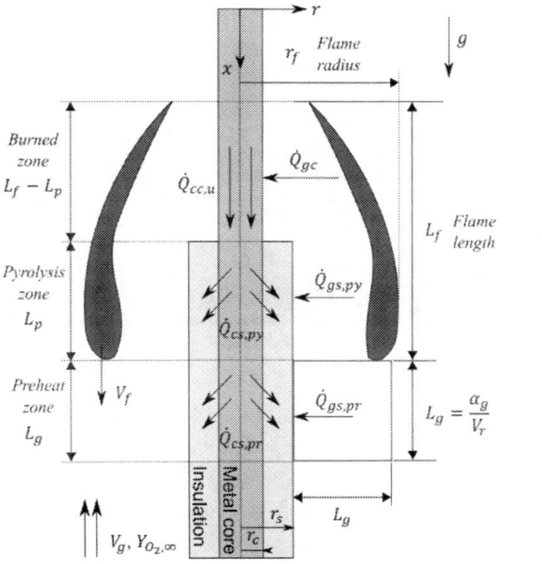

Figure 4: Conceptional description of downward spreading flame over an electric wire.

4.2 Flame spread rate (V_f)

The notable point of flame spread problem for electric wire is the heat conduction through the metal core plays significant role. From the heat balance equation in preheat zone, flame spread rate can be expressed as follow

$$V_f \sim \frac{\dot{Q}_{gs,pr} + \dot{Q}_{cs,pr}}{(A_s \rho_s c_s + A_c \rho_c c_c)(T_p - T_\infty)} \quad (4)$$

where c is the specific heat; T_p and T_∞ are pyrolysis temperature of insulation and ambient temperature; $\dot{Q}_{gs,pr}$ and $\dot{Q}_{cs,pr}$ are the heat conduction from the gas phase to the insulation and the heat conduction from the metal core to the insulation; subscript c shows the metal core. As for $\dot{Q}_{cs,pr}$, it depends on how much the metal core downstream is heated by the flame above the core. To estimate $\dot{Q}_{cs,pr}$, heat balance in burned zone is also considered and we applied the same assumption proposed by Hu et al. [8] that the wire surrounded by the flame is assumed as thermal isolation. This is a reasonable assumption under normal gravity condition because the flame tailing edge is always closed due to natural convection and it directly contacts flame, hence the maximum temperature in the metal core in axial direction may exist at the flame tailing edge location. Therefore, the heat balance in burned zone can be expressed as follow

$$\dot{Q}_{gc} \sim \dot{Q}_{cc,u} \quad (5)$$

where $\dot{Q}_{gc}$ and $\dot{Q}_{cc,u}$ are the heat conduction from the gas phase to the metal core in burned zone and the heat conduction inside the metal core in upstream axial direction. The $\dot{Q}_{gc}$ can be calculated by

$$\dot{Q}_{gc} = P_c(L_f - L_p)\dot{q}''_{gc} \quad (6)$$

where P_c, L_p, and $\dot{q}''_{gc}$ are the perimeter of the metal core, length of pyrolysis zone, and heat flux from flame to the metal core surface, respectively. Equation (6) shows that $\dot{Q}_{gc}$ increases as L_f becomes long.

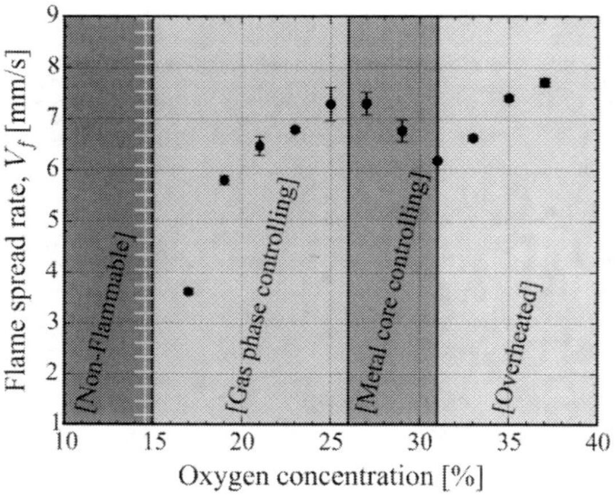

Figure 5: Three types of flame spread mode in various ambient oxygen concentration with high thermal conductivity metal wire (Cu wire).

To estimate the heat conduction through the metal core to the unburned insulation in preheat zone ($\dot{Q}_{cs,pr}$), the heat loss from the metal core to the insulation in the pyrolysis zone $\dot{Q}_{cs,py}$ is introduced. $\dot{Q}_{cs,py}$ can be calculated by the heat balance equation in the pyrolysis zone

$$\dot{m}L \sim \dot{Q}_{gs,py} + \dot{Q}_{cs,py}. \quad (7)$$

Eventually, $\dot{Q}_{cs,pr}$ can be calculated by

$$\dot{Q}_{gs,pr} \sim \dot{Q}_{gc} - \dot{Q}_{cs,py} \quad (8)$$

Substituting Eqs. (6) and (8) to Eq. (4), flame spread rate can be expressed as

$$V_f \sim \frac{\dot{Q}_{gs,pr} - \dot{Q}_{cs,py} + P_c(L_f - L_p)\dot{q}''_{gc}}{(A_s\rho_s c_s + A_c\rho_c c_c)(T_p - T_\infty)} \quad (9)$$

4.3 Role of oxygen concentration on flame length and flame spread for electric wire

The flame length (Eq. (3)) and flame spread rate formula (Eq. (9)) for electric wire were proposed in the former section. Equation (3) suggests the flame length L_f becomes short as oxygen concentration increases, while the conduction through the metal core to the unburned insulation $\dot{Q}_{cs,pr}$ decreases with flame length. Therefore, there is a strong interaction among oxygen concentration, L_f, and $\dot{Q}_{cs,pr}$.

Figure 5 shows the flame spread rate as function of oxygen concentration for Cu wire (the same plots in Fig. 3 (c)). For convenience purposes, the trend of flame spread rate for Cu wire is divided into three regions as shown Fig. (5) and we may call three regions as gas phase controlling region, metal core controlling region, and overheated region, respectively. In the gas phase controlling region, increasing flame temperature plays a dominant role on flame spread, hence flame spread rate monotonically increases with oxygen concentration. In metal core controlling region, flame length decreases with increasing oxygen concentration due to diffusion effects, hence $\dot{Q}_{cs,pr}$ in Eq. (9) decreases with increasing oxygen concentration, hence flame spread rate decreases with increasing oxygen concentration. In overheated region, gas phase temperature again starts playing a dominant role rather than flame length, because flame temperature is too high in this region.

5 Conclusions

Flame spread over the laboratory electric wire (LDPE insulated Cu and NiCr wire) was experimentally examined under various oxygen concentration. For NiCr wire, flame length became long and flame spread rate increased as oxygen concentration increased. On the other hand, for Cu wire, both flame length and flame spread rate showed the peak value for oxygen concentration, in addition there was a region where flame spread rate decreased with increasing oxygen concentration.

The theoretical expression for flame length (Eq. (3)) shows the flame length becomes short as oxygen concentration increases, while flame length becomes long as flame spread rate increases. The scaling model suggested that, dual effects of oxygen concentration on flame length plays a key role in downward flame spread over electrical wire. Particularly, when the conductivity of the metal core is large, three regimes controlled by different mechanisms appears for tested range of oxygen concentration. The first region is gas phase controlling region where flame spread increases monotonically as flame temperature increases. The second one is metal core controlling region where increasing oxygen concentration shortens flame length, hence the heat conduction through the metal core decreases, hence flame spread rate decreases. The third one is overheated region. This proposed model may help to understand the flame spread problem over electric wire. Further discussion would be necessary to validate the proposed scaling model in this work.

6 Acknowledgment

This research is supported by JAXA as a candidate experiment for the second phase utilization of JEM/ISS entitled "Solid Combustion" and third phase utilization of JEM/ISS entitled "FLARE".

References

[1] J. N. De Ris, Symp. Combust. 12 (1) (1969) 241–252.

[2] A. C. Fernandez-Pello, S. R. Ray, and I. Glassman, Symp. Combust. 18 (1) (1988) 579–589.

[3] S. Bhattacharjee, A. Simsek, F. Miller, S. Olson, and P. Ferkul, Proc. Combust. Inst. 36 (2) (2016) 2963-2969.

[4] S. Olson, P. Ferkul, Combust. Flame 180 (2017) 217–229.

[5] O. Fujita, K. Nishizawa, and K. Ito, Proc. Combust. Inst. 29 (2) (2002) 2545–2552.

[6] O. Fujita, M. Kikuchi, K. Ito, and K. Nishizawa, Proc. Combust. Inst. 28 (2) (2000) 2905–2911.

[7] Y. Nakamura, N. Yoshimura, H. Ito, K. Azumaya, and O. Fujita, Proc. Combust. Inst. 32 (2) (2009) 2559–2566.

[8] L. Hu, Y. Zhang, K. Yoshioka, H. Izumo, and O. Fujita Proc. Combust. Inst. 35 (3) (2015) 2607–2614.

[9] ISO4589-2, "Plastics -Determination of burning behavior by oxygen index Part 2: Ambient-temperature," (1996).

[10] S. Bhattacharjee, S. Takahashi, K. Wakai, and C. P. Paolini, *Proc. Combust. Inst.* 33 (2) (2011) 2465–2472.

11th Asia-Pacific Conference on Combustion,
The University of Sydney, NSW Australia
10th -14th December 2017.

Effects of slope and fuel bed width on fire spread: An experimental study

Han Li[1], Naian Liu[1,*], Xiaodong Xie[1], Linhe Zhang[1], Xieshang Yuan[1], Domingos X. Viegas[2]

[1]State Key Laboratory of Fire Science, University of Science and Technology of China, Hefei, Anhui 230026, PR China

[2]Department of Mechanical Engineering, University of Coimbra, 3030-788 Coimbra, Portugal

Abstract

In this work, a series of experiments were carried out on an inclinable combustion table to determine the effects of slope and fuel bed width on surface fire spread rate. Fires were ignited in plots with width ranging from 0.5 m to 2 m and slope from 0° to 30° in still air. It was observed that both the fuel bed width and slope terrain had a promotion effect on fire rate of spread, and flame length increased with slope significantly. Because of the unconfined lateral air entrainment, a V-shape fire line distortion was established from the initial linear ignition during the quasi-steady state when the propagation takes place under slope up to 20°. Flow pattern ahead of the fire front was identified, and the streamwise flow velocity decreased with the distance from fire front. Through the analysis of fuel consumption efficiency, the common assumption of a "thermally thin" fuel bed for steep slope fire propagations seems to be questionable.

1 Introduction

Each year, fires devastate forests from all continents causing casualties and irreparable environmental damages. Determining the parameters that control the spread of wildfires is therefore a major objective for the management and the prevention of these fires. The effects of slope, wind and vegetation properties on fire behavior have been extensively investigated at laboratory and field scale. In most studies, the researchers have focused on the rate of spread, the flame geometry, the temperature profile or the fuel consumption and mass loss. However, few attempts were made to test the effect of fuel bed width on fire behavior. In wind-driven grass fires, rate of spread was found to be dependent on the length of the ignition lines shorter than 50 to 75 m [1] and required longer acceleration time for higher winds. However, experiments by Wotton [2] showed no increase in flame radiation for ignition lines longer than about 2 m and there was a fuel bed width threshold of about 1 m, above which no effect of line width on fire spread rate was identified. Furthermore, Dupuy [3, 4] carried out experiments on pine needle beds and observed fuel bed width effects for upslope fires, but not so much quantitative analysis was made. With regards to theoretical studies, the main assumption of an infinite straight fire line [5, 6] became the shortcoming of modeling the curvature of upslope fire spread. Morandini [7] developed a two-dimensional model for slope fire, the fuel bed width effect was checked and pronounced to be a significant variable affecting the rate of spread, but parameters in the model were calculated from the experimental measured data and spatial variation of flame temperature was ignored, thus the prediction accuracy remained to be doubtful.

In this work, a set of experiments at laboratory scale was conducted to study the fire spread over excelsior fuel beds, in order to quantify the fuel bed width effect under slope condition, rate of spread, geometry of the fire front, fire line shape and fuel consumption efficiency were examined.

2 Materials and Methods

Experiments were carried out on a 6.5 m long and 3.2 m wide inclinable insulating board with painted distance marker on both side (Fig. 1). Table angle of inclination can vary between 0° and 30°, the area available for burning was 2.0 m wide and 6.0 m long. Four excelsior fuel bed widths (W = 0.5, 1.0, 1.5, 2.0 m) and five slope angles (θ = 10°, 20°, 25°, 30°) with a fuel load m_0 of 0.4 kg/m² were tested.

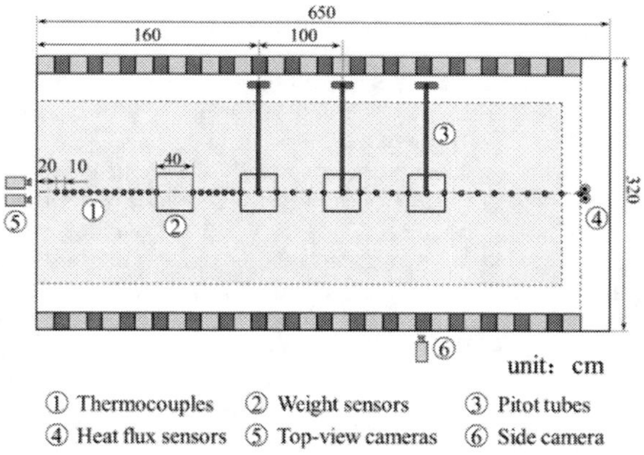

Figure 1: Schematic illustration of the experimental apparatus

The excelsior consisted of wood strands of fairly uniform rectangular crosssection (0.15 mm thick and 2 mm wide), but with different length, milled from pine wood boles. The excelsior was selected since it has relatively consistent size, is accessible, and is easy to be formed into a spatially uniform, randomly oriented and porous fuel bed. It also represents fine shrubs and grasses [8]. Adoption of this well-defined fuel bed facilitates good repetition.

Among all the tests, experimental conditions (ambient temperature and relative humidity) did not vary so much. To permit unobstructed natural convection, which may occur in practical situations, no side walls were attached. Ignition was initiated along the fuel bed width using a wool string soaked in n-heptane that would create a nearly instantaneous line of fire. During each experiment, a separate reference test was conducted at the same time on a small table with the same fuel

Corresponding author. Fax: +86-551-63601669
E-mail address: liunai@ustc.edu.cn

bed width under zero-slope condition, through this approach we obtained a reference fire rate of spread R_0.

In total 34 Type-K (chromel–alumel) thermocouples (with a bead diameter of 0.4 m and a wire diameter of 150 μm) were used to measure the gas temperatures over the fuel bed. All the thermocouples were installed 3 cm over the fuel bed, along its central axis, with a spacing of ~10 cm in the initial 2.4 m of the fuel bed and ~20 cm in the rest. Three calibrated temperature compensated bi-directional pitot tubes were positioned 7 cm above the fuel bed to measure the velocities of gas flow in the flame. Regarding to the actual fuel bed depth and flame length during the experiments, it was reasonable to believe that with such an experimental set, we would get a glimpse of the air flow conditions.

A couple of heat flux sensors (one total heat flux sensor and one radiant flux sensor) were installed 20 cm away from the end of fuel bed. The sensor surfaces, facing upwards, were positioned flush with the top surface of the fuel bed. A thermocouple was attached near the window of the flux sensor to measure the local air temperature. Four weight sensors measured the mass loss of a small plot (size 40.0 cm × 40.0 cm) of fuel along the central line of the fuel bed. The output signal of each load cell was calibrated under different slope angles. All the data outputs were recorded at 10Hz by a data acquisition system.

An infrared camera and a video recorder were placed at the rear of flame front to identify the fireline contours from top view, meanwhile, another video on the side, 4 m from the initial ignition line, was set to view the flame height and flame tilt angle.

3 Results and Discussion

3.1 Slope and fuel bed width effect on fire spread

Rate of spread was inferred from the output of streamwise distributed, near-bed-surface thermocouples. The moment when the temperature of each thermocouple reached 300°C [9] was identified as the fire front arrival. ROS shorted for rate of spread was obtained from the linear derivative of the curve 'flame front position versus times'. Fig. 2 showed the result for fuel bed width of 1 m under different slope. For slope angle higher than 20°, the spread undergoes an unsteady stage before the quasi-steady state, for higher slopes it would undergo longer acceleration stages. Apparently, the rate of spread was quasi-steady and increased with slope angle. When the slope angle was beyond 20°, the rate of spread suffered a prominent increase from 2.86 cm/s to 7.93 cm/s.

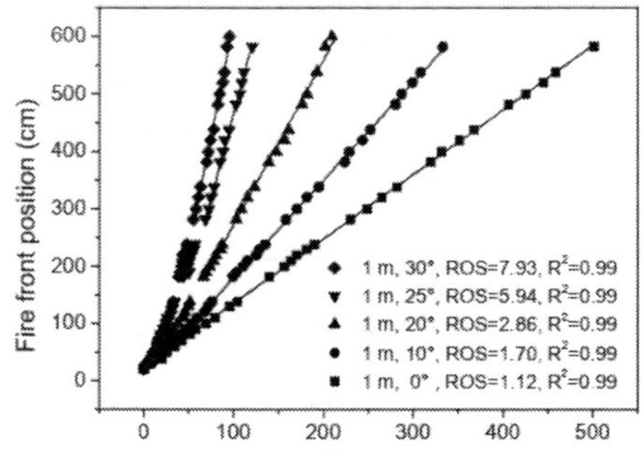

Figure 2: Flame front position *vs.* time for fuel bed width of 1 m. Line plots denote the corresponding linear regression lines.

Through the above linear fitting approach, quasi-steady rate of spread was obtained, thus non-dimensional rate of spread R/R_0 under slope conditions were presented in Fig. 3. For zero-slope condition, rate of spread remained under 2.0 cm/s and increased with fuel bed width. Under slope spread scenes, the effect of fuel bed width on rate of spread tends to increase with slope. During the 25° and 30°, the increasing trend of width effect on rate of spread seemed to be moderate but still showed an increase, no fuel bed width threshold [1, 10] was found because of the size restriction of the fuel bed.

Some empirical prediction models were tested, as for example the Canada fire model [11] shown in Fig. 3, as lines corresponding to:

$$R/R_0 = e^{3.533(\tan\theta)^{1.2}} \tag{1}$$

Fuel bed width effect on rate of spread is not explicitly included in the model, only the reference spread rate R_0 involves the width effect, the underestimation of ROS referred to a combined effect of fuel bed width and slope terrain.

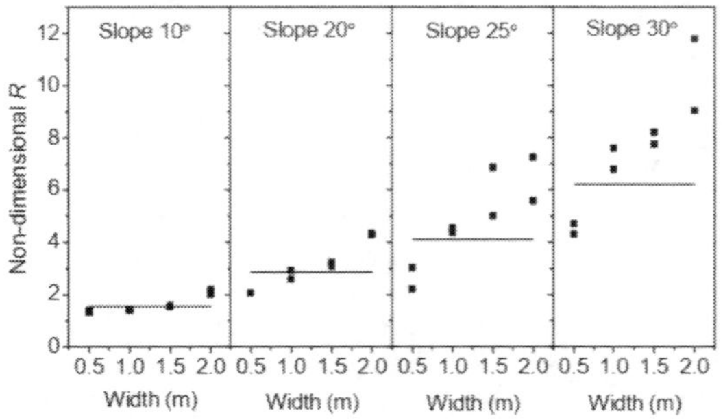

Figure 3: Non-dimensional R under slope conditions.

The analogy has often been invoked between wind and slope [12], suggesting that both factors caused similar effects, bringing flames closer to the unburned fuel, increasing heat transfer and modifying the air flow. Using this analogy, we can consider that the effect of slope on the rate of spread is equivalent to a given value of fire induced flow. According to the relationship between rate of spread R and ambient wind velocity U:

$$R \sim U^a \tag{2}$$

The maximum flow velocity measured by pitot tubes were averaged for each test and regarded as an indication of an effective wind. Rate of spread versus maximum flow velocity measured in the flame was plotted in Fig. 4. The rate of spread increased with maximum flow velocity in the flame with the correlation of $R = 1.17U_{fl}^{1.69}$, the subscript fl pertaining to flame. It is reasonable to believe that when the fire developed to an intense combustion phase, it would need to induce more oxygen from the ambient air, thus forming a great buoyancy-induced inflow along segments of flaming front.

Flame residence time is a measure of the dynamic properties of the fuel bed [13], and it can be characterized for example by the duration of the combustion reaction at a given place of the fuel bed. For each thermocouple, the duration of temperature above 300°C was identified, and the result showed that when the slope exceeded 20°, there was an increase stage of flame residence time corresponding to the acceleration rate of spread soon after the linear ignition, and then the residence time remained stable. A linear fitting with fixed slope of zero was done for the steady state of residence time versus flame arrival time, thereby the flame residence time for each test was determined.

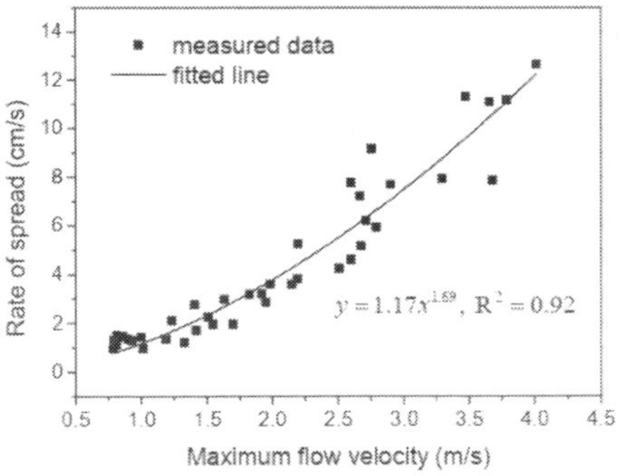

Figure 4: Rate of spread versus the maximum flow velocity measured in the flame.

As a result, the flame residence time increased with slope angle significantly, for fuel bed width of 1 m, the residence time varied from 12.3 s to 23.7 s when the slope increased from 0° to 30°. For different fuel bed widths under the same slope condition, flame residence time also showed a slight increase.

3.2 Flame characteristics

The preheating of virgin fuel is considered to depend strongly on flame properties. In this study, flame characteristics (flame length, flame angle, fire line shape) were systematically identified through video recorders at 25 fps and infrared camera.

With the evolution of fire spread, the fire line is generally considered to undergo a gradual change from the initial linear contour to a parabolic contour, and after that, the parabolic contour will keep steady till the end of flame spread. However, under steep slope above 20°, we observed that the fire line changed to a more pointed V-shape as delineated in Fig. 5. The included angle of V shape fire line was measured. The results show that the fire line was more pointed for higher slope but did not vary so much between different fuel bed widths.

Flame length and flame angle were determined from side-view camera. For each time period, a flame length was determined by using a vertical metallic grid (6×6 cm^2), 50 successive frames of the video were conducted and averaged to obtain a mean value. The maximum flame length showed a similar trend compared with the rate of spread. But for fast fire spread rate under 20° slope condition or above, the fire front was quite unsteady, and the flame tip fluctuated significantly. Therefore it is difficult to extract steady flame angle from the images.

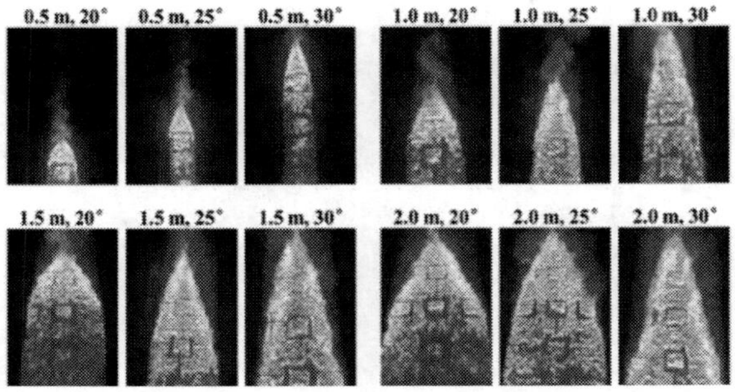

Figure 5: Fire line contours during steady spread state under different conditions.

3.3 Fire induced flow

Gas velocity profiles above the fuel bed were observed by pitot tubes. Regarding to four thermocouples ahead of each pitot tubes, when the fire front arrived at each thermocouple, air flow velocity measured by the pitot tube was recorded, thus we can identify the flow velocity distribution ahead of the fire front. As shown in Fig. 6, for a fuel bed width of 1 m, during the horizontal spread, the surface velocity remained relatively constant, only a weak reversal of flow was identified, thus only radiation from flame and combustion zone was the preheating source.

In 10° upslope fires, there was a small streamwise flow ahead of the fire front, such a flow may be caused by the turbulent diffusion of the flame edges. Under higher slope (> 20°) conditions, the flow velocity was streamwise and decreased with the distance from fire front. An obvious increase of flow velocity existed between slope 25° and 30°, which was associated with higher flame length and rate of spread. According to Anderson [8], such a streamwise air flow was also observed during wind-tunnel fires. The correlation raised here will benefit the modelling of convective heat transfer, since the convective heat transfer coefficient has a close relationship with local flow velocity.

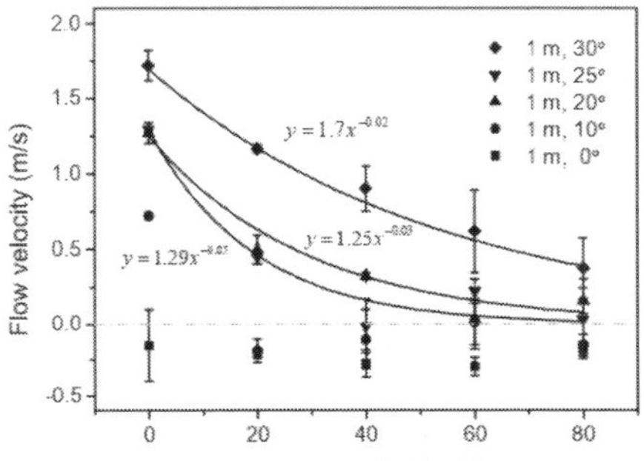

Figure 6: Local air flow velocity above the fuel bed as a function of distance from the fire front.

Through the video camera, V-shape fire lines were observed under steep slope higher than 20°, with vortices on the two flanks moving along to the fire front. These processes created a puffing like flame attachment to the fuel bed, which was accompanied with the streamwise flow. In the research of Dupuy [3], the vortices were systematically observed until 30° upslope fires, because our excelsior fuel bed is more porous and combustible than pine needle fuel bed in Dupuy's work. Under zero slope and 10° upslope conditions, the visible flames appeared to be stable, with no flame fluctuation observed.

3.4 Fuel Consumption

A reliable estimate of the total amount of fuel consumed in the flaming zone is important for calculating expected fire intensity and is useful for preparing prescriptions and guidelines for wildfire suppression and fuel reduction burning to reduce the hazard posed by wildfires. During the steady spread state, the fire line contour remained constant, thus completely consumed fuel mass on the weighing board must be m_0Rb, so the fuel consumption efficiency was calculated by using:

$$\eta_e = m'/(m_0Rb) \qquad (3)$$

m' represents for average value of fuel mass loss rate during the steady consuming state, b is the size of weighing board. As shown in Fig. 6, the fuel consumption efficiency was almost close to 1 in the zero-slope conditions, and didn't vary so much with fuel bed width, because of the low flame spread rate, all the fuel through the whole fuel bed depth was consumed completely during the flaming fire front spread.

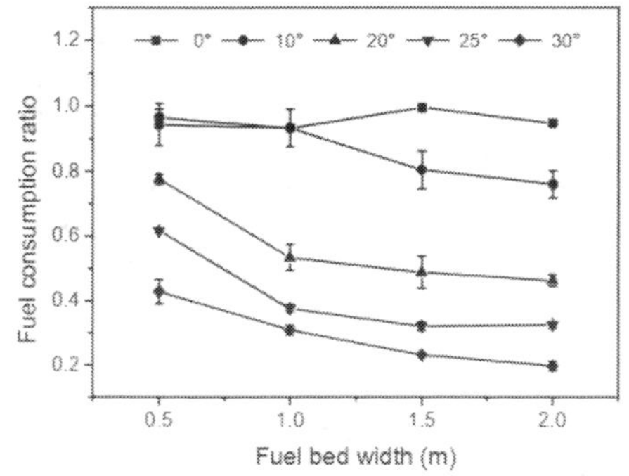

Fig. 6: Variation of fuel consumption efficiency with fuel bed width under different slope angle.

Under slope condition, the fuel consumption efficiency decreased with increasing slope angle and fuel bed width, regarding to the high rate of spread under such conditions. It is concluded that under high slope terrains, the fuel bed behaves as "thermally thick", the common assumption of a "thermally thin" [14] fuel bed remained to be doubted. Furthermore, it is interesting to note that such a decreasing trend of fuel consumption efficiency seems to ease when the fuel bed width increased to 2.0 m.

3.5 Heat Flux

Since the fuel bed width and slope angle had a significant effect on fire rate of spread, the heat transfer from the flame front to the unburnt fuels was enhanced. Fig. 7 presents the total and radiation heat flux measured. As shown, for each experimental condition, as the fire front approached the end of fuel bed, both heat flux underwent a sharp increase within a short period. When the slope is 30 °, the peak of total heat flux is much higher than that of radiation, no such apparent difference was found under zero slope, which suggests marked convective heating ahead of the flame front. Such a remarkable convective heating is also found for slope 20° and 25° when the fuel bed width is beyond 1.5 m.

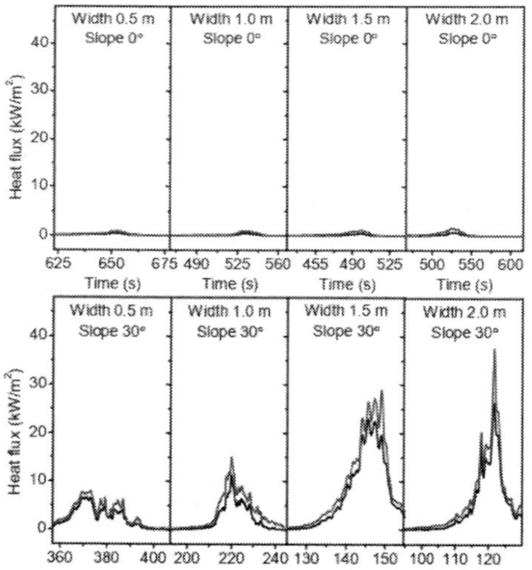

Fig. 7 Heat flux measured. Red line represents total heat flux, black line represents radiation heat flux.

4 Conclusions

This paper presents an experimental study of the fuel bed width and slope effects on surface fire spread. The major results are summarized as follows.

1) Fire rate of spread increases with fuel bed width and slope angle. For slope angle higher than 20°, the spread undergo an unsteady stage before the steady state, when the slope is higher, the acceleration distance will be longer. The relative effect of fuel bed width is relativelly larger for larger slopes.
2) The maximum flow velocity in the flame increased with the rate of spread with the correlation of $R = 1.17U_{fl}^{1.69}$, making a reasonable analogy between the effects of slope and ambient local wind.
3) Above 20°, the included angle of V-shape fire line was more pointed for higher slope but did not vary so much between different fuel bed widths. Meanwhile, the flow velocity ahead of the flame front was streamwise and decreased with the distance from fire front.
4) Under high slope terrain, the fuel consumption efficiency decreased with increasing slope angle and fuel bed width, and the fuel bed behaves as "thermally thick".

5 Acknowledgment

This research is funded by the National Key Research and Development Plan (No. 2016YFC0800100) and the National Natural Science Foundation of China (No. 51625602). This work was also supported by the USTC Fundamental Research Funds (No. WK2320000038).

References

[1] N. Cheney, J. Gould, Int J Wildland Fire 5(4) (1995) 237-247.

[2] B. Wotton, R. McAlpine, M. Hobbs, Int J Wildland Fire 9(4) (2000) 247-253.

[3] J.L. Dupuy, et al., Int J Wildland Fire 20(2) (2011) 272-288.

[4] J.L. Dupuy, J. Maréchal, Int J Wildland Fire 20(2) (2011) 289-307.

[5] P.J. Pagni, T.G. Peterson, Proc. Combust. Inst. (1973)

[6] A.M. Grishin, Publishing house of the Tomsk state university (1997).

[7] F. Morandini, P. Santoni, J. Balbi, Combuet Sci Technol 166(1) (2001) 67-90.

[8] W.R. Anderson, E.A. Catchpole, B.W. Butler, Int J Wildland Fire 19(3) (2010) 284-298.

[9] X. Silvani, F. Morandini, J.-F. Muzy, Int Commun Heat Mass 36(9) (2009) 887-892.

[10] P.M. Fernandes, iForest 7(2) (2014) 123.

[11] N. Liu, et al., Int J Wildland Fire 23(8) (2014) 1087-1096.

[12] D.X. Viegas, Int J Wildland Fire 13(1) (2004) 101-117.

[13] J. Mendes-Lopes, J. Ventura, J. Amaral. Forest Fire Research 14th Conference on Fire and Forest Meteorology (1998) 497-511

11th Asia-Pacific Conference on Combustion,
The University of Sydney, NSW Australia
10th -14th December 2017.

Development of Instantaneous Local Heat Flux Sensor for Internal Combustion Engine

Yuto Nakamura[1], Kazuhito Dejima[1], Tomohiro Tsuchiya[1], Osamu Nakabeppu[2]

[1]Graduate School of Science and Technology, Meiji University

1-1-1 Higashimita, Tama, Kawasaki, 214-8571, Japan

[2]Department of Mechanical Engineering, School of Science and Technology, Meiji University

1-1-1 Higashimita, Tama, Kawasaki, 214-8571, Japan

Abstract

We have developed a local three point heat flux measurement system with the MEMS (Micro-Electro-Mechanical-System) technology for the internal combustion engines. The heat flux sensor of three tiny thermometers of 315 μm square has measurement frequency of 10 kHz. The sensor was designed to observe turbulent heat transfer in the engine. The three tiny sensors enable to reproduce heat flux vector on the sensor surface flush to the engine inner wall. The developed sensor was composed of an Al alloy substrate, an Al_2O_3 insulation layer, Pt sensing layer and an Al_2O_3 passivation layer for improving the durability of the thin film sensors. The wall heat flux in the engine is derived by taking surface temperature change with the Pt sensor and analyzing the heat conduction inside the sensor with the temperature as the boundary condition. Performance test of the sensor was conducted in a laboratory engine of SIP, Cross-ministerial Strategic Innovation Promotion Program in Japan. The instantaneous heat flux with peak of 2.0 - 2.8 MW/m^2 with a noise level of 8.4 - 8.7 kW/m^2 was measured at 2000 RPM. Since the sensor can clearly measure the instantaneous value, remarkable variation among each cycle was confirmed. Simultaneous three point heat flux measurement also enable to visualize the local heat transfer in the three dimensional heat conduction analysis.

1 Introduction

Thermal efficiency of the internal combustion engines is required to be improved from the viewpoint of the environmental and energy saving issues. A cooling loss is one of the main factors of degrading the thermal efficiency and accounts for more than 20 % of the combustion energy. The thermal efficiency improvement needs a cooling loss mitigation technology. The SIP (Cross-ministerial Strategic Innovation Promotion Program in Japan) innovative combustion technology [1] has conducted a comprehensive research for realizing the maximum thermal efficiency of 50 %.

However, heat transfer mechanism in the combustion chamber of the engine has not been clarified yet. It is not easy to measure heat flux under the harsh environment in the engine. Some prior researches have been conducted. Enomoto et al. [2] developed a coaxial thermocouple type heat flux sensor and measured heat flux at the engine wall. Nakao et al. [3] developed an instantaneous heat flux sensor for corresponding to rapid change of temperature at the engine wall to investigate the effects of thermal insulation coating. Here, these sensors have a concern about the measurement accuracy due to the uncertainty of calibration.

The authors have been developing a high-performance heat flux sensor of the engine with the MEMS (Micro-Electro-Mechanical-System) technology in the SIP to understand the heat loss mechanism and to evaluate the heat loss mitigation method. Following target values were set at the beginning of the project for resolving the turbulent heat transfer; response over several kilo hertz, spatial resolution of a few hundred micrometer with close multipoint measurement, the maximum heat flux of 10 MW/m^2 and durability for the engine environment such as 10 MPa and 200 °C.

We have developed a multi-points instantaneous local heat flux sensor of thin film RTD type on a metal substrate for the engine with the MEMS technology. It is the first in the world. In this paper, we will report the developed heat flux sensor and its performance in a laboratory engine.

2 Heat flux measurement system

2.1 Sensor and driver

The three point thin film thermometer was fabricated on the metal substrate for heat flux measurement with high spatial and temporal resolution on the engine inner wall. Cross sectional image and thin film RTD pattern are shown in Fig. 1, and the specifications of layer structure are tabulated in Table 1. An aluminum alloy used as engine material is employed as the sensor substrate material for reproducing the heat transfer same as that in the engine wall. Al_2O_3 layers were used as an insulator layer and passivation layer. It is because the Al_2O_3 has thermal expansion coefficient relatively close to that of the aluminum alloy and reduction of thermal stress is possible. Higher thermal conductivity of the Al_2O_3 than that of SiO_2 also contributes to improve the thermal response.

Three RTD sensors of 315 μm square are located on the circumference of 900 μm diameter in three-hold symmetry. Distance between two sensors is 780 μm. The local three point heat flux measurement enables to observe heat flux vector at the center of the face. The thin film structure has a response frequency of several hundred kilo hertz.

Corresponding author. Fax: +81-44-934-7907
E-mail address: onakabep@meiji.ac.jp

Photo of the sensor is shown in Fig. 4. Sensor body is made compatible with a commercially available sensor, 'Medtherm' [4] heat flux sensor.

Temperature dependence of each RTD was measured by a box method. Figure 2 shows the resistance to temperature curves of a sensor taken before and after the lab-engine test. Although the absolute value showed a shift of 8 Ω which corresponds to 20 K, slope was kept in almost constant value. Since the heat flux is derived from temperature variation, the non-changing temperature dependence leads a reliable measurement.

The RTD sensor is driven under a constant current mode, and voltage drop across the RTD is amplified with lab-made 100-fold amplifier with a low pass filter. Frequency characteristics of the current source and amplifier are shown in Fig. 3. Cutoff frequency of the amplifier is over 100 kHz and current compensation factor is 23 kHz. Combined cut off turns out to 13 kHz. Since the sensor itself has the response frequency over 100 kHz, heat flux measurement is expected to have 10 kHz response at least.

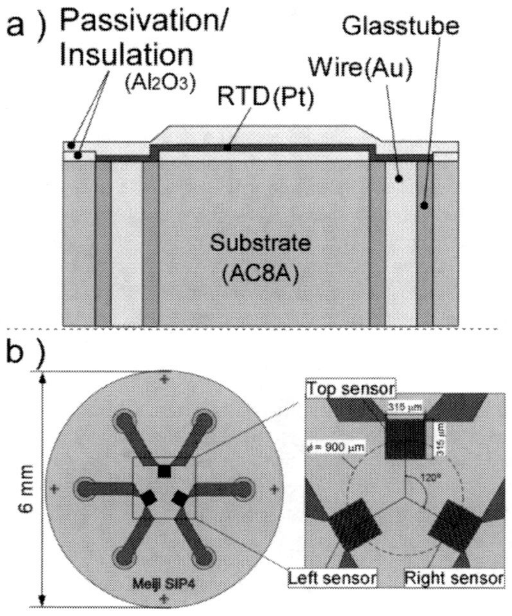

Figure 1: a) Cross sectional image of the sensor. The RTD is connected feedthrough electrode. b) The sensor pattern of three RTDs set on the circumference of 900 μm diameter. Thickness of the substrate was 4 mm.

Layer	Material	Thickness, μm
Passivation	Al_2O_3	2
Sensor	Pt	0.65
Insulation	Al_2O_3	2
Substrate	Al alloy	4000

Table 1: Specification of layers of the sensor.

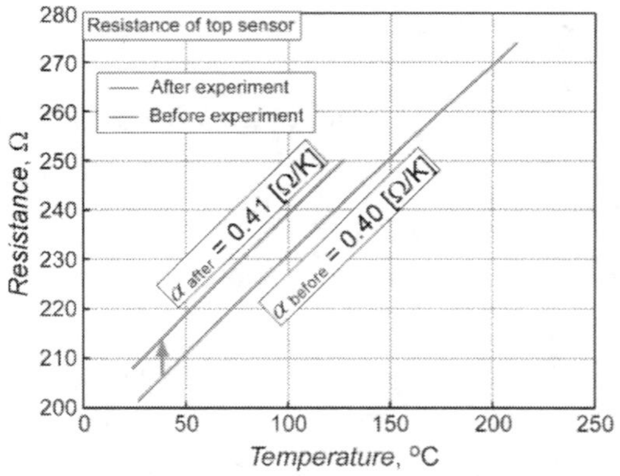

Figure 2: Temperature dependence of the sensor RTD. The slope does not change among before and after the heat flux measurement test in the lab-engine.

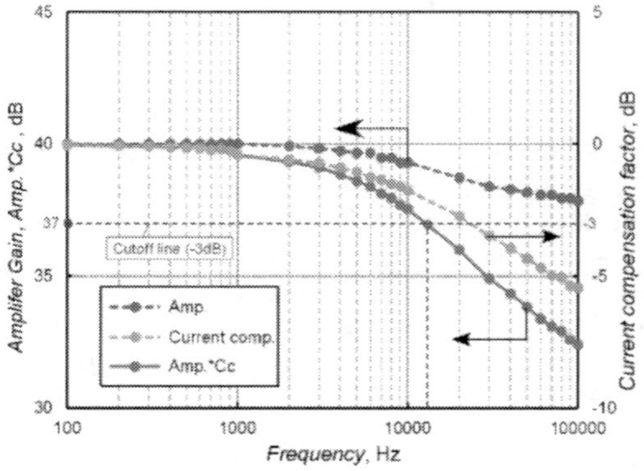

Figure 3: Cutoff frequency of the original driver circuit for heat flux measurement. Composite characteristics were evaluated form the constant current circuit and amplifier circuit.

2.2 Heat flux calculation

The heat fluxes are derived from 3D transient heat conduction analysis in the cylindrical coordinates system by using the measured temperature. In the numerical calculation, the three temperature data are given at the three measured points, and a consistent heat flux to the given temperature is applied at the surface of 120 deg. sector around each sensor as the boundary condition. Back face boundary condition of 4 mm height substrate is given by heat transfer coefficient of 1000 W/m^{2}K and the base environmental temperature.

Calibration of the analytical model was conducted by a self-heating of the RTD [5] to derive right heat flux. In general, there are unknown factors about thermal characteristics of the thin film structure of the sensor such as thermal properties, accurate length and interfacial thermal resistance. Therefor we tuned the response of the analysis to the known given heat flux at the thin film RTD. By using designed values and standard values for the analytical model, only the interfacial thermal resistance was set as a tuning parameter to adjust the measured heat flux to the excited heat flux. Fortunately, the decided interfacial resistance were valid for excitation heat flux of the wide frequency range of 200 Hz ~ 8 kHz.

The measurement error in the heat flux was estimated by following the law of propagation of errors. Theoretically, the wall heat flux of semi-infinite body with oscillating surface temperature at angular frequency ω and amplitude of T_0 is described as follows [6].

$$q_{w(t)} = \sqrt{\rho c\lambda\, \omega}\, T_0 \cos(\omega t + \tfrac{\pi}{4}) \quad (1)$$

The amplitude of the oscillating temperature $\tilde{T}_w$ is decided by measuring ac voltage $\tilde{V}$ across the RTD of base resistance of R_0 with current i, amplifier gain G, and converting with temperature coefficient α as following equation.

$$\tilde{T}_w = \frac{\tilde{V}}{G \cdot i \cdot \alpha \cdot R_0} \quad (2)$$

Further, interfacial thermal resistance R_i introduced to calibrate the analytical model has a dependence of 0.5 to 1 to the heat flux. By summing up main error factors, the error of the heat flux was evaluated to 13 % from (3). Here, error of R_i is 12 %, $\rho c\lambda$ of 5 %, α of 2.5 % and R_0 of 5 %.

$$\frac{\delta q_w}{q_w} = \sqrt{\left(\frac{\delta R_i}{R_i}\right)^2 + \left(\frac{1}{2}\frac{\delta \rho c\lambda}{\rho c\lambda}\right)^2 + \left(\frac{\delta \alpha}{\alpha}\right)^2 + \left(\frac{\delta R_0}{R_0}\right)^2} \quad (3)$$

3 Performance test of MEMS sensor

3.1 Conditions

The performance of the developed sensor was tested in terms of the noise level, durability against pressure and thermal strain in the lab-engine. Firing tests in a continuous 200-cycle run were conducted with tumble flow under IMEP = 400 kPa. The equivalent ratio was 1.0, and the engine speed was 2000 rpm. The ignition timing was 18 BTDC and data were sampled in every 0.1° CA, which corresponding to 120 kHz, and then, they were compressed into 24 kHz by 5-point average. The position of the sensor is indicated in Fig. 4. It is set through one of two pressure transducer ports. The top sensor was directed to the ignition plug.

3.2 Results of measurement

The ensemble average temperature and pressure in the experiment are shown in Fig. 5. This results indicated the developed sensor has the durability for high gas temperature of 2700 K at maximum, and high pressure of 3300kPa at maximum in the continuous 200-cycle run.

Instantaneous temperature and fluctuation component of 4.8 kHz ~ 24 kHz are shown in Fig. 6. The mean wall temperature rose by only 10 K from reference temperature and the fluctuation of 4.8 kHz ~ 24 kHz of the wall temperature is evaluated to 0.018 K (top), 0.018 K (left) and 0.019 K (right), respectively.

Wall heat fluxes calculated through the heat conduction analysis are shown in Fig. 7. The heat fluxes at the three points rapidly raised at a close timing, then took peaks ranging approximately 2.0 – 2.8 MW/m^2. In the decay process, rough trends were common but fluctuations showed difference each other. It is considered the sensor measures the heat flux corresponding to turbulence of the combustion gas with the scale of sub-millimeter.

The temperature distribution of the sensor can be calculated from the analysis and the vector of heat fluxes can be derived from the temperature distribution. The heat flux vector is also shown in Fig. 7. The heat flux in the in-plane direction probably reflects the gas flow and the turbulence. It is expected with the sensor to contribute to the investigation of the heat transfer mechanism in engines.

Furthermore, the noise of heat flux was estimated through the fluctuation over 4.8 kHz as shown in Fig. 8. According to the

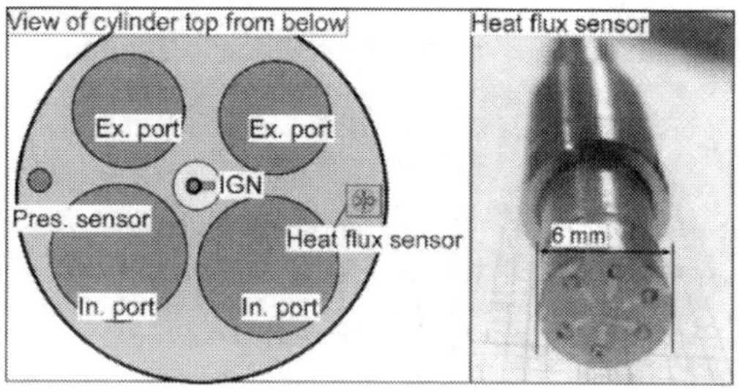

Figure 4: The position of the sensor in the cylinder top and the photo of the sensor.

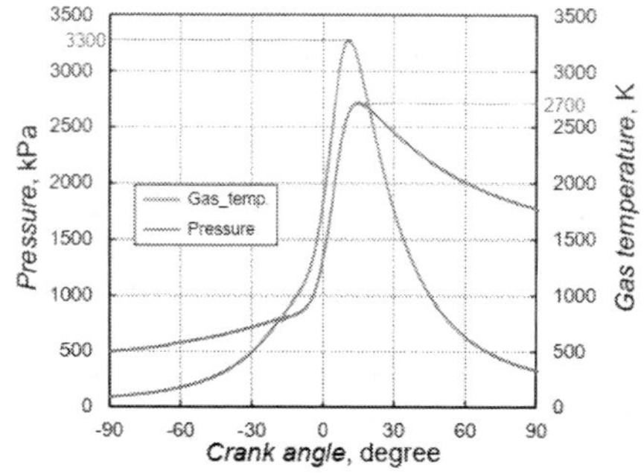

Figure 5: The ensemble averaged gas temperature and pressure in heat flux measurement test in the lab-engine.

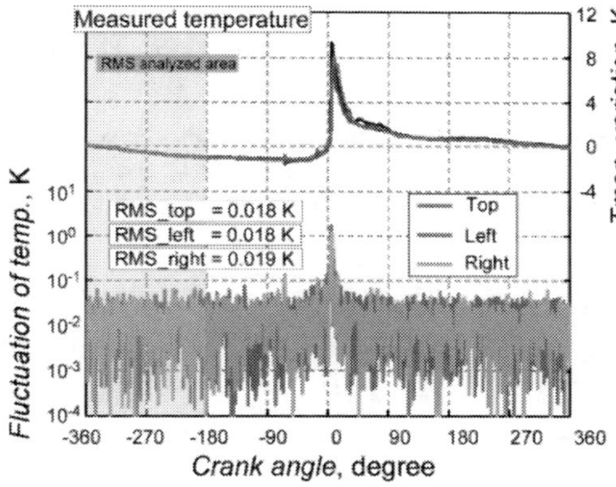

Figure 6: Instantaneous temperature variation and fluctuation component of 4.8 kHz ~ 24 kHz in the firing test.

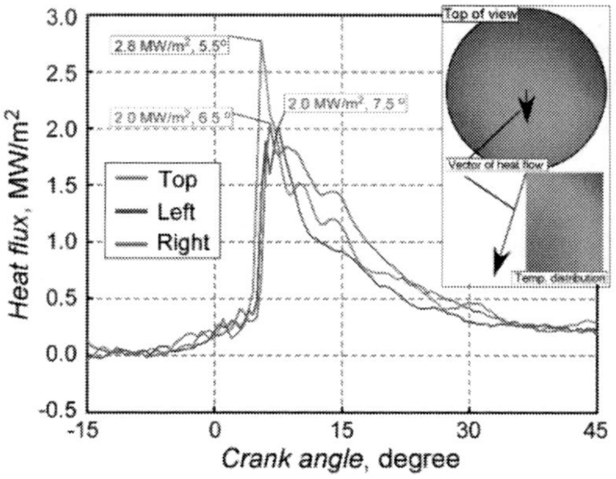

Figure 7: Instantaneous heat flux data and the vector of heat flow derived from temperature distribution.

results, the RMS of the three heat flux was about top: 8.5 kW/m^2, left: 8.4 kW/m^2, right: 8.7 kW/m^2, respectively. Since the RMSs of the three sensors are within 10 kW/m^2 despite of the noisy engine environment, the sensor enables to measure heat flux with enough low noise level to observe the instantaneous value and turbulent behavior.

From these results, it is demonstrated that the developed MEMS thin film sensor has enough durability in the harsh condition and is capable of measuring heat flux with low noise. The heat flux vector is expected to contribute to the investigation of the heat transfer mechanism in engines.

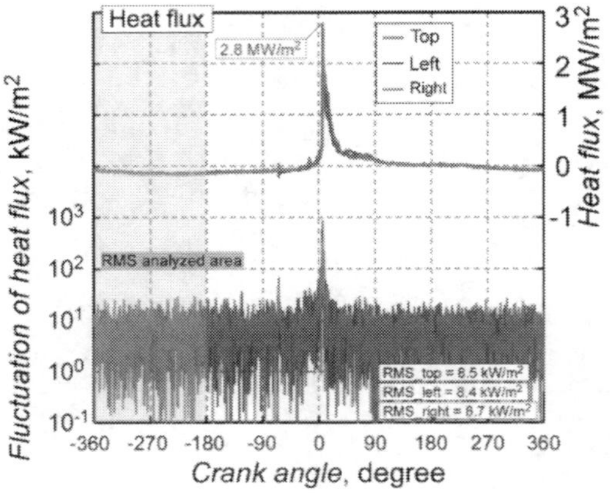

Figure 8: Heat flux in linear axis and its fluctuation over 4.8 kHz in logarithmic axis.

4 Conclusions

Following conclusions were obtained from the development of the thin film thermal sensor for instantaneous local heat flux measurement in the internal combustion engine.

- Thin film RTD sensors were developed on the Al alloy substrate to apply the MEMS technology to the heat flux measurement in the internal combustion engine. The measurement system has the response of 10 kHz, and enable to observe the heat flux vector at the center of the sensor surface with the simultaneous three point measurement with the tiny RTD of 315 μm square.
- It was experimentally demonstrated that the sensor can work under the harsh environment in the engine such as IMEP of 400 kPa, the average maximum gas temperature of 2700 K and the maximum pressure of 3300 kPa.
- Resultant heat flux showed the maximum of 2.0 - 2.8 MW/m^2 with the noise level of 8.4 – 8.7 kW/m^2. The measurement is capable with enough low level noise to observe the instantaneous value.
- The three sensors in submillimeter scale showed similar heat flux waves with small time lag. There is good possibility of extracting the information of turbulence from the simultaneous three point heat flux data.

5 Acknowledgment

This work was supported by Council for Science, Technology and Innovation (CSTI), Cross-ministerial Strategic Innovation Promotion Program (SIP), "Innovative Combustion Technology" (Funding agency: JST).

References

[1] Cabinet office, Government of Japan, Cross-ministerial Strategic Innovation Promotion Program (SIP), available at <http://www8.cao.go.jp/cstp/gaiyo/sip/index.html>, (2014), (accessed on 21 July.2017), (in Japanese)

[2] Y. Enomoto, et al., Study on Analysis of Instantaneous heat flux flowing into the combustion chamber wall of an internal combustion engine (Examination in the case of consideration of heat storage term and the temperature dependency of thermocouple's thermophysical properties), Trans. of JSME (B) Vol.57, No.539, (1991), p.2421-2427 (in Japanese)

[3] Y. Nakao, et al., Technology of Instantaneous Heat Flux Measurement on a High Response Heat Insulation Coating, Matsuda Technical Review, No.32, 2015, p.222-227 (in Japanese)

[4] Medtherm Corporation, available at <Medtherm.com>, (2017), (accessed on 29 July.2017)

[5] O. Nakabeppu, et al., Thin film resistance sensor for wall heat flux measurement in premixed combustion, Transactions of the Japan Society of Mechanical Engineers, Vol.82, No.840, 2016 (in Japanese)

[6] H.S. Carslaw & J.C. Jaeger, Conduction of Heat in solids, Oxford science publications,1959, p.64-70

11th Asia-Pacific Conference on Combustion,
The University of Sydney, NSW Australia
10th -14th December 2017.

Study on combustion and ignition characteristic of weakly flammable refrigerant R125 using a micro flow reactor with a controlled temperature profile

Shintaro Takahashi[1], Hisashi Nakamura[1], Takuya Tezuka[1], Susumu Hasegawa[1], Kaoru Maruta[1,2]

[1]Institute of Fluid Science, Tohoku University

2-1-1,Katahira, Sendai, Miyagi, 980-8577, Japan

[2]International Combustion and Energy laboratory, Far Eastern Federal University

Russky Island, Vladivostok, Russia

Abstract

A mixture refrigerant R410A is being used throughout the world. In order to develop a standard for the safe usage of R410A refrigerant, there is a need to understand the flammability properties of R410A components (R32 and R125). In this study, ignition characteristics of a stoichiometric C_2HF_5/air (R125) mixture were experimentally and computationally investigated by a micro flow reactor with a controlled temperature profile at atmospheric pressure. The experimental C_2HF_5 weak flame has two luminous regions in the case of a maximum wall temperature of 1300 K. The HRR profile of C_2HF_5/air flame has two heat release zones in computation with Linteris mechanism. As a results, the experimental and computational weak flame of a stoichiometric C_2HF_5/air mixture shows multi-stage oxidation at a maximum wall temperature of 1300 K although agreements between experimental luminosity profile and computational HRR profile are not satisfactory at this stage. Based on the obtained flame structure, a maximum wall temperature of 1300 K is found to be insufficient to attain the complete oxidation of a stoichiometric C_2HF_5/air mixture. Therefore, an additional computation with the maximum wall temperature of 2000 K was conducted. As a result, there exist three heat release zones in this case. First zone mainly corresponds to fuel decomposition less than 1350 K. Second zone is the region which produces CF_4 from CF_3. Third zone corresponds to CO oxidation. Reaction path analysis of key species was conducted. This analysis clarified dominant production and consumption reaction of key species, and reactions with heat release by each stage of oxidation process of C_2HF_5/air mixture.

1 Introduction

For air conditioning in buildings, vapor-compression refrigeration systems are widely used. Working fluids with high ozone depletion potential are used in these systems. The hydrofluorocarbons (HFCs) are promising candidates to replace conventional refrigerants such as chlorofluorocarbon (CFCs) and hydrochlorofluorocarbon (HCFCs). The refrigerant R410A, which is extensively used in residential air conditioners and heat pumps, is weakly flammable. This refrigerant consists of 50% of R32 (CH_2F_2) and 50% of R125 (C_2HF_5) in mass fraction. To facilitate their safe usage, it is essential to obtain fundamental knowledge on their combustion properties. In the past, measurements of laminar burning velocities for CH_2F_2 and C_2HF_5 were conducted by National Institute of Advanced Industrial Science and Technology (AIST) [1] and National Institute of Standards and Technology (NIST) [2]. Detailed chemical kinetic models for C_1-C_2 hydrofluorocarbons including CH_2F_2 and C_2HF_5 were developed and validated with experimental laminar burning velocities by NIST [3]. Figure 1 shows laminar burning velocities of premixed CH_2F_2/C_2HF_5/air mixtures as a function of equivalence ratio at atmospheric pressure and temperature [2]. These refrigerants have low laminar burning velocities compared with general hydrocarbons. For example, that of a stoichiometric CH_4/air mixture is around 38.0 cm/s at atmospheric pressure. Laminar burning velocities of CH_2F_2/C_2HF_5/air mixtures decrease with the increase of C_2HF_5 mass fraction. Although flame propagation characteristics of refrigerants have been investigated, there are very few studies on ignition characteristics of hydrofluorocarbons because it is difficult to experimentally evaluate ignition properties of weakly flammable fuels with conventional methods.

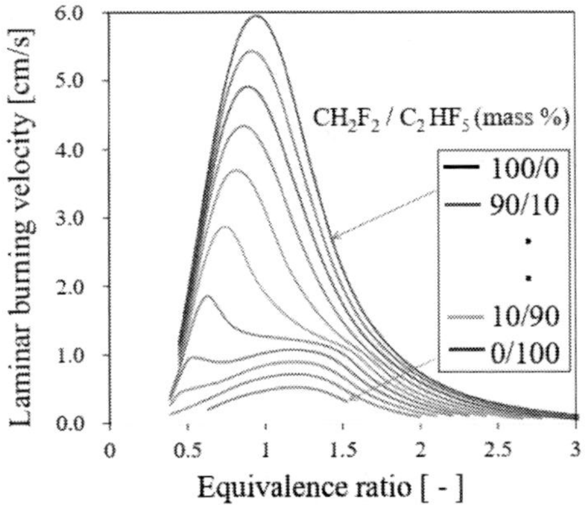

Figure 1: Laminar burning velocity of premixed CH_2F_2/air and C_2HF_5/air as a function of equivalence ratio [2].

In this study, we employed a micro flow reactor with a controlled wall temperature profile (MFR) [4] to evaluate ignition characteristics of C_2HF_5 since those of CH_2F_2 was investigated with MFR in our previous research [5]. A cylindrical quartz tube is employed as a reactor channel and the inner diameter of the tube is smaller than the ordinary quenching diameter. The tube is heated by an external heat source to give a stationary temperature profile along the inner surface of the tube wall. A premixed mixture is supplied to the reactor from the lower temperature region. Due to the small inner diameter of the tube, the gas-phase temperature strongly depends on the given wall-temperature profile. An inlet mixture is highly-preheated by the given wall-temperature at the small inner diameter and flames even for quite low reactivity are observed. In the previous study on CH_4 and C_3H_8 using MFR [4], three kinds of flames were observed depending on the inlet mixture flow velocity; normal flame, FREI (Flames with Repetitive Extinction and Ignition) and weak flames. At low flow velocity, weak flames, which are stable and with weak chemiluminescence, were observed. This study focused on weak flames in MFR since theoretical analysis [6] has shown that the weak flame branch is equivalent to the ignition branch of the Fendell curve [7]. Note that radical quenching has negligible effect on flame behavior under current conditions in MFR. [8] Thus, information on ignition property can be obtained by observing weak flame positions. In addition, since the system is well-defined 1-D system, investigations of reaction kinetics are possible by using weak flames.

In the present study, ignition characteristics of a stoichiometric C_2HF_5/air mixture were investigated using a MFR with a controlled temperature profile. One-dimensional computations with the detailed chemical kinetic models were conducted to investigate the weak flame structure in detail.

2 Experimental setup

Figure 2 shows a schematic of the experimental setup. A quartz tube with an inner diameter of 2 mm was employed as a reactor channel. The quartz tube was heated by a hydrogen/air flat-flame burner and a stationary wall temperature profile from ambient temperature (around 300 K) to 1300 K was formed along the inner surface of the tube. 1300 K is the temperature at which typical hydrocarbons are completely oxidized. A premixed fuel/air mixture was supplied to the reactor at an inlet mean flow velocity of 2 cm/s. The overall one-step reaction formula shown by Eq. (1) was used for the estimation of mixture compositions for a stoichiometric C_2HF_5/air mixture. This formula means that CO_2, HF, CF_2:O and CF_4 are final products when complete combustion of C_2HF_5/air mixture is attained. A digital still camera with a band-pass filter (transparent wavelength: 431.4 nm, half bandwidth: 6.4 nm) was used to observe weak flames. The experimental flame position was defined as the peak position in the chemiluminescence profile along the flow direction.

$$C_2HF_5 + O_2 = 0.5CO_2 + HF + CF_2{:}O + 0.5CF_4 \quad (1)$$

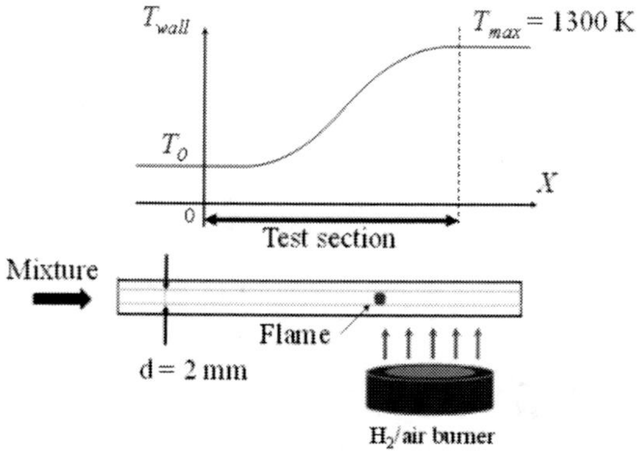

Figure 2: Schematic of a micro flow reactor with a controlled temperature profile.

3 Computational method

The flow in MFR was modelled as one-dimensional steady-state reactive flow without a boundary layer and computations were conducted using the PREMIX package in ANSYS CHEMKIN-PRO v17.2 [9]. To simulate the present micro flow reactor system, a heat convection term between the gas and the reactor wall was added to gas-phase energy equation (2).

$$\dot{M}\frac{dT}{dx} - \frac{1}{c_p}\frac{d}{dx}\left(\lambda A\frac{dT}{dx}\right) + \frac{A}{c_p}\sum_{k=1}^{K}\rho Y_k V_k c_{pk}\frac{dT}{dx} +$$

$$\frac{A}{c_p}\sum_{k=1}^{K}\omega_k h_k W_k - \frac{A}{c_p}\frac{4\lambda Nu}{d^2}\left(T_w - T\right) = 0 \quad (2)$$

In this equation, $\dot{M}$, T, x, C_p, A, ρ, λ and d are mass flow rate, gas-phase temperature, spatial coordinate, specific heat at constant pressure, cross-sectional area, density, thermal conductivity and channel diameter, respectively. Y_k, V_k, ω_k, h_k and W_k are mass fraction, diffusion velocity, rate of production, specific enthalpy and molecular weight of kth species, respectively. Nusselt number was selected to be constant (Nu = 4). The wall-temperature profile (from 300 to 1300 K), which was the same as the measured profile, was used in the computations. The mixture condition was stoichiometric and the flow velocity was 2 cm/s, as the same as in the experiments. The computations employed two detailed chemical kinetic models developed by NIST [2]. One is made by Linteris and Truett for fluorine inhibition of hydrocarbon flames [10]. This mechanism contains a hydrocarbon sub-mechanism and additional C_1 (200 reactions) and C_2 (400 reactions) fluorochemistry. It consists of 158 species and 1384 reactions (Linteris mechanism). The other is made by Babushok et al. for fire suppressant including fluorocarbons and brominated hydrocarbons which consists of 185 species and 1592 reactions (Babushok mechanism [11]). Both mechanisms used USC mechanism version II (USC II) which consists of 109 species and 784 reactions [12] as a hydrocarbon sub-mechanism. USC II was also used for computations of CH_4 weak flames as a reference. The flame position in the computations was defined at the maximum peak location of the heat release rate (hereafter HRR) profile. HRR was estimated by integration of relevant term in the energy equation. Chemiluminescence intensity in experiments and HRR in computations were normalized at the respective maximum values and the normalized profiles were compared.

Non-dimensional calculation was also conducted to investigate the chemical species composition of C_2HF_5 oxidation at equilibrium state. The AURORA package of ANSYS CHEMKIN-PRO v17.2 [13] was used. Computations were conducted for adiabatic and constant volume conditions for stoichiometric fuel/air mixtures at initial pressure of 0.1 MPa and initial temperature of 1000 K.

4 Results and discussion

4.1 Weak flame position of C_2HF_5/air

Figure 3 shows a weak flame image, normalized chemiluminescence intensity profile, and normalized HRR profiles of a stoichiometric C_2HF_5/air mixture at atmospheric

Corresponding author. Fax: +81-22-217-5296
E-mail address: s.takahashi@edyn.ifs.tohoku.ac.jp

pressure. The weak flame image and chemiluminescence intensity profile show two luminous zones. One is located in the wall temperatures from 1000 to 1270 K and another is in those higher than 1300 K. In the computation with Linteris mech., the two zones of heat release are seen in the HRR profile. One is located in the wall temperatures from 950 to 1260 K and shows a distinct peak in the HRR profile. Another is located at down steam side of maximum temperature and shows a small and broad HRR profile. On the other hand, Babushok mechanism shows single zone of heat release which is located from 980 to 1300 K. The rising position of the HRR profile with Babushok mechanism has a better agreement with that of the chemiluminescence intensity profile than that of the HRR profile with Linteris mechanism. However, Babushok mechanism does not show small and broad heat release in the down steam side of maximum temperature. Linteris mechanism reproduced two-stage oxidation as observed in experiments. In addition, computational HRR profile with Linteris mechanism for a stoichiometric CH_2F_2 /air mixture showed good agreement with the experimental chemiluminescence intensity profile [5]. Therefore, further analysis of the weak flame structure employed Linteris mechanism although correlation between experimental luminosity profile and HRR profile was still unsatisfactory at this stage.

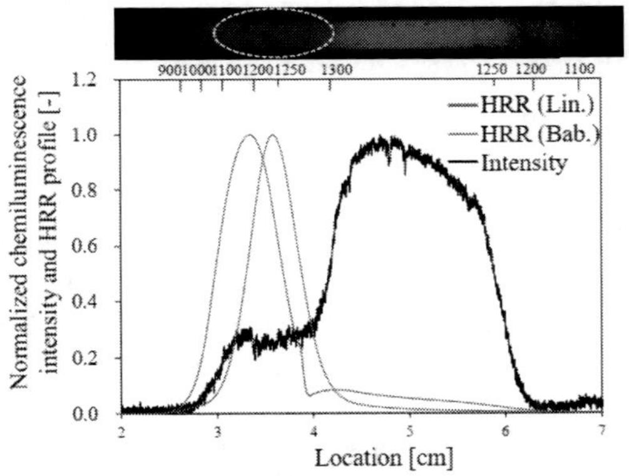

Figure 3: Experimental and computational weak flame positions.

4.2 Comparison of flame structure in the cases of high and low maximum wall temperatures

Figure 4 shows the computational weak flame structure for the C_2HF_5/air mixture with a maximum wall temperature of 1300 K. C_2HF_5 is consumed and CO, C_2F_6, CF_2:O, and HF are produced in the first heat release zone (around 1175 K). H_2O is an intermediate produced in this zone. In the second heat release zone (around 1300 K), C_2F_6 is partially consumed and CF_4 is formed. Oxidation of CO weakly proceeds in the second zone. However, large amount of CO and O_2 still exist at the end of reactor. It is reconfirmed that C_2HF_5 shows two-stage oxidation due to various species consumption and production at higher temperature zone.

O_2 (reactant) and CO (intermediate) were not completely consumed and CF_2:O and CF_4 (products) continued to increase in this computation. These resuilts indicate that the present maximum wall temperature of 1300 K is not enough to attain the complete oxidation of a stoichiometric C_2HF_5/air mixture. Therefore, an additional computation with higher maximum wall temperature at 2000 K was conducted. Figure 5 shows the computational weak flame structure of a stoichiometric C_2HF_5/air mixture with the maximum wall temperature of 2000 K. The HRR profile shows three zones of heat release: the first zone is located from 800 to 1350 K with the largest peak; the second one is located from 1350 to 1600 K with the second largest peak; and the third one is located from 1600 to 2000 K with small and broad heat release rate. As the same as in computations with the maximum wall temperature of 1300 K, first heat release zone corresponds to production of CO, C_2F_6, CF_2:O, and HF from fuel and second heat release zone corresponds to the zone where CF_4 is formed. The third zone corresponds to the formation of CO_2 by oxidation of CO. This mild CO oxidation at very high temperature would be due to less amount of H atom in the mixture since a C_2HF_5 molecule has single H atom. An H radical largely contributes to production of an OH radical which is quite important in the CO oxidation by $CO+OH=CO_2+H$. In C_2HF_5 weak flames, large amount of HF are formed in the first heat release zone and H atom is insufficient in the third HRR zone. Since the CO oxidation hardly proceed with O and O_2 without H atom by $CO+O_2=CO_2+O$ or $CO+O+M=CO_2+M$ which are very slow reactions even at temperatures higher than 1600 K. Gradual decrease of O_2 and CO in the third zone would be caused by these reaction with O_2 and O.

4.3 Reaction path analysis of C_2HF_5/air mixture

Reaction path analysis for major chemical species was conducted in order to investigate the reactions proceeded in each temperature zone. Schematic of reaction path dependence on wall temperature is shown in Fig. 6. Reactions of C_2HF_5/air mixture can be classified by wall temperature. First stage corresponds to the temperature zone lower than 1350 K where CO, C_2F_6, CF_2:O and HF are produced from fuel. Decomposition of C_2HF_5 starts by $C_2HF_5+F=C_2F_5+HF$ (R768) and $C_2HF_5+OH=C_2F_5+H_2O$ (R788). H_2O is consumed by F radical and produce OH radical which contribute for progress of reactions. C_2F_5 is also consumed with O_2 and O radicals, and CFO and CF_2:O are produced. Then, CFO is decomposed to CO and F radical by third-body reaction. CF_3 compose C_2F_6 at around 1300 K. Heat release at the first stage corresponds to the reactions of fuel decomposition with F radical and $2CF_3=C_2F_6$ (R390). Second stage mainly consists of reactions of CF_3 at the temperature lower than 1600 K. CF_3 produced from decomposition of C_2F_6 produced CF_4. Heat release at the second zone corresponds to this reaction ($CF_3+F=CF_4$, R195). Dominant reaction which release heat at the third zone is CO oxidation. From 1350 to 1800 K, dominant reaction of CO oxidation is $CO+OH=CO_2+H$ (R98), but it shifts to $CO+O_2=CO_2+O$ (R31) from 1800 K due to insufficient H radicals.

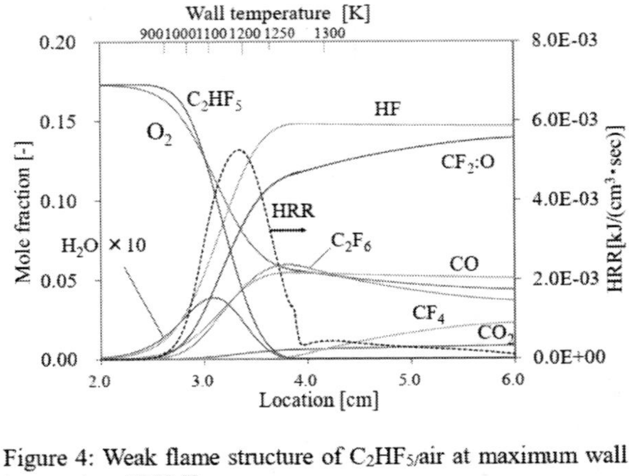

Figure 4: Weak flame structure of C_2HF_5/air at maximum wall temperature 1300 K.

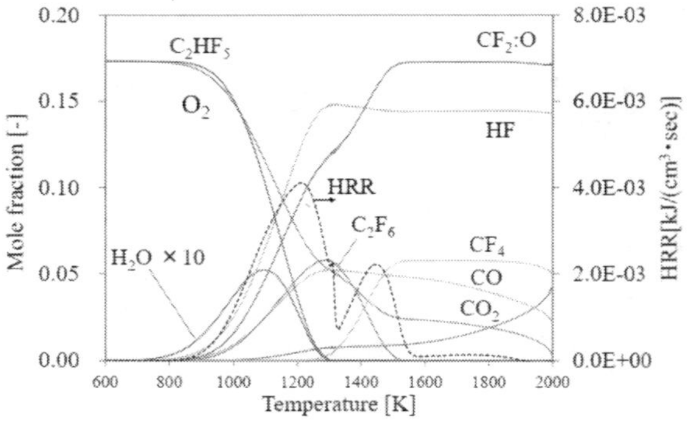

Figure 5: Weak flame structure of C_2HF_5/air at maximum wall temperature 2000 K.

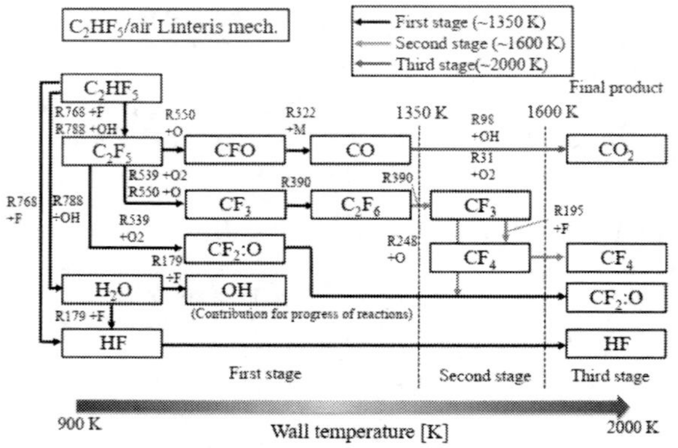

Figure 6: Schematic of reaction path dependence with a maximum wall temperature of 2000 K.

4.4 Time history of major species on C_2HF_5/air at 1000 K

Figure 7 shows the computational results of time dependent species mole fractions for C_2HF_5/air mixtures obtained by non-dimensional computations at 1000 K as the initial temperature. Intermediate products such as H_2O and C_2F_6 are completely consumed. CO does not react completely. Final products which consists of CO_2, HF, CF_2:O and CF_4 were produced after ignition. This tendency of chemical species profile of non-dimensional computation has a good agreement with the case of MFR at maximum temperature of 2000 K.

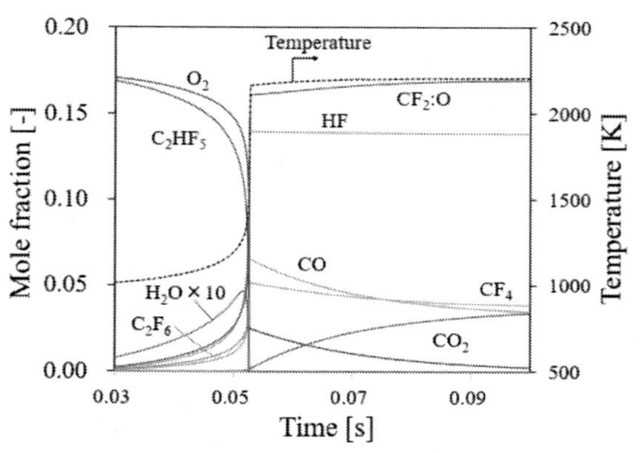

Fig. 7: Time history of major species of C_2HF_5/air mixture at initial temperature of 1000 K.

5 Conclusions

Ignition characteristics of stoichiometric C_2HF_5/air mixture were investigated using a micro flow reactor with a controlled temperature profile under atmospheric pressures.

· Two luminous regions of weak flame of C_2HF_5/air were observed experimentally, and Linteris mechanism showed the same number of heat release rate peaks, while Babushok mechanism only showed one heat release rate peak.

· From analysis of the flame structure of C_2HF_5/air at maximum wall temperature of 2000 K, HRR profile and reaction progress of C_2HF_5/air mixture are classified by wall temperature. First stage is mainly reaction of fuel decomposition at less than 1350 K. Second stage is production of CF_4 from CF_3. Third stage corresponds to CO oxidation.

· Reaction path analysis of major species were conducted. Reaction process of key intermediate species (C_2F_5, CFO, CF_3, C_2F_6, CO) of C_2HF_5/air mixture oxidation was indicated along the reactor wall temperature. In addition, this analysis showed that dominant reaction of CO oxidation change from R98 induced by OH radical to R31 induced by O_2 at wall temperature around 1800 K.

References

[1] K. Takizawa, A. Takahashi, K. Tokuhashi, S. Kondo, A. Sekiya, Combust. Flame, 141, 2005, 298-307.

[2] J. L. Pagliaro, G. T. Linteris, Spring Tech. Meeting East. States Section, Princeton, United States, 2016, 750.

[3] D. R. Jr. Burgess, M. R. Zachariah, W. Tsang, P. R Westmoreland, Prog. Energy Combust. Sci., 21, 1996, 453-529.

[4] K. Maruta, T. Kataoka, N.I. Kim, S. Minaev, R. Fursenko, Proc. Combust. Inst., 30, 2005, 2429-2436.

[5] S. Takahashi, H. Nakamura, T. Tezuka, S. Hasegawa, K. Maruta, The Ninth JSME-KSME Thermal and Fluids Engineering Conference, Okinawa, Japan, 2017, accepted.

[6] Y. Tsuboi, T. Yokomori, K. Maruta, Proc. Combust. Inst., 32, 2009, 3075–3081.

[7] S. Minaev, K. Maruta, R. Fursenko, R., Combust. Theor. Modeling, 11, 2007, 187–203.

[8] Y. Kizaki, H. Nakamura, T. Tezuka, S. Hasegawa, K. Maruta, Proc. Combust. Inst., 35, 2015, 3389–3396.

[9] R.J. Kee, J.F. Grcar, M.D. Smooke, J.A. Miller, PREMIX, Report No. SAND85-8240, 1993, Sandia National Laboratories.

[10] G. T. Linteris, L. Truett, Combust. Flame, 105, 1996, 15–27.

[11] V. I. Babushok, G. T. Linteris, P. T. Baker, Combust. Flame, 162, 2015, 2307–2310.

[12] USC Mech VersionII, available at <http://ignis.usc.edu/Mechanisms/USCMech%20II/USC_Mech%20II.html>

[13] E. Meeks, H. K. Moffat, J. F Grcar, R. J. Kee, AURORA, Report No. SAND96–8218, 1998, Sandia National Laboratories.

11th Asia-Pacific Conference on Combustion,
The University of Sydney, NSW Australia
10th -14th December 2017.

An Experimental Study on the Temperature Measurement Technique Using Ultrasonic Waves

Wuk Hwang[1], Hyun Jeong Seo[1], Jeong Min Jo[1], Min Chul Lee[1,*]

[1] Department of Safety Engineering, College of Engineering, Incheon National University,

Academy-ro 119, Yeonsu-gu, Incheon 22012, South Korea

Abstract

In this paper, air temperature was measured using ultrasonic non-contact method instead of conventional contact method such as thermocouples. Since high-speed and high-resolution temperature measurement is required to diagnose combustion status, especially combustion instability, in a gas turbine combustor, fast-responding ultrasonic wave measurement approach was adopted in this research. To validate the possibility of this approach, a basic experiment was conducted in a circular pipe, where the ultrasonic transmitter and the receiver was installed at both ends. A continuous 40 kHz sinusoidal ultrasonic wave was transmitted at one end and this sound was obtained at the other end at the rate of 1GHz so the time-of-flight in a circular tube was calculated and, then, sonic velocity was also obtained and finally air temperature was calculated from Boyle's law. For the reference, temperature using thermocouples was also measured at the same time. Measured temperatures from these methods were almost similar for the range of 28–65 °C. The temperature of the ultrasonic measurements was fluctuated during all tests due to the fast acquisition speed, but soon stabilized if the gas temperature kept constant. As increasing temperature, ultrasonic measurement method followed faster than the thermocouple method. In addition to this better dynamic characteristic, this ultrasonic methods has further strong points such as non-intrusive means to any combustor inside so it has excellent durability and flame stability, and easy installation at economical price. Therefore, this newly proposed methods would be the potential candidate of advanced temperature measurement method with high robustness, fast acquisition and short time constant.

1 Introduction

The combustor, which is one of the key components of the gas turbine, is the element that operates in the high temperature and high pressure environment in the whole engine [1]. It is mainly used for the contact type diagnosis such as the dynamic pressure sensor and the thermocouple temperature sensor to diagnose the combustion state in the combustor, the reliability of the measurement is deteriorated and the durability of the measuring apparatus is deteriorated and the loss frequently occurs [2].

The purpose of this study is to prevent and secure accidents and problems related to these gas turbine combustors by measuring the temperature inside the high temperature and high pressure combustor by using the non-contact diagnosis method using ultrasonic wave instead of the contact type diagnosis method [3].

Non-contact technology accurately diagnoses the flame temperature, Flame stability, Combustion instability mode and cause, and controls to the optimized combustion state so that the environment of the power plant (NOx and CO emission reduction), it is judged that it can contribute to high efficiency and high reliability management [4-6].

Corresponding author. Tel.: +82 32 835 8295
E-mail address: LMC@inu.ac.kr (M.C. Lee)

2 Theory

The speed of sound c, according to Boyle's law in an ideal gas at a constant pressure, is given by

$$C = \sqrt{\frac{\gamma R T}{M}} \quad (1)$$

Where R is the ideal gas constant, T the absolute temperature (in K), M the average molecular weight of the gas molecules. And γ =Cp/Cv, where Cp is the heat capacity of the gas at constant pressure and Cv is the heat capacity of the gas at constant volume [7, 8].

From Eq. (1) we know the temperature is

$$T = \frac{MC^2}{\gamma R} = \frac{M(\frac{L}{TOF})^2}{\gamma R} \quad (2)$$

Where T is in K, T (K) = T (°C) + 273.15, L is distance between ultrasonic sensors, TOF is ultrasonic wave time-of-flight [9, 10].

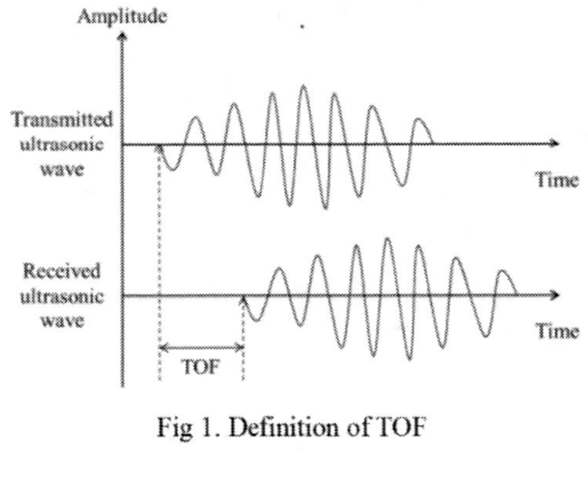

Fig 1. Definition of TOF

3 Experimental setup and method

The figure 2 shows a NI-DAQ (Data Acquisition) system for measuring temperature and an ultrasonic sensor transmitter and receiver both at a distance of 0.26m from the top of the frame. A K-type thermocouple temperature sensor was placed on the ultrasonic sensor and placed in a circular pipe to compare with the actual temperature. The air heated by the bottom electric heater passes through the transparent liner, raises the temperature of the air in the pipe, and measures the temperature inside the pipe using the ultrasonic sensor and the thermocouple temperature sensor. The measured data is compared and analyzed with the measured temperature of the ultrasonic sensor and the measured temperature of the thermocouple temperature sensor through LabVIEW. [11, 12].

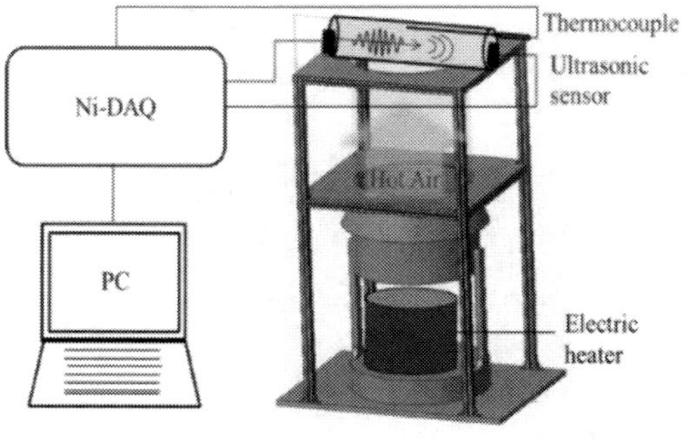

Figure 2: NI-DAQ system and temperature measurement setup

There are two methods for measuring the time-of-flight when the ultrasonic sensor receiver receives the frequency signal transmitted from the transmitter. The first threshold method is the most widely used method for detecting the time-of-flight of the transmitted ultrasonic signal from the receiver signal. The threshold level is specified for the amplitude of the received signal and the time is measured by judging the signal above the threshold value as a signal. However, since this method depends on the amplitude, it is difficult to measure accurately when the amplitude changes due to the distance between the ultrasonic sensors, the temperature, and the disturbance due to the flow of the fluid. If the signal is smaller than the noise, it is judged as noise and is recognized as a received signal [13].

Second, in the periodic recognition method, the received signal is divided into a noise signal and a received signal. The difference between the two signals is checked by the existence of the periodicity, and the period of the received signal is analyzed to find a period of continuous increase and the arrival time of the received signal is measured by determining the periodicity [14]. Since the received signal is usually a signal containing a noise signal, it is accurate to measure a stable signal after a lapse of time, but the signal of the first receiving period is affected by the weak noise signal. If the number of peaks in a period that is continuously increased is less than a specified number and is recognized as noise, it is recognized as a period of signals longer than the designated number, and has an error of one period. The difference of one cycle is 25 μs, and when the distance between ultrasonic waves is 0.26 m, 25 μs produces an error of about 17.5°C when converted to temperature [15-17].

4 Results and Discussion

Most ultrasonic sensors for distance measurement use sensors in the frequency range of 40 kHz. So in this experiment we used a sensor that generates ultrasonic waves at 40 kHz. Figure 3 shows the average flight time of the transmitter and receiver of the ultrasonic sensor before measuring the temperature using the ultrasonic sensor. This experiment shows measurement results up to 0.30 m in 0.01 m intervals to see the time-of-flight of ultrasonic sensors per distance.

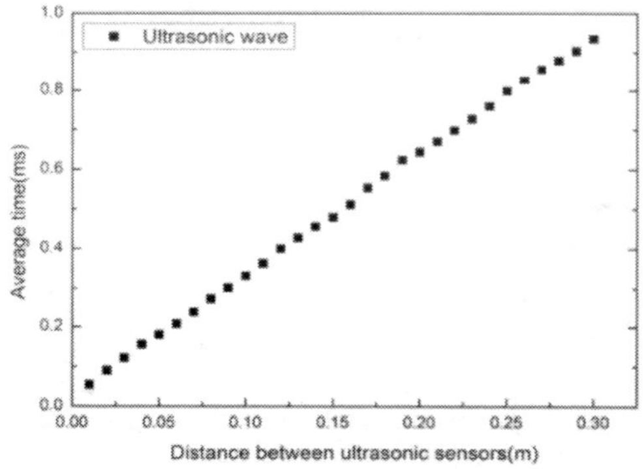

Figure 3: Average time of ultrasonic wave time-of-flight by distance

This figure 4 shows the measured values by combining the two methods to compensate for the problem of poor accuracy due to the amplitude change of the threshold detection method and the periodic error due to the noise signal of the period recognition method.

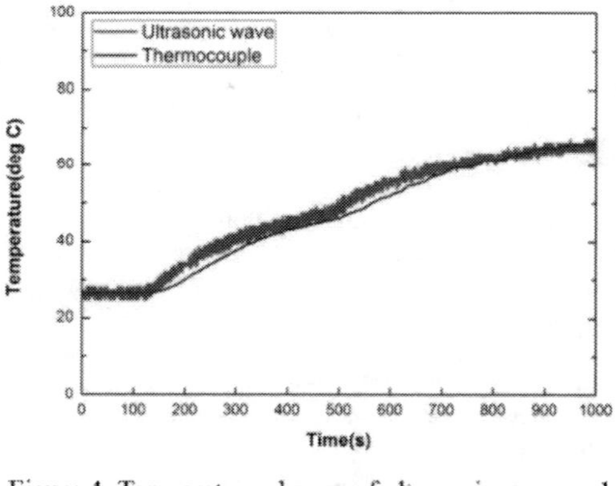

Figure 4: Temperature change of ultrasonic wave and thermocouple

As shown in the figure 5, the temperature measured by the ultrasonic sensor is more unstable than the temperature measured by the thermocouple temperature sensor. However, as the temperature increases with time, it is interpreted that the measured temperature of the ultrasonic sensor is relatively stable at a constant temperature range, so it is considered to be effectively used for measuring the temperature of the hot air. In addition, non-contact temperature measurements using ultrasonic sensors show that the response to a rise in temperature is faster than a thermocouple temperature sensor.

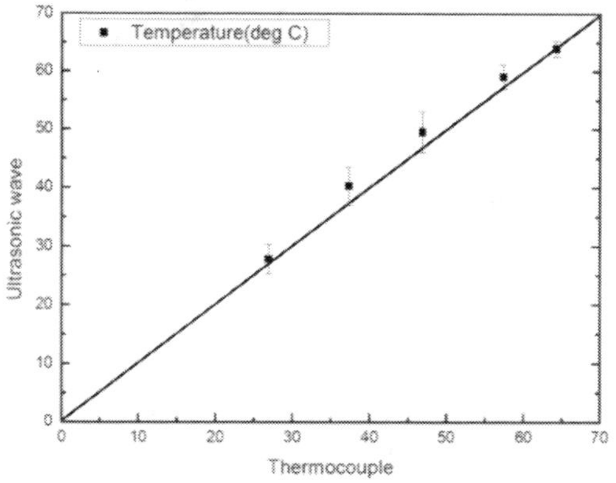

Figure 5: Error temperature of ultrasonic wave and thermocouple

To verify the results of the ultrasonic sensor and the results of the temperature sensor produced in this experiment, the correlation was analyzed using each temperature data. As shown in Table 1, the Pearson correlation coefficient of ultrasonic sensor and temperature sensors is highly correlated to 0.918, 0.957, 0.918, 0.918, 0.758, and 0.758.

Temperature	20°C ~ 30°C	30°C ~ 40°C	40°C ~ 50°C	50°C ~ 60°C	60°C ~ 65 °C
Correlation coefficient	0.918	0.957	0.918	0.918	0.758

Table 1. Pearson correlation coefficients among by temperature

4 Conclusions

In this study, an experiment was performed to measure the air temperature using an ultrasonic sensor. As a result, the temperature measured by the ultrasonic sensor appears to be more unstable than the measured temperature of the thermocouple temperature sensor. However, it is interpreted that as the measurement time increases, the temperature becomes relatively stable, and it is effectively used to measure the hot air temperature of the fast response. In addition, the non-contact temperature measurement using the ultrasonic sensor shows that the reaction according to the temperature rise is faster than the thermocouple temperature sensor. It is expected that high temperature combustion gas temperature measurement by ultrasonic wave is possible and measurement technology is developed and application to other fields is expanded (in boiler, rocket, aircraft, etc.). The noncontact temperature measurement using ultrasonic wave can be continuously measured, and in the future, it will improve the accuracy by minimizing the tremor and perform the analysis through experiment to enable the measurement at various temperatures higher than 100 °C.

5 Acknowledgment

This work was supported by the National Research Foundation of Korea (NRF) grant funded by the Korea government (MSIP) (No. NRF-2016R1C1B1006636). This research was supported by Korea Electric Power Corporation through Korea Electrical Engineering & Science Research Institute (grant number: R15XA03-13).

References

[1] J. C. Livengood, T. P. RonaJ, J. Baruch. Journal of the Acoustical Society of America, 26, 824 (1954);

[2] T. Y. Lee, S. H. Kim, S. R. Lee. Journal of the Korean Society for Precision Eng, 16 (1999) 196-202.

[3] J. B. Kim, K. Y. Jhang. Transactions of the Korean Society of Mechanical Eng, 38 (10) (2014) 1133-1137.

[4] Y. Choi., C. H. Yoon, H. S. Jeon. The Korean Society Mechanical Eng, (1998) 360-365.

[5] L. C. Lynnworth. Academic Press Inc, New York, (1989)

[6] C. H. Yoon., Y. Choi., H. S. Jeon. Transactions of the Korean Society of Mechanical Engs, 23 7, (1999) 893-900.

[7] W. Y. Tsai, H. C. Chen, T. L. Liao. Sensors and Actuators, A 132 (2006) 526–532.

[8] T. L. Liao, W. Y. Tsai, C. F. Huang. Sci. Technol, 15 (2004) 413-419.

[9] C. C. Tong, J. F. Figueroa, E. Barbieri. IEEE, 50 5 (2001).

[10] D. H. Lee, S. Y. Kim, K. S. Yoon, M. H. Lee. Journal of the Korean Society for Precision Eng, 24 8 (2007) 41-49.

[11] M. k. Choi, J. I. Park, J. H. PakThe Korean Institute of Electrical Eng, (2007) 144-145.

[12] A. B. Raine, N. Aslam, C. P. Underwood, S. Danaher. Sensors 15 5 (2015) 10705-10722

[13] T. S. Jin, Journal of the Korea Institute of Information and Communication Eng, 16 1 (2012) 41~47.

[14] D. Webster. IEEE Transactions on Instrumentation and Measurement, 43, 4 (1994) 578-582.

[15] M. S. Park, S. Y. Na. Journal of the Electronics Eng, 10 1 (1995) 25-28.

[16] M. Parrilla, J. J. Anaya, and C. Fritsch. IEEE, 40 4 (1991) 759-763

[17] D. Xiao, Q. Fan, C. Xu, X. Zhang. Ultrasonics 68 (2016) 150–154

11th Asia-Pacific Conference on Combustion,
The University of Sydney, NSW Australia
10th -14th December 2017.

Numerical Analysis of Upward Flame Spread Over Multiple Thin PMMA Slabs

H. R. Rakesh Ranga[1] and Vasudevan Raghavan[2]

[1] MS Scholar, Mechanical Engineering, Indian Institute of Technology Madras, Chennai, 600036, India

[2] Associate Professor, Mechanical Engineering, Indian Institute of Technology Madras, Chennai, 600036, India

Abstract

In this study, upward flame spread over multiple fuel slabs is numerically studied. The fuel slab made of Poly Methyl Methacrylate (PMMA) having 1 mm thickness has been used. The parameters varied are the number of slabs and the separation distance between them. Results obtained for the burning characteristics of multiple slabs are compared with those of a single slab. The burning rate, velocity field and temperature contours are studied for each case. It is found that for a given number of slabs, when the separation distance is small, the burning rate per slab is lesser than the single slab case. When the separation distance is large enough, the burning rate becomes similar to the case of single slab. This behavior is due the nature of air entrainment and heat transfer taking place with respect to a given configuration. It is also observed that for 1 mm thick slabs, if the separation distance is 8 mm, the burning rate tends towards that of single slab case. Further, an interesting behavior is observed for separation distances of 6 mm or more, for all the cases, except for the case of 3 slabs. The flame on the outer surface of the outermost slab does not exist for these cases and only the flame is contained in the gap between the outermost and adjacent slabs. However, experimental evidence is required to validate this aspect.

Keywords: Upward flame spread, PMMA, multiple slabs, separation distance.

1 Introduction

Study of flame spread over solid fuel surfaces is gaining a lot of attention as it is an important aspect of fire safety engineering. The flame spread rate and burning rate over solid fuel surfaces becomes all the more important if there are other surfaces that are also burning in their vicinity. Upward flame spread over solid fuel is rapid and highly transient in nature. Understanding the flame spread characteristics in situations, where the solid fuel is burning with interferences from other fuel surfaces burning in its vicinity, is not only important but also complex in nature. In this study, burning rate for the flame spread over multiple fuel slabs is analyzed. Temperature field, flow field and mass loss rates are presented to understand the effect of separation distance between, and the number of fuel slabs.

Various studies have been conducted on upward flame spread over solid fuels. Fernandez-Pello [1] developed a theoretical model for the upward propagation of laminar flame over vertical fuel surfaces. In the model, applicable to small scale fires, expression for the rate of flame spread was obtained as a function of thermo-physical properties of the fuel and oxidizer. Rangwala et al. [2] experimentally investigated the effect of width of the fuel sample on the burning characteristics in upward flame spread configuration. A decrease in flame height and the flame stand-off distance was observed for a decrease in the width of the sample. Orloff et al. [3] experimentally and theoretically investigated the upward flame spread over thermally thick condensed fuel surfaces. It was reported that in the laminar region, the heat transfer to the fuel was predominantly by convection whereas in the turbulent region, significant heat transfer occurred through radiation. Wang et al. [4] developed a detailed numerical model to predict the burning rate, convective and radiative fluxes in large scale vertical PMMA wall fires. This model was used to obtain the characteristics of the flame as a function of ratio of wall spacing to wall height. It was observed that when this ratio was large enough, the interference between the diffusion flame between two opposing walls become unimportant and the burning rate obtained was almost same as that of a single wall. Steady two dimensional vertically downward flame spread was experimentally studied by Kurosaki et al. [5]. A correlation for the flame spread rate, separation distance and the width of the fuel was reported. Avinash et al. [6] experimentally investigated the diffusion flame spread over thin parallel cellulose sheets (0.16 mm thick) in a natural convective environment. The effect of gravity on the flame spread rate was investigated by burning the fuel at various angles. It was found that if multiple burning sheets were close, due to lack of sufficient oxygen, the spread rates were lower. At a given separation, then due to the enhanced heat transfer from the neighboring flame, the spread rates obtained were high.

Little research has been done to study the flame spread behavior over multiple thin PMMA slabs. In the present study, burning of multiple PMMA slabs of 1 mm thick are numerically studied. The parameters varied are the number of slabs and the separation distance. The number of slabs are varied from 2 to 5. The separation distance is varied from 2 mm to 6 mm in steps of 2 mm. The result for each case is compared with that for a single slab.

2 Numerical Model

In this study, LES based computational fluid dynamics software, called Fire Dynamics Simulator (FDS) is used. It numerically solves Navier-Stokes equation appropriate for low-speed thermal gradient induced flow. Smagorinsky form of large eddy simulation is incorporated in the hydrodynamic model. The governing equations for the mass, momentum, species and energy conservation are solved in gas phase. A computational domain accounting for the symmetry in the thickness direction of the domain is employed (Fig. 1). Simulations are carried out considering a sample width of 5 mm. The number of cells used width-wise is unity. The thickness of the PMMA slab is 1 mm and the length of the slab is 75 mm. In the case of multiple slabs, all the slabs are simultaneously ignited using an ignition source, which is a hot

Corresponding author. Phone: +91-97-389-42653
E-mail address: hrrakeshranga@gmail.com

plate, at a temperature of 1200°C. The heating plates are of exact dimensions as that of the bottom surface of the PMMA slab and is placed 3 mm away from the surface. This ignition source is removed once the PMMA is ignited and a flame sustains over its surface.

The bottom of the domain is considered as an inert wall, which is at the ambient temperature. The bottom surfaces of PMMA slabs are 25 mm away from the bottom wall. Boundaries open to the ambient are indicated as OPEN in Fig. 1.

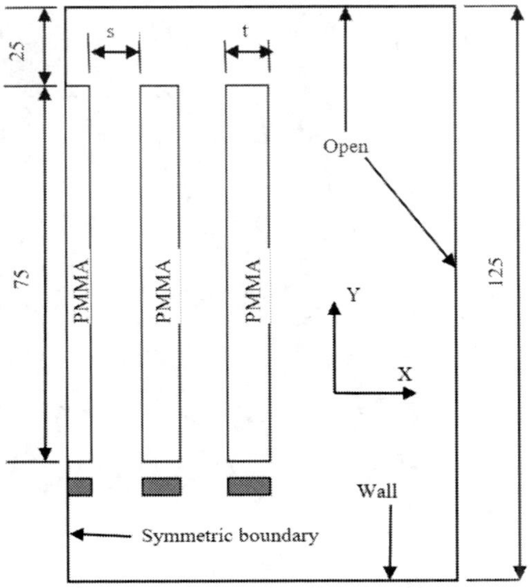

Figure 1: Computational domain for upward flame spread over 5 slabs. The grey colored boxes indicate the heater plates used as ignition sources.

Uniform meshing is provided in each direction. Along the thickness, 0.5 mm cells are used. The height of these cells, which correspond to mesh distribution along the height, is specified as 1 mm. In the width direction, keeping in mind that the whole width of the sample burns simultaneously, single cell is provided.

To model the pyrolysis, a sublimation type model is used. This involves a direct phase change from solid to vapor. The Stefan diffusion equation is used to estimate the mass gasification rate. Temperature and heat of pyrolysis are specified. Table 1 summarizes the physical and thermo-physical properties used for PMMA, in the numerical model.

Property	Units	Value
Density	kg/m^3	1190 [7]
Specific heat	kJ/kg.K	2.1 [7]
Conductivity	W/m.K	0.26 [7]
Emissivity	-	0.85 [8]
Absorption coefficient	m^{-1}	2700 [8]
Pyrolysis temperature	K	523 [7]
Heat of pyrolysis	kJ/kg	1620 [7]

Table 1: Physical and thermo-physical properties of PMMA

The combustion of the fuel vapors is modelled using infinite rate chemistry and a single step reaction, which is given as follows:

$$C_5H_8O_2 + 6(O_2 + 3.76N_2) \rightarrow 5CO_2 + 4H_2O + 22.56N_2$$

3. Results and Discussion

The effects of multiple slabs, held at varying separation distances, on burning characteristics are presented. As mentioned earlier, the parameters varied in this study are the number of slabs 'n' and the separation distance 's'. Figure 2 shows the variation of mass loss rate per fuel slab with respect to the separation distance. The ignition has been onset for the minimum required time. For all the cases, the heating plates have been kept in the domain until 15 s. For the case of 5 slabs each separated 2 mm away, ignition time has been 7 s. After the onset of ignition, simulation of burning under natural convective co-flow environment has been continued for 100 s. The cumulative mass lost during the process has been calculated from the mass loss rate data. This value has been divided by the number of slabs to arrive at the mass loss per slab.

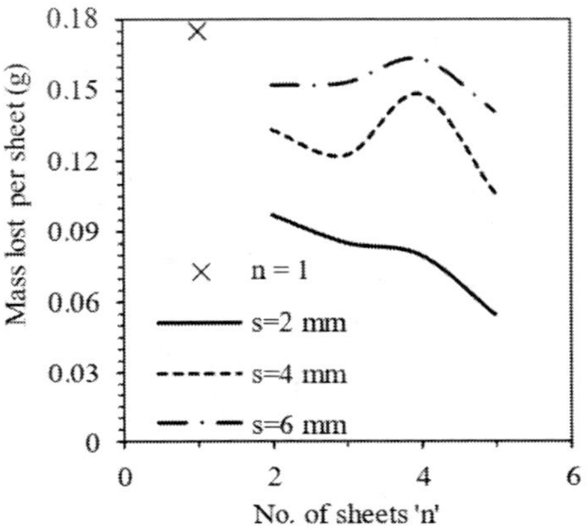

Figure 2: Variation of mass lost per fuel slab with respect to the number of fuel slabs for various separation distances (s in mm).

It is seen from Fig. 2 that for the least separation distance, i.e., 2 mm, the mass lost per slab for multiple slabs is the lowest. With an increase in number of slabs, the mass lost per fuel slab further decreases, with an exception of the case of n = 4, where the mass lost is higher than n = 3 and 5. This trend is observed in all separation distances. These aspects are also reflected in Fig. 3, where the mass lost per slab during 100 s burning is plotted as a function of separation distance for different n values. The reduction in the mass loss rate at lower separation distances is because of the fact that there is no proper air (oxygen) transport between the slabs resulting in lack of oxygen for burning. Hence, most of the burning takes place at the outer most slabs and inner slabs burn only after outer slabs regress. This results in an overall decrease in the mass lost per fuel slab. Further as the separation distance increases, the mass lost per fuel slab increases and approaches the value of n = 1. It is also demonstrated in Fig. 3 that for n = 2, at s = 8 mm, the mass lost per slab is almost similar as that of n = 1.

Two interference effects dictate the flame spread and mass loss rates. One, the oxygen supply towards the flame zone/fuel slabs and another, the heat transfer from the flame to the fuel surface. In the case of n = 2, at lower separation distances, there is an enhancement in the heat transfer of the inner most slab due to the burning of the outer slabs. However, oxygen is available only to the outer slabs and inner slab(s) starve for oxygen. This leads to pyrolyzate leaving through the small gaps and probably burn at the top portion of the sample. Further, burning in this configuration (low separation distance) is layer by layer. Once the outer layer regresses, the inner layer burn. Within 100 s, the scenario is that mass loss rate has not picked up in the inner layer as the outer layer have not regressed sufficiently. As the number of slabs are increased, the inner most layer starves for both oxygen supply as well as for heat transfer from the flame. For instance, in the case of n = 5, inner most slab is deprived of oxygen and heat from the flame. This aspect forms the reason for decreasing trend in mass loss rate with increasing number of slabs.

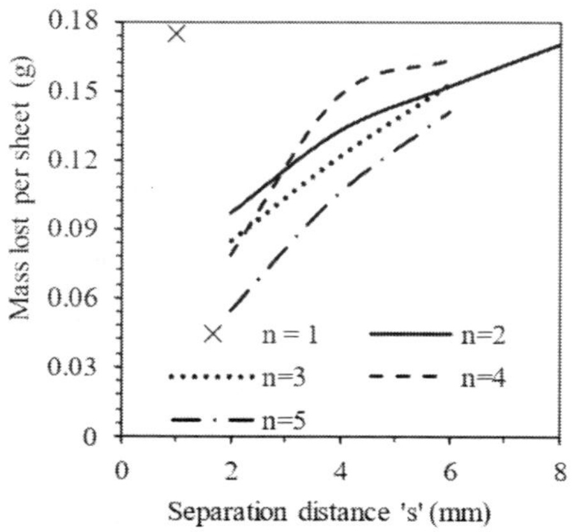

Figure 3: Variation of mass lost per slab (g) with respect to the separation distance (mm) for different number of slabs.

As the separation distance is increased, oxygen transport between the fuel slabs increases. In general, the mass loss rate per slab increases with increase in the separation distance. However, based on the number of slabs, the heat transfer between the flame and the slab varies and this aspect dictates the flame spread and mass loss processes. It is observed that even number of slabs, n = 2 or 4, have higher mass loss than the odd numbered cases, as seen in Figs. 2 and 3. In even number configuration Oxygen transport is enhanced because of entrainment from outer surfaces as well as that in between the fuel slabs. For n = 4, the outer most slab has good amount of air entrainment. It burns and heats up the adjacent inner slab. The inner slab receives heat from the outer slab as well as oxygen due to air entrainment from the gap between the inner and outer slabs. Therefore, due to enhancement in the oxygen and heat transport processes, mass loss rate increases for this case and becomes equivalent to that of n = 2. The cases, n = 3 and 5 have lesser mass loss rates than that of n = 4. In the case of odd number of slabs, for instance, 5 slabs, in the gap between two slabs located right or left to the central slab, oxygen flow depends upon the separation distance.

Figure 4 shows the temperature distribution along with velocity vectors for the separation distance of 4 mm, for the cases with different number of fuel slabs burning symmetrically in the domain. Velocity overshoot in the flame zone is clearly observed in Fig. 4. Further, the acceleration of the gases due to buoyancy driven flow is also quite apparent. The maximum value of the gas velocity within the slabs is clearly dictated by the separation distance. The flame anchored at the bottom of the regressing slabs clearly envelopes all the slabs. At this separation distance, for a given n, all the slabs have regressed almost to the same extent from the bottom.

For all the multiple slab cases with separation distance less than or equal to 4 mm, it is observed that recirculation zones are being formed above the slabs. The temperature of the gasses in these recirculation zones is around 650°C - 750 °C. The velocity of these gasses is around 0.12 m/s, which is much low when compared to the velocity of the combustion gasses that are buoyancy driven. These high velocity gasses engulf the low velocity hot gasses above the PMMA slabs, resulting in formation of recirculation zones. These recirculation zones heat up the top portion of the PMMA, thereby contributing in the pre-heating of the intermediate slabs. In certain cases, flames are observed on the top portion of the slabs as well.

Figure 5 presents the temperature contours and velocity vectors for n = 3 burning at different separation distances. Similar features are observed. However, it is interesting to note that at higher separation distance (s = 6 mm) inner slab regresses faster than the outer slabs.

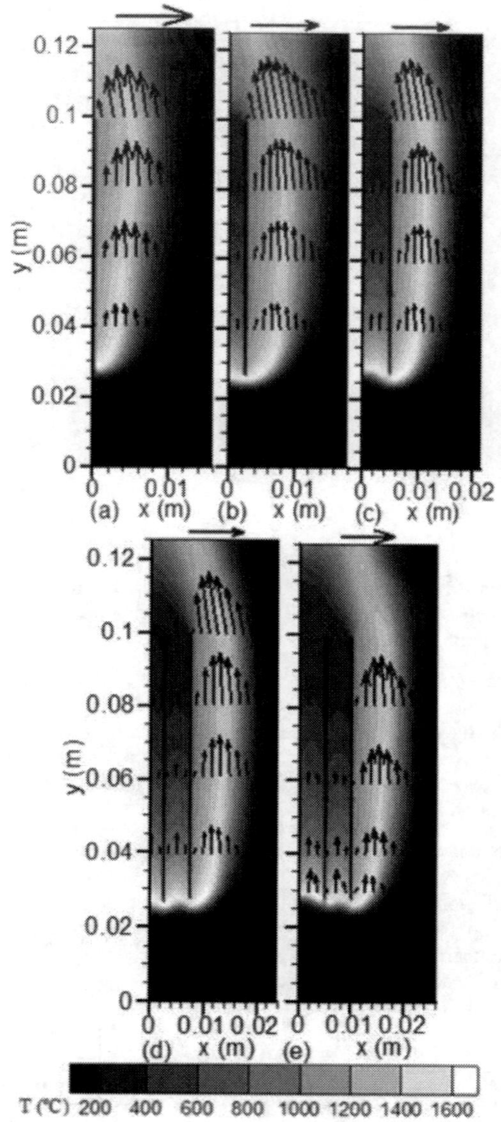

Figure 4: Temperature contours along with the velocity vectors for the cases with separation distance of 4 mm (a) n = 1, (b) n = 2, (c) n = 3, (d) n = 4 and (e) n = 5. The reference vector on top of each indicates a magnitude of 1 m/s.

Figures 6 and 7 show the temperature contours of n = 4 and 5, respectively, for separation distances of 2 mm, 4 mm and 6 mm. It is noted for s = 6 mm that the flame on the outer side of the outermost PMMA slab does not exists on its bottom portion, but it is established near the top portion of the slab. The outermost slab eventually shows regression due to the flame on its inner surface and top portion. The flames are established in the gaps between the slabs. This behavior needs experimental validation.

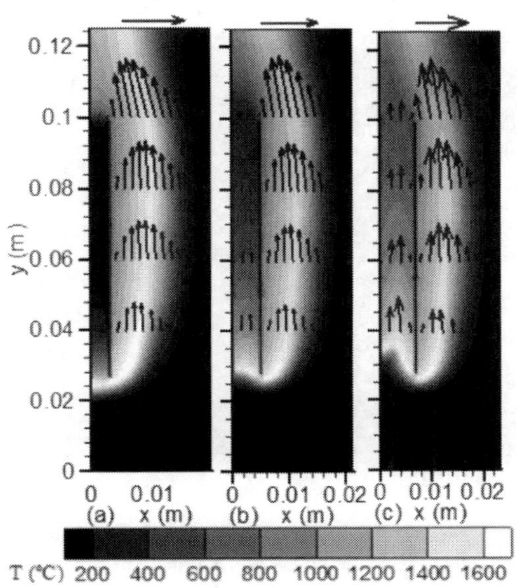

Figure 5: Temperature contours along with the velocity vectors for the cases with n = 3 (a) s = 2 mm, (b) s = 4 mm and (c) s = 6 mm. The reference vector on top of each indicates a magnitude of 1 m/s.

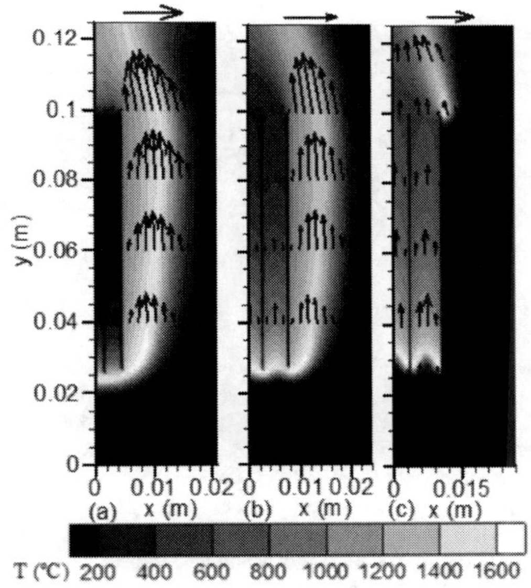

Figure 6: Temperature contours along with the velocity vectors for the cases with n = 4 (a) s = 2 mm, (b) s = 4 mm and (c) s = 6 mm. The reference vector on top of each indicates a magnitude of 1 m/s.

4. Conclusions

Symmetric cases are simulated to perform the above study. Within a burning time of 100 s, the multiple slab cases have lesser mass loss rate than the case of single slab. It is observed that, for a given number of slabs, the mass loss rate increases with increasing separation distance and finally tends to attain a value corresponding to that of a single slab. For a given separation distance, the general trend observed is that the mass loss rate decreases with increase in number of slabs, s = 4 mm being an exception, where the mass loss rate increases. At this separation distance, all slabs regress together. It is observed that even number of slabs, n = 2 or 4, have higher mass loss than the odd numbered cases. At separation distance of s = 6 mm, for n = 3, inner slab regresses faster than the outer slabs. For other n values, the flame on the outer side of the outermost PMMA slab does not exists on its bottom portion, but it is established near the top portion of the slab.

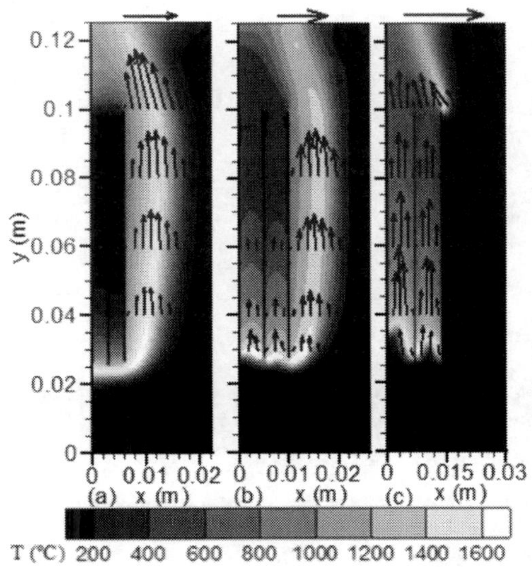

Figure 7: Temperature contours along with the velocity vectors for the cases with n = 5 (a) s = 2 mm, (b) s = 4 mm and (c) s = 6 mm. The reference vector on top of each indicates a magnitude of 1 m/s.

References

[1] A.C. Fernandez-Pello, A Theoretical Model for the Upward Laminar Spread of Flames Over Vertical Fuel Surfaces, Combust. Flame 31 (1978) 135-141

[2] A.S. Rangwala, S.G. Buckley, J.L. Torero, Upward Flame Spread on a Vertically Oriented Fuel Surface: The Effect of Finite Width, Proc. Combust. Inst. 31 (2007) 2607-2615.

[3] L. Orloff, J. De Ris, G.H. Markstein, Upward Turbulent Fire Spread and Burning of Fuel Surface, Fifteenth Symposium (International) on Combustion, The Combust. Inst. 15 (1975) 183-192

[4] H.Y. Wang, P. Joulain, J.M. Most, Modeling on Burning of Large Scale Vertical Parallel Surfaces With Fire Induced Flow, Fir Safety Journal 32 (1999) 241-271

[5] Y. Kurosaki, A. Ito, M. Chiba, Downward Flame Spread Along Two Vertical Parallel Slabs of Thin Combustible Solid, Symposium (International) on Combustion, 17 (1979) 1211-1220

[6] G. Avinash, A. Kumar, V. Raghavan, Experimental Analysis of Diffusion Flame Spread Along Thin Parallel Solid Fuel Surfaces in a Natural Convective Environment, Combust. Flame 165 (2016) 321-333

[7] R. Ananth, C.C. Ndubizu, P.A. Tatem, Burning Rate Distributions for Boundary Layer of a PMMA Plate in a Forced Flow, Combust. Flame 135 (2003) 35-55

[8] K. McGrattan, S. Hostikka, R. McDermott, J. Floyd, C. Weinschenk, K. Overholt, Fire Dynamics Simulator User's Guide, Natl. Inst. Stand. Technol. Spec. Publ., U.S., 2013

11th Asia-Pacific Conference on Combustion,
The University of Sydney, NSW Australia
10th -14th December 2017.

Influence of Tumble and Unsteady Effects on Near-Wall Heat Transfer in SI Engine

Ryosuke Okutani[1], Takeshi Yokomori[1], Masayoshi Matsuda[1], Hidetsugu Yamamoto[1], Norimasa Iida[1], Daijiro Ishii[2], Yuji Mihara[2]

[1]Keio University

3-14-1 Hiyoshi, Kohoku-ku, Yokohama, Kanagawa, 223-8522, Japan

[2]Tokyo City University

1-28-1 Tamadutsumi, Setagaya-ku, Tokyo, 158-8557, Japan

Abstract

Recently, elucidation of the near-wall heat transfer is strongly required for enhancing engine thermal efficiency. However, the previous wall heat flux models assuming the wall-function do not have enough prediction accuracy as for the recent engine technical development. This study focused on the influence of tumble and unsteady effects as the error factor and investigated them on the near-wall heat transfer in recent high compression ratio and high tumble SI (spark ignition) engine. In this paper, the measurements of heat flux and flow distribution in an optical SI engine, three kinds of 1-D calculation and estimation related to heat flux with previous models were performed. Finally, it is shown that four mechanisms regarding tumble and unsteady effects cause the previous models underestimation.

1 Introduction

Recently, global warming and energy security have received significant attention. In fact, the world vehicle fuel economy regulation has been strengthened about 4% annually. Thus an enhancing engine thermal efficiency is strongly required. In Fig. 1, it is assumed that the reduction of cooling heat loss and exhaust heat loss is one of effective way for further enhancing thermal efficiency. In recent engine technical development for this purpose, elucidation of near-wall heat transfer becomes important. However, previous models [1,2] do not have enough prediction accuracy. As for the cause, wall-function defect in a SI engine is suggested [3]. This paper investigates the influence of tumble and unsteady effects on near-wall heat transfer in a recent high compression ratio and high tumble SI engine.

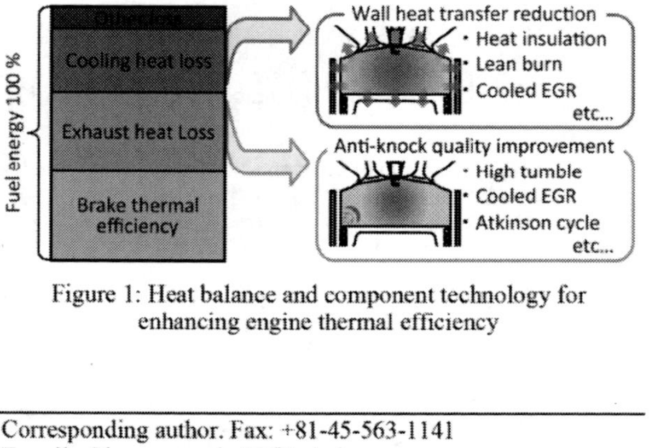

Figure 1: Heat balance and component technology for enhancing engine thermal efficiency

Corresponding author. Fax: +81-45-563-1141
E-mail address: ryosuke00@keio.jp

2 Methods

Figure 2 shows the measurement field and calculation domains in this study. Regarding experimental method, the flow velocity, in-cylinder pressure and temperature at the piston-top and 4 mm in-depth were determined in the combustion chamber. Regarding mathematical methods, the near-wall heat transfer analysis (gas-side domain) and solid heat conduction analysis (wall-side 4 mm domain) were performed in the each domain.

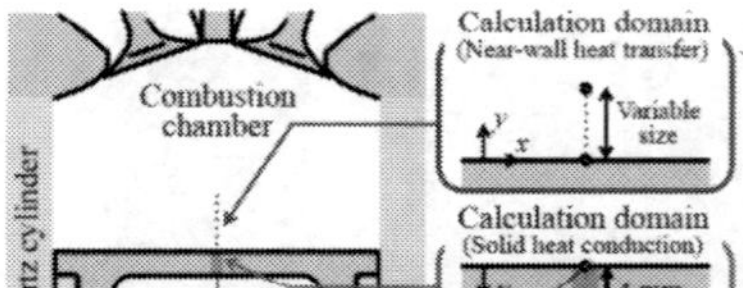

Engine parameters	Value
Bore	75 mm
Stroke	112.5 mm
Connecting rod	250 mm
Compression ratio	13
Intake valve open	330 deg.ATDC
Intake valve close	-86 deg.ATDC
Exhaust valve open	109 deg.ATDC
Exhaust valve close	-332 deg.ATDC

Figure 2: Measurement field and calculation domains

2.1 Experimental Setup

The in-cylinder pressure, flow velocity and temperature at the piston-top and 4 mm in-depth were measured with an optical SI engine. In Fig. 3, the flow velocity was determined with PIV (particle image velocimetry) measurements. Regarding this system, the double-pulsed laser sheet was irradiated to the engine by using two lasers (Lee Laser, LDP100MQG) and cylindrical lens. Here, tracer particles (SiO_2, 2.7 μm) were fed through the intake port and captured per 1 °CA with a high-speed camera (vision research, Phantom v 2011). In-cylinder pressure and temperature (piston-top and 4 mm in-depth) were determined per 0.1 °CA with a pressure sensor (AVL, GH15D) and coaxial thermocouples (MEDTHERM, K-type). Those data were recorded with a data station (Ono Sokki, DS-3000).

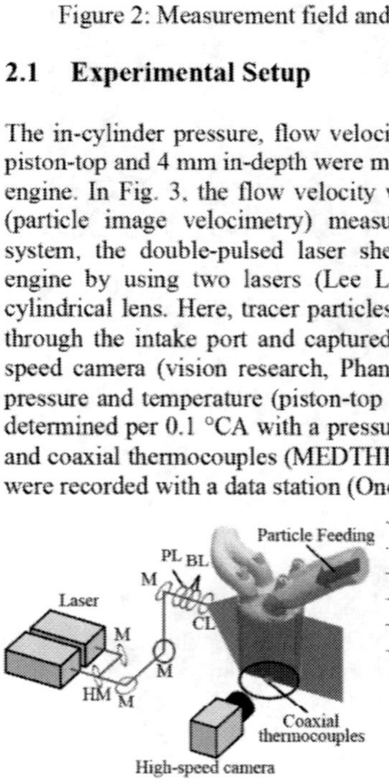

M	Mirror
HF	Half Mirror
PL	Plano-Convex Lens
BL	Bi-Convex Lens
CL	Cylindrical Lens

Figure 3: Schematic of PIV measurement system

2.2 Near-Wall Heat Transfer (Gas-Side) Analysis

In this analysis, 1-D calculation of the four governing equations took the unsteady effects into account was performed in the piston-top-center gas-side domain (see Fig. 2) since the wall-vertical quantities gradient are much larger than the wall-parallel one near the wall. Equation (1) is for solving the heat balance, but for calculating this, the eddy viscosity μ_t becomes necessary. Therefore, the momentum transfer equation (2) and two SST k-ω models equations (3) and (4) [4] were solved simultaneously. Regarding the turbulent model selection, Ma et al. [3] performed the in-cylinder μ-PIV measurements and confirmed that k-ω model can predict the friction velocity well.

$$\bar{\rho}C_p\frac{D\overline{T}}{Dt} = \frac{\partial}{\partial y}((\lambda + \frac{\mu_t C_p}{Pr_t})\frac{\partial \overline{T}}{\partial y}) + \frac{d\overline{P_m}}{dt} \tag{1}$$

$$\bar{\rho}\frac{D\bar{u}}{Dt} = \frac{\partial}{\partial y}((\mu + \mu_t)\frac{\partial \bar{u}}{\partial y}) - \frac{\partial \bar{P}}{\partial x} \tag{2}$$

$$\bar{\rho}\frac{D\bar{k}}{Dt} = \frac{\partial}{\partial y}((\mu + \sigma_k\mu_t)\frac{\partial \bar{k}}{\partial y}) + \mu_t(\frac{\partial \bar{u}}{\partial y})^2 - \beta^*\bar{\rho}\bar{\omega}\bar{k} \tag{3}$$

$$\bar{\rho}\frac{D\bar{\omega}}{Dt} = \frac{\partial}{\partial y}((\mu + \sigma_\omega\mu_t)\frac{\partial \bar{\omega}}{\partial y}) + \bar{\rho}\gamma(\frac{\partial \bar{u}}{\partial y})^2 - \beta\bar{\rho}\bar{\omega}^2 + 2(1 - F_1)\bar{\rho}\sigma_{\omega 2}\frac{1}{\bar{\omega}}\frac{\partial \bar{k}}{\partial y}\frac{\partial \bar{\omega}}{\partial y} \tag{4}$$

where subscript m means measured values and over-bar means ensemble-averaged values. $Pr_t = 0.85$ is the turbulent Prandtl number and $\partial\overline{P_m}/\partial t$ is the pressure work. Pressure gradient $\partial\bar{P}/\partial x$ is calculated from measured flow velocity $\overline{u_m}$ nearest to the piston-top-center outside boundary layer by Eq.(5) assuming $\partial\bar{P}/\partial y = 0$ and that the shear stress and advection can be neglected. Thermal conductivity λ (calculated with modified Eucken correction), constant pressure specific heat C_p [5] and viscosity v [6] were calculated from the gas temperature.

$$\frac{\partial \bar{P}}{\partial x} = -\bar{\rho}\frac{\partial \overline{u_m}}{\partial t} \tag{5}$$

Table 1 shows the calculation conditions for this analysis. The mesh is reconstructed so that the continuity is satisfied at each time step. As the result, the wall-vertical advection caused by the density change can be calculated. So as to achieve both low calculation cost and high-resolution mesh, the linear mesh size gradient was introduced so that the initial size at gas-side endpoint becomes 10 times larger than that at wall-side one. In order to investigate the influence of tumble, two kinds of boundary conditions were set. In the case of Neumann b. c., the turbulent physical quantities transfer from the outside of the boundary layer caused by tumble transformation and intake turbulent flow are not considered since these gradient are set to zero as the gas-side boundary condition. When the measured values were used, the turbulent energy was calculated from Eq.(6) and N_c =100 cycles data were used. However, the turbulent energy value used for the boundary condition is likely to be overestimated due to the influence of the cycle fluctuation. This is a future subject for detailed analysis.

$$\overline{k_m} = \frac{\overline{u_m'}^2}{2} = \sum_{t=1}^{N_c}\frac{|u_m^t - \overline{u_m}|^2}{2N_c} \tag{6}$$

Table 1: Conditions for near-wall heat transfer analysis

Condition	Configuration	Condition	Configuration
Initial domain size	25 mm	Initial condition T	358 K
Mesh number	200	Initial condition u	Measured
Spatial discretization	2 dim. central difference	Initial condition k	Measured
Time length	1 cycle	Initial condition ω	1000
Time step number	1500000	Boundary condition T	Gas: Neumann b. c. / Wall: Measured
Temporal discretization	4 dim. Runge-Kutta	Boundary condition u	Gas: Measured / Wall: 0
		Boundary condition k	Gas: Measured or Neumann b. c. / Wall: 0
		Boundary condition ω	Gas: Neumann b. c. / Wall: $60v/\beta_1\Delta y^2$ [4]

2.3 Solid Heat Conduction (Wall-Side) Analysis

In this analysis, 1-D calculation of the governing equation (7) was performed in the piston-top-center wall-side 4 mm domain (see Fig. 2) since the wall-vertical temperature gradient is much larger than the wall-parallel one. The wall heat flux q_w obtained from Eq.(8) in this analysis was used for validation of the near-wall heat transfer analysis shown in previous part.

$$\frac{\partial \overline{T}}{\partial t} = \alpha_w\frac{\partial^2 \overline{T}}{\partial y^2} \tag{7}$$

$$q_w = \lambda_w\frac{\partial \overline{T}}{\partial y}|_{y=0} \tag{8}$$

where α_w is the constant thermal diffusion coefficient and λ_w is the constant thermal conductivity.

Table 2 shows the calculation conditions for this analysis. The temperatures used as the boundary conditions were removed 400 Hz or higher frequency components and were eliminated the synchronized noise assuming that the 4 mm in-depth value is constant.

Table 2: Conditions for solid heat conduction analysis

Condition	Configuration	Condition	Configuration
Domain size	4 mm	Initial condition	Determined with iterative calculation so that the residual ratio is less than 0.001%
Mesh number	1000	Boundary condition	Piston-top: Measured 4 mm in-depth: Measured
Spatial discretization	2 dim. central difference		
Time length	1 cycle		
Time step number	10000		
Temporal discretization	2 dim. Crank-Nicolson		

3 Results and Discussions

3.1 Measurement Results

Figure 4 shows the measurement results at 2000 rpm. Wall-parallel flow velocity $\overline{u_m}$ and velocity fluctuation component $\overline{u_m'}$ are values at the measurement point nearest to the piston-top-center outside of the boundary layer. From Fig. 4 (d), the increasing of the velocity fluctuation component in the intake stroke and compression stroke after -30 °ATDC was shown. Regarding this, the intake flow and tumble transformation is considered to be the cause. In Eq.(6), this value was regarded as the turbulent velocity fluctuation component and used.

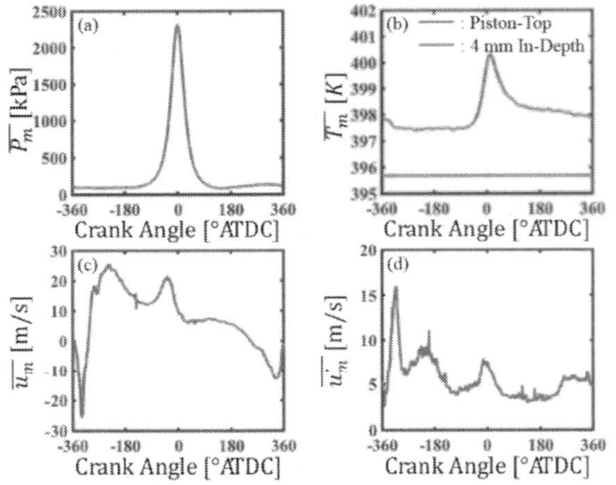

Figure 4: Measurement Results at 2000 rpm

3.2 Wall Heat flux

Figure 5 shows the comparison of the piston-center-top wall heat flux between the wall-side analysis, gas-side analysis and previous models calculation results. Here, the previous models are calculated with the shear velocity and temperature obtained from the gas-side analysis. Regarding the gas-analysis with Neumann b. c., the turbulent transfer from the outside of the boundary layer is not included to measured b. c. condition. In addition, Rakopoulos model [1] assumes the narrow-sense ideal gas, eddy viscosity wall-function [2] (solid line in Fig. 6) and wall-parallel flow. Moreover, in Han and Reiz model [2], the pressure work inside the boundary layer is not considered.

At first, the comparison between results of Han and Reiz and others shows that the pressure work relates to the behavior of the wall heat flux change. From the comparison of Rakopoulos and Neumann b. c. results, the underestimation of the former after -30 °ATDC is shown. As for the cause, in Fig. 6, the defects of the eddy viscosity wall-function, which may be caused by the remains of the turbulence generated until then, is shown. The comparison between two kinds of gas-side analysis results shows that measured b. c. is much larger than Neumann b. c. after -30 °ATDC. As the cause of this, Fig. 4 (d) shows the tumble transformation induced after -30 °ATDC, so the turbulent transfer to the boundary layer from the outside region is considered to be a main factor and also shown in Fig. 6.

Finally, in Fig. 5, the gas-side analysis result with measured b. c. shows a good agreement with the wall-side analysis result. This matching indicates that the tumble transformation, pressure work and turbulence remains are important for accurate near-wall heat transfer prediction in a recent SI engine.

3.3 Gas Temperature Distribution

Figure 7 shows the gas temperature distributions at each crank angle obtained from the gas-side analysis with measured b. c. for the turbulent energy. In Fig. 7, the prominent temperature gradient in 50 μm from the wall at TDC is shown. And the thicker temperature boundary layer is formed in the course of the compression and expansion stroke. As an interesting point, it is shown that a unique U-shaped temperature distribution in the Fig. 8. As the cause of this, the pressure work exceeding the heat transfer induces the specific temperature drop. Thus, this unique phenomenon is considered to occur in the immediate near-wall region where the turbulent thermal conductivity is low at the timing when the turbulence is weak and the pressure change speed is high.

3.4 Momentum Wall-Function

Figure 9 shows the comparison of the shear velocity between results with two kinds of gas analysis and log-law wall-function. The results show the difference at -180 °ATDC between measured b. c. and others. Regarding this difference, remains of turbulence generated by intake flow shown in Fig. 4 (d) seems to be the cause. At -90 °ATDC, the log-law result was different from others. This behavior was induced from the wall-vertical advection by density change. In Fig. 10, this influence affects the whole of boundary layer. At -30 °ATDC, a good agreement of the three results was indicated. As for the cause, the density change was low and, tumble transformation have not occurred until then. And the difference of the result with measured b. c. to others after -30 °ATDC was induced by tumble transformation. And Fig. 10 indicates that this phenomenon influences only log-low region. These results suggest a defect of the momentum wall-function in a recent SI engine.

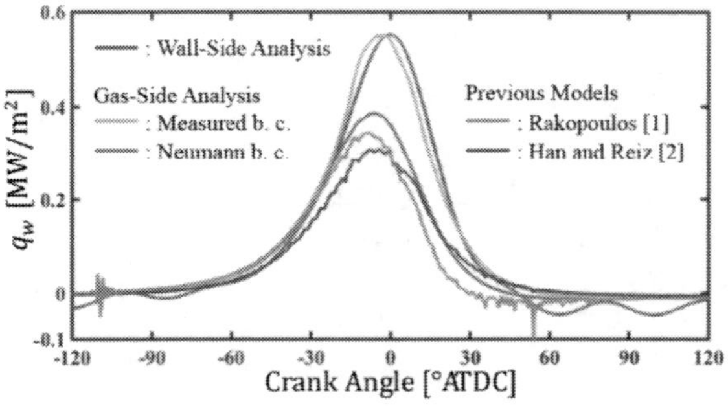

Figure 5: Comparison of piston-center-top wall heat flux q_w obtained from each calculation

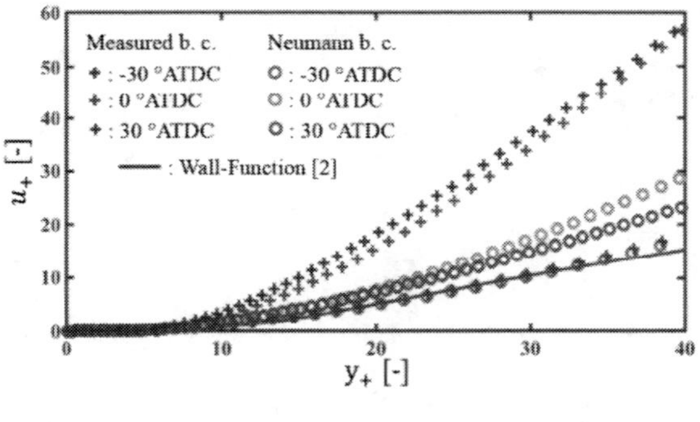

Figure 6: Eddy viscosity ratio $\nu_+=\nu_t/\nu$ at each crank angle

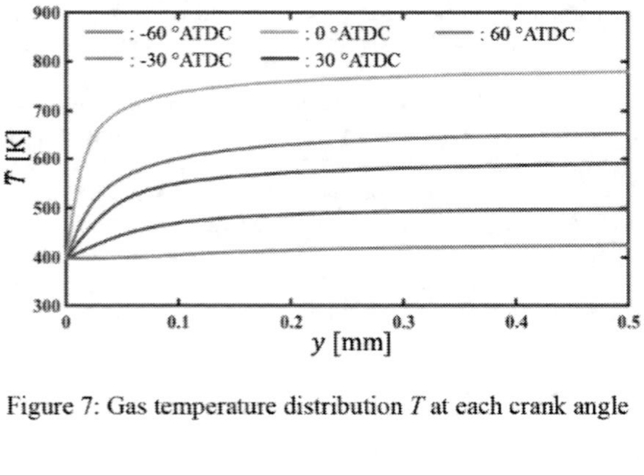

Figure 7: Gas temperature distribution T at each crank angle

4 Conclusions

This study revealed the influence of the tumble and unsteady effects on near-wall heat transfer in a high compression ratio and high tumble SI engine. The findings are as follows:

- Pressure work: This effect was shown to heavily relate to the behavior and underestimation of the wall heat flux and induce unique U-shape temperature distribution.
- Remains of turbulence: This effect was shown to increase the wall heat flux from the quasi-steady calculation results. The remaining turbulence generated until then was suggested to promote heat transfer.
- Tumble transformation: This effect was shown to induce the defect of the momentum wall-function and underestimation of the wall heat flux after tumble transformation timing. The transfer of the turbulence generated outside the boundary layer to log-low region was indicated.
- Wall-vertical advection: This effect shown to cause the defect of the momentum wall-function when the piston speed is high. This effect was suggested to influence not only log-law region but also the viscous sublayer.

Finally, considering all these phenomena, it was suggested that the wall heat flux could be well predicted even in a recent high compression ratio and high tumble SI engine.

5 Acknowledgment

This work was supported by Council for Science, Technology and Innovation (CSTI), Cross-ministerial Strategic Innovation Promotion Program (SIP), "Innovative Combustion Technology" (Funding agency: JST).

References

[1] Z. Han and R. D. Reitz, "A temperature wall function formulation for variable-density turbulent flows with application to engine convective heat transfer modeling", *Int. J. Heat Mass Transf.* vol. 40, pp. 613-625 (1997).

[2] C. D. Rakopoulos et al., "Critical evaluation of current heat transfer models used in CFD in-cylinder engine simulations and establishment of a comprehensive wall-function formulation", *Appl. Energy*, vol. 87, pp. 1612-1630 (2010).

[3] P. C. Ma et al., "Development and analysis of wall models for internal combustion engine simulations using high-speed micro-PIV measurements", *Flow Turb. Combust.*, vol. 98, pp. 283-309 (2017).

[4] F. R. Menter, "Two-Equation Eddy Viscosity Turbulence Models for Engineering Applications", *AIAA J.*, Vol. 32, pp. 1299-1310 (1994).

[5] NASA Thermo Build : available at <https://www.grc.nasa.gov/WWW/CEAWeb/ceaThermoBuild.htm>

[6] R. J. Kee et al., "A Fortran Computer Code Package for the Evaluation of Gas Phase, Multicomponent Transport Properties", SAND86-8246B (1998).

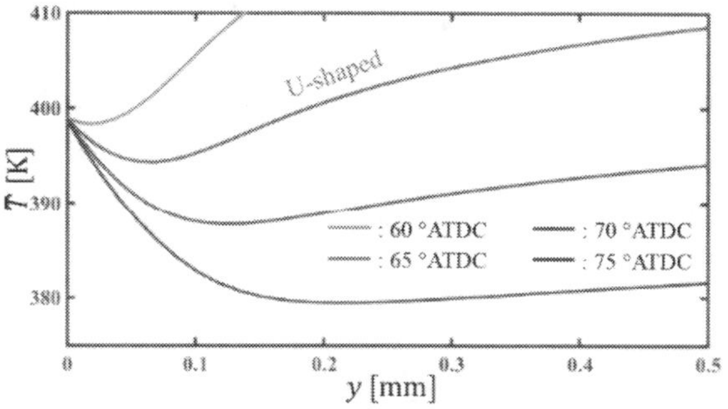

Figure 8: Gas temperature distribution T in expansion stroke

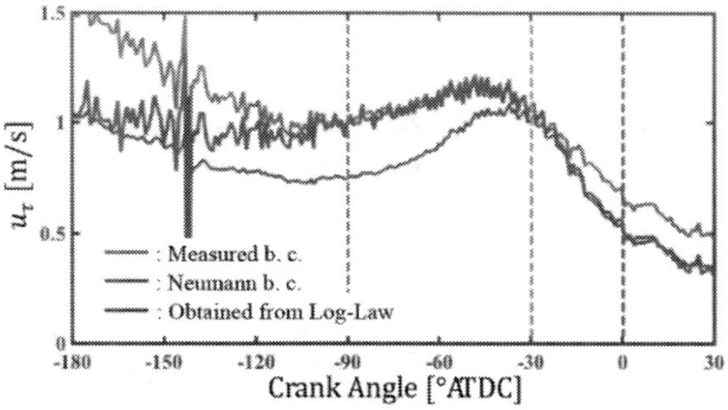

Figure 9: Comparison of shear velocity u_τ obtained from each calculation

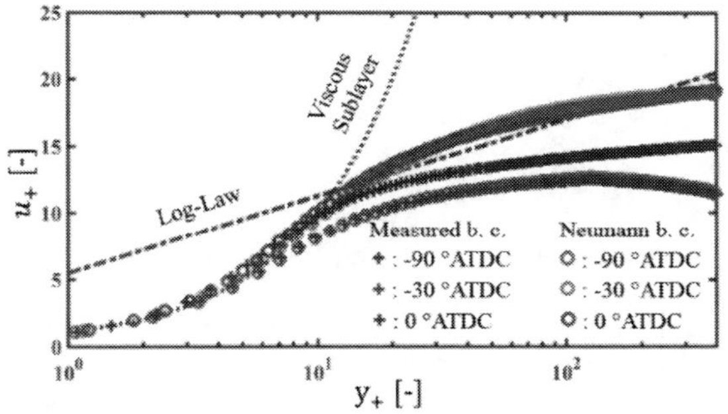

Figure 10: Comparison of non-dimensional velocity $u_+ = u/u_\tau$ at three lines crank angle in Fig. 9

11th Asia-Pacific Conference on Combustion,
The University of Sydney, NSW Australia
10th -14th December 2017.

Suitable pyrolysis model for physics-based bushfire simulation

Rahul Wadhwani[1, 2], Duncan Sutherland[1, 2, 3], Khalid Moinuddin[1, 2]

[1]Centre for Environmental Safety and Risk Engineering, Victoria University

Melbourne, VIC-3030, Australia

[2]Bushfire and Natural Hazards CRC

Melbourne, VIC-3002, Australia

[3]Department of Mechanical Engineering, University of Melbourne

Melbourne, VIC-3052, Australia

Abstract

Models for the thermal degradation (or pyrolysis) of solid fuel are fundamental to the physics-based simulation of grassfires. The pyrolysis process affects the combustion process and therefore the simulated flame, which defines the fire front. There are two competing models: a simple linear parameterisation and a non-linear Arrhenius model. The present work appraises these two models for Lucerne hay (a cured herbaceous fuel) to test their suitability for bushfire simulation. Thermogravimetric analysis and differential scanning calorimetry of Lucerne hay is conducted to measure the parameters required for the simulation. Simulations of pyrolysis are carried out using both the linear and single effective Arrhenius models, and compared with the experimental results. For this fuel, the linear model provides better agreement with the experimental data than the Arrhenius model. Hence, the linear model would be more suitable for large-scale wildfire simulation.

1 Introduction

Wildfire (or bushfire) is a global natural disaster which causes economic and ecological damage of worth billions of dollars each year. Australia is dominated by grass pasture lands which are significantly different from forest litter fuels. To study how fire propagates in a bushfire, physics-based fire models [1] are ideal for research purposes. They simulate the behaviour of the fire by solving mass, energy, and momentum conservation equations. Pyrolysis is an integral part of physics-based models to simulate wildfires [1] and combustion which represents the fire front [2]. Both forest and grass fuels exhibit variation in material thermophysical and chemical properties. They show complicated multiple step reaction kinetics of thermal degradation [3], variation in fuel properties with locations and season [4], and intra-species variation [5]. To understand the fire propagation the pyrolysis of forest fuels must be modelled [2]. There are two competing models for pyrolysis: linear [2,6], and Arrhenius [7]. These models appraised using the pyrolysis of Lucerne hay (LuH) (also called Alfalfa, Medicago satvia) in this study. LuH is one of the animal feedstock crops grown in Victoria, Australia. LuH selected to minimise the effect of site variation and increase reproducibility of this experimental work.

Corresponding author.
E-mail address: rahul.wadhwani@live.vu.edu.au

2 Numerical Model

Wildland-urban-interface Fire Dynamics Simulator (WFDS)-an extension of Fire Dynamics Simulator (FDS) is an open-source physics-based CFD fire model to simulate fire propagation as a thermally driven flow [8]. Currently, it is integrated as a vegetation sub-model in FDS [8]. Mell *et al.* [2,9] showed the capability of WFDS to simulate grassfire and crown fire. WFDS has two ways to define vegetation fuels: (a) the boundary fuel method, where fuel is defined in bulk layer e.g. grassland and is smaller than the grid size [2]; (b) the fuel element method, where fuel such as Douglas trees are defined by particles in the same size of the numerical grid elements [9]. For large-scale fire modelling, such as grassland it is appropriate to use the boundary fuel method. To use the boundary fuel method, a pyrolysis model is still required. The pyrolysis may be modelled as the linear model or the Arrhenius model. The two competing pyrolysis model are discussed below.

2.1 Linear model

The simple linear model [6] divides the mass loss rate of vegetation into three sections: (a) a step-change mass loss of moisture at 373K, (b) a linear rate of pyrolysis reaction ($\dot{w}_{pyr\_Lin}$) (Equation 1), and (c) a negligible char mass loss rate. The important pyrolysis step is

$$\dot{w}_{pyr\_Lin} = \frac{Q_{net}}{h_{pyr}} \frac{(T-T_0)}{(T_f-T_0)}, \tag{1}$$

where Q_{net} is the net input heat to the material (kJ/kg.K), h_{pyr} is the heat of pyrolysis reaction (kJ/kg), T_0 and T_f are the initial and final temperature of the pyrolysis range, and T is sample temperature at any time t. T_0 and T_f estimated from the thermogravimetric analyser (TGA), and h_{pyr} estimated from differential scanning calorimetry (DSC) for a particular material. Q_{net} requires measurement of specific heat capacity (C_P) of the sample using a hot disk analyser (HDA).

2.2 Arrhenius model

The Arrhenius model uses an Arrhenius equation to define the kinetics of the pyrolysis reaction (Equation 2). The default version of the model implemented in WFDS uses a single reaction kinetics based on Morvan & Dupuy [7]. A complex Arrhenius model [3] representing each component of the sample i.e. cellulose, hemicellulose, and lignin, is not considered because a single effective model for pyrolysis found to be suitable for large-scale fires [7]. Furthermore, the scale of grassfires is in kilometres, and the grid size used for simulations for a 200x200m plot grassfires is typically of order centimetres [10] to metres [2]. This requires computation time from days to weeks in conjunction to huge computational resources. There is a requirement to reduce the computational time which further compels the use of a single effective Arrhenius model in simulation studies. The Arrhenius model requires the estimation of kinetic parameters: Arrhenius frequency factor (A (s^{-1})), the activation energy of the reaction (E (kJ/mol)), and reaction function ($f(\alpha)$) [11]. The reaction function used in default model is reaction order model [11]. Vyazovkin *et al.* [11] suggested that, it is required to estimate the reaction function for the sample tested in TGA to choose appropriate kinetic parameters. Thus, the Arrhenius model is

$$\dot{w}_{pyr_{Arr}} = Ae^{\frac{-E}{RT}}f(\alpha) \quad . \qquad (2)$$

A non-isothermal method in which sample heated at a constant heating rate chosen to estimate the kinetic parameters. Furthermore, this technique has more relevance in fire science than the isothermal method [12]. Following the ICTAC recommendation [11], the isoconversional technique using the Kissinger-Akahira-Sunose (KAS) equation [13] is used to estimate the activation energy of the sample. A plot of heating rate (β) vs. 1/T at various heating rate gives the activation energy at each isoconversional slope [11,14]. After the activation energy measurement, a master YZ plot [11] is used to select the appropriate $f(\alpha)$, and then estimate the value of A.

3 Sample preparation

The LuH sample divided into three sections: leaf section (LuL) (representing the top part of hay consisting only the leaf, blades, seeds); stem section (LuS) (representing the bottom part of hay consisting only the stem, and nodes); and a mixture section (LuM) (which contains a 50-50 wt.% of LuL and LuS). The samples before any test conditioned at 27°C and 50% relative humidity for more than 36h. The initial mass of sample used in TGA and DSC kept constant to 7.5±0.01mg and evenly distributed. A constant initial sample mass allows the application of Hoffman and Pan's method [15], using DSC observations, to estimate heat of pyrolysis reaction [15].

A Mettler Toledo TGA/DSC 1 and a Mettler Toledo DSC 1, was used for the TGA and DSC experiments. The sample was tested from 30-900°C in an inert atmosphere of nitrogen at 20ml/min gas flow rate for TGA, for a wide range of heating rate (5, 7.5, 10, 20, 50, 100 K/min). The large range of heating rate considered here was chosen to match the heating rate expected in wildfires. The effect of material size and flow rate of the carrier gas found to be negligible on the TGA results [14]. The DSC experiments were conducted at similar conditions to the TGA, except the temperature range was 30-500°C, due to the limitations of the equipment and the sample holder. The Mettler Toledo DSC was calibrated using a standard sapphire disk supplied by Mettler Toledo.

A Thermtest TPS 500 HDA was used to estimate bulk material specific heat capacity using the transient plane source method [15]. Material specific heat is required for the WFDS simulation using the boundary fuel method. A powdered sample of LuM size 0.2-0.6mm with porosity 0.39 [16] was used to ensure repeatability of the experiment and minimise the effect of the random size of pore spaces in the sample.

4 Results

4.1 Experimental Results

Figures 1 & 2, show the mass loss thermogram and rate of fractional conversion for LuL, LuS, and LuM at a representative heating rate of 20 K/min. From Fig. 2, it is visible that there is the presence of two shoulder peaks: 190-240°C (associated with hemicellulose) and 370-490°C (associated with lignin). T_0 and T_f required for the linear model are estimated by visual inspection of Fig. 1 & 2. The T_0 and T_f are observed to be 393K and 633K respectively.

The presence of these shoulder peaks relative to the main peak of cellulose for LuL is found to be similar to the eucalyptus leaves, and pine needles studied in [14]. KAS methodology [13] is used to estimate the activation energy of the species as a function of conversion [11,14]. The activation energy as a function of fractional conversion (α) in the pyrolysis region shown in Fig. 3. It is visible that the activation energy varies with the fractional conversion, which is quite different from timber [11,14] but like litters materials [5,14]. The variation in LuH components suggests a multi-step reaction model corresponding to each component (cellulose, hemicellulose, and lignin). The activation energy for LuS and LuM is linearly increasing with fractional conversion. However, the variation in LuL is different to previous observations for leaves [5,14]. Braga *et al.* [17] observed the change in activation energy for elephant grass in the range of 186-279 kJ/mol. They also observed a similar variation of activation energy with fractional conversion as observed for LuL. The range of variation in activation energy ~100kJ/mol is of the same order as obtained for elephant grass and forest litter materials [5,14,17]. However, for a grassfire simulation, it is reasonable to assume the applicability of a single effective pyrolysis model, due to constraint such as computational grid size, temporal and spatial variability in the vegetation, and a desire to reduce computation speed.

Finally, after estimating activation energy, the reaction model ($f(\alpha)$) and corresponding Arrhenius frequency factor (A) are estimated (Vyazovkin *et al.* [11] describe this process in detail). We applied a truncated Sestak-Berggren model [10] and obtained John-Mehl-Avrami (JMA, with exponent $n = 4$) [11,14] model to be most appropriate for our data. That is,

$$f(\alpha)_{JMA(n=4)} = 4(1-\alpha)[-\ln(1-\alpha)]^{3/4} \quad . \qquad (3)$$

The KAS method is also used to estimate activation energy of moisture evaporation using the default reaction model [7]. Table 1 details the estimated kinetic parameters for LuH components. Furthermore, DSC was used to estimate the heat of pyrolysis required in the linear model. The experiment was performed with the same initial sample mass used in the TGA experiments so as to apply Hoffman & Pan's method [15]. The area between the data and calibration curves in the pyrolysis temperature range (obtained from Fig. 2) gives an estimate for the heat of the pyrolysis reaction. This is shown as a shaded region in Fig. 4, a positive value shows endothermic process. The Fig. 4 shows the second of the two-endothermic process: (1) moisture loss, (2) pyrolysis process; the curve is not horizontal due to inclined calibration curve.

The estimated values of the h_{pyr} are susceptible to the small inter-species variation of the samples affecting the computed area under the curve. Hence, the experiments are repeated ten times for consistent and repeatable observations. Most of the computed area under the curve from Fig. 4, estimates the heat of pyrolysis in the range of 525-575 kJ/kg. However, values in the range of 400-720 kJ/kg were also observed showing a probable variation in the estimation. Hence, a sensitivity analysis for the heat of pyrolysis on the simulations is conducted later to understand the impact of the variation.

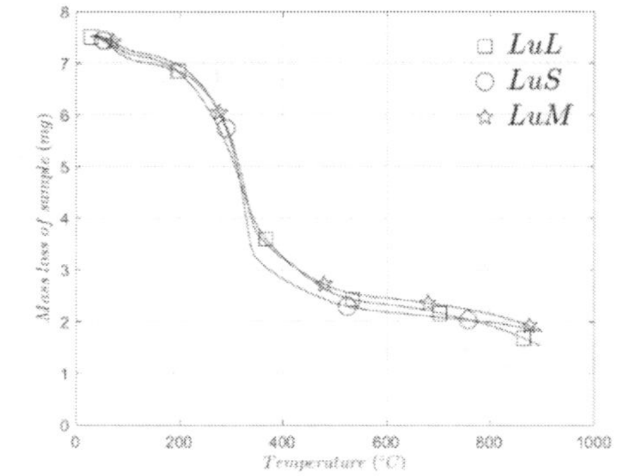

Figure 1: Mass loss thermogram for LuH components observed in the inert atmosphere of nitrogen in TGA at 20 K/min

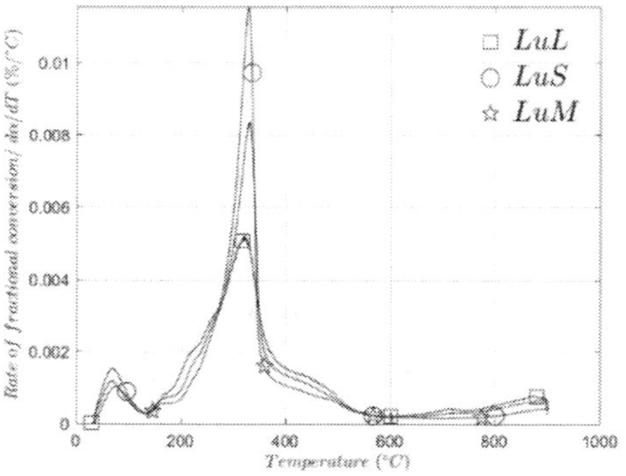

Figure 2: Rate of fractional conversion for LuH components in the inert nitrogen atmosphere at 20 K/min

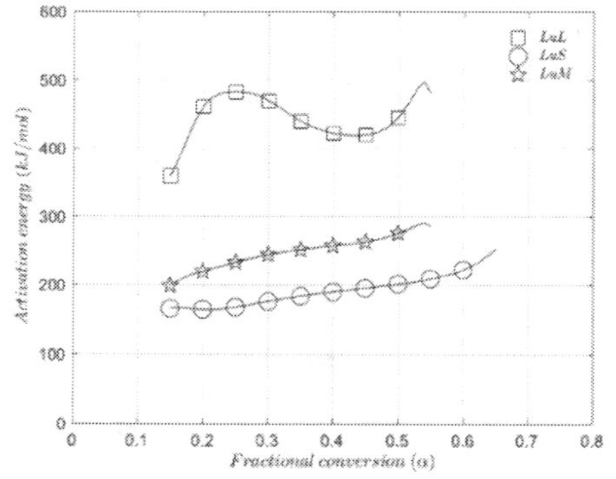

Figure 3: Activation energy as a function of fractional conversion (α) for LuH components by KAS

Variable	LuL	LuS	LuM
Density (kg/m^3)	397.88	497.43	447.66
A_{H_2O} (K$^{1/2}$s^{-1})	1.17e+8[a]	4.45e+6[a]	1.62e+7[a]
E_{H_2O} (kJ/mol)	61.67	48.60	55.20
A_{spec} (s^{-1})	2.3e+34[a]	2.65e+16[a]	1.27e+22[a]
E_{spec} (kJ/mol)	418.33	218.19	270.02

[a] estimated at 20 K/min

Table 1: Estimated kinetic parameters for LuH components

Heat capacity is measured as a function of temperature in the range of 30-150°C. A linear relation of C_P (kJ/kg.K) observed with temperature (T_{HDA} (K)) with r^2 =0.883. Discrepancies in the fit are due to changes in C_P after the moisture loss at 100°C. This is similar to the moisture effects observed by Bakar [14] for pine timber when tested from 30-225°C. Also, similar linear relations of heat capacity with temperature were observed for pine needles [16], for biomass [17], and for wood [18]. The linear equation of C_P at initial relative humidity of 50% is

$$C_p = 0.0096T_{HDA} - 1.802 \ . \tag{4}$$

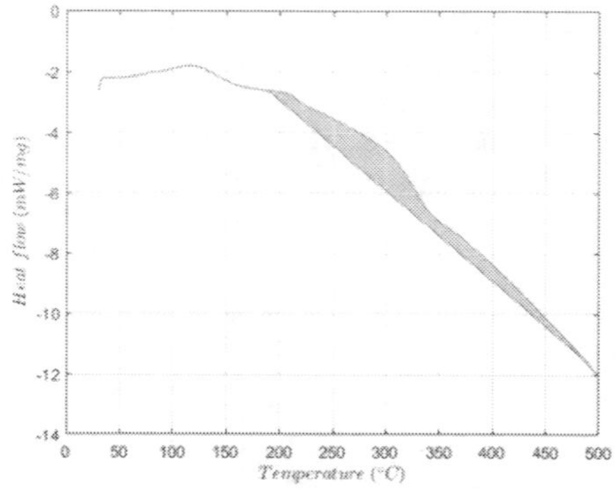

Figure 4: Heat flow of LuL observed in the inert nitrogen condition in DSC at 20 K/min

4.2 Simulated results

Simulations of the thermogravimetric experiments are carried out using Wildland-urban interface Fire Dynamic Simulator (WFDS) version 6.0.0. The simulation domain is x = -0.005, 0.005, y = -0.005, 0.005, z = 0, 0.01 (m) and the vegetation patch on z = 0 is x= -0.0025, 0.0025, and y = -0.00125, 0.00125 (m). The boundary of the domain is assumed to be radiant panels which heats from 30-900°C to match the heating rate of experiment. The fuel loading (without moisture) is 0.5411 kg/m^2; the vegetation height is 0.004 m; the vegetation density is defined in Table 1.

The linear model simulates the moisture evaporation section as a step change at 100°C [6] and hence, thermal degradation simulated from 120-900°C. Figure 5 shows the comparison between the experimental fractional mass loss observed in TGA and simulated fractional mass loss using WFDS with the linear model. Figure 6 shows a similar comparison using the Arrhenius model. It is quite visible from Fig. 5 & 6 that the linear model simulates the pyrolysis section 220-360 °C accurately.

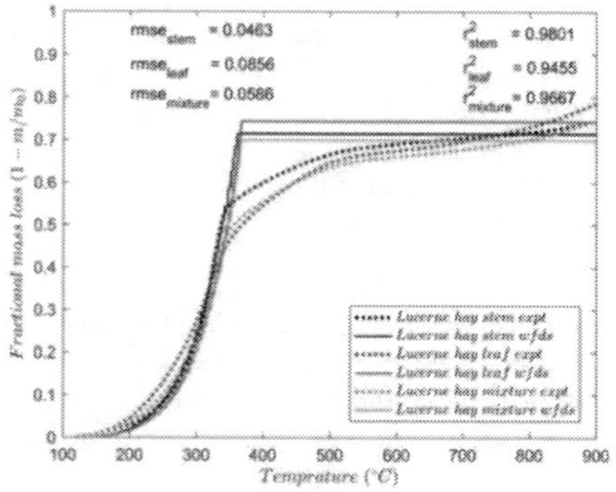

Figure 5: Fractional mass loss comparison for experimental & simulated case for LuH using the linear model at β=20 K/min

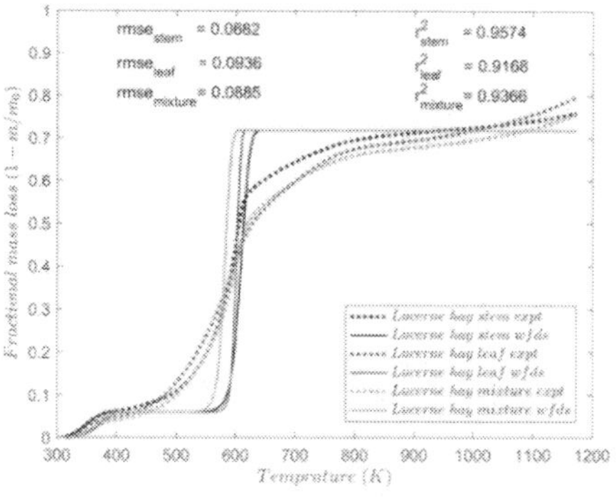

Figure 6: Fractional mass loss comparison for experimental & simulated case for LuH using the Arrhenius model at β=20 K/min

Similarly, in the Arrhenius model there is a good prediction for moisture section. While in the pyrolysis section, the Arrhenius model does a poor job due to the use of a single best-fit model for a complex multi-step thermal degradation reaction [11,12]. Hence, the value of combined effect results similar value of root mean square error (rmse).

Figure 7 shows the sensitivity of the linear model with the variation in h_{pyr} for LuM at β=20 K/min. The result show a maximum rmse = 0.0728 when h_{pyr} varied in ±30% while the rmse is 0.0546 when the mean h_{pyr} is 550 kJ/kg, obtained from DSC. The variation in h_{pyr} observed in DSC does not show significant impact on the accuracy of the linear model.

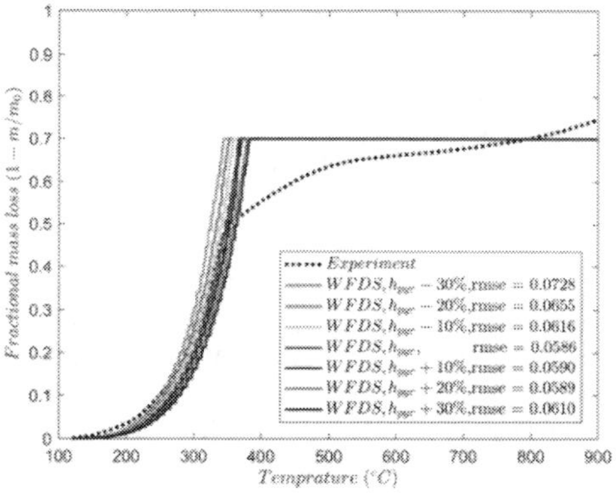

Figure 7: Sensitivity analysis for heat of pyrolysis for LuM at β=20 K/min

5 Conclusions

The present work appraises two competing models, namely the linear and Arrhenius models, for pyrolysis in physics-based fire simulations of fires in herbaceous fuels. The linear model shows perform better than the single step Arrhenius model for pyrolysis at microscale (e.g. TGA). The results also show that linear model is fairly independent of the estimation and inter-species variation observed in DSC measurement. Further, it also shows that it is independent on LuH components, which suggests the usage of one h_{pyr} value to represent LuH vegetation. However, further verification and validation of these models at medium- to larger-scale samples, i.e. cone calorimeter, room-scale, are required to test the model applicability for large-scale bushfire simulation. Also, it will be required to test the applicability of these models for forest litters which are significantly different from herbaceous fuels like LuH.

6 Acknowledgment

We would like to acknowledge the Bushfire and Natural Hazard Cooperative Research Centre (BNHCRC), Melbourne, Australia for providing us financial support to carry out this work. We also like to thank Mr. Lyndon Macindoe and Mr. Philip Dunn for technical support and sample collection.

References

[1] A.L. Sullivan, Int. J. Wildland Fire 18 (2009) 349-68.

[2] W. Mell, M. A. Jenkins, J. Gould, P. Cheney, Int. J. Wildland Fire 16 (2007) 1-22.

[3] H. Niu, N. Liu, Effect of Particle Size on Pyrolysis Kinetics of Forest Fuels in Nitrogen, Fire Saf. Sci., 11, 2014, p. 1393-405.

[4] D.H. Ashton, Aust. J. Bot. 23 (1975) 413–33.

[5] O.P. Korobeinichev, A.A. Paletsky, M.B. Gonchikzhapov, I.K. Shundrina, H. Chen, N. Liu, Procedia Eng. 62 (2013) 182-93.

[6] D. Morvan, J. Dupuy, Combust. Flame 138 (2004) 199-210.

[7] D. Morvan, J. Dupuy, Combust. Flame 127 (2001) 1981-94.

[8] K. McGrattan, S. Hostikka, J. Floyd, R. McDermott, C. Weinschenk, K. Overholt, Fire Dynamics Simulator Technical Reference Guide Volume 1: Mathematical Model, National Institute of Standards and Technology, Special Publication 1018, Gaithersburg, MD, 2013, p. 149.

[9] W. Mell, A. Maranghides, R. McDermott, S.L. Manzello, Combust. Flame 156 (2009) 2023-41.

[10] D. Morvan, S. Meradji, W. Mell, Fire Saf. J. 58 (2013) 195-203.

[11] S. Vyazovkin, A.K. Burnham, J.M. Criado, L.A. Pe´rez-Maqueda, C. Popescu, N. Sbirrazzuoli, Thermochim Acta. 520 (2011) 1-19.

[12] F. Richter, G. Rein, Fire Saf. J. 91 (2017) 191-9

[13] H.E. Kissinger, Anal. Chem. 29 (1957) 1702-6

[14] R. Wadhwani, D. Sutherland, K.A.M. Moinuddin, P. Joseph, J. Therm. Anal. Cal. (2017) (doi: 10.1007/s10973-017-6512-0)

[15] A.S.A. Bakar, *Characterization of Fire Properties for Coupled Pyrolysis and Combustion Simulation and Their Optimised Use*, PhD Thesis, Victoria University, Melbourne, Australia, 2016.

[16] N. Ouchiyama, T. Tanaka, Ind. Eng. Chem. Fundam 23(4) (1984) 490-3

[17] R. Braga, D. Melo, F. Aquino, J. Freitas, M. Melo, J. Barros, M. Fontes, J. Therm. Anal. Cal. 115(2) (2014) 1915-20.

[18] R. A. Susott, Forest Sci. 28 (1982) 839-51.

[19] W.C.R. Chan, M. Kelbon, B. B. Krieger, Fuel, 64 (1985) 1505-13.

[20] K. Radmanović, I. Đukić, S. Pervan, Drvna industrija, 65 (2014) 151-7.

11th Asia-Pacific Conference on Combustion,
The University of Sydney, NSW Australia
10th -14th December 2017.

The Influence of Reynolds Number on Soot Concentration in Turbulent Non-premixed Jet Flames

S. M. Mahmoud [1,3], G. J. Nathan [1,3], Z. T. Alwahabi [2,3], Z. W. Sun [1,3], P. R. Medwell [1,3], and B. B. Dally [1,3]

[1]School of Mechanical Engineering

[2]School of Chemical Engineering

[3]Centre for Energy Technology

The University of Adelaide, South Australia, 5005, Australia

Abstract

Soot volume fraction (*SVF*) was measured in six attached turbulent non-premixed jet flames of a C_2H_4-H_2-N_2 fuel mixture using the Laser-Induced Incandescence (LII) technique. The six flames comprise two sets with exit strain rates of 4100 and 7500 s^{-1}, and one individual flame with an exit strain rate of 12,900 s^{-1}. Within each set, the exit Reynolds number was changed both by varying the jet diameter and the fuel exit velocity. Measurements of the mean and radially-integrated *SVF* reveal a weak inverse dependence on the exit Reynolds number. The total soot yield was also found to be a strong function of both the exit strain and the flame volume, but to be almost independent of the exit Reynolds number. A non-negligible effect of buoyancy on *SVF* was also deduced from the global correlations.

1 Introduction

Soot formation and evolution in a flame is a complex process that is controlled by parameters such as pressure, temperature, fuel type, mixture fraction and mixing rate [1,2], most of which are interdependent in turbulent flows. Amongst these dependencies, the relationship between soot concentration and residence time, and consequently the mixing characteristics within a flame, is of crucial importance. For example, sufficiently fast mixing in diffusion flames can be effective in suppressing soot formation through altering the soot residence time in fuel rich and high temperature zones in the flames [3], while the soot volume fraction in turbulent flames scales inversely with the global residence time [4]. Additionally, the dependence of soot on turbulent mixing is due to the overlap in the timescales of soot formation and oxidation, with those of the turbulent mixing, typically at the order 10 milliseconds [5]. Therefore, these dependencies are non-linearly coupled in turbulent flames so that their resolution requires simultaneous measurements of all the controlling parameters. Also, the difficulty of performing these measurements in turbulent conditions has resulted in previous investigations being predominantly limited to characteristic, or global values, rather than local ones.

Corresponding author. Fax: +61 8 8313 4367
E-mail address: saleh.mahmoud@adelaide.edu.au

The interaction of soot and turbulence in jet flames have been the focus of several recent experimental studies. In one experimental study, Köhler *et al.* provided data from simultaneous soot and flow field measurements in turbulent diffusion jet lifted flames [6]. Despite the value of these data, it is desirable to perform measurements in attached flames, since lifted flames are more difficult to predict and strongly inhibit soot formation through partial premixing at the base. Qamar *et al.* performed comprehensive measurements of soot in piloted ethylene and natural gas turbulent jet flames [7], and investigated the effect of global mixing rate on soot concentration in Liquefied Petroleum Gas, LPG flames, in another publication [4]. However, the work of Qamar *et al.*, despite its significance, is not well suited to the development of validated models due to the use of different types of burners whose exit flow characteristics are both complex and difficult to characterise [4]. In addition, the kinetics of soot evolution for the fuel used is not as well understood as compared with ethylene, thereby introducing some challenges to model validation of their measurements.

On the other hand, the isolated effects of Reynolds number on soot evolution in turbulent jet flames has not been investigated explicitly. While Lee *et al.* [8] measured soot volume fraction in turbulent flames for various exit Reynolds number, their study did not isolate the Reynolds number from strain rate because the fuel exit velocity was varied for a constant burner diameter [8]. In order to isolate the separate influences of exit strain rate and Reynolds number, it is necessary to vary both the nozzle diameter and exit velocity [9]. However, the jet exit momentum flux, G, also varies with varying exit velocity, so that the ratio between jet momentum and buoyancy forces will also depend on whether the Reynolds number is varied at constant nozzle diameter or at constant velocity. Therefore, there is a need for new measurements that help to separate the effects of Reynolds number on soot from other parameters in turbulent flames.

In light of the points made above, the present study reports, through measurement of *SVF* in a set of turbulent non-premixed jet flames, the isolated effects of jet exit Reynolds number (Re_D) on the mean and integrated *SVF*, as well as the global correlations between soot and other controlling parameters. This study, along with the experimental database posted on the International Sooting Flame (ISF) website [10], aims to develop both a thorough understanding of soot evolution and predictive models of soot evolution in well-characterized turbulent non-premixed jet flames.

	Flame	Central jet diameter (D)	Mean jet exit velocity (U)	Exit strain rate (U/D)	Exit Reynolds number $Re_D=\rho UD/\mu$	Exit Froude number $Fr_e=U^2/gD$	Mean flame length L_f	Non-dim. flame length ξ_L	Flame Volume $V_f=\pi W_f^2 L_f/12$	Global mixing rate $(1/\tau_G)$
	error		(±2%)	(±2%)	(±2%)	(±4%)	(±8%)	(±10%)	(±24%)	(±24%)
	units	(mm)	(m/sec)	(s^{-1})	(-)	(-)	(mm)	(-)	(litres)	(s^{-1})
SET A	EHN-3	8	31.5	4,000	15,000	12,600	1,130	3.8	9.2	12.5
SET A	EHN-5	5.8	23.5	4,100	8,000	9,700	910	4.6	4.2	10.7
SET A	EHN-6	4.4	19	4,300	5,000	8,400	700	4.9	2.2	9.5
SET B	EHN-2	5.8	42.4	7,300	15,000	31,600	940	3.2	4.8	16.9
SET B	EHN-4	4.4	33.7	7,600	9,000	26,300	710	3.4	2.4	15.5
	EHN-1	4.4	56.8	12,900	15,000	74,700	710	2.4	2.4	26.1

Table 1. Flow conditions employed for the series of attached jet *EHN* ($C_2H_4/H_2/N_2$) flames.

2 Experimental Arrangement

2.1 Flames

The flames investigated in this paper are six attached turbulent non-premixed jet flames, each burning an identical mixture of ethylene-hydrogen-nitrogen, at a ratio of 40:41:19 by volume, and at ambient pressure and temperature (101.325 kPa and 21.1°C). These flames, termed "*EHN* flames", are classified into two sets, in addition to one individual flame, and were partly characterized in recent studies by the authors [9,11]. Set A comprises flames *EHN*-3, *EHN*-5 and *EHN*-6, which share an average exit strain rate of U/D= 4,100 s^{-1} but have different exit bulk mean Reynolds numbers. Set B comprises flames *EHN*-2 and *EHN*-4, which share an average exit strain of U/D= 7,500 s^{-1} and have different exit bulk mean Reynolds numbers. The sixth flame, *EHN*-1, has similar composition and an exit strain of 12,900 s^{-1} [9,11,12]. This arrangement was achieved through variation of both the jet diameter (D) and the fuel exit velocity (U) in each set. The exit strain rate, U/D, was chosen ahead of the global mixing rate, $1/\tau_G$, in the parametric study because it can be more accurately determined a-priori. W_f is measured to be the widest part of the flame. Experiments were conducted as carefully as possible to maintain constant exit strain in each set, although variations in flow meters and flow velocities resulted in slight deviations in strain rate, all within 5% of the average exit strain for each set. The exit Reynolds number is calculated based on the jet diameter with ρ, being the exit fuel density, of 0.725 kg/m³, and μ is the exit fuel viscosity of 1.21×10^{-5} kg/m/s. Full details of the flow conditions for *EHN* flames are provided in Table 1.

2.2 Burner Setup

The set of burners used in this study comprise three round aluminum tubes with inner diameters, D, of 4.4 mm, 5.8 mm, and 8 mm. All the pipes have a wall thickness of 1 mm, a length of 385 mm, and a tapered tip. All flames were surrounded by a stream of co-flowing air at a velocity of 1.1 m/s. A contraction with a square cross-section of dimensions 150 mm by 150 mm surrounded the burner and supplied the co-flowing air at ambient temperature. The burners were mounted on a traverse to enable measurements to be performed at various heights throughout the flame. An extraction hood was traversed with the burner to keep the distance between the flame tip and the hood constant and to avoid any influence of the exhaust extraction on the flames.

2.3 Laser-Induced Incandescence (LII)

The laser induced incandescence (LII) technique was used to measure soot volume fraction with an experimental setup, as reported in detail previously by the authors [9,11]. Soot incandescence was recorded with an ICCD camera and converted to *SVF* by calibration against a flat premixed ethylene-air flame stabilized on a McKenna burner. A pulsed Nd:YAG laser firing at 1064 nm was used to generated LII, while an ICCD camera was used to collect the incandescence through a 430-nm (10 nm bandwidth) interference filter. This arrangement was used to minimise interference from flame radiation and suppress the C_2 laser-induced emission [13]. The gate-width of the LII camera was set to 50 ns, with prompt timing relative to the LII excitation, because this has been proven to decrease the sensitivity of the signal to the soot particle size [14]. Soot extinction in the flat-flame was performed using a continuous-wave (CW) diode-pumped-solid-state laser (DPSS) operating at 1064 nm. It was assumed that the scattering contribution to beam extinction is negligible.

The minimum detection limit in the collected data is 1 parts per billion (ppb). The uncertainty of the measurements is 25%, which comprises of losses from the laser energy throughout the optics arrangement, uncertainties within the absorption function, and fluctuations in the laser power from the soot extinction measurements.

The LII laser sheet, with a height of approximately 24 mm, was aligned vertically with the burner axis and transmitted horizontally through the flame. The laser fluence was kept in the plateau region of the LII response curve at a fluence above 0.5 J/cm², to minimize the effect of beam "steering" and the influence on the LII signal of fluence variation from attenuation through the flame. Statistical data were obtained from the left half (laser-in side) of the flame only. Hot wire velocimetry measurements of the non-reacting flows were conducted at the exit plane to confirm the symmetry of the flow [11, 12]. Five hundred instantaneous images were collected at thirteen locations throughout the flame length. Accounting for the average thickness of the beam, the images were binned into arrays of 3×3 pixels so that the final volume of soot concentration measurements was a cube of 0.72 × 0.72 × 0.75 mm in length, width, and thickness, respectively.

3 Results

3.1 Axial distribution of *SVF*

The axial distribution of mean *SVF* for constant strain-rate (Set A) flames *EHN*-3, *EHN*-5, and *EHN*-6, corresponding to Re_D=15000, 8000 and 5000, respectively, normalized by their respective flame length is presented in Fig. 1. It can be seen that increasing Re_D leads to a lower mean *SVF* along the burner axis, with a 40 % reduction in the maximum mean SVF from Re_D = 5,000 to Re_D = 15,000. In addition, *SVF* profiles are found to be closely characterized by a Gaussian distribution, with a peak occurring in the range $0.58 < x/Lf < 0.64$. This trend is in agreement with those of the radiant heat profiles reported by Dong *et al.* [12] which highlights the interdependence of soot on flame radiation and temperature.

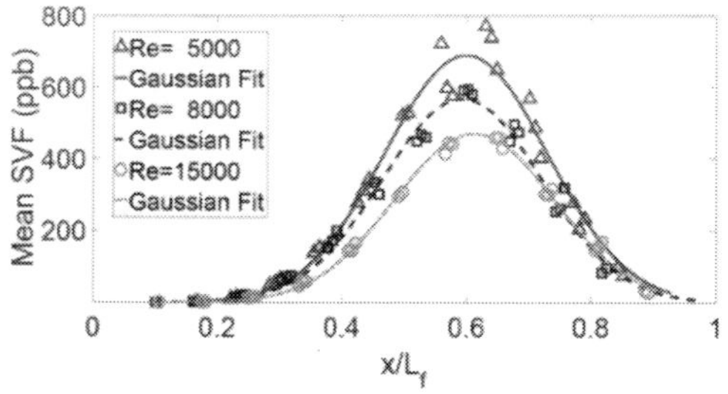

Figure 1. Axial profiles of centreline distribution of mean SVF as a function of the normalized length for Set A flames *EHN*-3 (Re=15,000), *EHN*-5 (Re=8,000), and *EHN*-6 (Re=5,000). Symbols represent experimental data, while the profiles are Gaussian fits to the experimental data.

Consistent with the observation of Lee *et al.* [8], the axial distance to soot inception here increases with Re_D. As shown in Table 1, the non-dimensional flames length for all five flames is in the range $2 < \xi_L < 10$, which implies that all flames are in the transitional regime between the momentum-driven and buoyancy-driven regimes, making them subject to both forces [15,16]. This deduction is also in agreement with Qamar *et al.*'s observation that the position of peak *SVF* moves further upstream for the buoyancy-dominated flame [4].

It should also be noted that the trends in Set B are consistent with those in Set A, although they exhibited lower mean *SVF* values overall, consistent with a higher exit *U/D* than those of Set A.

3.2 Radial distribution of *SVF*

The radial distributions of the mean *SVF* are presented at five different axial locations for Set A flames in Fig. 2. The Gaussian distribution provides a suitable fit to the radial profiles of mean *SVF* except for some locations where it yields a non-zero gradient at the axis.

Figure 2 shows that the radial *SVF* profiles exhibit a similar shape at a given normalized axial location, where the concentration of soot decreases with an increase in exit Re_D. An increase in the exit Reynolds number also leads to a broadening in the radial *SVF* profiles at a given axial position, in agreement with the observation by Copalle *et al.* [17] and Lee *et al.* [8] for turbulent jet flames. This normalized axial location of the soot inception, which is upstream from the location at which buoyancy becomes significant, can be explained by the constant residence time to the point of inception, which scales inversely with the constant exit strain for *EHN*-3, *EHN*-5 and *EHN*-6.

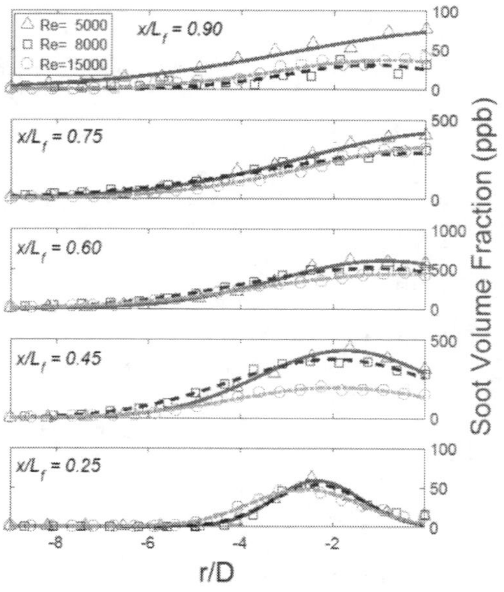

Figure 2. Radial profiles of the mean SVF at various downstream locations, for Set A flames. Data are reported only for the laser-in side of the flames. Symbols represent experimental data while the lines are best fits with Gaussian functions

3.3 Integrated *SVF*

Figure 3 present the axial profiles of the radially-integrated soot volume fraction, SVF_{rad}, for the flames in Set A. Here, SVF_{rad} represents the total volume of soot per unit length in mm², at a given normalized axial location. This was measured by integrating the mean *SVF* over the cross-section of the flame and over the image height of 18mm, at a given measured location i.e.

$$SVF_{rad} = \int SVF_{\text{mean}} \times 2\pi r dr \tag{1}$$

where *r* is the radial distance from the burner axis.

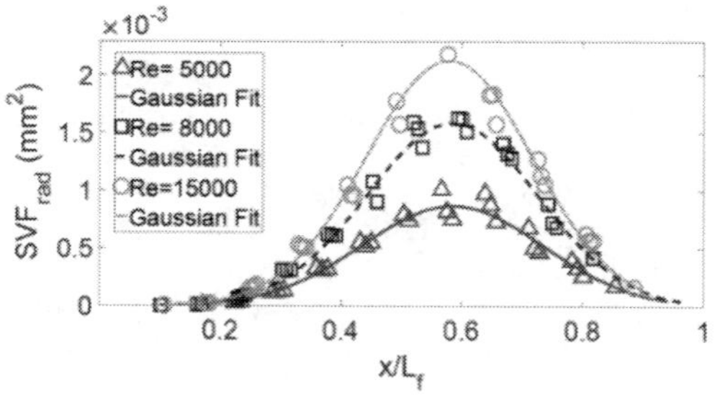

Figure 3. Axial profiles of radially integrated SVF (equation 1), in mm², over full flame length from Set A flames. Symbols represent experimental data, and profiles are Gaussian fits to the experimental data.

Similar to the mean axial and radial *SVF* profiles in Figs 1 and 2, SVF_{rad} profiles are well characterized by a Gaussian fit, with a correlation coefficient of ~97%. On the other hand, the radially integrated soot in Fig. 3 is found to scale directly with Re_D due to the increase in the flame volume with increasing flowrate, in contrast to the inverse relationship between Re_D and mean *SVF* observed in Figs. 1 and 2. Figure 3 also show that the location of the maximum radially integrated *SVF* for all flames occurs in the range of $0.58 < x/L_f < 0.62$, consistent with the observations for the centreline mean *SVF* in Fig. 1.

3.4 Global Correlations

The correlations between soot productivity and potential controlling parameters were assessed to provide more insight into which of these parameters best characterise the soot in these flames. Figure 4 presents the total soot volume in the flame, SV_{total}, as a function of the fuel flowrate for several values of U/D. Here, SV_{total} represents the volume-integrated SVF in mm^3, that is, the total soot volume in the flame obtained by integrating SVF_{rad} over the full flame length, using the equation

$$SV_{total} = \int SVF_{\mathrm{rad}} \cdot dL \qquad (2)$$

Two important points are evident. First, the soot productivity is independent of the Reynolds number, within the bounds of the error bars. Secondly, the data do not collapse with the Reynolds number, but retain instead a dependence on the strain rate. That is, each case of U/D results in a different normalized total soot volume. It can also be seen from Table 1 that the soot productivity increases with a reduction in Froude number, i.e. with an increase in the role of buoyancy.

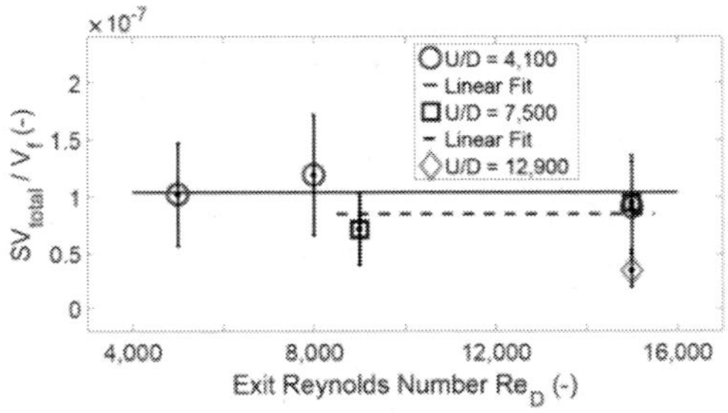

Figure 4. Total soot volume (equation 2) normalized by the flame volume as a function of exit Reynolds number for three values of exit strain, for the conditions reported in Table 1. The error bars correspond to the experiment uncertainty derived from the RMS of the scatter of the repeatability measurements

On the other hand, Fig. 5 presents the same soot productivity parameter as a function of Froude number, $Fr_e = U^2/gD$. This scaling does collapse the data, in contrast to Figure 4, while the soot productivity is also seen to decrease with Fr_e. That is, the soot productivity increases with an increased role of buoyancy, which is consistent with Fig. 4 and with the trends that have been deduced previously from measurements of flame temperature and NOx emissions [18,19].

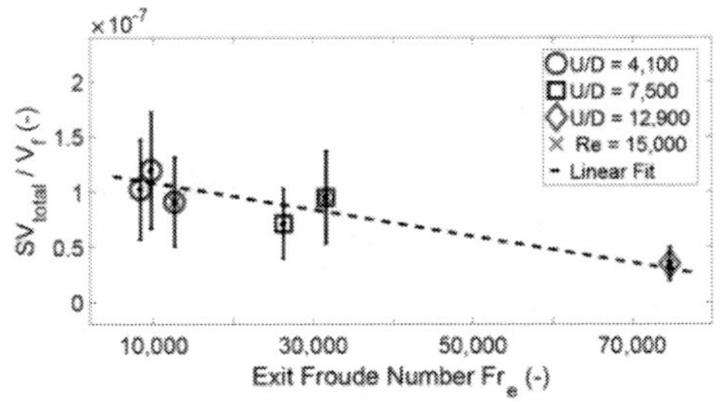

Figure 5. Total soot volume (equation 2) normalized by the flame volume as a function of exit Froude number for three values of exit strain, for the conditions reported in Table 1. The error bars correspond to the experiment uncertainty derived from the RMS of the scatter of the repeatability measurements.

4 Conclusions

The first detailed measurements of soot volume fraction in a series of turbulent non-premixed jet flames with constant exit strain rate and varying Reynolds number have been reported for a fuel mixture of C_2H_4-H_2-N_2. A weak and inverse relationship is found between the exit Reynolds number and mean SVF, which is consistent with these flames spanning the transitional regime between laminar and turbulent flames. Radial measurements of SVF show that soot concentration scales with a slight inverse dependence on exit Reynolds number in most regions of the flame.

The total soot in each flame is found to be a strong function of the flame volume, and the exit strain. In addition, the soot productivity (i.e. total soot volume normalised by flame volume) is almost independent of the exit Reynolds number, but decreases with an increase in Froude number. However, this does not necessarily imply that the soot concentration is independent of Reynolds number. This is because the flame volume itself may carry an implicit dependence on Re_D for these cases. In addition, it should be noted that the role of buoyancy in other applications should be assessed on a case-by-case basis, since buoyancy is typically not significant in larger, confined industrial flames.

5 Acknowledgment

The authors gratefully acknowledge the support of the Australian Research Council for their funding through Grant DP130100198.

References

[1] R.J. Santoro, H.G., Symp. (Int.) Combust. 20 (1985) 997–1006.
[2] R.J. Santoro, T.T. Yeh, J.J. Horvath, H.G. Semerjian, Combust. Sci. Technol. 53 (1987) 89–115.
[3] K.C. Lin, G.M. Faeth, Prop. Power 12 (1996) 691-698.
[4] N.H. Qamar, G.J. Nathan, Z.T. Alwahabi, D.K. King, Proc. Combust. Inst. 30 (2005) 1493–1500.
[5] M.E. Decroix, W.L. Roberts, Combust. Sci Tech. 160 (2000) 165-189.
[6] M. Köhler, K.P. Geigle, W. Meier, B.M. Crosland, K.A. Thomson, G.J. Smallwood, Appl. Phys. B 104 (2011) 409–425.
[7] N.H. Qamar, Z.T. Alwahabi, Q.N. Chan, G.J. Nathan, D. Roekaerts, K.D. King, Combust. Flame 156 (2009) 1339–1347.
[8] S.Y. Lee, S.R. Turns, R.J. Santoro, Combust. Flame 156 (2009) 2264–2275.
[9] S.M. Mahmoud, G.J. Nathan, Z. T. Alwahabi, Z. W. Sun, P. R. Medwell, and B. B. Dally, Proc. Combust. Inst. 36 (2017) 889–897.
[10] International Sooting Flame (ISF) Workshop. http://www.adelaide.edu.au/cet/isfworkshop/data-sets/turbulent.
[11] S.M. Mahmoud, G.J. Nathan, P.R. Medwell, B. B. Dally, Z. T. Alwahabi, Proc. Combust. Inst. 35 (2015) 1931-1938.
[12] X. Dong, G.J. Nathan, S. Mahmoud, P.J. Ashman, D.H. Gu, B.B. Dally, Combust. Flame 162 (2015) 1326–1335.
[13] C. Schulz, B.F. Kock, M. Hofmann, et al., Appl. Phys. B 83 (2006) 333–354.
[14] R.L. Vander Wal, Appl. Opt. 35 (1996) 6548–6559.
[15] M.G. Mungal, P.S. Karasso, A. Lozano, Combust. Sci. Technol. 76 (1991) 165–185.
[16] G.J.R. Newbold, G.J. Nathan, R.E. Luxton, Combust. Sci. Technol. 126 (1997) 71–95.
[17] A. Coppalle, D. Joyeux, Combust. Flame 96 (1994) 275–285.
[18] S.R. Turns, F.H. Myhr, Combust. Flame 87 (1991) 319–335.
[19] G.J.R. Newbold, G.J. Nathan, D.S. Nobes, S.R. Turns, Proc. Combust. Inst. 28 (2000) 481–487.

11th Asia-Pacific Conference on Combustion,
The University of Sydney, NSW Australia
10th -14th December 2017.

PDF-PBE for Particle Dispersion in a Turbulent Round Jet

Fatemeh Salehi[1], Matthew J. Cleary[1], Assaad R. Masri[1], Andreas Kronenburg[2]

[1] School of Aerospace, Mechanical and Mechatronic Engineering, The University of Sydney

Sydney 2006, NSW, Australia

[2] Institut für Technische Verbrennung, University of Stuttgart

70174 Stuttgart, Germany

Abstract

The probability density function form of the population balance equation (PDF-PBE) is employed in the large eddy simulation context to simulate the dispersion of droplets in a round turbulent jet. Test cases studied here include the injection of 60 and 90 μm droplets into a turbulent air jet with Re of 20,000 and 30,000. A newly developed *Stokes binning method* is used to account for the differential inertia of droplets of different sizes and to partition the PDF-PBE into classes based on the droplet Stokes numbers. The method leads to accurate modelling of the different trajectories that are followed by particles with different sizes in a computational tractable manner. A Langevin model is employed to reconstruct the subgrid scales in the carrier flows. The PDF-PBE model reproduces the radial dispersion and the mean droplet velocity fields very well. It is found that the direct effects of the subgrid scales on the dispersion of 60 and 90 μm droplets are insignificant in the present flow cases.

1 Introduction

This paper is part of a larger project aimed at developing reliable predictive capabilities for particles laden flows which are highly relevant for a wider range of applications that include particle dispersion in the atmosphere, pulmonary drug delivery, fire suppression systems and liquid spray fuel injection in diesel engines and gas turbines. Particles here refer to solid or liquid fragments suspended in a turbulent flow and undergoing complex reacting, deformation and dispersion processes while interacting with the surrounding field. Dispersion of inertial elements in turbulent carrier flows has a strong influence on these processes and hence must be validated separately. Droplets and solid particles mass density are, in general, vastly greater than that of the carrier phase and hence they cannot be assumed to simply follow the gas. The inertia effects lead to the development of inhomogeneities in the droplets and solid particles' spatial distribution and also dissipate the turbulence of the carrier.

In the concept of tow-phase flows, statistical approaches provide a probabilistic description of the dispersed phase in terms of droplet or particle properties and they have been widely adopted for modelling such flows [1-4]. Conventionally, when droplet or particle breakage or agglomeration is considered, the number of numerical particles grows or shrinks accordingly leading to uncontrolled computational load and it should be noted that with each new particles property (e.g. shape and morphology), the difficulty of controlling the computational load grows.

Several reviews on the use of the population balance equation (PBE) concept for a range of physical applications exist in the literature [5-8]. The PBE is mostly viewed as a stochastic approach and it capable to for modelling polydispersed droplets and particles in a computationally tractable way. The PBE presents evolution of a number distribution of a dispersed phase as a function of time and space as well as the physical properties, such as droplet size, introduced as internal coordinates of the number density function thus allowing representation of polydispersed distributions. Rigopoulos [5,6] extended the PBE method to turbulent flows by introducing the probability density function (PDF) of the PBE (termed PDF-PBE) for spherical solid particle cases where the relative velocity between the continuous and dispersed phases is neglected [5]. Salehi et al. recently developed a PDF-PBE approach for polysized and polyshaped droplets that, for the first time, explicitly addresses inertial effects [9]. An approach referred to as the method of *Stokes binning* was proposed to account for differential inertia of different sized droplets and particles in a computationally tractable way. In this approach, particles accelerate due to fluid dynamics forces associated with an averaged Stokes number in each bin.

The objective of this paper is to test the performance of the PDF-PBE model with Stokes binning against the experiment of Kennedy and Moody [10] for dispersion of polysized droplets in a turbulent round jet at different Reynolds numbers. In addition, the effects of subgrid scales on droplet dispersion are explored.

2 Methodology

This study uses a hybrid approach. Large eddy simulation (LES) is employed for the carrier phase and a PDF version of the PBE is chosen for the dispersed phase. The fundamental equations are presented here but more details can be found in [9].

2.1 The Population Balance Equation

The PBE governs the evolution of the particle number density function, n, in a phase space, R. The latter is decomposed into the coordinate space, $\boldsymbol{x} = (x_1, x_2, x_3)$, and the property (or state) space, $\boldsymbol{\xi} = (\xi_1, ..., \xi_M)$, which can include particle size, shape, surface area and so on.. $n(\boldsymbol{x}, \boldsymbol{\xi}, t) d\boldsymbol{x} d\boldsymbol{\xi}$ is defined as the

Corresponding author. Fax: +61-02-000-000
E-mail address: Fatemeh.salehi@sydney.edu.au

ensemble of entities having coordinates in the range $(d\boldsymbol{x}, d\boldsymbol{\xi})$ around $(\boldsymbol{x}, \boldsymbol{\xi})$ at a given time, t. Using this definition, the PBE is written as [11]

$$\frac{\partial n}{\partial t} + \nabla_x(\boldsymbol{u}n) + \nabla_\xi(\dot{\boldsymbol{\xi}}n) = w, \tag{1}$$

where $\boldsymbol{u}(\boldsymbol{\xi}; \boldsymbol{x}, t)$ and $\dot{\boldsymbol{\xi}}(\boldsymbol{\xi}; \boldsymbol{x}, t)$ are the velocities in the physical and the property spaces, respectively. The source term, w, is the net rate of change of number density due to discrete births and deaths of droplets or particles as a result of breakages and agglomeration. It is typically a highly non-linear function of the interactive forces between the carrier and dispersed phases. In the PDF-PBE, however, this non-linear term appears naturally in closed form. The third term on the left hand side of Eq. (1) accounts for growth and shrinkage due to condensation and evaporation leading to changes in droplet size and/or shape. In this study, which focuses on dispersion of spherical droplets without breakage, agglomeration or evaporation, the property space includes on the volume so that $\boldsymbol{\xi} = \xi_1 = V$ and w and $\dot{\boldsymbol{\xi}}$ are neglected.

Some form of averaging or statistical emulation of the turbulent field is required to solve Eq. (1) in turbulent flows. Here, a PDF version of the PBE is adopted for the dispersed phase and its turbulent distribution is represented discretely and stochastically by an ensemble of Lagrangian notional particles in a Monte Carlo simulation. In general, the number density function represents a range of droplet sizes with different slip velocities. The different trajectories of droplets with different inertia eliminates the possibility of representing all droplet sizes on each notional particle. To address this, the PDF-PBE is divide into classes, where each class represents the turbulent distribution of the number density function of droplets that have similar inertial interactions with the carrier phase. The notional particles in each class evolve according to stochastic differential equations [1]

$$d\boldsymbol{X} = \boldsymbol{U}dt \tag{2}$$

$$d\boldsymbol{U} = \boldsymbol{A}dt. \tag{3}$$

Here, the upper case symbols denote values associated with notional particles and are the sample space of the corresponding physical variables in Eq. (1). The acceleration, $\boldsymbol{A}$ is due to the drag force and hence Eq. (3) can be written as

$$d\boldsymbol{U} = \frac{1}{\langle\tau_p\rangle}(\boldsymbol{v} - \boldsymbol{U})dt, \tag{4}$$

Where $\boldsymbol{v}$ is the instantaneous carrier velocity and $\langle\tau_p\rangle$ is the mean (across the class of droplets) response time due to fluid-droplet interaction forces. In the concept of LES, $\boldsymbol{v}$ can be decomposed into the filtered velocity, $\bar{\boldsymbol{v}}$, and the subgrid velocity, $\acute{\boldsymbol{v}}$. If the droplet response time is large compared to the time scales of the turbulent flow, the direct effects of subgrid scales are insignificant and hence a subgrid model in the stochastic equations is not required [12]. In cases where the droplet response time is of the same order of magnitude as the flow time scales, a model is required for the unresolved velocity. Here, the Langevin model [1, 2] is employed to reconstruct the residual (or subgrid scale) carrier fluid velocity. Following Pozorski and Apte [13], the subgrid velocity is computed using

$$d\acute{v}_i = -\frac{\acute{v}_i}{\tau_L}dt + \sqrt{\frac{4K_{sgs}/3}{\tau_L}}d\omega_i, \tag{5}$$

where K_{sgs} is the subgrid kinetic energy, $d\omega$ is an increment of the Wiener process and τ_L is the characteristic time scale which differs in the directions parallel, $\tau_{L,||}$ and perpendicular, $\tau_{L,\perp}$ to the relative velocity. They are defined as

$$\tau_{L,||} = \frac{\tau_{sgs}}{\sqrt{1+\beta^2\zeta^2}}, \tau_{L,\perp} = \frac{\tau_{sgs}}{\sqrt{1+4\beta^2\zeta^2}}, \tag{6}$$

where τ_{sgs} is the time scale of subgrid fluid motions, ζ is $|\bar{\boldsymbol{v}} - \boldsymbol{U}|/\sqrt{2K_{sgs}/3}$ and β is the ratio of Lagrangian to Eulerian time scales. τ_{sgs} is defined as $C\Delta_f/\sqrt{2K_{sgs}/3}$ where C is the model constant and Δ_f is the filter size. Here, $\beta = 1$ and $C = 1$ are employed in the simulations as recommended by [2].

The droplet response time may be defined as

$$\frac{1}{\tau_p} = \frac{1}{2}C_D\frac{\rho}{\rho_d}\frac{A_d}{V_d}|\boldsymbol{v} - \boldsymbol{U}|, \tag{7}$$

where ρ is the carrier fluid density, ρ_d is the droplet density, C_D is the drag coefficient, A_d is a reference area (usually droplet projected area) and V_d is the droplet volume. The drag coefficient is modelled for spherical particles as [3]

$$C_D = \begin{cases} \dfrac{24}{Re}\left(1 + \dfrac{Re^{2/3}}{6}\right) & 0 < Re < 1000 \\ 0.424 & Re \geq 1000 \end{cases} \tag{8}$$

where Reynolds number is defined as $Re = \rho|\boldsymbol{v} - \boldsymbol{U}|d_p/\mu$.

2.2 The method of Stokes binning

A numerically efficient approach, called *Stokes binning*, was previously proposed to ensure a single notional particle can represent a class of the number density function without significantly compromising the accuracy of the droplets' inertial trajectories. The Stokes number defined as $St = \tau_f/\tau_p$ is commonly used to quantify the effects of inertia in which τ_f is the fluid characteristic time scale. The Stokes number expresses the response of an inertial droplet to changes in the carrier flow. Droplets with different sizes exhibit different accelerations as they experience different drag forces. In this method, specific Stokes number bins (i.e. intervals) of width ΔSt are defined in which inertial droplets are assumed to have approximately the same dynamic response to the carrier flow. The PDF-PBE classes are dynamically selected based on $dSt \leq \Delta St$ criterion where dSt represents the range of Stokes numbers of inertial droplets represented by a notional particle. The acceleration of a notional particle representing a turbulent realisation of a class of droplets is then determined by the average response time, $\langle\tau_p\rangle$, calculated over the droplets represented by that notional particle. To ensure that the criterion $dSt \leq \Delta St$ is met even as flow conditions change notional particles may be split during the simulation. The optimal value of ΔSt is determined by the desired accuracy and availability of computational resources. In line with our earlier findings, here we employ eight Stokes bins with $dSt = 0.38$ [9].

3 Test Cases Studied

The experiment studied dispersion of 60 and 90 μm spherical droplets in a turbulent round jet for a range of Re_{jet} using two different jet diameters [10]. Here, we consider two cases of the series with jet diameter of 7 mm and Re_{jet} of 20,000 and 30,000. The experiment injected hexadecane droplets with a velocity of 30 and 46 m/s for the low and high Re_{jet} cases, respectively. Droplets were more than 1000 diameters apart and hence, droplet interactions and effects of the droplets on the carrier phase were insignificant. Table 1 presents key parameters used in the simulations. The inlet Stokes number, defined as $St_{inlet} = \rho d_p^2 v_{jet}/18\mu D$, is also presented in Table 1. The the Stokes numbers are dynamically computed using, τ_p calculated from Eq. 7 and τ_f which is obtained using $\delta/\sqrt{2K_{sgs}/3}$ where δ is the radial distance at half maximum of the mean axial carrier flow velocity.

	$Re_{jet} = 20{,}000$	$Re_{jet} = 30{,}000$
v_{jet} (m/s)	43	64.5
D (mm)	7	7
U_0	30	46
ρ_d (kg/m^3)	750	750
d_p (μm)	60-90	60-90
St_{inlet}	53-122	77-178

Table 1: Parameters for test cases studied.

The evolution of the air jet at atmospheric pressure is simulated using a compressible LES code with a standard Smagorinsky subgrid model implemented in an open source C++ code known as OpenFOAM. The computational domain is a three-dimensional rectangular mesh of 50D in the axial direction and 14D in both transverse directions. Time-varying realistic turbulent velocity boundary conditions are applied for the inflow jet based on the measured turbulence intensity and an integral length-to-jet diameter ratio of 0.125. Fixed total pressure boundary conditions are considered for the domain side and the outlet whereas zero pressure-gradient is applied at the inflow. A uniform hexahedral mesh with 7.4M cells is employed here. The ratio between the resolved kinetic energy, K_{res}, and the total kinetic energy, $K = K_{res} + K_{sgs}$, is commonly employed to ensure the suitability of the computational meshes for LES computations. At least 80% of the total kinetic energy should be resolved in the LES [13]. Here, 98% of the domain have a value of K_{res}/K larger than 0.8 confirming a satisfactory mesh quality.

The experiment was conducted for two droplet diameters of 60 and 90 μm. However, in the PDF-PBE simulation, we assume that an injected notional particle initially represents a range of droplet size from 10 to 90 μm with a constant interval of 10 μm. A notional particle immediately splits into as many as required corresponding to the Stokes bin criterion. Since droplets do not alter the carrier phase velocities and there is no interaction between droplets, the considered condition does not affect the dispersion of the 60 and 90 μm droplets.

4 Results and Discussion

3.2 Jet evolution

Figure 1 presents the centerline profiles of axial mean carrier phase velocity, v, and its rms, v_{rms} normalised by v_{jet} for both Re_{jet}. Overall, the agreement between the LES and the experiment is satisfactory. It can be seen that the LES mean velocity is higher than the experiment at early stages for both Re_{jet} but this improves further downstream. This is consistent with the observation in Fig. 1b where the peak value of v_{rms} is overpredicted. One possible reason may be due to the assumption of a uniform turbulent intensity across the inlet jet but that aspect cannot be further investigated since the experimental radial velocity rms profiles are not available. The turbulence model also plays a significant role in predicting the jet breakup point and hence more advanced turbulence models may improve results. However, it is worth noting a similar level of agreement was reported by Sun et al [14] who employed a different turbulence model for simulation of the same jet. Although the accuracy of the predicted carrier phase has a significant impact on particle dispersion, we believe the LES accuracy is sufficient here since the disagreement in the carrier velocity rms is mainly observed at upstream locations where the particle dispersion is small. It can be seen that further downstream where most of the particle dispersion occurs, the LES prediction is satisfactory, particularly in the case with $Re_{jet} = 20{,}000$.

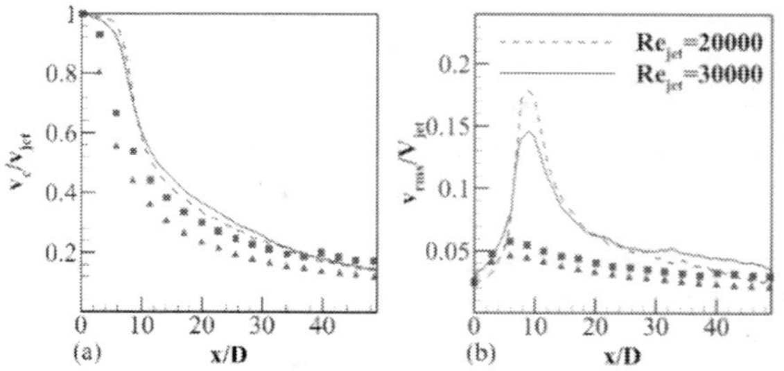

Figure 1: (a) Normalised mean carrier flow velocity, and (b) turbulence intensity along the centerline; dashed line, LES for $Re_{jet} = 20{,}000$; solid line, LES for $Re_{jet} = 30{,}000$; square symbol, experiment for $Re_{jet} = 20{,}000$; delta symbol, experiment for $Re_{jet} = 30{,}000$.

3.2 PDF-PBE results

The results presented here are computed as follows. The notional particles' quantities are first collected over three flow through times and then the mean Lagrangian values are computed at a selected axial location based on the azimuthally weighted average of those notional particles at the target axial location.

Figure 2 shows the dispersion of both 60 and 90 μm droplets as a function of axial location normalised by the jet nozzle diameter for both values of Re_{jet}. The radial dispersion, Θ, is defined as the mean square displacement of the droplets from the jet centerline at a given axial location. The dashed red line ($Re_{jet} = 20{,}000$) and solid blue line ($Re_{jet} = 30{,}000$) represent the PDF-PBE where the direct effects of the subgrid fluctuations of the carrier phase are modelled using Eq. (5). The results without the SGS model are also presented and are discussed in detail below. The PDF-PBE reproduces the droplet dispersion very well for both droplet sizes at $Re_{jet} = 20{,}000$. The radial dispersions for $Re_{jet} = 30{,}000$ are not as good as those for lows Re_{jet} but still satisfactory. At higher Re_{jet}, it seems that the predicted dispersion for 60 μm droplets agrees better with the experiment at $x/D = 35$ and 40. It should be noted that the experimental data shows the same level of dispersion (16 mm^2) for both sizes at $x/D = 35$ while the PDF-PBE model, as expected, predicts a higher dispersion for 60 μm droplets at this location. As reported by Kennedy and Moody [10], the calibration uncertainty is around ±12% for a dispersion of 25 mm^2 that may explain the observed behavior. Similarly, at $x/D = 40$, the measured dispersion is slightly higher for 60 μm droplets while the model reproduces a more remarkable difference in the dispersion of 60 and 90 μm droplets. At both Re_{jet}, there is a slight over-prediction at early stages that can be due to the discrepancy observed in the predicted carrier flow velocities.

Figure 3 presents the mean axial droplet velocities along the centerline. For both Re_{jet}, similar to the experiment, the droplet velocities initially increase, since they are injected at a lower velocity than the fluid, but later decrease as the droplets relax to the decaying carrier velocity. The agreement between the PDF-PBE and the experiment is excellent for both droplet sizes at $Re_{jet} = 20{,}000$ while the model underpredicts droplet velocities at $x/D < 20$ for $Re_{jet} = 30{,}000$. This is mainly due to the more noticeable discrepancy observed in predicted carrier phase velocities at upstream locations for higher Re_{jet}.

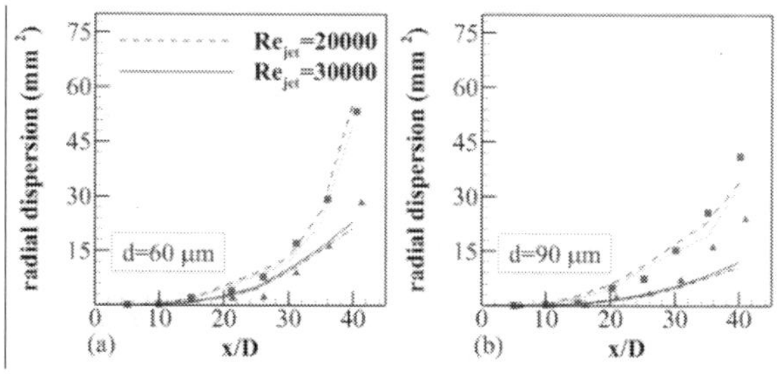

Figure 2: Radial dispersion; dashed (with SGS model) and dotted (without SGS model) red lines are for $Re_{jet} = 20{,}000$; solid (with SGS model) and dash dotted (without SGS model) blue lines are for $Re_{jet} = 30{,}000$; symbols, experimental data.

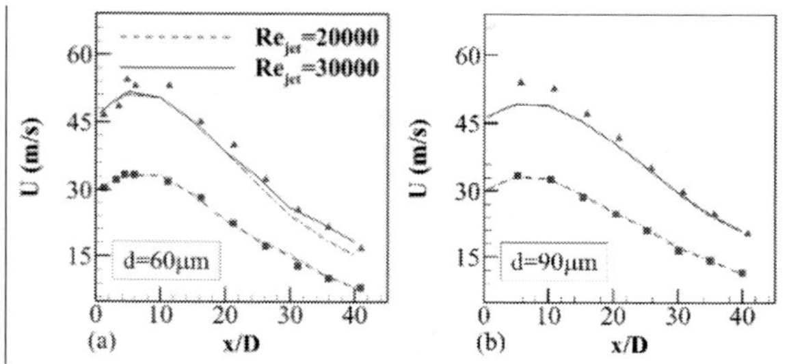

Figure 3: Mean droplet velocities. The legend is same as Fig. 2.

3.3 Subgrid effects

To investigate effects of the carrier phase subgrid model on predicted droplet dispersion, the PDF-PBE simulations are repeated with $v' = 0$. The results are presented in Figs. 2 and 3. As can be seen, the direct effects of the subgrid scale model on droplet dispersion and velocity are marginal for both Re_{jet}. This is attributed to the negligible effects of small scale carrier phase fluctuations on heavy droplets. However, we can still see that inclusion of the subgrid scale model results in slightly higher dispersion for both sizes.

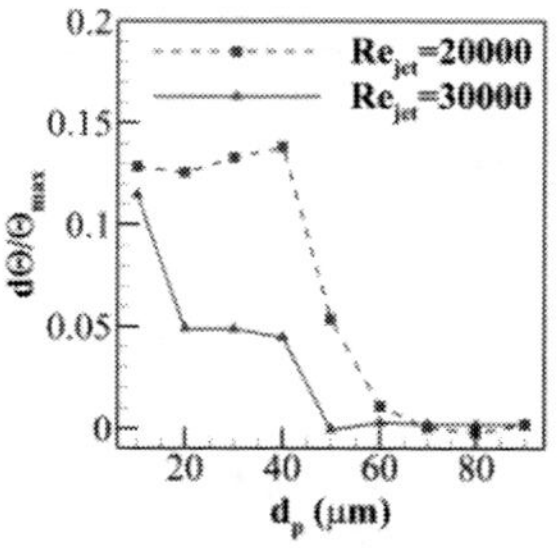

Figure 4: Normalised radial dispersion difference computed with and without SGS model at $x/D = 40$.

Effects of subgrid scales are also explored for all droplet sizes included in the simulations (10-90 µm). Figure 4 presents the difference between radial dispersion computed with and without SGS model, $\delta\Theta = \Theta_{\text{with SGS}} - \Theta_{\text{without SGS}}$, at $x/D = 40$ as a function of droplet size. $\delta\Theta$ is normalized by the maximum dispersion for each Re_{jet}. Several observations can be made here. Firstly, subgrid effects are negligible for droplets with $d_p \geq 60\ \mu m$ in both Re_{jet}. Secondly, for droplet with $d_p < 60\ \mu m$, normalised $\delta\Theta$ is larger at lower Re_{jet}, confirming importance of subgrid scales on the droplet dispersion when Re_{jet} and thus St decreases. Thirdly, $\delta\Theta$ does not linearly decrease as droplet size increases. The difference between the dispersion with and without SGS model is almost same for $d_p \leq 40\ \mu m$ and it suddenly decreases as the droplet size increases.

4 Conclusions

The recently developed PDF-PBE model with Stokes binning was validated against experimental data for the dispersion of 60 and 90 µm droplets in a round turbulent jet at two different Reynold numbers. The agreement between the PDF-PBE and the experiment was satisfactory for both Reynolds number. A Langevin model was employed to reconstruct the subgrid scales of the carrier flow and the effects of subgrid scales on droplet dispersion were also explored. It was found that the subgrid scale effects are insignificant for both 60 and 90 µm droplets but increase non-linearly for smaller droplets.

5 Acknowledgment

The work is funded by the ARC. FS, MJC and AK acknowledge the financial support provided by Universities Australia and Deutscher Akademischer Austausch Dienst (UA-DAAD). The authors acknowledge the HPC service at the University of Sydney for providing computing resources.

References

[1] J.-P. Minier, and E. Peirano, Phys. Rep. 352 (1-3) (2001) 1-214.

[2] J. Pozorski and S. V. Apte, Int. J. Multiphase Flow 35 (2) (2009).

[3] M. Bini and W. P. Jones, J. Fluid Mech. 614 (2008) 207-252.

[4] Fox, R.O., Ann. Rev. Fluid Mech. 44 (2012) 47-76.

[5] Rigopoulos, S., Chem. Eng. Sci. 62(23) (2007) 6865-6878.

[6] Rigopoulos, S., Prog. Energ. Combust. Sci. 36(4) (2010) 412-443.

[7] Y. Liao and D. Lucas, Chem. Eng. Sci. 64(15) (2009) 3389-3406.

[8] J. C. Lasheras, C. Eastwood , C. Martınez-Bazán, & J. L. Montanes, Int. J. Multiphase Flow 28(2) (2002) 247-278.

[9] F. Salehi, M. J. Cleary, A. R. Masri, submitted to J. Fluid Mech. (2017).

[10] I. M. Kennedy and M. H. Moody, Exp. Therm Fluid Sci. 18 (1) (1998) 11-26.

[11] F. Sporleder, Z. Borka , J. Solsvik, & H. A. Jakobsen, Rev. Chem. Eng., 28 (2012) 149–169.

[12] Q. Wang and K. D. Squires, Int. J. Multiphase Flow, 22 (4) (1996) 667-683.

[13] S. Pope, Prog. Energ. Combust. Sci. 11 (2) (1985) 119-192.

[14] G. Sun and J. C. Hewson and D. O. Lignell, Int. J. Multiphase Flow, 89 (2017) 108-122.

11th Asia-Pacific Conference on Combustion,
The University of Sydney, NSW Australia
10th -14th December 2017.

Dynamic behavior of temperature fluctuations during a turbulent fire

Kazushi Takagi[1], Hiroshi Gotoda[1], Isao T. Tokuda[2], Takaya Miyano[2]

[1]Department of Mechanical Engineering, Tokyo University of Science

6-3-1, Niijuku, Katsushika-ku, Tokyo, 125-8585, Japan

[2] Department of Mechanical Engineering, Ritsumeikan University

1-1-1 Nojihigashi, Kusatsu, Shiga 525-8577, Japan

Abstract

We numerically conduct a detailed study on the dynamic behavior of a buoyancy-induced turbulent fire using nonlinear time series analysis. Here, we focus on the complexity of temperature fluctuations which estimate the degree of randomness and disequilibrium in the underlying dynamics. The multiscale complexity-entropy causality plane (CECP) clearly shows the possible presence of two class of dynamics: low-dimensional chaos in the near-field dominated by the motion of large-scale vortices and high-dimensional chaos in the far-field forming a well-developed turbulent plume.

1 Introduction

Periodic and chaotic motion of flame front instability induced by natural convection is of fundamental importance in combustion and science researches. A substantial number of studies on a buoyancy-induced pulsation plume and pool fires have revealed the mean/instantaneous flow velocity and temperature distributions [1-10]. The characterization of the dominant oscillation frequency under a wide range of fuel flow velocities, burner diameters, and gravitational levels has been conducted using nondimensional numbers such as Froude number, Richardson number, Reynolds number, and Strouhal number [1-6].

Nonlinear time series analysis is becoming a useful tool for obtaining a more comprehensive understanding and interpretation of complex dynamics appearing in a wide spectrum of fields of combustion physics [11-20]. Recently, the permutation entropy [21] based on symbolic dynamics, which is the information entropy considering the probability distribution of possible existing patterns of rank order in a time series, has been applied to the analysis of flame front instability induced by buoyancy/swirl coupling [22], and thermoacoustic combustion oscillation [23,24]. It enables us to estimate the randomness of dynamics that cannot be extracted by power-spectrum analysis. Nevertheless, nonlinear time series analysis has yet to be considered in the area of fire physics. The main objective of this study is to reveal the dynamic behavior of temperature fluctuations during a buoyancy-induced turbulent fire from the viewpoint of symbolic dynamics. Note that the dynamic behavior of flow velocity fluctuations has been studied in our recent paper [25]. The dynamic behavior of a turbulent fire is obtained by using a three-dimensional large-eddy simulation. In this study, we consider the permutation entropy and complexity-entropy causality plane (CECP) proposed by Rosso et al. [26]. The latter enables us to quantify not only the randomness of dynamics but also the degree of correlational structures. In this study, we consider a multiscale CECP incorporating the variations in the time delay of the phase space constructed from temperature fluctuations.

Corresponding author. Tel: +81-03-5876-1717
E-mail address: gotoda@rs.tus.ac.jp

2 Governing equations and numerical simulation

To compute buoyancy-induced large-scale fire dynamics, we use the open-source Fire Dynamics Simulator (FDS) code developed by National Institute of Standard and Technology [27]. According to a precious study [10], the computational results obtained by the FDS are in good agreement with the experimental results for a buoyant turbulent fire with a diameter of 1 m. We incorporate the set of governing equations including the mass conservation equation, the momentum conservation equation, the energy conservation equation, and the chemical species equations, which is conducted a similar manner to that reported by Xin et al. [10].

$$\frac{\partial \rho}{\partial t} + \nabla \cdot \rho \mathbf{u} = 0 , \qquad (1)$$

$$\frac{\partial}{\partial t}(\rho Y_l) + \nabla \cdot \rho Y_l \mathbf{u} = \nabla \cdot \rho D_l \nabla Y_l + \dot{m}_l''' , \qquad (2)$$

$$\rho\left(\frac{\partial \mathbf{u}}{\partial t} + (\mathbf{u} \cdot \nabla)\mathbf{u}\right) = -\nabla p + \rho \mathbf{g} + \nabla \cdot \tau_v , \qquad (3)$$

$$\begin{aligned} &\frac{\partial}{\partial t}(\rho h) + \nabla \cdot \rho h \mathbf{u} \\ &= \frac{Dp}{Dt} + \dot{q}''' - \nabla \cdot \mathbf{q}_r + \nabla \cdot \lambda \nabla T + \sum_l \nabla \cdot h_l \rho D_l \nabla Y_l . \end{aligned} \qquad (4)$$

where ρ is the density, t is the time, $\mathbf{u}$ (= (u, v, w)) is the flow velocity vector, Y_l is the mass fraction of chemical species, D_l is the diffusivity of chemical species l, $\dot{m}_l'''$ is the production rate of chemical species l per unit volume, p is the pressure, $\mathbf{g}$ is the gravitational acceleration vector, τ_v is the viscous stress tensor, h is the enthalpy, $\dot{q}'''$ is the heat release rate per unit volume, $\mathbf{q}_r$ is the radiative heat flux vector, λ is the thermal conductivity, T is the temperature, and h_l is the enthalpy of chemical species l. The viscous stress in the momentum conservation equation is given by the Smagorinsky model. The heat release rate in the energy conservation equation is approximated by a mixture fraction combustion model. Similarly to in a precious work [10], the computational region in the x, y and z directions is set to 2.0 m, 2.0 m and 4.0m, respectively. The total number of cells with a uniform grid size of 2.5 cm is 1024000. The temporal resolution in the numerical simulation is 1 ms, and methane gas is supplied from a center area with the square shape of dimensions 1m × 1m. As shown in Figs. 1and 2, we focus on the extracted temporal evolution of temperature fluctuations along a centerline of the fire source.

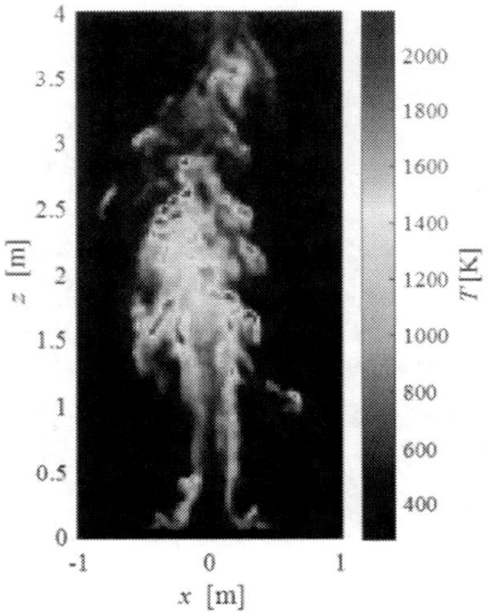

Figure 1: Instantaneous temperature field on the $x-z$ plane at t = 10.000 s

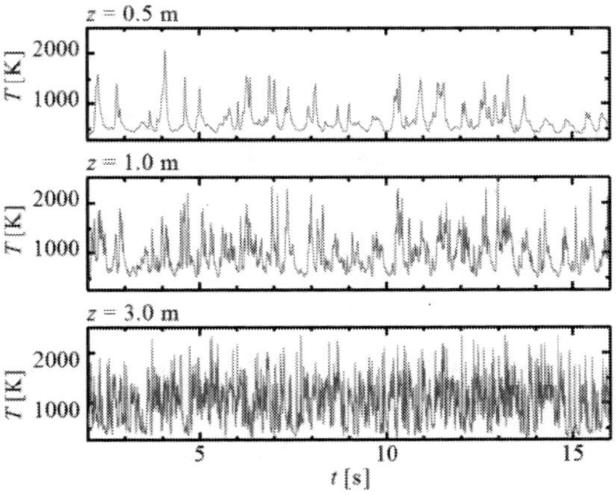

Figure 2: Temporal evolution of the centerline temperature T at different heights z.

3 Nonlinear time series analyses

3.1 Permutation entropy

Permutation entropy [21] in symbolic dynamics enables us to measure the degree of randomness from a sequence of ranks of the values of a time series. To compute the permutation entropy, we first construct D-dimensional phase space with an embedding time delay τ consisting of $\mathbf{T}(t) = (T'(t), T'(t+\tau), \ldots, T'(t+(D-1)\tau))$ from the temporal evolution of temperature fluctuations. After indexing all possible permutations ($D!$ permutations) of D data points taken from a time series, the relative frequency of each permutation pattern is estimated as $P(\pi)$. By substituting into the definition of information entropy, the normalized permutation entropy ranging from 0 to 1 is obtained in (5).

$$H_p = \frac{-\sum_{\pi} p(\pi) \log_2 p(\pi)}{\log_2 D!} \,. \tag{5}$$

3.2 Complexity entropy causality plane

The complexity entropy causality plane (CECP), which is a two dimensional plane with Jensen-Shannon statistical complexity and the permutation entropy, is useful for discussing the complexity of nonlinear dynamics. Chaotic dynamics is mainly influenced by the temporal resolution of the obtained time series. On this basis, we use a multiscale CECP [28] incorporating a temporally scale-dependent approach changing the embedding time delay. The importance of the multiscale CECP has been shown for flame front instability induced by buoyancy/swirling coupling [22]. As can be seen from (6) and (7), the complexity is described by considering both randomness and disequilibrium.

$$C_{JS}[\mathbf{P}] = Q_{JS}[\mathbf{P}, \mathbf{P_e}] H_p[\mathbf{P}] \,, \tag{6}$$

$$Q_{JS}[\mathbf{P}, \mathbf{P_e}] = \frac{\log_2 D!}{Q_{JS,max}} \{ H_p[(\mathbf{P} + \mathbf{P_e})/2] - H_p[\mathbf{P}]/2 - H_p[\mathbf{P_e}]/2 \}, \tag{7}$$

where C_{JS} is the Jensen-Shannon statistical complexity, Q_{JS} is the disequilibrium, $Q_{JS,max}$ is the maximum possible value of $Q_{JS}[\mathbf{P}, \mathbf{P_e}]$, and $\mathbf{P_e}$ is a uniform probability distribution. Q_{JS} becomes maximum when one of the components of $\mathbf{P}$ is unity and the other components are zero.

4 Numerical results and discussion

Figure 3 shows the variations in H_p of T' along the centerline of the fire source as a function of z. H_p is nearly 0.23 at z = 0.3 m corresponding to the formation region of unstable large-scale vortices, and significantly increases up to 0.82 in the far-field forming a well-developed turbulent plume. The value of H_p in the far-field is significantly higher than that in the near-filed. This clearly shows that the permutation entropy can capture a significant changes in the dynamics of a buoyancy-induced turbulent fire. Variations in H_p and C_{JS} of T' as a function of τ are shown in Fig. 4, together with the CECP involving $C_{JS,min}$ and $C_{JS,max}$. At z = 0.5 m, H_p monotonically increases with increasing τ and finally becomes constant. In contrast, C_{JS} takes a local maximum value at τ = 17 ms and decays with increasing τ. We observe a clear parabolic configuration in each CECP, and the trajectories of (H_p, C_{JS}) at z = 0.5 m moves from left to right with increasing τ. This trend corresponds to that of Lorenz chaos [29], which shows the possible presence of low-dimensional chaos in the region forming flame necking with unstable large-scale vortices. With increasing z, (H_p, C_{JS}) reaches the top right corner of the CECP, indicating high-dimensional chaos in the far-field. The time delay at which C_{JS} has a local maximum value corresponds to the minimal required sampling rate to capture all the information related to nonlinear correlations of the underlying chaotic dynamics [28]. On this basis, we extract the time delay τ_{Cmax} from the profile of C_{JS}. Figure 5 shows the variations in τ_{Cmax} for T' as a function of z. The value of τ_{Cmax} drastically deceases at around z = 0.8 m and finally takes constant value (τ_{Cmax} = 2 ms) in the far-field. This shows that the temporal scale required to capture the vortex structure in the near-filed is shorter than that in the far-field. The scale of convective vortices moving from the near-field becomes smaller in the far-field, and the smaller-scale vortex structure are dominant in the far-filed.

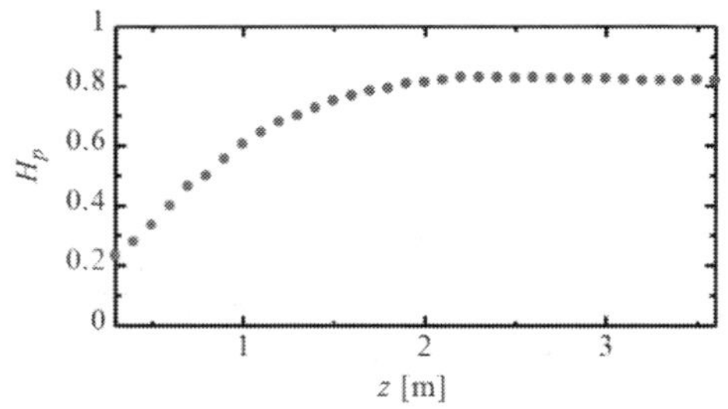

Figure 3: Variations in the permutation entropy H_p along the centerline of the fire source as a function of height z. Note that embedding time delay τ and embedding dimension D is set to 5 and 6, respectively.

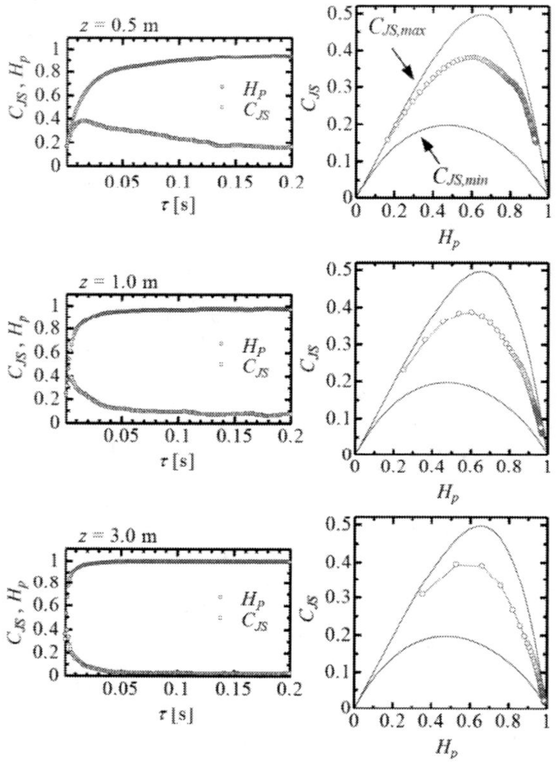

Figure 4: Variations in permutation entropy H_p and Jensen-Shannon statistical complexity C_{JS} of temperature fluctuations T' in terms of the time delay τ and height z, together with complexity-entropy causality planes involving $C_{JS,min}$ and $C_{JS,max}$ at different heights z.

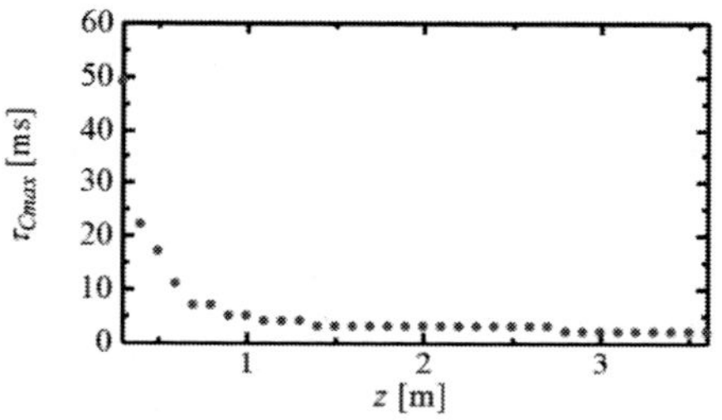

Figure 5: Variations in the time delay τ_{Cmax} for temperature fluctuations T' as a function of height z.

5 Conclusions

We numerically conduct a detailed study on the dynamic behavior of a buoyancy-induced turbulent fire from the viewpoint of symbolic dynamics and statistical complexity. The permutation entropy captures the significant changes in the randomness of temperature fluctuations. The multiscale complexity-entropy causality plane incorporating a scale-dependent approach shows a strong evidence of the presence of two classes of dynamics: low-dimensional chaos in the near-field dominated by the unstable motion of large-scale transverse vortex rings and high-dimensional chaos in the far-field forming a well-developed turbulent plume.

References

[1] A. F. Ghoniem, I. Lakkis, M. Soteriou, Proc. Combust. Inst. 27 (1996) 1531-1539.

[2] B. M. Cetegen, T. A. Ahmed, Combust. Flame 93 (1-2) (1993) 157-184.

[3] W. E. Mell, K. B. McGrattan, H. R. Baum, Proc. Combust. Inst. 26 (1996) 1523-1530.

[4] B. M. Cetegen, K. D. Kasper, Phys. Fluids 8 (11) (1996) 2974-2984.

[5] B. M. Cetegen, Y. Dong, M. C. Soteriou, Phys. Fluids 10 (7) (1998) 1658-1665.

[6] B. M. Cetegen, Y. Dong, Exp. Fluids 28 (2) (2000) 546-558.

[7] S. R. Tieszen, T. J. O'Hern, R. W. Schefer, E. J. Weck-man, T. K. Blanchat, Combust. Flame 129 (4) (2002) 378-391.

[8] P. E. DesJardin, T. J. O'Hern, S. R. Tieszen, Phys. Fluids 16 (6) (2004) 1866-1883.

[9] Y. Xin, J. P. Gore, K. B. McGrattan, R. G. Rehm, H. R. Baum, Combust. Flame 141 (4) (2005) 329-335.

[10] Y. Xin, S. A. Filatyev, K. Biswas, J. P. Gore, R. G.Rehm, H. R. Baum, Combust. Flame 153 (4) (2008) 499-509.

[11] L. Kabiraj, A. Saurabh, P. Wahi, R. I. Sujith, Chaos 22 (2) (2012) 023129.

[12] V. Nair, R. I. Sujith, Chaos 23 (3) (2013) 033136.

[13] L. K. B. Li, M. P. Juniper, Proc. Combust. Inst. 34 (2013) 947-954.

[14] K. Kashinath, I. C. Waugh, M. P. Juniper, J. Fluid Mech. 761 (2014) 399-430.

[15] L. Kabiraj, A. Saurabh, N. Karimi, A. Sailor, E. Mastorakos, A. P. Dowling, C. O. Paschereit, Chaos 25 (2) (2015) 023101.

[16] V. Nair, G. Thampi, R. I. Sujith, J. Fluid Mech. 756 (2014) 470-487.

[17] S. Balusamy, L. K. B. Li, Z. Han, M. P. Juniper, S.Hochgreb, Proc. Combust. Inst. 35 (2015) 3229-3236.

[18] M. Murugesan, R. I. Sujith, J. Fluid Mech. 772 (2015) 225-245.

[19] S. Suresha, R. I. Sujith, B. Emerson, T. Lieuwen, Phys. Rev. E 94 (4) (2016) 042206.

[20] R. Sampath, M. Mathur, S. R. Chakravarthy, Phys. Rev. E 94 (6) (2016) 062209.

[21] C. Bandt, B. Pompe, Phys. Rev. Lett. 88 (17) (2002) 174102.

[22] H. Gotoda, H. Kobayashi, K. Hayashi, Phys. Rev. E 95 (2) (2017) 022201.

[23] H. Gotoda, M. Amano, T. Miyano, T. Ikawa, K. Maki, S. Tachibana, Chaos 22 (4) (2012) 043128.

[24] H. Gotoda, H. Kinugawa, R. Tsujimoto, S. Domen, Y. Okuno, Phys. Rev. Appl. 7 (4) (2017) 044027.

[25] K. Takagi, H. Gotoda, I. Tokuda, T. Miyano, (2017) under review.

[26] O. A. Rosso, H. A. Larrondo, M. T. Martin, A. Plastino,and M. A. Fuentes, Phys. Rev. Lett. 99 (15) (2007) 154102.

[27] K. B. McGrattan, H. R. Baum, R. G. Rehm, A. Hamins,G. P. Forney, J. E. Floyd, S. Hostikka, Fire dynamics simulator, Report No. NISTIR6783, National Institute of Standards Technology, 2001.

[28] L. Zunino, M. C. Soriano, O. A. Rosso, Phys. Rev. E 86 (4) (2012) 046210.

[29] Y. Tang, A. Zhao, Y. Y. Ren, F. X. Dou, N. D. Jin, Physica A 449 (2016) 324-335.

11th Asia-Pacific Conference on Combustion,
The University of Sydney, NSW Australia
10th -14th December 2017.

Two-Dimensional Temperature Measurement of Burning Flammable Solid Along the Limiting Oxygen Index Test Using Fluorescent Two-Color Method

Ikuya Tanaka[1], Toshiki Kaneko[1], Takeshi Yokomori[1], Aki Hosogai[2], Yuji Nakamura[3]

[1]Keio University, 3-14-1 Hiyoshi, Kohoku-ku, Yokohama, Kanagawa, 223-8522, Japan

[2]Japan Aerospace Exploration Agency, 2-1-1 Sengen, Tsukuba, Ibaraki, 305-8505, Japan

[3]Toyohashi University of Technology, 1-1 Hibarigaoka, Tempaku-cho, Toyohashi, Aichi, 441-8580, Japan

Abstract

To reveal the heat transfer process inside a combustible solid, we investigated temperature distribution during flame spread along plate-formed specimens (PMMA, PS and PC) in the limiting oxygen index (LOI) test (ISO4589-2) for several oxygen concentration levels. Temperature was measured by phosphor thermometry based on the two-color intensity ratio obtained from up-conversion phosphor, Y_2O_3:Yb, Er, which can be used at wavelengths of 527 and 548 nm. Specimen dimensions were 100-mm long, 10-mm wide, and 2-, 4- or 10-mm thick, and phosphor was inserted at cross-sections in the width and diagonal directions of specimens. Results showed that temperature was higher near the surface than at the center of test specimens under high oxygen concentration, so that heat transfer progressed radially from the leading edge of the flame spread in all materials. Meanwhile, temperature distribution was uniform in the width direction under low oxygen concentration, especially near the extinction limit; therefore, heat transfer progressed uniformly only in the flame-spread direction in all materials. Flame-spread rate, V, increased with increasing heat flux value inside the specimen in the flame-propagation direction q_z, but the tendency varied depending on the type of material. These findings provide valuable insight into the characteristics of flame spread and flammability in the LOI test.

1 Introduction

Fire safety is one of the most important issues in manned space activity. Materials for use in spacecraft must be evaluated by NASA's fire safety standard STD-6001B, TEST1 [1]. However, this test method is pass-fail under conditions considered equivalent to those in the spacecraft. Thus, it is inconvenient that test results are applicable only to the exact conditions of the test. Therefore, the Japan Aerospace Exploration Agency (JAXA) proposed a new fire safety test method based on the Limiting Oxygen Index test (LOI test [2]) [3]. The LOI test, regulated by ISO 4589-2, evaluates the flammability of plastic materials by limiting oxygen concentration. This test has an advantage in that it can provide information on limiting flammability value. Therefore, test results can be used as numeric data to simplify current fire safety standards. However, as this evaluation method is based on empirical rules, close study from the viewpoint of combustion phenomena is necessary for the LOI test. In general, the flammability of a solid material and flame spread over the material surface depend on various environmental conditions, such as oxygen concentration, pressure, and gas-flow velocity. However it is thought that heat transfer from the flame and pyrolysis region to the unburned region is the primary controlling mechanism of flammability and flame spread [4]. Thus, an investigation of heat transfer into the unburned region is necessary for elucidation of flammability and flame spread. In order to discuss the heat transfer phenomena, measurement of the temperature distribution inside the material is effective. Thermocouples are generally used for measuring temperature distribution in materials [5-7]. However, point measurement lacks spatial resolution, and makes it difficult to determine the temperature gradient accurately. Our research group is developing a non-contact temperature measurement that involves a fluorescent two-color method using an up-conversion phosphor excited by near-infrared rays. We have already succeeded in measuring two-dimensional (2-D) temperature distribution, and determining heat flux distribution inside a transparent plastic with high spatial resolution. In this study, we measured the temperature inside a test specimen in the LOI test using the fluorescent two-color method with the up-conversion phosphor. The heat-transfer process inside test specimens was revealed to clarify flammability and flame spread in the material.

2 Experimental Setup and Procedure

2.1 Experimental Setup

Fig. 1 shows a schematic of the experimental apparatus. The LOI tester was designed to meet the requirements of ISO4589-2. A mixture gas of oxygen and nitrogen was supplied from the bottom of the chimney at 40 mm/s (mean velocity flowing through the cylinder). We varied oxygen concentration X_o by controlling the flow rate of both gases with mass-flow controllers. Polymethyl methacrylate (PMMA), polystyrene (PS) and polycarbonate (PC) were used as the test specimen; specimens were 100-mm long, 10-mm wide, and 2-, 4- or 10 mm in thick. Ignition was made by a methane-torch flame exposed at the top of specimen, and flame spread downward from there. Film-formed up-conversion phosphor was inserted at cross-sections in the width and diagonal directions of specimens, as shown Fig.1.

2.2 Temperature Measurement

We used the up-conversion phosphor Y_2O_3:Yb, Er, which is excited by near-infrared rays for temperature measurement, because PMMA has good transmissivity of such rays. This phosphor has strong emission intensity, and its two-color intensity ratio (527 nm and 548 nm wavelength) shows good temperature dependence in the range from ambient temperature to the pyrolysis temperature of each type of material. Excitation was achieved using a near-infrared laser (wavelength: 975 nm, laser power: 10 W). A laser diffuser was installed in the optical path of the irradiation system in order to eliminate the

Corresponding author. Fax: +81-45-563-1141
E-mail address: ikuya_tanaka@z3.keio.jp

unevenness of the laser intensity. The fluorescence was separated into two optical paths by a half mirror, and passed through band-pass filters with different center transmission wavelengths (W_1 = 527 nm and W_2 = 546 nm; the full width at half maximum of both filters was 10 nm). Therefore, the intensity distributions of the two wavelengths could be photographed separately with one CCD camera (PCO: Pixelfly usb). The camera had a 14-bit CCD with a resolution of 1392 pixels×1040 pixels. Image analysis software was used to correlate the locations of the two images photographed with different wavelengths, and the intensity ratio distribution was calculated. Finally, the temperature distribution was estimated using the calibration curve between the intensity ratio and temperature. Regarding to obtain the calibration curve, we installed a phosphor in a temperature-controlled furnace and photographed the fluorescence at each temperature by the measuring optical system. In addition, the intensity ratio was defined as equation (1).

$$IR = \frac{I_{546\ \text{nm, high temp.}}/I_{527\ \text{nm, high temp.}}}{I_{546\ \text{nm, room temp.}}/I_{527\ \text{nm, room temp.}}} \qquad (1)$$

where I is the intensity of the fluorescence modified by subtracting the disturbance of the flame photographed with the laser shutter closed. In order to eliminate variations in the light-receiving sensitivity of each CCD element, the intensity ratio was normalized by the value at room temperature. The shutter speed of the CCD camera was set to 0.5 s. The flame spread rate V was defined as the pyrolysis region of the specimen propagates vertically downward, and was calculated from the image recorded by the digital video camera. Flame spread reached steady state at temperature measurement region.

3 Results and discussion

3.1 Flame Spread Behavior and Flame Spread Rate of PMMA

Fig. 2 contains photographs of the downward-spreading flame over the PMMA specimen under six conditions: (a) δ = 2 mm, X_o = 20%; (b) δ = 4 mm, X_o = 20%; (c) δ = 10 mm, X_o = 20%; (d) δ = 2 mm, X_o = 17.6%; (e) δ = 4 mm, X_o = 18%; (f) δ = 10 mm, X_o = 18.5%. The flame was larger under high oxygen concentration than under low oxygen concentration. Pyrolysis and evaporation were occurring in the regions that appear white toward the top of the images. The shape of the pyrolysis and evaporation region was sharpened under high oxygen concentration, and flattened under low oxygen concentration. Fig. 3 shows the flame spread rate for PMMA, V_{PMMA}. Flame spread rate increased with increasing oxygen concentration. The flame temperature was high under high oxygen concentration, promoting pyrolysis and evaporation of the test specimen. Flame spread rate also became faster with thinner specimen. This tendency was more pronounced in the high oxygen concentration condition. On the other hand, in the near limiting oxygen concentration, the flame spread rate showed the same value, regardless of specimen thickness. Furthermore, in this experiment, limiting oxygen concentration became lower in thinner samples, and limiting oxygen concentrations of PMMA were 17.4%, 17.8%, and 18.5% at 2 mm, 4 mm, and 10 mm thickness, respectively.

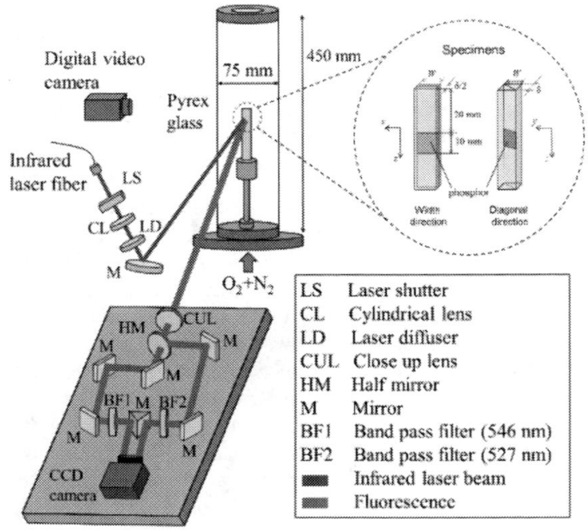

Figure 1: Schematic of experimental apparatus.

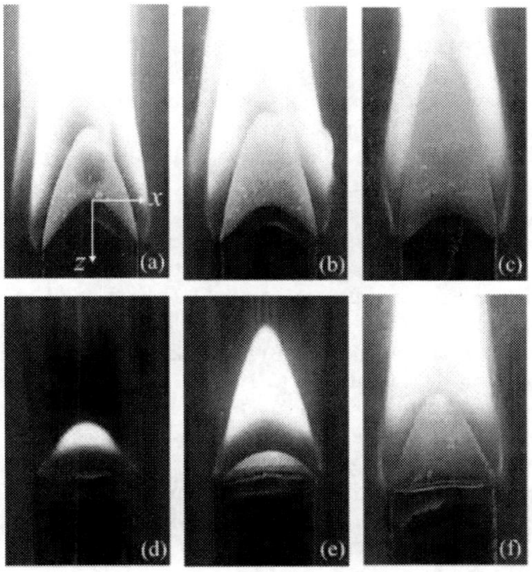

Figure 2: Photographs of flame spread in PMMA, (a) δ = 2 mm, X_o = 20%, (b) δ = 4 mm, X_o = 20%, (c) δ = 10 mm, X_o = 20%, (d) δ = 2 mm, X_o = 17.6%, (e) δ = 4 mm, X_o = 18%, (f) δ = 10 mm, X_o = 18.5%.

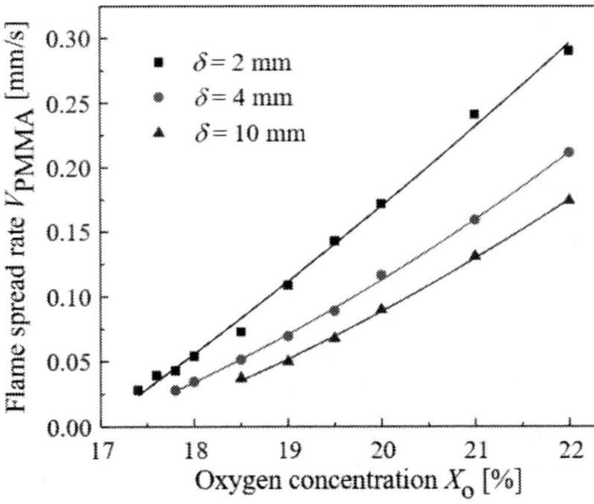

Figure 3: Flame spread rate for PMMA.

3.2 Two-Dimensional Temperature Distribution and the Heat Transfer Phenomena of PMMA

Fig. 4 shows 2-D temperature distributions and vector diagrams of the heat flux in the cross-sections in the width direction for PMMA. Fig. 5 shows 2-D temperature distributions and vector diagrams of the heat flux in the cross-sections in the diagonal direction. In these figures, x- and y-axes indicate the horizontal location in the width and diagonal direction, respectively, and the z-axis represents the vertical location in the downward direction. The origins of the x- and y-axes correspond to the center of the test specimen; that of the z-axis corresponds to the edge of the pyrolysis region, which was defined as the front position of the pyrolysis and evaporation region at the center of the test specimen. The arrows in the temperature distribution show the heat flux calculated from the temperature gradient.

For the high oxygen concentration condition shown in Figs. 4(a), (c), and (e), regardless of specimen thickness, the temperature and vector of the heat flux were larger near the surface than at the center of the specimen. In addition, the vector turned in a direction between the downward and width directions near the surface, so that the heat transfer progressed radially from the leading edge of the flame spread. This concurred with results showing that the propagation of the pyrolysis and evaporation region was faster near the surface than at the center under high oxygen concentration. As flame spread preceded near the surface of the test specimen under high oxygen concentration, it may be said that flame spread rate was dominated by the heat transfer near the surface. With thinner specimens, the temperature and the vector of the heat flux near the surface increased. For the thinner specimen condition shown in Fig. 2(a), flame spread preceded more near the surface so that the flame approached the measurement area, and temperature near the surface rose compared to thicker specimen. On the other hand, under low oxygen concentration near the extinction limit, temperature distribution was uniform in the width direction, and the vector of the heat flux turned only downward. Therefore, the heat transfer occurred only toward the flame spread direction, regardless of the position of the width direction. This was in accordance with our finding showing that the pyrolysis region spreading to the upper end of the test specimen became flat under low oxygen concentration [Fig. 2(d),(e),(f)].

With high oxygen concentration in the cross-section of the diagonal direction, shown in Fig. 5(a), the temperature and vector of the heat flux were larger near the surface, as in Fig. 4. On the other hand, under low oxygen concentration near the extinction limit, a different tendency was observed in the width direction where the temperature was larger near the surface than at the center of the specimen. In thicker specimen, such as 10-mm, the heat from the flame did not reach the inside of the specimen, so that the temperature gradient occurred in the width direction.

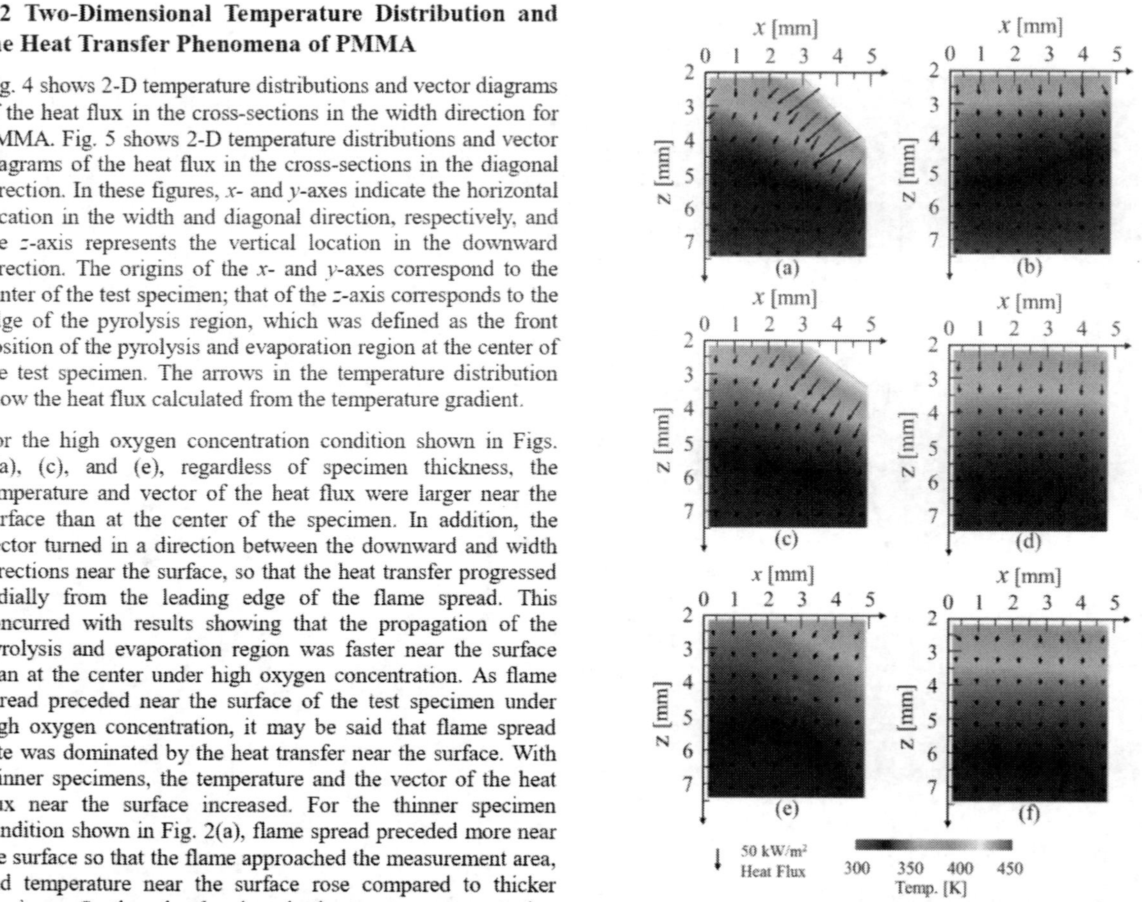

Figure. 4 Temperature and heat flux distributions in width direction for PMMA, (a) $\delta = 2$ mm, $X_o = 20\%$, (b) $\delta = 2$ mm, $X_o = 17.6\%$, (c) $\delta = 4$ mm, $X_o = 20\%$, (d) $\delta = 4$ mm, $X_o = 18\%$, (e) $\delta = 10$ mm, $X_o = 20\%$, (f) $\delta = 10$ mm, $X_o = 18.5\%$.

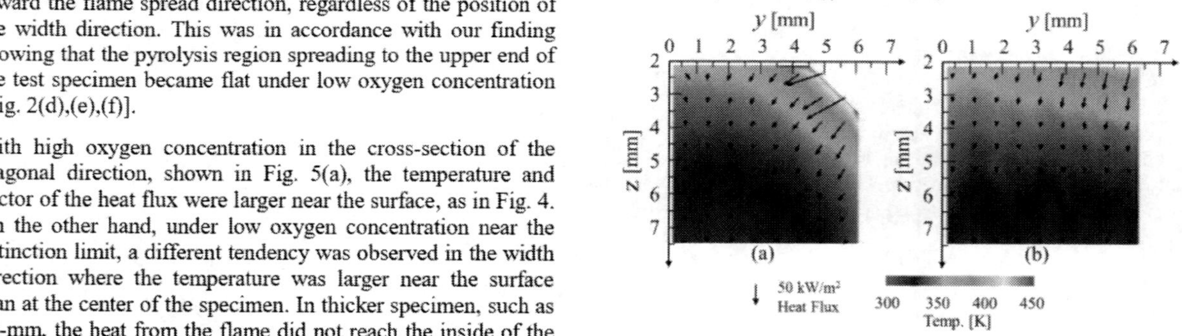

Figure. 5 Temperature and heat flux distributions in diagonal direction for PMMA, (a) $\delta = 10$ mm, $X_o = 20\%$, (b) $\delta = 10$ mm, $X_o = 18.5\%$.

3.3 Two-Dimensional Temperature Distribution and the Heat-Transfer Phenomena of PS and PC

Fig. 6 shows photographs of the downward-spreading flame over the PS and PS specimen under four conditions: (a) PS, $\delta = 2$ mm, $X_o = 20\%$; (b) PS, $\delta = 2$ mm, $X_o = 18\%$; (c) PC $\delta = 4$ mm, $X_o = 30\%$; (d) PC, $\delta = 4$ mm, $X_o = 28\%$. The green light inside the specimen is the fluorescence emission of the phosphor. In this experiment, limiting oxygen concentrations were 18% and 27.8% at the PS and PC specimen, respectively. Fig. 7 shows the 2-D temperature distributions and vector diagrams of the heat flux in the cross-sections of the width direction for PS and PC. For the high-oxygen-concentration condition shown in Figs. 7(a) and (c), the temperature and vector of the heat flux were larger near the surface than at the center of the specimen, and heat transfer occurred toward the inside, as it did in PMMA. Under low oxygen concentration near the extinction limit, temperature distribution was uniform in the width direction, and the vector of the heat flux was only downward for PS, as in PMMA. Despite near the extinction limit for PC, temperature distribution was slightly nonuniform in the width direction. As shown in Fig. 6(d), the propagation of the pyrolysis region was slightly faster near the surface than at the center, so that temperature was slightly higher near the surface. Although there was a difference depending on the type of material, the flame was formed at the upper end of the test specimen in the near limiting oxygen concentration, and the heat transfer from the entire upper end face appears conspicuously. Therefore, it can be suggested that one-dimensional heat transfer to the flame propagation direction is dominant in the near limiting oxygen concentration regardless of material.

3.4 Relationship between flame spread rate and heat flux

Fig. 8 shows the relationship between flame spread rate V and heat flux in the flame-propagation direction q_z in each material. Heat flux value at a position x = 4.5 mm and 0.5 mm away from the edge of pyrolysis region in the flame-propagation direction was defined as q_z. Propagation of the pyrolysis region was faster near the surface show in Fig. 2 and Fig. 6, it is considered that the point near the side surface has a greater influence on heat transfer, so we chose the point for the heat flux at the position of x = 4.5 mm. As shown in Fig. 8, flame spread rate increased with increasing heat flux value q_z in all the samples, but the increasing rate varied depending on the type of material. Flame spread rate of PMMA was relatively slow compared to that of other materials at high heat flux condition. In any case, it is thought that the heat flux in the vicinity of the pyrolysis region dominates the flame propagation, but further examination will be carried out based on the high accuracy of the measurement position of the heat flux and the relation with each physical properties of the material.

4 Conclusions

We conducted non-contact temperature measurement inside different types of test specimen with the LOI test using the fluorescent two-color method with up-conversion phosphor. Results of temperature and heat flux distributions revealed that the heat transfer process inside the test specimen varied depending on oxygen concentration, and that the heat transfer phenomenon is influenced by the type of material. The difference in heat transfer phenomena influenced flame spread and flammability.

References

[1] NASA-STD-6001B, Flammability, Offgassing, and Compatibility Requirements and Test Procedures (2011).

[2] ISO4589-2 Plastics-Determination of burning behavior by oxygen index Part 2: Ambient-temperature test.

[3] Fujita, O., "Solid combustion research in microgravity as a basis of fire safety in space", *Proc. Combust. Inst.*, vol. 35, pp. 2487-2502 (2015).

[4] Fernandez-Pello, A. C., Hirano, T., "Controlling mechanisms of flame spread", *Combust. Sci. Technol.*, vol. 32, pp. 1-31 (1983).

[5] Fernandez-Pello, A. C., Santoro, R. J., "On the dominant mode of heat transfer in downward flame spread", *Proc. Combust. Inst.*, vol. 17, pp. 1201-1209 (1979).

[6] Sibulkjn, M., Lee, C. K., "Flame propagation measurements and energy feedback analysis for burning cylinders", *Combust. Sci. Technol.*, vol. 9, pp. 137-147 (1974).

[7]Mamourian, M., Esfahani, J. A., Ayani, M. B., "Experimental and scale up study of the flame spread over the PMMA sheets", *Therm. Sci.*, vol. 13, pp. 79-88 (2009).

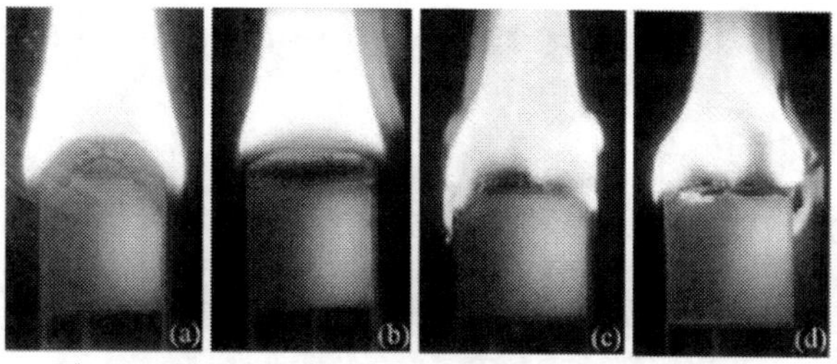

Figure. 6 Photographs of flame spread in PS and PC, (a) PS, δ = 2 mm, X_o = 20%, (b) PS, δ = 2 mm, X_o = 18%, (c) PC, δ = 4 mm, X_o = 30%, (d) PC, δ = 4 mm, X_o = 28% .

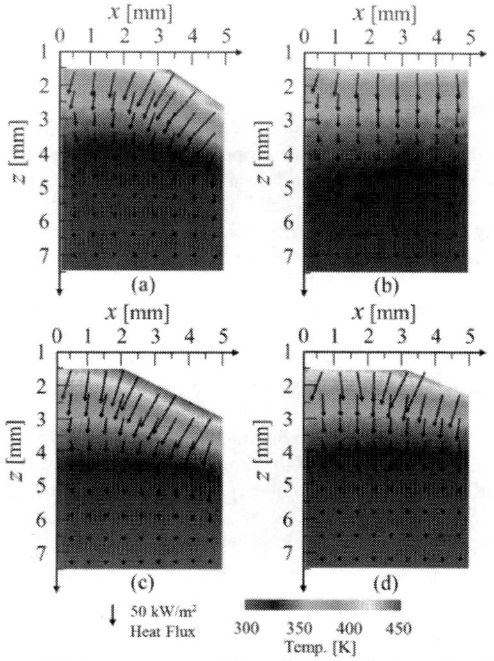

Figure. 7 Temperature and heat flux distributions in width direction for PS and PC, (a) PS, δ = 2 mm, X_o = 20%, (b) PS, δ = 2 mm, X_o = 18%, (c) PC, δ = 4 mm, X_o = 30%, (d) PC, δ = 4 mm, X_o = 28%.

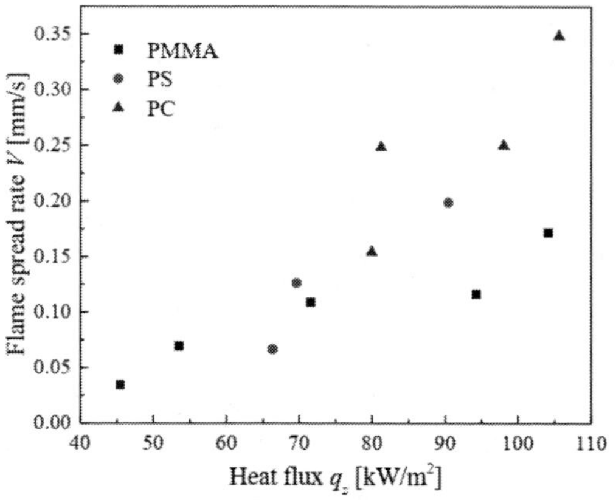

Figure. 8 The relationship between flame spread rate and heat flux.

11th Asia-Pacific Conference on Combustion,
The University of Sydney, NSW Australia
10th -14th December 2017.

On the enhancement of biodiesel droplet vaporization using electrostatic charge

Tushar Ahmed[1], Agisilaos Kourmatzis [2]

[1]Department of Engineering, Macquarie University

Sydney, NSW 2109, Australia

[2]School of Aerospace, Mechanical and Mechatronic Engineering, The University of Sydney

Sydney, NSW 2006, Australia

Abstract

Charge injection atomizers operate by introducing an electric charge into a bulk liquid jet such that Coulombic repulsion can assist in the liquid fragmentation process. These devices can not only help to atomize highly viscous liquids but can also reduce the total vaporization time of a fuel droplet. This is of particular relevance towards the implementation of highly viscous biodiesel fuels for combustion. In this contribution, the viability of using a charge injection atomizer with biodiesel fuels is investigated numerically. A previously published evaporation model which accounts for electrostatic charge has been extended to predict the evaporation time for individual fatty acid methyl esters (FAMEs), by accounting for the presence of electrical charge on the surface of the droplet and its breakup into a larger residual droplet and smaller "sibling" droplets. The numerical study is performed with a view towards determining the total vaporization time difference between the case of an uncharged biodiesel droplet and that of a charged droplet as a function of the ambient temperature (T_g), initial charge density, and droplet Reynolds number (Re). It is found that FAMEs with higher carbon chain lengths are more sensitive to charge for the chosen range (500~800 K) of temperature.

1 Introduction

Biodiesel is considered to be one of most promising alternatives to diesel fuel [1]. It has a similar chemical structure to the aliphatic hydrocarbons in petro-diesel and a higher cetane number than diesel fuel [1]. However, as with all other plant oils, biodiesel fuel has a higher viscosity. As a result, combustors face challenges in obtaining efficient atomization of biodiesel fuels [2]. In terms of mitigating this situation, a charge injection atomizer has significant potential. It allows for the introduction of electric charge into the liquid jet, thereby allowing for the presence of like charges to create enough of an electrical force (Coulombic) to exceed the surface tension force, resulting in atomization. Along with its usage as an atomizer in combustion applications, a charge injection atomizer could also be utilized in pesticide processing, industrial painting, film preparation, and ink-jet printing. It can work within a power range 2-250 mW and a supply pressure as low as 0.5 bar [3]. In addition, it allows a control over the droplet size, as well as droplet motions and trajectories by adjusting the applied electrical voltage. Furthermore, the presence of like charges creates a Coulombic repulsion that can prevent agglomeration or coalescence among droplets and produce a uniform charge distribution [2].

In the last three decades, considerable research has been carried out to develop charge injection atomizers working with insulating liquids such as diesel and kerosene. With the aim of increasing the spray specific charge, several modifications in nozzle design have been carried out [4-6]. The type of electrodes, the distance between the electrode and ground plate, and the orifice diameter have undergone optimization in order to maximize the spray specific charge and minimize the amount of current lost or "leaked" to the atomizer housing [7-9]. However, the use of bio-oils as a working liquid in a charge injection atomizer can rarely be found in the literature [2, 10], whilst in the case of biodiesels, which are largely electrically insulating dielectrics, and considering all of its constituent components, there have been no detailed publications to date focusing on the influence of electrostatic charge through a charge injection atomizer.

The aim of this work is to analyze the effect of electric charge on the vaporization time of FAMEs. The relevant key parameters used in this investigation are the ambient temperature, initial charge density, and Reynolds number, and the work is presented in this paper as follows: a brief description and assumptions of the evaporation model are given in section 2. Validation with experimental data and predictions from the model for a single biodiesel droplet are presented in Section 3.0. Finally, initial findings are discussed in Section 4.

2 Mathematical modeling

A previously published and validated equilibrium evaporation model [11] with a corrected Nusselt number (to account for the influence of evaporation on heat transfer) is extended to calculate the total vaporization time for a single droplet of individual FAME. The governing equations have been solved using a fourth order Runge-Kutta scheme. A detailed explanation can be found elsewhere [11]. For the reader's benefit, a brief discussion about the process of the charged droplet's breakup is presented here. As a droplet fragments due to electrostatic charge, it separates into a larger residual droplet and into an "n" number of smaller sibling droplets. In a charge injection atomizer, the surface energy which promotes atomization is obtained from the mutual repulsion of the net charges which are accumulated on the whole or partial surface of the droplets [2], and a single droplet with a diameter D [m] can hold a maximum surface charge given by equation 1 [11]

$$Q_{ray} = \pi(\varepsilon\sigma)^{1/2}(2D)^{3/2} \qquad (1)$$

Corresponding author. Tel: +61-(2)-9351-2928
E-mail address: akourmatzis@sydney.edu.au

where ε is the permittivity of vacuum [F/m], σ is the surface tension [N/m] and (1) is known as the Rayleigh limit. Through the vaporization process, the droplet diameter decreases, and as its charge (Q_0) approaches the Rayleigh limit, fragmentation occurs. In practice, individually charged droplets usually break up well below the Rayleigh limit [2] and therefore the fragmentation limit can be rewritten as follows:

$$Q_{ray} = \eta\left(\pi(\varepsilon\sigma)^{1/2}(2D)^{3/2}\right) \tag{2}$$

where η is the Rayleigh limit coefficient ($\eta \leq 1$), typically equal to 0.8 in this contribution unless otherwise specified.

Based on previous studies [11], for the whole simulation, it is assumed that after fragmentation, that the residual droplet has a mass of 97.5% of the mass of the core droplet and a charge of 80% of the charge of the core droplet (Q_{res}). The rest of the mass is distributed among the siblings and, depending on the number of siblings (n), the mass carried by an individual sibling can be calculated from equation (3). For the present study, it is assumed that n=5 and the simulation is terminated when the mass of the droplet reduces to 0.1% of the original mass. All the properties of FAMEs are calculated from [12], and the number of carbon atoms and double bonds for the FAMEs is shown as the left and right of the colon respectively. For instance, C=18:3 means that it has 18 carbon atoms and 3 double bonds.

$$m_{sib} = \frac{(1-0.975)\times mass\ of\ the\ core\ droplet}{number\ of\ siblings\ (n)} \tag{3}$$

3 Model Validation and Prediction

The performance of the equilibrium evaporation model with single component FAMEs presented in this study is evaluated by comparing its predictions to one of the experimental results by [13] which consisted predominantly of C18, as shown in Fig. 1 (for the case of no electric charge) thereby resulting in an excellent agreement with the model. Maximum deviation (7.51%) is found for C=18:0 and C=18:1 which likely occurs due to property estimation which is done using the 1/3 rule [11]. Further validation of the evaporation model may be found with simpler fuels in [11].

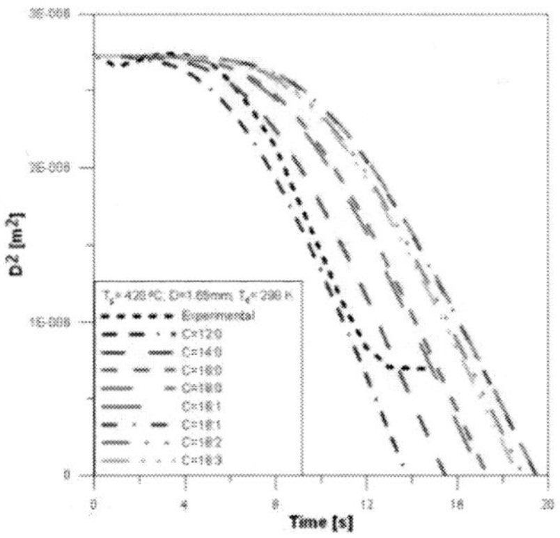

Figure 1: Comparison between the predictions of the present model and experimental findings [13] for the D^2 vs time history of FAMEs. Initial conditions are: T_g = 420 K, D = 1.62 mm, T_d = 25 K, Re= 0, and P_∞ =1 atm.

The model also predicts the total vaporization time for charged droplets of individual FAMEs by modifying the droplet diameter when it reaches the Rayleigh limit. Fig. 2 shows the evaporation history of a stagnant Stearate (C=18:0) droplet for three cases: evaporation without charge, evaporation with charge, and evaporation with charge and combustion. The model accounting for charge and combustion has been described in [11]. Briefly, in the case of combustion, it is assumed that the residual droplet evaporates using the equilibrium evaporation model described elsewhere [11], however the sibling droplets that are generated as a result of Rayleigh break-up are assumed to be enveloped by individual flames at the adiabatic flame temperature, and burn according to a D^2 law with the evaporation time for siblings, therefore, being calculated as:

$$t_{evap} = \frac{D^2}{k} \tag{4}$$

$$k = \frac{8k_g}{\rho_D C_{p,g}} \ln(1+B) \tag{5}$$

where D is the diameter of the siblings [m], k is the evaporation constant [m²/s], k_g is the thermal conductivity of the gas [W/(mK)] ρ_D is the density of the liquid [kg/m³], $C_{p,g}$ is the gas phase specific heat capacity [J/(kg K)], and B is the Spalding transfer number. The burning siblings subsequently enhance the vaporization rate of the charged residual droplet by increasing the ambient temperature around the residual. The Rayleigh fragmentation limit is seen as a step change in the droplet diameter on the droplet size time history as seen elsewhere [11].

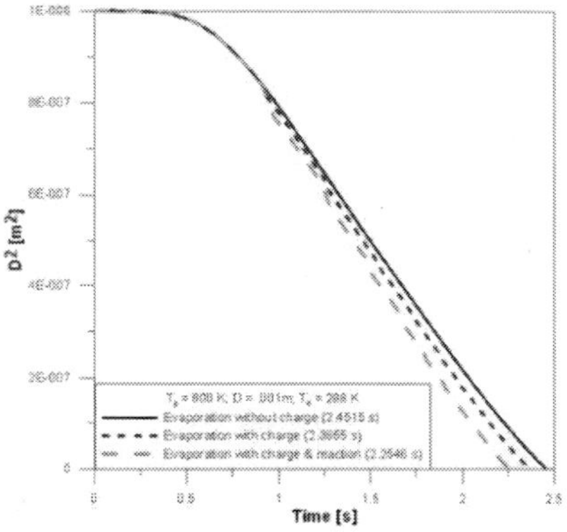

Figure 2: D^2 history for a Stearate (C 18:0) droplet with no charge, with charge and no reaction, and with charge and reaction for n = 5; Initial conditions are: D=0.001 m, T_g = 600 K, T_d = 298 K, Re= 0, and Q_i/Q_{ray} = 0.7

4 Results and Discussion

4.1 Effects of initial charge density

The initial charge (Q_i) on the droplet as a function of the Rayleigh limit, Q_{ray}, (Q_i/Q_{ray}) is an important key parameter with experimental relevance. The response of all the FAMEs has been tested with respect to initial charge density varying from Q_i/Q_{ray}=0.4~0.9 in increments of 0.1. Regardless of the ambient temperature and for any of the methyl esters, the same trend is observed, and thus the result for only C 18:0 is shown in Fig. 3. On the y axis, Δt represents the vaporization time difference for the following cases: (i) a core droplet and a core droplet with charge; (ii) a core droplet and a core droplet with charge and sibling droplet combustion. Throughout this paper, Δt will have the same meaning, and case (i), case (ii) will be noted as the charged case and the charge+combustion case respectively. From Fig. 3 it is clear that the effect of charge increases with increasing initial charge for the considered range with a significant impact up until a (Q_i/Q_{ray}=0.7). Above that level that the effect of charge diminishes due to the fixed Rayleigh limit coefficient, $\boldsymbol{\eta}$ = 0.8 which has been adopted, making this an important parameter to pre-define when designing an electrostatic atomizer.

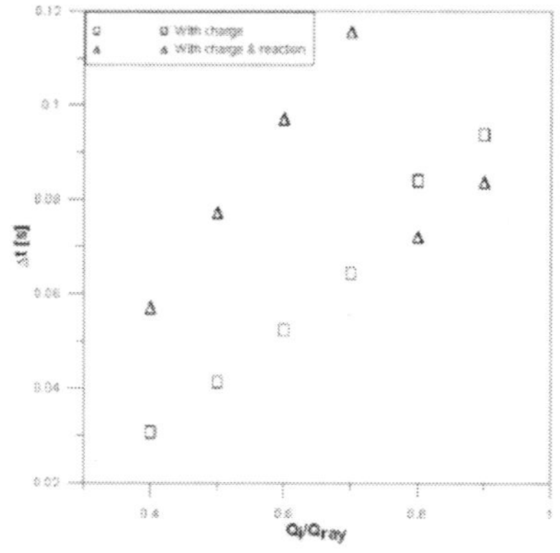

Figure 3: Vaporization time difference versus initial charge density factor (Q_i/Q_{ray}) for charge and reaction cases with n = 5. Initial conditions are: D=0.001 m, T_g = 700 K, T_d = 298 K, and Re = 0

4.2 Effects of ambient temperature

Evaporation with charge:

The amount of charge injected by an electrostatic atomizer per unit time is referred to as the total current, which is a combination of leakage current (lost to ground) and spray current (carried out by the liquid into the spray). The performance of an atomizer is assessed by measuring its spray current because the higher the spray current, the better the atomization. According to Shrimpton [14], for a highly viscous fluid, the residence time of an ion increases, resulting in an increase in the spray current with respect to the leakage current. Fig. 3 shows the effect of charge on FAMEs for the chosen range of temperatures. As expected, the maximum vaporization time difference is obtained for C=18:0, because, for FAMEs, increasing the chain length coincides with an increase in viscosity. However, a higher degree of unsaturation also lowers the viscosity, and the effect can be seen for C=18:2 and C=18:3 in Fig. 3. The effect of charge decreases for all FAMEs with increasing ambient temperature. At 800 K, the effect is at its minimum value and is almost similar for all the FAMEs, which can be explained using a mass fraction contribution as follows.

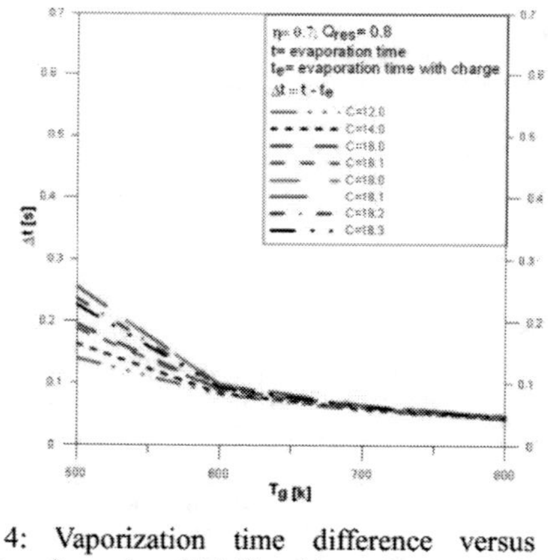

Figure 4: Vaporization time difference versus ambient temperature for various FAMEs with no charge and charge cases. Initial conditions are: D=0.001 m, T_d = 298 K, Re = 0 and Q_i/Q_{ray} = 0.7

In [11], from equations (8) and (9), it is clear that the evaporation rate is strongly related to the mass fraction. As the ambient temperature increases, the mass fraction of the vapor at the droplet surface, regardless of having charge or not, also increases, and at high temperatures, for all the FAMEs, the value becomes similar amongst fuels, therefore predicting almost identical evaporation times for both the conditions, a droplet with and without the charge. This suggests that the influence of charge in cold conditions may be more beneficial, which from a practical point of view would occur in the near-field of the spray.

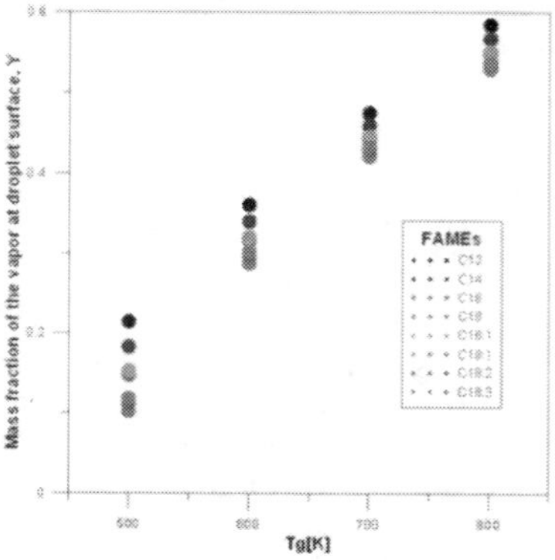

Figure 5: Mass fraction of the vapor at droplet surface versus ambient temperature for all the FAMEs. Initial conditions are: D=0.001 m, T_d = 298 K, Re = 0 and Q_i/Q_{ray} = 0.7.

Evaporation with charge and reaction:

The amount of heat generated from the combustion of siblings influences the vaporization time of the residual droplets and contributes to reducing the overall evaporation time further. Fig. 6 shows the effect of charge and reaction as a function of ambient temperature (T_g) for all the FAMEs. A reduction in sensitivity is again observed with increasing T_g. However, at 800 K, a variation is noted. Linoloeanate (C=18:3), although it was more sensitive than Palmitate (C=16:0) at 500 K, throughout the temperature range loses its sensitivity, and at 800 K, C=16:0 becomes more sensitive to charge regardless of having a lower carbon chain. The authors found this statement true for predicting their lifetime as well. A higher value of the evaporation constant (k) would possibly be responsible for this step down of C=18:3. As observed by Hashimoto et al. [15], the evaporation constant increases with increasing T_g and the higher value of k predicts a lower lifetime. From Fig. 7 it can be seen that, at 800 K, the value of k is higher for C=18:3 than for C=16:0, which causes this variation. However, this trend is not reflected by all the FAMEs here, as siblings carry only 2.5% of mass of the total core droplet.

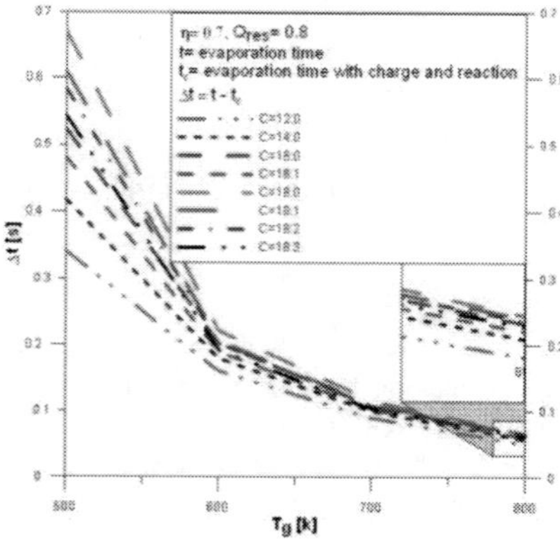

Figure 6: Vaporization time difference versus ambient temperature with charge and reaction cases for n = 5. Initial conditions are: D=0.001 m, T_d = 298 K, Re = 0 and Q_i/Q_{ray} = 0.7.

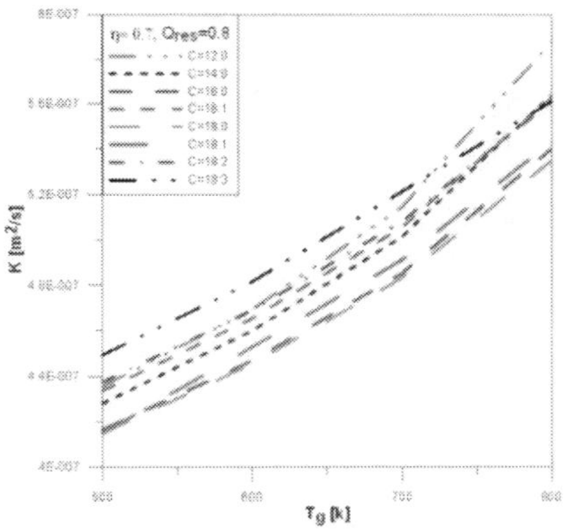

Figure 7: Evaporation constant versus ambient temperature for various FAMEs with no charge. Initial conditions are: D=0.001 m, T_d = 298 K, Re = 0 and Q_i/Q_{ray} = 0.7

4.3 Effects of Reynolds number

To predict the vaporization time of droplets for given operating conditions and fuel properties, the droplet Reynolds number is of great importance. Fig. 9 shows the effect of the droplet Reynolds number on the calculated time difference for both the cases. Here, the result for only C=18:0 is presented, as the same pattern is followed by the other fuels. It can be seen that the effect of charge on the total vaporization time decreases with increasing Re, while taking into account the sibling's combustion, the effect increases with increasing Re. This result is in agreement with previous work carried out with simpler fuels [11].

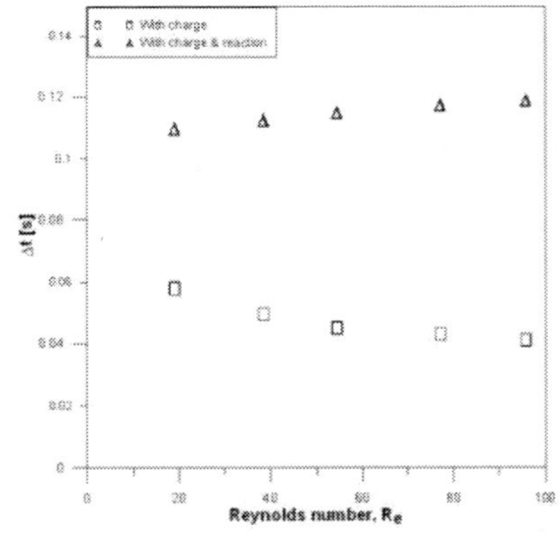

Figure 8: Vaporization time difference versus Reynolds number with charge and reaction cases for n = 5. Initial conditions are: D=0.001 m, T_g = 700 K, T_d = 298 K and Q_i/Q_{ray} = 0.7

5 Conclusions

The charged droplet evaporation model with and without combustion, originally presented in [11] has been extended to analyze the response of various FAMEs. Keeping practical parameters relevant to charge injection atomization in mind, the key parameters studied in this investigation are the initial charge on the droplet, ambient temperature, and Reynolds number. Numerical findings suggest that under the conditions tested in these simulations, an initial charge density of approximately 70% of the Rayleigh limit is advantageous towards reducing the vaporization time of the generated residual droplet. The effect of charge is found to be higher for highly viscous methyl esters throughout the range of temperatures examined. Increasing the Re results in a decrease in the influence of charge at the surface of the droplet, which is also in agreement with previous work. These initial simulations presented here form the first investigation of charge influence on the vaporization rate of electrically insulating biodiesel fuels and will form an invaluable tool for evaluating the performance of biodiesel fueled charge injection atomizers.

References

[1] T.P. Durrett, C. Benning, J. Ohlrogge, The Plant Journal 54(4) (2008) 593-607.

[2] Z.-T. Wang, A.M. Mitrašinović, J.Z. Wen, Energies 5(11) (2012) 4323-4339.

[3] A. Kourmatzis, *Pulsed Charge Injection Atomization and Turbulent Electrohydrodynamics*, PhD thesis, University of Southampton, Southampton, United Kingdom, 2011.

[4] M. Jido, J. Jpn. Soc. Chem. Eng. 40 (1986) 24-33.

[5] Kelly S. Robinson, Robert J. Turnbull, Kyekyoon Kim, IEEE Transactions on Industry Applications 1e-16(1980).

[6] A.R.H. Rigit, J.S. Shrimpton, Atomizationand Sprays 16 (2006) 401-419.

[7] J.S. Shrimpton, Y. Laoonual, International Journal for Numerical Methods in Engineering 67(8) (2006) 1063-1081.

[8] A.R.H. Rigit, J.S. Shrimpton, Experiments in Fluids 46(6) (2009) 1159-1171.

[9] A.R.H. Rigit, J.S. Shrimpton, Atomizationand Sprays 16 (2006) 421-442.

[10] J.S.S. G. AL-Ahmad, E.L. Ergene, and F. Mashayek, Atomization and Sprays 19 (2009) 547-566.

[11] A. Kourmatzis, Journal of Energy Engineering 143(3) (2016) 04016060.

[12] S.S. Sazhin, M. Al Qubeissi, R. Kolodnytska, A.E. Elwardany, R. Nasiri, M.R. Heikal, Fuel 115 (2014) 559-572.

[13] W.L.H. Hallett, N.V. Legault, Fuel 90(3) (2011) 1221-1228.

[14] J. Shrimpton, Charge Injection Systems, Springers, Verlag Berlin Heidelberg, 2009.

[15] N. Hashimoto, H. Nomura, M. Suzuki, T. Matsumoto, H. Nishida, Y. Ozawa, Fuel 143 (2015) 202-210.

11th Asia-Pacific Conference on Combustion,
The University of Sydney, NSW Australia
10th -14th December 2017.

Effective Parameters for Combustion Stability of Ultra-Low Emission CO_2 Generator using a Catalytic Combustor

Byungchul Choi[1], Kyungho Park[1], Pilsoo Jeong[2]

[1] School of Mechanical Engineering, Chonnam National University

[2] DaonRS Company

[1,2] 77 Yongbong-ro, Bug-gu, Gwangju, 61186, Republic of Korea

Abstract

The aim of this study is to develop a catalytic combustor for operation under ultra-lean burn conditions in order to suppress the formation of harmful gases such as NOx, CO, and HC, and to stably supply the CO_2 necessary for plant photosynthesis in a greenhouse. Toward this end, we optimized the parameters for achieving stable combustion in a catalytic combustor. The use of a mesh and porous material (SiC) that facilitated uniform supply of the mixture of air and fuel shortened the time required for the catalytic combustion device to reach the steady-state. The attachment of an ignition catalyst and combustion catalyst, and insulation of the catalytic combustor were effective for shortening the ignition time. The most favorable operating conditions for the CO_2 generator using the catalytic combustor are lambda = 3.5 and space velocity (SV) = 20,000 h^{-1}.

1 Introduction

The key issues in the development of combustors are energy savings and low pollution. To supply food to meet worldwide demand requires additional natural production of seasonally available foods as well as industrial-scale mass production systems that can produce crops under artificial conditions regardless of the season. For example, growing crops in a greenhouse can facilitate the production of a variety of crops, regardless of the season. The yield and quality of greenhouse crops may vary depending on proper supply of the carbon dioxide (CO_2) necessary for photosynthesis [1]. Currently, CO_2 is supplied to greenhouses as liquid CO_2, by sublimation of solid CO_2, or by using a flame burner [2]. The former two methods have expensive installation costs, the raw material supply is unstable, and the cost of the CO_2 supply is high. The latter method has low energy efficiency and generates harmful gases such as nitrogen oxides (NOx), hydrocarbons (HC), carbon monoxide (CO), and water vapor due to flame combustion. When water vapor emitted from the combustor is discharged to the greenhouse, the water vapor reportedly condenses on the leaves of the plants, thereby increasing the incidence of diseases in plants [3].

Catalytic combustion can expand the excess air-fuel ratio to lean operating conditions as compared with flame combustion, making it possible to increase the thermal efficiency of the combustor by achieving complete combustion at a low temperature; simultaneously, it is possible to eliminate the emission of HC and CO and the formation of thermal NOx [4]. However, catalytic combustion suffers from the disadvantage that it takes a long time to achieve stable combustion compared to the case of flame combustion.

The aim of this study is to develop a catalytic combustor that can operate under ultra-lean burn conditions in order to suppress the formation of harmful gases such as NOx, CO, and HC, and to stably supply the CO_2 necessary for plant photosynthesis in a greenhouse. We investigate the most effective parameters for achieving stable combustion in the catalytic combustor. Further, a heat exchanger is installed in the exhaust gas pipe of the catalytic combustor to minimize the water vapor in the exhaust gas and to preheat the air and fuel mixture by using the recovered waste heat, thereby improving the thermal efficiency and combustion stability of the catalytic combustor.

2 Experimental

2.1 Experimental apparatus

Fig. 1 shows the schematic diagram of the catalytic combustor.

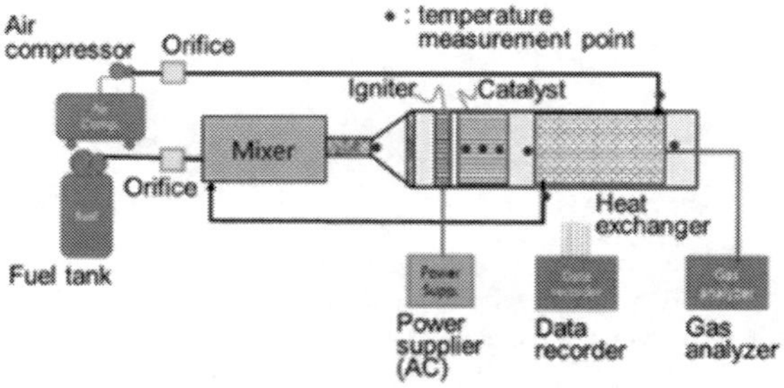

Fig. 1: Schematic diagram of the catalytic combustor.

The catalytic combustor of Fig. 1 consisted of a compressor for supplying air, a fuel tank (C_3H_8), a mixer for uniformly mixing the mixture, an igniter, a catalyst, a heat exchanger for exhaust heat recovery, temperature measuring thermocouples, and an exhaust gas analyzer. The flow rate was controlled by orifices. The air supply orifice was linearly adjustable from 5 to 150 L/min and the fuel supply orifice could be adjusted linearly from 1 to 5 L/min. Because the air and fuel are supplied in different tubes, a mixer was installed on the front of the catalytic combustor to facilitate homogeneous mixing. Inside the mixer, a spiral screw (114.3 (d) × 200 (L)) was installed in the longitudinal direction to promote the swirl of the fluid for thorough mixing. A diffuser was required to uniformly distribute

Corresponding author. Fax: +82-62-530-1689
E-mail address: bcchoi@jnu.ac.kr

the mixture from a 28.4 mm diameter tube to a 93-mm diameter catalyst. The angle of the diffuser was 60°. A metal mesh and porous ceramics (SiC) were also inserted into the diffuser for more uniform distribution in the mixer (Fig. 2).

The mixture on the catalyst was ignited by the catalytic igniter using an electric heater. The igniter is a substrate such as the combustion catalyst, which is coated with a Pt catalyst and passed through a nichrome wire. The nichrome wires were inserted into the substrate at regular intervals to achieve uniform temperature distribution. Ignition was performed by supplying 70 V AC power to the igniter by using a power supplier.

The substrate for combustion was 0.34 L cordierite (93 (d) × 50 (L) mm, 300 cpsi) coated with r-Al_2O_3 and about 3 wt% Pt catalyst (Fig. 3).

Generally, the emission from the CO_2 generator can be discharged directly into the greenhouse to supply CO_2 and heat the greenhouse. However, in this case, the large amount of water vapor contained in the exhaust gas may condense on the surface of the plant, thereby promoting fungal growth. Therefore, the water vapor must be condensed, and the temperature of the exhaust gas must be kept at about 100 °C or lower. A heat exchanger (7.72 (d) × 500 (L) mm, composed of 19 pipes) was installed in the exhaust pipe to condense the moisture in the exhaust and lower the exhaust temperature (Fig. 4). The exhaust gas from the catalytic combustor was passed into the heat exchanger tube, and the air supplied to the outside of the heat exchanger was preheated and sent to the mixer. All of the experimental apparatuses were coated with an insulator with a thickness of 3 cm.

2.2 Experimental parameters

The experimental parameters examined herein are the space velocity (SV), excess air ratio (lambda), presence of a distributer with metal meshes and porous ceramics, the distance between the igniter and the combustion catalyst, and insulation of the catalytic combustor. The SV range was set between 10,000 and 25,000 h^{-1}, the range of the excess air ratio was set to 3.5–5 in consideration of the thermal durability of the catalyst. The excess air ratio was calculated based on the CO_2 concentration.

Fig. 2: Diffuser with metals and ceramic porous materials (SiC).

Fig. 3: Catalyst for a catalytic combustor.

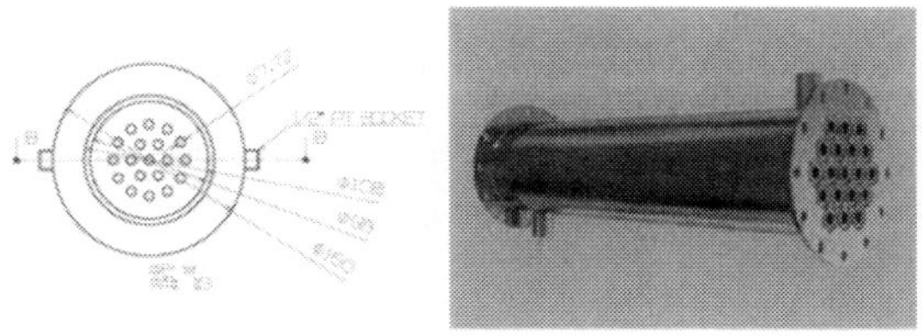

Fig. 4: Heat exchanger to recovery waste heat of exhaust gas.

3 Results and discussion

3.1 Effects of parameters on startability

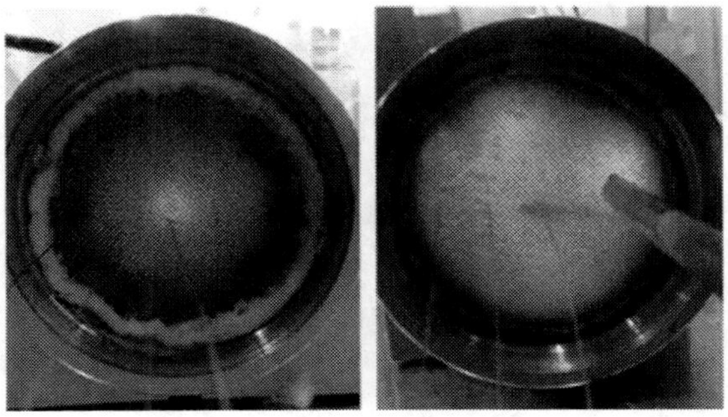

Fig. 5: Temperature distribution of radial direction without(left) or with(right) distributor.

Fig. 5 shows the temperature distribution in the radial direction without (left) and with (right) the distributor (mesh and SiC porous material). The experimental parameters were lambda = 3.5, SV = 20,000 h^{-1}. When the distributor was not inserted, the temperature was concentrated in the central region of the catalyst, and the temperature of the outermost portion of the catalyst did not increase. However, when the distributor was inserted, the air and fuel mixture spread evenly and the high temperature was widely distributed in the radial direction, and the catalyst temperature increased to steady-state within a short time (10 min).

When heat loss occurs to the outside of the catalytic combustor, the time for the catalyst temperature to reach the steady-state is increased. When the outside of the catalytic combustor case was not insulated, the ignition time was prolonged by about 30% as compared with the case when insulation was used, and the time required for the temperature to increase immediately after ignition was also extended.

The distance between the ignition catalyst and the combustion catalyst is also an important factor for achieving stable combustion. When the ignition catalyst and the combustion catalyst were attached, the ignition time was about 50% shorter than the case where the distance between the two catalysts was 2 cm.

Fig. 6 shows the temperature profiles of the catalytic combustor with the heat exchanger at lambda = 3.0, SV = 10,000 h^{-1}. The temperature in the center of the catalyst (CC IN1-2) increased sharply after ignition and rose to 725°C in 8 min. The outlet temperature of the heat exchanger (HE outlet) gradually increases from the value at 8 min to 309°C in 50 min. Due to the influence of the heat exchanger, the temperature of the feed air at the inlet of the catalyst (CC inlet) gradually increased from 24°C to 177°C in 50 min. The ratio of the heat recovery from

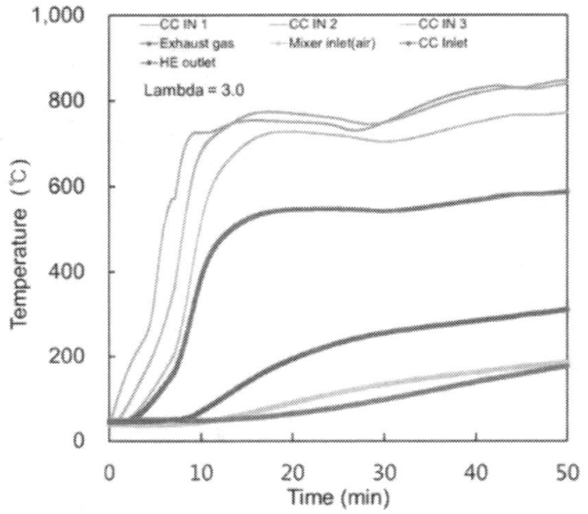

Fig. 6: Temperature profiles of the catalytic combustor with heat exchanger at lambda=3.0, SV=10,000 h-1.

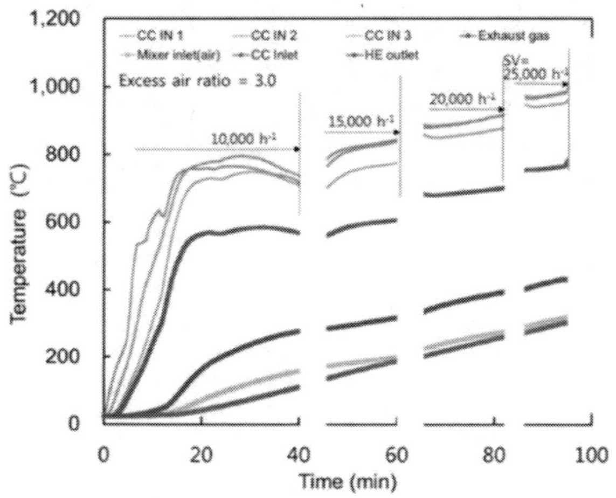

Fig. 7: Temperature profiles of the catalytic combustor with heat exchanger at lambda=3.0 and various with SV.

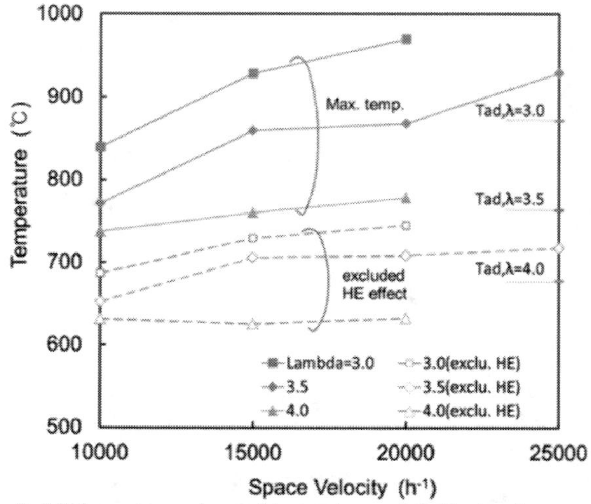

Fig. 8: Maximum temperature and temperature excluded heat exchange effect according to excess air ratio and SV.

the exhaust by the heat exchanger was about 51.1% at 50 min. The heat recovery ratio was about 29% based on the inlet temperature of the mixer.

Fig. 7 presents the temperature profiles of the catalytic combustor with the heat exchanger at lambda = 3.0 under various SV conditions. The SV in the catalytic combustor is an important factor for achieving stable combustion [5−7]. For lambda = 3.0, the temperature of the catalyst bed increased as the SV increased. This is because the outlet temperature of the heat exchanger that recovered the exhaust heat increased, and the heat supplied to the catalyst inlet increased. When the temperature of the catalyst bed reached or exceeded 980°C, the experiment was stopped to prevent thermal aging of the catalyst.

Fig. 8 presents the maximum temperature of the catalyst bed; the temperature excluded the heat exchanger effect based on the excess air ratio and SV. As shown in Fig. 7, the maximum temperature of the catalyst bed increased with an increase in the SV. For lambda = 3.0, the maximum temperature of the catalyst bed reached 970°C at SV = 20,000 h^{-1}. At this point, the maximum temperature was 98°C higher than the adiabatic flame temperature of 872°C. For lambda = 3.5, the maximum temperature of the catalyst bed increased to 929°C at SV = 25000 h^{-1}. At this point, the maximum temperature was 166 °C higher than the adiabatic flame temperature of 763°C. The concentrations of HC, CO, and NOx were non-zero in the above measurements. In the case of lambda = 4.0, the maximum temperature of the catalyst layer increased to 780°C at SV = 20000 h^{-1}; at this point, HC was emitted at 162 ppm because of incomplete combustion. Thus, for lambda = 4.0, the available SV range is below 20000 h^{-1}. The dotted line in the figure indicates the maximum temperature minus the temperature of the air and fuel mixture in the front region of the catalyst. This temperature excludes the effect of the heat exchanger at the highest temperature. In the absence of a heat exchanger, the temperature of the catalytic combustor was significantly lower than the adiabatic flame temperature, resulting in a greater heat loss for the catalytic combustor.

3.2 Stability of catalytic combustor

Fig. 9 shows the plot of the excess air ratio versus SV as an indicator of the stability of combustion in the catalytic combustor with the heat exchanger. Generally, a CO_2 generator using a catalytic combustor should operate under the following conditions: first, considering the thermal durability of the catalyst, the combustion temperature should be 900 °C or lower. Second, the CO_2 concentration should be at least 3%. Third, the HC, CO, and NOx emissions should be 0 ppm. In the figure, the shaded circles indicate the SV and lambda conditions where combustion is stable. The circles indicated by dotted lines indicate the conditions under which the combustion temperature does not exceed 900 °C. A CO_2 concentration of more than 3% was achievable when the lambda value was lower than 4.0. CO_2 generators should supply large amounts of CO_2 in a short time. Therefore, the most favorable operating conditions for the CO_2 generator using the catalytic combustor are lambda = 3.5 and SV = 20,000 h^{-1}.

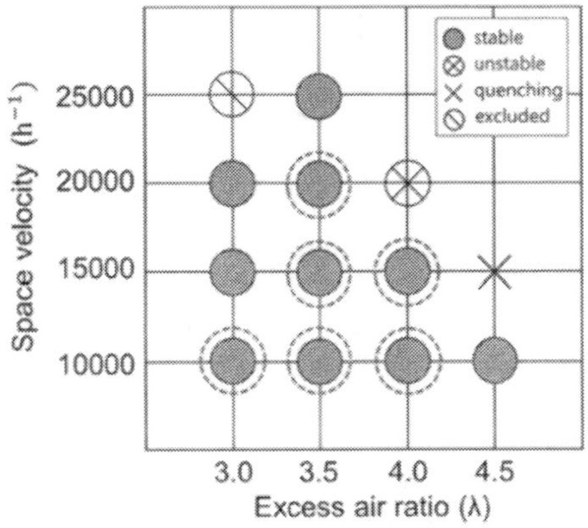

Fig. 9: Combustion stability of catalytic combustor with heat exchanger between excess air ratio versus SV.

4 Conclusions

From this attempt to develop a CO_2 generator using a catalytic combustor with zero emission and to stably supply CO_2 for plant photosynthesis in a greenhouse, we can draw the following conclusions:

(1) The use of a mesh and porous material (SiC) to uniformly supply the air and fuel mixture shortened the time required for the catalytic combustion device to reach steady-state.
(2) Attachment of the ignition catalyst to the combustion catalyst, and insulation of the catalytic combustor were effective for shortening the ignition time.
(3) The most favorable operating conditions for the CO_2 generator using the catalytic combustor are lambda = 3.5 and SV = 20,000 h^{-1}.

5 Acknowledgment

This study was supported by the 2017 Project of Technology Transfer Commercialization by Ministry of Science, ICT and Future Planning (DAONRS Co) and the Basic Research Lab (BRL) program (2015R1A4A1041746) funded by National Research Foundation of Korea. Authors thanks to Mr. W.B. Jo, Mr. C.H. Kim and Ms. J.S. Kim for cooperating in the preparation of the experimental equipment.
.

References

[1] O. Ghannoum, N. G. Phillips, J. P. Conroy, R. A. Smith, R. D. Attard, R. Woodfield, B. A. Logan, J. D. Lewis and D. T. Tissue, Exposure to preindustrial, current and future atmospheric CO_2 and temperature differentially affects growth and photosynthesis in eucalyptus, Global Change Biology, 16 (2010) 303-319.

[2] B.C. Choi, J.H. Lee, Emission and heat recovery characteristics of combustor-type CO_2 generator with an after-treatment device and regenerative heating device for Greenhouses, J. of Mechanical Sci. and Technol. 29 (2015) 1-9.

[3] C.P. Timothy, R.B. Richard, Biological control in greenhouse systems, Annual Review of Phytopathology, 39 (2001) 103-133.

[4] B.C. Choi, Combustion Engineering, Munundang, Seoul, 2016, pp. 477-521.

[5] M.S. Jung, C.M. Kim, M.Y. Kim, A parametric study on the catalytic combustion of gaseous methane, ethane and propane fuels, Journal of The Korean Society for Aeronautical and Space Sci. 40 (2012) 309-315.

[6] B.C. Choi, G.H. Park, D.S. Kim, Development of a catalytic combustor with thin-honeycomb substrate, 2nd World Congress on Momentum, Heat and Mass Transfer, 2017.

[7] W.C. Pfefferle, The catalytic combustor - An approach to cleaner combustion (by reaction control in gas turbine engines), J. Energy, 2(3) (1978) 142-146.

Contact point:

Corresponding author, Prof. Byungchul Choi, e-mail : bcchoi@jnu.ac.kr

11th Asia-Pacific Conference on Combustion,
The University of Sydney, NSW Australia
10th -14th December 2017.

Modelling of Flame Structure Using Thickened Flame Model Coupled With Multi-step Reaction Mechanism

Guo Shilong, Zhang Meng, Wang Jinhua*, Huang Zuohua

State Key Laboratory of Multiphase Flow in Power Engineering, Xi'an Jiaotong University

Xi'an, 710049, China

Abstract

Thickened flame (TF) model is one of the effective methods to simulate combustion phenomenon. Theoretically, it could be extended to multi-step chemical reaction mechanism. However, the results in the present work shown that TF model coupled with multi-step mechanism will underestimate burnt gas temperature, whether in confined or unconfined space. Meanwhile, some post-flame species (CO, OH) concentration are overestimated, which means the decrease in burnt gas temperature is mainly caused by enhanced species diffusion. Dynamic thickened flame (DTF) model coupled with multi-step mechanism performs well for predicting species concentration and combustion state. So it is not recommended to use TF model coupled with a multi-step chemical reaction mechanism. However, the dynamically thickened flame thickness is obviously thinner than the theoretical expectations for the dynamic thickening factor, which is always used to determine thickening factor for a certain mesh size. The main heat release rate zone couldn't be thickened as expected. A new DTF model used for multi-step reaction mechanism was proposed. The results show that it can both predict the flame structure accurately and get the flame thickness as expected.

1 Introduction

With the rapid improvement of parallel computing power in recent decades, the numerical simulation of combustion is becoming more and more extensive in scientific research and industry design. In practical numerical calculation of premixed flames, especially for three dimensional case, the thin flame front thickness (about 0.1-1mm) is generally smaller than the computational grid size. To get an accurate resolution of the thin flame front, approximately 5-8 grid points need to be specified inside the flame [1]. Namely, the grid size is about ten micrometer order. The calculation amount of simulation in such fine mesh size is far beyond the capacity of general workstation. To overcome this difficulty, besides flame tracking techniques, thickened flame (TF) model is also one of the effective methods [2].

The basic idea of the thickened flame model is assuming that there is a flame having the same fuel consumption speed but thicker than the real one. The chemical reactions inside thickened flame can be resolved using the Arrhenius law. Because of using the law of mass action, various combustion scene, such as flame-wall interactions, quenching and ignition phenomenon, could be directly calculated without other special sub-models. Moreover, thickened flame approach could be extended to multi-step chemical reaction mechanism in theory[2]. Initially, this approach could be applied in laminar flame simulation only, because the interaction between thickened flame and turbulence is changed which must be modeled. Colin er.al [2] proposed an efficiency function for large eddy simulations of turbulent premixed combustion to model the sub-grid scale flame wrinkling. Then a power law flame wrinkling model is developed by Charlette et.al [3-4]. Theoretically, both of these two model can be used in RANS with some parameters modification. The power law flame wrinkling model has been developed by EM2C [5-6] to get the more accurate efficiency function. Although thickened flame approach has many advantages, the thickened process changes the transport properties of burnt and unburnt mixture at the same time which may induce some undesirable results. A dynamic thickened flame (DTF) model was proposed by J. P. Legier et.al[7] to compute the thickened factor dynamically in space and time. A sensor is used to determine the reaction zone and control the thickening factor. In the reaction zone, the local thickening factor is larger than one, otherwise equals one.

In real configurations, tabulated approaches need a number of parameters to describe the flamelet. And premixed, part-premixed, even non-premixed combustion may occur at the same time, in particular with liquid fuel injection. In order to simplify the calculation, one or two step reaction mechanism coupled with thickened flame model were implemented to simulate turbulent combustion by many researchers[5-7]. But the mean temperature in numerical results could be overestimated[5] Because the global reaction mechanisms are fted to reproduce only global quantities of the flame, on a limited range of conditions, and might not respond correctly to the surrounding physics. Therefore, multi-step reaction mechanisms , especially for heavy fuels, are needed in calculating some complicated problems, such as pollutant formation mechanism during combustion.Although thickened flame model can be extended to detailed mechanism in theory [2], what happens to the flame structure people concerned (temperature or pollutant for example) is still not fully understood. In the present work, a thickened flame solver named TFReactingFoam was developed base on reactingFoam solver in open CFD source OpenFOAM. TFReactingFoam can be used to simulate both laminar and turbulent combustion base on the thickened flame model using multi-step chemical reaction mechanism. Here, we are interested in the flame structure calculated by thickened flame model coupled with multi-step chemical reaction mechanism, so the laminar flame was simulated. The flame structure calculated by TF model and DTF model are compared. A new dynamic thickened flame model base on previous research [7-8] is also introduced to determine thickened factor inside flame front more accurately. The constant TF model, DyTF model [8]and the new DyTF model are used to calculated confined and unconfined premixed combustion respectively.

Corresponding author: Jinhua Wang

E-mail address: jinhuawang@xjtu.edu.cn

2 Thickened flame model

The basic idea of TF model was originally proposed by Butler and Orourke [9]. Following the basic theories of premixed combustion [10], the unstrained laminar flame speed s_l^0 and the flame thickness δ_l^0 can be expressed as:

$$s_l^0 \propto \sqrt{D\bar{W}}, \quad \delta_l^0 \propto \frac{D}{s_l^0}, \tag{1}$$

where D is diffusivity coefficient and $\bar{W}$ the mean reaction rate. In TF model, the diffusivity coefficient is artificially increased by F times, whilst the reaction rate is divided by F. An artificial flame having the same unstrained laminar flame speed but F times thicker than the real one is obtained. If F is large enough, the artificial flame may then be resolved on coarser mesh.

Applying the TF model to the species mass balance equation

$$\frac{\partial \rho Y_F}{\partial t} + \nabla\cdot(\rho\, \mathbf{u}\, Y_F) = \nabla\cdot(\rho D \nabla Y_F) + A\, Y_F\, Y_O \exp\left(-\frac{T_a}{T}\right) \tag{2}$$

leads to

$$\frac{\partial \rho Y_F}{\partial t} + \nabla\cdot(\rho\, \mathbf{u}\, Y_F) = \nabla\cdot(\rho D F \nabla Y_F) + \frac{A}{F} Y_F\, Y_O \exp\left(-\frac{T_a}{T}\right) \tag{3}$$

A 74 step, 17 species multi-step chemical reaction mechanism developed by R. Sankaran et.al [11] is selected here for methane-air combustion. Species diffusivity expressed here calculated based on Hirschfelder and Curtiss approximation [12].

The dynamic thickened factor is determined as[8]:

$$\mathcal{F} = 1 + (\mathcal{F}_{max} - 1)\Omega, \quad \Omega = 16[c(1-c)]^2 \tag{4}$$

where c is the progress variable, 1 and 0 represent burnet and fresh gas respectively. Legier.J.P et.al[] used a "Arrhenius-like" expression to determine whether the flame should be thickened or not. We do not use this method here because of there are 17 species in the computational domain. The flame reagion is not related to only one specie.

2.2 1D Thickened Flame structure

In numerical simulation of combustion, people concern more about burnt gas temperature and species concentration, which are related to combustion efficiency and pollutants emission. Temperature, CO and OH here are considered as the representation of combustion state, pollutants emission and post-flame radical respectively. Fig. 1 compares the flame structures obtained from TF and DTF model. The results shows that the burnt gas temperature obtained from TF model decreases with the increases of thickened factor (F). CO and OH mass fraction increase with F. It indicates that using constant thickened factor may lead to the fuel cannot burnt completely and CO concentration is overestimate. This is because the intermediate components (CO, OH) diffusion velocity increase with the increase of diffusivity coefficient. Meanwhile, the increased diffusivity is sufficient to cool the burnt gases by diffusion towards the outlet boundary. Legier J P et.al [7] have used a single one-step scheme to describe propane-air combustion. They got same burnt gas temperature from TF and DTF model (F=20). It means TF model have little effect on burnt gas temperature and composition when using an overall reaction. Because there are no intermediate product to be diffused. The DTF model performs well for predicting temperature and post flame structure. However, there are two main drawbacks. First, the dynamically thickened flame thickness is obviously thinner than the theoretical expectations. Second, the flame inner structure, such as temperature and some species (CH_4, CO_2, H_2O, O_2), have been changed by the dynamic thickening factor.

The reason why dynamic thickened flame inner structure be changed and flame thickness unequal to theoretical expectation can be explained from Fig.2. The sensor inside flame front first increases up to unity (F= F_{max}) and then decrease eventually

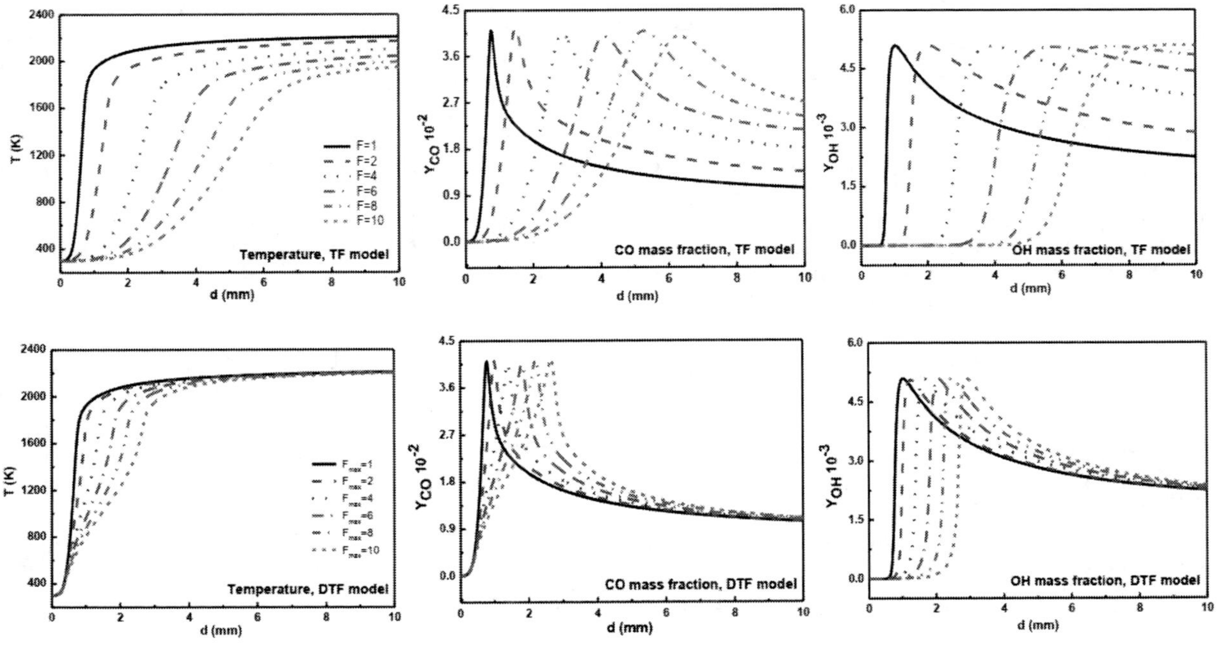

Fig.1 Flame structure information for different thickening factor

becoming stationary (F=1). As the sensor decrease, however, the HCO mass fraction (heat release rate) increases rapidly, which means that the main heat release zone have not been thickened effectively. Therefore, the slop of the temperature becomes steeper, and the same as other species.

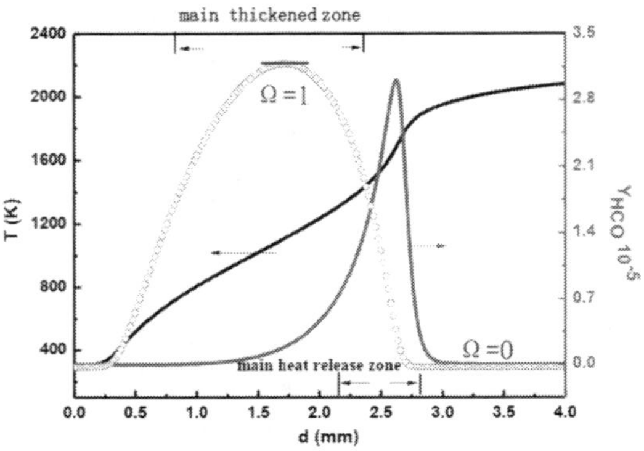

Fig.2 Flame inner structure (DTF model, F_{max}=10)

3. Unconfined and confined premixed flame

Based on the above analysis, it can be deduced that an ideal sensor should be constant unity inside flame front, and constant zero outside flame front. Hence, a new dynamic thickened flame model is proposed here base on the idea:

$$f(\Omega) = \begin{cases} 1 & (\Omega \geq a) \\ \frac{-2\Omega^3}{a^3} + \frac{3\Omega^2}{a^2} & (0 \leq \Omega < a) \end{cases} \quad (5)$$

$$F = 1 + (F_{max} - 1) f(\Omega) \quad (6)$$

where Ω is a sensor used to determine the flame front. We use the sensor suggested by Durand and Polifke [8] here, and it also can use the expression developed by J. P. Legier et.al [7] or other sensor expression. This piecewise function is continuous and derivable at Ω=0 and Ω=a. With detailed mechanism, there are multiple length scales associated with species and thermoal distribution. This piecewise function ensures that the whole flame thickeness, include preheated zone and reaction zone, could be thickened effective, as shown in Fig 3. So the thickened flame could be computated directly in coarse mesh.

As mentioned above, The increased diffusivity may cool the burnt gases by diffusion towards the atmosphere. In order to determine the impact of this factor, different thickened flame model is used to calculate a unconfined syngas laminar premixed Bunsen flame firstly. The burner geometry and discrete method are as same as Wang et.al[13]. The flame at stoichiometric ratio, Re = 1400, XH2=0.8 is simulated and compared with experiment. The details of the experimental and numerical setup can be found in Wang et.al[13]. The flame height obtained from four case, namely non-thickened (TF=1), constant thickened flame (F=4), Durand [12] dynamic thickened flame (F_{max}=4) and our new dynamic thickened flame (F_{max}=4), are quite close to the experiment measurement. Which implies the laminar flame speed keeps unchanged with different model. Because we are focus on the flame structure rather than flame speed here, the flame images are not shown here. Fig.3 presents a comparison of four numerical flame structures at flame tip. The zero point was chose at just ahead of the flame front on the unburnt side. The temperature and OH concentration profile shows that our new dynamic thickened flame model are good at predicting both flame thickness and flame structure. The main thickened zone covers the whole heat release zone rather than just the middle part of flame in Durand [8] DTF model, as show in Fig.3.

Secondly, a typical confined premixed flame propagation process is calculated. The combustion chamber is a closed rectangular duct 8 mm square by 50 mm long. The characteristic grid size is 250μm. A sequence of flame front (c=0.5) images is presented in Fig.4. The color represent 3D flame surface curvature which is calculated as ∇•n, where n is the flame surface unit normal vector defined as n = -∇c/|∇c|. Clanet and Searby [14] divided four stages during the flame propagation in confined duct, namely spherical flame (a), finger shaped flame (b), flame touching sidewalls (c-d) and tulip flame (e). These four stages are reproduced successfully using TF model and DTF model. Fig.5 presents the displacement speed of flame front as a function of positon. At the initial stage, the thickened flame speed is slightly less than that non-thickened one. It is because the increase in diffusivity makes the energy transferring more easily, and then the ignition becomes harder. Overall, the flame front evolution behaviors obtained from DTF model are in good agreement with those observed in TF model.

Fig.5 shows comparison of the flame temperature and OH massfraction structure from different thickened flame model. Thelocation of the zero point is defined at just ahead of the flame front on the unburnt side when the planar flame (Fig.4 e-d) appears. It is seen that, even though in the confined combustion case, the TF model perform poorly on predicting temperature and post-flame species concentration. The comparisons between Fig.3 and Fig.5 shows that the decrease in burnt gas temperature is mainly caused by enhanced species diffusion, which contain unreleased chemical energy. So it is not recommended to use TF model coupled with a multi-step chemical reaction mechanism, which would overestimate post-flame species concentration, then underestimate burnt gas temperature. Base on the previous analysis, it can be seen that the proposed DTF model is a combination of TF model and Durand's DTF. Therefore, it can both predict the flame structure more accurate and get the flame thickness as expected.

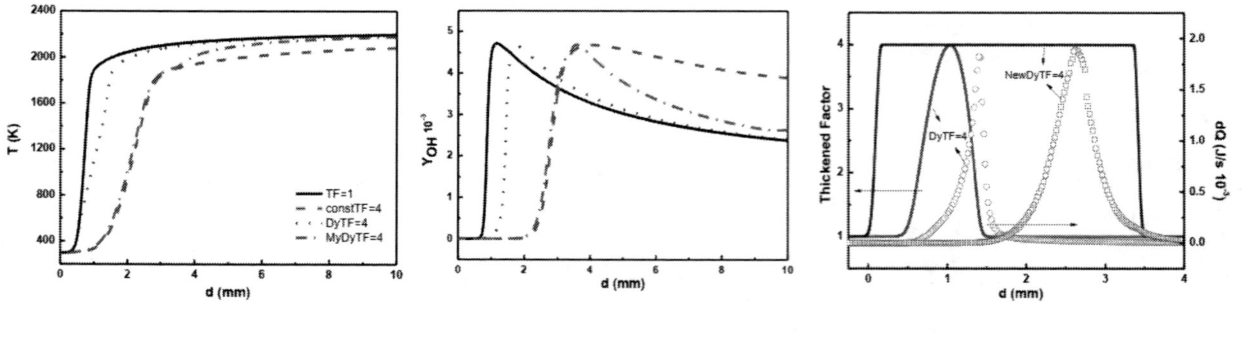

Fig.3 Unconfined premixed flame structure

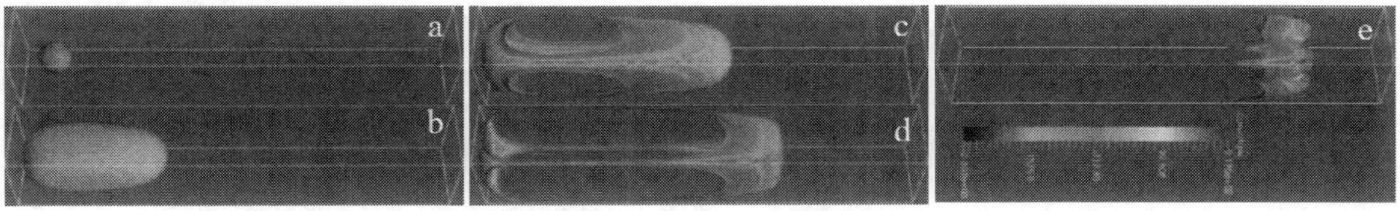

Fig.4 Flame shapes evolution (proposed DTF model)

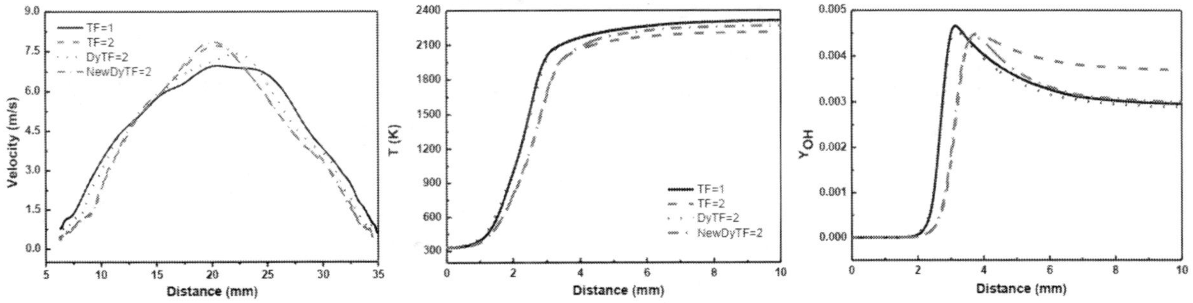

Fig.5 Confined flame speed and structure

4 Conclusions

In the present paper, we investigated the flame structure using thickened flame model coupled with multi-step reaction mechanism . Comparison shown that TF model coupled with multi-step mechanism will underestimate burnt gas temperature, no matter confined or unconfined case. At the same time, some post-flame species (CO, OH) concentrations are overestimated. which can be inferred that the decrease in burnt gas temperature is mainly caused by enhanced species diffusion. DTF model coupled with multi-step mechanism performs well for predicting combustion state. So it is recommended to use DTF model coupled with a multi-step reaction mechanism. However, the dynamically thickened flame thickness is obviously thinner than the theoretical expectations, which is always used to determine thickening factor for a certain mesh size. The main heat release rate zone also couldn't be thickened as expected. A new DTF model used for multi-step reaction mechanism was proposed, which can both predict the flame structure accurately and get the flame thickness as expected.

5 Acknowledgment

This work is supported by National Natural Science Foundation of China (No. 51376004). The authors also gratefully acknowledge Prof. Veynante. D for fruitful discussions.

References

[1] G. Lacaze, B. Cuenot, T. Poinsot, M. Oschwald, Combust. Flame 156 (2009) 1166-1180.

[2] O. Colin, F. Ducros, D. Veynante, T. Poinsot, Phys Fluids 12 (2000) 1843.

[3] F. Charlette, C. Meneveau, D. Veynante, Combust. Flame 131 (2002) 159-180.

[4] F. Charlette, C. Meneveau, D. Veynante, Combust. Flame 131 (2002) 181-197.

[5] G. Wang, M. Boileau, D. Veynante, Combust. Flame 158 (2011) 2199-2213.

[6] G. Wang, M. Boileau, D. Veynante, K. Truffin, Combust. Flame 159 (2012) 2742-2754.

[7] T.P. J. P. Legier, D. Veynante, Proceedings of the Summer Program, 2000. 157-169.

[8] L. Durand, W. Polifke. Int'l Gas Turbine and Aeroengine Congress & Exposition, 2007. 869-878.

[9] T.D. Butler, P.J. O'Rourke, Symposium on Combustion 16 (1976) 1503-1515.

[10] S R. Turns. An introduction to combustion : concepts and applications[M]. WCB/McGraw-Hill, 2000. p 219.

[11] R. Sankaran, E.R. Hawkes, J.H. Chen, T. Lu, C.K. Law, Proc. Combust. Inst. 31 (2007) 1291-1298.

[12] T. Poinsot, D. Veynante, R.t.edwards Inc (2001).

[13] J. Wang, Z. Wei, S. Yu, W. Jin, Y. Xie, M. Zhang, Z. Huang, Fuel 148 (2015) 1-8.

[14] C. Clanet, G. Searby, Combustion & Flame 105 (1996) 225-238.

11th Asia-Pacific Conference on Combustion,

The University of Sydney, NSW Australia
10th-14th December 2017.

Effect of radiation reabsorption on NO formation in premixed H_2/CO/air counterflow flames

Daoguan Ning, Bin Yao, Aiwu Fan*

State Key Laboratory of Coal Combustion, Huazhong University of Science and Technology

Luoyu Road,Wuhan, Hubei, 430074, China

Abstract

The effect of radiation reabsorption on flame temperature and NO formation of premixed H_2/CO/air counterflow flames was numerically investigated using modified OPPDIF code incorporated with the statistical narrow band model (SNB) and optical thin model (OTM). The results indicate that the effects of radiation reabsorption on flame temperature and NO formation both increases with the decrease of hydrogen proportion in the fuel. However, the relative errors between SNB and OTM are negligible for flame temperature, but become serious in predicting NO formation of H_2-lean syngas flames. Moreover, the effect of radiation reabsorption on NO formation strongly depends on equivalence ratio, namely, it is more significant for lean flames than rich flames. Kinetic mechanism analysis shows that the effect of radiation reabsorption has a strong effect on thermal NO formation, a moderate effect in the N_2O-route, and a weak effect in the NNH-route.

1 Introduction

Syngas is a clean and renewable fuel that consists essentially of CO and H_2[1]. Though it is a relatively environmentally friendly fuel, NOx emission is still a challenge because of high-temperature reaction of N_2. With increasingly stringent legislation on emissions, accurate prediction of NO formation of syngas combustion is important in investigating NO formation mechanism and it plays a significant role in the development of low-NOx combustion devices fueled with syngas.

NO emission characteristics of syngas combustion have gained numerous attention. Giles et al.[2] numerically investigated the effects of H_2-to-CO ratio and dilutions on NO emission. The results indicate that H_2O is more effective for mitigating NO emission than CO_2 and N_2. The influences of syngas components and strain rate on NO formation and combustion property were studied by Som et al. [3]. The results demonstrate that NO is generated primarily by the thermal route and the stable flame can be observed over wide ranges of strain rate and equivalence ratio. Park and Kim [4] numerically investigated the effects of N_2 dilution on NOx formation characteristics of CH_4/CO/H_2 syngas and air counterflow non-premixed flames. The computations manifest that the impacts of dilution on NO reduction become less effective with the increase of CH_4 proportion. Park et al. [5] studied the influence of radiation heat loss on flame characteristics of syngas diffusion flames diluted with CO_2 at a strain rate of $15s^{-1}$. The simulation results indicate that temperature difference between cases with and without considering radiation increases as CO and CO_2 mole fractions increase.

Literature survey showed that most of the numerical investigations adopted optical thin model (OTM) to calculate radiation heat loss, in which the radiation reabsorption was neglected. Although a few studies [6, 7] used statistic narrow band (SNB) model to investigate NO formation of syngas flames, quantitative results are still very scarce. Since the radiation reabsorption has a vital impact on NO formation of CH_4/air counterflow flame according to the reports of Wang and Niioka [8, 9], the similar quantitative investigation on syngas flame should be necessary. Therefore, the major objective of the present study is to numerically quantify the radiation reabsorption effect on flame temperature and NO emission of premixed H_2/CO/air counterflow flames over a wide range of fuel compositions and equivalence ratios.

2 Model description

The theoretical model considered here is a back-to-back counterflow premixed laminar flame, as schematically shown in Fig. 1. The distance between the two nozzles is d=30mm. The two mixture streams have the same initial temperature (300K) and velocity (15cm/s). The ambient temperature and pressure are 300K and 1atm, respectively. The global strain rate a is defined as:

$$a = 2|u|/d \qquad (1)$$

where u is the velocity at the nozzle exit. The magnitude of strain rate is fixed at $10s^{-1}$ in this study.

*Corresponding author. Fax: +86-27-87540724

E-mail address:faw@hust.edu.cn

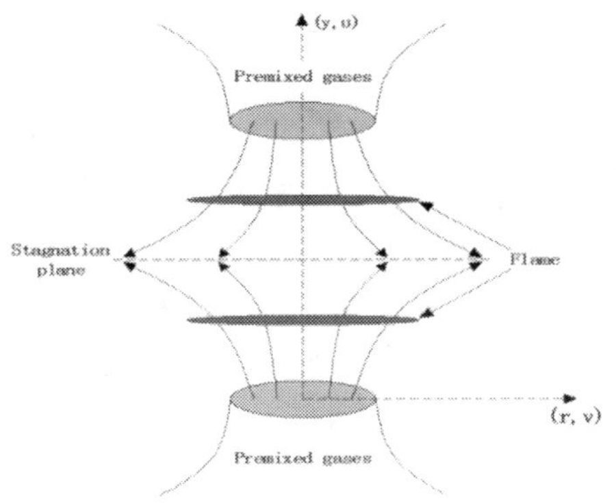

Figure 1: Schematic of the premixed counterflow flames.

The OPPDIF [10] code incorporated with different radiation models was used to simulate the premixed H_2/CO/air counterflow flames. The thermodynamic and transport properties were determined by using CHEMKIN II database [11] and TRANSPORT package [12], respectively. The detailed reaction mechanism USC_Mech_II [13] coupled with NOx reactions from GRI2.11 [14] was adopted for the present work. This reaction mechanism has been validated by Sahu and Ravikrishna [7]. In this study, two radiation models, namely, OTM and SNB, were used to compute radiation heat loss. The radiation participating species are CO, CO_2 and H_2O. In OTM, the heat flux of radiation heat transfer is calculated by [15]:

$$q_r = -4\sigma K_p(T^4 - T_\infty^4) \quad (3)$$

$$K_p = -\sum P_i K_i, i = CO_2, H_2O, CO \quad (4)$$

where σ is the Stefan-Boltzmann constant; K_p is the Planck mean absorption coefficient for the mixture. P_i and K_i are the partial pressure and Planck mean absorption coefficient for species i, respectively. Differing from OTM, radiation reabsorption is considered in the SNB model with exponential-tailed inverse line strength distribution. The spectral and temperature ranges and narrow band-width are 150~9300 cm^{-1}, 300-2900 K and 25 cm^{-1}, respectively. The radiative transfer equations were solved by the discrete ordinate method (DOM) with S6 quadrature scheme (detailed description about SNB model can be found in Ref.[8]). By comparing the results obtained by OTM and SNB, the effect of radiation reabsorption can be determined quantitatively.

3 Results and discussion

3.1 Effect of fuel compositions

To examine the effect of fuel compositions on maximum flame temperature and NO formation, stoichiometric H_2/CO/air mixtures were chosen. Figure 2 shows the maximum flame temperature (red and black lines) computed by using OTM and SNB for various fuel compositions. The relative errors are also presented in the same figure (green columns). It is seen that the maximum flame temperature modestly rises with a decreasing H_2 mole fraction because of different quantity and thermal properties of combustion products and heat release of different syngas (detailed clarification can be found in our previous study [16]). Moreover, the absolute discrepancy and relative error between the maximum flame temperatures predicted by these two radiation models increase with the decrease of H_2 content because of more reactant CO and product CO_2 in H_2-lean syngas flames, which both can enhance radiation reabsorption. In general, the effect of radiation reabsorption on the maximum flame temperature can be neglected because the largest relative error is only 3.1%.

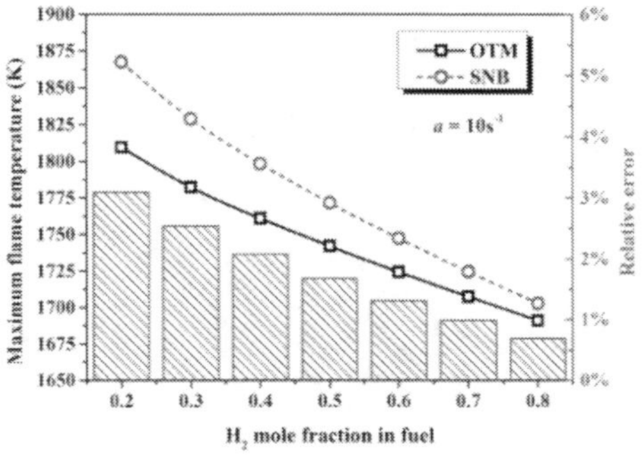

Figure 2: Maximum temperature of stoichiometric H_2/CO/air flames with different H_2 content.

Similarly, Fig. 3 demonstrates that the disparity between the peak NO mole fractions calculated by OTM and SNB increases with the decline of H_2 content, and more sharply at H_2-lean side (H_2 mole fraction<0.5). Furthermore, on the H_2-rich side, the relative errors (blue columns) between OTM and SNB are all below 5%. Thus, the effect of radiation reabsorption on NO formation of H_2-rich syngas combustion is negligible. In contrast, on the H_2-lean side, the relative errors are close to or even greater than 10%. The largest relative error reaches up to 19% at the 0.2H_2 condition. This demonstrates that radiation reabsorption can exert a significant effect on NO formation in H_2-lean syngas premixed combustion at small strain rates.

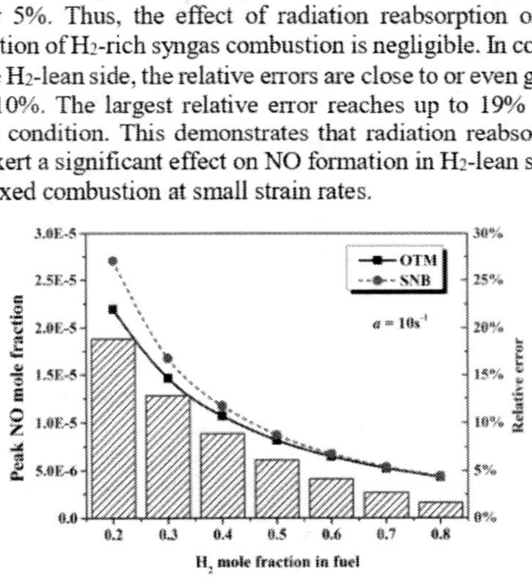

Figure 3: Peak NO mole fraction of stoichiometric H_2/CO/air flames with different H_2 content.

3.2 Effect of equivalence ratio

In this section, the effect of radiation reabsorption on flame temperature and NO formation will be discussed under various equivalence ratios. Two fuel compositions were selected to represent H_2-lean and H_2-rich syngas, namely, 0.2H_2+0.8CO and 08H_2+0.2CO.

Figure 4 shows the variation of maximum flame temperature and quantitative effect of radiation reabsorption (i.e., relative error) versus equivalence ratio. A well-known tendency is observed, namely, the maximum flame temperature first increases and then decreases with an increasing equivalence ratio and peaks at the stoichiometric mixture. Moreover, the temperature disparity shares the same tendency with the maximum flame temperature because the concentration of dominant radiation media (i.e., CO_2 and H_2O) is the highest at the equivalence ratio of unity. In addition, the histogram of the relative error exhibits an almost symmetrical distribution relative to the stoichiometric mixture.

This implies that the effects of radiation reabsorption have a roughly equal significance for both lean and rich mixtures. As the ability of radiation reabsorption of CO_2 is stronger than that of H_2O, the effect of radiation reabsorption on the flame temperature of H_2-lean syngas is more pronounced than that of H_2-rich syngas, which can be deduced from a comparison between Fig. 4(a) and (b).

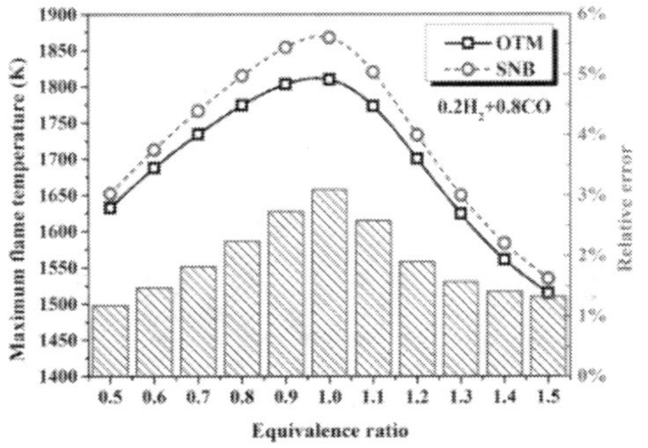

(a) $0.2H_2$+0.8CO

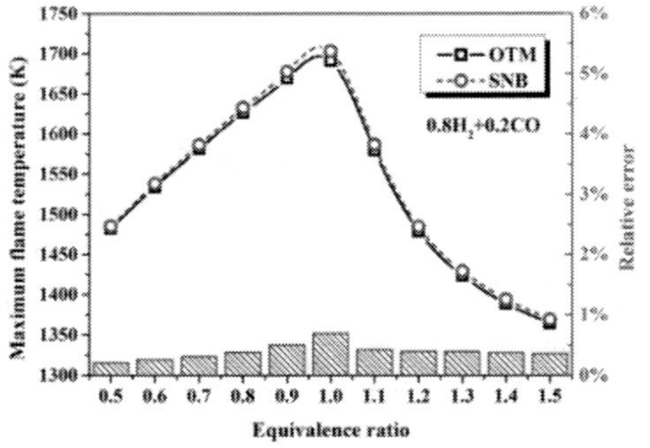

(b) $0.8H_2$+0.2CO

Figure 4: Variation of maximum flame temperature versus equivalence ratio.

Fig. 5 shows the profiles of peak NO mole fraction calculated by OTM and SNB as a function of equivalence ratio for both H_2-lean and H_2-rich syngas. It is seen that the profile of peak NO mole fraction versus equivalence ratio is similar with that of maximum flame temperature because the generation of NO is closely related to the maximum flame temperature and a higher flame temperature aggravates the NO formation. However, the relative errors between the NO predictions by OTM and SNB are apparently larger for lean mixtures as compared to rich mixtures, especially for H_2-lean syngas fuels (Fig. 5(a)). In other words, the effect of radiation reabsorption on NO formation in lean flames is more significant than rich flames. This is because more thermal NO is generated in lean flames due to the higher flame temperature.

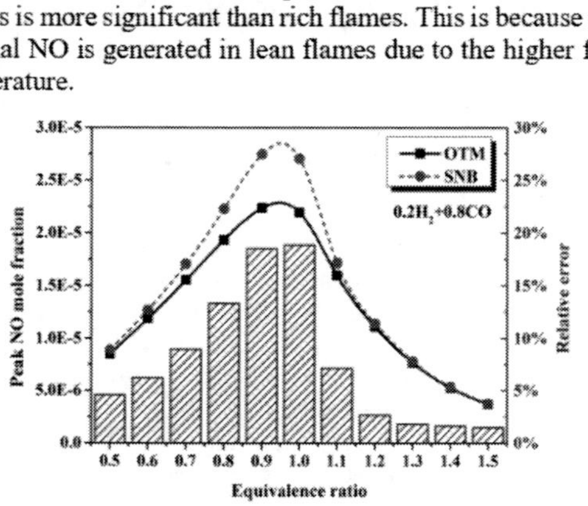

(a) $0.2H_2$+0.8CO

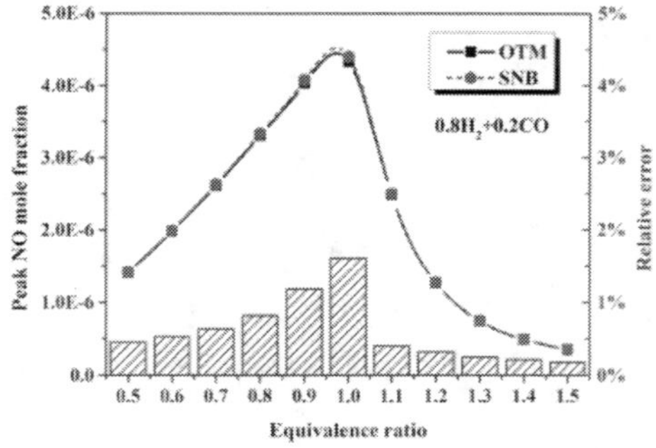

(b) $0.8H_2$+0.2CO

Figure 5: Variation of peak NO mole fraction versus equivalence ratio.

3.3 Kinetic mechanism analysis

Since the effect of radiation reabsorption has the most significant effect on the NO formation of stoichiometric ($0.2H_2$+0.8CO)/air flame, kinetic mechanism analysis was performed for this case. For pure H_2/CO/air flames, there are three types of NO formation routes, namely, thermal route, N_2O route and NNH route. Note that for the syngas that consists only of CO and H2 the prompt NO does not appear due to the absence of CH radical [6]. Figure 6 shows the relative error of integral net reaction rate calculated by OTM and SNB. Here, the integral net reaction rate is calculated from $\int_0^d R_i(y)dy$, where $R_i(y)$ is the net reaction rate of the elementary reaction i at position x. It is evident that the relative error of the thermal route is the largest, followed by the N_2O route, and the relative error of the NNH route is the smallest. Thus, neglecting radiation reabsorption will bring about a notable error in the prediction of thermal NO formation because the formation of thermal NO is extremely sensitive to temperature and its sensitivity increases with the rising of temperature.

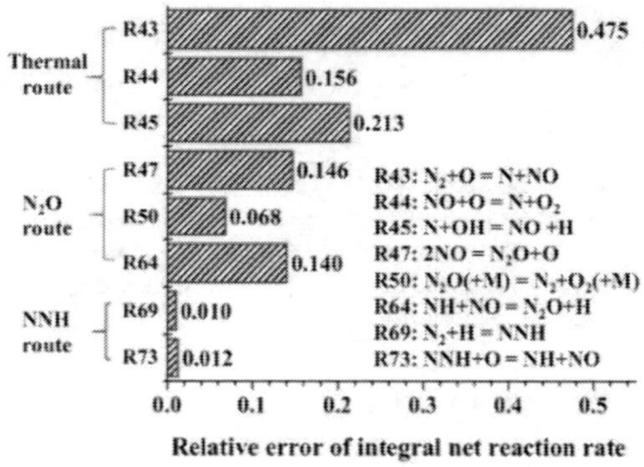

Figure 6: Relative error of integral net reaction rate of elementary reactions in various NO formation routes.

Figure 7 depicts the spatial distribution of NO production rate through each elementary reaction of the thermal route. As the flame structure is symmetrical about the stagnant plane, only the distribution on one side is plotted in Fig. 7. It is noted seen that the NO formation zone of R43 is remarkably wider than the other two reactions. This is because NO produced through R43 is generally considered to be formed in the post-flame gases [15] and the post-flame zone is wide at small strain rate. Moreover, there is a considerable discrepancy between NO production rates of R43 predicted by SNB and OTM in the post-flame zone.

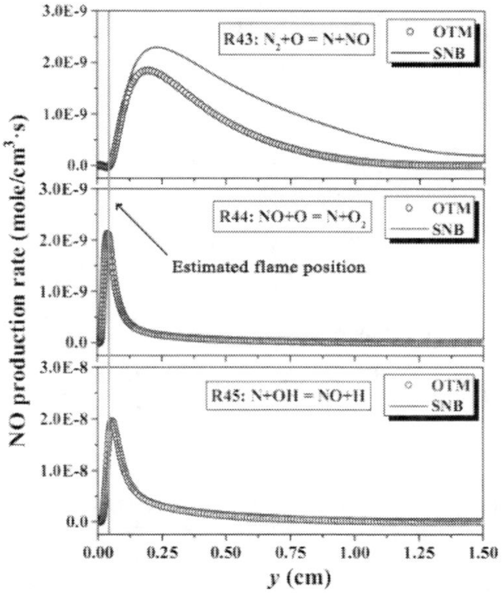

Figure 7: Spatial distribution of net production rate of thermal NO through each elementary reaction in the thermal route.

To reveal the underlying reasons for the above-mentioned phenomenon, Figure 8 illustrates the temperature profiles calculated by adopting SNB and OTM. It is manifested that OTM underestimates the temperature of post-flame gases because the flue gas is rich of CO_2 and H_2O, which have a strong ability of radiation reabsorption. As the activation energy for R43 is relatively larger (319,050 kJ/kmol), this reaction has a strong temperature dependency [17]. This explains why radiation reabsorption can significantly affect the NO production rate of R43.

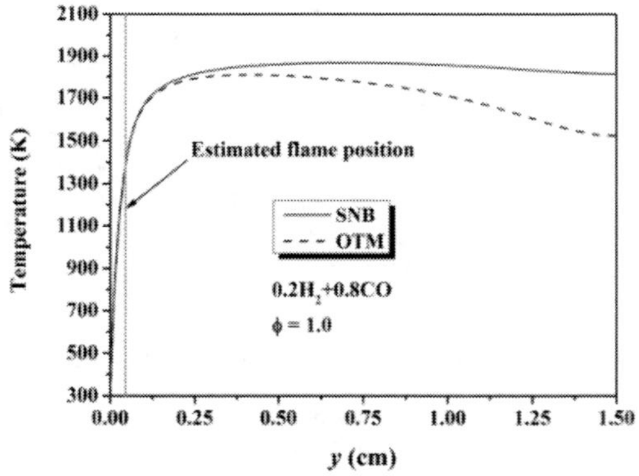

Figure 8: Temperature distribution of stoichiometric premixed (0.2H_2+0.8CO)/air flame.

4 Conclusions

The effect of radiation reabsorption on flame temperature and NO formation of H_2/CO/air counterflow premixed flames were numerically investigated over a wide range of equivalence ratio flames and H_2-to-CO ratio. Based on the numerical results obtained in this study, conclusions can be summarized as follows:

(1) The effect of radiation reabsorption can be neglected in predicting the temperature of premixed H_2/CO/air counterflow flames. However, it significantly affects NO formation of H_2-lean syngas/air flames with an equivalence ratio from 0.7 to 1.1.

(2) The effect of radiation reabsorption first increase and then decreases with the increase of equivalence ratio. Moreover, the effects on the maximum flame temperature are roughly equal for lean and rich flames, whereas it is much stronger in lean flames than rich flames on NO formation.

(3) Kinetic mechanism analysis indicates that radiation reabsorption has a strong effect on thermal NO formation, a moderate effect in the N_2O-route, and a weak effect in the NNH-route.

5 Acknowledgment

This work was supported by the National Natural Science Foundation of China (Grant Number: 51576084).

References

[1] T. Lieuwen, V. Yang, R. Yetter. Synthesis gas combustion: fundamentals and applications. CRC Press; 2009.

[2] D.E. Giles, S. Som, S.K. Aggarwal, Fuel 85 (12-13) (2006) 1729-1742.

[3] S. Som, A. Ramirez, J. Hagerdorn, A. Saveliev, S. Aggarwal, Fuel 87 (3) (2008) 319-334.

[4] S. Park, Y. Kim, Int. J. Hydrogen Energy 42 (16) (2017) 11945-11961.

[5] J. Park, D.S. Bae, M.S. Cha, J.H. Yun, S.I. Keel, H.C. Cho, T.K. Kim, J.S. Ha, Int. J. Hydrogen Energy, 33(23) (2008) 7256-7264.

[6] H.Y. Shih, J.R. Hsu, Combust. Flame 159(5) (2012) 1851-1863.

[7] A.B. Sahu, R. Ravikrishna, Combust. Flame, 173(2016)208-228.

[8] J. Wang, T. Niioka, Combust. Theor. Model, 5(3) (2001)385-398.

[9] J. Wang, T. Niioka, Proc. Combust. Inst., 29(2002)2211–2218.

[10] A. Lutz, R. Kee, J. Grcar, F. Rupley, OPPDIF: A Fortran Program for Computing Opposed Flow Diffusion Flames, Report No, SAND96-8243, Sandia National Laboratories, 1996.

[11] R. Kee, F. Rupley, J. Miller, Chemkin-II: A Fortran chemical kinetics package for the analysis of gas-phase chemical kinetics, Report No.SAND89-8009, Sandia National Laboratories, 1989.

[12] R.J. Kee, G. Dixon-Lewis, J. Warnatz, M.E, Coltrin, J.A. Miller, TRANSPORT: A Fortran computer code package for the evaluation of gas-phase multicomponent transport properties, Report No. SAND86-8246, Sandia National Laboratories 1986.

[13] H. Wang, X. You, A.V. Joshi, S.G. Davis, A. Laskin, F. Egolfopoulos, C.K. Law, available at <http://ignis.usc.edu/USC_Mech_II.htm>.

[14]C. Bowman, R. Hanson, D. Davidson, W. Gardiner Jr, V. Lissianski, G. Smith, D. Golden, M. Frenklach, M. Goldenberg,available at <http://www.me.berkeley.edu/gri-mech/>.

[15] C.L. Tien, Adv Heat Transfer, 5 (1969) 253-324.

[16] Ning D, Fan A, Yao H, Int. J. Hydrogen Energy, 42(15) (2017) 10466-10474.

[17] S. R. Turns. An Introduction to Combustion: Concepts and Applications, 3rd Edition, McGraw–Hill, New York, 2012.

11th Asia-Pacific Conference on Combustion,
The University of Sydney, NSW Australia
10th -14th December 2017.

Characteristics of Pure Oxygen/methane Flames in a Rapidly Mixed Tubular Flame Burner

Baolu Shi[1], Bo Li[1], Guoxing Wang[1], Xiaoyao Zhao[1], Jie Hu[2], Ningfei Wang[1]

[1]School of Aerospace Engineering, Beijing Institute of Technology

No.5 Zhongguancun South Street, Beijing, 100081, China

[2] Department of Mechanical Engineering, Kitami Institute of Technology

165 Koen-cho Kitami, Hokkaido, 090-8507, Japan

Abstract

Recently pure oxygen combustion has received keen interest from viewpoints of saving energy and reducing the emissions (NO_x and CO_2). To avoid potential hazards of flame flash back, an inherently safe technique of rapidly mixed tubular flame has been adopted in this study, in which fuel and oxygen are individually injected into the cylindrical burner. The flame structure and stability were experimentally investigated under various equivalence ratios. Detailed observations show that a steady tubular flame can be established close to the lean extinction limit; however, as the equivalence ratio was raised, steady combustion evolved into periodic oscillations; around stoichiometry the flame became steady, and thereafter evolved into unsteady state in the rich condition. Based on pressure fluctuation measurements and frequency spectral analyses, high frequency (around 2000 Hz) combustion-driven oscillation was observed. Steady tubular flame has been attained at the stoichiometric condition. Through comparing with the stoichiometric methane/air flame, the methane/oxygen flame illustrates remarkable advantages.

1 Introduction

Pure oxygen combustion is a promising approach for low-cost CCS combined with reduction of NO_x emissions [1, 2], since NO_x formation through the air containing nitrogen is completely inhibited, and in addition, in the exhausted gas carbon dioxide can be easily captured and sequestrated. Comparing with conventional air-firing combustion, pure oxygen combustion offers a number of technical advantages in terms of flame temperature, thermal efficiency, flame stability, heat transfer and pollutant emissions [1].

During the nineteenth and twentieth centuries, modern pure oxygen combustion techniques have been developed and they have been widely used in industries, including glass and ceramic manufactures as well as in the melting and/or cutting processes in steel-works [1]. However, due to very high burning velocity, destructive flame flash back even detonation may frequently occur if the combustion is conducted in a premixed mode. Hence most of studies adopted the non-premixed mode combustion, such as co-axial jet flames [3], counter-flow diffusion flame [4] and edge flame [5]. To avoid flame flash back, Shi et al. [6] adopted an inherently safe technique of rapidly mixed tubular combustion [7] to investigate the oxy-fuel combustion of methane under various oxygen contents. In this swirl tubular flame burner, the fuel and oxidizer were separately injected into the cylindrical combustion tube through their own slits to prevent flame flash back. Different from traditional swirling flame, the obtained tubular flame is usually a deterministic laminar flame.

In our previous studies, variation of flame structure with oxygen mole fraction was analyzed [6, 8]. And it was found that the fuel/oxygen mixing played an important role in determining the flame characteristics. The Damköhler number defined as the mixing to reaction time ratio has adopted to quantify the requirement for establishment of steady tubular combustion, which should be smaller than or around unity.

In the present study, to make a full utility of the remarkable merits of the tubular flame, an optimized tubular flame burner was adopted to attempt pure oxygen/methane combustion from lean to rich. The tangential inlet of the burner was reduced to very thin, and in addition, the width of fuel slit was reduced to half of the oxidizer slit according to the stoichiometry of methane/oxygen reaction. A high speed video camera was employed to record the temporal flame variation; a microphone was used to measure acoustic pressure fluctuations (APF), on which spectral analyses were made.

2 Experimental and Numerical Methods

The schematic of the burner used in the experiments is showed in Fig.1. The burner was made of stainless steel, which consists of four tangential slits and a downstream quartz tube. A quartz window was installed at the closed end to permit viewing and photographing. The inner diameter of the burner is 24 mm and the slit length is 12 mm. The length of the combustion tube was 75 mm.

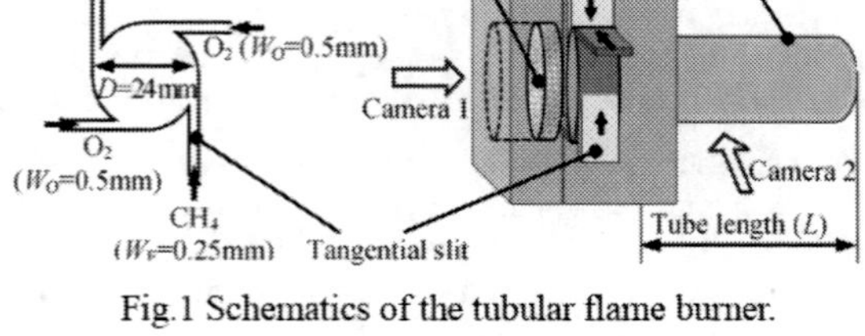

Fig.1 Schematics of the tubular flame burner.

Corresponding author: Baolu Shi. Fax: +86-10-6891-3623
E-mail address: shibaolu@bit.edu.cn

The fuel used was methane, which was vertically injected through two fuel slits with width of 0.25 mm; while the oxidizer was horizontally injected through two oxidizer slits (width of 0.5 mm). The equivalence ratio Φ is the overall value for the rapidly mixed combustion assuming complete mixing.

A high speed video camera Phantom v7.3 (VRI) was employed to record the flame structure. A frame rate of 10000 frames/s, a resolution of 512×512 pixels, and an aperture of 2.8 were utilized. To investigate the unsteady combustion, APF were recorded by a microphone MPA416 (BSWA TECH) controlled by a Labview software program at a sampling frequency of 50 kHz. The microphone was mounted perpendicular to the burner axis and was fixed 0.65 m away from the burner exit.

To make a further investigation on the acoustic resonant oscillation, the adiabatic flame temperature and sound speed were numerically analyzed using the CHEMKIN software [9, 10]. The GRI-MECH version 3.0 [11] was used as the reaction mechanism. Initial temperature of 298K, pressure of 1 atm, and composition of the unburned mixture were assigned to define the inlet boundary condition.

3 Results

At first, combustion was tested from lean to rich condition, in which oxygen flow rate Q_{O2} was maintained at 4 m^3/h. Figure 1 shows the results. At the very lean condition of Φ=0.1, a steady tubular flame with thin flame front (left picture of 1st row) was obtained. In the downstream, a blue flame with weak stripe was observed (the image is artificially intensified). By gradually increasing Φ, in the front view, two luminous flames (indicated by red arrows) were anchored at the exit of fuel slit (2nd row), which interrupted fuel and oxygen mixing. And the steady tubular flame evolved into periodic oscillations. Detailed analysis will be made in Figs. 4 and 5. By further increasing Φ to around unity, a steady tubular flame was obtained again, though the flame front was thick and illustrated non-uniform structure. By further raising Φ to the rich condition, the flame became unsteady again.

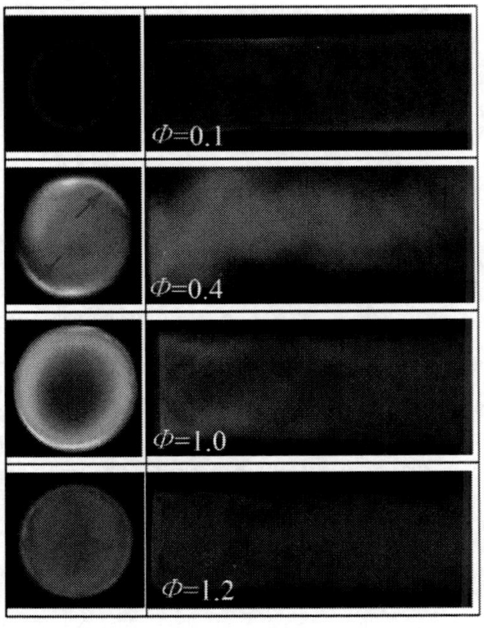

Figure 2 Flame appearances of CH_4/O_2 mixtures in the rapidly mixed tubular flame burner (Φ=0.1-1.2, Q_{O2}= 4 m^3/h).

To confirm the steady tubular flame at the stoichiometry, temporal variation of the flame structure was illustrated in Fig. 3. In the front view, beyond the center dark area, a circular reaction zone was formed though not uniform. The downstream flame was also non-uniform but steady. It is seen that a steady tubular flame was established at the stoichiometry under the pure oxygen condition, for which the adiabatic flame temperature can reach 3050 K. According to our previous observations on the mixing process in the rapidly mixed tubular flame burner, when the mean injection velocities of fuel and oxygen stream are equivalent, the mixing is best. In current case of Φ=1.0, injection velocities of fuel and oxygen are the same and very high (92.6 m/s) to attain the steady tubular flame.

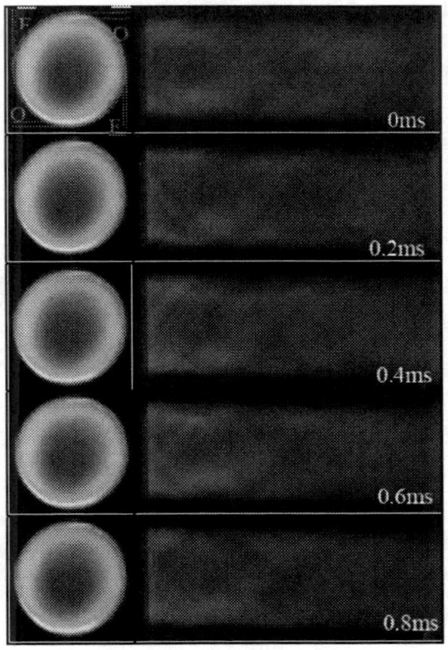

Fig.3 Steady tubular combustion at Φ=1.0 (Q_{O2}= 4.0 m^3/h, L=75 mm).

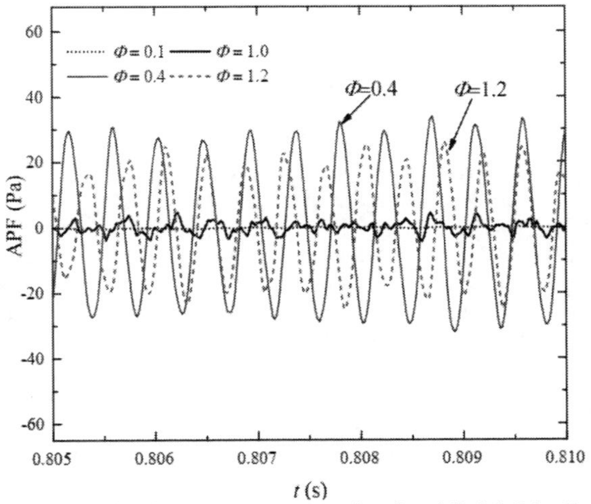

Fig.4 APF for the pure oxygen combustion (Φ=0.1-1.2, Q_{O2}= 4.0 m^3/h).

To make a further investigation, APF for the four tests in Fig. 2 were also examined. Figure 4 illustrates the results. At the very lean condition of Φ=0.1 (tubular flame), the amplitude of APF was almost around zero (dot curve); by increasing Φ to 0.4 which was laid in the oscillation region, the maximum amplitude reached around 35 Pa (blue solid curve); around the stoichiometry, the amplitude was almost within 5 Pa (bold solid curve), which certificates the establishment of steady tubular combustion at the stoichiometry; by further increasing Φ to 1.2 (dash curve), the amplitude of APF raised to a high value above 25 Pa, illustrating unsteady combustion.

A spectral analysis has been made on the two unsteady combustion tests, as shown in Fig.5. At the condition of Φ=0.4, an evident peak of large amplitude can be found at f =2269 Hz (Fig.5 (a)). Meanwhile, temporal variation of the flame structure was inserted in Fig. 5(a), which illustrates a periodic variation around 0.5 ms, particularly the anchored flame. It is seen that the cycle period of APF at f =2269 Hz coincides with variation period of flame structure. It may be inferred the anchored flame should be responsible for the periodic combustion-driven oscillation. In the rich condition (Fig.5 (b)), several peaks are distributed around f = 2584 Hz. The peak magnitude of APF reduced to half of that at Φ=0.4. Owing to difficulty in measurement for this high temperature flame, distinct variation in flame structure was not observed in the test.

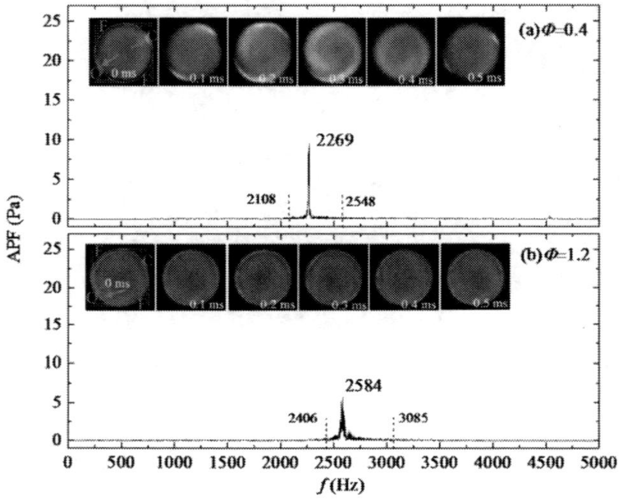

Fig.5 APF spectra (Φ=0.4 and 1.2, $Q_{O\text{-total}}$=4.0 m^3/h, L=75 mm).

To illustrate the advantage of the observed tubular flame under the pure oxygen condition, a comparison is made with the stoichiometric methane/air combustion in the same burner, in which pure oxygen is replaced with air with the same flow rate. Figure 6 shows the flame images, in which the flow rate of oxidizer is maintained at 6.0 m^3/h. As for the flame appearances, in the case of CH_4/air flame (upper row), a steady tubular flame with thin flame front is obtained, and inside the thin luminous zone is the burned gas with uniform temperature; for the CH_4/O_2 flame (lower row), a thick circular reaction zone is formed. The flame in the downstream illustrates helical structures, however, it is steady according to the images taking by the high speed video camera.

Based on numerical calculation with Chemkin, the adiabatic temperature is 2220 K for the stoichiometric CH_4/air flame. And the thermal conductivity of the combustion products is about 0.25 W/m·K according to ref. [12]. While the temperature reaches 3050 K for the CH_4/O_2 flame, and the thermal conductivity is 2.8 W/m·K [12]. Based on Fourier's law, for the CH_4/O_2 flame, heat exchange through thermal conduct becomes almost ten times higher than that of CH_4/Air. Furthermore, heat radiation is also much enhanced in the CH_4/O_2 flame, since the overall mole fraction of H_2O and CO_2 in the products is around 0.5. As for the net heat release rate, the peak value for premixed CH_4/O_2 flame is 2.54x10^5 J/cm^3·s, which is about 55 times higher than that of CH_4/Air flame (4.5x10^3 J/cm^3·s). In addition, according to fuel flow rate, the pure oxygen flame provides a large heat output of 27.3 kW (low caloric value), which is 4 times higher than that of air flame (5.5 kW). It is seen that the obtained steady pure oxygen flame shows much higher thermal efficiency and larger heat output than the air flame.

In addition, for the pure oxygen combustion, the overall mole fraction of the three radicals O, OH and H even approaches 0.2. In the heating process, when these active radicals reach the cold boundary, it will release much heat, referred to as thermochemical heat release [12].

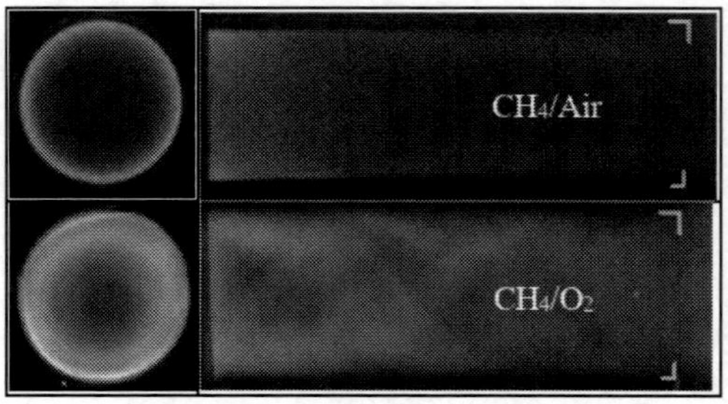

Figure 6 Stoichiometric CH_4/air (upper) and CH_4/O_2 (lower) flames (Φ=1.0, Q_{Air}= 6 m^3/h).

4 Discussion

The pressure fluctuation spectra in Figs. 4 and 5 show evident discrete characteristic peaks, indicating oscillatory combustion. Hence acoustic resonant frequencies in the tubular burner are roughly estimated by assuming that gas temperature is homogenous in the tubular flame burner. The frequencies can be calculated by assuming that the burner is a simple tube with one end closed and the other end open. According to the fundamental theory of the acoustic resonance in a cylindrical cavity with the boundary conditions of the present tubular flame burner, the natural frequencies can be given as [13, 14],

$$f = \frac{c}{2\pi}\sqrt{\left(\frac{\alpha_{mn}}{r}\right)^2 + \left(\frac{(2n_z+1)\pi}{2(l+\Delta l)}\right)^2} \qquad (1)$$

in which c (m/s) is the sound speed, m and n are the tangential and radial mode numbers (0, 1, 2. . .), respectively, α_{mn} is the nth solution of dJ_m/dr =0, where J_m is the mth order Bessel function, r (m) is the radius of the cylindrical cavity, nz is the axial mode number (0, 1, 2, . . .), l (m) is the length of the cylindrical cavity, and Δl (m) is the open end correction. The values of α_{mn} can be found in the literatures on mathematics and acoustics [12, 13].

By substituting α_{mn}=0 in Eq. (1), we obtain the pure axial model of natural frequency, which is most frequently observed in the resonant oscillatory combustion. The frequency can be obtained through the following relation,

$$f_z = (2n_z + 1)\frac{c}{4(l + \Delta l)} \quad (2)$$

In Eqs. (1 and 2), the sound speed depends on the actual temperature in the tubular flame burner. In our tests, due to difficulty in precisely measuring the very high flame temperature (above 2500 K), the adiabatic flame temperature based on Chemkin calculation is simply used to obtain the sound speed. This may lead a relative higher frequency in Eqs. (1 and 2), because of heat loss and inhomogeneous temperature distribution.

Here two combustion tests in the Fig.5 were analyzed. The harmonic frequencies based on the spectra of APF are summarized in the 2nd column; the frequency of flame structure variation calculated based on 300 flame images is also listed in Table 1 (3rd column). The calculated natural frequencies based on Eq. (1) exceed 20,000 Hz. In addition, sound speed at normal condition (340m/s) is also adopted, which gives the minimum cut-off frequency over 8000 Hz. These frequencies are much higher than the measured values. Hence the pure axial mode natural frequencies are calculated (based on Eq. (2)) and listed in the last column of Table 1.

Table.1 Measured oscillation frequencies and natural frequencies of axial mode oscillations.

Φ	Frequency of APF (Hz)	Frequency of flame structure variation (Hz)	Axial model acoustic resonant oscillation (Hz)
0.4	2269	2261	2413
1.2	2584	2567	3040

In the case of Φ=0.4, it can be seen that the measured APF frequency (2269 Hz) coincides well with the corresponding flame appearance variation frequency (2261 Hz), illustrating combustion-driven oscillation. The calculated axial mode natural frequency is 2413 Hz, which is a bit higher than APF frequency. The heat loss and particularly non-uniform temperature distribution should be the main causes, which yielding inaccurate sound speed in the calculation. In the case of Φ=1.2,APF frequency and frequency of flame variation are close to each other, however, they show a relative large difference with the axial mode natural frequency. In Fig. 5, the frequency, for which the amplitude is 2% of the peak amplitude, has also been marked by the vertical dash line offset the evident peak. The calculated axial mode frequency in the last column of Table 1 lie in the frequency range between two neighboring dash lines around the evident peak.

On the whole, the high frequency oscillation (above 2000Hz) observed in the test can be identified as the axial mode of acoustic resonant oscillation, i.e., the longitudinal acoustic resonant mode of the burner, which is different from that of tangential/radial mode observed in the large scale burners of propane/air premixed combustion [15].

5 Conclusions

Oxygen/methane combustion was experimentally investigated in a tubular flame burner with very thin inlet. To enhance mixing, the inlet width of fuel slit was reduced to half of the oxidizer slit according to methane/oxygen stoichiometry. Combustion was tested under various equivalence ratios. The following are specific conclusions.

(1) Various combustion modes were observed by varying Φ from the lean limit to the rich condition, from steady tubular flame to periodic oscillation, thereafter steady and unsteady combustion.

(2) Steady tubular flame around the stoichiometry has been obtained under the pure oxygen condition, which provides a (premixed-like) laminar flame with high temperature (adiabatic 3050 K) and thermal efficiency.

(3) The high frequency oscillations (above 2000 Hz) obersved were identified as axial mode acoustic resonant oscillation.

6 Acknowledgment

This work is supported by National Natural Science Foundation of China (No. 51676016).

References

[1] C.E. Baukal, Oxygen-Enhanced Combustion, CRC Press, Boca Raton, FL, 1998.
[2] T. Wall, Proc. Combust. Inst.31 (2007) 31-47.
[3] R.J. Harris, W. Horne and A. Williams, Combust. Flame 26 (1976) 311-321.
[4] A. Beltrame, P. Porshnev, W. Merchan-Merchan, et al., Combust. Flame 124 (2001) 295-310.
[5] S.K. Choi, J. Kim, S.H. Chung, J.S. Kim, Combust. Theor. Model.13 (1) (2009) 39-56.
[6] B. Shi, J. Hu, S. Ishizuka, Combust. Flame 162 (2015) 420-430.
[7] S. Ishizuka, T. Motodamari, D. Shimokuri, Proc. Combust. Inst. 31 (2007) 1085-1092.
[8] B. Shi, D. Shimokuri, S. Ishizuka, Combust. Flame 161 (2014) 1310-1325.
[9] R.J. Kee, J.F. Grcar, M.D. Smooke, J.A. Miller, A Fortran Program for Modeling Steady Laminar One-Dimensional Premixed Flames, in: Sandia National Laboratories (Ed.)Livermore, 1985.
[10] R.J. Kee, G. Dixon-Lewis, J. Warnatz, M.E. Coltrin, J.A. Miller, A Fortran Computer Code Package for the Evaluation of Gas-Phase, Multicomponent Transport Properties, in: Sandia National Laboratories (Ed.)Livermore, 1998.
[11] D.G. G. Smith, M. Frenklach, N. Moriarty, B. Eiteneer, M. Goldenberg, et al., GRI-Mech-An optimized detailed chemical reaction mechanism for methane combustion, Gas Research Institute, 1999.
[12] C.E. Baukal, B. Gebhart, Int. J. Heat Mass Transfer 39 (1996) 2989-3002.
[13] P.M. Morse, K.U. Ingard, Theoretical Acoustics, Princeton University Press, Princeton, 1987, p.902.
[14] D.R. Raichel, The Science and Applications of Acoustics, Springer, Verlag, 2006, p.144, 636.
[15] D. Shimokuri, Y. Shiraga, K. Ishii, H. Toh, S. Ishizuka, Combust. Flame 161 (2014) 2025-2037.

11th Asia-Pacific Conference on Combustion,
The University of Sydney, NSW Australia
10th -14th December 2017.

A Computational Study of Multiple Injection Strategies in Compression Ignition Engines at High Pressure

Gustav Nyrenstedt[1,*], Mohammed Jaasim [1], Eshan Singh[1], Rafig Babayev[1], Hong G Im[1], Mani Sarathy[1]

Bengt Johansson[1]

[1]King Abdullah University of Science and Technology

Thuwal, Makkah Province, 23955, Kingdom of Saudi Arabia

Abstract

Higher efficiency can be achieved by heat release control in order to get more ideal combustion and so extraction of maximum possible work. Concepts of multiple injectors are expected to increase the ability to control the heat release as well as decrease interference between the sprays. Computational Fluid Dynamics (CFD) studies on a Volvo D13 with altered compression ratio have been performed with n-heptane as fuel. The computations were used to derive methods that lead to constant pressure combustion utilizing three injectors with split injection strategies. Also the chemical mechanism used was validated, through Chemkin simulations, for the high pressure.

Promising CFD results with high pressures and control over the heat release were obtained. Isobaric combustion was achieved at pressures of up to 300 bar.

1 Introduction

New demands from customers as well as regulations from governments are expected to increase further in the future. These demands cover reduced emissions and so, better fuel economy. Especially the internal combustion engine (ICE) is being targeted when it comes to the problem of high emissions, such as particulates. European countries have set goals of zero pollution from vehicles in the future, which means that the internal combustion engine runs a risk of being banned.

One approach to reach higher fuel economy as well as lower emissions is to increase the engine efficiency. This is important for heavy-duty engines since fuel economy is more important when the driving distances are longer. A potential way of reaching higher efficiency in heavy duty diesel truck engines would be to use the double compression expansion engine (DCEE) concept that has shown potential of reaching 56 % indicated efficiency [1].

To further increase efficiency with this concept, more ideal thermodynamic cycles can be used [2]. This can be controlled by the rate of heat release. Isobaric combustion is one way of achieving high efficiency, which requires the timing of heat release to compensate for the pressure loss from expansion.

In order to reach improved control of the heat release, multiple fuel injectors can be used [3]. This is expected to give more degrees of freedom for the combustion control and make it easier to achieve ideal cycles at different load points.

However, it has not yet been evaluated if this concept is possible to combine with the high pressures used in for example the DCEE concept. The main focus of this study is to investigate how isobaric combustion can be achieved using three injectors at pressures close to 300 bar. In order to do so, due to the lack of experimental results, validation of chemical models at this high pressure is also needed. No experiments have been conducted for this study and as an initial effort, CFD simulations were performed to obtain an insight for the above said engine conditions.

*Gustav Nyrenstedt. ph: +966 560 458 328
E-mail Address: sven.nyrenstedt@kaust.edu.sa

2 CFD Study of Rate Shaping

To enhance understanding of how the heat release from combustion can be controlled, a 3D CFD study of different rate shaping strategies (control of the rate of injection) was performed using the three-dimensional computational CFD code CONVERGE. Also, a Chemkin study was performed to validate the chemical mechanism used at high pressures.

2.1 CFD Setup

The geometry studied here is one cylinder from a standard 4-stroke Volvo D13 engine with altered compression ratio (see Table 1). A constant engine speed of 1600 rpm was used for all different cases in order to compare the different injection strategies. All cases were run with n-heptane as fuel. Initial conditions at 180 CAD before top dead centre (BTDC) is shown in Table 2. Notable is that no initial swirl was set but instead an initial turbulence level was specified to enhance the mixing. Since the simulations started at 180 CAD BTDC, no intake stroke of air was performed and instead it was assumed that the combustion chamber was filled with air initially. The initial pressure was chosen as 10 bar.

Table 1: Geometry specifications of used cylinder

Parameter	**Magnitude**
Compression Ratio	11.5
Bore [m]	0.131
Stroke [m]	0.158
Displacement [l]	2.13
Con. Rod Length [m]	0.2675

The KH-RT spray breakup model was used to represent the spray breakup process [4] and the physical properties of diesel are incorporated into the code. Droplets were assumed to be spherical during the evaporation process and uniform thermal conductivity was assumed inside the droplet, i.e. the temperature distribution is homogeneous inside the droplet. Radius of the droplet was calculated using Frossling co-relation [5].

Evaporation of the droplet is governed by the mass transfer coefficient with quasi-steady state assumptions. The Rosslin-Rammler cumulative probability distribution was implemented to account for size distribution of droplets.

Table 2: Initial conditions at 180 CAD BTDC

Initial Conditions	Values
Pressure [bar]	10
Temperature [K]	500
Turbulent Kinetic Energy [m^2/s^2]	3.2

Turbulence is an important part of any CI engine and here a re-normalized group k-epsilon RANS turbulence model was used to account for in-cylinder turbulence. The chemical source terms are solved using multi-zone combustion sub-model by mapping the grid cells to the temperature bins of 5K and φ bins of 0.05 increments. The *Chalmers mechanism* for n-heptane reduced gas phase kinetic model was used to represent diesel combustion [6]. Other physical sub-models follow the standard built-in capabilities in CONVERGE v2.3 [7]. The simple Hiroyasu model was used for soot calculations [8] and the Zeldovich model was used for NOx calculations [9].

The mesh is generated in-situ, a base grid size of 4 mm was used in all three directions. Additional to this base mesh, embedding per injector was added to capture the flow near nozzles. A level 5 embedding was used for all the cases. Level 4 adaptive mesh refinement (AMR) was used based on criteria of temperature difference between the neighbouring cells (5K) inside the cylinder domain that refines the cells locally which is more economical than constant refinement of cells. Resulting new cell size in all three directions is calculated based on the standard formulation.

$$\text{New Cell Size} = \frac{\text{Base Grid}}{2^n} \quad (1)$$

where n is the level of refinement provided.

The CFD strategy is in general based on previous studies in a CI engine [10].

2.2 Injector Configuration

Three injectors were used in the simulations as shown by the representative sprays in Fig. 1. An injection pressure of 2500 bar along with nozzle holes with the geometric radius of 0.06 mm were used.

The reason for injection with two outer injectors is to gain additional degrees of freedom for the control of the heat release since all sprays are only using their own air. Another reason is to avoid the impingement on the wall which is increasing heat transfer. Additionally, little interference of the sprays occurs in this configuration. These last two features are expected to lower emissions and reduce heat losses without the impingement on the cylinder wall.

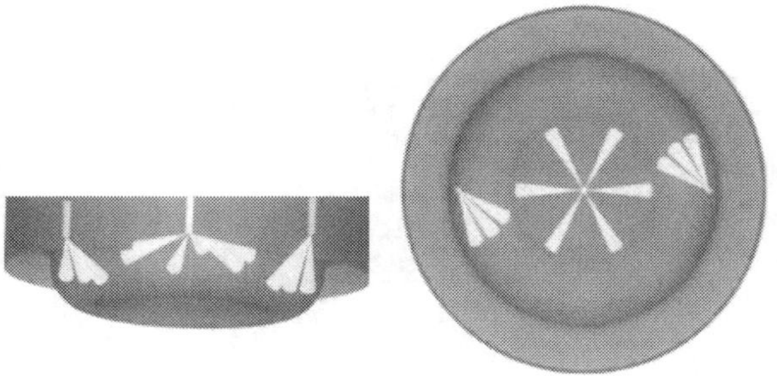

Figure 1: Representative view of the sprays in the three injector geometry.

2.3 Chemical Mechanism

At high pressures of around 300 bar, there are little or no experimental results to use for validation of chemical calculations models. Expected is that at high pressures the IDT (ignition delay timing) will not be much dependent on pressure change. This, since at high pressures chemical reactions such as combustion happens faster, which would decrease the IDT to very small values. A change in pressure would then not mean much.

To be able to see if a reduced chemical model, which can later be used in high-pressure CFD-simulations, 0-D calculations were performed using Chemkin™. This, since a reduced model, would be less computationally expensive. A detailed model was used in order to compare and investigate the accuracy of the reduced chemical model. Simulations with varying equivalence ratio as well as temperature were performed for pressures of 100, 200 and 300 bar. The detailed mechanism used is a mixture of iso-octane and n-heptane [11] but here, only the n-heptane part is used. The reduced model is an n-heptane mechanism, widely called the *'Chalmers mechanism'* [12].

The simulations were run in a constant volume homogenous batch reactor with n-heptane as fuel. The drawback of this is that the physical ignition delay, from evaporation and mixing etc., is not taken into account. This would definitely make the magnitude of the IDT inaccurate but it is here assumed that the trend will be the same for the real case. Combustion was considered achieved when reaching a temperature difference of 400 K between two calculation steps. The temperature sweep contained 25 equidistant points in the range 650-1250 K. This range was chosen since previous studies [13] have shown that there is little or no effect of varying pressure at either high or low temperatures. What could possibly be concluded from this is that the chemical models are still valid at high pressures, at least if the temperature is high enough.

The equivalence ratio was varied in order to see how the mechanisms behave at different engine load conditions. Only lean conditions were evaluated since a compression ignition engine is often run lean when diesel-like fuels are used.

3 Results

3.1 Ignition Delay Calculations

In general it can be said that the behaviour turned out to be quite as expected. Figure 2 shows that at lower temperatures there is little pressure dependence. From this figure, it is also realized that for high temperatures the ignition delay becomes so small that although there is some pressure dependence, it will not make much of a difference for the ignition delay (log- scale on y-axis). Notable is that, it was earlier said that the magnitude of IDT would not mirror the real case here, however it is assumed that it will be low enough at least when reaching temperatures that occur at TDC (around 1100 K).

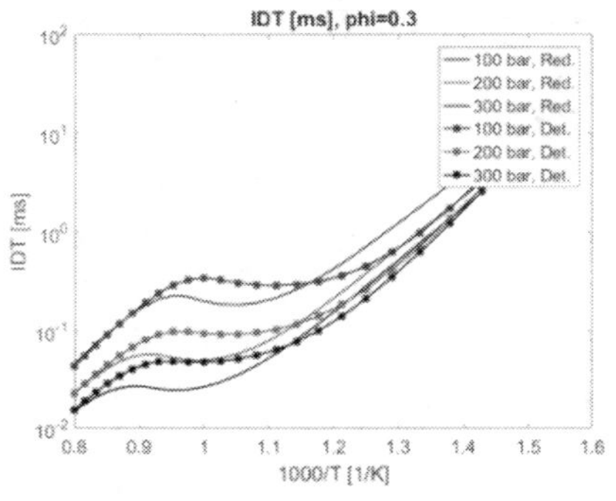

Figure 2: IDT vs temperature [K] for equivalence ratio of 0.3.

Two widely disparate mechanisms are responsible for ignition delay times at low, medium and high temperature zones. This results in a non-monotonically decreasing ignition delay time with respect to temperatures. At low temperatures, the molecular scission does not take place and atomic oxygen attached itself to the radical, followed by isomerization, and later beta-scissions. This leads to the chain branching step leading to OH, HO2 radical formation, which furthers the reaction at different rates (OH being a more reactive radical than HO2). This leads to a linear decrease of ignition delay at low temperatures (1000/T > 1.4). At high temperatures (1000/T <1), bond scission takes place readily leading to higher reactivity and lower ignition delay times with increasing temperature. In between the two regions, is a region where the sensitivity of ignition delay is relatively unaffected by changes in temperature. This region is called Negative Temperature Coefficient (NTC) region, due to the anomalous behaviour (as opposed to the Arrhenius behaviour of increasing reactivity with increasing temperature).

Reducing ignition delay times with increasing pressure is expected from previous literature. The extent of reduction is also in line with previous observations [14], which points to the fidelity of the simulation results. It is, as mentioned before, of interest to use a reduced mechanism to lower the expense of engine CFD calculations. Hence the assessment of the deviation was done compared to the detailed mechanism.

As expected, there is a big pressure dependence in the NTC region (see Figs. 2 and 3). This would then be the region of extra interest where it can be concluded if the reduced mechanism is valid or not. It can be seen that the absolute difference between the reduced and detailed mechanisms becomes bigger with increased pressure in the NTC region. This is, of course, not making the reduced mechanism more trustworthy for the high pressures. However, the error reduces in both low temperature and high temperature regions.

Comparing Fig. 2 and Fig. 3 we note that a higher equivalence ratio means shorter IDT in general which is expected due to addition of more fuel. The extra air acts as a cushion to the heat released from the chemical processes; hence the ignition delay of stoichiometric (usually a little rich mixtures) is lowest as there is no extra air (or fuel) to absorb the heat released. It is once again, in Fig. 3, noted that the increased pressure causes bigger differences in the NTC region.

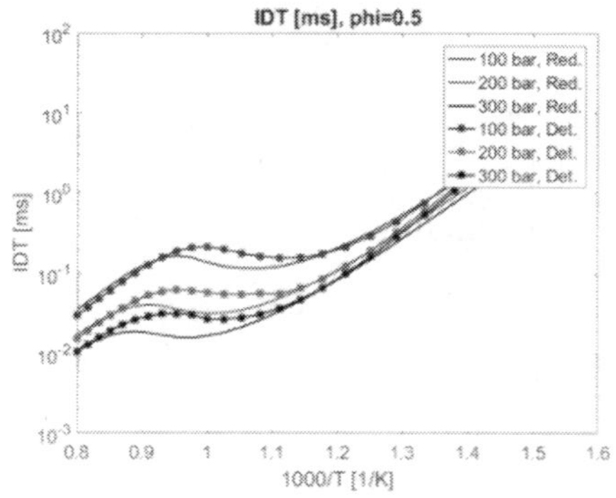

Figure 3: IDT vs temperature [K] for equivalence ratio of 0.5.

Concluded from all this is that the reduced mechanism is not very good in the NTC region for high pressures. The results at 300 bar for the two different equivalence ratios show monotonically increasing deviation, in the NTC region, from the detailed mechanism with increased pressure. However in many CI engine applications, such as used later in this study, the ignition starts close to TDC, which indicates high temperatures in general. For geometry and initial parameters later used in this paper the temperature at TDC is at values ~1100 K. This means that the reduced chemical model is trustworthy, since as has been seen in this 0-D study, there is smaller difference between reduced and detailed mechanisms for temperatures larger than 1000 K. Also the ignition delay would reach a smaller magnitude (in crank angle degrees) for high temperatures, which means that the absolute error will also be smaller than for low temperatures. With the engine speed of 1600 rpm, the IDT is less than one CAD for high temperatures.

Notable is that there might be local zones of varied temperature when the fuel is injected, however the temperature is assumed to be reasonably uniform since the pressure is high. High pressure gives shorter ignition delay since the temperature of the fuel is quickly rising and so the temperature distribution is quickly becoming uniform.

3.2 Isobaric Combustion

As mentioned before, for three injectors it is expected, that the heat release can be controlled in a simple way due to the many degrees of freedom. Many parameters such as in-cylinder pressure at the time of injection and turbulence in the cylinder during injection will play a major role in governing the heat release. The in-cylinder pressure as well as turbulence level will in general be different for changed injection timing. However, the studies performed here are concentrated in showing that it is possible to achieve constant pressure combustion at very high pressures.

Figure 4 shows a boot shaped split injection case and it is seen that isobaric combustion is indeed possible for high pressures when three injectors are used. The ignition delay is shorter at high pressures, which makes it easier to control the rate of heat release. This is seen in Fig.4 where the boot shaped rate of heat release, which is the desired profile, is accomplished. The split injection dictates the heat release so that the near constant pressure is achieved with the initial injection from the centrally placed injector. The pressure slightly increases and then decreases, although diminutive, before the second and third injections from the outer injectors are initiated. It is seen in Fig. 4 that the ignition delay is not long for these two later injections although much fuel is injected here. This is explained by the high pressure as mentioned before and also the multiple injectors used. The injector placement here means less interference and so there is air available for every single injection. It is understood that a small change of fuel amount or injection timing could reduce the small peak in cylinder pressure and a more perfectly isobaric combustion would be achieved.

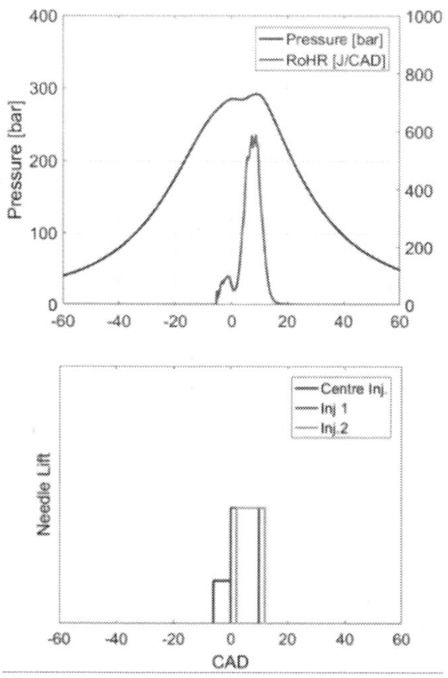

Figure 4: Pressure, Rate of Heat Release (RoHR) and Needle lift vs Crank angle degrees (CAD) using three injectors with split injection.

Since constant pressure combustion is now achieved and the benefits with multiple injectors in this matter discussed, future studies are interesting in the sense that many parameters are affecting the combustion. This means a large design of experiment and much possibilities as well as challenges. More injectors are expected to increase the production cost of the engine and therefore a study of the actual benefits in terms of efficiency is needed.

As a final note there is much more work to be done when it comes to chemical reactions at high pressures and this is only an initial study for future applications. Experimental results for validation would be very useful as well as more extensive computational studies.

4 Conclusions

- Reduced chemical mechanism trustworthy for high pressures when the temperature is also high.
- Three injectors provide good control of the heat release with increased degrees of freedom.
- It is possible to achieve isobaric combustion using three injectors at high pressures.
- Extended simulations as well as experiments are needed to determine possible potential of using multiple injectors with split injection at high pressures.

5 Acknowledgment

The simulations in this work were performed with the financial sponsorship of KAUST at the computing resources from KSL. Simulations in Converge were aided by Volvo Trucks.

References

[1] Lam, N., Tuner, M., Tunestal, P., Andersson, A. et al., "Double Compression Expansion Engine Concepts: A Path to High Efficiency," *SAE Int. J. Engines* 8(4):1562-1578, 2015, doi:10.4271/2015-01-1260.

[2] Bhavani Shankar, V., Lam, N., Andersson, A., and Johansson, B., "Optimum Heat Release Rates for a Double Compression Expansion (DCEE) Engine," SAE Technical Paper 2017-01-0636, 2017, doi:10.4271/2017-01-0636.

[3] T. Okamoto, Uchida, N., "New Concept for Overcoming the Trade-Off between Thermal Efficiency, Each Loss and Exhaust Emissions in a Heavy Duty Diesel Engine," *SAE Int. J. Engines 9(2):859-867,* 2016.

[4] R. Reitz, and Diwakar, R. , "Structure of High-Pressure Fuel Sprays," *SAE Technical Paper 870598,* 1987.

[5] A. A. Amsden, O'rourke, P. J., and Butler, T. D. , "KIVA-II: A computer program for chemically reactive flows with sprays (No. LA-11560-MS)," *Los Alamos National Lab., NM (USA),* 1989.

[6] Golovichev V., "Mechanism," 2016.

[7] K. J. Richards, Senecal, P. K., and Pomraning, E., "CONVERGE (Version 1.4.1) Manual," 2012.

[8] H. a. K. Hiroyasu, T., "Models for Combustion and Formation of Nitric Oxide and Soot in Direct Injection Diesel Engines," *SAE Technical Paper 760129,* 1976.

[9] C. Schwerdt, "Modelling NOx-Formation in "Combustion Processes " 2006.

[10] V. Raman. Vedharaj Sivasankaralingam, Mohammed Jaasim Mubarak Ali, and Adamu Alfazazi, "Experimental and Numerical Investigation of Ethanol/Diethyl Ether Mixtures in a CI Engine," *SAE Technical Paper 2016-01-2180,* 2016.

[11] G. K. Nour Atef, Samah Y. Mohamed, Mariam Al Rashidi, Colin Banyon, "A comprehensive iso-octane combustion model with improved thermochemistry and chemical kinetics" *Combustion and Flame,* vol. 178, pp. 111-134, 2017.

[12] http://www.tfd.chalmers.se/~valeri/MECH.html.

[13] M. P. Mehl, W. J.; Westbrook, C. K.; Curran, H. J. , "Kinetic modeling of gasoline surrogate components and mixtures under engine conditions," *Proc. Combust. Inst.,* 2011.

[14] Eshan Singh, Jihad Badra, Marco Mehl and S. Mani Sarathy, "Chemical Kinetic Insights into the Octane Number and Octane Sensitivity of Gasoline Surrogate Mixtures". *Energy and Fuels, ACS Publications,* 2017

11th Asia-Pacific Conference on Combustion,
The University of Sydney, NSW Australia
10th -14th December 2017.

Catalytic combustion of lean CO over CuO/TiO_2 nano-catalyst synthesized by flame spray pyrolysis (FSP)

Xin Chen [ξ], Zuwei Xu [ξ], Haibo Zhao [*]

State Key Laboratory of Coal Combustion, School of Energy and Power Engineering, Huazhong University of Science and Technology

1037 Luoyu Road, Wuhan 430074, China

Abstract

CuO/TiO_2 nanoparticles with different CuO contents of 2%, 8%, 12%, and 20% were synthesized by flame spray pyrolysis (FSP) method. The synthesized CuO/TiO_2 nanoparticles significantly increases CO combustion conversion at low temperature. The CuO/TiO_2 nanoparticles with the CuO content of 20% (20CT) exhibited excellent catalytic activity and achieved a complete catalytic oxidation of CO at a low temperature of 110 °C . The catalytic was characterized by N_2-physisorption isotherms, XRD, TEM, H_2-TPR and XPS. With the increase of CuO content, the specific surface area of nanoparticles decreased from $98m^2/g$ to $70m^2/g$. In addition, the crystal size and particles size increased with the increasing CuO content. The XPS show an oxygen defect of nanoparticles and a shift of Cu 2p binding energy value, which are due to support-metal oxide interaction between CuO and TiO_2. The TPR profiles of nanoparticles show two peaks which are attributed to stepwise reduction of different CuO species, and the reduction temperature of CuO is influenced by the CuO-TiO_2 interaction.

1 Introduction

Removal of lean CO is of importance in a variety of processes, such as cleaning exhaust gas and lowering automotive emissions, exhaust abatement in breathing apparatus, and CO preferential oxidation for proton exchange membrane fuel cells. Low-temperature catalytic oxidation of lean CO has attracted the growing interests of the scientific and industrial communities because it can provide a cheap and efficient air cleaning technology [1, 2]. Noble metal catalysts exhibit high catalytic activity and stability for CO catalytic combustion. Considering the limited resources and high cost of noble metals, more attention must be given to the catalysts with transition metals - based oxides as active component. Copper-based oxides catalyst has been found to be an excellent candidate for CO oxidation due to its relatively high catalytic performance and low cost. However, pure copper-based catalyst is less active and stable than the noble metal catalysts. Thus different metal oxide supports were introduced into the catalyst to improve the activity. Among them, TiO_2 shows an excellent effect on the activity of the catalysts due to its outstanding oxygen storage capacity (OSC) [3]. Metal nanoparticles supported on TiO_2 exhibit a strong metal-support interaction (SMSI), making the nano-catalyst remarkably stable with high dispersion even at elevated temperature [4].

CuO/TiO_2 catalysts were prepared in a variety of methods, such as impregnation, sol-gel, hydrothermal. But these preparation processes are relatively complicated and time-consuming. Many researchers have prepared a variety of high performance nano-catalysts by flame synthesis, which proved that synthesis of nanoparticles by flame synthesis is very appealing [5]. Among these flame synthesis methods, flame spray pyrolysis (FSP) is able to prepare nanoparticles by one-step synthesis with relatively low cost. Compared with other methods, the prepared material has high purity, good dispersion, controllable structure and composition, and easy separation and collection .

Here we reported catalytic performance of CuO–TiO_2 composite system for low-temperature oxidation of lean CO. FSP method was applied to prepare the nanoparticle catalysts. The morphology and phase composition of obtained catalysts were characterized by Brunauer-Emmett-Teller method (BET), X-ray diffraction (XRD), Transmission electron microscopy (TEM), X-ray photoelectron spectroscopy (XPS), and H_2 temperature programmed reduction (H_2-TPR).

2 Experimental

2.1 Catalyst preparation

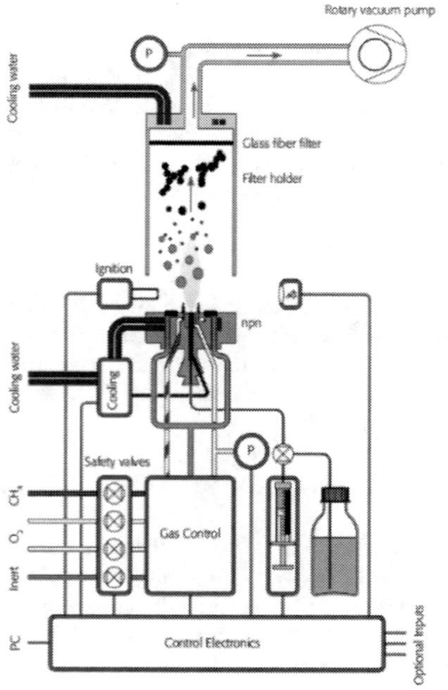

Figure 1: Schematic of flame spray pyrolysis synthesis system (Tethis NPS 10)

[ξ] X. Chen and Z.W. Xu contribute equally.
[*] Corresponding author. Tel: +86 -027-87545526-8208
E-mail: hzhao@hust.edu.cn (Zhao)

The CuO/TiO_2 catalysts were synthesized by a commercial laboratory-scale FSP system (NPS10, Tethis, Italy) [6]. A certain amount of $Cu(NO_3)_2 \cdot 3H_2O$ was dissolved in the ethanol. Then tetrabutyl titanate was added to the ethanol and reacted with free water to form the sol, so that the hydrophilic copper ions in the precursor solution were enriched around the sol. The concentration of tetrabutyl titanate in solution is 0.5mol/L. CuO/TiO_2 catalysts of different CuO contents were produced through varying the $Cu(NO_3)_2 \cdot 3H_2O$ concentration in the precursor solution. As shown in Fig. 1, the obtained precursor was injected with a 5 mL/min flow rate by a syringe pump through a stainless-steel capillary tube in a water-cooled nozzle, where it was dispersed into fine droplets by a coaxial gas flow of oxygen (5L/min) fed through an annular gap surrounding the capillary tube. The resulting spray was ignited by a pilot methane/oxygen annular flame (the flow rates of CH_4 and O_2 are 0.75 and 1.5 L /min, respectively), so that the organic part of the precursor solution was combusted, while the metal was oxidized and formed particles that grew by coagulation and sintering in the flame. The powders were collected on a glass microfiber filter supported on a stainless steel holder, located downstream the FSP burner. In the present work, we focused attention on four CuO/TiO_2 catalyst with different CuO contents, which were labeled as 2CT, 8CT,12CT and 20CT.

2.2 Catalyst characterization

Nitrogen sorption isotherms at −196 °C were obtained using a Micromeritics ASAP 2020 system. The surface area was calculated through the Brunauer–Emmett–Teller (BET) method. X-ray diffraction (XRD) patterns were obtained with a RU-200B diffractometer using Cu Kα1 radiation. Transmission electron microscopy (TEM) experiments were performed on a 2100F electron microscope (JEOL, Japan) with an acceleration voltage of 200 kV. XPS analysis was performed on a Thermo ESCALAB 250 high performance electron spectrometer, using Al Kα radiation (1486.6 eV) operating at an accelerating power of 150 W. All binding energies (BE) were referenced to the adventitious C 1s at 284.6 eV. H_2-TPR(temperature programmed reduction) was carried out on an automated chemisorption instrument (Micromeritics AutoChem II-2920). First, 100 mg of the sample was heated in N_2 (50 mL/min) from room temperature to 150 °C and held for 1 h, subsequently cooled to room temperature in a N_2 atmosphere and switched to the stream of 10 vol% H_2/Ar (50 mL/min) and held for 20 min Then the sample was heated from room temperature to 300 °C in H_2 at a heating rate of 10 °C/min.

2.3 Catalytic activity

The catalytic combustion of CO to CO_2 over CuO/TiO_2 was conducted in a fixed-bed quartz tube reactor (i.d = 8mm). The sample of each catalyst was loaded in the reactor and pretreated for 1 h by passing N_2 gas (100mL/min) through the samples heated at 120 °C to completely dry and remove adsorbed surface species. A feed steam of 1.5 vol% CO, 6 vol% O_2 and 93 vol% N_2 as diluent was delivered into the reactor at a space velocity (SV) of 60000 mL/(g·h). Online analyses of effluent gas were measured using a gas chromatograph (Synpec 3000B) equipped with a thermal conductivity detector (TCD).

The activity was expressed by the conversion of CO.

$$X_{CO} = \frac{[CO]_{in} - [CO]_{out}}{[CO]_{in}} \times 100\% \qquad (1)$$

whereas $[CO]_{in}$ and $[CO]_{out}$ are the CO concentrations in the feed gas and effluent gas, respectively.

3 Results and Discussion

3.1 Catalyst characterization

Table 1 reports the BET surface areas of CuO/TiO_2 samples. The surface areas of the nanoparticles present a decrease from 98.96 m^2/g of 2CT to 70.07 m^2/g of 20CT with the increase of CuO content. The same phenomenon has been reported in other literature [7]. This may be due to the addition of Cu, which promotes the sintering of the material. The sintering temperature of CuO is lower than TiO_2, so that the CuO component is easier to sinter in the process of particle formation and the crystallite size of the material is increased.

XRD patterns of CuO/TiO_2 are shown in Fig. 2, and the corresponding compositions and crystallite sizes are listed in Table 1. The rutile content increases with increasing of CuO content, as can be seen by the increasing intensity of the main rutile reflection (Fig. 2: $2\theta = 27.8°$), consistent with literature [8, 9]. At 20wt.% Cu a rutile content of 63.8 wt.% was obtained. It may be a result of the presence of Cu which increases the number of defects such as the oxygen vacancies in the material, thus contributing to the conversion of anatase to rutile [10]. The crystallite sizes of anatase and rutile both show a decrease with increasing CuO content, which is consistent with the change trend of particle size calculated from BET result. The particle size of 2CT is 15.0 nm and increase to 21.4 nm of 20CT.

To be noted, no $CuO/Cu_2O/Cu$ peaks are detected in the CuO/TiO_2 diffraction patterns even at the highest CuO content of 20%. This suggests that the CuO species are too small to be detectable by XRD. Thus, highly dispersed Cu species were successfully formed on the TiO_2 [11, 12].

Sample	CuO content [a]	BET surface area (m^2/g)	d_b (nm) [b]	Diameter of TiO_2(nm) [c]		Anatase (wt%)	Rutile (wt%)
				Anatase	Rutile		
2CT	2 wt%	98.96	15.0	25.12	29.96	72.3	27.6
8CT	8 wt%	83.56	18.0	32.20	32.36	51.0	49.0
12CT	12 wt%	77.28	19.4	33.54	33.71	37.9	62.3
20CT	20 wt%	70.07	21.4	40.25	35.17	36.2	63.8

a. Calculated by CuO wt%=$m_{CuO}/(m_{CuO}+m_{TiO_2})\times100\%$, where m is the mass of one component in catalyst ; b. the particle size was calculated by the formula: $d=6/(S_a \times \rho)$ (where S_a is the BET surface area, ρ is the density of particle); c. Calculated by the Scherrer equation from XRD results.

Table 1. XRD and BET data for the CuO/TiO_2 catalysts

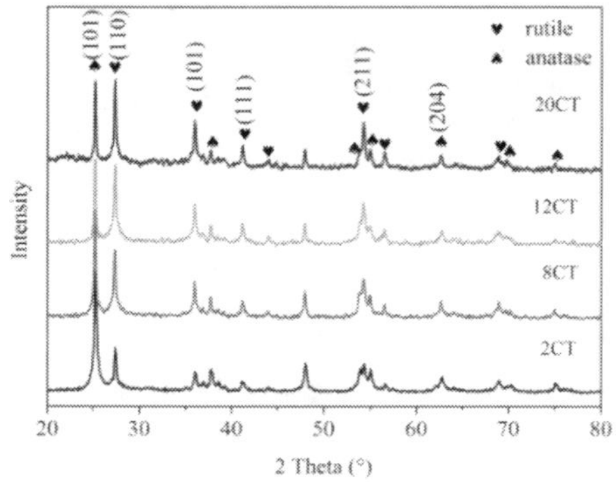

Figure 2: XRD patterns of the CuO/TiO_2 catalysts

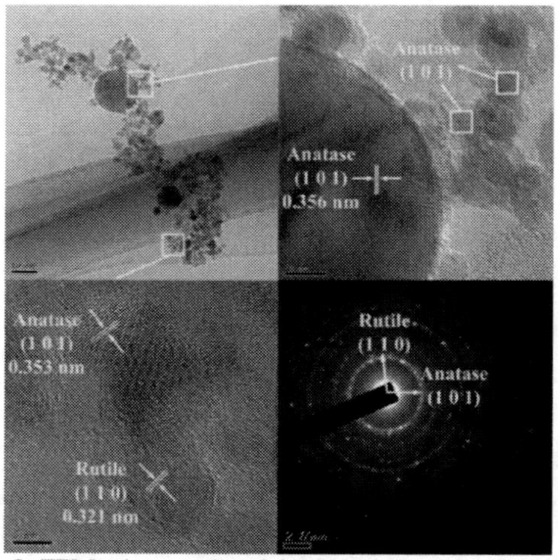

Figure 3: TEM micrographs of the FSP-derived Cu/TiO_2

The TEM micrographs and SAED (selected area electron diffraction) images of the catalysts are shown in Fig. 3. The catalyst exhibits spherical particles with a uniform particle size distribution (the average size is 15-25 nm), which agrees with d_b calculated by BET results. There is also a small amount of catalyst particles with a particle size of 50nm. It is assumed that the TiO_2 sol is formed by the hydrolysis of tetrabutyl titanate with the free water in the precursor solution, while the sol preferentially nucleated in the flame to produce large particles of TiO_2. It can be distinguished from Fig. 3 that the crystal size of anatase and rutile (the fringe of 0.321 nm is the characteristic of the (110) plane of rutile and the fringe of 0.353 nm denotes the (101) plane of anatase). The CuO and Cu_2O crystals are not found in the four samples which further confirmed the high dispersion of Cu species. The above results are in agreement with the XRD results.

As shown in Fig. 4 (XPS characterization), there are two sets of peaks for all the samples, corresponding to Cu $2p_{3/2}$. According to Papavasiliou et al. [13], the peak at 933.9 eV and the satellite peaks located at 938–946 eV represent Cu^{2+} and a peak at 932.9 eV is identified as the Cu^{1+}. The nonlinear curve-fitting process was performed using XPS peak41 software and the assignments were defined. The binding energy of Cu $2p_{3/2}$ are shown in Table 2. The binding energies of Cu $2p_{3/2}$ for the CuO/TiO_2 catalysts are 932.8–934.3 eV, which were slightly higher than those reported for the classic CuO and Cu_2O catalyst, indicating a stronger interaction between the metal and the support. A certain amount of Cu^{1+} exists on the surface of the four catalysts, which indicates that the decomposition of CuO occurs on the catalyst surface during the FSP synthesis process. At high flame temperatures, CuO releases oxygen to form Cu_2O. Moreover, high temperature flame also creates oxygen defect conditions, and rutile phases are produced under such conditions.

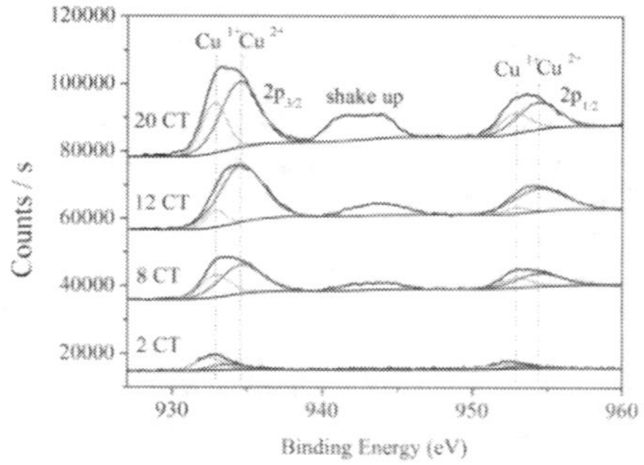

Figure 4: XPS results of the FSP-derived CuO/TiO_2

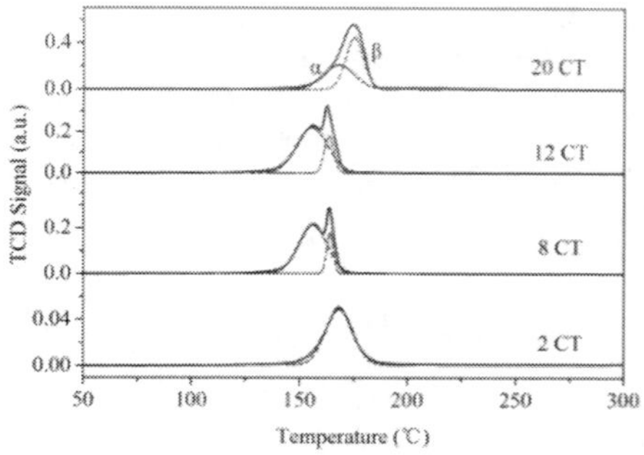

Figure 5: H_2-TPR profiles of the CuO/TiO_2 catalysts.

The TPR profiles of CuO/TiO_2 catalysts are shown in Fig.5. There are two reduction peaks at 140-190 °C corresponding to the stepwise reduction of different CuO species. The peak α corresponds to the reduction of finely dispersed particles which are easily reduced at low temperatures and the peak β is assigned to the reduction of relatively larger CuO species crystallites [14]. Table. 2 shows the reduction peaks area of four CuO/TiO_2 catalysts, and the peak area of peak β increases with the increasing CuO content which indicates the increase of CuO crystallites. As shown in Fig.5 and Table.2, when the CuO content is lower than 12%, with the increase of CuO content, the reduction temperature of CuO decreased and the binding energy value of Cu $2p_{3/2}$ increased, which perhaps indicates that the reduction temperature is influenced by the interaction between CuO and TiO_2 support.

Sample	BE (eV)	Cu^{1+}/Cu	α peak area (%)	β peak area (%)
2CT	932.79	0.65	100	--
8CT	933.65	0.35	83.62	16.38
12CT	934.29	0.12	80.38	19.62
20CT	933.62	0.33	45.01	54.99

Table 2: XPS and TPR data measured for CuO/TiO_2 catalysts

The catalytic combustion performance of the 1.5% CO is shown in Fig. 6a. CO begins to be oxidized at 50 °C. When the temperature is 50-90 °C, the conversion of CO increases slowly. And 12CT has the highest CO conversion rate at temperature of 40-80°C. This shows that 12CT has the best catalytic performance at lower temperatures, which is probably ascribed to its excellent oxidation properties. At a temperature of 90-100 °C, the CO conversion rate over 20CT increases rapidly and the complete oxidation of CO is achieved at a temperature of 110 °C. The other three catalysts achieved a conversion rate of 90% at 140 °C. Compared with the CuO/TiO_2 catalysts synthesized by other methods [11], the FSP-derived catalysts showed better performance to CO catalytic combustion. The Arrhenius curve of CO catalytic combustion is shown in Fig. 6b. A temperature range of 40-80 °C was selected to calculate the apparent activation energy, ensuring that the conversion of CO is below 20 % so that it can be assumed to be a differential reactor at this state [15]. 12CT has the lowest activation energy of 32.50 kJ/mol, which is consistent with its better catalytic performance at lower temperature. The reaction rate of CO on unit CuO content is shown in Fig. 6c. Different from the conversion rate, the reaction rate decreases with the increasing CuO content. 2CT has the highest unit reaction rate, and its reaction rate is four times that of other three catalysts at 110 °C. This may be attributed to 2CT having the highest specific surface area and CuO surface dispersion, thus providing a higher unit reaction area.

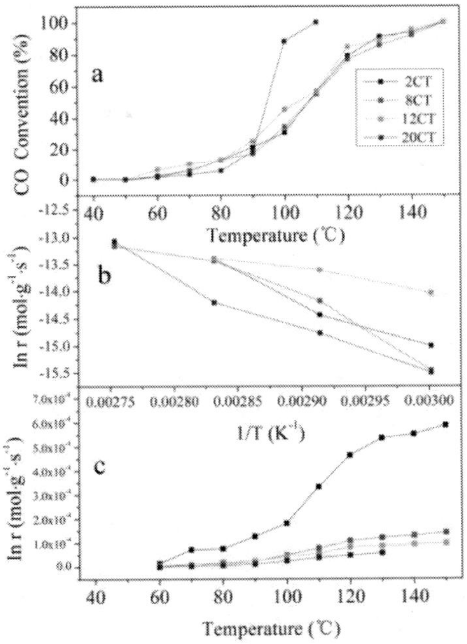

Figure 6: (a) CO conversion; (b) Arrhenius plots for CO combustion; and (c) CO oxidation rates per unit CuO mass as a function of temperature

Although 20CT has the lowest unit reaction rate and a relative poor catalytic activity at low temperature, it achieves complete CO convention at 110 °C. For CuO/TiO_2 catalysts, many researchers have demonstrated similar results that the rutile supported catalyst exhibits higher CO catalytic combustion activity than anatase supported catalyst [11, 16].

4 Conclusions

CuO/TiO_2 catalysts with different content of CuO were synthesized by flame spray pyrolysis, which has good catalytic performance for CO catalytic combustion at low temperature. A variety of physicochemical characterization methods were used to analyze and understand the relationship between catalytic activity and material composition/structure. The following conclusions were given here:

(1) With the increase of CuO content, the specific surface area of catalyst decreases, and the particle size and crystal size of catalyst increase, and the content of rutile of catalyst increases.

(2) It was found that 12CT (12 wt% CuO supported by TiO_2) presents the best oxidation ability through the TPR test. The reduction curve shows the stepwise reduction of dispersed CuO species and the reduction of CuO crystallites, respectively.

(3) 12CT presents the best catalytic performance at lower temperatures while 20CT (20 wt% CuO supported by TiO_2) has the lowest temperature to achieve complete CO convention, which may be related to a higher rutile content in the catalyst. 2CT (2 wt% CuO supported by TiO_2) has the highest unit reaction rate, which is related to the high dispersion of CuO on the active phase surface.

5 Acknowledgment

This research was funded by National Natural Science Foundation of China (51522603, 51606079), China Postdoctoral Science Foundation (2016M592331). The authors are grateful to the Analytical and Testing Center of HUST for XRD characterization, the Center for Materials Research and Testing of WHUT for XPS and TEM characterization.

References

[1] S. Royer, D. Duprez, J Cheminform 3 (1) (2011) 24-65
[2] B. White, M. Yin, A. Hall, D. Le, S. Stolbov, T. Rahman, N. Turro, S. O'Brien, Nano Lett 6 (9) (2006) 2095
[3] D. Widmann, R. J. Behm, Acc. Chem. Res 47 (3) (2014) 740-749
[4] D. Widmann, R. J. Behm, Angewandte Chemie 50 (43) (2011) 10241-10245
[5] F. E. Kruis, H. Fissan, A. Peled, J Aerosol Sci 29 (5) (1998) 511-535
[6] L. G. Bettini, M. V. Dozzi, F. D. Foglia, G. L. Chiarello, E. Selli, C. Lenardi, P. Piseri, P. Milani, Appl Catal B 178 (2015) 226-232
[7] R. Kydd, W. Y. Teoh, K. Wong, Y. Wang, J. Scott, Q. H. Zeng, A. B. Yu, J. Zou, R. Amal, Adv Funct Mater 19 (3) (2009) 369-377
[8] A. Teleki, N. Bjelobrk, S. E. Pratsinis, Sens Actuators B Chem 130 (1) (2008) 449-457
[9] C. Chaisuk, A. Wehatoranawee, S. Preampiyawat, S. Netiphat, A. Shotipruk, O. Mekasuwandumrong, Ceram Int 37 (5) (2011) 1459-1463
[10] M. S. P. Francisco, V. R. Mastelaro, Chem Mater 14 (6) (2002) 2514-2518
[11] M. Y. Kang, H. J. Yun, S. Yu, W. Kim, N. D. Kim, J. Yi, J Mol Catal A Chem 368–369 (1) (2013) 72–77
[12] . Wang, R. Büchel, A. Savitsky, M. Zalibera, D. Widmann, S. E. Pratsinis, W. Lubitz, F. Schüth, ACS Catal 6 (6) (2016)
[13] X. Yao, L. Zhang, L. Li, L. Liu, Y. Cao, X. Dong, Appl Catal B. s 150–151 (2014) 315-29.
[14] M. Y. Kang, W. Kim, B. J. Ji, N. D. Kim, H. J. Yun, J. Yi, Catal Letters 132 (3-4) (2009) 417-421
[15] W. Han, G. Zhang, G. Lu, Z. Tang, RSC Adv 5 (73) (2015) 59666-59676
[16] Y. Tang, L. Dong, C. Deng, M. Huang, B. Li, H. Zhang, Catal Commun 78 (2016) 33-36

11th Asia-Pacific Conference on Combustion,
The University of Sydney, NSW Australia
10th -14th December 2017.

Effects of Additional Diluents on Laminar Burning Velocities and Cellular instabilities in Outwardly Propagating Methane/Ethylene-Air Premixed Spherical Flame

H. J. Kim[1], K. H. Van[1], J. Park[1*], O. B. Kwon[1], Dae Keun Lee[2], Seung Gon Kim[2], Young Tae Ghauk[2], Dong soon Noh[2]

[1] Dep. Of Mechanical Engineering, Pukyong National Univ, Busan Korea

365 Sinseon-ro, Busan, Nam-gu, 48547, Republic of Korea

[2] Advanced Combustion Laboratory, Korea Institute of Energy Research, Daejeon Korea

152 Gajeong-ro, Daejeon, Yuseong-gu, 34129, Republic of Korea

Abstract

The present work investigated the laminar flame speeds and cellular instabilities in outwardly propagating methane/ethylene – air premixed flames diluted with carbon dioxide and helium. The diluents with the dilution ratio of 0.1 to 0.5 were added to methane/ethylene – air mixture to evaluate unstretched laminar burning velocities and Markstein lengths and cellular instabilities at elevated pressures up to 4 atm. The experimentally measured flame velocities were compared with numerical predictions from PREMIX and OPPDIF codes based on USC mechanism II. Also the effects of dilution ratio and elevated pressures on cell formation were examined through evaluating the effective Lewis number, flame thickness and thermal expansion ratio.

1 Introduction

Environmental regulation and increased energy consumption caused the overusing fossil fuels and depletion of energy sources. The concerns about above problems were required to search for reducing method of NO_x emissions, and conduct lean-premixed combustion. The combustion of lean hydrocarbon-air mixtures offered the potential of reduced flame temperatures and NO_x emissions according to the thermal mechanism [1]. Nonetheless, it had still remained occurrence of the instability of the flame decreases and extinction phenomena near the lean flammability limits. To understand the flame instability and flammability limits, several studies have been conducted to evaluate the impacts of the blended fuel or addition to diluents on the combustion experiment [1-4]. In CH_4/H_2-O_2/N_2 laminar premixed flames, lean flammable limit was greatly extended by additional H_2, extended flammable limits resulted in significant reduction of NO formation. The effect of reformate gas in flame extinction of CH_4-air was shown that NO formation was greatly reduced [5]. Gregory et al investigated the influence of added small amounts H_2 in CH4-air flame. The results showed that the improved lean flammability by adding H_2 tended to the increase flame speed with CH_4/H_2 mixtures in comparison to pure CH_4. However, the transport effects by decreased fuel Lewis number contributed to reducing sensitivity of the mixture flames [6]. Additionally, Vu et al studied effects of diluents (where CO_2, N_2, and He) about cellular instability in syngas-air premixed flames. And the results showed that the suppressant of diffusional-thermal instability was obtained for the syngas-air flame [7]. Also, Van et al as from their extended work carried out CH_4/C_2H_4/air premixed flame at high pressures up to 4 atm, and investigated cellular instabilities with increase of ethylene addition to the mixture [8]. Therefore, Because the effects of added diluents in CH_4/C_2H_4/air premixed flame have been hardly found in the literature, the present works was investigated by diluting CO_2, and He in the CH_4/C_2H_4/air premixed flame to secure the fuel flexibility particularly and evaluation of flame stability, it should be required to investigate unstretched laminar burning velocities, Markstein length, and cellular instability by adding diluents at elevated pressures (up to 4 atm).

Corresponding author. Fax: +00-00-000-000
E-mail address: jeongpark@pknu.ac.kr

2 Experimental setup

The experiments were conducted in a stainless steel, cylindrical constant volume chamber having 200 mm in an inside diameter and 220 mm in a length as shown in Fig. 1. For optical access, two quartz windows in 100mm diameter and 40 mm thickness were installed for optical access to mount oppositely. Opposed tungsten electrodes were connected to ignite combustible mixture with high voltage source (8 ~ 9 kV) at the center of the chamber. And also the electrodes were 0.5mm thickness and the gap between them was adjustable. Fuels (purity: methane-99.999%, ethylene - 99.999%) and Air (purity: 99.999 %) and Diluents (purity: CO_2-99.999%, He-99.999%) were supplied to the chamber at corresponding partial pressures using a pressure transmitter (PACE 1000, 25mbar to 1000bar, ± 0.005% FSO) to attain the total initial chamber pressure, P_u. A time duration was awaited about 15 minute to ensure complete mixing and attain steady condition. Once the voltage source was given, a flame kernel formed at the center of the chamber and started propagating outward spherically and quenched when it touched the wall of the chamber. The propagating spherical flame was visualized using a Schileren photography with a 300-W xenon light source and a pair of 150mm diameter concave mirrors, and recorded using a high-speed digital camera (Olympus i-Speed3), operating at 10,000 frames per second with 528×400 pixels of resolution. Operating the high-speed camera and measuring the instantaneous chamber pressure were triggered

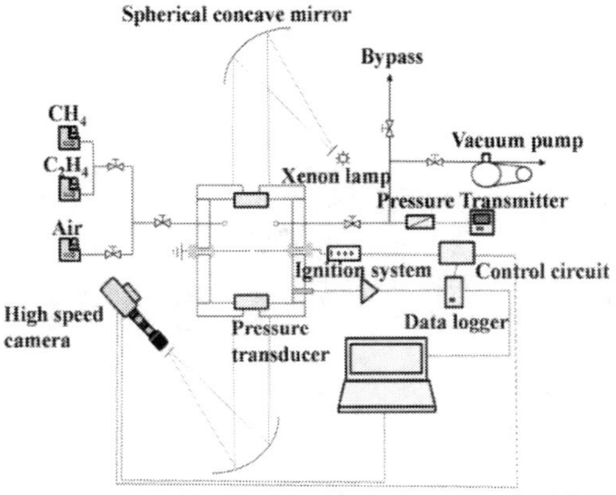

Figure 1. Schematic of experimental setup and schlieren system.

by the signal of the spark ignition source. The temporal chamber pressure during ignition and flame propagation was acquired by a dynamic pressure transducer (Kistler 601A, 0-250 bar) and a data logger (GRAPHTEC GL900). After combustion, the chamber was ventilated to the laboratory exhaust system and then purged using an air compressor to remove condensed water vapor prior to refilling for the next experiment.

3 Results and Discussion

For outwardly propagating spherical premixed flame, the burnt stretched flame propagation velocity, S_b, with respect to burned mixture was calculated by the instantaneous flame radius, R_f, measured from a Schlieren image, as $S_b = dR_f/dt$, where t denotes the time. The flame stretch rate, K, in an outwardly propagating spherical premixed flames were expressed as

$$K = \frac{1}{A}\frac{dA}{dt} = \frac{2}{R_f}\frac{dR_f}{dt} = \frac{2}{R_f}S_b \tag{1}$$

Based on previous study [8], experimental methodology was validated through comparison of a variety of flame radius ranges from experimental studies and three kinds of extrapolation model with numerical ones in methane-air flames on Gri v - 3.0. Based on reliable them, the most reliable detailed reaction mechanism to methane/ethylene-air premixed flame was examined through comparison among Gri v-3.0, USC Mech II, and Sung Mech to evaluate reliable detailed kinetic mechanisms. This results show that unstretched laminar burning velocities in CH4/C2H4/air premixed flames are best-fitted when the extrapolation model with nonlinear model [9, 10] is used at the flame radius range in 10-25mm. In the current study, The adopted USC Mech II and nonlinear model II of extrapolation in the following were represented:

$$\ln(S_b) = ln(S_b^0) - S_b^0 L_b \times (\frac{2}{R_f S_b}) \tag{2}$$

Experimental results for S_u^0 obtained with NM at 10-25 mm in CH_4/C_2H_4/air premixed flame reliable condition as previous validation with variation of diluents concentration were compared with numerical ones calculated attained with PREMIX code, and exhibited in Fig. 3. Here, the diluents concentration ($\Omega_{diluents} = X_{C_2H_4}/(X_{CH_4} + X_{C_2H_4} + X_{diluents})$) is defined by where X_i is the mole fraction of species i (i = C_2H_4, CH_4, CO_2 or He). As shown in Fig. 3, the peak discrepancy between experimental and numerical Su0 appears

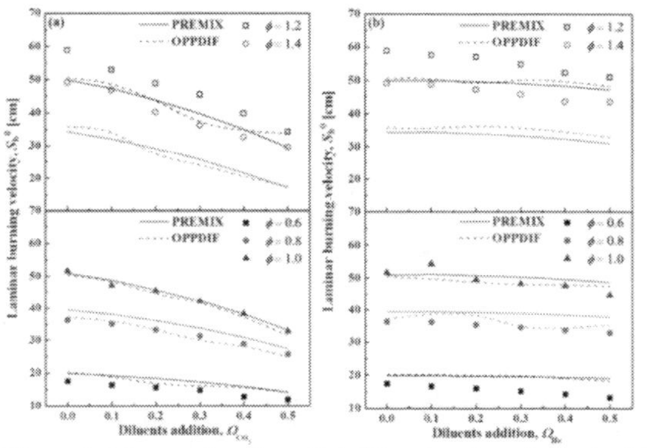

Figure 2. Measured and predicted Laminar burning velocities in CH_4/C_2H_4-air premixed flames diluted with CO_2 (a) and He (b) at P_u = 1 atm.

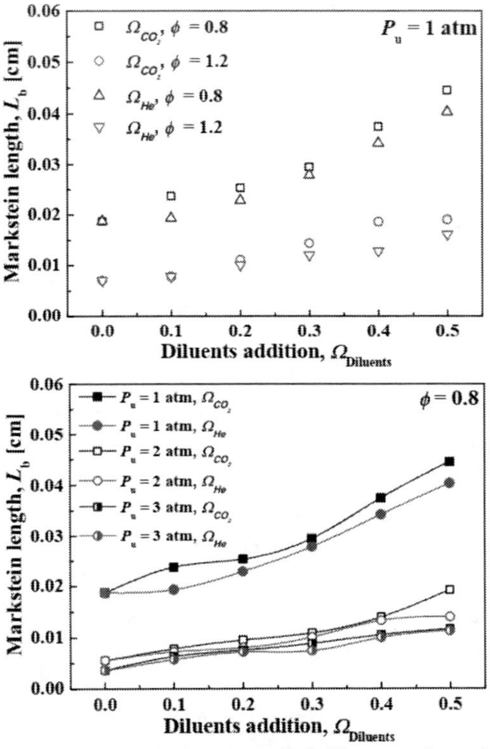

Figure 3. Marstein length against diluents volume fraction in CH_4/C_2H_4-air premixed flames diluted with CO_2 and He for ϕ = 0.8 and 1.2 at P_u = 1 atm (up) and at various initial pressures for ϕ = 0.8 (down).

at the equivalence ratio of 1.4 with diluents of CO_2 and He addition. It was seen that, at fuel-rich condition, the errors with simulation were larger than those with fuel-lean condition, further investigation could be required to clarify the reason on the discrepancy between experiment and simulation with diluents. Based on previous results, the OPPDIF [11]and PREMIX code [12] based on USC Mech II [13] were compared with experiment to give the priority between them in predicting S_u^0 in CH_4/C_2H_4/air diluted with CO_2 or He premixed flame, because of flame speed with unadiabatic and unstretched laminar burning velocity in OPPDIF. Then, the flame speed and strain rate which is defined as a = $(u_F + u_O)/2$ can be determined by the extrapolation of the linear correlation up to a = 0. Figure 3 shows that comparison of laminar burning velocity between experiment and prediction with PREMIX and OPPDIF based on USC Mech II. As shown in Fig. 3 (a), the

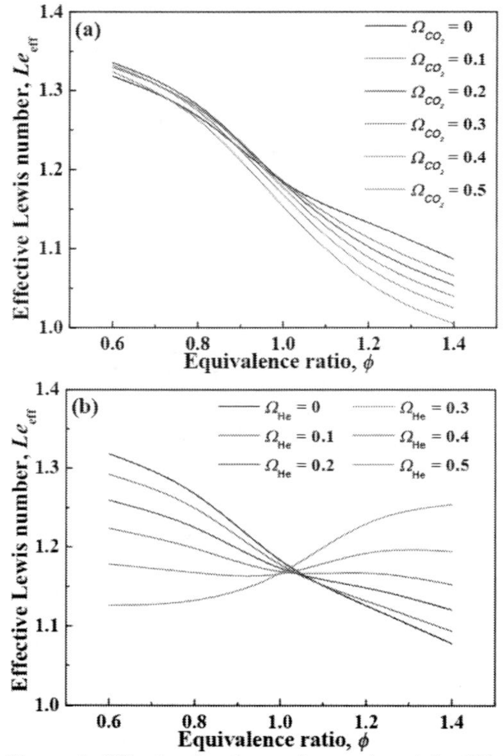

Figure 4. Effective Lewis number against ϕ in CH_4/C_2H_4-air premixed flames diluted with CO_2 (a) and He (b)

errors with the two numerical prediction were not sensitive at $0.6 \leqq \phi \leqq 1.0$. The prediction for OPPDIF well describes the CH_4/C_2H_4/air premixed flame with carbon dioxide addition. it increased in increase of in Ω_{CO_2}, e.g., 15.39 % for PREMIX at ϕ =0.6 and Ω_{CO_2} = 0.1. However, the discrepancy with PREMIX and OPPDIF increased as well, e.g., 59 % for OPPDIF at ϕ =1.4. For the He diluents cases in Fig. 3 (b), using PREMIX and OPPDIF, the errors were also insensitive for ϕ up to ϕ =1.0. Nonetheless, It was shown that the large discrepancy was represented in fuel-rich case. for ϕ = 1.4, the error with PREMIX are rather larger than those with OPPDIF (37.27 % for OPPDIF and 43.16 % for PREMIX at ϕ = 1.4 and Ω_{He} =0.1). The further confirmation to finding the reason on the discrepancy in the flame structure is the future work. Markstein length characterizes the effect of local heat release of a propagating flame on variation in the surface topology along the flame and the associated local flame front curvature. As mentioned in previous study, If the burnt Markstein length L_b > 0, the stretched flame speed decreases with the stable, and vice versa [1, 14, 15]. The measurement of L_b extracted from NM as a function of diluent concentration $\Omega_{diluents}$ for diluted CH_4/C_2H_4/air mixture at initial pressure of 1 atm for the ϕ = 1.4 are illustrated in Fig. 4. The tendency of L_b increase (decrease) with increasing diluent concentration (equivalence ratio). The both of diluents (CO_2 or He) results in reduction of laminar burning velocity dependence versus flame stretch. The Markstein length are positive, so no instability occurs during the atmospheric pressure for various ϕ. It was noted that The L_b decreases with the increase of the initial pressure. This indicate that further confirmation of evaluation on flame stability may be needed to the next work. Flame stability is another important characteristic of premixed flames. As previously mentioned, two instabilities of premixed flame were observed: diffusional-

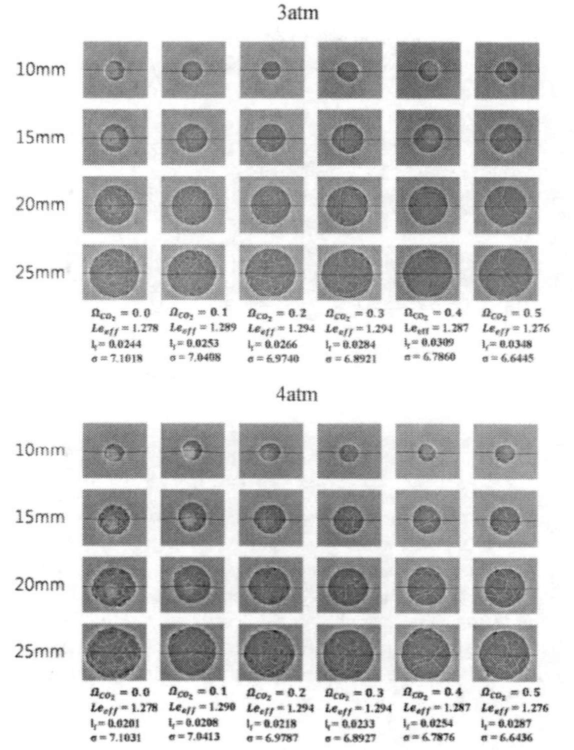

Figure 5. Effects of CO_2 addition on cellular instability in CH_4/C_2H_4/air premixed flame at ϕ = 0.8.

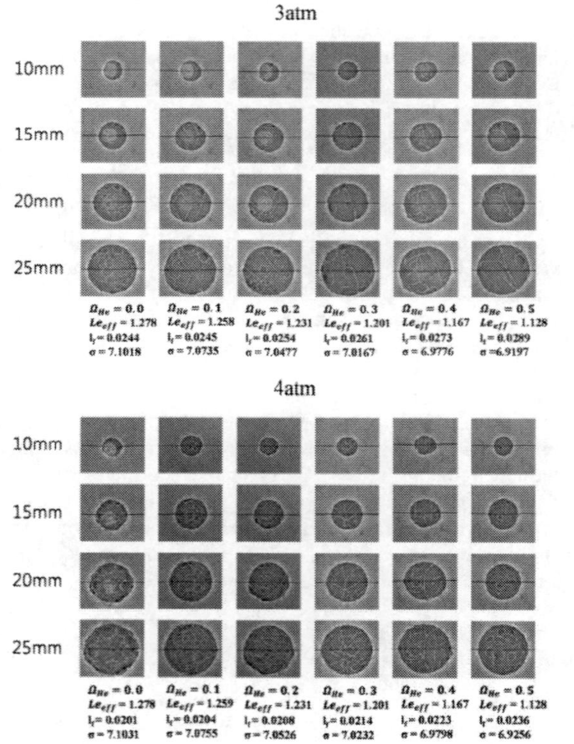

Figure 6. Effects of He addition on cellular instability in CH_4/C_2H_4/air premixed flame at ϕ = 0.8.

thermal instability and hydrodynamic instability. The effective Lewis number Le_{eff}, which represent the influence of diffusional-thermal instability, are analyzed. Figure 5 shows that the Le_{eff} of CH_4/C_2H_4/air with CO_2 (left) and He (right) as a function of with a variety of diluent concentration. This results show that of all the mixtures are larger than unity; therefore the diffusional-thermal instability can be sufficiently suppressed. As the content of CO_2 concentration increases in the fuel blends, the Le_{eff} increase (decrease) for fuel-lean (fuel rich), whereas the Le_{eff} decrease (increase) with increasing for lean mixture (rich mixture) with He diluent. The effects of initial pressure to destabilize the flame front are observed by capturing schlieren image, as shown in Fig. 5 and 6. And the Le_{eff}, the thermal expansion ratio, σ, and the flame thickness, l_f are indicated at the bottom of each photos the figure. This instability would be destabilized because of the significant increase in l_f and increase of in σ with increasing diluents concentration in both of diluents cases, while promoted with increasing initial pressure.

4 Conclusions

The present study was carried out to determine the effect of diluents on the laminar burning velocity, markstein length, cellular instability in CH_4/C_2H_4/AIR premixed spherical flame. The results showed that the increasing of dilution ratio caused the decrease of the flame propagation speed and the increase of markstein length. Also effective lewis number are always larger than unit, the cellular formations are not addressed to diffusive-thermal instability and the hydrodynamic instability was suppressed because σ (l_f) decreases (increases) in increase of diluents volume fraction. Consequently, the study was that flames were more stabilized by diluting N_2 and He.

5 Acknowledgment

This research was supported by SGER Program in National Research Foundation of Korea (NRF) founded by the Ministry of Education (2016R1D1A1A02937106).

References

[1] Halter. F, Chauveau. C, Djebaili-Chaumeix. N, Gokalp. (2005). Proc. Combust. Int. 30: 201.
[2] Jankson G.S, Sai. R, Plaia. J. M, Boggs. C. M, Kiger. K. T, Soloukhin RI. (2003). Combust. Flame. 132: 503.
[3] Gaudncheau J. L, Denet. D, Searby. G. (1998). Combust. Sci. Technol. 137:81.
[4] Ren J. Y, Qin W, Egolfopoulos F. N, Tsotsis T. T (2001). Combust. Flame. 124:717.
[5] Guo HS, Smallwood GJ, Gulder OL (2007). Combust. Flame. 121:41.
[6] Gregory S, Jackson, Roxanne Sai, Joseph M. P, Christopher M.B, Kenneth T. K (2003). Combust. Flame. 132:503.
[7] Vu TM, Park J, Kwon OB, Bae DS, Yun JH, Keel SI (2010). Int J Hydrogen Energy. 35:3868.
[8] Van K. H, Kim H. J, Park J, Kwon O. B, Lee D. K, Kim S. G, Guahk Y. T, Noh D. S. (2016). KOSCO.
[9] Kelley AP, Law CK (2009). Combust Flame. 156:1844.
[10] Bechtold JK, Cui C, Matalon M (2005). Proc. Combust. Int. 30:177.
[11] Lutz A. E, Kee R. J, Grcar J. F, Rupley F. M (1997). Sandia National Laboratories, USA, SAND96-8243 (1997).
[12] Kee R. J, Grcar J. F, Smooke M. D, Miler J. A, Meeks E (1985). Sandia National Laboratories, USA, SAND85-8240.
[13] Wang H, You X, Joshi AV, Davis SG, Salkin A, Egolfopoukos F, Law CK, USC MECH II <http://ignis.use.edu/USC_II.htm>.
[14] Vu TM, Park J, Kwon OB, Bae DS, Yun JH, Keel SI (2011). Int. J. Hydrogen Energy. 36:6914.
[15] Song W. S, Jung S. W, Park J, Kwon O. B, Kim Y. J, Kim T. H, Yun J. H, Keel S. I (2013). Int. J. Hydrogen Energy. 38:14102.

11th Asia-Pacific Conference on Combustion,
The University of Sydney, NSW Australia
10th -14th December 2017.

A Tomographic Background-Oriented Schieren Method for 3D Density Field Measurements in Heated Jets

C. Atkinson[1], S. Amjad[1], J. Soria[1,2]

[1] Laboratory for Turbulence Research in Aerospace and Combustion,
Department of Mechanical and Aerospace Engineering
Monash University, Victoria, 3800, Australia
[2] Department of Aeronautical Engineering, King Abdulaziz University,
Jeddah, Kingdom of Saudi Arabia

Abstract

Methodologies for the experimental measurement of three-dimensional instantaneous density fluctuations via tomographic background-oriented schlieren (TBOS) were assessed using synthetic background images, corresponding to experimental measurements of a heated turbulent jet. Filtered back projection and iterative algebraic reconstruction algorithms were explored. Results show a superior reconstruction when the solutions from filtered back projection were used as an initial solution to a masked and windowed iterative algebraic reconstruction. The influence of the number of cameras and the wavelength of density fluctuations are both investigated.

1 Introduction

To understand the structure of complex convective flows, it is essential to capture not only the velocity field but also the corresponding instantaneous density or temperature fields. Laser based optical measurement techniques, such as holographic or tomographic particle image velocimetry (PIV), are capable of quantifying instantaneous three-component three-dimensional (3C-3D) velocity fields [1,2], however the robust 3D measurement of density is less common. One means of quantifying the density of a flow is via an optical simplification to schlieren, known as background-oriented schlieren [3], which provides a measure of the integrated density gradients through a flow. To obtain the density gradients at discrete locations in the flow and hence enable the calculation of the density field, the distribution of density gradients along the integrated measurement must be determined. This represents an inverse Radon transform and if multiple projections of the integrated density gradients are available, corresponds to a tomographic reconstruction of the density gradient fields.

A handful of research groups report instantaneous 3D density measurements based on tomographic background-oriented schlieren (TBOS) [4,5], most using either a filtered back projection (FBP) [6] or an iterative algebraic reconstruction technique (ART) [7] to reconstruct the gradients of refractive index, which are related to the density gradients by the Gladstone-Dale relation. The validation of these approaches has generally been based on comparison with qualitative schlieren measurements, comparison of mean temperature fields or via the generation of synthetic background displacements based on density field data computed by Reynolds Averaged Navier-Stokes (RANS) simulations of specific flows.

In this paper we present and compare both FBP, ART and hybrid reconstruction techniques, combined with a random access iterative windowed and masked corrections, in order to enable the reconstruction of instantaneous 3D turbulent multi-scale density fields from simultaneous background-oriented schlieren projections. The ability of the TBOS technique to reconstruct these density fluctuations is assessed as a function of camera number and wavelength in order to aide in the planning of experimental measurements.

2 Principles of Background-Oriented Schlieren

Background-oriented schlieren (BOS) operates on the principles of a schlieren method and the relationship between the density of a fluid ρ and its refractive index n, as given by the Gladstone-Dale equation:

$$n - 1 = G(\lambda)\rho \tag{1}$$

where $G(\lambda)$ represents the Gladstone-Dale constant as a function of the wavelength λ of the incident light. If we consider the path followed by a light ray from a point in the image plane of a camera X_o to a point in the background pattern P_o (see Fig. 1), the variation in the refractive index along this path will result in the deflection of this ray due to refraction. If the volume is small with respect to the distance to the background Z_D then it can be assumed that the path followed by the ray remains unchanged and the refracted ray can be approximated by a small deflection of angle ε, with respect to the incident ray. If a background is placed behind the measurement volume then this deflection results in an apparent shift in the imaged background $X_o + \Delta X$, which is related to the deflection and optical setup as:

$$\Delta \mathbf{X} = f\left(\frac{Z_D}{Z_D + Z_A - f}\right)\varepsilon \tag{2}$$

where f is the focal length of the camera lens and Z_A the distance from the focal point to the centre of the measurement volume.

Corresponding author.
E-mail address: callum.atkinson@monash.edu

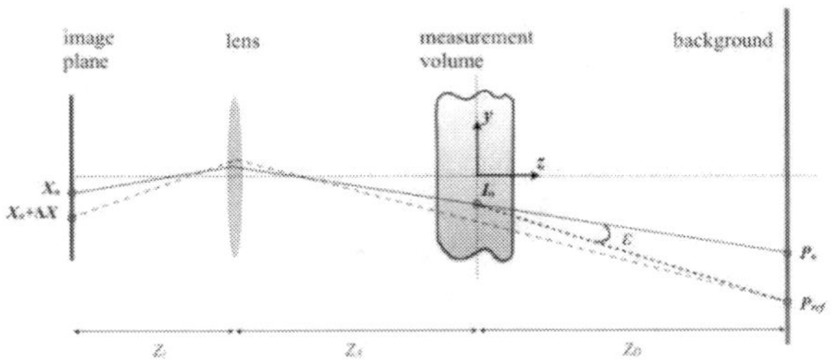

Figure 1: Schematic of the optical setup and parameters of BOS

For small angles, following the parallax assumption, the relative deflection associated with the diffraction of the ray relative to its incident angle can be approximated by:

$$\begin{aligned} \varepsilon_{z'} &= \frac{1}{n_o}\int \frac{\partial n}{\partial y'}\,\mathrm{d}x' \\ \varepsilon_{y'} &= \frac{1}{n_o}\int \frac{\partial n}{\partial z'}\,\mathrm{d}x' \end{aligned} \tag{3}$$

where $\boldsymbol{x'}$ represents the axis of the ray and $\boldsymbol{y'}$ and $\boldsymbol{z'}$ represent two axes orthogonal to the ray. The deflection angles between the ray $\boldsymbol{x'} = I_oP_o$ and $\boldsymbol{r} = I_oP_{\mathrm{ref}}$ can be determined by projecting rays from a point in the image X_o and from the point corresponding to the deflection of the background image $X_o + \Delta X$, which can be determined by standard cross-correlation particle image velocimetry algorithms. Assuming a small angle approximation, the deflections angles are given by the dot product of the local incident ray axes and the refracted ray $\boldsymbol{r}$:

$$\begin{aligned} \varepsilon_{x'} &= \mathbf{r}\cdot\mathbf{x}', \\ \varepsilon_{y'} &= \mathbf{r}\cdot\mathbf{y}', \\ \varepsilon_{z'} &= \mathbf{r}\cdot\mathbf{z}'. \end{aligned} \tag{4}$$

3 Tomographic BOS Reconstruction

In the present paper, tomographic reconstruction of the refractive index gradient fields $\partial n/\partial x_i\,(x,y,z)$ that correspond to the measured background displacements and associated ray deflections will be performed using either Fourier slice based filtered back projection (FBP) [6], iterative algebraic reconstruction techniques (ART) [7] or a combination of both.

3.1 Filtered Back Projection (FBP)

In the present implementation the reconstruction of each refractive index gradient is performed independently based on sinograms that represent the sum of each component of $\sum_{ray} \nabla n$ along the camera's axis at each position along the width of each camera. The sum of the gradients at each position in the image plane are determined by solving the following system of equations:

$$\frac{n_o}{\Delta x}\begin{bmatrix}\varepsilon_{x'}\\ \varepsilon_{y'}\\ \varepsilon_{z'}\end{bmatrix} = \begin{bmatrix}\mathbf{x}'^T\\ \mathbf{y}'^T\\ \mathbf{z}'^T\end{bmatrix}\sum_{ray}\nabla n \tag{5}$$

where n_o is the refractive index outside the measurement volume and Δx is the voxel width. Reconstruction is performed following the common practice of taking the inverse Radon transform of ramp filtered sinograms. A circular reconstruction domain is applied which corresponds to the common view of all cameras, outside which the density gradients are set to zero.

3.2 Algebraic Reconstruction Technique (ART)

The algebraic density gradient reconstruction is based on representing the defection angles $\varepsilon_{x'i}$, $\varepsilon_{y'i}$, $\varepsilon_{z'i}$ for the i-th ray as a projection of the gradients of the refractive index field ∇n where the contribution of each j-th point in the field is represented by a weighting w_{ij}. From equations (3) and (4) the contribution to the ray deflections can be expressed as:

$$\begin{aligned} \varepsilon^k_{x'i} &= L_i\frac{\sum_j w_{ij}\mathbf{x}'\cdot\nabla n^k_j}{n_o\sum_j w_{ij}} \\ \varepsilon^k_{y'i} &= L_i\frac{\sum_j w_{ij}\mathbf{y}'\cdot\nabla n^k_j}{n_o\sum_j w_{ij}} \\ \varepsilon^k_{z'i} &= L_i\frac{\sum_j w_{ij}\mathbf{z}'\cdot\nabla n^k_j}{n_o\sum_j w_{ij}} \end{aligned} \tag{6}$$

where k denotes the iteration of the reconstruction and L_i is the length of the path followed by the ray through the volume. A solution to the gradient field is computed by iteratively correcting a previous estimation of the field in order to minimise the difference between the projected deflection angles of each ray as calculated by equation (6) and the deflection angles estimated from the measured background displacements, equation (4). The required correction for each gradient component is determined by solving the following series of equations,

$$\frac{\lambda_j n_o w_{ij}}{L_i}\begin{bmatrix}\varepsilon_{x'i}-\varepsilon^k_{x'i}\\ \varepsilon_{y'i}-\varepsilon^k_{y'i}\\ \varepsilon_{z'i}-\varepsilon^k_{z'i}\end{bmatrix} = \begin{bmatrix}\mathbf{x}'^T\\ \mathbf{y}'^T\\ \mathbf{z}'^T\end{bmatrix}\left(\nabla n^{k+1}_j - \nabla n^k_j\right) \tag{7}$$

where λ_j is a relaxation parameter set to 2.0 through this paper.

3.3 Calculation of the Refractive Index Field

The refractive index fields $n(x, y, z)$ can be calculated from their reconstructed gradients by solving the Poisson equation:

$$\nabla^2 n = \frac{\partial^2 n}{\partial x^2} + \frac{\partial^2 n}{\partial y^2} + \frac{\partial^2 n}{\partial z^2} = q \tag{8}$$

The right-hand side is populated by taking a 2nd order central difference of the reconstructed density gradients with the solution obtained via an iterative successive over relaxation algorithm (SOR). In the present case the solution was terminated once the field had converged to 10^{-16}, to remove any influence of the convergence of the SOR algorithm on the assessment of the reconstruction methodology. As is often done in PIV a lower order discretisation is to reduce the influence of measurement and reconstruction noise associated with the gradient fields.

4 Generation of Synthetic Backgrounds

To assess the performance of the proposed reconstruction methods synthetic BOS images were created by tracing rays from each camera through a known refractive index distribution using Snell's law. A synthetic field consisting of a Gaussian air jet with a peak change in refractive index of $\Delta n_p = 1.5\times10^{-4}$ from $n_0 = 1.0$ was used, corresponding to a centreline temperature of approximately 368°C at standard atmospheric conditions, and a standard deviation σ = 9 voxels. The measurement domain of 65×3×65 voxels (43×2×43 mm) was sized based on a jet diameter of 10 mm and the use of a 1 Mpixel camera with a pixel size of 3.75×3.75 μm^2 equipped with f = 25 mm focal length lens with an aperture of f/22. The optical centre of the camera was positioned 275 mm from the centre of the volume and 575 mm from the background so that an interrogation window of 16 pixels would correspond to 16 voxels (0.66 mm) when projected to the volume centre and remain larger than the circle of confusion associated with the lens geometry of the BOS setup (see [8] for a discussion of the influence of finite aperture on BOS).

Synthetic background images were created for different camera numbers, with cameras evenly spaced in a 180° arc about the jet axis in the x-z plane (see Fig. 2). Such a camera configuration is relatively practical to setup, lends itself well to FBP and results in reconstruction that is almost independent of variation in the y*−axis*. This allows us to perform a more efficient comparison of the different reconstruction methods by limiting our region of interest to a thin slice normal to the jet axis.

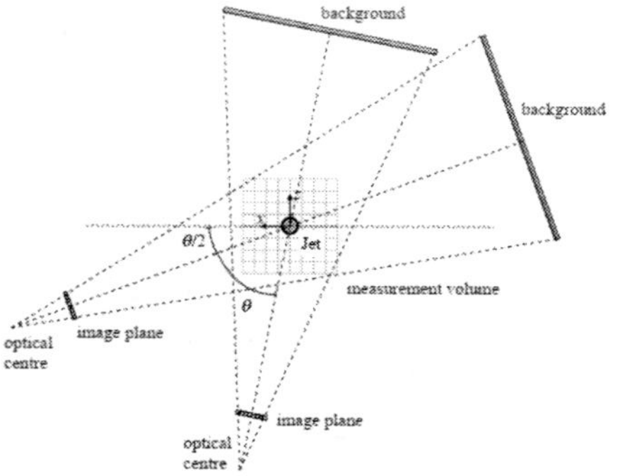

Figure 2: Schematic of the synthetic BOS configuration

To assess the ability of TBOS to resolve instantaneous turbulent density fluctuations, the Gaussian refractive index distribution was modulated using sinusoids in the x and z directions as given by:

$$n(x,y,z) = \frac{1}{2\pi\sigma^2} e^{-[x^2+z^2]/(2\sigma^2)} \left[1 + A\sin\left(\frac{2\pi}{\lambda_x}x\right)\sin\left(\frac{2\pi}{\lambda_x}z\right)\right] \quad (9)$$

where λ_x and λ_z are wavelengths of density fluctuations, A is the amplitude of the modulation, which was set to 0.25 such that the field consists of fluctuations up to 25% of the mean value, which decay towards the jet boundaries. This is similar to passive scalar fluctuations in a fully developed turbulent round jet [9]. The maximum background displacements in the image plane were 1 pixel.

5 Results

To assess the ability of the TBOS algorithms to resolve instantaneous density and corresponding refractive index fluctuations noiseless background displacement fields were generated for arrays of 6 to 22 evenly spaced cameras, with fluctuation wavelengths from $\lambda_x = \lambda_y = 4.6$ to 32.5 voxels or $L/14$ to $L/2$ where L is the measurement volume length.

Simulations showed that ART performed best when iterative corrections to the density field were performed by randomly sampling rays from each camera, which removes bias to any particular image, combined with a masking of the gradient field beyond a radius of 30 voxels and imposing a Hamming window to the correction, which similarly reduces the correction applied to the outer region of the reconstruction domain. As shown in Fig. 3, FBP provides a good reconstruction of gradients near the centre of the volume but also introduces considerable gradients outside the core of the

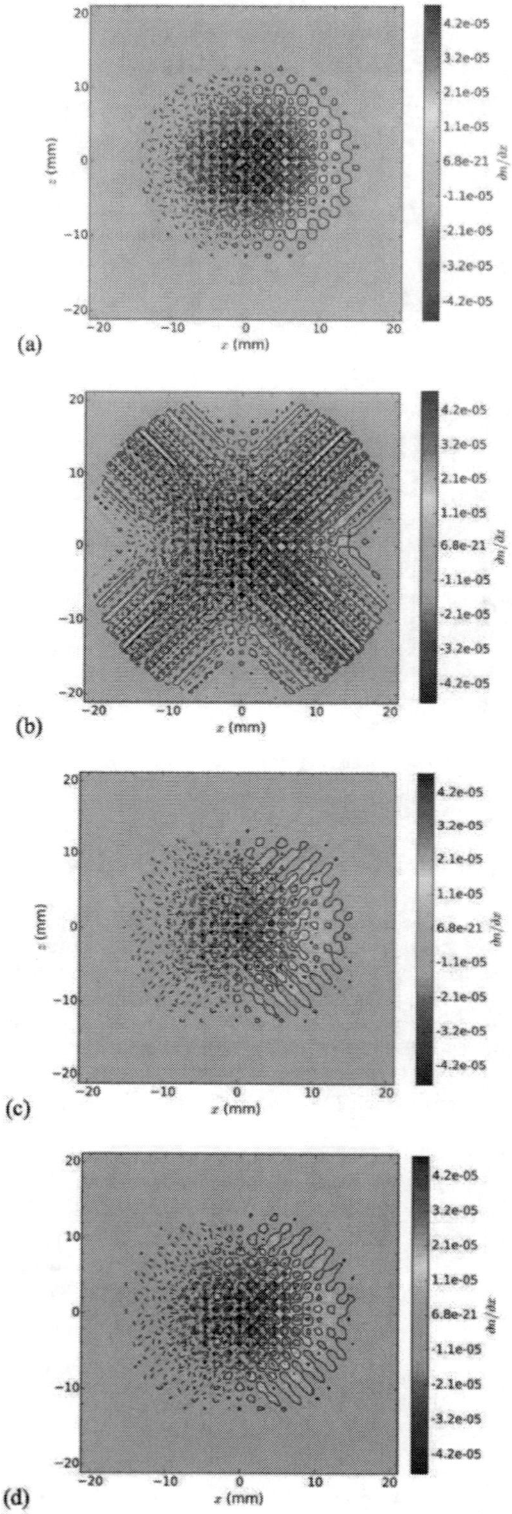

Figure 3: Contour plots for the reconstructed refractive index gradient $\partial n/\partial x$ for 14 cameras and $\lambda_{x,y} = L/14$: (*a*) synthetic field; (*b*) FBP; (*c*) ART 20 iteration; and (*d*) FBP+ART 10 iterations.

synthetic jet. Removing these regions then applying an ART reduces the sensitivity to the applied mask and in all cases converged to a solution better than that of ART alone by maintaining the strong gradients near the centre of the volume. ART with a null initial gradient field tends to under-predict the gradients near the centre of the volume and require approximately twice as many iterations to reach the same convergence.

The accuracy of a TBOS computed refractive index field depends not only on the reconstruction methodology but also on the number of the background views, the strength of the density gradients and the relative distance between the volume and the background. The influence of the number of cameras and the wavelength of the oscillation in the refractive index field for the FBP+ART reconstructions is demonstrated in Fig. 4 assuming exact calculations of the background displacement. A range of 6 to 22 cameras was considered as less camera results in significant errors and the benefit associated with the use of more cameras drops off significantly.

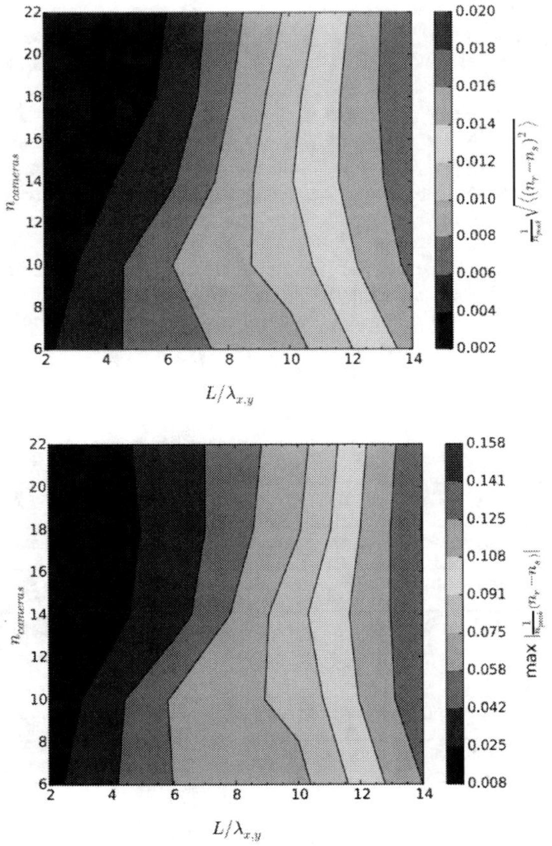

Figure 4: Contour maps of the RMS and peak errors between the synthetic and computed refractive index fields normalised by the peak refractive index gradient from Poisson solution with refractive index gradients computed by FBP+ART 10 iterations.

In all cases reducing the wavelength of the density oscillations significantly increases both the RMS and peak errors in the reconstructed fields when compared to the synthetic field. The extent to which the magnitude of these errors are influenced by spatial resolution is still under investigation. Comparison of the different reconstruction methods shows that FBP+ART provides the most accurate estimate of the refractive index field across the entire range of wavelengths. Increasing the number of cameras improves the reconstruction, however the benefit beyond the use of 18 cameras is almost negligible. This suggests that high quality TBOS measurements of turbulent flows will likely require between 16 to 18 cameras. It is important to note that unlike PIV cameras the cameras used for BOS do not have to operate in a double shutter mode, allowing for the use of lower cost machine vision cameras. While results indicate that the mean reconstruction error across the field is relative small, the reconstruction in unable to accurately predict the strongest small scale gradients resulting in peak errors of up to 15% of the peak.

4 Conclusions

The ability of a tomographic background oriented schlieren technique to reconstruct a fluctuating density field of differing scales and camera number is assessed using filtered back projection, algebraic reconstruction and a combination of the two. In all cases, using the filtered back projection as an initial solution to the algebraic solution was found to yield the best reconstruction of both the core of the heated jet and the flow boundaries. The accuracy of the reconstruction is strongly affected by the scale of the density fluctuations. Attempts to improve the effective resolution of such measurements is currently underway.

5 Acknowledgment

The support of Australian Research Council (ARC) for this work is gratefully acknowledged. Dr. Atkinson was supported by an ARC Discovery Early Career Researcher Award (DECRA) fellowship while undertaking this work.

References

[1] G. Elsinga,, F. Scarano, B. Wieneke, B. van Oudheusden, Experiments in Fluids 41(6) (2006) 933–947

[2] C. Atkinson, J. Soria, Experiments in Fluids 47(4) (2009) 553–568

[3] G. Meier, Experiments in Fluids 33 (2002) 181-187

[4] B. Atcheson, I. Ihrke, W. Heidrich, A. Tevs, D. Bradley, M. Magnor, H.P. Seidel, Proc. ACM transactions on graphics (TOG), ACM, vol 27 (2008) 132-

[5] E. Goldhahn, J. Seume, Experiments in Fluids 43(2-3) (2007) 241-249

[6] A. C. Kak and M. Slaney, Principles of Computerized Tomographic Imaging, IEEE Press, 1988

[7] G.T. Herman, A. Lent, Computers in biology and medicine 6(4): (1976) 273– 294

[8] H. Richard, M. Raffel, Measurement Science and Technology 12(9) (2001) 1576

[9] N. Panchapakesan, J. Lumley, Journal of Fluid Mechanics 246 (1993) 225–247

An Experimental Study of the Effect of Biodiesel on the Ignition and Combustion Characteristics of Single Droplets of Glycerol

Hendrix Y Setyawan, Mingming Zhu*, Zhezi Zhang, Dongke Zhang

Centre for Energy (M473), The University of Western Australia

35 Stirling Highway, Perth 6009, Australia

Abstract

The effect of residual biodiesel on the ignition and combustion characteristics of single droplets of glycerol was experimentally studied. Biodiesel was added into pure glycerol at concentrations ranging from 1 to 3v% to simulate biodiesel residue in glycerol as the by-product of trans-esterification of fatty acids. The single droplets of glycerol with and without biodiesel were suspended on the tip of a silicon carbide fibre and then ignited and combusted in a horizontal tube furnace maintained at 1023K. The ignition and combustion processes of the single droplets were recorded using a high-speed CCD camera. The droplet size, ignition delay time, burnout time, and burning rate were determined by analysing the history of the captured images. Biodiesel preferentially evaporated first, leading to the ignition of the droplet and the formation of a reddish flame. Subsequently, glycerol evaporated and burned with a greenish flame. As the biodiesel concentration increased, ignition delay and burnout times decreased while the burning rate of the droplets increased.

1 Introduction

The global production of biodiesel has continuously increased in the past decade, generating an increasing amount of crude glycerol as a by-product which cannot be absorbed by the existing glycerol market. In 2014, global biodiesel production was approximately 33.2 million tonnes, producing of approximately 3.3 million tonnes of crude glycerol [1]. Alternative technologies need to be developed to utilise this glycerol [2]. Burning it as a fuel for heat and power generation is perhaps one of the most economical and feasible options with potential to save fossil fuels and reduce biodiesel production cost [3]. However, using crude glycerol as a fuel remains challenging due to the properties and impurities [4]. Thus, understanding the role of impurities in the combustion process of glycerol will be beneficial in utilising crude glycerol as a fuel.

The properties of crude glycerol from the Wilmar Industry include 68.2 v% free glycerol with 2.2 v% methanol, 22.6 v% water, 1.3 wt % soap, 0.5 v% biodiesel, 1.8 v% free fatty acid (FFA), and 3 wt% ash impurities.

Our previous work has studied the effect of methanol [5], water [6], and soap [7] on the ignition and combustion characteristics of single droplets of glycerol. As a continuation of these studies, this paper reports the effect of biodiesel on the ignition and combustion characteristics of glycerol using the suspended single droplet experimental technique [8]. The ignition delay time, burnout time, and burning rate were investigated.

Corresponding author. Fax: +61-8-6488 7622
E-mail address: mingming.zhu@uwa.edu.au

2 Experimental

Mixtures of pure glycerol and biodiesel were used in this study. The pure glycerol was purchased from Sigma-Aldrich. The biodiesel was received from Wilmar International, Singapore. The key physical and thermodynamic properties of glycerol and biodiesel are listed in Table 1. Biodiesel were added into pure glycerol at concentrations of 1 v%, 2 v% and 3 v%, respectively. Since biodiesel and glycerol are immiscible, Tween 40 was used as a surfactant, added into each sample at a concentration of 0.1 v%. The biodiesel, glycerol, and Tween 40 were then stirred at 1500 rpm for 10 minutes to homogenise the mixture. Stirring was continued at 300 rpm throughout the experiment to retain its homogeneity. In this paper, the pure glycerol is denoted as PG and the mixtures are denoted as GBx, with x indicating the volume percentage of biodiesel in the glycerol. For instance, GB1 means pure glycerol with the addition of 1 v% biodiesel.

Fuel property	Glycerol	Biodiesel
Density @ 290K (kgm^{-3})	1261	878
Heating value ($MJkg^{-1}$)	18	42.2
Boiling point (K)	563	478 -716
Cetane number	5	51.7
Viscosity @313K (mPa.s)	1500	3.36 – 3.68
Auto-ignition temperature (K)	796	646
Specific heat capacity ($kJ\ kg^{-1}$)	2.43	2.2
Latent heat of vaporization at 373K ($kJ\ kg^{-1}$)	706	450
Vapour pressure at 298K (Pa)	0.01	135
Flash point (K)	433	373

Table 1: Key physical and thermodynamic property of glycerol and biodiesel [8, 9]

Figure 1 shows the experimental apparatus of the single droplet combustion experiments [8]. In brief, it consists of a horizontal tube furnace (600mm in length and 40mm in diameter) with temperature control providing a hot air environment, a droplet suspension system, a step motor for delivering the droplet into the furnace, and a CCD camera for measuring the combustion characteristics. Prior to the insertion of a single droplet, the furnace was heated to, and maintained at 1023K to represent typical combustion temperatures in diesel engines.

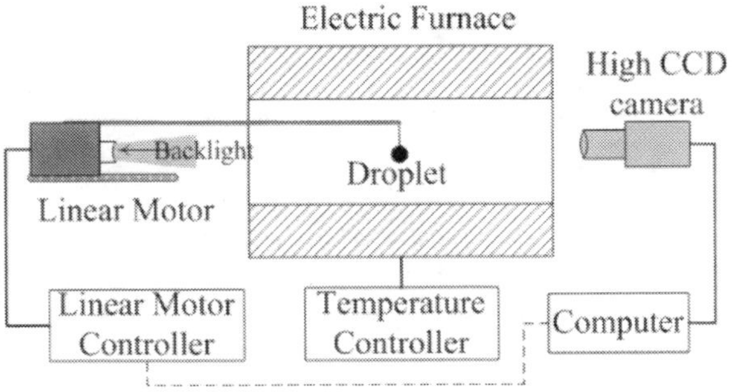

Figure 1: A schematic diagram of the apparatus for the droplet combustion experiments

The droplet was produced by a micropipette of 10 μL and was deposited onto the tip of a silicon carbide fibre of 0.142 mm in diameter. The actual shape of the suspended droplet was elliptical, and the droplet size was determined as the cubic root of the product of the droplet width squared and droplet length [10]. Aided by a step motor, the droplet was delivered to the centre of the furnace at a linear velocity of $1ms^{-1}$ or in 0.3s. A high-speed CCD camera (Basler PIA-210gc) was used to capture the images during the combustion process. In order to determine the droplet size, a 50 w lamp was used to backlight the burning droplet to assist the droplet size measurement. The recorded images of the backlit burning droplet were processed using Matlab to determine the evolution of the droplet size (d_s) during ignition and combustion.

The ignition delay time and burnout time of a burning droplet were approximated using the series of images taken by the CCD camera at 200 frames per second. The ignition delay time (t_i) was defined as the time difference between the time when the droplet arrived at the centre of the furnace and the time when the first visible flame was observed. The burnout time (t_b) was defined as the time difference between the ignition and the end of combustion. The burning rate was determined based on the classical d^2-law of droplet combustion [11]. From the d^2-t plot, the burning rate constant (k_b) was defined as $k_b = -\frac{d}{dt}\left(d_s^2\right)$.

3 Results and Discussion

3.1 Ignition and combustion phenomena of the single droplets

Figure 2 shows the time-sequenced images of burning glycerol droplets, glycerol – biodiesel mixture droplets, and neat biodiesel droplets of the initial droplet sizes of approximately 1mm. From the left to the right-hand side are: 1) the droplets arriving at the centre of the furnace; 2) ignition; 3-4) combustion of the droplets; and 5) end of droplet combustion. Upon arrival at the centre of the furnace, the glycerol droplets with and without biodiesel addition were of the same size (Image 1). At ignition (Image 2), flames were formed surrounding the droplets.

The flame of the pure glycerol droplet (PG) was greenish, suggesting a clean combustion with very little soot formation. The oxygen moiety in the glycerol may promote the combustion, and reduce soot formation. On the other hand, the flame from neat biodiesel was bright yellow, suggesting a high soot formation. At the end of the combustion of pure glycerol droplets, the fibre tip was clean, free of soot or ash deposit, while in the combustion of biodiesel droplets, soot formed and then quickly burnt.

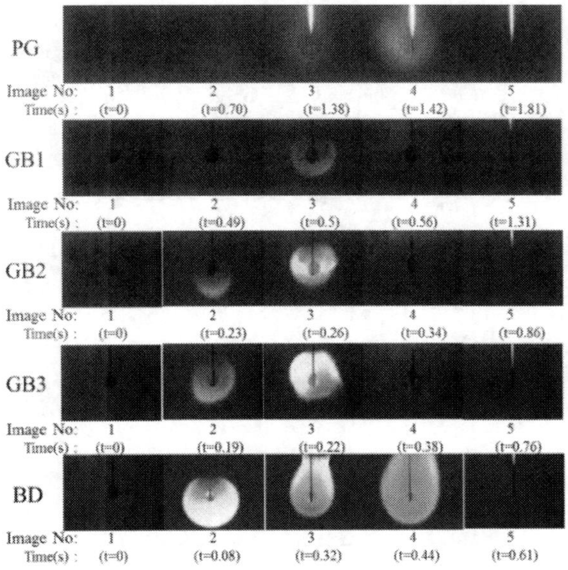

Figure 2: Typical time-sequenced images of burning glycerol droplets with and without biodiesel additions at 1023K. Pure glycerol (PG), pure glycerol – biodiesel mixture (GB), neat biodiesel (BD)

During the combustion of droplets of glycerol with biodiesel additions, a reddish flame was formed at ignition, lasting only for a few milliseconds, followed by a greenish flame until the end of the combustion. The reddish flame was mainly due to the combustion of biodiesel in the flame, and the duration of the reddish flame increased with biodiesel concentration. The reddish flame was mainly due to the combustion of biodiesel in the flame, which can be confirmed by the evidence that the combustion of neat biodiesel, which showed the same flame colour. The greenish flame is believed to be the combustion of pure glycerol.

Upon the heating, biodiesel would evaporate first, followed by the evaporation of glycerol. Compared to glycerol, biodiesel is a more volatile component due to its high vapour pressure. The latent heat of evaporation of biodiesel (250 kJ kg^{-1}) is also much lower than that of glycerol (706 kJ kg^{-1}). Biodiesel also has a higher cetane number (51.70) than pure glycerol (5). A higher cetane number means a shorter ignition delay time [14]. These suggest that when the ignition criterion was met, biodiesel was preferentially ignited and burned at the beginning of the droplet combustion. Once most of the biodiesel had been evaporated and burned, the glycerol evaporated and burned until completion of the combustion.

3.2 Droplet size, ignition delay time, burnout time, and burning rate

Figure 3 illustrates the temporal variation of the square of the normalised droplet diameters for glycerol droplets with various biodiesel additions after the ignition. Overall, the droplet sizes decreased linearly with time, implying that the combustion conforms to the d^2-law [11]. However, there were fluctuations in the droplet sizes before ignition, and the intensity of the fluctuation was stronger when the biodiesel concentration in the glycerol increased. When the biodiesel-glycerol droplet was heated, the droplet temperature increased. As previously discussed, the biodiesel evaporated first and glycerol concentrated at the droplet surface. The droplet temperature was therefore controlled by the more abundant, higher boiling point glycerol at the surface. When the droplet temperature of the droplet interior was higher than the boiling point of biodiesel, the biodiesel in the droplet interior would be superheated, resulting in the bubble formation. The bubbles would grow and collapse, leading to the observed fluctuations the shape and size of the droplet.

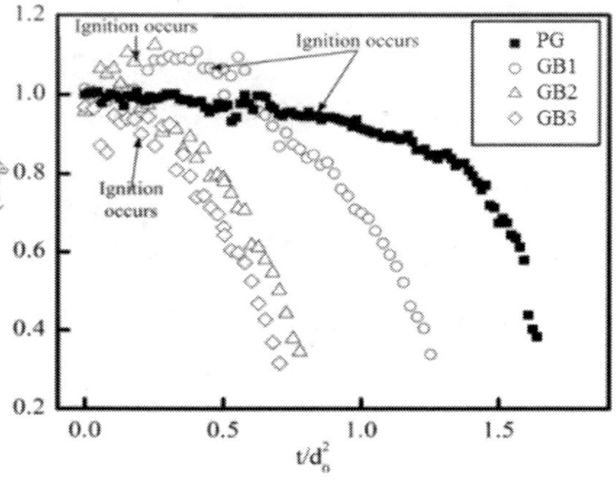

Figure 3: Temporal variation of the square of the normalised droplet diameters of glycerol with different biodiesel addition

Figure 4a shows the effect of biodiesel addition on the ignition delay of glycerol droplets. The ignition delay time decreased as the biodiesel in the mixture increased. The ignition delay time was influenced by droplet heating and evaporation, molecular diffusion and mixing with air, and the subsequent chemical reaction in the gas phase between the fuel and oxygen. As previously described, the ignition was initiated by the combustion of biodiesel. Biodiesel has a lower latent heat and boiling point and a much higher cetane number than glycerol. This explains why the ignition delay time became shorter as the biodiesel addition increased.

Figure 4b shows the burnout time of glycerol with various biodiesel additions. It is evident that the burnout time decreased as the biodiesel in the mixture increased. This suggests that the addition of biodiesel increased the droplet evaporation rate. Table 1 shows that the latent heat of evaporation of biodiesel (450 kJ kg^{-1}) is lower than that of glycerol (706 kJ kg^{-1}), suggesting that addition of biodiesel reduced the latent heat of the pure glycerol – biodiesel mixture.

Figure 4c shows the burning rates of droplets of glycerol with various biodiesel additions. It is clear that the addition of biodiesel increased the burning rate of the glycerol/biodiesel mixture. According to the classic combustion theory of droplets, the burning rate of fuel increases as the latent heat of evaporation, fuel boiling point, and density decrease [11]. As shown in Table 1, the addition of biodiesel to glycerol may reduce the heat of evaporation and density of glycerol, which consequently increased burning rates of the fuel mixture.

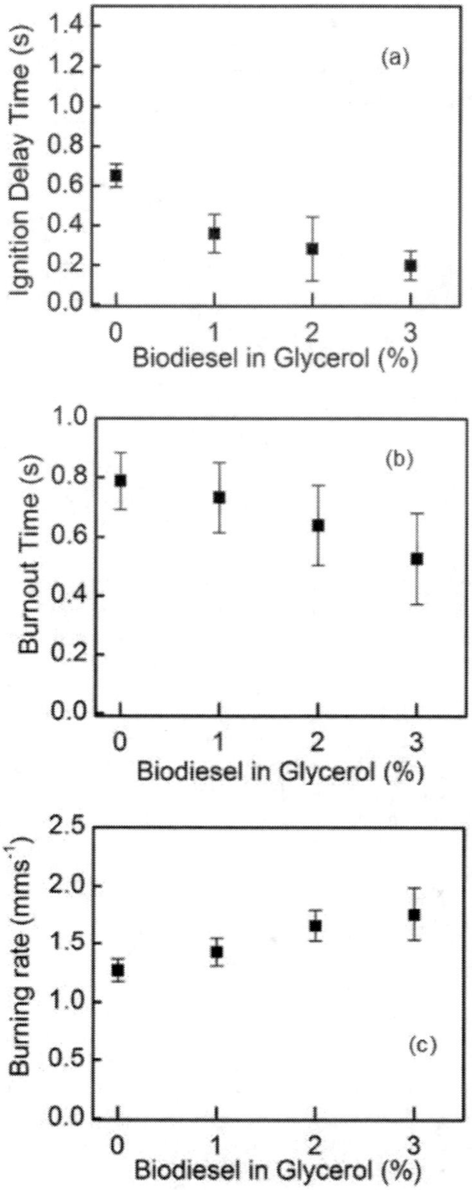

Figure 4: Effect of the biodiesel addition on (a) ignition delay time; (b) burnout time; and (c) burning rate of glycerol droplet

4 Conclusions

The effect of biodiesel addition on the ignition and combustion characteristics of single droplets of glycerol/biodiesel mixture was experimentally studied using the suspended single droplet combustion technique. It was found that biodiesel would preferentially evaporate first, leading to the ignition of the droplet and the formation of a reddish flame. Subsequently glycerol evaporated and burned with a greenish flame. The addition of biodiesel decreased the ignition delay times and burnout times while increasing the burning rates of the glycerol droplets.

5 Acknowledgements

Partial financial support has been received for this work from the Australian Research Council under the ARC Discovery

Project scheme (DP110103699) and the ARC Linkage Project scheme (LP100200135), BHP Billiton Iron Ore Pty Ltd and Ansac Pty Ltd. The biodiesel was provided by Wilmar International. Hendrix Setyawan also acknowledges the Australian Commonwealth Government for providing the Australia Awards Scholarship.

References

[1] OECD/FAO. OECD-FAO Agricultural Outlook 2014. OECD Publishing; 2014.

[2] J. McNeil, P. Day, F. Sirovski. Process Safety and Environmental Protection 90 (2012) 180-188.

[3] F.S.C. Junior FSC, F.S.A. Cavalcante, J.C.S. Alves, L.C. de Lima. International Transaction Journal of Engineering, Management, & Applied Sciences & Technologies 3 (4) (2012) 10.

[4] S. Hu, X. Luo, C. Wan, Y. Li. Journal of Agricultural and Food Chemistry 60(23) (2012) 5915-5921.

[5] H.Y. Setyawan, M. Zhu, W. Zhou. D. Zhang. Effect of methanol addition on combustion characteristics of single droplets of glycerol. Proceedings of the Australian Combustion Symposium, Australia, 2013, p. 332-335.

[6] H.Y. Setyawan, M. Zhu, Z. Zhang, D. Zhang, Energy Procedia 75 (2015) 578-583.

[7] H. Y. Setyawan, M. Zhu, Z. Zhang and D. Zhang, Combustion Science and Technology 189 (2017), p. 1540-1550.

[8] M. Zhu, Y. Ma, D. Zhang. Proc. Combust. Inst. 34 (1) (2013) 1537-1544.

[9] Glycerine Producers' Association. Physical properties of glycerine and its solutions. New York: Glycerine Producers' Association;1963.

[10] H.Y. Setyawan, M. Zhu, Z. Zhang and D. Zhang, Energy 113 (2016) 153-159.

[11] C.K. Law. Combustion Physics, Cambridge University Press, New York, 2006, p. 214 – 217.

[12] M. Lapuerta, J. Rodríguez-fernández, C. Estevez, N. Bayarri. Biomass and Bioenergy 76 (2015) 130-140.

[13] M.L. Botero, Y. Huang, D.L. Zhu, A. Molina, C.K. Law. Fuel 94 (0) (2012) 342-347.

11th Asia-Pacific Conference on Combustion,
The University of Sydney, NSW Australia
10th -14th December 2017.

Cu and Cu_2O oxidation in chemical looping processes: a first-principles theory study

Jie Cao[1], Haibo Zhao[1], Yongliang Zhang[2]

[1]State Key Laboratory of Coal Combustion, HuaZhong University of Science and Technology

1037 Luoyu Road, Wuhan, Hubei Province, 430074, P R China

[2]State Key Laboratory of Power Systems, Department of Thermal Engineering,

Tsinghua University, Beijing, 100084, P R China

Abstract

Copper oxide is widely utilized as oxygen carrier in many chemical looping processes such as chemical looping combustion (CLC) and chemical looping with oxygen uncoupling (CLOU). Understanding the oxidation processes of $Cu_2O + 1/2O_2 \rightarrow 2CuO$ (mainly in CLOU) and $2Cu + O_2 \rightarrow 2CuO$ (mainly in CLC) at a microcosmic level is critical for the exploration of intrinsic oxidation mechanism and kinetics, rational design of high-performance Cu-based oxygen carrier, and even CFD simulation, reactor design and scaling-up. In this work, the detailed information on Cu and Cu_2O oxidation was investigated by periodical density functional theory (DFT) calculations. First, we determined the most likely adsorption and dissociation pathways for O_2 on Cu (111) and Cu_2O (111) surfaces. The oxidation of Cu (111) surface favors the formation of CuO (NOT Cu_2O) at the initial stage, and so is the oxidation of Cu_2O (111) surface. Second, intrinsic mechanisms for the oxidation of inner Cu and Cu_2O were discussed. It was found that the outward diffusion of Cu cations presents smaller energy barriers than the inward diffusion of O anions, implying the regeneration of copper oxide bulk is mainly governed by the cations outward diffusion. Third, inner ions diffusion rather than surface reaction is identified as the rate-limiting step of Cu/Cu_2O oxidation, which is well consistent with the thermogravimetric analyzer (TGA) experiment. Finally, a whole mechanism for Cu/Cu_2O oxidation was elucidated.

1 Introduction

Chemical-looping combustion (CLC) has been raised to be a promising CCS (Carbon Capture and Storage) technology [1]. In this innovative combustion concept, air is replaced by oxygen carrier (OC) to combust fuels. In fuel reactor (FR), fuels is oxidized by active lattice oxygen in OCs (MeO_x). Ultra-pure CO_2 stream can be obtained through condensing H_2O, without extra energy involved for CO_2 separation. The reduced OCs are transported to air reactor (AR), where they are reoxidized by air to MeO_x, releasing a large amount of heat.

It is well-known that the oxygen carrier is a key component of CLC. Some metal oxides such as CuO are able to release gaseous O_2 at high temperature (800 −1000 °C) and oxygen-deficient atmospheres. The presence of O_2 will greatly prompt the combustion and conversion of solid fuels such as coal and the new chemical looping process is called as chemical-looping with oxygen uncoupling (CLOU) [2]. Cu-based oxygen carriers exhibit quite high lattice oxygen transport capacity within a moderate reaction temperature region, and therefore has been recognized as an excellent OC candidate for CLC/CLOU. There have been many publications focusing on the performance of Cu-based OC in FR [3-7]. This work focuses on the regeneration of OC, *i.e.*, the oxidation processes of Cu or Cu_2O, which is also of great importance to the whole chemical-looping system.

Till date, extensive experimental researches have been conducted to explore the oxidation mechanisms of CuO/Cu_2O/Cu system. At temperatures of interest for chemical looping process, Mimura et al. [8] deduced that the growth of Cu_2O predominates the oxidation in the range of 600~800°C, including lattice diffusion and grain boundary diffusion, while the lattice diffusion dominates the oxidation process at high temperature (850~1050°C) due to a fast disappearance of grain boundaries. Zhu et al. [9] reported that the oxidation of pure copper will produce both Cu_2O and CuO, and the surface is always covered by a thin layer of CuO before complete consumption of metallic copper. They conjectured [10] that at the initial stage of the oxidation, Cu reacts with air on the gas-solid interface, forming CuO directly; the later reaction is Cu + CuO→Cu_2O on the interface between Cu and CuO. However, these experimental observations can not *in situ* monitor the distribution of oxides and the migration of phase boundaries within the bulk. Therefore, the mechanisms proposed by Zhu et.al still needs to be confirmed at a microcosmic level.

With respect to the oxidation reaction kinetics, it is generally considered that the oxidation of pure copper at a relatively high-temperature region follows a parabolic kinetics, as most of the transition metals behave. However, the oxidation kinetics of Cu_2O→CuO is still disputable [11, 12], and it is still unclear whether inward diffusion of O^{2-} or outward diffusion of Cu^+ is responsible for the CuO growth [13].

The behavior of OC in AR is intrinsically dependent upon the electron density distributions and crystal structures of metal oxides. Hence, an atomic-scale modeling by quantum chemistry method will be very beneficial to reveal the intrinsic copper oxidation mechanism and kinetics with detailed geometric information and energy profile of every constituent elementary steps. In this study, periodical density functional theory (DFT) calculation was utilized to explore the surface Cu/Cu_2O oxidation reaction at the initial stage and the consequent oxidation of bulk Cu/Cu_2O. The most favorable O_2 adsorption site and O_2 dissociation pathway during the surface oxidation processes were determined, and the predominant inner ions migration/diffusion pathways during the bulk oxidation

Corresponding author. Fax: +86-27-87545526
E-mail address: hzhao@mail.hust.edu.cn (H. Zhao)

processes were identified. Meanwhile, an isothermal oxidation experiment was conducted in a TG analyzer to relate the DFT calculations about the rate-limiting step with some macroscopic observations. The calculated predictions agreed with the experimental results very well and gave an insight on the regeneration features of Cu-based OC in AR.

2 DFT calculation details

All DFT calculations were conducted in the CASTEP [14] program package of Materials Studio software. GGA-PW91 was chosen for the electron exchange correlation. The tested energy cutoff and k-point separation were 400 eV, 0.04 respectively. GGA+U method was used for the Cu 3d orbitals, and the reasonable U value was 7.5 eV [15].

The Cu_2O and Cu unit cells were optimized and the lattice parameters are listed in Table 1. Our previous work [6] used the CuO (111) surface to study the oxygen release characteristics of CuO, because it is the most stable low-index CuO surface which contains compact Cu–O layers under ambient conditions. Similarly, we chose the Cu_2O (111) and Cu (111) surfaces as the object surfaces according to others' experiment results [16].

Material	Calculation	Experiment
Cu_2O	a=b=c=4.296388 Å α=β=γ=90°	a=b=c=4.2685 Å α=β=γ=90° [17]
Cu	a=b=c=3.625647 Å α=β=γ=90°	a=b=c=3.604 Å α=β=γ=90° [18]

Table 1: Lattice parameters of Cu_2O and Cu crystal structures

In this study, the adsorption energy E_{ad} of O_2 molecule on the surface, was defined as

$$E_{\text{ad}} = E(\text{O}_2/\text{slab}) - [E(\text{O}_2) + E(\text{slab})] \qquad (1)$$

where $E(\text{O}_2)$ and $E(\text{slab})$ are the total energies of the free O_2 molecule and bare slabs, while the $E(\text{O}_2/\text{slab})$ is the total energy of the adsorption structure. When considering the reaction process, the complete LST/QST approach was used to find the transition state (TS) structure as well as to determine the energy barriers E_{b} of different steps. E_{b} was calculated as

$$E_{\text{b}} = E(\text{TS}) - E(\text{IS}) \qquad (2)$$

where $E(\text{TS})$ and $E(\text{IS})$ are the energies of the transition states and the initial states (IS) respectively.

3 Surface oxidation mechanism of Cu/Cu_2O

3.1 Cu_2O (111) surface oxidation process

The regeneration of Cu_2O was first studied. DFT calculations started with putting a free O_2 molecule upon the Cu_2O (111) surface (a 5 Å distance), and the optimized configuration was set as IS (as shown in Fig. 1(a)). Then the most favorable adsorption site was determined (see Fig. 1(b)), and the calculated adsorption energy is 2.09 eV, suggesting the surface is very active for the O_2 adsorption. In order to determine the most favorable dissociation pathway, we broke the adsorbed O-O bond, and enlarged the distance of the two O atoms. After the geometry optimization, a configuration with lowest overall energy was chosen to be the final state (FS) of the surface oxidation, as shown in Fig. 1(d). Finally, a TS search calculation was performed to identify the TS structure and the corresponding energy barrier. The most favorable pathway was shown in Fig. 1(c). The potential energy diagram for the surface oxidation is given in Fig. 2.

This surface oxidation process can be summarized as following: At the beginning, a free O_2 molecule adsorbs on the clean surface. Then, the two dissociated O atoms are motivated to migrate under high temperature environment, embedding in a stable position of Cu_2O (111) surface, and the energy barrier is 0.96 eV. It must be noted, in FS, one dissociated O anion is three fold-coordinated (O-3f) while another is four fold-coordinated (O-4f), which is similar to the atom configuration of CuO (111) surface. The transformation from Cu_2O phase to CuO phase indicates the dissociation process in our calculation is a reasonable surface oxidation pathway. The whole reaction energy is -2.63 eV.

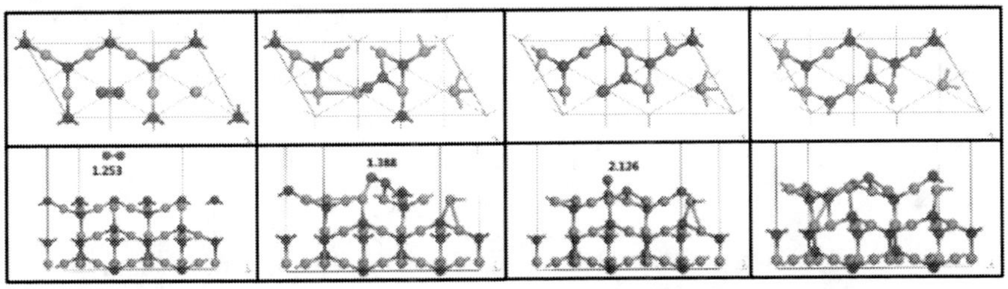

(a) Cu_2O+O_2 (g) (b) Cu_2O+O_2 (ad) (c) TS (d) $Cu_2O+_2O^{2-}$

Figure 1: O_2 molecule adsorption and dissociation processes on Cu_2O (1 1 1) slab, the blue balls present the adsorbing O atoms

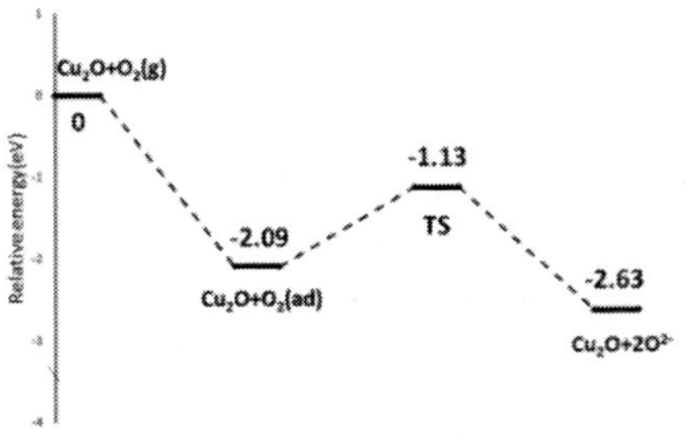

Figure 2: Potential energy diagram for the Cu_2O (111) surface oxidation process

3.2 Cu (111) surface oxidation process

On Cu (111) surface, the most favorable adsorption is that a horizontally directed oxygen molecule adsorbs at the top site of surface copper, releasing an energy of -1.16 eV (Fig. 3 Cu+O_2 (ad)). In the following process, two oxygen atoms dissociate on the surface driven by the high temperature, and bond with the surrounding Cu atoms. Meanwhile, the Cu cation on the adsorption site is drawn away from the compact surface gradually by the attraction force from the two oxygen anions, forming a horizontal O-Cu-O structure. The energy barrier of the dissociation process is only 0.06 eV, which is just required to break the O-O bond. Fig. 3 presents the structures and energy changes during the Cu (111) surface oxidation process.

To determine the product of Cu surface oxidation, a charge analysis was conducted. Based on the same set of optimization parameters, the Mulliken charge of the Cu cations in the Cu_2O (111) model ranges from 10.6 to 10.8 e, while that of CuO (111) model are 10.2-10.5 e. We calculated the involved ions charge in the final state (Fig. 3 FS-T) of Cu (111) oxidation. It was found that the Mulliken charge of the Cu cation bonding with the two O anions is 10.42 e, which means Cu^0 had been oxidized to Cu^{2+}. This findings is consistent with the research of Zhu et al [10], that is, the favorable reaction at the initial stage of Cu surface oxidation is $Cu+O_2 \rightarrow CuO$.

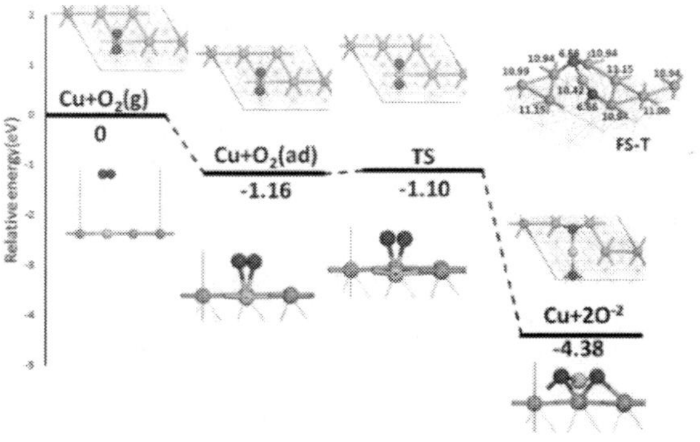

Figure 3: The surface oxidation process of Cu (111) slab, the red and green balls present oxygen atoms and Cu atom on the adsorption site respectively

4 Bulk oxidation mechanism of Cu/Cu_2O

The bulk oxidation of Cu/Cu_2O is dependent of the ions diffusion in the solid phase. There are two kinds of ions diffusion route: inward diffusion of anion and outward diffusion of cation. In this section, we used the final state of the surface reaction as the initial state of the bulk oxidation.

4.1 The ions diffusion during the oxidation of Cu_2O (111) slab

As shown in Fig. 4(a), there were two kinds of dissolved O anion, O-3f and O-4f respectively. O-3f is easier to diffuse because it possesses higher degree of freedom and bigger space of migration. It was moved vertically from the surface to the second layer, and the optimized structure was set as FS for this diffusion (Fig. 4(c)). At TS (see Fig. 4(b)), the O-3f migrates from the initial site to the middle position between surface and subsurface. The process is endothermic, and the energy barrier is 2.19 eV.

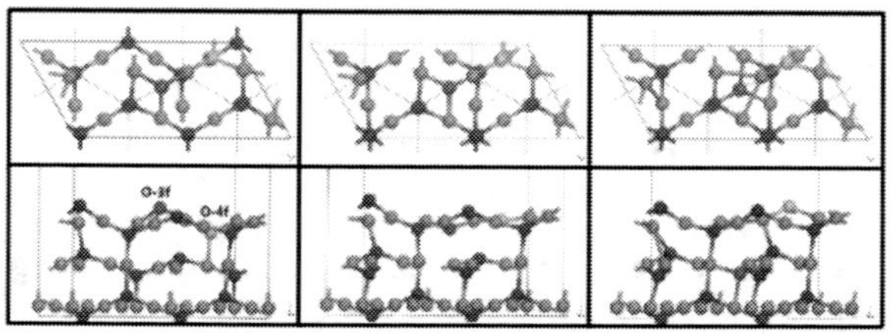

(a) IS (b) TS (c) FS

Figure 4: The structure of the IS, TS and FS for O anion inward diffusion in Cu_2O (111) slab. The blue and green balls present O-3f and O-4f respectively

Between two kinds of Cu cations (Cu-2f and Cu-3f), Cu-2f is more favorable to diffuse. The outward diffusion of Cu-2f was then calculated. TS and FS of this process were displayed in Fig. 5.The reaction energy of outward diffusion of Cu cation is 0.88 eV, and the energy barrier is 1.61 eV, both of which are significantly lower than those of O anion inward diffusion (see Fig. 6). It means that the outward diffusion of Cu cation requires less heat and is more kinetic favorable. Therefore, we deduced that the Cu cation outward diffusion dominates the growth of CuO during the bulk oxidation of Cu_2O slab.

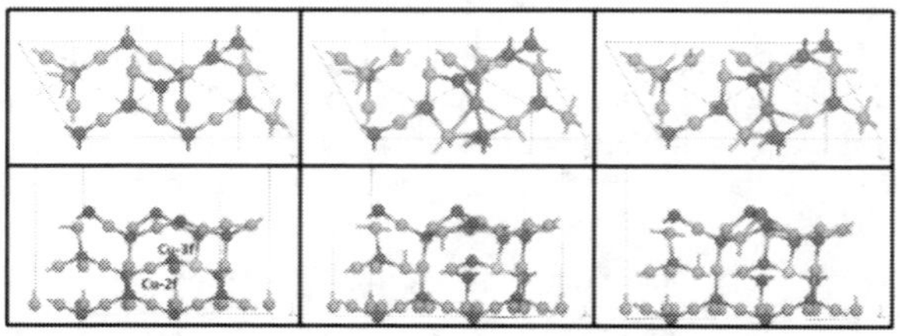

(a) IS (b) TS (c) FS

Figure 5: The structures of the IS, TS and FS for Cu cation outward diffusion in Cu_2O (111) slab. The blue and green balls present Cu-2f and Cu-3f respectively

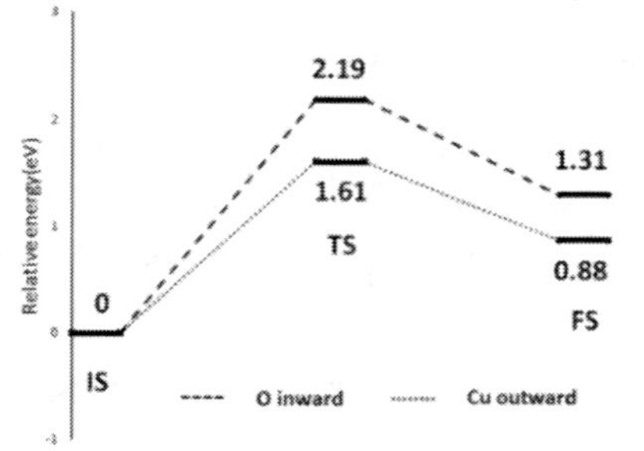

Figure 6: Potential energy diagram for the O anion and Cu cation diffusion processes in Cu_2O (111) slab

4.2 The ions diffusion during the oxidation of Cu (111) slab

Based on the final state of surface oxidation, the detailed information of O anion inward diffusion was first obtained, as shown in Fig. 7. It takes 1.81 eV to break the Cu-O bonds, and the reaction energy of this process is 1.72 eV.

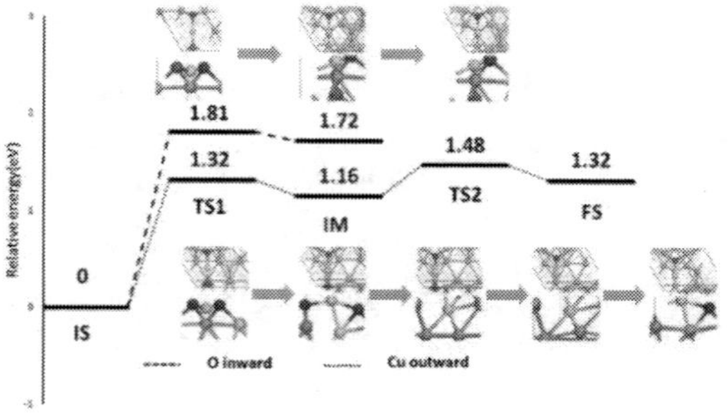

Figure 7: O anion inward diffusion and Cu cation outward diffusion during the bulk oxidation process of Cu (111) slab. The blue ball presents O anion, the light green and dark green balls present surface and subsurface Cu cations respectively

The outward diffusion of Cu cation is a two-step process. Firstly, surface copper vacancy (V_{Cu}) is occupied by the neighbor Cu atom (the light green ball in Fig. 7), creating a new vacancy (E_b=1.32 eV). The new bare copper vacancy is conductive to the outward diffusion of nearest subsurface Cu atom (Cu_{sub}) (the dark green ball in Fig. 7) with an additional energy of 1.48 eV. Clearly, the energy barrier for the cation outward diffusion in Cu (111) slab is still smaller than the anion inward diffusion, and the Cu cation outward diffusion is still predominant during the bulk oxidation of Cu (111).

5 Discussions

From the calculated barriers, it is noted that the surface oxidation reactions require quite smaller additional energy (0.96 eV for Cu_2O and 0.06 eV for Cu) than the bulk oxidation reactions (1.61 eV for Cu_2O and 1.32 eV for Cu). Therefore, it was deduced that the oxidation of Cu and Cu_2O is limited by the Cu cation outward diffusion rather than the surface process, which is totally different from the oxygen release process of CuO [6]. An isothermal oxidation experiment was conducted in a TGA (air atmosphere, 850°C) to validate the DFT calculations. The conversion and reaction rate are presented in Fig. 8. As shown, conversion rate achieves the maximum very rapidly, and then decreases gradually. This observation can be interpreted as that the surface reaction is completed within a short time period for a smaller barrier, but the bulk oxidation process (or the Cu cation outward diffusion) proceeds very slowly because of the large barrier, and become more terrible with the increase of migration distance. So here a good agreement between experiment observations and DFT calculations was obtained.

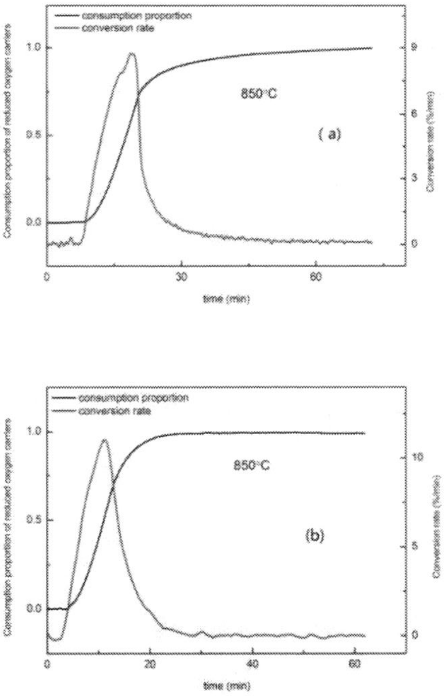

Fig. 8: The conversion (black line) and reaction rate (red line) of (a) Cu and (b) Cu_2O oxidation at 850°C.

6 Conclusions

Oxidation mechanisms of Cu_2O and Cu were investigated at an atomic level, and the major conclusions to be drawn from the work are as follows: (1) The most likely oxidation pathway for the low Miller index Cu_2O (111) and Cu (111) surface were obtained as well as the energy barrier of each step. Surface reaction of Cu_2O (111) leads to the transformation from Cu_2O phase to CuO phase. The oxidation of Cu (111) surface will produce a thin layer of CuO firstly, rather than Cu_2O. (2) The bulk oxidation process of Cu and Cu_2O are dominated by the Cu cation outward diffusion. (3) The rate-limiting step during the copper oxidation process is ions diffusion in the lattice instead of gas-solid reactions, which is quite different from the CuO decompositon process.

7 Acknowledgment

This work was supported by "National Key R&D Program of China (2016YFB0600801) and National Natural Science Foundation of China (51522603)".

References

[1] A. Nandy, C. Loha, S. Gu, P. Sarkar, M.K. Karmakar, P.K. Chatterjee, Renewable & Sustainable Energy Reviews 59 (2016) 597-619.

[2] T. Mattisson, A. Lyngfelt, H. Leion, International Journal of Greenhouse Gas Control 3 (2009) 11-19.

[3] D. Mei, A. Abad, H. Zhao, J. Adánez, Fuel Processing Technology 133 (2015) 210-219.

[4] S.Y. Chuang, J.S. Dennis, A.N. Hayhurst, S.A. Scott, Combustion and Flame 154 (2008) 109-121.

[5] P. Gayán, I. Adánez-Rubio, A. Abad, L.F. de Diego, F. García-Labiano, J. Adánez, Fuel 96 (2012) 226-238.

[6] Y. Zhang, H. Zhao, L. Guo, C. Zheng, Combustion and Flame 162 (2015) 1265-1274.

[7] H. Zhao, Y. Zhang, Y. Wei, J. Gui, Proceedings of the Combustion Institute 36 (2017) 4069-4077.

[8] K. Mimura, J.W. Lim, M. Isshiki, Y. Zhu, Q. Jiang, Metallurgical and Materials Transactions A 37 (2006) 1231-1237.

[9] Y. Zhu, K. Mimura, M. Isshiki, Materials Transactions 43 (2002) 2173-2176.

[10] Y. Zhu, K. Mimura, M. Isshiki, Corrosion Science 46 (2004) 2445-2454.

[11] Y. Zhu, K. Mimura, M. Isshiki, Oxidation of Metals 62 (2004) 207-222.

[12] Z. Grzesik, M. Migdalska, High Temperature Materials and Processes 30 (2012) 277-287.

[13] J.H. Park, K. Natesan, Oxidation of Metals 39 (1993) 411-435.

[14] J.P. Perdew, K. Burke, M. Ernzerhof, Physical Review Letters 77 (1996) 3865-3868.

[15] J. Hu, D. Li, J.G. Lu, R. Wu, The Journal of Physical Chemistry C 114 (2010) 17120-17126.

[16] Z. Gao, W. Huang, L. Yin, L. Hao, K. Xie, Catalysis Letters 127 (2009) 354-359.

[17] A. Kirfel, K. Eichhorn, Acta Crystallographica Section A 46 (1990) 271-284.

[18] G.M. Fehrenbach, H. Bross, Physical Review B 48 (1993) 17703-17714.

11th Asia-Pacific Conference on Combustion,
The University of Sydney, NSW Australia
10th -14th December 2017.

Bishomocubanes as Possible Binders in Composite Solid Propellants

Lovely Mallick[1], Hardik K. Thakker[2], Sohan Lal[3], Sudarshan kumar[1], Irishi N N Namboothiri[3], Arindrajit Chowdhury[2], Neeraj Kumbhakarna[2]

[1]Department of Aerospace Engineering

[2] Department of Mechanical Engineering

[3]Department of Chemistry

Indian Institute of Bombay, Mumbai, Mahrashtra, 400076, India

Abstract

High energy density materials (HEDMs) are an area of interest for synthetic chemists targeting new compounds and formulations to improve the performance parameters of existing propellants. Theoretical studies demonstrate that replacement of the traditionally used binder in composite solid propellants, hydroxy terminated polybutadiene (HTPB) can enhance the specific impulse as well as the density specific impulse of composite solid propellants. However, significant work needs to be done, especially on polymerization of such compounds before they can be used as effective replacements. As a first step, two high-nitrogen containing cage compounds such as DPTrizDMBHC and DTetzBHC were used as possible monomers by replacing 5% HTPB (by mass) in conventional solid propellant mixtures of AP and HTPB. The burn rates were found to be enhanced by 15 to 20% by such an addition.

1 Introduction

Researchers are on a constant lookout for new ingredients that would help achieve major parameters such as high energetics, high specific impulse, high density, low cost, and ease of storage and handling. On the basis of theoretical analysis it was concluded that cage compounds could possibly act as a replacement for HTPB as a binder in composite solid propellants [1]. The binders define mechanical properties as well as parameters for propellant processing and aging. Binder materials such as HTPB, carboxy terminated polybutadiene (CTPB), and nitrocellulose typically act as fuels, which get oxidized in the combustion processes.

High energy density materials (HEDMs) are appropriate in various volume-limited applications in rockets, gas generators, and explosives. Attractive properties of an ideal HEDM are high gravimetric and volumetric energy density, high positive HoF, high density, as well as low operational hazards such as low toxicity, high thermal stability, and low impact sensitivity. Eaton et al. [2] reported performance of the polycyclic cage hydrocarbon cubane. Cubane is a solid under standard conditions and can be used as a highly energetic additive in the solid propellant formulation. Unfortunately, multiple stages of synthesis restricted interest to continue research on cubane. However, homocubanes and bishomocubanes, with lesser strain energy, may be synthesized with fewer steps, and energetic groups may be attached easily with them. A comprehensive review of homocubanes, bishomocubanes, and trishomocubanes have been reported by Marchand [3]. In our previous theoretical studies [4,5], the energetic properties of high nitrogen substituted BHC compounds were examined. It was found that if HTPB in the propellant composition AP+HTPB is replaced by the BHC compounds considered in the present work, the specific impulse (I_{sp}) and density specific impulse of the propellant composition are higher as compared to those of the original composition [4,5]. In this work the effect of addition of DPTrizDMBHC and DTetzBHC was studied on the burn rate of composite solid propellants.

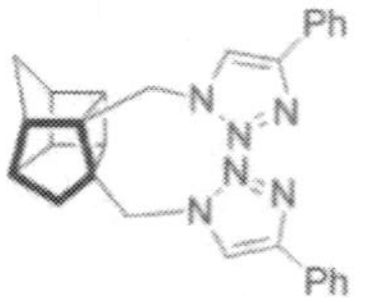

Diphenyltriazolo-dimethylbishomocubane (DPTrizDMBHC)

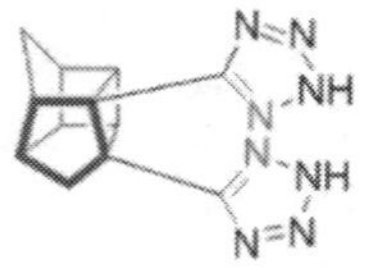

Ditetrazolobishomocubane (DTetzBHC)

2 Experimental Section

2.1 Particle Size Distribution (PSD) of Ammonium Perchlorate (AP)

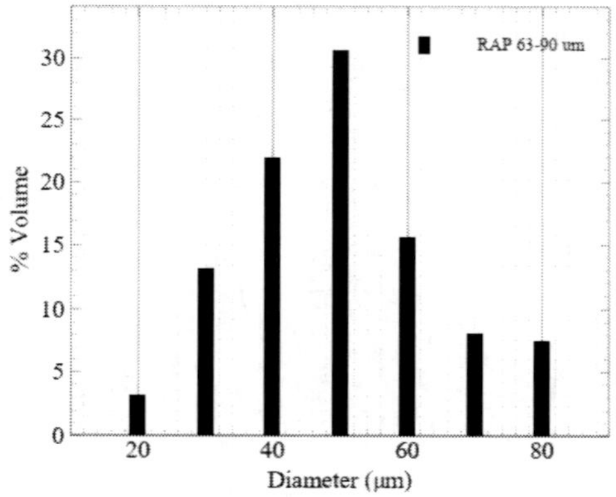

Figure 1: Particle size distribution of 63-90 μm AP.

The AP used in this study was provided by the High Energy Materials Research Laboratory, Pune, and was recrystallized after dissolving it in distilled water to remove the impurities. Owing to the hygroscopic and agglomerating nature of AP, the purified recrystallized particles were preheated to 60 °C in an oven to remove moisture, and subsequently sieved. The sieves were cleaned in an ultrasonic bath using deionized distilled water. The particles between the sieves with pore diameters ranging from 63–90 μm were chosen as the quantity of sieved AP was highest in that range. A digital microscope was used to acquire 30 images of approximately 160 particles, and the image processing steps such as segmentation, ellipsoid pattern recognition, and estimation of ellipsoid volume were carried out manually using a standard image processing software (GIMP) to determine the particle size distribution. The sizes of the AP particles were found to follow a Gaussian distribution, with the median occurring at around 50 μm, and is represented in Fig. 1.

Since the objective of this study was to estimate the efficacy of BHC-based compounds as possible binders, the base composition of the non-aluminized formulations were chosen such that the ratio of AP to HTPB was 80:20. The other two mixtures were formulated by replacing 5% HTPB by the BHC monomer. Thus the ratio of AP to DPTrizDMBHC or DTetzBHC to HTPB was 80:5:15. The composition of the CSPs is represented in Table 1.

Table 1: Composition of composite solid propellants.

Ingredients		**Weight (%)**	
Oxidizer	AP (Ammonium perchlorate)	80	80
Fuel Binder	HTPB (Hydroxyterminatedpolybutadiene)	15.92	11.94
Plasticizer	DOA (Dioctyladipate)	3	2.25
Curing Agent	IDPI (Isophoronediisocyanate)	1.074	0.81
Additive	BHC (Bishomocubane) cage compounds	---	5

Figure 2: Propellant casting setup with individual loads on each slurry strand of the propellants.

The conventional method of preparation of CSPs, as proposed by several authors [6-11], has been followed in this work. The preparation of CSPs follows three major steps: mixing, casting, and curing. Powdered AP was placed in the oven for 12 hours at 60°C to remove moisture. The ingredients for the propellant were measured using a digital measuring scale with a least count of 0.02 mg. HTPB and DOA were weighed in a glass beaker first and mixed using a glass rod for 15 minutes, while being heated at 60°C. After mixing for 15 minutes, the mixture was degassed in vacuum at 60°C for 15 minutes. Subsequently, dry powdered AP was weighed and added to the HTPB and DOA mixture in three batches with a time interval of 10 minutes between each batch. After adding all the AP powder in the mixture, the slurry was again mixed for 15 minutes and degassed in vacuum for 15 minutes. The required amount of IPDI was added to the mixture and mixed for 10 minutes. After adding IPDI the heating plate temperature was reduced to 40°C. The slurry was degassed again in vacuum for 10 minutes.

The final slurry was poured into the mold and transferred to the oven. The slurry was pressurized with a weight of 0.5 kg on top of each strand. A high vacuum pump was connected to the mold and the degassed for 45 minutes, while maintaining a temperature of 60 °C to reduce viscosity of the slurry and to facilitate the escape of trapped gases. Finally the slurry was allowed to cure in the oven for 7 days at 60°C. The cured propellant was removed carefully from the molds to avoid cracks or any damage and was stored in the desiccator. To avoid voids or gaps occurring due to the small amounts of propellants being handled, AP/HTPB and AP/DTetzBHC/HTPB propellants were further pressed in a pellet press at 1.5 ton for 10 minutes. Conversely, AP/DPTrizDMBHC/HTPB propellants were not pressed because the cured form of the propellants were well-compacted.

The length and diameter of each strand were approximately 10mm and 5mm, respectively. The cured strands were coated with Araldite to prevent lateral burning. A thin film of HTPB was coated on top of each strand to aid flame spread during ignition. The theoretical mean densities (TMD) of AP/HTPB, AP/DPhTrizBHC/HTPB, and AP/DTetzBHC/HTPB propellants were calculated to be 1.602 g/cm3, 1.632 g/cm3, and 1.658 g/cm3, respectively. However, the experimental densities were approximately 98%, 101% and 93% of the TMDs respectively. The values for the two BHC-based propellants require further verification, since their experimental crystal densities are unknown, and densities estimated using quantum chemical calculations were utilized in this study.

The linear strand combustion rates were determined by a strand burner setup. The design of the strand burner setup used for the solid propellant has been detailed in our previous work [12]. The thermal decomposition of AP/HTPB and AP/BHC/HTPB propellants were studied using a thermogravimetric analyzer (TGA), as reported in our previous work [13,14]. A piece of propellant weighing approximately 1 mg was placed in an open alumina crucible and heated from 30 °C to 500 °C at a heating rate of 5 Kmin-1.

3 Result and Discussion

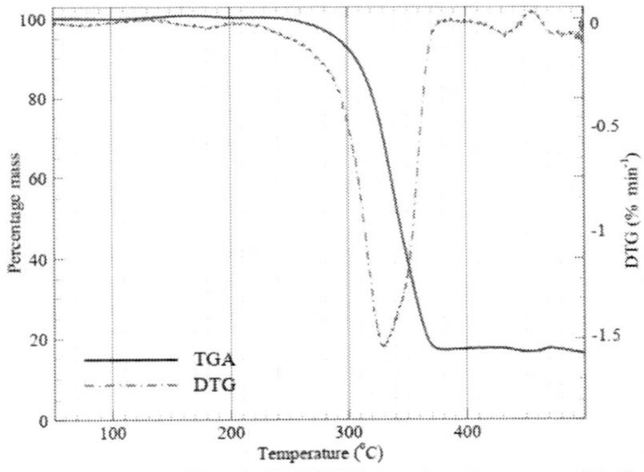

Figure 3: TGA-DTG of AP/HTPB at a heating rate of 5 $K.min^{-1}$

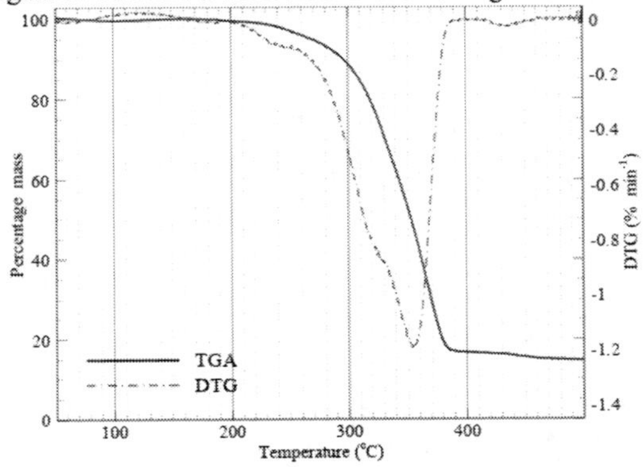

Figure 4: TGA-DTG of AP/DPTrizDMBHC/HTPB (80:5:15) at a heating rate of 5 $K.min^{-1}$

Thermal characterization of the propellants involved in this work was undertaken using a thermogravimetric analyzer to comprehend the thermal stability of the cured propellants during transportation, storage, and handling. The TGA and differential thermogravimetry (DTG) signals for AP/HTPB, AP/DPhTrizBHC/HTPB, and AP/DTetzBHC/HTPB propellants are shown in Figs. 3-5 respectively. 15% residual mass remained after decomposition of all compositions. The TGA curve in Fig. 3 shows that mass loss occurred in a single stage for AP/HTPB propellants, with an extrapolated onset temperature of 277 °C, an extrapolated end temperature of 371 °C, and a single strong DTG peak at 326 °C.

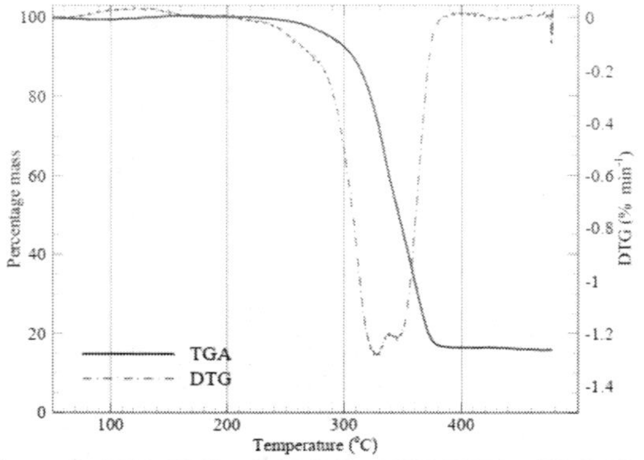

Figure 5: TGA-DTG of AP/DTetzBHC/HTPB (80:5:15) at a heating rate of 5 K.min-1

Similarly, the TGA signal from AP/DPTrizDMBHC/HTPB in Fig. 4 shows a three stage mass loss. The first hump (195 °C - 275 °C), which was more prominent compared to the AP/HTPB mixture, was indicative of the first stage mass loss during decomposition of AP. The onset temperature of the second and third stages, which were fairly inseparable, was 261 °C, which is 100 °C lower than pure DPTrizDMBHC. The extrapolated end temperature was 383 °C, and the largest DTG peak was at 351 °C. Figure 5 shows the mass loss information for AP/DTetzBHC/HTPB propellant, which was found to be similar to that of AP/DPTrizDMBHC/HTPB. The second and third stages had an onset temperature was 276 °C, an end temperature of 377 °C, and the largest DTG peak at 335 °C. Pure DTetzBHC is known to exhibit two step decomposition with mass loss of approximately 26% in the first stage. The onset, peak, and end temperatures of the pure components as well as the propellants are listed in Table 2.

Table 2: Thermal decomposition study of solid propellants and their ingredients.

Compounds/Propellants	Onset Temp (°C)	Peak Temp (°C)	End Temp (°C)
$AP_{63\text{-}90/LTD\text{-}HTD}$	260	315/401	426
HTPB	333	460	493
DPTrizDMBHC	357	382	397
DTetzBHC	245	251	263 (1st stage)
AP/HTPB Propellant	277	326	371
AP/DPTrizDMBHC/HTPB propellant	261	351	383
AP/DTetzBHC/HTPB propellant	276	335	377

Although trizole and tetrazoles based compounds were expected to be more energetic in nature, the presence of the phenyl group in DPTrizDMBHC along with the BHC structure itself stabilized the structure, and delayed the onset of its decomposition compared to pure HTPB. DTetzBHC was found to be more reactive than the triazole-based compound, as evidenced by the 100 °C drop in onset temperatures. But it was encouraging to observe that the onset and end temperatures of all the propellants were reasonably similar, thus ensuring that the new compositions are not highly unstable.

The experimental investigation of combustion characteristics of the propellant formulations was primarily constrained to measurement of the linear burn rates. The range of pressures considered were 10 to 50 bar. The nature of the flame at various stages of combustion is represented in Figure 6.

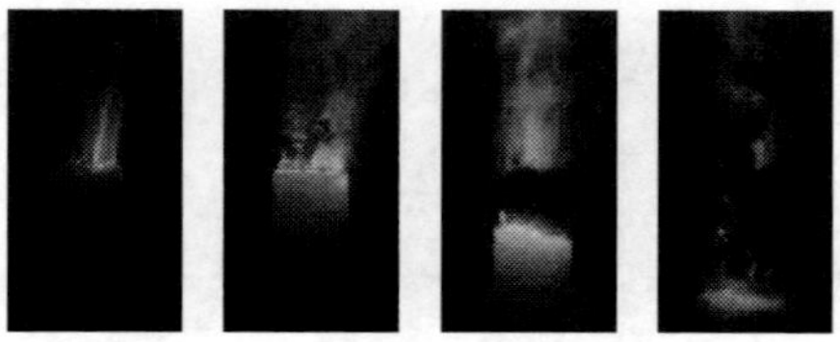

Figure 6: Various burn rate stages of AP/HTPB solid propellants at 20 bar.

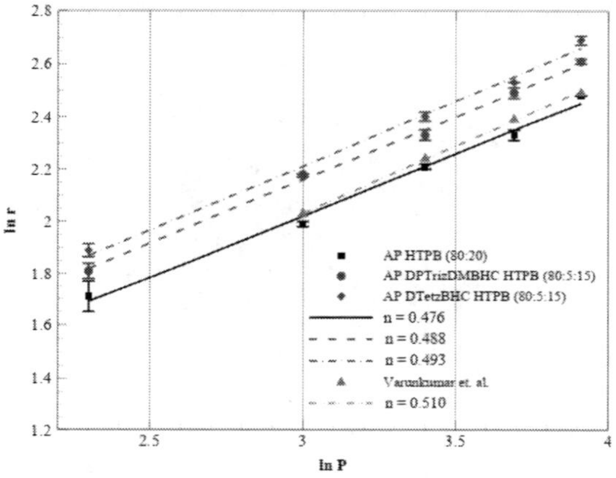

Figure 7: Burn rate of BHC added solid propellants.

Table 4: Characteristics and properties of propellants during combustion

Propellants	TMD (g cm^{-3})	Expt. Density (g cm^{-3})	Burn rate w.r.t AP/HTPB	Pressure Index (n)	ln a
AP/HTPB*	1.60	98 %	Baseline	0.48	0.59
AP/DPTrizDMBHC/HTPB#	1.63	101 %	15 %	0.49	0.69
AP/DTetzBHC/HTPB*	1.66	93 %	21 %	0.49	0.73

*pressed; and #unpressed

Figure 7 shows the comparison of burn rates of propellants with BHC-based formulations with that of the conventional AP/HTPB based non-aluminized formulation. The burn rates of AP/DPTrizDMBHC/HTPB, and AP/DTetzBHC/HTPB propellants were determined to be 15% and 21% higher than the conventional AP/HTPB propellants [15-19], respectively. The pressure index (n) of AP/HTPB, AP/DPTrizDMBHC/HTPB, and AP/DTetzBHC/HTPB propellants remained relatively unchanged by the addition of the compounds. In order to validate the burn rates of the AP/HTPB propellant, a combustion model HeQu-1D [20] was utilized and was found to provide an excellent match with the experimental burning rates. The model depends on various input parameters, such as particle size distribution and composition of propellants to determine the burn rates.

The higher burning rates of the BHC-based propellants were expected to be due to the higher flame temperatures owing to the higher heats of formation of the compounds. Hence, the adiabatic flame temperatures were calculated for the propellant formulations by using NASA CEA (Chemical Equilibrium and Applications) utility [21], and displayed in Table 4. The adiabatic flame temperatures of the BHC-based propellants were significantly higher compared to those of the AP/HTPB formulation. However, the values for the triazole-based formulation and the tetrazoles-based formulation were found to be reasonably similar, thus insinuating that the increment in burn rates should not be explained from this viewpoint alone. The effect of various physico-chemical parameters, such as enthalpy of sublimation, specific heats, conductivity, surface temperature, gas phase and condensed phase reaction rates, etc. should be taken into consideration. Additionally, decomposition of both DTetzBHC and DPTrizDMBHC [4] produces reactive byproducts such as HN3 and NH3. The presence of such reactive species in the condensed phase may lower the surface activation energies due to rapid oxidation by oxidizers produced by AP.

Table 5: Adiabatic temperatures of solid propellants using NASA CEA.

P (bar)	Tad of AP/HTPB (K)	Tad of AP/DPTrizDMBHC/HTPB (K)	Tad of AP/DTetzBHC/HTPB (K)
10	2339.49	2898.67	2900.26
20	2344.08	2949.44	2950.99
30	2346.18	2977.73	2979.25
40	2347.46	2997.08	2998.57
50	2348.34	3011.62	3013.09

4 Conclusions

Two new nitrogen rich cage compounds DPTrizDMBHC and DTetzBHC were used to replace 5% in AP-HTPB (80-20) based composite solid propellants. The propellant formulations were successfully cured into strands using conventional methods of preparation and combusted in a strand burner setup. The thermal stability was estimated using a thermogravimetric analyzer. The compositions with DPTrizDMBHC and DTetzBHC showed an improvement of 15% and 21% in burn rates, respectively. The pressure index (n) of AP/HTPB, AP/DPTrizDMBHC/HTPB, and AP/DTetzBHC/HTPB propellants were reasonably unchanged at 0.48 for all the formulations. The increased burn rates were determined to be partially due to the high adiabatic flame temperatures of the BHC-containing propellants. The thermal stability studies showed that all the formulations were equally stable. The study is an important step in the search for a replacement of HTPB as a binder.

5 Acknowledgment

We would like to thank High Energy Materials Research Laboratory, Pune, India, for providing ammonium perchlorate samples. The financial grant for the TGA–FTIR–MS setup, provided by the Industrial Research and Consultancy Centre, IIT Bombay, through the RIFC scheme, is gratefully acknowledged.

References

[1] S. Lal, S. Rajkumar, A. Tare, S. Reshmi, A. Chowdhury and I.N.N. Namboothiri, Chem. Asian. J., 9 (2014) 3533-3541.

[2] P. E. Eaton and T. W. Cole,J. Am. Chem. Soc., 1964,86, 3157-3158.

[3] A.P. Marchand, Z. Liu, D. Rajagopal, V.D. Sorokin, F. Zaragoza and A. Zope, Proceedings of the 7th ONR Propulsion Meeting, State University of New York, Buffalo, NY, 1994.

[4] S. Lal, L. Mallick, S.Rajkumar, O.P. Oommen, S. Reshmi, N. Kumbhakarna, A. Chowdhury, and I.N.N. Namboothiri, J. Mater. Chem. A, 3 (2015) 22118-22128.

[5] L. Mallick, S. Lal, S. Reshmi, I.N.N. Namboothiri, A. Chowdhury, and N. Kumbhakarna, New J. Chem., 41 (2016) 920-930.

[6] S.B. Haska, E. Bayramli, F. Pekel, and S. Ozkar, Mechanical properties of HTPB–IPDI-based elastomers, Inc. CCC 0021-8995/97/122347-08, John Wiley & Sons, 1997, 2347-2354.

[7] Ö. Hocaoğlu, Ö. Tülay, F. Pekel, S. Oskar, J. Appl. Poly. Sci., 79 (2001) 959-964.

[8] P.R. Patil, V.N. Krishnamurthy, S.S. Joshi, Propellants Explos. Pyrotech., 31 (2006) 442-446.

[9] S. Behera, Effect of RDX on elongation properties of AP/HTPB based case bonded composite propellants, DRDO Science Spectrum 2009, 31-36.

[10] K.R. Kreitz, *Catalytic nanoparticle additives in the combustion of AP/HTPB composite solid propellant*, Thesis, B.S., Texas A&M University, December 2010.

[11] R. Kurva, S.N. Jaywalker, S. Sac deva, Melilla, B. Bhattacharya, Cent. Eur. J.Energ. Mater., 9 (2012) 59-76.

[12] A. Ambekar, L. Mallick, R. Gandhi, A. Chowdhury, D. Radhakrishna, S. Challa, (AIAA 2013-3782), 49th AIAA/ASME/SAE/ASEE Joint Propulsion Conference, 2013, 10.2514/6.2013-3782.

[13] L. Mallick, S. Kumar and A. Chowdhury, Thermochim. Acta, 610 (2015) 57-68.

[14] L. Mallick, S. Kumar and A. Chowdhury, Thermochim. Acta, 653 (2017) 83-96.

[15] Y.L. Zhu, H. Huang, H. Ren, Q.J. Jiao, J. Energ. Mater., 32 (2014) 16-26.

[16] S. Vyazovkin, and C.A. Wight, Chem. Mater., 11 (1999) 3386-3393.

[17] M.A. Stephens, E.L. Petersen, D.L. Reid, R. Carro, and S. Seal, J. Propul. Power, 25 (2009) 1068-1078

[18] I. Kumar and P.A. Ramakrishna, J. Propul. Power, 30 (2014) 876-882.

[19] I. Kumar and P.A. Ramakrishna, J. Propul. Power, 30 (2014) 277-284.

[20] Vrunkumar, M. Zaved, H.S. Mukunda, Combust. Flame, 173 (2016) 411-424.

[21] S. Gordon and B. J. McBride, Computer Program for Calculation of Complex Chemical Equilibrium Compositions and Applications, in NASA Reference Publication 1311, 1996.

A Preliminary Study Modelling NO Emission by Subset Selection using a Genetic Algorithm and In-cylinder Parameters

T Bodisco[1,*], K Fang[2], A Zare[3], RJ Brown[3], Z Ristovski[3]

[1]School of Engineering, Deakin University

75 Pigdons Rd, Waurn Ponds, Victoria, 3216, Australia

[2]School of Engineering, University of Liverpool

The Quadrangle, Brownlow Hill, L69 3GH, United Kingdom

[3]School of Chemistry, Physics and Mechanical Engineering, Queensland University of Technology

2 George St, Brisbane, Queensland, 4000, Australia

Abstract

Introduced in this paper is the application of a genetic algorithm to perform subset selection to reduce the number of input parameters into a time history dependent model for the estimation of NO emission. For this work, a bespoke cycle, denoted as a sweep test, was utilised to provide the data for training the model. Input parameters into this model are in-cylinder parameters: indicated mean effective pressure, engine speed, peak pressure, peak pressure timing and the maximum rate of pressure rise, in addition to: intake air flowrate, instantaneous fuel consumption and boost pressure. Shown was that these input parameters allowed a high correlation between the estimated NO emission and the measured NO emission on the NRTC. A key advantage of subset selection is in being able to interpret the model itself to gain a physical understanding of what input parameters influence NO emission. A significant outcome from this work was in identifying that, for the engine under investigation, a time history of 8.5 s is needed to accurately estimate NO emission.

1 Introduction

With applications ranging from vehicle and emissions monitoring to engine, vehicle and after-treatment system calibration, there is an increasing need to be able to accurately estimate exhaust emissions from available data. To this end, there have been numerous empirical methods published in the literature exploring this, including: neural networks [1-3], linear regression [3] and empirical models [4]. This paper will explore using a genetic algorithm to perform subset selection in a basic linear model with time dependent data to estimate NO emission.

Exploring the literature indicates that the current method of choice among many researchers for the estimation of emissions from engine data is in the form of neural networks. Neural networks have the advantage of being fast to implement, can model non-linear processes and tend to produce models with excellent predictive power [5,6]. The existence of high quality libraries and toolboxes also make the use of this method attractive to many.

Roy, Banerjee and Bose [7] used neural networks to accurately predict the brake specific fuel consumption, brake thermal efficiency and carbon dioxide, oxides of nitrogen and particulate matter emissions of a common-rail diesel engine with exhaust gas recirculation (EGR). This study used input parameters of: load, fuel injection pressure, EGR (%) and fuel injected per cycle. The model was trained and validated on steady-state data. Roy, Banerjee and Bose also included a short discussion on the importance of careful selection of input parameters on subsequent model performance.

Ho et al. [5] used neural networks to predict the emissions from a hydrogen powered car. Input parameters for their implementation were: engine speed, throttle position, mass airflow, manifold air pressure, fuel pulse width, engine power, ignition advanced, engine temperature, air temperature, air fuel ratio and exhaust gas temperature. Ho et al. also trained and validated their model against steady-state data.

In an earlier study, Traver et al. [1] developed a neural network model to predict the emissions on a transient cycle (Federal Test Procedure or FTP) using in-cylinder parameters: peak pressure, combustion duration, indicated mean effective pressure, location of mass fraction burned (50%) and ignition delay. The training occurred using steady-state data. The final model was able to accurately predict the general trends of the emissions over the FTP cycle, but was unable to deliver a high degree of correlation. Despite this, validating against transient data is important in terms of establishing model utility.

Although neural networks are seen as having replaced conventional empirical modelling (polynomial and linear regression) [8], the utility of these conventional techniques should not be dismissed. An attraction to linear regression techniques is in the ability to interpret the model itself [9]. This is particularly true in implementations that involve some form of parameter reduction. Eliminating or reducing parameters that do not add value to the model improves the computation efficiency of using them, allows the analyst to better interpret the model itself for fundamental understanding and can improve their predictive accuracy [10].

Meta-heuristic algorithms, such as the genetic algorithm, are used extensively in optimisation problems [11], including feature selection [10]. The genetic algorithm provides an efficient means of exploring the parameter space for solutions which optimise a given cost function. In the example shown here, a genetic algorithm is employed to optimise basic linear

*Corresponding author.
E-mail address: t.bodisco@deakin.edu.au

regression models designed to estimate NO emission from in-cylinder pressure parameters.

This study uses transient data for both the training and the validation phases. For training, a bespoke engine cycle, denoted a sweep test, has been employed and the subsequent model validated against the nonroad transient cycle (NRTC). Another key feature in this implementation is the incorporation of time history. Most previous studies have trained and validated against steady-state data, time history in those samples is not as important owing to the engine being in thermal equilibrium. However, for a model to have true utility, it must be able to accurately predict transient emissions. Therefore, this paper will introduce the genetic algorithm as a method of subset selection, using a simple linear model to predict transient emissions with input parameters including those from a time history up to 20 s before the emission event.

2 Experimental Setup

The experimental work detailed in this paper was conducted on a EuroIII, common-rail, turbo-charged, six-cylinder Cummins diesel engine (ISBe220 31) coupled to an electronically controlled water brake dynamometer. The dynamometer has an advanced control system capable of being programmed to run a predetermined drive cycle (based on engine speed and engine torque).

In-cylinder pressure was sampled using a Kistler piezoelectric transducer (6053CC60) and the crank angle was determined by a Kistler crank angle encoder (type 2614) with a resolution of 0.5 degrees. The in-cylinder pressure and crank angle encoder signal were sampled at 200 kHz. Intake air flow rate and the instantaneous fuel consumption were sampled from the CANbus at 4 Hz. Nitrogen oxide (NO) emission was sampled at 10 Hz with a CAI 600 Series CLD analyser. For this work, all experimental data was time aligned and then resampled to 2 Hz. For more details on the experimental setup, please refers to Refs. [12,13]

3 Drive Cycles

Two drive cycles were utilised in this study. For training, a bespoke cycle, named a sweep test, was utilised, see Fig. 1. This cycle was designed to show a wide range of transient engine operation. The sweep test is 61 minutes in length.

For model confirmation, a standard drive cycle was selected. Used internationally for type approval, the nonroad transient cycle (NRTC) was deemed a suitable choice to measure how well the model would perform on unseen data. Figure 2 shows the NTRC drive cycle. For this work, both drive cycles were run from a hot engine. Figures 1 and 2 are shown as the engine ran them in this study, which has some natural deviation from the input to the dynamometer, this deviation can be considered minimal.

4 Modelling Approach

In matrix form, a basic linear model can be defined as:

$$y = \boldsymbol{X}\beta + \varepsilon \qquad (1)$$

where y is the scalar response (in this case the NO emission values), $\boldsymbol{X}$ is the design matrix (populated by the input paramters), β is a vector of the unknowns (coefficients of the model) and ε is a vector of the residuals. The vector parameters (y, β, ε), are all of length n, the number of data points, and the design matrix is $n \times p$, where p is the number of input parameters. In its basic form, this represents a very powerful and simple predictive model. Assuming that (1) is over-determined, solutions for β can be easily found using QR decomposition. As such, basic linear models are used in most fields of engineering and science extensively.

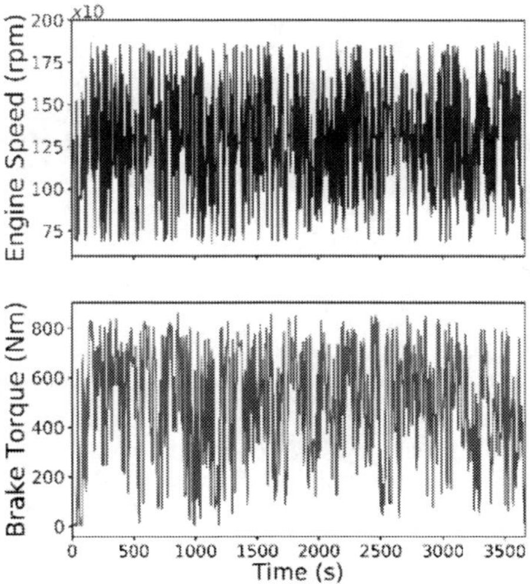

Figure 1: Engine speed and engine torque with respect to time for the bespoke sweep test as ran on the test engine

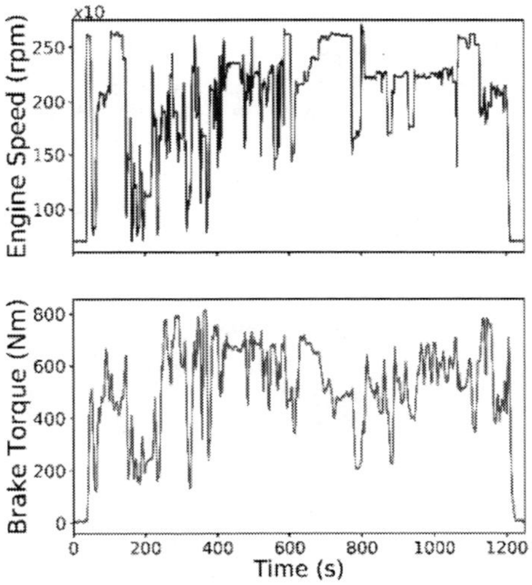

Figure 2: Engine speed and engine torque with respect to time for the NRTC engine cycle as ran on the test engine

A drawback of the basic linear model is that whilst they do tend to model with low bias, they can have large variance [9]. Another key issue is in the selection of parameters used to model the scalar response. Not all of the input parameters to the design matrix will necessarily be appropriate; the model itself may actually lose predictive power from the inclusion of parameters that are not linearly related to the scalar response. This is not a new problem and some sound solutions to it already exist in the literature: ridge regression, subset selection and the *lasso* (least absolute shrinkage and selection operator) [9]. The *lasso* method is effectively a combination of ridge regression and subset selection.

A key advantage of using a method which involves subset

selection is that it will reduce the number of parameters being used. In cases that may have many input parameters, this opens up the option to also interpret the ensuing model to gain additional physical understanding. This paper will present a simple subset selection method where the selection process is achieved with a genetic algorithm.

The input parameters used for this model are: intake air flowrate, instantaneous fuel consumption, boost pressure, indicated mean effective pressure, engine speed, peak pressure, peak pressure timing and the maximum rate of pressure rise. These parameters were chosen owing to their influence on NO emission. As NO is thermally generated, previous combustion events will have a direct influence on the emission formation. Therefore, it is important that any sound model for NO emission takes hysteresis into account. What is currently unknown is how much time history is required to properly explain instantaneous NO.

Based on the sweep test data, the design matrix contains the 8 parameters at the time-point the emission occurred (time aligned to match as best as possible with the combustion event) and the values of the parameters at 0.5 s intervals prior. For the purposes of this study, the model was resolved to include the previous history in 0.5 s steps up to 20 s prior to the emission event (i.e. 41 models were produced). Each model had 8 more input parameters than its predecessor, with the last model having 328 input parameters.

A genetic algorithm was employed to find the most appropriate subset of the input parameters to explain the NO emission. This was achieved by generating a Boolean array the same length as the number of input parameters. This Boolean array would then be used to produce the design matrix, the effectiveness of the design matrix would then be tested against a random 15% of the sweep test data (this 15% was kept fixed during the subset selection for each model, but was determined randomly for each model before the selection process). The fitness test was a simple square difference summation:

$$fitness = \sum_{i=1}^{n}\left(y_{i,model} - y_{i,data}\right)^2 \qquad (2)$$

Genetic algorithms are a form of numeric iteration. In this implementation, there were 50,000 iterations (each denoted as a generation) and within each generation there was a population of 20 (a population here is represented by the Boolean arrays, or strands in this context, where a strand is a single Boolean array). For the case with no time history, a strand would be formed by 8 genes. Each gene represents an input parameter, when a gene has a value of 1 it is included in the design matrix, when it has a value of 0 it is not. Being an iterative method, there is significant computation time required to optimise the subset selection. On a basic desktop computer, the most extreme case (Model 41, with 328 input parameters) takes approximately 4 days to converge on the optimal solution.

The initial population was randomly generated. At each generation, the 20 strands in the population were used to generate a design matrix for the linear model (1), each model was then evaluated using (2). At each generation, the 3 strands which yielded the smallest $fitness$ value were retained and the other 17 discarded. New strands were then *bred* from the remaining strands. This *breeding* was performed by sampling 2 random strands from the population (i.e. as the population regrows, the strands to potentially sample from increase). For each gene, the child gene is created from the corresponding parent genes. If the parent genes are different (i.e. one parent has 1 and the other a 0) the child is randomly selected with equal probability of being a 1 or a 0, if both parents have the same gene then the child has an 85% chance of inheriting that gene (when the child has a gene different to that of the parents it is known as a mutation, in this implementation the mutation rate was set as 15%). Having a reasonable population size ensures that some of the strands are somewhat random (this helps stop the process converging onto a local, rather than an absolute, solution), similarly the process also allows for a thorough exploration of the area local to the where the current *best* solution is. Given sufficient generations, the genetic algorithm should find the absolute solution (based off the given fitness function). The number of generations required to converge on an absolute solution is usually unknown, as a general rule-of-thumb the greater the number of input parameters and the more complex the fitness function, the greater the number of generations required. Also, the greater the mutation rate the greater the number of generations required to converge on a solution; however, a low mutation rate risks converging on a local rather than an absolute solution. Shown in Fig. 3 is the $fitness$ value against the number of generations, for the model with 20 s time history (328 input parameters), it can be seen that the process converged on the final value after 26,560 generations.

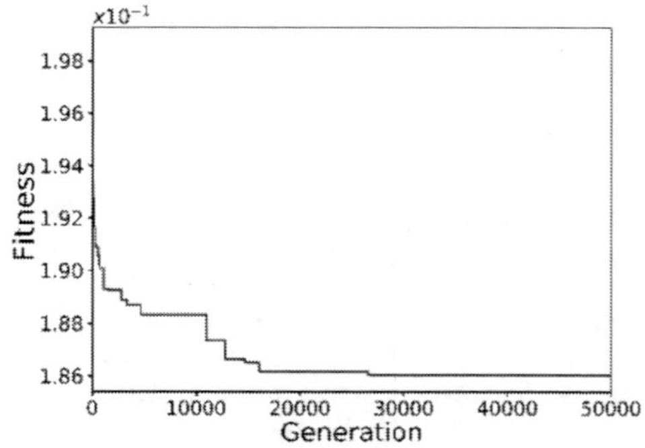

Figure 3: Progress of the genetic algorithm, the fitness value against the generation for the model with 20 s time history

4 Results and Discussion

After the training process, the 41 models were evaluated against the NRTC. An r^2, the coefficient of determination, value was determined between the experimentally measured NO against that predicted by each model by squaring the Pearson product-moment correlation. These results are shown in Fig. 4, there are a few interesting features in this figure. Clearly, increasing the time history does have a positive impact on the models ability to accurately predict the NO emission. However, what is also clear is that even without the time history, the indicated parameters in-conjunction with the air intake flow and instantaneous fuel consumption allowed for a high degree of predictive power. The genetic algorithm showed that the optimal fit occurred when the boost pressure was excluded.

Arguably the most interesting feature of Fig. 4 is the steep rise in predictive power finishing at 8.5 s time history. Increasing the time history from 8.5 s does continue to increase the predictive power, but the gain is minimal. Therefore, it can be concluded that most of the engine-out NO can be explained by the engine operation over the 8.5 s prior to the emission event. This time history is too long to be explained by phenomena such as turbocharger lag and is more likely to be explainable by thermal inertia of the combustion chamber, inlet manifold and after-cooler. In a study on a homogenous charge compression ignition (HCCI) engine, it was shown that the time for the wall temperature to become steady was in the order of 110 s [14]. This long settling time is indicative that an engine being run under normal transient conditions is never stable; therefore, steady-state tests are not indicative of emissions from a transiently run engine.

A further interesting feature of Fig. 4 is the decrease in predictive power seen by increasing the time history to 4.5 s, with the notable exception of the spikes at 1.5 s and 3.5 s time history. The spikes could be explained by the random selection of the 15% of the sweep test the genetic algorithm uses in the optimisation process. However, it is unlikely that the decreasing trend can be so easily dismissed. Intuitively, especially given that the models with greater time history contain all of the input data that their predecessors do, it would be expected that increasing the time history could only result in an increase in predictive power. A potential explanation could be that the sweep test contains smooth changes in engine speed and engine load, whereas the NRTC does not. The short time history of the sweep test, therefore, may be a mismatch with the transients seen in the NRTC. Suggestive that the training drive cycle may require a greater degree of erratic operation to be a fully effective cycle for training purposes. Additionally, the engine speed range of the NRTC is much greater. Therefore, the NO emission prediction in the engine speed ranges above that of the sweep test are extrapolations from the sweep test data used to develop the model. It would be expected that the overall predictiveness of this method would be substantially improved if the sweep test itself covered more of the operating area of the NRTC.

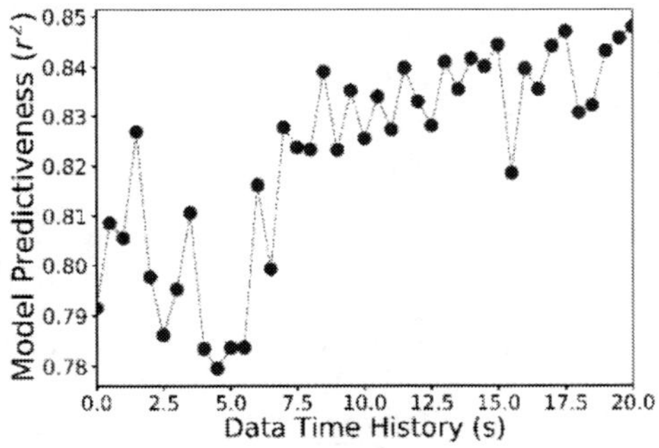

Figure 4: The coefficient of determination between the experimentally measured NO emission and that predicted by the linear model against the time history of the input data

4 Conclusions

This paper has introduced a method of subset selection using a genetic algorithm, with the application of modelling NO emission from a common-rail, turbo-charged, six-cylinder Cummins diesel engine. This method utilised a bespoke drive cycle, denoted as a sweep test, which was designed to show a wide range of engine operation for the purpose of training the model. The models developed using this process were subsequently employed on data collected by running the engine on the NRTC.

Using this method, 41 models were evaluated. The first model contained only 8 input parameters, each subsequent model contained not only all of the parameters of the previous models but also the parameters from a 0.5 s time step earlier. The 41st model had a time history of 20 s (328 input parameters).

Shown was that the method has merit and that indicated parameters yield a high degree of predictive power (with respect to NO emission). A further key outcome of this modelling exercise was in establishing that most of the NO emission can be explained by the data from a time history of 8.5 s from the emission event. Beyond the 8.5 s time history, the improvement in model predictiveness is not as pronounced.

5 Acknowledgment

The authors would like to acknowledge the work of Mr. Noel Hartnett in maintaining the Biofuel Engine Research Facility. Without his help and effort, the experiments for this work would not have been completed. Additionally, the authors would like to acknowledge the support of Mr. Mohammad Jafari for his support during the experiments.

References

[1] M. Traver, R. Atkinson, C. Atkinson, SAE Technical Paper 1999-01-1532 (1999).

[2] H. Taghavifar, H. Taghavifar, A. Mardani, A. Mohebbi, S. Khalilarya, S. Jafarmadar, J. Clean. Prod. 112 (2) (2016) 1729-1739.

[3] E. Tosun, K. Aydin, M. Bilgili, Alex. Eng. J. 55 (4) (2016) 3081-3089.

[4] S. d'Ambrosio, R. Finesso, L. Fu, A. Mittica, E. Spessa, Appl. Energ. 130 (2014) 265-279.

[5] T. Ho, V. Karri, D. Lim, D. Barret, Int. J. Hydrogen Energ. 33 (14) (2008) 3837-3846.

[6] M. Kiani Deh Kiani, B. Ghobadian, T. Tavakoli, A. Nikbakht, G. Najafi, Energ. 35 (1) (2010) 65-69.

[7] S. Roy, R. Banerjee, P. Bose, Appl. Energ. 119 (2014) 330-340.

[8] M. Canakci, A. Ozsezen, E. Arcaklioglu, A. Erdil, Expert Syst. Appl. 36 (5) (2009) 9268-9280.

[9] R. Tibshirani, J. R. Statist. Soc. B 58 (1) (1996) 267-288.

[10] F. Tan, X. Fu, Y. Zhang, A. Bourgeois, Soft Comput. 12 (2008) 111-120.

[11] M. Tayarani-N., X Yao, H Xu, IEEE T. Evolut. Comput. 19 (5) (2015) 609-629.

[12] T. Bodisco, R. Brown, Energ. 52 (2013) 55-65.

[13] M. Rahman, A. Pourkhesalian, M. Jahirul, S. Stevanovic, P. Pham, H. Wang, A. Masri, R. Brown, Z. Ristovski, Fuel 134 (2014) 201-208.

[14] C. Wilhelmsson, A. Vressner, P. Tunestål, B. Johansson, G. Särner, M. Aldén, SAE Technical Paper 2005-01-3731.

Experimental and CFD Investigation of Methane Partial Oxidation Above a Multiport Burner

Yael Bornstein, David F. Fletcher, Brian S. Haynes

School of Chemical and Biomolecular Engineering, The University of Sydney

Sydney, NSW 2006, Australia

Abstract

Partial oxidation of methane (POX) over a multiport burner is a potential route to syngas production. Experimental and Computational Fluid Dynamics (CFD) studies have been performed to investigate the post-combustion gas composition for a commercially-available burner. Reasonable agreement between the two approaches is obtained given the geometric simplification and uncertainty in thermal boundary conditions in the simulations.

1 Introduction

Hydrogen and syngas are important precursors to many industrial chemical processes, such as the Fischer-Tropsch process, production of ammonia and methanol amongst other organic compounds [1-3]. While there are a number of methods to produce hydrogen and syngas, at present 50% of the world's hydrogen is produced by steam methane reforming (SMR) which is energy intensive and hence expensive, bulky and requires long start up times [4,5]. There is an additional problem with using the syngas produced by this process for many of the downstream applications as these require a molar H_2:CO ratio of 2:1 not 3:1 (see Table 1).

Name	Reaction	ΔH_{298} (kJ/mol)
Steam methane reforming (SMR)	$CH_4+H_2O_{(g)} \rightleftharpoons CO+3H_2$	206
Partial oxidation (POX)	$CH_4+\frac{1}{2}O_2 \rightarrow CO+2H_2$	-36
Complete oxidation	$CH_4+2O_2 \rightarrow CO_2+2H_2O_{(g)}$	-802

Table 1: Key reactions in methane oxidation and syngas production.

Other methods of reforming have the potential to address these problems and can be used either in conjunction with SMR to tailor the production for specific applications, or to supplement supply. Specifically, a methane reformer using partial oxidation (POX) looks promising. Firstly, the process can produce a molar H_2:CO ratio of up to 2:1 which is suitable for downstream use. Secondly, due to the mildly exothermic nature of the partial oxidation reaction it can be brought about in a more compact reactor. This leads to the ability to build a reactor on site to produce hydrogen and syngas in situ, relying on the distribution network for natural gas which is well established [4]. Its high reaction rate also reduces the start-up time [5].

This method of production has its own challenges. To ensure that the partial oxidation reaction is the preferred pathway with significant yield of hydrogen and carbon monoxide, the combustion reaction needs to be run under rich conditions, ideally at equivalence ratios of 1.5 or higher. This leads to issues with stability as these conditions are at or outside the flammability limit of methane in air. As methane is in excess, and combustion produces a range of species, the final product needs to be purified, and the conversion can be low. These rich conditions have a propensity to soot which clogs the reactor, along with reducing yield. In addition, complete combustion (secondary oxidation of the product CO and H_2) is favoured, which also affects yield.

POX can be achieved either via catalytic combustion or via the thermal pathway, the latter of which is studied here. Some of the challenges associated with POX can be addressed through appropriate burner design. The burner used in this study has interacting laminar diffusion flames, which overcomes the stability issue. While the Hencken burner is one of several geometries which can achieve interacting diffusion flames, a configuration consisting of many coaxial ports spaced closely together is used in this study. This configuration will cause the port anchored flames to interact. These multiport burners have been studied extensively for heat transfer applications [6-8], but investigation of this burner concept for syngas production has been limited, with only one study found which used a Hencken burner operated at conditions far from those examined here [9].

In this paper we present experimental data obtained in the downstream region of a multiport burner and compare these with results from CFD simulations.

2 Experimental Study

Experimental investigations were performed using a commercially-available multiport burner and focused on obtaining temperature and major species compositions downstream of the combustion region. Figure 1 shows the burner in operation with a wire gauze present to stabilize the flame and the quartz probe used to take gas samples. The clear blue flames anchored to each of the ports are clearly visible.

Corresponding author. Fax: +61-2 9351 2854
E-mail address: david.fletcher@sydney.edu.au

Figure 1: Experimental setup showing the burner operation with the quartz sample probe present.

2.1 Experimental apparatus

A schematic diagram of the experimental setup is shown in Figure 2. The stainless steel burner (Carlisle PK-1885-D, catalogue number 13A105-0001) consists of 45 co-annular ports, arranged in a 3.8 mm (centre-to-centre) square grid, within an area of 1 in^2. This is located within a housing of 60 mm outer diameter. Cooling water is supplied and returned through stainless steel tubes ¼″ in diameter, mounted on opposite sides of the burner.

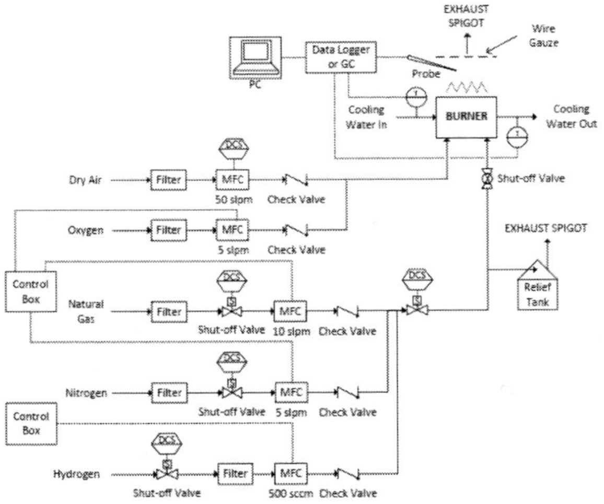

Figure 2: Schematic of the experimental apparatus.

Town gas was scrubbed of sulphur and used as the fuel. Air was dried and mixed with oxygen to provide the desired O_2:N_2 ratio as the oxidant. Details of the mass flow controllers and experimental operating procedure are given elsewhere [10].

2.2 Measurements

A gas sample was extracted from the flame at a point of interest using a 0.11 mm ID quartz probe. The sample was first passed through a cold finger trap to remove the water, followed by a particulate filter to remove soot and any other particulates. Suction was provided by a diaphragm vacuum pump which passed the sample through a dryer to a GC. The GC consisted of two columns. Channel 1 was a molecular sieve 5A (MS5A) column which detected H_2, O_2, N_2, CH_4 and CO. Channel 2 was a PoraPLOT U (PPU) column which detected CH_4, CO_2, C_2H_2, C_2H_4 and C_2H_6. The carrier gas was argon grade 5.0 which allows accurate detection of hydrogen.

The water vapour concentration was estimated by assuming that the water-gas shift reaction is in partial equilibrium, as indicated by the CFD results (see later).

A magnesia-coated 0.97 mm bead diameter S-Type thermocouple (Pt/Pt 90%-Rh 10%) was used to obtain temperature measurements in the post flame zone. Note the thermocouple was coated with magnesia to eliminate slow reversible changes to the thermocouple response that were observed to occur if the surface was untreated. A radiation correction was applied to the measured temperatures and the estimated temperature uncertainty is ~100 K.

2.3 Experimental Conditions

Here we focus on a base case with rich conditions in which the equivalence ratio ϕ = 3.0, the total volumetric gas flow rate $\dot{V}_{tot}$= 10 L/min and the mole fraction of oxygen in the oxidiser X_{O_2} = 0.4. Data were extracted at a number of positions above the burner at points located midway between two burner ports, as well as on horizontal scans across the burner.

The elemental ratio from the experimental measurements at the inlet and outlet are given in Table 2.

Ratio	Inlet	Outlet (incl. H_2O)
N:O	1.5	1.322
N:H	0.5	0.501
C:H	0.25	0.253
C:O	0.75	0.667
C:N	0.5	0.504

Table 2: Comparison of inlet and outlet elemental ratios.

There is excellent agreement between measured inlet and outlet data except for the ratio containing oxygen. It is believed these differences are due to estimation of the water mole fraction.

3 CFD Study

3.1 Model Setup

Simulations were performed to investigate the mixing and reaction behaviour for a single jet in the burner using a 2D axisymmetric geometry. Interaction between adjacent ports cannot be modelled directly in 2D but is accounted for here through the introduction of a symmetry plane at the outer edge of the computational domain. Details of the burner dimensions and inlets used are shown in Figure 3. The computational domain was extended 20 mm downstream of the burner face, the fuel inlet tube was given a length of 30 mm and that of the oxidizer inlet was 2 mm, the estimated thickness of the burner plate. The burner was made of stainless steel and the tube walls and burner face were included in the simulation by solving a conjugate heat transfer problem. In the experiments the burner is water cooled and a heat balance on the cooling water showed that 3.9 W per fuel channel was removed. In the CFD simulations the gas inlet temperatures and walls were set to 294 K and this resulted in a heat loss of 2.5 W. We attribute the difference to use of a simplified 2D axisymmetric geometry.

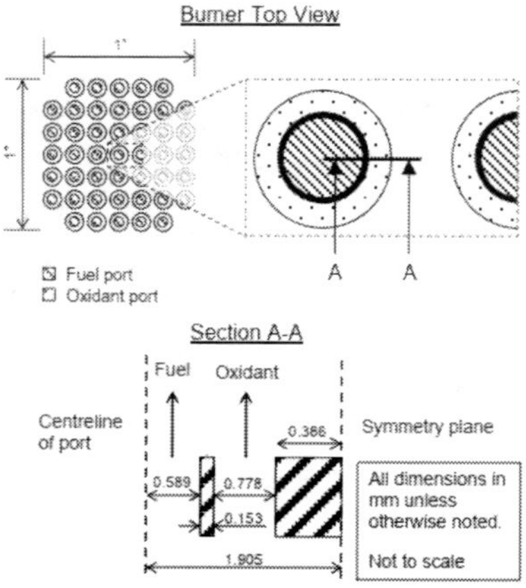

Figure 3: Details of the burner hole layout and dimensions.

Based on the jet Reynolds numbers (~100) the flow was determined to be laminar. Multi-component diffusion was modelled. Radiation heat transfer was ignored, as it was demonstrated to have a negligible effect on the results in the chosen geometry, as were buoyancy effects.

The reaction mechanism needs to be appropriate for very fuel rich conditions and after an extensive literature review we decided on the mechanism developed by Cuoci et al. [11], which has been used extensively for fuel rich conditions. The mechanism comprises 84 species and 1698 reactions.

3.2 Solution Method

The resulting model equations were solved using the commercially-available software package ANSYS Fluent (V 18.1). A multistep process similar to that applied in our earlier validation study of the stiff chemistry solver applied to a CH_4 flame [12] was applied here. First an isothermal flow was simulated, then the option to relax the chemistry to equilibrium was applied and then finally the finite rate chemistry solution was calculated using the CHEMKIN stiff chemistry solver, with ISAT being used to speed up the simulations. It proved necessary to suppress chemical reactions in the entry section of the burner as the very low reaction rates caused issues with the stiff chemistry solver in ANSYS Fluent.

A structured quad mesh was generated with mesh refinement in the region of the burner. The final mesh used had 29,000 cells and its distribution was a result of extensive mesh refinement studies.

The QUICK scheme was used for discretization of the convective derivatives and the pressure-based coupled solver with a cell-based Courant number of 200 for pressure-velocity coupling. Simulation took a very large number of iterations to converge, typically in the region of 20,000. Residual convergence was never achieved, so convergence was judged based constancy of exit-averaged variables, such as temperature and primary species mole fraction being constant.

4 Results

Results are presented here for the case described in Section 2.3. The experimental data showed that the chemical transformations were essentially complete at a distance of ~10 mm downstream of the burner so we confine our interest to these post combustion zone conditions that are of interest in the POX application.

Figure 4 shows results from the CFD simulation. The flame is seen to sit above the oxidant inlet. The incomplete combustion of the methane and the rapid development of uniform conditions across the burner are evident.

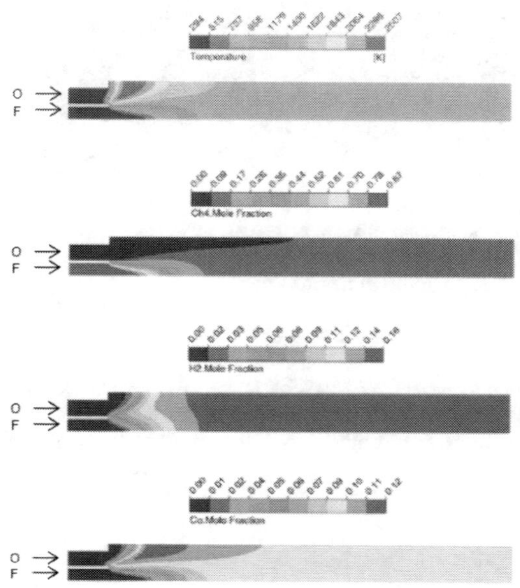

Figure 4: CFD results for the temperature and some major species.

The simulated temperature on an axial line situated midway between two adjacent burner ports is compared with the experimental data in Figure 5. It is clear that after a distance of ~3 mm in the experiments and ~7 mm in the CFD the temperature become essentially flat. The computed temperatures are approximately 120 K cooler than measured – this difference exceeds the estimated experimental uncertainties and must be related in part to the extent and product speciation of methane combustion.

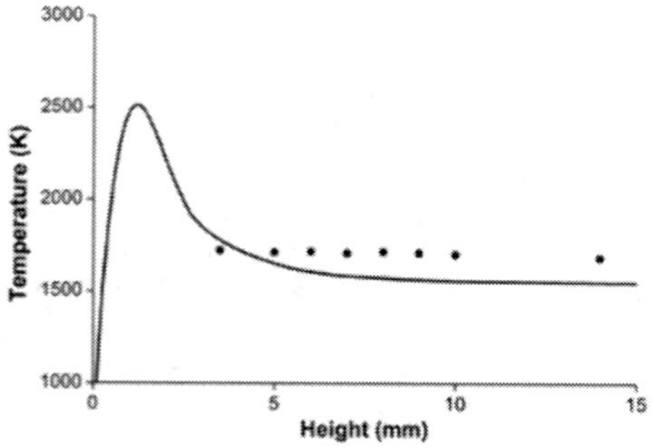

Figure 5: Experimental and CFD temperature on a line midway between the burner ports. Experimental data are shown by symbols and the CFD results by a line.

Figure 6 shows the equilibrium constant for the water gas shift reaction on the same line as used in Figure 5. It is evident that both the CFD and experimental data plateau around ~7 mm above the burner and that the experimental data show the reaction is close to equilibrium. The higher plateau value for the CFD results reflects the fact that the calculated temperatures are lower than those measured.

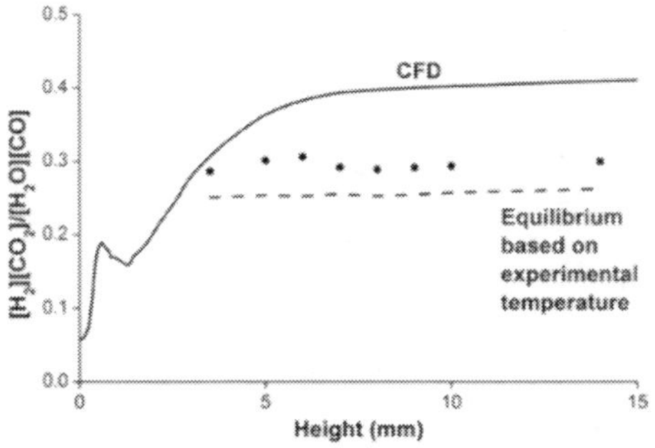

Figure 6: Reaction quotients and equilibrium constant for the water-gas shift reaction on an axial line midway between the burner ports. Experimental data are shown by symbols and the CFD results by a continuous line.

The experimental data for the major carbon species together with the CFD results are shown for the same location in Figure 7. It shows that the CFD predicts slightly more methane consumption and almost complete reaction of ethane. However there is generally good agreement with all measured species.

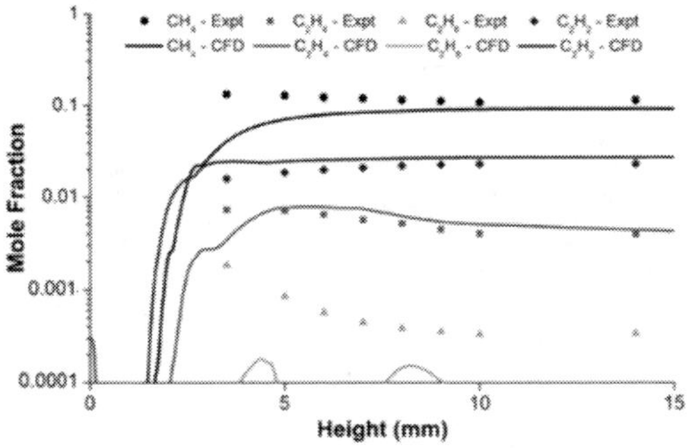

Figure 7: Experimental and CFD mole fractions of CH_4, C_2H_2, C_2H_4 and C_2H_6 along a line midway between the burner ports.

Finally, the species involved in the water gas shift reaction are shown in Figure 8. Again agreement is good, with the main difference being the hydrogen production is too high in the CFD simulations due to the greater conversion of methane.

5 Conclusions

Experimental studies have shown that combustion is readily stabilised at an equivalence ratio of 3.0 using a multiport burner. Reaction occurs very close to the burner, within typically ~10 mm of the burner and then more or less constant temperatures and species composition are reached. The water-gas shift reaction is found to be near to equilibrium downstream of the flame zone.

2D axisymmetric CFD simulations using a detailed chemistry mechanism appropriate for rich flames captures all the trends and gives acceptable agreement with the experimental data. Uncertainties in the heat loss and possible 3D effects could potentially explain the differences.

The coaxial burner studied here has potential for application to POX of methane for the production of syngas.

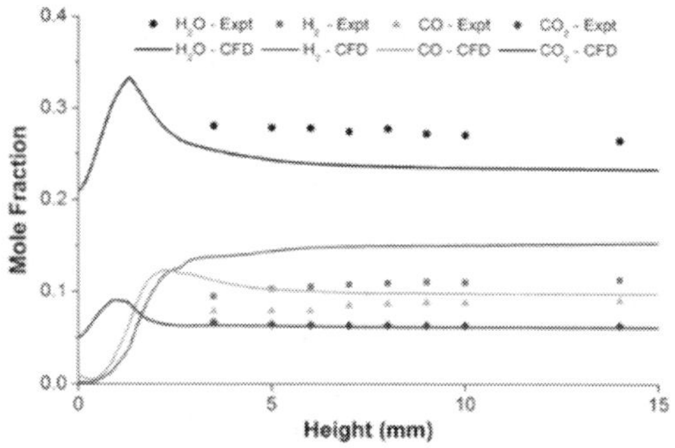

Figure 8: Experimental and CFD mole fractions of CO, CO_2, H_2 and H_2O along a line midway between the burner ports.

6 Acknowledgment

The authors acknowledge the University of Sydney HPC Service at The University of Sydney for providing High Performance Computing resources. DFF thanks Dr. Pravin Nakod of ANSYS India for considerable assistance in resolving issues with the stiff chemistry solver.

References

[1] M.A. Peña, J.P. Gómez and J.L.G. Fierro, Renewable and Sustainable Energy Reviews, 67 (1996) 597–611.

[2] Y. Xu, Z. Dai, C. Li, X. Li,. Z. Zhou, G. Yu. and F. Wang, J. Hydrogen Energy, 39 (17) (2014) 9149–9157.

[3] P.D.F. Vernon, M.H.L. Green, A.K. Cheetham and A.T. Ashcroft, Catal. Lett., 6 (2) (1990) 181–186.

[4] L. De Rogatis, T. Montini, A. Cognigni, L. Olivi and P. Fornasiero, Catal. Today, 145 (1–2) (2009) 176–185.

[5] R. Chaubey, S. Sahu, O.O. James and S. Maity, Renewable and Sustainable Energy Reviews, 23 (2013) 443–462.

[6] S. Chander and A. Ray, Energy Conversion and Management, 46 (18–19) (2005) 2803–2837

[7] V. Hindasageri, R.P. Vedula and S.V. Prahbu, Int. J. Heat Mass Tranf., 79 (2014) 914–925.

[8] S.T. Tillman, S. Kuchibhatla, K. Annamalai, J.A. Caton and D. Ranjan, J. Combust., (2012) Article ID 716050.

[9] Y.Q. Wang, Y.X. Shi, X.K. Yu, N.S. Cai, S.Q. Li, J. Electrochem. Soc., 160 (11) (2013) F1241–F1244.

[10] Y. Bornstein, *Numerical and Experimental Investigation of Partial Oxidation of Methane over a Multiport Burner*, MER Thesis, University of Sydney, Australia, 2017.

[11] A. Cuoci, A. Frassoldati, T. Faravelli and E. Ranzi, Combust. Flame, 45 (5) (2003) 870–886.

[12] D.F. Fletcher, C.T. Bowman, B.S. Haynes, CFD Simulation of a Confined Axisymmetric Laminar Methane-Air Diffusion Flame, Proc. Mediterranean Combustion Meeting, Çeşme, Izmir, Turkey, September 8-13, 2013.

11th Asia-Pacific Conference on Combustion,
The University of Sydney, NSW Australia
10th -14th December 2017.

Burn Rate Characterization of an Alternative Monopropellant –Hydroxyl Ethyl Hydrazinium Nitrate

Umakant Swami[1], K. Jayaraman[2], Arindrajit Chowdhury[1],

[1]Department of Mechanical Engineering,
Indian Institute of Technology Bombay, Mumbai, 400076, India
[2]DOLP, Defence Research Development Laboratory,
Hyderabad, Telangana, 500058, India

Abstract

The combustion characteristics of an energetic ionic liquid hydroxyethyl hydrazinium nitrate (HEHN) as liquid monopropellant is investigated. Strand combustion experiments were performed in an optically accessible combustion chamber with HEHN and Otto fuel-II (OF-II). The linear burning rates of liquid strands of HEHN and OF-II were extracted in a quiescent atmosphere of air with ambient pressures in the range of 10 to 50 bar. The regression surface profiles and burn rates were studied for HEHN and OF-II. The burn rates of HEHN strand increased with a pressure, from approximately 0.47 mm/s at 10 bar to 3.28 mm/s at 50 bar, and were found to be significantly higher than those of OF-II. The theoretical performance of HEHN as a monopropellant was found to be comparable to that of OF-II when evaluated using the NASA CEA code. The specific impulse of HEHN was evaluated to be 1.7% less than OF-II, while density specific impulse was approximately 13% higher than OF-II, rendering the fuel suitable for volume limited applications.

1 Introduction

Monopropellants are compounds or mixtures of compounds that decompose thermally or catalytically generating energy in the process [1]. The exothermic decomposition produces hot gases that are typically utilized for propulsion as well as power-generation in oxygen-deprived scenarios. Multi-component monopropellant formulations are amongst the prime candidates for replacing hydrazine as the mainstream propellant. These next generations of green monopropellants are intended to mitigate the various hazards posed by hydrazine and its derivatives. Various ionic liquid propellant formulations based on hydroxylammonium nitrate (HAN) as well as ammonium dinitramide (ADN) are at the forefront of this effort. Heterogeneous monopropellant formulations containing nanoparticles are also being explored. Historically, Otto fuel II (OF-II) has been a well-known liquid monopropellant formulation which was successfully utilized for several applications [2], consisting of 76% propylene glycol dinitrate (PGDN) as the monopropellant, 22.5% DBS as the desensitizer, and 1.5% 2-nitrodiphenylamine as the stabilizer (by weight). Although monopropellants were originally used for space-based applications, they have also been envisioned to be useful in various other applications such as underwater power sources, high-altitude unmanned aerial vehicles, robotic actuators, liquid-fueled guns, and as fuels in conventional reciprocating engines.

The combustion characteristics of these liquid monopropellants such as burn rates, flame temperature and structure, and flame standoff distances etc. are routinely extracted by conducting strand combustion studies in order to elucidate the variation of linear burning rates with pressure, temperature, and composition of surrounding gases. Strand combustion studies of various multi-component liquid monopropellants such as OF-II [3], HAN-based propellants [4], and pure liquid monopropellants such as hydrazine [5], isopropyl nitrate [6] and nitromethane [7] have been reported in the literature.

In the past few decades, ionic liquids (ILs) have been one of the most scintillating scientific discoveries in synthetic chemistry. Due to some unique properties such as low vapor pressure, low toxicity, high bulk density, high thermal stability, wide liquidus range, ionic conductivity and the ability to dissolve a wide range of chemical species, ILs have brought a green revolution to chemistry and chemical engineering [8]. Energetic ionic liquids (EILs) essentially are a new class of ILs, which have thermodynamically high energy yields on decomposition or combustion. In the past decade, significant efforts have been devoted to the research and development of EILs. A review by Zhang and Shreeve [8] details the synthesis of a large number of EILs as fuels and explosives. A review was recently published by Chowdhury and Kumbhakarna [9] on the mechanisms of pyrolysis of various EILs.

In the present study, a green EIL hydroxyethylhydrazinium nitrate (HEHN) was chosen as a monopropellant due to its low vapor pressure, reduced toxicity, high density, and enhanced stability. HEHN $[HO{-}CH_2{-}CH_2{-}NH_2{-}NH_2]^+[NO_3]^-$ has been synthesized at Defense Research and Development Laboratory, Hyderabad, by dropwise addition of nitric acid to 2-hydroxyethylhydrazine [10]. The salt was envisioned as a possible fuel [11,12] in bipropellant formulations, as well as a monopropellant for gas generators using iridium as the catalyst [13]. The combustion of HEHN was studied by the classical strand burner method, and the burning rates were extracted from 10 to 50 bar. The thermal decomposition characteristics were also analyzed using a thermogravimetric analyzer (TGA).

2 Experimental approach

The experimental setup utilized for analyzing the strand combustion of liquid monopropellants has been described in detail previously [6]. The current study utilized transparent quartz tubes of 5 mm internal diameter, thickness of 1 mm, and height of 40 mm to create the liquid strands. The combustion process was recorded by a Canon 550D DSLR camera at 60 frames per second, and the videos were converted to a sequence of frames for further processing. The position of the liquid meniscus was tracked through each image using the image processing software Spotlight 16 [14] yielding the rate of consumption of the liquid propellant in the tube.

Corresponding author. Fax: +91-22-25726875
E-mail address: arindra@iitb.ac.in

Ignition was achieved by electrically heating a nichrome wire dipped in the liquid strand. Immediately after ignition, the power supply to nichrome wire was cut off. The energy deposited by the heated nichrome wire was found to be insufficient for the experiments with OF-II strands. In these experiments, the nichrome wire was threaded through a small piece of composite solid propellant, measuring 2×2×1 mm^3, and comprised of an approximately stoichiometric mixture of ammonium perchlorate (AP) and hydroxyl terminated polybutadiene (HTPB). The solid propellant block was positioned approximately 1 to 3 mm above the tip of the tube and slightly offset from the tube axis in order to avoid contamination of the liquid by the combustion products of the solid propellant. It is notable that OF-II strands could not be ignited at the pressure of 10 bar despite repeated trials.

3 Results and Discussion

3.1 Combustion study of monopropellants

Various phenomena observed during the combustion of strands of HEHN and OF-II strands are discussed first. The combustion of ionic liquid HEHN strand at 20 bar chamber pressure was observed to possess a narrow, luminous, and yellow bipropellant flame at the tip of the tube, as shown in Fig. 1. A flickering bipropellant flame was observed at the tip of the tube and a reddish yellow monopropellant flame was observed above the liquid surface, as shown in Fig. 1 (c). The monopropellant flame was not visible clearly in Fig. 1 (d) due to soot deposited on the wall of the quartz tube. The combustion at 10 bar also exhibited a similar behavior. The combustion of HEHN was observed to be unsustainable after ignition at atmospheric pressure.

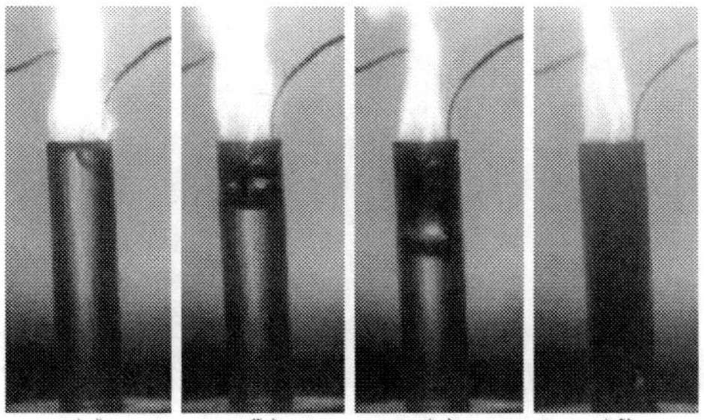

(a) (b) (c) (d)

Figure 1: Combustion of HEHN at 20 bar at (a) t = 0 s, (b) t = 8 s, (c) t = 13 s, and (d) t = 29 s

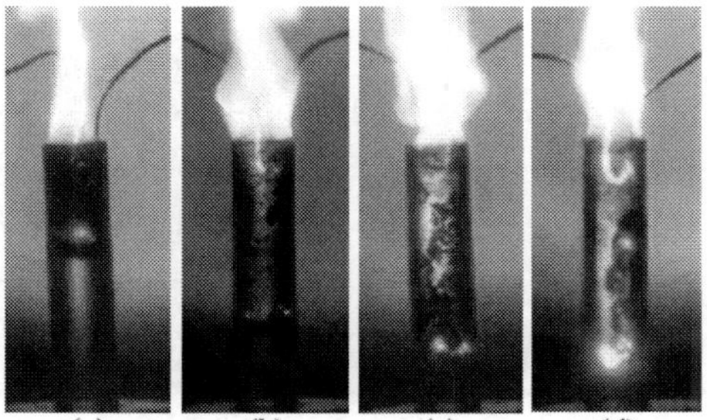

(a) (b) (c) (d)

Figure 2: Combustion of HEHN at (a) P = 20 bar, (b) P = 30 bar, (c) P = 40 bar, and (d) P = 50 bar

Figure 2 shows the selected frames of HEHN strand combustion from 20 to 50 bar pressure. The luminosity of monopropellant flame increases as the chamber pressure increases, as does the flickering nature of bipropellant flame.

The liquid monopropellant Otto fuel-II at 10 bar chamber pressure was not ignited after repeated tests. The combustion of OF-II strand at 20 bar chamber pressure was observed to demonstrate a narrow, luminous, and yellow bipropellant flame at the tip of the tube, as shown in Fig. 3. However, as opposed to HEHN, the monopropellant flame was not observed for 20 and 30 bar. OF-II displayed an initial flame at the tip of the tube which was more luminous and persistent than HEHN. After ignition of OF-II, a large number of bubbles were formed and continued to form at the liquid surface, during the entire combustion event. After combustion of the top layer of liquid OF-II, the flame traveled inside the tube assuming crescent-shaped flame as shown in Fig. 3 (c). The flame contained within the tube at 20 bar was observed to exhibit a pulsating nature.

(a) (b) (c) (d)

Figure 3: Combustion of OF-II at 20 bar at (a) t = 0 s, (b) t = 10 s, (c) t = 19 s, and (d) t = 96 s

Figure 4 shows selected frames of OF-II strand combustion at 20 to 50 bar in the presence of air as surrounding medium. The combustion of OF-II at 40 bar and 50 bar was accompanied by two distinct luminous flames. The monopropellant flame was observed to be bright yellow in color and more steady at higher pressures. The bipropellant flame at 40 and 50 bar remained at the tip of the tube and complete combustion of the strand was observed. The relative weakness of the monopropellant flame was apparent from a layer of dark decomposing liquid layer indicative of slower reactions and longer liquid residence times.

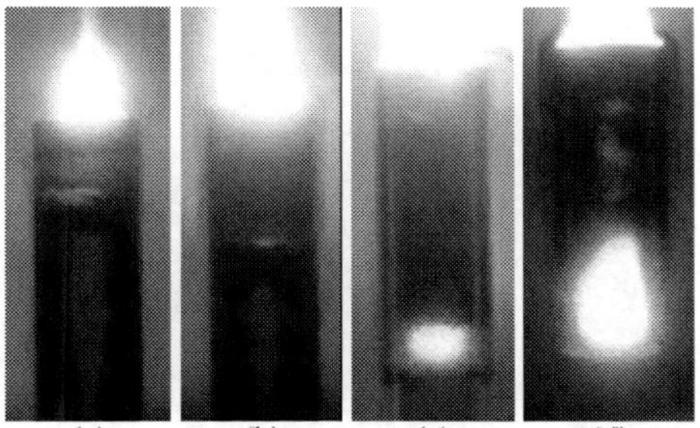

(a) (b) (c) (d)

Figure 4: Selected frames of OF-II strand combustion at (a) P = 20 bar, (b) P = 30 bar, (c) P = 40 bar, and (d) P = 50 bar

Figure 5 compares the regression profiles for the ionic liquid HEHN strand at various pressures. The monopropellant strand was not fully consumed at 10 bar. At chamber pressures of 20 bar and above complete combustion of all monopropellants was observed. The slope of the regression profile was found to increase with increase in chamber pressure. The data reported have been averaged over three tests to ascertain the repeatability of the event.

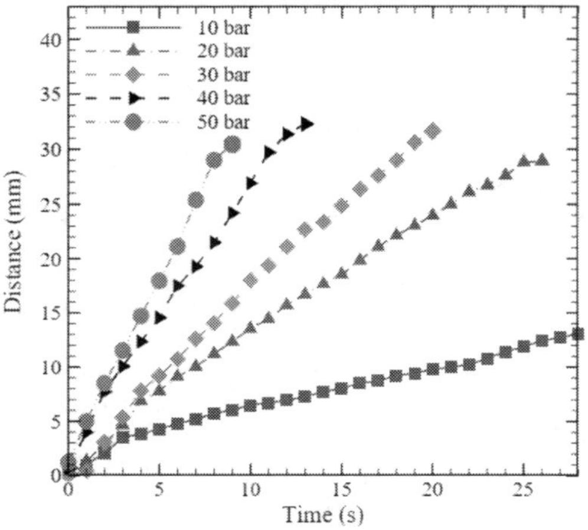

Figure 5: Regression profiles of HEHN at 10 to 50 bar pressure

Figure 6 shows the regression profiles of Otto fuel-II from 20 to 50 bar chamber pressure. At chamber pressures of 20 and 30 bar, Otto fuel-II initially produced linearly regressing surfaces as the bipropellant flame situated at the tip of the tube was close to the liquid surface and resulted in a larger heat feedback to the liquid. However, after a particular duration, the slope of the regression profile decreases because of the distance between the bipropellant flame and the liquid surface increases, and after some distance the flame collapses within the tube. At chamber pressures of 40 and 50 bar, linear regression profiles persisted throughout the entire combustion event owing to the dominant monopropellant flame situated at the liquid surface. The slope of the linear regression profiles increased with pressure.

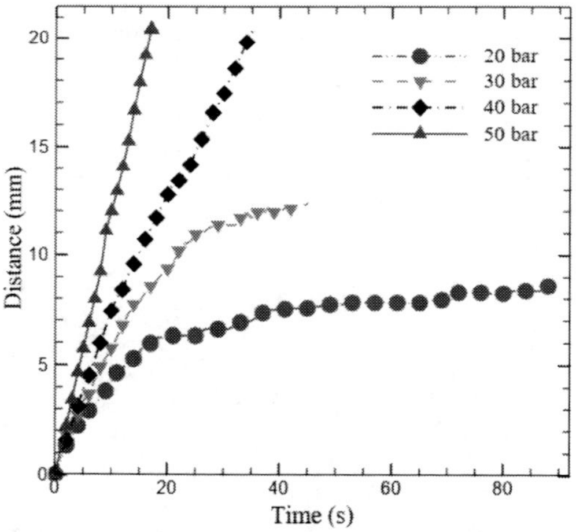

Figure 6: Regression profiles of OF-II at 20 to 50 bar

Figure 7 shows the comparison of burning rates at various pressures for HEHN and OF-II. The burn rates of HEHN and OF-II were calculated using the slope of regression profile, considering the entire combustion event at each pressure. However, the burn rates at 20 and 30 bar pressures for OF-II were evaluated by accounting for the initial slope of regression profile. The burn rate of isopropyl nitrate, another nitrate ester, previously measured by Ambekar et al. [6] was plotted on the same graph for comparison. The vertical error bars denote the repeatability of the measurements at each chamber pressure. The linear burning rate of HEHN was found to increase from 0.47 ± 0.06 mm/s at 10 bar to 3.28 ± 0.05 mm/s at 50 bar chamber pressures. A sudden increase in the burning rate of HEHN and OF-II were observed at 50 bar due to the possible presence of a steady monopropellant flame. The burning rate of OF-II was found to be less compared to HEHN in the pressure range of 20 to 50 bar..

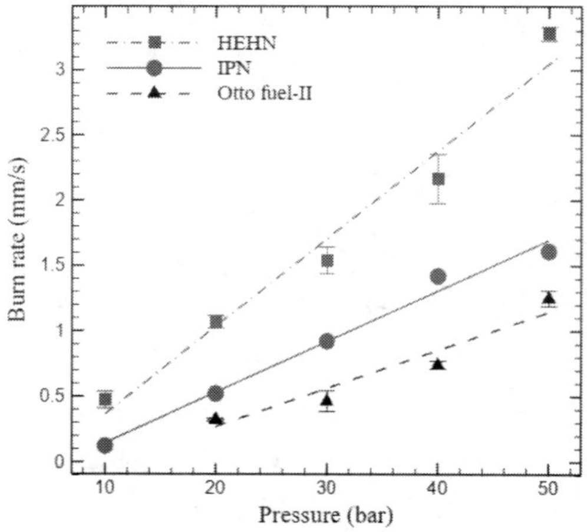

Figure 7: Burn rate compression of HEHN, IPN, and OF-II

The variation of the linear burning rate of a propellant with pressure may be defined using the mathematical form as $\dot{r} = aP^n$, where P is chamber pressure in bar, a is an empirical coefficient and n is the pressure exponent reported in [1]. Equation (1) and (2) show the linear burning rate for HEHN from 10 to 50 bar and OF-II from 20 to 50 bar chamber pressure respectively. The pressure exponent for pure IPN in the pressure range of 10 to 50 bar, shown in (3), has been reported to be 1.54 [6]. HEHN was found to have a lower value of pressure exponent compared to the other monopropellants, providing more stability to the combustion.

$$\dot{r}(mm/s) = 0.0328 \times [P(bar)]^{1.15} \quad (1)$$

$$\dot{r}(mm/s) = 0.0047 \times [P(bar)]^{1.40} \quad (2)$$

$$\dot{r}(mm/s) = 0.0046 \times [P(bar)]^{1.54} \quad (3)$$

In order to comprehend the reasons behind the higher burning rates of HEHN over OF-II, the adiabatic flame temperatures were calculated for both monopropellants shown in Table 1 at the pertinent pressures using the NASA CEA routines [15].

Pressure (bar)	HEHN (K)	OF-II (K)
10	1514.7	1462.5
20	1515.1	1472.4
30	1515.7	1484.0
40	1516.4	1495.4
50	1517.4	1506.1

Table 1: Monopropellant flame temperatures of HEHN and OF-II

As an input to the code, the heat of formation of HEHN was estimated to be -92.9 kcal/mol with the help of Gaussian 09 [16]. Although the monopropellant flame temperatures of HEHN were found to be marginally higher compared to OF-II at all pressures, the increment was not large enough to justify the rapid rise in burning rates over OF-II. Thus, it may be conjectured that the physical properties as well as chemistry of decomposition of HEHN play a significant role in enhancing the burning rates over OF-II. Efforts are currently devoted to better understand the combustion phenomenon of HEHN.

The performance of HEHN as monopropellant was also compared with that of OF-II based on specific impulses calculated using the NASA CEA code [15]. The chamber pressure and the expansion ratio for this hypothetical rocket problem were assumed to be 20 bar and 50 respectively. The vacuum specific impulse of OF-II and HEHN were found to be 256.6 s and 252.2 s respectively. However, the density specific impulse of HEHN was calculated to be 358.1 s as against a value of 298.2 s for OF-II, because of the higher density of HEHN (1.42 g/cm^3) in comparison with OF-II (1.23 g/cm^3).

The thermal decomposition of HEHN has been studied by thermogravimetric analysis (TGA). The TGA plots of HEHN subjected to a heating rate of 10 Kmin^{-1} is shown in Fig. 8. The curves are shown for three experimental runs with similar masses. The mass loss curve demonstrated that the decomposition process of HEHN may be divided into two regimes – the first decomposition regime, occurring from 40 °C to approximately 80 °C, accompanied by approximately 10% mass loss, and the second decomposition regime, occurring from 170 °C to 240 °C, accompanied by the remaining 80% mass loss. The first stage is associated with loss of water, while the second stage is expected to be an exothermic mass loss due to reactions in the condensed phase. The enhanced burn rates and reasonable thermal stability of HEHN may be enough to establish it as a frontrunner in power generation in oxygen-deficient conditions.

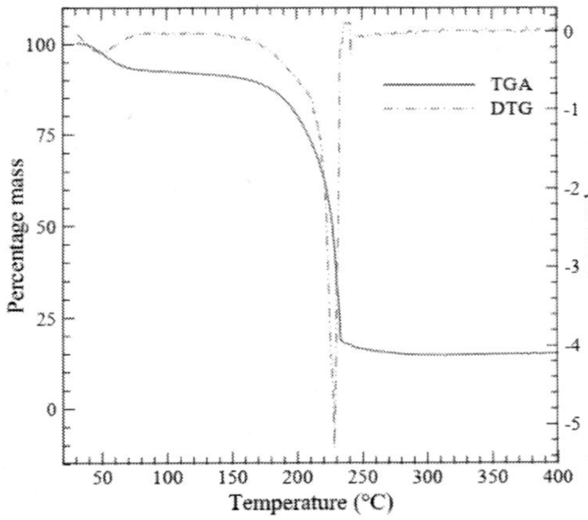

Figure 8: Thermal decomposition of HEHN at 10 Kmin-1

4 Conclusion

Combustion characteristics of HEHN and OF-II were investigated through strand combustion experiments conducted in the atmosphere of air with varying ambient pressure. The performance of HEHN was compared against the well-known monopropellant formulation OF-II. The combustion process of the propellant was affected by the ambient pressure as indicated by variations in burn rate, flame intensity, and the bubbly nature of regressing liquid interface. HEHN exhibited a distinct monopropellant flame visible at and above 10 bar chamber pressure. OF-II exhibited a hybrid combustion process at 20 and 30 bar. However, at 40 bar and 50 bar a distinct monopropellant flame was observed. The burn rate of HEHN and OF-II was observed to increase with pressure. The burn rates of HEHN at 10 bar and 50 bar chamber pressure were 32% and 62% higher than those of OF-II. The theoretical performance of HEHN was found to be comparable to that of OF-II when evaluated using the NASA CEA code. The specific impulse of HEHN was 252.2 s, while density specific impulse was approximately 358.1 s which is higher against OF-II. The gains in density specific impulse coupled with enhanced burn rates, and reasonable thermal stability is enough to establish HEHN as a frontrunner in power generation in oxygen-deficient conditions.

5 Acknowledgment

The author would like to acknowledge Defence Research and Development Laboratory (DRDL), Hyderabad for supporting this project.

References

[1] G.P. Sutton, "Rocket Propulsion Elements," Seventh Ed, John Wiley & Sons, 2001, p. 428.

[2] S.K. Schultheis, No. NUC-TP-276., Nav. Undersea Res. Dev. Center, San Diego, CA (1972).

[3] W.F. McBratney, Report No. AD-A086106, US Army Ballistic Research Laboratory, Aberdeen Proving Ground, Maryland, 1980.

[4] Y.P. Chang, K.K. Kuo, J. Propuls. Power, 18 (5) (2002) 1076-1085.

[5] C.B. Allison, Report No. NASA CR-13454, NASA Lewis Research Center, 1974.

[6] A. Ambekar, S. Sreedhara, A. Chowdhury, Combustion and Flame, 162(3), (2015) 836-845.

[7] E. Boyer, K.K. Kuo, Proc. Combust. Inst. 31 (2) (2007) 2045-2053.

[8] Q. Zhang, J.M. Shreeve, Chemical Reviews, 114 (20), (2014) 10527-10574.

[9] N. Kumbhakarna, A. Chowdhury, Curr. Org. Chem. 20 (2016) 2514.

[10] J.L. Shamshina, M. Smiglak, D.M. Drab, T.G. Parker, H.W.H. Dykes, R. Di Salvo, A.J. Reich, R.D. Rogers, Chem. Commun. (Camb). 46 (47) (2010) 8965-8967.

[11] A.J. Brand, & G.W. Drake, U.S. Patent No. 6,218,577, (2001).

[12] J. Smith, Brown, M. Frisby, S. Torabzadeh, US patent No. 0287824, 2010.

[13] R. DiSalvo, H.W. Dykes, R. D. Rogers, J. Shamshina, & M. Smiglak, U.S. Patent No. 8,636,860, (2014).

[14] R. Klimek, & T. Wright, Spotlight-8 image analysis software, (2006).

[15] B.J. McBride, S. Gordon, NASA Reference Publication 1311 (1996).

[16] M.J. Frisch, G.W. Trucks, H.B. Schlegel, G.E. Scuseria, M.A. Robb, J.R. Cheeseman, G. Scalmani, V. Barone, B. Mennucci, G.A. Petersson, Gaussian 09, 2009.

11th Asia-Pacific Conference on Combustion,
The University of Sydney, NSW Australia
10th -14th December 2017.

Combustion characteristics of demineralized coal pyrolysis char

Shaozeng Sun*, Wenda Zhang, Yijun Zhao, Pengxiang Wang

School of Energy Science and Engineering, Harbin Institute of Technology, 92 West Dazhi Street, Harbin, Heilongjiang Province, 150001, China

Abstract

In the oxy-coal combustion steam system (OCCSS), the demineralized coal combusts with gaseous O_2 and steam being recycled from condenser and injected stage by stage in order to produce a flue gas consisting predominantly of steam and CO_2. The flue gas is then expanded through an advanced turbine, and then separated in a condenser to produce high-purity CO_2. Stable demineralized coal combustion at high steam concentrations is crucial to the development of the OCCSS technique. In this study, Zhundong coal was used to prepare demineralized coal. Pyrolysis chars were prepared from the raw coal and the demineralized coal in a high-temperature drop tube furnace at 1200 °C. The effects of the reaction atmosphere (O_2/H_2O and O_2/N_2) and temperature (1100–1400 °C) on the combustion characteristics of the two chars were investigated. The results show that at 1200 °C, steam gasification of the demineralized coal pyrolysis char was slow because of alkali metal removal, which cannot compete with char O_2 combustion. The effects of steam on char combustion are mainly physical. Demineralized coal pyrolysis char combustion in an O_2/H_2O atmosphere accelerated significantly with increasing temperature. Steam gasification of demineralized coal pyrolysis char significantly promotes char conversion at 1300 °C. The effect of O_2 concentration on the demineralized coal pyrolysis char combustion was mild. When the O_2 concentration was higher than 40%, the combustion rate of the demineralized coal pyrolysis char increased significantly.

1 Introduction

Coal-fired power stations are the main source of CO_2 emissions. It is therefore important to investigate novel coal-based power generation techniques that simultaneously achieve energy-efficient conversion and CO_2 emission reduction. Recently, many new oxy-fuel combustion techniques using steam as the diluent have been reported. Clean Energy Systems Inc.[1-2] has designed and demonstrated techniques for oxy-natural gas power generation. Water or steam is injected into an O_2–natural gas combustion flame at high temperature in a high-pressure atmosphere; water or steam is directly converted to high-temperature steam. Salvador [3] and Seepana [4] investigated atmospheric-pressure combustion in an O_2/H_2O atmosphere. Hydrocarbon fuels are burned in O_2/H_2O atmospheres, and H_2O is heated indirectly through the water wall.

Since China is abundance of coal but lack of natural gas, Sun et al. proposed an oxy-coal combustion steam system (OCCSS) [5]. In this system, demineralized coal is obtained by removing ash from raw coal [6-8], which can combust as hypercoal without particle product. The demineralized coal combusts with gaseous O_2 and steam being recycled from condenser and injected stage by stage in order to produce a flue gas consisting predominantly of steam and CO_2. The flue gas is then expanded through an advanced turbine. The working fluid is easy to be separated in a condenser to produce high-purity CO_2. This system can be used for low-cost CO_2 capture. Steam is heated directly in the O_2–demineralized coal combustion flame. An advantage of the OCCSS is that the fuel chemical energy is directly converted into working fluid enthalpy in the same space. In this system, steam regulates the flame temperature and also participates in the combustion reaction; steam is also the working fluid in the turbine. This system removes the indirect heating method using in traditional boilers, and avoids superheater and reheater material requirements in high-efficiency supercritical boilers. This technique is approach to realizing coal clean utilization. However, achieving stable demineralized coal combustion at high steam concentrations is crucial for the development of OCCSS techniques.

The combustion characteristics of demineralized coal differ significantly from those of raw coal because demineralized coal gives almost no ash. Currently, much attention is focused on demineralized coal combustion and gasification reactivity in air. Rubiera [9,10] reported that demineralization greatly changes the morphology and reactivity of char, and improves the char combustion reactivity. Uzun's [11] results show that the reactivity of demineralized coal char is slightly reduced because of ash removal. Zhang [12] investigated the ignition temperature and burning rate of demineralized anthracite coal. Namkung[13] investigated low-temperature steam catalytic gasification of demineralized coal. Wang simulated [14] demineralized coal char catalytic gasification. Yoo reported that the steam gasification rate of demineralized coal is lower than that of raw coal. However, the combustion characteristics of pyrolysis char from demineralized coal in an O_2/H_2O atmosphere have not been reported.

The aim of this study was to investigate the effects of the reaction atmosphere and temperature on the combustion characteristics of pyrolysis char from demineralized coal in an O_2/H_2O atmosphere.

2 Experimental

2.1 Demineralized coal preparation

HF and HCl are generally used to remove coal ash. Many salts in coal, such as carbonates and phosphates, can be dissolved in HCl. Mineral materials such as silica in coal can be removed using HF, although HF is a weak acid. In this study, demineralized coal was prepared by washing Zhundong sub-bituminous coal with HF and HCl. The proximate and ultimate analytical results for the raw coal and the demineralized coal are shown in Table 1.

Corresponding author. Fax: +86-451-86412528
E-mail address: sunsz@hit.edu.cn

Table 1 Analyses of demineralized coal and raw coal

Coal	Proximate analyses (%)				Ultimate analyses (%)				
sample	M_{ad}	V_{ad}	A_{ad}	FC_{ad}	C_{daf}	H_{daf}	N_{daf}	S_{daf}	O_{daf}
ZD-raw	7.02%	27.59%	4.05%	61.34%	80.72	4.66	2.04	0.20	12.38
ZD-d	1.81%	29.84%	0.20%	68.15%	80.31	4.04	1.95	0.18	13.51

2.2 Pyrolysis char preparation

Chars were prepared from the raw coal and the demineralized coal in a high-temperature drop tube furnace (DTF). The DTF had six parts: a gas supply system, fuel feeding system, gas-preheating furnace, the main body of the DTF, a sampling system, and a cooling system. The reactor was an alumina tube of inner diameter 50 mm and length 1400 mm; it was heated electrically. The pyrolysis temperature was 1200 °C and the fuel feed rate was 12–15 g/h. The residence time of the coal particles was kept at 1 s by adjusting the carrier gas flow.

2.3 Pyrolysis char combustion experiments

Combustion experiments were performed on the pyrolysis chars from the raw coal and the demineralized coal in a DTF. The effects of reaction temperature (1100–1400 °C) and atmosphere (O_2/H_2O and O_2/N_2) were investigated. In the experiments, the mixture gas volume ratios were as follows: O_2/N_2 = 21/79, O_2/H_2O = 21/79, 30/70, 40/60, and 50/50. Steam was entrained into the heating furnace by an O_2 flow, and was heated to 360 °C. The steam was further heated to 500 °C in the second preheating furnace. The flow rate of the O_2–steam mixture was 5.5 NL/min. After the experimental temperature was reached, the char particles were entrained into the DTF by argon; the char feeding tube was cooled to 120 °C by recycling oil to prevent steam condensation on the outside of the fuel feeding tube. The char feeding rate was 6–8 g/h.

2.4 Calculation of char conversion rate

The char conversion rate is usually calculated based on the ash balance method. However, there is almost no ash in demineralized coal pyrolysis char. The ash balance method therefore cannot be used to calculate the char conversion rate of demineralized coal. Snape [15] reported a silica ash tracing method for accurately determining high-temperature volatile matter yields. Quartz is stable and resistant to high temperatures, and has no catalytic effect. In this study, the char conversion rate during combustion of demineralized coal pyrolysis char was calculated using the quartz sand tracing method. The char from demineralized coal pyrolysis was mixed with quartz sand, which was used as ash in normal coal. The mass of quartz sand in the mixture was assumed to be constant. The char conversion rate was calculated based on the char content of the mixture. The experimental results obtained using the quartz sand tracing method had good repeatability when the quartz sand particle size was 25–45 μm, and the mass ratio of char to quartz sand in the mixture was 7:1.

3 Results and discussion

3.1 Effects of atmosphere on char combustion

The addition of steam changes heat transfer and composition diffusion on the char particle surface compared with those in conventional pulverized coal oxy-fuel combustion and air combustion. The char steam gasification reaction competes with the char-O_2 reaction, which alters the pathway in char particle consumption. The effects of two different atmospheres (O_2/H_2O = 21/79 and O_2/N_2 = 21/79) on the conversion rates of raw coal pyrolysis char and demineralized coal pyrolysis char at various residence times during char combustion at 1200 °C were investigated. Figure 1 shows that when the residence time is less than 0.3 s, the conversion of raw coal pyrolysis char in an O_2/H_2O atmosphere is faster than that in an O_2/N_2 atmosphere. After 0.5 s, the char conversation rates in the two atmospheres are similar. When the residence time reaches 0.8 s, char combustion is almost complete. The char pore structure is relatively well developed, and the char skeleton contains large amounts of branched chains and small molecules in the initial stage of the reaction, and these favor char steam gasification. Steam addition results in the formation of large amounts of CO and H_2 on the char particle surfaces, and this promotes ignition and combustion of char particles. In the later stage of the reaction, the char pore structure collapses, and the structure becomes more graphitic. Char steam gasification is difficult, therefore the effect of steam on char consumption is not significant in the later stage of the reaction.

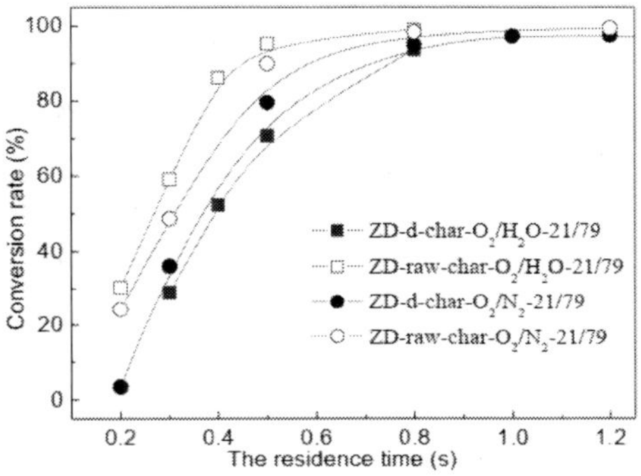

Figure 1: Char conversion rates in O_2/H_2O = 21/79 and O_2/N_2 = 21/79 atmospheres at 1200 °C.

For demineralized coal pyrolysis char, steam has the opposite effect on char conversion. The conversion rate of demineralized coal pyrolysis char in O_2/H_2O is slower than that in O_2/N_2. At 1200 °C, the steam gasification rate of demineralized coal pyrolysis char is greatly reduced because of alkali-metal removal. Char steam gasification cannot compete with char- O_2 combustion. At this point, the main effects of steam on the combustion of demineralized coal pyrolysis char are physical. The specific heat of steam is greater than that of N_2. The temperature of the char particle surface is therefore slightly lower at high steam concentrations, which results in a decrease in the char-O_2 combustion rate.

Figure 1 shows that the combustion rates of raw coal pyrolysis char in the two atmospheres (O_2/H_2O and O_2/N_2) are greater than those of demineralized coal pyrolysis char. The main reason is that Zhundong coal contains large amounts of sodium, potassium, and other alkali metals. In Zhundong coal, alkali metals are mainly present as water-soluble sodium, potassium, and organic compounds. The presence of alkali metals in the form of water-soluble and organic materials has a significant catalytic effect on the combustion and gasification of pulverized coal. The conversion rate of raw coal pyrolysis char is clearly higher than that of demineralized coal pyrolysis char in an O_2/H_2O atmosphere. The conversion rate of raw coal pyrolysis char at 0.5 s is more than 90%. However, the residence time needed to achieve the same demineralized coal pyrolysis char conversion rate is significantly longer. Another influencing factor is that the specific surface area of the raw coal pyrolysis char is significantly higher than that of the demineralized coal pyrolysis char. The specific surface area of the char particles is an important factor in the char combustion rate.

3.2 Effects of O_2/H_2O ratio on char combustion

High-concentration steam is an important characteristic of an OCCSS. The effects of steam on char combustion include endothermic and gasification reactions. The effect of the O_2/H_2O ratio on the combustion of raw coal pyrolysis char and demineralized coal pyrolysis char were studied.

Figure 2(a) shows the char conversion rates during raw coal pyrolysis char combustion at various O_2/H_2O ratios at 1200 °C. The char conversion rate gradually increases with increasing O_2 concentration and decreasing steam concentration. When the O_2 concentration is greater than 30%, the char conversion rate is more than 90% at a residence time of 0.3–0.4 s.

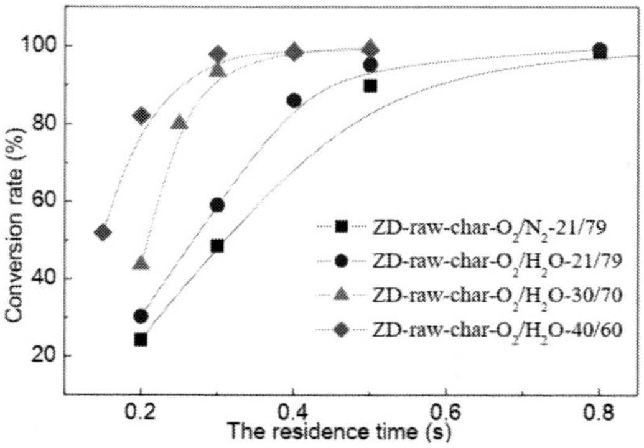

(a) the raw coal pyrolysis char

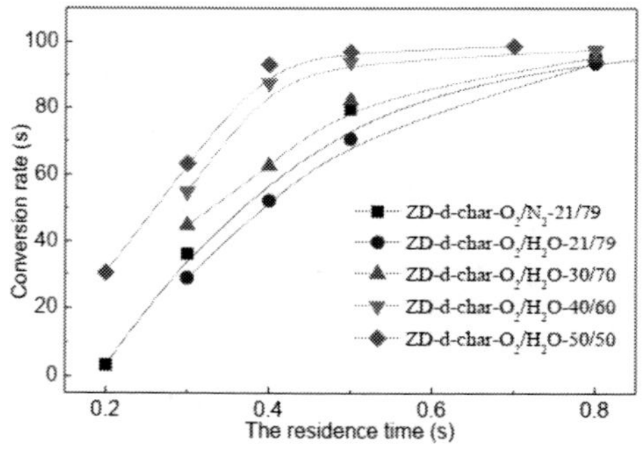

(b) the demineralized coal pyrolysis char

Figure 2: Char conversion rates during pyrolysis char combustion at various O_2/H_2O ratios at 1200 °C.

Figure 2(b) shows the char conversion rates during demineralized coal pyrolysis char combustion at various O_2/H_2O ratios at 1200 °C. As reported in section 3.1, steam addition inhibits demineralized coal pyrolysis char combustion when the O_2/H_2O ratio is 21/79. However, when the O_2/H_2O ratio is 30/70, the increased O_2 concentration promotes demineralized coal pyrolysis char combustion, and the decreased steam concentration reduces heat absorption on the char particle surfaces [14]. In this atmosphere, the combustion rate of demineralized coal pyrolysis char is slightly higher than that in O_2/N_2 = 21/79. When the O_2/H_2O ratio is 40/60, the combustion rate of demineralized coal pyrolysis char increases. The char conversion rate reaches 90% in 0.5 s. When the O_2 concentration is more than 40%, no clear increase in the char combustion rate is observed. This shows that the effect of the O_2 concentration on combustion is not significant.

A comparison of Figure 2(a) and (b) shows that the degrees to which the O_2/H_2O ratio affects the combustions of the raw coal pyrolysis char and the demineralized coal pyrolysis char are different. Clearly, O_2 promotes raw coal pyrolysis char combustion. The combustion rate of raw coal pyrolysis char increases rapidly when the O_2 concentration is 30%. Char combustion is complete in 0.4 s. However, the effect of the O_2 concentration on demineralized coal pyrolysis char combustion is milder. At an O_2 concentration of 40%, the char combustion rate clearly increases, and char combustion is complete in 0.7 s. This difference is mainly caused by the catalytic effect of the ash in the raw coal pyrolysis char. The specific surface area of the raw coal pyrolysis char is larger than that of the demineralized coal pyrolysis char, therefore the former provides more active sites and a higher reaction area for combustion.

3.3 Effects of reaction temperature on char combustion

The temperature significantly affects combustion and gasification. The effect of temperature is related to the reaction conditions. The activation energies of different reactions differ, therefore the sensitive temperature ranges for different reactions differ. For the char-O_2 combustion reaction, the activation energy is low, therefore the effect of temperature is significant at low temperatures. However, for char steam gasification, the activation energy is higher, therefore the temperature plays a role at higher temperatures. The O_2 diffusion rates in different atmospheres differ, therefore the effects of temperature on combustion in different atmospheres differ. The effects of the reaction temperature (1100–1400 °C) on the combustion characteristics of the raw coal pyrolysis char and demineralized coal pyrolysis char in two different atmospheres (O_2/H_2O = 21/79 and O_2/N_2 = 21/79) were investigated.

Figure 3(a) shows the effects of reaction temperature on the char conversion rate during raw coal pyrolysis char combustion in these two atmospheres. An increase in temperature promotes char combustion in the O_2/N_2 atmosphere. The residence time required to achieve a char conversion rate of 50% is 0.4 s at 1100 °C. When the reaction temperature increases to 1200 °C, the time decreases to 0.3 s. When the temperature is increased to 1400 °C, the time decreases to 0.2 s. These results show that raw coal pyrolysis char combustion in an O_2/N_2 atmosphere is significantly affected by the temperature. In an O_2/H_2O atmosphere, steam addition clearly promotes char combustion at 1100 °C. These results are in agreement with those reported by Zou [18]. However, increasing the reaction temperature has less effect on the combustion characteristics of raw coal pyrolysis char in an O_2/H_2O atmosphere. The gas coefficients calculated using Fuller's equation show that the diffusion coefficient of O_2 in N_2 is larger than that in steam. The effect of reaction temperature on raw coal pyrolysis char combustion in an O_2/H_2O atmosphere is therefore less than that in O_2/N_2.

Figure 3(b) shows the effect of reaction temperature on the char conversion rate during demineralized coal pyrolysis char combustion in O_2/H_2O = 21/79.

The combustion of demineralized coal pyrolysis char in an

O_2/H_2O atmosphere accelerates with increasing temperature. This is in agreement with the results reported by Wang et al.. In an O_2/H_2O atmosphere, the conversion rate of demineralized coal pyrolysis char reaches 75% in 0.7 s at 1100 °C. However, the char conversion rate reaches 80% in just 0.3 s at 1400 °C. The discussion in section 3.1 shows that in an O_2/H_2O atmosphere the steam gasification of demineralized coal pyrolysis char is not significant at 1200 °C. Figure 3(b) shows that the combustion rate of demineralized coal pyrolysis char clearly increases with increasing temperature from 1200 to 1300 °C. It is inferred that the steam gasification of demineralized coal pyrolysis char promotes char conversion at 1300 °C. When the reaction temperature increases to 1400 °C, the combustion rate of demineralized coal pyrolysis char increases slightly. This indicates that the reaction is in the diffusion-control zone.

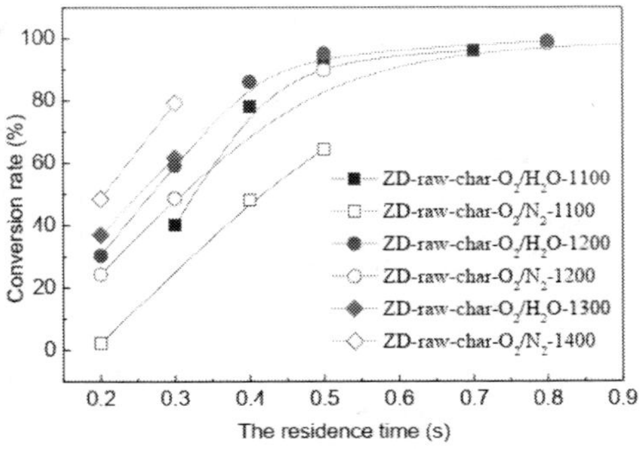

(a) the raw pyrolysis char

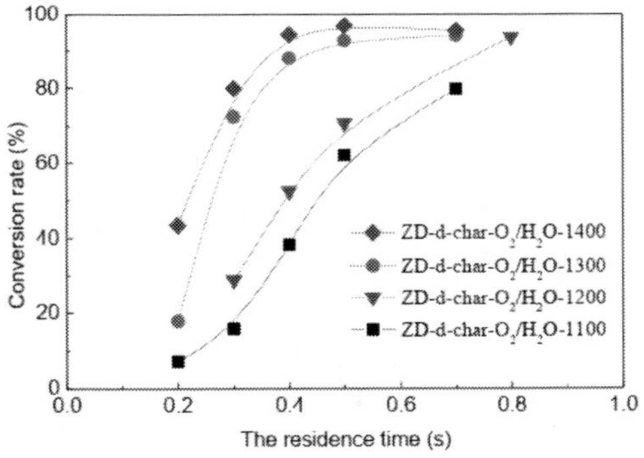

(b)the demineralized coal pyrolysis char

Figure 3: Effects of reaction temperature on char conversion rates during pyrolysis char combustion.

A comparison of Figure 3(a) and (b) shows that the effects of temperature on the combustion of different chars in an O_2/H_2O atmosphere differ. The combustion rate of demineralized coal pyrolysis char is slower than that of raw coal pyrolysis char at lower temperatures. However, at 1300 °C, the combustion rates of demineralized coal pyrolysis char and raw coal pyrolysis char are similar. The main reason is that when the temperature is increased, the catalytic action of ash is weakened. However, because of the low content of ash in demineralized coal pyrolysis char, ash does not block the internal diffusion of char particles. The demineralized coal pyrolysis char has a large amount of macropores, which is favorable for char steam gasification. The reactivity of the demineralized coal pyrolysis char is therefore higher at high temperatures in an O_2/H_2O atmosphere.

4 Conclusions

The ash in Zhundong coal was removed to obtain demineralized coal for use in the OCCSS, which gives CO_2 near-zero emissions. The effects of the reaction atmosphere and temperature on char combustion were investigated to study the combustion characteristics of demineralized coal pyrolysis char. Compared with raw coal pyrolysis char, the demineralized coal pyrolysis char had a larger number of macropores. However, the surface area of the demineralized coal pyrolysis char was smaller than that of the raw coal pyrolysis char. At 1200 °C, the steam gasification rate of the demineralized coal pyrolysis char was slow because of alkali-metal removal, and could not compete with char O_2 combustion. The effects of steam on char combustion were mainly physical. Demineralized coal pyrolysis char combustion in an O_2/H_2O atmosphere accelerated significantly with increasing temperature. Steam gasification of demineralized coal pyrolysis char promoted char conversion at 1300 °C. The effect of O_2 concentration on the demineralized coal pyrolysis char combustion was mild. When the O_2 concentration was higher than 40%, the combustion rate of the demineralized coal pyrolysis char clearly increased.

5 Acknowledgment

The authors acknowledge the financial support for this research provided by the National Natural Science Foundation of China (Grant No. 51536002) and the Foundation for Innovative Research Groups of the National Natural Science Foundation of China (Grant No. 51421063).

References

[1] R. Anderson, C. Hustad, P. Skutley. Energy Procedia 63(0) (2014) 511-523.
[2] R. Anderson, H. Brandt, H. Mueggenburg. Green Gas Control Technol (1999) 59-64.
[3] K .Zanganeh, C. Salvador, M. Mitrovic. 2nd International oxy-combustion Research Network. Windsor, CT, USA, 2007.
[4] S. Seepana, S. Jayanti. Energy Convers. Manage 51(10) (2010) 1981-1988.
[5] S. Sun, S. Meng, H. Xu, P. Qiu, R. Sun, Y. Zhao. One kind of coal-based near-zero emissions power generation system and method [P]: CN, 102628401A. 2012.08.08.
[6] K.M. Steel, J.W. Patrick. Fuel 80(14) (2011) 2019-2023.
[7] W. Niken, K. Teck, L. Zhang. Fuel Processing Technology 92 (11) (2011)2127-2137.
[8] E. Jorjani, H. Ghasemi Chapi, M. Tayebi Khorami. Fuel Processing Technology 92(10) (2011) 1898-1904.
[9] F. Rubiera, A. Arenillas, B. Arias, J. Pis, I. Sua´rez-Ruiz, K. Steel, J. Patrick. Fuel 82(15-17) (2003)2145-2151.
[10] F. Rubiera, A. Arenillas, C. Pevida, R. Garcı´a, J. Pis, K. Steel, J. Patrick. Fuel Processing Technology 79(3) (2002) 273-279.
[11] D. Uzun, S. ÖZDOGAN. Energy Sources, Part A 33(4) (2010) 283-297.
[12] S. Zhang, Z. Chen, X. Chen, X. Gong. Journal of Fuel Chemistry and Technology42(2) (2014) 166-174.
[13] H. Namkung, X. Yuan, G. Lee, D. Kim, T. Kang, H. Kim. Journal of the Energy Institute 87 (3) (2014) 253–262.
[14] J. Kopyscinski, J. Lam, C. Mims, J. Hill. Fuel 128(15) (2014) 210-219.
[15] T. Farrow, C. Sun, C. Snape. Journal of Analytical and Applied Pyrolysis113(5) (2015) 323-331.

11th Asia-Pacific Conference on Combustion,
The University of Sydney, NSW Australia
10th -14th December 2017.

Modeling of Ammonia/air Non-premixed Turbulent Swirling Flames in a Gas Turbine like Combustor

K.D.K.A. Somarathne[1], A. Hayakawa[1], N. Iki[2], O. Kurata[2], H. Kobayashi[1]

[1]Tohoku University

2-1-1, Katahira, Aoba-ku, Sendai, Miyagi, 980-8577, Japan

[2]National Institute of Advanced Industrial Science and Technology (AIST)

Tsukuba, Ibaraki, 305-8564, Japan

Abstract

In the present study, burner-stabilized non-premixed turbulent ammonia (NH_3)/air flames were succeeded using swirling flows, and moreover, the present study found that NO emission is significantly decreased with an increase in pressure in either non-premixed or premixed flames. Because, at high pressure, third body reaction of OH + H + M ⇔ H_2O + M has high sensitivity, and thereby, OH concentration reduction in high pressures leads to the reduction of NO generation, because OH radicals play an important role when amidogen (NH_2) is oxidized to NO through nitroxyl (HNO) in the reaction sequence. However, in non-premixed flames, especially in rich conditions, NO emissions are slightly higher than that of premixed flames, because NO production is depended on the local equivalence ratio of the flame region in the non-premixed flames. Thus, there is a significant NO production in the lean flame regions, in which near to the wall boundaries, owing to the high concentration of OH even in the overall-rich flame conditions. In order to avoid this distress, the present study suggests to use split NH_3 streams around the air stream, and thus, minimum NO emission was resulted when volumetric ratio, between inner and outer NH_3 swirlers, is 0.6: 0.4.

1 Introduction

Green energy resources are nowadays playing an important role in order to prevent the detrimental effects of fossil fuels on the environment, and among those, ammonia (NH_3) has recently been identified as a potential fuel source not only owing to the its carbon-free nature but also because of higher hydrogen capacity of 17.8% in weight. However much lower laminar burning velocity (7 cm/s, at stoichiometric condition, 300 K) and larger fuel NO generation than that of conventional hydrocarbon fuels [1] hinder the direct NH_3 combustion. However recent studies on NH_3/air turbulent premixed flames by Somarathne et al. [2-3] and Hayakawa et al. [4], numerically and experimentally, respectively, illustrated that by introducing a swirl flow, and thereby making a recirculation flow near the downstream of the swirler, NH_3/air premixed turbulent flames successfully achieved a stable combustion even in high turbulent intensities and elevated pressures. The studies of [2-4] found that there is a specific equivalence ratio, ϕ_{sp}, in rich flame condition in which NO and unburnt NH_3 emissions are minimal and in a same order such that ϕ_{sp} = 1.2, at the initial mixture temperature, T_0 of 500 K and operating pressure, P_0 of 0.1 MPa with an adiabatic wall conditions [2], whereas ϕ_{sp} was 1.05 at T_o and P_0 of 300 K and 0.1 MP, respectively, with non-adiabatic wall condition [4]. In addition, the study of [3] found that the pressure effect is very significant in order to reduce the NO and unburnt NH_3 emission, and the increase in pressure from 0.1 MPa to 0.5 MPa results the reduction of minimal NO and unburnt NH_3 emissions in the order from 700 ppm to 200 ppm of mole fraction at ϕ_{sp} = 1.2. The effect of pressure on NO reduction was related to the third body reaction of OH + H + (M) ⇔ H_2O + M, which has higher sensitivity in OH concentration reduction of flame region at high pressures [5], and subsequently leads to the NO reduction because OH radicals play an important role when amidogen (NH_2) is oxidized to NO through nitroxyl (HNO) in the reaction sequence [6] such as NH_2 + H(OH) ⇔ NH + H_2 (H_2O), NH + OH ⇔ HNO +H, HNO + OH ⇔ NO+ H_2O. Besides, the study of [3] illustrated that the decrease in chemical characteristic time with increase in operating pressure leads to the increase in over-all Damköhler number, which paves the way for reduction of unburnt NH_3 at high pressures in rich flame conditions.

However, most of gas turbine manufactures are like to have non-premixed flames in gas turbine combustors, because non-premixed flames are simpler to design and build owing to perfect reactant mixing in specific proportion is not required. In addition, non-premixed flames are safer as they do not propagate, and thereby do not flash back or auto-ignite in undesired locations, even inlet temperature is so high. Nevertheless, according to the best knowledge of authors there is no any comprehensive study on combustion and emission characteristics of NH_3/air non-premixed turbulent flames in a confined environment. By the way, NH_3/air combustion power generation has been successfully realized using a 50 kW micro gas turbine at the National Institute of Advanced Science and Technology (AIST), Japan by Kurata et al. in 2017 [7]. However, in that study, NO emission is still higher than that of government regulation such that 70 ppm of mole fraction at 16 % O_2 concentration. Therefore, the present study is dedicated to understand the combustion and emission characteristics of turbulent non-premixed NH_3/air swirl flames at various pressures and compare with corresponding premixed flame characteristics, which were already published in the studies of [2] and [3]. Accordingly, a relative simple test cases were chosen as affordable tests to evaluate the feasibilities of more complex gas turbine combustors, and in the present study, large eddy simulation (LES) with a finite rate chemistry is performed in a three-dimensional computational domain in order to discuss a detailed resolution of reacting flow field and emission characteristics in terms of overall and local equivalence ratio.

Corresponding author. Fax: +81-22-217-5323
E-mail address: kunkuma@flame.ifs.tohoku.ac.jp

2 Numerical Modelling

In the present study, spatially filtered Navier-Stokes equations were used, and the reacting flow simulations were conducted using open-source code, OpenFOAM. In the momentum equation, the unresolved subgrid-scale tensor is modeled by wall adapting local eddy viscosity (WALE) subgrid scale model, and unresolved fluxes of species and sensible enthalpy of the mass fraction and sensible enthalpy equations were described as a gradient assumption. Even though the laminar burning velocity estimations have a slightly over predictions than experimental measurements [1], the Miller's chemical kinetics [6], which consists of 23-species and 98-elementry reactions, for NH_3 oxidation was selected in this study by considering the computational burden of other detailed mechanisms even that have more closer laminar burning velocity estimations. In addition, *in-situ* adaptive tabulation algorithm library was used to calculate the composition changes through chemical reactions, and thereby reduce the computational cost. A Partially-Stirred-reactor (PaSR) model was used to account the turbulence-chemistry interactions.

In the first part of the present study, the unstructured three-dimensional computational domain consists of the cylindrical combustion chamber, which was installed vertically, and two annulus swirlers, which are used for the NH_3 and air streams and installed at the center of bottom of the combustion chamber. The chamber dimensions are same as the studies of [2] and [3], such that 72 mm in the inner diameter and 150 mm in the height. In addition, the outer swirler, which is used for the air stream in the present study, has inner and outer diameters of 14 mm and 24 mm, respectively, and the dimensions are same as the swirler dimensions, which was used in the premixed flame studies of [2] and [3]. Moreover, NH_3 stream passes through an inner swirler, which was located inside to the air swirler, and has inner and outer diameters of 8 mm and 13 mm, respectively. In the present study, the dimensions of swirlers are determined based on the studies of turbulent stratified flames [8]. The swirler angles of both swirlers are approximately 40 degree, and then, the swirler numbers of air and NH_3 swirlers are 0.68 and 0.69, respectively.

In all the cases, the temperature at the inlet flow, T_0 was 500 K, total volumetric flow rate, V is 0.009 m^3/s and the relative turbulent intensity, I of 30% was set for all inlet flows in order to retain a higher turbulent intensity in the flow field. These conditions are same as the studies of premixed flames in [2] and [3]. In addition, LES characteristics length in the flow field is approximately 0.63 mm, which corresponds to an approximately cubic root of a computational cell volume of 0.25 mm^3. The operating conditions, in terms of operating pressure, P_0, equivalence ratio, ϕ, and mean velocities of air and NH_3 streams, U_{air} and U_{NH3}, respectively, are shown in the Table 1. The origin of the coordinates was set at the center of bottom of the combustion chamber and the axial and radial coordinates are indicated by z and r, respectively.

P_0 [MPa]	0.1, 0.5			
overall ϕ	1.0	1.1	1.2	1.25
U_{air}[m/s]	30.53	29.93	29.33	29.03
U_{NH3}[m/s]	31.27	33.45	35.60	36.70

Table 1. Initial operating conditions

P_0[MPa]	0.1		
overall ϕ	1.2		
R	0.67: 0.33	0.60:0.40	0.55:0.45
U_{air}[m/s]	29.33		
$U_{NH3(innermost)}$[m/s]	23.73	21.36	19.60
$U_{NH3(outermost)}$[m/s]	4.53	5.47	6.10

Table 2: Operating conditions with the split NH_3 flow streams

In the second part of the present study, another annulus swirler was added into the outside to the air swirler for a NH_3 stream, which has inner and outer diameters of 25 mm and 30 mm, respectively. This swirler also has an angle of 40 degree and consequently, swirler number was 0.77. In this case also, T_0, V, and I are same as the first part of the study. However, NH_3 flow was split between the innermost and outermost swirlers according to the volumetric flow ratios, R as shown in Table 2.

3 Results and Discussion

3.1 Reacting flow characteristics

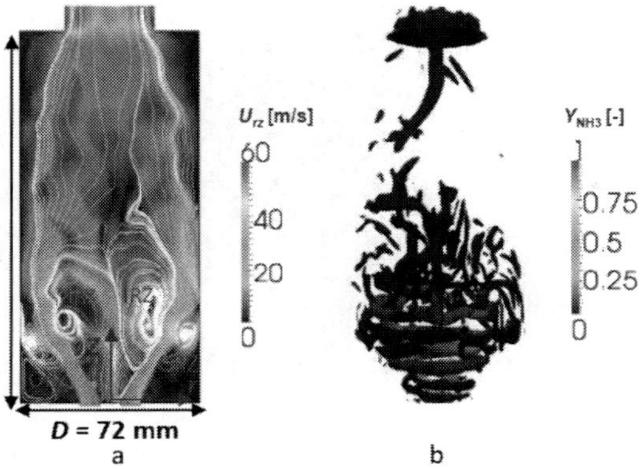

Figure 1. Instantaneous reacting flow field of turbulent non-premixed NH_3/air flames at overall ϕ =1 and P_0 = 0.1 MPa: (a) streamlines and velocity distribution in a vertical plane; (b) iso-surfaces of the second invariant of the velocity gradient tensor, Q, of 3 × 10^7, shaded with mass fraction of NH_3, Y_{NH3}.

The instantaneous reacting flow field of non-premixed turbulent NH_3/air flames at ϕ = 1 and P_0 = 0.1 MPa is shown in the Figs. 1 (a) and (b) by means of the streamlines in a vertical plane of the combustion chamber and iso-surfaces of the second invariant of the velocity gradient tensor, Q, of 3 × 10^7 (as the median value), respectively. In Fig. 1(a), inner recirculation zones (IRZ), which associate with the vortex breakdown bubble owing to the lower pressure of backsides in the swirling flow, bring unburnt NH_3, free radicals and hot burnt gases to the upstream of the combustion chamber and lead to maintain a continuous chemical reactions. Moreover, the rotational flow structures of the air and NH_3 swirling flows, in near downstream of the swirlers, are clearly shown in Fig. 1(b), and it helps to determine the flame length indirectly.

3.2 Characteristics of flame structures

Temperature distributions of non-premixed and premixed flames are shown in Figs. 2 (a) and (b) in terms of ϕ (overall ϕ) and P_0. In non-premixed cases also flames were successfully stabilized on the buff-body similar to the premixed flames. However, there is a lower temperature region in the center region of the flame in non-premixed cases unlike to the premixed cases at rich flame conditions. Moreover, center region temperature further decreases with increase in overall ϕ, and subsequently, flame was blow-off at the overall ϕ of 1.25 (not shown here).

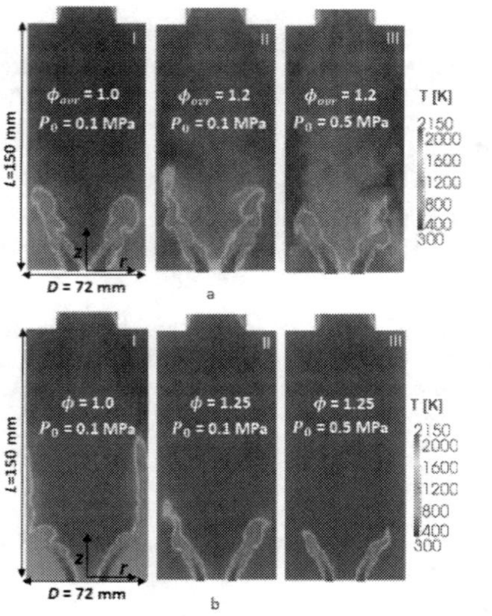

Figure 2. Temperature distributions in terms of ϕ (overall ϕ) and P_0: (a) non-premixed flames; (b) premixed flames.

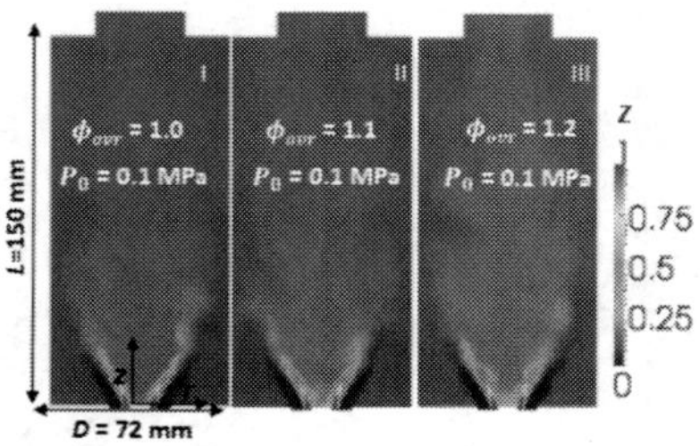

Figure 3. Mixture fraction variation in non-premixed NH_3/air flames in terms of overall ϕ at P_0 = 0.1 MPa.

This is possibly because of much richer flame conditions in the central region of the flame as shown in Fig. 3, using the mixture fraction, Z, distributions. In the present study, mixture fraction was formulated, as shown in (1), based on the Bilger's mixture fraction formulation for methane/air flames. The mixture fraction at the stoichiometric condition of NH_3/air flame, Z_{st} was 0.142.

$$Z = \frac{\frac{1}{4}M_H^{-1}[Z_H - Z_{H,2}] + \frac{3}{4}M_N^{-1}[Z_N - Z_{N,2}] - M_O^{-1}[Z_O - Z_{O,2}]}{\frac{1}{4}M_H^{-1}[Z_{H,1} - Z_{H,2}] + \frac{3}{4}M_N^{-1}[Z_{N,1} - Z_{N,2}] - M_O^{-1}[Z_{O,1} - Z_{O,2}]} \qquad (1)$$

where Z_H, Z_N, and Z_O are the elemental mass fractions of the hydrogen, nitrogen and oxygen elements, respectively; M_H, M_N, and M_O are the corresponding atomic masses; 1 and 2 digits correspond to the fuel (NH_3) and oxidizer (air) streams. In addition, similar to the premixed flames, in non-premixed flames also the flame length is decreased with increase in pressure as shown in Fig. 2 (a) II and III. This is associate with the significant decrease of chemical characteristic time scale with increases in operating pressure [3].

3.3 NO emission characteristics

Figures 4 and 5 show the NO and OH concentration distributions, X_{NO} and X_{OH}, respectively, of non-premixed and premixed flames in terms of ϕ (overall ϕ) and P_0, and it is clearly shows that NO production is corresponded to OH concentration. Moreover, in non-premixed flames, NO production is depended on the local equivalence ratios of the flame region, and there is a significant NO production in the lean flame regions around the wall boundaries owing to higher concentration of OH even in the overall-rich flame conditions. However, in premixed flames,

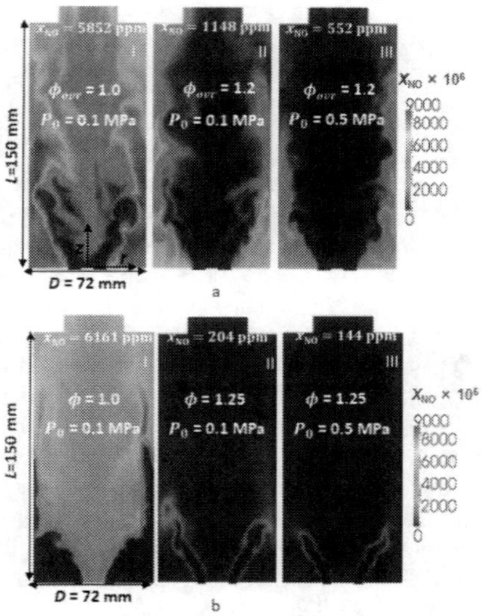

Figure 4. NO distributions in terms of in terms of ϕ (overall ϕ) and P_0: (a) non-premixed flames; (b) premixed flames.

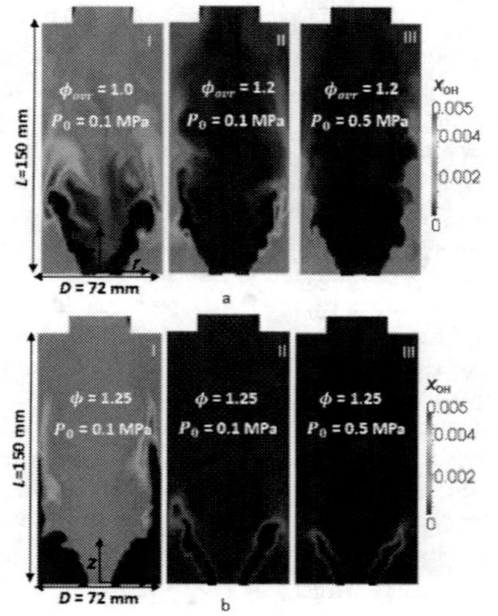

Figure 5. OH distributions in terms of in terms of ϕ (overall ϕ) and P_0: (a) non-premixed flames; (b) premixed flames.

there is no such locally-high OH concentration regions in the flame, and consequently, NO distribution is uniform in the flame. In addition, the space and time averaged emission (STAE) of NO, x_{NO}, was calculated at the exit of the combustor, and NO emissions decreases with the increase in overall ϕ and P_0 as similar to the premixed flames [3] as shown in Fig.4. Moreover, hereafter, X_k and x_k denote the mole fraction of species k in the combustion chamber and STAE of k, respectively.

3.4 Effect of pressure and overall equivalence ratio on NO/unburnt NH_3/H_2 emissions.

Figures 6 (a) and (b) show the STAE of NO, unburnt NH_3 and H_2 in terms of ϕ and P_0 for non-premixed flames and premixed flames, respectively. The present study verified that the effect of pressure is very significant in order to emission reduction of NO and unburnt NH_3 irrespective to the combustion type. NO emission reduction is related to the high sensitivity of OH + H + M ⇔ H_2O +M in high pressures [5], whereas the increase in over-all Damköhler at high pressures is caused to NH_3 emission reduction. Even though NO emissions of

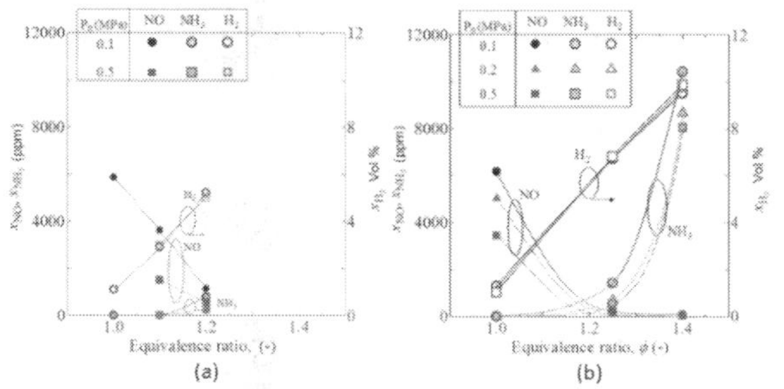

Figure 6. STAE of NO, unburnt NH_3, H_2 emissions in terms of ϕ and P_0: (a) non-premixed flames; (b) premixed flames.

of non-premixed flame, at $\phi = 1$ and $P_0 = 0.1$ MPa, is slightly lesser than that of premixed flame, it is importantly observed that, at rich flame conditions, STAE of NO of non-premixed flames is slightly higher than that of premixed flames such that at $\phi = 1.2$ and $P_0 = 0.5$ MPa, STAE of NO of non-premixed flame is in the order of 500 ppm of mole fraction, whereas, in premixed flames, STAE of NO is in the order of 200 ppm of mole fraction. This should be highly related to the local equivalence ratio variation in non-premixed flames. However, unburnt NH_3 emissions of non-premixed and premixed flames are in the same order of 200 ppm in mole fraction at $\phi = 1.2$ and $P_0 = 0.5$ MPa. In addition, there is no effect of pressure on H_2 emission.

3.4 NO emission characteristics with split NH_3 flows

In order to reduce the high NO generation in the region of near wall boundaries, the present study suggests to the split NH_3 stream around air stream using another swirler, which is located outside to the air swirler. The configuration of swirlers are described under the section two, and, in this section of the study, all calculations were carried out at overall $\phi = 1.2$ and $P_0 = 0.1$ MPa. Figures 7 (a) and (b) show the NO and OH concentration distributions, respectively, for the NH_3 spitting ratios (between innermost and outermost swirlers) of I: Non-split, II: 0.67:0.63, III: 0.60: 0.40, and IV: 0.55: 0.45. However, in the case of NH_3 flow split ratio of 0.50: 0.50, the flame was not stable, and this is possibly related to the steep velocity gradient, and consequently larger shear stress between inner NH_3 stream and air stream. Eventually, STAE of NO was recorded for the cases of I-IV as 1148 ppm, 1066 ppm, 608 ppm, and 637 ppm, respectively, and minimum STAE of NO was resulted in the case of III such that 608 ppm in mole fraction. This value is in the

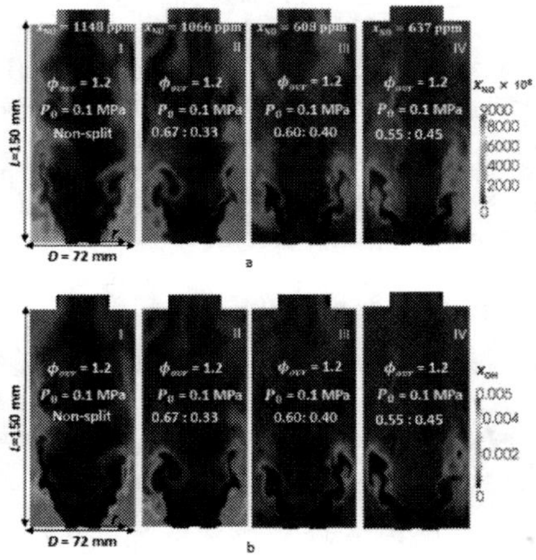

Figure 7. (a) NO; (b) OH concentration distributions in non-premixed flames at $\phi = 1.2$ and $P_0 = 0.1$ MPa in terms of NH_3 splitting ratio

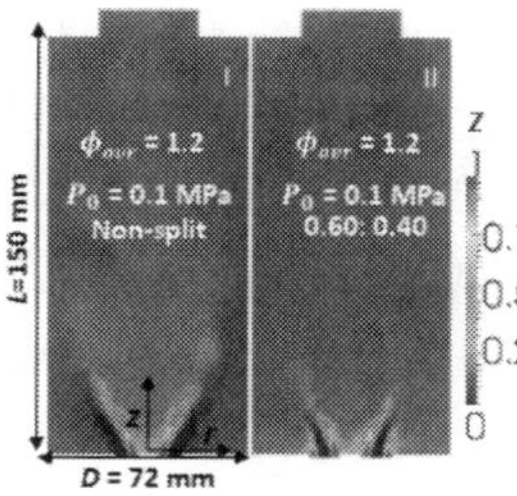

Figure 8. Mixture fraction variation in non-premixed flames at $\phi = 1.2$ and $P_0 = 0.1$ MPa with and without NH_3 splitting

same order of STAE of NO of premixed flame of $\phi = 1.2$ and $P_0 = 0.1$ MPa [2]. On the other hand, when NH_3 flow rate of innermost swirler is further reduced, local equivalence ratio of central region of combustor can be decreased and consequently, NO emission is again increased. Moreover, it was observed that in the case of III, locally-high OH concentration regions are disappeared in the combustor, and it should be related to the almost uniform local equivalence ratio distributions in the combustor, as shown in Fig. 8 by mixture fraction analysis.

4 Concluding remarks

The present study found that NO emission is decreased with an increase in pressure irrespective to the combustion type. However, it was observed that in the non-premixed flames, NO emissions are slightly higher than that of the premixed cases in overall rich flame conditions owing the variation of local equivalence ratio within the combustor. The study suggests to use split NH_3 streams around the air stream in order to achieve almost uniform local equivalence ratio distribution, and thereby NO generation reduction.

Acknowledgement

This research is supported by the Council for Science, Technology and Innovation (CSTI), the Cross-ministerial Strategic Innovation Promotion Program (SIP), "Energy Carriers" (Funding agency: The Japan Science and Technology Agency).

References

[1]A. Hayakawa, T. Goto, R. Mimoto, Y. Arakawa, T. Kudo, H. Kobayashi, Fuel 159 (1) (2015) 98-106.

[2]K.D.K.A. Somarathne, A. Hayakawa, H. Kobayashi, J.Fluid Sci. Tech. 11 (4) (2016) 16-00126.

[3]K.D.K.A. Somarathne, S. Hatakeyama, A. Hayakawa, H. Kobayashi, Fifty-fourth Symposium (Japanese) on Combustion, Combustion Society of Japan (2016) B-214.

[4]A.Hayakawa, Y. Arakawa, R. Mimoto, K.D.K.A. Somarathne, T. Kudo, H. Kobayashi, Int. J. Hydro. Energy, 42 (19) (2017) 14010-14018.

[5]A. Hayakawa, T. Goto, R. Mimoto, T. Kudo, H. Kobayashi, Mech. Eng. J. 2 (1) (2015) 14-00402.

[6]J.A. Miller, M.D. Smooke, R.M. Green, R. J. Kee, Combust. Sci. Tech. 34 (1983) 149-176.

[7]O. Kurata, N. Iki, T. Matsunuma, T. Inoue, T. Tsujimura, H. Furutani, H. Kobayashi, A. Hayakawa, Proc. Combust. Inst. 36 (3) (2017) 3351-3359.

[8]M.S. Sweeney, S. Hochgreb, M.J. Dunn, R.S. Balow, Combust. Flame 159 (9) (2012) 2912-2929.

11th Asia-Pacific Conference on Combustion,
The University of Sydney, NSW Australia
10th -14th December 2017.

Combustion characteristics of CH_4/CO flame with the variation of H_2 addition

Yu-Ting Chen [1], Tsarng-Sheng Cheng [2], Yueh-Heng Li [1], Fang-Hsien Wu[3], Jian-Syun Wu[1]

[1] Department of Aeronautics and Astronautics, National Cheng Kung University

Tainan, 70101, Taiwan

[2] Department of Mechanical Engineering, Chung Hua University

Hsinchu, 30012, Taiwan

[3] Research Center for Energy Technology, National Cheng Kung University

Tainan, 70101, Taiwan

Abstract

The study is aimed to investigate the effect of hydrogen addition on syngas combustion characteristics. The fuel composition of syngas, including methane, carbon monoxide, and hydrogen, was specified with identical heating value. The adiabatic flame temperature, laminar flame velocity, and extinction strain rate were determined by Chemkin collection 3.5 with GRI-Mech 3.0. The results show that the concentration of H radical is increased along with the hydrogen addition. During the combustion process, H radical which could inhibit the main reaction step of the production of carbon dioxide (R99), and produce OH radical with R38 and R46. The concentration distribution and the reaction rate at near extinction strain rate has been investigated to study the chemical effect with the hydrogen addition.

1. Introduction

Syngas, which consists of varying amounts of hydrogen, carbon monoxide, and small amount of methane, have gained great interest due to its wide varieties of sources. On the other hand, a new technology of integrated gasification combined cycle (IGCC) power generation has been developed to generate electricity by using syngas as the primary fuel and further reduce emission [1-3]. Fortunately, the detailed kinetic mechanisms for hydrogen and carbon monoxide have been tremendously studied due to their importance in fossil fuel combustion [1-15]. Scholte et al. [4]showed that hydrogen addition in fossil fuel leads to an increase of laminar burning velocity. Yu et al. [6] and Ren et al. [7] used opposed-jet method to investigate the effect of hydrogen addition on laminar flame speeds and strain rates, respectively. In additional to the increase in laminar burning velocity, a higher extinction strain rate (increased by 60%) has also been found. Vagelopoulos and Egolfopoulos [8] further investigated the H_2/CO/air and CH_4/CO/air flames by using experiments and model predictions with detailed chemical kinetic mechanisms. They concluded that the increase in laminar burning velocity and extinction strain rate were caused by the enhancement of H radicals, which could increase branching chain reactions and accelerate the CO oxidation reaction. Yetter et al. [9] summarized a comprehensive reaction mechanism for H_2/CO/Air system by comparing their experimental data obtained from shock tube with those measured from various types of reactors under a wide range of temperature (823-2870 K) and pressure interval (0.3-2.2 atm) conditions. At high temperature (above 2000K) regime, the reaction was dominated by H, O, and OH radicals. At low temperature regime (below 1200K), the reaction was dominated by HO_2 and H_2O_2. However, the concentrations of intermediates of H, O, OH, HO_2 and H_2O_2 are nearly the same order of magnitude in the intermediate temperature regime. Kim et al. [5]investigated the CO reaction under a wide range of pressure (1-9.6 atm) and low temperature (960 – 1200 K) conditions, and the pressure dependency of each reaction has also been studied. In the air-fuel condition, nitrogen can be decomposed to minor species and produce NO_x at high temperature [14]. Mueller et al. [16] added NO_x species to the H_2/O_2 and $CO/H_2O/O_2$ systems to investigate the catalytic effect at the temperatures below 1200 K. In both systems, similar results were found to show that the oxidation reaction has been catalysed by the third body reaction ($H + O_2 + M \leftrightarrow HO_2 + M$) at the atmosphere pressure. Meriem and Fouzi [17] studied the syngas flame with numerical method under oxygen-enriched condition, and found that the increase of oxygen addition accelerates the heat radiation and enhances the combustion efficiency. The production of NO_x is also reduced at certain strain rate.

Li et al. [11] used 1,3,5-trioxane, which could decompose to produce formaldehyde, with a variable pressure flow reactor to investigate the detailed reaction of formaldehyde and further revised the rate correlation for $CO + OH \leftrightarrow CO_2 + H$ and $HCO + M \leftrightarrow H + CO + M$. As a result, a comprehensive mechanism for hydrogen and carbon monoxide had been updated.

In this study, the extinction strain rates and laminar flame velocities with various hydrogen addition in CO/CH_4 combustion are systematically investigated.

2. Numerical methodology

The simulations were performed by using Chemkin collection 3.5 along with GRI-Mech 3.0. Equil-code is used for adiabatic flame temperature calculation under isobaric and isentropic conditions. One-dimensional laminar flame speed and normalized sensitivity are determined by Premix-code at steady state. Oppdif-code, which is based on the opposed-jet method, is used for flame temperature, species concentration, reaction rate, and heat release rate calculations. The initial temperature and pressure is set at 300K and 1atm, respectively. The distance

Corresponding author. Fax: +886-6-238-9940
E-mail address: yueheng@mail.ncku.edu.tw

between two opposed-jet nozzles is set at 2cm for simulation. GRI-Mech 3.0 including 53 reacting species and 325 reaction steps provide detailed chemical reaction mechanisms for this study. The variation of hydrogen addition used in the present study is shown in Table. 1 for stoichiometric condition. Same compositions of fuel were used for each side of the Oppdif-code. The simulations were performed at equivalence ratios of 0.6, 0.8 and 1.

Flame NO.	$H_2/CO/CH_4$ Fuel composition ratio (%)			$H_2/CO/CH_4$ Mole fraction(vol%)			Air (vol%)	Fuel (vol%)
	H_2	CO	CH_4	H_2	CO	CH_4	$\phi = 1$	
1	0	20	5	0	16.63	4.16	79.21	20.79
2	2	20	5	1.57	15.75	3.94	78.74	21.26
3	4	20	5	2.99	14.95	3.74	78.32	21.68
4	6	20	5	4.27	14.23	3.56	77.94	22.06
5	8	20	5	5.43	13.58	3.39	77.59	22.41
6	10	20	5	6.49	12.98	3.25	77.28	22.72
7	12	20	5	7.46	12.44	3.11	76.99	23.01
8	14	20	5	8.35	11.94	2.98	76.73	23.27
9	16	20	5	9.18	11.47	2.87	76.48	23.52
10	18	20	5	9.94	11.04	2.76	76.26	23.74
11	20	20	5	10.65	10.65	2.66	76.05	23.95

Table 1. H_2 addition matrix.

3. Results & Discussion

3.1. Adiabatic flame temperature

According to Table 1, the concentration of syngas (fuel) increases with the hydrogen addition, which could lead to a higher heating value and a higher adiabatic flame temperature of the mixtures. However, the adiabatic flame temperature only slightly increases with the increasing hydrogen addition at each quivalence ratio ($\phi = 0.6$, 0.8, and 1), as shown in Fig 1. Since the adiabatic flame temperature has no obvious change at the fixed composition of the present case, the discussion of thermal effect, which could affect the reaction rate and thermal expansion, is excluded in the study.

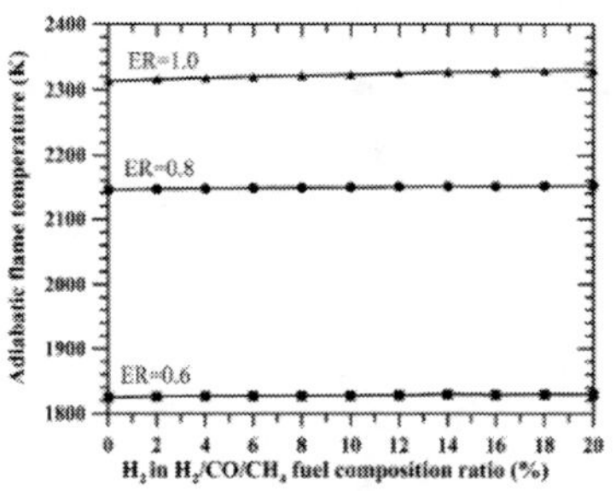

Figure 1. Adiabatic flame temperature relative to H_2 addition.

3.2. Laminar burning velocity

As indicated in Fig. 2, hydrogen addition has an effect to increase the reaction rate. It is expected that the higher concentration of H radical would be produced and accelerate the reaction along with the hydrogen addtion, as discussed in the previous studies [4, 6-9]. On the other hand, despite the lower equivalence ratios lead to lower laminar burning velocities, the increment of burning velocities is not proportional to the increment of equivalence ratio and hydrogen addition. It is to note that there is no significane of change in adiabatic flames with an increase of hydrogen addtion, as shown in Fig. 1. The increase of laminar burning velocity with hydrogen addition is contributed via chemical effect, instead of thermal effect. However, laminar burning velocity is mainly dominated by chemical reaction rate (RR) [18]. Furthermore, the chemical reaction rate is related to the species concentration and reaction constant.

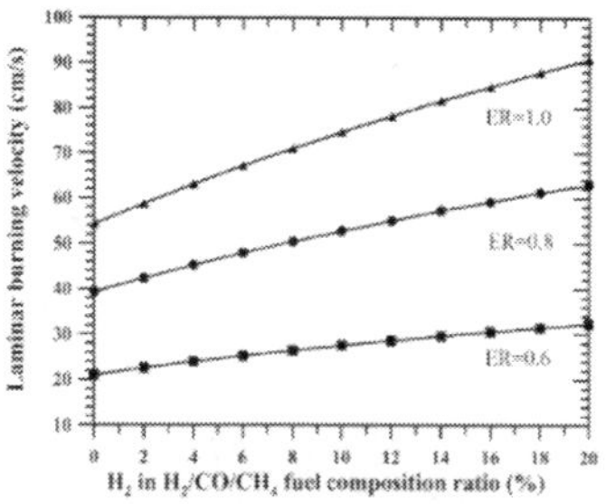

Figure 2. Laminar flame velocity relative to H_2 addition.

3.3. Sensitivity Analysis

To investigate the the influence of chemical reaction effect on laminar flame speed of the stoichiometric flame, a normalized sensitivity coefficient of flame speed was computed by Premixed-code, as shown in Fig. 3. When there is no hydrogen within the fuel, R99 plays the dominating role in the combustion process. However, the dependency of R99 drops along with the increasing hydrogen, then R35 and R46 become more important. The growing dependency of R35 showed that the production of H radical has been promoted, and the higher contents of H radical could promote R46 to produce OH radical and inhibit R99 to produce carbon dioxide. The effect of the hydrogen addition along with the chemical reaction will be further discussed below.

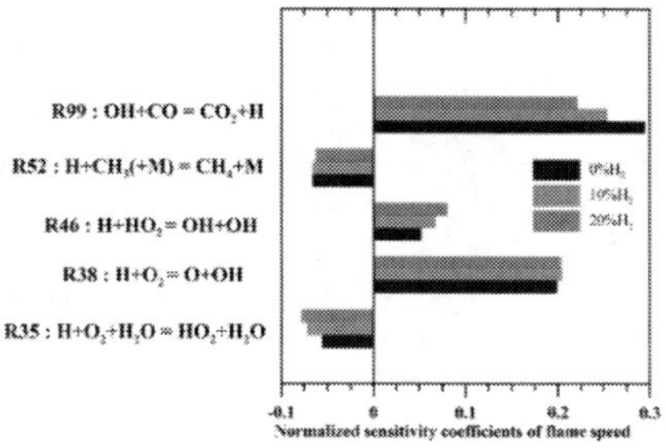

Figure 3. Normalized sensitivity coefficients of flame speed.

3.4. Extinction strain rate & Mole fraction

Physically, as the strain rate increases with the velocity, the flame thickness will be suppressed and become thinner. In the study, strain rate (α, s^{-1}) and flame position (x, cm) are defined by (1) and (2).

$$\alpha = \frac{2|V|}{L}\left[1 + \frac{V_1\sqrt{\rho_2}}{V_2\sqrt{\rho_1}}\right] , \quad (1)$$

$$x = 0.5(T_{int} + T_{final}) , \quad (2)$$

where L (cm) is the separated distance for each burner; V (cm/s) and ρ (g/cm^3) are the jet velocity and density of the mixtures came out from the burner, subscript 1 and 2 represent for each burner. T_{int} (K) and T_{final} (K) represent for initial and final temperature, respectively.

Figure 4 demonstrates the extinction strain rates of CO/CH_4 flames with different H_2 addition. The extinction strain rate could be numerically detected while the flame temperature suddenly drops to surrounding temperature. Therefore, the near extinction strain rates are 3700s^{-1}, 8450 s^{-1}, and 9450 s^{-1} for hydrogen composition ratio of 0%, 10%, and 20% at stoichiometric condition, respectively.

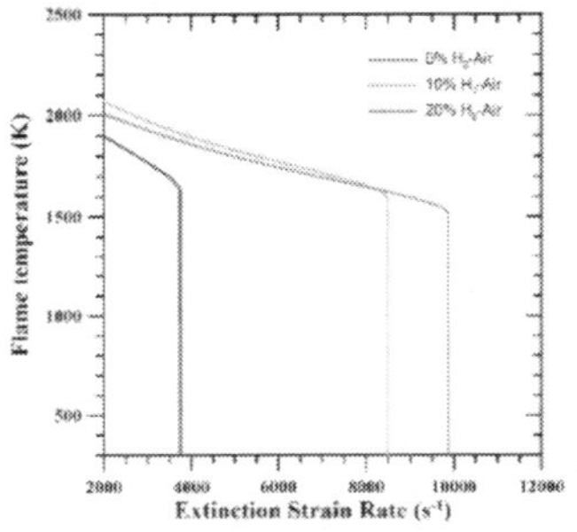

Figure 4. Extinction strain rate relative to H_2 addition.

Meanwhile, the flame temperature drops to around 1600K for each case. The detailed mechanism of the increasing extinction stain rate along with the hydrogen addition, as shown in Fig. 5, should be considered with the concentration distribution. The flame position extended backward to 0.97cm without hydrogen addition. In the case, methane will be decomposed to species such as H_2, CH_3, and C_2H_2 at the pre-heat zone. In the post-flame zone, the concentrations of the product, which consist of water and carbon dioxide, are lower than the normal condition (strain rate at $200s^{-1}$). On the other hand, the concentrations of H, O, and OH radicals raise from pre-heat zone to the reaction zone, and their concentrations are quite low due to the insufficient time for reaction. Further, H radical is dominated by the consumption of methane, and the concentrations of O and OH radicals are extremely low compared to the normal condition. During the reaction, carbon monoxide is consumed and produced by reaction of methane simultaneously. Also, the production of H radical promotes R46 to create more OH radical, which could further affect R99 to produce H radical.

The flame position extended backward to 0.98cm with the ratios of hydrogen composition raised to 10% and 20%, indicating that the flame thickness is thinner compared to the case without hydrogen addition. Also, the variations of methane and carbon monoxide become much smooth. At the same time, the concentration of C_2H_2 increases obviously. As to product concentration, water increased and carbon dioxide decreased. Accordingly, the addition of hydrogen tends to reduce the gradient of the consumption of methane and carbon monoxide. The concentration of H radical grows with R84 ($OH+H_2 \leftrightarrow H+H_2O$) and is up to 0.013, which is far from the concentration without hydrogen addition (0.008) from the pre-heat zone. The concentration of OH radical is slightly increased and the concentration of O radical is slightly decreased. It is worth to note that the local peak of OH at pre-heat zone is contributed by R46, and the other local peak occurred at the post flame zone is contributed by R38. As shown in Fig. 5, the concentration of hydrogen drops prior to carbon monoxide and methane, then slightly decreases after the reaction zone. It was anticipated that methane consumption produces hydrogen, resulting in the reduction of total consumption rate of hydrogen. Since H radical undergoes the process of consumption and production during the reaction, the chain branching of R46 and R52 may take place with the retained H radical. The higher contents of the H radical, has greatly affected the combustion process.

There is a great similarity between two different ratios of hydrogen composition. The concentration of each radical varies within a small interval. However, the more water and less carbon dioxide have been produced with the higher contents of hydrogen, and reaction of hydrogen has been strongly enhanced as well. R53 ($H+CH_4 \leftrightarrow CH_3+H_2$) and R99 have been both inhibited by the higher concentration of H radical, which is promoted by R3 ($O+H_2 \leftrightarrow H+OH$). The inhibition of R99 is also contributed by the lower concentration of OH radical, which could promote R84 and produce more H radical

3.5. Net reaction rate (near extinction strain rate)

Without hydrogen addition, OH radical is the dominating species, which would be produced by R38, R46and greatly affect the main reaction of R52, R98 ($OH+CH_4 \leftrightarrow CH_3+H_2O$), and R99 in the pre-heat zone. The hydrogen produced by R53 would react with the oxygen through R3 and generate H and OH radicals. These radicals could promote R46 and R84 to provide OH and H radical, further, R38 and R99 would be accelerated. However, R38 takes the leading role in the post-flame zone.

With the addition of hydrogen, pre-heat zone is dominated by R46, the reactants are provided by R45 ($H+HO2 \leftrightarrow O2+H2$) and R99 for the production of hydrogen and the H radical. As the concentration of H radical increases with the hydrogen addition, R3 has an obvious enhancement during the reaction zone, and R53 raises the concentration of hydrogen. The OH radical provided by R3 would react with carbon monoxide and hydrogen produced by R53 through R99 and R84, and H radical provided

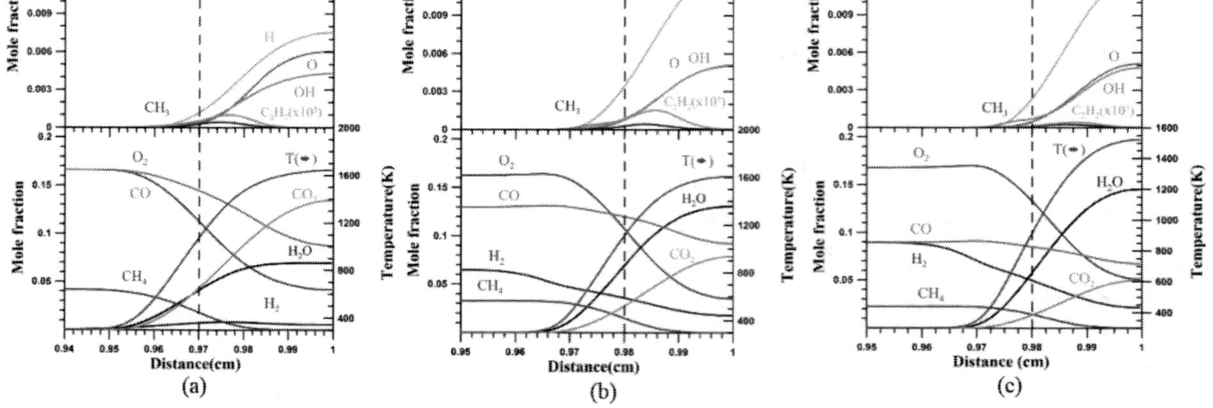

Figure 5. Concentrations distribution of species at near extinction strain rate (a) without H_2 addition; with (b) the ratio of 10% of H_2 in $H_2/CO/CH_4$ fuel composition ; (c) the ratio of 20% of H_2 in $H_2/CO/CH_4$ syngas.

by R3 would react with oxygen through R38, these reactions dominated the burning process at the reaction zone. A further increase in hydrogen will promote R35 to raise the concentration of HO_2, and makes R46 easier to produce OH and then accelerates R84. R3 has greatly enhanced and even greater than R99. Through R3 and R38 produce abundant OH, however, the reduction of the concentration of carbon monoxide has inhibited R99, in contrary, R38 and R84 have been further enhanced.

R3	$O+H_2 \leftrightarrow H+OH$
R35	$H+O_2+H_2O \leftrightarrow HO_2+H_2O$
R38	$H+O_2 \leftrightarrow O+OH$
R45	$H+HO_2 \leftrightarrow O_2+H_2$
R46	$H+HO_2 \leftrightarrow 2OH$
R52	$H+CH_3(+M) \leftrightarrow CH_4(+M)$
R53	$H+CH_4 \leftrightarrow CH_3+H_2$
R84	$OH+H_2 \leftrightarrow H+H_2O$
R98	$OH+CH_4 \leftrightarrow CH_3+H_2O$
R99	$OH+CO \leftrightarrow H+CO_2$

Table. 2 Chemical reactions.

4. Conclusions

Since the adiabatic flame temperature remains within a small interval, the increase of laminar burning velocity along with hydrogen addition is mainly contributed by chemical reaction rate. The flame thickness has been reduced along with the increasing strain rate, as well as residence time. From the analysis of concentration distribution and net reaction rate, the detailed mechanism of the dominating reactions (R38, R84 and R99) are greatly affected by the H radical, which plays a key role with the hydrogen addition. The concentration of H radical will increase with the increasing hydrogen addition. The hydrogen addition would inhibit the production of carbon monoxide, and hence R99 has been greatly inhibited by the reduction of carbon monoxide and OH radical.

5. Acknowledgment

This work was supported by the Ministry of Science and Technology, Taiwan, under contract MOST 105-2628-E-006-005-MY3.

References

[1] C.-J. Sung, and C. K. Law, Combust. Sci. Tech 180 (6) (2008) 1097-1116.

[2] K. J. Whitty, H. R. Zhang, and E. G. Eddings, Combust. Sci. Tech 180 (6) (2008) 1117-1136.

[3] M. Chaos, and F. L. Dryer, Combust. Sci. Tech 180 (6) (2008) 1053-1096.

[4] T. Scholte, and P. Vaags, Combust. Flame, (3) (1959) 511-524.

[5] T. J. Kim, R. A. Yetter, and F. L. Dryer, Proc. Combust. Inst. 25 (1994) pp. 759-766.

[6] G. Yu, C. Law, and C. Wu, Combust. Flame 63 (3) (1986) 339-347.

[7] J.-Y. Ren, W. Qin, F. Egolfopoulos, and T. Tsotsis, Combust. Flame 124 (4) (2001) 717-720.

[8] C. Vagelopoulos, and F. Egolfopoulos, Proc. Combust. Inst. 25 (1994) 1317-1323.

[9] R. A. Yetter, F. L. Dryer, and H. Rabitz, Combust. Sci. Tech 79 (1-3) (1991) 97-128, 1991.

[10] S. G. Davis, A. V. Joshi, H. Wang, and F. Egolfopoulos, Proc. Combust. Inst. 30 (1) (2005) 1283-1292.

[11] J. Li, Z. Zhao, A. Kazakov, M. Chaos, F. L. Dryer, and J. J. Scire, Int. J. Chem.Kinet. 39 (3) (2007) 109-136, 2007.

[12] H. Sun, S. I. Yang, G. Jomaas, and C. K. Law, Proc. Combust. Inst. 31 (1) (2007) 439-446.

[13] X. Li, X. You, F. Wu, and C. K. Law, Proc. Combust. Inst. 35 (1) (2015) 617-624.

[14] P. S. Veloo, Y. L. Wang, F. N. Egolfopoulos, and C. K. Westbrook, Combust. Flame 157 (10) (2010) 1989-2004.

[15] A. M. Starik, N. S. Titova, A. S. Sharipov, and V. E. Kozlov, Combust. Explos. Shock. Waves 46 (5) (2010) 491-506.

[16] M. Mueller, R. Yetter, and F. Dryer, Int. J. Chem.Kinet 31 (10) (1999) 705-724.

[17] M. Safer, F. Tabet, A. Ouadha, and K. Safer, Int. J. hydrogen energy 40 (6) (2015) 2890-2898.

[18] K. K. Kuo, Principles of combustion, John Wiley & Sons, 1986, p. 285.

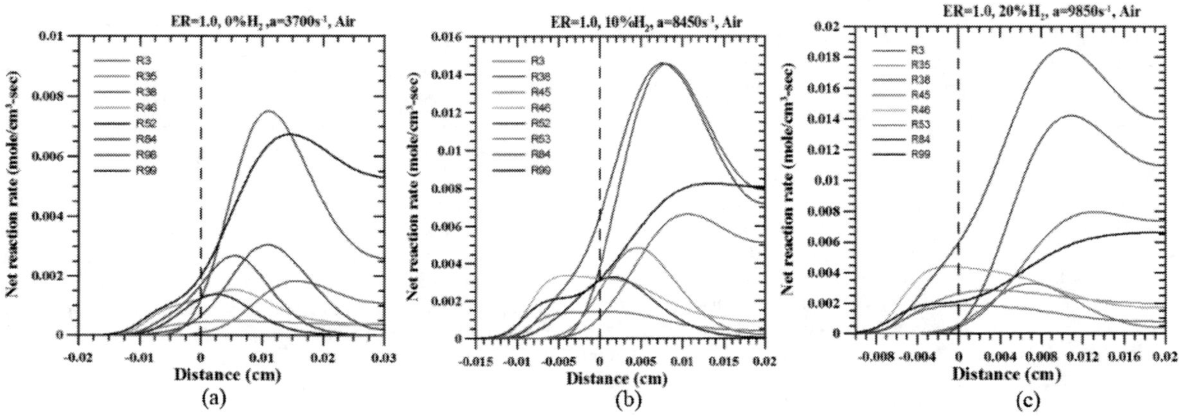

Figure 6. Reaction rates of species at near extinction strain rate (a) without H_2 addition; with (b) the ratio of 10% of H_2 in $H_2/CO/CH_4$ fuel composition ; (c) the ratio of 20% of H_2 in $H_2/CO/CH_4$ syngas.

11th Asia-Pacific Conference on Combustion,
The University of Sydney, NSW Australia
10th -14th December 2017.

Investigations of the low-temperature oxidation of acetaldehyde in a Jet Stirred Reactor

Tao Tao[1], Wenyu Sun[1], Nils Hansen[2*], Ahren W. Jasper[3], Kai Moshammer[2, 4], Bingjie Chen[5], Zhandong Wang[5], Bin Yang[1*]

[1]Center for Combustion Energy and Department of Thermal Engineering, Tsinghua University
Haidian, Beijing, 100084, PR China

[2]Combustion Research Facility, Sandia National Laboratories
Livermore, CA, 94551, USA

[3]Chemical Sciences and Engineering Division, Argonne National Laboratory
Argonne, IL, 60439, USA

[4]Physikalisch Technische Bundesanstalt (PTB), Bundesallee 100
Braunschweig, 38116, Germany

[5]King Abdullah University of Science and Technology (KAUST), Clean Combustion Research Center (CCRC)
Thuwal, 23955-6900, Saudi Arabia

Abstract

Acetaldehyde is not only an observed emission species but also a key intermediate produced during the combustion or low-temperature oxidation of fossil fuels as well as biofuels. Investigations into the low-temperature oxidation mechanisms for acetaldehyde are essential to better understanding some interesting phenomena like auto-ignition or cool flames. Therefore, experiments in a Jet Stirred Reactor were conducted with a fixed residence time around 2.7 s at low-temperature range (528 to 946 K) at 700 Torr. To our knowledge, this is the first set of detailed experimental results of the low-temperature oxidation of acetaldehyde. The mole fractions of 30 species were measured as functions of temperature by employing a molecular beam mass spectrometer using synchrotron generated vacuum ultra-violet photons as ionization source. Theoretical calculations were conducted to help with the identification and quantification of the peroxided intermediates (CH_3O_2 and $C_2H_4O_3$ isomers), providing valuable information for the low-temperature oxidation of acetaldehyde. Based on the measured concentration profiles of various species, especially for the oxygenated intermediates, suggestions for further improvement on the mechanisms are proposed. This experimental study provides an opportunity to advance our knowledge of the low-temperature oxidation of acetaldehyde.

1 Introduction

Over the past decades, there has been an increasing interest in biofuels, which are considered as long-term CO_2 neutral and clean. With higher oxygen content than fossil fuels, most biofuels can reduce the amount of some pollutants, for example soot, but could also increase the emissions of other toxic species, such as acids and aldehydes. It has been reported that acetaldehyde is not only an observed emission species but also a key intermediate produced during the combustion or low-temperature oxidation of fossil fuels as well as biofuels, especially for ethanol [1, 2]. A fundamental research about acetaldehyde low-temperature oxidation can be helpful to the validation of the low-temperature performance of the core mechanism, which plays a vital role in the hot auto-ignition procedure responsible for the knock in the spark ignition engine [3].

Over the past decades, there are a few experimental works about the oxidation of acetaldehyde. Flames fueled by acetaldehyde were investigated by several groups [4-6]. The oxidation and pyrolysis experiments of acetaldehyde were also conducted in shock tubes (ST) by Kern et al. [7], Won et al. [8] and Yasunaga et al. [9], analyzed by different analysis apparatuses. However, all the experiments were conducted under temperature higher than 1000 K. Dagaut et al. [10] measured the concentrations of some intermediates produced in the oxidation of acetaldehyde between 950 K and 1250 K in a Jet Stirred Reactor (JSR). A pressure-temperature ignition diagrams declaring five different regimes was summarized by Gray et al. [11] based on their experimental results in a well stirred reactor at low pressure under the temperature between 400 K and 620 K. Kaiser et al. [12] carried out the acetaldehyde oxidation experiments in a static and closed 1-L Pyrex reactor coupled to a mass spectrometer. A few of stable intermediates concentrations were measured as a function of time at only two temperature points (553 K and 713 K), which provided limited but valuable information for the development and validation of the mechanisms describing the low-temperature oxidation of acetaldehyde. Recently, low-temperature (500-700 K) oxidation experiments were performed by Barari et al. [13] at low pressure (8 Torr) in a flow reactor. Unfortunately, only the signals of some intermediates were reported as functions of time without further quantification information.

Improving the low-temperature oxidation mechanisms for

Corresponding author: nhansen@sandia.gov (Nils Hansen)
Corresponding author: byang@tsinghua.edu.cn (Bin Yang)

acetaldehyde is vitally important for the development of core mechanism. However, experimental data about the low-temperature oxidation of acetaldehyde are limited. The major aim of this paper is to quantify several important intermediates produced during the low-temperature oxidation of acetaldehyde in a jet stirred reactor (JSR) for the first time employing a synchrotron vacuum ultra violet photoionization molecular beam mass spectrometer (SVUV-PI-MBMS). Under the guidance of the quantified information, directions for the future modeling development are pointed out.

2 Experimental details

The experiments were performed at the Chemical Dynamics Beamline of the Advanced Light Source (ALS) at the Lawrence Berkeley National Laboratory. Details of this setup can be found in ref [14, 15], thus no more introduction is given herein. Temperatures were measured by a thermocouple (K-type, Thermocoax) placed near the tip of quartz nozzle. Similarly to our previous work [15], a linear assumption, was adapted to correct the deviation caused by the thermocouple position and surroundings. The temperature uncertainty was estimated to be less than 5%.

The experiments were operated at 31 different temperature points under low temperature ranges varying from 528 K to 946 K (corrected temperatures) with fixed pressure of 700 Torr. The flow rates of acetaldehyde, oxygen and argon were separately controlled by calibrated mass controllers to get a fixed residence time (2.7 s) at different temperature. The equivalence ratio was kept at 0.5 with the dilution ratio of argon fixed at 85%. The mass spectra were collected at each temperature point under several selected photoionization energies. Besides, energy-dependent mass spectra, which are known as photoionization efficiency (PIE) curve, were obtained between 9.5 eV and 12.0 eV at 638 K (corrected temperature) to help with the species identification.

3 Modeling and theoretical description

AramcoMech 2.0 [16-22], which is a widely used mechanism and recently updated to improve its prediction ability at low-temperature range, was employed to reproduce our experimental results. Another mechanism developed by Veloo et al. [23] for the low-temperature oxidation of butyraldehydes and propioaldehyde was also adopted as a reference for the following analysis and discussion. These two mechanisms will be referred as M1 and M2 for short.

All simulations were performed with the Chemkin Pro Software [24] using the recorded flow rates and corresponding corrected temperature as input, without fixing the residence time. In addition, rates of production (ROP) and sensitivity of reactions were analyzed to help with the following discussion.

The adiabatic ionization energies (IEs) and photoionization cross sections were calculated through the methods which have been validated in previous work [14, 15]. The adiabatic IEs for $C_2H_4O_3$ isomers and $C_2H_3O_3$ isomers were calculated at the M06-2X/aug-cc-pVTZ level of theory. The uncertainties for calculated adiabatic IEs were evaluated to be 0.2 eV.

4 Results and discussion

4.1 Major species

Since these two mechanisms have been developed for decades and validated by numerous experimental results of different fuels [25-27], especially for hydrocarbons [28], it is not surprising if both of the adopted mechanisms can well reproduce the mole fractions of major species and hydrocarbons. The comparisons between the experimental results and the predictions of M1 and M2 for major species show that the values and variation trend of the mole fractions under relative high temperatures are well captured, but both mechanisms under-predict the reactivity of acetaldehyde oxidation at low temperature. To avoid cluttering, only the results of M1 are shown here in Fig.1. The experimental results of fuel, CO_2 and H_2O clearly indicate the existence of a negative temperature coefficient (NTC) region approximately between 650 and 850 K, while the mechanisms fail to predict this important characteristic, which suggests that the low-temperature reactivity for acetaldehyde oxidation need to be enhanced.

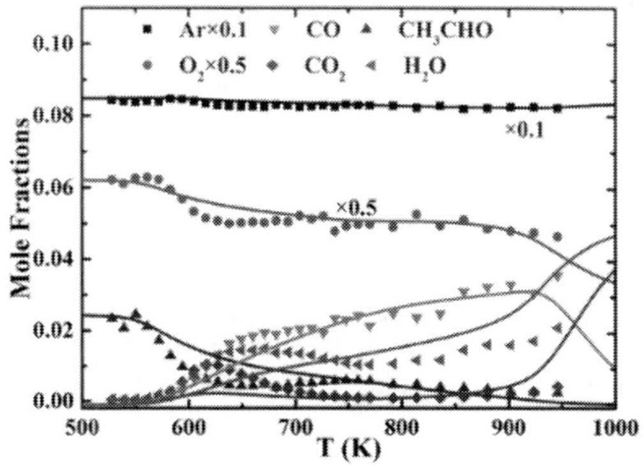

Figure 1: Experimental mole fraction profiles (symbols) of major species are compared with the modeling predictions of M1 (line). The experimental and modeling results of Ar and O_2 are divided by 10 and 2 respectively.

4.2 Fuel-related minor species

The first step of CH_3CHO oxidation is hydrogen abstractions by OH, CH_3CO_3 and CH_3O_2, which produces two different C_2H_3O isomers (CH_3CO and CH_2CHO). It will be helpful if the production ratio of these two isomers can be measured, because these two different structures will result in different oxygenated pathways. Unfortunately, the mole fractions of these two isomers, whose peaks are predicted as the order of 1~10 ppb by M1, are below the detection limit because of their high reactivity. However, our previous studies on the low-pressure premixed flames fueled by acetaldehyde [6] could give some valuable information for the validation of production ratio. Although the C-H bond energy on methyl functional group is higher than that on aldehyde group [29], typically the concentration of CH_2CHO radial will be higher than that of CH_3CO radicals.

The addition of molecular oxygen to the C_2H_3O isomers results in the formation of two $C_2H_3O_3$ isomers, then sequentially forms the other pair of $C_2H_3O_3$ isomers by internal H-abstraction or the two isomers of $C_2H_4O_3$ by various H-addition reactions. However, these four $C_2H_3O_3$ isomers cannot be quantified separately in the present work because that only an IE of 10.65 eV was observed in the PIE results, which doesn't match any calculated values. The vertical IEs were calculated here considering the zero point energies. The calculated IE for HO_2CH_2CO is 8.91 eV, while the results for CH_3CO_3, CH_2CO_3H and O_2CH_2CHO are similar (10.98 eV, 10.59 eV, and 10.72 eV separately). The IEs for these three species are close to the measured IE (10.60 eV), which makes it difficult to separate these four isomers.

The IE of HO_2CH_2CHO was calculated as 9.83 eV, while the IE was determined as 10.24 eV for CH_3CO_3H. In the PIE curve of $C_2H_4O_3$ isomers, only CH_3CO_3H could be observed in our experiments with the observed IE around 10.30 eV. The HO_2CH_2CHO's concentration is below the detection limit, which is consistent with the modeling results. Therefore, the mole fraction profiles of four $C_2H_3O_3$ isomers are derived as a group without separation, while the $C_2H_4O_3$ signals are assigned as CH_3CO_3H and quantified with calculated PICS.

In M1, separate descriptions are given for those two peroxided isomers discussed above except for CH_2CO_3H. The internal H-abstractions on CH_3CO_3 are ignored in M1. Although M1 contains HO_2CH_2CO and HO_2CH_2CHO, their predicted concentration are pretty low (< 0.01 ppm). In addition, none of the reactions containing HO_2CH_2CHO is directly connected with O_2CH_2CHO, which means that the HO_2CH_2CHO in M1 is irrelevant with the peroxide pathways. Therefore, the results for HO_2CH_2CO and HO_2CH_2CHO are not shown here. As depicted in Fig. 2 (a) and (b), the predicted concentration of O_2CH_2CHO in M1 is extremely lower than CH_3CO_3. Taking account of the high uncertainties of the experimental mole fractions, M1 slightly over predicts the total mole fraction of $C_2H_3O_3$ isomers, but the prediction for CH_3CO_3H is pretty higher than the experimental result. M2 only considers CH_3CO_3 and CH_3CO_3H, ignoring their counterparts O_2CH_2CHO and HO_2CH_2CHO. The predictions of CH_3CO_3 and CH_3CO_3H by M2 are unsatisfying in respect of the appearance temperature, but their peak concentrations are surprisingly closer to the experimental results than M1.

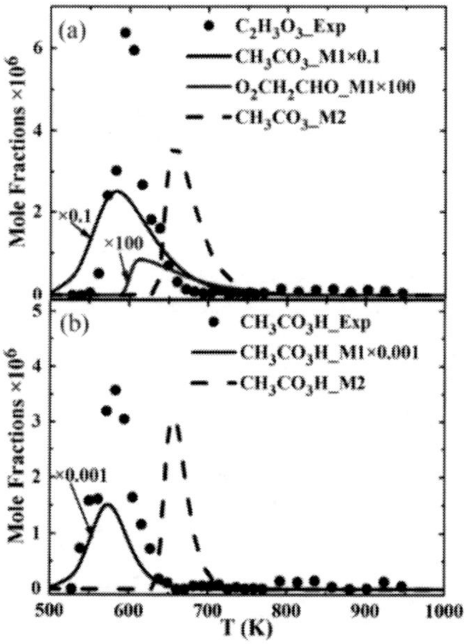

Figure 2: Experimental mole fraction profiles (symbols) of C2 oxygenated species are compared with the modeling predictions of M1 (line) and M2 (dash line). Predictions of CH_3CO_3 in both mechanisms as well as O_2CH_2CHO and CH_3CO_3H in M2 are scaled by different factors.

The sensitivity analysis of these two mechanisms gives a probable explanation for the divergence in the predictions of these two mechanisms. Reaction R1 and R2 are marked as the most two sensitive reactions for CH_3CO_3 and CH_3CO_3H consumption by the sensitivity analysis of M1. It appears that R1 has no direct relation with the formation or consumption of CH_3CO_3, however, it is actually a competitive reaction to reaction R2, which produces almost all CH_3CO_3. These two reactions are fighting for the supply of CH_3CO, which makes R1 sensitive for the consumption, instead of formation, of CH3CO3 and CH_3CO_3H. Enhancing the reactivity of R2 or weakening the reactivity of R1 will increase the concentrations of CH_3CO_3 and CH_3CO_3H. Tracing back to the mechanism, we can find that the reaction parameters of R1 were obtained via theoretical calculation [49], while those of R2 were only estimated without any specific validation, which may require further investigation.

$$CH_3CO(+M) = CH_3 + CO(+M) \quad \text{(R1)}$$

$$CH_3CO + O_2 = CH_3CO_3 \quad \text{(R2)}$$

There are some other possible reasons for the over-predictions of CH_3CO_3 and CH_3CO_3H in M1. According to the ROP analysis results, CH_3CO_3 will 100% transfer to CH_3CO_3H via the hydrogen addition in M1, while there is another pathway (R3) for the consumption of CH_3CO_3 in M2.

$$CH_3CO_3 \Rightarrow CH_2CO + HO_2 \quad \text{(R3)}$$

This reaction consumes large part of CH_3CO_3 in M2, which could explain why the peak concentrations for CH_3CO_3 as well as CH_3CO_3H are lower than those in M1 and closer to experimental results. However, this vital reaction is not even included in M1, which indicates that more consumption pathways for CH_3CO_3 like R3 need to be included to optimize M1's prediction ability.

4.3 Other oxygenated intermediates

The observation of other oxygenates, such as methyl peroxy (CH_3O_2), acetic acid (CH_3COOH) and so on, are also helpful for the validation and development of mechanisms at the low-temperature range. As shown in Fig. 3, CH_3O_2 can be well predicted by both mechanisms, while the CH_3COOH sub-mechanism need to be included in the mechanism.

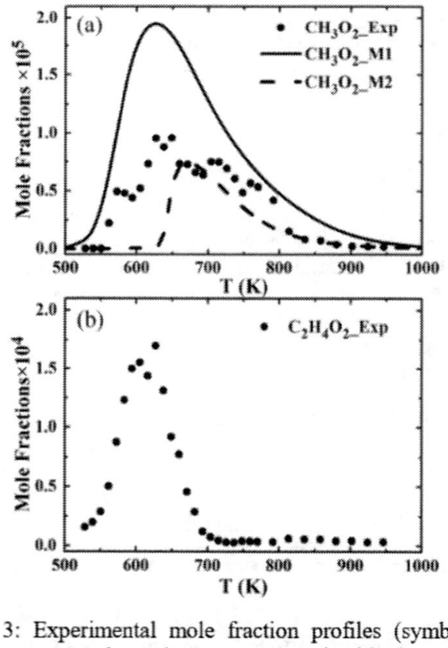

Figure 3: Experimental mole fraction profiles (symbols) of several oxygenated species are compared with the modeling predictions of M1 (line) and M2 (dash line).

Further analysis will be conducted based on the experimental results, and an updated mechanism will be developed to give a better explanation on the NTC behavior of acetaldehyde low-temperature oxidation.

5 Conclusions

The low-temperature oxidation of acetaldehyde was experimentally investigated for the first time by employing a jet stirred reactor coupled with a molecular beam mass spectrometer using synchrotron generated vacuum ultra-violet photons as ionization source. The mole fractions of 30 species, including 6 major species and 24 intermediates, were measured as functions of temperature at 700 Torr with fixed residence time around 2.7 s. The present study fill a gap in the experimental research exploring the low-temperature oxidation of acetaldehyde.

Two mechanisms are adopted to reproduce the experimental results and compared with each other. The mole fraction profiles for small hydrocarbons, CH_3O_2 and CH_3O_2H are well captured by AramcoMech 2.0, but the predictions for many other intermediates like $C_2H_3O_3$ and $C_2H_4O_3$ isomers are unsatisfactory, which indicates that fuel-related low-temperature pathways need to be optimized. With the help of rate of production analysis and sensitivity analysis, some guidance for further improvement on AramcoMech 2.0 are proposed. The rate coefficient parameters of the reactions discussed above, such as $CH_3CO+O_2 = CH_3CO_3$, need further investigation via kinetic experiments or theoretical calculations. More pathways for some intermediates like CH_3CO_3 and CH_3COOH should be included.

6 Acknowledgment

We really appreciate the productive discussion with Dr. Philippe Dagaut, as well as the expert technical assistance from Paul Fugazzi. This work is supported by the National Natural Science Foundation of China (No. 91541113). Tao want to thank for the support from China Scholarship Council (Grant No. 201506210349). This research used resources of the Advanced Light Source, supported by the Director, Office of Science, Office of Basic Energy Sciences, of the U.S. Department of Energy under Contract No. DEAC02-05CH11231. Sandia is a multi-mission laboratory operated by Sandia Corporation, a Lockheed Martin Company, for the National Nuclear Security Administration under contract DE-AC04-94-AL85000.

References

[1] S.M. Sarathy, P. Oßwald, N. Hansen, K. Kohse-Höinghaus, Prog. Energ. Combust. Sci. 44 (4) (2014) 40-102.

[2] N. Leplat, P. Dagaut, C. Togbé, J. Vandooren, Combust. Flame 158 (4) (2011) 705-725.

[3] J. Cavanagh, R.A. Cox, G.R. Olson, Combust. Flame 82 (1) (1990) 15-39.

[4] N. Leplat, J. Vandooren, Combust. Sci. Technol. 182 (4-6) (2010) 436–448.

[5] M. Christensen, M.T. Abebe, E.J.K. Nilsson, A.A. Konnov, Proc. Combust. Inst. 35 (1) (2015) 499-506.

[6] T. Tao, W. Sun, B. Yang, N. Hansen, K. Moshammer, C.K. Law, Proc. Combust. Inst. 36 (1) (2017) 1287-1294.

[7] R.D. Kern, H.J. Singh, K. Xie, Aip Conference Proceedings 208 (1) (1990) 487-492.

[8] S.J. Won, J.C. Ryu, J.H. Bae, Y.D. Kim, J.G. Kang, B. Korean Chem. Soc. 21 (5) (2000) 487-492.

[9] K. Yasunaga, S. Kubo, H. Hoshikawa, T. Kamesawa, Y. Hidaka, Int. J. Chem. Kinet. 40 (2) (2008) 73-102.

[10] P. Dagaut, M. Reuillon, D. Voisin, M. Cathonnet, M. McGuinness, J.M. Simmie, Combust. Sci. Technol. 107 (4-6) (1995) 301-316.

[11] P. Gray, J.F. Griffiths, S.M. Hasko, P.G. Lignola, Proc. R. Soc. Lond. A 374 (1758) (1981) 313-339.

[12] E.W. Kaiser, C.K. Westbrook, W.J. Pitz, Int. J. Chem. Kinet. 18 (6) (1986) 655-688.

[13] G. Barari, B. Koroglu, A.E. Masunov, V. Subith, J. Energy Resour. Technol. 139 (1) (2017) 012203.

[14] K. Moshammer, A.W. Jasper, D.M. Popolan-Vaida, A. Lucassen, P. Dievart, H. Selim, A.J. Eskola, C.A. Taatjes, S.R. Leone, S.M. Sarathy, Y. Ju, P. Dagaut, K. Kohse-Höinghaus, N. Hansen, J. Phys. Chem. A 119 (28) (2015) 7361--7374.

[15] K. Moshammer, A.W. Jasper, D.M. Popolan, Z. Wang, L. Ruwe, C.A. Taatjes, P. Dagaut, N. Hansen, J. Phys. Chem. A 120 (40) (2016) 7890-7901.

[16] C. Zhou, Y. Li, E. O'Connor, K.P. Somers, S. Thion, C. Keesee, O. Mathieu, E.L. Petersen, T.A. DeVerter, M.A. Oehlschlaeger, G. Kukkadapu, C. Sung, M. Alrefae, F. Khaled, A. Farooq, P. Dirrenberger, P. Glaude, F. Battin-Leclerc, J. Santner, Y. Ju, T. Held, F.M. Haas, F.L. Dryer, H.J. Curran, Combust. Flame 167 (3) (2016) 353-379.

[17] Y. Li, C. Zhou, K.P. Somers, K. Zhang, H.J. Curran, Proc. Combust. Inst. 36 (1) (2017) 403-411.

[18] U. Burke, W.K. Metcalfe, S.M. Burke, K.A. Heufer, P. Dagaut, H.J. Curran, Combust. Flame 165 (5) (2016) 125-136.

[19] S.M. Burke, U. Burke, R. Mc Donagh, O. Mathieu, I. Osorio, C. Keesee, A. Morones, E.L. Petersen, W. Wang, T.A. DeVerter, M.A. Oehlschlaeger, B. Rhodes, R.K. Hanson, D.F. Davidson, B.W. Weber, C. Sung, J. Santner, Y. Ju, F.M. Haas, F.L. Dryer, E.N. Volkov, E.J.K. Nilsson, A.A. Konnov, M. Alrefae, F. Khaled, A. Farooq, P. Dirrenberger, P. Glaude, F. Battin-Leclerc, H.J. Curran, Combust. Flame 162 (2) (2015) 296-314.

[20] S.M. Burke, W. Metcalfe, O. Herbinet, F. Battin-Leclerc, F.M. Haas, J. Santner, F.L. Dryer, H.J. Curran, Combust. Flame 161 (11) (2014) 2765-2784.

[21] W.K. Metcalfe, S.M. Burke, S.S. Ahmed, H.J. Curran, Int. J. Chem. Kinet. 45 (10) (2013) 638-675.

[22] A. Kéromnès, W.K. Metcalfe, K.A. Heufer, N. Donohoe, A.K. Das, C. Sung, J. Herzler, C. Naumann, P. Griebel, O. Mathieu, M.C. Krejci, E.L. Petersen, W.J. Pitz, H.J. Curran, Combust. Flame 160 (6) (2013) 995-1011.

[23] P.S. Veloo, P. Dagaut, C. Togbé, G. Dayma, S.M. Sarathy, C.K. Westbrook, F.N. Egolfopoulos, Combust. Flame 160 (9) (2013) 1609-1626.

[24] R.J. Kee, F.M. Rupley, J.A. Miller, M. Coltrin, J. Grcar, E. Meeks, H. Moffat, A. Lutz, G. Dixon-Lewis, M.D. Smooke, Reaction Design Inc., San Diego, CA (2008).

[25] P. Dagaut, C. Togbe, Energ. Fuel. 22 (5) (2008) 3499-3505.

[26] P. Dagaut, S.M. Sarathy, M.J. Thomson, Proc. Combust. Inst. 32 (1) (2009) 229-237.

[27] C. Togbe, P. Dagaut, A. Mzé-Ahmed, P. Dievart, F. Halter, F. Foucher, Energ. Fuel. 24 (11) (2010) 5859-5875.

[28] P. Dagaut, Phys. Chem. Chem. Phys. 4 (11) (2002) 2079-2094.

[29] J. Berkowitz, G.B. Ellison, D. Gutman, J. Phys. Chem. 98 (11) (1994) 2744-2765.

11th Asia-Pacific Conference on Combustion,
The University of Sydney, NSW Australia
10th -14th December 2017.

Effect of Miller Cycle and Fuel Injection Strategy on Performance of Marine Diesel Engine

Xiuxiu Sun[1,3], Xingyu Liang[1], Peilin Zhou[3], Yuehua Qian[2], Teng liu[2], Bo Liu[2]

[1]Affiliation

1. State key Laboratory of Engines, Tianjin University, Tianjin 300072, China

[2]Affiliation

2. China Shipbuilding Power Engineering Institute Co. Ltd, Shanghai 200120, China

[3]Affiliation

3. Dept. of Naval Architecture, Ocean &Marine Engineering, University of Strathclyde, Glasgow G4 0LZ, UK

Abstract

Computational fluid dynamics (CFD) is used to investigate the performance of a large two-stroke marine diesel engine. The simulated model is validated with experimental data. The in-cylinder pressure of the simulated model is in agreement with the experimental data. The errors of NO_x and CO_2 emissions are also within the accepted range. The effect of Miller cycle, injection sequence and pilot injection on combustion and emissions are investigated using this model. The results show that the in-cylinder pressure decreases with deeper Miller cycle level. However, NO_x emissions are reduced only slightly to 8.95 g/kWh. This decrease in NO_x emissions does not satisfy the requirements of Tier III. We also found that the injection interval angle between two injectors decreases the combustion pressure. However, the indicated specific fuel consumption is 7.3 g/kWh higher than the base value, when the injection interval angle is 8 °CA. Appropriate pilot injection strategy can decrease NO_x emissions and indicated specific fuel consumption, such as P10I5. However, NO_x emissions are not reduced sufficiently to meet the requirements of Tier III.

1 Introduction

Environmental protection has recently attracted increasing attention. Marine diesel engine emissions are strictly limited. The International Maritime Organization Tier III limits NO_x emissions to 3.4 g/kWh in the emission control areas, which is a 76% reduction compared with the 14.4 g/kWh stipulated in Tier II [1]. Because Tier III [2] is stricter against NO_x emissions, more measures are required in marine diesel engines to reduce these emissions. Various measures can be employed, such as exhaust gas recirculation (EGR), selective catalytic reduction (SCR), improved injection strategy and the use of the Miller cycle.

[1] Corresponding author, Tel/fax:+86 22 2789 1285
E-mail address: lxy@tju.edu.cn (XY. Liang)

Numerical simulation technology is widely used because it is efficient in terms of human and financial resources. Kontoulis et al. [3] used the KIVA-3 code as a modeling platform to study the effects of advanced injection strategies. By adding an appropriately timed pilot injection, they achieved fuel savings of approximately 1.7%, without increasing NO_x emissions. Verschaeren et al. [4] investigated NO_x emissions in a medium-speed heavy-duty diesel engine using exhaust-gas recirculation and Miller timing. An injection advance during low-load operation raised combustion pressures and shortened combustion duration, lowering CO and PM emissions and increasing efficiency. Chen et al. [5] studied pilot injection for improvement of combustion characteristics in a heavy-duty diesel engine. Their results demonstrated that with the advance of pilot injection timing, the peak in-cylinder pressure is lowered, the ignition delay of the main combustion is shortened, and NO_x and soot emissions are reduced. Injection strategy has large effect on marine engine performance and emissions. Thus, it is necessary to investigate different methods for reducing NO_x emissions.

This paper analyzes the effect of Miller cycle, injection sequence and pilot injection on marine diesel engine performance and emissions. The computational fluid dynamics (CFD) model is validated with experimental data. The in-cylinder pressure and emissions are compared for different measures. The potential to reduce NO_x emissions is analyzed. This paper provides guidance for researchers in choosing appropriate methods for reducing NO_x emissions.

2 Numerical Model

2.1 Marine engine CFD model

The present CFD simulation is conducted on a model of the 6S35ME-B9 marine diesel engine. The structural parameters can found in Table 1. The CFD model was constructed using CONVERGE 2.3 software.

cylinder number	6
Bore (mm)	350
Stroke (mm)	1550
Displacement (L)	149
Connecting rod length (mm)	1550
Speed (r/min)	142
Power (kWh)	3575

Table1: 6S35ME-B9 test engine specifications

Figure 1 shows the three-dimensional structure of model. This model includes a scavenge box, cylinder and exhaust port. The k-ε two-equation mode is used as turbulence model. Kelvin Helmholtz-Rayleigh Taylor model is applied to the simulation of spray breakup. The NTC collision model is implemented to simulate the fuel collision. The wall splash model is used in the O'Rourke model based on Weber number, film thickness and viscosity. The detailed chemical solver combustion model is used to simulate the fuel ignition. For the emissions generate models, the Extend Zeldovich NO_x model is applied to simulate the production of NO_x. Hiroyasu-NSC Model is used to simulate soot production respectively. Those models are chosen in CONVERGE. The injected liquid temperature is set to 345 K. The injection duration is 15.36°CA and the total injection mass is 0.0126 kg, which was obtained from experimental data.

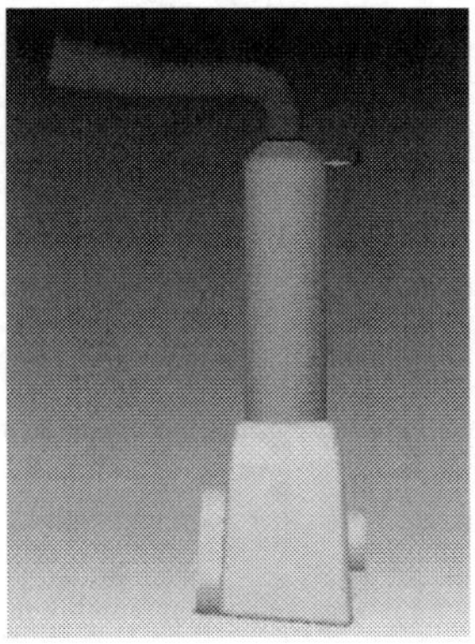

Figure 1:The three-dimensional geometric structure

The CFD model is based upon continuity, momentum and energy conservation laws [6]. Mesh grid size is decided based on the desired calculation time and calculation accuracy. Grid control includes a base grid, adaptive mesh refinement and fixed embedding. The effect of the base grid and the adaptive mesh refinement are shown in Fig. 2. As seen in Fig. 2 (a), the compression pressure of base grid 0.04 m is in agreement with that of base grid 0.03 m. The calculation accuracy increases gradually with decreasing base grid size. If the adaptive mesh refinement is set to 4, the calculated pressure is in good agreement with the experiment data, as shown in Fig. 2 (b). However, the calculation time also increases. Setting the base grid to 0.04 m and the adaptive mesh refinement to 4 can ensure the accuracy of the simulation.

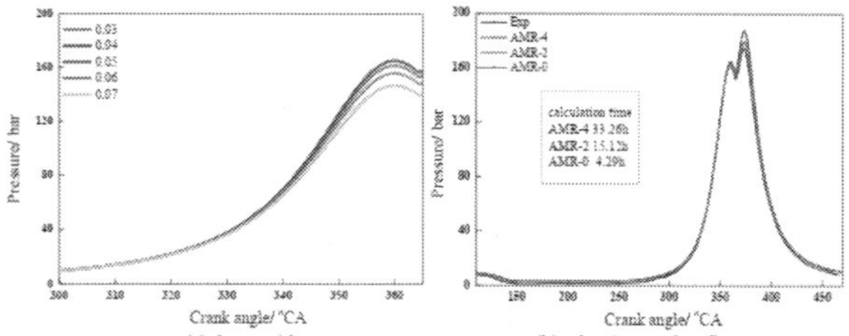

Figure 2: The effect of grid on the pressure: (a) base grid (b) adaptive mesh refinement

The CFD model will be used in the simulation. The simulated value was validated with the experimental data.

2.2 Model Validation

The in-cylinder pressure and emissions were tested when the engine speed was 142 r/min and the operating load was 100%. The calculated value was validated with the experimental data. Stratsianis et al. [7] studied the effects of post-injection on marine diesel engine performance. They also validated simulated values with experimental data under 100% load.

Figure 3 shows the comparison of in-cylinder pressure between the simulation and the experiment. The errors of maximum pressure during compress and combustion process are 0.8% and 1.5%, respectively. The delay in pressure change is 0.846°CA for the simulation model. The in-cylinder pressure is in good agreement with the experimental data. The model accurately predicts the pressure in the marine diesel engine.

The NO_x and CO_2 emissions are also compared with experimental data, to validate the accuracy of model, as shown in Figure 4. For NO_x emissions, the error of simulation pressure lower 2.8% than the experimental data, which is acceptable for predicting NO_x emissions. The error for CO_2 emissions is larger than that for NO_x emissions. The single component n-tetradecane is used as an alternative fuel to simulate the combustion process. The new n-tetradecane mechanism is used to decrease the calculation time. Some reactions are removed for the CO_2 production mechanism. NO_x emissions are close to the experimental data; that is, the CFD model can accurately predict NO_x emissions using n-tetradecane as an alternative fuel.

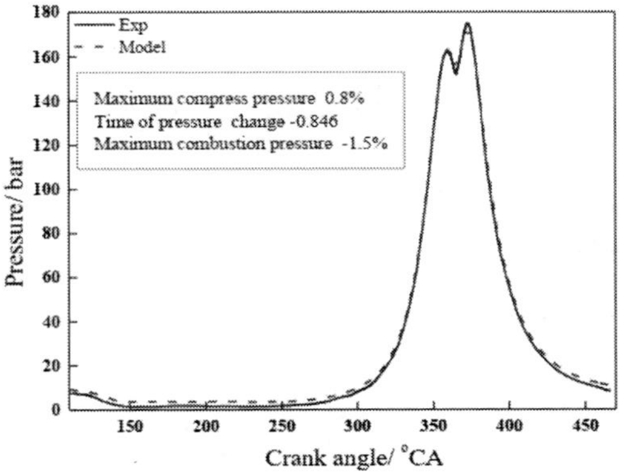

Figure 3: Comparison of in-cylinder pressure for the simulation and experiment

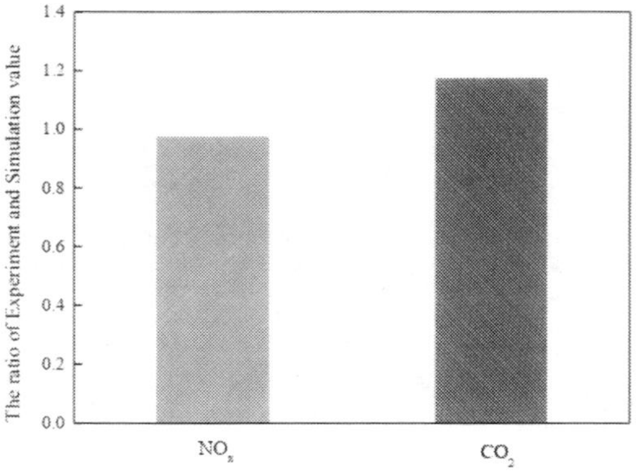

Figure 4: Comparison of NO_x and CO_2 for the simulation and experiment

3 Results and Discussion

3.1 Effect of Miller Cycle

Decreased inlet gas temperature lowers NO_x emissions. The Miller cycle lowers the in-cylinder gas temperature. This result can be realized by delaying the close time of the exhaust valve for two-stroke marine diesel engines

Different Miller cycle patterns were simulated to investigate their effects on marine diesel engine performance and emissions. The exhaust valve close time delay 5°CA, 10°CA, 15°CA and 20°CA compared with the base tine, which were named as M5, M10, M15 and M20. The scavenging pressure was improved to offset the loss of fresh charge. Figure 5 shows the changes of intake pressure and air-fuel ratio under different Miller cycles. The intake pressure increases with increased exhaust valve close time delay, which makes the equivalence ratio almost the same for the Miller cycle.

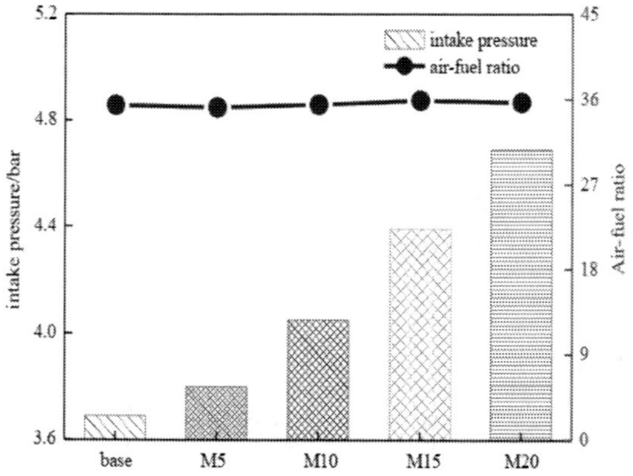

Figure 5: The changes of intake pressure and air-fuel ratio using different Miller cycle

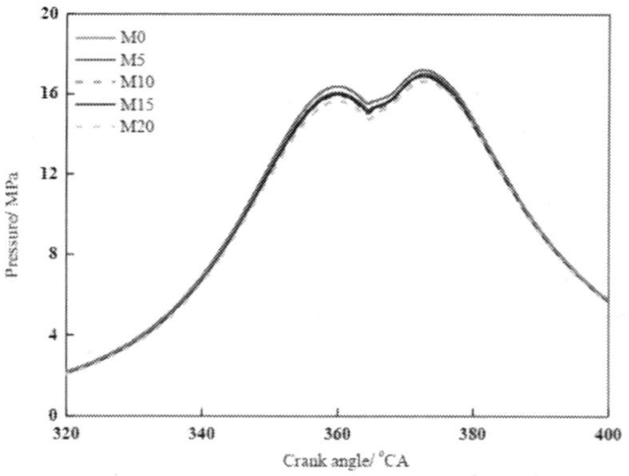

Figure 6: The changes of in-cylinder pressure using different Miller cycle

Figure 6 shows changes of the in-cylinder pressure with crank angle using different Miller cycles. It can be seen that the in-cylinder pressure decreases with increased exhaust valve close time delay. The higher intake pressure increases the in-cylinder pressure to higher than that of the original engine near the bottom dead center. Because, the equivalence ratios of different Miller cycles are almost same.

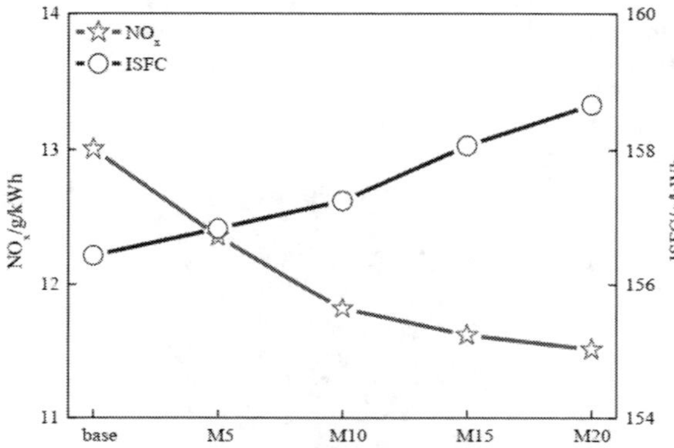

Figure 7: The changes of NO_x and ISFC using different Miller cycle

The changes in NO_x emissions and indicated specific fuel consumption (ISFC) are shown in Figure 7. The in-cylinder temperature decreases with increased exhaust valve close time delay. The fresh charge reduces residual gas temperature. Greater NO_x emissions are produced at higher temperatures and under oxygen-rich conditions. The range of high temperatures narrows gradually along with the increased exhaust valve close time delays. The NO_x emissions are 11.5% lower than those of the original engine. However, the achieved emissions still do not meet the requirements of Tier III. The ISFC increases with increased exhaust valve close time delay. The pressure decreases during the compression and power strokes (Figure 6), reducing the indicated power. However, the injection fuel mass remains constant. Together, these increase the ISFC.

3.2 Effect of injection sequence

The effects of different injection sequences on marine diesel engine performance were also analyzed. Two injectors were distributed uniformly on the cylinder head. The two injectors injected simultaneously in the original engine. Different interval angles between two injectors were investigated. The interval angle between the two injectors was set to 2°CA, 4°CA,

6°CA and 8°CA.

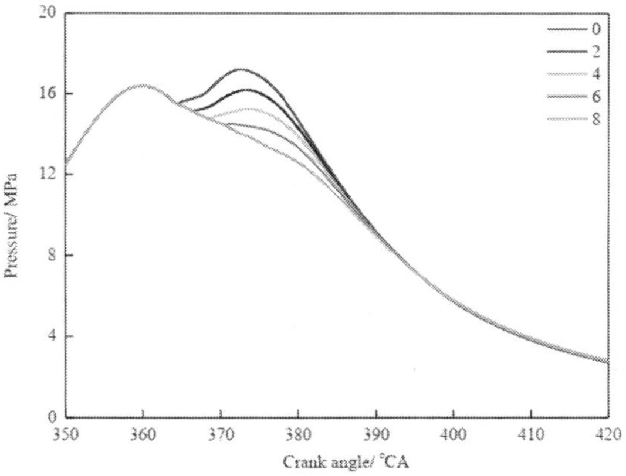

Figure 8: The changes of in-cylinder pressure using different injection sequence

The variations of in-cylinder pressure are shown in Figure 8 for the different injection sequences. It is obvious that the maximum pressure decreases with increasing interval angle between the two injectors during power stroke. The increasing pressure is caused by fuel combustion after top dead center. The maximum pressure is not rise during power when the injection interval angle is set to 8°CA. because, less fuel is combusted after top dead center.

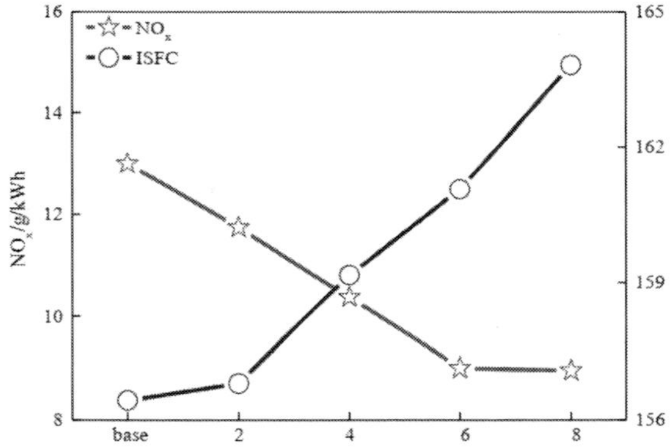

Figure 9: The changes of NO_x and ISFC using different injection sequence

Figure 9 shows the variations of NO_x emissions and ISFC using different injection sequences. NO_x emissions tend to decrease with increasing interval angle between the two injections. NO_x emissions are reduced by 4.05 g/kWh when the interval angle is increased to 8°CA. More fuel underwent incomplete combustion as interval angle was increased. The changes in in-cylinder pressure can explain this. The lower temperature reduces NO_x emissions. The injection mass is constant. The ISFC is inversely proportional to the power. The lower combustion pressure also decreases the power. Thus, ISFC increases with increasing interval angle.

In conclusion, increasing the interval angle between two injections reduces NO_x emissions. Changing injection sequence is more effective than changing the Miller cycle in reducing NO_x emissions; however, the resulting increase in ISFC is also large. Thus, using injection sequence is not advised for reducing NO_x emissions.

3.3 Effect of pilot injection

Pilot injection is widely used to reduce the NO_x emissions. Bae et al. [8] studied the effect of pilot injection on combustion and emission characteristics in a marine diesel engine. Their results showed that pilot injection timing affects auto ignition and heat release rate. Huang et al. [9] also invested the effect of pilot injection on low-temperature combustion in a diesel engine under a medium exhaust-gas recirculation rate. Decreasing the interval between the pilot and the main injection (pilot–main injection interval) can greatly reduce NO_x emissions and ISFC. Thus, pilot injection quantity and pilot–main injection interval were investigated in the present study for reducing NO_x emissions. Table 2 shows the investigated conditions. The pilot injection quantity is 5% and the pilot–main injection interval is 5°CA. This condition will be named P5I5. This naming pattern is also used for other conditions.

Pilot injection quantity	Interval time (°CA)		
5 %	5	10	15
10 %	5	10	15
15 %	5	10	15

Table2: the pilot injection quantity and pilot interval

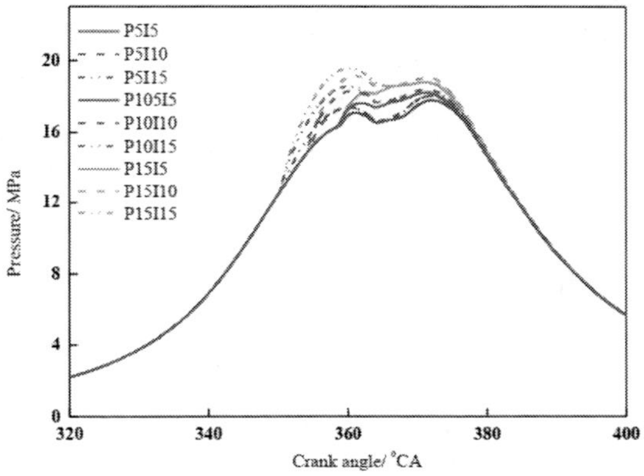

Figure 10: The changes of in-cylinder pressure using pilot injection strategy

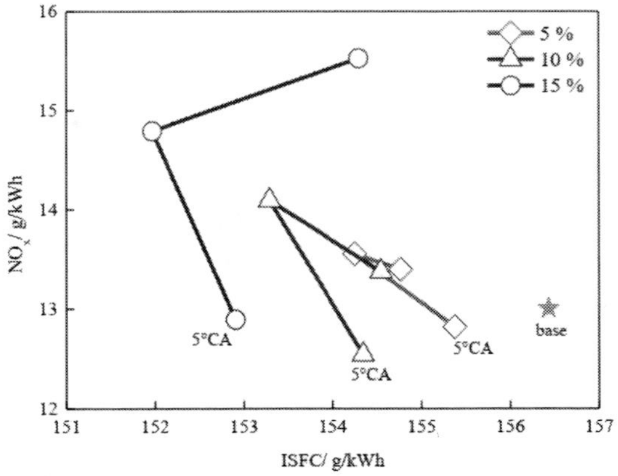

Figure 11: The changes of NO_x and ISFC using pilot injection strategy

Comparisons of in-cylinder pressure are shown in Figure 10. It can be seen that in-cylinder pressure increases with increasing pilot injection quantity and pilot–main injection interval. The fuel was combusted during the compression stroke, which increased the in-cylinder pressure. Owing to the lower quantity of fuel in the power stroke, the in-cylinder pressure remains constant for different pilot injection quantities and pilot–main injection intervals.

Figure 11 shows the relationships between NO_x emissions and ISFC using the pilot injection strategy. It can be shown that ISFC decreases with increasing pilot–main injection interval. The ISFC is lowest when pilot–main injection interval is 15°CA. The P10I5 condition has minimum NO_x emissions, but the corresponding ISFC is not minimum. However, the simulated value of NO_x emissions and ISFC are lower than the base value for the P10I5 condition. So, P10I5 is an optimum condition when considering NO_x emissions and ISFC, which are both lower than the base values.

In conclusion, pilot injection can be used to reduce NO_x emissions for marine diesel engines. However, the reduced of NO_x emissions and ISFC are lower than those using the Miller cycle or different injection sequences. That is, pilot injection needs to be used in conjunction with other strategies to reduce NO_x emissions to meet the requirements of Tier III.

4 Conclusions

A CFD model of a marine diesel engine was established for the present paper. The model was validated with experimental data. Then, the effects of Miller cycle, injection sequence and pilot injection were investigated on the combustion and emissions of a marine diesel engine. The main results are as follows:

1. In-cylinder pressure and emissions calculated by CFD model were in agreement with the experimental data.
2. In-cylinder pressure decreases with increasing exhaust valve close time delay. NO_x emissions also decrease, but the decrease in NO_x emissions is insufficient to meet the requirements of Tier III.
3. In-cylinder pressure and NO_x emissions decrease with increasing injection interval angle between two injectors. NO_x emissions decrease to 8.95 g/kWh when the injection interval angle set to 8°CA.
4. The P10I5 condition achieved the minimum value of NO_x emissions. However, pilot injection must be used with other strategies to meet the requirements of Tier III.

5 Acknowledgment

This work was supported by National Natural Science Foundation of China (No. 51376136 and No. 91641111) and National Sci-Tech Support of China (2016YFC0205304). The author also wants to appreciate the support of high-tech Ship Research Program of MIIT.

References.

- http://www.DieselNet.com/standards, Emission Standards, IMO Marine Engine, 13.07.2017(2017).
- Herdzik, J., Emissions from marine engines versus IMO certification and requirements of tier 3. Journal of KONES Powertrain and Transport, 18(2011)
- C.C.L.K. P. Kontoulis, The international symposium on diagnostics and modeling of combustion in intern al combustion engines 1(2008).
- R. Verschaeren, W. Schaepdryver, T. Serruys, M. Bas tiaen, L. Vervaeke, S. Verhelst, Energy 76(2014) 614-621.
- L. Chen, H. Zeng, X.B. Cheng, Applied Mechanics a nd Materials, 651-653(2014)866-874.
- A.A. Amsden, P.J. Orourke, T.D. Butler, Nasa Sti/rec on Technical Report N, 89(1989)
- Stratsianis, V., P. Kontoulis and L. Kaiktsis, Effects of Fuel Post-Injection on the Performance and Pollut ant Emissions of a Large Marine Engine. 2(2016).
- M.Bae1.D SikHan.Y J. Chang.J H Song.C H Jeon,,Jo urnal of the Korean Society of Marine Engineering,(2 010)37-45.
- HZ Huang, QX Wang, C Shi, QS Liu, CZ Zhou, Ap plied Energy, 179 (2016) 1194–1208

JOURNAL TITLE

11th Asia-Pacific Conference on Combustion

MANUSCRIPT TITLE

Effect of Miller Cycle and Fuel Injection Strategy on Performance of Marine Diesel Engine

Dear editor, reviewer

Thank you very much for your useful comments and suggestions on our manuscript. We have modified the manuscript accordingly, and the detailed corrections are listed below point by point:

Response to the editor and reviewers

- ✓ The whole manuscript has been revised carefully. The revisions and corrections have been highlighted in red color according to your requirement.
- ✓ Besides, the language in manuscript has been revised by professional.

Response to reviewer

Reviewer #1:

I have tried to read this manuscript a couple of times, and unfortunately, because of the poor English and poor sentence structure the meaning of many of the sentences is lost to me. This unfortunately is making it very difficult to provide a sufficient judgement on the quality of the paper. For now, I am recommending it be revised, as with a very significant editing of the manuscript it is possible it can ultimately be at an acceptable level, otherwise if the authors do not make significant changes I recommend it is rejected.

- ✓ Dear reviewer, we had revised the manuscript carefully. The manuscript has also been proofread by English speaker. Thank you very much.

I would recommend that the authors have the paper very thoroughly proofread by an English speaker. Generally, the paper is poorly structured which makes it very difficult to go from one section to the next. The knowledge gap and new contribution is also very unclear. There are also insufficient references (and they should be numbered, not listed as bullet points).

- ✓ Dear reviewer, we had revised the manuscript carefully. Three methods have been researched to reduce the emissions of marine diesel engine in this paper. The in-cylinder pressure and emissions of marine diesel engine have been analyzing in every section. This provides direction for researchers to reduce the NOx meeting the requirement of Tier III.

- ✓ We also added the number of reference to descript the significant of this paper. However, the references have been listed using bullet point. Because, this is requirement of the ASPACC2017.
- ✓ Thank you for your advices.

Examples of poor sections of the paper:

1. In the abstract it is stated "…the injection interval angles augment of two injectors makes the combustion terrible". This is very far from technical and accurate writing. What exactly does it mean if "combustion is terrible"? Reference to "terrible combustion" is also present in section 3.2, i.e. "The increment of interval angles causes the combustion terrible".

- ✓ Dear reviewer, the sentence has been advised. The unreasonable expressions have also been revised in manuscript. Thank you very much.

2. Much of section 3.3 is unfortunately difficult to follow. Particularly after Fig. 8, and this is again due to poor English. Sentences like "the in-cylinder compresses pressure increase" do not make sense, and this is only one example, many others throughout the manuscript, thereby making it too difficult for the reader to comprehend the technical value of sections of the paper.

- ✓ Dear reviewer, we added the figure of in-cylinder variations in manuscript. The Figure. 10 shows the variations of in-cylinder pressure using pilot injection strategy. It can be seen that the in-cylinder pressure increases with the pilot injection quantity augment.
- ✓ In this paper, the relationship of NOx and indicated power is key point. The effect of technology methods has been investigated on the emission and power. The potential of these methods has also been analyze in the end of every sections. So, this paper provides direction for researchers to reduce the emission of marine diesel engine.
- ✓ Thank you for your advices.

3. Additionally, In section 2.2 the authors state that the errors of "maximum compress pressure and combustion are 0.8% and 1.5%, respectively". What is the "error in combustion", this is extremely vague, but this may just be poor sentence structure? I cant be sure. I believe the authors mean they are comparing maximum in-cylinder pressure during combustion???

- ✓ Dear reviewer, we had revised the unreasonable expression. In section 2.2, the means of this sentence is that the errors of maximum in-cylinder pressure are 0.8% and 1.5% during the compress and power stroke.
- ✓ Thank you very much.

Other issues:

1. It is understood that there is little space in 4 pages to elaborate on a CFD model, but the authors have presented almost nothing, including no references to the modeling approaches,

which is not acceptable if they are not going to present their methodology here.

- ✓ Dear reviewer. We had added the reference of model approaches. We also added the structure of the CFD model (Fig.1). Thank you very much.

2. There is no information on the mesh resolution, how it was chosen, what the timestep is…The only information provided is that it is a "kappa-epsilon" two equation model. This in my opinion, is not to the standard required. Another passing statement on how soot production is modeled is also provided, but given that the entire paper is based on CFD, I cant possibly recommend acceptance without further information in this section.

- ✓ Dear reviewer, we had added the choose of mesh resolution. The effect of grid on the pressure were shown in Fig.2, which includes the base grid and adaptive mesh refinement. The detailed of mesh resolution has also added in manuscript, which can be found on Page 2.
- ✓ Thank you for your advice.

3. That being said, the authors do present some validation with experimental data later, which seems to be reasonable at least for the data presented however this is only done for one specific engine condition. I am not sure what the justification for this is. Is this a common testing condition? This must be explicitly defined here. Can we expect other conditions to yield similar comparisons to CFD, why or why not? With the lack of references as well, these questions remain unanswered.

- ✓ Dear reviewer, the in-cylinder pressure and emissions were validated with the experimental data when the load of marine diesel engine is 100 %. The errors of in-cylinder pressure and emission are in accept range. So, the CFD model can be used at the 100 % load. The researches of technology methods were based on the CFD model in manuscript.
- ✓ The CFD model needs to validate for other conditions. Because, many parameters are changes for other conditions, such as injection rate, fuel injection quantity and speed.
- ✓ Stratsianis et al. researched the post-injection on the effect of marine diesel engine performance. The simulated value also validated with the experimental data under the 100 % load.
- ✓ The CFD model is reasonable in this paper.
- ✓ Thank you very much.

4. The y-axis of Fig. 2 is not sufficient. It needs to be explicitly defined what is being normalized here. a y-axis label of "Experiment/Simulation" as a vertical dimensionless number is not acceptable.

- ✓ Dear reviewer, we had revised the define of Y-axis label for Fig.4. Thank you for your advice.

5. A reference is needed to the "Tier III" requirements.

- ✓ Dear reviewer, we had added the reference for the requirement of Tier III. Thank you for your advice.

11th Asia-Pacific Conference on Combustion,
The University of Sydney, NSW Australia
10th -14th December 2017.

Cold Plasma Methane Reforming

Amit Kumar[1], Anand M. S[1], L Rao[1], Dasappa S[1]

[1]Center for Sustainable Technologies, Indian Institute of Science

Karnataka, 560 012, India

Abstract

The article reports results from dry reforming of methane using cold plasma. An instrumented dielectric barrier discharge reactor has been setup to reform Methane using cold plasma and parametric analysis addressing the effects of gas residence time and plasma power on Methane conversion and product gas Hydrogen concentration. The results suggest improvement in Methane conversion and enhanced product gas Hydrogen concentration on independently increasing residence time and plasma power. At 70 W plasma power conditions and 130 ms gas residence time, a maximum CH_4 conversion of 16 % with an outlet H_2 concentration of 27 mole % was achieved.

1 Introduction

Natural gas ranks as the third most used energy resource in the world after crude oil and coal [1]. The principal component of natural gas is methane and is mainly being used for home and industrial heating and for the generation of electricity. Apart from a being used as a combustible fuel, NG also finds application in Hydrogen generation through the steam Methane reforming route. An alternative approach to steam reforming is the dry reforming of methane using non thermal plasma which has gained significant interest over the years.

Use of non-thermal plasmas for reforming of methane has advantages such as near room temperature and atmospheric pressure operations and no need for steam/water. Non thermal plasmas provide the necessary activation energy by electron impact excitation, ionization and dissociation reactions. Furthermore, higher conversions and selectivity's can be obtained without problems of soot formation [2, 3]. The other advantages of non-thermal plasmas is its high operational flexibility i.e., plasmas can be turned on and off just as a switch and plasma systems can be very compact.

The current study reports on the use of non-thermal dielectric barrier discharge (DBD) plasma based reforming of Methane. All reforming experiments were conducted at atmospheric pressure and temperature conditions.

2 Materials and Methods

Figures 1 and 2 depict the DBD reactor setup used in the current experiments. A close ended quartz tube having a length of 240 mm, inner diameter of 12 mm and a wall thickness of 2 mm was used as the plasma DBD chamber. The quartz tube consisted of two side ports of diameter 6 mm which were 150 mm apart along its length. An 8 mm metallic rod placed along the axis of the quartz tube acted like an inner plasma electrode. Teflon tape wrapped around the inner electrode positioned it along the axis of the plasma reactor and also provided the necessary gas sealing. Aluminum foil (ground electrode) 100 mm length wrapped on the outside of the quartz tube between the two side ports formed the other plasma electrode. A gas flow controller was used to adjust the residence time inside the reactor. A Perkin-Elmer make Gas Chromatograph (Claurus 680, Column Carboxen 1010) was used to analyze the composition of the feed and the outlet gas.

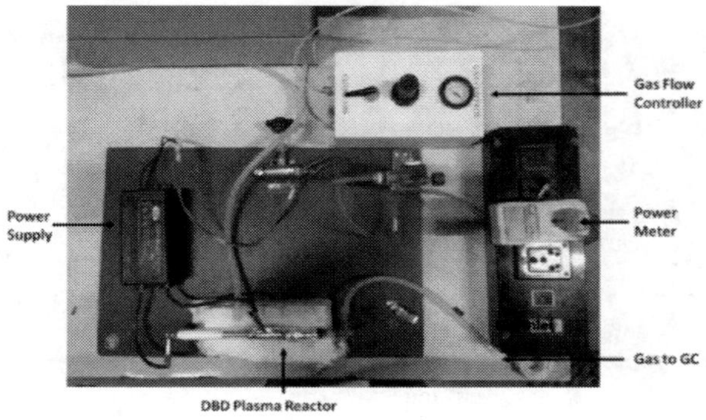

Figure 1: Dielectric barrier discharge non-thermal activation setup

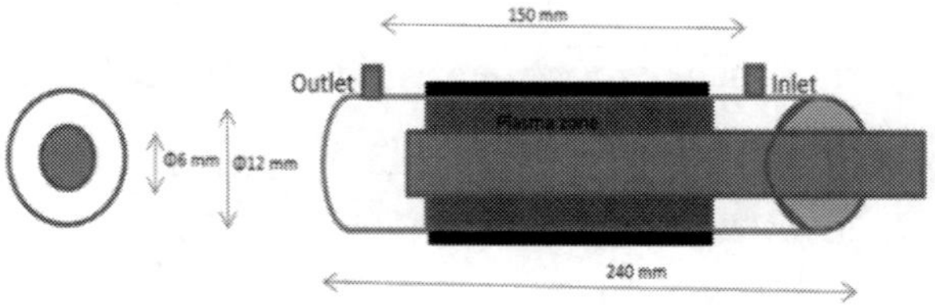

Figure 2: Picture of the Co-axial DBD reactor

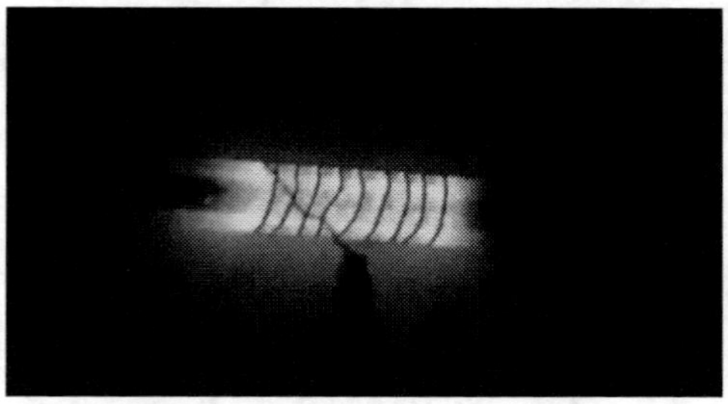

Figure 3: Picture of plasma discharge.

Corresponding author. Fax: +91-08-2360-1692
E-mail address: narayana@astra.iisc.ernet.in

A commercially available neon power supply system along with a power meter (Model MECO) has been used for supply and measurement of input power to the high voltage supply. Investigations to evaluate the fraction of the power delivered to the plasma have not been undertaken. All subsequent studies are solely based on varying the input power. Figure 3 shows the plasma discharge glow during a typical dry reforming operation.

Towards estimating the breakdown voltage and electric field strength, a high voltage waveform was captured by oscilloscope (Tektronix TDS2002B) through the use of high voltage probe (Tektronix P015A,1:1000). The oscilloscope waveform is as indicated in figure 4. The breakdown voltage and electric field strength was 3.37 kV_rms and 11.23 kV/cm.

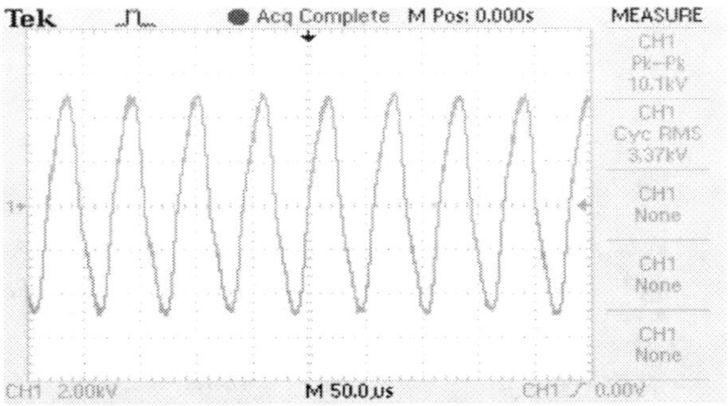

Figure:4 High voltage waveform captured by the oscilloscope during a typical DBD run.

A feed gas mixture from a cylinder having a composition as shown in Table 1 was used as the inlet gas. The gas flow controller was used to select the necessary gas residence time. Once the gas flow through the setup was stabilized, typically after about 15 to 20 mins from the start of the experiment, as confirmed by the steady GC signal, the plasma power supply was turned on and the power dissipated was noted. The outlet gas was analyzed for its composition and experiment was continued till the composition of the gas reached a steady state. Experiments were conducted at three gas residence times of 100 ms, 110 ms and 130 ms and at two plasma power values of 65 W and 70 W.

Species	H_2	N_2	CO	CO_2	CH_4
mole%	0	0.9±0.2	1.1±0.3	3.0±0.7	95.0±0.7

Table 1: Feed Gas Composition.

The product gas analysis results from a gas chromatogram are indicated in figure 5(a) and 5(b) for feed gas (without plasma) and plasma reforming gas respectively. It may be noted that very minor spikes are visible for trace gases like N_2, CO, CO_2 which are in trace quantities as against prominent peaks for those of CH_4 and H_2,.

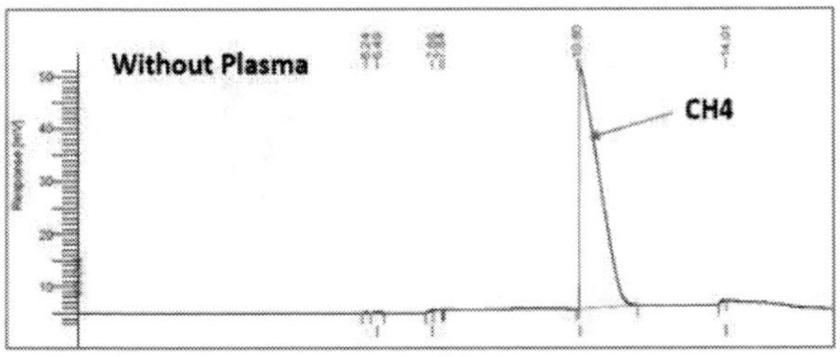

Figure:5(a) Gas chromatogram of feed gas

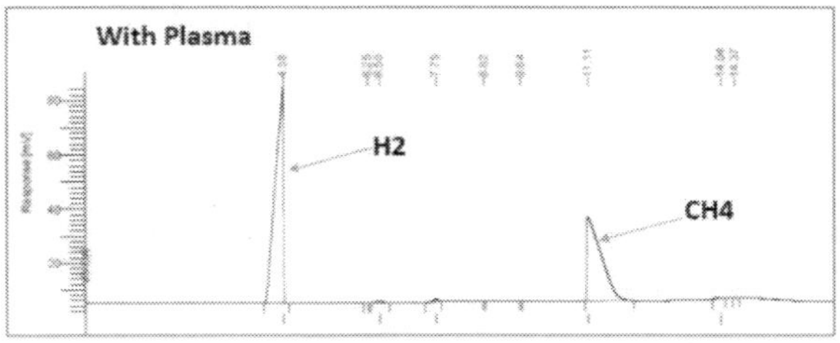

Figure : 5(b) Gas chromatogram of plasma reforming gas

For every experiment, the CH_4 conversion % as defined by "(1)" was calculated and the H_2 mole% as defined as the area under the H_2 peak of the GC signal was noted.

$$\left(\frac{moles\ of\ CH_4 converted}{moles\ of\ CH_4\ fed}\right) \times 100 = CH_4\ \%\ conv. \quad (1)$$

For each experimental setting a minimum of three individual trails were conducted and the average and standard deviation were calculated. The standard deviation is reported as error bars.

3 Results and Discussions

3.1 Effect of Gas Residence Time

The gas residence time in the reactor is primarily controlled by controlling the gas mass flow rate with the residence time being inversely proportional to the flow rate. The influence of gas residence time in the reactor at the two input power ratings is indicated in figure 6 and 7 respectively. Along expected lines, increasing the gas residence time in the plasma chamber increases the H_2 mole % and CH_4 conversion for both the power levels tested. A 30% increase in gas residence time from 100 ms to 130 ms, results in 48% and 67% increase in H_2 mole% for 65 W and 70 W plasma power respectively. That is, at 65 W plasma power H_2 mole % increases from 15.2% to 22.6% whereas at 70 W plasma power the H_2 concentration increase from 16.6% to 27.7%.

Similarly, as can be seen from Fig. 7, increase in gas residence time increase CH_4 conversion. A 30% increase in gas residence time, increases the CH_4 conversion by 57 % and 73% at 65 W and 70 W plasma power respectively. This result is as expected and is similar to earlier studies [3.4]. Higher gas residence time means lower gas flow rate and hence more time for the energetic electrons to activate the gas and thus increase conversion. Further exploration towards the limit condition is in progress.

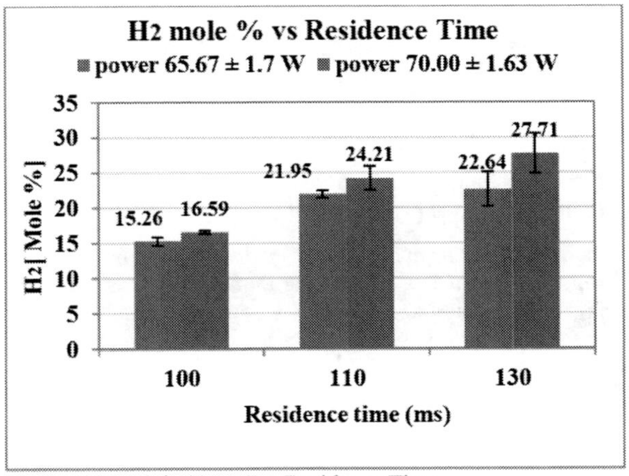

Figure 6: H_2 mole % vs Gas Residence Time

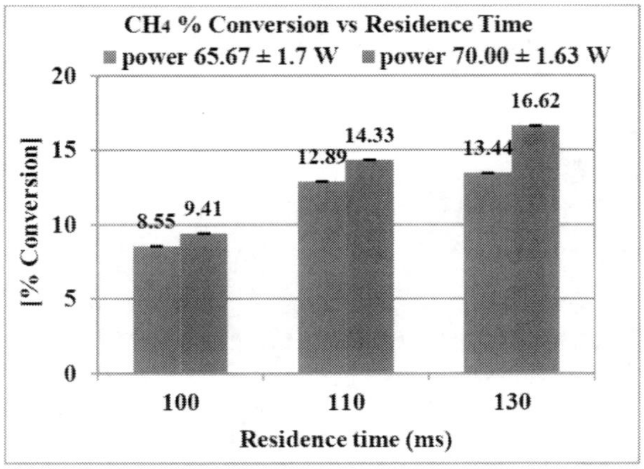

Figure 7: CH_4 % Conv. vs Gas Residence Time

In respect of the residence time, it is important to note that CH4 conversion occurs basically at the streamer (micro scale filament discharges) in time scales of nano seconds as evident from figure 8. The influence of residence time arises in terms of the contact time of CH4 with the establishing streamers. As such, improving the CH4 selectivity in terms of streamer establishment is a key factor in relation to physical flow path enhancement.

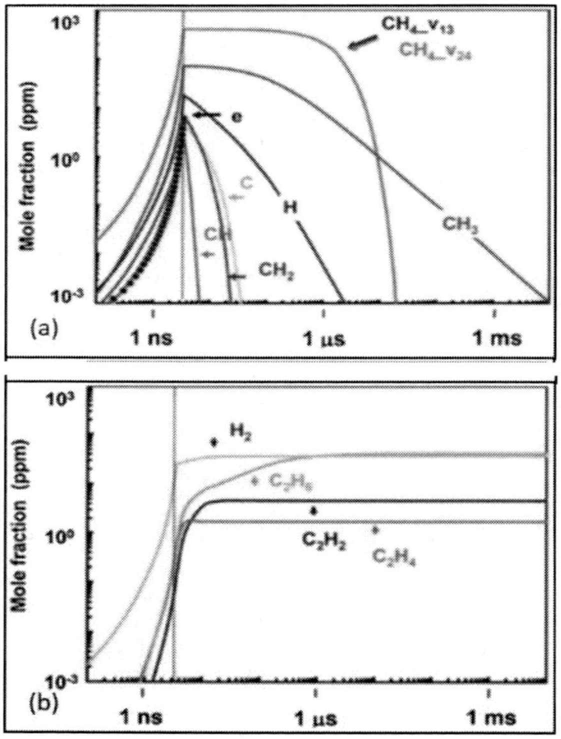

Figure 8. Radical (a) and specie (b) conversion timescales. From Nozaki et al [5]

In respect of the improvement in the conversion with power supply, literature suggests a logarithmic increase as indicated in Figure 9. It could again primarily be attributed to the increase in streamer density within the generic discharge regime.

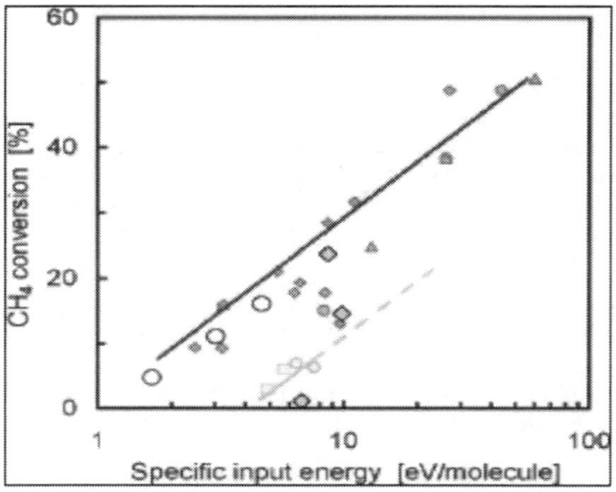

Figure 9. Methane conversion fraction with energy input. From Nozaki et al [6]

3.2 Effect of Plasma Power(W)

Figure 10 and 11 shows the H_2 mole % and CH_4 conversion as a function plasma power for all three gas residence time studied. It is very evident from these figures, that increasing the plasma power increase CH4 conversion and hence H2 mole% in the outlet gas increases. This trend was true for all the three gas residence times studied. At 130 ms gas residence time, increasing the plasma power by 8% (i.e., from 65 W to 70 W), increased the CH4 by 21% (i.e., from 13.45% to 16.26%) and H2 concentration by 22% (i.e., from 22.6% to 27.7%).

This result is again explained by the fact that increasing the plasma power yields more and higher energy electrons, which in turn results higher activation of the gas. These increase in the the number density of electrons and also their higher energy content due to higher plasma power dissipated causes higher conversion of the gas.

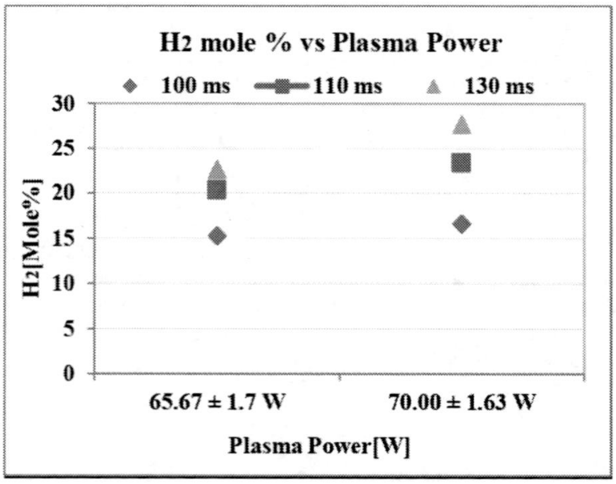

Figure 10: H_2 mole % vs Plasma Power

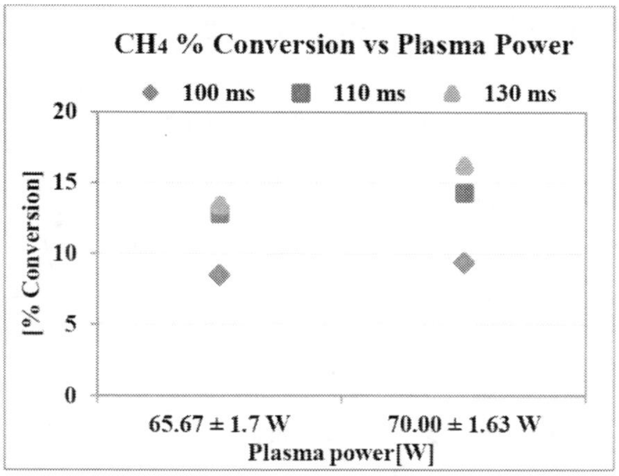

Figure 11: CH_4 % Conv. vs Plasma Power

3.3 Effect of Specific Energy Input(J/L)

Figure 12 and 13 shows the dependence of H_2 mole % and CH_4 conversion on the specific energy input. At 130 ms residence time, increasing the specific energy input by 6%(i.e., from 3200 J/L to 3400 J/L), increased the CH_4 by 21% (i.e., from 13.45% to 16.26%) and H_2 concentration by 22% (i.e., from 22.6% to 27.7%). At 100 ms residence time, increasing specific energy input by 6% (i.e., from 2615 J/L to 2769 J/L), increasing the CH_4 by 10% (i.e., from 8.55% to 9.41%) and H_2 concentration by 9% (i.e., from 15.26 to 16.59)

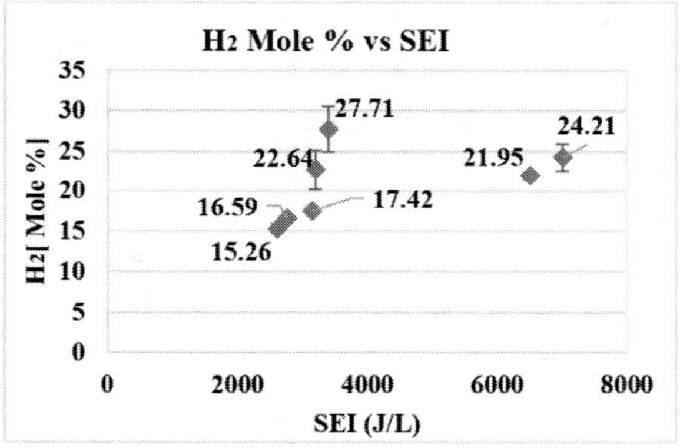

Figure 12: H_2 mole % vs SEI

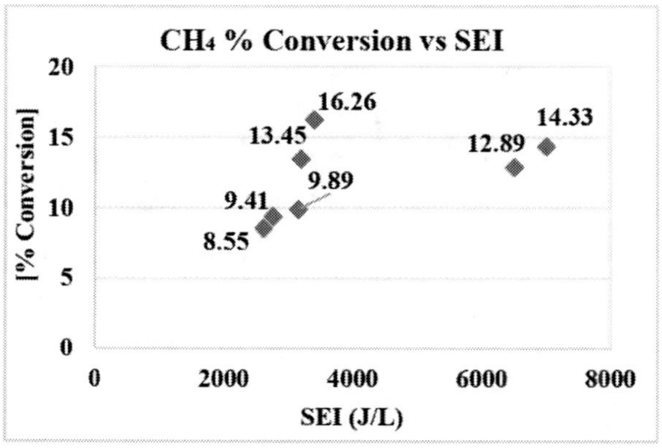

Figure 13: CH_4 % Conv. vs SEI

4 Conclusions

The current article has addressed the effect of gas residence time and the plasma power on dry reforming of methane to hydrogen using a dielectric barrier discharge plasma chamber under atmospheric pressure conditions. The results obtained suggests that

- Increasing the gas residence time increases the CH_4 conversion and hence the H_2 concentration in the outlet gas
- Operating at higher plasma power results in higher CH_4 conversion and hence higher H_2 concentration in the outlet gas.

5 Acknowledgments

The authors would like to acknowledge Indian Institute of Science for their financial assistance.

References

1. C. De Bie, B. Verheyde, T. Martens, J. van Dijk, S. Paulussen, A. Bogaerts, Plasma Process Polym 2011(8) 1033–1058
2. X. Tao, M. Bai, X. Li, H. Long, S. Shang, Y. Yin and X. Dai, Prog. Energy Combust. Sci., 2011(37) 113–124
3. R. Snoeckx, Y. X. Zeng, X. Tu, A. Bogaertsa, RSC Adv. 2015(5) 29799-29808
4. N. S. Matin, et. al., Plasma Chem Plasma Process 2010 (30) 333–347
5. Nozaki, Tomohiro, and Ken Okazaki. "Non-thermal plasma catalysis of methane: Principles, energy efficiency, and applications." Catalysis today 211 (2013): 29-38.
6. Nozaki, Tomohiro, et al. "Dissociation of vibrationally excited methane on Ni catalyst: Part 1. Application to methane steam reforming." Catalysis Today 89.1 (2004): 57-65.

Combustion Measurements of Waste Cooking Oil Biodiesel

C. Ming[1], I.M. Rizwanul Fattah[1], Q.N Chan[1], P.R. Medwell[2], S. Kook[1], E.R. Hawkes[1], G.H. Yeoh[1]

[1]School of Mechanical and Manufacturing Engineering

UNSW Sydney, NSW 2052, Australia

[2]School of Mechanical Engineering

The University of Adelaide, SA 5005, Australia

Abstract

An experimental investigation was performed to assess the combustion characteristics of a waste cooking oil (WCO) biodiesel fuel under simulated compression-ignition engine conditions. A conventional diesel was used as a base fuel for comparison purpose. The fuels were injected into the quiescent steady environment inside a constant-volume combustion chamber (CVCC) of 19.4 kg/m^3 ambient density, 6 MPa ambient pressure, and 1100 K bulk temperature. A range of optical diagnostics were performed, to compare the fuels over ambient O_2 concentrations of 10-21 vol.%, and injection pressures of 70-130 MPa. The results reveal that for the investigated test conditions (i) the lift-off lengths and the first soot distances of the biodiesel are consistently longer than that of diesel; (ii) the peak soot level of the flame increases with the relative distance between the first soot formation and flame lift-off; and (iii) the WCO biodiesel has a lower soot formation propensity than diesel. The fuel liquid lengths were also measured to be shorter than their lift-off distances, indicating no interaction between their spray and combustion processes.

1 Introduction

Biodiesels are mixtures of fatty acid methyl esters (FAMEs) with differing chain lengths and degrees of saturation, which can vary significantly depending on their feedstocks [1]. Biodiesel fuels contain oxygen, which have been reported to provide an effective route to enhance combustion, and to reduce soot formation in compression-ignition engines [2]. Nonetheless, biodiesels have physical and chemical properties that are different to that of conventional diesel. These differences in fuel properties can have a complicated impact on the fuel atomization, evaporation, combustion and soot formation processes, and hence, the ensuing engine performance [3].

This work is aimed at assessing the combustion properties of a biodiesel from the transesterification of waste cooking oil (WCO) under simulated compression-ignition engine conditions in a constant-volume combustion chamber (CVCC). Experiments are performed under three different ambient O_2 concentrations of 10, 15 and 21 vol.%, which are used to simulate different exhaust gas recirculation levels. The experiments are performed under fixed ambient density, pressure and temperature values of 19.4 kg/m^3, 6 MPa and 1100 K, respectively. The experiments are also assessed over three injection pressures of 70, 100 and 130 MPa. OH chemiluminescence and high-speed flame natural luminosity imaging are used to assess impact of these varied parameters on the air entrainment rate occurring upstream of the lift-off length, and the downstream soot formation and combustion processes of the biodiesel jet flame. To provide context to the examination, comparative measurements using a conventional diesel are also performed.

2 Experimental Details

Experiments were performed in an optically-accessible CVCC under simulated, quiescent diesel engine conditions. The CVCC has a 114 × 114 × 114 mm cubical combustion chamber, six window ports, and eight access ports for instrumentations and valves. For this work, sapphire windows were installed at four sides of the chamber to enable optical access. A diesel injector was mounted at the centre of a metal side port so that the fuel spray was directed into the centre of the chamber. An agitator was mounted at the top of the CVCC and was used to maintain a spatially uniform temperature environment. A schematic of the CVCC configuration along with the diagnostic set-ups used is shown in Fig. 1. The operation of the chamber was initiated by filling up the chamber with a premixed, lean combustible charge (*e.g.* acetylene, hydrogen, nitrogen and oxygen). The mixture was subsequently ignited with a spark plug to create a high temperature and pressure environment inside the chamber. Following the premixed combustion, the combustion products cooled over a relatively long time (~8 s) due to heat transfer to the chamber walls, the vessel pressure, therefore, decreased correspondingly. When the targeted pressure was reached, the injector was triggered and the fuel injection, auto-ignition and combustion processes ensued. The test conditions for the experiments are summarized in Table 1. The fuel injection system used comprised of a low pressure fuel

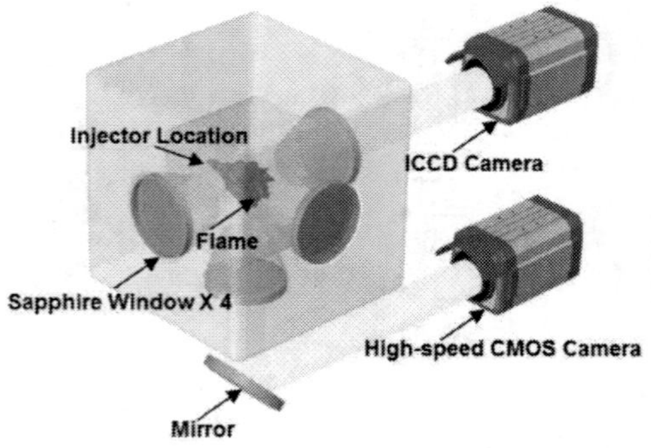

Figure 1: Schematic of experimental arrangement including the combustion vessel, OH-chemiluminescence and high-speed natural luminosity imaging setups.

pump a high pressure fuel pump, a Bosch commercial common-rail system (CP4) and an axially drilled single-hole solenoid injector with an orifice diameter of 105 μm. A common-rail PCV driver (Zenobalti, ZB-1200) was used to regulate and maintain the injection pressure, whilst an injector driver (Zenobalti, ZB-5100) was used to control the injection system to deliver the required fuel mass. A summary of the injection conditions used is provided in Table 1. For this work, comparative measurements were performed between a WCO biodiesel and a conventional diesel. The WCO biodiesel has higher fuel density and viscosity values when compared with diesel. It is noted that for the injection conditions used in this work, the difference in the actual injected mass of the biodiesel and diesel, when measuring with a Bosch tube type injection rate meter, was found to be within experimental errors of the injection rate meter across the varied conditions. The same injection parameters were therefore applied to both fuels, for all experiments. Gas-chromatography and mass-spectrometry (GCMS) was performed on the fuels. WCO biodiesel, is found to comprise mainly of FAMEs with long carbon chains such as oleic, linoleic and linolenic acids, and is oxygenated (11 wt%). The biodiesel therefore has a lower heating value and a higher stoichiometric mixture fraction than diesel. From the fuel specifications data that were provided by the fuel supplier, the biodiesel is also reported to have a higher cetane number than diesel. A summary of the main fuel properties of WCO biodiesel and diesel is provided in Table 2.

The optical combustion diagnostics used for this work, including intensified OH chemiluminescence imaging using an intensified CCD (ICCD) camera, and flame natural luminosity imaging with a high-speed CMOS camera. The cameras were arranged for viewing through the side or bottom windows of the CVCC, as depicted in Fig. 1, and were applied simultaneous to image the same injection event. For the OH-chemiluminescence imaging, the light emission from flame was detected through a 10 nm bandwidth interference filter (centred at 310 nm), using the ICCD camera (Andor iStar) that was fitted with a 50-mm f#3.5 UV lens. The camera was triggered to open for 2.5 ms after the quasi-steady lift-off length of the flame was established, and was stopped before the end of injection. A typical OH-chemiluminescence image recorded is shown in Fig. 2. In this figure, the injector orifice is located at the left edge of the image. The diesel fuel was injected from left to right at 100 MPa injection pressure and 15% ambient oxygen concentration by an injector equipped with a nozzle

Test conditions	
Ambient oxygen concentration (vol.%)	10, 15, 21
Ambient gas density (kg/m^3)	19.4
Ambient pressure (MPa)	6
Bulk temperature (K)	1100
Injection conditions	
Nozzle diameter (μm)	105
Injection pressure (MPa)	70, 100, 130
Injection quantity (mg)	10

Table 1: Summary of test and injection conditions.

Main properties		
Properties	WCO	Diesel
Density (kg/m^3)	883	832
Viscosity (@ 293 K, mm^2/s)	4.6	3.2
Oxygen content (wt%)	11	0
Higher HV (MJ/kg)	40.8	43.4
Stoichiometric air-fuel ratio (AFR) by mass	12.73	14.5
Cetane number	53.2	51

Table 2: Main fuel properties of waste cooking oil (WCO) biodiesel and diesel.

with an orifice diameter of 105 μm. The ambient pressure and temperature were maintained to be 6 MPa and 1100 K during the injection. For the high-speed flame luminosity imaging, a CMOS camera (Photron SA5) equipped with a 85-mm f#1.8 lens, was operated at a frame rate of 15,000 frames per second (fps) was used. The CMOS camera was synchronized to the start of injection of the injector, and 100 images were captured for each run. A typical instantaneous flame luminosity image captured for the same diesel jet flame presented in Fig. 3, but at 2 ms after start of injection (aSOI), is shown in Fig. 4. The diesel fuel was injected from the top to bottom in this image. It is noted that at least five runs were performed for each measurement to ensure good repeatability.

3 Results and Discussion

Flame structural information, including flame lift-off and first soot distance from the nozzle, during its mixing-controlled combustion phase has been shown to be critical towards the understanding of soot formation process [4]. The flame lift-off length, which refers to the distance from the fuel injector nozzle to where the flame stabilizes, is indicative of the location where intense combustion reaction of the flame initiates [5], and is widely used to characterize the amount of fuel-air premixing occurring upstream of the lift-off [6]. The first soot distance, on the other hand, refers to the distance from the injector to the location where intense flame luminosity is first detected and is used to indicate the initial onset of soot in the flame [6]. The lift-off lengths of the flames were derived from their time-averaged OH chemiluminescence images, following the approaches that were detailed in Ref. [5]. In brief, the lift-off length value was determined by finding the distance between the injector and first axial location where chemiluminescence intensity of the flame exceeded ~50% of the levelling-off value, as is shown in Fig. 2. The first soot distances, on the other hand, were derived from the high-speed natural flame luminosity images that were captured using the high-speed camera. The images were first corrected for background, dark charge and detector attenuation, and were subsequently binarized to enable logical operations [7, 8]. The first soot distance was determined by measuring the minimum distance from the injector to the location where half of the pixels on an arc with a spreading angle of $\Theta/2$ are occupied by

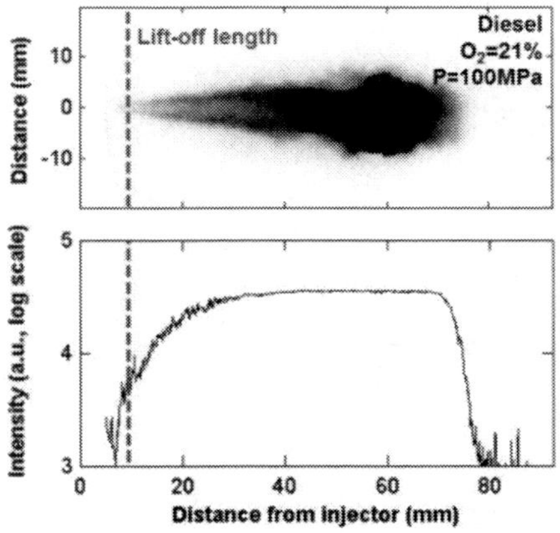

Figure 2: An example of the time-averaged OH-chemiluminescence emission image at 310 nm (top) and its axial intensity profile (bottom). Diesel was injected at 15 vol.% ambient oxygen concentration and 100 MPa injection pressure by a fuel injector with 105 μm nozzle orifice diameter.

the binarized flame area. This definition was used to minimize the sensitivity of the first soot distance calculations to the threshold setting used or potential fluctuations in the binarized flame area. It is noted that the flame central axis was taken as the line from the injector tip through to the centroid of the flame area, whilst flame spreading angle was defined as the average flame spreading angle between a starting and ending points along the flame central axis. The starting and ending points were set to be 45% and 55% of the instantaneous flame length for this work.

The averaged lift-off lengths and the first soot distances of the diesel and WCO biodiesel are plotted as a function of injection pressure, for the test conditions with ambient O_2 concentration of 10, 15 and 21 vol.%, in Fig. 4. From Fig. 4, it can be seen that lift-off lengths of the flames are found to increase with increasing injection pressure or decreasing ambient O_2 concentration. It is noted that previous studies have shown that an increase in injection pressure would result in higher velocity in spray, which would push the initial combustion zone of the flame downstream [9]. A reduction in ambient O_2 concentration, on the other hand, has been shown to necessitate a greater entrainment of total ambient gas with smaller fraction of oxygen (hence, longer mixing time) before combustion reactions can take place [10]. The current results are therefore consistent with the previously observed trends in lift-off lengths with respect to injection pressure and oxygen concentration effects. From Fig. 4, it can be observed that that the lift-off lengths of the WCO biodiesel flames are consistently higher than that of diesel flames. Previous studies have demonstrated flame lift-off is affected by the ignition quality of its fuel, which in turn, is correlated to its physical and chemical properties. From Table 2, it can be seen that the cetane number of the WCO biodiesel (*i.e.* 53.2), albeit slightly higher, is comparable to that of diesel (*i.e.* 51). The viscosity of the WCO biodiesel, on the other hand, is greater than that of diesel. Previous studies have reported that a high viscosity can impact on the atomization, vaporization and fuel-air mixing processes of the fuel. This can lead to a prolonged time between the beginning of fuel injection and the start of combustion process [11], which would subsequently impact on its ignition quality. Further investigation, however, is required to elucidate and extend the observed results. It should be noted that the liquid lengths of the WCO biodiesel and diesel were measured to be ~8.7 mm and ~7.7 mms, respectively, when diffused back illumination technique was applied by the authors. The longer liquid length observed for WCO biodiesel is expected because of its higher heat of vaporization due the presence of FAMEs with longer carbon chains [12]. The shorter liquid lengths of the fuels, when compared with their lift-off lengths, imply that there is no interplay between the spray and combustion processes for the fuels, under the investigated conditions.

In previous studies, it has been shown the soot inception time of a jet flame can be approximated from the observed location of the first soot formation relative to the flame lift-off [6]. This distance of lift-off length to soot formation region, which is termed as soot formation distance in the subsequent discussion, has been shown to depend strongly on fuel type, such that a longer distance is observed for fuel with lower soot formation tendency [4, 6]. In Fig. 5, the soot formation distances of the WCO biodiesel and diesel jet flames are plotted as a function of injection pressure and ambient O_2 concentration. The peak spatially integrated natural luminosity (SINL) values of the WCO and diesel flames under different test conditions are also computed and are presented in Fig. 5. It is acknowledged that the SINL signal, which was calculated by summing up the intensities of all pixels of each image, is a complex function of the soot concentration, properties and temperature [13]. Nonetheless, the peak SINL has been widely shown to be useful in providing a relative measure of the soot content of sooty flames, and is therefore used here to rank-order the soot level of the diesel and WCO biodiesel flames. In Fig. 5, it can be seen that a higher peak SINL value/soot content is observed for flame with shorter soot formation distance. From the figure, it can also be observed that under the same test and injection

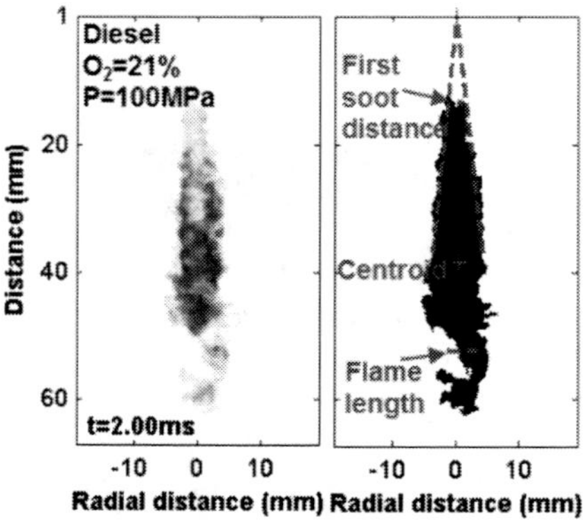

Figure 3: A typical instantaneous flame luminosity image of a diesel spray flame, recorded at 2 ms aSOI (left). The same flame image after background, dark charge and detector attenuation correction (right). The parameters of interest of the flame were derived from the boundary of the resulting image. Definitions of centroid, flame length and first soot distance are shown in the figure.

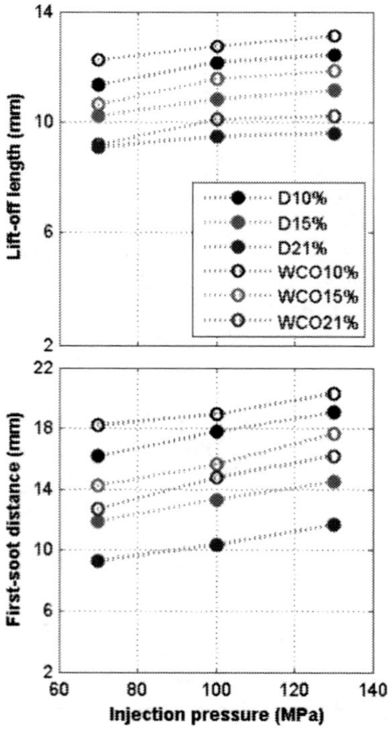

Figure 4: Comparisons of the lift-off lengths (top) and the first soot distance (bottom) of diesel and WCO jet flames. The lift-off lengths and the first soot distances are plotted as a function of injection pressure for three selected ambient oxygen concentrations of 10, 15 and 21 vol.%.

conditions, the soot formation distance of WCO biodiesel flame is consistently longer than that of diesel. As is noted in Table 2, the WCO biodiesel is oxygenated whereas the diesel fuel is not. The oxygen content of the biodiesel increases the critical mixture fraction for soot formation limit of biodiesel in relation to diesel [4]. It can therefore be expected that a smaller percentage of the biodiesel fuel jet mixtures would exceed its soot formation limit, and therefore, less downstream soot would form. The fuel molecular structure and composition of the WCO biodiesel are also less prone to soot formation than diesel. The WCO biodiesel contains carbon-oxygen bonds that are harder to break. This will hinder the formation of carbon-carbon double and triple bonds that can lead to the formation of soot precursors, and ultimately soot [4]. From the GCMS analysis that was performed (not shown here), the diesel used was found to contain aromatic compounds such as toluene, xylene and ethyl-toluene, which have strong tendencies to form soot during combustion, whereas the WCO used is aromatic-free. The fuel oxygenation, molecular structure and fuel composition effects of the WCO biodiesel, in combined with its longer observed lift-off length, would therefore contribute towards its observed lower peak SINL/soot content, when compared with diesel.

From Fig. 5, it can be seen that the soot formation distance for both WCO biodiesel and diesel increase with increasing injection pressure for both fuels. Different trends, however, are observed for the fuels across the varied ambient O_2 concentration. Specifically, the soot formation distance of diesel is found to decrease monotonically with increasing ambient O_2 concentration. The soot formation distance of WCO biodiesel, on the other hand, is observed to first decrease and

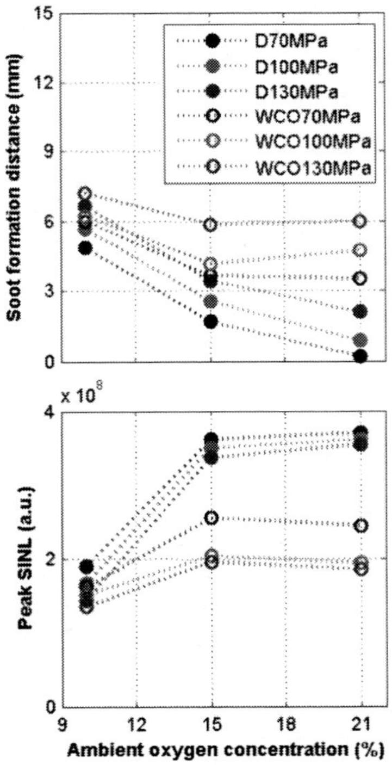

Figure 5: Comparisons of the soot formation distances (top) and the peak SINL (bottom) of diesel and WCO jet flames. The soot formation distances and the peak SINL are plotted as a function of ambient oxygen concentration for three selected injection pressures (70, 100 and 130 MPa).

then increase as the ambient O_2 concentration increases from 10% to 21%. It is noted that similar trends were also observed in a previous study [14] that was performed to examine the effects of ambient oxygen concentration on WCO biodiesel and diesel spray combustion. In their study, it was suggested that the soot formation of the WCO biodiesel was maintained without sufficient oxidation, when the ambient O_2 concentration decreases from 21% to 15%, which resulted in an increased soot yield for the biodiesel. Further investigation, however, is required to confirm this.

4 Conclusion

Experiments were performed in an optically-accessible constant-volume combustion chamber (CVCC) at ambient conditions that are representative of compression-ignition engine (19.4 kg/m^3 ambient density, 6 MPa ambient pressure, 1100 K bulk temperature, 10-21 vol.% ambient O_2 concentrations, 70-130 MPa injection pressure), comparing waste cooking oil (WCO) biodiesel to conventional diesel. The results reveal that for the investigated experimental conditions, the lift-off lengths and the first soot distances of the biodiesel were consistently longer than that of diesel. The peak soot level of the flame was observed to correspond to the distance from the lift-off length to the location of the first soot formation. The WCO biodiesel flames were found to have lower soot content than diesel, under the same test and injection conditions. The liquid lengths of both fuels were measured to be shorter than their lift-off lengths; implying interaction between the spray and combustion processes of the fuels was not significant.

References

[1] C.J. Chuck, C.D. Bannister, J.G. Hawley, M.G. Davidson, I. La Bruna, A. Paine, Energy 23 (2009) 2290-2294.

[2] X. Wang, Z. Huang, O.A. Kuti, W. Zhang, K. Nishida, Proc. Combust. Inst 33 (2011) 2071-2077.

[3] S.K. Hoekman, A. Broch, C. Robbins, E. Ceniceros, M. Natarajan, Renew. Sustain. Energy Rev 16 (2011) 143–169.

[4] J. G. Nerva, C. L. Genzale, S. Kook, J. M. Garcia-Oliver, L. M. Pickett, Int. J. Engine Res. 14 (2012) 373-390.

[5] B. Higgins, D.L. Siebers, SAE Paper 2001-01-0918 (2001).

[6] L.M. Pickett, D.L. Siebers, Int. J. Engine Res 7 (2005) 103-130.

[7] Q. N. Chan, P. R. Medwell, G. J. Nathan, Exp. Fluids 55 (2014).

[8] N.H. Qamar, G.J. Nathan, Z.T. Alwahabi, Q.N. Chan, Combust. Flame 158 (2011) 2458-2464.

[9] D.L. Siebers, B. Higgins, SAE Paper 2001-01-0530 (2001).

[10] D.L. Siebers, B. Higgins, L.M. Pickett, SAE Paper 2002-01-0890 (2002).

[11] M. Shahabuddin, A.M. Liaquat, H.H. Masjuki, M.A. Kalam, M. Mofijur, Renew. Sustainable Energy Rev. 21 (2013), 623-632.

[12] D.L. Siebers, SAE Paper 980809 (1998).

[13] C.J. Polonowski, C.J. Mueller, C.R. Gehrke, T. Bazyn, G.C. Martin, P.M. Lilo, SAE Paper 2011-01-1812 (2011).

[14] J. Zhang, W. Jing, W. L. Roberts, T. Fang, Energy 57 (2013) 722-732.

11th Asia-Pacific Conference on Combustion,
The University of Sydney, NSW Australia
10th -14th December 2017.

Effects of Sulphur and Vanadium Contents in Diesel Fuel on Engine Performance and Emissions: Principal Component Analysis (PCA)

Nicholas Surawski[1], Thuy Chu Van[2a, 3], Zoran Ristovski[2b], Nho Luong Cong[3], Huong Nguyen Lan[3], Chung-Shin Jonathan Yuan[4], S. M. Ashrafur Rahman[2a], Farhad M. Hossain[2a], Yi Guo[2b], Andelija Milic[2b], Thomas Rainey[2a], Vikram Garaniya[5], Richard J. Brown[2a]

[1]University of Technology Sydney, 81 Broadway, Ultimo NSW 2007, Australia

[2a]Biofuel Engine Research Facility (BERF) –

[2b]International Laboratory for Air Quality and Health –

Queensland University of Technology, 2 George St, Brisbane City, Queensland, 4000, Australia

[3]Vietnam Maritime University, 484 Lach Tray St, Hai Phong City, 180000, Vietnam

[4]National Sun Yat-sen University, 70 Lian-Hi Road, Kaohsiung 804, Taiwan

[5]Australian Maritime College, 1 Maritime Way, Newnham, Tasmania, 7248, Australia

Abstract

Marine diesel engines using Heavy Fuel Oil (HFO) produce emissions of toxic compounds that have a negative impact on the environment and human health. A very limited number of on-board ship emission measurement studies have been undertaken due their logistical and regulatory complexity. An alternative way to investigate some issues relating to HFO fuels is by the use of a proxy fuel for HFO in a laboratory based diesel engine. Sulphur (S) and vanadium (V) are two elements of particular interest in HFO because of their relationship to particle formation and corrosive salt properties, respectively. An experimental engine campaign has been conducted on a heavy duty high speed six-cylinder turbocharged and after-cooled diesel engine with a common rail injection system. Principal Component Analysis has been applied in this study to investigate the relationships between: (i) measured engine performance and emissions variables and (ii) fuel S and V content and engine load.

1 Introduction

Shipping is considered one of the most fuel efficient means of transportation [1], it accounts for over 90% of world trade by some 90,000 marine vessels [2]. However, exhaust emissions from ships have a negative impact on the environment and consequently on human health [3-10] and have become of global concern over the last decade [11]. To make the matters worse, these ships also burn low quality Heavy Fuel Oil (HFO) owing to its economic benefit [5]. Sulphur (S) and many trace metals such as silicon (Si), nickel (Ni), iron (Fe), lead (Pb), aluminum (Al), calcium (Ca) and vanadium (V) are observed in HFO, so HFO combustion consequently results in different compounds like sulphates, organic carbon (OC), black carbon (BC), ash and heavy metals in emitted particles [3, 7, 12], most of which pose a high toxicity risk [6]. In particular, shipping-related fine particle ($PM_{2.5}$) emissions alone can account for nearly 60,000 cardiopulmonary and lung cancer deaths each year [10]. Quantitative and qualitative research on ship emissions are needed for a deeper understanding for law makers and regulators [1], and is becoming more important [8]. However, a very limited number of on-board measurement studies have been undertaken [8, 13]. On-board ship emission measurements are an extremely complex task that need the participation of a wide range of institutions and modern instruments. Spiking fuel with S and V as a proxy for these elements in HFO in a non-marine diesel engine is an alternative approach to investigate these issues under controlled laboratory conditions.

Another issue caused by burning HFO is that it can cause corrosion and damage to engine parts. The main ash elements observed typically in HFO consist of Al, Fe, Ni, Ca, Si, Na, and V in which V and Na being the highest contributors. V contained in the oil is in a soluble form and its levels depend mainly on the crude oil source. Na presented in the oil is mainly associated with sea water contamination. Vanadium oxide (VO), vanadium dioxide (VO_2), and then vanadium pentoxide (V_2O_5) are formed during combustion due to oxidation of V. It is well known that V_2O_5 has a low melting point, so the condensed particles with V_2O_5 on the outer layers become semi-liquid and sticky, and then easily adhere to engine parts such as exhaust valves, piston crowns, and turbochargers [14]. Water vapour formed during combustion combines with Na to generate sodium hydroxide (NaOH) that reacts with the sulphur dioxide (SO_2) present in the exhaust gas to yield sodium sulphate (Na_2SO_4). This compound condenses and adheres to surfaces that already coated with V_2O_5. The V-Na-S compounds will dissolve the exhaust valve surface, which is a ferric oxide (Fe_2O_3) layer, thus exposing the under-lying steel surface to further oxidation attack and subsequent corrosion [15]. These melting points are strongly dependent on the ratios of V/Na. Therefore, finding the link between fuel ash elements with engine wear related particle components emitted is an interesting research topic that is relevant for this shipping industry.

The aim of this study is to investigate the effects of spiked diesel fuels with different S and V contents on diesel engine performance and emissions with a focus on nanoparticles, number size distributions and chemical components. A multivariate statistical analysis will be used to investigate relationships between these variables.

Corresponding author. Fax: +61-7-3138-1516
E-mail address: thuy.chuvan@qut.edu.au

2 Methodology

2.1 Fuel Preparation

Fuel S contents selected were 0.5% and 0.1% (by weight) that relate to S regulations issued by the International Maritime Organization (IMO) through Annex VI of the Marine Pollution Convention (MARPOL) applying in Emission Control Areas (ECAs) by 2015 and globally by 2020 respectively. The most abundant S compounds existing in diesel fuels are benzothiophenes and dibenzothiophenes [16]. V is contained in diesel fuels in a soluble form which is present as vanadium-porphyrin complexes [17]. Fe-based fuel-borne catalysts have been widely used in Europe as fuel additives [18]. Chemicals related to S, V and Fe contents therefore have been purchased and used to spike the diesel in order to make chemically altered fuels. Further details of the chemicals used in spiking fuels are in Table 1. The properties of fuel blends can be seen in Table 2.

Table 1: Chemicals used for fuel spiking

Chemical	CAS Number	Formula
Dibenzothiophene	132-65-0	$C_{12}H_8S$
Bis(cyclopentadienyl)vanadium (II)	1277-47-0	$V(C5H5)_2$
Bis(η5-cyclopentadienyl)iron or Ferrocene	102-54-5	$Fe(C5H5)_2$ $C_{10}H_{10}Fe$

Table 2: Properties of spiked fuels

Prop.	Units	Diesel	Fuel N01	Fuel N02	Fuel N03	Fuel N04
Density	kg/L	0.8376[a] 0.8444[b]	0.8456[b]	0.8472[b]	0.8486[b]	0.8498[b]
C	%mass	87.103[b]	87.781[b]	88.715[b]	88.648[b]	87.249[b]
N	%mass	0.0536[b]	0.0426[b]	0.0406[b]	0.0456[b]	0.0390[b]
S	%mass	$6.1x10^{-4a}$ $6.1x10^{-4a}$	0.102[b]	0.109[b]	0.475[b]	0.513[b]
V	mg/kg	0	5	15	5	15
Fe	mg/kg	0	10	10	10	10
HHV*	MJ/kg	45.64[b]	45.16[b]	45.31[b]	44.34[b]	44.68[b]

[a] from CALTEX, [b] tested at QUT; * High heating value

2.2 Engine Specifications and Test Setup

This experimental investigation used a heavy duty, six-cylinder, turbocharged and after-cooled diesel engine with a common rail injection system located at the Biofuel Engine Research Facility (BERF). Further details of the engine specifications can be seen in a previous study [19]. Figure 1 shows a schematic diagram of the experiment. The first sampling point was used for raw exhaust measurements by a Testo 350 XL for gaseous concentration including sulphur dioxide (SO_2), nitrogen oxides (NO_x), carbon monoxide (CO), carbon dioxide (CO_2), oxygen (O_2), and unburned hydrocarbons (HCs). The raw hot-exhaust was also directed to the DMS 500 2-stage dilution system, CAI NO_x and CO_2 analyzers from the second sampling point. Particle number size distributions in the size range of 5nm – 1.0μm were analysed with a sampling frequency of 1 Hz using a DMS 500 Fast Particulate Spectrometer (CAMBUSTION, Cambridge UK) through a heated sample line, and two dilution stages. The third sampling point was firstly diluted through the dilution tunnel and then divided into two paths. The first path was used for measurements with a Micro-Orifice Uniform Deposit Impactor (MOUDI 125R), a DustTrak™ and a Sable CA-10 carbon dioxide analyser, while another path sampled with an Aerosol Mass Spectrometer (AMS; Aerodyne Research USA) and a Sable CA-10 carbon dioxide analyser after an ejector diluter (Dekati, Kangasala, Finland) that used HEPA-filtered compressed air at room temperature for dilution. The MOUDI 125R is a 13-stage cascade impactor that works on the principle of inertial impaction combining with PTFE Teflon filters for metal measurements in particles emitted. A DustTrak™ II Aerosol Monitor 8530 (TSI), which is a light-scattering laser photometer giving real-time aerosol mass readings, was used to measure mass concentrations of PM_{10}, $PM_{2.5}$, and $PM_{1.0}$. An AMS was used to measure a real-time, non-refractory, size-resolved particulate chemical composition. Measured aerosol species were sulfate (SO_4), nitrates (NO_3), ammonium (NH_4), chlorides (Cl), and organics (ORG).

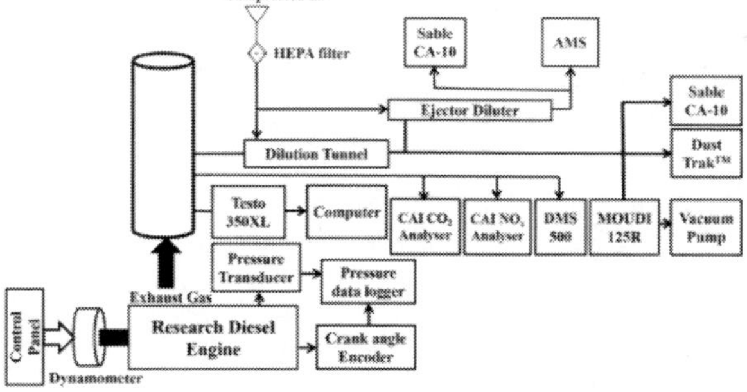

Figure 1: Schematic diagram of test setup

2.3 Principal Component Analysis

Principle Component Analysis (PCA) [20] was used as a quantitative tool to explore relationships between the (i) measured engine performance and emissions variables and (ii) fuel S and V content and engine load. In PCA, new variables called Principal Components (PCs) are found which are a linear combination of the existing variables. The PCs are found in such a way that they are mutually orthogonal, with the direction of the first co-ordinate axis capturing the most of the variation in the original dataset X. The utility of PCA stems from being able to map a high dimensional space to a low dimensional space, whilst at the same time capturing most of the variance in the original dataset. The algorithm to perform PCA relies on finding the Singular Value Decomposition (SVD) of X. The SVD produces two orthogonal matrices (U and V^T) which arise from the left and right singular vectors, as well as a square matrix (D) that contains the singular values as given by:

$$X = UDV^T. \tag{1}$$

It is possible to decompose X into what are called the loadings (P) and scores (T) matrices via:

$$X = (UD)V^T = TP^T. \tag{2}$$

T and P both have simple interpretations. The loadings matrix captures the contribution of the original variables to a particular PC. Elements of P with a higher magnitude indicate that a particular variable strongly influences the resulting PC. Alternatively, the scores matrix provides co-ordinates for data in the new co-ordinate space as found by the PCA. In this study, the number of experimentally measured variables was quite large relative to the number of engine test conditions. This makes PCA a useful technique to perform dimensionality reduction and to promote inferences related to the original experimental dataset. The PCA was performed in RStudio version 1.0.136 using the

FactoMineR [21] and psych [22] packages.

3 Results and Discussion

The factor loading diagram (Figure 2) indicates the quantitative relationship between variables measured in the engine performance and emissions characterisation and the resulting principal components. Black arrows indicate that the variable is positively correlated with the relevant principal component with red arrows indicating anti-correlation. The absolute value of the factor loading indicates the strength of the correlation. The factor loadings for PC1 are heavily influenced by gas/particle phase emissions of Na and V. The coupled nature of Na/V chemistry in the emissions makes a dominant contribution to PC1 and demonstrates empirical support for corrosive ternary salt formation processes in shipping exhaust.

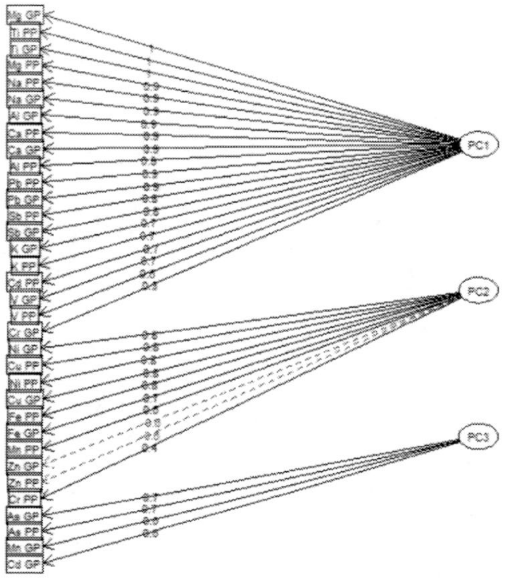

Figure 2: Factor loading diagram for the first three principal components (denoted PC1, PC2 or PC3). PP refers to a particle-phase element detected with ICP-MS and GP refers to a gas-phase element.

Scores plots for different fuel S and V contents are presented in Figures 3 and 4, respectively. Figure 3 shows a clear effect of S content in fuels. In particular, zero S contents (black open square) and low S (red filled circle symbols) are close to each other and behave similarly, while high S contents (blue open triangle) behave differently by being much more scattered in the 3-PC directions. A similar trend is observed for the V case and shows a clear V effect (Figure 4). High V contents (5 ppm in red filled circles and 15 ppm in blue open triangle symbols) behave totally differently to un-spiked fuel (0 ppm in open square symbols). From Figures 3 and 4, there the combustion of the spiked fuels leads to different emissions products from elements that are prominent in the lubricating oil (e.g. Ca, Mg).

The loadings plot presented in Figure 5 shows the relationship between variables. Correlated variables lie close to each other in the graph (± 45^0), anti-correlated variables lie in roughly opposite directions (135-225^0), whilst independent or un-correlated variables, lie in a roughly orthogonal direction (45-135^0) in the graph. For example, NO_x and CO_2 variables (both in the red square) located in the top left quadrant of Figure 5 are significantly correlated, while the NO_x and SO_4_AMS (orange ellipse) variables show a strongly anti-correlated trend. This may be due to higher S levels in the fuel lowering its HHV that may reduce combustion temperature. Consequently, NO_x emissions will be reduced.

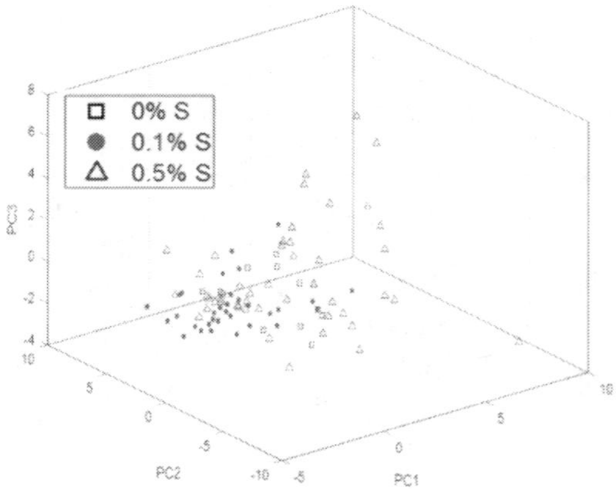

Figure 3: Scores plot for different fuel S contents

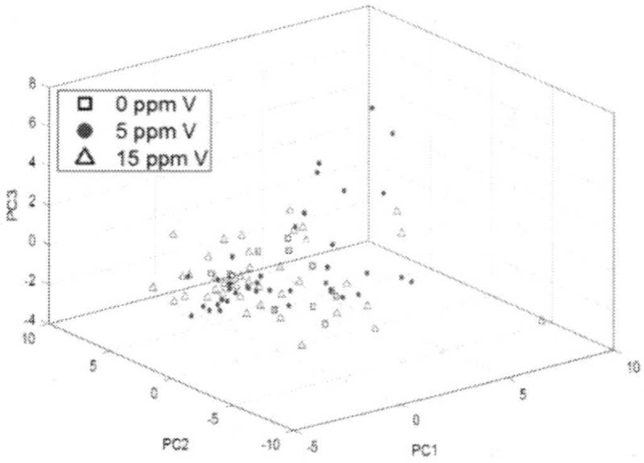

Figure 4: Scores plot for different fuel V contents

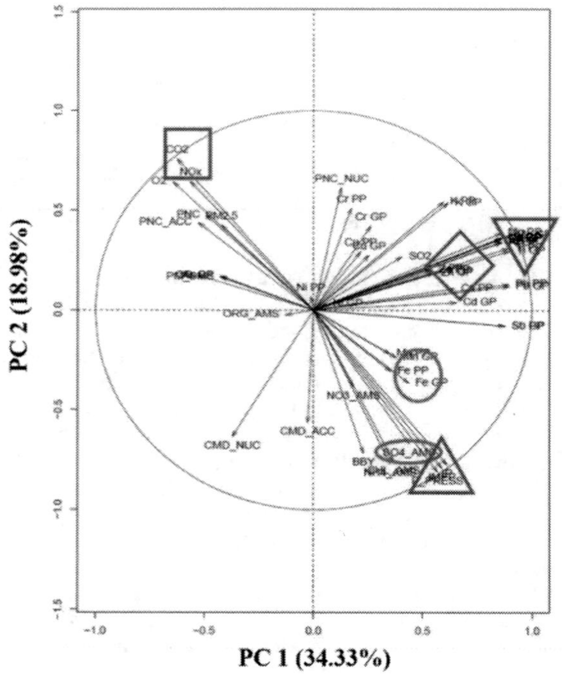

Figure 5: Loadings plot of variables

There is also a trend of increased vanadium content (purple diamond) being closely related to emitted Fe (green circle) and Al (red down triangle) in particles (Figure 5). This may be due to the oxidation attack and corrosion from engine parts caused by the availability of V in fuels used.

It is also clearly evident that higher in-cylinder pressure (higher engine loads) resulted in higher piston ring/liner loadings which can accelerate bore wear rates. Consequently more Fe can be created, so Fe (green circle) is highly correlated to engine pressure (blue up triangle) as presented in the bottom right quadrant of Figure 5.

4 Conclusions

This study has presented detailed measurements of gases and particle emissions from a non-marine diesel engine fueled with different S and V spiked fuels. Principal Component Analysis has been applied in this study to explore relationships between the measured engine performance and emissions variables and specific engine test conditions. Results showed that clear S and V effects have been presented by major strong evidences: an increase in SO_4, Al and Fe portions of emitted particles. Strong effect of engine loads have been investigated.

5 Acknowledgment

The second author gratefully acknowledges the financial support from the Vietnam Government for a PhD scholarship. The authors would also like to acknowledge: the software developer, Mr. Andrew Elder from DynoLog Dynamometer Pty Ltd; laboratory assistance from Mr. Noel Hartnett; QUT staff and colleagues for their support and guidance.

References

1. J.J. Corbett, Updated emissions from ocean shipping. Journal of Geophysical Research 108 (D20) (2003).

2. V. Eyring, Emissions from international shipping: 2. Impact of future technologies on scenarios until 2050. Journal of Geophysical Research, 110 (D17) (2005).

3. H. Winnes, J. Moldanová, M. Anderson, E. Fridell, On-board measurements of particle emissions from marine engines using fuels with different sulphur content. Proceedings of the Institution of Mechanical Engineers Part M: Journal of Engineering for the Maritime Environment 230 (1) (2016) 45-54.

4. A.A. Reda, J. Schnelle-Kreis, J. Orasche, G. Abbaszade, J. Lintelmann, J.M. Arteaga-Salas, B. Stengel, R. Rabe, H. Harndorf, O. Sippula, T. Streibel, R. Zimmermann, Gas phase carbonyl compounds in ship emissions: Differences between diesel fuel and heavy fuel oil operation. Atmospheric Environment 112 (2015) 370-380.

5. L. Mueller, G. Jakobi, H. Czech, B. Stegel, J. Orasche, J.M. Arteaga-Salas, E. Karg, M. Elsasser, O. Sippula, T. Streibel, J.G. Slowik, A.S.H. Prevot, J. Jokiniemi, R. Rabe, H. Harndorf, B. Michalke, J. Schnelle-Kreis, R. Zimmermann, Characteristics and temporal evolution of particulate emissions from a ship diesel engine. Applied Energy 155 (2015) 204-217.

6. F. Di Natale, C. Carotenuto, Particulate matter in marine diesel engines exhausts: Emissions and control strategies. Transportation Research Part D: Transport and Environment 40 (2015) 166-191.

7. M. Anderson, K. Salo, A.M. Hallquist, E. Fridell, Characterization of particles from a marine engine operating at low loads. Atmospheric Environment 101 (2015) 65-71.

8. J. Blasco, V. Durán-Grados, M. Hampel, J. Moreno-Gutierrez, Towards an integrated environmental risk assessment of emissions from ships' propulsion systems. Environment International 66 (2014) 44-47.

9. Z.D. Ristovski, B. Miljecviv, N.C. Surawski, L. Morawska, K.M. Fong, F. Goh, I.A. Yang, Respiratory health effects of diesel particulate matter. Respirology 17 (2) (2012) 201-212.

10. J.J. Corbett, J.J Winebrake, E.H. Green, P. Kasibhatla, V. Eyring, A. Lauer, Mortality from Ship Emissions: A Global Assessment. Environmental Scinece & Technology 41 (24) (2007) 8512–8518.

11. T. Chu-Van, T. Rainey, Z.D. Ristovski, A.M. Pourkhesalian, V. Garaniya, R. Abbassi, L. Yang, R.J. Brown, Emissions from a marine auxiliary diesel engine at berth using heavy fuel oil, The 10th AHMT Conference, QUT-Brisbane, Australia, 2016.

12. J. Moldanová, E. Fridell, O. Popovicheva, B. Demirdjian, V. Tishkova, A. Faccinetto, C. Focsa, Characterisation of particulate matter and gaseous emissions from a large ship diesel engine. Atmospheric Environment 43 (16) (2009) 2632-2641.

13. H. Agrawal, Q.G.J. Malloy, W.A. Welch, J.W. Miller, D.R. Cocker III, In-use gaseous and particulate matter emissions from a modern ocean going container vessel. Atmospheric Environment 42 (21) (2008) 5504-5510.

14. Vanadium Sulphur in Marine Fuels, the International Bunker Industry Association, available at http://www.ibia.net/ibia-publications-and-downloads/vanadium-sulphur-in-marine-fuels.pdf

15. Heavy Fuel Oil, American Bureau of Shipping, available at http://www.eagle.org/eagleExternalPortalWEB/ShowProperty/BEA%20Repository/Rules&Guides/Current/31_HeavyFuelOil/Pub31_HeavyFuelOil

16. P. Saiyasitpanich, M. Lu, T.C. Keener, F. Liang, S.J. Khang, The effect of diesel fuel sulfur content on particulate matter emissions for a nonroad diesel generator. Journal of the Air and Waste Management Association 55 (7) (2005) 993-998.

17. H.N. Dunning, J.W. Moore, H. Bieber, R.B. Williams, Porphyrin, Nickel, Vanadium, and Nitrogen in Petroleum. Journal of Chemical & Engineering Data 5 (4) (1960) 546-549.

18. Z-H. Zhang, R. Balasubramanian, Effect of cerium oxide and ferrocence nanoparticles addition as fuel-born catalysts on diesel engine particulate emissions: environmental and heath implications. Environmental Science & Technology, 51 (8) (2017) 4248-4258.

19. A. Zare, M.N. Nabi, T.A. Bodisco, F.M. Hossain, M.M. Rahman, T. Chu-Van, Z.D. Ristovski, R.J. Brown, Diesel engine emissions with oxygenated fuels: A comparative study into cold-start and hot-start operation. Journal of Cleaner Production 162 (2017) 997-1008.

20. R. Wehrens, Chemometrics with R. Multivariate Data Analysis in the Natural Sciences and Life Sciences, Springer, New York, USA, 2011.

21. F. Husson, Multivariate Exploratory Data Analysis and Data Mining, version 1.3.6. 2017.

22. W. Revelle, Procedures for Psychological, Psychometric and Personality Research, version 1.7.5. 2017.

11th Asia-Pacific Conference on Combustion,
The University of Sydney, NSW Australia
10th-14th December 2017.

ZND Structure of Cool Detonation in Dimethyl Ether-Oxygen-Carbon Dioxide Mixtures

R. Mével[1,2,*], J. Melguizo-Gavilanes[3] and M. Radulescu[4]
[1]Center for Combustion Energy, [2]Department of Automotive Engineering, Tsinghua University
30 Shuang Qing road, 100084 Beijing, China
[3]California Institute of Technology, Pasadena, USA
1200 E. California Blvd, MC 205-45, Pasadena, California 91125, USA
[4]Department of Mechanical Engineering, University of Ottawa
Ottawa, Ontario K1N 6N5, Canada, Canada

Abstract
Steady one-dimensional numerical simulations were performed to investigate the dynamics of the ZND detonation propagating in dimethyl ether-oxygen-carbon dioxide (DME-O_2-CO_2) mixtures. For dilution level as low as 70%, the chemical pathways characteristic of low-temperature and intermediate-temperature hydrocarbon fuels chemistry were activated. This leads to complex energy release profiles comprising up to three peaks in the thermicity profiles. A number of analyses have been performed including energy release rate per reaction as well as sensitivity and reaction pathway analyses. Additionally, 1D simulations including the effect of flow divergence suggest that, given the low energy release during the first oxidation step, momentum losses within the boundary layer may be sufficiently important to quench a cool detonation in DME-O_2-CO_2 mixtures.

1 Introduction

The detonation phenomenon has been historically associated with high-temperature chemistry [1]. For hydrocarbon fuel-air mixtures at normal conditions, the leading shock front of a detonation wave propagating at the Chapman-Jouguet velocity instantaneously brings the fresh mixture to temperature and pressure above 1500 K and 2-3 MPa, respectively. Under such thermodynamic conditions, the oxidative pathways which dominate in the low-temperature and negative temperature coefficient region (NTC), are too slow to compete with pyrolytic chemistry and monotonous energy release is observed. However, Liang et al. [2] recently showed that for high dilution with carbon dioxide, low-temperature chemistry could be activated at the von Neumann state for *n*-heptane-oxygen mixtures. Using a semi-empirical model, they estimated the detonation cell size to be on the order of 4 to 60 m for X_{CO2}=0.85. Because of its higher reactivity and its lower chemical complexicity as compared to normal alkanes, dimethyl ether (DME) is one of the most promising candidate for realizing experimental observation or multi-dimensional simulation of "cool detonation". A "cool detonation" may be defined as a detonation which propagation is driven by low-temperature chemistry.

The goal of the present study was to evaluate the chemical length scales and dominant chemical pathways of cool detonation propagating in DME-O_2-CO_2 mixtures using steady one-dimensional numerical simulations with detailed chemistry.

2 Modeling approach

The ZND model developed by Shepherd [3] and modified by Borzou and Radulescu [4] to account for lateral mass divergence was presently employed. The dynamics of the system is described by

$$\frac{dv}{dx'} = -\frac{1}{\rho v}\frac{dp}{dx'} = \frac{\dot{\sigma} - \dot{\sigma_A}}{1 - M^2}, \quad (1)$$

$$\frac{d\rho}{dx'} = -\frac{\rho}{v}\left(\frac{dv}{dx'} + \dot{\sigma_A}\right), \quad (2)$$

$$\frac{dY_i}{dx'} = -\frac{\omega_i}{\rho v}, \quad (3)$$

*Corresponding author: mevel@mail.tsinghua.edu.cn

with

$$\dot{\sigma} = \sum_{i=1}^{N} \left(\frac{\bar{W}}{W_i} - \frac{h_i}{C_P T} \right) \frac{\omega_i}{\rho} \tag{4}$$

and

$$\dot{\sigma_A} = (D - v) \frac{d \ln A}{dx}. \tag{5}$$

In Equation 1 to Equation 5, x' is the spatial coordinate in the detonation frame, v is the flow speed in the detonation frame, $x = Dt - x'$ is the laboratory frame coordinate, D is the detonation front in the laboratory frame, t is the time, p is the pressure, σ is the thermicity, M is the Mach number, ρ is the density, Y_i is the mass fraction, ω_i is the source term, N is the number of species, $\bar{W}$ is the mean molar mass, W_i is the molar mass, h_i is the enthalpy, C_P is the heat capacity at constant pressure, T is the temperature, $d \ln A/dx$ represents the rate at which the front expands laterally as well as the curvature of the front, and the subscript i represents the ith species.

All calculations were performed using the reduced reaction model developed by Pan et al. [5] for DME combustion. It consists of 29 species and 66 reactions and includes a detailed sub-mechanism for H_2-CO chemistry and a reduced model for C_1-C_2 chemistry.

3 Results and discussion

3.1 ZND structure of DME-O_2-CO_2 mixtures

The effect of carbon dioxide addition has been investigated using ZND simulations. The temperature and thermicity profiles obtained for different dilution levels are shown in Figure 1 and Figure 2, respectively. For CO_2 mole fraction as low as 70%, up to three inflection points in the temperature profiles can be observed, which result in several peaks in the thermicity profiles. The amplitude difference and distance between the peaks of thermicity can be as high as two orders of magnitude. The smallest length scale (distance to thermicity peak) is observed for a dilution of 75% and is on the order of few centimeters. For the same conditions, the third peak of thermicity is located at a distance of approximately 1.5 m. For comparison, CO_2 dilution of 85% was required to activate low-temperature chemistry in *n*-heptane-oxygen mixtures with smallest length scales on the order of 10 cm. In addition, only two peaks of thermicity were observed in the case of n-heptane-based mixtures.

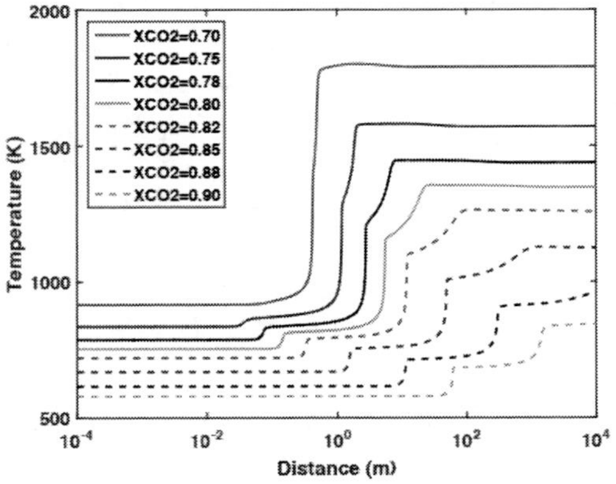

Figure 1: Temperature profiles of ZND detonation propagating in DME-O_2-CO_2 mixtures. Conditions: Φ=0.5; P_1=100 kPa; T_1=298 K.

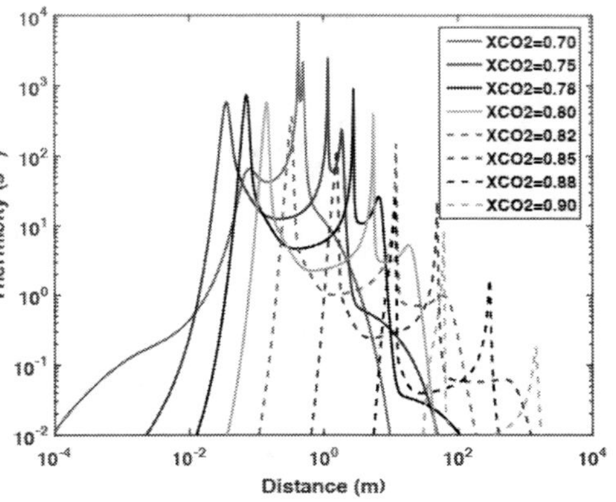

Figure 2: Thermicity profiles of ZND detonation propagating in DME-O_2-CO_2 mixtures. Conditions: Φ=0.5; P_1=100 kPa; T_1=298 K.

To explain the dynamics of the ZND structure of DME-O_2-CO_2 mixtures, a number of analyses has been performed.

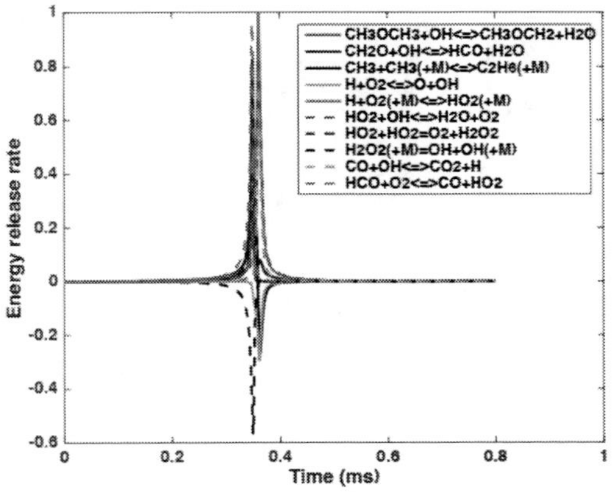

Figure 3: Energy release rate profiles in a ZND detonation wave propagating in a DME-O_2-CO_2 mixture. Conditions: Φ=0.5; X_{CO2}=0.60; P_1=100 kPa; T_1=298 K.

The energy release rate profiles are shown in Figure 3 and Figure 4 for 60% and 75% CO_2 dilution, respectively. For a dilution of 60%, the energy release is dominated first by CH_2O+OH=HCO+H_2O and HCO+O_2=CO+HO_2, and second by

H+O_2(+M)=HO_2(+M), HO_2+OH=H_2O+O_2 and CO+OH=CO_2+H. The two consecutive sequences are close enough to result in a monotonous energy release profile. Heat consumption is dominated by hydrogen peroxide decomposition and the chain branching reaction H+O_2=OH+O. For a dilution of 75%, three steps of energy release are observed. The first one is driven by the sequence: CH_3OCH_3+OH=CH_3OCH_2+H_2O and CH_3OCH_2+O_2=$CH_3OCH_2O_2$. The second one is due to the sequence: CH_2O+OH=HCO+H_2O and HCO+O_2=CO+HO_2. Finally, the third step is controlled by the sequence: H+O_2(+M)=HO_2(+M) and HO_2+OH=H_2O+O_2. Significant energy consumption is seen only during the second step of energy release and is mainly due to H_2O_2 decomposition.

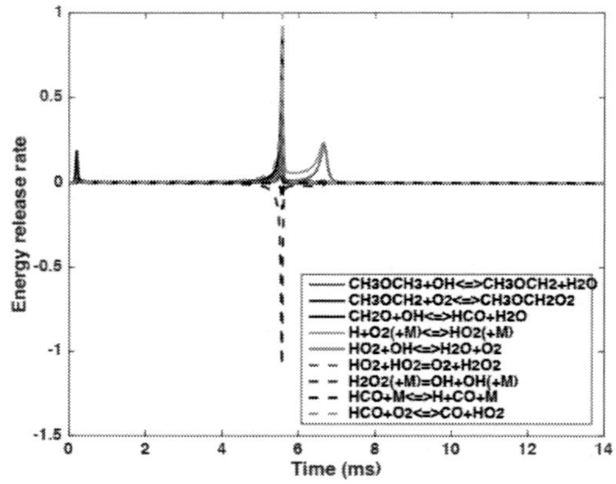

Figure 4: Energy release rate profiles in a ZND detonation wave propagating in a DME-O_2-CO_2 mixture. Conditions: Φ=0.5; X_{CO2}=0.75; P_1=100 kPa; T_1=298 K.

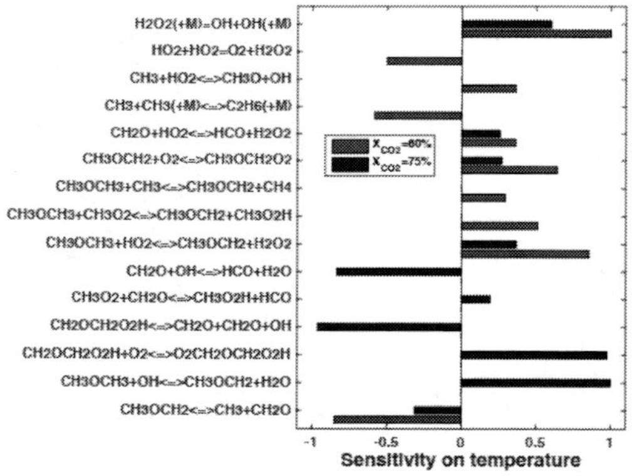

Figure 5: Sensitivity coefficient on temperature in a ZND detonation wave propagating in DME-O_2-CO_2 mixtures. Conditions: Φ=0.5; X_{CO2}=0.60 and 0.75; P_1=100 kPa; T_1=298 K.

The sensitivity coefficients on temperature are shown in Figure 5. For a dilution of 60%, the decomposition of H_2O_2 is the most sensitive reaction. Reactions that produce or consume H_2O_2 or its precursors also appear sensitive. The recombination of methyl radical exhibit an important negative sensitivity coefficient. For a dilution of 75%, the decomposition of hydrogen peroxide is still important but demonstrates a lower sensitivity coefficient. The reactions which form ($CH_2OCH_2O_2H$=CH_2O+CH_2O+OH) and consume (CH_2O+OH=HCO+H_2O) formaldehyde exhibit very high negative sensitivity coefficients. On the contrary, the formation of hydroperoxyketone ($O_2CH_2OCH_2O_2H$) and H-abstraction on DME by OH radical demonstrate high positive coefficients.

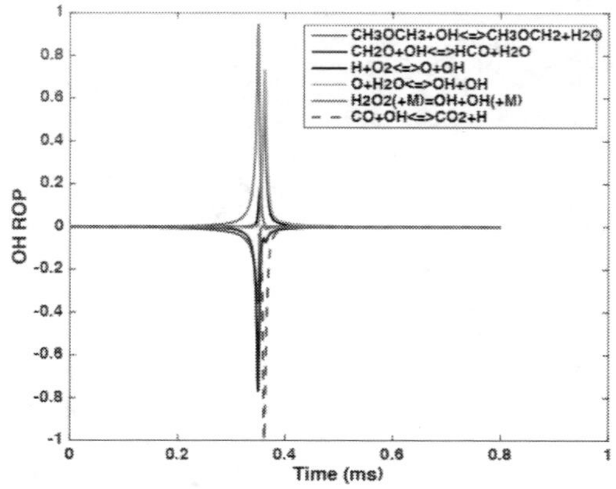

Figure 6: OH rate of production in a ZND detonation wave propagating in a DME-O_2-CO_2 mixture. Conditions: Φ=0.5; X_{CO2}=0.60; P_1=100 kPa; T_1=298 K.

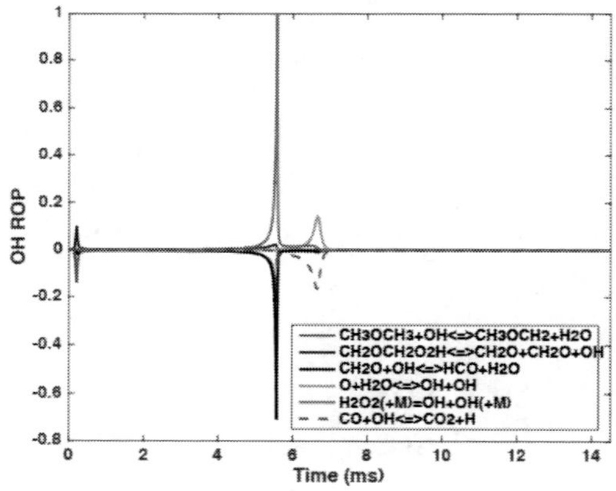

Figure 7: OH rate of production in a ZND detonation wave propagating in a DME-O_2-CO_2 mixture. Conditions: Φ=0.5; X_{CO2}=0.75; P_1=100 kPa; T_1=298 K.

The rates of production for OH are shown in Figure 6 and Figure 7. For a dilution of 60%, the production of OH is dominated by H_2O_2(+M)=2OH(+M), H+O_2=OH+O and O+H_2O=2OH whereas the consumption is due to CH_2O+OH=HCO+H_2O and CO+OH=CO_2+H. For a dilution of 75%, clear separation between the three steps of OH radical production is seen. During the first step, the production is due to $CH_2OCH_2O_2H$=CH_2O+CH_2O+OH and the consumption to CH_3OCH_3+OH=CH_3OCH_2+H_2O. Between the two first steps, accumulation of H_2O_2 and CH_2O is observed until the temperature is high enough to trigger the rapid decomposition of H_2O_2 and the consumption of the resulting OH radical by CH_2O. During the third stage, the production of OH is due to O+H_2O=2OH and its consumption to CO+OH=CO_2+H.

3.2 Effect of flow divergence

The generalized ZND model recently developed by Borzou and Radulescu [4] was employed to characterize the response to flow divergence of detonation propagating in DME-oxygen mixtures diluted with different amount of CO_2. Figure 8 shows the D_n-κ curves obtained for X_{CO2}=0, 0.5 and 0.75.

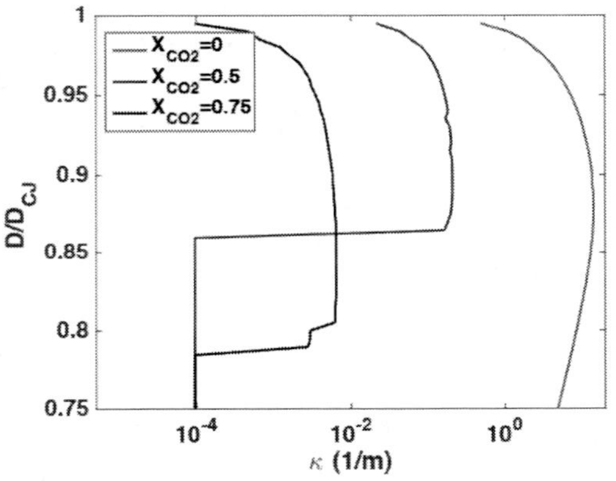

Figure 8: Detonation speed as a function of curvature for ZND detonation wave propagating in DME-O_2-CO_2 mixtures. Conditions: Φ=0.5; X_{CO2}=0, 0.5 and 0.75; P_1=100 kPa; T_1=298 K.

For the mixture without dilution, the evolution of the normalized detonation velocity as a function of curvature follows a typical evolution with a maximum curvature around 15 m^{-1} at D/D_{CJ} $\approx$0.85. As the dilution is increased to 50 and 75%, the maximum curvature is decreasing respectively to 0.21 and 0.064 m^{-1} for D/D_{CJ} in the range 0.8-0.9. Such values for the maximum curvature are unusually low and seem to indicate that cool detonation are particularly sensitive to flow divergence. Borzou and Radulescu [4] found that the lateral displacement due to boundary layer behind a detonation is on the order of 5.5 m^{-1} in a 19 mm wide channel. This result, along with our calculations, suggest that detonation propagating in highly CO_2 diluted DME-O_2 mixtures may not be able to sustain momentum losses induced by hydrodynamic boundary layer growth. Nevertheless, this information need to be considered with caution since the generalized ZND model is a one-dimensional simplification of a three-dimensional phenomenon. Various processes including ignition behind transverse shock waves [6], turbulent mixing, flow instability, and delayed burning of unreacted gas pockets behind the detonation front [7] have been pointed out previously as important features for detonation propagation and are believed to be responsible for extending the range of conditions in the D_n-κ space under which detonation propagation is possible [6, 7].

4 Conclusions

The structure of detonation propagating in dimethyl ether-oxygen-carbon dioxide mixtures has been studied using steady one-dimensional numerical simulation performed with detailed chemistry. For CO_2 dilution level above approximately 70%, the thermodynamic conditions behind the leading shock favor low-temperature chemistry and the energy release occurs through two or three successive steps. Simulations demonstrated the very high sensitivity of these mixtures to flow divergence. Given the low content of energy released during the oxidation step associated with low-temperature chemistry, momentum losses within the boundary layer may be sufficiently important to altogether quench the cool detonation in DME-O_2-CO_2 mixtures.

Acknowledgements

RM was supported by a start-up fund of the Center for Combustion Energy of Tsinghua University.

References

[1] Lee J.H.S., The Detonation Phenomenon, Cambridge University Press (2008).

[2] Liang W., Mével R. and Law C., Proceedings of the US Sections Meeting of the Combustion Institute, 10 (2017).

[3] Shepherd J.E., Progress in Astronautics and Aeronautics, 106 (1986) 263–293.

[4] Borzou B. and Radulescu M., Journal of Fluid Mechanics, (Submitted).

[5] Pan L., Kokjohn S. and Huang Z., Fuel, 160 (2015) 165–177.

[6] Xiao Q., Chang J., La Fleche M., Wang Y. and Radulescu M., Proceedings of the International Colloquium on the Dynamics of Explosions and Reactive Systems, (2017).

[7] Maxwell B., Bhattacharjee R., Lau-Chapelaine S., Falle S., Sharpe G. and Radulescu M., Journal of Fluid Mechanics, 818 (2017) 646–696.

11th Asia-Pacific Conference on Combustion,
The University of Sydney, NSW Australia
10th -14th December 2017.

High Temperature Unimolecular Decomposition of Cyclopentanone

Binod Raj Giri[1], Mohammed AlAbbad[1], John R. Barker[2] and Aamir Farooq[1]

[1]Clean Combustion Research Center, Division of Physical Sciences and Engineering, King Abdullah University of Science and Technology (KAUST), Thuwal, 23955-6900, Saudi Arabia

[2]Department of Climate and Space Sciences & Engineering, University of Michigan, Ann Arbor, Michigan 48109-2143, United States

Abstract

This work reports thermal decomposition of cyclopentanone behind reflected shock waves over the temperature range of 1150 – 1590 K and pressures between 750 and 1800 Torr. Carbon monoxide is one of the main reaction products and its formation was monitored using quantum cascade laser at wavelength near 4.559 μm. Our results suggest that cyclopentanone undergoes decomposition *via* reaction channels that solely produce CO under our experimental conditions. A recent *ab initio* study reveals that cyclopentanone may either open up the ring forming a di-radical which eventually leads to the formation of CO and ethylene or decompose *via* a concerted mechanism to directly produce CO and ethylene. However, at lower temperatures, the keto-enol tautomerism might play a significant role. To discriminate among the various channels, pressure and temperature dependence of the rate coefficients $k(T, p)$ for this multi-channel reaction system were calculated by solving time-dependent master equation.

1 Introduction

Among various potential biofuels, ketones present significant interest to the combustion community. Several researchers have demonstrated the idea of using chemical and biochemical conversion processes technology for biofuels production.[1] One of the technology involves the use of endophytic fungi to decompose lignocellulosic biomass, agro-industrial residues and wastewater into a whole array of volatile organic compounds (VOCs) such as saturated and unsaturated hydrocarbons, ketones, esters, acids, and alcohols.[1-4] These oxygenated compounds constitute advanced biofuels which can promote cleaner combustion and be environmental friendly due to emission reduction of the greenhouse gases and particulate matter. Ketones are also found as intermediates during the low temperature oxidation of large hydrocarbons or oxygenated fuels such as alcohols and esters. Therefore, the reactions kinetics involving ketones is important to understand the combustion behavior of these fuels.

Cyclopentanone (CPO) is a five- member ring cyclic ketone that can be efficiently produced from hydrogenation of biomass derived furfural[5], pyrolysis of biomass[6] and nylon[7]. Cyclopentanone is a high auto-ignition resistance fuel that can potentially be attractive for fuel blending applications in spark-ignition internal combustion engines. Also, it can be a promising feedstock for the synthesis of polycyclic alkanes[8,9] which are high-density fuels.[10] Therefore, these can be used as aviation fuels or as fuel blends to increase the volumetric heating value of conventional bio-jet fuels. Despite its importance in combustion, very few studies[11-17] are rendered in understanding the kinetic behavior of cyclopentanone.

The decomposition of cyclopentanone has been studied experimentally by using photochemical[11,12] and thermal[13,14] methods; and also using ab initio/master equation computational method[15]. The first study dates back to 1935 where Saltmarsh and Norrish[11] carried out photochemical decomposition of n-membered ($5\leq n \leq7$) ring cyclic ketones. They concluded that these cyclic ketones primarily decompose *via* simultaneous rupture of both C-C bonds adjacent to the carbonyl group forming CO and cyclic alkanes (n-1) as final products. However, in the case of cyclopentanone, they identified ethylene, butene isomers (C_4H_8) and CO as the decomposition products. They hypothesized that the decomposition of cyclic ketones readily produces cyclic hydrocarbons whenever there is a possibility of the formation of six or five carbon membered ring hydrocarbons. In a similar study, Benson and Kistiakowsky[12] suggested that the experiments of Saltmarsh and Norrish[11] who identified C_4H_8 as butene isomers was actually cyclobutene. They questioned the experimental techniques employed by Saltmarsh and Norrish[11] and further stated that the photolysis of cyclic ketones are much more complex than the simple hypothesis postulated by Saltmarsh and Norrish[11]. Johnson and Walters[13] studied the thermal decomposition of cyclopentanone in a cylindrical pyrex reaction vessel over the temperature range of 761 – 816 K and p = 99 – 314 Torr and reported the major products as CO, C_2H_4, H_2, 1-butene, 2-cyclopenten-1-one from the early stage of the reaction. Delles et al. [14] conducted similar experiments for T = 805 – 854 K and p = 11 – 30 Torr. They observed similar major products as that seen by Johnson and Walters[13], but they additionally identified 4-pentenal as one of the major products. Delles et al.[14] reported a rate expression ($k(T) = 1.5 \times 10^{12} \exp(-29642.7/T)$) for the thermal unimolecular decomposition of cyclopentanone for their experimental conditions. Both studies[13,14] have indicated that the kinetics for the decomposition of cyclopentanone is not simple. All of these experimental studies are confined to low temperatures $T \leq 850$ K only.

Recently, Zaras et al.[15] employed G3B3/master equation-RRKM computational methods to understand the importance of various decomposition pathways for cyclopentanone decomposition. The authors[15] were able to map out seven different reaction routes that led to the major decomposition products observed previously in the pyrolysis experiments. Their calculations showed that the reaction route leading to C_2H_4 and CO is the dominant channel and this channel was found to be significantly faster than the keto-enol tautomerism particularly at high temperatures. The authors further reported that enolization is less favored in cyclopentanone than in cyclohexanone due to large difference in the barrier heights of

Corresponding author. Fax: +00-00-000-000
E-mail address: binod.giri@kaust.edu.sa

the corresponding reactions. Their calculated values of the rate coefficients under-predicted the only existing low temperature experimental rate data of Delles et al.[14] by at least one order of magnitude. There are no other data available in the literature to validate the theoretical predictions of Zaras et al. [15] An additional study is, therefore, warranted to fully ascertain the kinetics of cyclopentanone decomposition. Our objectives behind this work are: (i) experimental measurements to deliver reliable rate coefficients for the thermal decomposition of cyclopentanone at high temperatures, (ii) statistical rate theory calculations to predict temperature- and pressure- dependence of the rate coefficients $k(T, p)$ to resolve the existing discrepancy, (iii) quantification of the role of keto-enol tautomerization channel.

2 Experimental Setup

Thermal unimolecular decomposition of cyclopentanone (CPO) was investigated in the low-pressure shock tube (LPST) facility at King Abdullah University of Science and Technology (KAUST). As the details of LPST can be found elsewhere[18], we provide here only a brief description. The LPST is made up of stainless steel which has 9 m long driver and driven sections with an inner diameter of 14.2 cm. The length of the driver section can be varied to achieve a reasonable reaction time behind reflected shock waves. This work used a 3 meter long driver section to achieve test times of about 1.5 ms. The driven section was separated from the driver section by a polycarbonate diaphragm. To ensure high purity of the shock tube, prior to each experiment, the driven section of the shock tube was pumped down to less than 10^{-5} Torr using turbo-molecular pump, whereas the driven section was evacuated to 10^{-3} Torr. Thereafter, the driven and driver sections were filled with cyclopentanone-argon mixture and helium gas, respectively. Shock waves were generated by gradually increasing the helium pressure in the driver section until diaphragm bursts. The incident shock wave speed was measured by a series of five PCB 113B26 piezoelectric pressure transducers (PZTs) placed in the last 1.3 m of the driven section. All experiments were conducted behind reflected shock waves, and the post-shock conditions (T_5, p_5) were calculated, with a maximum uncertainty of 1%, using standard shock-jump relations [19] embedded in the Frosh code. [20] Optical windows for CO laser diagnostic and a Kistler PZT were located at 2 cm from the shock tube endwall.

Carbon monoxide was monitored using the R(13) absorption line near 2193.36 cm^{-1} (4.56 μm) in the v_1 fundamental vibrational band of CO. A fixed-wavelength direct absorption strategy was used, where the mid-IR light was generated by a cw quantum cascade laser (QCL) supplied by Alpes Lasers. The mole fraction of CO was obtained using Beer-Lambert law: $I/I_0 = \exp(-S\phi x_i P_5 L)$ where I and I_0 are the transmitted and incident laser intensities, respectively; S (cm^{-2} atm^{-1}) is the line-strength of the transition, ϕ (cm) is the line-shape function, x_i is the mole fraction of the absorbing species, P_5 (atm) is the total pressure of the gas behind reflected shock wave and L (cm) is the absorption path length (internal diameter of the shock tube). The argon-broadening parameters for the R(13) transition were taken from Ren et al.[21], whereas line-strength was obtained from the HITRAN database.[22]

The gas mixtures of cyclopentanone (≥99% Sigma-Aldrich – further purified through freeze-pump-thaw cycles), diluted in Argon (99.999%, AHG) were prepared manometrically in a 24 litre teflon-coated stainless steel vessel equipped with a magnetically driven stirrer. Prior to mixture preparation, the vessel was pumped down to a pressure below 10^{-5} mbar. The mixtures were allowed to homogenize for at least two hours before use. The concentrations of cyclopentanone was varied between 0.2 – 1.2%.

3 Results and Discussion

The major reaction pathways for the unimolecular decomposition of cyclopentanone are shown in Fig. 1. The energetics for these pathways were taken from a recent theoretical work of Zaras et al. [15] who used G3B3 method to explore the potential energy surface. As can be seen, the energy barriers for the reaction pathways *via* TS4, TS5, TS6 and TS7 are very high compared to the channels proceeding *via* TS1, TS2 and TS3. The reported values of the rate coefficients by Zaras et al.[15] predicted that the contribution to these channels to the total decomposition rate of cyclopentanone is essentially zero.[15] This would mean that these energetically high lying channels can safely be ruled out for the analysis of our kinetic data. Consequently, there are three reaction pathways that are energetically accessible for the cyclopetanone decomposition under our experimental conditions. Among the three pathways, the pathway leading to cyclopentenol (keto-enol tautomerization) is the lowest energy path requiring an energy barrier of only 304 kJ/mol. Nevertheless, this value of the barrier height for enolization is found to be significantly higher than that of the similar process (cyclohexanone → cyclohexenol $\Delta E_0^{\neq} = 276$ kJ/mol)[23]. Through consecutive reaction sequences after enolization (R1), cyclopentanone yields various products, e.g., CO, H_2, C_2H_2, C_2H_4, 1,3-butadiene (see Fig. 4 of ref. 15 for reaction sequences).

CPO → cyclopentenol → → products (R1)

The reaction pathways via TS1 and TS2 lead to the same products, namely CO and C_2H_4. However, these two pathways differ in their identity in the potential energy surface. Cyclopentanone undergoes decomposition via TS2 in a concerted manner to produce CO and C_2H_4 (reaction R2); whereas in the other case, cyclopentanone first forms a diradical (•$CH_2(CH_2)_3C$•=O) through a barrierless process which eventually splits off via TS1 to finally produce CO and C_2H_4 (reaction R3).

CPO → CO + $2C_2H_4$ (R2)

CPO → •$CH_2(CH_2)_3C$•=O → CO + $2C_2H_4$ (R3)

R2 and R3 are accompanied by entropic gain, $\Delta S_{298.15K}^{\neq}$ = 18-38.5 and 40.2 J mol^{-1} K^{-1}, respectively, as opposed to the entropic loss $\Delta S_{298.15K}^{\neq}$ = -3.3 J mol^{-1} K^{-1} for reaction R1. This makes the reaction R1 least favored though its threshold energy is the least amongst the three channels. These analyses discussed above are based on the energetics and findings of the theoretical work by Zaras et al.[15]

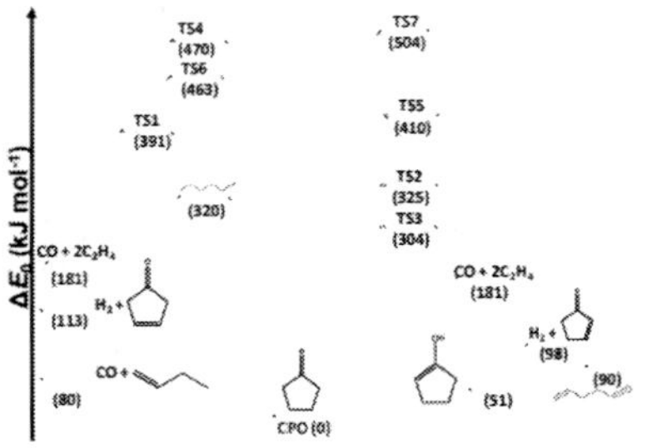

Figure 1: G3B3 energies for the major decomposition pathways of cyclopentanone (taken from reference 15).

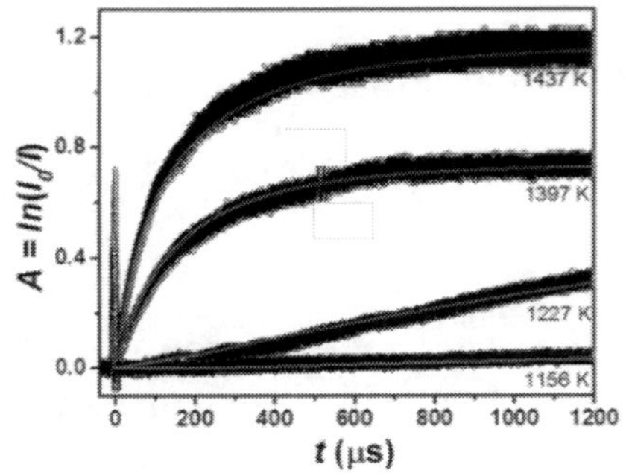

Figure 2: Representative absorbance-time-profiles for the formation of CO from the thermal decomposition of cyclopentanone. Solid lines indicate the best fit of kinetic simulations at the specified temperatures.

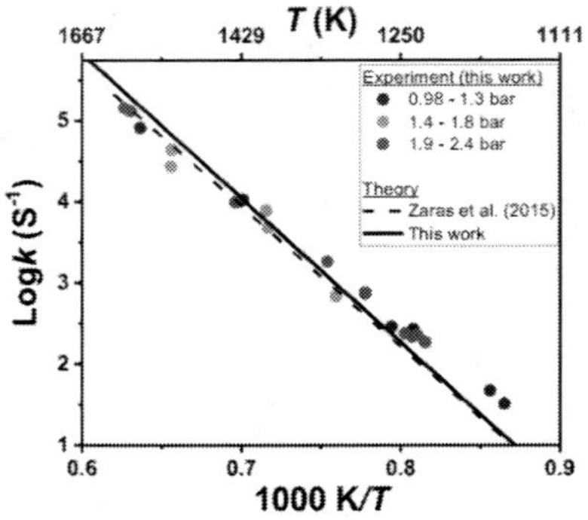

Figure 3: Arrhenius plot for the experimental and theoretical rate coefficients for the thermal decomposition of cyclopentanone. (filled circles): experimental data – this work; (—) G3B3/RRKM-master equation calculations – this work at 760 Torr; (----) G3B3/RRKM-master equation calculations – reference 15 for 760 Torr.

As discussed above, the most favored decomposition channels, R2 and R3, lead to the formation of CO and C_2H_4. Even the enolization channel (R1) eventually leads to the formation of energetically most favored channel forming CO and 1,3-butadiene. Because CO is one of the major products, we monitored the evolution of CO near 2193.36 cm^{-1} (4.56 μm) in our experiments to investigate the reaction kinetics of the thermal decomposition of cyclopentanone. As stated earlier, the experiments were carried out behind reflected shock waves over the temperature range of $T = 1150 - 1590$ K and $p = 750 - 1800$ Torr. The reaction rate coefficients $k(T, p)$ were extracted by iteratively varying the rate coefficients for R2 and R3 until the best fit for the measured absorbance-time- profiles of CO was obtained. The detailed kinetic model was adopted from Thion et al.[16] and the rate coefficients for R1, R2 and R3 were taken from Zaras et al.[15] as initial guess. Representative measured CO traces and best-fit kinetic simulations are shown in Fig. 2 and the extracted rate coefficients are plotted in Fig. 3. We note here that CO yield reaches unity ($[CO]_x/[CPO]_0 \approx 1$) at high temperatures and/or long reaction times. Moreover, we observed the evolution of CO starting at the onset of reaction at higher temperatures. This clearly indicates that reactions R2 and R2 are the major decomposition pathways. This observation is in line with the theoretical prediction of Zaras et al.[15]. As expected, our measured rate coefficients for pressures $p = 750 - 1800$ Torr did not exhibit any discernible pressure dependence. This would mean that our measured data are close to the high-pressure limit. As can be seen in Fig. 3, our experimental data compare very well with the theoretical predictions of Zaras et al.[15] However, there is a noticeable difference between the experimental and theoretical data as far as temperature dependence of the rate coefficients is concerned. The experimental data exhibit weaker T-dependence as opposed to the theoretical predictions. At the lower temperature end, our data show faster rate at least by a factor of two. This may indicate that the energy barrier for low lying channels may have been overestimated. In an attempt to understand the kinetics of cyclopentanone better, we computed pressure- and temperature-dependence rate coefficients using MultiWell Master Equation Program Suite[24]. For these calculations, we used the G3B3 parameters reported by Zaras et al.[15]. Argon was used as bath gas, and the energy transfer was treated using the conventional exponential-down model with $\langle \Delta E \rangle_d = 650\ cm^{-1}$. The Lennard-Jones parameters used were: $\sigma = 3.47$ Å and $\varepsilon/k_b = 114$ K for Ar and $\sigma = 5.8$ Å and $\varepsilon/k_b = 350$ K for CPO. Our calculated values of the rate coefficients agree well with the experimental data predicting that the decomposition of cyclopentanone leads exclusively to CO and C_2H_4 (≥98%) . However, the low temperataure data are not well captured by the theory and temperature dependence of the calculated rate coefficients did not improve either. Further improvement of the theoretical kinetic model is currently underway.

4 Conclusions

The kinetics for the thermal decomposition of cyclopentanone was studied behind reflected shock waves over $T = 1150 - 1590$ K and $p = 750 - 1800$ Torr. Our measured rate coefficients did not show any pressure effect indicating that the measured data are close to the high-pressure limit. Our measured yield for CO was found to be ~ 1 at high temperatures, suggesting that reactions R1 and R2 are the major channels. Our RRKM-Master Equation results match well with the experimental results. However, the weak T-dependence exhibited by the experimental data are not well predicted by the theory.

5 Acknowledgment

Research reported in this work was funded by King Abdullah University of Science and Technology (KAUST).

6 References

1. Luque, R.; Lin, C. S. K.; Wilson, K.; Clark, J., *Handbook of biofuels production Processes and Technologies*. Woodhead Publishing: Amsterdam u.a, 2016.

2. Strobel, G. A.; Knighton, B.; Kluck, K.; Ren, Y.; Livinghouse, T.; Griffin, M.; Spakowicz, D.; Sears, J., The production of myco-diesel hydrocarbons and their derivatives by the endophytic fungus Gliocladium roseum (NRRL 50072). *Microbiology* **2008,** *154* (11), 3319-3328.

3. Yousuf, A.; Sannino, F.; Addorisio, V.; Pirozzi, D., Microbial Conversion of Olive Oil Mill Wastewaters into Lipids Suitable for Biodiesel Production. *Journal of Agricultural and Food Chemistry* **2010,** *58* (15), 8630-8635.

4. Chen, X.-f.; Huang, C.; Xiong, L.; Chen, X.-d.; Chen, Y.; Ma, L.-l., Oil production on wastewaters after butanol fermentation by oleaginous yeast Trichosporon coremiiforme. *Bioresource Technology* **2012,** *118*, 594-597.

5. Hronec, M.; Fulajtarová, K., Selective transformation of furfural to cyclopentanone. *Catalysis Communications* **2012,** *24*,

100-104.
6. Demirbas, A., The influence of temperature on the yields of compounds existing in bio-oils obtained from biomass samples via pyrolysis. *Fuel Processing Technology* **2007,** *88* (6), 591-597.
7. Ballistreri, A.; Garozzo, D.; Giuffrida, M.; Montaudo, G., Mechanism of thermal decomposition of nylon 66. *Macromolecules* **1987,** *20* (12), 2991-2997.
8. Yang, J.; Li, N.; Li, G.; Wang, W.; Wang, A.; Wang, X.; Cong, Y.; Zhang, T., Synthesis of renewable high-density fuels using cyclopentanone derived from lignocellulose. *Chemical Communications* **2014,** *50* (20), 2572-2574.
9. Wang, W.; Li, N.; Li, G.; Li, S.; Wang, W.; Wang, A.; Cong, Y.; Wang, X.; Zhang, T., Synthesis of Renewable High-Density Fuel with Cyclopentanone Derived from Hemicellulose. *ACS Sustainable Chemistry & Engineering* **2017,** *5* (2), 1812-1817.
10. Chung, H. S.; Chen, C. S. H.; Kremer, R. A.; Boulton, J. R.; Burdette, G. W., Recent Developments in High-Energy Density Liquid Hydrocarbon Fuels. *Energy & Fuels* **1999,** *13* (3), 641-649.
11. Saltmarsh, O. D.; Norrish, R. G. W., 98. Primary photochemical reactions. Part VI. The photochemical decomposition of certain cyclic ketones. *Journal of the Chemical Society (Resumed)* **1935,** (0), 455-459.
12. Benson, S. W.; Kistiakowsky, G. B., The Photochemical Decomposition of Cyclic Ketones. *Journal of the American Chemical Society* **1942,** *64* (1), 80-86.
13. Johnson, E. R.; Walters, W. D., The Thermal Decomposition of Cyclopentanone1,2. *Journal of the American Chemical Society* **1954,** *76* (24), 6266-6271.
14. Delles, F. M.; Dodd, L. T.; Lowden, L. F.; Romano, F. J.; Daignault, L. G., The Pyrolysis of Cyclopentanone. *Journal of the American Chemical Society* **1969,** *91* (27), 7645-7647.
15. Zaras, A. M.; Thion, S.; Dagaut, P., Computational Kinetic Study for the Unimolecular Decomposition of Cyclopentanone. *International Journal of Chemical Kinetics* **2015,** *47* (7), 439-446.
16. Thion, S.; Togbé, C.; Dayma, G.; Serinyel, Z.; Dagaut, P., Experimental and Detailed Kinetic Modeling Study of Cyclopentanone Oxidation in a Jet-Stirred Reactor at 1 and 10 atm. *Energy & Fuels* **2017,** *31* (3), 2144-2155.
17. Scheer, A. M.; Welz, O.; Vasu, S. S.; Osborn, D. L.; Taatjes, C. A., Low temperature (550-700 K) oxidation pathways of cyclic ketones: dominance of HO2-elimination channels yielding conjugated cyclic coproducts. *Physical Chemistry Chemical Physics* **2015,** *17* (18), 12124-12134.
18. Badra, J.; Khaled, F.; Giri, B. R.; Farooq, A., A shock tube study of the branching ratios of propene+ OH reaction. *Physical Chemistry Chemical Physics* **2015,** *17* (4), 2421-2431.
19. Bradley, J., Shock waves in physics and chemistry. *Methuen, London* **1962**.
20. Campbell, M. F.; Haylett, D. R.; Davidson, D. F.; Hanson, R. K., AEROFROSH: a shock condition calculator for multi-component fuel aerosol-laden flows. *Shock Waves* **2015**, 1-19.
21. Ren, W.; Farooq, A.; Davidson, D. F.; Hanson, R. K., CO concentration and temperature sensor for combustion gases using quantum-cascade laser absorption near 4.7 μm. *Applied Physics B* **2012,** *107* (3), 849-860.
22. Rothman, L. S.; Gordon, I. E.; Babikov, Y.; Barbe, A.; Chris Benner, D.; Bernath, P. F.; Birk, M.; Bizzocchi, L.; Boudon, V.; Brown, L. R.; Campargue, A.; Chance, K.; Cohen, E. A.; Coudert, L. H.; Devi, V. M.; Drouin, B. J.; Fayt, A.; Flaud, J. M.; Gamache, R. R.; Harrison, J. J.; Hartmann, J. M.; Hill, C.; Hodges, J. T.; Jacquemart, D.; Jolly, A.; Lamouroux, J.; Le Roy, R. J.; Li, G.; Long, D. A.; Lyulin, O. M.; Mackie, C. J.; Massie, S. T.; Mikhailenko, S.; Müller, H. S. P.; Naumenko, O. V.; Nikitin, A. V.; Orphal, J.; Perevalov, V.; Perrin, A.; Polovtseva, E. R.; Richard, C.; Smith, M. A. H.; Starikova, E.; Sung, K.; Tashkun, S.; Tennyson, J.; Toon, G. C.; Tyuterev, V. G.; Wagner, G., The HITRAN2012 molecular spectroscopic database. *Journal of Quantitative Spectroscopy and Radiative Transfer* **2013,** *130*, 4-50.
23. Zaras, A. M.; Dagaut, P.; Serinyel, Z., Computational Kinetic Study for the Unimolecular Decomposition Pathways of Cyclohexanone. *The Journal of Physical Chemistry A* **2015,** *119* (28), 7138-7144.
24. J. R. Barker, T. L. N., J. F. Stanton, C. Aieta, M. Ceotto, F. Gabas, T. J. D. Kumar, C. G. L. Li, L. L. Lohr, A. Maranzana, N. F. Ortiz, J. M. Preses, J. M. Simmie, J. A. Sonk, and P. J. Stimac *MultiWell-2017 Software Suite*, 2017.

11th Asia-Pacific Conference on Combustion,
The University of Sydney, NSW Australia
10th -14th December 2017.

Characteristics of Gasoline/Methane Dual-Fuel Combustion in a Spark-Ignited Engines

Nan Li, Haiqiao Wei, Jiaying Pan*, Jianxiong Hua, Gequn Shu

State Key Laboratory of Engines, Tianjin University

No.92 Weijin Road, Tianjin, 300072, China

Abstract

Methane as an attractive alternative fuel offers the most potential in clean combustion and low CO_2 emissions. In this work, combustion characteristics of gasoline/methane dual-fuel have been investigated in a spark-ignited engine with port-injection of methane and direct-injection of gasoline, which allows for the variations of methane addition and excess air coefficient. The experimental results show that under low load conditions, as methane mass rate is raised, gasoline/methane dual-fuel combustion is promoted, and this becomes more obvious at lean condition. Similar observations can still be observed when the spark-ignited engine is operated at intermediate load conditions, but a prolonged combustion duration is observed with methane addition. It seems that methane addition plays an important role in the combustion duration under low load conditions, while excessive air coefficient is more dominant under intermediate load conditions. Present research provides a useful result to obtain the optimal ratio of gasoline/methane dual-fuel combustion in a spark-ignited engine.

1. Introduction

Nowadays the increasing concerns on fossil fuel storage and environment pollutions require more suitable solutions for reducing fuel consumption and exhaust emissions of internal combustion (IC) engines. As an attractive alternative fuel, methane shows the most potential to achieve large-scale applications in IC engines thanks to its advantages in energy saving and pollutant emissions [1].

Up to now, there are two main ways for using methane as fuel in IC engines. The first one is the gas engine where methane (or compressed natural gas, CNG) is either introduced into intake port before entering combustion chamber or directly injected into engine cylinder [2]. It was found that unless direct-injection strategy is adopted, gas engines with port fuel injection will suffer from a loss in volumetric efficiency over 10% in comparison to the same sized gasoline engine. Moreover, higher valve and seat wear rates are also observed due to the absence of liquid fuel spray acting as a cooling agent and lubricant [3-4]. Another popular way without significantly technological modifications is the dual-fuel concept, where most energy comes from gas fuel and minimum amount of liquid fuel such as diesel acts as ignition source [5].

To explore the synergy effect of gaseous methane and liquid gasoline fuel, Iorio et al. [6] carried out optical SI engine experiments under low load conditions, with port-injection of gasoline and direct-injection of methane. They interestingly found that the flame speed of gasoline/methane dual-fuel is higher than those of pure fuel scenarios, and possible reasons were attributed to the mixture stratification caused by fuel direct-injection. A recent laminar flame speed experiment by Wei et al. [7] showed that even under well homogeneous fuel/air mixture conditions, the higher dual-fuel flame speed still exists, especially under low pressure and lean conditions. However, this flame acceleration behavior is not observed in the experiments performed by Baloo et al. [8]. In addition, it is generally recognized that methane has a slow burning velocity and large cycle-by-cycle variation [9], but Genchi et al. [10] found that combustion instability can be effectively improved when gasoline/methane dual-fuel strategy is adopted. These results indicate that there are still many ambiguities in the combustion mechanism of gasoline/methane dual-fuel in actual combustion systems. Therefore, further investigations on the combustion characteristics of gasoline/methane dual-fuel under different conditions are crucially important before extensive applications of gasoline/methane dual-fuel mode into modern SI engines.

In current work, the combustion characteristics of gasoline/methane dual-fuel have been explored on a turbocharged direct-injection SI engine, with allowing for the variations in methane addition and excessive air coefficient. Port-injection of methane and direct-injection of gasoline strategy was adopted, and engine combustion was addressed under the low-to-intermediate load conditions. Current research results shall give insights into combustion optimization and product design of gasoline/methane (or gasoline/natural gas) dual-fuel SI engines.

2. Experimental apparatus and procedure

2.1 Experimental setup

The gasoline/methane dual-fuel experiment was performed on a four-cylinder, turbocharged, direct-injection, spark-ignited engine, further detailed technical parameters can be found in Table 1. Engine speed was controlled by an electric eddy current dynamometer at 1600 per minute (rpm). Fuel injection timing, injection duration and spark-ignition timing were controlled through a semi-open ECU manufactured by INCA system. Air-fuel ratio of mixture was measured by a wideband lambda sensor with a resolution of 0.001. Excessive air coefficient was closed-loop regulated by the fuel injection duration. The spark plug installed at cylinder 4# was a spark-plug typed piezo-electric pressure transducer, Kistler 6118BFD. The pressure signals measured were sent to a charge amplifier (Kistler 5018) and finally acquired by a SKLE combustion analyzer with an automatic TDC identifier. Pressure sampling with a 0.1 CAD resolution was obtained through a photoelectric encoder connected to the crankshaft. Intake air mass was measured by a flow sensor, and fuel mass rate was measured by a fuel

Corresponding author: Jiaying Pan
E-mail address: jypan@tju.edu.cn

consumption meter with an uncertainty of ±0.2%.

Engine type	Four cylinder, 4 stroke
Bore × stroke	84 mm × 90 mm
Displacement	2000 cc
Compression ratio	10:1
Spark plug timing	22 CAD bTDC
Rated torque	290 N•m
Intake valve opening /closing timing	384 CAD bTDC/ 120 CAD bTDC
Exhaust valve opening/ closing timing	558 CAD bTDC/ 306 CAD bTDC

Table 1: Specifications for the direct-injection SI engine

Considering large amounts of methane required in this experiment, high-purity natural gas (CH_4-92.6%, C_2H_6-3.69%, C_3H_8-0.66%, N_2-1.35%, CO_2-1.41%, etc.) was adopted here to represent methane fuel. It was found that small amounts of other components do not have significant impact on methane combustion [11]. The direct injected 92# gasoline was supplied by Shell in China. Initially methane fuel was compressed within a high-pressure bottle with 120 L in volume. In the experiments, the pressure of compressed methane was reduced to 0.3 MPa. The gaseous fuel injector was open-loop controlled by a Freescale Microcontroller. Methane mass rate was regulated through a Freescale Microcontroller, and a gas consumption instrument (ToCeiL-CMF010) was adopted to measure the amounts of methane fuel with an uncertainty of ±0.35%.

2.2 Experimental procedure

In this experiment, engine speed was fixed at 1600 rpm and two torque points (85 and 150 N·m) were selected to respectively explore the combustion characteristics of gasoline/methane dual-fuel SI engines under low and intermediate load conditions. intake temperature was kept at 27 ℃, coolant water temperature 82 ℃, lubricant oil temperature 87 ℃ before each engine experiment. 200 continuous engine cycles were carried out at each operation point and the average results were utilized to analyze dual-fuel combustion characteristics.

In this work, injection duration of gasoline and methane was simultaneously regulated to obtain the required dual-fuel mixture and excessive air coefficient. Excessive air coefficient λ>0 means lean combustion and methane addition is evaluated through an energy ratio defined in Equation 1:

$$\eta = \frac{M_{CH_4} \cdot LHV_{CH_4}}{M_{CH_4} \cdot LHV_{CH_4} + M_{gasoline} \cdot LHV_{gasoline}} \quad (1)$$

where η is the energy ratio of methane in dual-fuel mixture, M_{CH_4} the methane mass fraction, LHV_{CH_4} the methane lower heating value, $M_{gasoline}$ the gasoline mass fraction and $LHV_{gasoline}$ the methane lower heating value. The correlations of energy ratio and methane mass rate at low load conditions are presented in Table 2. It shows that as methane mass rate is raised, there is a linear increase in the energy ratio of methane in dual-fuel mixture. When methane mass rate reaches 2.5 kg/h, its corresponding energy ratio is beyond 80%, suggesting the dominance of methane combustion in SI engines. Meanwhile, this phenomenon is especially obvious for lean combustion conditions.

Mass rate of methane ψ_{CH4} (kg/h)	Energy ratio of Methane in mixture (%)			
	λ=1.0	λ=1.1	λ=1.2	λ=1.3
0.0	0.0	0.0	0.0	0.0
0.5	17.93	16.85	16.63	16.21
1.0	35.66	33.95	33.67	35.21
1.5	46.77	50.25	50.75	53.13
2.0	64.05	65.01	67.57	69.22
2.5	79.12	80.82	83.02	85.19

Table 2: Energy ratio of methane in dual-fuel mixture at low load conditions

3. Results and discussions

To illustrate the combustion characteristics of gasoline/methane dual-fuel under low load conditions, Fig. 1 shows the in-cylinder pressure and corresponding heat release rate with the variations of excessive air coefficient and methane mass rate under torque output of 85 N•m and spark advance angle of 22 deg. bTDC conditions. It is observed that under the same excessive air coefficient conditions, there is an obvious increase in the peak of in-cylinder pressure and advance in combustion phasing as methane mass rate is raised. This variation tendency becomes more obvious at higher excessive air coefficient conditions, e.g. when methane mass rate is raised from 0.0 to 2.5 kg/h at λ=1.0, the peak of in-cylinder pressure increases from 3.3 to 3.7 MPa, while it is from 2.4 to 3.3 MPa at λ=1.3 conditions. Under the same methane mass rate conditions, as the increases of excessive air coefficient, the peak of in-cylinder pressure decreases and combustion phasing is retarded due to the combustion instability at lean conditions. Corresponding heat release rate shows a similar variation tendency to in-cylinder pressure at different operation conditions. But it is also noted that for λ=1.0 case, there is no obvious variations in the peak of heat release rate as methane mass rate is raised, while the peak of heat release rate increases obviously for λ>1.0 cases. These observations suggest that methane addition can promote dual-fuel combustion, especially under lean combustion conditions.

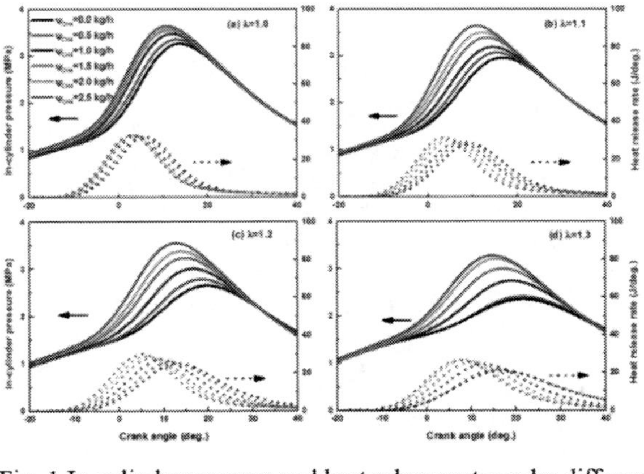

Fig. 1 In-cylinder pressure and heat release rate under different methane mass rates and excessive air coefficients conditions

To quantify the effects of methane addition on gasoline/methane dual-fuel combustion, Fig. 2 shows the ignition delay (CA0-CA10), combustion duration (CA10-CA90) and CA50 location under different methane mass rates and excessive air coefficients conditions. It is observed that under the same methane mass rate conditions, ignition delay time is prolonged and CA50 location is delayed as the increases of excessive air coefficient. Under the same excessive air coefficient conditions, ignition delay time is linearly shortened and CA50 location is linearly advanced as methane mass rate is raised from 0.0 to 2.5 kg/h. This variation tendency becomes more obvious at higher excessive air coefficient conditions. In terms of combustion duration, it shows an increasing tendency as methane mass rate is raised within certain range for λ=1.0 case, while it shows a decreasing tend when lean combustion strategy (λ>1.0) is adopted. Meanwhile, the larger excessive air coefficient is, the more methane mass rate is needed to satisfy a shorter combustion duration. These observations indicate that the improvement of methane addition in dual-fuel combustion is caused by the shortage in ignition delay time and combustion duration and the advance in CA50 location, especially at lean combustion conditions.

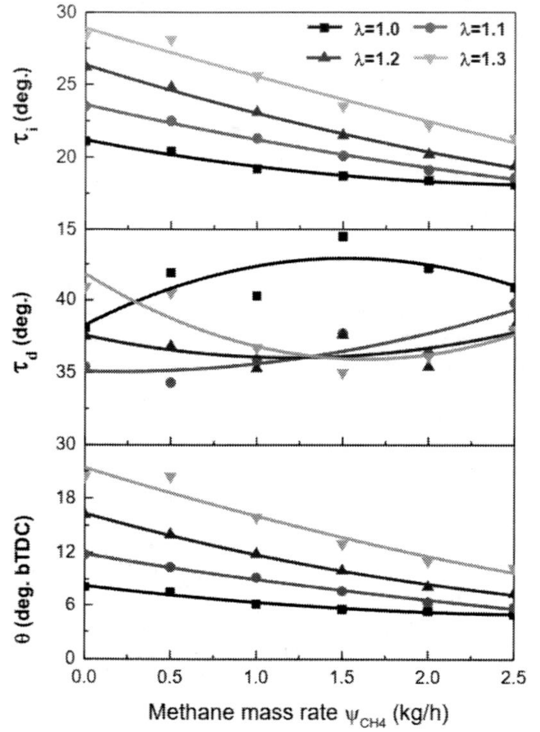

Fig. 2 Ignition delay (τ_i), combustion duration (τ_d) and CA50 location (θ) under different methane mass rates and excess air coefficients conditions

To reveal the underlying reasons for why methane addition promoting combustion observed, Fig. 3 shows the evolutions of in-cylinder pressure and heat release rate at different operation conditions. It is observed from Fig. 3(a) that as the increases of excessive air coefficient, much more air will be taken into engine cylinder, such that there is some increases in in-cylinder pressure before combustion. However, this improvement in thermodynamic state does not promote the peak increase of in-cylinder pressure because of lean combustion. Fig. 3(b) shows that for given excessive air coefficient conditions, as the increases of methane mass rate, there is initially no evident regularity for the in-cylinder pressure trace, e.g. the crank angle intervals of CA=-120~-116 deg.. As piston moves up and reaches CA=-60~-56 deg., in-cylinder pressure begins to vary orderly depending on methane mass rate, and this is more obvious when the piston is located at CA=-16~-12 deg. and later. This is because that for given torque output, increasing methane mass rate will reduce the amounts of gasoline fuel, such that the endothermic heat by liquid fuel evaporation decreases and thus in-cylinder thermodynamic state increases during compression stroke. The improvement of charge thermodynamic state will facilitate initial flame kernel formation. Besides, it is observed from the in-cylinder pressure evolutions at CA=-16~10 deg. that the case with methane mass rates of 0.5 kg/h shows a faster flame kernel development in comparison to that of 0.0 kg/h case. This may be due to the chemical property of gaseous methane fuel.

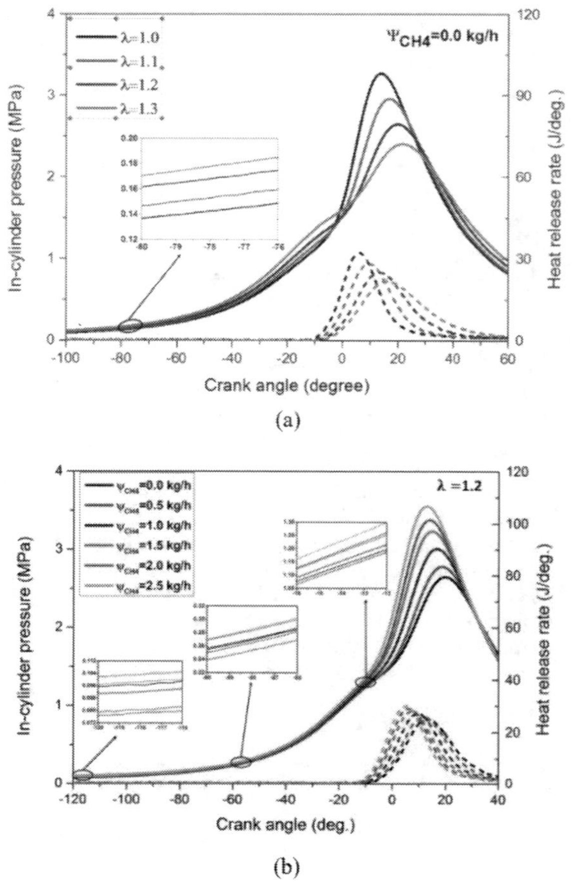

Fig. 3 Evolutions of in-cylinder pressure and heat release rate of (a) different methane mass rates at λ=1.2 and (b) different excess air coefficients at methane mass rates of 0.0 kg/h

Further combustion characteristics of gasoline/methane dual-fuel under intermediate load conditions have been studied, where torque output of 150 N•m and spark advance angle of θ_{sp}=16 deg. bTDC has been adopted. Fig. 4 presents the results of ignition delay time, combustion duration and CA50 location under different methane mass rates and excess air coefficients conditions. It is observed that as the increases of methane mass rate and excessive air coefficient, ignition delay time and CA50 location globally show a similar variation trend to that of low load conditions, even though there is an obvious decrease in ignition delay time. Meanwhile, different from the scenarios at low load conditions, combustion duration shows an increase with methane mass rate and a decrease with excessive air coefficient. In terms of the impact from methane addition, this may be due to that the Markstein length of methane fuel at high ambient pressure will become lower and thus its flame velocity is relatively slow compared with gasoline fuel [11-12]. To reveal the influence mechanism from excessive air coefficient, Fig. 5 further shows the accumulated heat release of the cases with methane mass rates of 2.0 and 4.0 kg/h cases at both stoichiometric and lean conditions. It is observed that despite the delayed of CA10 location at higher excessive air coefficient, the CA90 location of all cases seems almost located at the same crank angle degree. This suggests that the late stage of combustion is promoted under lean conditions and thus the combustion duration is decreased consequently.

From above results, it can be obtained that methane addition shows different influences on gasoline combustion, depending on SI engine load output. Under low load conditions, as the increases of methane mass rate, there is an improvement in gasoline/methane dual-fuel combustion, characterized by the advancement of CA50 location and the shortage of ignition delay time and combustion duration. Similar observations are still observed under intermediate load conditions, but a prolonged combustion duration is founded as the methane addition. Meanwhile, it seems that methane addition plays an important role in the combustion duration under low load conditions, while excessive air coefficient is more dominant under intermediate load conditions. But further increases in engine torque output to full load regime with wide open throttle, the promotion of methane addition in gasoline/methane dual-fuel combustion are not observed anymore.

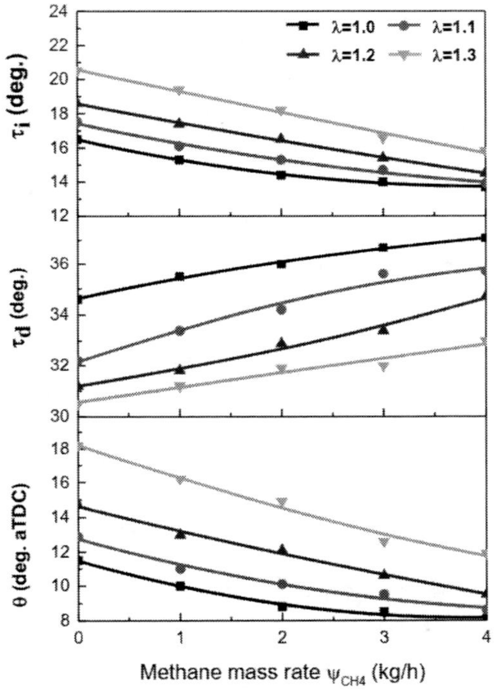

Fig. 4 Ignition delay (τ_i), combustion duration (τ_d) and CA50 location (θ) under different methane mass rates and excess air coefficients conditions

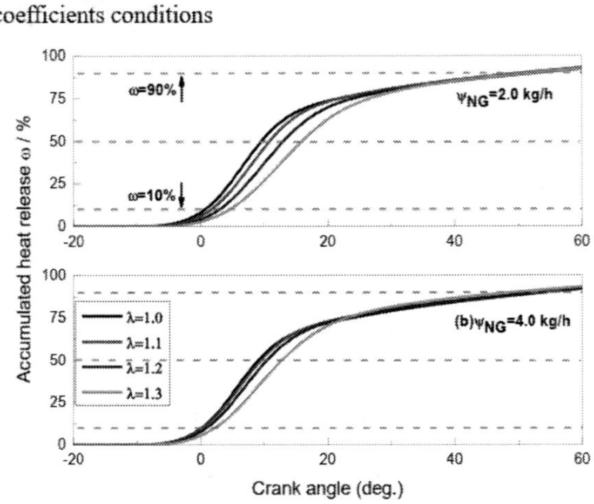

Fig. 5 Accumulated heat release of the cases with methane mass rates of 2.0 and 4.0 kg/h cases at stoichiometric and lean combustion conditions

4. Conclusions

Combustion characteristics of gasoline/methane dual-fuel were experimentally explored on a SI engine with port-injection of methane and direct-injection of gasoline in current work. Operation conditions with the variations in engine load, methane addition and excessive air coefficient were considered. The conclusions obtained are as follows:

First, under low load conditions, an improvement in gasoline/methane dual-fuel combustion is observed as the increases of methane mass rate, especially under lean combustion conditions, characterized by the increases in the peak of in-cylinder pressure and heat release rate. Further combustion analysis shows that there is a shortage of ignition delay time and combustion duration and an advanced of CA50 location with methane addition.

Second, the promotion of methane addition in gasoline/methane dual-fuel combustion can be still observed at intermediate load. But it is further found that combustion duration shows a decrease with excessive air coefficient and an increase with methane mass rate. The reasons can be ascribed to the enhancement of late-stage combustion under lean conditions and the fuel properties (e.g. Markstein length), respectively.

Third, methane addition seems to play an important role in the combustion duration under low load conditions, while excessive air coefficient is more dominant under intermediate load conditions. But further increases in engine torque output to full load regime with wide open throttle, the promotion of methane addition in gasoline/methane dual-fuel combustion are not observed anymore.

5 Acknowledgment

This work was supported by Major Research Plan of National Natural Science Foundation of China (91641203), National Natural Science Foundation of China (Grant No. 51476114, 51706152), China Postdoctoral Science Foundation (2016M590201), Tianjin Natural Science Foundation (17JCZDJC31500), Tianjin Science & Technology Supporting Program (13ZCZDGX04400).

References

[1] Z. Zakaria, S.K. Kamarudin, Direct conversion technologies of methane to methanol: An overview, Renew Sust Energ Rev, 65 (2016) 250-261.

[2] M. Choi, J. Song, S. Park, Modeling of the fuel injection and combustion process in a CNG direct injection engine, Fuel, 179 (1) (2016) 168-178.

[3] V. Delpech, J. Obiols, D. Soleri, L. Mispreuve, et al., Towards an Innovative Combination of Natural Gas and Liquid Fuel Injection in Spark Ignition Engines, SAE Int J Fuels Lubr, 3(2) (2010) 196-209.

[4] B. Yadollahi, M. Boroomand, The effect of combustion chamber geometry on injection and mixture preparation in a CNG direct injection SI engine, Fuel, 107 (2013) 52-62.

[5] L. Wei, P. Geng, A review on natural gas/diesel dual fuel combustion, emissions and performance, Fuel Process Technol, 142 (2016) 264-278.

[6] S. Di Iorio, P. Sementa, B. Vaglieco, F. Catapano, An

experimental investigation on combustion and engine performance and emissions of a methane-gasoline dual-fuel optical engine, SAE Technical Paper 2014-01-1329, 2014.

[7] H. Wei, D. Gao, L. Zhou, S. Petrakides, R. Chen, D. Feng, J. Pan, Experimental study on laminar flame characteristics of methane-PRF95 dual fuel under lean burn conditions, Fuel, 185 (2016) 254-262.

[8] M. Baloo, B. Dariani, M. Akhlaghi, I Chitsaz, Effect of iso-octane/methane blend on laminar burning velocity and flame instability, Fuel 144 (2015) 264–273

[9] S. Petrakides, R. Chen, D. Gao, H. Wei, Experimental Investigation on the Laminar Burning Velocities and Markstein Lengths of Methane and PRF95 Dual Fuels, Energ Fuel, 30(2016) 6777-6789.

[10] G. Genchi, E. Pipitone, Octane Rating of Natural Gas-Gasoline Mixtures on CFR Engine, SAE Int. J. Fuels Lubr., 7 (3) (2014) 1041-1049.

[11] C.K. Law, Combustion physics, Cambridge University Press, 2006.

[12] Y. Chen, Y. Wang, R. Raine, Correlation between cycle-by-cycle variation, burning rate, and knock: A statistical study from PFI and DISI engines, Fuel, 206 (15) (2017) 210-218.

[13] D.R. Derek, Lean combustion: technology and control, Academic Press, 2011.

11th Asia-Pacific Conference on Combustion,
The University of Sydney, NSW Australia
10th -14th December 2017.

Computation of gases radiation heat transfer for oxy-char combustion with weighted sum of gray gases model

Shiquan Shan[1], Liping Chen[1], Longzhen Guo[1], Licheng Shao[1], Zhijun Zhou[1*]

[1] State Key Laboratory of Clean Energy Utilization, Zhejiang University

38, Zhejiang University Road, Hangzhou, Zhejiang, 310027, P.R. China

Abstract

In this paper, a weighted sum of gray gases (WSGG) model was proposed for the flue gas radiation characteristics calculation of oxy-char combustion, and the coal char could come from coal pyrolysis poly-generation. Compared with the previous oxy-fuel WSGG model for coal or other fuels, new model could more accurately calculate the emissivity of oxy-char combustion flue gas. In several one-dimensional cases, the calculation results of the new model were similar to those of benchmark model. Therefore, the new model can be used in the numerical calculation of the actual combustion furnace; and it has great significance to the design and calculation of oxy-char combustion furnace.

1 Introduction

Coal is the main source of energy in China and many developing countries. Coal-based poly-generation technology can improve the energy utilization efficiency and resource utilization efficiency and is a very promising technology [1]. The poly-generation technology based on the coal pyrolysis could produce coal gas, tar and coal char, so as to further processing or use and achieve the graded utilization of energy and resources [2, 3]. The coal char produced by coal pyrolysis poly-generation can be used directly for power generation; what's more, the oxy-fuel combustion technology proposed in recent years can enrich CO_2 in combustion process and achieve CO_2 control in power plants [4-6]. When the pyrolysis poly-generation coal char is used for power generation with oxy-fuel combustion, not only the coal will be graded utilized, but also the carbon dioxide will be controlled, therefore, clean and efficient utilization of coal could be achieved.

For oxy-fuel combustion flue gas, its radiation characteristics have changed tremendously due to the increase in the molar fraction of CO_2 and H_2O [7], moreover, the radiation heat transfer is very important for the design of the boiler. However, there are few studies on oxy-char combustion, and there is little research on its radiation heat transfer characteristics. In many gas radiation characteristic models, the weighted-sum-of-gray-gases (WSGG) model is accurate and efficient; thus it is suitable for numerical calculation [8]. Yin [9] and Johansson [10] have proposed WSGG models for oxy-fuel combustion conditions; however, they are mainly for pulverized coal and other fuels combustion conditions. There will be very high CO_2 concentration of flue gas in oxy-char conditions, its radiation characteristics are different from those of other fuels, therefore, the previous models are no longer suitable for oxy-char combustion.

The main purpose of this paper is to propose a new WSGG model for oxy-char combustion conditions, which is based on the statistical-narrow-band (SNB) model as a benchmark, and the new model is suitable for the path-length range of 0.001-60 m and the temperature range of 500-2500 K. On this basis, the model proposed in this paper and the models in other publications were used to calculate the emissivity and the results were compared with the benchmark to verify the applicability of different models in oxy-char conditions. Finally, the new model was applied to a number of one-dimensional cases; the results were compared with those of the SNB benchmark model to verify the accuracy of the new model. The final results indicated that the new model had very good accuracy and could be used for the calculation of the flue gas radiation characteristics in oxy-char combustion.

2 Models and methods

2.1 SNB model

The statistical narrowband model is one of the narrowband models, it assumes that the spectral intensity in a narrow band is in terms of statistical distribution, and the bandwidth is usually chosen in the wavenumbers range of 5-50 cm^{-1}. The results of SNB model are closest to those of line-by-line method; it is generally believed that the SNB model given by EM2C laboratory [11] has the most accuracy [12], and its bandwidth is 25cm^{-1}. Therefore, the SNB model of EM2C laboratory was selected as a benchmark in this research.

2.2 WSGG model

The WSGG is an overall calculation model for gas radiation characteristics. It is widely concerned for its high calculation speed and adequate calculation accuracy. The WSGG model takes the gas emissivity as the calculation object. It assumes that the non-gray gas mixture is a combination of a number of gray gases and a white gas; it represents the actual non-gray gas emissivity with a weighted sum of emissivity of every gray gases. The non-gray gas emissivity expressed by n gray gases is as shown in equation (1).

$$\varepsilon = \sum_{i=1}^{n} w_i \left[1 - exp(-\kappa_i YPL)\right] \quad (1)$$

where w is the weight of the gray gas; κ_i is the absorption coefficient; P is the pressure, which is 1 for atmospheric pressure condition; and L is the path length. The subscript i represents the type of the gray gas. Y is the volume fraction of the gas mixture.

For the oxy-fuel combustion conditions, there are generally 4 gray gases; thus, n=4 in this paper. The weight coefficient w_i of

Corresponding author. Fax: +86-0571-87952889
E-mail address: zhouzj@zju.edu.cn

each gray gas is related to the temperature of gas mixture T, which can be expressed as:

$$w_i = \sum_{j=0}^{m} c_{i,j} \left(\frac{T}{T_{ref}}\right)^j \quad (2)$$

where w_i is the m-th polynomial of temperature T, we choose m as 4 in this paper. T_{ref} is the reference temperature, which makes the temperature term in Eq.(2) dimensionless to reduce the error, the recommended reference temperature in literature [13] is 2000K, thus, we selected this value in our paper.

For most of poly-generation coal char, the carbon content is more than 80%, while the hydrogen content is usual no more than 1%, what's more, considering that the oxy-fuel combustion atmosphere contains more than 70% of the CO_2; thus, the water content of combustion flue gas is very low, the molar ratio of H_2O to CO_2 is about 0.01. Therefore, the molar ratio MR is set as 0.01 to represent the oxy-char combustion condition. The oxy-fuel WSGG models proposed in [9] [10] [13] [14] all have a molar ratio MR above 0.125, these can represent coal and other gaseous fuels, but not suitable for coal char conditions.

The WSGG model coefficients for oxy-char combustion mainly include the κ_i and $c_{i,j}$. We first determined the emissivity of gas mixture using the benchmark SNB model in the path length range of 0.001 m-60 m and the temperature range of 500-2500 K. It is generally considered that such path length range and temperature range can include the most conditions of burners and boilers, which have practical significance. The WSGG model coefficients κ_i and $c_{i,j}$ were then determined using the optimization method so that the sum of the difference between the emissivity calculated through the WSGG model and that calculated through the SNB model is minimal [14].

2.3 Investigated cases

In order to verify the accuracy of the proposed new model, we used the WSGG model and the benchmark SNB model to calculate the gas radiation characteristics in several one-dimensional cases, and examined the radiation heat flux and the radiation source term to verify the accuracy of the WSGG model proposed in this paper. The whole process obtains the radiation intensity I by solving the radiation transfer equation (RTE), and then the radiant heat flux q and the radiation source $\dot{q}$ were calculated.

$$q = \int_{\Omega=4\pi} I(\vec{s})\vec{s}d\Omega \quad (3)$$

$$\dot{q} = \nabla \cdot q \quad (4)$$

Where $\vec{s}$ represents the unit vector for a given direction. The RTE was solved through the discrete coordinate method (DOM) and the specific procedures can refer to the literatures [10], [13] and [14].

In this paper, the one-dimensional cases includes four conditions: case 1 represents the isothermal and homogeneous condition, case 2 represents the isothermal and non-homogeneous condition, case 3 represents non-isothermal and homogeneous condition, and case 4 represents non-isothermal and non-homogeneous condition, respectively. The temperature distribution of non-isothermal condition is shown as follows [10]:

$$T = 1400 - 400\,cos(\tfrac{2\pi x}{X})\ (K) \quad (5)$$

The distribution CO_2 molar fraction in non-homogenous conditions is as follows:

$$Y_{CO2} = 0.9 - 0.05\cos(\tfrac{2\pi x}{X}) \quad (6)$$

Where x is the position coordinate and X is the length between the two walls. In all the cases, the molar ratio MR is 0.01, the length between the two walls was 1 m, and the temperature of the two blackbody walls was 1000 K. The specific information of each case is shown in Table 1:

Case	T(K)	Y_{CO2}
1	1200	0.9
2	1200	Eq.(6)
3	Eq.(5)	0.9
4	Eq.(5)	Eq.(6)

Table 1: Specific information of each case.

3 Results and discussions

3.1 WSGG model correlations for oxy-char combustion

The WSGG model coefficients are shown in Table 2 below, which are suitable for coal-pyrolysis poly-generation coal char in oxy-char combustion conditions. The new model was fitted through the results of SNB model as described in Section 2.2.

i	1	2	3	4
κ_i	0.029495	0.315046	4.2899	146.868
i	1	2	3	4
$c_{i,0}$	0.723645	0.122231	0.15737	-0.00202
$c_{i,1}$	-1.93588	0.928079	-0.53144	0.496947
$c_{i,2}$	3.359291	-2.74761	1.32711	-1.03386
$c_{i,3}$	-2.72808	2.498911	-1.29222	0.752951
$c_{i,4}$	0.792223	-0.75889	0.417128	-0.18816

Table 1: New WSGG model coefficients.

3.2 Emissivity results

New WSGG model proposed in this paper was used to calculate the emissivity at different temperatures of 800 K, 1200 K and 1800 K and the results were compared with the benchmark. the aim of this section is to verify accuracy of new model when calculating the emissivity of oxy-char combustion flue gas, what's more, the WSGG models proposed by Johansson (WSGG-1) [10] and Yin (WSGG-2) [9] were also used to and verify the applicability of the new model in oxy-char combustion condition.

Figure 1 shows the results comparison of new WSGG model and the benchmark SNB model when calculating the emissivity of the oxy-char combustion flue gas at different temperatures of 800 K, 1200 K, and 1800 K. It can be seen that the new model results were consistent with the benchmark; it indicates that the new model is suitable to calculate the flue gas emissivity. Moreover, the higher the temperature is, the lower the emissivity of flue gas will be. Figure 2 shows the emissivity results of different WSGG models. It can be seen that the

WSGG model proposed in this paper was more suitable for the calculation of oxy-char combustion, while the model proposed by other researchers were not suitable. The reason of this phenomenon is that these two previous oxy-fuel WSGG models are based on coal or other gaseous fuels and they have a minimum molar ratio MR of 0.125, while for the oxy-char combustion condition with high CO_2 concentration, such a molar ratio is still high; thus, there is large error.

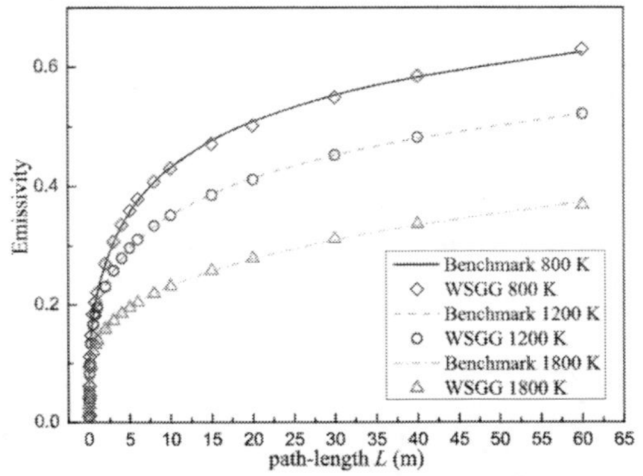

Figure.1 Comparison of emissivity calculated by the new WSGG model at different temperatures

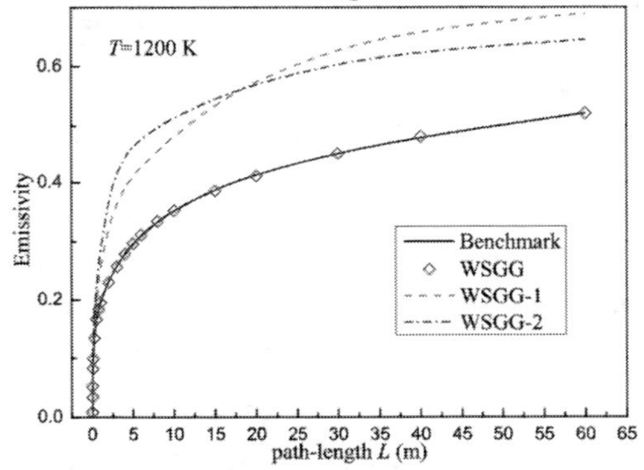

Figure.2 Comparison of emissivity calculated by different WSGG models

3.3. One dimensional cases results

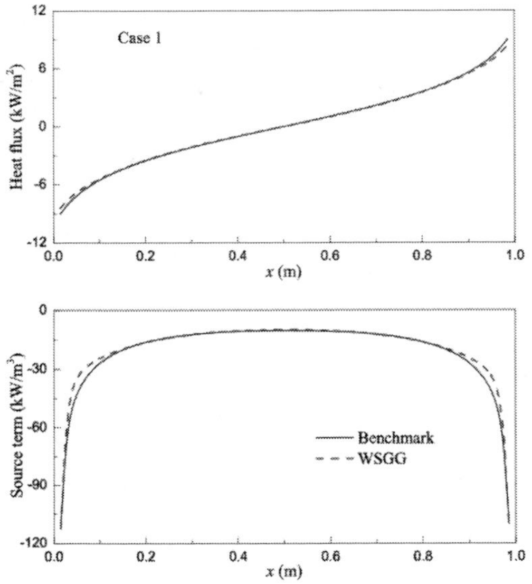

Figure.3 isothermal homogeneous condition results (case 1; Up: Radiation heat flux; Down: Radiation source term)

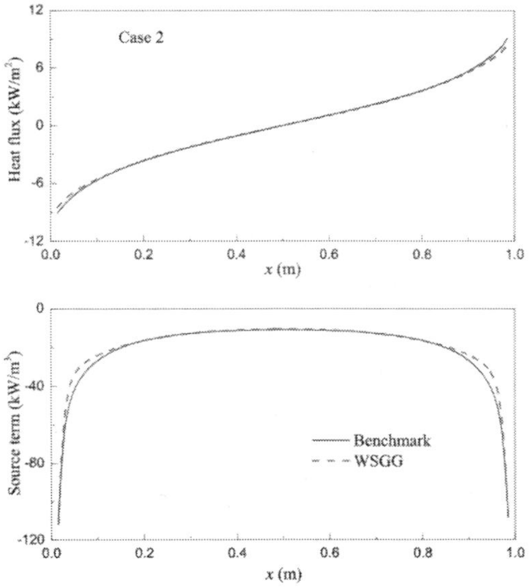

Figure.4 isothermal non-homogeneous condition results (case 2; Up: Radiation heat flux; Down: Radiation source term)

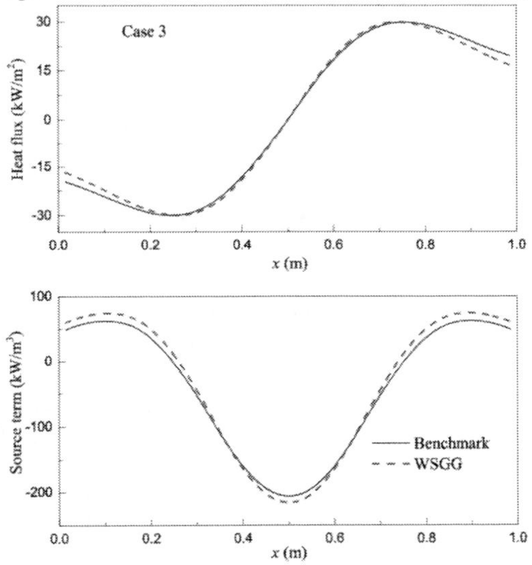

Figure.5 non-isothermal homogeneous condition results (case 3; Up: Radiation heat flux; Down: Radiation source term)

The results of the one-dimensional cases are shown in Fig. 3-Fig. 6. It can be seen from these figures that the radiant heat flux and the radiation source results calculated by the new WSGG model coefficients were all in accord with those of the benchmark SNB model in these four oxy-char combustion cases. What's more, Table 3 shows the average errors in these four oxy-char conditions, where $\overline{\delta q}$ denotes the average heat flow error and $\overline{\delta \dot{q}}$ denotes the average source term error.

	Case 1	Case 2	Case 3	Case 4
$\overline{\delta q}$(%)	3.03	3.02	4.73	4.71
$\overline{\delta \dot{q}}$(%)	6.59	6.72	10.96	10.89

Table 3: Average errors in four oxy-char conditions.

Figure 3 and Figure 4 are the results of isothermal homogeneous condition (case 1) and isothermal non-homogenous condition (case 2). It can be seen that the calculated results were in good agreement with the benchmark under isothermal conditions. Moreover, we can also see that the average errors of radiation heat flux and radiation source error were about 3% and 6% respectively from Table 3.

Figure 5 and Figure 6 are the calculation results non-isothermal homogeneous conditions (case 3) and non-isothermal non-homogeneous conditions (case 4) respectively; Table 3 also shows the average errors of radiation heat flux and radiation source term were about 4.7% and about 10% respectively. Compared with the isothermal condition, the errors in non-isothermal conditions were slightly more; however, the model was still possible to predict the radiant heat transfer of the oxy-char combustion flue gas under non-isothermal conditions.

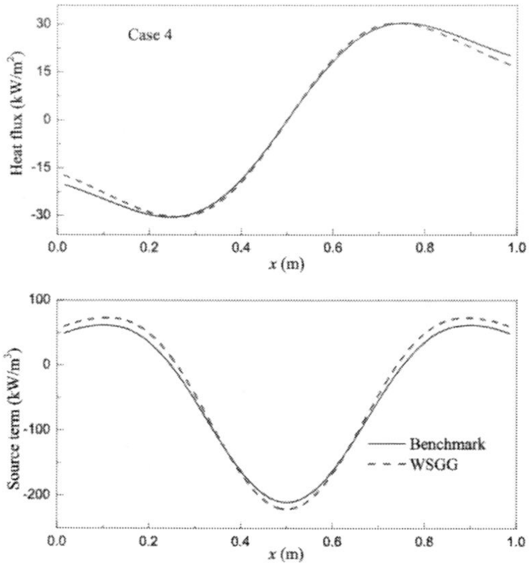

Figure.6 non-isothermal non-homogeneous condition results (case 4; Up: Radiation heat flux; Down: Radiation source term)

We can see that errors in non-isothermal conditions were slightly more than that of isothermal conditions; it indicates the accuracy of the WSGG model becomes lower when the change of temperature field is greater. It because the weight of WSGG model is a fitting polynomial of the temperature, and the calculation accuracy will decrease with the increase of the temperature field range. Comparing homogeneous conditions with non-homogeneous conditions, it can be seen that the non-homogeneous conditions had little effect on the calculation results.

4 Conclusions

In this paper, the WSGG model coefficients for the calculation of the flue gas radiation characteristics of the oxy-char combustion were proposed based on the statistical narrow band model. The results were verified in one-dimensional cases. Following conclusions are obtained:

(1) WSGG model is suitable for the calculation of heat transfer characteristics of flue gas radiation in oxy-char combustion for small calculation efforts.

(2) Due to the high concentration of CO_2 in the oxy-char combustion flue gas, the former WSGG models suitable for coal and gas fuels are not effective and it needs to develop new model correlations.

(3) The new WSGG model is in good agreement with the benchmark when used in one-dimensional cases, and the maximum average error is about 10%.

(4) The new WSGG model can be used to calculate the radiant heat transfer characteristics of oxy-char combustion conditions based on the poly-generation coal char, it can be used for design and CFD calculation of oxy-char combustion furnace.

5 Acknowledgment

The support of this work by the National Key Technology Support Program of China (No.: 2015BAA04B03) is gratefully acknowledged.

References

[1] M. Ni, C. Li, M. Fang, Q. Wang, Z Luo, K. Cen, Int. J. Coal. Sci. Technol 1(1) (2014) 39-45.

[2] Z. Guo, Q. Wang, M. Fang, Z. Luo, K. Cen, Chem. Eng. Technol 38(3) (2015) 463-472.

[3] Z. Guo, Q. Wang, M. Fang, Z. Luo, K. Cen, Applied. Energy 113 (2014) 1301-1314.

[4] M.A. Habib, H.M. Badr, S.F. Ahmed, R. Ben-Mansour, K. Mezghani, S. Imashuku, G.J. la O', Y. Shao-Horn, N.D. Mancini, A. Mitsos, P. Kirchen, A.F. Ghoneim, Int. J. Energy Res 35(9) (2011) 741-764.

[5] Z. Zhou, X. Hu, Z. You, Z. Wang, J. Zhou, K. Cen, Thermochim. Acta 553 (2013) 54-59.

[6] Z. Zhou, Z. You, Z. Wang, X. Hu, J. Zhou, K. Cen, Environ. Prog. Sustain 33(3) (2014) 993-999.

[7] T. Wall, Y. Liu, C. Spero, L. Elliott, S. Khare, R. Rathnam, F. Zeenathal, B. Moghtaderi, B. Buhre, C. Sheng, R. Gupta, T. Yamada, K. Makino, J. Yu, Chem. Eng. Res. Des 87(8) (2009) 1003-1016.

[8] P. Nakod, G. Krishnamoorthy, M. Sami, S. Orsino, Appl. Therm. Eng 54(2) (2013) 422-432.

[9] C. Yin, L.C.R. Johansen, L.A. Rosendahl, S.K. Kær, Energ. Fuel 24(12) (2010) 6275-6282.

[10] R. Johansson, K. Andersson, B Leckner, H. Thunman, Int. J. Heat. Mass. Tran 53(1) (2010) 220-230.

[11] A. Soufiani, J. Taine, Int. J. Heat. Mass. Tran 40(4) (1997) 987-991.

[12] H. Chu, F. Liu, H. Zhou, Int. J. Heat. Mass. Tran 54(21) (2011) 4736-4745.

[13] T. Kangwanpongpan, F.H.R. França, R.C. da Silva, P.S. Schneider, H.J. Krautz, Int. J. Heat. Mass. Tran 55(25) (2012) 7419-7433.

[14] R. Johansson, B. Leckner, K. Andersson, F. Johnsson, Combust. Flame 158(5) (2011) 893-901.

11th Asia-Pacific Conference on Combustion,
The University of Sydney, NSW Australia
10th -14th December 2017.

Two typical effects of microsecond plasma on flow rate disturbance premixed swirl flames

Wei Cui, Yihua Ren, Shuiqing Li*

Key Laboratory for Thermal Science and Power Engineering of Ministry of Education, Department of Thermal Engineering, Tsinghua University

Beijing, 100084, China

Abstract

The paper designed a flow rate disturbance system on a swirl burner. The heat release rate fluctuation and lean blowout limit variation have been analyzed in rich and lean disturbance. Microsecond pulsed plasma discharge was applied to determine the effect to the flame with flow rate disturbance. The results showed that the plasma plays double-edged effects to the flame due to the time delay with the flow disturbance. The aerodynamic actuation may influence the combustion stability, and the re-ignition may extend the lean blowout limit. The novel phenomenon provide a new method and strategy for combustion direct control.

1 Introduction

As the modern civil aircraft development, the jet engines operating envelops are much wider than their predecessors, and more prone to blowout near the lean blowout limit or during the transient operation [1,2]. The flow rate, density, throttle, power output, and other transient operation occurs during the aircraft maneuver [3]. In an investigation for a jet engines, more than half flameout conditions occurred in the descent operations [4]. To research the these kinds of engine transient characteristics, Rosfjord and Cohent developed a transient combustor testing facility based on a bypass system [5], and analyzed the combustor performance under the air flow rate oscillation. These researches showed the flow rate oscillation triggering the combustion instabilities via alternating the mixing characteristics and flow structure.

To mitigate the combustion instability, non-equilibrium plasma have been adopted in many combustion dynamics researches [6,7]. The nanosecond repetitively pulsed plasma shows great effect on the heat release rate of premixed flame[8]. Most plasma assisted combustion works apply loudspeaker to generate the continuous acoustics perturbation. To comprehensively simulate the transient working condition, the flow rate oscillation method should be considered.

The present paper presents a novel experimental research of combustion instability affected by pulsed plasma. This study design an individual rich or lean flow disturbance system on a premixed swirl burner. Based on the novel device, the microsecond repetitively pulsed plasma discharges are presented in the premixed flame area. The flame characteristics altered by the flow disturbance and plasma are investigated. According to the experiment, the plasma hydrodynamics and ignition effects are also discussed..

Corresponding author Tel:+861062773384,
Fax:+861062794068
E-mail address: lishuiqing@tsinghua.edu.cn

2 Experimental set-up

The experimental setup diagram presented in Fig. 1. The air and fuel mix in the vane swirler burner upstream. A bypass way with fast-acting valve is located in the air supply pipe to manipulate the air flow. The power supply connected with high voltage electrode, the center body is the ground electrode.

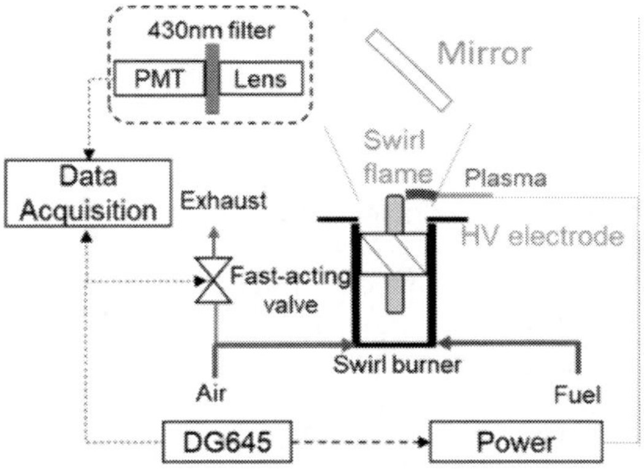

Figure 1: Experimental setup

2.1 Premixed swirl burner

A swirler-stablized burner with premixed gas of methane/propane and air is adopted in this research. The gases are controlled by Sevenstar D07 mass flow controllers, and mixing upstream a flash-back arrestor before flowing into the burner. The diameter of the outlet is 18mm. The swirler burner design is similar to the previous work fulfilled by Wang et al. [9] and located at the 22mm upstream position of the outlet. The swirler blade inclination angle is 45 degree and the swirl number is about 0.6. A 8mm diameter steel center body is installed in the swirler as a ground electrode, and is 22mm sticking from the outlet. A 5mm quartz plate placed at the exit flange of the burner as an insulator. .

2.2 Bypass system

A triplet bypass pipe is connected with the main air supply pipe. A flash-back arrestor is also inserted in the bypass way, to balance the flow damping with the main flow. A MAC35A-ACA direct solenoid fast-acting valve is selected to switch the bypass way on or off. An electric relay synchronized by a DG645 digital delay generator controls the valve, the open rising slope time is 6 ms and close drop slope is 2 ms. Changing the duty cycle of the valve opening time, could provide rich or lean disturbance for the flame. A Dantec Laser Doppler Anemometry and Shinagawa W-NK-10A flowmeter are applied to measure the main and bypass flow rate. As shown in Fig. 2, we set the valve working frequency as 5Hz, and the duty cycle 0.1, it is the average bypass flow rate versus the total air flow rate. When the bypass valve opening, about 10% air is exhausted via the bypass way, The sub image is the flow rate at the burner outlet. When the valve acting, the air flow decreases rapidly, so the velocity and equivalence ratio vary simultaneously, the premixed gas goes to rich, it is called the rich disturbance. After the acting period, the outlet flow rate back to the normal level. Likewise, the duty cycle 0.9 generates lean disturbance. The flow maximum deviation is about 10%. The fuel flow rate keep constant while the air changing.

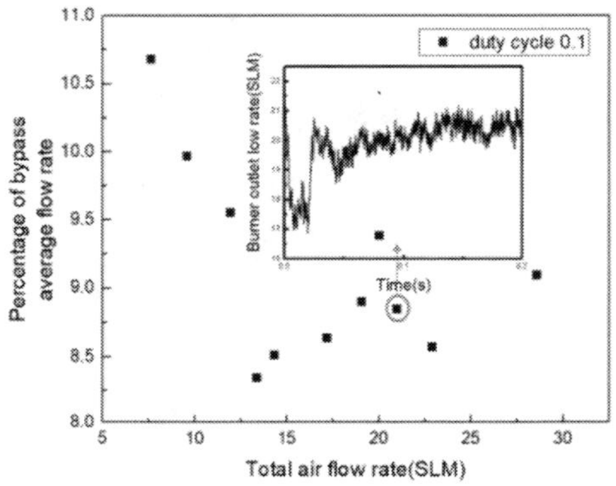

Figure 2:Average bypass flow rate

2.3 Discharge and diagnostic system

In previous works, we found that the nanosecond repetitively pulsed generator with steep rise-edge induce strong EMI for the devices especially for the electric relay and valve. So a microsecond pulse generator CMPC-40D design by Institute of Electrical Engineering of the Chinese Academy of Sciences is adopted to this research[10,11].This power supply could provide enough energy and generate non-equilibrium plasma in the flame field with less EMI and high reliability. The microsecond pulse generator output a TTL signal to synchronize the solenoid valve and diagnostic devices. A stainless steel pin located 20mm above the outlet quartz plate and at set a 5mm gap to the grounded center body. It is a slight influence to the swirl flame induced by the electrode.

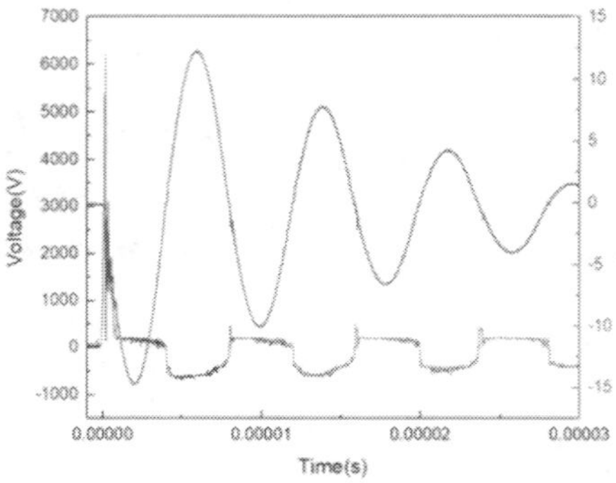

Figure 3: Voltage and current curve

A Tektronix DPO2024B oscilloscope with Tektronix P6015A high voltage probe and a Pearson current monitor 4100 is adopted to record the voltage and current curves. When the flame existing, the breakdown voltage is about 6kV because of the high gas temperature. The voltage and current curve shows in Fig3. The rising edge before breakdown is about 300ns of the microsecond-duration pulse generator. The total energy deposited in plasma is estimated by the circuit voltage and current amplitude. There is 8mJ energy deposited in the plasma at the first 4us after breakdown, 44mJ of the whole duration.

The 430nm chemiluminescence of excited CH* radicals is selected to diagnose the heat release rate fluctuations of the swirl flame. It is a well developed technique for heat release rate measurement of premixed gas flame and combustion dynamics research[12,13].A Hamamatsu CH-253 photomultiplier tube(PMT) is connected with NI USB-6356 to capture the PMT signal. A 45degree flat mirror is placed above the burner, a Nikkor 50mm lens and a 430nm bandpass filter is located before the PMT to collect the CH* chemiluminescence emission. A SRI DG645 digital delay generator synchronizes the entire system.

Because of the air discharge plasma, the nitrogen spectrum range covers the 430nm[14], it cannot be blocked by the 430nm bandpass filter. To determine whether the discharge emission and electro magnetic interference will interfere the heat release rate fluctuations recording, a validation test have been done without flame. The validation shows the discharge emission last about 500us. When we set a optic barrier before the PMT, there is also a 100us PMT signal affected by the EMI. Due to combustion dynamics time scale with rich or lean disturbance is about millisecond level, it could be recognized that the PMT signal induced by discharge emission and EMI interfere the flame signal slightly.

3 Results and discussion

3.1 Lean blowout limit characteristics

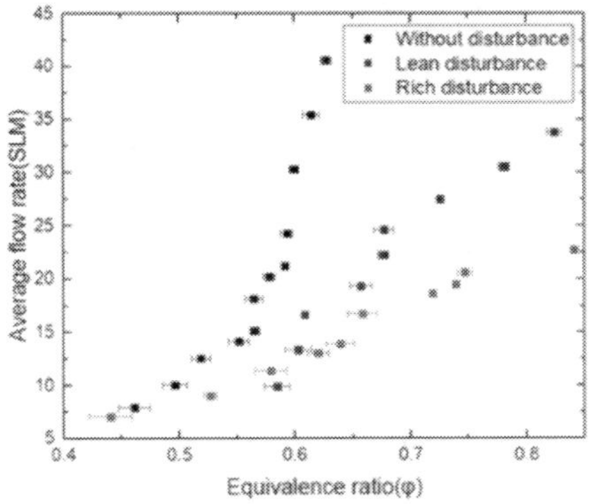

Figure 4: Lean blowout limit under the rich disturbance

To research the lean blowout limit variation under the rich or lean flow rate disturbance, two kinds of fuel was utilized. Because of the flow rate changing is time-variant and periodical, we use the flowmeter to determine the average flow rate at the burner outlet, and decrease the fuel flow rate to find the blowout point. In the following part, we define the average equivalence ratio at the lean blowout limit as the fuel flow rate versus the average oxidant flow rate over the stoichiometric ratio. The methane and propane are applied to analyze the lean blowout limit characteristics.

From the Fig.4, both of the lean and rich flow rate disturbance deteriorate the lean blowout limit. But the lean blowout limit curve of rich disturbance is below the one of lean disturbance both.

3.2 Plasma effect on the rich disturbance

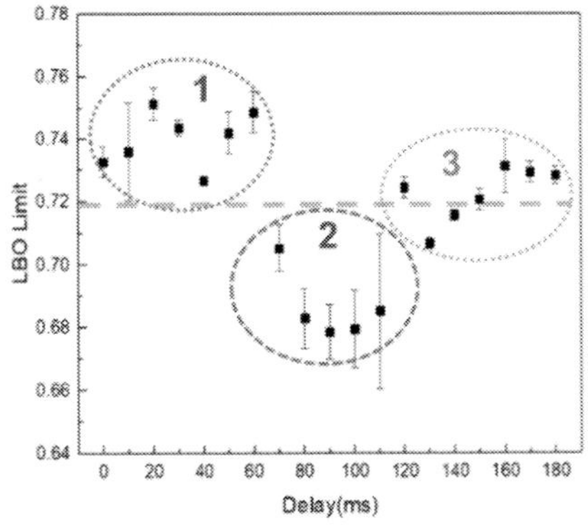

Figure 5: Lean blowout limit with different delay

Based on the previous study, we chose an operating point of the total flow rate is 20.4SLM with a 5Hz repetition rich disturbance, under this condition, the average flow rate at the burner outlet is 18.6SLM measured by the flowmeter, and average lean blowout limit equivalence ratio 0.72. A 5Hz repetitive high voltage pulse was applied on the electrode to generate the plasma as mentioned in part 2.3. We change the time delay between the signal for valve opening and the discharge from 0ms to 180ms. The time-variant flow rate at the outlet and the PMT signal representing the heat release rate fluctuation are synchronized at the same start time. We determined the lean blowout limit in every conditions.

The lean blowout limit equivalence ratio is 0.72 without plasma, we found a novel phenomenon that the effects of plasma on the rich disturbance could be divided as three regions on the Fig.5. The first part is from 0ms to 60ms, the lean blowout limit equivalence ratio is worse than the 0.72. The second part extends the limit, and the three part is nearby the 0.72.

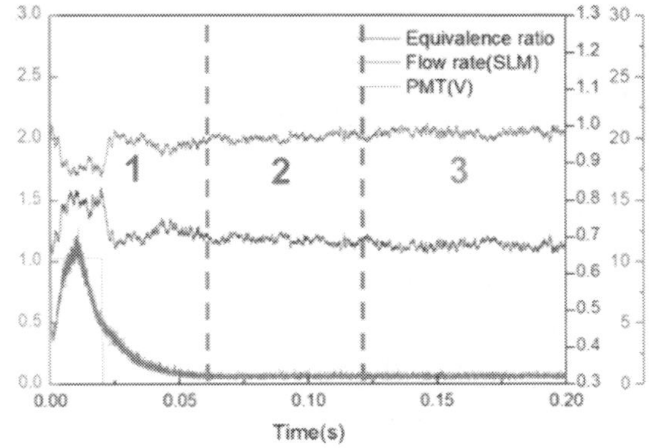

Figure 6: Regions of time delay

As seen from the Fig. 6, the flow rate, equivalence ratio and heat release rate change dramatically in the first time delay region as the Fig 5. In this region, the blowout limit is worsen by the discharge. In this moment, the shape of swirl recirculation zone was changed due to the simultaneous variation of velocity and equivalence ratio. The flame stability was weaken by the fluctuation. In the previous works, it is find that this pin to pin layout plasma actuator generates supersonic shockwave and strong aerodynamic actuation while discharge [15,16]. This aerodynamic actuation could the affect flow field fiercely, and the intensified the recirculation zone disturbance. So the discharge in region 1 made the LBO limit worse than the other regions. The detail mechanism has been undergoing a simulation research.

In the region 2, the lean blowout limit is obviously extended by plasma. We estimate the minimum ignition energy in this condition as (1). Because the discharge gap is a 5mm long tube area, the diameter of plasma area is about 1mm calculated from schlieren image in quiescent air.

$$E_{ign} = 4\pi dP\left(\frac{\alpha}{S_L}\right)^2\left(\frac{c_p MW_b}{R_u}\right)\left(\frac{T_b - T_u}{T_b}\right) \tag{1}$$

The minimum ignition energy is 2.8mJ while the unburnt gas temperature is 300K at the equivalence ratio 0.72. From the Fig.3, we could conclude that the energy deposited in the plasma is about 8mJ at the first 4us, and 44mJ totally deposited in the whole 500us. It means the discharge provides enough critical energy for the igniting the unburnt gas, and keep heating for 500us. Then the microsecond plasma re-ignited the flame on the verge of extinction and extended the lean blowout limit.

In region 3, the LBO limit was same as the condition without plasma. Because in this region, the flow fluctuation attenuated, and the PMT signals showed that the flame was also stable. Because the plasma played little effect on the stable flame propagation. Cathey also observed this phenomenon in the research[17].

4 Conclusions

The characteristics of laminar premixed flames under the flow rate disturbance has been investigated by analyzing the heat release rate fluctuation and lean blowout limit variation. Microsecond pulsed plasma discharge was applied to determine the effect to the flame with flow rate disturbance. The results showed that the plasma plays double-edged effects to the flame due to the time delay with the flow disturbance. The aerodynamic actuation may influence the combustion stability, and the re-ignition may extend the lean blowout limit.

In the plasma active combustion control and plasma assisted combustion development, comparing to the continuous discharge, the temporal control strategy becomes more smart and attractive. A well-designed plasma discharge strategy should consider and balance the double-edged effects to achieve high efficiency and reliability. Further quantitative and simulation investigation will be done to clarify the mechanism of plasma assisted flow disturbance combustion.

5 Acknowledgment

This work is funded by National Natural Science Foundation of China (No. 91641204). The authors are particularly grateful to researcher Tao Shao and associate researcher Cheng Zhang at the Institute of Electrical Engineering of the Chinese Academy of Sciences (IEECAS) for their helpful in experiment..

References

1 Muruganandam T M, Nair S, Scarborough D, et al. Journal of Propulsion and Power, 2005, 21(5): 807-814.

2 Nair S, Lieuwen T. Journal of Propulsion and power, 2007, 23(2): 423-429.

3 Cohen J M, Bennett Jr J C. AIAA Paper, 1996 (96-0322).

4 Richter B A, Ridenour-Bender M, Tsao M. Aircraft Turbine Engine Reliability and Inspection Investigations. SCIENCE APPLICATIONS INTERNATIONAL CORP SAN ANTONIO TX LOGISTICS TECHNOLOGY DIV, 1993.

5 Rosfjord T J, Cohen J M. Journal of Propulsion and Power, 1995, 11(3): 497-504.

6 Moeck J P, Lacoste D A, Laux C O, et al. Control of combustion dynamics in a swirl-stabilized combustor with nanosecond repetitively pulsed discharges. Proc. 51st AIAA Aerosp. Sci. Meeting. 2013: 1-11.

7 Ćosić B, Terhaar S, Moeck J P, et al. Combustion and Flame, 2015, 162(4): 1046-1062.

8 Lacoste D A, Moeck J P, Roberts W L, et al. Proceedings of the Combustion Institute, 2017, 36(3): 4145-4153.

9 Wang J, Li S, Yan W, et al. Proceedings of the Combustion Institute, 2011, 33(2): 1925-1932.

10 Shao T, Zhang D, Yu Y, et al. IEEE Transactions on Plasma Science, 2010, 38(7): 1651-1655.

11 Zhang C, Shao T, et al. Applied Physics Letters, 2014, 105(4): 044102.

12 Hurle I R, Price R B, Sugden T M, et al. Sound emission from open turbulent premixed flames.Proceedings of the Royal Society of London A: Mathematical, Physical and Engineering Sciences. The Royal Society, 1968, 303(1475): 409-427.

13 Lacoste D A, Xiong Y, Moeck J P, et al. Proceedings of the Combustion Institute, 2017, 36(3): 4183-4192.

14 Galley D, Pilla G, Lacoste D, et al.AIAA paper, 2005, 1193: 2005.

15 LI J, WU W, SONG H, et al. High Voltage Engineering, 2013, 7: 004.

16 Huimin S, Yinghong L, Qiaogen Z, et al. Plasma Science and Technology, 2011, 13(5): 608.

17 Cathey, Charles, et al. Combustion and Flame 154.4 (2008): 715-727.

Combustion of bioslurry and its fractions: different contributions to PM_{10} emission

Chao Feng, Wenran Gao, Hongwei Wu*

Department of Chemical Engineering, Curtin University

Kent Street, Perth, Western Australia, 6102, Australia

Abstract

Bioslurry and its three fractions, i.e. filtered-oil, filtered-tar, and filtered-char, were sent into a lab-scaled drop-tube-furnace (DTF) system for combustion and PM_{10} collection, respectively. The filtered-char, filtered-oil and filtered-tar fractions contributes 51, 30 and 19% of the total PM_{10} during bioslurry combustion. The filtered-char fraction contributes to 68% of the total PM_{1-10} emission while the filtered-oil fraction contributes to 35% of the total PM_1 emission and 30% of the total PM_{1-10} emission during bioslurry combustion.

1 Introduction

Bio-oil/biochar slurry (i.e. bioslurry) has high energy density, low production cost, low energy and carbon footprints, suitable for transport, and adaptable to conventional stationary combustion applications [1-3]. Particulate matter (PM) emission, especially those with aerodynamic diameter size less than 10 μm (hereafter referred to as PM_{10}), are notorious due to its difficulty for capture and adverse impact on environment. Previous investigations were carried out on PM emissions from the combustion of coal [4-8], biomass [9-13], bio-oil [14] and biochar [15]. There is little report on PM emission from the combustion of bioslurry and its fractions. This study focuses on the PM_{10} emission from combustion of different fractions of bioslurry via series of combustion experiments through DTF system in air at 1400 °C.

2 Experimental section

A bio-oil (fast pyrolysis of pine wood at 500 °C) was supplied by a company. The biochar was produced from pine wood at 500 °C using a fast pyrolysis reactor [15]. The bioslurry (10% biochar loading) were prepared via suspending fine biochar particles into bio-oil and stored for 1 week at room temperature prior to use [16]. Bioslurry was filtered and separated into three fractions, i.e. filtered-oil, filtered-tar, and filtered-char, accounting for 55, 35, and 10% of the bioslurry, respectively.

Experiments of combustion were carried out in a drop-tube-furnace (DTF, detailed elsewhere [17]) in air at 1400 °C. For filtered-char, an entrained-flow feeder (description given elsewhere [15]) was used. For bioslurry, filtered-oil, and filtered-tar, the fuel was fed into the DTF via a spray nozzle set (detailed elsewhere [14]). The atomisation conditions were adjusted spray the fuels into droplets of similar mean droplet size (~46 μm) measured using a previous method [18]. The values of λ (the ratio of the actual air/fuel ratio to the stoichiometric air/fuel ratio) for all experiments were greater than 8.0 and the complete combustion was achieved. PM_{10} samples were collected using a PM collection system (detailed elsewhere [17]). All experiments were carried out at least in triplicates.

*Corresponding author. Fax: +61-8-92662681
E-mail address: h.wu@curtin.edu.au

The proximate analysis and elemental compositions of all samples are reported in Table 1. The water contents and viscosities of filtered-oil and bioslurry were also determined (detailed elsewhere [16]). The concentrations of inorganic elements, i.e. Na, K, Mg, Ca, Al, Si and Fe were measured via method used previously [19]. The contents of Cl and S were quantified via an improved Eschka method [20]. The lower heating value was calculated based on fuel chemistry [21]. Figure 1 gives the distribution of the major inorganic elements in the three fractions of bioslurry. It shows that: the Na, K, S, Mg, Ca, and Si are mainly in filtered-char fraction, while the Cl, Al, and Fe are mainly in filtered-oil fraction. Further calculation gives that 65% of the inorganic elements are in filtered-char fraction, while, 63% of the volatile elements (Na, K, Cl and S) are in filtered-oil fraction, 65% of the refractory elements (Mg, Ca, Al, Si and Fe) are in filtered-char fraction. The mass-based particle size distribution (PSD) of PM_{10} was determined by measuring the mass of PM_{10} collected in each stage of DLPI and backup filter using a Mettler MX5 microbalance (accuracy of 0.001 mg). The inorganic elements (Na, K, Mg, Ca, Al, Si and Fe) in PM_{10} were also measured.

Samples	Filtered-oil	Filtered-tar	Filtered-char	Bioslurry
Proximate analysis (wt%, ar[a])				
Water/moisture	30.0	5.0	2.0	20.2
Ash	0.1	0.004	1.8[b]	0.3
Volatile matter	75.9	69.5	14.0[b]	61.6
Fixed carbon	1.7	25.4	82.2[b]	17.9
Elemental analysis (wt%, ar[a])				
C	36.39	53.39	82.99[d]	45.60
H	7.75	5.41	3.83[d]	6.55
N	0.08	0.15	0.12[d]	0.11
O[c]	55.78	41.01	13.01[d]	47.73
Viscosity (20 °C, mPa s)	40.0	nd[f]	nd[f]	148.4
LHV[e] (MJ/kg)	18.4	16.7	27.6	18.3

[a] as-received basis for filtered-oil, filtered-tar, and bioslurry; [b] dry basis; [c] by difference; [d] dry and ash-free basis; [e] lower heating value; [f] not determined

Table 1: Properties of filtered-oil, filtered-tar, filtered-char and bioslurry

3 Result and discussion

3.1 PSDs and yields of PM_{10} emission from combustion of bioslurry and its fractions

Figure 2 gives the particle size distributions (PSDs) and yields of PM_{10} emission from combustion of filtered-oil, filtered-tar, filtered-char, and bioslurry. The PM_{10} via sum of "filtered-oil + filtered-tar + filtered-char" was also presented. Figure 2a shows that the PSD curves of filtered-oil, filtered-tar, bioslurry, and "filtered-oil + filtered-tar + filtered-char" all display bimodal distributions each with a fine mode at ~0.043 μm and with a coarse mode at ~2.4 μm, while filtered-char shows a unimodal curve with a coarse mode at ~2.4 μm. Fig.2a shows that the filtered-oil has the highest fine mode peak, while the filtered-char has the highest coarse mode peak. Among the three fractions, filtered-oil has the highest PM_1 yield (0.08 mg/g), followed by filtered-tar (0.03 mg/g) and filtered-char (0.02 mg/g), while filtered-char has the highest PM_{1-10} yield (0.41 mg/g), followed by filtered-oil (0.16 mg/g) and filtered-char (0.06 mg/g). Therefore, the filtered-oil fraction plays an important role in PM_1 emission, while filtered-char fraction plays a significant role in PM_{1-10} emission. This distribution is consistent with the previous researches that biochar mainly produces PM_{1-10} emission during combustion [15], while bio-oil can produce both PM_1 and PM_{1-10} [14].

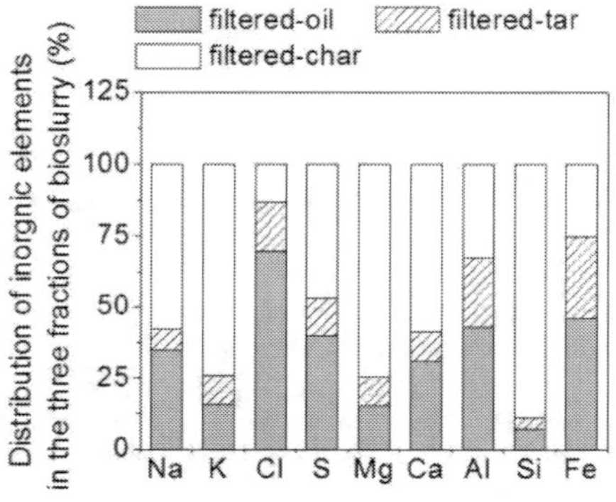

Figure 1: Distribution of the inorganic elements in filtered-oil, filtered-tar, and filtered-char fractions of bioslurry

3.2 Elemental PSDs and yields of PM_{10} emission from combustion of bioslurry and its each fraction

Figure 3 and 4 further gives the detailed elemental PSDs and yields of PM_{10} emission from combustion of the three fractions, bioslurry, and the sum of "filtered-oil + filtered-tar + filtered-char". It could be found that all the elements follow unimodal distributions however with different mode positions. The volatile elements all have a fine mode at ~0.043 μm, while the refractory elements show coarse modes at ~2.4 μm. Among filtered-oil, filtered-tar, and filtered-char, the peaks of the fine modes of volatile elements from filtered-oil are much higher than that from the other two fractions. As a result, the yields of the volatile elements in PM_1 from filtered-oil is higher than that of filtered-tar and filtered-char, indicating that during bioslurry combustion, filtered-oil fraction contributes more significantly to these elements emission in PM_1 than the other two fractions.

Similarly, the peaks of the coarse modes of refractory elements from filtered-char fraction are much higher than that from filtered-oil and filtered-tar, as a result, the yields of refractory elements in PM_{1-10} from filtered-char fraction is much higher than that of the other two fractions, indicating the significant contribution of filtered-char fraction to the refractory elements in PM_{1-10} emission during bioslurry combustion. These results are consistent with the distributions of these elements in the three fractions of bioslurry (see in Fig. 2), and further explained the higher PM_1 contribution from filtered-oil fraction, due to the higher emission of volatile elements in PM_1 via vaporization, nucleation and/or condensation [9,22], and the higher PM_{1-10} contribution from filtered-char fraction to the total PM_{10} emission due to the higher emission of refractory elements via catalyzed carbonate sintering effect [23] and coagulation to form aluminasilicates [5] during bioslurry combustion.

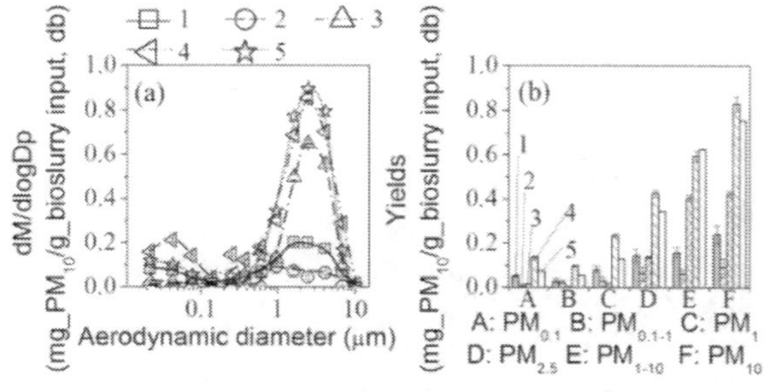

Figure 2: PSDs (a) and yields (b) of PM_{10} from combustion of 1.filtered-oil, 2.filtered-tar, 3.filtered-char, 4.bioslurry, and 5."filtered-oil + filtered-tar + filtered-char", respectively.

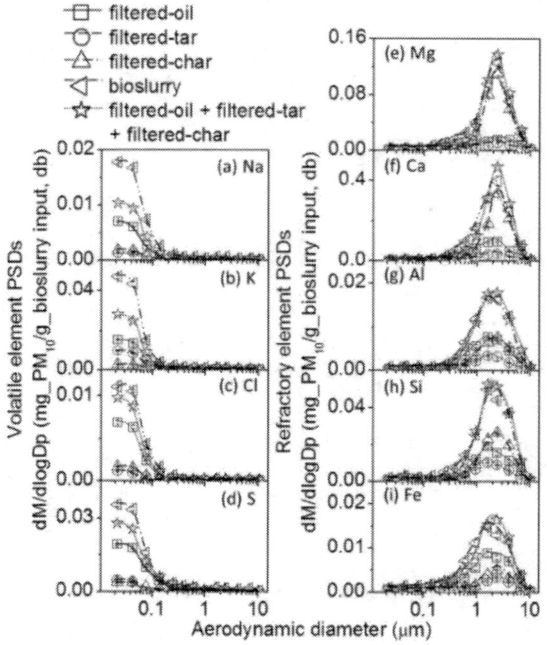

Figure 3: Elemental PSDs of PM_{10} from combustion of filtered-oil, filtered-tar, filtered-char, bioslurry, and "filtered-oil + filtered-tar + filtered-char"

3.3 Further discussion

Figure 2 illustrates that the fine mode of bioslurry is much higher than that of "filtered-oil + filtered-tar + filtered-char". As a result, the PM_1 yield from bioslurry combustion is much higher than that from "filtered-oil + filtered-tar + filtered-char", indicating that there must be some synergy happened during bioslurry combustion that contributes to the intense increase of PM_1; however, this mechanism is unknown. One of the possible reason for this phenomenon is the "volatile-char interactions" which would increase the volatilization of some inorganic species from filtered-char fraction during bioslurry combustion and is known to increase the PM_1 emission [24]. This could be further evidenced by the fact that the volatile elements in PM_1 from bioslurry combustion are obviously higher than that relevant from "filtered-oil + filtered-tar +

filtered-char" (see in Fig.5) correlating well with the fact that the "volatile-char interactions" has strong influence on volatile elements. The detailed mechanisms on the synergy during bioslurry combustion need to be further studied in the future. Thus, in this study, the different contributions of the three fractions of bioslurry and the synergy effect to the PM_1, PM_{1-10} and PM_{10} emission during bioslurry combustion could be calculated out (see in Fig. 5). As a result, the filtered-char fraction contributes more PM_{10} during bioslurry combustion than the other two fractions, accounting for 51% of the total PM_{10} from bioslurry combustion, followed by filtered-oil fraction (30%). Specifically, filtered-char fraction contributes 68% of the total PM_{1-10}, higher than filtered-oil and filtered-tar, while, filtered-oil fraction contributes 35% of the total PM_1, higher than filtered-tar and filtered-char, however, lower than that from the synergy effect (45%).

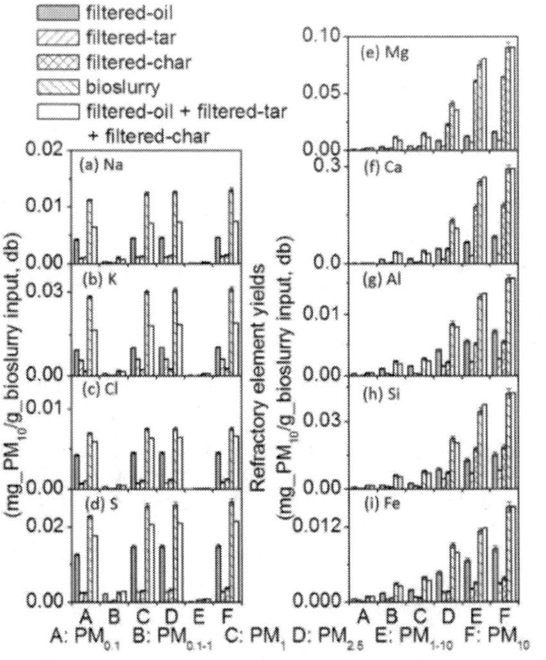

Figure 4: Elemental yields of PM_{10} from combustion of filtered-oil, filtered-tar, filtered-char, bioslurry, and "filtered-oil + filtered-tar + filtered-char"

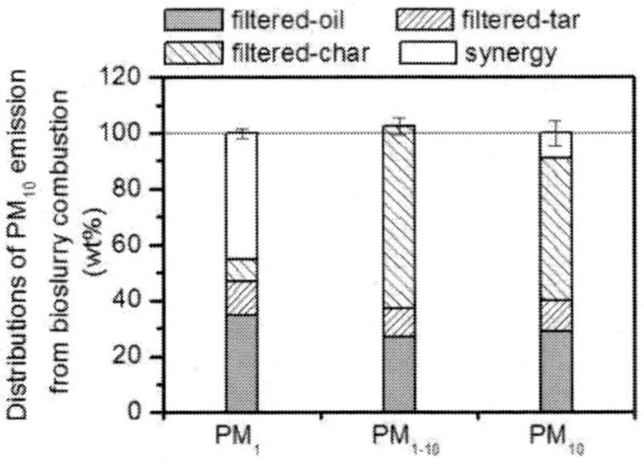

Figure 5: Contribution to PM_1, PM_{1-10} and PM_{10} from the combustions of bioslurry and its fractions and potential synergy

4 Conclusions

Bioslurry and its three fractions, i.e. filtered-oil, filtered-tar, and filtered-char, were sent into DTF for combustion and PM_{10} collection, respectively. From the results, the contribution of filtered-char fraction to PM_{10} emission during bioslurry combustion is more significant than the other two fractions, accounting for 51% of the total PM_{10}, followed by filtered-oil fraction (30%). In details, the filtered-char fraction contributes to 68% of the total PM_{1-10} emission from bioslurry combustion, while the filtered-oil fraction contributes to 35% of the total PM_1 emission and 30% of the total PM_{1-10} emission.

5 Acknowledgment

Partial support from the Australian Research Council through its Discovery Projects schemes is acknowledged. C.F. is grateful to the Chinese Scholarship Council (CSC) and Curtin University for providing scholarships to partially support his PhD study, and a provider (who has chosen to remain anonymous) for supplying the bio-oil sample, and WA&J King Pty Ltd (Western Australia) for supplying the pine wood sample.

References

[1] Wu H, Yu Y, Yip K, Energy & Fuels 2010;24:5652-5659.
[2] Yu Y, Wu H, Energy & Fuels 2010;24:5660-5668.
[3] Abdullah H, Mourant D, Li C-Z, Wu H, Energy & Fuels 2010;24:5669-5676.
[4] Neville M, Quann RJ, Haynes BS, Sarofim AF, Symposium (International) on Combustion 1981;18:1267-1274.
[5] Helble JJ, Sarofim AF, Combustion and Flame 1989;76:183-196.
[6] Carbone F, Beretta F, D'Anna A, Energy & Fuels 2010;24:6248-6256.
[7] Xu M, Yu D, Yao H, Liu X, Qiao Y, Proceedings of the Combustion Institute 2011;33:1681-1697.
[8] Kazanc F, Levendis YA, Maffei T, Energy & Fuels 2013;27:4984-4998.
[9] Christensen KA, Livbjerg H, Aerosol Science and Technology 1996;25:185-199.
[10] Valmari T, Kauppinen EI, Kurkela J, Jokiniemi JK, Sfiris G, Revitzer H, Journal of Aerosol Science 1998;29:445-459.
[11] Fernandes U, Costa M, Energy & Fuels 2013;27:1081-1092.
[12] Ruscio A, Kazanc F, Levendis YA, Energy & Fuels 2014;28:685-696.
[13] Gao X, Rahim MU, Chen X, Wu H, Proceedings of the Combustion Institute 2017;36:3313-3319.
[14] Feng C, Gao X, Wu H, Proceedings of the Combustion Institute 2017;36:4061-4068.
[15] Gao X, Wu H, Energy & Fuels 2011;25:2702-2710.
[16] Zhang M, Liaw SB, Wu H, Energy & Fuels 2013;27:7560-7568.
[17] Gao X, Wu H, Energy & Fuels 2010;24:4571-4580.
[18] Ghezelchi MH, Garcia-Perez M, Wu H, Energy & Fuels 2015;29:8058-8065.
[19] Yip K, Tian F, Hayashi J-i, Wu H, Energy & Fuels 2010;24:173-181.
[20] Rahim MU, Gao X, Wu H, Energy & Fuels 2013;27:6992-6999.
[21] Sheng C, Azevedo JLT, Biomass and Bioenergy 2005;28:499-507.
[22] van Lith SC, Jensen PA, Frandsen FJ, Glarborg P, Energy & Fuels 2008;22:1598-1609.
[23] Xiong Y, Kamal Akhtar M, Pratsinis SE, Journal of Aerosol Science 1993;24:301-313.
[24] Chen X, Liaw SB, Wu H, Combustion and Flame 2017;182:90-101.

11th Asia-Pacific Conference on Combustion,
The University of Sydney, NSW Australia
10th -14th December 2017.

Dynamic Adaptive Acceleration of Chemical Kinetics with Consistent Error Control

Wenwen Xie[1], Zhen Lu[2], Zhuyin Ren[1,2*], Graham Goldin[3]

[1]School of Aerospace Engineering, Tsinghua University

Beijing, 100084, China

[2]Center for Combustion Energy, Tsinghua University

Beijing, 100084, China

[3]Simens CD-Adapco

Lebanon, NH, 03766, USA

Abstract

The incorporation of detailed chemistry in combustion simulations is challenging due to the large number of chemical species and the wide range of chemical timescales. The performance of acceleration methods such as tabulation/retrieval strategies may deteriorate dramatically when large variation in the accessed composition space is present. In this study, a dynamic adaptive approach is proposed to accelerate chemistry calculations, in which the *in situ* adaptive tabulation (ISAT) or dynamic adaptive chemistry (DAC) is dynamically selected based on the encountered composition inhomogeneity. The principle component analysis is employed to identify a low-dimensional subspace, in which the composition inhomogeneity of the computational cells is quantified through reconstructing the histogram of composition. ISAT is invoked for cells being in composition regions with high cell/particle numbers to avoid unnecessary tabulations. DAC is employed for the remaining ones by invoking on-the-fly reduction to generate small skeletal mechanisms for local thermo-chemical conditions and therefore accelerates the chemistry integration. The consistent error control of ISAT and DAC is also addressed such that a single error tolerance is employed to control the local incurred error in composition for these two methods. The proposed approach is validated in model engine simulations and the performance is analyzed.

1 Introduction

Over the past several decades, the exploding development of detailed kinetics for hydrocarbon fuels facilitates the reliable prediction of near-limit flame phenomena such as local extinction/ re-ignition and emission characteristics in turbulent combustion simulations [1-5]. However, the time-intensive nature of integrating detailed chemistry due to the large number of chemical species and the wide range of chemical timescales involved poses a great computational challenge for multi-dimensional CFD, which may be further aggravated by the highly transient nature of the combustion processes such as those in internal combustion engines (ICE) [5].

Significant progress has been made in the development of numerical schemes and acceleration methodologies to efficiently accommodate detailed chemistry in combustion simulations [6-16]. For example, operator splitting schemes [14] have been employed to separate the integration into reaction and transport fractional sub-steps. In the reaction sub-step, the thermo-chemical state of each computational cell can be described by the n_s-sized species mass fraction $\boldsymbol{Y}$, the temperature T and the pressure P. Given a fixed pressure, the (n_s+1)-sized column vector $\boldsymbol{\phi} \equiv \{\boldsymbol{Y}, T\}$ evolves according to a set of ordinary differential equations (ODEs)

$$\frac{d\boldsymbol{\phi}(t)}{dt} = \boldsymbol{S}(\boldsymbol{\phi}(t)), \qquad (1)$$

where $\boldsymbol{S}$ is the reaction source term given by chemical kinetics. With operator splitting, skeletal mechanism [6], dynamic adaptive chemistry (DAC) [7-9], reaction mapping approximation [10-13] and fast ODE solvers [15, 16] can be employed to accelerate chemistry integration. Recently, the combination of *in situ* adaptive tabulation (ISAT) [10, 11], which expedites the integration of Eq. 1 by reusing the reaction mappings from similar initial composition, and DAC, accelerating the chemistry integration by generating on-the-fly small skeletal mechanisms for local thermo-chemical conditions, have gained significant interest for combustion simulations with relatively large time steps, i.e., 10^{-6} s to 10^{-4} s. For example, the integrated ISAT-DAC approach (TDAC) [17] has been explored for ultimately efficient ICE simulations, in which DAC is invoked to expedite the expensive chemistry integrations for adding new entries or failed retrieval attempts in ISAT. TDAC has been demonstrated to achieve a speedup factor of hundreds to thousands for near-homogenous engine simulations [17]. However, similar to ISAT, its performance deteriorates dramatically under realistic engine conditions with large pressure variation and composition inhomogeneity because the new entries added during the combustion stage cannot be reused so many times to pay off the computational overhead [18]. To address this challenge, a static adaptive chemistry acceleration approach has recently been explored to improve the computational efficiency of ICE simulations [5]. Instead of employing a fixed acceleration method throughout the entire simulation, acceleration method e.g., ISAT or DAC is employed at different combustion stages based on the encountered composition inhomogeneity and reaction activity. Specifically, ISAT with dynamic pruning is employed during the compression and post-combustion stages in which composition

Corresponding author.
E-mail address: zhuyinren@mail.tsinghua.edu.cn

inhomogeneity is in general not significant, and DAC is employed during the combustion stage. For the simulated HCCI combustion, the static adaptive approach achieves an overall speedup factor of more than five for chemistry calculation with a 171-species mechanism while maintaining accurate predictions of pressure, temperature and species concentration. It is three times more efficient than TDAC, demonstrating the great potential of adaptive acceleration for chemistry calculations. However, the specification of acceleration methods in the static method is empirical and is hard to extend to general combustion simulations. In addition, similar to TDAC, two separate error control parameters without any correlation, e.g., DAC reduction threshold and ISAT error tolerance are employed to control the simulation accuracy. This requires significantly extra efforts to determine the appropriate acceleration settings for a particular combustion simulation.

In this study, a dynamic adaptive acceleration method (DAAM) is developed. It dynamically selects the optimal acceleration method based on the encountered composition inhomogeneity, which is quantified through reconstructing the histogram of composition in a low-dimensional subspace determined by principle component analysis. ISAT is invoked for computational cells in composition regions with high cell/particle numbers to avoid unnecessary tabulations that would be rarely reused. DAC is employed to accelerate chemistry calculations for the remaining ones by invoking on-the-fly reduction to generate small mechanisms for local thermo-chemical conditions. A consistent error control strategy of ISAT and DAC is proposed so that a single error tolerance is employed to control the local incurred error in composition.

The rest of paper is organized as follows. In section 2, the methodology of DAAM is elaborated with emphasis on characterization of composition inhomogeneity and consistent error control. In Section 3, its performance is analyzed in ICE model simulations with various levels of composition inhomogeneity. Conclusions are in Section 4.

2 Methodology

In this work, the ISAT algorithm, ISAT5, with most recently used (MRU) pruning strategy is employed [11]. For DAC, both the DRG-based [6] and TSRA-based [8] DAC are considered to demonstrate the universality of the error control strategy.

2.1 Characterization of composition inhomogeneity

The principle component analysis (PCA) [19] is employed to identify the representative subspace of dimension D, in which the composition inhomogeneity is characterized. To obtain the accurate coordinate axes of the subspace and reduce the computational cost, some representative computational cells are selected with thermo-chemical states given by

$$\boldsymbol{\Phi} \equiv [\boldsymbol{\phi}^1 \cdots \boldsymbol{\phi}^k \cdots \boldsymbol{\phi}^{n_{sample}}], k = 1,2,\cdots,n_{sample}, \quad (2)$$

where n_{sample} is the number of representative cells. Each column of $\boldsymbol{\Phi}$ represents a sample point and each row represents a state variable in the chemical system. Let $\boldsymbol{W}$ denote the covariance matrix of $\widehat{\boldsymbol{\Phi}}$, obtained by scaling $\boldsymbol{\Phi}$ with the maximum value and mean centering for each attribute. The D-dimensional subspace for characterizing composition inhomogeneity is taken to be the subspace spanned by matrix **U**, which consists of the eigenvectors corresponding to the first D largest eigenvalues of $\boldsymbol{W}$. The dimension of the subspace can be obtained with a user-specified tolerance on the eigenvalues and it depends on the nature of the application.

The transformation $\boldsymbol{\Phi}^p = \boldsymbol{U}^T\boldsymbol{\Phi}$ projects cell compositions to the identified low-dimensional subspace, in which a uniform grid is generated to construct the histogram to quantify composition inhomogeneity. In each direction of the subspace, mass fraction vectors of the lower and upper boundary points are $\boldsymbol{Y}_{l,i}$ and $\boldsymbol{Y}_{u,i}, i = 1,2,\cdots,\mathrm{D}$, respectively. Assuming the gradient of species mass fraction is uniform in each direction of the subspace, one has

$$\boldsymbol{Y}_{u,i} - \boldsymbol{Y}_{l,i} = \frac{\partial \boldsymbol{Y}}{\partial x_j}\Delta x_{i,j}, i,j = 1,2,\cdots,\mathrm{D}, \quad (3)$$

where $\Delta x_{i,j}$ is the distance in j-th direction for boundary points in i-th direction. The grid interval in each direction is determined by

$$\Delta \mathrm{i} = \frac{\theta Y_{ref}}{|\partial \boldsymbol{Y}/\partial x_i|}, i = 1,2,\cdots,\mathrm{D}, \quad (4)$$

where θ is a model parameter, and Y_{ref} is the reference mass fraction value, which is specified according to the equilibrium state corresponding to the initial state of global equivalence ratio, i.e., $Y_{ref} = \sqrt{\sum_{i=1}^{n_s} Y_{eq,i}^2}$. The number of cells fall into each bin in the low-dimensional subspace is counted up to construct the histogram.

2.2 Dynamic adaptive specification of acceleration methods via computational time estimation

After the reconstruction of the low-dimensional histogram, ISAT is invoked for regions with high cell/particle numbers to avoid unnecessary tabulations that would be rarely reused. DAC is invoked for the remaining ones. Assuming that cell compositions to be evaluated in each bin in the reduced subspace can be retrieved from its boundary points when ISAT is invoked for this bin, the value of critical cell number n_c to switch between ISAT and DAC is given by

$$n_c t_{DAC} = 2^D t_F + (n_c - 2^D)t_R, \quad (5)$$

where t_{DAC} is the average CPU time for each integration with DAC, t_F represents the average CPU time to perform an operation invoking direct evaluation with ISAT, including the add, growth and replacement operation, and t_R denotes the average CPU time for the retrieval operation. ISAT is invoked for computational cells falling into the bins with more than n_c computational cells, while DAC is employed for the remaining ones.

2.3 Consistent error control in reaction mapping approximation

Considering the variety of the DAC methods [7-9] and the obscure connection between the reduction threshold and the incurred local error in composition, it is preferred to dynamically adjust the reduction threshold of DAC, ε_{DAC}, so that its incurred error is compatible with ISAT error tolerance ε_{ISAT} that effectively controls the local incurred error.

During a chemistry sub-step, DAAM classifies computational cells into ISAT-resolved and DAC-resolved ones. ISAT rigorously controls the difference between the linear approximation and the direct integration value, namely the local error

$$\varepsilon \equiv |\boldsymbol{R}_a(\boldsymbol{\phi}^q) - \boldsymbol{R}(\boldsymbol{\phi}^q)| < \varepsilon_{ISAT}, \quad (6)$$

where $\boldsymbol{R}_a(\boldsymbol{\phi}^q)$ is the linear approximation of $\boldsymbol{R}(\boldsymbol{\phi}^q)$. Similarly, ε_{DAC} can be dynamically adjusted such that the difference in results between DAC and direct integration satisfies Eq. 6, and thus one achieves the same error control for DAC and ISAT. Figure 1 illustrates the heuristic adjusting algorithm of ε_{DAC}, in which the induced local error from DAC $\varepsilon = |\boldsymbol{R}_{DAC}(\boldsymbol{\phi}) - \boldsymbol{R}(\boldsymbol{\phi})|$ is compared with ε_{ISAT}. The error tolerance ε_{DAC} will be decreased by a factor of β if $\varepsilon > \varepsilon_{ISAT}$, while be increased by a factor of β if $\varepsilon < \varepsilon_{ISAT}$. The adjusting iteration continues until the critical ε_{DAC} is ascertained. The iteration occurs once ISAT invokes direct integration and provides $\mathbf{R}(\boldsymbol{\phi})$. For all of cases in this work, β is set to be 2. In practice, the user-defined lower and upper bounds of ε_{DAC} help to confine its value to a realistic region. Note that chemistry integrations invoking DAC during this process bring additional computational costs.

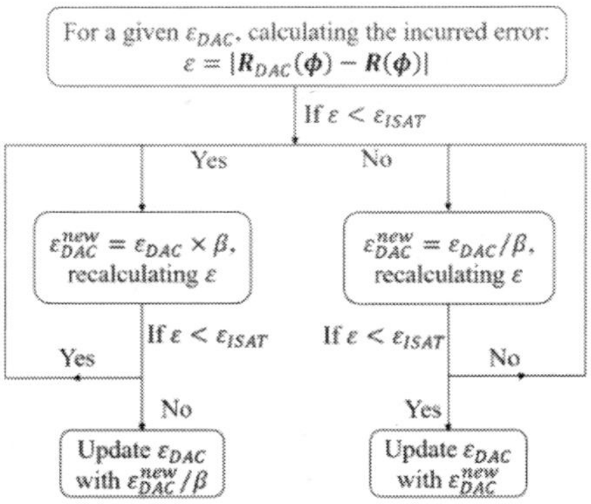

Figure 1: The heuristic algorithm for adjusting the DAC reduction threshold in DAAM.

Note that the computational overhead would be unacceptable if the above adjusting procedure is invoked for each time step. To resolve this issue, the adjusting procedure is invoked only when the largest eigenvalue of the covariance matrix changes more than 50% of its existing value, which is called the assessment operation. Until the next assessment operation, ε_{DAC} is kept as a constant, which is selected to be small enough to guarantee 90% error control effective in this time step, similar to the probability of effective error control in ISAT [11].

3 Results and Discussions

The DAAM method is validated in the ICE model cases for premixed, non-premixed and partial premixed combustion of n-heptane/air with the 88-species skeletal mechanism [20]. The implicit ODE solver DDASAC [21] is employed for chemistry integration with absolute and relative error tolerance set to be 1×10^{-12} and 1×10^{-9}, respectively.

The ICE model is similar to an individual grid cell in a probability density function (PDF) simulation of a turbulent reacting flow, consisting of n particles at any time. With Δt being the specified time step, the particle compositions evolve due to mixing and chemical reaction, which are treated in separate fractional steps. The mixing sub-step for each particle is governed by the modified curl model, which is used to model the micro-scale molecular diffusion process. With this ICE model, test cases of various levels of inhomogeneity and computational complexity can be easily designed through varying the initial distribution of particles' composition state.

For the ICE model simulations, the initial composition for each particle is set to be a mixture of pure fuel of 800 K and air of 900 K with a certain mixture fraction. The mixture fractions of those $n = 500$ particles are subjected to a Beta distribution, with the mean value $\tilde{Z}$ fixed to be 0.014. The root mean square of the mixture fraction $\sigma_0(Z)$ varies from 0, corresponding to the premixed cases, to 0.117, corresponding to the non-premixed cases. For the partial premixed cases, $\sigma_0(Z)$ is taken to be 5%, 20%, 40%, 60%, 80% of the non-premixed case, respectively. A pressure profile of $P = 50 + 30\sin(200\pi t - 0.5\pi)\ atm$ is used for all ICE model simulations. The mixing time scale is taken to be $\tau_{mix} = 2 \times 10^{-3}$ s, and the time step $\Delta t = 0.01\tau_{mix}$. All the cases are simulated for the time duration of 8 ms. The maximum size of the ISAT table is set to be 500 MB for all the cases. The starting species for DRG-based DAAM are fixed to be n-heptane, H and the two most abundant species with the upper and lower bounds of reduction threshold to be $\varepsilon_{DRG,u} = 0.2$ and $\varepsilon_{DRG,l} = 0.001$, respectively. And those for TSRA-based DAAM are set to be $\varepsilon_{TSRA,u} = 1 \times 10^{-2}$ and $\varepsilon_{TSRA,l} = 1 \times 10^{-5}$. For the low-dimensional manifold, the dimensionality of two is specified, together with a model parameter $\theta = 1 \times 10^{-3}$ for quantifying composition inhomogeneity.

An example of DRG-based DAAM without assessment operation, which means that the adjusting algorithm for reduction threshold is invoked for each chemistry time step, is analyzed to illustrate the details of the control strategy. Figure 2 (a) shows the scatterplot of ε_{DRG} with $\varepsilon_{ISAT} = 2 \times 10^{-4}$ for the case of $\sigma_0 = 0.117$. Multiple values of ε_{DRG} at a certain time instant show that the adjusting iteration happens more than once, relying on the frequency of ISAT invoking direct integration during this time interval. Temperature profiles of several particles together with the mean temperature profile are shown in Fig. 2 (b), which reveals that the inhomogeneity of particle composition leads to various reaction rates, further resulting in the need of adjusting ε_{DRG} to guarantee the consistent error with ISAT at each time instant.

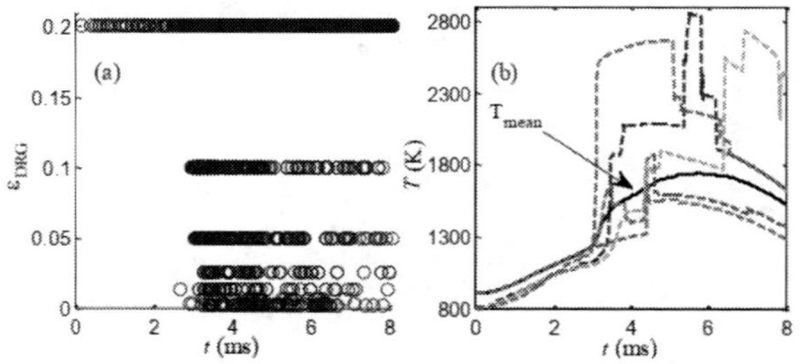

Figure 2: (a) The scatterplot of ε_{DRG} from DRG-based DAAM without assessment operation and (b) the temperature profiles of several selected particles together with the mean temperature profile with $\varepsilon_{ISAT} = 2 \times 10^{-4}$ for the case of $\sigma_0 = 0.117$.

Figure 3 shows the evolution of (a) ε_{DRG} and (b) ε_{TSRA} from DRG-based and TSRA-based DAAM invoking the assessment operation for the same case. During the time step invoking the assessment operation, the heuristic adjusting iteration occurs once ISAT invokes direct integration and an error tolerance of DAC is obtained, represented by the red symbols in Fig. 3. The value of ε_{DAC} which guarantees 90% error control effective at this time step is selected, represented by the blue line in Fig. 3, and is kept unchanged for DAC integrations until the next assessment operation. Besides, the selected reduction threshold of DAC decreases as expected at the fast combustion stage. Comparison of Fig. 2 (a) and Fig. 3 also demonstrates the significance of the assessment operation in alleviating the computational overhead.

Table 1 shows the percentage of the add, growth, retrieval and DAC-resolved operations over the whole simulation time of 8 ms. Both DRG-based and TSRA-based DAAM replace some unnecessary add operations with DAC integration, thus reduce the computational cost for the update of ISAT table. Consequently, the percentages of growth and retrieval decrease at the same time.

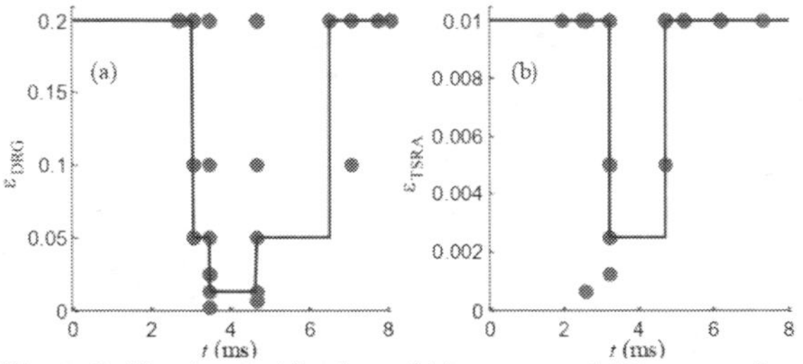

Figure 3: The threshold value of (a) ε_{DRG} and (b) ε_{TSRA} from DRG-based and TSRA-based DAAM invoking assessment operation with $\varepsilon_{\mathrm{ISAT}} = 2 \times 10^{-4}$ for the case of $\sigma_0 = 0.117$.

Percentage (%)	Add	Growth	Retrieval	DAC
ISAT	0.60	5.55	93.85	0
DAAM-DRG	0.39	3.63	89.93	6.05
DAAM-TSRA	0.43	4.01	91.13	4.43

Table 1: The percentage of operations add, growth, retrieval and DAC over the whole simulation of 8 ms.

Quantified relative errors of mean temperature ε_T against the corresponding speedup factors (sf) with $\sigma_0 = 0.117$ and $\sigma_0 = 0.047$ for $\varepsilon_{\mathrm{ISAT}} = 5 \times 10^{-4}, 2 \times 10^{-4}, 1 \times 10^{-4}, 5 \times 10^{-5}$ are shown in Fig. 4. Results from the other initial distributions are similar. For most of cases, DAAM method shows advantage over the standalone ISAT with higher speedup factors and lower relative error in composition. DAAM improves the acceleration performance by 44% at most than ISAT for these simple ICE model cases. The error control strategy facilitates the single input of error control parameter and achieves consistent error control of ISAT and DAC. For the homogeneous cases, DAAM degenerates to the standalone ISAT as expected.

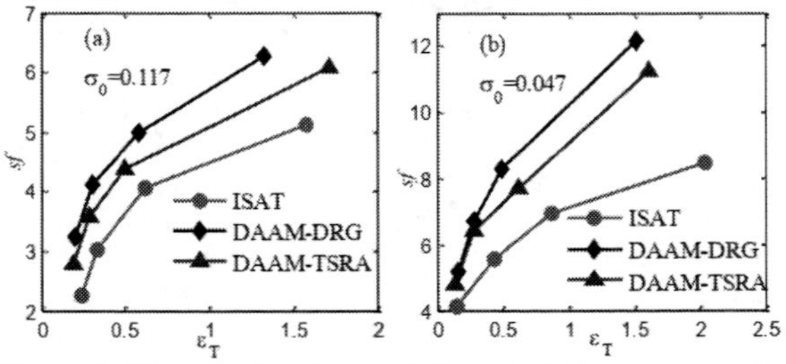

Figure 4: The speedup factors (sf) against the corresponding relative error of temperature ε_T with $\sigma_0 = 0.117$ and $\sigma_0 = 0.047$ for $\varepsilon_{\mathrm{ISAT}} = 5 \times 10^{-4}, 2 \times 10^{-4}, 1 \times 10^{-4}, 5 \times 10^{-5}$, respectively.

4 Conclusions

A dynamic adaptive acceleration method (DAAM) has been developed, in which the DAC and ISAT is dynamically selected based on the encountered composition inhomogeneity, described by the reconstructed composition histogram in a low-dimensional subspace identified by PCA. DAAM can dynamically adjust the DAC reduction threshold based on the user-defined ISAT error tolerance, which enables the single error control parameter for the combined use of these two methods and more importantly enables the rigorous local error control.

The DAAM method has been successfully demonstrated for simple ICE model cases with different initial inhomogeneity. It turns out that DAAM can improve the acceleration performance by 44%. The error control strategy facilitates the single input of threshold value and achieves the consistent error control of ISAT and DAC.

It is also suggested that the biggest potential to improve the acceleration performance lies on the fast calculation of sensitivity matrix $\boldsymbol{A}$, thus reducing the computational costs for add and replacement operations.

5 Acknowledgment

The work is supported by National Natural Science Foundation of China 91441202. Simulations are performed with the computational resources of the Tsinghua National Laboratory for Information Science and Technology. We are grateful to Prof. Lei Zhou for his help on the engine simulations.

References

[1] H. Wang, E.R. Hawkes, J.H. Chen, Combust. Flame 180 (2017) 110-123.

[2] R.R. Tirunagari, S.B. Pope, Proc. Combust. Inst. 36 (2) (2017) 1919-1927.

[3] H. Zhou, S. Li, Z. Ren, D.H. Rowinski, Combust. Flame 181 (2017) 136-148.

[4] R.P. Lindstedt, S.A. Louloudi, E.M. Váos, Proc. Combust. Inst. 28 (1) (2000) 149-156.

[5] H. Wei, L. Zhou, H. Zhou, Z. Lu, Z. Ren, G.M. Goldin, Combust. Sci. Technol. 189 (4) (2017) 623-642.

[6] T. Lu, C.K. Law, Proc. Combust. Inst. 30 (1) (2005) 1333-1341.

[7] L. Liang, J.G. Stevens, J.T. Farrell, Proc. Combust. Inst. 32 (1) (2009) 527-534.

[8] W. Xie, Z. Lu, Z. Ren, L. Hou, Proc. Combust. Inst. 36 (1) (2017) 645-653.

[9] O.O. Oluwole, Z. Ren, C. Petre, G. Goldin, Combust. Flame 162 (5) (2015) 1934-1943.

[10] S.B. Pope, Combust. Theory Model. 1 (1) (1997) 41-63.

[11] L. Lu, S.B. Pope, J. Comput. Phys. 228 (2) (2009) 361-386.

[12] A. Babajimopoulos, D.N. Assanis, D.L. Flowers, S.M. Aceves, R.P. Hessel, Int. J. Engine Res. 6 (5) (2005) 497-512.

[13] G. Goldin, Z. Ren, S. Zahirovic, Combust. Theory Model. 13 (4) (2009) 721-739.

[14] B. Sportisse, J. Comput. Phys. 161 (1) (2000) 140-168.

[15] A. Imren, D.C. Haworth, Combust. Flame 174 (2016) 1-15.

[16] Y. Gao, Y. Liu, Z. Ren, T. Lu, Combust. Flame 162 (2) (2015) 287-295.

[17] F. Contino, H. Jeanmart, T. Lucchini, G. D'Errico, Proc. Combust. Inst. 33 (2) (2011) 3057-3064.

[18] G. D'Errico, T. Lucchini, A. Onorati, G. Hardy, Int. J. Engine Res. 16 (1) (2014) 112-124.

[19] H. Abdi, L.J. Williams, Wiley Interdiscip. Rev. Comput. Stat. 2 (4) (2010) 433-459.

[20] C.S. Yoo, T. Lu, J.H. Chen, C.K. Law, Combust. Flame 158 (9) (2011) 1727-1741.

[21] M. Caracotsios, W.E. Stewart, Comput. Chem. Eng. 9 (4) (1985) 359-365.

11th Asia-Pacific Conference on Combustion,
The University of Sydney, NSW Australia
10th -14th December 2017.

Pressure Wave and Flame Front Positions Originating from Methane-Air Explosions in a 1m^3 Vessel with Circular Duct

Daniel Eschebach, Jafar Zanganeh*, Behdad Moghtaderi

Priority Research Centre for Frontier Energy Technologies & Utilisation, Discipline of Chemical Engineering, School of Engineering, Faculty of Engineering & Built Environment, University of Newcastle, Callaghan, NSW 2308, Australia

Abstract

Mitigation of an explosion propagating in a duct requires accurate pressure wave and flame velocity data to actuate a safety device preventing flame and pressure travel through a duct system. This study examines the rate of pressure rise along a duct system and provides correlations between pressure wave and flame front velocities with methane-air concentrations within the flammable range ignited by a 1 kJ pyrotechnic chemical ignitor under initially turbulent conditions. The experimental study was conducted in a 1 m^3 spherical explosion vessel with a high pressure injection chamber and 9.8 m 146 mm ID duct connected with a check valve. Piezoelectric pressure transducers and 200-1100 nm photodiodes were utilised to characterise the deflagration explosion pressure profiles and flame (by emitted light) velocities. High turbulence scenarios were investigated by using high pressure gas injection system moments before ignition. The maximum rate of pressure rise was higher inside the duct (194 bar.s^{-1} at 4.0 m from the ignition source) than inside the vessel (173 bar.s^{-1}) for 9.5% methane.

1 Introduction

Ventilation air methane abatement (VAMA) systems break down low methane concentration air into lower global warming potential (GWP) product gases (i.e. CO_2 and water vapour). In order to capture and direct VAM to an abatement unit a complex duct system must be designed which can safely deal with all explosion scenarios generated by methane (>90% methane composition of flammable gas, <1% flammable gas concentration in VAM) air mixtures. Methane has a LFL of 4.6±0.3%, and UFL of 15.8±0.8% and produces the maximum explosion pressure at the stoichiometric ratio condition of 9.5% [1]. Therefore the possibility of generating an explosive atmosphere in a VAMA system is high and the potential consequences could be catastrophic. The main type of hazard responsible for casualties in underground coal mining is explosions [2]. Accidental explosions almost always occur in unpredictable circumstances and therefore worst case scenarios need to be adequately addressed in safety control measures of design. Safety control measures are classified into active measures and passive measures. Active measures include fast acting isolation valves and powder spray suppression systems. Activation of active suppression systems rely on accurate knowledge of worst case explosion parameters within the system in order to locate the sensing elements and suppression or isolation device correctly and to fine tune the trigger detection limits for sensors so as to activate only in explosion scenarios. Detection systems are commonly dynamic pressure transducers and light sensing devices. A full data set of possible explosion parameters in different sized vessels and ducts is required in order to mitigate both minor and worst case scenarios. Furthermore, ducts can introduce an additional hazard by secondary explosion caused by ejection of oxygen depleted mixture into an oxygen rich atmosphere.

Literature has focused on initially laminar explosion characteristics of methane mixtures in closed vessels [3,4] and one end open straight ducts [5,6]; However, most published sources focus on small vessel size (20 L) and small pipe diameter [7-10]. Furthermore, the correlation between flame position and pressure wave profile is not well understood in ducts of varying size. Even if explosion parameters such as cloud size, fuel concentration, ignition point and vessel size are known, it is difficult to accurately predict the flame speed or pressure wave speed. Explosions in ducts are not scalable between different vessel size, for example, a smaller vessel has greater vessel wall to volume ratio which effectively acts as a heat sink on the explosion reducing explosion pressure. Furthermore, explosions in VAM ducting systems would typically occur while a gas stream is flowing rather than stationary, resulting in a more severe explosion. To address the lack of experimental data, experiments measuring pressure profile and flame position will be completed for a 1 m^3 vessel with 146 mm ID one end open duct.

The objective of this paper is to correlate pressure wave velocity to flame front velocity in a one end open duct, provide time delay data between the pressure wave and flame front and also the rate of pressure rise along the tube profile to elucidate the hazard of secondary explosion in ducts with respect to secondary pressure rate rise as compared to inside the vessel itself. To achieve the objectives of this study variable methane concentrations will be ignited in a 1 m^3 explosion vessel with 8.5 m connected one open duct. The experimental study followed and pressure plus light signal analysis along the tube length was completed at the Fire and Explosion Laboratory at the University of Newcastle- Australia.

2 Experimental

The explosion test apparatus consisted of a 1 m^3 explosion vessel, modified by connection of a 9.8 m length duct and 0.146 m internal diameter propagation tube, as shown below in Fig. 1.

E-mail address: jafar.zanganeh@newcastle.edu.au

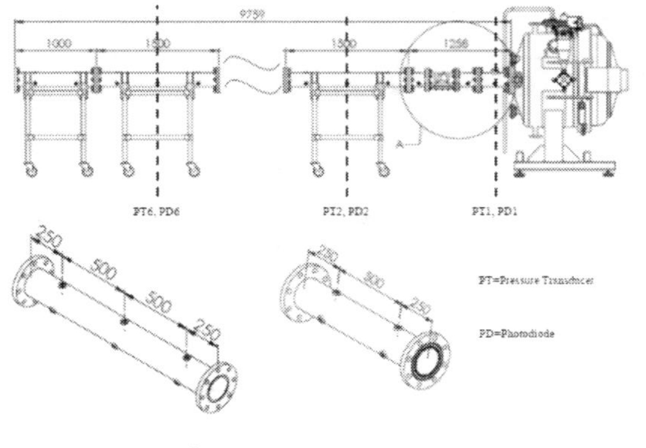

Figure 1: The 1 m^3 explosion test vessel with 146 mm ID duct.

The vessel and tube were connected by a bar flex check valve (model-FCVC-400, Process Systems Private Limited) which allowed isolation of the tube from the vessel by maintaining a small vacuum pressure in the vessel during setup. The valve was mounted 180° to allow gravity opening and opened just prior to ignition after the vessel reached near atmospheric pressure. Setup began with opening the vessel and loading the 1 kJ ignitor into the centre of the 1 m^3 vessel. The vessel was then closed and sealed with a manual lock and pin interlock safety system. The bar flex check valve was manually closed during vessel evacuation until a stable vacuum pressure was reached and filled with methane gas to the desired methane concentration. The initial vacuum was adjusted in order to achieve a final pressure of 96 kPa(a), maintaining a slight vacuum pressure to maintain check valve closure. The 4.5 L side chamber, which provided turbulent methane air mixture injection into the vessel, was filled initially with methane and then instrument air with a final methane-air mixture pressure of 21 MPa. Methane concentrations in the side chamber were 6.5%, 8.0%, 9.5%, 11.5%, 12.5% and 14%. The side chamber was injected into the vessel at time 0 ms, the bar flex check valve opened at 50 ms. The pre-loaded 1 kJ chemical ignitor was energized at 0 ms. The internal vessel pressure was measured by two pressure transducers (Kistler 701A 14 ns) and the duct length was monitored by six pressure transducers (STS 0-10 bar(g), 1 ms) and six photodiodes (DET10A, Thorlabs, 200-1100 nm, 1 ns). One pressure transducer and one photodiode were each mounted along the duct at distances of 0.5 m, 2.5 m, 4.0 m, 5.5 m, 7.0 m and 8.5 m. The photodiode viewing angle was 10° preventing downstream or upstream light capture. Pressure and photodiode data were collected by 10 MHz and 20 kHz data acquisition systems for the 1 m^3 vessel and duct, respectively, synchronized by the trigger signal. .

2.3 Equations

Data smoothing was utilized to reduce the number of data points and reduce the natural variation in sensor signal. The unweighted arithmetic mean function

$$f(x_i) = \frac{1}{2W+1}\sum_{k=i-W}^{k=i+W} x_k,\ m - W > i > W \quad (1)$$

was used to smooth the data, where m is the channel length and W is the smoothing width. For this function, the first and last W data points remain unsmoothed.

The pressure wave velocity and flame front velocity were determined using a fixed initial reference point method, which is the standard method for calculating explosion data. The following equation describes this calculation:

$$V_i = \frac{D_i}{t_i - t_0} \quad (2)$$

where V_i is the velocity at sensor, D_i is the distance of the sensor from the ignition source; t_i is the sensor detection time and t_0 is the time of ignitor ignition. Since open vessel data is more complex than closed vessel data, due to negative pressures and pressure wave reflections, the pressure wave arrival time was taken as the first point of pressure or voltage increase of the sensor, (11 kPag for pressure and 0.1 V for flame front). These values were selected based on the minimum value which just exceeded sensor noise.

Maximum rates of pressure rise along the tube profile were determined from integrated pressure-time profiles and maximum values $(dp/dt)_{max}$ are reported.

3. Results and Discussion

Pressure profiles of the methane explosions were generated both inside the 1 m^3 vessel and along the duct. The rate of pressure rise was measured at each pressure transducer location, Fig.1.

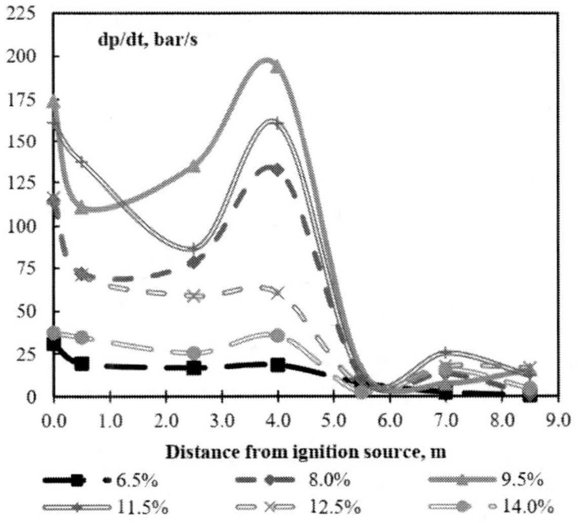

Figure 1: Maximum rate of pressure rise $(dp/dt)_{max}$ at distances along the duct length for methane gas explosions.

In the above figure, zero distance corresponds to the vessel wall of the 1 m^3 vessel. At methane concentrations near stoichiometry from 8.0% to 11.5% the maximum rate of pressure rise occurred in the duct at 4 m distance from the ignition source rather than at the ignition source. At methane concentrations far from the stoichiometric ratio, <8.0% and >11.5%, the maximum rate of pressure rise occurred inside the vessel at the ignition source. At greater than 5.5 m distance the rate of pressure rise was limited to <25 $bar.s^{-1}$ for all flammable methane concentrations. The rate of pressure rise is representative of the rate of the combustion reaction. At stoichiometric ratios of air and fuel mixture, the rate of combustion is at a maximum and most of the combustion reaction occurs within the vessel its self. Below the stoichiometric ratio (low air to fuel), an excess of fuel and limited oxygen supply cause a slow combustion reaction which ejects uncombusted fuel into the duct. However, at low stoichiometric ratios, this fuel is starved of oxygen until it reaches the end of the tube where a secondary explosion occurs as the fuel is mixed with the atmospheric oxygen. The rate of pressure rise in the tube is therefore low throughout this reaction. This was confirmed by visual observation of a secondary explosion occurring outside the vessel, by infrared video image. An external explosion was observed at methane concentrations significantly lower than the stoichiometric ratio (12.5% and 14%). At 6.5% methane no external explosion was observed, as expected at high stoichiometric ratio where excess oxygen allowed complete consumption of fuel within the vessel. The highest $(dp/dt)_{max}$ of 175 bar.s^{-1} within the 1 m^3 vessel portion, 0 m in the above figure, occurred at 9.5% methane. The highest $(dp/dt)_{max}$ of 194 bar.s^{-1} within the duct occurred for 9.5% methane at 4.0 m for from the ignition source.

The above observations provide insight into correlations between pressure wave and flame front velocities, Fig. 2.

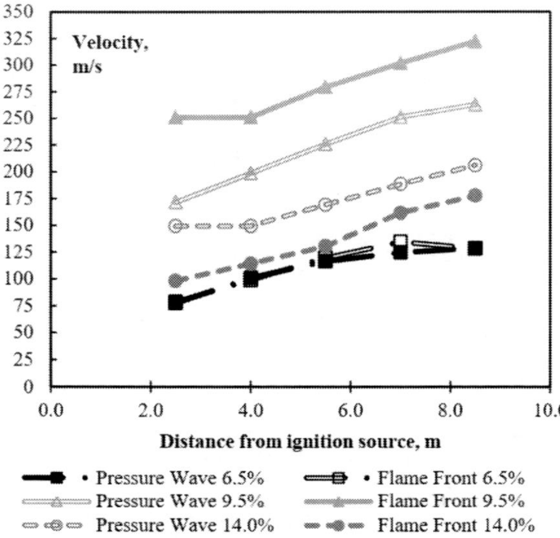

Figure 2: Pressure wave and flame front velocities for explosions of methane gas propagating through a 146 mm ID duct.

The flame front and pressure wave velocities were approximately equal when the fuel was quickly consumed at 6.5% methane. At 9.5% methane, the flame front velocity was faster than the pressure wave velocity and at 14.0% methane the flame front velocity was lower than the pressure wave velocity. At methane concentrations near stoichiometry, the continued combustion increased the pressure wave velocity significantly and the higher pressure along with optimum fuel-oxygen ratios accelerated the flame front velocity. The flame front velocity is a measure of the rate of combustion reaction which, for methane concentration at stoichiometry (9.5%), was very fast and caused significant acceleration of the flame front velocity above that of the pressure wave velocity. Correlations for the pressure wave and flame front velocities were calculated for Figure 2 data and are shown below in Table 1.

Table 1: Pressure wave and flame front velocity (m.s^{-1}) correlations for y=mx+b.

Methane concentration	m	±m	b	±b	Correlation coefficient
6.5%	17	3.4	15.7	17.1	0.83
6.5%	16	2.8	13.1	14.0	0.87
8.0%	31	4.0	13.1	19.7	0.92
8.0%	14	0.6	-0.4	2.8	0.99
9.5%	25	3.8	14.8	19.0	0.90
9.5%	15	2.0	7.6	10.2	0.91
11.5%	26	4.1	14.2	20.5	0.89
11.5%	14	2.1	8.0	10.5	0.90
12.5%	21	3.7	10.9	18.7	0.86
12.5%	28	6.8	35.0	33.9	0.77
14.0%	12	0.8	1.2	3.8	0.98
14.0%	24	5.9	30.1	29.2	0.77

Most of the pressure wave and flame front velocity correlations were strong with correlation coefficients above 0.90 and all above 0.77. It should be noted that in all cases, the flame front was behind the first pressure wave. The arrival time difference between the flame front and pressure wave is shown below in Fig. 3. Explosion delay time is the time between activation of the ignitor and onset of combustion. Similar to the rate of pressure rise (Fig.1) explosion delay time is at a minimum at 9.5% methane and at maximums near the UFL and LFL, [13]. In this study we define flame delay time as the time between arrival of the pressure wave and flame front at the same sensor. The delay times along the duct length were calculated at different lengths along the duct for each methane concentration tested, Fig. 3.

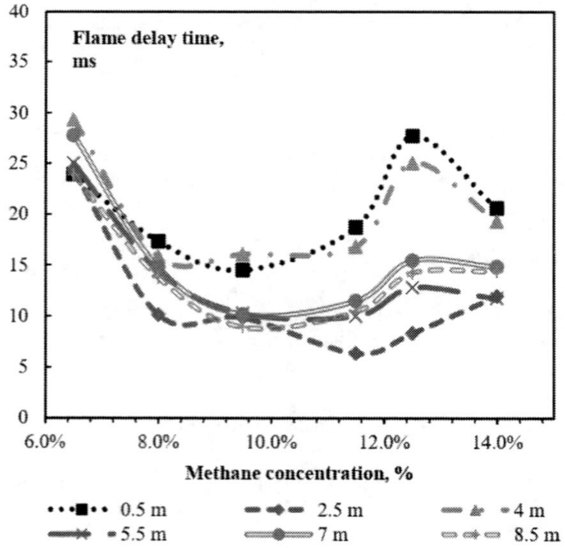

Figure 3: Flame front delay time from pressure wave at varying distances from the ignition source for explosions of methane

gas propagating through a 146 mm ID duct.

In the above figure, the flame delay time had a parabolic relationship between methane concentration and flame delay time, highly similar to explosion delay time [13]. A minimum exists in the range of 9.5% methane concentration and the shape of the flame front delay time is largely independent of the measurement distance along the duct length. Explosion profiles are dependent on the rate of reaction which can be expressed by the Arrhenius equation, demonstrating the exponential dependence of reaction rate on dependant on temperature but only proportional dependence on concentration or number of collisions. At higher concentrations of methane, the overall heat capacity of the mixture-air mixture increases. Therefore, the rate of temperature rise and hence the rate of reaction is slower at increased methane concentration as compared to decreased methane concentration about the equivalence point.

4 Conclusions

Observation of the rate of pressure rise within the vessel allowed observation of three pathways of combustion, **(I)** complete internal combustion, **(II)** partial internal combustion with secondary combustion in the pressure relief duct and **(III)** partial combustion with external combustion outside the duct. Pathway **(II)** resulted in more severe dp/dt explosion in the duct than at the ignition source. The highest $(dp/dt)_{max}$ within the 1 m^3 vessel was 175 $bar.s^{-1}$ and occurred at 9.5% methane. The highest $(dp/dt)_{max}$ within the duct was 194 $bar.s^{-1}$ and occurred for 9.5% methane at 4.0 m distance for from the ignition source. Although the flame front always followed the pressure wave, the flame front velocity was shown to exceed the pressure wave velocity at the stoichiometric fuel ratio of 9.5%. The flame delay time was at a minimum near stoichiometry and 2.5 m distance from the ignition source and at a maximum at both 0.5 ma and 4.0 m from the ignition source and at methane concentrations far from stoichiometry (6.5% and 14.0%). The pressure wave velocity varied from a 34-85 $m.s^{-1}$ at 14% to 9.5%, respectively. The flame front velocity was just slower the pressure wave velocity, having a range of 77 $m.s^{-1}$ at 9.5% methane to 32 $m.s^{-1}$ at 14% methane.

5 Acknowledgment

The authors wish to acknowledge the financial support provided to them by Australian Coal Association and Low Emission Technology (ACALET), Australian Department of Industry, the University of Newcastle (Australia).

References

[1] S. Kundu, J. Zanganeh, B. Moghtaderi, A review on understanding explosions from methane–air mixture, J. Loss Prev. Process Ind., 40 (2016) 507-523.

[2] D.M. Msiza, The review of the flammable gas and coal dust explosions, J Mine Ventilation Soc South Africa 56 (2003), 119–122.

[7] M.J. Ajrash, J. Zanganeh, B. Moghtaderi, Effects of ignition energy on fire and explosion characteristics of dilute hybrid fuel in ventilation air methane, J. Loss Prev. Process Ind., 40 (2016) 207-216.

[8] P.R. Amyotte, K.J. Mintz, M.J. Pegg, Y.H. Sun, The ignitability of coal dust-air and methane-coal dust-air mixtures, Fuel, 72 (1993) 671-679.

[9] M.J. Ajrash, J. Zanganeh, B. Moghtaderi, Methane-coal dust hybrid fuel explosion properties in a large scale cylindrical explosion chamber, J. Loss Prev. Process Ind., 40 (2016) 317-328.

[10] Q. Liu, Y. Hu, C. Bai, M. Chen, Methane/coal dust/air explosions and their suppression by solid particle suppressing agents in a large-scale experimental tube, Journal of Loss Prevention in the Process Industries, 26 (2013) 310-316.

[11] R. Kasmani, G. Andrews, H. Phylaktou, Experimental study on vented gas explosion in a cylindrical vessel with a vent duct, Process Saf. Environ. Protect., 91 (2013) 245-252.

[12] V.V. Molkov, Venting of defiagrations: dynamics of the process in systems with a duct and receiver, Fire Saf. Sci., 4 (1994) 1245-1254.

[13] G. Ferrara, S. Willacy, H. Phylaktou, G. Andrews, A. Di Benedetto, E. Salzano, G. Russo, Venting of gas explosion through relief ducts: Interaction between internal and external explosions, J. Hazard. Mater., 155 (2008) 358-368.

[14] S. Kundu, J. Zanganeh, B. Moghtaderi, Explosion Characteristics of Methane-Air Mixtures in a Spherical Vessel Connected with a Duct, Process Saf. Environ. Protect., Accepted 19 Jun 2017.

[15] EN 13673-1, Determination of maximum explosion pressure and maximum explosion pressure rise—Part I: maximum explosion pressure (2003).

[16] D. Bjerketvedt, J.R. Bakke. K. Wingerden._Gas Explosion Handbook, J. Hazard. Mater., 52 (1) (1997), p. 41.

[17] M. Gieras, R. Klemens, Experimental studies of explosions of methane air mixtures in a constant volume chamber, Comb. Sci. and Tech., 181 (2009) (4).

[18] A.M. Klimov, Premixed turbulent flames-interplay of hydrodynamic and chemical phenomena, Sov. Phys. Dokl. 20 (1975) (168).

[19] S.P. Sharma and C. Mohan, Fuels and Combustion, Tata McGraw-Hill Co. Ltd., (1984).

Development of SCR System with Optimized DEF Dosing Strategy to Meet BS-VI Emission Norms

Dhanyakumar K[1], Prachetas K[2], Swapnil S[2], Amit P[2], Brijesh P[3]

[1]ARAI Student

Kothrud, Pune, Maharashtra, 411038, India

[2]Cummins India Technical Centre

Balewadi, Pune, Maharashtra, 411045, India

[3]ARAI Faculty

Kothrud, Pune, Maharashtra, 411038, India

Abstract

The reduction of oxides of nitrogen (NO_X) without altering particulate maters (PM) and fuel consumption is challenging with in-cylinder emission control strategies in the diesel engines. Hence, selective catalyst reduction (SCR) aftertreatment technique is widely preferred for the reduction of diesel NO_X emissions. In this work, SCR system was developed for the 4-stroke, 6 cylinder, diesel engine and diesel exhaust fluid (DEF) dosing strategy was proposed to meet Bharat stage VI (BS-VI) emission requirements. A selection of SCR catalyst material and design of SCR and ammonia oxidation (AMOx) catalyst were done using appropriate analysis tools. The hot world harmonized steady cycle (WHSC) and hot world harmonized transient cycle (WHTC) tests were carried out with the optimized DEF dozing strategy. About 98% NO_X reduction ($DeNO_X$) efficiency was achieved with an average NH_3 slip of 8.5 ppm for hot WHSC with an average ammonia to NO_X ratio (ANR) of 1.3. While for hot WHTC with an average ANR of 1.2, $DeNO_X$ efficiency up to 96% with an average NH_3 slip of 4.5 ppm were achieved. Hence, SCR aftertreatment system with the optimized DEF dosing strategy and AMOx catalyst helped to meet BS-VI emission norms for NO_X and average NH_3 slip.

1 Introduction

Air pollutants are substances that adversely affect health of human being, their properties such as agricultural crops, man-made structures etc. and the environment by interfering with climate, the physiology of plants, animal species and entire ecosystems. In order to comply with the increasingly stringent emission standards, diesel engine manufacturers are forced to reduce NO_X and PM by about 88% and 50% respectively (from BS-IV to BS-VI engines). There are mainly two approaches: in-cylinder solutions and aftertreatment techniques to mitigate these emissions. Calibration of various parameters of in-cylinder solution is facing challenge of reduce NO_X and PM simultaneously. With conventional diesel combustion, trade-off between NO_X and PM has been observed. The use of exhaust gas recirculation (EGR) offered significantly lower levels of NO_X, but penalty on fuel consumption and PM was noticed [1].

Another way of reducing NO_X is to use of exhaust gas aftertreatment device mainly selective catalyst reduction (SCR) technique with diesel engines. Many of diesel engine manufacturers in the European market have chosen tuning of the engine for improved fuel economy and low PM emissions and then use of SCR catalyst to reduce the system out NO_X ($SONO_X$) as an overall strategy to meet stringent norm [2-3]. The diesel exhaust fluid (DEF), an aqueous urea solution made with 32.5% urea and 67.5% deionized water, is injected to the hydrolysis catalyst where urea will decompose thermally in to ammonia (NH_3) and carbon dioxide (CO_2). Various factors such as catalyst material and structure, SCR bed temperature, density, distribution of mass fraction and absolute pressure, adsorption of NH_3 on washcoat wall etc. define the NO_X reduction ($DeNO_X$) efficiency [4]. High $DeNO_X$ efficiency depends on accurate and controlled dosing of DEF, proper evaporation of DEF after mixing with exhaust gases and distributing of DEF over maximum surface area of catalyst [5-6]. DEF dosing depends on dosing rate, DEF spray angle and DEF droplets. While decomp tube length and decomp mixture design governs the mixing of DEF with exhaust gases. Distribution of mixture over catalyst surface depends on open area and uniformity index [6-7]. It is recommended that flow distribution on catalyst shall be more than 90% for best conversion performance and optimization of catalyst [8-10].

The study of Jungmin [11] showed that the use of SCR with engine offered reduction in fuel consumption and the volume of particulate filter which gave higher benefit than the cost of added urea. It was noticed that $DeNO_X$ efficiency was increased with increasing inlet ammonia concentration but high ammonia slip was observed too [12]. Hence, the oxidation catalyst is generally used to avoid any secondary emission of ammonia. If catalyst temperature is lower, catalyst is not active and urea becomes crystal. Hence, study of Jing et al. [13] indicated that urea can be injected when the catalyst temperature exceeds 200°C. Study of Grigorios [14] and Ehsan and Amir [15] also showed that catalyst temperature and space velocity were very critical and important factors for $DeNO_X$ efficiency.

In this work, selection of SCR catalyst material and size of SCR and AMOx catalyst were done using appropriate tools. Also, the DEF dosing strategy was optimized to meet BS-VI emission norms for NO_X and average NH_3 slip.

2 Methodology

2.1 Selection of catalyst material and size: The Pugh matrix tool was used for comparison of three catalyst materials i.e. Copper (Cu), iron (Fe), Copper Zeolite (CuZ) which are widely used for reduction reactions and selection of it for the study. Cu SCR catalyst has a high NO_X reduction performance at low temperatures, but it gives adverse performance at temperatures above 660°C [5, 16]. This may possible if catalyst oxidizes some of the NH_3 to NO at very high temperatures. The

Corresponding author. **Brijesh P**; Fax: +91-20-3023-1104

E-mail address: patel.pga@araiindia.com

Fe SCR catalyst has a significantly lower conversion than the Cu SCR catalyst at temperatures below 400°C. Based on Pugh matrix analysis, Copper Zeolite (CuZ) catalyst has been selected for the study due to its advantages in low temperature performance, high temperature stability and cost advantage.

After defining material, size of catalyst has been selected for the engine under study. One dimensional (1D) simulation analysis has been carried out using AVL-Boost for identifying optimized catalyst size. 1D model, as shown in Fig. 1, was developed. Engine out information like exhaust flow, exhaust temperature, NO_x mole fraction and air to fuel ratio were considered as an initial boundary conditions and has been modelled into ATB1 model. Exhaust pipes from turbo out and tail pipe have been modelled as ATP1 and ATP3 respectively (see Fig. 1). ANR input has been given to EI1 and FI1 models. DEF dosing strategy for constant ANR has been used to understand the aftertreatment (AT) performance. DEF dosing has been modelled as ATI1. ATP2 has been modelled to account decomposition process. Uniform index (UI) and flow distribution index (FDI) for this study have been assumed to be 1. Information for SCR and ammonia oxidation (AMOx) catalyst like catalyst type, washcoat, catalyst size, density, cells per square inch (CPSI), thermal conductivity and specific heat related to catalyst have been modelled into CAT1 and CAT2 respectively. The model was calibrated for thermal calibration only, since the kinetic boost models used for the analysis had been validated already.

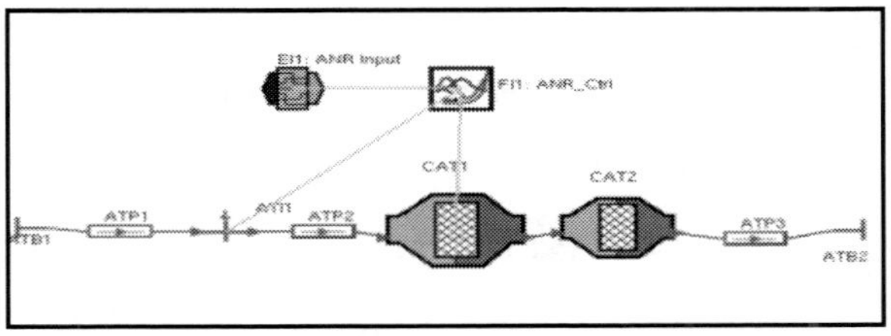

Figure 1: 1D model for selection of catalyst size

To meet 95% $DeNO_x$ efficiency, size of CuZ SCR catalyst has been finalized for the engine under study using validated 1D model and it was found 10.5 inch (D) x 6 inch (L). Also, AMOx catalyst size has been finalized as 10.5 inch (D) x 3 inch (L) to meet the requirement of average NH_3 slip below 10 ppm.

2.2 Testing methodology: Aftetreatment devices, with configuration selected from the analysis as discussed in Section 2.1, were installed with 6 cylinder, 4-stroke and common rail direct injection diesel engine (Note: Because of confidential concern, engine specifications are not provided). SCR has been installed at 2 meter from turbocharger out in order to maximize the NO_x conversion through minimizing temperature drop trough exhaust tubing. Aftertreatment system with measuring instruments is shown in Fig. 2. Tests were carried out for hot WHSC and hot WHTC with an appropriate DEF dosing strategy to meet the BS-VI emission norms with 15% engineering margin.

Figure 2: Representative test cell set up

Test data were acquired and analyzed with respect to catalyst bed temperature and space velocity (SV), as suggested in a literature study, for understanding of DEF dosing need.

3 Results and discussion

Steady state SCR catalyst performance study has been carried out. Figure 3 shows $DeNO_x$ efficiency and NH_3 slip Vs ANR for various catalyst temperatures and space velocity (SV). About 90% $DeNO_x$ efficiency with almost zero NH_3 slip is observed with 0.9 ANR for catalyst temperature above 265°C. NO_x reduction target to meet BS-VI emission norms is more than 95% which can be met with selected catalyst type and size at ANR above 0.95 as shown in Fig. 3. Also, maximum ~80% $DeNO_x$ efficiency may be achieved if engine is running with low catalyst temperature (less than 200°C).

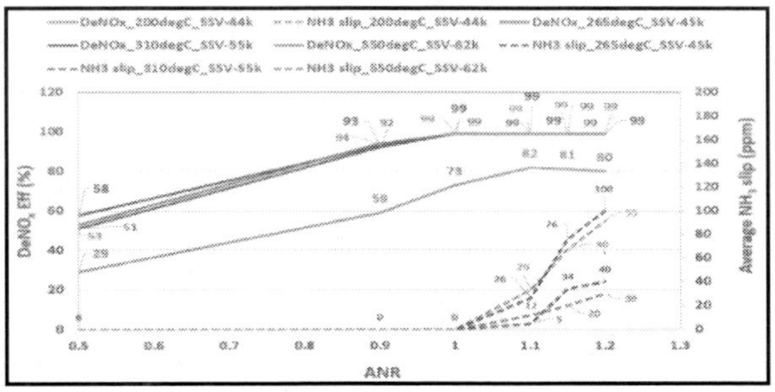

Figure 3: Steady state performance curve of CuZ catalyst

3.1 DEF dosing strategy for hot WHSC: SCR bed temperature is found above 200°C and 250°C for full WHSC and 97% of WHSC time respectively, as shown in Fig. 4. As can be seen in Fig. 5, space velocity (SV) is falling from 0k to 30k range with WHSC. As ~45% of cycle time is spent in the SV of 5k-10k, more time for catalyst reaction is available with it (see Fig. 5).

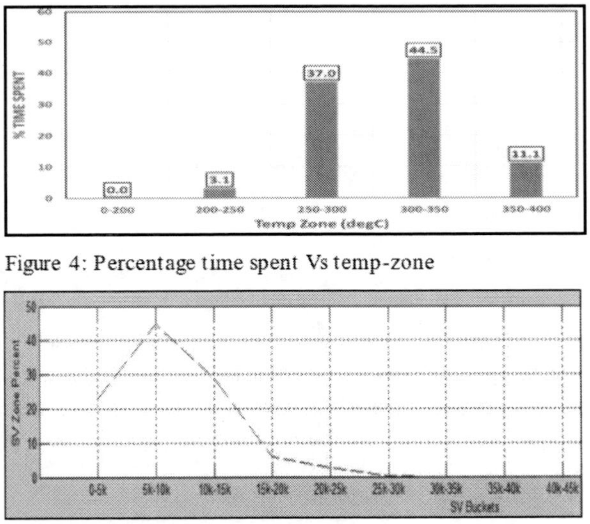

Figure 4: Percentage time spent Vs temp-zone

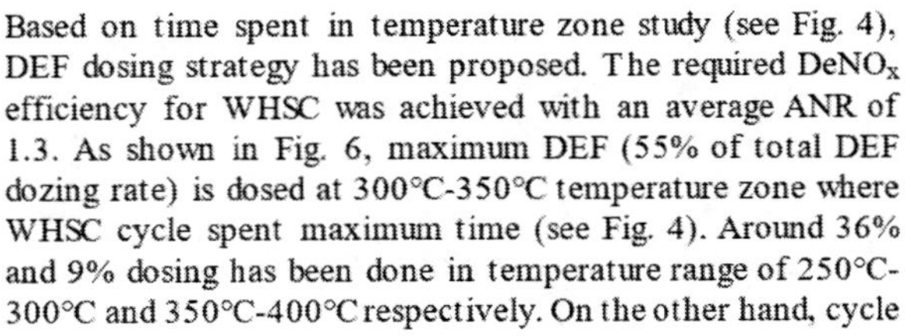

Figure 5: Percentage time spent Vs SV-zone

Based on time spent in temperature zone study (see Fig. 4), DEF dosing strategy has been proposed. The required $DeNO_x$ efficiency for WHSC was achieved with an average ANR of 1.3. As shown in Fig. 6, maximum DEF (55% of total DEF dozing rate) is dosed at 300°C-350°C temperature zone where WHSC cycle spent maximum time (see Fig. 4). Around 36% and 9% dosing has been done in temperature range of 250°C-300°C and 350°C-400°C respectively. On the other hand, cycle spent time about 3% at temperature zone below 250°C, hence no NH_3 dozing strategy was chosen. With selected dozing strategy (see Fig. 6), more than 98% $DeNO_x$ efficiency is achieved at temperature range of 250°C-350°C (where cycle spent time about 86% of total time, refer Fig. 4) as shown in Fig. 7. NH_3 storage capacity of catalyst helped in NO_x conversion at idle speed (200°C-250°C temperature zone) even though there is no NH_3 dosing at that temperature zone as shown in Figs. 6 and 7.

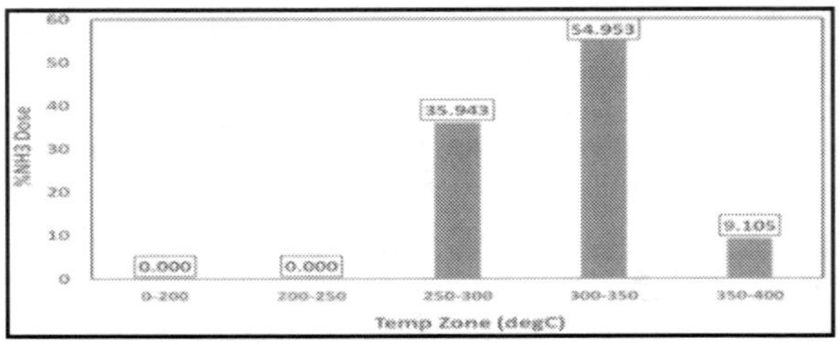

Figure 6: Percentage NH_3 dosing Vs temp-zone

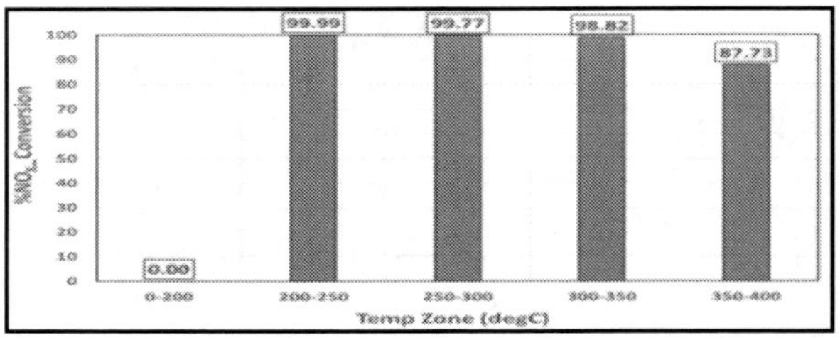

Figure 7: Percentage NO_x conversion efficiency Vs temp-zone

With the proposed DEF dosing strategy, average NH_3 slip was found 8.5 ppm. Percentage NH_3 slip with various temperature zone is plotted in Fig. 8. As can be seen in Fig. 8, 300°C-350°C temperature zone contributed maximum in total NH_3 slip.

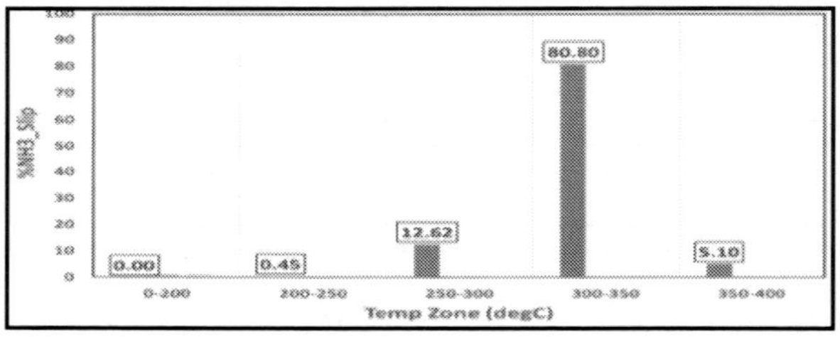

Figure 8: Percentage NH_3 slip Vs temp-zone

Summary of hot WHSC results is shown in Fig. 9. Engine out NO_x during WHSC was found about 12g/kWh. SCR with optimized DEF dosing strategy reduced system out NO_x to 0.22 g/kWh with average NH_3 slip of about 8.5 ppm. It can be seen in Fig. 9 that regulation limit was met with SCR with proposed DEF dosing strategy and AMOx catalyst without any modifications in engine calibration. However, a slight higher NH_3 slip was noticed than the set engineering limit which may possible to meet by reducing dosing amount at 300°C-350°C temperature zone.

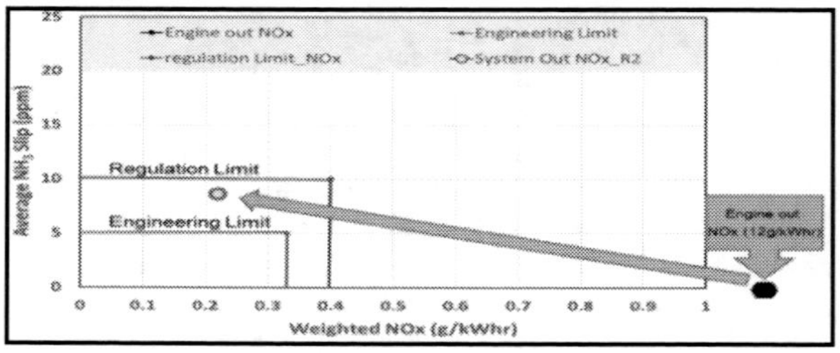

Figure 9: Summary of hot WHSC results

3.2 DEF dosing strategy for hot WHTC: Unlike hot WHSC, SCR bed temperature of ~39% of cycle is found below 250°C (see Fig. 10) with hot WHTC. As can be seen in Fig. 11, 70% of the cycle time is spent in the SV of 5k-10k which offer more time for catalyst reaction with that SV zone.

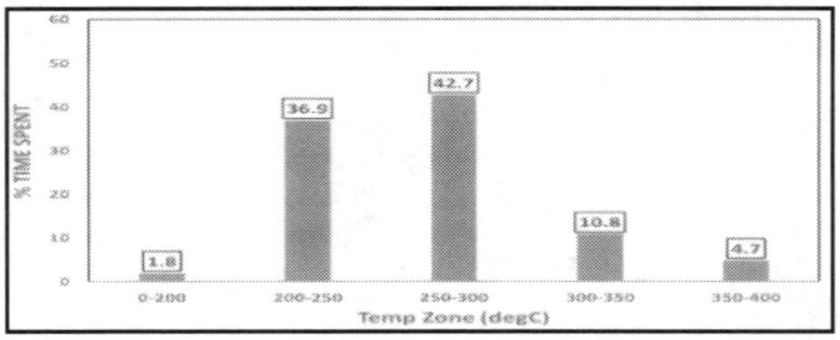

Figure 10: Percentage time spent Vs temp-zone

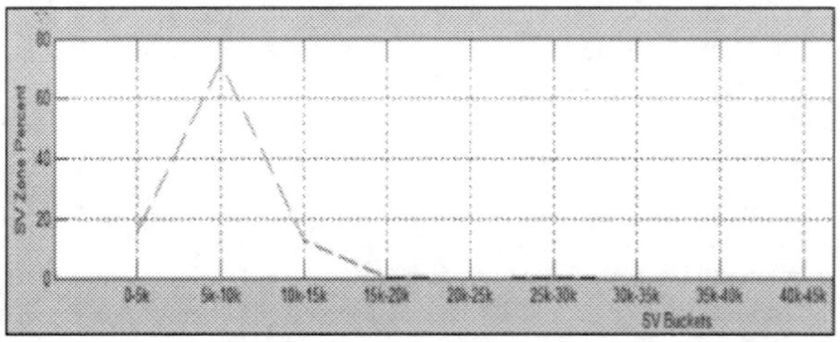

Figure 11: Percentage time spent Vs SV-zone

NH_3 dosing strategy for hot WHTC with temperature zone is plotted in Fig. 12. As compared to hot WHSC, WHTC spent maximum time in lower temperature zone (200°C-300°C), hence it is required to doze NH_3 for lower SCR bed temperatures too (see Fig. 12). The overall $DeNO_x$ efficiency for WHTC was achieved with an average ANR of 1.2.

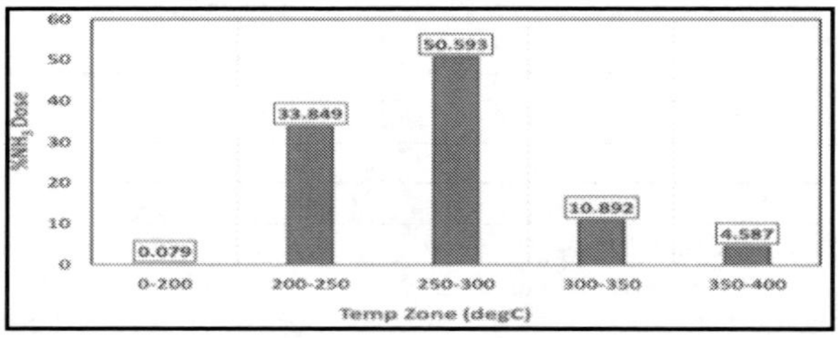

Figure 12: Percentage NH_3 dosing vs temp-zone

With selected dozing strategy (see Fig. 12), more than 96% $DeNO_x$ efficiency, as can be seen in Fig. 13, is achieved at temperature range of 250°C-400°C. On the other hand, even though a significant amount (~34%) of NH_3 is dozed below 250°C, $DeNO_x$ efficiency is achieved below 95%. It indicates that thermal management strategy (to improve SCR bed temperature above 250°C) may be needed to improve overall $DeNO_x$ efficiency during hot WHTC cycle.

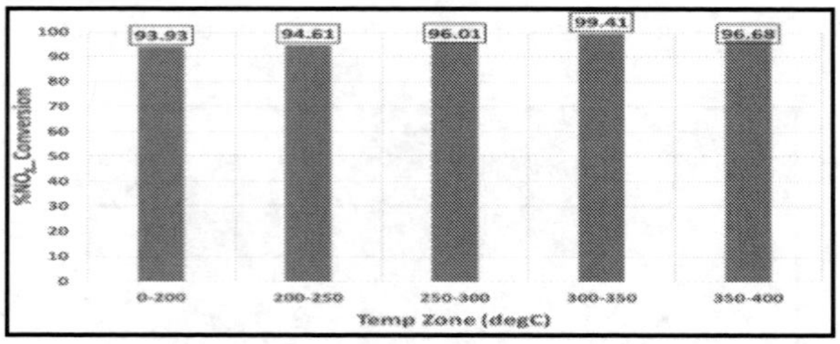

Figure 13: Percentage NO_x conversion efficiency vs temp-zone

Percentage NH_3 slip with various temperature zone is plotted in

Fig. 14. Results showed that 250°C-350°C temperature zone contributed maximum in total NH_3 slip. However, with the proposed DEF dosing strategy and AMOx catalyst, average NH_3 slip was observed about 4.5 ppm.

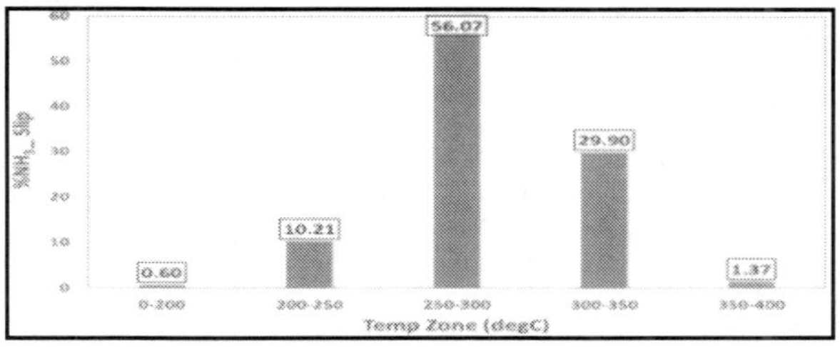

Figure 14: Percentage NH_3 slip vs temp-zone

Hot WHTC results are summarized in Fig. 15. Engine out NO_x during hot WHTC was found about 9.5 g/kWh (with 0 ppm NH_3 slip). SCR with optimized DEF dosing strategy reduced system out NO_x to 0.37 g/kWh with average NH_3 slip of about 4.5 ppm. It can be seen in Fig. 15 that regulation limit was met with SCR with proposed DEF dosing strategy and AMOx catalyst without any modifications in engine calibration. However, a slight higher NO_x was noticed than the set engineering limit which may possible to meet by improving overall $DENO_x$ efficiency using thermal management strategy.

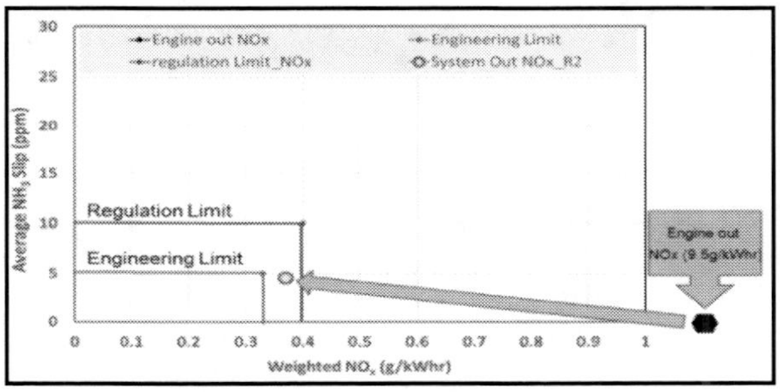

Figure 15: Summary of hot WHTC results

4 Conclusions

In this work, SCR aftertreatment system for the 4-stroke, 6 cylinder, and diesel engine was developed and DEF dosing strategy was proposed to meet Bharat stage VI (BS-VI) emission requirements. A selection of SCR catalyst material and the design of SCR and ammonia oxidation (AMOx) catalyst were done using Pugh matrix and 1D simulation using AVL-Boost respectively. The hot world harmonized steady cycle (WHSC) and hot world harmonized transient cycle (WHTC) tests were carried out with the optimized ammonia dozing strategy. Steady state SCR catalyst performance has also been studied. CuZ SCR catalyst with size of 10.5 inch (D) x 6 inch (L) has been finalized to meet 95% $DeNO_x$ efficiency. About 98% $DeNO_x$ efficiency was achieved with an average NH_3 slip of 8.5 ppm for hot WHSC with an average ANR of 1.3. On the other hand, $DeNO_x$ efficiency up to 96% with an average NH_3 slip of 4.5 ppm were achieved for hot WHTC with an average ANR of 1.2. SCR aftertreatment system with the optimized DEF dosing strategy and AMOx catalyst helped to meet BS-VI emission norms for NO_x and average NH_3 slip. However, a slight higher NH_3 slip and NO_x were noticed than the set engineering limit (15% lower than regulatory limit). The engineering limit for WHSC and WHTC can be met by a slight tuning of DEF dosing rate at 300°C-350°C temperature zone and thermal management respectively.

5 Acknowledgment

Authors would like to thanks Cummins India Technical Centre for allowing to publish work in the conference. Also, authors are thankful to FEV, India for providing their test facility.

References

[1] B Hollauf, B Breitschadel, Model Based NO_x Conversion Control, SAE Technical Paper, 2011-26-0042.

[2] Dieselnet.com, SCR system for mobile engines. www.dieselnet.com/cat_SCR_mobile_html, June2005.

[3] Indian Ministry of Road Transport and Highways (MoRTH), "2016-02-19 GSR 187 BS VI Draft", February2016.

[4] C Ming, W Shazam, Modelling and Optimisation of SCR systems, SAE Technical Paper, 2005-01-0969.

[5] C Mario, C Jonathan, C Sougato, Cu- and Fe-Based SCR Emission Control Systems, SAE Technical Paper, 2011-01-1329.

[6] T Joseph, Optimized for SCR Light-off, SAE Technical Paper, 2009-01-0901.

[7] A Munnannur, Development of Flow Uniformity, SAE Technical Paper, 2011-01-1239.

[8] W Johann, "Multi-Scale SCR Modelling, 1D Kinetic Analysis and 3D System Simulation", SAE Technical Paper 2005-01-0948.

[9] K Manoj, P Mohanan, S Nagraj, Numerical Study of SCR system Efficiency, International Journal on Theoretical and Applied Research in Mechanical Engineering, 1 (1), 2012.

[10] H Weltens, Optimisation gas flow distribution by CFD, SAE Technical Paper, 930780.

[11] S Jungmin, Aftertreatment Package Design for SCR, SAE Technical Paper, 2011-01-1135.

[12] P Walker, P Blakeman, B Magnusson, F Stunnenberg, Highly Durable SCR Catalyst Systems, SAE Technical Paper, 2004-01-1289.

[13] H Jing, Z Yan-guang, Z Yun-long, S Shi-Jin, W Jian-Xin, Closed-loop Control Strategy for Urea-SCR, SAE Technical Paper, 2011-01-1324.

[14] K Grigorios, F Pavlos, S Zissis, SCR System Optimization by Simulation Tools", SAE Technical Paper, 2013-01-1075.

[15] M Ehsan, H Amir, Development SCR catalyst Multi Objective Genetic Algorithm, SAE Technical Paper, 2011-01-1332.

[16] KI Akash, D Deodatta, VD Sunil, Selective Catalytic Reduction Technique, International Journal of Current Engineering and Technology, (7) (2017) 206-210.

AUTHOR INDEX

AUTHOR INDEX

AUTHOR INDEX

AUTHOR INDEX

AUTHOR INDEX

AUTHOR INDEX

AUTHOR INDEX

AUTHOR INDEX

AUTHOR INDEX

Combustion Institute China Section
c/o Section Chair: Fei Qi, Shanghai Jiao Tong University
fqi@sjtu.edu.cn

ISBN 978-1-5108-5646-2

CURRAN ASSOCIATES INC.
proceedings.com

11th Asia-Pacific Conference on Combustion 2017

Sydney, Australia
10 - 14 December 2017

Volume 2 of 2

11th Asia-Pacific Conference on Combustion 2017

Sydney, Australia
10 - 14 December 2017

Volume 2 of 2

ISBN: 978-1-5108-5646-2

Printed from e-media with permission by:

Curran Associates, Inc.
57 Morehouse Lane
Red Hook, NY 12571

Some format issues inherent in the e-media version may also appear in this print version.

Printed by Curran Associates, Inc. (2018)

Combustion Institute China Section
c/o Section Chair: Fei Qi, Shanghai Jiao Tong University

fqi@sjtu.edu.cn

Additional copies of this publication are available from:

Curran Associates, Inc.
57 Morehouse Lane
Red Hook, NY 12571 USA
Phone: 845-758-0400
Fax: 845-758-2633
Email: curran@proceedings.com
Web: www.proceedings.com

TABLE OF CONTENTS

VOLUME 1

VOLUME 2

11th Asia-Pacific Conference on Combustion,
The University of Sydney, NSW Australia
10th -14th December 2017.

DNS of cool flame propagation in n-dodecane spray autoignition

Taotao Zhou, Taohong Ye*, Minming Zhu, Wenlong Cheng

Department of Thermal Science and Energy Engineering, University of Science and Technology of China

Jinzhai Road, Hefei, Anhui, 230027, China

Abstract

Two dimensional DNS of n-dodecane spray autoignition at low temperature is conducted. The general autoignition process and the propagation characteristic of cool flame are analyzed. Results show that the cool flame spots firstly occur in the fuel lean regions and propagate into the regions where the mixture fraction is larger. The ignition instant is identified by the time derivative of maximum temperature of the flow field. The chemical reaction rate and heat release rate of second stage are much more intense than that of first stage. There exist occurrence, mergence and quenching during the propagation process of cool flame. The evolution curve of the front speed exhibits a general 'U' shape except several instants, which correspond to mergence and quenching of cool flame. Thus the general 'U' shape of speed curve is split into several little 'U' shapes. The temperature gradient at the reaction front is negatively correlated with the front speed of cool flame.

1 Introduction

The autoignition process of high-carbon fuel is essential for many combustion devices, such as aeroengine, diesel engine and newly developed HCCI engine. Experimental investigations [1-3] have demonstrated that multi-stage ignition and multi combustion regime exist during high-carbon fuel autoignition process due to the intermediate and low temperature chemical reactions. For liquid fuels, the combustion process becomes more complicated since there are complex interactions among atomization, evaporation, fuel/gas mixing and chemical reaction. Direct numerical simulation (DNS) is the most precise numerical methods, which is a useful tool for researchers to investigate these basic processes of spray autoignition and optimize the combustion chamber.

Because of the enormous computational requirements, most DNS studies are restricted to simple and small configuration. Borghesi [4] studied the effects of turbulence, global equivalence ratio and droplet diameter distribution on n-heptane autoignition at intermediate temperature and high pressure.Liu [5] investigated the effects of strain rate and initial temperature on n-heptane autoignition in a counterflow configuration. Ignition was found delayed for larger scalar dissipation rate, and the second stage ignition would not happen if the scalar dissipation rate was larger than a critical value. With the same configuration, Bansal et al. [6] studied the effect of oscillations of scalar dissipation on n-heptane autoignition. Mukhopadhyay [7] conducted a two-dimensional DNS to study the effect of compositional stratification on n-heptane autoignition at different initial temperature. Yoo [8] analyzed the influence of turbulent intensity and temperature stratification on the autoignition of n-heptane in their two-dimensional DNS. Krisman A[9] investigated the low temperature chemistry affected dimethyl ether lifted flame by two-dimensional DNS. The flame structure varies significantly at different initial temperature. Dai [10] analyzed the modes of propagating combustion wave which was caused by a cool spot and pointed out that the cool spot may also generate knock in engines. Zhao P [11] investigated the occurrence and propagation of cool flame as well as its transition to hot flame in a one-dimensional laminar premixed flame. Krisman A [12, 13] conducted a two-dimensional DNS for the atuoignition of dimethyl ether at low temperature. Four kinds of combustion regimes were distinguished in their study.

The researches stated above didn't consider the influence of droplets' influence. So far, there are no studies on low temperature chemistry affected spray autoignition. In the present research, a two dimensional DNS of n-dodecane spray autoignition at low temperature is conducted. N-dodecane is a surrogate for diesel fuel and shows obvious negative temperature coefficient (NTC) behavior[14]. The present work aims to study the propagation characteristic of cool flame and its influence on the second stage ignition.

2 Physical model and numerical methods

The numerical methods have been described in our previous paper [15], so only a brief overview is presented here. Low Mach assumption is adopted for gas phase governing equations. The density is solved through gas equation of state. Viscous heating, radiation and gravity are not considered in the simulation. Soret effect and Dufour effect are also neglected here. Lagrangian point-source method is used to track droplets' position, velocity, mass and temperature. Two-way coupling between gas phase and liquid phase is considered. The effects of liquid phase on gas phase are represented by sources in the governing equations of gas phase.

SIMPLE algorithm is adopted for the solution of gas phase governing equation. The momentum equation is discretized by sixth-order central difference scheme, and the scalar equations are discretized by fifth-order WENO scheme. Second-order semi-implicit Crank-Nicolson scheme is adopted for gas phase governing equation to advance in time and Second-order Runge-Kutta scheme is adopted for that of the liquid phase governing equations. A skeletal mechanism [14] containing 54 species and 269 element reactions is adopted for n-dodecane combustion reactions. The chemical reactions are integrated through DVODE solver. A sixth-order explicit filter [16] is applied to remove the aliasing error.

The computation configuration is shown in Fig.1, which is a square of length $L = 4mm$. A grid 256×256 is used and the resultant grid scale is $15.6\,\mu m$. Droplets are initially distributed in a slab layer $3/8L\sim5/8L$ of the domain. There is

*Corresponding author. Fax:
E-mail address: thye@ustc.edu.cn

no mean flow and the initial fluctuation velocity is 0.4m/s, given according to the prescribed turbulent energy spectrum [17]. The resulted Kolmogorov length scale and the turbulent integral length is 45 μ m and 0.26mm respectively, and the velocity fluctuations decay with time. The initial temperature of the oxidizer is 750K, and the initial droplet temperature is 300K. Periodic boundary conditions are adopted for all directions of the computation domain.

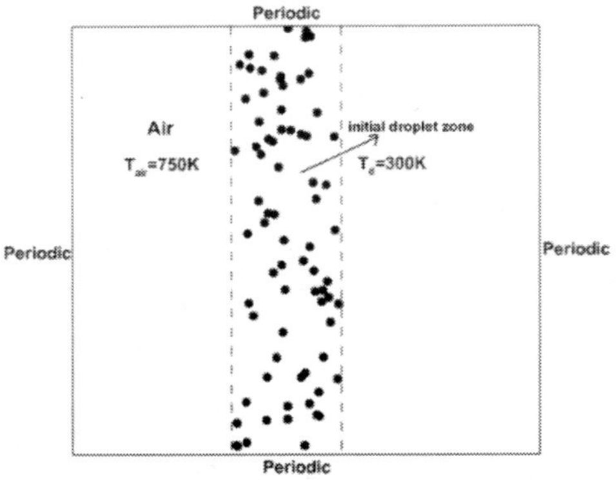

Figure 1: Computational domain and initial droplets distribution.

3 Zero dimensional (0-D) simulation

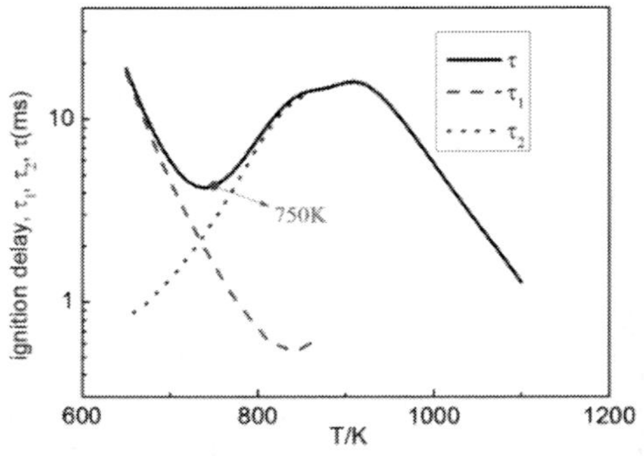

Figure 2: First-stage, second-stage, and total ignition delays of n-dodecane/air mixture under 6bar.

Before the conduction of two dimensional cases, homogenous ignition delays of n-dodecane/air mixture under 6bar are calculated by the software CHEMKIN. The ignition delay is defined by the peak value of time derivative of maximum temperature, dT_{max}/dt. As shown in Fig.2, the first-stage ignition delay generally decreases as the temperature rises up, while the second-stage ignition delay increases with temperature. The temperature range of NTC is about 735K~910K.

The temperature 750K is chosen as the initial temperature of oxidizer in our two dimensional cases, and the reasons are as followings:

(1)The temperature is in the NTC range, at which the interaction between low temperature chemistry and high temperature chemistry is intense;

(2) The second-stage ignition delay is relatively longer, thus there is a window time to study the propagation of cool flame;

(3) The total ignition delay is much shorter, which reduces the computational cost.

4 Two dimensional (2-D) case

4.1 General evolution

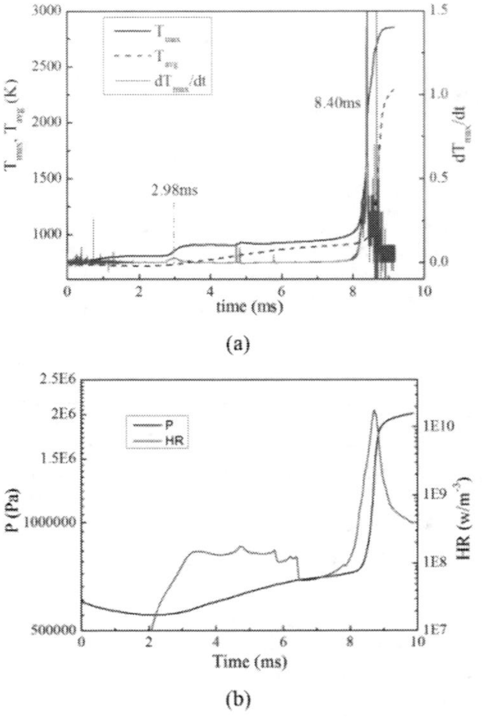

Figure 3: (a)The evolutions of averaged temperature, T_{avg}, maximum temperature, T_{max}, and the time derivative of maximum temperature, T_{max}/dt.(b)

Figure 3(a) shows the evolutions of averaged temperature, T_{avg}, maximum temperature, T_{max}, and the time derivative of maximum temperature, dT_{max}/dt. The ignition delay is defined by the peak value of time derivative of maximum temperature here. The first stage ignition delay τ_1 is 2.98ms, and the second stage ignition delay τ_2 is 8.40ms. The evolutions of pressure and volume averaged heat release rate are plots in Fig.3(b). At the initial time, there is a little decrease of averaged temperature, which is caused by the evaporation cooling of droplets. Meanwhile, the pressure gradually decreases with temperature. After the occurrence of first stage ignition, the maximum temperature rises quickly to about 900K. The averaged temperature and pressure gradually increase, which corresponds to the propagation of cool flame. The chemical reaction rate and heat release rate of second stage are much more intense than that of first stage, which can be seen from the value of heat release rate in Fig.3(b). And the hot flame propagates much faster than cool flame. The averaged temperature and maximum temperature rise quickly to above 2000K after the occurrence of second stage ignition. From Fig.3, there is a residence time window of about 5ms for the presence of cool flame, which enable us to study its propagation and influence on the whole flow field.

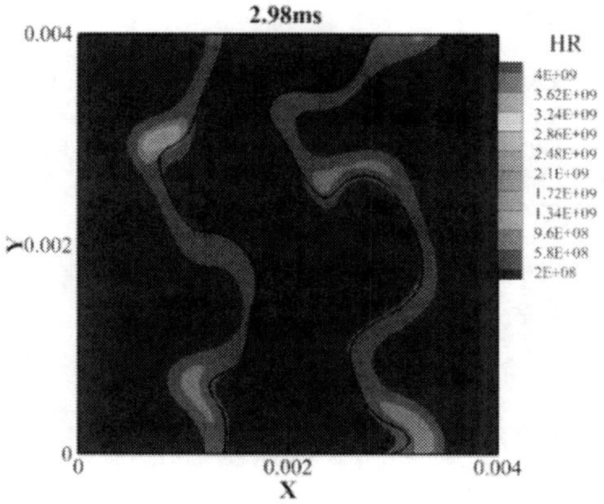

(a)

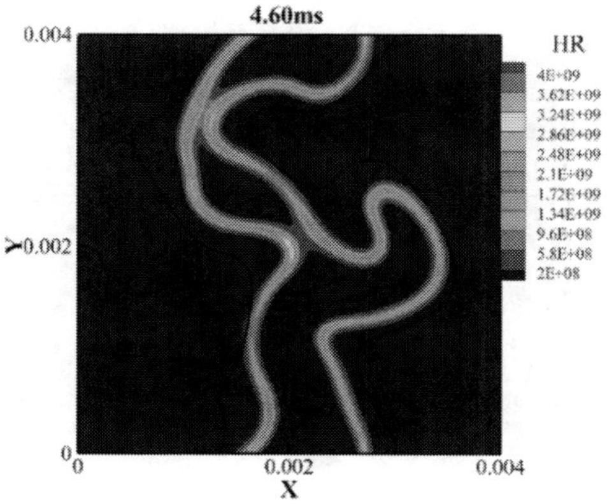

(b)

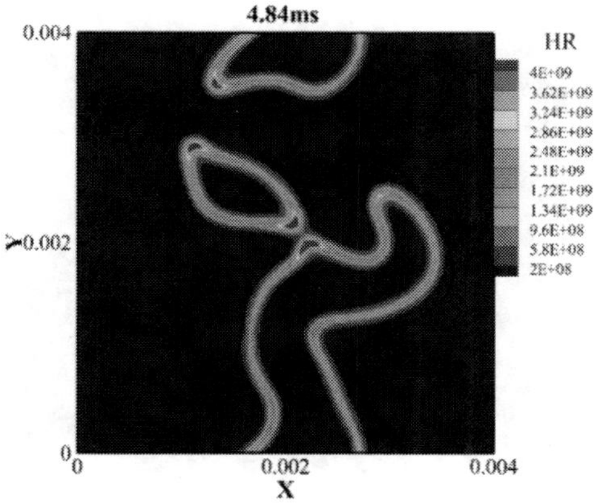

(c)

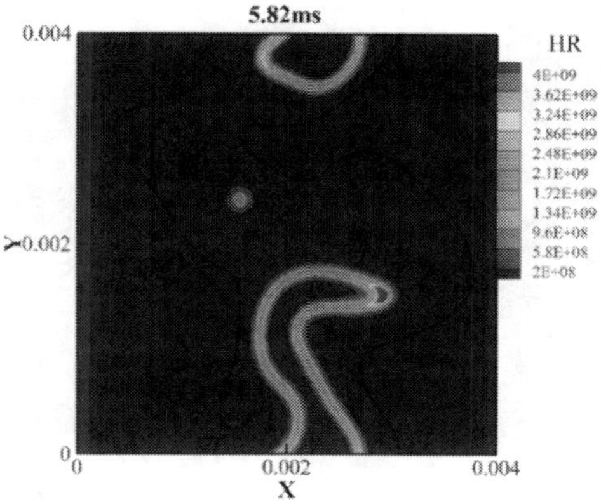

(d)

Figure 4: Isocontours of heat release rate (HR) at four instants (2.98ms, 4.60ms, 4.84ms, 5.82ms), the black line represents the stoichiometric mixture fraction, η_{st}.

Instantaneous views of heat release rate (HR) at four instants are displayed in Fig.4, and the black line in the figure represents the stoichiometric mixture fraction, η_{st}. The middle regions of the two black lines are fuel rich while the outside regions are fuel lean. The irregular shape of mixture fraction distribution is caused by droplets' evaporation under turbulence condition. At 2.98ms, the cool flame spot occurs in the fuel lean regions and propagate into the middle regions where the mixture fraction is larger. At about 4.60ms, the two cool flames meet each other and start to merge at two spots. At 4.84ms, three cool flames are formed after the mergence between initial two cool flames. The cool flames continue to propagate to the unburned regions, and the quenching of the middle circled cool flame happens at about 5.82ms due to the exhausted of 'fuel' for low temperature chemistry reaction. Exist of occurrence, mergence and quenching makes the propagation of cool flame a complex problem.

4.2 Reaction front speed of cool flame

The reaction front speed of flame is a key parameter during combustion process, which is relevant to the flame stabilization, combustion efficiency, quenching as well as soot production. The density-weighted displacement speed [8,18,19], S_d^*, is adopted here to investigate the propagation characteristic of cool flame:

$$S_d^* = \frac{1}{\rho_u|\nabla Y_k|}\left(\omega_k + \frac{\partial}{\partial x_i}\left(\rho D \frac{\partial Y_k}{\partial x_i}\right)\right) \quad (1)$$

where Y_k is the specie mass fraction, ω_k is net production rate of specie k and D is the specie diffusion coefficient. ρ is the density of gas mixture and ρ_u is the density corresponds to the unburned gas state. The isocontour $Y_k = Y_{OC12H23OOH} = 0.05$ is adopted for the evaluation of displacement velocity here. Specie OC12H23OOH is formed during the first stage combustion and vanishes before the happening of second stage ignition. The chosen isocontour coincides approximately with the location of maximum heat release rate, HR.

Temporal evolutions of mean front speed of cool flame, $\overline{S_d^*}$, and the mean magnitude of the temperature gradient at the surface location, $\overline{|\nabla T|}$, are displayed in Fig.5. Firstly, the speed curve of the whole time series exhibit a general 'U' shape except for several instants, and the temperature gradient at the surface location, $\overline{|\nabla T|}$, is negatively correlated with the front speed of cool flame, $\overline{S_d^*}$. These characteristics are similar with previous study of the front speed of hot flame [8, 20]. There exist large magnitude of the front speed at the initial and finish time of the propagation process, because $|\nabla Y_k|$ vanishes in the spots where ignition occurs or the remaining charge is burning out[8,20]. As demonstrated in Fig.4, mergence and quenching of cool flame will happen during the propagation of cool flame, and the occurring instants are consistent with the time when fluctuation happens in the speed curve in Fig.5. The general 'U' shape of speed curve is split into several little 'U' shapes by these fluctuations.

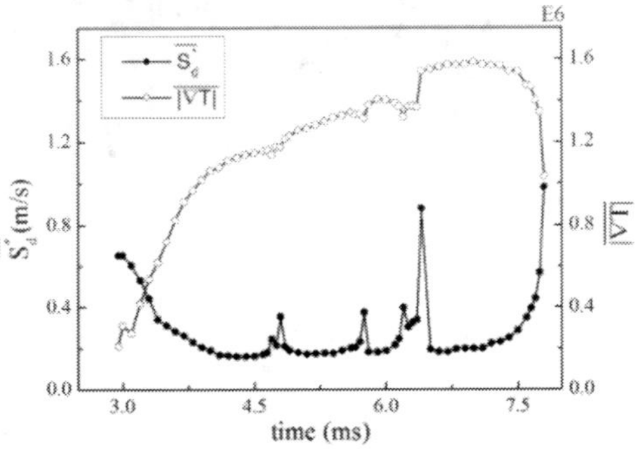

Figure 5: Temporal evolutions of mean front speed of cool flame, $\overline{S_d^*}$, and the mean magnitude of the temperature gradient at the surface location, $\overline{|\nabla T|}$.

5 Conclusion

A two dimensional DNS of n-dodecane spray autoignition at low temperature is conducted. The two stage ignition is identified by the time derivative of maximum temperature of the flow field. The chemical reaction rate and heat release rate of second stage are much more intense than that of first stage. After the propagation of cool flame, the temperature rises to above 900K.A residence time window of about 5ms enable us to study the propagation of cool flame. The cool flame spot occurs in the fuel lean regions and propagate into the middle regions where the mixture fraction is larger. There exist occurrence, mergence and quenching during the propagation process of cool flame. The speed curve of the whole time series exhibit a general 'U' shape except for several instants, and these instants correspond to the mergence and quenching of cool flame. Thus the general 'U' shape of speed curve is split into several little 'U' shapes. The temperature gradient at the surface location is negatively correlated with the front speed of cool flame.

6 Acknowledgment

This work was supported by the National Natural Science Foundation of China (Grant No. 51576182 and Grant No. 91441117). The numerical calculations in this paper have been performed on the supercomputing system in the Supercomputing Center of University of Science and Technology of China.

References

[1] Dec J E, SAE Technical Paper, 1997.

[2] Idicheria C A, Pickett L M, SAE Technical Paper, 2006.

[3] Musculus M P B, Miles P C, Pickett L M, Progress in Energy and Combustion Science 39(2) (2013)246-283.

[4] Borghesi G, Mastorakos E, Cant R S, Combustion and Flame 160(7) (2013): 1254-1275.

[5] Liu S, Hewson J C, Chen J H, Combustion and flame 137(3) (2004) 320-339.

[6] Bansal G, Im H G, Lee S R, Proceedings of the Combustion Institute, 2009, 32(1): 1083-1090.

[7] Mukhopadhyay S, Abraham J, Combustion and flame 158(6) (2011)1064-1075.

[8] Yoo C S, Lu T, Chen J H, Combustion and Flame 158(9) (2011) 1727-1741.

[9] Krisman A, Hawkes E R, Talei M, Proceedings of the Combustion Institute, 2015, 35(1): 999-1006.

[10] Dai P, Chen Z, Chen S, Proceedings of the Combustion Institute, 2015, 35(3): 3045-3052.

[11] Zhao P, Liang W, Deng S, Fuel 166(2016):477-487.

[12] Krisman A, Hawkes E R, Talei M, Proceedings of the Combustion Institute, 2015, 35(1): 999-1006.

[13] Krisman A, Hawkes E R, Talei M, Proceedings of the Combustion Institute, 2016.

[14] Yao T, Pei Y, Zhong B J, Fuel 191(2017) 339-349.

[15] Zhou T, Ye T, Zhu M, Fuel 187(2017)137-145.

[16] Kennedy C A, Carpenter M H, Applied Numerical Mathematics 14(4)(1994)397-433.

[17] Batchelor G K, Townsend A A. Decay of turbulence in the final period[C]//Proceedings of the Royal Society of London A: Mathematical, Physical and Engineering Sciences. The Royal Society, 194(1039) (1948) 527-543.

[18] Echekki T, Chen J H, Combustion and Flame 118(1) (1999) 308-311.

[19] Poinsot T, Veynante D, RT Edwards, Inc., 2005.

[20] Chen J H, Hawkes E R, Sankaran R, Combustion and flame 145(1) (2006)128-144.

11th Asia-Pacific Conference on Combustion,
The University of Sydney, NSW Australia
10th -14th December 2017.

Effects of Di-n-butyl Ether Addition to Turbulent Toluene Sooting Flames

Jingjing Ye[1], Stephan Kruse[2], Paul Medwell[1], Zhiwei Sun[1], Bassam Dally[1], Heinz Pitsch[2]

[1]The University of Adelaide

School of Mechanical Engineering, Adelaide, SA, 5005, Australia

[2]RWTH Aachen University

Institute for Combustion Technology, Aachen, 52056, Germany

Abstract

This paper presents an investigation of the effects on soot formation and oxidation when adding di-n-butyl ether (DBE) to toluene flames. Experiments were performed on a Jet in Hot Coflow (JHC) burner, where a prevaporised toluene and DBE mixture, carried by nitrogen at varying volumetric ratios, are injected into a 1544-K vitiated coflow with 7.5% oxygen by volume. Four flames with different DBE to toluene ratios at constant Reynolds number of 10,000 and coflow conditions are investigated in this study. Centreline temperature using a thermocouple and planar soot volume fraction using laser-induced incandescence (LII) are reported and analysed. It is found that the addition of DBE significantly affected soot inception, growth, and oxidation. The mean soot volume fraction and total soot loading values decrease nearly linearly with DBE addition while the maximum peak soot volume fraction values reduction trend is second order polynomial.

1 Introduction

Ethers have received considerable attention as second-generation biofuels [1]. They can be produced from lignocellulosic biomass, thus not directly competing with food production [1]. Di-n-butyl ether (DBE) is an attractive ether for applications such as diesel engines, due to its high cetane number (100) and comparable volumetric energy density (31.6 MJ/L) to that of petroleum fuels (36 MJ/L) [1]. Combustion of DBE is also expected to produce lower soot emissions than conventional diesel due to the presence of the C-O-C functional group.

Previous studies have primarily focused on short-chain ethers, and the combustion characteristics of DBE are not as widely investigated [1]. Beatrice *et al.* [2] studied the emission behaviour of DBE as a pure fuel in DI diesel engines, and reported that soot emissions were reduced considerably compared to those from combustion of n-tetradecane. However, there is a paucity of information on the impact of DBE as an additive on the sooting propensities of fuel blends.

This study investigates the effects of the addition of DBE to toluene on soot. Toluene is chosen in this study as it is a simple aromatic compound, and has been used as a surrogate for aromatics in diesel and gasoline fuels [3]. Aromatic compounds are important components in gasoline, diesel, and kerosene fuels, particularly from the perspective of pollutant formation.

Corresponding author. Phone: +61 (0)8 831 35460
E-mail address: jingjing.ye01@adelaide.edu.au

In this study, experiments are conducted with turbulent jet flames of prevaporised toluene and DBE carried by nitrogen, stabilised on a Jet in Hot Coflow (JHC) burner. The jet centreline temperature was measured with a thermocouple and soot volume fraction was determined from planar laser-induced incandescence (LII). The LII measurements were performed simultaneously with soot primary particle sizing via time-resolved planar laser-induced incandescence (TiRe-LII), laser-induced fluorescence of the OH radical (OH-LIF), and temperature from non-linear excitation regime two-line atomic fluorescence (NTLAF), although only soot volume fraction is reported in this paper.

2 Experimental Details

2.1 Experimental Setup and Operating Conditions

The JHC burner consists of a central insulated fuel jet (ID = 4.6 mm) surrounded by an annular coflow (ID = 108.2 mm). The hot coflow was produced from combustion of a lean mixture of natural gas and air in a secondary porous-bed burner. The resultant volumetric concentration of oxygen in the coflow was 7.5%. The hot coflow also included 6.1% CO_2 and 12.1% H_2O, balanced with N_2. The averaged coflow temperature was measured to be 1544 K by a thermocouple, after correction for radiation. The burner was surrounded by an additional outer coflow (ID = 152.4 mm), comprising air at room temperature.

Toluene/DBE blends were vaporised prior to entering the burner using a similar setup to a previous study [4]. A controlled evaporator and mixer (CEM) was used to preheat and premix the fuel and carrier gas (N_2). The pipelines conveying the mixture between the CEM and JHC burner were heated and insulated to prevent condensation. The temperature at the jet exit plane was measured to be approximately 508 K for the various cases, which is higher than the boiling points of both toluene (384 K) and DBE (416 K). Other relevant fuel properties are listed in Table 1.

Fuel	Formula	Y_c	T_{ad} (K)	S_L (cm/s)
toluene	C_6H_5-CH_3	0.91	2343	54
DBE	C_4H_9-O-C_4H_9	0.74	2294	58[1]

Table 1: Fuel properties: Y_c is the carbon mass fraction in the molecule; T_{ad} is the maximum adiabatic flame temperature for fuel-air mixture (Φ=1.05) at room temperature; S_L is the maximum laminar flame speed at 373 K for (Φ=1.1) at 1atm.

Four jet conditions were investigated during the experiments, all with toluene as the parent fuel and DBE mixed at differing volumetric ratios, namely: 0% (D00, pure toluene), 25% (D25), 50% (D50) and 75% (D75). The stoichiometric mixture fraction (Z_{st}) for the D00, D25, D50, D75 cases is 0.0986, 0.0997, 0.101, and 0.1024, respectively. Pure DBE flame was not investigated in this study. The bulk jet Reynolds number was kept constant at 10,000. The total mass flow rate of the fuel or the fuel blend was kept at 0.204 g/s, and the mass flow rate of carrier gas (N_2) was kept at 0.611 g/s. All the jet flames appeared attached and yellowish, but no smoke was detected from these flames.

2.2 Optical Details

Planar laser-induced incandescence (LII) was performed with the fundamental output (1064 nm) from an Nd:YAG laser. The LII signal was captured with an intensified CCD camera. Each LII instantaneous image was corrected for background and detector attenuation. After correction, the prompt LII signal was converted to soot volume fraction using a calibration factor, determined by performing LII of a steady and laminar ethylene/air flame stabilised on a Santoro-type burner with well-known soot volume fraction [5].

3 Experimental Results

3.1 Jet Centreline Temperature Profile

Figure 1 displays the mean temperature measured by a 75-µm type-R thermocouple along the jet centreline as a function of height above the burner (HAB) for three of the cases studied. Jet centreline temperature was also measured for the D25 case, however there are some issues with the measurement, and NTLAF images are currently being processed to cross-check those issues.

For non-premixed flames, the actual flame length is generally defined as the axial distance where the stoichiometric condition occurs at the jet centreline. As the peak temperature in non-premixed flames generally occurs close to the stoichiometric mixture fraction, the height where the maximum temperature occurs can be considered as the actual flame length. Figure 1 shows that the maximum jet centreline temperature is almost the same for various cases, suggesting that the flame length is similar for all the cases.

For the D75 case, the jet centreline temperature increases steadily with the axial distance until it reaches its peak. However, there is a dip in the temperature profile approximately between HAB = 150 to 260 mm in the D00 and the D50 cases. Such a dip was reported in sooting flames [6] and is caused by heat loss from the flame due to radiation by soot particles. As an aromatic compound, toluene has a stronger tendency to form soot than DBE, which explains the absence of a dip in the D75 case and a more profound dip in the D00 case than the D50 case. The measured temperature was found to decrease rapidly with time as soot started to deposit on the thermocouple after a certain period of time in this region. Similar phenomenon has also been reported in laminar sooting flames by McEnally and Pfefferle [7]. Hence, only the readings from the first few seconds were used to calculate the mean temperature. The thermocouple was cleaned after measurement at each height.

3.2 Soot Volume Fraction

Figure 2 displays a collage of time-averaged images of soot volume fraction, determined from the LII signal of toluene/DBE flames at various heights above the burner. The same colour scale is applied to all the soot volume fraction images to emphasise low f_v values, causing some images to appear saturated. These mean images are not strictly symmetrical, with higher signal on the left-hand-side of the image ($r < 0$ mm). The laser beam propagated from left to right. Similar asymmetry has been reported in previous studies, and was attributed to beam distortion and the local laser fluence redistribution, as a consequence of the variation in the refractive index field [8]. The refractive index of the gas mixture is determined by the local gas composition and temperature [8], which vary considerably from the hot coflow at the lean side to the cold fuel at the rich side.

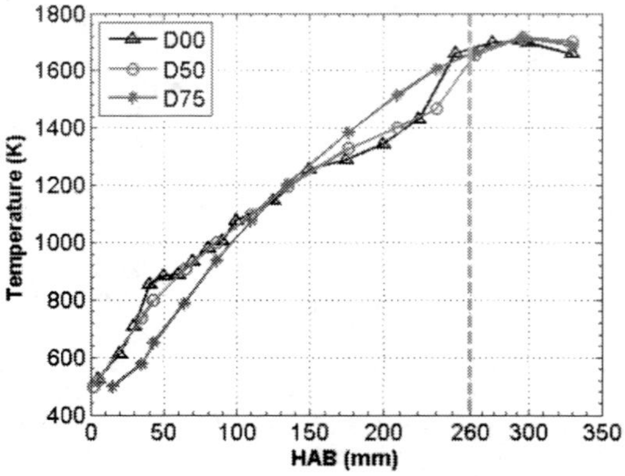

Figure 1: Mean jet centreline temperature profile. The green line indicates where the maximum mean f_v was measured along the jet centreline.

For the D00 case, soot inception occurred below HAB = 60 mm, as revealed in Fig. 2(a). The soot volume fraction gradually increases and reaches its maximum mean value of 0.8 ppm, followed by a decrease due to oxidation. As expected, the soot volume fraction decreases towards the tip of the flame at HAB = 315 to 350 mm, where oxidation takes place.

Soot inception takes place further downstream when toluene was blended with DBE. This change is most profound in the D75 case, where the f_v in the mean image is below the detection limit until HAB = 240 mm. The maximum mean f_v in this case was measured to be 0.1 ppm at HAB = 315 mm.

For the D00, D25, and D50 cases, the soot volume fraction is formed on the rich side of the reaction zone, on both sides of the flame, up to HAB ≈ 200 mm. Beyond this height, soot is transported inwards and the maximum f_v was measured near the jet centreline. Only the radial distribution of soot volume fraction at HAB ≈ 300 mm is presented in Fig. 3 for brevity. It demonstrates that the radial distribution of f_v did not change with the increasing concentration of DBE.

As shown in Table 1, toluene is an aromatic compound with a higher mass fraction of carbon than DBE. In addition to dilution and thermal effects, blending DBE with toluene affects soot formation and distribution mostly through influencing the chemical pathway, particularly the radicals and intermediates necessary for soot inception. These dilution, chemical, and thermal effects have been investigated previously for H_2 and N_2 [5,9]. In laminar diffusion ethylene flames, the radial location of maximum f_v was found to shift from the wings to the jet centreline as ethylene was diluted with more H_2 at a constant total flow rate [5]. Similarly, Smooke et al. [9] reported that the radial location of maximum f_v changed from the wings to the centreline as ethylene became more diluted with N_2 in laminar flames. Their numerical analysis revealed that this radial shift resulted from changes in the residence time within the soot formation region, and relative rates of inception and surface growth [9]. At the transition point of the radial shift in soot profiles, Z_{st} decreased from 0.0614 to 0.0602 in the ethylene diluted by H_2 case [5], while Z_{st} increased from 0.1017 to 0.1452 in the ethylene diluted by N_2 case [9]. In the current study, Z_{st} increases from 0.0986 in the D00 case to 0.1024 in the D75 case. The discrepancies observed in the radial distribution of soot are likely to result from different chemical pathways caused by the addition of DBE.

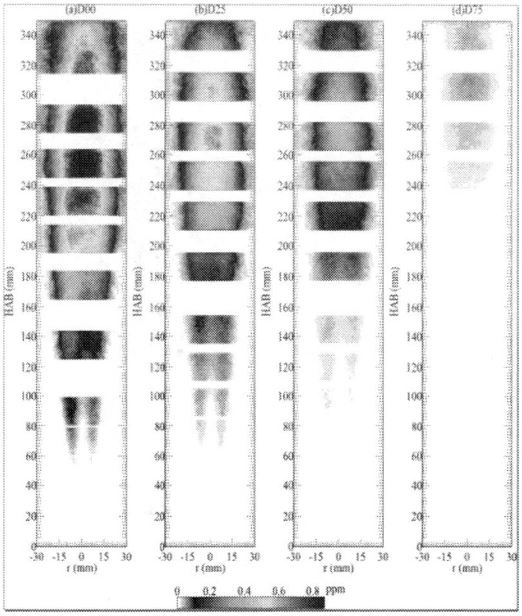

Figure 2: Collage of time-averaged images of soot volume fraction.

Figure 4 shows mean soot volume fraction along the jet centreline, averaged from 500 instantaneous images at each height. The peak f_v value along the jet centreline was calculated for each instantaneous image, and the maximum value amongst the 500 shots for each case is presented in Figure 5. Overall, both the time-averaged mean and the maximum peak f_v values were reduced as the amount of DBE in the fuel jet increased. This demonstrates that the soot reduction effect of DBE increases with its concentration.

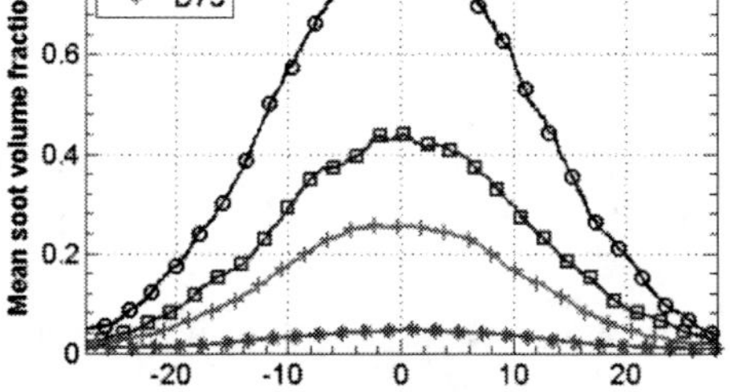

Figure 3: Time-averaged mean soot volume fraction as a function of radial distance with image centred at HAB=285 mm for the D00 case, and image centred at HAB=306 mm for other cases.

The time-averaged mean soot volume fraction along the jet centreline peaked approximately at the same height (HAB = 260 mm) for all the cases except for the D75 case (HAB ≈ 300 mm). The location of peak f_v is determined by the competition of the local soot formation rate, the oxidation rate, and soot transportation. These rates are dependent on the species concentration, mixing rate, and temperature. As shown in Fig. 1, the jet centreline temperature in the D75 case was close to that measured in other cases at HAB = 260 mm. For turbulent jet flames governed by momentum, the mixing rate is determined by the total mass flow rate of the fuel jet, which was kept constant for all cases. Hence, the shift in the axial location of peak f_v is suspected to be caused by differences in the species concentration, as a consequence of altered chemistry with the addition of large amounts of DBE.

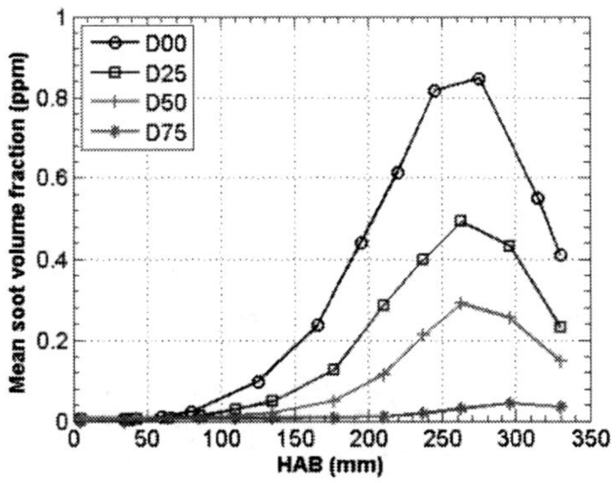

Figure 4: Time-averaged mean soot volume fraction.

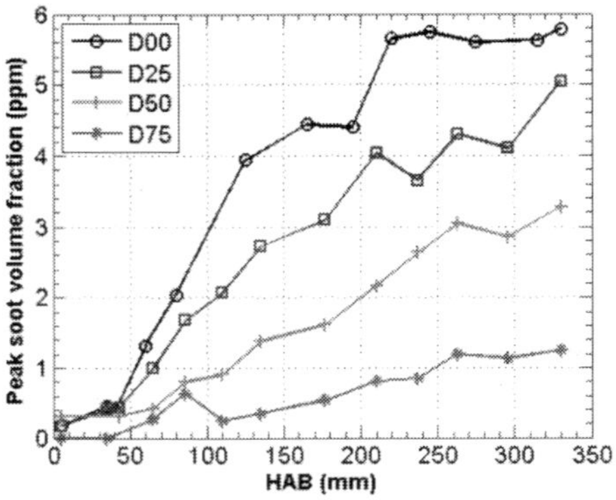

Figure 5: The maximum peak soot volume fraction along the jet centreline in instantaneous images.

The maximum peak f_v values in the instantaneous images are an order of magnitude higher than the time-averaged values of mean f_v. For instance, at HAB = 85 mm in the D75 case, no soot was observed in the mean image in Figure 2(d), while the maximum peak f_v was measured to be 0.8 ppm. This points to the high intermittency of soot formation in both space and time. Indeed, instantaneous images of soot volume fraction revealed convoluted structure of soot in isolated pockets, particularly in low-sooting regions.

Figure 2 and 5 have shown an interesting phenomenon where in the soot oxidation region (HAB= 280 – 340mm) the mean f_v reduces with the axial distance, and the maximum peak f_v value in instantaneous images increases with distance. For example, for the D25 case the recorded maximum peak f_v value is 4 ppm at HAB = 300 mm and 5 ppm at HAB = 340 mm. Similar behaviour is also seen in other cases. This suggests that, in the soot burnout region, the occurrence of detectable soot is less frequent as the flame propagates further downstream. Histograms of mean soot volume fraction at various heights support this hypothesis, which are not shown here for brevity.

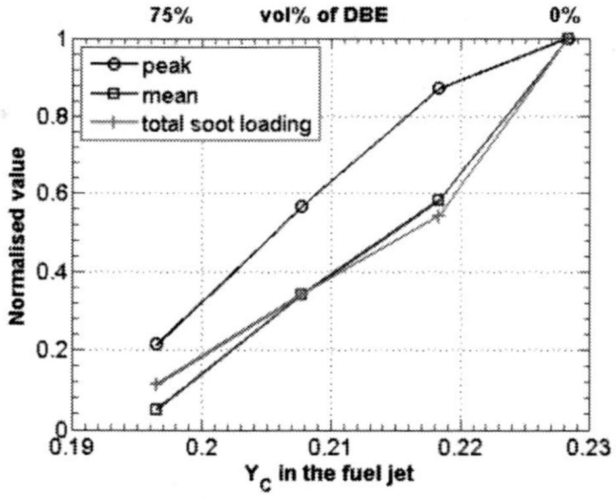

Figure 6: Normalised total soot loading, maximum peak and mean soot volume fraction as a function of the mass fraction of carbon (Y_c) in the fuel jet

In order to better demonstrate the soot reduction effect of DBE, normalised total soot loading (integrated over the full volume of the flame), maximum peak and mean soot volume fraction values are presented in Fig. 6 as a function of the mass fraction of carbon in the fuel jet. When the volumetric ratio of DBE increased from 0% to 75%, Y_c in the fuel jet decreased from 0.228 to 0.197. Worth noting is that the mean soot volume fraction and total soot loading values decrease nearly linearly with DBE addition while the maximum peak soot volume fraction values reduction trend is second order polynomial.

4 Conclusions

A systematic experimental study was conducted to investigate the impact of blending DBE with toluene on the soot formation of a turbulent flame stabilised on a JHC burner. Laser-induced incandescence was employed to measure the soot volume fraction distribution, while a thermocouple was used to measure the centreline mean temperature.

Time-averaged images of soot volume fraction indicate that the addition of DBE drastically reduced the soot formation in a toluene flame. Moreover, the addition of DBE shifted the soot inception to downstream position.

For all the cases investigated, the peak f_v values in the instantaneous images are an order of magnitude higher than the time-averaged f_v values. The axial location of peak f_v along the jet centreline was found to be constant amongst all the cases except for the D75 case which produced very little soot. This discrepancy is suspected to be caused by chemical effects. The radial location of maximum f_v was relatively constant as the concentration of DBE increased. This is different from previous studies [5,9] on laminar ethylene diffusion flames, in which a shift from the wings to the centreline was observed. Further studies are required to better understand the different radial distribution of soot in the present paper.

5 Acknowledgments

This research is funded by the China Scholarship Council and The University of Adelaide through a joint postgraduate scholarship. Funding from the Australian Research Council (ARC), United States Air Force Asian Office of Aerospace Research and Development (AOARD), and Deutscher Akademischer Austauschdienst (DAAD) is gratefully acknowledged. This work is also part of Cluster of Excellence "Tailor-Made Fuels from Biomass", funded by the Excellence Initiative of German federal and state governments.

References

[1] L. Cai, A. Sudholt, D. J. Lee, F. N. Egolfopoulos, H. Pitsch, C. K. Westbrook, S. M. Sarathy, Combust. Flame 161 (3) (2014) 798-809.

[2] C. B. C. Beatrice, J. D'alessio, N. Del Giacomo, M. Lazzaro, P. Massoli, Combust. Sci. Technol. 120 (1996) 335-355.

[3] L. Cai, H. Pitsch, Combust. Flame 162 (2015) 1623–1637.

[4] J. Ye, P. R. Medwell, M. J. Evans, B. B. Dally, Combust. Flame 183 (2017) 330-342.

[5] Z. Sun, B. Dally, G. Nathan, Z. Alwahabi, Combust. Flame 175 (2017) 270-282.

[6] S.M. Mahmoud, G.J. Nathan, P.R. Medwell, B.B. Dally, Z.T. Alwahabi, Proc. Combust. Inst. 35 (2015) 1931-1938.

[7] C.S. McEnally and L.D. Pfefferle, Combust. Flame 121 (4) (2000) 575-592.

[8] Z. W. Sun, Z. T. Alwahabi, D. H. Gu, S. M. Mahmoud, G. J. Nathan, B. B. Dally, Appl. Phys. B 119 (4) (2015) 731-743.

[9] M. D. Smooke, M. B. Long, B. C. Connelly, M. B. Colket, R. J. Hall, Combust. Flame 143 (4) (2005) 613-628.

11th Asia-Pacific Conference on Combustion,
The University of Sydney, NSW Australia
10th -14th December 2017.

Droplet combustion studies on an RP-1 surrogate and its constituent fuels

Anand Sankaranarayanan, Arindrajit Chowdhury, Neeraj Kumbhakarna

Department of Mechanical Engineering, IIT Bombay

Powai, Mumbai, Maharashtra, 400076, India

Abstract

The combustion characteristics of an RP-1 surrogate fuel (RP-1-s), obtained by mixing three pure compounds – methyl cyclohexane (MCH), n-dodecane, and isocetane in the molar ratio 59:32:9 was evaluated through droplet combustion studies under normal gravity. The burn rate data of the surrogate mixture was found to satisfactorily match that of RP-1 reported in a previous study, when tested under similar conditions. Fuel droplets in the size range 1-1.5 mm were used to study the combustion characteristics of RP-1-s and its constituents. The burn rate constants for the pure compounds followed the order isocetane < n-dodecane < MCH, while the sooting tendencies followed the order n-dodecane < isocetane < MCH. Flame height was found to be a good indicator for identifying the occurrence of micro-explosions. Burn rates, flame dimensions, flame stand-off ratios, and, the sooting behaviour of RP-1-s was found to lie between those of its constituent fuels.

1 Introduction

Surrogate fuels are commonly used to represent or study specific characteristics of complex fuels. They cannot be formulated to match all the properties of an actual compound [1]. Hence they are often designed to match specific parameters of interest. The usefulness of a surrogate fuel stems from the fact that the actual fuel under question may not be readily available for testing, as in the case of jet fuels or rocket propellants. In addition to this, each of these complex fuels is usually a mixture of numerous constituent fuel components. As a result, the batchwise variability in fuel properties can often be quite high for these compounds [2]. Studies aimed towards incremental improvements in fuel characteristics would often be inconclusive in such scenarios. Using a surrogate fuel for experimentation can help achieve the required consistency in test results. Surrogate fuels also present a channel for developing reaction mechanisms and predicting thermophysical properties, needed to model the combustion behaviour of complex fuels whose reaction chemistry can otherwise be difficult to formulate [3, 4].

RP-1 belongs to the general class of compounds known as kerosenes [3], and is commonly used as a propellant. Surrogates proposed for RP-1 can be found in previous works [1, 3]. The present study is intended to serve as a precursor to investigate the combustion characteristics of novel energetic propellants [5-8] , which can serve as additives to RP-1 for performance enhancement. Due to the difficulties in synthesis, most of these energetic compounds are typically made only in milligram or sub-milligram scales, which constrain the evaluation of their combustion behaviour to droplet combustion studies. Thus the main objective of the present work is to utilize droplet combustion studies under normal gravity to evaluate and establish the combustion behaviour of the RP-1 pyrolysis surrogate (RP-1-s) comprising of a mixture of methyl cyclohexane (MCH), n-dodecane and isocetane in the molar ratio 59:32:9 [9]. This blend was preferred ahead other surrogates due to the following reasons: the chosen surrogate closely matched the behaviour of RP-1 in terms of decomposition rates, compound class, ethylene yields and H/C ratio; the constituents used for preparation of the surrogate were easily available and well documented; and in practical combustion scenarios, the fuel (RP-1) is used for cooling the engine and hence gets pyrolyzed before it is actually combusted. Any surrogate that does not mimic the pyrolysis behaviour of RP-1 would have a different composition as compared to that of the actual pyrolyzed fuel, at the inlet to the combustion chamber. Regardless of the fact that it is preferable to subject RP-1-s to pyrolysis ahead of droplet combustion studies, this step was avoided to simplify the experimental procedure and allow for a comparison of RP-1-s with its constituent components– MCH, n-dodecane, and isocetane in terms of combustion characteristics. These compounds are commonly used constituents in jet surrogate fuels and the authors hope that the results can be beneficial for development of new surrogate fuels in this area. The suitability of droplet combustion experiments for developing surrogate fuels has been previously discussed by Avedisian [10]. Though the arguments presented therein pertained to spherically symmetric combustion, the general argument of matching burn rate characteristics of the surrogate fuel with that of the actual compound for a combustion surrogate was extended in this study to droplet combustion under normal gravity.

2 Experimental Work

2.1 Experimental Setup & Methodology

A schematic of the experimental setup used for the present work is shown in Fig. 1. The fuel droplet was deposited onto the tip of a quartz rod of size 0.27 mm, housed within the combustion chamber, with the help of a microliter syringe. The ignition of the fuel drop was achieved by means of a spark, generated with the help of an ignition transformer. The average spark energy provided for ignition was measured using the methodology outlined by Lee et.al [11] and was found to be approximately 6-12 mJ for spark durations between 0.1-0.2 s. Two DSLR cameras were simultaneously utilized for filming the experiments. Backlighting was provided to track the regression of the droplet surface in order to evaluate burn rates. Burn rate data was recorded using a CANON EOS 550D camera with 100 mm macro-lens, and flame characteristics

Corresponding author. Fax: +91-22-25726875
E-mail address: neeraj_k@iitb.ac.in

were captured with the help of a NIKON D5200 camera with an AF-S NIKKOR lens and additional zoom lens fittings. The soot deposited over the quartz rod was cleaned, and the air inside the combustion chamber was evacuated with the help of a vacuum pump after each trial. Additional videos of a rod of known dimensions were recorded after the completion of the experiments, under similar settings to determine the scaling factors. Further details of the combustion chamber design, ignition transformer, and vacuum pump used for the present study were outlined in a previous work [12].

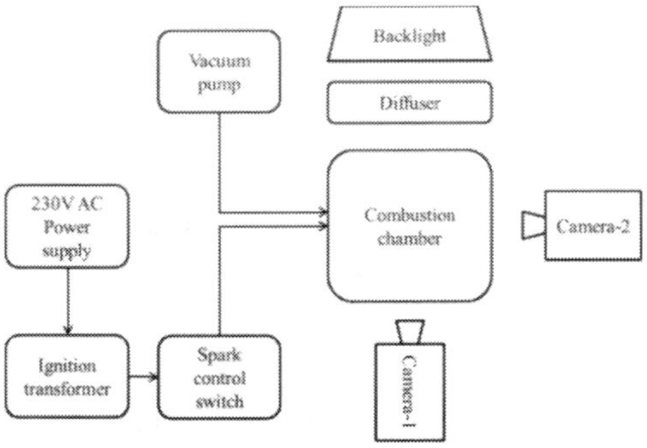

Figure 1: Schematic of experimental setup

2.2 Procedure for Data Analysis

Data was acquired at a rate of 50 frames per second for both the cameras. Synchronization between the two cameras was achieved based on the commencement of the sparking event, which was taken as t=0 s. The droplet and flame boundaries were tracked using the image processing software Spotlight [13]. These were then used as inputs to an in-house MATLAB code for recovering droplet size and flame characteristics as a function of time. The droplet was assumed to be ellipsoidal and the equivalent diameter was calculated as suggested by Kobayasi [14]. Flame diameter (D_f) and flame height (h_f) were defined to be equivalent to the instantaneous maximum width and length of the flame, defined by the luminous region, respectively. Each experiment was treated as a separate case as the initial conditions could not be exactly replicated. The estimated error in evaluation of burn rate constants was around ± 4% and that for flame characteristics was around ± 8%.

3. Results and Discussions

3.1 Comparison of burn rate behaviour of RP-1 surrogate with actual compound

For the first set of experiments, intended to compare the burn rate characteristics of the surrogate fuel with that of the actual compound, the experiments and data processing were conducted differently from the above mentioned procedure and in line with the available published data on RP-1 [15]. The results are shown in Fig. 2. In this case, the RP-1-s droplets were suspended using a quartz rod of size 0.67 mm as opposed to the 0.635 mm rods used for RP-1 in the baseline tests. The tests were done in open air and the droplets were ignited using a match stick. Despite the slightly larger tip diameter of the quartz rod for the present set of experiments, it was impossible to suspend a 2 mm diameter drop probably due to differences in viscosity and interfacial surface tension between the actual and surrogate fuel. As a result, the suspended fuel droplets were around 1.75 mm in diameter whereas those reported in literature are around 2 mm. The burn rate constants were evaluated for the time interval where $(D/D_0)^2$ reduced from 0.8 to 0.4. These were found to lie between 0.782-1.083 mm^2/s for RP-1 and 0.773-0.827 mm^2/s for RP-1-s. The reference work defines the maximum width of the droplet as the droplet diameter at any instant. This causes significant variations in burn rate as the shape of the droplet keeps changing during the course of combustion. Overall, the burn rate characteristics of the surrogate fuel were found to be fairly representative of the actual compound.

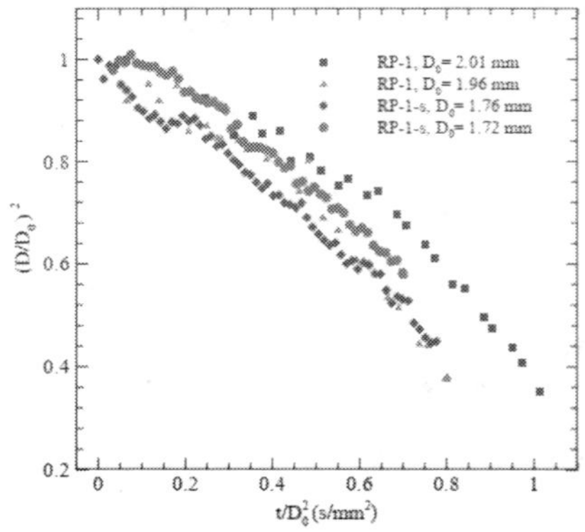

Figure 2: Comparison of burn rate behaviour of RP-1-s with RP-1

3.2 Burn rate and flame characteristics of RP-1-s

The burn rate characteristics of RP-1-s for various initial diameters are presented in Fig. 3. The effect of heat feedback from the quartz rod to the combusting fuel droplet increases as the droplet size keeps reducing. Hence the data has been presented only till a point where the droplet diameter reaches a value of about three times the suspension tip diameter. The results suggest that the surrogate fuel burns fairly smoothly.

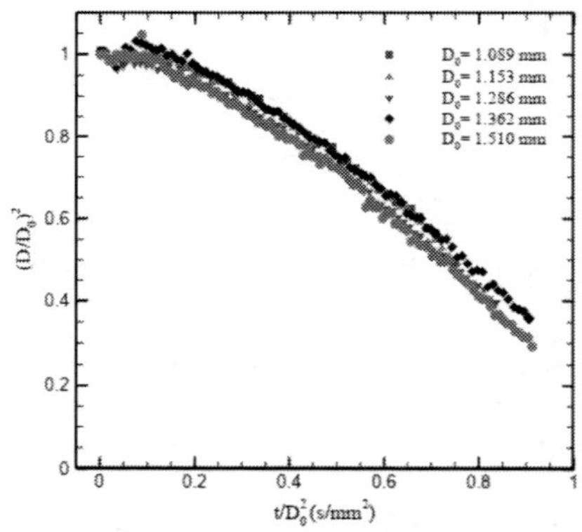

Figure 3: Variation of non-dimensionalized droplet diameter with scaled time for RP-1-s

The flame heights and flame diameters for the cases corresponding to the results in Fig. 3, are shown in Fig. 4, and, the flame stand-off ratios for the same are presented in Fig. 5. The flame geometry was tracked from the completion of sparking till the instant at which the droplet diameter was about 0.81 mm. Ignition using a spark causes the surrounding air to heat up and rise upwards due to buoyancy. This effect causes a stretching of the flame which continues approximately for 0.1 s, after the completion of sparking and can be seen in Fig. 4. It can also be discerned that small micro-explosions were readily visualized as disturbances in flame heights and that their occurrence was more frequent in droplets of larger size. This phenomenon was attributed to the higher retention time of the trapped volatiles within the droplet, on account of a smaller surface area to volume ratio, for a droplet of larger diameter.

The effect of initial droplet diameter on flame height was more prominent that on flame diameter. This was also observed in Fig. 5, where the ratio of instantaneous flame diameter to the instantaneous droplet diameter (Flame stand-off ratio–FSR) is plotted as a function of time. FSR signifies the relative position of the flame as compared to that of the droplet and is an important parameter for combustion modeling. Smaller droplets typical have larger FSR's than larger droplets. It can be also be seen that the FSR gradually increases with time as the flame diameter varies very little in the time interval studied, while the droplet diameter keeps reducing at an approximately constant rate. This observation is contradictory to the constant FSR predicted by the classical theory due to inherent limitations of the D^2 law, which yields a flame position that does not accurately account for thermal/transport parameters and neglects fuel vapour accumulation effects [16].

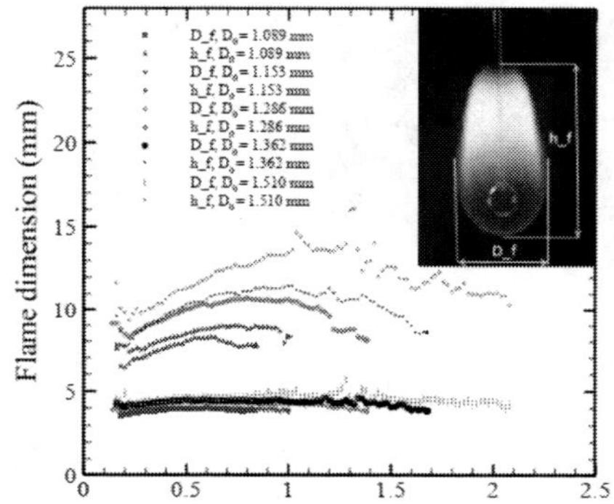

Figure 4: Variation of flame dimensions with initial droplet diameter for RP-1-s

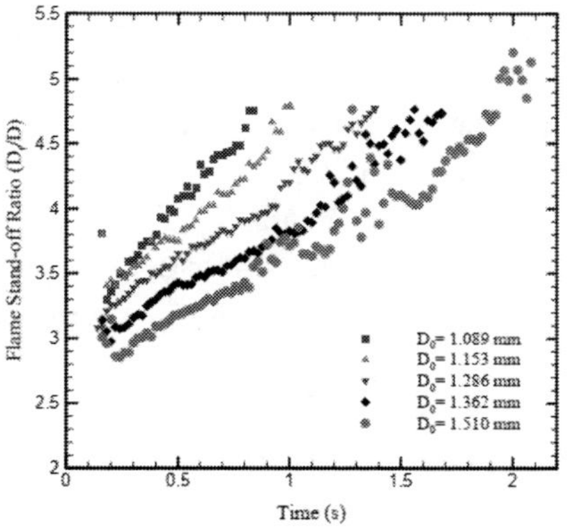

Figure 5: Variation of FSR's with initial droplet diameter for RP-1-s

3.3 Comparison of combustion characteristics of RP-1-s with its constituent fuels

The results for variation in burn rate constant as a function of initial droplet diameter for the chosen surrogate and its constituent fuels are presented in Fig. 6. The burn rate constant has been evaluated during a time interval starting from an instant where t/D_0^2 equals 0.2, till a point at which the droplet diameter becomes approximately thrice the suspension tip diameter. The burn rate constants for tested fuels were found to lie between 0.75-1 mm^2/s for initial droplet diameters between 1-1.5 mm. Based on the overall trend observed for all compounds, it can be concluded that the burn rate constant increases with an increase in initial droplet size, for droplet combustion under normal gravity. The effect is due to heat transfer enhancement resulting from greater buoyant forces on account of the larger initial droplet diameter [17]. Results suggest that isocetane has the lowest burn rate, followed by n-dodecane, RP-1-s, and MCH. The burn rate characteristics for droplets of similar size were compared for further understanding, as presented in Fig. 7. The slopes of n-dodecane, RP-1, and isocetane were found to be quite similar while that of MCH was distinctly higher than the other three.

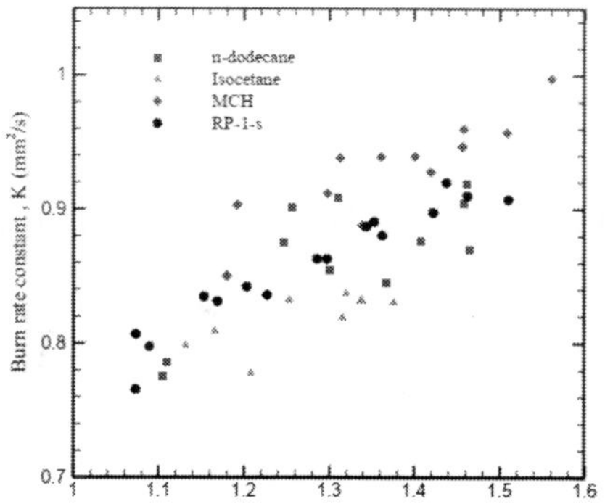

Figure 6: Variation of burn rate constants with initial droplet diameter for RP-1-s and its constituent fuels

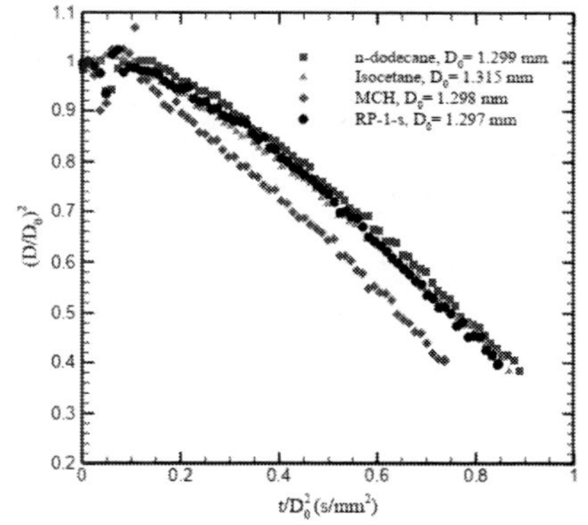

Figure 7: Comparison of burn rate behaviour of RP-1-s and its constituent fuels

The variation of flame height, flame diameter and FSR's for similar droplet sizes have been presented in Figs. 8 & 9, respectively. The sooting propensity of the tested fuels, as discerned from the deposits left over on the quartz rod, after completion of the combustion event, followed the order: n-dodecane < RP-1-s < isocetane < MCH. Higher burn rate for MCH was thought to be responsible for a rapid fall in flame height. MCH also displays a slightly larger initial flame diameter as compared to the other three compounds shown in Fig. 8. This coupled with faster burn rates, result in larger FSR's for MCH as compared to n-dodecane, isocetane and RP-1-s.

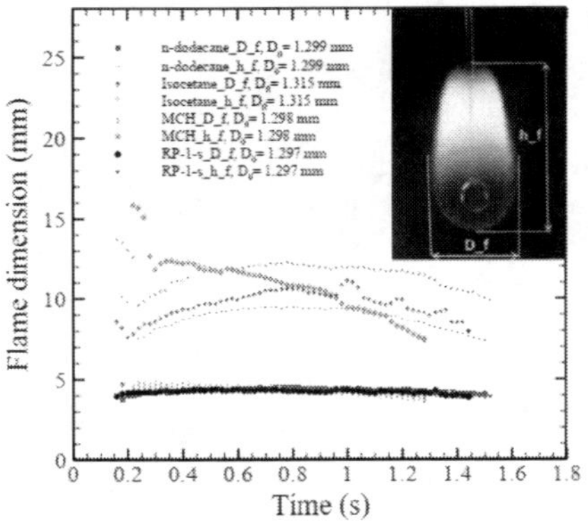

Figure 8: Comparison of flame dimensions for RP-1-s and its constituent components

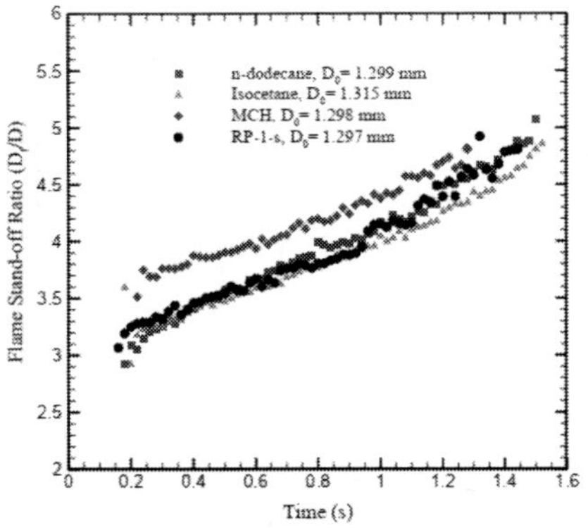

Figure 9: Comparison of FSR's for RP-1-s and its constituents

4 Conclusions

A surrogate mixture comprising of MCH, n-dodecane, and isocetane in the molar ratio 59:32:9 was found to fairly representative of RP-1 in terms of burn rate characteristics. Despite apprehensions regarding the lower boiling point of MCH in comparison with isocetane and n-dodecane, the occurrence of micro-explosions in the surrogate fuel was found to be minimal and mostly inconsequential for small size droplets, resulting in steady burn rate behaviour of the surrogate fuel, RP-1-s. The burn rate constants for tested fuels were found to lie between 0.75-1 mm^2/s for initial droplet diameters between 1-1.5 mm. For the time interval where the burn rate constant was evaluated, the flame diameters and flame heights were found to vary between 3.5-5 mm and 8-17 mm respectively. Flame height was found to be a good indicator for identifying the occurrence of micro-explosions. The results show that the combustion behaviour of the surrogate fuel, as assessed though the evaluation of burn rates, flame characteristics and sooting tendencies, lies between those of its constituent components.

5 Acknowledgment

The authors would like to acknowledge the grants from Industrial Research & Consultancy Centre (IRCC), IIT Bombay and Indian Space Research Organisation (ISRO) for supporting their project.

References

[1] M.E. MacDonald, *Decomposition Kinetics Of The Rocket Propellant RP-1 And Its Chemical Kinetic Surrogates*, PhD thesis, Stanford University Stanford, CA, USA, 2012.

[2] A. Violi, S. Yan, E. Eddings, A. Sarofim, S. Granata, T. Faravelli, E. Ranzi, Combustion Science and Technology, 174 (2002) 399-417.

[3] M.L. Huber, E.W. Lemmon, L.S. Ott, T.J. Bruno, Energy & Fuels, 23 (2009) 3083-3088.

[4] M. Colket, T. Edwards, S. Williams, N.P. Cernansky, D.L. Miller, F. Egolfopoulos, P. Lindstedt, K. Seshadri, F.L. Dryer, C.K. Law, in: 45th AIAA Aerospace Sciences Meeting and Exhibit, 2007, pp. 1-21.

[5] S. Rajkumar, R.S. Choudhari, A. Chowdhury, I.N. Namboothiri, Thermochimica acta, 563 (2013) 38-45.

[6] S. Lal, S. Rajkumar, A. Tare, S. Reshmi, A. Chowdhury, I.N. Namboothiri, Chemistry–An Asian Journal, 9 (2014) 3533-3541.

[7] S. Lal, L. Mallick, S. Rajkumar, O.P. Oommen, S. Reshmi, N. Kumbhakarna, A. Chowdhury, I.N. Namboothiri, Journal of Materials Chemistry A, 3 (2015) 22118-22128.

[8] L. Mallick, S. Lal, S. Reshmi, I.N. Namboothiri, A. Chowdhury, N. Kumbhakarna, New Journal of Chemistry, 41 (2017) 920-930.

[9] M.E. MacDonald, D.F. Davidson, R.K. Hanson, W.J. Pitz, M. Mehl, C.K. Westbrook, Fuel, 103 (2013) 1051-1059.

[10] C.T. Avedisian, Developing Surrogates for Liquid Transportation Fuels: The Role of Spherically Symmetric Droplet Combustion, in: Novel Combustion Concepts for Sustainable Energy Development, Springer, 2014, pp. 379-402.

[11] J.J. Lee, J.E. Shepherd, Spark Ignition Measurements in Jet A: part II, Explosion Dynamics Laboratory Report FM 99-7, California Institute of Technology, Pasadena, CA 91125, 2000.

[12] A.K. Yadav, A. Chowdhury, A. Srivastava, International Journal of Heat and Mass Transfer, 111 (2017) 871-883.

[13] R. Klimek, T. Wright, Spotlight-8 Image Analysis Software, NASA/TM-2006-214084, NASA Glenn Research Center, Cleveland Ohio, 2004.

[14] K. Kobayasi, in: Symposium (International) on Combustion, Elsevier, 1955, pp. 141-148.

[15] O.W. Dykema, S.A. Greene, Progress in Astronautics and Rocketry: Liquid Rockets and Propellants, (1960) 299-349.

[16] C. Law, S. Chung, N. Srinivasan, Combustion and flame, 38 (1980) 173-198.

[17] A. Ambekar, A. Chowdhury, S. Challa, D. Radhakrishna, Fuel, 115 (2014) 697-705.

11th Asia-Pacific Conference on Combustion,
The University of Sydney, NSW Australia
10th -14th December 2017.

Determination of Pressure Waveform in a T-burner Based on Standing Wave Ratio

Anchen Song[1], Junwei Li[1,*], Bingbing Sun[1], Xinjian Chen[1], Ningfei Wang[1]

[1] School of Aerospace Engineering, Beijing Institute of Technology,

5 South Zhongguancun Street, Beijing 100081, China

Abstract

In this paper, numerical simulation method based on T-burner test has been used to study the pressure wave propagation after a pulse, aiming to explore the pressure waveform characteristics after pulse and provide a theoretical guidance for future experiments. After the validity of the numerical method has been verified, the waveforms of pressure wave in T-burner are divided into five stages. Although different stages of waveforms can be initially distinguished by pressure contours and axis of pressure value, it is not intuitive enough for follow-up analysis and engineering applications. Thus, the concept of standing wave ratio in theoretical acoustics is introduced to determine the waveform in the T-burner. It can be more intuitive to determine the waveform by standing wave ratio, and more convenient in the application of the engineering project. In addition, the application of standing wave ratio also has an important significance in further understanding of energy characteristics of the T-burner, which is widely focused on by many combustion instability researchers.

1 Introduction

For more than half a century, the research on combustion instability is essential for a wide range of technologies, such as gas turbine [1], liquid rocket engine [2, 3], solid rocket motor [4-7] and some other burners [8, 9]. In the history of related research, pressure coupled response [10] has been thought of as a key factor in solid rocket motor combustion instability. It is also a characteristic parameter for predicting the stability. Pressure coupled response function can be measured by different methods [5, 11, 12]. Among these methods, T-burner is proved to be economical, simple, and effective, which makes it the preferred one in related research.

Compared with traditional double-base propellant, the widely used high energy aluminized AP/HTPB composite propellant is prone to have more difficulties in self-sustained oscillations [5] For composite propellant, the oscillation in the T-burner is usually caused by external pulser.

What researchers know about the coupled response function is largely based upon the assumption that harmonic oscillations and standing waves have been developed in the combustion chamber totally [10, 13]. Along with the growth in the use of pulsers, there is increasing concern over their effect on the test. Rezaiguia [14] made an analysis of the flame of a small solid propellant sample subjected to a longitudinal acoustic wave, trying to find a way to get particular oscillation mode by specific pulse. Yan [15] made experimental and theoretical study on characteristics of pulse excitation in T-burners to get the changing laws of pressure pulse and mean pressure. Song [4] revealed the existence of pressure wave propagation in a pulsed T-burner by numerical method.

The abovementioned studies have provided some important information on the role that pulse played in combustion instability tests. The first thing to note again is that, despite of using pulsers, these researches mostly based on a hypothesis of the successful establishment of standing wave in the combustion chamber [5, 14]. Only in this situation, the correlation theory, sound mode analysis for example, can be used. However, the researches of Yan [15] and Song [4] have revealed that the hypothesis may not always be established.

Therefore, a method of judging the waveform in the combustion chamber is necessary. In this paper, a judgment of waveforms of pressure wave based on standing wave ratio is presented and verified by numerical simulation based on the commercial programs (FLUENT®). It can be more intuitive to determine the waveform in the chamber, and more convenient in the application of the engineering project by this method.

2 Computational Model and Verification

The computational model is based on burning surface doubled and secondary attenuation test [5]. The T-burner of experimental setup is shown in Fig. 1 and the special grain structure is shown in Fig. 2. Because of the length of the paper, the process of the experiment can be found in Su's paper [5].

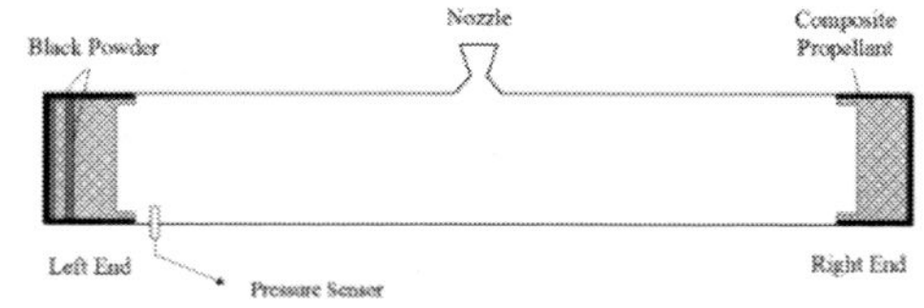

Figure 1: Schematic of T-burner with two propellant grains.

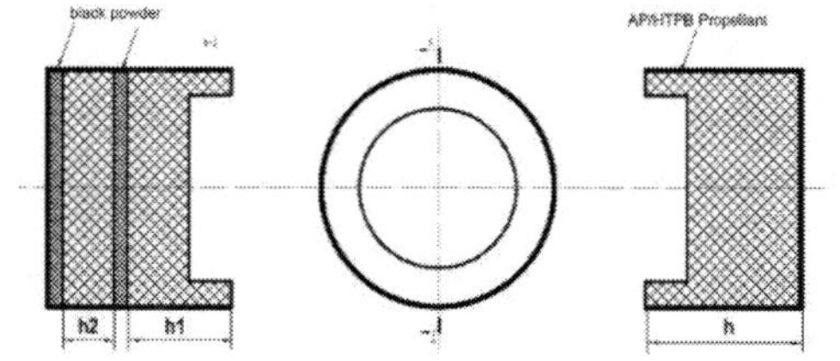

Figure 2: Grains of cup shape structure.

In the numerical simulation of this paper, the computational model domain contains about 2,059,210 grid nodes. The grid close to the nozzle wall was refined in order to reflect the drastic

*Corresponding author. Fax: +86-010-6891-3123
E-mail address: david78lee@gmail.com

changes of the flow more satisfactorily. The boundary condition is illustrated in Fig. 3. The length of the main body is 3.4m. The throat diameter is 0.0039m and the burning surface has a uniform inner radius of 0.070m. In the simulation, the coupled solution method which solves the governing equations of continuity, momentum, energy simultaneously is used. The N-S equations is closed by the RNG k–ε turbulence model. The second-order implicit formulation is used in an unsteady solver. Five pressure monitors are evenly distributed on the axis of symmetry. A more detailed description of the geometric model and boundary conditions can be found in Song's paper [4].

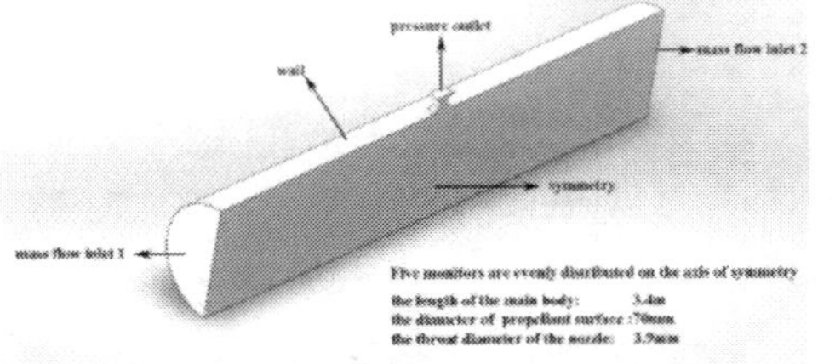

Figure 3: Computational geometry and boundary conditions.

For the mass flow inlet, the steady iterations are firstly carried out in the numerical calculation with a good convergence. After that, the two pulses are implemented via UDF (user defined functions) in the unsteady calculation process. The APN model is used to express the burning rate, as described in (1):

$$\dot{m} = \dot{m}_b / A_{in} = \rho_p \cdot a \cdot p^n \cdot A_b / A_{in} \qquad (1)$$

Where $\dot{m}$ is mass flux of the inlet, $\dot{m}_b$ is mass flow rate of burning surface of one grain, A_b is the area of it, ρ_p is the density of the propellant. The subscript a and n identify the burning rate coefficient and exponent, p is the pressure in T-burner, A_{in} is the area of the inlet. In the simulation, A_b/A_{in}=1. As for the APN model, the pressure in T-burner is independent of temperature. So it is set as a constant 2800K.

The verification of the numerical simulation has been done in the previous work [4]. It has been done by comparisons of numerical simulation results' P'–t curve, fluctuating pressure and pressure oscillation spectrums with experimental results'. Due to the length of the paper, it is not all repeated here. Only illustrate the pressure oscillation spectrums after the first and the second pulse in Fig. 4 and Fig. 5 for example.

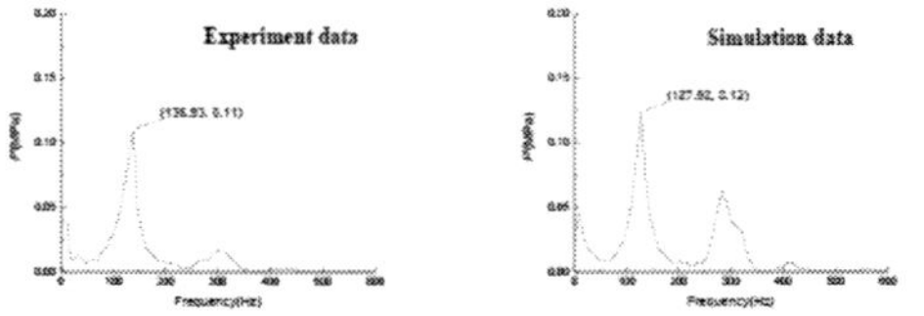

Figure 4: The pressure oscillation spectrums (the first pulse)

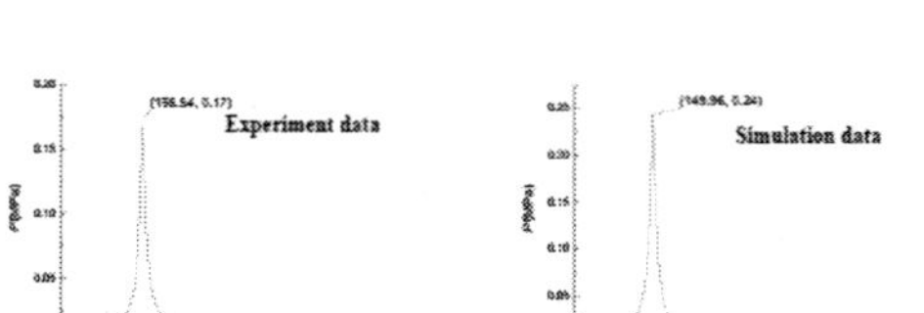

Figure 5: The pressure oscillation spectrums (the second pulse)

From the results above, the validity of the simulation results can be explained.

After a FFT (fast Fourier transform) method, pressure spectrum of the data from experiment and numerical simulation is shown. It is obvious that there is a peak at the frequency of 136.93Hz in the experiment, which is 156.54 in the numerical simulation. The frequencies of simulation are approximately 15% higher than the experimental frequency. The most likely causes of this discrepancy is the assumption of adiabatic walls, which results in a higher chamber temperature and thus a higher sound speed. The findings of a recent technical note by U.S. Air Force Research Laboratory [16] show a similar result. It indicates that the first axial acoustic mode has been excited by the pulse.

By method validation, the results suggest that the numerical simulation used in this paper is reasonable. After proving the effectiveness and practicability of the numerical method, a smaller T-burner modal is used to help us save both the time and the storage space, whose length of main body is 0.34m instead of 3.4m. The theoretical oscillation frequencies of the smaller model are 1560Hz, so the time step is set to 5×10-6s to meet the sampling frequency requirements.

3 Standing Wave Ratio

3.1 Concept Description

Standing wave ratio is commonly used in theoretical acoustics to characterize the sound absorption capacity of material. It can also characterize energy radiation in signal transmission process. The standing wave ratio is defined as the ratio of the partial standing wave's amplitude at an antinode (maximum) to the amplitude at a node (minimum) along the line. The standing wave ratio can be described as (2):

$$s = \frac{p_{a\,max}}{p_{a\,min}} \qquad (2)$$

Where s is standing wave ratio, p_{amax} is standing wave's amplitude at an antinode (maximum), p_{amin} is amplitude at a node (minimum). While s=1, it means that antinode and node of standing wave have the same amplitude value. There is no obvious node or antinode in the pressure wave line and pressure waveform in the chamber is travelling wave. While s tends to infinity, it means that a node amplitude in pressure wave line has been formed. In other word, it's an ideal standing wave.

3.2 Test Data Processing

Calculating the standing wave ratio in the T-burner according to the definition requires to know the pressure distribution on the axis. There is nothing difficult for numerical simulation. The pressure of the axis can be read.

However, as for the test result, it not an easy work to get the axis pressure directly. Because it is impossible to install enough pressure sensors to get the axis pressure whether for structural, safety, or cost considerations. In the usual experiment, the number of sensors will be no more than five, which is far from knowing the axis pressure directly. Therefore, an effective method of data processing, which can get the standing wave ratio with sensors as few as possible is necessary.

In the case of a T-burner, the first-order mode is dominant when the standing wave is formed in the chamber. Both numerical and experimental pressure oscillation spectrums can illustrate this point, which provides a possibility of knowing the node and antinode position without having to measure all the axis pressure. Two sets of p-t curve from the sensors in node and antinode position respectively are sufficient. (In this paper, they are the curve of the middle and the end of T-burner respectively). The next step is to take the pressure data for the period to be analyzed, and get the maximum and minimum values of the absolute value respectively. And the average value is taken as the node and antinode pressure value, and the standing wave ratio is obtained. Therefore, the test data processing can be described as:

$$s = \frac{p_{a\,max}}{p_{a\,min}} = \frac{U_{max}}{U_{min}} \tag{3}$$

In (3):

$$U_{max} = \frac{\left|max\{u_{a,1}, \ldots, u_{a,n}\}\right| + \left|min\{u_{a,1}, \ldots, u_{a,n}\}\right|}{2}$$
$$U_{min} = \frac{\left|max\{u_{n,1}, \ldots, u_{n,n}\}\right| + \left|min\{u_{n,1}, \ldots, u_{n,n}\}\right|}{2} \tag{4}$$

Where $u_{a,n}$ represent the antinode pressure and $u_{n,n}$ represent the node pressure respectively.

4 Results and Analysis

As mentioned above, a set of numerical simulation has been made with processed result to verify the standing wave ratio is effective for determining the pressure waveform in the T-burner. In contrast to the experimental method, the numerical method is much easier to get data in the calculation domain. Another advantage of using computer simulations is that it allows the essential simplification, which may help us better understand the core elements of the problem. Aiming at describing the variation of the waveform in the T-burner after pulse, a series of cases are simulated. A case is chosen here to conduct this exploratory study. The pulse time of this case is 1.6×10^{-4}s (A quarter of the T-burner cycle time), and the total pulse mass of the pulse is 2g. After application of the pulses (at the 0.01000s), the pressure oscillations in the T-burner began to occur. Then the contours of static pressure and temperature in different period are compared. It can be found that there is a significant difference between the pressure and the temperature distribution in the T-burner at different time after pulse. Five typical waveforms emerged from the analysis. The contours of the waveforms are illustrated in Fig. 6. The waveforms after pulse were divided into five groups according to the pressure distribution of different contours, they will be named as follows: the pulse stage, the transitional stage from pulse to traveling wave, the traveling wave stage, the transitional stage from traveling wave to standing wave, and the standing wave stage.

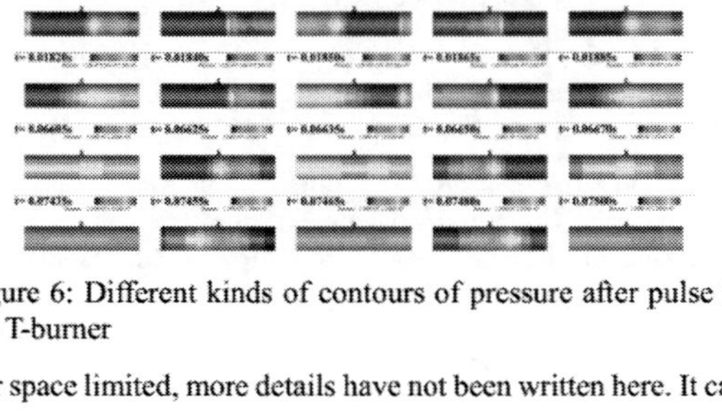

Figure 6: Different kinds of contours of pressure after pulse in the T-burner

For space limited, more details have not been written here. It can be partially found in Song's paper[4]. Based on the contours, therefore, the standing wave ratio of the numerical results will be calculated.

In this research, the pressure from 2000 points are taken evenly along the axis of the T-burner, and the axis pressure distribution curve is intercepted. In each cycle after the pulse, 15 different moments are chosen evenly, and then pressure on the axis are processed to get oscillation pressure. The axial oscillation pressure of the first eight cycles after the pulse is applied as shown in Fig. 8:

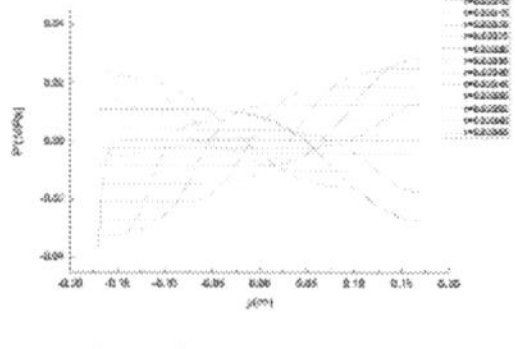

(a) The 1st cycle

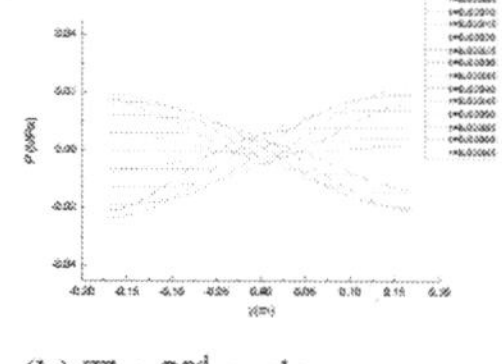

(b) The 2nd cycle

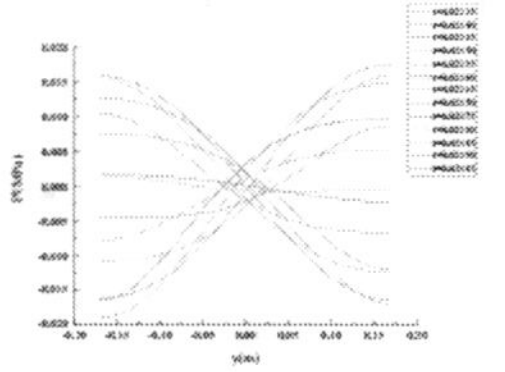

(c) The 3rd cycle

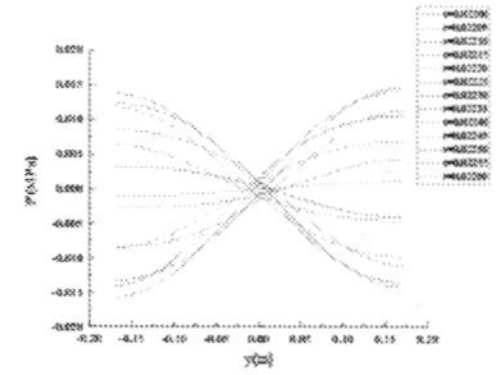

(d) The 4th cycle

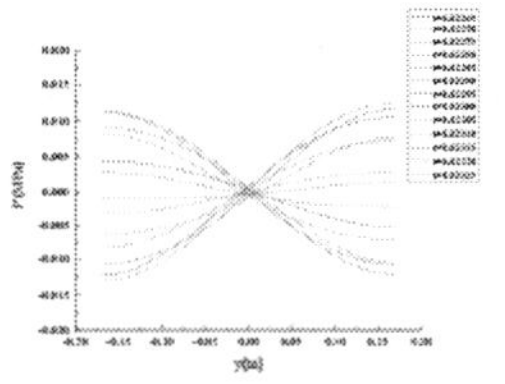

(e) The 5th cycle

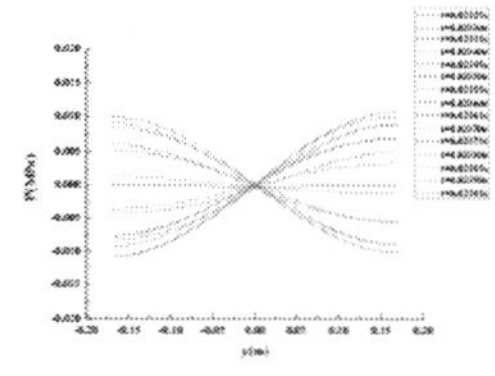

(f) The 6th cycle

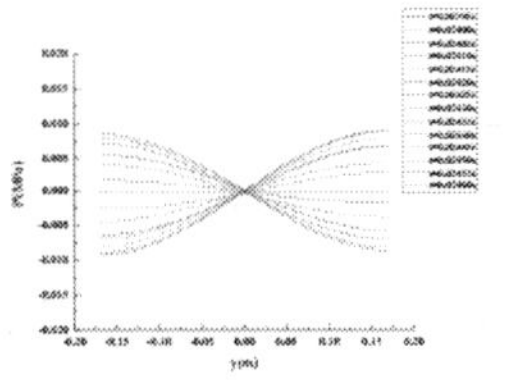

(g) The 7th cycle

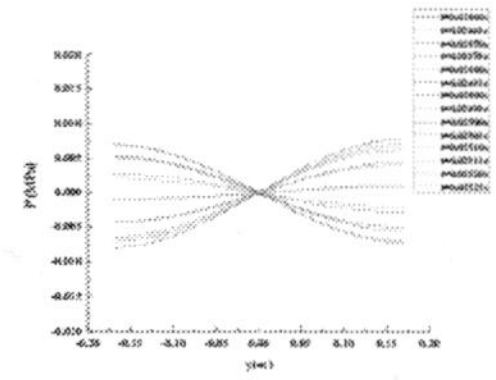

(h) The 8th cycle

Figure 7: The oscillation pressure distribution of the axis on the first eight cycles

It can be seen from the axis pressure in Fig. 8 that there is processes of waveform conversion in the first eight cycles. It is important to note that there is no need to read all lines to get the conclusion. The key point is the aggregation degree in the antinode and the node. In the first cycle, the pressure distribution in the whole T-burner is traveling waveform. There is no node, antinode or other characteristics of standing waveform. The axis pressure in this cycle presents a typical traveling wave characteristics instead. The transformation from traveling wave to standing wave starts at the second cycle. Though the waveform of T-burner has begun to superimpose which is no longer a typical traveling wave, the node has not yet formed at this time. The error of direct application of theory standing wave related cannot be ignored. The T-burner is basically a standing wave from the fifth cycle. The node is located half the axis.

In order to verify the effectiveness of the application of standing wave ratio in combustion instability, the standing wave ratio for each cycle is calculated separately. The calculation method is shown in (3), and the results are shown in Table 1.

Cycle No.	U_{max}(Mpa)	U_{min}(Mpa)	s
1st	0.02817	0.01155	2.44
2nd	0.02132	0.00520	4.10
3rd	0.01735	0.00297	5.84
4th	0.01459	0.00159	9.18
5th	0.01227	0.00080	15.34
6th	0.01046	0.00040	26.15
7th	0.00884	0.00019	46.53
8th	0.00758	0.00010	75.80

Table 1: The standing wave ratio of the combustion chamber in each cycle

As mentioned above, while s=1, the waveform of the pressure is travelling wave instead of standing wave; While s tends to infinity, it's an ideal standing wave. It can be seen from the axis pressure in Table 1 that the standing wave ratio s in the combustion chamber is increasing with cycles after oscillation. It can be seen that the standing wave ratio is 2.44 in the first cycle, and there is obvious traveling wave characteristic. As for the second to the forth cycle, the peak pressure is 3 times to10 times the trough pressure with the node gradually producing.

It belongs to the transition stage from the traveling wave to the standing wave this time. After the fifth cycle, most of the energy is reflected. It can be considered that the T-burner is already standing wave mode at the eighth cycle for the peak pressure is already 75.8 times the trough pressure. This conclusion is consistent with that shown in Fig. 8. Therefore, it is feasible to characterize the waveform in the combustion chamber by using the standing wave ratio.

5 Conclusions

The main goal of the current study was to find a parameter that characterizes the feature of pressure waveform in the combustion chamber. Based on T-burner test, the commercial program (FLUENT®) were used to study the effectiveness of the application of standing wave ratio in combustion instability. The results are as follows:

a) The waveforms conversion process after pulse in the combustion chamber has been proved.

b) The concept of standing wave ratio is commonly used in theoretical acoustics, which can be used in the research of combustion instability as well. While s=1, the waveform of the pressure is travelling wave instead of standing wave; While s tends to infinity, it's an ideal standing wave.

c) Typically the standing wave ratio s in the combustion chamber is increasing after pulse. The use of relevant conclusions or research methods need to determine the waveform of the combustion chamber first.

6 Acknowledgment

This work is supported by the National Science Foundation of China (No. 11572042). Mi Yan is greatly acknowledged for her kindly help on this paper.

References

[1] J.E. Temme, P.M. Allison, J.F. Driscoll, Combustion instability of a lean premixed prevaporized gas turbine combustor studied using phase-averaged PIV, Combustion and Flame 161(4) (2014) 958-970.

[2] A. Urbano, L. Selle, G. Staffelbach, B. Cuenot, T. Schmitt, S. Ducruix, S. Candel, Exploration of combustion instability triggering using Large Eddy Simulation of a multiple injector liquid rocket engine, Combustion and Flame 169 (2016) 129-140.

[3] P.P. Popov, A. Sideris, W.A. Sirignano, Low-Probability Events Leading to Rocket Engine Combustion Instability, AIAA Journal 55(3) (2017) 919-929.

[4] A. Song, J. Li, M. Yan, B. Sun, N. Wang, Propagation of Pressure Wave in a Pulsed T-burner, 53rd AIAA/SAE/ASEE Joint Propulsion Conference, Atlanta, GA, 2017.

[5] W. Su, N. Wang, J. Li, Y. Zhao, M. Yan, Improved method of measuring pressure coupled response for composite solid propellants, Journal of Sound and Vibration 333(8) (2014) 2226-2240.

[6] A.D. French, M. Panelli, G. Di Lorenzo, A. Schettino, F. Paglia, Combustion Instability and Pressure Oscillation Numerical Simulation in a Solid Rocket Motor, 53rd AIAA/SAE/ASEE Joint Propulsion Conference, Atlanta, GA, 2017.

[7] B. Sun, S. Li, W. Su, J. Li, N. Wang, Effects of gas temperature on nozzle damping experiments on cold-flow rocket motors, Acta Astronautica 126 (2016) 18-26.

[8] S. Chen, D. Zhao, H.K.H. Li, T.Y. Ng, X. Jin, Numerical study of dynamic response of a jet diffusion flame to standing waves in a longitudinal tube, Applied Thermal Engineering 112 (2017) 1070-1082.

[9] S. Li, D. Zhao, Heat flux and acoustic power in a convection-driven T-shaped thermoacoustic system, Energy Conversion and Management 75 (2013) 336-347.

[10] F.E.C. Culick, Stability of High-Frequency Pressure Oscillations in Rocket Combustion Chambers, AIAA Journal 1(5) (1963) 1097-1104.

[11] R.S. Brown, J.E. Erickson, W.R. Babcock, Combustion Response Function Measurements by the Rotating Valve Method, AIAA Journal 12(11) (1974) 1502-1510.

[12] L.D. Strand, K.R. Magiawala, R.P. McNamara, Microwave Measurement of the Solid-Propellant Pressure-Coupled Response Function, Journal of Spacecraft and Rockets 17(6) (1980) 483-488.

[13] W. Cai, P. Thakre, V. Yang, Transient Combustion Response of AP/HTPB Composite Propellant to Acoustic Oscillations in a Rocket Motor, Combustion Science and Technology 181(4) (2009) 597-617.

[14] H. Rezaiguia, P. Liu, T. Yang, Flame response of solid propellant AP/Al/HTPB to a longitudinal acoustic wave, International Journal of Spray and Combustion Dynamics (2017) 175682771769583.

[15] M. Yan, N. Wang, J. Li, A. Song, Y. Ma, Experimental and theoretical study on characteristics of pulse excitation in T-burners, Acta Astronautica 134 (2017) 278-290.

[16] M.E. Harvazinski, D.G. Talley, V. Sankaran, Influence of Boundary Condition Treatment on Longitudinal-Mode Combustion Instability Predictions, Journal of Propulsion and Power 32(2) (2016) 529-532.

11th Asia-Pacific Conference on Combustion,
The University of Sydney, NSW Australia
10th -14th December 2017.

The effect of cyclic variations on in-flame particulate morphology in a direct-injection petrol engine

D. Kim[1], Y. Zhang[1], Y. Gao[2], S. Kook[1,*], Q.N. Chan[1], and M. Xu[2]

[1]School of Mechanical and Manufacturing Engineering, The University of New South Wales

[2]School of Mechanical Engineering, Shanghai Jiao Tong University

Abstract

A new in-flame particulate sampling method has been developed and implemented in a direct-injection petrol engine to better understand the original causes of engine-out particulate emissions that have recently become a subject of regulations. The sampling probe has been designed to hold a transmission electron microscope (TEM) grid slightly above the piston top where wall-wetting induced pool fire and diffusion flame was expected. The particulates deposition was expected due to thermophoretic force between hot particulates within the flame and the cold TEM grid surface. The sampled particulates were imaged using a TEM and the images were post-processed to determine particulate size and fractal dimension. Of particular interest is how inherent engine cyclic variations impact the particulate morphology as the first evaluation of the newly developed sampling method. Results show that the increasing in-cylinder pressure conditions tend to increase the particulate number counts, especially small and compact ones. The measured mean primary particle diameters were ranged from 24.09 nm to 24.93 nm, and the aggregate radius of gyration was spanned from 61.34 nm to 68.47 nm depending on the test runs. The fractal dimensions were measured at 1.59 ~ 1.75. However, none of these fluctuations show a good correspondence with the measured in-cylinder pressure for various engine runs, suggesting that the engine cyclic variations do not make a significant impact on key morphology parameters of the sampled particulates.

1. Introduction

Petrol engines have been a favorable option in the automotive market for a lower level of engine noise and emissions than diesel engines. In the past decade, petrol engines have been advanced with spark-ignition direct-injection (SIDI) technology achieving better fuel economy and performance with its capability of more precise fuel injection control and charge cooling effect over conventional port-fuel injection (PFI) engines [1,2]. Despite the significant advantages of SIDI engine, the future of SIDI engine is skeptical with its problematic particulate matter (PM) emissions [3] which has negative effects on human health [4] and the environment [5]. The Euro 6 which regulates both particle mass and number emissions, applies a more strict threshold on the current SIDI engine PM emissions [6], which requires significantly improved fundamental understating of particulates formation occurring inside the cylinder of the engine.

Two main sources of PM emissions in SIDI engine are known to be fuel wall-wetting induced pool fire smoke [7,8] and in-flame soot formation occurring in locally rich mixtures [2]. Since SIDI engines possess wider adjustability for engine operating conditions than conventional PFI engines (e.g. fuel injection timings), the soot formation environment is more complex. For instance, previous studies reported that the advanced injection timing leads to increased wall wetting on the piston surface and the retarded injection causes limited mixing time, both of which increase engine-out particulate number and mass emissions [8–10]. This requires a very careful selection of fuel injection timing over various engine operating conditions.

The size distribution and morphology of engine-out particulates are extensively investigated through a sampling and transmission electron microscopy (TEM) approach [11–16]. However, the exhaust particulates are only a product of complex formation and oxidation processes occurring inside the cylinder of the engine. The lack of details about particulates nucleation, growth, and agglomeration [17] as well as oxidation [18] could be better understood by performing in-flame particulates sampling in a running engine. To fill this knowledge gap, we have developed a new in-flame particulates sampling system that is applicable to a working SIDI engine. The present study focuses on the sampled particulate TEM image analysis and its dependency on engine cyclic variations.

2. Experiments

The engine and particulate sampling system are illustrated in Fig.1 and the specifications of the engine and operating conditions are tabulated in Table 1. The sampling experiments were carried out in an optical engine based on a 0.5-litre single-cylinder engine with double-overhead cam (DOHC) and 16 valves. The bore/stroke is a square configuration of 86 mm. The cylinder has a compression ratio of 8.5 with the sampling system to secure marginal gap between extruded sampling probe and fast valve movement at around top dead centre. A conventional 95 RON petrol fuel was supplied with a constant pressure at 15 MPa to a side-mounted 6-hole injector (Continental DI XL2). The amount of fuel injected was maintained for stoichiometric condition with fixed injection duration of 2.4 ms. The fuel injection timing was intentionally selected for a wall-wetting condition with an advanced timing of 320 °CA bTDC, which corresponds to high particulate emissions conditions in various studies [12,13,16]. The spark ignition timing was fixed at 17 °CA bTDC. In order to keep a consistent in-cylinder thermal condition, the coolant water was heated to 90 °C and circulated through the cylinder engine, liner, and engine block. The intake air was supplied in the wide-open throttle position at 100 kPa absolute pressure (i.e. natural aspiration), and the intake air temperature was measured at

A/Prof S. Kook
E-mail address: s.kook@unsw.edu.au

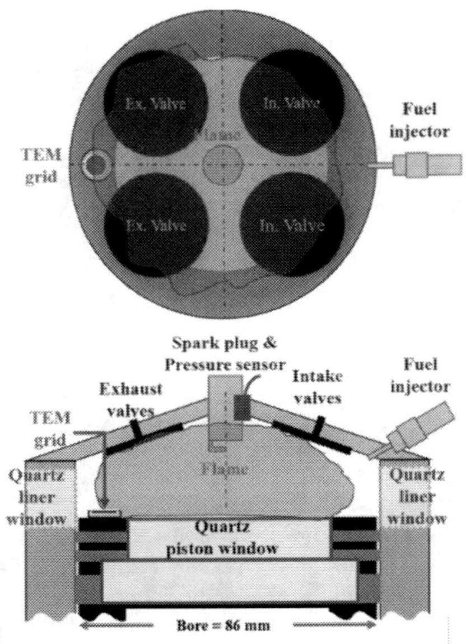

Figure 1: Illustration of the optical SIDI petrol engine and soot sampling system in top (top) and side view (bottom).

Table 1: Engine specifications and operating conditions

Displacement	0.5 L
Bore/Stroke	86mm/86mm
Compression Ratio	8.5
Valvetrain	DOHC, 4 valves
Intake Valve Timing	353 ~ 113 °CA bTDC
Exhaust Valve Timing	136 ~ 360 °CA aTDC
Fuel Injector	Continental DI XL2, 6 hole
Equivalence Ratio (ϕ)	1
Injection Pressure	15 MPa
Injection Timing	320 °CA bTDC
Spark Timing	17 °CA bTDC
Engine Speed	1200 RPM
Coolant Temperature	90 °C
Intake Air Temperature	30 °C
Intake Pressure	100 kPa (absolute)
Number of Firing Cycle per Sampling Run	5

30°C throughout the experiments.

The engine speed was held constant at 1200 RPM using a 36-kW AC motor and a large flywheel. The engine was fired once in every 10th cycle (i.e. 10 skip-firing) to avoid thermally over-loaded optical quartz windows, and to completely expel the residual gas from a previous cycle before the subsequent firing. For each TEM grid, five firing cycles were applied (*i.e.* 5-cylinder engine run) to have sufficiently high number of particulates while avoiding the over loading of particulate on the TEM grid. This was found from the trial-and-error based preliminary tests. Then the engine run was repeated for four different times to discuss the cyclic variation effects. The injection and spark timing were controlled via a universal timing control unit (Zenobalti ZB-9013P) based on the reference signal of the crankshaft position produced from a rotary encoder (Autonics E40S8). The in-cylinder pressure was recorded using a pressure transducer integrated onto the spark plug (Optrand C822J6-SP), thus enabling to calculate the apparent heat release rate (aHRR).

The particulate sampling probe was located on the opposite side of the fuel injector and between exhaust valves as illustrated in Fig 1. This was to reduce the collision risk with the exhaust valves, and more importantly to collect particulates formed due to pool fire developing around the region [19]. The height of sampling probe was set at 1 mm above the piston top to allow for sufficient exposure of the grid to the flame before it reaches the piston surface, thus enhancing in-flame soot sampling capability. A 3-mm diameter TEM grid (Emgrid CF400-Cu) with a standard carbon film supported by copper grid was contained in the sampling probe tip. The temperature gradient between hot flame and cold TEM grid surface would induce the thermophoresis and particulates deposition on the carbon film. It is expected that the reaction quenching would be immediate for the particulates due to high temperature difference, which would preserve the original particulate morphology [20].

The sampled particulates on the TEM grid were imaged using a TEM (JEOL 1400) with a point resolution of 0.38 nm and an accelerating voltage of 100 kV. The TEM images were recorded as digitised images via a CCD camera with a resolution of 11 mega pixels. Our previous study that conducted diesel in-flame soot sampling suggested various grid locations imaging to overcome soot morphology variations depending on the TEM grid location [21]. Therefore, a total of 30 TEM images from various on-grid locations were capture for each sample. The TEM images were post-processed via in-house developed Matlab code for particulate morphology data such as radius of particulates, primary particle diameter, and the fractal dimension. Detailed procedures of the Matlab-based image processing and determination of each morphology parameters can be found in our previous studies [22,23].

3. Results and Discussion

To investigate in-cylinder combustion conditions during each sampling run, averaged pressure traces and aHRR from 5 firing cycles for each sampling run have been plotted for various crank angles as shown in Fig 2. The peak pressures were varied from 2.8 to 3.2 MPa in the 4 sampling runs of the present study. The more rapid flame propagation and combustion are expected for run 3 and 4 as they have high peak pressure and more advanced CA50 than other sampling runs. This should be mainly due to the variation of mixtures in the vicinity of spark plug at the spark timing between firing cycles [24]. For ease of interpretation, run 1 to 4 were named for their increasing order of peak in-cylinder pressure and aHRR.

Selected TEM images of each sampling run are shown in Fig 3. On each image, the total number of particulates captured in TEM images and processed for size and fractal dimension analysis is noted. The sampled petrol in-flame particulates show typical chain-like fractal structures, similar to diesel soot particles. However, small and compact particles co-exist, suggesting significant variations in both the number and size of petrol particulates. A noticeable trend from Fig. 2 and 3 is the higher pressure and aHRR runs tend to show higher number of petrol particulates. This is largely led by the increased small and compact particles, which suggests the higher in-cylinder pressure and aHRR conditions cause higher formation of small soot particles (i.e. increased inception rate) that however do not appear to agglomerate with each other. It is understood that the increased temperature in pre-mixed flames leads to decreased soot formation due to the oxidation of soot nuclei [25], which is not the case for Fig. 3. Therefore, it is thought that the dominant source of particulates shown in the TEM images is diffusion flame that would show increasing soot formation with increasing temperature [26].

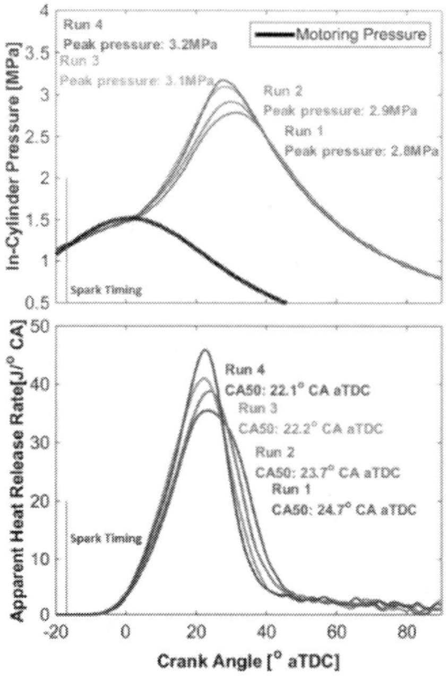

Figure 2: In-cylinder pressure and apparent heat release rate of the four 5-cycle sampling runs.

The characteristics of soot particle size have been determined by measuring primary particle diameter and radius of gyration of particulates, and their probability density functions (pdf) were plotted in Fig 4 top and middle, respectively. The fractal dimension was also calculated based on the mean primary particle diameter and radius of particulates in each processed particulate as shown in Fig 4 bottom. For the morphology analysis, at least 63 particulates and 845 primary particles were processed each of the four sampling runs. Figure 4 shows that the primary particle diameter is ranged from 10 nm to 60 nm and its mean values are varied from 24.09 nm to 24.93 nm, which is consistent with primary particle size distribution in the exhaust stream [11,13,14,16]. It is noticeable that both the mean primary particle size and the distribution do not vary with the level of run-to-run variations of the present study.

The aggregate size and fractal dimension, however, show marked variations. Figure 4 shows that the radius of gyration (RoG) of particulates is distributed from 20 nm to 275 nm, and the variation of mean values spans wider than that of the mean primary particle diameter (i.e. 61.34 nm to 68.47 nm). This was due to the increased soot inception rate for run 3 and 4 as discussed in the previous section, resulting in increased number of smaller particulates. The fractal dimension of particulates (D_f) falls into the 1.59 ~ 1.75, which is also a significant variation. However, the close inspection of the in-cylinder conditions (see Fig. 2) and the RoG and D_f results indicates there is no correspondence. The increasing in-cylinder pressure and aHRR, which showed the increased number of particulates (Fig. 3), does not simply increase the RoG or decrease the fractal dimension as one might expect. The distribution of RoG also shows no monotonic trend. Therefore, it could be concluded that the particulate number is very sensitive to the variations in peak in-cylinder pressure and aHRR; however, the size and fractal dimension of particulate aggregates show no

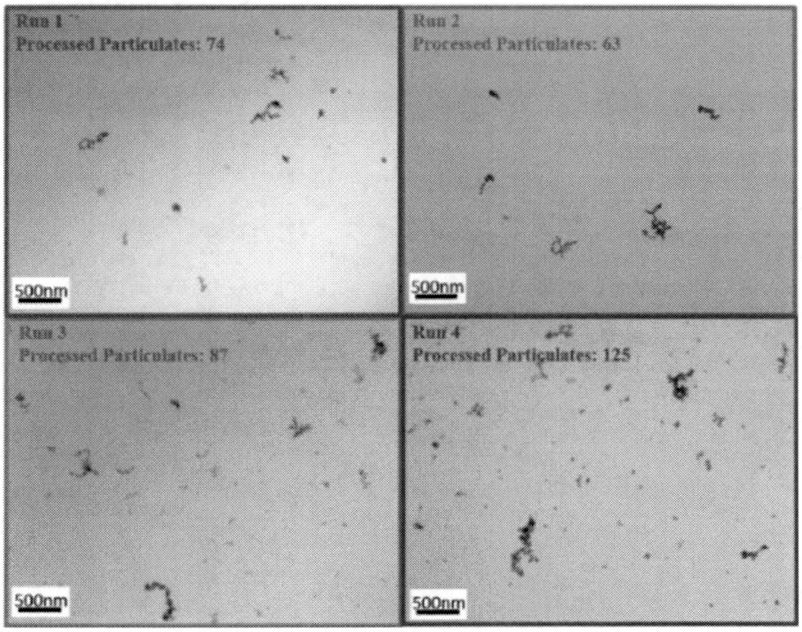

Figure 3: TEM images of the sampled particulates from 4 different engine runs. The number of processed particulates for each sample and the scale bar of 500 nm are shown on the top-left corner.

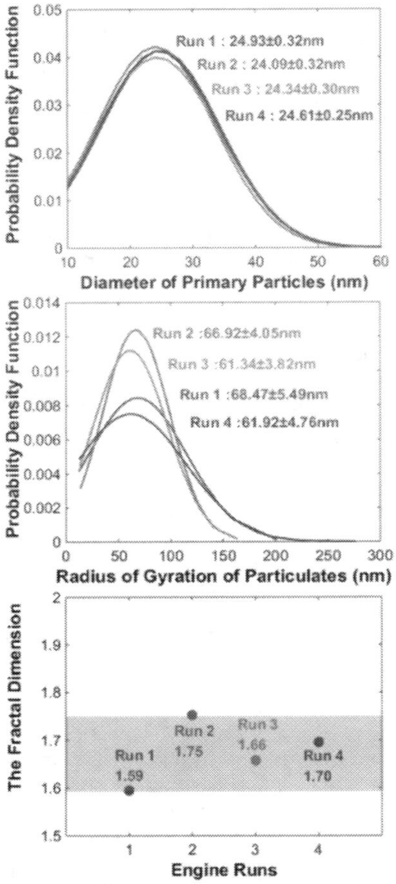

Figure 4: Probability density functions of the primary particle diameter (d_p, top), the particulate radius of gyration (R_g, middle), and particulate fractal dimension, D_f, for the four sampling runs.

significant dependency on the level of cyclic variations in the in-cylinder pressure conditions.

4. Conclusion

A new in-flame soot sampling method in direct-injection petrol engine has been applied to investigate morphology of particulates formed in combustion flame using TEM image. Four different engine runs were tested to better understand the influence of cyclic variations in the in-cylinder pressure conditions on particulate morphology. From the experiments, it is found that the particulate number counts increase with increasing in-cylinder pressure and aHRR within the tested run-to-run variations range. However, the primary particles forming larger aggregates show almost identical size distributions regardless of the run-to-run variations. Moreover, the aggregate size distribution and fractal dimension show no correspondence with the in-cylinder pressure and aHRR variations associated with the run-to-run variations, suggesting the cyclic variations do not make a significant impact on these key particulate morphology parameters.

Acknowledgment

Experiments were performed at the UNSW Engine Research Laboratory, Sydney, Australia. Support for this research was provided by the Australian Research Council via the Linkage Project scheme.

References

[1] Drake MC, Haworth DC. Proc Combust Inst (31);(2007):99–124.

[2] Zhao F, Lai M, Harrington DL. Prog Energy Combust Sci (25);(1999):437–562.

[3] Liang B, Ge Y, Tan J, Han X, Gao L, Hao L, et al. J Aerosol Sci (57);(2013):22–31.

[4] Kang M, Lim C, Han J. Toxicol Res (29);(2013):121–7.

[5] Pöschl U. Angew Chemie - Int Ed (44);(2005):7520–40.

[6] European Commission. Off J Eur Union (2008):136.

[7] Stevens E, Steeper R. SAE Tech Pap (2001-01-12);(2001).

[8] Drake MC, Fansler TD, Solomon AS, Szekely GA. SAE Tech Pap 2003-01-0547 (2003).

[9] He X, Ratcliff MA, Zigler BT. Energy & Fuels (26);(2012):2014–27.

[10] Hedge M, Weber P, Gingrich J, Alger T, Khalek I. SAE Int J Engines (4);(2011):650–66.

[11] Wu Z, Song C, Lv G, Pan S, Li H. Fuel (185);(2016):709–17.

[12] Miyashita K, Fukuda Y, Shiozaki Y, Kondo K, Aizawa T, Yoshikawa D, et al. SAE Tech. Pap. 2015-01-1872, (2015).

[13] Barone TL, Storey JME, Youngquist AD, Szybist JP. Atmos Environ (49);(2012):268–74.

[14] Gaddam CK, Vander Wal RL. Combust Flame (160);(2013):2517–28.

[15] Seong H, Lee K, Choi S. SAE Int (2013):10.

[16] Lee KO, Seong H, Sakai S, Hageman MD, Rothamer D. SAE Int (2013-24-01);(2013):9.

[17] Glassman I. Symp Combust (22);(1989):295–311.

[18] Stanmore BR, Brilhac JF, Gilot P. Carbon N Y (39);(2001):2247–68.

[19] Stojkovic BD, Fansler TD, Drake MC, Sick V. Proc Combust Inst (30 II);(2005):2657–65.

[20] Dobbins R a., Megaridis CM. Langmuir (3);(1987):254–9.

[21] Kook S, Zhang R, Szeto K, Pickett LM, Aizawa T. SAE Int J Fuels Lubr (6);(2013):80–97.

[22] Zhang Y, Zhang R, Rao L, Kim D, Kook S. Fuel (194);(2017):423–35.

[23] Zhang R, Kook S. Combust Flame (162);(2015):2720–8.

[24] Clark LG, Kook S, Chan QN, Hawkes ER. SAE Tech Pap 2016-01-2162 (2016).

[25] Böhm H, Hesse D, Jander H, Lüers B, Pietscher J, Wagner HGG, et al. Symp Combust (22);(1989):403–11.

[26] Hageman MD, Sakai SS, Rothamer DA. Proc Combust Inst (35);(2015):2949–56.

11th Asia-Pacific Conference on Combustion,
The University of Sydney, NSW Australia
10th -14th December 2017.

Kinetic study of multi-component pyrotechnics: observations and interpretations of the DSC curves

Anirudha Ambekar[1], Juyoung Oh[1], Yoocheon Kim[1] and Jai-ick Yoh[1]

[1] Department of Mechanical and Aerospace Engineering, Seoul National University,

Seoul, 151-742, Korea

Abstract

This study reports an experimental investigation into the chemical kinetics of several commercial pyrotechnic compositions. A differential scanning calorimeter was utilized to elucidate the thermo-kinetic characteristics of five multicomponent pyrotechnic compositions. The combustion process of typical pyrotechnics is primarily driven by condensed phase reactions. These reactions include processes like phase change, decomposition as well as oxidation reaction. Although the flame temperature of the pyrotechnics is much higher, the decomposition of various components as well as ignition reactions are known to occur well below 600 °C. The multicomponent nature of practical pyrotechnics results in a complex interaction between the various components when heated. The thermo-kinetic study was carried out to simulate the heating experienced by the pyrotechnics before the combustion zone. The physical process occurring within these temperature limits provide important insight into the rate of overall combustion. Thus, various observation made during the DSC study provided an insight into the phenomenology of the combustion process of energetic pyrotechnics. The chemical kinetic parameters such as overall activation energy (*Ea*) and pre-exponential factor (*A*) corresponding to these reactions are essential for a priory estimation of the burning rate of these pyrotechnics. The values of *Ea* and *A* were extracted from the experimental measurements.

1 Introduction

Pyrotechnics are fuel-rich energetic compositions that are closely related with other energetic materials such as explosives and propellants. Through the process of combustion, pyrotechnics may act as a source of gas, heat, light, smoke, sound, or any combination of thereof along with significant amount of solid phase combustion products. Typical pyrotechnics compositions consist of fine powders of oxidizer, fuel, as well as various additives, which are thoroughly mixed and compressed to form a porous solid charge. The additives may act as binders, stabilizers, catalysts, coloring agents, smoke dyes, and so on depending on the intended application. These granular heterogeneous porous composite energetic materials find utility in a range of applications such as chemical oxygen generators, pyrotechnic fasteners, automotive airbag inflators, thermites, material synthesis through solid-solid reactions, and fireworks.

Corresponding author. Fax: +82-2-882-1507
E-mail address: jjyoh@snu.ac.kr

The overall rate of combustion is an important characteristic that determines the performance and safety characteristics of pyrotechnics. The factors that determine this rate of pyrotechnics have been discussed in detail by Conkling and Mocella [1]. In summary, the combustion process is controlled by three fundamental aspects viz. the chemical activation energy, heat of reaction, and the fraction of the generated heat that reaches the unreacted propellant. Thus, underscoring the importance of the chemical kinetic parameters.

These parameters may be measured directly through a differential calorimetric study. Several studies [2-8] reporting combustion of binary pyrotechnic compositions or pure components have been reported in the literature. The current study aims to investigate a set of commercial multicomponent pyrotechnic compositions in order to elucidate their thermo-chemical behavior and extract the thermo-kinetic parameters. These parameters are typically related to the linear burning rate of the pyrotechnic composition and are of a practical interest. The nature of DSC curve for such complex propellants are expected to provide an insight into multiple parallel chemical reactions occurring in such propellants.

2 Experimental study

DSC experiments were carried out with a Mettler Toledo DSC 3 instrument utilizing the standard 40 μL aluminum pans with a purge flow of nitrogen maintained at 80 mL/min. The aluminum pans were sealed with an aluminum lid with a perforation of approximately 0.5 mm dimeter. The pans sealed in such a way allowed continuation of the chemical reactions in a self-generated atmosphere without interference of the surrounding atmosphere while allowing a safe vent for the gases generated during the process.

Furthermore, this study follows the ICTAC Kinetics Committee recommendations for collecting experimental data [9] as well as thermal analysis and data processing [10]. The pyrotechnics were tested under non-isothermal experimental conditions with heating rates of 5, 10, 20, 40, and 50 °C/min. The optimal sample mass was pre-determined according to the ICTAC Kinetics Committee recommendations in order to prevent self-heating of the sample while providing sufficient sensitivity for various peaks. This initial sample mass was found to be between 1.2 to 1.5 mg. The weight of the crucible before and after the experiments was measured in order to ascertain the mass-loss through escaping gas during DSC experiment. The data obtained through these experiments was processed using the AKTS thermo-kinetics software.

The mass based composition of the pyrotechnics being studied has been reported previously [11,12] and reproduced in Table 1.

These pyrotechnic compositions have been designated according to the color of the flame generated by them. These compositions are typically utilized in display pyrotechnics.

Component	Red	Green	Blue	Yellow	Silver
$KClO_4$	45 %	16 %	48 %	40 %	58 %
Mg	20 %	18 %	2.5 %	18 %	5 %
$SrCO_3$	20 %	-	-	12 %	-
$(C_2H_3Cl)_n$	8 %	8 %	6 %	7 %	6 %
$C_{10}H_{11}Cl_{17}$	7 %	8 %	8 %	5 %	-
CuO	-	-	23 %	-	10 %
$Ba(NO_3)_2$	-	50 %	-	-	14 %
S	-	-	12.5 %	-	7 %
Na_3AlF_6	-	-	-	18 %	-

Table 1: Composition of pyrotechnics used in current study.

3 Results and Discussion

3.1 Characteristic DSC curves

Typical calorimetric study reports the heat flow associated with the chemical reaction occurring within the sample. DSC analysis often elucidates the physical characteristics of material such as phase transition, melting and decomposition as well as combustion events. The characteristics DSC curves obtained at a heating rate of 10 °C/min for pyrotechnics under study are shown in Fig. 1.

Although, every pyrotechnic compositions studied here utilizes potassium perchlorate as the primary oxidizer and magnesium as the primary fuel, a significant variation in the nature of the thermo-chemical response curve can be seen. The DSC curves shown in Fig. 1 are a superposition of several simultaneous interacting thermo-chemical processes. Certain characteristics distinguishing the component processes can be identified from this figure.

The first endothermic peak was observed to occur close 115 °C for all compositions except red color pyrotechnic. This peak represents melting of sulfur or loss of the trace amount of water within the composition. The volatilization of sulfur is also manifested at approximately 455 °C for silver propellant while blue propellant does not show such a peak.

The DSC curve for pure $KClO_4$ is known [2] to show an endothermic phase transition close to 300 °C. Same peak may be observed for each pyrotechnic compositions except green and silver. A sudden change in baseline of a DSC curve indicates change in the sample mass. Such a change was observed for green and silver propellant close to the phase transition temperature of potassium perchlorate. In case of the green pyrotechnic the baseline deviation occurs around the $KClO_4$ endothermic peak while in case of silver propellant the endothermic peak is broadened due to the effect of transition occurring at 256 °C. Additionally, DSC curve for silver pyrotechnic shows an endothermic peak at 132 °C source of which could not be identified.

The onset of the primary exothermic peak was observed to be at approximately 340 °C. However, this peak may be considered as a superposition of the various exothermic and endotherm peaks related to the combustion and decomposition reactions.

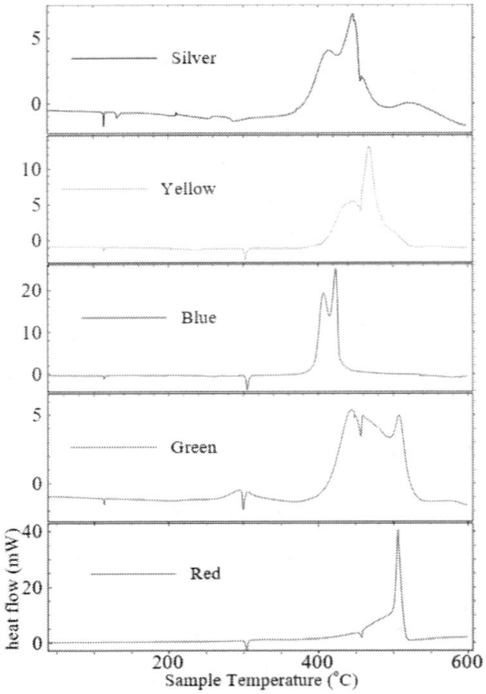

Figure 1: Characteristic DSC thermograms for pyrotechnics at 10 °C/min.

The normalized DSC exothermic peak area, nominal peak temperature, and nominal onset temperature for exothermic peak of various pyrotechnics at 10 °C/min are listed in Table 2.

	Normalized peak area (J/g)	Peak temperature (°C)	Onset temperature (°C)
Red	-2519	508	350
Green	-2748	450	360
Blue	-2449	390	335
Yellow	-2484	455	345
Silver	-1672	448	325

Table 2: Normalized peak area, peak temperature, and onset temperature.

3.2 Reaction kinetics

Chemical kinetic parameters for each pyrotechnic composition were evaluated from DSC experiments at multiple heating rates using ASTM standard method [13]. Figure 2 illustrates the variation of reaction rates of various pyrotechnics with temperature for different heating rates.

Owing to the multicomponent nature of the pyrotechnics, the overall reaction of each composition may be construed to be a mixture of multiple sequential and parallel reactions. Onset temperature of each of these constituent reactions may change by a different value with the change in the heating rate. This variation results in a non-similarity in consecutive thermograms and relatively poor fit in the Arrhenius plot indicated by low R-squared value for red, green, and silver compositions.

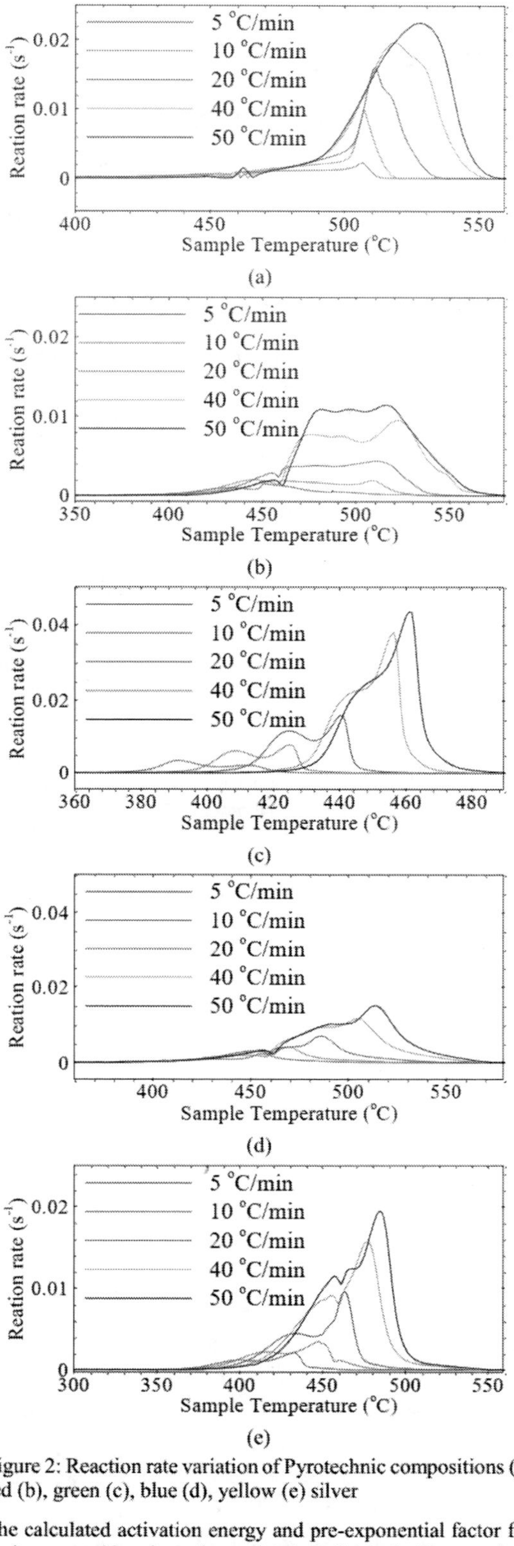

Figure 2: Reaction rate variation of Pyrotechnic compositions (a), red (b), green (c), blue (d), yellow (e) silver

The calculated activation energy and pre-exponential factor for each composition have been listed in Table 3. These values represent the overall chemical kinetics of the multicomponent combustion process where magnesium may be considered the primary fuel while $KClO_4$ acts as the primary oxidizer. Kang et al. [3] report the chemical kinetics of stoichiometric mixtures of Mg and $KClO_4$ in an atmosphere of air where the values of activation energy and pre-exponential factor for the ignition process and afterburning process were found to be 153.6 kJ/mol, 21.50 1/s, and 289.5 kJ/mol, 37.99 1/s, respectively. Similarly, the activation energy and pre-exponential factor for a Mg-$KClO_4$ mixture in a nitrogen atmosphere has been reported [4] to be 212.86 kJ/mol and 29.98 1/s respectively. The departure of the kinetic parameters of pyrotechnics in the current study from the values reported for pure Mg-$KClO_4$ mixtures was attributed to the reactions corresponding to the remaining components in a composition. Precise resolution of the effects due to additional components would need further study.

	Ea (kJ/mol)	*A* (1/s)
Red	536.04	78.00
Green	113.45	14.11
Blue	125.18	17.68
Yellow	185.74	25.71
Silver	186.84	26.91

Table 3: Activation energy and pre-exponential factor for the pyrotechnic compositions.

3.3 Mass-loss during DSC experiments

Typical DSC experiments provide the heat flow to/from the sample which can be normalized based on the sample mass. Subsequent analysis assumes the sample mass to remain unchanged during the experiment. However, the sample mass in the case of pyrotechnics was observed to change during the DSC experiment. This mass-loss was recorded by measuring the initial and final mass of the sealed aluminum pan, yielding a percentage mass-loss ratio. This reduction in sample mass must be taken into account while utilizing the analytical results for practical purposes. The experimentally measured values of averaged mass-specific heat released could be corrected with a factor based on the measured mass-loss. Equation (1) to (6) give the method of calculation for corrected mass-specific heat release.

$$\Delta h_E = \frac{E_R}{m_i} \tag{1}$$

$$\Delta h_A = \frac{E_R}{m_A} \tag{2}$$

$$m_A = \frac{m_i + m_f}{2} \tag{3}$$

$$m_{LR} = \frac{m_i - m_f}{m_i} \tag{4}$$

$$m_A = \frac{m_i}{2}\left(2 - m_{LR}\right) \tag{5}$$

$$\Delta h_A = \frac{2\Delta h_E}{2 - m_{LR}} \tag{6}$$

Where, Δh_E indicates the experimentally obtained averaged mass-specific heat release, E_R is the heat of reaction, m_{LR} is the mass-loss ratio, m_i and m_f are initial and final sample mass respectively, while Δh_A and m_A indicate the corrected averaged mass-specific heat release and corrected sample mass respectively. Table 4 shows the average values of initial and final mass, percentage change in sample mass, experimental mass-specific heat release, and corrected mass-specific heat release observed in this study for each pyrotechnic composition.

	m_{LR}	Standard deviation	Δh_E (J/g)	Δh_A (J/g)
Red	0.29	0.03	-2635.24	-3082.16
Green	0.33	0.04	-2592.46	-3104.75
Blue	0.53	0.03	-2271.91	-3091.04
Yellow	0.29	0.02	-2267.59	-2652.16
Silver	0.30	0.03	-1600.24	-1882.65

Table 4: Mass-loss during DSC experiments.

Additional observations could be made concerning the chemical kinetic parameters and the mass-loss ratio. The linear burning rates [12] for these compositions have been reported previously and corresponding theoretical estimates of the chemical kinetic parameters [14] have been reproduced in Table 5.

	Ea_T (kJ/mol)	A_T (1/s)
Red	88.34	7.13
Green	31.34	3.40
Blue	80.77	7.47
Yellow	76.44	6.42
Silver	97.39	8.07

Table 5: Theoretical activation energy and pre-exponential factor for the pyrotechnic compositions.

Although the experimental and theoretical values were found to be different, the ratio between theoretical and experimental values is fairly well-matched with the values and trends shown by the mass-loss ratio as shown in Fig. 3.

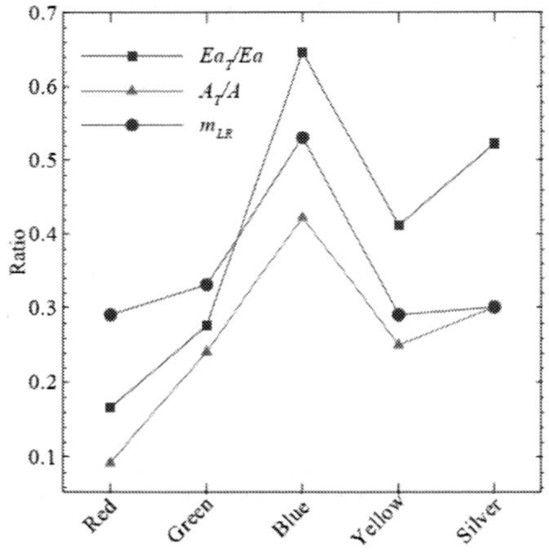

Figure 3: Comparison between trends of mass-loss and chemical kinetic parameters.

4 Conclusions

The study reports a calorimetric study of five multicomponent energetic pyrotechnic compositions. The pyrotechnic compositions with colored flame used for display pyrotechnics were selected. The DSC data was utilized to gain an insight into the combustion mechanism of the pyrotechnics. The DSC curve for each pyrotechnic was observed to be an approximate superimposition of the thermo-chemical response of each pure component. This study provides an insight into how the pure component thermal behavior is reflected in the behavior of the mixture. The non-isothermal DSC experimental technique was utilized to evaluate the apparent chemical kinetics parameters for these propellants. The *Ea* and *A* values evaluated through such a technique would be instrumental in theoretical estimation of the burning rate of the pyrotechnics. Furthermore, the overall mass loss during the experiments wherein the sample was heated from 30 °C to 600 °C was measured. Although, DSC analysis assumes a constant mass, the sample mass was noted to be decreasing significantly during the experiments. Appropriate scaling of the heat of reaction as well as kinetics parameters must be carried out in order to incorporate the effects of this mass loss.

5 Acknowledgment

This work was financially supported by BK21 PLUS program at the Department of Mechanical and Aerospace Engineering, Seoul National University. Additional support came from Advanced Research Center Program (NRF-2013R1A5A1073861) contracted through Advanced Space Propulsion Research Center at Seoul National University.

References

[1] J.A. Conkling, C. Mocella, Chemistry of Pyrotechnics: Basic Principles and Theory, CRC Press, 2010.

[2] J.S. Lee, C.K. Hsu, Thermochim. Acta. 367–368 (2001) 367–370.

[3] X. Kang, J. Zhang, Q. Zhang, K. Du, Y. Tang, J. Therm. Anal. Calorim. 109 (2012) 1333–1340.

[4] M. Fathollahi, H. Behnejad, J. Therm. Anal. Calorim. 120 (2015) 1483–1492.

[5] B. Berger, E.L. Charsley, J.J. Rooney, S.B. Warrington, Thermochim. Acta. 255 (1995) 227–239.

[6] S.G. Hosseini, A. Eslami, J. Therm. Anal. Calorim. 101 (2010) 1111–1119.

[7] M. Shamsipur, S.M. Pourmortazavi, S.S. Hajimirsadeghi, Combust. Sci. Technol. 183 (2011) 575–587.

[8] M. Shamsipur, S.M. Pourmortazavi, M. Fathollahi, J. Energ. Mater. 30 (2012) 97–106.

[9] S. Vyazovkin, K. Chrissafis, M.L. Di Lorenzo, N. Koga, M. Pijolat, B. Roduit, N. Sbirrazzuoli, J.J. Suñol, Thermochim. Acta.

590 (2014) 1–23.

[10] S. Vyazovkin, A.K. Burnham, J.M. Criado, L.A. Perez-Maqueda, C. Popescu, N. Sbirrazzuoli, Thermochim. Acta. 520 (2011) 1–19.

[11] A. Ambekar, M. Kim, J.J. Yoh, Appl. Therm. Eng. 110 (2017) 1066–1074.

[12] A. Ambekar, M. Kim, W.-H. Lee, J.J. Yoh, Appl. Therm. Eng. 121 (2017) 761–767.

[13] ASTM Standard test methods for Arrhenius kinetic constant for thermally unstable materials, Part 41, E698-79, 1982, p. 1012.

[14] A. Ambekar and J.J. Yoh, Appl. Therm. Eng. Under Review.

11th Asia-Pacific Conference on Combustion,
The University of Sydney, NSW Australia
10th -14th December 2017.

Droplet combustion studies of bis(nitratomethyl)-1-3-bishomocubane and its mixture with an RP-1 surrogate fuel

Anand Sankaranarayanan[1], Sohan Lal[2], I.N.N.Namboothiri[2], Sasidharakurup Reshmi[3], Arindrajit Chowdhury[1], Neeraj Kumbhakarna[1]

[1] Department of Mechanical Engineering, IIT Bombay
Powai, Mumbai, Maharashtra, 400076, India
[2] Department of Chemistry, IIT Bombay
Powai, Mumbai, Maharashtra, 400076, India
[3] Polymers and Special Chemicals Group, Vikram Sarabhai Space Centre,
Thiruvananthapuram, 690052, India

Abstract

The combustion behaviour of a novel energetic cage compound, bis(nitratomethyl)-1-3-bishomocubane (DNMBHC), and its mixture with an RP-1 surrogate fuel has been evaluated through droplet combustion experiments, under normal gravity. An RP-1 surrogate (RP-1-s), comprising of a mixture of methyl cyclohexane (MCH), n-dodecane, and isocetane in the molar ratio 59:32:9 was utilized for the purpose. The burn rate data of the surrogate mixture was found to satisfactorily match that of RP-1 reported in a previous study, when tested under similar conditions. Combustion studies on DNMBHC and its 12% mixture with RP-1-s indicated liquid phase pyrolysis of DNMBHC. The gases released during pyrolysis resulted in puffing and severe micro-explosions which caused the parent droplet to fragment into multiple smaller components. This phenomenon is usually desirable as it helps atomization of fuel droplets and promotes better mixing between fuel and oxidizer. Both DNMBHC and its 12% mixture with RP-1-s were found to be highly sooty under the conditions tested, but are expected to perform better in this regard, in an actual scenario. The potential of the compound to serve as an additive for performance enhancement is worth exploring.

1 Introduction

Cage hydrocarbons possess high energy densities owing to their highly strained C-C-C bond angles, which deviate significantly from the standard tetrahedral configuration of 109.5° [1]. The bishomocubane family represents one such category of compounds [2]. Theoretical studies on their performance show considerable promise for their use as possible additives to existing rocket propellant formulations, covering both solid and liquid variants [3]. As liquid propellants, these compounds generally have much higher densities as compared to their straight chain counterparts, making them attractive replacements in volume limited applications. However, due to the difficulties in synthesis, most of these compounds are typically synthesized only in milligram or sub-milligram scales. This constrains the evaluation of their combustion behaviour to droplet combustion studies [4, 5].

In the present work, the droplet combustion behaviour of the bishomocubane based cage hydrocarbon compound, bis(nitratomethyl)-1-3-bishomocubane (DNMBHC) [1] and its mixture with an RP-1 surrogate fuel [6] has been evaluated through droplet combustion experiments under normal gravity, in ambient atmosphere. The chosen RP-1 surrogate fuel (RP-1-s) is a blend of methyl cyclohexane (MCH), n-dodecane, and isocetane in the molar ratio 59:32:9 and has demonstrated an excellent match with the actual fuel in terms of compound class, decomposition rates, and ethylene yields [7]. The surrogate specifically targets to mimic the pyrolysis behaviour of RP-1. This is of significance in practical combustion scenarios where the fuel is used for cooling the engine and hence gets pyrolyzed before it is actually combusted. The suitability of droplet combustion studies for developing surrogate fuels to represent complex compounds has been previously discussed by Avedisian [8]. Though the arguments presented therein pertained to spherically symmetric combustion, the general argument of matching burn rate characteristics of the surrogate fuel with that of the actual compound for a combustion surrogate was extended in this study to droplet combustion under normal gravity.

The present work is aimed at understanding the combustion behaviour of DNMBHC and its mixture with RP-1-s. The burn rate and flame characteristics of RP-1-s are also explored.

2 Experimental Work

2.1 Experimental Setup & Methodology

A schematic of the experimental setup used for the present work is shown in Fig. 1. The fuel droplet was deposited onto the tip of a quartz rod of size 0.27 mm, housed within the combustion chamber, with the help of a microliter syringe. The ignition of the fuel drop was achieved by means of a spark, generated with the help of an ignition transformer. The average spark energy provided for ignition was measured using the

Corresponding author. Fax: +91-22-25726875
E-mail address: neeraj_k@iitb.ac.in

methodology outlined by Lee et.al [9] and was found to be approximately 6-12 mJ for spark durations between 0.1-0.2 s.

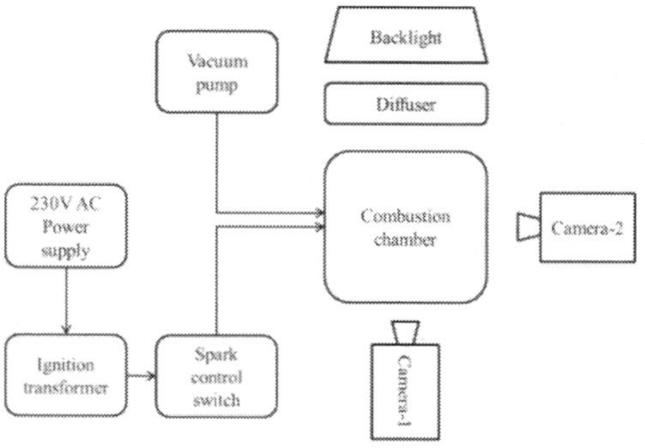

Figure 1: Schematic of experimental setup

Two DSLR cameras were simultaneously utilized for filming the experiments. Backlighting was provided to track the regression of the droplet surface in order to evaluate burn rates. Burn rate data was recorded using a CANON EOS 550D camera with 100 mm macro-lens, and flame characteristics were captured with the help of a NIKON D5200 camera with an AF-S NIKKOR lens and additional zoom lens fittings. The soot deposited over the quartz rod was cleaned, and the air inside the combustion chamber was evacuated with the help of a vacuum pump after each trial. Additional videos of a rod of known dimensions were recorded after the completion of the experiments, under similar settings to determine the scaling factors. Further details of the combustion chamber design, ignition transformer, and vacuum pump used for the present study were outlined in a previous work [10].

2.2 Procedure for Data Analysis

Data was acquired at a rate of 50 frames per second for both the cameras. Synchronization between the two cameras was achieved based on the commencement of the sparking event, which was taken as t=0 s. The droplet and flame boundaries were tracked using the image processing software Spotlight [11]. These were then used as inputs to an in-house MATLAB code for recovering droplet size and flame characteristics as a function of time. The droplet was assumed to be ellipsoidal and the equivalent diameter was calculated as suggested by Kobayasi [12]. Flame diameter (D_f) and flame height (h_f) were defined to be equivalent to the instantaneous maximum width and length of the flame, defined by the luminous region, respectively. The estimated error in evaluation of burn rate constants was around ± 4% and that for flame characteristics was around ± 8%.

3. Results and Discussions

3.1 Burn rate characteristics of RP-1 surrogate, DNMBHC and their mixtures

The burn rate characteristics of RP-1-s for various initial diameters are presented in Fig. 2. The effect of heat feedback from the quartz rod to the combusting fuel droplet increases as the droplet size keeps reducing. Hence the data has been presented only till a point where the droplet diameter reaches a value of about three times the suspension tip diameter. The results suggest that the surrogate fuel burns fairly smoothly. The flame heights and flame diameters for the cases corresponding to the results in Fig. 2, are shown in Fig. 3. The flame geometry was tracked from the completion of sparking till the instant at which the droplet diameter was about 0.81 mm. Ignition using a spark causes the surrounding air to heat up and rise upwards due to buoyancy. This effect causes a stretching of the flame which continues approximately for 0.1 s, after the completion of sparking and can be seen in Fig. 3. It can also be discerned that small micro-explosions were readily visualized as disturbances in flame heights and that their occurrence was more frequent in droplets of larger size. This phenomenon was attributed to the higher retention time of the trapped volatiles within the droplet, on account of a smaller surface area to volume ratio, for a droplet of larger diameter.

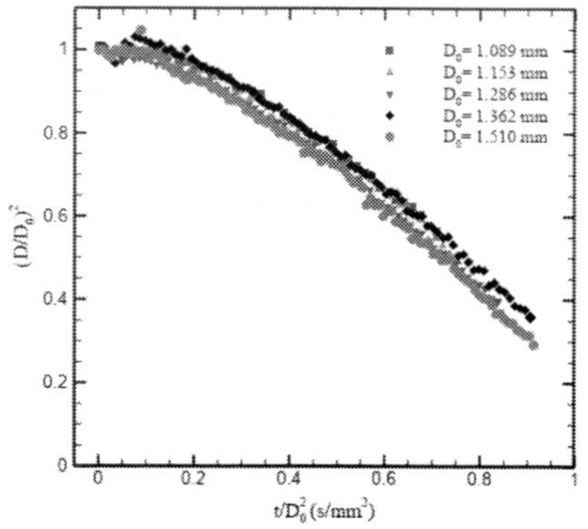

Figure 2: Variation of non-dimensionalized droplet diameter with scaled time for RP-1-s

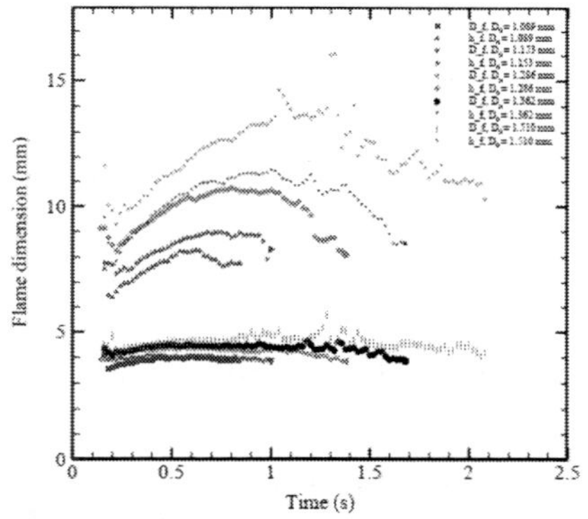

Figure 3: Variation of flame dimensions with initial droplet diameter for RP-1-s

The burn rate characteristics of pure DNMBHC and a 12% mixture of DNMBHC by mole fraction with RP-1-s are shown in Fig. 4, and Fig. 5, respectively. In the case of combustion of pure DNMBHC, the droplet size grows to a value of approximately 1.7 times that of the initial droplet diameter due to puffing of the droplet as a result of accumulation of gases in the condensed phase, from decomposition of the pure compound. Release of the trapped gases usually occurs in the form of vigorous micro-explosions which fragments the parent droplet into multiple smaller droplets, causing a decrease in the droplet diameter size. The results are indicative of liquid phase pyrolysis of DNMBHC. Similar results were also observed for the chosen mixture of DNMBHC and RP-1-s. A larger number of puffing events were seen in this case as compared to the pure compound. This was attributed to the slower rate of combustion, which allows for an intermittent release of trapped gases, resulting in the occurrence of more such events as compared to that of pure DNMBHC. The droplet mixtures in this case were also found to swell up to approximately 1.9 times the initial droplet size.

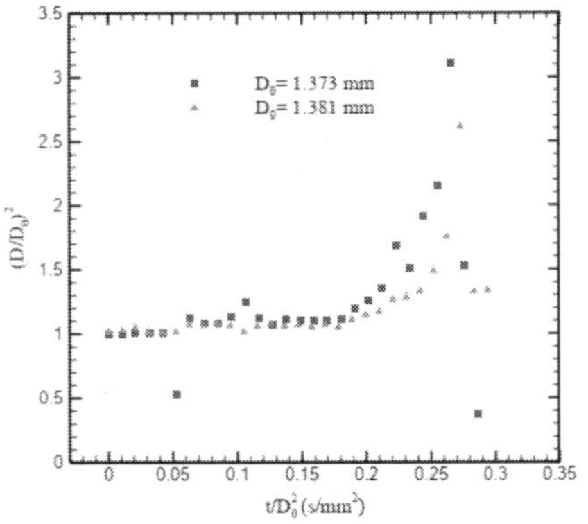

Figure 4: Variation of non-dimensionalized droplet diameter with scaled time for DNMBHC

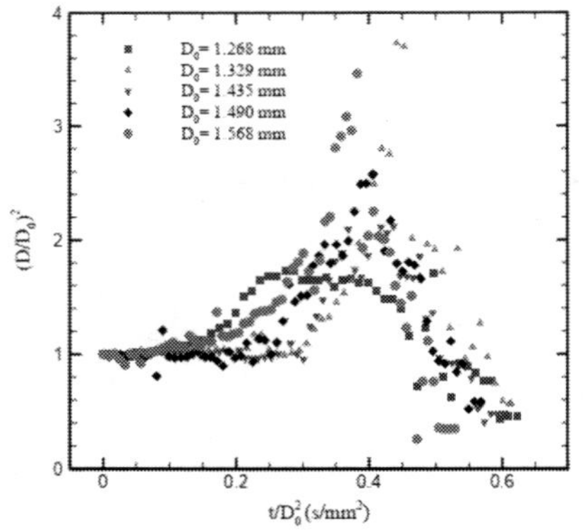

Figure 5: Variation of non-dimensionalized droplet diameter with scaled time for a mixture of 12% DNMBHC and 88% RP-1-s

3.3 Visualization of the combustion event

Sample images from the combustion event of DNMBHC and that of its 12% mixture with RP-1-s are presented in this section. The images of a DNMBHC droplet during different phases of combustion are shown in Fig. 6.

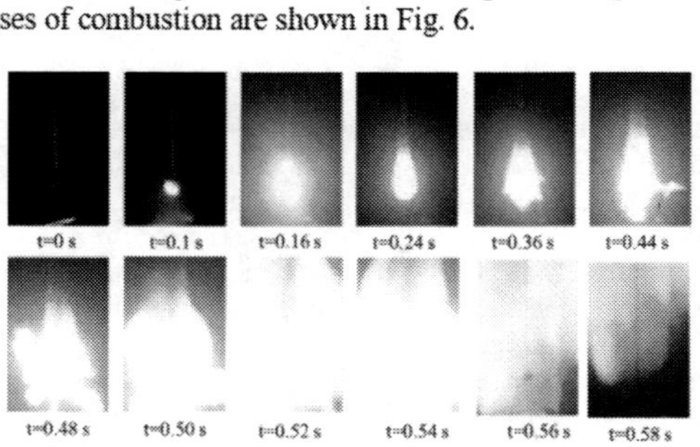

Figure 6: Selected frames from combustion of pure DNMBHC

The first 4 images respectively show the inception of spark, the first occurrence of a flame, the established frame front at the end of the spark, and, a steady self-propagating flame. Vigorous micro-explosions characterized by fragmentation of the parent droplet into multiple smaller droplets can be observed in the next 3 frames. The micro-explosions result in sudden and sustained expansion of the flame front. The culmination of the combustion event is captured in frames from t=0.5 to 0.58 s.

The backlit images of the same sequence of events are shown in Fig. 7. The soot residue left over after completion of combustion can be seen in the last 3 frames. Radiative heat losses due to the increased flame size plays a role in premature flame extinction, leaving behind large quantities of sooty residue. The increase in flame size also results in a larger separation between the flame front and the droplet. This in turn reduces the heat feedback to the droplet, causing a reduction in the condensed phase temperature, which hinders evaporation and further rapid pyrolysis. Significant quantities of suspended particulate matter were observed inside the combustion chamber, immediately after completion of combustion. The highly sooting nature of the compound can also be understood by observing the flame intensity in the backlit images. This behaviour was anticipated, as DNMBHC, a cage compound is expected to produce polyaromatic hydrocarbons (PAH's) during combustion, which are known soot precursors.

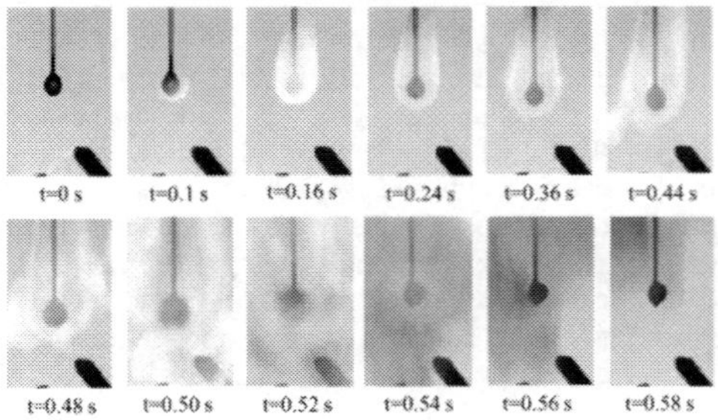

Figure 7: Backlit images of selected frames, highlighting combustion of DNMBHC

Significant events during the combustion of a mixture of DNMBHC and RP-1-s are shown Figs. 8 & 9. The initial phase of combustion is rather smooth. The frequency and strength of micro-explosions gradually increase, causing the flame front to grow in size, to a point where the flame extinguishes as explained previously. A sooty residue was seen to be deposited over the quartz rod in this case as well, though the quantity was considerably less as compared to pure DNMBHC. The puffing of the droplet during combustion can be clearly seen in the backlit image, in the frame corresponding to t= 0.88 s. Burn rate constants could not be quantified for either pure DNMBHC or its 12% mixture with RP-1-s due to puffing of the droplets and uneven combustion.

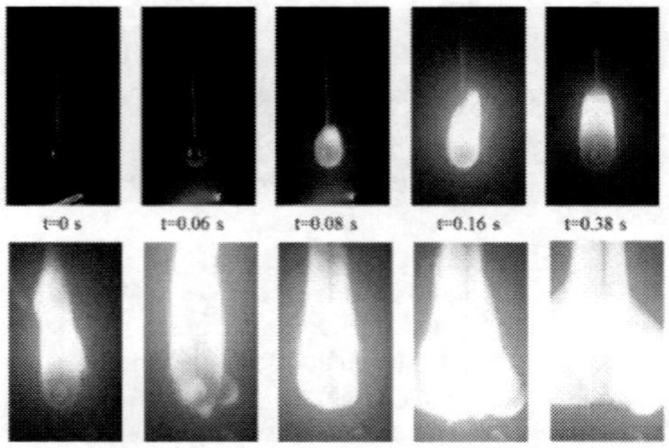

Figure 8: Selected frames from combustion of a mixture of 12% DNMBHC and 88% RP-1-s

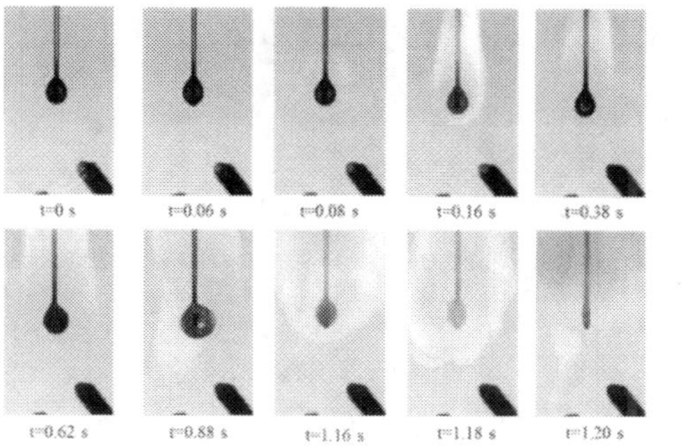

Figure 9: Backlit images of selected frames, highlighting combustion of a mixture of 12% DNMBHC and 88% RP-1-s

The time period for combustion of a pure droplet of DNMBHC, of initial diameter 1.381 mm was found to be 0.56 s as compared to a value of around 2 s for a similar size fuel droplet of RP-1-s. In the case of a 12% mixture of DNMBHC with RP-1-s, the combustion duration for an initial droplet size of 1.49 mm was found to be around 1.20 s as against a value of 2.22 s for a similar size fuel droplet of RP-1-s. Comparing the time required for combustion of the parent droplet provides a rough estimate of the enhancement in burn rate, if it is assumed that the fragmented droplets from the micro-explosions also undergo complete combustion within the considered time interval, and the time required for combustion of the left over soot residue is neglected. This method of quantification would roughly translate to an increase in burn rate of approximately 2.15 times for pure DNMBHC and 1.78 times for a mixture of 12% DNMBHC with 88% RP-1-s, as compared to RP-1-s, for initial droplet diameters of 1.381 mm and 1.49 mm respectively. Comparisons can be drawn to a similar study by Segal et.al [5] on methylated pentacycloundecane alkene dimer (MPCU), where a fivefold increase in burn rate constants for a 50% w/w mixture of MPCU with JP-10, as compared to that of pure JP-10 was reported. The burn rate constants were evaluated in this case through droplet combustion.

In an actual combustion scenario, flame temperature would be higher on account of higher ambient oxygen concentrations and the flame size would be lower because of smaller droplet sizes. This would in turn would promote a higher heat feedback to the combusting droplet and reduce the probability of premature flame extinguishment due radiative heat losses. Hence it is expected that the sooting tendency of both pure DNMBHC and its mixture with RP-1-s would be lower than what is currently observed.

4 Conclusions

A surrogate mixture comprising of MCH, n-dodecane and isocetane in the molar ratio 59:32:9 (RP-1-s) was found to satisfactorily represent RP-1 in terms of burn rate characteristics. Combustion studies on DNMBHC and its 12% mixture with the RP-1-s indicated liquid phase pyrolysis of DNMBHC. The gases released during pyrolysis resulted in puffing and severe micro-explosions which caused the parent droplet to fragment into multiple smaller components. This phenomenon is usually desirable as it helps atomization of fuel droplets and promotes better mixing between fuel and oxidizer in the combustion chamber, resulting in faster combustion, which directly translates to higher rates of energy release. For the range of initial diameters tested, the droplets were found to swell upto approximately 1.7 to 1.9 times the initial droplet size for pure DNMBC and its 12% mixture with RP-1-s, respectively. Combustion duration was used as a rough measure for quantifying burn rate enhancement. An increase in burn rate of approximately 2.15 times for pure DNMBHC and 1.78 times for a mixture of 12% DNMBHC with 88% RP-1-s, as compared to RP-1-s was obtained for fuel droplets of initial diameter 1.381 mm and 1.49 mm, respectively. The above method for quantifying burn rate enhancement is rather crude and future work in this area would be directed towards designing a new test setup to characterize the combustion behaviour of compounds that undergo pyrolysis in the condensed phase during combustion. Both DNMBHC and its 12% mixture with RP-1-s were found to be highly sooty under the conditions tested, but are expected to perform better in this regard, in an actual scenario. The potential of the compound to serve as an additive for performance enhancement is worth exploring.

5 Acknowledgment

The authors would like to acknowledge the grants from Industrial Research & Consultancy Centre (IRCC), IIT Bombay and Indian Space Research Organization (ISRO) for supporting their project.

References

[1] S. Rajkumar, R.S. Choudhari, A. Chowdhury, I.N. Namboothiri, Thermochimica acta, 563 (2013) 38-45.

[2] A.P. Marchand, Chemical Reviews, 89 (1989) 1011-1033.

[3] L. Mallick, S. Lal, S. Reshmi, I.N. Namboothiri, A. Chowdhury, N. Kumbhakarna, New Journal of Chemistry, 41 (2017) 920-930.

[4] T. Li, D. Zhu, C. Law, Journal of propulsion and power, 14 (1998) 45-50.

[5] C. Segal, S. Pethe, K. Williams, Combustion science and technology, 163 (2001) 299-244.

[6] M.E. MacDonald, D.F. Davidson, R.K. Hanson, W.J. Pitz, M. Mehl, C.K. Westbrook, Fuel, 103 (2013) 1051-1059.

[7] M.E. MacDonald, *Decomposition Kinetics Of The Rocket Propellant RP-1 And Its Chemical Kinetic Surrogates*, PhD thesis, Stanford University Stanford, CA, USA, 2012

[8] C.T. Avedisian, Developing Surrogates for Liquid Transportation Fuels: The Role of Spherically Symmetric Droplet Combustion, in: Novel Combustion Concepts for Sustainable Energy Development, Springer, 2014, pp. 379-402.

[9] J.J. Lee, J.E. Shepherd, Spark Ignition Measurements in Jet A: part II, Explosion Dynamics Laboratory Report FM 99-7, California Institute of Technology, Pasadena, CA 91125, 2000.

[10] A.K. Yadav, A. Chowdhury, A. Srivastava, International Journal of Heat and Mass Transfer, 111 (2017) 871-883.

[11] R. Klimek, T. Wright, Spotlight-8 Image Analysis Software, NASA/TM-2006-214084, NASA Glenn Research Center, Cleveland , Ohio, 2004..

[12] K. Kobayasi, in: Symposium (International) on Combustion, Elsevier, 1955, pp. 141-148.

11th Asia-Pacific Conference on Combustion,
The University of Sydney, NSW Australia
10th -14th December 2017.

Verification of Beating Phenomenon by Measuring of Pilot Flame Behavior in a Dual Swirl Combustor

Jaehyeon Kim[1], Munseok Jang[1], Keeman Lee[2]

[1]Department of Aerospace Engineering, Sunchon National Univ.

255, Jungang-ro, Suncheon-si, Jeollanam-do, 57938, Republic of Korea

[2]School of Mechanical and Aerospace Engineering, Sunchon National Univ.

255, Jungang-ro, Suncheon-si, Jeollanam-do, 57938, Republic of Korea

Abstract

Lean-premixed combustion systems have the advantage of lower NOx emission than diffusion combustion systems by low flame temperature, but have a serious problems of combustion instability. Controlling combustion instability is very important to efficiency and durability of combustor. In previous our studies, researching on thermo-acoustic instability in dual swirl combustor, the frequency of sound pressure is divided into two regimes with thermal power. In regime of low thermal power, the frequency is nearly constant regardless of change of combustor length unlike with be well-known resonance frequency equation. In this regime, it is confirmed by using governing equation of beating phenomenon that beating phenomenon occurs. Also, it is confirmed through direct image of flames that the behavior of main and pilot flames are different each. However, the possibility that subjective opinion affect the process of analysis images exists.

This study is performed to verify that beating phenomenon occurs by the different behavior of main and pilot flames through measuring only the pilot flame behavior. As the result, when beating phenomenon occurs, they are different respectively. The additional study is necessary to determine the obvious causes of beating phenomenon, especially in the regime of low thermal power.

1 Introduction

Diffusion combustion systems have been used in gas turbine, but recently lean premixed combustion systems are developing to replace them.[1,2] In the case of lean premixed combustion systems, combustion reaction occurs at low equivalence ratio, which have the advantage of reducing thermal NOx because of a lower temperature distribution than diffusion combustion systems. However, lean premixed combustion systems also have many problems, especially the trouble of combustion stability. Controlling combustion instability is very important performance to efficiency and durability of combustor.

The typical trouble of unstable flames generated in combustion chamber is thermo-acoustic instability. When the frequencies of the sound pressure fluctuations and the heat release fluctuations are identical with a same phase, they interact and generate a resonance phenomenon. In this case, if (1) called Rayleigh integral is satisfied, thermos-acoustic instability becomes to occur.[3]

$$RI = \int_V \int_T P'(x,t) * Q'(x,t)dtdv \qquad (1)$$

$$\geq \int_V \int_T L_i(x,t)dtdv$$

Until now, most of the studies on thermo-acoustic instability show that pressure fluctuations and heat release fluctuations are equal within the error range.[4,5] Also, in previous our studies on a dual swirl combustor, it is checked that sound pressure fluctuations and heat release fluctuations are same at the condition with combustion instability, and detected beating phenomenon through the measurement of sound pressure with the various conditions. Also, it is assured that main and pilot flames behave with different periods by analyzing high-speed camera images. [6,7] However, the possibility that subjective opinion affect the process analyzing images is still existed. The object of this study is clearly verification of fact that beating phenomenon occurs due to different behaviors of main and pilot flames through measuring of pilot flame behavior.

2 Experimental set up and method

The premixed combustor used in this study consists of respectively axial swirler and tangential swirler at inner and outer. In this case, due to swirl number of the main burner can't represent the axial direction momentum, it is calculated by (3) based on paper of Syred. Also, swirl number of the pilot burner is obtained by (2). [7, 8]

$$S_P = \frac{3}{2}[\frac{1-\left(\frac{R_1}{R_2}\right)^3}{1-\left(\frac{R_1}{R_2}\right)^2}]\tan\theta \qquad (2)$$

$$S_M = \left[\pi\left(R_0^2 - R_p^2\right)\left(R_0^2 - \frac{t}{2}\right)\right]/[(8 * t * h) * R_0] \qquad (3)$$

Figure 1 shows the dual swirl combustor schematically and the supply flow channel of the fuel and oxidizer. The fuel and the oxidizer were used respectively methane, which represents LNG, and compressed air. As main parameters, in this study, thermal powers were fixed at 4&7 kW, and lengths of combustion chamber were employed 400 and 600 mm. The gas flow rate was precisely controlled by a mass flow meter calibrated with a bubble meter.

Corresponding author. Fax: +82-61-750-3820
E-mail address: kmlee@scnu.ac.kr

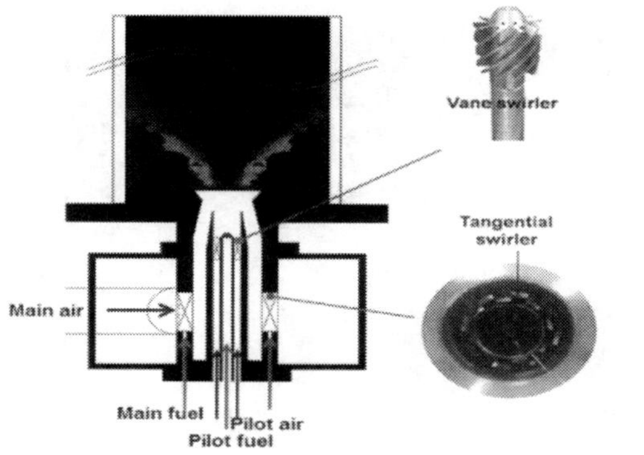

Figure 1: The configuration of dual swirl combustor schematically and the supply flow channel of fuel and oxidizer.

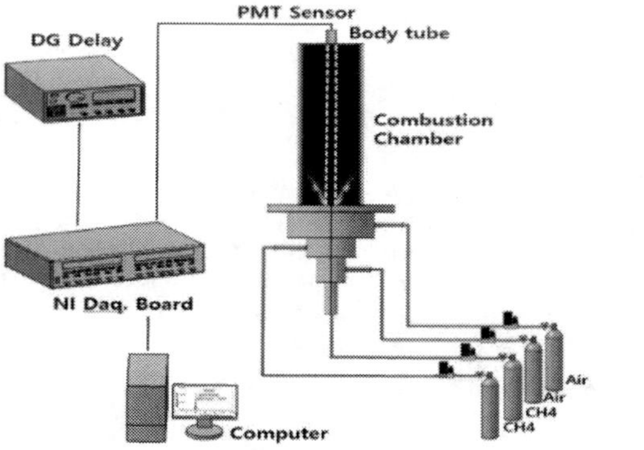

Figure 2: Experimental and measurement system.

Figure 2 turns up experimental and measurement system. For measuring the heat release fluctuations of the pilot flame, PMT sensor connected with optical fiber is placed in body tube located at the center of the combustion chamber downstream. So, the PMT sensor can only capture the signals of pilot flame behavior excluding the main flame. The equipment and position for measure of sound pressure fluctuations was explained in previous study detiledly.[7] It is possible that the frequency of main flame and pilot flame are separated by comparing the frequency of sound pressure fluctuation and heat release fluctuation. This help to judge that the measurement of heat release fluctuations of pilot flame used body tube is appropriate. Also, a trigger signal of a function generator (Agilent Co. 33220A) was used for signal synchronization. A data collector (Daq board. NI 9234 module) was used to acquisition and storage of data.

3 Results and discussion

Figure 3 represents the value of flame instability frequency with the various thermal powers and combustor lengths, which was acquired from the previous our study, and the resonance frequency equation is briefly shown in (4).

$$f_0 = \frac{(2n-1)c}{4L_c}, \mathrm{n} = 1,2,3, \ldots \quad (4)$$

In the (4), c is sound speed, which increases with rise of gas temperature, and L_c is the length of combustion chamber. Figure 3 tells that the frequency characteristics of flame instability can be divided in two regimes with thermal power. The frequency in Regime 1 of the relatively large thermal power

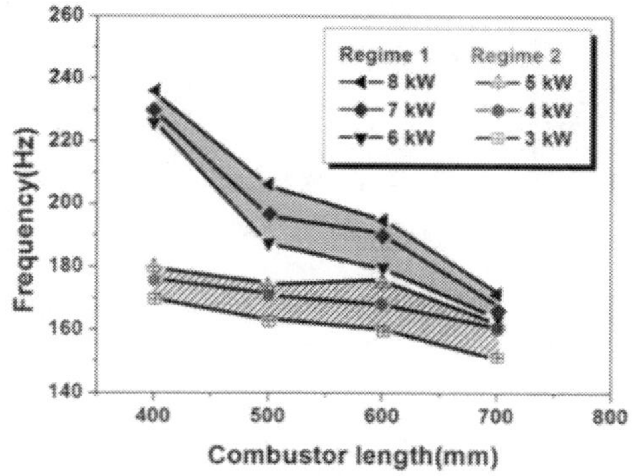

Figure 3: Experimental results of the frequency with the various thermal powers and combustor lengths.

is lower with longer the length of combustor and smaller thermal power.

This trend of instability frequencies corresponds with resonance frequency equation. The frequency in Regime 2 is higher with large of thermal power, however, is almost constant with longer the length of combustion chamber. In this case, it is guessed that beating phenomenon occurs in lower thermal power and longer the length of combustion chamber. This result is seemed to be a plausible through the insert of the frequencies acquired from experimental results in a governing equation of beating phenomenon.[7] In order to verify that main and pilot flame behave differently when the beating phenomenon occurs, the heat release frequency of pilot flame and the sound pressure frequency of flames are compared.

The result, which compared the heat release frequency of the pilot flame and the frequency of sound pressure, is shown in Fig. 4-7. The sound pressure frequency can consider as the behavior of main and pilot flame. When two signals with similar periods are synthesized, beating phenomenon occurs. Therefore, when the main and pilot flames behave simultaneously, beating phenomenon does not occurs. However, if the main and pilot flames behave similar but differently, beating phenomenon may occur.

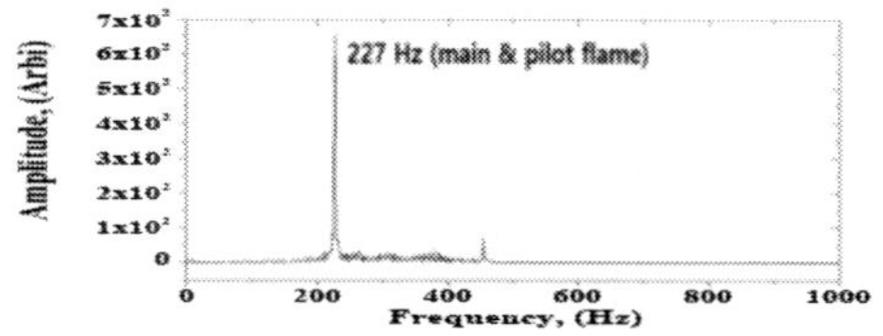

(a) FFT frequency of sound pressure.

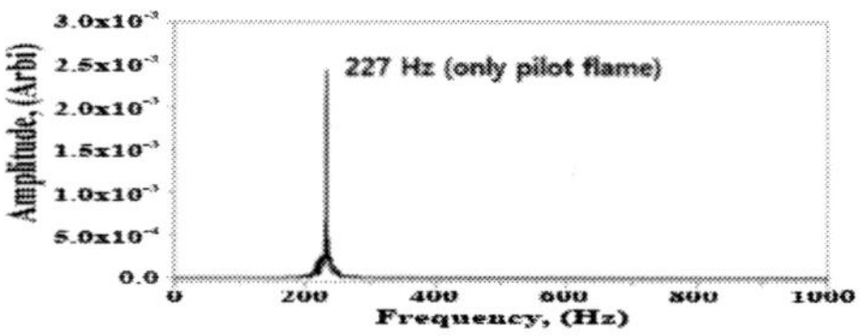

(b) FFT frequency of heat release.

Figure 4: FFT frequency of sound pressure and heat release. $(\dot{Q} = 7\ kW, L = 400\ mm)$

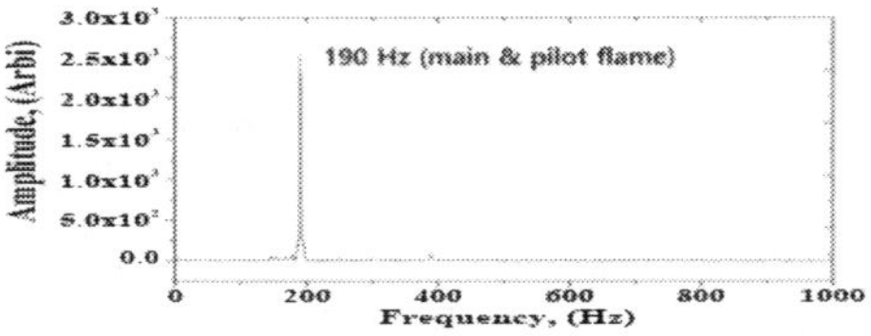

(a) FFT frequency of sound pressure

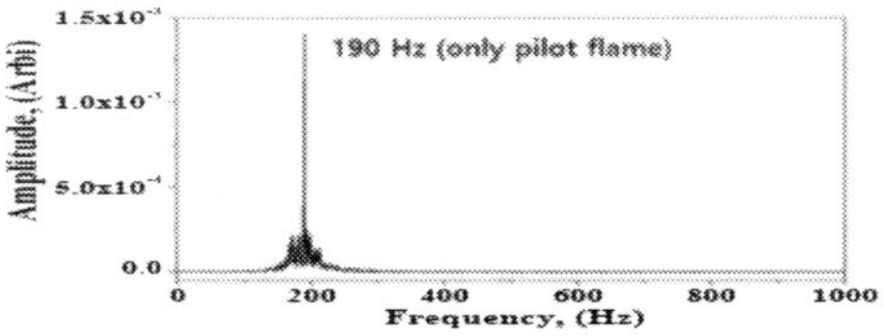

(b) FFT frequency of heat release

Figure 5: FFT frequency of sound pressure and heat release. ($\dot{Q} = 7\ kW, L = 600\ mm$).

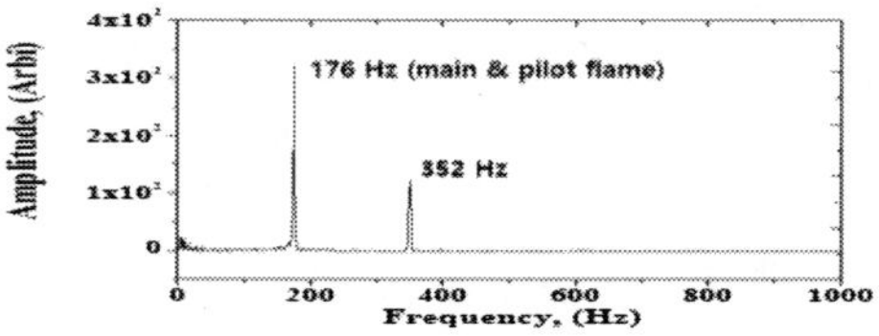

(a) FFT frequency of sound pressure

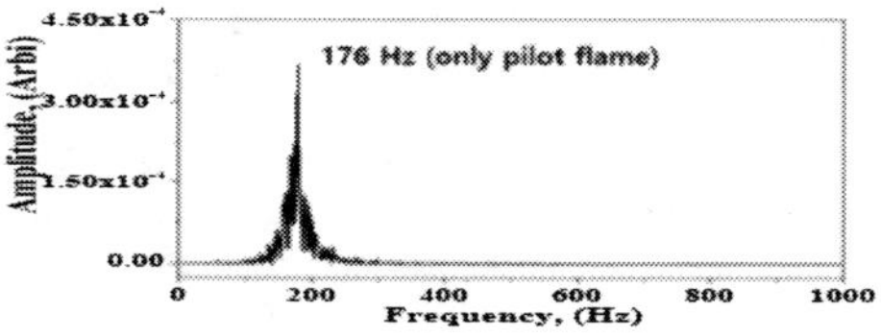

(b) FFT frequency of heat release

Figure 6: FFT frequency of sound pressure and heat release. ($\dot{Q} = 4\ kW, L = 400\ mm$)

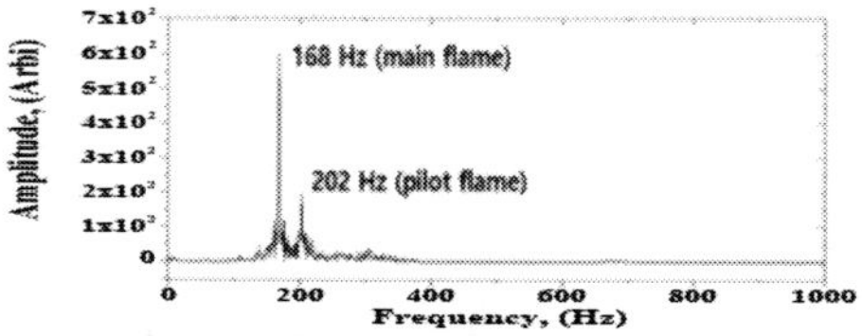

(a) FFT frequency of sound pressure

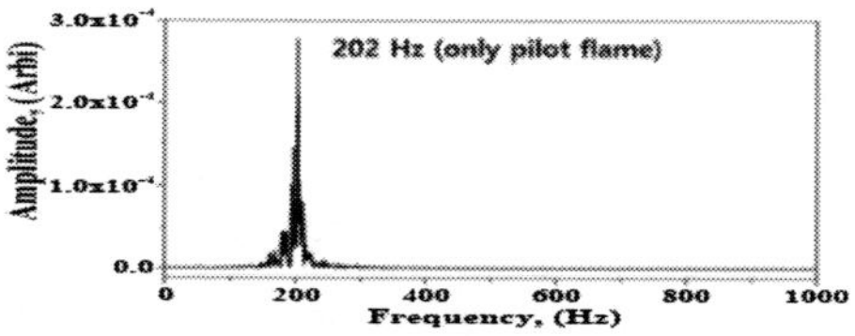

(b) FFT frequency of heat release

Figure 7: FFT frequency of sound pressure and heat release. ($\dot{Q} = 4\ kW, L = 600\ mm$)

In 7 kW as a typical thermal power of regime 1 with the length of combustion chamber of 400 mm, 600 mm, without beating phenomenon, it confirmed by Fig. 4-5 that the heat release frequency of pilot flame is same the frequency of sound pressure (i.e. main and pilot flames behave simultaneously). In case of 4 kW as a typical thermal power of regime 2, beating phenomenon not occurs in the length of combustion chamber of 400 mm and it is confirmed by Fig. 6 that the heat release frequency of pilot flame is same the frequency of sound pressure.

In case of Fig. 7, in the length of 600 mm with beating phenomenon occur, the heat release frequency of pilot flame is identical to the sound pressure frequency of pilot flame.

By measuring the behavior of the pilot flame, it is clearly considered that beating phenomenon occurs by behavior of main and pilot flames. Also parameters, which affect the different behavior of flames, are expected that the stratified combustion due to the difference of equivalence ratio of flames, and the resonant phenomenon with the alternative frequency change of main and pilot flames by combustor length and thermal power.

4 Conclusions

In previous study, in case of that the main frequency appears only one, the main flame and the pilot flame behavior same. But in case of that the main frequency and the second frequency occur, the main flame and the pilot flame behavior similar but differently. For reference, the second frequency shown in Jae is ignorable because is harmonic frequency as the value of integer multiple of the main frequency.

In this study, it is precisely verified by measuring the heat release fluctuations of pilot flame that beating phenomenon occurs due to different behavior of main and pilot flames in dual swirl combustor. Also, it is possible that the sound pressure frequency of main flame and pilot flame are separated by comparing the frequency of sound pressure fluctuation and heat release fluctuation. The causes of beating phenomenon are considered that the stratified combustion by the difference of equivalence ratio of main and pilot flames, and the resonant phenomenon by the frequency change of main and pilot flames with the combustor length and thermal power. The continuous study about the reason of beating phenomenon is necessary.

5 Acknowledgment

This research was supported by Basic Science Research Program through the National Research Foundation of Korea (NRF) funded by the Ministry of Education (NO.2015R1D1A1A01058610)

References

[1] S.K. Dhanuka, J.E. Temme, J.F. Driscoll, Lean-limit combustion instabilities of a lean premixed prevaporized gas turbine combustor. ScienceDirect of Proceedings of the Combustion Institute 33 (2011) 2961—2966.

[2] J.I. Jo, T.J. Park, C.H. Hwang, K.M. Lee, A Study on the Flame Shape and the Interaction between Pilot and Main Flames in a Dual Swirl Combustor. The Korean Society of Propulsion Engineers 18 (4) (2014) 33-42.

[3] L. Rayleigh, The explanation of certain acoustical phenomena. Notices of the proceedings meetings members of the Royal Institution of Great Britain 8 (1878) 536-542.

[4] Y. Huang, V. Yang, Dynamics and stability of lean-premixed

swirl-stabilized combustion. Progress in Energy and Combustion Science 35 (2009) 293-364.

[5] A.M. Steinberg, I. Boxx, M. Stohr, C.D. Carter, W. Meier, Flow-flame interactions causing acoustically coupled heat release fluctuations in a thermos-acoustically unstable gas turbine model combustor. Combustion and Flame 157 (2010) 2250-2266.

[6] I.C. Choi, K.M. Lee, M. Juddoo, A.R. Masri, A Study of Combustion Instability Mode in Dual Swirl Gas Turbine Combustor by PLIF and Chemiluminescence Measurement. The Korean Society of Combustion 19 (1) (2014) 29-38.

[7] M.S. Jang, K.M. Lee, Combustion Instability Characteristics due to the Beating Phenomenon in the Dual Swirl Gas Turbine Model Combustor. The Korean Society of Propulsion Engineers 20 (6) (2016) 61-69.

[8] N. Syred, M. Abdulsada, A. Griffiths, T. O'Doherty, P. Bowen, The effect of hydrogen containing fuel blends upon flashback in swirl burners. Applied Energy 89 (2012) 106-110.

11th Asia-Pacific Conference on Combustion,
The University of Sydney, NSW Australia
10th -14th December 2017.

Effect of reforming catalyst physicochemical properties on the reaction of hydrocarbon fuel

Xiaoxiong Zhang[1], Lingyun Hou[1], Dingrui Zhang[1]

Tsinghua University, Beijing, 100084, China

Abstract

Catalytic steam reforming of hydrocarbon fuel is a promising method for active cooling of scramjet. Catalyst is important for the successful application of this method. Three kinds of nickel/alumina (Ni/Al2O3) commercial catalysts doped with different contents of potassium, abbreviated as C1, C2 and C3 are compared via their reaction activities and effects on heat sink. C3 is demonstrated as the optimal catalyst with the best heat sink. Due to high specific surface area and large nickel atoms' molar fraction, gas yield of hydrogen in C3 is much more than those of C1 and C2 and contents of aromatics are much less. C3 owns the least average pore diameter, so large molecular poly-aromatics will not pass into the pore and cover the surface of active site which is beneficial for inhibiting the catalyst's deactivation. In view of the heat sink and reaction activity. C3 is the optimal catalyst for the application in the cooling channel.

1 Introduction

Due to the urgent demand for hypersonic flight, innovative propulsion should be presented. Till now, scramjet has been regarded as one of the most promising hypersonic aircraft [1,2]. However, plenty of aerodynamic heat will be generated during the hypersonic flight, which results in extremely high temperature on the wall of aircraft and even it will surpass the melting point of most materials. Therefore, active cooling of the wall by the fuel itself is necessary, which means the endothermic ability of fuel should be developed to a great extent [3,4]. The fuel's endothermic ability, i.e. heat sink, comes from its heating enthalpy and endothermic reaction under high temperature (>800K). When catalytic reforming reaction is introduced into the cooling channel, two main endothermic processes are thermal-cracking (equation (1)) and catalytic steam reforming reactions (equation (2)) of hydrocarbon fuel. Catalytic steam reforming reaction has more advantages over thermal-cracking reaction during the cooling, such as stronger endothermic ability, reactive products with more content of hydrogen which will be beneficial for subsequent combustion, and lower coke deposition on the wall of cooling channel, which can prolong the flight time [5-8].

$$C_nH_m \rightarrow aH_2 + bCH_4 + cC_2H_4 + dC_2H_6 + eC_3H_6 + \cdots \quad (1)$$

$$C_nH_m + nH_2O \xrightarrow{catalyst} nCO + \left(n + \frac{m}{2}\right)H_2 \quad (2)$$

As a result, attempts to obtain high levels of hydrogen production and a strong endothermic capacity by employing catalytic reforming of hydrocarbon fuels has recently attracted significant attention. Steam reforming of hydrocarbons are widely used in many fields, such as industrial apparatus of hydrogen generation [9], transportation and solid oxide fuel cell (SOFC) [10-12]. Catalyst are important for the steam reforming reaction of hydrocarbon. Nickel/Alumina are widely used in the process of industrial hydrogen generation [13], but many disadvantages such as coke deposition will restrict Ni/Alumina catalyst's further application in SOFC. To reduce the rate of coke deposition on the surface of Ni-based catalyst, some new methods are applied in the catalyst preparation. Tribalis et al. [12] improved the behavior of Ni/Alumina catalyst by modifying the support (alumina) doped with La-, Ba-, (La+Ba)- and Ce elements, then the coking rate was suppressed in the steam reforming of diesel. Tuan et al. [13] promoted the catalyst by adding Cu into Ni, and then the activity of Ni-Cu/Al2O3 catalyst for gasoline steam reforming was upgraded significantly. Although coke formation was inhibited greatly, the activity for steam reforming was reduced.

In order to obtain the complete conversion of hydrocarbons, all the catalysts for steam reforming of hydrocarbon fuel are chiefly used under the relatively high molar ratio of steam to carbon (S/C ratio) condition. Korabelnikov et al. [14] focused on the catalytic conversion and output characteristics of methane, heptane and decane at the S/C ratios of 2 and 4. But it was shown that under the 0.1 S/C ratio, the reaction temperature could not be higher than 500℃ because of large amounts of coke deposited on the catalyst surface and clogged the reactor. Hou et al. [6] has confirmed that a relatively low content of steam (S/C≈0.1) added into the feedstock could aggravate the total heat sink to a large extent. Hence, application of catalytic steam reforming of hydrocarbons as a cooling method is considerable different from that as a generation method of hydrogen in the industrial process which steam ratio are much higher (molar ratio S/C≈2~4)[9].

To apply the steam forming of kerosene as an available method for active thermal protection of engine successfully, the content of water is limited by the energy of products in the combustor. The original catalyst is appropriate for the condition of high S/C ratio. Some reactivity challenges of catalyst was presented at the low S/C ratio. This paper aimed at the choice of catalysts for steam reforming of kerosene in a micro-channel reactor at a low S/C ratio. A new method to evaluate the quality of catalysts by their effects on the heat sink and reaction activity during cooling process was proposed.

2 Experimental setup and catalytic coating

2.1 Experimental setup

The experiments were all conducted in the apparatus demonstrated in figure 5. The detailed description of the an be seen in Ref. [7]. Cycloalkanes and paraffins were the main components of the kerosene and constituted up to 99.5 wt%. Water and kerosene were added as a certain mass ratio, together

Lingyun Hou. Fax: +86-010-627-72157
E-mail address: lyhou@mail.tsinghua.edu.cn

with 0.3 wt% emulsifier, to form an emulsion fuel using a gear pump.

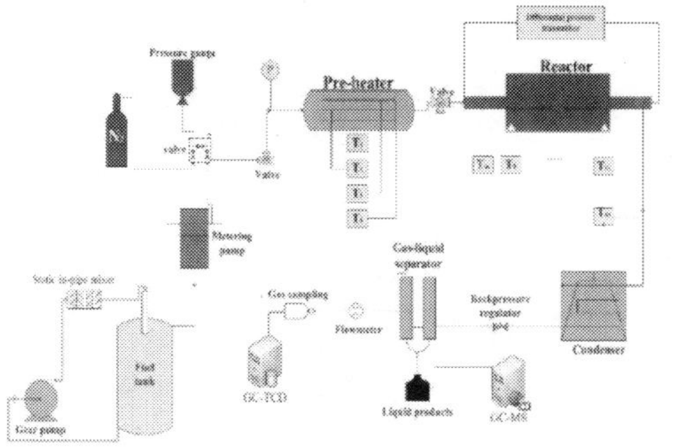

Figure 1: Experimental setup

Contents of gas production were detected by Gas Chromatography-Thermal Conductivity Detector (Agilent 7820A-GC, Agilent Technology, USA). H2, CO, CO2, CH4, C2H4, C2H6, C3H6, 1-C4H8 and 1-C4H10 were detected as the main gas compounds. These gases were discriminated by comparing their residence time with standard gas. The content of gas component i in gas products, ω_i(vol%), is determined as follows,

$$\omega_i = \frac{A_i}{A_{i,s}} \times \omega_{i,s} \quad (3)$$

Where A_i is peak area of gas compositon i in gas products; $\omega_{i,s}$ is content of gas composition i in standard gases, vol%; $A_{i,s}$ is peak area of gas compositon i in standard gases.

From ω_i, the gas yield of each non-condensable gas, Y_i (L/kg), is determined as follows,

$$Y_i = (\omega_i \times V_{gas}/m_f) \times (1000/60) \quad (4)$$

Where V_{gas} is the volume flow rate of non-condensable gases, L/min, which is measured by a flowmeter as shown in figure 1; m_f is the mass flow rate of the emulsified fuel, which is controlled as 1g/s.

The Gas Chromatography-Mass Spectrometery (GC-MS) system was used for the quantitative analysis of liquid products. Identification of each compound was based on retention times and matching the mass spectrum recorded with those in the spectral library (NIST-11). Peak area ratio of each compound reflect its mass ratio to some extent.

As referred in the introduction, heat sink (ΔH_{sink}, MJ/kg) plays an important role in the process of cooling process. It could be determined as follows,

$$\Delta H_{sink} = \frac{P_e - P_{loss}}{Q_m} \quad (5)$$

Where P_e is the electric power, kW; P_{loss} is the heat loss of the heating system as shown in figure 1, kW; Q_m is the mass flow rate of emulsified fuels, g/s.

2.2 Properties of catalysts

Three kinds of commercial catalysts which are provided by Qilu Petrochemical Corporation (a company of Sinopec). The three catalysts are abbreviated as C1; C2; C3. Their properties will be shown as follows:

Atoms composition of the catalyst surface was detected by a X-ray Photoelectron Spectroscopy (XPS) (Thermo Fisher/ ESCALAB 250Xi, USA). XPS operated at a constant pass energy of 30 eV with AlKα radiation (1486.6 eV). Table 1 summarizes the surface element components from the quantitative analysis of the XPS data.

Table 1: Element surface components of three catalysts

Catalyst	Average surface composition (atoms%)					
	Ni	Al	O	C	Fe	K
C1	4.68	21.74	47.15	20.9	2.1	3.39
C2	2.71	19.2	51.1	25.68	1.28	-
C3	5.97	30.29	45.99	15.03	2.29	0.39

The specific surface area of the catalysts was determined via a specific surface analyzer (QuadraSorb SI 5.06, Quantachrome, USA). Total pore volume and average pore size were measured using the Barrett-Joyner-Halenda (BJH) method and the nitrogen desorption curve. The specific surface area, total pore volume and average pore diameter of the three different catalysts are shown in table 2. It is found that C3 catalyst has the largest specific area (76.84 m^2/g) and pore volume (6.55E-02 cm^3/g), while its pore diameter is the minimum (1.46nm).

Table 2. Specific surface area and pore structure of three catalysts

Catalysts	S_{BET}	Total Pore Volume	Average pore diameter
	m^2/g	cm^3/g	nm
C1	22.46	5.99E-02	12.99
C2	15.77	6.17E-02	3.8
C3	76.84	6.55E-02	1.46

3 Results and discussions

3.1 Heat sink

To compare the effects of different catalysts on the heat sink of fuel. The heat sink at various outlet temperature of the three catalysts are measured in the experimental setup shown in figure 1, and the results are shown in figure 2. All of the experiments are conducted under the same condition, i.e. mass flow rate of 1g/s, water content of 10 wt% (S/C molar ratio is about 0.087), and reaction pressure of 3MPa.

The heat sink under both of the three catalysts condition increase with the increasing temperature. Before 600 ℃, compare the heat sink and its increasing rate under three catalysts condition, it can be found that under the same temperature, C3 owns the maximum heat sink and increasing rate, and heat sink and increasing rate under C1 catalyst condition is higher than C2 catalyst condition. This may be due to the higher catalytic steam reforming reaction activity for C3 catalyst than the other two catalysts under relatively low temperature. When the temperature surpass 600℃, increasing rate of the heat sink under three catalysts are both higher than that in the lower temperature condition less than 600℃. That is because the reactivity of the three catalysts are both upgraded due to the high temperature.

When the temperature comes up to 720℃, in the presence of C1 and C2 catalyst, the cooling channel is clogged by coke deposits on the inner wall surface very soon. While under C3 catalyst condition, the fluids could maintain its flowing as relatively stable, and its heat sink reaches as high as 5.01 MJ/kg.

More potassium (K) content seems well for metal-catalyzed coke inhibition, but has little suppression effects on condensation polymerization coke originated from unsaturated hydrocarbons. Therefore, C1 owns much more potassium content (3.39 mol%) than C3 (0.39%), but anti-coking capability of C1 is weaker than C3. This can be due to the coke inhibition mechanism of potash, K has more inhibition effects on metal-catalyzed coke, C3 has better suppression of condensation polymerization coke than C1.

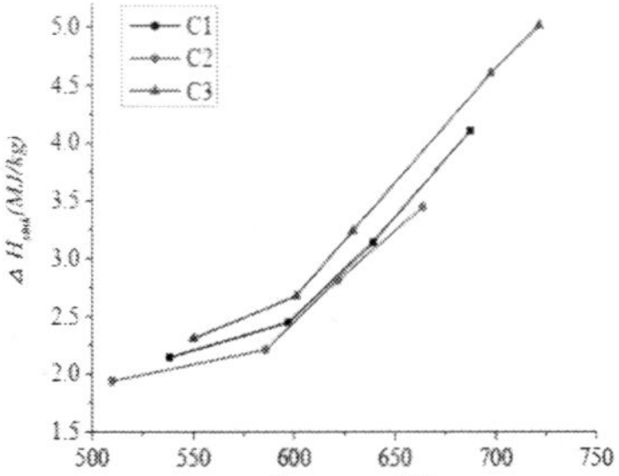

Figure 2: Heat absorption capacity versus temperature

3.2 Catalytic reforming reaction

Figure 3 demonstrates the gas yield of different gases in the gaseous products. Because of the highest specific surface area and Ni atoms content, C3 has the optimal activity to catalyze the steam reforming of hydrocarbons. At the same time, Fe atoms content (2.29 mol%) of C3 is the largest. Although Fe reduces the catalytic activity partially, it is beneficial for the hydrogen product selectivity. Hence, under C3 catalyst condition, the yield of hydrogen and carbon oxides are the largest among the three catalysts at various temperatures. From figure 3, under the relatively low temperature (less than 665℃), it can be found that no obvious differences of the gas yield of alkanes and olefins from the thermal cracking reaction among the three catalysts.

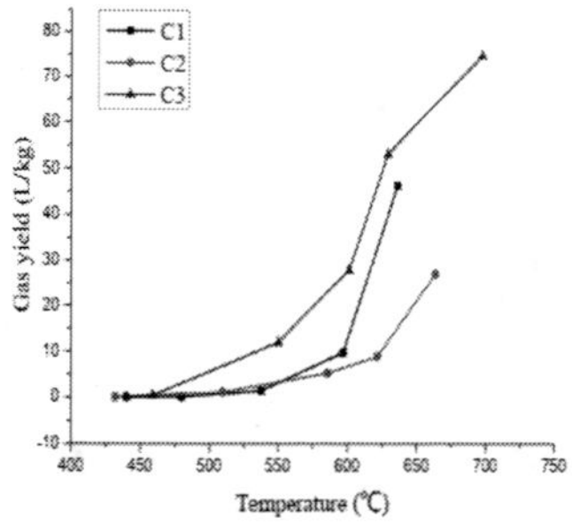

(a) H2

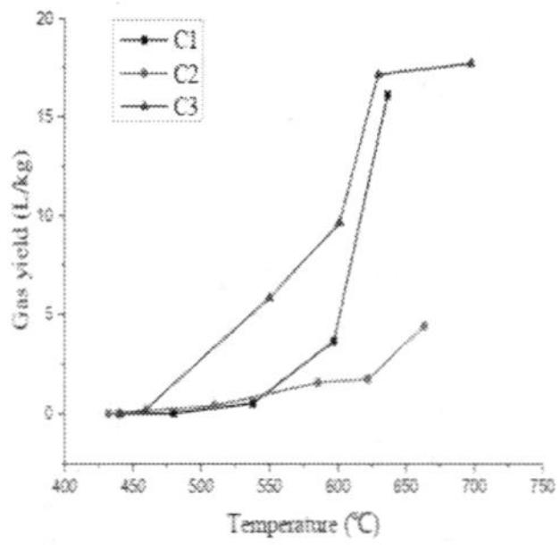

(b) CO+CO2

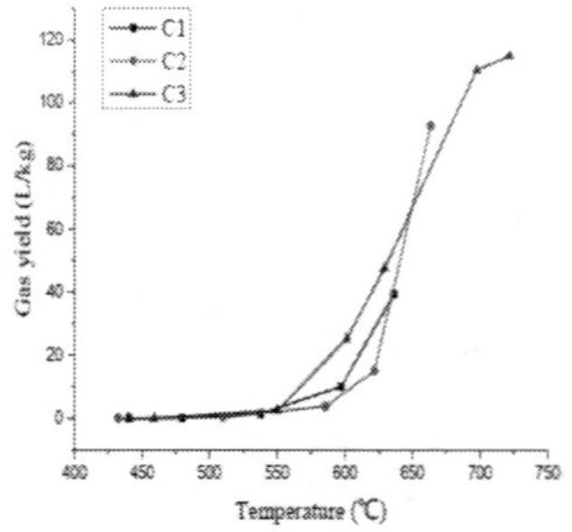

(c) Alkanes

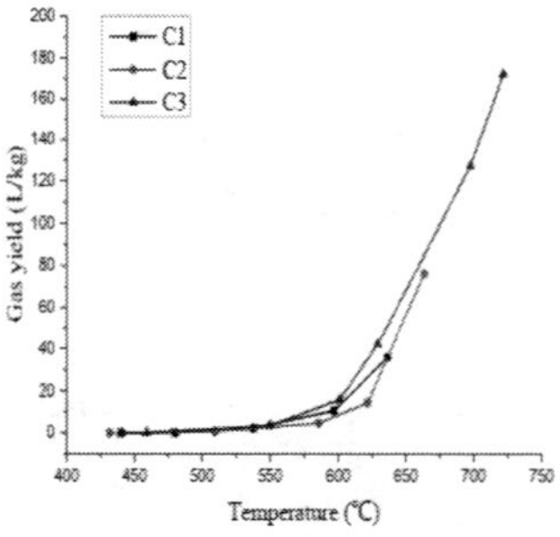

(d) Olefins

Figure 3: Gas yields with different gas products

Figure 4 demonstrates the contents of different hydrocarbons in the liquid product. With the increase of temperature, the contents of alkanes decrease gradually under any catalyst condition; contents of olefins increase for C3 and C2 catalysts, while decrease for C1 catalyst, and aromatics increase under any catalyst conditions. When the outlet temperature below 600℃, contents of aromatics under C1 and C2 conditions are both lower than 5 wt%, while there is no aromatics identified under C3 condition. As the temperature above 600℃, contents of aromatics increase to 23.28 wt% at 636℃ for C1 catalyst; contents of aromatics increase to 65.21 wt% at 626℃ and 72.34 wt% at 679℃ for C2 catalyst. But under C3 catalyst condition, the contents of aromatics is still below 1wt%. Until the temperature excess 700℃ (up to 722℃), the contents of aromatics only reach 5.17 wt%.

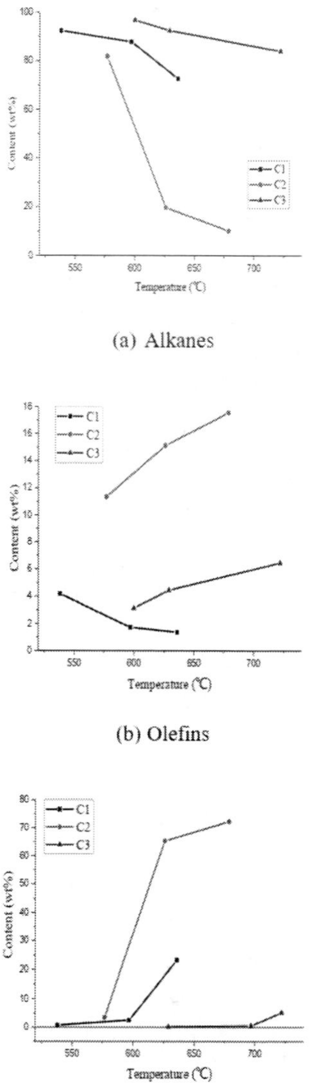

(a) Alkanes

(b) Olefins

(c) Aromatics

Figure 4. Contents of liquid products.

From table 1 and table 2, C3 catalyst has the largest specific surface area (3~5 times larger than C1 and C2) and the maximum molar ratio of active Ni (5.97 mol%), hence, catalytic activity of C3 is the best among the three catalysts. At the same time, both of oxygen and carbon molar ratio are the least in C3 catalyst, hence, less hydrogen or CO will be consumed for the reduction reaction of catalyst. While C2 has the most oxygen content, which exists as nickel oxide. A part of hydrogen is consumed to restore nickel oxide of the catalyst, therefore, gas yield of hydrogen is the least among the three catalysts.

4 Conclusions

Three kinds of catalysts, i.e. C1, C2, and C3, are compared via their reaction activities and effects on heat sink of the cooling system. Some conclusions should be presented as follows:

1) The heat sink of C3 was stronger than C1 and C2, which due to its high reactivity for catalytic steam reforming;

2) C3 catalyst had higher specific area and larger molar fraction of Ni than C1 and C2, which lead to a better reaction activity for steam reforming of kerosene. There were more hydrogen production in the gas products and much less aromatics in the liquid products because hydrogen in the gaseous products could suppress the condensation polymerization of unsaturated hydrocarbons;

3) Due to the small average pore diameter of C3, large molecular poly-aromatics would not enter into the pore, so the catalyst would not lose its activity because of active site covered by coke deposits.

5 Acknowledgment

The financial assistances provided by the National Natural Science Foundation of China (91641114) were gratefully acknowledged.

References

[1] J. M. Hank, J. S. Murphy, R. C. Mutzman, The X-51A Scramjet Engine Flight Demonstration Program, AIAA, Dayton, Ohio, USA, 2008, p. 2540.

[2] B. Palaszewski, L. S. Ianovski, P. Carrick, J. Propul. Power 14 (5) (2012) 641-648.

[3] J. D. Anderson, Hypersonic and high temperature gas dynamics, AIAA, Reston, VA, USA, 2006, p.711.

[4] H. Huang, D. R. Sobel, L. J. Spadaccini, Endothermic Heat-Sink of Hydrocarbon fuels for Scramjet Cooling, AIAA: Indiana, USA, 2002, p. 3871.

[5] H. Huang, L. J. Spadaccini, D. R. Sobel, J. Eng. Gas. Turb. Power. 126 (2) (2002) 284-293.

[6] L. Y. Hou, N. Dong, Z. Y. Ren, B. Zhang, S. L. Hu, Fuel Proc. Technol 128 (128) (2014) 128-133.

[7] L. Y. Hou, X. X. Zhang, Z. Y. Ren, Fuel Proc. Technol 154 (2016) 117-122.

[8] L. Y. Hou, N. Dong, D. P. Sun, Fuel 103 (1) (2013) 1132-1137.

[9] G. Bridger, W. Wyrwas, Chem. Eng. Process 48 (9) (1967) 101.

[10] S. Paloumbis, ACS Symposium Series, (1982) 489-497.

[11] M. Krupelt, T. R. Krause, J. D. Carter, J. P. Kopasz, S. Ahmed, Catal. Today 77 (1) (2002) 3-16.

[12] A. Tribalis, G. Panagiotou, K. Bourikas, L. Sygellou, S. Kennou, S. Ladas, A. Lycourghiotis, C. Kordulis, Catalysts 6 (1) (2016) 11.

[13] L. Tuan, N. Luong, K. Ishihara, Catalysts 6 (3) (2016) 45.

[14] A. V. Korabelnikov, A. L. Kuranov, Hypersonic Flight Vehicle Heat Protection using Chemical Heat Regeneration, AIAA, Reno, NV, USA, 2002, p. 913.

11th Asia-Pacific Conference on Combustion,

The University of Sydney, NSW Australia
10th -14th December 2017.

Experimental Determination of Ignition Delay for Hypergolic Bipropellants

Mahesh Dalwani[1], Umakant Swami[1], Arindrajit Chowdhury[1],

[1]Mechanical Engineering Department,

IIT Bombay, Mumbai, Maharashtra, 400076, India

Abstract

An experimental investigation of hypergolicity and ignition delays of hydrazine, unsymmetrical dimethyl hydrazine (UDMH), tetramethylethylenediamine (TMEDA), furfuryl alcohol, triethylamine (TEA), and 2,6-xylidine with white fuming nitric acid (WFNA) and red fuming nitric acid (RFNA) is presented. Results of ignition delays measured using drop test coupled with a high-speed camera and optoelectronic circuit are shown. Physical and chemical ignition delays of all fuels with RFNA are shown. All the fuels were found to be hypergolic in nature with WFNA and RFNA except for furfuryl alcohol, which displayed hypergolicity only with WFNA. Global ignition delays of all the fuels except for TEA were found to be less with RFNA as compared to WFNA. All the fuels showed global ignition delays less than 50 ms, which is the upper limit to classify the hypergols as useful. Amongst the compounds tested, TMEDA shows the lowest ignition delay with WFNA and RFNA and is seen as the best candidate to replace hydrazine based hypergolic compounds

1 Introduction

Hypergolic bipropellants are typically fuel-oxidizer pairs that ignite spontaneously when brought in contact with each other, rendering other ignition systems redundant, thus simplifying engine design and operation [1]. Ignition delay is defined as the time elapsed between propellants first contact and ignition. High ignition delays results in accumulation of fuel inside the combustion chamber which can lead to an explosion upon ignition known as a 'Hard-start' [2]. The upper limit set for ignition delay is 50 milliseconds to avoid 'Hard-start' [3]. The drop test is an apparatus in which a single drop of fuel is dropped onto an oxidizer pool in order to study the hypergolicity of propellants. The ignition delay values measured by a drop test setup are usually larger than a spray test because of poor mixing. Total ignition delays can be separated into physical and chemical ignition delays. The physical delay time is the time required for mixing of propellants and the chemical ignition delay quantifies the time period required for occurrence of ignition kernel after mixing has achieved [3].

Commonly used oxidizers for hypergolic applications are nitrogen tetroxide (N_2O_4/NTO), hydrogen peroxide (H_2O_2), red fuming nitric acid (RFNA), and white fuming nitric acid (WFNA). WFNA contains 98% of nitric acid and 1-2 % water, RFNA contains 5-20% of nitrogen tetroxide dissolved in 80-95% nitric acid [4]. Historically hydrazine (N_2H_4) and its derivatives like monomethylhydrazine (MMH), unsymmetrical-hydrazine (UDMH) and Aerozine50 (50% N_2H_4+50% UDMH), UH25 (25% N_2H_4+75% UDMH) have been widely used as hypergolic fuels in the thrusters of air-crafts like Titan2, Apollo, and space shuttle orbital maneuvering systems [5]. Hypergolic propellant can be stored for long durations because of their high boiling points. However, hydrazine and its derivatives are toxic in nature, have high vapor pressure, and are carcinogenic in nature [6].Therefore, the research community is on a constant lookout for hypergolic fuels having less toxic, optimum ignition delay, and comparable specific impulse to hydrazine based hypergolic fuels.

The current work explores some of the fuels which have gained considerable attention in the recent past as replacement for hydrazine based propellants. Furfuryl alcohol is hypergolic in nature in combination with nitric acid, also it is non-toxic [5]. N,N,N',N'-tetramethylenediamine (TMEDA) is also a promising candidate as it is less toxic than MMH and has shown hypergolicity with WFNA and RFNA [7]. Trimethylamine and 2,6-xylidene are alkyl amine based compounds which have also shown hypergolicity and are less harmful as compared to hydrazine based compounds.

2 Experimental Description

A drop test setup was developed to study the physical phenomena as well as chemical reactions between various pairs of hypergolic fuels and oxidizers. A schematic of the drop test setup consisting of a transparent cuvette, a syringe, a non-captive linear stepper motor, and a high-speed camera connected to the computer is shown in Fig. 1. The experiments were conducted under ambient conditions. A precise amount of the oxidizer was placed at the bottom of a 10×10 mm^2, 45 mm tall glass cuvette with the help of a micropipette (Finn-pipette F2). Approximately 20 to 25 µl of the fuel was loaded in a 50 microliter syringe (Hamilton 750RN).

The syringe was inserted between the stainless steel retaining plates. The location of the tip of the needle of the syringe was changed by fixing the retaining plates at different locations on the vertical guiding rods. A transparent acrylic casing was used to encompass the whole setup to prevent exposure to harmful gases evolved during the experiment. The fuel droplet was dispatched from a height of approximately 63 mm from the surface of the oxidizer pool. A high-speed camera (Photron FASTCAM SA5) was used to capture the combustion event. A Nikon lens (AF-S Nikkor, 18-55 mm), fitted with 4 mm and 2 mm close up lens, was used to focus on the surface of oxidizer pool. A rack and pinion mechanism, operated by a non-captive linear stepper motor was used to actuate the piston of the microliter syringe and control the rate of formation and descent of the fuel droplet. The drop of fuel released from the syringe fell freely into the pool of the oxidizer. The pre-ignition and post-ignition events were captured by the high-speed camera at

Corresponding author. Fax: +91-22-25726875

E-mail address: arindra@iitb.ac.in

a frame rate of 4000 fps. An LED light source (30W) was used as backlight for the experiments. The captured video was converted into individual frames using Photron FASTCAM Viewer. The ignition delay was measured as the time taken from the frame in which the fuel droplet impacted the oxidizer pool to the frame in which an ignition kernel was observed.

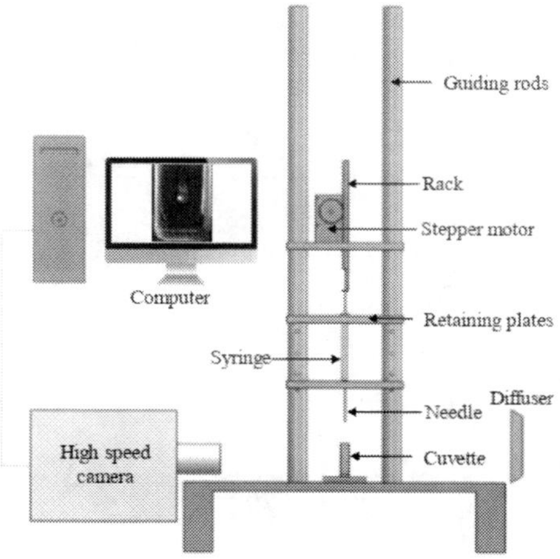

Figure 1: Schematic of the drop test setup

The schematic of the drop test setup with the optoelectronic circuit as shown in Fig. 2 (b) has almost the same configuration as that with the high-speed camera. A He-Ne laser (Holmarc model DL-R-5, wavelength 650 nm), 2 photodiodes (Photodiode 1, model BPW24R, and Photodiode 2, model TSAL6100), and an optoelectronic circuit were used in place of the high-speed camera. A 5 V regulated DC power supply (Aplab LD 3205) was used to power the optoelectronic circuit. A laser beam of 1 mm diameter passed through the cuvette and illuminated the first photodiode initially. The laser beam incident on the first photodiode first detects the droplet and subsequently the gases emerging from the liquid-droplet interaction. Generally, the voltages detected by the photodiode were found to drop suddenly when the path of the laser was obstructed by the droplet. A second photodiode was placed at the upper edge of the cuvette to detect ignition and the corresponding flame.

Figure 2 (a) shows the positive terminal of the first photodiode which was connected to a resistor of 10 kΩ. The negative terminal of same was connected to the positive source of a regulated DC power supply. An RC filter was used to eliminate the electrical noise generated by the linear stepper motor, which clouded the signal detected by the optoelectronic circuit. The values of resistance (150 Ω) and capacitance (0.1 μF) were fixed based on the desired cutoff frequency. The output was measured across the capacitor C1 and C2 corresponding to the first and the second photodiode respectively. The data acquisition system Easy Logger DS1M12 was used to record the data at a rate of 10^5 Hz.

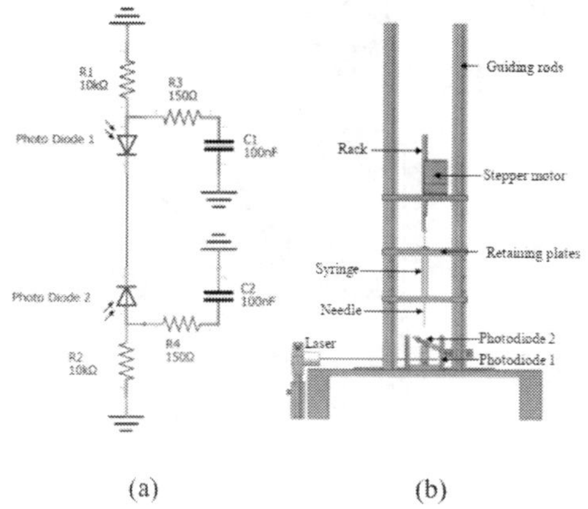

Figure 1: Schematic of drop test setup with optoelectronic circuit

3 Results and discussions

The ignition delay of different hypergolic fuels measured with the use of a drop test setup, coupled with a high-speed camera (HSC) is shown in Table 1. White fuming nitric acid ((WFNA), $HNO_3 \geq 99\%$, $H_2O \sim 0.5\text{-}1\%$) and red fuming nitric acid ((RFNA), $HNO_3 \sim 73.1\%$, $NO_2 \sim 17.5\text{-}22.5\%$, H_3PO_4 ~1-1.3%, HF~ 0.5-0.7%, $H_2O \sim 3.5$ wt %), were the oxidizer used for these tests.

Compounds	Chemical formula	Ignition delay (ms)	
		WFNA	RFNA
TMEDA	$C_6H_{16}N_2$	10 ± 0.5	8 ± 1.2
Furfuryl alcohol	$C_5H_6O_2$	19.8 ± 0.88	No ignition
TEA	$C_6H_{15}N$	17 ± 0.5	27 ± 2
2,6-xylidine	$C_8H_{11}N$	48 ± 1	21 ± 4
Hydrazine	N_2H_4	39 ± 10	13 ± 0.7
UDMH	$C_2H_8N_2$	1.63 ± 0.18	2 ± 0.43

Table 1. Hypergolic fuel tested

3.1 Reactions with white fuming nitric acid (WFNA)

Selected frames captured during the reaction of a furfuryl alcohol droplet, impacting on a 100 μl oxidizer pool is shown in Fig. 3. The onset of ignition delay is marked by the time instant at which the fuel droplet touches the oxidizer pool, as shown in Fig. 3 (a). Formation of soot from chemical reactions was observed in Fig. 3 (b). The first occurrence of an ignition kernel was observed in the upper portion of the cuvette at t=19 ms, as shown in Fig. 3 (c). A luminous flame was established after t=24 ms and persisted throughout the combustion event. A sample image representative of the same is shown in Fig. 3 (d). The flame extinguished at t =150 ms.

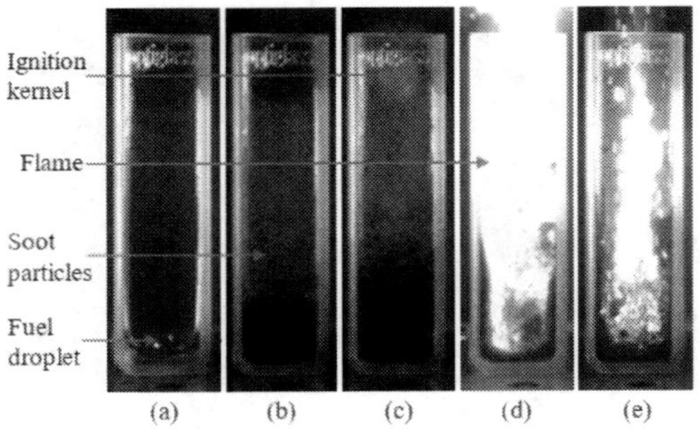

Figure 3: Selected frames for Furfuryl Alcohol into 100 µl of WFNA pool, (a) t=0 ms, (b) t=13.5 ms, (c) t=20.5 ms, (d) t=32.5 ms (e) t=136 ms

Figure 4 shows selected frames from the reaction of a 2,6-xylidine fuel droplet with the oxidizer pool. The impact of droplet on the pool surface, formation of gaseous cloud, occurrence of an ignition kernel, and a luminous flame are shown in Figs. 4 (a), (b), (c), and (d) respectively. The soot particles generated during the reaction were found to adhere to the side walls, obscuring visibility. The time period for the combustion event was found to be 77 ms.

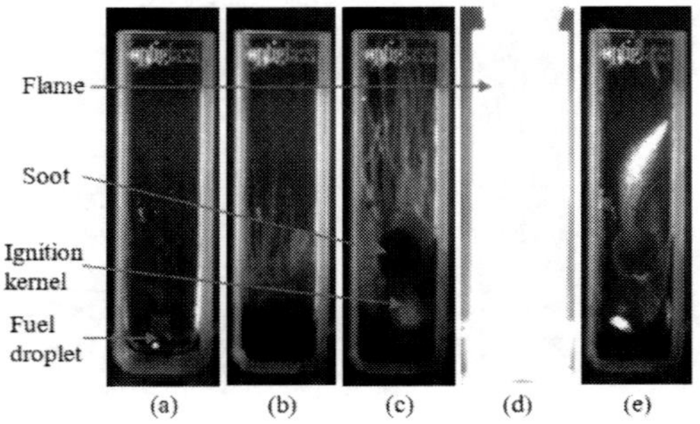

Figure 4: Selected frames for 2,6-xylidene into 100 µl of WFNA pool, (a) t= 0 ms, (b) t=13.75 ms, (c) t=48 ms (d) t=52 ms (e) t=76 ms

Similarly, combustion events of all fuels with WFNA were studied.

3.2 Reactions with red fuming nitric acid (RFNA)

Important combustion events resulting from the impact of a hydrazine fuel droplet on an oxidizer pool has been captured in Fig. 5. The fuel droplet rebounds from the oxidizer pool upon impact, reaches a certain height and then falls back to the pool surface, where it stabilizes and undergoes complete combustion. A flame front was found to be established almost instantaneously upon impact. Buoyant forces resulting from the hot combustion gases exert an upward force on the droplet and is thought to be responsible for lifting the fuel droplet from the pool surface. The fuel droplet at the end explode at time t=1.06 ms, shown in Fig. 5 (e). The hydrazine fuel droplet combustion period larger compared to other compound.

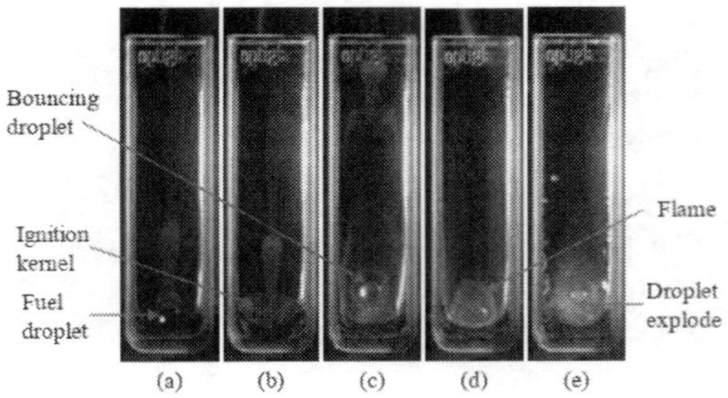

Figure 5: Selected frames for Hydrazine into 200 µl of RFNA pool, (a) t= 0 ms, (b) t=13 ms, (c) t=49 ms (d) t=363 ms (e) t=1.06 s

Figure 6 shows the selected frames of TEA droplet impacting on a 200 µl of an oxidizer pool. Gases were found to accumulate in the cuvette even before the droplet was released from the syringe shown in Fig. 6 (a). The flame was observed to be relatively non-sooty in nature. The flame extinguished at t =164 ms. Similar experiments for various fuels were performed with RFNA as an oxidizer.

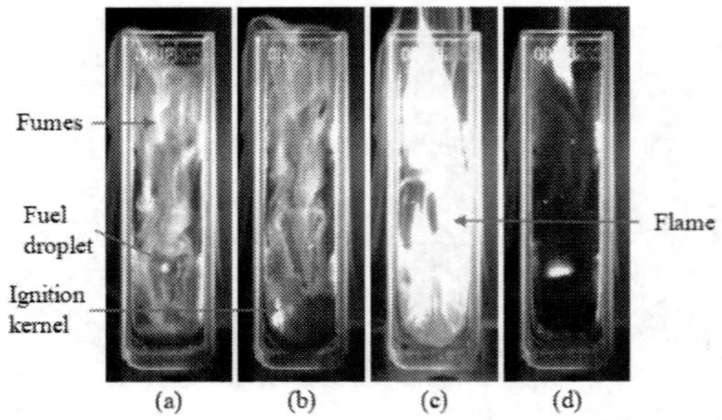

Figure 6: Selected frames for Triethylamine into 200 µl of RFNA pool, (a) t= -9 ms, (b) t=27 ms, (c) t=37.5 ms (d) t=156 ms

3.3 Ignition delay measurement using an optoelectronics circuit (OC)

Table 2 shows the ignition delay measured by an optoelectronic circuit for TMEDA, TEA, 2,6-xylidine, hydrazine, and UDMH with RFNA as the oxidizer. The ignition delays measured by the optoelectronic circuit were comparable to the results obtained using a high-speed camera as shown in Fig.7.

Fuel	Physical ignition delay (ms)	Chemical ignition delay (ms)	Global ignition delay (ms)
TMEDA	6.77 ± 0.33	1.22 ± 0.26	8.24 ± 0.72
TEA	25.93 ± 2.53	1.10 ± 0.03	26.49 ± 2.0
2,6-Xylidine	-	-	22.22 ± 2.5
Hydrazine	18.81	1.76	16.35 ± 3.7
UDMH	2.54 ± 0.18	0.75 ± 0.03	3.29 ± 0.16

Table 2: Ignition delays measured by the OC

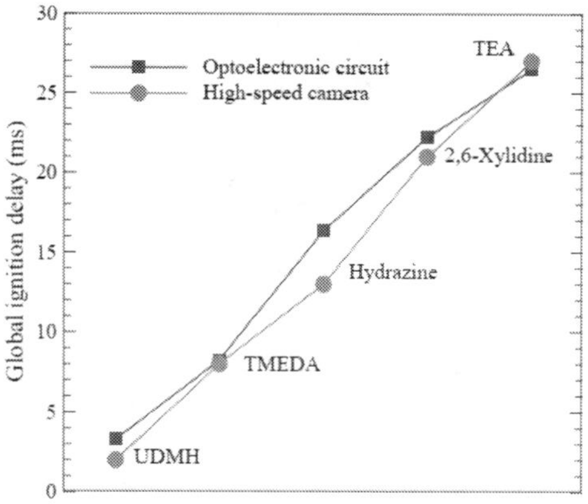

Figure 7: Global ignition delay by High-speed camera and optoelectronic circuit

Ignition delay for TMEDA with RFNA as an oxidizer, measured using an optoelectronic circuit is shown in Fig. 8. A laser beam of 1 mm diameter was passed through a height of 5.5 mm above the oxidizer pool. The photodiode 1 detects a dip in voltage signal after point A due to the obstruction of the laser beam by the falling droplet. The point B, at which the droplet touches the pool, marks the onset of ignition delay. The impact velocity of fuel droplet was calculated to be 1.07 m/s at point A. The time required for the droplet to travel from point A to point B was estimated as 5.01 ms. Point C, at which voltage falls again because the vapor coming from the pool which blocks the laser beam represents the start of chemical delay. The first visual occurrence of a flame front is captured as an increase in voltage signal at point D. The time periods between points B and C & C and D are representative of the physical and chemical ignition delays respectively for the reaction. The same was measured to be 6.77 ms and 1.25 ms for the reaction under consideration. The global ignition delay for the reaction is obtained by adding the physical and chemical ignition delays.

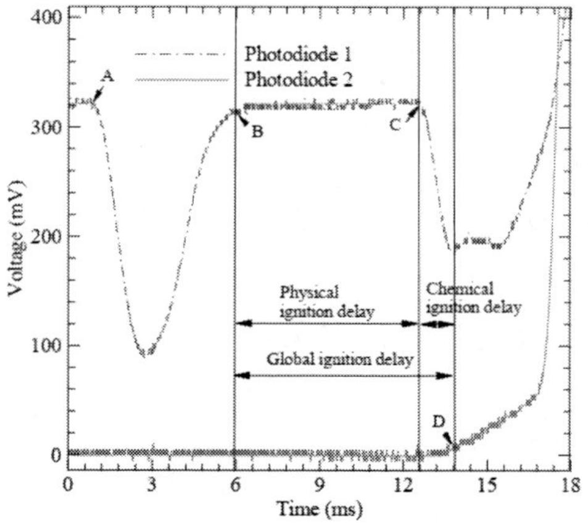

Figure 8: Signals of hypergolic ignition for TMEDA as fuel droplet into RFNA pool

Figure 9 shows the ignition delay determination using an optoelectronic circuit for a UDMH fuel droplet with an RFNA oxidizer pool. The voltage signal at point B does not reach its initial value as the signal is obstructed by the tail of gases formed behind the droplet. The point C, at which voltage drops because of gas clouds formed from the reaction, represents the start of chemical delay. The physical and chemical ignition delays were measured to be 2.3 ms and 0.78 ms respectively.

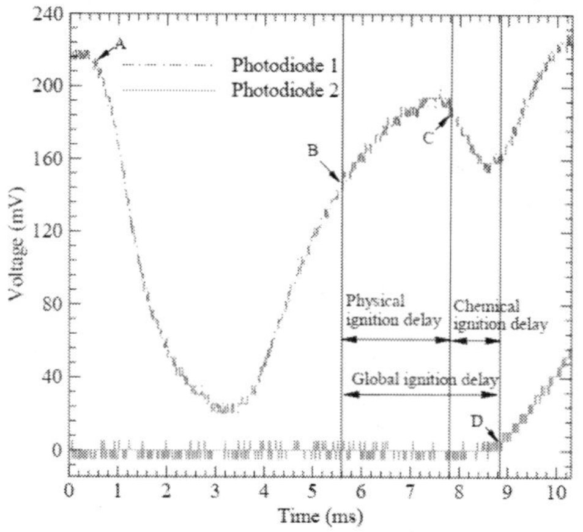

Figure 9: Signals for hypergolic ignition for UDMH and 200 µl RFNA

4 Conclusions

Hydrazine, UDMH, TMEDA, TEA, and 2,6-xylidine are hypergolic in nature with WFNA and RFNA. Furfuryl alcohol does not show hypergolicity when comes in the contact of RFNA. Global ignition delays for TMEDA, 2,6-xylidine, and hydrazine with WFNA were found to be 10 ms, 48 ms, and 39 ms respectively, which reduced to 8 ms, 21 ms, and 13 ms respectively with RFNA. For furfuryl alcohol with WFNA, the same was measured as 19 ms while for UDMH it was 2 ms with both WFNA and RFNA. Global ignition delay for ignition of TEA with RFNA was less as compared to WFNA since the hypergolicity of tertiary amines varies as RFNA < WFNA. All the fuels showed ignition delays less than 50 ms, which is the upper limit to classify the hypergols as useful. Amongst the compounds tested, TMEDA shows the lowest ignition delay with WFNA and RFNA and is seen as the best candidate to replace hydrazine based hypergolic compounds.

5 Acknowledgment

The author also wishes to acknowledge Defense Research and Development Laboratory (DRDL), Hyderabad for funding this project.

References

[1] G.P. Sutton, "Rocket Propulsion Elements," Seventh Ed, John Wiley & Sons, 2001.

[2] T. Kubal, E. Dambach, S. Son, W. Anderson, and T. Pourpoint, 46th AIAA/ASME/SAE/ASEE Jt. Propuls. Conf. & Exhib., no. July, pp. 1–18, 2010.

[3] U. C. Durgapal, P. K. Dutta, G. C. Pant, M. B. Ingalgaonkar, V. Y. Oka, and B. B. Umap, Propellants, Explos. Pyrotech., vol. 12, pp. 149–153, 1987.

[4] S. M. Davis and N. Yilmaz, Adv. Aerosp. Eng., vol. 2014, no. 1, p. 9, 2014.

[5] John D. Clark, An Informal History of Liquid Rocket Propellants, Rutgers University, New Jersey, U.S.

[6] W. G. Liu, S. Dasgupta, S. V. Zybin, and W. A. Goddard, J. Phys. Chem. A, vol. 115, no. 20, pp. 5221–5229, 2011.

[7] S. Wang, S. T. Thynell, and A. Chowdhury, Energy and Fuels, vol. 24, no. 10, pp. 5320–5330, 2010.

11th Asia-Pacific Conference on Combustion,
The University of Sydney, NSW Australia
10th -14th December 2017.

Experimental Study of Main-Post Injection under Compression-Ignition Engine Conditions, Part I: Effects of Dwell Time

C. Ming[1], Q.N. Chan[1], X.P. Pham[2], S. Kook[1], E.R. Hawkes[1], G.H. Yeoh[1], S. Yang[3], A.R. Masri[2]

[1]UNSW School of Mechanical and Manufacturing Engineering
UNSW Sydney, NSW, 2052, Australia

[2]School of Aerospace, Mechanical and Mechatronic Engineering
The University of Sydney, Sydney, NSW, 2006, Australia

[3]Beijing Hongshine New Energy Science and Technology Co Ltd
Zhichun Road, Haidan, Beijing 100086, China.

Abstract

Post-injection has shown significant benefits in the reduction of combustion emissions and soot formation from compression-ignition engines. However, there is a need to better understand the mechanisms underpinning the pollutant emission reduction, so that this injection strategy can be fully optimized. An experimental investigation was performed to understand the impact of dwell time on the interaction between the main and post injections for main-post injection cases. The measurements were performed in the quiescent steady environment inside a constant-volume combustion chamber (CVCC) of 19.4 kg/m^3 ambient density, 5.1 MPa ambient pressure, 900 K bulk temperature and 15 vol.% O_2 ambient conditions. A mass ratio of the main to post injection of 80% to 20% and a fixed injection pressure of 100 MPa were used. Main-post combustion properties were characterized from the high-speed flame luminosity imaging of the combustion cases. The results indicate that for the investigated test conditions, the ignition delay and location of the post injection increased with increasing dwell time. The flame luminosity of the post injection was observed to vary, depending on the dwell time used. The results indicate that the changes in the ignition and combustion characteristics were mainly driven by the extent of the interactions of the post injections with the combustion products originating from the mains.

1 Introduction

Multiple-injection is widely implemented in modern compression–ignition engines. The advantages of multiple-injection are well documented, with many studies reporting reduced combustion noise, decreased pollutant emissions and improved engine control [1-3]. Post-injection is a multiple-injection strategy that involves the injection of a small quantity of fuel after the main combustion event. Post-injection strategy is generally used as a particulate matter control measure to reduce soot emissions, or to enable exhaust gas after treatment technologies [2]. For post-injection, there remains much discrepancies in the understanding of the mechanisms through which reduced pollutant emissions is achieved, and whether interactions between the combustion of main and post injections contributed to the observed outcome [2, 4]. Previous studies have demonstrated that the mechanisms and the interactions between the injections can vary significantly, depending on the injection schedule (injection duration and dwell time) and ambient conditions under which they occurred. A more comprehensive understanding of the mechanisms involved can enable this fuel injection strategy to be refined and advanced [5]. This study focuses on the impact of dwell time on the interactions between the main and post injections for selected main-post injection schedules. A canola-based biodiesel fuel, which has been extensively studied by the authors [6], is injected into a quiescent ambient condition with 19.4 kg/m^3 ambient density, 5.1 MPa ambient pressure, 900 K bulk temperature and 15 vol.% O_2 concentration, inside an optically-accessible constant volume combustion chamber (CVCC). The mass ratio of the main to post injection is controlled at 80% to 20% for all injection cases. The dwell time between the injections, however, are varied from 1.5 to 2.5 ms. A high-speed digital camera is used to record the flame luminosity images from the combustion events. A detailed analysis is performed on the flame images recorded to characterize the main-post combustion properties, including ignition delay, ignition location, flame height and flame lift-off lengths, and to provide insights to their interactions.

2 Experimental Details

The experiments were conducted in an optically-accessible, constant-volume combustion chamber at UNSW Sydney, under simulated compression-ignition engine condition. A schematic of the chamber, along with the high-speed flame luminosity imaging setup, is provided in Fig. 1. The vessel has a cubical combustion chamber (114 by 114 by 114 mm), with optical access provided by side-port windows (102 mm diameter). A single-hole fuel injector was mounted at the center of a metal side-port of the CVCC. An agitator was mounted at the top of the CVCC and was used to maintain a spatially uniform temperature environment in the chamber. The operation of the CVCC facility is briefly described here. The compression-ignition engine condition was simulated by metering acetylene, hydrogen, oxygen and nitrogen gases into the CVCC. The combustible gas mixture was subsequently spark-ignited to

Corresponding author.
E-mail address: qing.chan@unsw.edu.au

create a high temperature and pressure environment. Following the premixed combustion, the combustion products inside the chamber were allowed to cool-down through heat transfer to the cooler chamber wall. The pressure inside the chamber was monitored with a pressure sensor (Kistler 6052C) during the cool-down period. The fuel injector was triggered once when the targeted condition was reached. For this study, an ambient condition with gas density, pressure and temperature of 19.4 kg/m^3, 5.1 MPa and 900 K, respectively, was targeted. The metering of the reactant gas was also controlled so that an environment with 15 vol.% oxygen concentration can be generated after the premixed combustion process. The experiments were therefore performed under low-temperature combustion (LTC), which is a combustion mode that is widely implemented in compression-ignition engines to reduce emissions [3]. A summary of the ambient experimental condition for this study is listed in Table 1. A canola-based biodiesel fuel (C1875), which have been extensively studied by the authors, was used in this study. A detailed description of the fuel can be found in Ref. [6] and therefore, only a short summary the main compositions and properties of the biodiesel is provided in Table 1. In brief, the biodiesel fuel has an oxygen content, iodine, saponification and cetane numbers of 10.83 wt%, 105, 185 and 59, respectively. The large iodine and saponification numbers indicate that the fuel is almost unsaturated and has long carbon chain length. The high cetane number, on the other hand, implies that the fuel has a short ignition delay.

For this study, a Bosch common-rail (CP4) system with a single-hole solenoid injector (105 μm) was used as the fuel injection system. A common-rail PCV driver (Zenobalti, ZB-1200) was used to regulate and maintain the injection pressure at 100 MPa. In addition, an injection driver (Zenobalti, ZB-5100) was used in conjunction with a digital delay generator (DG 535) to control the triggering of the fuel injection system. The single-hole injector was used to avoid the complexity that can arise from jet-jet interaction. In this study, the multiple-injection case was tailored to represent a 'main plus post' injection schedule, with a total injected fuel mass of 10 mg. The fuel mass ratio of the main to post injections was fixed as 80% to 20%. Dwell time (DT), which is defined here as the time between the end of electronic command of the main and the actual start of injection for the post, was set as 1.0, 1.5 and 2.5 ms for the main-post cases. A single-injection case, with its injection duration adjusted so that it has a total injected fuel mass of 10 mg (same as the multiple-injection case), is used as reference. The rate of injection (ROI) profiles of the main-post and single-injection cases, which were measured using a Bosch tube type injection rate meter, are shown in Fig. 2. A summary of the fuel injection conditions is also listed in Table 1.

The time-resolved flame luminosity images of the combustion of multiple and single-injection cases were captured using a high-speed camera (Photron SA-5), operating at 15,000 frames per second (fps) with an exposure time of 60 μs. The camera was equipped with a f#1.4 Nikkor lens. The pixel resolution of the image was ~0.2 mm per pixel. The camera was synchronized to the start of injection of the injector, and 120 images were captured for each run. At least five runs were performed for each measurement to ensure good repeatability. Typical time-sequenced flame luminosity images recorded for a main-post injection cases, with a dwell time of 1.5 ms, are presented in Fig. 3.

3 Results and Discussion

The spatial and temporal flame luminosity was calculated [5]

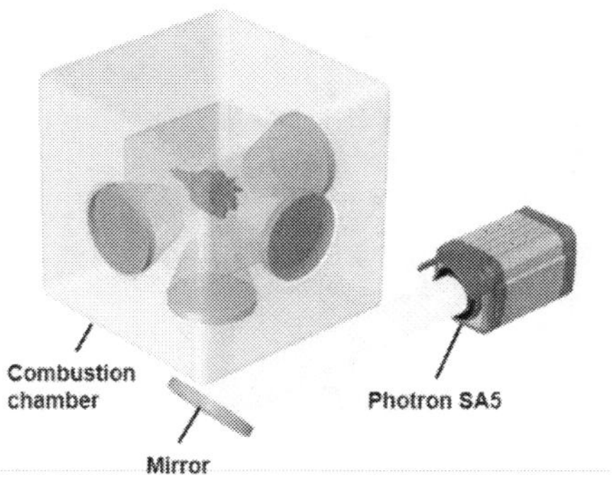

Figure 1: Schematic of experimental arrangement, including the combustion chamber and high-speed flame luminosity imaging setup.

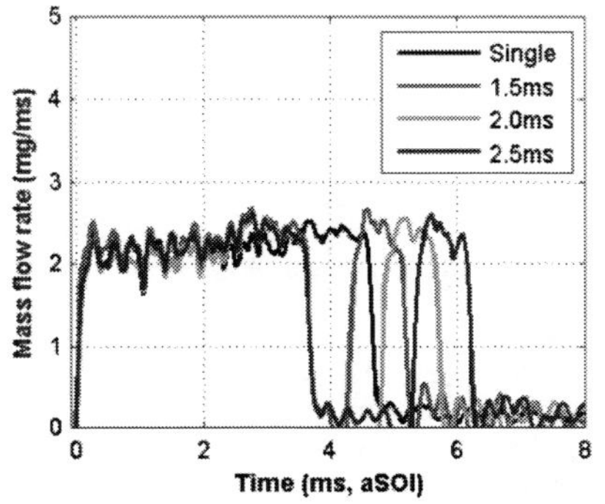

Figure 2: Injection rate profiles for main-post injection cases with dwell-timings (DTs) of 1.5, 2.0 and 2.5 ms, and single-injection.

Test conditions	
Ambient density (kg/m^3)	19.4
Ambient pressure (MPa)	5.1
Bulk temperature at injection (K)	900
Oxygen level (vol.%)	15
Fuel fatty acid profiles (wt%)	
Palmitic, C16:0	4.3
Stearic, C18:0	2.2
Oleic, C18:1	63.5
Linoleic, C18:2	18.9
Linolenic, C18:3	9.2
Eicosanoic, C20:0	0.4
Eicosenoic, C20:1	1.1
Fuel properties	
Fuel abbreviation	C1875
Oxygen content (wt%)	10.83
Iodine number	105
Saponification number	185
Cetane number	59
Fuel injection conditions	
Nozzle diameter (μm)	105
Injection pressure (MPa)	100
Injection quantity (mg)	10 (single) 8:2 (main-post)
Dwell time (ms)	1.5, 2.0 and 2.5

Table 1: Summary of test conditions, fuel composition, properties and injection conditions.

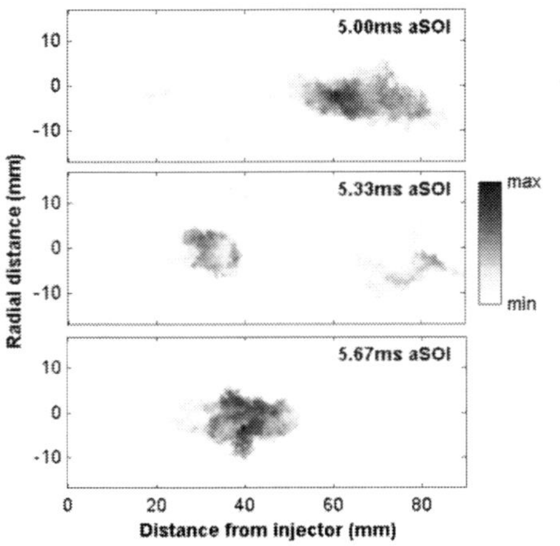

Figure 3: Typical time-sequenced flame luminosity images for a main-post injection case, with a dwell time (DT) of 1.5 ms. The fuel is injected from left to right in these images.

by performing a radial integration of the flame luminosity at each axial location of each frame, for the multiple and single-injection cases, as shown in Fig. 4. The ensuing contour plot, which is also termed Intensity-aXial-Time (IXT) plot in the literature [5], is subsequently used to derive key combustion characteristics of the injection cases, including ignition location, ignition delay, flame lift-off length, flame height and internal flame structure information. It is noted that ignition location and ignition delay time are defined here as the distance from nozzle and the elapsed time after the start of injection (aSOI) until the first detection of flame luminosity for the injection, respectively. The flame lift-off length and flame height refer to the average quasi-steady distance values of the flame base and tip from the injector nozzle.

In Fig. 4, for the multiple-injection cases, the two 'islands' on the plots represent the main (larger) and post (smaller) injection events. From the figure, it can be seen that whilst the overall shapes of the 'islands' on the contour plots are similar, some differences in the flame height, start of ignition, ignition location and intensity distribution can be observed for the post injections. In Fig. 4, the maximum flame height values of the post injections are found to be lower than their corresponding main injections, and the single-injection reference case. This is to not surprising, given that the post-injections were controlled to have smaller injection quantities (20% of the total injected mass) and hence, durations, than the mains in this study. The short injection durations therefore prevented the post-injection flames from reaching their maximum steady-state development. From Fig. 4, it can also be observed that both the ignition locations and the flame lift-off lengths of the post injections are lower than their mains and the single-injection reference case. In addition, when comparing the intensity distribution within the contour plots in Fig. 4, the peak luminosity values for both the main and post injection combustion events are found to be similar for all DTs. The post injections, however, have marginally smaller areas with higher intensity, with the smallest area observed when the shortest DT (*i.e.*, 1.5 ms) was used.

To further understand the changes in the ignition and combustion characteristics of the post injection events, the ignition delays, ignition locations and the flame lift-off lengths of the post injections are plotted against the averaged values of

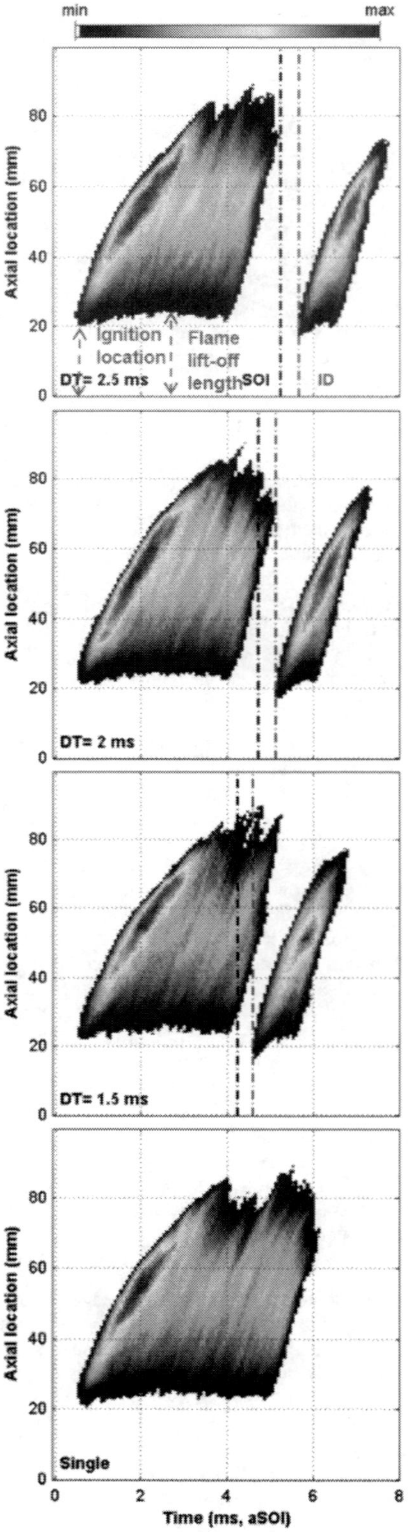

Figure 4: Averaged radial integration of flame luminosity at each various axial location from injector, as a function of time after start of injection, for the multiple and single-injection cases. The start of injection timings and ignition delays of the post injections are indicated with dashed lines.

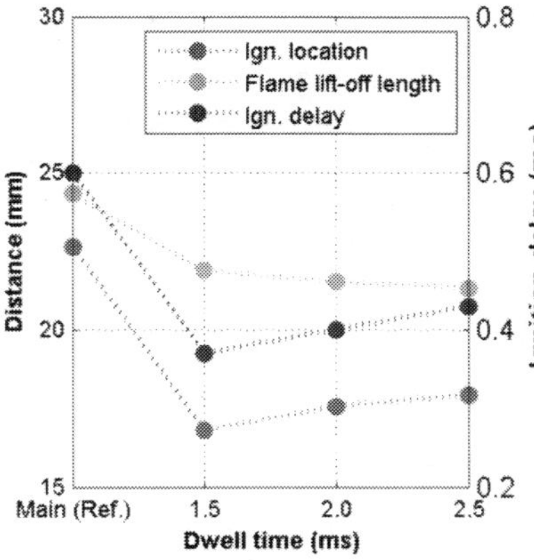

Figure 5: Ignition delay, ignition location and average flame lift-off length of main and post injection flames for the multiple-injection cases with dwell times (DTs) of 1.5, 2.0 and 2.5 ms.

their mains in Fig. 5. Dashed lines through the data are included to help visualize the trends. From the figure, it can be seen that the values of the ignition delays, ignition locations and flame lift-off lengths of the post injections are consistently shorter than those of the same case. In addition, there is also a clear trend of a shortening in the ignition delays and ignition locations with decreasing DT, whilst the flame lift-off lengths appear to reach an asymptote with longer DT. Previous studies have demonstrated that changes in ignition delay impact on the time available for fuel-air mixing. This will directly impact the heat release profiles of the premixed and the mixing-controlled combustion processes, which in turn, will affect the soot and temperature distribution of the flame [5]. Given that flame natural luminosity of a sooty flame mainly comprised of hot soot incandescence [7], which is a function of the local soot concentration and temperature [8], the observed changes in the intensity distributions in the IXT plots (Fig. 4) with varied dwell times can therefore be expected. It is also noted that a previous study by Cung *et al.* [5] has noted that when sufficiently short DT is used, the combined momentum of the closely-coupled injections can push the ignition location and flame lift-off length of the later injection downstream from the nozzle. A near linear correlation between the changes in the ignition delays and ignition locations, in addition to relatively invariant lift-off lengths of the flames across the varied dwell times, however, is observed here. These trends imply that for the test conditions of this study, the DTs used were sufficiently long such that the momentum from the main injections may to have decayed adequately.

From the time-sequenced flame luminosity images for the main-post injection with DT=1.5 ms (Fig. 3), it can be seen that whilst the combustion of the post injection was progressing towards the tail of the main, the combustion events were sufficiently separated. This observation is also reflected in the IXT plots presented in Fig. 4. From the figure, in the case of DT=1.5 ms, it can be seen that whilst the combustion of the post-injection initiated when the main was active, the contours associated with the two injection events are still clearly distinguishable. When the DT became longer, as in the cases of DT=2.0 and 2.5 ms, it can be also inferred from the IXT plots that the post combustion events only started after the mains had extinguished. Together, these findings imply that, for the test conditions of this study, no direct flame interaction occurred between the main and post combustion events, even when the shortest DT was used. The observed changes in the ignition and combustion characteristics of the post injections investigated in this study, was therefore likely to be driven by the extent of their interaction with the high-temperature products from the combustion of mains [9, 10, 11], but not through direct flame interaction. Further investigation is required for more definitive conclusion.

4 Conclusion

An experiment was performed in an optically-accessible, constant-volume combustion chamber to evaluation the effects of varying dwell time between the main and post injections. The ambient density, bulk temperature and oxygen concentration were maintained to be 19.4 kg/m^3, 900 K and 15 vol.% O_2 respectively, and the injection pressure was fixed at 100 MPa. A mass ratio of the main to post injections of 80% to 20% was used for this study. A high-speed digital camera was used to image the flame luminosity of the combustion cases. Main-post combustion properties, including ignition location, ignition delay, flame lift-off length, flame height and internal flame structure information were then characterized from the high-speed flame luminosity images. The results indicate that for the test conditions of in this study, the ignition and the combustion characteristics of the post injections were affected across the varied dwell times. The modifications in the ignition and combustion behaviours of the post injections appeared to be driven by the extent of their interactions with the combustion products of the main injections.

5 Acknowledgment

The authors wish to acknowledge the financial support of Australia Research Council (ARC) and Australian Academy of Technological Sciences and Engineering Priming Grant. The authors would also like to acknowledge the assistance of Brad Moody and I.M. Rizwanul Fattah.

References

[1] S.K. Chen, SAE Technical Paper 2000-01-3084 (2000).

[2] O'Connor, J., Musculus, M., SAE Int. J. Engines 6 (2013), 400-421.

[3] Stringer, V.L., Cheng, W.L., Lee, C.F., Hansen, A.C., SAE Paper 2008-01-1388 (2008).

[4] M. Bobba, M. Musculus, W. Neel, SAE Int. J. Engines 3 (2010), 496-516.

[5] K. Cung, A. Moiz, J. Johnson, S.-Y. Lee, C.-B. Kweon, A. Montanaro, Proc Combust Inst 35 (2015), 3061-3068.

[6] X.P. Pham, *Influences of Molecular Profiles of Biodiesels on Atomization, Combustion and Emission Characteristics*, PhD thesis, The University of Sydney, Sydney, Australia, 2015.

[7] X. Wang, O.A. Kuti, W. Zhang, K. Nishida, Z. Huang, Combust Sci Technol 182 (2010) 1369-1390.

[8] H. Zhao, N. Ladommatos, Prog. Energy Combust. Sci., 24 (1998) 221–255.

[9] M. Bolla, M.A. Chishty, E.R. Hawkes, S. Kook, Int J Engine Res 18 (2017), 6-14

[10] J. Lim, S. Lee, K. Min, Combust Sci Technol 183 (2010), 180-201.

[11] S. Skeen, J. Manin, L.M. Pickett, SAE Int J Engines 8 (2015), 696-715.

11th Asia-Pacific Conference on Combustion,
The University of Sydney, NSW Australia
10th -14th December 2017.

Effect of Water Mist Mass Fraction on Extinguishment of Counterflow Diffusion Flame in the Turbulent Air Flow

Ryohei Matsuo[1], Hiroyoshi Naito[2], Akira Yoshida[1]

[1]Tokyo Denki University

5 Senju-Asahicho, Adachi-ku, Tokyo, 120-8551, Japan

[2]Fire and Disaster Management Agency

2-1-2 Kasumigaseki, Chiyoda-ku, Tokyo, 101-8508, Japan

Abstract

Water mist is one of the effective fire suppression agents. In the present investigation, the effect of polydisperse water mist on extinguishment of the methane/air counterflow diffusion flame in the turbulent air flow is investigated experimentally. Burner consists of a 30-mm-diameter porous cylinder, from which methane is ejected uniformly. The turbulence is formed by two kinds of turbulence producing grids respectively. Polydisperse water mist is generated by single-fluid pressure type atomizers and is mixed with air far upstream of the burner. Size distribution of the water mist is measured by a phase Doppler particle analyzer (PDPA) and is found to be lognormal for all cases tested. Sauter mean diameter (SMD) is 20 μm and the mass fraction of water mist in air flow, Y_0, is changed from 0.00 to 0.05. For turbulent air flow, the critical stretch rate is found to be the sum of the bulk stretch rate and that induced by turbulence. The critical total stretch rate at extinguishment with water mist, $a_{T,ext,w}$, is normalized by that without water mist $a_{T,ext}$. $a_{T,ext,w}/a_{T,ext}$ decreases monotonously with Y_0 and all the data points collapse on one curve roughly independently of the turbulent intensity.

1 Introduction

Halon 1301 (trifluoroboromomethane, CF3Br) was a typical fire suppressant among halogenated hydrocarbons and was used worldwidely [1], but the production of fire suppressants containing bromine was banned by Montreal protocol because of their depletion potential of stratospheric ozone [2]. As a replacement for Halon 1301, water mist with the size less than 100 μm is considered, which can be suspended in air for much longer periods than large droplets generated by water sprinklers. Water mist has favorable suppression characteristics; water mist has high heat capacity and high latent heat of evaporation, which can absorb a significant quantity of heat from flames. Additionally, it expands 1700 times when it evaporates to vapor, which results in the dilution of the surrounding oxygen in air. Furthermore, water mist has favorable properties for fire suppression, as water mist is inexpensive, ubiquitous, non-electrically conductive and nontoxic.

Using counterflow diffusion flame, the behavior of water mist in laminar flames and the mechanism of extinguishment of laminar flame were investigated, within the limited range of water mist mass loading and mist size [3-6], and the critical stretch rate at extinguishment was found to depend strongly on the mist droplet surface area parameter. The effect of water mist on extinguishment of turbulent flame needs to be investigated because real fire is subjected to turbulent surroundings. In the present investigation, the effects of polydisperse water mist on extinguishment of counterflow diffusion flame in turbulent air flow was investigated experimentally.

Corresponding author. Fax: +81-3-5284-5469
E-mail address: 16kmk22@ms.dendai.ac.jp

2 Experimental Apparatus

In the previous studies, ultrasonic atomizers and twin-fluid atomizers were used to generate water mist [3-5]. However the difficulties were encountered, when the diameter and the mass fraction of water mist were varied independently. In the present investigation, single-fluid pressure type atomizers were adopted to generate water mist. An electric valve is equipped in the atomizer. The valve opening and closing were controlled by a pulse width modulation (PWM) controller. The mist diameter and the mass loading in air flow can be controlled independently by both the water supply pressure to atomizer and the duty ratio of electric valve opening.

The experimental apparatus is shown in Fig. 1 The experimental apparatus consist of blower, water mist generation system, wind tunnel, turbulence producing grid, combustion chamber, fuel supply system, phase Doppler particle analyzer (PDPA), laser Doppler velocimeter (LDV) and hot-wire anemometer. Air is supplied from a blower and the flow rate is measured using an appropriate orifice. Water mist generated by single-fluid pressure type atomizers is added to air flow in the wind tunnel, which consists of a diffuser, a settling chamber and a converging nozzle. The settling chamber contains honeycomb straightener and wire gauzes of 3 mesh, 5 mesh and 8 mesh, respectively. The combustion chamber is mounted on the converging nozzle. 10-mm-thick quartz windows are fitted on both sides of the combustion chamber to permit viewing, photographic observations and recording. The turbulence is formed by a turbulence producing grid which is installed between the converging nozzle and the combustion chamber. Two kinds of turbulence grids (*G1*, *G2*) are used. *G1* consists of the metal round rods with a cross section of φ = 2.0 mm. Rods are set 5.5 mm apart from each other. *G2* consists of the metal square rods with a cross section of 3.0 mm × 3.0 mm and rods are set 7.0 mm apart from each other. In the combustion chamber, an uncooled porous cylinder of 30 mm in diameter and 30 mm long is installed, the center being 110 mm downstream of the combustion chamber inlet. The downstream side of the cylinder surface is coated with heat-resisting paint in order to prevent ejection of fuel into the wake. The fuel used is methane (99.8 % pure) and the fuel flow rate is controlled by a mass flow controller. Size distribution of the water mist is measured by a PDPA, and turbulence intensity of the air flow is measured by an LDV. The autocorrelation of velocity fluctuation is calculated from the velocity measurements by a hot-wire anemometer. All measurements are carried out at the center of the converging nozzle exit and 45 mm upstream of the cylinder surface.

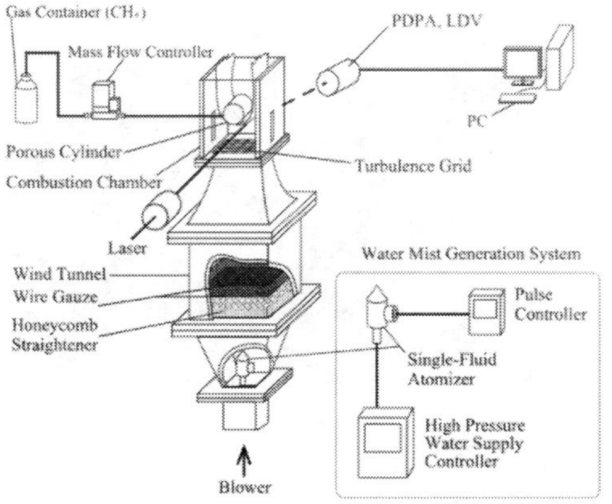

Figure 1: Experimental apparatus.

The schematic of high pressure water supply system is shown in Fig. 2 High-purity water is used to generate water mist. Size distribution of the water mist can be varied by changing nozzle tips and water pressure. Two types of the nozzle tips of which orifice diameter are 0.3 mm and 0.1 mm were used in the present study. By an electric valve of which opening and closing are controlled by a PWM, water mist mass loading is controlled. Water was recirculating in the closed loop supply system with an atmospheric reservoir tank. Water is pressurized by a plunger pump up to 10 MPa. Supply pressure is controlled by closing the return valve to the reservoir tank. An accumulater absorbs pressure fluctuation and a heat exchanger keeps the water temperature constant.

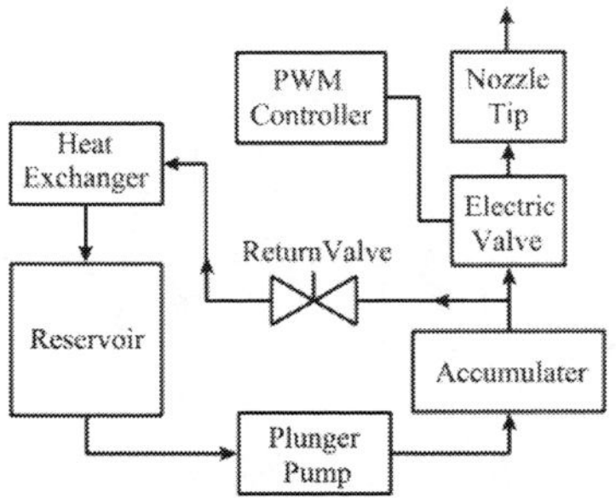

Figure 2: High pressure water supply system and atomizer.

Table 1 shows the specifications of used turbulence producing grids and characteristics of turbulence for the mean velocity V = 1.0 m/s. Hence, v' is turbulence intensity and L is the integral length scale of the turbulent air flow. In the laminar air flow, effect of turbulence on extinguishment is not considered because relative intensity v'/V is less than 1 %. In the present investigation, L is expressed, using Taylor's hypothesis, as

$$L = L_t \times V. \tag{1}$$

Here, L_t is integral time scale. L_t is calculated from the autocorrelation of velocity fluctuation. Typical autocorrelation is shown in Figure 3 for *G2* at V = 1.0 m/s and v' = 0.10 m/s. The vertical axis shows the autocorrelation coefficient and the horizontal axis shows the time interval. When time interval is large enough, the autocorrelation function is close to 0.

Grid no.	Cross section [mm]	Distance between bars [mm]	v' [m/s]	L [mm]
None	-	-	0.01	-
G1	φ 2.0	5.5	0.06	2.8
G2	3.0 × 3.0	7.0	0.10	3.7

Table 1: Turbulence producing grids and turbulence characteristics for V = 1.0 m/s.

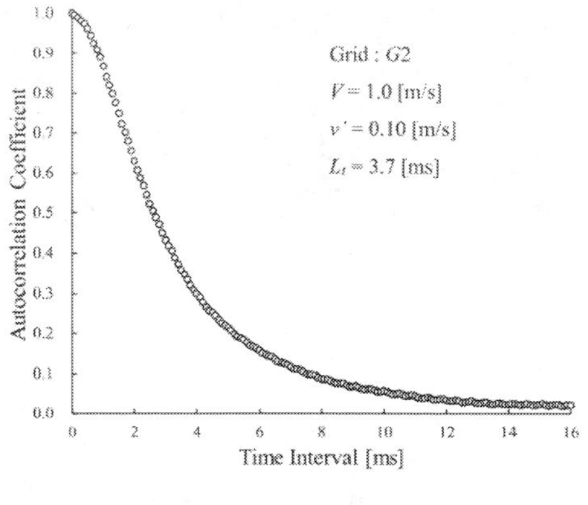

Figure 3: Autocorrelation for *G2* at V = 1.0 m/s, v' = 0.10 m/s.

3 Results and Discussion

3.1 Schlieren Images of Laminar and Turbulent Flames

Figure 4 shows schlieren images of (a) laminar and (b) turbulent methane diffusion flames for mean air flow velocity V = 1.5 m/s and fuel ejection velocity v_w = 0.21 m/s. For (b), the turbulence intensity was 0.21 m/s. The exposure time for laminar flame is 1/1000 s, and that for turbulent flame is 1/8000 s. For turbulent flame, large-scale distortions with small amplitude are generated on the apparently laminar diffusion flame.

3.2 Size Distribution of Water Mist

Typical size distribution of water droplets in the present study is shown in Fig. 5 for SMD = 20 μm at V = 1.9 m/s. The vertical axis shows the number ratio at a given diameter and the horizontal axis shows the mist diameter. In the present study, droplets sizes of water mist are less than 50 μm. SMD of the water mist is about 20 μm and the size distribution of water mist is log-normal.

3.3 Image of Mist Behavior

The behavior of the water mist near the flame zone is rather complicated. If all the mist droplets are small enough, they can evaporate in the flame zone and the diameter distribution should have only a secondary influence on the extinguishing behavior. However, if the diameter distribution includes some large droplets, they can go through the flame zone [7-9].

Figures 6 (a) and (b) show the traces of water mist during 1/125 s in laminar and turbulent air flows for $V = 1.0$ m/s and $v_w = 0.21$ m/s. For (b), the turbulence intensity v' is 0.14 m/s. Mist behavior is visualized by using 0.6-mm-thick laser sheet, which passes through the slit of the side walls of the combustion chamber. In the present study, almost all the water droplets evaporate in the flame zone. The water mist is uniformly dispersed in the air for the laminar flow case, but in the turbulent flow case, the traces become spotted and dark region in which mist droplets are not present appears due to the large velocity gradient around the turbulent eddies.

(a) Laminar flame.

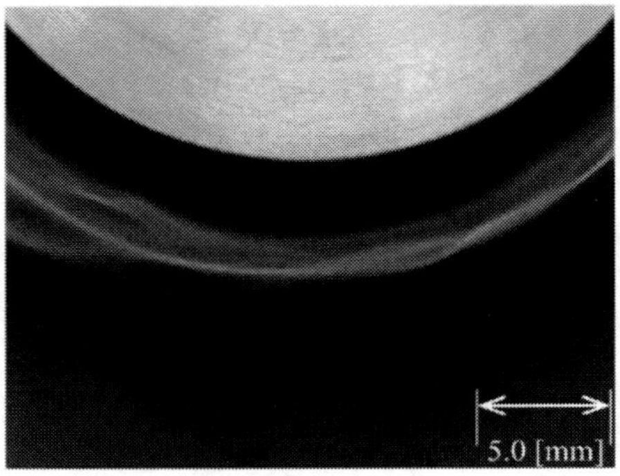

(b) Turbulent flame, $v' = 0.21$ m/s.

Figure 4: Schlieren images of laminar and turbulent methane diffusion flames for $V = 1.5$ m/s. (a) Laminar flame, exposure time = 1/1000 s, (b) turbulent flame, $v' = 0.21$ m/s, exposure time = 1/8000 s.

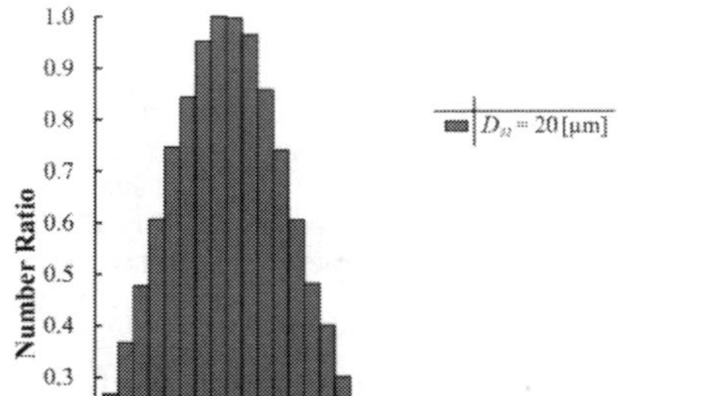

Figure 5: Water droplet size distribution for $D_{32} = 20$ μm

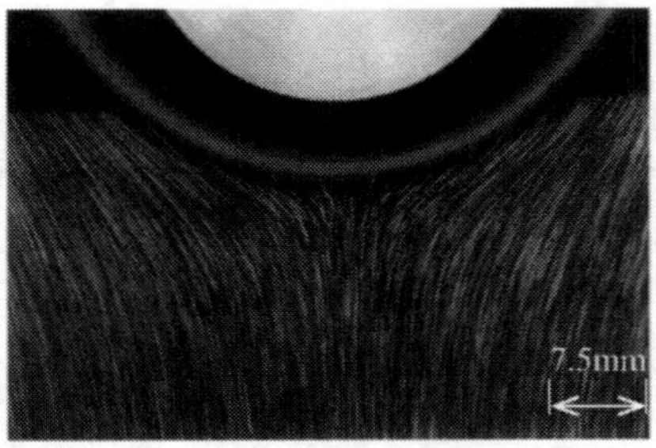

(a) Laminar air flow.

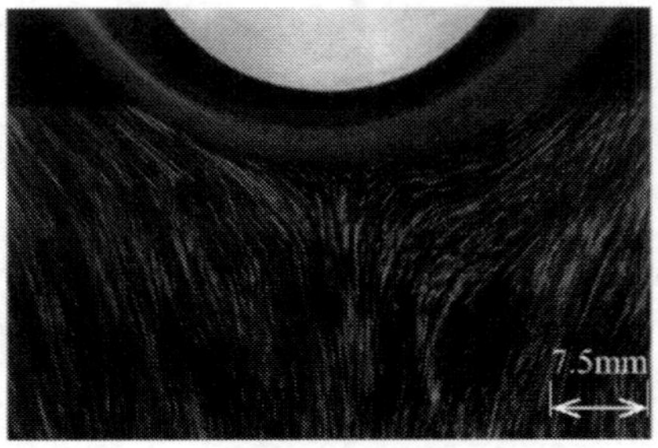

(b) Turbulent air flow, $v' = 0.10$ m/s.

Figure 6: Images of mist behavior for $V = 1.0$ m/s and $v_w = 0.21$ m/s. (a) Laminar air flow, (b) Turbulent air flow, $v' = 0.10$ m/s.

3.4 Limit of Flame Extinguishment

A flame stretch is dominant factor for extinguishment of the diffusion flame. Therefore, the effect of water mist can be estimated by the reduction of the critical stretch rate at extinguishment by water mist. In the previous study [10], methane or propane was ejected from a porous cylinder in the counterflowing turbulent air stream, and the extinguishment process due to turbulence was investigated. Here, the turbulence should act an additional stretch a_t on the flame zone in addition to bulk stretch a_b induced by mean flow. The total stretch rate a_T is estimated as the sum of the bulk flow stretch rate, a_b, and the turbulent stretch rate, a_t. a_b near the forward stagnation region of a cylinder was expressed by $2V/R$ and a_t was assumed to be equal to the inverse of the Kolmogorov time scale [11].

$$a_T = a_b + a_t. \tag{2}$$

$$a_b = 2V/R. \tag{3}$$

$$a_t = (\varepsilon/\nu)^{1/2}. \tag{4}$$

Hence, ν is the kinematic viscosity and ε is the dissipation rate. The dissipation rate ε can be related to the turbulence intensity, v', and Taylor microscale, λ for isotropic turbulence [12] by

$$\varepsilon = 15\nu(v'^2/\lambda^2). \tag{5}$$

Dryden [13] has used experimental data and shown that

$$\lambda^2 = 48.64(L\nu/v'). \tag{6}$$

Therefore,

$$\varepsilon = 0.3084(v'^3/L). \tag{7}$$

$$a_t = 0.555(v'^3/\nu L)^{1/2}. \tag{8}$$

In the present study, a_T is estimated from Equation (2).

Critical total stretch rate at extinguishment is shown in Figure 7. The vertical axis shows the non-dimensional fuel ejection parameter, $-f_w$ [5], and the horizontal axis shows a_T. Here, Y_0 is mass fraction of water mist. For the case without water mist, when $-f_w$ is large enough, the stretch rate at extinguishment is independent of $-f_w$ and the stretch rate at extinguishment in the laminar air flow agrees well with that for the turbulent air flow. This result is confirmed in Ref. 10. The critical stretch rate at extinguishment is due to chemical limitations on the reaction rate in the flame zone, and represents the extinguishment specific to the diffusion flames. When the water mist is added to the air flow, it is clear that the total stretch rate at extinguishment decreases. In addition, the critical stretch rate at extinguishment is independent of $-f_w$. Therefore, the effect of turbulence on the extinguishment of counterflow diffusion flame can be included in the total stretch rate through turbulent stretch rate.

3.5 Effect of Mass Fraction of Water Mist on Extinguishment

Heat capacity of water vapor and the latent heat of evaporation of water mist are proportional to the mass of the water mist. We investigated the effect of Y_0 on the critical total stretch rate at extinguishment. The result is shown in Fig. 8. The vertical axis shows $a_{T,ext,w}/a_{T,ext}$ and the horizontal axis shows the water mist mass fraction Y_0. Figure 8 also shows the results of numerical simulations for laminar flame [14]. In Ref. 14, the critical mass fractions of water droplets of D_{32} = 20 μm at extiguishment are simulated in methane/air counterflow diffusion flame in the laminar air flow and the size distribution of water mist is assumed to consist of 6 bins (8, 14 , 20, 28, 36 and 44 μm) of diameter range.

Figure 8 shows a quantitative agreement in trend that the critical stretch rate decreases monotonously with the mass fraction of water mist, and all the data points collapse on one curve roughly independently of the turbulent intensity. Therefore, even though spots in which water mist is sparse appear in turbulent air flow, the critical total stretch rate at extinguishment is expressed as a function of Y_0, which coincides with that for laminar flame. The concept of total stretch rate could be applied to the large scale real fire, even though the integral scale of turbulence should be much larger, and corresponding turbulent stretch rate should be smaller.

4 Conclusions

In the present study, the effect of water mist mass fraction on extinguishment of a methane-air counterflow diffusion flame in the turbulent air flow is investigated experimentally. Conclusions of the present study are summarized as follows.

1. The water mist is uniformly dispersed in the air for the laminar flow case, but for the turbulent flow case the water mist dispersion becomes spotted due to the large velocity gradient around turbulent eddies.
2. For the case with water mist addition, when the non-dimensional fuel ejection parameter $-f_w$ is large enough, the critical stretch rate at extinguishment is independent of $-f_w$.
3. The critical stretch rate decreases monotonously with the mass fraction of water mist and all the data points collapse on one curve independently of the turbulence intensity.
4. The total stretch rate can include appropriately the turbulence effect on extinguishment of the counterflow diffusion flame.

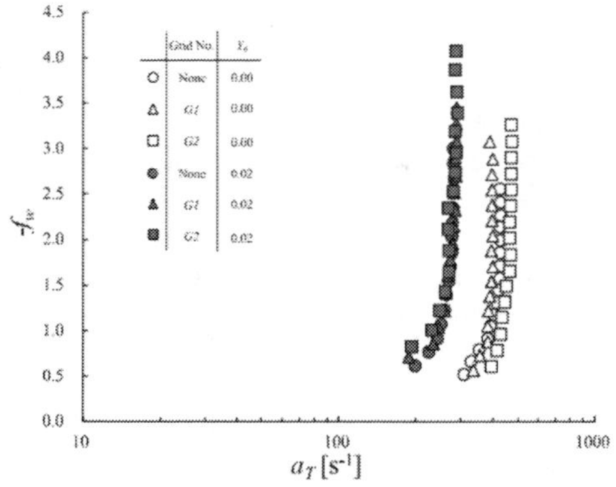

Figure 7: Critical total stretch rate at extinguishment for Y_0 = 0.00 and Y_0 = 0.02.

References

[1] R. E. Tapscott, T. A. Moore, J. D. Mather, J. A. Vitali, Halon Options Technical Working Conference (1998) 17–22.

[2] R. Hirst, K. Booth, Fire Technol.13 (1977) 296–315.

[3] H. Naito, T. Uendo, Y. Saso, Y. Kotani, A. Yoshida, Proc. Combust. Inst. 33 (2011) 2563-2571.

[4] A. Yoshida, R. Takasaki, K. Kashiwa, H. Naito, Y. Saso, Combust. Flame 160 (2013) 1357-1363.

[5] A. Yoshida, K. Kashiwa, S. Hashizume, H. Naito, Fire Safety J. 71 (2015) 217-225.

[6] S. Hashizume, H. Naito, A. Yoshida, Proc. 10th ASPACC (2015).

[7] A. M. Lentati, H. K. Chelliah, Combust. Flame 115 (1998) 158-179.

[8] H. K. Chelliah, A. K. Lazzarini, P. C. Wanigarathne, G. T. Linteris, Proc. Combust. Inst. 29 (2002) 369-376.

[9] H. K. Chelliah, Proc. Combust. Inst. 31 (2007) 2711-2719.

[10] H. Tsuji, A. Yoshida, N. Endo, Proc. Combust. Inst. 25 (1994) 1191-1197.

[11] L. W. Kostiuk, K. N. C. Bray, T. C. Chew, Combust. Sci. Technol. 64 (1989) 233-241.

[12] G. I. Taylor, Proc. R. Soc. A. 151 (1935) 421-444.

[13] H. L. Dryden, Q. Appl. Math. 1 (1943) 7-42.

[14] M. Kanno, A. Yoshida, M. Nishioka, Proc. Japanese Symp. Combust. (in Japanese) 52 (2014) 112-113.

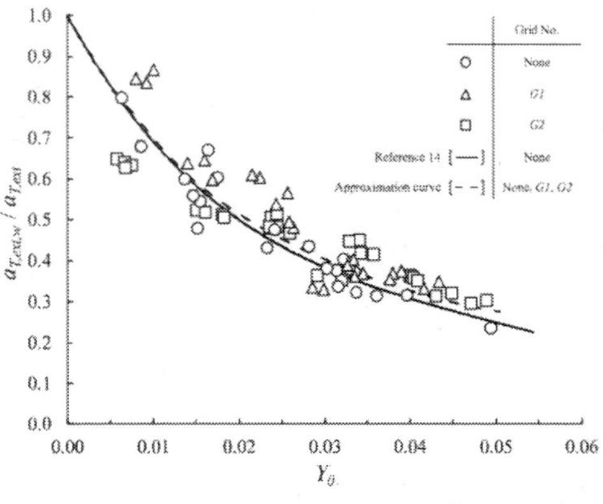

Figure 8: Effect of water mist mass fraction on the critical stretch rate at extinguishment.

11th Asia-Pacific Conference on Combustion,
The University of Sydney, NSW Australia
10th -14th December 2017.

Effect of Dilution by Carbon Dioxide on the Laminar Burning Velocity of *n*-Heptane/Air Mixtures

Mengni Zhou, Zunhua Zhang, Haiqi Ding, Junjie Liang, Gesheng Li

School of Energy and Power Engineering, Wuhan University of Technology

Wuhan, Hubei 430063, PR China

Abstract

Laminar burning velocities of *n*-heptane/air mixtures under carbon dioxide dilution rate from 0 to 15% at 0.1 MPa and 358 K were experimentally and numerically investigated via the outwardly expanding spherical flames and the simulation of one-dimensional planar flames. The physical effects of carbon dioxide were separated from its chemical effects by designing a type of fictitious carbon dioxide in the simulation. Results show that the laminar burning velocities *n*-heptane/air mixtures decrease with the increasing carbon dioxide dilution rate. The peaks value of mole fractions of the main radical species and net reaction rates of the main elementary reactions in the planar flames decrease with the increasing carbon dioxide dilution rate. Both physical and chemical effects of carbon dioxide suppress laminar burning velocities, the production of the radicals and the reaction rates of the elementary reactions of *n*-heptane flames, and the former dominates. In addition, the chemical effect has much more remarkable influence on the reaction rates of the elementary reactions with positive sensitivity coefficients than on those with negative sensitivity coefficients.

1 Introduction

Diesel engines are widely applied because of its high thermal efficiency and high power performance. On the other hand, high emissions of nitrogen oxides (NOx) from engine diesel have been noticed as major problems. Hence, how to control the NOx emission for diesel engine to meet the stringent requirement of emission regulation has become one of the key researches in the field of diesel engines. Exhaust Gas Recirculation (EGR) is an effective technique to reduce and control the NO_x emission from diesel engines [1].

The premixed diesel/air/exhausted gases mixtures, formed in ignition delay period, are combusted first during the combustion process of diesel engines with EGR. Considering the significant influence of premixed combustion on the parameters, such as the maximum heat release rate and the maximum rise rate of the cylinder pressure, premixed combustion characteristics for recycled exhausted gas and diesel mixtures need to be taken into account to design and optimize the diesel engines with EGR. Moreover, the premixed laminar flame plays an important role in the study of turbulence combustion and the validation of the chemical kinetic mechanism. Hence, it is necessary to study the effects of EGR gas on the premixed laminar diesel/air flames.

As a kind of typical diesel surrogate, the *n*-heptane laminar burning velocities have been reported in some open literatures [2-4]. N_2 and CO_2 are the two major components in EGR gas. Some researchers have studied the effects of N_2 addition on the laminar burning velocity [5-6]. They found that the thermal and dilution effects of N_2 on the laminar burning velocity play the leading role, and the chemical effect is marginal. However, as the best of authors' knowledge, few studies have been focused on the effect of the addition of CO_2 on *n*-heptane laminar burning velocity.

Therefore, the objective of the present study is to determine the dilution effect of CO_2 on the laminar burning velocities for *n*-heptane /air mixtures at 0.1 MPa, 358 K and equivalence ratios ranging from 0.8 to 1.5 by employing outwardly propagating spherical flames. In addition, freely propagating one-dimensional planar flames of *n*-heptane /air/ CO_2 were simulated at above initial conditions to analysis the kinetic effect of CO_2 on the laminar burning velocities of *n*-heptane/air flame. Moreover, a fictitious species F-CO_2 having the same thermal and transport properties with CO_2 was defined to separate the physical effect of CO_2 on *n*-heptane /air mixtures from its chemical effect.

2 Experimental setup and modeling method

In the present study, the experimental setup of outwardly propagating spherical flame was employed to perform the related experiments. Details about this experimental setup and the error analyses can be found in references 7 and 8. The diameters of optical windows are 120 mm. The purities of *n*-heptane, CO_2, N_2, and O_2 used in the experiments are 99.95%, 99.99%, 99.999% and 99.999%, respectively. The CO_2 dilution rate in the mixtures is defined as $R_{co_2} = n_{co_2} / (n_{co_2} + n_{N_2} + n_{O_2}) \times 100\%$. The experimental initial temperature and pressure is fixed at 358 K and 0.1 MPa, respectively, and the CO_2 dilution rate ranges from 0 to 15%. In the present study each experimental data was repeated at least three times, and a good repeatability was observed. The standard deviation of experimental data with 95% confidence interval was introduced to estimate the uncertainty of the laminar burning velocity, and the overall uncertainty is within 5%.

A computer program in our laboratory [9] was used to process the schlieren flame pictures obtained through the above experimental setup to derive the parameters such as stretched flame speed and laminar burning velocity. Considering the effect of ignition energy and the confinement of the constant vessel, the flame radii within 7–35 mm were used. The nonlinear extrapolation model based on the quasi-steady flame

Corresponding author. Fax: +86-27-8652-3305
E-mail address: gsliwh@gmail.com

propagation [10] was adopted, $\left(S_b / S_b^0\right)^2 \ln\left(S_b / S_b^0\right)^2 = -2L_b K / S_b^0$.

PREMIX code in CHEMKIN-PRO software [11] and LLNL detailed *n*-heptane mechanism [12] were employed to simulate the freely propagating one-dimensional planar flames of *n*-heptane /air/ CO_2 mixtures. The effect of thermal diffusion is considered, and the transport properties are evaluated by multi-component formulation. To ensure the results are grid-independent, the minimum gradient and curvature are set as 0.03 and 0.05 respectively, and all grid numbers in the simulation are above 500.

3 Experimental results and discussions

3.1 Effect of Addition of CO_2 on the Laminar Burning Velocities of *n*-heptane Flames

Figure 1 shows the experimental values and simulated values of laminar burning velocities of *n*-heptane/air mixtures at 358K and 0.1 MPa, different CO_2 dilution rates and equivalence ratios. As shown in the figure, the experimental values of laminar burning velocities have a reasonably good agreement with the modeled values. At the fixed CO_2 dilution rate, the laminar burning velocity of *n*-heptane/air mixture increases first and then decreases with the increase of equivalence ratio, and reaches the peak value at the equivalence ratio of about 1.1.

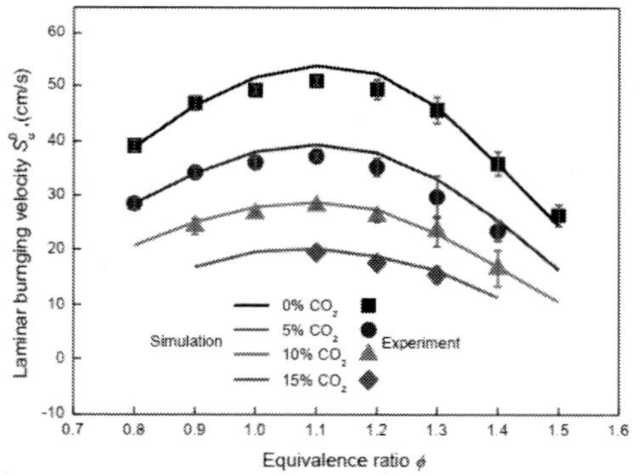

Figure 1: Variations of the laminar burning velocity of *n*-heptane flame at different CO_2 dilution rates with the equivalence ratio

In addition, as the CO_2 dilution rate increases, the laminar burning velocity of the *n*-heptane/air mixture decreases and the decrease rate gradually reduces. In the laminar flames, the laminar burning velocity has a positive relationship with the square root of the multiplication of thermal diffusion coefficient and chemical reaction rate [13], $S_u^0 \propto \sqrt{\alpha \cdot RR}$. According to the Arrhenius formula, the laminar burning velocity also has a positive correlation with adiabatic flame temperature. The additions of CO_2 to the mixtures increase the specific heat capacity of the mixtures, and decrease the concentration of oxygen in the oxidizer, which change the physical parameters of mixtures and reduce the heat release. As shown in Fig. 2, the thermal diffusion coefficients and adiabatic flame temperatures of *n*-heptane flames decrease with the increase of CO_2 dilution rate, therefore it is reasonable that the laminar burning velocities decrease with the increasing CO_2 dilution rate.

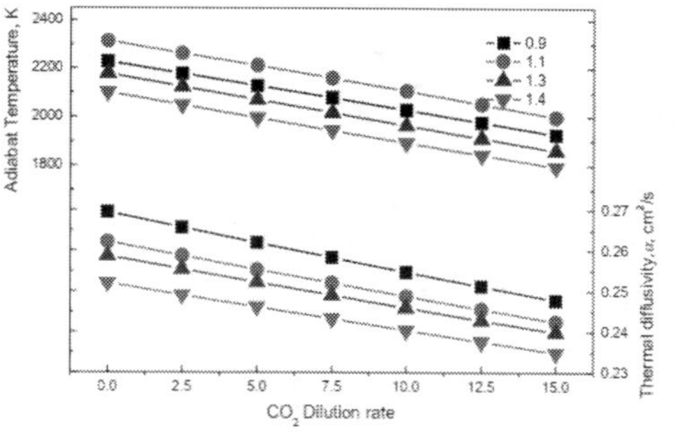

Figure 2: Variations of the thermal diffusion coefficients and adiabatic flame temperatures of *n*-heptane flames with the CO_2 dilution rate.

3.1.1 Sensitivity Analysis of the Laminar Burning Velocity

The sensitivity analysis of laminar burning velocities of stoichiometric *n*-heptane /air mixtures at different CO_2 dilution rates was conducted, as shown in Fig. 3. From the figure, it is seen that laminar burning velocities are sensitive to the elementary reactions involving oxidation of the H atom. The reactions that produce H atoms such as R1, R24, R11, R3 and R2 have positive coefficients, tending to promote the combustion. Although the R8 elementary reaction also produces the H atoms, its net reaction rate is negative, suggesting that the H atoms are consumed to form the H_2O. Thus its sensitivity coefficient is negative. In addition, as the chine termination reactions, R9、R98 and R31 elementary reactions consume H and OH atoms, reducing the free radical concentrations of the whole reaction. So those elementary reactions have negative sensitivity coefficients, tending to inhabit the combustion.

When the dilution ratio is increased, the sensitivity coefficient of R1 has the most significant influence on the laminar burning velocity, and its influence increases greatest. The sensitivity coefficient of R105, R163, R9 and R98 also increases gradually, however the sensitivity coefficients of R24 and R8 decrease.

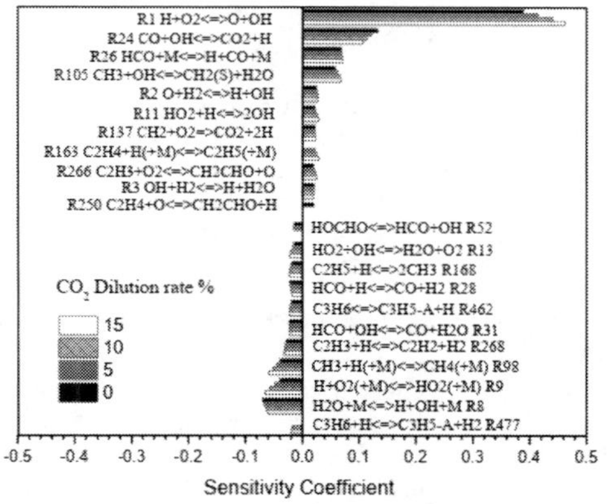

Figure 3: Sensitivity analysis of the laminar burning velocities for the stoichiometric *n*-heptane /air flames at different CO_2 dilution rates

3.1.2 Analysis of the concentration of radical species and the net reaction rate of elementary reaction

According to the modeled result of one-dimensional planar flames, the first five radical species which have the maximum mole fraction peaks are OH, H, O, CH_3, and HO_2 in order, and these peak values of mole fraction at different CO_2

dilution rate are shown in Fig. 4a. As seen in the figure, the peaks of mole fractions of OH, H, and O are greater than those of CH_3 and HO_2 by an order of magnitude. When elevating the CO_2 dilution rate, the peaks of the mole fractions of all of these five radical species decrease. In other words, to add CO_2 to *n*-heptane/air mixtures suppresses the radical production.

There is a strong influence of the dilution of CO_2 on the net reaction rates of the elementary reactions, as shown in Fig. 4b. Seven reactions are selected from Fig. 3 to show the influence. Here, regardless of the dilution rate, the peak net reaction rates of the reactions that have positive sensitivity coefficients are greater than those of the reactions that have negative sensitivity coefficients, thus enabling the radical pool to increase gradually. Then the reaction can be sustained. As the CO_2 dilution rates elevate, the peak net reaction rates of the reactions that have negative sensitivity coefficients almost keep steady. For the reactions that have positive sensitivity coefficients R1, R24 and R26, those peak net reaction rates remarkably decline.

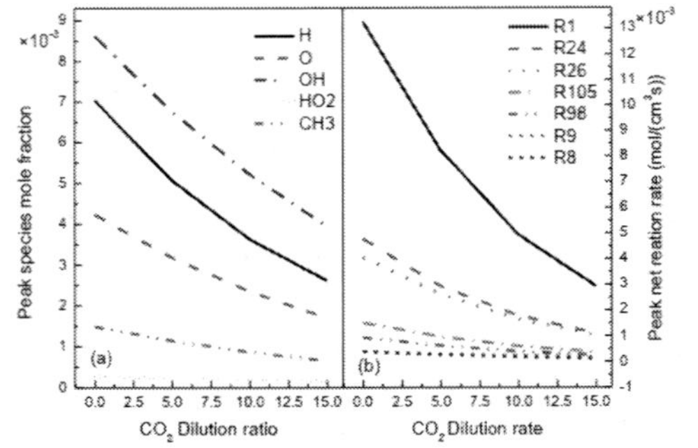

Figure 4: Peak values of the mole fraction of five radicals and the net reaction rates for the stoichiometric *n*-heptane/air flames at different CO_2 dilution rates

During the combustion process of hydrocarbon fuels, the concentration of H atom has a significant influence on the combustion reaction rate, and the effect of H atom mainly is reflected though the R1 reaction. R1 $H+O_2=O+OH$ is the most important chain branching reaction. This reaction consumes one H atom, then produces O and OH two radicals which can contribute to the rapid increase of the concentrations of free radicals. R24 $CO+OH=CO_2+H$ is the most significant chain propagation reaction, which consumes an OH atom and produce a H atom. It is also the main pathway of the oxidation of CO into CO_2. As above mentioned, with the increase of CO_2 dilution rate, the peak mole fraction of H atom and the net reaction rate of R1 and R24 in the *n*-heptane flames remarkably decrease, which is the reason that the laminar burning velocities decrease with the increase of CO_2 dilution rates.

3.2 Physical and Chemical Effects of the Dilution of CO_2

The overall effect of dilution of CO_2 on the laminar burning velocities includes the physical effect and chemical effect. In order to differentiate these two effects, a type of fictitious carbon dioxide F-CO_2 was designed in the present study. The F-CO_2 has the same thermodynamic and transport properties as normal CO_2. One-dimensional planar flames of *n*-heptane/air mixtures are simulated at CO_2 dilution rate of 5% and F- CO_2 dilution rate of 5%. The overall effect of CO_2 can be derived from the comparison of the modeled results between *n*-heptane/air mixtures without CO_2 and *n*-heptane/air mixtures with 5% CO_2 dilution rate, and the physical effect of CO_2 can be derived from the comparison between *n*-heptane /air mixtures without CO_2 and *n*-heptane/air mixtures with 5% F-CO_2 dilution rate. The chemical effect is equal to the overall effect minus the physical effect. Fig. 5 shows the variations of the physical and chemical effects of CO_2 on laminar burning velocities for *n*-heptane/air flames at the CO_2 dilution rate of 5% with the equivalence ratio. It can be seen that both physical and chemical effect suppress the laminar burning velocity over the range of the equivalence ratio and the most significant suppression effect is reached at the equivalence ratio of about 1.1. The physical effect of CO_2 is much stronger than the chemical effect on the laminar burning velocity.

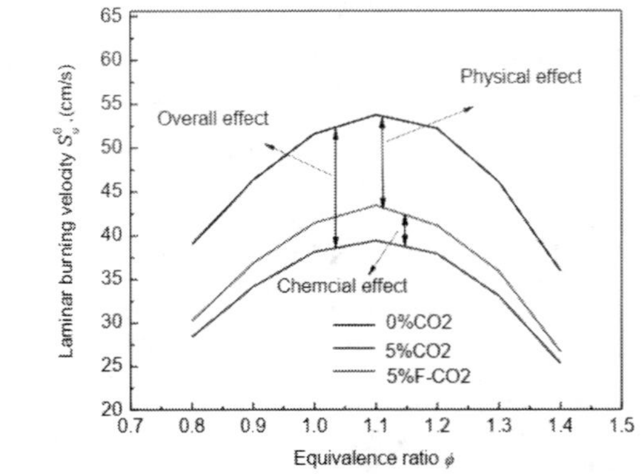

Figure 5: Variations of the physical and chemical effects of CO_2 on laminar burning velocities for *n*-heptane/air flames at the CO_2 dilution rate of 5% with the equivalence ratio

Shown in Fig. 6 are the physical and chemical effects of CO_2 on the radical production. From the figure, it is observed that both physical and chemical effects of CO_2 always show inhibition on the production of all five radical species. The physical and chemical effects of CO_2 on the production of the radicals H is at the same order of magnitude, while on the production of OH, O and CH_3, the physical effects of CO_2 are much stronger than the chemical ones. And the chemical effect of CO_2 on the production of HO_2 is marginal.

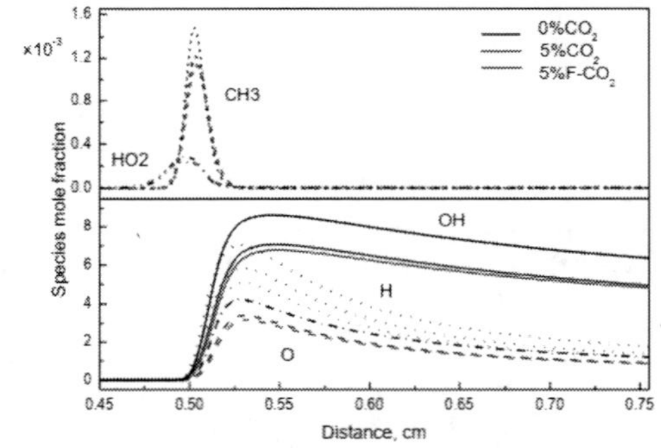

Figure 6: Physical and chemical effects of CO_2 dilution on the production of the radical species for the stoichiometric *n*-heptane/air flames

The behaviors of the physical and chemical effects of CO_2 dilution on the net reaction rates of the main elementary reactions are shown in Fig. 7. These elementary reactions have the three largest positive sensitivity coefficients and the three smallest negative coefficients. The physical and chemical effects of CO_2 always slow down the net reaction rates of the elementary reactions. In addition, the chemical effect of CO_2 on the elementary reactions with positive sensitivity coefficients is much stronger than the one on the elementary reactions with negative sensitivity coefficients.

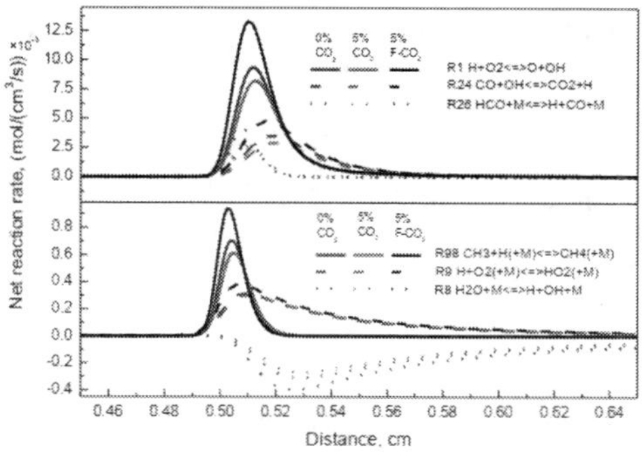

Figure 7: Physical and chemical effects of CO_2 on the net reaction rates of the selected elementary reactions in the stoichiometric *n*-heptane/air flames

4 Conclusions

The major conclusions of the study are as follows: (1) As the dilution rate of CO_2 elevates, the laminar burning velocities for *n*-heptane/air mixtures gradually decrease. The experimental values of laminar burning velocities have a reasonably good agreement with the modeled values. (2) The peak values of mole fractions of the main radical species and net reaction rates of the elementary reactions decrease with the increase of CO_2 dilution rate. (3) According to the sensitivity analysis of the laminar burning velocity, the sensitivity coefficient of R1 has the most significant influence on the laminar burning velocity, and most elementary reactions are influenced by the CO_2 dilution on the sensitivity. (4) Both physical and chemical effects of dilution by CO_2 on the *n*-heptane/air mixtures suppress the laminar burning velocity, but the former dominates. The physical and chemical effects of CO_2 always inhabit the production of the radical species mentioned here. The physical and chemical effects of CO_2 on the production of the radicals H is at the same order of magnitude, while on the production of OH, O and CH_3, the physical effects of CO_2 are much stronger than the chemical ones. And the chemical effect of CO_2 on the production of HO_2 is marginal. (5) The physical and chemical effect of CO_2 always inhibits the net reaction rates of the elementary reactions, and the chemical effect of CO_2 has a much weaker influence on the reaction rates than the physical effect.

5 Acknowledgment

This work was financially supported by the National Key Research and Development Program of China under Grant 2016YFC0205600, and the National Natural Science Foundations of China under Grant 51479149 and 51509198. The authors sincerely acknowledge "Fuels & Combustion Research Group in Wuhan University of Technology" for the assistance with setting up and testing experimental facilities.

References

[1]M. Zheng, G.T. Reader, J.G. Hawley. Energy Convers. Manage. 45(6) (2004) 883-900.

[2] J.P.J. Van Lipzig, E.J.K. Nilsson, L.P.H. De Goey, et al. Fuel 90(8) (2011) 2773-2781.

[3] S.P. Marshall, S. Taylor, C.R. Stone, et al. Combust. Flame 158(10) (2011) 1920-1932.

[4] L. Sileghem, V.A. Alekseev, J. Vancoillie, et al. Fuel 115 (2014) 32-40.

[5] I.Z. Syed, Y. Yeliana, A. Mukherjee, et al. SAE International Journal of Engines, 3(2010-01-0620) (2010) 517-528.

[6] K. Kumar, J.E. Freeh, C.J. Sung, et al. J. Propul. Power 23(2) (2007) 239-251.

[7] Z. Zhang, G. Li, L. Ouyang, et al. Int. J. Hydrogen Energy 36(20) (2011) 13194-13206.

[8] G. Li, Z. Zhang, J. Liang, et al. Int. J. Hydrogen Energy 37(5) (2012) 4490-4501.

[9] G. Li, J. Liang, Z. Zhang, et al. Energy Fuels 29(7) (2015) 4549-4556.

[10] A.P. Kelley, C.K. Law. Combust. Flame 156(9) (2009) 1844-1851.

[11] R.J. Kee, J.F. Grcar, M.D. Smooke, J.A. Miller, E. Meeks, PREMIX: A Fortran Program for Modeling Steady Laminar One-Dimensional Premixed Flames, Report No.SAND85-8240, Sandia National Laboratories, 1998.

[12] M. Mehl, W.J. Pitz, C.K. Westbrook, et al. Proc. Combust. Inst. 33(1) (2011) 193-200.

[13] K.K. Kuo, Principles of Combustion, 2nd ed, Wiley, New York, 2005, p. 457.

11th Asia-Pacific Conference on Combustion,
The University of Sydney, NSW Australia
10th -14th December 2017.

Experimental Study of Main-Post Injection under Compression-Ignition Engine Conditions, Part II: Two-color Pyrometry

P. Rorimpandey[1], C. Ming[1], Q.N. Chan[1,*], X.P. Pham[2], S. Kook[1], E.R. Hawkes[1], G.H. Yeoh[1], S. Yang[3], A.R. Masri[2]

[1]UNSW School of Mechanical and Manufacturing Engineering

UNSW Sydney, NSW, 2052, Australia

[2]School of Aerospace, Mechanical and Mechatronic Engineering

The University of Sydney, Sydney, NSW, 2006, Australia

[3]Beijing Hongshine New Energy Science and Technology Co Ltd

Zhichun Road, Haidan, Beijing 100086, China.

Abstract

A two-color pyrometry technique was applied to acquire spatially and temporally resolved soot temperature and soot distribution information of a main-post injection case, with a main to post injection mass ratio of 80% to 20% and a dwell time value of 1.5 ms, respectively. The measurements were performed in a quiescent steady environment of 19.4 kg/m^3 ambient density, 5.1 MPa ambient pressure, 900 K bulk temperature and 15 vol.% O_2 ambient concentration, inside a constant-volume combustion chamber (CVCC). A canola-based biodiesel was used for the experiments. The results reveal that for the test conditions of this study, the post injection is found to interact with the hot combustion gases at the tail of the main injection event, to ignite with a shorter ignition delay, and at a spatial location that is closer to the fuel injector. The combustion of the post injection event is observed to initiate and develop more rapidly, and at higher temperatures, when compared with that of the main injection event at equivalent injection timing. The soot concentration values of the post injection are also found to be lower and have a narrower distribution, when compared with the main injection.

1 Introduction

Laser-based techniques, including laser-induced incandescence [1, 2] and two-line atomic fluorescence [3, 4], have been used to provide the combustion research community with in situ, temporally and spatially resolved measurements that are important for understanding soot processes in flames. Producing experimental data sets in heavily sooty flames using laser-based techniques, however, has been challenging [5, 6], as significant interferences, such as absorption and scattering, which are associated with the presence of soot, can prevent many laser diagnostic tools from being reliably applied. Two-color pyrometry is a mature optical technique that has been applied to a multitude of fundamental and applied combustion environments. The technique is widely used by the combustion research community as it is relatively simple to implement. The technique can also be used to measure temporal and spatial evolution of soot, even under heavy soot load conditions where most laser-based techniques would fail.

In this study, two-color pyrometry technique is used to obtain spatially and temporally resolved soot temperature and concentration details for a main-post injection case. The measurements are carried out in a quiescent ambient condition inside an optically-accessible constant volume combustion chamber (CVCC). The main objective of this work is to use the two-color technique to analyze a main-post injection case, which has been presented in a companion paper [7].

2 Experimental details

The experimental layout used has been detailed in a companion paper [7]. In brief, an optically-accessible CVCC was used to simulate the operation condition of a compression-ignition engine (19.4 kg/m^3 ambient density, 5.1 MPa ambient pressure, 900 K bulk temperature and 15 vol.% ambient O_2 concentration). For this study, a double-injection case that was tailored to represent a 'main plus post' injection schedule was investigated. A total injected fuel mass of 10 mg, with a fixed fuel mass ratio of the main to post injections of 80% to 20%, was used. The dwell time (DT), which is defined here as the time between the end of electronic command of the main and the actual start of injection for the post, between the injections, was set as 1.5 ms. In this study, a canola-based biodiesel was utilized and was injected at an injection pressure value of 100 MPa into the CVCC [7].

2.1 Two-color Pyrometry

A classical two-color pyrometry technique was used to provide the two-dimensional temperature and soot distributions of the main-post injection case. A high-speed CMOS digital color camera (Photron SA5), which was equipped with a f#1.4 Nikkor lens, was used to implement this diagnostics technique. The high-speed camera was operated at 15,000 frames per second (fps) with an exposure time of 60 μs. The pixel resolution of the image captured was ~0.2 mm per pixel. The camera was synchronized to the start of injection of the injector, and ~120 images were captured for each run. All experiments were repeated at least five times to ensure good repeatability.

*Corresponding author.
E-mail address: qing.chan@unsw.edu.au

The mathematics of the two-color technique has been described in various literatures [8, 9]. A complete treatment is therefore not provided here. In essence, the signal from an optical detector viewing a radiating soot cloud can be described using a Planck's blackbody function that is modified by wavelength-dependent soot emissivity model, as well as the combined wavelength-dependent responsivity of the detection system used (optics train, filter and CMOS detector). The optical response of the CMOS detector used, for each of its color channel, is shown in Fig. 1. The pixel output, S, when observing soot irradiation at wavelength, λ, can therefore be described as:

$$S_\lambda = \int \epsilon_{soot} \frac{C_1 Q(\lambda)}{\lambda^5[\exp(\frac{C_2}{\lambda T_{soot}})-1]} d\lambda. \quad (1)$$

In Eq. 1, $Q(\lambda)$ is the optical system response coefficient of each wavelength, C_1 and C_2 are the Planck's first and second constants, and T_{soot} is the soot temperature. The soot emissivity, ϵ_{soot}, can be modeled with the empirical correlation of Hottel and Broughton [10]:

$$\epsilon_{soot} = 1 - \exp\left(-\frac{KL}{\lambda^\alpha}\right), \quad (2)$$

where α in Eq. (2) is given as [8]:

$$\alpha = 1.22 - 0.246 \times \ln(\lambda). \quad (3)$$

The K and L in Eq. (2) refer to the wavelength-dependent absorption coefficient and the gas layer thickness, respectively. Their product, KL, is widely used to characterize the soot content in various combustion systems [8, 9]. From Eq. 1, it can be seen that when the optical response coefficient, $Q(\lambda)$, is known, the S_λ will become a unique function of two unknown variables, T_{soot}, and KL. Given two unknowns, any two of the three color channels can therefore be used to produce two equations to solve for T_{soot}, and KL. In this study, only the red and green channels were used for the two color calculations, as these two channels have higher signal-to-noise ratios than blue (SNRs of 47, 21 and 9, for red, green and blue channels, respectively). The optical response coefficients of the detection system were determined through calibration with a tungsten filament light source of known spectral emission properties.

To reduce the computational time during image processing, two universal surfaces, which map the domains of the red and green pixel intensities to soot temperature and soot KL surfaces, were generated. By matching the pixel intensities recorded for the red and green channels to the parameter spaces of the universal surfaces, the implicit solutions to Eq. 1 can be determined in a time-efficient manner. It should be noted that the two-color technique is a line-of-sight method and therefore, its accuracy is susceptible to the spatial gradients in soot properties along the optical path [9]. It is therefore necessary to consider this limitation, when interpreting the two-color data obtained.

3 Results and Discussion

The rate of injection profile of the main-post injection case, measured using a Bosch tube type injection rate meter, is shown in Fig. 2. The two-color data of the main and post injections are compared at t_1=0.4 ms vs t_2=4.7 ms, t_1=0.6 ms vs t_2=4.9 ms and t_1=0.8 ms vs t_2=5.1 ms, with t_1 and t_2 corresponding to the time after start of injection (aSOI), for the main and post injections, respectively. A shift of 4.3 ms between the two injections has been selected, so that the same amount of fuel has been injected for the main and post injections at all time instants. The time instants have also been selected from the quasi-steady injection period, so that the

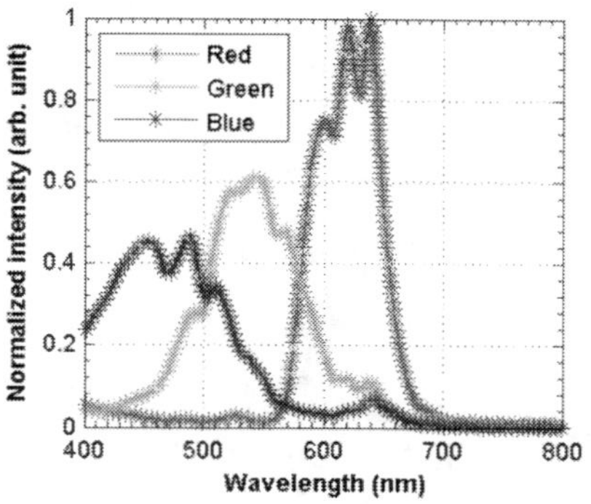

Figure 1: Normalized optical response of the CMOS camera.

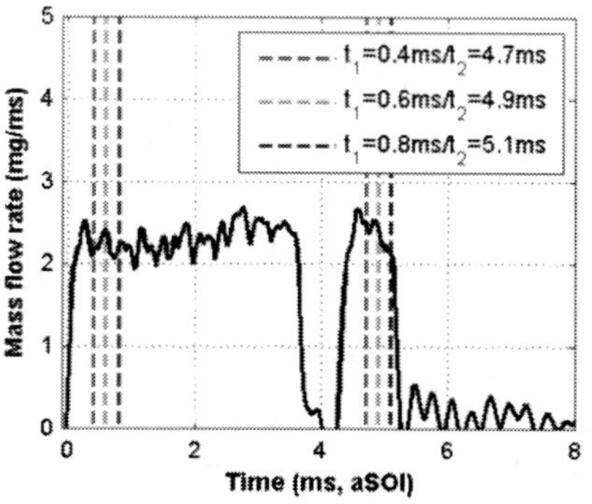

Figure 2: Injection rate profile for the main-post injection case, with a fuel mass ratio of main to post injections of 80% to 20%, and a dwell time (DT) of 1.5 ms. t_1 and t_2 represent the time after the start of injection for the main and post injections, respectively. The two-color data for the main and post injections are compared at t_1=0.4 ms vs t_2=4.7 ms, t_1=0.6 ms vs t_2=4.9 ms and t_1=0.8 ms vs t_2=5.1 ms.

findings of this study are not affected by the transient behaviors of the injections at the start and end of injection periods.

Figure 3 presents the soot temperature and soot KL distributions for the main-post injection case, at the three selected time instants. To facilitate direct, qualitative comparison, the soot temperature and soot KL images are divided into two, with the main and post injections presented at the left and right halves, respectively. The same color scaling is also applied to all images to ease qualitative comparison. For the soot temperature and soot KL images that were produced for the first time instant, it can be seen for the post injection event at t_2=4.7 ms, flame is detectable at an axial distance of ~20 mm, and the soot region is found to extend to an axial distance of ~28 mm from the injector nozzle. No data, however, is observed at t_1=0.4 ms, which indicates that the combustion of the main injection has not commenced at this time instant. For the images that were generated for second time instant, it can be seen that the flame region becomes detectable for both the main and post injections, at an axial distance of ~24 mm from nozzle. The flame region of the post injection at t_2=4.9 ms, however, is observed to have extended to an axial location of ~41 mm, which is downstream than that of the main injection at t_1=0.6 ms (~31 mm). For the third time instant, the same trends are observed to persist, such that the soot regions of both main and post injections at t_1=0.8 ms and t_2=5.1 ms become observable at similar axial distances from the nozzle (~25 mm).

The span of the soot region of the post injection at t_2=5.1 ms is also found to be longer than that of the main injection at t_1=0.8 ms. From the images, it can also be seen at all three of the selected time instants, the combustion of the main injection is found to be still active, when the combustion of the post injection is progressing. Nonetheless, the main and post injection combustion events are observed to be spatially separated and clearly distinguishable during these time instants. Together, these results indicate that there is no direct interaction between the main and post combustion events. Nonetheless, the interaction between the fuel injected from the post injection, with the hot combustion products of main, appear to be sufficient to advance the ignition of the post injected fuel, and subsequently, the location of the flame base. The flame zone of the post injection event is also observed to develop faster and farther than the main injection at the corresponding injection time. The faster flame penetration/development observed may be attributable to the post injection entering a reduced local density environment caused by presence of hot combustion products of the main injection, as well as slipstream effects [11].

The soot temperature and soot KL distributions images recorded at the three selected time instances are also used to analyze their correlations. Figure 4 presents the joint probability density functions (PDFs) between the soot temperature and soot KL for the three selected time instants, with the pseudo color contour representing the probability of the data set. From the PDFs, it can be seen that the span and the measured soot KL values decrease as the measured soot temperature values increase. It is important to note that the measurement results were generated using the two-color pyrometry technique, which is a line-of-sight method. The two-color signal recorded is therefore effectively weighted averages along its optical path. In previous studies (*e.g.*, [12]), optical diagnostics have shown that diesel jets have a relative cooler inner jet core that is surrounded by a thin diffusion flame, but with a comparatively uniform soot distribution. The two-color data obtained from a diffusion flame would therefore be more weighted towards the soot present in the hotter diffusion flame sheath, than in the flame core. In addition, as flame temperature increases, the weight of the soot signal from the flame sheath would also increase in relation to the soot signal from the flame core, such that the two-color data obtained becomes more representative of the soot properties in the flame edge only. All these can contribute towards the lower measured soot KL and span at higher flame temperatures observed in the PDFs.

For the PDFs that are generated for the first time instant, no data point is detected at t_1=0.4 ms, which is to be expected. The measured soot temperature values at t_2=4.7 ms, on the other hand, are found to be distributed between 1900 K and 2500 K. It should be noted that the observed distribution at t_2=4.7 ms is representative of the burnout of the main injection, which is the dominant combustion event at this time instant. From the PDFs that are produced for the second and third time instants, it can be seen that the soot temperatures measured at t_2=4.9 ms and 5.1 ms, are distributed between 1900 K and 2700 K, which are not only higher than the temperature values of the main injection combustion event observed at t_1=0.6 ms and 0.8 ms (between 1600 K and 2380 K), but also for the combustion process of the main injection at t_2=4.7 ms. Together, these results imply the combustion of the post injection event is occurring at higher temperature values. It should also be noted, however, the two-color pyrometry technique has been reported to have a temperature uncertainty of 150 K [9]. From the PDFs presented in Fig. 4, it can be seen that the most probable soot KL factors of the post injection is observed to have a narrower distribution and lower value than the main. It is possible that

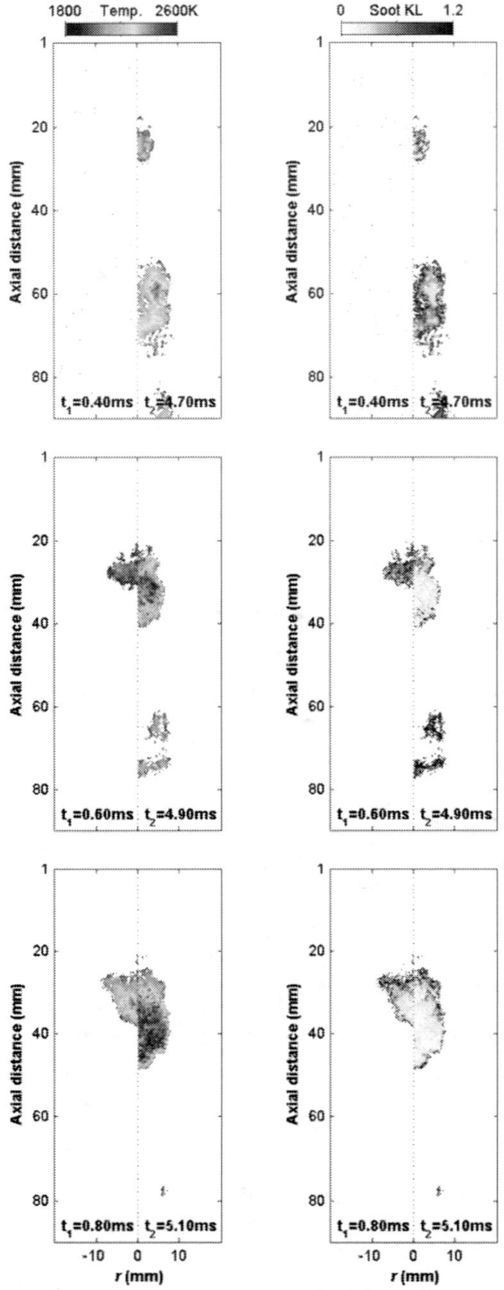

Figure 3: Comparisons of spatial distributions of typical, instantaneous soot temperature and soot KL of the main (left half) and post injection (right half), at selected time steps. The fuel is injected from the top to bottom, in these images.

the lower soot KL values observed could be a result of the main-post injection schedule used. Nonetheless, it is also likely that lower soot KL values observed may be a result of the two-color data becoming more biased towards the soot properties at the flame surface, due to the higher temperature values of the post injection combustion process. Further investigation is therefore required to elucidate and extend the current temperature and soot KL results. It should be noted that from Fig. 3, it can be seen that whilst the combustion of the main injection is still active at t_2=4.9 ms and 5.1 ms, the combustion

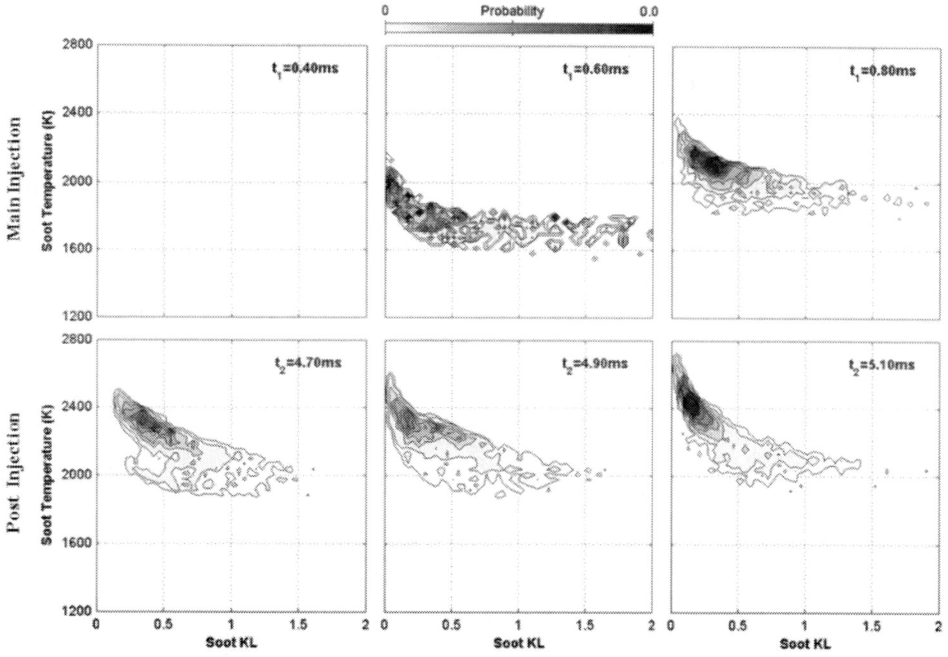

Figure 4: Comparisons of joint PDFs of soot temperature and soot KL of the main and post injections at the three selecte d time steps. All time instants have the same color scale.

regions that are associated with the main injection are small, and is assessed to not significantly influence the overall shapes and trends of the PDFs at these time instants.

4 Conclusion

Two-color pyrometry technique was applied to a main-post injection case, under a compression-ignition engine environment of 19.4 kg/m^3 ambient density, 5.1 MPa ambient pressure, 900 K bulk temperature, and 15 vol.% O_2 ambient concentration. The mass ratio of the main to post injection was specified as 80% to 20%, and the dwell time between the injections was fixed at 1.5 ms. A canola-based biodiesel was used. The results indicated that for the test conditions of this study, the fuel from the post injection interacted with hot combustion gases at the tail of the main injection event, to ignite at a shorter ignition delay and at a spatial location that is closer to the injector. When compared with the combustion processes of the main injection at equivalent injection timing, the interaction between the main and post was found to cause (1) the flame zone of the post injection to develop rapidly at higher temperatures after ignition, and (2) the distribution of the most probable soot KL factors for the post injection to be narrower, and have lower soot content values. Further studies, however, are required to confirm the observed soot concentration trends.

Acknowledgment

The authors wish to acknowledge the financial support of Australia Research Council (ARC) and Australian Academy of Technological Sciences and Engineering (ATSE) Priming Grant. The authors would also like to acknowledge the contributions of Brad Moody and I.M. Rizwanul Fattah.

References

[1] W. Jing, Z. Wu, W. L. Roberts, T. Fang, Fuel 181 (2016) 718-728.

[2] N. H. Qamar, Z. T. Alwahabi, G. J. Nathan, Q. N. Chan, Combust Flame 158 (2011) 2458-2464.

[3] P. R. Medwell, Q. N. Chan, P. A. M. Kalt, Z. T. Alwahabi, B. B. Dally, G. J. Nathan, Appl Opt 48 (2009) 1237-1248.

[4] P.R. Medwell, Q.N. Chan, B.B. Dally, S. Mahmoud, Z.T. Alwahabi, G.J. Nathan, Proc Combust Inst 34 (2013) 3619-3627.

[5] Q. N. Chan, P. R. Medwell, Z. T. Alwahabi, B. B. Dally, G.J. Nathan, Appl Phys B 104 (2011) 189-198.

[6] G.J. Nathan, P.A.M. Kalt, Z.T. Alwahabi, B.B. Dally, P.R. Medwell, Q.N. Chan, Prog Energy Combust Sci 38 (2011) 41-61.

[7] C. Ming, Q.N. Chan, X.P. Pham, A.R. Masri, S. Kook, E.R. Hawkes, G.H. Yeoh, S. Yang, Asia-Pacific Conference on Combustion, University of Sydney, Australia, 2017.

[8] K. Zha, R. -C. Florea, M. Jansons, Proceedings of the ASME 2011 Internal Combustion Engine Division Fall Technical Conference, West Virginia University, USA, 2011, ICEF2011-60146.

[9] K. I. Svensson, A. J. Mackrory, M. J. Richards, Dale R. Tree, SAE Paper 2005-01-0648 (2005).

[10] H. Hottel,F. Broughton, Ind Eng Chem 4 (1932) 166-175.

[11] S. Skeen, J. Manin, L.M. Pickett, SAE Int. J. Engines 8 (2015), 696-715.

[12] J. Dec, SAE Paper 970873 (1997).

11th Asia-Pacific Conference on Combustion,
The University of Sydney, NSW Australia
10th -14th December 2017.

Experimental Analysis of the Effect of Fuels on the Flame Describing Function of a Swirl-Stabilized Premixed Flame

Francesco Di Sabatino[1], Thibault F. Guiberti[1], Jonas P. Moeck[2], William L. Roberts[1], Deanna A. Lacoste[1]

[1]Clean Combustion Research Center, King Abdullah University of Science and Technology,
Thuwal, 23955-6900, Saudi Arabia

[2]Institut für Strömungsmechanik und Technische Akustik, Technische Universität Berlin,
Berlin 10623, Germany

Abstract

A study of the fuel-dependence of the Flame Describing Function (FDF) of a swirl-stabilized premixed flame is carried out. A swirl-stabilized burner, a hot wire to measure the velocity and a loudspeaker system to generate acoustic perturbations compose the experimental apparatus. The optical diagnostics include a photomultiplier tube to collect CH* chemiluminescence and an ICCD camera to acquire phase-locked images of OH* chemiluminescence to analyze the flame dynamics. A propane/air mixture with an equivalence ratio (Φ) equal to 0.7 and a methane/air mixture with an equivalence ratio equal to 0.67 are considered. The equivalence ratios are slightly different for the two fuels to maintain similar flame shape and size. Five different levels of forcing amplitude, from 5% to 25% of relative velocity fluctuation, are examined. The gain of the two FDFs shows the characteristic behavior for this type of flame: a local minimum followed by a local maximum. Moreover, increasing the forcing amplitude, the value of the local maximum of gain decreases. This is verified for the two considered fuels. Analyzing the phase-locked images of OH* chemiluminescence, it is possible to conclude that the size of the vortex roll-up at the tip of the flame plays an important role in the reduction of the value of the local maximum of gain when methane replaces propane.

1 Introduction

Modern low-emission combustor technology for stationary gas turbines and aero-engines suffers from the occurrence of combustion instabilities. This dynamic phenomenon effectively limits the operating range of the engine and may even lead to structural failure in extreme cases [1]. Combustion instabilities arise when the unsteady combustion process couples with the acoustic modes of the combustor [2]. The Flame Describing Function, which relates heat release rate fluctuations to velocity fluctuations in premixed flames, is generally used to analyze and model this interaction [3]. Commonly, a swirler is utilized to stabilize the turbulent flame in the combustor [4]. Therefore, the study of the response of a swirled flame to acoustic excitation is highly relevant. In the study of *Palies et al.* [5], the response of the flame to increasing level of forcing amplitude is analyzed in terms of the FDF. They highlighted that increasing the forcing amplitude decreases the value of the local maximum of gain. This nonlinear saturation is essential for the formation of limit cycle oscillations and determines the final pulsation amplitude. Furthermore, the characteristic shape of the gain of the flame describing function was shown to exhibit a local minimum followed by a local maximum. This strong variation in the gain response of swirl flames has been attributed to interference effects between axial velocity fluctuations associated with the acoustic wave and axial vorticity fluctuations generated by the interaction of the former with the swirler [5, 6, 7]. Other studies [8, 9] investigated the effects of swirl number, swirler geometry, and equivalence ratio on the FDF. However, studies on the effect of fuel on the flame describing function of swirled flames are scarce, and there is a lack of understanding of the influence of this parameter on the response of flames to acoustic modulation. This is the purpose of the present study. Flames of propane/air and methane/air are investigated in terms of gain and phase of the FDF. Moreover, the flame motion is analyzed using phase-locked images of OH* chemiluminescence.

2 Experimental Setup

The experimental apparatus is composed of a swirl-stabilized burner with a swirl number of 0.39 and a FDF measurement system. The burner is composed of a plenum and of a convergent section followed by an injection tube. A radial swirler is located upstream of the injection tube. The 4 kW flames are confined by a quartz tube of 70 mm of inner diameter and 100 mm length. A detailed description of the burner can be found in *Lacoste et al.* [10]. For perfectly premixed flames, the heat release rate can be assumed proportional to the CH* chemiluminescence intensity as shown in [11, 12]. The FDF system (see Fig. 1) comprises a photomultiplier tube (PMT) (Hamamatsu H10721-110) with a band-pass filter centered at 430 nm to collect heat release rate fluctuations through CH* chemiluminescence and a hot wire (HW) (Dantec miniCTA) to measure velocity perturbations upstream of the swirler. This measurement location will not introduce significant changes compared to a velocity measurement at the flame base, as shown in *Palies et al.* [5]. The experimental setup also includes a loudspeaker (Beyma 10LW30/N) connected to a high-fidelity amplifier (QSC GX5). The speaker is driven by a signal generator (NF WF1973) and generates the velocity fluctuations needed for the evaluation of the flame describing function. All data are recorded with an oscilloscope (Agilent Technologies Infiniium 2.5 Ghz). An ICCD camera (Princeton Instruments PI-MAX1) equipped with a band-pass filter centered at 310 nm is used to collect phase-locked images of OH* chemiluminescence. These images are used to analyze the flame motion. Synchronization between image acquisition and acoustic forcing is facilitated by connecting a function generator to the triggering system of the

Corresponding author.
E-mail address: francesco.disabatino@kaust.edu.sa

camera.

The methane/propane and air flows are set by thermal mass flow controllers (Brooks SLA 58 series) and are adjusted to provide an equivalence ratio of 0.7 for propane and of 0.67 for methane. A thermal power of about 4 kW is established for both fuels. A Reynolds number of 7800 for methane and of 7500 for propane is fixed considering a nozzle diameter of 18 mm.

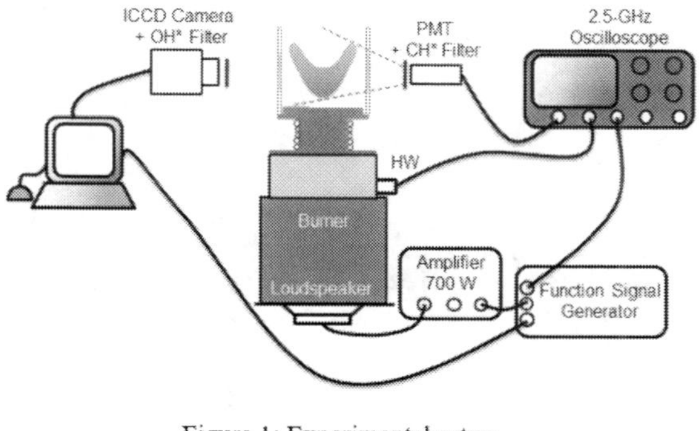

Figure 1: Experimental setup.

3 Flame Describing Function

The FDF is defined in the frequency domain as:

$$FDF(\omega, U') = \frac{\dot{Q}'/\bar{\dot{Q}}}{U'/\bar{U}} \quad (1)$$

where $\dot{Q}'$ is the heat release rate fluctuation, $\bar{\dot{Q}}$ is the mean heat release rate, U' is the axial velocity fluctuation and $\bar{U}$ is the mean axial velocity. Therefore, Eq. 1 represents the ratio of the relative heat release rate fluctuation over the relative axial velocity fluctuation. To evaluate the FDF, the incoming flow is perturbed by the loudspeaker actuated with a sinusoidal voltage signal. The frequency of the perturbations examined varies from 32 Hz to 400 Hz with increments of 16 Hz. The level of forcing amplitude is increased from 5% to 25% with increments of 5%, and it corresponds to the rms perturbation level determined at the hot wire location. The axial velocity and CH* chemiluminescence signals are recorded during 10 seconds for each excitation frequency.

4 Results and Discussion

4.1 Considerations on the Flame Shape

The equivalence ratio of the two fuels is selected to provide a similar flame shape and size as shown in Fig. 2.

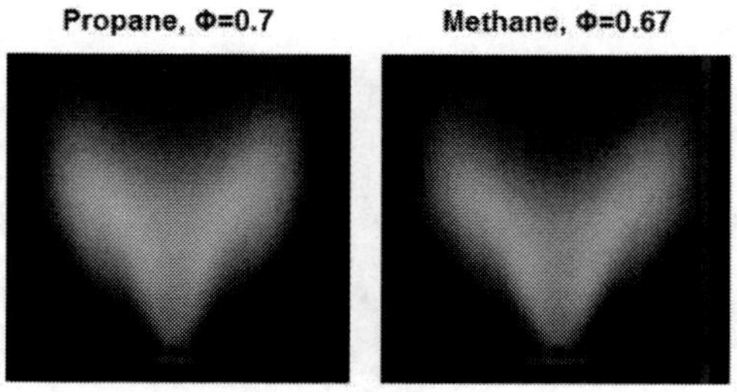

Figure 2: Comparison of the flame shapes for propane (left) and methane (right).

The two pictures are collected with a Nikon DSLR camera (D810) with a f-number of 4, an exposure time of 0.625 seconds and an ISO number of 250. Analyzing Fig. 2, it is possible to observe the characteristic "V" shape of this type of flames. The two flames are symmetric with respect to the central rod and are very similar to each other.

4.2 Flame Describing Function Evaluation

Using signals that are recorded with the hot wire and photomultiplier, it is possible to evaluate the FDFs for the two fuels in terms of gain and phase (see Figs. 3 and 4). The presence of a local minimum of gain followed by a local maximum of gain is observed for both fuels. This behavior is the result of the competition between two phenomena [5]: the oscillation of the flame base angle due to swirl number fluctuations and the vortex roll-up at the tip of the flame due to vortex shedding. When these two phenomena are out of phase, the top and the bottom part of the flame oscillate in phase opposition, resulting in a gain minimum. When these two phenomena are in phase, the top and the bottom part of the flame oscillate in phase generating larger gain values. Figures 3 and 4 also show that the value of the gain at the local minimum (near 100 Hz) is fairly independent of the forcing amplitude while the value of the gain at the local maximum (near 160 Hz) decreases when the forcing amplitude is increased. This behavior agrees with previous studies using a similar burner configuration [5, 8]. Up to the largest forcing amplitude considered in this study, the FDF phase remains essentially unaffected.

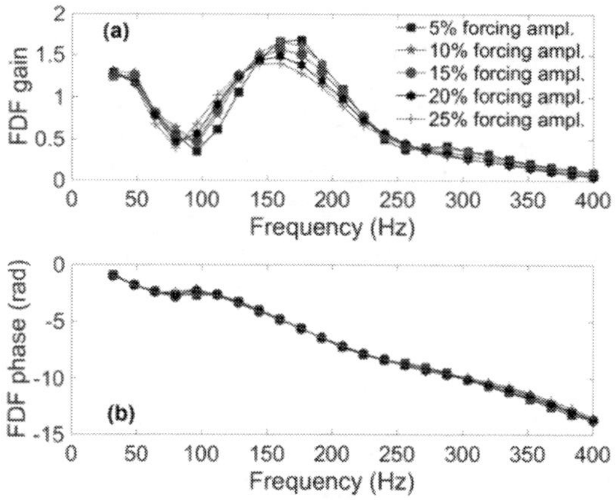

Figure 3: Gain (a) and phase (b) of the FDF for propane.

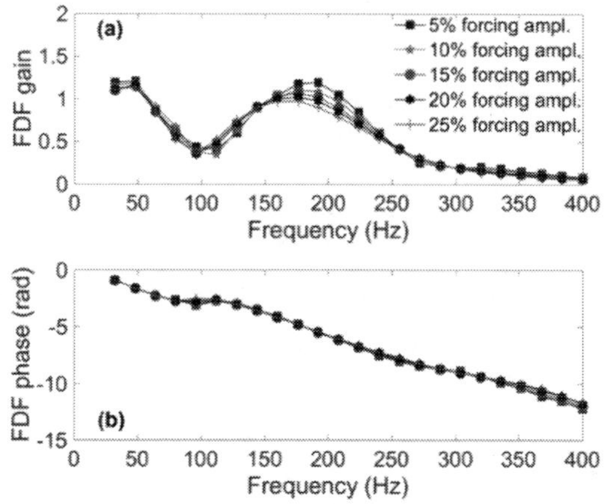

Figure 4: Gain (a) and phase (b) of the FDF for methane.

The FDF recorded for a forcing amplitude of 25% for propane and methane fuels are repeated in Fig. 5 for comparison purposes.

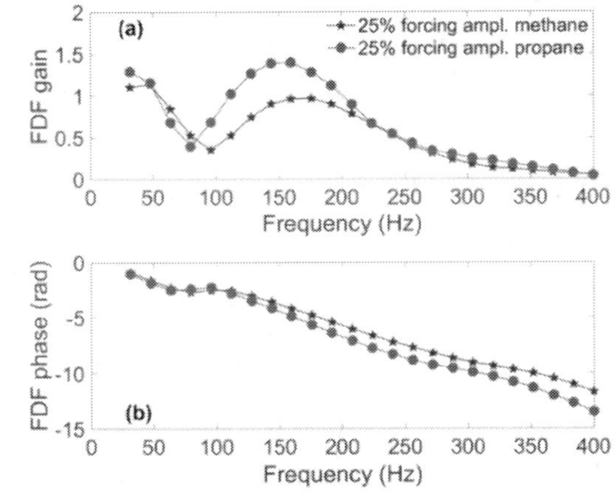

Figure 5: Comparison of the gain (a) and phase (b) of the FDF at 25% of forcing amplitude for propane and methane fuels.

It is shown that the value of the gain at the local maximum (near 160 Hz) is larger for propane than for methane. Conversely, the value of the gain at the local minimum (near 100 Hz) is very close for both fuels. To understand the different gain values obtained near 160 Hz, an analysis of the flame motion is carried out at this frequency using phase-locked images of OH* chemiluminescence.

4.3 Flame Motion Analysis

Ten phases, one each 36°, in the oscillation period (see Figs. 6 and 7) are considered for the flame motion study. The phase is defined with respect to the velocity signal. The gate width of the ICCD camera is selected equal to 300 μs and 480 images are averaged for each phase. In Figures 6 and 7, the mean signal does not fit perfectly the sample one because the mean is evaluated for the whole signal (10 seconds) while the sample is an instantaneous datum.

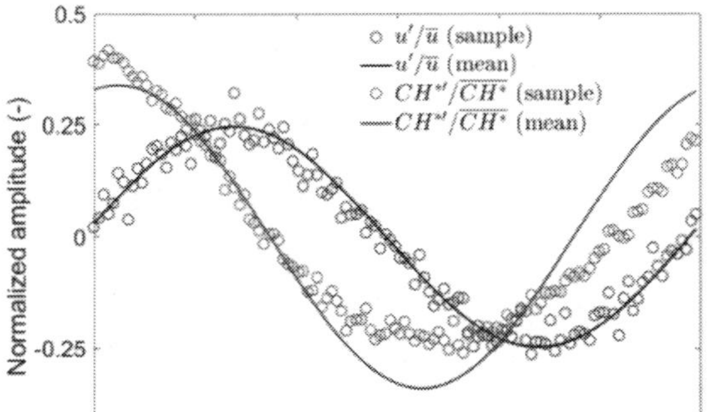

Figure 6: Normalized velocity (blue) and chemiluminescence intensity (red) fluctuation amplitude as a function of the phase at a forcing frequency of 160 Hz and a forcing amplitude of 25% for propane.

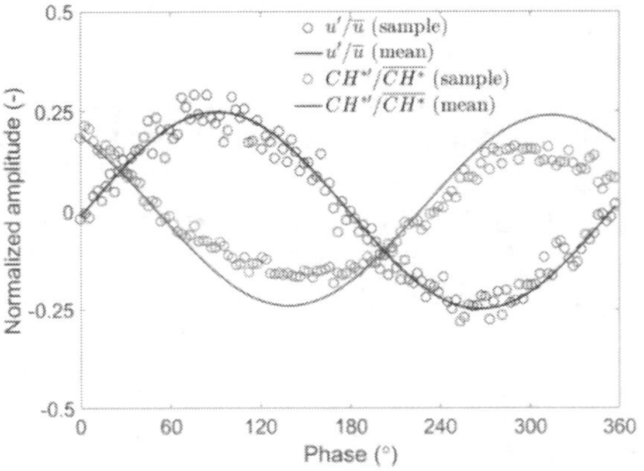

Figure 7: Normalized velocity (blue) and chemiluminescence intensity (red) fluctuation amplitude as a function of the phase at a forcing frequency of 160 Hz and a forcing amplitude of 25% for methane.

As previously stated, the value of the gain at the local maximum (near 160Hz) is linked to the relative oscillation of the top part of the flame with respect to the bottom part. In Figure 8, the normalized OH* chemiluminescence intensity for the top and bottom parts of the flame is plotted as a function of the phase for propane. Similar results are obtained for methane. The top and bottom parts are defined, for each fuel, by the threshold line shown in Fig. 9 that correspond to the position where the vortex roll-up starts.

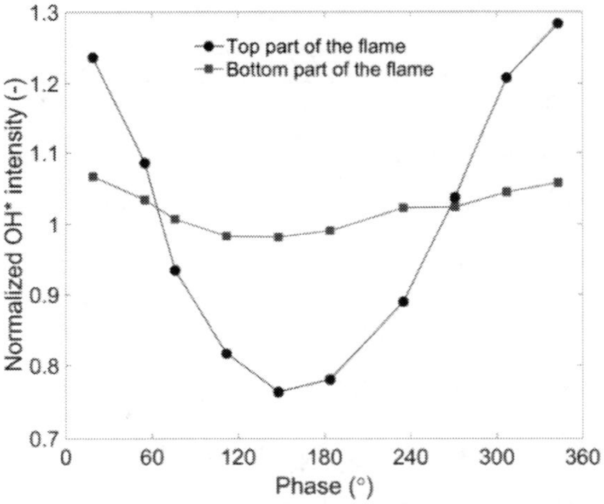

Figure 8: Normalized OH* intensity of the top and the bottom parts of the propane flame at a forcing frequency of 160 Hz and a forcing amplitude of 25%.

The top and bottom parts of the flame are oscillating almost in phase (Fig. 8). However, the amplitude of the oscillation of the top part is larger than the one of the bottom part, highlighting that the dynamics of the top part is predominant. Due to the axisymmetric nature of the flames, the Abel-inverted phase-locked images of OH* chemiluminescence (see Fig. 9) clarify why the motion of the top part of the flame is predominant.

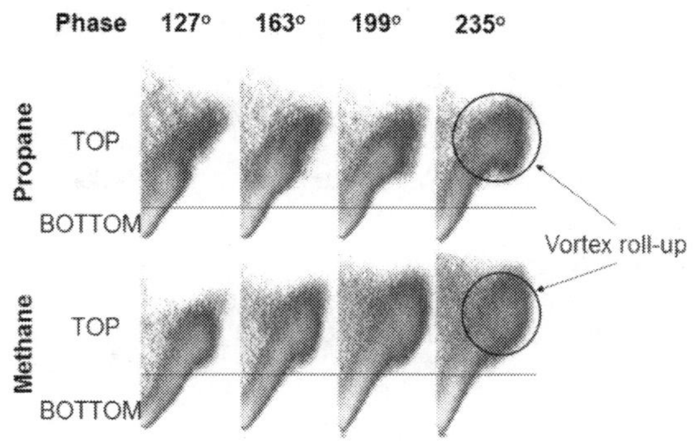

Figure 9: Phase-locked Abel-inverted images of OH* chemiluminescence at a forcing frequency of 160 Hz and a forcing amplitude of 25% for propane and methane.

From Fig. 9, it is possible to observe that the bottom part of both flames does not move significantly during the oscillation cycle while the top one does due to vortex roll-up. This phenomenon highlights that the vortex roll-up plays a fundamental role in the definition of the value of the gain at the local maximum near 160 Hz. In fact, comparing the maximum dimension of the vortex roll-up at the tip of the flame between propane and methane, a bigger size is observed for propane than for methane. Therefore, the difference in gain between the two fuels can be related to the difference in dimension of the respective vortex roll-ups. An estimate of the size of the vortex shows 24 mm for propane and 17.5 mm for methane.

5 Conclusions

An experimental analysis of the effect of different fuels on the flame describing function of a swirl-stabilized premixed flame has been carried out. Two different fuels have been examined: propane with $\Phi = 0.7$ and methane with $\Phi = 0.67$. Two different equivalence ratios have been selected in order to obtain similar flame shape and size for the two fuels. Five different levels of forcing amplitude, from 5% to 25%, have been considered.

The characteristic behavior of the gain of the FDF, a local minimum followed by a local maximum, has been observed for both fuels. A reduction of the value of the gain at the local maximum with increasing forcing amplitude has also been highlighted. Changing from propane to methane, the gain for the frequencies around this local maximum decreases. Using phase-locked images of OH* chemiluminescence, the flame dynamics at the local maximum near 160 Hz has been analyzed for both fuels. The predominance of the dynamic of the top part of the flame with respect of the bottom part has been underlined. The vortex roll-up at the tip of the flame has been recognized as an important factor in the definition of the gain. In particular, the larger gain of the FDF for propane with respect to methane can be associated to a bigger size of the vortex roll-up.

6 Acknowledgment

This work is partly funded by the Center of Competitive Funding from King Abdullah University of Science and Technology (Grant number 1975).

References

[1] T.C. Lieuwen, V. Yang, Instabilities in Gas Turbine Engines, vol. 210 of Progress in Astronautics and Aeronautics, AIAA, Reston, VA, 2005.

[2] S. Candel, Proc. Combust. Inst. 29 (2002) 1-28.

[3] S. Candel, D. Durox, T. Schuller, J.F. Bourgouin, and J.P. Moeck, Annu. Rev. Fluid Mech. 46 (2014) 147-173.

[4] Y. Huang, V. Yang, Prog. Energy Combust. Sci. 35 (2009) 293-364.

[5] P. Palies, D. Durox, T. Schuller, S. Candel, Combust. Flame 157 (2010) 1698-1717.

[6] T. Komarek and W. Polifke, J. Eng. Gas. Turbines Power 132 (2010) 061503 (7 pages).

[7] K.T. Kim, D.A. Santavicca, Combust. Flame 160 (2013) 1441-1457.

[8] P. Palies, D. Durox, T. Schuller, S. Candel, Combust. Sci. Technol. 183 (2011) 704-717.

[9] R. Balachandran, B.O. Ayoola, C.F. Kaminski, A.P. Dowling, E. Mastorakos, Combust. Flame 143 (2005) 37-55.

[10] D.A. Lacoste, J.P. Moeck, D. Durox, C.O. Laux, T. Schuller, J. Eng. Gas Turbines Power (2013).

[11] I.R. Hurle, R.B. Price, T.M. Sugden, A. Thomas, Proc. Roy. Soc. A. 303 (1968) 409-427.

[12] R.B. Price, I.R. Hurle and T.M. Sugden, Symp. (Int.) Combust. 12 (1971) 1093-1102.

11th Asia-Pacific Conference on Combustion,
The University of Sydney, NSW Australia
10th -14th December 2017.

Experimental Study on the Effect of Lubricant Oil on Ignition Characteristics of Hydrocarbons Using a Rapid Compression Machine

Yuusuke Wachi[1], Kazuki Iwakura[1], Kotaro Tanaka[1], Mitsuru Konno[1], Ying Jiang[2], Yasuyuki Sakai[2]

[1]Ibaraki University

4-12-1 Nakanarusawa, Hitachi, Ibaraki, 316-8511, Japan

[2] University of Fukui

3-9-1 Bunkyo, Fukui, Fukui, 910-8507, Japan

Abstract

Downsizing direct injection spark ignition (DISI) engines are one of the most effective and practical ways to achieve good fuel economy. Abnormal combustion, known as pre-ignition, is an obstacle to enhancing the gain of DISI engines. It is considered that this phenomenon occurs due to the ignition of lubricant oil droplets diluted by fuel.

In this study, in order to investigate the influence of lubricant oil and lubricant oil components on the ignition characteristics of fuel, ignition delay times of four types of fuels, isooctane, ethyl tert-butyl ether (ETBE), toluene, and 2,5-dimethylfuran (DMF), with/without lubricant oil and lubricant oil components, were measured using a rapid compression machine.

The ignition delay time of all the fuels with the lubricant oil were shorter than those without the lubricant oil. It was found that the factor contributing to the enhancement of the ignition depends on the fuel; in the case of isooctane, calcium included in the lubricant oil had the strongest influence on the fuel ignition, and, in the case of toluene, the lubricant base oil constituent had the strongest influence on the fuel ignition.

1 Introduction

In recent years, the depletion of fossil fuels and global warming, caused by carbon dioxide emissions from fossil fuel combustion, as well as air quality problems, have become more serious. For internal combustion engines, further reduction of carbon dioxide emissions and fuel consumption, through higher efficiency, is urgently required. With this motivation, the concept of engine downsizing has been the focus of much attention. Downsizing is undertaken to achieve higher efficiency by reducing engine displacement, and simultaneously compensating for the loss of torque/power by turbocharging. Generally, downsized engines use gasoline direct injection (GDI) for mitigating knocking by the mixture cooling effect, so that the downsizing ratio and compression ratio can be set higher. With the development of highly turbocharged direct-injection gasoline engines, low-speed pre-ignition (LSPI) was frequently observed [1]. LSPI is known to induce autoignition of stronger intensity, compared to normal knocking under conditions of low engine speed and high boost pressure, which needs to be suppressed.

Several types of LSPI have been reported and verified by many researchers [1–5]. Based on precise visualization, it has been reported that solid particles, generated or exfoliated by pressure oscillations in the first abnormal cycle, can subsequently pre-ignite due to self-heating in a normal cycle [1]. It has been also reported that engine lubricants or fuel-lubricant mixtures enter the combustion chamber and act as a pre-ignition source through a local autoignition event [2, 3, 4]. In the latter case, it is considered that calcium present in lubricant oil may cause LSPI [5].

In this study, in order to investigate the influence of lubricant oil and lubricant oil components on the ignition characteristics of fuel, ignition delay times for four types of fuels, isooctane, ethyl tert-butyl ether (ETBE), toluene, and 2,5-dimethylfuran (DMF), were measured using a rapid compression machine (RCM) at an equivalent ratio of 1.0, with/without lubricant oil and lubricant oil components, at 700 K and 3.0 MPa.

2 Experimental

2.1 Rapid compression machine (RCM)

Ignition delay times were measured using an RCM. The RCM comprises a driver piston, reactor piston, and driving air tank, as shown in Fig. 1. Specifications of the RCM are shown in Table 1. A pressure transducer (Kistler 6061BU20) was mounted on the endwall of the reactor cylinder, in order to measure the pressure in the cylinder. The lubricant oil was injected into the reactor cylinder using an injector mounted on the endwall of the cylinder. The reactor cylinder is covered by a heater, which is used to control the initial gas temperature. The gas mixtures were introduced into the preheated reactor cylinder from a stainless-steel vessel, where the homogeneous gas mixtures were prepared, and compressed in one stroke without rebound.

Bore × Stroke [mm]	100 × 130
Displacement [cm^3]	1021
Compression ratio	14.01
Compression time [ms]	33 ± 3

Table 1 Specifications of RCM

Kotaro Tanaka. Fax: +81-294-38-5080
E-mail address: kotaro.tanaka.comb@vc.ibaraki.ac.jp

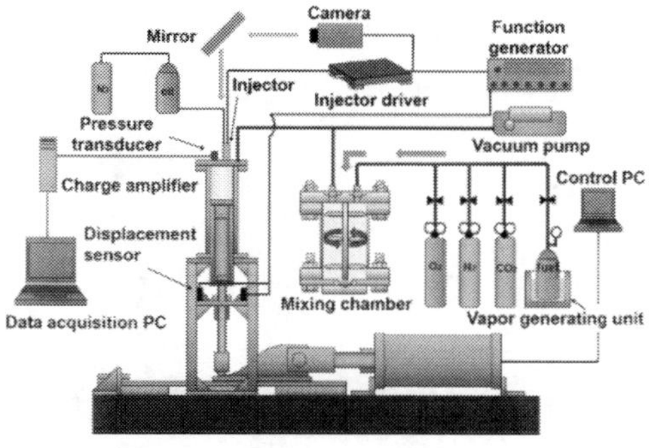

Fig. 1 Schematic diagram of rapid compression machine.

2.2 Experimental conditions

The desired compressed conditions can be obtained by changing the initial pressure (P_0) and preheat temperature (T_0), that were varied to attain the desired conditions at the end of compression, in terms of compressed pressure (P_C), and compressed temperature (T_C). The compressed temperature (T_C) is deduced from the measured pressure trace, using the adiabatic core hypothesis [6] :

$$\ln\left(\frac{P_C}{P_0}\right) = \int_{T_0}^{T_C} \frac{\gamma}{\gamma - 1}\frac{dT}{T}$$

where γ is the ratio of specific heat. The error of compressed temperature was ± 2 K which was derived from the initial temperature, pressure measurements and premixed gas composition.

The fuels used were isooctane, ethyl tert-butyl ether (ETBE), toluene, and 2,5-dimethylfuran (DMF). RON and MON of the selected fuels are summarized in Table 2 [7, 8]. Isooctane and ETBE have the low temperature oxidation, while toluene and DMF do not have it. Homogeneous fuel/oxidizer gas mixtures were prepared in a stainless steel vessel using heaters and stirrers. Oxygen (O_2), nitrogen (N_2), and carbon dioxide (CO_2) were used for the gas mixtures. Vaporized hydrocarbons were introduced manometrically into the vessel from the preheated fuel tank. After the introduction of the fuel, the other components (O_2, N_2, and CO_2) were also added monometrically, and the gas mixtures were mixed for 1 h, using stirrers to ensure homogeneity.

Species	RON	MON
Isooctane	100	100
ETBE	118	102
Toluene	124	112
DMF	106	98

Table 2 RON and MON of the selected fuels.

The lubricant oil and the components of the lubricant oil, icosane (one of the lubricant base oils) and calcium sulfonate (one of the lubricant additives), were injected into the reactor cylinder 26 ms after the start of the compression (corresponding to -30° ATDC). This injection timing was determined by the precise visualization of the practical DISI engine cylinder, where the lubricant oil, or the lubricant oil diluted by fuel flying, at approximately -30° ATDC (operating at 909 rpm) led to the pre-ignition [9]. The amount of the injected oil was set at 4 μl, which was the minimum injected-volume of the injector used. As the lubricant oil and their components had high viscosity, the injection system was heated to 327 K. The injection pressure was set to 5 MPa.

Ignition delay times of the fuels with the lubricant oil, or the components of the lubricant oil, were measured at a compressed pressure of 3 MPa, and a compressed temperature of 700 K, where pre-ignition typically occurred in the practical DISI engine.

2.3 Definition of ignition delay time

In this study, the ignition delay time was defined as the time interval between the end of compression (derivative of pressure equals zero) and the occurrence of the maximum rate of pressure rise. In the case of two-stage ignition, the first- and second-stage ignition delays were defined similarly. Figure 2 shows the definition of total ignition delay time (τ), and first-stage ignition delay time (τ_{LTO}). The units of both τ and τ_{LTO} were reported in the units of milliseconds. The maximum error of the ignition delay time was approximately ± 16% of the ignition delay time.

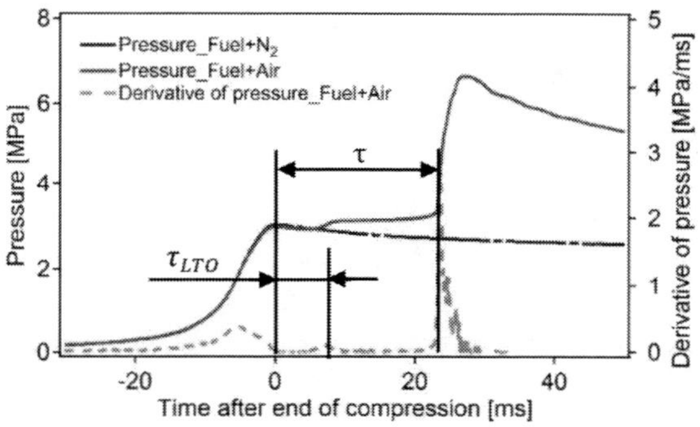

Fig. 2 Definition of the ignition delay time.

3 Results and discussion

Figure 3 shows the pressure profiles of isooctane, isooctane with lubricant oil, and isooctane with lubricant oil diluted by isooctane (lubricant oil : isooctane = 1 : 1 volumetric ratio). In Fig. 3, the ignition delay time of isooctane with lubricant oil was shorter than that of isooctane, and the ignition delay time of isooctane with lubricant oil diluted by isooctane was the shortest of the three. Figure 4 shows the pressure profiles of a series of ETBE experiments. In the case of ETBE, ignition did not take place. When the lubricant oil, or the lubricant oil diluted by ETBE (the same dilution ratio as that of isooctane), was injected, the ETBE gas mixture ignited. The ignition delay time of ETBE with the lubricant oil diluted by ETBE was shorter than that of ETBE with the lubricant oil.

Figures 5 and 6 show the pressure profiles for toluene and DMF, respectively. Under these experimental conditions, neither fuel ignited. When the lubricant oil was injected, the gas mixtures of both fuels ignited, and the tendencies were the same as that of ETBE. On the other hands, in the case of toluene and DMF, the ignition delay times of the fuels with the lubricant oil were shorter than those of the fuels with the lubricant oil diluted by the fuels, which was different from that of ETBE. From these experimental results, it is found that the lubricant oil enhanced the ignition of the fuels, whose tendencies depended on the ignition characteristics of the fuels.

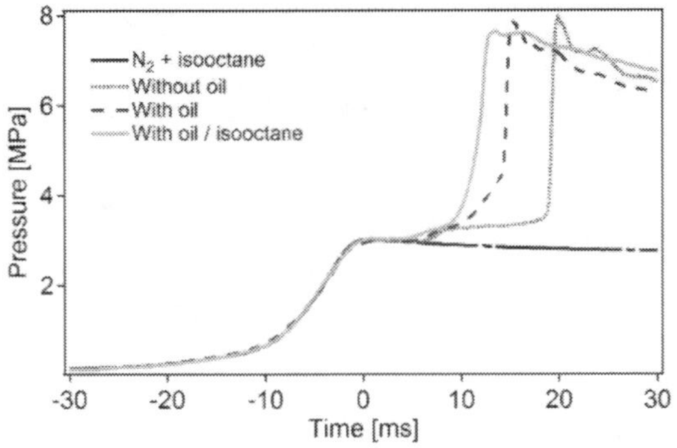

Fig. 3 Pressure profiles of isooctane premixed gas, isooctane premixed gas with lubricant oil, and isooctane premixed gas with lubricant/isooctane mixture. Pressure profile of isooctane/N_2 mixture is also shown in the graph.

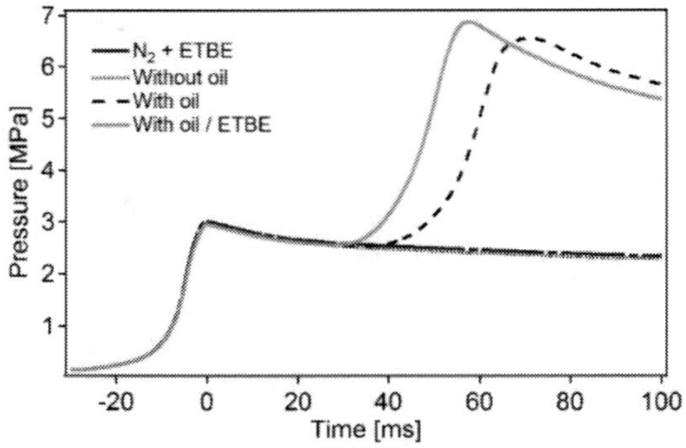

Fig. 4 Pressure profiles of ETBE premixed gas, ETBE premixed gas with lubricant oil, and ETBE premixed gas with lubricant/ETBE mixtures. Pressure profile of ETBE/N_2 mixture is also shown in the graph.

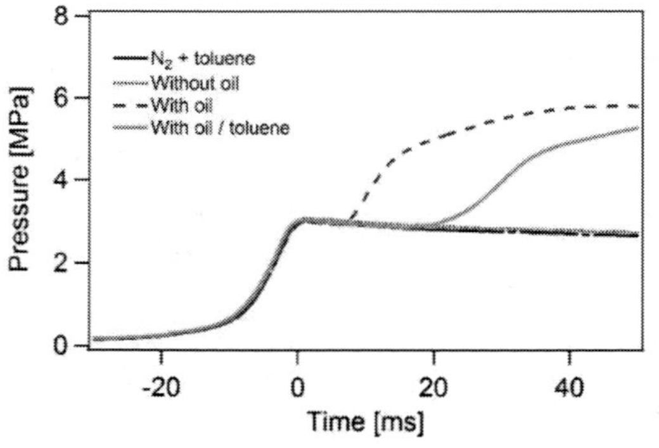

Fig. 5 Pressure profiles of toluene premixed gas, toluene premixed gas with lubricant oil, and toluene premixed gas with lubricant/toluene mixtures. Pressure profile of toluene/N_2 mixture is also shown in the graph.

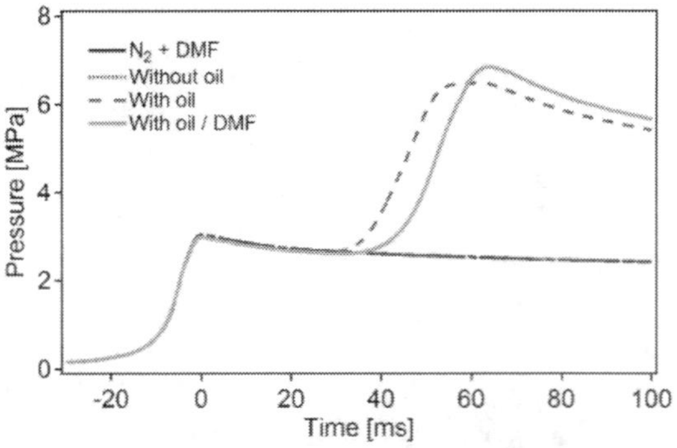

Fig. 6 Pressure profiles of DMF premixed gas, DMF premixed gas with lubricant oil, and DMF premixed gas with lubricant/toluene mixtures. Pressure profile of DMF/N_2 mixture is also shown in the graph.

To investigate the factor of the effect of lubricant oil on the ignition characteristics of the fuels, the two representative components included in the lubricant oil, icosane and calcium sulfonate, were also injected under the same conditions as those that the lubricant oil was injected, and the ignition delay times of the selected fuels were measured. Each component was diluted by the fuels (volumetric ratio of 1 : 1). The fuels used were isooctane and toluene. The effect of icosane included as the lubricant base oil, or calcium sulfonate included as the lubricant additive, on the ignition delay times of isooctane and toluene are shown in Figs 7 and 8, respectively. In Fig. 7, the ignition delay times of isooctane, with icosane diluted by isooctane, or calcium sulfonate diluted by isooctane, were shorter than those of isooctane, while they were longer than that of isooctane with lubricant oil diluted by isooctane. Although difference between the ignition delay times of isooctane with icosane diluted by isooctane and those of isooctane with calcium sulfonate diluted by isooctane was within the measurements error of the ignition delay times, the average values had clear difference between them. The ignition delay time of isooctane with calcium sulfonate was slightly shorter than that of isooctane with icosane, which indicates that in the case of isooctane, both icosane and calcium sulfonate enhance ignition. In the case of toluene, toluene with icosane ignited under these experimental conditions, while toluene with calcium sulfonate did not ignite. The ignition delay time of toluene with icosane was shorter than that of toluene with lubricant oil. It is found that icosane enhances the ignition of toluene.

From the abovementioned experimental results, enhancement of the ignition of the fuels by icosane is independent of the fuels used in the experiment, while with calcium sulfonate it depends on the fuel. The simulated ignition delay times of isooctane, toluene, and icosane are shown in Fig. 9. Simulation was performed in the temperature range 606–1667 K, a pressure of 3 MPa, and an equivalence ratio of 1.0 under adiabatic conditions. The detailed chemical kinetic model of isooctane and toluene in Ref [10], and the detailed chemical kinetic model of icosane constructed by KUCRS [11.12.13] were used. The simulations were performed using the CHEMKIN-PRO package [14]. The ignition delay times of icosane were shortest among three species, icosane, isooctane, and toluene. Hence, lubricant base oil is considered to enhance the ignition of isooctane and toluene. It is inferred that, when the fuels have a longer ignition delay time than that of icosane, ignition of the fuels is enhanced by the addition of lubricant oil. In the case of calcium sulfonate, the detailed mechanism of the enhancement of the ignition of the fuel is not well understood at present. Further study on the mechanism of the effect of the calcium sulfonate on the fuel ignition is required.

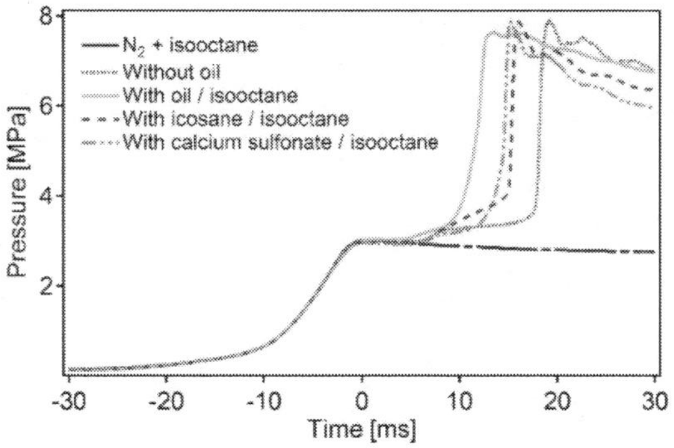

Fig. 7 Pressure profiles of isooctane, isooctane with isooctane/lubricant oil, isooctane with isooctane/icosane, and isooctane with isooctane/calcium sulfonate mixtures.

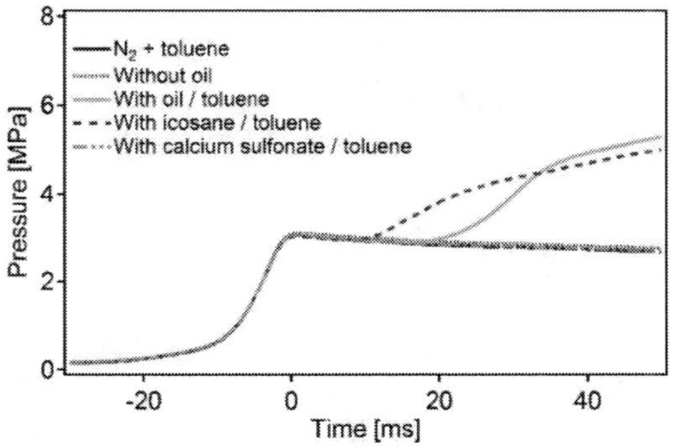

Fig. 8 Pressure profiles of toluene, toluene with toluene/lubricant oil, toluene with toluene/icosane, and toluene with toluene/calcium sulfonate mixtures.

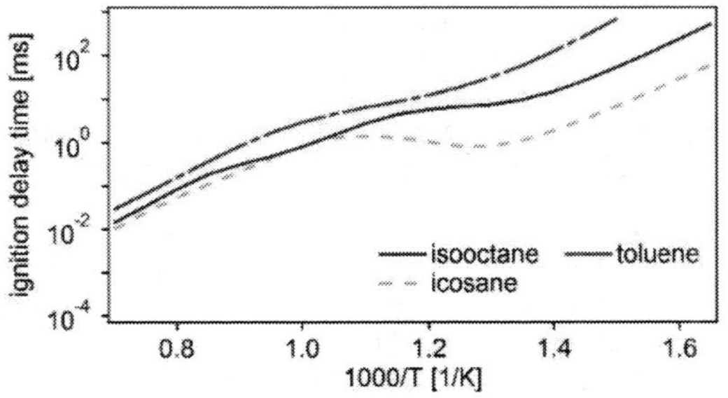

Fig. 9 Calculated ignition delay times of isooctane, toluene, and icosane in the temperature range 606-1667 K. Pressure is 3 MPa and an equivalence ratio is 1.0.

4 Conclusions

Ignition delay times of fuel premixed gas with lubricant oil were measured using an RCM at 3 MPa, 700 K, and an equivalence ratio of 1.0. Fuels selected for the experiment were isooctane, ETBE, toluene, and DMF. Lubricant oil was injected into the combustion chamber in the RCM. The lubricant oil was injected into the fuel/oxidizer mixtures in the RCM 26 ms after the start of compression, corresponding to a 30° BTDC. Ignition of the fuel was enhanced by the lubricant oil regardless of the type of fuel. As it was found that the lubricant oil enhanced the ignition of the fuels, the factor contributing to the enhancement of the fuel ignition was investigated. The contributing factor, that was independent of the type of fuel, was considered to be a lubricant base oil. Calcium sulfonate included in the lubricant oil as an additive also had an enhancing effect on some of the fuels, such as isooctane, while the ignition of some fuels, such as toluene, was not enhanced by calcium sulfonate.

References

[1]. Y. Okada, S. Miyashita, Y. Izumi, Y. Hayakawa, *SAE Int. J. Engines* 7 (2) (2014) 584-594.

[2]. M. Kassai, K. Torii, T. Shiraishi, T. Noda, T. K. Goh. K. Wilbrand, S. Wakefield, A. Healy, D. Doyle, R. Cracknell, M. Shibuya, SAE Technical Paper 2016-01-2292.

[3]. C. Dahnz, K. Han, U. Spicher, M. Magar, et al., *SAE Int. J. Engines* 3 (1) (2010) 214-224.

[4]. S. F. Dingle, A. Cairns, H. Zhao, J. Williams, O. Williams, R. Ali, SAE Technical Paper 2014-01-1222.

[5]. T. Kuboyama, Y. Moriyoshi, Journal of the Combustion Society of Japan 58 (183) (2016) 20-26.

[6]. G. Mittal, J. Sung, *Combust. Sci. Technol.* 179 (2007) 497-530.

[7]. American Petroleum Institute Research Project 45, Knocking Characteristics of Pure Hydrocarbons, American Society for Testing Materials, Philadelphia, 1916,p. 25, p. 27.

[8]. H. Hamid, M. A. Ali, Handbook of MTBE and Other Gasoline Oxygenates, Marcel Dekker, New York, 2004, p. 44

[9]. M. Kassai, H. Hashimoto, T. Shiraishi, A. Teraji, et al., SAE Technical Paper 2015-01-1867.

[10]. K. Hashimoto, M. Koshi, A. Miyoshi, Y. Murakami, T. Oguchi, Y. Sakai, H. Audo, K. Tsuchiya, SAE Technical Paper 2013-01-0887.

[11]. A. Miyoshi, KUCRS softoware library, revision 20170705, available at

<http://akmys.com/KUCRS/for update infomation>.

[12] E. R. Ritter, J. W. Bozzelli, *Int. J. Chem. Kinet.* 23 (1991) 767-778.

[13] A. Miyoshi, *J. Phys. Chem. A.* 115 (2011) 3301-3325.

[14]. ANSYS Chemkin-Pro, available at

<http://www.ansys.com/de-DE/products/fluids/ansys-chemkin-pro>.

11th Asia-Pacific Conference on Combustion,
The University of Sydney, NSW Australia
10th -14th December 2017.

Experimental Study on the Minimum Ignition Energy of Propane - Air Mixture with Various Water Mist Loading

Keisuke Nakahara[1], Chihong Liao[2], Hiroyoshi Naito[3], Akira Yoshida[1]

[1]Department of Mechanical Engineering, Tokyo Denki University

5 Senju-Asahicho, Adachi-ku, Tokyo, 120-8551, Japan

[2]Technical Laboratory, Morita Holdings Corporation

3-25-31 Nishishinbashi, Minato-ku, Tokyo, 105-0003, Japan

[3]Fire and Disaster Management Agency

2-1-2 Kasumigaseki, Chiyoda-ku, Tokyo, 101-8508, Japan

Abstract

Water mist is attracting attention as a substitute for water discharge from a hose or halon fire suppressants. The minimum ignition energy (MIE) is an important property for the ignition process of combustible mixture. In the present investigation, we investigated experimentally the relationship between minimum ignition energy density and water mist mass fraction in propane/air mixture with changing the equivalence ratio of the mixture. The water mist was generated by ultrasonic water mist generators. In the present study, instead of electric spark discharge and laser induced spark discharge, the mixture was ignited by a fine Nichrome wire which is heated electrically. The equivalence ratios of water mist added mixture φ are 0.8, 1.0, 1.3 and 1.6, and the mass fractions of water mist Y_0 are 0.00, 0.09 and 0.13. When water mist was added to the mixture, the ignition probability decreases and the minimum ignition energy density increases for all equivalence ratios. The ignition inhibition effectiveness of water mist is large for lean and rich mixtures, whereas it is small for near stoichiometric mixtures.

1 Introduction

When a fire occurs, an extinguisher is used to extinguish the fire and to inhibit the fire spreading. The water discharge from a hose is a representative method to extinguish a fire by water, but a large part of water discharged is not used to extinguish a fire and the lost water will cause the secondary damage. In addition, most of the production of Halon extinguishers used in the past was banned by the Montreal protocol because of their depletion potential of the ozone layer. Then the water mist attracts our attention as a substitute for those extinguishers including Halons.

Water mist has a high potential of fire suppression, because it has cooling effect, suffocation-dilution effect and reaction inhibition effect. For mitigation of explosion, the effects of water mist on the laminar flame speed and flame temperature have been investigated experimentally and numerically [1, 2]. Additionally, the water mist was found to reduce the explosion pressure [3]. To ignite the mixture including water mist, a new method of ignition was established and the minimum ignition energy of propane/air mixture with water mist was measured [4]. Furthermore, parameters controlling the ignition process were investigated [5].

In the present study, we measured the ignition probability of propane/air mixtures. The aim of this study is to measure the minimum ignition energy density by changing the equivalence ratio of the mixture and mass loading of water mist.

2 Experimental Apparatus and Method

2.1 Experimental Apparatus

Figure 1 shows the experimental apparatus. The experimental apparatus consists of combustion tube and water mist generator. Fuel used in the present study was propane, and propane and air were mixed far upstream of the combustion tube. The ignition experiments were performed in a vertical acrylic tube of 85 mm in diameter and 836 mm long. Since the initial flame kernel of which appearance was used as the ignition criterion is sufficiently smaller than the diameter of the combustion tube, it is considered that there is no effect of the dimension of the ignition tube on ignition criterion. The upper end is open to atmosphere and the flame propagates at nearly atmospheric pressure. A conical chimney was set up at the upper end of the combustion tube to prevent dilution of the mixture due to inflow of ambient air. A shutter valve was equipped at the upstream end of combustion tube to cut off the mixture supply.

The water mist was generated by an ultrasonic water mist generator and was mixed with unburned mixture. Since the mist generation rate by a piezoelectric atomizer is sensitive to the water temperature, it was kept constant at 40 degree C by an electric heater. The droplet size of water mist was measured by a phase Doppler particle analyzer and was found to range from 1 to 44 μm with a Sauter mean diameter of 20 μm. The homogenized mixture including water mist was supplied to combustion tube during 30 s with the mean velocity of 10 cm/s in the tube. Subsequently, the shutter valve was closed and we waited for 10 s before mixture settled inside the tube. Then, the mixture was ignited by electrically heated wire which is located at 254 mm from the tube exit. The terminal velocity by Stokes' law is a few cm/s and is so small that the mist stratification along the combustion tube was negligible. We assumed that the water mist was dispersed homogeneously within the combustion tube.

Corresponding author. Fax: +81-3-5284-5469
E-mail address: 16kmk16@ms.dendai.ac.jp

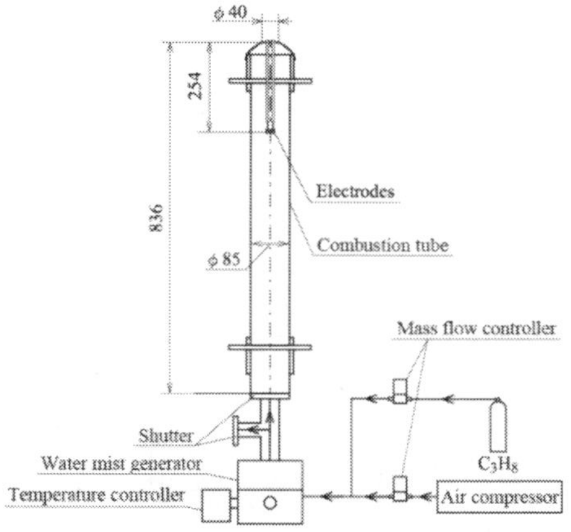

Figure 1: Experimental apparatus.

2.2 Ignition System

Figure 2 shows the ignition system used in the present study. Electric energy was supplied to Nichrome wire from a capacitor charged by a variable voltage DC power supply from 0 to 240 Volts. A capacitor bank consists of capacitors from 10 μF to 12600 μF. The current and voltage of the electric discharge were measured with a current probe (Iwatsu, SS-240A) and a voltage probe (Iwatsu, SS-103R), respectively, and were acquired by a digital oscilloscope (Tektronix, TDS3032B). The discharge duration was determined from the voltage trace of oscilloscope. The discharge duration used in this study was varied from 100 μs to 500 μs by changing the supply voltage and capacitance. Terminals were equipped at the end of the electrodes to cramp the Nichrome wire and a wire was fastened on the terminals by screws to reduce the resistance between Nichrome wire and terminals. The diameter of Nichrome wire ranged from 30 μm to 150 μm, and the length was 3.1 mm. The melting point of Nichrome wire is 1673 K. Just after melting, an arc discharge was produced between broken wires. The ignition probability was determined by ten trials for one condition.

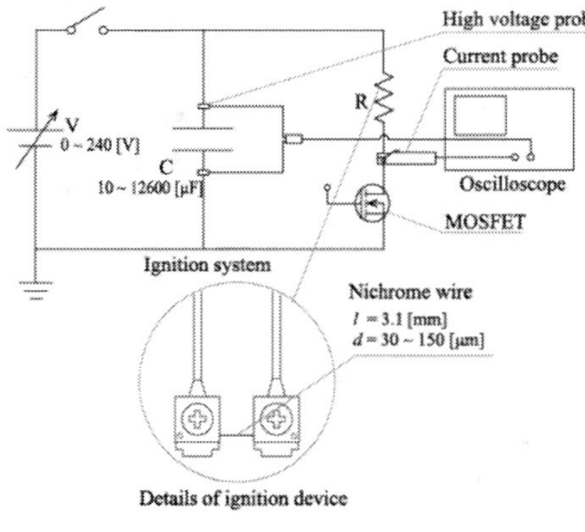

Figure 2: Ignition system.

2.3 Net Discharged Energy

The net discharged energy E_i was delivered to the flame kernel when the Nichrome wire was melted electrically. When V_c is the initial charging voltage to the capacitor, the charged energy E_c is $CV_c^2/2$. After discharge, the charging voltage decreased to V_r which is the voltage of remaining charge in the capacitor. The remaining energy in the capacitor E_r is $CV_r^2/2$. ΔE is the difference between the charged energy E_c and the remaining energy in the capacitor after discharge E_r. Then the net discharged energy E_i is expressed as follows.

$$\Delta E = E_c - E_r = E_i + E_h + E_m \qquad (1)$$

$$E_i = \Delta E - (E_h + E_m) \qquad (2)$$

where E_h is the sensible enthalpy to heat up the wire from ambient temperature to the melting point, E_m the latent heat of fusion of Nichrome wire.

2.4 Energy Density

In the past study, the minimum ignition energy without water mist was investigated experimentally and numerically [6, 7]. Electric spark discharge and laser induced spark have been used as ignition source for the flammable mixture [6, 8]. However, when water mist added to the mixture, it will be considered that electric charge will escape along the wetted electrodes for an electric spark discharge, and water mist will absorb the laser light for laser induced spark. Under these considerations, in the present study, we used a fine Nichrome wire which was heated electrically instead of the electric spark discharge and laser induced spark discharge.

The ignition device using the Nichrome wire was established by Ebina et al. [5]. They investigated the effect of water mist on the ignition probability of propane/air mixture, and determined the dominant parameter for ignition probability of the combustible mixture.

It can be expected that the energy density q_i is dominant parameter for ignition probability [9], instead of net discharged energy and hereafter we use the energy density as the dominant parameter for ignition defined as follows.

$$q_i = E_i / V_n \qquad (3)$$

where E_i and V_n are the net discharged energy and the volume of flame kernel [9], respectively. In the present study, the volume of the flame kernel V_n was assumed to be equal to the volume of the Nichrome wire, because the volume of the initial flame kernel is considered to be almost equal to the volume of the Nichrome wire.

2.5 Ignition Probability

Ignition is a statistical phenomenon and the ignition probability P_i (q_i) can be expressed by the cumulative probability distribution function [10],

$$P_i(q_i) = \frac{1}{1 + exp^{-\beta_0 - \beta_1 q_i}} \qquad (4)$$

where β_0 and β_1 are coefficients estimated by maximizing the likelihood function L . The likelihood function is given by the following expression.

$$\mathrm{L} = \prod_{i=1}^{n} P_i(q_i)^{y_i}(1 - P_i(q_i))^{1-y_i} \qquad (5)$$

where y_i is ignition probability for i^{th} test. The minimum ignition energy density (MIED) is defined at 1% ignition probability from using Eq. (4).

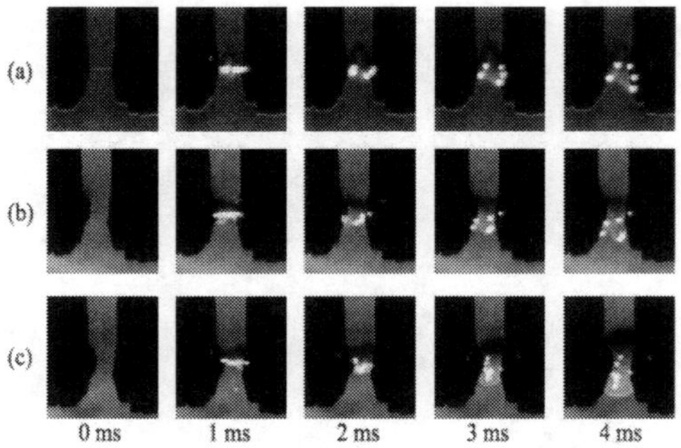

Figure 3: Time sequences of flame kernel evolution.

3 Results and Discussion

3.1 Ignition Process

Figure 3 shows the time sequences of the ignition process without water mist by schlieren photographs. (a) shows the Nichrome wire melting process in pure air, (b) shows non-ignition process of propane/air mixture for $\varphi = 1.3$, and (c) shows ignition process for $\varphi = 1.3$. The Nichrome wire is heated up within 1 ms in all conditions. The pure air or the mixture around the Nichrome wire is heated by heat transfer from the wire. After that, the temperature of the Nichrome wire rises up to the melting point. Then arc discharge occurs between the broken wires. The small flame kernel appears in Fig. 3(b) and (c), which is caused by heating of Nichrome wire and an arc discharge. On the other hand, Fig. 3(a) shows only the blurry front of the thermal wave. At $t = 2$ ms, the front of the thermal wave is more blurred in Fig. 3(a), and the front of the flame kernel becomes blurry in Fig. 3(b), whereas the front of the flame kernel remains sharply in Fig. 3(c). At $t = 3$ ms, only the thermal wave appears in Figs. 3(a) and (b), and the front of the thermal wave becomes more blurry. However, a definite flame front is formed in Fig. 3(c). At $t = 4$ ms, the front of thermal wave disappears due to the thermal diffusion in Figs. 3(a) and (b). The flame propagates spherically in Fig. 3(c). Thus, the ignition occurs due to the combined effect of heating by wire and an arc discharge.

3.2 Ignition Probability for $\varphi = 0.8$

Figure 4 shows the inhibition effect of water mist for equivalence ratio $\varphi = 0.8$. Error bars are shown in each plot. In this figure, Y_0 is the mass fraction of water mist in the mixture, and the curve shows the ignition probability. Solid line shows the ignition probability of mixture without water mist, dashed line shows that with water mist for $Y_0 = 0.09$, and dot-dashed line that for $Y_0 = 0.13$. β_0, β_1 are the fitting parameters of Eq. (4).

When $Y_0 = 0.00$, the ignition probability increases as the energy density increases. The ignition probability for $Y_0 = 0.09$ decreases comparing with that for $Y_0 = 0.00$, if the energy density is constant. Additionally, the minimum ignition energy density increases. When Y_0 is increased to 0.13, ignition probability further decreases and minimum ignition energy density increases significantly. The effect of water mist on the minimum ignition energy density is rather small if Y_0 is smaller than 0.09. However, when Y_0 is increased to 0.13, the minimum ignition energy density increases prominently.

This is because water mist absorbs the thermal energy from the heated Nichrome wire and an arc discharge and that the chemical reactions in the ignition process should depend nonlinearly on the temperature.

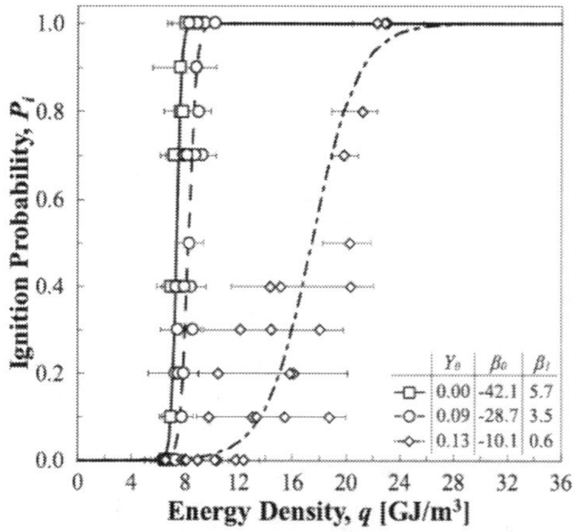

Figure 4: Ignition probability for $\varphi = 0.8$.

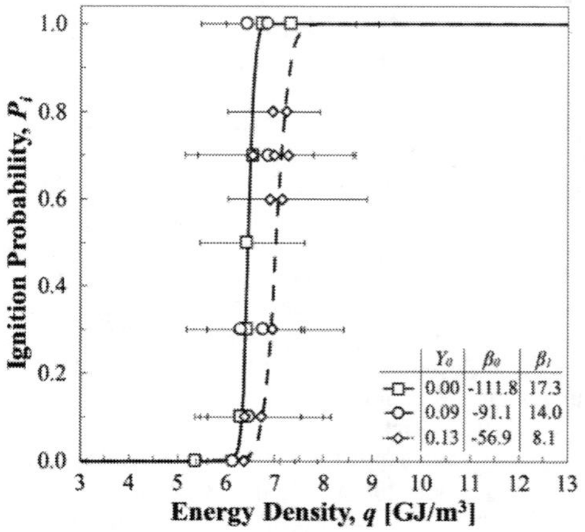

Figure 5: Ignition probability for $\varphi = 1.0$.

3.3 Ignition Probability for $\varphi = 1.0$

Figure 5 shows the effect of water mist for $\varphi = 1.0$. Compared to the cases for $Y_0 = 0.00$ and 0.09, we can see no meaningful difference and the inhibition effect of water mist on ignition is small. Even when Y_0 is increased to 0.13, inhibition effect is rather small. At the same time, it was confirmed by visual observation that the propagation speed of the flame decreases significantly as the mass fraction of water mist increases.

3.4 Ignition Probability for $\varphi = 1.3$

Figure 6 shows the effect of water mist on ignition probability for $\varphi = 1.3$. The ignition probability decreases as the mass fraction of water mist increases. For the case without water mist, the minimum ignition energy of propane/air mixture is the lowest for $\varphi = 1.3$ as shown by the investigation of Lewis and Elbe [6]. From this figure, it is found that a small amount of water mist has the effect to increase the minimum ignition energy density compared to that for the mixture of $\varphi = 1.0$.

3.5 Ignition Probability for $\varphi = 1.6$

The effect of water mist on ignition probability for $\varphi = 1.6$ is shown in Fig. 7. For this equivalence ratio, the ignition probability decreases as the water mist mass fraction increases, and the minimum ignition energy density increases significantly.

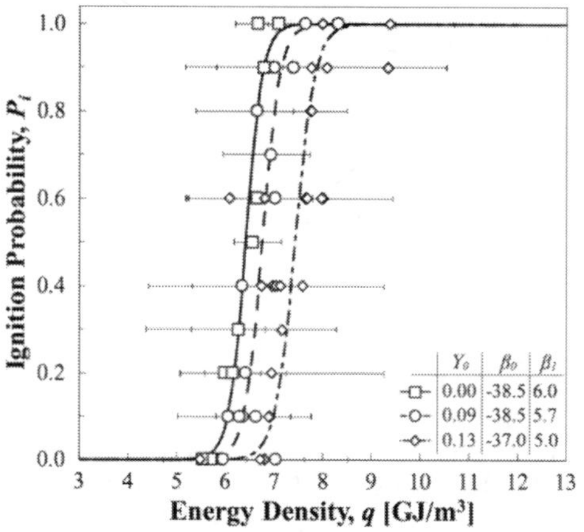

Figure 6: Ignition probability for $\varphi = 1.3$.

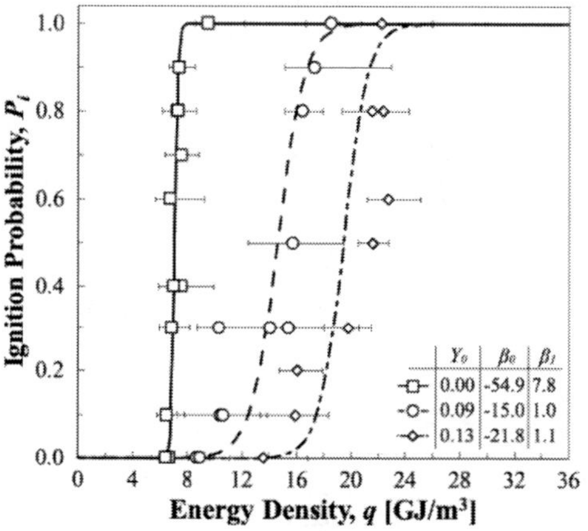

Figure 7: Ignition probability for $\varphi = 1.6$.

Even if ignition occurs and the small flame kernel is established, flame sometime does not propagate. From this result, for $\varphi = 1.6$, even a small amount of water mist is effective to decrease the ignition probability and the minimum ignition energy density increases even for $Y_0 = 0.09$. The equivalence ratio of $\varphi = 1.6$ is considered to be close to the rich limit of flame propagation. Therefore, the flame temperature is rather low compared to those for $\varphi = 1.0$ and $\varphi = 1.3$. Thus, the nonlinearity of temperature dependence should be small.

3.6 Minimum Ignition Energy Density

Figure 8 shows the minimum ignition energy density of each equivalence ratio. At all equivalence ratios, the minimum ignition energy density increases with increasing water mist mass fraction. This result qualitatively agrees with Ref. [11]. However, the inhibition effectiveness is different depending on the equivalence ratio of the mixture. For lean and rich mixtures ($\varphi = 0.8$ and 1.6), water mist acts effectively as suppressant. However, for the near stoichiometric mixtures ($\varphi = 1.0$ and 1.3), the inhibition effectiveness is rather small. These results are extremely important from the practical viewpoint to apply the water mist for extinguishment of the real fires. To prevent the explosion of hazardous combustible gases, it is important to supply water mist at the early stage of leakage of combustible gas to utilize as much as possible the suppression potential of water mist.

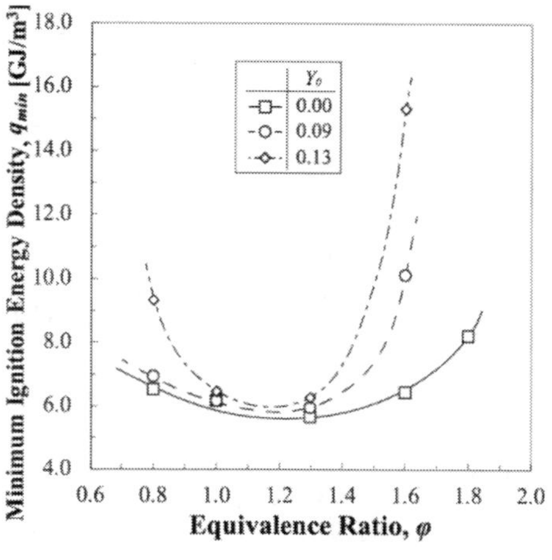

Figure 8: Relationship between q_{min} and φ.

4 Conclusions

The following conclusions were derived from the present study in which the ignition probability was measured for propane/air mixtures with various mass fractions of water mist.

(1) When the water mist mass fraction is increased, the ignition probability decreases and the minimum ignition energy density increases for all equivalence ratios.

(2) The suppression effectiveness of water mist is significantly large for lean and rich mixtures, whereas for near stoichiometric mixtures, it is rather small.

References

[1] A. Yoshida, T. Udagawa, Y. Momomoto, H. Naito, Y. Saso, Fire Saf. J. 58 (2013) 84-91.

[2] A. Yoshida, T. Okawa, W. Ebina, H. Naito, Combust. Flame 162 (2015) 1772-1777.

[3] P. Zhang, Y. Zhou, X. Cao, X. Gao, M. Bi, Saf. Sci. 62 (2014) 1-7.

[4] W. Ebina, C. Liao, H. Naito, A. Yoshida, Proc. Combust. Inst. 36 (2017) 3271-3278.

[5] W. Ebina, C. Liao, H. Naito, A. Yoshida, 10th ASPACC (2015)

[6] B. Lewis, G. von Elbe, Combustion, Flames and Explosions of Gases, 2nd Ed., Academic Press, (1961) 333-335.

[7] J. Han, H. Yamashita, N. Hayashi, Combust, Flame 157 (2010) 1414-1421.

[8] D.H. McNeill, Proc. Combust. Inst. 30 (2005): 2913-2920.

[9] T. Sloane, Combust. Sci. Tech. 73 (1990) 351-365.

[10] R.K. Eckhoff, M. Ngo, W. Olsen, J. of Hazard. Mater. 175 (2010) 293-297.

[11] W. Zhang, X. Gou, Z. Chen, Fuel 187 (2017) 111-116.

11th Asia-Pacific Conference on Combustion,
The University of Sydney, NSW Australia
10th-14th December 2017.

Experimental Study of the Nozzle-Ejector Effect on the Performance of a Pulse Detonation Engine

Zhiwu Wang[1], Yaqi Wang[1], Xing Liu[1], Hongwei Li[1], Kun Zhang[2], Longxi Zheng[1]

[1]School of Power and Energy, Northwestern Polytechnical University

[2]Avic Xi'an Aero-Engine (group) LTD

Abstract

In order to improve the propulsive performance of a Pulse Detonation Engine (PDE), several nozzle-ejector combinational configurations were designed and manufactured. A nozzle-ejector combinational configuration consisted of a nozzle and a straight ejector. There are three types of the nozzles in total, including straight nozzle(S nozzle), convergent nozzle(C nozzle)and convergent-divergent nozzle(C-D nozzle). The thrust of a gasoline-air PDE with three types of nozzle-ejector combinational configurations was measured experimentally. The nozzle-ejector configuration was installed positions X/D=-1,0,1 respectively(where X is the distance from detonation tube end to the head of ejector and D is the diameter of detonation tube.). The negative value represents the end of the detonation tube is fixed in the ejector. The thrust of every position and structure was conducted at 15Hz, 20Hz and 25Hz. The results indicated that the thrust augmentation of the PDE was dependent on the axial position of ejector, nozzle type and operating frequency. The thrust augmentation of the same nozzle-ejector combinational configuration changed with the operating frequency. Base on the present operating condition, the PDE with a straight nozzle-straight ejector combinational configuration had the best performance at the ejector position of X/D=0 at 15Hz.The highest thrust augmentation of the PDE with a convergent nozzle-ejector appears at position X/D=+1 at 20Hz. The PDE with a convergent-divergent nozzle-ejector got the best performance at the ejector position of X/D=-1 at 20Hz, and the thrust augmentation of 54.1% was the highest thrust augmentation of all these situations.

1 Introduction

Pulse detonation is a kind of combustion method with high thermal efficiency which utilizes the pulse detonation waves to produce burnt gas with high temperature, high pressure and high speed[1-2].A pulse detonation engine is an unsteady propulsion system which utilizes the impulse generated by the burnt gas to produce thrust [3-4]. PDEs have the potential advantages of high thermal cycle efficiency, simple structure, low specific fuel consumption, wide operating range and so on [5-6]. Specific impulse is an important index to character the propulsive performance of a pulse detonation engine. When an ejector is installed at the outlet of PDE, then fresh air can be ejected into the engine and the specific impulse increased [7-11].

Allgood et al.[12]investigated the performance of ejectors on pulse detonation engine by experiments. They found that the thrust augmentation of a pulse detonation engine with an ejector decreased at the increased filling factor and more thrust augmentation was generated from the increased ejector ratio of the length to diameter. Caldwell et al.[13] found that the detonation wave was weak and the velocity of leading shock wave was low under the low filling ratio of PDE through shadowgraph. Hoke et al[14]conducted experiments and found that when a single detonation tube in the center of the radial deviation from the ejector shaft was 4 times to the detonation tube diameter, the axial deviation from the optimum position ejector was 1.14 times to the detonation tube diameter range, ejector thrust augmentation could be maintained unchanged. Hoke et al.[15]validated that the detonation tube deviated from the center of the ejector had little effected on the ejector thrust augmentation, and the main cause of thrust augmentation was secondary flow of ejector and the secondary flow through the ejector lip. M.R.Nalim et al.[16]studied the wave rotor PDE ejector and found that an ejector could improve the engine performance. Huang et al. [17-18] conducted experiment and discovered that the axial position of ejector impacted the thrust augmentation significantly. They found that ejector could evidently improve the thrust of PDE, and the ejector inlet located upstream of the outlet PDE position have higher thrust augmentation compared to the downstream, as the higher frequency can generated higher thrust augmentation too. Wang et al. [19] obtains conclusions by experiments as bellows: the DDT distance of main detonation tube under the ejector mode is evidently shorter than regular electrical initiation of two-phase PDE, lengthened main detonation chamber and added convergence segment to the end section could improve detonation performance of ejector mode.

There are lots of factors affected the ejector performance of a PDE. Researchers have carried out much research on the PDE ejector, but the conclusions were different and even contradictory. So far, the combinational effect of nozzle-ejector including the nozzle type and ejector position on the PDE performance was not studied systematically. This paper focused on the experimental studied of the ejector position effect on the specific impulse of the PDE based on the combinational nozzle-ejector configurations. Three types of nozzles including the straight nozzle, convergent nozzle and convergent-divergent nozzle and one straight ejector were designed and manufactured. The effects of ejector position, nozzle type and operating frequency on the PDE ejector performance were investigated. Some useful results were obtained, which provided the reference to the optimal design of the PDE nozzle and ejector.

2 Experimental Setup

The experiment system of PDE-ejector was consisted of PDE,

Corresponding author. Fax: +029-88431116
E-mail :address:malsoo@mail.nwpu.edu.cn

nozzle, straight ejector, test bench, fuel and air supply system, ignition system, and data collection and control system. A PDE consisted of a mixing chamber, ignition chamber and detonation chamber. Three types of nozzles (including the straight nozzle, convergent nozzle with convergence angle of 9°and convergent-divergent nozzle with convergence angle of 5° and divergent angle of 5°) were mounted at the PDE outlet by flange. PDE, nozzle and ejector were installed on the moving frame by corresponding flanges. The relative position of the ejector and nozzle was adjustable. The experiments of nozzle-ejector combinational configure effect on the PDE performance were conducted with the three types of nozzle-ejector configurations at three different ejector positions of X/D=1, 0, + 1. X represents the distance between the nozzle outlet and the ejector inlet. D represents the inner diameter of the detonation tube. A schematic of the experimental setup is shown in Fig. 1. Gasoline and air were used as the fuel and oxidizer respectively. The liquid gasoline and air were supplied to PDE by adaptive control without valve. The operating frequency of PDE was controlled by the ignition frequency driven by a signal generator.

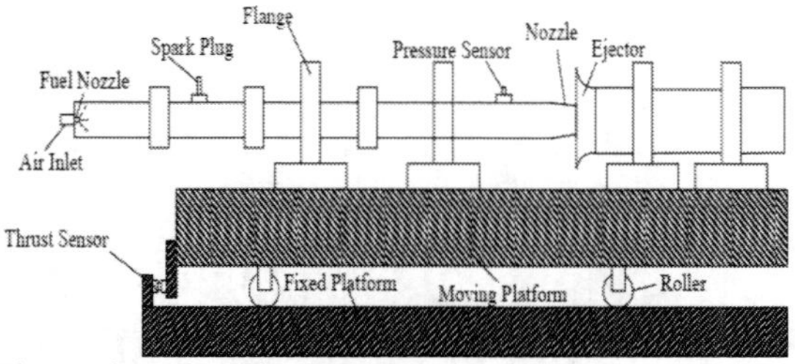

Fig. 1 Schematic of the experimental setup

The inner diameter of the PDE used in this paper is 60mm. The total length of the PDE is 1200mm, wherein the mixing chamber length is 250mm, the ignition chamber length is 300mm. The spark plug is located at the position with the distance of 150mm from the thrust wall (left end of PDE). The Shchelkin spiral is welded in the detonation chamber to accelerate the deflagration to detonation transition (DDT) and reduce DDT distance. All of the three nozzles have the same length of 300mm. The inner diameter of the straight ejector is 180mm and the length is 400mm.

A piezoelectric pressure transducers (Type: CA-YD-205) was installed at the detonation chamber tail with a distance of 1200mm from the thrust wall to measure the detonation pressure, represented by P1. The output signal from the pressure transducer was sampled with the DEWE3020 high-speed data acquisition and processing system. Sampling rates was 200 Ksamples/s and sampling time is 1s. The voltage value and the pressure value had the relationship of 1 V equaling 1 MPa. The calibration of pressure transducers was applied to convert the raw values to pressure values. The uncertainty of pressure measurement is about ± 72.5mV/MPa, and the inherent frequency is 200kHz.

The PDE thrust is measured by the Kistler force sensor with the sensitive degree of 3.678 PC/N and the measurement range of ±10000N .The PDE thrust measurement system is shown in Fig. 1. The moving frame and the fixed frame are connected by the force sensor.

3Experimental Methods and Results Analysis

All experiments were carried out with an initial pressure of 1 atm and initial temperature of 20℃. PDE was operated at three different frequencies of 15Hz, 20Hz and 25Hz. The gasoline/air equivalence ratio was about 1.0. In order to study the nozzle-ejector combinational configurations effect on PDE performance, the PDE performance without an ejector but a straight nozzle, convergent nozzle, or convergent-divergent nozzle was chosen as benchmark. Then detail analysis and discussion were conducted by comparing the PDE performance with the three different nozzle-ejector structures at three different ejector positions with that of the benchmark.

According to the theoretical analysis and experimental tests[20-22], the thrust of multi-cycle PDE can be summarized as Formulas (1)-(3), where F represents the theoretical average thrust by time, K is the thrust constant, and U_{CJ}, P_{CJ} represent the Chapman-Jouguet (CJ) gaseous detonation velocity and pressure respectively, V represents the PDE volume, f represents the PDE operating frequency, γ represents the specific heat ratio of burnt gas, P_0 represents the standard atmospheric pressure.

$$\overline{F} = K\frac{\Delta P_k}{U_{CJ}}Vf \quad (1)$$

$$\Delta P_k = \left(\frac{r+1}{2r}\right)^{\frac{2r}{r-1}} P_{CJ} - P_0 \quad (2)$$

$$K = 3.8687 + 1.48496e^{-0.10729f} \quad (3)$$

The thrust constant K at the different operating frequency can be obtained based on Formula (3), as shown in Table 1. The thrust K decreases with the increasing of PDE operating frequency. When the operating frequency is higher than 35Hz, K is nearly the constant 3.9.

P_{CJ} and U_{CJ} can be calculated by Chemical Equilibrium and Applications (CEA) code developed by NASA Lewis Research Center. For stoichiometric aero-gasoline and air at the initial pressure of 1atm and initial pressure of 20℃ (293.15K), P_{CJ}=18.685atm, U_{CJ}=1804.8m/s, γ=1.8127.According to Formulas (1)-(3), the PDE thrust at the operating frequencies of 15Hz, 20Hz and 25Hz were 59.2N, 77.1N and 94.2N, respectively.

f/Hz	5	10	15	20	25	30	35
K	4.74	4.37	4.16	4.06	3.97	3.92	3.9

Table1 Thrust constant K at the different PDE operating frequencies

The thrust augmentation is an important index to character the performance of the ejector and it is defined as follows:

$$T_{aug} = \frac{T_e - T_b}{T_b} \times 100\% \quad (4)$$

in which, T_{aug} is the thrust augmentation of the PDE, T_e is the thrust of PDE with ejector and T_b is the benchmark thrust of PDE without ejector.

The results of the experiment were shown in Fig. 2,Fig. 3 and Fig. 4.The thrust augmentation of straight nozzle at different frequency is similar. The optimal position of straight nozzle is position 0 where the thrust augmentation is the highest. However, the thrust augmentation decreased to 28.21% at

25Hz.This perhaps resulted from the decreasing of ejected air. When the operating frequency increased, the ejected time decreased and the air that was ejected into the ejector decreased.

The thrust augmentation of convergent nozzle is far away different at the different operating frequencies. The optimal performance of convergent nozzle-ejector at 15Hz was obtained at position 0,while the optimal performance at 20Hz and 25Hz was observed at position 1.The highest thrust augmentation obtained at position 1 at 20Hz was 47.54%. The highest augmentation decreased to 33.33% and 30.77% at the operating frequencies of 15Hz and 25Hz,respectively.

The performance of convergent-divergent nozzle reached optimal at position -1 at each of the three different frequencies. The thrust augmentation at ejector position 0 and 1 decreased sharply at all of the three frequencies, e.g. when the operating frequency is 20Hz,the thrust augmentation at ejector position 0 is 21.31% while the value of position -1 is 54.10%.Above all, when the operating frequency changed, the straight nozzle had better performance than other two nozzles at position 0, the convergent nozzle and convergent-divergent nozzle had the better performance at position 1 and position -1 respectively. Furthermore, the thrust augmentation increased with the increasing frequency. Fig. 5 shows the pressure and thrust histories of a PDE with a convergent-divergent nozzle-ejector at position -1.As shown in the figure 5, the peak pressure of each cycle is higher than 2MPa and the tendency of the thrust is similar for every cycle.

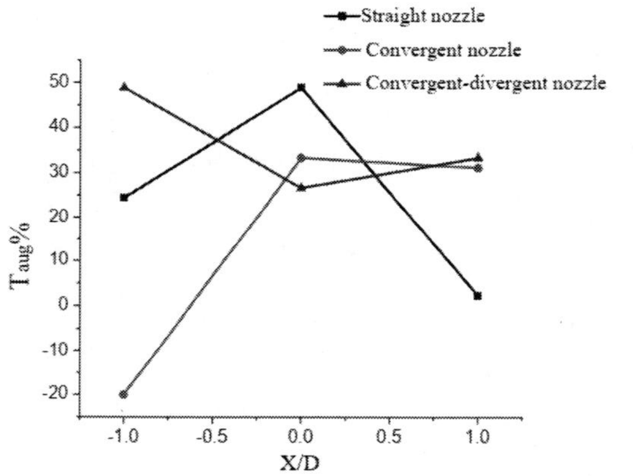

Fig. 2 Nozzle type effect and ejector position effect on PDE thrust augmentation at the operating frequency of15 Hz

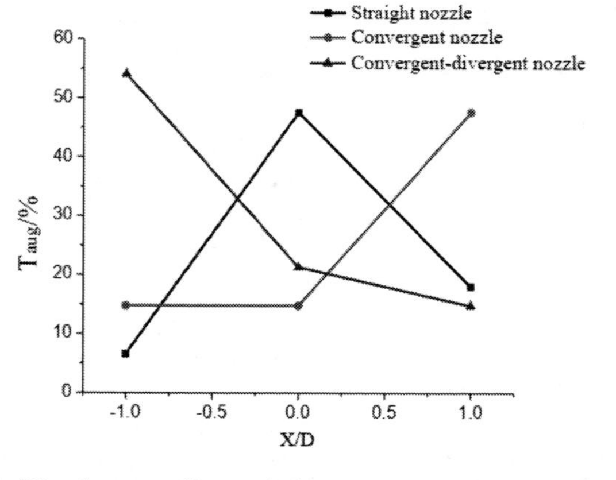

Fig.3Nozzle type effect and ejector position effect on PDE thrust augmentation at the operating frequency of 20 Hz.

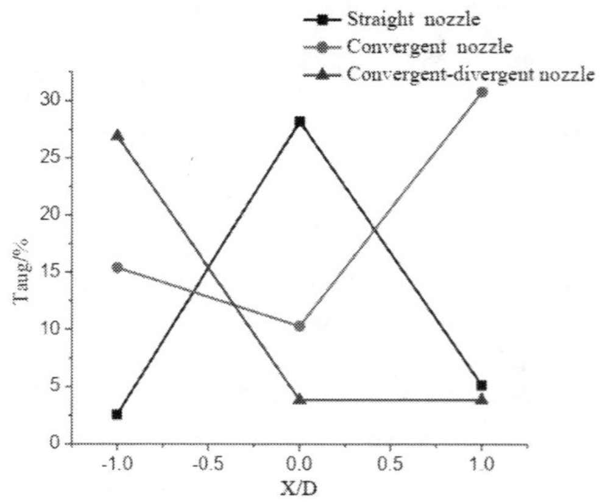

Fig.4 Nozzle type effect and ejector position effect on PDE thrust augmentation at the operating frequency of 25 Hz.

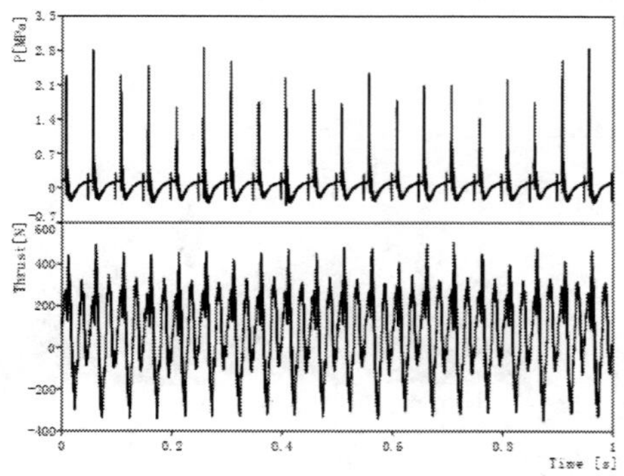

Fig. 5 Pressure and thrust histories of a PDE witha convergent-divergent nozzle-ejector at -1 ejector position at 20Hz.

4 Conclusions

Base on the pulse detonation engine with the inner diameter of 60mm, three different kinds of nozzle-straight ejector combination structure were experimental studied under different ejector installation position(X/D=-1, 0 and +1). The major conclusions of this paper are shown as follows:

(1) The thrust augmentation of PDE was influenced by both the axis position of ejector and the nozzle type.

(2) The axial position of ejector had different effect on the thrust performance of PDE with different nozzle. PDE with a straight nozzle-ejector combinational structure obtained the best propulsive performance at the ejector position of X/D= 0,while the PDE with a convergent nozzle-ejector combinational structure obtained the best propulsive performance at the ejector position of X/D= +1. For a PDE with a convergent-divergent nozzle-ejector combinational structure, the best propulsive performance was obtained at the ejector position of X/D= -1.

(3) The nozzle-ejector thrust augmentation was influenced by the operating frequency. The highest thrust augmentation of straight nozzle, convergent nozzle and convergent-divergent nozzle appears at 15Hz, 25Hz and 20Hz, respectively. Furthermore, the maximum thrust augmentation of convergent-divergent nozzle-ejector combinational structure achieved54.10%

at the operating frequency of 20 Hz at position -1.

5 Acknowledgment

This work was financially supported by the National Natural Science Foundation of China through Grant No. 51306153, the Natural Science Foundation of Shanxi Province of China through Grant No.2017JZ011 and the Fundamental Research Funds for the Central Universities through Grant No. 3102017jg2009.

References

[1] J.M.Austin, J.E.Shepherd. Detonations in hydrocarbon fuel blends [J]. Combustion and Flame, 2003, 132(1/2):73-90.

[2] M. Ken, E. Motoki, B. I. Ken, K. Jiro, M. Akiko, F. Ikkoh. Optical and thrust measurement of a pulse detonation combustor with a coaxial rotary valve [J]. Combustion and Flame, 2012,159(3):1321-1338.

[3] W. Fan, C. Yan, X. Huang, Q. Zhang, L. Zheng. Experimental investigation on two-phase Pulse Detonation Engine [J]. Combustion and Flame, 2003,133(4):441-450.

[4] C. Yan, W. Fan, X. Huang, Q. Zhang, L. Zheng.Exploratory research of new concept Pulse Detonation Engine [J]. Progress in Natural Science, 2002,12(10):1021-1025. (in Chinese)

[5] C. Yan, L. He, W. Fan, H. Lei, X. Huang, Q. Zhang.Research and development of Pulse Detonation Engine [J]. Aerospace Power, 2001,16(03):212-217. (in Chinese)

[6] Q. Zhang, C. Yan, W. Fan, X. Huang, Z. Wang. Conceptual design of multi-cycle Pulse Detonation Engine [J]. Propulsion Technology, 2003,24(06):500-504. (in Chinese)

[7] K. Landry, R. Shehadeh, N. Bouvet, S. Y. Lee, S. Pal, R. J. Santoro. Effect of operating frequency on PDE driven ejector thrust performance [C].41st AIAA/ASME/SAE/ASEE Joint Propulsion Conference & Exhibit, Tucson, Arizona, USA, 2005.

[8] G. Canteins, F. Franzetti, E. Zoclonska, B. A. Khasainov, R. Zitoun. Experimental and numerical investigations on PDE performance augmentation by means of an ejector [J]. Shock Waves, 2006, 15(2): 103—112.

[9] J. Wilson. A simple model of pulsed ejector thrust augmentation [R]. Hanover, MD, USA, NASA CR 2003-212541, 2003.

[10] Z. Gao, W. Fan, Y. Mu, Y. Yan, W. Chen. Effect of the axial position of tapered ejector on the Pulse Detonation Rocket Engine performance [J]. Propulsion Technology, 2012,33(05):826-830. (in Chinese)

[11] X. Huang, C. Yan, Z. Wang, W. Fan, B. Jiang. Experiment of the effect of detonation frequency on Pulse Detonation Engine ejector performance[J]. Aerospace Power, 2008,23(03):401-405. (in Chinese)

[12] D. Allgood, E. Gutmark, J. Hoke, R. Bradley, F. Schauer. Performance Studies of Pulse Detonaiton Engine Ejectors.Propulsion and Power. Vol.24,No.6 2008.

[13] N. Caldwell, E. Gutmark, J. Hoke, R. Bradley, F. Schauer. Investigation of Fundamental Processes Leading to Pulse Detonation Engine/Ejector Thrust Augmentation.46th AIAA Aerospace Sceinces Meeting and Exhibit.AIAA 2008-116, 2008.

[14] J. L. Hoke, R. Bradley, F. Schauer. Single-Ejector Augmentation of a Multi-Pulsed Detonation Engine.46th AIAA-Aerospace Sciences Meeting and Exhibit.AIAA 2008-115, 2008.

[15] J. L. Hoke, A. G. Naples, L. P. Goss, F. R. Schauer. Schlieren Imaging of a Single-Ejector, Multi-Tube Pulsed Detonation Engine. 47th AIAA Aerospace Sciences Meeting Including The New Horizons Forum and Aerospace Exposition. AIAA 2009-1065, 2009.

[16] M. R. Nalim, Z. A. Izzy, P. Akbari. Rotary wave-ejector enhanced pulse detonation engine [J]. Shock Waves, 2012, 22 (1):23-38

[17] X. Huang, C. Yan, J. Deng. Experiment of the effect of ejector axial position on the Pulse Detonation Engine ejector performance[J]. Propulsion technology, 2009,30(2):219-223. (in Chinese)

[18] X. Huang, Y. Li, L. Chen. Experiment studies of Pulse Detonaiton Engine ejectors performance of increase thrust[J]. Northwestern Polytechnical University Journal, 2013, 31(1):55-59. DOI:10.3969/j.Issn. 1000-2758. 2013. 01.011. (in Chinese)

[19] Y. Wang, W. Fan, Z. Gao. Experiment studies of Pulse Detonaiton Engine ejectors mode detonate [J]. Propulsion technology, 2013,34(8):1147-1152. (in Chinese)

[20] R. Zitoun, D. Desbordes. Propulsive Performances of Pulsed Detonations [J]. Combustion Science and Technology, 1999;144 (2): 93-114.

[21] T. Chao, E. Winenberger, J. E. Shepherd. On the Design of Pulse Detonation Engines[R]. GALCITT Reprot FM00-7, Graduate Aeronautical laboratories, Callifornia Institute of Technology,Pasadena,CA91125. 2001.

[22] Q. Zhang, C. Yan, W. Fan, X. Huang, Z. Wang. Multi Cycle Pulse Detonation Engine Conceptual Design[J]. Propulsion technology, 2003;24(6):500-504. (in Chinese)

• ***Personal communication and unpublished data*** is to be cited in parenthesis within the text

11th Asia-Pacific Conference on Combustion,
The University of Sydney, NSW Australia
10th -14th December 2017.

Influence of gas expansion on the interaction between spatially periodic shear flow and premixed flame

Ruixue Feng[1,2], Hongtao Zhong[2], Damir Valiev[1,2,3]

[1]Center for Combustion Energy, Tsinghua University,
Beijing 100084, China

[2]Department of Thermal Engineering, Tsinghua University,
Beijing 100084, China

[3]Applied Physics and Electronics, Umeå University,
Umeå 90187, Sweden

Abstract

It has been previously shown that thermal expansion may have a role in a boundary layer flashback of premixed turbulent flames [Gruber et al., J Fluid Mech 2012], enhancing the flame velocity through flame corrugation. The current study focuses on the two-dimensional numerical investigation of the interaction of a periodic shear flow and a laminar premixed flame. The periodic shear is a simplified model for the oncoming prolonged velocity streaks with alternating regions of high and low velocities found in a turbulent channel flow close to the wall. The parametric study focuses on the amplitude and wavelength of the periodic shear flow, and gas expansion ratio. It has been found that with the increase of the amplitude of the periodic shear flow, as well as the gas expansion, the curved flame velocity increases monotonically. The flame velocity dependence on the periodic shear wavelength is non-monotonic, which is consistent with previous theoretical studies of the curved premixed flame velocity.

1 Introduction

It has been recently demonstrated that the thermal expansion of burnt gas plays a role in the boundary layer flashback of premixed turbulent flames [1], enhancing the flame speed through the onset of Darrieus-Landau (DL) instability and associated flame corrugation. Recent experimental investigations revealed the existence of the flow reversal regions in both swirling and non-swirling configurations [2, 3, 4], with flame velocities approximately twice larger than the laminar flame speed. Three-dimensional simulations of Ref. [1] confirmed the creation of near-wall regions of flow reversal ahead of the flame sheet. It was shown that the flame in the boundary layer behaves in an almost laminar fashion, but its shape is modified by weak, highly anisotropic velocity fluctuations. In Ref. [1] it was shown that the flame acquires a corrugated shape with the characteristic time and length scales that can be correlated with the structures in the boundary layer. The oncoming prolongated velocity streaks with alternating regions of relatively high and low velocities interact with the initially planar flame. The observed flame shape evolution suggests that the streaky structure of the flow triggers the corresponding unstable wavelength of Darrieus-Landau (DL) instability. In Ref. [5] it was found that a planar flame that is subjected to oncoming periodic shear flow can propagate with increased velocities in the absence of thermal expansion. There still remain an open question regarding the influence of gas expansion on the flame shape and velocity subjected to external forcing by a spatially periodic shear flow for flames with realistic thermal expansion. A two-dimensional numerical parametric study of the interaction between planar premixed H_2/air flame and spatially periodic shear flow has been conducted in Ref. [6]. In this study it was found that shear flow-induced flame shape may be metastable, replaced by Darrieus-Landau-instability dominated configuration. The present study is methodologically similar to the investigation presented in Ref. [6], however, here we employ one-step irreversible Arrhenius reaction in order to study the effect of thermal gas expansion in isolation, with the other non-dimensional problem parameters fixed. This approach is not feasible with the use of complex chemistry and without affecting other fuel mixture or flame parameters, such as Lewis number or laminar burning velocity.

We perform a detailed investigation of the role of gas expansion on the regime of the curved flame front propagation subjected to external forcing by a spatially periodic shear flow. The parametric study is focused on the variation of flame propagation speed dependent upon the thermal gas expansion, as well as the amplitude and spatial scale of the oncoming periodic shear flow.

2 Details of the numerical simulation

We perform computational simulations of hydrodynamic and combustion equations including transport processes (thermal conduction, diffusion, and viscosity) and the Arrhenius chemical kinetics. The details about the system of equations can be found in Refs. [7, 8]. We take the Lewis number $Le = 1$ (representing a neutrally stable mixture to avoid thermal-diffusion instability), Prandtl and Schmidt numbers $\Pr = Sc = 0.75$ (a model value representing a typical gas mixture), and dynamic viscosity $\mu = 1.7 \times 10^{-5}\,\mathrm{Ns/m^2}$. The initial density, pressure and temperature were taken as $\rho_0 = 1.16\,kg/m^3$, $P_0 = 10^5\,Pa$, $T_0 = 300K$, respectively. The thermal expansion factor is taken in the range

Corresponding author.
E-mail address: dvaliev@tsinghua.edu.cn

$\Theta \equiv \rho_b / \rho_0 = 5 - 14$, where ρ_b is the burnt gas density. A characteristic constant of time dimension τ_R in the irreversible Arrhenius reaction is adjusted to obtain a particular value of the laminar flame speed S_L by solving the associated eigenvalue problem for the range of Θ. The burning zone thickness is conventionally described by the parameter of length dimension often identified as a characteristic flame thickness $\delta_L \equiv \mu / \mathrm{Pr}\,\rho_0 S_L$. We have employed an activation energy $E_a / \Theta R_u T_0 = 5$, an initial flame propagation Mach number $Ma = S_L / c_0 = 0.001$, where c_0 is the sound speed in the fuel mixture. Channel width varied in the range $D = (20-400)\delta_L$ (we use non-dimensional scaling by δ_L). We employed translational periodic boundary condition at the top and the bottom boundaries, and the non-reflecting inflow/outflow boundary conditions at the ends of the channel, see Fig.1.

We used sinusoidal inlet velocity profile. The inlet velocity varied with vertical coordinate x as

$$u = S_L(1 - A\cos(2\pi kx)).$$

Here, A is the amplitude of the periodic shear, k is the wavenumber. Amplitude A varied in the range 0.125 – 1.0. Wavenumber k was kept constant, k = 2. Variation of the wavelength of the periodic shear inflow was attained through the variation of the domain width D.

The initial flame structure was set according to Zeldovich-Frank-Kamenetskii solution.

A two-dimensional, fully compressible Navier-Stokes solver was used in the current study. It has been previously successfully employed in the laminar premixed flames studies, hydrodynamic combustion instability studies, as well as the investigation of flame acceleration [7, 8] and phenomena related to the aero-acoustic applications [9, 10]. The solver uses a cell-centered, finite-volume numerical scheme, with second-order accuracy in time and fourth-order in space for the convective terms, and second order in space for the diffusive terms. In this particular work, a uniform grid with the quadratic cells of size $0.25\delta_L$ was used in the subdomain surrounding the flame front to ensure the isotropic propagation of the curved flame in the all directions. We performed a grid independence study to ensure that the chosen grid spacing is sufficient. The longitudinal size of calculation domain changes dynamically, following the flame propagation. Splines of the third order are used for re-interpolation of the flow variables during periodic grid reconstruction to preserve the second order accuracy of the numerical scheme. The grid size in the flame subdomain varied in the range $(0.125-1.0)\delta_L$.

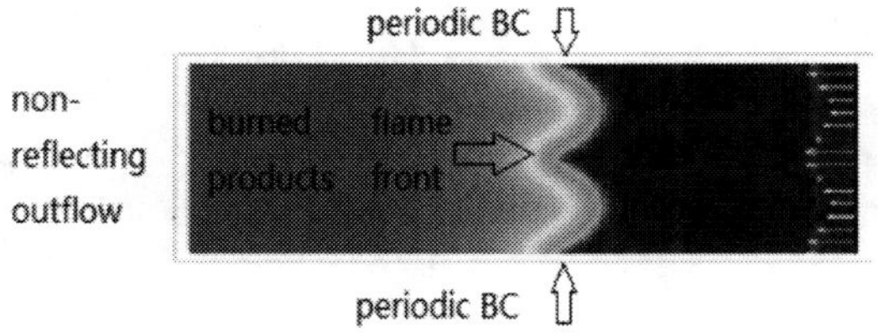

Figure 1: Sketch of the numerical setup.

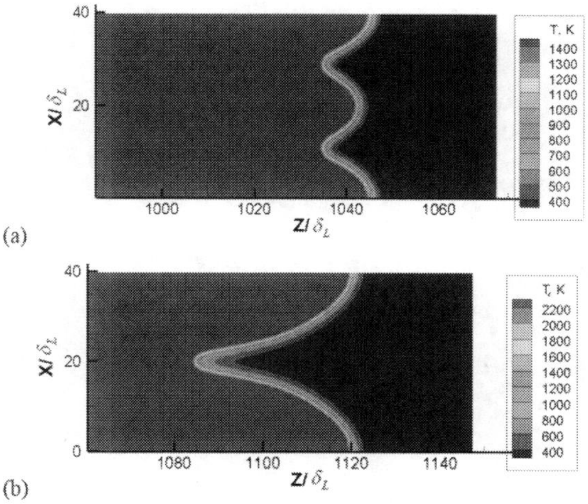

Figure 2: Temperature field, A=1, k=2, θ=8, (a) scaled time $tS_L / D = 0.888$, (b) scaled time $tS_L / D = 1.775$.

3 Results and discussion

We performed several sets of simulations for a range of expansion ratios Θ, inlet velocity variation amplitudes A and domain widths D (we keep the wavenumber of the inlet velocity variation constant, k=2, and change its wavelength by changing D).

In line with the results of Ref. [6], it was observed that the flame shape can experience changes during a simulation run, first changing from a planar one to a two-hump configuration, and then to a one-hump configuration, as illustrated by a typical example presented in Fig. 2 (a,b). We see how a metastable shear-flow-induced flame configuration can be replaced by a stable Darrieus-Landau-instability induced configuration.

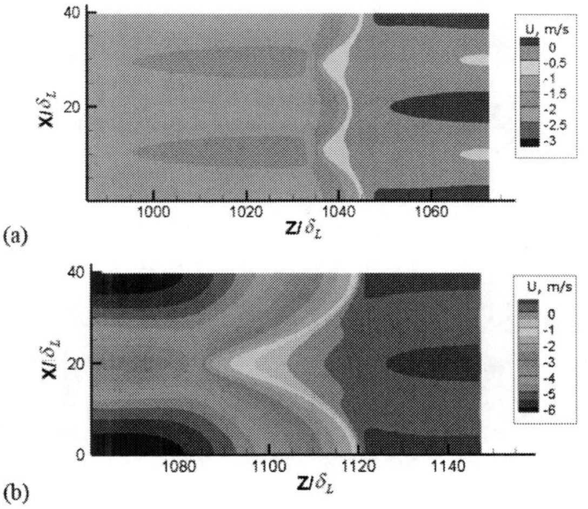

Figure 3: Velocity profile A=1, k=2, θ=8, (a) scaled time $tS_L / D = 0.888$, (b) scaled time $tS_L / D = 1.775$.

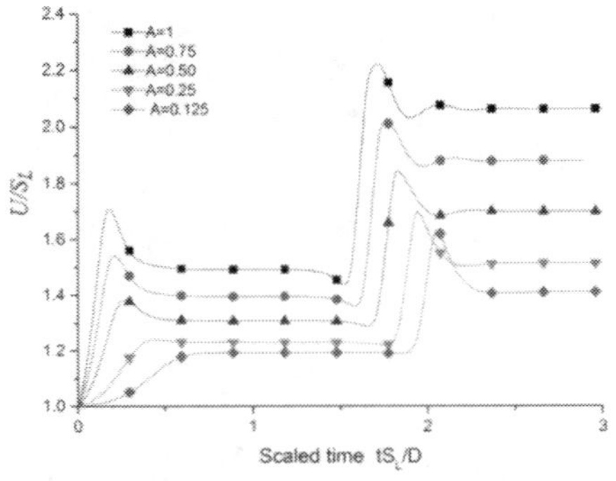

Figure 4: Temporal evolution of the scaled flame velocity U/S_L (proportional to a scaled front length) for different amplitudes A (non-dimensional), k=2, θ=8.

Figure 3 (a, b) shows the velocity fields corresponding to the snapshots of the Fig. 2. It is seen that the velocity field in Fig 3 (a) is correlated with the flame shape in Fig 2 (a), suggesting that the two-hump flame shape is induced by the sinusoidal inflow velocity profile with k = 2. On the contrary, the flame shape in Fig. 3 (b) is no longer dominated by the inflow velocity profile. Similar phenomenon was observed in Ref. [1], where it was found that boundary layer streaks determine the flame front shape initially; however, at the later stages of the boundary layer flashback the Darrieus-Landau instability starts to dominate the flame front shape.

Each simulation in the parametric study was running until a stationary one-hump flame propagation was achieved. However, In the current study, we focus on the quantifying the curved flame velocity at the first, shear-flow induced metastable stage, denoted as U_1. It is assumed that the increase of the curved flame velocity is proportional to the increase of the flame surface.

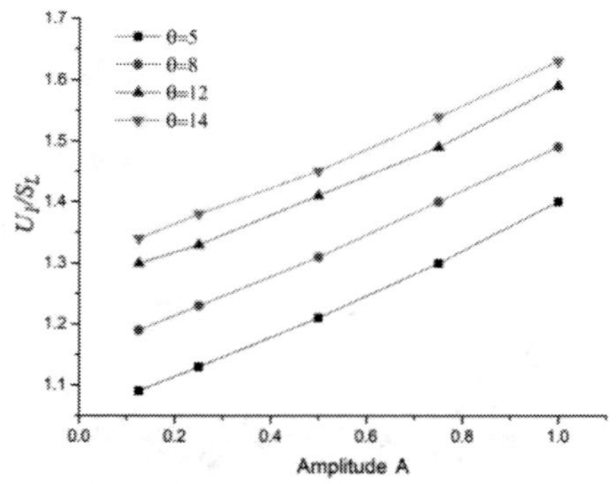

Figure 5: Scaled velocity U_1/S_L vs amplitude plots at different gas expansions θ.

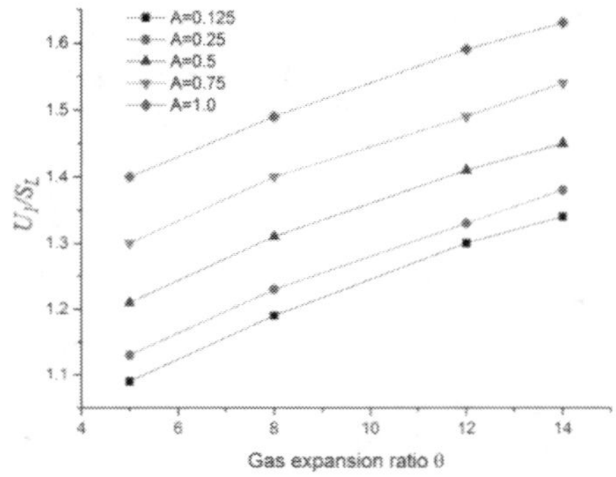

Figure 6: Scaled curved flame velocity U_1/S_L vs gas expansion θ ratio at different amplitudes A (non-dimensional).

Figure 4 shows the scaled flame velocity evolution for different inlet shear flow amplitudes. It is seen that both the velocity of the first (metastable) two-hump flame configuration and of the second stable DL-induced one-hump flame configuration is increasing with amplitude A. Note that the scaled time instant for the flame shapes at Fig. 2(a) corresponds to the first metastable velocity plateau at the Fig. 4, see the line with square markers (A = 1). The flame shape for the Fig. 2 (b) corresponds to the second velocity plateau, respectively.

It is noted that from Fig. 4 it is seen that the shear-flow induced velocity can exceed the maximum velocity of the curved laminar flame corrugated due to the Darrieus-Landau instability [11]. Indeed, the theoretical prediction for the maximum flame velocity curved due to Darrieus-Landau instability can be predicted as [11, 12]

$$U_m = S_L\left[1+\frac{\Theta(\Theta-1)^2}{\Theta^3+\Theta^2+3\Theta-1}\right]. \qquad (1)$$

For example, for θ = 8 the velocity increase predicted by Eq. (1) is ~ 1.65, which is lower than the velocity of the one-hump flame at the second stationary stage at Fig. 4 for the inlet shear flow amplitudes A higher than 0.5.

The variation of the one-hump flame velocity with the inlet shear flow amplitude could be attributed to the flame-flow interaction and the diversion of the flow ahead of the highly curved flame, as seen in Fig. 3 (b). This effectively changes the flow ahead of the flame into the shear flow with modified wavelength.

Figure 5 shows the scaled shear-flow induced flame velocity versus inlet shear flow amplitude for different expansion ratios. It is seen that the shear flow-induced velocity grows monotonically with the inlet velocity variation amplitude for all expansion ratios. The result is in line with findings of Brailovsky and Sivashinsky [5] for the flames with zero expansion, as well as the results of Ref. [6].

Figure 6 shows the scaled shear-flow induced flame velocity versus expansion ratio for a range of shear flow amplitudes θ = 5 - 14. Similarly, it is observed that the shear flow-induced

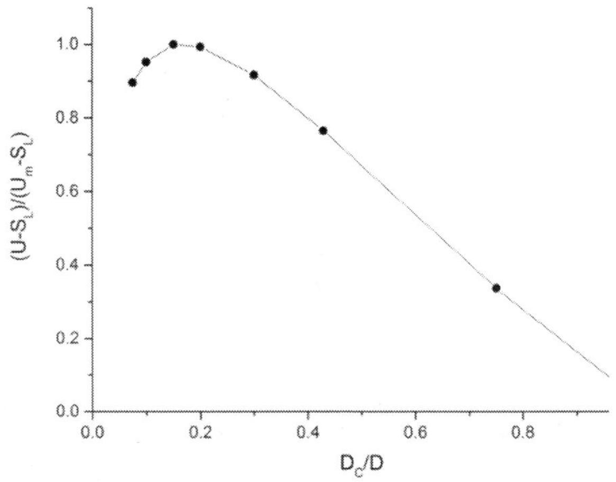

Figure 7: Scaled velocity of a curved stationary flame $(U_l-S_L)/(U_m-S_L)$ vs channel width, A=1, θ=8.

velocity would also grow monotonically with the expansion ratio for all inlet velocity variation amplitudes.

Figure 7 shows the scaled velocity of a shear-flow induced curved stationary flame $(U_l-S_L)/(U_m-S_L)$ versus the inverse channel D_C/D width for inlet shear flow amplitude A=1.0 and gas expansion θ=8. Here, D_C is the critical channel width, below which the Darries-Landau instability would be suppressed [13]:

$$D_C = \frac{2\pi\delta_L}{\Theta - 1}\left[\Theta - 1 + \Theta \ln \Theta(\Theta + 1)\right].$$

For θ=8, the critical channel width can be estimated as $D_C \approx 40\delta_L$.

It is observed that with the increase of the channel width U_l first grows, and then decreases. Similar behavior is observed for the laminar curved stationary flames corrugated due to Darrieus-Landau instability only [13]. This indicates that in fact the curved flame velocity at both the first metastable stage and the second stationary stage may be influenced by both Darrieus-Landau instability and shear flow, but in different proportions.

4 Conclusions

The present study is focused on the influence of gas expansion on the stationary curved flame velocity subjected to external forcing by a spatially periodic shear flow. It was found that the curved flame velocity increases with increasing gas expansion ratio and with the increase of the amplitude of the periodic shear flow. The shear-flow-induced curved flame velocity can exceed the maximum velocity of the curved laminar flame corrugated by the Darrieus-Landau instability only. The curved flame velocity dependence on the periodic shear wavelength is non-monotonic.

5 Acknowledgment

We thank Andrea Gruber for useful discussions and Florian Schmidt for the help with computations. The work is sponsored by the National Science Foundation of China (NSFC) through the Thousand Young Talents program and by the Stiftelsen Lars Hiertas Minne Grant No. FO2015-0916. Computational resources of the High Performance Computing Center North (HPC2N) available under project SNIC 2016/1-565 were used.

References

[1] A. Gruber, J. H. Chen, D. Valiev, C. K. Law, Journal of Fluid Mechanics 709 (2012) 516-542.

[2] C. Heeger, R. L. Gordon, M. J. Tummers, T. Sattelmayer, A. Dreizler, Exp. Fluids 49 (2010) 853-863.

[3] C. Eichler, T. Sattelmayer, ASME J. Engng Gas Turbines Power 133 (2011) 011503.

[4] D. Ebi, N. T. Clemens, AIAA paper 2013-2604 (2013) DOI:10.2514/6.2013-2604.

[5] I. Brailovsky, G. Sivashinsky, Physical Review E 51(2) (1995) 1172-1183.

[6] D. Valiev, A. Gruber, C.K. Law, J.H. Chen, Numerical study of interaction between Darrieus-Landau instability and spatially periodic shear flow, 25th International Colloquium on the Dynamics of Explosions and Reactive Systems (ICDERS), August 2 – 7, 2015, Leeds, UK.

[7] D. Valiev, V. Bychkov, V. Akkerman, L.-E. Eriksson, M. Marklund, Physics Letters A 372 (2008) 4850-4857.

[8] D.M. Valiev, V. Bychkov, V. Akkerman, L.-E. Eriksson, Physical Review E 80 (2009) 036317.

[9] N. Andersson, L.-E. Eriksson, L. Davidsson, Int. J. Heat Fluid Flow 26 (2005) 393-410.

[10] C. Wollblad, L. Davidson, L.-E. Eriksson, AIAA Journal 44 (2006) 2340-2353.

[11] V. Bychkov, Phys. Fluids 10 (1998) 2091–2098.

[12] A. Gruber, A. R. Kerstein, D. Valiev, H. Kolla, J. H. Chen, Proceedings Combustion Institute 35 (2014) 1485–1492.

[13] V. Bychkov, M. Liberman, Physics Reports 325(4) (2000) 115-237.

11th Asia-Pacific Conference on Combustion,
The University of Sydney, NSW Australia
10th -14th December 2017.

A Practical Gasoline Homogeneous Charge Compression Ignition (HCCI) Engine

Anand Lalwani, Swapnil Awate, Arindrajit Chowdhury, Sheshadri Sreedhara

IIT Bombay

Powai, Mumbai, Maharashtra, 400076, India

Abstract

HCCI combustion has been found to be an excellent technology for achieving cleaner emissions along with economical operation. It exploits the benefits of both gasoline and diesel engines simultaneously through compression ignition of a lean homogeneous charge, thus producing ultra-low nitrogen oxides (NO_x) and particulate matter (PM) emissions. Intake air heating and port fuel injection (PFI) were employed to achieve stable HCCI combustion at part load conditions using gasoline as fuel. Load based rpm control system based on a proportional integral derivative (PID) control algorithm was developed and utilized to run the engine at a constant rpm in the HCCI mode. Preliminary simulations were carried out using a chemical kinetics software (CHEMKIN) which provided a basis for the selection of the experimental parameters. Combustion characteristics, performance characteristics, and emission characteristics were analyzed and compared for the gasoline HCCI mode and the conventional diesel mode. Low NO_x emissions with higher brake thermal efficiencies (BTE) than the diesel mode were observed in the HCCI mode. Though the hydrocarbon (HC) and the carbon monoxide (CO) emissions were high in the HCCI mode, economical after-treatment systems can be used for their management.

1 Introduction

Internal combustion (IC) engines play a crucial role in various applications in the field of transportation and captive power generation such as automobile engines, marine engines, aircraft engines, generator sets, etc. Such a wide array of applications of IC engines as a source of energy ensures that a slight improvement in the efficiency would result in significant overall fuel savings. Additionally, the upcoming stringent regulations are expected to lower the emission norms of carbon dioxide, placing even more focus on engine efficiencies. It would also be necessary to further reduce the emission norms of NO_x, CO, etc. noxious gases. This has led to increasing focus towards research in the field of newer combustion technologies for IC engines.

IC engines have conventionally been classified into two categories based on the method for ignition: spark ignition (SI) and compression ignition (CI). The compression ratio (CR) in SI engines is limited by knocking and is between 8 and 12, thus reducing efficiency. Though the bulk gas temperature is low, the high temperature flame-front leads to significantly high NO_x emissions in SI engines [1]. Alternatively, in CI engines, the fuel rich zones give rise to unburned hydrocarbon (UHC), CO and PM emissions. NO_x emissions are a result of the high-temperature combustion occurring in the near stoichiometric zone. The available after-treatment systems of these emissions, particularly those for NO_x and PM, are expensive.

Modifications in these conventional principles can aid in improving engine performance. HCCI combustion is one such technology which provides a unique solution, promising to exploit the benefits of SI and CI engines, while avoiding the peak temperature zones that lead to higher NO_x formation. HCCI combustion is termed homogeneous as auto-ignition is expected to occur simultaneously at numerous locations within the combustion chamber, leading to efficient combustion of the fuel-air mixture without formation of a distinct flame front as in SI or locally inhomogeneous mixtures as in CI. This reduces the local peak temperatures inside the chamber. NO_x formation increases exponentially above 1800 K, and the capability of containing the peak combustion temperature below this in HCCI ensures ultra-low NO_x emission [2]. Also as the mixture is uniform and lean, the formation of particulate matter is suppressed. During HCCI combustion, oxidation of CO to CO_2 is expected to be suppressed due to lower overall combustion temperature which suggests higher CO emissions [3].

A PFI system, wherein the fuel is injected into the intake manifold and a homogeneous mixture enters the combustion chamber during the intake stroke, has been widely used for homogeneous mixture preparation to run the engine in HCCI mode [2, 4, 5]. Aoyama et al. [4] observed the in-cylinder combustion of the premixed lean mixture using a high speed camera for a 17.4 CR engine with gasoline injected into the inlet port. Multi-point compression ignition was confirmed with no evidence of the flame propagation in the shadow images. Oakley et al. [2] achieved gasoline HCCI combustion in a 4 stroke VCR engine with CR 11.5 and elevated intake temperature (593 K). They established an operating range with respect to air-fuel ratio and exhaust gas re-circulation (EGR) and studied their effect on engine performance and emissions. Yang et al. [5] built an HCCI-SI dual mode engine and optimized the engine performance by using the waste thermal energy in the exhaust gases and cam phasing for controlling valve overlap and effective compression ratio.

Part load operation of HCCI using gasoline as fuel has been found to be better than both SI and CI systems in terms of NO_x and PM emissions. Efficiencies better than SI engines and comparable to CI engines can also be achieved using gasoline HCCI. Building on the previous work [6, 7] this study presents successful employment of a load-based rpm control system on a gasoline HCCI engine with a PFI system and compares its performance and emissions characteristics with conventional diesel mode of combustion.

2 Numerical Analysis

Preliminary numerical investigations were carried out using CHEMKIN to determine the values of parameters such as

Corresponding author
E-mail address: swapnil.awate@iitb.ac.in

compression ratio, inlet air temperature, and equivalence ratio that would lead to successful implementation of HCCI combustion. A homogeneous zero degree reactor was considered for simulations, which idealizes HCCI combustion as a variable volume fixed mass reactor model. The engine parameters utilized in the simulations were identical to the experimental set-up being used. A combination of a CR of 14, an inlet temperature of 420 K, and an equivalence ratio of 0.25 was selected to be the starting point for the experiments of gasoline HCCI combustion based on these simulations.

3 Experimental Set-up

Experiments were performed on an agricultural single cylinder 4-stroke single speed diesel engine manufactured by Kirloskar Oil Engines Ltd. The engine specifications are stated in Table 1.

Sr. No.	Parameter	Value
1	Make	Kirloskar Oil Engines Ltd.
2	Model	AV1
3	Bore	80 mm
4	Stroke	110 mm
5	Displacement	553 cc
6	CR	16.5:1
7	Rated Power	5 HP at 1500 rpm
8	Cooling	Water cooled

Table 1: Engine specifications

The engine has a design CR of 16.5. It was modified into a variable compression ratio (VCR) engine by M/S Legion Brothers and the CR was set to 14.4 for the experiments. The engine was loaded by an adjustable eddy current dynamometer (Powermag). In-cylinder pressures were measured by a pressure transducer (Cityzen) with an operating range of 0-200 bar. Crank angles were measured by using an encoder (Kübler, Sendix 5000) with an operating range of 0-6000 rpm. The pressure-crank angle data was logged for 50 consecutive cycles and processed appropriately for analysis. A calibrated multi-gas analyser (AVL) was used to measure exhaust gas emissions such as NO_x, CO, CO_2, O_2 and UHC. Multiple modifications were made to run the engine in HCCI mode as shown in Fig. 1.

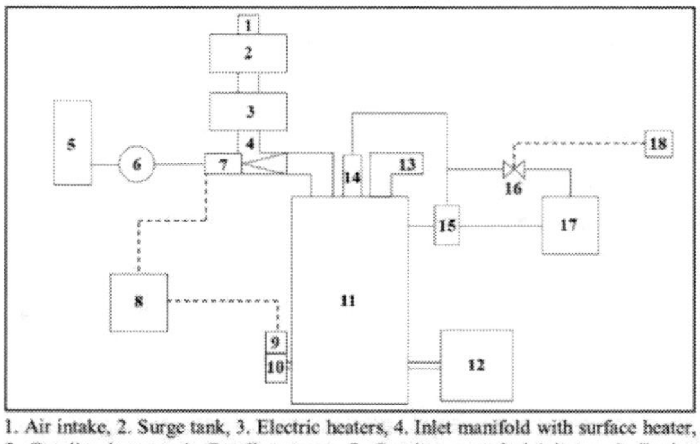

1. Air intake, 2. Surge tank, 3. Electric heaters, 4. Inlet manifold with surface heater, 5. Gasoline burette, 6. Gasoline pump, 7. Gasoline port fuel injector, 8. Engine control unit (ECU), 9. Magnetic sensor, 10. Modified gear on camshaft, 11. Kirloskar CI engine, 12. Dynamometer, 13. Exhaust manifold, 14. Diesel injector, 15. Diesel pump, 16. Diesel by-pass with solenoid valve, 17. Diesel storage tank, 18. Switch to operate by-pass valve

Figure 1: Diesel engine modified for HCCI operation [8]

A heating system consisting of 2 electric heaters, 3.5 kW each, was installed to pre-heat the intake air to the desired value. This temperature was selected to 433 K for a setting of CR 14.4 based on the preliminary results obtained from the analysis using CHEMKIN as discussed in Section 2. A surface heater was installed just after the injector to avoid wall wetting. The temperature of the surface heater was set to 393 K.

A separate fuel delivery system for gasoline was installed for the operation in HCCI mode. Pressurized gasoline was injected by a solenoid-operated injector (Valeo, 348002) into the port, where it was expected to evaporate and mix with the heated intake air to form a homogeneous mixture. Both the gasoline pump and the gasoline injector were powered by a regulated power supply (RPS). To ensure no diesel was injected into the combustion chamber during the HCCI mode, a high-pressure solenoid operated directional poppet valve (Spica, Y04147-002, DTDA-MCN-212) was installed in the diesel supply line between the fuel pump and the injector to by-pass the diesel back to the diesel storage tank. When this solenoid valve was actuated, fuel followed the least resistance path which was open to atmospheric pressure and thus no fuel was injected by the diesel injector.

The injection timing and the injection duration of the gasoline injector were controlled by a custom-made ECU. A detailed description of the development process of this ECU was reported by Jeeva et al. [9]. The objective of load-based rpm control in a constant rpm engine is to control the amount of fuel injected to maintain the rpm when the load is changed. This was achieved by modifying the ECU program to perform proportional integral derivative (PID) control. The parameters, K_P, K_I and K_D, were tuned appropriately through experiments to ensure smooth operation of the engine along with minimum fluctuations in the rpm.

4 Results and discussions

During an experiment, readings were taken twice for every load and three such experiments were conducted to address repeatability. The data from the HCCI mode was compared with the data from an optimized Kirloskar diesel engine with a stroke volume of 661 cc, CR of 14 and, other parameters same as that of the engine used for experiments in this study.

4.1 Combustion characteristics

The variation of in-cylinder pressure with crank angle for different loads is plotted in Fig. 2.

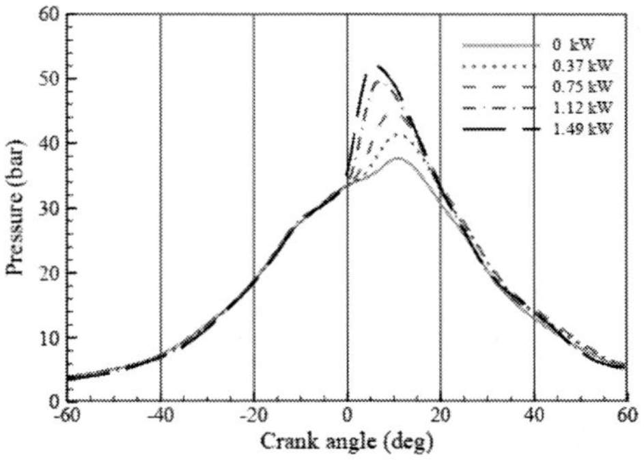

Figure 2: Variation of in-cylinder pressure trace with load

It was observed that the peak pressure increased with load in agreement with the results of the simulations, where the increase in the equivalence ratio was representative of the increase in the load. The selection of inlet air temperature as 433 K was found to be an appropriate choice as the start of combustion (SOC) occurred near the top dead centre (TDC) for every load. In-cylinder peak pressure (P_{max}) was observed to be higher at every load in HCCI mode compared to the conventional diesel mode. The higher rate of in-cylinder pressure rise ($dP/d\theta_{max}$) owing to faster rates of combustion in the HCCI mode caused the P_{max} to be higher than the conventional diesel mode. The $dP/d\theta_{max}$ values for every load were less than the knocking criterion stated by various researchers [10, 11]. For the values of loads above 1.49 kW, which corresponds to 40% of the maximum load, the intensity of engine noise was found to be unacceptable.

Figure 3 shows the effect of load on the ROHR plotted against crank angle.

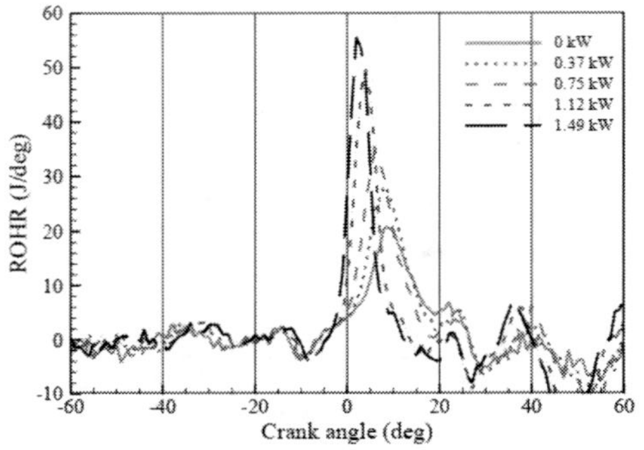

Figure 3: Behavior of ROHR with varying load

It was observed that the peak heat release rate increased with load owing to the increase in P_{max} and $dP/d\theta_{max}$. Also, the crank angle at which peak heat release occurred was found to advance as the load increased. A single stage heat release was observed as significant heat release from low temperature reactions is typically not observed for a fuel such as gasoline with a high octane number [12].

4.2 Performance characteristic

Figure 4 shows the variation of brake thermal energy (BTE) against load.

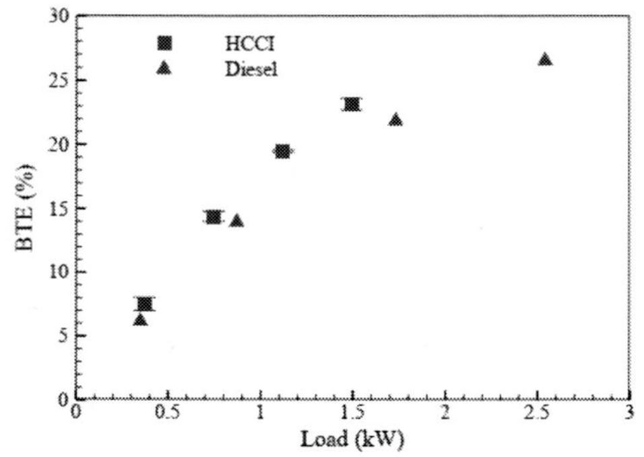

Figure 4: Variation of BTE with load

It was observed that at any given load, a higher BTE was achieved in the HCCI mode than the diesel mode during the experiments. Near constant volume combustion similar to Otto cycle is expected in HCCI combustion. For the same compression ratio, Otto cycle exhibits higher efficiencies than diesel cycle, providing an explanation for the observed trend. As expected, BTE was found to increase with an increase in load for the range of operation under consideration. Though

4.3 Emission characteristics

Pertinent emissions such as NO_x, HC, CO, and CO_2 were measured and compared for the HCCI mode and conventional diesel mode. Owing to the fundamentally less sooty nature of the gasoline fuel as compared to the diesel fuel, the comparison of PM emissions was omitted in this study, following various researchers working on gasoline HCCI [4, 6, 7].

Figure 5 depicts a comparison of NO_x emissions between HCCI mode and diesel mode with increasing load.

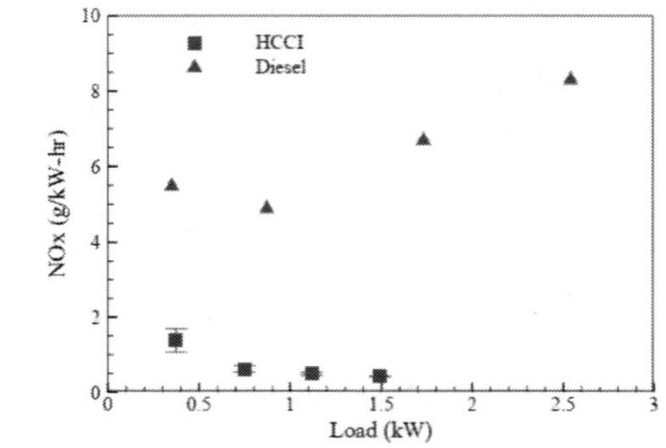

Figure 5: Variation of NO_x emissions with load

It was observed that NO_x emissions were lower in the HCCI mode as compared to the diesel mode owing to the lower in-cylinder temperatures which is a characteristic of HCCI combustion. Although the brake thermal NO_x values were found to be decreasing with load for HCCI mode, the absolute NO_x values were found to be increasing with load for both the modes. NO_x in IC engine emission is primarily thermal NO_x which is strongly dependent on in-cylinder temperatures [2]. The NOx emissions were found to increase with load owing to the increase in the peak in-cylinder temperature.

Figure 6 shows the variation of HC emissions against load.

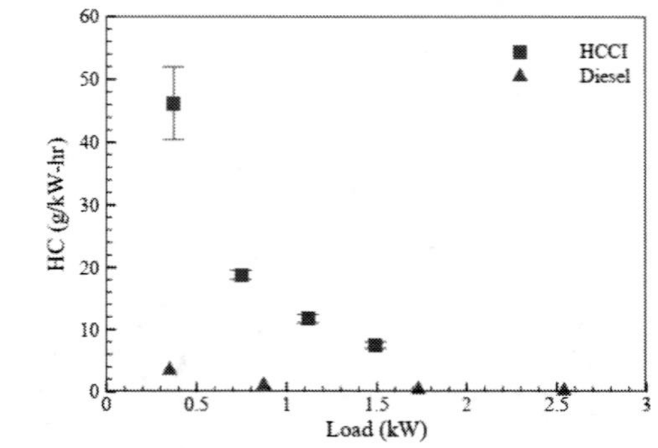

Figure 6: Variation of HC emissions with load

It was observed that the amount of HC emissions at all loads were much higher in the case of the HCCI mode as compared to the diesel mode. It was conjectured that in HCCI mode a PFI system was used, which typically leads to higher HC emissions, most of which come from the crevices [13]. Reduction of the HC emissions with load in both the modes was because of the higher in-cylinder temperature during the exhaust stroke which promoted the oxidation of the HC.

Figure 7 shows a comparison between the CO emissions of the HCCI mode and the conventional diesel mode.

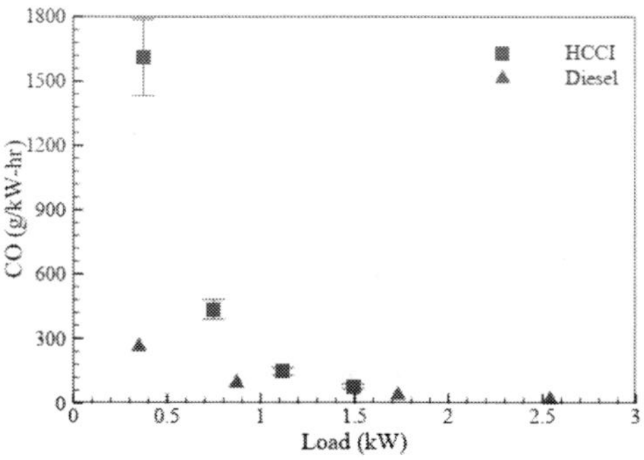

Figure 7: Variation of CO emissions with load

It was observed that the CO emissions were higher for the HCCI mode as compared to the conventional mode at all the loads. It was also observed that the difference between the emission values of the two modes reduced at higher loads. CO emissions reduced with load for both the modes which was in agreement with higher in-cylinder temperatures expected at higher loads promoting the oxidation of CO.

CO_2 emissions for the HCCI mode and the diesel model are shown against load in Fig. 8.

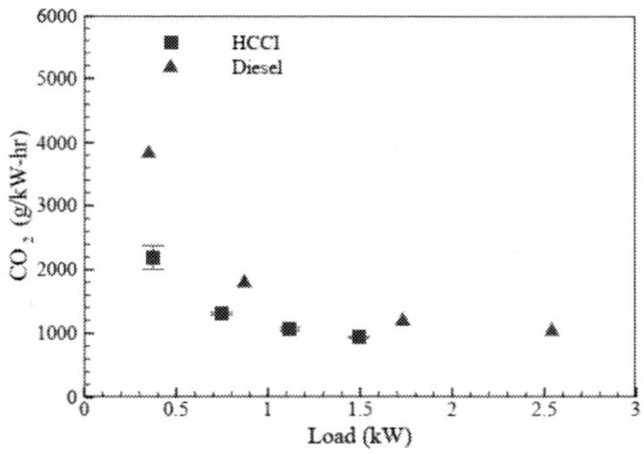

Figure 8: Variation of CO_2 emissions with load

It was observed that the amount of CO_2 emissions at all loads is lower in the HCCI mode as compared to the conventional diesel mode owing to the combustion of lesser amount of fuel. The amount of brake specific CO_2 emission was found to reduce for both the modes as the load increased which is in congruence with the increasing BTE with load.

5 Conclusions

Experiments were conducted to run a single cylinder CI engine in HCCI mode using gasoline as fuel and stable operation was achieved under part-load conditions. Load-based rpm control was implemented using PID algorithm to run the engine at constant rpm, thus enabling comparison between conventional diesel and gasoline HCCI mode at different loads. The combustion and performance characteristics such as in-cylinder pressures, ROHR and BTE were compared for both the modes. It was found that BTE values are higher for the operation in HCCI mode. Additionally, exhaust gas emissions were measured and compared for both the modes. It was found the HCCI mode has lower NO_x emissions but higher HC and CO emissions than the conventional diesel mode.

Thus, in order to reduce overall emissions and maintain high fuel economy, operating the engine in HCCI mode for part load conditions and switching to conventional mode for higher loads is found to be a sustainable solution.

6 Acknowledgments

Authors would like to acknowledge the Department of Science and Technology (DST), India for funding this project. Authors are thankful to Mr. Jeeva Baalu, Mr. Dharmvir Singh, Mr. Abhishek Sinha and Mr. Yashraj Gurumukhi for their contributions in the experimental work.

References

[1] T. W. Ryan III and T. J. Callahan, SAE Paper No. 961160, 1996.

[2] A. Oakley, H. Zhao, N. Ladommatos, and T. Ma, SAE Paper No. 2001-01-1030, 2001.

[3] J. E. Dec, SAE Paper No. 2002-01-1309, 2002.

[4] T. Aoyama, Y. Hattori, J. Mizuta, and Y. Sato, SAE Paper No. 960081, 1996.

[5] J. Yang, T. Culp, and, T. Kenney, SAE Paper No. 2002-01-2832, 2002.

[6] S. Awate, H. Kalyankar, B. Jeeva, A. Chowdhury, and S. Sreedhara, Proceedings of the 10th Asia-pacific conference on Combustion, Beijing, China, 2015.

[7] S. Awate, B. Jeeva, A. Chowdhury, and S. Sreedhara, Proceedings of the 23rd National conference on IC Engines and combustion SVNIT, Surat, India, 2013.

[8] S. Sreedhara, A. Chowdhury, B. Jeeva, S. Awate, Patent No.3787/MUM/2014 A, 2014.

[9] B. Jeeva, S. Awate, J. Rajesh, A. Chowdhury, and S. Sheshadri, Proc. 2014 2nd Int. Conf. on Emerging Technol. Trends Electron. Commun. Networking, SVNIT, Surat, India, 2014.

[10] A. Bhave, M. Kraft, F. Mauss, A. Oakley, and H. Zhao, SAE Paper No. 2005-01-0161, 2005.

[11] T. Aroonsrisopon, V. Sohm, P. Werner, and D. E. Foster, SAE Paper No. 2002-01-2830, 2002.

[12] M. Sjöberg and J. E. Dec, SAE Paper 2003-01-3173, 2003.

[13] M. Christensen, B. Johansson, and A. Hultqvist, SAE Paper No. 2006-01-0631, 2006.

11th Asia-Pacific Conference on Combustion,

The University of Sydney, NSW Australia
10th -14th December 2017.

Engine Performance Affected by Direct Injection Timing in a Dual Ethanol Injection SI Engine

Nizar F.O. Al-Muhsen, Guang Hong

University of Technology Sydney

15 Broadway, Ultimo, NSW, 2007, Australia

Abstract

Ethanol fuel, as a bioproduct, has become an alternative option to gasoline fuel in SI engines. Dual ethanol injection strategy (DualEI) is a new technology under development to maximise the benefits of ethanol's properties to the SI engine use. The direct injection timing could significantly affect the DualEI engine performance and fuel economy. In this study, four direct injection timings were set after the intake valve closed to investigate the combustion and emissions characteristics and specific fuel consumption of the DualEI engine. Experiments were conducted at a light load (Engine torque 6.5-7.5 N.m) and four-selected spark timings at a fixed direct injection ratio. The results showed that the IMEP and engine thermal efficiency improved with advanced direct injection timing from TDC, and the indicated specific fuel consumption was reduced. As the direct injection timing works interactively with spark timing, advanced spark timing could improve the engine performance when the direct injection timing was retarded. The maximum improvement to the IMEP was 7.61%, and the indicated specific fuel consumption reduction was 7.07%. This was attributed to the improved mixture and combustion quality caused by the longer time available when the direct injection timing was advanced. The greater in-cylinder pressure and temperature could play a significant role in the engine performance improvement when the direct injection timing was retarded, and the spark timing was advanced. Consequently, the indicated specific nitric oxide (ISNO) was sharply increased because it is directly related to the burned gas temperature. On the other hand, because of the cooling effect, ISNO decreased with the retarded timing of the direct injection. The indicated specific carbon monoxide was increased with advanced direct injection and spark timings. This was attributed to the lean mixture quality caused by the decreased equivalence ratio with the advanced direct injection timing.

1. Introduction

Emissions reduction and the fuel economy improvement are the most common drivers of the internal combustion engine development. Moreover, alongside with modern SI engine development such as GDI engines, more tightened regulations were published, which aimed to reduce emissions produced by IC engines. Compared to Euro 3, Euro 6 was tightened the emissions standards of the NO, CO and HC for the light duty cars by about 60%, 56.5% and 50% respectively [1]. These legislations are evident instances that highlight the significance of the environmental protection issue. Recently, several new technologies have been implemented to reduce the emissions of the vehicles. Ethanol fuel has been used widely as an alternative fuel to fossil fuel in engines to address the issue of sustainability [2]. Ethanol is a renewable fuel produced from a variety of biological resources. Compared to gasoline, ethanol has a greater flame speed, octane number and heat of vaporisation and it provides a cleaner combustion [3]. Nevertheless, ethanol has lower heating value, which consequently increases the specific fuel consumption compared to gasoline. Slow evaporation rate is also reported to ethanol fuel at low ambient temperature [4]. Blended fuel, gasoline or diesel plus ethanol, is one technology which is extensively used in the engine development research [5, 6]. However, blended gasoline or diesel with ethanol does not provide a fully utilised to ethanol fuel properties. Dual injection strategy can provide a superior utilising to the ethanol fuel properties such as the latent heat of vaporisation. Liu et al. [19] investigated the combustion characteristics using ethanol port injection plus gasoline direct injection to a naturally aspirated SI engine. Their results showed that using ethanol fuel could significantly mitigate the engine knock and improve the combustion efficiency. Taniguchi et al. from Toyota Motor Company reported that using pure ethanol fuel to GDI engine improved the thermal efficiency by enhancing the volumetric efficiency at high engine load [7]. It was reported that the combustion performance significantly improved when the dual injection strategy was used. Zhuang et al. [8, 9] investigated the merits of the ethanol direct injection strategy (EDI) to the gasoline port injection (GPI) engine performance. Huang et al. analysed the combustion process numerically to the same project [10, 4]. Their results showed that the engine performance could slightly improve due to injecting a certain amount of ethanol directly into the combustion chamber. The greater laminar flame speed combined with an oxygen content of ethanol helped in enhancing the combustion quality. A significant reduction in the NO emission occurred due to the ethanol lower adiabatic flame temperature and the cooling effect of the EDI strategy. However, increasing the amount of ethanol directly injected led to combustion performance deterioration, which resulted in greater HC and CO emissions were produced. This was attributed to the ethanol film formed on the combustion chamber walls caused by the ethanol impingement. Zhu et al. [11] tested the effect of different dual fuel, ethanol plus gasoline, injection techniques on the combustion characteristics. Gasoline port injection plus ethanol direct injection (GPI+EDI) and ethanol port injection plus gasoline direct injection

(EPI+GDI) strategies were included. The best combustion performance was recorded at GPI+EDI that the indicated mean effective pressure (IMEP) was increased by 2% at light engine load. The effect of EDI timing on the performance of a GPI engine was investigated [8], and the effect of the ethanol port injection timing on the GDI engine performance was experimentally investigated [12]. The results of both investigations showed that the injection timing significantly affected the engine performance and knock mitigation. Using biofuel such as ethanol to the SI engines applications presents new challenges to the engine development. Dual ethanol injection strategy (DualEI) is a new technology under development aiming to exploit the benefits of using pure ethanol to SI engines application. This paper aims to investigate the effect of the direct injection timing (DIT) associated with the spark timing (ST) on the combustion and emissions characteristics of a naturally aspirated SI engine equipped with a DualEI system.

2. Experimental Setup

A modified Yamaha YBR250 motorcycle engine was used in the experiments. It is a single cylinder four-stroke air-cooled naturally aspirated SI engine. The original engine was modified by Hents Technology to meet the research needs. The engine modification included the installation of a direct fuel injection system and an electronic control unit (ECU). The ECU was used to adjust the throttle position, direct injection timing, mass of the injected fuel per cycle and direct injection pressure. The direct fuel injection system comprised of a returnless high-pressure pump and a six-hole high-pressure injector. A Super Micro Flow Mate fuel flow meter was used to measure the fuel flow rate of the fuel injected. The fuel flow meter accuracy is ±3%. An eddy current dynamometer was used to set the required engine speed and measure the engine torque. A Kistler 6115B cylinder pressure transducer with a Kistler 5011 charge amplifier was used to record the in-cylinder pressure. K-type thermocouples were used to measure the cylinder head temperature and exhaust gas temperature with a resolution of 0.1 °C and an uncertainty of 0.35%. A MEXA-584L Horiba exhaust gas analyser was used to measure the exhaust gas emissions of CO, CO2, HC, NO and lambda (λ). The intake airflow was stabilised by using an 80L intake buffer tank. The intake airflow rate was measured by using a ToCeiL20N thermal air-mass flow meter. More details about the research engine and its specifications can be found in [13].

3. Experimental Procedure

It is difficult to use pure ethanol for cold starting in SI engines without using an auxiliary system with gasoline fuel [14]. In the experiment, the engine was started and warmed up by using GPI only until the cylinder head temperature became stable at around 200°C. After that, the GPI was switched to ethanol port injection, and the mixture was kept at the stoichiometric condition. The engine operating conditions and the main engine specifications are listed in Table 1. The volumetric ratio of the EDI to ethanol port injection was kept at around 50% during the experiments. The spark timing was swept from ST28 to ST34. The direct ethanol injection was started after the intake valve closed. At each spark timing, the DIT was swept from 60 to 120 CAD bTDC (DIT60 to DIT120) at light engine load as shown in Table 1. The throttle opening was set at 23%, and the engine speed was fixed at 3500 RPM. The equivalence ratio (λ) was kept at around the stoichiometric condition (λ=1). To ensure that the value of λ was consistent during the experiments, a Horiba exhaust gas analyser and a Bosh Wide-band lambda sensor were used independently to ensure the measured value of λ. A LabVIEW code was written to record the engine data at each tested engine operation condition. Five samples of data were collected at a rate of one record each second, and the average values were used in the calculations and analysis. The in-cylinder pressure data was recorded at 0.5 crank angle degree resolution with 100 consecutive cycles in each sample. The ensemble average of the cylinder pressure data was used in analysing the combustion and emissions characteristics.

Corresponding author. Nizar F.O. Al-Muhsen
E-mail address: Nizar.Al-Muhsen@student.uts.edu.au

Table 1. Engine specifications

Engine type	Four stroke, SI engine
Engine displacement	249 cm^3
Tested engine Spark Timings	ST28, ST30, ST32,ST34
Tested port fuel injection timing	410 [CAD bTDC]
Tested direct fuel injection timing	DIT60, DIT80, DIT100 and DIT120
Timing of intake valve [CAD bTDC]	Opening: 282.2 Closing: 126.2
Timing of exhaust valve [CAD bTDC]	Opening: 594.6 Closing: 340.7

4. Results and Discussions

The effect of the direct injection timing on the IMEP, indicated thermal efficiency and indicated specific fuel consumption (ISFC) will be presented and discussed first, then on the combustion characteristics and finally on emissions of the indicated specific nitric oxide (ISNO) and indicated specific carbon monoxide (ISCO).

4.1 Effect of DIT on Engine Performance

Figure 1 shows the effect of the direct injection timing on the IMEP of the DualEI engine at four-selected spark timings. The port injection timing was 410 CAD bTDC which was set originally by the manufacturer. As shown in Fig. 1, the IMEP increases significantly with advanced direct injection timing from DIT60 to DIT100 due to the improved mixture and combustion quality. This is because the fuel impingement due to EDI strategy reduces significantly with the advanced direct injection timing from DIT60 to DIT100. However, the IMEP slightly decreases when the timing of direct injection is further advanced from DIT100 to DIT120. This is possibly due to the lower in-cylinder pressure and temperature at DT120 comparing to DIT100. Compared with the DIT60, the maximum IMEP improvement is 7.61% at DIT100 and 5.92% at DIT120 when the spark timing is ST34. This is caused by the increase of the combustion pressure and temperature when the spark timing is advanced from ST28 to ST34 [15]. This indicates that the spark timing and the direct injection timing interactively work together when the DualEI is used. As shown in Fig. 2, the indicated thermal efficiency increases with the advanced injection timing from DIT60 to DIT100 at all spark timings. On the other hand, the indicated thermal efficiency slightly decreases when the direct injection timing is further advanced from DIT100 to DIT120. The greatest indicated thermal efficiency improvement is recorded at DIT100 and ST34 caused by the IMEP improvement. Figure 3 shows a significant reduction of the ISFC when the direct injection timing is advanced from DIT60 to DIT100, but it slightly increased when the direct injection timing is further advanced from DIT100 to DIT120. This performance is consistent with the improvement of IMEP and thermal efficiency. Compared with the DIT60, the maximum reduction of ISFC is 6.87% and 7.07% at spark timing ST32 and ST34 respectively and for DIT100. This may be attributed to the combustion performance improvement, which is caused by the time available to the ethanol directly injected to be homogeneously mixed with air before the combustion starts. Moreover, the greater in-cylinder pressure and temperature at the time of injection may play a significant role in increasing ethanol evaporation rate, which leads to improving the combustion quality and then better fuel economy performance.

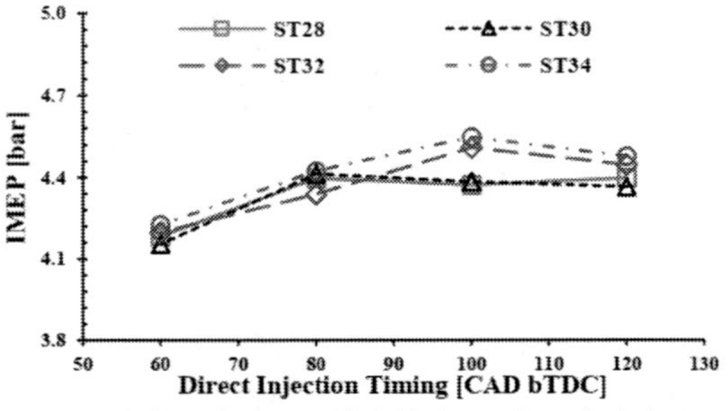

Figure 1 Variation of IMEP with injection and spark timings.

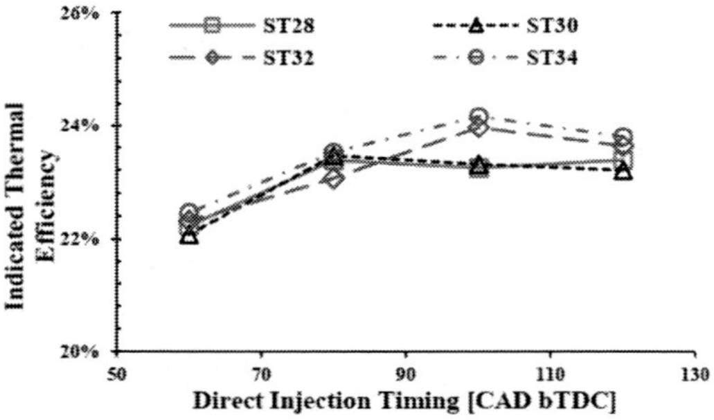

Figure 2. Variation of ITE with injection and spark timings.

4.2 Effect of DIT on Combustion

Figure 4 shows the variation of the CA0-10% with the direct injection timing at four-selected spark timings. The CA0-10% is defined in CAD which is calculated from the spark timing to the timing when 10% of the fuel mass is burnt [16]. As shown in Fig. 4, the CA0-10% is shorter when the direct injection timing is advanced from DIT60 to DIT120. This shows the significant improvement in combustion stability by advancing the direct injection timing [17]. This can be attributed to the longer time available to the ethanol-air mixture to be homogeneously mixed before the combustion starts. As shown in Fig. 4, the CA0-10% at late spark timing ST28 is shorter than that in early spark timing ST34. This indicates to the significant effect of the time available to the mixture on the CA0-10%. Figure 5 shows the variation of the CA10-90% with the timing of direct injection at four-selected spark timings. The CA10-90% is defined in CADs which is calculated from 10% to 90% of the fuel mass fraction burnt. As shown in Fig. 5, in general, the CA10-90% is shorter when the direct injection timing is advanced from DIT60 to DIT80, but the CA10-90% increases significantly when the direct injection timing is further advanced to DIT120. However, at the same injection timings, the IMEP and the indicated thermal efficiency were increased as shown in Figs. 1 and 2. This can be attributed to the combustion speed reduction associated with the cooling effect of EDI strategy. On the other hand, the early spark timing ST34 reduces the CA10-90% due to the greater in-cylinder pressure and temperature. Increasing the combustion temperature may promote faster evaporation rate for ethanol and thus better mixture and combustion quality.

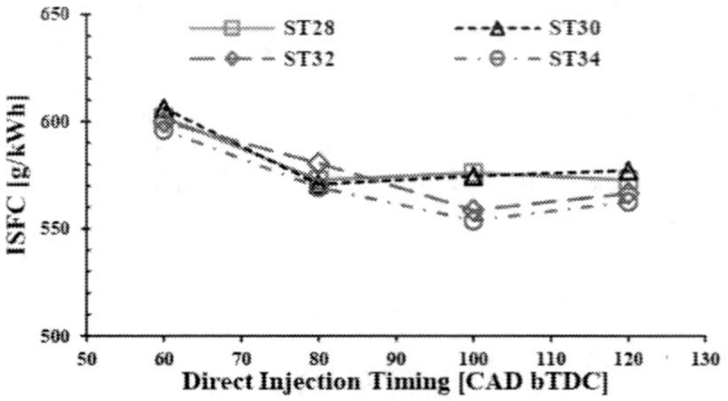

Figure 3. Variation of ISFC with injection and spark timings.

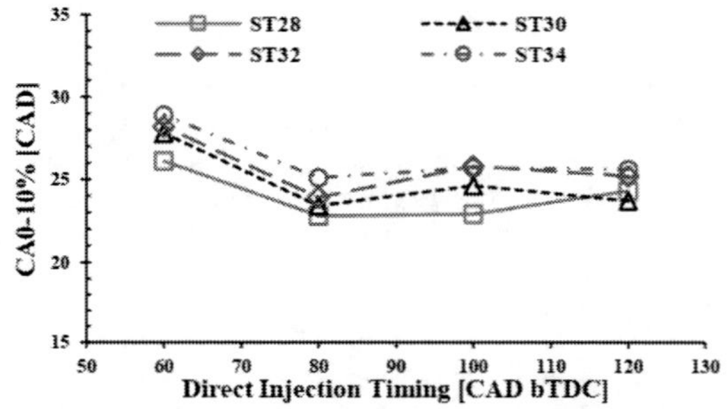

Figure 4. Variation of CA0-10% with injection and spark timings.

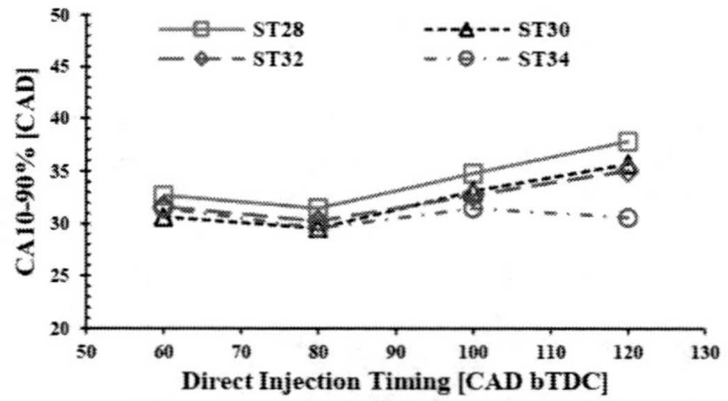

Figure 5.Variation of CA10-90% with injection and spark timings.

4.3 Effect of DIT on engine Emissions

Figure 6 and 7 show the variation of the ISNO and ISCO with the timing of the direct injection at different spark timings. As shown in Fig. 6, the ISNO increases noticeably with the advanced direct injection and spark timings. The maximum increase of the ISNO is 60.36% at DIT120 and ST34. This is because the ISNO is directly related to the combustion temperature [3]. Consistently with the causes behind the indicated thermal efficiency improvement and the ISFC reduction as shown in Figs. 2 and 3, more heat is released per crank angle degree resulting in greater combustion pressure and temperature. Furthermore, greater combustion pressure and then the temperature can be achieved when the spark timing is advanced from ST28 to ST34 due to the change in the combustion phase. As shown in Fig. 7, in general, the ISCO increases with advanced direct injection timing from DIT60 to DIT120 at all the spark timings. This is because the equivalence air-fuel ratio (lambda) increases significantly when the direct injection timing is retarded from DIT60 to DIT120 as shown in Fig. 8. This may provide more air to the fixed amount of injected fuel leading to reduce the ISCO.

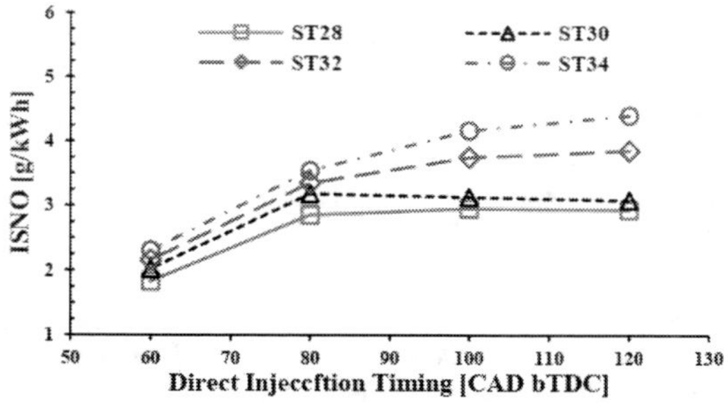

Figure 6. Variation of ISNO with injection and spark timings.

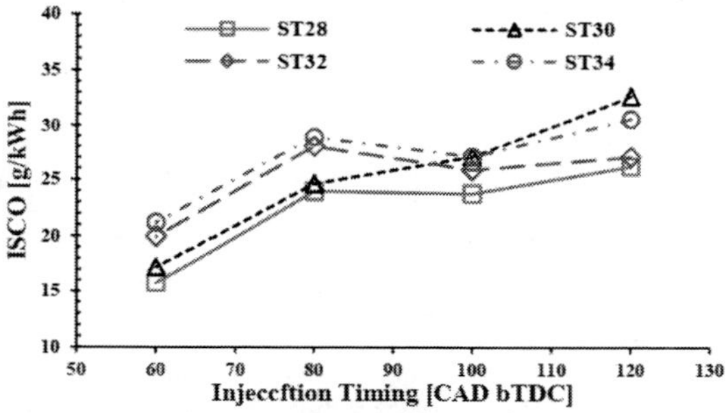

Figure 7. Variation of ISCO with injection and spark timings.

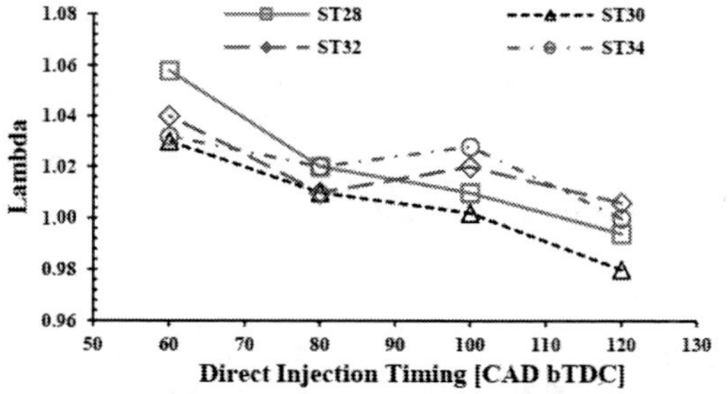

Figure 8. Variation of lambda with injection and spark timings.

5. Conclusions

Experiments were carried out to investigate the effect of direct injection timing on the performance of a 250Cc single cylinder SI engine equipped with a DualEI system. At fixed engine speed and load, the direct injection timing was varied from 60 to 120 CAD bTDC at an interval of 20 CAD. At each direct injection timing, the spark timing was varied from 28 to 34 CAD bTDC at an interval of 2.0 CAD. The results of this study can be concluded as follows.

(1) The engine performance was improved when the direct injection timing was advanced from DIT60 to DIT120. The IMEP was improved by up to 7.61%, and the ISFC was reduced by up to 7.07%. This could be attributed to the improved mixture quality at the time of ignition associated with the in-cylinder pressure and temperature increased by the advanced spark timing.

(2) The initial combustion duration reduced with the advanced direct injection timing. This could be attributed to the longer time available to the fuel directly injected to be homogeneously mixed with air. Furthermore, the larger volume of the combustion chamber at the time of injection could reduce the fuel impingement. However, the major combustion duration increased when the direct injection timing was advanced from DIT80 to DIT120. This was because the equivalence ratio was increased with advanced injection timing.

(3) The ISNO sharply increased with advanced direct injection. This was attributed to the increased combustion pressure and temperature by better combustion quality. In general, the ISCO increased with advanced direct injection timing. This was attributed to the equivalence ratio reduction when the timing of direct injection was advanced.

6. Acknowledgment

The scholarship provided by the Iraqi Ministry of Higher Education & Scientific Research is gratefully appreciated.

References

1.law, EUR-Lex/ Access to European Union. 2007. EUR-Lex/ Access to European Union law.

2.Bajpai, Pratima. 2013. 2013 ed. India: Springer New Delhi Heidelberg New York Dordrecht London. Book.

3.Masum, B. M., H. H. Masjuki, M. A. Kalam, I. M. Rizwanul Fattah, S. M. Palash, and M. J. Abedin. 2013. *Renewable and Sustainable Energy Reviews* 24:209-222. doi: 10.1016/j.rser.2013.03.046.

4.Huang, Yuhan, Guang Hong, and Ronghua Huang. 2015. *Energy Conversion and Management* 92 (0):275-286.

5.Turner, Dale, Hongming Xu, Roger F. Cracknell, Vinod Natarajan, and Xiangdong Chen. 2011. *Fuel* 90 (5):1999-2006. doi: 10.1016/j.fuel.2010.12.025.

6.Bielaczyc, Piotr, Andrzej Szczotka, and Joseph Woodburn. 2011.10.4271/2011-24-0177.

7.Taniguchi, Satoshi, Kaori Yoshida, and Yukihiro Tsukasaki. 2007.10.4271/2007-01-2037.

8.Zhuang, Yuan, and Guang Hong. 2014. *Fuel* 135 (0):27-37.

9.Zhuang, Yuan, and Guang Hong. 2013. *Fuel* 105:425-431.

10.Huang, Yuhan, Guang Hong, and Ronghua Huang. 2015. *Applied Energy* 160:244-254.

11.Zhu, G, D Hung, and H Schock. 2010. *Proceedings of the Institution of Mechanical Engineers, Part D: Journal of Automobile Engineering* 224 (3):387-403.

12.Kim, Namho, Seokwon Cho, and Kyoungdoug Min. 2015. *Fuel* 158:725-732.

13.Al-Muhsen, N, J Wang, and Guang Hong. 2016. 20th Australasian Fluid Mechanics Conference.

14.Huang, Yuhan, and Guang Hong. 2016. *Energy Conversion and Management* 123:338-347.

15.Al-Muhsen, Nizar F.O., and Guang Hong. 2017. SAE International Conference, 2017-01-2230

16.Heywood, John B. 1988. McGraw-Hill, Inc.

17.Pulkrabek, Willard W. 1997. United States of America: Prentice-Hall, Inc. book.

11th Asia-Pacific Conference on Combustion,
The University of Sydney, NSW Australia
10th -14th December 2017.

Multi-Environment PDF Modeling for Diesel Spray Combustion Processes using Tabulated Chemistry

Namsu Kim, Kiyoung Jung, Yongmo Kim*

Hanyang University

Haengdang-Dong, Sungdong-Ku, Seoul 133-791, Korea

Abstract

The Eulerian Probability Density Function approach coupled with a tabulated chemistry model has been developed to realistically predict the auto-ignition and combustion processes in the constant-volume spray combustion system (ECN spray A). Computations are made for four different ambient oxygen level conditions including 0%, 13%, 15% and 21%. Numerical results are in the reasonably good agreement with experimental data in terms of mixture fraction distribution, ignition delay, and flame lift of length. These numerical results clearly indicate that the present FGM-based MEPDF approach has the capability to realistically predict essential features of the diesel combustion processes..

1 Introduction

To comply with the progressively stringent emission regulations for diesel engines, systematic researches are persistently required to develop advanced compression-ignition combustion technologies with lower pollutant emission and higher engine performance. The next-generation diesel combustion strategies involve the optimal modifications for pressure, temperature, levels of Exhaust-Gas Recirculation (EGR), injection parameters including injection pressure and timing, and in-cylinder flow. Thus, reliable comprehensive combustion models are needed to realistically simulate the modified engine combustion processes as well as to identify the optimum combustion condition for advanced compression-ignition engines [1-4].

Diesel combustion involves complex physical processes such as liquid atomization, droplet breakup, vaporization, turbulent mixing, autoignition, flame propagation, flame lift-off, local quenching and pollutant formation. The issues associated with temporal evolution of spray combustion are quite important in the practical applications because these essentially transient combustion phenomena highly influence combustion efficiency and pollutant formation. In the compression-ignition engines, the high-speed liquid jets injected into high-temperature environment in the combustion chamber generate an intense turbulent flow field which plays an important role in the mixing and combustion processes. Since the ratio of the characteristic chemical and mixing time scales changes significantly during the injection of liquid fuel, the turbulence-chemistry interaction (TCI) highly influences autoignition, subsequent multi-mode combustion processes and pollutant formation. Moreover, the detailed chemistry for a large hydrocarbon has a number of intermediate chemical species and elementary reactions, the precise reaction pathways are directly linked to ignition delay [5] and pollutant formation. Therefore, TCI and detailed chemical mechanism should be appropriately considered in the diesel spray combustion modeling.

The objective of the current work is to develop a robust and efficient numerical framework for the two-phase mixing and combustion in diesel engines. For gas mixture, a deterministic Eulerian PDF transport method coupled with a tabulated chemistry method is used to account for the turbulence chemistry interaction. Meanwhile, liquid phase is described by traditional Lagrangian parcel approach with physical submodels. Interactions between two phases are modelled with interphase source terms. Due to the inherent advantages of computational efficiency and compatibility with conventional CFD codes, the multi-environment probability density function (MEPDF) approach [6] has been adopted to analyze combustion process and structures in the diesel-like spray combustion conditions. In this work, we implement the RANS based MEPDF turbulent spray combustion model together the Flamelet Generated Manifolds (FGM) procedure [5,7] in OpenFOAM [8]. Then, the model has applied to numerically investigate the ECN n-dodecane non-reacting and reacting cases called Spray A [9]. Numerical results are compared with experimental data in terms of ignition delay and flame lift-off length for three ambient oxygen concentrations. Based on numerical results, the detailed discussions are made for the essential features of the n-dodecane fueled spray jet flames under the diesel-like condition.

2 Mathematical modelling

2.1 Multi-Environment Probability Density Function

In MEPDF method, the shape of the PDF is assumed to be sum of multiple delta functions with different weights,

$$f_{\phi}(\boldsymbol{\psi}; \boldsymbol{x}, t) = \bar{\rho}(\langle\boldsymbol{\phi}\rangle, \boldsymbol{x}, t) \sum_{n=1}^{N_e} p_n(\boldsymbol{x}, t) \prod_{i=0}^{N_s} \delta[\psi_i - \phi_{i,n}](\boldsymbol{x}, t), \quad (1)$$

where $\bar{\rho}$ is the Reynolds averaged density, N_e is the number of environments, $p_n(\boldsymbol{x}, t)$ is the mass weight of environment n, and $\phi_{i,n}$ is abscissa i nth of environment. The integration of the transport PDF equation [10] in the composition space yields N_e sets of transport equations of weight and weighted abscissa for each environment,

$$\frac{\partial(\bar{\rho} p_n)}{\partial t} + \nabla \cdot (\bar{\rho} \tilde{\boldsymbol{u}} p_n) - \nabla \cdot (\Gamma_{eff} \nabla p_n) = a_n, \quad (2)$$

Corresponding author. Fax: +82-2-2297-3432
E-mail address: ymkim@hanyang.ac.kr

$$\frac{\partial(\bar{\rho}p_n\phi_{i,n})}{\partial t} + \nabla\cdot(\bar{\rho}\tilde{\boldsymbol{u}}p_n\phi_{i,n}) - \nabla\cdot\left(\Gamma_{eff}\nabla(p_n\phi_{i,n})\right) = b_{a,n} + p_n S_i(\boldsymbol{\phi_n}) + \Omega_m p_n(\langle\phi_i\rangle - \phi_{i,n}), \quad (3)$$

where $\tilde{\boldsymbol{u}}$ is the Favre averaged velocity, $p_n\phi_{i,n}$ is the weighted abscissa i of the nth environment, Γ_{eff} is the sum of the moelcular diffusivity and the turbulent diffusivity, and Ω_m is the micro mixing frequency. Note that the Interaction by Exchange with the Mean (IEM) model is used to close the conditional diffusion term. The source terms a_n and $b_{a,n}$ are determined using direct quadrature method of moments [6] to accurately reconstruct the high-order moments. Detail descriptions of the model could be found in [6,11].

2.2 Flamelet Generated Manifolds

In the present study, The igniting flamlets [5] are used to construct the FGM to account for the auto ignition of the flame. The temporal evolution and structure of the igniting flamelets are obtained from solutions of unsteady flamelet equations with constant strain rate (300 s-1). According to the numerical results, the ignition delay is not significantly changed if the strain rate is under the certain value [5]. Thus, the constant strain rate is used for the unsteady flamelet calculations. The rest of the FGM library is based on the collection of steady flamelet solutions with strain rates lower than 300 s-1. The unsteady counterflow flame is described by one-dimensional equations of mass, flame stretch rate, species, and energy. The flamelet solutions are computed using in-house code based on Cantera [12] and the 130-speices/2395-reactions mechanism by Ranzi et al. [13]. The unitary Lewis number assumption is used to simplify the equations. The progress variable is defined as [5]:

$$Y_c = \frac{Y_{CO}}{MW_{CO}} + \frac{Y_{CH2O}}{MW_{CH2O}} + \frac{Y_{CO2}}{MW_{CO2}}, \quad (4)$$

where Y_i is the mass fraction of the species i and MW_i is the molecular weight of the species i. The Bilger's definition of mixture fraction [14] and the progress variable are used as the low-dimensional space parameters. In the FGM library, the major quantities of nth environment, including density, temperature, reaction source term of the progress variable, and species mass fractions, are represented by a function of two control variables.

2.3 MEPDF for a diesel spray combustion processes

Evolution of liquid phase in the space and time is described by the Lagrangian parcel. In this method, many identical droplets are indicated as a parcel. The mass transfer from the liquid phase to the gas phase due to evaporation is modeled using correlation of Frössling and the heat transfer between two phase including heating of liquid droplet and evaporation is modeled using Ranz-Marshall correlation. A detail description of the equations and submodels can be found elsewhere [1].

Coupling between gas phase and liquid droplet particle equation is done by introducing interphase source terms in the conservation equation of mass and momentum in conjunction with the MEPDF approach. Since the PDF transport equation is derived for a single-phase flow, additional interphase source terms are required to account for the interphase exchange of heat, mass, and momentum.

The weight in the MEPDF formulations represents the relative mass (probability) of typical environment existed in a computational cell. Thus, we will describe the spray combustion system as a two-stream combustion. One stream is the evaporated fuel and the other stream is the ambient gas. Therefore, it is natural to employ the two-environment MEPDF approach to treat this two-phase problem and the source terms from evaporated mass should be zero for ambient gas weight. Then, the weight transport equation of fuel (p_{Fuel}) and ambient gas(p_{Atm}) can be derived as follows:

$$\frac{\partial\bar{\rho}p_{Fuel}}{\partial t} + \nabla\cdot(\bar{\rho}\tilde{\boldsymbol{u}}p_{Fuel}) = \nabla\cdot(\Gamma_{eff}\nabla p_{Fuel}) + \bar{S}_\rho, \quad (5)$$

$$\frac{\partial\bar{\rho}p_{Atm}}{\partial t} + \nabla\cdot(\bar{\rho}\tilde{\boldsymbol{u}}p_{Atm}) = \nabla\cdot(\Gamma_{eff}\nabla p_{Atm}). \quad (6)$$

In this MEPDF formulation, the sum of fuel weight transport equation (4) and ambient gas weight transport equation (6) with the constraint $p_{Fuel}(\boldsymbol{x},t) + p_{Atm}(\boldsymbol{x},t) = 1$ exactly reconstructs the mass conservation equation with interphase source term,

$$\frac{\partial\bar{\rho}}{\partial t} + \nabla\cdot(\bar{\rho}\tilde{\boldsymbol{u}}) = \bar{S}_\rho. \quad (7)$$

The transport equations of the weighted mixture fraction and the weighted progress variable can be also derived in a similar fashion,

$$\frac{\partial\bar{\rho}pZ_{Fuel}}{\partial t} + \nabla\cdot(\bar{\rho}\tilde{\boldsymbol{u}}pZ_{Fuel}) = \nabla\cdot(\Gamma_{eff}\nabla pZ_{Fuel}) + p_{Fuel}\Omega_m(\tilde{Z} - Z_{Fuel}) + \bar{S}_\rho, \quad (8)$$

$$\frac{\partial\bar{\rho}pZ_{Atm}}{\partial t} + \nabla\cdot(\bar{\rho}\tilde{\boldsymbol{u}}pZ_{Atm}) = \nabla\cdot(\Gamma_{eff}\nabla pZ_{Atm}) + p_{Atm}\Omega_m(\tilde{Z} - Z_{Atm}), \quad (9)$$

$$\frac{\partial\bar{\rho}pY_{C,Fuel}}{\partial t} + \nabla\cdot(\bar{\rho}\tilde{\boldsymbol{u}}pY_{C,Fuel}) = \nabla\cdot(\Gamma_{eff}\nabla pY_{C,Fuel}) + p_{Fuel}\Omega_m(\widetilde{Y_C} - Y_{C,Fuel}) + p_{Fuel}\bar{S}_{Yc,Fuel}, \quad (10)$$

$$\frac{\partial\bar{\rho}pY_{C,Atm}}{\partial t} + \nabla\cdot(\bar{\rho}\tilde{\boldsymbol{u}}pY_{C,Atm}) = \nabla\cdot(\Gamma_{eff}\nabla pY_{C,Atm}) + p_{Atm}\Omega_m(\widetilde{Y_C} - Y_{C,Atm}) + p_{Atm}\bar{S}_{Yc,Atm}. \quad (11)$$

Here, $\bar{S}_{Yc}$ is the reaction source term of the progress variable and it can be expressed as a function of mixture fraction and progress variable, $\bar{S}_{Yc,n}(Z_n, Y_{c,n})$. The Favre averaged mixture fraction and progress variable can be computed by $\tilde{Z} = pZ_{Fuel} + pZ_{Atm}$ and $\widetilde{Y_C} = pY_{C,Fuel} + pY_{C,Atm}$ respectively.

3 Simulation Results

To validate the present FGM-based MEPDF spray combustion model, computations are made for n-dodecane non-reacting and reacting spray jets (Spray A) in a diesel-like condition. Experimental data for Spray A are available at the ECN online website [9]. In the measurements, the n-dodecane liquid jet is injected into the pre-heated constant volume chamber. Among all test conditions of Spray A, we have chosen four cases with four different ambient oxygen concentrations (0, 13, 15 and 21%) as the validation cases. The ambient gas conditions for four test cases are summarized in Table 1. The ambient gas temperature is 900K and the chamber pressure is about 6MPa (It varies with ambient oxygen concentration). The injector diameter is 0.9μm and the injection pressure is 150MPa. The temperature of liquid n-dodecane is 363K and the injection velocity is 590m/s.

O2%	N2%	CO2%	H2O%	T[K]	P[MPa]
0	89.71	6.52	3.77		
13	77.1	6.26	3.64	900	6
15	75.15	6.23	3.62		
21	69.33	6.11	3.56		

Table 1: Ambient gas composition for four test cases [6].

3.1 Non-reacting Spray-A

Before we get into the simulation of diesel-fueled spray flames, the evaporating and non-reacting spray case with zero oxygen level is numerically analyzed to evaluate the prediction capability of spray submodels as well as to verify the MEPDF formulations developed for a two-phase flow problem. The atomization is modeled by initial droplet size distribution which follows Rosin-Rammler distribution. The Kelvein-Helmholtz (KH) and Rayleigh-Taylor (RT) model are used to predict the droplet breakup. The present study has utilized the model constants suggested by Wehrfritz et al. [5]. The standard k-epsilon model is used to describe turbulence.

Fig. 1 shows the predicted and measured liquid and vapor penetration for the non-reacting spray A [9]. The liquid and vapor penetration are defined as the distance from the nozzle exit to the axial location where the liquid volume and vapor mass fraction is 0.001. In terms of liquid and vapor penetration length, the present spray model yields the good conformity with measurements [9]. Since the liquid penetration and vapor penetration are directly linked to the mixing field, the accurate prediction of these quantities is quite crucial for analyzing the spray flame fields.

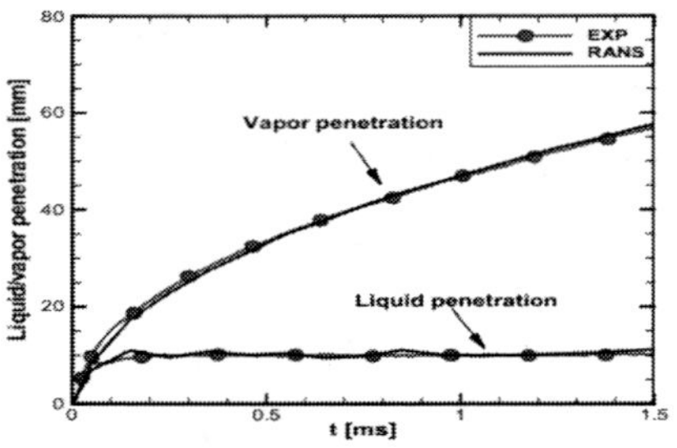

Figure 1: Liquid and vapor penetration versus time for the non-reacting spray jet.

Fig. 2 shows the radial profiles of mean and rms mixture fraction at two axial stations (x=25, 40mm). As identified in Fig. 2, the levels of mean and rms mixture fraction are noticeably overestimated at the upstream (x=25mm). However, these deviations are within the range of experiment errors. At the downstream region (x=40mm), the agreement with measurements is quite good. Even if there exist the marginally deviations between experimental data and numerical results, the numerical results obtained in this study are comparable to those of the previous RANS-based works [15-16]. These results suggest that the proposed MEPDF approach together with spray submodels has the capability to realistically predict the non-reacting two-phase flow mixing field.

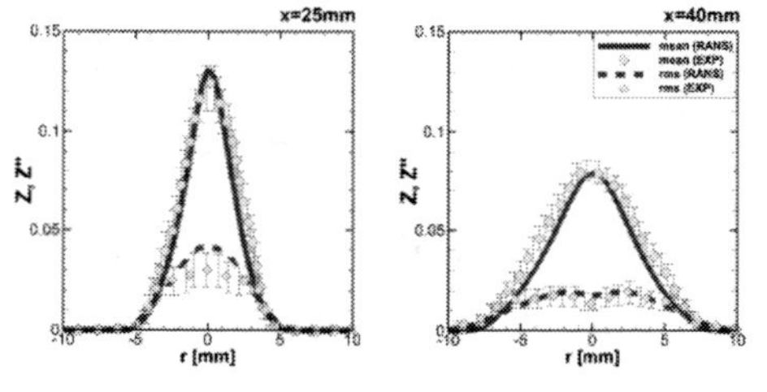

Figure 2: Radial profiles of mean(solid line) and rms(dashed line) mixture fraction at two axial positions(left: x=25mm, right: x=40mm). Symbols are experimental data [6].

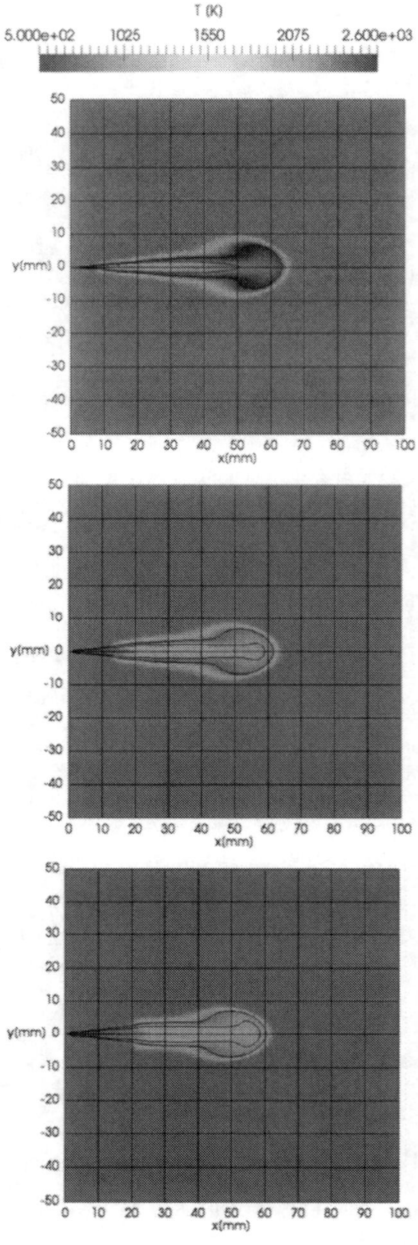

Figure 3: Temperature distributions for reacting Spray-A jets with three ambient oxygen levels. Oxygen concentration is 21%(top), 15%(middle), and 13%(bottom).

3.2 Reacting Spray-A

Next, the present RANS-based FGM-MEPDF approach has been applied to investigate the effects of ambient oxygen level on a diesel spray combustion processes. Fig. 3 displays the predicted contours of mean temperature at t=1.5ms after the start of injection (ASOI) for three different oxygen levels (21%, 15%, and 13%) from left to right respectively. These numerical results indicate that the higher oxygen concentration leads to the higher peak temperature and the shorter flame lift-off length. To precisely identify the fuel-rich mixing field for three reacting spray flames, the iso-mixture fraction lines of Z=2Zst and Z=Zst are plotted in Fig. 3 (Zst=0.045). The inner and outer black curves in the spray-combusting field are corresponding to Z=2Zst and Z=Zst, respectively. The peak temperature region is generated near the stoichiometric mixture fraction line and the low-temperature oxidation mainly is observed in the fuel-rich region with the equivalence ratio higher than 2.0. Although the precise flame characteristics are sensitively varied with the oxygen level, these three spray jet flames have the similar flame pattern which basically resembles a conceptual model of diesel combustion proposed by Dec [17].

In the conceptual model, the spray jet flame consists of liquid fuel jet, low temperature oxidation zone, and high temperature oxidation zone. As shown in Figure 4, The low-temperature oxidation zone near the centerline is dominantly controlled by the fuel-rich premixed combustion while the high-temperature oxidation zone is mainly influenced by the near-stoichiometric non-premixed combustion.

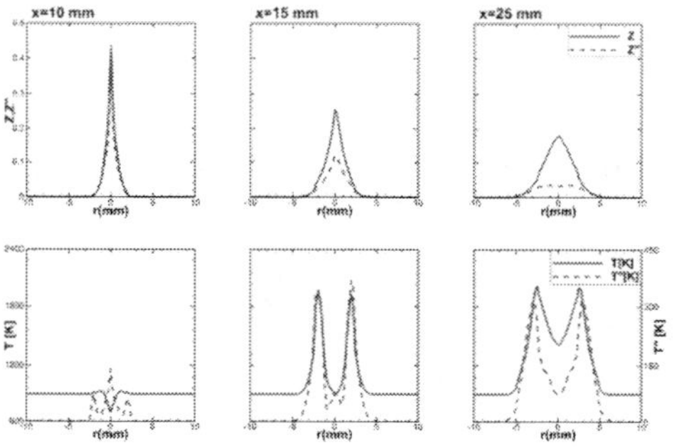

Figure 4: Radial profiles of mean and rms of mixture fraction and temperature for O2=15% case at x=10mm, 15mm, and 25mm.

Figure 5 shows the predicted and measured ignition delay and FLOL versus ambient oxygen concentration. Overall agreements between the measurements are good. Both the ignition delay and FLOL decreases by increasing the ambient oxygen level in the chamber. In terms of ignition delay and FLOL, numerical results are reasonably well agreed with experimental data [6]. The higher temperature field elevated by the higher oxygen level possibly amplifies the strength of interphase coupling between liquid fuel and ambient gases by progressively increasing heat and mass transfer. Especially at the flame lift-off region for the highest oxygen concentration case (21%), there exists the much stronger interphase coupling at the upstream flame stabilization zone near the injector due to the much shorter flame lift-off length. In this aspect, it is speculated that the spray model constants adjusted for non-reacting sprays could be optimized for reacting cases for the better agreement. Since the flame stabilization process of a turbulent diesel spray jet is inherently non-stationary, the RANS-based turbulent combustion model has basically the limitations to correctly reproduce the dynamics of flame lift-off process.

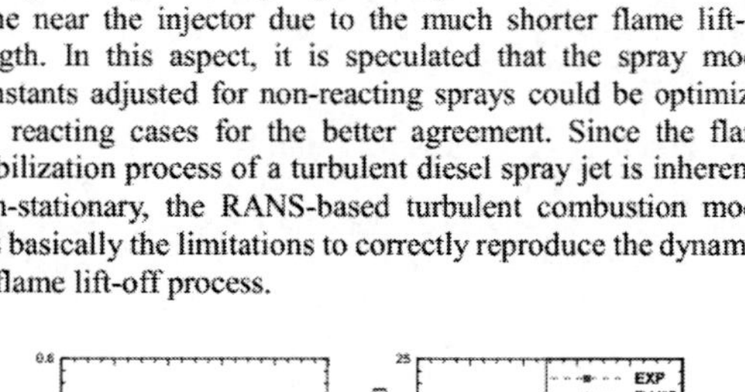

Figure 5: Ignition delay and Flame lift-off length versus ambient oxygen concentration.

4 Conclusions

The multi-environment probability density function approach coupled with a tabulated chemistry model has been developed to realistically predict the auto-ignition and combustion processes in the constant volume chamber spray combustion system (ECN spray A). Numerical results obtained in this study are reasonably well agreed with experimental data in terms of mixture fraction distribution, ignition delay, and flame lift-off length. These numerical results clearly indicate that the present FGM-based MEPDF approach has the capability to realistically predict the essential features of the diesel combustion processes.

5 Acknowledgment

This work was supported by the Development of 300_MW class Korean IGCC demonstration plant technology of the Korea institute of Energy Technology Evaluation and Planning (KETEP) grant funded by the Korea government Ministry of Knowledge Economy.

References

[1] C. Baumgarten, Mixture formation in internal combustion engines, Springer, 2006

[2] J. E. Dec, Proceed. Inst. Combust. 32(2009) 2727-2742.

[3] A Workshop to Identify Research Needs and Impacts in Predictive Simulation for Internal Combustion Engines (PreSICE), Office of Science and Office of Energy and Renewable Energy, US Department of Energy, 3 March 2011, 2011. <http://www1.eere.energy.gov/vehiclesandfuels/pdfs/presice_rpt.pdf>.

[4] R.D. Reitz, Combust. Flame 160 (2013) 1–8.

[5] A. Wehrfritz, O. Kaario, V. Vuorinen, B. Somers, Combust. Flame, 167 (2016) 113-131.

[6] R.O. Fox, Computational Models for Turbulent Reacting Flows, Cambridge University Press, Cambridge, 2003.

[7] J.A. van Oijen, L.P.H. de Goey, Modelling of Premixed Laminar Flames using Flamelet-Generated Manifolds, Combust. Sci. Tech. 161 (2000) 113-137.

[8] OpenCFD Ltd, OpenFOAM user guide, version 2.2.0, 2011.

[9] <http://www.ca.sandia.gov/ecn>

[10] S.B. Pope, Prog. Energy Combust. Sci. 11 (1985) 119-192.

[11] N. Kim, Y. Kim, Combust. Flame 182 (2017), 190-205.

[12] D.G. Goodwin, Cantera C++ user's guide, California Institute of Technology, California, U.S.A (2002).

[13] E. Ranzi, A. Frassoldati, A. Stagni, M. Pelucchi, A. Cuoci, T. Faravelli, Int. J. Chem. Kinet. 46 (9) (2014) 512–542

[14] R.W. Bilger, Symp. (Int.) Combust. 22(1989) 475-488.

[15] G. D'Errico, T. Lucchini, F. Contino, M. Jangi, X. S. Bai, Combust. Theory. Mod. 18 (2014) 65-88.

[16] Y. Pei, E. R. Hawkes, S. Kook, G. M. Goldin, T. Lu, Combustion. Flame, 162 (2015) 2006-2019.

[17] J.E. Dec, SAE Paper 970873, 1997.

11th Asia-Pacific Conference on Combustion,
The University of Sydney, NSW Australia
10th -14th December 2017.

Multi-Environment Probability Density Function Approach for Turbulent Partially-Premixed Methane/Air Flames

Namsu Kim and Yongmo Kim*

Hanyang University

Haengdang-Dong, Sungdong-Ku, Seoul 133-791, Korea

Abstract

The present study has systematically assessed the capability of a multi-environment PDF approach using non-premixed and premixed tabulated chemistry to predict the fundamental characteristics of turbulent partially-premixed flames. Validation cases include Sydney pilot jet flame with inhomogeneous inlet and partially premixed flame under the moderate and intense low oxygen dilution condition with the strong flow reversal. Numerical results for these two flames have demonstrated that the FGM-based multi-environment PDF approach has the certain capability to realistically predict the essential features of the turbulent partially premixed flames.

1 Introduction

New generation gas turbine and internal combustion engine will operate under complex flame regimes with lean premixed or partially premixed mixtures to comply with strict environmental regulations. Recently, Meares et al. [1,2] and Barlow et al. [3] experimentally investigated the stability and structure of turbulent partially-premixed flames with a broad range of mixture inhomogeneity in a jet piloted burner. Their results indicated that flame stabilization is sensitively influenced by the degree of inhomogeneity in the mixture. The inhomogeneous partially premixed mixtures are also found in Moderate and Intense Low oxygen Dilution (MILD) combustion condition. In MILD combustor, to achieve the low pollutant emission and the elevated flame stability, the fuel and air mixture effectively mixed before they reach main reaction zone. Thus, understanding essential features of partially-premixed flames with an inhomogeneous mixture is desirable to develop high efficient and ecofriendly power technology.

In this study, the Multi Environment Probability Density Function(MEPDF) approach is adopted to systematically analyze combustion processes and precise structures in turbulent partially-premixed flames. This model can describe the evolution of joint composition PDF with relatively low computational costs compare to stochastic PDF transport approaches that utilize a much larger number of Lagrangian particles or Eulerian fields. Moreover, the present MEPDF approach adopted tabulated chemistry to effectively reduce the numerical burden for the large number of transported scalars involved in the detailed chemistry.

Herein, a novel combustion model that combines the MEPDF approach with FGM [4] is used to predict turbulent partially-premixed flames. The model applied to the Sydney pilot flames with inhomogeneous inlet [1-3] and a lab scale MILD combustor with strong flow reversal [5,6]. Based on our numerical results, discussions are made for the essential features of turbulent partially-premixed flames.

2 Mathematical modelling

2.1 Multi-Environment Probability Density Function

In MEPDF method, the joint composition PDF is assumed to be sum of multiple delta functions with different weights,

$$f_{\phi}(\boldsymbol{\psi};\boldsymbol{x},t) = \bar{\rho}(\langle\boldsymbol{\phi}\rangle,\boldsymbol{x},t)\sum_{n=1}^{N_e} p_n(\boldsymbol{x},t)\prod_{i=0}^{N_s}\delta[\psi_i - \phi_{i,n}](\boldsymbol{x},t), \quad (1)$$

where N_e is the number of environments, $p_n(\boldsymbol{x},t)$ is the mass weight of environment n. The integration of the transport PDF equation [7] in the composition space yields N_e sets of transport equations of weight and weighted abscissa for each environment,

$$\frac{\partial(\bar{\rho}p_n)}{\partial t} + \nabla\cdot(\bar{\rho}\tilde{\boldsymbol{u}}p_n) - \nabla\cdot(\Gamma_{eff}\nabla p_n) = a_n, \quad (2)$$

$$\frac{\partial(\bar{\rho}p_n\phi_{i,n})}{\partial t} + \nabla\cdot(\bar{\rho}\tilde{\boldsymbol{u}}p_n\phi_{i,n}) - \nabla\cdot\left(\Gamma_{eff}\nabla(p_n\phi_{i,n})\right) = b_{a,n} + p_n S_i(\boldsymbol{\phi_n}) + \Omega_m p_n(\langle\phi_i\rangle - \phi_{i,n}), \quad (3)$$

where p_n is the mass weight, $p_n\phi_{i,n}$ is the weighted scalar i of the nth environment, the source terms a_n and $b_{a,n}$ are determined using direct quadrature method of moments [8] to accurately reconstruct the high-order moments, and Ω_m is the micro mixing frequency. Detail descriptions of the model could be found in [4,8].

2.2 Flamelet Generated Manifolds

The Flamelet Generated Manifolds(FGM) is based on the assumption that a turbulent flame can be described as an ensemble of laminar flamelets [9]. In this regard, one-dimensional laminar flame solutions could represent complex turbulent flame structure with appropriate parameters. For adiabatic simulations, two parameters are used in this study. One is the mixture fraction and the other is the reaction progress variable. The mixture fraction is known as the most important parameter for the turbulent non-premixed flame and the progress variable is widely used in the premixed combustion model to represent overall progress of the reaction. The Bilger's definition of mixture fraction [10] is used and the progress variable is defined as:

Corresponding author. Fax: +82-2-2297-3432
E-mail address: ymkim@hanyang.ac.kr

$$Y_C = \frac{Y_{H_2}}{M\,W_{H_2}} + \frac{Y_{H_2O}}{M\,W_{H_2O}} + \frac{Y_{CO_2}}{M\,W_{CO_2}}. \qquad (4)$$

Here, Y_i is species mass fraction and $M\,W_i$ is molecular weight of species i. Meanwhile, for non-adiabatic simulation, an additional parameter enthalpy deficit (H) is added in the library dimension to account for heat loss effects. The enthalpy deficit, difference of enthalpy between non-adiabatic enthalpy and adiabatic enthalpy, can be obtained from the enthalpy transport equation with heat sink term [11] and its definition:

$$\frac{\partial \bar{\rho}\tilde{h}}{\partial t} + \nabla\cdot\left(\bar{\rho}\tilde{u}\tilde{h}\right) - \nabla\cdot\left(\Gamma_{eff}\nabla\tilde{h}\right) = \tilde{q}_{sink} \quad , \qquad (5)$$

$$H = \left(\tilde{h}_{ad} - \tilde{h}\right)/\left(1 - \tilde{Z}\right). \qquad (6)$$

Here $\tilde{h}_{ad}$ is the adiabatic enthalpy, $\tilde{h}$ is the Favre-averaged enthalpy, $\tilde{q}_{sink}$ is the heat sink term, and $\tilde{Z}$ is the Favre-averaged mean mixture fraction.

In the present study, non-premixed FGM and premixed FGM are used. Non-premixed FGM is constructed from the one-dimensional counter flow diffusion flame solutions and premixed FGM is constructed from the freely propagating flames. CANTERA[12] and GRI-3.0[13] chemistry mechanism are used for all one-dimensional flame calculations. For the non-adiabatic FGM-MEPDF model, we combine three different premixed FGM libraries which vary the temperature of the oxidizer temperature with 773, 523, and 297 K.

3 Results and discussion

To validate the present MEPDF approach tabulated chemistry, we chose two target cases. The first target case is the Sydney pilot jet flames with inhomogeneous inlet (FJ200-5GP-Lr75-57). Figure 1 shows the schematic configuration of the burner used in these measurements [2,3]. The main jet inlet consists of two tubes surrounded by an annular pilot jet. Upstream of the recess, fuel is injected through the central jet and air is injected through the co-flow jet. The recession length of the main jet controls the fuel-air mixing and the inhomogeneity of the inlet. Detailed information for the boundary condition is available in the previous work[4]. The second target case is a lab scale MILD combustion burner which was experimentally and numerically studied by Rebola et al [5,6]. The combustor schematic is displayed in Figure 2. Fuel is injected through the center of jet, and heated air is injected through the co-flow of jet. The combustor operating conditions used as the measurements and the computations are given in Table 1.

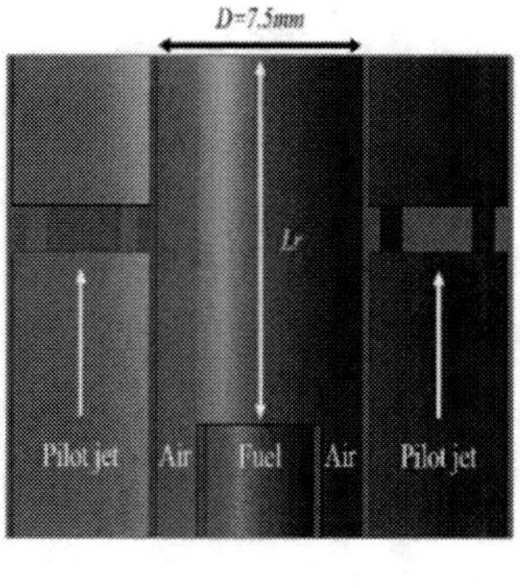

Figure 1: Schematic configuration of the Sydney pilot jet flame burner used in measurements [2,3].

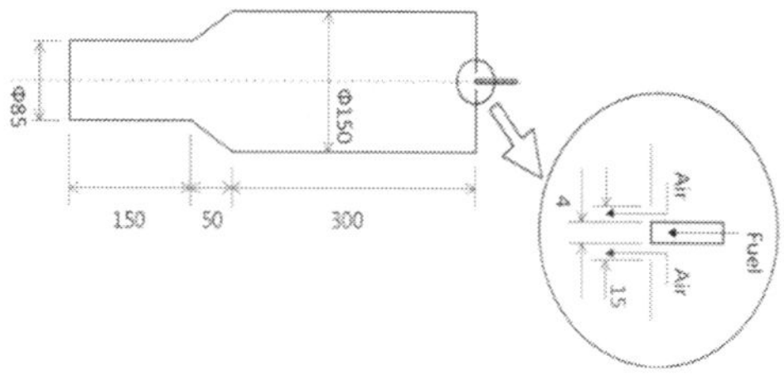

Figure 2: Schematic configuration of the MILD combustor [5].

Injection air temperature (K)	773
Inlet air mass flow rate (g/s)	6.9
Inlet air velocity (m/s)	102.1
Injection methane temperature (K)	293
Inlet methane mass flow rate (g/s)	0.2
Inlet methane velocity (m/s)	24.4
Wall temperature (K)	1173

Table 1: Operating condition of the MILD combustor [5].

3.1. Sydney pilot jet flame with Inhomogeneous inlet

Figures 3 and 4 present the radial profiles of the mean and rms mixture fraction and temperature for the inhomogeneous case at four axial locations. In terms of the mean and rms mixture fraction, the numerical results obtained from the premixed and non-premixed manifolds agreed reasonably well with the measured profiles at all axial locations, even if some slight deviations exist. The predicted and measured temperature profiles indicate that the inhomogeneous case produced a slightly higher peak temperature and a relatively broader hot-flame zone than the homogeneous case. These flame characteristics enhance the flame stability.

Figure 5 presents the radial profiles of the mean CO2 and CO mass fraction for the inhomogeneous case. At the upstream region (x/D=1) with multiple combustion modes, the non-premixed manifold accurately reproduced the measured CO peak and distribution, while the premixed manifold underestimated the CO level at the outer pilot region. At relatively downstream locations (x/D=10 and 15), the non-premixed manifold showed better conformity with the measured peak CO mass fraction, while the premixed manifold considerably overestimated the CO mass fraction at the fuel-rich side.

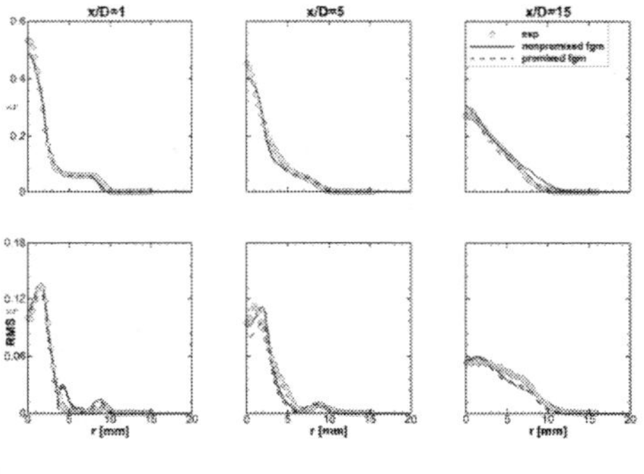

Figure 3: Radial profiles of the mean and rms mixture fraction in the inhomogeneous case (symbols: experimental data [3]; solid lines: non-premixed FGM; dashed lines: premixed FGM).

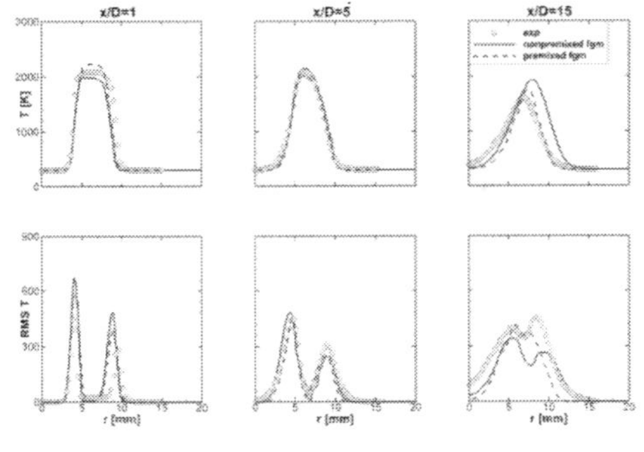

Figure 4: Radial profiles of the mean and rms temperature in the inhomogeneous case (symbols: experimental data [3]; solid lines: non-premixed FGM; dashed lines: premixed FGM).

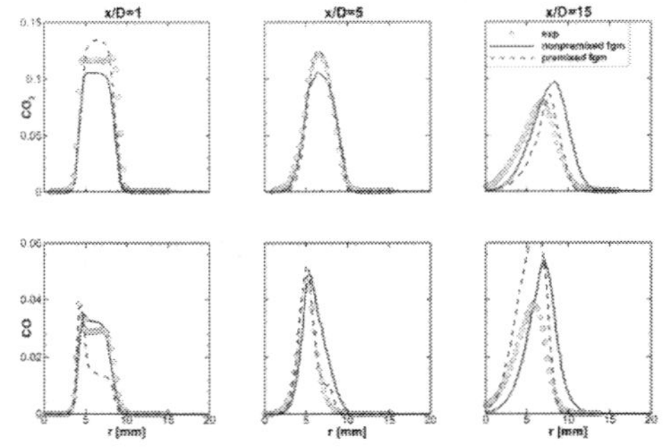

Figure 5: Radial profiles of the CO_2 and CO mass fraction in the inhomogeneous case (symbols: experimental data [3]; solid lines: non-premixed FGM; dashed lines: premixed FGM).

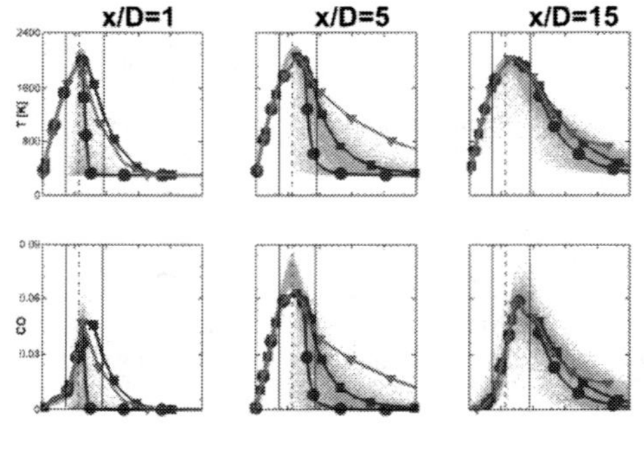

Figure 6: Environment-conditioned profiles of temperature, and CO predicted by the non-premixed FGM-MEPDF approach at x/D=1, 5, and 15. The fuel stream environment (red square), pilot jet environment (green triangle), and air co-flow environment (blue circle) are compared with experimental data [3] (orange dots).

As shown in Figures 6, the profiles of the environment-conditioned values predicted by the MEPDF-FGM approach reasonably well reproduced the overall distribution of the measured conditional scatters at all axial stations. The peak CO levels occurred at the slightly rich near-stoichiometric condition; this trend is directly tied to the stratified-premixed flame structure. At axial locations shortly after the flame base (x/D=5, 10, and 15), the peak CO levels were progressively moved to the richer side. This could be a clear indication of the flame transition from a premixed flame to a diffusion-dominated flame. Near the flame base (x/D=1) with multiple combustion modes, the two manifold approaches have limited capabilities to realistically predict the measured conditional statistics.

3.2. Lab scale MILD combustor

Figure 7 presents the streamlines and the mean temperature distribution in the MILD combustor. As would be expected, this flameless combustion situation creates the large recirculation zone with the enhanced temperature uniformity. In this recirculation zone, the fluid particles experience the long residence time. Hence, the convective heat loss though the wall is not negligible even if the peak temperature is substantially lower compared to the conventional methane/air non-premixed flame. Since the precise flame structure is highly influenced by the recirculated flow, it is actually necessary to assess the prediction capability for the velocity field. However, the detailed assessment is not feasible for the flow fields owing to the lack of experimental data for the velocity distribution.

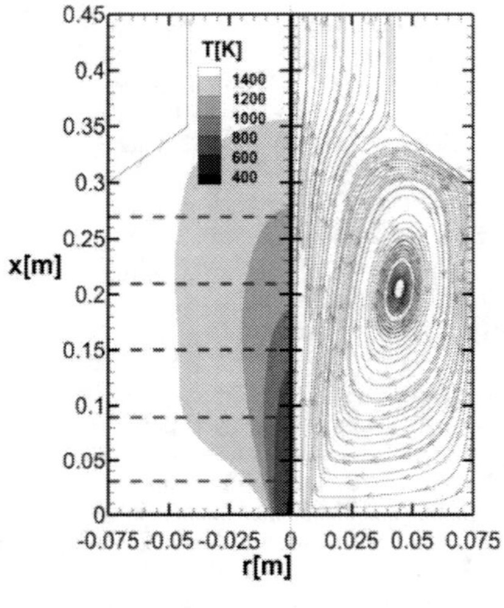

Figure 7: Contour of the streamlines and mean temperatures in the recirculation zone. Dotted lines indicate five measurement locations.

Figure 8 and 9 show the predicted and measured radial profiles of the mean temperature, and major species (O2, CO2 and CO) mole fraction at five axial locations (x = 30 mm, 90 mm, 150 mm, 210 mm, and 270 mm). Generally, numerical results obtained by the present non-adiabatic MEPDF approach are in good agreement with the experimental data [5]. However, it can be clearly seen that the temperature level is considerably underestimated at the relatively upstream center core region (x < 150 mm). The similar deviations were found in numerical results previous obtained by the EDC approach [6]. This discrepancy could be attributed mainly to the limitations of the RANS based turbulence model as well as partly to the experimental uncertainties.

At the central zones in the relatively upstream location (x < 150mm), the non-adiabatic MEPDF approach yields the overestimated O2 mole fraction and the underestimated CO2. mole fraction. These trends are quite apparent at the upstream region (x = 30mm). These tendencies are directly tied with the underestimation of the mean temperature in the upstream locations. However, at the whole flame field excluding the upstream central region, numerical results obtained by the non-adiabatic MEPDF approach are reasonably well agreed with measurements [5]. In terms of the mean temperature, O2 and CO2 mole fraction, the present non-adiabatic MEPDF approach yields numerical results comparable to those of the previous EDC model [6] even if the slight differences exist. Unlike the profiles of mean temperature, O2 and CO2 mole fractions, there exist the distinctly distribution for the CO mole fraction. The non-adiabatic MEPDF approach reasonably well predicts the measured profiles of the CO mole fraction.

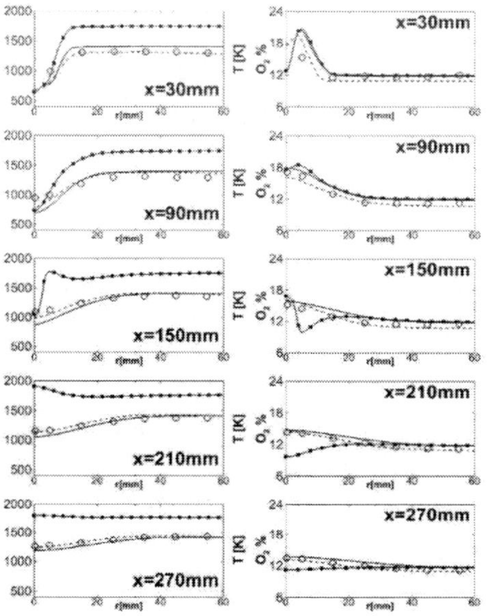

Figure 8: Radial profiles of temperature and O2 mole fraction (Circle: experimental data [5], solid line: non-adiabatic MEPDF, solid line with ■: adiabatic MEPDF, dashed line: EDC [6]).

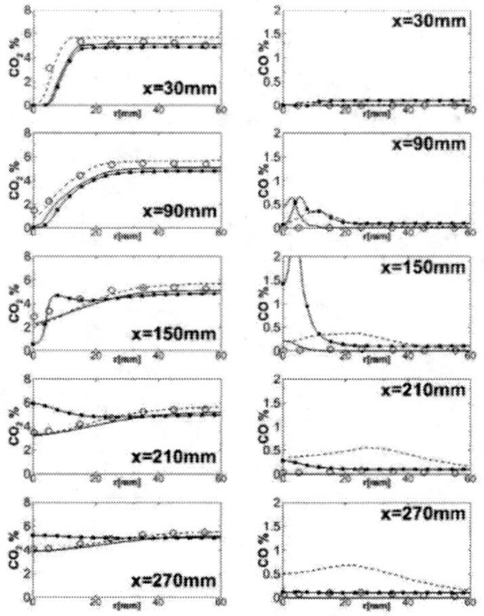

Figure 9: Radial profiles of CO2 and CO mole fraction (Circle: experimental data [5], solid line: non-adiabatic MEPDF, solid line with ■: adiabatic MEPDF, dashed line: EDC [6]).

4 Conclusions

The RANS-based multi-environment PDF approach, combined with a FGM library, was applied to predict two turbulent partially-premixed flames. For the Sydney pilot jet flame with inhomogeneous inlet, even if the premixed and non-premixed manifolds have the limited ability to reproduce the measured distribution, especially near the flame base with multiple combustion modes, numerical results in the whole flame regions are in overall agreement with experimental data. For the lab scale MILD combustor, the non-adiabatic MEPDF approach yields the reasonably good conformity with measurements in terms of temperature and major species mole fractions. Moreover, in terms of CO mole fraction, the present approach shows the much better prediction capability, compared to the EDC model.

References

[1] S. Meares, A.R. Masri, Combust. Flame 161 (2) (2014)484-495.

[2] S. Meares, V.N. Prasad, G. Magnotti, R.S. Barlow, A.R. Masri, Proc. Combust. Inst. 35 (2) (2015)1477–1484.

[3] R.S. Barlow, S. Meares, G. Magnotti, H. Cutcher, A.R. Masri, Combust. Flame 162 (10) (2015) 3516-3540.

[4] N. Kim, Y. Kim, Combust. Flame 182 (2017) 190-205.

[5] A. Rebola, M. Costa, and P. J. Coelho, Experimental evaluation of the performance of a flameless combustor, Applied Thermal Engineering, 50 (1) (2013) 805-815.

[6] A. Rebola, P. J. Coelho, and M. Costa, Assessment of the Performance of Several Turbulence and Combustion Models in the Numerical Simulation of a Flameless Combustor, Combust. Sci. Tech., 185 (2013) 600-626.

[7] S.B. Pope, Prog. Energy Combust. Sci. 11 (1985) 119-192.

[8] R.O. Fox, Computational Models for Turbulent Reacting Flows, Cambridge University Press, Cambridge, 2003.

[9] J.A. van Oijen, L.P.H. de Goey, Combust. Sci. Tech. 161 (1) (2000) 113-137.

[10] R.W. Bilger, Symp. (Int.) Combust. 22(1989) 475-488.

[11] M. Ihme, H. Pitsh, Modeling of radiation and nitric oxide formation in turbulent nonpremixed flames using a flamelet/progress variable formulation, Phys. Fluids 20, 055110 (2008).

[12] D.G. Goodwin, Cantera C++ user's guide, California Institute of Technology, California, U.S.A (2002).

[13] G.P. Smith, D.M. Golden, M. Frenklach, N.W. Moriarty, B. Eiteneer, M. Goldenberg, C.T. Bowman, R.K. Hanson, S. Soonho, W.C. Gardiner, Jr., V.V. Lissianski, Z. Qin, GRI mechanism, version 3.0.; 1997. <http://www.me/berkeley.edu/gri_mech>.

11th Asia-Pacific Conference on Combustion,
The University of Sydney, NSW Australia
10th -14th December 2017.

Firewhirl dynamics:

what we know and what we don't

A. Y. Klimenko[1*], F. A. Williams[2]

[1]The University of Queensland, SoMME,

Qld, 4072, Australia

Street, City, State, Zip-code, Country

[2]Dept. of Mechanical and Aerospace Engineering, University of California, San Diego,

9500 Gilman Drive MC 0411, La Jolla, USA

Abstract

This paper analyses qualitative aspects of interaction of rotational motion and density change due to heat release in firewhirls. The mechanisms of firewhirl stabilisation, elongation and, possibly, breakdown are discussed in general and in the context of significant density change, which affects firewhirl dynamics and can delay or inhibit the breakdown. The details of these mechanisms are not understood well and need further investigation. This paper also presents a qualitative theory of vortex breakdown for firewhirls and other tornado-like flows.

Figure 1. Firewhirl near Curtin Springs (NT, Australia) filmed by Chris Tangey of the Alice Springs Television (https://wordlesstech.com/fire-tornado-in-australia/)

*Corresponding author. E-mail: a.klimenko@uq.edu.au

1 Introduction

The phenomenon of firewhirls combines the intense rotation of tornadoes with the fierce heat release of large fires [1-3]. The rotation strengthens and stabilises the fire while intense heat stimulates updraft that, in turn, intensifies the rotation—the combination of these effects makes firewhirls extremely dangerous and difficult to combat. During the 1923 Great Kantō earthquake in Japan a single gigantic firewhirl, which engulfed the Army Clothing Depot in downtown Tokyo, killed tens of thousands of people. Firewhirls have repeatedly been observed in Australia (see Figure 1). The structure of a firewhirl can be seen in Figure 2, where intense heat release and rotation are clearly visible.

Figure 2: Firewhirl filmed by the U.S. Fish and Wildlife Service (Wikipedia).

The properties of firewhirls have been examined in a number of publications [4-7] under different assumptions. A buoyancy-controlled inviscid solution for velocity components near the axis was suggested by Battaglia et al. [4], although this solution does not determine the length of the flame in firewhirls and does not apply at low Rossby numbers. Kuwana et al. [5] used the Burke-Schumann formulation of the problem [8,9] and investigated the flame length while taking diffusion into account but assuming constant density. Klimenko and Williams [6] have generalised this approach and linked it to the theory of the compensating regime in vertical flows [10]. Yu and Zhang [7] generalised the solution of Klimenko and Williams [6] and investigated some of the compressibility effects. The present work briefly reviews the dynamics of firewhirls and concludes that firewhirls represent a complex phenomenon, which needs a combination of analytical, numerical and experimental approaches to its investigation.

2 Inclined firewhirls

Firewhirls over inclined surfaces are of great interest for any study of the phenomenon, since they are an opportunity to observe the relative significance of rotational dynamics and buoyancy [11]. Figure 3 illustrates this point: the direction of the flame near the surface is perpendicular to the surface. This indicates that rotational dynamics dominate the flow at these scales even if there is a very large change of density within the flame. At larger scales, however, the direction of the flame becomes vertical, which points to the predominant influence of the buoyancy higher up the flame. While it is buoyancy that creates the overall updraft motion and stimulates rotation due to conservation of the angular momentum, the region of intense rotation is controlled by vortex dynamics.

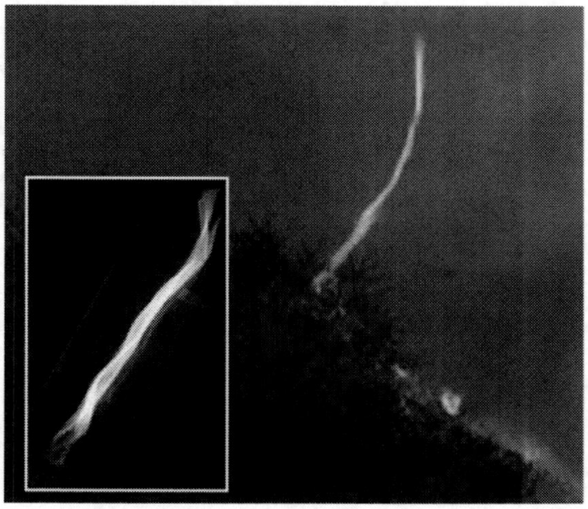

Figure 3: Firewhirl over inclined surface in California (right) compared with the lab experiment (left) conducted by Chuah et. al [11].

3 Strong vortices in atmosphere

The dynamics of tornado-like strong vortices (which includes large atmospheric vortices such as tornadoes, hurricanes, firewhirls, dust devils etc.) necessarily involve generation of tangential vorticity in a developed vertical flow [10]. This effect is shown in Figure 4. Figure 4 (left) illustrates a tornado-like vortex without tangential vorticity. This might be the case in the early stages of vortex formation, but a developed stable vortex must have updraft-stimulating tangential vorticity, as shown in Figure 4 (right).

The generation of tangential vorticity is controlled by the the vertical swirl ratio K

$$K \equiv \frac{1}{\mathrm{Rs}} \equiv \frac{(\gamma\omega_z)^{1/2}}{V_{tz}} \qquad (1)$$

which is the reciprocal of the *swirling Rossby number* Rs. Here, V_{tz} is translational velocity, $\gamma = rV_\theta$ is circulation and ω_z is axial vorticity. Understanding that, generally, this parameter must be balanced in a realistic developed vortex leads to the 4/3 power law – see reference [10] for details and analysis. This corresponds to a compensating regime that under normal conditions has stream function

$$\psi \sim r^\alpha z \qquad (2)$$

and axial vorticity

$$\omega_z \sim \frac{1}{r^\alpha}, \qquad (3)$$

where α = 4/3 (more accurately, varies between 4/3 and 3/2 depending on conditions). The expected values of the compensating exponent α have been confirmed by measurements in hurricanes and tornados [10].

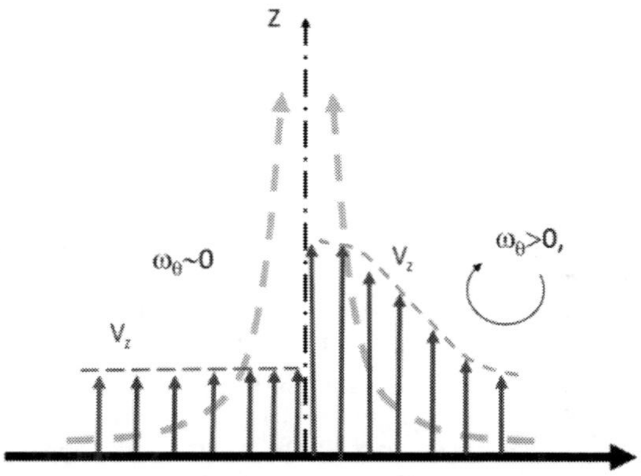

Figure 4. A tornado-like flow. Left: without tangential vorticity. Right: with tangential vorticity as in a realistic developed tornado-like vortex.

4 Vortex breakdown

While the compensating mechanism ensures the general stability and persistence of tornado-like flows, gradual increase of circulation can lead to excessively high values of the vertical swirl ratio parameter K in the core of the flow and to vortex breakdown. The breakdown mechanism is illustrated in Figure 5. Strong vorticity stimulates the near-surface centripetal flow (due to the Ekman effect). This reverses the sign of radial shear near the axis, inducing negative tangential vorticity that stimulates a downdraft. The radius of the core of the flow increases, reducing the intensity of rotation and decreasing K.

Essentially, vortex breakdown is a part of the compensating mechanism preventing extreme values of K in the rest of the flow and keeping most of the vortex intact. Vortex breakdowns have been observed in large and intensive tornadoes, and in the strongest hurricanes (causing hurricane eyes to appear).

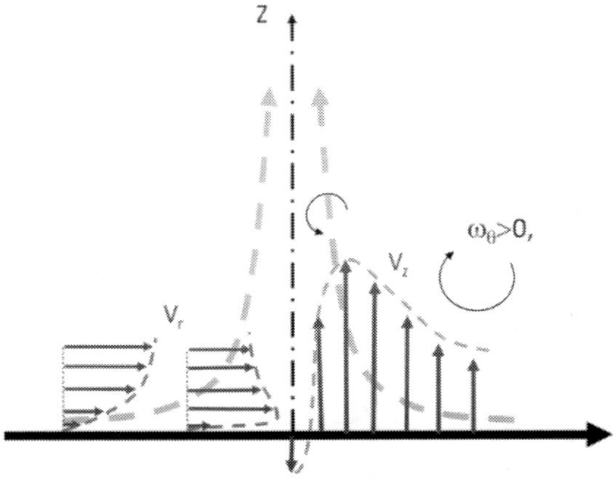

Figure 5. Evolution of the radial velocity (left) leading to vortex breakdown and axial downdrafts (right).

5 Stretching the firewhirls

For the Burke-Schumann model, the exponents of the compensating regime lead to flame stretching [6]. As shown in Figure 6, this is indeed in good agreement with experimental data [11]: the presence of rotation corresponds to elongated flames with the exponent α taking values from the compensating range $4/3 < \alpha < 3/2$, The flames without rotation correspond to the potential value of the exponent $\alpha = 2$. The problem seems resolved but there is, however, one important point in the theory [6] that needs clarification: the analysis is conducted for constant density, while density is very low within the flame due to high temperature. Klimenko and Williams [6] understood this and restricted validity of the compensating exponents to the outer sections of the flow where the density is constant. This needs to be done since, even without density change, the compensating exponents cannot be extended to the core of the flow [10]. However, this approach does not resolve the physics of density/vorticity interactions.

Yu and Zhang [7] have suggested to use an analogue of the Howarth-Dorodnitsyn transformation and eliminate density from the Burke-Schumann model for the scalar transport by using transformed coordinates. This undoubtedly represents a step forward compared to the analysis of Klimenko and Williams [6], but does it resolve the main physical problem of how density change affects the vortex? It appears that the answer is negative. First, the new analysis [7] introduces only secondary effects into the expression for flame length: variation of molecular diffusivity with temperature and differential diffusion. These effects are rotation-independent and not likely to be relevant to real firewhirls, where the flow is turbulent and the influence of molecular diffusivity is insignificant. Most importantly, the improved treatment does not resolve the outstanding physical issues that are discussed in the following section.

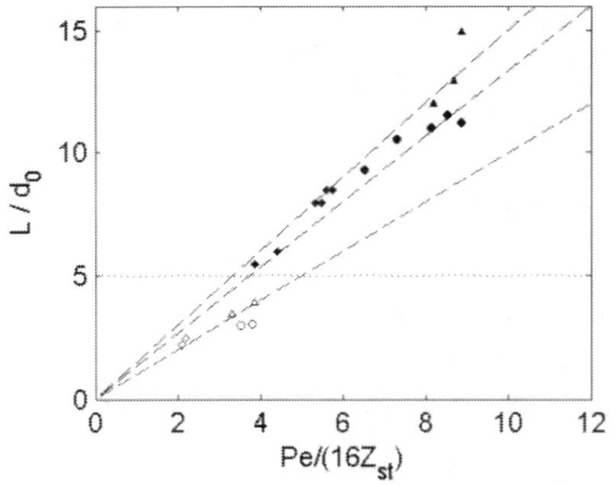

Figure 6. Evaluation of the flame length [6] in comparison with experiments [11] (solid symbols -- with rotation; open symbols – without rotation). The dashed lines corresponds to the following values of the compensating exponent α = 2, 3/2, 4/3 (from the bottom to the top)

6 Effect of density change on strong vortices

Transformations of coordinates of the Howarth-Dorodnitsyn type can remove density from the scalar transport equation in boundary layers but, generally, cannot do the same for dynamic components since the interaction between the density gradient and the pressure gradient produces new physical effects. Specifically, the boundary layer equations still involve the density-dependent term

$$\frac{DV_z}{Dt} = -\frac{1}{\rho}\frac{\partial p}{\partial z} + \ldots \qquad (4)$$

even after the Howarth-Dorodnitsyn transformation. Since the pressure in this flow is r-independent (omitting the centrifugal component of the pressure), the lower density would stimulate a stronger updraft near the axis. The same effect can be considered from the perspective of vorticity generation; in variable-density flows this involves the density gradient term

$$\frac{D\vec{\omega}}{Dt} = (\vec{\omega}\cdot\nabla)\vec{V} - \vec{\omega}(\nabla\cdot\vec{V}) + \frac{\nabla\rho\times\nabla p}{\rho^2} + \ldots \qquad (5)$$

Note that the pressure has axial gradient $\partial p/\partial z<0$ due to 1) hydrostatic reduction of pressure by gravity and 2) axial acceleration of the flow (since $V_z = 0$ at the surface $z = 0$ and V_z increases for larger z). The reduction of density near the axis (i.e. the radial density gradient $\partial\rho/\partial r>0$) in combination with the axial pressure gradient generates the tangential vorticity ω_θ that stimulates the updraft (see Figure 4, right). These vorticity-generating effects are physical and cannot be eliminated by a mere transformation of coordinates. Therefore, the effect of generation of tangential vorticity by transformation of the axial vorticity (which is present in all developed tornado-like flows) would overlap and interfere with the effect of vorticity generation by the density gradient (which can be expected to be significant in the core of firewhirls).

Since the region of significant variations of density is localised near the axis, its effect should generally be localised but can also be expected to enhance stability of the vortex [6]. Indeed, generation of updraft-stimulating tangential vorticity ω_θ should counterbalance dynamic generation of downdraft-stimulating vorticity by the vortex-breaking mechanism illustrated in Figure 5. This should delay vortex breakdown, allowing the possibility of larger values of the vertical swirl ratio *K*. Hence, intensive heat release in firewhirls should lead to extension of the vortex stability region to faster rotations and to further elongation of the flame. While vortex breakdowns are less likely in firewhirls than in tornado-like vortices with constant density, breakdowns are still possible in firewhirls when the mechanisms of vorticity generation by buoyancy and axial pressure gradient are overpowered by intense rotation. In this case the flame should shorten. These conclusions seem logical and have been confirmed by initial computer simulations [12-14].

7 Conclusion

The dynamic processes, which need to be considered in firewhirls, involve a complex balance of various factors that may have opposite effects on the flow. The present paper discusses basic features of the firewhirl dynamics and offers a general qualitative theory for the breakdown mechanism of tornado-like vortices. Further progress in understanding firewhirls will require careful evaluation of the factors discussed here by a combination of theoretical analysis with well-planned computer simulations and experiments. These simulations and experiments can be performed with gradually increasing level of vorticity to determine the point of vortex breakdown in presence or absence of various factors including density change and buoyancy.

References

[1] H. W. Emmons, S. J. Ying, Proc. Combust. Inst 1 (1928) 2–11.

[2] F. A. Williams, Prog. Energy Combust. Sci. 8 (1982) 317–354.

[3] K. Kuwana, K. Sekimoto, K. Saito, F. A. Williams, Fire Safety Journal 43 (2008) 252–257.

[4] F. Battaglia, R. G. Rehm, H. R. Baum, Physics of Fluids 12 (2000)

[5] K. Kuwana, S. Morishita, R. Dobashi, K. H. Chuah, K. Saito, Proc. Combust. Inst. 33 (2011) 2425 – 2432.

[6] A.Y. Klimenko , F.A. Williams , Combust. Flame 160 (2013) 335

[7] D. Yu, P. Zhang, Combust. Flame 182 (2017) 36–47

[8] S. P. Burke, T. E. W. Schumann, Proc. Combust. Inst 1 (1928) 2–11.

[9] F. A. Williams, Combustion Theory, Addison-Wesley, Reading, MA, 2nd edition, 1985 pp.38-47.

[10] A.Y. Klimenko, J. Fluid Mech. 738 (2014) 268

[11] K. H. Chuah, K. Kuwana, K. Saito, F. A. Williams, Proc. Combust. Inst. 33 (2011) 2417–2424.

[12] C. Dowd, BE thesis, SoMME, The University of Queensland, 2013.

[13] A.N. Kamitsis, BE thesis, SoMME, The University of Queensland, 2015.

[14] C. Murphy, BE thesis, SoMME, The University of Queensland, 2016.

11th Asia-Pacific Conference on Combustion
The University of Sydney, NSW Australia
10th-14th December 2017.

A Flame Index for the Lagrangian Probability Density Function Method

Zhen Lu[*1], **Zhuyin Ren**[2], and **Hong G. Im**[1]

[1]Clean Combustion Research Center, KAUST, Thuwal, 23955, Saudi Arabia
[2]School of Aerospace Engineering, Tsinghua University, Beijing, 100084, China

Abstract

To separate the nonpremixed and premixed modes in partially premixed flame simulations with the transported PDF method, a flame index is proposed based on the subgrid mixing rate between Lagrangian particles. The mixing rate on local equivalence ratio and progress variable is compared to measure the combustion mode. With a renormalized progress variable, the proposed flame index is 0 for equilibrium solution of the nonpremixed flame, and 1 for the premixed flame with homogeneous cold flow. The index is validated on a series of constructed turbulent combustion subgrid particles and the partially stirred reactors with different mixing models. Tests demonstrate that the flame index for PDF method works well for nonpremixed flames with variant scalar dissipation rate. Furthermore, the index shows a consistent trend at different parameters. Results also confirms that the localness of the EMST model make it predict the physical process well.

1 Introduction

Partial premixing plays a significant role in practical combustion devices [1]. Due to the inhomogeneity of compositions, different combustion modes, such as premixed, nonpremixed, and auto-ignition, may co-exist.

Modeling approaches for such flames are developing [2, 3, 4, 5]. One way is to combine the developed modeling methods for nonpremixed and premixed combustion, with a flame index to switch [2]. The other way is applying a "universal" model for partially premixed flame. The transported probability density function (PDF) method [6], originally designed for the nonpremixed combustion, has shown potential to handle the premix mode [3, 4, 5].

A typical method to check the mode of combustion in partially premixed flame is through the scattering plots of temperature or progress variable versus mixture fraction at a plane. This kind of plot shows the statistical behaviour of the flame. In addition to the analysis on global behavior, transient and local information would contribute to the understandings of underlying turbulence flame interactions. For the partially premixed flame, flame index [7, 8] is a useful tool. The Takeno index [7] is firstly proposed based on the orientations of fuel and oxidizer gradients, which is applied to the analysis of direct numerical simulaiton (DNS) data. It is then extended to the large eddy simulation (LES) with modeled subgrid information from the resolved field [8]. Knudsen and Pitsch [2] proposed a flamelet transformation to yield a "combustion regime index" for the partially premixed combustion. It basically compares the contributions from transport between different flamelets and along one flamelet on the governing equation of progress variable.

Existing indexes more or less apply resolved gradient to model subgrid flux, and neglecting some subfiltered terms. Althogh the transported PDF method with Lagrangian particle description need to model the subgrid mixing, a suitable mixing model could keep some insights of the subgrid physical processes. In the present work, a flame index based on composition transport between particles is proposed.

2 PDF method and flame index

2.1 PDF model

In the transported PDF method, the transport equation [9]

$$\begin{aligned}\frac{\partial F_L}{\partial t} + \frac{\partial F_L\widetilde{v_i}}{\partial x_i} + \frac{\partial F_L\widetilde{v_i'|\psi}}{\partial x_i} = \\ -\frac{\partial}{\partial \psi_\alpha}\left[F_L\overline{\frac{1}{\bar{\rho}}\frac{\partial}{\partial x_i}\left(\rho D_{(\alpha)}\frac{\partial \phi_\alpha}{\partial x_i}\right)|\psi}\right] \\ -\frac{\partial F_L S_\alpha}{\partial \psi_\alpha},\end{aligned} \quad (1)$$

*Corresponding author
Email address: zhen.lu@kaust.edu.sa

is solved for the filtered mass density function

$$F_L(\psi; \boldsymbol{x}, t) \equiv \int G(\boldsymbol{x} - \boldsymbol{x}') \rho(\boldsymbol{x}', t) f'(\psi; \boldsymbol{x}, t) d\boldsymbol{x}', \tag{2}$$

with mixture-averaged diffusivity model in the form of Fick's law. Due to the high dimensions in composition space, a Monte Carlo method with Lagrangian particles [6] is typically adopted to solve the transport equation of PDF. Note that for the PDF transport equation, the chemical source term is in a closed form. The filtered governing equations can be obtained by calculating the zeroth and first moment of the FMDF.

While the reaction source term is closed elegantly in PDF method, the conditional convection and diffusion terms in Eq. 1 require modeling, which are the third and fourth terms, respectively. There widely employed approaches to model the conditional diffusion term, i.e. the micro mixing are listed in the followings.

1. Interaction by exchange with the mean (IEM) [10] or linear mean-square estimation (LMSE) [11]
   IEM model treat the conditional diffusion term as,
   $$\frac{d\phi}{dt} = -\frac{\phi - \tilde{\phi}}{2\tau_{mix}}, \tag{3}$$
   where $\tilde{\phi}$ is the Favre-averaged composition. τ_{mix} is the mixing time scale obatained with
   $$\tau_{mix} = \frac{\widetilde{Z''^2}}{\tilde{\chi}}, \tag{4}$$
   with Z stands for the mixture fraction and $\tilde{\chi} = 2D_t \widetilde{\nabla Z \nabla Z}$ the filtered scalar dissipation rate. It is clear that the IEM makes the particle mix with the mean of the subgrid, which is indicated in the name.

2. Modified Curl (MC) [12]
   MC applies stochastic particle interaction on the subgrid particles. For an ensemble of N particles, $N \times \frac{dt}{\tau_{mix}}$ pairs of particles are randomly picked. Mixing in these pairs is done as
   $$\begin{aligned} \phi^{p,new} &= \phi^p + \frac{1}{2}a(\phi^q - \phi^p), \\ \phi^{q,new} &= \phi^q + \frac{1}{2}a(\phi^p - \phi^q), \end{aligned} \tag{5}$$
   where a is a random number in $(0, 1)$ with a uniform distribution.

3. Euclidean minimum spanning tree (EMST) [13]
   EMST first constructs the Euclidean minimum spanning tree in composition space. Then the mixing pairs are selected based on the proximity of particle pairs in composition space along the edges. Therefore, EMST guarantees the localness in composition space of the mixing process.

2.2 Flame index

Regarding the use of flame index, it should imply a clear separation for nonpremixed and premixed combustion, or two limits corresponding to these two modes respectively. Furthermore, it is expected that the index show a consistent trend varying from the non-premixed limit to the premixed limit

Consider the Burke-Schumann solution, all points in the diffusion flame have the progress variable $C = 1$, while mixture fraction Z or local equivalence ratio Φ differes. Thus,

$$\delta C = 0, \quad |\delta Z| \geq 0. \tag{6}$$

Unfortunately, due to the reversible chemistry, the compositions in a nonpremixed flame cannot reach the fully burnt states, and the limit of progress variable is obtained at the equilibrium. A renormalized progress variable [14]

$$C = \frac{C_O}{C_{O,eq}}, \tag{7}$$

is employed to make $C = 1$ for the nonpremixed limit, where C_O is the progress variable based on oxygen mass fraction, subscript eq stands for the value at equilibrium state. Consequently, Eq. 6 still holds for the renormalized progress variable at equilibrium solution.

As for a premixed flame with homogeneous cold flow, mixture fraction is same for all particles, while C changes through the locations with respect to the flame, leading to

$$|\delta C| \geq 0, \quad \delta Z = 0. \tag{8}$$

Based on the above observation, a flame index based on mixing rate of fluid particles is proposed as

$$\mathrm{FI} = \frac{|\delta C|/C}{\beta|\delta\Phi^*|/\Phi^* + |\delta C|/C}, \tag{9}$$

where

$$\Phi^* = \frac{\Phi}{\Phi + 1} \tag{10}$$

is the normalized equivalence ratio [15] for each particle, β is a scaling factor for the mixing on mixture fraction, defined as

$$\beta = \max\left(\left|\frac{\partial C_{O,eq}}{\partial \Phi^*}\right|_{\Phi^*=0.5^-}, \left|\frac{\partial C_{O,eq}}{\partial \Phi^*}\right|_{\Phi^*=0.5^+} \right). \tag{11}$$

In Eq. 9, $|\delta C|/C$ and $|\delta\Phi^*|/\Phi^*$ stand for particle mixing rate of reaction progress and mixture fraction respectively. Φ^* is chosen instead of Z to avoid over-estimating the importance of mixing between streams with similar compositions. The scaling factor β is designed to deal with the "dip" of the progress variable near the stoichiometric condition. Although the slope of C is smoothed by the renormalization [14], β can still represent the scale of $|\partial C/\partial\Phi^*|$ for the slope of the "dip". With such a scaling factor, it is expected that the nonpremixed combustion has FI < 0.5, while close to 0.5 means the nonpremixed flame is away from the equilibrium state, i.e. close to extinction. Note the the two limits of equilibrium nonpremixed solution with FI = 0 and homogeneous premixed flame with FI = 1 do not change with the scaling factor. Furthermore, the renormalization guarantees the limits for nonpremixed and premixed modes with different definitiona of the progress variable.

3 Validation of the flame index for PDF

For simplicity, hydrogen is taken as the fuel in the present work. The fuel stream is composed of hydrogen and nitrogen, with the volume ratio $H_2 : N_2 = 1 : 1$, $T = 300$ K and the oxidant stream is air with $O_2 : N_2 = 1 : 3.76$, $T = 300$ K. With such two inlet streams, the scaling factor β is 4.3. A 10 species, 20 reactions hydrogen oxidation mechanism [16] is applied with assistance of the *in situ* adaptive tabulation (ISAT) [17, 18] algorithm to accelerate calculations. The progress variable used in the present study is defined based on oxygen [14]. C_O for hydrogen oxidation is defined as the ratio between the mass of O bound to fuel atoms in the form of H_2O and that would be bound in the fully burnt states, and $C_{O,eq}$ is the value of C_O when the mixture is at equilibrium.

3.1 Construction of turbulent non-premixed flames with flamelet solution

Although the limit of FI on equilibrium solution is guaranteed with the definition of the renormalized progress variable, the application of FI on nonpremixed flames away from the equilibrium state remains concerned. For the nonpremixed with substantial stretching, the progress variable is not a constant over Z from 0 to 1. Therefore, there will be changes on C with Z, i.e. $\partial C/\partial Z \neq 0$.

In the present study, subgrid particles constructed with flamelet solutions are applied to check the behavior of FI with respect to stretching. Assuming a beta PDF distribution of mixture fraction, with a set of $\tilde{Z}$ and $\widetilde{Z''^2}$, N_p particles on the flamelet solution [19] corresponding to a χ_{st} can be seleted as a group of particles in one cell of non-premixed turbulent flame simulations. Increasing χ_{st} makes the nonpremixed flame move from equilibrium solution to extinction. The mixing models are applied for one step to extract the particle mixing information, with the mixing timescale obtained with $\tau_{mix} = \widetilde{Z''^2}/\tilde{\chi}$. The FI in Eq. 9 is then calculated to check its validity for non-premixed flames. Time step for mixing is set to be $1/25$ of τ_{mix}.

For the cases shown in the followings, $N_p = 100$ particles are picked from the flamelets with χ_{st} varies from 2 (1/s) to 1700 (1/s). $\tilde{Z} = Z_{st} = 0.304$ and $\widetilde{Z''^2}/(\tilde{Z} \times (1 - \tilde{Z}))$ varies from 0.2 to 0.8. Results on the mean of flame index $\langle FI \rangle$ obtained with EMST are shown in Fig. 1. Considering not all particles reside in the "dip" regime, $\langle FI \rangle$ should be less than 0.5. For all the three mixing models, $\langle FI \rangle$ is below this limit, while three mixing models have different behaviors against χ_{st} and $\widetilde{Z''^2}$. EMST presents result with $\langle FI \rangle$ not apparently related with variance. In addition, the $\langle FI \rangle$ increases with χ_{st} approaching the extinction limit. It indicates that the localness of EMST makes particles mix in subgrid like a 1D flamelet solution.

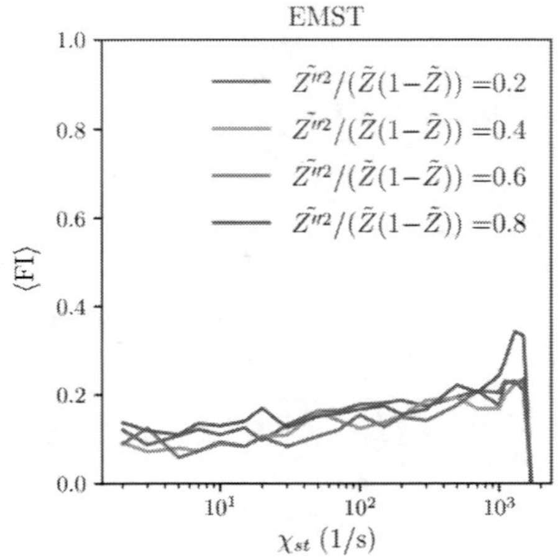

Figure 1: Means of the flame index $\langle FI \rangle$ obtained with the EMST model at different χ_{st}, with particles samples from the corresponding flamelet solution.

3.2 PaSR

The partially stirred reactor (PaSR) has two inlet streams for fuel and oxidizer respectively. Inside the reactor, reaction occurs and the mean thermochemical properties are assumed to be statistically spatially homogeneous, but the fluid is imperfectly mixed at the molecular level, having a mixing timescale of τ_{mix}. An outflow with a mass flow rate equals the sum of inlets make the mass in the reactor constant, related with a residence time τ_{res}. PaSR is similar to a single cell in the PDF modeling of turbulent combustion. A splitting scheme for mixing and reaction on particles are applied for the calculations of the mixing reaction processes [20].

The global equivalence ratio $\Phi_g = 1$ is controlled by the flow rate. Pressure is set to be 1 atm. Time step for calculations is the minimun between $\tau_{res}/10$ and $\tau_{mix}/25$, with $N_p = 1000$. Results are taken after the PaSR reaches the statistical stationary.

When τ_{mix} goes to 0, the PaSR should reach the PSR limit, which would behave like the premixed combustion. Therefore, $\langle FI \rangle$ should increase with the decrease of τ_{mix}. Results obtained with $\tau_{res} = 4 \times 10^{-3}$ s and $\Phi_g = 1$ is shown in Fig. 2. It is observed that for both IEM and EMST, ⟨FI⟩ rises with the decrease of τ_{mix}, while results from MC do not change much.

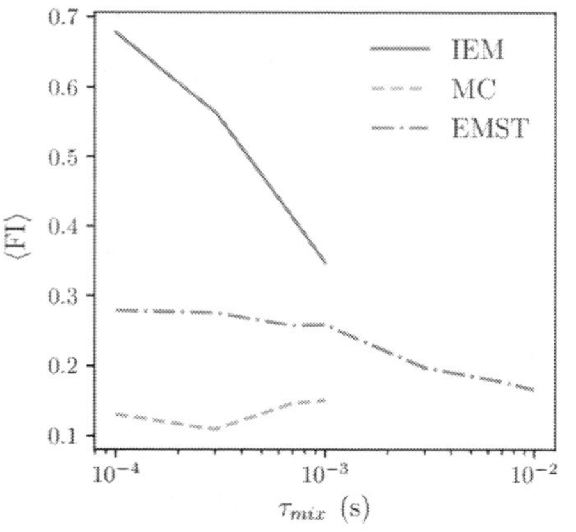

Figure 2: Means of the flame index ⟨FI⟩ for different mixing time scales τ_{mix} with $\tau_{res} = 4 \times 10^{-3}$ s.

4 Conclusions

In the present work, a flame index based on the mixing rate of fluid particles has been proposed for the transported PDF method. The flame index is designed to show the local combustion mode to assistant the post analysis on LES/PDF simulations. Definition of the index guarantees it is 0 for the nonpremixed flame at equilibrium state and 1 for premixed flame with homogeneous cold flow. A scaling factor for mixing on mixture fraction and progress variable makes the nonpremixed mode has a flame index less than 0.5. Flamelet solutions with different scalar dissipation rate are employed to construct nonpremixed flames approaching to extinction. The flame index is then validated based on the particle data. It is observed that the localness of EMST makes it predict the mixing process physically. PaSR calculations with different mixing time scales, equivalence ratio, and partial premixing extent are conducted, with the flame index showing a consistent behavior.

5 Acknowledgement

This research work was sponsored by King Abdullah University of Science and Technology (KAUST). The simulations utilized the computing resources at KAUST Supercomputing Laboratory.

References

[1] A. Masri. Proc. Combust. Inst. 35 (2015) 1115–1136.

[2] E. Knudsen, H. Pitsch. Combust. Flame 156 (2009) 678–696.

[3] H. Wang, P. Zhang. Proc. Combust. Inst. 36 (2017) 1693–1703.

[4] M. Kuron, E. R. Hawkes, Z. Ren, J. C. Tang, H. Zhou, J. H. Chen, T. Lu. Proc. Combust. Inst. 36 (2017) 1987–1995.

[5] H. Zhou, S. Li, Z. Ren, D. H. Rowinski. Combust. Flame 181 (2017) 136–148.

[6] S. B. Pope. Prog. Energy Combust. Sci. 11(2) (1985) 119–192.

[7] H. Yamashita, M. Shimada, T. Takeno. Proc. Combust. Inst. 26 (1996) 27–34.

[8] P. Domingo, L. Vervisch, K. Bray. Combust. Theory Model. 6 (2002) 529–551.

[9] F. A. Jaberi, P. J. Colucci, S. James, P. Givi, S. B. Pope. J. Fluid Mech. 401 (1999) 85–121.

[10] J. Villermaux, J. C. Devillon. In Proc. 2nd Int. Symp. on Chemical Reaction Engineering. New York, pp. 1–13.

[11] C. Dopazo, E. E. O'Brien. Acta Astronaut. 1 (1974) 1239–1266.

[12] J. Janicka, W. Kolbe, W. Kollmann. J. NonEquilib. Thermodyn. 4(1) (1979) 47–66.

[13] S. Subramaniam, S. Pope. Combust. Flame 115(4) (1998) 487–514.

[14] R. S. Barlow, G. Magnotti, H. C. Cutcher, A. R. Masri. Combust. Flame 179 (2017) 117–129.

[15] C. K. Law. Combustion Physics. Cambridge University Press, New York, USA, 1st ed. (2006).

[16] Z. Hong, D. F. Davidson, R. K. Hanson. Combust. Flame 158 (2011) 633–644.

[17] S. B. Pope. Combust. Theory Model. 1(1) (1997) 41–63.

[18] L. Lu, S. B. Pope. J. Comput. Phys. 228(2) (2009) 361–386.

[19] H. Pitsch. FlameMaster, A C++ computer program for 0D combustion and 1D laminar flame calculations (2017).

[20] Z. Ren, S. B. Pope. Combust. Flame 136(1) (2004) 208–216.

11th Asia-Pacific Conference on Combustion,
The University of Sydney, NSW Australia
10th -14th December 2017.

Flame temperature measurement using color-ratio pyrometry with a consumer grade DSLR camera

Anand Sankaranarayanan, Umakant Swami, Arindrajit Chowdhury, Neeraj Kumbhakarna

Department of Mechanical Engineering, IIT Bombay

Powai, Mumbai, Maharashtra, 400076, India

Abstract

A CANON EOS-550D DSLR camera with a 100 mm zoom macro-lens was calibrated for measurement of whole field temperature using color-ratio pyrometry (CRP). The spectral response curve of the camera-lens system was obtained with the help of a laser power meter and a tunable laser source. The response curves thus obtained were used to generate look up tables for black body as well as soot radiation. The black body look up table was validated with the help of temperature measurements in the range 750-1500 °C. The assumption that the emissivity of the soot particles varied as $\lambda^{-1.38}$ was utilized to generate soot lookup tables. The result obtained using CRP for a candle flame was compared against thermocouple measurements. These results indicate that CRP applied using a consumer grade camera is a suitable technique for obtaining flame temperatures greater than 1000 K with reasonable accuracy, for sooty flames. The method allows 2D, non-intrusive temperature measurement at very low costs.

1 Introduction

Measurement of flame temperature forms one of the cornerstones of combustion diagnostics. It is important to determine the temperature distribution with the combustion zone from a perspective of efficiency as well as design and safety. Flame temperatures have a direct impact on combustion efficiency and emission control, including mechanisms of soot formation and destruction [1]. Thus it is imperative to accurately estimate the temperature distribution for the combustion phenomenon being studied.

Various tools and techniques for measurement of temperature has been developed over the past century [2]. The techniques may be broadly categorized as intrusive and non-intrusive in nature. The simplest intrusive technique is the use of thermocouples. These are typically used for point measurements and are restricted by the maximum temperature that can be measured, owing to limitations placed by the material of construction of the thermocouple. In addition to this, the accuracy of flame temperatures measured using thermocouples is quite limited due to several reasons. Often, the uncertainty in measurements may be of the order of 10% or more, owing to heat losses from bead, errors in estimating flame emissivity, and soot deposition on the thermocouple bead. Non-intrusive techniques benefit from their ability to determine temperatures using optical means without affecting the thermal and flow field. Laser based techniques such as Rayleigh scattering, Coherent Anti Raman Spectroscopy and other methods [3] typically provide higher degrees of accuracy. Their sensitivity comes at the expense of high initial investments and the requirement of extensive care during handling such instruments. In addition concerns regarding safety and space limitations limit their usage in certain environments [4].

Temperature estimation methods based on detecting the intensity of light emitted by soot particles in a flame is conventionally known as soot pyrometry. This technique can be classified into different categories based on its implementation. The most popular among these are the two-color pyrometry method and the color-ratio pyrometry (CRP) method. The wavelengths used for detection may be narrow band or broad band. Soot pyrometry in various forms has been previously utilized to measure the flame temperatures in several applications such as burner flames [5, 6] , diesel engines [7], and, coal fired boilers [8]. Kuhn et.al [6] demonstrated that DSLR cameras may be suitably calibrated as pyrometers to acquire whole field temperature distribution. The present work discusses the steps involved in calibrating a consumer grade DSLR camera to evaluate flame temperatures, using detection of light emitted by soot particles in the visible spectrum, based on CRP. Once calibrated, the method provides a cost-effective and easy-to-implement solution for measuring 2D temperature profiles of sooty flames with reasonable accuracy. A candle flame was characterized and the measured thermal field was compared with thermocouple measurements. Results are presented for a candle flame.

2 Theory behind calibration of a DSLR camera

2.1 Working of a DSLR

A DSLR camera usually consists of a Charge Coupled Device (CCD) sensor or a Complementary Metal-Oxide-Semiconductor (CMOS) sensor, coated with a Color Filter Array (CFA). The CFA in turn consists of an array of three distinct filter elements designated as Red (R), Green (G) and Blue (B) filters, which selectively allow only a fraction of the visible radiation emitted by the source to reach the sensor. One of the most popular arrangements of the filter elements in the CFA is known as the Bayer pattern [9]. It consists of repetitive units of a 2x2 pattern of square pixels with the R, G, and B elements appearing at least once in each unit. In order to recreate the image of an object under consideration, each pixel must have its own set of RGB values. This is obtained by interpolating the information from the neighbouring pixels through demosaicing. The built-in processor within the camera also applies several other corrections to the input signal for ease of processing and to render the processed image visually appealing to the human eye [10].

2.2 Image processing details

One of the basic requirements of CRP is that the output signal generated by the detector of the camera must be directly proportional to the input signal received. This requires the elimination of unwanted image processing steps, such as white balancing and gamma correction. As it is often difficult to determine the sequence and extent of post-processing performed on the input data by the camera's processor, it is advisable to record the images in the RAW mode for pyrometry applications. The RAW mode is a proprietary format which functions as a digital negative for the camera. The steps to be followed for generating a tiff image from an image in the RAW format for analyzing the required temperature distribution using the open source code (dcraw) and MATLAB is shown in Fig. 1.

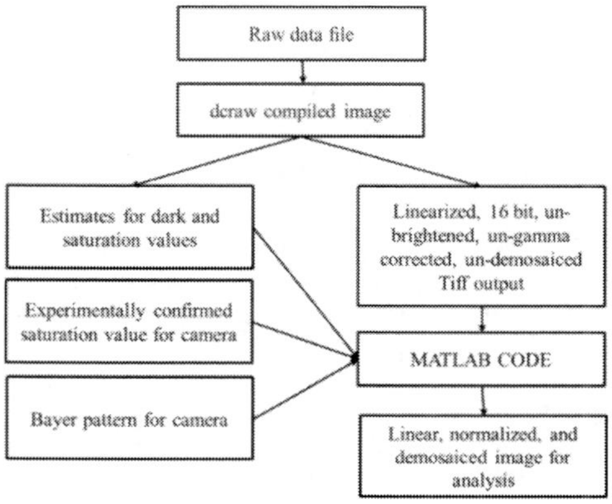

Figure 1: Image processing steps in CRP

2.3 Theory behind CRP

Soot pyrometry measures the temperature of soot particles in the flame. These particles are usually very small and attain thermal equilibrium with the surrounding combustion gases within 1 to 10 μs [11]. Hence the soot temperature may be safely approximated to be the same as the surrounding gas temperature. The intensity of light (*I*) emitted by an object at a given wavelength (λ) is a function of temperature (*T*) and is given by (1).

$$I(\lambda,T) = \varepsilon(\lambda)\frac{2\pi hc^2}{\lambda^5(e^{hc/\lambda kT}-1)} \tag{1}$$

where $\varepsilon(\lambda)$ is the emissivity of the object at the given wavelength, h is Planck's constant, c is the velocity of light, and k is the Boltzmann constant. The signal received by the detector of the camera (S_F) for a given exposure time (τ), after the light has passed through the lens and CFA is given by (2)

$$S_F = 2\pi hc^2\tau\int_{\lambda_1}^{\lambda_2}\frac{\varepsilon(\lambda)\eta(\lambda)}{\lambda^5(e^{hc/\lambda kT}-1)}d\lambda \tag{2}$$

where $\eta(\lambda)$ accounts for combined efficiency of the detector, the lens array, the transmission efficiency of the CFA as well as a geometric factor [6]. In this case, λ_1 and λ_2 are the limits of visible wavelength range, whose values are of 400 and 700 nm respectively. The ratio of signals received by each pixel of the camera can then be found using the relation provided in (3)

$$\frac{S_{F_1}}{S_{F_2}} = \frac{\int_{\lambda_1}^{\lambda_2}\eta_{F_1}(\lambda)\frac{\varepsilon(\lambda)}{\lambda^5}\left[\exp(\frac{hc}{\lambda kT})-1\right]^{-1}d\lambda}{\int_{\lambda_1}^{\lambda_2}\eta_{F_2}(\lambda)\frac{\varepsilon(\lambda)}{\lambda^5}\left[\exp(\frac{hc}{\lambda kT})-1\right]^{-1}d\lambda} \tag{3}$$

where F_1 and F_2 represent filters 1 and 2 respectively (either of R,G, and B). For a black or gray body, the emissivity may be eliminated from the numerator and denominator. Equation (3) is integrated numerically to generate look up tables for the color-ratio values, for a given range of temperatures. For generating soot look up tables, the soot emissivity is assumed to vary as $\lambda^{-1.38}$ [6].

3 Experimental Work

3.1 Experiments for characterizing the spectral response of a DSLR camera

A schematic of the experimental setup used to characterize the spectral response of the camera is given in Fig. 2.

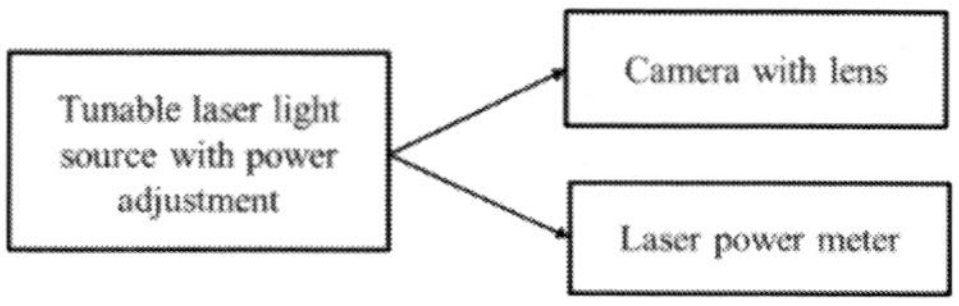

Figure 2: Schematic of the experimental setup used for characterizing the camera's spectral response

A tunable laser source, Fianium Whitelase SC400, whose wavelength and intensity could be varied, was the most important component of the setup. The emission from the laser was imaged onto the camera sensor through the chosen lens system. The aperture, ISO setting, and f-stop value of the camera were kept unchanged during the course of calibration. The laser power settings used for the experiments were typically less than 0.5 μW and was measured with the help of a FieldMate laser power meter. In order to ensure that the power readings remain unaffected by the ambient light, the calibration experiments were conducted in a dark room. The images were shot in the RAW mode and the ISO setting was kept at the minimum value possible to reduce ambient noise. The inbuilt image processing by the camera was minimalized by avoiding all image enhancement options or by setting them to their default values.

3.1.1 Detector linearity check

The linearity of the CMOS detector was confirmed at a fixed wavelength by holding all the camera and laser settings constant, except shutter speed which was varied gradually. The RGB values of the resulting images were obtained with the help of MATLAB, using the procedure outlined in the section 2.2. The results presented in Fig. 3, show that the response of the detector is indeed linear for the camera-lens system utilized.

Corresponding author. Fax: +91-22-25726875
E-mail address: neeraj_k@iitb.ac.in

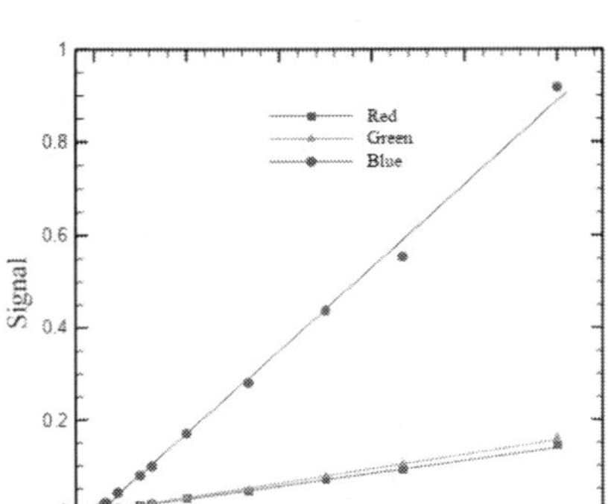

Figure 3 : Detector linearity at a wavelength of 400 nm

3.1.2 Characterization of spectral response

For characterizing the spectral response of the system, experiments were conducted from 400 to 700 nm at intervals of 10 nm. The laser power setting for each trial was noted and the values were measured by a laser power meter. The images were processed as explained in section 2.2, to obtain the RGB data for each pixel. The ratio of intensities R/B and R/G were also computed separately. The response function for the red filter was calculated in an arbitrary scale by (4).

$$\eta_{R\_arb}(\lambda) = \frac{R}{P \times \tau} \quad (4)$$

where, R is the normalized value of the red signal between 0 and 1, averaged for a central zone inside the imaged laser spot, P is the measured power value of the laser beam for the given setting, and τ is the time interval of data acquisition.

This was then normalized to a scale between 0-1 using (5)

$$\eta_{R\_Scaled}(\lambda) = \frac{\eta_{R\_arb}(\lambda)}{MAX(\eta_{R\_arb}(\lambda))} \quad (5)$$

where, $MAX(\eta_{R\_arb}(\lambda))$ represents the maximum value of $\eta_{R\_arb}(\lambda)$, within the range of considered wavelengths. The R/B and R/G ratios were then used to derive corresponding scaled values for blue and green signals as given in (6) and (7).

$$\eta_{B\_Scaled}(\lambda) = \frac{\eta_{R\_Scaled}(\lambda)}{(R/B)} \quad (6)$$

$$\eta_{G\_Scaled}(\lambda) = \frac{\eta_{R\_Scaled}(\lambda)}{(R/G)} \quad (7)$$

where, B and G are the normalized values of the blue and green signals, defined in a manner similar to R.

A final normalization was implemented using the highest of these scaled values at all measured wavelengths ($MAX(\eta_{R/G/B\_SCALED}(\lambda))$, to obtain the normalized spectral response curve for the camera, shown in Fig. 4. The equations used for the same are as follows:

$$\eta_R(\lambda) = \frac{\eta_{R\_Scaled}(\lambda)}{MAX(\eta_{R/G/B\_SCALED}(\lambda))} \quad (8)$$

$$\eta_G(\lambda) = \frac{\eta_{G\_Scaled}(\lambda)}{MAX(\eta_{R/G/B\_SCALED}(\lambda))} \quad (9)$$

$$\eta_B(\lambda) = \frac{\eta_{B\_Scaled}(\lambda)}{MAX(\eta_{R/G/B\_SCALED}(\lambda))} \quad (10)$$

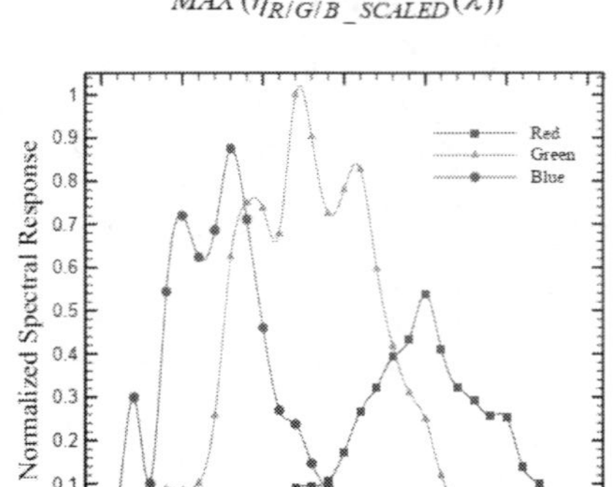

Figure 4: Normalized spectral response curve for the camera

3.2 Validation of response curve with black body calibration results

Once the response curves for the camera's CFA were obtained, the lookup tables were generated as discussed previously. These were compared against black body calibration data from 750 to 1500 °C. The black body calibration was carried out using a CALsys 1500BB calibration source. Readings were obtained in steps of 50 K and the results are presented in Fig. 6. The spectral response curves for R, G and B were modified slightly (< 4%), by multiplication with a scaling factor, for a better agreement with the blackbody results [6]. The scaling factors used were 1.04, 1.01 and 1 for R, G and B respectively. The curves were found to be reasonably monotonous from 1000 to 3000 K, ensuring unique values of temperature from each pixel.

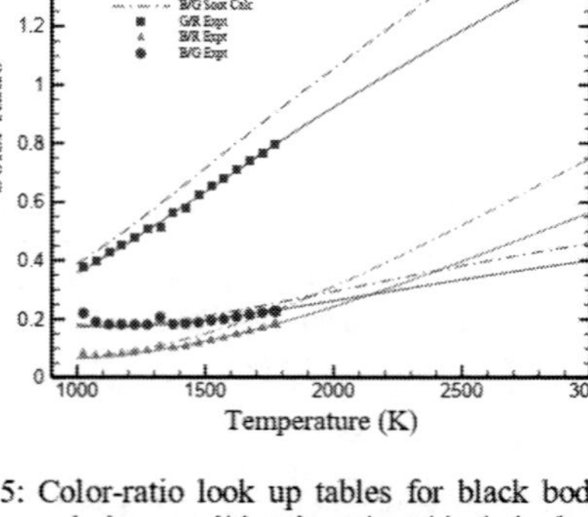

Figure 5: Color-ratio look up tables for black body and soot radiative emissions, validated against black body calibration data

4. Results and Discussions

The validity of the generated lookup tables was estimated by comparing the whole-field thermal data for a candle flame obtained using CRP with point measurements from an S-type thermocouple. The thermocouple wires were 80 µm in diameter, with a bead size of approximately 150 µm. The RAW image was captured with the thermocouple located within the flame, to identify its location in the 2D temperature profile obtained for the rest of the flame. The uncorrected thermocouple readings are shown in Fig.6. The 2D temperature contour retrieved from the flame, based on the generated soot lookup tables is shown in Fig. 7. It was observed that the temperature of the gas surrounding the thermocouple was approximately around 1700-1800 K for pixels in the vicinity of the bead, while that measured by the thermocouple showed an average value of about 1650 K at steady state. Since the temperature correction due to radiative losses is expected to be of the order of 100 K, the data obtained using soot pyrometry was determined to be in excellent agreement with the expected values.

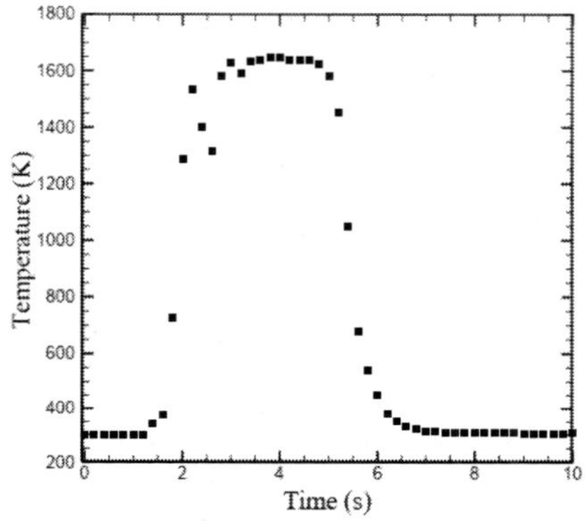

Figure 6: Uncorrected thermocouple readings for candle flame

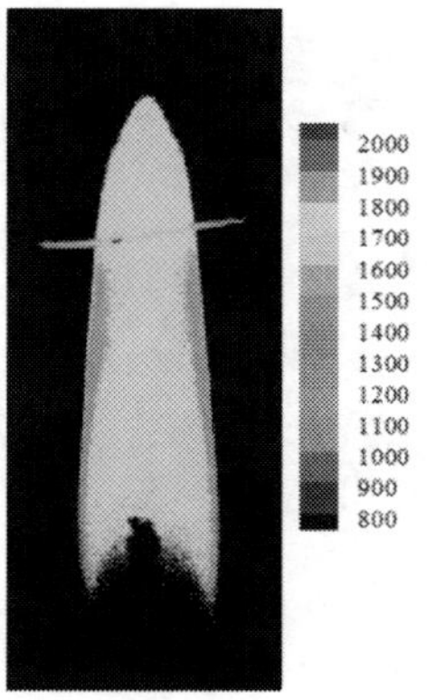

Figure 7: 2D temperature distribution of a candle flame (K)

The axisymmetric nature of a candle flame was captured well in the temperature distribution obtained. It should be noted that the CRP technique works well only for yellow sooty flames and fares poorly when applied to bluish flames.

4 Conclusions

The spectral response curve for the camera's CFA was characterized using a tunable laser source and a laser power meter. The images were shot in the RAW mode with the settings chosen so as to minimize the camera's inbuilt processing steps. The open source code (dcraw) was used to obtain estimates for dark and saturation values of each image and convert the RAW image to a linearized, 16 bit, un-brightened, un-gamma corrected and un-demosaiced tiff output which could be read in MATLAB. An in-house MATLAB code which used inputs of dark and saturation values of the recorded image and the bayer pattern for the camera's CFA, was used to generate a linear, normalized and demosaiced image for analysis. The spectral response curves thus obtained were used to generate color-ratio lookup tables for black body and soot radiation. The black body lookup table was validated against black body calibration data from 750-1500 °C. Soot look up tables were validated against S-type thermocouple data for a candle flame. It was observed that the temperature of the gas surrounding the thermocouple was approximately around 1700-1800 K for pixels in the vicinity of the bead, while that measured by the thermocouple showed an average value of about 1650 K at steady state. Since the temperature correction due to radiative losses is expected to be of the order of 100 K, the data obtained using soot pyrometry was determined to be in excellent agreement with the expected values. Once calibrated, a consumer grade DSLR can prove to be cost effective yet useful tool for non-intrusive measurement of 2-D temperature data for sooty flames.

5 Acknowledgment

The authors would like to thank Dr. Parinda Vasa and Dr. B.P. Singh, from the department of Physics, IIT Bombay, for allowing us to utilize their facility in order to characterize the spectral response of our camera. The assistance provided by Ms. Geeta towards conducting the experiments is also appreciated.

References

[1] T. Panagiotou, Y. Levendis, M. Delichatsios, Combustion and Flame, 104 (1996) 272-287.

[2] P. Childs, J. Greenwood, C. Long, Review of scientific instruments, 71 (2000) 2959-2978.

[3] J. Zetterberg, Z. Li, M. Afzelius, M. Aldén, Applied spectroscopy, 62 (2008) 778-783.

[4] A.K. Yadav, A. Chowdhury, A. Srivastava, International Journal of Heat and Mass Transfer, 111 (2017) 871-883.

[5] H. Guo, J.A. Castillo, P.B. Sunderland, Applied optics, 52 (2013) 8040-8047.

[6] P.B. Kuhn, B. Ma, B.C. Connelly, M.D. Smooke, M.B. Long, Proceedings of the Combustion Institute, 33 (2011) 743-750.

[7] T. Fu, Z. Wang, X. Cheng, Journal of Heat Transfer, 132 (2010) 051602.

[8] C. Lou, H.-C. Zhou, P.-F. Yu, Z.-W. Jiang, Proceedings of the Combustion Institute, 31 (2007) 2771-2778.

[9] J.M. Densmore, M.M. Biss, K.L. McNesby, B.E. Homan, Applied Optics, 50 (2011) 2659-2665.

[10] R. Sumner, Processing RAW Images in MATLAB, Department of Electrical Engineering, University of California Santa Cruz, USA, 2014.

[11] H. Ma, R. Stevens, R. Stone, in, SAE Technical Paper, 2006.

11th Asia-Pacific Conference on Combustion,
The University of Sydney, NSW Australia
10th -14th December 2017.

Flow Reactor Study on Benzene Pyrolysis and Oxidation

K. Kashiwa[1], T. Kitahara[2], M. Arai[2], and Y. Kobayashi[2*]

[1]Graduate School of Advanced Science and Technology, Tokyo Denki University

[2]Graduate School of Engineering, Tokyo Denki University

Senjyu-Asahi-cho 5, Adachi-ku, Tokyo, 120-8551 Japan

Abstract

Nanometer size Particulate Matter (PM) formation from hydrocarbon combustion field was troublesome problem for environmental emission and human health. It was formed from fuel rich combustion field under a certain temperature condition. The mechanism of PM formation is fuel pyrolysis and its partial oxidation. In this study, pyrolysis and oxidation processes in a benzene mixture were investigated using a flow reactor system. As the result, it was conformed that benzene pyrolysis with zero O_2 condition formed SOF like PM, whereas dry soot PM was formed under high temperature and O_2 exist condition. SOF like PM under low temperature condition increased with an increase of O_2 concentration in the mixture. Much of dry soot PM was formed under high temperature condition but it was oxidized with an increase of O_2 concentration. Lowest PM formation was observed at 1273 K (medium temperature condition).

1 Introduction

PM contained in combustion exhaust and Polycyclic Aromatic Hydrocarbons (PAHs) which is PM precursors, have adversely effect on the human body. Then their reduction is strongly required. Soot is the main component of PM formed by incomplete combustion field, and temperature and oxygen concentration in the field are dominant factors of PM formation.

There are many researches on soot formation. For example, formation behavior of PAHs and soot in a laminar diffusion flame have been reported, including temperature effect [1] and oxygen concentration effect [2] in transition region from PAHs to soot. Pyrolysis and oxidation of hydrocarbon fuel were also investigated using a shock tube [3, 4], and relation between oxygen concentration and soot formation is reported there. However, there are many unclear points such as the temperature and oxygen concentration, relating formation mechanism and characteristics of PM formed under different conditions.

Control of temperature and oxygen concentration was important to investigate transition behavior from PAHs to soot, but it was difficult to control artificially flame temperature and oxygen concentration. Therefore, in this study, benzene pyrolysis and formation behavior of PM were investigated using a flow reactor. Obtained results reported here are fundamental relationship between PM formation and field conditions such as temperature, O_2 concentration and residence time in the heated flow reactor.

Yoshihiro Kobayashi, Fax: +81-3-5284-5466
E-mail address: ykoba@mail.dendai.ac.jp

2 Flow reactor and experimental method

2.1 Experimental setup

Experimental setup for benzene pyrolysis and oxidation is illustrated in Fig.1. It consists of a fuel supply system and a flow reactor with heating furnace. Liquid benzene was regulated by a tube pump, and was supplied to an evaporation tank. N_2 and air mixed with vaporized benzene in the tank. Here benzene flow rate was 0.36 g/min and total flow rate of the mixture was 5.4 L/min for the main experimental condition. It corresponded 1.9±0.2 vol% of benzene mixture. Oxygen concentration was varied by changing the supply ratio of N_2 and air.

A quartz tube of 26 mm dia. and 2420 mm length was used as a flow reactor. Heating zone located in the middle part of the reactor, and its length was 300 mm. Pyrolysis and oxidation reactions were progressed in this heating zone. Point A and B were the sampling points of the mixture before reaction and Point C and D were for the mixture after reaction.

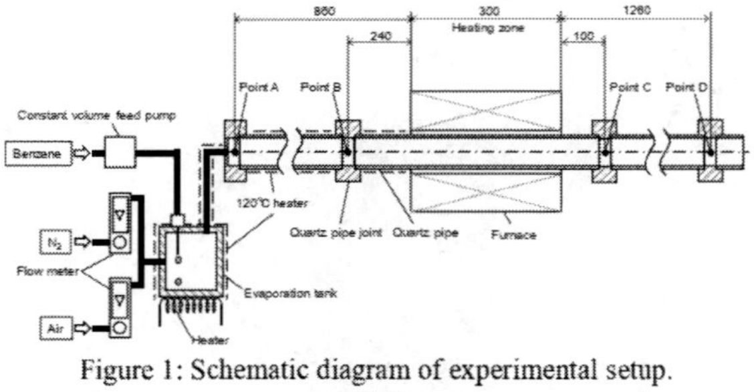

Figure 1: Schematic diagram of experimental setup.

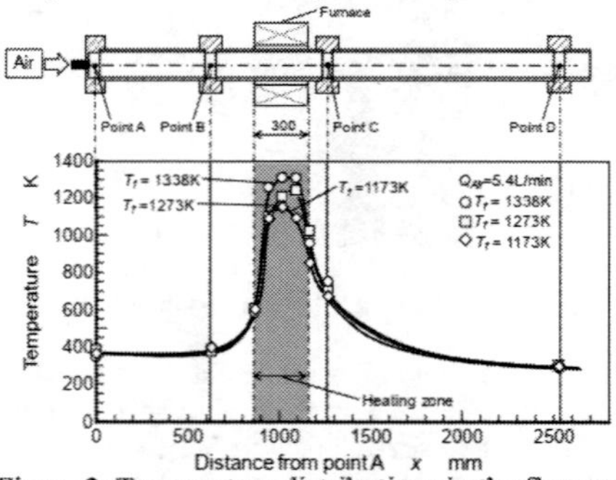

Figure 2: Temperature distributions in the flow reactor.

2.2 Temperature distribution along the flow reactor

Typical temperature distributions in the reactor are shown in Fig. 2. Temperature of the mixture was kept at around 400 K before the heating zone and heated up to regulated temperature, after then it was cooled down to room temperature. The regulated furnace temperatures (T_f) in the figure are 1173 K, 1273 K, and 1338 K. Since the maximum temperature of the mixture in the heating zone consisted with the regulated furnace temperature, hereafter, this furnace temperature was used as a representative temperature in the heating zone.

2.3 Residence time in the flow reactor

Since pyrolysis and oxidation reactions were progressed in the heating zone, it was considered that the residence time in the heating zone had substantial influence on these reactions. Therefore, the flow speed and the residence time in the heating zone were varied by inserting a quartz bar into the furnace. The conditions of flow rate, flow speed and residence time in the heating zone is listed in Table 1. As listed in the table, the residence time could be varied from 0.031 sec to 1.8 sec by changing the diameter of the quartz bar.

	Flow rate L/min (20°C, 1atm)	Furnace Temp. K	Flow velocity in furnace m/s	Residence time in furnace sec
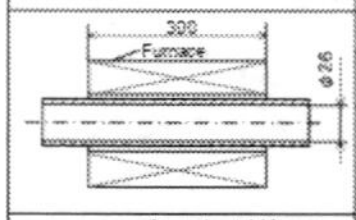	2.2	1173	0.27	1.1
	2.7	1173	0.34	0.88
	5.4	293 1173 1273 1338	0.17 0.68 0.74 0.78	1.8 0.44 0.41 0.39
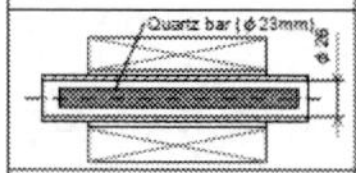	5.4	293 1273	0.78 3.4	0.38 0.089
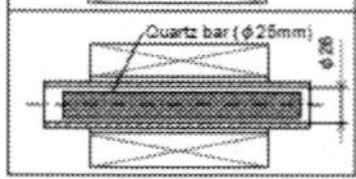	5.4	293 1273	2.3 9.8	0.13 0.031

Table 1: Residence time in the heating zone.

2.4 Gas composition and PM analysis

For gas composition analysis, sampling was performed at points A, B, C, and D of the reactor. Figure 3 shows the schematic diagram of the gas composition analysis. Sampling probe was inserted, and gas in the reactor was collected in the sampling bag by a suction pump. It was analyzed by Gas Chromatography (detector: TCD), O_2 analyzer (magnetic wind-type), and CO_2 analyzer (NDIR).

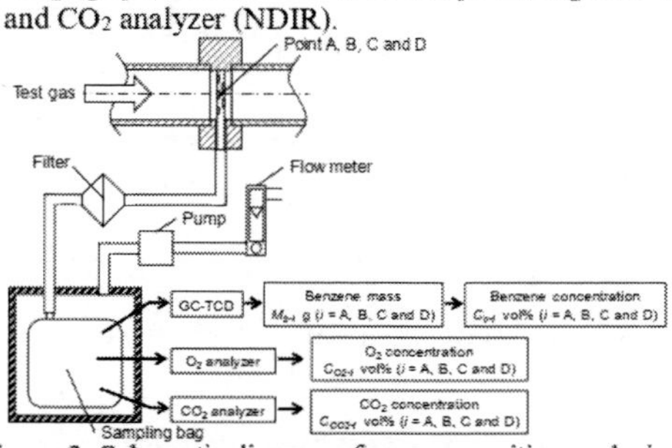

Figure 3: Schematic diagram of gas composition analysis.

The PM sampling probe was also inserted at point D, and PM was captured by a quartz fiber filter. PM sampling and mass analysis scheme are shown in Fig. 4. Owing to less amount of PM, sampling time was around one hour. First, sampled mass of PM was directly measured by an electronic microbalance (measurement accuracy is ±5%). Its mass was called "Wet PM". Since "Wet PM" contained moisture, it was removed by an oven at 120 °C for 150 min. Moisture removed PM was measured as the next step, and its mass was called "Dry 120 °C PM". After then, "Dry 120 °C PM" was dried up again at 240 °C for 60 min to remove low-boiling volatile component. Its mass was measured again, and it was called "Dry 240 °C PM". Further, the PM after dried up at 240 °C was analyzed for separation of Soluble Organic Fraction (SOF) and Dry Soot by using a combustion type PM analyzer (HORIBA MEXA-1370PM).

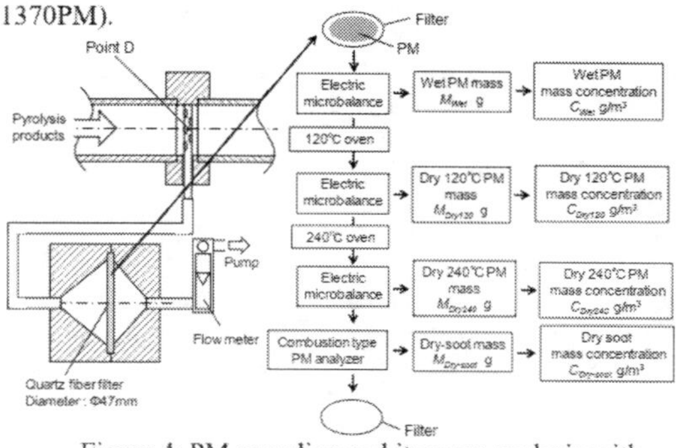

Figure 4: PM sampling and its mass analysis with drying treatment.

3 Pyrolysis and Oxidation of Benzene

3.1 Overall behavior in a flow reactor

Overall reaction behavior along the reactor at benzene concentration $C_{b\text{-}IN}$ = 1.9 vol%, an oxygen concentration $C_{O2\text{-}IN}$ = 2.3 vol%, and a furnace temperature T_f = 1273 K is shown in Fig. 5. The upper part of the figure shows concentration changes of benzene, oxygen and carbon dioxide, the middle part shows the formation of "Dry 120 °C PM", and the lower part shows the temperature distribution in the flow reactor.

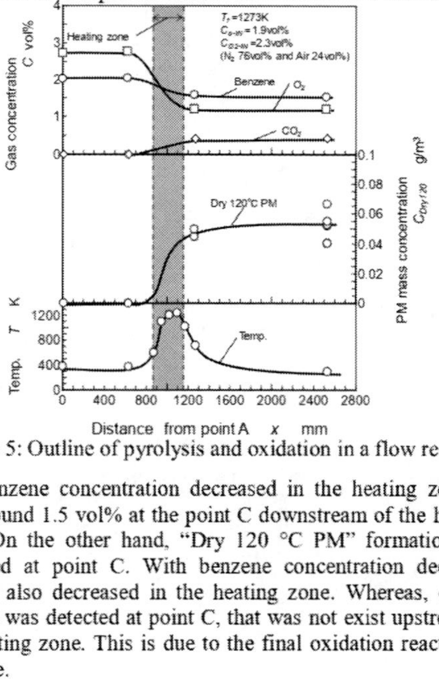

Figure 5: Outline of pyrolysis and oxidation in a flow reactor.

The benzene concentration decreased in the heating zone. It was around 1.5 vol% at the point C downstream of the heating zone. On the other hand, "Dry 120 °C PM" formation was observed at point C. With benzene concentration decrease, oxygen also decreased in the heating zone. Whereas, carbon dioxide was detected at point C, that was not exist upstream of the heating zone. This is due to the final oxidation reaction of benzene.

There were no differences between gas concentrations and PM at point C and D. It meant that all the reactions might be stopped by the cooling down effect in the process from point C to D, or all reactions might be ceased by achieving the final equilibrium state. Residence time in the heating zone was considered as the critical parameter of these two cases.

3.2 Residence time and reaction progress

Residence time effect on the final benzene concentration measured at point D are shown in Figs. 6 and 7. The furnace temperature were 1173 K and 1273 K, respectively. The final benzene concentration measured at point D ($C_{b\text{-}D}$) was normalized by the supplied benzene concentration ($C_{b\text{-}IN}$).

When the heating zone temperature was low as shown in Fig. 6, no concentration change was observed for the benzene mixture under zero O_2 concentration (benzene and N_2 mixture). It meant that no benzene pyrolysis was induced in 1173 K. Whereas the final concentration of benzene decreased under a certain concentration of O_2. Since some pyrolysis processes were needed before oxidation of benzene, it suggested that O_2 assist pyrolysis might be induced in the mixture of benzene with O_2 and N_2.

The final benzene concentration decreased with increasing the residence time until 0.8 sec under the conditions with O_2. When the residence time increased to 0.9 sec or more, the benzene concentration stopped decreasing. Therefore, when the residence time was shorter than 0.8 sec, the reaction might stop in the midway of the reaction by the cooling effect during the process from point C to D. In other words, it is considered that when the residence time was sufficiently long, it become the equilibrium state after pyrolysis and oxidation.

When the heating zone temperature was elevated to 1273 K, on the contrary, benzene pyrolysis under zero O_2 condition was clearly observed by the evidence of the final benzene concentration decrease in Fig. 7. It meant that benzene pyrolysis was induced independently with either exist or absence of oxygen. The final concentration of benzene reached to the constant levels when the residence time was over 0.2 sec. Further visible flame was observed at 10.1 vol% O_2 condition because of high oxygen condition near stoichiometry (ϕ=1.4, explained later). ϕ=1.4 for 10.1 vol% O_2 is the special case, and it suggests that strong oxidation becomes the dominant parameter of benzene reaction.

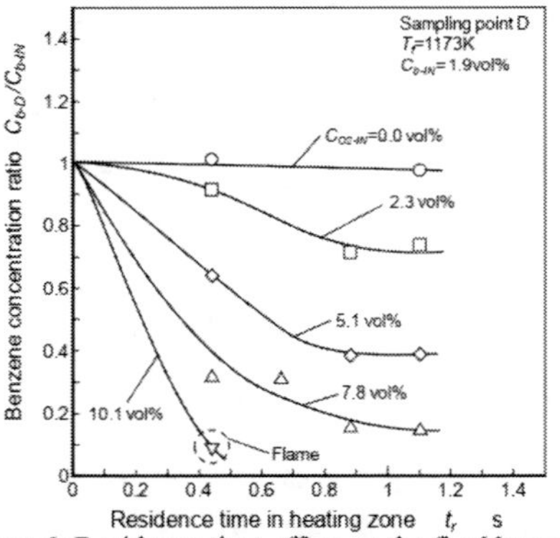

Figure 6: Residence time effect on the final benzene concentration (T_f = 1173 K).

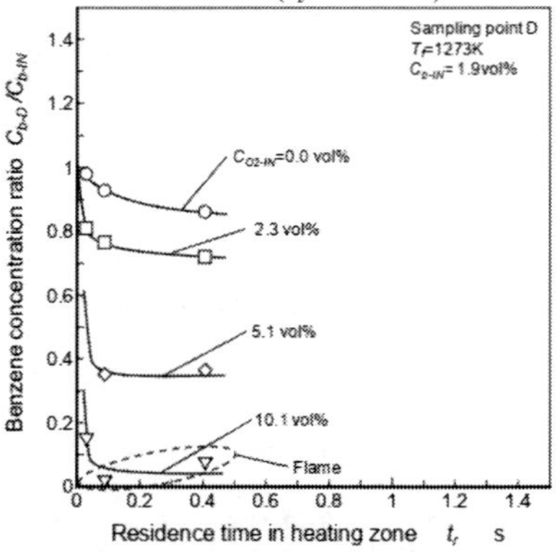

Figure 7: Residence time effect on the final benzene concentration (T_f = 1273 K).

When the residence time was longer than a certain residence time, the final concentration of benzene became constant level and pyrolysis and oxidation reactions might be ceased. Higher the temperature and higher the concentration of O_2, shorter the residence time for sufficient reaction was enough. From the comparison of the results shown in Figs. 6 and 7, the temperature effect on this critical residence time (reaction rate) was meaningful.

Final products of pyrolysis and oxidation with enough residence time seemed to be the new equilibrium state after reaction progressed under heating zone temperature. Figure 8 shows the residence time effect on the combustion products including PM. Heating zone temperature was 1173 K as same as Fig. 6. Gas concentration and PM mass measured at point D (C_{b-D}) were normalized by the supplied benzene concentration (C_{b-IN}).

The results of benzene and oxygen are shown in the upper row. The oxygen concentration decreased with an increase of the residence time until 0.8 sec and decreasing tendency was ceased when residence time was beyond 0.8 sec. Increasing tendency of CO_2 in lower part of the figure was also ceased there. Since few quantities of benzene and O_2 were remained there, reactants and products seemed to be an equilibrium state after reaction when the residence time was longer than 0.8 sec.

Mass of Dry 120 °C PM shown in the lower part increased with the residence time and its increasing tendency was also ceased by the enough residence time such as 0.8 sec. It was clear that PM of 0.01% level of the benzene was formed by benzene pyrolysis under zero O_2 condition

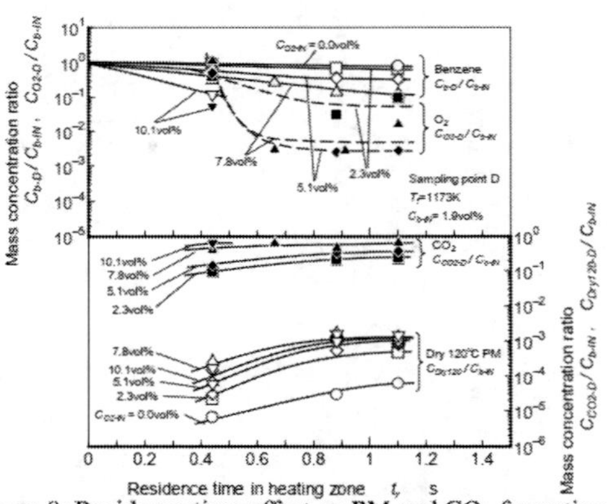

Figure 8: Residence time effect on PM and CO_2 formation (T_f = 1173 K).

In addition, it was observed that higher the oxygen concentration, the higher mass concentration of carbon dioxide and PM. This is probably because the oxygen in the mixture activates PM formation as same as pyrolysis and oxidation of benzene.

Fuel equivalence ratio ϕ to O_2 was changing with O_2 concentration. Benzene and N_2 mixture with zero O_2 meant ϕ= ∞, 2.1 vol% oxygen corresponded ϕ=6.1, 5.1 vol% was ϕ=2.8 and 7.8 vol% was ϕ=1.8. Then all mixture conditions in this report were fuel rich conditions. Since the fuel rich reaction was maintained here, fuel benzene remaining was quite reasonable. However, some amount of O_2 was also remained after pyrolysis and oxidation in the heating zone. It meant that new thermal equilibrium state after reaction was taken place under long residence time reaction.

3.3 Temperature effect on PM formation

In the equilibrium state after reaction, the PM formation was dominated by heating zone temperature and O_2 concentration. Therefore, the temperature effects on PM formation were summarized. Here PM was characterized as Dry 120 °C PM, Dry 240 °C PM and Dry Soot. These were corresponding to the PM sampled from point D with enough residence time. Dry 120 °C PM with remained benzene is shown in Fig. 9. Since the concentration of the remained benzene after equilibrium depended on the O_2 concentration but did not depend on the temperature, the equilibrium state after reaction was mainly dominated by O_2 concentration.

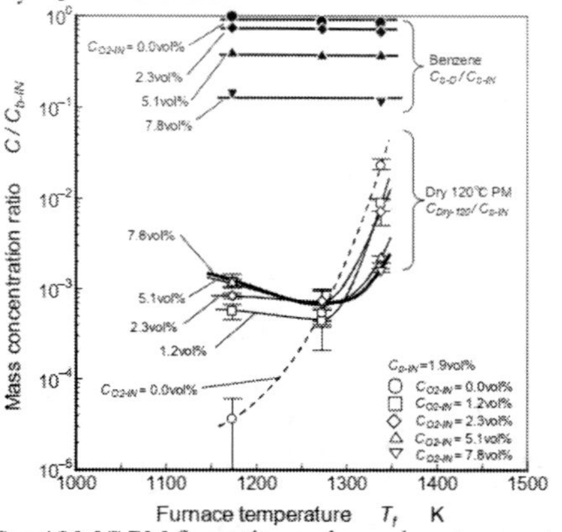

Figure 9: Dry 120 °C PM formation under various temperature and various O_2 concentration conditions.

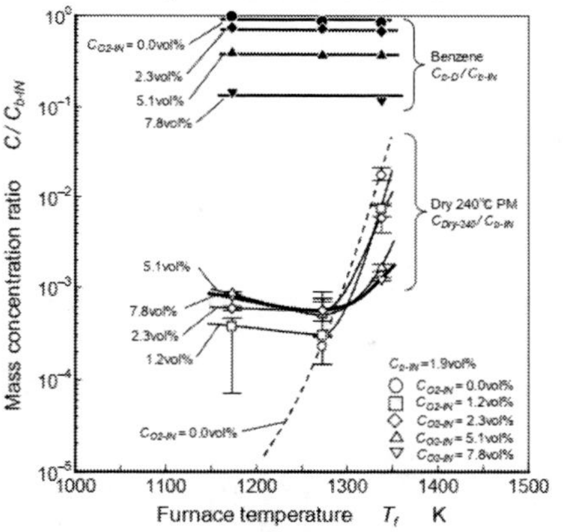

Figure 10: Dry 240 °C PM formation under various temperature and various O_2 concentration conditions.

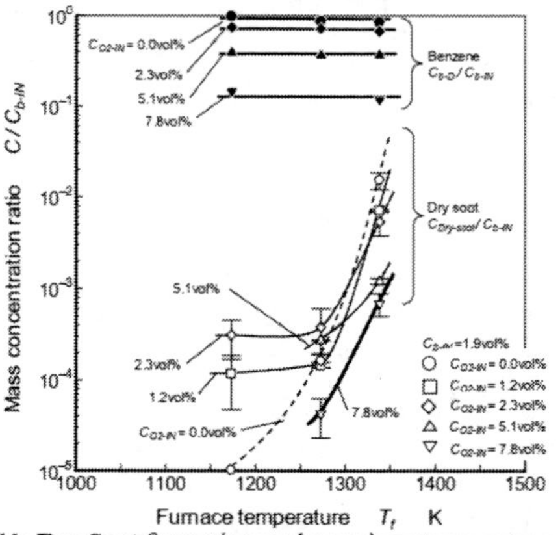

Figure 11: Dry Soot formation under various temperature and various O_2 concentration conditions.

As the result of Dry 120 °C PM, the PM mass concentration significantly increased from 1273 K to 1338 K. PM formation was promoted by high temperature. Also, since the PM mass concentration decreased with an increase of the oxygen concentration, it was considered that the oxidation reaction was active at 1338 K. At 1173 K, on the other hand, the PM mass concentration became higher as higher the oxygen concentration. It is considered that the oxidation reaction is weak at low temperature, and the thermal pyrolysis reaction becomes active.

The dashed line in the figure shows the pour pyrolysis with zero O_2 concentration. PM concentration at 1338 K was higher than those PMs with O_2 exist conditions. These results can lead following discussions. When the heating zone temperature is low, PM is hardly formed in zero O_2 condition but it is formed under O_2 exist conditions. On the other hand, when the heating temperature is higher than a certain temperature, PM formation under zero O_2 condition is activated strongly but PM oxidation is simultaneously taken place.

Dry 240 °C PM shown in Fig. 10 indicated almost same manner as Dry 120 °C PM. However, the absolute mass of PM was reduced by the dry up treatment between 120 °C to 240 °C. PM mass reduction ratio from 120 °C to 240 °C was obvious under low temperature conditions. It meant that PM formed in low temperature condition involved SOF of low boiling point.

Above tendency was observed more clearly in Dry Soot behavior shown in Fig. 11. One order less Dry Soot was detected under low temperature condition when it was compared with Dry 120 °C PM. It indicated that benzene pyrolysis under pyrolysis dominant temperature condition could hardy produce Dry Soot. On the other hand, as for PM produced under high temperature condition, most of PM was dry PM.

4 Conclusions

Benzene pyrolysis and formation behavior of PM were investigated using a flow reactor. The main results are summarized as follows:

1. Thermal equilibrium state after pyrolysis and oxidation was only achieved with long residence time reaction in the flow reactor.
2. SOF like PM was formed at low temperature condition even though the absence of oxygen. It increased with an addition of oxygen
3. Dry Soot PM was formed under high temperature conditions. It decreased with an increase of oxygen concentration.

References

[1] Y. Kobayashi, T. Furuhata, K. Amagai, M. Arai, Combust. Flame 154 (2008) 346-355.

[2] Y. Zama, K. Shimizu, M. Tsukui, T. Furuhata, M. Arai, Transact. Jpn. Soc. Mech. Eng. Ser. B 79 (2013) 1147-1154, No. 802.

[3] M. Frenklach, Physical Chemistry Chemical Physics, 2002, 20, 4, 2028-2037.

[4] M. Frenklach, M. K. Ramachandra and R. A. Matula, Proc. Combust. Inst., 1984, 20, 871-878.

11th Asia-Pacific Conference on Combustion,
The University of Sydney, NSW Australia
10th -14th December 2017.

Morphological Evolution of Flexible Polyurethane Foam during Pyrolysis and Effects on Smoldering Establishment and Propagation

Fei YOU[1*], Shiqiang HU[1,2]

[1]College of Safety Science and Engineering & Institute of Fire Science and Engineering, Nanjing Tech University

30 South Puzhu Rd., Nanjing, Jiangsu, 211800, China

[2]Siemens Numerical Control Ltd.

18 Siemens Rd., Nanjing, Jiangsu, 211100, China

Abstract

The pyrolysis properties and induced apparent and micro-morphologies of Flexible Polyurethane Foam (FPUF) were investigated by Thermogravimetry (TG), Digital Camera and Scanning Electron Microscope (SEM). Results show that the mass loss and pyrolysis stages lie mainly between 250 ℃ and 350 ℃ under a heating rate of 20 K/min and accounts for 83.0 % of total mass loss. Isocyanate and polyol compounds are produced, further degraded to carbonaceous residues and chars and oxidized in the process to contribute primary heat sources to maintain self-sustained smoldering propagation and carbon source to form charred skeletons. Meanwhile, distinctive apparent and micro-morphologies evolution shows up: the skeleton structure melts and converts from original open-cell to membrane-like bodies at 250 ℃, fiber-like edges at 300 ℃, needle-like fibers at 400 ℃, until loses its shape at 500 ℃ and turns into ash powders completely after 600 ℃. Such physical and chemical evolution of carbon-based skeletons promotes formation of balanced conditions among heat, fuel and oxygen. Smoldering has finally established at about 328 ℃. Increasing heating rates to 50 K/min accelerates pyrolysis rates and promote balanced conditions of smoldering propagation.

1 Introduction

Smoldering is a special flameless combustion process in which heterogeneous chemical reactions involved are relatively slow, released heat is relatively low, more toxic intermediate gaseous products are produced. Fuel, heat and oxygen are dynamically and slowly balanced and maintained in steady propagation process of smoldering. These leads to difficult detection and long-term latency. It is widely seen in porous micro-environments like those of forest fires, underground mine fires and housing fires with peat depositions, coal particles, polyurethane foam mattress as fuels. Once interactive balance is destroyed, transition from smoldering to flame is inevitably induced and exaggerates fire spreads. Hereby, it is important to figure out how such a balance is established and kept, i.e., formation mechanism of smoldering establishment and propagation. Also, it can be seen smoldering is inherently an interlocking chain-like process called "smoldering chain", say, includes initial incubation, establishment, sustaining, propagation, transition, flaming and ultimate fire formation, among which pyrolysis process is crucial for establishment of smoldering. Induced critical heat and fuel from pyrolysis and oxygen introduced externally meet in this dynamically balanced stage.

As mentioned above, rigid and porous carbonaceous or even char structures generated by pyrolysis has an important role in the smoldering establishment and propagation. It is beneficial for oxygen diffusion to the very frontal surface where heat derived from exothermic oxidation[1] of carbonaceous residues and condensed and gaseous fuels are provided. In this process, the micro-morphological evolution of pyrolyzed bulk material prior to establishment and propagation of smoldering is much important to explore facts inherently.

As a typical smoldering material, pyrolysis characteristics of FPUF have been much addressed. Lucas et al [2-5] found that heating rate, end temperature and atmosphere had significant effects on distributions of pyrolysis products. Rein et al[6] provided preliminary research regarding micro-morphology changes of FPUF during smoldering propagation. Branca et al[7] focused on rigid polyurethane foam and found that heat radiation flux affected pore sizes, altered proportions of three heat transmission means and corresponding smoldering propagation. However, the apparent and micro-morphology evolution of FPUF during the pyrolysis and its effects on establishment and propagation of smoldering are seldom concerned.

In this work, the pyrolysis characteristics of FPUF were studied by Thermogravimetry (TG) under three different heating rates of 10, 20, 50 K/min in a mixed atmosphere in which O2 and N2 accounts for 20 vol% and 80 vol%. Carbonaceous residues at end temperatures of 200, 250, 300, 400, 500, 600, 700 and 800 ℃in pyrolysis were characterized by Digital Camera and SEM. Morphological evolution and its affects on establishment and propagation of smoldering were further analyzed.

2 Materials and methods

Corresponding author. Fax: +86-25-8323-9949
E-mail address: yfei@njtech.edu.cn

2.1 Material and instrument

The experimental FPUF materials were purchased from Sinomax Polyurethanes (Shanghai) Co., Ltd. SDT (simultaneous DSC-TGA) 600 of TA Instrument (USA) and Scanning Electron Microscope (SEM) TM 300 of Hitachi were used for characterizations.

2.2 Methods

FPUF was cut into portions of about 10mg and squeezed into a sample crucible. The tested foams were heated to a series of end temperature (200, 250, 300, 400, 500, 600, 700 and 800 ℃) under three different heating rates (10, 20, 50 K/min) respectively with same inflow gas velocity of 100 ml/min and mixed atmosphere (N_2:O_2=8:2, vol%: vol%). Apparent and micro-morphology of virgin and pyrolyzed samples at 7 end temperatures were characterized by Digital Camera and SEM.

3 Results and Discussions

3.1 Pyrolysis process and apparent morphology evolution of FPUF

Generally, the pyrolysis process of FPUF consists of complex physical and chemical evolution under various conditions. It is accepted that C-O bond is broken and isocyanate and polyol are generated firstly, then decomposition and carbonization of polyol are introduced at higher temperature[8]. In general, multiple reactions like oxidation and decomposition are carried out simultaneously and usually parallel during pyrolysis. Figure 1 shows the TG and DTG profiles of FPUF. The mass loss rate and apparent morphology of FPUF corresponding to different end temperatures are shown in Table 1 and Figure 2.

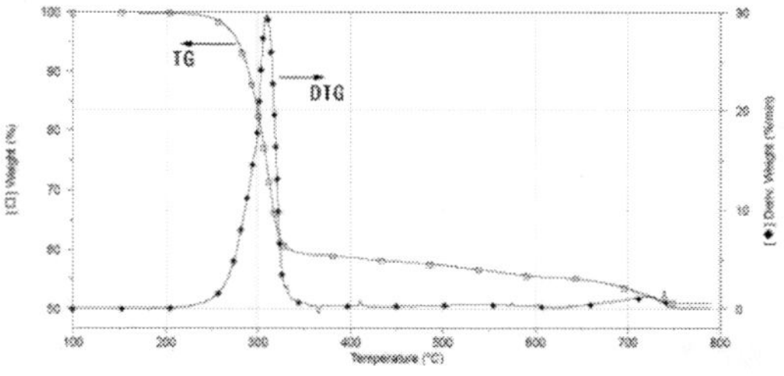

Figure 1: TG and DTG profiles of FPUF (20 K/min, N_2:O_2=8:2 in vol%)

The DTG curve (Figure 1) shows the temperature of main mass loss stage lie in between 250℃ and 350℃ under 20 K/min, accounting for 83% of total mass loss (table 1). The total mass loss is defined as that of FPUF at 800℃, i.e. 49.1%). In this temperature range, isocyanates and polyols are produced, carbonized and oxidized at higher temperature as mentioned above.

Table 1: The mass loss rate of FPUF at 5 different temperatures (20 K/min, N_2:O_2=8:2 in vol%)

Temperature (℃)	200	250	300	350	400
mass loss rate (%)	0.2	1.0	17.0	40.7	41.5
Percentage of total mass loss (49.1%)	0.4	0.2	35.0	83.0	85.0

The DTG curve in Figure 1 also shows that mass loss rate of FPUF at 200℃ is close to zero and consumes up only to 0.2% (Table 1), indicating that prevailing mass loss reactions have not yet occurred. This can also be proved by Figure 2b which shows both apparent morphology and color do not change much at 200℃. Some chemical bonds like biuret and allophanate break down above 110 ℃ whose decomposition reactions are endothermic. In this case, no released heat and combustible fuels can be accumulated and it is difficult for self ignition, say, smoldering can not be initiated.

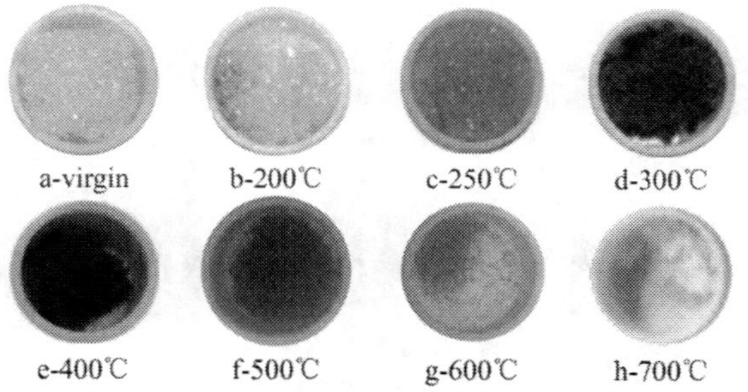

Figure 2: Apparent morphologies of FPUF heated to different end temperatures (20 K/min, N_2:O_2=8:2 in vol%)

However, there occur obvious changes at 250℃ (Figure 2c). The apparent color turns brown gradually mainly due to the volatilization of isocyanate. The DTG profile shows that the mass loss rates of FPUF increased rapidly after 250 ℃ (with a range of 3.1-29.6 %/min until 300 ℃) (Figure 1) although absolute mass loss around this point is only about 1.0 % (Table 1).These all indicate that rapid thermal degradation is triggered at 250℃and isocyanate (highly toxic chemical) and polyol compounds are produced. Part of isocyanate compounds further break down into diamino toluene and NO_x before escape from interior bulk material due to air flow inside [9,10]. This is why fumes from smoldering of FPUF would kill persons even without flaming fires. The heat released by the further oxidation of polyols at this temperature constitutes primary heat sources to maintain self-sustained smoldering propagation [1], and polyols are also the main carbon source to form charred skeletons.

The color of tested foam turns black at 300℃ from brown at 250℃ gradually (Figure 2d), meanwhile exposed material begins to shrink but still keep an original skeleton structure completely. The mass loss of FPUF is about 17.0 % at 300℃ that accounts for 35% of total mass loss (Table 1). Oxidation or (and) carbonization between 250 ℃ and 300 ℃ of volatile isocyanate and polyols cause the mass loss. It is exactly the carbonization of polyols that makes tested foam turns black.

The color of tested foam turns black completely at 400℃ (Figure 2e), and the volume of exposed material reduces significantly, but a relatively complete skeleton structure is still discernible. The mass loss of FPUF is about 41.5% at 400℃, accounting for 85% of total mass loss (Table 1). This is mainly due to decomposition of polyols and oxidation of carbonaceous residues between 300 ℃ and 400 ℃. It is found that transition from smoldering to flaming occurs not at the interface but inside the hot char region below smoldering front[11]. The char region corresponds to the partially pyrolyzed and oxidized residues after 300 ℃. There are two reasons to explain why the flaming easily occurs inside char region: First, thermal decomposition of carbonaceous materials generate combustible gaseous products (including intermediates and aerosols) at high temperature; Second, the temperature of hot char region is highest in smoldering structure due to aftermath exothermic oxidation of carbonaceous residues and chars.

The tested foam still keep a basic skeleton structure at 500℃ (Figure 2f) with lowered strength. It is easy to collapse in smoldering propagation. After 600℃, the skeleton structure is completely destroyed and turns into powders. The color of powder changes from black to white gradually (Figure 2g and Figure 2h). The remaining white powders are inorganic additives and ashes.

3.2 Micro-morphological Evolution of FPUF

Micro-morphologies of FPUF correspond at different end temperatures in pyrolysis are shown in Figure 3.

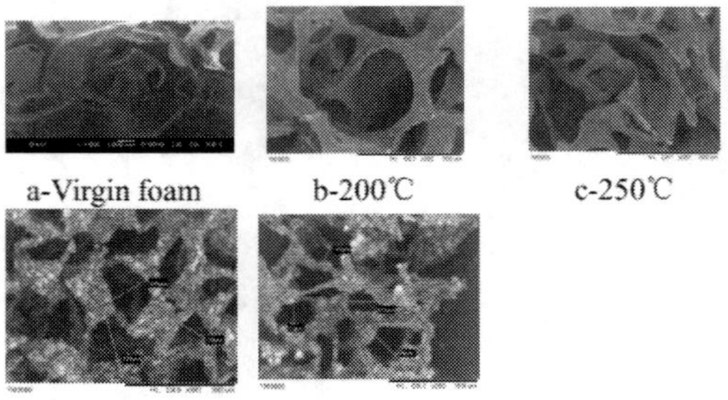

a-Virgin foam b-200℃ c-250℃

d-300℃ e-400℃

Figure 3: Micro-morphologies of FPUF corresponding to different end temperatures (20 K/min, N_2:O_2=8:2 in vol%)

The micro-morphology of virgin foam is shown in Figure 3a, which present a football-like porous skeleton structure. Most parts of bulk materials are open (see the pore marked 1 in Figure 3a). Only a few ones are closed-pores due to original processing technique (see the pore marked 2 in Figure 3a).

When tested foam is heated to 200 ℃, the micro-morphology basically has no change compared to that of virgin foam, a skeleton structure is kept completely and the surface is smooth (Figure 3b), indicating that the thermal degradation has not yet occurred before 200 ℃. This corresponds exactly to the apparent morphology of Figure 2b.

When tested foam is heated to 250 ℃, the skeleton structure is melt and membrane-like frames are resulted, pore size is getting smaller (Figure 3c). This phenomenon is in agreement with previous research[6]. The tested foam begins to decompose at 250 ℃ and release isocyanate that causes the material to become brown, but the mass loss of FPUF is only about 1% (Table 1). So there is no obvious evidence of decomposition from micro-morphology of FPUF as shown in Figure 3c, and tested foam remained elasticity as shown in Figure 2c.

When tested foam is heated to 300℃, membrane-like bodies of FPUF disappear and convert to fiber-like edges (Figure 3d). , Compared to virgin foam, rougher surfaces due to carbonization and migration of inorganic additive particles from interior bulk are formed, pore sizes become smaller, skeleton structure is getting thinner. Tested foam loses elasticity and becomes more fragile because of carbonization at 300 ℃. As also can be seen from Figure 2d, pores are getting smaller because of the melting and shrinkage of foam during heating process, the volume of exposed material decreases by more than 50 %. Thus permeability and carrying capacity of oxygen inflow and resulted effluent gas outflow in tested foam are all reduced with decreased pore sizes. However, since porosity of virgin foam is up to 95%, the decreasing pore sizes in early pyrolysis stage play minor role in smoldering propagation. Smoldering has finally established at about 328 ℃ for FPUF[12], then the heat required for melting and carbonization is basically equal to that of exothermic oxidation at this temperature.

When tested foam is heated to 400 ℃, skeleton structure of tested foam is further thinned, some even becomes needle-like fibers without a uniform structure any longer, and some bleaching points (Figure 3e) show up. These phenomena are mainly caused by further oxidation of carbonaceous residues and chars. The mass loss of FPUF is about 41.5 % at 400 ℃, accounting for 85.0 % of the total mass loss (Table 1) as mentioned in 3.1, which indicates that carbon oxidation is almost complete. Exothermic reaction due to such oxidation finishes at 405 ℃ under same heating rate and atmosphere conducted in previous study, which can also prove that oxidation of carbonaceous residues and chars can be finished basically at 400 ℃. Exothermic carbon-based oxidation mainly occurs between 300 ℃ and 400 ℃ under designed conditions that is consistent with that in 3.1. Such released heat is the primary energy to maintain a steady smoldering. The exposed material is further embrittled and overall bearing capacity is reduced dramatically once carbon-based skeleton structure has been oxidized to the largest extent. The smoldering structure is easily to collapse into ashes.

3.3 Influences of Heating Rate on Micro-morphology Evolution

The tested foam heated to around 300 ℃ were selected for further investigations. In order to make the results comparable, the samples were pyrolyzed at same mass loss rates, then the micro-morphologies were characterized by SEM. Maximum pyrolysis temperatures are set to be 300 ℃, 313 ℃ and 329 ℃ under 10 K/min, 20 K/min and 50 K/min, respectively. At these temperatures the mass loss ratios are all close to 71.0 % (Figure 4).

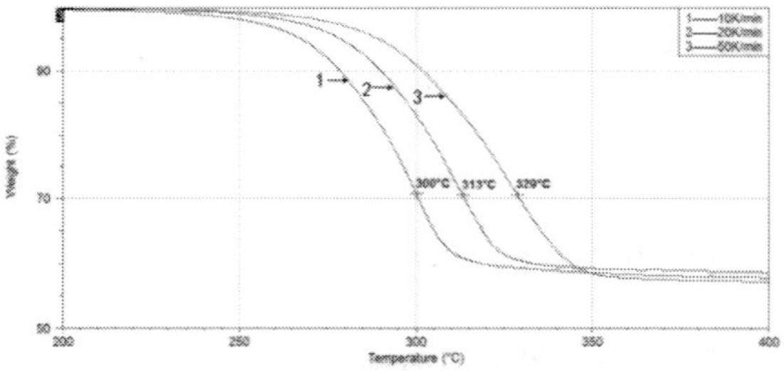

Figure 4: Selected temperatures under three heating rates

Figure 5 is the micro-morphology of FPUF at 300 ℃, 313 ℃ and 329 ℃ under three kinds of heating rate (10 K/min, 20 K/min and 50 K/min), respectively.

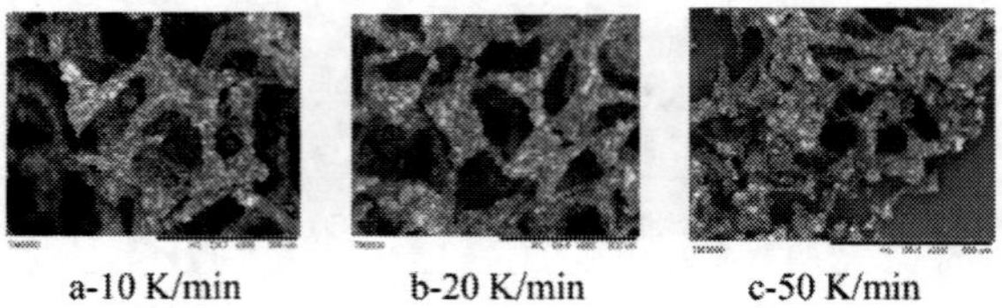

a-10 K/min (300℃) b-20 K/min (313℃) c-50 K/min (329℃)

Figure 5: The influence of heating rate on micro-morphology of FPUF

Figure 5a and Figure 5b clearly show that the tested foam has an uniform structure under heating rate of both 10 K/min and 20 K/min, and their pore sizes have no significant differences. Therefore, the heating rates increased from 10 K/min to 20 K/min show little effects on micro-morphologies of FPUF.

When the heating rate is increased to 50 K/min, the tested foam still keeps a skeleton structure, but no more uniform structure exists. The foam melts and aggregates into blocks at connection points (Figure 5c). This is mainly due to the rapidly raised temperature and strong gradient occurs. This coincides with the phenomenon of burning behavior of FPUF in a fire, say, melting, collapsing, dropping, bubbling and burning, when the heating rate is about 200 K/min to 300 K/min[9]. Heating rate in fire is several times higher than that in this work, so the tested foam can still keep complete skeleton structure and only partial is melting and shrinking into blocks.

From another perspective, increasing heating rates could accelerate pyrolysis rates and promote balanced conditions of smoldering propagation. Therefore, there exist an appropriate heating rate in the establishment and propagation stage of smoldering, i.e. the temperature of heat source to ignite smoldering or the temperature of smoldering should keep in a suitable range. When heating rate or temperature is too high, smoldering material can not maintain balance conditions among heat, fuel and oxygen and then introduce a transition to flaming; When heating rate or temperature is too low, balance conditions can not be guaranteed at a lower heat value due to more heat losses, or less oxygen and fuels, thus smoldering can not be established[13].

4. Conclusions

(1) The temperature of main mass loss stage is between 250℃ and 350℃ under the heating rate of 20 K/min and accounts for 83% of total mass loss. In this temperature range, FPUF breaks down into isocyanate and polyols firstly and then carbonization and oxidation of polyols occur at higher temperatures.

(2) For both apparent morphologies and micro-morphologies, FPUF shows no obvious physical and chemical changes before 200 ℃; when heated to 250 ℃, the skeleton structure melts to membrane-like bodies and results in smaller pore sizes and lower permeability for both oxygen and effluent gases; tested foam presents apparent charring features at 300 ℃, resulting in smaller pore sizes, thinner and more fragile skeleton structures, reduced volume accommodation capability, bearing capacity and permeability, which is further intensified at 400 ℃; exothermic oxidation of carbonaceous residues and chars mainly occurs between 300 ℃ and 400 ℃. Skeleton structure loses its shape at 500 ℃ and turns into powders completely after 600 ℃.

(3) The heating rates increased from 10 K/min to 20 K/min show little effects on micro-morphologies of FPUF. The tested foam still keeps a skeleton structure, but no more uniform structure exists. The foam melts and aggregates into blocks at connection points. Also, increasing heating rates could accelerate pyrolysis rates and promote balanced conditions of smoldering propagation.

5 Acknowledgment

The study is financially supported by the National Natural Science Foundation of China (Grant No. 50906039 and 51376089).

Reference

[1] C.Y.H. Chao, J.H. Wang, Combust. Flame 127 (4) (2001) 2252-2264.

[2] L. B. Valencia, T. Rogaume, E. Guillaume, Fire Safety J. 44 (7) (2009) 933-940.

[3] Y.H. Zhang, Z.B. Xia, H. Huang, J. Analytical and Applied Pyrolysis 84 (1) (2009) 89-94.

[4] D. Włodarczak, J. Applied Poly. Sci. 36 (2) (1988) 377-386.

[5] L. Xue, Y. Yang, Q.Y. Xie, J. Applied Fire Sci. 17 (2) (2007) 167-175.

[6] G. Rein, C. Lautenberger, A.C. Fernandez-Pello, Combust. Flame 146 (2006) 95-108.

[7] C. Branca, C.D. Blasi, A. Casu, V. Morone, C. Costa, Thermochimica Acta 399 (1-2) (2003) 127-137.

[8] S.Q. Hu, F. You, Procedia Engineering 52 (2013) 145-151.

[9] R.H. Krämer, M. Zammarano, G.T. Linteris, Poly. Degrad. Sta. 95 (6) (2010) 1115-1122.

[10] M. Ravey, E. M. Pearce, J. Applied Polymer Science 63 (1) (1997) 47-74.

[11] S.D.Y. Tse, *An Experimental Investigation of Two-Dimensional Smoldering and the Transition to Flaming in Flexible Polyurethane Foam*, PhD thesis, University of California, Berkeley, U.S.A., 1996.

[12]Y. Lei, D. Liang, S.J. Shan, Fire Sci. Tech. 30 (10) (2011) 881-885.

[13]C. Lu, L. Peng, J.J. Zhou, J. Combust. Sci. Tech. 11 (6) (2005) 487-492.

11th Asia-Pacific Conference on Combustion,
The University of Sydney, NSW Australia
10th -14th December 2017.

Genesis and Evolution of Premixed Flames in Turbulence

Himanshu L. Dave[1], Abinesh M.[1], Swetaprovo Chaudhuri[1]

[1]Department of Aerospace Engineering, Indian Institute of Science,
Bangalore, Karnataka, 560012, India

Abstract

In this work, we address: from where and how do the complex topology and physico-chemical state of a turbulent premixed flame generate and evolve in time. Direct Numerical Simulation (DNS) of H_2-air turbulent premixed flames is combined with a newly developed computational methodology called the Backward Flame Particle Tracking (BFPT). Uniformly distributed flame particles that entirely span isotherms at time t_f are tracked backward to an earlier time t_i ($t_i < t_f$). On backtracking, these flame particles form multiple clusters in the leading locations of the corresponding isotherms. Since Zeldovich, such leading locations or leading points have remained an enigmatic concept in combustion literature inducing strong hypotheses without concrete proofs on their role. The direction of smallest principal curvature at these points is preferentially aligned with vorticity and the most extensive direction of the left Cauchy-Green strain rate tensor, causing the leading regions to stretch and disperse generating the complete surface. The local flame displacement speed of the leading points is also shown to control the global turbulent flame speed S_T, albeit with a finite time-lag. Finally, the flame particles are found to follow Batchelor's pair dispersion law which yields a ratio of generation to annihilation timescales for turbulent premixed flames.

1 Introduction

Turbulence-flame interactions are ubiquitous in nature and engineering devices. Owing to the evolving topology and state, turbulent premixed flames propagate faster than the corresponding laminar flames. Understanding these are critical for developing efficient land and aero-propulsion engines, preventing industrial explosions, and deciphering Supernovae Ia [1]. A premixed flame could be considered as an ensemble of propagating isoscalar surfaces. In an Eulerian investigation, flame-flow properties are extracted on such surfaces, usually described by a progress variable, and averaged over the complete surface to investigate flame-turbulence interactions. A Lagrangian approach [2] to investigate local dynamics of non-reacting turbulence is well known to be insightful. For turbulent combustion, a Lagrangian method could provide an instantaneous description of the flame surface along with the time history of specific portions of a flame surface. In Flame Particle Tracking (FPT), spatio-temporally evolving points (Flame Particles) defined on reactive isoscalar surfaces which propagate relative to the local fluid velocity are studied [3]. It has been developed as a computational tool based on the rigorous foundations developed by Pope [4], and offers a new paradigm for investigating turbulence-flame interactions. A flame particle is uniquely identified by $\boldsymbol{X}^F(\psi_o)$. $\boldsymbol{X}$ is the initial position, ψ_o is the isovalue of the scalar ψ, and superscript F identifies the corresponding parameter at a flame particle. The evolution of the position of a flame particle is governed by [3,4]:

$$\frac{\partial \boldsymbol{x}^F(\boldsymbol{X}^F(\psi_o),t)}{\partial t} = \boldsymbol{v}^F(\boldsymbol{X}^F(\psi_0),t) = \boldsymbol{u}^F + S_d^F\boldsymbol{n}^F \qquad (1)$$

Here, $\boldsymbol{x}^F$ is the current position; $\boldsymbol{u}^F$ is the local fluid velocity; S_d^F is the local flame displacement speed; $\boldsymbol{n}^F$ is the local surface normal pointing towards unburnt reactants; and $\boldsymbol{v}^F$ is the resultant velocity in a fixed frame of reference.

A premixed flame interacts with turbulence over multitude of length and time scales causing certain regions of the flame to annihilate or extinguish, resulting in the reduction of burning area [5]. But, for a statistically steady configuration, the flame must be continuously generated to balance the annihilation processes. However, the locations on the flame surfaces that would generate the new surfaces are not known *a priori*. Forward tracking of an entire flame surface would mostly lead to its annihilation at ellipsoidal flame islands [6]. To resolve this problem, the backward tracking algorithm for non-reacting fluid particles [7] is modified and complemented with the FPT algorithm. This enables tracking the flame particles backward in time from t_f to t_i, addressing the following specific questions concerning turbulent premixed flames.

1. Where do the fully developed isoscalar surfaces evolve from? What are the special flame-flow properties of flame particles that characterize these generation locations?
2. What governs the expansion of these generation locations towards a fully developed state? What could be a suitable mathematical description?
3. Could the global turbulent flame speed S_T, at time t_f, be obtained as a function of the local property S_d at an earlier time?
4. Is there any law that governs the dispersion of these flame particles? If so, could that law be used to find out the ratio of generation to annihilation timescales?

2 DNS Database

A DNS of H_2-air premixed flame at an equivalence ratio of $\phi = 0.81$ in an inflow-outflow configuration was performed using the open-source Pencil code [8]. Detailed kinetics of 9 species and 19 reactions was used [9]. The computations are 6th order accurate in space and 3rd order accurate in time. A cuboid of size 1.918cm×0.48cm×0.48cm was initialized with a planar laminar flame. Along X, Y, and Z direction, $960 \times 240 \times 240$ grid points were used. Navier-Stokes Characteristic Boundary Condition (NSCBC) along X direction and periodic boundary condition along Y and Z directions were specified. Pre-computed isotropic turbulence at 310K, superimposed with a mean flow, was issued from the inlet YZ plane of the cuboid. $\Delta x/\eta \approx 1.25$. Δx is the grid spacing and $\eta = (\nu^3/\langle\epsilon\rangle)^{1/4}$ is

Corresponding author.
E-mail address: schaudhuri@aero.iisc.ernet.in

the Kolmogorov length scale. ν and $\langle\epsilon\rangle$ are the kinematic viscosity and mean dissipation rate in the unburnt reactants, respectively. $\delta_L/\Delta x \approx 18$. δ_L is the thermal thickness of an unperturbed planar laminar flame. These parameters indicate sufficient spatial resolution of the turbulence and the premixed flame, respectively. The Taylor scale Reynolds number, $Re_\lambda = 85$. After a fully developed turbulent flame was realized, the DNS snapshots for the interval $t_f - t_i = 250\mu s$, were saved at time intervals $\Delta t = 1\mu s$. Kolmogorov time scale $\tau_\eta = 12\mu s$. The saved DNS data were processed using the BFPT algorithm.

3 Backward Flame Particle Tracking

This section presents the details of the BFPT algorithm. While tracking the flame particles backward in time (1) cannot be solved by usual numerical integration techniques because of the implicit nature of the equation [7]. This difficulty is resolved in the BFPT algorithm, which is implemented in two stages: estimation stage and the correlation stage. In the estimation stage, a RK-2 type predictor-corrector integration step advects the flame particle from the current position $\boldsymbol{x}^F(t)$ to $\boldsymbol{x}^{F^*}$ at $t - \Delta t$. The particle tracking is done using the FPT algorithm detailed in [3]. Since, the Eulerian fields, from which interpolations to particle positions are made, are read in reverse, $\boldsymbol{x}^{F^*}$ requires correction which is done in the correlation stage. The correlation stage introduces a set of dummy points around $\boldsymbol{x}^{F^*}$ followed by interpolation of $\boldsymbol{u}$, $\boldsymbol{n}$, and S_d at them. From these values the velocity $\boldsymbol{v}$ is computed and stored in an array A. This array therefore contains the "available" velocities that would advect the dummy points forward from time $t - \Delta t$ to t. A second array B, is formed by dividing the difference of the position vectors of the dummy points and $\boldsymbol{x}^F(t)$ by Δt. Thus, B contains the velocity vectors "required" by dummy points to advect to $\boldsymbol{x}^F(t)$. Choice of the best approximation among dummy points is obtained by seeking the highest correlation D_i between the two arrays A and B. The correlation coefficient $D_i = \boldsymbol{A}_i.\boldsymbol{B}_i/\max\{|\boldsymbol{A}_i||\boldsymbol{B}_i|\}^2$. The subscript i is an index that runs from 1 to total number of dummy points. $D_i \in [-1,1]$. $D_i = 1$ represents perfect correlation. This condition is possible only when the ith dummy point is actual flame particle position at time $t - \Delta t$, within interpolation and temporal errors. Due to finite number of dummy points D_i may not be exactly 1. In such cases, the dummy point which gives the maximum correlation among all the dummy points is selected. To validate the BFPT algorithm, backtracked flame particles at t_i were tracked forward in time using FPT. The average difference between the backward and forward trajectories was found to be much less than δ_L.

4 Results and Discussions

4.1 Flame Particle Clusters and their Flow-Flame Properties

Using BFPT, 2000 flame particles uniformly distributed on each of the wrinkled isotherms (T_o= 350K, 500K, 800K, 1100K) at time t_f were tracked back to t_i. Figure 1 shows the states of the flame particles on T_o= 350K and 1100K surfaces at time t_i, intermediate reference time t_c, and final time t_f ($t_i < t_c < t_f$). Results for other isotherms are similar. This clearly illustrate that the complete flame surface at t_f evolves from the multiple sets of flame particles clustered at t_i, and that these multiple clusters constitute the leading locations of the convoluted surface. Portions of a surface which are not covered by flame particles at t_i annihilate before t_f. Therefore, the DNS-BFPT methodology enables us to identify the locations on a surface at a time that generate new surface elements at later times. The isotherm corresponding to T_o= 350K is referred to as the *leading surface*; the clustered flame particles on it are the *leading flame particles*.

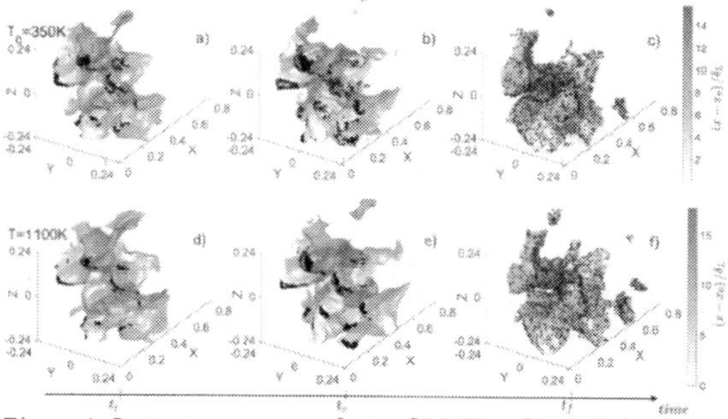

Figure 1: Instantaneous snapshots of 350K and 1100K isotherms with flame particles at t_i, t_c, and t_f. Axes unit is cm. Colorbar represent the normalized axial distance.

The leading flame particles defined here have properties similar to the leading points defined in [10,11]. Therefore, in this work, the terms leading flame particles and leading points will be used interchangeably. Figure 2 clearly illustrate that at t_i the clustered flame particles are spatially leading towards reactants, have low fluid speeds, and are located predominantly in the positively curved regions of the leading surface. Positively (negatively) curved surfaces are convex (concave) toward reactants. As time progresses, the leading points disperse to span the entire leading surface. Consequently, the probability density function (pdf) of the properties at leading flame particles approach the pdf of the leading surface at t_f. To our knowledge, these are the first observations of multiple clusters of leading points generating the complete surface at a later time, in partial consonance with Zeldovich's conjecture of a singular "pilot point in a non-stationary flame" [10]. Recent observations from DNS suggest that leading points contribute to the turbulent flame speed of turbulent premixed flames [11]. The pdfs of the constituents of stretch rate, $d[ln(\delta A)]/dt = K = a_T + 2S_d\kappa_m$, for the leading surface (without superscript) and leading points (with superscript) are also shown in Fig.2. δA represents an infinitesimal area element. a_T is the tangential strain rate while $2S_d\kappa_m$ is stretch rate due to propagation of curved surfaces. Initially, the mean curvature at leading points, κ_m^F, is predominantly positive and along with S_d^F contributes towards surface generation. At all times from t_i to t_f, the contribution of a_T^F to K^F is largely positive and hence helps in generating new surface elements. In fact, the pdf of a_T^F coarsely approximates the pdf of a_T, except around t_c, where a_T^F attains larger values to shift its pdf to the right of the pdf of a_T. This suggests a possibility of maximum in the history of stretch rate at the leading points.

Figure 3 shows the variation of $\langle S_d^F\rangle$ vs. $\langle K^F\rangle$ among the different isotherms. It is found that the ensemble-averaged stretch rate $\langle K^F\rangle$ for leading points and flame particles on the isotherms of 350K, 500K, 800K, and 1100K attain corresponding maximum values around t_c followed by a near-linear increase in the ensemble-averaged flame displacement speed $\langle S_d^F\rangle$. These maximum mean stretch rates on comparison with the extinction strain rate of a planar, strained, premixed, H_2-air laminar flame with $\phi = 0.81$, in a counter-flow configuration show that leading points are not critically stretched, on average. Inset of Fig.3 shows that $\langle S_d^F\rangle$ remains a linear function of temperature, at all times t_i, t_c and t_f. Therefore, propagation characteristics of these parts of the flame can be unambiguously described in terms of flame particles on an arbitrary reference surface.

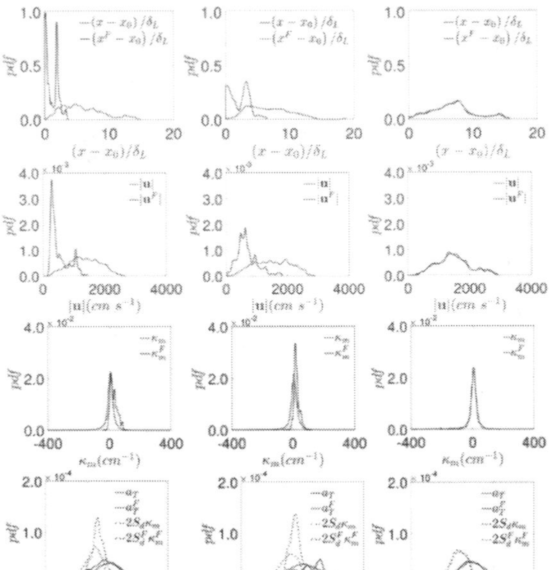

Figure 2: First row: pdfs of 'x' component of position vector; x_0 is the 'x' component of most leading point of the surface. Second row: pdfs of fluid speed. Third row: pdfs of mean curvature. Fourth row: pdfs of constituents of stretch rate. Blue lines correspond to complete leading surface; black lines correspond to leading points.

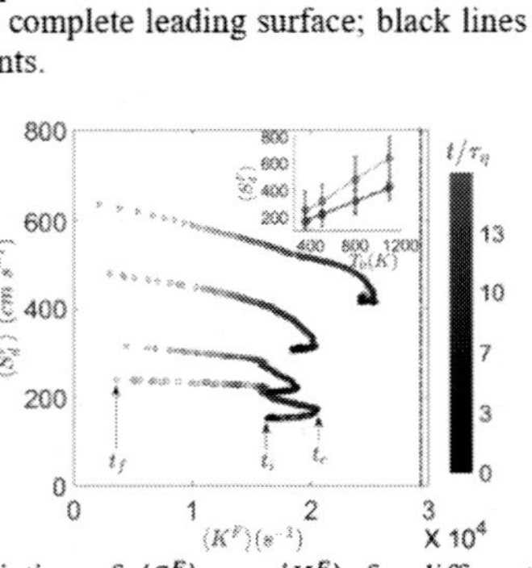

Figure 3: Variation of $\langle S_d^F \rangle$ vs. $\langle K^F \rangle$ for different isotherms. Colorbar represents time normalized by τ_η. The dash-dotted line corresponds to the extinction strain rate of a planar, strained, H_2-air laminar flame. Inset shows linear variation of $\langle S_d^F \rangle$ vs. T_0 of the isotherms at t_i, t_c and t_f. Error bar in the inset represents one standard deviation. o-350K; □-500K; ◊-800K; ▷-1100K

4.2 Finite Strain Study on Leading Points

Surface generation from the leading points is a finite time process. Here we use FST to quantify the deformation of leading points from their reference configuration at t_i to the deformed configuration at t_f. To compute $F_{ij}(t) = \partial x_i^F / \partial X_j^F$ - the deformation gradient tensor, the flame velocity gradients $A_{ij}(t) = \partial v_i^F / \partial x_j^F$ are interpolated at the leading points at each time instant. On integrating $\partial F_{ij}/\partial t = A_{ik} F_{kj}$, subjected to the initial condition $F_{ij}(0) = \delta_{ij}$, $F_{ij}(t)$ is obtained. δ_{ij} is the Kronecker delta function. The A_{ij} tensor defined here captures the effect of straining due to the variations in fluid velocity $\boldsymbol{u}^F$ and those due to S_d^F and κ^F as well, hence is different from the A_{ij} defined in [12]. Using $\boldsymbol{F}$, the left Cauchy-Green deformation tensor $\boldsymbol{C} = \boldsymbol{F}.\boldsymbol{F}^T$ is constructed. The statistics of the alignment of the unit eigenvectors $\hat{\boldsymbol{l}}_j$ of $\boldsymbol{C}$ in a co-ordinate system $(\hat{\boldsymbol{e}}_i)$, local to the leading points are computed to understand the surface generation. Here, $(\hat{\boldsymbol{e}}_3)$ is the local normal (identical to $\boldsymbol{n}$ defined in (1)). $\hat{\boldsymbol{e}}_1$ and $\hat{\boldsymbol{e}}_2$ are the directions of maximal and minimal principal curvatures ($|\kappa_1| > |\kappa_2|$), respectively. The eigenvalues of $\boldsymbol{C}$ are sorted as $\Lambda_1 > \Lambda_2 > \Lambda_3$. Thus, $\hat{\boldsymbol{l}}_1$ is the direction of maximum extension. The statistics of Λ_i (not shown) indicate that stretching dominates along $\hat{\boldsymbol{l}}_1$ and $\hat{\boldsymbol{l}}_2$, while compression dominates along $\hat{\boldsymbol{l}}_3$. Figure 4a show the pdfs of the alignment of $\hat{\boldsymbol{\omega}}$ with $\hat{\boldsymbol{e}}_2$ and $\hat{\boldsymbol{l}}_1$ at time $t_i + \Delta t$, t_c, and t_f. It shows that the vorticity is preferentially aligned with the least curved principal direction as well as with the most extensive principal strain. Figure 4b show the pdfs of alignment of $\hat{\boldsymbol{l}}_1$ with $\hat{\boldsymbol{e}}_1$ and $\hat{\boldsymbol{e}}_2$. The preferential alignment of $\hat{\boldsymbol{l}}_1$ with $\hat{\boldsymbol{e}}_2$ shows that, overall, the leading points are preferentially strained along the direction of the least curvature. However, the shape of the leading points initially is close to cylindrical and the pdfs of $2S_d\kappa_m$ from Fig.2 show that positively curved surfaces undergo stretch due to propagation as well. Thus, the leading points undergo stretching due to two different, but coupled, mechanisms acting in two different directions.

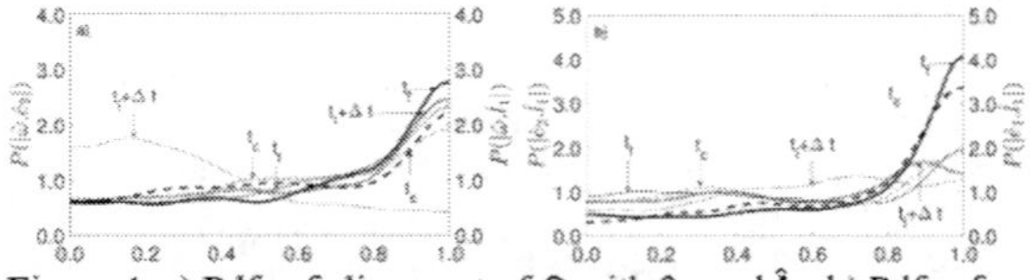

Figure 4: a) Pdfs of alignment of $\hat{\boldsymbol{\omega}}$ with $\hat{\boldsymbol{e}}_2$ and $\hat{\boldsymbol{l}}_1$; b) Pdfs of alignment of $\hat{\boldsymbol{l}}_1$ with $\hat{\boldsymbol{e}}_2$ and $\hat{\boldsymbol{e}}_1$. Solid lines: t_f, Dashed Lines: t_c, and dotted lines: $t_i + \Delta t$. Black lines: left Y-axis and blue lines: right Y-axis

Along the most curved direction the stretching is primarily due to propagation. But this stretching effect is less pronounced compared to the stretch along the least curved direction, which is accounted in a_T. The source of this stretching can be imagined to be the action of vortices due to their alignment with the surface. Since, the leading surface is very close to the unburnt reactants it can be expected that the turbulence is not strongly altered due to combustion and that the contribution to a_T is primarily due to the turbulence in unburnt region.

4.3 S_T as a Function of S_d of Leading Points

The Lagrangian framework is used to connect the current global state of the flame to a previous reference state of the leading points. The mass flow rate at t_f into the surface of area $a(t_f)$ is given by $\dot{m}_{in}(t_f) = -\int_{a(t_f)} \rho \boldsymbol{u}_r^F.\boldsymbol{n}^F da$ where ρ is the density and $\boldsymbol{u}_r^F = \boldsymbol{u}^F - \boldsymbol{v}^F = -S_d^F \boldsymbol{n}^F$ by using (1). Therefore,

$$\dot{m}_{in}(t_f) = \rho S_T(t_f) A_L = \int_{a(t_f)} \rho S_d^F(t_f) da \approx \rho \langle S_d^F(t_f) \rangle \int_{a(t_f)} \boldsymbol{n}^F.\boldsymbol{n}^F da$$

Using $\langle S_d^F(t_f) \rangle \approx \langle S_d^F(t_c) \rangle + l_{M,t} \langle K_c^F \rangle$ from Fig.3 and Nanson's theorem [13] $\boldsymbol{n}^F da = J\boldsymbol{F}^{-T}\boldsymbol{N}^F dA$, we get:

$$S_T(t_f) = A_L^{-1}[\langle S_d^F(t_c) \rangle + l_{M,t} \langle K_c^F \rangle] \int_{A(t_c)} \boldsymbol{n}^F.(J\boldsymbol{F}^{-T}\boldsymbol{N}^F) dA \qquad (3)$$

In (3), A_L is the cross-sectional area of the YZ plane. da, $\boldsymbol{n}$ are the area element and unit normal, respectively, in the deformed configuration at t_f. $dA, \boldsymbol{N}$ are the area element and unit normal, respectively, in the reference configuration at t_c. $A(t_c)$ corresponds to the area of only the leading regions at time t_c and not the complete surface area. $l_{M,t}$ is turbulent Markstein length, J is determinant of $\boldsymbol{F}$. (3) therefore shows that $S_T(t_f)$ is related to S_d at the leading points from an earlier time t_c, as empirically observed in [11].

4.4 Dispersion of Flame Particles and Ratio of Timescales

The final question on the time-scales for generation and destruction of the flame surfaces could be addressed from the dispersion characteristics of two distinct sets of flame particles. Set-G (Set-Generatrix) contains flame particles clustered at time t_i on the 350K, 500K, 800K, and 1100K isotherms, obtained from BFPT. Set-D (Set-Destructrix) contains flame particles uniformly distributed on the isotherms at t_i and tracked forward using the FPT algorithm [3], which cluster in the negatively curved regions before annihilation. Only intra-cluster pairs with separation distance equal and above 50η are considered for dispersion statistics. $\langle\epsilon\rangle$ and τ_η are conditioned on the respective isotherms as $\langle\epsilon_T\rangle$ and $\tau_{\eta,T}$, respectively. As shown in Fig.5a and Fig.5b, the pair dispersion statistics for Set-G and Set-D follow Batchelor's t^2 dispersion law irrespective of the temperature value, but with modified constants: $\langle|\boldsymbol{\Delta}^F - \boldsymbol{\Delta}_o^F|^2\rangle = \frac{11}{3}C(\epsilon\Delta_o^F)^{\frac{2}{3}}t^2$. Validity of this dispersion law for Set-D flame particles is known [14], but applicability of the same t^2 law for the leading points allows estimation of the flame surface generation and destruction time-scales: τ_G and τ_D, respectively. τ_G could be estimated as time taken for particle pairs in Set-G to grow from a clustered state with initial separation of 50η to a final separation around L_o, such that $\langle|\boldsymbol{\Delta}_G^F - \boldsymbol{\Delta}_{G,o}^F|^2\rangle = (L_o - 50\eta)^2$, yielding

$$\tau_G = \left((L_o - 50\eta)^2 / \left[C_G \frac{11}{3}(50\eta\langle\epsilon_T\rangle)^{\frac{2}{3}}\right]\right)^{1/2} \quad (4)$$

τ_D is the time required for Set-D particle pairs to shrink to around 50η from an initial separation of L_o such that $\langle|\boldsymbol{\Delta}_D^F - \boldsymbol{\Delta}_{D,o}^F|^2\rangle = (50\eta - L_o)^2$.Consequently,

$$\tau_D = \left((50\eta - L_o)^2 / \left[C_D \frac{11}{3}(L_o\langle\epsilon_T\rangle)^{\frac{2}{3}}\right]\right)^{1/2} \quad (5)$$

Therefore,$\tau_G/\tau_D \approx (L_o/\eta)^{1/3}(C_D/C_G)^{1/2} \sim Re_{L_o}^{1/4}$. The above expression reveals that, statistically, the surface takes more time for generation than annihilation. Hence, the total flame surface area or turbulent flame speed will fluctuate in time [11,15].

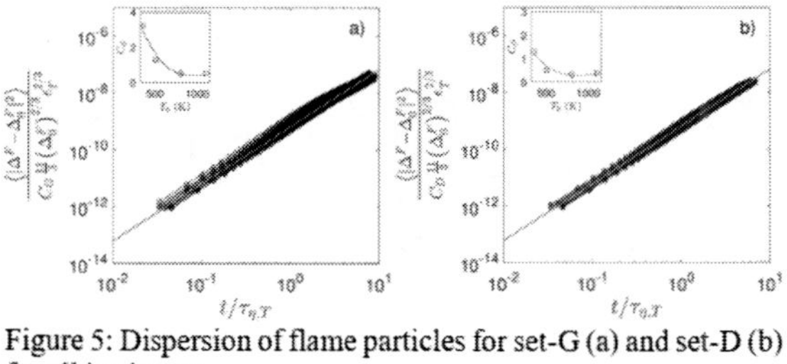

Figure 5: Dispersion of flame particles for set-G (a) and set-D (b) for all isotherms

4 Conclusions

In this work, Backward Flame Particle Tracking and Finite Strain Theory are applied in turbulent combustion to show the following for the first time.

1. Fully developed turbulent flame surface evolves from multiple clusters of flame particles resident on the corresponding surface at earlier times. These flame particles are characterized as leading, positively curved, and have low fluid speed.
2. The least curved direction $\hat{\boldsymbol{e}}_2$ of these leading points preferentially aligns with the vorticity $\hat{\boldsymbol{\omega}}$ and the most extensive principal direction $\hat{\boldsymbol{l}}_1$ of the left Cauchy-Green strain rate tensor. The leading points are strained in the most curved direction primarily due to propagation and along the least curved direction due to the tangential strain rate.
3. S_d of the leading points indeed control S_T, albeit with a finite time lag, as shown in (3)
4. The ratio of the generation to destruction time-scales of the turbulent flame surface scales as $Re_{L_o}^{1/4}$.

5 Acknowledgements

Discussions with Dr. Narayana Sundaram of IISc are gratefully acknowledged. HD and AM acknowledge the scholarship support from the Ministry of Human Resource and Development, India

References

[1] E.S. Oran, Proc. Combust. Inst. 35 (2015) 1–35.
[2] K.R. Sreenivasan, J. Schumacher, Philos. Trans. A. Math. Phys. Eng. Sci. 368 (2010) 1561–1577.
[3] S. Chaudhuri, Proc. Combust. Inst. 35 (2015) 1305–1312.
[4] S.B. Pope, Int. J. Eng. Sci. 26 (1988) 445–469.
[5] A. Lipatnikov, Fundamentals of Premixed Turbulent Combustion, 2012.
[6] H.A. Uranakara, S. Chaudhuri, H.L. Dave, P.G. Arias, H.G. Im, Combust. Flame 163 (2016) 220–240.
[7] a. Nahum, A. Seifert, Phys. Rev. E 74 (2006) 1–10.
[8] N. Babkovskaia, N.E.L. Haugen, A. Brandenburg, J. Comput. Phys. 230 (2011) 1–12.
[9] J. Li, Z. Zhao, A. Kazakov, F.L. Dryer, Int. J. Chem. Kinet. 36 (2004) 566–575.
[10] A. Amato, M. Day, R.K. Cheng, J. Bell, T. Lieuwen, Proc. Combust. Inst. 35 (2015) 1313–1320.
[11] S.H. Kim, Proc. Combust. Inst. 0 (2016) 1–8.
[12] R. Ni, N.T. Ouellette, G. a. Voth, J. Fluid Mech. 743 (2014) R3.
[13] L.E. Malvern, Introduction to the Mechanics of Continuous Medium (1969).
[14] S. Chaudhuri, Phys. Rev. E 91 (2015) 21001.
[15] A.Y. Poludnenko, Phys. Fluids 27 (2015).

11th Asia-Pacific Conference on Combustion,
The University of Sydney, NSW Australia
10th -14th December 2017.

Global Sensitivity Analysis of Chemical Kinetic Reaction Mechanisms

Shengqiang Lin, Xie Ming, Weixing Zhou

School of Energy Science and Engineering, Harbin Institute of Technology

Harbin, Heilongjiang, 150001, China

Abstract

In this work, the FAST method is upgraded for sensitivity analysis of chemical reaction mechanisms, and a criterion is presented for assigning appropriate frequencies to input parameters so that the interference from the input parameters can be effectively avoided. In addition, the new FAST method can successfully identify whether a reaction has a positive or negative effect on the products. In order to identify the important reactions in the entire reaction process, we employed the FAST method to obtain time-varying sensitivity indices since the model describing a chemical reaction process is always time-dependent. The performance of the new FAST method was evaluated by a detailed mechanism of H_2/air combustion model. The results show that the sensitivity indices obtained from the new FAST method can not only reflect the positive or negative contribution of a reaction to the products, but also quantitatively measure the effect of reactions on the products. Thus, this method can be a remarkably effective tool for the modelling and diagnosis of complex chemical reaction mechanisms..

1 Introduction

The chemical process of practical fuel combustion can be described by detailed kinetic reaction mechanisms, which involve hundreds or even thousands of reactions. However, there is a large uncertainty in the reaction rate coefficients of many reactions, even for the best-known ones. Nevertheless, the combustion models used for predicting products need precise input parameter values. Thus, for reaction mechanism modelling, much effort should be directed at finding the appropriate rate coefficients for large reaction mechanisms and determining the effects of important reactions on the products. In the modelling of chemical reaction mechanisms, sensitivity analysis is a routine diagnostic tool for the investigation of chemical kinetic behaviors. The sensitivity analysis method ascertains the dependence of model output on the input factors in order to find most important input factors [1].

There are two main classes of sensitivity analysis techniques: local sensitivity analysis (SA) and global sensitivity analysis [2-4]. Local sensitivity analysis is a valuable tool to quickly achieve sensitivity coefficients for nominal parameter set of model which is regarded to be the best knowledge of model. However, input parameters of chemical kinetic mechanisms are not know with low uncertainty. Many reports indicate that local sensitivity analysis is inaccurate for complex nonlinearity models, or becomes impractical as the number of parameters subjected to variation increases [5-10]. Thus, many GSA methods have been developed to analyze complex models, such as FAST method [11-13], Sobol' method [14], HDMR method [5, 15-18] and ordinary least squares approach [19]. And these method are applied successfully in complex models such as reaction mechanisms construction [20-23], chemical reaction mechanisms analyzation [5, 18, 24], experimental analyzation [9]. However, relatively few studies are available on evaluate the positive or negative effect of an increase in the input factor value on the product.

The application of the FAST method in studying reaction mechanisms started with the work of Cukier et al. [11, 13]. They introduced a classic FAST method to evaluate simple reaction systems with large uncertainty in reaction rate coefficients. In this work, the FAST method was chosen partly because it can evaluate complex nonlinear models with small sample sizes and can determine whether increasing the value of input factors will have a positive or negative effect on the product. In the FAST method, each input parameter is explored with a different integer frequency. The sampling techniques of the FAST method are based on selecting N design points over a particular space-filling curve in the kth dimensional input space. A classic FAST method establishes the relationship between a subset of the Fourier coefficients and $\partial c_i / \partial k_l$, where c_i is the concentration of species i and k_l is reaction rate of lth reaction. Hence, a large Fourier coefficient corresponds to a large sensitivity index and a small Fourier coefficient corresponds to a small sensitivity index. Moreover, an important advantage of the FAST method is that it can distinguish whether a reaction plays a positive or negative role in products. Importantly, this feature can be greatly useful to modelling detailed mechanisms since we know how to enhance or weaken combustion characteristic parameters such as ignition delay time and products.

However, assigning frequencies to input factors in the FAST method is challenging since this requires a complex algorithm and becomes more and more difficult with increasing numbers of input factors. Moreover, the interference of input parameters in the set of frequencies should be avoided because the interference can lead to the overestimation of the sensitivity index, S_i. Thus, the FAST method may not accurately estimate the sensitivity index if the frequencies are not properly assigned [27]. In this work, assigning frequencies to input factors so that the sensitivity indices can be accurately estimated by the FAST method. In addition, the performance of the new FAST method was evaluated by a H_2/O_2 combustion reaction mechanism [28]

2 Methodologies

2.1 FAST method

A classic FAST method was introduced by Cukier et al [11-13]. as a diagnostic tool to investigate chemical reaction mechanisms with a large uncertainties in rate coefficients. The effect of the reaction rate coefficients on the products can be

Corresponding author. Fax: +86-0451-86403142
E-mail address: fujianlsq@163.com

written in the form:

$$\frac{dy_i}{dt} = f(x_j) \tag{1}$$

where x_j is the jth rate coefficient, and y_k is product determined by chemical reaction mechanisms and reactor model, i is the number of products. y_i can be concentration of species, temperature, and ignition delay time. Perfectly Stirred Reactor (PSR) is selected in this work as reactor model to investigate the uncertainty in the rate coefficients effect on products. This reactor model is commonly used by other researchers for global sensitivity analysis of complex reaction mechanisms [7, 18].

When FAST method is applied to estimate contribution of input factors to products, we must firstly choose a set of r integer frequencies, $\{\omega_1, \omega_2, \cdots, \omega_r\}$, to input factors where r is the number of independent reaction rate coefficients. In order to evaluate the importance of input factors, the FAST generates sampling points in the kth dimensional input space by space-filling curve. In this work, we use a space-filling curve provided by Saltelli et al [29] since it can provide more a uniformly distributed sample. Such space-filling curve defined as follow:

$$x_j(s) = \frac{1}{2} + arcsin\left(sin(\omega_j s)\right), \quad \forall j = 1, 2, \ldots \tag{2}$$

where $x_j(s)$ is a probability density function (pdf) of the jth input parameter, s is the parametric variable varying in $(-\pi, \pi)$ which is sampled over its range using N points, and ω_j is the integer frequency. It is noted that the frequencies should avoid interferences up to a given order M since reactor model is run at each sample points and the Fourier spectrum is calculated at specific frequencies $\{\omega_j, 2\omega_j, \cdots, M\omega_j\}$ to estimate the sensitivity index of factor x_j. y_i is expanded in Fourier series as:

$$y_i(s) = \frac{A_0^{(i)}}{2} + \sum_{j=1}^{r}\left(A_{\omega_j}^{(i)}\cos(\omega_j s) + B_{\omega_j}^{(i)}\sin(\omega_j s)\right) \tag{3}$$

where the Fourier amplitude, $A_{\omega_j}^{(i)}$ and $B_{\omega_j}^{(i)}$, can be written as follow:

$$\begin{cases} A_{\omega_j}^{(i)} = \frac{1}{2\pi}\int_{-\pi}^{\pi} y_i(s)\cos(\omega_j s)\,ds \\ B_{\omega_j}^{(i)} = \frac{1}{2\pi}\int_{-\pi}^{\pi} y_i(s)\sin(\omega_j s)ds \end{cases} \tag{4}$$

The sensitivity indices measured in the classic FAST method for input factor x_j are written in the form:

$$S_{\omega_j}^{(i)} = \sum_{p=1}^{M} B_{p\omega_j}^{(i)} \tag{5}$$

A classic FAST method only considers Fourier amplitude $B_{\omega_j}^{(i)}$ as a sensitivity index of factor x_j. Such a sensitivity index can be positive or negative; the positive sign of Fourier amplitude indicates increase in products with increasing x_j and the negative sign for Fourier amplitude indicates increase in products with increasing x_j since $B_{\omega_j}^{(i)}$ is proportional to $\frac{dy_i}{dt}$. It is note that interferences in FAST method may occur, such that $p_1\omega_1 = p_2\omega_2$. Thus, we may overestimate the sensitivity indices for input factor x_1 and x_2 since such sensitivity indices involve extra contribution.

Due to the symmetry properties of space-filling curve and reduce the number of model evaluations, equations (4) are upgraded by Koda et al. [30] and written as follows:

$$A_{\omega_j}^{(i)} = \begin{cases} 0 & \text{if } \omega_j \text{ odd} \\ \frac{1}{\pi}\int_{0}^{\pi/2} y_i(s)\cos(\omega_j s)ds & \text{if } \omega_j \text{ even} \end{cases} \tag{6}$$

$$B_{\omega_j}^{(i)} = \begin{cases} 0 & \text{if } \omega_j \text{ even} \\ \frac{1}{\pi}\int_{0}^{\pi/2} y_i(s)\sin(\omega_j s)ds & \text{if } \omega_j \text{ odd} \end{cases} \tag{7}$$

Koda et al. also introduced ANOVA to FAST method, and sensitivity indices measure in such variance-based FAST method can be written in the form:

$$S_{\omega_j}^{(i)} = \frac{\sigma_{\omega_j}^2}{\sigma_i^2} \tag{8}$$

where $S_{\omega_j}^{(i)}$ is referred to as the first order sensitivity index, σ_i^2 is the total variance of product y_i, and $\sigma_{\omega_j}^2$ is a part of total variance σ_i^2 and indicates the contribution of input factors x_j to the product y_i since frequency ω_j are associate with x_j. The variance σ_i^2 and $\sigma_{\omega_j}^2$ can be written respectively as follows:

$$\sigma_i^2 = 2\sum_{l=1}^{\infty}\left[\left|A_l^{(i)}\right|^2 + \left|B_l^{(i)}\right|^2\right] \tag{9}$$

$$\sigma_{\omega_j}^2 = 2\sum_{p=1}^{M}\left[\left|A_{p\omega_j}^{(i)}\right|^2 + \left|B_{p\omega_j}^{(i)}\right|^2\right] \tag{10}$$

Note that $S_{\omega_j}^{(i)}$ is a normalized sensitivity index in the variance-based FAST method so that the $S_{\omega_j}^{(i)}$ may be ordered with respect to y_i. Although a normalized sensitivity index can reflect the contribution of the input factor to product at particular time, it may not greatly reflect the effect of input factor on the entire reaction process. Moreover, chemical reaction mechanisms modelling are very sensitivity to the positive or negative effect of reaction on products; however, the variance-based GSA method may inaccurate estimate the sensitivity indices since it considers the negative contribution of input factors to product as a positive contribution. In the present work, we will compare sensitivity indices with or without normalization.

2.2 Choice of the frequencies

As mentioned above, we introduce two types of FAST method to measure the sensitivity indices of a model with uncertainty in multiple parameters, and explain the problem of interferences. Since it is difficult to assign frequencies to models with large number of input factors, we propose a criterion for selecting appropriate frequencies for input factors so that the new FAST method can effectively avoid the problem of interferences, while being able to identify the positive or negative effects of input factors on products. The criterion for frequency selection for input factors can be written in the form:

$$\begin{cases} \omega_j > M \times r + 2(j-1) + 1 \\ \omega_j \text{ is odd} \end{cases} \tag{11}$$

where M is the interference factor, r is the number of input factors. For example, we can set input frequencies as $\{13, 15, 17\}$ which meet minimum requirement for model with three factors with the condition $M = 4$. Such an equation provides a simple algorithm to select frequencies for model with many input factors, and can accurately estimate sensitivity index since it effectively avoid the interferences from $p_i\omega_j = p_k\omega_l$ for $\forall i,j,k,l$ and $j \neq l$ where $i,k \leq M$ and $j,l \leq r$. Moreover, we can simplify equations (6) and (7) since ω_j is odd and the simplified equations can be written in the form:

$$A_{\omega_j}^{(i)} = 0 \quad (12)$$

$$B_{\omega_j}^{(i)} = \frac{1}{\pi}\int_0^{\pi/2} y_i(s)\sin(\omega_j s)ds \quad (13)$$

where $B_{\omega_j}^{(i)}$ contains all contributions of input factors on the output. Since $B_{\omega_j}^{(i)}$ is proportional to $\frac{dy_i}{dt}$, we can obtain sensitivity indices that can identify the positive or negative effect of the increasing of input factors on products. Thus, we can obtain two types of FAST method though equation (8) and (13) respectively. In the present work, sensitivity indices from equation (13) are discussed in detail since the some features of this FAST method cannot be found in other GSA methods, and a cases is applied to illustrate the performance of this FAST method.

3 Results and Discussion

3.1 Detailed mechanisms

In this section, we study the effect of uncertainty in the reaction rate on the ignition delay time and other products based on detailed mechanisms. Ignition delay time is commonly defined as the time when the temperature increases by 400 K from the initial temperature; this definition has been applied in this paper. Since the ignition delay time is closely associated with temperature, we can study the effect of uncertainty in the reaction rate on the ignition delay time by observing the sensitivity of reaction rate coefficients to temperature. In this section, we validate the criterion for frequency selection and the performance of FAST method again using the H_2/air combustion model developed by Konnov [28]. This model was selected to illustrate the performance of the new FAST method for several reasons. Firstly, the H_2/O_2 combustion model is a well-studied system because of which some reactions are relatively well defined. Secondly, this model was developed by evaluating the uncertainty of each rate coefficient, which rendered it suitable for investigating the effect of uncertainty in rate coefficients on products. In this work, this test is employed to study the effect of uncertainty in the reaction rate on the ignition delay time and other products, which were calculated with the Senkin code [30] or a stoichiometric H_2/air mixture at a temperature of 1000 K and pressure of 1.59 bar.

To use the FAST method for global sensitivity analysis of a chemical detailed mechanisms of hydrogen combustion which involves 33 chemical reactions, we firstly determine the values of some parameters for FAST method: the numbers of input factors i=33, interference factor, M=4, and time-points, NT=1000, uniformly distributed in [0 0.0005] for generating the profiles of sensitivity indices. Thus, the frequencies $\omega_i = 135 + 2(i-1), i = 1, 2, \ldots, 33$ are assigned to the reaction rate based on the equation (11). Hence, FAST method needs 1593 samples to study the sensitivity of the products to uncertainties in the reaction rates; these results shown in Fig. 1. It is seen that reaction 16 (H + O_2 = OH + O) is the major contributor to products including temperature and H_2, O_2, and H_2O, and plays a positive role in outlet temperature. Therefore, ignition delay time can decrease with increase in the rate of reaction 16 which accelerates the chemical reaction process, and thus, ignition state occurs early. It is further seen in Fig.1 that the consumption of H_2, O_2 and the production of H_2O can be effectively enhanced by increasing of the rate of reaction 16 as it accelerates the H_2/air combustion reaction, thereby facilitating the consumption of H_2 and O_2 and the production of H_2O. These results of FAST method are in good agreement with what could be predicted on the basis of chemical knowledge.

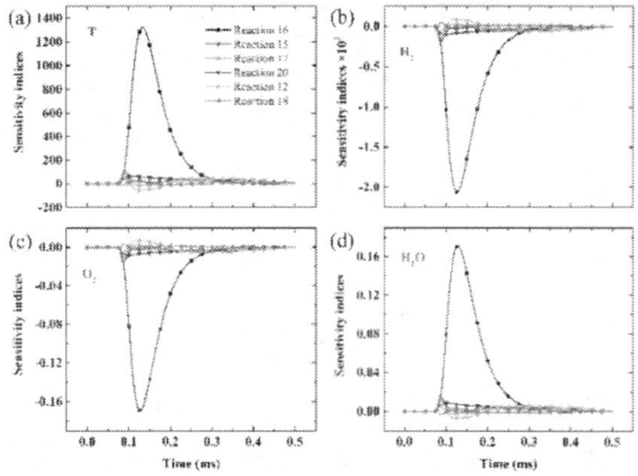

Fig. 1: The profiles of sensitivity indices for detailed mechanisms of H_2/air combustion model using the FAST method with 1593 samples at P=1.59 bar, T=1000 K and stoichiometric mixtures of H_2/air. We list the important reactions which significantly effect on the major products involving temperature (a), mass fraction of H_2 (b), mass fraction of O_2 (c) and mass fraction of H_2O (d)

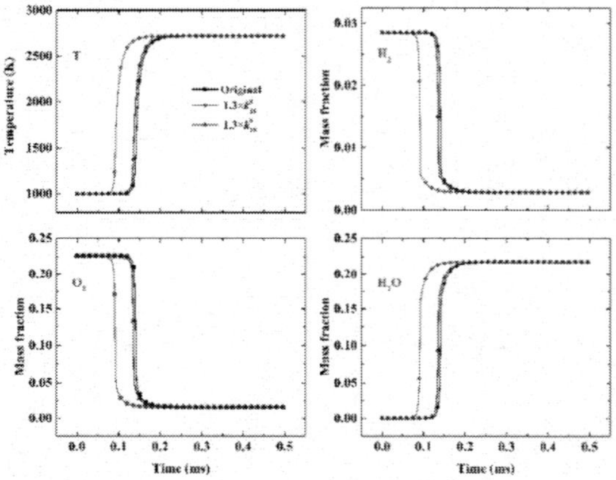

Fig. 2: The prediction results comparisons of original combustion model and two validated models. $1.3 \times k_i^0$ indicates that the rate of the reaction I of this validated model are calculated with original reaction rate k_i^0 multiplied by 1.3, and other reaction rates of the validated model is all derived from original reaction rates

3.2 validated model

In order to validate the aforementioned results of the FAST method, two validated models of H_2/air were designed. For the two validated models, all the reaction rate coefficients were derived from the original model, with the exception that the rate of reaction 16 of the original model, $k_{16}^{(0)}$, was replaced with $1.3 \times k_{16}^{(0)}$ for the first validated model and that of reaction 15 of the original model, $k_{15}^{(0)}$, was replaced with $1.3 \times k_{15}^{(0)}$ for the second validated model, where $k_{15}^{(0)}$ and $k_{16}^{(0)}$ were calculated with the original reaction rate coefficients. Fig. 2 shows the comparison of products predicted using the two validated models and detailed mechanisms. It can be seen that the ignition delay time decreases with increase in the rates of reactions 15 and 16 since these reactions can positively affect the combustion of H_2/air. Hence, increasing the rates of reactions 15 and 16 can increase the consumption of H_2 and O_2 and the production of H_2O. Moreover, the products are more sensitive to the change in reaction 16 than to the change in reaction 15. Thus, these results confirm the superior performance of the FAST method in the diagnosis of chemical reaction models. Therefore, the FAST method can be an effective tool for modelling detailed mechanisms and diagnosing combustion reaction kinetic models.

4 Conclusions

In the present study, we proposed a criterion for assigning appropriate frequencies to input parameters to simplify the frequency selection process of the FAST method. The new FAST method can quantitatively analyze the positive or negative effect of a reaction on the products. We also extended this GSA method to obtain a set of time-varying sensitivity indices to gain a better understanding of combustion reactions. In order to validate the performance of the new FAST method, a detailed mechanism of H_2/air combustion, were employed. Furthermore, the validated models were developed to illustrate the sensitivity predictions of the FAST method. The analyses of new FAST methods revealed that the sensitivity index profiles could determine the important reactions that greatly affect the chemical reaction processes, and that the value of sensitivity index of a reaction reflects its importance in the chemical reaction process. The new FAST method can be a powerful tool for the modelling and diagnosis of chemical reaction mechanisms.

5 Acknowledgment

This work was supported by the National Natural Science Foundation of China (No. U1432130, and 51676056).

References

1. T. Turányi; A. S. Tomlin, Analysis of Kinetic Reaction Mechanisms, Springer Berlin Heidelberg, 2014.
2. H. Wang; D. A. Sheen, Prog. Energ. Combust. 47 (2014) 1-31.
3. T. Turányi, Comput. Chem 14 (3) (1990) 253-254.
4. T. Nagy; T. Turányi, int. J. Chem. Kinet. 43 (7) (2011) 359–378.
5. T. Ziehn; A. S. Tomlin, Int. J. Chem. Kinet. 40 (11) (2008) 742–753.
6. J. Zádor; I. G. Zsély; T. Turányi, Reliab. Eng. Syst. Safe. 91 (10) (2006) 1232-1240.
7. S. Li; B. Yang; F. Qi, Combust. Flame 168 (2016) 53-64.
8. T. Ziehn; K. J. Hughes; J. F. Griffiths; R. Porter; A. S. Tomlin, 13 (4) (2009) 589-605.
9. Y. Pei; M. J. Davis; L. M. Pickett; S. Som, combust. Flame 162 (6) (2015) 2337-2347.
10. M. Koda; G. J. Mcrae; J. H. Seinfeld, Int. J. Chem. Kinet. 11 (4) (2010) 427-444.
11. R. I. Cukier; C. M. Fortuin; K. E. Shuler; A. G. Petschek; J. H. Schaibly, j. Chem. Phys. 59 (8) (1973) 3873-3878.
12. J. H. Schaibly; K. E. Shuler, J. Chem. Phys. 59 (8) (1973) 3879-3888.
13. R. I. Cukier; J. H. Schaibly; K. E. Shuler, J. Chem. Phys. 63 (3) (1975) 1140-1149.
14. I. M. Sobol, Math. Comput. Simulat. 55 (1-3) (2001) 271-280.
15. G. Li; C. Rosenthal; H. Rabitz, J. Phys. Chem. A 105 (33) (2001) 7765-7777.
16. G. Li; S. Wang; H. Rabitz, J. Phys. Chem. A 106 (37) (2002) 8721-8733.
17. G. Li; S. W. Wang; H. Rabitz; S. Wang; P. Jaffé, Chem. Eng. Sci. 57 (21) (2002) 4445-4460.
18. M. J. Davis; R. T. Skodje; A. S. Tomlin, J. Phys. Chem. A 115 (9) (2011) 1556-1578.
19. M. J. Davis; W. Liu; R. Sivaramakrishnan, J. Phys. Chem. A 121 (3) (2017) 553-570.
20. L. Xing; S. Li; Z. Wang; B. Yang; S. J. Klippenstein; F. Zhang, Combust. Flame 162 (9) (2015) 3427-3436.
21. A. S. Tomlin, proc. Combust. Inst. 34 (1) (2013) 159-176.
22. S. J. Klippenstein; L. B. Harding; M. J. Davis; A. S. Tomlin; R. T. Skodje, Proc. Combust. Inst. 33 (1) (2011) 351-357.
23. C. F. Goldsmith; A. S. Tomlin; S. J. Klippenstein, proc. Combust. Inst. 34 (1) (2013) 177-185.
24. D. D. Zhou; M. J. Davis; R. T. Skodje, J. Phys. Chem. A 117 (17) (2013) 3569-84.
25. I. M. Sobol, Matem. Mod. 1 (4) (1993) 407-414.
26. I. M. Sobol, Matem. Mod. 2 (1) (1990) 112–118.
27. A. Saltelli; M. Ratto; T. Andres; F. Campolongo; J. Cariboni; D. Gatelli; M. Saisana; S. Tarantola, Global sensitivity analysis: the primer, John Wiley & Sons, 2008.
28. A. A. Konnov, Combust. Flame 152 (4) (2008) 507-528.
29. A. Saltelli; P. S. Chan, Technometrics 41 (1) (2012) 39-56.
30. M. Koda; G. J. Mcrae; J. H. Seinfeld, Int. J. Chem. Kinet. 11 (4) (1979) 427-444.
31. R. J. Kee; J. F. Grear; M. D. Smooke; J. A. Miller, Chemkin-II, A Fortran Chemical Kinetics Package for the Analysis of Gas Phase Chemical Kinetics, SAND-89-8009, 1989.

11th Asia-Pacific Conference on Combustion,
The University of Sydney, NSW Australia
10th -14th December 2017.

A Five-Equation Model for the Simulation of Miscible, Compressible Fluids Including Molecular Species Transport.

Michael Groom[1], David Youngs[2], Ben Thornber[1]

[1]The University of Sydney

Sydney, NSW, 2006, Australia

[2]Strathclyde University

Glasgow, G11XQ, United Kingdom

Abstract

When simulating compressible multispecies flows using interface-capturing finite-volume methods, the typical approach of using transport equations for species mass fractions can produce errors in pressure and/or temperature across material interfaces under certain circumstances. Adding quasi-conservative transport equations for species volume fractions, such as the five-equation model of Allaire *et al.* [J. Comput. Phys. 181 (2002) 577-616], has been shown to remove these errors in inviscid flows and here we extend that approach to include the effects of species diffusion as well as viscosity and thermal conductivity. A second order accurate finite-volume method is derived to solve the new set of equations and formal convergence studies are presented to demonstrate the benefits with respect to the typical four-equation approach based on mass fractions.

1 Introduction

Compressible turbulent reacting flows, such as those encountered in aeronautical applications, involve a wide range of scales and complexities. Typical modelling strategies rely on Reynolds Averaged Navier-Stokes (RANS) approaches to compute ensemble averaged fields, however the appearance of unsteadiness and inhomogeneities in these flows limits the precision of such approaches [1]. Increased precision can be obtained with Large Eddy Simulation (LES) approaches, where the turbulent length scales in the flow are explicitly calculated, or even Direct Numerical Simulation (DNS), where every length scale in the flow is resolved. These approaches involve simulating either the filtered or fully instantaneous governing equations, which for single-phase multifluids consist of the Navier-Stokes equations augmented with additional scalar transport equations that determine the composition of the mixture [2].

As molecular transport effects must be explicitly modelled in reacting flows, the typical approach is to use species mass fractions in the scalar transport equations, resulting in a system that is fully conservative [3]. However, it has been well documented that when using this set of equations to simulate flows where the ratio of specific heats γ varies with mixture composition, spurious oscillations in pressure and/or temperature are generated across material interfaces, in particular when the interface is advected through the computational mesh [4,5,6]. While such errors theoretically disappear in the DNS limit, the computational efficiency will certainly be reduced, and in general will still be present in a LES. Various approaches exist for removing such errors, all of which sacrifice the fully conservative nature of the governing equations [6].

A popular approach due to Allaire *et al.* [7] that produces very small conservation errors is to have an additional transport equation for the species volume fraction, written in advection form. With suitable thermodynamic closure, this five-equation model is suitable for simulating any equation of state while also maintaining pressure/temperature equilibrium across material interfaces. The non-conservative terms may be hard to discretise however. Despite originally only being presented for inviscid flows, subsequent studies have included the effects of viscosity and thermal conductivity [8]. However, a necessity for DNS/LES of both reacting and non-reacting flows is the molecular diffusion of species, yet the inclusion of these effects into the five-equation volume faction model has eluded previous efforts. This motivates the current work presented here, in which the five-equation model of Allaire *et al.* is extended to include species diffusion, thus enabling accurate and efficient DNS and LES of arbitrary gas mixtures.

2 Extension of the Five-Equation Model

2.1 Derivation

We start with the original five-equation model of Allaire *et al.*:

$$\frac{\partial \rho_k z_k}{\partial t} + \nabla \cdot (\rho_k z_k \boldsymbol{u}) = 0 \tag{1}$$

$$\frac{\partial \rho \boldsymbol{u}}{\partial t} + \nabla \cdot (\rho \boldsymbol{u} \times \boldsymbol{u}) = -\nabla \cdot p\mathrm{I} \tag{2}$$

$$\frac{\partial \rho e}{\partial t} + \nabla \cdot [(\rho e + p)\boldsymbol{u}] = 0 \tag{3}$$

$$\frac{\partial z_k}{\partial t} + \boldsymbol{u} \cdot \nabla z_k = 0 \tag{4}$$

Refer to [7] for the full details. Note that in the above system, (4) only needs to be solved for N-1 species. The following derivation is for non-reacting flows with two species, however both the inclusion of reactions and the extension to an arbitrary number of species is possible. First we note that as mixing occurs at the molecular level, both pressure and temperature

Corresponding author. Fax: +61-2-9351-4841
E-mail address: ben.thornber@sydney.edu.au

equilibrium is expected in the diffusive flux. As the two gases are at the same pressure and temperature, according to Avogadro's hypothesis [9], they also have the same number density. Therefore each molecule in the mixture occupies the same volume, and volume fractions z_k may be equated with number density fractions n_k. In addition, as each molecule for a given species has the same mass, the number-weighted, volume-weighted and mass-weighted mean velocities for species k are all the same and are denoted $\boldsymbol{u}_k$. The evolution of total number density N is given by:

$$\frac{\partial N}{\partial t} + \nabla \cdot (N\bar{\boldsymbol{u}}) = 0 \tag{5}$$

where $\bar{\boldsymbol{u}} = \frac{\sum_k n_k \boldsymbol{u}_k}{N}$ is the number-weighted mean velocity of the mixture. Noting that $z_k = \frac{n_k}{N}$ leads to:

$$\frac{\partial N z_k}{\partial t} + \nabla \cdot (N z_k \boldsymbol{u}_k) = 0 \tag{6}$$

$$N\frac{\partial z_k}{\partial t} - z_k \nabla \cdot (N\bar{\boldsymbol{u}}) + \nabla \cdot (N z_k \boldsymbol{u}_k) = 0 \tag{7}$$

$$N\frac{\partial z_k}{\partial t} + N\bar{\boldsymbol{u}} \cdot \nabla z_k - \nabla \cdot (z_k N \bar{\boldsymbol{u}}) \ldots$$

$$\ldots + z_k \boldsymbol{u}_k \cdot \nabla N + N \nabla \cdot (z_k \boldsymbol{u}_k) = 0 \tag{8}$$

$$\frac{\partial z_k}{\partial t} + \bar{\boldsymbol{u}} \cdot \nabla z_k + \nabla \cdot [z_k(\boldsymbol{u}_k - \bar{\boldsymbol{u}})] + z_k(\boldsymbol{u}_k - \bar{\boldsymbol{u}})\frac{\nabla N}{N} = 0 \tag{9}$$

For binary mixtures, if Fickian diffusion [9] is assumed then:

$$z_k(\boldsymbol{u}_k - \bar{\boldsymbol{u}}) = -D_{12} \nabla z_k \tag{10}$$

The volume fraction equation then becomes:

$$\frac{\partial z_k}{\partial t} + \bar{\boldsymbol{u}} \cdot \nabla z_k = \nabla \cdot (D_{12} \nabla z_k) + D_{12} \nabla z_k \cdot \frac{\nabla N}{N} = 0 \tag{11}$$

The number-weighted mean velocity is related to the mass-weighted mean velocity by:

$$\bar{\boldsymbol{u}} = \boldsymbol{u} - \frac{\sum_k M_k z_k (\boldsymbol{u}_k - \bar{\boldsymbol{u}})}{\sum_k M_k z_k} = \boldsymbol{u} + \frac{M_1 - M_2}{M_1 z_1 + M_2 z_2} D_{12} \nabla z_1 \tag{12}$$

where M_k is the molecular weight of species k. The assumption of Fickian diffusion (i.e. Soret and Dufour effects are ignored) means that the diffusive fluxes of $\rho_k z_k = \rho Y_k$ can be written in terms of mass fraction gradients, as per the standard four-equation mass fraction model. The contributions due to species diffusion in the energy equation are the same as for the four-equation model [10], as are the contributions due to viscosity and thermal conduction. The full set of equations is given below:

$$\frac{\partial \rho_k z_k}{\partial t} + \nabla \cdot (\rho_k z_k \boldsymbol{u}) = \nabla \cdot (\rho D_{12} \nabla Y_k) \tag{13}$$

$$\frac{\partial \rho \boldsymbol{u}}{\partial t} + \nabla \cdot (\rho \boldsymbol{u} \times \boldsymbol{u}) = -\nabla \cdot \mathrm{S} \tag{14}$$

$$\frac{\partial \rho e}{\partial t} + \nabla \cdot [(\rho e + p)\boldsymbol{u}] = -\nabla \cdot (\mathrm{S} \cdot \mathbf{u} + \boldsymbol{q}_c + \boldsymbol{q}_d) \tag{15}$$

$$\frac{\partial z_k}{\partial t} + \left(\boldsymbol{u} + \mathcal{M} D_{12} \nabla z_1 - D_{12}\frac{\nabla N}{N}\right) \cdot \nabla z_k = \nabla \cdot (D_{12} \nabla z_k) \tag{16}$$

where $\mathrm{S} = p\mathrm{I} - \lambda_b(\nabla \cdot \boldsymbol{u})\mathrm{I} - \mu[\nabla \boldsymbol{u} + (\nabla \boldsymbol{u})^T]$, $\lambda_b = -\frac{2}{3}\mu$, $\boldsymbol{q}_c = -\kappa \nabla T$, $\boldsymbol{q}_d = \sum_k \rho D_{12} \nabla Y_k h_k$, $\mathcal{M} = \frac{M_1 - M_2}{M_1 z_1 + M_2 z_2}$, $N = \frac{P}{k_B T}$.

Thus there are three processes that modify the volume fraction; advection with the number-weighted mean velocity, diffusive mixing and pressure-temperature equilibration.

2.2 Discretisation

The governing equations have been implemented in the University of Sydney compressible CFD code Flamenco [11], which has an existing implementation of the inviscid volume fraction model of Allaire *et al.* The existing numerical algorithm is a Godunov-type finite volume method in a structured, curvilinear, multiblock framework. The code employs a fifth order spatial reconstruction for the inviscid component [12], the inviscid flux is computed with the HLLC approximate Riemann solver [13], and the viscous/diffusive fluxes are computed using second order central differences. Temporal integration is performed using a second order TVD Runge-Kutta method [14], and the algorithm has been modified to provide reduced dissipation at low Mach number [15].

The modifications made to the existing algorithm for the new set of equations are now given. We begin by writing the LHS of the volume fraction equation, shown here for the solution in the x-direction, as:

$$\frac{\partial z_k}{\partial t} + \mathcal{U} \cdot \nabla z_k = \frac{\partial z_k}{\partial t} + \frac{\partial \mathcal{U} z_k}{\partial x} - z_k \frac{\partial \mathcal{U}}{\partial x} \tag{17}$$

where $\mathcal{U} = (\boldsymbol{u} + \mathcal{M} D_{12} \nabla z_1 - D_{12}\frac{\nabla N}{N})$. Following [4], the intercell flux is computed as:

$$F_{i+\frac{1}{2}} = (z_k \mathcal{U})^{RS}_{i+\frac{1}{2}} - z_k \mathcal{U}^{RS}_{i+\frac{1}{2}} \tag{18}$$

where $(.)^{RS}$ denotes a term arising from the solution of the Riemann problem at the cell interface. Note that compared to the original inviscid five-equation model, the upwind direction of the advection of volume fraction is determined by not only the mass-weighted mean velocity but also the gradients of number density and volume fraction itself. These gradients are computed using second order central differences, while $\mathcal{M}$ and N are computed as the average of the interface reconstructed values. The volume fraction fluxes are calculated by modifying the signal speeds of the HLLC approximate Riemann solver to incorporate the additional diffusion velocities. Recent advances in finite-volume methods could permit an arbitrarily high-order discretisation, which will be the focus of future work.

The RHS terms of the volume fraction equation are also discretised using second order central differences. Another modification must be made, this time to the discretisation of the RHS of (13) as the assumption of pressure-temperature equilibrium must be enforced to ensure consistency with the terms on the LHS of (16). If this is not enforced in the calculation then errors will be introduced in the species densities and temperatures. Within the current framework this is addressed by re-writing the RHS of (13) as:

$$\nabla \cdot (\rho D_{12} \nabla Y_k) = \nabla \cdot \left(\rho D_{12} \frac{M_k \nabla X_k}{M}\right) \tag{19}$$

where M is the molecular weight of the mixture and X_k is the mole fraction of species k, which is equal to volume fraction under the assumption of pressure-temperature equilibrium.

3 Convergence Studies

This section considers two fundamental test cases in one-dimension, in order to demonstrate the advantage of the new five-equation model over the standard four-equation model for simulating multispecies flows. By considering the simple case of (approximately) incompressible diffusion between two quiescent gases, the availability of an analytical solution allows us to assess the accuracy of both approaches by performing formal grid convergence studies. Galilean invariance of the governing equations also allows us to analyse the effects of advection by performing the same grid convergence study but with an additional velocity applied to the interface.

3.1 Diffusion of Air and Methane

The first test case considers the diffusion of air and methane on a domain of $0 \leq x \leq 1$, with reflective boundary conditions applied and grid sizes ranging from 32 to 2048 cells. The ideal gas equation of state is used for both gases, which have a fixed ratio of specific heats of $\gamma_{air} = 1.4$ and $\gamma_{CH_4} = 1.32$ respectively. The initial pressure and temperature are $P = 100000$ Pa and $T = 300$ K, and the molecular weights are taken to be $M_{air} = 28.964$ g/mol and $M_{CH_4} = 16.0425$ g/mol, resulting in a density ratio of 1.805. The analytical solution for the volume fraction distribution is given by:

$$z_1 = \frac{1}{2}[1 - \mathrm{erf}(\frac{x-x_0}{\sqrt{4D_{12}t+h_0{}^2}})] \qquad (20)$$

where $x_0 = 0.5$, $h_0 = 0.02$ and the diffusion coefficient is artificially fixed to be $D_{12} = 0.01$. The velocity field is initialised by assuming that the volume-weighted mean velocity is initially zero, which is equivalent to stating that the divergence of the velocity field must be zero. This leads to the following expression for the initial mass-weighted mean velocity:

$$u = \frac{-D_{12}}{\rho}\frac{\partial \rho}{\partial x} \qquad (21)$$

The grid convergence study is performed by comparing the computed and analytical volume fraction profiles for each grid resolution. The expected order of accuracy of the algorithm is second order. Table 1 gives the errors norms for the five-equation volume fraction model, while Table 2 gives the errors norms for the four-equation mass fraction model.

N_x	L^1 norm	L^∞ norm	$O(L^1)$	$O(L^\infty)$
32	6.25E-04	2.32E-03	-	-
64	1.59E-04	5.96E-04	1.98	1.96
128	3.99E-05	1.49E-04	1.99	2.00
256	1.00E-05	3.72E-05	2.00	2.00
512	2.51E-06	9.32E-06	1.99	2.00
1024	6.40E-07	2.34E-06	1.97	2.00
2048	1.71E-07	5.88E-07	1.90	1.99

Table 1: Error norms for Case 1 at $t = 0.5$ for the volume fraction model.

As can be seen in Table 1 and Table 2, the mass fraction model is actually more accurate for this particular test case. There are two reasons for why this is the case. Firstly, the majority of the errors that arise in the mass fraction approach are produced when the material interface is advected through the computational mesh [5,6]. Secondly, the current discretisation is derived for advection-dominated flows, as there is no feedback of the source terms (i.e. viscosity, diffusion) back

N_x	L^1 norm	L^∞ norm	$O(L^1)$	$O(L^\infty)$
32	4.70E-04	1.61E-03	-	-
64	1.27E-04	4.47E-04	1.89	1.85
128	3.19E-05	1.13E-04	1.99	1.98
256	7.99E-06	2.83E-05	2.00	2.00
512	2.01E-06	7.08E-06	1.99	2.00
1024	5.13E-07	1.78E-06	1.97	1.99
2048	1.39E-07	4.47E-07	1.88	1.99

Table 2: Error norms for Case 1 at $t = 0.5$ for the mass fraction model.

onto the inviscid flux component. This affects the volume fraction model more severely as part of the diffusion of species volume fraction is calculated in the inviscid flux and the remainder calculated in the viscous flux. Even though the amount of diffusion in this test case has been amplified by a few orders of magnitude, this is still not an ideal approach and future work will be aimed at mitigating such errors by discretising the advection and source terms together in a unified framework.

Figure 1 displays the error norms contained in Table 1 and Table 2 as a log-log plot, showing that second order accuracy is obtained for both the mass fraction and volume fraction models. The slightly lower order of convergence shown for both models at the finest grid resolution is because the analytical solution is for incompressible mixing, whereas very small compressibility effects are becoming resolved in the simulations.

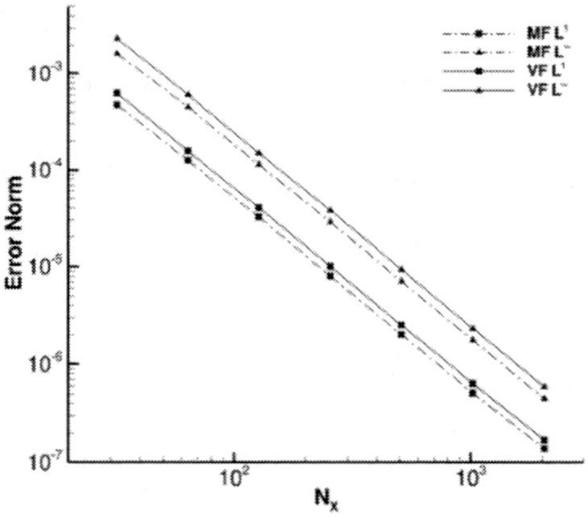

Figure 1: Comparison of error norms for the mass fraction (MF) and volume fraction (VF) models for Case 1.

3.2 Advection-Diffusion of Air and Methane

The second test case considers the same process of diffusion of air and methane, but this time with an additional mean velocity of 4 m/s applied, with the domain now $0 \leq x \leq 2$ and the boundary conditions changed to periodic. In other words, the previous case has been mirrored about $x = 1$ and is advected until the interface returns to its starting position at $t = 0.5$. The initial conditions are the same as for Case 1, and the grid sizes used in the convergence study are also doubled so as to maintain the same grid spacing. Table 3 gives the errors norms for the volume fraction model, while Table 4 gives the errors norms for the mass fraction model.

N_x	L^1 norm	L^∞ norm	$\mathcal{O}(L^1)$	$\mathcal{O}(L^\infty)$
32	3.28E-03	5.04E-03	-	-
64	2.04E-04	3.31E-04	4.01	3.93
128	7.55E-05	1.48E-04	1.43	1.16
256	1.98E-05	3.77E-05	1.94	1.97
512	5.01E-06	9.35E-06	1.98	2.01
1024	1.28E-06	2.34E-06	1.97	2.00
2048	3.42E-07	5.88E-07	1.90	1.99

Table 3: Error norms for Case 2 at t = 0.5 for the volume fraction model.

With the additional advection of the interface the advantage of the five-equation model becomes clear, with the results close to an order of magnitude more accurate. This highlights the motivation for developing the new model, as usually some advection will be present in a turbulent mixing calculation.

N_x	L^1 norm	L^∞ norm	$\mathcal{O}(L^1)$	$\mathcal{O}(L^\infty)$
32	7.91E-03	1.74E-02	-	-
64	2.15E-03	4.32E-03	1.88	2.01
128	5.33E-04	1.14E-03	2.01	1.92
256	1.33E-04	2.86E-04	2.00	2.00
512	3.33E-05	7.14E-05	2.00	2.00
1024	8.35E-06	1.79E-05	2.00	2.00
2048	2.11E-06	4.47E-06	1.99	2.00

Table 4: Error norms for Case 2 at t = 0.5 for the mass fraction model.

The results shown in Table 3 and Table 4 and visualised in Figure 2 show that for the same error, the volume fraction model can be run on somewhere between a half to a quarter of the grid resolution. Even though the five-equation model is roughly 1.5x more computationally expensive, this result indicates that it can still produce equivalent results for somewhere between 10% to 40% of the cost of the mass fraction model in one dimension (assuming a CFL restriction).

4 Conclusions

In this paper we have derived both a new set of governing equations for compressible multispecies flows, as well as a discretisation of the new equations in the form of a second

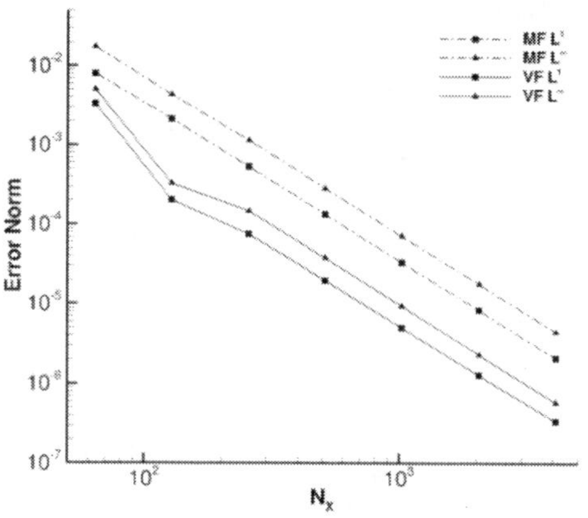

Figure 2: Comparison of error norms for the mass fraction (MF) and volume fraction (VF) models for Case 2.

order accurate finite volume method. Two fundamental test cases were used to demonstrate that the new set of equations can produce equivalent results while requiring as little as 10% of the computational cost compared to the existing approach.

5 Acknowledgment

This research was supported under the Australian Research Council's Discovery Projects funding scheme (project number DP150101059).

References

[1] L.Y.M. Gicquel, G. Staffelbach, T. Poinsot, Prog. Energy Combust. Sci. 38 (2012) 782-817.

[2] F.A. Williams, Combustion Theory, Westview Press, 1985.

[3] B. Thornber, R.W. Bilger, A.R. Masri, E.R. Hawkes, J. Comput. Phys. 230 (2011) 7687-7705.

[4] R. Abgrall, J. Comput. Phys. 125 (1996) 150-160.

[5] P. Jenny, B. Muller, H. Thomann, J. Comput. Phys. 132 (1997) 91-107.

[6] R. Abgrall, S. Karni, J. Comput. Phys. 169 (2000) 594-623.

[7] G. Allaire, S. Clerc, S. Kokh, J. Comput. Phys. 181 (2002) 577-616.

[8] V. Coralic, T. Colonius, J. Comput. Phys. 274 (2014) 95-121.

[9] S. Chapman, T.G. Cowling, The Mathematical Theory of Non-Uniform Gases, Cambridge University Press, 1970.

[10] A.W. Cook, Phys. Fluids 21 (2009) 055109.

[11] B. Thornber, Phys. Fluids 28 (2016) 045106.

[12] K. Kim, C. Kim, J. Comput. Phys. 208 (2005) 570-615.

[13] E.F. Toro, Riemann Solvers and Numerical Methods for Fluid Dynamics, Springer-Verlag, 1997.

[14] R. Spiteri, S. Ruuth, SIAM J. Num. Anal. 40 (2002) 469-491.

[15] B. Thornber, D. Drikakis, D. Youngs, R. Williams, J. Comput. Phys. 227 (2008) 4873-4894.

11th Asia-Pacific Conference on Combustion,
The University of Sydney, NSW Australia
10th -14th December 2017.

High-temperature rate constant measurement for the reaction of GVL with OH

D. Liu, F. Khaled, A. Farooq*

Clean Combustion Research Center, Physical Science and Engineering Division, King Abdullah University of Science and Technology, Thuwal 23955-6900, Kingdom of Saudi Arabia

Abstract

γ-Valerolactone (GVL) is a promising bio-based platform molecule. Hydrogen (H) abstraction is the main consumption pathway for GVL at intermediate to high temperature conditions. Overall reaction rate coefficients are measured in a shock tube for the reaction of GVL with hydroxyl radicals over 990 – 1294 K and pressures near 1.2 atm. Ultraviolet OH absorption diagnostic is used to measure the consumption of OH radicals. This work presents the first direct measurements of the titled reaction. Results are fitted in Arrhenius expression:

$$k = 4.87 \times 10^{-10} exp(-4440\ K/T)\ cm^3 molecule^{-1} s^{-1}$$

The data from this work are compared with GVL isomer and show reasonable agreement with the experimental values of methyl butanoate.

1 Introduction

There is growing interest in biofuels in search of alternative energy from fossil fuels. Recently γ-valerolactone (GVL) is proposed as a promising bio-derived platform molecule [1]. Several synthesis pathways have been studied from cellulose [2, 3] as well as hemicellulose [4, 5]. Some physical properties of GVL make it favored as a gasoline additive, such as high boiling point (480 K), low vapor pressure (0.28 Torr at 296 K) [6], and relatively large content of oxygen (reduced soot emission) [7].

Horváth et al. [2] tested the behavior of GVL as a fuel additive, by comparing the property of gasoline/ethanol (10 vol%) with gasoline/GVL (10 vol%) blends. The results showed that the data of gasoline/GVL blend were comparable with those of gasoline/ethanol blend. But GVL has lower vapor pressure than ethanol, and the production process of GVL demands less energy than ethanol. Bruycker et al. [8] studied the pyrolysis of GVL experimentally in a tubular continuous flow reactor from 873 to 1113 K and theoretically calculated the thermodynamic and kinetic data of relevant species at the CBS-QB3 level. Based on those calculations, a detailed kinetic model was developed. In their work, hydrogen abstraction and isomerization to 4-pentanoic acid were found to be the dominant GVL consumption channel. Sudholt et al. [9] extended the above mechanism for oxidation, by investigating the flame temperature as well as species concentration of the stoichiometric methane/GVL low-pressure flame both experimentally and numerically. Different from pyrolysis condition, hydrogen abstraction was observed to be the sole dominant GVL consumption pathway at low-pressure flame oxidation condition.

Despite the importance of H abstraction reaction, there is no direct measurement for this reaction. In this work, we will carry out the first direct measurement of the overall rate constant for H abstraction by OH from GVL.

*E-mail address: aamir.farooq@kaust.edu.sa

2 Experimental method

Rate constant measurement is conducted in the low-pressure shock tube (LPST) facility at King Abdullah University of Science and Technology (KAUST). A detailed description of this setup could be found elsewhere [10], here only a brief introduction is given. The KAUST LPST is made up of 9 m driver part and 9 m driven part. Both parts have same inner diameter of 14.2 cm. The driver section length is adjustable according to requirement. The incident shock speed was measured by five piezoelectric PCB transducers, which were linearly placed along the last 1.3 meter of the driven section. All measurements were performed behind the reflected shock waves. The temperature and pressure of reflected shock wave were calculated from the thermodynamic data of the gas mixture and incident shock velocity, utilizing standard shock-jump equation. The uncertainty in calculated pressure and temperature was about ±1% and ±0.7%, respectively. The shock tube facility is equipped with a 24-litre magnetically stirred mixing tank. Before each experiment, a gas mixture of GVL/tert-butyl-hydroperoxide (TBHP)/argon was prepared and left to mix for 3 hours to reach homogeneity. Helium was used as the driver gas. A rough pump and turbomolecular pump were used to evacuate the shock tube to ~10^{-5} Torr level between successive runs. The shock tube is also equipped with a heating system, for those molecules with low vapor pressure, the whole setup could be heated up to 100°C.

A 532 nm Millenia-Prime laser is used to pump the 614 nm ring-dye cw red light laser. An external frequency doubler is used after the 614 nm red light to generate the 306.69 nm ultraviolet (UV) light which is in the center of the well-studied $R_1(5)$ OH line in the A2 Σ+ ← X2 Π transition band. The OH mole fraction is calculated from Beer-Lambert Law, $I/I_o = \exp(-X_{OH} k_{OH} PL)$. In this formula, I and I_o are transmitted and incident laser intensities, respectively; P and L are pressure and absorption length; K_{OH} is the absorption coefficient; X_{OH} is the OH mole fraction. The uncertainty in pressure and temperature generates ±3% uncertainty in OH mole fraction measurement.

High purity GVL (99%) was purchased from Sigma Aldrich. A 30% TBHP in water solution was used as clean OH precursor. For all experiments, much large amount of GVL than TBHP was used in the gas mixture so that the pseudo-first-order condition ([GVL] >> [OH]) was always satisfied. Variation of GVL/TBHP ratio was used to isolate the influence of secondary reactions. The concentration of GVL ranged from 200 – 300 ppm while TBHP was 10 – 30 ppm. CHEMKIN [11] simulations were conducted to determine rate constants from the measurements. Kinetic mechanism from Sudholt et al. [9] was used to run the constant UV simulation.

3 Result and Discussion

Three different gas mixtures were used in the measurements. The first mixture contained 300 ppm GVL and 30 ppm TBHP, and second mixture 300 ppm GVL and 15 ppm TBHP, while the third mixture contained 200 ppm GVL and 10 ppm TBHP, all with balance argon. These mixtures were shock-heated to 990 – 1294 K at a pressure near 1.2 atm by shock waves. Sensitivity analysis was conducted using CHEMKIN to understand the relevant importance of reactions in our reaction system. The definition of sensitivity is as follow: $S_{OH} = (\partial X_{OH} / \partial k_i) \times (k_i / X_{OH})$, where k_i is the i_{th} reaction, and X_{OH} is OH mole fraction. One representative result presented in **Fig. 1** at 990 k and 1.23 atm shows that the OH mole fraction is most sensitive to the GVL + OH reaction. Some minor interference from CH_3 is observed, but these secondary reaction rates are taken from from Pang et al. [12] and Oehlschlaeger et al.[13], and added to the GVL mechanism. A representative measured OH mole fraction (black solid line) and best-fit simulation value (red solid line) is plotted in Fig. 2, with perturbation curves of +50% (dash-dot line) and -50%(dot line). Uncertainty in the determined rate constant mainly comes from gas composition (±5%), the uncertainty of fitting the profile (±5%), OH fraction value (±3%), temperature (±0.7%), the UV beam frequency (±0.002cm^{-1}), and zero time in fitting (±0.5 us). Each uncertainty source contribution is calculated independently, and the overall uncertainty is obtained via the root-sum-squared method. The total uncertainty for titled reaction at 990 K is calculated to be 21%, which is shown in Fig. 3 as the error bar.

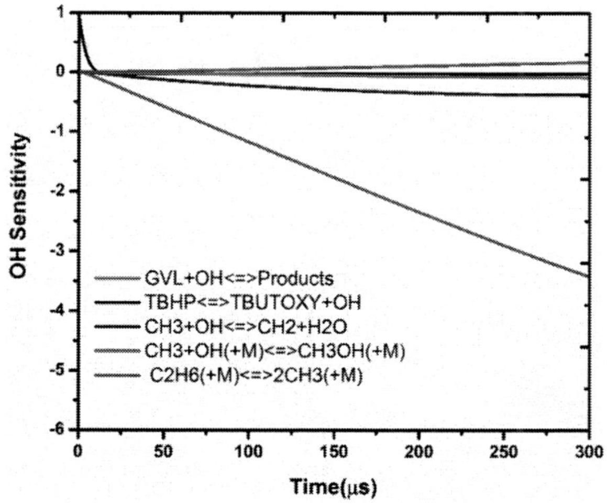

Fig. 1. Sensitivity analysis for hydroxyl radical in the reaction of GVL+OH at T = 990 K and P = 1.23 atm. Gas mixture : 200 ppm GVL + 6 ppm TBHP (14 ppm H2O) in Argon.

The measured rate constants are listed in Table 1 with corresponding reflected shock temperatures and pressures. These rates are fitted in an Arrhenius expression:

$$k = 4.87 \times 10^{-10} exp(-4440\ K/T)\ cm^3 molecule^{-1} s^{-1}$$

P_5 atm	T_5 K	k cm^{3}molecule^{-1} s^{-1}
1.12	1259	1.49E-11
1.18	1179	9.97E-12
1.16	1113	8.30E-12
1.12	1191	1.08E-11
1.17	1294	1.66E-11
1.17	1093	8.31E-12
1.23	990	6.64E-12

Table 1: Summary of GVL + OH rate constant values with temperature and pressure.

Methyl butanoate (MB) and ethyl propanoate (EP) share the same chemical formula ($C_5H_{10}O_2$) as GVL. Thus rate constant values of MB + OH and EP + OH are plotted together with our measured GVL + OH data in Fig. 3 for comparison. Solid lines in Fig. 3 denote theoretical values, and dots indicate experimental value. For methyl butanoate, Huynh et al. [14] theoretical value underestimates experimental value presented by Lam et al. [15]. At 1100 K, the experimental value is 15 times larger. Lam et al.'s [15] measured rate constant value is close to our experiment data, though our data exhibit larger activation energy. Huynh et al. [14] theoretical prediction shows similar activation energy with our data. Hou et al. [16] calculation for ethyl propanoate agrees with Huynh et al. [14] calculation for methyl butanoate at high temperatures. The measured rates for GVL is slower than those of MB, which is reasonable considering the cyclic structure of GVL.

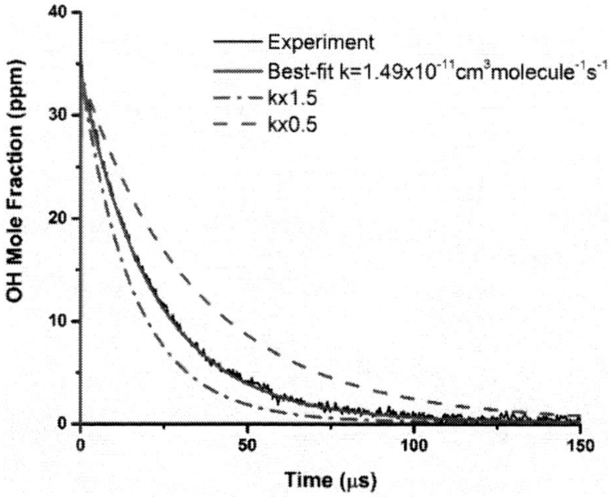

Fig. 2. Representative OH time-history profile for GVL + OH reaction at 1259 K and 1.12 atm. Gas mixture: 300 ppm GVL + 33 ppm TBHP. The best-fit simulated profile and perturbation of best-fit rate by ±50% are also plotted.

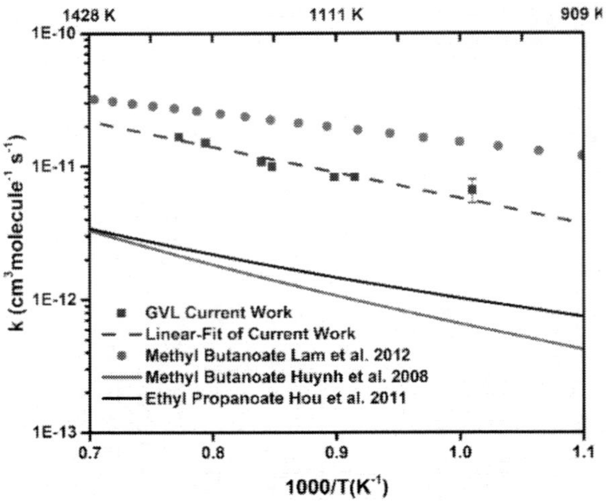

Fig. 3. Arrhenius plot of rate constant for GVL + OH ⇒ products, in comparison with methyl butanoate (theoretical and experimental data) and ethyl propanoate (theoretical data).

4 Conclusions

Overall rate constants were measured for GVL + OH reaction over 990 – 1294 K and pressures 1.2 atm. Measured rate constants were fitted in Arrhenius expression and compared with two isomers of GVL, namely methyl butanoate and ethyl propanoate. Due to cyclic structure, GVL exhibits different reactivity and different rate constants for H-abstraction in comparison to methyl butanoate. The measured data give insights about GVL high-temperature reactivity and will be very useful for detailed kinetic mechanisms.

.

References

[1] W. Li, J.-H. Xie, H. Lin, Q.-L. Zhou, Highly efficient hydrogenation of biomass-derived levulinic acid to γ-valerolactone catalyzed by iridium pincer complexes, Green Chemistry 14 (2012) 2388.

[2] I.T. Horváth, H. Mehdi, V. Fábos, L. Boda, L.T. Mika, γ-Valerolactone—a sustainable liquid for energy and carbon-based chemicals, Green Chem. 10 (2008) 238-242.

[3] X. Tang, X. Zeng, Z. Li, L. Hu, Y. Sun, S. Liu, T. Lei, L. Lin, Production of γ-valerolactone from lignocellulosic biomass for sustainable fuels and chemicals supply, Renewable and Sustainable Energy Reviews 40 (2014) 608-620.

[4] L. Bui, H. Luo, W.R. Gunther, Y. Roman-Leshkov, Domino reaction catalyzed by zeolites with Bronsted and Lewis acid sites for the production of gamma-valerolactone from furfural, Angew Chem Int Ed Engl 52 (2013) 8022-8025.

[5] R. Xing, W. Qi, G.W. Huber, Production of furfural and carboxylic acids from waste aqueous hemicellulose solutions from the pulp and paper and cellulosic ethanol industries, Energy & Environmental Science 4 (2011) 2193.

[6] D.R. Stull, Vapor pressure of pure substances. Organic and inorganic compounds., Industrial & Engineering Chemistry 39 (1947) 517-540.

[7] C.K. Westbrook, William J. Pitz, and Henry J. Curran., Chemical Kinetic Modeling Study of the Effects of Oxygenated Hydrocarbons on Soot

Emissions from Diesel Engines, The journal of physical chemistry A 110 (2006) 6912-6922.

[8] R. De Bruycker, H.-H. Carstensen, M.-F. Reyniers, G.B. Marin, J.M. Simmie, K.M. Van Geem, An experimental and kinetic modeling study of γ-valerolactone pyrolysis, Combustion and Flame 164 (2016) 183-200.

[9] A. Sudholt, R. Tripathi, D. Mayer, P.-A. Glaude, F. Battin-Leclerc, H. Pitsch, The oxidation of the novel lignocellulosic biofuel γ -valerolactone in a low pressure flame, Proceedings of the Combustion Institute 36 (2017) 577-585.

[10] J. Badra, A.E. Elwardany, F. Khaled, S.S. Vasu, A. Farooq, A shock tube and laser absorption study of ignition delay times and OH reaction rates of ketones: 2-Butanone and 3-buten-2-one, Combustion and Flame 161 (2014) 725-734.

[11] i.C.P.R.e. Reaction Design, San Diego, CA, 2010.

[12] G.A. Pang, R.K. Hanson, D.M. Golden, C.T. Bowman, High-temperature rate constant determination for the reaction of OH with iso-butanol, J Phys Chem A 116 (2012) 4720-4725.

[13] M.A. Oehlschlaeger, D.F. Davidson, R.K. Hanson, High-temperature ethane and propane decomposition, Proceedings of the Combustion Institute 30 (2005) 1119-1127.

[14] L.K.H.a.A. Violi, Thermal Decomposition of Methyl Butanoate: Ab Initio Study of a

Biodiesel Fuel Surrogate, The Journal of organic chemistry 73.1 (2008) 94-101.

[15] K.-Y. Lam, D.F. Davidson, R.K. Hanson, High-Temperature Measurements of the Reactions of OH with Small Methyl Esters: Methyl Formate, Methyl Acetate, Methyl Propanoate, and Methyl Butanoate, The Journal of Physical Chemistry A 116 (2012) 12229-12241.

[16] C. Hou, C.-G. Ci, T.-Y. Jin, Y.-X. Wang, J.-Y. Lium, Mechanism and Kinetics of the Ch3ch2c(O)Och2ch3 + Oh Reaction: A Theoretical Study, Journal of Theoretical and Computational Chemistry 10 (2011) 691-709.

11th Asia-Pacific Conference on Combustion,
The University of Sydney, NSW Australia
10th -14th December 2017.

High temperature chlorination of PbO and CdO by NaCl during incineration

Xinye Wang[1], Rong Du[1], Guilin Piao[1], Jubing Zhang[1], Min Chen[1], Yaji Huang[2]

[1] School of Energy and Mechanical Engineering, Nanjing Normal University

78 Bancang Street, Nanjing, Jiangsu, 210042, China

[2] School of Energy and Environment, Southeast University

2 Sipailou, Nanjing, Jiangsu, 210096, China

Abstract

NaCl could chlorinate PbO and CdO (low volatility) to $PbCl_2$ and $CdCl_2$ (high volatility) via SiO_2 and/or Al_2O_3 directly or indirectly. The indirect chlorination means NaCl first releases Cl-containing gas which then chlorinates metal oxide. With H_2O participation, the indirect chlorination occurred initially at 900-1000°C lower than that (1000-1100°C) with O_2 participation. The indirect chlorination could occur without O-containing gas participation initially at 1000-1100°C. The direct chlorination needed NaCl turn to vapor. Below the melting point (801°C), the slow sublimation of NaCl caused the low volatilization rate of Pb and Cd. Above the melting point, the relatively much faster evaporation of NaCl caused the much faster volatilization rate. PbO was easier to chlorinate directly by NaCl than CdO. The kinetics parameters of direct chlorination were obtained at last.

1 Introduction

Lead (Pb) and cadmium (Cd) within solid waste have been raising severe concern due to their high toxicity and elevated concentrations of 200-2000 mg/kg and 10-1000 mg/kg, respectively [1-3]. During solid waste incineration, the combustion temperature is typically above the melting points of $PbCl_2$ and $CdCl_2$ (~500°C), but close to the boiling points of PbO and CdO (~900°C) [4-6]. Consequently, due to evaporation-condensation, the semi-volatile heavy metals Pb and Cd migrate from solid fuel to bottom ash, fly ash, and even submicron aerosol, which cannot be effectively captured by conventional dedusting equipment such as electrostatic precipitators and bag filters [7-10]. Submicron Pb and Cd can directly be inhaled into the lungs and subsequently permeate into the blood stream, therefore making submicron aerosol the most harmful form of both metals [11].

During incineration, chlorine (Cl) can bond with Pb and Cd, forming $PbCl_2$ and $CdCl_2$, which have much lower melting points compared to PbO and CdO. The main Cl-donators in solid waste are PVC and NaCl supplying organic Cl and inorganic Cl, respectively [12]. It has been confirmed that PVC and NaCl could enhance the volatilization of Pb and Cd in most incineration experiments [13-15]. There are two possible paths for PVC to chlorinate metal oxides: direct chlorination and indirect chlorination. The direction chlorination means metal oxide reacts with Cl-donator directly to form metal chloride, while the indirect chlorination means Cl-donator releases Cl-containing gas (HCl, Cl_2 etc.), and then metal oxide reacts with Cl-containing gas to form metal chloride. We found PbO and CdO were chlorinated by PVC indirectly, in agreement with the conclusion from Wang et al. [13, 16]. Kosuda et al. found that ZnO was chlorinated by PVC directly, and then proposed an interesting mechanism [17].

The metal oxides chlorination by NaCl were much more complex than that by PVC. Whether direct or indirect path, some other reactants are indispensable to involve in the chlorination by NaCl. However, PVC releases HCl by its decomposition, or reacts with metal oxides without other reactants' participation. Matsuda et al. considered that NaCl could release HCl with the participation of SO_2+O_2 or SO_2+H_2O or SO_2+O_2+H_2O according to the thermodynamic calculation [18]. Besides these gases, the ashes could also cause Cl-containing gas release. It was considered that NaCl could react with silicon (Si) and/or aluminum (Al) with the participation of H_2O or O_2 [19, 20]. The direct chlorination of PbO by NaCl with the participation of SiO_2 and Al_2O_3 has been confirmed in the research of electric arc furnace dust thermal treatment [21].

In the previous research, we found the chlorination action of NaCl to PbO and CdO by the participation of SiO_2 and Al_2O_3, which represented the simplified ash contents during incineration. Like other previous researches, it was still not well understood that how NaCl chlorinated PbO and CdO during incineration. In this paper, we focused on the indirect chlorination and direct chlorination by NaCl via SiO_2 and Al_2O_3. These two paths were investigated separately to obtain the corresponding temperature windows and different characteristics between Pb and Cd. The results would facilitate the understanding of volatilization enhancement by NaCl during waste incineration.

2 Materials and Methods

The direct and indirect chlorination were investigated through two sets of experiments. The first is indirect chlorination experiments. The main step of indirect chlorination is Cl release from NaCl, so the indirect chlorination research focused on the effect of temperature and atmosphere on Cl release. The second is direct chlorination experiments which simulated the incineration condition.

Corresponding author. Fax: +86-25-85481124
E-mail address: xinye.wang@njnu.edu.cn

2.1 Experimental Materials

The proposed reactions between NaCl and the mixture of SiO_2 and Al_2O_3 are shown in (1) and (2) [16, 22].

$$2NaCl+2SiO_2+Al_2O_3+H_2O=Na_2O\cdot 2SiO_2\cdot Al_2O_3+2HCl \quad (1)$$

$$2NaCl+2SiO_2+Al_2O_3+O_2=Na_2O\cdot 2SiO_2\cdot Al_2O_3+Cl_2 \quad (2)$$

According to the equation coefficients, the mole ratio of $NaCl/SiO_2/Al_2O_3$ is 2:2:1. However, the melting of NaCl caused the fusion of residue and corundum ark during the initial attempt, so that the residue could not be moved out to for sample analysis. To avoid the extensive melting, the mole ratio was reduced to 1:2:1.

The proposed direct chlorination is shown in (3). Pb and Cd are represented as M.

$$2NaCl+2SiO_2+Al_2O_3+MO=Na_2O\cdot 2SiO_2\cdot Al_2O_3+MCl_2 \quad (3)$$

To simulate the incineration condition, the ratio of reactants did not follow the equation coefficients, but the mass fraction in municipal solid waste. In the simulated waste, the mass fractions of combustible matters and incombustible matters were 70% and 30%, respectively. Incombustible matters were composed of SiO_2 and Al_2O_3 with same mole amount. NaCl addition fraction is 1% of simulated waste. PbO and CdO were added and counted to both 1500 ppm in simulated waste. Our previous research found that combustible matters produced reducing atmosphere which promoted the volatilization of Pb and Cd significantly [16]. Therefore, combustible matters were screened out in the experiments.

2.2 Experimental Methods

The Cl release experiments were carried out using a horizontal tube furnace (Φ50mm×500mm) made of quartz. The mixtures of $NaCl+SiO_2+Al_2O_3$ were loaded in the corundum ark and placed in the middle of furnace with the heating rate of 10°C/min, the heat preservation of 30 min and the gas flow rate of 500 mL/min. Whether NaCl released Cl was judged by X-ray diffraction (XRD) detection of residues that if new compound of Na with Si and/or Al was found, Cl was considered to release out. In the experiments, there were four reaction temperatures (900°C, 1000°C, 1100°C and 1200°C) and three kinds of atmosphere taken into consideration (N_2, O_2 and O_2+H_2O). O_2+H_2O means 500 mL/min O_2 passes through 55°C water and becomes wet gas.

The direct chlorination experiments were carried out using the same furnace with the same heating rate. We supposed that the direct chlorination rates of Pb and Cd were far lower than the volatilization rates of $PbCl_2$ and $CdCl_2$. Besides that, the extremely weak volatilizations of PbO and CdO without NaCl addition were demonstrated that the total volatilization fractions were less than 1%. Therefore, the volatilization rates of Pb and Cd could be considered as the direct chlorination rates in the experiments.

In the previous researches, the volatilized Pb and Cd were collected by filtering or washing exhaust gas outside furnace for easy operation [16, 23]. However, the metal recovery was not high enough that a large part of volatilized metals deposited on the walls between corundum ark and metal collector. Therefore, we designed the second generation sampling device which worked inside furnace where volatilized metals kept vapor. Once the metals volatilized, the vapor mixed with cooling air and condensed as particles depositing on the sampling wall inside partially and the rest was captured by filter. To obtained the dynamic volatilization data, the sampling device was replaced quickly by a new one after a certain time.

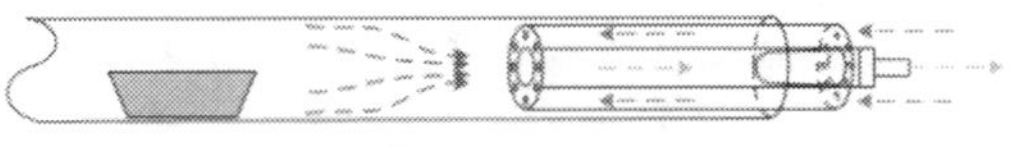

Figure 1: Diagram of dynamic sampling device and gas flow paths.

2.3 Kinetic model

The apparent kinetics were established on the basis of Arrhenius equation. The kinetic parameters were calculated using Coats-Redfern method [24].

3 Results and Discussions

3.1 Indirect chlorination

The Cl release reaction ranged from 900°C to 1200°C. At each temperature, the mixture of $NaCl/SiO_2/Al_2O_3$ was calcined at N_2, O_2 and O_2+H_2O atmosphere. The XRD patterns of residue under different reaction conditions were shown in Figure 1. In the raw materials, NaCl and SiO_2 were crystal while Al_2O_3 was amorphous. In all 900°C conditions, NaCl diffraction peaks weakened for the volatilization of NaCl. However, no new patterns were found at three atmospheres, so no reaction occurred. In 1000°C/O_2+H_2O condition, an unkown peak appeared at 2θ=22°. Then, this unknown peak appeared in all 1100°C and 1200°C conditions. It was very weak in the condition of 1100°C/N_2, and then enhanced when the temperature is 1200°C. The unkown peak responded to an unkown matter. According to the changes of patterns with temperature and atmosphere, it was considered that higher temperature was favorable to the production of this unkown matter and the positive effect of atmospheres followed the sequence of $O_2+H_2O > O_2 > N_2$.

These regularities met the expectation of Na silicate, aluminate or aluminosilicate, so this unkown matter was considered as the compound of Na which indicated Cl release and indirect chlorination. Thus, the participation of H_2O was most favorable for NaCl to release Cl (HCl) via SiO_2 and/or Al_2O_3, with the initial temperature between 900-1000°C. O_2 was corresponding to 1000-1100°C. The previous researches usually considered that the Cl release from NaCl via SiO_2 and/or Al_2O_3 needed the participation of oxygen-containing gas [16, 18, 22]. Nevertheless, the Cl release at N_2 atmosphere above 1000°C broke this consideration.

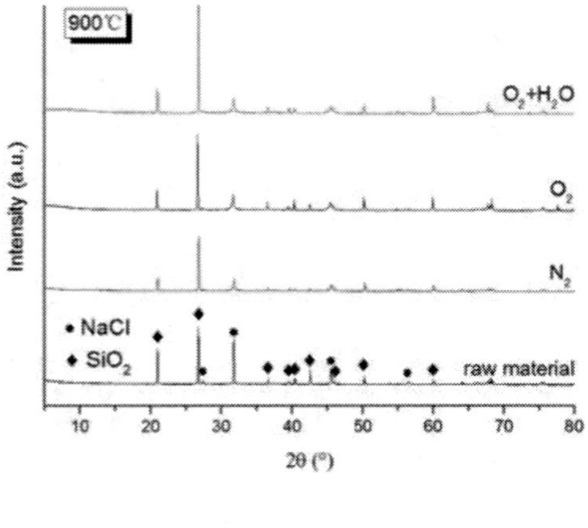

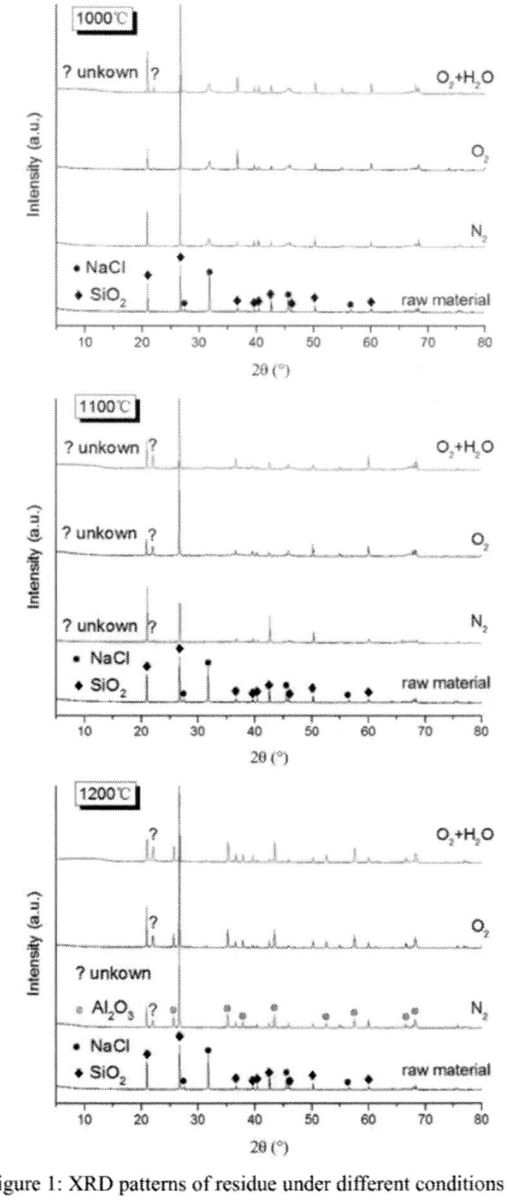

Figure 1: XRD patterns of residue under different conditions

3.2 Direct chlorination

To avoid the interfere of indirect chlorination, the direction chlorination were investigated at 600-1000°C under N_2 atmosphere so that indirect chlorination would not occurred. The evaporation fraction at different temperature range is shown in Figure 2. During the cooling step (1000-20°C), there were still considerable Pb and Cd volatilizing. It indicated that the volatilization fraction and rate were not lowered by the lack of reactants.

The weak volatilization of Pb and Cd was found at the initial temperature of 650-700°C. The significant volatilization enhancement was at 800-850°C, higher than the melting point of NaCl (801°C). Matsuda et al found that NaCl volatilization occurred at 627-727°C and suggesting that NaCl underwent a partial melting from temperature far below the nominal melting temperature [18]. The initial temperature of NaCl sublimation was close to that of direct chlorination, so the direct chlorination was considered to occur when NaCl turned to vapor. At 650-700°C, sublimation was so weak that NaCl vapor production was slow, so the direct chlorination was weak as well. At 800-850°C, evaporation was much stronger than sublimation that NaCl vapor production got much faster, so the direct chlorination was enhanced significantly.

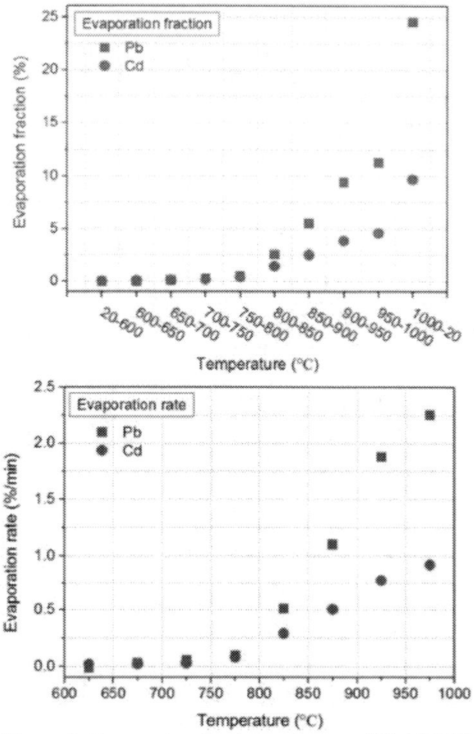

Figure 2: Evaporation fraction and rate at 600-1000°C

The thermogravimetric analysis (TGA) has shown that the volatilization rates of $PbCl_2$ and $CdCl_2$ were almost the same, as shown in Figure 3. Therefore, the higher evaporation rate of Pb than that of Cd indicated that Pb was easier to chlorinate directly than Cd.

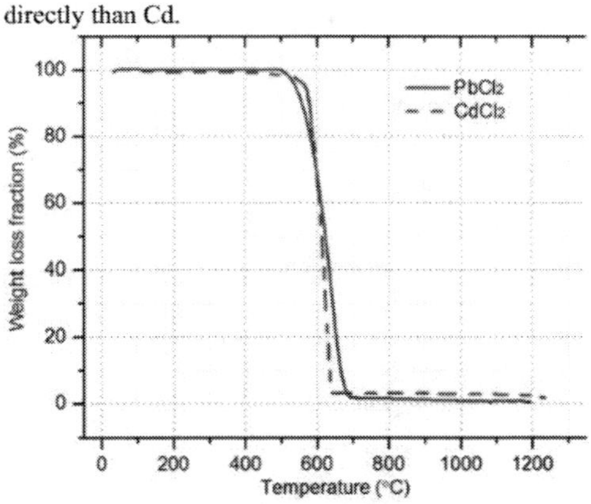

Figure 3: Weight loss curves of $PbCl_2$ and $CdCl_2$ obtained by TGA in N_2 atmosphere with heating rate of 10°C/min

According to Coats-Redfern method, the mechanism function with the maximum linearly dependent coefficient is $f(\alpha)=(1/3)(1-\alpha)^4$ for both Pb and Cd volatilization. The activation energies are 207 KJ/mol and 148 KJ/mol, respectively, while the pre-exponential factors are 1.6×10^8 min-

^{1} and 1.1×105 min^{-1}, respectively. Yoo et al investigated the direct chlorination of PbO by NaCl in electric arc furnace dust. The activation energy was 175 kJ/mol and the mechanism function was $f(\alpha)=(3/2)(1-\alpha)^{4/3}[(1-\alpha)^{-1/3}-1]^{-1}$ [21]. The differences indicates that the ash composition has significant effect on direct chlorination kinetics.

4 Conclusions

The direct chlorination and indirect chlorination were investigated separately. With H_2O participation, the indirect chlorination occurred initially at 900-1000°C lower than that (1000-1100°C) with O_2 participation. The indirect chlorination could occur without O-containing gas participation initially at 1000-1100°C. The direct chlorination needed NaCl turn to vapor. Below the melting point (801°C), the slow sublimation of NaCl caused the low volatilization rate of Pb and Cd. Above the melting point, the relatively much faster evaporation of NaCl caused the much faster volatilization rate. PbO was easier to chlorinate directly by NaCl than CdO. The kinetics parameters of direct chlorination were obtained at last. The kinetics parameters were decided by ash composition.

5 Acknowledgment

The authors gratefully acknowledge the support from the National Natural Science Foundation of China (No. 51476031 and No. 51676040).

References

[1] D. Poole, B.B. Argent, V.N. Sharifi, J. Swithenbank, Fuel 87 (7) (2008) 1318-1333.

[2] H. Zhou, A.H. Meng, Y.Q. Long, Q.H. Li, Y.G. Zhang, Renew. Sust. Energ. Rev. 36 (2014) 107-122.

[3] A.P.G. Angela Cristina Kasper, Erich Lopes Braitback de Oliveira, Nichele Cristina de Freitas Juchneski, Hugo Marcelo Veit, Electronic Waste Recycling [M], Springer International Publishing, 2015.

[4] H. Cheng, Y. Hu, Bioresource Technol. 101 (11) (2010) 3816-3824.

[5] H.M.Z. Hossain, Q.H. Hossain, M.M.U. Monir, M.T. Ahmed, Renew. Sust. Energ. Rev. 39 (2014) 35-41.

[6] B. Leckner, Waste Manage. 37 (2015) 13-25.

[7] J. Pagels, M. Strand, J. Rissler, A. Szpila, A. Gudmundsson, M. Bohgard, L. Lillieblad, M. Sanati, E. Swietlicki, 34 (8) (2003) 1043-1059.

[8] W.P. Linak, J.O.L. Wendt, Prog. Energ. Combust. 19 (2) (1993) 145-185.

[9] H. Yi, X. Guo, J. Hao, L. Duan, X. Li, J. Air Waste. Manage. 56 (9) (2006) 1243-1251.

[10] H. Yi, J. Hao, L. Duan, X. Tang, P. Ning, X. Li, Fuel 87 (10-11) (2008) 2050-2057.

[11] A. Seaton, W. Macnee, K. Donaldson, D. Godden, Lancet 345 (8943) (1995) 176-178.

[12] G.Y. Chen, N. Zhang, W.C. Ma, V.S. Rotter, Y. Wang, Fuel 140 (2015) 317-327.

[13] S.J. Wang, H. Zhang, L.M. Shao, S.M. Liu, P.J. He, Chemosphere 117C (2014) 353-359.

[14] B. Zhang, X.Y. Yan, K. Shibata, T. Uda, M. Tada, M. Hirasawa, Mater T Jim 41 (10) (2000) 1342-1350.

[15] B. Nowak, S. Frías Rocha, P. Aschenbrenner, H. Rechberger, F. Winter, 179 (2012) 178-185.

[16] X. Wang, Y. Huang, C. Liu, S. Zhang, Y. Wang, G. Piao, 192 (2017) 1-9.

[17] T. Kosuda, T. Okada, S. Nozaka, Y. Matsuzawa, T. Shimizu, S. Hamanaka, S. Mishima, 97 (4) (2012) 584-591.

[18] H. Matsuda, S. Ozawa, K. Naruse, K. Ito, Y. Kojima, T. Yanase, 60 (2) (2005) 545-552.

[19] S.S. A Jakob, RPWJ Struis, Environ Sci Technol 30 (11) (1996) 3275-3283.

[20] W.A. Punjak, F. Shadman, 2 (1988) 702-708.

[21] J.M. Yoo, B.S. Kim, J.C. Lee, M.S. Kim, C.W. Nam, Mater Trans 46 (2) (2005) 323-328.

[22] M. Uberoi, F. Shadman, 30 (4) (2002) 624-631.

[23] Y. Zhang, Y. Chen, A. Meng, Q. Li, H. Cheng, Journal of hazardous materials 153 (1-2) (2008) 309-319.

[24] L. Zhang, F. Duan, Y. Huang, Bioresource technology 175C (2014) 174-181.

11th Asia-Pacific Conference on Combustion,
The University of Sydney, NSW Australia
10th -14th December 2017.

Lean methane combustion over palladium loaded on alumina and HBETA zeolite – the role of the support on water vapour inhibition

Hadi Hosseiniamoli [1], Eric M. Kennedy[1], and Michael Stockenhuber[1]

[1]The University of Newcastle

1 University Drive, Newcastle, NSW, 2308, Australia

Abstract

Catalytic combustion of ventilation air methane is a potential solution for abatement of this greenhouse gas (GHG) source. A leading hypothesis for the observed catalyst deactivation of palladium-based catalysts when operating at low temperature is water poisoning. In this research, high silica BETA and alumina solids were used as supports for the Pd-based catalyst. The long term stability of methane conversion is significantly improved over HBETA compared to alumina supported Pd. TEM and chemisorption analyses confirms a greater level of dispersion of Pd in the HBETA compared to alumina support. TPD results shows that the Pd/Alumina catalyst adsorbs more water vapour at lower temperatures compared to Pd/HBETA resulting in an enhancement in the extent of water inhibition. Evidence for the presence of stored oxygen was observed in oxygen desorption results, where oxygen and water were co-adsorbed over the samples. This feature is also consistent with the supposition that Pd/HBETA has greater lattice oxygen mobility and a higher potential storage capacity compared to that of Pd/Al_2O_3. In addition, it was found that at lower temperatures, the effective surface coverage of oxygen was higher for Pd/HBETA compared to Pd/Alumina which was confirmed by TPD analysis.

1 Introduction

Catalytic combustion of VAM is a potential solution for reducing the net level of GHG fugitive emissions from mining [1]. Catalyst deactivation of palladium based catalysts during methane oxidation is a major challenge associated with this catalytic processes. At temperatures below 450ºC, a leading hypothesis for catalytic deactivation is poisoning of the catalyst by water vapour, which results in the formation of surface $Pd(OH)_2$ species. This catalyst inhibition due to water vapour is dependent on the type of support especially when oxygen mobility on the support has been found to influence water inhibition effect on the catalytic reaction [2]. The presence of strong polar groups such as hydroxyl groups or metal-oxygen vacancies on the surface of the catalyst are the primary source of hydrophilicity of the conventional heterogeneous materials such as alumina and zeolites [3]. For example, high Si/Al in a zeolite is equivalent to a reduction in the aluminum concentration and in turn a reduction in the concentration of bridging OH groups and a more neutral framework. This reduction in OH groups results in a decrease in the propensity of zeolites to interact with water molecules, i.e., a higher hydrophobicity and even greater thermal stability [4].

In this research, HBETA and alumina were used as supports for Pd. The activity and stability of these catalysts were evaluated under VAM oxidation conditions, and the role of support on catalytic combustion of methane in the absence and presence of water vapor in the feed and generated during reaction was studied. The physical and chemical properties of Pd/HBETA was investigated in order to have better understanding of the performance of high silica Beta zeolite in comparison with a conventional Pd/Alumina catalyst for VAM oxidation.

2 Experimental

2.1 Catalyst Preparation

HBETA (Si/Al_2=300) was purchased from Zeolyst International, Kansas, USA and alumina spheres (phase, purity) of diameter 1.5-2 mm were purchased from Sasol (South Africa) and ground via mortar and pestle to a powder form. Reagent grade palladium (II) nitrate was purchased from Sigma-Aldrich. 1.2 wt% palladium was loaded on calcined samples using incipient wetness impregnation, using Pd(II) nitrate solution (10 wt% in 10 wt% nitric acid). Catalysts were oven dried at 110°C for approximately 20 h. The dried catalyst was then crushed and sieved to 250 – 400 µm. The sieved catalyst particles were calcined in a tubular fixed-bed reactor at 500°C for 2 h under air.

2.2 Characterization Techniques

Surface area measurements were performed by using Micromeritics Tristar surface area analyser via nitrogen adsorption and desorption at 77 K. Micromeritics Vacprep 061 was used for degassing of the samples. A JEOL 2100 Transmission Electron Microscope (TEM) with EDS was used for imaging nano-sized Pd particles on the surface of the support. A purpose built Chemisorption apparatus, with volumetric glass apparatus, was used to measure metal dispersion and active particle size by hydrogen chemisorption at 35ºC. The quantity of Pd loaded on catalysts was measured using a Varian 715-ES inductively coupled plasma optical emission spectrometer (ICP-OES). A purpose built Temperature-programmed desorption (TPD) apparatus with a Pfeiffer Prisma quadrupole mass analyser was used for detection the H_2O and O_2 desorption.

2.3 Catalytic Activity Measurements

Catalytic combustion of lean methane was performed in a tubular stainless steel fixed bed micro reactor. The inlet methane concentration was set at 0.7% in air with a total gas hourly space velocity (GHSV) of 100 000 h^{-1}. Methane conversion was monitored using a gas chromatograph (Shimadzu GC-2014) with packed Hysep and Molecular Sieve dual column, and a thermal conductivity detector (TCD). The

Corresponding author. Fax: +61-49-854-410
E-mail address: hadi.hosseiniamoli@uon.edu.au

temperature of the bed was measured by a thermocouple placed into the catalyst bed. An average carbon mass balance of 97% was obtained for all experiments. For the saturated feed experiment, water vapour was introduced into the feed stream. The average water vapour concentration was determined by a humidity probe and determined to be constant, with approximately 85% relative humidity (RH) at ambient temperature (25°C ± 3°C).

3 Results and discussion

3.1 Catalyst Stability

The stability of Pd/HBETA and Pd/Al_2O_3 under VAM conditions, for 24 h continuous reaction on stream, was assessed. As plotted in Fig .1, the stability of both catalysts was assessed under dry conditions during the early period of each experiment. Following this period, 3.4% water vapour, balanced in air, was introduced into the feed stream. This resulted in a sudden drop in catalyst activity for both catalysts. During time on stream analysis, the Pd/HBETA catalyst was shown to rapidly return to its initial activity while the Pd/Al_2O_3 catalyst continued to deactivate over the 24 h period of reaction. This highlights a start difference in the influence water vapour has on the stability of the catalyst in terms of the use of different supports. Pd/HBETA displayed a resistance to water poisoning in comparison to Pd/Al_2O_3, which deactivated even under dry feed conditions.

It has been reported that the presence of water vapour in the inlet feed stream is the primary factor for Pd based catalyst deactivation, greater than catalyst deactivation through the production of water vapour during the reaction [5]. The reason for continued deactivation of Pd/Al_2O_3 catalyst by water inhibition may be related to the presence and role of hydroxyl groups on the surface of the alumina. Hydroxyl groups originating from water vapour accumulate on the active sites of Pd/Al_2O_3 by formation of $Pd(OH)_2$ species and these species in turn block access to the PdO sites for methane dissociation. This interpretation is consistent with that proposed by others in the literature [5].

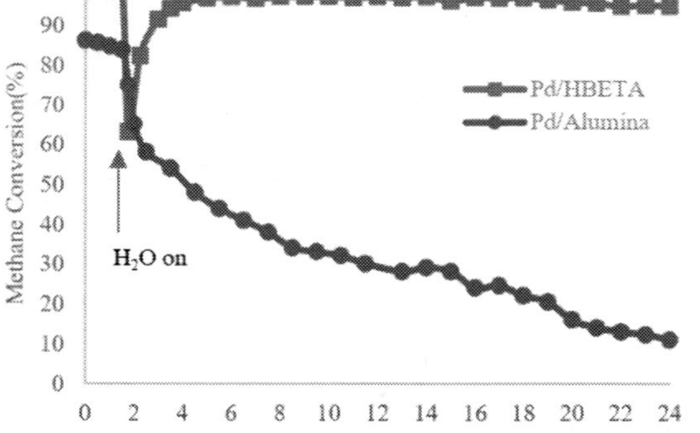

Figure 1: Methane conversion as function of time on stream at a constant bed temperature 400°C, GHSV 100 000 h^{-1}.

Stability tests were performed at 400°C, a temperature at which water inhibition was found to be significant over Pd/Al_2O_3. Fig. 2 shows the experimental results for methane conversion over Pd/Al_2O_3 to achieve 90% conversion and how this temperatures is needed to be altered in order to maintain this conversion level. In order to maintain a methane conversion level of 90%, the temperature was increased to over 500°C progressively during the 50 h period of time on stream. It is evident that the catalyst stability after 50 h of reaction at the temperature over 500°C, presumably at a temperature which is sufficient to eliminate the inhibiting effect of water causing deactivation of the catalyst. It can be concluded that the water inhibition effect is most notable at temperatures less than 500°C. Note that there is still slow deactivation occurring at temperatures higher than 500°C.

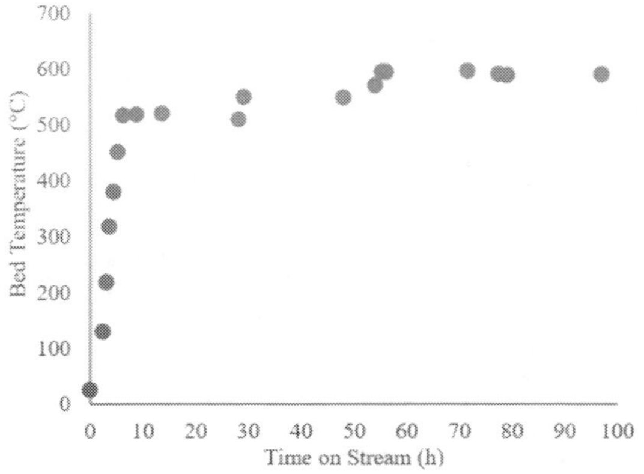

Figure 2: Methane conversion over Pd/Al_2O_3 to achieve approximately 90% conversion as a function of time on stream, GHSV 100 000 h^{-1}

3.2 Physical and Chemical Properties of Catalyst

The ICP result showed the concentrations of palladium on the surface of Pd/HBETA and Pd/Al_2O_3 were 1.18 wt% and 1.06 wt%, respectively. Surface area of Pd/HBETA and Pd/Al_2O_3 were determined by N_2 physiosorption. Not surprisingly, Pd/HBETA has much higher surface area compared to Pd/Alumina because of the nature of pore micropores in zeolite HBETA. The BETA zeolite framework provides approximately 20.5 % accessible volume [6]. The gas adsorption analysis was used to estimate the Langmuir surface area of 694 m^2g^{-1} and 247 m^2g^{-1} for Pd/HBETA and Pd/Al_2O_3, respectively. Since the surface area of Pd/HBETA is more than double that for Pd/Al_2O_3, it may provide a more accessible surface for Pd particles dispersion. Therefore, it is suggested that the nature of BETA support appears to engender relatively high level of Pd particle dispersion in comparison to an alumina support, consistent with the TEM images shown in Fig. 3.

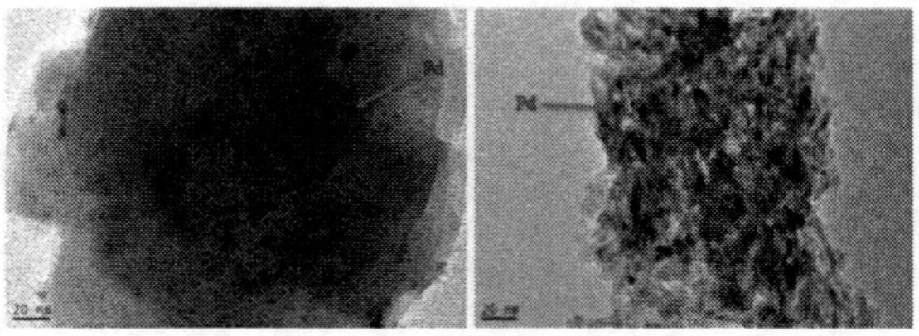

Figure 3: TEM images for Pd/HBETA (left) and Pd/Al_2O_3 (right)

Scanning Transmission Electron Microscopy (STEM) was used to study Pd dispersion on the surface of Pd/HBETA and Pd/ Al_2O_3 catalysts. Fig. 3 shows a typical TEM image and on the surface of these catalysts, respectively. STEM images disclose a smaller average Pd particle size, with a higher level of dispersion on the surface of HBETA compared to that of Al_2O_3. Statistical analysis shows the distribution of Pd particles in the range of 1 nm to 10 nm on both catalyst supports. Based on TEM analysis, most of the PdO particles on the Pd/ HBETA support are in the average range of 2-3 nm compared to 4-5 nm for Pd/Al_2O_3. It can be concluded that the nature of support influences the distribution and morphology of Pd nanoparticles, and this finding has been reported by others in literature [7, 8]. The trend in the differences in Pd dispersion has also been confirmed by H_2-chemisorption, with Pd dispersion from H_2-chemisorption at 35°C suggests a Pd dispersion of 24.6 % for Pd/HBETA and 17.4% for Pd/ Al_2O_3.

The surface properties of Pd/HBETA in comparison with Pd/ Al_2O_3 were also investigated by Temperature Program Desorption (TPD) analysis. The effect of water vapour on catalyst activity was examined using this technique in order to understand the interactions between water, support and active sites of catalyst. Figure 4 shows the results of TPD analysis for desorption of water vapour from active sites of fresh catalysts Pd/HBETA and Pd/Alumina. Prior to water vapour adsorption, fresh catalysts were calcined for 1 h at 500 °C (rate 5°C/min^{-1}), followed by H_2O adsorption at 110 °C. This adsorption temperature was used in order to reduce the extent of physical adsorption of water vapour on the catalyst surface [9]. Water desorption was then performed from room temperature up to 700°C. Low temperature peaks were observed for Pd/Alumina between 200°C and 400°C and a high temperature peak, observable between 550°C and 650°C, was detected from both catalysts. It is suggested that the higher temperature peaks are related to the desorption of surface hydroxyl groups of the alumina solid lattice [10, 11]. Zeolites structure can contains different types of silanol groups, which may result in preferential adsorption sites for water, and by increasing the temperature, silanol dehydroxylation can occur with the formation and detection of additional water molecules [12]. In the present study, involving the use of high silica HBEA, the reduced number of silanol groups in the framework, encountered with dehydroxylation, results in the formation and detection of additional desorbing water molecules as is exhibited in TPD spectra at around 600°C.

The area of the water desorption peak at low temperatures for Pd/Alumina is much larger than the area associated with the higher temperature peak. This suggests that, in our study, the Pd/Alumina catalyst adsorbed more water at lower temperatures resulting in an enhancement in the extent of water inhibition on the rate of combustion. This is consistent with the stability data shown in Fig. 1 and Fig. 2. In contrast, the Pd/HBETA catalyst, with its significantly reduced water adsorption capacity at lower temperature meant reduced influence of water vapour. It has been reported that the inhibiting effect of water vapour, present in the reaction mixture, increases with a lower dispersion of Pd and a higher concentration of water vapour in the feed [13]. Subsequently, the inhibitory effect of water vapour on the Pd/Alumina catalyst, which has a lower level Pd dispersion compared to Pd/HBETA catalyst is predicted and was observed in the present investigation.

In order to understand the role of support oxygen capacity and mobility involved in catalytic activity in presence of water vapour, O_2 and H_2O were co-adsorbed over Pd/HBETA and Pd/Al_2O_3 catalysts. In order to prepare catalysts for this aspect of the study, catalyst samples were heated under vacuum at 500°C for 1 h to remove any pre-adsorbed compounds. Following exposure of the catalyst to oxygen and H_2O(v) at 110°C, the samples were heated under vacuum to 800°C. Fig. 5 shows that the maximum rate of oxygen desorption was observed below 400°C for Pd/HBETA and above 600°C for Pd/Alumina, the trend be the opposite to that observed for H_2O desorption (see Fig. 5). These observations would suggest that the oxygen adsorption capacity of Pd/HBETA is higher than Pd/Alumina. It is reported in literature that high Pd dispersion on the support can lead to higher oxygen adsorption capacity [13], and this is consistent with the results of TEM analysis where Pd/HBETA has a higher level of Pd dispersion compared to Pd/Alumina. Additionally, the support can play an important role in influencing Pd oxygen adsorption. It is generally accepted that there is a strong correlation between the oxygen adsorption capacity of noble metals and their catalytic activity, and the influence of support on the Pd adsorption of oxygen [13, 14]. This role can be attributed to the creation of oxygen vacancies in the support and increased lattice oxygen mobility, which would suggest that the desorption peak at around 400°C in Pd/HBETA (attributed to surface oxygen adsorbed on PdO species) leads to a higher catalytic activity at this temperature, which was consistent with the data acquired in the present study. In addition, the peak that appears above 600°C could be assigned to the decomposition of a PdOx species at high temperatures [15, 16] and this occurred along with a change in color of the recovered catalyst from TPD apparatus caused by Pd reduction at high temperatures.

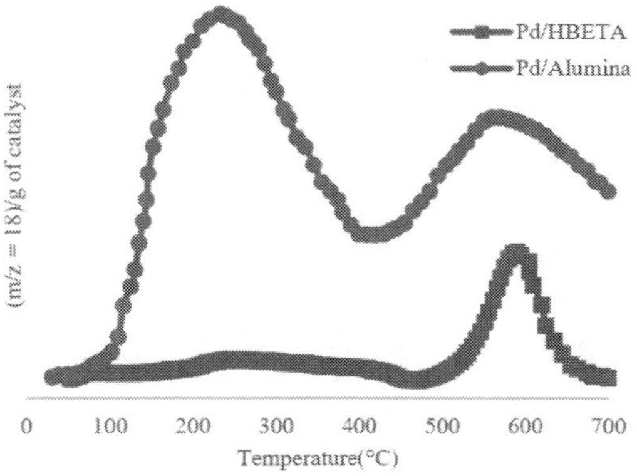

Figure 4: TPD spectra of water desorption from Pd/HBETA and Pd/Alumina, Water adsorbed at 110°C.

The TPD data indicates that varying the type of support, the oxygen adsorption capacity and potentially lattice oxygen mobility of the catalyst can lead to a change in the inhibiting effect of water over the catalyst. Pd/HBETA has enhanced oxygen storage capacity compared to Pd/Al_2O_3, and these attributes are supported by TPD data analysis. Oxygen storage and lattice oxygen mobility of a catalyst is involved in limiting water inhibition [17, 18]. Evidence for the presence of stored and mobile oxygen was observed in oxygen desorption TPD results displayed in Fig.5, where oxygen and water were co-adsorbed over the samples. It was found that at lower temperatures, the effective surface coverage of oxygen was higher for Pd/HBETA compared to Pd/Alumina. Moreover, competitive adsorption between H_2O and O_2 led to lower oxygen surface coverage over the Pd/Al_2O_3 catalyst which resulted in a lower stability. However, the relative coverage of H_2O and O_2 reduced when increasing the reaction temperature, because the heat of water adsorption is lower than oxygen adsorption on Pd-based catalysts [19, 20]. This suggests that the high silica zeolite, with their particular adsorption characteristics, would be a more suitable support compared to alumina for Pd metal catalyzed methane combustion at low temperatures.

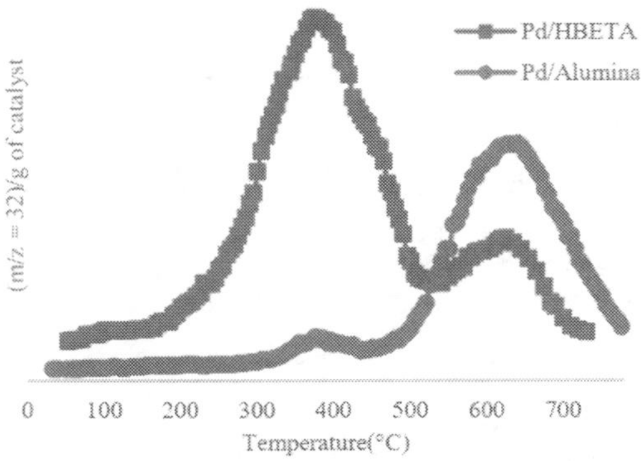

Figure 5: TPD spectra of O_2 (m/z = 32) desorbed from catalysts. Oxygen was co-adsorbed with H_2O at 110°C.

4 Conclusions

Lean methane conversion over Pd loading on HBETA and alumina was investigated. Methane conversion, as a function of time on stream under humid conditions, results in an improvement in the stability of HBETA compared to alumina support. TEM and chemisorption analyses confirm a greater level of dispersion of Pd in the HBEA compared to alumina support. The high Si/Al ratio in Pd/HBETA (and corresponding reduced aluminium concentration) leads to a more neutral framework which results in greater hydrophobicity and higher thermal stability. This, in turn, resulted in decreasing catalyst potential for oxygen adsorption due to competition of active sites with water vapour, confirmed by TPD results. This feature is also consistent with the supposition that Pd/HBETA has greater oxygen storage capacity compared to that of Pd/Al_2O_3, which has again been confirmed by TPD analysis.

5 Acknowledgment

We would like to acknowledge ACARP for their generous financial support of the project. We thank Hongwei Liu for assistance with TEM analysis. We are grateful to acknowledge financial support provided by University of Newcastle. Also, we would appreciate Jenny Zobec and Huiming Zhang from the X-ray Unit at UON for their help in TEM analysis.

References

[1] P. Gélin, M. Primet, a review. Applied Catalysis B: Environmental, 2002. 39(1): p. 1-37.

[2] D. Ciuparu, M. Lyubovsky, E. Altman, L. Pfefferle, A. Datye, Catalysis Reviews, 2002. 44(4): p. 593-649.

[3] A. Sah, H. Castricum, A. Bliek, D. Blank, J. Ten, Journal of membrane science, 2004. 243(1): p. 125-132.

[4] N. Chen, The Journal of Physical Chemistry, 1976. 80(1): p. 60-64.

[5] A. Setiawan, J. Friggieri, E. Kennedy, B. Dlugogsrski, Mi. Stockenhuber, Catalysis Science & Technology, 2014. 4(6): p. 1793-1802.

[6] Coster, D., A. Blumenfeld, J. Fripiat, The Journal of Physical Chemistry, 1994. 98(24): p. 6201-6211.

[7] Y. Liu, S. Wang, T. Sun, C. Zhang, Applied Catalysis B: Environmental, 2012. 119: p. 321-328.

[8] L. Wenge, , G. Deyong, X. Xin, China Petroleum Processing & Petrochemical Technology, 2012. 14(3): p. 1-9.

[9] A. Setiawan, J. Friggieri, H. Hosseiniamoli, E. Kennedy, B. Dlugoohorski, A. Adesina, M. Stockenhuber, Physical Chemistry Chemical Physics, 2016. 18(15): p. 10528-10537.

[10] M. Zamora, A. Cordoba, The Journal of Physical Chemistry, 1978. 82(5): p. 584-588.

[11] M. Egashira, M. Nakashima, S. Kawasumi, T. Selyma, The Journal of Physical Chemistry, 1981. 85(26): p. 4125-4130.

[12] D. Serrano, G. Calleja, J. Botas, F. Gutierrez, , Separation and purification technology, 2007. 54(1): p. 1-9.

[13] B. Stasinska,A. Machocki, K. Antonaik, M. Rotko, F. Goncalves, Catalysis Today, 2008. 137(2–4): p. 329-334.

[14] J. Nilsson, P. Carlsson, S. Fouladvand, N. Martin, J. Gustafson, M. Skogundh, ACS Catalysis, 2015. 5(4): p. 2481-2489.

[15] Y. Lou, J. Ma, W. Hu, L. Wang, X. Cao, P. Hu, ACS Catalysis, 2016.

[16] L. Simplício, S. Sales, L. Lietti, F. Bozon., Applied Catalysis B: Environmental, 2006. 63(1): p. 9-14.

[17] J. Au-Yeung, K. Chen, A. Bell, E. Iglesia, Journal of Catalysis, 1999. 188(1): p. 132-139.

[18] C. Bozo, N. Guihaume, M. Primet E, Catalysis Today, 2000. 59(1–2): p. 33-45.

[19] Li, G., P. Xiao, and P. Webley, Langmuir, 2009. 25(18): p. 10666-10675.

[20] G. Zakumbaeva, S. Artamonov, Reaction Kinetics and Catalysis Letters, 1979. 10(2): p. 183-186.

11th Asia-Pacific Conference on Combustion,
The University of Sydney, NSW Australia
10th -14th December 2017.

The structure and modelling analysis of a high Ka CH_4/air stratified premixed jet flame using DNS

Haiou Wang[1*], Bruno Savard[1], Evatt R. Hawkes[1,2], Jacqueline H. Chen[3]

[1]School of Mechanical and Manufacturing Engineering, The University of New South Wales, NSW 2052 Australia

[2]School of Photovoltaic and Renewable Energy Engineering, The University of New South Wales, NSW 2052 Australia

[3]Sandia National Laboratories, Livermore, CA 94550, USA

Abstract

In the present study, a direct numerical simulation (DNS) of a high Karlovitz number (Ka) CH_4/air stratified premixed jet flame is performed and anlysed. The central jet has an equivalence ratio of 0.4, which is considerably smaller than the pilot equivalence ratio of 0.9. A reduced chemical mechanism for CH_4/air combustion with NO_x based on GRI-Mech3.0 was used. More than 5 billion grid points were employed in the DNS. The effects of high Ka were significant and the flame structure was rather distributed due to the effects of small-scale turbulent eddies. The modelling of the highly turbulent stratified premixed flame was examined in the framework of tabulated chemistry models. It was found that stratified premixed flamelets based on a reactant-to-product counterflow laminar flame configuration are able to predict the conditional means of the DNS accurately by accounting for the interactions of different mixture fractions/progress variables, particularly when the assumption of unity Lewis number is employed.

1 Introduction

Many combustion applications, including gas turbines, furnace and lean-burn internal combustion engines, operate in a lean premixed mode [1] to reduce pollutant emission and improve combustion efficiency. However, lean premixed combustion is difficult to achieve in highly turbulent environments due to operational issues such as combustion instability, extinction, and emissions of CO and unburned hydrocarbons. Stratified premixed combustion has been shown to extend the flammability limits and improve the overall-lean combustion stability [2–6]. Improved understanding of stratified premixed combustion is therefore desirable to enable improved design of efficient and clean combustion devices.

In stratified premixed flames, a reaction front propagates through a range of equivalence ratios and the flame structure is influenced by mixture stratification. Earlier studies on stratified premixed combustion focused on laminar flame configurations. [7-11]. Recently, an increasing number of experiments have been performed to understand turbulent stratified premixed flames in various configurations [12-19]. Both large-scale stratification and small-scale stratification have been found to influence the stratified premixed flame structure.

Direct numerical simulation (DNS) resolves all relevant physical and chemical scales both in space and time, and it plays an increasingly important role in understanding turbulence/flame interactions of turbulent flames [20]. However, only few DNS of turbulent stratified premixed flames have been reported so far and none of them was performed in conditions of high Karlovitz numbers (Ka). The Karlovitz number measures the ratio of flame characteristic time to Kolmogorov time scale [21]. In many practical lean combustion systems, the relatively weak lean stratified premixed flames interact with intense turbulence, resulting in a high Ka.

Unfortunately, it is not feasible to perform DNS of turbulent stratified premixed flames in practical combustors, even with the most powerful supercomputing resources. Accurate prediction of these flames relies significantly on combustion sub-models for turbulent stratified premixed flames. In the literature, there have been a number of studies of turbulent stratified premixed flames in the context of Reynolds-averaged Navier-Stokes (RANS) and large-eddy simulation (LES), among which tabulated chemistry methods, such as flamelet generated manifolds (FGM) and non-premixed flamelets, were widely used [6, 22-25]. However, FGM does not account for the flux in the mixture fraction space while non-premixed flamelets can not capture the flame propagation [26]. It is therefore desirable to develop/validate more suitable combustion sub-models for turbulent stratified premixed flames.

Based on the above discussion, the objective of the paper is to perform the first DNS of a high Ka stratified premixed jet flame, present some preliminary results of this flame, and discuss the modelling implications. The paper is organised as follows. Section 2 describes the DNS configuration and numerical methods employed in the present work. The DNS results are shown and discussed in Section 3. Finally, some conclusions are drawn in Section 4.

2 Numerical methods

Recently, Wang et al. [27-30] studied a high Ka experimental premixed flame using DNS, in which the jet diameter was D = 1.5 mm, the jet bulk velocity, U_b, was 110 m/s, and the jet Reynolds number, Re, based on D and U_b was 10,500. The jet operated at atmospheric pressure and a temperature of 300 K. The jet equivalence ratio was $\phi_j = 0.7$. A pilot flame of CH_4/air mixture at $\phi_c = 0.9$ was established surrounding the burner to provide a hot co-flow to assist with flame stabilization. In the present work, the same configuration is employed. However, the jet bulk velocity is increased to U_b = 165 m/s, and the jet Reynolds number reaches Re = 15,764. Also, the jet has an

Corresponding author. Fax: +02-93-857-762
E-mail address: haiou.wang@unsw.edu.au

equivalence ratio of ϕ_j = 0.4, which is smaller than the equivalence ratio of the pilot (ϕ_c = 0.9), so that the DNS features mixture stratification.

In the DNS, the mean jet axial velocity profile is specified using a power law velocity profile for an axisymmetric fully developed turbulent pipe flow. A turbulence field is obtained by generating an auxiliary homogeneous isotropic turbulence field based on a prescribed Passot-Pouquet energy spectrum [31]. The turbulent velocity is 4 m/s and the integral length scale is 0.75 mm, which are consistent with those reported in [27, 28]. The isotropic turbulence field is filtered outside of the jet then added to the mean inlet velocity using Taylor's hypothesis. The filtering function has the following top-hat form:

$$u'(r)/u'_{\max} = \frac{1}{2} \times \left(1 + \tanh(\frac{D/2 - r}{\delta_u})\right) \quad (1)$$

Where the shear layer width δ_u is specified as 0.025D.

The Karlovitz number is evaluated as Ka = τ_L/τ_η, where τ_η is the Kolmogorov time scale defined as $\tau_\eta = (\tilde{\nu}/\tilde{\epsilon})^{1/2}$ and τ_L is the flame time scale. Here, $\tilde{\nu}$ is the Favre averaged viscosity and $\tilde{\epsilon}$ is the Favre averaged kinetic energy dissipation rate. Since the flame is stratified, there is no unique flame time scale. A reference flame time scale from a laminar premixed flame with an equivalence ratio of 0.7 is selected for the analysis, which results in τ_L = 3.4 ms. The variations of Ka at different axial locations along the isoline of $\tilde{c}$ = 0.9 are estimated based on this time scale as shown in Table 1, where c is the progress variable and the tilde represents Favre averaging. It is obvious that the Ka are extremely high and the flame is located in the broken reaction zones regime of the premixed combustion regime diagram [21].

x/D	$\tilde{c}$ = 0.9
8	887
16	580
24	430
32	312

Table 1: The variations of the Ka at different axial locations along the isoline of $\tilde{c}$ = 0.9.

The temperature and species mass fraction profiles are specified using the following expression [27, 28]:

$$\psi(r) = \psi_c + \frac{\psi_j - \psi_c}{2}\left(1 + \tanh(\frac{D/2 - r}{\delta_s})\right) \quad (2)$$

where ψ_j and ψ_c denote the values of ψ in the jet and the pilot, respectively. The radial distance is denoted as r and δ_s is a characteristic thickness for the scalar variations. ψ_c is determined from a chemical equilibrium calculation of a CH_4/air mixture with ϕ = 0.9 at a defined temperature of 1,800 K

The physical domain for the simulation is $L_x \times L_y \times L_z = 48D \times 36D \times 36D$ in the streamwise, x, and lateral directions, y and z, respectively. A uniform grid spacing of Δx = 20 μm is used in the streamwise direction x, while algebraic stretching is used in the lateral directions y and z outside of the shear zone. The grid in the y (z) direction is uniform with Δy (Δz) = 20 μm in the region between y/D (z/D) = -5 and 5. The grid spacing of the DNS is chosen to adequately resolve both the flame and turbulence structures. In the present high Ka case, the resolution of turbulence is more demanding than that of the flame. The Kolmogorov length scale, denoted as $\eta = (\tilde{\nu}^3/\tilde{\epsilon})^{1/4}$, has a minimum of 8 μm in a narrow region near the potential core where the flame does not overlap. The criterion, $\eta/\Delta x > 0.5$ [32], is satisfied elsewhere. Thus, the smallest scales of the turbulent flow are reasonably resolved in the flame region. The resultant number of grids is $N_x \times N_y \times N_z = 3600 \times 1200 \times 1200$. To remove artefacts from initialisation, the simulation was first advanced for $8\tau_j$ on a half-resolved grid, where τ_j is the flow through time defined as $\tau_j = L_x/U_b$. The results were then mapped to the fine grid, and the solution was advanced for another $3\tau_j$ to provide stationary statistics.

The DNS were carried out using the code S3D [33], which was originally developed at Sandia National Laboratories. It solves the compressible governing equations of mass continuity, momenta, energy, and mass fractions of chemical species including chemical reactions using finite difference methods. The equation sets are solved in a fully coupled way including the two-way interaction of heat release and flow.

A previously published reduced chemical mechanism for premixed CH_4/air flames with NO_x, derived from GRI-Mech3.0, was employed [27, 34]. The skeletal mechanism contains 268 elementary reactions and 44 species, of which 28 species were transported on the DNS grid [34]. Constant Lewis numbers, determined from a fit to mixture-averaged transport properties in a premixed flame, are used for the species. The Lewis numbers were presented in Ref. [27].

The simulation was performed on Titan at Oak Ridge Leadership Computing Facility (OLCF) on 192,000 Opteron processor cores for approximately 5 days, requiring 25 million CPU hours, and on Edison at National Energy Research Scientific Computing Center (NERSC) on 48,000 Intel Ivy Bridge processor cores for 3 days, requiring 3.5 million CPU hours.

3 Results and discussion

Figure 1 shows the instantaneous distributions of heat release rate and mass fractions of CH_4, CH_2O and OH. It is noted that, for methane/air combustion, CH_4 is attacked by radicals such as OH and H in the fuel consumption layer, and forms CH_3, which further produces CH_2O, H_2, CO and other intermediates. The species CH_2O characterises the preheat zone of the flame, heat release rate the reaction zone and OH the oxidation zone. One notable feature of Fig. 1 is that the flame is relatively weak in the near field. Heat release rate is enhanced in the downstream region, where the radicals have higher levels of concentration. Most of the heat is produced in the primary reaction zone, where all the fuel is consumed. However, it is seen that non-negligible heat release rate also takes place in the high temperature region, behind the primary reaction zone. This region is denoted as the secondary reaction zone, similar to that discussed in [35]. The existence of the secondary reaction zone is mainly attributed to the fuel fragments (CO, H_2, etc.) in the high temperature region, and involves exothermic reactions, including $CO+OH \Leftrightarrow CO_2+H$ (R94), $H_2+OH \Leftrightarrow H_2O+H$ (R79), etc.

It is also seen from Fig. 1 that the flame structure is influenced by turbulence significantly. As mentioned in the previous section, the Ka of the present flame is extremely high. Therefore, strong interactions between turbulence and the flame are expected. The broadening of the flame due to turbulence is evident. Particularly, turbulent eddies can enter into the preheat zone and reaction zone, the thickness of which increases progressively downstream. The detailed mechanism for the flame thickening in high Ka jet flames has been explored in Wang et al. [28]. For the sake of brevity, it is not repeated here.

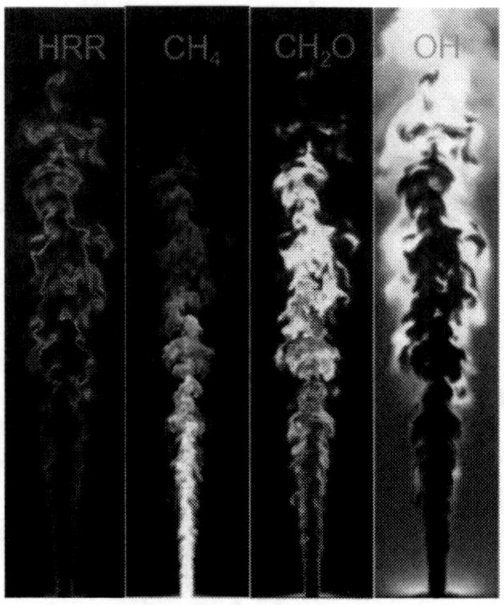

Figure 1: Instantaneous distributions of heat release rate and species mass fractions of CH_4, CH_2O and OH.

Next, the modelling of the current turbulent stratified premixed flame is discussed in the framework of tabulated chemistry models. In particular, a reactant-to-product counterflow flame configuration is employed to generate the flamelets using FlameMaster [36]. Following the work of Wang et al. [30], the reactants and products in the counterflow flame calculations are taken locally from the DNS. In particular, the temperature and composition in the reactants and products of the laminar flames are determined from the corresponding values from the DNS at various axial locations. Therefore, stratified premixed flamelets are generated, which is able to account for the interactions of different mixture fractions/progress variables across the flame.

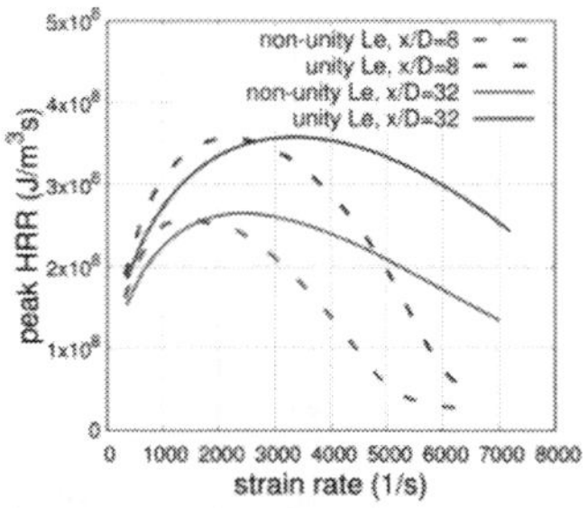

Figure 2: Peak HRR as a function of strain rate in counterflow laminar flames with both unity and non-unity Lewis numbers, and with the reactants taken from the centerline at two axial locations in the DNS.

Various strain rates are applied to the flamelets and the response of the peak heat release rate (HRR) to the strain rate is revealed in Fig. 2. The results of the flamelets with unity and non-unity Lewis numbers (Le) are included. It is seen that the sensitivity of HRR to the strain rate is reduced with increasing axial location, as a result of the different boundary conditions in the flamelets extracted from the DNS, but the maximum peak HRR is not influenced. Differential diffusion significantly reduces the amplitude of the peak HRR, and this is consistent both at $x/D = 8$ and 32.

The comparison of the flame structure at various axial locations between the turbulent stratified flame and counterflow laminar flames are shown in Fig. 3. The strain rate in the flamelets is chosen such that the magnitude of the peak HRR matches as closely as possible that of the conditional mean from the DNS (see Fig. 2), which is consistent with the approach taken in Wang et al. [30]. It is encouraging to observe that the flamelets are in good agreement with the conditional means of the DNS. Overall, the unity Le flamelets approximate the DNS better. These results indicate the tabulated chemistry method using stratified premixed flamelets with unity Le assumptions has the potential to predict this highly turbulent stratified premixed flame accurately, which will be tested *a-posteriori* in future work.

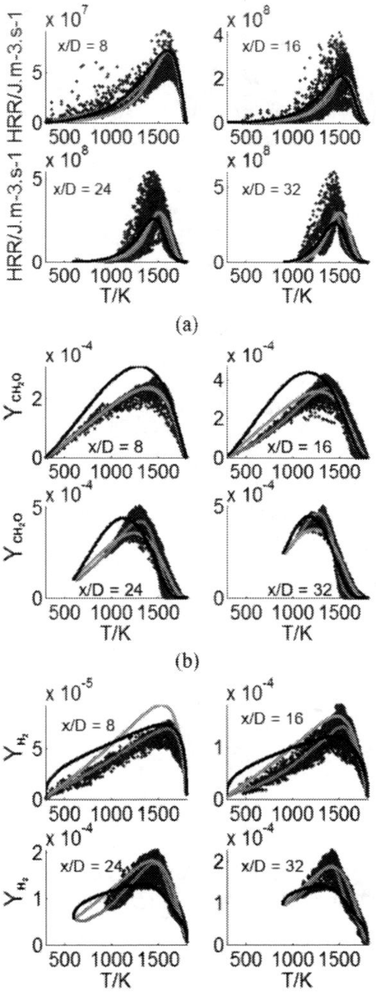

Figure 3: Comparison of the flame structure at four axial locations between the turbulent stratified flame and counterflow laminar flames for (a) HRR, (b) Y_{CH2O} and (c) Y_{H2}. Red line: DNS results; green line: results from unity Le flamelets; black line: results from non-unity Le flamelets.

4 Conclusions

Direct numerical simulation (DNS) of a high Ka CH_4/air stratified premixed jet flame was performed for the first time. The DNS data were analysed and it was found that the instantaneous flame structure is influenced by turbulence significantly. Turbulent eddies can enter into the preheat zone and reaction zone, the thickness of which increases progressively downstream. The conditional means of heat release rate, and species mass fractions of CH_2O and OH were compared with the results from stratified premixed flamelets. It was observed that the flamelet results are in good agreement with the conditional means of the DNS, with the unity Le flamelets approximating the DNS better.

5 Acknowledgment

This research used resources provided by the National Energy Research Computing Center (NERSC) which is supported by the Office of Science of the US DOE under contract No. DE-AC02-05CH11231, the Oak Ridge Leadership Computing Facility at the Oak Ridge National Laboratory which is supported by the Office of Science of the U.S. DOE under contract No. DE-AC05-00OR22725, and the Pawsey Supercomputing Centre with funding from the Australian Government and the Government of Western Australia. This work was supported by the Australian Research Council. The work was also supported by computational resources at Pawsey awarded through the National Computational Merit Allocation Scheme. The work at Sandia National Laboratories was supported by the Division of Chemical Sciences, Geosciences and Biosciences, the Office of Basic Energy Sciences, the US DOE. Sandia National Laboratories is a multiprogram laboratory operated by Sandia Corporation, a Lockheed Martin Company, for the US Department of Energy under contract DE-AC04-94-AL85000.

References

[1]. D. Dunn-Rankin, Lean Combustion, Academic Press 2007.

[2]. T. Kang, D.C. Kyritsis, Proc. Combust. Inst. 31 (2007) 1075–1083.

[3]. Y.M. Marzouk, A.F. Ghoniem, H.N. Najm, Proc. Combust. Inst. 28 (2000) 1859–1866.

[4]. R. Sankaran, H.G. Im, Proc. Combust. Inst. 29 (2002) 77–84.

[5]. A.R. Masri, Proc. Combust. Inst. 35 (2015) 1115–1136.

[6]. A.N. Lipatnikov, Prog. Energy Combust. Sci. 62 (2017) 87-132.

[7]. A. Pires Da Cruz, A.M. Dean, J.M. Grenda, Proc. Combust. Inst. 28 (2000) 1925–32.

[8]. E.S. Richardson, V.E. Granet, A. Eyssartier, J.H. Chen, Combust. Theory Model. 14 (2010) 775–792.

[9]. R. Zhou, S, Hochgreb, Combust. Flame 160 (2013) 1070–1082.

[10]. X. Shi, J.Y. Chen, Combust. Flame 184 (2017) 233–245.

[11]. B. Coriton, M.D. Smooke, A. Gomez, Combust. Flame 157 (2010) 2155–2164.

[12]. G. Kuenne, F. Seffrin, F. Fuest, T. Stahler, A. Ketelheun, D. Geyer, D. Geyer, J. Janicka, A. Dreizler, Combust. Flame 159 (2012) 2669–2689.

[13]. S.K. Dhanuka, J.E. Temme, J.F. Driscoll, Proc. Combust. Inst. 33 (2011) 2961–2966.

[14]. M.S. Sweeney, S. Hochgreb, M.J. Dunn, R.S. Barlow, Combust. Flame 160 (2013) 322–334.

[15]. C. Galizzi, D. Escudié, Combust Flame 157 (2010) 2277–2285.

[16]. P.C. Vena, B. Deschamps, G.J. Smallwood, M.R. Johnson, Proc. Combust. Inst. 33 (2011) 1551–1558.

[17]. P. Anselmo-Filho, S. Hochgreb, R.S. Barlow, R.S. Cant, Proc. Combust. Inst. 32 (2009) 1763–1770.

[18]. N. Pasquier, B. Lecordier, M. Trinité, A. Cessou, Proc. Combust. Inst. 31 (2007) 1567–1574.

[19]. S. Meares, A.R. Masri, Combust. Flame 161 (2014) 484–495.

[20]. J.H. Chen, Proc. Combust. Inst. 33 (2011) 99–123.

[21]. N. Peters, Turbulent combustion, Cambridge University Press, 2000.

[22]. E. Knudsen, H. Pitsch, Combust. Flame 2012 (159) 242-264.

[23]. P. Auzillon, O. Gicquel, N. Darabiha, D. Veynante, B. Fiorina, Combust. Flame, 159 (2012) 2704–2717.

[24]. F. Proch, A.M. Kempf, Combust. Flame 161 (2014) 2627–2646.

[25]. B.A. Perry, M.E. Mueller, A. R. Masri, Proc. Combust. Inst. 36 (2017) 1767–1775.

[26]. P. Dguyen, L. Vervisch, V. Subramanian, P. Domingo, Combust. Flame 157 (2010) 43–61.

[27]. H. Wang, E.R. Hawkes, B. Zhou, J.H. Chen, Z. Li, M. Aldén, Proc. Combust. Inst. 36 (2017) 2045–2053.

[28]. H. Wang, E.R. Hawkes, J.H. Chen, B. Zhou, Z. Li, M. Aldén, J. Fluid Mech. 815 (2017) 511–536.

[29]. H. Wang, E.R. Hawkes, J.H. Chen, Phys. Fluids 28 (2016) 095107.

[30]. H. Wang, E.R. Hawkes, J.H. Chen, Combust. Flame 180 (2017) 110–123.

[31]. T. Passot, A. Pouquet, J. Fluid Mech. 118 (1987) 441–466.

[32]. S.B. Pope, Turbulent flows, Cambridge University Press, 2000.

[33]. J.H. Chen, A. Choudhary, B. de Supinski, et al., Comput. Sci. Discov. 2 (2009) 015001.

[34]. T.F. Lu, C.K. Law, Combust. Flame 154 (2008) 761–774.

[35]. C. Jimenez, B. Cuenot, T. Poinsot, D. Haworth, Combust. Flame 128 (2002) 1–21.

[36]. H. Pitsch, Flamemaster: a C++ computer program for 0D combustion and 1D laminar flame calculations, 1998.

An investigation of high power laser pulses on soot using an IR pump UV probe approach

Hamdy A. Ahmed[1, 2], Matthew J. Dunn[1], Daniel Bartos[1], Assaad R. Masri[1]

[1]The University of Sydney

Sydney, NSW, 2006, Australia

[2]Zagazig University

Zagazig, Sharkia, 44519, Egypt

Abstract

A comprehensive understanding of the mechanisms and interaction between light and solid particles is crucial for the optimal development of many applications including concentrated solar radiation and laser light-driven reactions. In this study, we investigate the interaction of intense, high-power pulsed infrared (IR) laser-light with particles produced in a sooting ethylene–air laminar diffusion flame using an IR pump and ultraviolet (UV) probe technique. The IR pump pulse preferentially heats highly carbonaceous soot particles, whereas the UV probe pulse is used to monitor scattering and fluorescence from molecules and particles over a range of delay times. In the low sooting region of the flame, the IR pulse does not show any detectable interaction with the nanoparticles for fluence levels up to 1.7 J/cm^2. In the sooting region of the flame at IR fluence levels sufficient for sublimation processes to occur, the temporal and spectral response of the signal generated from the UV probe signal indicates a significant decrease in the fluorescence signal. The IR pump pulse causes up to a 50% drop in scattering from the UV probe pulse, indicating a significant change in the size and morphology of the largest scattering particles. This study indicates that although IR light does not interact directly with fluorescent species and particles in flames, there is a significant indirect interaction mechanism at high fluence levels.

1 Introduction

The interaction of high-intensity light and solid particles is relevant to many current and emerging material synthesis and energy transformation processes, especially those are potentially powered by solar radiation. Solar radiation has been used to drive solar gasification [1] and hybrid combustion systems [2], where the interaction of the light with particles and droplets is the primary coupling mechanism of the solar energy to the system. Solar pumped lasers have been shown to be able to produce highly concentrated continuous wave light [3] as well as pulsed light at high instantaneous power levels [4]. These solar pumped lasers can be then utilised to generate temperatures and enable chemical reactions that can be used in applications such as magnesium reduction [3], ablation based nanoparticle production [5] and ablation plumes to deflect asteroids [6].

Although there are many current and future applications that rely on a thorough understanding of light-particle interactions, the actual mechanisms of high-intensity light-particle interactions are not thoroughly understood [5]. One of the challenges of understanding light-particle interactions is due to the complex particle, chemical and optical interaction mechanisms that can vary for different conditions and materials [5]. One area where there has been significant progress in understanding the interaction of high-intensity light-particle interactions is laser-induced incandescence (LII) of soot in combustion systems [7]. In LII, a high-power pulsed laser rapidly heats highly carbonaceous soot particles to temperatures close to 4000 K. The subsequent incandescence emitted from the heated soot is then used to determine information on the local soot fraction [7], mean size and morphological distribution [8].

As the primary purpose of LII is a diagnostic technique to determine information on the soot from the incandescence signal, only a modest number of studies have focused on the properties of the soot and the probed medium after the LII process. The LII heating process causes the soot to form multi carbon onion shells [9], increases particle annealing and graphitisation as well as causing size and mass loss due to fragmentation, sublimation and other mass loss mechanisms [7]. As would be expected from such potentially dramatic structural changes during and after LII, the resultant soot particles optical properties are also modified, as determined using IR pump IR probe experiments [8, 10] as well as IR pump, near IR and visible wavelength line of sight absorption probe measurements [11].

While a reasonable body of knowledge exists on the evolution of the structure of the soot particles properties during and after LII, the fate of the lost mass from the soot particles during the LII process is not well understood. A consensus has formed in the literature that sublimation is the dominant process that limits the maximum soot temperature in LII [12]. For fluence levels larger than the plateau initiation fluence, C_2 and C_3 emission will start to interfere with the blackbody emission spectrum from the incandescing soot [7]. However, Gloay et al [13] have found that C_2 and C_3 emission in high fluence LII is not in local thermal equilibrium, as would be expected from a sublimation process but rather in a higher excited state. The excited C_2 and C_3 state emission indicates that either or both, non-thermal particle ejection process from the soot particle and heterogeneous C reactions on the soot particle surface are important in the formation of carbon clusters. These small carbon clusters can then condense back onto the original soot particles [14], or through homogeneous nucleation or heterogeneous nucleation onto fragments form new particles [15]. The importance and occurance of carbon cluster formation and subsequent cluster nucleation is supported by the

Corresponding author. Fax: +61-2-9351-7060
E-mail address: hamdy.ahmed@sydney.edu.au

results of Stipe et al. [16]. Stipe et al. [16] report complete soot particle photofragment and subsequently generate large amounts of carbon fragments by a high fluence (14.7 J/cm^2) 193 nm excimer laser and a new particle with a regular size distribution and a much smaller diameter than the original particles. Recent measurements by [11] also seem to support the idea of the nucleation of the carbon clusters as an important mechanism after LII, as they report an evolution of the local absorption properties 50-800 ns after the LII pulse. The change in absorption properties has been hypothesised to be attributed to sublimated soot material forming light-absorbing gaseous species.

The work presented in this paper focuses on furthering the understanding of the nature, mechanism and evolution of the lost mass from soot particles following intense pulsed heating (LII). Understanding the lost mass from rapidly heated particles, while relevant to many applications, is particularly relevant to the generation of tailored and controlled particle structures from pulsed lasers. Specifically, in this study, we employ a 1064 nm IR pump pulse followed by a 266 nm UV probe beam pulse approach. The use of a UV probe beam enables a broad range of optical interactions to be probed including fluorescent nanoparticles [17], carbon clusters such as C_2 and C_3 as well as soot. The signal generated from the UV probe beam is spectrally dispersed into four spectral regions, with temporal evolution of the signal in each of the four regions monitored and recorded by a fast response photomultiplier tube (PMT). Results are recorded for a range of probe beam delays up to 1 µs and pump beam fluences from 0.06-1.7 J/cm^2. Measurements are recorded in a laminar sooting ethylene diffusion flame at two locations, a very low sooting location and a high sooting location.

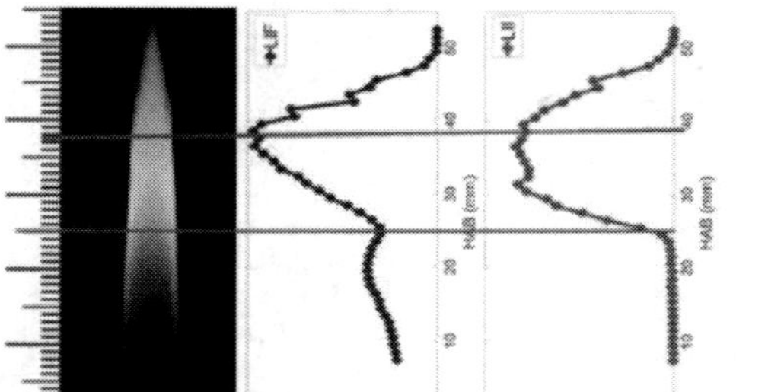

(a) (b) (c)

Figure 1: Overview of the probed flame structure, visible flame emission (a), 266 nm UV excited fluorescence emission at 350 nm (b), LII signal which is proportional to soot volume fraction (c).

2 Experimental Setup

A duplicate of the Yale laminar jet diffusion flame burner [18] is utilised in this study as a steady and repeatable source of soot particles. The burner features are 4 mm inner diameter jet surrounded concentrically by a 74 mm outer diameter coflow of air. The jet stream is a mixture of 60% ethylene diluted with 40% nitrogen by volume. The mean velocities of the jet and coflow air are maintained at 35 cm/s, resulting in a visible laminar flame height of 55 mm as shown in Fig. 1a. Measurements are acquired at two heights above the burner (HAB), on the centerline of the flame, a very low sooting but nanoparticle producing region at HAB = 25 mm and in a highly sooting region at HAB = 37.5 mm. To give context to the selection of the measurement locations relative to the flame structure, the centreline axial profiles of nanoparticle fluorescence and LII are shown in Fig. 1 b and c. The results shown in Fig. 1b) and c) were measured using an optical diagnostic setup used in previously reported measurements investigating nanoparticle and soot formation in flames [19] and are included in more details in another paper which is under review.

The optical diagnostic setup utilised in this paper is presented in Fig. 2, and is similar to that used in [19]. The main difference in this paper to previous studies is the timing of the UV and IR beams is varied and a range of IR beam energies are utilised. For the IR pump beam, the fundamental of an Nd:YAG 10 Hz pulsed laser (Spectra-Physics Quanta-Ray Pro-350) at 1064 nm and 8 ns FWHM is focused to a spot size diameter of 450 µm. The UV probe beam is generated by the frequency quadrupled output of a picosecond mode-locked Nd:YAG 10 Hz pulsed laser (Ekspla PL2251) with a 80 ps pulse duration, forming a spot size of 250 µm diameter at the probe volume. Three IR pump beam fluence levels are examined in this study, 0.06, 0.72, and 1.7 J/cm^2, while the UV probe beam fluence is held constant at 1.1 J/cm^2. The delay between the pump and probe beam is generated by a digital delay generator (SRS DG535) for a range of delays up to 1 µs. The signal generated from the pump and probe beams from the probe volume is collected and collimated by an $f_\#$=3.4 off-axis parabolic mirror and subsequently focused onto the 20 µm entrance slit of the 0.5 m spectrometer by another off-axis parabolic mirror. The spectrometer spectrally disperses the collected light and temporal evolution of four spectral windows (PMT1 @ 266±15 nm, PMT2 @ 350±15 nm, PMT3 @ 445±15 nm, PMT4 @ 575±15 nm) are monitored by PMTs (3x Hamamatsu H10721-210 and 1xH10721-20). The signals from the four PMTs are recorded by a fast response oscilloscope (4 GHz bandwidth, Tektronix DSA70404C) at a sample-rate of 25 GS/s. All results presented are based on an average of 500 realisations to minimise the effects of noise.

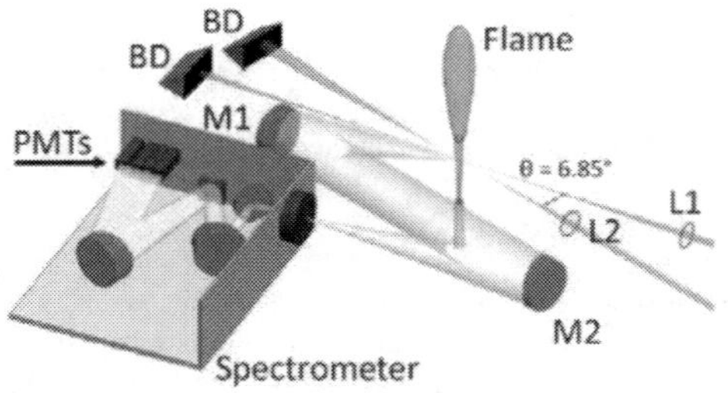

Figure 2: An overview of the experimental setup including the IR pump beam, UV probe volume, collection optics and PMT based spectrometer. M1 and M2: off-axis parabolic mirrors, BD: beam dump, L1 is the 266 nm laser focusing lens, and L2 is the 1064 nm laser focusing lens.

3 Experimental Approach and Data Reduction

By delaying the UV probe beam relative to the IR pump beam the evolution of the soot and lost mass from the soot during the LII process has been investigated over a broad range of times, up to 1 µs. However, as a result of the IR pump beam, an LII signal is generated that is still present when the UV probe beam arrives at the probe volume. As such, in the signal produced by the UV probe beam, there is some level of interferences from the residual IR probe beam LII process. The signal generated by the UV probe is corrected for LII interferences by subtracting a reference case for the corresponding delay with no UV probe beam. By doing so, we are inherently assuming an independence of the LII signal and the signal generated by the probe beam. This assumption of independence of LII and probe beam signal, whilst sufficient for the two UV based PMTs, will need to be investigated further in future studies for PMT4 centered at 575 nm. The LII interference on the UV probe beam signal and its correction with the final signal used for processing is shown in Fig. 3.

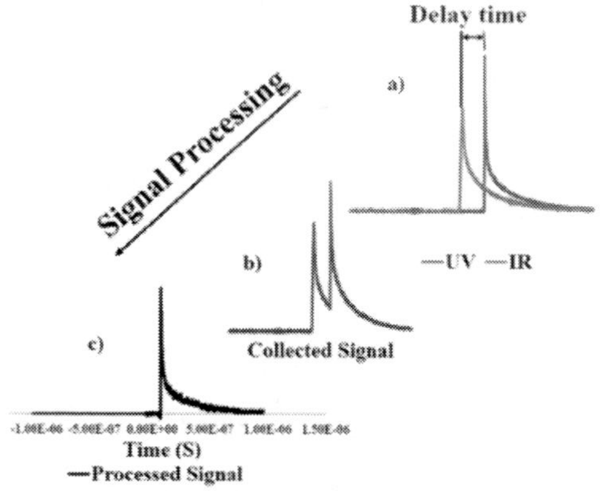

Figure 3: Signal processing example at HAB=37.5 mm for PMT4 (575 nm), pure IR and UV signals without interaction (a), measured interacting IR and UV (b), measured UV with IR subtracted (c).

4 Experimental Results and Discussions

The results for two positions on the centreline of the studied flame are presented here, HAB=25 mm and HAB=37.5 mm. The results are processed by a background subtraction and a residual LII correction, the peak value of the resultant signal is used for subsequent analysis. As the signal generated by the probe beam resembles an exponential decay signal, other parameters such as the decay rate and integrated signal are possible to extract, but analysis of these measures is left for a future publication. For the results at HAB=25 mm in Fig. 4, the differences in the signal levels for the four PMTs are within the uncertainty of the measurement system (~2%) for all delays times and IR pump intensities. It is shown in Fig. 1 at this location that there is no significant soot from the LII and some fluorescence, we infer that there are some fluorescent nanoparticles in this region as per previous measurements [17] but no large soot. As such the conclusion here is that the IR pump does not perturb or significantly interact with the nanoparticles in the flame.

The results for the high sooting region, HAB = 37.5 mm, are presented in Fig. 5. The scattering levels at 266 nm on PMT1 at zero delay in Fig. 5a) are much larger than those reported in HAB=25 mm in Fig. 4a) due to the presence of large particles of soot which is further backed up by the plot of relative soot volume fraction reported in Fig. 1 c) illustrating the high levels of soot at HAB = 37.5 mm vs. HAB=25 mm. For the low fluence level of 0.06 J/cm^2, there is no significant signal change with probe beam delay for any of the wavelength regions reported in Fig. 5. However, for the higher fluence measurements, 0.72 and 1.7 J/cm^2, there is a significant evolution of the interaction with the probe beam delay. Fig. 5 a) indicates that for high fluences after 25 ns delay, the scattering signal is decreased by up to 50%, indicating that the soot particles have undergone a significant change in structure to decrease the scattering level, through a combination of sublimation, fragmentation and particle ejection. At least within the delay window that we have examined, heterogeneous nucleation and condensation of the lost mass from the soot particles to recover a structure similar to the original structure before the LII process can be ruled out. An additional contributing factor in the scattering signal reduction is the change in the structure of the soot particles due to processes such as annealing. However, it is difficult to argue that these physical structural changes alone, and not particle mass loss during LII, causes the 50% drop in the scattering signal.

Without the IR pump pulse, the signal generated on PMTs 2, 3 and 4 (UV and visible) at HAB=37.5 mm is a combination of nanoparticle fluorescence, UV generated LII and UV photofragmentation of the soot particles. For the high IR pump fluence level cases, a significant decrease in signal with delay

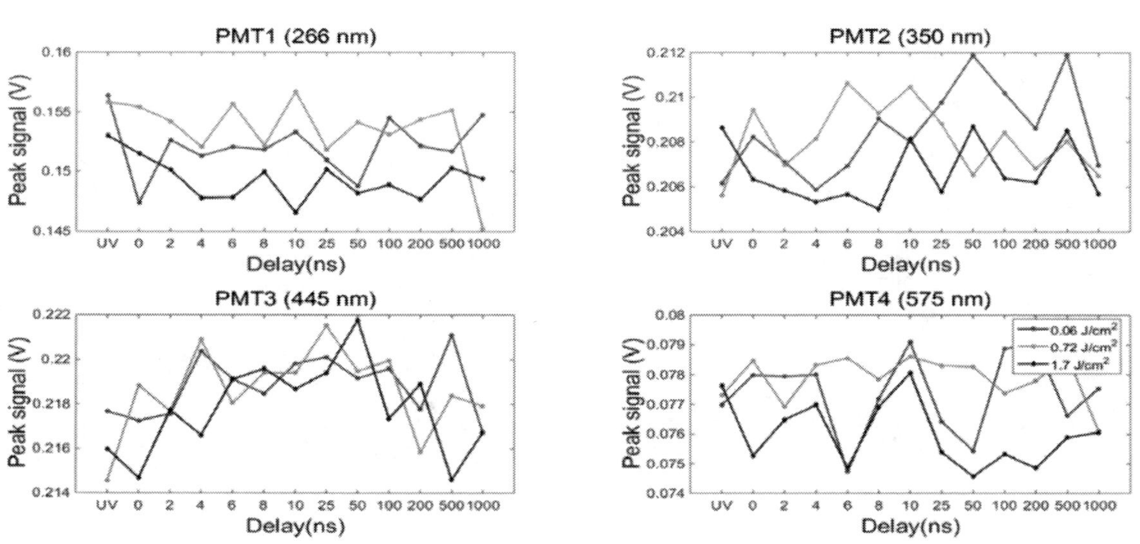

Figure 4: Peak signals for the 4 PMTs at 25 mm HAB for 1064 laser fluences of 0.06, 0.72, and 1.7 J/cm^2.

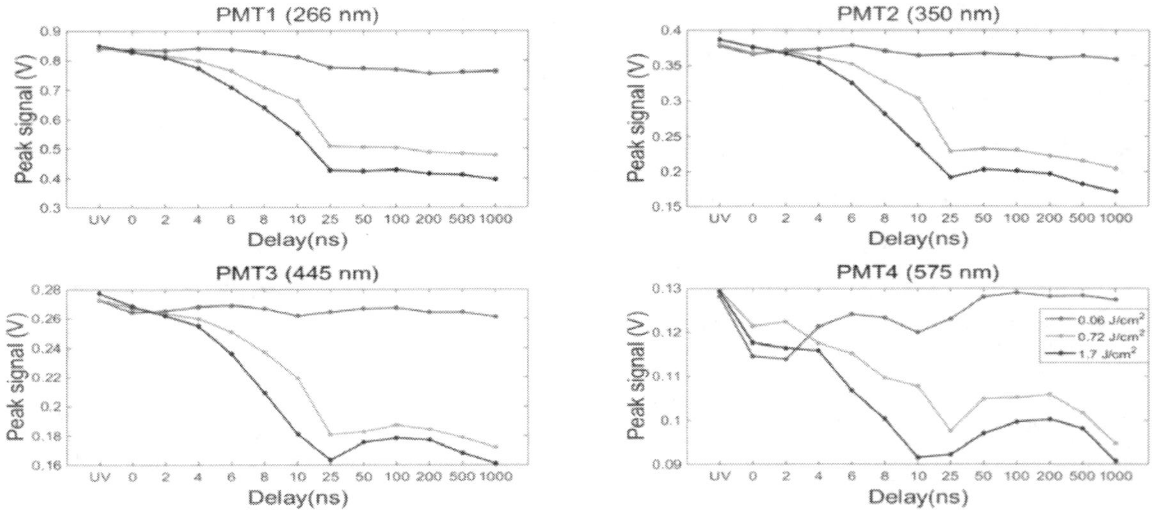

Figure 5: Peak signals for the 4 PMTs at 37.5 mm HAB for 1064 laser fluences of 0.06, 0.72, and 1.7 J/cm^2.

time is observed, which for PMT2 is close to a 50% decrease. Although the unambiguous interpretation of these results requires further work, some observations can be made. The fluorescence yield from nanoparticles coated via heterogeneous reactions and condensation of with the carbon clusters and fragments lost by the soot particles will be reduced. Significant modification of the structure of the nanoparticles by oxidisation due to local gas heating from the LII is unlikely in the relatively short delay times employed. The mass loss, soot particle fragmentation and particle structure modifications such as annealing will decrease the efficiency of the UV based photofragmentation process of the soot. However, the signal in the UV and visible decreases at high pump fluence levels where it might be expected that carbon cluster fragments and sublimation products from the soot would generate C_2, C_3 and larger carbon clusters which would be expected to have high quantum yield efficiencies and increase the signal on the UV and visible PMTs for 266 nm excitation.

5 Conclusions

In this study, we have investigated the effect of high power laser pulses on carbonaceous nanoparticles and soot in laminar ethylene flames using an IR pump and UV probe approach. The IR pump beam does not show any significant photochemical interaction with the nanoparticles up to high fluence levels (1.7 J/cm^2). The IR pump beam interaction with soot creates strong interactions with the local optical properties that evolves over time causing a reduction of up to 50% in the local scattering and UV signal from the UV probe beam. Further study is needed to get a clear understanding of the possible interactions happened during the period from 25 to 1000 ns and beyond after the excitation of the IR pulse.

6 Acknowledgment

This research is supported by the Australian Research Council. We acknowledge the financial support from the Egyptian government in the form of a scholarship for the student.

References

1. A. Z'Graggen; A. Steinfeld, Chemical Engineering and Processing: Process Intensification 47 (4) (2008) 655-662
2. X. Dong; Z. Sun; D. Gu; P. J. Ashman; Z. T. Alwahabi; B. B. Dally; G. J. Nathan, Combustion and Flame 171 (2016) 103-111
3. M. Oliveira; D. Liang; J. Almeida; C. R. Vistas; F. Gonçalves; R. Martins, Solar Energy Materials and Solar Cells 155 (2016) 430-435
4. M. Lando; Y. Shimony; Y. Noter; R. M. J. Benmair; A. Yogev, Applied Optics 39 (12) (2000) 1962-1965
5. C. B. Stipe; J. H. Choi; D. Lucas; C. P. Koshland; R. F. Sawyer, Journal of Nanoparticle Research 6 (5) (2004) 467-477
6. A. Gibbings; M. Vasile; I. Watson; J.-M. Hopkins; D. Burns, Acta Astronautica 90 (1) (2013) 85-97
7. H. A. Michelsen; C. Schulz; G. J. Smallwood; S. Will, Progress in Energy and Combustion Science 51 (2015) 2-48
8. E. Cenker; W. L. Roberts, Applied Physics B 123 (3) (2017) 74
9. A. Inoue; T. Seto; Y. Otani, Carbon 50 (3) (2012) 1116-1122
10. R. L. Vander Wal; T. M. Ticich; A. B. Stephens, Applied Physics B 67 (1) (1998) 115-123
11. M. Saffaripour; K.-P. Geigle; D. R. Snelling; G. J. Smallwood; K. A. Thomson, Applied Physics B 119 (4) (2015) 621-642
12. C. Schulz; B. F. Kock; M. Hofmann; H. Michelsen; S. Will; B. Bougie; R. Suntz; G. Smallwood, Applied Physics B 83 (3) (2006) 333
13. F. Goulay; L. Nemes; P. E. Schrader; H. A. Michelsen, Molecular Physics 108 (7-9) (2010) 1013-1025
14. F. Memarian; F. Liu; K. Thomson; K. Daun; D. Snelling; G. Smallwood, Applied Physics B: Lasers and Optics 119 (4) (2015) 607-620
15. H. A. Michelsen; A. V. Tivanski; M. K. Gilles; L. H. Van Poppel; M. A. Dansson; P. R. Buseck, Applied Optics 46 (6) (2007) 959-977
16. C. B. Stipe; D. Lucas; C. P. Koshland; R. F. Sawyer, Applied Optics 44 (31) (2005) 6537-6544
17. M. Commodo; F. Ossler; C. de Lisio; A. D'Anna; P. Minutolo, Combustion Science and Technology 184 (7) (2012) 916-928
18. M.D. Smooke, M.B. Long, B.C. Connelly, M.B. Colket, and R.J. Hall, Soot formation in laminar diffusion flames, Combust. Flame 143 (2005) 613-628.
19. M. Sirignano; D. Bartos; M. Conturso; M. Dunn; A. D'Anna; A. R. Masri, Combustion and Flame 176 (2017) 299-308

11th Asia-Pacific Conference on Combustion,
The University of Sydney, NSW Australia
10th -14th December 2017.

Ion chemistry investigation in rich methane premixed flames

Haoyi Wang [1], Bingjie Chen[1], Jie Han [1,3], Heng Wang [1], Nils Hansen [2], S. Mani Sarathy [1]

[1] Clean Combustion Research Center (CCRC), King Abdullah University of Science and Technology (KAUST)
Thuwal, 23955-6900, Saudi Arabia

[2] Combustion Research Facility, Sandia National Laboratory
Livermore, CA, 94551, USA

[3] Media Group (Current address)
Foshan City, Guangdong Province, China

Abstract

Applying external electric field show great performance on emission control, spark ignition, and combustion efficiency improvement in internal combustion engines and gas turbines. To understand the reason why soot, NOx and other pollutants can be suppressed by electric field, ion chemistry needs to be studied and developed. In this work, positive ions were measured in low pressure (P=60 Torr), burner stabilized, rich premixed methane/oxygen/argon flames. Detailed ion profiles were measured and identified as signals versus height-above-burner by utilizing a quadrupole molecular beam mass spectrometry (MBMS). Three equivalence ratios (ϕ=1.5, 1.8 and 2.0) were applied to investigate ion chemistry. Top three measured ions were identified as H_3O^+, $C_3H_3^+$ and $C_2H_3O^+$, along with some minor ions. When flames were slightly rich (ϕ=1.5), H_3O^+ had largest signal in the flame zone; however, when flames were very rich (ϕ=1.8 and 2.0), dominant ions changed to $C_3H_3^+$, suggesting a different ion chemistry. Neutrals and radicals in the same flame (ϕ=1.5) were additionally measured by Photo Ionization Time-Of-Flight Molecular Beam Mass Spectrometry (PI-TOF-MBMS) to further investigate neutral-ion reactions. Correlations between neutrals and ions indicated major ion structures and ion reaction channels. This work can complete experimental data for methane flames combined with previous research, and offer new insights for future ion chemistry development.

1. Introduction

Ions in flames are attracting the attention of future combustion enhancement. They are normally produced by chemi-ionization reactions, with varying mole fraction from 10^{-7} to 10^{-9} [1]. Ion chemistry has been gradually more appealing in the recent decade due to the applications of external electric field and plasma assisted combustion, showing great prospect on suppressing soot formation [2], increasing flame stability [3], and improving the overall combustion efficiency. Ion sensors have been applied for spark ignition [4], knock control [5] and in-cylinder pressure regulation [6] in internal combustion engines. Detailed ion chemistry needs to be developed in order to understand and utilize electric field application on combustion. Investigations are required on reaction rates of neutral-ion initiation, neutral-ion propagation and ion-electron recombination.

Ion measurements by mass spectrometry within the flames helped neutral chemistry studies in flame decades ago [7-11]. Since ions have very low mole fractions in flames, it is difficult to apply laser diagnostics. Molecular beam mass spectrometry (MBMS) was used to provide ion measurements with higher accuracy [1], compared to other detection techniques such as Langmuir probe. Ions of different flames have been detected using MBMS: Goodings *et al.* [8, 9] and Hayhurst *et al.* [12] for methane flames; Deckers *et al.* [9, 10] for acetylene flames; Colcate *et al.* [7] for ethylene flames. Prager *et al.* [13] developed the first ion chemistry model on methane flames based on the experimental results from Goodings *et al.* [9]. Alquaity *et al.* [14, 15] recently validated this model in premixed flames. Nevertheless, Hayhurst *et al.* [12] notified the ion chemistry in rich and sooty methane flames is different from lean and stoichiometry conditions, so ion chemistry for rich methane flames still needs further research.

Therefore, a systematic investigation on ion chemistry of rich methane flames was conducted in this work. Flame temperatures and ion profiles of rich methane premixed flames under three equivalence ratios were measured by B-type thermocouple and quadrupole MBMS, respectively. Neutral species were measured by Photo-Ionization Time-Of-Flight MBMS (PI-TOF-MBMS). After that, the identification and quantification on the major ion $C_3H_3^+$ in rich flames were performed, followed by predicted initiation pathways of rich flames.

2. Experimental methods

2.1 Quadrupole MBMS

A quadrupole MBMS at KAUST was used to measure the ion profiles, as described in previous work [14]. A stainless water-cooled McKenna burner with 6cm diameter was utilized for burner stabilized premixed methane/oxygen/argon flames. The McKenna burner was mounted with a high resolution translation stage (SMC, LEY- 25RC-350G-S5), with uncertainty of ± 0.01 mm. Different flame positions could be measured from pre-flame zone to post flame zone. A Hiden HPR-60 quadrupole MBMS was used to measure ion profiles as a function of height above burner. A 24mm long nickel cone of 57 degree was used to sample ions directly from flames and reduce the cone effect on flames. After sampling, a molecular beam was formed by applying three stages of high vacuum

Corresponding author. Bingjie Chen
E-mail address: bingjie.chen@kaust.edu.sa

pumping. Different electric fields were applied on the second and third stages to help the desired ions reach the detector. Premixed flames in this work were all performed at low pressure (P=60 ± 0.3 Torr), with cooling water temperature at 25.0 C. Gas flow rates used for each flames were tabulated in Table 1, which were controlled by calibrated MKS mass flow controller, within uncertainty as 1% of the full range (5 SLPM). The overall uncertainty of measurements were estimated to be 20%.

Table 1: Gas flow rates for three rich flames (in sccm).

Flames	phi	CH_4	O_2	Ar
A	1.5	1607	2143	1250
B	1.8	1776	1974	1250
C	2	1875	1875	1250

2.2 Photo-Ionization Time-Of-Flight MBMS

PI-TOF-MBMS was applied to measure the neutral species in flame A at Chemical Dynamic Beamline 9.0.2 in Advanced Light Source, Lawrence Berkeley National Laboratory. The experimental setup was the same as described in previous work [16]. A quartz probe was used to sample the products from the methane flame. Then, similar to quadrupole MBMS, a molecular beam was formed by applying three stages of high vacuum pumping. Different from quadrupole MBMS, for PI-TOF-MBMS, a tunable vacuum ultraviolet beam generated by synchrotron radiation was used to soft ionize the neutrals without breaking their chemical bonds. Ionized neutrals were then measured and identified by time-of-flight MBMS. Thus, neutral signals were measured as a function of height-above-burner, and their mole fractions within the flame were calculated. In this work, 12 photo-ionization energies (9.5, 9.8, 10.0, 10.5, 11.0, 11.5, 12.3, 13.2, 14.35, 15.4, 16.2, 16.65 eV) were implemented to determine mole fractions of different species. The overall uncertainty of measurements was estimated to be 20%. In order to differentiate the potential isomers for neutral species, Photo-Ionization Energy (PIE) scan was introduced with specific ionization energies (from 9.0 to 11.5 eV) at HAB=4 mm to identify different isomers due to their different photo-ionization energies.

2.3 Temperature measurement

Flame temperature was measured by a SiO_2 coated B-type thermocouple (Omega) with bead size of 0.6 mm . The coating layer had the approximate thickness of 0.01 mm. The temperature was corrected for radiation by corrected relation from Shaddix [17]. The corrections range from 100K to 200K, along with corrected flame temperature varying from 1000K to 2000K.

3. Results and discussions

3.1 Ion and neutral pool distributions in flames

According to the measurement by quadrupole MBMS, the major ions and the temperature profile for rich flames were shown in Fig 1 as H_3O^+, $C_3H_3^+$ and $C_2H_3O^+$. Seven minor ions were also observed as HCO^+, CH_3O^+, CH_5O^+, $C_3H_5^+$, $C_4H_5^+$, $C_2H_5O^+$ and $C_7H_7^+$. The comparison of the signals clearly showed that observed major ions and minor ions corresponded to the results of rich methane flames from Goodings *et al.* [9] and Hayhurst *et al.* [12], both using Benson burner for their works. Extrapolated from Fig 1, at slightly rich flame (ϕ=1.5), H_3O^+ was the dominant ion with the largest signal among all the ions. However, when the equivalence ratios increased to 1.8 and 2.0, $C_3H_3^+$ became the dominant ion instead of H_3O^+. This different trend indicated different chemistry in very rich flames from lean, stoichiometric, and slightly rich flames.

For neutrals and radicals, the major neutral species observed were CH_4, O_2, Ar, CO, CO_2 and H_2O. Some minor neutral species were also detected such as C_2H_2, C_2H_4, CH_2O, CH_3CHO, C_3H_4, C_4H_4, CH_2CO, C_4H_6, and their isomers. The main detected radicals were $C_3H_3\cdot$ and $CH_3\cdot$. Relating the measured data for ions, neutral species, and radicals, specific major ion $C_3H_3^+$ structures and neutral-ion reactions within rich flames were derived and revealed. The conversion method from signal to mass fraction is still under investigation.

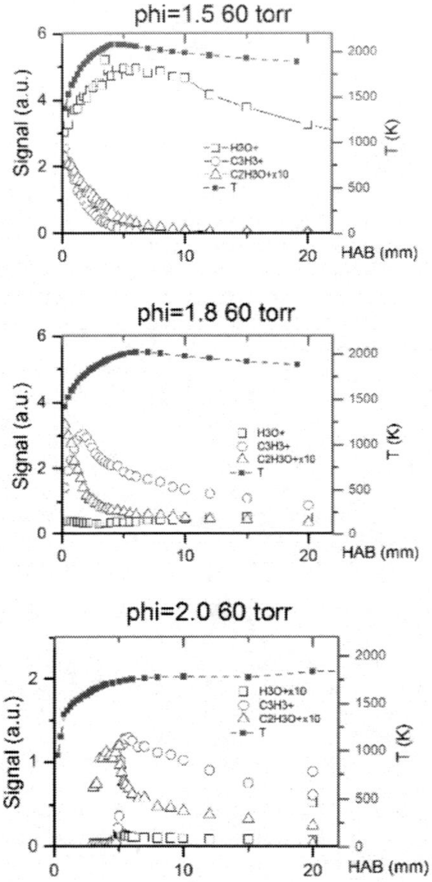

Figure 1: Major ion profiles in three rich methane premixed flames.

3.2 Major ion $C_3H_3^+$ structure

Three possible structures for $C_3H_3^+$ were shown in Fig 2 as propargyl, cyclopropenyl and allyl. In order to find the dominant ion structure, a reasonable assumption was made that correlated ion and neutral profiles can indicate their similar structures because $C_3H_3^+$ and C_3H_3 radical were likely to be generated at the same time during reaction. For example, $C_3H_3^+$ ions should have the similar structure as dominant C_3H_4 neutrals and C_3H_3 radicals. Thus the ion structure could be deducted if the corresponded structure of neutrals or radicals was known.

Therefore, the structures of neutrals and radicals was identified first. Photo-ionization energy scan, as referred in the previous section, was introduced to differentiate isomers of neutrals due to their different photo-ionization energies. For C_3H_4 neutrals, PIE signal shown in Fig 3 started at energy of 9.7 eV, which corresponded to allene structure with the ionization energy of 9.65eV. The signal drastically increased at the ionization energy of 10.25 eV, which corresponded to propyne structure with the ionization energy of 10.2 eV. Thus, both allene and propyne structure were recognized as possible structures for C_3H_4, and propyne was the dominant one. Similarly, for C_3H_3 radical shown in Fig 3, the photo-ionization energy scan showed that the initial signal appeared at ionization energy of 8.5 eV, consisted with propargyl structure. Thus, C_3H_3 radical was assigned as the propargyl structure.

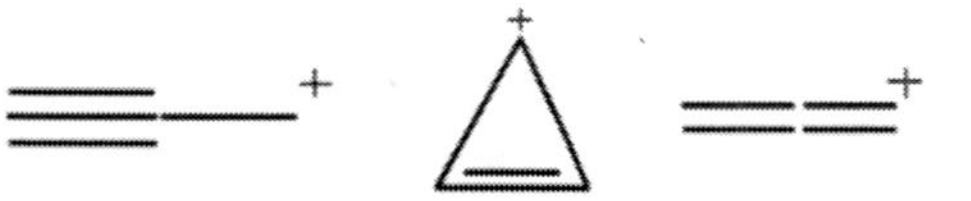

Figure 2: Three possible structures for $C_3H_3^+$. Left: propargyl; Middle: cyclopropenyl; Right: allyl.

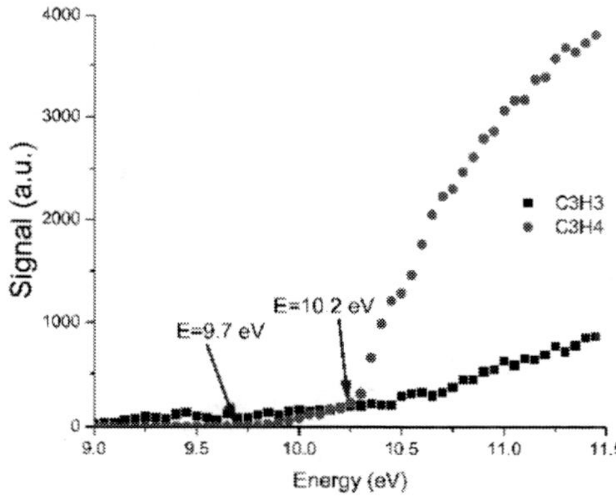

Figure 3: PIE scan for C_3H_4 neutral and C_3H_3 radical.

After the neutral structures being identified, corresponded neutrals and radicals were put together with $C_3H_3^+$ ions. Neutrals were plotted in mole fractions versus height above burner, whereas ion as signals versus height above burner. In order to correlate the positions between ions and neutrals due to different cone sizes between MBMS in KAUST and PI-TOF-MBMS in Lawrence laboratory. Both ions and neutrals were measured by MBMS at KAUST as a reference. After neutrals being measured by PI-TOF-MBMS in Lawrence laboratory, the profile could be shifted according to the reference. As a result, neutral profile was shifted 4 mm towards the burner in order to compensate the cone effect on ion measurements and normalized to the same positions as the neutrals. As a result, strong correlations among $C_3H_3^+$, and propargyl radical and propyne were observed in Fig 4. This indicated $C_3H_3^+$ should have similar structure with propargyl, because when $C_3H_3^+$ was produced, propargyl radical and propyne can also be produced in parallel channels. Thus, it can be deduced that the dominant structure for $C_3H_3^+$ was propargyl structure, as the first structure in Fig 2.

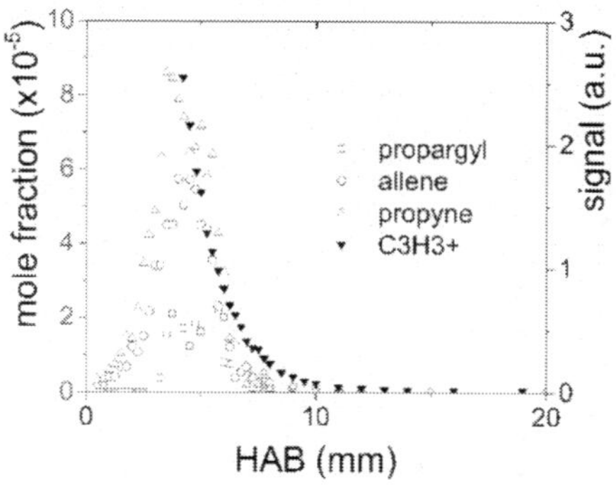

Figure 4: Correlations among ions and their close neutrals. The ion profile normalized to the same position as the neutrals.

3.3 Initiation reactions in rich flames

According to previous studies on lean, stoichiometry and slightly rich methane flames [9, 14, 15], the initiation reaction was considered as (1). However, in very rich flames, $C_3H_3^+$ was reported as dominant ions. Same observation was found by Hayhurst *et al.* [12]. Therefore, for very rich flames, due to the high concentration of acetylene, the initiation reaction was proposed to be (2), which was also consisted with the results from [18] .

$$CH\ (\text{or } CH\cdot) + O \Leftrightarrow HCO^+ + e^- \quad (1)$$

$$CH\ (\text{or } CH\cdot) + C_2H_2 \Leftrightarrow C_3H_3^+ + e^- \quad (2)$$

From Fig 1, when the flame became very rich (ϕ=1.8 and 2.0), H_3O^+ ion became a minor ion, instead of major ion in not very rich flames. According to the model derived by Prager *et al.* [13], the main production reaction for H_3O^+ was (3). Low concentration of H_3O^+ ions indicated the low concentration of HCO^+ in very rich flames. In very rich flames, acetylene concentrations were much higher than oxygen in flame zone, driving initiation reaction from (1) to (2). Moreover, from Fig 5, correlated profiles of $C_3H_3^+$, $C_3H_3\cdot$, and C_2H_2 suggested their correlations and the initiation reaction in this case should be switched from (1) to (2).

$$H_2O + HCO^+ \Leftrightarrow H_3O^+ + CO \quad (3)$$

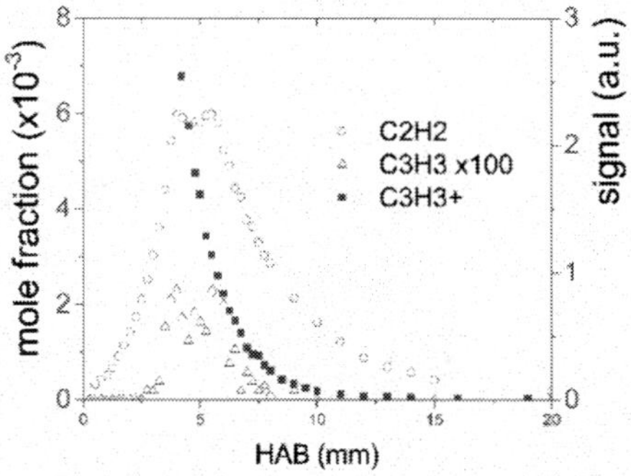

Figure 5: Correlations among $C_3H_3^+$, acetylene and propargyl radicals. Ion profile was shifted away from the burner 4 mm.

Therefore, when H_3O^+ was the dominant ion in the flames, the concentration of the oxidizer was comparably enough, (1) should be the main initiation reaction, which was applicable in lean, stoichiometry and slightly rich flames. When the flames became very rich, high concentration of acetylene and low concentration of oxidizer would make (2) as main initiation reaction, and resulting in $C_3H_3^+$ as dominant ion. Thus, (2) was considered as ion initiation reaction in very rich methane flames.

4. Conclusions

In this work, ion profiles were measured in three premixed rich methane flames by quadrupole MBMS. Major ions were found to be H_3O^+, $C_3H_3^+$ and $C_2H_3O^+$, with several minor ions also identified and quantified. Additionally neutral measurements were performed by PI-TOF-MBMS to further investigate neutral-ion reactions. Correlations among ions and close formula neutrals were observed. Most possible structures of measured ions were deduced. After ion register, ion initiation reactions and neutral-ion propagation reactions were proposed and discussed, revealing different ion chemistry in very rich methane flames from Prager's ion chemistry model developed on lean, stoichiometry and slightly rich methane flames. Updates on ion chemistry model will be attempted by implanting more neutral-ion reactions. Numerical simulations on the updated ion chemistry model will also be performed in order to find correspondence of the dominant ions as observed with experiments. For the future model development, more reaction pathways need to be considered in slight rich region, and soot precursor ions ($C_3H_3^+$, $C_4H_5^+$, $C_7H_7^+$) require further study.

5. Acknowledgment

Primary support of this work was from Clean Combustion Research Center in King Abdullah University of Science and Technology, Thuwal, Saudi Arabia. Neutral measurements were done with Sandia Flame Team collaboration at Advanced Light Source in Lawrence Berkeley National Laboratory, Berkeley, USA. ALS is supported by the Director, Office of Basic energy Sciences, of U.S. Department of Energy, under contract No. DEAC02-05CH11231. Sandia is a multi-program laboratory, operated by Sandia Corporation, a Lockheed Martin Company, for the National Nuclear Security Administration under contract No. DE-AC04-94-AL85000.

References

[1] A. B. Fialkov, Prog. Energy Combust. Sci. 23 (1997) 399-528.
[2] M. Saito, T. Arai, M. Arai, Combust. Flame 119 (1999) 356-366.
[3] H. F. Calcote, R. N. Pease, Ind. Eng. Chem. Res. 43 (1951) 2726-2731.
[4] M. Hellring, T. Munther, T. Rögnvaldsson, N. Wickström, C. Carlsson, M. Larsson, J. Nytomt, SAE Technical Paper (1999) No. 1999-01-1162.
[5] M. Asano, T. Kuma, M. Kajitani, M. Takeuchi, SAE Technical Paper (1998) No. 980162.
[6] A. Gazis, D. Panousakis, R. Chen, W.-H. Chen, Int. J. Engine. Res. 7 (2006) 271-282.
[7] H. Calcote, J. Reuter, J. Chem. Phys. 38 (1963) 310-317.
[8] J. M. Goodings, D. K. Bohme, C.-W. Ng, Combust. Flame 36 (1979) 45-62.
[9] J. M. Goodings, D. K. Bohme, C.-W. Ng, Combust. Flame 36 (1979) 27-43.
[10] J. Deckers, A. Van Tiggelen, Nature 181 (1958) 1460-1460.
[11] J. Deckers, A. Van Tiggelen, Nature 182 (1958) 863-863.
[12] H. R. Jones, A. N. Hayhurst, Combust. Flame 166 (2016) 86-97.
[13] J. Prager, U. Riedel, J. Warnatz, Proc. Combust. Inst.31 (2007) 1129-1137.
[14] A. B. Alquaity, B. Chen, J. Han, H. Selim, M. Belhi, Y. Karakaya, T. Kasper, S. M. Sarathy, F. Bisetti, A. Farooq, Proc. Combust. Inst. 36 (2017) 1213-1221.
[15] A. B. Alquaity, J. Han, M. Chahine, H. Selim, M. Belhi, S. M. Sarathy, F. Bisetti, A. Farooq, Combust. Sci. Technol. 189 (2017) 575-594.
[16] N. Hansen, T. A. Cool, P. R. Westmoreland, K. Kohse-Höinghaus, Prog. Energy Combust. Sci. 35 (2009) 168-191.
[17] C. R. Shaddix, Correcting thermocouple measurements for radiation loss: a critical review, Report No. CONF-990805--, Sandia National Laboratories, 1999.
[18] J. Cancian, B.A.V. Bennett, M.B. Colket,, M.D. Smooke, Combust. Theor. Model. 17.2 (2013): 294-315.

11th Asia-Pacific Conference on Combustion,
The University of Sydney, NSW Australia
10th -14th December 2017.

The Control of Mixture Homogeneity in a Double-Injection Gasoline Compression Ignition (GCI) Engine

H. Goyal[1], S. Kook[1,*], E.R. Hawkes[1] and Q.N. Chan

[1]School of Mechanical and Manufacturing Engineering, The University of New South Wales, Sydney, NSW 2052, Australia

Abstract

The use of low ignition quality fuel in gasoline compression ignition (GCI) engines has exhibited significantly improved engine efficiency and reduced NO_x and smoke emissions owning to extended ignition delay and mixture homogeneity. The partially premixed charge that avoids uHC and CO emissions while keeping the close coupling between the fuel injection event and combustion phasing is a real benefit of GCI engines for practical applications. In the present study, the first injection proportion of double-injection GCI combustion is varied to understand its influence on premixed charge preparation at fixed engine operating conditions of 2000 rpm and 940 kPa net indicated mean effective pressure (IMEP). Specifically, a conventional gasoline fuel with 91 research octane number (RON) was tested at various first injection proportions of 20~40% while the combustion phasing was held constant by varying the second injection timing. The results show that the increased first injection proportion causes a more stretched apparent heat release rate (aHRR) profile with lower initial burning rate due to increased mixture homogeneity of the first-injection charge. This leads to increased net indicated efficiency because of the proper positioning and longer lasting of the combustion event. The increased mixture homogeneity also causes reduced engine-out NO_x emissions.

1. Introduction

In order to develop next-generation diesel engines meeting future emission regulations, challenging steps much be taken to reduce the regulated emissions of nitrogen oxides (NO_x) and particulate matters (PM) with no fuel economy penalty. One near-term approach to address this concern is gasoline compression ignition (GCI) combustion in which a low cetane number (or high octane number) fuel is directly injected into the combustion chamber using conventional common-rail fuel injection system in a diesel engine [1, 2]. As the injection timing of GCI combustion is close to top dead centre (TDC), the extended ignition delay is controlled predominantly by high auto-ignition resistance of gasoline [3]. Because of the relatively shorter premixing time as compared to other premixed-charge-based combustion regimes such as homogeneous charge compression ignition (HCCI) and stratified charge compression ignition (SCCI), the mixture is neither homogeneous nor fully pre-mixed but partially pre-mixed [4]. This partial premixing makes the combustion phasing closely coupled with the fuel injection event [4-6], and effectively overcomes the high-load limit of HCCI combustion by avoiding spontaneous auto-ignition of mixtures [7]. From various tests, it was reported that the partially premixed charge combustion achieves stable engine operation across a wide range of engine loads and speeds [8, 9] with PM and NO_x emissions below the current regulations limit [7].

One particular concern of the GCI charge preparation is an optimised level of mixture homogeneity. From many previous investigations on HCCI engines [10, 11], it is well known that increased mixture homogeneity can reduce engine-out emissions of PM and NO_x while increased unburnt hydrocarbon (uHC) and carbon monoxide (CO) become an issue. In partially premixed charge realised in GCI engines, however, it is unclear how the overall mixture homogeneity impacts the engine efficiency and engine-out emissions in what ways. The present study implements the double-injection-based GCI combustion in a small-bore, single-cylinder diesel engine to study the impact of first injection proportion that is expected to control the mixture homogeneity. The in-cylinder pressure traces and net indicated mean effective pressure (IMEP) were measured for various first injection proportions. The apparent heat release rate (aHRR) was also calculated using the ensemble-averaged in-cylinder pressure traces, which together with the measured injection rate were used to evaluate the net indicated power, net indicated engine efficiency and indicated specific fuel consumption (ISFC). The aHRR was further analysed to derive the combustion phasing, which was characterized by measuring the crank angle location corresponding to 50% heat release (i.e. CA50).

2. Experiments

The schematic diagram of a single-cylinder common-rail diesel engine for GCI combustion applications is shown in Fig.1. The direct injection light-duty diesel engine shares the production engine head but three of the four cylinders were deactivated. As listed in table 1, the engine has a single-cylinder displacement volume of 497.8 cm^3 with bore and stroke of 83 mm and 92 mm, respectively. The geometric compression ratio is 17.7 and the swirl ratio is 1.4 according to manufacturer's specification. The shape of the piston crown is cylindrical bowl as depicted in the cross-sectional view of the combustion chamber in Fig.1. In order to reduce the pressure fluctuations in the intake and exhaust manifolds, two 60-litre surge tanks are positioned on the intake and exhaust sides. A conventional common-rail fuel injection system (Bosch CP3) was used to provide 50 MPa of injection pressure for the direct injection of gasoline via a solenoid-type injector. The injector nozzle had seven holes with nominal hole diameter of 134 μm, the included angle of 150°, and the Bosch K-factor of 1.5. The nozzle is mini-sac type with

Corresponding author: Sanghoon Kook; Fax: +00-00-000-000
E-mail address: s.kook@unsw.edu.au

a convergent, hydro-eroded orifice and a discharge coefficient of 0.86 (i.e. KS1.5/86 according to Bosch specifications). The hydraulic flow rate (HFR) was 400 cm^3 and is measured for time duration of 30 s. The common-rail pressure, injection timing, the number of injections and injection duration were accurately controlled by an electronic injection control unit (Zenobalti ZB-9013P).

The engine was connected to Eddy Current (EC) dynamometer (FroudeHoffmann AG-30HS) for constant speed tests of 2000 rpm. The GCI engine operating conditions are also shown in table 1. The intake air temperature was fixed at 80°C. The temperature of the coolant that was circulated through the cylinder head and the liner was maintained at 90°C using a water heater/circulator (ThermalCare Aquatherm RA Series). The experiments in this study were performed using a conventional gasoline with 91 RON, which is widely available in Australian petrol stations. The physical properties of the used gasoline fuel are summarised in table 1. A double injection strategy was implemented with the early first-injection timing fixed at 170°CA bTDC to provide partially premixed charge condition [12] while the late near-TDC second injection was executed to trigger the combustion [3]. The proportion of the fuel mass delivered in the first injection was varied from 20~40% to test various charge premixing and mixture homogeneity levels. Accordingly, the second injection timing was adjusted between 11 to 7°CA bTDC to maintain a fixed CA50 of 10°CA aTDC. While the injection conditions were varied, the net IMEP was measured at about 940 kPa. The injected fuel mass was measured using a Bosch-tube type injection rate meter [13]. During GCI combustion tests, the in-cylinder pressure traces were recorded using a pressure transducer integrated into the glow plug adaptor (Optrand PSI-glow-A). Simultaneously, the exhaust emissions of smoke (Horiba Opacimeter MEXA-600S) and NO_x (Ecotech 9841 AS) were measured. The unburnt hydrocarbon (uHC) and carbon monoxide emissions were also measured using a HC/CO analyser (Horiba MEXA-584L).

3. Results and Discussion

Figure 2 shows the in-cylinder pressure and apparent heat release rate (aHRR) traces for various first injection proportions ranging between 20~40% at fixed total injected mass of 22 mg. The CA50 was also fixed at 10°CA bTDC by adjusting the second injection timing as discussed previously in table 1. From the figure, a clear trend is found such that increased first injection proportion leads to decreased peak in-cylinder pressure and peak of aHRR. While both the start of combustion determined by the rise of in-cylinder pressure and aHRR and the combustion phasing are held constant, the magnitude of in-cylinder pressure and aHRR decreases as the injection proportion is increased. This trend could be explained with the increased mixture homogeneity for higher first injection proportion, similar to HCCI combustion [14]. It was implied that the increased first injection proportion played a significant role on GCI combustion employing the early first injection such that well-distributed fuel-air mixture decreased the combustion temperature as well as the local equivalence ratio [15]. This is well reflected in the decreased in-cylinder pressure and aHRR. It was also notable that for higher first injection proportion, the initial burning rate was decreased because of the increased mixture homogeneity of the first-injection charge while the overall aHRR profile appears to be stretched out with higher aHRR measured at later crank angles. The observed trends confirm that the mixture homogeneity is predominantly controlled by the first injection proportion of the double-injection-based GCI combustion.

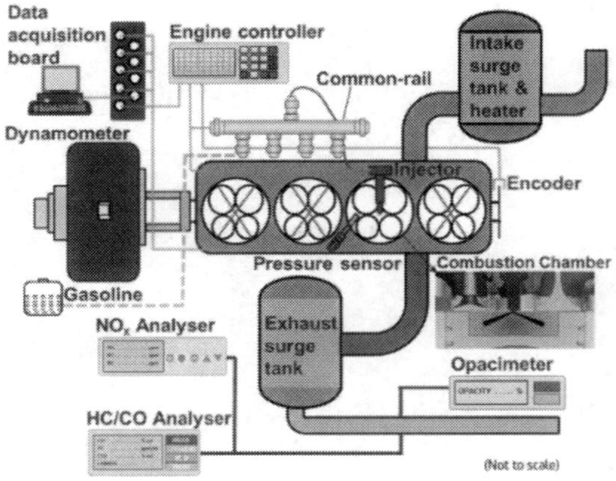

Figure 1: Schematic diagram of a single-cylinder, common-rail diesel engine setup for GCI combustion experiments

Engine specifications	
Displacement	497.8 cm^3
Bore	83 mm
Stroke	92 mm
Piston	Cylinder bowl
Number of valves	2 intake and 2 exhaust
Compression ratio	17.7
Swirl ratio	1.4
Injection system	7-hole Bosch common-rail Nominal hole diameter: 134 µm Included angle: 150° K-factor: 1.5 Discharge coefficient: 0.86 HFR: 400 cm^3 for 30 s
Operating conditions and fuel properties	
Engine speed [rpm]	497.8 cm^3
Intake air temperature [°C]	80
Coolant temperature [°C]	90
Common-rail pressure [MPa]	50
Net IMEP [kPa]	940 ± 40
Fuel	Gasoline (91 RON)
Fuel density at 15°C, kg/m^3	730
Low heating value, MJ/kg	44.4
Kinematic viscosity at 40°C, mm^2/sec	0.75
Fuel injection strategy	Double
1st injection timing [°CA bTDC]	170
2nd injection timing [°CA bTDC]	11~7
Combustion phasing, CA50 [°CA aTDC]	10 ± 0.5
1st injection proportion [%]	20~40
Total injected fuel mass [mg]	22
Overall equivalence ratio [ϕ]	0.58

Table 1: Engine specifications, operating conditions and fuel properties

The calculated net IMEP and coefficient of variation of IMEP using the measured in-cylinder pressure for various first injection proportions is shown in Fig. 3. For fixed combustion phasing conditions realised by the variations of the second injection timings, the net IMEP shows a monotonically increasing trend with an increase in the first injection proportion. As it was shown in the apparent heat release traces (Fig. 2), the increased mixture homogeneity would result in lower initial heat release with higher late heat release i.e. more stretched heat release, which leads to higher IMEP. The coefficient of variation of IMEP (CoV of IMEP) indicates a slight increase in the cyclic fluctuation at higher first injection proportion, which however is all within 4%, indicating stable GCI engine operations.

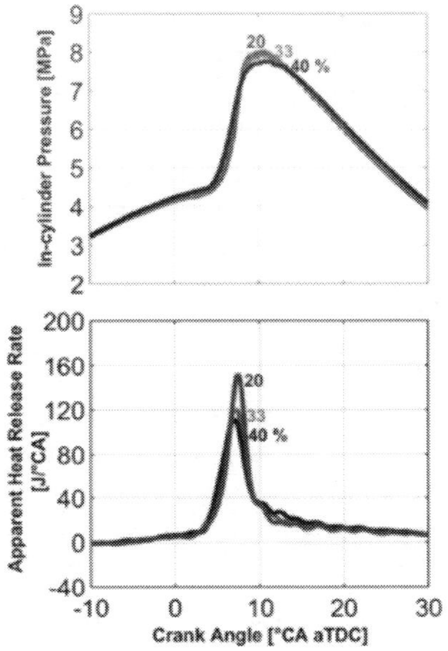

Figure 2: Effect of first injection proportion on in-cylinder pressure and apparent heat release rate (aHRR) of GCI engine combustion at fixed combustion phasing (CA50). The first injection timing was fixed at 170°CA bTDC

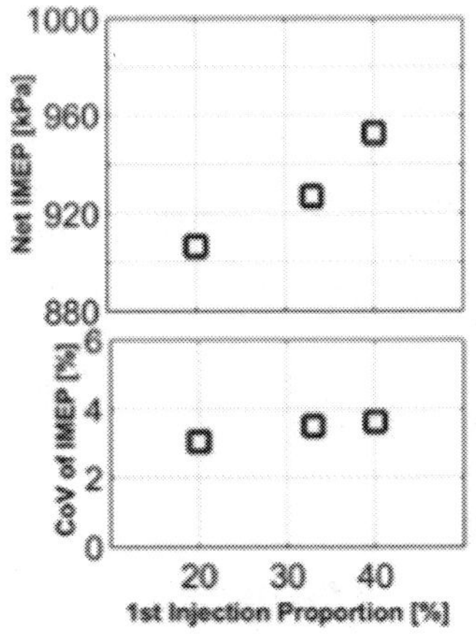

Figure 3: Effect of first injection proportion on net IMEP, CoV of IMEP of GCI combustion

Fig. 4 shows the net indicated power, estimated net indicated engine efficiency and indicated specific fuel consumption (ISFC) which follow a consistent trend with net IMEP (Fig. 3) as the total energy was fixed. It is clearly seen that the engine efficiency increases with the increasing first injection proportion as the heat release occurs for a longer period of time (Fig. 2). This means much lower fuel consumption as shown in the ISFC plot. It was thought that the combustion of partially premixed charge with the optimised mixture conditions and proper positioning of the combustion phasing (i.e. neither too advanced nor too retarded) resulted in higher engine efficiency and lower fuel consumption. Fig. 4 shows that high net indicated efficiency of 49% and low ISFC of 166 g/kWh could be achieved at 40% first injection proportion. In comparison, the diesel baseline case of this engine [13] shows 41% efficiency and 210 g/kWh ISFC. This means 19.5% increase in the net indicated efficiency and 21.9% decrease in ISFC can be simultaneously achieved using the GCI combustion implementing a double injection strategy.

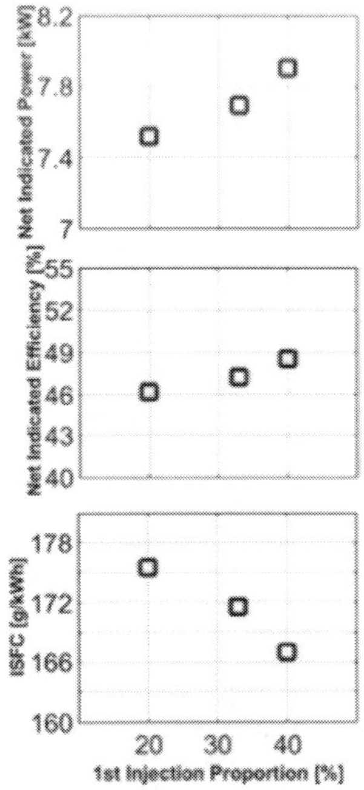

Figure 4: Effect of first injection proportion on net indicated power, indicated efficiency and ISFC of GCI combustion

Fig. 5 shows the measured engine-out emissions for various first injection proportion. The observed trend of increased net indicated specific emissions of uHC and CO with increased first injection proportion were expected considering the decreased in-cylinder pressure and peak of aHRR for higher first injection proportion. Because of the overall lean charge and the reaction quenching at relatively cool chamber surfaces, the combustion temperature could be lower resulting in incomplete combustion products in the exhaust, similar to HCCI combustion [16, 17]. However, NO_x emissions show an inverse trend to uHC/CO emissions – i.e. the increased first injection proportion leads to reduced NO_x emissions. This is because the well-mixed homogeneous charge allows more parcels of fuel-air mixture to reach a lower local equivalence ratio and hence the flame temperature was lower, which leads to lower thermal NO formation [14-17]. A typical smoke-NO_x trade-off characteristics can be seen – i.e. the smoke opacity increases with an increase in first injection proportion. From Fig. 5, uHC/CO/NO_x emissions show a good agreement with mixture homogeneity, however, the smoke trend raises a question as the increased mixture homogeneity should have decreased smoke emissions, as in HCCI combustion [18, 19]. It should be noted that each of the double injection used in the present study can make a significant impact on engine-out smoke emissions, despite having a significant level of homogeneous mixture. Given that 20 to 40% of fuel has been already premixed during the first injection and injecting the rest of the fuel during the second injection in the existing mixtures lead to different overall ambient conditions of the combustion chamber. That is, because of the second fuel injection, the ambient temperature and pressure is strongly influenced by the first injection proportion. It is thought that due to the combustion of the first injected fuel, the higher first injection proportion leads to a shorter time between the start of the second injection and the start of main combustion, which can increase the smoke emissions. Therefore, the double-injection-based GCI combustion in this study performs differently to HCCI combustion but similar to conventional diesel combustion.

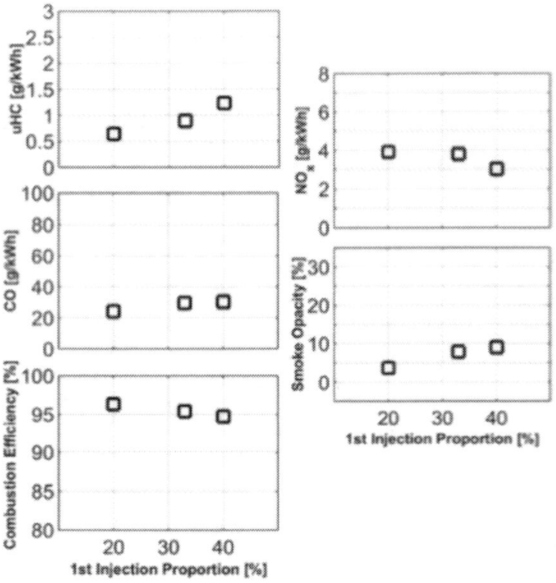

Figure 5: Effect of first injection proportion on engine-out emissions of uHC, CO, smoke opacity and NO_x of GCI combustion

4. Conclusion

The performance and emissions of double-injection-based gasoline compression ignition (GCI) combustion has been successfully tested in a single-cylinder light-duty diesel engine equipped with a conventional common-rail fuel injection system at medium load conditions. The results show that increased first injection proportion leads to increased mixture homogeneity of the first-injection charge as evidenced by a decreased initial burning rate but a stretched heat release profile as the high aHRR continued through the later crank angles. The latter appeared to make a more significant impact, resulting in improved engine efficiency and lower ISFC. The engine-out emissions of NO_x show a decreasing trend due to the increased mixture homogeneity and hence decreased peak of aHRR; however, the uHC/CO/smoke emissions increase with higher first injection proportion. This trade-off characteristics indicates that GCI combustion implementing the double injection strategy of the present study behaves similar to conventional diesel combustion. However, compared to the diesel baseline case, the GCI operation exhibits significantly lower NO_x and smoke emissions with higher engine efficiency, thereby clearly demonstrating the advantages of controlling the mixture homogeneity using the fuel delivered during the first injection.

Acknowledgment

Experiments were performed at the UNSW Engine Research Laboratory, Sydney, Australia. Support for this research was provided by the Australian Research Council.

References

[1] G.T. Kalghatgi, P. Risberg and H.E. Ångström, SAE Technical Paper No. 2006-01-3385; 2006.

[2] G.T. Kalghatgi, P. Risberg and H.E. Ångström, SAE Technical Paper No. 2007-01-0006; 2007.

[3] V. Manente, C.G. Zander, B. Johansson, P. Tunestal and W. Cannella, SAE Technical Paper No. 2010-01-2198; 2010.

[4] L. Hildingsson, B. Johansson, G.T. Kalghatgi and A.J. Harrison, SAE Int. J. Engines 3 (1) (2010), pp. 440-460.

[5] M. Sellnau, M. Foster, K. Hoyer, W. Moore, J. Sinnamon and H. Husted, SAE Int. J. Engines 7 (2) (2014), pp. 835-851.

[6] M.C. Sellnau, J. Sinnamon, K. Hoyer, J. Kim, M. Cavotta and H. Husted, SAE Technical Paper No. 2013-01-0272; 2013.

[7] M.C. Sellnau, J. Sinnamon, K. Hoyer and H. Husted, SAE Int. J. Engines 5 (2) (2012), pp. 300-314.

[8] V. Manente, P. Tunestal and B. Johansson, SAE Int. J. Engines 2 (2) (2009), pp. 71-88.

[9] V. Manente, P. Tunestal, B. Johansson and W.J. Cannella, SAE Technical Paper No. 2010-01-0871; 2010.

[10] K. Epping, S. Aceves, R. Bechtold and J. Dec, SAE Technical Paper No. 2002-01-1923; 2002.

[11] M. Yao, Z. Zheng and H. Liu, Prog Energy Combust Sci 35 (5) (2009), pp 398-437.

[12] C. Woo, S. Kook and E.R. Hawkes, Fuel 180 (0) (2016), pp. 9-19.

[13] H. Goyal, S. Kook, E. Hawkes, Q.N. Chan, S. Padala and Y. Ikeda, SAE Technical Paper No. 2017-01-0742; 2017.

[14] H. Suzuki, N. Koike and M. Odaka, SAE Technical Paper No. 980509; 1998.

[15] S. Kook, S. Park and C. Bae, Energy and Fuels 22 (1) (2008), pp. 331-337.

[16] A. Wimmer, H. Eichlseder, M. Klell and G. Figer, Int. J. Veh. Des. 41 (1-4) (2006), pp. 32-48.

[17] R.K. Maurya and A.K. Agarwal, Appl. Energy 88 (4) (2011), pp. 1169-1180.

[18] J.T. Kashdan, N. Docquier and G. Breuneaux, SAE Technical Paper No. 2004-01-2945; 2004.

[19] S.M. Aceves, D.L. Flowers, J. Martinez-Frias, J.R. Smith, C.K. Westbrook, W.J. Pitz, R. Dibble, J.F. Wright, W.C. Akinyemi and R.P. Hessel, SAE Technical Paper No. 2001-01-1027; 2001.

11th Asia-Pacific Conference on Combustion,
The University of Sydney, NSW Australia
10th -14th December 2017.

The progressive ash formation and deposition dynamics in a 25 kW coal combustor

Qian Huang[1,2], Qiang Yao[1], Shuiqing Li[1,*]

[1]Key laboratory of Thermal Science and Power Engineering of Ministry of Education,
Department of Thermal Engineering, Tsinghua University, Beijing 100084, China
[2]Zhou Pei-Yuan Center for Applied Mathematics, Tsinghua University, Beijing 100084, China

Abstract

The ash deposition process in practical boilers is dependent upon the coal property and the combustion condition, which complicates the prediction and precaution of the problem. In this paper we investigate the progressive ash formation and deposition behaviors in a 25 kW self-sustained pulverized coal combustor for three types of coal samples. The Si/Al-dominated Hami lignite experiences an increase in the ultrafine $PM_{0.2}$ quantity during the char burning regime, opposite to the sodium-rich ZD lignite, and that trend is attributed to the contribution of refractory elements. As for the deposition characteristics, the deposit formed by HAF-2 bituminous is remarkably strengthened when the temperature goes up to 950-1150 °C. Moreover, the deposit morphology shows the presence of coarse mode particle only within the central windward region. The incident velocity distribution reveals that the oblique impaction is an essential factor that hinders the particle sticking on the lateral sides.

1 Introduction

Ash deposition onto the heat transfer tubes is the major technical challenge encountered in modern coal-fired power plants [1]. Among the various damages, the deposits serve as additional heat resistances, blockage agents of the flow passage and triggers to the fire-side tube corrosion/erosion problems, jeopardizing the efficient and safe operation of boilers [1,2]. The remedies, or even precautions to meet the urgent industrial needs definitely require thorough understandings on the mechanisms and controlling factors of the ash deposition processes at both qualitative and quantitative levels, to which a large number of field and lab investigations have dedicated [1-6]. Generally, the molten deposit in the radiation zone is termed as slag, whereas the unmelted fly ash in the downward convective region causes fouling problems [3]. The two cases exhibit distinct properties in terms of particle-wall interaction and subsequent deposit growth behaviors [3]. This diversity in mechanisms ultimately leads to a complex situation of ash deposition in different locations within the boiler, which is indeed relevant to the coal properties and the combustion conditions. For instance, typical low rank lignite that is often rich in alkali and alkaline elements (AAEM) has been widely noticed in recent years for the severe fouling propensity in convective zones [4]. Besides, the role of combustion condition in deposition behaviors is expected to manifest itself in next-generation ultra-supercritical boilers with higher operation temperatures.

Faced with these complexities, it is hence desirable to investigate the ash deposition behaviors in a progressive manner. Numerous studies have demonstrated the applicability of the kW-scale combustor to attack the deposition problems [4-6]. Featured with similar time-temperature history of coal particle to that in practical boilers, this type of combustor enables a comprehensive investigation on the ash formation and deposition behaviors [4,5]. It is thus expected that, with the simulated environments of different furnace locations, the deposition behaviors can be quantitatively characterized, which will shed light on the deposition mechanisms, the dependence on coal properties, and then help address the fouling/slagging problems in power plants.

In this work, we conducted a combined characterization of fly ash and deposit collected at different locations for three types of coal samples. The progressive ash formation and deposition dynamics are discussed thereafter.

2 Experimental

2.1 Coal properties

Three coal samples were used in this work, as coal properties listed in Table 1. ZD lignite is a low rank coal with low ash content but high concentrations of Ca, Na and S. In contrast, the HAF-2 bituminous has an ash content dominated by Si and Al. Hami lignite has moderate content of Na and the lowest coal rank among the three samples.

	ZD lignite	Hami lignite	HAF-2 bituminous
Proximate analysis (wt %, dry basis)			
Volatile	30.58	33.00	22.51
Ash	5.88	26.60	22.32
Fixed C	63.54	40.40	55.17
HHV (MJ/kg)	28.83	20.74	27.19
Ultimate analysis (wt %, dry and ash free basis)			
C	71.60	60.28	85.44
H	3.16	2.55	5.18
N	0.78	1.08	1.42
S_{total}	0.52	0.41	0.35
O (by difference)	23.94	35.68	7.61
Ash composition (wt %)			
SiO_2	5.31	57.04	69.08

Corresponding author. Fax: +86-10-6279-4068
E-mail address: lishuiqing@tsinghua.edu.cn

Al_2O_3	5.25	14.12	22.04
Fe_2O_3	2.30	4.80	1.66
CaO	36.00	5.60	0.58
MgO	6.90	0.91	0.53
SO_3	34.75	3.61	0.39
K_2O	0.34	1.79	2.17
Na_2O	5.76	1.92	0.49

Table 1: The coal properties.

2.2 The 25 kW combustor setup

Figure 1 shows the schematic view of the 25 kW self-sustained down-fired combustor. Detailed descriptions are referred to Ref. [4,7]. Four sampling ports, P1-P4, are available for ash formation and deposition investigation purposes. During the experiments, the stable combustion condition is definitely a prerequisite. As shown in Fig. 1, the temperature profiles along the furnace were kept similar for the three coal samples in the experiments. Table 2 then lists the simulating conditions of ports 2-4, which were used in the experiments.

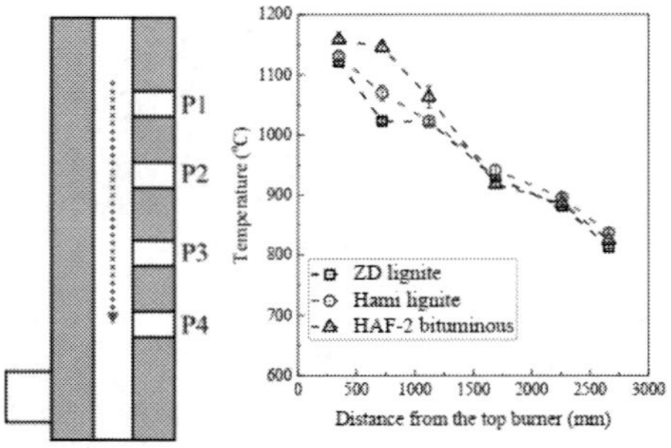

Figure 1: The schematic view of the 25 kW down-fired pulverized coal combustor and the temperature profiles for the several coal samples.

Sampling port	P2	P3	P4
Residence time (s)	0.4-0.5	~1.0	~1.5
Temperature (°C)	~ 1100	900-950	800-850
Simulation condition	Furnace exit	Super-heater	Reheater

Table 2: The simulating conditions for sampling port 2-4.

The deposit samples were collected for various elapse times using a removable coupon with outer dimeter of 20 mm and surface temperature mutually controlled [4,8]. The coupon was then measured for the deposit mass, and was also solidified for SEM observations on the deposit structure. Besides, fly ash particulate was sampled using an aspiration probe with two-stage dilution [4]. The Dekati Electrical Low-Pressure Impactor (ELPI) was applied to measure the particle size distribution (PSD) of the fine particulate PM_{10}.

3 Results and discussion

3.1 The particulate formation characteristics

The progressive evolutions of the fine PM_{10} for the three coal samples are shown in Fig. 2. The case for bituminous is referred to Ref. [7], in which a soot oxidation zone is recognized from P1-P4. Hence, in Figs. 2a&b we investigated the PM_{10} evolution for low rank lignites. Note that the y-axis is set as the PM quantity per unit entrained ash. Opposite trends are observed for the ultrafine $PM_{0.2}$ mode from P2 to P4 between the two samples. A slight decrease in the $PM_{0.2}$ concentration is present for the sodium-rich ZD lignite. This trend is consistent with the Na-dominated gas-to-particle conversion pathway which contributes to the majority of the ultrafine particles formed during ZD combustion [4,9]. In contrast, the Si/Al-dominated Hami lignite experiences a dramatic increase by 2-4 times for the ultrafine $PM_{0.2}$, which is then attributed to the intensive "reduction/re-oxidation" process of the refractory elements (Si, Al, Fe, etc) during char combustion [10]. A comparison among the three coal samples clearly shows the greatest ability of ultrafine PM formation for the sodium-rich ZD lignite. Nevertheless, the formation abilities of micron modes at P4 are similar. It is noticed that both lignites exhibits a slight increase of $PM_{0.2-10}$ from P2-P4, which is a consequence of the continuous mineral particle shedding and breakage during coal/char combustion [11].

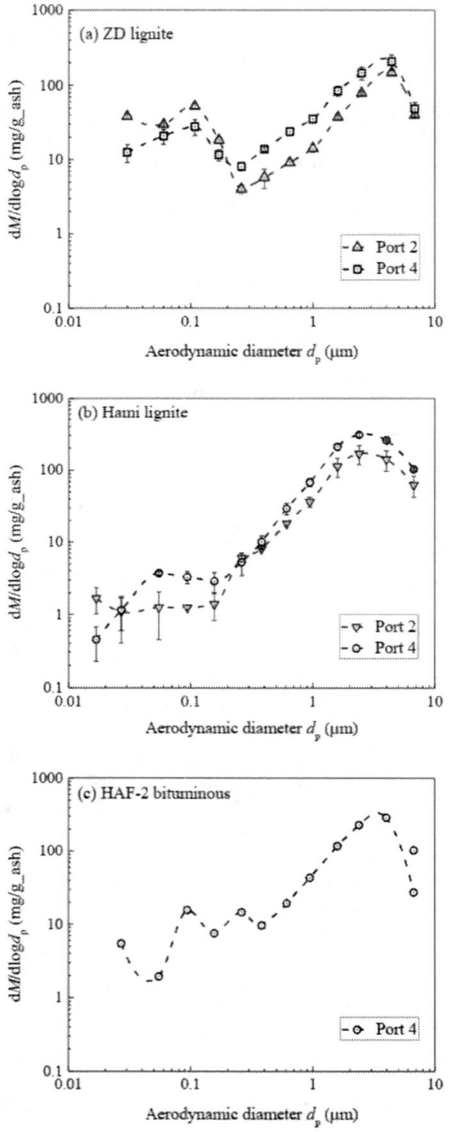

Figure 2: The progressive evolutions of the fine PM_{10} for the

three coal samples.

3.2 The ash deposition characteristics

The progressive ash deposition dynamics for the three coal samples are shown in Fig. 3. It should be noted that the surface temperatures of the probe coupon are controlled at different values (600 °C at P4, 700 °C at P3 and 850 °C at P2), similar with the cases in practical boilers. First, during the combustion of ZD lignite, the deposit grows linearly with time at all the sampling ports 2-4, with higher absolute mass in the higher temperature region. This trend indicates no ash shedding in the initial 1 h period, and that the ash sticking probability is larger under higher temperatures, most likely due to the partially molten salts condensed on the particle surface that enhance the particle stickiness. Similar trends are observed for Hami lignite. The deposits collected in P3 have larger mass than those in P4. Massive ash shedding occurs during 20-30 min and 60-90 min for the deposit in P4, whereas the P3 deposit only shows a shedding process during 60-90 min.

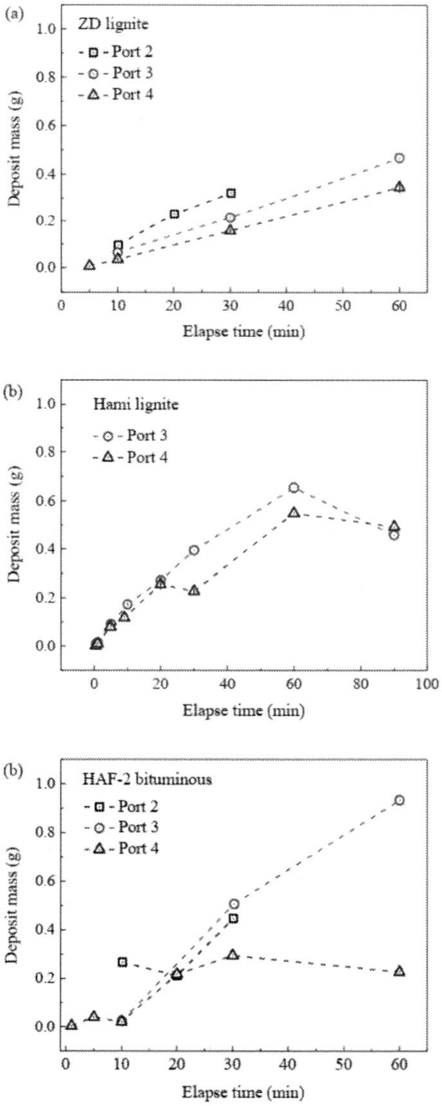

Figure 3: The progressive ash deposition dynamics for the three coal samples.

In contrast, the HAF-2 bituminous exhibits a different behavior. The P4 deposit is low in mass and also strength. The deposit reduces its mass during 5-10 min and 30-60 min. However, the deposit in P3 shows a very different growth curve with continuous mass accumulation during the 1 h period, resulting in a deposit mass ~5 times of that in P4. The result implies that the temperature increase from P4 to P3 (~100 °C) greatly strengthens the deposit structure such that no shedding emerges. This trend is readily preserved in the deposit formed at P2. The 10-min deposit at P2 is heavier than those in P3 and P4 by an order. After a following mass decrease during 10-20 min, the deposit mass increases again during 20-30 min. The severe slagging propensity of HAF-2 bituminous is also directly observed during the operation of the 25 kW combustor, as shown in Fig. 4. The slags are formed and then grow in the inner furnace tube in the regions near P2 and P3. It is believed that the high-temperature induced partial melting and sintering are essential to this strength reinforcement and that care should be taken when burning this bituminous in practical boilers.

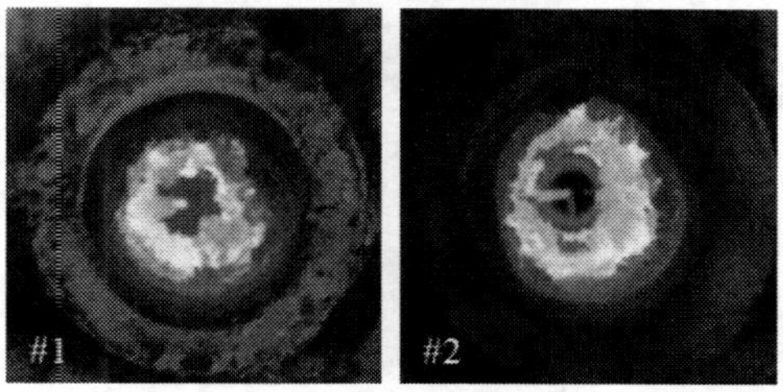

Figure 4: Top view of the slag formed in the combustor tube after burning HAF-2 bituminous in separate runs. The visible slags are mainly formed near P2 and P3 with temperatures 1000-1150 °C.

3.3 The ash deposition dynamics

Figure 5 shows a further attempt to characterize the deposit cross-sectional morphology aiming at uncovering the deposit formation dynamics. It is first noted that the coarse mode ash particles larger than 10 or even 100 µm are located within a narrow azimuthal region in the windward direction. A closer look into the side region shows the detailed structure of the deposit layer at azimuthal angles of 90° and 180°, in which only micron-sized particles are present.

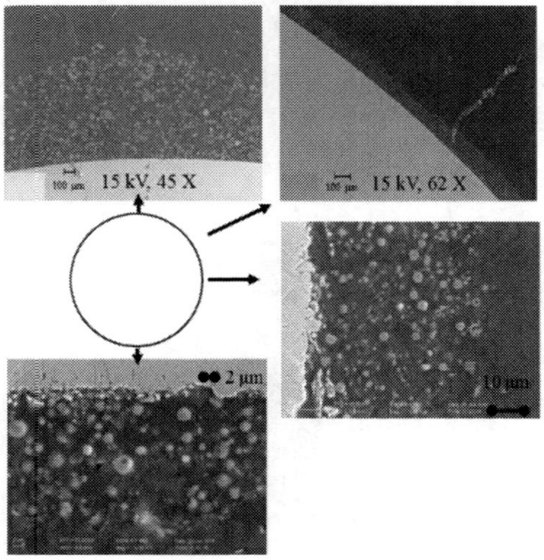

Figure 5: The cross-sectional morphology of the ash deposit collected in P3 and formed by Hami lignite during 1 h period.

The dynamics of ash deposition should be accounted for when trying interpreting the deposit property and structure shown in Figs. 3&5. The particle impaction is the first step that decides whether the particle resides or rebounds subsequently. Hence, in Fig. 6a we present the numerical result of the particle incident velocity distribution along the azimuthal angle of the coupon for different particle sizes. Larger particles exhibit larger incident velocity, because larger particles have more kinetic energy to penetrate through the flow boundary layer near the probe coupon surface. A local minimal velocity is present at the windward stagnation point (θ=0), whereas the maximal values occur in the azimuthal angle range of 30-50°. It is thus clear that the oblique impaction with larger incident particle hinders the sticking of impinging particle, leading to the deposit morphology shown in Fig. 5 in which coarse mode particles can only reside through vertical impaction in the central windward region [12]. Interestingly, it is further revealed in Fig. 6b that the oblique impaction with largest incident velocity has induced remarkable erosion on the lateral sides of the coupon surface. The impact craters are visible when the deposit is carefully erased.

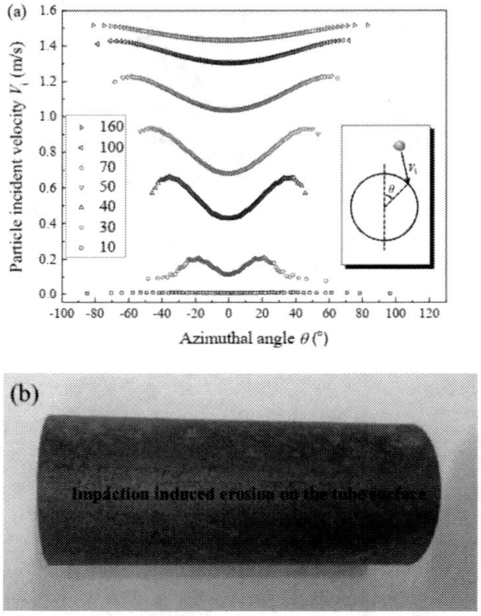

Figure 6: (a) The particle incident velocity as a function of azimuthal angle; (b) The impaction induced erosion of the probe coupon.

4 Conclusions

In this paper we investigated the progressive ash formation and deposition dynamics for three representative types of coal samples. The main conclusions are drawn below:

(1) The evolution of the ultrafine $PM_{0.2}$ in the char burning zone from sampling port P2 to P4 are different for the two lignites. The sodium-rich ZD lignite exhibits a slight decrease, while the Si/Al-dominated Hami lignite experiences an increase of 2-5 times in the $PM_{0.2}$ concentration. In contrast, the $PM_{0.2\text{-}10}$ concentrations slightly increase from P2 to P4 for the both lignites.

(2) The deposit masses of ZD and Hami increase moderately with temperature. Nevertheless, the deposit formed by HAF-2 bituminous is significantly strengthened when the temperature goes up to 950-1150 °C, despite the loosely packed deposit formed in P4 (850 °C).

(3) The deposit morphology always shows the presence of coarse mode particle only within the central windward region. The incident velocity distribution reveals that, in addition to gravitational shedding, the oblique impaction is another factor that hinders the particle sticking on the lateral sides.

5 Acknowledgment

This work was supported by the National Natural Science Fund of China (No. U1361201).

References

[1] E. Rassk, Mineral impurities in coal combustion: Behavior, problems, and remedial measures, Hemisphere Publishing Corporation, New York, 1985.

[2] L.L. Baxter, R.W. Desollar. Fuel 72 (1993) 1411-1418.

[3] R. Bryers, Prog. Energy Combust. Sci. 22 (1996) 29-120.

[4] G. Li, S.Q. Li, Q. Huang, Q. Yao, Fuel 143 (2015) 430-437.

[5] Z. Zhan, J.O.L. Wendt, Energy Fuels 31 (2017) 2198-2202.

[6] I. Naruse, D. Kamihashira, Khairil, Y. Miyauchi, Y. Kato, T. Yamashita et al, Fuel 84 (2005) 405-410.

[7] J.K. Zhuo, S.Q. Li, Q. Yao, Q. Song, Proc. Combust. Inst. 32 (2009) 2059-2066.

[8] Q. Huang, Y.Y. Zhang, Q. Yao, S.Q. Li, Powder Technol. 317 (2017) 449-457.

[9] Q. Huang, S.Q. Li, G. Li, Q. Yao, Combust. Flame 182 (2017) 313-323.

[10] R.J. Quann, A.F. Sarofim, Proc. Combust. Inst. 19 (1982) 1429-1440.

[11] L. Yan, R.P. Gupta, T.F. Wall, Fuel, 81 (2002) 337-344.

[12] A.G. Konstandopoulos, Powder Technol. 37 (2006) 292-305.

11th Asia-Pacific Conference on Combustion,
The University of Sydney, NSW Australia
10th -14th December 2017.

Transported PDF simulations of turbulent CH_4/H_2 flames under MILD combustion conditions

Hu Wang[1], Hua Zhou[1], Zhuyin Ren[1,2*]

[1]Center for Combustion Energy, Tsinghua University, Beijing, 100084, China

[2]School of Aerospace Engineering, Tsinghua University, Beijing, 100084, China

Abstract

Transported probability density function (TPDF) simulations have been carried out for turbulent non-premixed CH_4/H_2 flames of jet-into-hot-coflow (JHC) burner, a typical model to emulate moderate or intense low oxygen dilution (MILD) combustion condition. Two cases with different O_2 levels in the coflow stream, namely HM1 and HM3, are simulated with the differences between MILD combustion and hot-temperature combustion being analyzed. With a detailed 25-species mechanism and the EMST mixing model with $C_\phi = 2.5$, the TPDF simulations can well predict the temperature and species distributions for both cases, demonstrating the great potential of TPDF simulations for investigating the physio-chemical process in kinetics-controlled and low-temperature combustion. The effects of composition boundary condition distribution and mixing model constant on the predicted flame characteristics are also investigated and quantified.

1 Introduction

Moderate or intense low oxygen dilution (MILD) combustion has received more attention in this decade regarding its advantages of high efficiency, low pollutant emission, distributed reaction zone and high reliability [1]. MILD combustion is also regarded as representative process of kinetics-controlled and low-temperature combustion systems, including autoignition and premixed systems in which flames are stabilized by vitiated and hot coflows [2]. Therefore, it is essential to gain more insight into physical and chemical aspects of MILD combustion from both the fundamental research and practical application perspectives.

Different model burners have been designed and utilized to study flame structure and flame characteristics under MILD combustion condition [3]. For example, Dally et al have performed systematic experimental investigation of a turbulent jet flame in an unconfined hot coflow burner – JHC burner, which has served as a guidance to explore MILD combustion in turbulent conditions [3].

To complement the experimental study, numerical simulations, especially those employing Reynolds-averaged Navier-Stokes formulation (RANS), have been widely employed to reveal flame structure, flame stabilization mechanism and detailed physio-chemical interaction under MILD combustion conditions. Christo & Dally implemented EDC combustion model in RANS simulations to predict JHC flame and validate the importance of molecular diffusion [7]. The effects of EDC model parameters on flame predictions are further quantified for the optimal tuning to avoid over-prediction of temperature in certain flame extinction region [8]. Multi-environment probability density function approach [11] and perfectly stirred reactor model [12] have recently also been utilized to simulate JHC flames and obtained reasonable prediction. To simulate the kinetics-controlled MILD combustion process, the transported PDF method has great advantage for capturing the finite rate kinetics processes since it provides an exact description of the mean chemical source term in turbulent flame without any modelling, a capability not available in other combustion models.

In this study, for the first time, the TPDF method in the context of RANS is applied to simulate two representative JHC burner flames, namely HM1 and HM3, with the emphasis on studying the finite-rate reaction process in MILD combustion and on revealing the difference in flame characteristics between MILD and hot-temperature combustion. In addition, the effects of composition boundary condition and scalar mixing rate on the flame characteristics are also analyzed through a parametric study. In the remainder of this paper, the RANS-PDF methodology for simulating JHC burner flames is first described, followed by an analysis of the preliminary results.

2 RANS-PDF simulations

A hybrid RANS-TPDF approach is utilized to simulate the CH_4/H_2 flame in the JHC burner, in which a finite-volume solver solves the conservation equations of mean mass, momentum, energy, turbulent kinetic energy $\tilde{k}$ and dissipation rate $\tilde{\epsilon}$, and a particle-based Monte Carlo method solves the transport equation of one-point, one-time joint composition PDF [13]. Each particle carries the full set of fluid properties, e.g. velocity U, position x, mass m, and composition ϕ, and the evolution of the particles in physical and composition space is governed by the following stochastic differential equations [16],

$$\frac{d\phi_i}{dt} = S_i(\boldsymbol{\phi}) + C_\phi M(\phi_i) \quad (1)$$

$$dx_k = U_k dt \quad (2)$$

$$dU_k = -\frac{1}{\langle\rho\rangle}\frac{\partial\langle p\rangle}{\partial x_k}dt - \left(\frac{1}{2}+\frac{3}{4}C_0\right)\Omega\left(U_k - \widetilde{U_k}\right)dt + \left(C_0\tilde{k}\Omega\right)^{\frac{1}{2}}dW_k \quad (3)$$

$$d\omega = -C_{\omega 3}\Omega(\omega - \tilde{\omega})dt - S_\omega\Omega\omega dt + (2C_{\omega 3}C_{\omega 4}\tilde{\omega}\Omega\omega)^{1/2}dW \quad (4)$$

where the simplified Langevin model (SLM) [18] is employed for velocity and the stochastic frequency model [19] is employed for turbulence frequency. In the equations, S_i is the reaction source term for the ith species ϕ_i, M is a mixing model representing the rate of change in composition due to molecular diffusion, C_ϕ is the mixing model constant, ρ and p are fluid density and pressure, W_k is the kth component of an isotropic vector-valued Wiener process, W is an independent Wiener process, $\tilde{k}$ is the turbulent kinetic energy,

Corresponding author. Fax: +86-10-62797371
E-mail address: zhuyinren@tsinghua.edu.cn

Ω is the conditional mean turbulence frequency which is defined as

$$\Omega \equiv C_\Omega \frac{\langle \rho\omega | \omega \geq \tilde{\omega} \rangle}{\langle \rho \rangle} \quad (5)$$

S_ω is the source term for ω defined as

$$S_\omega \equiv C_{\omega 2} - C_{\omega 1} \frac{P}{k\Omega}, P \equiv -\widetilde{u_k u_l} \frac{\partial \tilde{U}_k}{\partial x_l} \quad (6)$$

C_0, C_Ω, $C_{\omega 1}$, $C_{\omega 2}$, $C_{\omega 3}$, $C_{\omega 4}$ and C_ϕ are model constants, which are identical to those in [20].

The finite-volume solver provides the mean velocity and turbulent diffusion coefficient to the particle solver, while the particle solver feeds the finite-volume solver with the mean density computed from the thermochemical state. The pseudo time averaging of the statistics over a number of iterations is taken after the solution reaches a statistically stationary state, to ensure that sufficient temporal convergence is attained. The hybrid method has been assessed by previous studies [21] and the details of the method can be found in [24].

2.1 Numerical settings

For the JHC burner simulated [3], a central jet of methane/hydrogen mixture is issued into a hot and diluted coflow. The inner diameter of fuel pipe is $D_j = 4.25$mm, and the outer diameter of coflow pipe is $D_c = 82$mm. The burner is surrounded by air, which has the same speed as hot coflow. The details of the two simulated flames HM1 and HM3 are listed in Table 1. Note that HM1 has only 3% oxygen in the high temperature coflow, which favors MILD combustion. In contrast, HM3 has three times more oxygen in the coflow and it corresponds to high temperature combustion.

Table 1: The stream properties of HM1 and HM3 flames.

Case		HM1	HM3
Central jet	Re	10000	10000
	$U/m \cdot s^{-1}$	71.457	71.457
	T/K	305	305
	$V(CH_4)$: $V(H_2)$	1:1	1:1
High-T coflow	$U/m \cdot s^{-1}$	3.2	3.2
	T/K	1300	1300
	$Y(O_2)$	0.03	0.09
	$Y(N_2)$	0.85	0.79
	$Y(H_2O)$	0.065	0.065
	$Y(CO_2)$	0.055	0.055
Air	$U/m \cdot s^{-1}$	3.2	3.2
	T/K	300	300

The 2-D, axisymmetric computational domain extends $80D_j$ and $20D_j$ in the axial and radial directions, respectively. In the axial direction, 97 grid nodes are uniformly distributed, while 50 gird nodes are arranged in the radial direction to guarantee gird-independent solution. In the RANS-TPDF simulations, 30 particles per cell are employed and the statistics are time averaged for 5 flow-through times.

A skeletal mechanism of 25 species is used to describe the methane oxidation, which yields similar mean and rms results as the detailed GRI-Mech 3.0 [25]. The EMST mixing model [26] is employed with the mechanical-to-scalar timescale ratio model constant, C_ϕ, being varied from 1 to 3.5 to investigate the effect of scalar mixing rates on the flame characteristics.

The significance of composition boundary conditions for numerical predictions has been emphasized in [2]. In this work, two types of composition boundary conditions, e.g., the nominal value for the composition at the inlet, and the reconstructed distribution of mean composition from experimental measurement, are considered and compared. For the inlet mean velocity and turbulence quantities, they are taken from a separate pipe flow simulation.

3 Results and Discussions

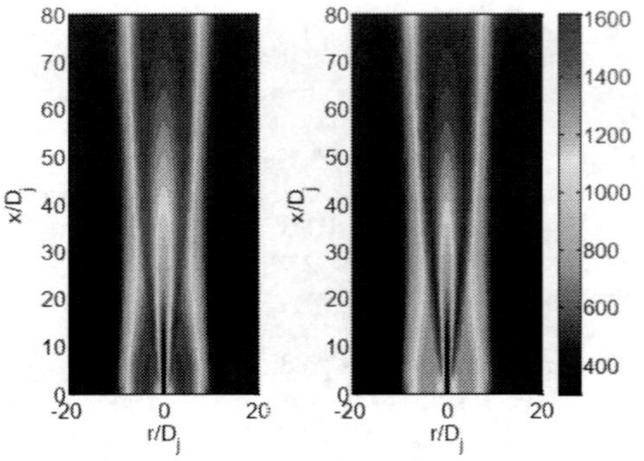

Figure 1: The predicted temperature contours with $C_\phi = 2.5$ and reconstructed distribution from experimental measurement for boundary conditions. Left, HM1; right: HM3.

Figure 1 shows the predicted temperature contours for both cases. As shown, due to the decreased oxygen concentration in the coflow, the maximum flame temperature has been significantly reduced from 1700K for HM3 to around 1400K for HM1. Accordingly, for HM3, significant heat release occurs at the very upstream in the shear layer between fuel jet and hot coflow (close to the inlet), while for HM1, significant heat release is obvious only at the very end of the flow domain i.e., after about x/D_j=60. According to experimental analysis, the influence of coflow persists about x/D_j =23.5 downstream of inlet plane. The hot-temperature combustion observed in HM1 is due to the combustion of the remaining fuel with the oxygen supplied from the surrounding air stream. In HM1, for the region mainly influenced by the interaction between the jet and coflow streams, e.g., x/D_j <23.5, it is kinetics-controlled low-temperature combustion.

The difference between MILD combustion and hot-temperature combustion is further analyzed in Fig. 2, which illustrates the radial flame structure at x/D_j=14.1, using hydroxyl (OH) as the flame marker. It is evident that chemical reaction occurs in HM1 as indicated by the accumulated OH concentration, but is much weaker than that in HM3. At this particular axial location, the peak OH concentration in HM1 is only about one third of that in HM3, which is well correlated to the reduced oxygen concentration in the coflow stream. In addition, as shown, the peak OH position shifts slightly towards the oxidizer side in HM1. The weaker heat release and weakened reaction activity in HM1 are direct signatures of MILD combustion.

More importantly, as shown, for both flames, the RANS-TPDF simulations well capture both the peak and radial distributions of OH radical, demonstrating the great potential of the TPDF method for modelling MILD combustion. The observed discrepancy between simulation and experimental data may arise from the specification of composition boundary conditions, mixing model constant or kinetics mechanism, which are subject to further study.

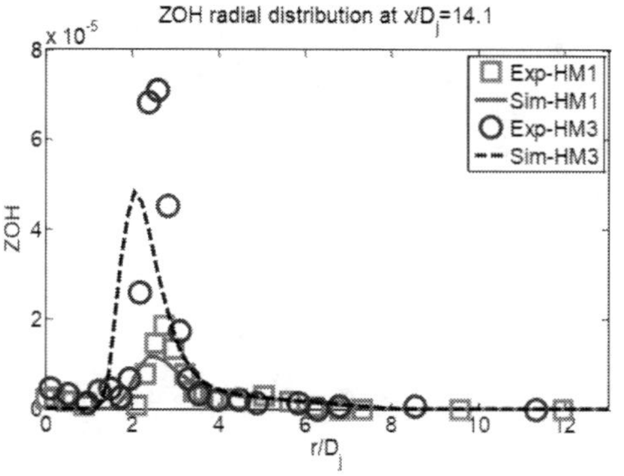

Figure 2: The predicted OH radial profiles at the height of x/D$_j$=14.1 along the axis with $C_\phi = 2.5$ and reconstructed distribution from experimental measurement for boundary conditions. The symbols are experimental data from [9].

3.1 Effects of composition boundary conditions

The effects of composition boundary conditions are quantified using two different specifications, i.e., the nominal value specification and the reconstructed distribution from experimental measurement. As shown in Fig. 3, close to the inlet, different scalar distributions do not lead to significant difference around the reaction region. However, at further downstream location (see Fig. 4), temperature around the reaction region is significantly increased about 200K if the nominal value distribution is employed. Actually, the nominal value distribution would specify higher temperature in the outside region of coflow, and correspondingly shorten the ignition delay time for the mixture, and promote the reaction extent. This is consistent with the observation by Ihme and See [2], that the influence of composition boundary condition distribution will be preserved among the entire flame domain.

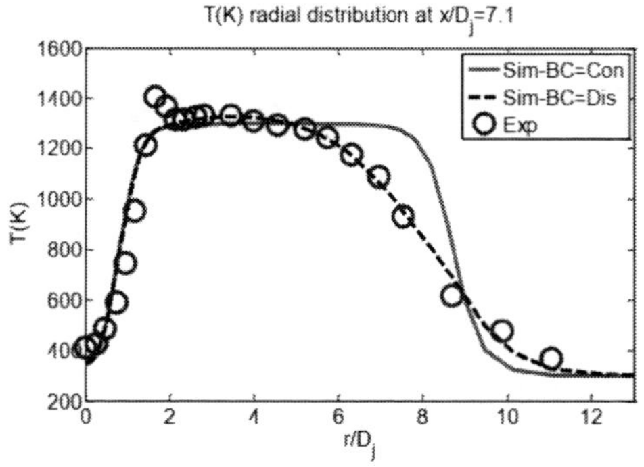

Figure 3: The predicted temperature radial profiles in HM1 obtained with different specification of composition boundary condition.

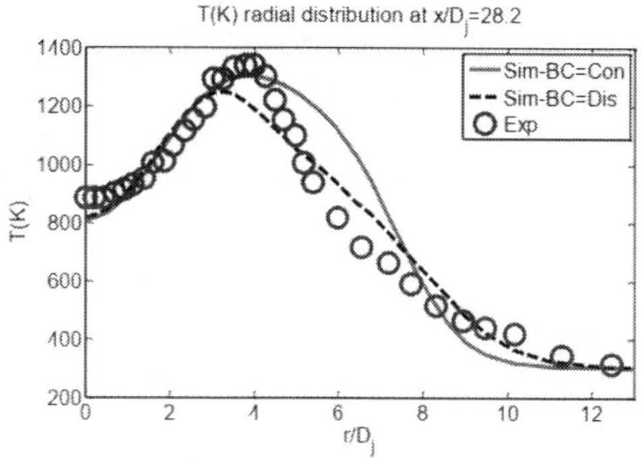

Figure 4: The predicted temperature radial profiles in HM1 obtained with different specification of composition boundary condition.

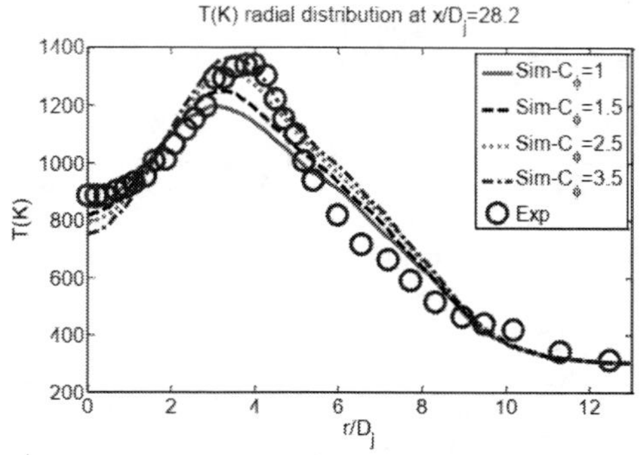

Figure 5: The predicted temperature profiles in HM1 obtained with different values of mixing model constant C_ϕ.

3.2 Effects of scalar mixing rates

In TPDF simulations, while the mean chemical reaction source term is treated exactly, the molecular diffusion term is unclosed in the one-point one-time PDF representation and is modelled by mixing models. Each mixing model consists of a specific mixing formulation describing the manner in which mixing occurs, coupled with the specification of the scalar mixing timescale. Previous studies found that the mixing models have a crucial impact on the computed flame characteristics e.g., local extinction/reignition or turbulent burning velocities [27]. Here the effects of scalar mixing rates on MILD combustion characteristics are investigated through a parametric study of varying mixing model constant C_ϕ. Note that the constant C_ϕ is the mechanical-to-scalar timescale ratio, e.g., $C_\phi = \tau_{turb}/\tau_s$ and is typically taken to be around 2.0 based on studies of passive scalar mixing in shear flows [28].

As shown in Fig. 5, the change in mixing rate has significant impact on the predicted temperature under MILD combustion condition. The peak temperature at the selected axial location has been increased from about 1200K with $C_\phi = 1$, to 1400K with $C_\phi = 3.5$. Meanwhile, with the increase of C_ϕ, the temperature on the axis is decreased. Inside the shear layer between fuel jet and hot coflow, the change of scalar mixing rate make notable difference in the mixture compositions, which directly changes the ignition delay time. A higher mixing rate yields a higher peak temperature and a corresponding shift of the radial position of peak temperature toward the hot coflow. As illustrated by the temperature profiles, a proper specification of scalar mixing rate is crucial for predicting the MILD combustion characteristics, and $C_\phi = 2.5$ provides the best prediction in terms of temperature profiles. More analysis of the effects of scalar mixing rates will be performed with particle-level sensitivity analysis as future work.

4 Conclusions and future work

Transported PDF simulation has great advantage for capturing the finite rate kinetics processes in MILD combustion since the mean chemical source term is treated exactly. For the first time, the TPDF method in the context of RANS has been applied to simulate two representative JHC burner flames, namely HM1 and HM3, to study the finite-rate reaction process in MILD combustion and reveal the differences in flame characteristics between MILD and hot-temperature combustion.

Preliminary results show that with a detailed 25-species mechanism and the EMST model with $C_\phi = 2.5$, the TPDF simulations yield excellent predictions of the temperature and species distributions for both cases.

For the two different inlet scalar boundary conditions considered, although they do not lead to significant difference around the reaction region close to the inlet, temperature at further downstream locations could differ by 200K, indicating the crucial impact of inlet boundary conditions. The parametric study of scalar mixing rate indicates that a proper specification of scalar mixing rate is crucial for predicting the MILD combustion characteristics. A higher mixing rate leads to a higher maximum peak temperature and a corresponding shift of the radial position of peak temperature towards the hot coflow

In this work, the differences between MILD combustion and general combustion are evident based on direct and detailed comparison of mean species composition and temperature profiles. For future work, Lagrangian tracking method and sensitivity analysis [29, 30] will be employed to investigate the relative importance of reaction, mixing and boundary conditions in these flames, and to enhance the understandings of physio-chemical process in MILD combustion.

5 Acknowledgements

The work is supported by National Natural Science Foundation of China 91441202 and 51476087. Simulations are performed with the computational resources of the Tsinghua National Laboratory for Information Science and Technology.

References

[1] A. Cavaliere, M. de Joannon, Prog. Energ. Combust. 30 (4) (2004) 329-366.
[2] M. Ihme, Y. C. See, Proc. Combust. Inst 33 (1) (2011) 1309-1317.
[3] B. B. Dally, A. N. Karpetis, R. S. Barlow, Proc. Combust. Inst. 29 (1) (2002) 1147-1154.
[4] E. Oldenhof, M. J. Tummers, E. H. Van Veen. D. J. E. M. Roekaerts, Combust. Flame 158 (8) (2011) 1553-1563.
[5] C. M. Arndt, M. J. Papageorge, F. Fuest, J. A. Sutton, W. Meier, M. Aigner, Combust. Flame 167 (2016) 60-71.
[6] P. R. Medwell, P. A. M Kalt, B. B. Dally, Combust. Flame 148 (1) (2007) 48-61.
[7] F. C. Christo, B. B. Dally, Combust. Flame 142 (1) (2005) 117-129.
[8] J. Aminian, C. Galletti, L. Tognotti, Fuel 165 (2016) 123-133.
[9] A. Parente, M. R. Malik, F. Contino, A. Cuoci, B B Dally, Fuel 163 (2016) 98-111.
[10] A. Mardani, Fuel 191 (2017) 114-129.
[11] J. Lee, S. Jeon, Y. Kim, Combust. Flame 162 (4) (2015) 1464-1476.
[12] Z. Chen, V. M. Reddy, S. Ruan, N. A. K. Doan, W. L. Roberts, N. Swaminathan, Proc. Combust. Inst. 36 (3) (2017) 4279-4286.
[13] M. Muradoglu, P. Jenny, S.B. Pope, D.A. Caughey, J. Comput. Phys. 154 (1999) 342-371.
[14] P. Jenny, S.B. Pope, M. Muradoglu, D.A. Caughey, J. Comput. Phys. 166 (2001) 218-252.
[15] M. Muradoglu, S.B. Pope, D.A. Caughey, J. Comput. Phys. 172 (2001) 841-878.
[16] S.B. Pope, Prog. Energ. Combust. 11 (1985) 119-192.
[17] S.B. Pope, Annu. Rev. Fluid. Mech. 26 (1994) 23-63.
[18] D.C. Haworth, S.B. Pope, Phys. Fluids. 29 (1986) 387-405.
[19] P. Van Slooten, S. Pope, Phys. Fluids. 10 (1998) 246-265.
[20] D.H. Rowinski, S.B. Pope, Phys. Fluids. 25 (2013) 105105.
[21] R.R. Cao, S.B. Pope, A.R. Masri, Combust. Flame 142 (2005) 438–453.
[22] R.R. Cao, H. Wang, S.B. Pope, Proc. Combust. Inst. 31 (2007) 1543–1550.
[23] M. Muradoglu, K. Liu, S. Pope, Combust. Flame 132 (2003) 115–137.
[24] M. Muradoglu, S.B. Pope , D.A. Caughey. Comput. Phys. 172 (2001) 841–878.
[25] V.V. Lissianski, Z. Qin, available at <http://www.me.berkeley. edu/gri-mech/>.
[26] S. Subramaniam, S.B. Pope, Combust. Flame 115 (1998) 487–514.
[27] H. Wang, S.B. Pope, Combust. Theor. Model. 12 (2008) 857-882.
[28] C. B. BN R, I. Dekeyser, B.E. Launder, Phys. Fluids. 21 (1978) 307-310.
[29] Z. Ren, S.B. Pope, Proc. Combust. Inst. 32 (2009) 1629–1637.
[30] H. Zhou, S. Li, Z. Ren, and D.H. Rowinski, Combust. Flame 181 (2017) 136-148.

Reviewer #1:

General Feedback:

In general, the paper is well written and very interesting. However, some radial and centreline temperature profile comparisons between the HM1 and HM3 cases involving both the simulated and experimental results will help to strengthen the main objective.

Detailed Comments:

The title of Table 1 should be directly above the table.

In Fig.1, use Dj instead of D to achieve consistency.

Both Fig.4 & 5 show an under prediction of peak temperature and over prediction of flame span in the radial direction regardless the use of different BC profile or the mixing constant. Any comment on that?

Please clarify the meaning of the first sentence in section 3.2, "In TPDF simulations, while the nonlinear chemical reaction terms is treated exactly... ".

The text style of the title of sub-section 3.3 should be aligned with other. The content of sub-section 3.3 does not fit its title. The most part of this sub-section is detailing the way of conducting sensitivity test, rather than an actual discussion.

In section 4 "Conclusion", paragraph 2, the reasoning of selecting 2.5 as the final mixing constant (C_ϕ) cannot be found in the paper.

Reference number 31, the name of journal should be "Combust. Theor. Model.".

In Fig. 3 – 5, try to follow the same style as Fig. 2. Use symbols for experiment data, and lines with different styles/colours for different simulation results.

Reply:

1. Thanks for the suggestions. The table title, notations, line styles and references have been updated in the revised manuscript.

2. By increasing the mixing model constant over 2.5, the peak temperature is reasonably predicted. The overprediction of flame span may result from the deficiency of the predicted turbulent transported scalar flux, which can be improved using mixing models local in velocity space.

3. In TPDF method, the mean chemical reaction rates can be treated exactly.

4. Thanks for the suggestions, the sensitivity analysis previously described in discussion part is move to the conclusions and future work part in the revised manuscript.

5. Explanation for selecting 2.5 as the final mixing constant has been added to the revised manuscript.

Ignition Delays of Blended Unsymmetrical Dimethyl Hydrazine with an Energetic Ionic Liquid

Umakant Swami[1], Mahesh Dalwani[1], Krishna Mohan[2], Arindrajit Chowdhury[1]

[1]Mechanical Engineering Department,
IIT Bombay, Mumbai, Maharashtra, 400076, India
[2]DOLP, Defence Research Development Laboratory,
Hyderabad, Telangana, 500058, India

Abstract

An experimental investigation of hypergolicity and ignition delays of blends of hydroxyethylhydrazinium nitrate (HEHN) and unsymmetrical dimethyl hydrazine (UDMH) with white fuming nitric acid (WFNA) and red fuming nitric acid (RFNA) has been presented. HEHN by itself was found to be non-hypergolic in combination with RFNA and WFNA under ambient conditions. However, UDMH-HEHN blends containing a minimum of 30% and 40% UDMH by weight were found to be hypergolic with the same oxidizer. Tests were conducted to measure the ignition delays in a drop test setup equipped sequentially with a high-speed camera as well as an optoelectronic circuit. The effects of UDMH concentration on ignition delays in HEHN-UDMH blends were studied. The physical and chemical ignition delays were segregated with the use of the optoelectronic circuit. The ignition delays and viscosities were found to reduce as the percentage of UDMH in HEHN was increased. The hypergolic blend containing 50% UDMH was found to be the best candidate for replacement of UDMH owing to its low ignition delay with RFNA. The density of this blend was approximately 40% higher than UDMH, making it advantageous for volume limited applications.

1 Introduction

Hypergolic bipropellants are typically fuel-oxidizer pairs that ignite spontaneously when brought in contact with each other, rendering other ignition systems redundant, and thus simplifying engine design and operation [1]. The ignition delay is defined as the time between the physical contact of fuel and oxidizer to the appearance of an ignition kernel. The ignition delay should be less than ten milliseconds to avoid excessive accumulation of propellant in the combustion chamber to avoid dangerous pressure spikes known as 'hard starts', which could rupture the engine [2]. Engines using hypergolic propellants are widely used in the orbital-maneuvering and attitude control in various spacecraft and satellite. Hypergolic ignition mainly relies on exothermic reactions both in the condensed phase as well as gas phase to initiate combustion throughout the chamber. The vaporization of propellants as they enter the confined chamber space and the initial chamber pressure rise is one of the factors influencing ignition. The other factors are vapor pressure of the fuel and oxidizer, viscosity, surface tension, and density of the fluids involved, as well as condensed phase reaction rates and gas phase reaction rates [3].

In the past two decades, ionic liquids (ILs) have been one of the most scintillating scientific discoveries. Due to several unique properties such as low vapor pressure, low toxicity, high density, high thermal stability, liquidity over a wide temperature range, ionic conductivity, and the ability to dissolve a wide range of chemical species, ILs have bought a green revolution to chemistry and chemical engineering [4]. "Energetic ionic liquids (EILs)" are a class of ILs that generate energy on decomposition. In the past decade, significant efforts have been devoted to the research and development of EILs [5]. In contrast to the relatively long history of EILs as explosives, the history of hypergolic EILs is much shorter. In 2008, the use of dicyanamide-based ILs as replacements for hydrazine in propellant formulations was proposed [6]. This was the first report on the investigation of EILs as hypergolic EILs, which encouraged researchers to search for more environmentally benign hypergolic ILs and to explore their potential as non-toxic propellants.

The energetic ionic liquid, hydroxyethylhydrazinium nitrate (HEHN), is expected to be a suitable candidate for the replacement of hydrazine-based hypergolic fuels [7]. Although HEHN is not a hypergolic fuel at room temperature, it was rendered hypergolic by adding acetone and adding catalysts to the oxidizer by Smith et al. [8]. However, HEHN may also be expected to be converted to a hypergolic fuel by the addition of other hypergolic fuels, such as hydrazine, monomethyl hydrazine, and unsymmetrical dimethyl hydrazine (UDMH). Blending allows us to attain desired ignition delays along with lower toxicity levels as compared to commonly used hypergolic fuel. However, the blending of two such fuels also modifies the physical properties such as viscosity, density, and vapor pressure [9].

Due to the lack of a robust model to predict the ignition delay of various fuel-oxidizer combinations, experiments are typically conducted to measure ignition delays [2]. The drop test is a method primarily used for screening of propellants, by determining the hypergolicity of fuel. A drop test setup coupled with phototransistors has been utilized by Farmer et al. [10] to measure chemical ignition delays of hypergolic fuels. The global ignition delays have also been measured using the drop test coupled with laser diagnostic technique [11].

In a present study, HEHN is blended with the known hypergol, UDMH. HEHN was chosen as a green propellant owing to its low vapor pressure, low toxicity and high density while UDMH was chosen due to its low ignition delays and miscibility with HEHN. HEHN was mixed in UDMH in the proportion of 20 to 80 % and the ignition delays of these fuel mixtures with white fuming nitric acid (WFNA) and red fuming nitric acid (RFNA) oxidizers were measured experimentally.

Corresponding author. Fax: +91-22-25726875
E-mail address: arindra@iitb.ac.in

2 Experimental approach

2.1 Drop test setup with high-speed camera

A drop test setup was developed to study the physical phenomena as well as chemical reactions between various pairs of hypergolic fuels and oxidizers. A schematic of the drop test setup consisting of a transparent cuvette, a syringe, a non-captive linear stepper motor, and a high-speed camera connected to the computer is shown in Fig. 1. The experiments were conducted under ambient condition. A precise amount of the oxidizer was placed at the bottom of a 10×10 mm^2, 45 mm tall glass cuvette with the help of a micropipette (Finn-pipette F2). Approximately 20 to 25 µl of the fuel was loaded in a 50 microliter syringe (Hamilton 750RN).
The syringe was inserted between the stainless steel retaining plates. The location of the tip of the needle of the syringe was changed by fixing the retaining plates at different locations on the vertical guiding rods. A transparent acrylic casing was used to encompass the whole setup to prevent exposure to harmful gases evolved during the experiment. The fuel droplet was dispatched from a height of approximately 63 mm from the surface of the oxidizer pool. A high-speed camera (Photron FASTCAM SA5) was used to capture the combustion event. A Nikon lens (AF-S Nikkor, 18-55 mm), fitted with 4 mm and 2 mm close up lens, was used to focus on the surface of oxidizer pool. A rack and pinion mechanism, operated by a non-captive linear stepper motor was used to actuate the piston of the microliter syringe and control the rate of formation and descent of the fuel droplet. The drop of fuel released from the syringe fell freely into the pool of the oxidizer. The pre-ignition and post-ignition events were captured by the high-speed camera at a frame rate of 4000 fps. An LED light source (30W) was used as backlight for the experiments. The captured video was converted into individual frames using Photron FASTCAM Viewer. The ignition delay was measured as the time taken from the frame in which the fuel droplet impacted the oxidizer pool to the frame in which an ignition kernel was observed.

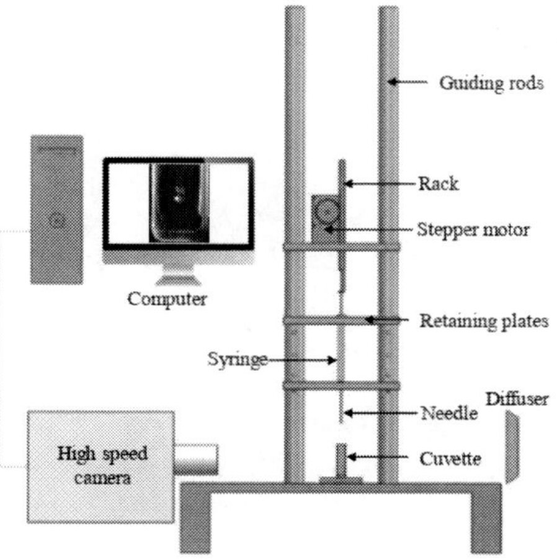

Figure 1: Schematic of the drop test setup with the high-speed camera

2.2 Drop test setup with optoelectronic circuit

The schematic of the drop test setup with the optoelectronic circuit as shown in Fig. 2 (b) has almost the same configuration as that with the high-speed camera. A He-Ne laser (Holmarc model DL-R-5, wavelength 650 nm), 2 photodiodes (Photodiode 1, model BPW24R, and Photodiode 2, model TSAL6100), and an optoelectronic circuit were used in place of the high-speed camera. A 5 V regulated DC power supply (Aplab LD 3205) was used to power the optoelectronic circuit. A laser beam of 1 mm diameter passed through the cuvette and illuminated the first photodiode initially. The laser beam incident on the first photodiode first detects the droplet and subsequently the gases emerging from the liquid-droplet interaction. Generally, the voltages detected by the photodiode were found to drop suddenly when the path of the laser was obstructed by the droplet. A second photodiode was placed at the upper edge of the cuvette to detect ignition and the corresponding flame.

Figure 2 (a) shows the positive terminal of the first photodiode which was connected to a resistor of 10 kΩ. The negative terminal of same was connected to the positive source of a regulated DC power supply. An RC filter was used to eliminate the electrical noise generated by the linear stepper motor, which clouded the signal detected by the optoelectronic circuit. The values of resistance (150 Ω) and capacitance (0.1 µF) were fixed based on the desired cutoff frequency. The output was measured across the capacitor C1 and C2 corresponding to the first and the second photodiode respectively. The data acquisition system Easy Logger DS1M12 was used to record the data at a rate of 10^5 Hz.

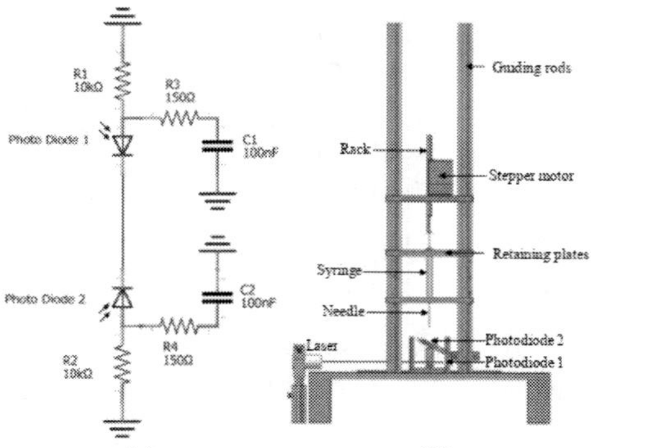

(a) (b)

Figure 2: Schematic of drop test setup with optoelectronic circuit

3 Results and Discussion

The oxidizers used in the present study were WFNA ($HNO_3\geq$ 99%, and H_2O~1%) and RFNA (HNO_3~73.1%, NO_2~17.5-22.5%, phosphoric acid~1-1.3%, hydrofluoric acid ~ 0.5-0.7%, and H_2O ~3.5%). HEHN, UDMH, and RFNA were supplied by DRDL Hyderabad. WFNA was procured from Sigma Aldrich.

3.1 Tests with white fuming nitric acid (WFNA)

Figure 3 shows the selected frames of a UDMH fuel droplet with a diameter of 2.38 mm interacting with a WFNA pool. The volume of the oxidizer pool was kept constant across all experiments at approximately 200 µl. The velocity of the droplet at the time of impact was estimated to be 1.11 m/s. At t=0 ms the fuel droplet touched the pool, which marked the start of ignition delay measurement, shown in Fig. 3 (a). Chemical reactions were found to be initiated at the interface of the droplet and pool leading to evolution of gases from the pool. The gaseous products were found to react further above the pool, providing heat feedback to the condensed phase. The reaction rates in condensed phase increased further due to increasing temperatures owing to exothermic reactions as well as heat feedback. Figure 3 (b) shows the ignition kernel in the gas phase at t=2 ms. However, combustion was not sustained owing to the spillage of the liquid outside the cuvette shown in Fig. 3 (c). A secondary flame envelope was observed surrounding a smaller droplet of UDMH falling back into the cuvette, due to vigorous reactions between UDMH and WFNA as shown in Fig. 3 (d).

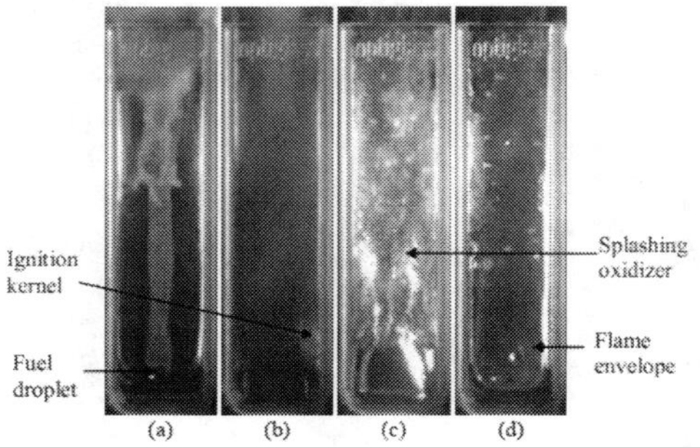

Figure 3: Selected frames for combustion of UDMH in 200 µl WFNA, Frame (a) t= 0 ms, (b) t= 2 ms (ignition), (c) t= 8 ms, and (d) t= 175 ms

The notation UHX is used in this work to denote the composition of mixtures of UDMH and HEHN, containing X% of HEHN in UDMH by weight. The blends UH20, UH30, UH40, and UH50 were observed to ignite spontaneously with WFNA, with average ignition delays of 4.5, 16.75, 33, and 46.9 ms respectively. The ignition delays of UH60 and UH70 with WFNA were observed to be 119 ms and 176 ms respectively, which were significantly higher than UH50, due to the lower reactivity of HEHN. The UHX mixtures combusted slower than UDMH owing to their lower reaction rates with WFNA.

3.2 Tests with red fuming nitric acid (RFNA)

The combustion of UDMH fuel droplet with RFNA is shown in Fig. 4. UDMH was found to react more vigorously with RFNA than WFNA, although ignition delays were of the same order. The spillage of mixture outside cuvette was observed due to the vigorous reaction of UDMH with RFNA. The combustion sustained till the UDMH droplet was fully consumed at time t = 92 ms.

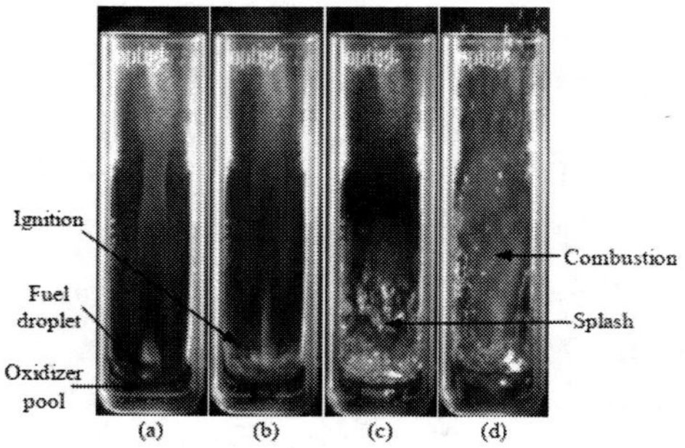

Figure 4: Selected frames for combustion of UDMH in 200 µl RFNA, Frame (a) t= 0 ms, (b) t= 1.5 ms (ignition), (c) t= 4 ms, and (d) t= 12 ms

The blends UH20, UH30, UH40, and UH50 showed the ignition delay of 2.5 ms, 3.3 ms, 5.8 ms, and 7.8 ms respectively. The ignition delay of UH40 with RFNA was found to be approximately five times lower as compared to that with WFNA. The difference was attributed to the higher percentage of N_2O_4 in RFNA, which reacted more readily with HEHN than HNO_3.The combustion was also found to take place without any splashing. However, the overall duration of the combustion event was 245 ms, which was longer as compared to that with WFNA. The UH50 blend combusted in a fashion similar to that of UH40. Ignition delay with RFNA was approximately one-fifth of the value observed with WFNA.

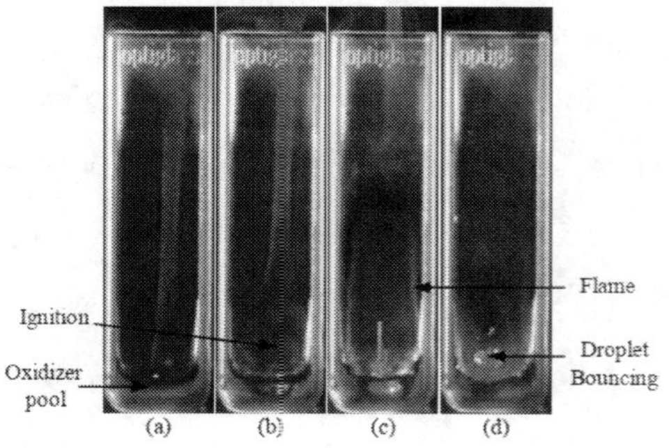

Figure 5: Selected frames for combustion of UH40 fuel in 200 µl RFNA, Frame (a) t= 0 ms, (b) t= 4 ms (ignition), (c) t= 12.5 ms, and (d) t= 130 ms.

The variation of ignition delay with the percentage of HEHN by mass in UDMH is depicted in Fig. 6. For both RFNA and WFNA, the ignition delay was found to rise gradually, up to 50 percent of HEHN by mass in UDMH. A rapid increase in ignition delay was observed after this, in the case of WFNA. There was no ignition for UH80 with WFNA as an oxidizer and UH70 with RFNA as an oxidizer. The ignition delay of UH60 with RFNA was not found to be repeatable and varied from 9 to 341 ms in different tests. Unlike the test with WFNA, the UHX mixtures were found to combust with much lower ignition delays with RFNA, possibly owing to the presence of N_2O_4.

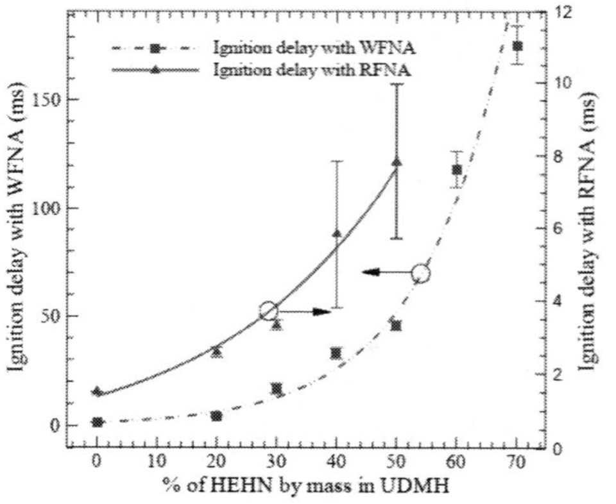

Figure 6: Variation of ignition delay with the percentage of HEHN in UDMH.

3.3 Tests with the optoelectronics circuit

The results discussed in sections 3.1 and 3.2 were obtained using a high speed camera. The same set of experiments was carried out using an optoelectronic circuit. The method was used to segregate the physical and chemical ignition delays. The global ignition delay measured by an optoelectronic circuit was comparable to that obtained using the high-speed camera. Details of the results have been summarized in this section.

Figure 7 shows the measurement of ignition delay by an optoelectronic circuit for UH50 with 200 µl of RFNA. Photodiode first detects a drop in voltage signal after point A, when the path of a laser is obstructed by the droplet. At point B the drop touches the pool, which marks the start of ignition delay measurement. The velocity of the droplet at the time of impact was estimated to be 1.11 m/s and time required for the droplet to travel from point A to point B was estimated as 3.54 ms. The voltage did not rise back to baseline value after the passage of the droplet past the laser beam at B owing to the presence of a translucent wake formed by the reaction of UDMH vapors with RFNA vapors. From C to D, the signal suffered a second drop due to the evolution of gases from the condensed phase. The second photodiode was found to detect the ignition kernel at D. The global ignition delay was measured to be 5.6 ms from point B to D, the physical ignition delay was measured to be 4.53 ms from point B to C, and the chemical delay was measured to be 1.07 ms from point C to D.

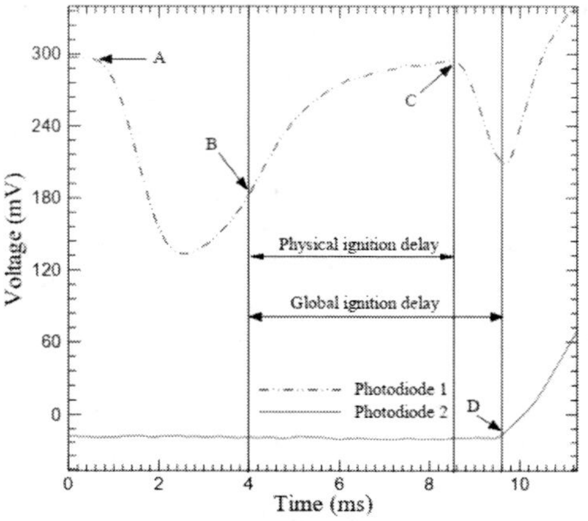

Figure 7: Ignition delay measurement for UH50 droplet into 200 μl of RFNA

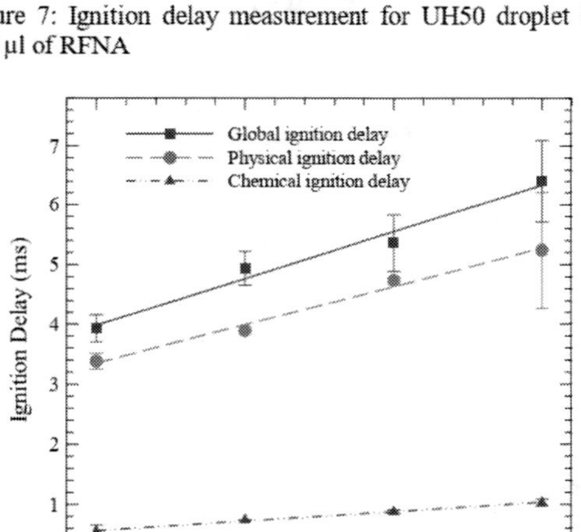

Figure 8: Variation of Ignition delay for UH20, UH30, UH40, and UH50 with RFNA

Figure 8 shows the separation of global, physical, and chemical ignition delays using the optoelectronic circuit for UH20 to UH50 with RFNA. The other parameters such as the volume of oxidizer (200 μl of RFNA), the distance of the needle from the surface of the oxidizer (63 mm), and temperature were kept constant. It was interesting to note that while the global and physical ignition delays increased linearly with increase in the percentage of HEHN in UDMH, the chemical ignition delay increased at a lower rate.

4 Conclusions

HEHN is a highly viscous and non-hypergolic compound at ambient condition. Its viscosity was reduced by blending it with known hypergols with less viscosity namely UDMH, which also imparted hypergolicity to HEHN. The drop test method was used for screening of hypergolic fuels. The physical and chemical ignition delays were segregated with the use of an optoelectronic circuit. The hypergolic fuel mixtures UH50 and UH40 were found to be good candidates for replacement of UDMH owing to their low ignition delays of 8 ms and 6 ms with RFNA respectively. The density of the UH50 was 1.08 g/cm^3, which was approximately 40% higher than UDMH. This improvement in density is expected to be useful in volume limited applications.

5 Acknowledgment

The author also wishes to acknowledge Defence Research and Development Laboratory (DRDL), Hyderabad for funding this project.

References

[1] G.P. Sutton, "Rocket Propulsion Elements," Seventh Ed, John Wiley & Sons, 2001.

[2] J. Forness, T.L. Pourpoint, S.D. Heister, 49th AIAA/ASME/SAE/ASEE Jt. Propuls. Conf. (2013).

[3] D.K. Huzel, Modern Engineering for Design of Liquid-Propellant Rocket Engines, 1962.

[4] M. Smiglak, A. Metlen, R.D. Rogers, Acc. Chem. Res. 40 (2007) 1182.

[5] Q. Zhang, J.M. Shreeve, Chem. Rev. 114 (2014) 10527.

[6] S. Schneider, T. Hawkins, M. Rosander, G. Vaghjiani, S. Chambreau, G. Drake, Energy & Fuels 22 (2008) 2871.

[7] J.L. Shamshina, M. Smiglak, D.M. Drab, T.G. Parker, H.W.H. Dykes, R. Di Salvo, A.J. Reich, R.D. Rogers, Chem. Commun. (Camb). 46 (2010) 8965.

[8] J.E. Smith, G.E. Ogden, C.J. Brown, P.M. Frisby, S.A. Torabzadeh, (2010).

[9] V.K. Bhosale, S.G. Kulkarni, P.S. Kulkarni, Chem. Sel. 1 (2016) 1921.

[10] M.J. Farmer, L.O. Mays, C.S. Hampton, J.E.S. Jr, J. Propuls. 20 (2004) 372.

[11] L.O. Mays, M.J. Farmer, J.E.J. Smith, Combust. Sci. Technol. 134 (1998) 127.

11th Asia-Pacific Conference on Combustion,
The University of Sydney, NSW Australia
10th -14th December 2017.

Improving ignition and combustion performance through burner retrofit for a supercritical down-fired boiler

Pengyuan Liu[1], Hai Zhang*[1], Yang Zhang[1], Yuxin Wu[1], Junfu Lv[1]

1. Key Laboratory for Thermal Science and Power Engineering of the Ministry of Education, Department of Thermal Engineering, Tsinghua University, Beijing 10084, China

Abstract

To improve the ignition and combustion performance of a 600MW supercritical down-fired boiler (DFB) equipped with slot burners, some retrofits on the burners and associated concentrators were conducted. Comparative measurements on the performance of concentrators, and temperatures in the zones of burner and hopper were measured before and after the retrofit under different operation settings. Results revealed that before retrofit though in general the fuel-rich coal/air mixture could be ignited within the distance of 2.5m, the ignition of the fuel-lean coal/air flows was poor. For all burners, ignition was postponed with the increasing secondary air. After retrofit, carbon content in fly ash and slag decreased by improving the primary air separation devices, air distribution and nozzle structure design. The ignition distance of the fuel-rich flows and the depth of flame traveling path became remarkably shorter, and flame temperature in the hopper region decreased. The combustion stability and efficiency of the boiler were improved.

Keywords: Down-fired boiler, combustion, ignition, anthracite, burner retrofit

1 Introduction

Down-fired boilers (DFBs), sometimes called as arch-fired boilers, have been widely used to burn anthracite coals around the world for about three decades in power generation[1-4].DFBs are commonly featured with the configuration of an enlarged lower furnace formed by the front and rear water membrane walls. The connections between the lower and upper furnaces are called the arches. On the arches, a number of burners are arranged, shooting the primary air (PA, the mixture of air and coal particles) downwards to create a W-shaped flame or two U-shaped flames to prolong the residence times of coal particles in the furnace. The secondary air (SA) is injected into the furnace from the arches and the vertical walls in the lower and/or upper furnaces. The lower furnace is covered mostly with thermal insulation materials to create a high temperature burning zone.

According to their burner configuration, DFBs can be divided into two types. One uses round burners and the other uses rectangular slot burners. Though the round burner type DFBs dominate the market, the slot burner DFBs have some advantages such as lower pressure drop of pipeline, higher momentum in down-shooting jets and thereby higher usage efficiency of lower furnace. The development of DFBs still caught great attention.

China imported the first slot burner DFB, in subcritical pressure, made by Mitsui Babcock Energy Limited (MBEL) in late 1990s. During its operation, some problems were found such as delayed coal ignition, high carbon content in fly ash and slag, and high NOx emission (reaching levels of 1600 mg/m^3 at 6% O_2)[5]. Hence, some modifications were made in the latest built slot-burner type 600MW supercritical DFB, including using louver concentrators and new configuration of the slot burners[6]. The operational data showed that supercritical DBF realized relatively low NOx emission (~850 mg/m^3 at 6% O_2) and alleviated asymmetric combustion. However, there are still some problems, including high loss of ignition (LOI) and tube burst in the hopper region, seriously impacting the cost and safety of the boiler.

To investigate the causes of these problems, a series of on-site measurements was conducted, including the performance of the concentrators, ignition distance of two selected nozzles and temperature fields in the burner zone and the hopper region. Based on the evaluation, some retrofits were proposed and conducted. In this paper, the details of the retrofit and thereafter comparative measurements were introduced. The performance of anthracite ignition and combustion were significantly improved.

2 Experimental

2.1 Boiler furnace

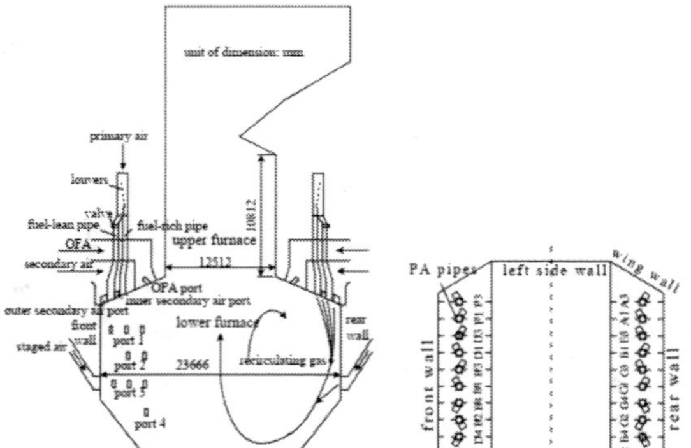

a. vertical furnace cross-section and measuring port layout in experiment

b. horizontal furnace cross-section

Figure 1: Schematic of furnace of a 600MW supercritical DFB

Figure 1 shows the schematic of the 600MW supercritical DFB. On the front and rear arches, 12 groups of burners were symmetrically positioned. Before retrofit, each group of burner was connected to a pair of louver concentrators. The louver

Corresponding author: Hai Zhang, Tel: 010-62773153
E-mail address: haizhang@tsinghua.edu.cn

concentrator divided PA into fuel-rich and fuel-lean streams, and their ratio was adjusted by the air dampers installed on each fuel-lean pipe.
The fuel-rich PA flow was injected vertically down around the middle of the arch while the fuel-lean PA flow was injected near the membrane water walls. The SA bellows symmetrically installed on the arches were divided into two layers. The secondary air (SA) is injected into the furnace from the arches and the vertical walls in the lower and/or upper furnaces.

2.2 Fuel

The fuel used was an anthracite coal and its properties are listed in Table 1.

Proximate analysis, wt.% (as received)				
Volatile matter	Ash	Moisture	Fixed Carbon	Net Heating Value(MJ/kg)
7.47	35.90	7.8	48.83	20.21
Ultimate analysis, wt.% (as received)				
Carbon	Hydrogen	Oxygen	Nitrogen	Sulfur
47.96	2.38	1.13	0.79	4.04

Table 1: Coal properties used in the experiments

2.3 Measurement conditions

Temperature measurements were conducted under 5 cases. Case 1 was the baseline case with full boiler load and the dampers openings of 55%, 75%, 50%, and 100% for the SA, staged air, OFA and vent air respectively. Cases 2-4 were set up to assess the effects of each air stream, while Case 5 was used to assess the performance after the retrofit. The damper of fuel-lean air and outer SA was individually changed. To maintain the flow rate of the total combustion air, the staged air damper openings were adjusted simultaneously with SA damper openings to keep the pressure in the SA wind box constant.

Case #	Before retrofit				After retrofit
	1	2	3	4	5
SA (%)	55	55	25	80	55
Velocity of SA(m/s)	38	38	25	49	33
Staged air (%)	75	75	100	50	75
Fuel-lean air (%)	100	70	100	100	100
velocity of fuel-rich flow (m/s)	17	18	17	17	13
O_2 at the furnace exit (%)	3.4				3.4
Exhaust gas temperature (℃)	142				135

Table 2: Air damper openings and operation parameters for the on-site measurements at full load

2.4 Aerodynamics field measurements

The PA flow rates were measured by Pitot tubes mounted on each PA pipe. Calibrations of the mass flow rate of SA at various damper openings were conducted in the real furnace before the on-site measurements. The flow rates of SA, staged air and OFA were measured at their exits using a hot-wire anemometer. The data deviation was less than 10%. The fireworks were tied at the PA and staged air ports to check the air flow direction.

2.5 Temperature measurements

Temperatures in the burner zone and along the flame travel path (vertically through the nozzle) were measured, using a 10 m long K-type thermocouple with ϕ0.3 mm wire and ϕ4mm stainless steel sheath, similar to that used by Costa[7-9]. The measurement points are shown in Fig. 2, similar to the one as we reported in previous study [10].

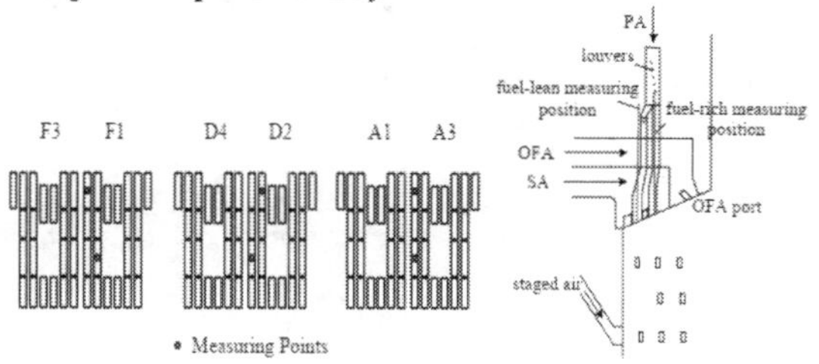

Figure 2: Measuring points and position for primary coal/air jet temperature measurements

2.6 Concentration characterization of concentrator

The separation efficiency of the concentrators was measured using the isokinetic sampling technique using a probe with a flat-head isokinetic device[10]. The air mass flow rates were measured simultaneously. The total sampling time for every test was 6 minutes.

2.7 Retrofit of the boiler

It was found that the original louver concentrators had unsatisfied performance. Thus, higher efficient separators (i.e. cyclones) were used to reduce the amount of coal in fuel-lean flows. The momentum of jet flows from the arches was reduced to weaken the flame penetration depth. As shown in Figure 3, the aspect ratio of fuel rich and fuel lean nozzles was increased from 4.9 to 5.9 and 6.6, respectively.

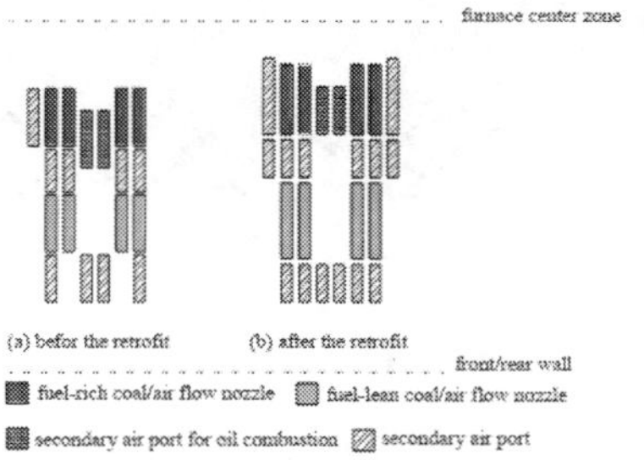

Figure 3: Nozzle arrangement in the burner

3 Results and Discussion

3.1 Aerodynamics field

Table 3 gives the velocities of 12 burners on the front and rear arches, including those of SA, OFA and staged air measured at cold state for Cases 1 and 4 before retrofit. It can be found that the velocity deviation of PA flow could be controlled within 10% by adjusting the opening valve installed on the pulverized coal pipeline. The momentum of SA and staged air was almost equal between the front and rear walls, forming a symmetrical cold flow field in the furnace. For the same each burner, the exit velocity sequence was SA > OFA > staged air. The momentum ratio of SA and staged air was 5.64 in baseline case, indicating that the SA injected from the arches would lead to too deep penetration. The results indicated that the lift effect of staged air was not obvious and the staged air could turn upward near the hopper region.
Figure 4 shows the aerodynamics field illustrated by fireworks before and after retrofit. The dashed line represents the location of staged air. Before retrofit, PA and SA had higher momentum. The fireworks were carried by high speed air flow and the jets traveled near the staged air. The direction of flow did not

obviously change near the staged air even at Case 3. The main reason was that the high velocity and momentum of SA lead too deep penetration. After retrofit, for the same burner, the exit velocities of each air were equal. The momentum of the jets from the arches reduced. At Case 5, the depth of jet flow became shorter. At the location of the staged air, air flow turned upward significantly.

Burners	A1, A3		F1, F3		D2, D4	
Case #	1	4	1	4	1	4
velocity of SA(m/s)	16.7	19.9	17.4	19.9	18.5	22.9
velocity of staged air (m/s)	10.6	13.4	8.3	11.9	6.1	13.4
velocity of OFA(m/s)	11.9		13.1		14.1	
flow rate of SA(kg/s)	17.8	21.3	18.5	21.0	19.5	24.2
flow rate of staged air (kg/s)	16.1	10.4	16.6	8.2	13.2	6.0
flow rate of OFA (kg/s)	6.6		7.2		7.8	

Table 3: The velocities and mass flow rates of different burners

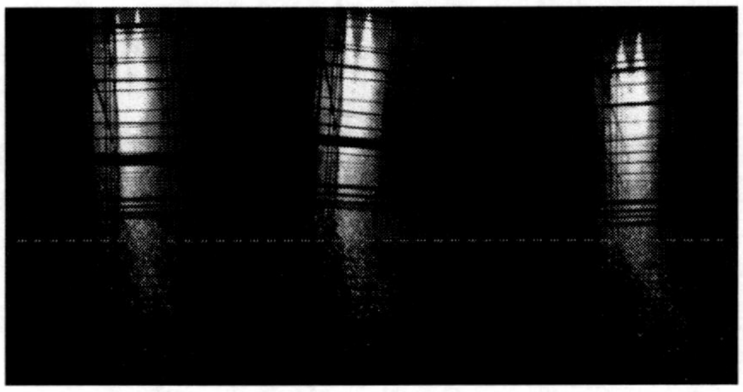

(a) Case 1 (b) case 3 (c) case 5

Figure 4: Fireworks tracing of aerodynamics field

3.2 Performance of concentrators before and after retrofit

Pipes		Louver concentrator		Cyclone	
		Fuel-rich flow	Fuel-lean flow	Fuel-rich flow	Fuel-lean flow
D2	coal	45.6	54.4	96.6	3.4
	air	47.5	53.5	39.3	60.7
B4	coal	62.1	37.9	97.8	2.2
	air	47.2	52.8	36.6	63.4
A3	coal	57.3	42.7	95.4	4.6
	air	46.9	53.1	35.7	64.3

Table 4: Mass percentages of pulverized coal and air flow in fuel-rich and fuel-lean streams with different concentrators

The ignition of PA jets strongly depended on separation performance of the separators. Tables 3 give the measured results of the particle mass and air flow fractions in the fuel-rich and fuel-lean flows. The results reveal that the highest separation efficiency of the louver concentrators was only 62.1%, occurring in the fuel-rich pipe of B4 burner. For the fuel-rich flow of D2 burner, it was even below 50%. On the contrary, cyclones had high separation efficiency. The mass of pulverized coal in fuel-rich flow were more than 95% and the air mass flows were below 40% for each pipe. The retrofit of the concentrators kept a large amount of coal particles in fuel-rich flows, which was benefit for coal ignition.

3.3 Temperatures along the fuel-rich and fuel-lean PA streams

Figure 5 shows the PA temperature profiles along the traveling paths of the fuel-rich and fuel-lean streams for A3 and D2 burners at different damper opening combinations. It can be seen that near the both burners, as the flame proceeded, the fuel-rich PA temperature increased rapidly, while fuel-lean PA temperature increased much more gradually. From the measured temperature results, the ignition distance of the PA jet, which is significant for the safety and efficiency of DFB was determined. The ignition distance was expected to be 0.3 m - 2.5 m, since too short distance could cause burn-out of the nozzles and heavy slagging, while too long distance could lead to too short residence time of coal in the furnace, thereby high carbon content in fly ash. For the turbulent PA jet in a utility boiler, it is difficult to accurately determine its ignition temperature and distance. Alternatively, based on previous studies[11], 900℃ was considered as the ignition point to determine the ignition position for the anthracite coal used in this utility boiler.

Determined by the above method, at the baseline case (Case 1 in Table 2) for fuel-rich PA jets, ignition occurred when x ≥ 2.5m for A3 burner and when x ≥ 1.1 m for D2 burner. However, no ignition was detected for fuel-lean PA jets in the measurement range of 4.6 m. The maximum temperatures measured through fuel-lean nozzles were only 394℃ for A3 burner and 650℃ for D2 burner.

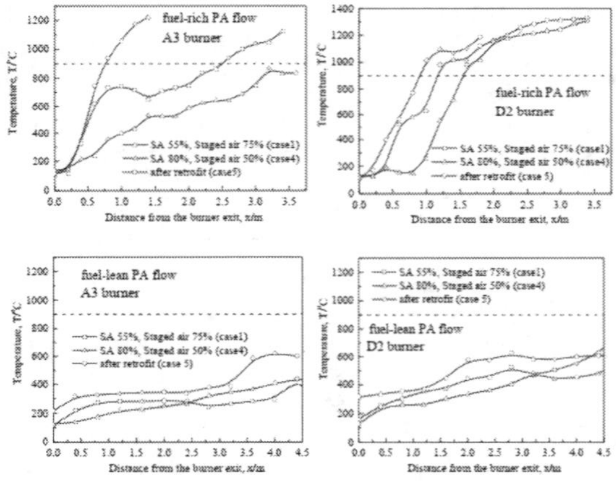

Figure 5: Temperature profiles for the fuel-rich and fuel-lean PA flows at different cases

By comparison, the ignition distance was longer for the fuel-rich flow from A3 burner than that from D2 burner, and temperature of the fuel-lean flow from A3 burner was lower than that from D2 nozzle at the same distance. The major cause could be attributed to the different locations of A3 and D2 burners. A3 burner was positioned near the low-temperature side wall while D2 burner was positioned near the middle of the furnace. Thus, the PA from D2 burner absorbed more radiation heat and mixed more intensively with the surrounding hot recirculation gas.

As the SA damper opening increased from baseline setting, i.e., changed from 55% to 80% accompanied by the staged air damper opening decreased from 75% to 50%, temperatures of the fuel-rich PA flows from both A3 and D2 burners became lower and ignition was delayed (even impossible for A3 burner).

Two factors were counted for this change: (1) more relatively cold SA was mixed into the fuel-rich PA flow; (2) the increase of the vertical momentum induced higher entraining capability and deeper penetration of PA flow.

The temperatures of the fuel-lean streams from A3 and D2 burners were below 900℃ when SA air damper opening was 55% (normal case) and 80%. The foremost reason was the burner configuration. Shown in Fig. 3, the inner SA acted as a restraint to the heat transfer between fuel-rich and fuel-lean PA flows. As a result, the fuel-lean flow was mainly heated up by the hot recirculation gas from two sides of the burners and radiation heat from furnace, which was not enough for its successful ignition. When the inner air damper opening was large, the restraint became stronger, further delaying the ignition of the fuel-lean PA stream. The other reason was the poor concentration effect of louver concentrators that kept large amount of coal particles in the fuel-lean flows.

After retrofit, velocities of both PA and SA were reduced and the concentrations of pulverized coal in the fuel-rich flows increased significantly. The results revealed that adjacent to the burner exits, temperature of the fuel-rich PA stream increased rapidly for both burners. The ignition distance of PA streams from fuel-rich nozzles of A3 and B2 burner became shorter. Ignition occurred when $x \leq 1$ m for both burners. Primary air in the fuel-rich flow decreased and the heat needed for coal ignition reduced. Meanwhile, the increase of concentration of pulverized coal in fuel-rich flows would be benefit for the coal ignition. More pulverized coal particles are injected into the furnace through the fuel-rich nozzles and more likely to be ignited and burnt in furnace center. Higher pulverized coal concentration is benefit for combustion stability. Moreover, for the fuel-lean stream, the gas temperature near the burner nozzle does not rise too much. But less coal particles are in the fuel-lean stream and there would not be too much effect of poor ignition on combustion efficiency.

3.3 Gas Temperatures near the hopper region

Figure 6 showed the gas temperatures near the hopper region under different damper openings. Before retrofit, as the flame proceeded, the gas temperatures surged to 1280℃ within 0.2 m no matter how the damper openings were adjusted. In fact, as the thermocouples were injected further, they were burned out quickly, indicating the flame temperature could be beyond its upper limit of 1500℃.

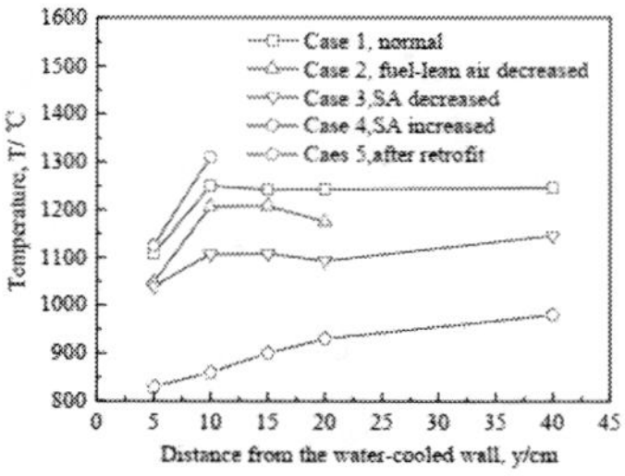

Figure 6: Gas temperatures near the hopper region

The temperatures in the hopper region increased as the SA damper opening increased or the fuel-lean air damper opening decreased. In addition, increasing the boiler load also increased the flame temperature. Too high flame temperature close to the water wall would be no doubt a safety threat to boiler operation. These phenomena could be accounted by multiple factors. (1) The PA had larger vertical momentum and penetrated deeper than the PA in conventional DFBs; (2) The boiler was not in favor of fuel-lean PA ignition due to the air nozzle arrangement and unsatisfied separation performance of the louver concentrators. After the retrofit, the momentum of jet flows from the arches reduced and the temperature decreased about 200℃. This was because the flame center moved upwards, which was helpful for the tube safety in hopper region.

After the retrofit, measurements showed the carbon content in fly ash and slag decreased from 7% to 4% and 12% to 3%, respectively.

4 Conclusions

To improve the ignition and combustion performance of a 600MW supercritical down-fired boiler (DFB) equipped with slot burners, some retrofits on the burners and associated concentrators were conducted. Before retrofit, although the fuel-rich coal/air mixture could be ignited within the distance of 2.5 m, the ignition of the fuel-lean coal/air flows was poor as their temperatures were below 900℃ within 5 m. The ignition distance became longer with the increasing secondary air. Poor ignition of fuel-lean flows resulted in high LOI. Gas temperatures measured near the hopper region further showed that too deep penetration occurred for the coal/air flows and this phenomenon barely changed with the damper opening settings of SA air and staged air. Modifications of the burner and air distribution, improvement of solid separators, and reduction of momentum of jet flows from the arches were implemented on the boiler. The pulverized coal concentration of fuel-rich flows increased. The ignition distance of the fuel-rich flows and the depth of flame traveling path became remarkably shorter, and flame temperature in the hopper region decreased. The combustion stability and efficiency were improved.

5 Acknowledgment

The financial support from MOST project (No. 2015BAA04B01) is appreciated.

References

[1] Fan, W. D., Lin, Z. C., Li, Y. Y., et al. Energy Fuels 23 (2009) 111−120.

[2] Lee, J. M., Kim, D. W., Kim, J. S. Energy 36 (2011) 5703−5709.

[3] Steer, J., Marsh, M., Griffiths, A., Malmgren, A., Riley, G. Energy Convers Manage 66 (2013) 285−294.

[4] Zhang D.L., Meng C.W., Zhang H., et al. Applied Thermal Engineering 103 (2016) 264-273.

[5] Kuang, M., Li, Z. Q., Xu, S. T., Zhu, Q. Y. Environ. Sci. Technol 45 (2011) 3803−3811.

[6] Kuang, M., Li, Z. Q., Liu, C. L., Zhu, Q.Y. Environ. Sci. Technol 47 (2013) 4850−4858.

[7] Costa, M., Silva, P., Azevedo, J. L.T. Combust. Sci. Technol 175 (2003) 271−289.

[8] D. S. De. J. Inst. Energy 77 (1981) 113−116.

[9] Costa, M., Azevedo, J. L.T. Combust. Sci. Technol 179 (2007) 1923-1935.

[10] Liu, P.Y., Gao, J.J., Zhang, H., Zhang, D. L., Wu, Y.X., Zhang, M., Lu, J.F.. Fuel Processing Technology 158 (2017) 172-179.

[11] Cui K., Liu B., Zhang H., Wu Y. X., Matsumoto K., Takeno K. Energy Fuels 27 (2013) 2246-2254.

11th Asia-Pacific Conference on Combustion,
The University of Sydney, NSW Australia
10th -14th December 2017.

Experimental Study of Spherical Turbulent Flame Propagation of Pulverized Coal Particle Cloud in O_2/N_2 Atmosphere

Khalid Hadi[1], Ryo Ichimura[1], Nozomu Hashimoto[1], Osamu Fujita[1]

[1]Division of Mechanical and Space Engineering, Hokkaido University

Kita 13 Nishi 8, Kita-ku, Sapporo-city, Hokkaido, 060-8628, JAPAN

Abstract

To clarify the effect of the turbulence intensity on the flame propagation of the coal particle cloud, a new experimental apparatus, in which coal particles can be dispersed homogeneously in the turbulent flow field formed by two fans, was developed. By using the new experimental apparatus, experiments of spherical turbulent flame propagation of pulverized coal particle cloud in spherical constant volume chamber in various turbulence intensity were conducted. Flame propagation velocities were obtained based on the flame propagation images taken by the high-speed camera. As a result, it was found that turbulence has positive effect on flame propagation velocity since the flame propagation velocity increases as the turbulence intensity increases. This paper is the first paper in the world that is reporting the effect of the turbulence intensity on spherical pulverized coal cloud flame propagation.

1 Introduction

Coal remains as an extremely important fuel and promising energy source besides of other fossil fuels because of its most abundant source. The pulverized coal combustion research has been conducted for a long time to gather fundamental information for using it in energy generation and for an understanding pulverized coal fire disaster. The mechanism of pulverized coal combustion is a complex phenomenon compared to gaseous or liquid fuel because devolatilization and chemical reactions of volatilized fuel and residual char with oxidizer occurs simultaneously.

In previous, lots of pulverized coal combustion study have been conducted experimentally and numerically. Fujita et al. [1], Kiga et al. [2] and Suda et al. [3] examined the quiescent pulverized coal cloud combustion characteristics in microgravity condition. Other previous works studied the turbulent pulverized coal combustion in jet flames [4]. Hayashi et al. [5] and Hashimoto et al. [6] explored about soot formation characteristics in a lab-scale turbulent pulverized coal flame. However, the effect of the turbulence intensity on the flame propagation of the pulverized coal particles has not been clarified yet. Therefore, it is necessary to conduct new experiments to obtain the data for the fundamental understanding of turbulent pulverized coal combustion.

As for gaseous fuels, spherical propagating premixed flames in a constant volume combustion chamber have been investigated to clarify the fundamental characteristics of premixed flames. Smallbone et al. [7], Kitagawa et al. [8] and Hayakawa et al. [9] studied spherically propagating flame for laminar and turbulent to clarify turbulent burning velocity in various conditions. Tse et al. [10] investigate propagation phenomena of spherically premixed flames in various conditions up to 60 atm. The most recent work of Goulier et al. [11] on laminar and turbulent flame speed of premixed hydrogen/air mixtures suggested that a homogeneous and isotropic turbulence is generated using several fans.

The objective of the present study was to investigate the effect of turbulence intensity to the spherical flame propagation velocity of pulverized coal particle cloud. The experiments were conducted at atmospheric pressure under turbulent field generated by fans in a closed spherical vessel.

2 Experimental apparatus and procedures

Experiments were carried out using a constant volume spherical combustion chamber shown in Fig. 1. The inner diameter of the chamber is 200 mm and the height is 280 mm. The total volume of the chamber was approximately 6.19×10^{-3} m^3. A spark igniter composed of two stainless steel electrodes with diameter of 1.8 mm was mounted inside the chamber. Spark gap was set to 2 mm, where the gap within 1 to 3 mm is the best [12]. A capacitor discharge ignition (CDI) circuit was adopted in the spark ignition. A series of capacitors with total 50 μF was charged by 469 VDC in the CDI circuit. To ensure the ignition of pulverized coal particles, the total spark energy of 5.5 J was discharged to the ignition coil. Turbulence was generated in the chamber by counter-rotation of two identical seven bladed fans located vertically and symmetrically. The fan was coupled to Graphite Brushes, 150W motor equipped with encoder connected to ESCON 50/5 controller to control the rotation of fan separately.

Flame propagation of pulverized coal particles cloud was observed by direct photography at sample rate of 1800 pps using a high-speed camera (Miro C with UV lens) through 50 mm diameter quartz glass window. Pressure history inside combustion chamber was measured using the Valcom VPRTF-A4 pressure sensor and recorded by Hioki 8870 data logger. Another pressure sensor (Valcom VPRTF-A4) and F37N-S Valcom digital panel meter was employed for monitoring the pressure of gas mixture in the dispersion tank.

Common bituminous coal of 48 μm average diameter on mass basis with properties as shown in Table 1 was used as the fuel. The 40% O_2 and 60% N_2 used as the ambient gas in order to obtained the flame propagation. Coal concentration inside the combustion chamber in this study was set as 1.3 kg/m^3. According to work by Suda et al. [3], the higher flame propagation velocity of pulverized coal particles is around 1.3 kg/m^3 of coal concentration. Dispersion gas of 40% oxygen and 60% nitrogen were mixed in dispersion tank with the total

Corresponding author. Fax: +81-11-706-7841
E-mail address: hadi.psa@gmail.com

pressure of 300 kPa. Before the coal particle dispersion, coal particles were placed in four filter cups and tubes. Subsequently, the coal particles were dispersed into the chamber through four inlets within 0.7 seconds using the dispersion gas. After 0.3 seconds dispersion end, the mixture was ignited at the center of chamber at 0.1 MPa. The fan rotation was started approximately 3 minutes prior to dispersion and kept running during the flame propagation. Fans rotation speed was confirm using a digital tachometer. Minimum of three experimental data for each condition were used for analysis. Flame propagation velocities have been determined from the time histories of the flame radius. Keyence KV 1000 was employed as sequence controller.

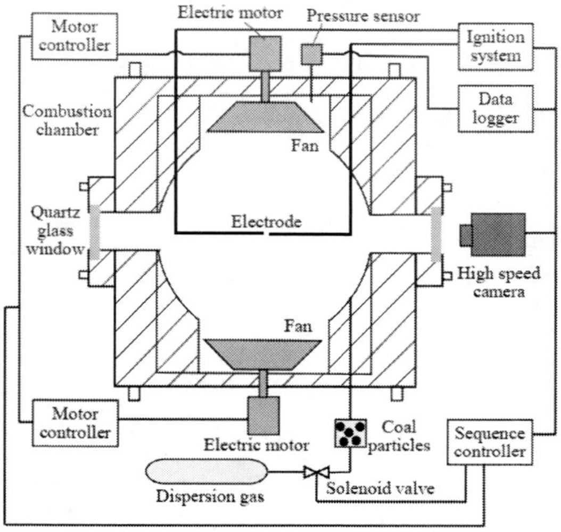

Figure 1: Experimental apparatus schematic diagram.

Analysis	Common bituminous coal
Proximate [wt%]	
Moisture	0.7
Ash	14.2
Volatile matter	33.5
Fixed carbon	52.3
Ultimate [wt%, dry]	
Carbon	70.5
Hydrogen	4.64
Nitrogen	1.66
Oxygen	8.59
Sulfur	0.46
Heating value [MJ/kg]	27.8

Table 1: Physical and chemical analysis of coal.

2.1 Turbulence intensity measurements

A correlation between fan rotational speed, N and turbulence intensity, u' was obtained from Particle Image Velocimetry (PIV) measurements carried out by Seika Digital Image Co., Ltd. The PIV measurement system consists of a CCD camera of 1600 x 1200 pixels resolution, a double pulsed Nd:YAG laser of 532 nm wavelength with delay 10 μs, a timing controller, and seeding generator. Figure 2 shows the combustion chamber and equipment of PIV measurement. Oil mist of 1 μm in size was used as particle tracing seed. Various scale eddies exist in a turbulent field, therefore MicroNikkor105mm f/2.8G and Nikkor28mm f/1.8 lens were employed to image the turbulent flow in small and large measurement field, respectively. Analysis window was set to 32 x32 pixels with 50% overlap for both measurement field. Small measurement field of 8 mm window with 7 μm pixel size in flow for high resolution, whereas 60 mm window size with 54 μm pixel size in the flow of large field measurement.

Figure 3 shows the relationship between the fan rotational speed and the turbulence intensity. It is shown that the turbulence intensity is proportional to the fan speed. The turbulence was considered homogeneous with no regular bulk motion in the center of the combustion vessel [7]. The integral length scale of turbulence was determined by utilizing the data obtained from this PIV experiment.

Figure 2: View of the equipment of PIV measurement.

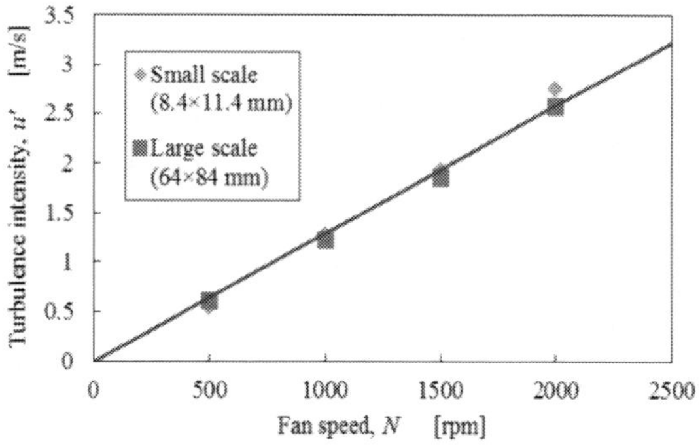

Figure 3: Turbulence intensity, u' as a function of fan speed, N.

The result of large measurement field was adopted to estimate the longitudinal integral length scale. In this study, distribution of turbulence intensity analysis area of 47 x 34 mm was adopted. The longitudinal velocity correlation coefficients, $R_{11}(x)$ were calculated using fluctuation component of velocity obtained from the PIV measurement. Hence, $R_{11}(x)$ was approximated by an exponential function as shown in (1).

$$R_{11}(x) \approx exp(-px^q) \tag{1}$$

Where x is the separation distance, as shown in Fig. 4. The constant $p = 7.80 \times 10^{-3}$ and $q = 1.54$ were determined from least-square fitting. Figure 4 shows that the solid curve is approximated curve of $R_{11}(x)$ by (1). The longitudinal integral length scale, L_f is defined as the integration of $R_{11}(x)$ with respect to x from zero to infinity. Therefore, the integration can be determined using Gamma function, Γ, as expressed in (2).

$$L_f = \int_0^\infty R_{11}(x)dx = p^{-1/q} \cdot \Gamma\left(1 + \frac{1}{q}\right) \tag{2}$$

Longitudinal integral length scale, L_f of 20.9 mm was calculated regardless of the turbulence intensity. The approach follows that of Smallbone et al. [7].

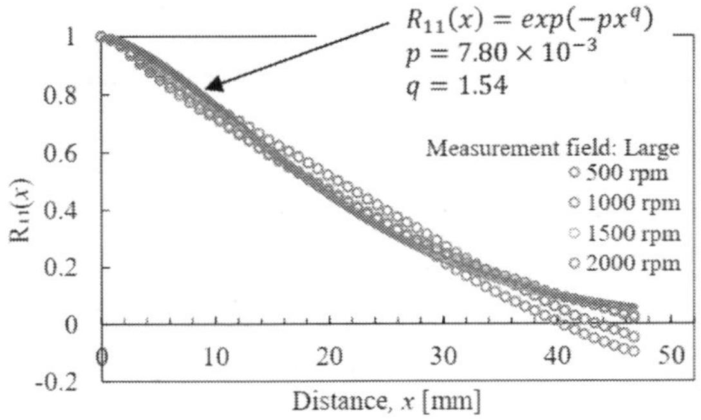

Figure 4: Variation of longitudinal velocity correlation coefficient, $R_{11}(x)$, with distance, x.

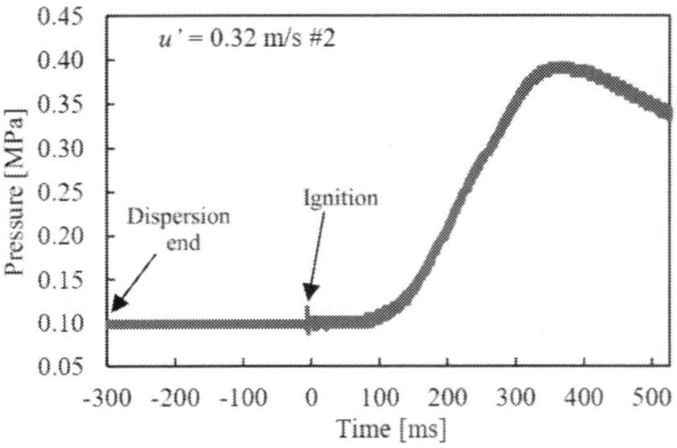

Figure 6: Pressure history during flame propagation.

3 Experimental results

3.1 Flame observation

Figure 5 shows sequential direct images of the spherical turbulent pulverized coal cloud flame propagation in 40% O_2 and 60% N_2 gas mixture from ignition until 20 ms with turbulence intensities of 0.32, 0.65, 0.97 and 1.29 m/s. Although the partially irregular shapes of flame front were observed, on the whole, the flame propagated almost spherically for all cases. The irregular shape of the flame front maybe influenced by the heat loss to the electrodes, in the horizontal direction, especially in lower turbulence intensity. In addition, that is possibly caused by the convection of the whole flame or interaction with large turbulent eddy as the flames grow [13]. The flame diameter was determined by measuring the distance between the farthest flame fronts. As shown in Fig. 5, the flame diameter at a same elapsed time from the ignition increases as turbulence intensity increases.

Figure 6 illustrates the history of the pressure inside the chamber at the turbulence intensity of u'= 0.32 m/s. The pressure is within ± 5% of the atmospheric pressure after the dispersion of coal particles for all cases and remains almost constant during flame propagation up to 50 ms from the onset of ignition. Therefore, the pressure in the chamber during flame propagation was assumed to be constant within the observation range.

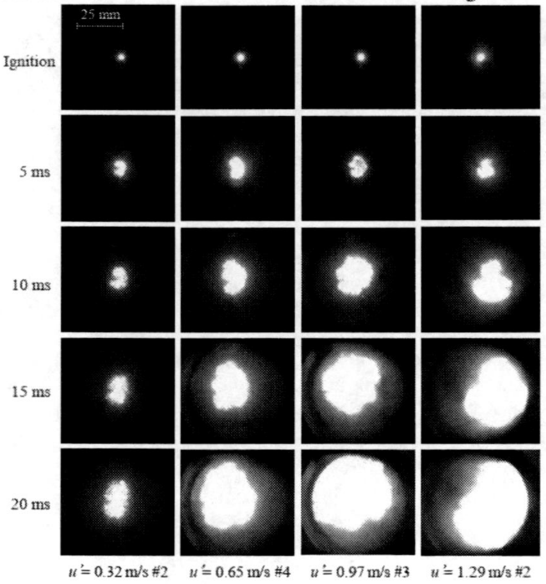

Figure 5: Flame propagation in 40% O_2 and 60% N_2 mixture at atmospheric pressure.

3.2 Flame propagation velocity

Shown in Fig. 7 are the relationship between flame radius and time elapsed from 5 ms after the onset of spark ignition with various turbulent intensities. Spark images appear within the first 5 ms from the onset of the ignition thus concluded as ignition affected period. After 5 ms, the flame propagation was assumed to be not affected by the ignition energy. The ignition affected period was not considered in the calculation of flame propagation velocity. As shown in Fig. 7, the flame radius increments rate increases as the turbulence intensity increases for all turbulence intensities in this study. The flame radius was measured until the flame front reaches the edge of the window.

Figure 8 shows the relationship between flame propagation velocities and mean flame radius. Only flame propagation velocities at 5 ms after ignition and onwards were plotted in Fig. 8. It is shown that the flame propagation velocity increases as the flame radius expand for all cases. The flame propagation velocity increased because of the flame stretch rate decreased. The similar trends can be seen in the gaseous fuel combustion [14].

Figure 9 shows the effect of turbulence intensity on the flame propagation velocity at a mean flame radius of 10.45 mm. The flame radius of 10.45 mm as shown in Fig. 8 was chosen because it is equivalent to the flame diameter of 20.9 mm, which is the same length as the longitudinal integral length scale, L_f, calculated in the analysis of the turbulent field inside the chamber. This approach follows that of Mandilas et al. [15] and Kitagawa et al. [8].

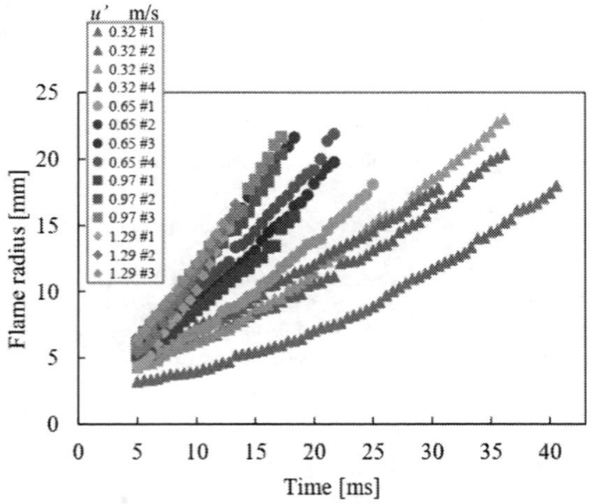

Figure 7: Expansion of flame radius as a function of time elapsed.

It is shown from Fig. 9 that the flame propagation velocity increases as turbulence intensity increases. In the research of gaseous fuel flame propagation conducted by Kitagawa et al. [8], it was suggested that turbulence makes the wrinkling of the flame front. Consequently, the flame area is increased and thus burning velocity is also increased.

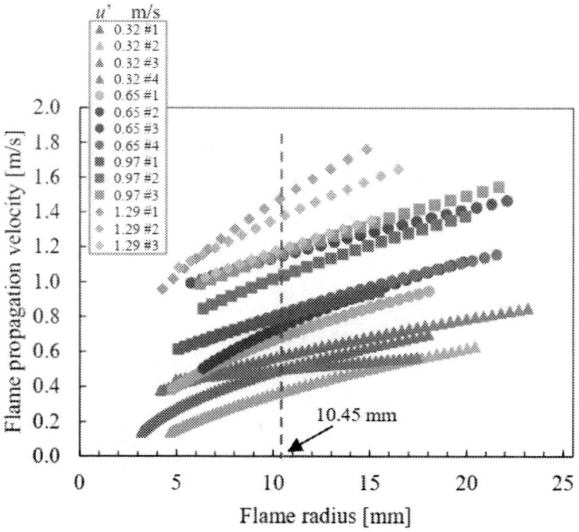

Figure 8: Relationship between flame propagation velocity and mean flame radius.

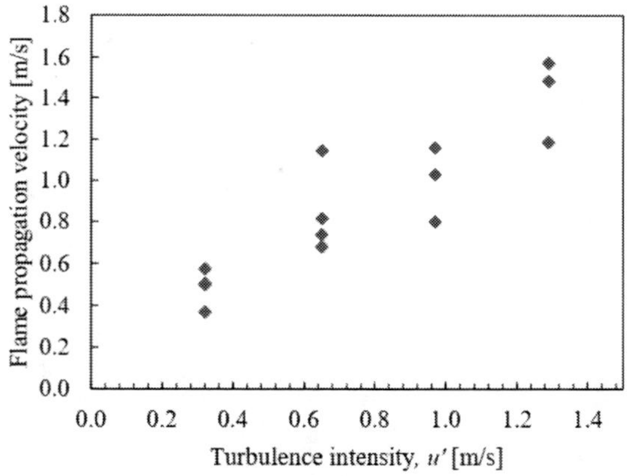

Figure 9: Flame propagation velocities at various turbulence intensities, *u'*.

4 Conclusions

For the first time, the spherical turbulent pulverized coal cloud flame propagation have been studied experimentally. The effect of turbulence intensity on the flame propagation velocity of pulverized coal particle cloud at atmospheric pressure was clarified. Results show that nearly spherical shape of flame was observed during flame propagation. The flame propagation velocity increases with increasing turbulence intensity. This result revealed that the turbulence has a positive effect on flame propagation velocity of pulverized coal particles.

The results were expected to be useful for our further understanding of spherical turbulent pulverized coal cloud flame propagation in various turbulence intensities. Therefore it leads to the improvement of pulverized coal combustion at atmospheric pressure.

5 Acknowledgment

Part of this work was supported by JST research promotion program Sakigake (PRESTO) "Creation of Innovative Core Technology for Manufacture and Use of Energy Carriers from Renewable Energy" Grant Number JPMJPR1542. The authors are indebted to Dr. Suda of IHI Co., Prof. Kitagawa of Kyushu Univ., Prof. Hayakawa of Tohoku Univ. for their helpful advices and discussion. The authors also indebted to the Central Research Institute of Electric Power Industry for providing coal particle samples.

References

[1] O. Fujita, K. Ito,T. Tagashira, J. Sato, Heat Transfer in Microgravity 269 (1993).

[2] T. Kiga, S. Takano, N. Kimura, K. Omata, M. Okawa, T. Mori, M. Kato, Energy Convers. Manag. 38 (1997) S129–S134.

[3] T. Suda, K. Masuko, J. Sato, A. Yamamoto, K. Okazaki, Fuel 86 (2007) 2008-2015.

[4] S. M. Hwang, R. Kurose, F. Akamatsu, H. Tsuji, H. Makino, M. Katsuki, Energy and Fuels 19 (2) (2005) 382–392.

[5] J. Hayashi, N. Hashimoto, N. Nakatsuka, H. Tsuji, H. Watanabe, H. Makino, F. Akamatsu, Proc. Combust. Inst. 34 (2013) 2435–2443.

[6] N. Hashimoto, J. Hayashi, N. Nakatsuka, K. Tainaka, S. Umemoto, H. Tsuji, F. Akamatsu, H. Watanabe, H. Makino, J. Therm. Sci. Technol. 11 (3)(2016) 16-00387.

[7] A. Smallbone, K. Tsuneyoshi, T. Kitagawa, J. Therm. Sci. Technol. 1 (1) (2006).

[8] T. Kitagawa, T. Nakahara, K. Maruyama, K. Kado, A. Hayakawa, S. Kobayashi, Int. J. Hydrogen Energy 33 (20) (2008) 5842–5849.

[9] A. Hayakawa, Y. Miki, Y. Nagano, T. Kitagawa, J. Therm. Sci. Technol. 7 (4) (2012) 507–521.

[10] S. D. Tse, D. L. Zhu, C. K. Law, Proc. Combust. Inst. 28 (2000) 1793–1800.

[11] J. Goulier, N. Chaumeix, F. Halter, N. Meynet, A. Bentaïb, Nucl. Eng. Des. 312 (2017) 214–227.

[12] S. S. Shy, Y. W. Shiu, L. J. Jiang, C. C. Liu, S. Minaev, Proc. Combust. Inst. 36 (2017) 1785–1791.

[13] D. Bradley, M.Z. Haq, R.A. Hicks, T. Kitagawa, M. Lewis, C.G.W. Sheppard, R. Woolley, Combustion and Flame 133 (2003) 415-430.

[14] A. Hayakawa, T. Goto, R. Mimoto, Y. Arakawa, T. Kudo, H. Kobayashi, Fuel 159 (2015) 98-106.

[15] C. Mandilas, M. P. Ormsby, C. G. W. Sheppard, R. Woolley, Proc. Combust. Inst. 31 (2007) 1443–1450.

11th Asia-Pacific Conference on Combustion,
The University of Sydney, NSW Australia
10th -14th December 2017.

Electrospray driven wet cyclone electrostatic precipitator for ultrafine particles collection inside a combustion flue gas

Jong Won CHOI[1], Dae Hun CHUNG[1], Young Chan CHOI[1], Joeng Guen KIM[1], Wook Hyun LEE[1], Youngmin WOO[1], Min Hye YOON[1]

[1]Energy Efficiency Research Division, Korea Institute of Energy Research
Yuseongdae Ro 1312, Dae Jeon, 34101, Korea

Abstract

The global regulations subject to the ultrafine particles concentration inside a combustion flue gas become stricter due to the harmful effects on both atmosphere and human body. While previously the coal fired power plant industry employed the dry electrostatic precipitator (EP), nowadays it is inevitable to install the additional treatment unit to satisfy the stricter regulations for ultrafine dusts. One of the promising systems may be a wet EP. It however have a few problems such as non-uniform water film formation on the wall, large system size, excessive water and energy consumption, and thereby large operation costs. To improve such drawbacks, we herein propose a novel wet EP which includes the cyclone and electrospray technology. In this work, we present the design, the fabrication and the sub-1㎛ particles (PM1.0) collection test of the electrospray driven wet cyclone EP. The experimental results showed the maximum collection efficiency was 94% for the electrical field strength of 4.0 kV/cm with the water amount of 12 ml/min for every single nozzle.

1 Introduction

The wet electrostatic precipitator (wet EP) is an effective control system to remove ultrafine particles or water soluble gases inside a combustion flue gas. However, it has been reported to consume large amount of water and energy compared with the dry EP even though its dust collection efficiency is much higher than the dry EP. Furthermore, the mal-distributed electric field sometimes leaded to the fire on the spot-agglomerated dusts when the water film is not formed uniformly on the collecting electrode surface [1]. Therefore, we modified the wet EP system by means of substituting the simple pressurizing spray by the electrospray technology, which can provide the smaller but highly charged droplets and more uniform water film formation even with much lower energy consumption.

Electrospray technique is used in many industrial applications such as analytical chemistry [2], nanoparticle production [3], uniform coatings [4], and space propulsion [5]. In electrospray driven EP, water flows directly through the high voltage (HV) needle, breaks down to much smaller and highly charged droplets and then fly away to the ground electrode. In more detail, once the charged droplets are sprayed from the nozzle, they follow the electrical field formed between the high voltage needle and the ground plate, and finally reaching the ground plate. During this travel, most of ultrafine particles inside a flue gas meet the charged droplets by the electrostatic attraction. After then, the water and dusts sludge is naturally flowing down in the gravitational direction. The water electrospray characteristics depend on the applied voltage, nozzle geometry, liquid properties such as conductivity, viscosity and surface tension [6-7]. In this study, we used the tap water as a spray solution with the conductivity of about 500 $\mu S \cdot cm^{-1}$. The electrospray motion between the needle tip-to-plane electrodes with the interspace d=74 mm was observed to decide the desirable nozzle size.

In the meantime, the cyclone separators are industrially employed to remove slightly larger particulates from an exhaust air prior to the electrostatic precipitators, of which collection mechanism includes the vortex momentum difference and gravity. Larger particles in the rotating air have too much inertia to follow the tight curve, and therefore collides the cyclone body wall. The rotational speeds of larger particles become lower than the air and finally fall to the bottom of the cyclone cone [8]. Among the fine particles removal systems, the cyclones owing to swirling motions and longer residence time of the gases have been reported to show about 10 times larger specific treatment capacity (defined by the amenable gas treatment capacity per unit system volume) than that of general EPs, which is as shown in Table 1. Hence, we have started designing a novel wet EP based on the high efficiency cyclone geometry in order to reduce the whole system size because the economical construction costs is also one of the most important issues.

Accordingly, we herein proposed a novel electrostatic precipitator to remove ultrafine particles less than 1 μm with high efficiency. This work scope includes the design and fabrication of the cyclone combined with the electrospray technology and dust collection performance evaluation. Applying the negative potential to the nozzles, the highly charged- and sub-50 ㎛ water droplets are sprayed and travels to the counter polar collector along the electric field. During a travel, the charged water droplets pull the ultrafine particles by electrostatic interaction, of which collecting mechanism was quite promising for next generation wet EP system.

Table 1: Specifications comparison among various industrial combustion gas treatment units

Types	Specific treatment capacity ($Nm^3/h\ m^3$)	Pressure drop (mmH_2O)	Removal efficiency (%)	99% removal particle size (μm)
Cyclones	**2,340**	**20~70**	**90.0**	**25**
Bag filters	440	60~230	99.9	1.0
Scrubbers	340	60	95.4	1.0
EPs	220	12.5	99	5

2 Design and Experimentals

Prior to designing the electrospray driven wet cyclone EP system, the basic characteristics of electrospray between a single needle and a SS304 plate with the distance d=80 mm was visually investigated as shown in Fig. 1. In order to visualize the spray motion, the pictures were taken by the digital camera Canon 5D mark II and 70-200 mm lens. We observed the spraying motion in the dependence of nozzle size, the applied voltage and water feed flow rate.

Testing the visualization experiments with a nozzle size ranging from 0.1 to 1.3 mm, applied voltage from 0 to 40 kV, and water feed from 5 to 50 ml/min, it was found that the water droplet size decreases in a narrower nozzle, in a higher voltage, and in smaller water feed conditions (not shown), which are attributed that i) the droplet size in nature decreases in a narrower nozzle, ii) at higher voltage condition, the water droplets become more charged and thereby have stronger repulsive force among them to break into smaller water droplets, and iii) the larger amount of water feed is too hard to be sufficiently charged and broken in, respectively. In addition, the electrical field strength between a nozzle and a plate is inversely proportional to the square root of nozzle size. Out of preliminary results, one may opine that the narrower nozzle may be beneficial to generate much smaller and highly charged water droplets for the application to the electrostatic precipitator in terms of high electrical field strength and high specific droplet surface area. However, a narrower nozzle is easier to be clogged and the smaller and thereby lighter droplets tend to be blown away by following a flue gas without a collection onto the EP body. Therefore, in this study, we prepared 0.45 mm-nozzles (inner diameter) for the electrospray inside a cyclone for the practical and stable application. The proposed electrospray wet cyclone EP is designed on the modified Stairmand cyclone, of which dimension is summarized in Fig. 2 [9]. It simply consists of 4 water feed tubes with the electrical insulator on top and 16 nozzles on each water feed tube as shown in Fig. 3. The distance between a nozzle tip and the cyclone body was measured by 74.4 mm.

To analyze the dust collection performance, a square wind tunnel (0.6m×0.6m×5m) was prepared for testing the fine particle collection performance. As shown in Fig. 4, the velocity of air in the wind tunnel is adjusted by controlling the speed of the electric blower using the variable frequency drive.

We also fabricated a particle condensation system working with Hexamethlydisilazane solution ((CH3)3SiNHSi(CH3)3, HMDS, Sigma Aldrich) to generate and supply the ultrafine dusts. The bubbler tank with HMDS solution was set in a temperature controlled vessel (35°C). Pure nitrogen gas (8 $Lmin^{-1}$) was fed into the HMDS bottle, carrying the HMDS vapor and was mixed with air (30 $Lmin^{-1}$) before entering the electric furnace with the temperature kept at 750°C. As HMDS

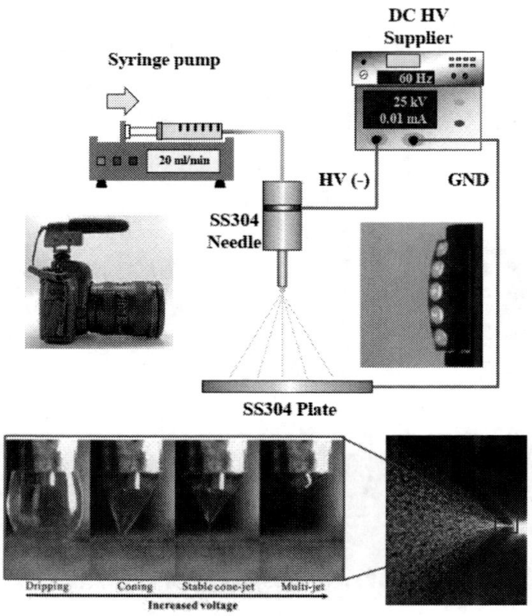

Figure 1: Experimental setup of an electrospray motion of a single nozzle with tap water (500 $\mu S \cdot cm^{-1}$ in conductivity)

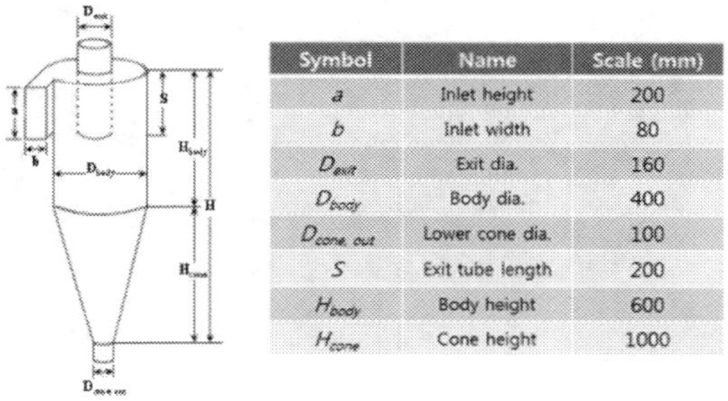

Figure 2: Dimension of modified Stairmand cyclone

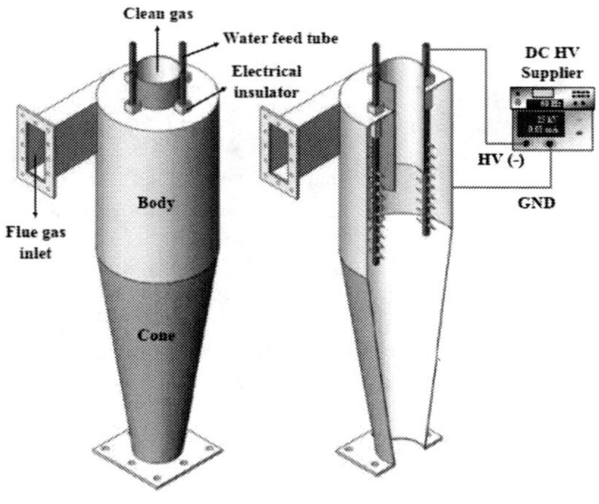

Figure 3: Inner structure of electrospray wet cyclone EP

Figure 4: Pictures of the electrospray wet cyclone EP (upper) and the collection efficiency test equipments (lower)

vapor flows into the electric furnace, it was immediately oxidized, and then turned into the stable oxides. After nucleation, condensation and coagulation, the either stable silicon dioxide (SiO2) or the silica tetrahedron (SiO4) particles were finally generated. A DC power supply (Model HM200-50K-SP, Daesung Tech., Korea) with the capacity of up to -50 kV was prepared to apply the negative DC voltage to the water feed tubes. The inlet gas velocity was measured using a hot wire anemometer (TSI, Model 9515). The average gas velocity flowing into the wet cyclone was measured as 10 m/s, which is equivalent to 9.6 Nm^3/min. The collection efficiency was calculated for different voltages ranging from 0 to -40 kV. The particle size and its mass concentration distribution were measured by using a commercial optical particle sizer (Model: TSI3300, TSI instruments, USA). The sampling was carried out more than 10 times at every single condition and then averaged. Finally, the mass removal efficiency (η) of the wet cyclone EP was calculated as follows:

η = (mass concentration at EP_{off} – mass concentration at EP_{on}) / mass concentration at EP_{off}

Prior to testing the collection performance, the degree of particle deposition along the wind tunnel is investigated by comparing the particle number concentrations at between upstream and downstream of the EP without any voltage application, which was negligible.

3 Results and Discussion

Differently from other dust collecting systems, the cyclone has an inherent flaw, that is, high pressure drop. Fig. 5 shows the pressure drops of the wet cyclone EP designed in this study. With increasing the inlet air velocity, the pressure drop also increases. In addition, the pressure drop in electrospray mode is also slightly higher (about 10%) due to the water droplet disturbance inside a cyclone. While the pressure drop was approximately five times higher than that in commercial electrostatic precipitator, the specific treatment capacity of the wet cyclone EP (about 3,840 $Nm^3.h^{-1}.m^{-3}$ in this study) was calculated 15 times larger than the commercial EPs, meaning that the system size can be reduced down to 7% for the same flue gas treatment [10].

Fig. 6 is a plot of the particles size distribution generated from the home-made particle generator. We also measured the particles concentration with the In-stack cascade impactor (Model: DLPI, Dekati, Finland) with the 47 mm PTFE-T filter. After collecting the ultrafine particles onto the filter, it was maintained in the oven at 60°C for 3 hrs, and then the mass variation was measured with the electronic precise balance (Model: Auw220d, Shimadzu, Japan) with the accuracy of 10^{-5} g. The averaged dust concentration was measured as 18.3 $mg.m^{-3}$ with the deviation of ± 0.6 $mg.m^{-3}$. The main particle size is located in-between 0.2 and 1.2 ㎛, of which size has been known too hard to remove [11].

Finally, Fig. 7 shows the particle collection efficiency as the applied voltage increases when the supplying air flow is 9.6 CMM and the water feed flow rate through 64 nozzles is totally measured by 770 $ml.min^{-1}$ (in other words, about 12 $ml.min^{-1}$ for every single nozzle). The collection efficiency without a doubt increases with increasing the voltages, which were in detail 54, 78, 90, 92 and 94% for 2.0, 2.5, 3.0, 3.5 and 4.0 kV/cm in electrical field strength, which is attributed that the higher voltage produces the finer water droplets, and thereby leads to increasing spray angle, spray velocity, collision

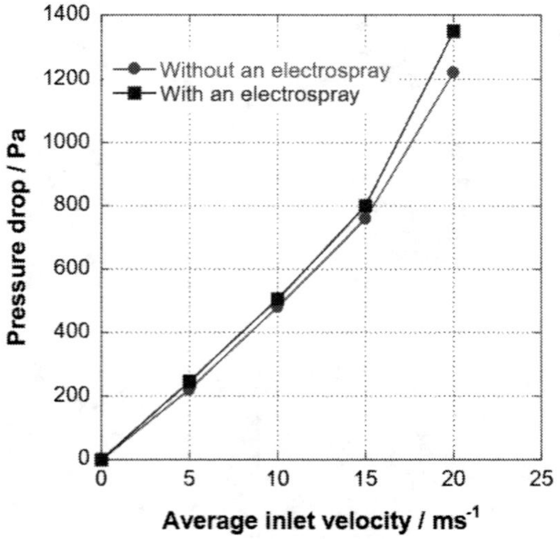

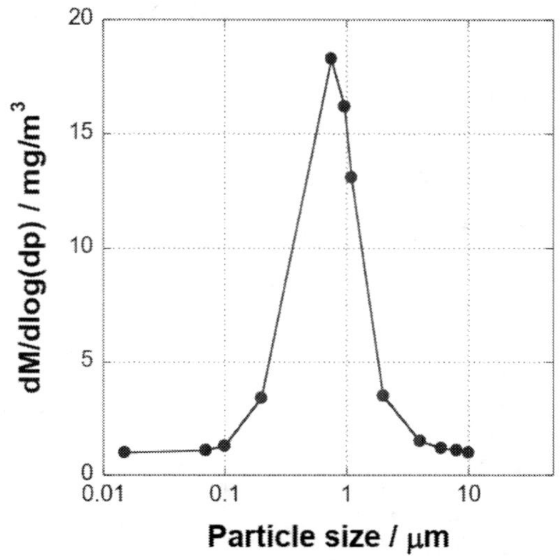

Figure 6: SiOx particles size distribution inside a supplying air

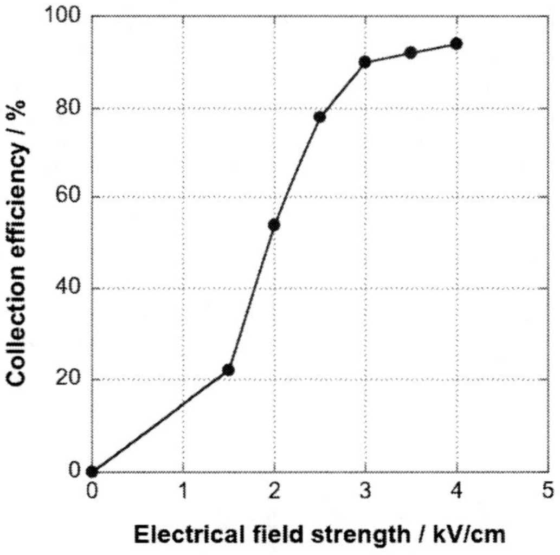

Figure 7: PM 1.0 particles collection efficiency of the electrospray driven wet cyclone EP

probability and electrostatic attraction with the incoming dusts.

4 Conclusions

We herein proposed the application of electrospray technology to the wet cyclone EP to both enhance the collection efficiency and reduce the system size. From visualization to collection tests, we obtained a few useful conclusions as below.

1) With increasing the applied voltage as well as decreasing the nozzle size and water feed amount, the finer water droplets are produced in electrospray system.

2) Although the pressure drop was measured about five times higher than that in the commercial EP, our electrospray wet cyclone showed the 15 times larger specific treatment capacity.

3) The sub-1μm particles collection efficiency increases as the applied voltage increases, which is attributed that the smaller droplets (in other words larger number of charged water droplets) enhance the spray speed, screening area, collision probability and electrostatic attraction with fine particles.

Further experiments for various nozzle sizes, inflowing dust concentrations and water feed amount will give comparative information of energy and water consumption with other industrial electrostatic precipitators, which remains on-going researches.

5 Acknowledgment

This work was supported by the Korea Institute of Energy Technology Evaluation and Planning (KETEP) granted financial resource from the Ministry of Trade, Industry & Energy, Republic of Korea (No. 20161120200090).

References

[1] Y. Liu, B. Hu, L. Zhou, Y. Jiang, L. Yang, Energy & Fuel 30 (10) (2016) 8441-8447.

[2] J. Fenn, M. Mann, C. Meng, S. Wong, C. Whitehouse, Science, 246 (4926) (1989) 64-71.

[3] A. Rulison and R. Flagan, Journal of the American Ceramic Society, 77 (12) (1994) 3244-3250.

[4] A. Seaver and K. Eckhardt, 1988, U.S. Patent, No. 4,748,043.

[5] C. Bartoli, H. Rohden, S. Thompson, J. Blommers, J. Phys. D: Applied Physics, 17 (12) (1984) 2473.

[6] J. Borra, P. Ehouarn, D. Boulaud, J. Aerosol Sci., 35 (11) (2004) 1313–1332.

[7] J. Lopez-Herrera, A. Barrero, A. Boucard, I. Loscertales, M. Marquez, J. Amer. Soc. Mass Spectrom., 15 (2) (2004) 253–259.

[8] J. Dirgo, D. Leith, Jounal of Aerosol Science and Technology, 4 (4) (1985) 401-415.

[9] D. Leith, D. Mehta, Atmospheric Environment, 7 (5) (1973) 527-549.

[10] A. Jaworek, A. Krupa, T. Czech, J. Electrost., 65 (3) (2007) 133-155.

[11] M. Najafabadi, H. Tabrizi, A. Aramesh, M. Ehteram, J. Electrost., 72 (5) (2014) 402-411.

11th Asia-Pacific Conference on Combustion,
The University of Sydney, NSW Australia
10th -14th December 2017.

Combustion characteristics of multiple *n*-heptane pool fires

Yan Jiao, Wei Gao, Naian Liu*, Jiao Lei, Linhe Zhang, Cong Ji

State Key Laboratory of Fire Science, University of Science and Technology of China,
Hefei, Anhui, 230026, China

Abstract

In this work, we present a comprehensive experimental investigation on the combustion characteristics of *n*-heptane MPF (multiple pool fire). A series of fire tests were conducted to explore the fire interaction and its effect on the combustion characteristics of MPF. Analyses indicate that the two dominating mechanisms of heat feedback enhancement and air entrainment restriction have competing effects on the combustion characteristics of MPF, depending on the ratio of pool interval S and pool diameter D. It is found that MPF merges at a critical range of S/D = 3.63~4.18 (D = 5.5 cm, S = 20~23 cm in this work). For the non-merged fires, the time-averaged burning rate, axial velocity and temperature increase steadily with reducing S/D. However, for the merged fires, the time-averaged burning rate, axial velocity and temperature first increase sharply and then decrease with reducing S/D, and the peak values appears when S/D = 2.27~2.73 (D = 5.5 cm, S = 12.5~15 cm in this work). A modified heat release rate correlates the axial velocity and temperature of the fully merged flames reasonably well, and accords well with the classic McCaffrey line.

1 Introduction

Among previous studies, one single pool fire has attracted much attention, while relatively few studies touched on the multiple fires burning. Multiple fires close to each other naturally interact with each other, which may promote heat transfer and restrict air entrainment, thereby the burning process and characteristics may be influenced.

Putnam and Speich [2] pioneered a work of multiple jet fires. Thomas et al. [3] and Baldwin [4] suggested that different pool intervals S and pool diameters D may lead to very different burning behaviours of multiple fires, and fire merging criteria and flame height scaling models were proposed. Weng et al. [5, 6] and Delichatsios [7] introduced the dimensionless HRR (heat release rate) to correlate the flame height of merging fires.

In fire accidents, especially in industrial fires, MPF (multiple pool fires) has drawn more and more attention due to the serious consequence [8]. The knowledge of combustion characteristics of MPF has great importance for fire prevention. Huffman et al. [9] performed nine-burner and thirteen-burner circular pool fire tests using methanol, acetone, hexane, cyclohexane and benzene as fuels. They suggested that the burning process of MPF was determined by heat feedback rate, based on which a simplified burning rate correlation was testified. However, air entrainment condition was not incorporated. Sugawa and Takahashi [10] presented experimental data and correlations of the flame height of hexane MPF with fire source configuration of two line pools, three and four circular pools. Nevertheless, the burning rate, velocity and temperature distribution of those fires were not studied. Our previous work [11-13] analyzed the effect of heat feedback enhancement and air entrainment restriction on the BOT (burn-out time) of MPF, by which the spatial fluctuations of the two interaction effects were interpreted. However, the BOT analysis method does not allow one to explore the instantaneous burning process, and is able neither to capture the combustion characteristics during the quasi-steady period.

In this work, we present a comprehensive experimental study on the combustion characteristics of *n*-heptane MPF. The effects of fire interaction on the fire merging behavior and burning process are examined. The instantaneous and time-average burning rate, and the velocity and temperature distribution under the fire interaction are also analyzed.

2 Experimental

Figure 1 shows the experimental design. Fire tests were performed over a square fire board platform elevated 50 cm above the ground.

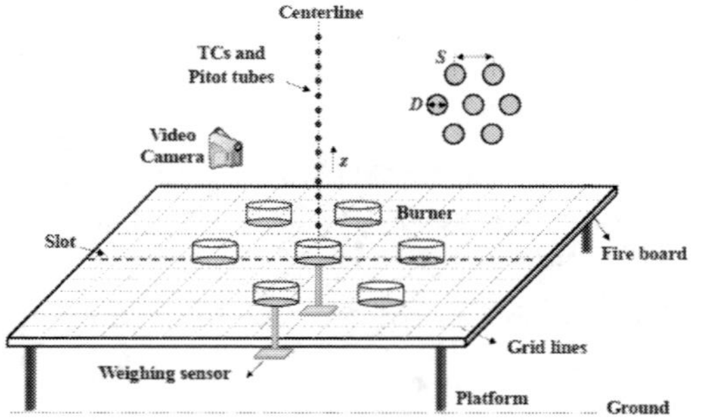

Figure 1: Schematic of experimental design of the present work.

A seven-burner fire pattern was used, among which the outer six burners formed a regular hexagon. All burners are made of iron and for all, the fuel of *n*-heptane has an initial identical lip height of 2.5 cm. The bottom of burners was positioned flush with the top surface of platform. Three slots made it easy to adjust the outer pools along the diagonal lines of regular hexagon. The pool interval S is defined as the distance between the center of the two adjacent pools, which could be determined with the help of grid lines drawn on the platform. In this work, two sets of fire tests were conducted. For one set, we set pool interval S to be 10 cm and used six identical outer pools (D = 5.5 cm), but varied the center pool size. For this set, three tests were performed: 1) with

Naian Liu. Fax: +86-551-63601669
E-mail address: liunai@ustc.edu.cn

a center pool of D = 10.0 cm; 2) with a center pool of D = 5.0 cm; 3) without a center pool. For another set, fire tests with identical D (5.5 cm) and different pool intervals S (8 cm, 9 cm, 10 cm, 12.5 cm, 15 cm, 20 cm, 23 cm, 25 cm, 30 cm, 33 cm, 35 cm, 40 cm and 45 cm) were performed.

The instantaneous burning rate of the center burner was recorded and compared simultaneously by an electronic balance and a weighing sensor, while six weighing sensors were used for that of the outer six burners. Along the centerline of the center burner, ten Pitot tubes (series 160S, Dwyer Instruments, Inc.) were located at an identical interval (≈10 cm) to measure the axial velocity. The lowest and highest Pitot tubes were respectively positioned 86 mm and 992 mm higher than the top of burner lip. Twenty K-type thermocouples (with a bead diameter of 0.5 mm and a wire diameter of 0.15 mm) were used for axial temperature measurement, for which ten thermocouples were located at the points where the velocity was measured by Pitot tubes. The lowest and highest thermocouples were respectively positioned 49 mm and 1016 mm higher than the top of burner lip. A video camera was used to record the burning process.

3 Results and discussions

3.1 Experimental observation

Figure 2 presents several typical experimental photos (D = 5.5 cm, S = 20 cm). At the initial stage after ignition, fires burned slowly, and the fire interaction was insignificant since the flames were almost vertical (Fig. 2a). In the meantime, the burning rate curves of all fires were almost consistent. Several minutes later, with continuous heat feedback, the fuel temperature substantially rose (with no boiling, though), thus leading to a higher burning rate and a lengthened flame height (Fig. 2b). During this period, the center fire competed with the outer fires for the entrained air, as witnessed by the inward tilting of the outer fires. Soon after that, the fuel inside the pools reached the boiling stage, evidenced by sharp increase of burning rate. At the same time, the flame height was further lengthened and the flame tips began to touch each other (Fig. 2c). Subsequently, the heat feedback intensity and burning rate reached peak values and the fires were entirely merged (Fig. 2d).

Figure 2: Typical experimental photos of MPF of the present work (D = 5.5 cm, S = 20 cm). The time after ignition of each sub-figure was also presented.

As observed, the merging flames were periodically puffing and oscillating (Fig. 2e), which may be due to the Rayleigh-Tayler instability of natural convection. This period only lasted for about one minute when the fuel was consumed quickly. After that, the flame of the center burner firstly burnt out (Fig. 2f). As compared with the single pool fire, the BOT of MPF was obviously shortened, which was consistent with our previous observations [11-13].

In addition, it was observed that, the occurrence of fire merging is dependent on S/D. A critical criterion of S/D = 3.63~4.18 (S = 20~23 cm) was obtained, as the flame tips of fires were closely touched, and the flames showed a steady merging behavior. For a larger S/D, fires tended to tilt inward and touch each other sometimes, but did not merge.

3.2 Burning rate

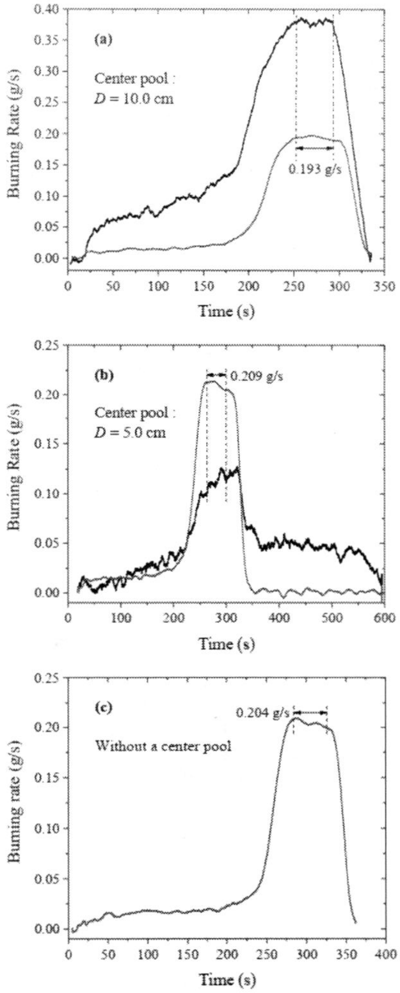

Figure 3: Instantaneous burning rate curve of MPF in the present work. (black line: the center fire; red line: the outer fire)

The interaction of MPF will be reflected in the burning rate variation during burning. In the present work, we firstly performed fire tests with the set of pool interval S = 10.0 cm, and used six identical outer pool (D = 5.5 cm) with different center pool diameters. As shown in Fig. 3a and Fig. 3b, when the flames began to merge (around 200 seconds in the figure), the burning rates of the center and outer flames show synchronous sharp increase, implying that the flames mutually promoted heat feedback. Comparing with the single n-heptane pool fire, for which the time-average burning rate is 0.11 g/s, 0.05 g/s and

0.03g/s for D = 10 cm, D = 5.5 cm and D = 5.0 cm, the MPFs showed nearly fourfold burning rate increase due to flame merging.

On the other hand, however, the experimental data reveal that fire interaction may also have a mechanism resulting in reduction in the burning rate of MPF due to air entrainment restriction. As shown in the figures, for the time-averaged burning rate of the outer fires during fire-merging, although the MPF with center fire of D = 10.0 cm received the highest heat feedback, the burning rate (0.193 g/s) was still lower than that of the MPF with center fire of D = 5.0 cm (0.209 g/s) and without a center fire (0.204 g/s). This was due to the fact that, the center fire needed to consume the entrained air and would reduce the air supply available for the outer fires, thus leading to decrease of burning rate. Therefore, for the burning rate of MPFs, the heat feedback enhancement and the air entrainment restriction indeed constitute two competing mechanisms.

Figure 4 presents the time-averaged burning rate of MPF with identical D (5.5 cm) and different flame intervals S. As a whole, the burning rate of flames first increased and then decreased with reducing S, among which the peak value appeared when S/D = 2.27~2.73 (S = 12.5~15 cm).

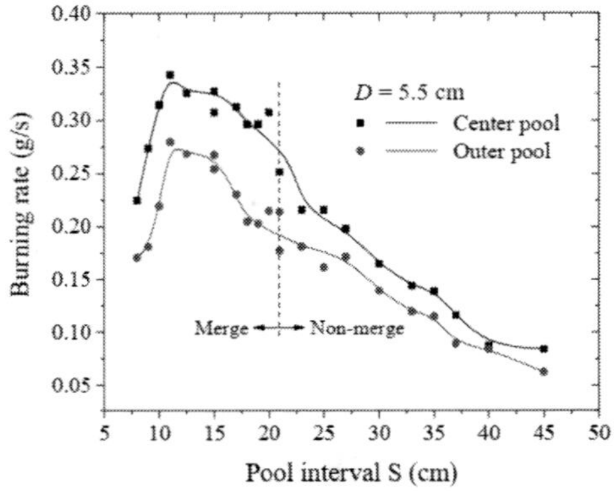

Figure 4: Time-averaged burning rate of MPF with identical D (5.5 cm) and different flame intervals S in the present work.

As mentioned in Section 2, fire merging occurred at the critical point of S = 20~23 cm. For this, we compared the burning rate of S = 23~45 cm between that of S = 12.5~20 cm, and the latter obviously has a sharper increase, which reflects the marked influence of heat feedback enhancement due to fire merging. However for S = 8~12.5 cm, although significant fire merging occurred, the burning rate of MPF appears to decrease with decreasing S, implying that the effect of air entrainment restriction had exceeded the effect of heat feedback enhancement. Therefore, the effect of the two competing mechanisms on the burning rate of MPF is dependent on S.

3.3 Axial velocity and temperature distribution

In previous works, the fire merging behavior and flame height of MPF drew the majority of attentions. However, the velocity and temperature distributions of MPF have not been fully studied despite of their importance. Fire interaction exerts notable impact on the burning rate, thus will naturally affect the heat release rate $\dot{Q}$, which dominates the axial velocity and temperature. Thus, fire interaction is supposed to affect the axial velocity and temperature distributions of MPFs.

Figure 5 and Figure 6 present the time-averaged axial velocity and temperature versus height z of MPF with identical D (5.5 cm) and different flame intervals S. The origin of z-coordinate was defined at the top of burner lip. As a whole, for the cases of non-merged fires, the axial velocity and temperature were comparable to that of a single fire, and increased steadily with decreasing S/D. However, for the merged fire tests, the time-averaged axial velocity and temperature were considerable higher than that of a single fire, and first increased sharply then decreased with decreasing S/D. This variation tendency with S/D is similar to that of time-averaged burning rate, which is closely related to the two competing effects caused by fire interaction.

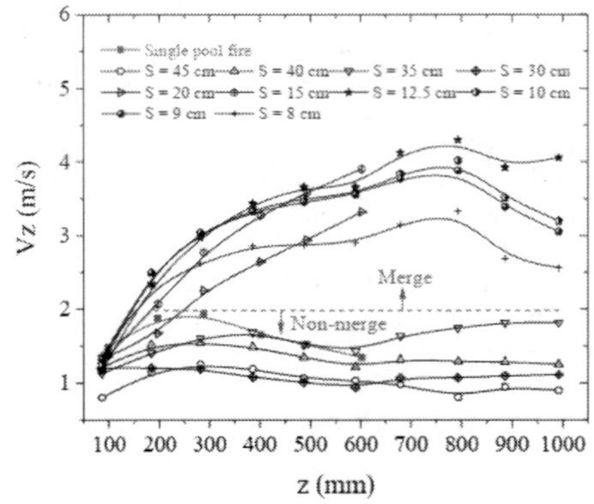

Figure 5: Time-averaged axial velocity of MPF with identical D (5.5 cm) and different flame intervals S, with comparison to that of a single pool fire in the present work.

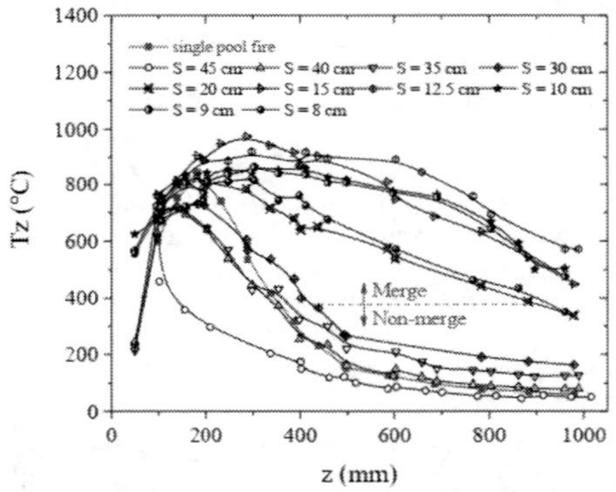

Figure 6: Time-averaged axial temperature of MPF with identical D (5.5 cm) and different flame intervals S and the comparison with that of a single pool fire in the present work.

3.4 Correlations of dimensionless axial velocity and temperature of the fully merged fires

In fire studies, the heat release rate $\dot{Q}$ is commonly used to correlate the axial velocity and temperature distributions of flame. The classic correlations are proposed by McCaffrey [1]:

$$\left(\frac{V_z}{\dot{Q}^{1/5}}\right) \propto \left(\frac{z}{\dot{Q}^{2/5}}\right)^n \begin{cases} n = 1/2 & \text{for continuous flame} \\ n = 0 & \text{for intermittent flame} \\ n = -1/3 & \text{for buoyant plume} \end{cases} \quad (1)$$

$$T_z \propto \left(\frac{z}{\dot{Q}^{2/5}}\right)^n \begin{cases} n = 0 & \text{for continuous flame} \\ n = -1 & \text{for intermittent flame} \\ n = -5/3 & \text{for buoyant plume} \end{cases} \quad (2)$$

where $\dot{Q}$ equals to $\eta \dot{m} \Delta H_c$, with η, $\dot{m}$ and ΔH_c respectively representing combustion efficiency, burning rate and heat of combustion.

For the fully merged MPF with very small S/D, it may be reasonable to treat it as a single pool fire which has a self-similarity in axial velocity and temperature distribution. The axial velocity and temperature depend on $\dot{Q}$, in which the burning rate variation is incorporated. In this way, $\dot{Q}$ should be modified as:

$$\dot{Q}_{total} = \eta \dot{m}_{center} \Delta H_c + 6\eta \dot{m}_{out} \Delta H_c \tag{3}$$

where $\dot{Q}_{total}$, $\dot{m}_{center}$ and $\dot{m}_{out}$ respectively represent the total heat release rate of MPF, burning rates of the center and the outer fire. The term $6\eta \dot{m}_{out} \Delta H_c$ denotes the heat release rate of the six outer fires, among which the coefficient six denotes the number of fires. In this work, only MPFs with $D = 5.5$cm and $S = 8\sim15$cm are treated as fully merged MPF based on experimental observations and the data are used to correlate.

As shown in Figure 7 and Figure 8, all the axial velocity and temperature data correlates well, and accords well with the classic McCaffrey line, indicating that it may be reasonable to apply the McCaffrey correlations to the fully-merged MPFs.

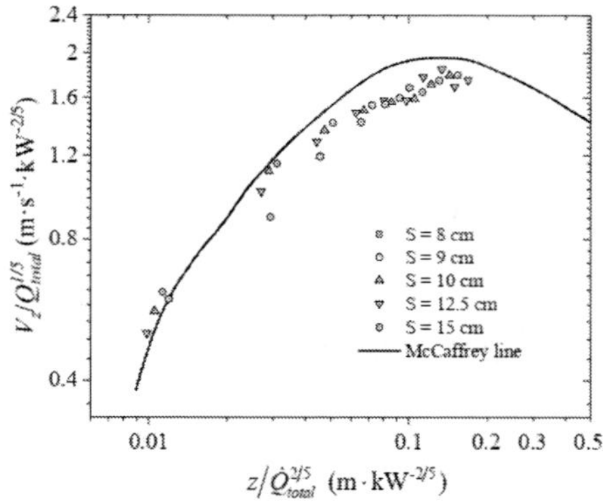

Figure 7: Correlated time-averaged axial velocity of MPF with identical D (5.5 cm) and various S and the McCaffrey line.

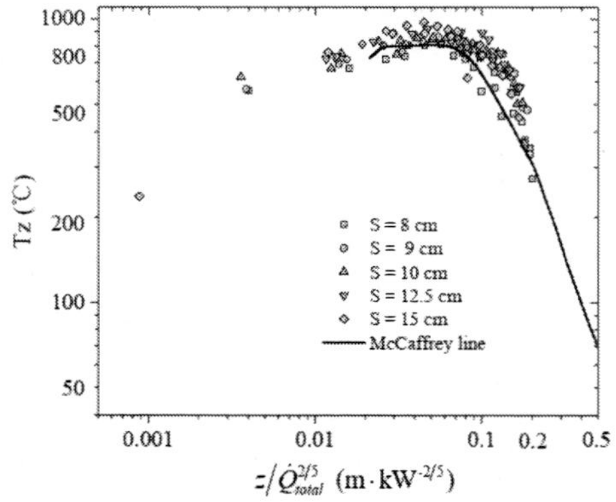

Figure 8: Correlated time-averaged axial temperature of MPF with identical D (5.5 cm) and various S and the McCaffrey line.

4 Conclusions

In the present work, we present a comprehensive experimental investigation on the combustion characteristics of n-heptane MPF. The effects of fire interaction on the combustion characteristics of MPF are discussed. The major findings are summarized as follows.

1) MPF interacts during burning. The critical criterion of fire merging is S/D equals to 3.63~4.18.
2) For the non-merged fires, the time-averaged burning rate, axial velocity and temperature increase steadily with reducing S/D. However, for the merged fires, the time-averaged burning rate, axial velocity and temperature first increase sharply and then decrease with reducing S/D, and the peak values appears when S/D equals to 2.27~2.73. These nonlinear variations are attributable to the two competing mechanisms of heat feedback enhancement and air entrainment restriction.
3) A modified heat release rate correlates the axial velocity and temperature of the fully merged fires reasonably well, and accords well with the classic McCaffrey line.

5 Acknowledgment

This research is funded by the National Key Research and Development Plan (No. 2016YFC0800100) and the National Natural Science Foundation of China (No. 51625602). This work was also supported by the USTC Fundamental Research Funds (No. WK2320000038).

References

[1] B.J. McCaffrey, Purely buoyant diffusion flames: some experimental results. Technical Report NBSIR 79-1910. (1979) 1-44.

[2] A.A. Putnam, C.F. Speich, Proc. Combust. Inst. 9 (1) (1963) 867-877.

[3] P.H. Thomas, R. Baldwin, A.J.M. Heselden, Proc. Combust. Inst. 9 (1) (1965) 983-996.

[4] R. Baldwin, Combust. Flame 12 (4) (1968) 318-324.

[5] W. Weng, D. Kamikawa, Y. Fukuda, Y. Hasemi, K. Kagiya, Combust. Sci. Technol. 176 (12) (2004) 2105-2123.

[6] W. Weng, D. Kamikawa, Y. Hasemi, Proc. Combust. Inst. 35 (3) (2015) 2597-2606.

[7] M.A. Delichatsios, Fire Sci. Technol. 26 (1) (2007) 1-8.

[8] S. Vasanth, S.M. Tauseef, T. Abbasi, S.A. Abbasi, J. Loss Prev. Process Ind. 29 (2014) 103-121.

[9] K.G. Huffman, J.R. Welker, C.M. Sliepcevich, Fire Technol. 5 (3) (1969) 225-232.

[10] O. Sugawa, W. Takahashi, Fire Mater. 17 (3) (1993) 111-117.

[11] N. Liu, Q. Liu, Z. Deng, S. Kohyu, J. Zhu, Proc. Combust. Inst. 31 (2) (2007) 2589-2597.

[12] N. Liu, Q. Liu, J.S. Lozano, L. Shu, L. Zhang, J. Zhu, Z. Deng, S. Kohyu, Proc. Combust. Inst. 32 (2) (2009) 2519-2526.

[13] N. Liu, Q. Liu, J.S. Lozano, L. Zhang, Z. Deng, B. Yao, J. Zhu, S. Kohyu, Proc. Combust. Inst. 34 (2) (2013) 2555-2564.

11th Asia-Pacific Conference on Combustion,
The University of Sydney, NSW Australia
10th -14th December 2017.

Numerical Investigation into Natural Gas–Diesel Dual-Fuel Engine Configuration

Baocong Jing[1], Andrew P. Wandel[2]

[1]School of Mechanical and Electrical Engineering, University of Southern Queensland
West St, Toowoomba, Queensland, 4350, Australia

[2]Computational Engineering and Science Research Centre, School of Mechanical and Electrical Engineering,
University of Southern Queensland
West St, Toowoomba, Queensland, 4350, Australia

Abstract

Natural gas is used as an additive to a Diesel engine in a dual-fuel configuration. The natural gas is port injected with the charge, while the Diesel is directly injected. The effect of increasing the energy content contributed by the natural gas (to 11% and 22%) is tested for two engine speeds (2000 and 2500 rpm) using ANSYS Forte with *n*-heptane as the surrogate fuel for Diesel and methane as the surrogate fuel for natural gas. It was found that the main combustion phase was retarded by additional amounts of natural gas, with 22% methane approaching misfire at 2500 rpm. The thermal efficiency for the other methane cases was higher than Diesel at the corresponding engine speeds, with 11% methane the highest at both engine speeds. Increasing the amount of natural gas significantly reduced the emissions of CO and NO_x. Because the natural gas is separately regulated, in practice the amount can be dynamically controlled by the ECU to satisfy the desired performance/emissions levels. The current results suggest that small amounts of natural gas can have a significant positive influence on the emissions profile.

1 Introduction

The transportation sector in Australia produced 93 Mt CO_2-e in the year to June 2016, which accounted for 18% of the nation's greenhouse gas emissions [1]. Of this figure, 37% is due to trucks, busses and light commercial vehicles [2]. Improvements in Diesel engine efficiency can contribute significantly to the improvement of these figures, reduce operating costs and extend fossil fuel reserves.

A promising alternative is to use natural gas with Diesel in dual-fuel operation because this can reduce NO_x and soot, but could result in increased CO and unburned CH_4. The conventional configuration is to maintain direct injection of the Diesel, with the natural gas port injected to premix with the charge. Various studies have investigated this concept. One study used CONVERGE CFD to study the effect of Exhaust Gas Recirculation (EGR) for a light-duty VW TDi engine [3]. They found that increasing the natural gas content retarded the main combustion phase so that once the energy content of the natural gas exceeded a certain level, no reactions occurred. The NO_x emissions also decreased, which can be partially attributed to the decrease in power generation, while the unburned hydrocarbon (UHC) and CO levels were approximately constant until the misfire point. Advancing the injection timing substantially reduced the NO_x, with corresponding significant increases in UHC and CO. Finally, splitting the Diesel injection into separate pulses increased the NO_x and decreased the UHC and CO the more mass there was injected into the second pulse.

Another study used experiments and a two-zone model (where the expanding "burning zone" is conical during fuel injection and spherical during combustion) to study the performance of a single-cylinder research engine [4]. The Diesel injection timing was close to TDC (5°, 10° and 15°BTDC) and the consistent ignition delay time resulted in all conditions being close to misfire. Increasing the energy content supplied by the natural gas mitigated the misfire conditions. Conditions closer to misfire had decreased NO and soot production but increased CO and specific energy consumption.

A modified single-cylinder Caterpillar 3400-series heavy-duty engine was studied at 25% load both experimentally and using the AVL FIRE v2014 software coupled with CHEMKIN [5]. A fixed energy content of 75% natural gas was used with different injection timings in the range 10–50°BTDC. At 10°BTDC, the unburned methane and CO were highest, but NO_x was low. At 30°BTDC, the NO_x was at its maximum and beyond this injection timing, unburned methane and CO emissions were approximately the same.

The current study investigates the effect of engine speed and energy content of natural gas for a single-cylinder engine at full load.

2 Test conditions

ANSYS Forte was used to simulate the engine with test conditions listed in Table 1. The RNG k-e model was used for the turbulence; a 35-species 74-reaction *n*-heptane chemical kinetics scheme was used. Pure Diesel (modelled as *n*-heptane) was considered along with natural gas (modelled as methane) contributing 10% and 20% of the mass of fuel (which equates to 11% and 22% of the energy based on the LHV of the fuels).

Corresponding author. Fax: +61-7-4631-2526
E-mail address: andrew.wandel@usq.edu.au

Bore	8.25 cm
Stroke	11.6 cm
Compression Ratio (CR)	17
Injection start–end	30°BTDC–7°ATDC
Equivalence ratio λ	3.31
Engine speeds	2000, 2500 rpm
Load	Full

Table 1: Engine parameters for Results in Sect. 4.

3 Validation

A qualitative validation was performed against Diesel experiments [6]. In analysing the effect of engine speed (Figs. 1 and 2), similar effects can be observed in that increasing engine speed retards the location and magnitude of the peak pressure, eventually leading to the onset of misfire when there are effectively two local maxima of pressure. The heat release rate shows the two separate instances of heat release, with the first release always occurring at essentially the same timing, with the major heat release becoming retarded and weaker with increasing engine speed. At the highest engine speed tested, the major heat release has two distinct increases, so the final heat release is curtailed as misfire begins at higher engine speeds.

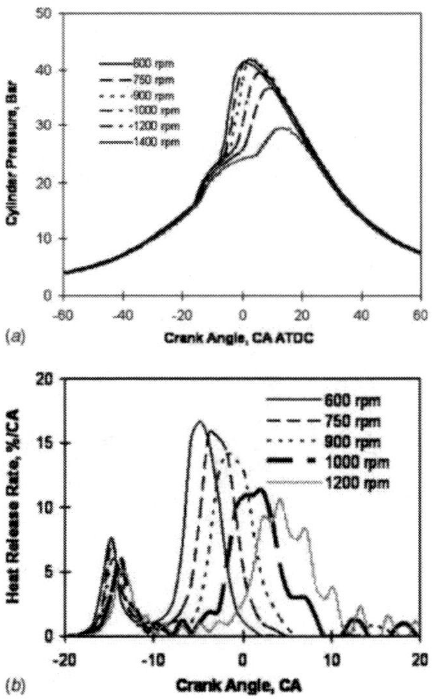

Figure 1: Effect of engine speed on (a) cylinder pressure and (b) heat release rate at CR = 10, AFR = 50, T_{in} = 40°C [6].

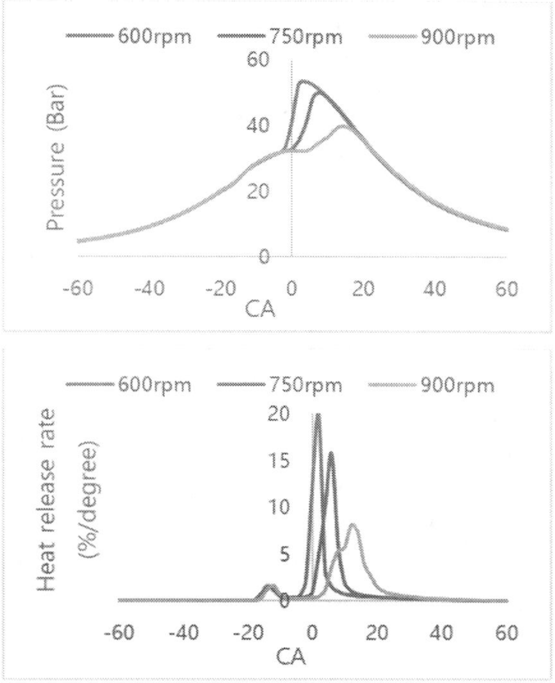

Figure 2: Simulation results corresponding to Fig. 1.

4 Results

4.1 2000 rpm

The pressure trace for the different fuels is shown in Fig. 3. In line with other studies [3,4], the main combustion phase is retarded by the addition of the methane. The fundamental reason for this is that the autoignition temperature of methane is not reached by TDC for the motored pressure: it requires the main combustion phase of the Diesel to reach the autoignition temperature. Therefore the port-injected methane acts as a diluent during the ignition delay of the Diesel. For the methane to autoignite before TDC requires extraordinary levels of charge pressure and temperature.

The corresponding heat release rate is shown in Fig. 4. The initial combustion phase is unaffected by the addition of methane, but the main combustion phase is slightly delayed, takes longer and reaches a lower peak heat release rate with increasing amounts of methane, with a corresponding decrease in accumulated heat (Table 2).

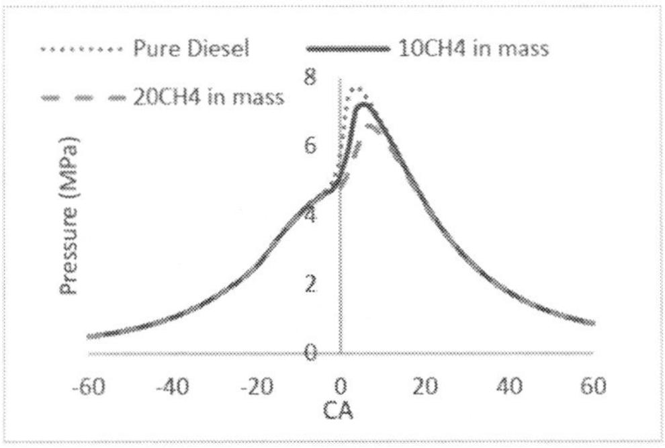

Figure 3: Pressure trace for 2000 rpm.

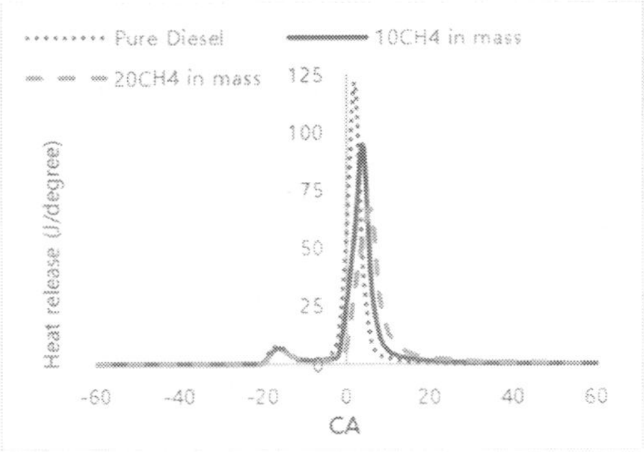

Figure 4: Heat release rate for 2000 rpm.

Engine speed	Diesel	10% CH_4	20% CH_4
2000 rpm	567 J	558 J	540 J
2500 rpm	586 J	546 J	498 J

Table 2: Accumulated heat for different cases tested.

4.2 2500 rpm

The pressure trace for the different fuels is shown in Fig. 5. Because of the faster engine speed, the duration shown in Fig. 5 is only 80% of the duration in Fig. 3, therefore it is unsurprising that the main combustion phase is significantly retarded, with the 20% methane approaching misfire. This outcome is reflected in the heat release rate (Fig. 6), where the initial combustion phase is unaffected, but the main combustion phase has two stages, which are separated in the case of 20% methane. Because the main combustion commences after TDC, the peak heat release rate is reduced by more than the 20% implied by the increased engine speed. The total energy released actually slightly increases for Diesel, but decreases for the methane cases, with a marked loss of power for 20% (Table 2).

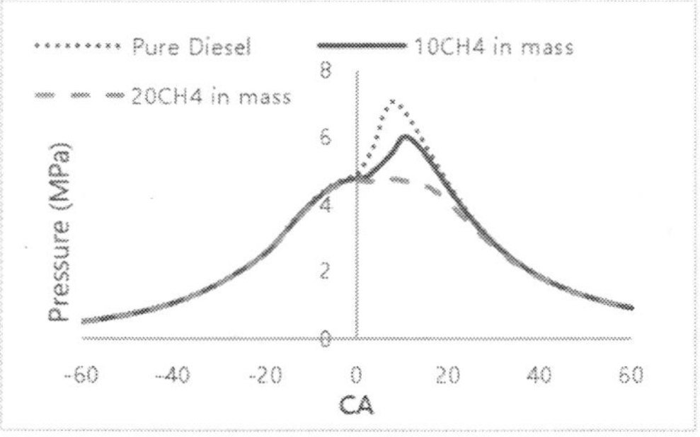

Figure 5: Pressure trace for 2500 rpm.

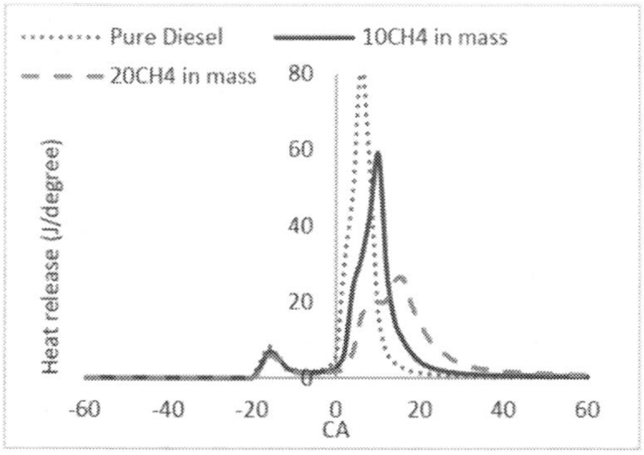

Figure 6: Heat release rate for 2500 rpm.

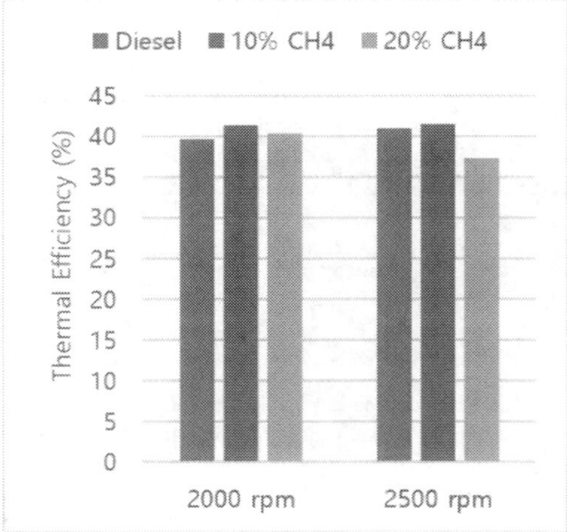

Figure 7: Thermal efficiency for all cases.

4.3 Performance and emissions

The thermal efficiency for the various cases is reported in Fig. 7. The 10% methane case had the highest value at both engine speeds, with 20% methane higher at 2000 rpm, but significantly lower at 2500 rpm due to its closeness to misfire. The CO and NO_x levels (Figs. 8 and 9 respectively) both reduced with increasing methane content. During the initial combustion phase, small amounts of methane are produced by the Diesel, so the additional methane aids the combustion process approach completion. The reduced cylinder pressure (and corresponding temperature)—due to the methane dilution retarding the moment of peak heat release—results in conditions that are not favourable for NO_x production.

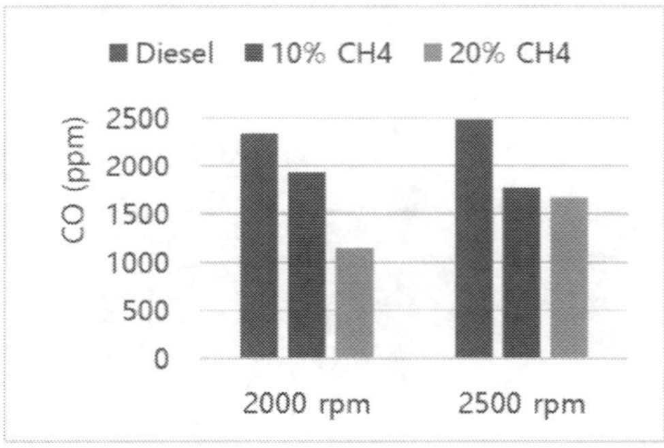

Figure 8: CO emissions for all cases.

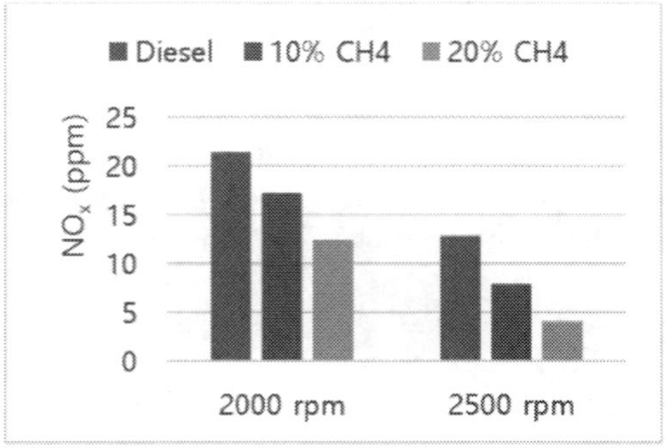

Figure 9: NO_x emissions for all cases.

5 Conclusions

The effect of natural gas operating in a dual-fuel configuration in a Diesel engine has been investigated via a simulation study. For the limited range of settings, it was found that increasing the fraction of energy content supplied by natural gas retards the main combustion phase without impacting on the initial combustion phase. At the higher engine speed, 20% CH_4 by mass was close to misfire; the thermal efficiency was slightly improved for the other cases with methane. Increasing the amount of natural gas increasingly reduced the emissions of CO and NO_x. Because the ECU would be able to dynamically control the amount of natural gas injected, based on the current results, it would appear that the best control strategy would be to reduce the amount of natural gas with increasing engine speed. Further research will seek to quantify such a control strategy and also investigate the impact of Diesel injection timing.

6 Acknowledgment

The current study was performed for a research project as part of a coursework Master's degree.

References

[1] Commonwealth of Australia, Quarterly Update of Australia's National Greenhouse Gas Inventory: June 2016, 2016.

[2] Commonwealth of Australia, Tracking to 2020: An interim update of Australia's greenhouse gas emissions projections, December 2015.

[3] K. Poorghasemi, R.K. Saray, E. Ansari, B.K. Irdmousa, M. Shabakhti and J.D. Naber, Applied Energy 199 (2017) 430–446.

[4] R.G. Papagiannakis, S.R. Krishnan, D.C. Rakopoulos, K.K. Srinivasan and C.D. Rakopoulos, Fuel 202 (2017) 675–687.

[5] A. Yousefi, M. Birouk and H. Guo, Fuel 203 (2017) 642–657.

[6] H. Guo, W.S. Nell, W. Chippior, H. Li and J.D. Taylor, J. Eng. Gas Turb. Power (2010) 1–10.

11th Asia-Pacific Conference on Combustion,
The University of Sydney, NSW Australia
10th-14th December 2017.

Effect of air coflow and swirl on droplet dispersion in reacting sprays

Mallela Jagadeesh[1], Manish M[1], Srikrishna Sahu[1]

[1]Department of Mechanical Engineering, IIT Madras, Chennai, 600036, INDIA

Abstract

The aim of this paper is to study the effect of air coflow and air swirl on droplet dispersion in reacting sprays. Different optical techniques are used to characterize the spray droplets as well as to visualize the spray flame. The flame characteristics such as height, width and liftoff are found to be significantly affected due to the air co-flow, which strongly modifies the local droplet number density within the spray. Though, imparting swirl on the air coflow again redistributes the droplets resulting in wider and shorter flame. Since droplet often burn as a group rather than individually, characterization of droplet groups or clusters is important. This is achieved by identifying droplets on the Mie-scattering images followed by a statistical method in order to quantify the degree of clustering and estimate length scales of the droplet clusters.

1 Introduction

The ever-increasing demands in energy and transportation sectors and stricter norms for emission regulation necessitate thorough investigation of spray combustion process for optimal performance as well as emission reduction. The dispersion of fuel droplets significantly influences the structure of the spray flame. Hence, detail understanding on the factors influencing distribution of droplets within combustion chamber, for example, air coflow around the spray and/or air swirl, is crucial. In addition, it is well known that the turbulent dispersion of droplets may lead to droplet clusters, which burn as groups. The group combustion of droplets may affect the flame location and distributions of temperature, fuel vapor and oxygen [1, 2, 3]. Thus, characterizing droplet dispersion, in particular droplet clustering in reacting sprays, and understanding the respective influences of air coflow and air swirl is the goal of the present research work. For this purpose, a coaxial burner is considered with provision for imparting swirl on the coflowing air around the central fuel spray. A pressure swirl atomizer is considered for fuel injection, which finds application in gas turbine engines apart from several other industrial applications.

2 Flow arrangement and instrumentation

The experimental setup is as shown in the Figure 1. A central stainless-steel tube within the burner supplies kerosene to a Delavan pressure swirl atomizer (0.5 GPH, 70^0 cone angle), which generates a hollow cone spray. The fuel supply pressure was maintained at 6 bar such that the fuel flow rate is same for all experiments. The coflowing air around the spray is supplied through the annular passage between the central fuel tube (23 mm outer diameter) and a concentric metallic tube (48 mm inner diameter). The volume flow rates of the fuel and air flow are measured with respective rotameters and regulated using control valves. An air swirler fabricated by rapid prototyping (eight twisted blades of 45^0 blade angle and geometric swirl number, S = 0.77) is placed 15 mm below the injector exit within the annular passage (as shown in Fig. 1). The experiments were conducted for three different co-flowing air flow rates, 0 LPM, 200 LPM and 400 LPM, while the fuel flow rate is maintained same. These flow rates of air correspond to Reynolds number of co-flowing air, Re_{co} equal to 0, 3250, 7500, respectively. Experiments were conducted both in absence and in presence of the air swirler. The droplet measurements were carried out using different laser based techniques (as explained below) for different axial locations, z = 3 cm, 5 cm, 7 cm, 10 cm and 14 cm from atomizer exit and for four radial locations i.e., r = 0 cm, 2 cm, 4 cm and 6 cm from the injector axis (as shown in Fig 1). A CANON EOS 70D DSLR camera was used to capture spray flame images to characterize the flame (height, width, lift-off) for different operating flow conditions. Figure 2 demonstrates some of the flame images, which highlights the differences in spray dispersion and flame structure depending on the operating conditions. The droplet number density and droplet velocity were measured by application of the particle image velocimetry (PIV) technique. For this purpose a Nd:YAG laser (Quantel, EverGreen: 145 mJ/pulse at 532 nm; 5 mm beam diameter) was used to illuminate the burning spray. The scattered light from droplets was collected through a lens (105 mm; f/2.8G ED Nikon lens) in conjunction with a suitable band-pass optical filter centred on 532 nm. The images of focused droplets were captured through a camera (PCO Pixelfly: 14 bit, 1,040 x 1,392 pixels^2) placed at an angle of 90° with respect to the plane of the laser sheet. The field of view was about 1.4 cm x 1.8 cm such that the spatial resolution was around 13 μm/pixel. In order to identify the droplets on the PIV images, while rejecting the contribution due to background flame which sometimes appears as bright patches, a continuous wavelet transform (CWT) based on Mexican hat wavelet was applied to each row of the image. Since the length scale of intensity variation in the flame region was different in comparison to that for the droplets, thus by selecting an appropriate wavelet scale, it was possible to suppress the flame signal and identify only the droplets.

The Interferometric Laser Imaging for Droplet Sizing (ILIDS) technique was used to size the spray droplets. ILIDS is a planar defocusing technique based on detecting the reflected and the first order refracted light scattered from a droplet, which, at a specific forward scattering angle, interfere to produce parallel fringes on a defocused plane [4, 5]. The characteristic interferogram is observed with a far field arrangement of receiving optics. The number of fringes present in each of the recorded fringe patterns is proportional to the droplet diameter. Introduction of a pair of cylindrical lenses between the objective and the CCD camera leads to optical compression of the circular fringe pattern for each droplet in the vertical direction only such that each droplet is imaged as a rectangular region with a superimposed fringe pattern [6, 7]. The camera was set at an angle of 74° forward scattering angle. Some examples of the instantaneous ILIDS images of the reacting spray are presented in Fig. 3 for the condition Re_{co} = 7500, S = 0. The droplet number density is high at the center of the spray, which makes fringe counting difficult. At r = 4 cm location the flame is also visible. The fringe patterns at r = 6 cm correspond to the droplets which escaped from the flame.

Corresponding author.
E-mail address: ssahu@iitm.ac.in

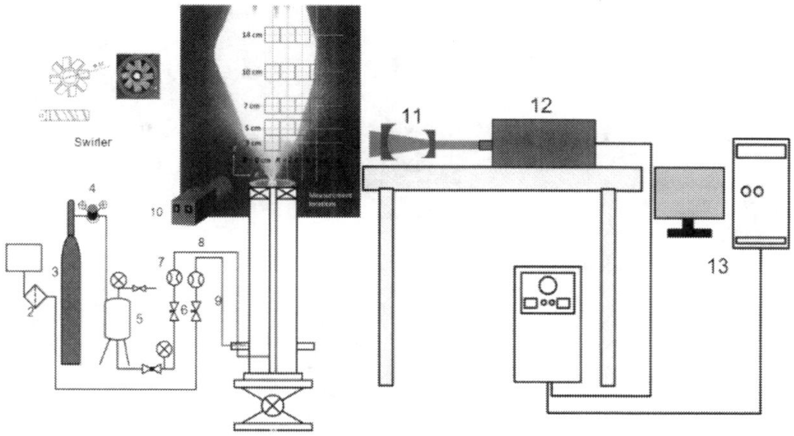

1. Compressor 2. Filter Regulator Lubricator 3. Nitrogen cylinder 4. Pressure regulator 5. Pressure vessel 6. Needle valve 7. Flowmeter 8. Kerosene line 9. Air line 10. Camera 11. Optics 12. Nd-Yag Laser 13. Data acquisition system

Figure 1: Experimental setup and optical arrangement for studying the spray flame

$Re_{co} = 0$ $Re_{co} = 7500, S = 0$ $Re_{co} = 7500, S = 0.77$

Figure 2: Spray flame visualization in the reacting spray for different flow conditions.

Figure 3: Instantaneous ILIDS images at z = 7 cm for different radial locations for the operating condition $Re_{co} = 7500$, S = 0

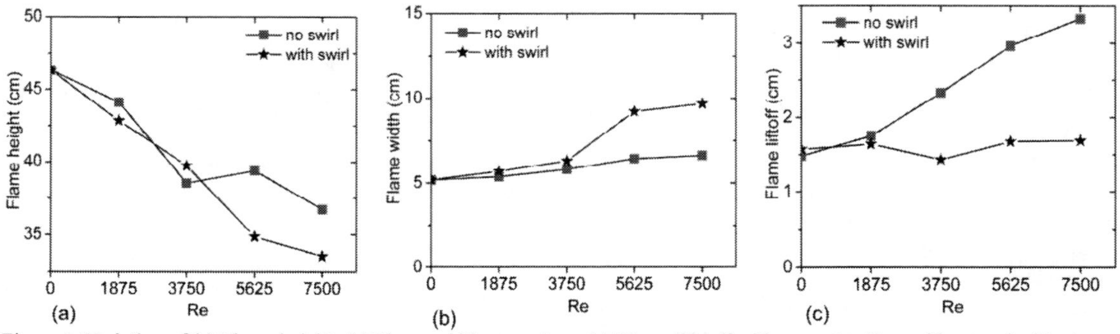

Figure 4: Variation of (a) Flame height, (b) Flame width at z = 5 cm (c) Flame lift-off with respect to Re_{co} without and with air swirl.

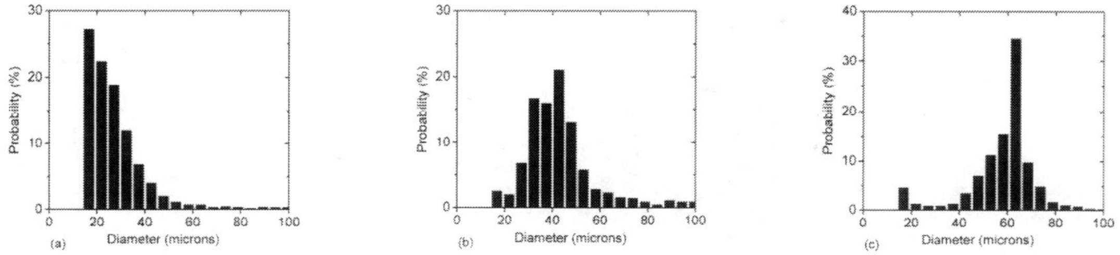

Figure 5: Probability distribution of droplet diameter at z = 7 cm axial location with Re = 7500, S = 0 along the radial locations (a) 0 cm (b) 4 cm and (c) 6 cm

3 Results and discussions

3.1 Flame characteristics

As shown in in Fig. 4a the flame height decreases with increasing the coflow rate of air. This is also evident in Fig. 2. However, the influence of air swirl is realized only for higher coflow rates. The width of the flame and the flame lift-off are also important parameters in designing combustion chambers. The flame width presented here was calculated at an axial distance of z = 5 cm from the nozzle exit. At higher air flow rates imparting swirl on the coflowing air leads to air recirculation close to the burner exit. Thus, part of the air surrounding the central spray reverses its direction towards the injector. As a consequence, the flame width is nearly doubled (Fig 4b) as compared to the flame without coflowing air. In addition, the flame stand-off distance reduces substantially since the flame can be stabilized closer to the burner exit. The flame stand-off distance increases with increasing co-flow (without swirl) in agreement with [9]. Interestingly, for the condition corresponding to maximum coflow rate and air swirl (Re_{co} = 7500, S = 0.77), a mild central flame is observed within the spray (Fig. 2), which indicates low velocity region within the recirculation zone. It was also observed that the soot from kerosene flame reduced with increase in co-flow. This is due to better availability of oxidizer which is supplied during co-flow. We note here that these macroscopic flame characteristics depend on the flow rates and velocity of liquid and air at the burner exit and are expected not be influenced by the spray microstructures such as droplet clustering, which correspond to much smaller length scales (as will be discussed later).

3.2 Droplet size

Fig 5 presents the histograms of droplet size distribution for different radial locations at z = 7 cm. It can be observed that away from the spray center the size distribution is skewed towards larger droplets. This is due to the hollow-cone spray generated by the simplex injector. Since larger droplets travel at larger velocity near the boundary of the spray, the flame is stabilized inside the spray boundary instead of enveloping the whole spray

3.3 Droplet number density

The mean and the root mean square (rms) of fluctuations of droplet number density are calculated for different measurement locations. The results are presented in Fig. 6, where the values are normalized with the average number density at z = 3 cm, r = 0 cm location (denoted as $\overline{N_0}$). In general, for all cases the mean droplet number density decreases away from the injector exit due to droplet evaporation (Fig 6a), which though depends on the flame lift-off distance. For instance, for no air coflow the flame begins very close to the injector exit. Hence, in that case the droplet count reduces rapidly from z = 3 cm to 5 cm locations. However, for the case Re_{co} = 7500 the flame stands away from the injector exit (around z = 5 cm), so the reduction in droplet count is less pronounced. The droplet number density at locations away from the injector exit was found to increase with increasing air coflow in general. However, the introduction of air swirl significantly alters the dispersion. For example, no droplets were observed along the injector axis in contrast to the cases without air swirl. Hence, the corresponding the number density is nearly zero in Figure 6. This is due to two reasons, first, centrifuging of droplets towards the periphery of the spray. In addition, due to the central vertical flame this region is devoid of droplets due to evaporation. In general, the rms of droplet number density fluctuations decrease away from the injector exit similar to the mean. However, for higher air coflow and away from the injector the fluctuations are considerable as compared to respective mean values.

3.4 Droplet clustering

The degree of droplet clustering is statistically characterized by the *D* parameter ([8]). In this approach, each PIV image is divided in to certain number of uniformly sized cells. The number of droplets inside each cell is counted. The probability of given number of droplets per cell is obtained considering the ensemble of images. For a purely random droplet distribution this probability is represented by a Poisson distribution. This is because we seek for either success or failure of an event (i.e. a given number of droplets per cell), which is given by the Binomial distribution, which tends to a Poisson distribution when the number of droplets and cells tend to be large. The *D* parameter is calculated as the normalized difference between the standard deviations of the measured probability distribution (σ) and that of the Poisons distribution (σ_p). The magnitude of *D* indicates the degree of clustering corresponding to a given cell size. The *D* parameter is calculated for cells of different sizes (Fig 7). It can be observed that the *D* parameter increases and reaches a maximum value and then decreases. The maximum value of *D* signifies the degree of droplet clustering and the cell size corresponding to D_{max} provides a statistical measure of the length scale of clustering (L_c). Figure 8 shows axial variation of D_{max}, which increases away from the injector exit for Re_{co} = 7500 . This trend is opposite to that for the spray flame with no-coflow. This means the tendency of the droplets to burn as a group is higher towards the injector exit in absence of coflow (this is qualitatively observed in the spray flame image in Fig 2), while in presence of coflow droplet grouping occurs away from the injector.

Though not shown here, the characteristic dimension of the droplet clusters, L_c was found to scale with the size of the large turbulent eddies (L), which is approximated as the spray half width (obtained from droplet velocity measurement). This is in agreement with earlier literature [8]. Since L is about 14 mm this means L_c, is nearly equal to the dimensions of the image viewing area (1.4 cm x 1.8 cm). Also, it does not to change significantly along the axis of the spray and for different operating flow conditions. Measurement of droplet clustering in reacting sprays has been rarely reported earlier. Nevertheless, more investigation is required to conclude on role of droplet clustering on spray flame propagation.

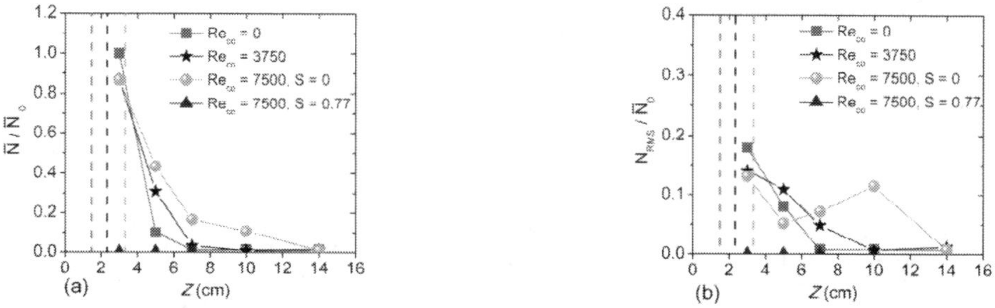

Figure 6: Variation of (a) normalized average no.density (b) normalized RMS of no.density along the axis of the spray. (dashed lines represent the flame lift off location for corresponding colour series)

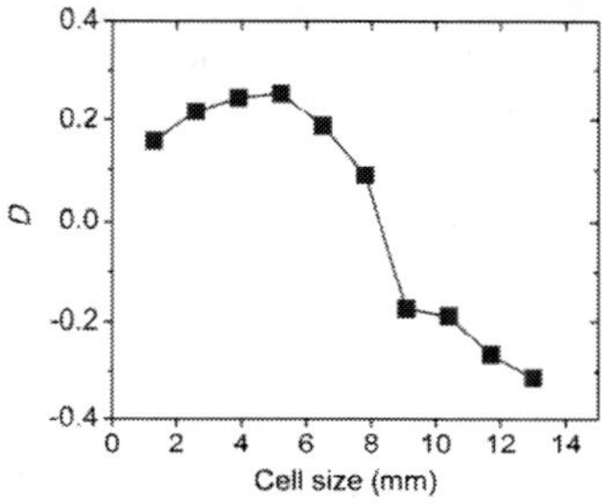

Figure 7: Variation of D with cell size at z = 3 cm, r = 0 cm for Re = 7500.

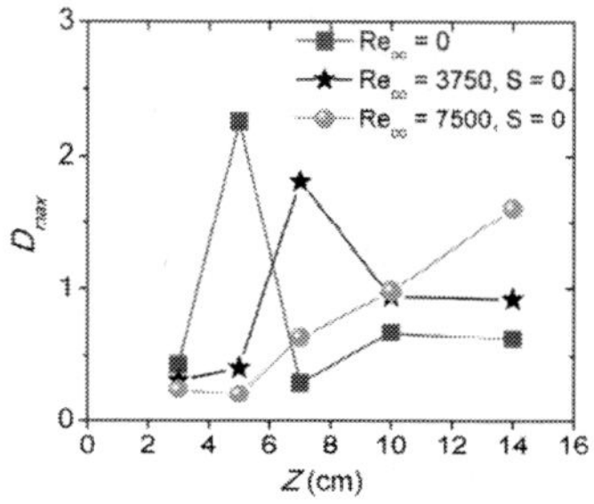

Figure 8: Variation of D_{max} along the spray axis

4 Conclusions

The present work mainly focused on flame and droplet characterization in a reacting kerosene spray using different optical diagnostic techniques. The goal of the present work is to study the effect of coflowing air (in presence and absence of air swirl) on droplet dispersion. The coflowing air either with or without swirl strongly influences the droplet dispersion in the spray as confirmed by the visualization images of the spray flame. The overall effect of air swirl is observed in terms of significant reduction in flame height, wider flame width and shorter flame stand-off distance. The coflow results in significant increase in local mean as well as fluctuations of number density of droplets. However, detail characterization of droplets droplet clusters are required for further understanding on group burning of droplets. The degree of droplet clustering was characterized by calculating a statistical quantity known as D parameter based on droplet number probability within an interrogation cell on the PIV images. The D parameter was found to increase along the spray axis in presence of coflow around the spray, while the trend is opposite in absence of coflow.

References

[1]. Hardalupas, Y., Liu, C. H. and Whitelaw, J. H. (1994) Experiments with disk stabilized kerosene-fuelled flames. Comb. Sci. and Tech. 97, 157-191.

[2]. S. Tsushima, H. Saitoh, F. Akamatsu, and M. Katsuki, "Observation of combustion characteristics of droplet clusters in a premixed-spray flame by simultaneous monitoring of planar spray images and local chemiluminescence," Proc. Combust. Inst. 27, 1967 1998.

[3]. F. Akamatsu, Y. Mizutani, M. Katsuki, S. Tsushima, and Y. D. Cho, "Measurement of the local group combustion number of droplet clusters in a premixed spray stream," Proc. Combust. Inst. 26, 1723 1996.

[4]. A. Glover, S. Skippon, and R. Boyle, Interferometric laser imaging for droplet sizing: a method for droplet-size measurement in sparse spray systems," Applied Optics 34, 8409-8421 (1995).

[5]. N. Damaschke, H. Nobach, T. I. Nonn, N. Semidetnov, and C. Tropea, "Multi-dimensional particle sizing techniques," Experiments in fluids 39, 336–350 (2005).

[6]. M. Maeda, Y. Akasaka, and T. Kawaguchi, "Improvements of the interferometric technique for simultaneous measurement of droplet size and velocity vector field and its application to a transient spray," Experiments in Fluids 33, 125–134 (2002).

[7]. Hardalupas, Y., Sahu, S., Taylor, A. M., Zarogoulidis, K., 2010. Simultaneous planar measurement of droplet velocity and size with gas phase velocities in a spray by combined ilids and piv techniques. Experiments in fluids 49 (2), 417–434.

[8]. John Fessler, Jonathan D Kulick, John K Eaton, Preferential concentration of heavy particles in a turbulent channel flow, Physics of fluids 6, 3742 (1994).

[9]. Mahendra V Reddy, Darshan Trivedi, Sudarshan Kumar, Experimental Investigations on Lifted Spray Flames for a Range of Coflow Conditions, Combustion Science and Technology Vol. 184 , Iss. 1,2012.

Effects of pressure and oxygen concentration on a stabilized premixed combustion

Jean-Michel Most and Michel Champion

CNRS - Pprime UPR 3346, ISAE-ENSMA, Université de Poitiers
1, avenue Clément ADER, BP 40109, 86961 FUTUROSCOPE Cedex, France
jean-michel.most@ensma.fr; michel.champion@ensma.fr

Abstract

In the context of thermal mechanisms induced by a hot spot at the surface of a fuel material, the objective of this work is to describe the conditions of the development, over a long period, of its self-heating. On a practical point of view, a better understanding of the phenomena involved during the potential transition from a steady flameless combustion process (subcritical regime) to auto-ignition (supercritical regime) should help to prevent fire.
An original set-up has been designed to study the behavior of a gasoline vapors/oxygen enriched air premixed combustion stabilized at the surface of a grid. The effects of the ambient pressure and oxygen concentrations on the stabilization of the flameless combustion has been studied. This allows to identify the conditions (fuel consumption, flame temperature, chemical species concentrations, etc.) under which auto-ignition and flaming of the material becomes possible.
Using this experimental configuration, the preliminary steps of a heterogeneous combustion of a fuel solid (or liquid) material, or combustion in a porous medium has been investigated. This experiment is supposed to represent the self-heating of a material leading either to complete extinction or to flaming. This situation can be found in industrial, domestic or wild fire scenarios after the deposit of energy on the surface of a fuel material by natural or human actions.

1 Generalities and objectives

The global objective of this work is to contribute to the understanding of the mechanism involved in the heterogeneous combustion of either a solid or a liquid fuel surface. This mechanism can be observed to be at the origin of fire scenarios when a hot spot is formed consecutively to the impingement of heat flux to the surface of a fuel material. Frank-Kamenetskii [1,2,3] developed a combustion theory to explain why, after a time delay which can reach several hours of smoldering regime, eventually flaming can occur. The main steps, presently studied, include (1) a **self-heating** of the material with an increase in temperature due to exothermicity of internal reactions, (2) a possible **thermal runaway**, where the self-heating rapidly accelerates to high temperatures, and finally, **(3)** a spontaneous auto-ignition process. In some practical situations these three steps may lead to flaming.
The theory of the self-heating is based on the generation of heat occurring through the internal substance which is proportional to the fuel mass (or volume) and also depends on the ambient conditions as temperature. This situation of self-heating is illustrated in Figure 1 based on the concept of the Semenov theory of ignition [1,4] which is briefly reviewed below for a heat production occurring within a material.

Considering Figure 1, let us suppose that the material is initially at temperature T_1, a point where the heat gains and losses are not balanced, then, since the heat generation is over the heat losses, temperature increases up until reaching point A where the energy equilibrium is obtained at temperature of point A. A second thermal equilibrium is obtained at point C. The examination of the shapes of the curves shows that at point A, the thermal equilibrium is stable, the slope of heat generation being lower than the one of heat losses; if a small increase in temperature takes place, the heat losses become larger than the gains and temperature goes down to point A. Conversely, a small decrease of temperature leads to a process for temperature to return to point A. On the other hand, at point C, the situation is different, a small rise of temperature cause heat generation to exceed heat losses and the system runs away to ignition. Conversely a small fall of temperature in the system leads back to point A. The equilibrium at point C is unstable.

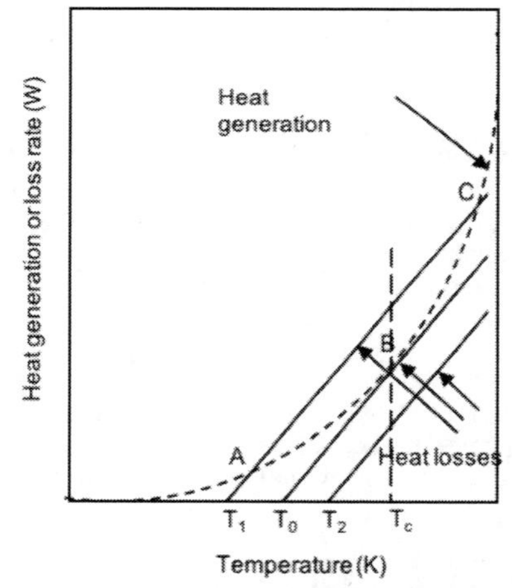

Figure 1: Heat losses and gains, as represented in the Semenov theory

If we consider now Figure 1 at a higher room temperature T_2, it can be seen that there will be no place for the heat gains and the losses to be in equilibrium: the system will just keep rising in temperature until ignition occurs, it is the **supercritical regime**; a system which starts at T_1 is **subcritical.**
The starting point B represents an unstable equilibrium condition. Any ambient temperature lower than T_0 will lead to a stable, non-igniting behavior and it can be deduced that the material will stay indefinitely at the same temperature below T_c.

Several parameters control the probability of flaming such as the material type, the porosity [], the shape, the homogeneity, the surrounded conditions (temperature, pressure, available oxygen) and finally the heat losses to the environment, though data are missing on the conditions leading to the self-heating and flaming phenomena as well as on the behavior of combustion under overpressure and oxygenation conditions [5]. Then, the basis of the Semenov theory of self-heating of a solid or a porous system can be applied to a configuration simulating a gaseous combustion stabilized in contact of a beforehand heated metallic grid (called also burner). The exothermic combustion zone is supplied by the diffusion of degradation products of a gaseous mineral gasoline and the surrounding oxidant medium.
In real fire situations, the thermal runaway may or may not occur, while the final outcome, if thermal runaway occurs, may be the spontaneous combustion with fire propagation through the environment with impacts on persons and properties. Moreover, in some cases, pressure [6,7,8], and oxygen concentrations are over the normal conditions (hyperbaric enclosure, medical walk-in chamber, ocean deep chamber, over normal oxygen condition consecutively to an oxygen leak, aeronautical engines, etc.), so, these modified conditions can lead to either a safety situation or a catastrophic fire event. This studied combustion configuration can be partly recovered in ramjet engines where the ignition delay is a major parameter.
The system, considered in this study, partly simulates a heterogeneous combustion of a solid fuel in an oxidant medium. A material (materialized by a metallic grid), previously heated by an external heat source, transfers a heat flux, by diffusion and radiation processes, to a fuel medium. Under this heat effect, fuel is gasified inducing a mass flow of gaseous degradation fuel products towards the grid, counter flow the heat flux. A premixed combustion zone is stabilized at the grid surface.
For this purpose, an original set-up [8] was designed in order to stabilize a premixed flameless combustion (subcritical regime) at the burner surface to investigate the mechanism of self-heating, thermal runway and spontaneous combustion with flaming as function of the ambient pressure, oxygen concentration and heat losses.

2 Experimental device

The combustion system (Figure 2) is 0.07m wide (W), 0.09m high (H) and 0.015m deep (D). It is made of a fuel tank filled with cotton fibers soaked with a liquid fuel (mineral Zippo gasoline: 64% n and iso-paraffin, 35% of cyclo-paraffin and 1% of aromatic hydrocarbons, heat of combustion: 48400 kJ/kg, heat of gasification: $320 kJ/kg_{fuel}$). These values correspond to the standard hydrocarbon characteristics. The metallic grid (the burner) has a surface $S = 0.025 \times 0.018 \ m^2 = 4{,}5.10^{-4} m^2$, and is located 0.02 m over the cotton surface. A O_2-N_2 mixture flows around the combustion system which is confined in a pyrex box to ensure the oxidant mixture composition at the grid level. The complete set-up is placed in a 330 liter pressurized enclosure equipped with silica windows to visualize the behavior of the system.
Beforehand heated, the grid acts as a flame holder and a quenching device stabilizing an extremely diluted gasoline/O_2-N_2 premixed flameless combustion on the metallic surface. The steady exothermic chemical reaction, located at the external part of the grid, is supplied by the fuel mass transfer from the gasoline bed to the reaction zone in the opposite direction of the radiated and diffused heat transfer.
Tests were performed at various volumic fraction of oxygen FO_2 (0.23, 0.3, 0.4, 0.5 and 0.75) and pressure P (1.0 to 2.8 atm). The heat losses can be also changed by wrapping the combustion system with a thermal insulator and by modifying the burner-soaked cotton distance (these results are not presented in this work). The test protocol consists in heating the burner surface with a pilot flame and to flow the known oxidant mixture (given FO_2 value) around the device body. The working conditions (vessel pressure, oxidant flow rate (the gas mixture composition are fixed a priori by adjusting the gas partial pressures in a gas bottle) are measured respectively with a pressure sensor and a mass flow meter. Insulated K-type thermocouples are located at the surface of the burner and inside the cotton-gasoline bed, close to its surface. The fuel weight loss, due to the fuel gasification, is measured by a load cell to determine both the fuel velocity burning rate (kg/s) and the flame speed (cm/s). It is assumed that the fuel combustion through the reacting zone is quasi-complete.
Moreover, due to the low burning rate, the fuel diffusion velocity by capillarity in the cotton allows to keep a constant fuel concentration at the exposed face of the cotton surface, at least during the first period of the trial.
The system (gasoline burning rate and heat release in the flame) is controlled by the energy balance at the fuel bed which depends on the head released at the grid level (heat source from the exothermic chemical reactions), on the heat flux from the combustion zone to the cotton surface, on the energy of the fuel liquid, on the heat losses in the system body, and, finally, on the heat conduction inside the tank.

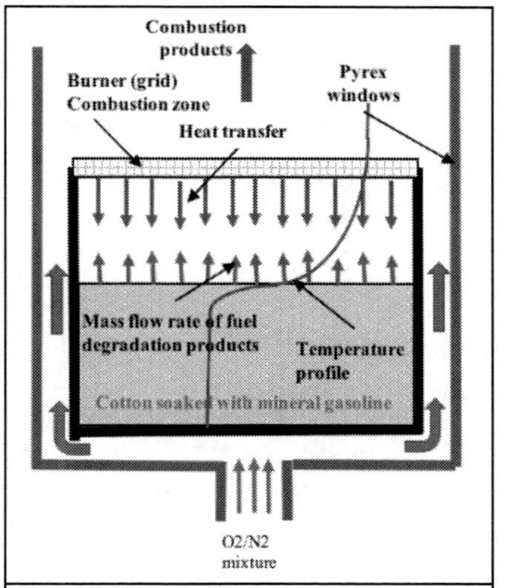

Figure 2: Scheme of the experimental setup

3 Main results

Two types of information are sought after: the thermo-chemical characterisation of the system in over-atmospheric pressure and over-oxygenated medium, and the thermo-chemical conditions leading to the transition from a flameless surface reaction towards an open fire.

3.1 The combustion regimes

Depending on the test conditions of pressure and oxygen partial pressure, the obtained experimental results exhibit two different burning regimes: a flameless regime where the chemical and the heat source are located completely at the grid surface (subcritical combustion mechanism), or a transition to flaming (called supercritical) where both the liquid gasification rate and the heat release rate drastically increase up to a complete ignition of the fuel. A diagram describing these phenomena will be presented later.

To characterize the combustion type, the gas concentrations are measured upstream, on, and downstream of the grid (Table 1). Observations show rather a premixed combustion mode with the formation of a fuel/air mixture upstream the reaction position due to the fact that heat diffusion by the grid is balancing the chemical heat release.

Table 1 presents the gas composition (gas between the soaked fuel cotton surface and the burner, at the burner surface and over it). The results show a diffusion of oxygen through the reaction zone positioned at the burner surface.

Volumic fraction F_i	Upstream the grid	At the grid surface	Over the grid
Hydrogen	Traces	Traces	Traces
Oxygen	13.96	11.28	15
Nitrogen	68.8	69.6	69
Carbon monoxide	0	0	0
Carbon dioxide	4.8	8.35	3.25
Methane and ethane	Traces	Traces	Traces
Heavy Hydrocarbons	9.85	2	1.49

Table 1: Gas composition expressed in volumic fraction $F_i=V_i/V$ (where V is the total volume). Gas sampling and analysis by gas chromatography

3.2 The chemical zone temperature

Temperatures at the grid T_{grid} and near the gasoline soaked cotton provide the main information on the process. Evolution of the grid temperature with pressure exhibits an increase for both the reaction zone (grid zone) and the gasoline up to a plateau where an equilibrium heat balance is reached.

The T_{grid} value are always lower (600 to 900°C) than in a standard flame due to the heat losses by the grid.

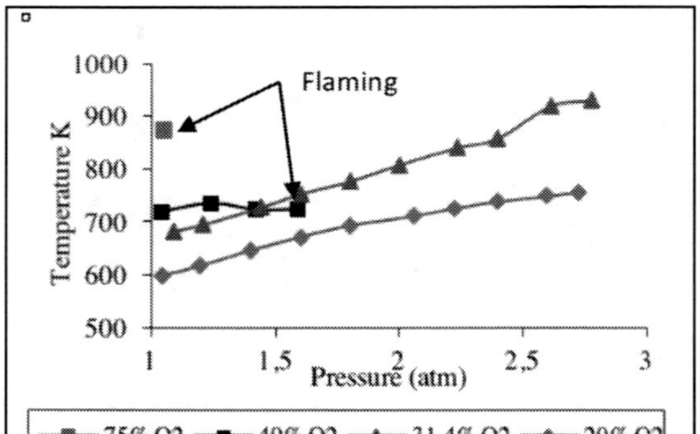

Figure 3: Influence of pressure on temperature

Figure 3 shows, for O_2 volumic fraction (FO2) smaller than 50%, a constant increase of the T_{grid} temperature with pressure at the grid level. Oxygen volumic fractions larger than this value, i.e. FO2≥0.5, lead to flaming. This observation predicts an increase of the fuel burning rate with these parameters, because, an increase of temperature intensifies the heat transfer to the soaked cotton surface, then an increase of the gasified fuel flow rate at equivalent heat losses in the pressurized enclosure.

Figure 4 represents the grid temperature variations, supposed to be close to the chemical reaction temperature, versus the ambient oxygen partial pressure (P_{O2}).

In Figure 4, the T_{grid} evolutions for all oxygen volumic fractions are superposed. Flaming is always observed as soon as $P_{O2} \geq 0.75$ atm. Then, for $F_{O2} = 0.75$, flaming is instantaneous at normal pressure and no thermal equilibrium can be accurately measured to determine the grid temperature.

Different behaviors of the grid temperature variations are observed for various pressure evolutions. For a constant pressure ramp (0.55 bars/min), the grid temperature increases up to flaming which always occurs at an oxygen volumic fraction of 0.75. For a pressure evolution by steps of 0.2 bars, the grid temperature increases rapidly before decreasing slowly towards the thermal equilibrium. Finally, for a stiff increase of pressure of 1.0 bar, flaming appears for $F_{O2} > 0.3$.

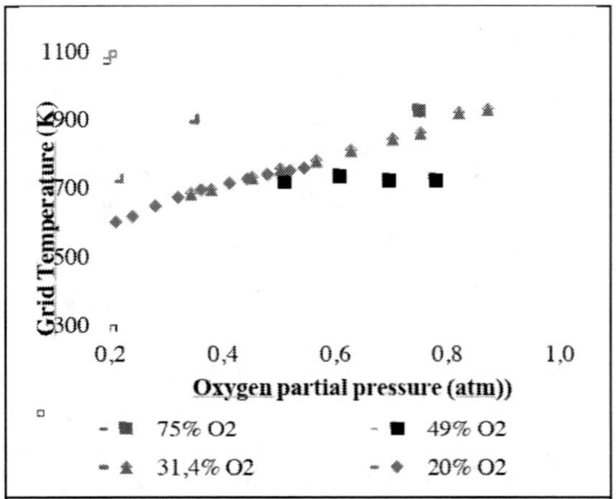

Figure 4: Burner temperature variation with the oxygen partial pressure

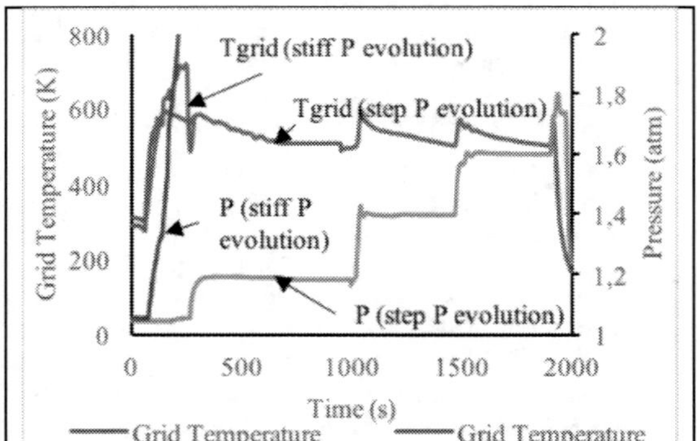

Figure 5: Grid Temperature variations with pressure evolutions for $F_{O2} = 0.5$

A few seconds before flaming, a red spot is always observed on the grid surface. It can be supposed that radiative and diffusive heat fluxes from the flameless combustion increase, so that temperature inside the soaked cotton increases up to the fuel gasification temperature; this leads to a large mass flow of gaseous gasoline, which is instantaneously ignited at the hot point location.

3.3 The burning rate

The burning rate Vc (g/s) is given in term of an equivalent laminar flame speed S_l (m/s). This quantity is determined from the measured combustor mass loss (g/s). The proportionality coefficient between V_c and S_l is evaluated to be 220 cm/g. Figure 6 presents the variation of the laminar flame speed for the studied working conditions. We observe an increase of the laminar burning rate with both the enclosure pressure and oxygen volumic fraction F02. Self-ignition of the mixture between the solid surface and the grid is never reached under the studied conditions, namely for $F_{O2} = 0.2$ and 0.3 and under pressure varying in the range 1.0-2.7 bars. For $F_{O2} = 0.5$, ignition appears for $P \geq 2.5$ atm; for $F_{O2} = 0.75$, ignition is always observed, even at atmospheric pressure.
Ignition seems to be correlated to the initial slope of the laminar flame speed: the ignition phenomenon, related to the characteristics of the heat and mass transfers between the grid and the fuel surface (cotton soaked gasoline), mainly depends on the thermodynamic properties of the system.
These findings can be extrapolated to the self-heating of a solid, where the ignition probability depends on the system via the chemical reaction rate, the heat released and losses rate.
However, no simple correlation between S_l and P was found to deduce a global reaction order.

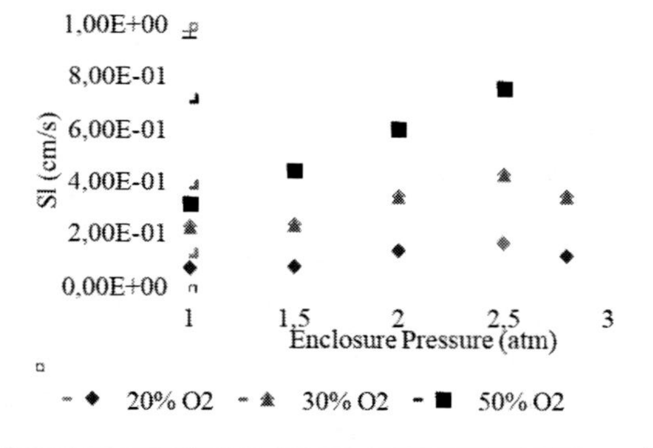

Figure 6: Flame speed (gasoline burning rate) variation vs ambient pressure

The small decrease of the equivalent flame velocity (fuel burning rate) observed for the largest values of the pressure is interpreted as resulting from the lower diffusion velocity of the gasoline in the porous medium due to the pressure effect on the cotton surface.
The thermal output power of the system is 13.8W with air in normal conditions of pressure.
Experimental results providing the ignition curve in term of pressure and oxygen volumic fraction are shown in Figure 7. Clearly the domain under the curve contains states where combustion is limited to the grid (subcritical regime) whereas the upper part contains states where self-ignition of the mixture occurs (supercritical regime).

4 Conclusion

We have presented an experimental investigation of the effect of pressure and oxygen fraction on the existence of a combustion zone stabilized by a grid located two centimetres over a permeable wall supplying gaseous fuel. At the grid, steady combustion is obtained as a result of a competition between chemically induced heat release and heat losses within the grid.

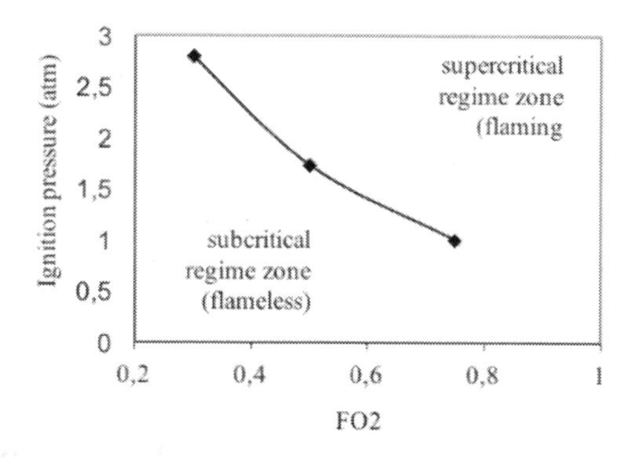

Figure 7: Subcritical (flameless) and supercritical (flaming) regimes in self-heating configuration

It was shown that the existence of such a steady combustion region, i.e. self-heating combustion, is depending on the pressure and the oxygen volumic fraction: when either the pressure or the presence of oxygen between the wall and the grid is large enough, self-ignition of the mixture occurs, this leading essentially to the so called flaming conditions.
From a practical point of view, under similar combustion conditions, this work is a contribution to the understanding of the mechanism responsible (as well as the origin) for a fire to happen in a hyperbaric chamber. From a fundamental point of view, results presented in this work allowed the identification of the thermo-physical parameters, in particular pressure and local oxygen concentration, responsible for the transition from a stable combustion on a combustible solid surface to an unstable combustion leading eventually to the ignition of the fuel reserve.

References

1. Babrauskas, V. (2003). Ignition Handbook, Fire Science Publishers and Society for Fire Protection Engineers, 1116 pp.
2. Frank-Kamenetskii, D. A., Diffusion and Heat Transfer in Chemical Kinetics, 2nd ed, Plenum Press, New York, (1969)
3. Quintiere J. G., (2006) Fundamentals of fire phenomena, Wiley Authenticity Guarantee
4. Jones J. C., Journal of Fire Sciences (2008), Volume 26 issue : 3 ; pages 197-211.
5. Read, A. J., Harrison, W. J., and Walker I. K., New Zealand J. Sci. 10, 964-978 (1967).
6. Corlett R.C., Luketa-Hanlin A. (2008), Saito K. (eds) Progress in Scale Modeling. Springer, Dordrecht
7. McAllister S., Fernandez-Pello C., Urban D., Ruff G., *Combustion and Flame*", 157 (2010) 1753-1759.
8. Most J-M., Mandin P., Chen J., Joulain P., Durox D., Fernandez-Pello A. C., Symposium Int. on Combustion 26 (1), 1311-1317.

11th Asia-Pacific Conference on Combustion,
The University of Sydney, NSW Australia
10th -14th December 2017.

Effect of In-Cylinder Injection Strategies on Performance and Emissions of Gasoline Direct Injection Engine using Simulation Tools

Patel Kamlesh Hirji[1], Sourabh Deshpande[2], Brijesh P[3]

[1]ARAI Student
Kothrud, Pune, Maharashtra, 411038, India
[2]Application Engineer, CEI
Bavdhan, Pune, Maharashtra, 411021, India
[3]ARAI Faculty
Kothrud, Pune, Maharashtra, 411038, India

Abstract

Significant progress has been made in development of direct injection internal combustion engines. This could possibly help in eliminating fuel short circuiting loss in two-stroke spark ignition (2S-SI) engine. The subject of the study is 2S, single cylinder, gasoline direct injection (GDI) engine with spark plug located nearby exhaust port and centrally located injector. The aim of this work was to investigate the effect of various operating parameters such as injection pressure, start of injection and injection duration on performance and emission of 2S-SI GDI engine. Some initial inputs required for simulation in Converge computational fluid dynamics (CFD) were generated by carrying out simulation in one dimensional (1D) Ricardo Wave. The effect of increase in injection pressure while maintaining the same overall equivalence ratio throughout the simulation was studied. The simulations were carried out with injection pressure for 80 bar, 130 bar and 180 bar. The soot was reduced with increasing injection pressure. But, with 180 bar injection pressure, the equivalence ratio required nearby spark-plug for combustion was found inadequate resulted in incomplete combustion/misfire situation and hence produced very high hydrocarbon. Soot with 180 bar injection pressure was found lower as the soot formation was found less because of lower in-cylinder temperature.

1 Introduction

A significant research over the time has been carried out with spark-ignition engines to overcome its drawbacks specially on improvement of part load efficiency by changing load control strategy [1]. For same GDI strategy is proposed which has an ability to drive the stratified lean mode at part load. Thereby like a diesel engine, pure air is sucked un-throttled into the combustion chamber and fuel injection takes place late on in the compression stoke [2]. If a homogenous charge is formed in stratified lean mode then the mixture won't be ignitable as it will exceed the lean limit. To avoid this phenomenon, an ignitable (approximate stoichiometric) mixture is generated only close to the spark plug region with the predominant part of combustion chamber filled with ultra-lean mixture or pure air.

In place of port injection, Yamagishi et al. [3] injected gasoline directly in cylinder using mechanical injector and achieved 52% reduction in hydrocarbon (HC) emissions. Vielledent [4] also achieved 3 fold reduction in HC emissions and 40% reduction in fuel consumption using low pressure electronic injection of gasoline through the cylinder barrel. Douglas and Blair [5] reported a significant improvement in fuel economy (by 30%) from direct injection (DI) gasoline inside the cylinder with an overall penalty of 20% on power. Similar results were also observed by Leighton and Ahern [6]. Air assisted direct injection strategy proposed by them with 2S-SI engine offered about 30% improvement in fuel economy. Badami et al. [7] had compared effects of DI and manifold injection (MI) strategies in a 2S-SI engine for various engine speeds and full load condition while maintaining air-fuel ratio at stoichiometry. Maximum improvement of 39% in power at low speeds was reported and on an average 90% and 60% reduction in HC and oxides of nitrogen (NO_x) emissions were observed respectively with DI system. Schmidt et al. [8] studied the effect of injection timing with GDI on fuel consumption and emissions using 3D CFD tool. With late injection, 75% and 15% reduction in fuel consumption at idling and full load respectively and 80% reduction in NO_x+HC emissions were observed. Mattarelli et al. [9-10] did study with novel 2-stroke GDI single-cylinder engine, rated at 30 kW at 4500 rpm. Results showed benefit with GDI engine over other prime mover.

Adepalli and Mallikarjuna [11] studied evaluation of the effect of engine parameters on the characteristics of a GDI engine using CFD analysis. The analysis was carried out for three engine speeds, three compression ratios and three fuel injection pressures (200 bar, 300 bar and 400 bar). The fuel injection pressure was found to play a vital role in obtaining combustible mixture near the spark plug at the time of spark. Effect of injection pressure and timing on performance and emissions were studied by Dube and Ramesh [12]. Result of the study showed that end of injection (EOI) timing was the most significant parameter with the spark plug location near the exhaust port being the most suitable in the 2S-GDI mode which exhibited significantly higher efficiency, lower HC and carbon monoxide (CO) emissions for the same power output as compared to 2S-MI operation.

Based on literature outcome, in this work, effect of various injection parameters on performance and emissions have been investigated using 1D and 3D simulation tools.

2 Simulation Setup and Validation

The engine selected for this simulation was 2S-SI single cylinder direct injection air-cooled engine. The specifications of the engine used are given in Table 1 [12].

Corresponding author. Brijesh P; Fax: +91-20-3023-1104
E-mail address: patel.pga@araiindia.com

Bore × Stroke	61 mm × 68.2 mm
Compression Ratio	7.7:1
Displacement	200 cc
Max. Torque	16.5 Nm @ 3000 RPM
Rated Power	6.2 kW @ 4500 RPM

Table 1: Engine specifications [12]

The 3D CFD simulation was carried out using Converge CFD, while 1D simulation was carried out with Ricardo Wave. Engine model was validated with the in-cylinder pressure traces, performance and emissions data of engine operating with equivalence ratio of 0.7 at 70% throttle and 3000 rpm. Base model of 1D and 3D both were validated with 130 bar fuel injection pressure, end of injection (EOI) 50 crank angle degree (CAD) and 167 CAD spark timing. CAD in this paper is defined with respect to "suction top dead centre (TDC)".

2.1 Ricardo Wave Numerical Setup and Validation

Ricardo Wave is one dimensional engine analysis software based on empirical relations for all the activities taking place in engine [13]. The cylinder of the engine is present at the centre. To the left of the cylinder is the air intake system (transfer ports) whereas to the right is the exhaust system. All the ducts present in the model are of the same dimensions as that present on the test bed. As model is prepared for GDI engine, injector, as shown in Fig. 1, is directly placed over the engine. The combustion model and emission model used for NO_x prediction were SI Wiebe combustion model and Extended Zeldovich model respectively.

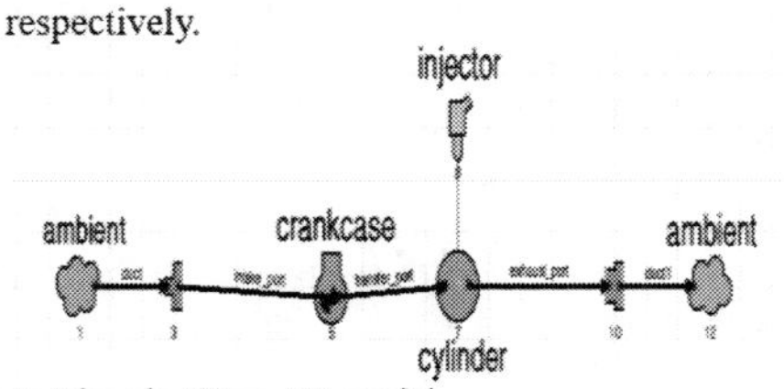

Figure 1: Ricardo Wave 1D model

The Ricardo Wave code for described model has been validated against the experimental results of Dube and Ramesh [12]. Simulated in-cylinder pressure traces are compared with same of experimental traces (which are available only for a few crank position) in Fig. 2.

Figure 2 shows a very good correlation between simulated and experimental in-cylinder pressure traces. Also, experimental and simulated values of brake power (BP), brake thermal efficiency (BTE), NO_x and HC are compared in Table 2 and found very good matching (within 8% variation). Hence, the model can be considered as validated and used for further simulations. The model constants, solver settings, surface temperatures were kept same for the remaining simulations.

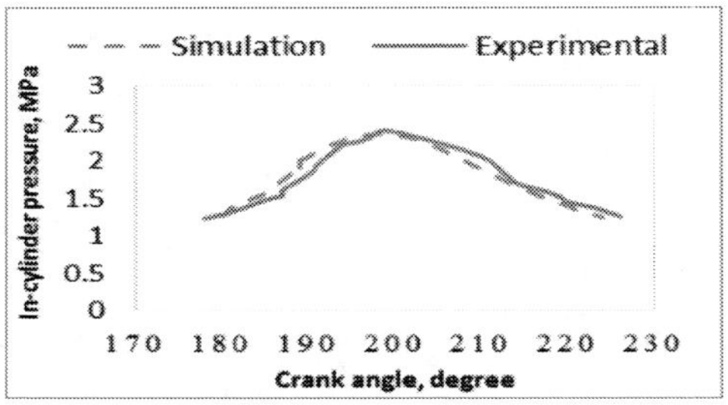

Figure 2: In-cylinder pressure vs crank angle comparison between experimental and 1D Ricardo Wave simulation.

Parameters	Experimental	Ricardo Wave
BP (kW)	4.4	4.23
BTE (%)	0.2598	0.2392
NO_x (ppm)	681	704
HC (ppm)	412	391

Table 2: Comparison between experimental and simulation data

2.2 Converge Numerical Setup and Validation

The Converge CFD-tool which is very attractive to simulate internal combustion (IC) engine problems was used for 3D CFD simulation. For more information please see Ref. [14]. A full cylinder, open cycle model was used for the numerical analysis. The engine under study [12] was modelled in Catia V5 and .stl file was exported to Converge. The geometry of 2S GDI engine with intake, transfer and exhaust port along with their respective boundary condition can be seen in Fig. 3.

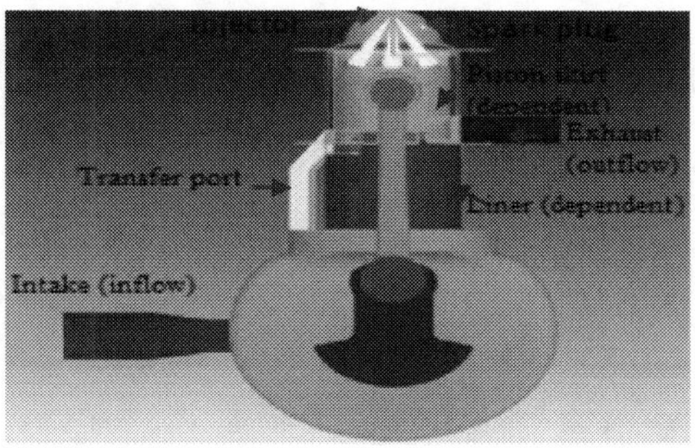

Figure 3: Converge surface .stl file

Piston skirt is a dependent region as it is a moving boundary. It corresponds to cylinder region when it is at top dead centre (TDC) and to crankcase region when at bottom dead centre (BDC). Liner is also a dependent boundary as it has traces in both cylinder and crankcase region. With two-stroke engine open cycle simulation, if sealing feature of Converge has not been used then computational inaccuracies became a problem due to significant crevice volumes between the liner and the piston skirt [14]. Hence, following boundaries have been sealed in this work.

- Connecting rod seal from wristpin (Moving)
- Crankcase Seal from Crankshaft (Moving)
- Liner (Moving)
- Piston Skirt (Stationary)

The maximum distance between the two boundaries for appropriate sealing should be less than or equal to 10^{-6} m [14].

A full cylinder Converge code was run for open cycle. Since the detailed chemistry of gasoline fuel is not available, iso-octane has been used to simulate the spray and combustion. The SAGE detailed chemistry model (having 48 species and 152 reactions) has been used to model the combustion. The Kelvin-Helmholtz and Rayleigh-Taylor (KH-RT) model with breakup length has been used for spray modelling with the recommendations set of constant for moderate flash boiling with GDI. The Extended Zeldovich NO_x model has been used to predict NO_x. The Hiroyasu soot formation model and Nagle and Strickland-Constable (NSC) soot oxidation model have been used to compute the soot emissions.

As mentioned earlier, in-cylinder experimental data were available for limited crank angle, hence the Converge code for described model has been validated against the 1D Ricardo Wave. A very good matching between both the traces can be seen in Fig. 4. Hence, code is considered as validated and used to study the effect of various fuel injection pressure and duration on performance and emissions.

Effect of higher and lower fuel injection pressure with step size of 50 bar than base fuel injection pressure (i.e. 130 bar) has been investigated. Hence, three simulations with 80 bar, 130 bar and 180 bar injection pressure were carried out. The value of injection duration were varied according to the injection pressure for keeping equivalence ratio and EOI equal to base operating condition i.e. 0.7 and 50 CAD respectively. The spark timing was also kept constant (167 CAD) for all three cases.

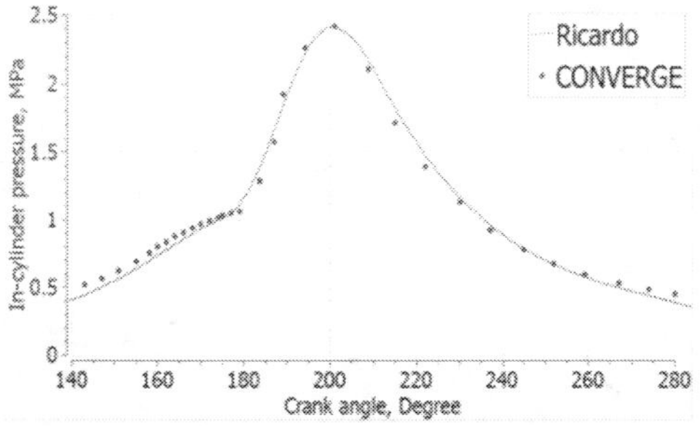

Figure 4: In-cylinder pressure vs crank angle comparison between experimental and Converge simulation

3 Results and Discussions

3.1 Effect of Injection Pressure on In-cylinder Pressure

The fuel injection pressure plays an important role in the mixture formation which in turn affects peak in cylinder pressure traces of the engine. Figure 5 shows the variation of in-cylinder pressure with crank angles for the three fuel injection pressures viz. 80 bar, 130 bar and 180 bar. It can be noticed from Fig. 5 that at the fuel injection pressures of 80 bar and 130 bar, the peak in-cylinder pressures are about 19.1 bar and 24.1 bar respectively which indicates increasing in-cylinder pressure (by about 26%) with increased fuel injection pressure (by about 62%). However, with the fuel injection pressure of 180 bar, the combustion did not take place at all. The occurrence of the peak pressure marginally shifts toward the TDC at the fuel injection pressure of 130 bar as compared to 80 bar (see Fig. 5), which is mainly because of the reduced ignition delay due to better atomization of the injected fuel.

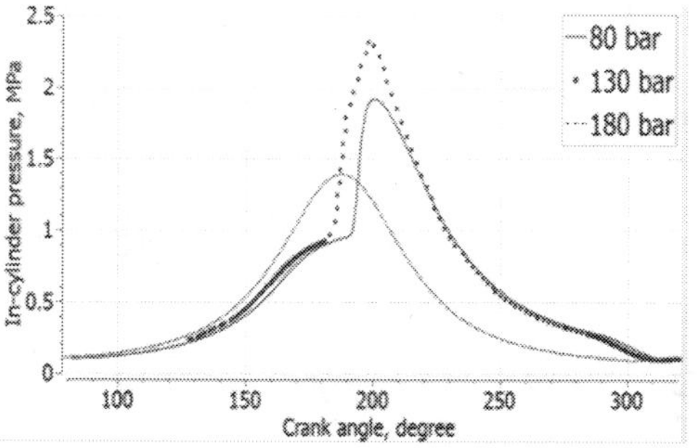

Figure 5: Comparison of in-cylinder pressure traces for various injection pressures

Figure 6 shows the contours of equivalence ratio for the various fuel injection pressures at close to spark timing event. With fuel injection pressure of 180 bar, the ignitable air-fuel mixture doesn't exist near the location of the spark plug, as can be seen in Fig. 6c (region is highlighted using black colour circle). Thereby, there won't be any pressure rise due to combustion, and thus in-cylinder pressure curve simply looks like motoring curve (see Fig. 5).

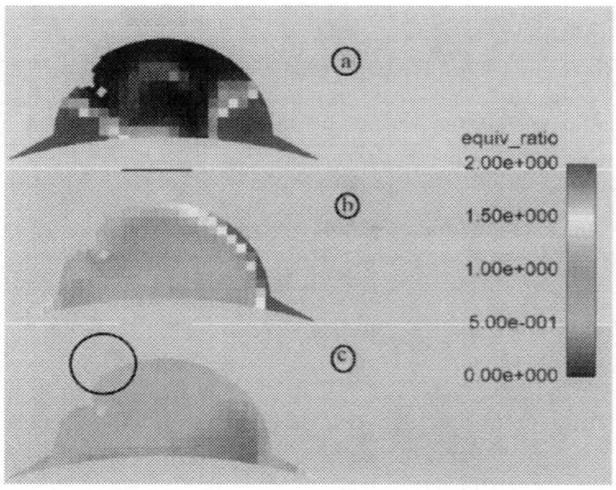

Figure 6: Contour of equivalence ratio for a) 80 bar, b) 130 bar and c) 180 bar fuel injection pressure at 170 CAD

It is also evident from Fig. 6 that a homogeneous air-fuel mixture is seen for injection pressure of about 180 bar due to fine atomisation whereas for the lower pressure a charge stratification pattern is visible. A reason behind this might be increase in the turbulent kinetic energy (TKE) with the increased injection pressure. Effect of injection pressure on tumble ratio was analysed and found insignificant.

3.2 Effect of Injection Pressure on NO_x Emissions

Figure 7 shows the variation of NO_x formation with crank angles for various fuel injection pressures. From Fig. 7, it is observed that the formation of NO_x increases steadily from the time of spark (167 CAD). This is because of increase in the gas temperature beyond 1600 K during combustion. Afterward, their formation is frozen and concentration remains constant up to the exhaust port open (EPO). Usually the gas temperature decreases during the expansion stroke, which inhibits the NO_x formation during expansion.

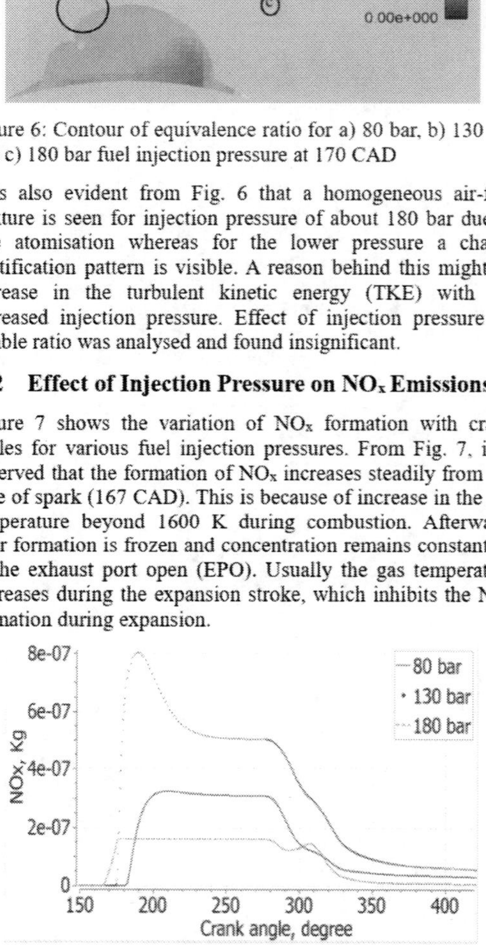

Figure 7: Comparison of NO_x for various injection pressures

In-cylinder temperature contours of 80 bar and 130 bar fuel injection pressure in Figs. 8a and 8b respectively indicate the direct co-relation of NO_x level with in-cylinder temperature during combustion. However with fuel injection pressure of 180 bar (see Fig. 8c), because of misfiring, low in-cylinder temperature and hence low NO_x formation is observed (see Fig. 7. Note: The value of NO_x for 180 bar injection pressure in Fig. 7 is multiplied by 10^3 on Y-axis to magnify the face value). It is also found that the NO_x emissions from the engine are about 58% higher when the fuel injection pressure increases from 80 bar to 130 bar.

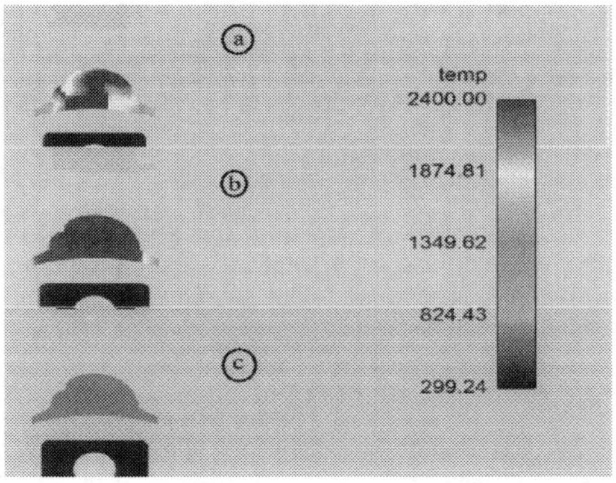

Figure 8: Contour of in-cylinder temperature for a) 80 bar, b) 130 bar and c) 180 bar fuel injection pressure at 205 CAD

3.3 Effect of Injection Pressure on HC Emissions

HC emissions are found very low, as shown in Fig. 9, for 80 bar and 130 bar fuel injection pressure. However, at the fuel injection pressure of 180 bar, because of misfiring, the HC emissions shoot up.

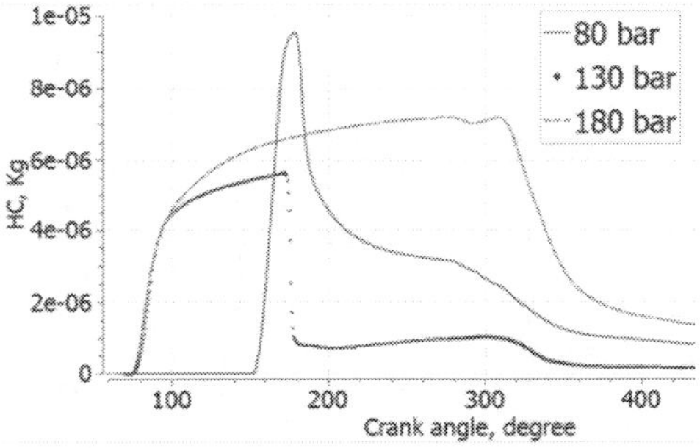

Figure 9: Comparison of HC for various injection pressures

3.4 Effect of Injection Pressure on Soot Emissions

It is observed from Fig. 10 that the soot levels are higher at lower fuel injection pressures. As the fuel injection pressure increases fuel is finely atomised which helps in formation of homogeneous mixture and reduces soot. Soot for 180 bar injection pressure is low compared to other two cases because of partial combustion and soot formation was very low, whereas in case with 130 bar injection pressure, soot formation is lower (as compared to 80 bar injection pressure) due to better fuel atomisation. (Note: The value of soot at 180 bar injection pressure in Fig. 10 is multiplied by 10 on Y-axis to magnify the face value).

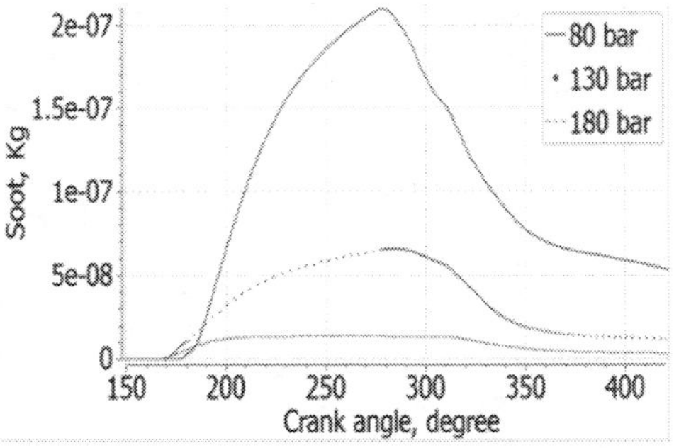

Figure 10: Comparison of soot for various injection pressures

4 Conclusions

In this work, the effect of fuel injection pressure on the performance and emission characteristics of a two-stroke GDI engine was investigated using Converge CFD analysis. A full engine model was created and open cycle simulation was carried out for the study. 1D simulation was also carried out to generate initial conditions for Converge. Both the code were validated with published experimental data and found very good matching. Increase in in-cylinder pressure and NO_x was observed with increase in fuel injection pressure. On the other hand, HC and soot were found lower with increased fuel injection pressure. However, with higher fuel injection pressure (180 bar), the combustion did not take place at all. As a result, very high HC emissions were noticed with 180 bar fuel injection pressure. It is concluded from the study that the moderate fuel injection pressure (~130 bar) for engine configuration under study is required to get an ignitable mixture near the spark plug location at the time of spark. Further study with effect of spark timing and injection rate shape may be planned to optimized GDI engine further.

5 Acknowledgment

The authors thank Computational Engineering India Pvt. Ltd, Pune, for providing opportunity and availing their valuable resources for simulations required for the project at their esteemed organization. Authors would also like to thank the Ricardo India Pvt. Ltd. for providing free academic license to ARAI academy.

References

[1] Spicher, U., Reissing, J., Kech, J.M. and Gindele, J., SAE Technical Paper No. 1999-01-2938.

[2] Eichlseder, H., Baumann, E., Müller, P. and Rubbert, S., SAE Technical Paper No. 2000-01-0248.

[3] Yamagishi, G., Sato, T. and Iwasa, H., SAE Technical Paper 720195.

[4] Vieilledent, E., SAE Technical Paper 780767.

[5] Douglas, R. and Blair, G, SAE Technical Paper 820952.

[6] Leighton, S. and Ahern, S., SAE Technical Paper 2003-260019.

[7] Badami, M., Marzano, M., Millo, F. and Nuccio, P., SAE Technical Paper 1999-01-3311.

[8] Jajcevic, D., Almbauer, R., Schmidt, S., Glinsner, K. et al., SAE Technical Paper 2010-32-0014.

[9] Mattarelli, E., Rinaldini, C. and Cantore G., SAE Technical Paper 2013-24-0085, 2013, doi:10.4271/2013-24-0085.

[10] Mattarelli, E., Rinaldini, C. and Baldini, P., SAE Technical Paper 2014-01-1672.

[11] Addepalli S.K. and Mallikarjuna J.M, Alexandria Eng. J. (2016), http://dx. doi.org/10.1016/j.aej.2016.10.007.

[12] Dube, A. and Ramesh, A., SAE Technical Paper 2016-01-1052.

[13] http://software.ricardo.com/products/wave-rt. Accessed on 1st July 2017.

[14] Richards K.J., Senecal P.K., Pomraning E., CONVERGETM (Version 2.2.0) Manual, Convergent Science Inc., 2016.

11th Asia-Pacific Conference on Combustion,
The University of Sydney, NSW Australia
10th -14th December 2017.

Optimizing the Operating Parameters of Torrefaction Process on the Combustion Enhancement of Coal/Biochar Blended Fuels

Kai-Lin Xiao[1], Chao-Wei Huang[2], Yueh-Heng Li[1*], Guan-Bang Chen[3]

[1]*Department of Aeronautics and Astronautics, National Cheng Kung University, Taiwan*
[2]*Department of Chemical and Material Engineering, National Kaohsiung University of Applied Science, Taiwan*
[3]*Research Center for Energy Technology and Strategy, National Cheng Kung University, Tainan*
*Corresponding Author, Email: yueheng@mail.ncku.edu.tw

Abstract

Although the biomass is regarded as a suitable alternative fuel, its fuel properity and combustion characteristics are much different from fossil fuel. In order to make the properties of biomass more similar to coal, torrefaction was used to improve these disadvantages of biomass. However, operating parameters of torrefaction process are associated with the fuel property of biochar and coal/biochar co-combustion. Therefore, the Taguchi method was utilized to reduce the experimental runs and search for the optimum condition of torrefaction process. The produced biochar would blend with pulverized coal at 50% on gravimetric basis. Comparing the combustion behavior among the pulverized coal, biochar, and blended fuels, a thermogravimetric analyzer was used to observe weight loss of fuel from ambient temperature to high temperatures at fixed heating rate. The results showed the ignition temperature and burnout temperature of blended fuel would occur earlier than pulverized coal. In addition, the fuels were compressed as pellets and carried out combustion experiment in a practical system. The blended fuel has longer volatile combustion due to more volatile matter contained, while pulverized coal contained more fixed carbon results in persisting longer time during char combustion. The results showed that the coal/biochar blended fuel can effectively enhance the fuel conversion rate and combustion characteristic index. Furthermore, the results can be applied to co-firing in large scale boilers in the future.

Keywords: Torrefaction, Miscanthus biochar, Co-firing, Thermogravimetric analysis, Single pellet combustion

1 Introduction

With the air pollution and greenhouse effect getting worse and worse, a lot of countries have drawn attention on renewable energy. In the renewable energy, biomass energy utilization has increased attention because of their rapid growth, low mineral content and good carbon balance [1]. In general, the biomass could be divided into the following types: agricultural, forest, municipal waste, energy crops and biological [2]. Miscanthus is regarded as inedible energy crop which is a woody, C4 perennial and rhizomatous grass. It is well known to its wide availability, strong adaptability, and good combustion properties [3]. Miscanthus also possesses high carbon fixation efficiency, so that it is suitable for being biomass fuel.

Although biomass is a good candidate for alternative fuel, it has some disadvantages, for example, high moisture content, low energy density, low grindability and nonhomogeneous chemical properties [4]. Therefore, biomass must enhance the combustible properties by essential pretreatment. Torrefaction is usually regarded as a method of fuel pretreatment.

It is a thermochemical process in an inert or limited oxygen environment where biomass is slowly heated to a temperature ranging from 200°C to 300°C [5]. The hydroxyl groups are removed during the heating process, namely producing hydrophobic material [6]. For this reason, the properties of biochar are more coal-like.

In order to optimize the fuel pretreatment, the Taguchi method is used to find the optimization routes and identify the important parameters. The Taguchi method is versatile and employed in various field. Chen et al. [7] employed Taguchi method to find the optimum co-gasification operation of torrefied biomass and coal in an entrained flow gasifier.

Biomass contains lesser amounts of carbon dioxide and sulfur than those in coal. Therefore, through co-firing could effectively reduce the emission of CO_2 and SO_2. In addition, many biomasses contain calcium, which is very effective in absorbing the sulfur released from the coal during co-firing. Some researches had proven these phenomena by experiment [2,8]. Furthermore, Fahlstedt et al. [9] mentioned that the co-firing can slightly enhance the combustion efficiency. Muthuraman et al. [10] used municipal solid waste to blend with different rank coals and observed the co-combustion can improve devolatilazation properties of coal. From these results, we can find that biomass blend with coal to carry out the co-firing not only reduce the harmful gas emission but enhance the combustible behavior of coal.

Nowadays, some researchers discussed about pretreatment methods [6,7]. However, the optimal operating parameters of pretreatment method of biomass are rarely discussed. So, this study is going to create a composite index which combines the properties and combustion characteristics of biochar. Eventually, the Taguchi method was used to find out the optimal operating parameters of torrefaction.

2 Experimental methods

2.1 Sample preparation

Biomass is not unfavorable for co-firing. Therefore, torrefaction was employed as pretreatment method to convert the chemical and physical properties of biomass. The feedstock of this experiment was air-dried Miscanthus floridulus obtained from the Chiayi Agricultural Experiment Branch in Taiwan. During the torrefaction process, approximately 10 g of Miscanthus powder was packed in a quartz cylindrical holder. The holder was placed in a tubular furnace and filled with nitrogen. the flow rate was set to 50, 100 and 150 c.c./min. The temperature in the tubular furnace was heated to three temperature ranges of 200, 250 and 300 °C and then maintained for 60, 90 and 120 min. After that, the raw Miscanthus would convert into Miscanthus biochar.

Corresponding author. Fax: +886-6-238-9940
E-mail address: yueheng@mail.ncku.edu.tw

2.2 Thermogravimetric analysis

In order to observe the properties and combustion behavior of fuel, a thermal analyzer (PerkinElmer, STA 8000) was used to simultaneously perform thermogravimetric analysis. In each case, approximately 15 mg of fuel at particle sizes smaller than 74 μm was put into the crucible and heated to 1000°C from 30°C with fixed heating rate of 20 °C/min. The carry gas was nitrogen and air at fixed 50 ml/min for the pyrolysis and oxidation process. In the oxidation process, the ignition temperature was obtained by intersection method, and the burnout temperature was defined as the temperature at which the fuel conversion reaches 99%. The results obtained were used to determine the combustion characteristic index (S). A high S value implies that the fuel possesses better combustion performance. This index is defined as follows: [11]

$$S = \frac{(\frac{dW}{d\tau})_{max} \cdot (\frac{dW}{d\tau})_{mean}}{T_i^2 \times T_b} \quad (1),$$

where (dW/dt)max and (dW/dt)mean represent the maximum and average mass-loss rates, respectively.

2.3 Single pellet combustion

To investigate the combustion behavior of fuel. Fuel powder was compressed a fuel pellet whose weight was nearly 800 mg, and then, used the free-drop furnace to operate the pellet combustion experiment. The temperature in the furnace was set to 600 and 800 °C with a 220V/2kW power supply. Figure 1 showed the schematic diagram of experimental apparatus. Fuel pellet was dropped through the transmission tube to the stainless mesh platform. The weight change was measured by an electronic balance in real time (once per second). The video camera set in front of furnace observation window was used to record the combustion process. Gas analyzer (MRU VarioPlus) was utilized to measure the flue gas emission (including O_2, CO, CO_2 and NOx). In the experiment, the ignition delay time, total combustion time, and flame patterns of fuel were determined and the characteristics of fuel could be further specified.

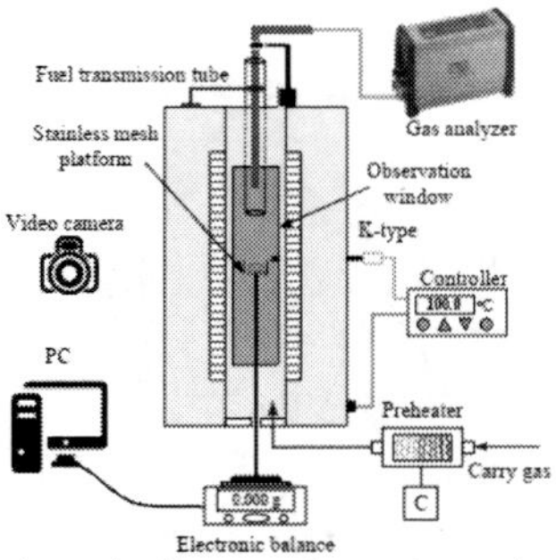

Fig. 1. Schematic diagram of experimental apparatus.

3 Results and discussion

3.1 Optimization of torrefaction condition

In the torrefaction process, temperature, residence time, and N_2 flow rate are the three important control factors. Those parameters would affect the characteristics of produced fuel. Taguchi method was used to optimize torrefaction condition, heating value (HV) and Energy yield (EY), by minimal number of experiments. The control factors were assumed to be independent of each other, and each factor had three levels, such as Table 1. The original 27 experiments reduce to 9 experiments by Taguchi method.

NO.	Temperature (°C)	N_2 flow rate (sccm)	Residence time (min)
1	200	50	60
2	200	100	90
3	200	150	120
4	250	50	90
5	250	100	120
6	250	150	60
7	300	50	120
8	300	100	60
9	300	150	90

Table 1. Experimental layout using an L9(3^3).

The larger-the-better (LTB) characteristics were used to obtain the optimal condition of torrefaction for highest HV and EY in the Taguchi method. The characteristics of LTB is defined as: [12]

$$\frac{S}{N_{LB}} = -10\log\left[\frac{1}{n}\sum_{i=1}^{n}\frac{1}{y_i^2}\right] \quad (2),$$

where n is the number of tests and yi is the value of the EY or the HV. Table 2 showed that the S/N ratio was calculated by Eq. (2). Based on the analytical results, the optimal conditions could be obtained. For the highest EY, the torrefaction temperature, residence time and N_2 flow rate are 200 °C, 60 min, and 100 sccm, respectively. Moreover, the optimal torrefaction condition for the highest HV set the torrefaction temperature: 300 °C, residence time: 120 min and nitrogen flow rate: 50 sccm. The large difference of the S/N rates implies the importance of parameters in the torrefaction process. The torrefaction temperature appears to be the most significant variable, followed by the residence time and finally the nitrogen flow rate.

NO.	HV (MJ/Kg)	EY (%)	S/N ratio for HV	S/N ratio for EY
1	17.17	94.41	72.26	-0.499
2	16.65	91.29	71.99	-0.792
3	17.34	95.93	72.35	-0.361
4	17.86	85.83	72.60	-1.327
5	17.87	87.52	72.61	-1.158
6	17.87	87.93	72.61	-1.117
7	21.18	69.74	74.09	-3.13
8	20.70	73.16	73.89	-2.715
9	19.78	69.15	73.49	-3.204

Table 2. HV and EY of biochar with the corresponding S/N ratio.

3.2 Thermogravimetric analysis

Table 3 presented the basic characteristics of Australia coal and Miscanthus biochar. From the proximate analysis, the moisture, volatile matter, fixed carbon and Ash can be determined. Among them, the volatile matter and fixed carbon are two important parameters for combustion behavior and gas emission. Australia coal possess higher ratio of fixed carbon (50.5%). Therefore, it has greater heating value. In contrast, Miscanthus possesses lower fixed carbon ratio (5.9%) which may cause the reduction of the energy efficiency, but its high volatile matter content (82.2%) makes it easily being ignited. When the Miscanthus carry out torrefaction, the hydroxyl groups are removed during the heating process. Therefore, H/C and O/C ratio gradually reduce. As the temperature higher, the properties of Miscanthus biochar would be more similar to Australia coal. No matter biochar samples under the best-EY or best-HV torrefied conditions, the properties of biochar were between Australia coal and raw Miscanthus.

Fuel	Proximate analysis (wt. %)				Elemental analysis (wt. %)			HHV (MJ/kg)
	M	VM	FC	Ash	C	H	O	
Australia coal	1.4	29.6	50.5	18.5	73.3	4.2	5.3	25.6
Raw Miscanthus	4.9	82.2	5.9	7.0	44.9	6.6	48	16.0
Biochar (best EY)	1	77.7	8.4	12.9	46	6.2	47.6	16.9
Biochar (best HV)	1.1	52.8	31.5	14.6	58.1	3.6	38	21.2

Table 3. Physical and chemical properties of Australia coal, raw Miscanthus, and biochar.

The pyrolysis profiles of the studied fuels are shown in Fig. 2, which illustrates the mass loss of the fuel samples and their corresponding time derivative of the mass loss (DTG) with regard to increasing temperature. There are two main mass losses in the pyrolysis process. The first mass loss occurred due to moisture evaporation at below 140 °C. The second mass loss appeared between 200-660 °C. In this temperature region, fiber composition of biochar underwent thermal degradation. Such as Fig. 3b, because of the biochar contain great amount fiber content (cellulose, hemicellulose and lignin), they have larger mass loss than Australia coal and blended biochar. When Australia coal was heated to 1000°C, the maximum mass loss shifts to high-temperature zone due to its high carbon contents and the lowest volatile contents. To observe the influence of blended fuel, the biochar blended with Australia coal at 50% on the weight basis. From the pyrolysis process, it can be found that the mass loss of coal-blended biochar has linearly superposing with the mass loss of coal and torrefied biochar. This indicates that no chemical reaction occurred between the coal and torrefied biochar when the blended fuel was subjected to pyrolysis.

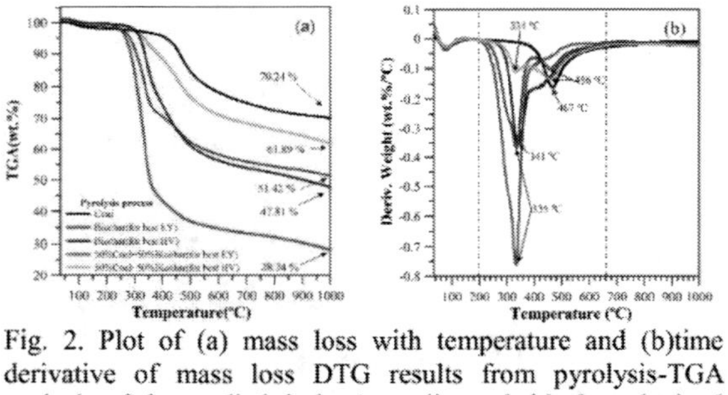

Fig. 2. Plot of (a) mass loss with temperature and (b)time derivative of mass loss DTG results from pyrolysis-TGA analysis of the studied fuels, Australia coal, biochar obtained under the best-HV and best-EY torrefaction, and 50% biochar–50% coal under nitrogen atmosphere.

Figure 3 showed the oxidation process (or combustion process) for the studied fuels under air atmospheres. The maximum mass loss of Australia coal occurred at 594°C, which is higher than those of the other studied fuels, because it has higher fixed carbon content. When the environmental temperature reached 1000°C, the residual mass rate was 19.44% in Fig. 3a. Figure 3b showed the blended biochar with best EY. There were regions of main mass losses presented at 220-360°C and 360-500°C. The peak temperatures in the two temperature ranges were 303 and 386 °C. When the surrounding temperature reached 360 °C, the residual mass rate was 41.42%. These results suggest that more than half of the fuel was released as volatile gas. The combustion process was accomplished at 500 °C and the corresponding residual mass rate was 13.54%, which is lower than that of Australia coal. In addition, the blended biochar with best HV main mass losses occurred in the temperature ranges of 240–320 and 320–560 °C which a little bit higher than the blended biochar with best EY. This is because that the rising of torrefaction temperature results in an increase in the carbon content of torrefied fuels. Figures 3d and 4e demonstrate the TGA and DTG curves of coal blended with biochar with various torrefaction parameters.

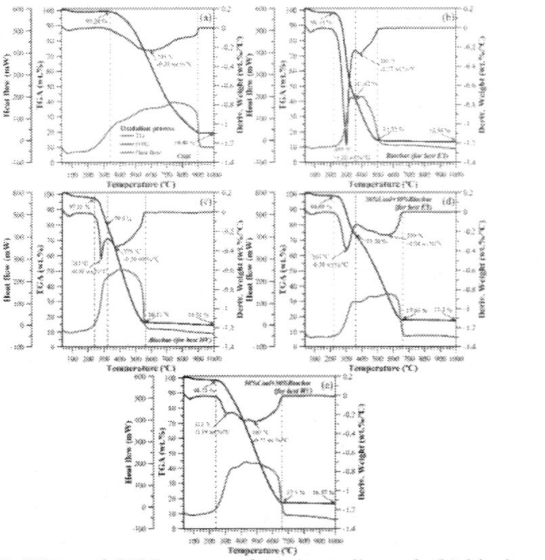

Fig. 3. TGA and DTG curves of (a) Australia coal, (b) biochar with best-EY torrefaction, (c) biochar with best-HV torrefaction, (d) 50% coal–50% biochar with best-EY torrefaction, and (e) 50% coal–50% biochar with best-HV torrefaction in ambient air.

To compare the combustion behavior of studied fuels, the combustion characteristics index (S) were calculated by TGA. Table 4 presented the biochar with best EY have optimal combustion characteristics. In contrast, the Australia coal has the lowest S index. The high S index implies the fuel contain more volatile gas lead to higher reactivity or lower ignition temperature.

Fuel	T_i (°C)	T_b (°C)	S x 10^7
Coal	446.8	887.3	0.871
Biochar (best EY)	265.5	488.8	49.7
Biochar (best HV)	272.9	504.8	25.9
50%Coal+50%Biochar (best EY)	272.7	639.7	6.8
50%Coal+50%Biochar (best HV)	317.2	644.9	3.63

Table 4. Ignition temperature, burnout temperature, and combustion characteristic index for five studied fuels.

3.3 Single pellet combustion

The study chose three solid fuels, namely Australia coal, 50% coal–50% biochar with the best-EY torrefaction condition, and 50% coal–50% biochar with the best-HV torrefaction condition, to investigate combustion behavior of fuel. When the fuels were dropped into high surrounding temperature in a flash. The solid fuels will undergo these processes including heating, drying, pyrolysis/ devolatilization, volatile burning, char burning, and ash formation. These stages were affected by fuel's characteristics. As shown in Fig. 4, the biochar with the best-EY torrefaction condition was easy to ignite and had the longest flame length due to the higher volatile content. By contrast, the combustion behavior of the biochar with the best-HV torrefaction condition was similar to that of Australia coal because of further carbonization during pretreatment.

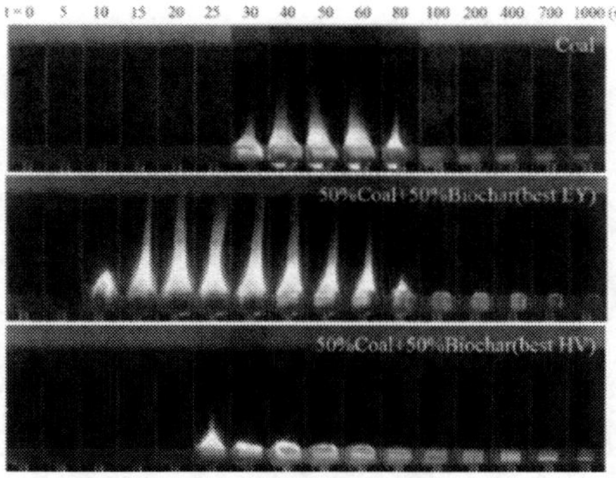

Fig. 4. Combustion process of three solid fuels at a surrounding temperature of 600 °C.

The flue gas emission was measured by Gas Analyzer (GA) when it operated the single pellet combustion. The results showed that reduction in oxygen and an increase in carbon monoxide before the gas phase ignition. This stage carried out the heterogeneous combustion reaction ($2C+O_2 \rightarrow 2CO$). Following the gas flame formation, CO carried out homogeneous reaction with the surrounding oxygen molecules to form CO_2 ($CO+0.5O_2 \rightarrow CO_2$). When this stage finishes, the heterogeneous reaction remains dominant. The CO_2 concentration is rarely detected, but the CO concentration is abundant. Until the char combustion finished, CO concentration sharply decreases and O_2 concentration recedes back to 21%.

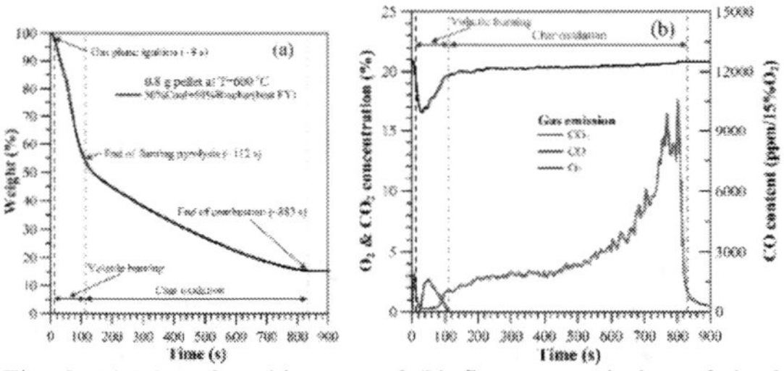

Fig. 5. (a) Mass-loss history and (b) flue-gas emission of single pellet of 50% coal–50% biochar with best-EY torrefaction at a surrounding temperature of 600°C.

T (°C)	Fuel	t_i (s)	t_{flame} (s)	t_{char} (s)	t_{total} (s)	Ash (wt. %)
	A	30	73	1489	1592	20.97
600	B	9	103	721	833	15.50
	C	24	88	1087	1199	18.84
	A	4	70	1255	1329	18.67
800	B	1	82	658	741	15.29
	C	2	80	1024	1106	18.01

Table 5. Characteristic time and ash content of three studied fuels (A: Australia coal, B: biochar with best-EY torrefaction, C: biochar with best-HV torrefaction) at surrounding temperatures of 600 and 800 °C.

4 Conclusions

To find the optimal torrefaction condition by Taguchi method, the Miscanthus biochar and Australia coal were subject to thermogravimetric analysis to observed the combustion behavior of studied fuels. In addition, these fuels were pelletized as tablets and dropped into a single-pellet furnace to investigate the combustion process and gas emission of the single pellet at various surrounding temperatures. The following results were obtained:

1. In the thermal pyrolysis process, Miscanthus biochar main engaged in the thermal degradation of hemicellulose and cellulose content. Australia coal have high carbon content, the main mass loss of coal tended to occur in a higher temperature range. In addition, observing the pyrolysis features of the blended fuels revealed linear superposition of coal and the biochar with specific torrefaction conditions.
2. By calculated S index, it can be found that biochar with best-EY torrefied condition had a relatively low ignition temperature and a low burnout temperature, resulting in a higher S index value.
3. Fuel pellet with high volatile matter was easily ignited. However, the volatile gas combustion process was longer. By contrast, a fuel pellet with high fixed carbon content had long-lasting char burning. In general, when the surrounding temperature increases, the thermal pyrolysis process and devolatilization can be accelerated.

5 Reference

[1]. Saxena, R.C., Adhikari, D.K., Goyal H.B., "Biomass-based energy fuel through biochemical routes: a review." Renew Sustain Energy Rev(2009), 13(1):167-178.

[2]. Basu, P., "Biomass Gasification, Pyrolysis and Torrefaction-Practical Design and Theory." Published by Elsevier Inc. (2013)

[3]. Brosse, N., Dufour, A., Meng, X., Sun, Q., Ragaukas, A., "Miscanthus: a fast growing crop for biofuels and chemicals production." Biofuels Bioprod Biorefin (2012), 6: 580–598.

[4]. Jeguirim, M., Dorge, S., Loth, A., Trouve, G., "Devolatilization kinetics of Miscanthus straw from thermogravimetric analysis." Int J Green Energy(2010), 7:164-173.

[5]. Bergman, P.C.A., Boersma, A.R., Zwart, R.W.R., Kiel, J.H.A., "Torrefaction for biomass co-firing in existing coal-fired power stations." Report ECN-C-05-013, ECN(2005a)

[6]. Wilk, M., Magdziarz, A., "Hydrothermal carbonization, torrefaction and slow pyrolysis of Miscanthus giganteus." Energy (2017), 1-13.

[7]. Chen, W.H., Chen, C.J., Hung, C.I., "Taguchi approach for co-gasification optimization of torrefied biomass and coal." Bioresource Technology (2013), 144: 615-622.

[8]. Hein, KRG., Bemtgen, J.M., "Co-combustion of coal and biomass." Fuel Process Technol (1998), 54: 159-169.

[9]. Fahlstedt, I., Lindman, E., Lindberg, T., Anderson, J., "Co-firing of biomass and coal in a pressurized fluidised bed combined cycle." Proceedings of the 14th International Conference on Fluidized Bed Combustion in Vancouver (1997), P.295-299.

[10]. Muthuraman, M., Namioka, T., Yoshikawa K., "Characterisrics of co-combustion and kinetic study on hydrothermally treated municipal solid waste with different rank coals: A thermogravimetric analysis." Applied Energy (2010), 87: 141-148.

[11]. Ren, X., Meng, J., Moore, A.M., Chang, J., Gou, J., Park, S., "Thermogravimetric investigation on the degradation properties and combustion performance of bio-oil." Bioresource Technology (2014), 152: 267-274.

[12]. Mcneill, I. C., "15-Thermal Degradtion." Comprehensive Polymer Science and Supplements, 6 pp.451-501, 1989.

11th Asia-Pacific Conference on Combustion,
The University of Sydney, NSW Australia
10th -14th December 2017.

LES on Knocking Combustion and End-gas Auto-ignition Based on A Downsized Spark-ignited Engine

Jiaying Pan, Haiqiao Wei* , Gequn Shu,

State Key Laboratory of Engines, Tianjin University

No.92 Weijin Road, Tianjin, 300072, China

Abstract

A great challenge for highly boosted/downsized spark-ignited engines is conventional knocking and super-knock, where sporadic auto-ignition with localized origin devotes to the nature of abnormal combustion. Nowadays Large Eddy Simulation (LES) shows more attractive because of its capability of providing local time-resolved information and reproducing the sporadicity of abnormal combustion. Using a LES framework coupling detailed chemistry solver, this work quantitatively investigates knocking combustion and end-gas auto-ignition by means of adjusting spark-ignition timing (ST) in a downsized spark-ignited engine. Firstly, current LES results show that as the advance of ST, normal combustion to knocking combustion with various intensities are observed, agreeing well with engine experiments. Secondly, knocking intensity increases proportionally with the ST in conventional knocking regime, but super-knock event is not observed even at very early ST. Moreover, the linear relationship between knocking intensity and the burned mass fraction (BMF) by end-gas auto-ignition is confirmed by current LES, and meanwhile, a second regime of knocking intensity is identified when the BMF reaches a threshold. Finally, corresponding to the knocking intensity at different STs, various auto-ignition scenarios are obtained from end-gas region, which show that local auto-ignition can be induced by single and multiple hot-spots, or occur with a homogeneous behavior. However, the traces of thermodynamic state show that the origin of strong pressure wave during knocking combustion is mainly contributed by the homogeneous auto-ignition or the auto-ignition initiated by multiple hop-spots, instead of the early one by single hot-spot.

1. Introduction

Downsized spark-ignited (SI) engines with turbocharging technology have become increasingly attractive because of their increased thermal efficiency and low CO2 emissions. However, severe thermodynamic conditions at low-speed and high-load may induce the occurrence of abnormal combustion like conventional knock or super-knock [1]. Theoretically, both combustion phenomena are resulted from local auto-ignition, encouraged by single or multiple hot-spots and followed by the initiation and propagation of premixed flame fronts [2]. Although numerous investigations have been performed during the last decade, knocking issues are still the main barriers for SI engine development, largely ascribing to the lack of comprehensive understanding on the complex combustion mechanism [3].

When knocking occurs, apparent pressure waves (or even shock waves) are induced by local auto-ignition at end-gas region before the arrival of main flame. These aerodynamic waves propagate back and forth in cylinder and result in intense pressure oscillations, which severely damage engine body and structure [4]. Knocking combustion highly depends on flame propagation velocity as well as on species composition and temperature fluctuations in cylinder, which all influence auto-ignition delay time, making knocking a recurrent but non-cyclic phenomenon. As a result, experimentally investigates this combustion is not an easy task in practical. Nonetheless there have been several pioneer studies of knocking experiments, which have shed light on the fundamental physicochemical properties for knocking combustion [5-10].

In comparison to engine experiments, it has shown that Computation Fluid Dynamics (CFD) is able to mimic practical combustion process and to offer interesting physicochemical information of the combustion process in an enclosed combustion chamber [11]. Zhou [12] et al. numerically investigated knocking phenomena in a Reactivity Controlled Compression Ignition (RCCI) engine fuelled with biodiesel and methanol. Using a tabulation approach, Knop et al. [13] studied end-gas auto-ignition scenarios during flame propagation in SI engines based on Reynolds-Averaged Navier–Stokes (RANS) equation formalism. Above CFD results are well consistent with engine experimental data under different operation conditions. However, RANS simulation classically emphasizing the prediction of mean cycles may be not necessarily subjected to the knocking combustion with a sporadic and non-cyclic behavior [14].

Alternatively, Large Eddy Simulation (LES) appears attractive because it allows reproducing the sporadic abnormal combustion and accessing to more accurate localized information of individual cycles. There have been several LES frameworks with reduced kinetic scheme or tabulated kinetics, which attempt to investigate the abnormal combustion subjected to knocking events [15-16]. Lecocq et al. [17] developed a new LES framework coupling flame surface density model and tabulated kinetics to study knocking and pre-ignition in a SI engine. The results showed that the LES can well capture the sporadic and localized auto-ignition features. Using dual heat transfer and two-step reduced scheme, Misdariis et al. [18] carried out a LES work on engine knocking to investigate the effect of temperature field on local auto-ignition. They found that the occurrence of local auto-ignition and subsequent knocking combustion largely depend on the inhomogeneous temperature distributions, e.g. the temperature difference between intake and exhaust valves. Above investigations basically indicate the possibility of using LES approach to accurately predict knocking issues.

Despite above investigations relevant to engine knock, there are still many ambiguities associated with the key physical-chemical mechanisms, such as the interactions between multiple

Corresponding author: Haiqiao Wei
E-mail: whq@tju.edu.cn

hot-spots [19], reaction front and pressure wave interactions [20], deflagration to detonation transition (DDT) [21] and the role of chemistry [6]. The main objective of current LES work is to further explore knocking combustion and end-gas auto-ignition, especially addressing the variation of knocking intensity and auto-ignition scenarios affected by low-temperature combustion. Current LES framework has coupled detailed chemistry solver and one-equation viscosity model, and different knocking intensities have been achieved by adjusting spark-ignition timing (ST).

2. Methodology and Specifications

2.1 Computation models

Theoretically end-gas auto-ignition is a combustion phenomenon controlled by chemical kinetics during combustion. Accurately predicting this sporadic combustion need detailed chemical kinetics allowing for both low- and high-temperature chemistry. To solve this issue, the SAGE detailed chemistry solver [22] integrated into CONVERGE CFD [23] code has been utilized in this work. In calculation, the SAGE calculates reaction rate for each elementary reaction while CFD code solves transport equations. Along with Adaptive Mesh Refinement (AMR), an accurate mechanism can be used for modeling many combustion regimes, including ignition, premixed, mixing-controlled, etc. Therefore, The SAGE is perfect for the simulation relevant to end-gas auto-ignition and knocking combustion. A LES sub-model, one-equation eddy viscosity model [25-26], has been used in present work to solve a sub-filter-scale (SFS) kinetic energy transport equation, and turbulent viscosity is modeled by employing the SFS turbulent kinetic energy and filter size. Because of the better performance in adjusting various grid levels of computation domain, this model has been extensively used to resolve the flow with complex geometries such as internal combustion engines.

2.2 Model Specifications

The current knocking experiments have been carried out in a Port Fuel Injection (PFI), single-cylinder turbocharged engine with four valves. The main geometrical specifications and operation conditions for the engine can be found in Table 1. In simulation, current LES work is performed through CONVERGE CFD code [23], which automatically generates real-time and high-quality orthogonal hexahedral meshes. To achieve the balance of computation efficiency and computational accuracy, the AMR strategy has been utilized to automatically adjust mesh number, which depends on the real-time distributions of local velocity, temperature and species mass fraction. During combustion, the cell size is dx=0.1~0.2 mm around spark plug and end-gas auto-ignition region. For the rest of time, it is close to dx=0.5 mm for the cell at above regions and dx=1.0 mm at other regions. The total mesh number for whole computational domain varies from 1.0 million at Bottom Dead Center (BDC) to 0.8 million at Top Dead Center (TDC).

A reduced chemical kinetic mechanism developed by Liu et al. [27] has been selected rather than detailed chemical kinetics in order to improve computation efficiency. This Primary Reference Fuel (PRF) mechanism has been well validated and tested against a wide range of SI engine conditions [28-31]. To accurately predict flame propagation and end-gas auto-ignition, the SAGE detailed chemistry solver and one-equation eddy viscosity model have been integrated into current LES framework. Meanwhile, isothermal conditions are adopted for the treatment of wall heat transfer, and the wall temperatures for cylinder head, piston, intake and exhaust valves are 450, 450, 639 and 784 K, respectively. Further, the maximum Courant-Friedrichs-Lewy (CFL) number is reduced from 50 to 1.75 once combustion starts to guarantee high resolution of pressure waves during knocking combustion. Eight monitor points (P1-P8) have been set up at edge of cylinder head, with an interval of 45 degree mutually, to obtained more information of local combustion. The geometry for current SI engine and monitor points are shown in Fig. 1. Considering that for a given cycle, the flow and combustion have a low impact on the next cycle in SI 4-stroke engines due to the uncoupling of intake and exhaust phases [32], a strategy is proposed here. As the cases with different STs have the same operating conditions in terms of engine speed and load output, 8 LES consecutive cycles with a reference ST have been performed to provide independent initial conditions, including the fields of velocity, energy and mass fraction, for the other STs. Consequently, only the combustion phases of each cycle are simulated by adjusting the ST, which makes the CPU cost much cheaper.

Parameters	value
Compression ratio	10.6
Bore/ Stroke [mm]	77/85.6
Connection rode length [mm]	132
Engine speed [rpm]	1800
Intake pressure [atm]	1.80
Intake temperature [K]	305
Fuel	IC8H18
IVO/IVC [CAD]	353/-162

Table 1: Main specifications for the experimental SI engine

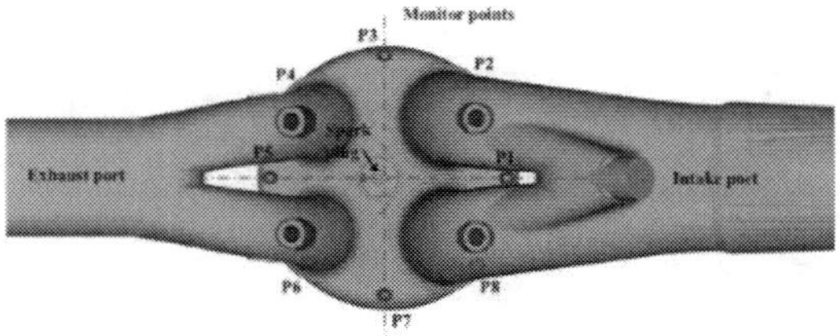

Fig. 1 Schematic of SI engine model and monitor points for local auto-ignition during knocking combustion

3. Results and discussions

3.1 Validation for LES Model

The experimental pressure is from 500 engine cycles with a ST sweep from ST=-2 to 8 crank angle degree (CAD), while the LES results are obtained from a ST sweep from ST=-8 to 8 CAD to extend research scope. Fig. 2 shows the results of in-cylinder pressure traces by both engine experiments and simulations. It is observed that under current downsizing level, engine combustion process seems very sensitive to the ST. At a later ST (e.g. ST=4 CAD), in-cylinder pressure curves for both experiments and simulations are relatively smooth without obvious pressure oscillations, which can be regarded as a normal combustion event. As the advance of ST (e.g. ST=0 CAD), pressure peak and pressure rise rate are significantly increased, and the advance of combustion phasing is followed by apparent pressure oscillations on in-cylinder pressure traces, suggesting the occurrence of knocking combustion. Compared with the pressure peak of P_{max}=5.7 MPa at ST=4 CAD, it is increased up to P_{max}=8.6 MPa at ST=-1 CAD, with a growth rate of approximate 51%. On the whole, current LES results agrees well with experimental data in terms of combustion phasing and pressure level, demonstrating the good performance of current LES in predicting knocking combustion.

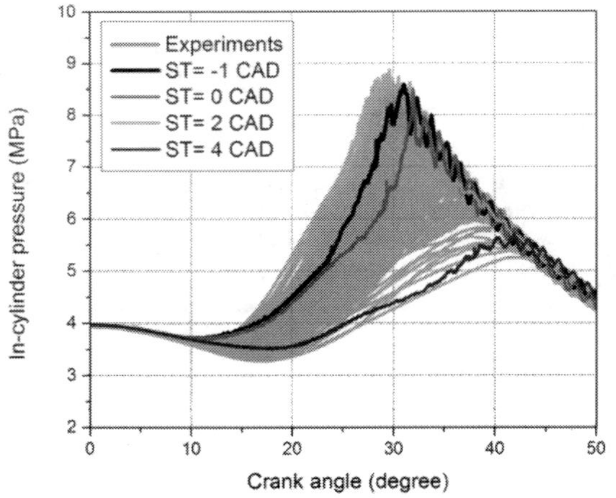

Fig. 2 Comparison of in-cylinder pressure traces between engine experiments and simulations

The experimental pressure is from 500 engine cycles with a ST sweep from ST=-2 to 8 crank angle degree (CAD), while the LES results are obtained from a ST sweep from ST=-8 to 8 CAD to extend research scope. Fig. 2 shows the results of in-cylinder pressure traces by both engine experiments and simulations. It is observed that under current downsizing level, engine combustion process seems very sensitive to the ST. At a later ST (e.g. ST=4 CAD), in-cylinder pressure curves for both experiments and simulations are relatively smooth without obvious pressure oscillations, which can be regarded as a normal combustion event. As the advance of ST (e.g. ST=0 CAD), pressure peak and pressure rise rate are significantly increased, and the advance of combustion phasing is followed by apparent pressure oscillations on in-cylinder pressure traces, suggesting the occurrence of knocking combustion. Compared with the pressure peak of P_{max}=5.7 MPa at ST=4 CAD, it is increased up to P_{max}=8.6 MPa at ST=-1 CAD, with a growth rate of approximate 51%. On the whole, current LES results agrees well with experimental data in terms of combustion phasing and pressure level, demonstrating the good performance of current LES in predicting knocking combustion.

To further illustrate the complex pressure characteristics during knocking combustion, the instantaneous in-cylinder pressure of different monitor points at ST=0 CAD has been shown in Fig. 3. It is observed that for the combustion stage prior to 30.5 CA, the instantaneous pressure history for different locations almost varies along the same track, representing the stage of normal combustion. When the piston moves to 30.5 CA, the pressure curve corresponding to point P2 begins to deviate from others and to show obvious pressure fluctuations. This indicates that local auto-ignition may firstly occur around the location adjacent to point P2 and result in local over-pressure. Subsequently, the pressure wave induced propagates forward or new auto-ignition is initiated, which further induce over-pressure at other locations, as shown by the apparent pressure fluctuations of other monitor points in Fig. 3. In this case, the maximum of instantaneous in-cylinder pressure reaches approximately P_{max}=8.5 MPa (from point P7). The variations in pressure traces of different locations basically suggest the complexity of knocking combustion, as shall be discussed in Part 3.3.

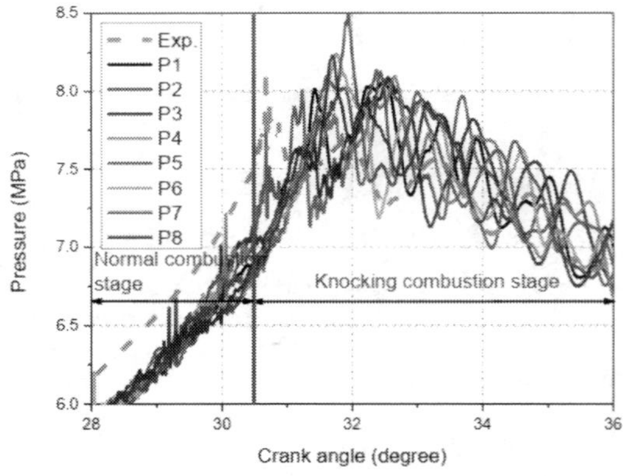

Fig. 3 Instantaneous in-cylinder pressure of different monitor points at ST=0 CAD

3.2 LES for Knocking Characteristics

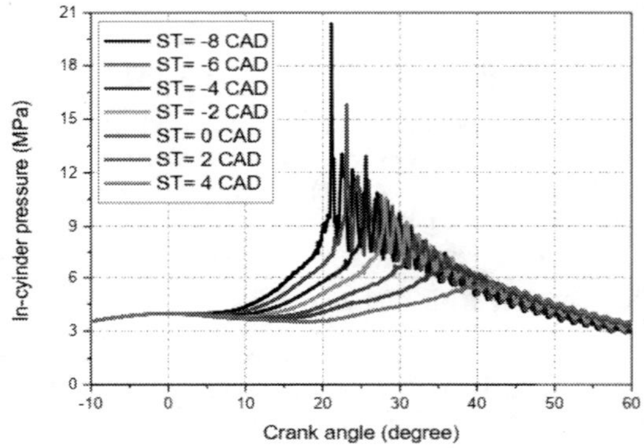

(a)

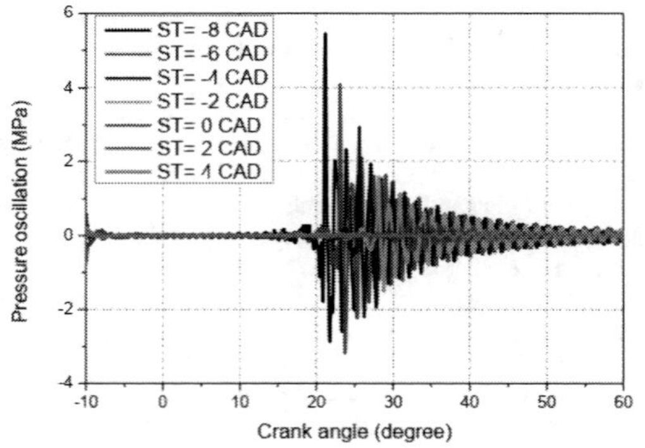

(b)

Fig. 4 (a) Instantaneous in-cylinder pressure and (b) pressure oscillations of different spark-ignition timings

Fig. 4 shows the instantaneous in-cylinder pressure and corresponding pressure oscillations at different STs. The pressure oscillations obtained have been processed by the Fast

Fourier Transform (FFT) with a band frequency of 5~40 kHz. There are many evaluation methods for knocking characteristics. In current work, the knocking onset is defined as the instant of sharp increase in the first pressure derivative, and knocking intensity is evaluated by the maximum amplitude of pressure oscillations (MAPO), as previous work done [33]. It is observed from Fig. 4(a) that as ST is advanced, pressure rise rate and pressure peak are significantly increased, with the appearance of obvious high-frequency pressure oscillations superimposing on mean pressure traces. Fig. 4(b) shows that at very later ST (e.g. ST=4 CAD), the amplitude of pressure oscillations is below 0.2 MPa. When it is advanced to ST= 0 CAD, the amplitude of pressure oscillations reaches 0.5 MPa, representing moderate knocking combustion. Further advance in the ST (e.g. ST=-6 CAD), the amplitude of pressure oscillations is increased up to 4.0 MPa, which can be classified as a heavy knocking event. Meanwhile, the increases in knocking intensity are accompanied by the advance of knocking onset, e.g. it is changed from 32 CA at ST=0 CAD to 20 CA at ST=-8 CAD. However, different from the LES work in [16], the super-knock events induced by pre-ignition have not been observed through varying ST in current work.

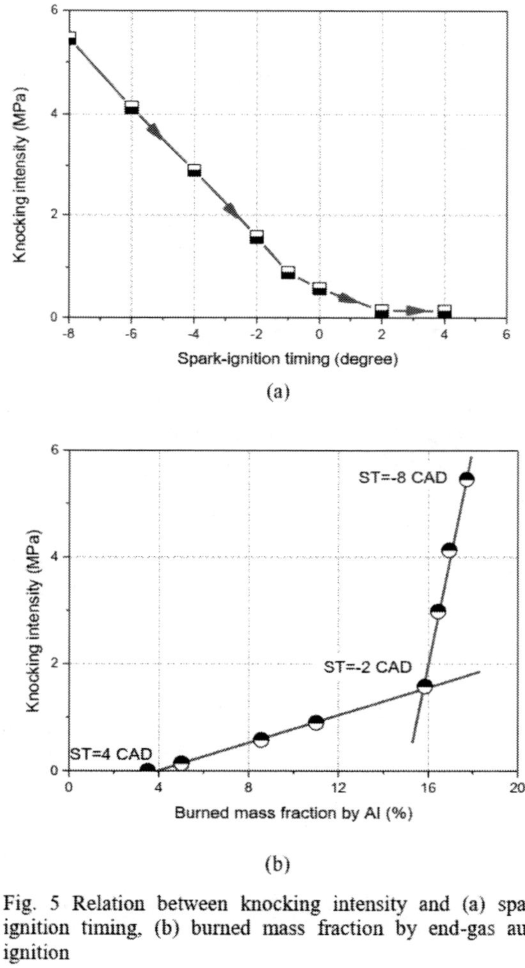

Fig. 5 Relation between knocking intensity and (a) spark-ignition timing, (b) burned mass fraction by end-gas auto-ignition

A linear correlation has always been utilized by engine manufactures to correlate knocking intensity with the burned mass fraction (BMF) by end-gas auto-ignition [1,34]. To validate this point of view, Fig. 5 shows the variation of knocking intensity as a function of ST and the BMF. Noted that the mixture in cylinder is homogeneous, the burned volume fraction has been used to represent the BMF by end-gas auto-ignition. Fig. 5(a) shows that as the advance of ST, there are three levels in the variation of knocking intensity. The ST later than ST=2 CAD corresponds to the normal combustion without pressure oscillations. For the ST between ST=-1 and 2 CAD, knocking intensity is proportional to the ST with a medium slope, representing the regime of moderate knocking combustion. When the ST is advanced earlier than ST=-1 CAD, knocking intensity increases proportionally as ST with a much higher slope, corresponding to heavy knocking combustion. It is observed from Fig. 5(b) that the linear correlation is confirmed by current LES, with a knocking intensity proportional to the BMF by end-gas auto-ignition. Meanwhile, when the BMF exceeds approximate 15%, there is a faster increase in knock intensity, suggesting a second regime for knocking intensity, which appears approximately for the ST before the TDC.

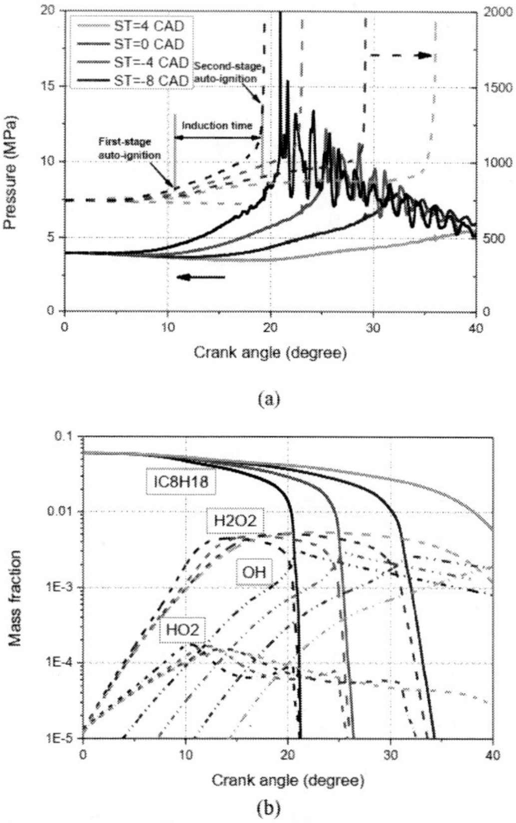

Fig. 6 Evolutions of (a) in-cylinder thermodynamic state and (b) corresponding species mass fraction of the point with first auto-ignition at different spark-ignition timings.

In order to further demonstrate the correlation between knocking combustion and end-gas auto-ignition, evolutions of in-cylinder thermodynamic state and species mass fraction from the point with first auto-ignition at different STs are shown in Fig. 6. It is observed that before the occurrence of knocking combustion, end-gas mixture experiences a series of low-temperature combustion stages. Taking ST=-8 CAD for example, when the piston moves to 5 CA, the temperature and pressure and begin to deviate the compression-expansion traces, characterizing the emergence of low-temperature combustion, as shown in Fig. 6(a). When it is approximate 12 CA, the first auto-ignition event takes place, followed by an induction time (also termed the second-stage ignition delay). This can be identified by the exponential decreases in IC8H18 and increases in OH, CH2O and H2O2 from Fig. 6 (b). Although it is relatively slow for the low-temperature combustion, the thermodynamic state gradually is increased in this duration. When the piston moves to 18 CA, end-gas main auto-ignition occurs with sharp increases in temperature and decreases in IC8H18 and H2O2, but the in-cylinder pressure does not behave oscillating behavior immediately. This indicates that strong pressure waves or shock waves may not directly initiated by the first hot-spots auto-ignition during knocking combustion. Instead, the strength of the resulting pressure pulse which determines knock intensity may depend on the subsequent severe auto-ignition events or the interactions of the auto-ignition front with the ignition delay gradient which determines the evolution of the pressure wave, as pointed out in [19]. Meanwhile, it is observed that the later spark-ignition timing, the longer time low-temperature combustion experiences before final main auto-ignition, and thereby low knocking intensity. Therefore, knocking combustion not only depends on the premixed flame propagation, but on the progress of low-temperature combustion and the scenarios of auto-ignition initiation and development.

3.3 LES for End-gas Auto-ignition

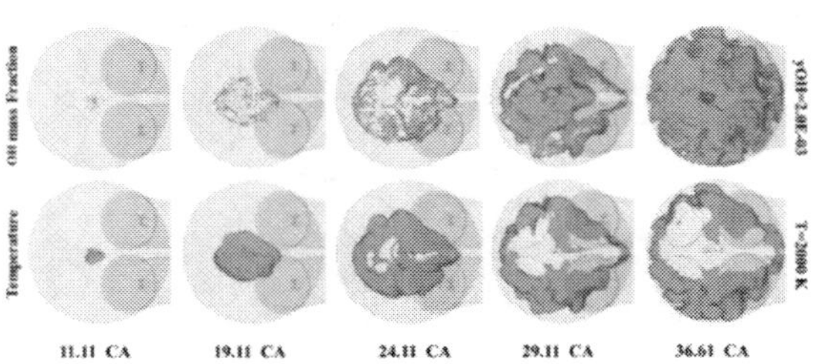

Fig. 7 Isosurfaces of OH mass fraction and temperature during premixed flame propagation at ST= 4 CAD

Before discussing different auto-ignition scenarios during knocking combustion, it is necessary to show the main premixed flame initiation and propagation under normal combustion conditions. Fig. 7 shows the isosurfaces of OH mass fraction of y_{OH}=2.3E-3 and temperature of T=2000 K at ST= 4 CAD. It is observed that during the formation of steady flame kernel, main premixed flame front develops in a smoothly spherical shape, characterizing a laminar flame behavior. As the combustion is progressed, the smooth flame front begins to be stretched and wrinkled by the disturbances of multi-scale turbulent flow in cylinder. Eventually, the main flame front reaches end-wall region and consumes all the remaining mixture without auto-ignition occurrence. Obviously, such wrinkled flame feature cannot be captured by previous RANS work [36].

To illustrate the different auto-ignition scenarios at end-gas region during knocking combustion, Fig. 8 presents the evolutions of flame propagation and end-gas auto-ignition at ST=-4 CAD, which are represented by the isosurfaces of T=2000 K and OH mass fraction of y_{OH}=0.002 as well as the contours of H2O2 mass fraction. It is observed that before the main flame arrival, there is an auto-ignition center initiated from single hot-spot at 23.11 CA at the location adjacent to point P8. Consequently, the auto-ignition front produced rapidly propagates outwards and consumes all surrounding mixture within approximate 2 CAD, corresponding to a flame speed of dozens of meters per second. Subsequently, an auto-ignition event induced by multiple hot-spots occurs at 25.11 CA in the location adjacent to point P2. Due to the interactions between multiple hot-spots, the auto-ignition reaction front consumes surrounding mixture with much shorter time (within about 0.5 CAD). For the contours of H2O2 mass fraction, it shows that before main flame reaching end-wall, there are obvious accumulations in H2O2 mass fraction, suggesting the progress of low-temperature chemical reactions experienced by end-gas mixture. When the main auto-ignition is initiated by local hot-spots, there is a fast decrease in H2O2 mass fraction around auto-ignition center. These auto-ignition scenarios agree well with the variations of temperature and species mass fraction discussed in Part 3.2.

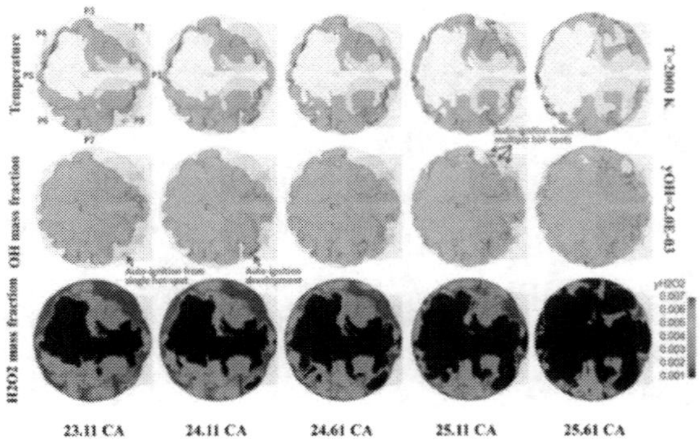

Fig. 8 Evolutions of flame propagation and end-gas auto-ignition represented by the variation of temperature and species mass fraction at ST=-4 CAD

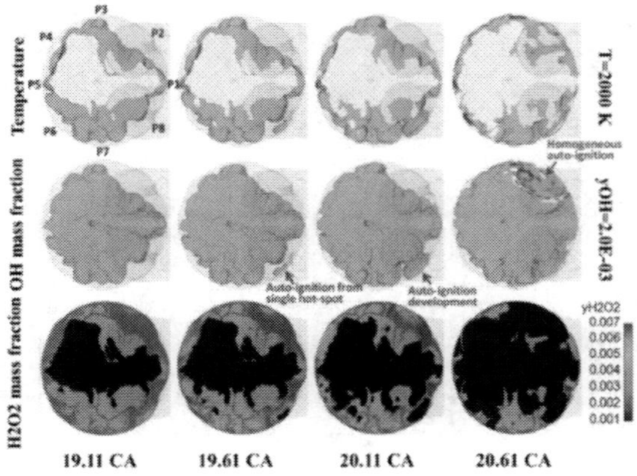

Fig. 9 Evolutions of flame propagation and end-gas auto-ignition represented by the variation of temperature and species mass fraction at ST=-8 CAD

Fig. 9 further presents the results of ST=-8 CAD case, which shows that similar to ST=-4 CAD case, an auto-ignition is induced by single hot-spot at the location around point P8 when the piston moves to 19.11 CA. Consequently, the auto-ignition front quickly develops outwards and consumes surrounding mixture with approximate 1.0 CAD. The propagation velocity for this auto-ignition reaction front is much higher, beyond one hundred meters per second. Furthermore, when the piston moves to 20.61 CA, there is a homogeneous auto-ignition occurring at the location around point P2. This is obviously different from the auto-ignition scenarios induced by multiple hot-spots at ST=-4 CAD case. This homogeneous auto-ignition consumes surrounding mixture with a behavior of quasi constant-volume combustion and releases amounts of chemical energy within microseconds, leading to sharp increases in local pressure. It is observed from Fig. 10 that the local over-pressure induced by the homogeneous auto-ignition reaches a pressure amplitude beyond 20 MPa, and the propagation velocity of auto-ignition front reaches around 700 m/s. However, this auto-ignition reaction front does not develop into detonation due to insufficient mixture remained, low flame speed and pressure peak, consisting with the conclusion of knocking intensity variations as the ST discussed before.

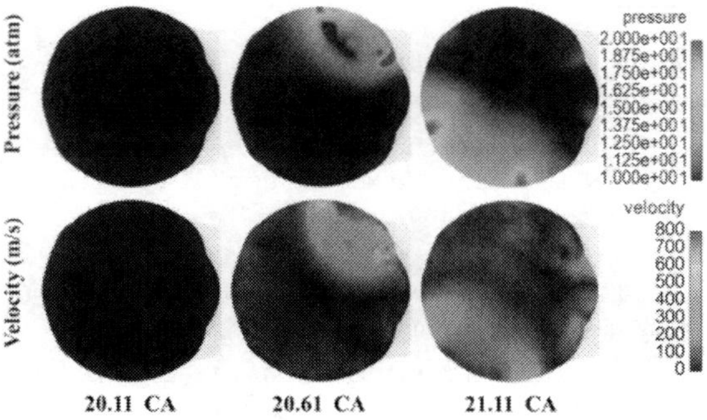

Fig. 10 Evolutions of pressure and velocity field at different crank angles at ST=-8 CAD

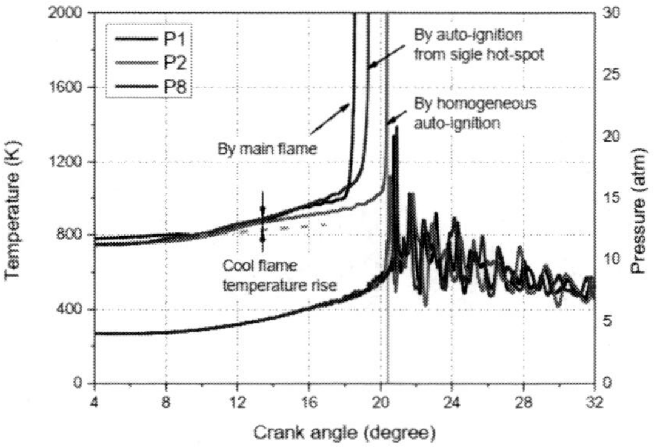

Fig. 11 (a) In-cylinder temperature and (b) pressure traces at different locations for ST=-8 CAD case

Although different auto-ignition scenarios have been discussed, the origin of strong pressure waves during knocking combustion needs to be further clarified. Fig. 11 shows the temperature and pressure curves from three different locations (P1, P2 and P8). Here point P1 denotes the event by main flame combustion, while P2 and P8 represent the one by end-gas auto-ignition. It is observed that before the completion of combustion, the end-gas mixture for all cases experiences low-temperature chemical reactions, characterized by obvious cool flame temperature rise in temperature traces, further validating the conclusion obtained from Fig. 6. For point P1 case, main flame consumes end-gas mixture before auto-ignition occurrence, so no local pressure wave is generated. For point P8 case, there is an auto-ignition induced by single hot-spot, however, this auto-ignition event does not result in local over-pressure and thus fails to initiate strong pressure wave. When the homogenous auto-ignition occurs at point P2, it is observed that obvious pressure oscillations appear in pressure traces, demonstrating the generation of strong pressure waves. Therefore, it is concluded that the early auto-ignition induced by single hot-spot just changes the thermodynamic of surrounding mixture, and knocking combustion is mainly contributed by the subsequent homogeneous auto-ignition or the auto-ignition from multiple hop-spots, as pointed out in knocking experiments [37]. These observations in current work will shed light on the fundamental mechanism for knocking combustion in both experimental and simulation work in further.

4. Conclusions

In current work, knocking combustion and end-gas auto-ignition have been investigated through a LES work coupled with detailed chemistry solver and one-equation eddy viscosity model. 8 LES consecutive cycles with reference spark-ignition timing have been calculated, and knocking intensity has been achieved by adjusting the spark-ignition timing.

First, compared with experimental data, current LES can predict knocking combustion with good performance in terms of combustion phasing and pressure level. As spark-ignition timing is advanced, pressure peak and pressure rise rate are increased, and the advance of combustion phasing is followed by apparent pressure oscillations on in-cylinder pressure traces, suggesting the occurrence of knocking combustion.

Second, knocking intensity is very sensitive to spark-ignition timing at current downsizing level. As the advance of spark-ignition timing, three levels of knocking intensity have been identified, i.e. normal combustion, moderate knocking and heavy knocking. Furthermore, the linear correlation between knocking intensity and the burned mass fraction (BMF) by end-gas auto-ignition has been confirmed, and meanwhile, a second regime for knocking intensity is found when the BMF reaches a threshold, characterized by a greater slope with the BMF. However, the super-knock event induced by pre-ignition is not observed in current LES work.

Third, knocking occurrence is closely relevant to end-gas chemical reactions. During main flame propagation, the increases of end-gas thermodynamic state result in a series of chemical reactions, including the low-temperature combustion characterized by accumulations of typical species and cool flame temperature rise and the final high-temperature auto-ignition initiated by local hot-spots. During knocking combustion, the local auto-ignition can be initiated by single and multiple hot-spots, or occur with a homogeneous behavior. However, local thermodynamic traces show that it is the homogeneous auto-ignition or the one by multiple hot-spots that results in strong pressure wave, rather than the one by single hot-spot.

5 Acknowledgment

This work was supported by Major Research Plan of National Natural Science Foundation of China (91641203), National Natural Science Foundation of China (Grant No. 51476114), China Postdoctoral Science Foundation (2016M590201), Tianjin Natural Science Foundation (17JCZDJC31500), Tianjin Science & Technology Supporting Program (13ZCZDGX04400).

References

[1] Pan JY, Shu GQ, Zhao P, Wei HQ, Chen Z. Interactions of flame propagation, auto-ignition and pressure wave during knocking combustion. Combust Flame 2016; 164: 319–328.

[2] Zhen XD, Wang Y, Xu SQ, Zhu YS, Tao CJ, Xu T, Song MZ. The engine knock analysis – An overview. Appl Energy 2012; 92: 628–636.

[3] Pan JY, Shu GQ, Wei HQ. Interaction of Flame Propagation and Pressure Waves During Knocking Combustion in Spark-Ignition Engines, Combust Sci Technol 2014; 186: 192–209.

[4] Wei HQ, Feng DQ, Pan MZ, Pan JY, Rao XK, Gao DZ.

Experimental investigation on the knocking combustion characteristics of n-butanol gasoline blends in a DISI engine. Appl Energy 2016; 175: 346–355.

[5] Pöschl M, Sattelmayer T. Influence of temperature inhomogeneities on knocking combustion. Combust Flame 2008; 153: 562–573.

[6] Kawahara N, Tomita E, Sakat Yo. Auto-ignited kernels during knocking combustion in a spark-ignition engine. P Combust Inst 2007; 31: 2999–3006.

[7] Qi YL, Wang Z, Wang JX, He X. Effects of thermodynamic conditions on the end gas combustion mode associated with engine knock. Combust Flame 2015; 162: 4119–4128.

[8] Bates L, Bradley D, Paczko G, Peters N. Engine hot spots: Modes of auto-ignition and reaction propagation. Combust Flame 2016; 166: 80–85.

[9] Kawahara N, Tomita E. Visualization of auto-ignition and pressure wave during knocking in a hydrogen spark-ignition engine. Int J Hydrogen Energ 2009; 34: 3156–3163.

[10] Pan J, Sheppard C, Tindall A, Berzins M et al. End Gas Inhomogeneity, Autoignition and Knock. SAE Technical Paper 1998-98-2616; 1998.

[11] Drake MC, Haworth DC. Advanced gasoline engine development using optical diagnostics and numerical modeling. P Combust Inst 2007; 31: 99–124,

[12] Zhou DZ, Yang WM, An H, Li J. Application of CFD-chemical kinetics approach in detecting RCCI engine knocking fuelled with biodiesel/methanol. Appl Energy 2015; 145: 255–264.

[13] Knop V, Michel JB, Colin O. On the use of a tabulation approach to model auto-ignition during flame propagation in SI engines. Appl Energy 2011; 88: 4968–4979.

[14] Linse D, Kleemann A, Hasse C. Probability density function approach coupled with detailed chemical kinetics for the prediction of knock in turbocharged direct injection spark ignition engines. Combust Flame 2014; 161: 997–1014.

[15] Robert A, Richard S, Colin O, Martinez L, Francqueville LD. LES prediction and analysis of knocking combustion in a spark ignition engine. P Combust Inst 2015; 35: 2941–2948.

[16] Robert A, Richard S, Colin O, Poinsot T. LES study of deflagration to detonation mechanisms in a downsized spark ignition engine. Combust Flame 2015; 162: 2788–2807.

[17] Lecocq G, Richard S, Michel JB, Vervisch L. A new LES model coupling flame surface density and tabulated kinetics approaches to investigate knock and pre-ignition in piston engines. P Combust Inst 2011; 33: 3105–3114.

[18] Misdariis A, Vermorel O, Poinsot T. LES of knocking in engines using dual heat transfer and two-step reduced schemes. Combust Flame 2015; 162: 4304–4312.

[19] Bradley D, Kalghatgi GT. Influence of autoignition delay time characteristics of different fuels on pressure waves and knock in reciprocating engines. Combust Flame 2009; 156: 2307–2318.

[20] Terashima H, Koshi M. Mechanisms of strong pressure wave generation in end-gas auto-ignition during knocking combustion. P Combust Flame 2015; 162: 1944–1956.

[21] Wang Z, Qi Y, He X, Wang J, Shuai S, Law CK. Analysis of pre-ignition to super-knock: Hotspot-induced deflagration to detonation. Fuel 2015; 144: 222–227.

[22] Senecal PK, Pomraning E, Richards KJ. Multi-Dimensional Modeling of Direct Injection Diesel Spray Liquid Length and Flame Lift-off Length using CFD and Parallel Detailed Chemistry. SAE Technical Paper 2003-01-1043; 2003.

[23] Richards KJ, Senecal PK, Pomraning E. CONVERGE Users Guide & Reference Manual (Version2.1.0), Convergent Science, Inc., Middleton, WI, 2013.

[24] Turns SR. An Introduction to Combustion, McGraw-Hill, Inc., 1996.

[25] Yoshizawa A, Horiuti K. A Statistically-Derived Subgrid-Scale Kinetic Energy Model for Large Eddy Simulation of Turbulent Flows. J Phys Soc Jpn 1985; 54: 2834–2839.

[26] Menon S, Yeung PK, Kim WV. Effect of Subgrid Models on the Computed Interscale Energy Transfer in Isotropic Turbulence. Comput Fluids 1996; 25: 165–180.

[27] Liu Y, Jia M, Xie M, Pang B. Improvement on a skeletal chemical kinetic model of iso-octane for internal combustion engine by using a practical methodology. Fuel 2013; 103: 884–891.

[28] Raju M. Wang M, Dai M, Piggott W., et al. Acceleration of Detailed Chemical Kinetics Using Multi-zone Modeling for CFD in Internal Combustion Engine Simulations. SAE Technical Paper 2012-01-0135; 2012.

[29] Givler S, Raju M, Pomraning E, Senecal P., et al. Gasoline Combustion Modeling of Direct and Port-Fuel Injected Engines using a Reduced Chemical Mechanism. SAE Technical Paper 2013-01-1098; 2013.

[30] Liu YD, Jia M, Xie MZ, Pang B. Enhancement on a skeletal kinetic model for Primary Reference Fuel oxidation by using a semi decoupling methodology. Energ Fuel 2012; 26: 7069–7083.

[31] Park S, Furukawa T. Validation of Turbulent Combustion and Knocking Simulation in Spark-Ignition Engines Using Reduced Chemical Kinetics. SAE Technical Paper 2015-01-0750; 2015.

[32] Vermorel O, Richard S, Colin O, Angelberger C, Benkenida A, Veynante D. Towards the understanding of cyclic variability in a spark ignited engine using multi-cycle LES. Combust Flame 2009; 156: 1525–1541.

[33] Shu GQ, Pan JY, Wei HQ. Analysis of onset and severity of knock in SI engine based on in-cylinder pressure oscillations. Appl Therm Eng 2013; 51: 1297–1306

[34] Xiaofeng G, Stone R, Hudson C, Bradbury I. The Detection and Quantification of Knock in Spark Ignition Engines. SAE Technical Paper 1993-93-2759; 1993.

[35] Law CK. Combustion physics. Cambridge University Press, 2010.

[36] Pan JY, Shu GQ, Wei HQ. Research on in-cylinder pressure oscillation characteristic during knocking combustion in spark-ignition engine. Fuel 2014; 120: 150–157.

[37] Rudloff J, Zaccardi JM, Richard S, Anderlohr JM. Analysis of pre-ignition in highly charged SI engines: Emphasis on the auto-ignition mode. P Combust Inst 2013; 34: 2959-2967.

11th Asia-Pacific Conference on Combustion,
The University of Sydney, NSW Australia
10th -14th December 2017.

Large eddy simulation of turbulent partially premixed flames with inhomogeneous inlets using the DSMC model

Yun Bai, Kun Luo, Wenjian Bi, Jianren Fan

State Key Laboratory of Clean Energy Utilization, Zhejiang University, Hangzhou, 310027, China

Abstract

A recently developed turbulent combustion model, namely the dynamic second-order moment closure (DSMC) model is used to simulate the Sydney piloted partially premixed flames with inhomogeneous inlets in the framework of LES. Although a simple four-step global reaction mechanism is applied, good agreement is observed for time-averaged radial profiles, scatter plots and conditional profiles for mixture fraction and temperature, implying the capability to reproduce the partially premixed flames with the DSMC model.

1 Introduction

Partially premixed flame modeling is becoming an outstanding challenge in large eddy simulation (LES) of turbulent combustion. Recently, an experiment-scale configuration, where the fuel and air are inhomogeneously mixed at the burner exit, namely Sydney piloted flame series with inhomogeneous inlets, has been extensively studied [1]. In these flames, premixed, non-premixed and partially premixed combustion co-exist along with local extinction and reignition phenomena [1], showing a great challenge for turbulent combustion models. To simulate these flames, a sparse-Lagrangian multiple mapping conditioning model (MMC) [2], a multi-regime flamelet model (MRF) [3] and a revised two mixture fraction flamelet/progress variable ($2\xi - FPV$) model [4], including the *PDF-like* and *flamelet-like* models, have been performed.

In the present paper, a recently developed turbulent combustion model, namely the dynamic second-order moment closure (DSMC) model [5] is used to simulate the Sydney piloted flame series with compositionally inhomogeneous inlets. In the previous study [5], two typical non-premixed flames, Sandia Flame D and Sydney bluff-body stabilized swirling non-premixed flame, have been simulated with simple one-step and two-step chemical reaction mechanisms. The LES results showed good agreement with experimental data, demonstrating the capability of the DSMC model to reproduce non-premixed flames. In the present study, the applicability of the DSMC model for partially premixed flame is further checked.

2 Experimental Conditions

The target flames were performed using a modified Sydney piloted jet burner which consists of two concentric tubes surrounded by a pilot stream which has an inside diameter of 18mm. The inner diameters of the concentric tubes are 4 mm and 7.5 mm respectively. Cold coflow air is provided through a wind tunnel at 15 m/s. Fuel is injected through the inner tube and air flows the outer annulus. The inner tube can be recessed within the outer tubes with a recession distance L_r to vary the degree of premixing between the two streams. When $L_r = 0$ mm, the fuel/air mixture is fully non-premixed, while completely homogeneous mixing occurs when $L_r = 300$ mm. The experimental studies observed that the recession length or rather the mixture inhomogeneity at the burner exit has significant influence on flame stability.

Corresponding author. Fax: +86-571-87953687
E-mail address: zjulk@zju.edu.cn

In the present work, two flame cases, FJ200-5GP-Lr300-59 (Flame H) and FJ200-5GP-Lr75-80 (Flame I), with the homogeneous inlet and inhomogeneous inlet respectively, are selected for simulations. The pilot is a 5-gas pilot (5GP) burning mixture of C_2H_2, H_2, CO_2, N_2 and air, which matches the adiabatic equilibrium temperature and composition of stoichiometric CH_4/air mixture. In experimental study, both flames are at 70% of the blow-off velocity at the corresponding value of L_r.

3 Numerical Approach

3.1 Dynamic Second-order Moment Closure Model

The dynamic second-order moment closure model for turbulent combustion has been recently developed [5]. The main idea is that the filtered chemical reaction rate can be directly calculated in the form of Arrhenius law using the reacting scalars instead of mixture fraction, avoiding the separate treatments for premixed/non-premixed flame regimes. The sub-grid unresolved reaction rate is modeled with a second-order moment closure model. A brief description is presented below.

Considering an arbitrary reaction j in a chemical system with multi-step reactions, the forward or reverse progress rate of this reaction can be expressed as (here we take a two-reactant kinetics for simplification):

$$\vartheta_j = A_j \exp(-E_j / RT)\left(\frac{\rho Y_1}{W_1}\right)^m \left(\frac{\rho Y_2}{W_2}\right)^n , \quad (1)$$

where A_j and E_j denote the pre-exponential factor and the activation energy in reaction j respectively. R is the universal gas constant. The subscripts *1* and *2* represent the reactants and superscripts m and n are the concentration exponents. W is molecular weight. The filtered reaction progress rate can be expressed as:

$$\overline{\vartheta_j} = \overline{A_j \exp(-E_j / RT)\left(\frac{\rho Y_1}{W_1}\right)^m \left(\frac{\rho Y_2}{W_2}\right)^n} = \overline{B_j \exp(-E_j / RT) R_1^m R_2^n} , \quad (2)$$

where $R_1 = \rho Y_1$, $R_2 = \rho Y_2$, $B_j = A_j W_1^{-m} W_2^{-n}$. To avoid the complexity from the Taylor series expansion of temperature function, the whole temperature function term is treated as a single variable K, namely $K = exp(-E_j/RT)$. By assuming smaller species variances and ignoring the third-order correlation term, the filtered reaction rate becomes:

$$\overline{\vartheta}_j = B_j\left(\overline{KR_1}^m\overline{R_2}^n + mn\overline{KR_1}^{m-1}\overline{R_2}^{n-1}\overline{R_1'R_2'} + m\overline{R_1}^{m-1}\overline{R_2}^n\overline{K'R_1'} + n\overline{R_1}^m\overline{R_2}^{n-1}\overline{K'R_2'}\right). \quad (3)$$

For these unresolved second-order variance correlation terms, $\overline{R_1'R_2'}$, $\overline{K'R_1'}$ and $\overline{K'R_2'}$, their transport equations can be derived following the similar deriving method of the transport equations of Reynolds stress, and can be concluded using the dimensional terms C_1, C_2 and C_3 in the form as [5]:

$$\frac{\partial\overline{\phi'\varphi'}}{\partial t} + \frac{\partial\left(\overline{u}_j\overline{\phi'\varphi'}\right)}{\partial x_j} = C_1\frac{\partial}{\partial x_j}\frac{\partial\overline{\phi'\varphi'}}{\partial x_j} + C_2\frac{\partial\overline{\phi}}{\partial x_j}\frac{\partial\overline{\varphi}}{\partial x_j} + C_3\overline{\phi'\varphi'}, \quad (4)$$

which means that the change of the correlation term can be determined by its convection, diffusion, dissipation and chemical production terms. In theory, these unresolved terms can be calculated by solving their transport equations. For simplification, however, we assume a steady state and balance of the convection and diffusion terms, i.e. the chemical production term is equilibrium with the local dissipation term. That is to say, in resolved space and time scales, the production of unresolved second-order variance correlation term by chemical reaction is mainly consumed by dissipation, and this balance assumption has scale similarity. Then a simple algebraic closure method is obtained as:

$$\overline{\phi'\varphi'} = C_{\phi\varphi}\Delta^2\frac{\partial\overline{\phi}}{\partial x_j}\frac{\partial\overline{\varphi}}{\partial x_j}. \quad (5)$$

The coefficient can be determined by a dynamic procedure.

For the filtered highly non-linearity exponential function, we apply a fluctuation averaged closure method to model $\overline{K}$:

$$\overline{K} = \frac{1}{2}\left\{\exp\left[-\frac{E_j}{R\left(\tilde{T} + \widetilde{T''T''}^{1/2}\right)}\right] + \exp\left[-\frac{E_j}{R\left(\tilde{T} - \widetilde{T''T''}^{1/2}\right)}\right]\right\}, \quad (6)$$

where the temperature variance $\widetilde{T''T''}$ is modeled using the sub-grid variance model with the dynamic procedure to determine the coefficient [6]:

$$\overline{\rho}\widetilde{T''T''} = C_{TT}\overline{\rho}\Delta^2\left|\nabla\tilde{T}\right|^2. \quad (7)$$

These closure methods for the second-order correlation terms and $\overline{K}$ have been validated a priori and a posteriori in the previous study [5].

3.2 Numerical Implementation

The DSMC model has been implemented into a LES solver where the variable-density, low-Mach number transport equations for all scalars are solved using a staggered space-time, conservative discretization scheme. The dynamic Smagorinsky model is applied for turbulence model. A four-step methane/air global reaction mechanism, rewritten by Andersen et al. [7] based on Jones and Lindstedt four-step mechanism, is used. This mechanism can perform well for both fuel-lean and moderately fuel-rich mixtures [7].

The homogeneous/inhomogeneous inflow boundary conditions for velocity and scalars are obtained from the separate non-reacting simulations of the upstream nozzle performed by Perry et al. [4], which show good agreement with the experimental boundary conditions.

The simulation domain is taken as a cylindrical domain with an axial length of 40D and a radius of 4D, and discretized by 188 × 67 × 64 grid points in the axial, radial and azimuthal directions, with the total grid number of 806 144. The statistical data are obtained through sampling over about five more flow-through periods.

4 Numerical Results and discussions

4.1 Temperature Scatters and Conditional Profiles

First of all, the combustion modes of the two flames are checked by presenting the scatter plots of temperature versus mixture fraction in **Figures 1-2**. The mixture fraction ξ is defined based on element mass fractions:

$$\xi = \frac{2\left(Y_C - Y_{C,2}\right)/W_C + \left(Y_H - Y_{H,2}\right)/2W_H + \left(Y_O - Y_{O,2}\right)/W_O}{2\left(Y_{C,1} - Y_{C,2}\right)/W_C + \left(Y_{H,1} - Y_{H,2}\right)/2W_H + \left(Y_{O,1} - Y_{O,2}\right)/W_O}. \quad (8)$$

Pure CH_4 is taken as fuel and the stoichiometric mixture fraction is $\xi_{st} = 0.055$.

For Flame H as shown in **Fig. 1**, the dominant non-premixed flame structure is obviously evident and the DSMC model performs very well. At the location of $x/D = 15$, the local extinction is detected by LES. Further downstream at $x/D = 30$, the local extinction appears to be over-predicted.

The scatter plots for Flame I shown in **Fig. 2** reveal a very different flame structure, especially near the nozzle. At the location of $x/D = 1$, the points on the rich side of the flame remain around 300K, corresponding to the cold mixing within the fuel/air jet. Close to the stoichiometric mixture fraction, a steep rise in temperature from cold unburnt to hot burnt conditions, indicating a stratified-premixed combustion mode, is well-predicted. Similar to the measurement, the transitions from the premixed to the non-premixed mode at $x/D = 5$ are also predicted, though the transition is not that quick along with a delayed ignition. At $x/D = 15$, the flame has completed the transition to the non-premixed mode. The re-ignition can also be predicted, while the local extinction still is over-predicted at $x/D = 30$. According to these scatter plots for both cases, the current simulations are all in good agreement with the experiments by predicting the transitions from the premixed to non-premixed mode albeit with a little wider shapes downstream, indicating the capability to simulate the partially premixed flame with the DSMC model.

Figures 3-4 present the conditional mean and rms temperature as function of mixture fraction for two flames. In the present study, the results calculated from the MMC model [2], MRF model [3] and $2\xi - \mathrm{FPV}$ model [4], are included in the comparisons once available. It should be noted that the detailed reaction mechanisms are used in these models. In general, the conditional mean and rms values computed by DSMC model show a good agreement with the experiment, in spite of some discrepancies. In Flame H, the mean temperature near the stoichiometric mixture fraction is always under-predicted. For Flame I, it is found that the conditional temperature at the fuel-lean side has an excellent agreement with experiment. At the location of $x/D = 1$, the dip on the lead side, which has not been predicted by MMC model or MRC model, is reproduced very well by the present DSMC model, showing the capability to simulation the stratified-premixed mode around stoichiometric conditions. Due to the absence of the radicals' prediction using the current global mechanism, the ignition event is not captured accurately under the premixed condition and the ignition on the rich side is postponed which can result in the under-prediction of the conditional mean temperature on the rich side at $x/D = 5, 10$. After the successful ignition, the DSMC model, even though with a simple chemistry mechanism, can have a good prediction downstream.

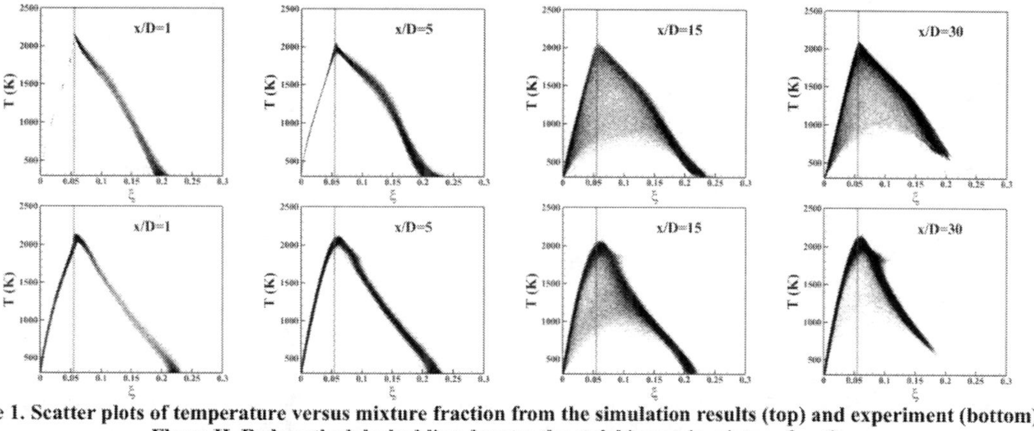

Figure 1. Scatter plots of temperature versus mixture fraction from the simulation results (top) and experiment (bottom) for Flame H. Red vertical dashed line denotes the stoichiometric mixture fraction.

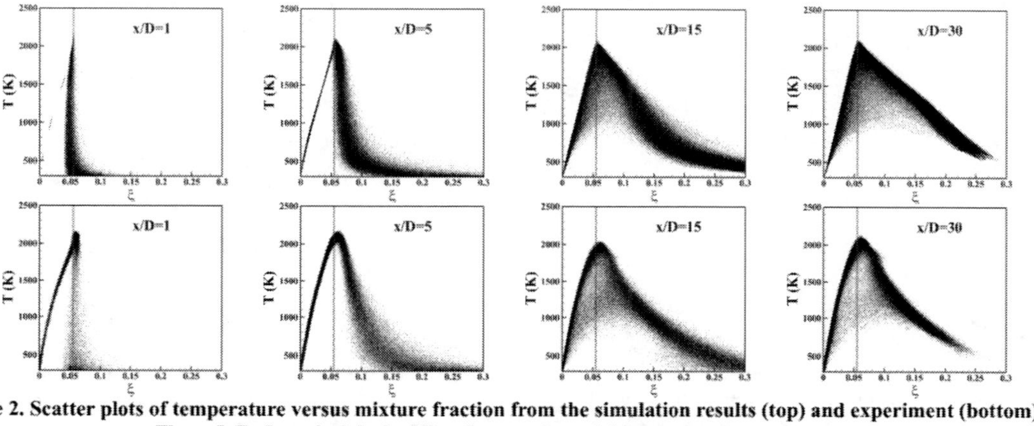

Figure 2. Scatter plots of temperature versus mixture fraction from the simulation results (top) and experiment (bottom) for Flame I. Red vertical dashed line denotes the stoichiometric mixture fraction.

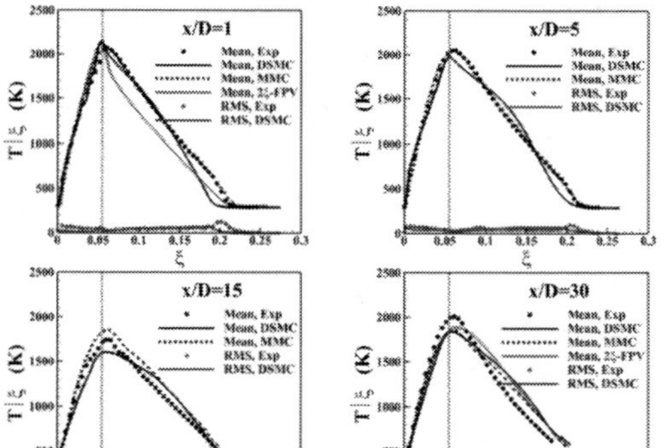

Figure 3. Comparison of experimental and simulated conditional mean and rms of temperature for Flame H.

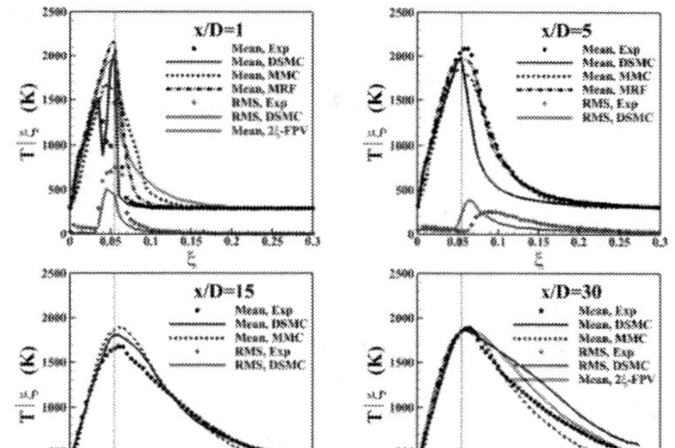

Figure 4. Comparison of experimental and simulated conditional mean and rms of temperature for Flame I.

4.2 Radial Profiles

Figures 5-6 display radial profiles of the mean and rms values of mixture fraction and temperature for Flame H. The simulation results by other models are also imposed. It is obvious that both mean and rms values are generally in good agreement with the experimental data. Due to the same inflow boundary condition that has an overestimated mixture fraction profile near the centerline, the mixture fractions obtained by the DSMC model and $2\xi - \text{FPV}$ model are both over-predicted near the centerline upstream. Despite the simple reaction mechanism, the mixture fractions predicted by DSMC model are very close to those from the $2\xi - \text{FPV}$ model with a detailed mechanism. Compared with the results from the MMC model, the current calculated temperature profiles are similar. At $x/D = 30$, both the DSMC and MMC models overestimate the mixture fraction at the centerline and under-predict the temperature in the outer region. The predicted rms values show similar trend as those of the mean values. The temperature profiles show that the calculated flames narrow down compared to the experimental flame. These comparisons indicate that the present DSMC combustion model can simulate the homogeneous case as well as the MMC model, even employing a simple four-step global chemical mechanism.

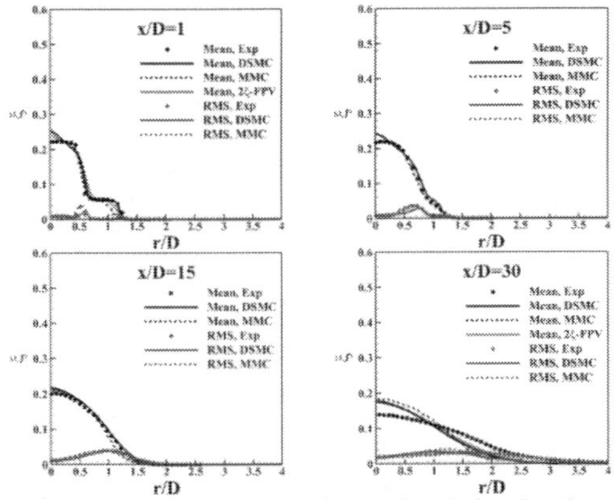

Figure 5. Comparison of experimental and simulated mean and rms results of mixture fraction for Flame H.

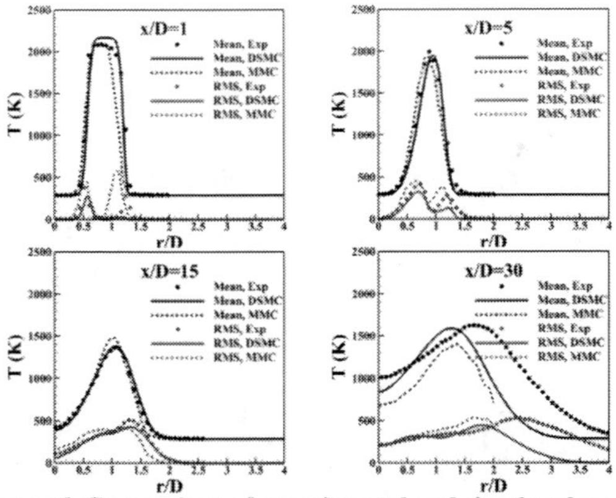

Figure 6. Comparison of experimental and simulated mean and rms results of temperature for Flame H.

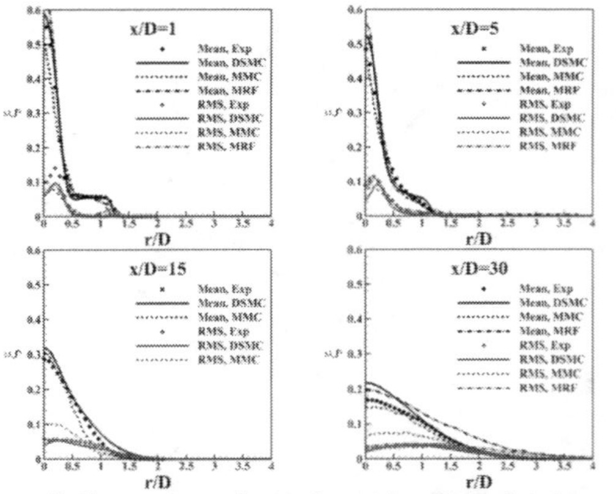

Figure 7. Comparison of experimental and simulated mean and rms results of mixture fraction for Flame I.

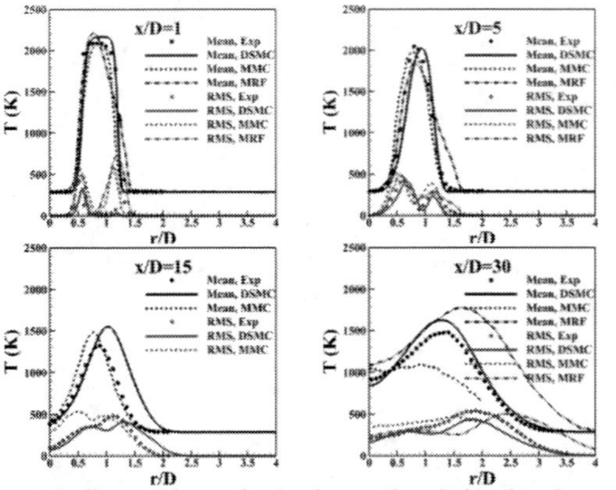

Figure 8. Comparison of experimental and simulated mean and rms results of temperature for Flame I.

Radial profiles for Flame I are presented in **Figures 7-8**. Compared to Flame H, Flame I has a similar exit profile of mean temperature but different mixture fraction profile at $x/D = 1$. For Flame H, the mixture fraction is nearly homogeneous, while Flame I has near-stoichiometric reactants at the jet edge ($r/D = 0.5$) and very rich mixture near the jet centerline with a steep gradient. Because of the ignition delay on the rich side, the calculated mean mixture fraction is higher near the centerline, and is still over-predicted at $x/D = 30$. Meanwhile the radial location of the temperature peak is shifted outwards, indicating that the predicted flame is broadened, and the predicted temperature outside of the radius of $r/D = 1$ is overestimated at the location of $x/D = 15$. At $x/D = 30$, a good agreement is obtained with the DSMC model while larger discrepancies can be found in the rest two models. In addition, excellent agreement is observed for rms values at all measurement locations, better than the rest models. These radial profiles show that the present DSMC model with four-step reaction mechanism can reasonably predict the partially premixed flame well and even better at some locations than the other two models.

5 Conclusions

In the present work, the Sydney turbulent partially premixed flames with both homogeneous and inhomogeneous fuel/air inlets are simulated using the dynamic second-order moment turbulent combustion model in the LES framework. The simulation results are compared with the experimental data and the results calculated by other combustion models. Although the ignition event is not captured accurately due to the absence of radicals' predicting with the current simple four-step global reaction scheme, the calculated results by the DSMC model are in reasonably good agreement with the measurements generally for both flames and comparable or even better at some locations than other models with detailed chemical mechanisms, indicating the capability of DSMC model to reproduce the partially premixed flames.

6 Acknowledgment

This work was supported by the National Nature Science Foundation of China (NO. 51390493).

References

[1] R.S. Barlow, S. Meares, G. Magnotti, H. Cutcher, A.R. Masri. Combust. Flame 162 (2015) 3516-3540.

[2] S. Galindo, F. Salehi, M.J. Cleary, A.R. Masri. Proc. Combust. Inst. 36 (2017) 1759-1766.

[3] K. Kleinheinz, T. Kubis, P. Trisjono, M. Bode, H. Pitsch. Proc. Combust. Inst. 36 (2017) 1747-1757.

[4] B.A. Perry, M.E. Mueller, A.R. Masri. Proc. Combust. Inst. 36 (2017) 1767-1775.

[5] K. Luo, J.S. Yang, Y. Bai, J.R. Fan. Fuel 187 (2017) 457-467.

[6] C.D. Pierce. *Progress-variable approach for large-eddy simulation of turbulent combustion*. PhD thesis, Stanford University, 2001.

[7] J. Andersen, C.L. Rasmussen, T. Giselsson, P. Glarborg. Energy Fuels 23 (2009) 1379-1389.

11th Asia-Pacific Conference on Combustion,
The University of Sydney, NSW Australia
10th-14th December 2017.

Large Eddy Simulation of Turbulent Flame with Inhomogeneous Inlets

T.Z. Hou[1], P. Wang[2], Q. Yu[1], C.J. Wang[1]

[1] School of Energy and Power Engineering, UJS
Xuefu Road 301, Zhenjiang, Jiangsu, 212013, China
[2]Institute for Energy Research, UJS
Xuefu Road 301, Zhenjiang, Jiangsu, 212013, China

Abstract

Partially premixed combustion is often found in practical combustion devices, however, large eddy simulations (LES) of it is still a challenge. In the present work, a REDIM-PFDF sub-grid combustion model, based on the reaction-diffusion manifold (REDIM) and presumed filtered density function (PFDF), is used to calculate the partially premixed flame FJ200-5GP-Lr75-57 with inhomogeneous inlets measured jointly by the University of Sydney and Sandia Laboratory. In order to keep consistent with the experimental state, a periodic inflow condition is used to obtain fully developed turbulence inside the jet pipes. The flame and flow field structure of the studied state are analyzed, and LES results are compared with the experimental data as well. The comparison shows that the LES predicted temperature, CO_2 and CO mass fractions are in good agreement with the experimental data. The overall good agreement demonstrates the capability of the REDIM-PFDF combustion model for simulating partially premixed flame.

1 Introduction

It is difficult for fuel and air to fully mixed in practical combustion devices, and there will be partially premixed phenomenon basically in them. But the academic researches over the past 10 years are basically focused on the study of complete premixed flame and diffusion flame, researches for partially premixed flame need to be further in-depth. The historical classification of flame into premixed and non-premixed is gradually becoming less relevant in practice with the advent of "mixed mode" combustors. Practical engineering applications include low temperature diesels (LTD), direct injection stratified charge engines (DISC) and even homogenous charge compression ignition engines (HCCI) [1], therefore, there is an urgent need to develop deep research for partially premixed flame. Current understanding as well as modeling capabilities of turbulent non-premixed, premixed, and homogeneous partially premixed flame have improved significantly, particularly with respect to turbulence-chemistry interactions and associated transient processes such as extinction and re-ignition [2,3]. However, such capabilities are not yet sufficient to meet the demands for accurately predicting partially premixed and stratified flames that involve a high degree of inhomogeneous in the mixing fraction.

To establish turbulence models, a variety of problems need to be solved: how to couple the chemical mechanism, and considering the interaction between turbulence and chemical reactions. Modeling approaches that are now capable of accounting adequately for finite-rate chemistry effects in turbulent premixed and non-premixed flames include laminar flamelets [4], conditional moment closure [5], and probability density function methods [6], more details can be found in the review paper [3]. The immediate outstanding challenge of numerical simulation is to improve the capabilities of computing finite-rate chemistry effects of the flame with a range of flow complexities and intense levels of stratification.

In the present study, REDIM [7] method is coupled with PFDF model to construct a REDIM-PFDF sub-grid combustion model. This model has a good performance in the large eddy simulation of the lean premixed flame [8] and the turbulent stratified flame [9]. The coupling model is used in this study to simulate the partially premixed reference flame of University of Sydney. The behavior of the flame is studied, at the same time, the feasibility of the model is verified by comparing simulation results with the experimental data.

Corresponding author. Fax: +86 511 88799500
E-mail address: pingwang@ujs.edu.cn (P. Wang)

2 Numerical Method

In the present study, REDIM-PFDF sub-grid model is implemented into the in-house Finite-Volume code LESOCC2C [10]. The above code solves the Low-Mach number version of the compressible Navier-Stokes equations on body-fitted curvilinear block-structured grids using second-order central schemes in space and a three-step Runge-Kutta method of second order in time, and is powerful in dealing with turbulence combustion.

2.1 Reaction-Diffusion Manifold (REDIM)

N. Peters et al. proposed a simplified method that the quasi-steady state approximation is used for intermediate product in chemical reaction, with partial equilibrium assumptions for some rapid chemical reactions [11]. Maas and Pope suggested an alternative method called intrinsic low-dimensional manifolds (ILDM) [7] based on the above method. This tabulation technique select several progress scalars (α_1, $\alpha_2 \ldots \alpha_n$) as the independent variable and the other variable such as species, rate of reaction, temperature and pressure (Φ) can be regarded as a function of the selected scalar that can interpolated in the look-up table.

$$\Phi = f(\alpha_1,\ \alpha_2 \cdots \alpha_n) \tag{1}$$

Bykov and Maas developed a new low-dimensional manifold (REDIM) technology based on ILDM because of the deficiencies of it. REDIM abandons the hypothesis introduced in the ILDM method, and takes into account the coupling process of chemical reaction and molecular transport in generating low-dimensional manifold, which makes it more accurate and applicable.

In this paper, REDIM is used to generate a two-dimensional manifold. As shown in Fig. 1, the two independent variables of the REDIM table are the mass fraction of CO_2 and N_2, respectively, and any thermodynamics scalar can be obtained in this two-dimensional table by interpolation. Figure. 1 shows the two-dimensional REDIM lookup table with the production rate of CO_2 as the dependent variable.

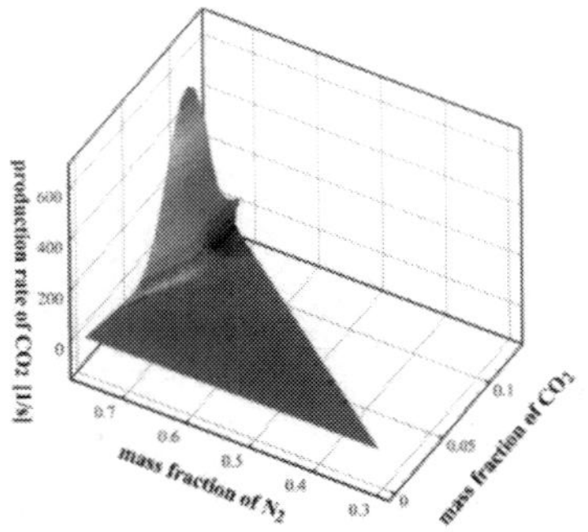

Figure 1: The used REDIM chemistry table for production rate of CO_2 as a function of mass fractions of CO_2 and N_2.

2.2 REDIM-PFDF Combustion Model

The low-dimensional manifold obtained by REDIM uses the simplified method to represent the detailed data of the laminar flame. Therefore, the above REDIM table cannot describe the scalar fluctuation in the turbulent flame and cannot be directly invoked in the LES calculation of the turbulent flame. Appropriate methods should be used to combines the above low-dimensional manifold with turbulence.

In the present study, the PFDF method is coupled with the REDIM manifold to construct a REDIM-PFDF sub-grid combustion model. The PFDF method is used to describe the distribution of the mass fraction of CO_2 and N_2 in the turbulence field of the sub-grid control volume, and coupling low-dimensional manifold with turbulence thereby. A chemical thermodynamics lookup tables are calculated and plotted by LESOCC2C, all required thermodynamic variables can be interpolated in this lookup table.

3 Calculation Settings

In order to investigate the structure, behavior and other thermodynamic information of partially premixed flames and provide a benchmark flame for the development of highly applicable combustion models, University of Sydney and Sandia Laboratory did a series of benchmark experiments based on different initial conditions and boundary conditions with the same partially premixed burner [12,13].

3.1 Inhomogeneous Jet Flame by University of Sydney

The burner consists of two concentric tubes (inner and main outer tube) surrounded by a pilot with an inside diameter of 18mm and wall thickness 0.2mm. The inner tube, which can be retracted upstream of the burner exit plane, has an inside diameter of d=4.0mm and wall thickness of 0.25mm. The main outer tube has an inside diameter of D=7.5mm and wall thickness of 0.25mm. The burner assembly is centered in a wind tunnel which had a square cross section of 25×25cm supplying a co-flowing air stream at a fixed velocity, U_c=15m/s. The burner was operated in two configurations, one with fuel injected from the inner tube (FJ) and air from the main outer tube, and the other with fuel injected from the main outer tube (FA) and air from the inner tube.

The flames investigated by University of Sydney had a volumetric air to fuel ratio of V_A/V_F=2.0. The fuel used in experiments was compressed natural gas, which consisted of 88% CH_4, 7.8% C_2H_4, 1.9% CO_2, and 1.2% N_2 by volume with the remaining balance of 1.1% consisting of H_2, H_2O and other hydrocarbons. The pilot used in experiments was a mixture of C_2H_2, H_2, CO_2, N_2 and air in proportions to match the composition of stoichiometric CH_4/air. The flame of this state do not exist flame-out phenomenon.

3.2 Grid

The mesh used is a body-fitted multi-block structured grid in which the whole domain is divided into 349 blocks and contains 5.25 million cells, it is generated with O-type grid method using ICEM-CFD software. The radial size of grid in the inflow pipe and its downstream is $\Delta r \approx 0.11$mm, while $\Delta x \approx 0.4$mm along the axial direction. The structure of the whole burner and local mesh in the inflow pipe are shown in Fig. 2a and 2b respectively.

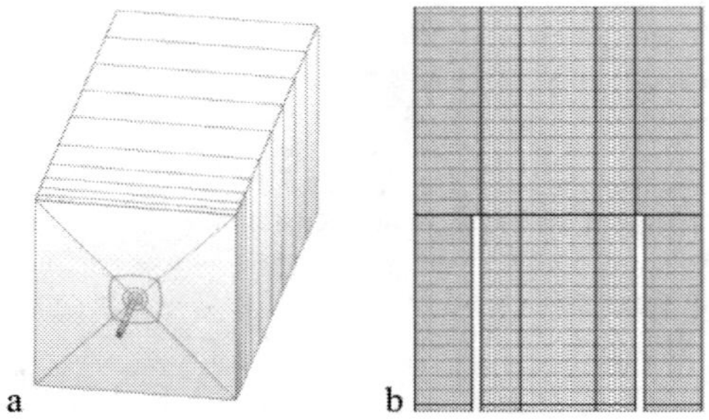

Figure 2: a: The structure of the whole burner; b: Local mesh in the inflow pipe.

This paper select FJ200-5GP-Lr75-57 reference flame to carry out LES study, whose main parameters is shown as follows:

FJ: fuel injected from the inner tube and air from the main outer tube;

5GP: using 5 gases (C_2H_2, H_2, CO_2, N_2 and air) in the pilot substitute for the stoichiometric CH_4/air;

L_r: recession distance of inner tube with respect to jet exit plane is 75mm;

U_J: bulk velocity at jet exit plane is 57m/s;

U_{bo}: bulk velocity at jet exit when blow off occurs is 114.3m/s;

U_A: air bulk velocity is 59.5m/s;

U_F: fuel bulk velocity is 67.0m/s.

4 Results and Discussion

4.1 Flame Structure

An important prerequisite for predict turbulent flame is predict the turbulence field accurately. In order to obtain the fully developed turbulence field corresponding to the experiment, the periodic boundary condition is used in the upstream of the inner and the main outer tube. A fine turbulent flow field in the inflow pipe (inner and main outer tube) can be clearly observed in Fig. 3a, which shows the instantaneous distribution of the axial velocity of the flow field.

In the present study, pure fuel and air are injected from the inner and main outer tube respectively, and mixed downstream of the main outer tube. Figure. 3b shows the distribution of φ (equivalent ratio) in the combustion chamber. It can be found that air at the end of the main outer tube is not fully mixed with fuel and continuously mixed after entering the combustion chamber due to the turbulence, while the stoichiometric CH_4/air evenly premixed gas is injected into the combustion chamber from the pilot.

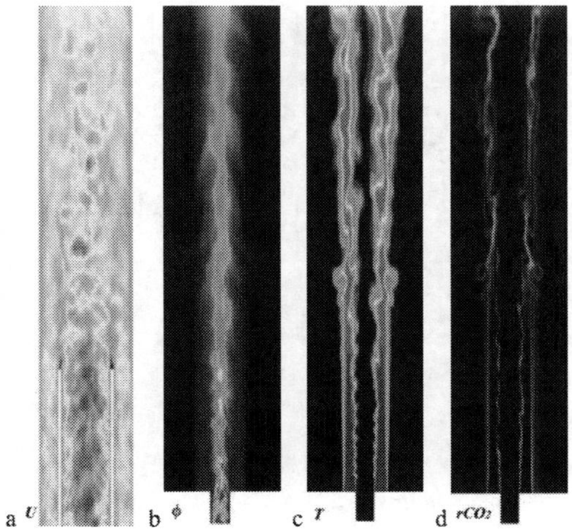

Figure 3: a: Distribution of the axial velocity in local area of inflow pipe; b: Distribution of φ in the combustion chamber; c: Distribution of temperature in the flow field; d: Distribution of the production rate of CO_2.

It can be found in Fig. 3c that a stable and long high temperature zone is created in the pilot exit plane and its downstream. Therefore, the temperature increases abruptly with radius r and then falls back to ambient temperature. The drop of the temperature occurs because of the small difference of velocity between the pilot and wind tunnel. The same, the small difference inhibit the occurrence of shear force, thus the flame is slim and long shown in Fig. 3c, and also, the flame do not become a "V"-shaped flame in the further downstream.

Production rate of CO_2 can be used to indicate the location of the flame front. But it can be found that there are two peaks at the downstream of the pilot from the video of production rate of CO_2. Analysis shows that the outer peak is caused by the reversibility of the CO_2 reaction process. The direct contact of the high temperature exhaust gas with the ambient cold air at the position of this peak stimulates the dissociation and combination of the CO_2, therefore, the production rate of CO_2 downstream the pilot wall is bigger than other zone of the burner. Figure 3d shows the instantaneous distribution of the production rate of CO_2

4.2 Statistics of Temperature and Species

It can be seen from the Fig. 4, which shows the mean and rms fluctuation of temperature along the radial direction at different axial distances, that the simulation results have a good agreement with the experimental data, especially in the upstream (x = 7.5mm), which shows that the model used in this paper can be used to simulate partially premixed flame, and also can predict flame behavior under different conditions. But it still have deviation between them in the downstream. The reason may be as follows: First of all, the calculation have not taken enough time, statistical averaging have not been complete, therefore, the predicted value cannot show the most realistic results. In addition, due to the rapid combustion of the jet flame in the experiment, it must have some deviation when measuring the experimental data, which is also a factor of the deviation.

CO_2 is the main product of CH_4 combustion, whose mass fraction represents the reaction process, and is obtained by solving the Favre filter transport equation of the mass fraction. Figure. 5a and 5b show the mean and rms fluctuation of the CO_2 mass fraction along the radial direction at different axial distances, their general trend are basically the same as that of in Fig. 4, and also can agree well with the experiment.

CO is the intermediate product of CH_4 combustion, and the prediction result of CO is an important criterion to verify the feasibility of the model. Mass fraction of CO is smaller than most that of other species, and it is difficult to describe its distribution accurately in experiment because several elementary reactions involve CO to make its value more active, therefore, the experimental data have deviation in most time. But it shows that the prediction data of CO agree well with the experimental value in Fig. 6, therefore, it can be assumed that the REDIM-PDF model can predict partially premixed flame of CH_4 very well.

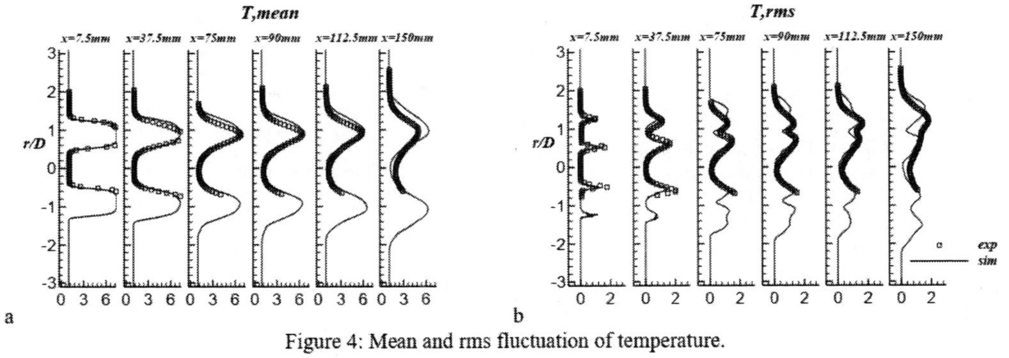

Figure 4: Mean and rms fluctuation of temperature.

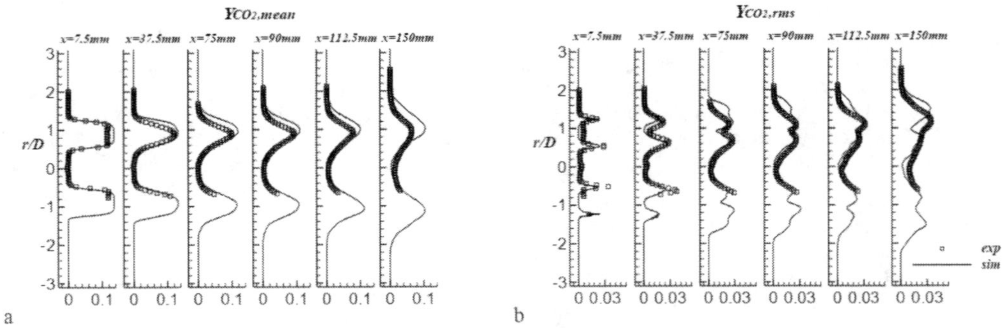

Figure 5: Mean and rms fluctuation of fluctuation of CO_2 mass fraction.

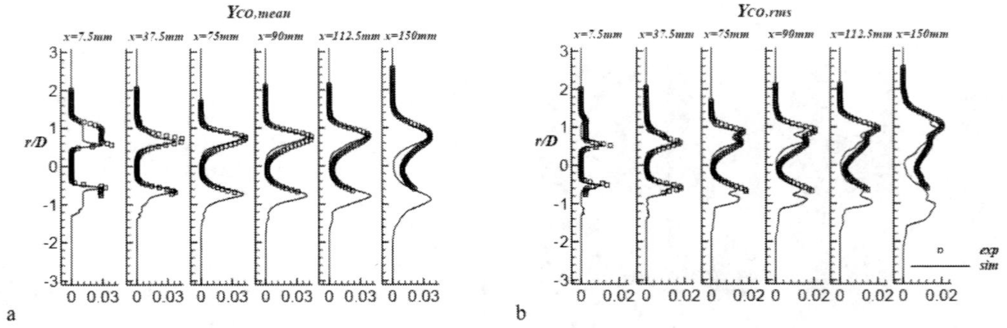

Figure 6: Mean and rms fluctuation of fluctuation of CO mass fraction.

5 Conclusion

In the present paper, LES of the inhomogeneous jet flame FJ200-5GP-Lr75-57, measured jointly by the University of Sydney and Sandia Laboratory, is carried out by using the REDIM-PFDF sub-grid combustion model. REDIM is a novel method for reducing detailed chemical reaction mechanism, which takes into account the coupling of chemical reaction and molecular transport. The chemical reaction table generated by it is suitable for describing the premixed, non-premixed and partially premixed combustion modes. In order to obtain the real mixing state of fuel and air inside the jet pipes, the jet pipes are included in the computation domain and a periodic inlet boundary condition is used in the upstream of the inlet pipes. The LES predicted temperature and species distributions are in good agreement with the experimental data, which demonstrates the good performance of the REDIM-PFDF model in simulating partially premixed combustion.

6 Acknowledgement

The support of National Natural Science Foundation of China through project 51576092, the support of Natural Science Foundation of Jiangsu Province through project BK20151344, and the support of PA PD project funded by Jiangsu Higher Education Institution are gratefully acknowledged. The authors also thank Prof. U. Maas (KIT, Germany) and Prof. A.R. Masri (University of Sydney, Australia) for providing the REDIM chemistry table and experimental results, respectively. All computation is performed on the High Performance Computing Platform of Jiangsu University.

Reference

[1] J.E. Dec, Proc. Combust. Inst 32 (2) (2009) 2727-2742.

[2] R.W. Bilger, S.B. Pope, K.N.C. Bray, J.F. Driscoll, Proc. Combust. Inst 30 (1) (2005) 21-42.

[3] S.B. Pope, Proc. Combust. Inst 34 (1) (2013) 1-31.

[4] E. Knudsen, H. Pitsch, Combust. Flame 159 (1) (2012) 242-264.

[5] Y. Ge, M.J. Cleary, A.Y. Klimenko, Proc. Combust. Inst 34 (2013) 1325-1332.

[6] R.P. Lindstedt, S.A. Louloudi, E.M.Vaos, Proc. Combust. Inst 28 (1) (2000) 149-156.

[7] V. Bykov, U. Maas, Proc. Combust. Inst 32 (1) (2009) 561-568.

[8] P. Wang, F. Zieker, R. Schießl, N. Platova. J. Fröhlich. U.Maas, Proc. Combust. Inst 34 (1) (2013) 1269-1280.

[9] C.J. Wang, S.J. Lu, P Wang, J. Chem. Ind. Eng (China) 66 (12) (2015) 4948-4959.

[10] P. Wang, J. Fröhlich, V. Michelassi, W. Rodi, Int. J. Heat. Fluid. Fl 29 (3) (2008) 654-664.

[11] N. Peters, Reducing. Mechanisms (1991): 48-67.

[12] R.S. Barlow, S. Meares, G. Magnotti, et al. Combust. Flame 162 (10) (2015) 3516-3540.

[13] S. Meares, V.N. Prasad, G. Magnotti, R.S. Barlow, A.R. Masri, Proc. Combust. Inst 35 (2015) 1477-1484.

11th Asia-Pacific Conference on Combustion,
The University of Sydney, NSW Australia
10th -14th December 2017.

Influence of Injection Timing on In-Cylinder Flow and Combustion Performance in a Spark-Ignition Direct-Injection (SIDI) Engine

L.G. Clark[1], S. Kook[1], Q.N. Chan[1], and E.R. Hawkes[1]

[1]School of Mechanical and Manufacturing Engineering, The University of New South Wales

Abstract

Particle image velocimetry (PIV) was applied to an optical spark-ignition direct-injection (SIDI) engine in order to analyse various turbulent properties of the flow-field. The engine was motored at 1200 RPM with an intake pressure of 100 kPa, while PIV images were acquired on the horizontal (swirl) plane. The flow was decomposed into mean and fluctuating components via spatial filtering with an 8-mm cut-off length. Five dual-injection timing strategies were analysed, ranging from two injections early in the intake stroke to two injections during the compression stroke. The in-cylinder pressure traces, indicated mean effective pressure (IMEP), and combustion phasing data suggested that engine performance is incredibly sensitive to relatively minor changes in injection timing. An expected global trend of increasing turbulence with retarded injection timing was clearly observed. Relationships between turbulent kinetic energy (TKE) and burn rate were present, but not as obvious as anticipated, as total kinetic energy (KE) and variations in large-scale flow structures were also found to play a significant role.

1. Introduction

The combustion process within an engine is greatly influenced by the in-cylinder flow field. Spark-ignition (SI) engines are particularly sensitive to the intensity and scale of turbulent flow fluctuations, which have been shown to directly affect flame propagation speed and thereby the performance, efficiency, and emissions output of an engine [1–3]. It is necessary to continually improve the knowledgebase of these in-cylinder processes to meet the demands of ever-tightening regulations on pollutant formation and the increased expectations of consumers.

It is well known that there is a near-linear relationship between engine speed and burn rate in SI engines, which allows them to operate over a large speed range. Multiple studies utilising methods such as hot wire anemometry (HWA), laser Doppler velocimetry (LDV) and particle image velocimetry (PIV) have shown that the reduction in time available for combustion at high engine speeds is compensated for by increased turbulence intensity, which in-turn increases the flame propagation speed of the mixture [1–4].

Direct-injection (DI) adds further complexity to in-cylinder flow behaviour. The introduction of high-momentum fuel sprays, injected at pressures upwards of 10 MPa, can have a significant impact on both large-scale bulk motion and small-scale turbulent fluctuations [5–9].

Lewis Clark
E-mail address: lewis@unsw.edu.au

This study aims to improve the understanding of the effects of fuel-injection events on bulk in-cylinder flow, small-scale turbulence, burn rate, and indicated output. Experiments on variations in injection strategy and timing are conducted in an optically accessible research engine operating at low speed (1200 RPM) and high load (wide-open throttle). The tools used to analyse the performance of each condition include in-cylinder pressure measurements to calculate parameters including combustion phasing (CA10/CA50) and indicated mean effective pressure (IMEP), and high-speed particle image velocimetry (HSPIV) to resolve in-cylinder flow features.

2. Experiments

2.1. Optical Engine Specifications and Operating Conditions

The experiments were carried out in an optically accessible single cylinder spark-ignition direct-injection (SIDI) engine with a side mounted injector and flat-topped piston [10,11], depicted schematically in Figure 1. The engine head is based on a production car double-overhead cam 16 valve inline-4 cylinder engine which has been modified to remove cylinders 2, 3 and 4. The cylinder has a geometric compression ratio of 10.5 and is of square configuration with a bore and stroke of 86 mm, resulting in a displacement volume of 500 cm^3.

In order to replicate a thermally stable warmed-up operating condition, 90°C water was constantly circulated through the engine using a water heater/circulator (ThermalCare Aquatherm RA-series). The throttle was fixed in the wide-open position which resulted in an intake air pressure of 100 kPa

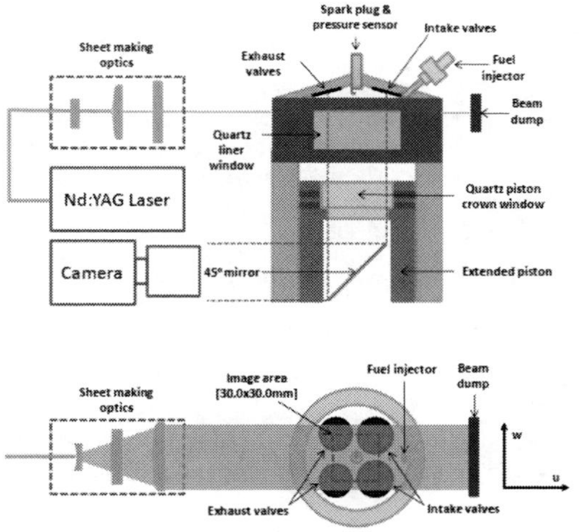

Figure 1: Optical Engine & Imaging Configuration.

absolute, and while not specifically controlled the intake air temperature was recorded consistently at 30°C throughout the experiments. An AC motor and a large flywheel maintained a stable engine speed of 1200 RPM.

The engine was controlled via a universal timing unit (Zenobalti ZB-9013P) which uses a rotary encoder (Autonics E40S8) on the crankshaft as its reference. Synchronisation of the laser and camera was achieved through the combination of a delay generator (Stanford DG-535) and two frequency dividers (Pulse Research Lab PRL-240A). Crank-angle resolved in-cylinder pressure traces were recorded via a pressure transducer integrated into the spark plug (Optrand C822J6-SP).

Conventional 95 RON petrol fuel was supplied at a constant pressure of 15 MPa to a side mounted 6-hole injector (Continental DI XL2) placed between the intake valves. A 1:1 ratio double-injection strategy was implemented for this study, based on our previous work conducted in the same optical engine [11]. Injection timings were as shown in Table 1 – ranging from two injections early in the intake stroke to two injections during the compression stroke. Ignition timing was fixed at 14°CA bTDC, corresponding to the minimum advance for best torque (MBT) for the best performing injection timings.

Injection Strategy	1:1 split ratio, double-injection
Fuel Pressure	15 MPa
Injection Duration	1.15 ms per injection
Injection Timings	300°CA bTDC & 270°CA bTDC 300°CA bTDC & 210°CA bTDC 300°CA bTDC & 110°CA bTDC 260°CA bTDC & 110°CA bTDC 140°CA bTDC & 110°CA bTDC

Table 1: Fuel-injection strategy and timings.

2.2. Image Acquisition

The intake flow was seeded with olive oil droplets from a four-jet atomiser, with a quoted nominal diameter of ~1 μm. The beam from a frequency doubled 532 nm dual oscillator Nd:YAG laser (Quantronix Hawk-Duo) was formed into a sheet using three cylindrical lenses – two controlling the breadth and one to gradually reduce the thickness of the sheet to roughly 1 mm at the region of interest. Image pairs are captured with a high-speed CMOS camera (VisionResearch Phantom v7.3) with a pulse separation (Δt) of 20 μs, as this was found to provide a suitable dynamic range for all velocities encountered in this study, while maintaining a sufficiently high signal to noise ratio. The images were acquired at 28°CA bTDC, slightly earlier than ignition timing, due to the limitation of the piston blocking the field of view.

Figure 1 depicts the orientation of the camera and laser for the swirl plane imaging, whereby the laser sheet was introduced horizontally through the top of the quartz window liner 5 mm below the fire deck, and the camera captured the images through the quartz piston crown via the 45-degree mirror in the extended piston. A Micro Nikkor 200 mm f/4 lens provided a 30.0 mm x 30.0 mm region of interest at a resolution of 468 x 468 pixels, corresponding to 64 μm per pixel. The aperture of the lens was stopped down to f/8 to increase the depth of field and improve image sharpness.

2.3. PIV Analysis

Image pairs were processed with PIVlab, a Matlab-based open-source code [12]. 4 passes of a FFT window deformation cross-correlation algorithm were performed with 50% overlap, and a Gaussian 3-point sub-pixel estimator was employed for peak finding. The first two passes were performed with an interrogation size of 48 x 48 pixels, and the last two with an interrogation size of 24 x 24 pixels – corresponding to 1.54 x 1.54 mm. This resulted in a vector spacing of 0.77 mm and a dynamic velocity range of 19.2 m/s according to the ¼ window displacement rule.

2.4. Flow-Field Post-Processing

A Reynolds decomposition was performed on the flow-fields in order to separate the mean and fluctuating components:

$$U(\theta,x,y,i) = \bar{U}(\theta,x,y,i) + u_{hf}(\theta,x,y,i) \qquad (1)$$

Where:
- $U(\theta,x,y,i) =$ Instantaneous velocity.
- $\bar{U}(\theta,x,y,i) =$ Mean velocity.
- $u_{hf}(\theta,x,y,i) =$ High frequency fluctuating velocity.

A low frequency fluctuating velocity component, which gives an insight into cycle-to-cycle variation, was also calculated:

$$u_{lf}(\theta,x,y,i) = \bar{U}(\theta,x,y,i) - \bar{U}_{EA}(\theta,x,y) \qquad (2)$$

Where:
- $\bar{U}_{EA}(\theta,x,y) =$ Ensemble average velocity.

Fluctuation intensities u'_{hf} and u'_{lf} were defined as the root-mean-square (RMS) of the fluctuating velocity for crank angle θ of each individual cycle i across the spatial domain (similarly for u_{lf}):

$$u_{hf}'(\theta,i) = \sqrt{\frac{1}{n \times m}\sum_x^n \sum_x^m u_{hf}(\theta,x,y,i)^2} \qquad (3)$$

Turbulent kinetic energy (TKE) was then calculated, ignoring the 3rd directional component:

$$TKE(\theta,i) = \frac{1}{2}\left(u_{hf}'^2 + w_{hf}'^2\right) \qquad (4)$$

The mean flow $\bar{U}$ was calculated through applying a low-pass spatial filter based on a selected cut-off length. A 2-D fast Fourier transform (2-D FFT) was performed on each flow field to convert it into the spatial frequency domain, where a high-order low-pass Butterworth filter was applied. In accordance with Equation 1, the high-frequency fluctuating velocity u_{hf} was then calculated by subtracting the low-pass filtered mean flow $\bar{U}$ from the instantaneous velocity U. To account for filtering edge effects the turbulent flow field was cropped by 4 data points along each side.

3. Results and Discussion

3.1. Effect of Injection Timing on In-Cylinder Pressure, Indicated Output, and Combustion Phasing

Engine performance marked variations with changes in injection timing. In Figure 2 the mean pressure trace of each operating condition is shown, where it is clear that the 300 & 110°CA bTDC and the 140 & 110°CA bTDC cases exhibit the highest peak in-cylinder pressure and earliest combustion phasing. Surprisingly, the 260 & 110°CA bTDC case performs similarly to the two early injection cases, despite only a minor change in injection timing from the 300 & 110°CA bTDC case.

The indicated mean effective pressure (IMEP) of each case is

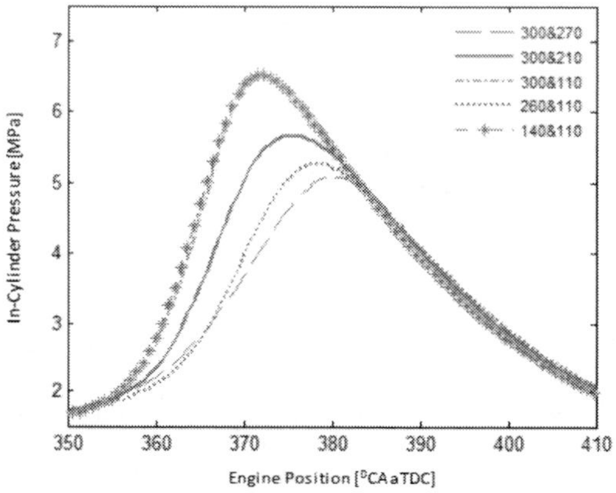

Figure 2: Mean in-cylinder pressure traces for each case.

shown in Figure 3, where the centre-line represents the median value, the box limits are the 25th and 75th percentiles, and outliers are noted as an orange +. As expected based on the pressure traces, the 300 & 110°CA bTDC case results in the highest IMEP; however, the 140 & 110°CA bTDC case lags behind – likely due to the lack of mixture preparation time. The 300 & 270°CA bTDC and the 260 & 110°CA bTDC cases both possess a larger spread than the other three, indicating a higher level of cyclic variation.

Early flame development is investigated in Figure 4 through the analysis of CA10, which describes the time taken (in crank angles) for the first 10% of mass fraction burned.

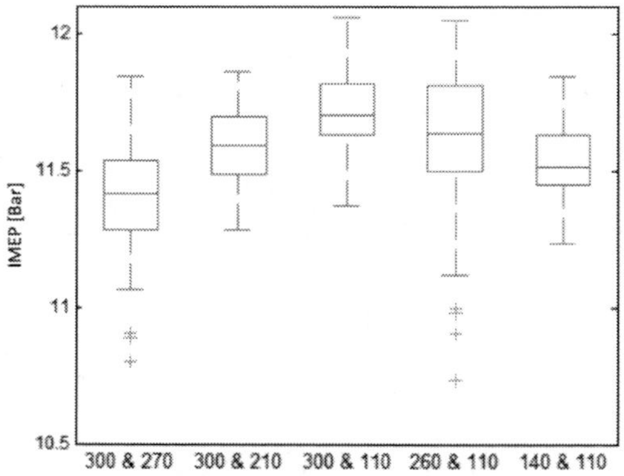

Figure 3: IMEP for various injection timings

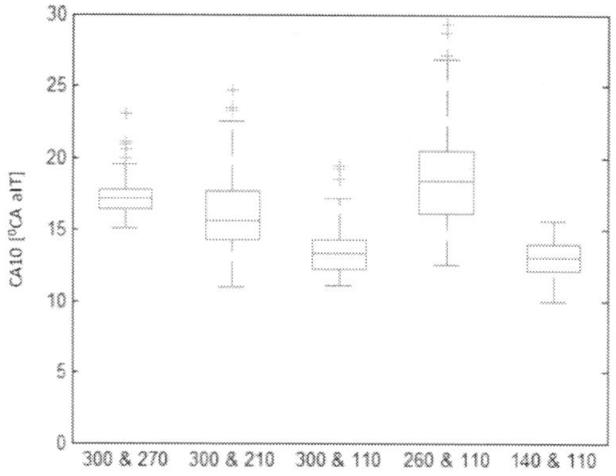

Figure 4: Combustion Phasing (CA10: crank angle corresponding to 10% heat release) for various injection timings.

Unsurprisingly, the trends in the median values are similar to those observed in the mean pressure traces in Figure 2. As with the IMEP values, the 260 & 110°CA bTDC case exhibits the largest spread, although in this instance the spread of the 300 & 270°CA bTDC case is much tighter. This is an interesting observation, as cyclic variation is often linked to behaviour in early flame development.

3.1. Effect of Injection Timing on In-Cylinder Flow

The results discussed above can mostly be described according to existing pre-mixed turbulent combustion models, which state that an increase in turbulent fluctuation intensity will result in a corresponding increase in heat release rate [4]. Retarded injection timings should enhance turbulence intensity at the point of ignition as there is less time available for the turbulence to decay through viscous effects. The TKE of each fuel-injected condition (and the non-injected case) is shown in Figure 5, where an expected overall trend of increasing turbulence with retarded injection timing is observed. Despite the lower heat release rates (i.e. delayed combustion phasing) displayed in Figure 4, the 260 & 110°CA bTDC case shows slightly higher TKE than the 300 & 110°CA bTDC case. In addition, the spread of TKE values for these two cases is very similar, suggesting that the high level of variation in CA10 for the 260 & 110°CA bTDC case is not as a result of inconsistent turbulent intensity.

Analysing TKE alongside total kinetic energy (KE) in Figure 6 reveals a greater insight into the observed combustion phasing

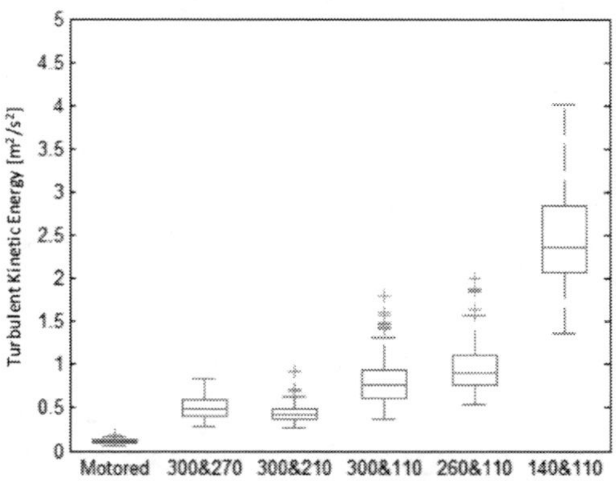

Figure 5: TKE for various injection timings

behaviour. The 300 & 210°CA bTDC and 300 & 110°CA bTDC cases each have higher KE than the 300 & 270°CA bTDC and 260 & 110°CA bTDC cases respectively, which aligns with the CA10 trends in Figure 4. This suggests that the larger scales of flow also have a noticeable effect on heat release rate.

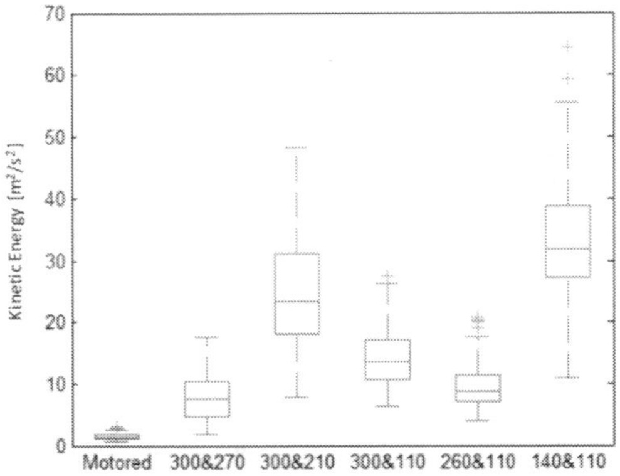

Figure 6: KE box plot.

A further explanation of the cycle-to-cycle variation in combustion phasing and IMEP may be obtained through investigating the low-frequency velocity fluctuation behaviour. In Figure 7, the ratio of low-frequency fluctuation intensity to the mean flow speed of each case is shown. Comparing this to Figure 3, it is apparent that there is a correlation between the cycle-to-cycle variation of bulk flow structures and IMEP, with the 300 & 270°CA bTDC and 260 & 110°CA bTDC cases exhibiting the widest IMEP spread and highest normalised fluctuation intensity.

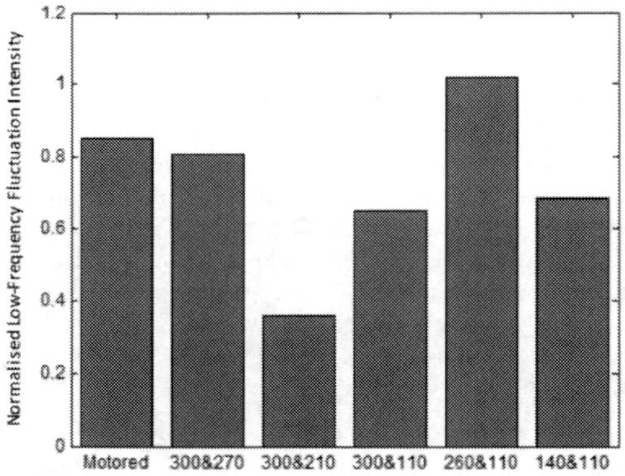

Figure 7: Normalised low-frequency fluctuation intensity.

4. Conclusion

Particle image velocimetry (PIV) was performed in an optical spark-ignition direct-injection engine to determine the effect of injection timing on in-cylinder flow and combustion performance. The results suggest that optimal performance is achieved through balancing mixture formation time and turbulence enhancement. Trends in heat release rate / combustion phasing were not completely explained through interpreting TKE results, as total KE was also found to play a role. Similarities between cycle-to-cycle variation in flow structures and IMEP were observed, suggesting that injection timings which lead to less repeatable bulk flow behaviour are likely to result in less stable engine output.

Acknowledgment

Experiments were performed at the UNSW Engine Research Laboratory, Sydney, Australia. Support for this research was provided by the Australian Research Council via the Linkage Project scheme.

References

[1] D. Bradley, M.Z. Haq, R.A. Hicks, T. Kitagawa, M. Lawes, C.G.W. Sheppard, R. Woolley, Turbulent burning velocity, burned gas distribution, and associated flame surface definition, Combust. Flame. 133 (2003) 415–430. doi:10.1016/S0010-2180(03)00039-7.

[2] L. Gillespie, M. Lawes, C.G.W. Sheppard, R. Woolley, Aspects of Laminar and Turbulent Burning Velocity Relevant to SI Engines, SAE Tech. Pap. 2000-01-0192. (2000). doi:10.4271/2000-01-0192.

[3] M.J. Hall, F.V. Bracco, A Study of Velocities and Turbulence Intensities Measured in Firing and Motored Engines, SAE Tech. Pap. 870453. (1987). doi:10.4271/870453.

[4] N. Peters, The turbulent burning velocity for large-scale and small-scale turbulence, J. Fluid Mech. 384 (1999) 107–132. doi:10.1017/S0022112098004212.

[5] J.E.T. Rimmer, E.J. Long, C.P. Garner, G.K. Hargrave, D. Richardson, S. Wallace, The Influence of Single and Multiple Injection Strategies on In-Cylinder Flow and Combustion within a DISI Engine, SAE Tech. Pap. 2009-01-0660. (2009). doi:10.4271/2009-01-0660.

[6] S.H.R. Müller, S. Arndt, A. Dreizler, Analysis of the In-Cylinder Flow Field / Spray Injection Interaction within a DISI IC Engine Using High-Speed PIV, SAE Tech. Pap. 2001-01-1288. (2011). doi:10.4271/2011-01-1288.

[7] W. Zeng, M. Sjöberg, D. Reuss, Using PIV Measurements to Determine the Role of the In-Cylinder Flow Field for Stratified DISI Engine Combustion, SAE Int. J. Engines. 7 (2014) 2014-01-1237. doi:10.4271/2014-01-1237.

[8] P.G. Aleiferis, M.K. Behringer, Modulation of integral length scales of turbulence in an optical SI engine by direct injection of gasoline, iso-octane, ethanol and butanol fuels, Fuel. 189 (2017) 238–259. doi:10.1016/j.fuel.2016.10.087.

[9] H. Zhuang, V. Sick, H. Chen, Impact of Fuel Sprays on In-Cylinder Flow Length Scales in a Spark-Ignition Direct-Injection Engine, SAE Int. J. Engines. 10 (2017) 2017-01-0618. doi:10.4271/2017-01-0618.

[10] Q.N. Chan, Y. Bao, S. Kook, Effects of injection pressure on the structural transformation of flash-boiling sprays of gasoline and ethanol in a spark-ignition direct-injection (SIDI) engine, Fuel. 130 (2014) 228–240. doi:10.1016/j.fuel.2014.04.015.

[11] L.G. Clark, S. Kook, Q.N. Chan, E.R. Hawkes, Influence of Injection Timing for Split-Injection Strategies on Well-Mixed High-Load Combustion Performance in an Optically Accessible Spark-Ignition Direct-Injection (SIDI) Engine, SAE Tech. Pap. 2017-01-0657. (2017). doi:10.4271/2017-01-0657.

[12] W. Thielicke, E.J. Stamhuis, PIVlab – Towards User-friendly, Affordable and Accurate Digital Particle Image Velocimetry in MATLAB, J. Open Res. Softw. 2 (2014). doi:10.5334/jors.bl.

11th Asia-Pacific Conference on Combustion,
The University of Sydney, NSW Australia
10th -14th December 2017.

Molecular Characterization of Lignin Pyrolysis Tar by High-Resolution Orbitrap Mass Spectrometry

Chunjinag Liu, Yizun Wang, Chaoqun Zhou, Hao Ma, Xiamin Chen, Fei Qi, Zhongyue Zhou*

Shanghai Jiao Tong University

800 Dongchuan Rd., Shanghai, 200240, China

Abstract

The molecular composition of lignin slow pyrolysis tar was characterized by using positive- and negative-ion electrospray ionization (ESI) orbitrap mass spectrometry. The results of positive- and negative-ion ESI mass spectra show complementarity of the two ionization mode. The detailed analysis with DBE versus carbon number plots shows that monomers and dimers in O_2 and O_3 class compounds are the major products.

1 Introduction

As the most abundant and renewable carbon source, lignocellulosic biomass holds promising potential to be used to produce alternative fuels and fine chemicals [1,2]. Pyrolysis is one of widely adopted strategies to directly convert biomass into bulk bio-oils. Due to the cellulose, hemicellulose and lignin components of lignocellulosic biomass, bio-oils contain a high amount of oxygen, which make it corrosive and unstable [3,4]. Before used as practical fuels, bulk bio-oils is usually upgraded through complicated strategies. Thus, accurate understanding of the basic composition of bio-oil is of great importance for upgrading of the crude bio-oil and improvement of the pyrolysis processes. As one of the main components of lignocelluloic biomass, lignin has a significant influence on the pyrolysis behavior of biomass.

So far, various analytical methods, such as Fourier transform infrared spectroscopy (FT-IR), thermogravimetric analysis (TGA), gas chromatography-mass spectrometry (GC-MS), gel permeation spectroscopy (GPC), nuclear magnetic resonance (NMR), were employed for analyzing the composition of bio-oil. Only those compounds that are found from the spectral database are unequivocally identified, complete characterization of specific bio-oil is still challenging due to its extreme complex composition, diversity of sources and limitations of analytical instruments. Ultrahigh resolution mass spectrometry, such as Fourier transform ion cyclotron resonance mass spectrometry (FT-ICR MS) and orbitrap mass spectrometry have been successfully used for detailed characterization of bio-oils.[5-17] For example, Jarvis et al.[7,8] analyzed the pyrolysis oil from red pine, pine pellets and peanut hulls employing negative-ion ESI FT-ICR MS. Smith et al. identified over 100 molecular compounds in a bio-oil from pyrolysis of a loblolly pine tree,[6] and characterize over 800 chemical compositions in a bio-oil samples from fast pyrolysis of red oak , and most of them are previously known in conventional GC–MS.[5] Liu et al.[11] studied the distribution of heteroatoms in the solvent fractions of the red pine fast pyrolysis bio-oil. The high mass resolution and mass accuracy enable unique assignment of each peaks in a single mass spectrum of complex samples. ESI is usually used because of its "soft" and "selective" ionization of polar, heteroatom-containing compounds, such as highly oxygenated species in bio-oils. Negative-ion ESI is usually employed as the most abundant polar oxygenated molecules in bio-oil are easily deprotonated to form ions.

In this paper, we present the detailed characterization of tar from lignin pyrolysis by ESI orbitrap mass spectrometry.

2 Experimental Section

2.1 Sample

The lignin brown powder used in our work was purchased from Sigma-Aldrich and used as received. Prior to the pyrolysis, the raw lignin was characterized by elementary analysis and FTIR.

The amounts of C, H, N and S were determined with an Elementar elemental analyzer (vario EL cube), while the amount of oxygen was obtained as the difference of 100% - (CHNS). Elemental analysis results shown in Fig. 1 reveal that the heteroatom content is comprised of mostly oxygen (29.64%), with very low sulfur (1.49%) and nitrogen (0.83%) contents.

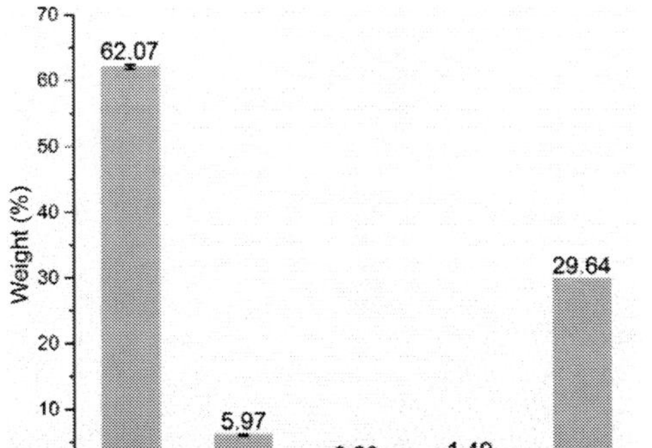

Figure 1. Element analysis of kraft lignin. The amount of oxygen was obtained as the difference of 100% - (CHNS). Error bars are ± the standard deviation of four parallel analyses.

The FTIR spectrum of lignin was recorded at room temperature using a FTIR spectrometer (Nicolet 5700, Thermofisher Scientific) with a resolution of 4 cm^{-1} and 32 scans in the range from 4000 to 400 cm^{-1}. Before the measurement, lignin was diluted with KBr (1%, w/w) and pressed into a wafer. The

*Corresponding author.
E-mail address: zhongyue.zhou@sjtu.edu.cn

FTIR spectra are presented in Fig. 2. The corresponding peaks are identified according to previous reports [18-22], and assignments are listed in Table 1.

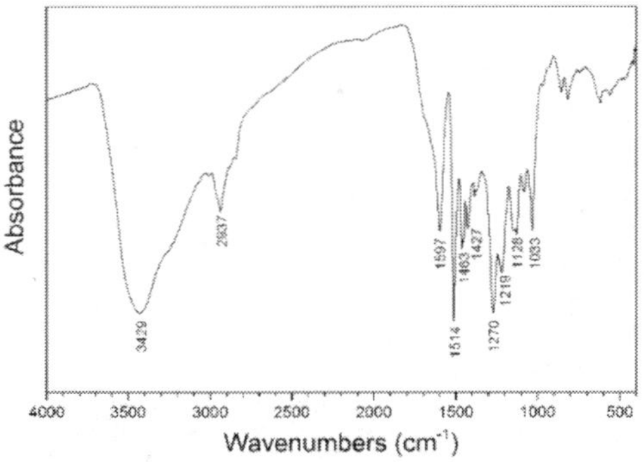

Figure 2. FTIR analysis of the kraft lignin.

Table 1. Peak assignments for the FTIR spectra of kraft lignin

Wavenumbers (cm^{-1})	Assignment
3428	O-H stretching vibration
2937	C-H stretching vibration
1597	C=O stretching conjugated to the aromatic ring
1513	Aromatic ring vibration
1463	C-H deformations, asymmetric in $-CH_3$ and $-CH_2$-
1427	Aromatic ring vibrations combined with C-H in plane deformation
1270	C-O stretching vibration in guaiacyl ring
1219	C-O stretching vibration in guaiacyl ring
1128	C-H in plane deformation of syringyl units
1033	C-O deformation in primary alcohols

2.2 Preparation of lignin pyrolysis tar

A laboratory scale fixed-bed reactor facility was used for collecting lignin pyrolysis tar. The facility includes a pyrolysis reactor and a condensing system for tar collection. A condenser with 30 mL methanol (HPLC grade, Thermofisher) was bathed in ice water (0 °C). Before pyrolysis, the reactor was stabilized at 500 °C and the reactor system was flushed with nitrogen (~200 mL/min) for 20 min. Then, the flow of nitrogen was switched to 50 mL/min, which kept the residence time of the volatiles to approximately 2 s. The quartz boat with 100 mg lignin was inserted into the center of the reactor. The average heating rate was measured as 300 °C/min. The pyrolysis tar was collected using condenser. The non-condensed gases were not collected in present experiment. The methanol dissolved with pyrolysis products was used as crude samples for mass spectrometry analysis.

2.3 Mass Spectrometry Analysis

The collected tar solutions were analyzed by an orbitrap mass spectrometer (Orbitrap Fusion, Thermo Scientific) with both positive- and negative-ion ESI modes. The bulk solutions were further diluted in methanol (HPLC grade, Thermofisher) to create 1 mL/mL ESI stock solutions. During all the experiments, a syringe pump delivered the samples to the ionization source at a rate of 5 μL/min. Nitrogen was used as a sheath gas (8 L/min) and auxiliary gas (8 L/min). the ESI needle was floated with 4500 V and 3500 V for positive- and negative-ion ESI, respectively. In order to aid ionization efficiency, 1% (v/v) formic acid (HPLC grade, Thermofisher) was added to the stock solutions for positive-ion ESI, while 1% (v/v) was added for negative-ion ESI.

For each spectrum, 200 scans were averaged to improve the signal-to-noise (S/N) ratio. Full scan mass spectra in the mass range from 100 to 500 were obtained with a mass resolving power of 500000 (at m/z 200). The mass spectra were externally calibrated using the calibration solutions. The peaks greater than 0.1% relative abundance were chosen for data analysis. An elemental formula $C_{0\text{-}30}H_{0\text{-}60}N_{0\text{-}2}O_{0\text{-}20}S_{0\text{-}2}$ and mass tolerance of ±0.1 mmu were used for molecular assignment. In order to interpret the mass spectra, data are visualized by plots of double bond equivalents (DBE, number of rings plus double bonds) versus carbon number for each heteroatom class.

3 Results and Discussion

3.1 ESI Mass Spectra of the lignin pyrolysis tar

Figure 3 shows the broad band mass spectra of kraft lignin pyrolysis tar acquired by positive- and negative-ion ESI orbitrap mass spectrometer. The mass spectra exhibit a dramatic difference between two ionization modes. For the position-ion ESI mode, the peaks ranged from m/z 100 to 350 and was centered around m/z 140. In comparison, more peaks are observed with the negative-ion ESI. The resolving power ($m/\Delta m_{FWHM}$) is calculated to be over 500 000 at m/z 183 in negative-ion ESI mass spectrum, which ensures a unique assignment of elemental composition for each peak in the mass spectrum.

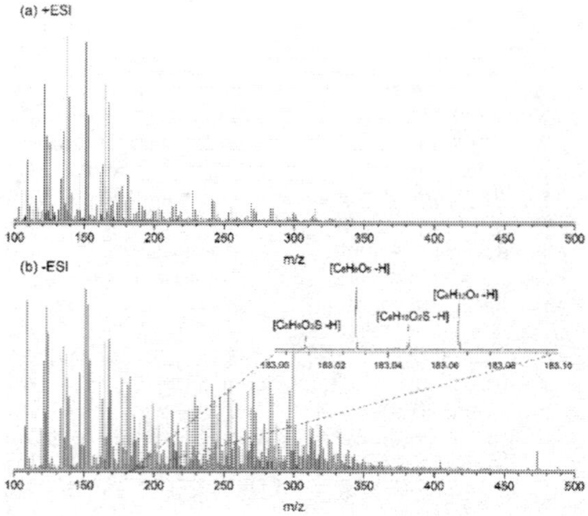

Figure 3. Broadband mass spectra of kraft lignin pyrolysis tar: (a) positive-ion ESI spectra and (b) negative-ion ESI spectra.

Figure 4 shows a comparison of the distributions of heteroatom compositions obtained by both positive- and negative-ion ESI.

The relative abundance each specie is a corresponding percentage of the sum of all detected signals. Ionization with negative-ion ESI is particularly suitable for polar compounds in bio-tar, as the O_x compounds are easily deprotonated to form ions. The oxygenated O_1-O_4 class compounds accounted for more than 70% of the total intensity and the O_3 class species were the most abundant group.

The O_x class species were found to have 2-21 DBE, with 4 and 5 DBE dominating. Lignin is an aromatic polymer consisting of hydroxy- and methoxy-substitued phenylpropane units linked through ether and condensed types of linkages. Considering a benzene ring contributes four DBE, it can be concluded that the most abundant species in the bio-tar are phenolic compounds with hydroxyl or carboxyl functionalities. Moreover, as shown in Fig. 4, the groups with DBE value of 6-10 in O_2-O_4 classes account for about 30% of respect heteroatom class. These oxygenated compounds with high unsaturation degree leads to a high reactivity and instability of bio-oil.

The rest minor amounts belonged to the O_xS and O_xN, which is consistent with the bulk elemental analysis of kraft lignin shown in Fig. 1. Obviously, more S- and N-containing compounds are detected using negative-ion ESI.

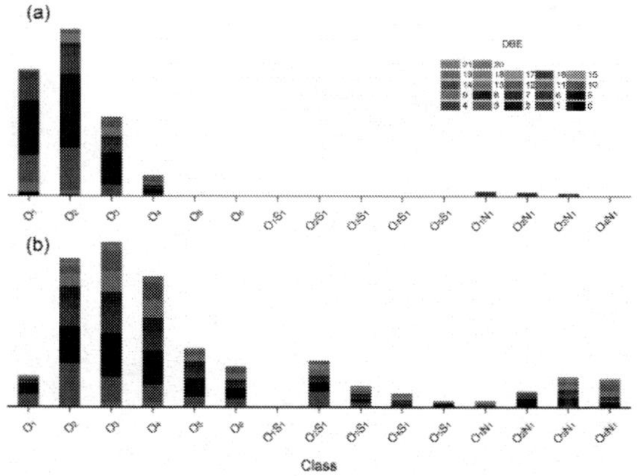

Figure 4. Heteroatom class distributions for lignin pyrolysis tar obtained from (a) positive- and (b) negative-ion electrospray orbitrap mass spectra.

It must be pointed out that parallel comparison for a specific class between Fig. 4 (a) and Fig. 4(b) is infeasible because that (i) ESI has a large difference in the ionization efficiency, (ii) the sum of the signal intensity depends upon the tuning conditions and (iii) no common internal standard was used for either mode.

3.2 DBE versus Carbon Number Plots for O_x class compounds

Typically, ESI is featured a soft ionization technique for polar species in complex mixtures. In the case of bio-oil, negative-ion ESI is usually employed as the most abundant O_x species in bio oil are easily deprotonated to form ions [23]. The DBE versus carbon number plots of O_2 and O_3 class compounds in lignin pyrolysis tar are presented in Fig. 5. The DBE versus carbon number plot act as a fingerprint of each heteroatom class. The notable difference between the positive- and negative-ion spectra shows complementarity of the two ionization mode. Specifically, more compounds with low degree of unsaturation (low DBE range) is detected with positive-ion ESI, especially in the DBE range of 2-4, whereas unsaturated compounds were identified with negative-ion ESI.

As lignin is a phenolic macromolecule with a large number of cross linking between the phenylpropane units including p-coumaryl, coniferyl and sinapyl structures, phenols and their derivaives are major components. As shown in Fig. 5, the compounds with DBE=5-8 are the most abundant O_2 and O_3 classes, which should be monomers of lignin.

Furthermore, the plots in Fig. 5 exhibited a proportionate increase in DBE with carbon number, in other words, the aromaticity increases with carbon addition. The distribution of O_2-O_3 classes present a bimodal distribution, namely, the primary O_2-O_3 classes contain compounds with DBE values of 4-6 and carbon numbers of 6-10, while the secondary groups lie in the DBE range of 8-11 and carbon number range of 13-20. On the basic of the carbon number and DBE value, the main compounds in the secondary group are likely lignin-derived dilignols, which is consistent with lignin result in previous report.

The plot of DBE vs carbon number for O_3 class exhibits similar compositional space coverage as O_2 classes. The most abundant O_3 class species are found in the DBE range of 4-6 with a broad carbon number range (6-10).

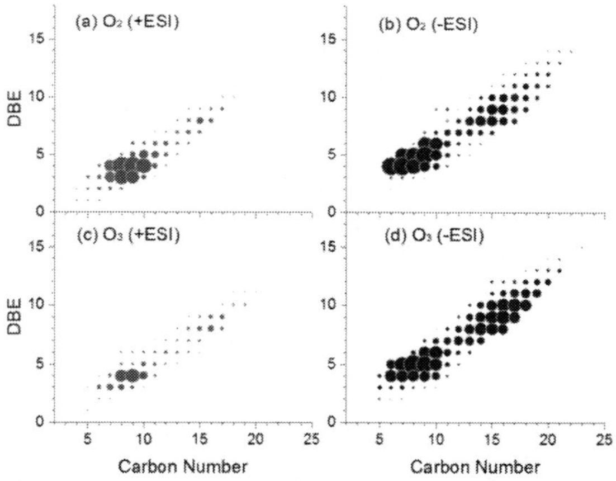

Figure 5. Comparison of O_2 and O_3 classes detected by positive- and negative-ion ESI mass spectra.

4 Conclusions

The chemical composition of kraft lignin slow pyrolysis tar was characterized with a high-resolution orbitrap mass spectrometer equipped with ESI ion source. Although the kraft lignin has relatively simpler structure compared to practical bulk biomass, the results show that the slow pyrolysis tar of lignin is a very complex mixture. Detailed analysis with DBE versus carbon number plots shows that monomers and dimers in O_2 and O_3 class compounds are the major products.

5 Acknowledgment

This work was supported by the National Natural Science Foundation of China (11675111, 91641205) and the Science and Technology Commission of Shanghai Municipality (17142202000).

References

[1] J. Hill, E. Nelson, D. Tilman, S. Polasky, D. Tiffany, Proc. Nat. Acad. Sci., 103 (30) (2006) 11206-10.

[2] D. Tilman, J. Hill, C. Lehman, Science, 314 (2006) 1598-1600

[3] A. V. Bridgwater, Chem. Eng. J. 91 (2003) 87−102.

[4] F.-X. Collard, J. Blin, Renew. Sustain. Energy Rev. 38 (2014) 594–608.

[5] E. A. Smith, S. Park, A. T. Klein and Y. J. Lee, Energy & Fuels, 26 (2012) 3796-3802.

[6] E. A. Smith and Y. J. Lee, Energy & Fuels, 24 (2010) 5190-5198.

[7] J. M. Jarvis, D. S. Page-Dumroese, N. M. Anderson, Y. Corilo and R. P. Rodgers, Energy & Fuels, 28 (2014) 6438-6446.

[8] J. M. Jarvis, A. M. McKenna, R. N. Hilten, K. C. Das, R. P. Rodgers and A. G. Marshall, Energy & Fuels, 26 (2012) 3810-3815.

[9] C. R. Smith, R. L. Sleighter, P. G. Hatcher and J. W. Lee, Environmental Science & Technology, 47 (2013) 13294-13302.

[10] Y. Liu, Q. Shi, Y. Zhang, Y. He, K. H. Chung, S. Zhao and C. Xu, Energy & Fuels, 26 (2012) 4532-4539.

[11] F. Stankovikj, A. G. McDonald, G. L. Helms, M. V. Olarte and M. Garcia-Perez, Energy & Fuels, 31 (2017) 1650-1664.

[12] N. S. Tessarolo, R. V. S. Silva, G. Vanini, A. Casilli, V. L. Ximenes, F. L. Mendes, A. d. R. Pinho, W. Romao, E. V. R. de Castro, C. R. Kaiser and D. A. Azevedo, J. Anal. Appl. Pyrol., 117 (2016) 257-267.

[13] M. Stas, J. Chudoba, D. Kubicka and M. Pospisil, Energy & Fuels, 29 (2015) 3233-3240.

[14] J. M. Santos, L. O. dos Santos, F. F. Silva, M. N. Eberlin and A. Wisniewski, Jr., Bioenergy Research, 8 (2015) 1938-1945.

[15] T. Kekalainen, T. Venalainen and J. Janis, Energy & Fuels, 28 (2014) 4596-4602.

[16] M. Ghidotti, D. Fabbri, O. Mašek, C. L. Mackay, M. Montalti and A. Hornung, Environmental Science & Technology, 51 (2007) 6580-6589.

[17] D.-D. Zhang, Z.-M. Zong, J. Liu, J.-H. Lv, T.-M. Wang, J. Gui, M. Qu, L.-L. Guo, Z.-H. Feng and X.-Y. Wei, Fuel Processing Technology, 142 (2016)167-173.

[18] D. Shen, J. Zhao, R. Xiao, S. Gu J. Anal. Appl. Pyrol. 111 (2015) 47-54.

[19] S. Wang, H. Lin, B. Ru, W. Sun, Y. Wang, Z. Luo, J. Anal. Appl. Pyrol. 108 (2014) 78-85.

[20] O. Faix, Holzforschung – Int. J. Biol. Chem., Phys. Technol. Wood 45(1991) 21–28.

[21] C.E.L. Pasquali, H. Herrera, Thermochim. Acta 293 (1997) 39–46

[22] J. Hu, R. Xiao, D. Shen, H. Zhang, Bioresour. Technol. 128 (2013) 633–639.

[23] J.M. Jarvis, D.S. Page-Dumroese, N.M. Anderson, Y. Corilo, R.P. Rodgers, Energy Fuel, 28 (2014) 6438-6446.

11th Asia-Pacific Conference on Combustion,
The University of Sydney, NSW Australia
10th -14th December 2017.

The Effect of After Injection on In-Cylinder Soot Particulates in a Small Bore Diesel Engine

L. Rao[1], Y. Zhang[1], D. Kim[1], S. Kook[1,*], K.S. Kim[2], and C.-B. Kweon[2]

[1]School of Mechanical and Manufacturing Engineering, The University of New South Wales
[2]US Army Research Laboratory

Abstract

After injection, which introduces a short injection following the main injection, is proven to be effective in reducing diesel engine soot emissions. The present study aims to investigate the in-cylinder soot reduction mechanism triggered by after injection in a small-bore optical diesel engine with a particular emphasis on the morphology of soot particles. Accordingly, in-bowl soot sampling based on the thermophoresis principle was performed for the direct collection of soot from the reacting diesel flame. The subsequent transmission electron microscope imaging enabled the structural analysis of soot particles, which provides fundamental insight regarding the formation and oxidation processes of in-bowl soot particles. The results show that, with the introduction of after injection, a downsizing trend of soot particles is found regardless of the sampling locations, which was explained by the increased in-cylinder bulk gas temperature and thus higher oxidation rate. A significant after-injection-induced decrease in soot aggregate size also suggests an accelerated oxidation of in-bowl soot particles through the fragmentation of large soot aggregates.

1. Introduction

While diesel engines are favoured for fuel economy and high torque delivery, adverse impacts of soot emissions on the environment and human health are yet to be resolved. Within the past decades, new technologies for reduced in-cylinder soot and exhaust filtering have been developed and successfully implemented. As one of the most widely used exhaust after-treatments, diesel particulate filter (DPF) effectively reduces the engine-out soot emissions; however, the backpressure issues and increased costs are still unavoidable. Therefore, the in-cylinder reduction of soot is crucial, which can be achieved using high-pressure common-rail fuel delivery systems capable of multiple fuel injections.

Among all multiple injection strategies, after injection, or in some cases been referred as close-coupled post-injection, has been proven to be effective in reducing soot emissions. Previous studies reported that the benefits of after-injection are attributed to the enhanced oxidation of in-cylinder soot particles by either a mixing mechanism through interaction between the two separate injection events [1,2] or a thermal mechanism with elevated in-cylinder temperature from the after-injection-induced heat release [3,4]. On the other hand, some studies showed that the reduced soot is due to the suppressed soot formation [5]. This is primarily caused by decreased injection duration of the main injection such that the fuel jet head is less replenished by the shortened fuel injection leading to lower soot formation caused by reduced local equivalence ratio of the mixture [5].

The present study aims to visualise the structural impact of the after injection on in-cylinder soot particles, and thus better understand the soot reduction mechanism. Currently, the particular interest is to study after injection in a small-bore single-cylinder optical diesel engine since by brining the wall closer to injector, the interaction between the main soot and after-injection jet will be significantly different to that in large-bore, heavy-duty engines upon which most of the previous investigations were performed [6]. For the present study, the thermophoresis-based sampling of soot particles and analysis using a transmission electron microscope (TEM) has been conducted for two different in-bowl locations. With main injection only case used as a baseline, the structural change of soot particles due to after injections is visualised, and detailed information towards the in-cylinder mechanism of soot reduction is discussed.

2. Experiments

The sampling technique was performed in a single-cylinder, light-duty optical diesel engine modified from a conventional 2-litre, four-cylinder diesel engine. The simplified engine configuration is shown in figure 1 and the engine specifications as well as operating conditions are summarised in table 1. The engine has a displacement volume of 497.5 cm^3 with 83 mm bore and 92 mm stroke, a geometric compression ratio of 15.5, and a swirl ratio of 1.4. Throughout the experiments, the engine speed was held constant at 1200 rpm.

The engine was kept with the cylinder head and liner wall temperature of 363 K and an intake air temperature of 303 K to simulate warmed-up and thermally stable operating conditions. A piezoelectric transducer (Kistler 6056A1) is installed next to the centrally mounted common-rail fuel injector (Bosch CP3) on the conventional flat engine head between two pairs of intake and exhaust valves. The original 7-hole injector nozzle with 150° included angle was laser welded with only one hole left open for the present study such that the complex jet-jet interaction phenomenon is avoided. The nozzle hole size is 134 μm in diameter (nominal). The common-rail pressure was fixed at 100 MPa. Methyl decanoate was selected as the diagnostics fuel since previous laser diagnostic imaging of it has fully revealed the in-cylinder development of its combustion process with no beam attenuation issues owing to its low sooting propensity [7]. Therefore, methyl decanoate was used as an excellent reference for the purpose of designing sampling locations and validating sampling results. The main injection

Corresponding author. Fax: +61 (0)2 9663 1222
E-mail address: s.kook@unsw.edu.au

timing was fixed at 11°CA aTDC with duration 1.2 ms on which a short after injection at 25°CA aTDC with duration of 0.3 ms was added.

As shown in Figure 2, a 3-mm TEM grid was installed at the tip of the sampling probe, which was installed on the piston-bowl wall. The probe tip had a 2-mm diameter hole that provides a direct exposure of the TEM grid to the soot-laden reacting gases. The restrictive passage was to protect the carbon layer of the grid from excessive heat. The high temperature gradient between the hot soot particles and the cold TEM grid surface induced thermophoresis that forced soot particles to deposit on to the grid surface with immediate reaction quenching [8]. For reference, a laser diagnostic image of the combustion process under same operating conditions is shown in Figure 2 (bottom left), where soot-PLII signal is coloured in red and OH-PLIF signal coloured in blue. The red dashed box highlights the after-injection fuel jet, meanwhile the main-combustion soot and high-temperature reaction zone have travelled to the back of the piston bowl region as shown. This illustrates no physical interaction between the after-injection-induced combustion region and the main-combustion soot cloud. Hence two sampling locations, noted as A and B were selected to capture only main-injection-generated soot particles, which was isolated from the impact of after injection. The two sampling locations were achieved by using two sampling probes with different protrusions into the bowl (shown in Figure 2 top left, noted the short and long probes). Further to note, the soot sampling experiment was conducted for a total of 10 firing cycles with a 10 skip-fired mode before the engine was stopped for the removal and storage of the grid. The 10 firing cycles, selected based on the trial-and-error-based preliminary tests, were needed for the sufficient number of soot particles for the statistically meaningful data analysis, which however was low enough to avoid possible soot overloading issues [9]. The sampled soot particles were imaged using a TEM (JEOL 1400). To avoid the potential bias in result, a total of 30 TEM images were taken at various on-grid locations for each sample. The TEM images were post-processed using an in-house developed automatic Matlab code to obtain key morphology parameters.

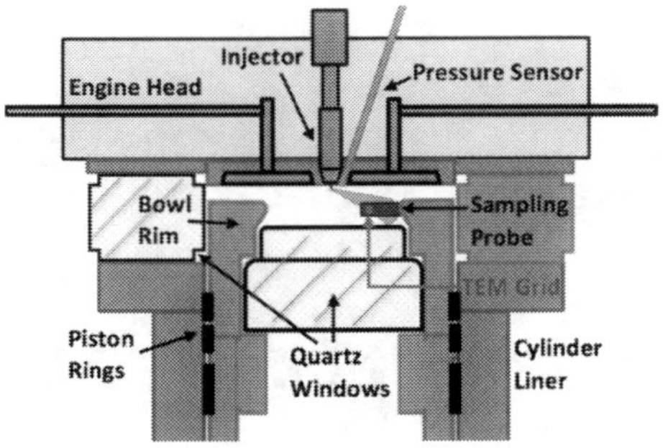

Figure 1: Schematic diagram of the optical diesel engine and soot sampling setup.

Displacement (single-cylinder)	497.5 cm³
Bore	83 mm
Stroke	92 mm
Compression ratio	15.5 (geometric)
Engine speed	1200 rpm
Swirl ratio	1.4
Wall (coolant) temperature	363 K
Intake air temperature	303 K
Injector type	Common-rail (Bosch CP3)
Number of holes	1
Nozzle type	Hydro-grounded, K1.5/0.86
Nozzle diameter	134 μm
Included angle	150°
Rail pressure	100 MPa
Fuel	Methyl Decanoate
Lower heating value	37.7 MJ/kg
Cetane Number	52
Main injected fuel mass per hole	11.5 mg
Main injection signal duration	1.2 ms
Main injection signal timing	11°CA bTDC
After injection signal timing	25°CA aTDC
After injection fuel mass per hole	2.9 mg
After injection signal duration	0.3 ms

Table 1: Engine specifications and operating conditions

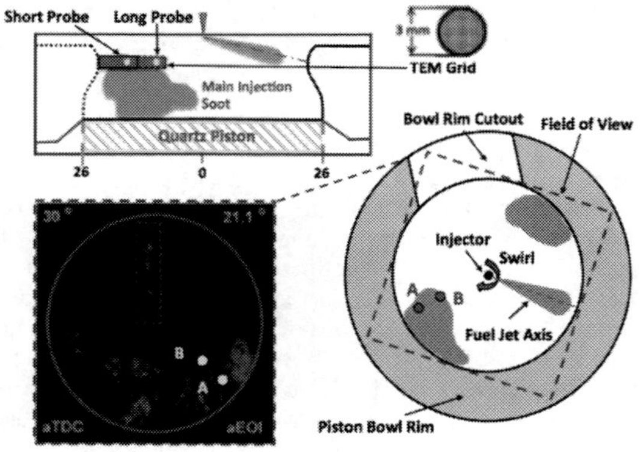

Figure 2: Illustration of the soot sampling system from the top view (right) and side view (top) of the piston bowl. Laser diagnostic image of the after-injection-assisted combustion (bottom left).

3. Results and Discussion

The ensemble-averaged traces of in-cylinder pressure and apparent heat release rate (aHRR) together with the estimated bulk gas "in-cylinder" temperature are shown in Figure 3. The after injection case is colored in blue meanwhile the main-only injection data are plotted using a black line as reference. It is apparent that the three characteristics of in-cylinder phenomenon are well matched in terms of the overall combustion phasing and pressure/aHRR magnitudes between the two cases. Hence, it can be concluded that the effects of after injection on in-cylinder soot particles are effectively isolated from the thermodynamic conditions. Shown in figure 3, shortly after the start of injection event at 27°CA aTDC, a slight drop in aHRR can be spotted, indicating a short ignition delay period for the after injection. This is then followed by a distinct spike in aHRR as an indicator of a second combustion event. Accordingly, the in-cylinder pressure for the after-injection case rises from the reference main-only injection case. The pressure deviation is highlighted in the insert plot for the close-up view of this crank angle range. The influence of the after injection on in-cylinder conditions is also observed in the estimated in-cylinder temperature, which is roughly 30 K higher compared with the main-only injection case.

The selected TEM images in Figure 5 show large and chain-like soot aggregate composed of many large soot primary particles at location A for using main injection only. Also noticed is the low amount for large aggregate as barely one or two are found in each TEM image at this sampling location, meanwhile many small soot structures are also noticed. By contrast, soot particles sampled at location B are larger with more clustered structure, albeit smaller in number. In right

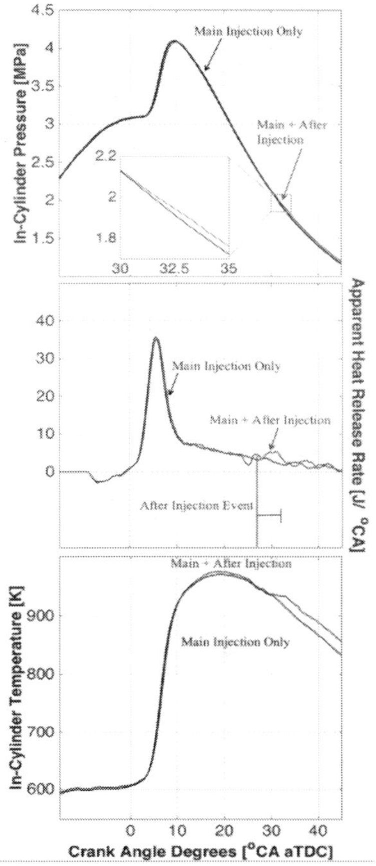

Figure 3: Ensemble-averaged in-cylinder traces, apparent heat release traces and in-cylinder bulk gas temperature traces for the main-only injection and main + after-injection cases.

panel of Fig. 4, the after-injection case was found for showing a consistent trend that soot particles at location A are large in number but smaller in size and less compacted in structure than location B. This is explained that location B was where a high rate of soot oxidation was expected due to the soot interaction with surrounding OH radicals and the formation of turbulent ring vortex in the wall-jet head region, and only those large and highly agglomerated soot aggregates survived the intensive oxidation. Care should be taken to the after-injection case, since those soot particles sampled at the two locations are mostly generated from main injection while showing apparently smaller size, more compact structure and larger number, as compared to those of main-injection only. This is because the oxidation of in-cylinder soot was enhanced by the after injection as large soot aggregates from the main injection broke down into smaller particles/aggregates, which is a common pathway for the oxidation of in-cylinder soot particles [10]. With the previous study confirming an interaction-free environment between the after and main injections, the enhanced oxidation is suspected to be caused solely by the elevated in-cylinder temperature as shown in Figure 3.

Figure 5 shows the probability density functions (pdf) of soot primary particle diameter (d_p) and aggregate radius of gyration (R_g) in the top and middle rows respectively, for sampling location A (left column) and B (right column). The pdfs of the after-injection case is plotted in red to distinguish them from the main only injection case (black lines). The logarithmic plots for fractal dimension (D_f) are also plotted in the bottom row of figure 8. At both sampling points, the primary particle diameters between the two injection strategies are similar in terms of both mean value and pdf distribution. This indicates that the impact of after injection on the soot primary particle is minor – *i.e.* insignificant effect with respect to surface-oxidation from the after injection. However, a drastic impact of after injection can be clearly seen as the dramatic decrease in R_g despite sampling locations, which is also consistent with the visual inspection of the TEM images. This confirms the previous statement that large soot particles decompose into smaller ones via enhanced oxidation due to elevated temperature caused by after injection. Such efficacy is further illustrated by the fractal structure of sampled soot particles as higher fractal dimensions are found at both sampling locations when after injection is added to the main injection.

4. Conclusion

This study implements the in-bowl soot sampling technique in a single-cylinder small-bore optical diesel engine with an emphasis on the morphological variation of in-flame soot particles under the introduction of after injection. It is apparent that the after-injection triggers a second combustion apart from the main combustion that increases in-cylinder temperature. With after injection, the sampled soot particles exhibit a downsizing trend in aggregates with more compact structures. This is due to enhanced oxidation of in-cylinder soot particles through the fragmentation of large soot aggregates, which is caused by the increased temperature.

Acknowledgement

Experiments were performed at the UNSW Engine Research Laboratory, Sydney, Australia. Support for this research was provided by the US Army RDECOM ITC-PAC and in part by the Australian Research Council.

References

[1] Hotta Y, Inayoshi M, Nakakita K. Achieving Lower Exhaust Emissions and Better Performance in an HSDI Diesel Engine with Multiple Injection. R&D Rev Toyota CRDL 2002;37:9–16. doi:10.4271/2005-01-0928.

[2] Yun H, Reitz RD. An Experimental Investigation on the Effect of Post-Injection Strategies on Combustion and Emissions in the Low-Temperature Diesel Combustion Regime. J Eng Gas Turbines Power 2007;129:279. doi:10.1115/1.2180812.

[3] Bobba M, Musculus M, Neel W. Effect of Post Injections on In-Cylinder and Exhaust Soot for Low-Temperature Combustion in a Heavy- Duty Diesel Engine. SAE Int J Engines 2010;3:496–516. doi:10.4271/2010-01-0612.

[4] Mancaruso E, Merola SS, Vaglieco BM. Study of the multi-injection combustion process in a transparent direct injection common rail diesel engine by means of optical techniques. Int J Engine Res 2008;9:483–98. doi:10.1243/14680874JER01308.

[5] Han Z, Uludogan A, Hampson GJ, Reitz RD. Mechanism of Soot and NOx Emission Reduction Using Multiple-injection in a Diesel Engine. SAE Tech. Pap. Ser., vol. 960633, 1996. doi:10.4271/960633.

[6] O'Connor J, Musculus M. In-Cylinder Mechanisms of Soot Reduction by Close-Coupled Post-Injections as Revealed by

Imaging of Soot Luminosity and Planar Laser-Induced Soot Incandescence in a Heavy-Duty Diesel Engine. SAE Int J Engines 2014;7:673–93. doi:10.4271/2014-01-1255.

[7] Le MK, Zhang R, Rao L, Kook S, Hawkes ER. The development of hydroxyl and soot in a methyl decanoate-fuelled automotive-size optical diesel engine. Fuel 2016;166:320–32. doi:10.1016/j.fuel.2015.11.006.

[8] Dobbins R a., Megaridis CM. Morphology of flame-generated soot as determined by thermophoretic sampling. Langmuir 1987;3:254–9. doi:10.1021/la00074a019.

[9] Kook S, Zhang R, Szeto K, Pickett LM, Aizawa T. In-Flame Soot Sampling and Particle Analysis in a Diesel Engine. SAE Int J Fuels Lubr 2013;6:80–97. doi:10.4271/2013-01-0912.

[10] Echavarria CA, Jaramillo IC, Sarofim AF, Lighty JS. Burnout of soot particles in a two-stage burner with a JP-8 surrogate fuel. Combust Flame 2012;159:2441–8. doi:10.1016/j.combustflame.2012.03.011.

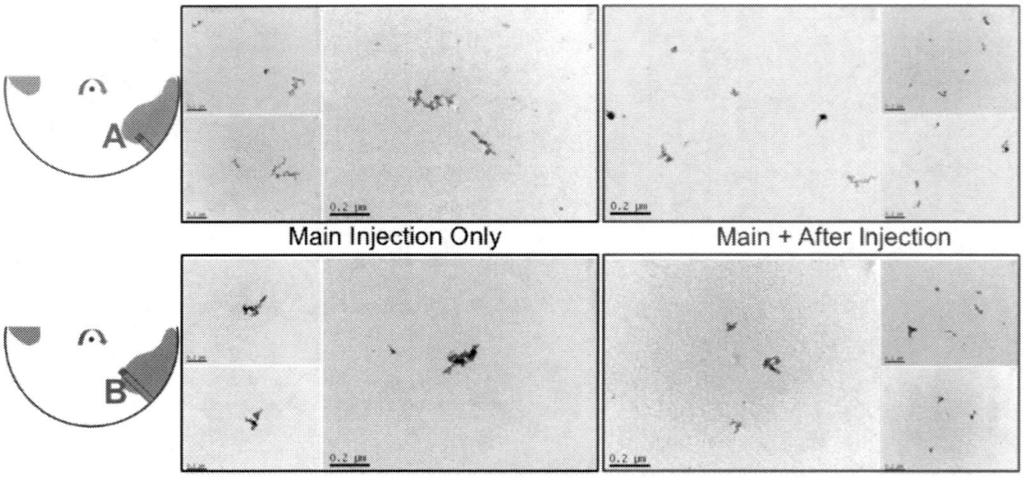

Figure 4: Transmission electron microscope (TEM) images of the sampled soot particles from location A (top) and location B (bottom) for the main injection only case and the main with two different after-injection timings.

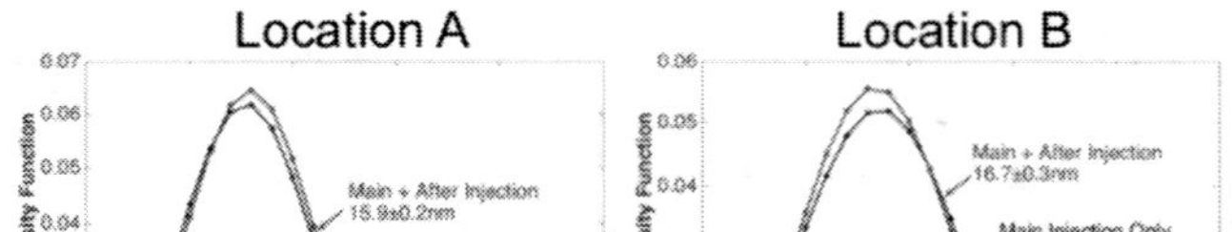

Figure 5: Size distributions of soot primary particles and aggregates and the fractal dimension analysis of the sampled soot particles from location A (left) and location B (right) for the main injection only case and the after-injection case.

11th Asia-Pacific Conference on Combustion,
The University of Sydney, NSW Australia
10th -14th December 2017.

Ignition Improvement and NOx Reduction with Local Oxygen-Enriched Coal-fired Burner

Hai Zhang*, Kai Cui, Yang Zhang, Yuxin Wu, Qing Liu, Junfu Lv, Guangxi Yue

Key Laboratory for Thermal Science and Power Engineering of the Ministry of Education, Department of Thermal Engineering, Tsinghua University, Beijing 10084, China

Abstract

In this paper, an experimental study on the High Temperature Air Combustion (HTAC) with local oxygen (O_2) enrichment for pulverized coal was conducted, with an 8MW model burner. A small amount of diluted O_2 stream with O_2 concentration of 25%-50% (in volume) was injected into a pre-heating chamber located at the exit of the burner. Experimental results showed that with a small amount of low O_2 concentration the flame stabilization of the burner could be significantly improved. Furthermore, the combination of local O_2 enrichment with HTAC could be also effective to reduce NOx formation. For bituminous coal, ~50% reduction was achieved while for the low volatile coals, ~20% NOx was reduced against the air combustion. The simulation results illustrated that high temperature zone outside the burner was enlarged and flame temperature increased with O_2 enrichment. Also, there was optimal VO_2 for NOx reduction in OE-HATC burner. For present configuration, the optimal $VO_2 \approx 50\%$, at which reduction rate is largest while the production rates for fuel NOx and thermal NOx formations are rather weak. This study showed that with a proper burner design, and organization of oxygen enrichment, a new low-NOx combustion technology for pulverized coal could be established.

Keywords: HTAC (high temperature air combustion); Pulverized coal; Burner; Oxygen enrichment; NOx reduction

1. Introduction

High Temperature Air Combustion (HTAC), also named as Flameless Oxidation (FLOX), Low NOx Injection (LNI) or MILD Combustion, is a proven advanced technology for gaseous fuel combustion due to its merits in energy saving, flame stability enhancement and NOx emission reduction [1-7]. Conventionally, it is featured with high temperature secondary air of a temperature up to 1273 K heated by the recuperative or regenerative heat exchangers installed outside the furnace. Laboratory experiments found that HTAC significantly enhanced ignition and remarkably reduced NOx formation of a pulverized coal jet. However, for industrial application, the external heat exchangers used for gas combustion are not suitable for coal combustion because of the ash slagging and fouling at high temperature. To overcome this problem, special designed burner [8-11] or high speed air jets [12-15] for proposed for HTAC of pulverized coal.

Obviously, increasing oxygen concentration (VO_2) is also in favor of flame stabilization [17-20]. Thus, O_2 enriched flows (with $VO_2 > 21\%$) were used for ignition assistance or even long term operation to enhance combustion of low grade coals [13, 21]. However, two main concerns exist for O_2-enriched combustion. One is the cost; namely, the consumption of O_2. The other is the possible increase of NOx formation. On one hand, increasing VO_2 will introduce a zone with higher temperature and stronger oxidizing atmosphere. On the other hand, increasing VO_2 is benefit for the faster and more volatile matter (VM) release of coal particles, which is in favor of NOx formation reduction [9, 22]. Given such an effect could be dominant with a small amount of O_2 enrichment and mild VO_2 increase, a new low-NOx combustion technology for pulverized coal could be established. Consequently, the feasibility of integrating the HTAC with O_2-enriched combustion for pulverized coal firing was experimentally and numerically assessed in the present study.

2. Experimental

Experiments were conducted based on the HTAC burner used in previous study [9], as schematically shown in Figure 1. The primary air (air-coal particle flow) is off-centrally injected into a preheating chamber, generating an inner recirculation zone. By the secondary air swirling or other means, a bulk reverse flow of high temperature flue gas is generated at the same time. When heat and mass of flue gas is continuously supplied into the preheat chamber, rapid heat up of the primary air stream is realized. In addition to the existing HTAC setup, an O_2-enriched stream is also supply to the preheating chamber.

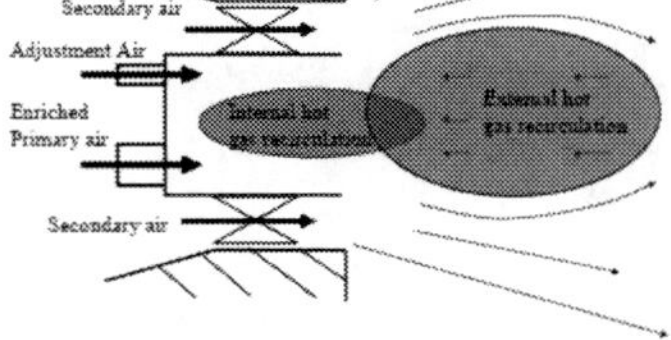

Figure 1: The schematic of the oxygen-enriched high temperature air combustion (OE-HTAC) burner

The thermal output of the burner was 8 MW. The experiments were conducted at ambient temperature. The total flow rate of the air was ~ 9600 Nm^3/h. The designed velocities for the primary air V_1 = 16-20 m/s and V_2 =18-22 m/s.

The experimental system was shown in Fig. 2. The furnace was insulated with refractory material. The length of the furnace was ~5.2 m, followed by a section of water jacket to cool down the flue gas. Five windows were installed along the furnace for flame observation. Exhaust gases were measured with gas analyzer before they are sent to the chimney.

*Corresponding author. Fax: +86-10-6278-1743
E-mail address: haizhang@tsinghua.edu.cn

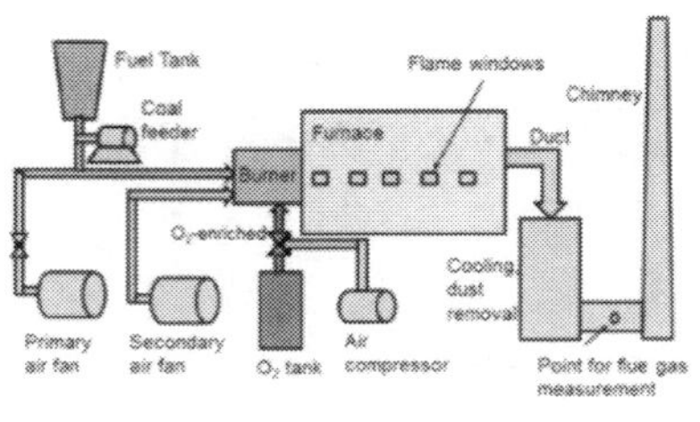

Figure 2: Schematic diagram of the experimental system.

Three Chinese coals with different volatile matter content were tested. The proximate and ultimate analyses of the tested coals were listed in Table 1.

Table 1: Properties of tested coals

	SB	CZ	HC
Ultimate (dry) (%)			
C	66.54	46.60	54.82
H	3.60	1.70	2.75
O	5.19	1.91	4.42
N	0.82	0.69	0.85
S	1.48	3.76	2.74
Proximate (%)			
Dry loss	2.73	1.45	1.05
Ash	19.64	43.89	33.37
Volatile (dry, ash free)	37.28	12.94	20.07
Low Heating Value (MJ/kg)	26.52	16.77	21.72
Ignition temperature (K, by TGA)	703	783	753

The O_2-enriched flows were generated by mixing the air stream supplied by an air compressor with the pure O_2 stream from O_2 tanks. Besides air condition, three O_2-enriched cases were tested with VO_2 of 30%, 40% and 50%. The O_2-enriched stream supplied from a stainless tube with inner diameter of 25 mm was placed in the middle of the primary air stream. The flow rates of the O_2-enriched streams are given in Table 2.

Table 2: The settings of the O_2-enriched

Item	Unit	Case1	2	3	4
O_2 concentration in O_2-enriched flow	%	21	30	40	50
Flow rate of O_2-enriched	Nm^3/h	26.5	26.5	26.5	26.5
Pure O_2 consumption	Nm^3/h	0	3.0	6.4	9.7
Air flow rate from air compressor	Nm^3/h	26	23.5	20.1	16.8

3. Numerical Approach

The numerical simulation using the same configuration of the tested model burner was based on an Eulerian-Lagrangian formulation of the 3D gas-solid two phase flow, using FLEUNT. The computational domain and the gridding are illustrated in Fig. 3. The hexahedral grid type gridding was used because it could result in not only more stable calculation, but also less mesh number than other types. Around the burner, meshes were denser. Independency of grid number was checked by monitoring the variation of temperature at a selected point located 50 cm away from the burner exit with the grid number. The temperature barely changed when the grid number was more than 400,000. Thus the simulations used 500,000 grids and the discrepancy introduced by the grid number was kept to less than 1%.

The RSM model was selected to simulate the turbulence flow with strong swirling to improve the precision and numerical convergence [23]. EDC model was selected for the VM combustion to account for the chemical kinetics of key intermediate species. VM reactions were described with a skeletal mechanism with 16 species and 41 reactions, reduced from GRI [24, 25]. Details regarding to the EDC model applied for the coal combustion can be referred to our previous study [25]. The initial temperature of the air streams and O_2-enriched stream was 300 K. The mass flow rates of the primary air, secondary air, O_2-enriched stream and pulverized coal are kept at 0.628, 3.36, 0.0143 and 0.391 respectively. VO_2 in O_2-enriched streams varied from 21% -100%, and its volumetric flow rate was about 2.2%-8.5% of the primary air. The furnace temperature was set to 673 K, approximately equaled to the experimental measurement.

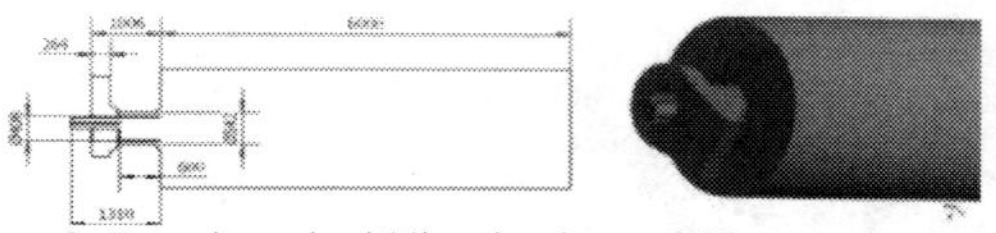

Figure 3: Domain and gridding for the model burner simulation

The modeling of NOx for the coal firing boilers is still under development [26, 27], and its accuracy depends on that of the sub-models. Some parameters used in the modeling are difficult to be obtained by laboratory experiments or theoretical analyses. In the present study, the models used in the FLUENT were adopted to individually calculate the formations of thermal NOx, fuel NOx, and the reduction NOx. The key model parameters were determined by the measured NOx concentrations at two given conditions through the try-by-error method. The key determined parameters are listed in Table 3. More details of the NOx modeling methodology can be referred to Ref. 28.

Table 3: Parameters for NO_x modeling

Source	Item	Value/description
Thermal NOx	[O]	instantaneous results from EDC modelling
	[OH]	
Fuel NOx from volatile-N	Mass % of N in volatile	2.48
	Volatile-N conversion ratio	100%
	Intermediate species of volatile-N	$HCN/NH_3/NO$
	HCN % in volatile-N	100%
	NH_3 % in volatile-N	0
	NO % in volatile-N	0
Fuel NOx from Char-N	Mass % of N in char	0.83
	Char-N conversion ratio	100%
	Intermediate species of char-N	$HCN/NH_3/NO$
	HCN % in char-N	80%
	NH_3 % in char-N	20%
	NO % in char-N	0
	BET inner surfaces,	25000 m^2/kg
NO_x reduction model	Reburn model	partial-equilibrium
	Reburn fuel species	CH_4 & CH_3
	Equivalent fuel type	CH_4

4. Results and Discussion

The flame stabilization performance of the OE-HTAC burner was assessed by using the minimum flow rate of the ignition-assisted heavy oil.

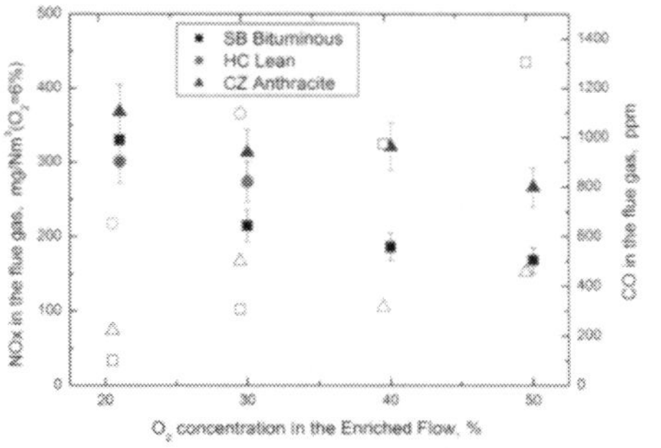

Figure 4: Measured NOx and CO concentrtaions in the flue gas under different O_2 enrichment degrees for three tested coals (solid symbol: NOx; open symbol: CO)

Figure 4 showed the NOx concentration measured at the exhaust section of the flue gas with various local O_2 enrichments for three coals with different VMs. The results revealed that NOx concentration decreased rather than increased with the increasing VO_2. This demonstrated that OE-HTAC burner was benefit to further decrease NOx emission for pulverized coal boilers.

Moreover, the results illustrated that as VO_2 increased to 30%, the effect of O_2 enrichment is obviously. When VO_2=50%, NOx was reduced for 50% for the high VM SB bituminous coals, and 20% for the low volatile HC and SZ coal. For the low VM coals, the additional advantage of OE-HTAC burner was the enhancement of flame stabilization, which can be used for ignition assisted oil saving.

Figure 5 depicts the calculated temperature contours under O_2 enrichments. Results showed that the high temperature zone enlarged and the flame became wider as a small extra amount of O_2 was supplied to the burner.

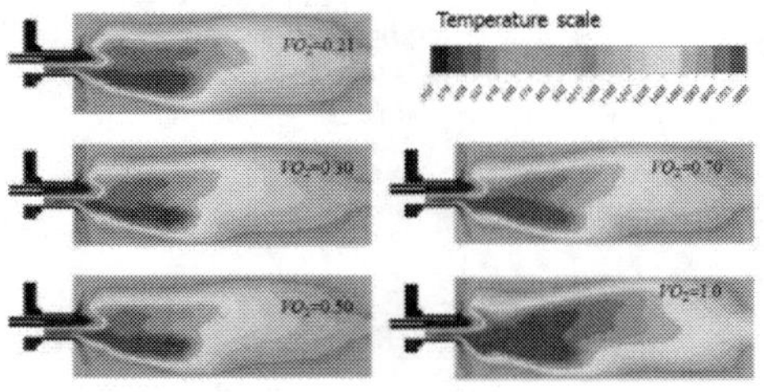

Figure 5: Flame temperature contours under various O_2 concentrations in the O_2-enriched flows

The average temperature across the exit plane of the preheating chamber can be used as an index for the effective HTAC [9, 23]. The averaged exit temperature was above 1170K, indicating that higher temperature flue gas can effectively heat up the primary air. When VO_2 < 50%, the average exit temperature barely increases with VO_2. However, when VO_2 varies over 21%-100%, the exit temperature still only increases ~25 K. Thus, the O_2 enrichment using present configuration keeps the effectiveness of original HATC implementation.

Ignition occurrence was defined on the isothermal curve of 1173 K. Figure 6 shows the ignition curve on the central horizontal plane of the burner when VO_2 in the enriched flow varied from 0.21 to 1.0. The area around the turning points was a place with concentrated coal particles, thus a main combustion zone. When O_2 enrichment was enhanced, the ignition location moved towards the burner center line, indicating ignition was accelerated.

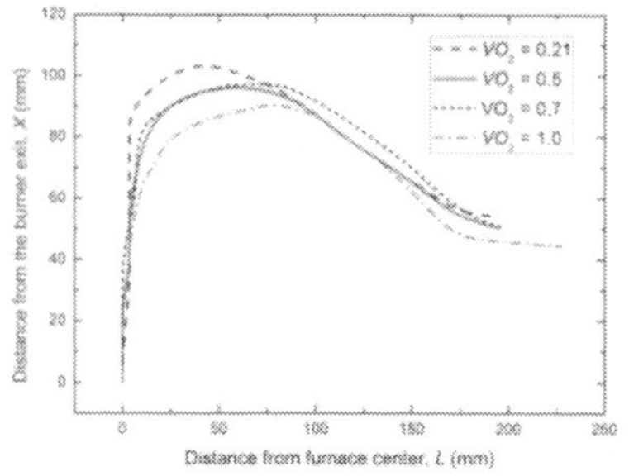

Figure 6: Variation of ignition location with O_2 concentration in the O_2-enriched flow

Simulation results show roughly, every 10% increment of VO_2 raises T_{max}, ~11 K. Given T_{max} was already ~1900 K, the thermal NOx counts for a large portion in total NOx formation. Any remarkable augment of T_{max} could significantly increase thermal NOx formation. On the other hand, VM release was enhanced at a higher T_{max}. The changes of net NOx formation depend on the competition between these two mechanisms.

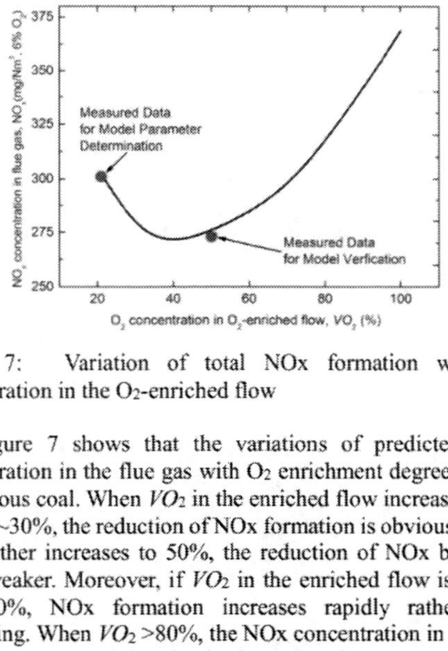

Figure 7: Variation of total NOx formation with O_2 concentration in the O_2-enriched flow

Figure 7 shows that the variations of predicted NOx concentration in the flue gas with O_2 enrichment degree for SB bituminous coal. When VO_2 in the enriched flow increases from 21% to ~30%, the reduction of NOx formation is obvious. When VO_2 further increases to 50%, the reduction of NOx becomes much weaker. Moreover, if VO_2 in the enriched flow is higher than 50%, NOx formation increases rapidly rather than decreasing. When VO_2 >80%, the NOx concentration in the flue gas becomes even higher that in air combustion.

Figure 8 depicts the variations of predicted furnace-averaged thermal and fuel NOx production rates, and NOx reduction rate with VO_2 in the O_2-enriched flow. The production rate of fuel NOx is 3~4 times faster than thermal NOx. When VO_2 <50%, the rate of NOx reduction is greater than that of thermal NOx formation. There is a peak in difference when VO_2 = 50%-70%. However, as VO_2 further increases, thermal NOx formation rate approaches to the NOx reduction rate. When pure O_2 stream is applied, production rate of thermal NOx increases more than 50%. However, the formation of fuel NOx is insensitive of VO_2 in the O_2-enriched flow. NOx reduction rate first decreases with VO_2 when VO_2 < 30%, and increases with VO_2 when VO_2 = 30% -70%; and a peak value exist at VO_2 = 50%-70%. There was an optimal VO_2 existed for NOx reduction in OR-HATC burner. For present configuration, the optimal VO_2

≈ 50%, at which reduction rate is largest while the production rates for fuel NOx and thermal NOx formations are rather weak.

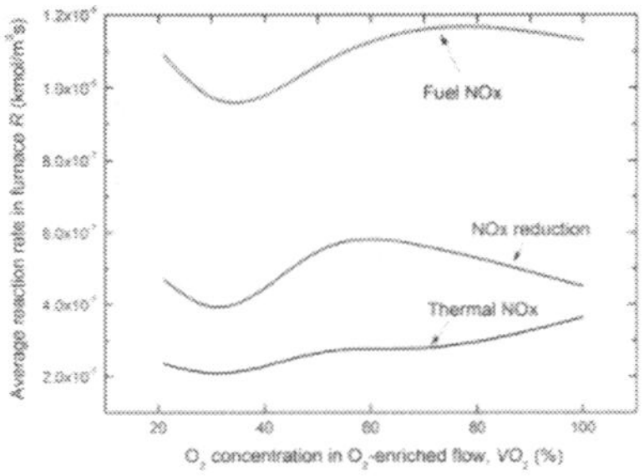

Figure 8 Variation of NOx formation rates and reduction rate with O_2 concentration in the O_2-enriched flow

4 Conclusions

Oxygen-enriched High Temperature Air Combustion (OE-HTAC) was proposed aiming at further improving flame stabilization and reducing NOx formation in the existing HTAC and a model burner was designed to implement such a combustion technology. The combustion performance of the OE-HTAC model burner was evaluated using an 8MW test rig and 3D numerical simulation. Experimental results showed that OE-HTAC burner remarkably improved flame stabilization even with a small extra amount of O_2 consumption, and rather low O_2 concentrations in the O_2-enriched flow. Combined O_2 enrichment with HTAC could be also effective to reduction NOx formation as well. For bituminous coal, ~50% reduction was achieved while for the low volatile coals, ~20% NOx was reduced against the air combustion. The simulation results illustrated that high temperature zone outside the burner was enlarged and flame temperature increased with O_2 enrichment. The temperature at the burner slightly increased with O_2 enrichment. Simulation with EDC model for gas phase combustion improved NOx formation and reduction prediction. The predicted NOx emission agreed with the data measured from the model burner experiments. In addition, simulation showed that there was optimal VO_2 for NOx reduction in OE-HATC burner. For present configuration, the optimal $VO_2 \approx$ 50%, at which reduction rate is largest while the production rates for fuel NOx and thermal NOx formations are rather weak.

Overall, present study confirmed that with proper burner design and organization of O_2 enrichment, O_2-enriched HTAC technology can be a practicable, efficient and clean technology for pulverized coal boilers.

Acknowledgment

The financial support from MOST project (No. 2015BAA04B01) is appreciated.

References

[1] Weinberg, F. J., *Combust. Sci. Technol.*, 121 (1996) 3–22.

[2] M. Katsuki, T. Hasegawa, *Proc. Combust. Inst.* 27(1998), 3135 - 3146.

[3] T. Fujimori, D. Riechelmann, J. Sato, *Proc. Combust. Inst.* 27(1998), 1149-1158..

[4] J. Yuan, I. Naruse, *Energy Fuels* 13 (1999) 99-106.

[5] R. Weber, S. Orsino, N. Laliemant, A Verlaan, *Proc. Combust. Inst.* 28(2000), 1315–1321.

[6] S. Sugiyama, N. Suzuki, Y. Kato, K. Yoshikawa, A. Omino, T. Ishii, K. Yoshikawa, T. Kiga *Energy* 30 (2005) 399–413.

[7] R. Weber, J. P. Smart, W. Kamp, *Proc. Combust. Inst.* 30 (2005) 2623-2629.

[8] T. Suda, M. Takafuji, T. Hirata, M. Yoshino, J. Sato, *Proc. Combust. Inst.* 29(2002),503–509.

[9] H. Zhang, G. X. Yue, J. F. Lu, Z. Jia, J. X. Mao, T. Fujimori, T. Suko, T. Kiga, *Proc. Combust. Inst.* 31 (2007), 2779-2785

[10] N. Schaffel-Mancini, M. Mancini, A. Szlek, R. Weber, *Energy*, 35(7), (2010) 2752-2760.

[11] N Schaffel, M Mancini, R Weber, *Combust Flame* 156(9) , (2009)1771-1784.

[12] C. Galletti, A. Parente, L. Tognotti, *Combust Flame*, 151(4), (2007) 649-664.

[13] G. G. Szegö, B. B. Dally, G. J. Nathan. *Combust Flame*, 156, (2009) 429-438。

[14] P. Li, J. Mi, B. B. Dally, R. A. Craig, F. Wang. Energy Fuels, 25 (7), (2011) 2782–2793。

[15] E. Abtahizadeh, J. van Oijen, P. de Goey. *Combust Flame*,159(6), (2012) 2155-2165.

[16] M. Derudi, R. Rota. *Proc. Combust. Inst.* 33 (2011), 3325-3332

[17] L. Chen, S.Z. Yong, A.F. Ghoniem, *Progress in Energy and Combustion*, 38(2), (2011) 3325-3332

[18] A. Molina, C.R. Shaddix, *Proc. Combust. Inst.* 31(2), (2007) 1905–1912

[19] C. R. Shaddix, A. Molina, *Proc. Combust. Inst.* 32(2), (2009) 2091-2098.

[20] B. Liu, Z. Zhang, H. Zhang, H. Yang, D. Zhang. *Fuel*, 116(15), (2014) 77-83

[21] P. F. Li, J. C. Mi, B. B. Dally et al. *Science China Technological Sciences*, 54(2), (2011), 255-269.

[22] A. Molina, E. G. Eddings, D. W. Pershing, A. F. Sarofim, *Combust Flame*,136 (2004) 303–312.

[23] K. Cui, H. Zhang, W.L. Wang, Y.X. Wu, H.R Yang, J.F. Lu. *J. of Engr. Thermophysics* 33(11), (2012) 2006-2009. (in Chinese)

[24] S. B. Pope, *Combust. Theor. Model.* 1, (1997), , 41−63.

[25] K. Cui, B. Liu, H. Zhang, Y. X. Wu, K. Matsumoto, K. Takeno, *Energy Fuels* 27, (2013) 2246−2254

[26] E. H. Chui, H. Gao. Fuel, 89(10), (2010) 2977-2984.

[27] S. R. Gubba, D. B. Ingham, K. J. Larsen K J, et al. Fuel Processing Technology, 104, (2012),181-188.

[28] K. Cui, Ph.D. dissertation, 2013, Tsinghua University.

11th Asia-Pacific Conference on Combustion,
The University of Sydney, NSW Australia
10th -14th December 2017.

Modelling the effects of ethanol fumigation on engine performance and emissions in a six-cylinder, common rail diesel engine

F Lodi[1], T Bodisco[2,*], N Surawski[3], RJ Brown[4], Y Yousef[1]

[1]College of Engineering, Alfaisal University

Alzahrawi Street, Almaather, Riyadh, 11533, Saudi Arabia

[2]School of Engineering, Deakin University

75 Pigdons Road, Waurn Ponds, VIC, 3220, Australia

[3]Faculty of Engineering and IT, University of Technology, Sydney

81 Broadway, Ultimo, NSW, 2007, Australia

[4]Biofuel Engine Research Facility, Queensland University of Technology

2 George Street, Science and Technology Faculty, Brisbane 4000, Australia

Abstract

This paper describes a one-dimensional thermodynamic model developed using AVL BOOST with the objective of analysing the performance, combustion parameters and NOx emissions of a Euro III, six-cylinder turbocharged Cummins diesel engine. The model was validated against experimental data obtained from the same engine run at a constant speed of 2000 rpm at varying load conditions (full, three quarter (3Q) and half load) using low sulphur diesel fuel (D100), as well as fumigated ethanol at 10% (D90), 20% (D80) and 30% (D70) substitutions (by energy). The results for D100, D90, D80 and D70 were found to be in good agreement with the experimental data. The percentage variation for engine performance parameters such as: brake power (BP), indicated power (IP), indicated torque (IT) and mean effective pressure (MEP) for D100 have been found to be approximately in the range of -5% to 1.5% for all loads, whereas, the fuel energy variation was only 0.33% for all loads. With increasing ethanol fumigation, a rise in peak pressure of the cycle, more rapid initial heat release rate and a reduction in the NOx emissions were observed in this study.

1 Introduction

Alcohol is a form of renewable energy which can be produced from carbon-based agricultural feed stocks, locally grown crops and even waste products [1]. Sugarcane residue is another renewable energy source of alcohol production [2]. In the last decade, an increasing trend of alcohol fuel production from renewable sources has been noted globally [3]. Ethanol is considered to be one of the more promising alternative fuels. Compared to methanol, ethanol has a lower enthalpy of vaporisation and auto-ignition temperature, hence better ignition characteristics [4, 5].

Broukhian et al. [6] found that knock at high loads limited the proportion of ethanol which could be fumigated. At nearly all engine operating conditions, modest gains in thermal efficiency were observed, while NOx decreased at all load conditions. In fact, with the use of fumigated alcohol, if efficient air utilisation takes place, then it can result in even greater power output. Although displacement of diesel by ethanol can be done to about 50%, it has limitations, which have been discussed in detail by [4, 7, 8]. Furthermore, as the evaporation of alcohol in the intake manifold lowers the temperature of the incoming air, which results in an increase in density, more air, and hence oxygen, could be made available for combustion [8, 9]. Ghadikolaei et al. [10] found that fumigating alcohols in diesel engines leads to a reduction of NOx and CO_2 in most tests, and PM in all cases.

Fumigated ethanol has been shown to cause a significant increase in the rate of pressure rise, relative to operation using diesel fuel alone [11]. The increased rate of pressure rise has been attributed to the auto-ignition of the homogeneous fumigated charge after initial combustion reactions had already raised the cylinder temperature and pressure [10]. It has been reported that this effect is worsened by an increase in ignition delay, associated with the cooling effect of the fumigated alcohol [10]. Similar findings associated with an increase in ignition delay with ethanol fumigation have also been reported [4, 5, 8, 9]. Bodisco et al. [12] utilised combustion resonance as a means of determining the onset of ignition. The experiments comprised of fumigated ethanol substitutions up to 50% at full load, 3Q load and half load. The results have demonstrated that at full load there is a decrease in ignition delay with increasing ethanol substitutions, whereas at half load there is an increase in ignition delay with increasing ethanol substitutions. Bodisco et al. [12] suggested that this conflicting result is a consequence of the auto-ignition of ethanol. This has also been reported in other literature sources [11].

As ethanol is fumigated into the engine, higher values of peak pressure are obtained owing to more rapid combustion [13]. Other combustion parameters, which are affected by ethanol fumigation, are ignition delay and combustion duration [13]. Fuels having a low cetane number (such as ethanol) will typically have a longer ignition delay, as ignition occurs sufficiently late in the expansion process, owing to the high latent heat of vaporisation of ethanol [13]. This has been reported in the literature by various authors [12, 13]. With an increase in load, a decrease in ignition delay is observed, attributable to a richer fuel supply [13]. Ignition timing will affect most of the in-cylinder parameters including in-cylinder pressure, maximum rate of pressure rise, peak cylinder pressure

* Corresponding author.
E-mail address: t.bodisco@deakin.edu.au

and temperature, hence affecting NOx emissions. Combustion duration has also been observed to decrease by increasing the quantity of ethanol fumigated. This has been attributed to enhanced mixing of fuel and air and an increased availability of oxygen in the cylinder, due to the changes in physico-chemical properties of the air-fuel mixture, namely; combustion temperature, oxygen concentration, latent heat of vaporisation, fuel distribution, ignition delay and cetane number [13]. The peak heat release rate (HRR) is also observed to increase as the quantity of injected ethanol is increased. In comparison to diesel fuel only (with no ethanol), the crank angle at which the peak heat release rate occurs is also retarded by 1-4° with ethanol fumigation [13]. This is due to ethanol being an oxygenated fuel, the HRR increases due to the availability of more oxygen and thus the combustion is improved [13].

Owing to the complex combustion issues associated with ethanol fumigation and the potential gains in terms of reducing global dependence on non-renewable fuels, this paper will detail an AVL BOOST engine model for investigating the effects of ethanol fumigation in terms of performance and emissions. This model can be utilised in future studies to understand the behaviour of ethanol during cold-start conditions, improve our fundamental understanding of the effects of ethanol fumigation and to develop strategies for the control of engines utilising ethanol as a secondary fumigated fuel.

2 Development of the AVL BOOST Model

2.1 Engine Model

A six-cylinder engine model has been built using AVL BOOST software, replicating the geometry and dimensions of the engine for generating and connecting different elements used for model design, such as: cylinders, engine, turbocharger, intercooler, ECU (engine control unit) and the engine interface for the ECU to communicate to the engine [14]. The intake and exhaust manifolds, turbocharger and intercooler are connected in the workspace through pipes, having measuring points at the ends, wherever required, and the pipe flow model has been developed as a one-dimensional thermodynamic model based on the first law of thermodynamics. The elements are defined according to the physical parameters of the engine and the physical geometries such as pipes, manifolds, etc. have been defined after being measured accurately on the engine.

The model is based on the fuel consumption data and start of injection data from the engine running at 2000 rpm (at different load conditions) from the experiments conducted by Bodisco et al. [5].

2.2 Combustion Model Setup

The combustion model used in the engine is a zero-dimensional (two-zone spatial discretisation), with the energy conservation between the burned and unburned zones being maintained. The two-zone model allows for the calculation of emissions such as NOx, CO and soot. The double vibe function, used in the combustion model, is defined by two single vibe functions, i.e. the start of combustion and the combustion duration, defined by a parameter '*m*' (denoted as the shape parameter). The following form of equations can be used to describe this single vibe function and the rate of heat release approach [14].

$$\frac{dQ_B}{d\alpha} = Q_B \frac{a}{\Delta\alpha_c}.(m+1).y^m.e^{-a.y^{(m+1)}} \qquad (1)$$

where $y = \frac{\alpha-\alpha_0}{\Delta\ \alpha_c}$, Q_B is fuel heat input, α is the crank angle in degrees, α_0 is the angle of the start of combustion in degrees, $\Delta\alpha_c$ is the angle of combustion duration in degrees, m is the shape parameter and a is the vibe parameter [14]. Here, the value of the vibe parameter, a, is set to 6.9, in order to obtain complete combustion. The values of the shape parameter m, the vibe parameter a, combustion start and combustion duration are specified as an input to the combustion model. These values can be specified either as constant values or as a function of engine speed (rpm) and engine load. To obtain an estimate of the required combustion duration and selection of the variables, a standard table can be referred to in the manual. [14]. The choice of these parameters in the model for dual fuel (diesel/ethanol) was possible because of the availability of experimental fuel consumption and combustion start data for this paper.

2.3 Engine Model Simulation

The model was run at 2000 rpm (full, 3Q and half load) for diesel only fuel. An engine operating speed of 2000 rpm was selected as rated power is achieved at this speed, for this engine. The model is further reformed to include a secondary ethanol injector to inject ethanol in varying proportions in fumigation mode. The location of the ethanol injector is selected to be after the intercooler (before the intake manifold). In the cases of ethanol injection, a portion of the diesel fuel, on an energy basis, is replaced by ethanol fuel, as 10% (D90), 20% (D80) and 30% (D70), with the fuel data used from the experiments conducted by [5].

3 Results and Discussion

3.1 D100 (Diesel only)

The results are presented in tabulated form for engine performance parameters. Rows 3 and 4 of Table 2 show the results for D100 compared with experimental results. Figure 1 shows the comparison between the pressure vs crank angle curve obtained from the experiment and the model for full load, with very good agreement reached between the experiment and the model.

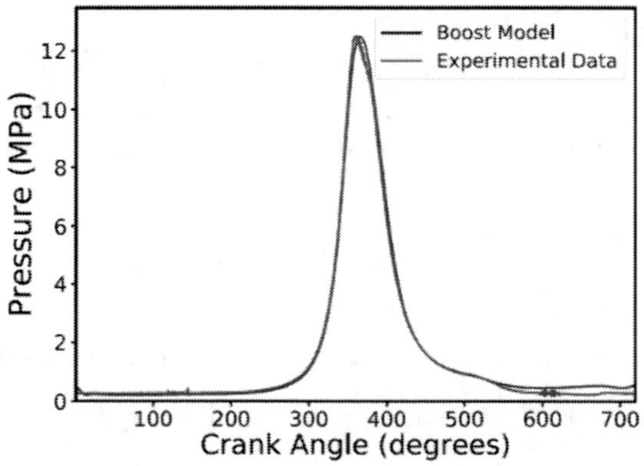

Figure 1: Comparison of in-cylinder pressure for Cylinder 1 for D100, full load, 2000 rpm between the BOOST model and the experimental data

3.2 D90, D80 and D70 (ethanol fumigation)

Rows 6 and 7 of Table 2 show the model results for engine performance parameters for D90 (10% ethanol fumigation), for 2000 RPM, full load, compared with experimental results. Tables for D80 and D70 have been omitted owing to space issues. However, the D90 results shown in Table 2 are reflective of the overall fit. Owing to space constraints, tabulated results 3Q and half load for D100 and D90 have been omitted. Similarly, the D80 and D70, 2000 RPM at full, 3Q and half load conditions have not been shown. However, the discussion of the results will be all-encompassing.

3.3 Discussion

As the ethanol fuel is injected before the intake manifold, it gets pre-mixed with the incoming charge of air and enters as a homogeneous mixture of ethanol and air. As diesel is injected, it initiates the combustion, igniting the homogeneous mixture of ethanol and air. This results in faster combustion, with a higher rate of heat release for the same volume of fuel, especially in the early stages of combustion, consequently resulting in higher peak pressure.

At full load, more rapid pre-mixed combustion and higher peak pressures are observed, with increasing ethanol fumigation, as also reported in the literature [12, 13], which would result in better combustion, with the peak pressure occurring closer to TDC. An important factor that would contribute to the high peak pressure would be the auto-ignition temperature and the lower cetane number of ethanol, compared to diesel. This pattern is seen in the present model, which is the same as that observed in the experiments at full load (refer to Table 2 for full load readings and Fig 2). At half load, there is an increase in the ignition delay and decreased combustion duration, as also reported in the literature by [12, 13]. This suggests that a decrease in combustion duration leads to higher peak pressures, but this effect will be compensated by the increase in ignition delay, and hence the values of peak pressure only see a limited increase with increased ethanol fumigation. However, as the quantity of fumigated ethanol is further increased, this effect becomes more significant and there is a decrease in the peak pressure values, but due to more oxygen being available for combustion with increasing ethanol, the combustion is improved and hence, we see an improvement in the performance parameters of the engine. This is the pattern observed in the present model and the same is observed from the experimental data.

At 3Q load, it is observed in the model that there is hardly any change, or very minimal change, in the performance parameters of the engine. At the other loads, with an increase in ethanol fumigation, the performance parameters are enhanced. This pattern can be observed in the model and in the experiments. This could be due to no change in the ignition delay and combustion duration at the 3Q load condition, and the cooling effects of ethanol would cause a decrease in the temperature of the inlet manifold and then an overall temperature decrease in the combustion chamber, leading to lower peak pressure.

There is a slight improvement in the performance of the engine with ethanol fumigation in the present model; with the brake thermal efficiency increasing from 36% with D100 to 38% with D90 (10% ethanol fumigation), at full load. The increase in the brake thermal efficiency can be attributed to the high laminar flame speed of ethanol, which would result in increased pre-mixed combustion—which is evident from the higher rate of pressure rise and increased in-cylinder pressure. The increased pre-mixed combustion, due to the availability of more homogeneous air-ethanol mixture at higher loads, is also verified by the heat release diagram, Fig 2, with ethanol fumigation of 30% (D70) having the highest HRR. In the D100 (diesel only fuel) case, the model performance parameters for 3Q and half load are very close to the experimental performance parameters, with the variations approximately around 1.5% (approximately 7% for full load). However, the variations slightly increase in all cases of ethanol fumigation for lower loads, but decrease at full load. In the D90 (10% ethanol) case, the variations in the performance parameter are in the range of approximately 4% to 5% for full load and approximately 5% to 6% for 3Q and half load. For D80 (20% ethanol), the variations improve with increasing ethanol fumigation and are in the range of 0.5% for full and 3Q load, and 2% for half load conditions. For D70 (30% ethanol), the variations are in the range of approximately 2% to 3% for full and 3Q load, while it is approximately 8% to 9% for half load. This could be due to the limitations of the combustion model in AVL BOOST when it handles the complexity of the incoming charge, as a fumigated homogeneous mixture of ethanol-air and then pockets of heterogeneous diesel fuel, as diesel is injected into the cylinder to initiate combustion. However, as the ethanol fumigation further increased and a more homogeneous mixture is available for the engine, the combustion (in the model) seems to become more stable, with the variations in the performance parameters coming down for higher ethanol fumigation cases, as compared to D90 (10% ethanol).

Figure 2 shows the heat release rate in the model for different fuel setups (D100, D90, D80 and D70) for full load. D70 shows the highest rate of heat release with the highest in-cylinder peak pressure. The limitations with the HRR diagram can be attributed to two main causes. Firstly, the combustion model used in AVL BOOST has limited capability to handle the complexity of the charge when ethanol is fumigated. Secondly, the combustion duration and other combustion parameters such as the 'shape parameter' – 'm' (m = 6.9 to have complete combustion), with ethanol fumigated cases have been kept the same as D100 in the model. The combustion start timings were slightly advanced for ethanol fumigation cases in the model, according to the experimental data, however, the combustion duration was not changed, as compared to D100. The trend shown in Fig 2 is quite satisfactory, showing the increase in heat release (more rapid pre-mixed combustion phase) with increasing ethanol fumigation. With the availability of a faster burning homogeneous mixture of ethanol-air, the combustion is faster and earlier. Cycle-by-cycle variations and ignition delay could be two very important parameters to investigate the combustion process and its limitations observed in this paper.

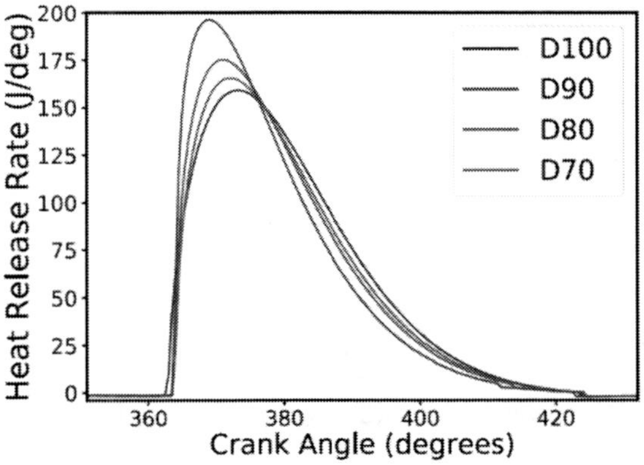

Figure 2: Heat release rate for increasing ethanol fumigation (D100, D90, D80 and D70) for full load, 2000 RPM, as determined by the BOOST model for Cylinder 1.

As the model is developed according to the rate of flow of fuel in the experiments, with ethanol replacing the diesel fuel on the basis of substitution by energy, the fuel energy variation for D100, as well as for all ethanol fumigation cases, between the model and experiments, comes out to be approximately 0.33%.

The combustion model used in the engine model developed in AVL BOOST has limited capability to accurately calculate the NOx emissions. Rather, it gives the trend of the emissions at various operating points, which is also influenced by the model. The multipliers used in the model to influence the emissions have been tuned up for the D100 case, and then maintained to be approximately the same for ethanol fumigation cases to estimate the trend of the NOx emissions. NOx results are shown in Table 1, the NOx trend in the model has been found to be similar to the experiments.

Load	D100 Model	D100 Exp	D90 Model	D90 Exp	D80 Model	D80 Exp	D70 Model	D70 Exp
	(ppm)	(ppm) *	(ppm)	(ppm)	(ppm)	(ppm) *	(ppm)	(ppm) *
Full	72.7		61.5	60.9	62.9	59.8	59.5	59.4
3Q	53.5		47.9	50.9	43.8		45.1	48.6
Half	26.3		32.3	36.5	23.8		23.1	

Table 1: Boost model and experimental NOx results
*some experimental data was not available

4 Conclusions

This modelling work was undertaken to understand the effects of ethanol fumigation in a Euro III, six-cylinder diesel engine running under steady state conditions, at 2000 RPM, under varying load conditions. The model discussed in this paper has been found to be in good agreement with the experimental results in terms of engine performance, combustion parameters, and NOx emissions. It has provided insight into the behaviour of ethanol, when supplied in the fumigated mode, for the engine running under steady state, at various load conditions. The variation of engine performance parameters, combustion parameters and NOx in the model have been well contained within 5% as compared to the experimental data, except for the case of D70, at half load, where the engine performance parameters are more than 5%. The model can serve as a useful tool for further potential research involving more complex transient running conditions of the engine, as well as to investigate the behaviour of fumigating other secondary fuels with diesel. The limitations of the combustion model can be a subject for further study, as there are changes in the combustion process with the use of different fuels, owing to chemical and physical differences.

Acknowledgement

The authors would like to express their appreciation to AVL List GmbH for supplying the AVL BOOST software through their University Partnership Program. Special thanks are also due to Caltex for donating the ethanol used in the original experimental campaign. Lastly, the authors would like to express their gratitude to Mr Noel Hartnett and Dr Svetlana Stevanovic for their support.

Parameter	Engine Speed	NOx	Brake Power	Indicated Power	Indicated Torque	Brake Mean Effective Pressure	Indicated Mean Effective Pressure	Brake Thermal Efficiency	Peak Firing Pressure	Fuel Energy (Diesel)	Fuel Energy (Ethanol)	Total Fuel Energy
Unit	rpm	ppm	kW	kW	Nm	kPa	kPa	%	kPa	kJ	kJ	kJ
D100 Experiment	2006	N/A	157.2	161.6	771.6	1600.4	1648.1	39.03	12418	24.2	0.0	24.2
D100 Model	2000	72	146.9	152	725.8	1498	1550.3	36.27	12656	24.3	0.0	24.3
% variation	-0.3	-	-6.9	-6.3	-6.3	-6.8	-6.3	-7.5	1.9	0.3	0.0	0.3
D90 Experiment	2003	61	158.5	162.2	774.5	1617.5	1641.7	40.20	12471	21.7	1.9	23.6
D90 Model	2000	61.5	151.6	156.6	747.7	1547	1597.1	38.57	12697	21.7	1.9	23.6
% variation	-0.2	1.0	-4.5	-3.5	-3.5	-4.5	-2.7	-4.2	1.7	0.0	0.3	0.1

Table 2: Comparison of engine performance data for D100 and D90 between the model and the experiment at 2000 RPM, full load

References

[1] M. Balat, Energ. Source. 27 (6) (2005) 569-577.

[2] E. Bocci, A. Carlo, and D. Marcelo, Energ. 34 (2009) 689-698.

[3] R. Gilbert, and A. Perl, Energy and Transport Futures, University of Calgary, (2005) 1-96.

[4] M. Qudais, O. Haddad, and M. Qudaisat, Energ. Convers. Manage., 41 (2000) 389-399.

[5] T. Bodisco, S. Low Choy, and R.J. Brown, Appl. Ther. Eng. 60 (1-2) (2013) 79-87.

[6] E. Broukhiyan, and S. Lestz, SAE Technical Paper 811209 (1981).

[7] B. Chauhan, N. Kumar, S. Pal, and J. Du, Energ. 36 (2) (2011) 1030-1038.

[8] G. Prakash, A. Ramesh, and A. Shaik, SAE Technical Paper 1999-01-0232 (1999).

[9] Z. Sahin, and O. Durgun, Energ. Convers. Manage., 48 (7) (2007) 1952-1964.

[10] M. Ghadikolaei, Renew. Sust. Energ. Rev., 57 (2016) 1440-1495.

[11] T. Hayes, L. Savage, R. White, and S. Sorenson, SAE Technical Paper 880497 (1988).

[12] T. Bodisco, P. Trondle, and R.J. Brown, Energ., 84 (2015) 186-195.

[13] D. Jamuwa, D. Sharma, and S. Soni, Energ. Convers. Manage., 115 (2016) 221-231.

[14] AVL BOOST, User guide, Edition, 2010.

11th Asia-Pacific Conference on Combustion,
The University of Sydney, NSW Australia
10th -14th December 2017.

Large eddy simulation of reacting flows in a scramjet combustor

Majie Zhao[1], Taohong Ye[1*], Hong Ye[1]

[1] Department of Thermal Science and Energy Engineering, University of Science and Technology of China
Hefei, Anhui, China

Abstract

Large eddy simulation (LES) has been performed to investigate transverse hydrogen jet mixing and combustion process in a scramjet combustor model with a compression ramp at inlet to generate shock train. Partially Stirred Reactor (PaSR) sub-grid combustion model with a skeleton of 19 reactions and 9 species hydrogen/air reaction mechanism was used. Comparisons of statistical results show that the present simulations are capable of predicting flow and combustion profiles. Intense combustion consists of two regions: one is in the near field upstream due to the intense bow shock compression heating and the second region locates the downstream flow field(x/D>12). The combustion process is dominated by non-premixed mode in the mixing layer and is dominated by premixed mode near the wall.

1 Introduction

Transverse jet in supersonic cross-flow (JISCF), as a basic structure of the scramjet combustor and its high fuel/air mixing efficiency, has been widely studied through experiments and numerical simulations[1,2,3]. Mixing process of fuel and oxidizer which is mainly affected by complex large scale coherent structures and shock structures is the major point of interests in the previous non-reacting case of JISCF combustor[4]. interactions between those coherent structures and the complex three dimensional shock waves in the non-reacting case have important effect on the mixing process, and thus affect the combustion process in the reacting case of JISCF combustor. Although some work[5,6] have been explored, the mechanism of turbulent mixing and combustion process effected by shock waves are still not clear completely, and further research is needed.

Early studies limited by the measurement techniques of supersonic transverse jet combustion were focus on hydrogen jet ignition and flame stabilization[7]. In recent years, the transverse hydrogen injection system has been conducted successfully on the shock tube experiment and the ground based flight test, such as Gamba combustor [8], Hyshot II scramjet engine [9], X-43A scramjet. HyShot II scramjet configuration[9] was designed to provide benchmark experimental data including wall pressure and wall heat flux distribution on supersonic combustion, which was critical for the later researchers [10,11] to study the interaction between the wall and the flame. Hanson et al[8] of Stanford University studied the effect of jet to cross-flow momentum flux ratio (J=0.3-5.0) on flame stabilization mechanism and ignition position by changing the jet inlet pressure, and suggested that the flame stabilization was supported by self-ignition. NO-PLIF and PIV were used by Mai et al[12] to study the interaction between an oblique shock wave and supersonic transverse jet combustion. However, because of the extremely complicated measurement and huge investment of experimental study, computational fluid dynamics (CFD) has attracted increasing attention on the development of the efficient, high-performance scramjet engine. Compared to the RANS method, which could only provide some mean field information without the turbulent characteristics of complex unsteady supersonic flow, large eddy simulation (LES) directly resolves the largest turbulent structures and can provide more unsteady phenomena, which has become a prosperous tool to study supersonic transverse jet turbulent mixing and combustion process.

In the present research, LES-PaSR has been performed to investigate interactions of fuel jet flame with the shock train structure that develops along the constant area combustor due to the compression ramp of 10° in the Gamba combustor[13]. LES has a strong potential in the study of the turbulent mixing and combustion process of supersonic transverse jet combustion. In our previous study[4], large eddy simulations has been performed to investigate the flow structures and mixing characteristics of a high-enthalpy hydrogen JISCF combustor. Moreover, based on the open source code OpenFOAM, LES-Flamelet model and LES-PaSR model have been adopted to study the turbulent mixing and combustion process in scramjet[14,15]. The paper which explores this study is structured as follows. Section 2 presents physical model and numerical methods, and LES results are summarized in section 3. Finally, the paper finishes with conclusions in section 4.

2 Physical model and numerical methods

2.1 Description of JISCF combustor

Computational domain of the JISCF combustor is similar to the experimental setup at Stanford University[13]. The side-view schematic of the chamber is shown in Figure 1. The injection orifice with a diameter of D = 2 mm is located at 70 mm downstream the cross-flow inlet and the width is 75 mm in the span-wise z-direction. The flow conditions of the incoming air stream and the hydrogen jet are given in Table 1.

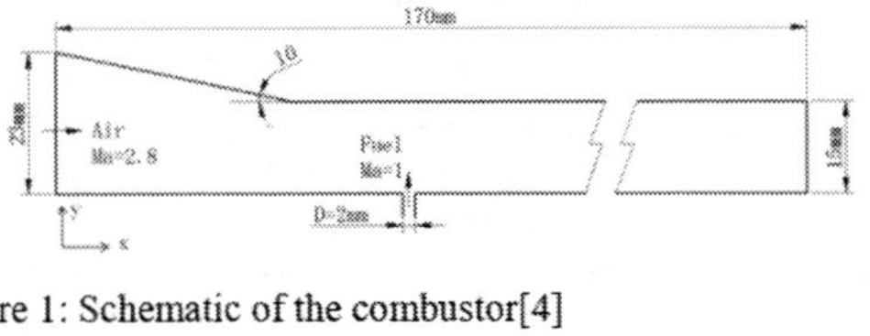

Figure 1: Schematic of the combustor[4]

Corresponding author. Fax: +0086-551-63603048
E-mail address: thye@ustc.edu.cn

J	M_a	P_a/kP_a	T_a/K	$Y(H_2)$	$Y(O_2)$	$Y(N_2)$
2.11	1.0	668.25	250	1.0	0	0
Air	2.8	40	1200	0	0.232	0.768

Table. 1 Inflow conditions of the hydrogen jet and incoming air stream

2.2 Computational methods

The filtered LES compressible governing equations are given by

$$\frac{\partial \bar{\rho}}{\partial t}+\frac{\partial \bar{\rho}\tilde{u}_j}{\partial x_j}=0 \tag{1}$$

$$\frac{\partial \bar{\rho}\tilde{u}_i}{\partial t}+\frac{\partial \bar{\rho}\tilde{u}_i\tilde{u}_j}{\partial x_j}=-\frac{\partial \bar{P}}{\partial x_i}+\frac{\partial}{\partial x_j}\left(\bar{\tau}_{ij}-\tau_{ij}^{sgs}\right) \tag{2}$$

$$\frac{\partial \bar{\rho}\tilde{Y}_m}{\partial t}+\frac{\partial \bar{\rho}\tilde{u}_j\tilde{Y}_m}{\partial x_j}=\frac{\partial}{\partial x_j}\left[\bar{\rho}D\frac{\partial \tilde{Y}_m}{\partial x_j}-\tau_{\Phi}^{sgs}\right]+\overline{\dot{\omega}_m}\ (m=1,\cdots,N) \tag{3}$$

$$\frac{\partial \bar{\rho}\tilde{E}}{\partial t}+\frac{\partial}{\partial x_j}\left[(\bar{\rho}\tilde{E}+P)\tilde{u}_j\right]=\frac{\partial}{\partial x_j}\left[\lambda\frac{\partial \tilde{T}}{\partial x_j}+\tilde{u}_j\bar{\tau}_{ij}-H^{sgs}-\sigma^{sgs}\right]+\overline{\dot{\omega}_E} \tag{4}$$

where superscript "$-$" denotes spatially-filtered parameters, and "~" denotes Favre-filtered parameters for LES. $\bar{\rho}$ is the filtered density. The filtered velocity vector is denoted by $\tilde{u}_{j(j=1,2,3)}$. $\bar{P}$ is the pressure. The filtered viscous stress tensor $\bar{\tau}_{ij}$ can be computed using the filtered strain rate tensor $\tilde{S}_{ij}$. $\tilde{Y}_m$ is the filtered mass fraction of species m and D is the molecular mass diffusivity. λ is the thermal diffusivity, $\lambda=\mu C_p/Pr$ and μ is the laminar viscosity. The filtered total energy $\tilde{E}$ is the sum of kinetic and internal energies $\tilde{E}=\tilde{e}+\frac{1}{2}\tilde{u}_j^2+k^{sgs}$. Perfect gas state equation (5) is assumed to link the relation between pressure and temperature,

$$\bar{P}=\bar{\rho}R(\tilde{Y}_m)\tilde{T} \tag{5}$$

where $R(\tilde{Y}_m)$ is the mixture gas constant.

In the present study, the sub-grid stress τ_{ij}^{sgs} is closed by the dynamic Smagorinsky model, and the sub-grid enthalpy flux H^{sgs} and the sub-grid viscous work σ^{sgs} in Eqs. (4) are closed as

$$H^{sgs}+\sigma^{sgs}=-\frac{\mu_t C_p}{Pr_t}\frac{\partial \tilde{T}}{\partial x_j}-(\mu_t+\mu)\frac{\partial k^{sgs}}{\partial x_j}+\tilde{u}_i\tau_{ij}^{sgs} \tag{6}$$

where the sub-grid kinetic energy and viscosity are modeled by $k^{sgs}=C_I\Delta^2(\tilde{S}_{ij}\tilde{S}_{ij})$ and $\mu_t=\bar{\rho}C_D\Delta^2(\tilde{S}_{ij}\tilde{S}_{ij})^{1/2}$ respectively. Δ is the filter width. The sub-grid scalar stresses are approximated using an eddy-diffusivity model, which is written as,

$$\tau_{\Phi}^{sgs}=\left(\bar{\rho}\widetilde{u_j Y_m}-\bar{\rho}\tilde{u}_j\tilde{Y}_m\right)=-\bar{\rho}\tilde{D}_t\nabla\tilde{\Phi} \tag{7}$$

where $\tilde{D}_t$ is the turbulent diffusivity modeled as $\bar{\rho}\tilde{D}_t=\mu_t/Sc_t$. The turbulent Schmidt number is set as $Sc_t=0.7$.

The Partially Stirred Reactor (PaSR) model is used to model the interactions between turbulence and combustion. the reacting structures shows some correlation with the dissipative structures of the flow and the spatial scales is typically smaller than Δ in high Re number turbulent combustion, in which the flame is highly wrinkled, perforated or fragmented. Therefore, it could be assumed that the reacting flow is composed of reacting fine structures and the surroundings interacting with the large scale structures as the PaSR model. The ratio of the fine reacting structures volume to the LES cell volume is modeled as

$$\kappa=\frac{\tau_c}{\tau_c+\tau_m} \tag{8}$$

where, τ_c and τ_m denote the chemical reaction time and subgrid turbulent mixing time, respectively. The reaction time scale τ_c is determined by a chemical kinetics model of 9 species (H_2, O_2, H_2O, H, O, OH, HO_2, H_2O_2 and N_2, where N_2 is regarded as inert species), 19 reaction steps for hydrogen[16], which has been adopted to study supersonic combustion by some researchers[17,18]. The subgrid turbulent mixing time is calculated as

$$\tau_m=c_m\sqrt{\tau_\Delta\tau_k} \tag{9}$$

where, c_m is the model constant based on specific flow configurations, $\tau_\Delta=\Delta/v'$ is the sub-grid time scale and $\tau_k=(\nu/\varepsilon)^{1/2}$ is the Kolmogorov time scale, where Δ is the computational cell scale, $v'=\sqrt{2k/3}$ is the sub-grid velocity fluctuation, k is the sub-grid turbulent kinetic energy, ν is the laminar kinematic viscosity and $\varepsilon=v'^3/\Delta$ is the sub-grid dissipation rate.

The filtered equations are solved with a fully explicit finite volume discretization, density based solver developed in OpenFOAM[19]. The convective fluxes are reconstructed using a second order TVD scheme and viscous diffusion fluxes discretization is implemented using the second order central differencing scheme. The marching time-step is adjusted according to Courant number under 0.25 to capture the unsteady flow field structure. In addition, the fine mesh in present LES is about 85 million.

3 Results and discussion

Comparisons of the predicted and experimental results of mean pressure distribution on the upper wall is presented in Fig. 2. The LES results show a fairly good agreement with the experiments of Gamba et al[13]. Fig. 2 also shows the logarithmic distribution of density gradients($\log\|\nabla\bar{\rho}\|$) in the central plane (z/D=0). Due to the compression ramp of 10°, the shock train that develops along the constant area combustor interacts with the fuel jet and walls. The pressure profile increases steeply on the upper wall where the interactions between the reflected shock and the upper wall take place. In addition, the grid sensitivity analysis is also presented in Fig. 2.

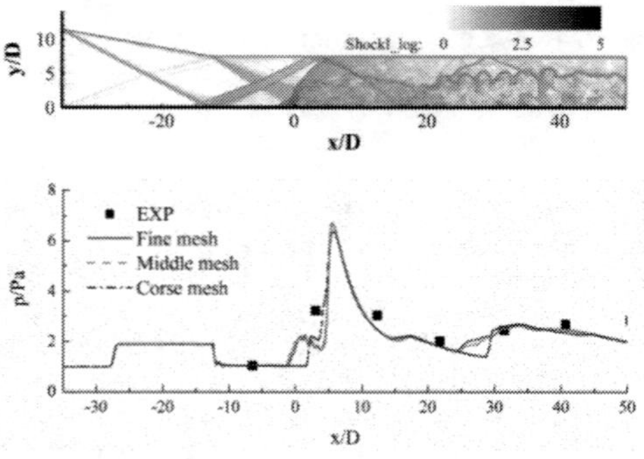

Figure 2: logarithmic distribution of density gradients($\log\|\nabla\bar{\rho}\|$) in the central plane (z/D=0), and the predicted and

experimental results of mean pressure distribution on the upper wall.

Figure 3 shows the predicted results of instantaneous distribution of OH mass fraction (down) in the central plane (z/D=0) and the OH-PLIF snapshot by Gamba et al[13] (top). Although the PLIF signal depends on OH number density not mass fraction, It is nearly the same by using OH mass fraction instead. The main characteristic combustion process such as the flame structure, interactions between shocks and flame, interactions between shocks and wall, could all be captured by the present LES and the subgrid combustion model. It could be seen that the flame is anchored in the recirculation zone upstream of the jet orifice, and the windward-side flames are wrinkled and stretched caused by the unstable characteristics of jet shear layer. The jet shear layer flame just after the bow shock is bright and enhanced due to the bow shock compression and heating.

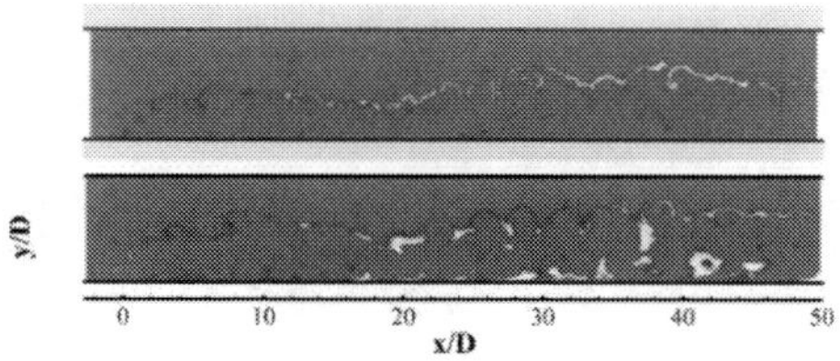

Figure 3: OH-PLIF snapshot by Gamba et al[13] (top) and the predicted results of instantaneous distribution of OH mass fraction (down) in the central plane (z/D=0).

Figure 4 shows the predicted isosurface of instantaneous OH mass fraction colored by Takeno Flame Index[20] and isosurface of Lamda2-criterion colored by instantaneous H2 mass fraction. Takeno Flame Index (TKI) is adopted for identification of premixed or non-premixed modes. Takeno Flame Index greater than zero indicates a premixed flame while Takeno Flame Index less than zero indicates a non-premixed flame. It could be found that intense combustion consists of two regions: one is in the near field upstream due to the intense bow shock compression heating and the second region locates the downstream flow field(x/D>12). Combining the instantaneous OH mass fraction distribution and Takeno Flame Index, it can be found that the combustion process is dominated by non-premixed mode in the mixing layer and is dominated by premixed mode near the wall especially the region where the shock wave interacts with the wall. After the interaction of reflected oblique shock and jet plume, the combustion process is promoted and enhanced due to the enhanced mixing by the complex large scale coherent structures and compression heating by the shock structures.

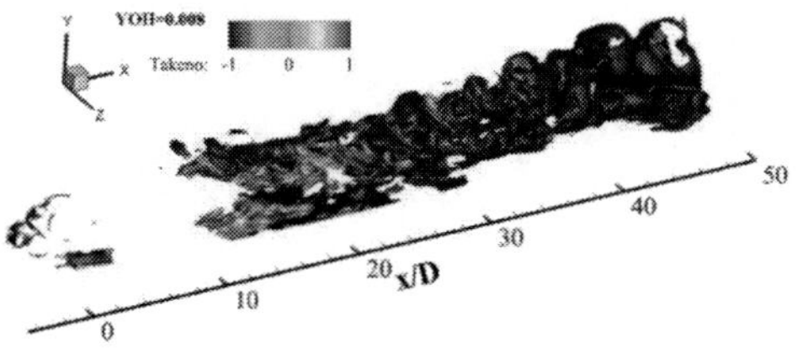

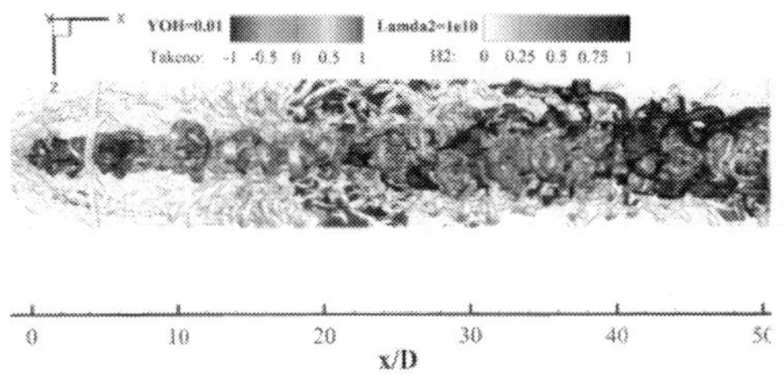

Figure 4: Isosurface of instantaneous OH mass fraction colored by Takeno Flame Index and isosurface of Lamda2-criterion colored by instantaneous H2 mass fraction.

4 Conclusions

LES-PaSR has been performed to investigate mixing and combustion process in the high-enthalpy hydrogen JISCF combustor. some conclusions could be summarized briefly as follows.

The numerical solver developed based on OpenFOAM is evaluated for supersonic reacting flows. The flame distribution and wall-pressure distribution predicted from the LES simulation show a fairly good agreement with the experiments. The flame structures identified by combining the Takeno Flame Index and OH mass fraction suggest that the combustion process is dominated by non-premixed mode in the mixing layer and is dominated by premixed mode near the wall especially the region where the shock wave interacts with the wall. Moreover, the complex large scale coherent structures and compression heating by the shock structures formed in the transverse jet in supersonic cross-flow with a compression ramp have important effect on the turbulent mixing and combustion process.

5 Acknowledgment

This work was supported by the National Natural Science Foundation of China (Grant No. 91441117 and Grant No. 51576182). The numerical calculations in this paper have been performed on the supercomputing system in the Supercomputing Center of University of Science and Technology of China.

References

[1] Huang W, Yan L. Progress in research on mixing techniques for transverse injection flow fields in supersonic crossflows[J]. Journal of Zhejiang University SCIENCE A, 2013, 14(8): 554-564.

[2] Mahesh K. The interaction of jets with crossflow[J]. Annual review of fluid mechanics, 2013, 45: 379-407.

[3] Huang W. Transverse jet in supersonic crossflows[J]. Aerospace Science and Technology, 2016, 50: 183-195.

[4] Zhao, M., Ye, T., Cao, C., Zhou, T., & Zhu, M. (2016). Study of sonic injection from circular injector into a supersonic cross-flow using large eddy simulation. *International Journal of Hydrogen Energy*, *41*(39), 17657-17669.

[5]Peterson D M, Boyce R R, Wheatley V. Simulations of mixing in an inlet-fueled axisymmetric scramjet[J]. AIAA journal, 2013.

[6] Gehre R M, Wheatley V, Boyce R R. Combustion Regimes in Inlet-fueled, Low Compression Scramjets[C]//20th AIAA

International Space Planes and Hypersonic Systems and Technologies Conference. 2015: 3507.

[7] Yoshida A, Tsuji H. Supersonic combustion of hydrogen in a vitiated airstream using transverse injection[J]. AIAA Journal, 1977, 15(4): 463-464.

[8] Gamba, M., & Mungal, M. G. (2015). Ignition, flame structure and near-wall burning in transverse hydrogen jets in supersonic crossflow. Journal of Fluid Mechanics, 780, 226-273.

[9] Smart M K, Hass N E, Paull A. Flight data analysis of the HyShot 2 scramjet flight experiment[J]. AIAA journal, 2006, 44(10): 2366-2375.

[10] Fureby C, Chapuis M, Fedina E, et al. CFD analysis of the HyShot II scramjet combustor[J]. Proceedings of the Combustion Institute, 2011, 33(2): 2399-2405.

[11] Chapuis M, Fedina E, Fureby C, et al. A computational study of the HyShot II combustor performance[J]. Proceedings of the Combustion Institute, 2013, 34(2): 2101-2109.

[12] Mai T, Sakimitsu Y, Nakamura H, et al. Effect of the incident shock wave interacting with transversal jet flow on the mixing and combustion[J]. Proceedings of the Combustion Institute, 2011, 33(2): 2335-2342.

[13] Gamba, M., Miller, V., Mungal, G., & Hanson, R. (2012, January). Combustion characteristics of an inlet/supersonic combustor model. In *50th AIAA Aerospace Sciences Meeting Including the New Horizons Forum and Aerospace Exposition* (p. 612).

[14] Cao, C., Ye, T., & Zhao, M. (2015). Large eddy simulation of hydrogen/air scramjet combustion using tabulated thermo-chemistry approach. *Chinese Journal of Aeronautics*, *28*(5), 1316-1327.

[15] Berglund, M., Fedina, E., Fureby, C., Tegnér, J., & Sabel'Nikov, V. (2010). Finite rate chemistry large-eddy simulation of self-ignition in supersonic combustion ramjet. *AIAA journal*, *48*(3), 540-550.

[16]C.J. Jachimowski, An Analytical Study of the Hydrogen - air Reaction Mechanism with Application to Scramjet Combustion, NASA-TP-2791, 1988.

[17]Gong, C., Jangi, M., Bai, X. S., Liang, J. H., & Sun, M. B. (2017). Large eddy simulation of hydrogen combustion in supersonic flows using an Eulerian stochastic fields method. *International Journal of Hydrogen Energy*, *42*(2), 1264-1275.

[18]Wang, H., Wang, Z., Sun, M., & Qin, N. (2015). Large eddy simulation of a hydrogen-fueled scramjet combustor with dual cavity. *Acta Astronautica*,*108*, 119-128.

[19] Jasak H, Jemcov A, Tukovic Z. OpenFOAM: A C++ library for complex physics simulations[C]//International workshop on coupled methods in numerical dynamics. IUC Dubrovnik, Croatia, 2007, 1000: 1-20.

[20] Yamashita, H., Shimada, M., & Takeno, T. (1996, January). A numerical study on flame stability at the transition point of jet diffusion flames. In*Symposium (International) on Combustion* (Vol. 26, No. 1, pp. 27-34). Elsevier.

11th Asia-Pacific Conference on Combustion,
The University of Sydney, NSW Australia
10th -14th December 2017.

Large eddy simulation of fuel injection under trans/supercritical conditions

Wu Wei, Maozhao Xie, Ming Jia, Hongsheng Liu, Liang Li

School of Energy and Power Engineering, Dalian University of Technology, Dalian 116024, China

Abstract

The n-heptane jet with subcritical and supercritical inlet temperatures injected into a supercritical environment filled with multi-component gas, has been numerically investigated using large eddy simulation. Mainly attentions are paid on the effects of density-gradient, thermodynamic and transport properties on the flow evolution. We find that large density-gradients are easier formed in transcritical injection and the larger specific heat has the effects of increasing the density gradients. While the diffusion coefficient decrease in the larger density gradient regions, suggesting that large density gradient has the effects of suppressing the flow diffusion mixing. In addition, when the fluid transits across the pseudo boiling temperature, the specific heat obtained a maximum value, while this value is suppressed by the dissolution effects, indicating that multi-component injection is different from the single species injection and a special investigation is needed.

1 Introduction

Many combustion devices, including internal combustion engines, gas turbines and liquid rockets, operate in high pressure and temperature regimes with significantly optimized fuel injection systems and increased mixing efficiency between the fuel and ambient gas, which will be of great importance to combustion thermal efficiency and pollution emission. As such, the elevated pressure and temperature in the combustion chamber may be above the critical pressure and temperature of the injected fuel, while the temperature of the injected fuel may be below or above its critical temperature, which is corresponding to supercritical and transcritical injection, respectively. However, the mixing behaviors under such conditions have been only limited understood, especially for diesel engines and gaining a deep insight into the fundamental mechanisms of the fuel jet evolution under trans/supercritical condition is a prerequisite for controlling and improving the injection and mixing process.

Research over the past decade has provided significant insights into the mixing process in the rocket engine combustion chamber. Recent experimental studied were conducted by Mayer, Oschwald and Cheroudi et al[1-3] with the injection temperature below and above injecting into a constant-pressure chamber. In their experiments, the classical breakup and atomization process in subcritical regime cannot be seen due to the decrease of surface tension between fuel jet and ambient gas. So drops and liquid ligaments are no longer detected, instead, a large number of finger-like structures emerge from the jet core. Driven by the space propulsion, recent experiment at internal combustion engine can be found in the review papers of Zoldak et al[4] and Engine Combustion Network[5]. In numerical simulation aspect, Zong et al [6,7] perform an investigation on cryogenic nitrogen injected into supercritical nitrogen gas environment. They suggest that the density-gradient play an important role on jet evolution and has the effects of stabilizing jet surface. Identically, Park[8] conduct a serials of cryogenic nitrogen injection with different turbulent model, and it is found that turbulent models have limited influence on jet evolution, while the real fluid equation of state plays an important role on the numerical simulation. For supercritical fuel injection investigations, Oefelein et al[9,10] using the n-heptane and dodecane inject into a constant-volume chamber filled with nitrogen or air. They suggest that fuels enter the chamber as a compressed liquid fluid, rather than a spray, and the real-fluid equation of state must be considered under supercritical condition.

In the present study, studies are focused on the density gradient effects on transcritical and supercritical injection. In addition, the disintegration features are analyzed.

2 Governing equations

The conservation equation of mass, momentum, and energy are Favre-filtered and solved by LES. These equations are derived by filtering the small-scale dynamics from the resolved scales over a set of spatial and temporal intervals, and can be express as:

$$\frac{\partial \bar{\rho}}{\partial t}+\frac{\partial(\bar{\rho}\tilde{u}_i)}{\partial x_i}=0 \tag{1}$$

$$\frac{\partial(\bar{\rho}\tilde{u}_i)}{\partial t}+\frac{\partial(\bar{\rho}\tilde{u}_i\tilde{u}_j+\overline{p\delta_{ij}})}{\partial x_j}=\frac{\partial(\tilde{\tau}_{ij}-\tau_{ij}{}^{SGS})}{\partial x_j} \tag{2}$$

$$\frac{\partial \bar{\rho}\tilde{e}_t}{\partial t}+\frac{\partial[(\bar{\rho}\tilde{e}_t+\bar{p})\tilde{u}_i]}{\partial x_i}=\frac{\partial[\tilde{u}_j\tilde{\tau}_{ij}+\bar{q}_i-H_i{}^{SGS}+\sigma_i{}^{SGS}]}{\partial x_i} \tag{3}$$

$$\frac{\partial \tilde{\rho}\bar{Y}_i}{\partial t}+\frac{\partial}{\partial x_j}(\tilde{\rho}\bar{Y}_i\bar{u}_j-\tilde{\rho}D_{i,m}\frac{\partial \bar{Y}_i}{\partial x_j}+\Phi_{j,m}^{SGS})=0 \tag{4}$$

Generally, these unclosed SGS terms are difficult to solve and models are needed to close them. Hence, Smagorinsky model[11] is used to model the unclose terms. Finally, the PR real equation of state is used to closed the group of governing equation. The viscosity and thermal conductivity are calculated by using the model of Chung et al[12]. For simplification, the mass diffusion coefficients of the individual components are computed using Kinetic Theory.

The density, constant-pressure specific heat, viscosity and thermal conductivity predicted by PR EOS and Chung et al are shown in Figure 1. In addition, the NIST[13] data are also plot for comparison. The density and specific heat agree well with NIST measurement except that the specific heat has a large deviation from NIST data at around the pseudo-boiling point, but it follows the trend well. For the viscosity and thermal

Corresponding author. Fax: +86-0411-8470-6302
E-mail address: xmz@dlut.edu.cn

conductivity, though there are some difference between the calculation and NIST data, but they follow the trend well, indicating that the calculation of the thermodynamic properties and transport properties is reliable

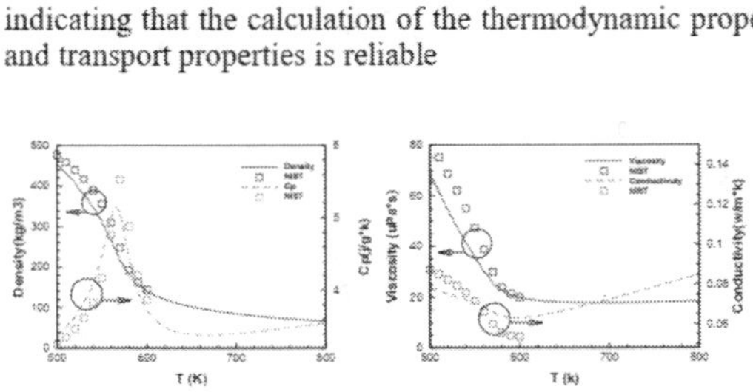

Fig1. Density, constan-pressure specific heat, viscosity and thermal conductivity calculated by PR EOS and compare with NIST data

3 Flow configuration and numerical method

The physical model in Hossain et al[7] is simulated, where n-heptane with initial temperature 573K is injected into the chamber filled with a mixture of water, oxygen and nitrogen, and the composition of species are listed in table 1. The initial temperature and pressure in the chamber are 770K and 4.0 MPa, respectively, which is well above the critical values (P_c=2.74MPa and T_c=540K) of the injected n-heptane. For comparison, a transcritical case with initial injection temperature 520K while other conditions do not change is also calculated, and the initial injection conditions of the two cases are shown in table 2. The physical model is cylindrical chamber with an injection diameter D_{inj}=0.15mm at axial direction. The cylindrical computational domain is 55 mm in the axial directions with a diameter of 20 mm, which is sufficient to eliminate the influence of far-field boundary conditions on the near-injector flow. The geometry has been discretized using an O-grid with a total number of cells about 4.8 million.

The filtered compressible Navier-Stokes equations are solved by means of finite-volume method. According to the method suggested by Park[11], the PISO algorithm is lightly modified to be capable of handling the real-gas fluid. For LES, in order to obtain high accuracy, the convective and viscous terms are discrete by means of second-order scheme, along with a second-order implicit scheme for temporal derivatives. To improve the computation efficiency, a multi-block domain decomposition technique with message passing interfaces (MPI) is used for parallel computing.

Table 1. Chamber initial condition

O_2(mole%)	H_2O(mole%)	N_2(mole%)	P_∞(Mpa)	ρ_∞(kg/m^3)
0.064	0.128	0.808	4.0	16.64

Table 2. Physical parameters for simulation

case	T_{inj}[K]	U_{inj}[m/s]	T_∞[K]	P[MPa]	ρ_{inj}/ρ_∞	Re_{inj}
1	573	348	770	4.0	13.7	432800
2	520	348	770	4.0	26.4	334200

4 Result and discussion

Figure 2 show the penetration length based on the species of n-heptane for the two cases, and the experimental data conducted by Kishwar et al[4] are given for comparison. Penetration depth for the simulation is the maximum axial distance of an iso-volume where the mass-fraction value is 1% of n-heptane. By comparing case 1 with measurement data, the trend of simulation result follows the experiment well, but there is some deviation because of the experiment data given in the experiment is uncertain as the supercritical jet penetrates further into the combustion chamber; The penetration length in case 2 is larger than case 1 because the larger density ratio has the effects of stabilizing the jet flow.

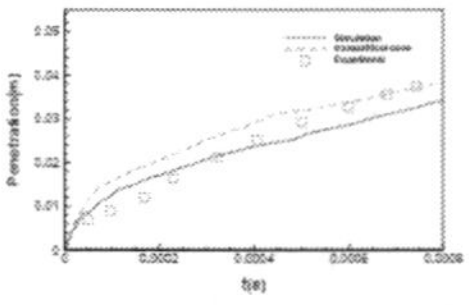

Fig.2 Measured and simulation penetration of heptane

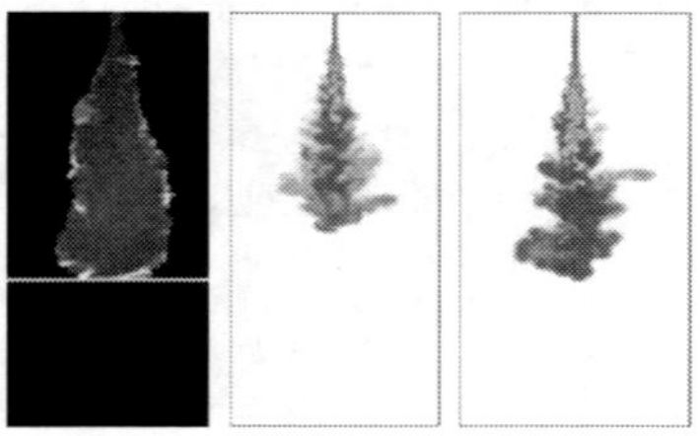

a) t=0.3 ms

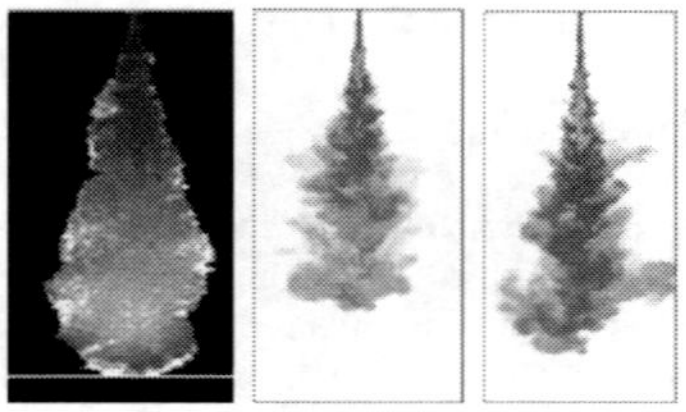

b) t=0.7 ms

Fig.3 Experimental plume and simulation contours based on mass-fraction

Figure 3 shows the contours of the mass-fraction of n-heptane and graph of visual characteristics obtained by Kishwar et al[7] for case 1 at t=0.3 ms and t=0.7 ms. As seen here, the simulation basically capture the shape of the experimental jet well. By close inspect, the jet surface is more easily disturbed and wrinkled in case 1, indicating that supercritical injection has the ability of increasing mixing efficiency. Hence, the potential core in case 2 is longer due the larger density ratio suppressing the development of flow into the radial direction.

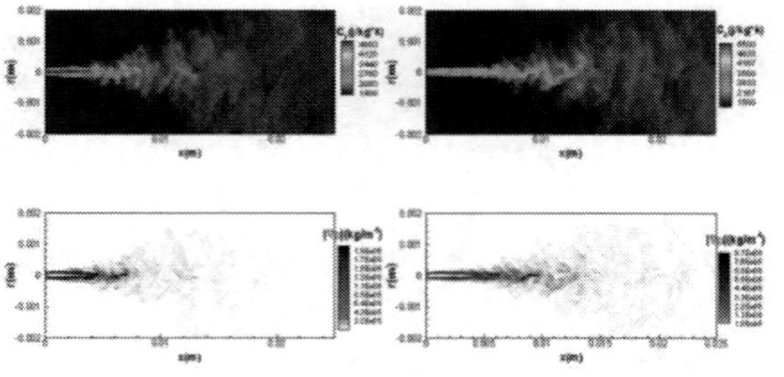

Fig.4 Snapshots of density gradient, constant pressure specific heat, for case 1(left column) and case 2(right column)

Figure 4 shows the snapshots of constant pressure specific heat and density gradient in the x-r plane at θ=0 for the two cases. For the supercritical injection of case 1, the largest constant pressure specific heat is concentrated on the potential core region, as the fuel jet mix with the surrounding gas, this value gradually decrease. However, different from case 1, the specific heat in the transcritical of case 2 reach the maximum values at the end of potential core districts, where the temperature transit across the pseudo-critical point. In the density gradient distributions, these values show much larger in case 2 due to the large density ratio effects and forming a longer potential core. Besides, the large density gradient distribute in the pseudo-critical regions, indicating that besides density ratio, the pseudo boiling temperature has some effects of forming high density gradient.

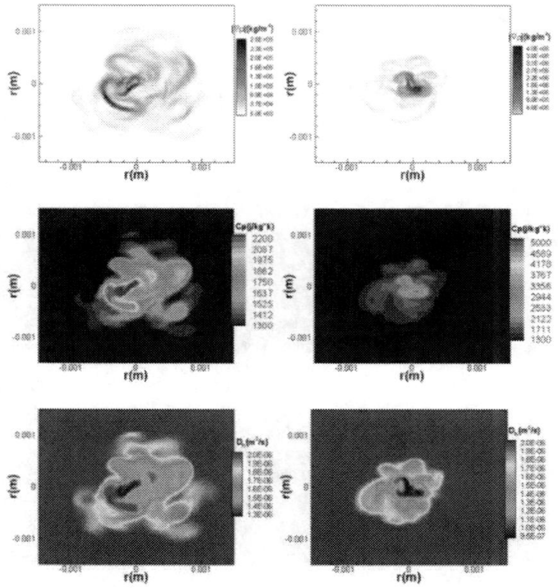

Fig.5 Snapshots of density-gradient, constant pressure specific heat and diffusion coefficient at x=0.008 for case 1(left column) and case 2(right column)

To further understand the supercritical and transcritical injection, the radial distribution of density gradient, constant pressure specific heat and diffusion coefficient at x=0.008m in Figure 5 are used to reveal the mixing feature. An analysis of the thermodynamic and transport properties distribution on the large density gradient is significant important to understand the mixing mechanism under supercritical injection. By comparing the two cases, the large density-gradient regions mainly concentrate on the center and the value is much larger in case 2. For this reason, the diffusion coefficient in case 1 is larger than case 2 and the distribution of thermodynamic and transport properties spread wider domain in radial direction. Based on Figure 4, the conditional averages for both cases are listed in table 3 and 4. By comparing the two cases, the constant pressure specific heat values are larger in case 2 at all cutoffs, which is consistent with the contour plots above. However, the diffusion coefficient shows the opposite result. The constant pressure specific heat increases with the elevation of cutoff value, indicating that the large specific heat has the effects of forming large density. While the diffusion coefficient decrease in the larger density gradient regions, indicating that large density gradient has the effects of suppressing the flow diffusion mixing.

Table 3. Conditional averages of specific heat and diffusion coefficient over regions where $|\nabla\rho|$>cutoff in case 1

Cutoff	average Cp(j/kg*k)	average D_h(m^2/s)
$0.1|\nabla\rho|_{max}$	1.733e+3	1.583e-6
$0.2|\nabla\rho|_{max}$	1.829e+3	1.522e-6
$0.3|\nabla\rho|_{max}$	1.882e+3	1.481e-6

Table 3. Conditional averages of specific heat and diffusion coefficient over regions where $|\nabla\rho|$>cutoff in case 2

Cutoff	average Cp(j/kg*k)	average D_h(m^2/s)
$0.1|\nabla\rho|_{max}$	2.494e+3	1.044e-6
$0.2|\nabla\rho|_{max}$	2.836e+3	1.001e-6
$0.3|\nabla\rho|_{max}$	3.087e+3	9.851e-7

The instantaneous data are collected to calculate the mean properties after the flow reach its stationary state. Figure 7 shows the mass fraction of n-heptane and the constant-pressure specific heat along the centerline. The n-heptane fraction profile remains unchanged in the potential core region, whose length is larger in the transcritical case. In constant-pressure specific heat profiles, it keeps a constant value followed by a smooth decrease in case 1. While this is of great difference in the transcritical case 2, where there is a peak value that is absent in the supercritical case 1. This is because the temperature transit across the pseudo boiling point, where more energy is consumed to increase its temperature. While the maximum value is downsized comparing to that of the n-heptane (seen in Fig.1), for the surrounding gas has dissolve into the jet and the critical properties of jet has been changed. Hence, the multi-component mixing jet is different from the single specie inject mixing and the dissolution effects play an important role on the jet mixing feature.

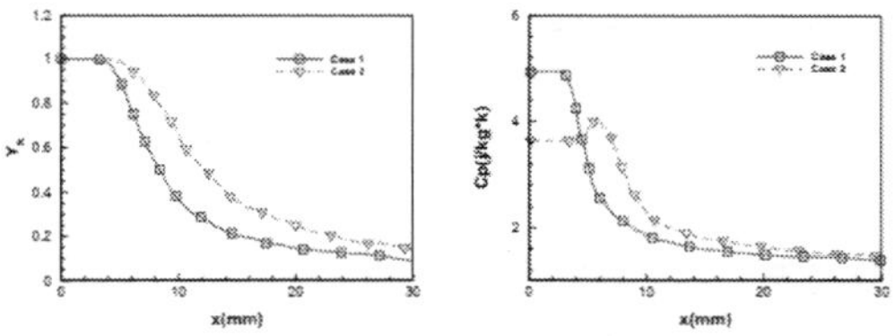

Fig7. Mass fraction of n-heptane and constant-pressure specific heat along the jet centerline

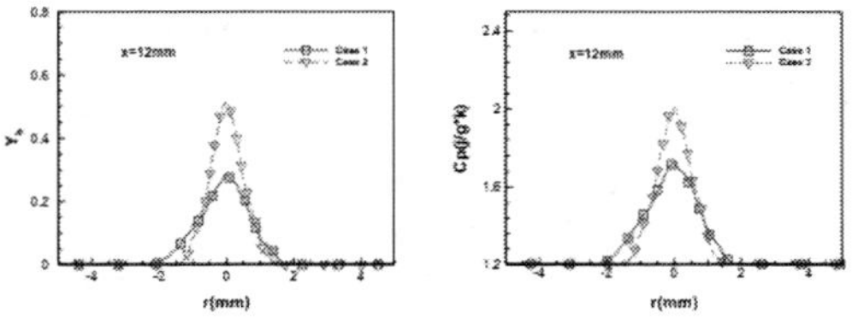

Fig8. Mass fraction of n-heptane, constant-pressure specific heat in radial distribution at x=12mm

To further understand the mixing behavior, the mass fraction of n-heptane and constant pressure specific heat at x=12mm are shown in Figure 8. It is notable that these properties show similar distribution, where a maximum value is obtained at the axial position and have a much larger value in the transcritical case 2 around the jet centerline. However, as the fluid jets develop far from the axis position, these properties are larger in the supercritical case 1, indicating that n-heptane develop deeper into the radial position.

The T-Tc and P-Pc are displayed in Figure 9. It is found that the pressures are all above its critical point, and the temperature is subcritical for the transcritical case while supercritical for the supercritical injection. To further understand the disintegration behavior in transcritical and supercritical injection, the injection fluid field can be divided into several parts: the length of the liquid-like core L_L: T-Tc<0, the length of the dense core L_D where the density-gradient reach to a maximal value, the end of the transitional region L_T which the density-gradient come to an unchanged region. According this definition, the length of each part is displayed in Figure 9, and these values are listed in table 5. One can recognize that liquid-like core is still exists, indicating that the surface tension may not disappear completely, while in the supercritical case, no liquid-like regions are detected, where intensive turbulent diffusion dominates the mixing process. Because of this reason, the length of the dense core (L_D) and the end of transition region (L_T) shows much higher value in the transcritical injection than in the supercritical injection. All these disintegration features indicate that the supercritical jet tend to be more easily mixed with the ambient gas and reach the self-similar region.

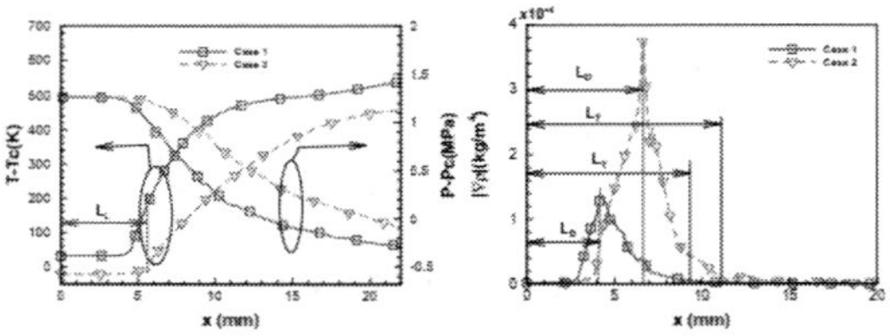

Fig.9 T-Tc, P-Pc distribution (left) and density-gradient (right) along the jet centerline

Table 5. The length of disintegration at each part for the transcritical and supercritical injection

Length type	Case 1	Case 2
L_L	---	5.82mm
L_D	4.1mm	6.52mm
L_T	9.22mm	11.0mm

4 Conclusions

Supercritical and transcritical multi-species injection mixing process are investigated using large eddy simulation. The thermodynamic and transport properties are calculated based on the real gas equation of state, and the PISO algorithm was slightly modified to handle the real fluid state.

The thermodynamic and transport properties are easily affected by the density-gradient. Larger density-gradient is formed in the transcritical case, therefore a longer potential core region is detected and the jet penetrates further downstream. The constant pressure specific heat increases with density-gradient, indicating that the large specific heat has the effects of forming large density-gradient. While the diffusion coefficient decrease in the larger density gradient regions, suggesting that large density gradient has the effects of suppressing the flow diffusion mixing. By comparing with heptane, the specific heat of mixture in the pseudo-critical temperatures is suppressed for the change of critical parameter, indicating that multi-component injection is different from the single species injection and a special investigation is needed.

Investigation is performed on the disintegration mechanism in transcritical injection and supercritical injection and we find that the jet disintegration process in transcritical injection is different from that in supercritical, and the thermal-dynamic and transport properties play an important role in the disintegration behavior.

5 Acknowledgment

This work is supported by National Natural Science Foundation of China (Grant No. 51376029. No. 51576029).

References

• Oschwald M, Schik A. Supercritical nitrogen free jet investigated by spontaneous Raman scattering. *Experiments in Fluids*. 1999;27(6):497-506.

• Chehroudi B, Cohn R, Talley D. Cryogenic shear layers: experiments and phenomenological modeling of the initial growth rate under subcritical and supercritical conditions. *International Journal of Heat & Fluid Flow*. 2002;23(5):554-563.

• Chehroudi B, Talley D, Coy E. Visual characteristics and initial growth rates of round cryogenic jets at subcritical and supercritical pressures. *Physics of Fluids*. 2002;14(2):850-861.

• Hossain K, Qiu J, Shetty S, Zoldak P, Boer CD. Transonic Combustion: Model Development and Validation in the Context of a Pressure Chamber. *Sae Technical Papers*. 2012.

• ECN. Engine Combustion Network. http://www.sandia.gov/ecn/. [Online].

• Zong N, Meng H, Hsieh SY, Yang V. A numerical study of cryogenic fluid injection and mixing under supercritical conditions. *Physics of Fluids*. 2004;16(12):4248-4261.

• Zong N, Yang V. Gryogenic fluid jets and mixing layers in transcritical and supercritical environments. *Combustion Science & Technology*. 2006;178(1):193-227.

• Park TS. LES and RANS simulations of cryogenic liquid nitrogen jets. *Journal of Supercritical Fluids the*. 2012;72(12):232–247.

• Oefelein JC, Yang V. Modeling High-Pressure Mixing and Combustion Processes in Liquid Rocket Engines. *Journal of Propulsion & Power*. 1998;14(5):843-857.

• Oefelin J. Analysis of transcritical spray phenomena in turbulent mixing layers. Paper presented at: 34th Aerospace Sciences Meeting and Exhibit, 1996.

• Smagorinsky J. General circulation experements with the primitive equations. *Monthly Weather Review*. 1963;91(3):99-164.

• Chung TH, Ajlan M, Lee LL, Starling KE. Generalized multiparameter correlation for nonpolar and polar fluid transport properties. *Industrial & Engineering Chemistry Research*. 1988;27(27):671-679.

• http://webbook.nist.gov/chemistry/fluid

11th Asia-Pacific Conference on Combustion,
The University of Sydney, NSW Australia
10th -14th December 2017.

Large Eddy Simulation of Dilute Spray Dispersion and Evaporation in Different Co-flow Temperatures

YANG Qiong, GONG Huifeng, HU Zongjie

School of Automotive Studies, Tongji University

4800 Cao An Road, Shanghai, 201804, P. R. China

Abstract

By combining Large Eddy Simulation (LES) model and Discrete Phase Model (DPM), the study investigates turbulent dispersion and evaporation of non-reacting dilute sprays. The results of computation are in conformity with measurements. Comparing two cases in different co-flow temperatures (313K and 433K), the study finds: Both gas and droplet axial mean velocities at central axial positions rise remarkably with the increase of the co-flow temperature. The Stokes number shows little difference in two co-flow temperatures. Dynamically steady state exists during the evaporation period, and the higher the co-flow temperature is, the less time it takes to reach the state. The droplet distribution space decreases and the n-heptane vapor concentration gradient becomes higher when the co-flow temperature increases.

1 Introduction

Spray combustion is widely applied in industries, including piston engines and gas turbines. How to improve their efficiencies and reduce toxic emissions has become a world-wide issue. Spray evaporation plays an important role in combustions as evaporation rates often directly affect combustion rates, efficiencies as well as product components.

Considering that real sprays in engines are too complicated to investigate and single droplet researches neglect turbulent dispersions, this study focuses on dilute sprays in which droplet intervals are much larger than droplet diameters and droplet-droplet interactions can be negligible. Crowe [1,2], Mashayek [3] and Santanu De [4] have done some comprehensive reviews of two-phase flow simulations, but the work has been based on Reynolds averaged approaches which cannot describe details of turbulent flows. Bini and Jones [5,6] did some LES simulations of dilute spray evaporations based on the experimental results of Masri [7], but they paid attention mainly on modeling rather than on spray evaporation. The present study aims to investigate dilute spray dispersion and evaporation, and to compare the effects of different co-flow temperatures based on LES.

2 Mathematical formulation

Lagrange method of the dispersed phase is combined with an Eulerian description of the continuous phase in simulations [8]. The continuous gas phase which includes the carrier air and the co-flow hot air is calculated by LES model firstly. After the gas flow calculation reaches steady state, droplets are injected into the computational domain and tracked by DPM model.

For continuous gas flows, the control equations of LES [9] are:

$$\frac{\partial \rho_g}{\partial t} + \frac{\partial}{\partial x_j}(\rho_g \bar{u}_j) = S_m \tag{1}$$

$$\frac{\partial}{\partial t}(\rho_g \bar{u}_i) + \frac{\partial}{\partial x_j}(\rho_g \bar{u}_i \bar{u}_j) = -\frac{\partial \bar{p}}{\partial x_i} + \frac{\partial}{\partial x_i}\left(\mu \frac{\partial \bar{u}_i}{\partial x_j}\right) - \frac{\partial \tau_{ij}}{\partial x_j} + S_i \tag{2}$$

$$\frac{\partial}{\partial t}(\rho_g \bar{h}) + \frac{\partial}{\partial x_j}(\rho_g \bar{h} u_j) = \frac{\partial}{\partial x_j}\left(\rho_g \alpha \frac{\partial \bar{h}}{\partial x_j}\right) - \frac{\partial q_{hi}}{\partial x_j} + S_h \tag{3}$$

$$\frac{\partial}{\partial t}(\rho_g \bar{Y}) \frac{\partial}{\partial x_j}(\rho_g \bar{Y} u_j) = \frac{\partial}{\partial x_j}\left(\rho_g D \frac{\partial \bar{h}}{\partial x_j}\right) - \frac{\partial q_{Yi}}{\partial x_j} + S_m \tag{4}$$

Equations (1)~(4) are the mass conservation equation, the momentum equation, the energy equation and the species transport equation of the gas phase respectively. Here, ρ_g is the air density, $\bar{u}$ is the gas velocity, μ is the viscosity coefficient, $\bar{Y}$ is the mass fraction of vapor, D is the binary diffusion coefficient, α is the thermal diffusion coefficient, $\bar{h}$=$\bar{Y}h_v + (1-\bar{Y})h_{air}$ is the enthalpy of the gas mixture. And the sub-grid scale stress tensor τ_{ij} is solved by the Dynamic Smagorinsky-Lilly model.

For discrete droplets, the equations [10] are:

$$\frac{dx}{dt} = u_p \tag{5}$$

$$\frac{du_p}{dt} = \frac{1}{\tau_p}(u - u_p) + a \tag{6}$$

$$m_p c_p \frac{dT_p}{dt} = hA_p(T_\infty - T_p) + \frac{dm_p}{dt} h_{fg} \tag{7}$$

$$\frac{dm_p}{dt} = k_c A_p \rho_\infty ln(1 + B_m) \tag{8}$$

Equations (5)~(8) are the mass conservation equation, the momentum equation, the energy equation and the species transport equation of the droplet phase respectively. Here, x is the droplet position, u_p is the droplet velocity, a is the resultant of accelerations of all forces except the drag force, T_p and T_∞ are the droplet temperature and the gas temperature, the convection coefficient h is calculated by $N_u = \frac{hd_p}{k_\infty} = 2 + 0.6Re^{\frac{1}{2}}Pr^{\frac{1}{3}}$, and the mass transfer number $B_m = \frac{Y_{i,s} - Y_{i,\infty}}{1 - Y_{i,s}}$.

Besides, interactions between the continuous phase and the discrete phase are considered in the source items. Therefore, the source items in the gas control equations can be calculated by:

$$S_m = -\frac{1}{V}\sum_n \frac{dm_p}{dt} \tag{9}$$

$$S_i = -\frac{1}{V}\sum_n \frac{d(m_p u_p)}{dt} \tag{10}$$

$$S_h = -\frac{1}{V}\sum_n \frac{d(c_p m_p T_p)}{dt} \tag{11}$$

where, m_p, u_p, c_p are the droplet mass, the droplet velocity, the heat capacities at constant pressure. V is the cell volume.

Corresponding author: Zongjie Hu. Fax: +86-21-69589978
E-mail address: zongjie-hu@tongji.edu.cn

3 Computational details and validations

3.1 Computational details

The simulation is based on Gong Huifeng's spray evaporation experimental results [11]. In the experiments, n-heptane droplet clusters (293K) atomized by a Sonotek nebulizer, are blown by the fast carrier air (293K) into the thermal atmosphere through a water cooling central tube which is long enough so that the droplet velocity at the tube exit is nearly the same as that of the carrier gas. The thermal atmosphere consists of the hot air (313~513K) which is heated by electrical heaters and is enclosed by several glass and metal walls to reduce heat transfer. PDPA system is used to measure droplet velocities and diameters at the same time while droplet clusters are dispersing.

The computational domain is displayed in Fig. 1. The tube inner diameter is 5mm, which is much smaller than the glass window scale so that droplets can seldom impinge on walls. Both the carrier air inlet and co-flow air inlet are velocity inlets. The co-flow air velocity is constant (0.5m/s) while the carrier air velocity follows a normal distribution in consideration of the viscous friction from tube walls, so the carrier gas velocity is initialized by a user defined function which is derived from velocities of droplets whose diameters are smaller than 5μm. In order to avoid reverse flows, a 300mm circular wall is added above the glass windows. No slip conditions are enforced at walls and all the walls are adiabatic. The outlet is set as outflow.

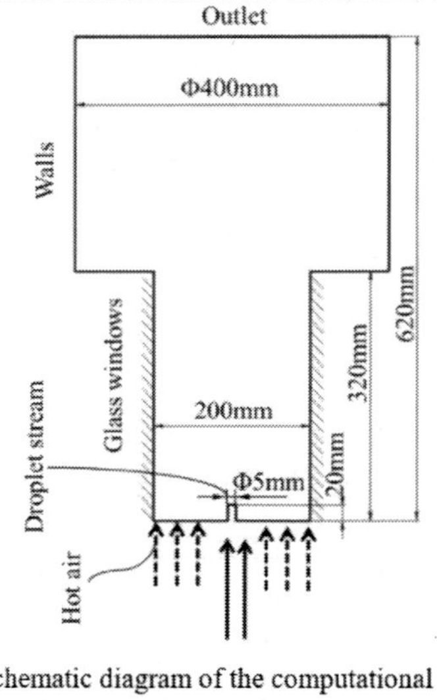

Figure 1: Schematic diagram of the computational domain

The computational grid is a mixture of structured meshes and O grids, both the axial and radical mesh distribution of which are uneven. Along the radical direction, meshes at droplet stream zone and shear layers are much denser. Along the axial direction, meshes become sparser as it goes far from the spray inlet. The tiniest mesh scale is around 20μm, appearing at the carrier air inlet boundary, and the total number of this computational domain is over 1.65×10^6.

The simulations are performed using 7 different classes of droplets. Figure 2 shows the mass flow rates of each droplet class respectively. The total mass flow rate of the spray is 1.25×10^{-4} kg/s and the droplet volume fraction is less than 0.02%. The droplet initial velocities are derived from experimental results. In the computation, each droplet is tracked until it completely evaporates or escapes from the computational domain, which yields more than 1.5×10^5 droplets in the domain at steady state.

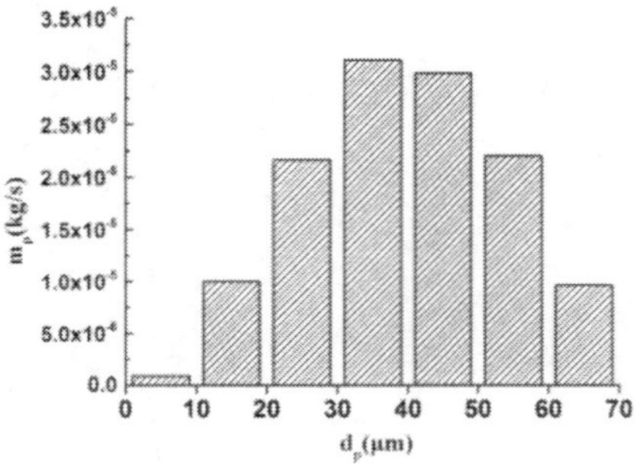

Figure 2: Mass flow rate distribution per class

3.2 Validation

Figure 3~Figure 5 compare the experimental and computational results of droplet mean velocities, droplet velocity fluctuations and droplet mean diameters at different axial distances respectively. It is obvious that the simulation results are in conformity with the experimental results. But at x/D=40 in Fig. 5 where the number of droplets is small, the statistics insufficiency causes a diameter discrepancy.

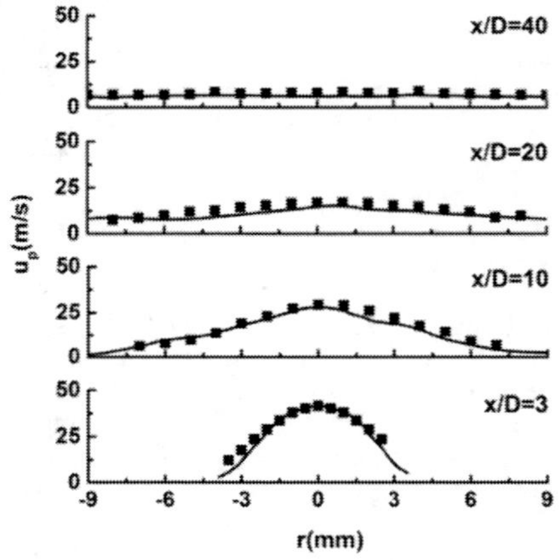

Figure 3: Droplet axial mean velocity at different axial distances

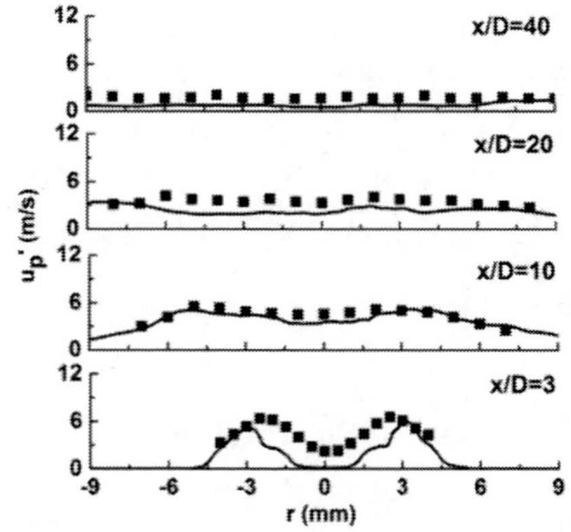

Figure 4: Droplet axial velocity fluctuations at different axial distances

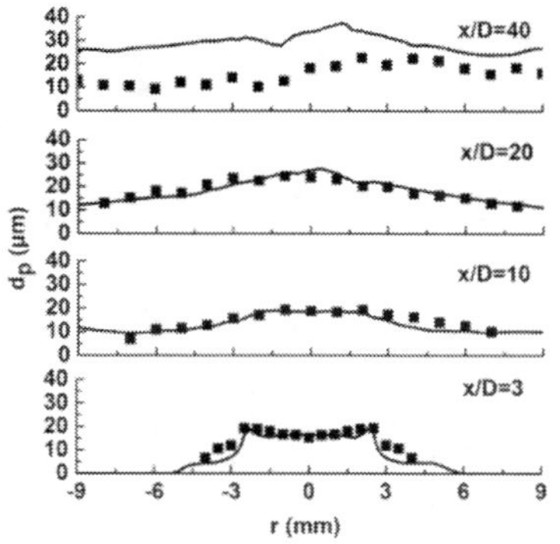

Figure 5: Droplet mean diameters at different axial distances

Based on the model, the turbulent dispersion and the evaporation of dilute sprays are investigated in two different co-flow temperatures while other computational details are completely the same. One hot air temperature is 313K (Case1) which is lower than n-heptane boiling point (371K), the other is 433K (Case2) which is higher than n-heptane boiling point.

4 Results

4.1 Spray dispersion

Figure 6 shows the gas and droplet axial mean velocities in 313K (Case1) and 433K (Case2) at different central axial positions. With the increase of the co-flow temperature, both gas and droplet axial mean velocities rise remarkably. The volume expansion in higher temperature causes the growth of the gas mean velocities, so droplet clusters which are carried by the gas phase run faster accordingly. In addition, for both cases, droplet velocities are higher than gas velocities because of droplets' inertia, which can be expressed by the Stokes Number quantitatively.

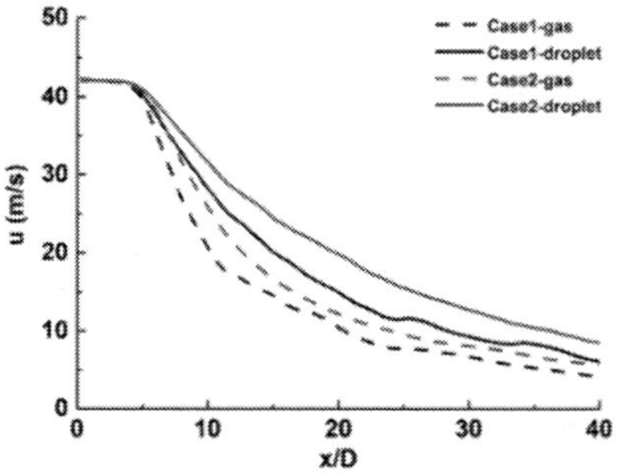

Figure 6: Comparison of gas and droplet axial mean velocities at different central axial positions

Stokes number St is defined as the ratio of droplet relaxation time and Kolmogorov time scale. The smaller the Stokes number for the droplets is, the more easily they can follow the gas fluctuations. Table1 displays the Stokes number for two different droplet diameters in different co-flow temperatures. At the same positions, the Stokes number for 5µm droplets is smaller than 1, indicating their excellent gas following abilities. Therefore, velocities of droplets whose diameters are less than 5µm can be regarded as the gas velocities. But much larger Stokes number for 55µm droplets means that they cannot response the gas turbulence fast enough, so these droplets' velocities can decay slower while the gas phase has decelerated, making droplet velocities go faster than the surrounding gas velocities. However, the little Stokes number difference between Case1 and Case2 illustrates that co-flow temperatures have no significant effects on droplet responsiveness to the gas turbulent fluctuations.

	Case1 (313K)		Case2 (433K)	
	5µm	55µm	5µm	55µm
x/D=3	0.085	2.624	0.091	2.707
x/D=10	0.296	4.433	0.387	6.516
x/D=20	0.071	1.080	0.072	1.026

Table 1: Stokes number calculated at different axial positions for droplet diameters 5µm and 55µm

Hence, Case1 is taken as an example to investigate the distribution difference between two diameter classes caused by different Stokes number in Fig. 7. Class1 droplets (d<10µm) with smaller Stokes number can follow the gas turbulence better and disperse into broader space while most Class6 droplets (50≤d<60µm) stay at central positions because of their weaker gas following abilities.

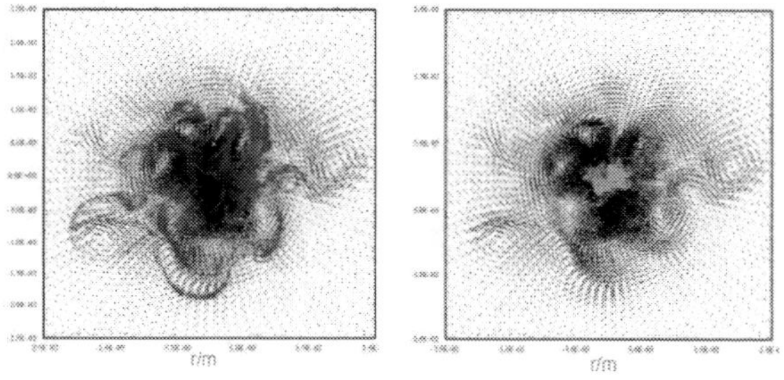

(a) Class1 droplets (d<10µm) (b) Class6 droplets(50≤d<60µm)

Figure 7: Droplet distributions in Case1 at x/D=20

4.2 Spray evaporation

Figure 8 shows the cumulative mass of n-heptane vapor and remaining droplets as a function of the injecting duration time. With droplets injected continually, the amount of n-heptane vapor accumulates in the computational domain. It is obvious to find that the vapor cumulative mass in Case2 (433K) grows faster than its counterpart, indicating that higher temperatures enhance the spray evaporation. In addition, for Case2, most n-heptane existing in the computational domain is liquid when the injecting duration time is less than 10ms, but after a 10-millisecond injection, the cumulative mass of vapor surpasses that of droplets, and the situation that the cumulative vapor mass increases linearly and the cumulative droplet mass remains constant represents that the evaporation reaches dynamically steady state. Meanwhile the higher the co-flow temperature is, the less time it takes to reach the steady state.

Figure 9 displays droplet distributions after the injection lasts 40ms. The number of droplets reduces and the distribution space narrows in Case2 due to the higher co-flow temperature and the stronger evaporation. But compared with the droplet axial penetration distance shown in Fig. 9(a), the radical penetration distance sees a larger decrease as indicated in Fig. 9(b) when the co-flow temperature increases from 313K to 433K. It is thought that bigger droplets have smaller evaporation rates and higher axial velocities, so the axial distance does not shrink remarkably although the stronger evaporation occurs in Case2. However, droplets at the edge of dilute sprays are mainly smaller droplets and they have larger evaporation rates, so these droplets vanish quickly in higher temperatures, leading to the narrow radical distribution.

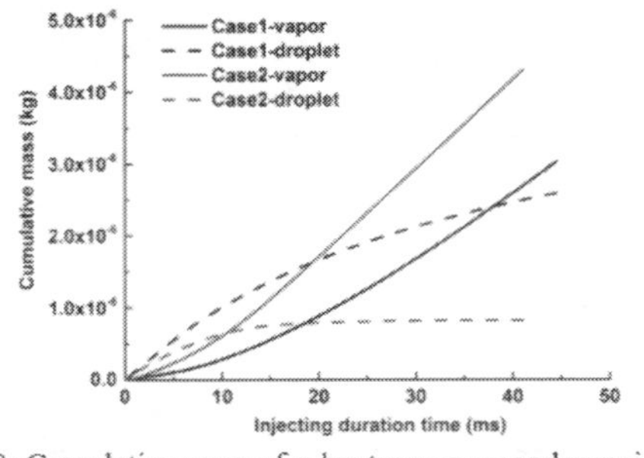

Figure 8: Cumulative mass of n-heptane vapor and remaining droplets as a function of injecting duration time

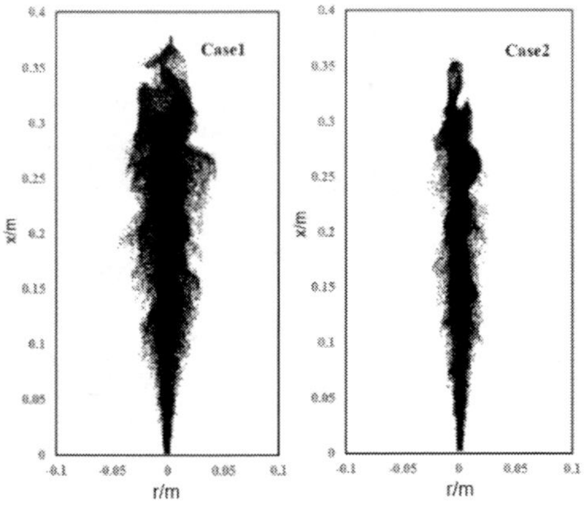

(a) Droplet axial distribution

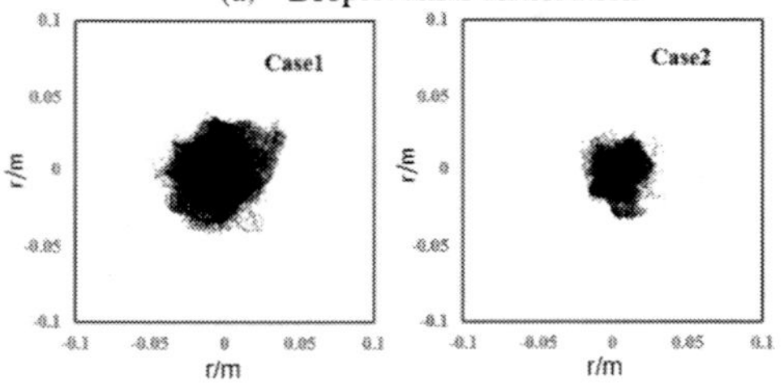

(b) Droplet radical distribution

Figure 9: Droplet axial and radical distributions in different co-flow temperatures (t=40ms)

Although the droplets at spray edges evaporate faster, the mean mass fraction of n-heptane vapor at center is much higher than that at edges for both cases, shown in Fig. 10. The mass fraction is defined as the ratio of vapor mass and mixture gas mass. It can be attributed to different vapor production rates of different droplet classes. Droplets whose volumes are bigger can release more n-heptane vapor during the same injection period in spite of their lower evaporation rates, and these bigger droplets are mainly located at center so that there is higher n-heptane vapor mass fraction in the central computational domain. Consequently, the n-heptane vapor concentration gradients in Case2 are much steeper, which means that the fuel mixture gas concentration in higher temperatures is more uneven.

5 Conclusions

By applying LES-DPM model and detailed boundary

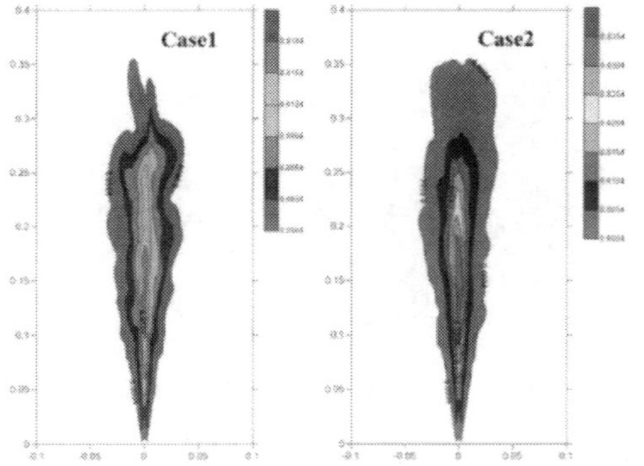

Figure 10: Contours of mean mass fraction of n-heptane vapor in different co-flow temperatures (t=40ms)

conditions, computational results agree well with experimental results.

With the increase of the co-flow temperature, both gas and droplet axial mean velocities at central axial positions rise remarkably. The Stokes number which indicates the droplet responsiveness to the gas turbulent fluctuations has little difference in two investigated co-flow temperatures. However, droplets in different sizes have different Stokes number so that their space distributions are distinctive, leading to different evaporation results. Droplet stream evaporation has dynamically steady state. The higher co-flow temperature, inducing the stronger the evaporation, can shorten the time it takes to reach the steady state. In addition, when the co-flow temperature increases from 313K to 433K, the spray radical distributions shrink more evidently than the axial distributions. The mean mass fraction of n-heptane vapor increases and the mixture gas distribution becomes more uneven at higher co-flow temperature.

6 Acknowledgment

This work is supported by the National Natural Science Foundation of China (NO. 51106113).

References

[1] Crowe C T, Fluids Engineering 104(3) (1982) 297-303.

[2] Crowe C.T., Troutt T.R., Chung, J.N., Fluid Mechanics 28(1) (1996) 11-43.

[3] Mashayek F, Pandya R V R., Prog. Energy Combust. Sci., 29(4) (2009) 329-378.

[4] De S, Lakshmisha K N., Spray and Combustion Dynamics 1(2) (2009) 169-198.

[5] Bini M, Jones W P., Heat and Fluid Flow 30(3) (2009) 471-480.

[6] Jones W P, Lyra S, Marquis A J., Heat and Mass Transfer 53(11) (2010) 2491-2505.

[7] Chen Y C, Stårner S H, Masri A R., Multiphase Flow 32(4) (2006) 389-412.

[8] Dukowicz J K. Computational Physics 35(2) (1980) 229-253.

[9] Galperin B. Large eddy simulation of complex engineering and geophysical flows, Cambridge University Press, 1993.

[10] Miller R S, Harstad K, Bellan J. Multiphase Flow 24(6) (1998) 1025-1055.

[11] Gong H.F., No. 2016-01-0843. SAE Technical Paper, 2016.

11th Asia-Pacific Conference on Combustion,
The University of Sydney, NSW Australia
10th-14thDecember 2017.

Mercury removal from coal-fired flue gas by modified clay minerals

Huan Liu, Lin Chang, Yongchun Zhao, Junying Zhang*, Jihua Qiu

State Key Laboratory of Coal Combustion, Huazhong University of Science & Technology

Luoyu Road, Wuhan, China, 430074

**Corresponding author. Email: jyzhang@hust.edu.cn*

Abstract

Four kinds of minerals including mordenite (Mor), bentonite (Ben), attapulgite (Atp), vermiculite (Ver), were processed into several modified adsorbents by different active substances and then were used to conduct the experiment via a bench-scale fixed-bed reactor system in simulated flue gas for the capture of elemental mercury (Hg^0). In addition, the effects of different flue gas components on mercury removal efficiency of samples were also studied. Furthermore, combined with a variety of characterization results, the physicochemical properties and the adsorption mechanisms of adsorbents were analyzed. Among the unmodified minerals, bentonite had the largest adsorption capacity while attapulgite and mordenite had the ability to oxidize Hg^0. Successfully, the mercury removal efficiencies of all the modified minerals were improved with different degrees. Specially, the modification with $CuCl_2$, $NaClO_3$ and KI could make minerals reach 90% Hg^0 removal. O_2 and HCl were beneficial to increasing mercury removal efficiency while SO_2 and NO played negative effects. During the adsorption process of $CuCl_2$-Atp, Hg^0 was oxidized to Hg^+/Hg^{2+} while Cu^{2+} was reduced to Cu^+, and part of Cu^+ combined with Cl and O to form new compounds. The adsorption for mercury should attribute to the physicochemical adsorption, but chemical adsorption dominated it.

1 Introduction

In recent years, toxic and bioaccumulative mercury has received increased attention because of its environmental and ecological threat. According to previous literature [1-3], the annual global emissions of mercury amounted to about 5000 tons and coal-fired power generation contributed approximately one-third of the anthropogenic emissions. Treacherously, all forms of mercury could be converted to methylmercury which can cause minamata disease [4]. Given these, the Environmental Protection Agency (EPA) issued related laws and standards on mercury control for new power plants to limit the emission of mercury [5, 6].

Mercury exists mainly in three forms: elemental mercury (Hg^0), davelient mercury (Hg^{2+}), particulate mercury (Hg_p) [2]. Hg^{2+} and Hg_p can be effectively removed by the existing air pollutant control devices (APCDs) of power plants, while Hg^0 is difficult to capture because of its volatile ability [3]. Accordingly, the academic field has started a wave of research on mercury adsorbents, such as carbon adsorbents [7, 8], fly ash adsorbents [9, 10], Ca-based adsorbents [11], metal or metal oxide adsorbents [12, 13], mineral adsorbents [2, 14] and so on. Among them, activated carbon injection technology (ACI) has been widely applied in coal-fired power plants in America recently. However, the high cost, large consumption, and adverse impact on the commercial value of fly ash limit its development [15]. Fly ash is cheap and avaliable, but the adsorbents will be affected by many factors. Ca-based adsorbents can effectively capture Hg^{2+}, but invalid for Hg^0. Metal or metal oxide adsorbents have high removal efficiency but expensive cost. Considering economic viability and high efficiency, mineral clays with unique physicochemical properties are worth studing. Some researchers have suggested that modified minerals could be attractive alternatives for Hg sorbents [16]. He et al. [17] synthesized TiO_2 pillared clay which was proved to have a high Hg^0 oxidation efficiency of 88.2% at 300 ℃ with 5% O_2 in the flue gas. Shi et al. [18] found that attapulgite modified with CeO_2 had more stable adsorption sites and the high activity in the wider rang. Cai et al. [2] indicated that the KI-clays performed much better than the KBr-clays in terms of Hg^0 removal under the same conditions. This work focused on the performance of modified clay minerals to remove mercury in different conditions, and sheding light on the possible mechanisms of mercury removal. The major goal of this study is to seek for mineral adsorbents with high mercury removal efficiency.

2 Materials and methods

2.1 Adsorbent preparation

Mordenite, bentonite, attapulgite and vermiculite with big specific surface area were chosen as adsorbent carriers, and active substances (S, $CuCl_2$, $NaClO_3$, KBr, KI) acted as modifying agents. All the reagents involved were analytically pure. As for S modification, 5ml 2 mol/L HCl diluted with 10ml deionized water and 2.5g minerals were mixed together, and then 0.75g sodium thiosulfate were added. After the process of stirring, filtering, drying and milling, the S-promoted adsorbents were obtained. As for other adsorbents,, impregnation was the main method. 0.25g $CuCl_2 \cdot 2H_2O$ and 2g minerals were added into 30ml deionized water. The subsequent procee was similar to S modification. $NaClO_3$, KBr, and KI modification were similar to $CuCl_2$ modification but their dosage was 0.5g. Finally, the samples obtained were sieved to 300 mesh for further application.

2.2 Adsorbent characterization

X-ray diffraction (XRD), X-ray fluoresent (XRF), and environment scanning electron microscope (ESEM) were

Corresponding author. Fax: +00-00-000-000
E-mail address:abc@abcde.com

conducted to test crystalline state, element content and morphology of samples. The specific surface area was represented by Brunauer-Emmett-Teller (BET) method. In addition, X-ray photoelectron spectroscopy (XPS, Thermo ESCALAB250Xi) was used to determine copper and halogen species covering the surface of adsorbents. The C1s binding energy value of 284.8 eV was taken as the reference for all binding energy calibration.

3 Results and discussions

3.1 Performance of unmodified minerals on mercury removal

From Fig. 1, four unmodified minerals had poor Hg^0 removal abilities, among which Mor had the maximum efficiency of 45% and Ver had the minimum efficiency of 13%. The BET of unmodified Atp, Ben, and Mor was 40.9, 6.8, and 153.6 m^2/g, respectively. Combined with the above results, the correlation between the mercury removal efficiency and BET of samples was not obvious, because the BET of attapulgite was six times of bentonite, but the efficiency of attapulgite was lower. Besides, the tendency of mercury online result was similar to that of digestion. Interestingly, Mor and Atp could oxidize Hg^0, which should be attributed to their structure and Ti in samples to form special active sites after combined with O_2. However, the adsorption ability of Ben was related to its stratified structure that had hydrone and cation.

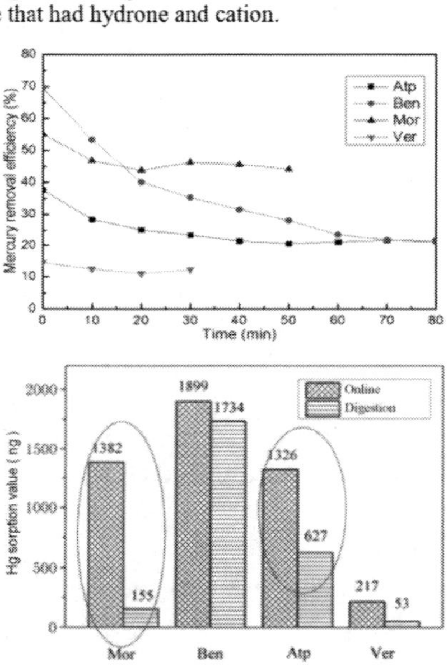

Figure 1: Mercury removal efficiency curves and disgestion of four unmodified minerals

3.2 Performance of modified minerals on mercury removal

Considering Ver had the worst performance, the other three minerals were further studied. The equation of mercury removal efficiency (E) was given in (1) below.

$$E = (1 - C_t/C_0) \times 100\% \qquad (1)$$

where C_0 and C_t represent the instantaneous Hg^0 concentrations in the inlet and outlet of the fixed-bed reactor, respectively.

From Fig. 2, the modifications were successful except for KBr modification, which was attributed to the poor KBr adsorbability of samples. After modified by elemental sulfur (S), the demercuration performance of Atp and Ben were enhanced significantly, while that of Mor was weakened. The negative effect was induced by S, which entered into the pore microstructure of Mor. For samples modified by $NaClO_3$, the efficiency of attapulgite was more than 90%, while the other was no below 50%. Combined with results of XRF, Fe and Cl played promoting roles. Ferric ion in adsorbents could form Fe^{3+}-O by attacking a not bonding electron in $NaClO_3$, which promoted the oxidation of mercury. Unlike KBr, after modified by KI, efficiencies of samples reached about 90%. Undoubtedly, I_2 was an indispensable reagent in the reaction process of samples modified by KI and mercury. Excitingly, efficiencies of Cu-Atp and Cu-Ben presented more than 80%.

Taking Cu-Atp for an example, the possible mechanism from the perspective of XPS spectra of samples was proposed. From Fig. 3, Cu^{2+} existed in different orbits for fresh sample, respectively. However, the peaks of Cu^{2+} became weak while the peaks of Cu^+ were formed in spent sample. So, there are some transformation. Besides, the peak site of Cl also moved during the reaction process. The high efficiency of Cu-Atp was the coefficient results of Cu and Cl. Similar reaction process was also put forward by Yang et al. [19] when they had a research on the mercury removal performance of fly ash.

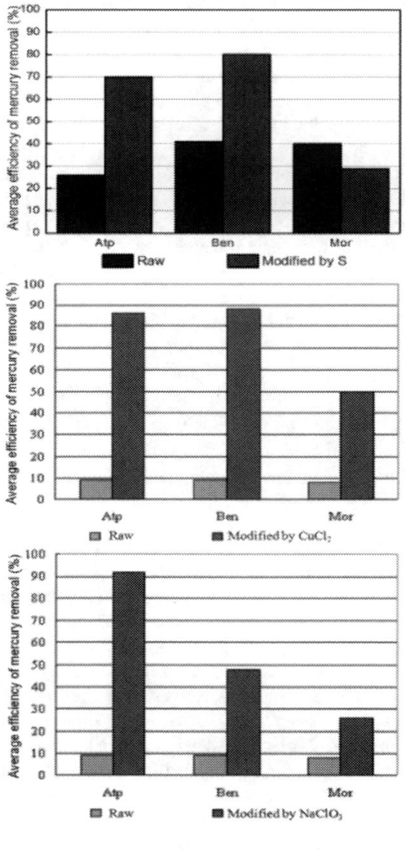

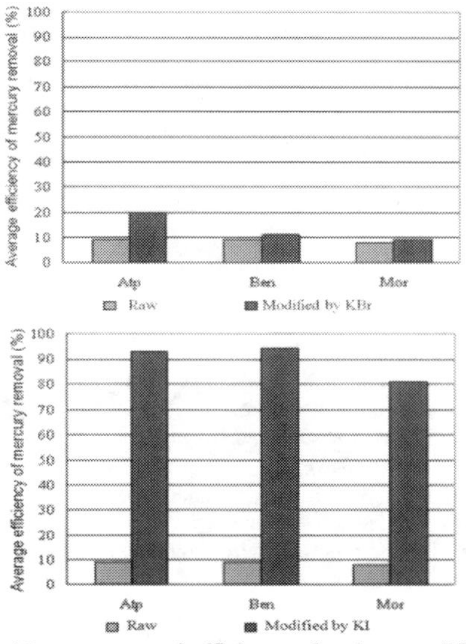

Figure 2: Mercury removal efficiency of various modified minerals

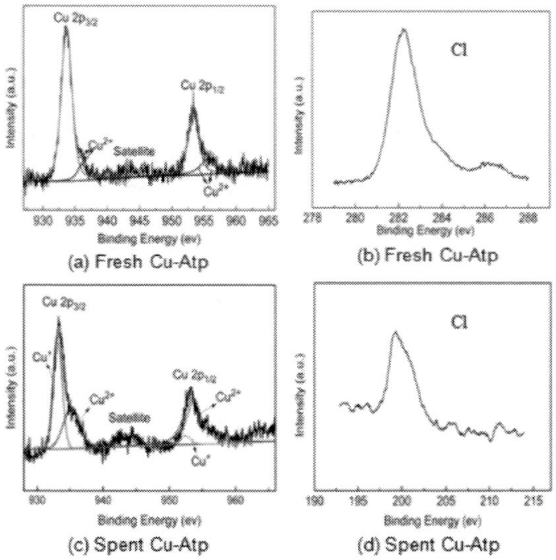

Figure 3: XPS spectra of fresh and spent Cu-Atp

3.3 Effect of individual flue gas composition on performance of samples

As is known to all, the actual working condition is very complex [20], especially the components of flue gas [21, 22]. Therefore, it is significant to have a deep research on the different flue gas atmosphere [23]. Here, effects of O_2, HCl, SO_2 and NO on mercury removal capacity over S-Atp /Ben and Cu-Atp /Ben were studied, respectively.

From Fig. 4, it's not hard to find that both O_2 and HCl contributed to raising mercury removal efficiency of the adsorbents, while SO_2 and NO played the impeding role. Compared with other gases, the influence of SO_2 was greater on S-Atp and S-Ben.

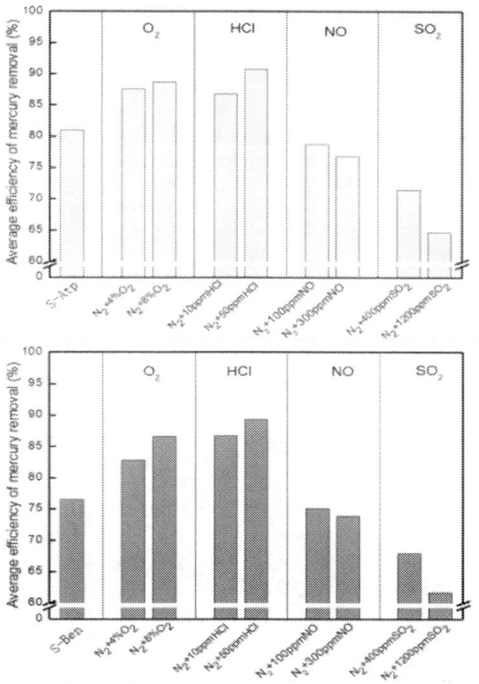

Figure 4: Effects of O_2, HCl, SO_2 and NO on mercury removal efficiency over S-Atp /Ben at 120℃

Figure 5 showed the effects of O_2, HCl, SO_2 and NO on mercury removal performance of Cu-Atp and Cu-Ben. The trends were similar to the effects on S-Atp and S-Ben. Differently, Cu-Atp and Cu-Ben always maintained good performance in different gas components with more than 80% mercury removal efficiency. Therefore, Cu-Atp and Cu-Ben were good choices for mercury capture, which could be equivalent to that of activated carbon.

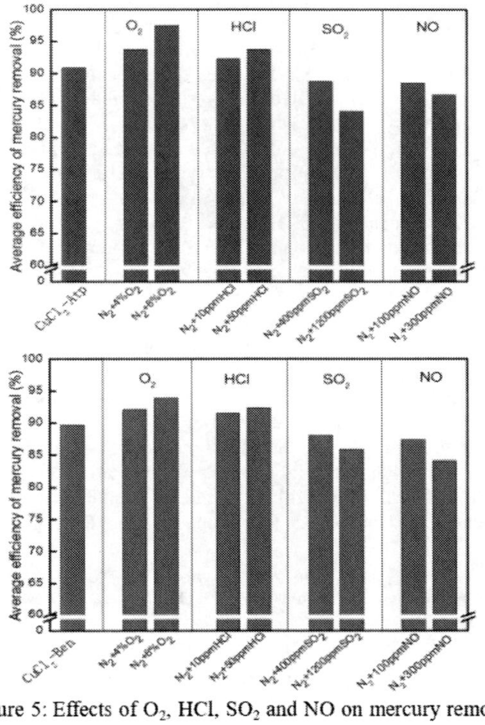

Figure 5: Effects of O_2, HCl, SO_2 and NO on mercury removal

efficiency over Cu-Atp /Ben at 120℃

4 Conclusions

The correlation between the mercury removal efficiency and BET of samples was not obvious. Among the four minerals, attapulgite and bentonite proved to be excellent adsorbent supporters for mercury removal from coal-fired flue gas, based on their structural characteristics. Attapulgite and mordenite can oxidize Hg^0, with some Ti in them forming. special active sites. The performance of modified samples had certain relation with the load of active substances. After modified by KI, the efficiency of all the samples kept above 80%, and the increase of reaction temperature was beneficial to mercury removal because I_2 was an indispensable reagent in the reaction process. Proper chemical modifications could improve the performance of adsorbents to remove mercury, especially for Cu-Atp which could be compared with activated carbon. That Cu-Atp performed best was the results of two kinds of oxidising active sites Cu^{2+} and Cl. Both O_2 and HCl acted as positive facilitators in mercury removal of samples, while SO_2 and NO had the opposite effect. In addition, Cu-Atp was influenced slightly by the different gas components.

5 Acknowledgment

The authors are grateful to the National Key Research and Development Program of China (Grant No. 2016YFB0600604), the National Nature Science Foundation of China (NSFC) (No.41672148, No.51506066) for their financial support. The authors also acknowledge the Analysis and Test Center of Huazhong University of Science & Technology. The authors would like to thank anonymous reviewers for their critical comments.

References

1. T.D. Brown, D.N. Smith, R.A. Hargis, W.J. O'Dowd, Mercury Measurement and its Control: What we Know, Have Learned, and Need to Further Investigate. Journal of the Air & Waste Management Association, 49 (9) (1999) 1-97.
2. J. Cai, B. Shen, Z. Li, J. Chen, C. He, Removal of Elemental Mercury by Clays Impregnated with KI and KBr. Chemical Engineering Journal, 241 (2014) 19-27.
3. H. Yang, Z. Xu, M. Fan, A.E. Bland, R.R. Judkins, Adsorbents for Capturing Mercury in Coal-Fired Boiler Flue Gas. Journal of Hazardous Materials, 146 (1-2) (2007) 1-11.
4. K.F. Lambert, D.C. Evers, K.A. Warner, S.L. King, N.E. Selin, Integrating Mercury Science and Policy in the Marine Context: Challenges and Opportunities. Environmental Research, 119 (2012) 132-142.
5. EPA, Air Toxics Standards for Utilities. http://www.epa.gov/airquality/powerplanttoxics/actions.htm, (accessed 27 July, 2011).
6. EPA, National Emission Standards for Hazardous Air Pollutants From Coal- and Oil-Fired Electric Utility Steam Generating Units and Standards of Performance for Fossil-Fuel-Fired Electric Utility, Industrial-Commercial Institutional, and Small Industrial-Commercial-Institutional Steam Generating Units; Revisions; Proposed Rule. Federal Register, 80 (31) (2015)
7. S.H. Lee, Y.J. Rhim, S.P. Cho, J.I. Baek, Carbon-Based Novel Sorbent for Removing Gas-Phase Mercury. Fuel, 85 (2) (2006) 219-226.
8. K. Reddy, A.A. Shoaibi, C. Srinivasakannan, Gas-Phase Mercury Removal through Sulfur Impregnated Porous Carbon. Journal of Industrial and Engineering Chemistry, (2014)
9. J.C. Hower, C.L. Senior, E.M. Suuberg, R.H. Hurt, J.L. Wilcox, Mercury Capture by Native Fly Ash Carbons in Coal-Fired Power Plants. Progress in Energy and Combustion Science, (2010)
10. J. Yang, Y. Zhao, J. Zhang, C. Zheng, Removal of Elemental Mercury From Flue Gas by Recyclable CuCl2 Modified Magnetospheres Catalyst From Fly Ash. Part 1. Catalyst Characterization and Performance Evaluation. Fuel, 164 (2016) 419-428.
11. Y. Zhuang, J.S. Thompson, C.J. Zygarlicke, J.H. Pavlish, Impact of Calcium Chloride Addition On Mercury Transformations and Control in Coal Flue Gas. Fuel, 86 (15) (2007) 2351-2359.
12. J.A. Hrdlicka, W.S. Seames, M.D. Mann, D.S. Muggli, C.A. Horabik, Mercury Oxidation in Flue Gas Using Gold and Palladium Catalysts On Fabric Filters. Environmental Science & Technology, (2008)
13. H. Li, C. Wu, Y. Li, J. Zhang, CeO_2 TiO_2 Catalysts for Catalytic Oxidation of Elemental Mercury in Low-Rank Coal Combustion Flue Gas. Environmental Science & Technology, 45 (17) (2011) 7394-7400.
14. C. He, B. Shen, J. Chen, Adsorption and Oxidation of Elemental Mercury Over Ce-MnOx/Ti- PILCs. Environmental Science & Technology, 48 (14) (2014) 7891−7898.
15. Y. Liu, D.J.A. Kelly, H. Yang, C.C.H. Lin, S.M. Kuznicki, Z. Xu, Novel Regenerable Sorbent for Mercury Capture from Flue Gases of Coal-Fired Power Plant. Environmental Science & Technology, 42 (16) (2008) 6205-6210.
16. L. Zhang, Y. Zhuo, W. Du, Y. Tao, C. Chen, X. Xu, Hg Removal Characteristics of Noncarbon Sorbents in a Fixed-Bed Reactor. Industrial & Engineering Chemistry Research, 51 (14) (2012) 5292-5298.
17. C. He, B. Shen, G. Chi, F. Li, Elemental Mercury Removal by CeO2/TiO2-PILCs Under Simulated Coal-Fired Flue Gas. CHEMICAL ENGINEERING JOURNAL, 300 (2016) 1-8.
18. D. Shi, Y. Lu, Z. Tang, F. Han, R. Chen, Q. Xu, Removal of Elemental Mercury From Simulated Flue Gas by Cerium Oxide Modified Attapulgite. Korean Journal of Chemical Engineering, 31 (8) (2014) 1405-1412.
19. J. Yang, Y. Zhao, J. Zhang, C. Zheng, Removal of Elemental Mercury From Flue Gas by Recyclable $CuCl_2$ Modified Magnetospheres Catalyst From Fly Ash. Part 2. Identification of Involved Reaction Mechanism. Fuel, 167 (2016) 366-374.
20. A.A. Presto, E.J. Granite, Survey of Catalysts for Oxidation of Mercury in Flue Gas. Environmental Science & Technology, 40 (18) (2006) 5601-5609.
21. H. Li, Y. Li, C. Wu, J. Zhang, Oxidation and Capture of Elemental Mercury Over SiO_2–TiO_2–V_2O_5 Catalysts in Simulated Low-Rank Coal Combustion Flue Gas. Chemical Engineering Journal, 169 (1-3) (2011) 186-193.
22. A. Zhang, W. Zheng, J. Song, S. Hu, Z. Liu, J. Xiang, Cobalt Manganese Oxides Modified Titania Catalysts for Oxidation of Elemental Mercury at Low Flue Gas Temperature. Chemical Engineering Journal, 236 (2014) 29-38.
23. S. Eswaran, H.G. Stenger, Z. Fan, Gas-Phase Mercury Adsorption Rate Studies. Energy & Fuels, 21 (2) (2007) 852-857.

11th Asia-Pacific Conference on Combustion,
The University of Sydney, NSW Australia
10th -14th December 2017.

Measurement of the distribution of temperature and emissivity of a candle flame using hyperspectral imaging technique

Li Ni[1], Shu Zheng[1*], Huawei Liu[1], Huaichun Zhou[1*]

[1]Key Laboratory of Condition Monitoring and Control for Power Plant Equipment of Ministry of Education, School of Energy, Power and Mechanical Engineering, North China Electric Power University

BeiNong No.2, Beijing, 102206, China

Abstract

A candle flame is experimentally studied using a hyperspectral imaging device in this paper. Then the distributions of temperature and wavelength-dependent emissivity are calculated using a Newton-type iterative method.The results show that the spectral radiation intensity increases from the outside to the inside in the candle flame, and at the same spatial point, the intensity increases as the wavelength increases. Then the results reveal that the highest temperature, which is about 2000 K, occurs close to the flame border, indicating the most vigorous reaction. They also show that the emissivity near the flame boundary is high, whereas near the center of the candle flame it is low because of the influence of the wick. Additionally, the radiation characteristics of the candle flame differ from a gray-body, and the emissivity of the flame decreases as the wavelength increases.
Keywords: candle flame, hyperspectral images, radiation temperature, flame emissivity

1 Introduction

A candle flame exemplifies the basic nature of a diffusion flame in the study of fire. It involves the evaporation of a liquid fuel from a porous surface, and unlike an uncontrolled fire, quickly reaches a steady state. Recently, research has been carried out on candle flames to study different applications and different conditions, including electric field effects [1], flame flickering [2], microgravity experiments [3,4], smoke point measurements [5], fire safety [6,7], flame shape analysis [8], and numerical models [7,9]. However, a systematic study on measurement of temperature and emissivity of candle flames has not been reported to date.

Many researchers have devoted their research efforts to temperature measurements in flames using nonintrusive optical techniques. Tunable diode laser absorption spectroscopy [10]–[12], laser-induced incandescence [13,14], and emission-based tomography [15]–[18]all have proved to be reliable tools. Additionally, due to the nonrequirement of a high-quality radiation source for transmission-based measurements, emission-based tomography based on line-of-sight intensity has received more and more attention [15]. De Iuliis et al. [16] and Snelling et al. [17]used a multiwavelength flame emission technique for high spatial resolution determination of soot temperature and soot volume fraction in axisymmetric laminar diffusion flames.

Measurement using a multiwavelength method is reliable from the standpoint of the spectral resolution. Recently, the hyperspectral imaging device was developed. Hyperspectral imaging is a promising technology, where each pixel measures radiation intensity at a large number of continuous wavelengths, providing radiation images at these wavelengths, which contain both spatial and spectral information [19]. It is widely used in remote sensing [20], astronomy, agriculture, medicine, food quality, safety applications [19,21], etc. For example, Liu et al. [22]used a hyper-spectral imaging device to measure an ethylene/air laminar diffusion flame. The spectral radiation intensities were used to estimate a two-dimensional(2D) distribution of temperature and wavelength-dependent emissivity in a sooty flame. The method did not involve spatial reconstruction, which is more concerned with combustion community. At present, the application of hyperspectral imaging in candle flame measurement and diagnostics is not yet available.

In this paper, firstly, a Newton-type iterative method can be used to solve the unknown coefficients in the polynomial function of emissivity varying with wavelength and the unknown temperature. Secondly, data processing and results from an experiment on a candle flame using the hyperspectral imaging technology will be presented and the corresponding discussion will be given. Additionally, some concluding remarks are provided.

2 Methodology

According to Wien's Law, the spectral intensity of a real emitter can be expressed as [23]:

$$I(\lambda,T)=\varepsilon(\lambda)\cdot I_b(\lambda,T)=\varepsilon(\lambda)\cdot\frac{c_1}{\pi\lambda^5 e^{c_2/\lambda T}} \tag{1}$$

Where, ε represents the wavelength-dependent emissivity which varies with the wavelength λ, c_1 and c_2 are constants, T represents the temperature. In the study of a series of radiative spectral intensities under various wavelengths for an object to be studied, the unknowns to be solved are the emissivities varying with the wavelengths and the temperature. In this case, according to the Taylor series, emissivity ε is generally expressed as a function of the wavelength λ [24]:

$$\varepsilon(\lambda)=a_0+a_1\cdot\lambda+a_2\cdot\lambda^2+...a_n\cdot\lambda^n \tag{2}$$

where a_1, a_2, … , a_n represent the unknown polynomial coefficients, then equation(1)can be rewritten as:

$$I(\lambda,T)=\left(a_0+a_1\cdot\lambda+a_2\cdot\lambda^2+...a_n\cdot\lambda^n\right)\cdot\frac{c_1}{\pi\lambda^5 e^{c_2/\lambda T}} \tag{3}$$

In such conditions, the question becomes how to solve a_1, a_2, … , a_n and T with radiative intensities $I(\lambda_1)$, $I(\lambda_2)$, … , $I(\lambda_m)$, whose total number is m.

The equations to be solved can be expressed as:

$$\alpha_j\cdot I(\lambda_j,T)-\alpha_j\cdot\left(a_0+a_1\cdot\lambda_j+a_2\lambda_j^2\cdot+...a_n\cdot\lambda_j^n\right)$$

*Corresponding author: Shu Zheng and Huaichun Zhou
E-mail address: shuzheng@ncepu.edu.cn, hczhou@ncepu.edu.cn

$$\cdot \frac{c_1}{\pi\lambda^5 e^{c_2/\lambda_j T}} = 0, j = 1,2,...,m. \qquad (4)$$

Where, α_1, α_2, ... , α_m are weight coefficients, which are the normalized device response values in this paper, and will be seen later. To some extent, the response values can represent the sensitivity of the device, and thus reflect the reliability of the measured data. Evidently, the equations to be solved are non-linear, accordingly, a Newton-type iterative method is used to solve equation (4) in this paper. Liu et al. [25] adopted a Newton-type iterative method to solve the unknown coefficients in the polynomial relationship between the emissivity and the wavelength, as well as the unknown temperature of the ethylene/air laminar diffusion flame.

3 Experiment

Radiation images were taken with a hyper-spectral imaging system(Type: SOC710-VP; Surface Optics Corporation, San Diego, CA, USA). The camera can take images of flame and create 128 sheets of images with 696 (h)×520 (v) pixels. Figure 1 is A photograph of the experimental set-up, composed of a candle, a hyper-spectral imaging system and a Laptop Notebook PC.

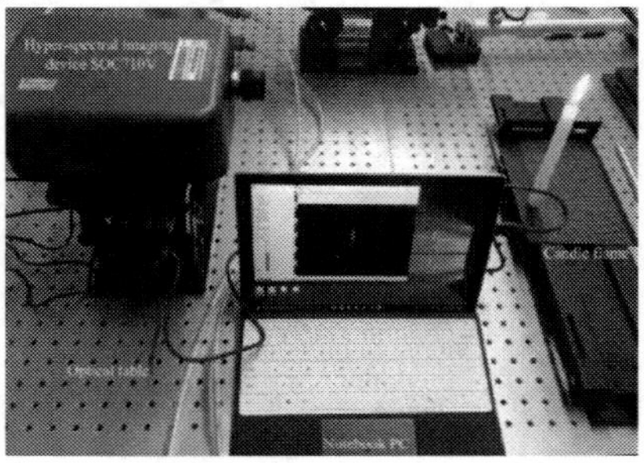

Figure 1. Photograph of experimental set-up.

The hyperspectral images shown in Figure 2 contain spatial and spectral information, from which both the spatial distribution characteristics and spectral radiation properties can be analyzed. The results of the analysis shown in Figure 2 reveal that the spectral radiation intensity increases from the outside to the inside in the candle flame. Also, at the same spatial location, the intensity increases as the wavelength increases.

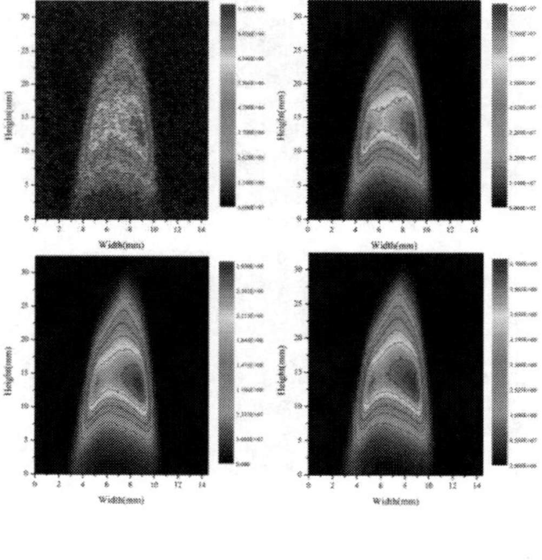

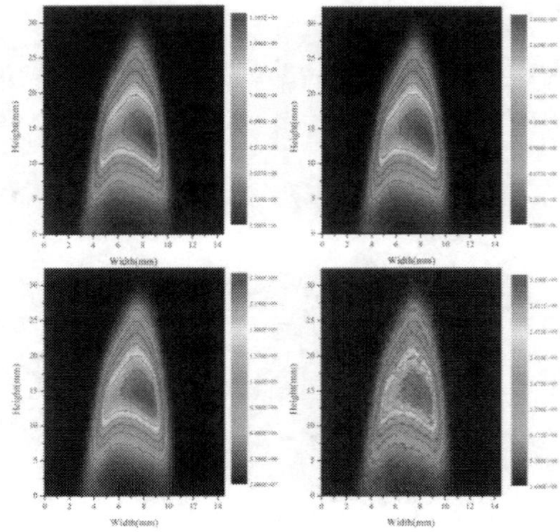

Figure 2. Hyperspectral images of the flame radiation intensity ($Wm^{-3}sr^{-1}$). Including 8 sheets of images from various available wavelengths, specially λ_1=447.66nm, λ_2=528.84nm, λ_3=611.10nm, λ_4=694.44nm, λ_5=778.86nm, λ_6=864.35nm, λ_7=950.93nm, λ_8=1038.60nm.

The 2D distribution of the radiation temperature and emissivity were estimated using a Newton-type iterative method. The estimated radiation temperature distribution is shown in Figure 3. The results in Figure 3 reveal that the highest temperature, which is about 2000 K, occurs near the flame border. M.C. Thomsen et al. [26] measured the peak temperature of the candle flame between 2000 and 2200 K, which proves that the result in Figure 3 is reasonable. Such a highest temperature location indicates that the most vigorous reaction is controlled by air amount and fuel content. For zones near the wick, air transported by the diffusion effect is insufficient, while for zones away from the wick, the fuel is inadequate due to consumption. Therefore, the most vigorous reaction occurs between the two zones, i.e. close to the flame border. Also, it can be seen that the trend of the change of the temperature was opposite to that of the emissivity. The 2D distribution of wavelength-dependent emissivity calculated by the polynomial coefficients obtained by this method is shown in Figure 4. From the data shown in Figure 4, it can be concluded that for the same detection position, emissivity decreases as the wavelength increases. Also, the highest emissivity occurs in the zone near the flame boundary, while the emissivity is very low near the flame center because of the wick.

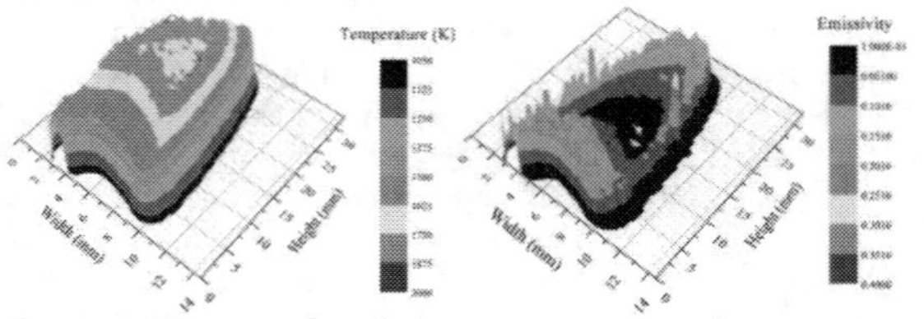

Figure 3. Image of radiation temperature and 528.84nm emissivity distribution.

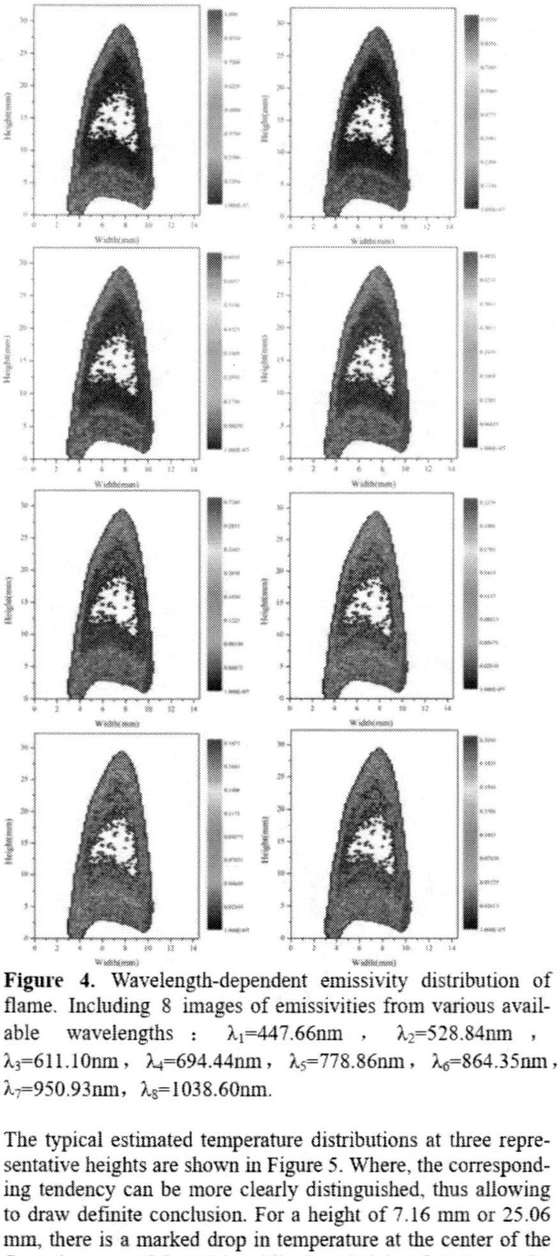

Figure 4. Wavelength-dependent emissivity distribution of flame. Including 8 images of emissivities from various available wavelengths : λ_1=447.66nm , λ_2=528.84nm , λ_3=611.10nm, λ_4=694.44nm, λ_5=778.86nm, λ_6=864.35nm, λ_7=950.93nm, λ_8=1038.60nm.

The typical estimated temperature distributions at three representative heights are shown in Figure 5. Where, the corresponding tendency can be more clearly distinguished, thus allowing to draw definite conclusion. For a height of 7.16 mm or 25.06 mm, there is a marked drop in temperature at the center of the flame because of the wick, while for a height of 17.9 mm, the temperature is the highest at the center of the flame because there is no wick influence. The emissivity distribution at 550 nm wavelength is shown in Figure 6. For a height of 7.16mm, the peak emissivity is 0.29 at a width of 7 mm, while for a height of 25.06 mm, the peak emissivity is 0.24 at a width of 6 mm. In contrast, for a height of 17.9 mm, the peak emissivity is 0.225 at a width of 4.8 mm, away from the center of the flame because of the influence of the wick.

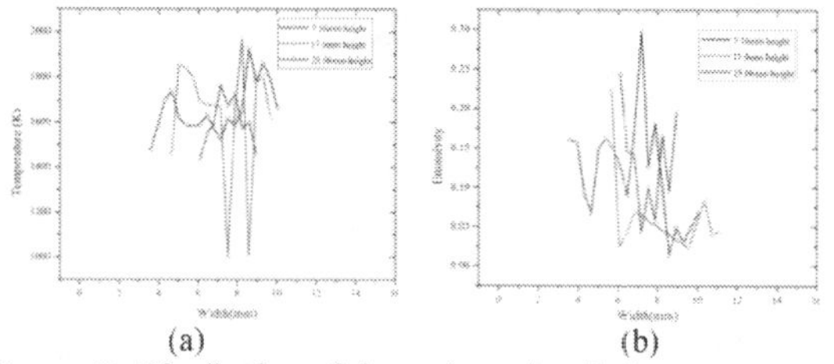

(a) (b)

Figure 5. Distribution of the estimated radiation temperature and the corresponding emissivities at 550 nm wavelength for three representative heights, specially 7.16, 17.9 and 25.06 mm. (a)Distribution of the estimated radiation temperature for the three representative heights. (b)Emissivity distribution at 550 nm wavelength for the three representative heights.

The distribution of the emissivity varying with wavelength for 3 characteristic points at the 3 representative heights mentioned above is shown in Figure 6. According to the calculated results, it can be roughly inferred that this is an approximately inverse relationship between the emissivity and the wavelength.

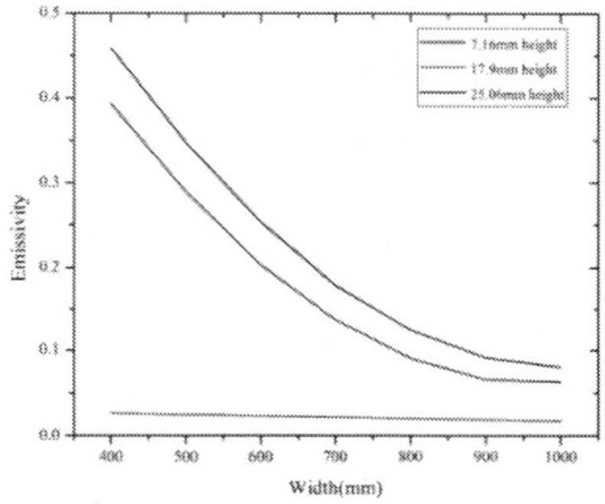

Figure 6. Emissivity varying with wavelength for 3 characeris tic points at three representative flame heights (7.16,17.9 and 25.06 mm) of flame.

4 Conclusions

A Newton-type iterative method can be used to solve the unknown coefficients in the polynomial function of emissivity varying with wavelength and the unknown temperature. Based on this method, an experiment using a hyperspectral imaging camera is conducted to study a candle flame. The concluding remarks are summarized below:

(1) The radiation characteristics of the candle flame differ from a gray-body, and the emissivity varies with the wavelength.

(2) The emissivity decreases as the wavelength increases, the data presented in Figure 8 reveals an approximate inverse relationship between the emissivity and the wavelength.

(3) The emissivity near the flame boundary is high, whereas near the center of the candle flame it is low because of the influence of the wick.

(4) The highest temperature, which is about 2000 K, occurs near the flame border, indicating the location of the most vigorous reaction.

(5) There is a marked drop in temperature at the center of the flame because of the influence of the wick.

5 Acknowledgment

This research was supported by the National Key R&D Program of China(No.2017YFB0601900), the National Natural

Science Foundation of China(No.51406095,51025622) and the Fundamental Research Funds for the Central Universities(No.2017ZZD005).

References

[1] F.B. Carleton, F.J. Weinberg, Electric field-induced flame convection in the absence of gravity, Nature 330 (1987) 635–638.

[2] T. Maxworthy, The flickering candle: transition to a global oscillation in a thermal plume, J. Fluid Mech. 390 (1999) 297–323.

[3] H.D. Ross, R.G. Sotos, J.S. T'ien, Observations of candle flames under various atmospheres in microgravity, Combust. Sci. Technol. 75 (1–3) (1991) 155–160.

[4] D.L. Dietrich, H.D. Ross, Y. Shu, P. Chang, J.S. T'ien, Candle flames in non-buoyant atmospheres, Combust. Sci. Technol. 156 (April 207) (2000) 1–24.

[5] K.M. Allan, J.R. Kaminski, J.C. Bertrand, J. Head, P.B. Sunderland, Laminar smoke points of wax candles, Combust. Sci. Technol. 181 (5) (2009) 800–811.

[6] M. Urban, D. Griffin, D. Gard, Comparative soot diagnostics: preliminary results, in: Proc. 4th International Microgravity Combustion Workshop,Cleveland, 1997, pp. 205–210.

[7] A. Hamins, M. Bundy, S. Dillon, Characterization of candle flames, J. Fire Protection Eng. 15 (4) (2005) 265–285.

[8] P. Sunderland, J. Quintiere, G. Tabaka, D. Lian, C.-W. Chiu, Analysis and measurement of candle flame shapes, Proc. Comb. Inst. 33 (2) (2011) 2489–2496.

[9] M.P. Raju, J.S. T'ien, Modelling of candle burning with a self-trimmed wick, Combust. Theor. Model. 12 (2) (2008) 367–388.

[10] S. Wagner, B. T. Fisher, J. W. Fleming, and V. Ebert, "TDLAS-based in situ measurement of absolute acetylene concentrations in laminar 2D diffusion flames," Proc. Combustion Inst., vol. 32, no. 1,pp. 839–846, 2009.

[11] C. Liu, L. Xu, Z. Cao, and H. McCann, "Reconstruction of axisymmetric temperature and gas concentration distributions by combining fan-beam TDLAS with onion-peeling deconvolution," IEEE Trans. Instrum. Meas.,vol. 63, no. 12, pp. 3067–3075, Dec. 2014.

[12] C. Liu, L. Xu, and Z. Cao, "Measurement of nonuniform temperature and concentration distributions by combining line-of-sight tunable diode laser absorption spectroscopy with regularization methods," Appl. Opt.,vol. 52, no. 20, pp. 4827–4842, 2013.

[13] C. Schulz et al., "Laser–induced incandescence: Recent trends and current questions," Appl. Phys. B, vol. 83, no. 3, pp. 333–354, 2006.

[14] F. Liu, K. A. Thomson, and G. J. Smallwood, "Effects of soot absorption and scattering on LII intensities in laminar coflow diffusion flames,"J. Quantitative Spectrosc. Radiat. Transf., vol. 109, no. 2, pp. 337–348,2008.

[15] Q.-X. Huang et al., "Reconstruction of soot temperature and volume fraction profiles of an asymmetric flame using stereoscopic tomography,"Combustion Flame, vol. 156, no. 3, pp. 565–573, Jan. 2009.

[16] S. D. Iuliis, M. Barbini, S. Benecchi, F. Cignoli, and G. Zizak, "Determination of the soot volume fraction in an ethylene diffusion flame by multiwavelength analysis of soot radiation," Combustion Flame, vol. 115, nos. 1–2, pp. 253–261, Oct. 1998.

[17] D. R. Snelling, K. A. Thomson, G. J. Smallwood, Ö. L. Gülder,E. J. Weckman, and R. A. Fraser, "Spectrally resolved measurement of flame radiation to determine soot temperature and concentration,"AIAA J., vol. 40, no. 9, pp. 1789–1795, Sep. 2002.

[18] M. M. Hossain, G. Lu, and Y. Yan, "Optical fiber imaging based tomo-graphic reconstruction of burner flames," IEEE Trans. Instru. Meas.,vol. 61, no. 5, pp. 1417–1425, May 2012.

[19] P. M. Mehl, Y.-R. Chen, M. S. Kim, and D. E. Chan, "Development of hyperspectral imaging technique for the detection of apple surface defects and contaminations," J. Food Eng., vol. 61, no. 1, pp. 67–81,2004.

[20] D. Manolakis, D. Marden, and G. A. Shaw, "Hyperspectral image processing for automatic target detection applications," Lincoln Lab. J.,vol. 14, no. 1, pp. 79–116, 2003.

[21] A. A. Gowen, C. P. O'Donnell, P. J. Cullen, G. Downey, and J. M. Frias, "Hyperspectral imaging—An emerging process analytical tool for food quality and safety control," Trends Food Sci. Technol.,vol. 18, no. 12, pp. 590–598, 2007.

[22] H. Liu, S. Zheng, H. Zhou, and C. Qi, "Measurement of distributions of temperature and wavelength-dependent emissivity of a laminar diffusion flame using hyper-spectral imaging technique," Meas. Sci. Technol.,vol. 27, no. 2, p. 025201, 2015.

[23] Yan W, Zhou H, Jiang Z, Lou C, Zhang X and Chen D 2013 Experiments on measurement of temperature and emissivity of municipal solid waste (MSW) combustion by spectral analysis and image processing in visible spectrum Energy Fuels 27 6754–62.

[24] Gardner J L, Jones T P and Davies M R 1981 A six-wavelength radiation pyrometer High Temp. High Press. 13 459–66.

[25]Liu H, Zheng S, Zhou H, et al. Measurement of distributions of temperature and wavelength-dependent emissivity of a laminar diffusion flame using hyper-spectral imaging technique[J]. Measurement Science & Technology, 2016, 27(2):025201.

[26]Thomsen M C, Fuentes A, Demarco R, et al. Soot measurements in candle flames[J]. Experimental Thermal & Fluid Science, 2017, 82:116-123.

11th Asia-Pacific Conference on Combustion,
The University of Sydney, NSW Australia
10th -14th December 2017.

Micro-explosion in Fuel Droplets of Hydrous Ethanol-Diesel Emulsion

Xiaoqing Zhang[1,2], Tie Li[1,2], Jianpeng Song[1], Yijie Wei[1]

[1]Collaborative Innovation Center for Advanced Ship and Deep-Sea Exploration, Shanghai Jiao Tong University
Dongchuan Rd., Shanghai, 200240, China

[2] State Key Laboratory of Ocean Engineering, Shanghai Jiao Tong University
Dongchuan Rd., Shanghai, 200240, China

Abstract

With the growing stringent emission regulations, the utilization of clean alternative fuel is becoming more urgent. Bio-ethanol, as a carbon-neutral fuel, is considered to have great potential for use in diesel engines as a clean alternative fuel. Micro-explosion has been studied to be beneficial on improving the secondary breaking of the spray. However, limited studies have been published in public with regard to the micro-explosion in fuel droplets of hydrous ethanol-diesel emulsions. In this study, Experiments are conducted to measure the boundary temperature and observation of the micro-explosion of hydrous ethanol-diesel emulsion. The experimental results indicate that the temperature boundary condition is approximately 170 °C. It is related only to the type and proportion of the light components, and is not related to the proportion of diesel fuel. The intensity of the micro-explosion is defined, and it meets the parabolic rule with the fraction of the light component. Micro-explosion is observed under the conditions of homogeneous nucleation and heterogeneous nucleation. The impacts of micro-explosion are investigated. The smaller the droplet diameter, the higher the emulsifier content, and the more difficult it is for a micro-explosion to occur.

1 Introduction

"Micro-explosion" is a phenomenon in the evaporation and combustion processes of multicomponent fuel droplets. Typically, the micro-explosion arises, because the droplets of fuel blends are of a homogeneous nucleation, and the whole droplets disintegrate into smaller droplets. [1, 2] Micro-explosion can improve the mixing of dispersed fuel spray by promoting secondary breaking of the mixture [2-5].The micro-explosion phenomenon has been observed and investigated in present literatures [3, 6-9] Most of these studies are focused on the emulsion with water in diesel fuel or bio-diesel fuel. Limited studies have been presented to study the micro-explosion in fuel droplets of hydrous ethanol-diesel emulsions.

Preliminary research studies on micro-explosion primarily investigate various impact factors such as temperature, pressure, droplet composition, and proportion. The intensity of micro-explosion are described[3, 10]. However, few studies are focused on the mathematical model of micro-explosion. There is still have difficult to find the accuracy boundary temperature conditions based on a theory model of micro-explosion in fuel droplets of hydrous ethanol-diesel emulsions.

Given the growing population and stringent emission regulations, the utilization of clean alternative fuel is becoming more urgent and attractive. Bio-ethanol, as a carbon-neutral fuel and renewable energy, is considered to have great potential for use in diesel engines as a clean alternative fuel [11-13]. Moreover, studies find that it can strike a better balance between energy gain and production cost for bio-ethanol [14, 15]. However, it also leads to the stratification of blends owing to the high hydrophilicity of ethanol. Surfactants are added into blends to form emulsions, which can maintain a stable state for a longer period of time [16, 17]. Thus, hydrous ethanol-diesel emulsion has received greater attention [11, 18-21].

The literatures indicate that hydrous ethanol-diesel emulsion is considered as an alternative fuel for traditional fossil energy. Micro-explosion can improve the mixing of dispersed fuel spray by promoting secondary breaking of the blends. However, there have been few studies on the micro-explosions of hydrous ethanol-diesel emulsion droplets, in particular, there is no general mathematical model of micro-explosion suiting for various kinds of emulsions. In this regard, some experiments are conducted to investigate the boundary temperature of hydrous ethanol-diesel emulsion; Two different type of micro-explosion phenomena are observed. Further, the impact factors on the micro-explosion are investigated.

2 Experimental setups

Figure 1 shows the experimental setup of the observation system for droplet micro-explosions. The hanging wire method is employed to observe the droplets. The droplets are placed in a constant volume vessel. The external copper blocks and a resistance wire are used to heat and create a high and even temperature field. The temperature of the droplets can be measured by thermocouples. The metal probe of the thermocouple is very close to the droplet. Therefore, the measured temperature can be considered as the temperature of the droplets. A high-speed camera is used to observe and record the developing process of the droplet. The temperature of the micro-explosion is recorded by observation.

The shadow-graph method [8] is employed to observe the micro-explosion of the droplet. Figure 2 shows the droplet suspension system. A microliter syringe with a range of 3 μL and a nickel wire with a diameter of 0.2 mm are used to form a droplet with a diameter of approximately 1.5 mm.

Corresponding author. Fax: +86-21-3420-8348
E-mail address: litie@sjtu.edu.cn

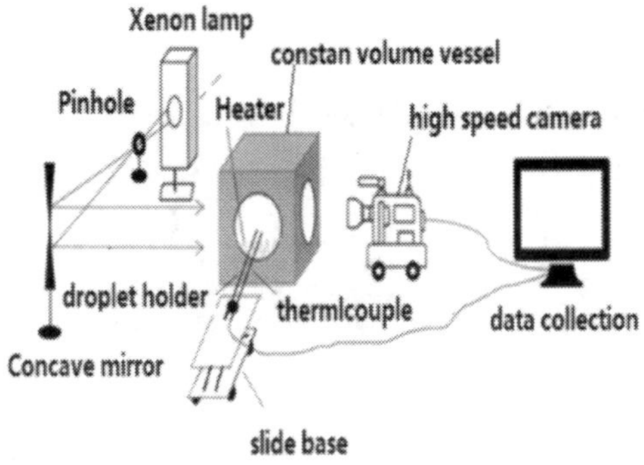

Figure 1: Experimental setup for observation and measuring system for micro-explosion of droplet

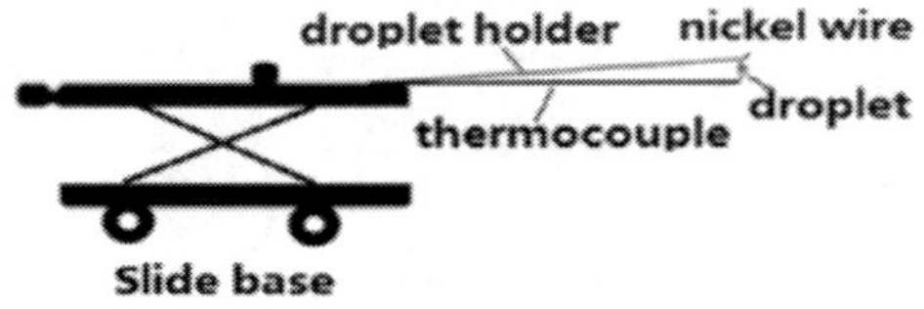

Figure 2: Droplet suspension system

In this study, HE10 represents the emulsion containing the hydrous ethanol by 10% in volume, and the naming rule is also used for HE20, HE30, *etc.* HE10, HE20, HE30, HE40, and HE50 emulsified fuels are developed by using Shah's method. Every test condition repeats five shots.

3 Definition intensity of micro-explosion

It should be noted that some works in the literature defined the intensity as the ratio of the droplet diameter after the micro-explosion to the initial diameter at the beginning of the explosion. However, in the experiments, some droplets exhibited multiple micro-explosions under heterogeneous nucleation. The droplet diameter may change slightly; thus, the intensity calculation may not reflect the facts. In this study, the intensity of the micro-explosion was defined as the ratio of the diameter of the droplet after the micro-explosion to the initial diameter of the droplet, as shown in (1):

$$\Phi = \frac{D}{D_0} \quad (1)$$

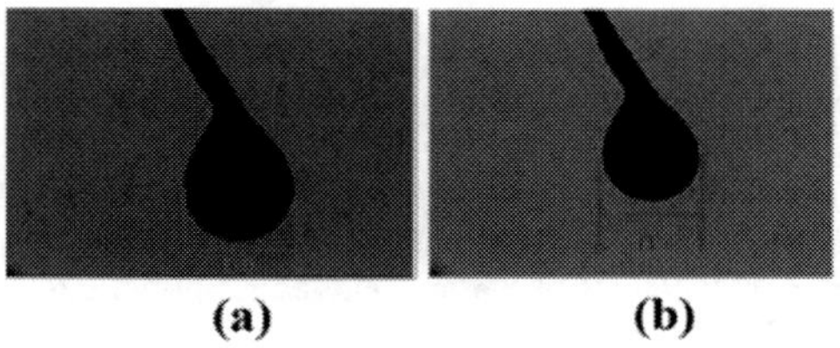

Figure 3: Definition of intensity of micro-explosion

where Φ is the micro-explosion strength, D is the diameter after the end of the micro explosion, and D_0 is the initial diameter, as shown in Fig.3.

4 Results and discussion

4.1 Temperature boundary condition

Figure 4 shows the relationship between the temperature boundary condition of the micro-explosion and the volume fraction of hydrous ethanol for various emulsions. At atmospheric pressure, the temperature boundary condition is almost keep the same as the different volume fraction of hydrous ethanol, it is approximately 170 °C. The fitting curve is almost a horizontal line, and the slope of the fitting curve is approximately 0.56. This means that the temperature increases by only 0.56 °C with the 10% increase in the volume fraction of hydrous ethanol. Therefore, the boundary temperature seems have no relationship with the proportion of the component in emulsions, it is just related with the type of component. Of course, the proportion may have impacts on the intensity of micro-explosion, this will discussion in later.

It should be noted that the temperature boundary condition is defined as the point where the primary micro-explosion begins. Here, the heterogeneous nucleation phenomena may occur, and some small fuel droplets may splash out at the beginning. However, since the main part of the droplet has not been overheated, it is not regarded as the start of the micro-explosion.

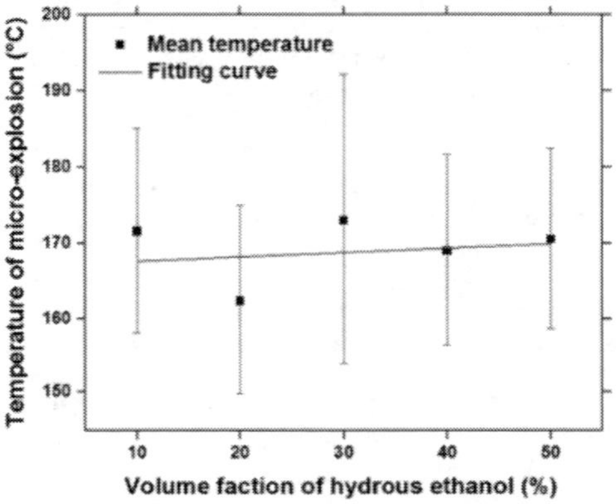

Figure 4: Boundary temperature for various hydrous ethanol-diesel emulsions

4.2 Observation of micro-explosion

In this study, the micro-explosions occur under homogeneous nucleation and heterogeneous nucleation are observed in the experiments. Figure 5 shows the development processes of the micro-explosions both under the homogeneous nucleation and heterogeneous nucleation.

The first stage is the nucleation and formation of the water-free layer [Fig.5(a)] as the light components rapidly escape from the surface of the droplet. It is different with a micro-explosion in that the burst component in the moment of the micro-explosion is composed of small fuel droplets, and it is a part of the emulsion. Once the nucleation is formed, the overheated molecules will soon be absorbed into it, and it expands rapidly as it heats up. Usually, a bright spot can be seen in the center of the droplet. This is the image of the light source through the droplet.

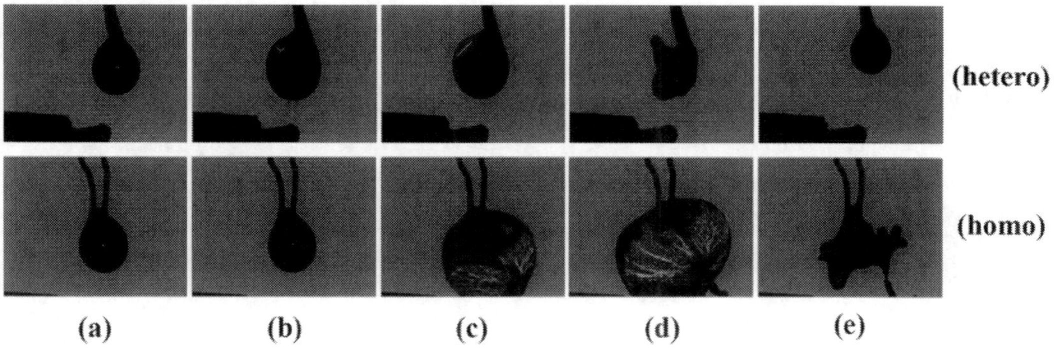

Figure 5: Development process of micro-explosion under heterogenous nucleation and homogeneous nucleation

The next stage is the process of bubble growth [Fig. 5 (b) and Fig.5(c)]. The gas in the bubble does not remain stationary, and the liquid outside the bubble keeps moving owing to the existence of a vortex, which makes the size and shape of the bubble change continuously. Zeng & Chia-fon[22] proposed an accumulation formula for the disturbance value to judge the occurrence of a micro-explosion through a critical value of perturbation.

The last stage is the explosion [Figure 5(d) and Fig.5 (e)]. Theoretically, the droplet will explode as the perturbation value in the droplet reaches the critical value. The difference is the micro-explosion delay.

As seen in Fig.5, the micro-explosion under heterogenous of a droplet is also observed in this study. It occurs as uneven heating or unstable liquid composition. The site of nucleation is not in the center, or the number of nucleation is more than one time. It is usually small and occurs earlier than the micro-explosion. It can be explained that the gas core is closer to the surface of the droplet, and it does not have sufficient time to grow considerably; therefore, the droplet being not completely broken is often the case. Part of the droplet bursts out, while the rest forms a small droplet again owing to the effect of surface tension. This small droplet usually evaporates rapidly. In some cases, two or more bursts may occur.

4.3 Effect of droplet size

Two sizes of HE30 droplet are investigated in this study: a droplet with a diameter of about 1.5 mm, and a droplet with a diameter of 1 mm. The results are listed in Table 1. The micro-explosion does not occur for the smaller size of HE30 droplet (D_1 = 1.0 mm), while a micro-explosion is observed for the larger-diameter droplet (D_2 = 1.5 mm). It can be explained that droplets with different diameters have different evaporation rates. For a small droplet, given its small surface area, the droplet may evaporate completely before the temperature in the droplet reaches the superheat limit. Then, the chance of a micro-explosion occurring is lost. the temperature boundary of the micro-explosion should be similar because their light components are the same. For further verification, the fuel blends, having the same components as the HE30 except without an emulsifier, are used, and the same experiments are conducted. The results (see Table 1) indicate that there is a slight micro-explosion in the small droplets (D_1= 1.0 mm). Therefore, the smaller the droplet diameter, and the higher the emulsifier content, the more difficult it is for a micro-explosion to occur. However, this is because the droplet evaporation rate affects the achievement of the heating limit, and there is no change in the temperature boundary condition.

	Intensity of micro-explosion(Φ)		
NO.	D_1< 1.0 mm	D_1^a< 1.0 mm	D_2> 1.5 mm
1	-	0.71	0.70
2	-	0.92	0.50
3	-	0.81	-
4	-	0.75	0.64
5	-	-	0.58
Ave	-	0.80	0.60

a. without emulsifier

Table 1: Intensity of micro-explosion for HE30 under $D1$=1.0mm and $D2$=1.5mm

4.4 Effect of fuel droplet composition

Figure 6 shows the relationship between the intensity of the micro-explosion and the fuel type. It meets the parabolic rule. The intensity of a micro-explosion reaches its maximum as the mass ratio of the light component and heavy component is 1:1. This is because the superheated steam of the light component can provide the energy to break the oil film. On the other hand, the heavy component is needed to maintain enough superheat surrounding it. Thus, both the light component and heavy component are required to form a micro-explosion.

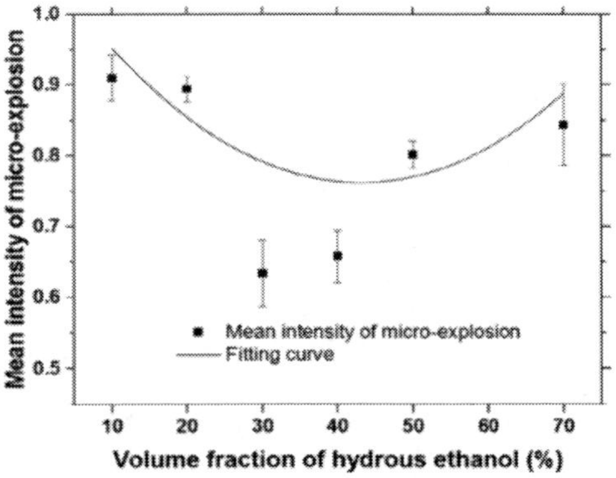

Figure 6: Intensity of micro-explosion under various emulsions

4 Conclusions

(1) The temperature boundary condition of a micro-explosion for hydrous ethanol-diesel emulsion is investigated. At atmospheric pressure, the temperature boundary condition is

approximately 170 ℃. The temperature boundary condition is mainly related to the type of the light components, and is weak related to the proportion of the components.

(2) the micro-explosion is observed. There are two kinds of micro-explosions occurred under homogeneous nucleation and heterogeneous nucleation. Micro-explosion under heterogeneous may occur multiple times.

(3) The intensity of a micro-explosion is defined. It meets the parabolic rule with the fraction of the light component, and the intensity of the micro-explosion reaches its maximum when the mass ratio of the light component and the heavy component is approximate 1:1.

(4) The smaller the droplet diameter, and the higher the emulsifier content, the more difficult it is for a micro-explosion to occur. However, as the droplet evaporation rate impacts the achievement of the heating limit, there is no change in the temperature boundary condition.

5 Acknowledgment

The support by the Natural Science Foundation of China (91541104) is gratefully acknowledged.

References

[1] Avulapati, M.M., et al., Puffing and micro-explosion of diesel–biodiesel–ethanol blends. Fuel 166 (2016) 59-66.

[2] Shinjo, J., et al., Physics of puffing and microexplosion of emulsion fuel droplets. Physics of Fluids 26(10) (2014) 103302(1-21).

[3] Hagos, F.Y., A.R.A. Aziz, and I.M. Tan. Water-in-diesel emulsion and its micro-explosion phenomenon-review. in Communication Software and Networks (ICCSN), 2011 IEEE 3rd International Conference on. 2011. IEEE.

[4] Chia-fon, F.L., K.-T. Wang, and W.L. Cheng, Atomization Characteristics of Multi-component Bio-fuel Systems under Micro-Explosion Conditions. 2008, SAE Technical Paper.

[5] Watanabe, H., et al., An experimental investigation of the breakup characteristics of secondary atomization of emulsified fuel droplet. Energy 35(2) (2010) 806-813.

[6] Lee, C.H., et al., Dynamic behavior and micro-explosion characteristics of impinging droplets on a high-temperature surface. Journal of Visualization 1(18) (2015). 1(18) 59-70.

[7] Califano, V., R. Calabria, and P. Massoli, Experimental evaluation of the effect of emulsion stability on micro-explosion phenomena for water-in-oil emulsions. Fuel 117 (2014) 87-94.

[8] Mura, E., et al., Study of the micro-explosion temperature of water in oil emulsion droplets during the Leidenfrost effect. Experimental Thermal and Fluid Science 43 (2012) 63-70.

[9] Rao, D.C.K., S. Karmakar, and S. Som, Puffing and micro-explosion behavior in combustion of butanol/Jet A-1 and acetone-butanol-ethanol (ABE)/Jet A-1 fuel droplets. Combustion Science and Technology (2017) (just-accepted).

[10] Mura, E., et al., Emulsion droplet micro-explosion: Analysis of two experimental approaches. Experimental Thermal and Fluid Science 56 (2014) 69-74.

[11] Kumar, B.R. and S. Saravanan, Use of higher alcohol biofuels in diesel engines: a review. Renewable and Sustainable Energy Reviews 60 (2016) 84-115.

[12] Tamilselvan, P. and N. Nallusamy, Reduced emissions using blends of diesel fuel and Chicha oil biodiesel. Energy Sources, Part A: Recovery, Utilization, and Environmental Effects, (2017) 1-6.

[13] Karthikeyan, S. and A. Prathima, Engine emission characteristics of algal biofuel with micro emulsion. Energy Sources, Part A: Recovery, Utilization, and Environmental Effects 38(24) (2016) 3661-3667.

[14] Shapouri, H., J. Duffield, and M. Wang, The energy balance of corn ethanol revisited. TRANSACTIONS-AMERICAN SOCIETY OF AGRICULTURAL ENGINEERS 46(4) (2003) 959-968.

[15] Martinez-Frias, J., S.M. Aceves, and D.L. Flowers, Improving ethanol life cycle energy efficiency by direct utilization of wet ethanol in HCCI engines. Journal of Energy Resources Technology 129(4) (2007) 332-337.

[16] Campos-Fernández, J., et al., A comparison of performance of higher alcohols/diesel fuel blends in a diesel engine. Applied energy 95 (2012) 267-275.

[17] Lapuerta, M., O. Armas, and R. Garcia-Contreras, Stability of diesel–bioethanol blends for use in diesel engines. Fuel 86(10) (2007) 1351-1357.

[18] Li, T., et al., Characteristics of non-evaporating, evaporating and burning sprays of hydrous ethanol diesel emulsified fuels. Fuel 191 (2017) 251-265.

[19] Ambrós, W., et al., Experimental analysis and modeling of internal combustion engine operating with wet ethanol. Fuel 158 (2015) 270-278.

[20] Breaux, B.B. and S. Acharya, The effect of elevated water content on swirl-stabilized ethanol/air flames. Fuel 105 (2013) 90-102.

[21] Asad, U., et al., Ethanol-fueled low temperature combustion: a pathway to clean and efficient diesel engine cycles. Applied Energy 157 (2015) 838-850.

[22] Zeng, Y. and F.L. Chia-fon, Modeling droplet breakup processes under micro-explosion conditions. Proceedings of the Combustion Institute 31(2) (2007) 2185-2193.

11th Asia-Pacific Conference on Combustion,
The University of Sydney, NSW Australia
10th-14thDecember 2017.

Experimental Study on the Performance of a Pulse Detonation Rocket Engine with Injected Flows

Qibin Zhang, Wei Fan*, Ke Wang, Wei Lu, Yongjia Wang

School of Power and Energy, Northwestern Polytechnical University

Xi'an, 710072, PR China

Abstract

To explore the impact of secondary flows on the pressure fluctuation of the exhaust flow and the thrust of the pulse detonation engine (PDE), experiments on the operation and the propulsive performance of a two-phase pulse detonation rocket engine (PDRE) with injected flows have been carried out in this study. Two types of injected flows and five types of transistor-transistor logic (TTL) control signals for supply and ignition were tested. The results indicated that the injected flows helped reducing the pressure fluctuation of the exhausted flow, and a distinct reduce was observed when the nitrogen injection was introduced. Oxygen-enriched injection can also promote the combustion of the residual fuel in the burned gas. Thrusts when the injected flows were employed have been improved compare to the cases with no injection, and the increases all exceed 50%. The propose of injecting secondary flows near the tube exit has been proved as an effective way for enhancing the propulsion performance of a PDRE.

1 Introduction

The engines based on detonations have more potential advantages in energy release rate and thermodynamic efficiency than the conventional propulsion concepts which utilized deflagrations [1-3]. In recent decades, the development of a competitive pulse detonation engine(PDE) has always been a hot research spot. Lots of attempts have been tried to turn these potentials into applications.

Nozzles are usually used to improve thrusts of jet propulsion devices. But the inherently unsteady flow field in a PDE makes it difficult for nozzle optimization, and it leads to thrust loss [4]. How to reduce the fluctuation of the exhausted flow becomes a great challenge. High frequency operation was tried for solving the problem. Wang et al. proposed a kind of valveless PDRE which utilized gasoline for fuel and oxygen-enriched air with 25% ~ 45% oxygen by volume for oxidizer, and high frequency operations up to 110 Hz were obtained [5]. The valveless operation mode was improved by Lu et al. [6]. By introducing purge gas into the detonation tube, the valveless PDRE can operated stably from 10 Hz to 160 Hz. Fully-developed detonation waves were proved to be obtained successfully at a maximum frequency of 130 Hz. There was also a high frequency operation research of PDE presented by Wu et al. [7], a micro-thruster based on gaseous pulse detonation was developed, and it has been successfully operated at 200 Hz. These high frequency operations partially reduced the pressure fluctuation of the exhausted flow in a PDE.

Another solution for the unsteady exhaust flow of a PDE is finding appropriate design of the optimum nozzle. An analytical model for quantifying the specific impulse of a single-cycle detonation tube fitted with diverging nozzles was developed by Barbour et al. [8]. It indicated that the optimized area ratio was shown to be mainly a function of the ambient/plateau pressure ratio. Experiments have also been carried out to explore the fitted nozzle for the PDE. Performance of a valveless PDRE with different nozzle configurations, e.g., converging, diverging, and converging-diverging nozzles has been studied in Ref. [9]. According to the impact of nozzles on the propulsive performance, converging nozzles are suggested to be used when the fill fraction is larger than one, while diverging nozzles are suggested when the fill fraction is smaller than one. In another creative research, a kind of fluidic nozzle for the PDE was proposed by Brophy et al. [10]. In order to avoid over-expended loss, secondary flows were introduced into the diverging section of the nozzle. So that it provided the dynamic ability for a nozzle to adjust its effective area ratio for the various nozzle pressure ratios during the cyclic operation of a PDE. And eventually it helped providing back-pressurization, accelerating the exhaust products, and generating adequate thrust coefficients. Injecting secondary flows was proved to be an effective way for cutting down the loss caused by unsteady exhaust flows and enhancing the propulsive performance of a PDE.

The secondary flows can also be introduced into the detonation tube near the exit. It is mainly employed to dilute the detonative mixture, and reduce the detonation intensity. The decrease of the pressure spikes resulted in the fluctuation reduction of the exhaust flow, which helps cutting down the thrust loss. On the other side, when oxygen-enriched flows are employed, it can promote the combustion of the residual fuel in the burned mixture especially when liquid fuel are utilized. Hence, it may help improving the propulsion performance of the PDE eventually. In this study, the impact of secondary flows which were introduced near the detonation tube exit on the pressure fluctuation of the exhausted flow and the PDRE thrust were investigated. A two-phase PDRE with bell-shaped converging-diverging nozzle was utilized, and the secondary flows included oxygen-enriched air and nitrogen.

2 Experimental setup

2.1 The supply system

The supply system utilized in the current study is shown

*Corresponding author. Fax: +86 29 88431116
E-mail address:weifan419@nwpu.edu.cn

schematically in Fig. 1. Kerosene, oxygen, and nitrogen were utilized for fuel, oxidizer, and the purge gas. Oxygen-enriched air with an oxygen volume percentage of 40% and nitrogen were employed for injected flows. A spark plug with an ignition energy of about 50 mJ was used to ignite the kerosene/oxygen mixture. A check valve was used to prevent flashback to the oxidizer passage. Solenoid valves were adopt for turning on/off the supplying of fuel, oxidizer, and injected flows before or after each operation. Filters, pressure gages, and regulators were all utilized in every supply passage.

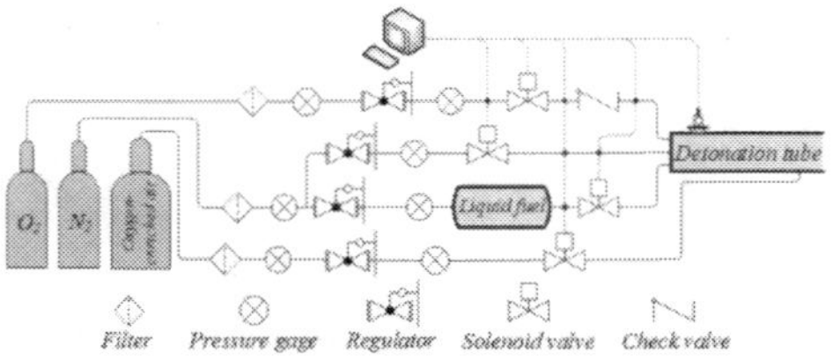

Figure 1: Schematic of the supply system

2.2 The detonation tube

Figure 2 shows the schematic of the detonation tube. The detonation tube employed in this study was a straight tube with one end closed, and the inner diameter was 30 mm. From the closed end to the open end, the tube consisted of three sections (the injection and ignition section, the deflagration-to-detonation transition (DDT) section and the secondary injection and measurement section). The length of these sections were 165 mm, 360 mm, and 285 mm respectively. Fuel and oxidizer were injected into the tube coaxially through the closed end. A pressure transducer (p_0) which was 20 mm away from the closed end was employed to record the pressure history near the close end of the detonation tube. The distance between the spark plug and p_0 were 100 mm. In order to accelerate the flame velocity during the DDT process, a typical Shchelkin spiral was introduced. An annular flow injector was introduced in the secondary injection and measurement section, and it was located in the middle of this section. Three pressure transducers (p_1, p_2, and p_3) were used to monitor the instantaneous pressure. The locations of p_1 and p_2 were 50 mm away up or down from the injector respectively, and p_3 was 92.5 mm downstream from p_2. The angle between the injected flow and the tube axis was 45°, and the gap width of the annular injector was 2 mm. A bell-shaped converging diverging nozzle was attached to the end of the detonation tube. The converging area ratio was 2.3 and the diverging area ratio was 10.0, while the lengths of the converging section and the diverging section were 10 mm and 64 mm, respectively.

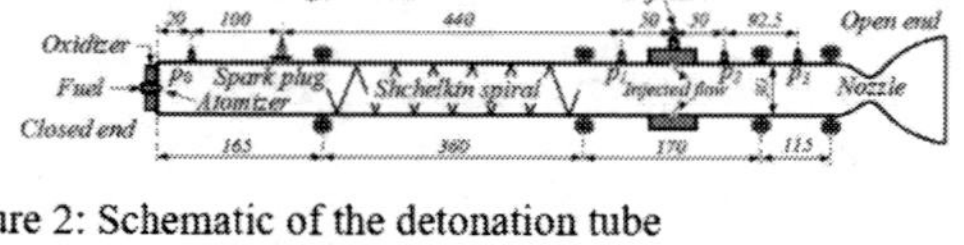

Figure 2: Schematic of the detonation tube

The measuring uncertainty caused by the pressure sensors employed in this investigation were within ±3%. The pressure signals obtained by these transducers were processed by a signal conditioner module, and then recorded by a multi-channel data acquisition system. A sampling rate of 200 kHz was utilized, and the thrust test bench of Ref. [11] was used.

2.3 Control signals

Based on how and how much of the injected flows would be introduced into the detonation tube, five types of the TTL control signals for the supply and ignition were tested in this study. It is presented in Fig. 3. In case 1, the injection of secondary flows starts with the beginning of the fuel/oxidizer supply and finishes with the ending of the fuel/oxidizer supply; in case 2, the injection starts with the beginning of the fuel/oxidizer supply and finishes with the beginning of the purge gas supply; in case3, the injection continues during the entire operation; in case 4, the injection starts with the ending of the fuel/oxidizer supply and finishes with the beginning of the purge gas supply; in case 5, the injection starts with the ending of the fuel/oxidizer supply and finishes with the ending of the purge gas supply.

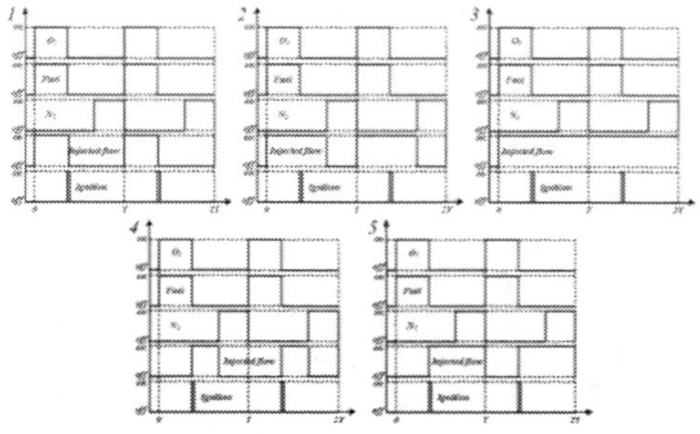

Figure 3: TTL control signals for supply and ignition

3 Results and discussion

3.1 Operations with injected flows under five types of TTL control signals

Operations of the PDE with injected flows under five types of the TTL control signals were presented at 5 Hz in this research. The initial pressure of fuel was 0.6 MPa, while 2.25 MPa for the oxidizer, 2.0 MPa for the purge gas, and 0.5 MPa for the injected flow. The equivalence ratio of the detonative mixture was 1.5. Pressure histories at p_1, p_2, and p_3 with no injection and with oxygen-enriched air under five types of TTL control signals were shown in Fig. 4.

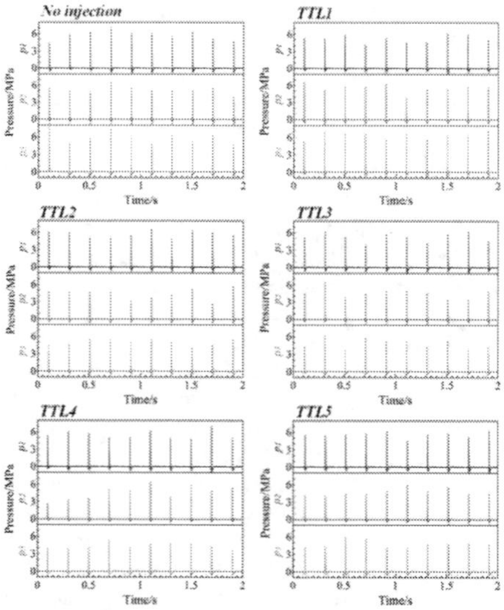

Figure 4: Pressure histories with and without injected flows

In these operations, the pressure spikes of the detonation wave all exceeded 4 MPa, and the velocities all exceeded 2000 m/s.

It means that fully developed detonations have been obtained with and without injections at 5 Hz.

3.2 Fluctuations of the pressure spikes

A main purpose of the introduction of the injected flows is to reduce the pressure fluctuation of the exhausted flow in the detonation tube. Then the impact of the injected flows on the pressure fluctuation of the exhaust flow was investigated. Pressure histories recorded by p_3 when oxygen-enriched air was introduced under five types of TTL control signals were shown in Fig. 5. Compared to the case which operated with no injection, the average of pressure spikes reduced when the injected flows were utilized in most of these cases. It was 9 % under TTL2, and 11%, 20%, and 13% under TTL3, TTL4, and TTL5 in sequence. When the oxygen-enriched air was injected into the detonation tube, the equivalence ratio of the kerosene/air mixture was decreased. Therefore, the detonation waves became weaker, and the pressure fluctuation of the exhausted flow was reduced accordingly.

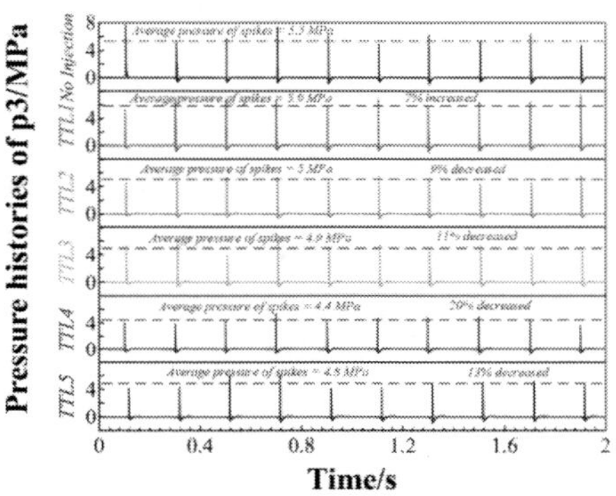

Figure 5: Pressure histories of p_3 with five types of TTL control signals

The difference between the impact of the oxygen-enriched air and the nitrogen on the pressure fluctuation of the exhausted flow was also studied. The pressure histories of p_3 with and without injected flows were illustrated in Fig.6. These cases were carried out with the TTL4 control signals. When the oxygen-enriched air was introduced, the average pressure spikes decreased by 20%, while it was 25% for the nitrogen flow. Compared to the oxygen-enriched air injection, when nitrogen was introduced, the equivalence ratio of the kerosene/oxygen mixture decreased more. Therefore, the average of the pressure spikes was getting lower. The results validate that it is an effective way to reduce the pressure fluctuation of the exhausted flow of a PDRE by injecting secondary flows into the tube near the exit.

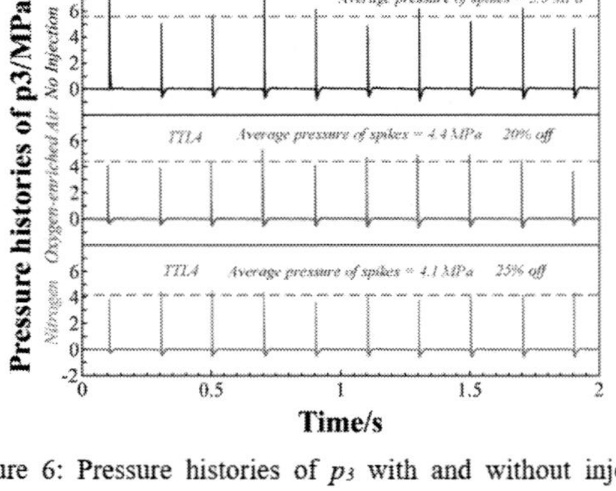

Figure 6: Pressure histories of p_3 with and without injected flows

3.3 Propulsive performance

The effect of the injected flows on the propulsion performance of the PDRE was also investigated. Both oxygen-enriched air and nitrogen injections were examined. The supply pressures of the injected flow were 0.5 MPa and 1 MPa. Thrusts of the PDE when the oxygen-enriched air was introduced as secondary flow were summarized in Table 1. Compared to the case with no injection, greater thrusts were obtained when oxygen-enriched air was introduced with all types of the TTL control signals. When the supply pressure of the injected flow was 0.5 MPa, the increase of the thrust all exceeded 68.1%. While it was 76.1% when the supply pressure of the injected flow was 1 MPa. Because the initial equivalence ratio of the kerosene/oxygen mixture was 1.5 in this study, there was a little residual fuel in the exhausted flows. Then a part of fuel was exhausted out. The oxygen-enriched air which was introduced into the detonation tube not only reduced the pressure fluctuation of the exhausted flow but also promoted the combustion of the residual fuel after the detonation, and eventually increased the thrust.

Supply Pressure/MPa	0	0.5	1
TTL1/N	12.55	25.4	26.4
TTL2/N		24.6	24.7
TTL3/N		24.3	23.7
TTL4/N		21.1	22.1
TTL5/N		23.1	24.3
Increase		≥68.1%	≥76.1%

Table 1: Thrusts when the oxygen-enriched air was utilized as the injected flow.

The influence of the nitrogen injection on the thrust was tested next, and the results with five types of TTL control signals were summarized in Table 2. Greater thrusts were also achieved in all the cases. When the supply pressure of the injected flow was 0.5 MPa, the increase of the thrust all exceeded 50.6%. While it was 65.7% when the supply pressure of the injected flow was 1 MPa. According to the results in Tables 1 and 2, it shows that the oxygen-enriched injection improves the thrust more efficiently.

Supply Pressure/MPa	0	0.5	1
TTL1/N	12.55	22.9	25.8
TTL2/N		25.2	25.6
TTL3/N		23.4	24.2
TTL4/N		18.9	20.8
TTL5/N		20.5	25.0
Increase		≥50.6%	≥65.7%

Table 2: Thrusts when the nitrogen was utilized as the injected flow.

4 Conclusions

Experiments on a PDRE with injected flows near the tube exit were carried out, and five types of TTL control signals were tested for the supply of injected flows. Kerosene, oxygen, and nitrogen were used as fuel, oxidizer, and the purge gas separately. The oxygen-enriched air and nitrogen were utilized as injected flows. It indicated that the injected flows reduced pressure fluctuations of the exhaust flow. In other word, it can cut down part of the performance loss caused by the unsteady periodical detonations. The nitrogen injection has an obvious effect on reducing the fluctuation, meanwhile, greater propulsion performance was achieved when the injected flows were employed. The thrust increases all exceed 50% when oxygen-enriched air and nitrogen injection were introduced. And it also revealed that different from the nitrogen injection, oxygen-enriched air flows can simultaneously promote the combustion of the residual fuel as well as reducing the pressure fluctuation. Therefore, it effectively enhanced the propulsion performance of a PDRE. In conclusion, injecting oxygen-enriched air into a detonation tube near the exit was validated a successful way for a higher propulsive performance PDE.

5 Acknowledgement

The authors wish to thank National Natural Science Foundation of China (91441201, 91641101, and 51376151) for financial support of this work. This work also be sponsored by Innovation Foundation for Doctor Dissertation of Northwestern Polytechnical University (CX201714).

References

[1] K. Kailasanath, AIAA J., 38 (9) (2000) 1698-1708.

[2] P. Wolanski, P. Combust. Inst., 34 (2013) 125–158.

[3] W. Fan, C. Yan, X. Huang, Q. Zhang, L. Zheng, Combust.

Flame, 133 (2003) 441–450.

[4] Z. C. Owens, R. K. Hanson, J. Propul. Power, 23 (2) (2007) 325-337.

[5] K. Wang, W. Fan, W. Lu, F. Chen, Q. Zhang, C. Yan, Energy, 71 (2014) 605-614.

[6] W. Lu, W. Fan, K. Wang, Q. Zhang, Y. Chi, P. Combust. Inst., 36 (2017) 2657–2664.

[7] M. Wu, T. Lu, J. Micromech. Microeng. 22 (2012) 105040 (10pp).

[8] E. A. Barbour, R. K. Hanson, J. Propul. Power, 25 (1) (2009) 162-172.

[9] Q. Zhang, W. Fan, K. Wang, W. Lu, Y. Chi, Y. Wang, Appl. Therm. Eng., 100 (2016) 1161–1168.

[10] C. M. Brophy, D. F. Dausen, L. R. Smith, R. G. Wright, Fluidic Nozzles for Pulse Detonation Combustors, AIAA Paper 2012-1035, 50th AIAA Aerospace Sciences Meeting including the New Horizons Forum and Aerospace Exposition, 09 - 12 January 2012, Nashville, Tennessee.

[11] K. Wang, W. Fan, W. Lu, Q. Zhang, F. Chen, C. Yan, Q. Xia, Energy, 79 (2015) 228-234.

11th Asia-Pacific Conference on Combustion,
The University of Sydney, NSW Australia
10th -14th December 2017.

Premixed flame structure of CH_4/air in a novel mesoscale combustor with a plate flame holder and preheating channels

Jianlong Wan, Haibo Zhao*

State Key Laboratory of Coal Combustion, School of Energy and Power Engineering, Huazhong University of Science and Technology, 1037 Luoyu Road, Wuhan, Hubei, 430074, China

Abstract

A novel mesoscale combustor with a plate flame holder and preheating channels was manufactured in this paper, and the laminar premixed flame structures were studied experimentally and numerically. The experimental results showed that this special structure has excellent performance for anchoring flame, and the flame can remain steadily symmetrical behavior at a very low equivalence ratio. The three-dimensional numerical simulation reproduced the flame structure in combustion chamber very well. The transverse and longitudinal characteristics of flame structure were displayed from different viewing angles and the transverse superimposed effect of flame front was revealed. For the transverse structure of flame front, the flame root shifts downstream and the flame top shifts downstream firstly and then shifts upstream when approaching the side wall. With respect to the longitudinal structure, the flame front near the center of combustion chamber is nearly straight, but the flame front near the side wall presents obvious bending.

1 Introduction

It is challenging to obtain steadily symmetrical flame with a wide flammability limit under some extreme combustion conditions (such as large heat loss due to a big surface-area-to-volume ratio, the short residence time of gaseous mixture and the fuel mixture of low concentration and low caloric value) [1, 2]. The above issues result in various dynamic flame behaviors in the combustors. For instance, Pizza *et al.* [3] through direct numerical simulation (DNS) found five dynamic flame types which include periodic ignition and extinction, symmetric V-shaped, asymmetric, oscillating and oscillating flames in mesoscale plane channels. Maruta *et al.* [4] found the flame with repetitive extinction and ignition (FREI).

Many technical approaches have been employed to extend the flammability limits and improve flame stability. Heat recirculation is a frequently adopted strategy to improve flame stability. For example, Kim *et al.* [5] and Kuo *et al.* [6] demonstrated that the flammability limits of fuel mixture can be significantly extended in the “Swiss-roll” configuration combustor. Veeraragavan [7] pointed out that the wall made of orthotropic thermal conductivity material can create "hot pockets" in the wall to stabilize the flame propagation in a plate burner. Moreover, the flow recirculation zone is another effective method to stabilize the flame. The cavity flame holder is usually used to improve the flame stabilization [8]. Sun *et al.* [8] pointed out that the cavity shear layer has significant effect on anchoring flame. This stabilization method was also adopted in mesoscale combustion [9]. Bluff body is another method which can generate flow recirculation to stabilize the flame [10]. Miguel-Brebion *et al.* [10] found that the temperature of flame holder has a significant effect on the flame topology. Wan *et al.*[11] found that the introduction of the bluff body is able to extend flame blow-off limits in a mesoscale combustor.

From the above investigations, it is known that flow recirculation and heat recirculation are beneficial for stabilizing the flame. In order to further improve the flame stabilization, we developed a novel micro combustor with a plate flame holder and preheating channels, and the two-dimensional numerical results revealed that lean premixed CH_4/air flames can be well stabilized in this combustor [12]. In this paper, a structurally similar but mesoscale combustor was manufactured and the CH_4/air premixed flame structures were experimentally and numerically studied. Note that the three-dimensional numerical simulation was conducted to better reproduce the flame structures in the real combustor.

2 Experimental device and method

2.1 Geometric model of combustors

Figure 1 schematically shows the cross-section of a symmetric combustor with a plate flame holder and preheating channels. Transparent quartz glass was used to manufacture this combustor. Two blue dashed lines at the inlet of combustion chamber are used to divide the whole channel of the combustor into the preheating channel and combustion chamber. The red dashed lines divide the combustor solid wall into four segments: the plate flame holder, one wall of the preheating channels and two preheating walls of combustion chamber. The wall thickness (δ) is 2.0 mm, and the channel gap and width of the preheating channel are 3.0 mm and 40.0 mm, respectively. Other geometrical dimensions are shown in Fig. 1.

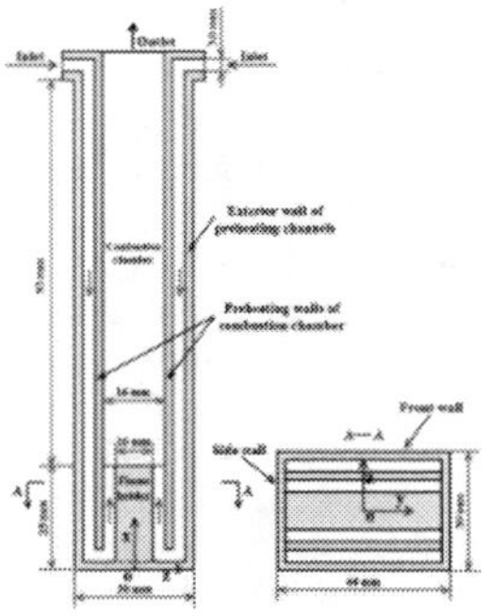

Figure 1: The schematic of the mesoscale combustor with a plate flame holder and preheating channels.

Figure 2a shows the whole geometrical structure of combustor. It can be seen that the combustor is symmetrical with respect to the horizontally and vertically centre parting planes (see Fig.

Corresponding author. Fax: +86-27-87545526
E-mail address: hzhao@mail.hust.edu.cn

2b). One of the four same parts of combustor was adopted as the computational region (see Fig. 2b and 2c). The two parting planes were set as symmetry boundary (see Fig. 2c).

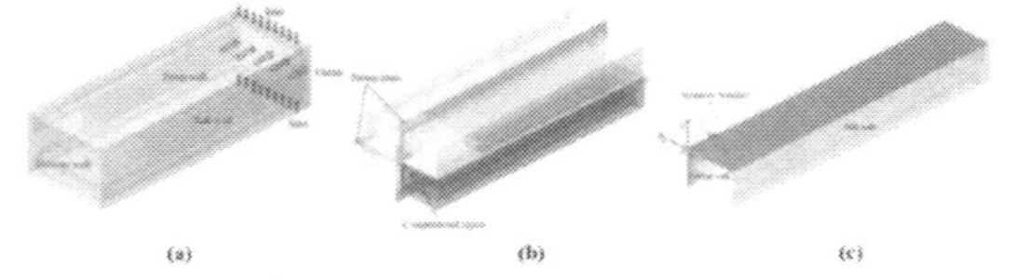

Figure 2: Three-dimensional schematics of the mesoscale combustor with a plate flame holder and preheating channels.

2.2 Experimental system

The experimental system is schematically presented in Fig. 3. Methane and air were fully mixed in a mixing tank. The equivalence ratio (ϕ) and inlet velocity (V_{in}) of CH_4/air premixed mixture were controlled by two mass-flow meters. The unburned mixture was ignited by a flame gun at the exit of the combustor. A flash-arrester was installed in the fuel mixture line. A temperature measurement system via dynamic transient method which was developed by Xu *et al.* [13] was installed at the center of exit to measure the exhaust gas temperature. A digital video camera was applied to take photographs. The present work mainly focused on the laminar flame structure characteristics under small inlet velocity and lean fuel.

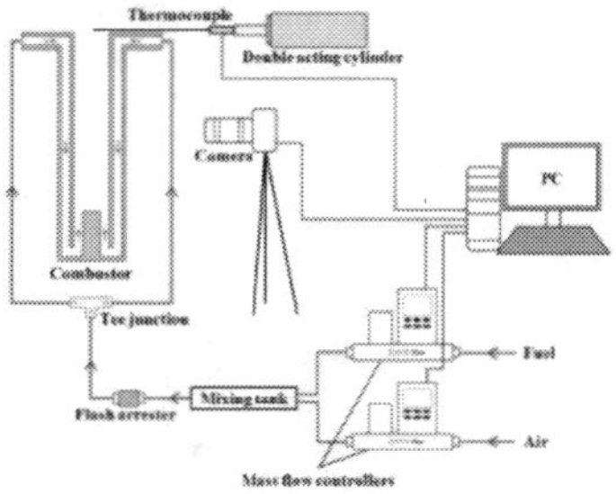

Figure 3: Schematic of the experimental system

The fresh fuel mixture was ignited firstly under a suitable operating condition, at which the flame can be stabilized in the combustor. Then under a constant inlet velocity the equivalence ratio was decreased with a step of 0.025 until flame blows off. In the present paper, three inlet velocities (V_{in}=0.6, 0.8 and 1.0 m/s, and the corresponding Reynolds numbers in the preheating channel are 120, 160 and 200, respectively) are adopted.

2.3 Experimental results

Figure 4 presents that the flame is nearly symmetrical. For quantitative description, the upstream part of flame front very near the flame holder is defined as the flame root, and the downstream part of flame front near the preheating wall of combustion chamber is defined as the flame top. Under an equivalence ratio, the strong anchoring effect of flame holder makes the flame root nearly rest on the same location with the increasing inlet velocity, but the flame top shifts downstream significantly. As a result, the height the flame front is increasing with an increasing inlet velocity. However, the flame root shifts downstream under a smaller equivalence ratio due to a lower temperature level of flame holder.

Figure 5 shows that the flame can still remain symmetrically stable under a very small equivalence ratio, which means that the current special combustor structure can significantly broaden the flammability limit. Moreover, the flame becomes slimmer and slimmer with the decrease of equivalence ratio, and the distance between the right and left flame fronts obviously decreases with the decreasing equivalence ratio. What is more, the right and left flame roots at ϕ=0.450 and 0.425 merged into one "U" shaped flame root, a constricted "waist" near the middle of flame front (see the white arrow in Fig. 5e) was observed. It should be pointed out that the photographs of stable flame in Fig. 4 and 5 are from the side views, and the horizontally (y direction) superimposed effect of flame front makes the flame seem thicker.

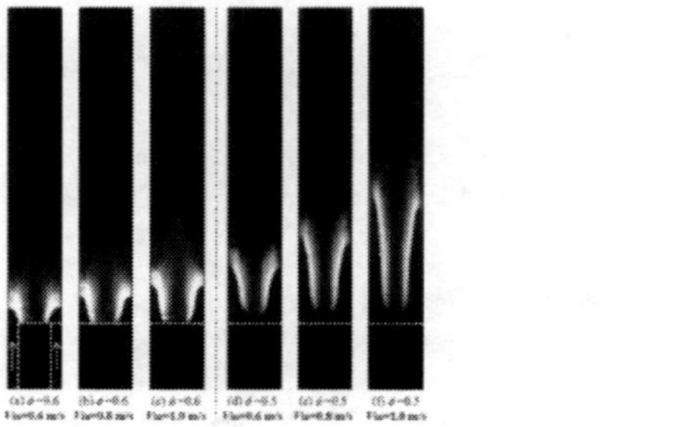

Figure 4: Side-viewing photographs of stable flame under different inlet velocity at ϕ=0.6 and 0.5 (the end wall of flame holder is marked by the horizontally dashed lines, and the yellow arrow indicates the flow direction of unburned mixture)

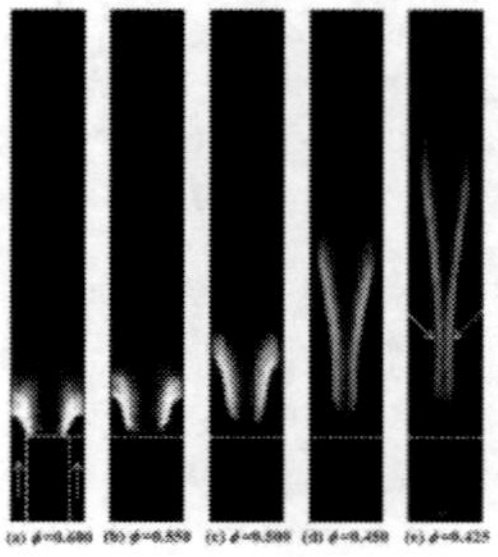

Figure 5: Side-viewing photographs of stable flame under different equivalence ratios at V_{in}=0.6 m/s

Figure 4 shows that the flame behaviors under an equivalence ratio for different inlet velocities are similar. Figure 5 shows that the brightness of flame under a bigger equivalence ratio (ϕ> 0.45) is lighter. ϕ> 0.45 is recommended in practical operation, which is also the main scope of the paper. The flame structures under the condition of under well operating conditions were further investigated numerically. Due to the space limit, the case of V_{in}=0.6 m/s and ϕ=0.55 was adopted as a typical example to illustrate the flame structure in this kind of combustor. The flame under a lower equivalence ratio which is near the extinguished limit (Near-limit Flame) ($\phi \leq 0.45$) is darker, which will be used to reveal the flame blow-off mechanisms in our future work.

3 Numerical analysis

3.1 Computation scheme

The quartz was used as solid material, whose density, specific heat capacity and thermal conductivity are 2650.0 kg/m^3, 750.0 J/(kg·K) and 2.0 W/(m·K), respectively. The heat transfer in the wall was considered in the computation [7]. The detailed C1 chemistry mechanism was adopted to simulate CH_4/air premixed combustion [14]. The thermodynamic and transport properties of the gaseous species were taken from the

CHEMKIN databases [15, 16]. The heat fluxes in the fluid and solid wall were computed using the Fourier's law. The inner wall surfaces were assumed to be chemically inert and no-slip. The CH_4/air mixture with uniform concentration and velocity distributions of 300 K were set at the inlet. A Neumann boundary condition was given for the exit. The heat loss from the outer wall surfaces included the natural convection and radiation. The computational fluid dynamics software Fluent was applied to solve the momentum, mass, energy and species conservation equations. Grid independency was conducted by using three sets of grid system ($\Delta x=\Delta y=\Delta z=250$ μm, 200 μm and 125 μm). When comparing the two key radicals (HCO and CH) distributive profiles, it was found that the cell size of 250 μm is sufficiently fine to capture the flame structure. Further refinement of the meshes near the plate flame holder was conducted, and a non-uniform grid system with 1,562,592 cells was employed.

3.2 Model validation

Figure 6 shows that the flammability limits under different inlet velocities are almost same, especially for the smaller inlet velocity. It is noted that the maximum relative deviation between predicted and measured flammability limits is 5.88 % at V_{in}=1.0 m/s. Furthermore, the following results will indicate that the flame shape and structure by simulation are very similar or almost same with experimental measurement. These confirm reasonable accuracy of the numerical model adopted in the present work.

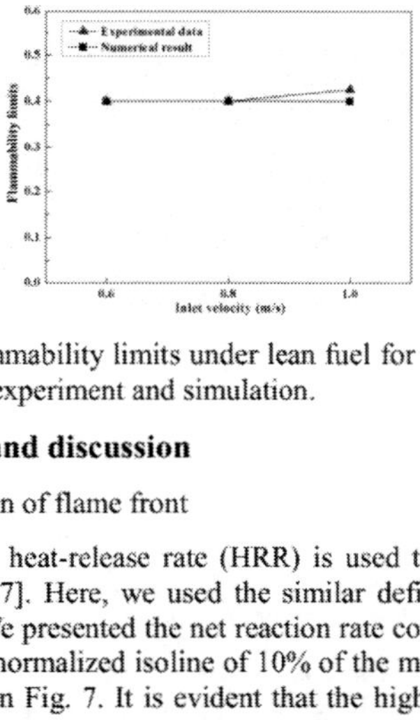

Figure 6: Flammability limits under lean fuel for different inlet velocities via experiment and simulation.

3.3 Results and discussion

3.3.1 Definition of flame front

Generally, the heat-release rate (HRR) is used to identify the flame front [17]. Here, we used the similar definition for the flame front. We presented the net reaction rate contours of CH_4 with overlaid normalized isoline of 10% of the maximum HRR at y=0.0 mm in Fig. 7. It is evident that the high reaction rate zone of CH_4 lies in the scope of normalized isoline of 10% of maximum HRR. Similar structures have also been observed in other cases. So we used the normalized 10% of maximum HRR isoline to visualize the flame front.

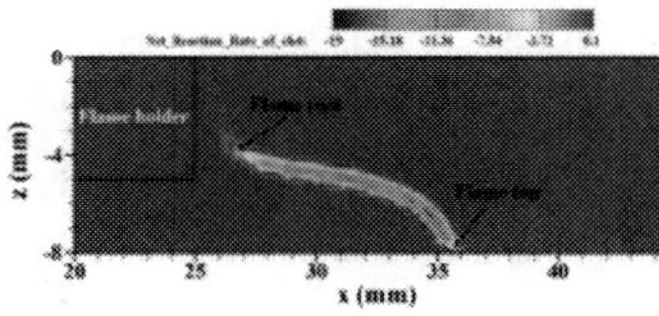

Figure 7: Net reaction rate contours of CH_4 with overlaid normalized isoline of 10% of maximum HRR (blue dashed line) at y=0.0 mm cross-section.

3.3.2 Three-dimensional flame shape

Figure 8 shows the three-dimensional temperature field with overlaid flame front. Figure 8a is the side-view of combustor from the same angle of the experimental observation, which is very similar with that of the experimental result (see Fig. 8e). The flame roots and flame tops obtained by experiment and simulation are almost at the same location, respectively, which further confirms the accuracy of the present numerical model. It seems that in Fig. 8e the flame shape between the experiment and simulation is a little different. This is mainly because that the observed flame behavior of experiment is the high temperature zone near the middle region of flame, and it is hard to capture the edge regions of flame near the side wall of combustor by the camera due to its lower temperature, which results in a flame front of "S" shape from the side-view. The superimposed shape of the middle region of flame front also demonstrates the "S" shape via simulation.

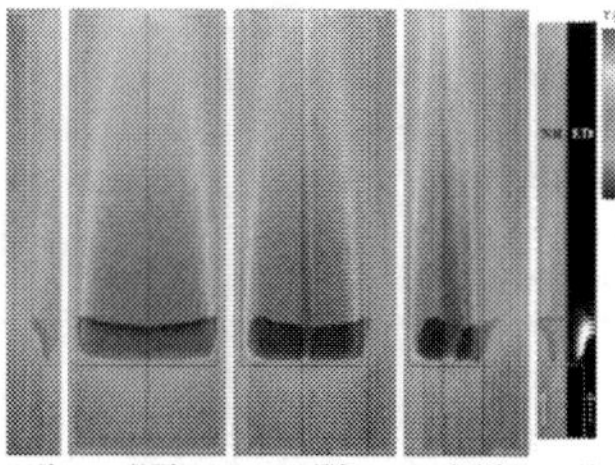

Figure 8: Three-dimensional temperature field with overlaid normalized 10% isosurface of maximum HRR (dark curved surface) for different angles rotating around the x axis

3.3.3 Transverse structure of flame front

Figure 9 presents that the shapes of flame front near the middle region (-18.0 mm ≤ y) are actually "S" patterns, and the "C" pattern only occurs very near the side wall (y ≤ -19.0 mm). Moreover, the exhaust gas temperature is higher in the middle region of combustor, and the exhaust gas temperature closer the side wall is lower. Furthermore, Fig. 9 indicates that the flame root is mainly anchored by the low flow velocity zone which is in the range of 0-0.2 m/s. The flame root shifts downstream with a decreasing distance to the side wall, and the flame root at y=-19.0 mm is near the 0.2 m/s isoline.

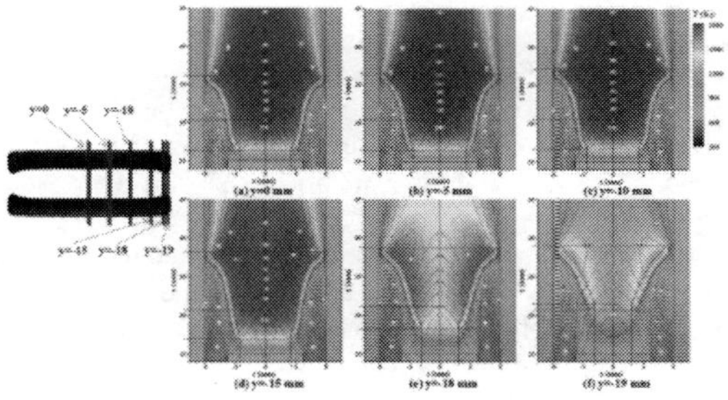

Figure 9: Temperature contours with overlaid x-direction isolines of flow velocity (black lines) and 10% maximum HRR isoline (blue dashed lines) for six transverse sections.

Figure 10 presents that, as it closes to the side wall (y from 0 to -19 mm), the height of recirculation zone increases, the flame root shifts downstream and flame top shifts downstream firstly and then shifts upstream. In such a way, the height of flame front increases firstly and then decreases.

3.3.4 Longitudinal structure of flame front

Fig. 11 shows the temperature contours with overlaid flame front for six longitudinal sections. This figure presents that the high temperature zone is larger for a more downstream longitudinal section, and the flame front is extending to inner

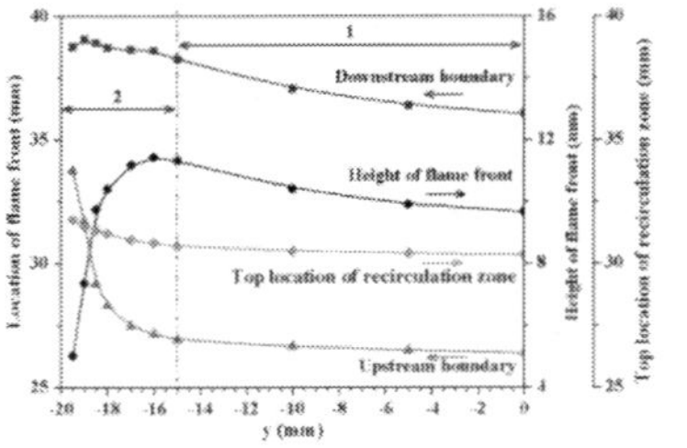

Figure 10: The variations of upstream and downstream boundary locations of flame front, the height of the flame front and the top location of recirculation zone for different transverse sections.

wall of combustor at the same time. Fig. 11 shows that the flame front near the center of combustor is almost straight, and the flame front near the side wall is obviously bending. In order to study this characteristic quantitatively, the flame front is divided into three zones: the central region that κ (κ= dz/dy, dz is the variation of flame front on z direction, dy is the variation of flame front on y direction) is smaller than 0.01 is defined as central zone of flame front (CZ), and the region of flame front near the side wall are defined as boundary zone (BZ).

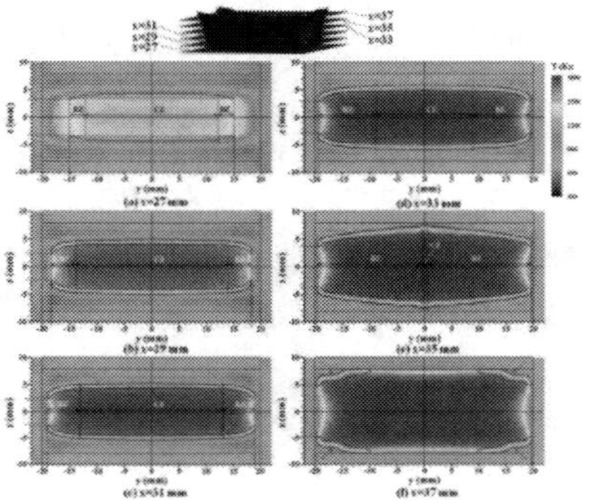

Figure 11: Temperature contours with overlaid 10% of maximum HRR isoline (blue solid lines) for six longitudinal sections (the vertically black dashed lines indicate the ranges of central zone and boundary zone).

Figure 12 shows that the total width of flame front is increasing with the downstream movement of longitudinal section, but the increasing amplitude is decreasing. The width of BZ is also increasing with the downstream movement of longitudinal section, and the increasing amplitude sharply increases at a more downstream location (~32.0 mm $\leq x$). However, with the downstream movement of longitudinal section, the width of CZ increases firstly and then decreases, and the decreasing amplitude sharply decreases at a more downstream location, and the percent of central zone in total flame front is decreasing.

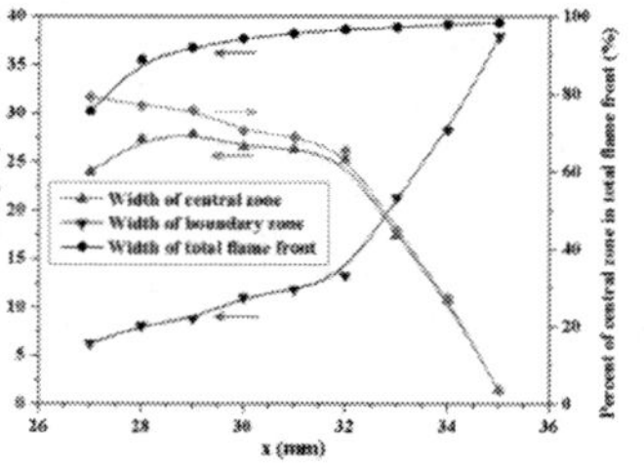

Figure 12: The variations of the widths of central zone, boundary zone and total flame front, and the percent of central zone in total flame front for different longitudinal sections.

4 Conclusions

The laminar premixed flame structures in a novel combustor with a plate flame holder and preheating channels were experimentally and numerically studied. This special structure demonstrated an excellent performance for anchoring the flame. The flame becomes more slender for a larger inlet velocity and shifts downstream with the decrease of equivalence ratio. Three-dimensional numerical simulation was used to re-appear and quantitatively investigate the flame structure. We firstly observed the flame structures from different viewing angles. Then, the transverse characteristics of flame structure are studied by making multiple transverse sections. As it is closer to the side wall, the recirculation zone increases, the flame root shifts downstream and the flame top shifts downstream firstly and then shifts upstream. The longitudinal characteristics of flame structure were studied by making multiple longitudinal sections. It can be known that the flame front near the center of combustion chamber is nearly straight, but the flame front near the side wall presents obviously bending. In order to quantitatively investigate this characteristic, the flame front at longitudinal section was divided into the central zone and boundary zone based on the bending degree, and their distribution characteristics were explored.

5 Acknowledgements

This work was supported by the Natural Science Foundation of China (Nos. 51706080, 51522603, 51390494) and the China Postdoctoral Science Foundation (2016M600591).

References

[1] Y. Ju, K. Maruta, Prog. Energ. Combust. Sci. 37 (2011) 669-715.

[2] S.J. Shanbhogue, S. Husain, T. Lieuwen, Prog. Energ. Combust. Sci. 35 (2009) 98-120.

[3] G. Pizza, C.E. Frouzakis, J. Mantzaras, A.G. Tomboulides, K. Boulouchos, Combust. Flame 155 (2008) 2-20.

[4] K. Maruta, T. Kataoka, N.I. Kim, S. Minaev, R. Fursenko, Proc. Combust. Inst. 30 (2005) 2429-2436.

[5] N. Kim, S. Kato, T. Kataoka, T. Yokomori, S. Maruyama, T. Fujimori, et al., Combust. Flame 141 (2005) 229-40.

[6] C.H. Kuo, P.D. Ronney, Proc. Combust. Inst. 31 (2007) 3277-84.

[7] A. Veeraragavan, Energ. 93 (2015) 631-40.

[8] M.B. Sun, H.Y. Wu, Z.Q. Fan, H.B. Wang, X.S. Bai, Z.G. Wang, et al., P I MECH ENG G-J AER 225 (2011) 1351-65.

[9] J.L. Wan, A.W. Fan, Y. Liu, H. Yao, W. Liu, X.L. Gou, et al., Combust. Flame 162 (2015) 1035-45.

[10] M. Miguel-Brebion, D. Mejia, P. Xavier, F. Duchaine, B. Bedat, L. Selle, et al., Combust. Flame 172 (2016) 153-61.

[11] J.L. Wan, A.W. Fan, H. Yao, W. Liu, Energ. 113 (2016) 193-203.

[12] J.L. Wan, A.W. Fan, H. Yao, Combust. Flame 170 (2016) 53-62.

[13] Z. Xu, X. Tian, H. Zhao, Proc. Combust. Inst. 36 (2017) 4443-51.

[14] R.J. Kee, J.F. Grcar, M.D. Smooke, J.A. Miller. Sandia National Laboratories Report, SAND85-8240. (1994).

[15] R.J. Kee, F.M. Rupley, J.A. Miller. Sandia National Laboratories Report, SAND87-8215B. (1990).

[16] R.J. Kee, J.F. Grear, M.D. Smooke, J.A. Miller. Sandia National Laboratories Report, SAND85-8240. (1985).

[17] B.J. Lee, H.G. Im, Proc. Combust. Inst. 36 (2017) 1569-76.

11th Asia-Pacific Conference on Combustion,
The University of Sydney, NSW Australia
10th -14th December 2017.

Effects of hydrogen enrichment on combustion and emissions formation processes in a gasoline rotary engine

Jinxin Yang, Changwei Ji*, Shuofeng Wang, Teng Su, Du Wang

Beijing University of Technology

Pingleyuan No.100, Chaoyang District, Beijing, 100124, China

Abstract

This study aims to numerically investigate the combustion and emissions formation processes in a SI rotary engine fueled with hydrogen-gasoline blends. The RNG k-ε turbulence coupled with a skeletal PRF mechanism were adopted to simulate the engine working process under 0%, 2% and 4% hydrogen volume fractions in CONVERGE software. Results showed that a mainstream flow field along with the rotor movement was formed during the compression stroke. The center of the burned zone would move in same direction with the mainstream flow. Meanwhile, due to the effect of the mainstream flow, the flame propagation in direction of the mainstream flow was expedited. While the flame propagation in contrary direction was retarded, consequently, the unburned mixtures at rear region of combustion chamber suffered incomplete combustion. The distributions of NO and CO emissions were also affected by the flow field. After hydrogen addition, the increased OH, H and O radicals concentrations combined with the intense flow flied accelerated the combustion process, which resulted in the improvement and advancement of in-cylinder pressure and temperature. The increased temperature and the changed chemical reaction path promoted the formation of NO emission. At last, CO emission was decreased with the increase of hydrogen-enrichment fraction.

1 Introduction

Compared with the conventional reciprocating engine, the rotary engine has many advantages, such as high power to weight ratio from the high operating speed, compact and simple design due to the less moving parts, less vibration and noise resulting from non-reciprocating components, etc. [1-2]. For these reasons, the rotary engine is a promising energy system for mobile applications. The disadvantages of rotary engine, however, cannot be ignored. Generally, it includes higher HC and CO emissions, poor fuel economy and low thermal efficiency [3]. Fuel property is a key factor for rotary engine performance.

Since it has a high flame speed, short quenching distance, high diffusive speed and wide flammability, etc., the hydrogen enrichment is a potential approach to overcome drawbacks in rotary engine by improving homogeneity and burning velocity of fuel-air mixtures. Ji et al. [4-7] investigated the effect of hydrogen enrichment on the combustion and emissions characteristics of a gasoline rotary engine at idle and part load conditions. The test results indicated that the brake mean effective pressure, in-cylinder temperature and thermal efficiency were increased and the HC emissions was decreased after the hydrogen enrichment. Thus, hydrogen enrichment is demonstrated to be a feasible way to improve the rotary engine performance. However, due to limitations of the experimental testing measures, it is hard to insight details of the flow field, flame development and propagation, emissions forming and distributions in combustion chamber by experiment.

The mathematical models have been widely used in investigating the combustion process and optimizing the design of a rotary engine. Fan et al. [8-11] systematically studied the influence of pocket shapes, injection strategies, ignition positions and timings on mixture formation, combustion process and emissions characteristics of a natural-gas-fueled rotary engine with and without hydrogen addition. According to their computed results, the optimal pocket shape, injection and ignition strategies were obtained. Besides, they also found that the peak pressure of natural-gas-fueled rotary engine could be increased by 29% with hydrogen enrichment. Current investigation established a three-dimensional dynamic simulation model for the hydrogen-enriched gasoline rotary engine. Models for the dynamic grid, turbulent and combustion were built using the software CONVERGE. The model accuracy was validated with the experimental results.

2 Mathematical model establishment and validation

2.1 Modeled rotary engine and computational mesh

The three-dimensional dynamic simulation model was built based on a side-ported and gasoline-fueled Z160F rotary engine. The specifications of testing engine are listed in Ref. [4-7]. The basic grid size (d_{base}) was set as 4.0 mm. Meanwhile, to achieve high mesh quality and acceptable calculation time, adaptive mesh refinement (AMR) was adopted. AMR will automatically refined the grid basing on the fluctuating and moving conditions. Three refinement levels (embed_scale) were chosen for the different computational domains to improve grid resolution. The refined grid size (d_{embed}) is calculated by the follow equation:

$$d_{embed} = d_{base} \cdot 2^{-(embed\_scale)} \quad (1)$$

The embed_scale was set as one for intake and exhaust ports, two for combustion chamber and three for spark plug region, boundary layers and fluctuating domains of velocity, temperature and species. The grid quantity of single combustion chamber changes from 50,000 to 250,000.

2.2 Turbulence, chemical kinetic and combustion models

The turbulence in the hydrogen-enriched gasoline rotary engine was described by the RNG k-ε model. Some researchers have already conducted studies on the rotary engine's flow field either

Corresponding author. Fax: +86 1067392126
E-mail address: chwji@bjut.edu.cn

in experiments by Particle Image Velocimetry or CFD simulations [9-11]. Their test results indicated that, the RNG k-ε model could better reflect the velocity and position of vortex in rotary engine. The parameters settings were selected as suggested by the CONVERGE user-guide in this paper [12].

A skeletal kinetic model for primary reference fuel (PRF) [13] oxidation was used in hydrogen-enriched gasoline combustion simulation. This chemical kinetic models is specially developed for CFD simulations and has been validated in shock tube, jet-stirred reactor, constant volume combustion bomb and internal combustion engines over a wide range conditions. Since, the hydrogen oxidation model is a fundamental sub model in all hydrocarbon fuels. This model also could be used in hydrogen-enriched gasoline combustion. To show NO formation inside the cylinder, a NOx sub model from Golovitchev was added [14]. The combustion process was computed by the SAGE model, which allows for any number of chemistry and reactions to be modeled.

2.3 Model validation

The original rotary engine was modified and the detailed information has been listed in Ref. [4-7]. The investigations were conducted at rotary engine speed (N) of 4500 rpm and MAP of 35 kPa. The spark timing was fixed at 25 °CA BTDC and the equivalence ratio (Φ) of hydrogen-gasoline-air mixtures was kept at 1.0. The hydrogen enrichment level (α_{H2}) is represented by the hydrogen volume fraction in the total intake gas. The conditions where α_{H2} = 0%, 2% and 4% were chosen to explore the combustion process in the hydrogen-enriched gasoline rotary engine.

The accuracy of established model is validated by comparing the in-cylinder pressure with the experimental results in the same conditions. Figure 1 illustrates the in-cylinder pressure trace. As it is shown in Fig. 1, the maximum absolute error in the peak in-cylinder pressure is lower than 0.1 bar, while the relevant crank angle for the peak in-cylinder pressure is lower than 2 °CA. The good agreement between model prediction and experimental results suggested that it could be used to predict the hydrogen-enriched gasoline rotary engine working process.

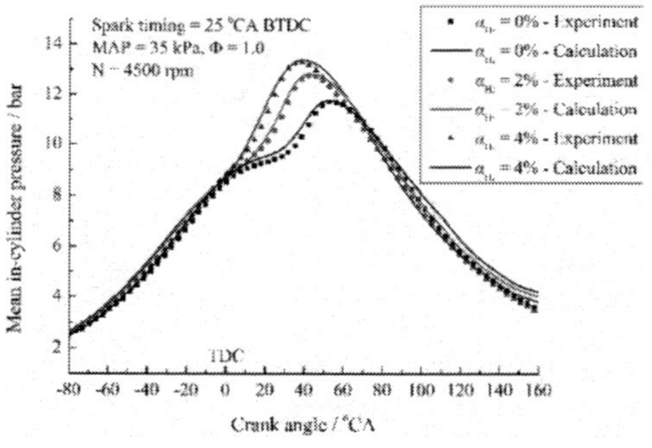

Figure 1: Model validation for in-cylinder pressure

3 Simulation results and discussion

3.1 Combustion process

The rotary engine has a special mechanical design and operation mode. Thus, a mainstream flow field exists during the compression stroke, whose direction is same to the rotor movement. The special flow field of the rotary engine is likely to play an important role in combustion process. The distribution and concentration of OH radical are commonly used in reflecting the combustion location and intensity. For flame development stage, Figures 2 describe the OH distribution with different hydrogen volume fractions in the combustion chamber at TDC. In Fig. 2, as the mainstream flow field exists in the combustion chamber, the flame propagation direction is deflected to the rotor movement direction. Meanwhile, a larger distribution area and higher concentration of OH radical are achieved at higher hydrogen volume fractions. This means that the combustion process is accelerated after the hydrogen addition. This is because the hydrogen enrichment increases the reaction rate of OH + H2 <=> H + H2O, and consequently helps form more H radical and promote the reaction rate of H + O2 <=> O + OH. And this reaction is the most important branching reaction in hydrocarbon combustion, which directly affects the burning velocity and accelerates the combustion process.

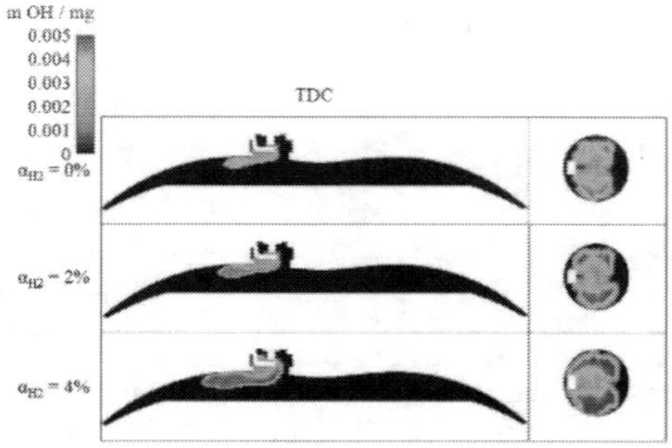

Figure 2: OH distribution under different α_{H2} at TDC

The mass distribution of CO_2 is depicted in Fig. 3 to show the flame propagation stage with different hydrogen volume fractions at 10, 20, 40 and 60 °CA ATDC. CO_2 is the main product of hydrocarbon fuel, which mainly distribute in the burned zone. Thus, the flame propagation could be represented by CO_2 mass distribution. As it is shown in Fig. 3, the volume of burned zone increases with the hydrogen enrichment levels. This means hydrogen addition accelerates the flame propagation velocity. Meanwhile, the increased burned zone after hydrogen enrichment is mostly located in front-end of the combustion chamber. This is because the flame propagation against the direction of rotor movement is retarded and the flame propagation in the direction of rotor movement is expedited by effect of the mainstream flow flied. Besides, the unburned mixtures are not burnt until it is pushed into the broad area of combustion chamber from the narrow rear region, nevertheless, the mixtures could not be burnt completely before the exhaust valve opening. Such behaviors would lead to the tremendous unburned hydrocarbon emissions.

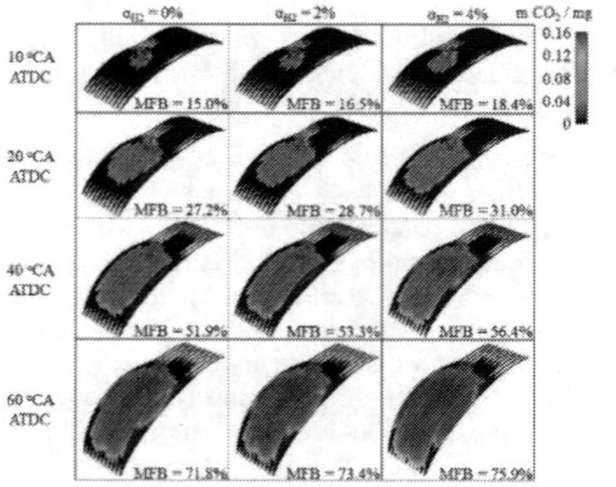

Figure 3: Flame propagation figured with CO_2 mass

Figure 4 illustrates the mean in-cylinder temperature of hydrogen-enriched gasoline rotary engine. Although hydrogen enrichment occupies the volume of intake air and decreases the total mass and energy of gasoline-hydrogen-air mixtures, the accelerated combustion process makes the heat release process is more concentrated. This could result in the higher and advanced peak in-cylinder temperature after the hydrogen addition. Meanwhile, the higher in-cylinder temperature also results in the improve in-cylinder pressure as described in Fig. 1. Compared with original gasoline case, the peak in-cylinder pressure is increased by 9.1% and 13.7% with 2% and 4% hydrogen blends, respectively. Besides, it can also be observed from Fig. 4 that the temperature of post-combustion is effectively reduced after hydrogen blends. The thermal efficiency is enhanced by the faster combustion process, which decreases the post-combustion.

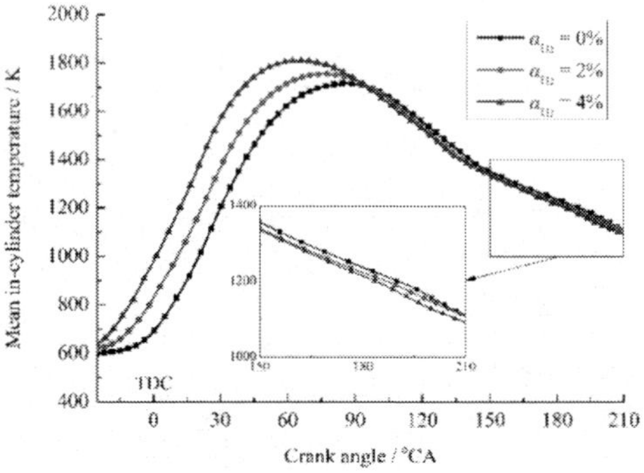

Figure 4: In-cylinder mean temperature versus crank angle

3.2 Emission formation and distribution

Figure 5 displays the distribution of NO emission in the combustion chamber with original gasoline case at 20, 60, 120 and 200 °CA ATDC.

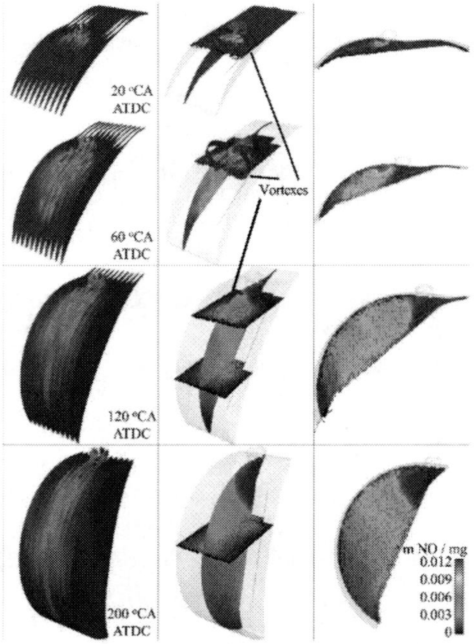

Figure 5: NO emissions distribution in the combustion chamber

As it is illustrated in Fig. 5, the highest NO concentration region is near the spark plug. This could be explained by the following reasons. Firstly, NO emission is mostly influenced by the local mixture temperature and the temperature in the burned zone is relatively high. Secondly, as the flame propagation is retarded by the mainstream flow field in rear region of the combustion chamber, the sustained combustion increases the temperature near the spark plug region. Therefore, a higher concentration of NO is obtained in this region. It can also be seen from Fig. 5, two NO enrichment regions appear near the spark plug region. This is because the mainstream flow impinges on the cylinder wall and forms two vortices with opposite rotation. The burned gas and nitrogen continuously mix up and form NO with high temperature inside the vortices. With the crank angle increases to 60 °CA ATDC, the distribution region of NO is enlarged and the concentration gets higher due to the increasingly rising in-cylinder temperature. With the further increasing of crank angle, NO emission is pushed toward to front region of the rotor housing by the mainstream flow and the concentration of NO emission is decreased. This is because the opposite rotation vortices are dissipating in bigger combustion chamber and the increased volume decreases in-cylinder temperature. Thus, the enrichment and formation of NO emission is restricted.

Figure 6 displays the distribution of CO emission in the combustion chamber with original gasoline case at 20, 60, 120 and 200 °CA ATDC. As it is shown in Fig. 6, the highest CO mass is distributed near the flame front. This is because CO is a intermedia component of the hydrocarbon fuel combustion. A large number of CO forms in the regions where the combustion intensity is relative high. Some of those CO would be further oxidized by the oxygen and release substantial heat. Therefore, a higher concentration of CO is achieved near the flame front. In Fig. 6, CO is also concentrated by the opposite rotation vortices and forms enrichment region. Because of the further complete oxidation of CO in this region, only one CO enrichment region is obtained in the burned zone. At 200 °CA ATDC, the unburned fuel-air mixtures are continuously pushed into the burned region to combust, which lead to a higher mass of CO is displayed in the rear of combustion chamber.

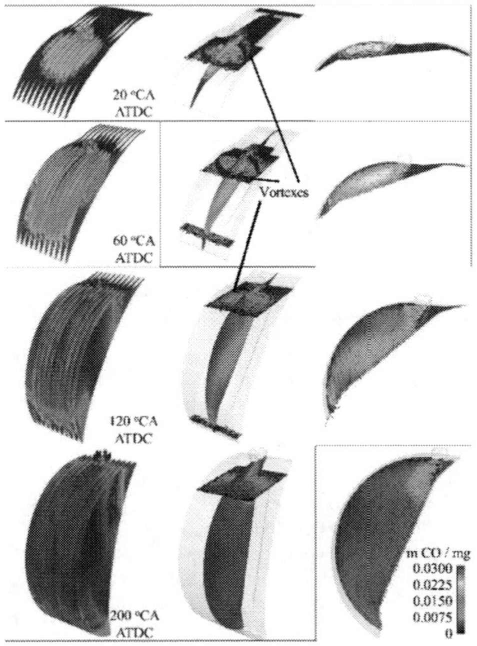

Figure 6: CO emissions distribution in the combustion chamber

Figure 7 describes curves of NO and CO mass varying in the rotary engine combustion chamber with different hydrogen enrichment levels. The peak of NO is obviously increased with hydrogen enrichment. This is because the higher combustion temperature of hydrogen increases the in-cylinder temperature. In addition, the increased concentration of active radicals also promotes the NO formation. Therefore, hydrogen enrichment provides a suitable condition for NO. When the exhaust valve is opened at 210 °CA ATDC , the masses of NO emission are 0.102, 0.113 and 0.148 mg respectively for hydrogen volume fractions of 0%, 2% and 4%. This indicates that hydrogen enrichment leads to more NO emission for gasoline rotary engine. The mass of CO under different hydrogen volume fractions is also described in Fig. 7. It can be seen from Fig. 7 that the peak of CO mass is decreased after hydrogen enrichment. This is because hydrogen addition occupies the volume of intake air and reduces the inhaled mass of gasoline. Therefore, the carbon content in the binary fuel decrease. While, CO + OH <=> CO2 + H is the dominate path for CO oxidation, the increased OH concentration also promotes CO transferring to CO_2. Moreover, the mass of CO with hydrogen blends decreases at post-combustion period. This is can be explained by the fact that the accelerated combustion process after hydrogen enrichment promotes the oxidation of CO and the short quenching distance of hydrogen reduces the unburned mixtures in rear region of the combustion chamber. Compared with original gasoline case, the masses of CO emission are reduced by 14.7% with 2% hydrogen blends and 33.0% with 4% hydrogen enrichment.

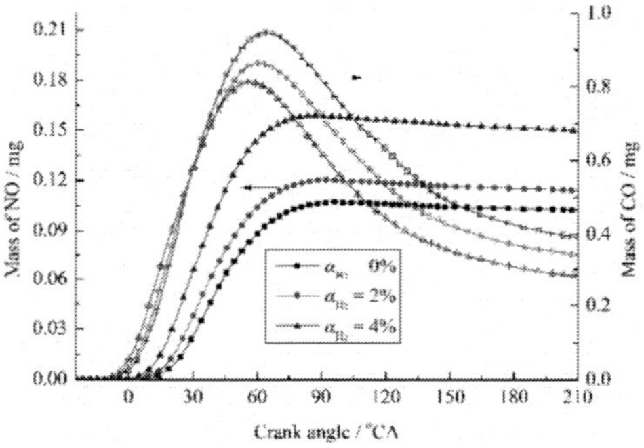

Figure 7: NO and CO mass versus crank angle

4 Conclusions

In the present study, a three-dimensional dynamic simulation model for a SI rotary engine fueled with hydrogen-gasoline blends was established using CFD software. Satisfying agreement between the numerical and experimental results was obtained for all the tested cases. The combustion and emission formation processes in the hydrogen-enriched gasoline rotary engine were numerically investigated. Conclusions are as follows:

(1) Due to the mainstream flow flied, the flame propagation against the direction of rotor movement is retarded and the flame propagation in the direction of rotor movement is expedited. Meanwhile, the unburned mixtures in the narrow rear region are not burnt until it is pushed into the burned zone, nevertheless, the mixtures could not be burnt completely before the exhaust valve opening. After hydrogen enrichment, the increased concentration of active radicals accelerated the combustion process and lead the heat release more concentrated. Meanwhile, the higher combustion velocity also resulted in the improvement of in-cylinder temperature and pressure. Besides, the post-combustion was also effectively reduced after hydrogen addition.

(2) NO was mainly distributed near spark plug of the burned zone where the local temperature was relatively high. Two NO emission enrichment regions were formed in the opposite rotation vortices from impinging of the mainstream on the cylinder wall. With the increase of crank angle, the distribution region of NO was enlarged until the vortices dissipated. The highest CO concentration was obtained near the flame front where the combustion intensity is relative high. CO was also concentrated by the opposite rotation vortices, but the further complete oxidation reduced its concentration in this zone. Before the exhaust valve opening, the combustion of unburned fuel-air mixtures in the rear regions resulted in a higher mass of CO. Compared with original gasoline case, a 45.1% increased NO emission and 33% decreased CO emission were obtained after 4% hydrogen enrichment.

5 Acknowledgment

This work was supported by National Natural Science Foundation (Grant No.51476002), National Key Basic Research Development Project (973) (Grant No.2013CB228403), Key Program of Sci & Tech Project of Beijing Municipal Commission of Education (Grant No.KZ201610005005), Beijing Municipal Commission of Science and Technology (Grant No. Z141100003814017) and Sci & Tech Project of Beijing Municipal Commission of Education (Grant No.KM201510005011).

References

[1] C. Hubmann, F. Beste, H. Friedl, W. Schoffmann, SAE. paper 2013-32-9132 (2013).

[2] J. Ribau, C. Silva, F. Brito, J. Martins, Energy. Convers. Manage 58 (2012) 120-133.

[3] K. Siadkowska, M. Wendeker, A. Majczak, G. Baranski, M. Szlachetka, SAE paper 2014-01-2611 (2014).

[4] C. Ji, T. Su, S. Wang, B. Zhang, M. Yu, X. Cong, Energy. Convers. Manage 121 (2016) 272-280.

[5] T. Su, C. Ji, S. Wang, L. Shi, J. Yang, X. Cong, Energy. Convers. Manage 148 (2017) 120-127.

[6] T. Su, C. Ji, S. Wang, L. Shi, J. Yang, X. Cong, Energy. Convers. Manage 42 (2017) 12696-12705.

[7] T. Su, C. Ji, S. Wang, L. Shi, J. Yang, X. Cong, Int. J. Hydrogen. Energy 42 (2017) 11893-11901.

[8] B. Fan, J. Pan, W. Yang, Y. Liu, S. Bani, W. Chen, Energy. Convers. Manage 133 (2017) 511-523.

[9] B. Fan, J. Pan, Z. Pan, A. Tang, Y. Zhu, H. Xue, Appl. Therm. Eng 89 (2015) 11-27.

[10] B. Fan, J. Pan, Y. Liu, Y. Zhu, Energy. Convers. Manage 103 (2015) 218-234.

[11] B. Fan, J. Pan, W. Yang, Y. Zhu, W. Chen, Int. J. Hydrogen. Energy 41 (2016) 4039-4053.

[12] Convergent. Science. Corp, CONVERGE Theory Manual, 2014.

[13] Y. Liu, M. Jia, M. Xie, B. Pang, Energ. Fuel 26 (2012) 7069-7083.

[14] V.L. Golovitchev, available at <http://www.tfd.chalmers.se/~valeri/MECH.html/>.

Reviewer Comments:

Reviewer #1:

The English still needs to be polished. For example, in the fourth section "Conclusions", the authors wrote "Conclusions as follows" can be better polished to "Conclusions are as follows".

Response: Thank you for the careful review and valuable comments! According to your suggestions, this manuscript has been revised.

11th Asia-Pacific Conference on Combustion,
The University of Sydney, NSW Australia
10th -14th December 2017.

Fire behaviour modelling of polypropylene under autoignition conditions

Long Shi, Guomin Zhang, Sujeeva Setunge

Civil and Infrastructure Engineering, School of Engineering, RMIT University, Melbourne, VIC 3001, Australia

Abstract

Fire behaviour modelling of solid fuel is the key to the success of a computational fluid dynamics (CFD) tool. Fire behaviour modelling of polypropylene (PP) under autoignition conditions was taken and validated by cone calorimeter experiments. Modelling considered both solid and gas phases, such as pyrolysis reactions, melting processes, gas and liquid transportation, in-depth radiation and gas phase combustion. Mass loss of PP samples is quite following typical non-charring polymers with three stages: a sharply increasing mass loss of ignition stage, a stably increasing period of molten stage, and finally a peak with boiling liquid state.

1 Introduction

Polymers are one type of frequently used materials in our daily lives and almost every industry. From 1960 to 2000, the consumption of polymers has increased about 30 times [1]. However, their fire risk is always threatening human beings and properties because they are based on organic and thus flammable. A large number of studies have been performed on their fire behaviours [2-5]. Fire threats of non-charring polymers, such as polypropylene (PP), are even bigger than other types of polymers because they can result in liquid pool fires after melting. A large volume of gas volatiles can be produced as almost all the polymers can change into gas volatiles, leaving no or very few residues [6-8].

Assumptions and simplifications in numerical modelling always compromise the complexity of a numerical model and define somehow its limitations [4, 9]. For example, few previous models enable to provide a good combination of solid and gas phase modelling, and challenges still exist to predict the burning behaviours of solids [10]: (*i*) applications of gas products were limited because of the difficulty in describing gas transportation inside solid materials, while gas volatiles were assumed to exhaust immediately after produce; (*ii*) no common agreements were given for some physical behaviours, such as evaporation and melting; and (*iii*) most of the previous pyrolysis models only considered solid phase.

Therefore, this study has focused on fire behaviour modelling of a typical non-charring polymer under autoignition conditions. The study is aiming to provide a useful tool for fire behaviour modelling under various conditions.

2 Mathematical model

2.1 Major assumptions

The major features of the numerical model include: (i) it models both gas and solid phases; (ii) consideration of physical/mechanical processes, as well as heterogeneous and homogeneous chemical reactions, makes it capable of describing water evaporation, polymer melting, pyrolysis and oxidation reactions; and (iii) consideration of permeability and porosity makes it capable of simulating liquid and gas transportation processes inside the solid phase.

Schematic of a computational domain is shown in Fig. 1. Several assumptions were adopted: (i) solid phase was considered as one-dimensional, and the gas phase as two-dimensional, due to the predominant (normal) direction of heat transfer in the solid phase and simplicity of the gas phase, respectively; (ii) homogeneous reactions in the solid phase were ignored, considered in the gas phase only; (iii) all gas species have the same thermal properties. Gas species in the solid phase were assumed to be ideal gases, and gas species in the gas phase were treated as incompressible ideal gas; (iv) influence of volume change in the solid phase on the gas phase was ignored; (v) failure processes such as cracking were ignored due to the complexity; and (vi) thermal feedback from flame to solid materials was assumed to be negligible compared to external radiation.

2.2 Governing equations

The major features and equations of this model are briefly introduced here, and more details such as initial and boundary conditions can be seen in our previous studies [11, 12]. For the solid phase, equations for energy conservation, consumption of solid species, and mass conservation for gas and liquid are given below, respectively:

$$\sum_{i=1}^{N_s+N_l+N_s} \frac{\partial \rho_i \phi C_{p,i} T_s}{\partial t} + \sum_{j=1}^{N_l+N_s} \frac{\partial \rho_j \phi C_{p,j} T_s u_j}{\partial x} = \frac{\partial}{\partial x}\left(\lambda \frac{\partial T_s}{\partial x}\right) + Q_{pyr} + \frac{\partial \dot{q}''_{rad}}{\partial x} \quad (1)$$

$$\frac{\partial \rho_i}{\partial t} = \Theta_i \quad (2)$$

$$\frac{\partial \rho_g \phi}{\partial t} + \frac{\partial \rho_g \phi u}{\partial x} = -(\Theta_i + \Theta_l) \quad (3)$$

$$\frac{\partial \rho_l}{\partial t} + \frac{\partial \rho_l u}{\partial x} = \Theta_l \quad (4)$$

where ρ is the density; ϕ is the porosity; C_p is the specific heat capacity; T is the temperature; t is the time; u and v are the velocities; λ is the thermal conductivity; x and y are the Cartesian coordinate; Q_{pyr} is the pyrolysis heat; q_{rad} is the radiation heat; Θ is the production or reaction heat; i and j represent species; N is the total number of species; and l and g indicate liquid and gas species, respectively.

Reaction terms are simulated as:

$$\Theta_i = \sum_j v_{1,s}^j k_j \rho_j^{n_j};\; k_j = A_j \exp(-E_j/RT);\; \Theta_l = A_l \cdot \exp(-E_l/RT)\rho_l \quad (5)$$

where k is the reaction rate constant; A is the pre-exponential factor; E is the activation energy; R is the Universal gas

Corresponding author. Fax: +61-3-9925 3215
E-mail address: long.shi@rmit.edu.au (L. Shi)

constant; and ν^j is the coefficient.

Heat generated or absorbed during pyrolysis reactions is:

$$Q_{pyr} = \sum_j \Delta h_j k_j \tag{6}$$

where h is the specific enthalpy of materials.

Absorption of radiation is described following the Beer-Lambert Law:

$$\dot{q}''_{rad} = \dot{q}''_{ext} \exp(-\kappa x) \tag{7}$$

where κ is the absorption coefficient; and *ext* means external.

Currently, no appropriate model has been found in describing the gas and liquid transportation inside non-charring polymer because of the complexity of problem. As the related siulation is impossible unless a pressure equation is invoked, therefore, following the Reference [13], gas and liquid transportations are estimated by the constitutive relations (Darcy's law) applying to flows through the matrix of the solid:

$$u_g = -\frac{\gamma_g}{\mu_g \phi}\frac{\partial P}{\partial x};\ u_l = -\frac{\gamma_l}{\mu_l}\frac{\partial P}{\partial x};\ P = \rho_g RT/M_g;$$
$$\phi = 1-(\rho_{theor})^{-1}\sum_{i=1}^{N_s}\rho_i;\ M_g = \left(\sum_{i=1}^{N_g}\frac{Y_i}{M_i}\right)^{-1} \tag{8}$$

where P is the pressure; γ is the permeability; Y is the mass fraction; and *theor* represents the theoretical value.

It should be mentioned that the adoption of Darcy's law will largely affect gas phase, but with limited influence on solid phase outputs. The simulation of gas transportation is the key to the gas phase, we will revise the model when there is an appropriate model available. The related validation will be taken in our future works.

Two-dimensional Navier-Stokes equations, supplemented with the energy, continuity, momentum and species conservation equations, were utilized for the gas phase, respectively:

$$\frac{\partial \rho C_p T}{\partial t} + \frac{\partial \rho C_p u T}{\partial x} + \frac{\partial \rho C_p v T}{\partial y} = \frac{\partial}{\partial x}\left(\lambda\frac{\partial T}{\partial x}\right) + \frac{\partial}{\partial y}\left(\lambda\frac{\partial T}{\partial y}\right) + Q_{reac} \tag{9}$$

$$\frac{\partial u}{\partial x} + \frac{\partial v}{\partial y} = 0 \tag{10}$$

$$\frac{\partial \rho u}{\partial t} + \frac{\partial \rho u u}{\partial x} + \frac{\partial \rho v u}{\partial y} = \frac{\partial}{\partial x}\left(\mu\frac{\partial u}{\partial x}\right) + \frac{\partial}{\partial y}\left(\mu\frac{\partial u}{\partial y}\right) - \frac{\partial P}{\partial x} \tag{11}$$

$$\frac{\partial \rho v}{\partial t} + \frac{\partial \rho u v}{\partial x} + \frac{\partial \rho v v}{\partial y} = \frac{\partial}{\partial x}\left(\mu\frac{\partial v}{\partial x}\right) + \frac{\partial}{\partial y}\left(\mu\frac{\partial v}{\partial y}\right) - \frac{\partial P}{\partial y} \tag{12}$$

$$\frac{\partial \rho Y_i}{\partial t} + \frac{\partial \rho u Y_i}{\partial x} + \frac{\partial \rho v Y_i}{\partial y} = \frac{\partial}{\partial x}\left(\rho D\frac{\partial Y_i}{\partial x}\right) + \frac{\partial}{\partial y}\left(\rho D\frac{\partial Y_i}{\partial y}\right) + \Theta_{reac} \tag{13}$$

where *reac* represents reaction process; μ is the dynamic viscosity; and D is the diffusivity coefficient.

Non-charring polymer burns out leaving very few or no residue. During the numerical modelling, it then becomes impossible to continue with the process when the size and mass of grids become zero after burning out. In order to avoid empty grids, very few char residues were assumed for non-charring polymers. As liquid water content inside polymer is very low [14], its presence in the polymer samples can be ignored. Reactions considered were:

$$[Polymer] \rightarrow [bPolymer] \tag{14}$$

$$[Polymer] \rightarrow [Fuel] + [Char\ (very\ little)] \tag{15}$$

$$[bPolymer] \rightarrow [Fuel] + [Char\ (very\ little)] \tag{16}$$

where *Polymer* is the polymer in its original state of solid and *bPolymer* is considered as the bubbled polymer after melting.

In the absence of detailed information on the chemical composition of pyrolysis products (which is quite complex and difficult to quantify precisely), the chemical reaction in the diffusion flame is written in a general form on mass basis:

$$[Fuel] + \nu_{O_2}[O_2] \xrightarrow{kf} \nu_{CO}[CO] + \nu_{CO_2}[CO_2] + \nu_{H_2O}[H_2O] + \Delta h_f \tag{17}$$

where *Fuel* is the general designation of gas volatiles from the pyrolysis reactions of solids. And the pyrolysis reaction rate constant and the heat of reaction are given by, respectively:

$$k_f = A_f \cdot \exp(-E_f/RT);\ Q_{reac} = k_f \rho^2 Y_F \cdot Y_{O_2} \cdot \Delta h_f \tag{18}$$

Sensitivity analysis of modelling was taken and the details can be seen in Reference [15]. Regarding the influence of char yield, the influence is relatively limited as very few char was considered during the numerical modelling. Sequential coupling technique was used for the solid and gas phases interaction. The boundary conditions of bottom inflow of gas phase were obtained by the gas flow from the solid phase, including both gas velocity and temperature.

3 Experimental and numerical methodology

The left part in Fig. 1 shows test set-up of cone calorimeter. During experiments, cone calorimeter was positioned under a bigger hood and the volume extraction rate was keeping at 24 L/s. The volume extraction provides a forced convection driven flow under the cone calorimeter. External heat fluxes were chosen to represent low, middle and high radiation, such as 25, 50 and 75 kW/m^2, respectively. Spark plug was not used, and autoignition of the samples was considered to occur when visible flame was first observed. Each experiment was repeated at least two times, with more runs taken if the repeatability was not good. More details of the experiments can be found in our previous studies [3, 16, 17].

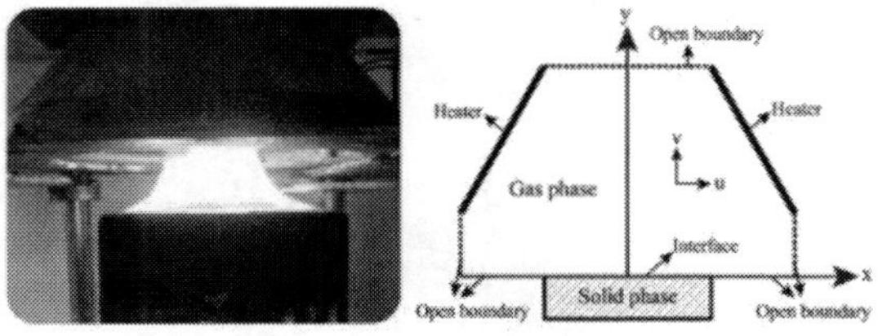

Fig. 1. Test set-up (left) and computational domain (right)

All the tested samples in this study were not fire-retarded, which were originally in sheet form and were cut to desired size of 100 mm × 100 mm × 30 mm (thickness).

Governing equations were solved by fully implicit Finite Volume Method (FVM). To avoid oscillations and convergence problems, staggered grids were utilized for the pressure field. As a high order differencing scheme, Quadratic Upstream Interpolation for Convective Kinematics (QUICK) was utilized to discretise all convective terms. Pressure Implicit with Splitting of Operators (PISO) was used to solve the pressure-velocity coupling problem.

4 Results and discussion

4.1 Mass loss rate

Table 1 shows the thermal properties of PP for modelling

inputs. PP is a non-charring polymer, which can melt and burn out with very few or no residues. Three chemical reactions were considered during modelling, as shown in Eqs. (14) to (16). It should be noticed that the melting process was also considered, and the melting rate dependent on temperature can refer to Reference [7]. Some inputs are very difficult to be obtained from the literature, such as gas permeability and diffusion coefficient. These data were assumed from similar materials or substance in the literature. Very few char are left after the burning, so the consideration of char during numerical modelling is to avoid empty grid. For those data difficult to be found in the literature, the properties form similar material were referred to for the numerical modelling.

Table 1. Thermal properties as modelling inputs

Property	Unit	PP/bPP	Char (nominal)
Pyrolysis reaction rate	1/s	3.17×10^{24}exp(-373,400/RT) [18]	-
Melting rate	1/s	2.90×10^{47}exp(-366,250/RT) [7]	-
Density	kg/m^3	938.6 [measured]	95
Heat of pyrolysis	kJ/kg	1750	-
Gas permeability	m^2	5.80×10^{-16}	5.80×10^{-16}
Water permeability	m^2	3.79×10^{-19}	3.79×10^{-19}
Diffusion coefficient of water	m^2/s	5.11×10^{-8}	5.11×10^{-8}
Diffusion coefficient of gas	m^2/s	1.85×10^{-9}	1.85×10^{-9}
Specific heat capacity	J/kg·K	1930 [3]	1350 [19]
Surface emissivity	-	0.97	0.8 [20]
Thermal conductivity	W/m·K	0.21 [3]	0.35
Absorption coefficient	1/m	960	0.5

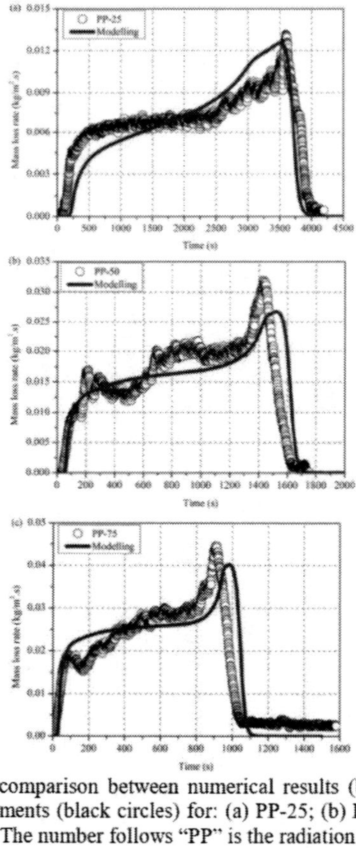

Fig. 2. A comparison between numerical results (blue lines) and experiments (black circles) for: (a) PP-25; (b) PP-50; and (c) PP-75. The number follows "PP" is the radiation heat flux

Fig. 2 shows the comparison of *MLR* of PP samples between numerical modelling and experiments. It shows the mass loss rate from the experiment start until the extinguishment, including the ignition. Ignition time can be seen in Reference [3]. It is observed that the predictions of PP samplers under various heat fluxes obey quite well with experiments. The burning processes are quite following the typical non-charring polymers mentioned by Tewarson [21]. They clearly show three stages. Firstly, the mass loss rate (*MLR*) increase sharply after the autoignition, reaching to the first peak. Then it goes through a stably increasing period, representing the molten stage. Finally, after all the sample melts, it undergoes the boiling liquid state, where another peak *MLR* is shown. After the peak, the *MLR* keeps decreasing under the continuing consumption of the melted sample, until burning out.

4.2 Temperature in the solid phase

Fig. 3 shows the predicted temperature inside the PP samples under various radiation heat fluxes. In these subfigures, the axes X, Y and Z represent the time, depth in the solid, and temperature. The colour bars on the right side show a temperature range of 300-800 K.

It is known from Fig. 3 that the surface temperatures increase quite fast since the beginning of modelling. The increase of the bottom temperature is relatively slower than that. This is because it needs time for the radiation heat transfer to the bottom. After the maximum temperature, the temperatures at various depths are decreasing until the end. This is because the consumption of solid material and the original space occupied by solid components is then filled by the hot air or fuel. The temperatures then keep decreasing.

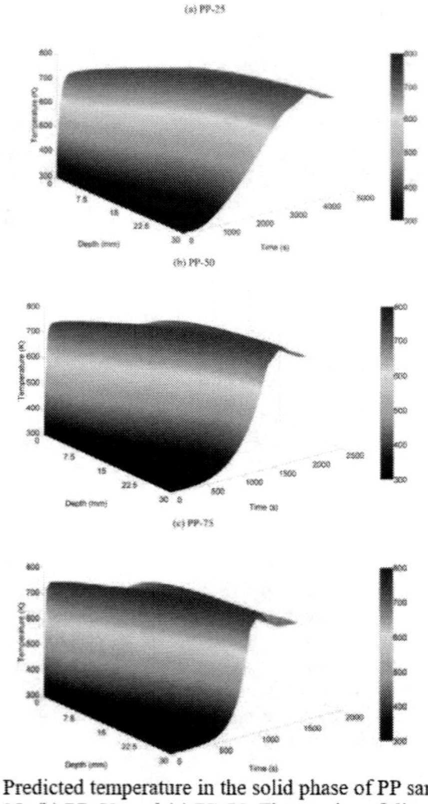

Fig. 3. Predicted temperature in the solid phase of PP samples: (a) PP-25; (b) PP-50; and (c) PP-75. The numbers follow PP is the external heat flux.

Comparing to the results from charring materials in our previous study [11], two differences can be observed: (a) the non-charring polymers show much lower temperatures in the solid phase; and (b) the surface temperature decrease after achieving the maximum for non-charring polymers, while the charring materials almost keep constant. These two phenomena are quite consistent with the characteristics of non-charring polymers, such as low absorption coefficient and very few residues.

4.3 Gas volatiles

Predictions of mass fractions of O_2 and Fuel are shown in Fig. 4. In this figure, the left half shows the mass fraction of O_2 and the right half is the Fuel. The colour bars in the left (within a range of 0-0.25) and right sides (within a range of 0-0.8) show the O_2 and Fuel, respectively. It is known that the range of mass fraction for Fuel is much higher than that of O_2. It is also observed that the mass fraction of O_2 in most part of the gas phase is close to ambient value, excepting the area above the slab surface. This is because this is the location happened with the combustion reactions, consuming large volume of O_2.

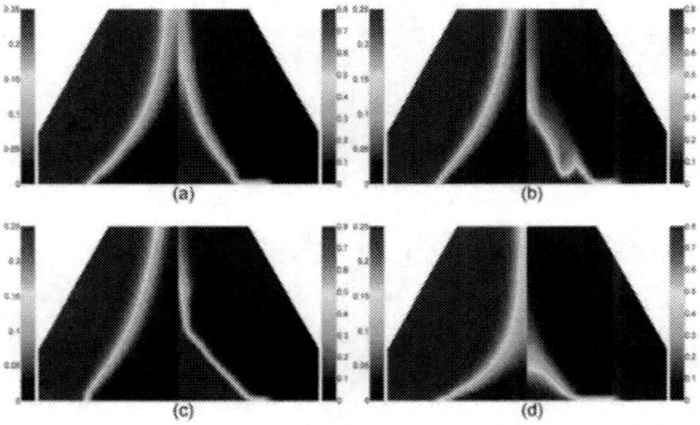

Fig. 4. Mass fraction of O_2 (left half) and Fuel (right half) in the gas phase for 30 mm PP samples under 75 kW/m^2 heat flux at: (a) 100 s; (b) 200 s; (c) 800 s; and (d) 1200 s

5 Conclusions

To expand the applications and improve the modelling accuracy of current computational fluid dynamics (CFD) tools, a numerical model was developed to predict the fire behaviours of solids slabs under various radiations fluxes. The model considers both solid and gas phases, including pyrolysis reactions, melting process, gas and liquid transportation, in-depth radiation and gas phase combustion. The predictions of burning rate of polypropylene (PP) were validated by cone calorimeter experiments. The modelling results of gas phase such as gas species were also quite consistent with the major heat transfer processes.

Based on both experiments and numerical modelling, the mass loss history of PP samples under autoignition is quite following typical non-charring polymers: namely a sharply increasing mass loss rate of ignition stage, a stably increasing period of molten stage, and finally a peak with boiling liquid state. The modelling accuracy of this model will be largely improved when some of the experimental data are available, such as melting rate, liquid and gas permeability, and some key properties for melted polymers. Our future efforts then enable the validation of gas volatiles release rate from the solid phase and gas phase combustion, and probably the development of empirical model to predict gas and liquid transportation inside non-charring polymers.

References

[1] Mouritz AP, Gibson AG, Fire properties of polymer composite materials, Springer, Netherlands, 2007.

[2] Sonnier R, Dorez G, Vahabi H, Longuet C, Ferry L, FTIR-PCFC coupling: A new method for studying the combustion of polymers, Combust. Flame, 161 (2014) 1398-1407.

[3] Shi L, Chew MYL, Fire behaviors of polymers under autoignition conditions in a cone calorimeter, Fire Saf J, 61 (2013) 243-253.

[4] Bal N, Rein G, Relevant model complexity for non-charring polymer pyrolysis, Fire Saf J, 61 (2013) 36-44.

[5] Pater P, Hull TR, Stec AA, Lyon RE, Influence of physical properties on polymer flammability in the cone calorimeter, Polym. Adv. Technol., 22 (2011) 1100-1107.

[6] Lyon RE, Solid-state thermochemistry of flaming combustion, in, Federal Aviation Administration, USA, 1999.

[7] Li J, Stoliarov SI, Measurement of kinetics and thermodynamics of the thermal degradation for non-charring polymers, Combust. Flame, 160 (2013) 1287-1297.

[8] Di Blasi C, Modeling and simulation of combustion processes of charring and non-charring solid fuels, Prog. Energy Combust. Sci., 19 (1993) 71-104.

[9] Di Blasi C, Modeling chemical and physical processes of wood and biomass pyrolysis, Prog. Energy Combust. Sci., 34 (2008) 47-90.

[10] Shi L, Chew MYL, A review of fire processes modeling of combustible materials under external heat flux, Fuel, 106 (2013) 30-50.

[11] Shi L, Chew MYL, Novozhilov V, Joseph P, Modelling the pyrolysis and combustion behaviours of non-charring and intumescent-protected polymers using 'FiresCone', Polymers, 7 (2015) 1979-1997.

[12] Shi L, Chew MYL, Novozhilov V, A model to predict fire behaviors of combustible materials under external heat flux, in: The 14th International Conference and Exhibition on Fire and Materials, San Francisco, USA, 2015, pp. 256-270.

[13] Lautenberger C, A generalized pyrolysis model for combustible solids, PhD Thesis, Department of Mechanical Engineering, University of California, Berkeley, 2007.

[14] MatWeb, Database of material properties, www.matweb.com, accessed in 2015.

[15] Shi L, Pyrolysis and combustion processes of combustible materials under external heat flux, PhD Thesis, Department of Building, National University of Singapore, 2014.

[16] Shi L, Chew MYL, Experimental study of carbon monoxide for woods under spontaneous ignition condition, Fuel, 102 (2012) 709-715.

[17] Shi L, Chew MYL, Influence of moisture on autoignition of woods in cone calorimeter, J. Fire Sci., 30 (2012) 158-169.

[18] Grammelis P, Basinas P, Malliopoulou A, Sakellaropoulos G, Pyrolysis kinetics and combustion characteristics of waste recovered fuels, Fuel, 88 (2009) 195-205.

[19] Alves SS, Figueiredo JL, A model for pyrolysis of wet wood, Chem. Eng. Sci., 44 (1989) 2861-2869.

[20] Ragland KW, Aerts DJ, Properties of wood for combustion analysis, Bioresour. Technol., 37 (1991) 161-168.

[21] Tewarson A, Generation of heat and chemical compounds in fires, in: SFPE Handbook of Fire Protection Engineering, National Fire Protection Association, Quincy, 2002, pp. 3.82-83.161.

11th Asia-Pacific Conference on Combustion,
The University of Sydney, NSW Australia
10th -14th December 2017.

An optical absorption method to deduce the temperature dependence of gas viscosity

Rongkang Gao[1], Sean O'Byrne[1]

[1]School of Engineering and Information Technology, University of New South Wales

Northcott Drive, Canberra, A.C.T, 2600, Australia

Abstract

This paper demonstrates a method to quantify the viscosity dependence on temperature by using the technique of tunable diode laser absorption spectroscopy (TDLAS). A hollow core photonic crystal fiber (HC-PCF), used as a gas channel, is surrounded by a furnace for temperature adjustment. The path-integral absorbance of gas inside fiber as a function of time during transient is measured by using TDLAS. These measurements under non-isothermal conditions are correlated to the temperature dependence of viscosity, which can be used to infer the constant of the Power Law. The measurement sensitivity depends on the high-temperature length fraction of the fiber. It is shown that 5m fiber length (93% in the furnace) can provide a reasonable measurement resolution based on numerical simulations.

1 Introduction

The measurement of gas viscosity at elevated temperatures is a challenge. Most of the techniques, such as capillary viscometers and spinning disk viscometers, involve moving components (a mercury bead in the capillary viscometer, a rotating disk in the spinning disk viscometer), which can be difficult to operate at elevated-temperature conditions. The mercury bead will be vaporized at a temperature of 630 K at atmospheric pressure, and the elastic modulus of the spinning disk assembly will change with temperature, as will the effect of loss mechanisms in the rotating system. Harvey et al. [1] illustrated a summary of viscosity measurement of CO2, showing a shortage of laboratory data at temperatures higher than 700 *K*. In addition, the available analytical solutions such as the Sutherland's Law and the Power Law is applicable to moderate temperatures. For the aforementioned reasons, we present the TDLAS measurement for inferring the temperature dependence of viscosity in a non-isothermal condition. Our system has no moving part and can be conducted at elevated temperature due to the high heat resistance of the HC-PCF used. The measurements of viscosity under ambient temperature were presented in a paper [2]. In this paper, due to the lack of the enough fiber length, we focus on the feasibility analysis of high-temperature measurements using this technique.

2 Experimental arrangement and methods

The experimental setup consists of two gas cells connected by a HC-PCF (NKT Photonics HC-2000-01). This setup is similar to that used for our previous research, but with a different laser source (EP200-DM-B01-PM) and a compatible current controller (Newport Model 505). This laser source is used because it provides operating wavelengths corresponding to strong CO_2 absorption lines around 2005 nm, compared with the much weaker lines near 1577 nm used in the previous study. The fiber is twined around an alumina tube in the furnace to maximize the high-temperature length fraction of the fiber. A transient gas flow is initiated by a pressure differential caused by two gas cells (cell 2 hold vacuum, cell 1 increased to target pressure instantaneously). The laser beam is coupled into the fiber, being absorbed by CO_2. The light attenuation as a function of time during the transient process is monitored by the photo detector. A detailed description of the measurement procedures and signal modulations is discussed in our previous paper[2]. The transition line used for experiments is R12e at 2005.09 nm.

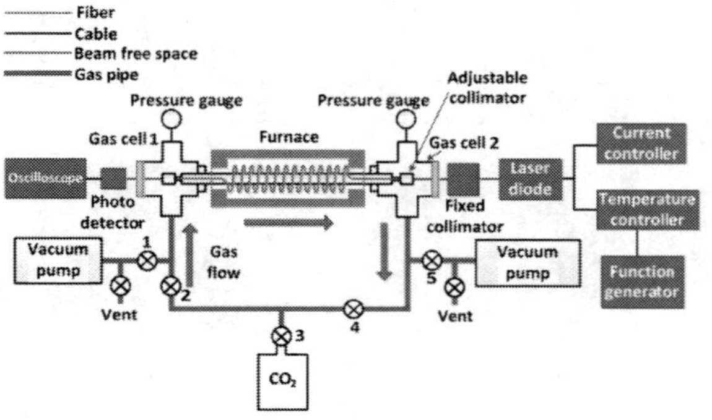

Figure 1: Experimental setup

2.1 Data processing

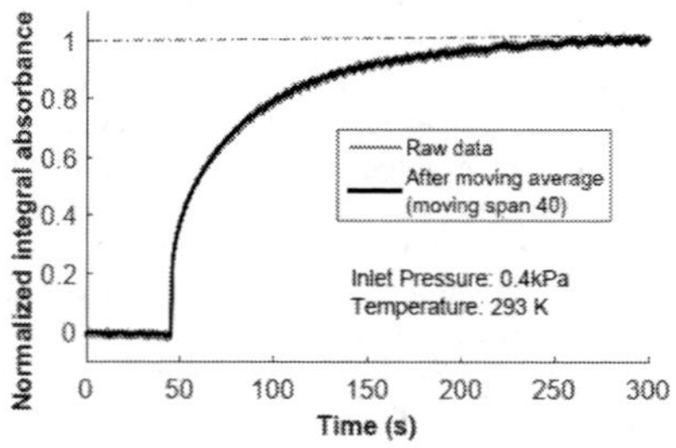

Figure 2: Normalized integral absorbance as a function of time

The raw data of integral absorbance, shown in Figure 2, is obtained from the transition in modulated signal in each scan. Gao et al. [2] cover a thorough description of the data processing to achieve the integral absorbance. The data presents a much higher signal-to-noise ratio than the previous data [2], due to the stronger transition used near 2005.09 nm.

2.2 Calibrated temperature profiles

The temperature distribution along the furnace length is

Corresponding author.
E-mail address: rong.gao@student.adfa.edu.au

calibrated by using a thermocouple and a translational stage for position adjustment. The temperature profiles inside the furnace and extra 10cm lengths outside are therefore obtained under different input values of temperature from 400 K to the highest operating temperature of furnace 1473 K, shown in Figure 3. Based on this, the temperature distribution along the fiber can be easily obtained with known twined length of fiber in the furnace.

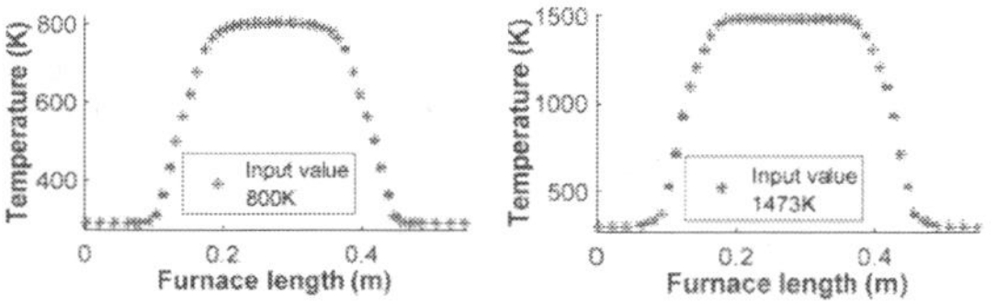

Figure 3: Temperature profiles along the furnace

2.3 Thermal expansion effect on measurement

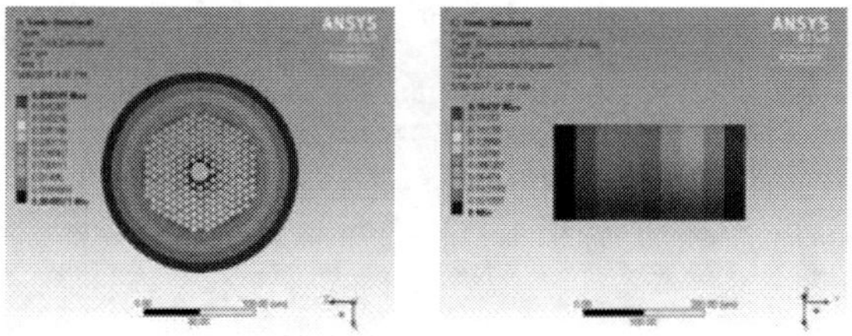

Figure 4: Deformation contour of HC-PCF in radial and longitudinal directions (temperature 1473 K, radius 6.5 um, length 300 um)

ANSYS Workbench is used for the finite element (FE) modelling to quantify the thermal expansion of HC-PCF in both radial and longitudinal directions. The deformation contours are shown in Figure 4. Rather than using actual fiber length (0.83 m), a small length (300 um) was used for the computational efficiency. The results are compared to the linear thermal expansion theory:

$$\Delta L = \alpha L_0 \Delta T \tag{1}$$

where α is the thermal expansion coefficient (5.5×10^{-5}/K for silicon), L_0 and ΔL refer to the original and expanded value of length or radius. It is shown in Figure 5 that the expansion results from FE modelling match perfectly with Equation (1). Therefore, Equation (1) can be used to infer expansion of actual length (0.83m) used in experiments. We take the expansion under highest temperature for the worst-case analysis (ΔL=0.054 m, ΔR=4.2×10^{-3} um). In measurements, the increase of radius will shorten the transient time for gas flow to reach a steady state. Because the quantity of time constant is used to calibrate viscosity, this decrease in time constants will cause a decrease of the measured value of viscosity. The expansion in length, on the contrary, will raise the time constant of gas flow, and increase the measured viscosity correspondingly. To analyze the coupled effect of expansions to the measured value of viscosity, the numerical curves of integral absorbance as a function of time with and without considering expansions were generated and depicted in Figure 6. The detailed algorithm of this simulation under isothermal conditions is covered in Gao et al. [2]. As shown in Figure 6, under the same operating conditions of temperature and pressure, the time taken for gas flow reaches a steady state with expanded radius and length (50.992s), is slightly higher compared with the time constant 50.898s in which case the thermal expansion is not considered. This indicates that the effect of expanded length is stronger than the effect of expanded radius on the time constant. To match the time constant (50.992s) under thermal expansion condition, the viscosity should increase from 5.19×10^{-5} Pa s to 5.1984×10^{-5} Pa s when the radius and length remain the initial value. This leads to a decline of 0.0084×10^{-5} Pa s (0.28%) in the measured value of viscosity, which is far below the uncertainty (0.025×10^{-5} Pa s) in previous viscosity measurements [2].

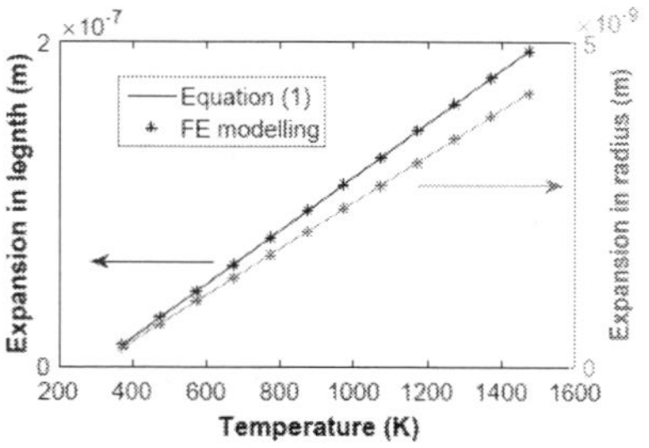

Figure 5: Thermal expansions of radius and length at different temperatures

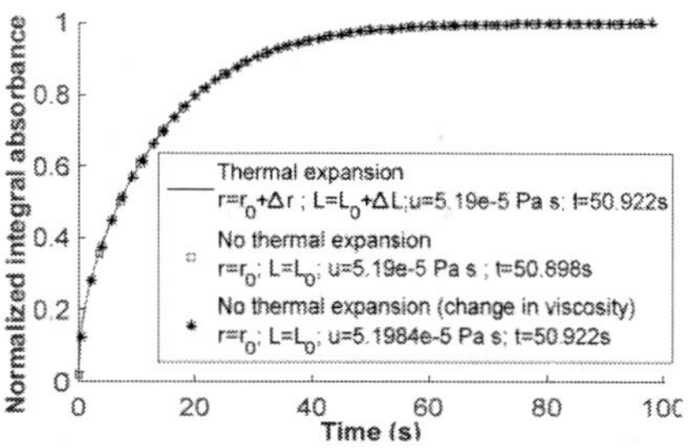

Figure 6: Normalized integral absorbance as a function of time with and without expansion

Power law of viscosity is expressed as $\mu = \mu_0 \left(\frac{T}{T_0}\right)^n$, where n refers to Power law's constant (n=2/3), μ_0 signifies the reference viscosity at the reference temperature T_0. The constant n can be expressed as n=f (u, T). To extrapolate the deviation of the constant of power law due to the thermal expansion, the total derivative of n is calculated as follows

$$\mathrm{d}n = \frac{\partial f}{\partial u}du + \frac{\partial f}{\partial T}dT = \frac{1}{ln\frac{T}{T_0}}\frac{du}{u} - \frac{ln\frac{u}{u_0}}{\left(ln\frac{T}{T_0}\right)^2}\frac{dT}{T} \tag{2}$$

we obtain Δn<0.00096 based on Δu. This result shows that if we ignore the expansion in length and diameter at experiments, the discrepancy of measured power law constant is smaller than 0.00096. Compared with the value of n, four decimal places deviation caused by thermal expansion is negligible.

3 Numerical modelling

3.1 Numerical scheme

The purpose is to solve the integral absorbance of gas flow, under non-isothermal conditions, as a function of time in a transient process. Based on the Beer-Lambert law, the integral absorbance is proportional to the number density N, line strength S and path length L, expressed as follows:

$$\int_{v_1}^{v_2} A(t)dv = N(t)SL \tag{3}$$

S is a function of temperature, describing the intensity of a particular transition, the value of which at each specified temperature can be achieved from the HITRAN database [3]. In non-isothermal conditions, line strength S(T) is also a function of space. Based on the temperature profile $T(x)$, the spatial distribution of line strength along the fiber length $S=S(T(x))$ is obtained and shown in Figure 8. Figures in this section are calculated with parameters: L_{fiber}=0.55 m; $T_{furnace}$=600 K; P_{in}=26.7 kPa.

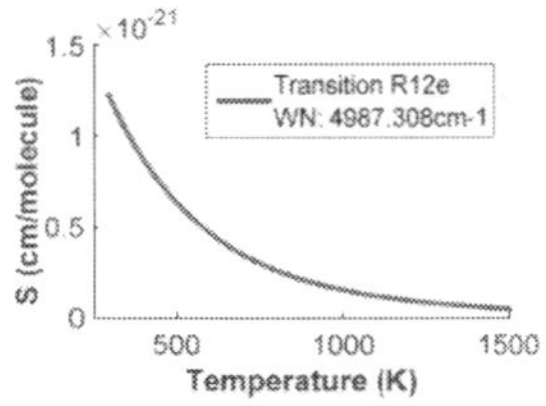

Figure 7: Line strength as a function of temperature

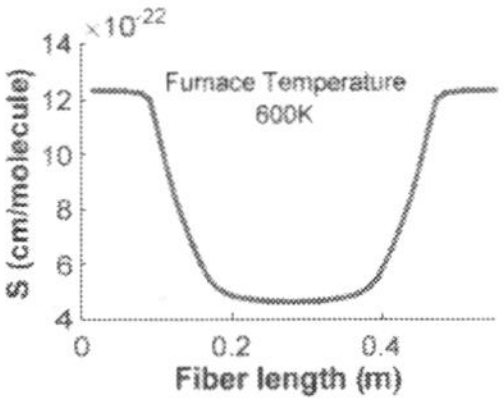

Figure 8: Line strength profile along the fiber

In an unsteady state, the number density is a function of time and space. The integral absorbance at any arbitrary point t in Equation (3) is discretized and expressed as a spatial integral of the product of line strength and number density the stream-wise direction as follows:

$$\int_{v_1}^{v_2} A(t)dv = \int_0^L N(x,t)S\big(T(x)\big)\,dx \qquad (4)$$

Based on the ideal gas law, the number density $N(x, t)$ is determined by the pressure $P(x, t)$. The local pressure evolution $P(x, t)$ is solved by using a model from Sharipov and Graur [4]:

$$\frac{\partial p}{\partial t} = T\frac{\partial}{\partial x}\frac{p}{\sqrt{T}}\left(G_P\frac{\partial \ln p}{\partial x} - G_T\frac{\partial \ln T}{\partial x}\right) \qquad (5)$$

p, T, x, t are the dimensionless quantities that are normalized respectively by the inlet pressure (P_0), ambient temperature (T_0), fiber length (L) and the characteristic time scale (τ), which is expressed as $\tau = 2L^2/av_m^2$, where a denotes the tube diameter and v_m refers to the most probable speed of gas.

G_P and G_t signify the Poiseuille and thermal creep coefficients respectively, both of which are functions of the rarefaction parameter δ, described as $\delta=ap/(\mu v_m)$. In an isothermal flow field, the most probable speed v_m ,which is a function of temperature, depicted as $v_m = \sqrt{2kT(x)/m}$, is therefore a function of space x. We incorporate the power law into this model to portray viscosity change with temperature. The viscosity coefficient μ and the most probable speed v_m vary with the stream-wise position x. As a result, the parameter G_P, G_T are functions of space x and time t. The expansion of Equation (5) is therefore expressed as follows.

$$\frac{\partial p}{\partial t} = T^{\frac{1}{2}}\frac{\partial p}{\partial x}\frac{\partial G_P}{\partial x} - T^{-\frac{1}{2}}\left(G_P + \frac{1}{2}G_T\right)\frac{\partial p}{\partial x}\frac{\partial T}{\partial x} + G_P T^{\frac{1}{2}}\frac{\partial^2 p}{\partial x^2} - pT^{-\frac{1}{2}}\frac{\partial T}{\partial x}\frac{\partial G_T}{\partial x} + \frac{3}{2}G_T p T^{-\frac{3}{2}}\left(\frac{\partial T}{\partial x}\right)^2 - G_T p T^{-\frac{1}{2}}\frac{\partial^2 T}{\partial x^2}$$
(6)

We use the finite difference method to determine the local pressure evolution, with time discretization and space discretization calculated as

$$\frac{\partial P}{\partial t} = \frac{P_i^{N+1}-P_i^N}{\Delta t},\ \frac{\partial P}{\partial x} = \frac{P_{i+1}^N-P_{i-1}^N}{2\Delta x},\ \frac{\partial^2 P}{\partial x^2} = \frac{P_{i+1}^N+P_{i-1}^N-2P_i^N}{\Delta x^2},$$
$$\frac{\partial T}{\partial x} = \frac{T_{i+1}-T_{i-1}}{2\Delta x},\ \frac{\partial^2 T}{\partial x^2} = \frac{T_{i+1}+T_{i-1}-T_i}{\Delta x^2},$$
$$\frac{\partial G_P}{\partial x} = \frac{G_{P_{i+1}}^N-G_{P_{i-1}}^N}{2\Delta x},\ \frac{\partial G_T}{\partial x} = \frac{G_{T_{i+1}}^N-G_{T_{i-1}}^N}{2\Delta x} \qquad (7)$$

Therefore, the partial differential Equation (6) is expressed in a finite-difference form using Equation (7). $P(x, t)$, $T(x)$, $G_P(x, t)$ and $G_T(x, t)$ are expressed by P_i^N, T_i , G_{Pi}^N *and* G_{Ti}^N , N and i denotes discrete time and discrete space, while Δt and Δx represent the time and space steps, respectively. Poiseuille and thermal creep coefficient G_P and G_T under different values of the rarefaction parameter δ can be found in the literature[5].

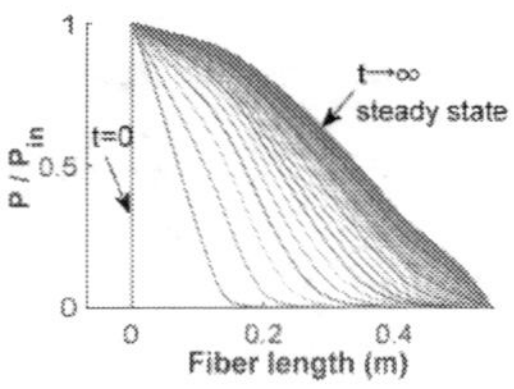

Figure 9: Pressure profile along fiber at different times(P_{in}: inlet pressure)

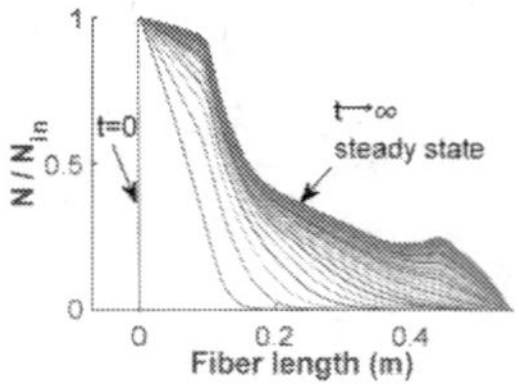

Figure 10: Number density profile along fiber at different times(N_{in}: inlet number density)

The local pressure distribution at any arbitrary time t is achieved and depicted in Figure 9, which is converted to local number density as a function of space in Figure 10, by using the profile of $T(x)$. Based on the line strength distribution $S(x)$ and local number density $N(x, t)$ at time point t The spatial distribution of the product of $N(x, t)$ and $S(x)$ is obtained and depicted in Figure 11. By integrating each curve in Figure 11 with respect to fiber length, the path-integral absorbance as a function of time is achieved and shown in Figure 12, where the absorbance is normalized by the steady state value.

This numerical calculation is carried out with different values of power law constant (n=1/3, 2/3 and 1). Figure 12 shows that three curves with different power law constants present very close tendencies, which makes it difficult for curve differentiation, indicating the integral absorbance is not sensitive to viscosity effect, under above conditions (fiber length is 0.55m). This is because the high temperature in the furnace bring down the values of both the line strength and number density (when pressure is fixed), leading to a lower value of the local absorbance in the furnace than the part out of the furnace, as shown in Figure 11. Besides, the fraction of fiber length (36%) in/out the furnace used in experiments is not high enough to compensate this effect on the path-integral absorbance. These result in a low influence of the high-temperature effect to the integral absorbance we measured, therefore a low sensitivity of the measurement to power law constants. However, the sensitivity can be improved by increasing the fiber length inside the furnace.

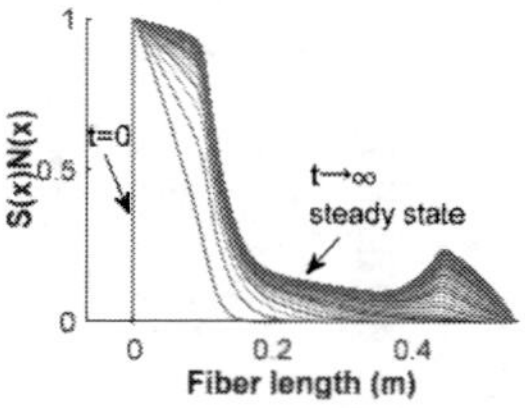

Figure 11: Spatial distribution of the product of line strength and number density

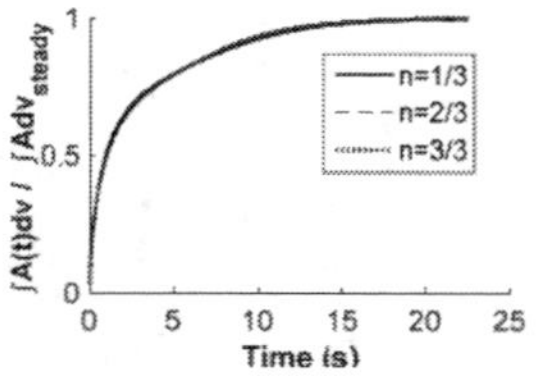

Figure 12: Normalized Integral absorbance as a function of time

3.2 Sensitivity analysis

This section is to evaluate the sensitivity of the integral absorbance to the parameters of fiber length and furnace temperature. Figure 13 presents the simulations conducted under the same conditions compared to the last section, but using 5 m fiber length to increase length fraction in the furnace. As shown in Figure 13, the gaps between curves, which is computed with different constants of the Power Law, increases significantly compared with Figure 12, making it feasible to differentiate numerical curves by comparing with the TDLAS measurements.

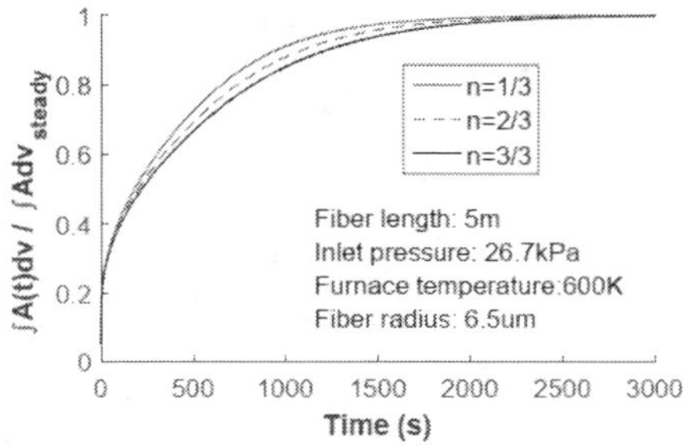

Figure 13: Normalized integral absorbance as a function of time (L_{fiber}=5m; $T_{furnace}$=600K)

Figure 14 shows the simulation conducted using 5m fiber under furnace temperature of 1473 K. The increase of furnace temperature converge the curves all together in the beginning, during which process the integral absorbance reaches to 80% of the steady state value. After that, the curves start to diverge from one another. The above tendency is due to the fact that the increase of temperature up to 1473 K significantly reduces the line strength and number density, causing the product of the $S(x)$ and $N(x)$ inside the furnace below two orders of magnitude compared to the part outside the furnace, as indicated in Figure 15. The integral absorbance reaches to a high value in the beginning is because the gas flow starts from fiber length outside the furnace, where $S(x)N(x)$ is much higher. Based on Figure 14, it can be concluded that the increase of the furnace temperature causes the calculation less sensitive to viscosity, and therefore reduces the measurement resolution.

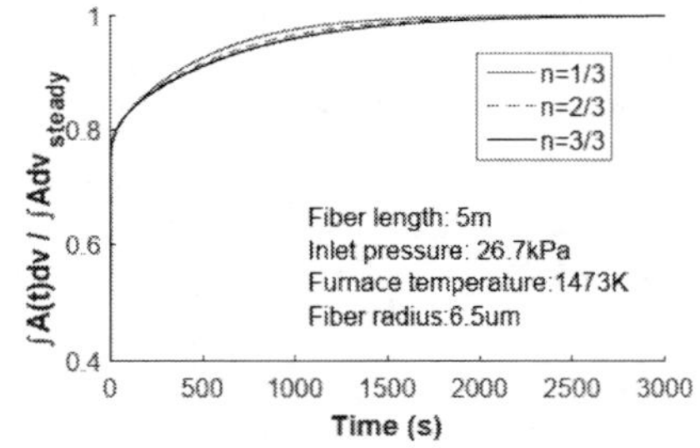

Figure 14: Normalized integral absorbance as a function of time (L_{fiber}=5m; $T_{furnace}$=1473K)

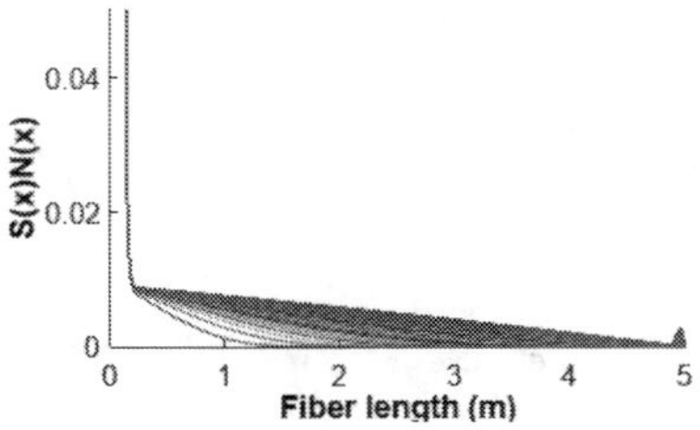

Figure 15: Spatial distribution of the product of line strength and number density ($T_{furnace}$=1473K)

Apart from the fiber length and temperature, the hollow core radius and the operating pressure are two sensitive parameters of the measurements. As mentioned in [6], larger radius and higher pressure increase the separation between numerical curves, thus increase the measurement resolution. Figure 16 shows the simulation results conducted with fiber core radius 25 um. The gaps between curves become significantly larger compared to 6.5um core radius. This presents an alternative way to increase the measurement resolution. However, different core radius corresponds to different operating wavelengths and therefore different transition lines, which needs to be taken into account based on gas species of interest.

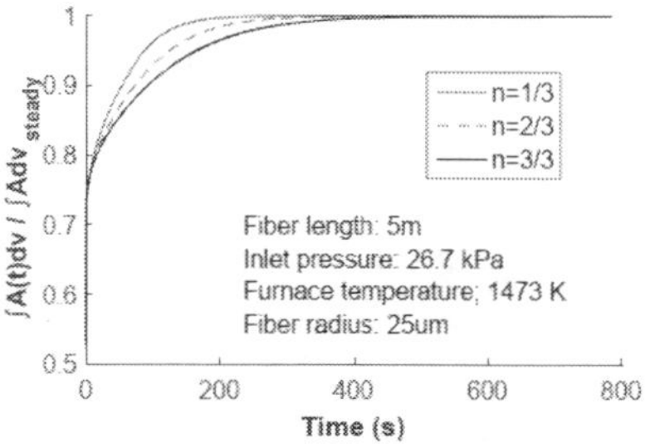

Figure 16: Normalized integral absorbance as a function of time (L_{fiber}=5m; $T_{furnace}$=1473K; R_{fiber}=25um)

4 Conclusions

An optical method using diode laser absorption spectroscopy to quantify the temperature dependence of viscosity up to 1500K is demonstrated in this paper. Our method can extend the higher temperature limit (700 K) for current experimental CO_2 viscosity data. The effect of thermal expansions in fiber radius and length were shown to be sufficiently small to be considered negligible to measurements. Low length fraction in furnace causes low measurement sensitivity to viscosity, due to the end effect where the temperature is non-isothermal. It is shown that 5m fiber length (93% in the furnace) provide a reasonable measurement resolution based on numerical simulations, demonstrating the feasibility of this technique in investigating the temperature dependence of gas viscosity. The numerical simulations also predict that the measurement resolution decreases with the increase of temperature.

References

[1] Harvey AH, Huber ML, Laesecke AR, Muzny CD, Perkins RA. Progress toward new reference correlations for the transport properties of carbon dioxide. Southwest Research Institute: The 4th International Symposium–Supercritical CO_2 Power Cycles: Citeseer; 2014.

[2] Gao R-K, Sheehe S, Kurtz J, O'Byrne S, Ketsdever A, Struchtrup H. Measurement of gas viscosity using photonic crystal fiber. AIP Conference Proceedings: AIP Publishing; 2016. p. 080008.

[3] Rothman LS, Gordon IE, Babikov Y, Barbe A, Benner DC, Bernath PF, et al. The HITRAN2012 molecular spectroscopic database. Journal of Quantitative Spectroscopy and Radiative Transfer 2013;130:4-50.

[4] Sharipov F, Graur I. General approach to transient flows of rarefied gases through long capillaries. Vacuum 2014;100:22-5.

[5] Sharipov F, Seleznev V. Data on internal rarefied gas flows. Journal of Physical and Chemical Reference Data 1998;27:657-706.

[6] Gao R, O'Byrne S, Sheehe S, Kurtz J, Liow J-L. Transient gas viscosity measurement using tunable diode laser absorption spectroscopy. Submitted to Journal of Experiments in Fluids 2017.

11th Asia-Pacific Conference on Combustion,
The University of Sydney, NSW Australia
10th -14th December 2017.

Simulation of Knock and Super-Knock in Spark Ignition Engines

M. Jaasim[1], F. E. Hernández Pérez[1], S. Vedharaj[1], V. Raman[1], R. W. Dibble[1], H. G. Im[1]

[1]King Abdullah University of Science and Technology

Thuwal, Makkah province, 23955, Saudi Arabia

Abstract

Pre-ignition in spark ignition engines is a critical issue that leads to severe knocking events and can damage the engine catastrophically. It is widely accepted that pre-ignition emanates from hot spots inside the combustion chamber. The location of the hot spot is expected to influence the knock intensity that may result from the pre-ignition event. In this study, full cycle engine simulations are conducted to investigate the effects of the location of the hot spot inside the cylinder on subsequent combustion behavior. The simulations are performed using CONVERGE, a three-dimensional computational fluid dynamics (CFD) code, incorporating reduced chemistry, turbulence modeling and moving structures (valves, piston). Gasoline direct injection (GDI) spray is represented by the Kelvin-Helmholtz/Rayleigh-Taylor (KH-RT) spray breakup model with the renormalization group k-epsilon turbulence model to describe the internal flow field. A G-equation model for flame tracking coupled with a multi-zone model is utilized to capture flame propagation, auto-ignition and subsequent combustion processes. The endgas temperature is higher for pre-ignition cycles when compared to normal spark combustion cycles, which favors the endgas auto-ignition. Strong formation of CH_2O, which is an indicator of cool flame, was found in the endgas region ahead of the flame front in pre-ignition cycles. A method to simulate pre-ignition cycles and techniques to analyze the pre-ignition results are proposed in the study.

1 Introduction

Regulations imposed by governments on automobile manufacturers are stricter each year in terms of emissions and efficiency. Particulate emissions are a major problem in compression ignition engines, which is alleviated in spark ignition (SI) engines. However, improving the efficiency of SI engines comparable to that of diesel engines is challenging due to various limitations. Downsizing of the SI engines is usually adopted to improve efficiency [1], which is also accompanied by the problem of pre-ignition leading to knock or super-knock when operated at boosted operating conditions. Ignition of mixture by an external source before firing the spark plug is widely called as pre-ignition and predominantly occurs at low speed and hence called as low speed pre-ignition. It is also called stochastic pre-ignition because the occurrence of pre-ignition is stochastic in nature and not limited only to low speed operation [2]. Pre-ignition leads to very high in-cylinder pressures accompanied with severe pressure oscillations called as super-knock. Super-knock cycles are capable of causing permanent damage to the moving parts of the engines. The exact source of pre-ignition in engines has not been clearly identified, but various reasons are mentioned in the literature [3] such as composition of the engine oil, oil droplet from the crevice region, impurities inside the cylinder, carbon deposits, spark plug temperature, fuel properties, etc. The source of pre-ignition can also be due to hotspots available inside the cylinder region, which can pre-ignite the mixture at undesirable times.

The initial smooth pressure rise in pre-ignition cycle is due to the initiation of the pre-ignition flame, followed by auto-ignition of the mixture, which in turn can generate a flame front with very high velocity and could lead to detonation. Various one-dimensional (1-D) direct numerical simulation studies were performed to explain the physics involved with pressure rise and oscillations. Endwall auto-ignition is predominant in cases when the temperature of the endgas falls into the negative temperature coefficient (NTC) regime. The endwall auto-ignition can generate an auto-ignition front propagating faster than the speed of sound and causing detonation, which has been demonstrated in the literature [4].

The location of the hotspot and time at which pre-ignition is initiated play a vital role in deciding the fate of the cycle. Early pre-ignition can lead to consumption of all fuel before the piston reaches top dead center (TDC) and the cycle will end up compressing the burnt gas. Late pre-ignition cycles may not effectively lead to super-knock. A parametric study of the above said factors will provide further insight into pre-ignition cycles that lead to super-knock. The energy of the hotspot plays a major role in causing detonation, which leads to super-knock as reported in the literature. The local temperature gradient of the hotspot and critical radius of the hotspot determines whether the hotspot will lead to detonation [5]. The pressure and temperature of the endgas are found to be important, which can significantly influence the nature of the cycle.

In this study, pre-ignition is induced by a hotspot of preset temperature and a pre-ignition flame front is induced inside the cylinder at known crank angle and cylinder location. One objective of the study is to obtain detailed insights about the events taking place inside the cylinder after pre-ignition and the local pressure and other conditions that influence the pre-ignition cycle to produce super-knock. The main focus will be on methods to simulate and analyze the pre-ignition cycle results using economical Reynolds averaged Navier-Stokes (RANS) simulations.

2 Computational Setup

The engine geometry and mesh generated in-situ for the study are shown in Figure 1. The initial cell size of the mesh was set to 4 mm in all directions which was refined with adaptive mesh

Corresponding author. Phone: +966-569-060-779
E-mail address: mohammedjaasim.mubarakali@kaust.edu.sa

refinement (AMR) based on the temperature and velocity gradient between the neighboring cells, with thresholds of 5 K and 1 m/s, respectively. The refinement of the cells locally reduced the cell size to 0.125 mm (new cell size = basegrid/2^n, where n defines the refinement level). For the sake of brevity the details about the governing equations are not presented here, but they can be found in [6]. The engine specifications and operating conditions for the present study are listed in Table 1. The initial and boundary conditions for the combustion simulations are obtained by matching the motoring peak pressure.

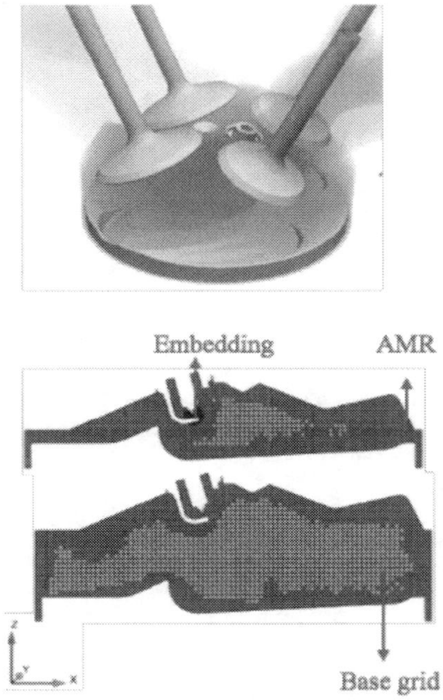

Figure 1: Geometry (top) of the engine and mesh generated for the study (bottom).

Parameters	Values
Bore	85 mm
Stroke	90 mm
Connecting rod length	138 mm
Compression ratio	9.5
Valves	2-intake, 2-exhaust valves
Spark plug energy	40 mJ
Spark time	-5 CAD, 0 CAD
RPM	1200
Intake temperature	300 K
Intake pressure	1 bar
Lambda (λ)	1
Fuel	*iso*-octane

Table 1: Specifications and operating conditions of the engine.

2.1 Spray modeling

The engine was operated in gasoline direct injection mode at full load with fuel injection during intake valve opening event (IVO) in the experiments. Proper modeling of the spray and its characteristics are important to obtain similar mixture distribution compared to experiments. The spray breakup was described by Kelvin Helmholtz-Rayleigh Taylor (KH-RT) model. The droplet is assumed to be spherical during the entire evaporation phase. Infinite heat conductivity is assumed inside the droplet, i.e., the droplet is assumed to be at uniform temperature. The evaporation of the droplet is governed by mass conservation and the radius of the droplet is calculated using the Frossling relation. The collision and coalescence of the droplet are modeled using the model proposed by Amsden et al. and the Rosslin-Rammer size distribution is used to represent the initial size distribution of the injected fuel droplets.

2.2 Combustion Model

Modeling of pre-ignition is a challenging task, which requires a model that can capture the flame propagation and auto-ignition inside the cylinder. The G-equation model was used for the initial pre-ignition flame tracking, coupled with a multi-zone model, which was used to describe the endgas auto-ignition during the entire cycle. The multi-zone model groups the cells based on temperature and equivalence ratio (5 K and 0.05 φ, respectively) to solve for the reaction rates in the endgas. A schematic of the combustion model representation is shown in Figure 2. The gas phase chemical kinetic model developed by Liu et al. was used in the simulations and the heat loss to the walls is accounted by the model developed by Han et al. The in-cylinder flow field is represented by a RANS approach with the renormalization group k-epsilon model. Further details about the models used in the simulations can be found in [6].

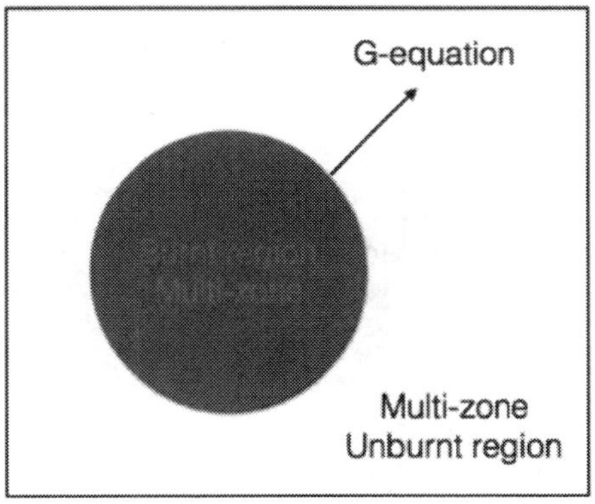

Figure 2: Schematic of the combustion model used for the simulations.

2.3 CFL limit

It is well known that CFL number ensures convergence and accuracy of the solution. Pre-ignition problems are unique due to flame propagation, auto-ignition and detonation which occurs simultaneously and requires special treatment of the numerical schemes and methods to handle the pressure oscillations generated inside the cylinder due to knocking (auto-ignition) in the cylinder. To capture the pressure oscillations the very small CFL Mach number should be used. This small CFL Mach number ensures to capture the pressure oscillations. CFL Mach number of 1 was used in the study, which has been also used in the previous studies [6,7].

$$CFL\ Mach\ number = a.(dt/dx) \quad (1)$$

Where 'a' is the speed of the sound in the system. The

convergence stability and grid converged result was ensured for the present grid size and CFL number (not shown).

3 Results and Discussions

3.1 In-cylinder Pressure

The experiments were conducted for the operating conditions specified in Table 1 and simulations were performed for the same conditions. For the purpose of validation, the in-cylinder pressure comparison between the experiments and simulations is shown in Figure 3, displaying good agreement. The results indicate the fidelity of the models used in the simulations.

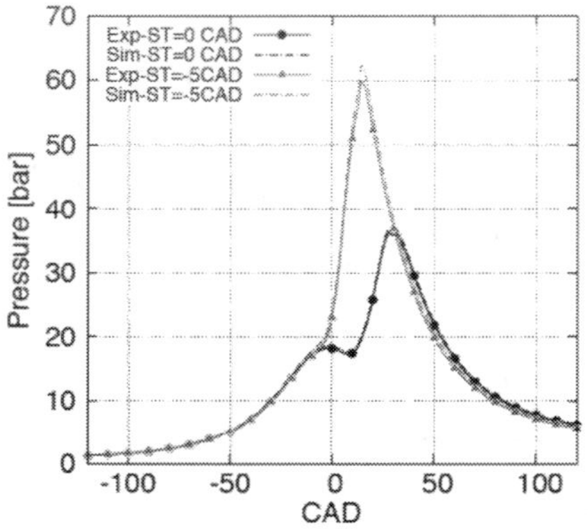

Figure 3: Comparison of the experimental in-cylinder pressure and numerical simulations (P_{in} = 1 bar, λ=1)[6].

Note that the simulation pressure reported in Figure 3 is a volume-averaged result. The pre-ignition was induced in the cylinder using a hotspot of 2000 K at the desired crank angle over the piston as illustrated in Figure 4. The value of 2000 K is somewhat arbitrary, but as mentioned before, the focus of the study is on the events after pre-ignition since the source that induces pre-ignition is not clearly known in the engines. The location of the hotspot was selected based on the previous studies by the authors [6] and it was noted that the location of the hotspot has influence over the knock intensity and peak pressure achieved in the cycle.

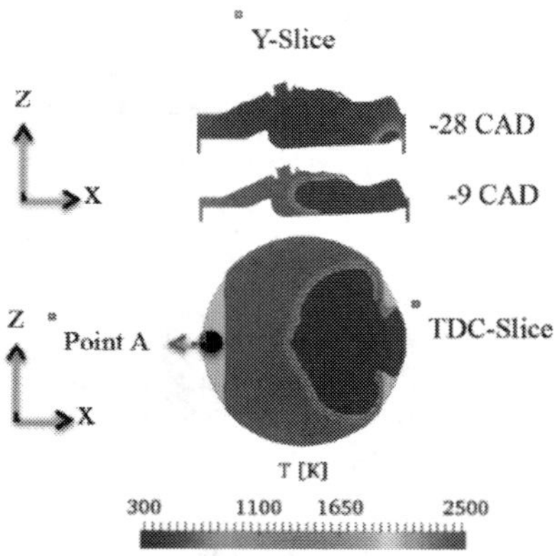

Figure 4: Pre-ignition flame initiation and flame propagation by a hotspot of 2000 K.

In the present study, pre-ignition was initiated at -20 CAD aTDC (P-20) with 2 bar intake pressure and λ=1 operating condition. To study and analyze pre-ignition, the local pressure variation was considered at 1 mm from the endwall (point A in the endgas region as shown in Figure 4). The above described models and methods are able to capture the pressure oscillations inside the cylinder as displayed in Figure 5, where pressure obtained for a normal cycle with spark timing at 0 CAD is compared against a pre-ignition cycle for which the flame was initiated at -20 CAD aTDC at surface of the piston near liner location. It can be seen from Figure 5 that the pre-ignition cycle (-20P) shows higher pressure values accompanied with large oscillations when compared with the normal combustion cycle (NC).

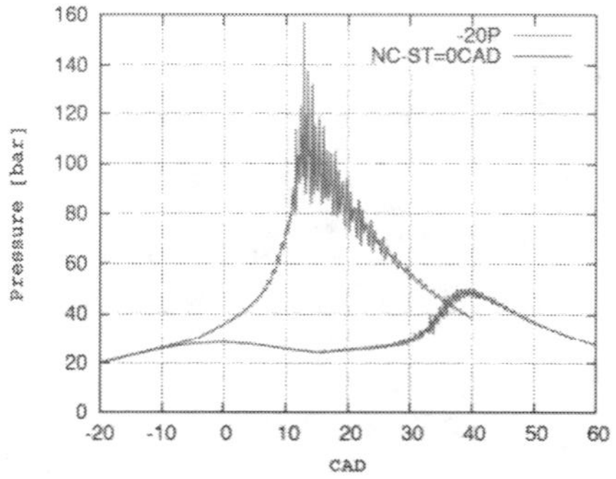

Figure 5: Comparison of local pressure at point A (Pin= 2 bar, λ=1).

In the previous work by the authors, the maximum pressure inside the cylinder was used to represent the pre-ignition cycle that leads to super-knock [6]. The pressure (P_A) exhibits a smooth pressure rise before TDC due to the pre-ignition flame propagation, followed by severe pressure oscillations. This behavior is not observed in the normal spark ignition cycle. The knock intensity calculated for the normal combustion and pre-ignition cycles using Equation (2) shows that the knock intensity is ~70 times larger than that of a normal combustion cycle. The allowed knock intensity limit above which the cycle is said to knock is 0.5. Knock intensity (KI) is difference between the maximum and minimum of the knock signal.

$$\text{Knock signal} = (P_i+P_{i+1}+P_{i+2})/3 - P_{mean} \quad (2)$$

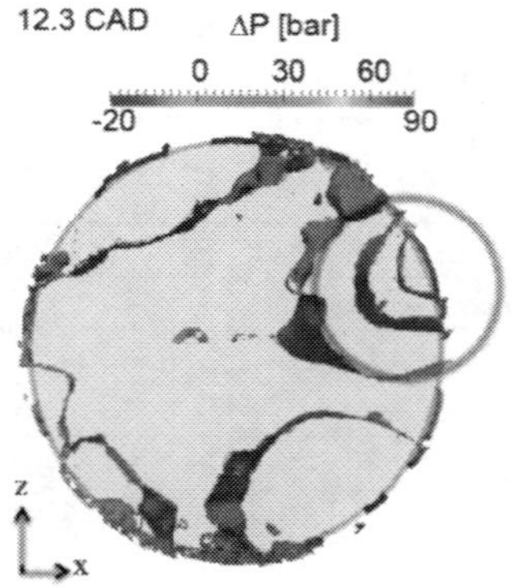

Figure 6: Iso-contours of pressure difference. The region with higher pressure is circled.

Next the in-cylinder pressure is further investigated to identify the high pressure regions inside the cylinder which can provide further insights into the origin of pressure fluctuations and consequences of the super-knock cycle. The plot in Figure 6 shows the iso-surface contours of difference between the local pressure and mean pressure ($\Delta P = P_i - P_{mean}$) between -20 bar to 90 bar. It can be inferred that very high pressure is observed near the endwall and the crevice region. This high pressure region moves back and forth causing a pressure oscillation which has also been observed in direct numerical simulations (DNS). The difference in pressure is higher than 90 bar, which is capable of tilting the piston and cause damage to the engine.

3.2 Flame Propagation and Endgas Auto-ignition

The flame propagation inside the cylinder is illustrated in Figure 7, where temperature contours are displayed. The propagation speed of the flame increases after ~40% of the injected mass is consumed by the pre-ignition flame. The pressure of the system is found to increase due to the pre-ignition flame, which is followed by the severe pressure oscillations. In the endgas region multiple auto-ignition zones are identified, which are the source of the observed pressure oscillations.

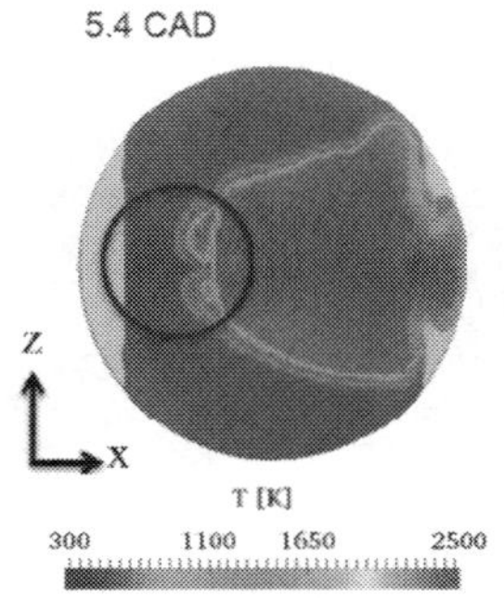

Figure 7: Flame propagation and endgas auto-ignition.

3.3 Endgas Temperature and Reactivity

Comparison of temperature of the endgas between the normal cycle and pre-ignition cycle is shown in the Figure 8. The endgas temperature is approximately 250 K higher for the pre-ignition cycle. Note that the temperature lies in the NTC range that favors endwall auto-ignition, which has been observed in the simulations and also has been reported previously in DNS studies [4]. The reactivity of the endgas is increased due to higher pressure and temperature. Strong formation of CH_2O is found in the endgas, which indicates low temperature flame [8] and increased probability of auto-ignition in the region as shown in Figure 9.

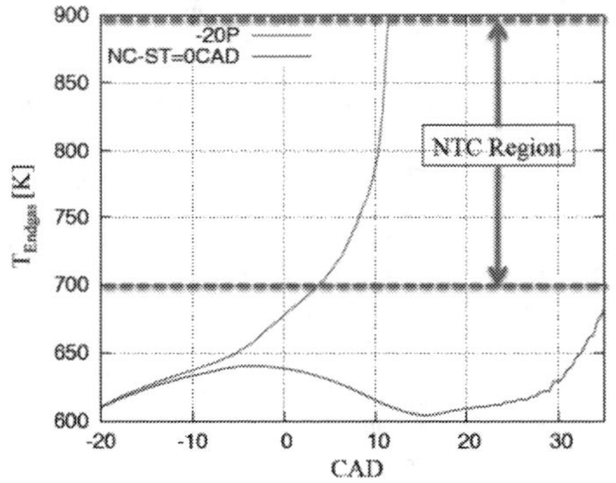

Figure 8: Comparison of endgas temperature history for pre-ignition and normal combustion cycles.

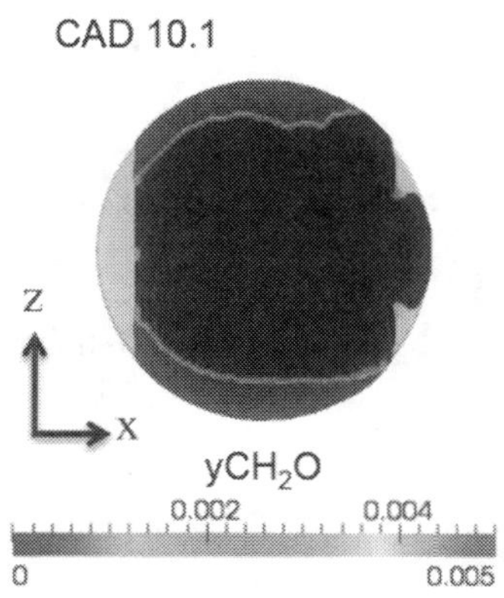

Figure 9: Mass fraction of CH_2O in endgas region ahead of the flame region.

4 Conclusions

Three-dimensional pre-ignition simulations were performed and the results were analyzed. In summary:

- The models and numerical methodology used in the study are capable of capturing the auto-ignition events in the endgas.
- Regions of high pressure are found near the endwall and crevice region. The high-pressure regions change location at very high speeds. The pressure difference is higher than 90 bar, which can damage the engine.
- The temperature and pressure of the endgas in a pre-ignition cycle favors auto-ignition of the endgas. Increased reactivity and cool flame characteristics are found in the endgas when compared to the normal combustion cycles.

5 Acknowledgment

This work was supported by King Abdullah University of Science and Technology (KAUST) with Saudi Aramco under the Fuelcom II program and made use of KAUST supercomputing resources. The authors thank convergent science for providing their code for the simulations.

References

[1] B. Lecointe, G. Monnier, SAE Technical Paper 2013-01-0542,2003, doi:10.427/2003-01-0542.

[2] E. Chapman, V. Costanzo, SAE Int. J. Engines, 9(1) (2016),107-142.

[3] G. Kalghatgi, K. Morganti, I. Algunaibet, M. Sarathy, SAE Technical Paper 2016-01-0702, 2016, doi:10.4271/2016-01-0702.

[4] H. Yu, Z. Chen, Combust. Flame 162 (2015) 4102-4111.

[5] G. T. Kalghatgi, D. Bradley, Int. J. Engine Res. 13 (4) (2012) 399-414.

[6] M. Mubarak Ali, F. Hernandez Perez, S. Vedharaj, R. Vallinayagam, H. G. Im, R. W. Dibble, SAE Technical Paper 2017-01-0686, 2017, doi:10.4271/2017-01-0686.

[7] S. Park, T. Furukawa, SAE Technical Paper 2015-01-0750, 2015, doi:10.4271/2015-01-0750.

[8] B. Bauerle, F. Hoffmann, F. Behrendt, J. Warnatz Proc. Combust. Inst. 25 (1994) 135-141.

11th Asia-Pacific Conference on Combustion,
The University of Sydney, NSW Australia
10th -14th December 2017.

Hypergolic Ignition Induced by Binary Collision of TMEDA and WFNA Droplets: Non-monotonic Effects of Impact Parameter

Dawei Zhang, Peng Zhang*

Department of Mechanical Engineering

The Hong Kong Polytechnic University, Kowloon, Hong Kong

Abstract

Ignition delay of hypergolic bipropellants of TMEDA and WFNA by their binary droplet collision has been experimentally studied, with the particular emphasis on the effects of impact parameter, B, which measures the deviation of the collision from the trajectory of head-on situation. The experimental results reveal a non-monotonic effect of the impact parameter. For B slightly larger than 0, the intrinsic deformation and internal flow pattern promote the mixing between the colliding droplets and hence reduce the ignition delay. If B increases to so that off-center separation happens, the hypergolic ignition would be prolonged owing to the less mass transfer.

1 Introduction

Being widely adopted for the space propulsion applications, the hypergolic bipropellant has been of great interests in the past decades[1-3]. Unlike the ignition of a homogenous mixture of non-hypergolic reactants, the auto-ignition of the hypergolic bipropellants essentially relies on the mixing process between the liquid fuel and oxidizer preceding the eventual gas-phase ignition[3].

Many experimental efforts have been made in order to quantify the influence of the liquid-phase mixing process on the hypergolic ignition[4-7]. In spite of the worthy knowledge obtained from these previous studies, the liquid-phase mixing effect remains quantitatively uncertain because of the existence of the "wall effect" from their experimental facilities.

Recently, an experimental apparatus with capability of eliminating all the "wall effect" was designed and established by the authors, through which the hypergolic ignition of N,N,N',N'–tetramethylethylenediamine (referred to as TMEDA hereinafter) and white fuming nitric acid (referred to as WFNA hereinafter) was investigated by the binary collision of 'freely' moving droplets[8]. The focus of the study was mainly placed on the head-on collisions at variable collision Weber numbers(*We*s) and size ratios(Δs). Hypergolic ignition delay of the bipropellants was thereby interpreted by using the previous knowledge of the mixing and interaction effects in non-reactive head-on droplet collisions[9-10]. Nevertheless, it is noted that the axisymmetric head-on collision is a rare event in the real situation and the almost sure events are off-center collisions.

To quantify off-center collisions, a dimensionless impact parameter, B, is defined as the ratio of the projection of the distance of two droplet centers to the direction of their relative velocity, χ, to the sum of the two droplet radii, $(D_S + D_L)/2$, and more specially $B = 2\chi/(D_S + D_L)$ [11]. The head-on collision corresponds to $B = 0$, off-center collisions to $0 < B < 1$, and a grazing collision to $B = 1$.

It has been found that B affects not only the outcomes of the binary collisions but also the mixing during the collision process [12]. Specifically, it has been numerically found that a non-monotonic variation of the mixing rate with B would exhibits at various collision *We*s[13-14]. Whether the hypergolic ignition induced by the off-center collision of TMEDA and WFNA would be in line with this trend is the question that merits studies.

Motivated by the foregoing concerns, the present study aims mainly to experimentally study the hypergolic ignition characteristics of bipropellants of WFNA and TMEDA by their off-center droplet collisions with intricate mixing patterns and fluid physics. We therefore present the study as follows. The experimental apparatus and the measurement methods are expatiated in Section 2. The results for four representative cases are presented in Section 3 to illustrate the hypergolic ignition processes. Dependence of ignition delay on the impact parameter is presented in Section 4, followed by concluding remarks in Section 5.

2 Experimental Specifications

The experimental specifications have already been expatiated in our previous paper [8] and therefore only a brief description is given herein. Droplets of WFNA and TMEDA were generated by two unlike and independently controlled droplet generators, each of which is fastened on a XYZ microscope stage through which the generator could be adjusted precisely in both the dispensing positions and angles. As with intensive corrosivity, the WFNA droplet was generated by using an in-house needle-shape nozzle, which was manufactured with the highly anti-corrosion material of Teflon (Polytetrafluoroethylene). With the carefully arranged connecting pipe and driving pressurized ultra-pure nitrogen gas, the WFNA droplet size was able to be fixed at 1.45mm at constant generation frequency. The electromagnetic micro-valve made by Fritz Gyger AG was employed to generate TMEDA droplets, whose sizes vary from 0.2mm to 1.5 mm. Temporally and spatially stable collisions between the TMEDA and WFNA droplets were achieved by precisely adjusting the tunable droplet generation time and the position of XYZ stage.

A high speed camera V711 from Phatom was adopted to record shadowgraph images of the collision and ignition processes with a fixed sampling rate of 5000fps and a fixed shutter time of 5μs. A MATLAB program was developed to help analyze the images in determining the diameters, the impact velocities and the impact parameter, B. The measurement of B, as shown in

* Corresponding author.

E-mail address: pengzhang.zhang@polyu.edu.hk

Fig.1, relies on the measureent of two velocity vectors at an appropriate temporal interval. The measurement error for droplet diameter and impact parameter are less than 7%, and the uncertainty of measured droplet velocity, which is determined by averaging the displacement of the droplet center during five successive images (i.e. 1.0ms), is about 3%-8%. The sphericity of the large WFNA droplet was ensured by the small downward velocity at the moment of colliding with TMEDA droplet.

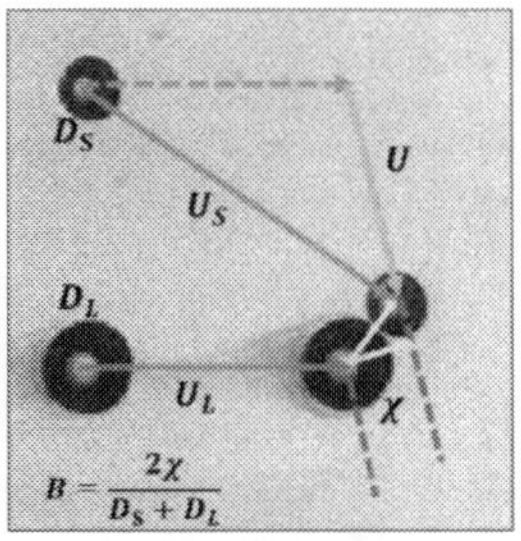

Figure 1. The procedure of the calculation of impact parameter B in experiment for hypergolic ignition by binary off-center collision of TMEDA and WFNA droplets.

A grayscale analysis of the shadowgraph images as proposed in our previous study also is applied to precisely determine the ignition delay time with an uncertainty of less than 0.2 ms. Such an analysis is based on that large difference of grayscale levels (0-256) can emerge in the shadowgraph images, where the darkest areas (levels are lower than 5) are occupied by droplets and their vapors; the brightest areas (levels are higher than 250), by luminous ignition kernels and flames. A photodetector and an infrared detector has been employed in our previous study to verify the reliability and accuracy of the grayscale analysis.

3 Phenomenological Description

3.1 Hypergolic Ignition

In order to characterize the off-center (impact parameter) effect in the present study, four representative collisions with various Bs of 0, 0.3, 0.6 and 0.9, but with fixed We = 60.9, Δ = 1.6 and $Oh = 2.5 \times 10^{-3}$ are presented, in which the Weber number, We, measurs the relative importance of droplet inertia compared with its surface tension and is with the definition by $We = \rho_O D_O U^2/\sigma_O$, where ρ_O, σ_O, and D_O are the density, the surface tension and the diameter of the WFNA droplet and U is the relative velocity of the two droplets; size ratio, Δ, is defined by the ratio of the diameter of the bigger droplet to that of the smaller one; impact parameter, B, is as defined above and Ohnesorge number, Oh, is adopted to measure the importance of viscous force relative to surface tension.

Figure 2 shows in the sequence of both the physical time and dimensionless time, $T = t/t_{osc}$ where $t_{osc} = \sqrt{\rho_o R_o^3/\sigma_o}$ is approximately the natural oscillation time of the WFNA droplet, the entire collision and ignition processes for the four representative cases. For a clear description of the entire ignition process, multiple length scales were adopted as indicated on the images. The initial moment, t = 0, is defined when the two droplets were just about to collide.

As shown in Fig.2, in the initial stage, the interaction between the colliding TMEDA and WFNA droplets for all the four cases are similar to those non-reactive collisions.[12] Reflexive deformation dominates in the collision of B = 0 and B = 0.3, while the two droplets tend to stretch away from each other at B = 0.6 and B = 0.9. Since the merged droplet was heated up by the exothermic liquid-phase reaction, opaque vapors and gaseous species were generated and started to expand outwardly at around 3.6ms for all the four collisions. The opaque gas-phase matters continued growing so that the surface of the merged droplet at B = 0 and 0.3 and the separated droplets with the satellite droplets at B = 0.6 and 0.9 are blurred or even concealed inside the dark vapor and gaseous products, as seen at 11.4ms. The flame, indicated by the luminous overexposure area on the shadowgraphs, can be observed at the moments of 31ms, 25ms and 33.0ms for collisions B = 0, B = 0.3 and B = 0.6 respectively. While no bright flame occurs throughout the whole period for collision B = 0.9, so it should be regarded as non-ignitable. Non-combustible products or unburned extra reactants could be observed along with or after the luminous flame, as indicated on the images.

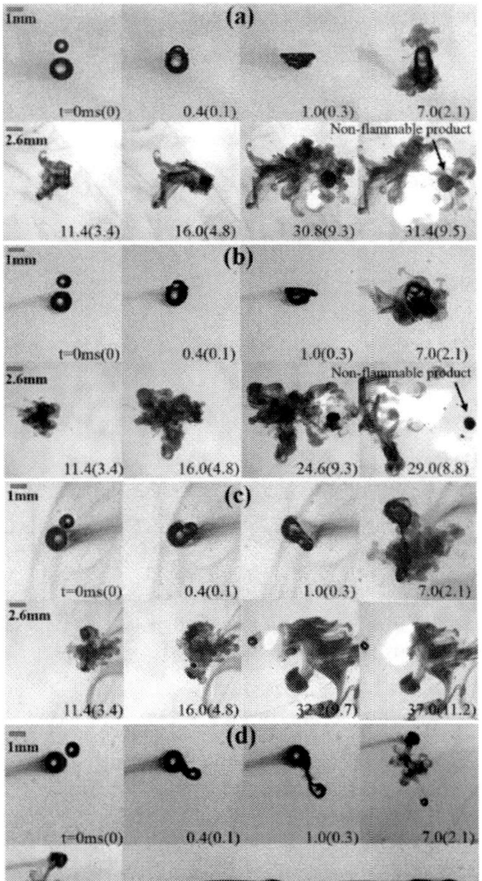

Figure 2. Shadowgraph images of the hypergolic ignition at selected times for a representative case with We=60.9 and Δ=1.6 and (a) B = 0; (b) B = 0.3; (c) B = 0.6; (b) B = 0.9.

3.2 Impact Parameter Effects on Mixing, Evaporation and Ignition

To systemically describe and analyze the evaporation and condensed phase chemical reaction processes, we quantified the pixel-area occupied by the opaque matters in the shadowgraph images by a normalized parameter of r_d. The time-dependent ratio of r_d is defined by [8]

$$r_d = N_d(t|G < G_{low})/N \quad (1)$$

where $N_d(t|G < G_{low})$ is the total number of pixels having the grayscale levels, G, lower than G_{low}; N is the total number of pixels in the image.

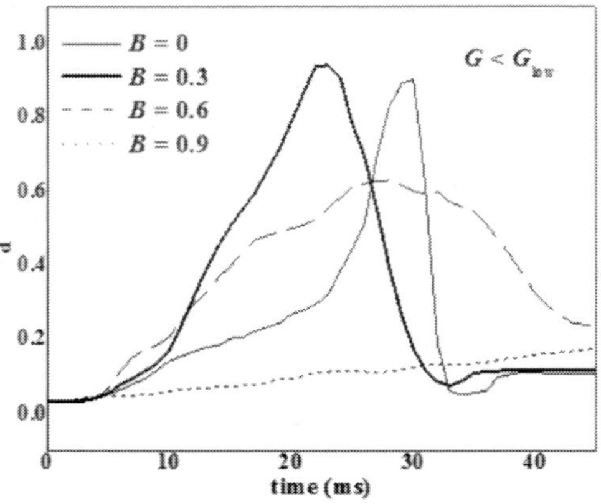

Figure 3. Grayscale level analysis of r_d for the representative cases shown in Figure 2.

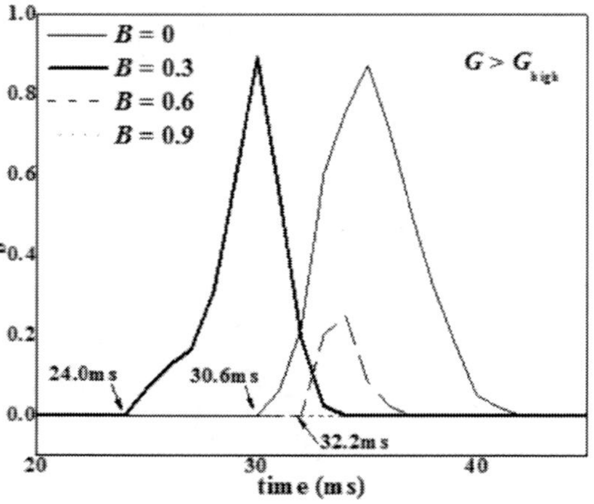

Figure 4. Grayscale level analysis of r_b for the representative cases shown in Figure 2.

As shown in Fig 3, r_d remains the lowest level at the beginning for almost the same time duration for all the four collisions and begins to rise up afterwards. In the following growth period of the opaque matters, the peak values of the occupying area at B = 0.6 and B = 0.9 are much smaller than those for B = 0.0 and B = 0.3. The underlying reason is that, since the stretching separation happens at B = 0.6 and B = 0.9, it diminishes the amount and time period of mass transfer between the droplets and hence impairs the evaporation and condensed phase chemical reactions.

Another interesting phenomenon is that the starting instant of the rapid evaporation stage comes about 10ms earlier at B = 0.3 than the head-on collision. The difference in the duration of the rapid evaporation stage is owing to the different mixing efficiencies for the two merging collisions. When the two droplets collide with non-zero impact parameters, both reflexive interaction and stretching will effect and a four-lobe-shape drop would therefore always be formed, as shown in the shadowgraph at 7.0ms for the collision at B = 0.3. The droplet internal under the competing effects would be more complicated than it does in the head-on collision under symmetric reflexive interaction. The intricate internal flow will thereby enhance the mixing and expedite the evaporation.

Based on the observation and analysis above, we found that the impact parameter would influence the mixing between the two droplets in two competing ways. At relatively small Bs, for example, B = 0.3, the reflexive interaction is dominant and it strengthens the mixing by the "jet-like" internal flow patterns [17]. While the mixing enhancement will be weakened due to the reduced time for mass transfer if stretching separation happens at higher Bs.

Affected by the mixing of the fuel and oxidizer and the accompanying evaporation, the hypergolic ignition delay is therefore related to the impact parameter effect. Another normalized parameter, r_b, whish is silimar to r_d but with a high threshold of grayscale level of G_{high}, representing the development of the luminous flame was employed to quantitatively identify the ignition delay. As shown in Fig. 4, the ignition delay times for the three ignitable representative collisions B = 0, B = 0.3 and B = 0.6 are 30.6ms, 24.0ms and 32.2ms, respectively. Such non-monotonic behavior hereto can be attributed to the two competing mechanisms on the mixing of the fuel and oxidizer droplets prior to ignition event in the off-center collisions.

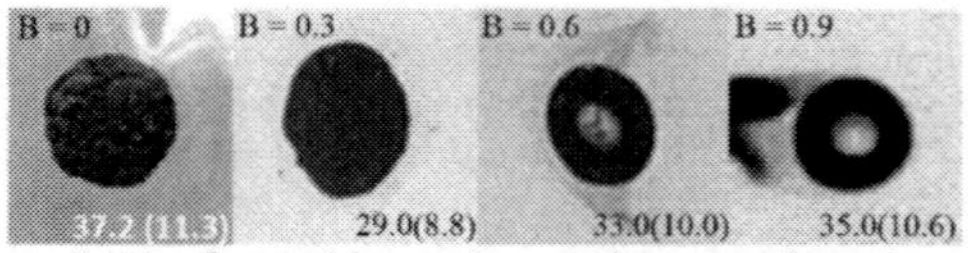

Figure 5. Nonflammable condensed-phase products from the representative cases shown in Figure 2.

Appearances of the final reaction products for the four collisions are another evidence of the non-monotonic behavior. As shown in Fig. 5, the solid-like surface appearance of the product for the cases of B = 0 and B = 0.3 implies a complete reaction, while the smooth surfaces for the cases of B = 0.6 and B = 0.9 suggest that the reduced mass transfer and hence chemical reaction result in a certain amount of unburned liquid propellant.

4. Non-monotonic Effect of Impact Parameter on Ignition Delay

The non-monotonic dependence of the ignition delay on B at We = 60.9 and Δ = 1.6 has been adequately investigated and it can be further confirmed by comparing the results at various Wes and Δs. As shown in Fig.6, the non-monotonic dependence of ignition delay on the impact parameter B can be observed. Specifically, for collisions with We = 41.0, Δ = 1.3; We = 60.9, Δ = 1.3; We = 60.9, Δ = 1.6; We = 83.0, Δ = 1.6; and We = 83.0, Δ = 2.0, the ignition delay is inversely proportional to B when B is slightly larger than 0. However, if B increases to a value about 0.3~0.4, the stretching separation happens and the time for ignition delay will be prolonged with increasing B.

Some interesting observations on the dependence of ignition delay on B can be found for large size ratios. For Δ = 2.0, the ignition delay monotonically decreases with B from 0.2 to 0.5 for We = 60.9 and from 0.4 to 0.6 for We = 41.0, as shown in Fig. 6. It will in fact turn into being non-ignitable when the collision becomes a head-on or grazing one. As reported by Zhang et al.[15], the dominant exothermic liquid-phase reaction between WFNA and TMEDA is

$$TMEDA + 2HNO_3 \rightarrow TMEDADN \quad (2)$$

whose overall equivalence ratio for Δ = 2.0 is as low as 0.07. Consequently, the deficiency of heat release in the extremely "fuel lean" reaction and the substantial heat lost to the surrounding environment from such small droplets make the ig nitability range of B is smaller with increasing the size ratio.

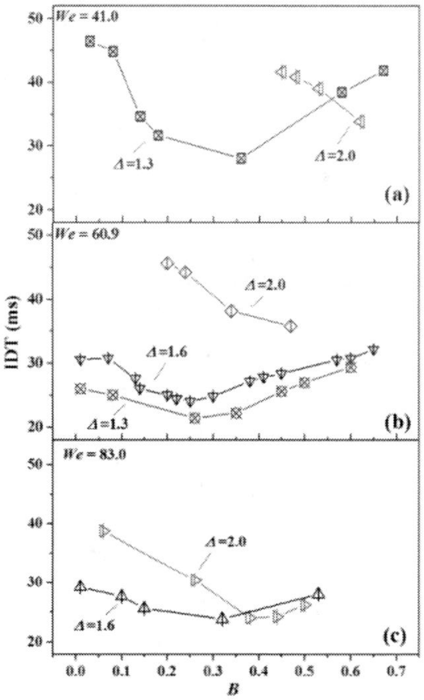

Figure 6, Dependence of ignition delay on B with (a)We = 41.0, Δ= 1.3 and 2.0; (2) We = 60.9, Δ = 1.3, 1.6 and 2.0; and (c) We = 83.0, Δ =1.6 and 2.0.

5 Conclusions

The hypergolic ignition induced by the binary collision by WFNA and TMEDA droplets was experimentally studied. Non-monotonic dependence of the hypergolic ignition delay on the impact parameter effect was observed and analyzed for various Weber number and size ratios. The ignition seems to be suppressed at small *Wes* and/or large size ratios. The ignitibility regime merits future study, particularly in consideration of heat lost and size effect have not been adequately studied.

6 Acknowledgment

This work was supported by the Hong Kong RGC/GRF (operating under contract numbers PolyU 152217/14E and 152651/16E) and partly by the Hong Kong Polytechnic University (G-UA2M and G-YBGA).

Reference

[1] G.P. Sutton, O. Biblarz, Rocket propulsion elements, 8ed., John Wiley & Sons, Hoboken, New Jersey, 2010.

[2] E.A. Fletcher, G. Morrell, Ignition in liquid propellant rocket engines, in: M.G. J. Ducerme, A. H. Lefebvre (Ed.), Progress in Combustion Science and Technology Volume I, Pergamon, Oxford. London. New York. Paris, 1960, pp. 183-215.

[3] R.J.M. Eric A. Hurlbert, Propellant Ignition and Flame Propagation, in: V. Yang, M. Habiballah, J. Hulka, M. Popp (Eds.), Liquid Rocket Thrust Chambers: Aspects of Modeling, Analysis, and Design, American Institute of Aeronautics and Astronautics, Inc., Virginia, 2004.

[4] M. Kilpatrick, L.L. Baker, Symp. (Int.) Combust, (1955) 196-205.

[5] S. Wang, S.T. Thynell, A. Energy & Fuels 24 (2010) 5320-5330.

[6] Forness, T.L. Pourpoint, S.D. Heister, 49th AIAA/ASME/SAE/ASEE Joint Propulsion Conference (2013) 3772.

[7] E.M. Dambach, Y. Solomon, S.D. Heister, T.L. Pourpoint, J. Propul. Power 29 (2013) 331-338.

[8] D. Zhang, P. Zhang, Y. Yuan, T. Zhang, Combust Flame 173 (2016) 276-287.

[9] C. Tang, J. Zhao, P. Zhang, C.K. Law, Z. Huang, J. Fluid Mech., 795 (2016) 671-689.

[10] C. Tang, P. Zhang, C.K. Law, Phys Fluids 24 (2012).

[11] J. Qian, C.K. Law, J. Fluid Mech. 331 (1997) 59-80.

[12] N. Ashgriz, J. Poo, J. Fluid Mech. 221 (1990) 183-204.

[13] T. Inamuro, S. Tajima, F. Ogino, International journal of heat and mass transfer 47 (2004) 4649-4657.

[14] X. Chen, D. Ma, P. Khare, V. Yang, 49th AIAA Aerospace Sciences Meeting Including the New Horizons Forum and Aerospace Exposition; 2011. p. 771.

[15] P. Zhang, L.D. Zhang, C.K. Law, Combust. Flame 162 (2015) 237-248.

[16] W.-G. Liu, S. Dasgupta, S.V. Zybin, W.A. Goddard III, The J. Phys. Chem. A 115 (2011) 5221-5229.

[17] C. Tang, J. Zhao, P. Zhang, C.K. Law, Z. Huang, J. Fluid Mech., 795 (2016) 671-689.

Lewis number effect on the flame height of circulation-controlled firewhirls

Dehai Yu[1] and Peng Zhang[1*]

[1]Department of Mechanical Engineering, The Hong Kong Polytechnic University

Hung Hom, Kowloon, Hong Kong

Abstract

A theoretical analysis was carried out on the circulation-controlled firewhirls, considering multiple physical effects, including the previously investigated variable physical properties and strong vortex, and with emphasis on the effect of non-unity Lewis number. By means of perturbation method, an analytical expression of the flame height, with all effects presented in explicit forms, was derived in the situation of both fuel and of oxidizer Lewis numbers being near unity. In leading order approximation, the non-unity Lewis number effect tends to change the flame height by a factor of Le_O/Le_F.

1 Introduction

Firewhirls, as a natural phenomenon widely occurring in wild and urban fire and holding potential to cause severe damages to lives and properties, have attracted numerous investigations involving both experimental and theoretical approaches in the past a few decades[1-19]. As the flame height of firewhirls increasing, the radiative heat transfer tends to facilitate the ignition remotely therefore expediting the spread of the fires. Hence, there is always a particular interest in the flame height of firewhirl [3-8, 15, 19].

In Chuah et al.'s experimental study[5], the vortical flow was set up inclined, resulting in a correspondingly inclined firewhirl. This experiment testifies that the flame height of firewhirl was determined by circulation, instead of the previously recognized buoyance [6, 11, 20], provided that the circulation is sufficiently strong. By assuming constant density and mass diffusivity, Burgers vortex, and unity Lewis numbers, Chuah et al. proposed an theoretical prediction for the flame height[5]

$$\frac{x_h}{d_0} = \frac{Pe}{16Z_{st}} \tag{1}$$

where the flame height is linearly proportional to the diameter of the fuel pool, d_0, to the Peclet umber, Pe, and inversely proportional to the stoichiometric mixture fraction, Z_{st}. Compared with their own experimental results, (1) always gives an underestimation on the flame height.

Klimenko and Williams[7] argued that the Burgers vortex was not strong enough to describe the actual circulation-controlled firewhirls[21]. By replacing the Burgers vortex by a strong vortex[21], Klimenko and Williams derived a revised flame height expression

$$\frac{x_h}{d_0} = \frac{2}{\alpha_v}\frac{Pe}{16Z_{st}} \tag{2}$$

where α_v characterizes the strong vortex, and assumes value less than 2. The equation (2) agrees well with the experimental results of Chuah et al.[5].

In the theories of both Chuah et al.[5] and Klimenko and Williams[7], density and mass diffusivity were assumed as constants. Such an assumption is rather questionable in firewhirl, where the temperature is significantly higher than that of the ambience. As a result, as temperature increases, the density decreases and so does the flow inertia, and therefore the fuel can be transported into higher altitude, rendering a larger flame height[22]. Retaining the unity-Lewis-number assumption and regarding the vortical flow as Burgers vortex, the authors carried out a theoretical analysis for the variable physical properties effect on the flame height[15]. Introducing a Howarth-Dorodnitsyn[23] like coordinate transformation, the governing equations can be converted into a simplified density- and mass diffusivity-free form. An analytically explicit flame height expression can also be obtained[15]

$$\frac{x_h}{d_0} = \left(\frac{T_m}{T_0}\right)^{2-\alpha_T}\frac{Pe}{16Z_{st}} \tag{3}$$

where T_m represents the specially defined "mean temperature", assuming value higher than the fuel pool temperature T_0, and α_T characterizes the temperature dependence of mass diffusivity[24] and it is always less than 2. The flame heights predicted by equation (3) also agree well with the experimental results of Chuah et al.[5].

Both variable physical properties and strong vortex tend to lengthen the flame height as indicated in (2) and (3). Because both effects are independent in physics, a simple combination of them in a theory must overshot the predictions on the flame height. Consequently, there must exist a mechanism that reduces the flame height. Inspired by the theoretical research on droplet combustion[22, 25], the authors[19] found that the discrepancy between the fuel and oxidizer mass diffusivities can be an effective loss-mechanism to the flame height, thereby a more general flame height expression was derived

$$\frac{x_h}{d_0} = \alpha_D\frac{2}{\alpha_v}\left(\frac{T_m}{T_0}\right)^{2-\alpha_T}\frac{Pe}{16Z_{st}} \tag{4}$$

where $\alpha_D = D_F/D_O$ is the ratio of fuel mass diffusivity to that of oxidizer. In Chuah et al.' experiments, the fuels adopted were methanol, ethanol, and 2-propanol, whose molecular weights are larger than that of air, therefore the fuel mass diffusivities are smaller than that of oxidizer[24], thus the values of α_D are less than unity, resulting in shorter flame height.

In all the above theoretical studies, the fuel and oxidizer Lewis numbers are assumed to be unity. However, in many combustion problems, such an assumption can seldom be exactly

Corresponding author. Fax: +852-27666664
E-mail address: pengzhang.zhang@polyu.edu.hk

satisfied[22, 25]. In this paper, utilizing the perturbation method and regarding the deviation of Lewis number from unity as small parameters[26], we proposed a theoretical analysis, attempting to reveal the non-unity Lewis number effect on the flame height of circulation-controlled firewhirls.

2 Mathematical Formulation

2.1 Governing Equations

Complete specification of the circulation-controlled firewhirl system requires three transport equations governing the fuel mass fraction, oxidizer mass fraction, and energy, all of which are inhomogeneous due to the presence of chemical reaction term. Considering the stoichiometry of the chemical reaction, the governing equations can be reduced to two equations, in which the reaction terms are formally removed[25, 26]. The specific mathematical forms are given as follows:

$$\begin{aligned}&\left(\tilde{\rho}\tilde{u}\frac{\partial}{\partial \tilde{x}}+\tilde{\rho}\tilde{v}\frac{\partial}{\partial \tilde{r}}\right)(\tilde{Y}_F+\tilde{T})\\&-\frac{2Le_F}{Pe}\frac{\partial}{\partial \tilde{x}}\left[\tilde{\rho}\tilde{D}_F\frac{\partial}{\partial \tilde{x}}\left(\frac{\tilde{Y}_F}{Le_F}+\tilde{T}\right)\right]\\&-\frac{2Le_F}{Pe}\frac{1}{\tilde{r}}\frac{\partial}{\partial \tilde{r}}\left[\tilde{\rho}\tilde{D}_F\tilde{r}\frac{\partial}{\partial \tilde{r}}\left(\frac{\tilde{Y}_F}{Le_F}+\tilde{T}\right)\right]=0\end{aligned} \tag{5}$$

$$\begin{aligned}&\left(\tilde{\rho}\tilde{u}\frac{\partial}{\partial \tilde{x}}+\tilde{\rho}\tilde{v}\frac{\partial}{\partial \tilde{r}}\right)(\tilde{Y}_F-\tilde{Y}_O)\\&-\frac{2Le_F}{Pe}\frac{\partial}{\partial \tilde{x}}\left[\tilde{\rho}\tilde{D}_F\frac{\partial}{\partial \tilde{x}}\left(\frac{\tilde{Y}_F}{Le_F}-\frac{\tilde{Y}_O}{Le_O}\right)\right]\\&-\frac{2Le_F}{Pe}\frac{1}{\tilde{r}}\frac{\partial}{\partial \tilde{r}}\left[\tilde{\rho}\tilde{D}_F\tilde{r}\frac{\partial}{\partial \tilde{r}}\left(\frac{\tilde{Y}_F}{Le_F}-\frac{\tilde{Y}_O}{Le_O}\right)\right]=0\end{aligned} \tag{6}$$

in which the nondimensional quantities are defined as $\tilde{\rho}=\rho/\rho_0$, $(\tilde{r},\tilde{x})=(r,x)/r_0$, $(\tilde{u},\tilde{v})=(u,v)/u_0$, $\tilde{D}_F=D_F/D_{F0}$, $\tilde{Y}_F=Y_F$, $\tilde{Y}_O=\tilde{Y}_O/\sigma_{FO}$, and $\tilde{T}=c_pT/q_c$, with the quantities having index 0 referring to their values at the ground $x=0$, σ_{FO} specifying the stoichiometry of the reaction, and q_c representing the combustion heat release by consuming unit mass of fuel. The nondimensional numbers are defined as $Le_F=\lambda/\rho c_pD_F$, $Le_O=\lambda/\rho c_pD_O$, and $Pe=u_0d_0/D_{F0}$.

Equations (5) and (6) satisfy the most general situation in combustion, including non-unity Lewis numbers, i.e., $Le_F\neq 1$ and $Le_O\neq 1$, and they reduce to the conservation equations for the conventional species-species and species-enthalpy coupling function, $\beta_S=\tilde{Y}_F-\tilde{Y}_O$ and $\beta_T=\tilde{Y}_F+\tilde{T}$, respectively as Lewis numbers being unity. The boundary conditions for (5) and (6) are specified as: BC(1) at $\tilde{r}=0$, $\partial\tilde{Y}_F/\partial\tilde{r}=\partial\tilde{Y}_O/\partial\tilde{r}=\partial\tilde{T}/\partial\tilde{r}=0$; BC(2) at $\tilde{r}=\infty$, $\partial\tilde{Y}_F/\partial\tilde{r}=\partial\tilde{Y}_O/\partial\tilde{r}=\partial\tilde{T}/\partial\tilde{r}=0$; BC(3a) at $\tilde{x}=0$ and $\tilde{r}\leq 1$,

$$\begin{aligned}\tilde{\rho}\tilde{u}\tilde{Y}_F-(1/Pe)\tilde{\rho}\tilde{D}_F(\partial\tilde{Y}_F/\partial\tilde{x})&=\tilde{\rho}\tilde{u}\\\tilde{\rho}\tilde{u}\tilde{T}-(2Le_F/Pe)\tilde{\rho}\tilde{D}_F(\partial\tilde{T}/\partial\tilde{x})&=\tilde{\rho}\tilde{u}(\tilde{T}_0-q_v/q_c)\\\tilde{\rho}\tilde{u}\tilde{Y}_O-(1/Pe)\tilde{\rho}\tilde{D}_F(\partial\tilde{Y}_O/\partial\tilde{x})&=0\end{aligned}$$

BC(3b) at $\tilde{x}=0$ and $\tilde{r}>1$, $\partial\tilde{Y}_F/\partial\tilde{x}=\partial\tilde{Y}_O/\partial\tilde{x}=\partial\tilde{T}/\partial\tilde{x}=0$; BC(4) at $\tilde{x}=\infty$, $\tilde{Y}_F=0$, $\tilde{Y}_O=\tilde{Y}_{O,\infty}$, $\tilde{T}=\tilde{T}_\infty$. BC(1) and BC(2) are the axisymmetric and radially far field conditions, respectively. BC(3) describes the Stefan flow of fuel evaporation induced by the heat transfer from the flame. BC(4) is the axially far field condition.

Introducing a Howarth-Dorodnitsyn like density-mass-diffusivity-weighted coordinate defined by[15, 19]

$$\xi=\frac{D_{F0}}{u_0r_0}\int_0^{\tilde{x}}\tilde{\rho}^2\tilde{D}_Fdx'=\frac{2}{Pe}\int_0^{\tilde{x}}\tilde{\rho}^2\tilde{D}_Fdx',\qquad \eta=\int_0^{\tilde{r}}\tilde{\rho}dr' \tag{7}$$

equations (5) and (6) are simplified by assuming large Peclet number, i.e., $Pe\gg 1$.

$$\left(\hat{u}\frac{\partial}{\partial\xi}+\hat{v}\frac{\partial}{\partial\eta}\right)(\tilde{Y}_F+\tilde{T})=Le_F\frac{1}{\eta}\frac{\partial}{\partial\eta}\left[\eta\frac{\partial}{\partial\eta}\left(\frac{\tilde{Y}_F}{Le_F}+\tilde{T}\right)\right] \tag{8}$$

$$\left(\hat{u}\frac{\partial}{\partial\xi}+\hat{v}\frac{\partial}{\partial\eta}\right)(\tilde{Y}_F-\tilde{Y}_O)=Le_F\frac{1}{\eta}\frac{\partial}{\partial\eta}\left[\eta\frac{\partial}{\partial\eta}\left(\frac{\tilde{Y}_F}{Le_F}-\frac{\tilde{Y}_O}{Le_O}\right)\right] \tag{9}$$

The corresponding boundary conditions are simplified to BC(I) at $\eta=0$, $\partial\tilde{Y}_F/\partial\eta=\partial\tilde{Y}_O/\partial\eta=\partial\tilde{T}/\partial\eta=0$; BC(II) at $\eta\to\infty$, $\partial\tilde{Y}_F/\partial\eta=\partial\tilde{Y}_O/\partial\eta=\partial\tilde{T}/\partial\eta=0$; BC(III-a) at $\xi=0$ and $\eta\leq 1$, $\tilde{Y}_F=\tilde{Y}_{F0}$, $\tilde{T}=\tilde{T}_0-q_v/q_c$, $\tilde{Y}_O=0$; BC(III-b) at $\xi=0$ and $\eta>1$, $\tilde{Y}_F=0$, $\tilde{T}=\tilde{T}_\infty$, $\tilde{Y}_O=\tilde{Y}_{O,\infty}$. The detailed derivation has been presented in[19].

Considering strong vortex effect, the vortical flow are characterized by the following stream function[19]

$$\psi=\begin{cases}s(\xi)\eta^{\alpha_v'}, & \eta<\eta_c\\ \eta_c^{\alpha_v'-\alpha_v}s(\xi)\eta^{\alpha_v}, & \eta\geq\eta_c\end{cases} \tag{10}$$

in terms of which the velocity components can be determined as

$$\hat{u}=\begin{cases}\alpha_v'\eta^{\alpha_v'-2}s(\xi), & \eta<\eta_c\\ \alpha_v\eta_c^{\alpha_v'-\alpha_v}s(\xi)\eta^{\alpha_v-2}, & \eta\geq\eta_c\end{cases} \tag{11}$$

$$\hat{v}=\begin{cases}-s'(\xi)\eta^{\alpha_v'-1}, & \eta<\eta_c\\ -\eta_c^{\alpha_v'-\alpha_v}s'(\xi)\eta^{\alpha_v-1}, & \eta\geq\eta_c\end{cases} \tag{12}$$

where η_c represents the radius of the vortex core, and α_v' and α_v characterize the vortical flow inside and outside the vortex core, respectively.

Defining the stream function coordinate as

$$\chi=\frac{\alpha_v}{2}\xi,\qquad \zeta=\sqrt{2\psi} \tag{13}$$

equations (8) and (9) are converted into

$$\frac{\partial}{\partial\chi}(\tilde{Y}_F+\tilde{T})=Le_F\frac{1}{\zeta}\frac{\partial}{\partial\zeta}\left[\zeta\frac{\partial}{\partial\zeta}\left(\frac{\tilde{Y}_F}{Le_F}+\tilde{T}\right)\right] \tag{14}$$

$$\frac{\partial}{\partial\chi}(\tilde{Y}_F-\tilde{Y}_O)=Le_F\frac{1}{\zeta}\frac{\partial}{\partial\zeta}\left[\zeta\frac{\partial}{\partial\zeta}\left(\frac{\tilde{Y}_F}{Le_F}-\frac{\tilde{Y}_O}{Le_O}\right)\right] \tag{15}$$

in which the convection and diffusion transports are completely separated. Correspondingly, the boundary conditions become: BC(i) at $\zeta=0$, $\partial\tilde{Y}_F/\partial\zeta=\partial\tilde{Y}_O/\partial\zeta=\partial\tilde{T}/\partial\zeta=0$; BC(ii) at $\zeta\to\infty$, $\partial\tilde{Y}_F/\partial\zeta=\partial\tilde{Y}_O/\partial\zeta=\partial\tilde{T}/\partial\zeta=0$; BC(iii-a) at $\chi=0$ and $\zeta\leq 1$, $\tilde{Y}_F=\tilde{Y}_{F0}$, $\tilde{T}=\tilde{T}_0'$, $\tilde{Y}_O=0$; BC(iii-b) at $\chi=0$ and $\zeta>1$, $\tilde{Y}_F=0$, $\tilde{T}=\tilde{T}_\infty$, $\tilde{Y}_O=\tilde{Y}_{O,\infty}$.

2.2 Perturbation Solution

In the problems with distinct fuel and oxidizer (non-unity) Lewis numbers, an exact solution to the fuel and oxidizer mass fractions as well as temperature requires the division of the whole domain into two regions, namely, the fuel region and the oxidizer region. The governing equations must be solved separately to yield general solutions being valid in either region, in terms of which the flame location as well as the flame temperature can be determined by matching the solutions at the flame location [22, 25]. However, the three-dimensionality of the firewhirl system invalidates the above approach for arbitrary non-unity Lewis number.

If both fuel and oxidizer Lewis numbers are near unity, equations (14) and (15) can be approximately solved by means of perturbation expansion. Following Chung and Law's approach[26], we define the Lewis number weighted coupling function by

$$\beta_S = \frac{\tilde{Y}_F}{Le_F} - \frac{\tilde{Y}_O}{Le_O}, \qquad \beta_T = \frac{\tilde{Y}_F}{Le_F} + \tilde{T} \tag{16}$$

which become the conventional coupling functions as Lewis numbers are unity. In the absence of unity Lewis number, the coupling functions defined by (16) have better mathematical property than conventional coupling function, since the latter are smooth across the flame whereas the former is continuous but not smooth[22, 26]. For near-unity Lewis numbers, we can introduce two small quantities,

$$\left(1 - \frac{1}{Le_i}\right) = l_i, \qquad i = F, O \tag{17}$$

in terms of which the distributions of fuel and oxidizer mass fractions and temperature can be expanded as

$$\tilde{Y}_F = \tilde{Y}_F^0 + l_F \tilde{Y}_F^1 + \cdots \tag{18}$$

$$\tilde{Y}_O = \tilde{Y}_O^0 + l_F \tilde{Y}_O^1 + \cdots \tag{19}$$

$$\tilde{T} = \tilde{T}^0 + l_F \tilde{T}^1 + \cdots \tag{20}$$

Substituting (18)-(20) into (16) gives the expansion of Lewis-number-weighted coupling functions, and further substitution of the latter into (14) and (15), gives the leading order equation, describing the conservation of the leading order Lewis-number-weighted coupling function, being valid in the whole field

$$\frac{\partial \beta_T^0}{\partial \chi} - Le_F \frac{1}{\zeta}\frac{\partial}{\partial \zeta}\left(\zeta \frac{\partial \beta_T^0}{\partial \zeta}\right) = 0 \tag{21}$$

$$\frac{\partial \beta_S^0}{\partial \chi} - Le_F \frac{1}{\zeta}\frac{\partial}{\partial \zeta}\left(\zeta \frac{\partial \beta_S^0}{\partial \zeta}\right) = 0 \tag{22}$$

The boundary conditions are given by: LBC(1) at $\zeta = 0$, $\partial\beta_S^0/\partial\zeta = \partial\beta_T^0/\partial\zeta = 0$; LBC(2) at $\zeta \to \infty$, $\partial\beta_S^0/\partial\zeta = \partial\beta_T^0/\partial\zeta = 0$; LBC(3a) at $\chi = 0$ and $\zeta \le 1$, $\beta_S^0 = \tilde{Y}_{F0}/Le_F$, $\beta_T^0 = \tilde{Y}_{F0}/Le_F + \tilde{T}_0'$; LBC(3b) at $\chi = 0$ and $\zeta > 1$, $\beta_S^0 = -\tilde{Y}_{O,\infty}/Le_O$, $\beta_T^0 = \tilde{T}_\infty$.

2.3 Flame Height

The flame contour expression can be determined by the equating (24) to zero, implying the complete consumption of fuel and oxidizer

$$\left(\frac{\tilde{Y}_{F0}}{Le_F} + \frac{\tilde{Y}_{O,\infty}}{Le_O}\right)\int_0^\infty J_0(\omega\zeta_f)J_1(\omega)\exp(-Le_F\omega^2\chi_f)d\omega = \frac{\tilde{Y}_{O,\infty}}{Le_O} \tag{25}$$

The flame height is the highest point on the flame contour:,

$$\left(\frac{\tilde{Y}_{F0}}{Le_F} + \frac{\tilde{Y}_{O,\infty}}{Le_O}\right)\int_0^\infty J_1(\omega)\exp(-Le_F\omega^2\chi_h)d\omega = \frac{\tilde{Y}_{O,\infty}}{Le_O} \tag{26}$$

where χ_h represents the flame height in stream function coordinate. The left-hand side of (26) can be integrated exactly, yielding

$$\int_0^\infty J_1(\omega)\exp(-Le_F\omega^2\chi_h)d\omega = 1 - \exp\left(-\frac{1}{4Le_F\chi_h}\right) \tag{27}$$

Substitution of (27) into (26) gives

$$\left(\frac{\tilde{Y}_{F0}}{Le_F} + \frac{\tilde{Y}_{O,\infty}}{Le_O}\right)\left[1 - \exp\left(-\frac{1}{4Le_F\chi_h}\right)\right] = \frac{\tilde{Y}_{O,\infty}}{Le_O} \tag{28}$$

Since the firewhirl is circulation-controlled, we can assume large flame height, i.e., $4Le_F\chi_h \gg 1$, and have

$$\exp\left(-\frac{1}{4Le_F\chi_h}\right) \sim 1 - \frac{1}{4Le_F\chi_h} \tag{29}$$

Substituting of (29) into (28) and recalling the definition of stoichiometric mixture fraction[22],

$$Z_{st} = \frac{\tilde{Y}_{O,\infty}}{\tilde{Y}_{F0} + \tilde{Y}_{O,\infty}} \tag{30}$$

the flame height expression can be derived as

$$\chi_h = \frac{Le_O}{Le_F}\frac{1}{4Z_{st}} \tag{31}$$

where all terms of order $O(l_F)$ and higher are neglected in leading order approximation. Inverting the stream function coordinates to density-mass-diffusivity-weighted coordinates, and then the latter to physical coordinates, the flame height expression becomes

$$\frac{x_h}{d_0} = \frac{2}{\alpha_v}\frac{Le_O}{Le_F}\left(\frac{T_m}{T_0}\right)^{2-\alpha_T}\frac{Pe}{16Z_{st}} \tag{32}$$

Equation (32) reveals three mechanisms that can change flame height. The strong vortex effect tends to intensify the axial stretching effect of the vortical flow, therefore increasing the flame height. The variable physical properties effect resolves the decreasing of density due to high temperature, resulting in the fuel being more readily transported to higher altitude, i.e., lengthening the flame height. The non-unity Lewis number effect modifies the flame height according to the specific combustion system. For those using hydrocarbons with large molecular weight, the fuel Lewis number is usually larger than that of air, thus the effect of non-unity Lewis number tends to decrease the flame height. All four factors in the flame height expression (32) refer to independent physical cause, thus are separable, and moreover can be reduced to the previous formulas (1) – (4) under their particular situation.

In generally, the flame contours can be plotted by numerically solving (25) under various pairs of Lewis numbers, as shown in Fig. 1. It is seen that the decreasing of fuel Lewis number tends to expand the flame contour, particularly in axial direction, while the oxidizer Lewis number exhibits the inverse effect on the flame contour. The Lewis numbers can in physics be interpreted as the ratio of thermal diffusivity to mass diffusivity. The smaller Lewis number of fuel than that of oxidizer implies that the fuel has larger mass diffusivity than the oxidizer, due to which the fuel has higher capability to be transported to larger altitude, i.e., extending the flame contour in axial direction. In opposite situation, i.e. Lewis number of oxidizer being lower than that of fuel, the larger mass diffusivity of oxidizer tends to squeeze flame contour, especially in the axial direction because the flame end close to the fuel pool is anchored at the rim of fuel pool due to flame sheet approximation.

In a special situation of equal Lewis numbers, the flame contours are presented in Fig 1(right), which are only moderately different from each other. It testifies that (32) that the flame height tends to rely on the variation of Lewis numbers by their ratio, which, according to respective influence of fuel and oxidizer Lewis numbers on the flame contour, must be Le_O/Le_F. In addition, the flame shape with larger Lewis numbers tend to be slimmer than that with lower Lewis numbers. The physical reason can be interpreted as follows. The larger Lewis numbers for both fuel and oxidizer mean lower mass diffusivities so that the axial convection tends to be more dominant than diffusion, resulting in a more stretched flame shape in the axial direction. Similarly, lower Lewis numbers for fuel and oxidizer mean enhanced mass diffusion towards all the directions, which tends to counteract the axial convection and to make flame shape stouter.

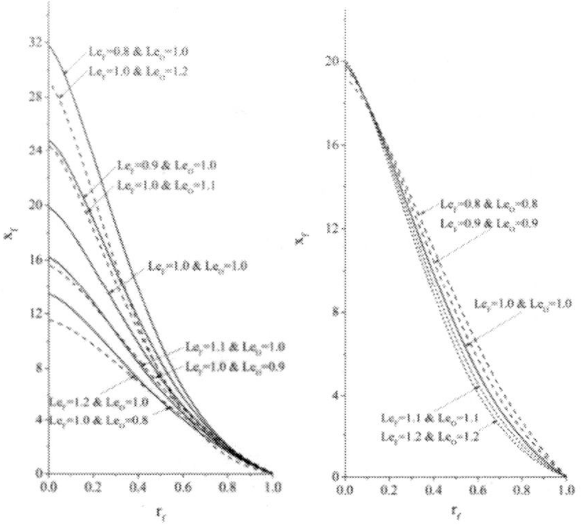

Fig 1. Flame contours with various pairs of Lewis numbers.

3 Conclusions

This paper presents a theoretical analysis to extend the existing theories on the circulation-controlled firewhirls to non-unity Lewis number. Utilizing the perturbation method, an analytical expression of flame height is derived in the situation of Lewis number being near unity. The results show that both strong vortex and variable physical properties effects tend to lengthen the flame height, but the non-unity Lewis number effect can reduce the flame height of firewhirls.

5 Acknowledgment

This work was supported by the Hong Kong RGC/GRF (operating under contract numbers PolyU 152217/14E and 152651/16E) and partly by the Hong Kong Polytechnic University (G-UA2M and G-YBGA).

References

[1] F. Battaglia, K.B. McGrattan, R.G. Rehm, H.R. Baum, Combust. Theory Modelling 4 (2000) 123-138.

[2] F. Battaglia, R.G. Rehm, H.R. Baum, Phys. Fluids, 12 (2000) 2859-2867.

[3] K.H. Chuah, G. Kushida, Proc. Combust. Inst. 31 (2007) 2599-2606.

[4] K.H. Chuah, K. Kuwana, K. Saito, Combust. Flame 156 (2009) 1828-1833.

[5] K.H. Chuah, K. Kuwana, K. Saito, F.A. Williams, Proc. Combust. Inst. 33 (2011) 2417-2424.

[6] H.W. Emmons, S.-J. Ying. The fire whirl. Symposium (International) on Combustion, 1967. p. 475-488.

[7] A.Y. Klimenko, F.A. Williams, Combust. Flame 160 (2013) 335-339.

[8] K. Kuwana, S. Morishita, R. Dobashi, K.H. Chuah, K. Saito, Proc. Combust. Inst. 33 (2011) 2425-2432.

[9] K. Kuwana, K. Sekimoto, K. Saio, F.A. Williams, Y. Hayashi, H. Masuda, Aiaa Journal 45 (2007) 16-19.

[10] K. Kuwana, K. Sekimoto, K. Saito, F.A. Williams, Fire Safety Journal 43 (2008) 252-257.

[11] J. Lei, N. Liu, K. Satoh, Proc. Combust. Inst. 35 (2015) 2503-2510.

[12] J. Lei, N.A. Liu, L.H. Zhang, K. Satoh, Combust. Flame 162 (2015) 745-758.

[13] J.A. Lei, N.A. Liu, L.H. Zhang, H.X. Chen, L.F. Shu, P. Chen, Z.H. Deng, J.P. Zhu, K. Satoh, J.L. de Ris, Proc. Combust. Inst. 33 (2011) 2407-2415.

[14] F.A. Williams, Progress in Energy and Combustion Science 8 (1982) 317-354.

[15] D. Yu, P. Zhang, Proc. Combust. Inst. (2017) 3097-3104.

[16] K. Zhou, N. Liu, K. Satoh, Fire Safety Science 10 (2011) 681-691.

[17] K. Zhou, N. Liu, L. Zhang, K. Satoh, Fire Technology 50 (2014) 1573-1587.

[18] R. Zhou, Z.N. Wu, J. Fluid Mech. 583 (2007) 313-345.

[19] D. Yu, P. Zhang, Combust. Flame 182 (2017) 36-47.

[20] P.B. Sunderland, J.E. Haylett, D.L. Urban, V. Nayagam, Combust. Flame 152 (2008) 60-68.

[21] A. Klimenko, Theoretical and Computational Fluid Dynamics 14 (2001) 243-257.

[22] C.K. Law, Combustion Physics, Cambridge University Press, 2006.

[23] P.M. Chung, Advances in Heat Transfer 2 (1965) 109-270.

[24] E.L. Cussler, Diffusion: mass transfer in fluid systems, Cambridge University Press, 2009.

[25] F. Williams, Combustion Theory 2nd. The Benjamin/ Cummings, 1985.

[26] S.H. Chung, C.K. Law, Combust. Sci. and Tech. 37 (1984) 21-46.

11th Asia-Pacific Conference on Combustion,
The University of Sydney, NSW Australia
10th -14th December 2017.

A Study on the Basic Combustion Characteristics in a Metal Fiber Burner

Jaehyeon Kim[1], Minsoek Han[1], Keunseon Sim[1] and Keeman Lee[2]

[1]Department of Aerospace Engineering, Sunchon National University

255, Jungang-ro, Suncheon-si, Jeollanam-do, 57938, Republic of Korea

[2]School of Mechanical and Aerospace Engineering, Sunchon National University

255, Jungang-ro, Suncheon-si, Jeollanam-do, 57938, Republic of Korea

Abstract

This study investigated the basic combustion characteristics in a metal fiber burner. Metal fiber burners have been used owing to high efficiency, low emission and the extension of flammability limit. Main parameters of this study are thermal power (204, 255, 306, 357, 408 kW/m^2) and equivalence ratio (0.4-1.3). The fuel and the oxidizer were used the propane with commercial grade and the compressed air. In this study, the flame mode is divided according to the shape of the flame. First, the blue flame mode appears relatively at lean conditions. Second, the radiation mode turns up near the stoichiometric ratio. Finally, the pool flame mode is shown relatively at rich conditions. In the case of the blue flame mode, the flame has a low surface temperature, efficiency and CO, NOx emission, while the flue temperature was high. For the radiation mode, the surface temperature and efficiency is high. But, it is confirmed that the increase of the CO, NOx emission is relatively larger than the blue flame mode. Finally, the pool flame mode represents relatively a lower temperature and efficiency compared to the radiation mode.

1 Introduction

According to rigid environmental regulations and demands of the development of combustion system with high efficiency and low emissions, the various combustion systems and burners have been developed and studied. Until now, Bunsen burner has been used in commercial boilers, but porous burner has been recently received attention for the above reasons. A boiler with a Bunsen burner couldn't meet environmental regulations and energy requirement because of problems related to thermal efficiency and NOx emission. On the other hand, a porous burner has advantages that complete combustion and uniform heating are possible. In addition, the flammability limit expands and CO emission decreases owing to preheating effect of unburned mixture by regenerative combustion. Also, thermal NOx decrease because the temperature of flame is low. The materials of porous burners include metal fiber, ceramic, stainless steel. Among them, the metal fiber offers advantages of low pressure drop and the durability to thermal stress.[1-5]

However, the study on the combustion characteristics of each flame shape is relatively insufficient with the combustion condition in premixed surface radiant burner. The object of this study is to investigate the radiant efficiency, exhaust gas emission and preheating effect of unburned mixture in a metal fiber burner. As a main parameter, thermal powers and equivalence ratios each were adjusted from 204 to 408 kW/m^2 and 0.4 to 1.3.

Corresponding author. Fax: +82-61-750-3820
E-mail address: kmlee@scnu.ac.kr

2 Experimental set up and method

Figure 1 shows experimental system schematically used in this study. A burner with a diameter of 50 mm and a metal fiber with a porosity of 86% of the FeCrAlloy fibre mat sintered were used in the experiment. Also a quartz with a diameter of 50 mm to simulate a combustor chamber was used to observe the behavior of flame.

The fuel and the oxidizer were used the propane with commercial grade and the compressed air. They were mixed enough for the premixed combustion in the mixing chamber before entering to the burner. Two-stage ceramic honeycombs of different sizes were positioned under the metal fiber to form a uniform flow field of premixed mixture.

To investigate combustion characteristics, a thermocouple of K-type was placed to measure the exhaust gas temperature at a distance of 300 mm above from the surface of burner. A thermocouple of R-type was used to measure the surface temperature of the burner. The surface temperature of the burner in this study was measured and averaged at the 9 points along surface for the detailed temperature measurement by two-dimensional traverse.[6, 7] Also, to investigate the preheating effect of the unburned mixture, a thermocouple of R-type was fixed at the center of the back side of the metal fiber by ceramics adhesive, and a thermocouple of K-type was positioned at a distance of 2 mm from the back side of metal fiber mat.

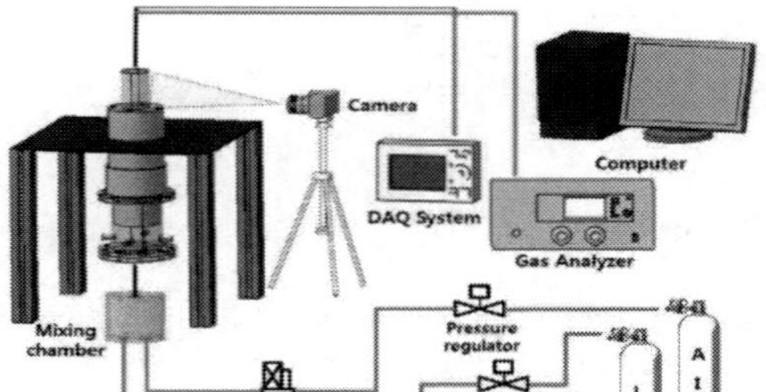

Figure 1: Experimental set-up.

The exhaust gas acquired by a probe at the same height as a thermocouple (i.e. 300 mm in the downstream of the combustion chamber) was analyzed to checked production of CO and NOx using a gas analyzer (Optima 7, Germany). To investigate shapes of flame, the direct flame images according to the respective combustion conditions were acquired through the DSLR camera.

As main parameter in this study, thermal power was fixed from 204 kW/m^2 to 408 kW/m^2 (i.e., 204 kW/m^2, 255 kW/m^2, 306 kW/m^2, 357 kW/m^2, 408 kW/m^2) and equivalence ratio was adjusted from 0.4 to 1.3 at intervals of 0.1.

Using the averaged burner surface temperature, the radiation efficiency of a metal fiber burner is defined as η_r by (1). Using temperatures of the back side of a metal fiber, the temperature of unburned mixture and the preheating efficiency is defined by the heat transfer equation as η_p described in (2).

$$\eta_r = \frac{Q_r}{Q_F} * 100 \quad [\%] \quad (1)$$

$$\eta_p = \frac{Q_p}{Q_{cond}} * 100 \quad [\%] \quad (2)$$

In (1) and (2), Q_F is the thermal power of the inlet fuel and Q_r, Q_p, Q_{cond} are the heat transfer rates of the radiation, the preheating and the conduction respectively. The related equations are defined following as in (3), (4), (5) and (6).

$$Q_F = \dot{m_f} * L.H.V \quad [W] \quad (3)$$

$$Q_r = \sigma\epsilon A(T_{sur}^4 - T_\infty^4) \quad [W] \quad (4)$$

$$Q_p = \dot{m}c_p(T_{mix} - T_\infty) \quad [W] \quad (5)$$

$$Q_{cond} = kA\frac{T_{sur} - T_{back}}{dx} \quad [W] \quad (6)$$

In (4), (5) and (6), T_∞ is the ambient temperature and $T_{sur}, T_{mix}, T_{back}$ are the temperature of the surface, the unburned mixture and the back side of the metal fiber respectively.

3 Results and discussion

3.1 The flame mode

In this study, the flame mode is divided according to the shape of the flame. First, the blue flame mode occurs at the lean equivalence ratios, the relatively long blue flame is formed on respectively many small porous holes of the metal fiber. Second, the radiation mode occurs near the stoichiometric equivalence ratio, the flame permeates into the metal fiber mat and heats it. In the pool flame mode appears in the rich conditions, the flame is formed widely on the metal fiber unlike the flame of the blue flame mode formed respective on the hole of the metal fiber.

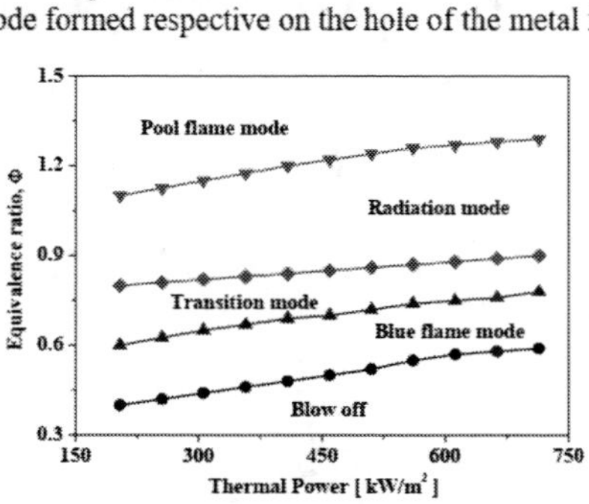

Figure 2: The flame mode map of the metal fiber burner according to thermal power.

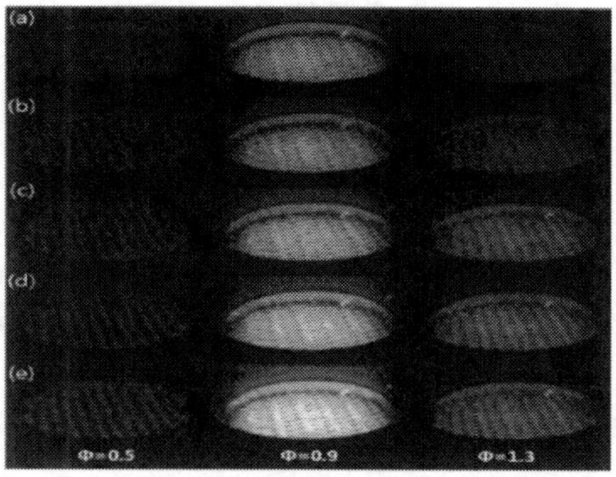

Figure 3: The flame images of the blue flame mode (Φ=0.5), the radiation mode (Φ=0.9) and the pool flame mode (Φ=1.1) with various thermal powers.(a: 204, b: 255, c: 306, d: 357, e:408 kW/m^2).

Also, the transition mode lies between of the blue flame mode and the radiation mode. Fig. 2 shows the flame mode map according to thermal power. And the flame images of the three mode (the blue flame mode, the radiation mode and the pool flame mode) is shown in Fig. 3 at the same equivalence ratio according to thermal power. It is confirmed that the combustion is even possible under the equivalence ratio 0.5 known as the flammability limit of propane as shown in Fig. 2. It is the feature of the surface radiant burners.

3.2 Thermal characteristics

The surface temperature of metal fiber burner is shown with the various thermal powers and equivalence ratios in Fig. 4. The surface temperature appears to peak at an equivalence ratio of 1.0. The cause is considered that the mat is heated more in the radiation mode than in the blue flame mode and the combustion reaction is more active. Also, the cause why the peak value increases with the rise of thermal power is that the combustion reaction becomes active by the increase of amount of fuel.

Figure 5 represents the exhaust gas temperature with thermal power and equivalence ratio. It exhibits the lowest point at an equivalence ratio of 1.0 oppositely to the surface temperature. This is regarded to be a phenomenon in which the heat of the flame used to heat the exhaust gas is used to heat the mat in the radiation mode. Also, due to the same cause with the increase of the surface temperature according to growing of thermal power, the exhaust gas temperature increases with the rise of the thermal power.

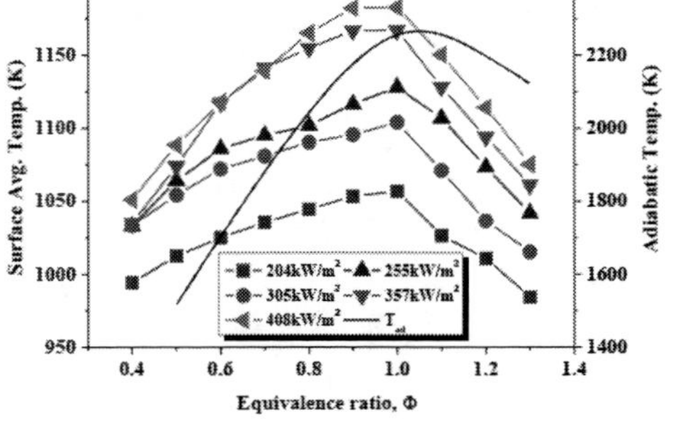

Figure 4: The averaged surface temperature of the burner with the various thermal powers and equivalence ratios.

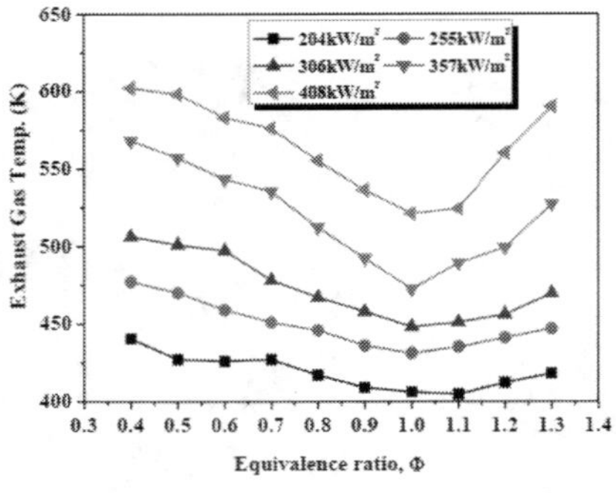

Figure 5: The exhaust gas temperature with the various thermal powers and equivalence ratios.

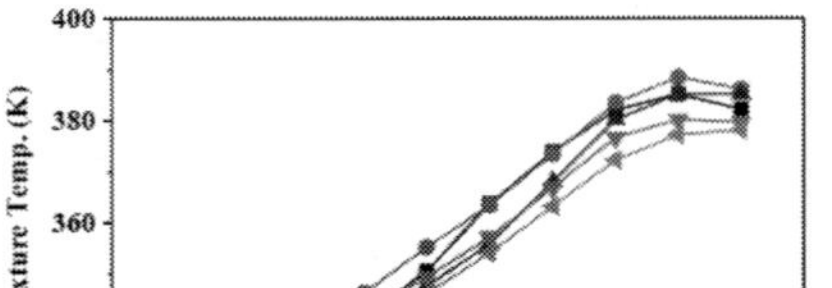

Figure 6: The unburned mixture temperature with the various thermal powers and equivalence ratios.

Figure 6 shows the unburned mixture temperature with the various thermal powers and equivalence ratios. The temperature increases with the rise of equivalence ratio. The reason is considered be a change in the position of the flame in the metal fiber and the increase of surface temperature. Due to the increase in unburned mixture gas, the temperature decreases slightly with the rise of thermal power.

3.3 Thermal efficiency

In this section, the radiation efficiency and the preheating efficiency as defined respectively by (1) and (2) are examined. The radiation efficiency with the thermal power and equivalence ration of burner is shown in Fig. 7. The maximum efficiency appears at an equivalence ratio of 1.0 and decreases with the

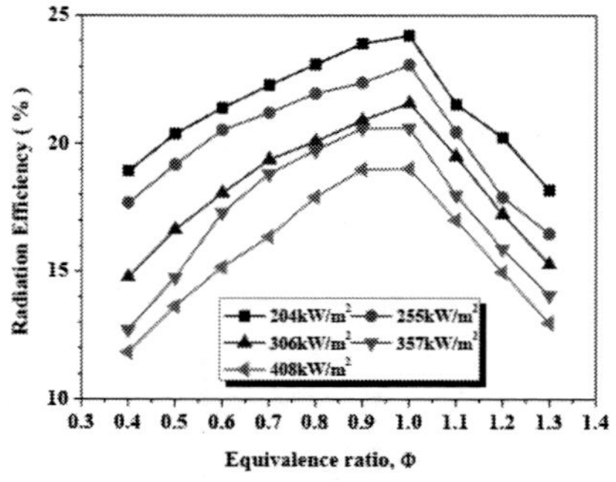

Figure 7: The radiation efficiency with the thermal power and equivalence ratios.

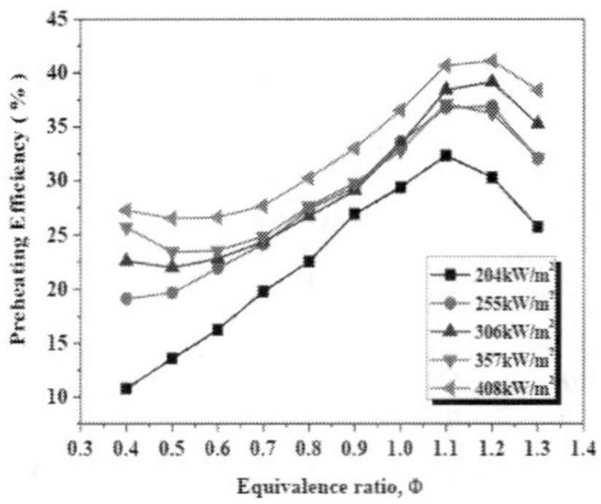

Figure 8: The preheating efficiency with the various thermal powers and equivalence ratios.

increase of thermal power. This is considered to be caused by the rise of the flame position due to the increase of the flow rate of the mixture by the rise of thermal power.[3, 4]

The preheating efficiency appears with the thermal power and equivalence ratios in Fig. 8. The maximum efficiency occurs at an equivalence ratio of 1.0 and increased with the rise of thermal power. This is considered that the effect of rise in numerator value by the increase of flow rate is larger than the effect of rise in denominator value by the increase of the surface temperature. (i.e. In equation (2) explained at the section 2, when the thermal power increases, the increase rate of Q_p is larger than the increase rate of Q_{cond})

3.4 Emission characteristics

Figure 9 shows CO emission performance with the thermal power and equivalence ratio. At an equivalence ratio of less than 0.9, the CO emission is very low. This is considered to be the effect of complete combustion, which is characteristic of the surface radiant burner. The cause of the rapidly increase of CO from the equivalence ratio of 1.0 is regarded that incomplete combustion due to rich equivalence ratio.

Figure 10 represents NOx emission performance with the thermal power and equivalence ratio. A peak of emission appears at an equivalence ratio of 1.0 and increases with the rise of thermal power like the trend of the surface temperature. This is considered as the increase thermal NOx with the increase of the surface temperature.

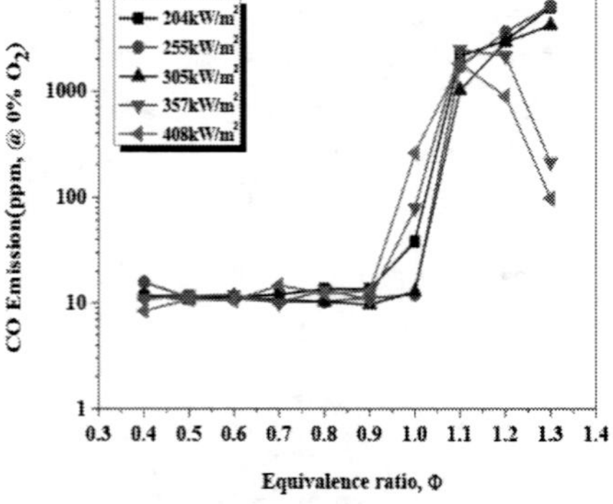

Figure 9: The CO emission of the exhaust gas with the various thermal powers and equivalence ratios.

4 Conclusions

The flame mode and the thermal characteristics according to main parameters in metal fiber burner of premixed combustion are experimentally studied. The flame image was used to define the blue flame mode, the radiation mode and the pool flame mode. And the flame was confirmed to exist under the lean

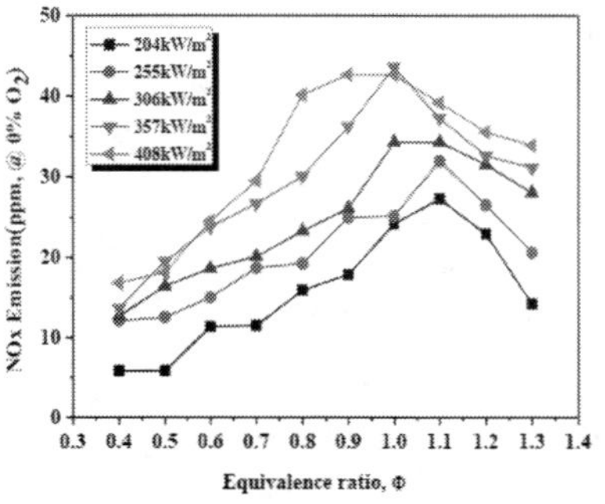

Figure 10: The NOx emission with the various thermal powers and equivalence ratios.

conditions of Φ=0.5 known as the flammability limit of propane by the flame mode map. In the case of the blue flame mode, the flame formed at a low equivalence ratio, the flame had low temperature and CO, NOx emission. In addition, since the flame is located on the outside of the metal fiber, efficiency is low but the exhaust gas temperature is high. In the case of the radiation mode, which appeared at an equivalence ratio of near 1.0, the flame is located in the metal fiber. Because of heating of the metal fiber, the surface temperature and efficiency were high. But, it was confirmed that the increase of the CO, NOx emission relatively than the blue flame mode. Finally, the pool flame mode, which appeared in the rich conditions, shows lower temperature, efficiency by forming long flames on the burner surface

5 Acknowledgments

This research was financially supported by the Ministry of Trade, Industry, and Energy(MOTIE), under the "Regional Specialized Industry Development Program" supervised by the Korea Institute for Advancement of Technology(KIAT).(Ref. number : R0005948)

References

[1] S.A. Leonardi, R. Viskanta, J.P. Gore, Radiation and thermal performance measurements of a metal fiber burner. Quantitative Spectroscopy & Radiative Transfer 73 (2002) 491-501.

[2] B.G. Yu, S.M. Kum, C.E. Lee, S.R. Lee, Combustion characteristics and thermal efficiency for premixed porous-media types of burners. Energy 53 (2013) 343-350.

[3] M.M. Abdelaal, M.K. El-Riedy, A.M. El-Nahas, Effect of oxygen enriched air on porous radiant burner performance and NO emissions. Experimental Thermal and Fluid Science 45 (2012) 163-168.

[4] K.W. Cho, K.W. Han, Y.K. Lee, D.S. Noh, H.M. Yoon, K.J. Riu, K.H. Lee, Premixed combustion of coke oven gas in a metallic fibre mat. Fuel 80 (2001) 1033-1036.

[5] K. Qiu, A.C.S. Hayden, Premixed gas combustion stabilized in fiber felt and its application to a novel radiant burner. Fuel 85 (2006) 1094-1100.

[6] C. Keramiotis, B. Stelzner, D. Trimis, M. Founti, Porous burners for low emission combustion: An experimental investigation. Energy 45 (2012) 213-219.

[7] S. wood, A.T. Harris, Porous burners for lean-burn applications. Progress in Energy and Combustion Science 34 (2008) 667-684.

11th Asia-Pacific Conference on Combustion,
The University of Sydney, NSW Australia
10th -14th December 2017.

Non-monotonic Cellular Instability of Methanol/Air Flames

Guoqing Wang[1,*], Lei Li[1], Yuyang Li[1], Fei Qi[1]

[1]School of Mechanical Engineering, Shanghai Jiao Tong University

Shanghai, 200240, P. R. China

Abstract

Cellular instability of spherical methanol/air flames was investigated in a constant-volume single-chamber cylindrical combustion vessel at initial temperature of 423 K, initial pressure of 1-10 atm and equivalence ratio of 0.7-2.2. The cellular instability on the flame surface was observed at 5-10 atm. Asymptotic analysis was calculated to determine the logarithmic growth rate of cellular instability. The critical Peclet number and critical flame radius were extracted from both experimental observation and asymptotic results. The cellular instability of methanol/air mixtures monotonically increases with initial pressure and non-monotonically varies with equivalence ratio. The non-monotonic flame thickness and monotonic critical Peclet number result in the non-monotonic critical flame radius as well as cellular instability over a wide range of equivalence ratio.

1 Introduction

Cellular instability is one of the intrinsic characteristics of laminar premixed flame. Typical cellular flames are subjected to two instable modes: hydrodynamic and diffusive-thermal instability. Both the two modes of cellular instability have been observed in spherical flames [1, 2]. Theoretical analysis of cellular instability in spherically expanding flame was first introduced by Istratov and Librovich [3]. Bechtold and Matalon [4] subsequently developed an explicit asymptotic analysis to account for both hydrodynamic and diffusive-thermal effects. This asymptotic analysis was corrected and adopted by Bradley et al. [5, 6] in explosion flame. Addabbo et al. [7] further improved the asymptotic analysis by considering variable transport properties. Based on the asymptotic analysis, the transition from stable to cellular flames can be calculated. An accurate asymptotic analysis of spherical flame instability depends on properly extracted flame thickness, global activation energy, Lewis numbers and so on [8].

Equivalence ratio (ϕ) is a critical factor for cellular instability because the combustible mixtures with a component of lower molecular weight reactants, such as in rich heavy hydrocarbon/air mixtures and in lean methane/air or hydrogen/air mixtures, easily become cellular [9]. Numerous studies have been performed for the cellular instability of heavy fuel/air mixtures. However, most of the previous studies focus on the lean and moderate rich flames ($\phi < 1.4$) [10-15]. Cellular instability under richer conditions ($\phi > 1.4$) are inadequate understanding. The great challenges of investigating very rich flames include two aspects. On the one hand, the premixed mixtures of fuel/air should be combustible at extreme rich conditions ($\phi \approx 2.0$). On the other hand, the flame front should propagate faster enough to weaken the buoyancy effect [16].

To obtain a fuel with wide flammability limits and large propagation velocity, methanol is selected as the target fuel. Methanol is an important alternative for transportation fuels considering its abundant production sources, low cost, high octane number and high oxygen content [17]. The flammable range of methanol/air mixtures is 6.7%-36% in mole (ϕ = 0.51-4.0), and the laminar burning velocity of methanol/air mixtures is also much larger than that of other alcohols/air and n-heptane/air mixtures [18]. Bechtold et al. [19] has investigated the Markstein numbers of several hydrocarbons and alcohols, and then found that the Markstein number of methanol/air mixtures change similarly to that of other heavy fuel/air mixtures. This indicates the cellular instability of methanol/air mixtures is analogous to that of heavy fuel/air mixtures.

The aim of this paper is to investigate the flame instabilities of methanol/air mixtures under extreme rich conditions. Firstly, the spherically expanding flames was measured in a high-pressure constant-volume cylindrical combustion vessel over a wide ϕ range. Secondly, the asymptotic analysis was performed for the investigated methanol/air mixtures to figure out the instability difference between extreme rich flames and other flames.

2 Experimental methodology

All the experiments were performed in a constant-volume single-chamber cylindrical combustion vessel. The combustion vessel is made of 304 stainless steel and has an inner volume of 2.77 L (inner diameter 150 mm, inner length 152 mm). Two quartz windows (diameter 100 mm) were clamped to the vessel flange, providing an optical access with 75 mm diameter clear apertures. During the experiments, partial pressure method was used for preparing reactant mixtures in the premixing vessel. The prepared mixtures were stirred by a magnetic stirrer for 10 minutes to ensure uniform mixing. Both the combustion vessel and the premixing vessel can be heated up to 500 K by flexible detachable heating jackets. The mixture of reactants in the combustion vessel was ignited by two tungsten electrodes. An outwardly propagating spherical flame was formed after the

*Corresponding author.

E-mail address: wgqing@sjtu.edu.cn

ignition and a schlieren system was used to record the flame propagation. Clear schlieren images were recorded by a high speed camera. The camera was operated at 10000 frames/sec and the spatial resolution was set at 480×480 pixels to map a region of 75×75 mm^2. In this work, all the experiments of methanol were performed at T_u = 373-423 K, P_u = 1-10 atm and ϕ = 0.7-2.2. Methanol (≥99.5%) were provided by Aladdin Reagent. Synthetic air consisting of 21% O_2 and 79% N_2 provided by Nanjing Special Gas Factory was used as the oxidizer. Each experiment was repeated 3-4 times.

3 Asymptotic analysis

For a spherically expanding flame, the instant flame front can be expressed by $r = R_f(t)$. When the flame suffers from cellular instability, the perturbed flame front can be expressed in the form of $r = R_f(t)[1 + A(t)\mathbb{S}_n(\theta, \varphi)]$, where A is the amplitude of the disturbance and $\mathbb{S}_n$ is the surface harmonics with wavenumber n [2]. The logarithmic growth rate $\bar{\mathcal{A}}$ of the disturbance can be calculated by (1) according to the asymptotic analysis in [4, 7].

$$\bar{\mathcal{A}}(n, Pe) = \frac{d\ln A}{d\ln Pe} = \omega - \frac{\Omega}{Pe} \qquad (1)$$

A negative value of $\bar{\mathcal{A}}$ indicates a stable flame while a positive value an unstable one. In (1), the first coefficient ω represents the destabilizing effect of hydrodynamic instability. The second coefficient Ω represents the influence of diffusive-thermal instability and its detailed expression is given as $Q_1 - [\beta(Le_{eff} - 1)/(\sigma - 1)]Q_2 + PrQ_3$ [7]. Peclet number Pe is a nondimensional flame radius normalized by laminar flame thickness (ℓ_f). In general, the two coefficients depend on density ratio $\sigma = \rho_u/\rho_b$, wavenumber n, effective Lewis number Le_{eff}, Zeldovich number $\beta = E_a(T_b - T_u)/(R^o T_b^2)$ and Prandtl number Pr. To accurately determine the stability of methanol flame, these flame parameters must be properly extracted. Le_{eff} is calculated according to definition given by Bechtold et al. [19] and the Lewis number Le corresponding to the deficient reactant is weighted more heavily. The nondimensional Le and Pr are calculated by $\lambda/(\rho_u c_p D)$ and $\mu c_p/\lambda$, respectively. The overall activation energy is calculated according to $E_a = -2R^0[\partial \ln f^0/\partial(1/T_b)]_p$. The more basic parameters, such as unburnt gas density ρ_u, burnt gas density ρ_b, thermal conductivity λ, heat capacity c_p, molecular diffusion coefficient D, kinetic viscosity μ, burnt gas temperature T_b, laminar burning flux f^0, are all calculated using the one dimensional simulation results of adiabatic planar flames in Chemkin-Pro software [20]. Most recently, Zhang et al. (SJTU model) [17] and Burke et al. (NUI model) [21] developed detailed kinetic models of methanol based on laminar burning velocities, low pressure premixed flame, ignition delay time, jet-stirred reactor and literature data. The models were adopted in this work. The one dimensional simulation was performed with the Premixed Laminar Flame-Speed Calculation module in the Chemkin-Pro software. Mixture-averaged transport was selected and Soret effect was considered. Ending axial position was set to be 10 cm and the maximum number of grid points was set to be 1000 in order to eliminate the grid effect. The solution control was set to be CURV 0.1 and GRAD 0.1.

4 Results and discussion

4.1 Flame Morphology

Figure 1 compares the schlieren images during the propagating process of methanol/air flames at P_u = 5 atm. Flame kernels are greatly curvature in the early phase. No cellular structure appeared on flame surface except for few cracks induced by ignition. As flame propagates, the initial cracks start to develop along regions of high curvature and secondary crack forms. An influence on the flame propagation velocity is not observed in this phase. When the flame radius is larger than a critical value (critical flame radius, $R_{f,c}$), large amount of cellular structures simultaneously appear over the whole flame surface as shown in Fig. 1. Both the flame surface area and flame propagation velocity significantly increase once the flame becomes cellular. It is also observed that methanol/air flame becomes more instable as ϕ increases at ϕ < 1.4. However, the flame become more and more stable as the premixed mixture becomes richer at ϕ > 1.4.

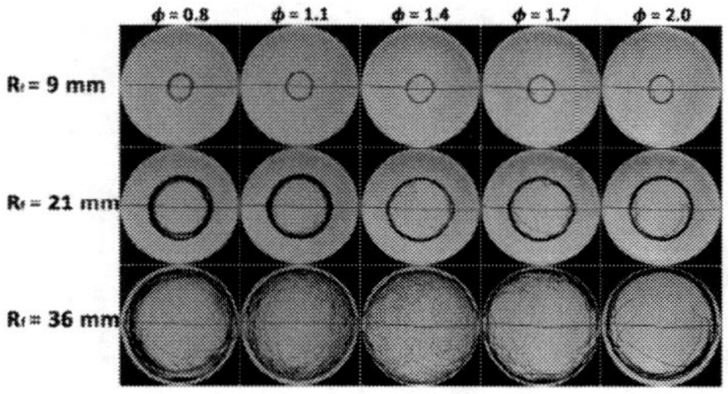

Figure 1: Schlieren images of methanol/air flames at T_u = 423 K and P_u = 5 atm

In order to further confirm the flame instability of methanol/air mixtures at extreme rich conditions, Figure 2 compares the recorded schlieren images at R_f ≈ 3.6 cm over wide ϕ and P_u ranges. The flame surfaces keep smooth and only few cracks appear at P_u = 1-2 atm. This indicates that methanol/air flames do not suffer from cellular instability at P_u ≤ 2 atm. As P_u increases, all the flames monotonically become more instable. More cellular structures are observed on the flame surface at higher pressures. Figure 2 also shows that the instability structures of methanol/air mixtures increase firstly and reach the most chaotic state at ϕ ≈ 1.4. Then the cellular instability is gradually suppressed with increasing ϕ. In other words, the investigated methanol/air flames become smoother and less cellular structure is observed in extreme rich flames.

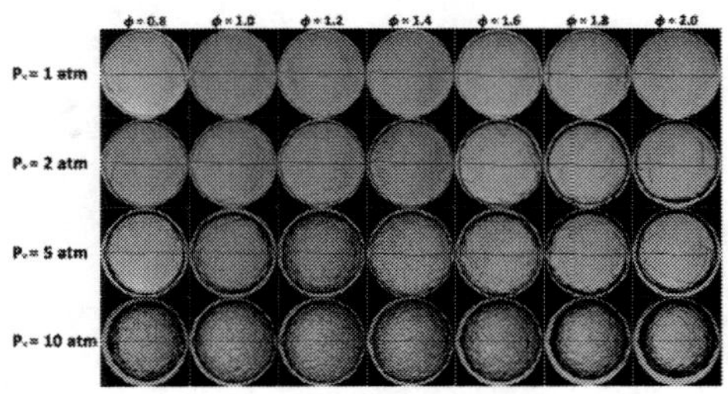

Figure 2: Schlieren images of methanol/air flames at T_u = 423 K and P_u = 1-10 atm

4.2 Critical Peclet Number

To interpret the non-monotonic cellular instability of methanol/air flames over a wide ϕ range, asymptotic analysis is calculated according to previous studies [4, 7]. The calculated $\bar{\mathcal{A}}$ of disturbance is plotted in Fig. 3 as an example. The light blue region in Fig. 3 represents stable region and the corresponding cellular structure does not occur in this region. When Pe number is small, the flame is stable and no cellular structure occurs. As flame propagates, there is a critical Peclet number (Pe_c) beyond which $\bar{\mathcal{A}}$ becomes positive. The nose of the peninsular instability determines the critical condition, namely Pe_c and n_c. Then the stretch effect becomes weaker and more cellular structures with different n appear on the flame surface. The cellular structure with moderate n has the largest $\bar{\mathcal{A}}$.

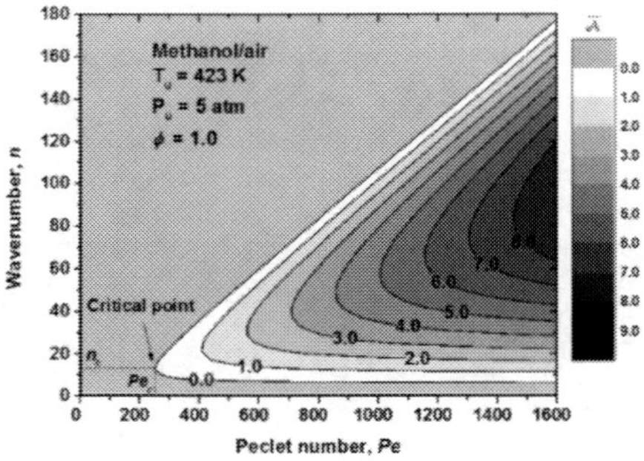

Figure 3: Calculated logarithmic growth rate of disturbance using the SJTU model [17] at different Pe and n. Negative values are not shown. The nose of the peninsular instability determines Pe_c and n_c.

The calculated Pe_c are plotted in Fig. 4 and their values decrease with increasing ϕ. Pe_c declines faster at $\phi < 1.4$ while the variation becomes very slow at $\phi > 1.7$. The nearly constant Pe_c at extreme rich conditions is very important for the non-monotonic cellular instability in Figs. 1-2. It is also observed that Pe_c is insensitive to pressure increase and its values at P_u = 1-10 atm are almost the same.

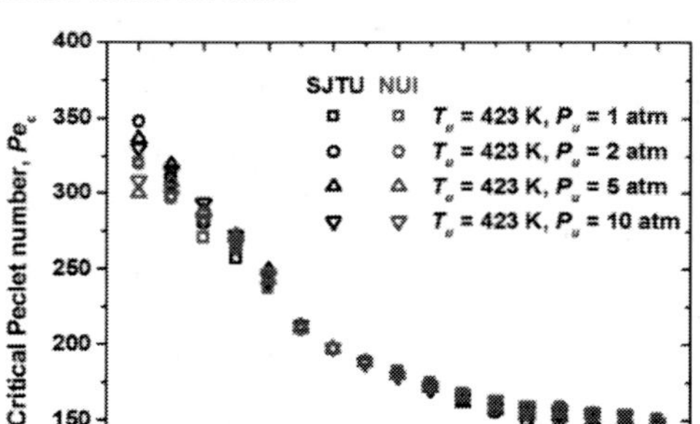

Figure 4: Critical Peclet number of methanol/air flames at T_u = 423 K, P_u = 1-10 atm and ϕ = 0.6-2.2.

4.3 Critical Flame Radius

Critical flame radius $R_{f,c}$ is the product of Pe_c and ℓ_f. As Pe_c is a nondimensional flame radius, the calculated $R_{f,c}$ is more valuable compared to Pe_c. In this work, the laminar flame thickness is calculated according to the definition $\ell_f = (T_b - T_u)/(dT/dx)_{max}$, where $(dT/dx)_{max}$ is the maximum temperature gradient across the flame front. Figure 5 shows that the calculated ℓ_f of methanol/air flames greatly decrease with increasing P_u. As ϕ increases, the values of ℓ_f firstly decrease to a minimum at $\phi \approx 1.3$, then they begin to increase. Figure 5 indicates that the ℓ_f of methanol/air flames changes non-monotonically. The product of the non-monotonic ℓ_f and monotonic Pe_c is also non-monotonic over a wide ϕ range.

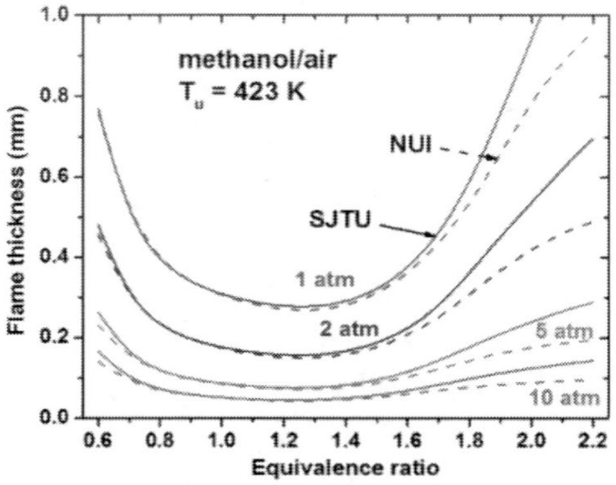

Figure 5: Calculated flame thickness of methanol/air mixtures at T_u = 423 K, P_u = 1-10 atm and ϕ = 0.6-2.2.

According to the neutral stability curves in Fig. 3, the critical flame radii are extracted and plotted in Fig. 6. The calculated $R_{f,c}^{cal}$ monotonically decrease with increasing P_u and non-monotonically change with increasing ϕ. There is a dramatic decrease of $R_{f,c}^{cal}$ on the lean side. The dramatic decrease results from the consistent decline of Pe_c and ℓ_f. At $\phi \approx 1.3$, $R_{f,c}^{cal}$ varies slowly because both Pe_c and ℓ_f are almost unchanged under these conditions. The minimum $R_{f,c}^{cal}$ locates at $\phi \approx 1.3$ which corresponds to the most instable flame. As methanol/air mixtures become more richer, Pe_c keeps the same but ℓ_f starts to enter its accelerating growth stage. The larger slope of ℓ_f at P_u = 1-2 atm determines that the atmospheric $R_{f,c}^{cal}$ increase faster. However, the increase of $R_{f,c}^{cal}$ becomes less apparent at pressures higher than 10 atm considering the small values of ℓ_f and its imperceptible variation.

According to the schlieren images in this work, the experimental critical flame radius ($R_{f,c}^{exp}$) can be extracted. If no cellular instability appears on flame surface, the flame propagation velocity vary relatively slowly and hence can be considered in quasi-steady phase [22]. Once the cellular structures fully developed on flame surface, a sharp increase of the recorded propagation velocity will be observed. This sharp increase can be easily recognized. It should be noted that $R_{f,c}^{cal}$ is the indicator at which flame first becomes unstable and cells start to grow. $R_{f,c}^{exp}$ is the indicator at which the formed cells start to increase flame surface area and propagating speed. Therefore, $R_{f,c}^{exp}$ should be larger than the one evaluated by the asymptotic analysis. Qualitative comparison is then performed for the asymptotic and experimental critical points. As might be expected, $R_{f,c}^{exp}$ is always larger than $R_{f,c}^{cal}$. The measured $R_{f,c}^{exp}$ also decreases with increasing P_u which agrees with the more unstable flame morphology of methanol/air flames at high pressures. The non-monotonic characteristic of $R_{f,c}^{exp}$ is also observed in Fig. 6 over a wide ϕ range. $R_{f,c}^{exp}$ decreases firstly to the minimum value at $\phi \approx 1.3$ and then increases with increasing ϕ. This is consistent with the smoother flame surface under extreme rich conditions.

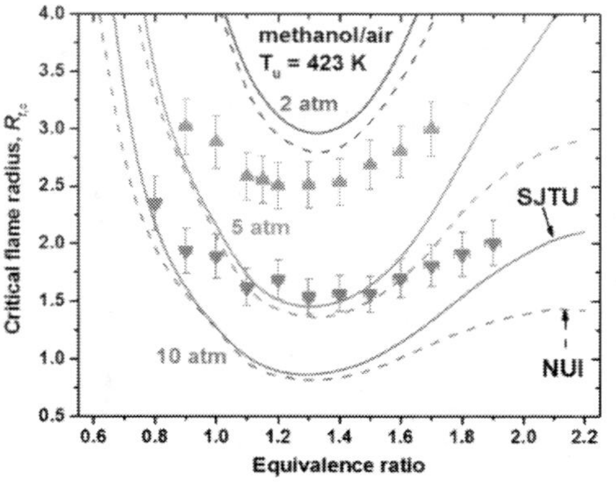

Figure 6: Comparison of experimental (symbols) and calculated (lines) critical flame radius of methanol/air flames at different pressures.

4 Conclusions

Considering the scarce studies on extreme rich fuel/air flames at elevated pressures, spherically expanding flames of methanol/air were investigated in a constant-volume single-chamber cylindrical combustion vessel at $T_u = 423$ K, $P_u = 1$-10 atm and $\phi = 0.7$-2.2. The flame morphology indicates that cellular instability structures monotonically increase with increasing P_u and non-monotonically vary over a wide range of ϕ. Asymptotic analysis was calculated to determine the logarithmic growth rate of instability and critical parameters. The decrease of flame thickness results in the decrease of critical flame radius and enhancement of cellular instability. The non-monotonic flame thickness and monotonic critical Peclet number result in the non-monotonic critical flame radius over a wide ϕ range.

5 Acknowledgment

The authors appreciate the funding supports from National Natural Science Foundation of China (91541201, 51622605, 91641205).

References

[1] G. I. Sivashinsky, Philos. T. Roy. Soc. A, 1990, 332 (1624): 135-148.

[2] M. Matalon, Annu. Rev. Fluid Mech., 2007, 39 (1): 163-191.

[3] A. G. Istratov, V. B. Librovich, Astronautica Acta, 1969, 14 (5): 453-467.

[4] J. K. Bechtold, M. Matalon, Combust. Flame, 1987, 67 (1): 77-90.

[5] D. Bradley, Philos. T. Roy. Soc. A, 1999, 357 (1764): 3567-3581.

[6] D. Bradley, C. Harper, Combust. Flame, 1994, 99 (3-4): 562-572.

[7] R. Addabbo, J. K. Bechtold, M. Matalon, Proc. Combust. Inst., 2002, 29 (2): 1527-1535.

[8] J. Beeckmann, R. Hesse, S. Kruse, A. Berens, N. Peters, H. Pitsch, M. Matalon, Proc. Combust. Inst., 2016:

[9] G. I. Sivashinsky, Combust. Sci. Technol., 1977, 15 (3-4): 137-145.

[10] O. C. Kwon, G. Rozenchan, C. K. Law, Proc. Combust. Inst., 2002, 29 (2): 1775-1783.

[11] C. Tang, Z. Huang, J. Wang, J. Zheng, Int. J. Hydrogen Energy, 2009, 34 (5): 2483-2487.

[12] X. Gu, Z. Huang, S. Wu, Q. Li, Combust. Flame, 2010, 157 (12): 2318-2325.

[13] X. Wu, Z. Huang, X. Wang, C. Jin, C. Tang, L. Wei, C. K. Law, Combust. Flame, 2011, 158 (3): 539-546.

[14] X. Zhang, C. Tang, H. Yu, Z. Huang, Energy Fuels, 2014, 28 (3): 2258-2266.

[15] M. Baloo, B. M. Dariani, M. Akhlaghi, M. AghaMirsalim, Fuel, 2016, 170: 235-244.

[16] L. Qiao, Y. Gu, W. J. A. Dahm, E. S. Oran, G. M. Faeth, Proc. Combust. Inst., 2007, 31 (2): 2701-2709.

[17] X. Zhang, G. Wang, J. Zou, Y. Li, W. Li, T. Li, H. Jin, Z. Zhou, Y.-Y. Lee, Combust. Flame, 2017, 180: 20-31.

[18] S. M. Sarathy, P. Oßwald, N. Hansen, K. Kohse-Höinghaus, Prog. Energy Combust. Sci., 2014, 44 (0): 40-102.

[19] J. K. Bechtold, M. Matalon, Combust. Flame, 2001, 127 (1–2): 1906-1913.

[20] (2009) Chemkin-Pro 15092, Reaction Design: San Diego.

[21] U. Burke, W. K. Metcalfe, S. M. Burke, K. A. Heufer, P. Dagaut, H. J. Curran, Combust. Flame, 2016, 165: 125-136.

[22] A. P. Kelley, C. K. Law, Combust. Flame, 2009, 156 (9): 1844-1851.

11th Asia-Pacific Conference on Combustion,
The University of Sydney, NSW Australia
10th -14th December 2017.

Microscopic Characterization of the Near-nozzle Field Diesel Sprays

Yijie Wei, Tie Li*, Bin Wang

Collaborative Innovation Center for Advanced Ship and Deep-Sea Exploration, State Key Laboratory of Ocean Engineering, Shanghai Jiao Tong University. 800 Dongchuan RD. Minhang District, 200240, Shanghai, China

Abstract

To further understand fundamental mechanism of mixture formation in diesel engines, the high speed CMOS camera with a long distance microscopy and the high frequency diode laser were employed to record the near-nozzle field injection processes under both non-evaporating and evaporating diesel-like conditions. At the room temperature condition, small droplets were observed at the spray tip before the end of injection and along the jet boundaries during the quasi-steady injection process. During the end-of-injection transients, while there were some poorly-atomized large droplets and ligaments with low penetration velocity, the mist-like tiny droplets suspending at the vicinity of the nozzle dominated the droplet number distribution. Under the 4.0 MPa and 900 K ambient condition, no droplets and ligaments at the spray tip and along the jet boundaries were clearly observed before the end of fuel injection. During the end-of-injection transients, owing to the low penetration velocity and sparse distribution of the fuel, some droplets were observed, indicating the surface tension confining the liquid to a sphere shape; at the same time, however, some cotton-like structures were also observed, implying the diminishment of the surface tension of the fuel.

1 Introduction

The well prepared mixture of fuel and ambient gases enable designers to achieve considerable improvements of the combustion process in terms of reduced fuel consumption and engine raw emissions[1, 2]. Establishment of a suitable model to predict the spray breakup and mixture formation is one of the most significant challenges. The primary breakup of the spray results in large ligaments and droplets that form the dense region near the nozzle. It is very important to study the atomization and mixture formation in the near nozzle field under different operating conditions.

A lot of experimental studies of the mixture preparation in the near-nozzle region was operated below the critical points of the liquid fuel. Nowadays, with the intake boosting technology application in the diesel engines[3], fuel may be injected into the combustion chamber at conditions close or even beyond the fuel critical point in the future. If a liquid fuel were injected into the systems where the working fluid exceeded the thermodynamic critical condition of the liquid phase under some conditions, the surface tension forces would started to diminish so that the classical low-pressure gas-liquid interface was replaced by a diffusion dominated mixing layer.[4] For supercritical injection, B. Chehroudi [5] carried out earlier experimental studies, the jets they investigated behave similar to an incompressible variable-density gas jet when the ambient pressure reaches and exceeds the critical pressure of the injectant. It has been well recognized in the liquid rocket community for years. Some researchers started to study the effects of supercritical phenomena in diesel engines.

Dahms et al.[4]injected n-dodecane into a constant volume combustion chamber. At constant chamber density and supercritical fuel pressures, liquid structures affected by surface tension such as ligaments and drops develop at low ambient temperatures. But at high-temperature conditions, a different, more diffusive mixture preparation process without measurable fuel drops was found. Contrary to conventional wisdom, they revealed that gas-liquid interfacial diffusion layers develop, not necessarily because of vanishing surface tension forces, but owing to the combination of a reduction in mean free molecular path and broadening interfaces, which then enter the continuum length scale regime.

Manin et al. [6] investigated the atomization and mixing processes of dodecane sprays injected into environment with progressively higher pressure and temperature and got similar results with Dahms.

Shinabuth[7] et al. injected a mixture into a rapid compression and expansion machine (RCEM) to determine the spray cone angle at vicinity of the nozzle. They found if the ambient temperature went beyond a critical temperature of the used fuel, the cone angle was considered independent of ambient pressure. They supposed this was possibly caused by surface tension diminishment.

From the above analysis we can see that researches on spray mixture formation under supercritical conditions were carried out using alternative fuels. In this work the diesel fuel (Commercial #0 diesel of China) was injected into a constant-volume combustion chamber at both non-evaporating and evaporating conditions to research the surface tension effects on the spray structure.

2 Experimental setup

The experiments of this work were conducted on a constant-volume combustion chamber with optical access through sapphire windows under diesel engine-like conditions. At the same time, a set of high-speed microscopic imaging system was established to take the near-field images of the spray.

2.1 Combustion chamber and microscopic imaging system

The combustion chamber used to simulate the desired ambient conditions employed two sapphire windows with a diameter of 90mm for optical access. The constant volume combustion chamber and the optical arrangements of near-nozzle field microscopic imaging system are shown in Fig. 1. Before spray

*Corresponding author. Fax: +86-21-3420-8348
E-mail address: litie@sjtu.edu.cn

injection, premixed mixture of ethylene (C_2H_4), nitrogen (N_2) and oxygen (O_2) with the proper fraction was charged into the combustion chamber, and then ignited by two spark plugs. After the mixture of the specified density was ignited, the pressure decreased gradually after a quick rise owing to the heat transfer through the chamber wall. A K type thermocouple was used to measure the mixture temperature prior to ignition and a piezo-electric transducer (KISTLER 6125C) coupled with a charge amplifier (KISTLER 5018A) was used to record the dynamic pressure in the combustion chamber after ignition. A mixing fan with a speed of 5000 rpm was adopted to reduce the temperature inhomogeneity in the chamber before fuel injection. The fuel injection pressure was provided by a motor-driven common rail system. When the targeted ambient pressure was reached, the fuel injection, high-speed camera was triggered by a precise in-house made mechanism based on the pressure feedback. [8]

In this paper, the spray images were taken by means of shadowgraph technique. As shown in Fig. 1, the microscopic imaging system consists of three main equipment, the pulsed-diode laser with a 50 ns pulse duration is used as light source to freeze the motion of spray. A high-speed CMOS camera coupled with a long-distance microscope is used to record the near-nozzle field spray images. The pulse laser beam generated by the puled-diode laser device passed through the sapphire windows and illuminated the spray injected vertically from the top of the combustion chamber. The microscope objective was configured for a magnification of 1.6×. Based on the magnification and the camera pixel size, the final optical resolution was 12.5 um/pixel. The imaging parameters used in this work are listed in Table 1.

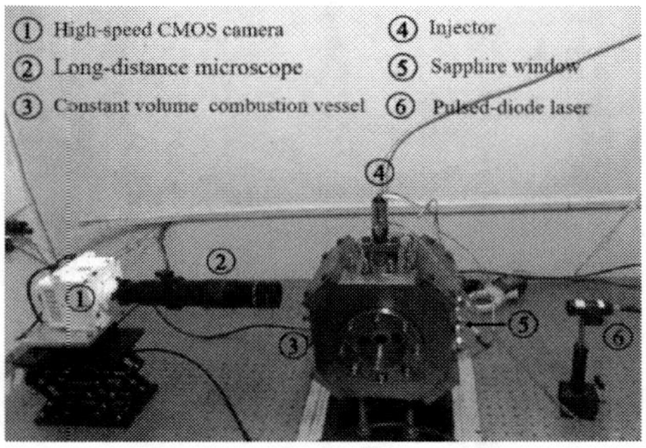

Fig. 1: Optical arrangements of near-nozzle field microscopic imaging system.

Light source	CAVILUX HF
Wave length	810 nm
Repetition rate	30 kHz, 60 kHz
Pulse duration	50 ns (FWHM)
High-speed camera	Phantom v710, v2012
Frame rate	30 kHz, 60 kHz
Exposure time	1μs [a]
Resolution	272*448, 384*720
Long-distance microscope	K2 DistaMax
Working distance	17 cm
Magnification	1.6×

a. Although the camera exposure time was set to 1 μs, the illumination system pulse duration determines the exposure timescale (50 ns).

Table 1: Imaging parameter settings.

2.2 Experimental conditions

At room temperature condition (1.4 MPa, 293 K and 15 kg/m³), the combustion chamber was charged with nitrogen to obtain specified density. At 4.0 MPa, 900 K and 15 kg/m³ ambient condition, to create a high ambient pressure and temperature environment at the beginning of spray injection, a premixed mixture of ethylene (C_2H_4), nitrogen (N_2) and oxygen (O_2) with the proper fraction was charged into the combustion chamber, and then ignited by two spark plugs. The volume fraction of C_2H_4, N_2 and O_2 was 3.5%, 10.5% and 86%, respectively. The chamber wall was heated up to 443K to avoid water contaminating on the windows. Table 2 shows the specifications for the injection system and the experimental conditions conducted in this work.

Injector type	Common-rail solenoid injector
Injection pressure [MPa]	120
Nozzle type	Mini sac
Orifice number	1
Orifice diameter [mm]	0.28
Injection duration [ms]	3
Ambient density [kg/m³]	15
Ambient temperature [K]	293K, 900K
Ambient oxygen concentration	0

Table 2: Specifications for injection system and experimental conditions.

3 Results and discussion

During the fuel injection, fuel enters the combustion chamber at a speed of several hundred meters per second, and the jet breaks up according to the mechanism of the atomization regime. Immediately after leaving the nozzle exit, the jet starts to spread out and break up into a conical spray, which is caused by the factors such as aerodynamic, cavitation, turbulence and liquid velocity relaxation. A dense spray filled with large ligaments and droplets is formed near the nozzle. The microscopic details of the ligaments or droplets can be observed at the end of injection or at the edge of spray.

3.1 Sequential images of near-nozzle field spray

Fig. 2 shows two sequential images of the fuel injection. As shown in Fig. 2 (a), the spray is injected into a room temperature ambient condition (Ambient pressure 1.4 MPa, ambient temperature 293 K and ambient density 15 kg/m³). For this low temperature and pressure condition, fuel was injected at low-subcritical pressures, atomization occurs forming a distinct spray. It can be seen that the spreading angle of high-pressure diesel sprays is wider at the start of injection than during the quasi-steady period. The relatively low lift of the injector needle during the start of injection transients throttles the fuel passage through the needle seat to the orifice. It is the same for the high temperature experiment in Fig.2 (b). Small droplets are observed at the spray tip at the end of injection and along the jet boundaries during the quasi-steady injection process. Particle size distribution near the nozzle during the end of injection is shown in Fig. 4. During the end-of-injection transients, while there are some poorly-atomized large droplets and ligaments with low penetration velocity, the mist-like tiny droplets with diameter around 14 micrometer suspending at the vicinity of the nozzle dominate the droplet number distribution, as shown in Fig. 3 and Fig. 4.

Fig. 2(a) shows that the effect of surface tension against the aerodynamic forces responsible for the breakup of jet. Large fuel ligaments can be seen leading to the formation of droplets with relatively large diameter (up to 200 μm).

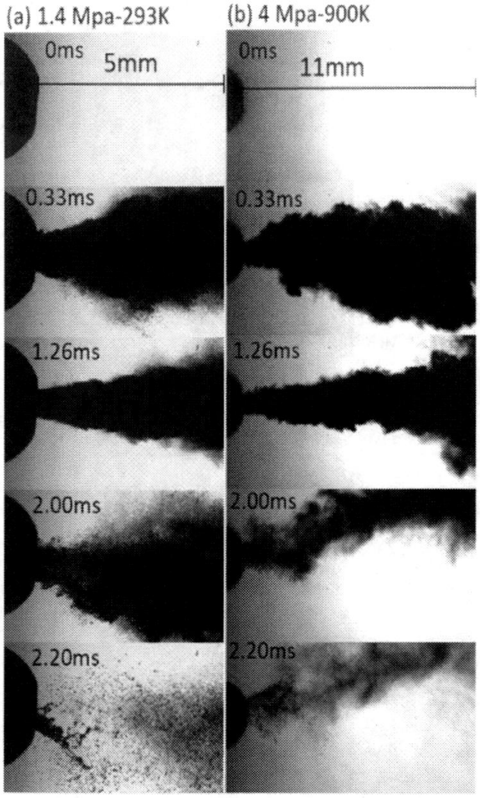

Figure 2: Sequence of spray images and time after start of injection.

As shown in Fig. 2(b), the spray is injected into the 4 MPa, 900K and 15 kg/m³ ambient condition, no droplets or ligaments at the spray tip and along the jet boundaries is clearly observed during the quasi-steady period. At the end of fuel injection, as can be seen from Fig. 2(b)-2.20ms and Fig. 5, there is no obviously gas-liquid interface at the edge of the droplet-like fuel, but a blurred mixing state. This may be caused by combustion chamber condition exceeding the critical point of the liquid fuel. For this situation, the injected diesel jet exhibits diffusion dominated mixing processes.

As shown in Fig. 5, during the end-of-injection transients, owing to the low penetration velocity and sparse distribution of the fuel, some droplets are observed, indicating the surface tension confining the liquid to a sphere shape; at the same time, however, some cotton-like structures are also observed, implying the diminishment of the surface tension of the fuel. Droplet-like fuel and cotton-like structures coexist in the combustion chamber is consistent with the droplets behavior in supercritical fluids. In the research of Givler [9] the surface tension was clearly still in effect during the mixing of fuel and oxidizer in liquid-rocket motors, despite the fact that the sprays were injected into an environment that exceeded the supercritical point with respect to the fuel. During the spray injection, the fuel with a relatively low temperature was injected into the high ambient temperature and pressure chamber, the fuel–ambient mixture temperatures will range from the liquid fuel temperature (below the critical point of the fuel injected) to ambient gases temperature (above the critical point the liquid fuel). Because of this transcritical nature of the mixing process, droplets have been observed in the high temperature condition as shown by Fig. 5.

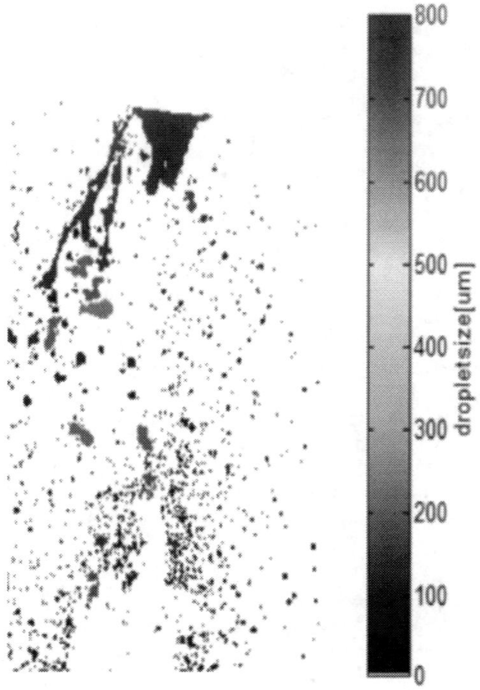

Figure 3: Droplet size distribution of 2.25 ms after start of injection under room temperature condition.

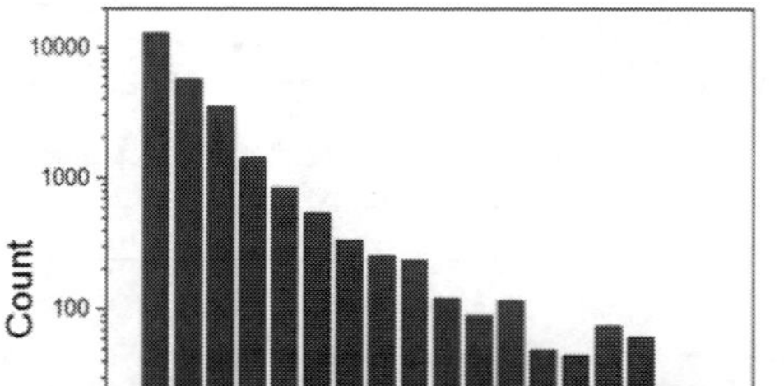

Figure 4: Particle size distribution near the nozzle during the end of injection.

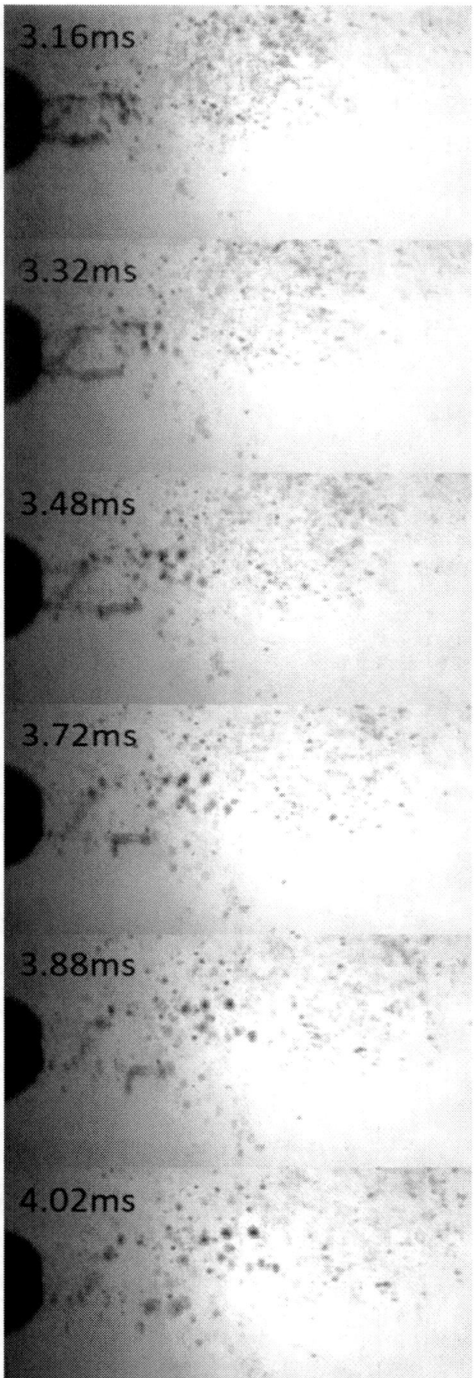

Figure 5: Sequence of spray images during the end of injection at the high temperature condition.

4 Conclusions

High-speed microscopic imaging experiments have been conducted on diesel sprays under both non-evaporating and evaporating diesel-like conditions using 0# commercial diesel fuel.

At the room temperature condition, small droplets were observed along the jet boundaries and at the end of injection, which confirmed the effects of surface tension and showed the effect of surface tension against the aerodynamic forces responsible for the breakup of jet. During the end-of-injection transients, while there were some poorly-atomized large droplets and ligaments with low penetration velocity, the mist-like tiny droplets suspending at the vicinity of the nozzle dominated the droplet number distribution.

At the high temperature condition, no droplets and ligaments at the spray tip and along the jet boundaries were clearly observed during the quasi-steady period, the diesel jet exhibited diffusion dominated mixing processes. During the end-of-injection transients, owing to the low penetration velocity and sparse distribution of the fuel, some droplets were observed, indicating the surface tension confining the liquid to a sphere shape; at the same time, however, some cotton-like structures were also observed, implying the diminishment of the surface tension of the fuel.

5 Acknowledgment

The Supports by Natural Science Foundation of China (51276115&91541104) are gratefully acknowledged.

References

[1] S. H. Park, S. H. Yoon, and C. S. Lee, "Effects of multiple injection strategies on overall spray behavior, combustion, and emissions reduction characteristics of biodiesel fuel," Applied Energy, vol. 88, no. 1, pp. 88-98, 2011.

[2] L. M. Pickett, J. Manin, C. L. Genzale, D. L. Siebers, M. P. Musculus, and C. A. Idicheria, "Relationship between diesel fuel spray vapor penetration/dispersion and local fuel mixture fraction," SAE International Journal of Engines, vol. 4, no. 2011-01-0686, pp. 764-799, 2011.

[3] H. Osada, Y. Aoyagi, K. Shimada, Y. Goto, and H. Suzuki, "Reduction of NOx and PM for a Heavy Duty Diesel Using 50% EGR Rate in Single Cylinder Engine," SAE Technical Paper0148-7191, 2010.

[4] R. N. Dahms, J. Manin, L. M. Pickett, and J. C. Oefelein, "Understanding high-pressure gas-liquid interface phenomena in Diesel engines," Proceedings of the Combustion Institute, vol. 34, no. 1, pp. 1667-1675, 2013.

[5] B. Chehroudi, D. Talley, and E. Coy, "Visual characteristics and initial growth rates of round cryogenic jets at subcritical and supercritical pressures," Physics of Fluids, vol. 14, no. 2, pp. 850-861, 2002.

[6] J. Manin, M. Bardi, L. M. Pickett, R. N. Dahms, and J. C. Oefelein, "Microscopic investigation of the atomization and mixing processes of diesel sprays injected into high pressure and temperature environments," Fuel, vol. 134, pp. 531-543, 2014.

[7] S. S. Shinabuth D, Kosaka H, "Diesel spray at Supercritical Conditions," JSAE20155272, 2015.

[8] T. Li, X. Q. Zhang, B. Wang, T. Guo, Q. Shi, and M. Zheng, "Characteristics of non-evaporating, evaporating and burning sprays of hydrous ethanol diesel emulsified fuels," Fuel, vol. 191, pp. 251-265, 2017.

[9] S. D. Givler and J. Abraham, "Supercritical droplet vaporization and combustion studies," Progress in Energy and combustion Science, vol. 22, no. 1, pp. 1-28, 1996.

11th Asia-Pacific Conference on Combustion,
The University of Sydney, NSW Australia
10th -14th December 2017.

Modelling and Analysis of Gas Feeding System Dynamics for Control of Constant Volume Combustion Rig Boundary Conditions

Vladimir Krivopolianskii[1], Sergey Ushakov[1], Vilmar Æsøy[1], Maximilian Malin[2], Eilif Pedersen[1]

[1]Norwegian University of Science and Technology (NTNU)
Otto Nielsens veg 10, Trondheim, 7046, Norway
[2]Graz University of Technology
Inffeldgasse 19/II, Graz, 8010, Austria

Abstract

For focusing on large-scale marine injection systems, a unique research facility has been developed based on existing Constant Volume Combustion Rig (CVCR) available in Marine Machinery laboratory of NTNU. In order to increase operation flexibility, associated with tolerance of wide pressure range during pre-burn phase, a number of modifications has been introduced, particularly those related to gas feeding process.

This study was intended to investigate a gas dynamics during the sequential gas feeding into a large-volume CVCR prior combustion. The main interest is in prediction of the gas composition in the rig by the end of gas feeding process. For these needs, a detailed bond graph model was created and simulation of the gas filling system has been performed. The validity of the model has been achieved by the tuning of the model using the obtained experimental data. The results showed the applicability of the bond graph approach for such systems and possibility to predict the gas composition in CVCR and eventually find the optimal operational profile regardless of used gas composition and initial conditions of the system.

1 Introduction

Operation of a pre-burn type of constant volume chamber implies burning of a suitable gas mixture to reach high-pressure and high-temperature ambient conditions. Examined fuel injection process occurs at certain time during the cooling down phase with the corresponding ambient thermodynamic property [1-3] .

The Constant Volume Combustion Rig, available in NTNU/SINTEF Marine Machinery laboratory is a large volume cell that is used to assess the performance of marine-type multi-hole nozzle injectors at high-temperature and high-pressure conditions. The first study using current CVCR was performed by Malin et. al[4] and showed chamber's research potential for testing of diesel-like fuels, however the experimental setup indicated certain pressure/density limitations (about 100 bar and 12 kg/m^3) that restricted applicability of the obtained results. To overcome this limitation a number of modifications of CVCR body were introduced enabling it to withstand enhanced load conditions, while expending its volume from 4.9 to 6.8 dm^3, and providing additional rig body heating. The enlargement of the CVCR was motivated by possible increase in a spray penetration length and was achieved by increasing the inner body length from 100 to 140 mm. At the same time, the reasons of heating system implementation include: reduction of heat losses to the chamber walls, reduction of temperature gradient after pre-combustion phase, mitigation of condensation on windows and resembling engine combustion chamber walls temperature. Moreover, in order to increase the efficiency of installed heating elements and to maintain desired temperature level of the rig body, an external insulation was additionally applied. Taking into account the increased maximum working pressure of CVCR (around 270 bar), the gas feeding procedure has been modified from fully premixed model to so-called hybrid mixing that combine both premixed and sequential gas feeding in order to increase the safety by keeping combustible and oxidizing gases in separate bottles before feeding them into the combustion rig. In its turn, the introduction of sequential gas filling processes together with preheating of the combustion cell created a new challenge associated with control of gas composition by the end of filling. According to the study of Baert, Frijters [1], chemically pre-heated combustion rigs with sequential gas filling model have significant drawback associated with relatively long time interval between the measurements. Figure 1 outlines the main components of CVCR and its subsystems.

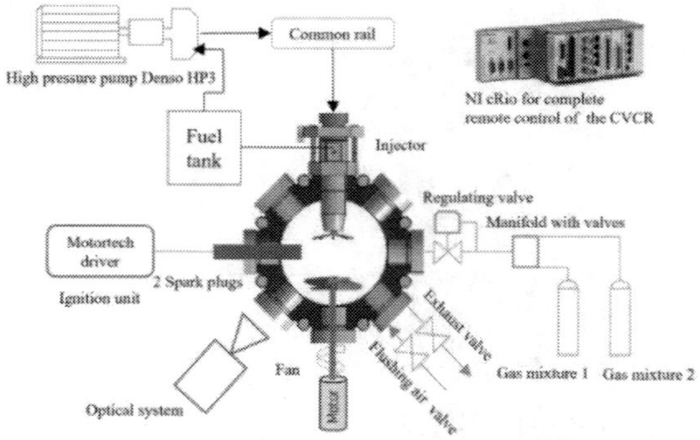

Figure 1. Schematic of the Constant Volume Combustion Rig

The main goal of current work was to solve the above-mentioned problem by a change in operation procedure of CVCR, namely by tuning of the gas feeding process, in order to reach the required gas composition without need to wait for adjustment of gas to rig walls temperature [1]. This is achieved by the detailed study of the gas filling process dynamics and finding appropriate control settings to reach high tests accuracy and repeatability.

2 Modelling of gas filling system

Referring to the paper of Baert, Frijters [1], that declares around 300-360 seconds per gas charging, and taking into account main

Corresponding author. Tel: +47-73-59-50-00
E-mail address: Vladimir.Krivopolianskii@ntnu.no

design differences between current CVCR and combustion cell in the reference, it becomes obvious that some operational changes are needed to keep the period of CVCR working cycle as short as possible, while retaining full control over the composition of a gas mixture. Solution to this problem was found in detailed studying of system thermodynamics and finding appropriate control settings of the gas filling process. The detailed simulation was performed using the bond graph methodology that ensures flexibility in modelling of physics in multidisciplinary domain, particularly in fluid mechanics and thermodynamics that are very relevant for the current study [5, 6]. Conceptual bond graph model of Pedersen [7], which is closely related to one of Karnopp [8], is used here as a basis for modelling the gas system feeding system of considered CVCR.

A simplified gas system flow diagram and block diagram of pseudo-bond graph model, that represents the physics during gas filling sequence, are presented respectively in Fig.2 and 3 (check valves, pressure regulators and filters are not shown). The influence of check valves and filters on system dynamics has been simulated by placing of anisotropic nozzle with corresponding flow characteristics, while regulators, that main purpose is to maintain a constant pressure in gas lines, were realized in the model by "infinite" big volumes of the gas bottles. In order to reflect phenomena of heat exchange between the compressed gases and air surrounding a pipe, empirical relations for convection both in laminar and turbulent conditions were implemented based on study of Gnielinski [9]. The simplified system consists of two 50 l gas bottles with oxidizing mixture ("OM") and combustible mixture ("CM") respectively; two ball valves to open/close gas lines according to an operation sequence; 10 mm diameter piping; a regulating needle valve (Hoke 2335F4Y) with electrical actuator (0121F2E) and CVCR.

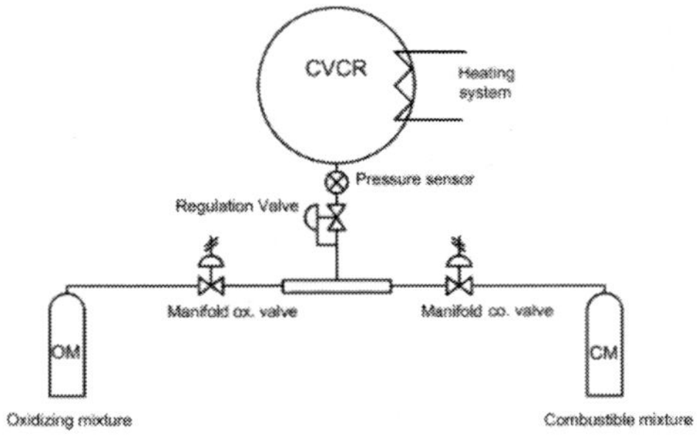

Figure 2. Simplified gas filling system

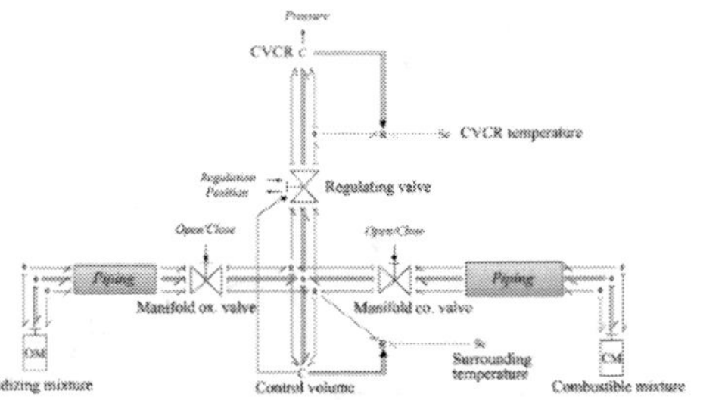

Figure 3. Physical model of the gas filling system

The operation of the gas system follows the subsequent control system rules:

- The same regulating valve is operated for feeding both combustible and oxidizing mixtures . During operation, pressure inside the CVCR increases in two equal increments i.e. 5 bar + 5 bar.
- The regulating valve is controlled by a digital actuating signal with rig set pressure - valve opening area correlation according to an interpolation of values presented in Table 2.
- A deadband for the regulating valve operation was specified as 0.4% of maximum opening of the valve.

An implementation of above-listed features of the control system could be seen in Fig.4.

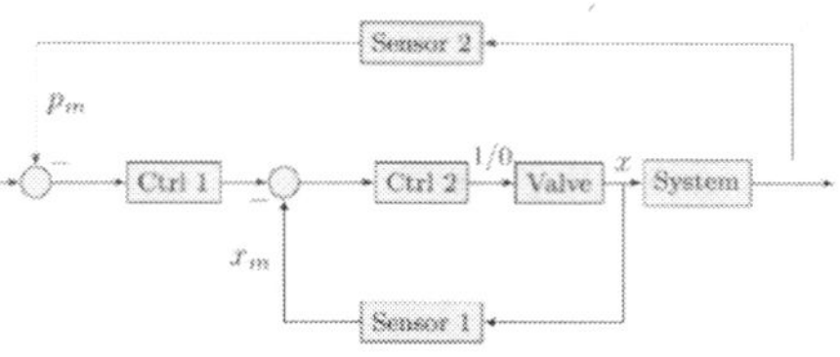

Figure 4. Control system of the gas system

Here "Ctrl1" block is used for the interpolation, "Ctrl2" is a deadband block, "Valve" takes into account the opening characteristics of the regulating valve, while "System" represents dependence of the regulating valve opening on pressure in CVCR. Signal delays have been fitted into "Sensor 1" and "Sensor 2" blocks. The pre-burn of the gas in the CVCR can described by the following global chemical reaction:

$$aCO + bO_2 + cN_2 \rightarrow aCO_2 + cN_2 + (b - 0.5a)O_2 \quad (1)$$

Reactants used in equation (1) allow avoiding possible water condensation on optical windows during experiments and ensure the required oxygen fraction in products. Considering 21% of oxygen in products of chemical reaction (1), the separation of combustible and oxidizing mixture should be: CO = 49.6 %v, N_2 = 50.4 %v in "CM" bottle and O_2 = 62.2 %v, N_2 = 37.8 %v in "OM" gas bottle.

3 Results

3.1 Tuning of the model

In order to take into account thermal and flow coefficients not only of CVCR, but also of gas pipes, check valves and filters, a preliminary model tuning was required. For this purpose, a study case with the following boundary and initial conditions has been selected:

- The pressure for gas regulators (near the gas bottles) was set to 29 bar and 18 bar for line with "OM" and "CM" respectively.
- "CM" gas bottle was filled with pure nitrogen, while "OM" gas bottle was filled with mixture containing 42 %v O_2 and 58 %v N_2 (by volume).
- The gas composition in the CVCR by the end of filling process (heating is not applied) is expected to be similar to that for air.

The list of instrumentation, used for measuring system's flow dynamics, is given in Table 1. The gas analyzer that measured O_2 content in a gas mixture after the filling process was also used to estimate the gas composition in CVCR.

Even though the gas regulators were set to hold constant pressure in the gas lines, hence reducing the cycle-to-cycle variability in initial conditions, the performed measurements for the tuning of the model have revealed certain differences in

pressure after gas regulators in both gas lines. The initial pressure appeared to be lower than the set pressure both in "CM" gas pipe 24.3 bar vs. 29 bar and in "OM" line, 16 bar vs.18 bar in the course of 11 tests (refer to Fig. 5). Therefore, calibration of the system has been made, referring to a cycle with approximately average conditions (based on 11 tests): 26.8 bar in "OM" gas line and 17 bar in "CM" line.
The comparison of simulation model results with those obtained from calibrating measurements is shown in Fig.6 for regulating valve opening area and pressure and temperature in the CVCR.

During the tuning procedure the rate of pressure increase in CVCR, the opening of the regulating valve and oxygen content in the chamber after the filling have been chosen as a main parameters. Here the "Deviation", given by equation (2), was used as criterion for tuning accuracy. The absolute values of temperature in CVCR were not taken into account; however, some considerations have been made regarding the temperature dynamics.

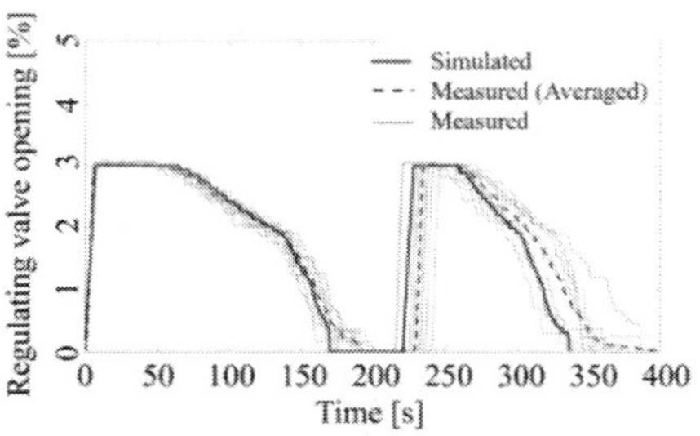

Figure 5. Tuning: Comparison between simulated regulating valve opening and 11 measured cycles.

The main argument here is spatial temperature variance inside CVCR and difficulties of bulk temperature estimation. Figure 6a shows perfect match of tuned regulating valve behavior with one acquired during experiment under the same boundary and initial conditions. The same could be addressed to the pressure in CVCR during the feeding process, while it is possible to notice some discrepancy between the temperatures. Measured oxygen content after gas feeding, with 95% confidence interval, is 21.27 ± 0.31, while in simulations it came to 21.1% with deviation of 0.2%.

$$Deviation = 1 - \frac{\int V.opening_{sim}}{\int V.opening_{mes}} \quad (2)$$

Pressure inside CVCR	Kulite HEM 375
Pressure after regulating valve	GE PM5076-TA-A3-CA
Pressure in oxidizing mixture line	PTX1400 (0-100 bar)
Pressure in combustible mixture line	UNIK 5000 (0-70 bar)
Gas analyzer	Horiba PG-250

Table 1. List of instrumentations

3.2 Validation of the model

The normal operation condition of the CVCR corresponds to a feeding gas pressure of 15 bar at 60 °C that yields in 15.6 kg/m^3 of ambient density. In contrast to the tuning work (performed before), here the combustion chamber was filled with combustible gas mixture and oxygen content was measured after the pre-burn phase. The procedure of oxygen measurement was changed due to limitations of the used gas analyzer (maximum O_2 = 25 %v). To verify the gas feeding model and to find optimum system settings, experiments and simulations were performed under 3 different regulating valve modes with 6 repetitions each (refer to Table 2).
The main results, where simulations are compared with the experimental data, are shown in Fig.7-8. Due to the fact that during "Very fast" valve opening mode, the pressure in the combustion rig increased significantly above the set point, the obtained results were not considered in this study. Regulating valve opening during "Slow" mode, shown in Fig.7, is characterized by certain delay of second phase, where oxidizing mixture is led to the combustion chamber. This occurred due to small swing in first phase when pressures were set into balance, since reference pressure sensor for filling process is mounted close to, but still outside of the heated CVCR.

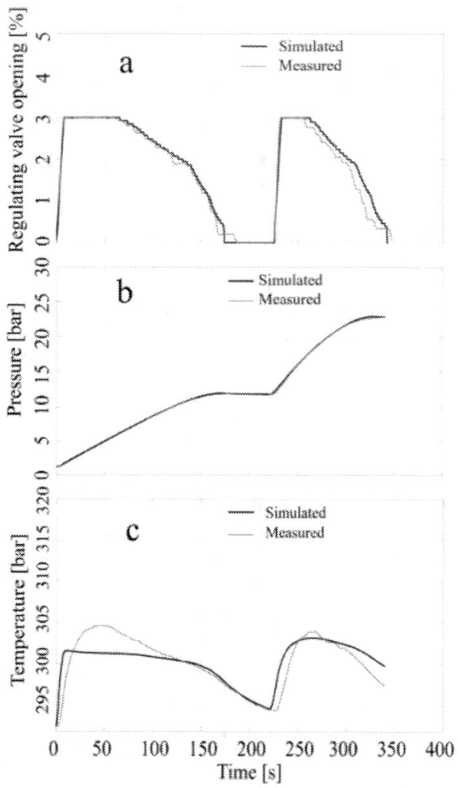

Figure 6. Tuning: Comparison between measured and simulated parameters (a is regulating valve opening; b is Pressure inside CVCR and c is Temperature inside CVCR)

Pressure difference, [bar]	"Slow" Valve opening, [%]	"Fast" Valve opening, [%]	"Very Fast" Valve opening, [%]
10	3	6	12
5	2.5	5	10
1	1.5	3	6
0.5	1	2	4
0.25	0.75	1.5	3
0	0	0	0

Table 2. Settings of the regulating valve for different modes

Moreover, a relatively well-differentiated regulating valve opening in "Slow" mode under small pressure deviation between set point and actual measured pressure inside CVCR (see Table 2) has being equilibrated until the valve deadband was not reached. As a result these delays led to only 5% deviation based on equation (2), while the entire filling cycle took not more 300 sec, which is similar to Baert, Frijters [1] when difference in rig volumes is taken into consideration. In contrast to "Slow" mode, the "Fast" mode, as seen, is not creating delays, however it is characterized by overshoot of the regulating valve opening, unfulfilment of predefined settings during the complete valve profile and a big discrepancy according to "Deviation" parameter. In addition to that, "Fast" mode did not provide an accurate partial pressure control during gas filling, as valve settings appeared to be relatively rough for small pressure differences.

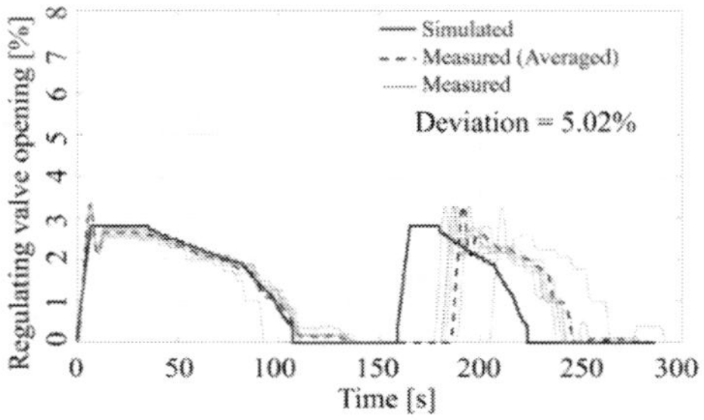

Figure 7."Slow" valve opening mode. Comparison between experiments and simulation

The values of the measured oxygen content after the gas combustion and simulated oxygen content by the end of a filling process (all with 95% CI) can be summarized as follows: "Slow" mode, measured is 20.8 ± 0.28, "Slow" mode, simulated is 31.11, "Fast" mode, measured - 20.36 ± 0.5, and "Slow" mode, simulated - 31.02.

Results of oxygen measurements show that change from "Slow" operating mode to the "Fast" one is associated with reduction of oxygen content in both experiments and simulation studies. Experimental results emphasize a relatively poor control of gas composition during "Fast" mode due to rough control settings and partial pressure overshoot during gas feeding. It also could be noticed that oxygen contents obtained from simulations are almost identical. This could be explained by idealized (tuned - like) behavior of the regulating valve.

5 Conclusion

Despite the fact that the considered simulation model demonstrates some deviation from experimental results, the study clearly shows its applicability for prediction of gas filling system dynamics for pre-burn-type constant volume combustion chambers. Moreover, the model with appropriate and comprehensive tuning can be very useful in achieving its main goal - estimation of the optimum settings for regulating valves and hence for reduction of period between the measurements. Finally, it should be highlighted that some further work is required (and is planned in future) to increase fidelity level of the existing model:

1. Implement detailed regulating valve dynamics that represents its entire operational range.
2. Study the conjugate heat transfer in the CVCR.
3. Implement more accurate flow models that take into consideration geometrical features of the used hardware.
4. Select more representative operating conditions for the tuning process of the model.
5. Install additional temperature sensors for more accurate representation of gas flow dynamics during the feeding sequence.

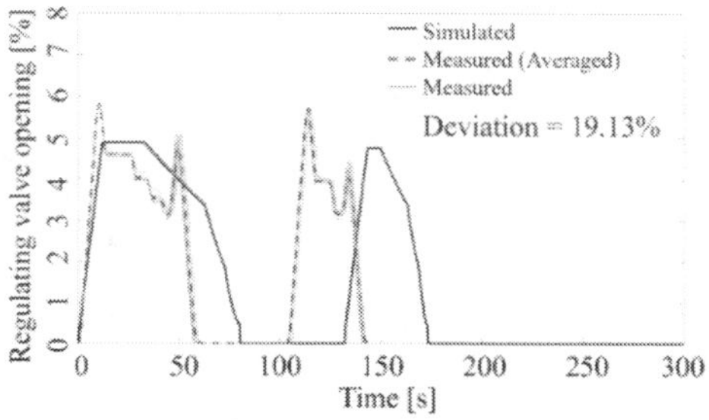

Figure 8."Fast" valve opening mode. Comparison between experiments and simulation

6 Acknowledgment

This work was supported by the project "SFI Smart Maritime (237917/O30) - Norwegian Centre for improved energy-efficiency and reduced emissions from the maritime sector" that is partly funded by the Research Council of Norway.

References

1. Baert, R.S., et al., Design and operation of a high pressure, high temperature cell for HD diesel spray diagnostics: guidelines and results. 2009, SAE Technical Paper.
2. Pickett, L.M., et al., Comparison of diesel spray combustion in different high-temperature, high-pressure facilities. SAE International Journal of Engines, 2010. 3(2010-01-2106): p. 156-181.
3. Meijer, M., et al., Engine combustion network (ECN): characterization and comparison of boundary conditions for different combustion vessels. Atomization and Sprays, 2012. 22(9).
4. Malin, M., et al., Soot Investigation on Fish Oil Spray Combustion in a Constant Volume Cell. SAE International Journal of Fuels and Lubricants, 2015. 8(2015-24-2479): p. 581-596.
5. Pedersen, E. and H. Engja, Mathematical modelling and simulation of physical systems: lecture notes in course TMR4275 Modelling, simulation and analysis of dynamic systems. 2008, Trondheim, Department of Marine Technology, Norwegian University of Science and Technology.
6. Karnopp, D.C., D.L. Margolis, and R.C. Rosenberg, System dynamics: modeling, simulation, and control of mechatronic systems. 2012: John Wiley & Sons.
7. Pedersen, E., Modelling thermodynamic systems with changing gas mixtures. SIMULATION SERIES, 1999. 31: p. 189-195.
8. Karnopp, D.C., State variables and pseudo bond graphs for compressible thermofluid systems. Journal of Dynamic Systems, Measurement, and Control, 1979. 101(3): p. 201-204.
9. Gnielinski, V., New equations for heat and mass transfer in the turbulent flow in pipes and channels. NASA STI/recon technical report A, 1975. 75: p. 8-16.

11th Asia-Pacific Conference on Combustion,
The University of Sydney, NSW Australia
10th -14th December 2017.

Generalisation of the eddy-dissipation concept for low turbulence, low Damköhler number jet flames

M.J. Evans[1], C. Petre[2], A. Parente[3]

[1]School Mechanical Engineering
The University of Adelaide, Adelaide, South Australia, 5005, Australia
[2]ANSYS Belgium
Avenue Pasteur, 4, 1300 Wavre, Belgium
[3]Ecole Polytechnique de Bruxelles, Aero-Thermo-Mechanics Laboratory
Université Libre de Bruxelles, Brussels, 1050, Belgium

Abstract

Moderate or intense low oxygen dilution (MILD) combustion has been the focus of a range of fundamental experimental and numerical studies. Reasonable agreement between experimental and numerical investigations, however, requires finite-rate chemistry models and, often, ad hoc model adjustment. To remedy this, an adaptive eddy dissipation concept (EDC) combustion model has previously been developed. This work presents a generalised version of the modified EDC model without the need for an assumed, single-step chemical reaction or ad hoc coefficient tuning. The results show good agreement with experimental measurements of two CH_4/H_2 flames in hot coflows, showing improvements over the standard EDC model as well as the previously published modified EDC model. The updated version of the EDC model also demonstrates the capacity to reproduce the downstream transition in flame structure of a MILD jet flame seen experimentally.

1 Introduction

Combustion in the moderate or intense low oxygen dilution (MILD) regime has been identified as offering improved efficiency and reduced pollutant emissions [1]. To better understand non-premixed MILD combustion, fundamental research has been undertaken using jet-in-hot-co-/cross-flow (JHC) burners [2, 3]. Despite extensive investigations, these experimental studies have proved a challenge for simulations, due to the relatively slow chemical time-scales and low turbulence in non-premixed MILD flames.

A characteristic feature of MILD combustion in JHC burners is strong turbulence-chemistry interactions resulting in a Damköhler number, *Da*, near unity [4]. As a result, computational modelling of MILD jet flames requires a finite-rate chemistry approach, with the eddy dissipation concept (EDC) model [5] previously demonstrating reasonable accuracy and computational cost [4, 6-9]. The EDC model uses the assumption that a reacting flow-field may be discretised into a network of perfectly-stirred reactors within "fine structures" at the Kolmogorov length scale [5]. The mean reaction rate of any given species in each reactor is derived from analytical and empirical arguments based on the turbulent energy cascade between different length scales [5].

It has been shown that the standard EDC model cannot be applied in flames where values of the turbulence Reynolds number, Re_T, are below 64 [4, 7, 9]. Under such conditions, the EDC model has required *ad hoc* adjustments of the empirically derived model coefficients to match experimental measurements and avoid discontinuities in calculated reaction-rates [8, 9]. This deficiency at low Re_T has prompted the extension of the EDC, with dynamic coefficients based on Re_T and *Da* [4]. This modified model introduces empirical blending coefficients for low Re_T however required a global estimate of reaction rates to estimate *Da* [4]. Physical interpretation of the adjusted EDC coefficients used with Reynolds-averaged Navier-Stokes (RANS) flow-fields can describe distinguishing features of MILD combustion. Previous work has suggested that, under MILD combustion conditions in a JHC burner, reaction zones are smaller than in conventional combustion [4, 7, 9] with longer residence times [4, 7, 9], which both serve to reduce the local mean reaction rate [4-9].

This work will present results of an updated version of the modified EDC model proposed by Parente *et al.* [4] to include a better estimate of chemical time-scales and reaction rates.

2 The Modified Eddy Dissipation Concept

The eddy dissipation concept (EDC) combustion model is a multi-species, finite-rate combustion model. The key assumption in the derivation of the EDC model is that all reactions are confined to "fine structures" [5]. These reactions are governed by a reaction rate derived on the assumption that all energy is transferred to, and dissipated at, the smallest length scales [5], although there is evidence to suggest that there may be energy transfer to larger scales in [premixed] reacting flows for *Da* on the order of unity [10]. The characteristic length, γ, and time, τ^*, of fine scales may be given as multiples of the Kolmogorov length and time scales [4, 5]:

$$\gamma = C_\gamma\,(\nu\varepsilon/k^2)^{1/4} \tag{1}$$

$$\tau^* = C_\tau\,(\nu/\varepsilon)^{1/2} \tag{2}$$

This leads to the mean reaction rate of species i, R_i [4]:

Michael J Evans. Fax: +61-8-8313-4367
E-mail address: m.evans@adelaide.edu.au

$$R_i = \rho\gamma^2/[\tau^*(1-\gamma^3)](Y_i - Y^*_i) \quad (3)$$

Where ρ is the mean density, ν the kinematic viscosity, ε the dissipation rate, k the turbulent kinetic energy and Y_i and Y^*_i are the mass fractions of species i surrounding, and within, the scales, respectively. The coefficients C_τ and C_γ are empirically derived and held constant in the original EDC model [5].

The derivation of the modified EDC combustion has been previously provided by Parente *et al.* [4] and is not reproduced here. This derivation reformulates the EDC model coefficients as functions of Re_T and Da. Based on the energy cascade assumed in the k-ε turbulence model, the modified EDC introduces a relationship for the model coefficients [4], such that:

$$C_\tau \propto [Da^2 (Re_T + 1)]^{-1/2} \quad (4)$$

$$C_\gamma \propto [Da\,(Re_T + 1)]^{1/2} \quad (5)$$

These forms are used for $Re_T > 5$, and chemical timescales τ_c, < 0.1 s. Results were found to be insensitive to these cut-off values, to within an order of magnitude. Finally, the coefficients are bound by those originally derived by Magnussen for a fully turbulent flow [5] and the values differing by an order of magnitude: C_τ is limited between 0.4083 [5] and 5.0, and C_γ to between 0.5 and 2.1377 [5]. Values of C_τ and C_γ which are respectively below and above the values estimated by Magnussen [5] tend towards infinitely fast chemistry, which is unsuitable for modelling combustion with Da near unity. Although values of C_τ below 0.4083 have been suggested for modelling MILD combustion in a JHC burner [11], permitting values in this range did not affect the results in the current study.

The value of Da was estimated by Parente *et al.* [4] using a global reaction rate [12] for one-step CH_4 combustion. This estimation chemistry cannot account for the finite-rate reaction chemistry controlled affected by minor species in different regions of the flame and is fuel-specific. There is hence a need to extend this to a general form without explicit dependency on the boundary conditions. The value of Da is calculated herein as the ratio of the flow timescale, τ_f, to τ_c, where:

$$\tau_f = (\nu/\varepsilon)^{1/2} \quad (6)$$

$$\tau_c = \max(Y_i/\omega_i) \quad (7)$$

Here, ω_i are the reaction rates (dY_i/dt) of oxygen, the fuels and major species (not including H_2O): CH_4, H_2, O_2, CO and CO_2. Finally, a characteristic Reynolds number for fine scales, Re^*, was defined by Parente *et al.* [4] and may be written as:

$$Re^* = 4/3\, C_\tau C_\gamma^2 \quad (8)$$

By the definition of Re^* in (8), the characteristic scale of the standard EDC model is 2.5 [4], and equal to 1 at the Kolmogorov length scale.

3 Numerical Setup

Previously studied CH_4/H_2 flames [2] are modelled using a modified EDC model [4] in ANSYS FLUENT 14.5. Both these flames have jet Re = 10,000 and have been previously studied numerically to better understand the MILD combustion regime.

The domain used to model the 1:1 CH_4/H_2 (by volume) flame [2] has been used previously to model the same flame cases with a modified EDC model [4]. This study focuses on two flames issuing into 1300-K coflows with 3 or 9% O_2 (by mass), referred to as HM1 and HM3 respectively. All aspects of the simulations, except the evaluation of Da, are the same as those used previously by Parente *et al.* [4] for direct comparison.

4 Results and Discussion

Radial profiles of temperature, Y_{CO}, Y_{OH} and Y_{H2O} for the HM1 and HM3 flames are presented in Figs. 1 and 2. Figures 1 and 2 show comparisons of the current, and previous modelling efforts (labelled "Parente 2016") [4], along with experimental data [2]. The experimental data include vertical estimates of a 95% confidence interval of the mean values [4], although these are essentially indistinguishable from the hollow, circular glyphs use to indicate the mean data.

Examination of Fig. 1 demonstrates the superior predictive qualities of the current modified EDC model compared to the standard EDC and the previous, *ad hoc* modified EDC model [4]. These results demonstrate excellent prediction of temperature and the peak concentrations CO and H_2O throughout the flame, as well as the concentration of the OH radical in the near-field, 30 mm from the jet exit plane. Results at 60 mm from the jet exit plane (not shown for brevity) showed similar predictive qualities to results taken at 30 mm. These results indicate a significant improvement in modelling the major features of the MILD HM1 flame [2], particularly for CO and H_2O concentrations. This flame has not previously been modelled with similar accuracy without case-specific coefficient tuning or a pre-set reaction time-scale.

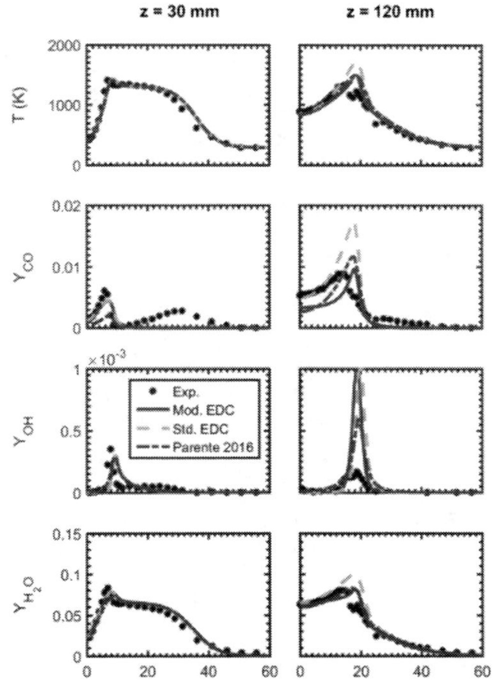

Figure 1: Comparisons of experimental [2] and predicted radial temperature, Y_{CO}, Y_{OH} and Y_{H2O} distributions for the HM1 flame at two different downstream locations. The current modelling is referred to as "Mod. EDC'", the standard EDC model is referred to as "Std. EDC" and the work of Parente *et al.* [4] is referred to as "Parente 2016".

The trends observed in Fig. 1 are also observed in Fig. 2. In both cases, the features of the flame structure are accurately predicted in the well-controlled, coflow-dominated region of the flame (< 100 mm from the exit [2]). Notably, none of the models can match the temperature field at both 30 mm and 120 mm from the jet exit plane, however the modified EDC demonstrates excellent agreement in the near-field and maintains reasonable agreement downstream. This demonstrates the effectiveness of the modified EDC combustion model in predicting the structure of the HM1 flame HM3 flames, which may be described as MILD and autoignitive respectively near the jet exit plane [13].

Beyond the coflow-controlled region, the results in both Fig. 1 and Fig. 2 show improved predictions of temperature and CO and H_2O mass fractions compared to the standard EDC model and the previous study by Parente *et al.* [4]. The discrepancies in peak OH concentrations have previously been attributed to intermittent flame extinction events and fluctuations in local mixture fraction in the downstream flame brush, suggesting a possible shortcoming in the approach to intermittency included in the standard EDC model [4, 6, 14]. These results are step towards validating the modified EDC model for use in low Re_T and low Da jet flames.

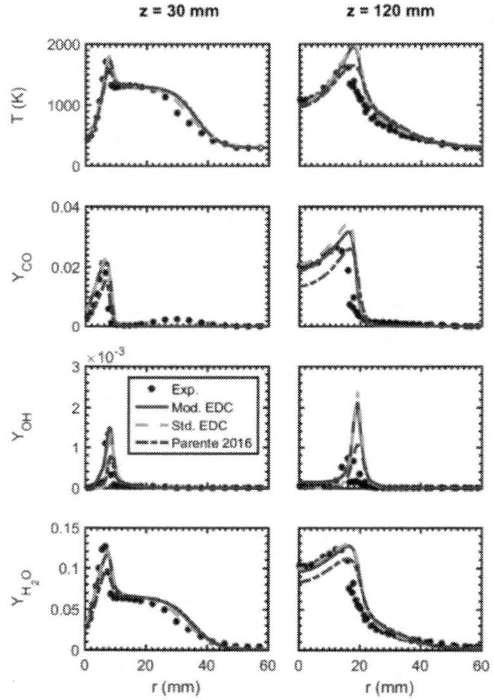

Figure 2: Comparisons of experimental [2] and predicted radial temperature, Y_{CO}, Y_{OH} and Y_{H2O} distributions for the HM3 flame at two different downstream locations. The current modelling is referred to as "Mod. EDC'", the standard EDC model is referred to as "Std. EDC" and the work of Parente *et al.* [4] is referred to as "Parente 2016".

Spatial distributions of Da, Re_T, and the EDC model variables, C_τ and C_γ, along with Re^*, may be used to provide further insight into structure of the HM1 and HM3 flames. These are presented in Figs. 3-7 respectively. The distribution of Da in the HM1 flame (Fig. 3) shows a distinct transition from a weak, thin reaction zone to a more intensely burning region with a change from $Da < 0.1$ to $Da > 0.1$. Although this is similar in appearance to the transitional structure seen in ethylene flames [3, 9], the transition in the HM1 flame occurs approximately 115 mm downstream, and hence is likely to be the effect of air entrainment from the surrounding wind tunnel. This structure appears similar to that seen from experimental photographs of the flame, which show a broad flame brush downstream of this point [2, 14]. In contrast to the sudden change in the Da profile of the HM1 flame, the distribution of Da in the HM3 flame suggests an attached flame without a noticeable discontinuity across the transition from the coflow-controlled near-field to the downstream flame brush. Although the change in the structure of the HM1 flame may be attributed to wind tunnel air stream, the low turbulence flow caused by the hot coflow persists downstream for both the HM1 and HM3 cases (Fig. 4). This is consistent with conclusions of previous work [9] and has a negative effect on the validity of the standard EDC model, which assumes a highly turbulent, isotropic energy cascade.

The motivation for an adaptive EDC model is supported by the distributions of C_τ and C_γ in Figs. 5 and 6. This results indicate that the greatest change in the modified coefficients, relative to the standard model values, occur at the transition point of the HM1 flame and near the jet centreline in the stabilisation region of the HM3 flame. In these regions, C_τ is increased whilst C_γ is reduced, analogous to reductions in the fine structure size with simultaneous increases in residence time. This results in the increased Re^* shown in Fig. 7 where reaction zones both upstream of the HM1 transition point, and on the rich side of both flames have characteristic dimensions up to eight times larger than that of the Kolmogorov scale. Interestingly, regions of low Re^* near the centreline of the HM3 flame correspond to reaction zones smaller than the Kolmogorov length scale, which are not seen in the HM1 case. In contrast, this quantitative analysis of the reaction zone scales relative to the underlying flow-field, suggests larger reaction zones in the MILD HM1 flame, near the jet shear layer.

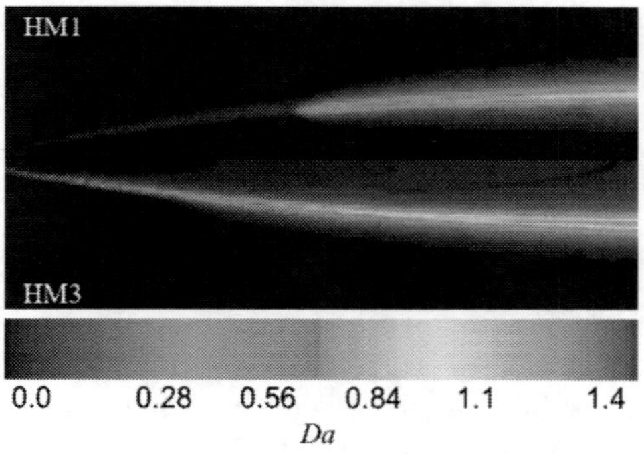

Figure 3: Damköhler number for the HM1 and HM3 flames, extending 250 mm downstream of the jet exit plane.

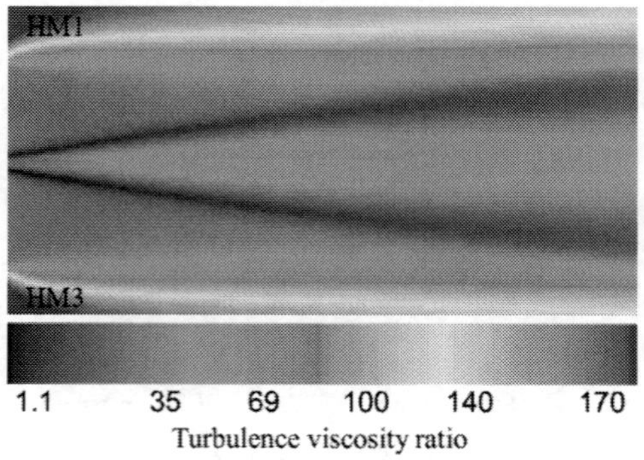

Figure 4: Turbulence viscosity ratio for the HM1 and HM3 flames, extending 250 mm downstream of the jet exit plane.

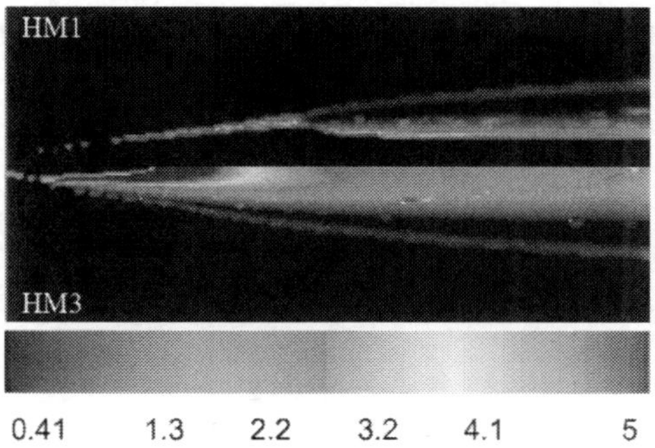

Figure 5: Modelled values of C_τ for the HM1 and HM3 flames, extending 250 mm downstream of the jet exit plane.

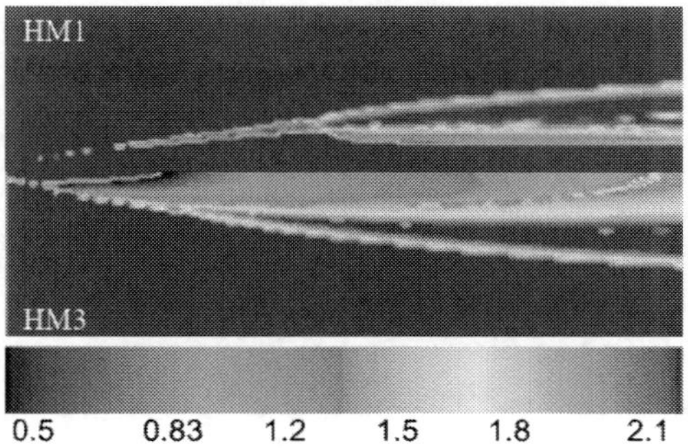

Figure 6: Modelled values of C_γ for the HM1 and HM3 flames, extending 250 mm downstream of the jet exit plane.

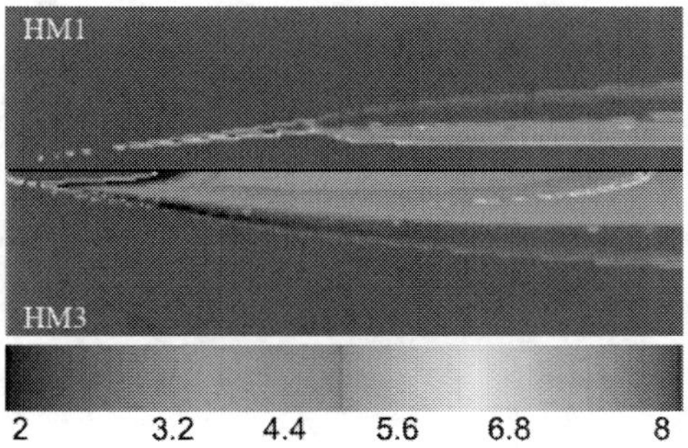

Figure 7: Modelled characteristic fine structure Reynolds number for the HM1 and HM3 flames, extending 250 mm downstream of the jet exit plane.

5 Conclusions

Experimental measurements of the HM1 and HM3 flames [2] have been used to demonstrate a modified EDC combustion model without the need for *ad hoc* coefficient tuning or an estimation of a representative, one-step chemical reaction. The results generated using this model show very good agreement with measured temperature and species concentration data in the low turbulence intensity region near the jet exit plane. Further downstream, the model captures the increase in *Da* corresponding to the HM1 flame brush seen experimentally [2, 14] and demonstrates the changes in characteristic Re^*, and reaction zone size relative to the Kolmogorov length scale, across the flame reaction zone. The results show that the HM3 flame features small reaction zones near the centreline of the jet, which are not present in the MILD HM1 flame, where stabilisation is controlled by reactions near the jet shear layer.

6 Acknowledgments

The authors thank A/Prof. Paul Medwell of the University of Adelaide for thought-provoking discussions on MILD combustion. Dr Evans gratefully acknowledges support from the University of Adelaide, the Australian Research Council (ARC) through the Discovery (DP and DECRA) grant scheme and the United States Asian Office for Aerospace Research and Development (AOARD), as well as eResearch SA for high-performance computing services.

References

[1] A. Cavaliere, M. de Joannon, Prog. Energy and Combust. Sci. 30 (2004) 329-366.

[2] B.B. Dally, R. Barlow, A.N. Karpetis, Proc. Combust. Inst. 29 (2002) 1147-1154.

[3] P.R. Medwell, P.A.M. Kalt, B.B. Dally, Combust. Flame. 152 (1-2) (2008) 100-113.

[4] A. Parente, R. Malik, F. Contino, A. Cuoci, B.B. Dally, Fuel. 163 (2016) 98-111.

[5] B.F. Magnussen, 19th AIAA aerospace science meeting, American Institute of Aeronautics and Astronautics, St. Louis, U.S.A., 1981, paper AIAA 81-0042.

[6] F.C. Christo, B.B. Dally, Combust. Flame 142 (2005) 117-129.

[7] A. De, E. Oldenhof, P. Sathiah, D. Roekaerts, Flow. Turb. Combust. 87 (4) (2011) 537-567.

[8] S.R. Shabanian, P.R. Medwell, M. Rahimi, A. Frassoldati, A. Cuoci, Appl. Therm. Eng. 52 (2) (2013) 538-554.

[9] M.J. Evans, P.R. Medwell, Z.F. Tian, Combust. Sci. and Tech. 187 (7) (2015) 1093-1109.

[10] J. O'Brien, C.A.Z. Towery, P.E. Hamlington, M. Ihme, A.Y. Poludnenko, J. Urzay, Proc. Combust. Inst. 36 (2017) 1967-1975.

[11] A. Mardani, Fuel 191 (2017) 117-129.

[12] C.K. Westbrook, F.L. Dryer, Prog. Energy and Comb. Sci. 10 (1984) 1-57.

[13] M.J. Evans, P.R. Medwell, H. Wu, A. Stagni, M. Ihme, Proc. Combust. Inst. 36 (2017) 4297-4304.

[14] J. Aminian, C. Galletti, L. Tognotti, Fuel 165 (2016) 123-133.

11th Asia-Pacific Conference on Combustion,
The University of Sydney, NSW Australia
10th -14th December 2017.

Mode coupling due to the non-uniformly distributed heat release in combustion instabilities

Lei Li[1*], Liangliang Xu[1], Guoqing Wang[1]

[1] School of Mechanical Engineering, Shanghai Jiao Tong University

Shanghai, 200240, P. R. China

Abstract

In the present paper a low-order model is developed to investigate the combustion instabilities in a chamber with circumferentially non-uniform heat release. It is shown that the mode coupling can be brought by the non-uniformly distributed heat release. From the results it can be found that the instability frequency is less influenced by the mode coupling. The unstable growth rate, however, can be significantly affected by the non-uniformity of the heat release. In the longitudinal mode dominant instability case, the azimuthal acoustic mode becomes significant for the sine shaped heat distribution. However, the effects of vorticity waves can be neglected, as the magnitude is weak enough compared to the acoustic waves.

1 Introduction

Thermoacoustic combustion instabilities may occur in the combustors of energy systems such as rocket engines, aero engines and gas turbines. It may lead to large pressure fluctuations and structural damages, even the failure of a combustor. Combustion instabilities are very difficult to be predicted and eliminated in the design stage, because the occurrence depends on many factors such as operating and environmental conditions, fuel properties or geometrical detail of combustor elements. Moreover, the possible mechanics driving combustion instabilities underly in the interactions of small disturbances and reacting turbulent flow. The scales of physical elements involved in combustion instabilities range from millimeter (turbulence scale) to several meters (acoustic wave lengths). Research on such a multi-physics and multi-scale problem are very difficult and always are big challenges to the researchers.

To understand the mechanisms of the combustion instabilities, it is important to study the response of the flame to the acoustics. Flame transfer functions, or flame describing functions, are therefore introduced. In a flame transfer function, the unsteady heat release is typically given as a function of the inlet velocity disturbance. The flame transfer functions can be determined experimentally or theoretically. In experiments, the emission signals of excited radicals OH* or CH* are used to characterize the heat release rate [1-3]. These experimental data also provide an understanding of the physical mechanism inducing the flame perturbation. Theoretical modelling of flame transfer function is usually developed based on the linearized G-equation. For example, the response of a laminar premixed flame to flow oscillations was theoretically investigated by Fleifil et al. [4] The authors neglected the variation of the flame speed and assumed that the unsteady heat release is induced by the flame area perturbation. Thus the leading order of the unsteady heat release rate was obtained. In fact, the laminar flame speed can be significantly changed by the stretch of the flame, due to the imbalance between the heat and mass transfer in combustion process. By considering such effects, the decay of the flame wrinkles along the flame surface for large frequency was explained in theoretical models [5, 6]. The flame dynamics of the turbulence flame can also be analyzed theoretically. For example, the flame transfer function of a premixed swirling flame was derived by using the G-equation derived by taking an ensemble average with respect to the turbulent fluctuations [7]. A good agreement was obtained between the experimental and theoretical results.

In the annular combustion chambers of gas turbines, thermoacoustic combustion instabilities may be longitudinal or azimuthal [8]. Longitudinal instabilities are characterized as axial acoustic mode and many works have been done on the interaction between the flame and the longitudinal acoustic waves [9, 10]. Azimuthal instabilities occur due to the coupling of the azimuthal acoustic mode and flames. Typically, in analysis of azimuthal instabilities, the annular geometry is usually assumed to be thin enough that the radial dependence is ignored and the pure azimuthal modes are considered [11, 12].

CFD methods can be used for the simulation of the combustion instabilities in combustor. Large eddy simulation (LES) has proven to be a powerful tool in simulation of swirl-burner behaviors [13-15] and azimuthal instabilities in combustion system [16]. LES is associated with many advantages, but it is always involved with high computation cost. To reduce computational cost, reduced-order models are needed in the analysis of combustion instabilities. Linear prediction methods are usually used [17-21]. Such simplified low-order model is associated with low computation cost. But it should be noted that the accuracy of the results depends strongly on the modelling of the turbulence combustion process [4, 22, 23] and the description of the system boundary conditions [24-26].

Typically, a flame transfer function should be included in a low-order model to predict the combustion instability of a system. For longitudinal instabilities, the flame transfer function can be given as a jump condition, in which the flame is assumed to concentrate on a zero thickness plane [17]. If the length of the flame is not small enough compared with the acoustic wave length, however, the axial distribution of the heat release should be considered [27, 28]. For example, a flame model with triangular distribution was introduced by Kato et al., for the prediction of combustion oscillation in a dry low emission combustor [27]. It was found that the triangular flame model gave a better prediction of the combustion instabilities compared to the simple flame sheet model.

Corresponding author. E-mail address: Lilei81@sjtu.edu.cn

The prediction of azimuthal instabilities is much more complicated than the longitudinal instabilities. To simplify the problem, zero flow Mach number is assumed and net-work model can be used, in which both the axial and radial directions of the annular chamber can be neglected and only circumferential modes are considered. For example, a simple analytical model was given by Parmentier et al. [11] The single annular chamber with N burners is modeled by using one-dimensional zero Mach number formulations. The flame transfer function of each burner is included as lumped parameter. This model can be used for the prediction of the combustion instabilities of the system with different types of burners. If the flow Mach number is not zero, however, the convection mode should be considered and new prediction models are needed. By using a two-dimensional low-order model and a liner Euler equation method, it was found that vorticity waves are generated from the interaction of acoustic waves and the flame, which may have significant effects on azimuthal instabilities [12].

The spatial distribution of the flame model may have crucial effects on the combustion instabilities, no matter for longitudinal or azimuthal modes. It has been proved that the method of mixing different types of burner in one annular chamber can be used to suppress the azimuthal instabilities [11]. In the previous work, a linear Euler Equation method was introduced and the combustion instability with azimuthally non-uniform heat release was studied [29, 30]. It was observed that the circumferential distribution of heat source really has effects on the longitudinal instabilities. The generation of vorticity wave was also observed for such conditions. However, in these work, the generated azimuthal modes have not been quantitatively determined.

In the present paper, a low-order model is developed. It will be seen that the coupling of longitudinal and azimuthal modes is caused by the azimuthal non-uniformity of the heat source. The generation of aizimuthal modes and vorticity waves will also be investigated in details. This paper is organized as follows. Section 2 gives the description of the problem. Section 3 presents the method and the results.

2 Problem description

Figure 1 shows the physical configuration. An annular duct containing a planar flame, which separates the upstream and downstream regions. We assume that the heat source is circumferentially non-uniform. To make the problem simple, the classical n-τ model is employed to model unsteady heat release. Here the fluctuations in heat release are taken as linearly proportional to the velocity perturbation with a time delay,

$$Q'(t) = k_q C_p (\bar{T}_{02} - \bar{T}_{01}) \bar{\rho}_1 u'_{x1}(t-\tau) \quad (1)$$

where Q' is the unsteady heat release, C_p is the specific heat of constant pressure, $\bar{\rho}_1$ is the mean density of the upstream region, u'_{x1} is the axial velocity perturbation at the upstream boundary of the flame, and τ denotes time delay. $\bar{T}_{01}$ and $\bar{T}_{02}$ are mean stagnation temperatures of upstream and downstream regions, respectively. We also assume that the heat source is azimuthally non-uniform, hence in the calculation of the heat release rate equation $\bar{T}_{02} = \bar{T}_{01}(k + \alpha\sin\theta)$ is used, where k is the ratio of azimuthally mean stagnation temperature and α is used to refer to the degree of the non-uniformity. If the heat source is non-uniform, it can be imagined that the flow in the downstream of the flame is also non-uniform. It should be noted that the non-uniform flow may induce flow instabilities which may interact with the thermoacoustic instabilities. In this work, however, we will focus on the effect of the non-uniform heat source on the combustion instabilities, therefor we assume that the flow in the downstream zone is uniform due to the quick mixing of the flow. It means the stagnation temperature of the flow after the flame is $\bar{T}_{02} = k\bar{T}_{01}$. The heat release form for the non-uniformity is quite simple here, however, a complicated azimuthally non-uniform heat source can be decomposed into the sum of many sine-type distributed heat source. Therefore, through the investigation of such simple case, we can gain insight into the effects of the non-uniform heat source on the instabilities.

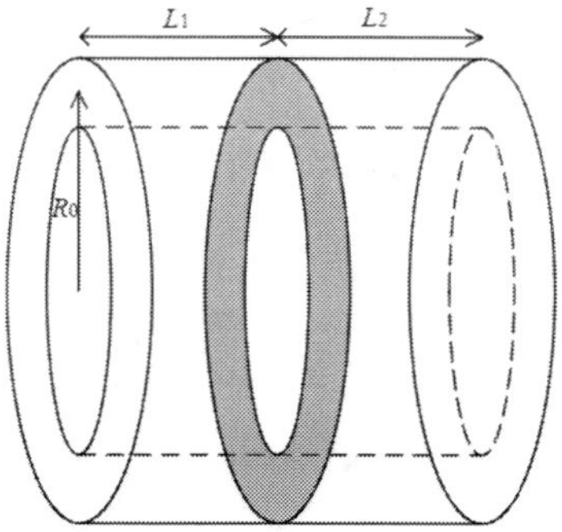

Figure 1. Annular chamber with planar flame

3 Method and results

3.1 Low-order model

In a LEE method is developed to investigate the effects of non-uniform heat source on the combustion instabilities [30]. In LEE method, linear Euler equations are discretized directly and the eigenvalues of a huge matrix should be calculated. A low-order model is used in this work, for which the matrix size is very small, e.g., less than 100*100. In the low-order method, two-dimensional analytical descriptions of the perturbations are given first. Across the flame there are conservation laws of mass, momentum and energy. Employing a linearization procedure, jump equations describing the correlations of the disturbance quantities across the flame will be obtained. For the present two-dimensional problem, we have one mass equation, two momentum equations and one energy equation. The details of the low-order method can be found in Ref. [12].

If the heat source is uniform in circumferential direction, each of the azimuthal modes is independent. However, the modes are coupled if the heat source is circumferentially non-uniform, because the integration of the production of heat source term for different modes with θ is no longer zero. In the following, the mode coupling of longitudinal and azimuthal modes will be investigated. It should be noted that we will focus on the problem of longitudinal dominant modes.

For simplicity, the left boundary condition is taken to be acoustic closed and the right boundary condition is given as acoustic open.

3.2 Results and discussion

Figure 2 shows the real and the imaginary parts of the angular frequency as a function of time delay for a uniform heat source. The frequencies correspond to the first longitudinal mode. We can observe the period variation of the frequency with the time delay, which can be explained by Rayleigh criterion.

The effects of the azimuthally non-uniform heat source on the oscillation frequencies are shown in Fig. 3. Both the LEE results and the low-order model results are listed. We can observe a good agreement between the results of the two methods. It can also be found that with the increase of the non-uniformity, the real part of the frequency (growth rate) changes from negative to positive, which shows that the system becomes stable from unstable state. These results are in agreement with that of Ref. [30].

The ratios of pressure amplitude of each mode to the zero mode are depicted in Fig. 4. From the results, it can be found that the first azimuthal mode becomes important with the increase of the non-uniformity. The pressure amplitude of the first azimuthal mode even reaches more than one half of the longitudinal mode. It can also be found that the second and third azimuthal modes are smaller enough and can be neglected.

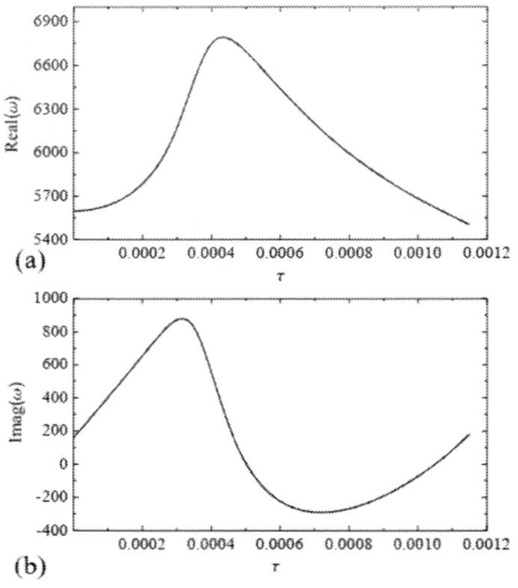

Figure 2. Real (a) and imaginary (b) part of frequency as a function of time delay for 0.02 Mach number, k=3, L_1=L_2=0.1m, k_q=0.2, α=0

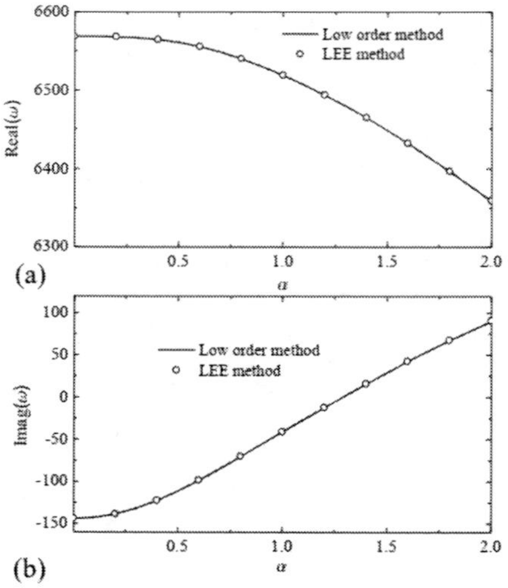

Figure3. Real (a) and imaginary (b) part of frequency as a function of α for 0.02 Mach number, k=3, L_1=L_2=0.1m, k_q=0.2, τ=0.00055s

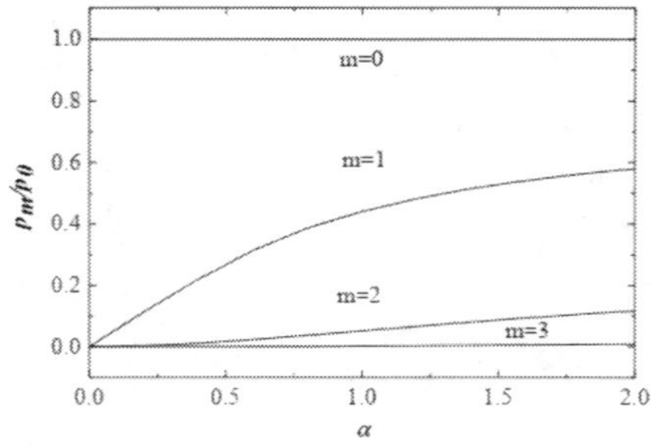

Figure 4. Ratio of pressure amplitude of each mode to the zero mode as a function of α for 0.02 Mach number, k=3, L_1=L_2=0.1m, k_q=0.2, τ=0.00055s

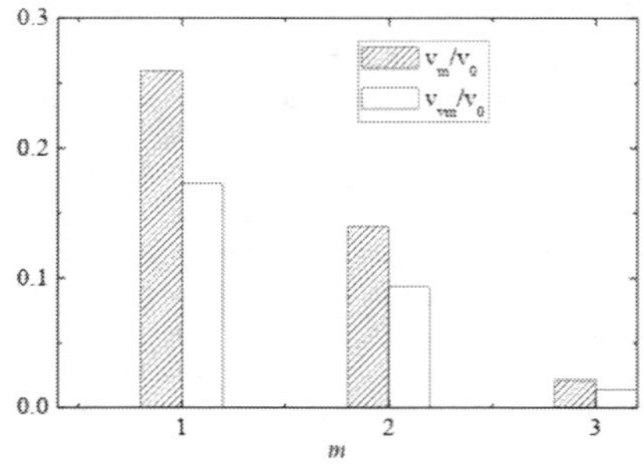

Figure 5. Ratio of circumferential velocity amplitude of each mode to zero mode axial velocity perturbation for 0.02 Mach number, k=3, L_1=L_2=0.1m, k_q=0.2, τ=0.00055s and α=2

Table 1. Ratio of axial velocity amplitude of each mode to zero mode axial velocity perturbation for 0.02 Mach number, k=3, L_1=L_2=0.1m, k_q=0.2, τ=0.00055s and α=2

m	0	1	2	3
V(acoustic)	1	7.83E-01	7.73E-01	2.55E-02
V(vorticity)		2.95E-03	3.18E-03	7.28E-04

Figure 5 shows the ratio of circumferential velocity amplitude of each mode to axial velocity perturbation of zero mode. Both the acoustic and vorticity components are listed. It can be found that the acoustic and the vorticity modes are important for circumferential velocity. However, for axial velocity, the vorticity modes are very small, less than 1% of that of acoustic induced velocity perturbation, as shown in Table 1.

It should be noted that the temperature ratio in these cases is given as k=3. In a real devise the temperature ratio may be larger than this value, but as shown in the previous work [29, 30], it does not change the conclusions. Moreover, as shown in figure2, the time delay τ determines the instabilities of the system, however, it has little effects on the coupling of the azimuthal and longitudinal modes.

4 Conclusions

In the present paper the circumferentially non-uniform heat release on the combustion instabilities has been investigated. From the results of the low-order model, it is found that the non-uniformity of the heat source can change the stability characteristics of a combustion system. The model coupling is caused by the azimuthally non-uniform heat release. And the calculated results show that the first azimuthal acoustic mode is important. However, the vorticity mode is not significant compared with acoustic mode.

5 Acknowledgment

This work was supported by the National Natural Science Foundation of China (51676126).

References

[1] Thumuluru SK, Lieuwen T. Characterization of acoustically forced swirl flame dynamics. Proceedings of the Combustion Institute. 2009;32:2893–900.

[2] Palies P, Durox D, Schuller T, Candel S. The combined dynamics of swirler and turbulent premixed swirling flames. Combustion and Flame. 2010;157:1698–717.

[3] Candel S, Durox D, Schuller T, Bourgouin J-F, Moeck JP. Dynamics of Swirling Flames. Annual Review of Fluid Mechanics. 2014;46:147-73.

[4] Fleifil M, Annaswamy AM, Ghoneim ZA, Ghoniem AF. Response of a laminar premixed flame to flow oscillations: A kinematic model and thermoacoustic instability results. Combustion and Flame. 1996;106:487-510.

[5] Wang HY, Law CK, Lieuwen T. Linear response of stretch-affected premixed flames to flow oscillations. Combustion and Flame. 2009;156:889-95.

[6] Preetham, Thumuluru SK, Santosh H, Lieuwen T. Linear Response of Laminar Premixed Flames to Flow Oscillations: Unsteady Stretch Effects. Journal of Propulsion and Power. 2010;26:524-32.

[7] Palies P, Schuller T, Durox D, Candel S. Modeling of premixed swirling flames transfer functions. Proceedings of the Combustion Institute. 2011;33:2967-74.

[8] Lieuwen T, Yang V. Combustion instabilities in gas turbine engines: operational experience, fundamental mechanisms, and modeling: American institute of Aeronautics and Astronautics; 2005.

[9] Stohr M, Sadanandan R, Meier W. Experimental study of unsteady flame structures of an oscillating swirl flame in a gas turbine model combustor. Proceedings of the Combustion Institute. 2009;32:2925-32.

[10] Patton M. Allison YC, Matthias Ihme, James F. Driscoll. Coupling of flame geometry and combustion instabilities based on kilohertz formaldehyde PLIF measurements. Proceedings of the Combustion Institute. 2015;35:3255-62.

[11] Parmentier JF, Salas P, Wolf P, Staffelbach G, nicoud F. A simple analytical model to study and control azimuthal instabilities in annular combustion chambers. Combustion and Flame. 2012;159:2374-87.

[12] Li L, Sun X. Effect of vorticity waves on azimuthal instabilities in annular chambers. Combustion and Flame. 2015;162:628-41.

[13] Huang Y, Yang V. Dynamics and stability of lean-premixed swirl-stabilized combustion. Progress in Energy and Combustion Science. 2009;35:293-364.

[14] Tay-Wo-Chong L, Bomberg S, Ulhaq A, Komarek T, Polifke W. Comparative Validation Study on Identification of Premixed Flame Transfer Function. Journal of Engineering for Gas Turbines and Power. 2012;134:021502.

[15] Poinsot T. Prediction and control of combustion instabilities in real engines. Proceedings of the Combustion Institute 2017;36:1-28.

[16] Staffelbach G, Gicquel LYM, Boudier G. Large eddy simulation of self excited azimuthal modes in annular combustors. Proceedings of the Combustion Institute. 2009;32:2909-16.

[17] Dowling AP. The calculation of thermoacoustic oscillations. Journal of Sound and Vibration. 1995;180:557-81.

[18] Nicoud F, Wieczorek K. About the zero Mach number assumption in the calculation of thermoacoustic instability. International Journal of Acoustics and Vibration. 2009;1:67-111.

[19] Dowling AP, Stow SR. Acoustic analysis of gas turbine combustors. Journal of Propulsion and Power. 2003;19:751-64.

[20] Miranda AC, Polifke W. Combustion Stability Analysis of Rocket Engines with Resonators Based on Nyquist Plots. Journal of Propulsion and Power. 2014;30:962-77.

[21] Schulze M, Kathan R, Sattelmayer T. Impact of Absorber Ring Position and Cavity Length on Acoustic Damping. Journal of Propulsion and Power. 2015;52:917-27.

[22] Ducruix S, Durox D, Candel S. Theoretical and experimental determinations of the transfer function of a laminar premixed flame. Proceedings of the Combustion Institute. 2000;28:765-73.

[23] Lieuwen T. Modeling premixed combustion-acoustic wave interactions: a review. Journal of Propulsion and Power. 2003;19:765-81.

[24] Marble FE, Candel SM. Acoustic disturbance from gas non-uniformities convected through a nozzle. Journal of Sound and Vibration. 1977;55:225-43.

[25] Stow SR, Dowling AP, hynes TP. Reflection of circumferential modes in a chocked nozzle. Journal of Fluid Mechanics. 2002;467:215-39.

[26] Duran I, Morgans AS. On the reflection and transmission of circumferential waves through nozzles. Journal of Fluid Mechanics. 2015;773:137-53.

[27] Katoa S, Fujimoria T, Dowlingb AP, Kobayashi H. Effect of heat release distribution on combustion oscillation. Proceedings of the Combustion Institute. 2005;30:1799-806.

[28] Kim KT, Lee JG, Quay BD, Santavicca DA. Spatially distributed flame transfer functions for predicting combustion dynamics in lean premixed gas turbine combustors. Combustion and Flame. 2010;157:1718-30.

[29] Li L, Yang L, Sun X. Effect of distributed heat source on low frequency thermoacoustic instabilities. Journal of Sound and Vibration. 2013;332:3098-111.

[30] Li L, Sun X, Lioi C, Yang V. Effect of Azimuthally Nonuniform Heat Release on Longitudinal Combustion Instabilities. Journal of Propulsion and Power. 2017;33:193-203.

11th Asia-Pacific Conference on Combustion,
The University of Sydney, NSW Australia
10th -14th December 2017.

Temperature imaging of gaseous *n*-heptane flames in hot vitiated coflows

M.J. Evans, P.R. Medwell, Z.W. Sun, B.B. Dally

School of Mechanical Engineering, The University of Adelaide

Adelaide, SA, 5005, Australia

Abstract

The scenario of fuel injected into hot surrounds is found in a range of practical combustion applications. These flame conditions have been emulated using a jet in hot coflow burner. Four lifted, turbulent flames, with a constant jet flow of a prevapourised *n*-heptane/N_2 mixture at 380 K into various hot and vitiated coflow conditions, are investigated in this paper. Three of these flames issued into coflows generated by the combustion of different mixtures of ethylene/air and one had a coflow from a natural gas/air flame. Gas sampling and the non-linear excitation regime two-line atomic fluorescence (NTLAF) technique were used to investigate the mixing between the hot coflow and the surrounding air, as well as the flame temperature for the different coflow configurations. A comparison of results for the turbulent *n*-heptane flames issuing into hot coflows is presented, suggesting that the hottest flame is not associated with the coflow containing the highest concentration of O_2. The results also highlight the low temperature of the jet along centreline in comparison previous studies using simpler fuels.

1 Introduction

Flames issuing into hot and vitiated environments may be found in the practical applications of combustion in gas turbines, furnaces and other devices featuring strong exhaust gas recirculation, such as HCCI engines. These environments may be emulated in laboratory-scale research burners by the use of the jet-in-hot-coflow (JHC) configuration, where a jet of fuel issues into a coflow of combustion products [1,2]. The JHC burner is similar to the vitiated coflow burner (VCB) [3,4]. Both configurations have previously measured soot-free flame regions [1,3] or investigated flame stabilisation in the coflow-controlled region of the burner [2-4,5]. These studies have not, however, been able to provide detailed, downstream measurements of non-premixed jet flames in regions of significant soot loading due to interference of the optical diagnostic techniques that were used.

The ignition and structure of flames in hot and diluted coflows have been studied extensively in the context of lifted flames [3-5] and moderate or intense low oxygen dilution (MILD) combustion [1,2]. Studies of these flames have often focused on the near-field, coflow-controlled, region of these flames [2]. This limited region of interest has arisen because of the focus on flame stabilisation mechanisms. Although detailed temperature and species data have been collected in the soot-free region of a range of flames under MILD combustion conditions [1,6] these datasets have tended to use scattering-based diagnostic techniques, and therefore cannot be applied to sooting or spray flames. Moreover, whilst these data have been used extensively for model validation, the boundary conditions of the burners have often not been fully characterised.

Michael J. Evans. Fax: +61 (0)8 8313 4367
E-mail address: m.evans@adelaide.edu.au

One of the key components of primary reference fuels and diesel surrogates is *n*-heptane (*n*-C_7H_{16}) [2]. Investigations of this alkane in hot and oxygen-diluted environments have demonstrated a significantly different behaviour to smaller hydrocarbons, which has been attributed to its multiple fuel pyrolysis/decomposition pathways [2]. The combination of different ignition pathways, and a distinct region of negative net heat release with highly diluted (3% O_2 vol./vol.) hot oxidants, warrants the further investigation of *n*-heptane as a surrogate for more complex, practical fuels.

This work explores the behaviour of *n*-heptane flames under different hot and vitiated coflow conditions, particularly the downstream differences resulting from the different coflow streams. This work, similar to previous studies [2], focuses on prevapourised *n*-heptane to eliminate the complex mechanism of droplet evaporation, and provide comparative data for future spray-flame studies. Additionally, the persistence and mixing of the coflow stream are reported for the first time, which aid interpretation of the NTLAF results and will be beneficial for future modelling of these types of flames.

2 Methodology

2.1 Burner Description and Flame Cases

Experiments were undertaken in flames stabilised on a JHC burner similar to one studied previously [1,2]. The burner features a 4.6 mm I.D. (6.35 mm O.D., tapered over 4 mm to 5.2 O.D. mm at the jet exit plane) central jet which protrudes 25 mm above a 110 mm I.D. porous bed burner, producing an annular coflow of hot combustion products. The total length of the central pipe was more than 100 times its diameter to ensure fully-developed flow at the jet exit plane. The porous bed burner was surrounded by a secondary annular coflow (130 mm I.D.) of air at approximately equal velocity to reduce fluctuations downstream caused by the mixing of the hot coflow and quiescent room air (at approximately 21°C).

The fuel jet stream contained 35% *n*-heptane and 65% N_2, by mass, and issued from the central jet at a temperature of 380 K to ensure completely prevapourised fuel. The bulk mean Reynolds number at the jet exit remained constant at 10,000. The *n*-heptane/N_2 was combined and heated using a Bronkhorst controlled evaporated mixer (CEM). Approximately 10% of the total volumetric flow (4 nLpm) of N_2 in the jet bypassed the CEM to carry indium produced by means of ablation [7,8]. The CEM and seeded streams were recombined in a heated pipe upstream of the burner assembly.

Four different coflows are investigated: three from the combustion of ethylene/air (Cases 1-3) and one from a natural gas/air flame (Case 4). The bulk velocities of the hot coflow streams were held at 1.1 m/s in all cases. The coflow conditions are summarised in Table 1, which include the calculated equilibrium molar concentration of oxygen (X_{O2}), carbon dioxide (X_{CO2}) and water vapour (X_{H2O}) in the coflow, as well as the measured temperature (T_{cofl}).

Case	X_{O2}	X_{CO2}	X_{H2O}	T_{cofl} (K)
1	0.108	0.063	0.063	1490
2	0.096	0.070	0.070	1590
3	0.085	0.077	0.077	1590
4	0.075	0.061	0.121	1520

Table 1: Summary of the coflow streams (wet basis).

2.2 Experimental Techniques

Planar thermography was performed using non-linear excitation regime two-line atomic fluorescence (NTLAF) of indium [7]. The optical set-up for the planar NTLAF thermography imaging was similar to a previous study [8].

Thermography through NTLAF is performed on atomic indium seeded into the fuel stream by laser ablation [9]. Laser sheets with wavelengths tuned to the indium transitions at 410.18 nm and 451.13 nm were produced by two Quantel TDL90 dye lasers, and used to pump the ground and thermally excited states (with electronic energy difference ΔE_{10}) of atomic indium. A total of 500 image pairs of Stokes and Anti-Stokes signals were recorded using a single f-number 1.2 lens through a custom-made dichroic beam splitter arrangement onto and two Princeton PI-MAX 4 ICCD cameras. After image matching, the spatial resolution is 150 µm/pixel in the region of interest. The relative strengths of the Stokes and Anti-Stokes signals, F_{21} and F_{20} respectively, are used to determine the gas-phase temperature, T, through (1):

$$T = \frac{\Delta E_{10}/k}{\ln\left(F_{21}\times\left(1+{}^{C_S}/_{I_{20}}\right)\right)-\ln\left(F_{20}\times\left(1+{}^{C_A}/_{I_{21}}\right)\right)+C_T} \qquad (1)$$

In (1), I_{20} and I_{21} are the energy of the incident Stokes and Anti-Stokes laser beams, k is the Boltzmann constant, and C_A, C_S and C_T are constants determined through daily experimental calibration. This technique is valid in fuel-rich mixtures with a single-shot accuracy within 100 K, between 800-2500 K [7]. The technique is not accurate below this temperature range owing to the low population density of thermally excited, indium which is probed using the Anti-Stokes beam, or in the fuel-lean reaction zone where atomic indium is oxidised [7]. Finally, it should be noted that as the Stokes single originates from atomic indium in its ground state, this signal may be used as a surrogate for rich, cool mixtures in the flow.

Laser sheets were nominally 20 mm high, with the central 15 mm presented herein. Data were collected at ten measurement locations for each case. The height of sheet centres ranged from 15 mm to 330 mm (spanning a total range of x/D from 2 to 72) above the jet exit plane to capture the temperature field from the flame base to the broadest part of the flame base. All flames were estimated to be approximately 500 mm long from images and visual observations.

The coflow temperature for all cases and downstream jet centreline profile of Case 2 were measured with an R-type thermocouple with a 75-µm-diameter bead, corrected for radiative heat transfer. Gaseous O_2, CO and CO_2 were measured (on a dry basis) using a calibrated Nova 7466K gas analyser sampled through a 3.2 mm O.D., 1.6 mm I.D. probe, with a length of 80 mm normal to the flow bulk velocity. The probe diameter was kept as small as possible by not cooling it in order to minimise the disturbance of the flow-field. Quenching of the products is not necessary because they are at equilibrium conditions at the coflow exit plane. The estimated absolute accuracy of the gas analyser (based on calibration) is ±0.1% for O_2 and ±0.2% for CO_2.

3 Results and Discussion

Presented in Figure 1 is the measured mean mole fraction of O_2, for Case 2, at the centre plane region above the jet. Data were collected at 15 radial locations spanning a total of 100 mm: the centreline, then every 5 mm until 20 mm, followed by every 10 mm until 50 mm either side of the centreline; at 5 heights including the jet exit plane up to 85 mm downstream. The central, hatched, region of the species field represents the fuel stream where O_2 concentrations could not be measured due to the high concentration of prevapourised liquid fuel. These results suggest that the limit to which flame may be considered coflow-controlled is approximately 60 mm downstream of the jet exit plane. Further downstream the mean concentration of O_2, immediately outside of the bands of low O_2 (and elevated CO_2 shown in Fig. 2), exceeds that of the coflow indicating mixing between quiescent room air and the coflow stream. A similar spatial distribution was recorded for CO_2 (shown in Fig. 2), with 7.5% CO_2 (vol./vol.) measured in Case 2 coflow. The apparent discrepancy between this dry-basis measurement and the value in Table 1 is owing to the calculation of equilibrium CO_2 on a wet-basis. Converting the wet-basis equilibrium of CO_2 in Table 1 to a dry-basis, gives a concentration of 7.53% (vol./vol.), in agreement with the measured value. The CO_2 concentration could not be measured in the central region of Fig. 2 for the same reasons described in the discussion of Fig. 1.

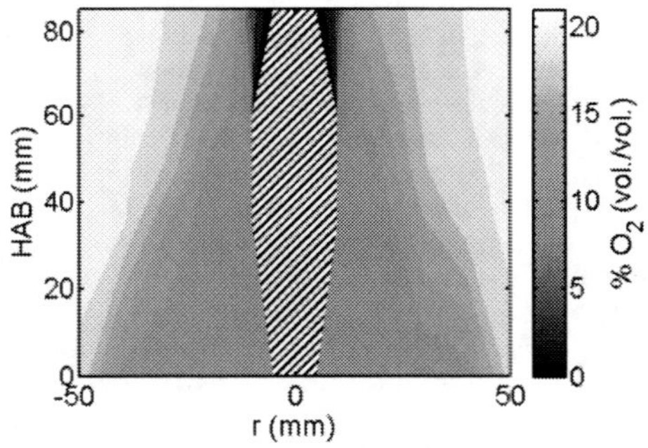

Figure 1: Measured O_2 mole fraction (dry basis), for Case 2, at the centre plane region above the jet.

The measured coflow temperature for Case 2 is shown in Fig. 3. The radial temperature profile at the jet exit plane shows a maximum difference of less than 10 K at equal radial locations, within 30 mm of centre of burner centre. Temperature distributions for the other coflows were similarly uniformly distributed and axisymmetric.

Understanding the extent and shape of the mixing field is imperative for interpreting the results of the flame appearance and laser-based NTLAF (among other) diagnostics. Without knowledge of the mixing field, the properties of the oxidant cannot be determined locally at the flame front and conclusions drawn about flame behaviour or structure may be erroneously attributed to the coflow. It is therefore critical that the mixing between the three fuel, coflow and air streams be understood for interpreting downstream observations and model validation [10]. Although some data from these fields are present in previous datasets of flames in hot coflows studied using scattering [1-5], the three-stream mixing cannot be measured using NTLAF (as the conditions are both cool and lean). The data presented in

Figs 1 and 2 may be used to provide verification for modelling the three-stream mixing effects in JHC burners using computational fluid dynamics, facilitating more accurate understanding of the fundamental combustion processes.

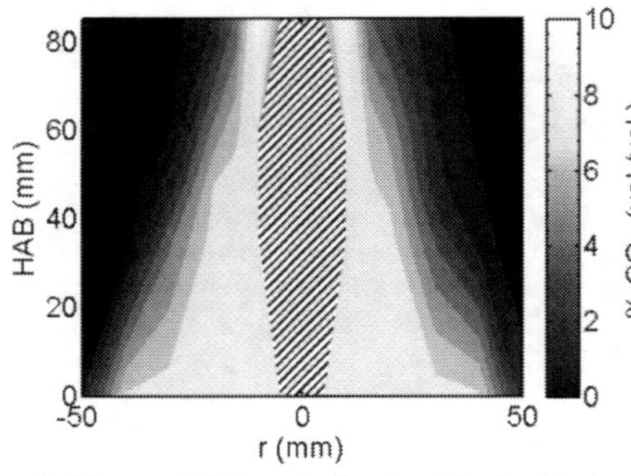

Figure 2: Measured CO_2 mole fraction (dry basis), for Case 2, at the centre plane above the jet.

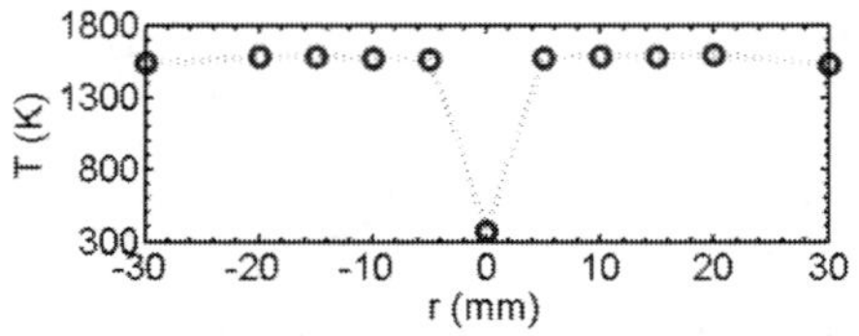

Figure 3: Mean radial profile of measured coflow temperature, for Case 2, at the jet exit plane.

Instantaneous images of the temperature field of all three flames were taken centred at heights ranging from 15 mm to 330 mm above the jet exit plane. No smoothing or filtering was performed on the quantified temperature images. The mean flame bases were lifted from 15 mm (Case 1) to 18 mm (Case 4), with height increasing with decreasing coflow O_2 concentration. Lift-off heights were based on a temperature threshold of 800 K from the averaged, calculated temperature fields. Images centred 15 mm above the jet exit plane (not shown for brevity) show isolated pockets of high temperatures, suggesting the presence of ignition kernels, upstream of lifted flame bases. This mechanism of flame stabilisation is typical of autoignitive turbulent flames [11], as is consistent with previous conclusions of prevapourised *n*-heptane flame stabilisation in hot and diluted coflows [2]. The lift-off heights of each flame are subsequently summarised in Table 2, along with the most probable temperature from PDFs from each ensemble of images centred about three different heights. Noticeably, the flame in Case 4 is the hottest at all heights despite being the most lifted.

Case	Mean lift-off height (mm)	Most probable temperature (K) centred at height of:		
		60 mm	110 mm	165 mm
1	15.0	1250	1250	1250
2	16.7	1300	1350	1350
3	17.1	1300	1350	1350
4	18.1	1500	1500	1600

Table 2: Summary of flame characteristics in all four cases.

Figure 5 shows typical instantaneous images for Cases 1 and 4, centred about 60 mm above the jet exit plane. Images show an area 15 mm high and 50 mm wide. These typical instantaneous images demonstrate that both cases (with the maximum and minimum O_2 concentration in their coflows) feature unbroken flame fronts of similar width in the coflow-controlled region of the flames. Comparison between PDFs of temperature in these sets of images, quantified in Table 2, however, demonstrate that the flame in Case 4 is hotter than in Case 1 (which, in turn, is of a similar temperature to Cases 2 and 3), despite having 3% less available O_2 (vol./vol.) in the coflow stream. This trend is consistent across all heights. As both coflows in Cases 1 and 4 are held at similar temperatures, this could demonstrate the effect of the coflow CO_2/H_2O ratio on *n*-heptane combustion through chemical or physical effects. This is consistent with simulations of laminar C_2H_4 flames with hot coflows where coflows with higher ratios of H_2O to CO_2 exhibited hotter flame fronts, particularly under MILD conditions [12]. Although most images of the temperature field at this height show unbroken flame fronts, intermittent 'holes' are evident in a small number of the temperature images centred about 60 mm above the jet exit plane. These 'holes' appear on both sides of the jet exit plane with similar frequency, but are not seen to occur on both sides simultaneously in any frame.

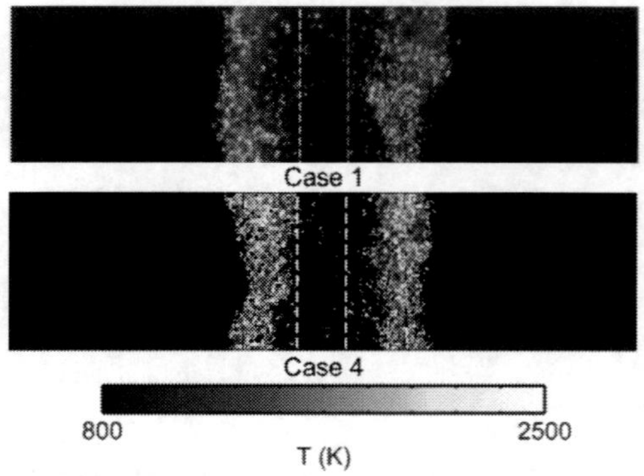

Figure 4: Instantaneous temperature fields centred 60 mm above the jet exit plane. Imaging area is 15 mm × 50 mm, and white lines represent the I.D. of the central fuel jet.

Downstream of the initial mixing between the coflow and quiescent air mixing region downstream, the instantaneous temperature fields of Cases 1 and 4 initially appear similarly affected by the turbulent mixing of the jet and quiescent air. In both cases, the high temperature regions are broadened and distorted by mixing with cold air.

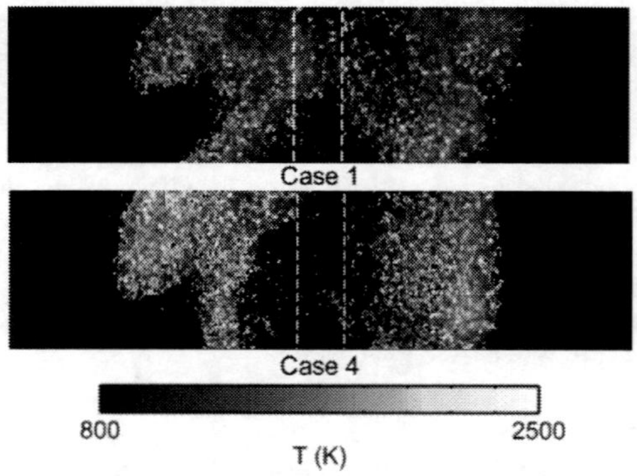

Figure 5: Instantaneous temperature fields centred 110 mm above the jet exit plane. Imaging area is 15 mm × 50 mm, and white lines represent the I.D. of the central fuel jet.

Instantaneous images of the temperature field are presented in Fig 7 to identify breaks in the flame front at 165 mm above the jet exit plane. The temperature images determined obtained through NTLAF are equivalent to conditioning on T ≥ 800 K in rich mixtures [7]. Therefore, without additional analyses, any 'holes' may be indicative of quenching due to high strain rate or a cool pocket of rich or lean mixture. Although NTLAF cannot be used to explicitly detect a lean flame front, this would produce a rich preheat zone which would be imaged through the NTLAF technique.

Both Cases 1 and 4 demonstrate a similar propensity for breaks in the rich, hot zone of the flame in images centred 110 mm downstream of the jet exit plane (with an example shown in Fig. 7). Comparison with the Stokes signal (fluorescence of ground-state indium), however, shows an absence of atomic indium, suggesting no cool, rich mixture. This implies that the mixture in this 'hole' is lean, and not indicative of a reacting mixture quenched due to high local strain rates.

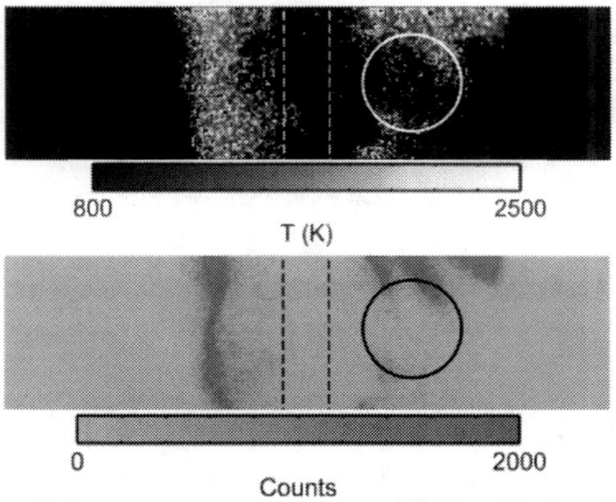

Figure 6: Instantaneous temperature and field and Stokes signal in Case 4 centred 110 mm above the jet exit plane with a break in the flame circled. Imaging area is 15 mm × 50 mm, and dashed lines represent the I.D. of the central fuel jet.

Typical images centred at 165 mm downstream of the jet exit plane are shown in Fig. 7. At this height, the flame sheet exhibits several 'holes' between all images taken of Cases 1 or 4. These 'holes' all occur in Case 4, which also demonstrates a cooler jet region than Case 1. These images suggest that the central jet region in both cases is near 800 K, close to the low temperature ignition temperature of *n*-heptane near 830 K [13]. Despite low instantaneous temperatures along the centreline in Fig. 7 being evident at this height in all cases, spatial fluctuations in the location of this cooler region result in a thermocouple measurement of 1250 K along the jet centreline at this height for Case 2. This mean centreline temperature, at approximately x/D of 36, is lower than in diluted, non-premixed natural gas/H_2 flames, which exhibit centreline temperatures in excess of 1200 K as close to the jet as $x/D = 20$ [7]. In Case 2, the mean temperature centreline at $x/D = 19$, measured using a thermocouple, is 925 K. This discrepancy is found despite both fuels (*n*-heptane and the natural gas/H_2 blend) having similar (undiluted) stoichiometric flame temperatures with air. These initial results suggest that the central jet region only begins to exceed 800 K close to 36 jet diameters from the jet exit plane, however the structure of the reaction zone at this location requires further investigation.

4 Conclusions

Temperature fields of turbulent, autoignitive, lifted, *n*-heptane flames issuing into four different hot and diluted coflows have been imaged using NTLAF thermography. These data extend from flame base region to over 330 mm downstream and are complemented by gas composition measurements of the mixing field between the fuel jet, hot coflow and quiescent air streams. Thermography using the NTLAF technique suggest that flames issuing into the natural gas/air coflow are hotter than those issuing into ethylene/air coflows. Whilst both cases examined in detail demonstrate 'holes' in the flame sheet in, and just beyond, only the flames issuing into the natural gas/air coflow demonstrate 'holes' at $x/D = 36$. At this height, the centreline temperature is close to that of the low temperature ignition temperature of *n*-heptane, warranting further investigation, and the mean temperature less than jet flames previously studied with the same diagnostic technique.

5 Acknowledgments

The authors thank Dr Alfonso Chinnici from the University of Adelaide, and Mr Qian Li from Zhejiang University for their assistance in data collection. Financial support was provided by through the Australian Research Council (ARC) Discovery (DP and DECRA) grant scheme, and the United States Asian Office for Aerospace Research and Development (AOARD).

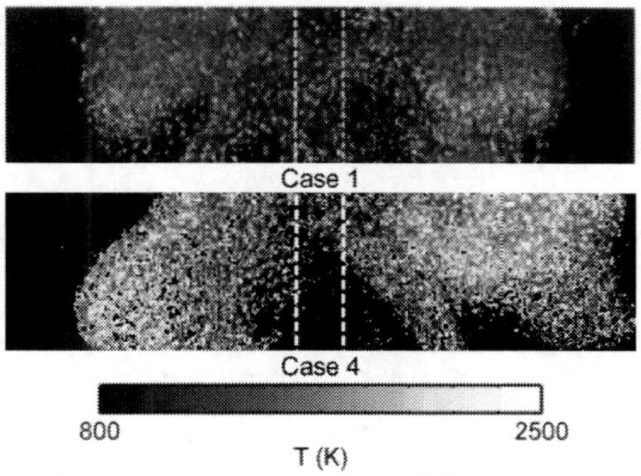

Figure 7: Instantaneous temperature fields centred 165 mm above the jet exit plane. Imaging area is 15 mm × 50 mm, and white lines represent the I.D. of the central fuel jet.

References

[1] B.B. Dally, A.N. Karpetis, R.S. Barlow, Proc. Combust. Inst. 29 (2002) 1147-1154.

[2] J. Ye, P.R. Medwell, M.J. Evans, B.B. Dally, Combust. Flame 183 (2017) 330-342.

[3] R. Cabra, J.Y. Chen, R.W. Dibble, A.N. Karpetis, R.S. Barlow, Combust. Flame 143 (4) (2005) 491-506.

[4] A.R.W. Macfarlane, M.J. Dunn, M. Juddoo, A.R. Masri, Proc. Combust. Inst. 36 (2017) 1661-1668.

[5] T.F. Guiberti, H. Cutcher, W.L. Roberts, A.R. Masri, Energy Fuels 31 (3) 2128-2137.

[6] P.R. Medwell, P.A.M. Kalt, B.B. Dally, Combust. Flame. 152 (2008) 100-113.

[7] P.R. Medwell, Q.N. Chan, B.B. Dally, S. Mahmoud, Z.T. Alwahabi, G. J. Nathan, Proc. Combust. Inst. 34 (2013) 3619-3627.

[8] D. Gu, Z. Sun, B.B. Dally, P.R. Medwell, Z.T. Alwahabi, G.J. Nathan, Combust. Flame 179 (5) (2017) 33-50.

[9] P.R. Medwell, Q.N. Chan, B.B. Dally, Z.T. Alwahabi, S. Mahmoud, G.F. Metha, G.J. Nathan, Appl. Phys. B 107 (3) (2012) 665-668.

[10] M. Ihme, Y.C. See, Proc. Combust. Inst. 33 (2011) 1309-1317.

[11] M.J. Evans, P.R. Medwell, H. Wu, A. Stagni, M. Ihme, Proc. Combust. Inst. 36 (2017) 4297-4304.

[12] M.J. Evans, P.R. Medwell, Z.F. Tian, J. Ye, A. Frassoldati, A. Cuoci, Combust. Flame 178 (2017) 297-310.

[13] N. Peters, G. Paczko, R. Seiser, K. Seshadri, Combust. Flame 128 (1-2) (2002) 38-59.

11th Asia-Pacific Conference on Combustion,
The University of Sydney, NSW Australia
10th -14th December 2017.

Modelling Turbulent Transport Effects on Kernel Formation

Linghan Kong, Lingyun Hou*, Zhuyin Ren

School of Aerospace Engineering, Tsinghua University,

Beijing, 100084, China

Abstract

As an important section of the working process of gas turbine engine, ignition play a significant role in the research of combustion and gas engine. For ignition kernel produced by electric spark, small-scale turbulence exists inside the post-flame zone and preheat zone could enhance the mixing and heat transmission process. The present work investigates the effects of small-scale turbulence on ignition and flame kernel propagation by means of the incorporation of turbulence induced diffusion in the preheat and post-flame zone of one-dimensional spherical flames for n-dodecane/air, simulated with a 54-species chemical mechanism at high-altitude gas conditions. It is found that the width of flammability limit decreases with increased turbulent intensity. Detailed investigation of effects of small-scale turbulence on flame structure and flame speed are presented and it shows that turbulence can broaden the flame front and enhance the heat release rate there, eventually increasing the absolute flame speed.

1 Introduction

It is becoming increasingly common to operate gas turbines under fuel-lean or diluted operating conditions due to stringent emission regulations and efficiency concerns, and the development of advanced low-emission aero-engine combustors demands more careful design of ignition devices and their placement in the combustion chamber to ensure reliable relighting in case of flame blow-out [1-2]. In particular, the relight capability at high altitudes is critical. There is a strong practical interest to develop a better understanding of physical processes involved and consequently improve numerical methods capable of predicting the physio-chemical processes involved.

Currently, electrical spark is the most common type of ignition systems in gas turbine engines. The success of the spark in initiating and propagating the flame is a complex function of all flow-field variables. Four stages are basically identified in the successful relight of a gas turbine [3-4]: (i) energy deposition, (ii) formation of a viable flame kernel, (iii) flame growth and propagation, and (iv) stabilization. The establishment of an ignition probability map for a real industrial burner by experimental means is both practically challenging and expensive. As for the simulation of ignition probability, a combination of low-dimensional ignition model and 3D CFD simulations is needed. For example, a one-dimensional ignition model is developed for modelling the ignition stage (i) and (ii) [5]. This model is used to build the ignition probability map with several simple criteria, and is introduced to provide initial conditions for the 3D flame propagation simulations. However, this model is less accurate for flame kernel formation at highly turbulent conditions in which turbulence-chemistry interaction has to be accounted for.

To simulate the effects of turbulence in one-dimensional model, a reduced order model for turbulent scalar transport is needed. Ren et al [6] developed a method that incorporate turbulence induced diffusion in 1-D unstretched premixed flames. The interaction between small-scale turbulence and chemistry is accounted for via enhanced diffusion in the preheat zone of a premixed flame. However, this model need to be modified to simulate the ignition process which is homogenous at the spherical coordinate, and a detailed model for turbulent diffusion coefficient is also needed.

The aim of this study is to develop a reduced order model, which accounts for the effect of turbulent transport on flame kernel formation through the incorporation of turbulence-induced diffusion in the flame kernel. As an outline of the paper, in Section 2, the one-dimensional spherical flame model with turbulent transport is described with the focus on the construction of turbulence-induced diffusivity and numerical implementation. In Section 3, the effects of turbulent transport on flame kernel formation and initial stage of flame kernel propagation at two representative conditions are investigated. Conclusions are in Section 4.

2 One dimensional flame kernel formation model with turbulent transport

In the model, spark energy is assumed to be uniformly deposited in a spherical region, two-phase flow and evaporation is not considered here. The one-dimensional governing equations in spherical coordinates are

$$\frac{\partial \rho}{\partial t}+\frac{1}{r^2}\frac{\partial}{\partial r}(r^2\rho u_r)=0 \tag{1}$$

$$\rho\frac{\partial Y_i}{\partial t}+\rho u_r\frac{\partial Y_i}{\partial r}-\frac{1}{r^2}\frac{\partial}{\partial r}\left(r^2\rho D_{i,g}^{*}\frac{\partial Y_i}{\partial r}\right)=\dot{\rho}_{i,\chi} \tag{2}$$

$$\rho c_p\frac{\partial T}{\partial t}+\rho u_r\frac{\partial h}{\partial r}-\frac{1}{r^2}\frac{\partial}{\partial r}\left(r^2\rho\lambda_g^{*}\frac{\partial T}{\partial r}\right)=\sum_{i\epsilon\chi}\dot{\rho}_{i,\chi}h_i \tag{3}$$

Where ρ, u_r, Y_i, T, h denote the density, velocity, mass fraction of species i, temperature and enthalpy, c_p is mass-averaged heat capacity, $\dot{\rho}$ denotes the mass source term and χ means the term is related to chemical reaction. The effective diffusivity $D_{i,g}^{*}$ and conductivity $\lambda_g^{*}$are given by:

$$D_{i,g}^{*}=D_{i,g}+D_{turb}(r,t) \tag{4}$$

$$\lambda_g^{*}=\lambda_g+\lambda_{turb}(r,t) \tag{5}$$

where $D_{i,g}$, λ_g are the laminar diffusivity and conductivity, D_{turb} , λ_{turb} are the turbulent diffusivity and conductivity varying with time and

Corresponding author. Lingyun Hou
E-mail address: lyhou@mail.tsinghua.edu.cn

position. A detailed model is needed to estimate the value of D_{turb}. With a unity turbulent Lewis number, λ_{turb} is computed as:

$$\lambda_{turb} = \rho c_p D_{turb} \tag{6}$$

In order to show the interaction of turbulence and combustion, the whole calculation area is divided into four regions: post flame, reaction, preheat and unburned zones, which is showed in Fig.1 in detail. The treatment of turbulence is different at four regions. For the post-flame zone, turbulences whose length scale is larger than the radius of ignition kernel cause the ignition kernel to be displaced and this effect cannot be simulated by the current model. For the reaction zone, it is considered that turbulence had no effects on this zone. For the preheat zone, turbulences whose length scale is larger than the thickness of preheat zone will cause the flame front to be wrinkled and this effect cannot be simulated by the current model. For the unburned zone, turbulence effects are ignored since there is no scalar gradient at this zone.

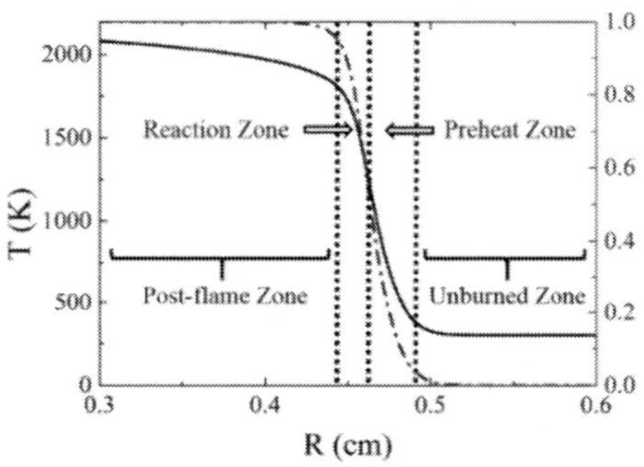

Figure 1: Temperature (solid line) and progress variable (dash dot) against radius in the laminar spherical flame, the boundary of four regions is shown.

From the discussion above, it is clear that the turbulent diffusivity varies with different zones. The turbulent diffusion coefficient is considered to be zero at reaction zone. For the unburned zone, turbulent diffusivity has no effects so its value is simply set as the value of the abutting preheat zone. Since the thickness of the preheat zone is far less than the ignition kernel radius, the turbulent length scales in the preheat and post flame zones are very different, and must be treated separately. However, there is something in common between the two zones: only turbulent length scales smaller than a given value is taken into consideration. So the Smagorinsky model of LES [7-9] can be used. From this model, the residual eddy viscosity can be calculated from the filter width Δ and the turbulent dissipation rate ε [10]:

$$v_t = \frac{2}{3\pi^{2/3}} C_0 \Delta^{4/3} \varepsilon^{1/3} \tag{7}$$

Here $C_0 \approx 1.5$ is the Kolmogorov constant, then the residual turbulent diffusivity can be calculated as:

$$D_{turb} = \frac{2}{3\pi^{2/3}} C_0 \Delta^{4/3} \varepsilon^{1/3} / Sc_t \tag{8}$$

Here Sc_t is turbulent Schmidt number. If we choose the filter width Δ as the ignition kernel radius R and the thickness of preheat zone δ_p, the turbulent diffusivity at the post-flame zone D_{t1} and at the preheat zone D_{t2} can be directly calculated from (8).

To accurately estimate the turbulent diffusivity, we need a detailed model for calculating the thickness of preheat zone, and the boundaries of the post-flame, reaction, preheat and unburned zones. We choose the mass fraction of oxidized carbon atoms (CO and CO2) in the whole carbon compounds as the progress variable,

$$C = \frac{Y(CO) + Y(CO_2)}{\sum Y(C_x H_y O_z)} \tag{9}$$

The boundary between post-flame and reaction zones is defined as C=95%, and the boundary between preheat and unburned zones is C=5%. C=50% is chosen as the boundary of preheat and reaction zones. Then the thickness of preheat zone could be defined as the distance between the positions of C=5% and C=50%, which was approximately 0.025mm in the laminar flame.

The progress variable C at every space grid is calculated every time step. In addition, when the progress variable at the space center C(0,t) is less than 95%, it is considered that there is no flame structures and the whole calculation area will be affected by the turbulence whose length scale is smaller than the ignition kernel radius R. The value of turbulent diffusion coefficient is calculated from (8) with a given turbulent dissipation rate and turbulent Schmidt number, kernel radius R and thickness of preheat zone δ_p is calculated at every time step so the turbulent diffusion coefficients also vary with time.

The computational domain is set to be [0,1] cm with a homogeneous distribution of unburnt mixture. The boundary conditions are fixed to be equilibrium products at R=0 cm and the fresh stoichiometric mixture at R=1 cm. A second-order, implicit Crank-Nicolson scheme with central differences and a first-order upwind scheme are employed for the diffusion and convection terms, respectively. Grid size is $5 \times 10^{-6} m$ and the time step is $1 \times 10^{-7} s$, which guarantees numerical convergence. The ignition spark energy released by the electrodes is assumed to be uniformly deposited in the region [0, 0.25] cm, and this process is treated as instantaneous. The spark energy is 150mJ, leading to an initial kernel temperature of about 1800K.

3 Results and Discussions

3.1 Effects of turbulent transport on kernel formation process

One-dimensional adiabatic premixed spherical flames of pre-vaporized n-dodecane/air mixtures are simulated under the conditions of high altitude with H=6km, where pressure is 0.5atm and temperature is 230K, together with the detailed mechanism by Yao et al [11] consisting of 54 species and 269 reactions. The turbulent intensities are characterized by the additional turbulent diffusion coefficients. The turbulent transport intensity is denoted by the value of initial additional turbulent diffusion coefficients at preheat zone, which is symbolized by D_t. The value of D_t is zero in the laminar flame, and 1.0, 2.0, 3.0 cm^2/s for turbulent flames, and will increase with time since the thickness of preheat zone also increases with time under the turbulent condition. Turbulent diffusion coefficient at post-flame zone is calculated by $D_t$with the known R and δ_p.

Figure 2 shows the flammability limits of kernel formation of laminar and turbulent ignitions under different turbulent transport intensities. Every cross denotes a calculation case of ignition failure, and the circle represents the successful case determined by the continual increase of the radius of ignition kernel in the whole simulation time (t=5ms). The flammability limits of kernel formation of turbulent ignition is much narrower than laminar ignition, table 1 shows the detailed value. It shows that turbulent ignition has a much harsher equivalence ratio condition than laminar ignition, and turbulence shows inhibition on the formation of ignition kernel.

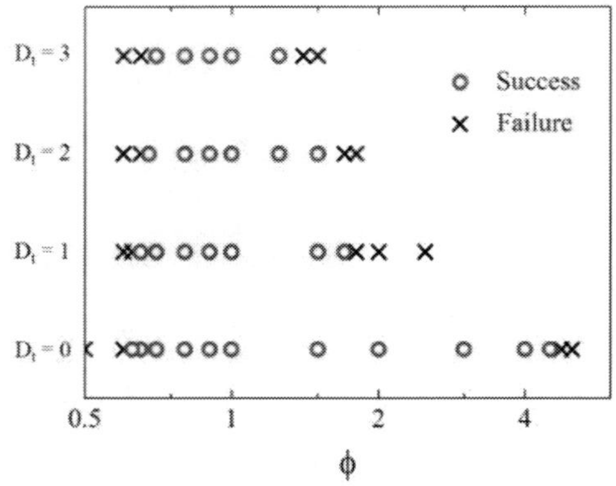

Figure 2: Flammability limits of laminar and turbulent ignitions.

Flame limits	Lean-burn	Rich-burn
$D_t = 0$	0.60	4.5
$D_t = 1$	0.64	1.75
$D_t = 2$	0.66	1.6
$D_t = 3$	0.7	1.3

Table 1: Flammability limits of different Dt.

3.2 Effects of turbulent transport on kernel propagation process

During the process of kernel propagation, Fig. 3 shows the comparison of the temperature change rate between diffusion and reaction terms in the energy equation under different turbulent intensities. From Fig.3 (a), kernel radius increases from 0.25cm to 0.6cm in 2ms and the heat dissipation rate at flame front is smaller than heat production rate, which means the ignition kernel can successfully propagate. On the contrary, Fig.3 (b) denotes a kernel propagation failure case with high turbulent intensity. In this case the ignition kernel formation process is successful and the flame front can be detected. However, ignition kernel cannot propagate because the heat dissipation rate at flame front is larger than heat production rate, the kernel radius decreases with time.

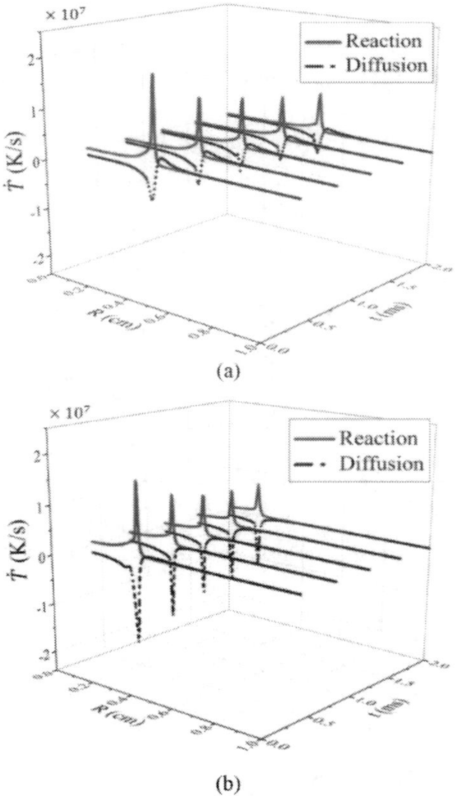

Figure 3: Reaction term and diffusion term in the energy equation with time and radius (a) $D_t = 1$ (b) $D_t = 3$. $\phi = 1.5$.

Figure 4 shows the temperature and species mass fraction of laminar ignition ($D_t = 0$) and turbulent ignition ($D_t = 3$) with the radius. It is clear that the added turbulent diffusion broadens the preheat zone based on the temperature profile. The distributions of radicals such as CHx are also thickened as greatly as temperature distribution. Note that turbulent diffusion can also affect the peak radical concentrations. For example, turbulent diffusion slightly reduces the peak value of CHx by 20%. In the post-flame zone, the added turbulent diffusion reduces the temperature and temperature gradient, and the distributions of hydroxyl radicals are also thickened and the peak value increases by 15%.

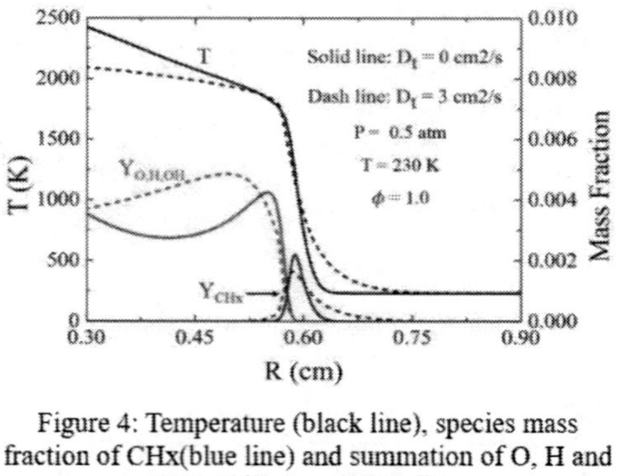

Figure 4: Temperature (black line), species mass fraction of CHx(blue line) and summation of O, H and OH(red line) against radius.

Figure 5 shows the spatial profiles of reduced chemical heat release rate at different ignition kernel radius with a comparison of laminar and turbulent conditions. Here heat release rate ω_F is divided by ρc_p so that its value is equal to the local temperature change rate. The peak value of heat release rate increases by approximately 50% at turbulent conditions and the increment doesn't change a lot with time. On the other hand, the thickness of heat release rate profiles is broadened by adding turbulent diffusion and this effect is strengthened with an increase in ignition kernel radius. Heat release rates at post-flame and preheat zones are enhanced and is more significant at post-flame zone.

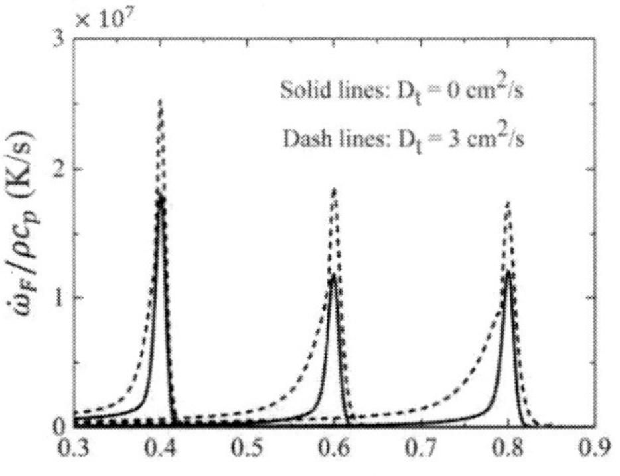

Figure 5: Reduced chemical heat release rate at different ignition kernel radius, $\phi = 1.0$.

Figure 6 shows the absolute flame speed as a function of stretch rate at different turbulent diffusion coefficients. Linear extrapolation is conducted to obtain the un-stretched flame speed in the range of flame radius from 0.5 to 1 cm. When the flame stretch is relatively large (it means that flame radius is small), the flame speed slightly increases with increasing turbulent diffusion. On the other hand, the slope of the extrapolation line, which corresponds to the Markstein length, and un-stretched flame speed S_{a0} increase with increasing turbulent intensity. The absolute flame speed rapidly increases with turbulent transport intensity and is 2.85 times of the laminar flame speed at $D_t = 3$.

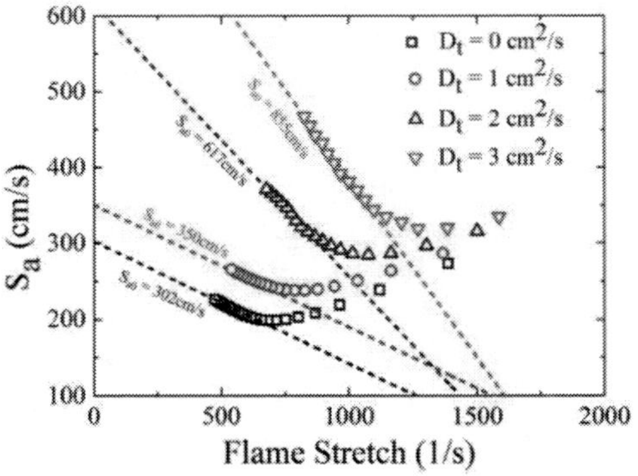

Figure 6: Absolute flame speed as a function of stretch rate under different turbulent diffusion coefficients, $\phi = 1.0$.

4 Conclusions

A detailed small-scale turbulent influencing method for one-dimensional ignition calculation was developed to predict the turbulent ignition characteristics by means of the incorporation of turbulence induced diffusion in spherical premixed flame. Flammability limits under different turbulent intensities were calculated, and the effects of small-scale turbulence on the ignition kernel propagation was presented. It is shown that the width of flammability limit decreases with an increase in turbulent intensity. In the stage of kernel formation, turbulent diffusion enhanced the heat dissipation at the flame front, and ignition failure occurs when heat dissipation is larger than the heat product at the flame front. In the stage of kernel propagation, turbulent diffusion broadens the thickness of preheat zone and spatial profiles of radicals, and then enhance the heat release rate, eventually increases the absolute flame speed, which is positive to ignition.

Acknowledgement

The work is supported by National Natural Science Foundation of China 51476087 and 91441202 and by Gas Turbine Initiatives at Tsinghua University 20141081218.

Reference

[1] I. Glassman, R. A. Yetter. Combustion (Fourth Edition), 2008, p. 379-408.
[2] C. Law, Combustion Physics, Cambridge University Press, 2006, p. 305-313.
[3] A.H. Lefebvre, Gas Turbine Combustion (Second Edition), Crc Press, 1998.
[4] E. Mastorakos, Prog. Energy Combust. Sci. 35 (1) (2009) 57-97.
[5] A. Bruyat, G. Linassier, P. Villedieu, Towards Fuel Spray Ignition in Aircraft Engine, ICLASS 2012, 12th Triennial International Conference on Liquid Atomization and Spray Systems, Heidelberg, Germany, 2012.
[6] Z. Ren, H. Yang, T. Lu, Fuel 115(1) (2014) 241-247.
[7] P. Sagaut, Measurement Sci. & Technol. 12(10) (2001) 1745-1746.
[8] S.B. Pope, Prog. Energy Combust. Sci. 11(2) (1985) 119-192
[9] S.B. Pope, Turbulent Flows 12(11) (2000) 806.
[10] S.B. Pope, Annual Review of Fluid Mechanics, 19(1) (1987) 237-270.
[11] T. Yao, Y. Pei, B. J. Zhong, Fuel, 191 (2017) 339-349.

11th Asia-Pacific Conference on Combustion,
The University of Sydney, NSW Australia
10th -14th December 2017.

Influence of spray impingement on fuel combustion and emissions

Tianyu Ma, Haifeng Liu, Hu wang, Lei feng, Mingfa Yao*

State Key Laboratory of Engines, Tianjin University

Weijin Road 92#, Tianjin, 22, China

Abstract

The effects of spray wall impingement on fuel combustion and emissions were numerically studied based on a newly developed spray impact model. The model shows good performance in spray impingement cases under constant volume system conditions. The numerical result indicates that the ambient temperature and wall temperature greatly affects the fuel film deposition and the combustion process. Higher ambient temperature could accelerate fuel evaporation and reduce fuel deposition. The peak soot production increase linearly with the deposited film mass. Different near wall combustion processes are found based on different deposited film mass. For insignificant fuel deposition case, the near wall mixture dominates the combustion process. Higher wall temperature accelerates the evaporation and products high temperature mixture in the near wall region, results in faster soot oxidation. For significant wall film case, a typical pool fire phenomenon can be observed. Higher wall temperature could advance the fuel film combustion process, and accelerates the production and oxidation of soot.

1 Introduction

Fuel injection technology has been improved a lot since the development of common rail electronically controlled fuel injection system. Although the spray/wall impact phenomenon can be partially avoided with multi-injection strategies, there is still chance of spray impingement in port injection, gasoline direct injection and diesel injection under high load conditions. Many researchers have found there is strong relationship between soot emission in the near wall region and spray wall impact [1-3], making it more and more important to study the mechanism of spray impact.

Early injected fuel is easily deposited on walls. Secondary atomization, fuel evaporation and near wall emissions are influenced by the wall film developing process, as shown in Fig.1. Previous experimental research has shown that the film fluctuation is much more violent due to the spray impingement [4]. Desoutter et al. [5] studied the turbulent boundary layer over an evaporating liquid film. Strong coupling effect was found among the film evaporation, boundary layer flow and heat transfer processes based on numerical results. Drake et al. [6] and Jiao et al. [2] studied the effects of piston gasoline film on the combustion process by experimental and numerical methods, respectively. Pool fire was observed on piston surface. A positive correlation was observed between soot emission and film mass. Liu et al. [1] explored the effects of early injection strategy on soot emission in a diesel engine. The soot emission reached peak with -55°CA ATDC injection timing. Most fuel film deposited on cylinder wall, which forms dense mixture in the squish region. Recently, with the development of reactivity-controlled compression ignition (RCCI), the injection timing is further advanced in order to achieve better mixture concentration and reactivity stratification controls.

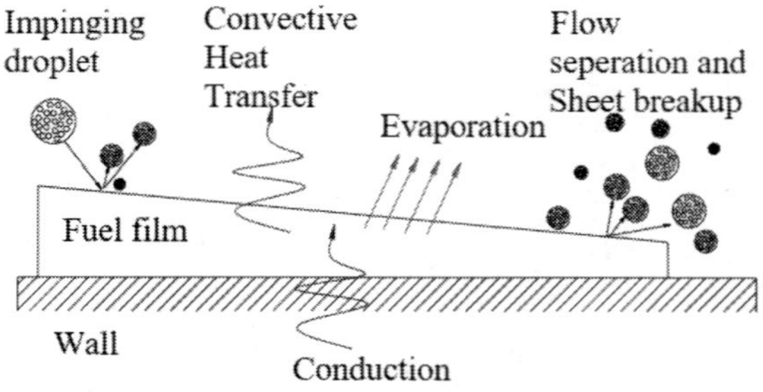

Fig.1 Diagram of wall film formation and evaporation.

Hereinafter, a new spray wall impingement model is developed and validated in spray impact cases in a constant volume combustion chamber to investigate the influence of deposited wall film on combustion and soot emission.

2 Wall impingement model

A new spray wall impingement model is developed based on droplet impact phenomenon [7]. Impact criteria and secondary droplet properties are described, wherein prompt splash and corona splash are added. On the micro level, both droplet incident conditions and wall conditions would influence the impact outcomes. A reduced primary reference fuel mechanism is used for the prediction of soot formation during near wall combustion [8].

2.1 Conversation equations

Mass transformation during droplet impact is an important process, which decides the fuel distribution in the near wall region. In a stick or spread case, all the fuel is added into the wall film part. On the contrary, all the fuel drop flies back into the combustion chamber again for reflect cases. The splashing outcomes are divided into corona splash and prompt splash, as shown in Fig.2. The morphology dynamics of these two splash modes are analyzed based on geometrical parameters.

a) b)

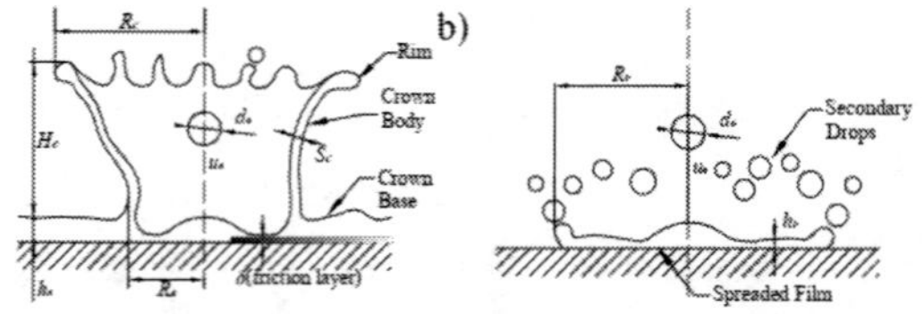

Figure.2 Diagrams of splash. a) corona splash. b) prompt

Corresponding author. Mingfa Yao. Fax: +86 22 27383363
E-mail address: y_mingfa@tju.edu.cn

splash.

The mass conversation equation for corona splash can be described as Equation (1):

$$\pi\rho\left(R_B+R_c\right)S_cH_c+\sum_{n=1}^{N}\sum_{i=1}^{n}m_{si}=\frac{\pi}{6}\rho d_0^3+\eta_m\left(\pi\rho R_B^2h_0\right) \quad (1)$$

Where R_B and R_C represent the radius of crown bottom and rim. Hc and h_0 are the heights of crown and crown base, respectively. Sc is the average thickness of crown. Here the entire rim and a part of crown are considered to transform into secondary droplets. The left-hand sides (LHS) of (1) are the mass of crown and secondary drops respectively, while the right-hand sides (RHS) are incident droplet mass and the splashed film mass. For prompt splash, the crown formation process is replaced by a thin film deposition. Its mass conversation equation is shown as:

$$\pi\rho R_P^2h_P+\sum_{n=1}^{N}\sum_{i=1}^{n}m_{si}=\frac{\pi}{6}\rho d_0^3 \quad (2)$$

where h_p is the residual film thickness. Both (1) and (2) are used for solving the mass of splashing droplets. While the energy conversation equation for splash cases can be expressed as Equation (3):

$$E_0=E_{diss}+E_\sigma+E_{ks}+E_{\sigma s} \quad (3)$$

where E_0 represents the total incident energy and the four terms on the RHS of (3) are dissipation energy, surface energy of crown (including lamella), kinetic energy of secondary droplets and surface energy of secondary droplets, respectively. The incident energy contains both kinetic energy and surface energy:

$$E_0=\frac{1}{12}\pi\rho d_0^3u_0^2+\sigma\pi d_0^2 \quad (4)$$

Here we assumed that the mass of secondary drops originates from both the incident droplet and the wall film. Therefore, the surface energy of wall film should be considered. However, the increased surface energy of wall film is transformed into the crown part and rim part, which is calculated as $E\sigma$, thus the original surface energy of wall film is offset on either side in (4).

2.2 Coupling in CFD code

Kiva 3vR2 is used as a CFD platform for model implementation. Firstly, the droplet incident conditions and the wall condition decide the outcome of the impinging drop. Secondly, the physical properties of secondary drops, such as mass, diameter, velocity, are calculated based on conversations equations shown above. Thirdly, a wall film would be created for dense spray impact cases.

3 Results and discussion

The model is validated and tested in a constant volume vessel for diesel spray impact cases. The shadow of the impinging spray is recorded and transformed into grey-scale maps to analyze the spray radius penetration and the combustion process. A 1/8 sector computational grid, which contains a 15mm refined block near the wall is used in the simulation. Totally 15665 cells are used for calculation. The spray impingement model is first validated in two non-reacting spray impact cases shown in Table 1. Then the influence of spray impingement on combustion and emissions is studied by two cases also shown in Table 1.

Diesel fuel cases	1	2	3	4
Impinging distance(mm)	35	35	35	35
Ambient temperature (K)	400	600	753	723
Wall temperature (K)	400	600	473-723	
Injection pressure (MPa)	60	60	60	60
Ambient pressure (MPa)	1	1	4	4
Fuel mass (mg)	23.9	23.9	23.9	23.9
Injection duration (ms)	5.5	5.5	5.5	5.5
Diameter of injection nozzle (mm)	0.14	0.14	0.14	0.14
Inclination angle of wall (deg.)	0	0	0	0
Number of parcel	3000	3000	3000	3000

Table 1. parameters of spray impingement cases

Figure 3 shows the radius penetration of impinging diesel spray for case 1 and 2. It can be seen that the spray development after impact is well captured under different ambient and wall temperature conditions. It is observed that the radius penetration is restrained at high wall temperature impact case. This is because the wall impingement greatly accelerates the droplet break up process, generating more micro drops which are easy to evaporate. High ambient and wall temperature heats the droplets rapidly, such that they could not penetrates far along the wall.

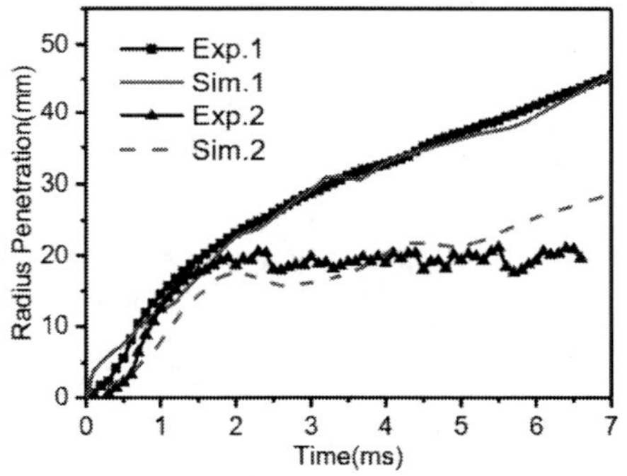

Figure. 3 Radius penetrations of unburnt impinging spray

The influence of wall temperature on spray impingement and combustion are studied by tracking the wall film mass, heat release rate (HRR) and emissions. Figure 4a shows the curve changing of vapor fuel mass and HRR. Although the fuel is consumed faster with the increase of wall temperature, the HRR show similar trends. This can be explained by the wall film mass shown in Fig. 4b. For all three cases, the wall film mass does not exceed 1.0mg, meaning that more than 96% of the injected fuel still remains above the wall and mixing with air. So, the combustible mixture is nearly the same for these three cases, which in turn depict similar HRR curves. However, results presented in Fig. 4b indicate that with the increase of wall temperature, the wall film mass and the soot production decrease together. On one hand, higher wall temperature accelerates the evaporation process in the near wall region, such that less fuel droplets contact the wall and deposits on the wall. On the other hand, High-Temperature mixture is formed in the neat wall region, which promotes soot consumption.

Figure 5 shows the influence of wall temperature on spray impingement and combustion under low ambient temperature (case 4). Nearly 1/4 injected fuel became wall film as shown in Fig. 5b. A typical 'pool fire' happens in the near wall region. It can be seen in Fig. 5a that with the increase of wall temperature,

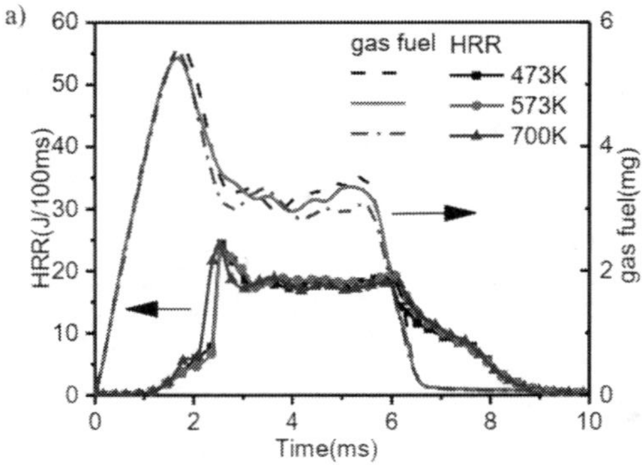

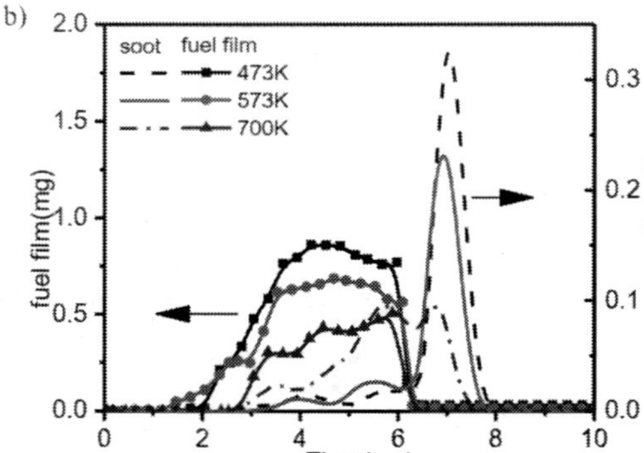

Figure. 4 Influence of wall temperature at 753K (case 3). a) Changing of gas fuel and HRR with time b) Changing of wall film and soot

the wall film evaporates and combusts faster, thus the fuel vapor reduces more quickly. This is verified by the HRR curves: lower wall temperature greatly postpones the heat release rate peak, which slows down the fuel vapor consumption. What is more, with the increase of wall temperature, the soot decreases. This can be attributed to higher combustion temperature in the neat wall region which accelerates soot oxidation.

Comparing the onset time of soot formation in Fig. 4b and Fig. 5b, it is found that the soot generates when the film begins to accumulate in Fig. 4b, while the soot appears after the film evaporates and combusts in Fig. 5b. This explains the difference between near wall mixture combustion and pool fire. These two-different combustion process are mainly distinguished by the near wall mixture temperature. High wall region temperature promotes the film evaporation and heats mixture rapidly. Thus, the combustion process is accelerated, resulting in faster soot oxidation.

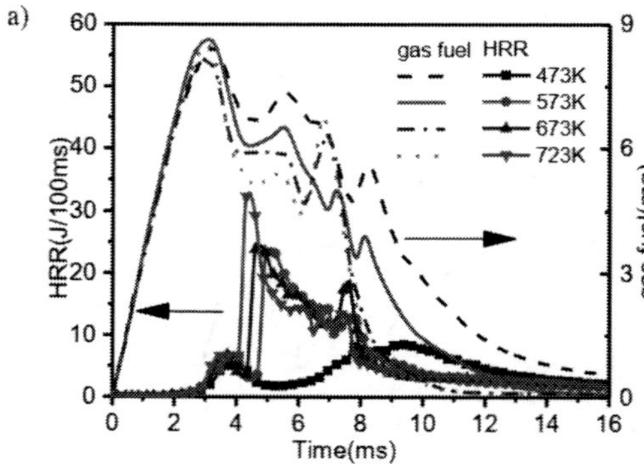

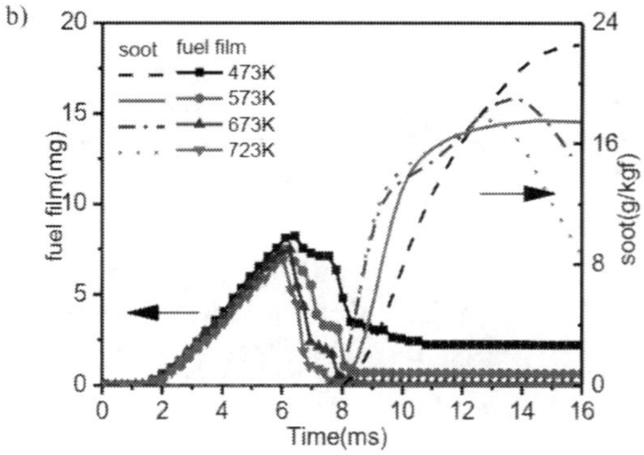

Figure. 5 Influence of wall temperature at 723K (case 4). a) Changing of gas fuel and HRR with time b) Changing of wall film and soot

Comparing all these cases above, it is found that the ambient temperature has a significant effect on film formation and combustion process. Hereinafter, images have been collected to analyze the difference in combustion process between case 3 and case 4. 3D slices of the spray field are used for better understanding the near wall temperature and soot variations. Fig. 6 shows the high-speed images of flame at 753K and the nephograms of temperature and soot. Local equivalence ratio is also marked by contour lines, shown in Fig. 6c. It is observed that the flame is relatively stable. An obvious combustion region is formed after spray impingement, shown in Fig. 6a and 6b. The temperature distribution also proved that the combustion region is stable, which does not move along the wall (Fig. 7c). This means that the spray fuel is well mixed in this region and the process is continuous during the entire impinging process. Thus, the soot can be quickly oxidized, resulting in less soot emissions (Fig. 6d).

However, when the ambient temperature decreases to 723K, the combustion region becomes unstable, as shown in Fig. 7a and 7b. The mixture in the spray outer vortex region firstly burns, the flame gradually spreads towards the inner region and ignites the wall film. Fig. 7c clearly shows the variation of temperature and mixture distribution. The mixture in the spray inner region does not burn due to lower ambient temperature. When the deposited fuel film being ignited by the outer flame, a typical pool fire phenomenon happens and results in a large amount of soot formation (Fig. 7d).

Moreover, the development of deposited fuel film is shown by Fig. 6e and Fig.7e. A high ambient temperature(753K) directly leads to neglectable wall film, which concentrates on the center of spray axis. In contrast, the film spreads a wider region under lower temperature(Fig.7e). This deposited fuel postpones the combustion time of the spray inner region, resulting in pool fire and raised soot production.

Comparing the two cases above, it is seen that the ambient temperature greatly influences the combustion process after spray impingement. The deposited film near the spray inner region is shown to be the main source of soot production. Thus the temperature of this region controls the combustion rate of fuel film. The evaporation rate increases with higher ambient temperature, thus accelerates the mixing process in the near wall region. This results in more stable combustion in the near wall region and consequently generates less soot.

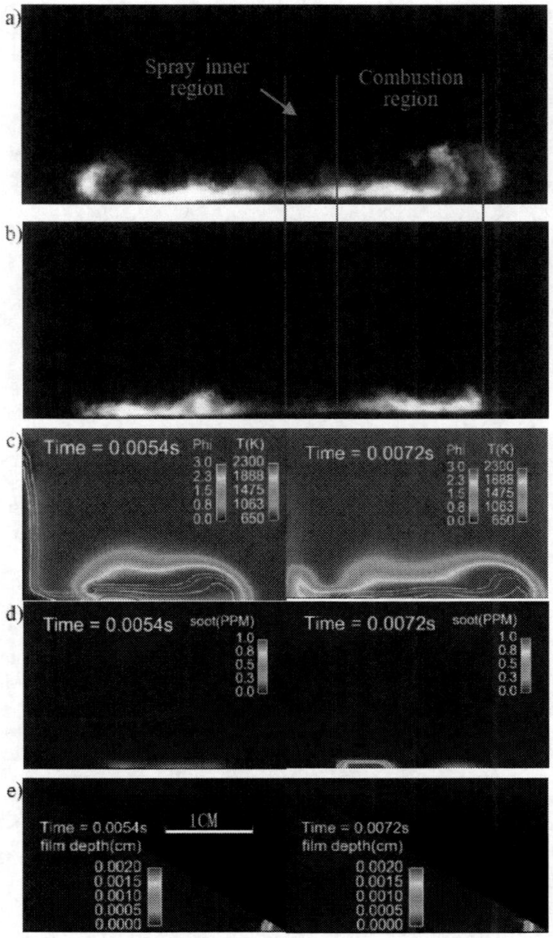

Figure. 6 Pictures of combustion field at 753K. a) Flame photo at 5.4ms b) Flame photo at 7.2ms c) Nephogram of temperature and phi d) Nephogram of soot e) Deposited film depth

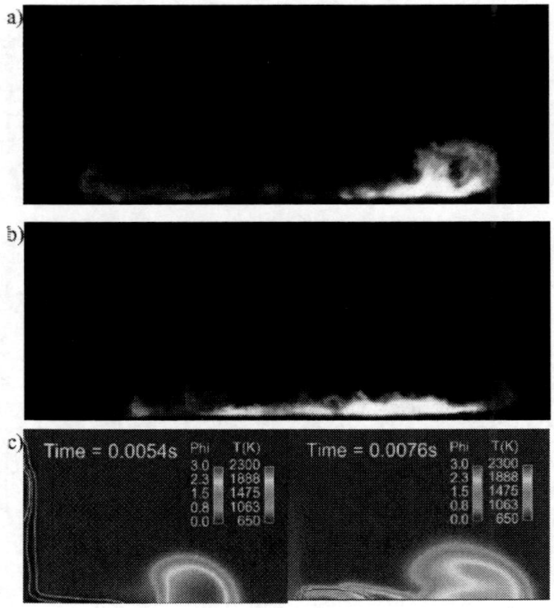

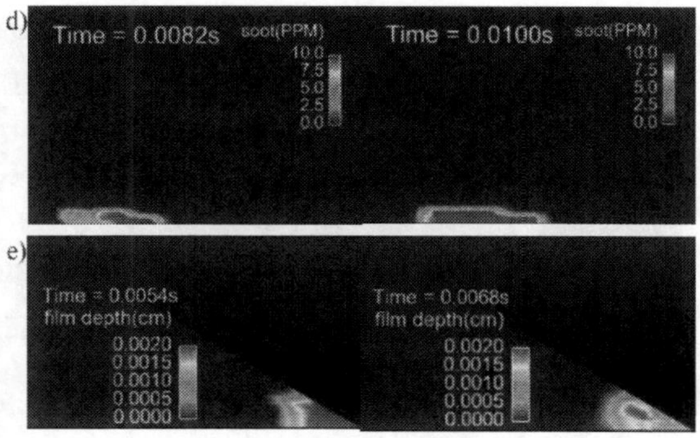

Figure. 7 Pictures of combustion field at 723K. a) Flame photo at 5.4ms b) Flame photo at 7.2ms c) Nephogram of temperature and phi d) Nephogram of soot e) Deposited film depth

4 Conclusions

A new spray wall impingement model is used to study the spray impingement phenomenon and following combustion process. The result shows that the spray impingement combustion process is sensitive to the ambient temperature. Higher ambient temperature accelerates the mixing process and generates stable combustion in the near wall region. In contrast, lower ambient temperature causes more fuel film deposition, which exacerbates the combustion process and generates more soot.

Wall temperature directly influences the variation of wall film and near wall mixture temperature. When there is large amount of wall film, a typical pool fire phenomenon is observed. Higher wall temperature could advance the fuel film combustion, and accelerate the production and oxidation of soot. However, when there is less fuel film, high wall temperature accelerates the evaporation and products higher temperature mixture in the near wall region, results in faster soot oxidation.

5 Acknowledgment

This work was supported by National Natural Science Found of China (NSFC) through the Projects of 91541111 and 51320105008.

References

• H.F. Liu, S.Y. Ma, Z Zhang, Z.Q. Zheng, M.F. Yao, Fuel 139: 472-481, 2015.

• Q Jiao, R.D. Reitz, International Journal of Engine Research, 2015, 16(8): 994-1013.

• B Wang, S Mosbach, S Schmutzhard, Applied Energy, 2016, 163: 154-166.

• D Kalantari, C Tropea, Fluid Dyn. Mater. Process, 2014, 10: 37-61.

• G Desoutter, C Habchi, B Cuenot, International Journal of Heat & Mass Transfer, 2009, 52(25-26):6028-6041.

• M.C. Drake, T.D. Fansler, A.S. Solomon et al, SAE transactions, 2003, 112(3): 762-783.

• T.Y. Ma, L Feng, H Wang, H.F. Liu, M.F.Yao, International Journal of Heat and Mass Transfer, 2017, 112: 401-412.

• H.Wang, M.F.Yao, R.D. Reitz, Energy & Fuels, 2013, 27(12): 7843-7853.

11th Asia-Pacific Conference on Combustion,
The University of Sydney, NSW Australia
10th -14th December 2017.

Online multi-point LIBS measurement on sodium release during the combustion of Zhundong coal and coal/additive blends

Yingzu Liu[1], Zhihua Wang[1], Kaidi Wan[1,*], Yu Lv[2], Yong He[1], Jun Xia[3], Kefa Cen[1]

[1]State Key Laboratory of Clean Energy Utilization, Zhejiang University, Hangzhou 310027, China

[2]Department of Mechanical Engineering, Stanford University, Stanford, CA 94305, USA

[3]Department of Mechanical, Aerospace and Civil Engineering & Institute of Energy Futures, Brunel University London, Uxbridge UB8 3PH, UK

Abstract

The retention performance of 5 different sorbent additives, including 2 pure additives, i.e., silica and alumina, and 3 typical natural mineral additives, i.e., kaolin, mica and pyrophyllite, on the release of sodium during the combustion of Zhundong coal is investigated using combined online- and offline-measurement techniques. An online multi-point Laser-Induced Breakdown Spectroscopy (LIBS) technique, upgraded from the single point LIBS, is used for the in-situ measurement of the time-resolved sodium release flux of the coal and coal/additives mixtures. On the other hand, offline inductively coupled plasma atomic emission spectrometer (ICP-AES) is applied to examine the individual behavior of each sodium class. The online and offline measurement results for the sodium release of Zhundong coal and the coal/additives mixtures are compared and verified with each other. The study suggests that all the additives can effectively reduce sodium release during the whole three stages of coal combustion, and the highest sodium retention efficiency is achieved in the ash cooking stage. The variation of the content of the four different classes of sodium indicates that the mineral additives reduce the global sodium release via reacting with the water-soluble sodium to form insoluble sodium.

1 Introduction

Coal, as an energy source, supports approximately 40% of the worldwide electricity [1]. It was known that the presence of sodium in coal can lead to fouling and corrosion of heat transfer surfaces of coal-fired boilers [2]. These sodium-induced issues significantly restrict the utilization of the low rank coal with a relatively high sodium concentration, e.g., Zhundong coal [3, 4].

Better understanding of the release characteristics of sodium plays a key role in the development of sodium-controlling technologies. In the past decades, research on sodium release during coal combustion mainly employs offline measuring techniques, e.g., sampling measurements by analyzing the bulk composition of fly ash and ash deposits [5]. According to the pioneering work in the field, the forms of sodium in coal can be categorized to four classes [6, 7]: (1) water-soluble sodium, which can be dissolved within moisture and exist as Na salts; (2) NH_4Ac-soluble sodium, which is organically bounded with carboxyl groups; (3) HCl-soluble sodium, which is organically bounded with nitrogen- or oxygen-containing functional groups; (4) insoluble sodium, which is bounded with clay minerals. Although offline measurements can detect the final amount and composition of sodium species at post-combustion stage, it cannot directly capture the dynamic sodium release process and obtain the time-resolved information of the sodium release rate at different coal combustion stages. To overcome these disadvantages of offline techniques, online measurements using advanced laser diagnostics, e.g., Laser-Induced Fragmentation Fluorescence (ELIF) [8], Tunable Diode Laser Absorption Spectroscopy (TDLAS) [9], Laser-Induced Breakdown Spectroscopy (LIBS) [10] and Planar Laser-Induced Fluorescence (PLIF) [11], have been recently implemented. However, to our best knowledge, the directly time-resolved measurement of total sodium flux from a burning coal particle via multi-point LIBS technique has not been reported.

Pulverized-sorbents can be injected into boilers along with pulverized-coal as additives to reduce sodium release from the coal during combustion. Since this method can be directly applied in the current combustion systems without any major reconfiguration, it has attracted a lot of attentions [12-19]. Previous studies on additives for the retention of sodium release found that silica (SiO_2) and alumina (Al_2O_3) are the two important and effective compounds [15]. However, these studies on additives for sodium retention employed offline measurements to monitor the amount of sodium released, and therefore there is no information about the temporal release characteristics of sodium for coal blended with additives. Moreover, the sodium retention performance of additives at different stages of coal combustion has rarely been revealed yet.

In the present study, the performance of different sorbent additives on the release of sodium during the combustion of Zhundong coal will be investigated by combined online- and offline-measurements. Online multi-point LIBS technique, upgraded from the single point LIBS [20], will be employed for the in-situ measurement of the time-resolved sodium release flux of the coal and coal/additives mixtures. Offline inductively coupled plasma atomic emission spectrometer (ICP-AES) will be applied to investigate the individual behavior of each sodium class. The offline ICP-AES results will serve as a reference to verify the online quantitative multi-point LIBS measurements. Based on these experiments, the advantage and disadvantages of different additives for sodium retention of Zhundong coal have been discussed.

Corresponding author. Fax: +86-571-87951616
E-mail address: wankaidi@zju.edu.cn

2 Experimental methods

2.1 Coal and mineral additives

A typical Chinese brown coal called Zhundong coal, which has appreciable sodium content, is considered in the present study. The results of proximate analysis and ultimate analysis, as well as the ash composition of the coal sample can be found in [20]. Different additives were blended with Zhundong coal to explore their effects on sodium retention. Specifically, typical natural mineral additives, i.e., kaolin, mica and pyrophyllite, and pure silica and alumina are employed. The physical and chemical properties of these additives have been summarized in [20]. All the additives are blended with the coal at a dosing ratio of 3% by weight, which is a typical dose for sorbents [20].

2.2 Identification of sodium classes

Different classes of sodium are identified according to their solubility in different solvents via a chemical fractionation method. 1 g of the coal sample was first added into 100 ml water at 333 K and stirred for 24 hours. The mixture is filtered afterwards. Inductively coupled plasmas atomic emission spectroscopy (ICP-AES) is used to analyze the filtrate to identify the content of first class water-soluble sodium like sodium sulfates and chlorides. The solid filter residue is then sequentially dissolved in 100 ml NH_4Ac (1 mol/L) and 100 ml HCl (1 mol/L), with the filtrate of each case analyzed using ICP-AES. This analysis is able to identify the second class of sodium organically bounded with carboxyl groups, and the third class organically bounded with nitrogen- or oxygen-containing functional groups. The final residue after HCl extraction is digested in strong acid and also analyzed via ICP-AES to obtain the information of the last class non-soluble sodium, such as sodium silicate.

2.3 Multi-point LIBS measurement system

The coal and additives are first grounded and sieved to a fine powder (<75 μm). Then they are blended at the desired mass ratio (97% coal and 3% additives). 50 mg of the coal/additive mixture is pressed into a 4 mm spherical pellet.

A spherical coal pellet is suspended on two ceramic rods (diameter of 1 mm) at a height of 10 mm above a heat flux burner. The burner produces a laminar premixed methane/air flame at an equivalence ratio of 0.8 with the flow rates of methane and air at 0.59 SL/min and 7.06 SL/min, respectively. Based on CHEMKIN simulation of the flame with GRI-3.0 mechanism, the temperature at the pellet height is estimated to be ~1892 K with the following primary gas composition: 3.9% O_2, 7.6% CO_2, 15.4% H_2O and 72.8% N_2.

An Nd:YAG laser (Spectra Physics, Model PRO-250) with a fundamental wavelength of 1064 nm is focused into the gas plume 10 mm above the burning coal pellet. The repetition rate, pulse duration and average laser energy are 10 Hz, 10 ns and 300 mJ, respectively. The LIBS signal is collected by an Ocean Optics spectrometer (Model USB 4000). The laser and spectrometer are synchronized via a digital pulse generator (Stanford Research System, DG535). The laser focusing lens and LIBS signal collection optics are installed on an electric translational platform to perform multi-point LIBS measurement (Fig. 1). The LIBS measuring point is moving periodically in the radius distance (r) of -12 mm, -9 mm, -6 mm, -3 mm, 0 mm, 3 mm, 6 mm, 9 mm, 12mm. The moving frequency of the platform is 10 Hz, while the LIBS measuring frequency at the same point is 1 Hz.

Based on the sodium concentration of all the measuring points, the flux of sodium ($Na_{flux,t}$) passing through the horizontal plane 10 mm above the burning coal pellet can be determined as:

$$Na_{flux,t} = \int_0^{2\pi}\int_0^{\infty} u \times C_{Na} \times r \times dr \times d\theta \quad (1)$$

where u is the axial velocity and θ is the radian. In the present study, the pellet is spherical and burned in a uniform gas flow provided by the burner. Hence, Eq. (1) can be simplified as:

$$Na_{flux,t} = 2\pi \times \int_0^{\infty} u \times C_{Na} \times r \times dr \quad (2)$$

The sodium flux $Na_{flux,t}$ at a given combustion time can be obtained by performing an integration of a function of r. More details about the procedure can be found in our previous study [Y.Z. Liu, Y. He, Z.H. Wang, K.D. Wan, J. Xia, J.Z. Liu, K.F. Cen, Measurement and kinetics of sodium release from a burning Zhundong coal particle using multi-point LIBS method, Combust. Flame, (2017) under revision].

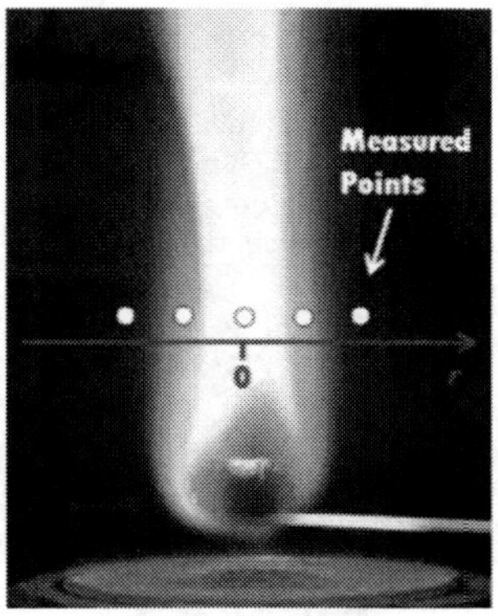

Figure 1: Multi-point LIBS measuring points.

3 Results and discussion

3.1 Global sodium retention characteristics of different additives

Pulverized Zhundong coal samples with different additives was burnt in a muffle furnace at 815 °C to produce ash according to the Chinese National Standard GB/T 1574-2007. By performing the ICP-AES measurements for the raw coal and the residual ash, the global amount of sodium released during coal combustion can be quantified.

The content of the different classes of sodium in Zhundong raw coal and the ash collected from different burnt coal samples are summarized in Table 1. The values given in the table are the mass of Na (mg) in a sample prepared from 1 g of Zhundong raw coal. Corrections have been made to account for the reduction of sodium in the sample due to the displacement of coal by sorbent additives. For Zhundong raw coal, the major sodium classes are water-soluble (66.8%) and insoluble (21.3%) sodium, while the content of the other two sodium classes, i.e., NH_4Ac-soluble and HCl-soluble sodium, are minor. Similar characteristics can also be observed in the ash prepared without additives. For coals with the 3% additives, the insoluble sodium (57.0%–85.9%) plays a dominant role in the 4 classes of sodium. For the sample with 3% pyrophyllite, its total sodium content is slightly higher than that in Zhundong raw coal, which could be attributed to experimental uncertainty.

Table 1: The content of different classes of sodium in Zhundong raw coal and the ashes burnt from the coal with different additives.

Sample	Na content (mg/1 g of raw coal)				
	Water-soluble	NH4Ac-soluble	HCl-soluble	Insoluble	Total
Zhundong raw coal	4.946	0.550	0.334	1.579	7.409
Na content in the ash					
No additive	1.894	0.095	0.310	1.593	3.892
3% silica	0.794	0.701	0.546	5.053	7.094
3% alumina	0.887	0.948	1.062	3.835	6.732
3% kaolin	0.454	0.258	0.918	5.678	7.307
3% mica	0.814	0.732	0.938	4.805	7.289
3% pyrophyllite	0.289	0.258	0.505	6.393	7.445

Comparing the sodium content in the ash and the raw coal, the variation of the content of different classes of sodium before and after combustion in Zhundong coal and coal/additive blends can be obtained, as shown in Fig. 2. The negative value means the consume of a sodium class, which should be either released into the gas phase or transformed into other sodium classes, while the positive value means the produce of a sodium class, which should be generated from other sodium classes. For the coal with no additive, it can be found that 47.5% (Table 1) of the total sodium is released during the combustion, of which is mainly the water-soluble sodium. The contribution of NH_4Ac-soluble, HCl-soluble and insoluble sodium to the total sodium release is minor. When the coal is burnt with additives, its sodium release characteristics are clearly affected. The amount of total sodium released during combustion has been significantly reduced after the coal is mixed with additives. Specifically, the released percentage of total sodium is 4.3%, 9.1%, 1.4%, 1.6% and 0.0% for the additive of silica, alumina, kaolin, mica and pyrophyllite, respectively. From the variation of the content of the 4 different classes of sodium, it can be deduced that the mineral additives reduce the global sodium release via reacting with the water-soluble sodium to form insoluble sodium, which is consistent to the sodium retention mechanism proposed by Kyi and Chadwick [15].

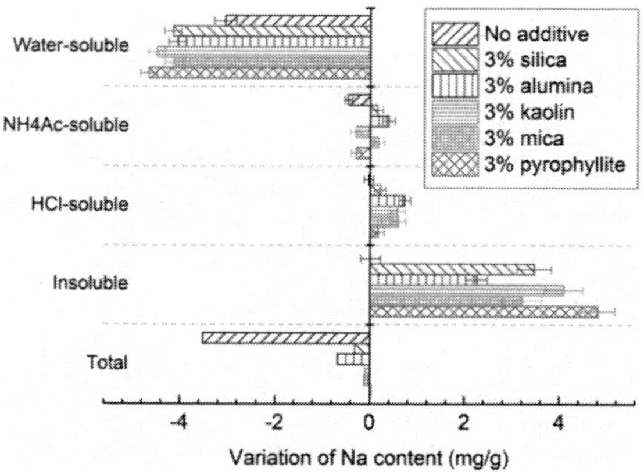

Figure 2: Variation of the content of the different classes of sodium before and after combustion in Zhundong coal and the coal/additive blends. Error bars indicate the statistical uncertainty of three measurements.

3.2 In-situ measurement of temporal sodium release during combustion

Multi-point LIBS technique introduced in Section 2.3 has been employed for the in-situ measurement of sodium concentration.

The obtained temporal release profiles of sodium during the combustion of Zhundong coal and coal/additive blends are shown in Fig. 3. Each measurement is repeated three times. The three characteristic stages of coal combustion can be clearly observed for the coal and coal/alumina samples: first, a narrow peak during the devolatilization stage; then, a left skewed large peak characterize the char burnout stage; finally, a slow decrease to baseline in the ash cooking stage. For other coal/additive blends, the sodium flux profiles exhibit slightly different characteristics: the second peak for the char burnout stage becomes less obvious.

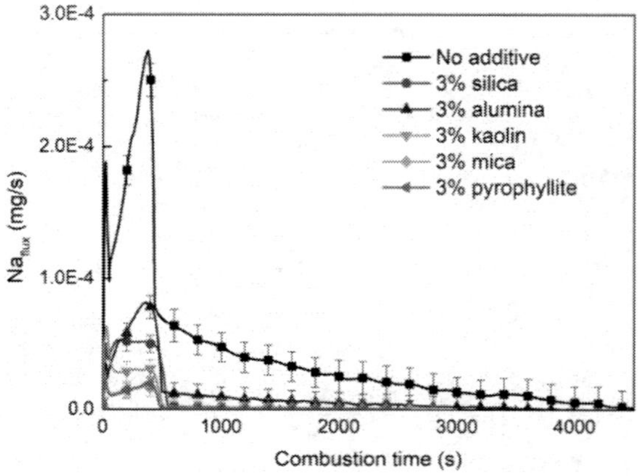

Figure 3: Temporal sodium release profile for Zhundong coal and coal/additive blends measured by multi-point LIBS. Error bars indicate the statistical uncertainty of three measurements.

By integrating the sodium flux with the combustion time, we can obtain the amount of sodium released at the three different stages of the pellet burning, as listed in Table 2. Note that the values in the table have been converted for 1 g of raw coal, in order to facilitate the comparison with the ICP-AES results in Section 3.1. Corrections have been made to account for the reduction of sodium in the sample due to the displacement of coal by sorbent additives. For the total amount of sodium released, the multi-point LIBS measurements agree well with the ICP-AES results within an error of 0.274 mg per 1 g of raw coal. The small deviation could be due to the difference of combustion configurations of the coal/additive blends. In multi-point LIBS experiments, the coal/additives were pressed into pellets and burnt in a premixed methane/air flame, while in ICP-AES experiments the coal/additives were burnt in a muffle furnace. By comparing the sodium released at the three stages for Zhundong coal, it can be found that the char burnout and ash stages contribute to 41.4% and 54.9% of the total sodium released, respectively, while the devolatilization stage only release a marginally amount (3.7%) because of its short duration. With the additives, the sodium release during the combustion of Zhundong coal is significantly reduced.

Figure 4 illustrates the sodium retention efficiency of different additives calculated using the results of pure coal (i.e., sample of "No additive" in Table 2) as a reference. It can be found that all the five additives show significant sodium retention effects with an overall retention efficiency larger than 70%. Silica is found to have a stronger sodium retention effect than alumina. For the natural mineral additives, all the three minerals, i.e., kaolin, mica and pyrophyllite, show a higher sodium retention efficiency than the two pure additives. Because the main compounds of the three natural minerals are silica and alumina, there could be some synergistic effects during the sodium retention of silica and alumina. Moreover, all the additives can effectively prohibit the sodium release during the whole three stages of coal combustion, and the highest sodium retention efficiency is achieved in the ash cooking stage.

Table 2: Sodium released at three stages of coal combustion for Zhundong coal and coal/additive blends.

Sample	Na released (mg/1 g of raw coal)			
	Devola-tilization	Char burnout	Ash	Total
No additive	0.139	1.540	2.042	3.722
3% silica	0.031	0.484	0.049	0.564
3% alumina	0.022	0.575	0.354	0.951
3% kaolin	0.076	0.239	0.006	0.322
3% mica	0.019	0.131	0.043	0.193
3% pyrophyllite	0.019	0.129	0.015	0.163

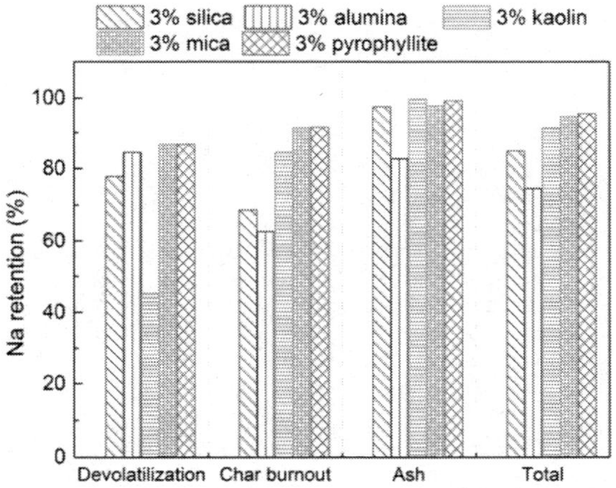

Figure 4: Sodium retention efficiency of different additives at three stages of coal combustion.

4 Conclusions

The sodium retention performance of five different sorbent additives, including two pure additives, i.e., silica and alumina, and three typical natural mineral additives, i.e., kaolin, mica and pyrophyllite, for the combustion of Zhundong coal has been studied by combined online- and offline-measurements. All the additives were blended with the coal at a dosing ratio of 3% by weight. The individual behavior of each sodium class and the time-resolved sodium release characteristics during the combustion of Zhundong coal and different coal/additives mixtures were investigated by offline ICP-AES and online multi-point LIBS techniques, respectively. The offline and online measurement results are compared and verified with each other, and the obtained online temporal sodium release profile is important for understanding the sodium release mechanism during coal combustion.

Generally, all the five sorbent additives show a significant sodium retention effect with an overall retention efficiency larger than 70%. It is found that silica has a stronger sodium retention effect than alumina. For the natural mineral additives, all the three minerals, i.e., kaolin, mica and pyrophyllite, show higher sodium retention efficiencies than the two pure additives. Moreover, all the additives can effectively reduce sodium release during the whole three stages of coal combustion, and the highest sodium retention efficiency is achieved in the ash cooking stage. The variation of the content of the four different classes of sodium indicates that the mineral additives reduce the global sodium release via reacting with the water-soluble sodium to form insoluble sodium.

5 Acknowledgment

This work was supported by the National Natural Science Foundation of China (51422605, 51621005) and the Fundamental Research Funds for the Central Universities (2016QNA4013).

References

[1] R. Khatami, Y.A. Levendis, Combust. Flame, 164 (2016) 22-34.

[2] Y. He, K.Z. Qiu, R. Whiddon, Z.H. Wang, Y.Q. Zhu, Y.Z. Liu, Z.S. Li, K.F. Cen, Sci. Bull., 60 (22) (2015) 1927-1934.

[3] X. Wu, X. Zhang, K. Yan, N. Chen, J. Zhang, X. Xu, B. Dai, J. Zhang, L. Zhang, Fuel, 181 (2016) 1191-1202.

[4] Z.H. Wang, Y.Z. Liu, R. Whiddon, K.D. Wan, Y. He, J. Xia, K.F. Cen, Combust. Flame, 176 (2017) 429-438.

[5] R.J. Quann, M. Neville, M. Janghorbani, C.A. Mims, A.F. Sarofim, Environ. Sci. Technol., 16 (11) (1982) 776-781.

[6] S.A. Benson, P.L. Holm, Ind. Eng. Chem., Prod. Res. Dev.;(United States), 24 (1) (1985).

[7] J. Zhang, C.-L. Han, Z. Yan, K. Liu, Y. Xu, C.-D. Sheng, W.-P. Pan, Energy Fuels, 15 (4) (2001) 786-793.

[8] F. Greger, K.T. Hartinger, P.B. Monkhouse, J. Wolfrum, H. Baumann, B. Bonn, Symp. (Int.) Combust., 26 (2) (1996) 3301-3307.

[9] E. Schlosser, T. Fernholz, H. Teichert, V. Ebert, Spectroc. Acta Pt. A-Molec., 58 (11) (2002) 2347-2359.

[10] Y. He, J. Zhu, B. Li, Z. Wang, Z. Li, M. Aldén, K. Cen, Energy Fuels, 27 (2) (2013) 1123-1130.

[11] P.J. van Eyk, P.J. Ashman, Z.T. Alwahabi, G.J. Nathan, Combust. Flame, 155 (3) (2008) 529-537.

[12] S.H.D. Lee, I. Johnson, Journal of Engineering for Power, 102 (2) (1980) 397-402.

[13] W.A. Punjak, F. Shadman, Energy Fuels, 2 (5) (1988) 702-708.

[14] T. Takuwa, I. Naruse, Proc. Combust. Inst., 31 II (2007) 2863-2870.

[15] S. Kyi, B.L. Chadwick, Fuel, 78 (7) (1999) 845-855.

[16] H.B. Vuthaluru, J.M. Vleeskens, T.F. Wall, Fuel Process. Technol., 55 (2) (1998) 161-173.

[17] A. Kosminski, D.P. Ross, J.B. Agnew, Fuel Process. Technol., 87 (12) (2006) 1051-1062.

[18] K.Q. Tran, K. Iisa, B.M. Steenari, O. Lindqvist, Fuel, 84 (2-3) (2005) 169-175.

[19] M. Uberoi, W.A. Punjak, F. Shadman, Prog. Energy Combust. Sci., 16 (4) (1990) 205-211.

[20] Y. He, R. Whiddon, Z.H. Wang, Y.Z. Liu, Y.Q. Zhu, J.Z. Liu, K.F. Cen, Energy Fuels, 31 (2) (2017) 1082-1090.

11th Asia-Pacific Conference on Combustion,
The University of Sydney, NSW Australia
10th -14th December 2017.

Measurement on evaporation characteristics of multi-component fuel spray

Wenyuan Qi[1], Yuyin Zhang[1] *, Shunhua Yang[2]

[1]School of Mechanical Engineering, Shanghai Jiao Tong University,

800 Dongchuan Road, Shanghai, 200240, China

[2]Science and Technology on Scramjet Laboratory of Hypervelocity Aerodynamics Institute, CARDC,

6 Second Ring South Road, Mianyang, Sichuan, 621000, China

Abstract

The volatility difference of each component in multi-component fuel spray have significant effect on evaporation. In this paper, a laser diagnostic based on Ultraviolet/Visible Laser Absorption Scattering(UV-LAS) was developed for measurement of vapor and liquid concentration distributions in multi-component fuel spray. Error estimation, which was conducted by comparing the total injection mass of fuel and the evaporated mass measured via UV-LAS when the fuel has been evaporated completely, indicated that the uncertainty of vapor mass measurement was below 10% under the pressure and temperature conditions.

1 Introduction

Fuel volatility has a great effect on its evaporation process and thus the mixture formation and combustion processes in internal combustion engines [1]. To date, however, many researchers have been using single-component fuel in their quantitative experiments or modeling, because it is too complicated to interpret the real physics behind the evaporation and combustion characteristics if using multi-component fuels. Fortunately, more and more researchers are challenging the multi-component evaporation modeling [2] and quantitative measurements. [4-6]

Planar laser-induced fluorescence (PLIF) is an effective method for experimental investigation on fuel concentration distributions in multi-component spray. Zhang [3] developed a two–tracer PLIF diagnostic technique and applied to visualize the in-cylinder evaporation and distribution of different volatility fuel components in an optical spray guided spark-ignition direct-injection (SIDI) engine. Yoon [4] analyzed the vapor-phase distribution and behavior of each component in multi-component fuel using PLIF with acetone, TMPD and anthracene as dopants. Itani [5] investigated the effects of preferential evaporation of multi-component fuels by detection of two aromatic fluorescence tracers with complementary evaporation characteristics matched to different components of a multi-component.

The ultraviolet/visible laser absorption-scattering (UV-LAS) imaging technique was also proved to be capable of imaging the vapor/droplet concentration distribution of spray simultaneously [7,8]. UV-LAS technique has several advantages, such as having greater S/N ratio, without oxygen-quenching, with high reliability when applied in high temperature environment and so on. Zhang [6] has made an attempt for quantitatively determining vapor phase concentration distributions in an evaporating multicomponent fuel spray using the UV-LAS imaging technique.

In this paper, UV-LAS imaging technique was developed to measure the vapor/liquid concentration distributions of fuels with components of different volatility. As a practice, p-xylene and fluorobenzene were selected as the substitutes for n-octane and n-hexane respectively in a two-component n-octane/n-hexane fuel spray. The vapor concentration distributions of each components in the spray were obtained and an error estimation was conducted by comparing the injection mass of fuel and the evaporated mass measured from UV-LAS when the fuel has been completely evaporated.

2 Principle and experimental setup

UV-LAS technique is based on the principle of the fuel vapor absorption at one wavelength and the droplets scattering at another wavelength. The ultraviolet and visible LAS technique consists of two measurements: a transmission measurement of visible light to detect the droplets optical thickness at the transparent wavelength and a transmission measurement of ultraviolet light to determine the joint optical thickness of vapor and droplets at the absorbing wavelength. The optical thickness(OT) at ultraviolet and visible light can be expressed as:

$$OT_{uv} = \log(I_0/I_T)_{\lambda_a} = \int_0^L \varepsilon(\lambda_a) \cdot C_{v(x)} dx + \int_0^L K_a^{ext} dx \quad (1)$$

$$OT_{vis} = \log(I_0/I_T)_{\lambda_T} = \int_0^L K_{\lambda_T}^{sca} dx \quad (2)$$

where I_0 is the incident light intensity, I_T is the transmitted light intensity, $\varepsilon(\lambda_a)$ is the molar absorption coefficient of fuel vapor at λ_a($m^3 \cdot m^{-1} \cdot mol^{-1}$), $C_{v(x)}$ is the vapor concentration ($mol \cdot m^{-3}$) at certain location, L is the optical path length (m), $K_{\lambda_A}^{ext}$ is the extinction coefficient for a cloud droplet (m^{-1}), $K_{\lambda_T}^{sca}$ is the scattering coefficient (m^{-1}).

2.1 Vapor/liquid concentration of fuel spray

From (1) and (2), the line-of-sight averaged vapor concentration can be derived as follow:

$$\overline{C_v} = \frac{1}{\varepsilon(\lambda_A) \cdot L}[\log(I_0/I_T)_{\lambda_A} - R \cdot \log(I_0/I_T)_{\lambda_T} \quad (3)$$

where R is the ratio of drop optical thickness at two wavelengths and defined as follow:

$$R = \left. \frac{\int_0^L K_{\lambda_A}^{ext} dx}{\int_0^L K_{\lambda_T}^{sca} dx} \right|_{droplet} \quad (4)$$

In fact, it has been experimentally confirmed that the optical

Corresponding author. Tel/Fax: +86(21)34204449
E-mail: yuyinzhang@sjtu.edu.cn

thickness of liquid phase at the two wavelengths closely approaches the same value, i.e. $R \approx 1$, when dropsize is big enough compared to the wavelength of the incident light [7], such as in the case of spray in the typical gasoline engines. When the spray can be regarded as axisymmetric, the spatial concentration distributions can be reconstructed based on the onion-peeling model [9]. Otherwise, when the spray could not be treated as axisymmetric, the concentration distribution projected into a plane can be calculated as:

$$M_v = \overline{C_v} \cdot L = \frac{1}{\varepsilon(\lambda_a)}\left[\log(I_0/I_T)_{\lambda_a} - R \cdot \log(I_0/I_T)_{\lambda_T}\right] \qquad (5)$$

The detailed depiction of the UV-LAS principle and the processing method is referred to the previous publications. [7, 8]

2.2 Temperature correction

For quantitative measurement, the molar absorption coefficients $\varepsilon(\lambda_a)$ at different pressure and temperature conditions are required in (4), which can be obtained by calibration experiment [7]. However, the parameter is related to the temperature and the temperature field of fuel-gas mixture when applied this method to an evaporated fuel spray, and the mixture is usually inhomogeneous in temperature, so it is necessary to correct the value of $\varepsilon(\lambda_a)$ due to temperature the field in spray for improving the accuracy of quantitative vapor concentration measurement. Based on energy equation, the temperature of mixture(T_{mix}) can be estimated by the bulk mean temperature of liquid and ambient gas as follows [10]:

$$T_{mix} = \frac{C_v \cdot c_{fv} \cdot T_{sat} + \rho_a \cdot c_{pa} \cdot T_a - C_{vi} \cdot H - C_v \cdot (T_{sat} \cdot - T_f) \cdot c_{fl}}{C_v \cdot c_{fv} + \rho_a \cdot c_{pa}} \qquad (6)$$

where, C_v is the concentration of fuel vapor(kg/m^3), ρ_a is the concentration of the ambient gas(kg/m^3), C_{fv} is specific heat of fuel vapor (J/(kg·K)), C_{pa} is specific heat of the ambient gas(J/(kg·K)), C_{fl} is specific heat of the liquid fuel(J/(kg·K)), T_a is the temperature of ambient gas(K), H is the latent heat of fuel vaporation, T_{sat} is the saturate temperature of the fuel (K), T_f is the liquid fuel temperature. The subscript (i,j) means the position in images. A detail flow-chart of the image processing in the UV-LAS technique [1,8] is demonstrated in Fig.1

2.3 Experimental setup and fuel component substitution

Based on the principle of UV-LAS technique, an imaging system was produced as shown in Fig. 2. A coaxial laser beam consisting of the second harmonic and the fourth harmonic of an Nd:YAG laser(90mJ/pulse) was separated into two, an ultraviolet beam (266nm) and a visible beam (532nm), by a dichroic mirror. The two beams were then expanded to a diameter of 100mm. After being attenuated due to absorption and/or scattering of the fuel vapor and droplets, the beams were re-separated into two, a visible one and an ultraviolet one. Then, a pair of collimating lens focused the two beams into the respective ICCD (1024x1024pixels, with 266nm bandpass filter) and CCD (1608x1208pixel, with 532nm bandpass filter). Apertures before the cameras cut off the optical noises beyond the incident wavelength and scattered light beyond the designed optical half-angle, respectively.

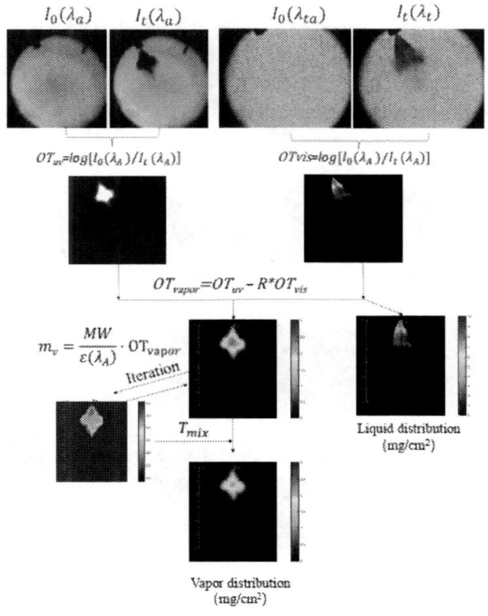

Figure 1: Flow-chart of the image processing in the UV-LAS technique

In this work, p-xylene and fluorobenzene were selected as the substitutes for n-octane and n-hexane respectively in a two-component n-octane/n-hexane fuel spray. P-xylene and fluorobenzene have strong absorption at 266nm wavelength light. Fig. 3 shows the absorption spectroscopes of p-xylene and fluorobenzene at different temperature. The former represents the high boiling point (HBP) component with higher volatility, and the latter represents the low boiling point (LBP) component with lower volatility. This is because p-xylene has similar

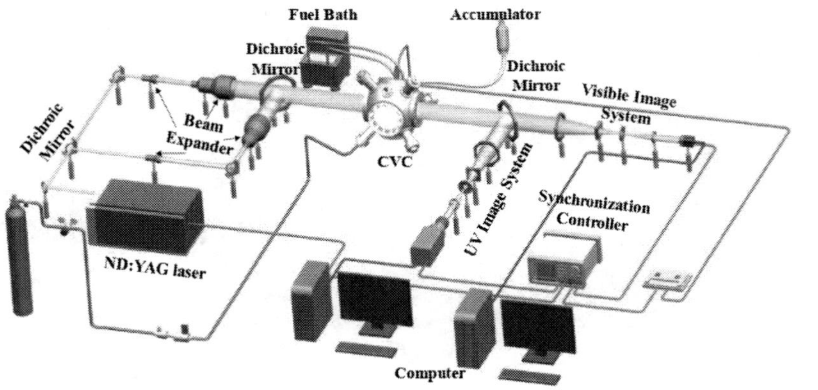

Figure 2: Imaging system of UV-LAS

thermal-physical properties to those of n-octane, and so does for fluorobenzene and n-hexane [6]. Therefore, for determining vapor of LBP, the fuel composes of fluorobenzene and n-octane (abbreviated as FB/n-oc); for determining vapor of HBP, the fuel composes of p-xylene and n-hexane (abbreviated as p-xy/n-he).

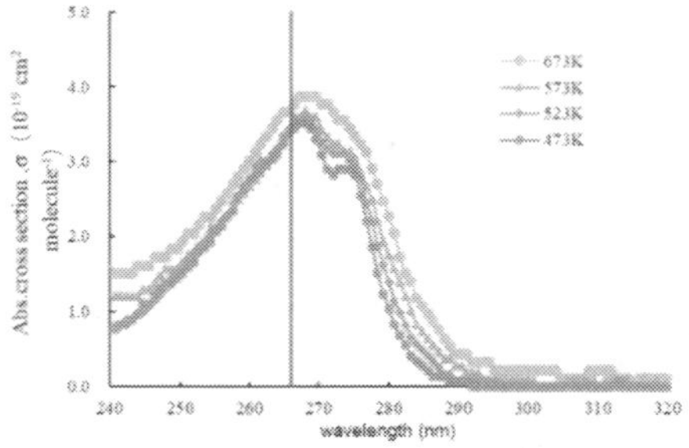

(a)

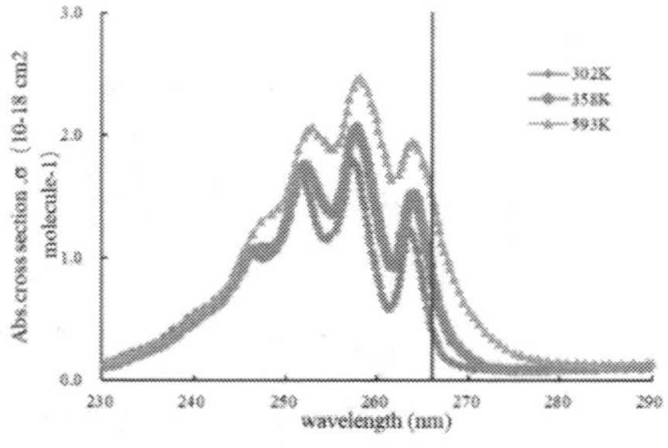

(b)

Figure 3: Absorption spectra of p-xylene and fluorobenzene (a): p-xylene, (b): fluorobenzene

3 Result and discussion

3.1 Liquid/vapor mass distributions in two-components fuel spray

The liquid/vapor mass distributions were obtained based on UV-LAS imaging technique. In spray of p-xylene/n-hexane binary component fuel, the liquid distribution of p-xylene/n-hexane and only p-xylene vapor distribution were given in Fig.4.

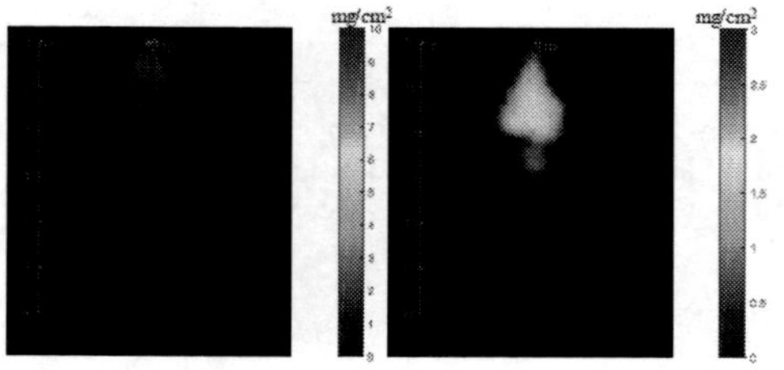

(a) (b)

Figure 4: Liquid/vapor mass distributions in p-xylene/n-hexane binary component spray projected on the plane, (a): p-xylene/n-hexane liquid phase (b): p-xylene vapor phase.

It should be noted that the liquid distribution includes p-xylene and n-hexane, however, vapor distribution only includes p-xylene component because only p-xylene vapor absorbs 266nm wavelength. So, the p-xylene vapor distribution in p-xylene/n-hexane can characterize the high boiling point component evaporation.

3.2 Difference in vapor mass distribution for different boiling-point components

Figure 5 shows a sequence of vapor mass distributions of LBP and HBP in binary component sprays in comparison to the pure fuels. The ambient temperature T_a =623K, the ambient pressure P_a=4.5bar, the injection pressure P_{inj}=3MPa, the fuel temperature T_f=298K, injection duration T_{inj} is 1ms. The fuel spray maintains a symmetrical shape at ASOF=1ms (during injection) and it becomes non-axisymmetric when ASOF>2ms (after end of injection). By comparing the p-xylene (HBP) vapor distribution in p-xylene/n-hexane spray to fluorobenzene (LBP) vapor distribution in fluorobenzene/n-octane spray, it can be observed that the spatial stratification due to volatility differences is insignificant for fluorobenzene (LBP component) and p-xylene (HBP component). It is likely that the evaporation rates of both LBP and HBP components liquid droplets are high at this high enough ambient temperature (T_a=623K) which is also considered to be the diffusion-controlled regime [3].

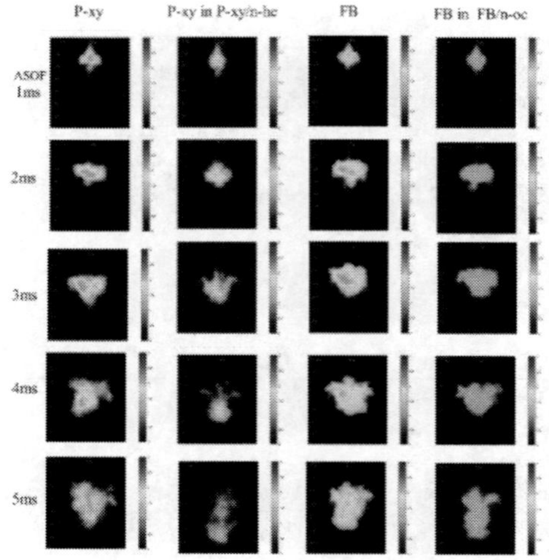

Figure 5: Vapor concentration of different boiling point component at T_a=623K (ASOF: after start of fuel)

3.3 Measurement accuracy

Figure 6 shows the temporal variation of the evaporated fuel mass measured by UV-LAS. As time elapses, the mass of vapor increases until fuel evaporated completely. And then it almost

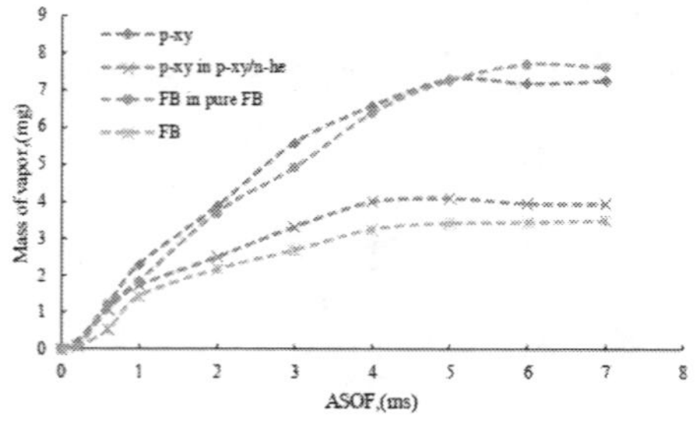

Figure 6: Temporal variation of evaporated mass of different components in the sprays

keeps constant. The measurement accuracy of UV-LAS can be estimated by comparing this constant value obtained by UV-LAS to the total fuel mass injected.

The injection fuel mass was controlled by changing the injection duration time from 0.4ms to1.4ms. The UV-LAS measurement error estimations were shown in Fig.7 for pure and binary component fuel. For the total evaporated mass measurement by UV-LAS, the error is less than 6.94% in pure component and is less than 9.85% in binary component measurement. The main sources of measurement error come from the calibration of molar absorption coefficient $\varepsilon(\lambda_a)$ and the temperature field inhomogeneity in spray. The error of binary component measurement is larger than the pure component, it is because that the UV-absorption spectra shift when mixing. It is also significant to improve the temperature correction method and verify the substitution effect on absorption spectra. It is noticed that the error is achieved when only vapor phase exists and there is no multiple scattering from dense droplets. It needs to more investigation for estimating accuracy where dense droplets exist.

4 Conclusions

In this work, the UV-LAS imaging technique system was developed to realize the measurement of liquid/vapor mass distributions in binary-component fuel spray. The measurement accuracy was estimated by comparing the vapor mass obtained from UV-LAS measurement and the total injection fuel mass. The errors are less than 6.94% and 9.85 % for pure and binary component, respectively. This LAS imaging technique has been proved to have great potential as a quantitative laser diagnostic for the vapor distributions of different volatile components in a multicomponent fuel spray.

5 Acknowledgment

The research was sponsored by National Natural Science Foundation of China (No. 51376120), International Science and Technology Cooperation Program of China (No. 2016YFE0127500), Key Laboratory of Hypersonic Ramjet Technology (No. STS/MY-KFKT-2014001).

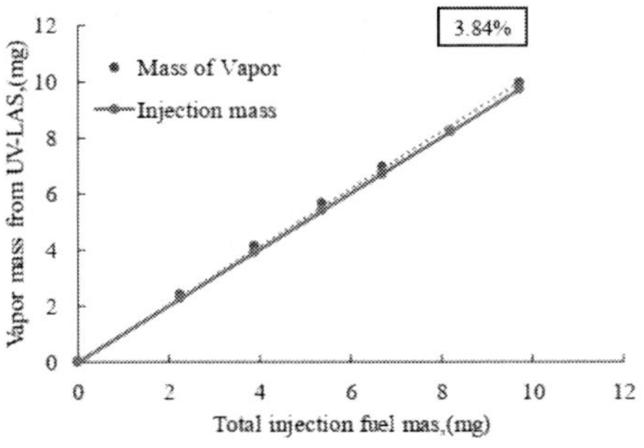

(a) Pure p-xylene spray

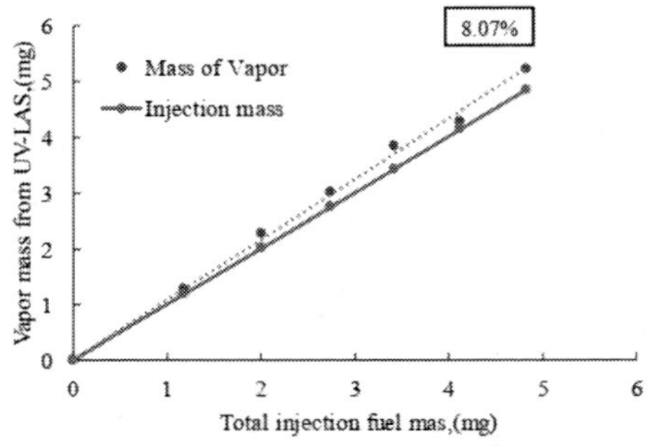

(b) p-xylene in p-xylene/n-heptane spray

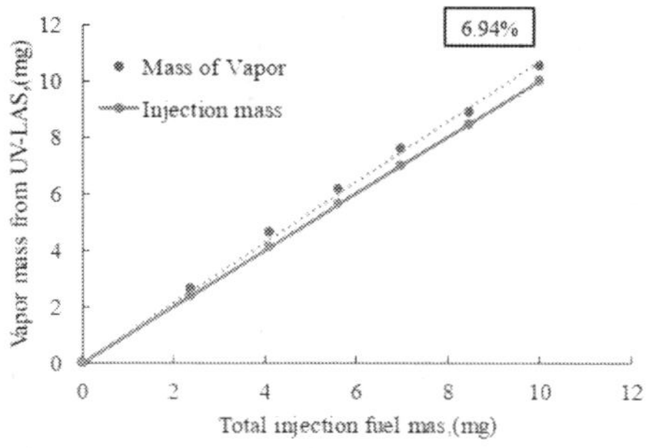

(c) Pure FB spray

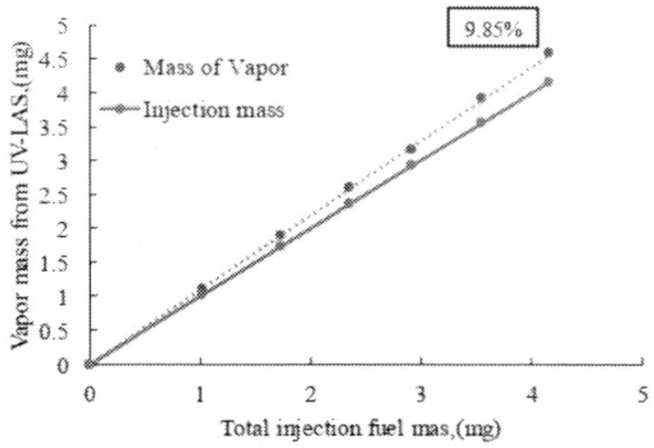

(d) FB in FB/n-octane spray

Figure 7: Error estimations for measurement of pure/binary component fuel spray by UV-LAS

References

[1] Jenny P, Roekaerts D, Beishuizen N. Progress in Energy and Combustion Science. 2012;38(6):846-887.

[2] Lippert A M, Reitz R D. Control Conference (CCC), 2014 33rd Chinese. IEEE, 1997:4175-4180.

[3] Zhang R, Sick V. SAE Technical Papers, 2007.

[4] Yoon J K, Myong K J, Senda J, et al. Journal of Mechanical Science and Technology, 2009, 23(9):2565-2573.

[5] Itani L M, Bruneaux G, Lella A D, et al. Proceedings of the Combustion Institute, 2015, 35(3):2915-2922.

[6] Zhang Y, Kotani Y, Yoshida A, et al. Combustion Science & Technology, 2007, 179(5):863-881.

[7] Zhang Y, Yoshizaki T, Nishida K. Applied Optics. 2000;39(33):6221-6229.

[8] Zhang Y, Li S, Qi W, et al. International Journal of Engine Research, 2016.

[9] Jr H D. Appl Opt, 1981, 20(3):493-499.

[10] Senda J, Kanda T, Kobayashi M, et al. SAE Technical Papers, 1997.

11th Asia-Pacific Conference on Combustion,
The University of Sydney, NSW Australia
10th -14th December 2017.

Numerical Study of Premixed and Stratified Combustion using Flame Surface D ensity

Zhou Yu [1], Taohong Ye[1*]

[1] Department of Thermal Science and Energy Engineering, University of Science and Technology of China

Jinzhai Road 96, Hefei, Anhui, 551, China

Abstract

Large eddy simulation was applied to investigate operating conditions SwB1 and SwB5 of Cambridge stratified swirl burner, using flame surface density (FSD) model to describe combustion process and chemistry tabulation to determine major scalars, with the purpose to research flame structures and develop suitable sub-grid scale (sgs) combustion models for the premixed/stratified conditions. The LES results show that the sgs model is well suit for the calculation of upstream, while it exists deviation in downstream, which may caused by the wrinkling model adapted underestimate the shape change of turbulent to flame. The recirculation zone formed behind the bluff body is closed to stable owe to the re-laminarisation caused by heat release of combustion. Compared with PPDF-FGM model, the FSD model show ability in preserving the flame progatation, especially for stratified flame.

1 Introduction

Combustion is in a position of vital significance for energy transition at nowadays. Its efficient utilization attracts much attention with purpose to reduce energy consumption and protect environment. Premixed combustion, especially lean premixed mode, is widely used in gas turbine and aero-engine, because of its even temperature field and low emission [1]. While there usually exists some adverse phenomenon in combustion equipment, suchlike backfire, blowoff, thermal acoustic instability, at lean premixed condition. Generally, ideal full premixed condition is difficult to achieve, the reactants are non-uniform, hence stratified combustion is more common in actual working conditions. Compared with lean premixed, stratified combustion catches more attention lately, as it could extend the combustion limits and improve combustion stability [2].

Recently, some fundamental experiments for premixed and stratified flames have been carried out. Among them, the Cambridge stratified swirl burner designed and operated by Sweeney et al [3] is a well benchmark case owe to its detailed measurements of the flow field and scalar field. In order to develop more suitable models for premixed and stratified combustion, massive efforts have been invested. Nambully et al [4] simulated the non-swirl cases based on LES using a novel filtered-laminar-flame PDF SGS model, meanwhile the effects of differential diffusion has also been discussed. Zhang et al [5] put forward a new dynamic thickened flame (DTF) model which considering the variances of characteristic scalars and simulated the premixed non-swirling and swirling conditions to evaluate the effect of variances. Brauner et al [6] used LES with transported PDF to calculate the most of cases in Cambridge flame series and obtained good agreements.

In premixed and stratified combustion, correct descriptions for its propagation and flame thickness are the important criteria for sgs models. Thereinto, flame surface density model, which represents combustion occurs on the flame front through a monotonically increasing characteristic scalar, namely progress variable, could guarantee the flame propagation, is a reasonable choice for the premixed and stratified combustion simulations [7,8]. While in terms of FSD model, the ignorance of the flame inner species information prevents analyzing flame structures meticulously. Among the mentioned models, the chemistry tabulation combined with presumed PDF is a very efficient way for turbulent combustion simulations by decoupling flow and reaction, with an assumption that the compositions changing through low-dimensional manifolds, but it fails to meet the requirements mentioned previously. Lecocq et al [9] took advantage of the two above models to proposed a new coupled model and applied it successfully in LES, the satisfactory results implies that the new couple model could not only locate the reaction zone accurately, but also predict the reaction process in flame front correctly.

In the present research, large eddy simulation has been applied to investigate operating conditions SwB1 and Swb5 of Cambridge stratified swirl burner, using FSD to describe combustion process and chemistry tabulation to determine major scalars, aiming at developing a suitable model for premixed and stratified combustion and analyzing flame structures. The present paper is made up by the follow sections. In section 2, a brief introduction for Cambridge stratified swirl burner and a detailed description of the present used sgs models, including FSD model, chemistry tabulation method, presumed probability density function closure (PPDF), et al are presented. The numerical results and flame structures are shown in section 3. Finally, the paper ends up with conclusions in section 4.

2 Physical model and numerical methods

2.1 Experiment setup

The exit geometry of the Cambridge stratified swirl burner is shown in Figure 1. It consists of two coaxial round pipes and a central bluff body. The experiments are carried out in a wind tunnel with a diameter being 196.8mm, in order to avoid unnecessary influences. The operating conditions SwB1 and

Corresponding author. Fax: +0086-551-63603048
E-mail address: thye@ustc.edu.cn

SwB5 are given in Table 1, the meaning of each variable could be found in [5].

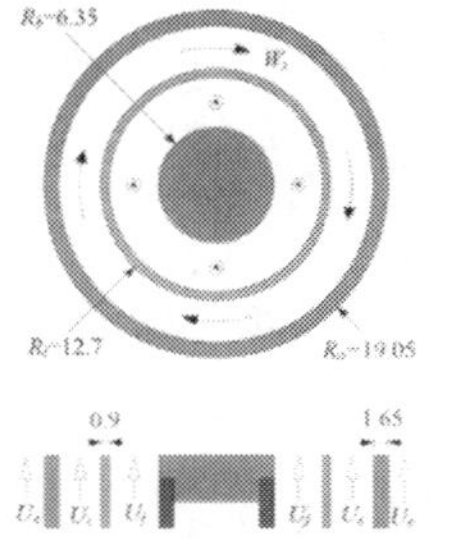

Figure 1: Schematic of the exit geometry of the Cambridge stratified swirl burner (SwB). All dimensions are to scale and in mm [3]

Case	U_j	U_s	U_e	ϕ_j	ϕ_s	ϕ_e
SwB1	8.31	18.7	0.4	0.75	0.75	0.0
SwB5	8.31	18.7	0.4	1.0	0.5	0.0

Table 1: Operating conditions of SwB1 and SwB5

2.2 Computational methods

2.2.1 Flame surface density model

In the present simulation, the distribution of the reactants is represented using mixture fraction Z defined by Bilger et al [10]. As mentioned above, in terms of FSD model, a normalized quantity called progress variable is required and generally expressed using mass fraction of fuel Y_F as,

$$c = \left(Y_F - Y_F^u(Z)\right) / \left(Y_F^b(Z) - Y_F^u(Z)\right) \tag{1}$$

where superscript u and b indicate unburned and burnt state, respectively.

With low Mach number and unity Lewis number assumptions, the Favre filtered transport equation for the normalized progress variable can be simplified as [8],

$$\frac{\partial(\bar{\rho}\tilde{c})}{\partial t} + \frac{\partial(\bar{\rho}\tilde{u}_i\tilde{c})}{\partial x_i} = \frac{\partial}{\partial x_i}\left(\bar{\rho}\tilde{D}\frac{\partial\tilde{c}}{\partial x_i}\right) + \overline{\omega}_c + \frac{\partial(\bar{\rho}\tilde{u}_i\tilde{c} - \overline{\rho u_i c})}{\partial x_i} + \overline{\frac{2\rho D}{Z}\frac{\partial c}{\partial x_i}\frac{\partial Z}{\partial x_i}} \tag{2}$$

the last term in RHS of (2) is cross scalar dissipation rate, and modeled as [11],

$$\overline{\frac{2\rho D}{Z}\frac{\partial c}{\partial x_i}\frac{\partial Z}{\partial x_i}} = \frac{2}{\tilde{Z}}\bar{\rho}\left(\tilde{D} + \tilde{D}_t\right)\frac{\partial\tilde{c}}{\partial x_i}\frac{\partial\tilde{Z}}{\partial x_i} \tag{3}$$

where $\tilde{D}_t$ is turbulent diffusion coefficient approximated by Germano dynamic process [12].

Combined with the FSD model and Weller sgs flux model, the first three terms in RHS of (2) can be modeled in the following manner [11],

$$\frac{\partial}{\partial x_i}\left(\bar{\rho}\tilde{D}\frac{\partial\tilde{c}}{\partial x_i}\right) + \overline{\omega}_c + \frac{\partial(\bar{\rho}\tilde{u}_i\tilde{c} - \overline{\rho u_i c})}{\partial x_i} = \rho_0 S_L \Xi|\nabla\tilde{c}| + \frac{\partial}{\partial x_i}\left(\bar{\rho}\tilde{D}_t\frac{\partial\tilde{c}}{\partial x_i}\right) \tag{4}$$

where ρ_0 is unburned gas density, S_L is laminar flame speed which is sensitive to chemistry mechanism. The wrinkling factor Ξ which represents the change of flame surface area caused by the effect of turbulent eddies is expressed using Charlette2 model,

$$\Xi = \left\{1 + \min\left[\frac{\Delta}{\delta_L}, \Gamma_\Delta\left(\frac{u'_\Delta}{S_L}\right)\right]\right\}^{\beta_1} \tag{5}$$

where Δ is filter width, δ_L is laminar flame thickness and Γ_Δ is an efficiency function. The expressions of each parameter in (5) could be found in [8]. The sgs velocity fluctuations is estimated using the Smagorinsky model, as recommended by Ma et al [8].

$$u'_\Delta = \frac{\nu_t}{C_S\Delta}, \nu_t = (C_S\Delta)^2\left|\frac{1}{2}\left(\frac{\partial\tilde{u}_j}{\partial x_i} + \frac{\partial\tilde{u}_i}{\partial x_j}\right)\right| \tag{6}$$

where the ν_t is turbulent viscosity. The coefficient C_S is determined through Germano dynamic process.

The Favre filtered transport equation for the normalized progress variable solved in present simulation is shown in 2.2.3. It should be mentioned that the flame inner species information is neglected in FSD model.

2.2.2 Chemistry tabulation and presumed PDF model

A series of one dimensional free propagation laminar premixed flames are calculated by FlameMaster [13] to construct the laminar chemistry table. The detailed process to built the table using characteristic scalars can be found in [5]. Briefly, the laminar chemistry table can be tabulated as mixture fraction and normalized progress variable, namely, $\varphi = \varphi(Z, c)$.

Considering the interactions between turbulence and combustion, the presumed PDF is employed to model sgs flame structures. The turbulent scalars can be obtained as follow,

$$\tilde{\varphi} = \iint \varphi(Z, c)\tilde{p}(Z, c)dZdc \tag{7}$$

where $\tilde{p}(Z, c)$ is the filtered joint PDF of mixture fraction and normalized progress variable. With assumption that the two scalars are statistic independence, the joint PDF can be approximated as,

$$\tilde{p}(Z, c) = \tilde{p}(Z) \times \tilde{p}(c) \tag{8}$$

The distributions of PDF for mixture fraction and normalized progress variable are both modeled using the beta PDF [5].

The turbulent chemistry table can be expressed as $\tilde{\varphi} = \tilde{\varphi}(\tilde{Z}, \tilde{Z}_v, \tilde{c}, \tilde{c}_v)$, the $\tilde{Z}_v$ and $\tilde{c}_v$ are the variances of mixture fraction and normalized progress variable respectively. In the following simulation, the turbulent scalars are determined by solving the transport equations for each characteristic scalar.

2.2.3 LES governing equations

The Favre filtered transport equations for mass and momentum are solved to describe the evolution of flow field, with the sgs stress term expressed using dynamic Smagorinsky model. Besides, four additional transport equations are required for the combustion simulation, namely, mixture fraction, normalized progress variable and their relevant variances. According to the method proposed by Domingo et al. [14], the related governing equations are expressed as follow,

$$\frac{\partial(\bar{\rho}\tilde{Z})}{\partial t} + \frac{\partial(\bar{\rho}\tilde{u}_i\tilde{Z})}{\partial x_i} = \frac{\partial}{\partial x_i}\left(\bar{\rho}\tilde{D}\frac{\partial\tilde{Z}}{\partial x_i}\right) + \frac{\partial(\bar{\rho}\tilde{u}_i\tilde{Z} - \overline{\rho u_i Z})}{\partial x_i} \tag{9}$$

$$\frac{\partial(\bar{\rho}\tilde{Z}_v)}{\partial t} + \frac{\partial(\bar{\rho}\tilde{u}_i\tilde{Z}_v)}{\partial x_i} = \frac{\partial}{\partial x_i}\left(\bar{\rho}\tilde{D}\frac{\partial\tilde{Z}_v}{\partial x_i}\right) + \frac{\partial(\bar{\rho}\tilde{u}_i\tilde{Z}_v - \bar{\rho}\widetilde{u_i Z_v})}{\partial x_i} + 2\left(\bar{\rho}\tilde{u}_i\tilde{Z} - \bar{\rho}\widetilde{u_i Z}\right)\frac{\partial\tilde{Z}}{\partial x_i} - \bar{\rho}\tilde{\chi}_Z^{res} \tag{10}$$

$$\frac{\partial(\bar{\rho}\tilde{c})}{\partial t}+\frac{\partial(\bar{\rho}\tilde{u}_i\tilde{c})}{\partial x_i}=\frac{\partial}{\partial x_i}\left(\bar{\rho}\widetilde{D}_t\frac{\partial\tilde{c}}{\partial x_i}\right)+\rho_0 S_L\Xi|\nabla\tilde{c}|+\frac{2}{\tilde{Z}}\bar{\rho}\big(\widetilde{D}+\widetilde{D}_t\big)\frac{\partial\tilde{c}}{\partial x_i}\frac{\partial\tilde{Z}}{\partial x_i} \tag{11}$$

$$\frac{\partial(\bar{\rho}\tilde{c}_v)}{\partial t}+\frac{\partial(\bar{\rho}\tilde{u}_i\tilde{c}_v)}{\partial x_i}=\frac{\partial}{\partial x_i}\left(\bar{\rho}\widetilde{D}\frac{\partial\tilde{c}_v}{\partial x_i}\right)+2(\widetilde{c\omega_c}-\tilde{c}\widetilde{\omega}_c)+\frac{\partial(\bar{\rho}\tilde{u}_i\tilde{c}_v-\overline{\rho u_i c_v})}{\partial x_i}+2(\bar{\rho}\tilde{u}_i\tilde{c}-\bar{\rho}\widetilde{u_i c})\frac{\partial\tilde{c}}{\partial x_i}-\bar{\rho}\tilde{\chi}_c^{res}+\sqrt{(\tilde{c}_v)}\overline{\frac{2\rho D}{Z}\frac{\partial c}{\partial x_i}\frac{\partial Z}{\partial x_i}} \tag{12}$$

The eddy diffusivity model is adopted for the sgs scalar fluxes, with the model coefficients determined using dynamic procedure. The sgs scalar dissipation rates $\tilde{\chi}_Z^{res}$ and $\tilde{\chi}_c^{res}$ are modeled using a linear relaxation hypothesis. The chemical source in transport equation for normalized progress variable variance is pre-computed and stored in the turbulent chemistry table.

In the present study, the widely used GRI3.0 mechanism (53 species, 325 reactions) [15] is employed to built the chemistry table. In order to balance computational accuracy and memory requirements, the turbulent chemistry table $\tilde{\varphi}(\tilde{Z},\tilde{Z}_v,\tilde{c},\tilde{c}_v)$ contains 150×25×150×25 grid points in the corresponding coordinate direction. The grid points are uniform along the $\tilde{Z}$, $\tilde{c}$ direction, and parabolic distribution at rest directions.

2.2.4 Numerical setup

The cylinder computational domain has 240mm and 80mm in length at axial and radial directions, and is discretized by structured grid. There are 186 and 197 non-uniform grid points in axial and radial directions, and 32 uniform grid points in circumferential direction, respectively. The mesh is refined around jet exit and shear layer region, respectively. The total number of grid points is almost 1.17 million. The non-slip boundary condition is chosen on the wall, the convective boundary condition is used for the domain exit.

The LES calculations are performed using an in-house Fortran code. The Favre-filtered governing equations are discretized using finite difference method. The detailed numerical algorithms could be found in [5]. The time step is adjusted dynamically to guarantee the Courant number less than 0.5. The statistic ensembling average is last for 10τ ($\tau=L/U_j$, where $L=$ 240mm and $U_j=$ 8.31m/s) after running over 10τ.

3 Results and discussion

Instantaneous axial velocity fields of non-reacting and reacting flow for premixed and stratified flames are shown in Figure 2. The black solid lines stand for the isoline that c = 0.5 to mark the flame front. The red solid lines stand for the isoline that axial velocity being 0 to mark the recirculation zone. One can see that, compared with non-reacting cases, the recirculation zone formed behind the bluff body is closed to stable owe to the re-laminarisation caused by heat release of combustion. Through momentum exchange and thermal expansion, both two recirculation zones for reacting cases are dissipated gradually at downstream. As for stratified combustion, a relative shorter recirculation zone and wider flame brush could be observed, because of the higher heat release.

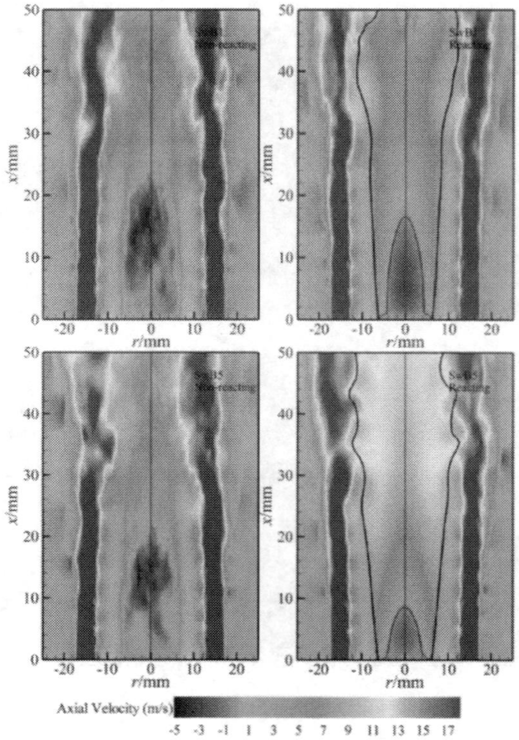

Figure 2: Snapshots of instantaneous axial velocity fields of non-reacting and reacting flow cases.

Figure 3 shows radial distributions of temperature, mass fractions of CH_4 and H_2O at different streamwise locations for premixed and stratified combustion. Firstly, for premixed flame, compared with PPDF-FGM model, almost even better results could be observed for FSD model. At x=10mm, the predictions of temperature using two models respectively are smiliar. While at rest locations, the FSD model show ability in preserving the flame progatation. Secondly, for stratified flame, much better results are achieved using FSD model. At x=10mm, the caculated temperature are overrated which may due to the ignorance of heat loss caused by the bluff body. At x=30mm, the LES results are in good agreement with experiment data. While the discrepancies at x=50mm are reflected in the inaccurate prediction of the location for the flame front.

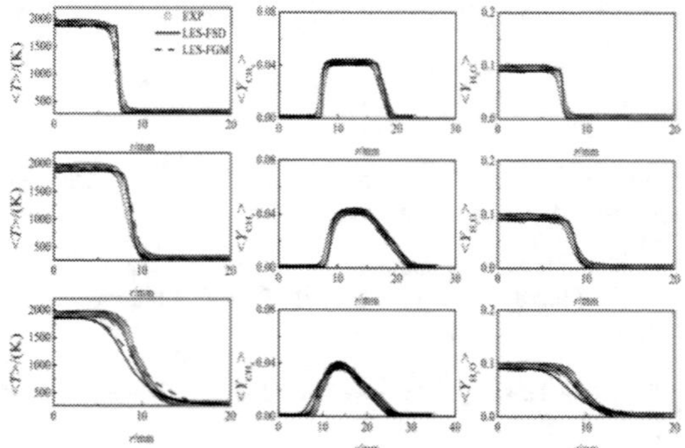

(1) SwB1

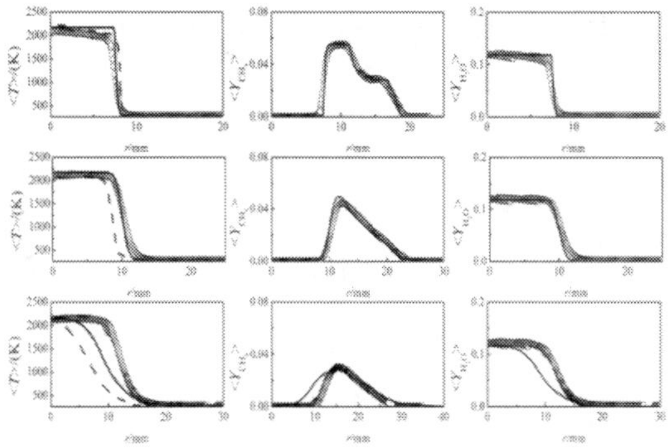

(2) SwB5

Figure 3: Radial distribution of temperature, mass fractions of CH_4 and H_2O at different streamwise locations of x=10, 30 and 50 mm from top to bottom. The results of PPDF-FGM are obtained from [5].

4 Conclusions

Large eddy simulation has been applied to investigate operating conditions SwB1 and SwB5 of Cambridge stratified swirl burner, using FSD to describe combustion process and chemistry tabulation to determine major scalars. Some conclusions can be obtained as follow.

Reasonable descriptions of the distributions of major species for premixed and stratified combustion could be achieved by using the chemistry tabulation which based on one dimensional free propagation laminar premixed flame. It should be noted that the tabulation method could also solve the ignorance of the flame inner species information, which is a major disadvantage for the traditional FSD model. The LES results show that the sgs model is well suit for the calculation of upstream, while it exists deviation in downstream, which may caused by the wrinkling model adapted underestimate the shape change of turbulent to flame. The recirculation zone formed behind the bluff body is closed to stable owe to the re-laminarisation caused by heat release of combustion. Compared with PPDF-FGM model, the FSD model show ability in preserving the flame progatation, especially for stratified flame.

5 Acknowledgment

This work was supported by the National Natural Science Foundation of China (Grant No. 91441117 and Grant No. 51576182).

References

[1] H. Zhang, C. Han, T. Ye, J. Turbul 17 (3) (2016) 327-355.

[2] X. Shi, J.Y. Chen, Z. Chen, Combust. Flame 163 (2016) 394-405.

[3] M.S. Sweeney, S. Hochgreb, M.J. Dunn, Combust. Flame 159 (9) (2012) 2896-2911.

[4] S. Nambully, P. Domingo, V. Moureau, Combust. Flame 161 (7) (2014) 1756-1774.

[5] H. Zhang, T. Ye, G. Wang, Flow. Turbul. Combust 98 (3) (2017) 841-885.

[6] T. Brauner, W.P. Jones, A.J. Marquis, Flow. Turbul. Combust 96 (4) (2016) 965-985.

[7] M. Boger, D. Veynante, H. Boughanem, Symposium on Combustion 27 (1) (1998) 917-925.

[8] T. Ma, O.T. Stein, N. Chakraborty, Combust. Theor. Model 17 (3) (2013) 431-482.

[9] G. Lecocq, S. Richard, O. Colin, Combust. Flame 158 (6) (2011) 1201-1214.

[10] R.W. Bilger, S.H. Stårner, R.J. Kee, Combust. Flame 80 (2) (1990) 135-149.

[11] D. Butz, Y. Gao, A.M. Kempf, Combust. Flame 162 (9) (2015) 3180-3196.

[12] C.D. Pierce, P. Moin, Phys. Fluids 10 (12) (1998) 3041-3044.

[13] H. Pitsch, available at <http://www.stanford.edu/group/pitsch>.

[14] P. Domingo, L. Vervisch, S. Payet, Combust. Flame 143 (4) (2005) 566-586.

[15] G.P. Smith, D.M. Golden, M. Frenklach, available at: <http://www.me.berkeley.edu/gri_mech>.

11th Asia-Pacific Conference on Combustion,
The University of Sydney, NSW Australia
10th -14th December 2017.

Numerical study on flow field and slagging property of SNG Gasifier

Yi SU[1], Min Yao[2], Chuntao LUO[2], Jianping KUANG[2], Yunhuan JING[2], Zitong ZHAO[1], Yinjian MA[2]

[1]Shanghai Marine Diesel Engine Research Institute
No. 3111 Huaning Rd., Shanghai, 201108, China
[2]SNCC Coal Chemical Industry Company, Shenhua Ningxia Coal Industry Group,
No. 168 Beijing Rd., Yinchuan, Ningxia, 750411, China

Abstract

A new SNG gasifier was developed by Shenhua Group(China) with 2000t/d capacity. A 3D numerical model coupled with ash slagging was built to investigate flow, temperature, species and slagging of the gasifier. The flow field is composed of four regions: center jet, upper recirculation, lower recirculationand and plug flow. The lower recirculation field help prolong coal particle residence time and also promote carbon conversion. Slagging properties are strongly associated with wall temperature, heat flux and particle movement. The results showed good agreement with real operation data.

1 Introduction

Coal gasification technology is developing fast in China in recent years. Shenhua Group(China) developed a SNG Gasifier. It is a top fired entrained flow type gasifier with a capacity of 2000t/d coal consumption. Since measurement of temperature inside coal gasifier is still a big technical challenge, numerical analyse is the best way to understand and optimize reaction inside gasifier.

For the past few years, many numerical models have been developed to predict velocity, temperature, species, etc. inside gasifier. Fletcher[1] used CFX software to calculate gaseous product composition and temperature at gasifier outlet. Choi[2] biult a 2D model for coal slurry gasifier based on modified vortex crushing model. The temperature field was calculated. hi[3], Slezak[4]and Watanabe[5] built 3D model for CRIEPI Gasifier. k-ε and modified EBU models were used to investigate gaseous turbulence flow property. Yu[6]and Liu[7, 8] investigated flow properties inside opposed four-nozzles coal slurry gasifier. Wu [9-11] used simplified PDF model to calculat temperature field of Texaco gasifier.

A 3D numerical model coupled with ash slaging was built in present work. Velocity, temperature and thickness of slag were calculated. The results showed good agreement with real operation data.

2 Physical model, mesh and numerical methods

2.1 Physical model

The SNG Gasifier is a top-fired, oxygen-blown entrained flow reactor. The operation temperature is controled above ash melting point, in order that slag flow long the membrane wall into the quench chamber. The SNG Gasifier operated with a capacity of 2000t/d coal consumption. The size of gasifier was shown in Fig 1. The control area is from top buner outlet to furnace slag outlet. Table 1 shows the operation parameter, and Table 2 shows the specification of coal.

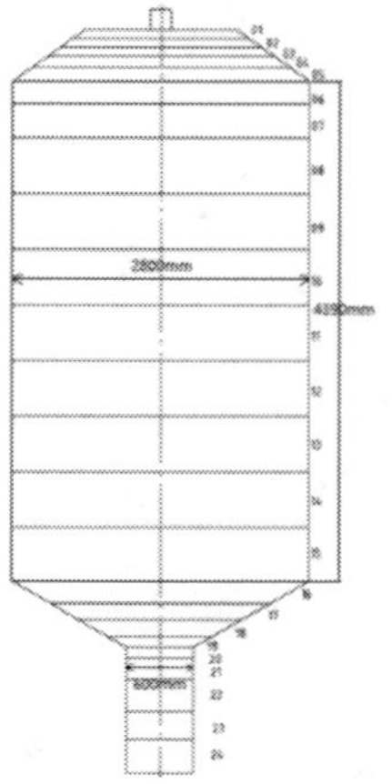

Figure 1: Schemetic of SNG Gasifier

	Coal	Oxygen	CO_2	stream
Pressure/Mpa	5.3	5.5	5.3	5.5
Temperature/°C	70	140	70	360
Flow rate/ t/h	91.477	67.54	20.329	1.62

Table 1: Gasifier operation parameter

Proximate Anlysis[a]				LHV	Particle size/μm	
w / %				MJ/kg	≥120	5%
M	A	FC	V	23.69		
4	16.44	52.25	27.31		40~120	85%
Ultimate Analysis[b] (Wdaf /%)						
C	H	N	S	O(diff.)	<40	10%
77.3	5.1	0.8	1.6	15.2		

a. As received base; b. dry ash free base

Table 2: Coal specification

2.2 Grid

Corresponding author. Fax: +86-21-51711742
E-mail address: colinsuyi@163.com

The fluid area was meshed by ANSYS ICEM software. Grid independece was studied with grid nymbers of 810000, 1230000, 1560000, 1830000 by comparing the temperature and distribution of CO concentration along the gasifier. It showed that when gird number exceeded 1230000, the results changed a little, therefor the grid of 1230000 was finally used(Fig. 2).

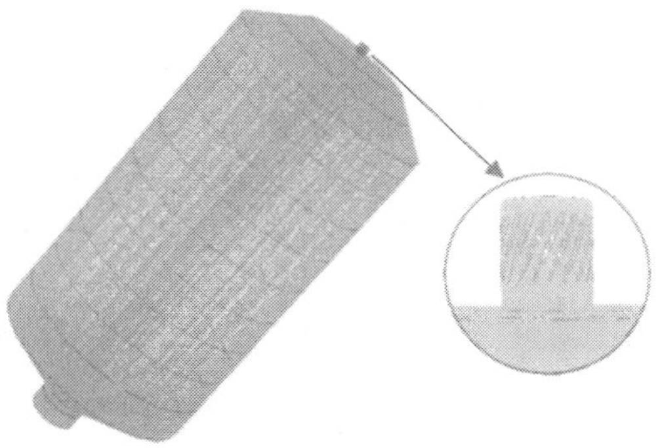

Figure 2: 3D Grid

2.3 Numerical methods

Since the flow inside gasifier is highly swirling, a realizable k-ε model was adopted because it takes into account the swirling effect[12]. Lagrangian tracking was used to describe the motion of particles because solid phase takes less than 10% volume fraction of the gasifier. Water evaporation and coal devolatilization process satisfies the theory of Wen and Chaung[13]. A simple one-step model developed by Badzioch and Hawksley[14]was used. Since the temperature and pressure are sufficiently high, the modeling was conducted on the basis of thermodynamic equilibrium, which could give close enough results[15]. The P-1 model was used to caldulate radiation flux.

ANSYS Fluent software was used for numerical simulation. SIMPLE algorithm was used to deal with the pressure-velocity coupling. PRESTO! Sheme was used for pressure diescretization. The pressure based implicite method was used to solve the discretized equations and second order upwind scheme was applied to other terms.

3 Results of flow field

3.1 Model verification

The comparison between real operation results and simulation results is listed in Table 3. Results from this study showed favorable agreement with operation parameter. The method used in this study is appropriate for SNG Gasifier.

	CO (V%)	H2 (V%)	CO2 (V%)	Syngas Temp.(K)
Real operation data[16]	69.6	24.1	3.5	1703
Simulation	71.2	22.1	4.1	1715

Table 3: Comparison between operation and simulation results

3.2 Velocity and particle movement

The velocity and path of coal particle movement is shown in Fig 3. It is clear that flow region inside the SNG gasifier is formed by four regious: center jet, upper recirculation, lower recirculationand plug flow. The oxygen ejected from burner with high swirlig at speed of 110m/s. The pulverized coal was soonly induced by oxygen at the burner outlet. Reaction of coal and oxygen happens simultaneously. There is a recirculation region at upper and lower dome, which prolongs the residence time of coal particles.

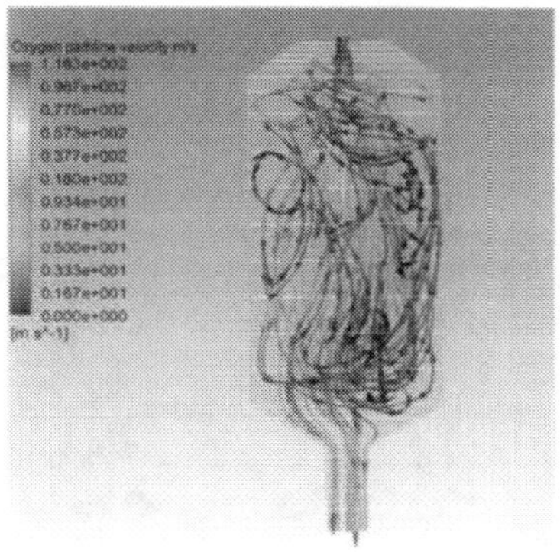

(a) Flow path

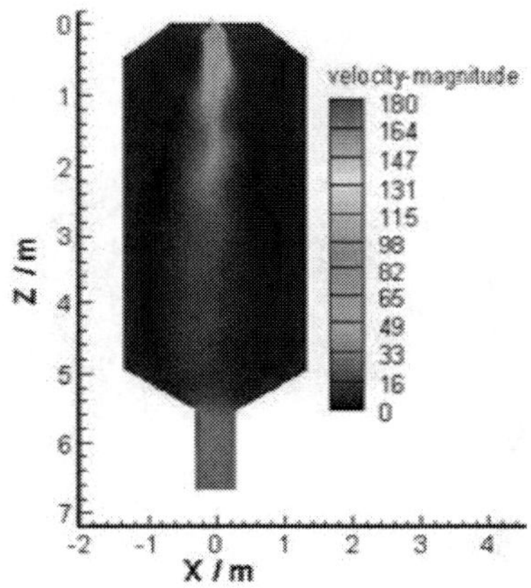

(b) Velocity distribution

Figure 3: Velocity distribution and path

3.3 Temperature and species

Temperature profile inside gasifier is presented in Fig. 4. The flame is conical shaped with wavy tail. The flame didn't starts from the burner outlet, therer is a "black area" . which caused by endothermic reaction, such as water evaporation and coal devolatilization. Fig. 5 shows that the highest devolatization rate accur at the outlet of burner. The devolatization process ended within 0.3m along the gasifier axis. This is accordance with the temperature field in Fig 4. Temperature at upper dome area is comparatively lower than other field. The high temperature region was distributed in the center jet region and lower dome region. Temperature at recirculation region and plug flow region were uniform.The averave temperature at gasifier outlet is 1715K. Heterogeneous oxidation can be reflected by carbon burnout rate shown in Fig 6. High burnout rate area released large amount of heat, which was absorbed by endothermic gasification reaction. After carbon burnout, the gaseous reaction is moderate and resulted in a uniform temperature distribution.

Figure 7 to Fig 9 shows the mole fractions of CO, H2 and CO2 across the gasifier. The concentration of CO and H2 were almost zero at the center jet area, because sufficient oxygen in this area converted CO and H2 into CO2 and H2O, CO2 distribution in

Fig 9 can be an anvidence. Although oxygen is dominated in center jet area, it is below equivalence ratio over the whole reaction area. Therefore, gasification reaction was dominated. As a result, CO and H2 concentration were high inside the gasifier, which take up to 93.3% of the total syngas at gasifier outlet. The simlation results are comparable with real operation data as shown in Table 3. The species concentration underwent a minor change out of the flame area, which indicated the mixing of particle and gaseous species was sufficient and reactions could reach equilibrium quickly.

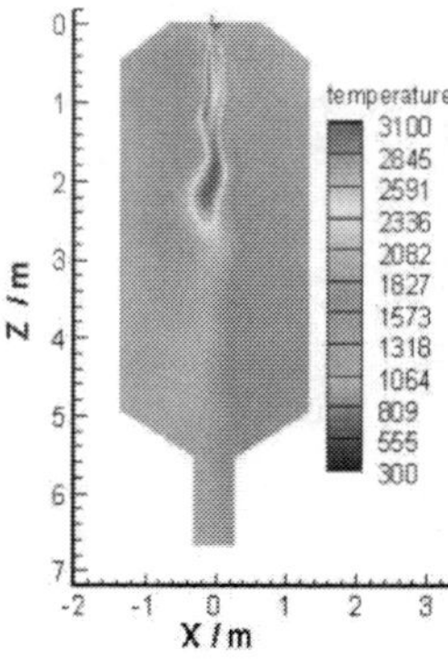

Figure 4: Temperature

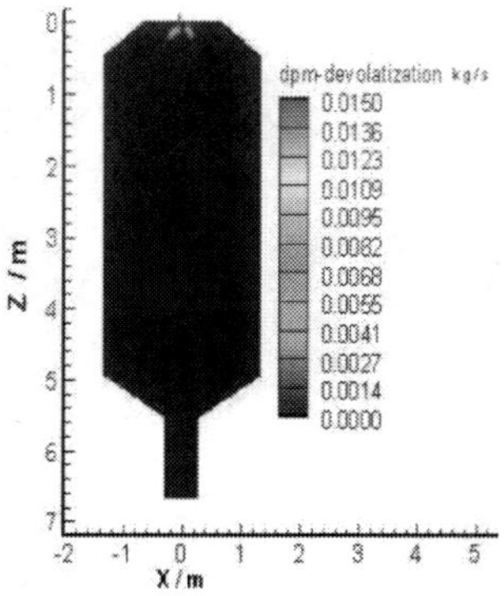

Fig 5 Coal devolatization rate

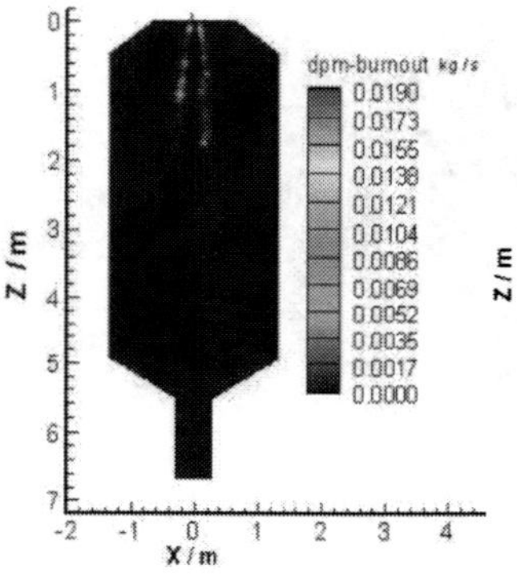

Figure 6: Carbon burnout rate

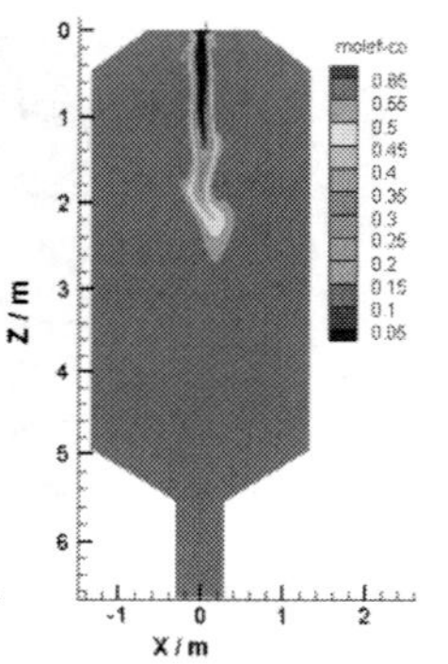

Figure 7: Mole fraction of CO

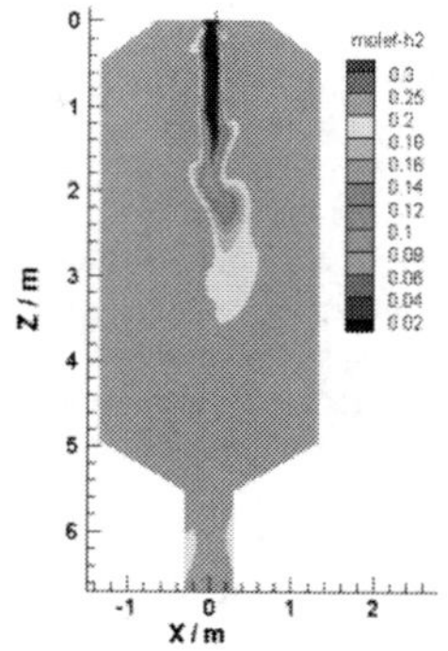

Figure 8: Mole fraction of H_2

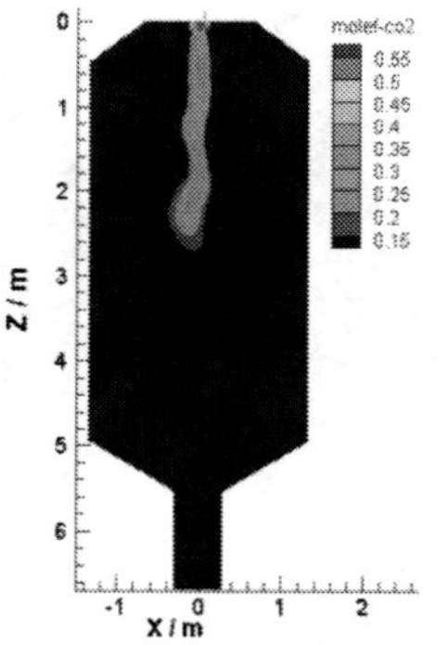

Figure 9: Mole fraction of CO_2

4 Slagging properties of SNG Gasifier

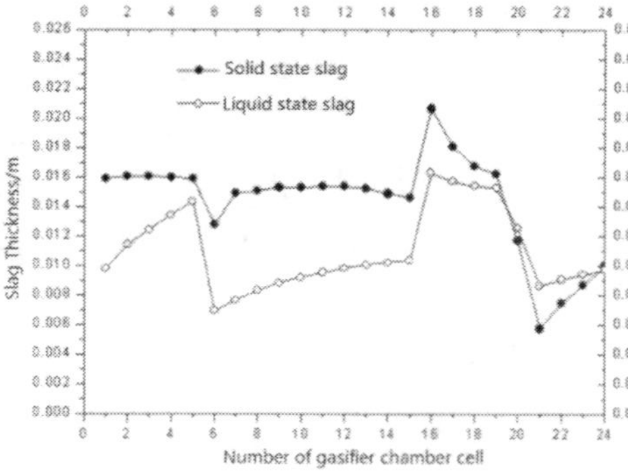

Figure 10: Predicted slag thickness along gasifier cells

The slagging properties along the gaisifer membrane was studied using a slagging model developed by Seggiani[17]. The model provides the slag deposit thickness (solid and liquid), the temperature across the deposit, and the heat flux to the membrane wall. The configuration of the gasifier chamber is represented in Figure 1. The chamber is subdivided into 24 cells in the axial. Temperature, mass of total slag (assumed equal to ash composition in Table 2) and heat released from membrane wall were inputted into the slagging model. The thichness of solid phase slag and liquid state slag within each cell can be calculated. Results were printed and shown in Fig. 10.

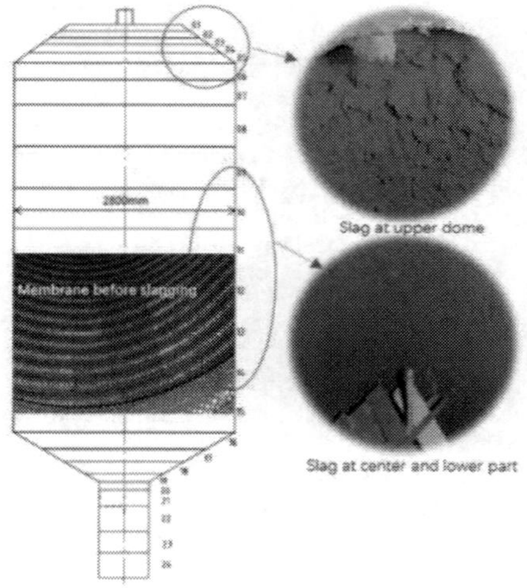

Figure 11: Slagging preoperty at different cells of gasifier

The solid and the liquid slag thicknesses are fractions of the total slag deposit thickness, and depend on the temperature profile in the layer. At upper dome, temperature is comparatively lower(cell 1-5), therefore, liquid state slag is thick, which leads to a fluffy deposit of ash(as shown in Fig. 11). While, temperature at straight part of gasifier(cell 6-10) is high (Fig 4). The heat fluxes to slag surfaces is also intense. The liquid slag viscosity decreases owing to the higher temperature and slag flow rate increases. Correponding to this, photo took at this area(Fig.11) shows a smooth and thick salg deposit, which inndicated that slage flow was quick. This results in an increase of both solid and liquid slag in lower dome part (cell 15-20)(shown in Fig. 10). These results confirm the importance of temperature control at bottom of the gasifier inorder to prevent tap blockage by solid slag.

5 Conclusion

Numerical analysis of SNG Gasifier was conducted. A slag building model was integrated with gasifier flow field model to analyse the slag behavior on the reactor wall.

The flow field of SNG Gasifier is composed of four regions: center jet, upper recirculation, lower recirculationand and plug flow. The flame is conical shaped with wavy tail. Temperature at upper dome area is comparatively lower than other field. The lower recirculation field help prolong coal article residence time and also promote carbon conversion.

Slaging properties are strongly associated with wall temperature, heat flux and particle movement. The predicted slag thichness is in agreement with measurement data. Slag is fluffy at upper dome due to low temperature and low circulation of particles. In contrast, slag is compact at center and lower dome of the gasifier.

At present, measurement of temperature and slag inside coal gasifier is still a big challenge. However, numerical analyse could provide a good reference for gasifier operator. Further consideration will be given to the influence of coal type variation.

6 Acknowledgment

The authors kindly acknowledge the funding support of Shanghai Rising-Star Program(16QB1404800), Shanghai Municipal Science and Technology Project Management Approach (16dz1206302), China Postdoctoral Science Foundation (2016M602946XB).

References

[1]. Letcher D F, Haynes B S, Christo F C, Applied Mathematical Modelling 24(3) (2000) 165-182.

[2]. Choi Y C, Li X Y, Park T J, Fuel 80(15) (2001) 2193-2201.

[3]. Shi S P, Zitney S E, Shahnam M, Journal of the Energy Institute 79(4) (2006) 217-221.

[4]. Slezak A, Kuhlman J M, Shadle L J, Powder Technology, 203(1) (2010) 98-108.

[5]. Watanabe H, Otaka M, Fuel, 85(12-13) (2006) 1935-1943.

[6]. Yuan H Y, Qu H G, Ren H, Journal of East China University of Science and Technology 31(3) (2005) 393-397.

[7]. Wang J. *Experimental and numerical study on slag flow at slag tap hole of gasifier*, PhD thesis, East China University of Sience and Technology, Shanghai, 2013.

[8]. Cheng X X, Hou G J, Liang Q F, Chemical engineering 40(3) (2012) 58-62.

[9]. Wu Y X. *Numerical Simulation Studies on Entrained Flow Coal Slurry Gasifiers and Modeling for the Particle-Eddy Interaction*, PhD thesis, Tsinghua University, Beijing, 2007.

[10]. Wu Y X, Zhang J S, Yue G X, Journal of Chemical Industry and Engineering 59(3) (2007) 537-543.

[11]. Wu Y X, Zhang J S, Wang M M, Proceedings of the CSEE 27(32) (2007) 57-62.

[12]. Shih T-H, Liou W W, Shabbir A, etc., Computers & Fluids 24(3) (1995) 227-238.

[13]. Wen C Y, Chaung T Z, Ind Eng Chem Process Des Dev 18(4) (1979) 684–695.

[14]. Badzioch S, Hawksley P G W, Industrial & Engineering Chemistry Process Design and Development 9(4) (1970) 521-530.

[15]. Barba D, Prisciandaro M, Salladini A, etc., Fuel 90(4) (2011) 1402-1407.

[16]. Kuang J P, Coal Science and Technology 43(2015) 102-106.

[17]. Seggiani M, Fuel 77(14) (1998) 1611-1621.

11th Asia-Pacific Conference on Combustion,
The University of Sydney, NSW Australia
10th -14th December 2017.

Numerical Investigation of Advanced Low Temperature Combustion Techniques in Diesel Engine

Nikhil Khedkar[1], P. Brijesh[1], S. Sreedhara[2]

[1] The Automotive Research Association of India
Pune, Maharashtra, 411038, India
[2]India Institute of Technology, Mumbai
Mumbai, Maharashtra, 400076, India

Abstract

Diesel engines, also known compression ignition (CI) engines, are more popular over gasoline engines due to their undisputed benefit of fuel economy. However, diesel engines primarily produce oxides of nitrogen (NO_x) and particulate matter (PM) which are toxic for human health as well as environmental. Hence, regulatory bodies are imposing stringent restrictions on these pollutants. Recent studies suggested that in-cylinder solutions like low temperature combustion (LTC) and multiple injection strategies are efficient techniques to reduce these emissions. In this work, *Converge* computational fluid dynamics (CFD) tool was used to simulate combustion for experimentally optimized LTC and optimum two-shot injection with LTC strategies to get a better understanding of in-cylinder processes and emission formation. A lower peak combustion temperature, lower percentages of oxygen (O_2) and less available time for the reactions with optimized LTC run and optimum two-shot injection strategies were the reason behind a significant reduction in NO_x (~98%) emissions with these strategies. Results of the study showed that a downward motion of piston for most of the period during fuel injection with optimized LTC strategy improved the mixing rate of fuel with surrounding air. As a result, lower PM emissions were achieved with optimized LTC strategy.

1 Introduction

In internal combustion (IC) engines, compression ignition engines, also known as diesel engines, are more popular due to its high fuel economy, high efficiency and high reliability over spark ignition (SI) engines. However, CI engines emit considerably higher oxides of nitrogen (NO_x) and particulate matter (PM) which are falling under stringent emission limits. In-cylinder solutions are efficient ways to reduce these emissions. Formation of NO_x is a function of three parameters: temperature, presence of oxygen and sufficient time for reactions [1]. On the other hand, inefficient droplet evaporation and improper mixing of fuel-air lead to a formation of fuel-rich pockets resulted in formation of PM. Raeie et al. [2] found that NO_x was decreased with retarded injections but at the same time PM was increased due to incomplete combustion of diesel fuel. A typical trend of NO_x and soot production during combustion process in diesel engines was shown on equivalence ratio (Φ)-T plane by Potter and Durrett [3]. It showed that NO_x and soot can be reduced simultaneously when combustion occurs at low Φ and low temperature regimes which was defined as low temperature combustion (LTC) regime. An exhaust gas recirculation (EGR) is generally used for achieving LTC. EGR increases the heat capacity along with reduction in oxygen content of an in-cylinder mixture which limits the peak combustion temperature [1]. A study by Zheng et al. [4] showed that NO_x and PM can be reduced simultaneously with EGR above 55%. An increase in ignition delay was a reason for the reduction in PM at higher EGR, which improved the mixing of fuel-air, resulted in lower local equivalence ratio. Kimura et al. [5] demonstrated modulated kinetics (MK) combustion mode by using retarded injection timing (near top dead centre (TDC)) and heavy EGR. A simultaneous reduction in NO_x and PM emissions was achieved over a low-to-moderate speed and load range. However, with retarded injection, there was an increase in fuel consumption due to a lower expansion ratio. Brijesh et al. [6] and Yin et al. [7] observed similar results with cooled EGR (below 75° C).

Corresponding author. Brijesh P; Fax: +91-20-3023-1104
E-mail address: patel.pga@araiindia.com

Recent developments in high-pressure electronically controlled common-rail injection systems give freedom to optimize the fuel injection schedule to reduce in-cylinder NO_x and PM formation as much as possible. Multiple injection strategies offer a better solution for reducing emissions with lower effect on engine performance [7, 8]. Mobasheri et al. [9] found that ~20 crank angle (CA) dwell was beneficial for a simultaneous reduction of NO_x and PM emissions. A similar study was done by Chen [10] in which maximum benefit in terms of reduction in NO_x and PM was observed with the small pilot (~13% of total injected fuel) fuel quantity and main injection timing near TDC. The combination of multiple injections with EGR was used by Sarangi et al. [11] which resulted in near zero NO_x emissions and a significant reduction in PM emissions.

In this work, experimentally optimized LTC and optimum two-shot injection with LTC strategies have been numerically investigated with *Converge* computational fluid dynamics (CFD) tool to have insight on in-cylinder processes and emission formation.

2 Methodology

The experimental work had been carried out on a naturally aspirated, 661 cubic centimetre (cc), variable compression ratio (VCR) diesel engine having provision for cooled EGR and multiple injection strategy. The detailed information of engine configuration, test rig and experimental analysis may be found in Ref. [12].

The CFD software used in this numerical study was *Converge*, which is a general purpose CFD code for simulating three-dimensional reacting flow simulations with stationary or moving boundaries. The cylindrical domain of the engine was divided into angular sectors such that one injector nozzle placed at the center of each sector. The injector has three nozzles separated by 120°. Figure 1 shows a 120° sector model having hemispherical combustion chamber (HCC) where faces 1, 2, 3 and 4 are piston surface, cylinder wall, cylinder head and front-back faces of 120° sector model respectively.

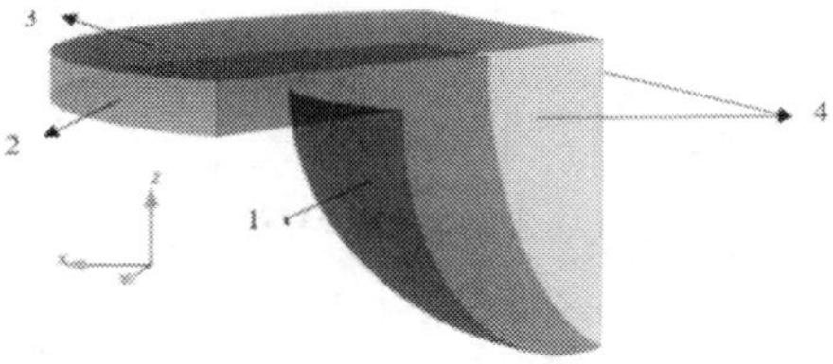

Figure 1: 120° sector model for HCC of the diesel engine

2.1 Simulation Setup and Code Validation

Before using the *Converge* CFD code for described model to study the in-cylinder combustion processes and emissions, it is necessary to validate its predictive ability. For this reason, numerical data was validated against the experimental results of Nikhil et al. [12]. The model was calibrated for 50% load at 1500 rpm engine speed. Injection parameters such as the start of injection (SOI) and injection pressure were fixed at -27 crank angle degrees after top dead center (CAD aTDC) and 240 bar respectively and EGR was kept 0%. In this work, this condition has been called as a base run. Close cycle simulations i.e. intake valve close (IVC= -144 CAD aTDC) to exhaust valve open (EVO= 144 CAD aTDC) were carried out. The RANS-based Re-normalization group (RNG) k–e model was used for the simulations in this work to model turbulence. Wall heat transfer was modelled using the Han and Reitz heat transfer model [13]. The spray was initiated using the blob injection method [14]. The modified Kelvin-Helmholtz (KH)-Rayleigh-Taylor (RT) breakup model along with No Time Counter (NTC) collision model suggested by Schmidt and Rutland [15] was used. Initial conditions at IVC, mass of fuel injected, injection rate shape and the SOI were given as inputs to code. The SAGE detailed chemistry model was used to model the combustion. The kinetic mechanism used in the simulation consisted of 42 species and 168 reactions. The extended Zeldovich NO_x model was used in this work to predict NO_x. The Hiroyasu soot formation and Nagle and Strickland-Constable (NSC) soot oxidation models were used to compute the soot emissions.

Figure 2 shows the comparison between experimental and simulation pressure traces. Close comparability of pressure traces for the majority of crank angles along with the similar start of combustion (SOC) timing can be observed from Fig. 2. Table 1 shows a good comparability between simulated and experimental NO_x and soot emissions. Overall, it may be concluded that the code was validated well with the experimental data for pressure, NO_x and soot. Further simulations were performed based on this calibrated model. Initial settings like model constants, surface temperatures, initial grid size and time-step were kept same for the remaining simulations.

Base run	NO_x (g/kg of fuel)	Soot (g/kg of fuel)
Experimental	37.53 (= 14.6 g/kWh)	1.69 (= 0.65 g/kWh)
Numerical	31.83	1.63

Table 1: Emissions comparison

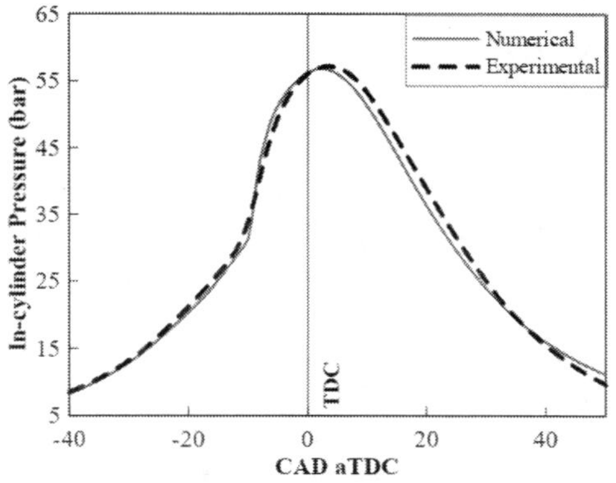

Figure 2: Comparison of simulated and experimental in-cylinder pressure traces

2.2 Implementation of Optimized LTC and Optimum Two-Shot Injection Strategies

The experimental study of Nikhil et al. [12] showed that optimized LTC (cooled EGR 30%, SOI -5 CAD aTDC, injection pressure 270 bar) offered a significant reduction in NO_x (~98%) and PM (~61%) with reduction in brake thermal efficiency (BTE) by 13% as compared to the base run. Optimum two-shot injection strategy (18 CAD dwell, 10% pilot fuel injection quantity, main injection at -5 CAD aTDC, injection pressure 270 bar) with LTC offered improvement in BTE by around 8% with a minor increase in NO_x and PM as compared to optimized LTC. More details about experimental work may be found in ref. [12]. Hence, these two optimized strategies were simulated using calibrated *Converge* code to understand physics of in-cylinder processes and reason behind a huge reduction in emissions. Input conditions were modified to cooled EGR 30%, SOI -5 CAD aTDC and injection pressure 270 bar to simulate optimized LTC run without changing any of combustion model constant. Similarly, spray and fuel inputs were modified to simulate optimum two-shot injection strategy. Figure 3 shows good comparability between simulated and experimental NO_x and PM trends for both the optimized strategies.

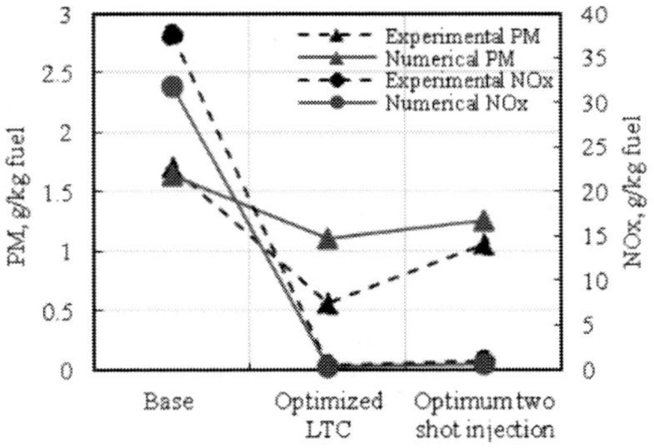

Figure 3: Simulated and experimental emission trend comparison for various strategies

3 Results and Discussion

A literature study highlighted that formation of NO_x is a function of high temperature, presence of oxygen (O_2) and sufficient time for the reactions. Hence, a maximum local peak temperature (T_{max}), duration for which local peak temperature (T) is higher than 2100 K during the in-cylinder combustion process and in-cylinder mass fraction of oxygen available at the start of combustion were analyzed for the base run, optimized LTC and optimum two-shot injection strategies and tabulated in Table 2. Study of Potter and Durret [3] showed that formation of NO_x starts in high temperature regime above 2100 K. Therefore, 2100 K was chosen as temperature limit for this study. T_{max} for optimized LTC and optimum two-shot injection strategy were found 2432 K and 2526 K respectively which were considerably lower than that of the T_{max} with the base run, as described in Table 2. Effect of cooled EGR and late heat release during expansion stroke can be seen on T_{max} in the combustion chamber. For the base run, local peak temperature during combustion process was observed higher than 2100 K for the duration of 75 CAD. On the other side, as provided in Table 2, the duration for optimized LTC run and optimum two-shot injection strategies are found 50 CAD and 63 CAD respectively. Use of EGR had reduced oxygen mass fraction available in the combustion chamber before the start of combustion for optimized LTC and optimum two-shot injection strategy as compared to that for the base run (see Table 2). Lower peak temperature, lower percentages of O_2 and less available time for reactions were observed for both optimized LTC and optimum two-shot injection strategies as compared to those with base run. As a result, a significant reduction in NO_x was observed for optimized LTC and optimum two-shot injection strategies as compared to the base run (see Fig. 3).

Run	Maximum local peak temperature (K)	Duration (CAD) for T_{max} > 2100 K	In-cylinder O_2 mass fraction at SOC
Base run	2858	75	0.23
Optimized LTC	2432	50	0.19
Optimum two-shot injection strategy	2526	63	0.19

Table 2: Parametric comparison of factors affecting on NO_x formation

Figure 4 shows PM traces for the base run, optimized LTC and optimum two-shot injection strategies. PM formation and oxidation are a function of fuel-air mixing and temperature. A high soot formation rate was observed with base run and optimum two-shot injection strategy as compared to that with the optimized LTC run (see Fig. 4).

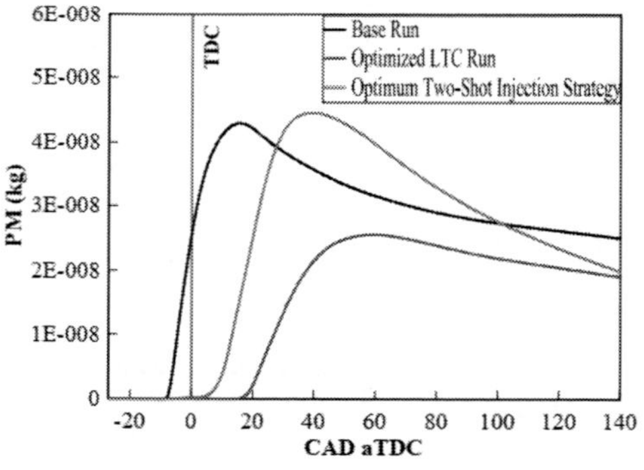

Figure 4: In-cylinder PM emission trends for various strategies

To understand this complex soot formation and oxidation processes, contours of equivalence ratio and in-cylinder temperature were plotted in Fig. 5 and analysed. Contours of equivalence ratio on a central plane and piston bowl surface are represented by Columns 1 and 2 respectively. Column 3 represents corresponding in-cylinder temperature contours on the central plane. Rows a, b and c represent base run, optimized LTC run and optimum two-shot injection strategy respectively. As soot formation instances were different for base run, optimized LTC and optimum two-shot injection strategies (see Fig. 4), contours were plotted for different instances for each case. For base run (a), optimized LTC run (b) and optimum two-shot injection strategy (c), contours were taken at 0 CAD aTDC, 43 CAD aTDC and 18 CAD aTDC respectively. These were the instances where equal amount of soot was already formed for above mentioned cases during respective soot formation period. Regions with higher equivalence ratio (~5) and high temperatures (>1500 K) are the regions where majority of soot forms [3]. Figures 5(a-1) and 5(a-2) show fuel rich pockets near to the base of the piston which is at a higher temperature (see Fig. 5(a-3)) which resulted in high soot formation with the base run. Based on flow analysis, it was observed that due to the upward motion of piston during fuel injection, high amount of fuel got accumulated near to base of the piston which was responsible for the formation of fuel rich pockets in that region (see Fig. 6a). On contrary, for optimized LTC run, piston motion was downward for most of the period during fuel injection. Therefore, instead of fuel getting accumulated near base of piston, majority fuel was pushed in upper part of combustion chamber as well as in squish region (see Fig. 6b), where fuel got mixed with surrounding air producing pockets with comparatively low equivalence ratio (~2). As an effect of LTC, combustion temperatures were also found lower (see Fig. 5 (b-3)). The majority of lower equivalence ratio mixture at relatively lower temperatures (1400 K-1700 K) resulted in lower soot formation for optimized LTC run [3].

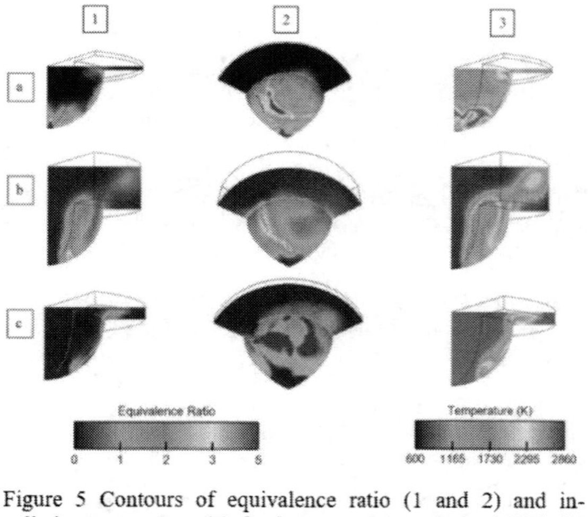

Figure 5 Contours of equivalence ratio (1 and 2) and in-cylinder temperature (3) for base run (a), optimized LTC (b) and optimum two-shot injection strategy (c) at 0, 43 and 18 CAD aTDC respectively

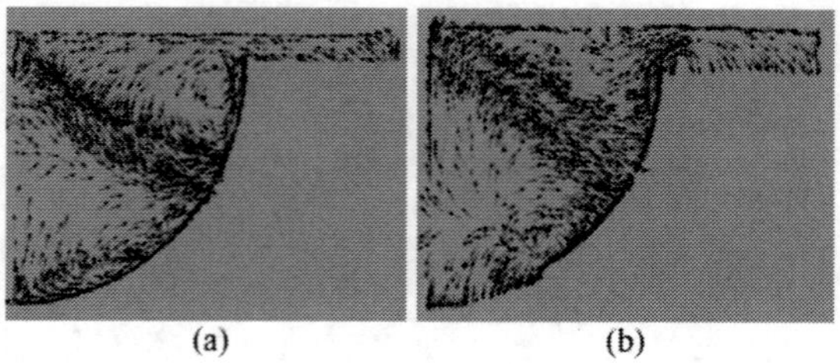

Figure 6: Velocity vectors for (a) base run (b) optimized LTC

Mean gas temperature traces were plotted for the base run, optimized LTC and optimum two-shot injection strategy (see

Fig. 7). Early combustion of pilot fuel resulted in higher mean in-cylinder gas temperature during main injection (at -5 CAD aTDC) for optimum two-shot injection strategy as compared to that for the optimized LTC run. Because of higher temperature, combustion of fuel injected in the main injection started earlier which can be seen in experimental heat release traces [12]. This caused insufficiency of time for fuel (injected in the main injection) to get mix with surrounding air and produced pockets of rich mixture near to piston bowl surface, that can be clearly observed in Figs. 5(c-1) and 5(c-2). Also, high temperatures (above 1500 K) were observed near to the rich mixture pocket (Fig. 5(c-3)). Hence, higher soot formation was observed with optimum two-shot injection strategy (see Fig. 4).

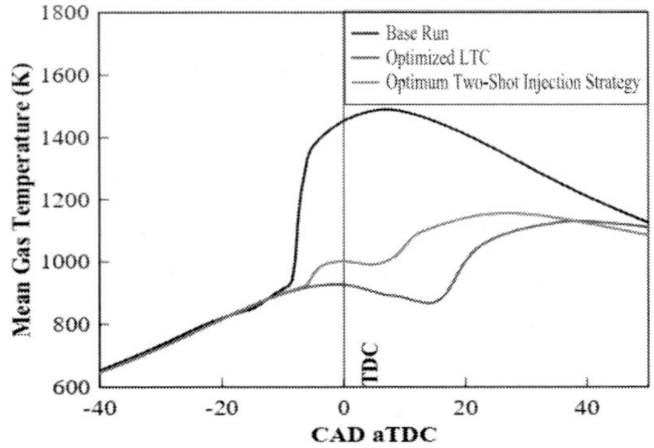

Figure 7: Mean gas temperature trends for various strategies

4 Conclusions

In this work, *Converge* CFD code for 120° sector model has been validated against the experimental results of Nikhil et al. [12]. Experimentally achieved optimized LTC and optimum two-shot injection strategies were simulated by modifying inlet conditions and injection parameters. Results of simulation were analysed to understand NO_x and PM emission formation processes.

- Lower local peak combustion temperatures and less time for the reactions were responsible for a significant reduction in NO_x emissions with optimized LTC and optimum two-shot injection strategies.
- Direction of piston motion during fuel injection plays major role in fuel-air mixing. Fuel injection during expansion stroke resulted into better fuel-air mixing for optimized LTC.
- Combustion of pilot fuel advanced SOC of main injection fuel which led to insufficiency of time to form fuel-air mixture resulted into high soot formation for optimum two-shot injection strategy.

Further improvement in LTC with two-shot injection strategy may be possible with modularized injection strategy.

5 Acknowledgment

The authors are grateful to the Industrial Research and Consultancy Centre (IRCC), IIT Bombay for providing numerical tool *Converge* for research work. Authors are also thankful to "Convergent Science" for providing *Converge* licenses with academic price to IIT Bombay.

References

[1] Heywood, J., Internal Combustion Engine Fundamentals. McGraw-Hill Education, New York, 1988.

[2] Raeie, N., Emami, S. and Sadaghiyani, O.K., Effects of Injection Timing, before and after Top Dead Center on the Propulsion and Power in a Diesel Engine. Propulsion and Power Research, 3(2)(2014) 59-67.

[3] Potter, M. and Durrett, R., High-Efficiency Clean Combustion Design for Compression Ignition Engines. 12th Annual Diesel Engine Emissions Reduction (DEER) Conference, Detroit, *USA* (2006).

[4] Zheng, M., Han, X. and Reader, G.T., Empirical Studies of EGR Enabled Diesel Low Temperature Combustion. Journal of Automotive Safety and Energy, 1(3)(2010) 219-228.

[5] Kimura, S., Aoki, O., Kitahara, Y. and Aiyoshizawa, E., Ultra-Clean Combustion Technology Combining a Low-Temperature and Premixed Combustion Concept for Meeting Future Emission Standards. SAE Technical Paper, (2006) (No. 2001-01-0200).

[6] Brijesh, P., Chowdhury, A. and Sreedhara, S., Effect of Ultra-Cooled EGR and Retarded Injection Timing on Low Temperature Combustion in CI Engines. SAE Technical Paper, (2013)(No. 2013-01-0321).

[7] Yin, B., Wang, J., Yang, K. and Jia, H., Optimization of EGR and Split Injection Strategy for Light Vehicle Diesel Low Temperature Combustion. International Journal of Automotive Technology, 15(7)(2014) 1043-1051.

[8] Brijesh, P. and Sreedhara, S., Experimental and Numerical Investigations of Effect of Split Injection Strategies and Dwell between Injections on Combustion and Emissions Characteristics of a Diesel Engine. Clean Technologies and Environmental Policy, 18(7)(2016) 2325-2334.

[9] Mobasheri, R., Peng, Z. and Mirsalim, S.M., Analysis of the Effect of Advanced Injection Strategies on Engine Performance and Pollutant Emissions in a Heavy-Duty DI-Diesel Engine by CFD Modeling. International Journal of Heat and Fluid Flow, 33(1)(2012) 59-69.

[10] Chen, S.K., Simultaneous Reduction of NO_x and Particulate Emissions by Using Multiple Injections in a Small Diesel Engine. SAE transactions, 109(3)(2000) 2127-2136.

[11] Sarangi, A.K., Garner, C.P., McTaggart-Cowan, G.P., Davy, M.H., Wahab, E. and Peckham, M., The Effects of Split Injections on High Exhaust Gas Recirculation Low-Temperature Diesel Engine Combustion. International Journal of Engine Research, 14(1)(2013) 68-79.

[12] Nikhil K., Vehzan R., Brijesh P. and Sreedhara S., Low Temperature Combustion with Two-Shot Injection Strategies for Simultaneous Reduction of NO_x and PM in Compression Ignition Engines. International Conference on Sustainable Energy and Environmental Challenges, 2017, p. 367.

[13] Han, Z. and Reitz, R.D., A Temperature Wall Function Formulation for Variable-Density Turbulent Flows with Application to Engine Convective Heat Transfer Modeling. International Journal of Heat and Mass Transfer, 40(3)(1997) 613-625.

[14] Reitz, R.D. and Diwakar, R., Structure of High-Pressure Fuel Sprays. SAE Technical Paper, (1987) (No. 870598).

[15] Schmidt, D.P. and Rutland, C.J., A New Droplet Collision Algorithm. Journal of Computational Physics, 164(1) (2000) 62-80.

11th Asia-Pacific Conference on Combustion,
The University of Sydney, NSW Australia
10th -14th December 2017.

Study of Combustion Behavior and Emission Formation of Compression Ignition Engine under Transient Operating Conditions Using Simulation Tools

Nikhil Mahajan[1], Brijesh P[2]

[1]ARAI Student

Kothrud, Pune, Maharashtra, 411038, India

[2]ARAI Faculty

Kothrud, Pune, Maharashtra, 411038, India

Abstract

Diesel engines need to be tested under transient conditions because most of the time, the driving schedule of the vehicles is transient. The need to test the diesel engine under transient operating conditions is to know about the areas which cause emission formation. The purpose of this work was to study the in-cylinder combustion analysis of diesel engine under transient conditions. A portion in the United States Environmental Protection Agency (US EPA) agricultural tractor transient cycle was simulated in Converge Computational Fluid Dynamics (CFD)-tool to study combustion under transient conditions. Some of the inputs required for simulation in Converge CFD-tool were taken from the simulation done in one-dimensional (1D) Ricardo Wave software for same transient conditions. In this work, the deterioration in combustion due to transient conditions was analysed and the effect of increase in injection pressure when transient conditions are prevailing was investigated too. The soot emission formation was also studied with the help of change in acetylene mass throughout the engine cycle. The results showed that with increase in injection pressure soot emissions were reduced but nitrogen oxide (NO_x) emissions were increased. The effect of cylinder wall temperatures on emissions formation was also predicted.

1 Introduction

The transient events taking place in a diesel engine can be broadly classified into three categories which are load increase transient event, speed increase transient event and cold starting. In load increase transient event, a torque deficit is created between engine torque and resisting torque. With this event, engine speed drops resulted in injection of more fuel making the fuel and air ratio high. This deteriorates the combustion leading to overshoot in particulate matter (PM) [1,2]. In speed increase transient event, the resisting torque is constant but the increase in torque demand by the driver will result in increase of fuel injection [1,2]. In cold starting, the in-cylinder temperatures are not enough to ignite the fuel. Hence, more fuel is injected leading to more soot emissions.

Yum et al. [3] compared the transient results with static loads and quasi-steady estimations experimentally. It was found that compared to static loads, the largest deviations for brake specific fuel consumption (bsfc) were 6.7% and 9.1% for lowest and medium load respectively. For quasi-steady estimations, the largest errors for bsfc were 2.9% and 4.3% for lowest and medium load respectively. Hagena et al. [4] also did experimental study with quasi-steady and transient events. Results showed increase in PM with increased load with transient event.

High fuel air ratios are generally found after an increase in speed or load producing higher combustion temperatures. NO emissions increase due to local high temperatures formed because of close to stoichiometric air-fuel mixtures [5]. Soot forms in the regions where fuel mass fraction and temperatures have higher values, i.e. in the richer and hotter regions [6]. There is a rapid increase in fuel injection pressure upon the onset of the instantaneous load step causing the penetration of the liquid fuel jet to increase. The rate of mixture preparation becomes slow due to liquid fuel impingement on the cool combustion walls, enhancing the heterogeneity of the mixture resulted in soot emissions [7].

Zhang et al. [8] had found that the accumulative NO_x was increased by 12.7% and 27.6% as the injection timing was advanced by 2 crank angle degree (CAD) and 4 CAD respectively with various transient events. The increase in injection pressure led to improve in the quality of atomization and evaporation leading to a shortened ignition delay period and an accelerated combustion rate. Effect on emissions while transition from conventional mode of diesel engine operation to low temperature combustion (LTC) mode was experimentally investigated by Sarangi et al. [9]. The results of study showed that the load transient pose occurs during mode change and resulted in higher carbon monoxide, total hydrocarbon and smoke emissions. Henein et al. [10] investigated the in-cylinder condition at the time of cold starting. It was found from the cold starting pressure diagrams that combustion failure is intermittent, leading to a series of firing and misfiring cycles. Effect of heat loss during cold starting transient event was investigated in various works [11-13]. It was observed that low heat loss with the highly insulated cylinder wall produced higher NO emissions due higher in-cylinder temperatures.

In general, a literature study showed that various transient events played a significant role in exhaust emissions of diesel engines. Hence, numerical study with transient event has been carried out in this work to understand in-cylinder combustion analysis and emission formation.

2 Simulation Setup and Validation

The 3D CFD simulation was carried out using Converge CFD-tool for understanding of transient combustion phenomena. Some of the inputs required for CFD were obtained from simulation done in 1D Ricardo Wave. The engine selected for this simulation was single cylinder, 4-stroke, direct injection, light-duty diesel engine. Please see Refs. 14 and 15 for details

Corresponding author. Brijesh P; Fax: +91-20-3023-1104
E-mail address: patel.pga@araiindia.com

of engine configuration required for simulation purpose. Both 1D and 3D models were validated against the experimental results of the operating condition (–27 crank angle degree after top dead centre (CAD aTDC) injection timing, 18 compression ratio, 240 bar injection pressure and 0% EGR) at 75% load of the light-duty diesel engine [14].

2.1 Ricardo Wave Numerical Setup and Validation

Ricardo Wave is an one dimensional engine analysis software based on empirical relations for all the activities taking place in the engine. Model prepared for the engine under study [14] is shown in Fig. 1. All the ducts/components used in the model are of the same dimensions as that used with the test bed. The combustion model and heat transfer model used were diesel wiebe combustion model and Woschni heat transfer model respectively. The Extended Zeldovich model [16] was used for prediction of NO_x. The Hiroyasu soot formation model [17] and Nagle and Strickland-Constable (NSC) soot oxidation model [18] have been used to compute soot emissions.

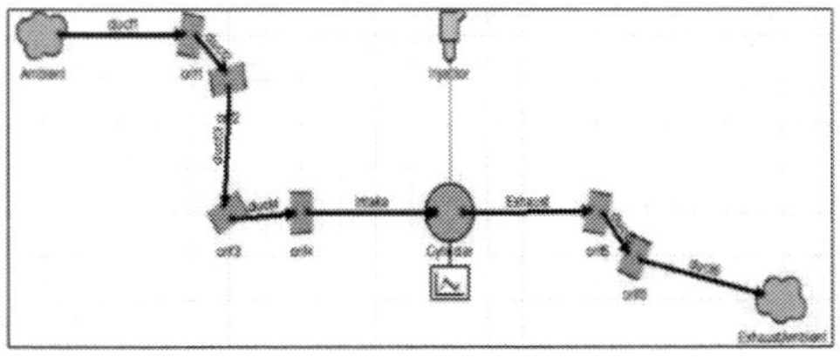

Figure 1: Ricardo Wave 1D model

The Ricardo Wave code for described model has been validated against the experimental results of Brijesh et al. [14]. Comparison between experimental and Ricardo Wave in-cylinder pressure traces is shown in Fig. 2. A good comparability between both traces is observed. Also code is capable to predict the values of NO_x, PM, brake power and bsfc very close with the experimental values. Based on the validation results, 1D model was found suitable to obtain various inputs required for 3D CFD.

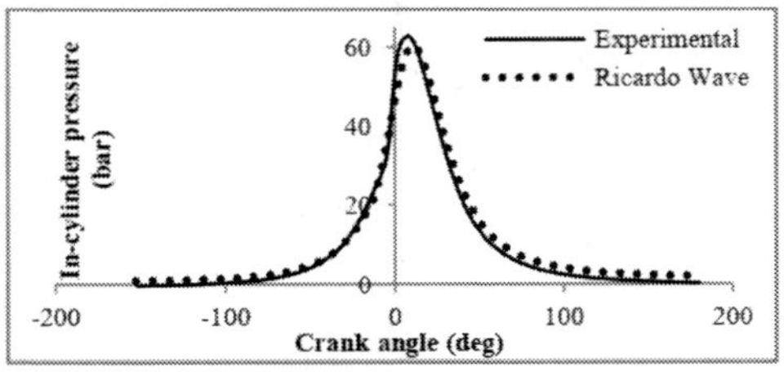

Figure 2: Comparison between experimental and Ricardo Wave in-cylinder pressure traces with crank position

2.2 Converge Numerical Setup and Validation

The Converge CFD-tool was used for 3D CFD simulation. Ease of gridding, adaptive mesh refinement (AMR) capabilities and ease of parallelization made Converge CFD-tool very attractive to simulate internal combustion (IC) engine problems. For more information please see Ref. [19]. In Converge, the cylindrical domain of the engine can be divided into angular sectors such that one injector nozzle falls at the centre of each sector. The engine under study has three nozzles. The injector location was axis-symmetric. Hence, a 120° sector model, as shown in Fig. 3, is created for the diesel engine having hemispherical combustion chamber and a flat cylindrical head. The simulations were carried out from inlet valve closing (IVC = –145 CAD aTDC) till exhaust valve opening (EVO = 145 CAD aTDC).

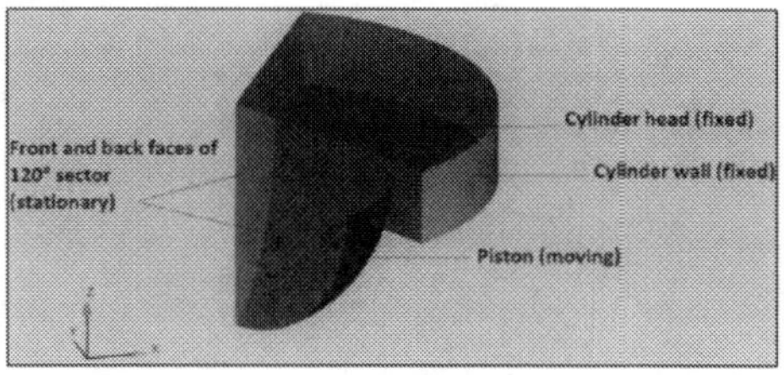

Figure 3: A 120° sector model with boundary conditions

Diesel fuel was used for liquid phase to simulate the spray. Since the detailed chemistry of diesel fuel is not available, n-heptane has been used to simulate the combustion. A variant of k-ε RNG model has been used for modelling turbulence, which takes into account the compressibility and spray-turbulence interaction and predicts large flame structures better. Wall heat transfer was estimated by using the heat transfer model suggested by O'Rourke and Amsden. Spray was initiated using the blob injection method suggested by Reitz and Diwakar [20]. In this method, initial parcel sizes were set equal to the nozzle diameter. The modified Kelvin-Helmholtz (KH)-Rayleigh-Taylor (RT) breakup model with breakup length along with No Time Counter (NTC) collision model suggested by Schmidt and Rutland [21] has been employed. The rebound/slide spray wall interaction model was used. Injection profile generated during experimental work [14] has been given as an input. Using the input values of injection duration and mass of fuel injected, Converge CFD-tool scaled the injection velocity profile to inject the specified mass of fuel in the specified crank angle duration. The solver used was a variable time step solver with initial time step as $1e^{-07}$. The change in time step occurs due to change in grid spacing. This is controlled by Courant-Friedrich's Lewy (CFL) numbers. The SAGE detailed chemistry model has been used to model the combustion [22]. Kinetic mechanism used in the simulation consisted of 42 species and 168 reactions. The extended Zeldovich NO_x model [16] has been used for NO_x prediction. The Hiroyasu soot formation model [17] and Nagle and Strickland-Constable (NSC) soot oxidation model [18] have been used to compute the soot emissions.

The Converge CFD code for described model has been validated against the experimental results of Brijesh et al. [14]. A very good comparability between experimental and Converge traces for a wide range of crank angles may be seen in Fig. 4. A very good comparability is observed between experimental and numerical values of exhaust NO_x and soot concentrations as described in Table 1. Hence, the model can be considered as validated and can be used for further simulations.

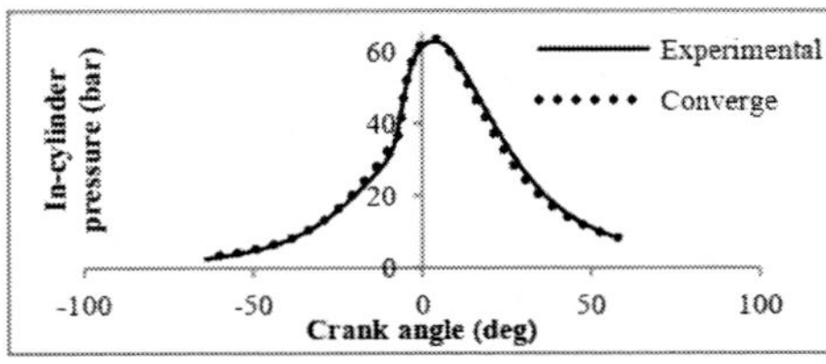

Figure 4: Comparison between experimental and Converge in-cylinder pressure traces with crank position

Base run	NO_x (g/kg of fuel)	Soot (g/kg of fuel)
Experimental	37.0	13.224
Converge	20.6910	14.513

Table 1: Comparison of emissions values

3 Testing Methodology Adopted

To study the in-cylinder processes during transient event, it was necessary to run the simulation in Converge under transient cycle. The transient cycle selected for this work was US EPA agricultural tractor transient cycle, as can be seen in Fig. 5. Only a certain portion in the whole cycle (highlighted with red dotted rectangle in Fig. 5) was selected for simulation due to lack of computation capabilities. The portion to be simulated was a spike consisting of increasing engine speed from 50% to 90% and engine torque from 35% to 90%.

The calculation was done for conversion of normalised values to actual values. Engine speed needs to be increased from 695 rpm to 971 rpm. For this requirement, engine torque should increase from 5.77 Nm to 14.84 Nm. The speed increase was decided to be achieved in two cycles of engine operation. Hence, the speed increase step of 276 rpm was divided into two equal halves. A simulation was run in 1D Ricardo Wave by giving the above speed increase in validation model to get the stated increase in torque, keeping all other inputs the same. The 1D simulation was done to find the initial pressure and temperature at −145 CAD aTDC which were required as inputs to 3D simulation. The change in injected mass from 8.9 mg to 16.2 mg during the speed increase event was also found from 1D simulation. Hence, injection duration and injection pressure need to be increased to 16 CAD and 186.65 bar respectively with Cycle 2 as compared to Cycle 1 (where injection duration and pressure were kept 10 CAD and 104.4 bar respectively).

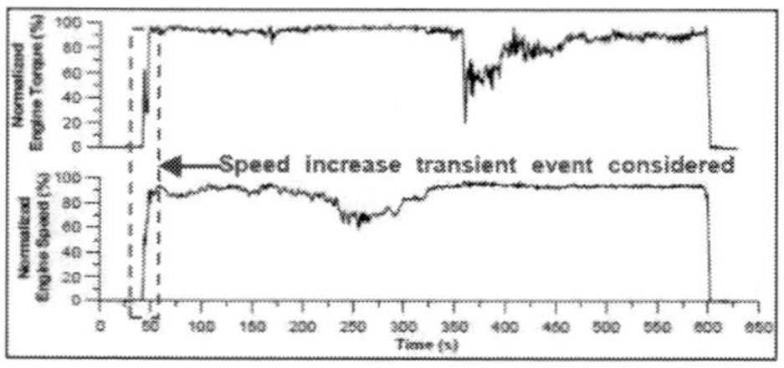

Figure 5: US EPA agricultural tractor transient cycle [1]

4 Results and Discussion

4.1 Effect of Increase in Injection Pressure on Air Fuel Mixing

The droplet velocity contours of the Cycles 1 and 2 at −20 CAD aTDC is shown in Fig. 6. It can be seen in Fig. 6 that the droplet velocity at the start of spray with Cycle 2 is almost three times of that with Cycle 1. The injection velocity increases as the injection pressure increases. More the droplet velocity more will be the penetration. The averaged turbulent kinetic energy (TKE) over entire domain in Cycles 1 and 2 is compared in Fig. 7. The high injection velocity aided with compression of the charge led to increase TKE with Cycle 2. The mixing of the charge was enhanced due to increase in turbulence resulted in reduced ignition delay by around 2 CAD with Cycle 2 than Cycle 1.

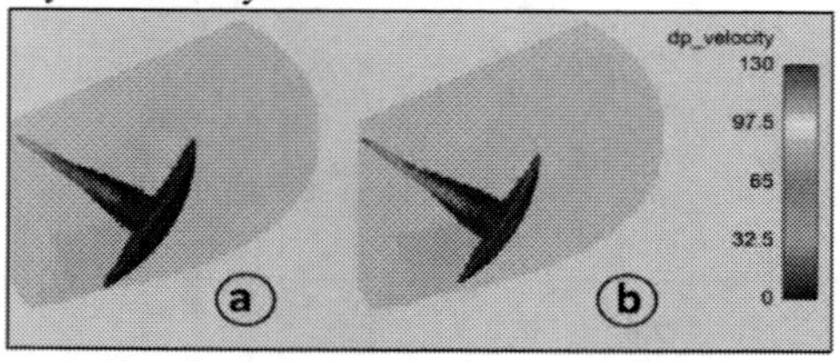

Figure 6: Droplet velocity contour comparison between a) Cycle 1 and b) Cycle 2 at −20 CAD aTDC

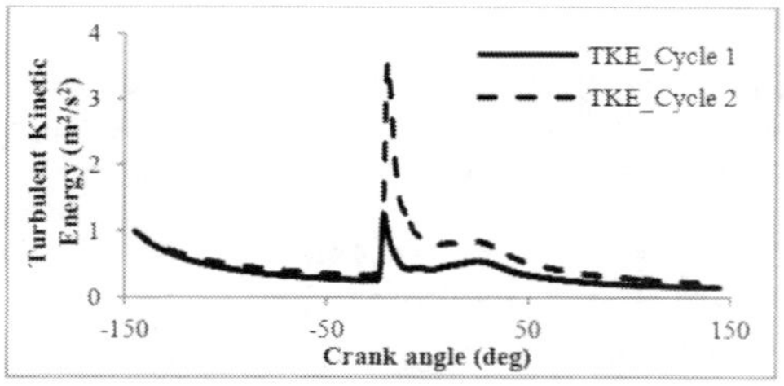

Figure 7: TKE plots for Cycles 1 and 2

4.2 Effect of Increase in Injection Pressure on Emissions

As injection pressure was increased with Cycle 2, reduction in soot was observed. Soot formation generally depends on the density in the centre of fuel spray and entrainment of air into it. The increased injection pressure caused good mixing, due to which complete combustion took place resulted in higher NO_x observed with Cycle 2. Table 2 gives the comparison of NO_x and soot emissions for both the cycles.

Cycle No	NO_x (g/kg of fuel)	Soot (g/kg of fuel)
Cycle 1	8.3733	22.5029
Cycle 2	34.12	15.9242

Table 2: Comparison of NO_x and soot for Cycles 1 and 2

Contours of oxygen content, temperature, NO_x and soot for Cycle 2 are plotted in Fig. 8. NO_x is formed in the common region where stoichiometric conditions are present and temperature is high. Soot is formed in regions having less oxygen content and moderately high temperatures. There is no soot formation in the regions where NO_x is formed. This shows the trade-off between NO_x and soot.

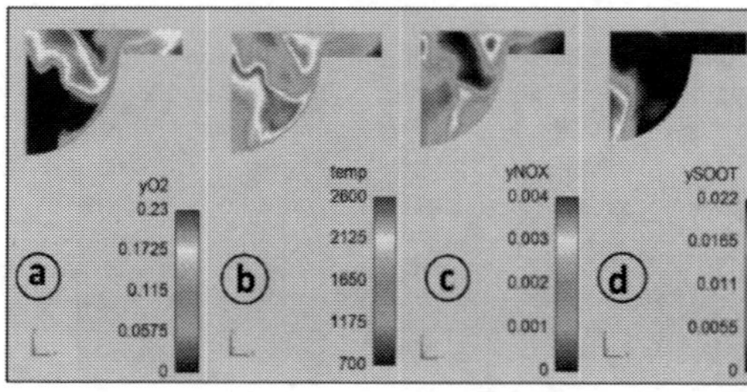

Figure 8: Contours of a) oxygen content, b) temperature, c) NO_x and d) soot for Cycle 2 at 25 CAD aTDC

Acetylene is believed to be one of the precursors of soot [23]. Concentration of acetylene (C_2H_2) and soot with respect to crank position for Cycle 2 are plotted in Fig. 9. Increasing trend of soot with decreasing trend of acetylene in Fig. 9 indicates the formation of soot from acetylene.

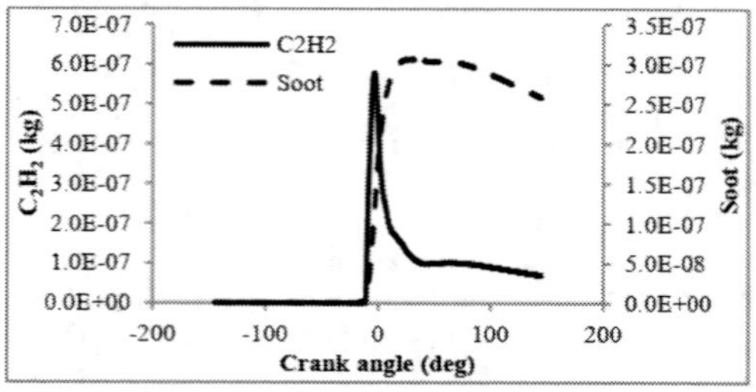

Figure 9: Acetylene (C_2H_2) and soot concentration with crank position for Cycle 2

4.3 Heat Entrainment due to Cylinder Wall Temperatures

Figure 10 shows the contours of in-cylinder temperature and the temperature on piston boundary. As can be seen in Fig. 10, the temperature on the surface of the piston (highlighted with a black circle) is higher than the in-cylinder temperature. Hence, heat transfer might be taking place from piston boundary to inside the cylinder. This may increases the temperature inside the cylinder leading to higher NO_x formation and enhances soot oxidation.

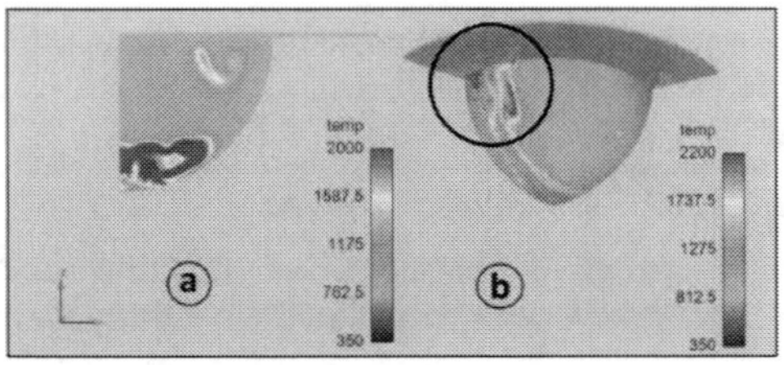

Figure 10: a) In-cylinder temperature and b) piston boundary contours at 5 CAD aTDC

5 Conclusions

The simulation for studying in-cylinder combustion analysis for speed increase transient event was done in Converge CFD software. The inputs required for the simulation in Converge CFD software were taken from the 1D simulation in Ricardo Wave software with the same operating conditions as given in Converge. Converge CFD code for 120° sector model has been validated against the experimental results for 75% load steady state conditions. The results for the speed increase transient events were extracted and studied thoroughly. Subsequent conclusions are derived from the analysis of the results.

- Improper mixing of air fuel mixture leads to combustion deterioration resulting in increase of soot and NO_x emissions.
- Increase in injection pressure led to proper mixing of air fuel mixture followed by proper combustion resulting in reduction in soot emissions but increase in NO_x emissions.
- The formation of soot from acetylene was noticed.
- Heat transfer may takes place inside the cylinder leading to increase in in-cylinder temperature resulting in high NO_x formation and enhance soot oxidation.

6 Acknowledgment

The authors thank "Convergent Science" and "Computational Engineering India Pvt. Ltd, Pune" for providing their valuable resources for simulations done. Authors are also thankful to "Ricardo India Pvt. Ltd." for providing free of cost Ricardo wave licenses to ARAI academy. Authors also recognize support provided by IC Engine lab members, Indian Institute of Technology (IIT), Mumbai in terms of sharing data related to engine and experimental results.

References

1. C.D. Rakopoulos, E.G. Giakoumis, Diesel Engine Transient Operation, Springer-Verlag, London, 2009.
2. C.D. Rakopoulos, E.G. Giakoumis, SAE Technical Paper 2006-01-0884.
3. K.K. Yum, N. Lefebvre, E. Pedersen, Applied Energy 185 (2017) 472–481.
4. J.R. Hagena, Z.S. Filipi, D.N. Assanis, SAE Technical Paper 2006-01-1151.
5. C.D. Rakopoulos, A.M. Dimaratos, E.G. Giakoumis, D.C. Rakopoulos, Energy Conversion and Management 50 (9) (2009) 2381–2393.
6. C.D. Rakopoulos, D.T. Hountalas, G.N. Taklis, E.I. Tzanos, International Journal of Energy Research 19 (1) (1995) 63-88.
7. H. Harndorf, H.W. Kuhnt, MTZ worldwide 56 (1995) 8-20.
8. L.P. Zhang, Z.C. Liu, J. Tian, Sci China Tech 57 (3) (2014) 480-488.
9. A.K. Sarangi, C.P. Garner, G.P. McTaggart-Cowan, M.H. Davy, E. Wahab, M. Peckham, Proceedings of the Institution of Mechanical Engineers Part D: Journal of Automobile Engineering 229 (7) (2014) 850–865.
10. N.A. Henein, A.R. Zahdeh, M.K. Yassine, W. Bryzik, SAE Technical Paper 920005, 1992.
11. C. Arcoumanis, X.G. Yao, Transient smoke and unburnt hydrocarbon emissions during cold-start in a turbo-charged DI diesel engine, Institution of Mechanical Engineers Seminar on 'Transient Performance of Engines' (1994) 43–60.
12. C.D. Rakopoulos, E.G. Giakoumis, D.C. Rakopoulos, Energy Conversion and Management 45 (17) (2004) 2627–2638.
13. E.G. Giakoumis, C.D. Rakopoulos, A.M. Dimaratos, M. Peckham, SAE Technical Paper 2010-01-1287.
14. P. Brijesh, A. Chowdhury, S. Sreedhara, *Experimental and numerical investigations of in-cylinder solutions to reduce emissions in compression ignition engines*, PhD Thesis, Indian Institute of Technology Bombay, India, 2014.
15. P. Brijesh, S. Abhishek, S. Sreedhara, SAE Technical Paper 2015-01-0402.
16. J.B. Heywood, Internal Combustion Engine Fundamentals, McGraw Hill, Inc., U.S.A, 1988.
17. H. Hiroyasu, T. Kadota, SAE Technical Paper 760129, 1976.
18. J. Nagle, R.F. Strickland-Constable, Oxidation of Carbon Between 1000-2000°C, Proceedings of the Fifth Carbon Conference, 1 (1962), p. 154.
19. K.J. Richards, P.K. Senecal, E. Pomraning, CONVERGE v2.2 Theory Manual, Convergent Science, Inc., Madison, WI (2016).
20. R.D. Reitz, R. Diwakar, SAE Technical Paper 870598, 1987.
21. D.P. Schmidt, C.J. Rutland, Journal of Computational Physics 164 (1) (2000) 62–80.
22. P.K. Senecal, E. Pomraning, K.J. Richards, SAE Technical Paper 2003-01-1043.
23. H. Omidvarborna, A. Kumar, D.S. Kim, Renewable and Sustainable Energy Reviews 48 (2015) 635–647.

11th Asia-Pacific Conference on Combustion,
The University of Sydney, NSW Australia
10th -14th December 2017.

Numerical Simulation on Process of Pulse Detonation Engine Driven Ejector

XiQiaoHuang[1], XiangZhouFeng[1],MiaoMiaoWang[2], QianXiangLi[3],BoTao[4]

[1] School of Power and Energy,

Northwestern Polytechnical University,

Xi'an 710072, China

Abstract

In order to study the ejecting mechanism of pulse detonation engine (PDE)-ejector, the paper presents numerical simulation on the flow field parameters along the PDE detonation chamber in the conditions of ejector with different structure or axial positions by using CFD. The entraining processes of ejector with different structure are simulated at a 2D axisymmetric model to completely reveal the ejecting mechanism. It is concluded that there are three basic parts of the PDE-ejector cycle by analyzing an entire process of numerical simulation: (1) the detonation wave exists from the PDE followed by a flow reversal out of ejector inlet, (2) the generation of positive entrainment at the ejector inlet, (3) steady entrainment accompanied with strong flow.

1 Introduction

The pulse detonation combustion is a type of intensive combustion method with high thermal efficiency, which is achieved by periodic detonation wave. The detonation wave produces the combustion products with high temperature, pressure and speed[1]. Pulse detonation engine is an unsteady system that provides the propulsion by impulse of combusted gas with high temperature and high pressure generated by pulse detonation wave[2]. Compared with the other usual propulsion system,there are two obvious characteristic: unsteady work and detonation combustion process. PDE has the potential advantages such as high thermal efficiency, simple structure, lighter and low sfc etc[3].

Thrust is an important performance parameter of pulse detonation engine. The methods of improving the average thrust of PDE include increasing the impulse of unit volume, increasing the scale of the engine and increasing the work frequency. In the research of single-cycle detonation, a nozzle is usually equipped at the end of the PDE to improve the impulse of PDE, the results of experiment and simulation both indicate that a nozzle can improve obviously the impulse of PDE[4]. But something should be reminded: the nozzle is filled by fresh air in single-cycle. However the improvement of impulse is caused by the fresh air in nozzle jet and the air from the PDE, it is very different from ones in multi-cycle. The method that can eject fresh air and improve the impulse in multi-cycle is equipping an ejector at the outlet of PDE (abbreviated as PDE-ejector)[5].

Ejectors utilize the turbulent diffusion function of jet to mix two different flows, causing the exchange of energy. There has been the rape theory of design and utilization of the steady ejector. Although some research of the unsteady ejector has been conducted, there has almost no real utilization. The recently study shows that the unsteady ejector can generate higher improvement of thrust compared to the steady ejector.

The research reports about the ejector of PDE is increasing year by year, mainly on the working frequency of the engine, ejector aspect ratio and inlet shape, and axial position of the ejector thrust augmentation effect, but the conclusion is often different and even contradictory. Therefore, a large number of research and analysis are needed for the study of the unsteady ejector.

2 Numerical method

2.1 Basic equations and turbulence models

The unsteady two dimensional axisymmetric N-S equations [6], 5 groups of single step irreversible finite rate model of Chemical reactions, $k-\varepsilon$ model of turbulence models and standard wall function methods near the wall are employed.[7].The governing equations are as follows:

$$\frac{\partial \overline{Q}}{\partial t}+\frac{\partial \overline{E}}{\partial x}+\frac{\partial \overline{F}}{\partial r}-\frac{\partial \overline{E}_v}{\partial x}-\frac{\partial \overline{F}_v}{\partial r}+\frac{\overline{S}}{r}=0 \qquad (1)$$

Where $\overline{Q}$ is dependent variable, $\overline{E}$ and $\overline{F}$ are non-viscous fluxes in two coordinate directions, $\overline{E}_v$ and $\overline{F}_v$ are viscous flux in two coordinate directions, and $\overline{S}$ is the source term.

2.2 The computational domain and the computational grid[8-9]

Computational domain included detonation chamber and ejector. Detonation chamber consisted of ignition region and detonation propagation region. Ejector had mixing zone and far field. The detonation tube is 1600mm long with an inner diameter of 60mm (D). The ignition section is located in the closed end of PDE with 4 mm width. Ejector device is located in the open end of pulse detonation engine. Ejector is 950mm long. Inner diameters are respectively 120mm, 150mm, 170mm. The ejector entrances with diameter of 150mm have two shapes. Convergent inlet contour had a quarter of circular arc which lip radius being 50mm, another rounded inlet contour with a quarter of circular arc lip which radius being 5mm. Various positions of the ejector entrance with respect to the detonation tube exit are attempted to characterize engine performance. To normalize the parameters, it referred to the overlap position as x/d, with x being the axial distance from the detonation chamber exit to the ejector inlet and d being the inner diameter of the detonation chamber. As a convention, it selected a positive x/d factor to represent

XiQiaoHuang
E-mail address:janehxq@nwpu.edu.cn

overlapping of the engine and the ejector in the models and negative x/d factor to represent separation. Computational grid is structured grid. The total number of cells is 53494 and the minimum face area is 9e-07 m^2 .Figure 1 shows the computational domain of PDE with the ejector which radius is 150mm.

Figure. 1. computational domain of detonation ejector

2.3 Initial and boundary conditions

Assuming that all the walls are thermal insulation and no slipping.The boundary of the far field is pressure inlet, and inflow of gas components is air which total pressure is 1atm and total temperature is 300K. Ejector inlet and exit are internal flow fields. The thrust wall boundary condition is variable. The mixed gas filling stage is set as the velocity inlet, and the filling velocity is 60m/s. The detonation tube is filled with stoichiometric ratio mixture of C_3H_8/O_2. When the mixture is completely filled, the thrust wall boundary condition is changed into wall condition. Then the energy is added to the ignition region. The ignition zone pressure is set at 20atm, and the temperature is 3000K. The detonation waves are soon formed[10]. The second cycle is repeated after 40ms. In the beginning of the second cycle the inlet valve opened again. However, air is firstly filled, and then the mixture is filled subsequently otherwise directly filling the mixture would occur preignition.

2.4 Numerical calculation method

The finite volume method is employed in calculation process. The second-order implicit scheme is used in the time advancing process. All parameters are discretized by second-order upwind scheme, and the simple algorithm is used for the coupling of velocity and pressure.

3 Simulation results

3.1 Working process simulation of ejector with different axial positions

The mass of entrainment secondary flow could change at different time of one cycle for the ejector driven by unsteady flow. However, the total pressure of primary flow exit is constant for steady-flow ejector, and there is a fixed transition surface between the primary flow and secondary flow so that the mass flow could be constant for steady flow. A large number of studies had examined that the enhanced performance of unsteady ejectors is typically associated with a more efficient energy transfer process between the primary flow and the secondary (entrained) flow through inviscid processes, while the steady ejector relies primarily on viscous shear mixing. In fact, some analytical studies had indicated that the best performance for unsteady ejectors could be attained for relatively short ejector lengths with a fluid pulse that is of high momentum, low mass and short duration.

In order to study pulse effect on PDE performance when the ejector is located at different axial position relative to the exit of the PDE, we calculated the entraining process of three typical ejectors with x/d factor of -8,0,2.

Figure 2 showed the temperature contour of PDE-ejector in position 0. The filling process is too long to be presented in the cycle of PDE-ejector. The development process of flow field is traced out when detonation exhaust product expelled out of detonation tube. The transient mass flow and the pressure at the exit of the detonation chamber, the inlet and outlet of the ejector are presented in figure 3 and 4.

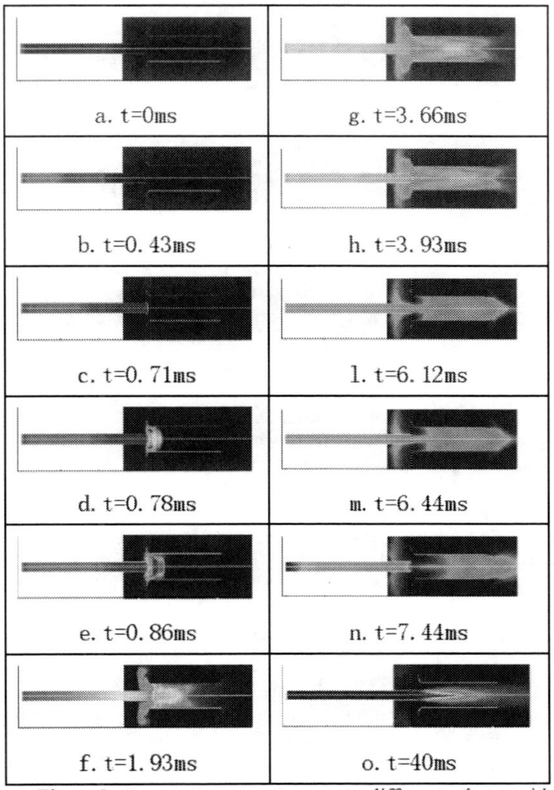

Figure.2 temperature contours at different time with D=150mm, r=50mm and x/d=0

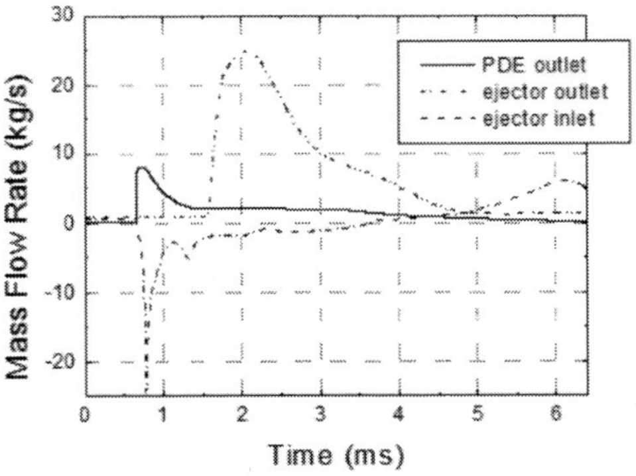

Figure.3 mass flow curves at different time with D=150mm, r=50mm and x/d=0

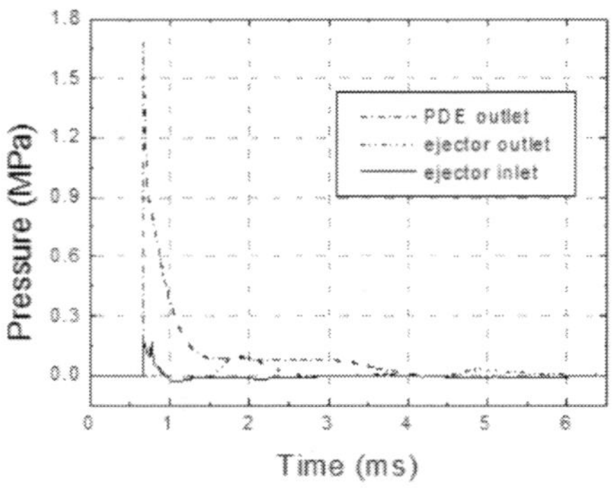

Figure.4 pressure curves at different times with D=150mm, r=50mm and x/d=0

The working processes when the ejector located different positions are similar. But the entraining effect is different. When the x/d factor comes to -8, the distance between the ejector inlet and detonation tube exit is farther than other conditions so that the reverse flow in the inlet of ejector appeared later while the positive flow started earlier in one cycle than other two conditions. The pressure peak of reverse flow is smaller, which is attributed to the weak reflect shock wave coming from the detonation tube exit due to the longer propagation path. The reverse flow would result in negative thrust. So the drastic expansion degree of shock wave propagating out of ejector inlet is higher than other conditions, which would form lower negative pressure and higher positive entraining mass flow. The high temperature gas expelled downstream from detonation tube exit and expanded drastically when the x/d=2. When it reaches the ejector inlet, the expanded gas covers the ejector inlet and impactes on the surface of ejector inlet, which resulte much more reverse mass flow. So the pressure peak of reverse flow is higher than other conditions.

3.2 Simulation of working process of ejector with different diameters

As are shown in figure.5-6, in addition to the ejector with a diameter of 150mm, we also simulated the entraining process of other two kinds of ejector, which is composed of D=120mm and D=170mm. The simulation results indicated that after the detonation wave coming out of the detonation chamber, it formed a spherical shock wave, which would reflect when it spread to the frontal surface of ejector inlet. Moreover, the strength of the reflected wave is related to the diameter of the ejector. Because of the larger flow area of larger diameter ejector, the entraining mass flow increased with the increase of the diameter of the ejector. However, it is not true that the longer diameter of the ejector is, the better it is. It is well known that the detonation wave expelled out of detonation tube would decay into a shock wave due to lacking energy supply by high temperature gas. The decayed shock would be reflected on the frontal surface of ejector inlet once it reached the ejector inlet. The reflect wave from smaller diameter ejector is nearly perpendicular to the centerline of PDE-ejector and is stronger. Conversely, the shock wave would fully expand and grew up because there is adequate space. The larger shock would cover the detonation tube exit as if there is not any ejector. So the reflect wave would take more time to spread a longer distance before reached to the centerline, which indicated that the strength of the reflected wave would decrease with the rising of ejector diameter and the thrust would decrease accordingly. Different diameters of the ejector had different entraining mass flow. However the pressure peaks had a slight difference because the shape tour of the ejector is same.

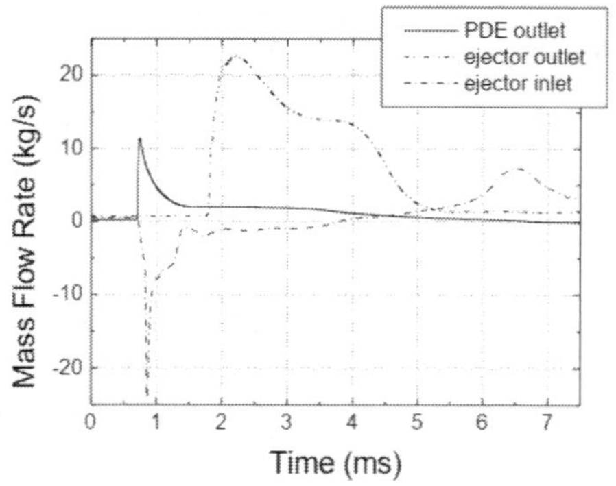

Figure.5 mass flow curves at different cross sections with D=170mm.

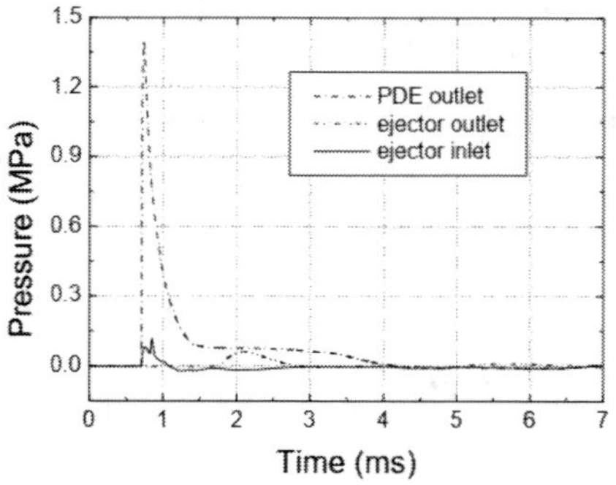

Figure.6 pressure curves at different cross sections with D=170mm.

3.3 Basic process of the detonation ejecting

In conclusion, the simulation results showed that the different ejector structures had different effects, but their basic processes are the same. Detonation ejecting can be roughly divided into three stages: (1) The PDE discharges the detonation wave, which blocks the inlet of the ejector completely where it comes into a severe reverse flow; (2) The inlet of the ejector has a positive ejecting; (3) Ejecting stability. In the following, the specific working processes of flow fields with the ejector inlet being at position of 240mm from the exit of PDE are described.

In the first stage, the high temperature and high pressure gas discharged from the detonation chamber which is rapidly expanding, and the secondary flow channel is completely blocked. Because there is not an explosive mixture in the ejector, the detonation wave decayed rapidly for a strong shock at the PDE exit. With the backward propagation of the shock, some exhaust gas from last cycle and the air in the mixing section of ejector are compressed and accelerated backwards. As is showed in figure.7a, the zone B2 is the exhaust gas of the detonation chamber. The zone B1 of the mixing section is filled with the high temperature gas pushed out of the detonation tube by the filling air in the last cycle. The zone C is full of the ejecting air that is compressed by the shock. And the uncompressed air is after the zone C. After the shock wave spreading out of mixing section, the original gas in the mixing section is compressed and is accelerated and expelled backward. At the same time, as is shown in figure.7b, the shock wave from detonation chamber reflected forward on the inner wall of the ejector. A severe reverse flow appeared in the inlet of the ejector. After that, the cycle came into the second stage. Figure.7c and 7d showed that the high temperature gas near the ejector inner wall formed a vortex ring. The gas in the front of vortex ring flowed backward between vortex ring and jet flow of the detonation chamber while the gas in the inlet of the ejector began to flow backward. At the third stage, the vortex ring at the exit of the detonation chamber moved backward and decayed while the fresh air filled in the ejector gradually, as is showed by figure.7e and 7f. The working mechanisms of the three stages are very different. The first stage indicates that the shock reflected forward mainly and blocked second-flow completely, which led to reflux at the inlet of the ejector; In the second stage, the entrainment effect of jet flow is dominant; In the third stage, the vortex ring gradually weakened in the ejector, and the bypass flow continued to be pushed back under the action of inertia and pressure difference.

NO.	Density contours
(a)	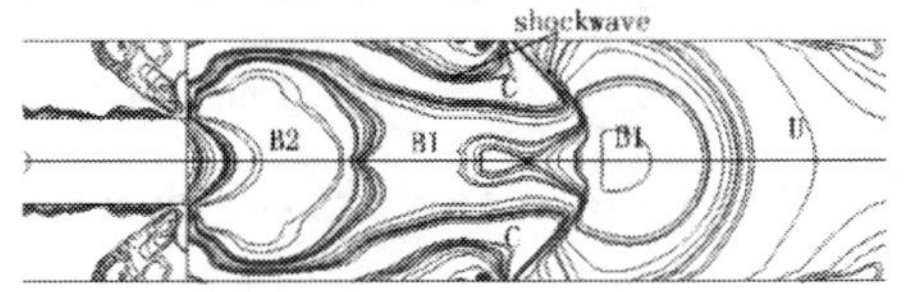
(b)	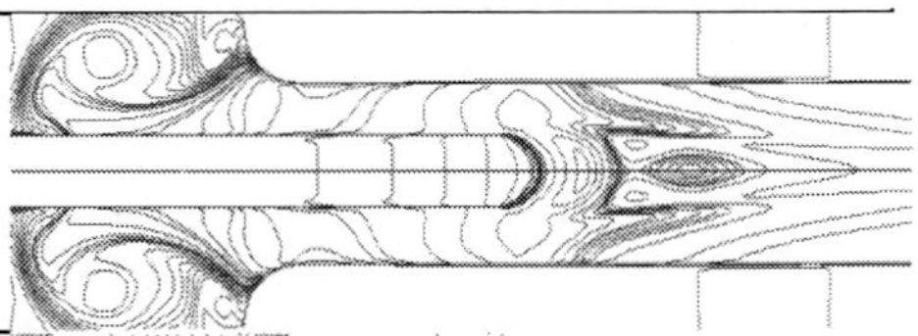
(c)	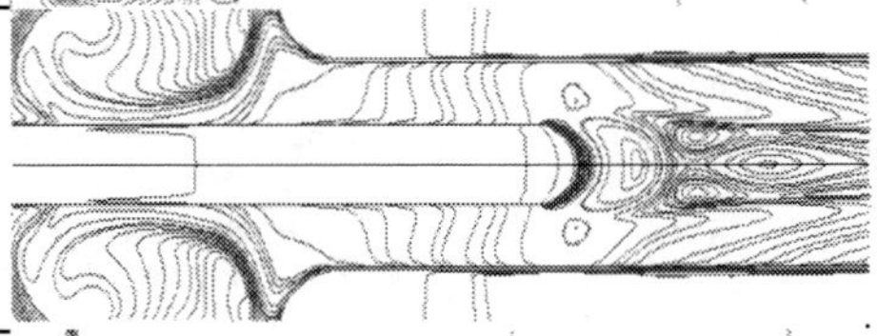
(d)	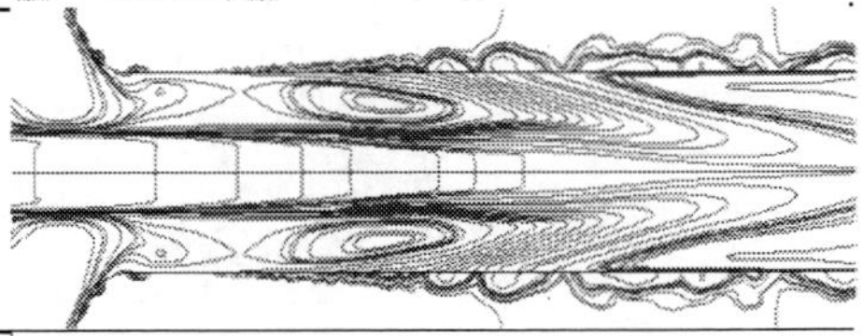
(e)	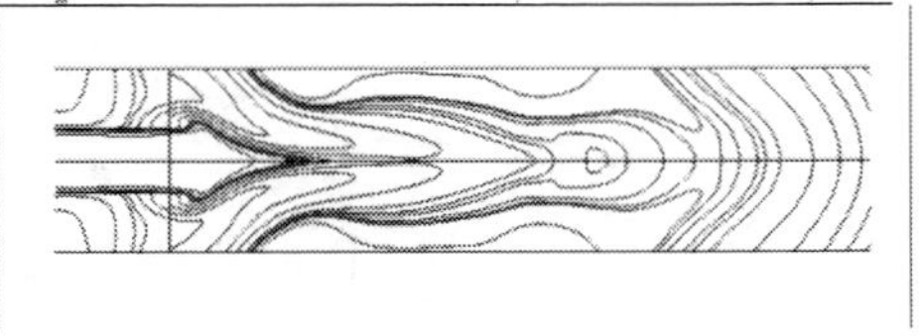
(f)	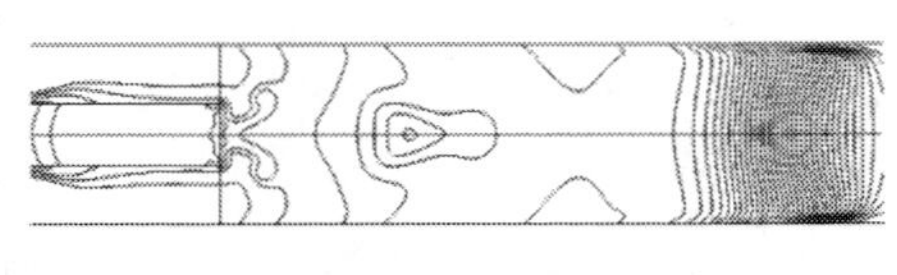

Figure.7 Density contours of the ejecting process

4. Conclusion

The numerical calculation of the working process of a pulse detonation engine ejector with different structures is carried out. The calculation results show that the different structures of the ejector have different ejecting effect , but no matter what kind of structure, the working process is basically the same. Detonation ejecting can be roughly divided into three stages: (1) After the detonation wave discharged from the PDE, completely blocked ejector inlet, at the entrance of ejector has a dramatic reverse flow; (2) The ejector inlet starts to generate a forward injection; (3) Ejector begins ejecting a stable and strong airflow. In addition, the outlet pressure of each type ejector is significantly lower than that of the detonation tube outlet pressure, which can effectively reduce the noise and change its frequency distribution. The numerical simulation results agreed with experimental results very well.

5 Acknowledgment

The authors would like to appreciate the financial supports of National Natural Science Foundation of China (51476135) and the Aviation Foundation (2015ZB53026)

References

[1] Ken Matsuoka, Motoki Esumi, Ken Bryan Ikeguchi, Jiro Kasahara, Akiko Matsuo, Ikkoh Funaki. Optical and thrust measurement of a pulse detonation combustor with a coaxial rotary valve [J]. Combustion and Flame, 2012,159(3):1321-1338.

[2] Ko Kawane, Satoshi Shimada, Jiro Kasahara, Akiko Matsuo. The influence of heat transfer and friction on the impulse of a detonation tube [J]. Combustion and Flame, 2011, 158(10):2023-2036.

[3] W. Fan, C. J. Yan, X. Q. Huang, Q. Zhang, L. X. Zheng. Recent advances in research on pulse detonation engine with new concept [J]. Aircraft Design, 2003,23(2):55-61.

[4] Chiping L, Kailasanath K. Performance Analysis of Pulse Detonation Engines with Partial Fuel Filling. AIAA 2002-0610. 2002.

[5] Canteins G, Franzetti F, Zoclonska E, Khasainov B A, Zitoun R. Experimental and numerical investigations on PDE performance augmentation by means of an ejector [J]. Shock Waves, 2006, 15(2): 103–112.

[6]. J. X. Zhao. Numerical simulation of combustion [M]. Beijing , The Science Publishing Company, 2002 : 403-410.

[7]. W. Q. Tao. Numerical Heat Transfer [M]. Xi'an, Xi'an Jiao Tong University Press. 2001 : 353-362.

[8]. F. J. Wang. Computational Fluid Dynamics Analysis [M]. Beijing : Tsinghua University Press, 2004.

[9]. Z. Z. Han, J. Wang, X. P. Lan. FLUENT Fluid Engineering Emulator Calculation Example and Application [M]. Beijing Institute of Technology press , 2004.

[10]. Z. J. Liang. Numerical Simulation on the Ejection and Secondary Detonation in a Pulsed Detonation Engine [D]. Xi'an: Northwestern Polytechnical University,2017 : 11-16.

11th Asia-Pacific Conference on Combustion,
The University of Sydney, NSW Australia
10th -14th December 2017.

Nanostructure Analysis of Soot Particles at Different Axial Locations in high-pressure diesel jet flames

Hao Jiang[1], Tie Li [1,2*], Yifeng Wang[1], Pengfei He[1]

[1]State Key Laboratory of Ocean Engineering

[2] Collaborative Innovation Center for Advanced Ship and Deep-Sea Exploration

Shanghai Jiao Tong University, 800 Dongchuan Road, Minhang District, Shanghai 200240, P.R. China

Abstract

For better understanding the formation and oxidation process of soot particles in high-pressure diesel jet flames, size and nanostructure are extracted from HRTEM (high-resolution transmission electron microscope) images of soot primary particles which were directly sampled by a thermophoresis probe at different axial locations from the injector nozzle tip (30, 40, 50, 60, 70 mm) and extracted by alcohol solution. High-pressure diesel jet flames were achieved in a constant volume combustion vessel (CVCV) under diesel-like conditions. Diameters of soot primary particles firstly increase in the upstream region (30 to 50 mm), exhibit a maximum value around the mid-stream region (50 mm), and then decrease in the downstream region (50 to 70 mm). Average values of lengths and endpoint distances of lattice fringes increase with the increasing of sampled distance, however, average lattice fringe tortuosities keep decreasing along the axial direction of diesel jet flames.

1 Introduction

In order to reduce emission of diesel particulate matter, it is important to understand the soot formation and oxidation process in a diesel jet flame. It is known that microstructures of soot particles closely associate with this process. Advanced analytical techniques have added increasingly detailed knowledge in our understanding on the morphology and reactivity of soot. Soot nanostructures are often studied by HRTEM. In recently years, many studies have demonstrated a direct relationship between nanostructure of particle and its oxidation activity. Specifically, the chaos and amorphous internal structure of the soot particles have much higher oxidative activity and lower activation energy than the particles with graphite structure [1,2]. Therefore, the study of nanostructure not only reveals the similarities and differences of soot particles, but also provides a better understanding of the formation mechanism for soot particles.

[1]Kook, Aizawa and Pickett, etc. have directly sampled soot particles via a sampler in high-pressure jet flame. Soot particles can be acquired by using the thermophoretic probe when the TEM grid is exposed to the high-temperature flame with relatively low fuel injection quantity [3,4,5]. It is noteworthy that higher combustion temperature and pressure may notably change the reactivity and the nanostructures of soot particles especially when the injection quantity is more than 20 mg [4,5]. Worse, severe damage may occur to the TEM grid when the combustion temperature is relatively high with a large fuel injection quantity. In this paper, soot particles are directly sampled based on the principle of thermophoresis at 5 different axial locations, from 30 mm to 70 mm in front of the injector nozzle, aiming at obtaining more detailed understanding of the mechanism of soot formation and oxidation process. Properties of soot particles, such as primary particle size, lattice fringe length, endpoint distance of lattice fringe and lattice fringe tortuosity are analyzed from the obtained TEM images.

Corresponding author: Prof. Tie Li, Fax: +86-21-3420-8348
E-mail address: litie@sjtu.edu.cn

2 Experimental apparatus and procedure

2.1 Experimental system of jet combustion

Here, diesel engine conditions were simulated in a constant-volume combustion vessel (CVCV) brought to high temperature and high pressure conditions by a premixed combustion prior to the fuel injection. A schematic of the experimental system for characterization of jet combustion processes is depicted in Fig. 1. The diesel-like high-temperature and high-pressure conditions were established by spark-ignited of a pre-mixture comprised of nitrogen (N_2), oxygen (O_2) and ethylene (C_2H_4). Two spark plugs were employed to ignite the mixture in the CVCV. The diesel fuel injection was enabled by a motor-driven common-rail injection system. The injector with one hole of 0.28 mm diameter was employed in this study. The chamber wall was heated up to 110 ℃ to keep the windows from being contaminated by water condensation [6].

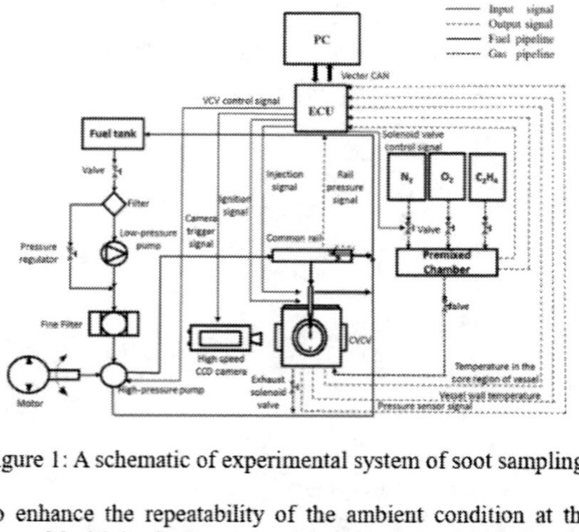

Figure 1: A schematic of experimental system of soot sampling.

To enhance the repeatability of the ambient condition at the start of fuel injection, a pressure feed-back control mechanism is developed to trigger the injection event. Fig. 2(a) shows the pressure trace after the spark-ignited of the pre-mixture during one test cycle. After quick rise caused by the burning of the pre-mixture, the pressure decreases gradually owing to the heat transfer through the chamber wall. When the targeted pressure of 3.96 MPa is reached, the fuel injection is activated as marked by the red arrow shown in Fig. 2(a). Fig. 2(b) shows the typical combustion process in the CVCV. The sampling probe is installed in the opposite position of the injector and is used for sampling soot particles in the diesel jet flame.

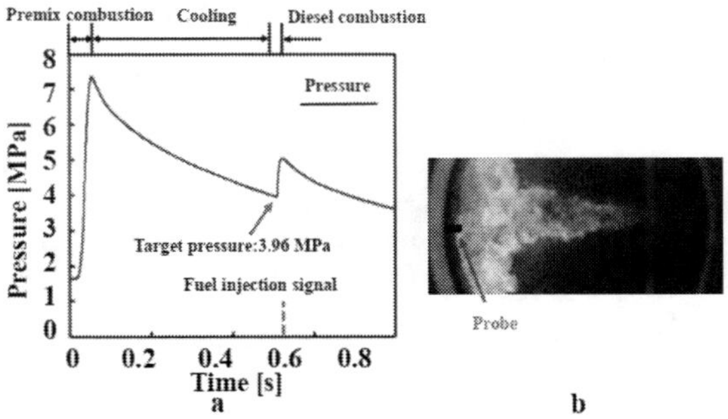

Figure 2: (a) Determination of fuel injection timing with the pressure feed-back control, (b) The sampling probe in the high-pressure diesel jet flame.

2.2 Soot sampler

In Fig. 3 (a), it's the schematic of the sampling probe. The probe volume is small enough to reduce the impacting on diesel jet flame. The diameter of the little circular groove is only 4.5 mm in the main body of the sampling probe which is installed on a window by a screw rod of diameter of 2 mm. To avoid the burn-off of the grid at high thermal load, fire-resistant carrier, which can bear the higher temperature and pressure for a longer time, should be employed. Soot particles are sampled on the carrier based on the principle of thermophoresis. The small cover plate with a small round hole which could protect the carrier from the impact of the combustion surrounding is fixed on the little circular groove by a screw. Soot samples from different spacial location in the CVCV can be easy to be trapped by the sampling probe with screw rods of different lengths. Fig. 3 (b) shows the installed position of the sampling probe in the CVCV. The sampling probe is installed in front of the injector. Soot particles are directly sampled in the diesel jet flame, which can effectively avoid influences of dilution, cooling and residence time during the sampling process.

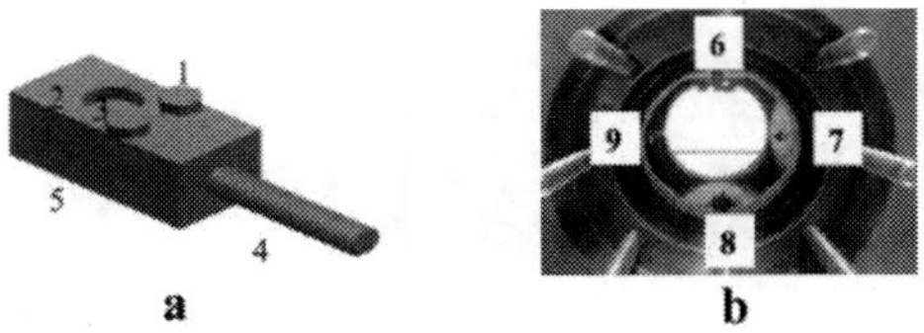

Figure 3: (a) The schematic of sampling probe, (b) Sampling probe installing position in the CVCV. 1. Screw; 2. Cover plate; 3. Carrier; 4. Screw rod; 5. Main body of sampling probe; 6. Fan; 7. Injector; 8. Sparking plug; 9. Probe.

2.3 Experimental conditions

Commercial diesel is used in this experiment. Different sampling locations along the axial direction impacting on the soot particle samples are studied with the fixed fuel injection pressures of 120 MPa and injection quantity of 30 mg. The experiment scheme is shown in Table 1. Images of primary particles are observed by using a high-resolution TEM (Talos F200X). The maximum magnification is up to ×1,500,000. The specified point resolution of the microscope is 0.12 nm and the electron acceleration voltage is 200 keV.

Environmental density	15 kg/m^3
Environmental temperature	900 K
Environmental oxygen concentration	21%
Injection pressure	120 MPa
Injection quantity	30 mg
Temperature of CVCV	383 K
Fuel injector type	Φ 0.28 mm
Distance from injector nozzle	30 mm–70 mm

Table 1: Experimental conditions and parameters

2.4 TEM images analysis

Primary particle size, lattice fringe length, endpoint distance of lattice fringe and lattice fringe tortuosity of primary particles are essential properties that can help to studying the formation and oxidation process of soot particles. The diameter of primary particle is acquired by a software from Fudan university. Lattice fringe length (L_a), endpoint distance of lattice fringe (D_s) and tortuosity (T_f) are extracted from TEM images by using a Matlab-based software.

3 Results and discussion

3.1 TEM images of soot particles

High quality soot samples are necessary for HRTEM observing and are used for analyzing morphology and nanostructures of primary particles. Monocrystalline silicon pieces with sampled soot particles are oscillated by an ultrasonic instrument in a small amount of ethanol for about 20 min, in the following, ethanol solution containing soot particles is dropt on a micro grid carbon film by an eppendof. Soot samples are finally dried in the atmosphere. In this method, high quality HRTEM samples were successfully obtained. Parameters extracted from the TEM images, dimensions and nanostructures of primary particles at different axial locations are investigated. Fig. 4 shows the micromorphology of soot aggregates which hangs at the edge of metal frame. In Fig.4 (b) and (c), they indicate that some almost spherical primary particles pack together and form soot aggregates.

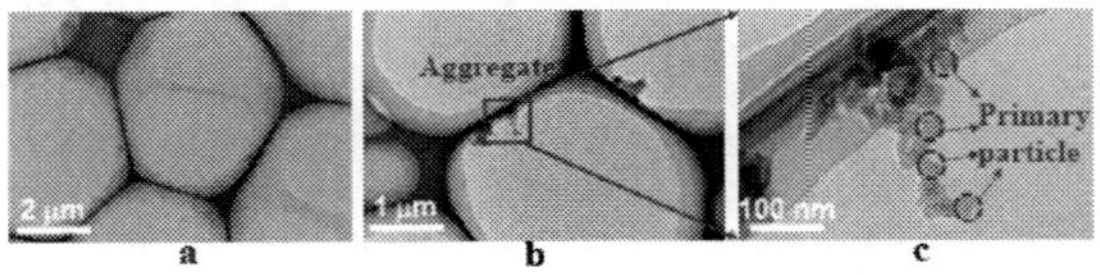

Figure 4: Micromorphology of soot particle aggregates in micro grid carbon film, ([O_2] = 21%, P_{inj} = 120 MPa, Q_f = 30 mg).

Fig. 5 shows TEM images of soot primary particles at axial locations of 30, 50 and 70 mm, respectively. Nanostructural parameters of primary particles can be clearly observed and accurately extracted from the HRTEM images at the magnification of x1,050,000. In Fig. 5(a), the fine particles with short and randomly oriented graphitic segments inside primary particle sampled in the upstream region of flame are surrounded by many shorter lattice fringes which are disorder and near to chaos, besides, outer shell are not good stratified structure. In Fig. 5(b), primary particles sampled in the mid-stream region present more mature structure with longer and ordered lattice fringe. Those short and curving lattice fringes are slowly stretch. In Fig. 5(c), comparing with other primary particles, the fine particles are surrounded by long and ordered lattice fringe, which have different and outstanding structure from outer shell and are easy to identify in the TEM images. There are longer lattice fringes with more straight structure. In another word, soot primary particles sampled in the downstream region of the flame are most mature and have highest graphite structure. Three TEM images intuitively reflect the formation and oxidation process of soot primary particles along the axial direction from the injector nozzle. Following the change of nanostructures, soot primary particles gradually become mature.

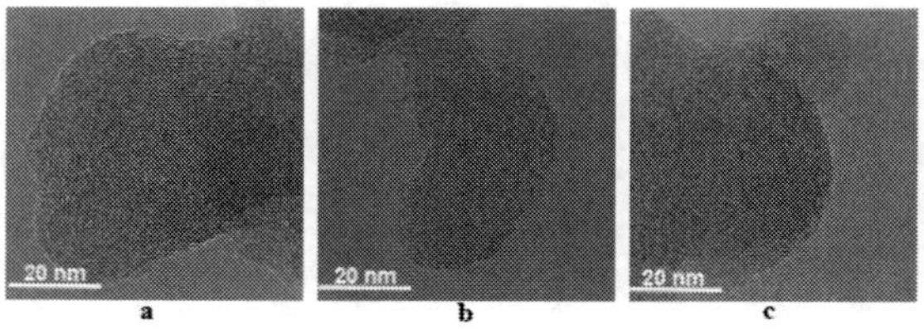

Figure 5: Nanostructure of primary particles at different axial locations (30, 50, 70 mm).

3.2 Diameter of primary particles

Diameters of primary particles are analyzed from the TEM images. Fig. 6(a) shows distribution curves of diameters of soot primary particles, which are similar to the normal distribution, sampled at 30, 40, 50, 60, 70 mm from the injector nozzle, respectively. Diameters of primary particles are measured based on TEM images at a magnification of x150,000, which is useful for identification of each primary particle within aggregates and at the same time the number of soot particles in the view field is statistically acceptable. In the present research, only apparent identifiable primary particles, including the separate existing ones and ones within the aggregates, are measured. The diameter measurements of primary particles are conducted by measuring the projected length for each primary particle. Numbers of primary particles are 478, 451, 433, 635 and 809, starting from 30 mm to 70 mm, respectively. Comparing with the diameter distribution of soot primary particles sampled at 30 mm location from the injector nozzle, it is easy to find that the distribution curve of diameter of soot primary particles sampled at 70 mm location is more concentrated in the smaller diameter region. Peak values of the distribution curves move to small values with the increasing of distances from the injector nozzle. Average diameters of primary particles starts at 29.15 nm at 30 mm, increases to 31 nm at 50 mm, and then decreases to 24.55 nm at 70 mm, as shown in Fig. 6(b) with error bars. From the upstream region to the mid-stream region, young soot particles gradually grow to mature with the bigger and bigger size by absorption PAH (polycyclic aromatic hydrocarbon). In the downstream region, diameters of soot particles will decrease slightly because outmost fringe layers are oxidized due to the ambient with a higher temperature. Since soot particles have become stable and less active, a further change in particle size becomes extremely difficult, so that diameters of soot particles will basically remain unchanged. The sizes of primary particles directly sampled in the diesel jet flame are similar to that sampled in the exhaust gas stream [7, 8], nevertheless, they are relatively bigger than that analyzed by Kondo, etc [9].

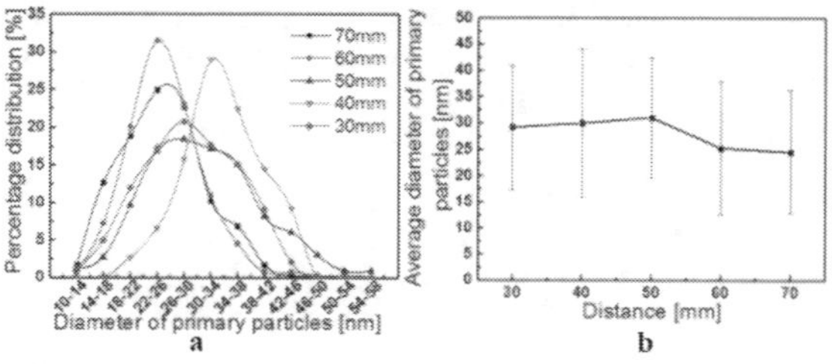

Figure 6: (a) Distribution curve of diameter of primary particles at different axial locations; (b) Average diameter of primary particles at different axial locations (30, 40, 50, 60, 70 mm).

3.3 Analysis of nanostructure parameters

To quantitatively characterize the nanostructural evolution of soot primary particles during the combustion process along the axial direction, more than 30 primary particles are randomly selected from different soot aggregates and studied at each sampled location. Lattice fringe lengths of soot primary particles are accurately extracted from the HRTEM images at the magnification of x1,050,000 which is useful to identify each soot layer fragment within primary particles. Fig. 7(a) shows that lattice fringe length falling within the range of 0.5 - 1 nm is more than 77.5%, 66.8% and 62.3% at 30, 50, 70mm sampled location, respectively. When gradually leaving away from the injector nozzle, crystallite grows in-plane via the incorporation of existing graphene layers and non-organized carbon and coalescence of crystallites with a side-by-side junction of the graphene layers due to high temperature and long residence time. Finally, the small crystallites grow to mature with the larger dimension. It means that there are much longer lattice fringe far away from injector nozzle. In Fig. 7(b), comparing among lattice fringe lengths in different axial locations, it indicates that lattice fringe lengths increase with the increasing of sampling distance from the injector nozzle. The graphite layers with planar dimensions are consisted of carbon atoms located at the basal plane and the edge-site positions, and reactivities of carbon atoms at the basal plane are far lower than that of carbon atoms at the edge-site positions [10,11]. Longer fringes corresponding to larger graphite layers mean a greater proportion of carbon atoms at the basal plane. Consequently, more soot particles with ordered structure and lower reactivity exist in the location far away from the injector nozzle.

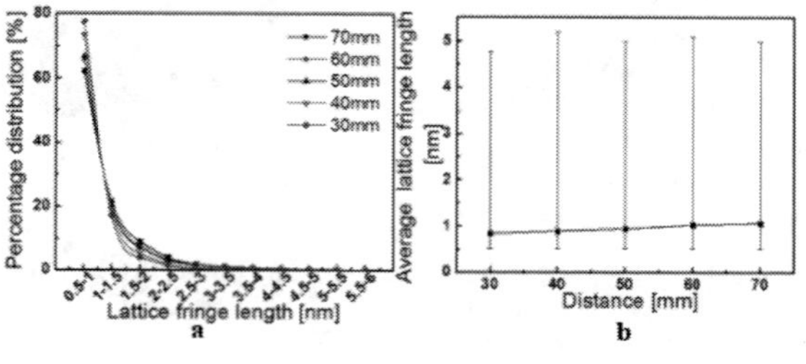

Figure 7: (a) Distribution curve of lattice fringe length at different axial locations; (b) Average lattice fringe length at different axial locations (30, 40, 50, 60, 70 mm).

Fig. 8(a) is distribution curves of endpoint distances of lattice fringes of soot particles at different sampled locations. As expected from the observation of the HRTEM images showed in Fig. 5, the endpoint distances of lattice fringes tend to increase. Increasing with the sampled distances, there are higher and higher frequencies of the longer endpoint distances of lattice fringes. Fig. 8(b) indicates that average endpoint distances of lattice fringes are 0.6844, 0.7132, 0.7495, 0.7785 and 0.8488 nm from 30 mm to 70 mm, respectively. The mean shifts toward bigger values when sampled locations are far away from the injector nozzle.

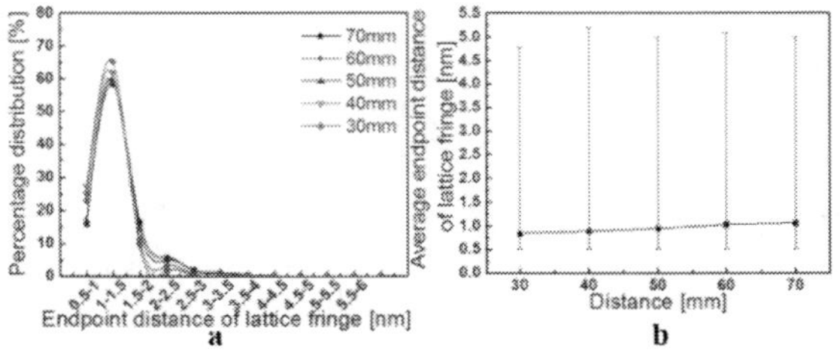

Figure 8: (a) Endpoint distance of lattice fringe at different axial locations; (b) Average endpoint distance of lattice fringe at different axial locations (30, 40, 50, 60, 70 mm).

Fig. 9(a) plots the distribution curves of the lattice fringe tortuosities. There are higher frequencies in the region of smaller values such as 1.0-1.2 at the longer sampled locations from the injector nozzle. This phenomenon means that graphite layers with planar dimensions gradually stretch opening and become straighter. As shown in Fig. 9(b), a progressive decrease of tortuosities accompanying the increasing of sampled distances is evidenced, with the peak value shifting from 1.277 nm to 1.22 nm. Further oxidized graphite fragments of soot particles sampled far away from the injector nozzle are speculated that high temperature and long residence time are responsible for the gradual decrease in lattice fringe tortuosities. Straighter graphite layers means that there are fewer odd-membered carbon rings with weak C–C bonds [10]. When tortuosities are decreasing, there will be carbon atoms of a greater proportion located at the basal plane and carbon atoms of a fewer proportion located at the edge-site positions. This reveals that the oxidation process causes the change from the amorphous structure to the higher graphitized structure. Higher tortuosity indicates that there are more odd-membered rings within the carbon planes and that there is an increase in the sp2/sp3 hybridization ratio, reflecting less electron resonance stabilization [10,12].

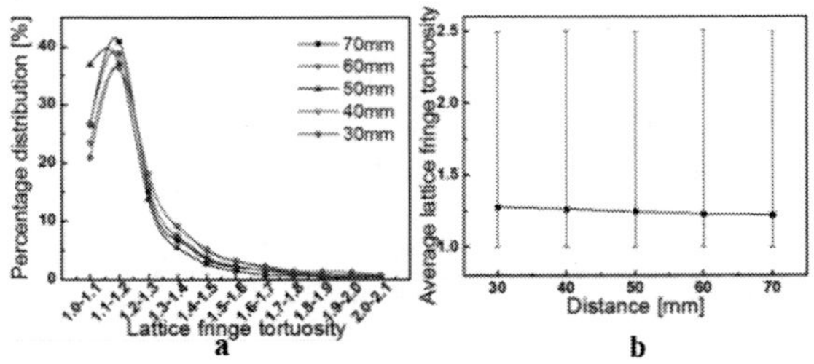

Figure 9: (a) Lattice fringe tortuosity at different axial locations; (b) Average lattice fringe tortuosity at different axial locations (30, 40, 50, 60, 70 mm).

4 Conclusions

In the diesel jet flame, soot particles were directly sampled by sampling probe with monocrystalline silicon pieces at different axial locations, 30 mm to 70 mm from the injector nozzle. After extracted in the alcohol solution, high quality soot samples were acquired and were observed by using the High-Resolution Transmission Electron Microscope (HRTEM). The diameters of primary particles, lattice fringe lengths, endpoint distances of lattice fringes and lattice fringe tortuosities of the sampled soot particles were analyzed. The obtained conclusions in this study are as follows:

1. HRTEM images of soot primary particles gradually present mature and graphitized structure with obvious lattice fringe and seldom find random oriented graphitic segments along the axial locations from the injector nozzle.

2. The sizes of primary particles directly sampled in the diesel jet flame are similar to that sampled in the exhaust gas stream, nevertheless, they are relatively bigger than that analyzed by Kondo, etc, who sampled soot particles in the constant volume combustion chamber and the diameters fell in the range of 12.6-20.2 nm.

3. Nanostructure parameters such as lattice fringe lengths and endpoint distances of lattice fringes are increasing with sampled location far away from the injector nozzle, whereas lattice fringe tortuosities are decreasing. These trends are considered to reflect the soot formation and oxidation process along the axial direction in a diesel jet flame.

5 Acknowledgment

The financial supports by the Youth Science Foundation of Natural Science Foundation of China (51506122) is gratefully acknowledged. The contributions by classmates and other students in the Combustion Engine Laboratory of NAOCE in SJTU are highly appreciated.

References

[1] M.P. Musculus, P.C. Miles, L. M. Pickett., Progress in Energy and Combust. Sci 39.2 (2013) 246-283.

[2] R.D. Reitz, Combust. Flame 1.160 (2013) 1-8.

[3] R. Zhang, S. Kook, Environmental science & technology 48.14 (2014) 8243-8250.

[4] K. Kondo, J. Takahashi, T. Aizawa, SAE International Journal of Fuels and Lubricants 7.2014-01-2637 (2014) 683-692.

[5] L.M. Pickett, Lyle M., and D.L. Siebers, Combust. Flame 138.1 (2004) 114-135.

[6] T. Li, X Q. Zhang, B. Wang, et al, Fuel 191 (2017) 251-265.

[7] K.O. Lee, R. Cole, R. Sekar, et al, Proceedings of the Combustion Institute 29.1 (2002) 647-653.

[8] A. Neer, Combust. Flame 146.1 (2006) 142-154.

[9] K. Kondo, T. Yamaguchi, et al, No. 2011-24-0068. SAE Technical Paper, 2011.

[10] Z. Li, C. Song, J. Song, et al, Combust. Flame 158.8 (2011) 1624-1630.

[11] R.L. Vander Wal., and A.J. Tomasek, Combust. Flame 136.1 (2004) 129-140.

[12] M. Alfè, B. Apicella, R. Barbella, et al, Proceedings of the Combustion Institute 32.1 (2009) 697-704.

Dr. Hao Jiang will make the verbal presentation.

1. Review questions

(1) Soot sampling under such conditions is extremely challenging and any insight into the process under such complex conditions provides valuable insight. What is the expected size limitations of the particles able to be captured by the TEM grid?

(2) Some of the figures are compressed and labelling is small, for example figure 8 the text on this figure is quite small. Figure 5 the size reference bar is impossibly small to read.

2. Revisions:

Dear reviewers, thank you for your heuristic suggestions for my paper. In the following, I am will to answer the questions according to your advice.

(1) The size of one grid on the TEM grid is about 5 μm, and the size of a soot aggregate is about 0.2-0.5 μm, so that the size of grid is much bigger than the soot aggregate size. When the ethanol solution containing soot particles is dropt on a TEM grid by an eppendof, soot sample containing a large number of soot aggregates is acquired and every soot aggregate is composed by several to dozens of primary particles. In addition, there is very large number of aggregates on the sample and the statistical data are acquired. Radius of gyration and fractal dimension of soot aggregate cannot be affected by this sampling method. Finally, in this paper, we have only analyzed the nano-structure parameters of soot primary particles, which are not affected by the sampling method, too.

(2) Figure 5 and figure 8 have been revised according to advice of reviewers.

11th Asia-Pacific Conference on Combustion,
The University of Sydney, NSW Australia
10th -14th December 2017.

Combustion characteristics of a methane jet flame in hot coflow of O_2/H_2O *vs*. O_2/N_2

Z. Shu, C. Dai, K. Cheong, J. Mi*

College of Engineering, Peking University, Beijing 100871, China

Abstract

Combustion characteristics of a CH_4 jet flame in hot coflow (JHC) of O_2/H_2O versus O_2/N_2 are investigated numerically. Both traditional and MILD oxy-fuel combustions are generated by diluting the coflow oxygen to f_{O_2} = 9%, 21% and 85% (mole fraction of oxygen in coflow). To distinguish physical and chemical effects, simulations of the JHC combustion are also conducted for the dilution of XH_2O (artificial H_2O) versus H_2O.

Results show that both the average temperature and size of the JHC reaction zone decrease when changing the diluent from N_2 to H_2O. These variations derive primarily from the physical effect. Specifically, a higher thermal capacity of H_2O than N_2 plays a dominant role in physically decreasing the reaction temperature. However, the dilution of H_2O enhances the chemical effect on the CO and OH formations so that the physical one becomes minor. Moreover, the H_2O dilution chemically delays the ignition and thus significantly increases the flame lift-off height for f_{O_2} = 9%. All the above reveal this: the MILD regime is more feasible in the combustion of $CH_4/O_2/H_2O$ than $CH_4/O_2/N_2$.

1 Introduction

The oxy-fuel combustion has been a promising technology of Carbon Capture and Storage (CCS) of CO_2 for decades [1]. It is established when oxygen is diluted by CO_2 or H_2O instead of N_2 and its concentration > 21%. As a result, the exhaust gas of this combustion contains mainly CO_2 and H_2O. The oxy-fuel combustion technologies have developed intentionally for the 'near zero-emission' of fossil-fuel combustion by CCS but faced some challenges. For example, air-ingress causes high NO_x emissions [2] and poor stability occurs due to delayed ignition [3]. However, those challenges may vanish when the oxy-fuel combustion operates under the MILD regime, i.e. using MILD oxy-combustion. The advantages of this flameless oxy-combustion technology has been well demonstrated by Linde in steel industry for practical furnaces such as walking beam furnace and pit furnace [4].

Previous works [5-12], among others, have revealed the operation limits of this new combustion technology. Li et al. [5] investigated by experiment the MILD oxy-combustion of gaseous fuels at furnace and made several significant findings. For instance, the stable MILD combustion was found to occur over a wider range of equivalence ratio for dilution by CO_2 than by N_2. Moreover, they found more uniform temperature distributions and lower emissions of NO_x and CO in the MILD than conventional oxy-combustion. Sabia et al. [6] experimentally examined auto-ignition delay times of methane in diluted and pre-heated conditions in a one-dimensional flow reactor, obtaining longer auto-ignition delays for MILD oxy-combustion.

Numerically, Mei et al. [7] simulated the JHC combustion of CH_4 diluted by N_2 and CO_2, and revealed that the MILD combustion of CO_2 dilution produces far larger reaction zone, lower temperature, and more CO. The fundamentals behind those differences were sought by Zhang et al. [8] through chemical kinetic calculations of well-stirred reactor (WSR). These authors revealed that high concentration of CO_2 from the dilution by CO_2 delays ignition, slows oxidations and reduces the temperature rise, thus facilitating the establishment of MILD combustion. All this results mainly from a greater heat capacity of CO_2 than N_2; relatively, the chemical effect is minor. However, oxy-combustion diluted by H_2O is quite different from the traditional combustion and O_2/CO_2 combustion because the physicochemical properties of H_2O is very unique. Recently, there are few studies paying attention on the performance of O_2/H_2O combustion. Wang et al. [9] studied effects of dilution of CO_2 and H_2O on a counter-flow methane flame. They found that both chemical and thermal effects of CO_2 dilution play equal roles to significantly reduce the flame temperature, yet, on which the H_2O dilution has little chemical effect. Moreover, Tu et al. [10] compared conventional combustion and MILD combustion of CH_4/H_2 jet flame in hot coflow diluted by N_2, CO_2, and H_2O, and claimed that the replacement of N_2 by either CO_2 or H_2O will lead to a conspicuous difference in reducing the maximum flame temperature under conventional combustion in contrast to MILD combustion. Zou et al. [11] investigated the steam's effect on temperature in methane oxy-H_2O combustion. They indicated that the chemical impact of H_2O increases the maximum combustion temperature. Hong et al. [12] implied that the presence of H_2O advanced the oxidation of CH_4 because of H_2O reactivity at high temperatures. Despite the above investigations [5-12], the oxy-combustion diluted by H_2O has been paid insufficient attention and so not understood as much as it diluted by CO_2.

To help address the above deficit, this work is designated to examine the combustion characteristics of a methane jet in hot coflow (JHC) of O_2/H_2O (instead of O_2/CO_2) versus O_2/N_2. The main objective is three-fold:

(1) To show the discrepancies in the JHC flame characteristics caused by N_2 and H_2O dilutions;
(2) To differentiate physical and chemical effects of changing the diluent from N_2 to H_2O on the JHC flame;
(3) To find reasons behind distinct characteristics of combustion due to using different diluents of H_2O and N_2.

To identify chemical and physical effects of H_2O on combustion characteristics, the oxygen dilution is changed from N_2 to H_2O and artificial species XH_2O. Note that XH_2O has the same thermodynamic coefficients as the real H_2O but does not participate in any chemical reactions. Therefore, only the chemical effect plays a role in making differences between JHC flames diluted by H_2O and those by XH_2O. And the differences between XH_2O and N_2 dilutions are due to the physical effect. Both traditional and MILD oxy-combustions are considered by taking the coflow oxygen molar fraction at f_{O_2} = 9%, 21% and

85%. To this end, a number of simulations through RANS modeling are performed using the Eddy Dissipation Concept (EDC) model with the detailed chemistry-reaction mechanism (GRI-Mech 3.0). Computational details (not presented herein) refer to our previous work [7]. Note also that the present model of JHC combustion is modified from the experimental combustor of Dally et al. [13]. Our previous study [7] has well validated the RANS modeling using the experimental data [13], which is not repeated here.

2 Results and discussion

Following Mei et al. [7], we define the contour of $R_{CO} = f_{CO}/f_{CO_{max}} = 0.01$ approximately as the border of the reaction zone, where f_{CO} and $f_{CO_{max}}$ are the concentration of CO and its maximum. The rationality is that carbon monoxide may be regarded as the last intermediate product of hydrocarbon combustion. This definition was experimentally confirmed by Mei et al. [7] to be effective for the methane JHC combustion at different modes.

Figures 1(a-c) show contours of the mean temperature T, together with reaction zones defined by $R_{CO} = 0.01$, respectively for $f_{O_2} = 9\%$, 21% and 85%. Correspondingly, Figures 2 and 3 display contours of the mean concentrations of CO and OH. Note that the log-scale is used to properly view low contours of the CO and OH concentrations. Results for dilution by XH_2O are also displayed to examine how differently chemical and physical factors play in the combustion characteristics. Moreover, Table 1 shows the volume-weighted averages of the temperature rise $\overline{\Delta T}$, thermal capacity $\overline{\rho C_p}$ and heat released per unit $\overline{\rho C_p \Delta T}$, as well as ΔT_{max}, for $f_{O_2} = 9\%$, 21% and 85%. A number of interesting observations can be made directly or indirectly from Figs. 1-3 and Table 1, which are detailed below.

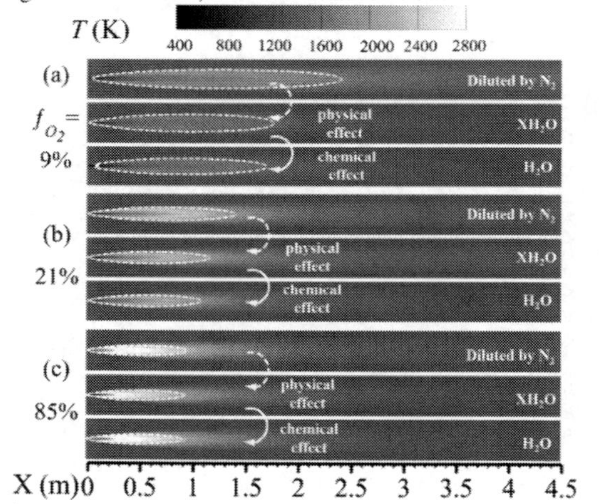

Figure 1: Contour distribution of temperature for $f_{O_2} = 9\%$, 21% and 85% (write dotted line: $R_{CO}=0.01$).

Influence of oxygen fraction in coflow: Comparisons of plots from Figures 1-3 reveal straightaway that: as f_{O_2} increases from 9% to 85%, the reaction zone decreases dramatically in size and simultaneously the distributions of the temperature and concentrations of CO and OH become more inhomogeneous. Indeed, see Table 1, ΔT_{max}, another indicator of the temperature inhomogeneity, increases significantly with varying f_{O_2} from 9% to 85% for all the diluents; e.g., ΔT_{max} grows from 498K to 1706K for the case of H_2O dilution. These discrepancies can be understood here: when a fuel jet issues and mixes with coflow at $f_{O_2} = 9\%$, combustion reactions occur more slowly, thus taking a larger volume in space to burn out, than for the cases with more oxygen at $f_{O_2} = 21\%$ and 85%. Of note, for the three cases with $f_{O_2} = 9\%$, ΔT_{max} is below the auto-ignition temperature of methane (≈ 810K), see Table 1. However, for either $f_{O_2} = 21\%$ or 85%, ΔT_{max} is greater than the auto-ignition temperature. The definition of MILD combustion [14] is hence met in the low oxygen cases but not in other cases. Namely, the MILD combustion occurs at $f_{O_2} = 9\%$ but not at $f_{O_2} = 21\%$ and 85%.

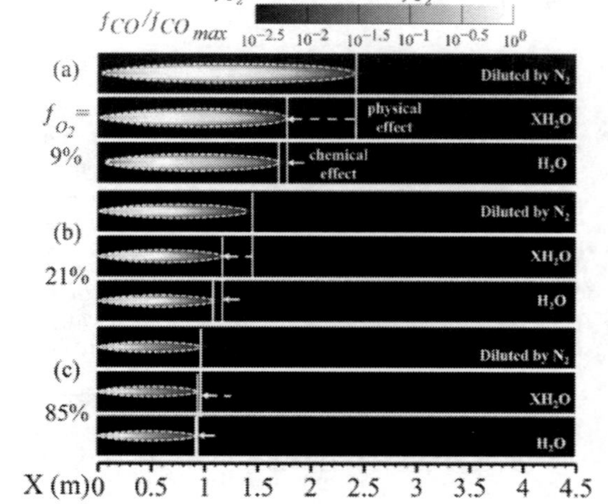

Figure 2: Normalized contours of the CO concentration for $f_{O_2} =$ 9%, 21% and 85%. Symbols: ----- R_{CO}=0.01; --→ physical effect; → chemical effect.

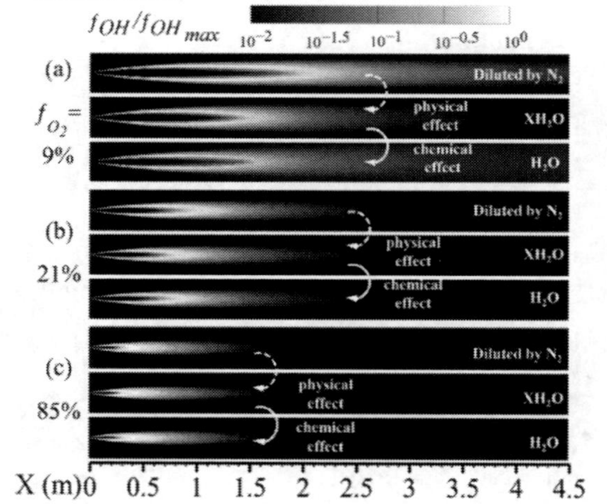

Figure 3: Normalized contours of H_2O on OH concentrations for $f_{O_2} = 9\%$, 21% and 85%.

Influence of diluent: Figures 2(a)-3(a) compares the CO/OH related reaction zones of the JHC flames diluted by N_2, H_2O and XH_2O for $f_{O_2} = 9\%$. It is observed immediately that the dilution by H_2O leads to the reaction zones significantly smaller than those by N_2. Highly interestingly, the maximal temperature rise for the former is $\Delta T_{max} = 498$K that is considerably smaller than $\Delta T_{max} = 652$K for the latter. This is perhaps unexpected, given that the reaction zone is larger in the former case, but can be explained below. Apparently, as expected, the above observations become less significant as the coflow oxygen is increased from $f_{O_2} = 9\%$ to $f_{O_2} = 85\%$, see Figs. 2(b,c)-3(b,c)

*Corresponding author. Fax: +86-10-62767074
E-mail address: jcmi_pku@hotmail.com

and Table 1.

Table 1. Data of ΔT_{max}, $\overline{\Delta T}$, $\overline{\rho C_p}$ and $\overline{\rho C_p \Delta T}$

f_{O2} (%)	diluent	ΔT_{max} (K)	$\overline{\Delta T}$ (K)	$\overline{\rho C_p}$ (J/m³/K)	$\overline{\rho C_p \Delta T}$ (J/m³)
9	N_2	652	461	260	116708
9	XH_2O	484	305	368	119141
9	H_2O	498	352	362	120250
21	N_2	1192	713	253	147960
21	XH_2O	996	664	318	149019
21	H_2O	1104	694	315	153610
85	N_2	1736	1015	251	234669
85	XH_2O	1728	951	263	230074
85	H_2O	1706	1032	251	235000

The smaller reaction zone for the case of H_2O dilution may be understood for two reasons. The first is associated with chemical reaction R59:

$CO + OH \Leftrightarrow CO_2 + H + heat$,

whose reaction rate depends on concentrations of both CO and OH. The oxidation and heat release for the methane combustion is predominately controlled by R59. When adding H_2O into the coflow for dilution, the production of OH radical will grow and so the forward (CO to CO_2) reaction rate will increase. Consequently, the fuel is burnt out, and thus *heat* releases, more rapidly. So, the reaction-occupied volume reduces.

The second reason is related with the entrainment of coflow fluid by the central fuel jet. Ricou and Spalding [15] investigated by experiment the entrainment of round turbulent jets including combustion cases and found that the jet entrainment rate (R_{ent}) can be formulated generically as

$$R_{ent} = \frac{m}{m_0} = 0.32\frac{x}{d_o}\left(\frac{\rho_{cof}}{\rho_o}\right)^{\frac{1}{2}}.$$

Here, m_0 and m represent the jet exit mass flux and entrained mass flux, respectively, while ρ_o and ρ_{cof} denote fluid densities of the center jet and surrounding coflow. It follows that the ratio of the volume of the entrained fluid (V_{ent1}) at ρ_{cof1} to that at ρ_{cof2} (V_{ent2}) is

$$\frac{V_{ent1}}{V_{ent2}} = \frac{m_{ent1}}{m_{ent2}} \cdot \left(\frac{\rho_{cof2}}{\rho_{cof1}}\right) = \left(\frac{\rho_{cof2}}{\rho_{cof1}}\right)^{\frac{1}{2}}.$$

When f_{O_2}= 9% and T_{cof} = 1300K, the coflow density of O_2/N_2 is ρ_{cof1} = 0.266 kg/m³ and that of O_2/H_2O is ρ_{cof2} = 0.18 kg/m³. It is thus estimated that the mole fraction of the entrained coflow of O_2/H_2O is about 22% higher than that of O_2/N_2. Equally, the entrained oxygen from the coflow of O_2/H_2O exceeds that of O_2/N_2 by 22%. Since the increased oxygen concentration accelerates chemical reactions, plus the first reason, the reaction zone of the H_2O-diluted JHC MILD flame is thus reduced significantly in size from the N_2 case, see Figs. 2(a) and 3(a).

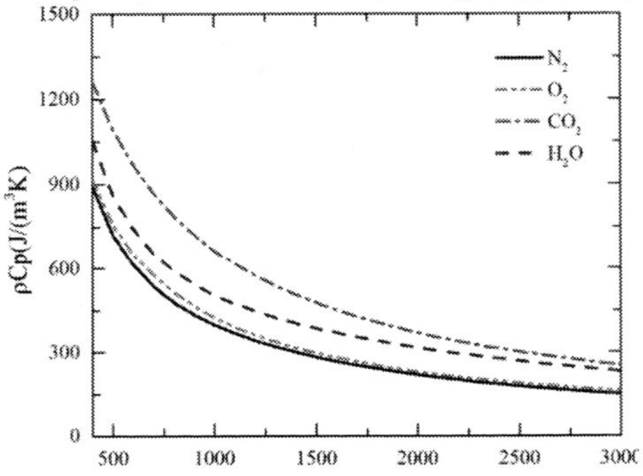

Figure 4: Thermal capacities of N_2, O_2, CO_2 and H_2O with temperature ranging from 400 K to 3000 K.

Now we explore the reasons behind the lower temperature for the H_2O dilution than for the N_2 case. Figure 4 compares the values of $\overline{\rho C_p}$, the thermal capacity, of gases N_2, O_2, CO_2 and H_2O with temperature ranging from 400 K to 3000 K. Evidently, the nitrogen $\overline{\rho C_p}$ is considerably higher than that of H_2O. This implies that the nearly same heat release, which should be approximately equal to $\overline{\rho C_p \Delta T}$, from the two cases will lead to a lower rise of the average temperature ($\overline{\Delta T}$) for the H_2O dilution than for the N_2 case. Indeed, Table 1 shows that $\Delta T_{max} \approx$ 498K and $\overline{\Delta T}$ ≈ 352K for the former whereas those two are approximately 652K and 461K for the latter.

Relative influence of chemical and physical factors: One objective of the present study is to identify and compare chemical and physical effects on combustion characteristics when changing the diluent from N_2 to H_2O. Then, the chemical factor of H_2O is marked in Figs. 2(a-c) by white arrows, which are measured from the end of reaction zone developed by XH_2O. The physical factor is represented by white dotted arrows, which are based on the end of reaction zone for the N_2 dilution.

It is immediately observed from Fig. 2 that both physical and chemical factors act consistently to size down the reaction zone when switching the diluent from N_2 to H_2O. At f_{O_2}= 9%, the contribution of physical factor (≈ 26%) is far higher than that of chemical factor (5%), both acting to decrease the flame length approximately by 31%, in the combustion diluted by H_2O. The total effect of the two factors is also reflected indirectly in $\overline{\rho C_p \Delta T}$ from Table 3. Given that the total heat release is H, the global volume of reaction zone may be expressed as $V \approx H/\overline{\rho C_p \Delta T}$, i.e. V is inversely proportional to $\overline{\rho C_p \Delta T}$; in other words, the greater the latter, the smaller the former. As seen in Table 1, at f_{O_2}= 9%, $\overline{\rho C_p \Delta T}$ ≈ 116708 and 120250 J/m³, respectively, for the dilutions of N_2 and H_2O. That is, the global volume of reaction zone should be smaller for the H_2O case, which is consistent with the results of Figs. 2-3.

Table 2. Combustion characteristics for different cases

f_{O2} (%)	diluent	L (m)	T_{max} (K)	f_{COmax} (%)	f_{OHmax} (%)
9	N_2	2.37	1952	3.0	0.15
9	XH_2O	1.75	1784	2.7	0.14
9	H_2O	1.63	1798	1.1	0.16
21	N_2	1.41	2492	10.2	0.17
21	XH_2O	1.17	2296	8.8	0.15
21	H_2O	1.08	2304	6.1	0.19
85	N_2	0.96	3036	27.2	8.76
85	XH_2O	0.93	3028	27.1	8.90
85	H_2O	0.91	3011	26.6	9.24

Table 2 reports the data of L and the maxima of the mean temperature and concentrations of CO (f_{CO}) and OH (f_{OH}) over the whole reaction zone, denoted by T_{max}, f_{COmax} and f_{OHmax}. Figure 5 compares physical and chemical effects of H_2O dilution on L, T_{max}, f_{COmax} and f_{OHmax} for f_{O_2}= 9%. It is demonstrated that the physical effect of H_2O dilution is considerably more important than the chemical one in reducing the reaction zone. Likewise, the variation of the flame temperature results primarily from the physical effect whereas the chemical effect is nearly negligible. On the other hand, the chemical effect is more significant on f_{OHmax} and particularly on f_{COmax}. Moreover, the maximum temperature is apparently lest sensitive, while the maximum concentration of CO is most sensitive, to the change of diluent. As the coflow f_{O_2} increases, both physical and chemical effects weaken, see Table 2, since the diluent fraction in coflow drops.

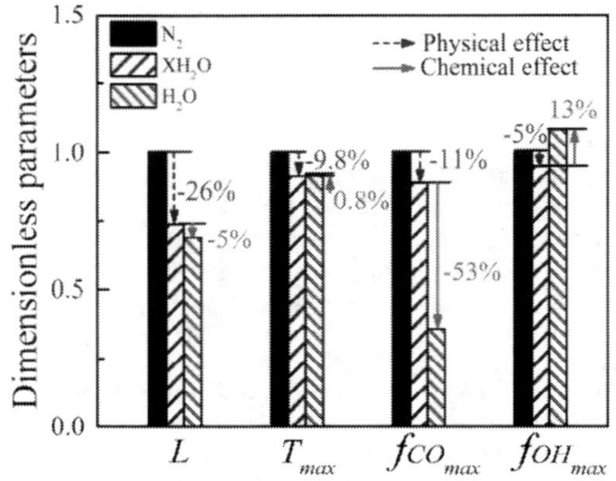

Figure 5: Comparison of physical and chemical effects of H_2O dilution on L, T_{max}, fCO_{max} and fOH_{max} for f_{O_2}= 9%.

Influence of H_2O and N_2 dilutions on JHC flame lift-off: Careful checks to Figs. 2(a) and 3(a) find that the JHC flame diluted by N_2 for f_{O_2}= 9% starts at some distance downstream from the jet nozzle exit. This indicates that the flame lift-off occurs in the N_2 case. The same figures also show the difference in the JHC flame lift-off due to varying the diluent from N_2 to XH_2O (physical effect) and then to H_2O (chemical effect). Relative to the N_2 case, the onset of MILD combustion diluted by H_2O move farther downstream from the jet nozzle. Indeed, the lift-off height H is greater in the latter case, i.e., H = 30 mm and 70 mm for N_2 and H_2O dilutions at f_{O_2}= 9%. It is interesting that there is nearly no lift-off to occur in the XH_2O case. That is, the physical mixing appears to have little influence on the ignition process. Accordingly, the flame lift-off results almost only from the chemical effect. The reason behind appears to be that the chemical effect of adding H_2O increase the ignition delay time [16] and so move the stable position of flame farther downstream.

3 Conclusions

The present simulations have been carried out to investigate the JHC combustion characteristics of CH_4 (central fuel jet) + O_2/H_2O versus O_2/N_2 (oxidant coflow). The artificial gas XH_2O is used to separate physical and chemical effects of changing the diluent from N_2 to H_2O. Based on the present work, several conclusions may be drawn below:

(1) When changing the diluent from N_2 to H_2O, physical and chemical effects act together to reduce the size of reaction zone and also decrease the reaction temperature. In the joint action, the physical effect plays a far more important role than does the chemical one. In particular, the temperature decrease results predominantly from a higher thermal capacity of H_2O than N_2.

(2) However, the dilution of H_2O significantly enhances the chemical effect on the CO and OH formations. As a result, the chemical effect is higher than the physical one on fOH_{max} and particularly on fCO_{max}.

(3) Relative to the N_2 dilution, the H_2O dilution chemically delays ignition and thus increases the flame lift-off height when the coflow oxygen fraction is sufficiently low, e.g., at f_{O_2}= 9%.

(4) In general, the MILD combustion should occur more probably in burning $CH_4/O_2/H_2O$ than $CH_4/O_2/N_2$, even though the MILD combustion is found to occur in the two cases at f_{O_2}= 9%.

4 Acknowledgment

The support of Nature Science Foundation of China (No. 51276002) is gratefully acknowledged.

References

[1] J. Lilliestam, J. M. Bielicki, A. G. Patt, Energy Policy 47 (2012) 447–455.

[2] Normann, F.; Andersson, K.; Leckner, B.; Johnsson, F. Prog. Energy Combust. Sci. 35 (5) (2009) 385–397.

[3] T. Suda, K. Masuko, J. Sato, A. Yamamoto, K. Okazaki, Fuel 86 (12) (2007) 2008–2015.

[4] J. von Schéele, P. Vesterberg, O. Ritzén, Invisible flames for clearly visible results. Nord Steel Min Rev Swed 2005:16.

[5] P. Li, B. B. Dally, J. Mi, F. Wang, Combust. Flame 160 (5)(2013) 933–946.

[6] P. Sabia, M. de Joannon, A. Picarelli, R. Ragucci, Combust. Flame 160 (1) (2013) 47–55.

[7] Z. Mei, J. Mi, F. Wang, C. Zheng, Energy & Fuels 26 (6) (2012) 3257–3266.

[8] J. Zhang, J. Mi, P. Li, F. Wang, B.B. Dally, Energy and Fuels 29 (7) (2015) 4576–4585.

[9] L. Wang, Z. Liu, S. Chen, C. Zheng, J. Li, Energy and Fuels 27 (2013) 7602–7611.

[10] Y. Tu, H. Liu, W. Yang, Energy & Fuels 31 (3) (2017) 3270–3280.

[11] C. Zou, Y. Song, G. Li, S. Cao, Y. He, C. Zheng, Int. J. Heat Mass Transfer 75(2014) 12–18.

[12] D. Hong, L. Liu, Y. Huang, C. Zheng, X. Guo, Energy and Fuels 30 (10) (2016) 8491–8498.

[13] B.B. Dally, A.N. Karpetis, R.S. Barlow, Proc. Combust. Inst. 29 (2002) 1147–1154.

[14] A. Cavaliere, M. de Joannon, Prog. Energy Combust. Sci. 30 (2004) 329-366.

[15] F. P. Ricou, D. B. Spalding, J. Fluid Mech. 11 (1961) 21-38.

[16] Z. Cao, T. Zhu, Chinese J. Chem. Eng. 20 (4) (2012) 701–709.

11th Asia-Pacific Conference on Combustion,

The University of Sydney, NSW Australia
10th -14th December 2017.

Numerical Study of a High Swirl Number Swirler

Majid Aghayari[1], Sadegh Tabejamaat[1], Bashir Ajjele[1]

Amirkabir University of Technology,424 Hafez Ave, Tehran, Iran, 15875-4413. +98 (21) 64540

Abstract

In this paper, swirl combustion flow is studied by numerical method. In this simulation, RANS method is used. Two turbulence model (Reynolds Stress Model and k- ε) are compared with experimental results which is obtained from Laser Doppler velocimetry (LDV) method. Then, EDC (Eddy Dissipation Concept) and FGM (Flamelet Generated Manifolds) model are compared to the combustion modelling. The results show good agreement of temperature and velocity with experiment results in this chamber. This means using a suitable combustion model like FGM and turbulence modeling like k- ε to predict the combustion flow behavior can give acceptable information. In cold case, results show there is small and outspread vortex in domain and they are grown and orderly when combustion occurs. There are vortex shedding in cold case but the vortex are moving very low in hot case.

1 Introduction

Swirling turbulent flow has important role to control combustion process and the performance of turbine. So, it needs more attention to predict characteristics of swirling flow in combustion to improve engine design. Flow in combustion chamber of gas turbine is high dependent to geometry and inlet conditions. By improving the using equipment in simulation, more accurate methods such as Large Eddy Simulation (LES) and Direct Numerical Simulation (DNS) has gained popularity among researchers. But with all the benefits, these methods in the simulation process has made complex problems and increase computational burden associated, the time to solve the problem and the cost of needed equipment for the simulation. To simulate the flow with lower computational cost, researchers are still appreciated RANS methods. Since RANS equations are inefficient to simulate the vortices, it has been created different models to simulate the vortex. Among these models, k- ε model are popular which a lot of researchers used it. But this model has some defection and thus the other models are presented also to more precise. Among the presented models to higher accuracy, Reynolds stress model can be noted.

Giridharan and Mongia [1] simulated swirlers' outflow of CFM56 combustion chamber. Li and et al. [2] simulated combustion and wall cooling of a combustion chamber of gas turbine by using a k- ε model. Shirinzadeh and Mahdavi Moghadam [3] studied the effects of some variables such as air to fuel ratio, swirl number, and fuel injection angle by using k- ε model. Mir Hosseini and et al. [4] simulated flow and heat transfer of an air turbine by using standard k- ε model. Poinsot and et al. [5] also simulated flow inside a combustion chamber by using LES model and studied combustion instability in it. Ajjeleh and et al. [6] studied methane combustion in swirling flow. They used k- ε model to simulate turbulence modelling. Wu and et al. [7] investigated a swirling flow in a combustion chamber by using LES method. Agbonzikilo and et al. [8] studied numerically and experimentally mixing of air and fuel in the gaps of radial swirler of a low NOx burner in gas turbine. In this simulation, they used several RANS model such as k- ε and concluded SST model revealed the closest answer to experimental results.

As stated above, previous researcher simulated the combustion chamber with k- ε model or LES. But in this study, Reynolds Stress Model (RSM) is used to compare with k- ε model. FGM and EDC model is used to combustion modeling. The previous study usually used EDC model. EDC is a good model when an appropriate kinetic mechanism is used, but solution time will be very high. So, in this study, FGM model is used that has a lower solution time and EDC model is compared with it. In EDC model, DRM22 mechanism is used. DRM22 seems be adequate for this simulation.

2 Geometry

The studied geometry is shown in figure 1 and 2. Figure 1 shows swirler which is placed on chambers that is shown in figure 2. It has 12 blade and flow passes radially. Length of chamber is 114 mm, outlet diameter is 40 mm, and main diameter of chamber is 85 mm. Flow passes through swirler and enter to chamber from a passage with 27.85 mm diameter.

Figure 1: swirler geometry

Corresponding author. Fax: +98-21-6695-020

E-mail address: sadegh@aut.ac.ir

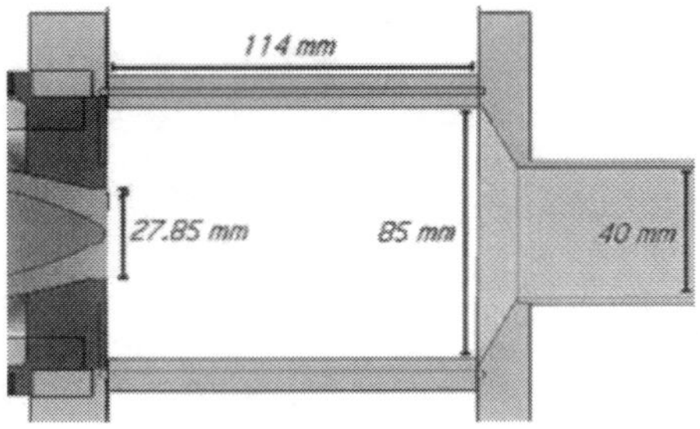

Figure 2: combustion chamber geometry

3 Numerical simulation

in this solution, swirler and chamber are simulated simultaneous. In outlet of chamber, due to discharge production to environment and do experiment in atmospheric mode, pressure outlet boundary condition has been used. Constant Adiabatic condition is used for wall. In table 1, all boundary condition are presented.

Name Of Boundary	Type of Boundary	Value
Air inlet	Mass flow inlet (g/s)	12.23
Fuel inlet	Mass flow inlet (g/s)	0.59
Outlet	Pressure Outlet	0
Operational Pressure	-	101325 (pa)
Wall	Adiabatic	-

Table 1: Simulation Boundary Condition

Inlet air is considered uniform. According to presented data in [5], inlet and outlet value are selected. Equivalence ration is 0.83. The mixing inlet temperature and chamber pressure are 300 k and 101325 pa, respectively. Since changing is low and there is excess air and major species in chamber are air, the properties of inlet and outlet fluid is considered incompressible ideal gas. heat capacity of participation species in reaction are considered polynomial function relative to temperature. The mesh is generated by Gambit and it is shown in figure 3.

Figure 3: Generated mesh by Gambit

3.1 Simulation Method

Simulation is done by Fluent. Boundary condition and setting is applied according to what was mentioned. It is used k- ε RNG and RSM for turbulence modelling and standard wall function is applied to it. Grid number are about 1968000. As well as, grid study is done by 1300000 and 2400000 cells. FGM (Flamelet Generated Manifold) and EDC (Eddy Dissipation Concept) is used for combustion modelling. In EDC model, a reduced model is used. This mechanism is known as DRM22. This mechanism has 24 species and 104 step reaction. In FGM model, mixture fraction equations is solved and then GRI 3 mechanism can be used for this model.

The basic idea of the FGM technique is that the flame structure is a combination of a preheating zone where detailed transport phenomena including stretch, curvature and preferential diffusion effects take place and a much thinner reaction layer in which only chemical reaction and molecular diffusion take place and balance each other. The chemical source term, perturbed and balanced by molecular diffusion effects, is the key information stored in the FGM database. This chemical source is not very sensitive to stretch and curvature, but major effects due to changes in conserved variables like pressure, enthalpy and elemental composition at the reaction layer must be taken into account. This is invoked by creating a multiple-dimensional manifold, generated by choosing a set of appropriate flamelets in which the same or similar perturbations appear. A generalized flame stretch theory, based on these ideas, has been developed to show this and to analyze the flame dynamics [9].

EDC is used for turbulent combustion under the finite rate chemistry assumption, whose chemical time scale is governed by Arrhenius law. EDC follows the hypothesis of Eddy Break up (EBU) model that the reaction zone is viewed as a collection of reactants pockets transported by turbulent eddies, in other words, the region where molecular mixing and reaction of reactants occurs only takes up a limited fraction of the total volume of fluid, which is related to the dissipation of turbulence, thus the model is named as "eddy dissipation" [10].

4 Result Discussion

4.1 Validation

Validation is done by comparing numerical results and experimental data which is gained by Laser Doppler Velocimetry (LDV) in [5]. According to that in combustion process, velocity of fluid, reaction and temperature changes should be simulated simultaneously, axial velocity and temperature profiles are studied.

Figure 4 shows comparison of axial velocity between k- ε model, RSM model, and experimental results. According to this figure, at the beginning, k- ε model presents more error than RSM model, but by keeping away from entrance, k-ε model has better behavior compared to RSM and is closer to experimental results. The reason of this problem, can be made due to high swirling of flow in the entrance and boundary layer interaction that the k- ε model couldn't simulate it. But by keeping away from inlet and vanishing of boundary layer, this model, simulated flow as well. So, It seems k- ε model is a better than RSM. Furthermore, in k- ε model (2equation), the equation number is less than RSM model (7 equation) and then time of solution is less. According to this figure, it can be said k- ε model has a good accuracy compared with experimental results.

In figure 5, temperature curves are compared together. According this figure, FGM model has better solution than EDC model. in FGM model, temperature of central of domain is simulated very good, but near the walls, especially at the end of the chamber, it has a little errors compared to actual value. The cause of it would be large scale mesh in that parts where the answers is slightly further away from the actual values. But EDC model has generated completely unrealistic across the field. So, it can be said EDC model is not appropriate model for this case.

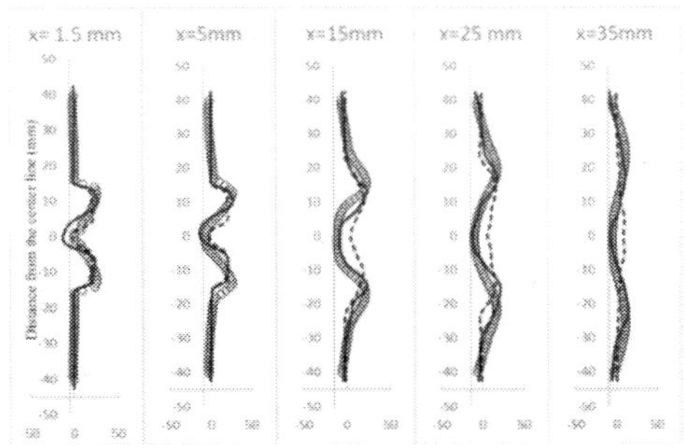

Figure 4: Axial velocity of experimental (o), k- ε (—), and RSM model (---)

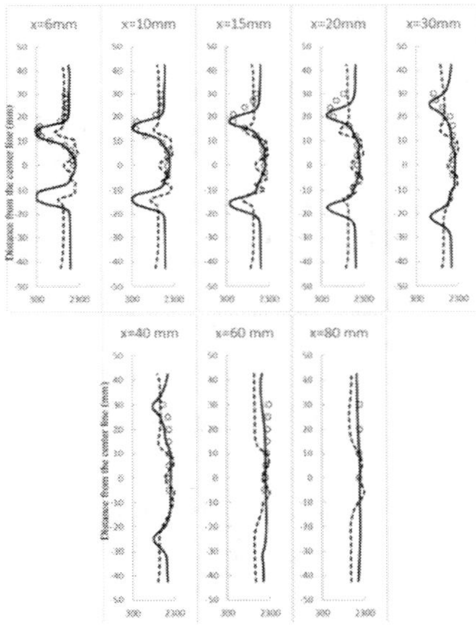

Figure 5: Temperature comparison in different locations along the longitudinal the chamber. experimental (o), k- ε (—), and RSM model (---)

4.2 Flame Structure

In figure 6, axial velocity contour has been shown in cold and hot case. As indicated in the figure, near the entrance to the chamber, because of the geometry and existence rotational speed, the flow has a tendency to diverge. So, in the entrance of chamber, flow vortex has generated. Combustion usually is stable in vortex zone. Therefore, in this case, flame forms in this zones. Near the wall of chamber, small vortex has generated due to geometry changes suddenly. Of course, these zones with central recirculation zone where they have hot species caused by combustion, made hot inlet flow and as a result flame has generated for inlet flow and flame has stabilized. A regarding temperature about 1500 k for flame, it can be seen in figure 7, in range of 0.05 m from inlet, flame has been completed. Center of flame (the most important reaction zone view of energy released) is the range of 0.09 m from inlet. Another thing that can be noted is small amount of flame getting into the swirler that is shown in figure 7. In other words, the flame will deal with the nose. This is due to the boundary layer which reduces the flow rate and thus the flame arrive into the canal and are dealing with nose.

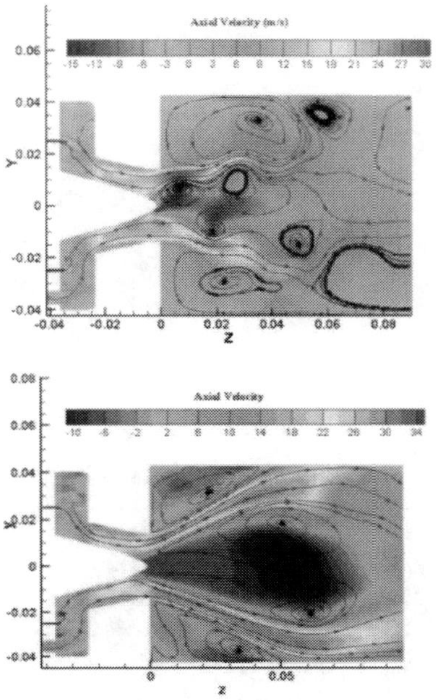

Figure 6: Axial velocity contour in cold (top) and hot (bottom) case

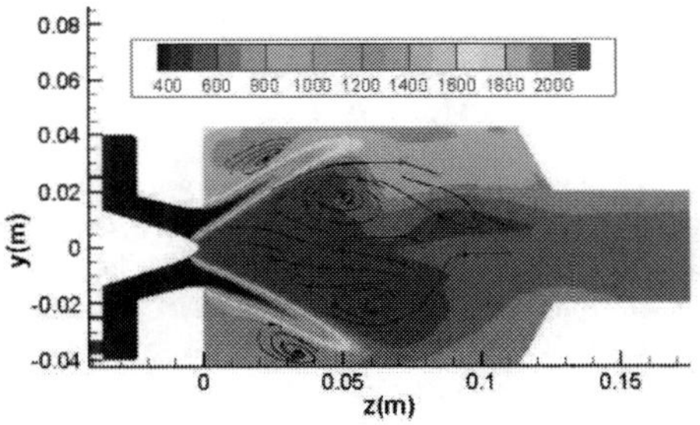

Figure 7: Contour of temperature in hot case

Another significant issue is the generated eddy in the center of domain. As it can be seen in figure 6, the vortex in cold case is very different with hot case. They are different in number and direction of rotation. In hot case, number of main vortex is 5, but in hot case, they are more. This issue represent different behavior of hot case and cold case.

Figure 8, shows the methane concentration distribution. Methane is injected to field of flow by some holes in the swirler and mixing of it and air begin. Inside of the passage, mixing increases to the end almost uniformly enters into the combustion chamber. Then, methane burns and concentration of it decreases. Near the outlet of the combustion chamber, there is almost no methane which represents the efficiency of the swirler and chamber for the fuel and air rate.

In the figure 9, mole fraction of HO_2 is shown. The radical is formed at the beginning of the methane-air chain reaction and is converted to OH radical quickly and is considered as a marker for edge flame or check the ignition. According to figure 9, it can be said that flame is V formed and reactions begin on the outer and inner surfaces. This means that the central recirculation zone and lateral recirculation zone cause an increasing in temperature of the reactants and reactions are started.

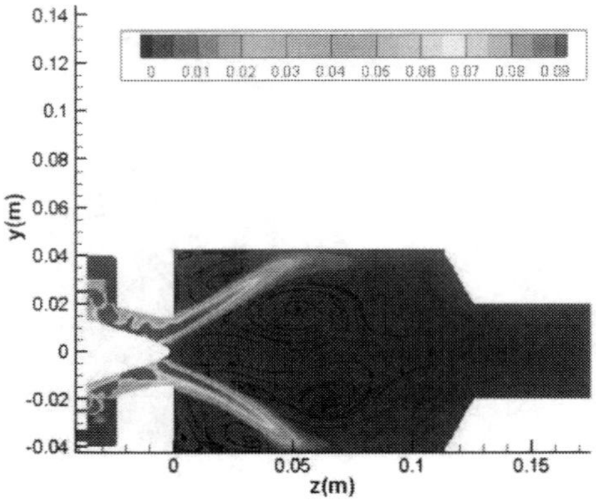

Figure 8: Methane mole fraction

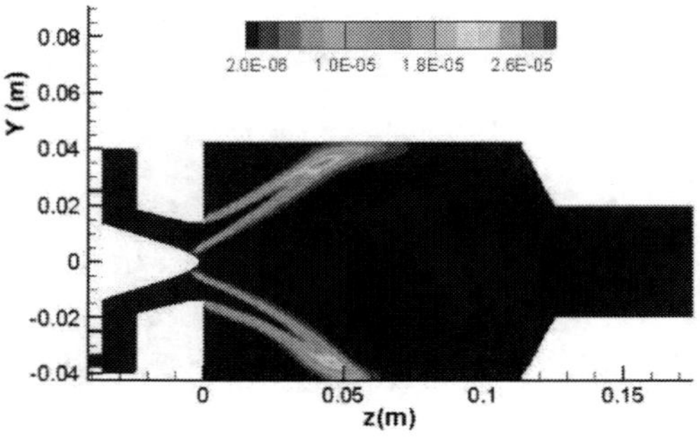

Figure 9: HO_2 mole fraction

6 Conclusion

In this paper, the swirling flow result from a swirler in burner, is simulated and studied in hot and cold case. In the non-reacting case, k- ε RNG and RSM model are selected and the average axial velocity in some section are validated with experimental results. k- ε RNG model simulated behavior of flow better than RSM and it can be concluded that the model is better in wall bounded and swirling flow simulation. Also in the hot case, the FGM and EDC model were selected and FGM model seems better than EDC model by comparing simulation temperature with experimental data. In this paper, it was also found behavior of cold and hot case are different. The cold flow vortices are different from the hot case in the number and the direction of rotation. Other important issues in the model geometry was the existence of boundary layer at the output of the swirler and thus entering the combustion into it.

References

[1] M. Giridharan, H. Mongia, S. M. Jeng, "SWIRL CUP MODELING – PART VIII: SPRAY COMBUSTION IN CFM-56 SINGLE CUP FLAME TUBE", AIAA-2003-0319, 2003.

[2] L. Li, X.F. Peng, T. Liu, Combustion and cooling performance in an aero-engine annular combustor, Applied Thermal Engineering, pp. 1771-1779, 2006.

[3] F. Shirinzade, H. Mahdavi moghadamm, "Numerical investigation and optimization of effective parameters in gas turbine combustion chamber by using Fluent", 1st conference of Mechanical Engineering, Tehran, RCME2011-05-032 ،1389.

[4] Sh.A. Mirhosseini, M.H. Moghadas, A.R. Mostofizade, "Study of liner heat transfer and distribution of the flow in the annular combustion chamber of J85-Ge21 Engine", IRANAPA2012-16789, 1390.

[5] L. Y. M. Gicquel, G. Staffelbach, T. Poinsot,"Large Eddy Simulations of gaseous flames in gas turbine combustion chambers," Toulouse : Elsevier, 2012.

[6] B. Ajjele, S. Tabejamaat, M.A. Attarzade, "Methane reaction study in swirling flow", 5th fuel and combustion conference of Iran, University of Science and Technology, 1392.

[7] Y. Wu, C. Carlsson, R. Szasz, L. Peng, L. Fuchs, X.S. Bai, "Effect of geometrical contraction on vortex breakdown of swirling turbulent flow in a model combustor," Fuel 170,2016, p.p.210–225.

[8] F. E. Agbonzikilo, I. Owen, J. Stewart, S. K. Sadasivuni, M. Riley, V. Sanderson, "Experimental and Numerical Investigation of Fuel–Air Mixing in a Radial Swirler Slot of a Dry Low Emission Gas Turbine Combustor," Journal of Engineering for Gas Turbines and Power, Vol. 138 / 061502-1, JUNE 2016.

[9] J.A. van Oijen , A. Donini, R.J.M. Bastiaans, J.H.M. ten, Thije Boonkkamp, L.P.H. de Goey, "State-of-the-art in premixed combustion modeling using flamelet generated manifolds", Progress in Energy and Combustion Science 57 (2016) 30–74.

[10] H. Bao, "Development and Validation of a New Eddy Dissipation Concept (EDC) Model for MILD Combustion", Delft University of Technology, (2017) Jan.

11th Asia-Pacific Conference on Combustion,
The University of Sydney, NSW Australia
10th -14th December 2017.

Numerical study on the laminar flame characteristics of acetylene-methanol blends

Qianqian Li[1*], Yemiao Zhang[1], Yingjia Zhang[1], Zuohua Huang[1]

[1] State Key Laboratory of Multiphase Flow in Power Engineering, Xi'an Jiaotong University

No. 28th, Xianning West Road, Xi'an, Shaanxi, 710049, People's Republic of China

Abstract

Laminar flame characteristics of the blends of acetylene and methanol were numerically studied at 0.1 MPa and 363 K. A model was adopted to do the simulations with the accuracy validated with published laminar flame speeds data of methanol and acetylene. Simulation results showed that the methanol addition into acetylene leads to the significant increase of laminar flame speed with the equivalence ratio corresponding the maximum value shifting to the rich side. The reasons behind were analyzed and found the thermal and chemical kinetic effects both contribute the laminar flame speed variation while the contribution of the diffusion effect can be ignored. Besides, the equivalence ratio shift is found to be closely related to the adiabatic temperature and OH concentrations.

1 Introduction

Alcohols have been considered to be promising alternative fuels around the world. Methanol is the simplest alcohol and were widely studied on engines. It is found the addition of methanol into gasoline or diesel can significantly decrease the harmful emissions of CO, HC and soot. However, methanol suffers many disadvantages like the hygroscopicity, the low energy cost and the corrosion problem, retarding the practical use of pure methanol fuel [1-4]. Nowadays, methanol is always added into the conventional fuels due to the low cost of methanol and the methanol-gasoline blends is being sold in the market. Therefore, it is of great significance to study the blends of methanol with different components of conventional fuels like alkanes, alkenes and aromatics. Actually, these fuels will decompose to be several small molecule species at high temperature and acetylene (C2H2) is one important intermediate specie. Thus, comprehensive study conducted on the blends of acetylene and methanol will be helpful to learn more details of the blends of methanol and the conventional fuel components.

Until now, considerable study has been conducted to focus on the variations of the emissions by blending alcohols into acetylene [5-7]. The combustion is a complex process involving the heat transfer, mass transfer, fluent and the chemical kinetics, and is highly affect the harmful emissions and the thermal efficiency. However, little attention has been paid on the combustion process of the acetylene-methanol blends.

Laminar flame speeds is a representative parameter to learn the fundamental combustion characteristics. It plays significant role in studying the turbulent flame. Besides, laminar flame speed is an important parameter to validate the mechanism accuracy. Laminar flame speeds of pure methanol and acetylene have been measured over wide initial condition range. Jomaas et al. [8] measured the laminar flame speeds of acetylene-air mixtures at wide initial pressures. Park et al. [9] conducted the comprehensive studies on the laminar flame speed and extinction strain rate of acetylene-oxygen-nitrogen with the counterflow configuration. Simulations results indicated that prediction ability of the existing model needs to be improved at rich mixtures. Zhang et al. [10], Li et al. [11] and Veloo et al. [12] all have measured the laminar flame speeds of methanol. There are other studies conducted the blends of acetylene with different fuels. Al-Hamamre et al. [13] studied the combustion chacteristics of acetylene-hydrogen blends and found that acetylene addition accelerate the flame speed however the concentration of H radical decreases.

This study aims to study the laminar flame characteristics of acetylene-methanol blends numerically. An accuracy model will be selected first and then laminar flame speed variation will be presented with the addition of methanol into acetylene. Finally, the reason caused the laminar flame speed variation will be illustrated in details.

2 Kinetic Modeling and Analysis Method

The model selected to do the simulations was published by Li et al. [11] which was developed for methanol-isooctane blends. Details of the model construction can be referenced the paper [11]. The prediction ability of the model on the laminar flame speeds of acetylene and methanol were examined. Fig. 1 shows the comparison between the simulations and the experimental results for pure acetylene and methanol at atmospheric pressure. It is seen the model yields well predictions for both acetylene and methanol at all given conditions. Therefore, it is reasonable to adopt the model to simulate the one-dimensional freely propagating flame of acetylene-methanol blends.

Model simulations were conducted using CHEMKIN II/Premix code [14-16], and the simulations considered the Soret diffusion effect and mixture-averaged transport. The initial temperature is 363 K and the initial pressure is 0.1 MPa. The equivalence ratio ranges from 0.6 to 1.6. The blending ratio of methanol by molecule ratio was set to be 0.0, 0.25, 0.5, 0.75, 1.0. The calculation domain was set from -2.0 cm to 10.0 cm corresponding to the upstream to the downstream. The adaptive grid control based on solution gradient and the curvature were all decreased to be 0.02.

Qianqian Li. Fax: +86-29-82668789
E-mail address: qianqianli@xjtu.edu.cn

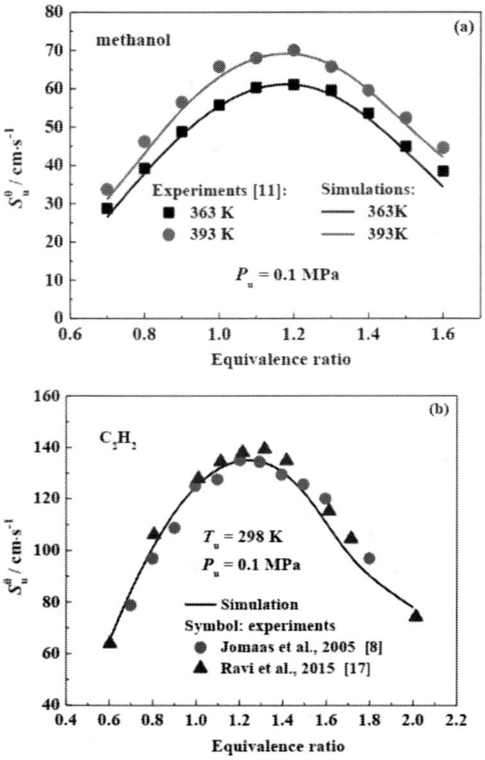

Fig. 1. Comparison between the simulations and the experimental results from the references.

3 Results and Discussion

3.1 Simulated Laminar Flame Speed

Fig. 2 shows the simulated laminar flame speeds of acetylene-methanol blends at 363 K, 0.1 MPa and different blending ratios. It is seen that, with the addition of methanol, the laminar flame speeds decrease significantly and an equivalence ratio shift to the leaner side is observed for the maximum laminar flame speed. However, the equivalence ratios corresponding to the peak laminar flame speeds for different blends is larger than 1.1, which is different from many hydrocarbon fuels.

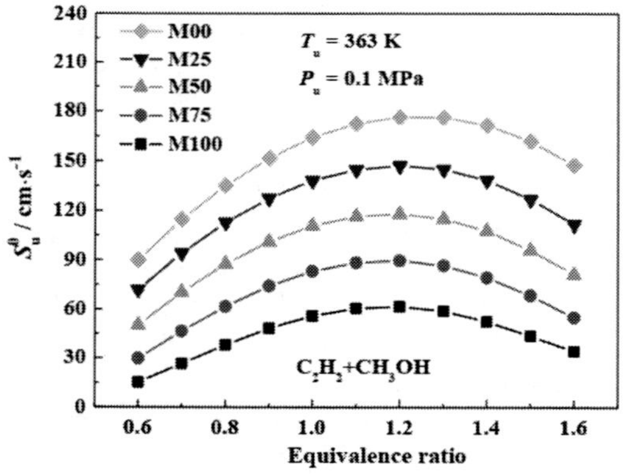

Fig. 2. Simulated laminar flame speeds of acetylene-methanol blends at 363 K and 0.1 MPa.

3.2 Discussion on the Variation of Laminar Flame Speed

Laminar flame speed is considered to be affected by three properties of diffusive, thermal and kinetic, according to the equation as [18],

$$S_u^0 \sim (\alpha Le)^{1/2} \exp(-E_a / RT_{ad}) \qquad (1)$$

where α is the thermal diffusivity and acquired through $\alpha = \dfrac{\lambda}{\rho C_p}$; λ is the thermal conductivity; ρ is the density of the unburned mixture; C_p is the specific heat; E_a is the overall reactivation energy; R is the universal gas constant. T_{ad} is the adiabatic temperature and calculated according to the thermal equilibrium theory. *Le* represents the Lewis number, and here we use the overall effective Lewis number. The thermal effect can be characterized by T_{ad}, and the diffusion effect can be indicated by *Le*. Higher adiabatic temperature and stronger diffusion will result in faster flame speed. Fig. 3 and Fig. 4 shows the effective Lewis number and adiabatic temperature of different blends. The overall effective Lewis number, Le_{eff}, can be deduced through [18],

$$Le_{eff} = \begin{cases} \dfrac{Le_O + (1-\tilde{\phi})Le_F}{2-\tilde{\phi}} & \phi < 1 \\[2ex] \dfrac{Le_F + (1+\tilde{\phi})Le_O}{2+\tilde{\phi}} & \phi > 1 \end{cases} \qquad (2)$$

in which $\beta = E_a(T_b - T_u)/RT_b^2$ and $\tilde{\phi} = \beta(\phi - 1)$. ϕ represents the equivalence ratio, β Zeldovich number. Le_F and Le_O respectively indicates the Lewis number of the thermal diffusivity of the mixture to the mass diffusivities of the fuel and oxidizer. The Lewis numbers of the blending fuels are calculated based on the mole fractions of the pure fuels. It is obviously seen that Le_{eff} and T_{ad} all decrease at lean and rich sides with the addition of methanol, which is consistent with the laminar flame speed variation. This indicates that the both thermal and diffusion effects contribute the the reduction of the laminar flame speed.

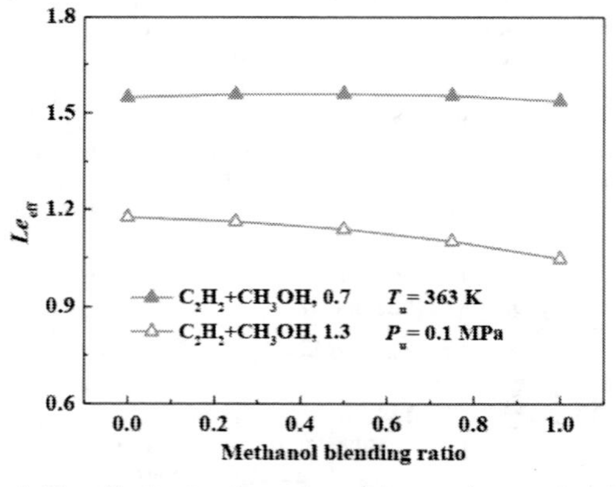

Fig. 3. The effective Lewis number of the acetylene-methanol blends

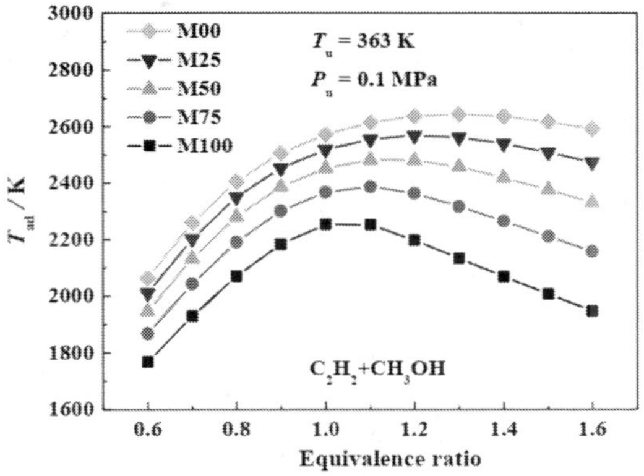

Fig. 4. The adiabatic temperature of the acetylene-methanol blends

Further anaylsis on the effects of different factors were developed. The two terms of equation 1 respectively represents the diffusion effect and the Arrhenius effect. Figure 5 shows the diffusion effect, Arrhenius effect and the combined effect at lean and rich mixtures. It is seen that, with the increase of methanol blending ratio, the diffusion effect yields negligible variation while the Arrhenius effect shows significant decrease and plays dominant role in the laminar flame speed variation.

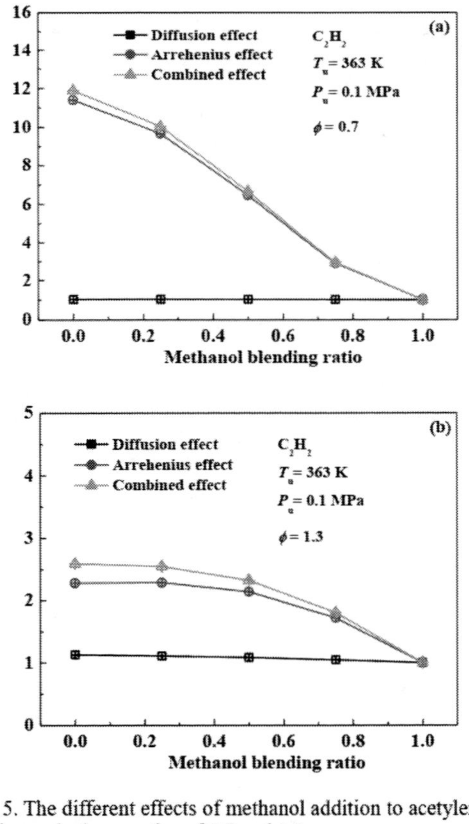

Fig. 5. The different effects of methanol addition to acetylene for the equivalence ratios of 0.7 and 1.3

The Arrhenius effect concludes the effects of both thermal and kinetic factors. In order to account the contribution of each factor, the laminar flame speeds were calculated at the adiabatic temperature of pure methanol by adjusting the ratio of nitrogen to remove the thermal effect, as shown in Fig. 6. It is seen that, with the increase of methanol addition, laminar flame speeds decrease significantly. Therefore, the kinetic effect also play significant effect on promoting the laminar flame speed. Besides, the equivalence ratio of the maximum laminar flame speed shift to the lean side, which is different from the result obtained in Fig. 2. Especially, the equivalence ratio of the maximum laminar flame speed is highly consistent with the equivalence ratio of the maximum adiabatic temperature for pure acetylene, indicating the effect of adiabatic temperature on the laminar flame speed variation with equivalence ratio.

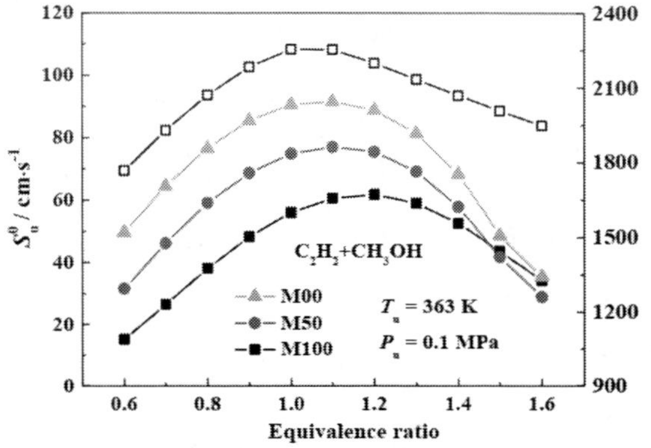

Fig. 6. Laminar flame speeds of three different blends at the same adiabatic temperature

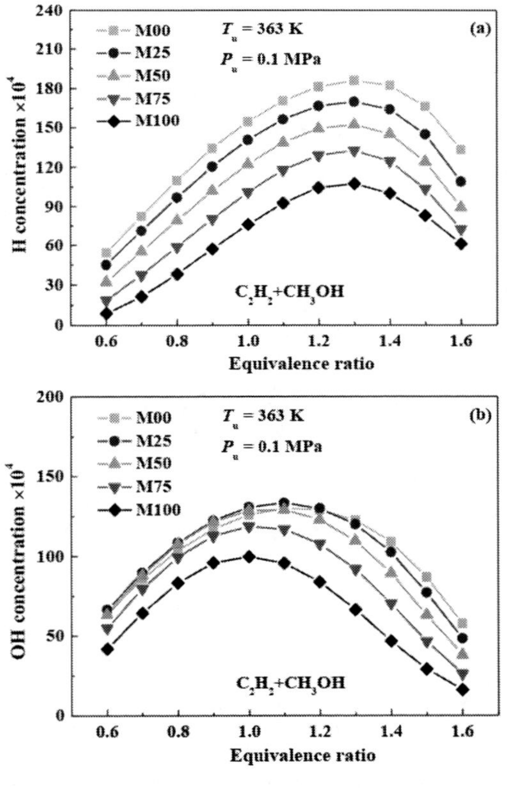

Fig. 7. Maximum concentrations of H and OH for different methanol-acetylene blends

H and OH species dominants the chain branching reactions in the flame propagation. Fig. 7 shows the maximum mole fractions of H and OH for different methanol-acetylene blends. Obviously, H and OH concentration show decrease tendency with the addition of methanol, which retards the whole reaction and finally slow the laminar flame speed. Especially, the equivalence ratio shift to the lean side is also observed for the variation of OH concentration, indicating the effect of kinetic effect on the equivalence ratio shift of laminar flame speed.

4 Conclusions

Laminar flame characteristics of acetylene-methanol blends were numerically studied at 363 K and 0.1 MPa. A model developed for methanol-isooctane blends were validated to be accurate to do the simulations. Simulation results indicate the methanol addition into acetylene leads to the significant decrease of laminar flame speed and the equivalence ratio of the maximum value shifts to the lean side. Further analysis show that laminar flame speed variation can be contributed by the thermal effect, the kinetic effect and the diffusion effect. Both the thermal effect and the kinetic effect dominated the laminar flame speed variation while the diffusion effect can be neglected.

5 Acknowledgment

This work is supported by the National Natural Science Foundation of China (Grant No. 51406159, 91441203 and 50876085), the National Basic Research Program (2013CB228406), the China Postdoctoral Science Foundation (2015T81025), and State Key Laboratory of Engines, Tianjin University (K2016-02).

References

[1]. Li J, Gong C, Wang E, Yu X, Wang Z, Liu X. Emissions of formaldehyde and unburned methanol from a spark-ignition methanol engine during cold start. Energ Fuel 2010; 24: 863-870.

[2]. Li L, Ge Y, Wang M, Li J, Peng Z, Song Y, Zhang L. Effect of gasoline/methanol blends on motorcycle emissions: Exhaust and evaporative emissions. Atmos Environ 2015; 102: 79-85.

[3]. Ninomiya JS, Golovoy A, Labana SS. Effect of methanol on exhaust composition of a fuel containing toluene, n-heptane, and isooctane. J Air Pollut Control Assoc 1970; 20: 314-317.

[4]. Qi DH, Chen H, Geng LM, Bian YZ, Ren XC. Performance and combustion characteristics of biodiesel–diesel–methanol blend fuelled engine. Appl Energy 2010; 87: 1679-1686.

[5]. Esarte C, Abián M, Ángela Millera, et al. Gas and soot products formed in the pyrolysis of acetylene mixed with methanol, ethanol, isopropanol or n-butanol. Energy, 2012, 43(1):37-46.

[6]. Esarte C, Millera A, Bilbao R, Alzueta MU. Gas and soot products formed in the pyrolysis of acetylene-ethanol blends under flow reactor conditions. Fuel Processing Technology, 2009, 90(4): 496-503.

[7]. Wu J, Song KH, Litzinger T, Lee SY, et al. Reduction of PAH and soot in premixed ethylene-air flames by addition of ethanol. Combustion and Flame, 2006, 144(4): 675-687.

[8]. Jomaas G, Zheng XL, Zhu DL, Law CK. Experimental determination of counterflow ignition temperatures and laminar flame speeds of C2–C3 hydrocarbons at atmospheric and elevated pressures. Proceedings of the Combustion Institute, 2005, 30(1): 193-200.

[9]. Park O, Veloo PS, Egolfopoulos FN. Flame studies of C2 hydrocarbons. Proceedings of the Combustion Institute, 2013, 34(1): 711-718.

[10]. Zhang ZY, Huang ZH, Wang XG, Xiang J, Wang XB, Miao HY. Measurements of laminar burning velocities and Markstein lengths for methanol-air-nitrogen mixtures at elevated pressures and temperatures. Combust Flame 2008; 155: 358-368.

[11]. Li Q, Zhang W, Jin W, et al. Laminar flame characteristics and kinetic modeling study of methanol-isooctane blends at elevated temperatures. Fuel, 2016, 184:836-845.

[12]. Veloo PS, Wang YL, Egolfopoulos FN, Westbrook CK. A comparative experimental and computational study of methanol, ethanol, and n-butanol flames, Combust Flame 2010; 157: 1989-2004.

[13]. Al-Hamamre Z, Yamin J. The effect of hydrogen addition on premixed laminar acetylene-hydrogen-air and ethanol-hydrogen-air flames. International Journal of Hydrogen Energy, 2013, 38(18): 7499-7509.

[14]. Kee RJ, Grcar JF, Smooke MD, Miller JA. PRMIX: a FORTRAN program for modeling steady laminar one-dimensional premixed flames. Sandia National Laboratory, SAND Report 85-8240. 1985.

[15]. Kee RJ, Miller JA. Computational Modeling of Flame Structure, Physica D 1984; 12: 198-211.

[16]. Kee RJ, Rupley FM, Miller JA. CHEMKIN-II: a Fortran chemical kinetics package for the analysis of gas-phase chemical kinetics. Sandia National Laboratory, Technical Report SAND89-8009, 1989.

[17]. Ravi S, Sikes TG, Morones A, Keesee CL, Petersen EL. Comparative study on the laminar flame speed enhancement of methane with ethane and ethylene addition. Proceedings of the Combustion Institute, 2015, 35(1): 679-686.

[18]. Law CK. Combustion Physics, first ed., Cambridge University Press, Cambridge, 2006.

11th Asia-Pacific Conference on Combustion,
The University of Sydney, NSW Australia
10th -14th December 2017.

OH Imaging in a Non-Uniform, Hydrogen-Fueled Scramjet Engine

Tristan Vanyai, Stefan Brieschenk, Timothy J. McIntyre

The University of Queensland

Brisbane, Queensland, 4068, Australia

Abstract

The hydroxyl radical, OH, acts as a convenient indicator of combustion within hypersonic engines. This paper presents results from OH imaging within a scramjet combustor in the T4 hypersonic shock tunnel. Three methods are presented, being time-integrated emission, time-resolved emission, and planar laser-induced fluorescence. All methods capture emission in ultraviolet wavelengths, around $\lambda = 310$ nm. The resulting 2D images are representative of OH density and hence combustion. These images show the dependence of combustion on the shock structure within the scramjet combustor.

1 Introduction

Supersonic combustion is an important subject of investigation in the development of hypersonic engines. As the flight velocity of a vehicle increases beyond approximately Mach 6, the process of slowing the incoming air to subsonic speeds becomes increasingly inefficient [1].

If the combustion process in such an engine were to be subsonic, the resulting combustor temperatures can easily reach several thousand Kelvin [2], even without considering the heat addition from combustion. This would lead to severe restrictions on the choice of materials suitable for these conditions, considering melting points and thermal stress.

More significantly, these high temperatures prevent the recombination of combustion products, water and carbon dioxide, significantly decreasing cycle efficiencies. As such, the temperature should remain below approximately 2500 K [1, 3].

The flow must therefore remain supersonic throughout the combustor. Supersonic Combustion ramjets (scramjets) are designed such that a fuel-air mixture will ignite and complete combustion within a practical length, whilst maintaining supersonic speeds throughout the entire engine.

Scramjet engines have been developed and tested in ground and flight tests for several decades. Whilst pressure and heat transfer measurements can be made easily at the flow boundaries, it is more difficult to examine the conditions within the core flow. Optical diagnostic techniques are therefore typically used to probe key thermodynamic properties without disturbing the flow.

2 Methodology

This paper presents results from an experimental campaign within the T4 Stalker Tube at The University of Queensland, Australia. T4 is a free-piston hypersonic shock tunnel, capable of producing flow at speeds between Mach 4 and Mach 10, with total enthalpies typically ranging between 2.5 and 15 MJ/kg.

2.1 Scramjet Model

The scramjet engine was designed to take advantage of a combustion process called "Thermal Compression" (see [4] for more details.) As such, the flow is highly three-dimensional throughout the engine.

Figure 1 shows a schematic of the engine model, with key measurements given in mm. The engine consists of a three-dimensional inlet, a constant area combustor, and a single ramp nozzle. The model has a constant width of 75 mm.

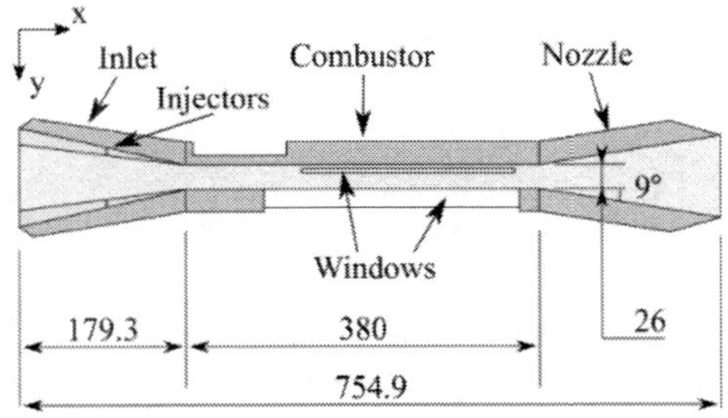

Figure 1: Scramjet engine model, adapted from [5].

As shown in Fig. 2, the intake ramp angle varies in the spanwise direction, z, but is constant in the streamwise direction, x. This varying angle was designed to produce a non-uniform flow field within the combustor. As such, the engine had high and low compression sides, where pressures and temperatures were higher and lower than the mean, respectively.

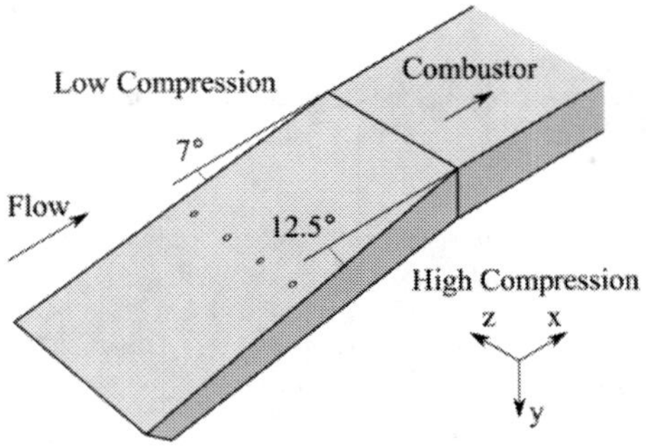

Figure 2: Scramjet model intake, adapted from [5].

Tristan Vanyai. Fax: +61-33-654-799
E-mail address: t.vanyai@uq.edu.au

This non-uniform flow field enabled ignition to occur on the high compression side, even though the mean flow conditions were unsuitable for robust combustion. Heat release from the high compression side would be transmitted to the low compression side through pressure waves (Thermal Compression) that enable combustion throughout the entire engine. For more details on these flow fields, see [5,6].

2.2 Experimental Condition

Experiments were conducted in an air test gas at the freestream conditions presented in Table 1. They correspond to the flow behind a hypersonic vehicle forebody with a turning angle of 9°, at a flight equivalent Mach 10 condition with a dynamic pressure of 50 kPa. This is typical of an expected hypersonic flight condition with relatively low dynamic pressure, resulting in low drag and minimizing structural stress.

Quantity	Units	Value
P_∞	kPa	4.20 ± 0.61
T_∞	K	408 ± 39
M_∞	–	7.27 ± 0.16
h_{Total}	MJ/kg	4.66 ± 0.34

Table 1: Experimental freestream conditions at the scramjet engine entrance.

Hydrogen fuel was injected in the intake through cylindrical injectors at sonic conditions producing a stoichiometric fuel-air ratio, i.e. an equivalence ratio of one.

2.2 Optical Arrangement

Three optical diagnostic methods were used to probe the distribution of OH radicals throughout the scramjet combustor. These are labeled as Time-Integrated Emission (TIE), Time-Resolved Emission (TRE) and Planar Laser-Induced Fluorescence (PLIF). All methods used a PI-MAX3 ICCD to view the scramjet through the window in the positive y-direction in Fig. 1.

The two emission methods simply capture the light emitted from excited OH radicals, OH*. This is primarily due to chemiluminescence, where OH radicals are produced in the excited state, before de-exciting to generate emission. Since emission is produced throughout the entire flow field, emission signals are line-of-sight integrated.

In the PLIF method, OH radicals are excited by a 2D laser sheet, passing through the combustor along the z-direction. This produces a much higher proportion of molecules in the excited state, thus the signals are typically much stronger than in equivalent emission methods. Only the molecules within the 2D sheet are excited, thus PLIF signals are not line-of-sight integrated.

To produce the laser sheet, an Nd:YAG laser was used to pump a tunable dye laser. OH molecules were excited using the $Q_1(9)$, $Q_2(8)$ doublet in the (1,0) vibrational band at 283.93 nm. Fluorescence was then observed in the (1,1) and (0,0) bands between approximately 305-320 nm.

In the TIE method, the ICCD is exposed for 1 ms, during which time a fluid particle is expected to travel several model lengths. As such, the TIE images represent the mean flow field, with no unsteady features resolved.

Both TRE and PLIF methods had exposure times of 300 ns, although the PLIF signal typically only lasts a time of the order of 50 ns. In this shortened exposure, the flow travels less than 1 mm. Thus finer, unsteady flow features can be resolved.

Since T4 is a short-duration test facility, the test condition lasted approximately 1 ms. Each method can therefore only capture one image per test. The TIE method is integrated over the entire duration of the test to maximize signal strength. The Nd:YAG laser used in the PLIF method has a frequency of 10 Hz, thus only one laser sheet can be produced within a given 1 ms timeframe. Additionally, the ICCD has a readout time of > 10 ms, thus obtaining multiple TRE images within the test time is not possible with the current arrangement.

Emission in the TIE and PLIF methods were filtered by narrow-band filter with a centre wavelength of 310 nm and FWHM of 11 nm, which captured emission in the (1,1) and (0,0) vibrational bands. Due to necessarily lower intensities in the TRE signal, this method was filtered using a combination of UG11 and WG305 filters, which resulted in a much larger bandwidth, with significant transmission between 300-385 nm. The peak transmission of this combination of filters was approximately three times greater than the narrow-band filter.

PLIF signals are additionally complicated by variation of laser intensity over the laser sheet. This was addressed by diverting a portion of the laser sheet onto a cell filled with laser dye. The resulting fluorescence was observed, and used to normalize the PLIF intensity, assuming a linear dependence.

3 Results

Images of the OH distribution through the scramjet combustor are shown in Fig. 3-5. The horizontal and vertical axes show streamwise and spanwise distances respectively, measured from the engine entrance and high compression wall. Note that each image corresponds to a separate test, thus flow conditions are not identical between Fig. 3-5.

As expected, the TIE method produces a "smeared" image, where time-integrating has removed unsteady features. Conversely, finer features are visible in the TRE image, however this method has a much lower signal-noise ratio.

Nonetheless, the distribution of signal, corresponding to distribution of OH density, is consistent between the two emission methods. The signal gets stronger along the streamwise direction, as combustion evolves.

The spanwise shock generated by the high compression side of the inlet can be seen impinging first the low compression wall at approximately (x, z) = (340,75), traversing the combustor, then hitting the high compression wall at (x,z) = (450,0). It is clear that OH production is coupled to this spanwise shock, with emission intensity increasing dramatically behind each reflection.

There is also a weaker spanwise shock that impinges the low compression wall at approximately (x,z) = (460,75), which was generated from the low compression side of the inlet. This shock structure has a less dramatic effect on combustion except at the wall impingement.

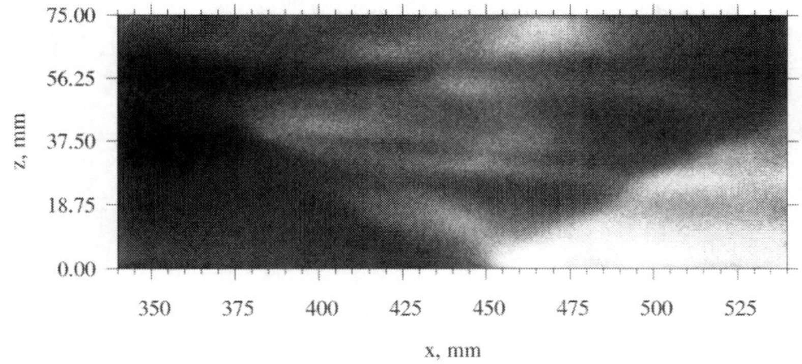

Figure 3: Resulting image using the Time-Integrated Emission (TIE) method, from [5].

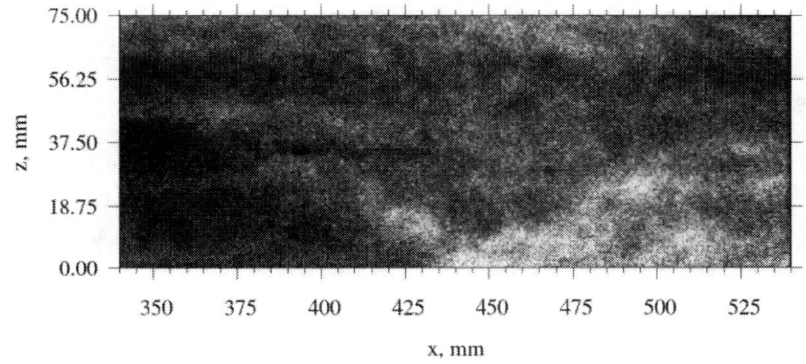

Figure 4: Resulting image using the Time-Resolved Emission (TRE) method, from [5].

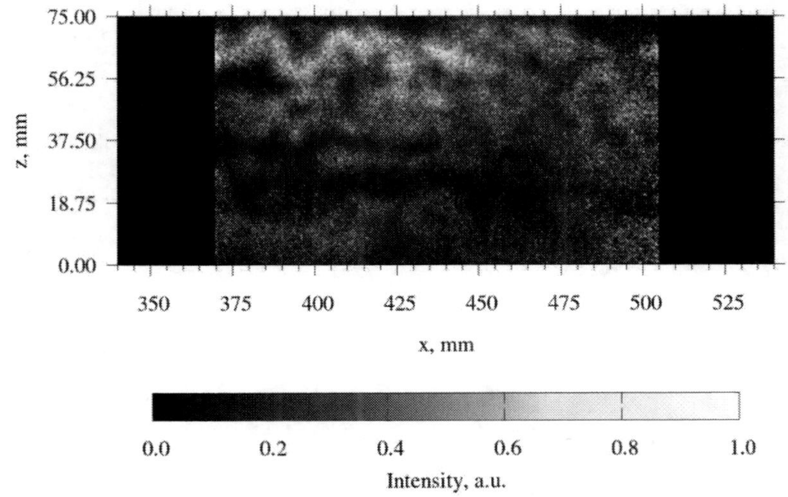

Figure 5: Resulting image using the Planar Laser-Induced Fluorescence (PLIF) method, from [5].

Finally, the PLIF image in Fig. 5 shows a much different depiction of the flowfield. In this case, unsteady features are clearly visible, with an oscillating feature seen near the low compression wall (z = 75 mm).

Compared with the TRE method, the PLIF image has a much stronger signal-noise ratio, as the fluorescence draws its power from the incident laser sheet. PLIF is thus a more useful, instantaneous diagnostic technique.

The laser sheet was located at a vertical location 6.5 mm from the both the combustor floor and the symmetry plane, in the y-direction. This corresponds to the slot window shown in Fig. 1. OH is typically found in a membrane of mixed fuel between the fuel rich wall and the lean core flow, hence the choice of laser sheet position. Since the flow is highly three-dimensional, this OH membrane passes through the laser sheet plane, producing regions of high signal.

However there are also regions where the membrane remains displaced in the out-of-plane direction (y), producing regions of low signal. This can be seen along the z = 18.75 mm line, where numerical simulations indicate the flow is significantly lean.

Interestingly, the bright feature visible in the other methods is not reproduced using PLIF. There are a number of possible causes for this difference: Lower OH density in this region, lower laser intensity, reduced population in the probed energy state, emission loss through quenching, and absorption of the fluorescence en route to the detector.

The most likely cause of the lack of this feature in the PLIF signal is simply that the laser sheet does not intersect this region. Since the TIE and TRE methods are line-of-sight integrated, emission is captured for all flow in the y-direction. This indicates that combustion in this region is not occurring within the plane of the laser sheet, and highlights a drawback of the PLIF methodology in this arrangement.

Emission from all OH in a cross-section could be obtained by aligning the laser sheet normal to the flow direction (x). However this arrangement would not capture the development of the flame along the combustor within a single shot.

A detailed examination of the other possible causes for the lack of this feature in the PLIF image can be found in [5].

The images produced by these methods are currently qualitative. Quantitative measurements would require in situ calibration with a well-defined flame profile, not included in the current arrangement. These images do, however, provide useful visualization of the OH density distribution which can be compared with numerical simulations.

4 Conclusions

Three optical techniques have been used to probe the OH distribution within the core flow of a scramjet engine combustor.

Each method has its own benefits and disadvantages: Time-integrated emission produces strong images, at the cost of temporal resolution; time-resolved emission results in weaker intensities, but can capture unsteady flow features; PLIF produced a strong image that clearly resolved unsteady features, but was not representative of the distribution of OH in the out-of-plane direction.

Combustion was thus inferred from the distribution of OH, and was found to be closely coupled to the shock structure. The resulting distributions will be compared to numerical simulations in future work.

5 Acknowledgments

The authors would like to acknowledge the contributions made by Tamara Sopek and Russell Boyce to enable the use of the optical diagnostic system. Thanks also to Augusto Fontan Moura for his help during the experimental campaign.

References

[1] M. Smart, AIAA Journal 50 (3) (2012) 610-619.

[2] A. Ferri, The Aeronautical Journal, 68 (645) (1964) 575-597.

[3] T. Vanyai et al., Scramjet Performance for Ideal Combustion Processes, manuscript submitted for publication.

[4] T. Vanyai et al., An Experimental Investigation of a Thermal Compression Scramjet with OH Imaging, 20th AIAA International Space Planes and Hypersonic Systems and Technologies Conference, AIAA, 2015-3610.

[5] T. Vanyai, *Experimental Investigation of a 3D Thermal Compression Scramjet Using Advanced Optical Techniques*, PhD thesis, The University of Queensland, Brisbane, Australia, Submitted.

[6] M. Bricalli, L. Brown, R. R. Boyce, AIAA Journal, 53 (7) (2015) 1740-1760.

11th Asia-Pacific Conference on Combustion,
The University of Sydney, NSW Australia
10th -14th December 2017.

Influence of NO_x in Combustion of Biomass Fuel

Ibukun Oluwoye[1], Bogdan Z. Dlugogorski[1], Jeff Gore[2], Mohammednoor Altarawneh[1]

[1]School of Engineering and Information Technology, Murdoch University

90 South Street, Murdoch, WA 6150, Australia

[2]Dyno Nobel Asia Pacific Pty Ltd

Mt Thorley, NSW 2330, Australia

Abstract

Utilisation of biomass in combustion technologies alleviates several pollution-related environmental impacts. This process often occurs in thermally reactive conditions. The present study highlights the cogent influence of nitrogen oxides (NO_x) in combustion of biomass (surrogate) fuel. At practical industrial fuel-rich condition, we investigated the thermal decomposition of biomass surrogate (morpholine) with oxides of NO_x. A laboratory-scale tubular laminar-flow reactor enabled the flexible optimisation of reaction parameters within the temperature range of 300 °C to 1100 °C at different He-O_2-NO_x reactive mixtures, with Fourier transform infrared (FTIR) spectroscopy, NO_x chemiluminescence and gas chromatography (µGC) serving to assess the major gas products as well as N-conversion profiles. The results offer two important remarks; (i) the presence of NO_x lowers the ignition temperature of the morpholine as a result of initial nitration processes, formulating insights towards the safety implications in a typical biomass operating plants; (ii) at elevated temperature above 800 °C, morpholine converts NO_x at approximately 80 % reduction efficiency, depending on oxygen abundancy of the system. These findings shed light on biomass-NO_x interactions occurring in various industrial processes, such as in thermal firing of solid biomass fuels, and particularly, in fuel stagging technology utilising biomass as the major reburn fuel.

1 Introduction

Biomass comprises a group of plant-derived organic matter, representing a valuable source of renewable energy worldwide [1-4]. Production of energy from biomass necessitates various conversion technologies (i.e., thermo-chemical processes) such as gasification, combustion, pyrolysis, torrefaction, liquefaction, and hydrothermal upgrading [5, 6]. In combustion systems, the use of biomass fuels reduces the emission of atmospheric pollutants, e.g., nitrogen oxides (NO_x). The NO_x formation includes the fuel, the thermal, and the prompt mechanisms. These channels serve as an active platform for NO_x species (and its precursors) to react with volatile hydrocarbons, as well as the parent biofuel [7-9].

Furthermore, biomass interacts with NO_x in many other industrial thermal processes. For instance, reburning technology can apply biomass residues to abate NO_x formation in coal-burning boilers and kilns [10-14]. The process basically involves the introduction of secondary fuel (in a reburning zone) to reduce the NO_x initially produced in a slightly lean-fuel condition (primary zone), into intermediate nitrogenous species and then into molecular nitrogen (N_2) in the final burnout zone [14, 15]. Among other alternatives, biomass fuels such as wood, straw, rice husk, bio-oil, sewage sludge, and carbonised municipal solid waste provide an effective means of reducing NO_x formation (50 – 75 %) in strategic reburning systems [16-25].

Thus, it is essential to understand the fundamental factors underpinning the thermal decomposition of biomass/biofuel in reactive NO_x-containing atmospheres. Accordingly, this study seeks to gain further insights on the influence of NO_x in combustion of biomass via experimental procedures. The result focuses on the ignition features of biomass surrogate in NO_x, and high-temperature N-conversion profiles. The conclusions drawn from the present work should apply equally to herbaceous, woody and/or chaff biomass.

2 Materials and Applied Methods

2.1 Biomass Surrogate

Biomass exhibits complex and varying structural properties. This imposes the use of representative model compounds, generally termed as surrogates, to gain an understanding of the thermo-chemical conversion of biomass-derived fuels. These model compounds include hetero-element species, embodying the well-defined N- and O-functionalities within their structures [4, 9, 26-30]. In this study, we employed morpholine as a multi-functional group cyclic biomass model. Morpholine (C_4H_9NO, 1-oxa-4-aza-cyclohexane) represents an oxygen and nitrogen-containing biofuel. Its N-, C-, and O-contents somewhat correspond to that revealed by ultimate analysis of typical biomass feedstocks, hence it is commonly deployed as a surrogate compound for biomass Furthermore, morpholine is widely used as an impregnating agent for fruit, cardboard, and paper, and as a fuel additive. Several investigations have described its decomposition mechanisms and products, combustion, ignition and flame characteristics [26, 31-37].

2.2 Experimental

Figure 1 illustrates the experimental apparatus. A digital syringe pump delivered the reactant fuel (morpholine; purity > 99.0 %, 2.6 – 1.1 µL/min) into the preheated combined stream of helium, NO_x and oxygen gases flowing through the system at

Corresponding author. Fax: (+) 61 8 9360 6346
E-mail address: B.Dlugogorski@murdoch.edu.au

combined rates of between 730 and 305 mL/min (at STP), resulting in a molar-based concentration of 1000 ppm. An electrically heated three-zone horizontal split furnace accommodated the quartz reactor tube, and ensured a well-defined isothermal reaction zone (cf. Fig. 1). The reactor operated at nominal Reynolds numbers of 7.8 (at 300 °C) – 1.8 (at 1100 °C), indicating a laminar flow regime with an average hydrodynamic entry length of approximately 0.9 % of the total reaction zone (300 mm). We maintained NO_x concentration at 620 ppm (600 ppm of NO and 20 ppm of NO_2), adjusting the flow rate of O_2 to sustain the fuel-oxygen equivalence ratio Φ of 1.25 (i.e., moderately fuel rich condition, also corresponds to oxygen-fuel equivalent ratio λ = 0.8) as per the following stoichiometric reaction:

$$C_4H_9NO + 5.75O_2 = 4CO_2 + 4.5H_2O + 0.5N_2 \qquad (1)$$

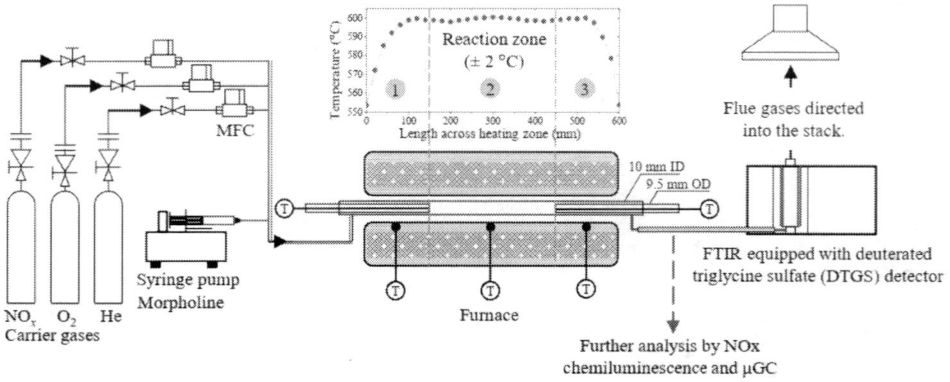

Figure 1: The schematic representation of experimental setup.

We set a constant residence time of 1.0 s in the reaction zone, at ambient pressure of 1.01 bar, by regulating the flowrate by the mass flow controllers (MFC, calibrated and operated at the room temperature, and adapting the gas expansion according to temperature-volume relationship. Similarly, we employed a mass balance to convert the flow rate of the fuel needed for the predefined concentration at the isothermal reaction temperature, into the corresponding set-values at room temperature. The volume of the preheating and cooling-down zones amounts to 4 % of the reaction zone.

Fourier transform infrared (FTIR) spectroscopy facilitated online monitoring of product species exiting the tubular reactor. The setup incorporated an electrically heated transfer line (3.175 mm ID × 300 mm) linking the reactor to a small volume (100 mL) 2.4 m path-length online gas sampling cell (Pike) placed inside the FTIR compartment. The spectrometer averaged 64 accumulated scans per spectrum at 1 cm^{-1} resolution. Both the sample line and the sampling cell operated at 140 °C, avoiding the condensation of volatile products prior and during the analysis. Likewise, the chemiluminescence NO_x analyser (Thermo Scientific model 42*i*-HL) and of μGC (Agilent 490 micro-gas chromatography, 20 m MolSieve-5A column, heated injection) measure NO and N_2 concentrations, respectively.

3 Results

3.1 Ignition characteristics of Morpholine in NO_x

Table 1 lists the infrared (IR) band frequencies of the vaporised fuels. Generally, characteristic vibrations with absorbance value below 0.8 are direct representation of product's concentration exiting the reactor. Due to the structural vulnerability of the heterocyclic ring of morpholine [31-36], we have selected one of the group frequencies at 1132 cm^{-1} to compute the fuel conversion:

$$X = \left(\frac{Abs_i - Abs_T}{Abs_i}\right) \times 100\,\% \qquad (2)$$

with respect to the initial concentration Abs_i at the reactor inlet (i.e., proportional to absorbance value at 1132 cm^{-1}), and the concentration of the fuel at the reactor outlet Abs_T.

Table 1: IR bands of vaporised morpholine, 1000 ppm in 2.4 m path-length gas cell.

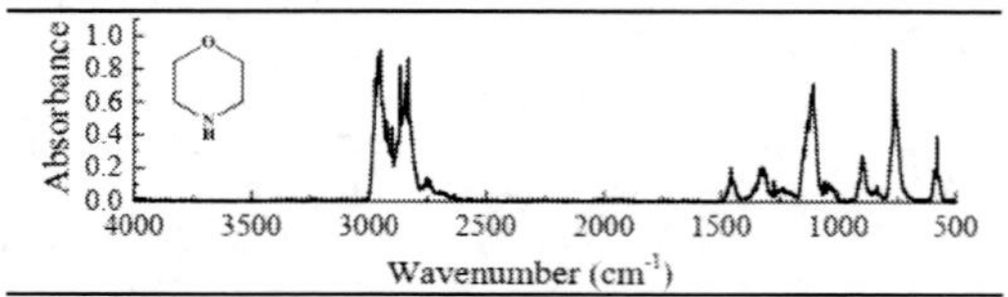

cm^{-1}	Vibration
2840 – 2955	CH_2 stretches
1455	CH_2 scissors
1321	CH_2 wagging
1237, 1062	Oxazines C-O-C symmetric stretches
1320	Asymmetric C-N-C stretches
1132	Selected group frequency
712	Aromatic N-H wag

In our previous study, we monitored the conversion of morpholine as a function of temperature and reactive gas composition [37]. Fig. 2 depicts the conversion trend of morpholine at different gas-mixture conditions, extrapolating the onset ignition temperature. The points have been connected by basis-spline function, as guide to the eye. The trend shows that, the presence of NO_x somewhat inhibits the decomposition of morpholine under pyrolytic conditions, due to the formation of highly stable nitroso and nitro compounds via the R· + NO → RNO and the R· + NO_2 → RNO_2 reaction channels [37]. However, in the presence of O_2, NO_x reduces the ignition temperature by approximately 150 °C due to sensitising activities of NO_x (see Section 4).

3.2 N-Conversion Profile

At elevated temperatures above 800 °C, FTIR confirms the formation of gaseous products such as acetylene (C_2H_2), ethylene (C_2H_4), methane (CH_4), formaldehyde (CH_2O), carbon monoxide (CO), carbon dioxide (CO_2), hydrogen cyanide (HCN), ammonia (NH_3) and nitric oxide (NO). Herein, as shown in Fig.3, we focus on the major nitrogenous species, namely, NO and N_2. Equation 3 defines the transient NO conversion in terms of NO concentrations at the inlet of the reactor $[\mathrm{NO}]_{i,\mathrm{inlet}}$, the outlet of the reactor $[\mathrm{NO}]_{i,\mathrm{outlet}}$, and the diminutive NO concentration in the decomposition product of morpholine $[\mathrm{NO}]_{i,\mathrm{fuel}}$. The subscript i denotes the gas composition, i.e., with and without O_2.

$$X_{\mathrm{NO}} = 1 - \frac{([\mathrm{NO}]_{i,\mathrm{outlet}} - [\mathrm{NO}]_{i,\mathrm{fuel}})}{[\mathrm{NO}]_{i,\mathrm{inlet}}} \times 100\ \% \qquad (3)$$

The maximum NO conversion amounts to approximately 83 %, at 900 °C, after which it decreases to about 80 % due to the activity of the reduction scheme at higher temperatures [10, 14, 38]. This was confirmed by global N_2 formation, increasing to maximum with respect to overall NO reduction.

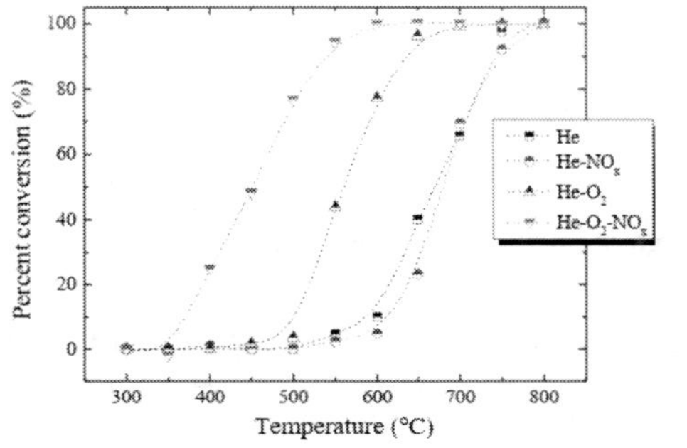

Figure 2: Conversion of morpholine, at the reactor outlet, at different temperature and atmospheres. The extrapolated onset ignition temperature amounting to 581 °C, 613 °C, 494 °C, and 356 °C for He, He-NO$_x$, He-O$_2$, and He-O$_2$-NO$_x$ conditions, respectively.

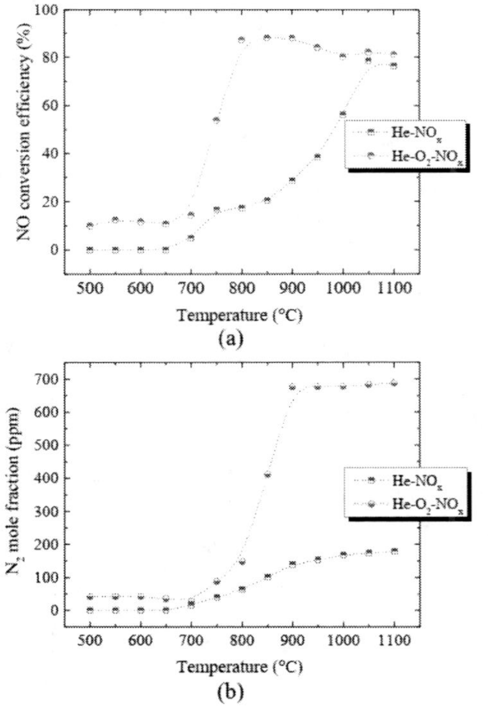

Figure 3: NO$_x$ conversion efficiency of morpholine (a), and formation of N_2 (b) during the thermal decomposition of the fuel at different experimental conditions.

4 Discussion

Evidently from Fig. 2, NO$_x$ enhances the ignition of biomass fuels. This implies that, with an adequate heat/ignition source, NO$_x$ can initiate premature combustion of biomass residues [15, 37]. Previous studies [39, 40] have identified the sensitisation reaction of NO$_x$. The presence of NO/NO_2 accelerates the oxidation of hydrocarbons by turning the usual low-temperature chain-terminating steps of formation of peroxyl (RO_2) radicals into a chain-propagating step [40-47]. Such oxidative nitration reactions can be summarised as follows:

$$\mathrm{R\cdot} + \mathrm{O_2} \rightarrow \mathrm{ROO\cdot} \qquad (4)$$

$$\mathrm{ROO\cdot} + \mathrm{NO} \rightarrow \mathrm{RO\cdot} + \mathrm{NO_2} \qquad (5)$$

$$\mathrm{ROO\cdot} + \mathrm{NO_2} \rightleftharpoons \mathrm{ROONO_2} \rightarrow \mathrm{RO\cdot} + \mathrm{NO_3} \qquad (6)$$

Our earlier investigation [37] employed the density functional theory (DFT) to compute the enthalpic requirements of (4) – (6), identifying their respective spontaneity, as well as other low-temperature channels involving hydroperoxyl radical (HO_2). As a practical industrial implication, the risk of detrimental fire incidences (e.g., in wood-processing plants) [48-54] could be minimised by avoiding trace concentration of NO$_x$ (e.g., from engine exhausts) within the vicinity of combustible dust sediments (e.g., on hot machineries) and suspensions.

Furthermore, in agreement with a former experimental study on furan (C_4H_4O) [55], the oxygen content of morpholine allowed low-temperature NO$_x$ reduction during pyrolytic conditions (cf. Fig. 3). However, O_2 in the dilution stream improves the NO$_x$ reduction efficiency. The maximum NO$_x$ conversion exceeds those recorded for actual biomass residues. For example, biomass fuels such as wood, straw, rice husk, bio-oil, sewage sludge, and carbonised municipal solid waste delivered a maximum NO$_x$ reduction efficiency of about 50 – 80 % [11, 16-25, 56, 57]. This discrepancy can be attributed to the unparalleled contribution of the cellulose, hemicellulose and lignin constituent of actual biomass. On this note, it should be noted that phenol (six-membered aromatic compound) delivered a maximum conversion of 5 – 60 % (from 900 – 1100 °C) only at similar equivalent ratio condition [58]. Hence the interaction of NO$_x$ with cellulose, hemicellulose and lignin content biomass is postulated to follow different mechanism.

5 Conclusions

The contribution reports the influence of NO$_x$ on combustion characteristic of biomass surrogate (morpholine). The results show that, at conditions pertinent to industrial applications, the sensitising effects of NO$_x$ can reduce the ignition temperature of biomass, e.g., by 150 °C for the case of morpholine under a characteristic fuel-rich condition. On the other hand, the reported NO$_x$ reduction efficiency elucidates morpholine as a good biomass surrogate with efficiency values similar to that of actual biomass residues. In presence of oxygen, the effectiveness of morpholine as NO reduction reagent compares well to that of high-importance fuels such as acetylene, ethylene, ethane, natural gas, methane, and biogas [11, 59-63]. The O-content of morpholine produces O/OH radicals that permit NO reduction at relatively lower temperature, making it an excellent reburning fuel in a very O_2-deficient environment.

6 Acknowledgment

This study has been funded by the Australian Research Council (ARC LP110201152 and LP160101169) and Dyno Nobel Asia Pacific.

References

[1] P. McKendry, Bioresour. Technol., 83 (2002) 37-46.
[2] M. Altarawneh, B.Z. Dlugogorski, Environ. Sci. Technol., 49 (2015) 2215-2221.
[3] M. Altarawneh, B.Z. Dlugogorski, PCCP, 17 (2015) 1822-1830.
[4] M. Altarawneh, K. Altarawneh, J. Phys. Org. Chem., 25 (2012) 431-436.
[5] P. McKendry, Bioresour. Technol., 83 (2002) 55-63.
[6] O. Senneca, R. Chirone, P. Salatino, J. Anal. Appl. Pyrolysis, 71 (2004) 959-970.
[7] S.C. Hill, L.D. Smoot, Prog. Energy Combust. Sci., 26 (2000) 417-458.
[8] P. Glarborg, Prog. Energy Combust. Sci., 29 (2003) 89-113.
[9] Q. Ren, C. Zhao, Renew. Sustainable Energy Rev., 50 (2015) 408-418.
[10] P. Glarborg, M.U. Alzueta, K. Dam-Johansen, J.A. Miller, Combust. Flame, 115 (1998) 1-27.
[11] L. Prada, J. Miller, Combust. Sci. Technol., 132 (1998) 225-250.
[12] H. Spliethoff, U. Greul, H. Rüdiger, K.R. Hein, Fuel, 75 (1996) 560-564.
[13] E.C. Zabetta, M. Hupa, K. Saviharju, Ind. Eng. Chem. Res., 44 (2005) 4552-4561.
[14] L.D. Smoot, S.C. Hill, H. Xu, Prog. Energy Combust. Sci., 24 (1998) 385-408.
[15] I. Oluwoye, B.Z. Dlugogorski, J. Gore, S. Vyazovkin, O. Boyron, M. Altarawneh, Environ. Sci. Technol., 51 (2017) 7714-7722.
[16] B. Adams, N. Harding, Fuel Process. Technol., 54 (1998) 249-263.
[17] C. Casaca, M. Costa, P Combust Inst, 32 (2009) 2641-2648.
[18] N.S. Harding, B.R. Adams, Biomass Bioenergy, 19 (2000) 429-445.
[19] A. Kicherer, H. Spliethoff, H. Maier, K. Hein, Fuel, 73 (1994) 1443-1446.
[20] P. Maly, V. Zamansky, L. Ho, R. Payne, Fuel, 78 (1999) 327-334.
[21] H. Rüdiger, U. Greul, H. Spliethoff, K.R.G. Hein, Fuel, 76 (1997) 201-205.
[22] R. Salzmann, T. Nussbaumer, Energy Fuels, 15 (2001) 575-582.
[23] C. Casaca, M. Costa, Combust. Sci. Technol., 177 (2005) 539-557.
[24] S.V. Pisupati, S. Bhalla, Biomass Bioenergy, 32 (2008) 146-154.
[25] P. Fang, Z.-J. Tang, J.-H. Huang, C.-P. Cen, Z.-X. Tang, X.-B. Chen, Fuel Process. Technol., 137 (2015) 1-7.
[26] K. Kohse-Höinghaus, P. Oßwald, T.A. Cool, T. Kasper, N. Hansen, F. Qi, C.K. Westbrook, P.R. Westmoreland, Angew. Chem. Int. Ed., 49 (2010) 3572-3597.
[27] A. Lucassen, K. Zhang, J. Warkentin, K. Moshammer, P. Glarborg, P. Marshall, K. Kohse-Höinghaus, Combust. Flame, 159 (2012) 2254-2279.
[28] M. Altarawneh, M.H. Almatarneh, A. Marashdeh, B.Z. Dlugogorski, Combust. Flame, 163 (2016) 532-539.
[29] M. Altarawneh, A.a.H. Al-Muhtaseb, B.Z. Dlugogorski, E.M. Kennedy, J.C. Mackie, The Journal of Physical Chemistry A, 114 (2010) 11751-11760.
[30] S. Rawadieh, I. Altarawneh, H.B. Alateyat, M. Altarawneh, Computational and Theoretical Chemistry, 1018 (2013) 45-49.
[31] P. Nau, A. Seipel, A. Lucassen, A. Brockhinke, K. Kohse-Höinghaus, Exp. Fluids, 49 (2010) 761-773.
[32] A. Lucassen, P. Oßwald, U. Struckmeier, K. Kohse-Höinghaus, T. Kasper, N. Hansen, T.A. Cool, P.R. Westmoreland, P. Combust. Inst., 32 (2009) 1269-1276.
[33] A. Lucassen, N. Labbe, P.R. Westmoreland, K. Kohse-Höinghaus, Combust. Flame, 158 (2011) 1647-1666.
[34] M. Altarawneh, B.Z. Dlugogorski, J. Phys. Chem. A, 116 (2012) 7703-7711.
[35] P.P. Dholabhai, H.-G. Yu, J. Phys. Chem. A, 116 (2012) 7123-7127.
[36] S. Li, D.F. Davidson, R.K. Hanson, N.J. Labbe, P.R. Westmoreland, P. Oßwald, K. Kohse-Höinghaus, Combust. Flame, 160 (2013) 1559-1571.
[37] I. Oluwoye, B.Z. Dlugogorski, J. Gore, P.R. Westmoreland, M. Altarawneh, Fire Saf. J., 91 (2017) 235-242
[38] M.U. Alzueta, P. Glarborg, K. Dam-Johansen, Combust. Flame, 109 (1997) 25-36.
[39] A.B. Bendtsen, P. Glarborg, K. Dam-Johansen, Combust. Sci. Technol., 151 (2000) 31-71.
[40] Y. Tan, C. Fotache, C. Law, Combust. Flame, 119 (1999) 346-355.
[41] P.G. Ashmore, B.P. Levitt, Symp. (Int.) Combust., 7 (1958) 45-52.
[42] C.L. Rasmussen, A.E. Rasmussen, P. Glarborg, Combust. Flame, 154 (2008) 529-545.
[43] P. Dagaut, F. Lecomte, S. Chevailler, M. Cathonnet, Combust. Sci. Technol., 148 (1999) 27-57.
[44] P. Dagaut, J. Luche, M. Cathonnet, Combust. Sci. Technol., 165 (2001) 61-84.
[45] G. Dayma, K.H. Ali, P. Dagaut, P. Combust. Inst., 31 (2007) 411-418.
[46] I. Oluwoye, M. Altarawneh, J. Gore, H. Bockhorn, B.Z. Dlugogorski, J. Phys. Chem. C, 120 (2016) 3766-3775.
[47] I. Oluwoye, M. Altarawneh, J. Gore, B.Z. Dlugogorski, Combust. Flame, 162 (2015) 3681-3690.
[48] T. Abbasi, S.A. Abbasi, J. Hazard. Mater., 140 (2007) 7-44.
[49] S. Callé, L. Klaba, D. Thomas, L. Perrin, O. Dufaud, Powder Technol., 157 (2005) 144-148.
[50] Y.S. Chin, L.I. Darvell, A.R. Lea-Langton, J.M. Jones, A. Williams, Energy Fuels, 30 (2016) 4398-4404.
[51] U. Krause, Fires in Silos: Hazards, Prevention, and Fire Fighting, John Wiley & Sons, 2009.
[52] J.M. Jones, A. Saddawi, B. Dooley, E.J.S. Mitchell, J. Werner, D.J. Waldron, S. Weatherstone, A. Williams, Fuel Process. Technol., 134 (2015) 372-377.
[53] Á. Ramírez, J. García-Torrent, A. Tascón, J. Hazard. Mater., 175 (2010) 920-927.
[54] M. Todaka, W. Kowhakul, H. Masamoto, M. Shigematsu, J. Loss Prev. Process Indust., 44 (2016) 538-543.
[55] F. Guarneri, E. Ikeda, J.C. Mackie, Energy Fuels, 15 (2001) 743-750.
[56] P. Dagaut, F. Lecomte, Energy Fuels, 17 (2003) 608-613.
[57] H.-S. Do, Y. Bunman, S. Gao, G. Xu, Energy Fuels, 31 (2017) 4499-4506.
[58] C.-Y. Liu, C.s. Zhang, R.h. Yin, Environ. Prog. Sustain. Energy, 34 (2015) 47-53.
[59] R. Bilbao, A. Millera, M.U. Alzueta, L. Prada, Fuel, 76 (1997) 1401-1407.
[60] J. Giménez-López, V. Aranda, A. Millera, R. Bilbao, M.U. Alzueta, Fuel Process. Technol., 92 (2011) 582-589.
[61] F. Lecomte, P. Dagaut, S. Chevailler, M. Cathonnet, Combust. Sci. Technol., 150 (2000) 181-203.
[62] J. Zhao, Q. Wang, L. Yu, L. Wu, Atmos. Environ., 132 (2016) 290-295.
[63] P. Dagaut, J. Luche, M. Cathonnet, Combust. Flame, 121 (2000) 651-661.

11th Asia-Pacific Conference on Combustion,
The University of Sydney, NSW Australia
10th -14th December 2017.

An Experimental and Numerical Study of $C_4H_8O_2$ Esters Combustion Chemistry

K. N. Osipova[1,2,*], D. A. Knyazkov[1,2], I. E. Gerasimov[1], A. M. Dmitriev[1,2], A. G. Shmakov[1,2], O. P. Korobeinichev[1]

[1]Voevodsky Institute of Chemical Kinetics and Combustion

Institutskaya str. 3, Novosibirsk, 630090, Russia

[2]Novosibirsk State University

Pirogova str. 2, Novosibirsk, 630090, Russia

Abstract

Fatty acids esters are the main biodiesel components. Experimental studies, especially at atmospheric and higher pressures, are of great importance for combustion models development. In this work, we studied near-stoichiometric and rich flames of two isomeric esters - ethyl acetate (EA) and methyl propanoate (MP). We did experiments to study $EA/O_2/Ar$ and $MP/O_2/Ar$ flame structure at atmospheric pressure. For numerical simulation, we used two published mechanisms for esters oxidation. Significant differences between experiment and modeling and between models arise for some intermediates. Analysis of fuels decomposition reaction paths shows MP to decompose mostly in radical reactions. Besides radical reactions, EA undergoes monomolecular decomposition with ethylene and acetic acid formation.

1 Introduction

Fossil fuels are the basis of the modern energy industry. Combustion of these fuels produces nitrogen and sulfur oxides, soot, carbon dioxide etc. Thus, a search of alternative energy sources is of crucial importance. Biodiesel is the most promising candidate. Its physical and chemical properties are close to those of oil-based fuels and it suits to existing diesel engines without design changes. Biodiesel is biodegradable, less toxic and contains no sulfur and aromatic species. Experimental and numerical study of biodiesel combustion and the development of its combustion models are an important task. Fatty acids esters are main biodiesel parts. However, the rate constants for many combustion reactions of esters are either unknown or known with insufficient accuracy. Therefore, combustion models needed to be tested against experiments over a wide range of conditions. The complex structure and high molecular weight of esters impede experiments. Hence the results for the oxidation and combustion of smaller esters are extrapolated to heavier ones. Most papers on biodiesel deal with methyl esters. Ethyl esters combustion is cleaner because they are produced from bioethanol instead of methanol. Osswald et al.[1] studied differences in combustion chemistry of two isomeric esters – methyl acetate (MA) and ethyl formate (EF). In the EF flame, the concentrations of benzene, C_2 – and C_4 – hydrocarbons are higher. Hydrogen atom elimination from the methoxy group of MA causes formaldehyde formation. Acetaldehyde forms via hydrogen atom elimination from the EF ethoxy group. A stoichiometric (stoich.) methane/air flame with ethyl acetate (EA) addition was studied by Gasnot et al. [2]. Intermediates of EA combustion include C_2H_4 and CH_3COOH. The high concentration of these species suggests their importance in the combustion mechanism. Yang et al.[3] determined the influence of fuel molecule structure on intermediates composition. They studied three esters: methyl butanoate, methyl iso-butanoate and ethyl propanoate. The maximum concentration of C_2 – hydrocarbons was higher in ethyl ester flame. Fuel molecule structure also influences on the ignition delay time. Esters with formula $C_5H_{10}O_2$ (ethyl propanoate - EP and methyl butanoate - MB) were studied in [4]. The ignition delay of EP was shorter because its combustion products – ethylene and propanoic acid are very reactive. Pitz and Mueller [5] did experiments with C_9 methyl and ethyl esters in a laboratory rapid compression machine. Ethyl esters showed higher reactivity. The main distinctive feature of ethyl esters is a primary reaction with a six-membered transition state formation and its decomposition into fatty acid and ethylene [3]. Data on hydrocarbons combustion suits to model esters combustion [6] because esters molecules have hydrocarbons fragments in alkyl chain. This fact was used to develop a chemical-kinetic mechanism of ethyl pentanoate oxidation in [7]. The proposed mechanism is an extended version of C_1-C_5-fuels oxidation model. The main reactions begin via thermal breaking of C-C, C-O and C-H in EPe molecule. EPe also decomposes into pentanoic acid and ethylene. Six primary radicals form in hydrogen atom elimination reactions. Reactions of the radical cross-isomerization were also added to the mechanism and a. submechanism for pentanoic acid was proposed. During EPe combustion, the same radicals as in the flames of ethyl butyrate, ethyl propanote and ethyl acetate form. Thus, the mechanism for EPe describes combustion of smaller fuels. Westbrook et al. [8] developed a kinetic mechanism for small alkyl esters - methyl formate (MF), methyl acetate (MA), ethyl formate (EF) and ethyl acetate (EA). They used the principle of functional groups similarity. Each ester molecule has a central O(C=O)- group and methyl or ethyl group, connected with O and H- atoms (for formates) or methyl radical (for acetates), connected with a C=O fragment. The bond strength of ethyl groups of EF and EA are almost the same, and rate of hydrogen elimination from them is also equal. The same is true for methyl groups of MA and MF, the acetyl groups of EA and MA and the formyl groups of EA and MF. This approach suits for development of oxidation mechanisms for heavier esters.

There is a lack of data on the structure of esters at atmospheric and higher pressures. In this work, we did experiments under

Corresponding author. Fax: +7(383)3307350
E-mail address: anna-amalia123@yandex.ru

atmospheric pressure, because experiments at higher pressures involve technical difficulties. This paper presents an experimental and numerical study of the combustion mechanism of two esters – ethyl acetate (EA) and methyl propanoate (MP) at atmospheric pressure. As EA and MP are isomers, the C/H/O ratio in their combustible mixtures is the same. We studied the influence of the esters functional group, because it mostly determines combustion chemistry [6].

2 Experiment

Laminar flames of premixed gaseous mixtures of EA/O_2/Ar and MP/O_2/Ar were stabilized on a Botha-Spalding flat burner [9] at atmospheric pressure. For each ester, near stoich. and rich flames were studied. Combustible blends composition and gas velocity on the burner surface (V) are in the Table 1.

Fuel	φ	Reagents mole fractions			V, m/sec
		[fuel]	[O_2]	[Ar]	
EA	1.14	0.0371	0.1629	0.8	0.1543
	1.73	0.0771	0.2229	0.7	0.094
MP	1.14	0.0371	0.1629	0.8	0.1543
	1.73	0.0771	0.2229	0.7	0.094

Table 1: Combustible mixtures composition and gas velocity on the burner surface.

Liquid fuel was supplied to evaporator with a syringe, moved by a step motor. The mass flow rates of the fresh gas mixture components were set with mass flow controllers (MKS Instruments Inc.). The burner temperature was maintained at 348 K with thermostatic water supply through a cooling jacket, soldered with the burner. Species mole fractions vs height above the burner were obtained with molecular-beam mass-spectrometry with soft electron impact ionization. The flames were sampled by a conic quartz probe with a 0.08 mm orifice diameter. A detailed set up description is given in [10]. Ionizing electron energy (IEE) was chosen individually for each species (Table 2). Mole fractions of the reactants and main combustion products were calculated from the corresponding mass-peak intensities by direct calibrations or by elements balance [10]. For intermediates we used the method of relative ionization cross-sections (RICS) [11, 12]. Measurements error for the reactants and the main combustion products is ±15%. For molecular hydrogen it is ±30% because of difficulties in its pumping out from the vacuum system. For other species the accuracy is ±50%. Temperature profiles were measured by S-type (Pt/Pt+10%Rh) thermocouples. A detailed description of thermocouple construction is in [10]. Radiation heat loss correction was calculated by the method, proposed in [13, 14]. Temperature measurements accuracy is ±40 K.

3 Modeling

To simulate flames structure, we used the PREMIX code [15] from the CHEMKIN-II package [16]. To consider perturbations, caused by probe sampling, the input data for modeling were the experimental temperature profiles, measured by thermocouple, placed near the probe orifice. The energy conversion equation was not solved. For modeling we used two mechanisms: the first (Model 1) has 522 species and 2719 reactions [17] and the second (Model 2) – 404 species and 2557 reactions [18].

4 Results and discussion

4.1 Temperature and mole fraction profiles

Figure 1 (a, b) presents profiles of temperature, reactants and main combustion products in the near stoich. and the rich EA flames. The post-flame zone temperature in the near stoich. flame is 1575 K, and flame zone width - 1.15 mm. For the rich flame post-flame zone temperature is slightly lower – 1539 K, but the flame zone width is significantly larger – 2.3 mm. Figure 1 (a, b) also shows comparison of experimental and simulation results. They agree within the experimental error. The same results for MP flames are on the Fig. 1 (c, d). The post-flame zone temperature for the near stoich. flame is 1545 and for the rich – 1530 K. Flame zone width of the near stoich. flame is 1.1 mm, for the rich - 1.7 mm. The discrepancies between experimental and numerical data are also within the experimental error - Fig. 1 (c, d). The post-flame zone temperatures are close enough to compare flames of different esters. Figure 2 shows intermediate mole fraction profiles and their modeling for EA flames. Both models adequately predict profiles shapes and maximum concentration position. For the near stoich. EA flame Model 1 better predicts CH_4, C_2H_2 and C_2H_4. Model 2 is better for CH_3, C_2H_6, C_2H_4O and C_3H_6 + CH_2CO. In the case of the rich EA flame Model 1 describes better CH_4, CH_2O, C_3H_8 and C_3H_6 + CH_2CO, Model 2 is better for CH_3, C_2H_6, C_2H_4O and CH_3COOH. The discrepancies between the models and experiments are likely due to imprecise values of rate constants of some reactions. The discrepancies between the models arise because they were developed based on different experimental data: Model 1 was tested against flame propagation speed, ignition delays and data of oxidation in a jet-stirred reactor; Model 2 was created based on flame structure data, obtained at low and atmospheric pressures. The maximal acetic acid concentration is in rich flame is ~ 2-3 times higher than that in the near stoich. The simulation qualitatively shows the same trend. Both mechanisms overestimate the maximal C_2H_2 concentration. It correlates with the data of [18].

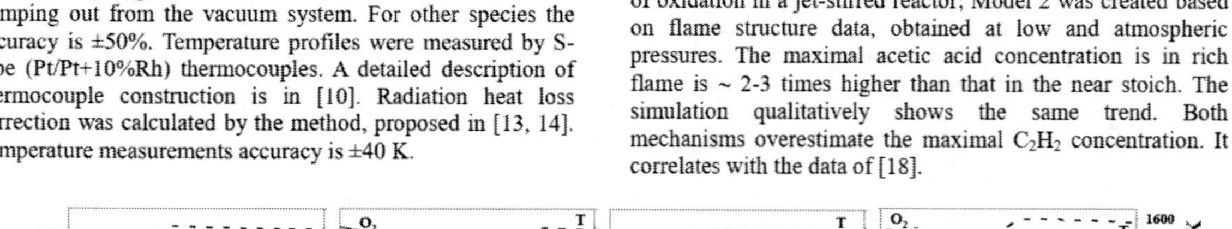

Figure 1: Temperature, reagents and main combustion products mole fraction profiles: EA (near stoich.) - a, EA (rich) - b, MP (near stoich.) - c and MP (rich) - d. Symbols - experiment, solid lines - Model 1, dashed – Model 2, dash-dot – temperature.

Intermediates profiles for MP flames are on the Fig. 2. In the near stoich. MP flame Model 1 better predicts C_2H_2 profile and Model 2 is better for C_2H_4. For the rich MP flame, the profiles, simulated using both models, differ only slightly. As the models give correct profiles shape and maximum position for intermediates, reaction paths analysis, based on the mechanisms, is possible.

4.2 Reaction paths analysis.

Schemes of EA conversion in the near stoich. and the rich flames (Model 1) are on the Fig. 3 (a, b). Numbers near arrows show fuel percentage, decomposing via the corresponding channel. EA has two types of conversion paths: radical reaction (H atom elimination and formed radical decomposition) and monomolecular decomposition reaction with formation of stable intermediates – acetic acid and ethylene. In the rich EA flame, contribution of monomolecular decomposition reaction increases from 42% to 88% In the near stoich. and in the rich MP flames relation of fuel conversion channels is almost equal - Fig. 3 (c, d). MP decomposes mainly in radical reactions. Therefore, the used mechanisms reveal connection of esters molecule structure with primary decomposition paths, despite they predict some intermediates only qualitatively.

m/z	Formula	Species	Ionization energy, eV	IEE, eV	Calibration method
1	H	Atomic hydrogen	13.6	16.65	Direct
2	H_2	Hydrogen	15.43	20	Balance H
15	CH_3	Methyl radical	9.84	13.2	RICS(CH_4)
16	CH_4	Methane	12.61	15.4	Direct
17	OH	Hydroxyl	13.02	16.65	Direct
18	H_2O	Water	12.62	15.4	Balance O
26	C_2H_2	Acetylene	11.4	12.3	RICS(C_2H_4)
28	C_2H_4	Ethylene	10.51	12.3	Direct
28	CO	Carbon monoxide	14.01	15.4	Direct / Balance C
30	H_2CO	Formaldehyde	10.88	11.5	RICS(CO)
30	C_2H_6	Ethane	11.52	12.3	RICS(CO)
32	O_2	Oxygen	12.07	14.35	Direct
40	Ar	Argon	15.76	16.2	Direct
42	C_3H_6+CH_2CO	Propene +Ketene	9.73/9.62	12.3	RICS(CO_2)
44	C_2H_4O	Acetaldehyde	10.23	10.9	RICS(CO_2)
44	C_3H_8	Propane	10.94	12.3	RICS(CO_2)
44	CO_2	Carbon dioxide	13.78	15.4	Direct / Balance C
60	CH_3COOH	Acetic acid	10.65	12.3	Direct
86	$C_4H_6O_2$	Vinyl acetate	9.2	12.3	Estimated
88	$C_4H_8O_2$	Ethyl acetate	10	20	Direct
88	$C_4H_8O_2$	Methyl propanoate	10	20	Direct

Table 2: Detected species, ionizing electron energy and calibration method.

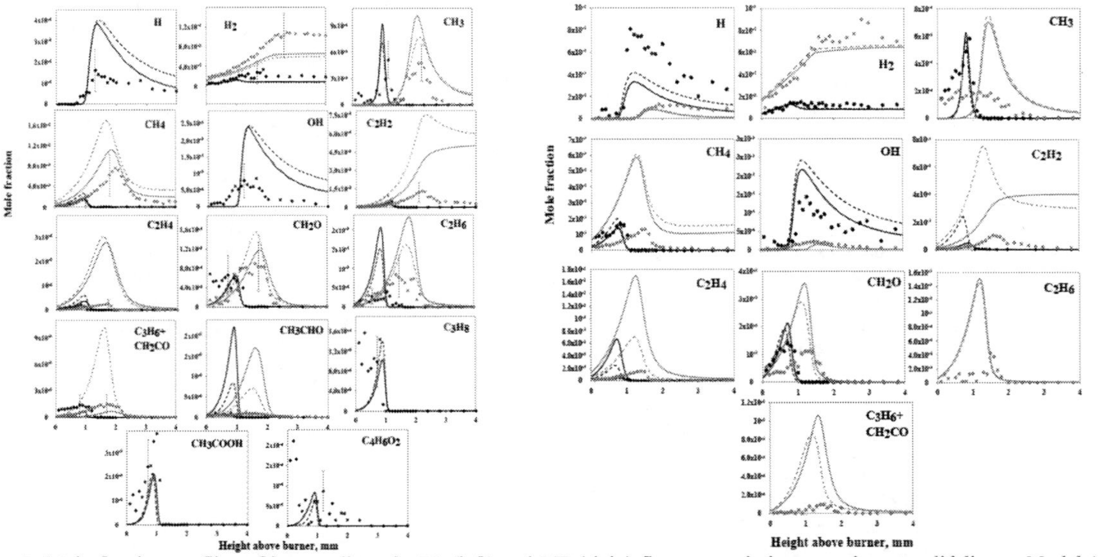

Figure 2: Mole fraction profiles of intermediates in EA (left) and MP (right) flames: symbols - experiment, solid lines - Model 1, dashed lines - Model 2. Black symbols and lines – near stoich, grey - rich.

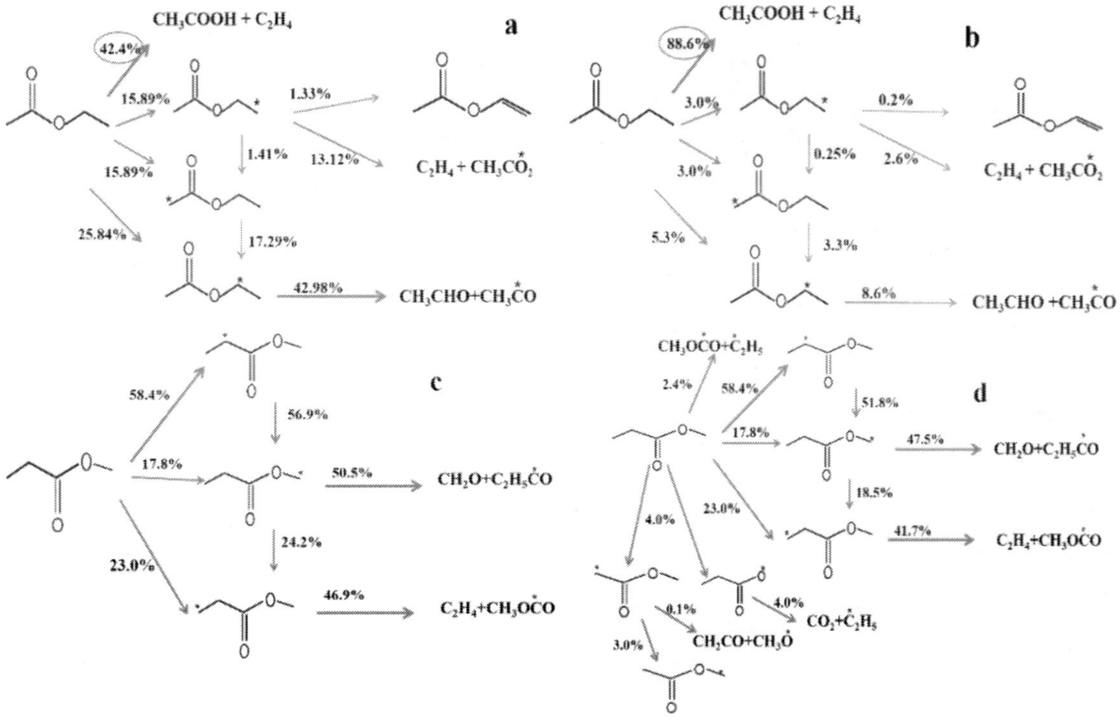

Figure 3: Schemes of EA and MP conversion: a - EA, near stoich.; b - EA, rich; c - MP, near stoich.; d - MP, rich flame.

5 Conclusions

1) Profiles of the temperature, reactants, main combustion products and intermediates in EA and MP burner-stabilized flames at atmospheric pressure were measured experimentally.
2) Flames structure was simulated with two published mechanisms for esters oxidation.
3) The used mechanisms describe quantitatively reagents and main combustion products mole fraction and qualitatively mole fractions of intermediates.
4) Based on numerical results we analyzed primary reaction paths of fuels conversion.
5) One of the decomposition paths of ethyl esters is reaction with ethylene and corresponding fatty acid formation. The contribution of this path rises as equivalence ratio increases.

References

1) P. Osswald, U. Struckmeier, T. Kasper, et.al, J. Phys. Chem.A 111(19) (2007) 4093-4101.
2) L. Gasnot, V. Decottignies, J. F. Pauwels, Fuel 83 (2004) 463-470.
3) B. Yang, C. K. Westbrook, T. A. Cool, et.al, Phys. Chem. Chem. Phys. 13 (15) (2011) 6901-6913.
4) W. K. Metcalfe, S. Dooley, H. J. Curran, et.al, J. Phys. Chem. A 111 (19) (2007) 4001-4014.
5) W. J. Pitz, C. J. Mueller, Prog. Energy Combust. Science 37 (3) (2011) 330-350.
6) C. K. Westbrook, Annu. Rev. Phys. Chem. 64 (2013) 201-219.
7) G. Dayma, F. Halter, F. Foucher, et.al, Energy Fuels, 26 (2012) 4735-4748.
8) C. K. Westbrook, W. J. Pitz, P. R. Westmoreland et.al, Proc. Combust. Inst. 32 (2009) 221-228.
9) J.P. Botha, D. B. Spalding, Proc. Roy. Soc. Lond A 225 (1954) 71.
10) O. P. Korobeinichev, S. B. Ilyin, V. V. Mokrushin, A. G. Shamakov, Combust. Sci. Technol. 116-117 (1-6) (1996) 51-76.
11) T.A. Cool, K. Nakajima, K. A. Taatjes, et.al, Proc. Combust. Inst. 30 (2005) 1681-1688.
12) I. E. Gerasimov, D. A. Knyazkov, S. A. Yakimov, et.al, Combust. Flame 159(5) (2012) 1840-1850.
13) W. E. Kaskan, Proc. Combust. Inst. 6 (1957) 134-141.
14) C. R. Shaddix, Proc. 33rd Nat.l Heat Transf. Conf. HTD99-282 (1999).
15) R. J. Kee, J. F. Grcar, M. D. Smooke, Sandia Nat. lab. Report SAND85-8240 (1985).
16) R. J. Kee, F. M. Rupley, J. A. Miller, Sandia Nat. lab. Report SAND89-8009 (1989).
17) G. Dayma, F. Halter, F. Foucher, et.al, Energy Fuels 26 (2012) 6669-6677.
18) O. P. Korobeinichev, I. E. Gerasimov, D. A. Knyazkov, et.al, Zeit. Phys. Chem. 229(5) (2015) 759-780.

11th Asia-Pacific Conference on Combustion,
The University of Sydney, NSW Australia
10th -14th December 2017.

Formation of Incipient Soot Particles from Polycyclic Aromatic Hydrocarbons: A ReaxFF Molecular Dynamics Study

Qian Mao[1,2], Adri C.T. van Duin[2], K. H. Luo[1,3*]

[1] Center for Combustion Energy, Key Laboratory for Thermal Science and Power Engineering of Ministry of Education, Department of Thermal Engineering, Tsinghua University, Beijing 100084, China

[2]Department of Mechanical and Nuclear Engineering, The Pennsylvania State University, University Park, PA 16802, USA

[3]Department of Mechanical Engineering, University College London, Torrington Place, London WC1E 7JE, UK

Abstract

In this study, we present the results from a series of ReaxFF molecular dynamics (MD) simulations to uncover the underlying mechanisms behind the nucleation and growth of incipient soot particles from polycyclic aromatic hydrocarbons (PAHs). PAHs, namely, naphthalene, anthracene, pyrene, coronene, ovalene and circumcoronene, are selected for ReaxFF MD simulations over a range of temperatures from 400 to 2500 K. Distinctive mechanisms of incipient soot formation are identified with respect to PAH mass and temperature. At low temperatures, all types of the above PAHs can nucleate into incipient soot particles in stacked structures due to physical interactions. With the increase of temperature, the possibility of physical nucleation decreases for each PAH. At moderate temperatures, it becomes difficult for these PAH monomers, except circumcoronene to grow into incipient soot particles. When the temperature increases to 2500 K, all the PAHs become chemically active, which not only leads to the formation of incipient soot particles but also takes the graphitization with the increase of the carbon-to-hydrogen (C/H) ratios in the particles. In addition to the formation of fullerene-like soot particles, stacked particles connected by 'carbon bridges' are also observed for large PAHs like coronene, ovalene and circumcoronene.

1 Introduction

Carbon materials, including zero-dimensional (0D) fullerenes, 1D nanotubes, 2D graphene and 3D graphite, are commercially produced in high-temperature aerosol environment by flame synthesis [1] and attract extensive attention in the materials science and condensed-matter physics [2]. At the same time, soot, composed of impure carbon, is generated from incomplete combustion of hydrocarbons and fossil fuels, which has a detrimental impact on the combustion efficiency [3], atmospheric visibility [4], and human health [5,6]. For both kinds of carbon-based materials, which are typically generated in high-temperature environments, it is accepted that nucleation is the starting point of the whole gas-to-particle conversion process [1,7]. Unfortunately, this nucleation process is currently poorly understood. Based on abundant numerous experimental studies, polycyclic aromatic hydrocarbons (PAHs) are widely accepted to be the precursors for fullerenes, graphite and soot [8–10]. Currently, there are three postulated soot nucleation paths [7], namely physical nucleation of PAHs into stacked clusters, chemical nucleation of PAHs into crosslinked three-dimensional structures and generation of fullerene-like structures. Though the above mechanisms all contribute to soot nucleation in flames, their relative importance or feasibility at different temperatures is still unclear.

From an experimental perspective, determination of the low number density and short-lived PAHs is very difficult via laboratory methods [7], let alone detailed in-situ observation of soot nucleation and further growth. The complexity of the chemical and physical reactions during soot nucleation, makes it both numerically and experimentally challenging. Previous studies of the physical nucleation of soot are performed utilizing atomic simulation methods, such as density functional theory (DFT) calculations [7,11,12], molecular dynamics (MD) simulations [13–19] and Monte Carlo (MC) simulations [20]. Compared to DFT, MD is more capable to cover a wider range of mass distribution and to shed light on issues involving formation of incipient soot particles from PAH monomers. Currently, MD and MC simulations are based on semi-empirical or empirical force fields that only describe physical interactions between PAHs and rigid connection between atoms in PAHs. ReaxFF MD [21] is such a method to describe the dynamic process involving both physical and chemical reactions for large systems at the atomic level with affordable computational cost. In this paper, we utilize the ReaxFF MD to investigate the soot nucleation and growth from PAHs of different masses at varied temperatures. By examining the sizes of soot particles and tracing the structural evolutions, we are able to divulge the underlying soot nucleation mechanisms of temperature and PAH mass dependence.

2 Methodology

2.1 Simulation method

The ReaxFF force field is based on bond order/bond length [21] and parameterized against QM-based training sets containing full reaction pathways of the target species, which is capable of describing chemical reactions entailed in bond formation and dissociation during MD simulations. Up to now, this ReaxFF C/H/O description has been successfully incorporated with Ni, Si and noble gas ion like He+, Ne+, Ar+, and Kr+ ions and applied in the study of carbon nanotube nucleation from hydrocarbon precursors [22], high-kinetic energy collisions of

Corresponding author. Fax: +44-20-7388-0180
E-mail address: K.Luo@ucl.ac.uk

nanoparticles and noble gas ions with graphene [23]. The similarity between graphene and PAH [24] indicates that this force field should be capable of describing thermal and chemical properties of PAHs. The parameter sets for C/H/O description can be found in the supplementary material.

2.2 Simulation procedures

According to the mass spectra of high-molecular weight species in flames, PAH monomers are detected with mass ranging from about 100 to 700 amu [25] with carbon-to-hydrogen (C/H) ratio of 1.4 - 2.5 [26]. Therefore, PAHs constituted of 2-, 3-, 4-, 7-, 10-, and 19- numbered aromatic rings are selected and investigated in this study. First of all, the structure of each PAH monomer is minimized via a conjugate gradient algorithm. Each of the minimized PAH monomers is duplicated 50 times and then randomly distributed in a cubic box of $100 \times 100 \times 100$ Å^3. Subsequently, a system energy minimization is performed. Prior to MD simulations, PAH monomers are required to equilibrate at the target temperatures. Firstly, the interactions between PAH monomers are eliminated and the vibrational energy of every monomer is equilibrated to the specified simulation temperature for 1×10^6 iterations with a time step of 0.25 fs, while the translational and rotational components of the monomer velocity are kept zero. In experiments, the well-distributed translational and rotational velocities come from sufficient collisions with surrounding molecules, which, however, is difficult to achieve within the limited durations in MD simulations. Thus, based on the principle of equipartition of energy, the translational and rotational velocities of PAH monomers are given in Maxwellian distributions of the specified temperature.

Afterwards, dynamic processes of soot nucleation from different types of PAH are performed in the constant temperature canonical ensemble (NVT). Simulation temperatures are chosen at 400, 800, 1200, 1600, 2000 and 2500 K. Each simulation takes 8×10^6 iterations with a time step of 0.25 fs, which accounts for a total simulation time of 2 ns. The Nosé-Hoover thermostat with a damping constant of 100 fs is adopted to control the system temperature. All the simulations are repeated for three times with different initial conformations to evaluate statistical relevance. MD simulations are performed using ReaxFF force field, which is implemented in the Large-scale Atomic/Molecular Massively Parallel Simulator (LAMMPS) package [27]. Snapshots and movies in this study are prepared by the Visual Molecular Dynamics (VMD) software [28].

2.3 Post-processing method

For incipient soot particles comprised of PAHs with chemical bonds, it is relatively easy to identify them from the atomic connection tables. However, for physical bonding, an appropriate criterion to distinguish soot particles from PAH monomers is desirable. Dimerization of PAH monomers leads to stacked, T-shaped, 'shifted graphite' as well as crossed conformations, though stacking the PAH molecules yields the most stable motif [16]. From experimental studies, the inter-layer plane spacing for the stacked PAH structures ranges from 3.5 to 3.9 Å depending on the degree of misalignment of the stacked layer planes [17]. Therefore, in terms of the above possible conformations of PAH dimers, interatomic distance (d) is measured as the nearest two atoms from different monomers shown in Fig. 1 (a) and compared to the interatomic distance criteria (D). That is, if the interatomic distance is less than the criterion, it is believed that the two monomers belong to one cluster. Here, for example, growth of an incipient soot particle from coronene monomers at 400 K is scrutinized by utilizing criteria of 3, 4 and 5 Å and the results are displayed in Fig. 1 (b). The monomer numbers in the soot particle are quite consistent when employing criteria of 4 and 5 Å compared to that of 3 Å. Hence, criterion of 4 Å is used for our later analysis to distinguish PAH monomers from soot particles or clusters.

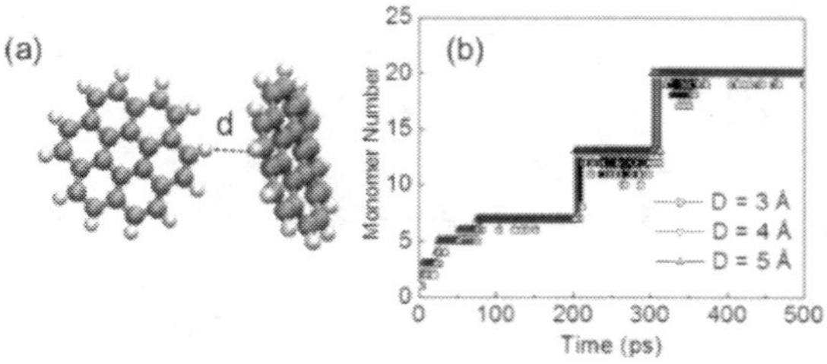

Figure 1. (a) Definition of the interatomic distance (d) between two PAH monomers. (b) Growth of monomer number of coronene in an incipient soot particle through physical nucleation at 400 K by changing the interatomic distance criteria (D) of 3, 4 and 5 Å.

3 Results and discussion

Based on the computational settings described in the previous sections, relationship between the PAH type/mass and temperature as well as evolutions from PAH monomers to incipient soot particles are discussed in this section. Due to the thermal stability of PAHs, decomposition of PAH may only happen at temperatures above 2000 K [29]. Below the decomposition temperature, two distinctive kinds of processes are observed, (1) physical nucleation of PAHs into incipient soot particles at low temperatures is discussed in Section 3.1 and, (2) formation of PAH dimers/trimers at elevated temperature is displayed in Section 3.2. Above the decomposition temperature, formation of incipient soot particles through chemical reactions is analyzed in Section 3.3. Finally, in Section 3.4, we summarize the soot nucleation mechanisms in terms of PAH masses and temperatures and compare them with literature data.

3.1 Physical nucleation

At low temperatures, detailed physical nucleation process of coronene at 800 K is displayed in Fig. 2. Through tracing the conformation evolution, nucleation of incipient soot particles by physical interaction can be well understood. Initially, randomly distributed coronene monomers collide with each other to form PAH dimers. Most of the dimers are in stacks, but a few T-shaped dimers are also observed at time instants b and c (Fig. 2). It is noteworthy that one-dimensionally aligned coronene stacks are composed of a maximum of 4 ~ 5 monomers at time instants d (Fig. 2). This is consistent with the HR-TEM images of soot particles extracted from diesel engine as the average number of coronene monomer in stacks was reported to be 3.85 [30]. Moreover, T-shaped conformations of 3 ~ 4 stacked monomers with one monomer perpendicular to them on the side are also found during this period. Then, at time instant e (Fig. 2), collisions between PAH stacks lead to the formation of incipient soot particles comprised of stacked clusters in three-dimensional shapes. All the above events occur within the initial 150 ps. With the increase of the particle size, the particle number density in the system decreases significantly, which leads to the decrease of collision frequency and consequently the rate of soot growth. Later, collisions between soot particles lead to the formation of much larger soot particles as shown from time instants h to i (Fig. 2). The final soot particle tends to be spherical with locally aligned stacked clusters, which is also described as a turbostratic structure [31].

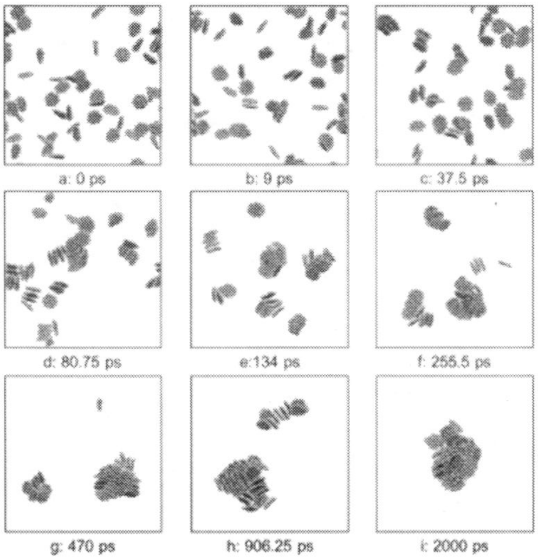

Figure 2. Snapshots of key stages of the physical nucleation of incipient soot particles from coronene monomers at 800 K. Cyan and white spheres represent carbon and hydrogen atoms, respectively.

3.2 No nucleation

When the temperature increases to 1600 K, the dynamic behaviors of the coronene monomers are recorded and displayed in Fig. 3. Initially, collisions between PAHs lead to the formation of some dimers and few trimers. However, the dimers or trimers are quite unstable and may dissociate through collisions with surrounding monomers. Throughout the simulation, none of them survives to grow further into incipient soot particles. Since the translational velocity and collision frequency between PAHs at 1600 K are higher than that at 800 K, the binding energy between PAHs is not strong enough to stabilize the dimer or trimer from evaporation. This consequently limits the growth from PAH dimers or trimers to incipient soot particles. According to the previous studies [7], dimerization is regarded as the first nucleation step. In our simulations, we do observe the formation of PAH dimers and trimers, but they cannot grow further into soot particles at elevated temperatures. Therefore, we propose that formation of relatively larger PAH clusters instead of PAH dimers is essential for the continuous soot nucleation and further growth.

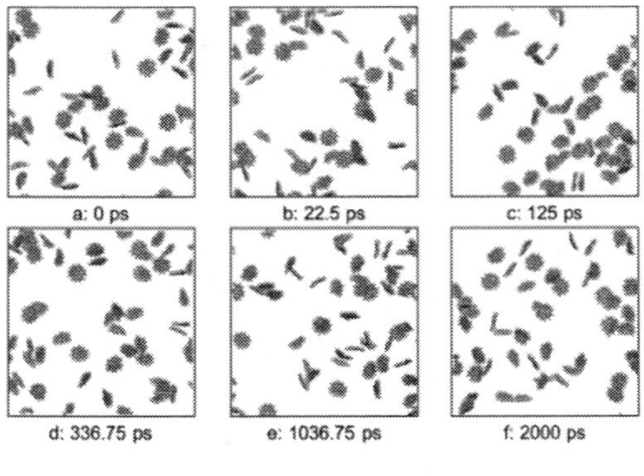

Figure 3. Snapshots of key stages of the dynamics of coronene monomers at 1600 K. Cyan and white spheres represent carbon and hydrogen atoms, respectively.

3.3 Chemical nucleation

When the temperature increases to above 2000 K, PAHs become thermodynamically unstable and decompose to small radicals or molecules. A detailed dynamic process for PAH fragmentation and soot nucleation, starting from circumcoronene monomers at 2500 K, is shown in Fig. 4. As displayed in Fig. 4, formation of the incipient soot particles through chemical nucleation can be roughly divided into three stages. Initially, from approximately 0 to 1000 ps, PAH dehydrogenation takes place to produce aryl radical from time instants a to b (Fig. 4) - the dissociated H is highlighted in yellow. The aryl radical is unstable and further reaction happens from steps c to d (Fig. 4) with the scission of C-C bond as the ring opening, which is similar to the β scission in alkanes. Meanwhile, pentagon/heptagon (5/7) pairs appear at the edges of the cracks as a result of edge rearrangements seen at time instant e (Fig. 4), which can also be found in Fig. 6 as the numbers of 5 C-member and 7 C-member rings increase simultaneously during the same period.

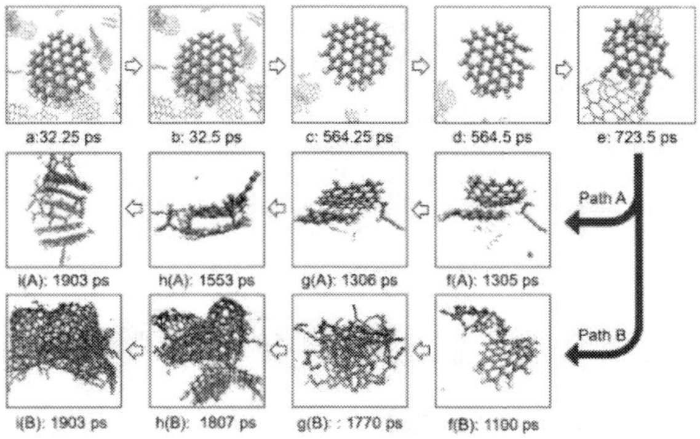

Figure 4. Snapshots of key stages of PAH fragmentation and soot inception from circumcorenene monomers at 2500 K in path A and path B. Cyan and white spheres represent carbon and hydrogen atoms, respectively. Important carbon and hydrogen atoms are highlighted by orange and yellow.

Afterwards, from approximately 1000 to 1700 ps, followed by successive C-C bonds splitting. Linear or branched polyacetylenic-like radicals or small molecules like C2H2 are generated from ring-opening reactions. These radicals help the growth of soot particles in later stages. Our simulation results are in good agreement with theoretical work and experimental results on PAH fragmentation. That is, H- and C2H2-loss are the lowest energy channels and all H atoms in a PAH monomer have similar binding energies with values almost independent of the PAH size [32,33]. Further growth from the carbon-based fragments to soot particles is observed in two separated ways as highlighted by path A and path B in Fig. 5. For path A, the polyacetylenic chain radicals connect to the aryl radicals from time instants f (A) to g (A) (Fig. 4). Subsequently, another PAH connects with the aryl radical by H abstraction. These PAH dimers, with the PAH planes parallel to each other, are connected by polyacetylenic chains - highlighted in orange at time instant h (A) (Fig. 4). Further growth from the PAH dimer to PAH cluster is shown at time instant i (A) (Fig. 4). The stacked structure is more like what have been observed during the physical nucleation at low temperatures but connected by 'carbon bridges' on the edge. The connecting covalent bonds are much stronger than the physical interactions, which prevents evaporation at high temperatures. The other nucleation path, indicated as path B in Fig. 4, is through self-assembly formation of a fullerene-like soot particle. Resulting from the ring-opening and fragmentation process at the initial stage, abundant polyacetylenic linear and branched chains are formed and connect to form long chains with branches as shown in g (B) (Fig. 4). Then the mixed chains connect to some intact PAHs or aryls which act as the nuclei for the reorganization or condensation of the 6-member rings as shown in h (B) (Fig. 4).

4 Conclusions

Temperature and PAH mass dependent homogenous soot nucleation mechanisms are summarized from ReaxFF molecular dynamics simulations. At low temperatures, e.g., 400 K, physical nucleation of incipient soot particles is applicable to all the six types of PAHs through step-wise addition of monomers or clusters and forming incipient soot particles comprised of stacks in different orientations. With the increase of temperature, the feasibility to form incipient soot particles decreases with the decrease of PAH mass. Specifically, naphthalene, anthracene and pyrene, with 2-, 3- and 4-numbered aromatic rings, cannot grow into incipient soot particles at temperatures above 800 K. Instead, predominant formation of dimers, trimers or tetramers is observed. When the temperature rises to 1200 K and above, coronene with 7-numbered aromatic rings can no longer form incipient soot particles. A similar phenomenon happens for ovalene at temperatures above 1600 K. Only circumcoronene, with 19-numbered aromatic rings, is found to form incipient soot particles even at 2000 K. At high temperatures (2500 K), PAHs are thermodynamically unstable and growth of incipient soot particles results from chemical reactions.

5 Acknowledgment

Support from Major Programmes of the National Science Foundation of China (Grant No. 51390493 and Grant No. 91441120), the China Scholarship Council and the Center for Combustion Energy at Tsinghua University is gratefully acknowledged. The simulations were performed on ARCHER funded under the EPSRC project "UK Consortium on Mesoscale Engineering Sciences (UKCOMES)" (Grant No. EP/L00030X/1). ACTvD acknowledges funding from NSF grant# 1462980. The authors are particularly grateful to Prof. Richard A Yetter at Penn State for helpful discussions.

References

[1] S. Li, Y. Ren, P. Biswas, S.D. Tse, Prog. Energy Combust. Sci. 55 (2016) 1–59.

[2] A.K. Geim, K.S. Novoselov, Nat. Mater. 6 (2007) 183–91.

[3] K. Okude, K. Mori, S. Shiino, T. Moriya, SAE Trans. 113 (2004) 1002–1013.

[4] H. Horvath, Part A. Gen. Top. 27 (1993) 293–317.

[5] K.S. Johnson, B. Zuberi, L.T. Molina, M.J. Molina, M.J. Iedema, J.P. Cowin, et al., Atmos. Chem. Phys. 5 (2005) 3033–3043.

[6] M. Matti Maricq, J. Aerosol Sci. 38 (2007) 1079–1118.

[7] H. Wang, Proc. Combust. Inst. 33 (2011) 41–67.

[8] J.B. Howard, Symp. Combust. 24 (1992) 933–946.

[9] P. Gerhardt, S. Löffler, K.H. Homann, Symp. Combust. 22 (1989) 395–401.

[10] M. Frenklach, Phys. Chem. Chem. Phys. 4 (2002) 2028–2037.

[11] H. Nagai, M. Nakano, K. Yoneda, R. Kishi, H. Takahashi, A. Shimizu, et al., Chem. Phys. Lett. 489 (2010) 212–218.

[12] S.H. Chung, A. Violi, Proc. Combust. Inst. 33 (2011) 693–700.

[13] P. Elvati, A. Violi, Proc. Combust. Inst. 34 (2013) 1837–1843.

[14] A. Violi, S. Izvekov, Proc. Combust. Inst. 31 (2007) 529–537.

[15] T.S. Totton, A.J. Misquitta, M. Kraft, Phys. Chem. Chem. Phys. 14 (2012) 4081–4094.

[16] T.S. Totton, D. Chakrabarti, A.J. Misquitta, M. Sander, D.J. Wales, M. Kraft, Combust. Flame. 157 (2010) 909–914.

[17] A. Raj, M. Sander, V. Janardhanan, M. Kraft, Combust. Flame. 157 (2010) 523–534.

[18] D. Chen, T.S. Totton, J.W.J. Akroyd, S. Mosbach, M. Kraft, Carbon N. Y. 67 (2014) 79–91.

[19] M. Rapacioli, F. Calvo, F. Spiegelman, C. Joblin, D.J. Wales, J. Phys. Chem. A. 109 (2005) 2487–2497.

[20] M. Sirignano, A. Collina, M. Commodo, P. Minutolo, A. D'Anna, Combust. Flame. 159 (2012) 1663–1669.

[21] A.C.T. van Duin, S. Dasgupta, F. Lorant, W.A. Goddard III, J. Phys. Chem. A. 105 (2001) 9396–9409.

[22] U. Khalilov, A. Bogaerts, E.C. Neyts, Nat. Commun. 6 (2015) 10306.

[23] K. Yoon, A. Ostadhossein, A.C.T. Van Duin, Carbon N. Y. 99 (2016) 58–64.

[24] K. Yoon, A. Rahnamoun, J.L. Swett, V. Iberi, D.A. Cullen, I. V. Vlassiouk, et al., ACS Nano. 10 (2016) 8376–8384.

[25] J. Happold, H.-H.H. Grotheer, M. Aigner, Combust. Gener. Fine Carbonaceous Part. Sci. Publ. (2009) 277–288.

[26] H.A. Michelsen, Proc. Combust. Inst. 0 (2016) 1–19.

[27] S. Plimpton, J. Comput. Phys. 117 (1995) 1–19.

[28] W. Humphrey, A. Dalke, K. Schulten, J. Mol. Graph. 14 (1996) 33–38.

[29] S.E. Stein, A. Fahr, Am. Chem. Soc. 89 (1985) 3114–3125.

[30] S. Mosbach, M.S. Celnik, A. Raj, M. Kraft, H.R. Zhang, S. Kubo, et al., Combust. Flame. 156 (2009) 1156–1165.

[31] R.D. Heidenreich, W.M. Hess, L.L. Ban, J. Appl. Crystallogr. 1 (1968) 1–19.

[32] R.H. Chen, S.A. Kafafi, S.E. Stein, J. Am. Chem. Soc. 111 (1989) 1418–1423.

[33] C. Lifshitz, Int. Rev. Phys. Chem. 16 (1997) 113–139.

Reply to Reviewers' Comments on Manuscript ID: 460

We sincerely appreciate the comments and suggestions by the two reviewers, which have been used to improve the quality and clarity of the revised manuscript. Detailed replies to the two reviewers' comments are as follows:

Reviewer 1:

Comments: *The overall structure is clear, which mainly talked about different mechanisms for PAHs under different temperature. The description of the methodology is also understandable even to people who are not in this specific area of research. One suggestion is about Figure 1, which is a bit unclear and unreadable due to the three colors' presenting. Also, I do not think the format for reference is right, but it is not as important as the main content.*

Author reply: We have corrected the size and the color of Figure 1, which is more clear and readable. The format of the reference is also corrected according to the format in the Template.

Reviewer 2:

Comment 1: *Abstract: correct:"...except circumcoronene grows into incipient soot particles." -> "...except circumcoronene to grow into incipient soot particles."*

Author reply: It has been corrected.

Comment 2: *Section 2.1 "The parameter sets for C/H/O description can be found in the supplementary material." There is no supplementary material for the paper. Either give here a reference to a website or other paper or remove.*

Author reply: Sorry for the mistake. The supplementary material is now attached.

Comment 3: *The simulations are conducted for a total of 2ns, a potential issue is that there is questions about if for incipient soot formation if dimerisation is stable enough at flame temperatures for these to survive long enough to make any meaningful contributing before breaking part or stronger connecting bonds are formed.*

Author reply: We found the physical PAH dimerization happens at low temperatures. In the combustor or engines, the temperature distribution ranges from about 400 to 2200 K. Moreover, the strong chemical bonds always form at much higher temperatures than the physical dimerization. If a PAH dimerization forms at a low temperature region and then passes to a high temperature region, the PAH dimer may break part. The configuration of nascent soot particles is reported to be core-shell structures with stacked PAHs in the center and aliphatic chains on the shell region[1].

Comment 4: *Is the present code capable of including (or currently include) growth mechanisms from smaller aromatic-aliphatic molecules as has been hinted at as an important mechanism from recent photoionisation studies?*

Author reply: We appreciate this advice. Additional MD simulations of PAHs with substituted PAH with aliphatic chains (4-vinyl-3a^1,5a^1-dihydropyrene [A4-C_2H_2]) is added as a comparison to A4. Their conformations and particle growth processes are shown in Fig. 1 and 2, respectively. Compared to A4, A4-C_2H_2

can physically nucleate into soot particles at temperature as high as 800 K.The size of soot particle is similar for A4 and A4-C_2H_2 monomers at 800 K. However, both PAHs initiate chemical nucleation at high temperatures (above 2000 K). Through this comparsion, we can conclude that the aliphatic chain as attached to the pericondensed PAH may help the physical nucleation at a relatively higher temperature compared to the corresponding pericondensed PAH but has little effect on the soot nucleation at typical flame temperatures (1600 K). Due to the fact that the masses and sizes of A4-C_2H_2 are larger than that of A4, it is consistent with our conclusion in the manuscript as '*Large PAH monomers have a higher possibility of growing into incipient soot particles than small PAHs due to stronger physical interactions.*'

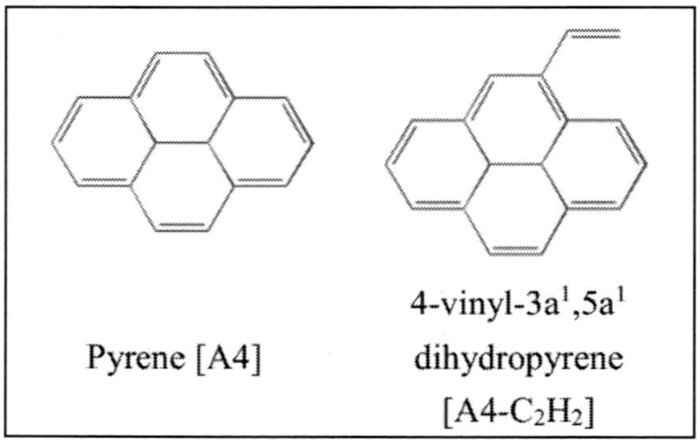

Figure 1. Structure of pericondensed PAH (A4), substituted PAH with aliphatic chains (A4-C_2H_2).

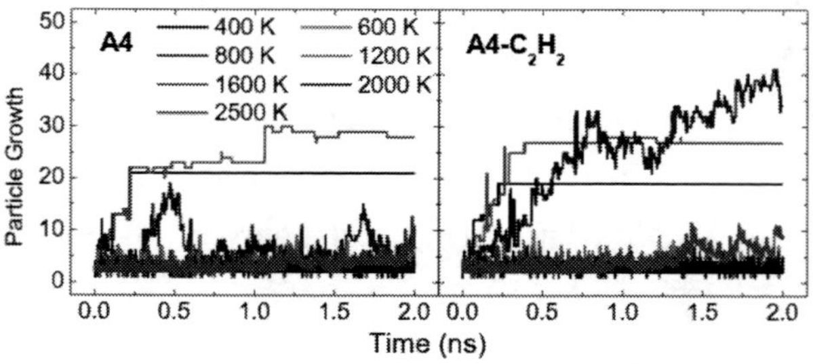

Figure 2. Growth of the maximum soot particles from pericondensed PAH (A4), substituted PAH with aliphatic chains (A4-C_2H_2) at temperatures ranging from 400 to 2500 K.

Comment 5: *Some of the text and labelling in the figures is very small and is need to be increased in size to be readable.*

Author reply: We have increased the text, labeling and size of the figures.

Comment 6: *Revise first line of conclusion: "Temperature and PAH mass dependent homogenous soot nucleation mechanisms summarize."*

Author reply: This sentence has been revised to 'Temperature and PAH mass dependent homogenous soot nucleation mechanisms are summarized from ReaxFF molecular dynamics simulations.'

Reference

[1] H. Wang, Formation of nascent soot and other condensed-phase materials in flames, Proc. Combust. Inst. 33 (2011) 41–67. doi:10.1016/j.proci.2010.09.009.

SUPPLEMENTARY MATERIAL

Reactive MD-force field: C/H/O

```
 39       ! Number of general parameters
   50.0000 !Overcoordination parameter
    9.5469 !Overcoordination parameter
   26.5405 !Valency angle conjugation
parameter
    1.7224 !Triple bond stabilisation
parameter
    6.8702 !Triple bond stabilisation
parameter
   70.0000 !C2-correction
    1.0588 !Undercoordination parameter
    4.6000 !Triple bond stabilisation
parameter
   12.1176 !Undercoordination parameter
   13.3056 !Undercoordination parameter
  -70.5044 !Triple bond stabilization energy
    0.0000 !Lower Taper-radius
   10.0000 !Upper Taper-radius
    2.8793 !Not used
   33.8667 !Valency undercoordination
    6.0891 !Valency angle/lone pair parameter
    1.0563 !Valency angle
    2.0384 !Valency angle parameter
    6.1431 !Not used
    6.9290 !Double bond/angle parameter
    0.3989 !Double bond/angle parameter:
overcoord
    3.9954 !Double bond/angle parameter:
overcoord
   -2.4837 !Not used
    5.7796 !Torsion/BO parameter
   10.0000 !Torsion overcoordination
    1.9487 !Torsion overcoordination
   -1.2327 !Conjugation 0 (not used)
    2.1645 !Conjugation
    1.5591 !vdWaals shielding
    0.1000 !Cutoff for bond order (*100)
    2.1365 !Valency angle conjugation
parameter
    0.6991 !Overcoordination parameter
   50.0000 !Overcoordination parameter
    1.8512 !Valency/lone pair parameter
    0.5000 !Not used
   20.0000 !Not used
    5.0000 !Molecular energy (not used)
    0.0000 !Molecular energy (not used)
    2.6962 !Valency angle conjugation
parameter
  3    ! Nr of atoms; cov.r;
valency;a.m;Rvdw;Evdw;gammaEEM;cov.r2;#

alfa;gammavdW;valency;Eunder;Eover;chiEEM;etaE
EM;n.u.
            cov r3;Elp;Heat
inc.;n.u.;n.u.;n.u.;n.u.
            ov/un;val1;n.u.;val3,vval4
 C    1.3674   4.0000  12.0000   2.0453
0.1444   0.8485   1.1706   4.0000
      9.0000   1.5000   4.0000  30.0000
79.5548   4.8446   7.0000   0.0000
      1.1168   0.0000 181.0000  14.2732
24.4406   6.7313   0.8563   0.0000
     -4.1021   5.0000   1.0564   4.0000
2.9663   0.0000   0.0000   0.0000
 H    0.8930   1.0000   1.0080   1.3550
0.0930   0.8203  -0.1000   1.0000
      8.2230  33.2894   1.0000   0.0000
121.1250   3.7248   9.6093   1.0000
     -0.1000   0.0000  61.6606   3.0408
2.4197   0.0003   1.0698   0.0000
    -19.4571   4.2733   1.0338   1.0000
2.8793   0.0000   0.0000   0.0000
 O    1.2450   2.0000  15.9990   2.3890
0.1000   1.0898   1.0548   6.0000
      9.7300  13.8449   4.0000  37.5000
```

116.0768 8.5000 8.3122 2.0000
0.9049 0.4056 59.0626 3.5027
0.7640 0.0021 0.9745 0.0000
-3.5500 2.9000 1.0493 4.0000
2.9225 0.0000 0.0000 0.0000
6 ! Nr of bonds;
Edis1;LPpen;n.u.;pbe1;pbo5;13corr;pbo6

pbe2;pbo3;pbo4;n.u.;pbo1;pbo2;ovcorr
1 1 80.8865 107.9944 52.0636 0.5218 -
0.3636 1.0000 34.9876 0.7769
6.1244 -0.1693 8.0804 1.0000 -
0.0586 8.1850 1.0000 0.0000
1 2 180.6309 0.0000 0.0000 -0.4794
0.0000 1.0000 6.0000 0.6281
12.2202 1.0000 0.0000 1.0000 -
0.0670 6.8158 0.0000 0.0000
2 2 153.3934 0.0000 0.0000 -0.4600
0.0000 1.0000 6.0000 0.7300
6.2500 1.0000 0.0000 1.0000 -
0.0790 6.0552 0.0000 0.0000
1 3 163.3110 83.9973 54.4316 -0.5220 -
0.3123 1.0000 10.2503 1.0000
0.3553 -0.3757 7.0000 1.0000 -
0.1331 4.6021 0.0000 0.0000
2 3 160.0000 0.0000 0.0000 -0.5725
0.0000 1.0000 6.0000 0.5626
1.1150 1.0000 0.0000 0.0000 -
0.0920 4.2790 0.0000 0.0000
3 3 142.2858 145.0000 50.8293 0.2506 -
0.1000 1.0000 29.7503 0.6051
0.3451 -0.1055 9.0000 1.0000 -
0.1225 5.5000 1.0000 0.0000

3 ! Nr of off-diagonal terms;
Ediss;Ro;gamma;rsigma;rpi;rpi2
1 2 0.1200 1.3861 9.8561 1.1254 -
1.0000 -1.0000
1 3 0.1347 1.8343 9.7934 1.3139
1.1498 1.1039
2 3 0.0283 1.2885 10.9190 0.9215 -
1.0000 -1.0000
18 ! Nr of
angles;at1;at2;at3;Thetao,o;ka;kb;pv1;pv2
1 1 1 74.9085 44.7514 0.9144 0.0000
0.0050 0.3556 2.5715
1 1 2 68.0294 13.4722 5.5819 0.0000
0.6849 0.0000 1.0031
2 1 2 68.4575 22.1235 1.2937 0.0000
3.0000 0.0000 1.5009
1 1 3 15.7798 9.0805 4.0304 0.0000
1.8785 70.0000 1.1737
2 1 3 65.0000 13.4505 1.8249 0.0000
1.5646 0.0000 1.2173
3 1 3 74.7266 45.0000 1.8020 -16.7178
2.6091 0.1000 2.3556
1 2 2 0.0000 0.0000 6.0000 0.0000
0.0000 0.0000 1.0400
1 2 1 0.0000 7.5000 5.0000 0.0000
0.0000 0.0000 1.0400
1 2 3 0.0000 45.0000 3.0000 0.0000
1.0000 0.0000 1.0400
2 2 2 0.0000 27.9213 5.8635 0.0000
0.0000 0.0000 1.0400
2 2 3 0.0000 8.5744 3.0000 0.0000
0.0000 0.0000 1.0421
3 2 3 0.0000 15.0000 2.8900 0.0000
0.0000 0.0000 2.8774
1 3 1 76.7840 44.2266 0.9343 0.0000
1.3483 0.0000 1.8301
1 3 3 63.9120 17.1680 0.8751 0.0000
0.0693 50.9415 3.0000
1 3 2 79.6413 28.6488 0.3789 0.0000
1.6776 0.0000 1.0010
2 3 2 85.8000 9.8453 2.2720 0.0000
2.8635 0.0000 1.5800
2 3 3 79.5453 45.0000 2.1630 0.0000
3.0000 0.0000 1.2391

3 3 3 80.7324 30.4554 0.9953 0.0000
1.6310 50.0000 1.0783
21 ! Nr of
torsions;at1;at2;at3;at4;;V1;V2;V3;V2(BO);vcon
j;n.u;n
1 1 1 1 2.1207 26.8713 0.5160 -
9.0000 -2.8394 0.0000 0.0000
1 1 1 2 1.4658 44.1251 0.4411 -
5.3120 -2.1894 0.0000 0.0000
2 1 1 2 1.4787 40.5128 0.4396 -
5.2756 -3.0000 0.0000 0.0000
1 1 1 3 0.9963 17.2365 0.1491 -
2.5000 -1.0000 0.0000 0.0000
2 1 1 3 1.5159 28.6602 0.7169 -
7.5489 -3.0000 0.0000 0.0000
3 1 1 3 -0.2000 23.6540 -1.0000 -
5.9155 -1.1552 0.0000 0.0000
1 1 3 1 1.8231 46.5696 -1.0000 -
3.0536 -3.0000 0.0000 0.0000
1 1 3 2 1.4836 80.0000 0.0363 -
4.7349 -1.0000 0.0000 0.0000
2 1 3 1 0.5983 49.5033 0.7210 -
3.4046 -1.6880 0.0000 0.0000
2 1 3 2 -0.2000 76.3511 1.0000 -
4.0709 -1.0000 0.0000 0.0000
1 1 3 3 -0.2000 5.0000 -1.0000 -
3.6523 -2.9000 0.0000 0.0000
2 1 3 3 2.5000 80.0000 1.0000 -
2.6071 -3.0000 0.0000 0.0000
3 1 3 1 -0.2000 80.0000 -1.0000 -
3.6863 -3.0000 0.0000 0.0000
3 1 3 2 2.5000 38.8954 -0.8368 -
4.6681 -2.9000 0.0000 0.0000
3 1 3 3 -0.2000 78.1766 0.0250 -
2.8895 -3.0000 0.0000 0.0000
1 3 3 1 2.5000 0.1000 1.0000 -
2.6905 -2.7573 0.0000 0.0000
1 3 3 2 0.5241 69.2788 -1.0000 -
4.4539 -2.8081 0.0000 0.0000
2 3 3 2 2.5000 0.1000 -0.4869 -
2.8372 -1.0000 0.0000 0.0000
1 3 3 3 2.5000 0.1000 1.0000 -
3.4298 -1.0000 0.0000 0.0000
2 3 3 3 -0.2000 0.1000 -1.0000 -
3.5698 -1.0000 0.0000 0.0000
3 3 3 3 -0.2000 0.1000 1.0000 -
3.8409 -1.0000 0.0000 0.0000
1 ! Nr of hydrogen
bonds;at1;at2;at3;Rhb;Dehb;vhb1
3 2 3 2.1200 -3.5800 1.4500 19.5000

11th Asia-Pacific Conference on Combustion,
The University of Sydney, NSW Australia
10th -14th December 2017.

Investigation of the performance of After-Treatment Devices on diesel and biodiesel particles

Yi Guo[1], Svetlana Stevanovic[1], Mohammad Jafari[1], Puneet Verma[1,2], Richard Brown[2],

Chiemeriwo Godday Osuagwu[1], Barbara D'Anna[3], Zoran Ristovski[1,2]

[1]International Laboratory of Air Quality and Health (ILAQH) –

[2]Biofuel Engine Research Facility (BERF) –

Queensland University of Technology, 2 George St, Brisbane City, Queensland, 4000, Australia

[3] CNRS, IRCELYON, Université de Lyon, Lyon, 69626, France

Abstract

The performance of after-treatment devices (ATDs) for diesel and biodiesel particulate matter were investigated by measurement and comparison of nanoparticle concentration before and after the ATD, which was mounted on a diesel engine. The engine was run with both diesel and coconut biodiesel. The loading of the diesel particulate filter (DPF) can be divided in two stages, (i) initial loading, beginning with a clean filter, and (ii) final loading, when the DPF has been loaded for 3 hours, and where the filtration efficiency was stable. The results show that, during the initial loading, the biodiesel particle number concentration (PN) was higher than the PN from neat diesel, while the particle mass concentration (PM) was lower than that of diesel. However, in the final loading process, the PN and PM of biodiesel particles were both lower than for neat diesel. The filtration efficiency of ATDs is different for diesel and biodiesel particles. The performance of the ATD was better with neat diesel fuel as compared to biodiesel. This would suggest that a more detailed study is needed to understand the interaction of ATDs with particles being produced by combustion of new and emerging biodiesel fuels.

1 Introduction

Vehicle emissions have been recognized as one of the most significant contributors to atmospheric pollution in urban areas [1].

After-Treatment Devices (ATD) are applied to mitigate the adverse health effects of diesel emissions on ambient air quality. It is believed that treated exhaust, combined with the usage of biodiesels, can reduce vehicle emissions even further. In particulate reduction, Diesel Oxidation Catalysts (DOCs) and Diesel Particulate Filters (DPFs) are the most common ATDs [2-5]. DPFs are devices that capture particles before they are released into the air. The DPF is most commonly made of ceramic wall-flow monoliths. The wall-flow filter is designed in such a way as to enable the exhaust flow to penetrate the walls to the adjacent channels which allow the passage of flow. This filter type has regular porosities which are plugged at each end. A presentation of the flow-through is shown in Figure 1.

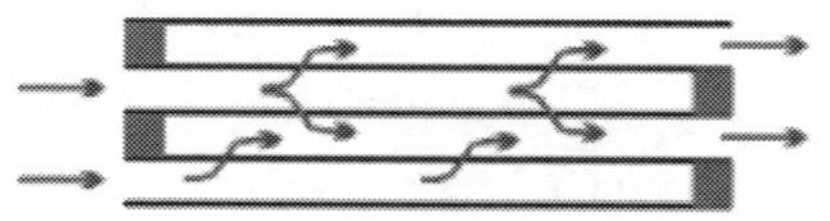

Figure 1: Flow Representation of the Wall-Flow Substrates [4]

From the figure above, it can be observed that the filter separates the airborne particles from the exhaust stream by retaining them on the collection surface. Separation is obtained by a porous barrier which retains the particulates.

Biodiesels have been recognized as one of the most promising alternative fuels, and are expected to be increasingly used to replace diesel in the near future [6], owing to their lower particulate emissions and sustainability.

Therefore, as two of the most effective alternatives in decreasing vehicular emissions, the interaction between ATDs and biodiesel exhaust particles needs to be investigated. It is vital to evaluate the performance of ATDs with particles originating from the combustion of biodiesel, and the change of physiochemical properties of the particles passing through the ATD. Here, we have investigated the particle number concentration (PN) and particle mass concentration (PM) of the particles after the ATD when using diesel, and one neat biodiesel from a coconut feedstock.

The primary aim of this research is to measure the properties of biodiesel particles and their impact on the performance of ATDs and the magnitude of this effect. These results will improve the understanding of particle emissions from diesel engines. In addition, it is expected that this study will help in the understanding of the performance of ATDs when used with biodiesels.

2 Methodologies

2.1 Experimental setup

The performance of the ATD with diesel and biodiesel particles was investigated using the setup as shown in Figure 2.

Corresponding author. Fax: +61 4 5259 7868
E-mail address: y8.guo@hdr.qut.edu.au

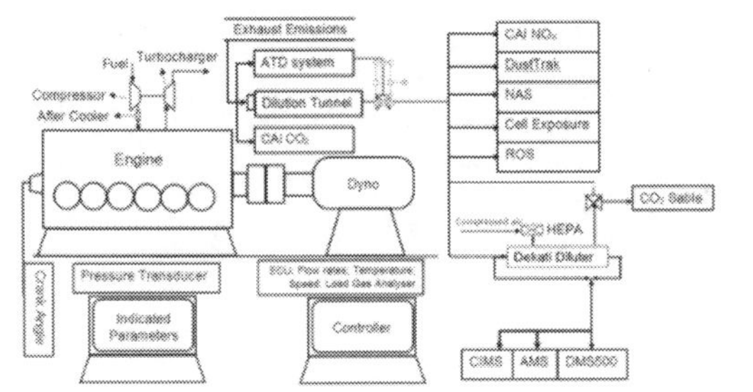

Figure 2: Schematic diagram of After-treatment Devices (ATD), showing experimental loading set up

The ATD was installed on the engine exhaust pipe. PM properties, including mass and number concentrations, as well as size distribution were monitored before and after the ATD system. Filtration efficiency was then calculated. During the DPF loading process, testing instruments were monitoring the emissions after the ATD. Once the emissions became stable, primary emissions were measured before the ATD by placing the equipment after the dilution system and before the ATD. For each fuel, the same test procedure and loading time were applied, to ensure comparable results for the different fuels.

2.2 Facilities and instruments

This experimental study was performed on a Euro III Cummins diesel engine coupled to a water brake dynamometer, which are both connected to an electronic control unit (ECU). Table 1 shows the specifications of the test engine. The engine was operated at 1500 rpm with 75% load and 100% load for initial loading and final loading respectively.

Model	Cummins ISBe220 31
Cylinders	6 in-line
Capacity (L)	5.9
Bore × Stroke (mm)	102 × 120
Maximum power (kW/rpm)	162/2500
Maximum torque (Nm/rpm)	820/1500
Compression ratio	17.3
Aspiration	Turbo-charged & after-cooled
Fuel Injection	Common rail
After-treatment devices	None
Emissions certification	Euro III

Table 1: Test engine specifications

The DOC and DPF media are housed in a filter system holder, which is made from stainless steel. The system has three connecting modules which are clamped together. The design of the system is illustrated in Figure 3.

Figure 3: Diesel Particulate Filter assembly (Left: inflow with the DOC, Mid-section: wall-flow substrates and DPF, Right: outflow with cap)

In the experiment, the filter system was connected to the engine exhaust pipe with a stainless steel tube. The outlet of the filter system was connected to the measurement system, which contains a number of other instruments for particle and gaseous characterisation. The flow rate through the ATD was kept constant at 17 lpm for every test to ensure the results are comparable. To monitor the performance of the ATD during the loading test, the pressure drop over the ATD system was continuously measured. The monotonous increase in pressure drop is a typical sign of the accumulation of soot inside the DPF.

As shown in Figure 2, the PN and PM were measured by a Fast Particulate Spectrometer DMS 500. The dilution tunnel was used to dilute raw exhaust with HEPA-filtered laboratory air. A second diluter (Dekati) was connected in series with the first stage dilution to further increase the dilution ratio and decrease the particulate concentrations. The purpose of the dilution was to bring down the temperature as well as the concentration of gases and PM within the measuring range of the instruments. Diluted exhaust was then sent to different gaseous and particle measuring instruments. A CAI 600 series CO_2 analyser was used to measure the CO_2 concentration directly from the raw exhaust. A second CO_2 meter (SABLE, CA-10) connected via a three-way valve between the two diluters was used to record the CO_2 concentration from the diluted exhaust. In addition, the flow rate of the exhaust inlet of the Dekati diluter was measured for the calibration of the dilution ratio results.

2.3 Fuel selection

Commercially available diesel was used as the baseline fuel, along with the coconut biodiesel.

Recently, there has been a significant growth in biodiesel usage [7] and it is predicted that its use will be increased in the near future[6]. The feedstock to produce the biodiesel is strongly dependent on national and local issues including: government policy settings, agriculture, prevailing crude oil prices and the age of the diesel engine fleet. Of the range of biodiesels in common use, coconut is the dominant feedstock in the Pacific Islands. In Europe, around 11% of biodiesel is sourced from coconut feedstock. Coconut biodiesel shares some common fatty acids with other biodiesels, however its distinctive is that it contains 50% methyl laurate and is around 90% saturated.

3 Results and Discussion

The initial loading and final loading results are analysed separately, and are shown in Figures 4 and 5, respectively. In both figures, Figure (a) is the mass and number concentration result, while Figure (b) is the filtration efficiency result. The filtration efficiency (E_f) calculation is defined as the following formula:

$$E_f = (C_{in} - C_{out})/C_{in} \quad (1)$$

C_{in} refers to the particle mass or number concentrations in the inlet of the ATD (see Table 2.), while the C_{out} refers to the outlet of the ATD.

Fuel	Engine operation	PN (#/cm^3)	PM (mg/m^3)
Diesel	75% load, 1500rpm	7.21E+07	111.49
B100*	75% load, 1500rpm	1.52E+07	6.49
	100% load, 1500rpm	1.05E+07	4.38

Table 2. Particle concentrations, mass and number, before the filter system. *B100 denotes 100% biodiesel.

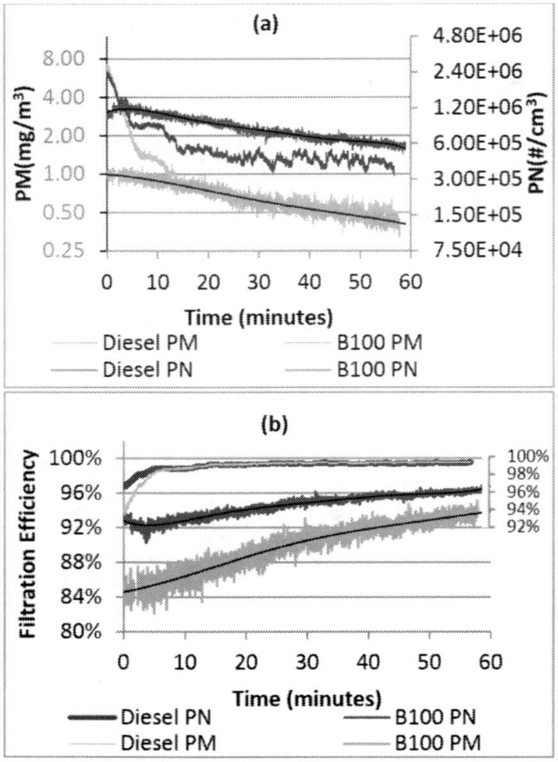

Figure 4. Initial loading results measured after the ATD at 75% load with 1500rpm. (a) mass concentration (PM) and number concentration (PN) after the filter; (b) filtration efficiency.

The particle concentrations after the filter, both mass and number, rapidly drop as the filter is loaded with neat diesel (Figure 4.(a)). This is a result of an increase in the filter efficiency (Figure 4.(b)) as the particle cake is rapidly formed on the filter. This brings filter efficiencies to the maximum for both number and mass in ~10 minutes.

With biodiesel, the initial change in the concentration, mass and number, after the filter is not as rapid (Figure 4.(a)), although a clear reduction is observed. The particle concentration before the filter with pure biodiesel is almost 8 times less than for diesel. This means that the DPF is initially loaded at a much lower rate than with pure diesel, and therefore the particle cake on the DPF is formed much more slowly. This can also be seen in the much slower increase of the filtration efficiencies for B100.

In the final stage of loading, the particle concentrations after the DPF reach a steady value (Figure 5 (a)) both for neat diesel and B100. It is worth noting that the particle concentrations, mass and number, for neat diesel are still higher than for B100. The filter efficiency for B100 is also lower in the final stage than for neat diesel. Also of note is the efficiency expressed in terms of particle mass for B100, being lower than the efficiency expressed in terms of particle number. This was not the case for neat diesel.

If the different filter efficiencies for B100 and neat diesel were only due to the differences in the rate the filters were loaded, and therefore the rate the filter cake is formed, one would expect that in the final loading for both fuels the same value would be reached. As the final filter efficiencies were different for the two fuels, other PM properties, such as particle size, soot structure or other issues could have had some influence.

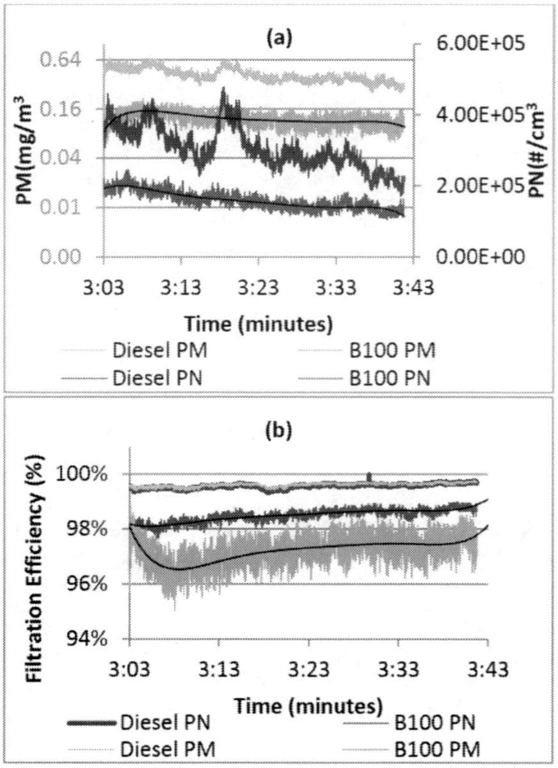

Figure 5. Final loading results measured after the ATD at 100% load with 1500rpm. (a) mass concentration (PM) and number concentration (PN) after the filter; (b) filtration efficiency.

To investigate why the filtration efficiency from different fuels varies, we measured the particle size distributions before and after the filters for all of the fuels analysed. The particle size distribution of the diesel before and after the ATD is shown in the figure 6a, and particle size distribution of biodiesel before and after the ATD on figure 6b.

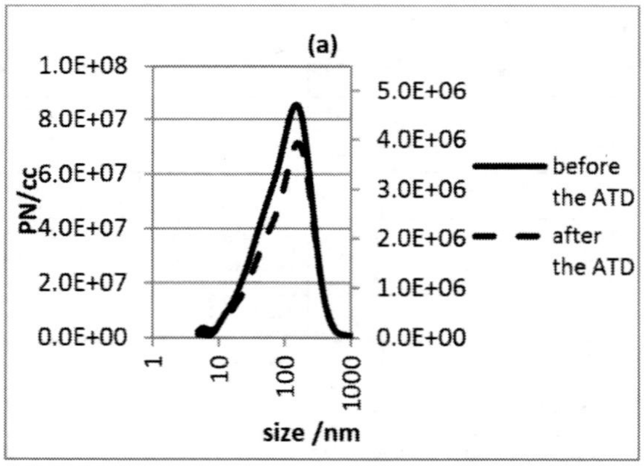

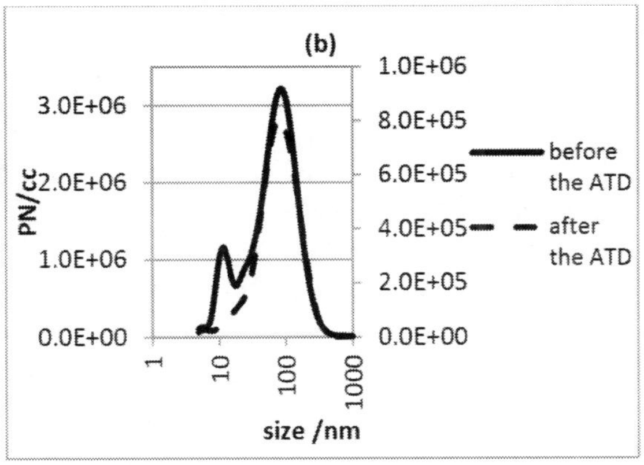

Figure 6. The particle size distribution of different fuels (The curve before the ATD applies the primary axis, while the curve after the ATD applies the secondary axis) From the graph 6a, we can see that the particle size distribution varies with different fuels combusted. Although both fuels gave a mono-mode distribution, the modes of the size distributions were different. Diesel particles peaks at 178nm, while the B100 peaks at 64.9nm.

According to the filtration mechanism, there are four basic main physical deposition mechanism. They are interception, inertial impaction, diffusion and gravitational settling[8]. In terms of removal, except for diffusion, all others are more effective for bigger particles than smaller particles. Comparing the size distributions before and after the filter it is clear that smaller particles are removed in the initial filtration process during the cake formation. It seems that diffusion processes play a bigger role in the initial cake formation process and all other mechanisms start playing an important role in the particulate filtration after the cake formation.

Other factors, like the particle density might impact the filtration process as well [9], but to which extent it contributes to the filtration requires further investigation.

4 Conclusions

DPF performance during the loading and final stage was measured for two different fuels, neat diesel and pure biodiesel (B100). Due to a significantly lower PM concentration being emitted from the diesel engine with B100, the loading and cake formation was significantly slower than for neat diesel. It is interesting that the final filter efficiency was also lower for B100. The observed lower filter efficiency for B100 could be a consequence of the biodiesel particles having not only lower concentrations, but also different properties such as particle shape and structure. This indicates that a more detailed study is needed to understand the interaction of DPFs with particles being produced by combustion of the new and emerging biodiesel fuels.

5 Acknowledgments

The author gratefully acknowledges the financial support from the Chinese Government for his PhD course at QUT. The author would also like to acknowledge: laboratory assistance from Mr. Noel Hartnett; QUT staff and colleagues for their support and guidance. Biodiesel supplied by Dr Doug Stuart of Suncoast renewables is gratefully acknowledged.

References

1. Liati, A., et al., Variations in diesel soot reactivity along the exhaust after-treatment system, based on the morphology and nanostructure of primary soot particles. Combustion and Flame, 2013. 160(3): p. 671-681.

2. Johnson, T.V., Diesel emissions in review. SAE International Journal of Engines, 2011. 4(2011-01-0304): p. 143-157.

3. Johnson, T.V. Diesel emission control in review. in 2008 SAE World Congress, SAE Technical Paper. 2008.

4. Majewski, W.A., Diesel particulate filters. 2011, Ecopoint Inc. p. 24.

5. Transport, E.C.D.-G.f.M., White Paper on Transport: Roadmap to a Single European Transport Area: Towards a Competitive and Resource-efficient Transport System. 2011: Publications Office of the European Union.

6. Directive 2009/139/EC of the European Parliament and of the Council. 2009, The European Parliament and the Council of the European Union: Strasbourg. p. 9.

7. Setting National Fuel Quality Standard, F.a.U.O.P. Section and D.o.t. Environment, Editors. 2008, Department of Environment and Energy of the Australian Government.

8. Hinds, W.C., Aerosol technology: properties, behavior, and measurement of airborne particles. 2nd ed. 1999, New York: Wiley.

9. Wakeman, R., The influence of particle properties on filtration. Separation and Purification Technology, 2007. 58(2): p. 234-241.

11th Asia-Pacific Conference on Combustion,
The University of Sydney, NSW Australia
10th -14th December 2017.

Effect of H_2 and CO Dilution on NO Formation in Coflow CH_4/air Diffusion Flames

Ye Y [1], Jianfei X [1], Zhongzhu G [1], Haiyang A [1], Xianpeng Z [1]

[1]School of Energy and Mechanical Engineering, Nanjing Normal University

78 Bancang Street, Nanjing, Jiangsu Province, 210042, China

Abstract

The effect of H_2 and CO addition on NO formation in coflow CH_4/air diffusion flames was investigated by numerical simulation with experimental verification. The volume fraction of dilution changed from 0.0 to 0.5. The simulation results agree well with experimental data. The results show that NO formation increases with H_2 and CO addition in coflow CH_4/air diffusion flames. Specifically, thermal NO formation increases with H_2 addition, and both thermal NO and prompt NO increase with CO addition.

1 Introduction

Energy and environment problems are global hot issues for now, and the utilization of clean energy has become an important research direction. Natural gas is a high-quality fuel and important industrial chemicals. Natural gas is mainly consist of methane, and contains a small amount of ethane, propane, butane and nitrogen. Compared with traditional fossil fuel, natural gas is a kind of clean gas fuel with abundant reserves.

IGCC[1] technology is a promising clean combustion technology. The fuel of IGCC system is produced through gasification of solid fuels, such as coal and biomass. H_2 and CO are the main products of the gasification progress. Although IGCC technology can reduce the emission of soot and SO_x, NO_x emission cannot be reduced effectively. As a gas pollutant, NO_x (mainly NO) has a great harm to human body and environment. Therefore research on the effect of H_2 and CO dilution on NO formation in coflow CH_4/air diffusion flames has practical significance for developing clean combustion technology.

There have been abundant studies concerning methane combustion. These studies are about reducing pollutants generated in methane flame by adding dilution [2,3](such as CO_2, N_2 and H_2O) and the effect of changing combustion mode on pollutants formation and combustion characteristics[4,5]. Besides, some scholars have studied reaction mechanism and ignition characteristics of methane combustion by numerical method [6]. This paper presents a detailed numerical study on the effect of H_2 and CO addition on NO formation in laminar CH_4/air diffusion flames with experimental verification. It help to reveal the variation trend and formation mechanism in diluted flames and provide theoretical basis of the pollution control for combustor design.

Corresponding author. Fax: +86-025-85481124
E-mail address: jianfeixi@njnu.edu.cn

2 Experimental Methodology

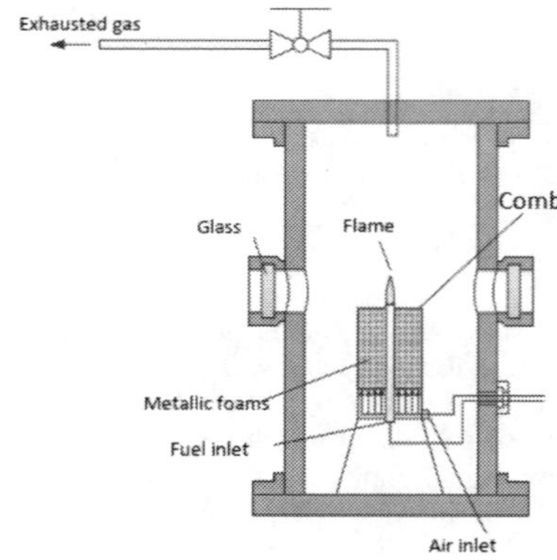

Figure 1: Schematic diagram of the combustion system.

The experiment was conducted in a coflow axisymmetric laminar diffusion flame burner, as shown in Fig. 1. The fuel stream issued from a 10.9-mm-inner-diameter vertical tube, and the oxidant (air) from the annular region between the fuel tube and a 101-mm-diameter concentric tube. Before exiting the annular region the air passed through packed beds of glass beads and porous metal disks, which are conducive to keeping the flame stable. The wall thickness of the fuel tube is 0.96mm. The base flame is coflow methane/air diffusion flame. During the experiment, the mass flow rates of methane and air were, respectively, 0.141 g/min and 4.724g/min at room temperature (298 K) and atmosphere pressure. A flame enclosure made of flexible steel mesh protects the flame from air flow in the room. A quartz glass window is set up on one side of the combustor to observe the flame shape. The concentration of NO in off-gas was measured by handheld gas analyzer (ecom-J2KN made in Germany). The gas analyzer consists of gas flow system and circuitry. It works by the following principle, gas is sent to gas cell of the sensor from the sampling tube by sampling pump, and the electrical signal which is transformed by the sensor is converted into the concentrations of measured gas.

3 Numerical model

The experimental flames above were modeled by numerical simulation. Low Mach number assumption was adopted. The governing equations were discretized using the finite volume method in axisymmetric cylindrical coordinates. The SIMPLE numerical scheme [7] was used to handle the pressure and velocity coupling. The diffusion terms in the conservation equations were discretized by the central difference method and the convective terms were discretized by the power law method [8]. To speed up the convergence process, the discretized governing equations of gas species were solved in a fully coupled fashion at each control volume. Those of momentum, energy and pressure correction were solved using the tri-diagonal matrix algorithm. OPT[9] model was utilized to calculate radiation heat loss, which not consider the reabsorption of gas.

The chemical reaction mechanism used is GRI-Mech 3.0[10], which contains 53 kinds of speices and 325 elementary reactions. GRI-Mech 3.0 has been shown to offer reasonable prediction of NO formation for various flames[11].

4 Results and discussion

In this paper, the fuel stream consists of CH_4 and dilution. The fraction of dilution is defined as:

$$\alpha_d = \frac{V_d}{V_d + V_{CH_4}} \tag{1}$$

Where V_d and V_{CH_4} are the diluent and methane volume flow, respectively.

The NO_x emission is described by the emission index, which is defined as the ratio of total formed NO_x to total heat release and calculated as：

$$EI_{NO} = \frac{V_{FG} \cdot M_{NO} \cdot \alpha_{NO}}{V_{CH_4} \cdot q_{CH_4} \cdot M_{CH_4} + V_d \cdot q_d \cdot M_d} \tag{2}$$

Where α_{NO} is the NO volume fraction, and V_{CH_4}, V_d, V_{FG} is the volume flow of methane, diluent and theoretical dry flue gas, respectively, in normal temperature and pressure. Quantities M_{NO_x}, M_{CH_4} and M_d are, respectively, the NO, methane and diluent molecular weight. Quantities q_{CH_4} and q_d is the methane and diluent heat value.

4.1 Experimental Verification of Mathematical Model

Fig.2 and Fig.3 display the variations of the NO mole fraction as the diluent fraction α_{H_2}, and α_{CO} change, respectively, and the comparison of the experimental and computational results in different diluent conditions. There are discrepancies between experiment data and computational results, this phenomenon may be due to inevitable factors in experiments. However, the relative errors are less than 15%, and the variation trend of experimental data is consistent with that of computational results, which can prove that the mathematical model is acceptable and can be used to the numerical research on NO formation in diluted coflow CH_4/air diffusion flames.

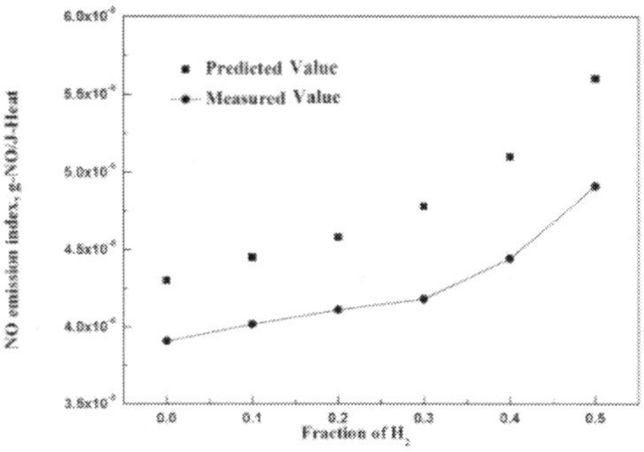

Figure 2: Variation of NO emission index at different H_2 fractions

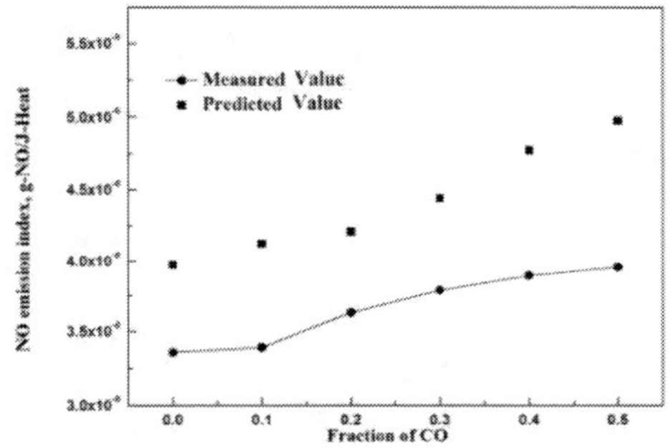

Figure 3: Variation of NO emission index at different CO fractions

4.2 Simulation of NO Formation

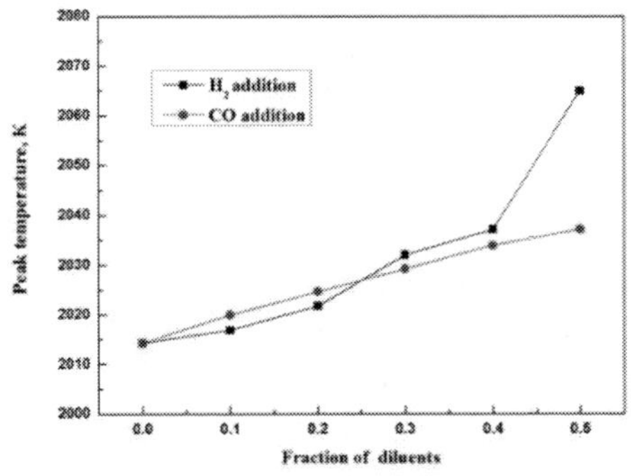

Figure 4: Variation of peak temperature

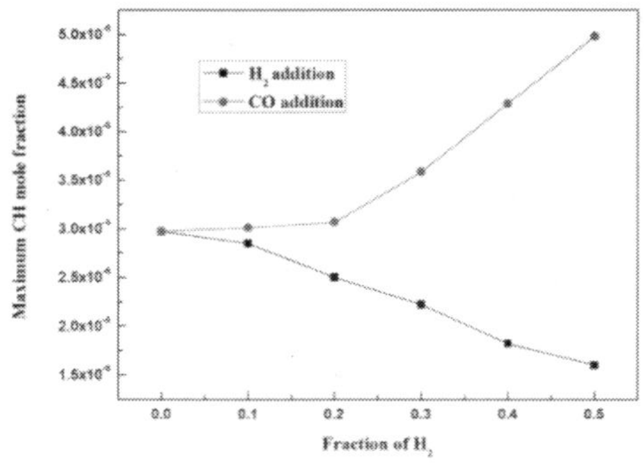

Figure 5: Variation of CH mole fraction at different fractions of dilution.

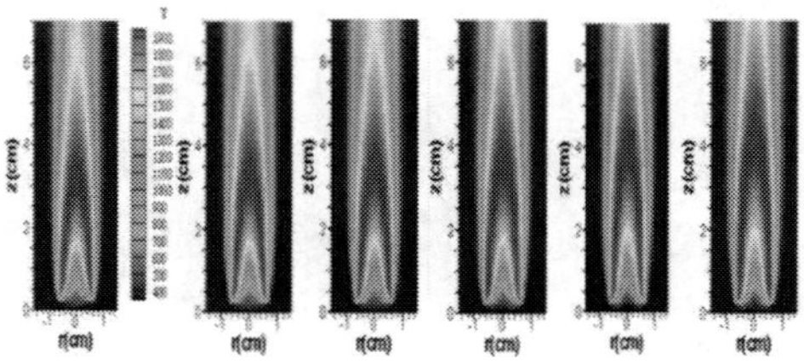

Figure 6(a): The temperature distributions with the H_2 fraction from 0 to 0.5

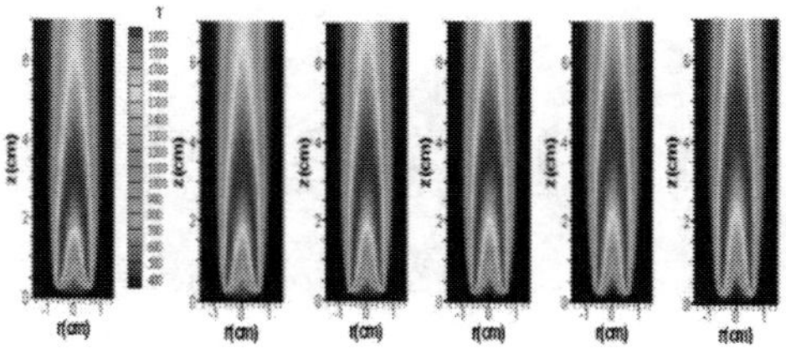

Figure 6(b): The temperature distributions with the CO fraction from 0 to 0.5

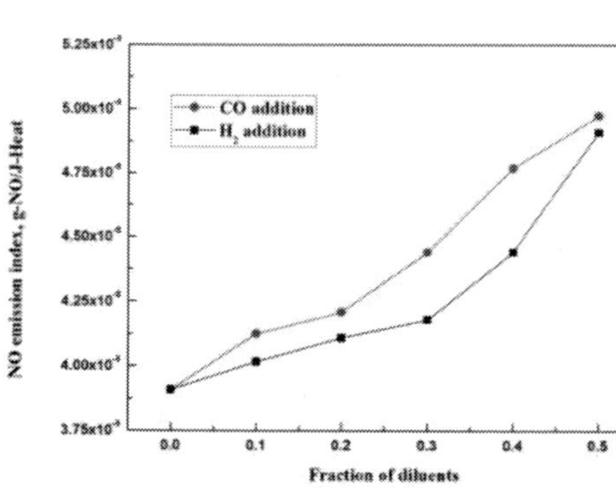

Figure 7: Variation of NO emission index at different fractions of dilution.

Fig.2 displays the variation of NO emission index as the diluent fraction α_{H_2} changes. As shown in the figure, when α_{H_2} increases, NO emission index increases monotonically. The variation above can be explained by the mechanism of NO formation. NO is mainly produced through the following two reaction in H_2/air flame, N + OH = NO + H and N_2 + O = NO + N. The previous research proves that the method to identify the mechanism of NO formation in a flame should not be based on how NO is finally formed, but on how molecular nitrogen is originally converted to atomic nitrogen or element nitrogen [12]. Molecular nitrogen is converted to atomic nitrogen through N_2 + O = NO + N, which is typical thermal route. Therefore, with α_{H_2} increasing, the formation of thermal NO in CH_4/air diffusion flame increases. It depicts that both thermal NO and prompt NO should be investigated to analyse the NO formation in CH_4/air diffusion flame with H_2 adding. Therefore flame temperature and CH mole fraction was investigated. As depicted in Fig.4 and Fig.5, when α_{H_2} increases, the peak temperature of CH_4/air diffusion flame increases obviously, but the maximum CH mole fraction decreases. It indicates that addition of H_2 promotes the formation of thermal NO and inhibits the formation of prompt NO through thermal effect and chemical effect. Considering the net variation of NO emission index, the promoting effect is stronger than the inhibition effect.

Fig.3 displays the variation of NO emission index as the diluent fraction α_{CO} changes. As shown in the figure, when α_{CO} increases, NO emission index also increases monotonically. The variation above can be explained by the chemical effect caused by CO. The initial oxidation of CO is OH + CO = H + CO_2. The addition of CO causes promoting effect on forward reaction of the oxidation, then more OH will be participated the reaction, meanwhile more H will be produced. The forward reaction of main chain reaction of the combustion (H + O_2 = OH + O) will be promoted with H increasing. During the combustion progress of coflow axisymmetric laminar diffusion flame, the residence time of CH_4 is not long enough for completed combustion. As a result, the reaction zone will be in oxidation atmosphere. The addition of CO causes promoting effect on forward reactions of two reactions mentioned above, leading to an increase in combustion intensity. CH mole fraction and flame temperature increase respectively (as shown in Fig.5 and Fig.4), with the combustion intensity increasing, leading to an increase in NO emission. Therefore, adding CO to the fuel side changes NO emission characteristics through thermal effect and chemical effect, resulting in an increase in NO emission index.

Fig.6(a) and 6(b) displays the variation of temperature distributions with H_2 and CO dilution. As shown in the figures, the height and width of flame increases with H_2 and CO adding, respectively. As shown in the figure, width of H_2 addition flame is wider than that of CO addition flame. This effect is presumably due to faster flame speeds and better diffusivity of H_2, which allows fuel to burn more completely through transport effect. Considering radiation effect, the absorption coefficient of CO is higher than that of H_2, which implies more radiation heat loss with CO addition, inhibiting the increase of temperature. Transport effect and radiation effect occur simultaneously and are intimately coupled, which leads to higher flame temperature of H_2 than that of CO as shown in Fig.4.

Fig.7 displays the comparison of effects of CO and H_2 dilution on NO emission index. As shown in the figure, the effect of CO dilution is more significant. The increase in both CH mole fraction and temperature causes the increase in both thermal NO and prompt NO with CO adding. However, the addition of H_2 promotes the formation of thermal NO and inhibits the formation of prompt NO. Therefore, NO emission with H_2 adding is slightly less than NO emission with CO adding.

5 Conclusions

The effect of H_2 and CO addition on NO formation in coflow CH_4/air diffusion flame was investigated by numerical simulation with experimental verification, the simulation data agree well with experimental result. The result shows that:

When adding CO as dilution, NO emission increases with α_{CO}. The reason is that, the increase in both OH mole fraction and temperature causes the increase in NO emission.

When adding H_2 as dilution, NO emission increases with α_{H_2}i. The reason is that the addition of H_2 promotes the formation of thermal NO and inhibits the formation of prompt NO, and the promoting effect is stronger than the inhibition effect, resulting in an increase in NO emission.

References

• F.Emun, M Gadall, T Majozi. Computers & Chemical Engineering 34 (3) (2010) 331-338

• Z Yang, X Yu, J Peng. Lee. Experimental Thermal & Fluid Science 81 (2017) 209-222

• Jeong Park, Seung-Gon Kim, Kee-Man Lee. Energy Res 26 (2002) 1141-1160

• L Ren. Russian Journal of Physical Chemistry A. 91 (5) (2017) 866-875

• SE Hosseini, S Salehirad, MA Wahid. International Meeting of Adcances in Thermofluids. 1440 (1) (2012) 1309-1312

• Park JS, Kim JS, Keel S. International Journal of Energy Research 31 (2007) 472–485.

• S.V. Patankar. Numerical Heat Transfer, Fluid Flow. Hemisphere, New York, 1980

• Guo H, Liu F, Smallwood GJ. Combustion and Flame, 145 (2006) 324-338

• Guo H, Y. Ju, K. Maruta, T. Niioka. Combust. Flame, 109 (1997) 639-646

• Smith G P, Golden G M, Frenklach M. http://www.me.berkeley.edu/ gri_mech, 1995

• Sahu AB, Ravikrishna RV. Hydrogen Energy, 39 (2014) 17358~70

• F. Liu, H. Guo and G. S. Smallwood. Combust. Flame, 138 (1-2) (2004) 136-154

11th Asia-Pacific Conference on Combustion,
The University of Sydney, NSW Australia
10th -14th December 2017.

A Comparison of the Performance of Uni-element Injectors for Laboratory Scale Gas-Liquid Rocket Engines

Tausif Shaikh[1], Rajat Sawanni[1], Arindrajit Chowdhury[1], Swami Umakant[1]

[1]Mechanical Engineering Department,

IIT Bombay, Powai, Mumbai, Maharashtra, 400076, India

Abstract

A quantitative study of the atomization of injector sprays is required to accurately predict and improve the performance characteristics of liquid propellant rocket engines. A laboratory scale setup of a Gas-Liquid Rocket Engine (GLRE) was designed to characterize the properties of a class of new fuels synthesized by the Defense Research and Development Laboratory, Hyderabad. An efficient atomization system was required for accurately predicting the I_{sp} values. Four injectors were designed and fabricated based on available literature. A preliminary spray characterization of the injectors under atmospheric conditions was conducted to analyse the trends of the Sauter Mean Diameter (SMD) with the radial and axial positions. The swirl coaxial injector was selected as the most suited design for the current engine design owing to its small breakup length, good atomization and good mixing. The shear coaxial injector provided better atomization but with a larger breakup length and higher velocities.

1 Introduction

Liquid fueled rocket engines are used in most space shuttle main engines, launch vehicles, and missiles where engines with high power to weight ratio are required for propulsion. Additionally, the research community is constantly trying to synthesize new fuels, thus requiring complete or partial redesign of the components. The testing and characterisation of these fuels is an important task that requires high accuracy saddled with low developmental costs. Thus, laboratory scale setups of liquid rocket engines are typically designed for testing and scrutinizing such new fuels. The efficiency of such systems is dependent on a plethora of component designs.

One such important component is the injector which serves the purpose of atomizing and mixing the propellants for better vaporization and combustion. The specific injector design usually takes into account the nature of the propellants. In a previous work, Shaikh et al. [1-3] studied the performance characteristics of gasoline-gaseous oxygen (GOX) lab-scale gas-liquid rocket engine (GLRE) with a design thrust of 10 kg. Initial tests showed lower specific impulse values as compared to theoretical predictions. The chamber dimensions dictate a shorter breakup length and good mixing and atomisation near the injector face. In this design, the GLRE uses a commercially available solid cone spray nozzle (Model B1/8GG-1 Full cone) with an orifice diameter of 0.79 mm. The atomisation with such a system is poor and needs significant modification to improve the efficiencies of the systems. Four such types of injectors were fabricated and studied for application in the rocket engine setup using hydrocarbon fuels. Shear Coaxial (SC) and Gas Centered Swirl Coaxial (GCSC) injectors were designed using guidelines from the literature [4-9]. A solid cone spray nozzle with surrounding oxygen jets at 90^0 to the injector face, designated as (SSC) and 60^0 to the injector face, designated as (ISC) were also designed and fabricated. An extensive study of these atomisers at atmospheric conditions is expected to provide qualitative as well as quantitative data on the atomisation in the chamber during a hot fire test. A direct correlation has not been determined for shadowgraphy results at atmospheric conditions with the hot fire results. However, literature suggests a positive correlation between the two with cold flow tests in high pressure chambers actively correlated to hot fire results [5, 10]. Hence, the quantitative values derived from this study would be utilised for computational modelling of the combustion chamber and prediction of various pertinent parameters.

2 Injector Design

The goal of designing the injector was to achieve the highest c^* efficiency while maintaining the durability and simplistic design of the injector. The efficiency is a result of better atomisation and mixing coefficients achieved by the injector.

2.1 Shear Coaxial Injector (SC)

The basic mechanism governing atomisation in SC injectors is the shear-stripping of the liquid stream (central jet) by the higher velocity gaseous stream (annular stream). It is initiated by the initial vortices at the stream boundaries which become further unstable to form surface waves and is reinforced by the Kevin-Helmholtz instability mechanism because of the higher velocity gaseous stream. This contributes to the primary breakup of the central liquid jet which is further broken down in the downstream region with the help of aerodynamic drag. A myriad of geometric and flow parameters account for the atomisation and mixing in such injectors. Based on the extensive experimental study done by Falk and Burick [4] on the characterisation of sprays from SC injectors, non-dimensional numbers were developed to reduce the number of parameters deciding the vaporisation (η_{c*_vap}) and mixing efficiency (η_{c*_mix}). It was devised that η_{c*_mix} and η_{c*_vap} were respectively the functions of two non- dimensional parameters as shown in Eq. (1a) and (1b).

$$\eta_c^*|_{mix} = f\left(\frac{(\rho_g V_g)^2}{MR.V_l}\right) \quad (1a)$$

$$\eta_c^*|_{vap} = g\left(\frac{V_g - V_l}{MR.V_l}\right) \quad (1b)$$

Corresponding author. Fax: +91-22-257-26875
E-mail address: arindra@iitb.ac.in

Where V_g represents the velocity of the gases, V_l is the velocity of the liquid phase, ρ_g is the density of the gas phase and MR is the mixture ratio. Broadly speaking, Falk and Burick et. Al. [4] emphasised that the atomisation is basically dependent upon the velocity difference between the gas and liquid stream, the mixing is dependent upon the momentum transfer by the gaseous flow.

In this case, the SC injector was designed to achieve an SMD of 100 microns and with the goal of maximizing the c^* efficiency of the combustor.

2.2 Gas Centered Swirl Coaxial (GCSC) Injector

In a GCSC injector, the gaseous oxidizer at high velocity enters directly through the center of the injector and is surrounded by a swirling liquid film injected along the periphery of the injection element. The liquid film is subsequently stripped and fragmented into drops by the high velocity gas stream. In order to effectively evaluate performance of atomisers, it is essential to precisely predict liquid film dynamics inside the injector. The liquid film thickness and length are a function of the injector recess length, and they affect the atomized drop size to a large extent. It is, therefore, noted that the characterisation of GCSC injectors involve thorough understanding of the variation of liquid film thickness, breakup length, and spray angle with various non dimensional parameters. The design and fabrication of the GCSC injector was conducted after careful evaluation of the geometric considerations from available literature. It has been made clear from [6-8, 11] that a higher value of momentum flux ratio (*M*) results in better atomisation which is characterized by shorter breakup lengths. A critical momentum flux ratio (M_c) is defined such that a further increase in M after M_c leads to no substantial decrease in breakup length. The effect of gaseous post recess length is also studied and a higher value of the same is needed for a lower value of M_c. The design of the injector was finalised to keep the momentum flux ratio beyond or atleast at the critical value. This assures a better atomisation and a lower breakup length of the GCSC injector. A higher recess length was used to provide enough space for the swirling liquid stream to mix and interact with the central gas stream before it was injected into the chamber [6].

The expected velocities of both the liquid and vapour components in the cold flow conditions were kept the same as in hot fire run. This was achieved by keeping the momentum flux difference to be the same as in hot fire case. The current design is a basic one which is neither converger, diverger, nor pre-filmer type as noted in [9]. Although a higher atomisation and mixing is known to be achieved for converger type of elements, the feasibility of fabrication was to be studied first and as a consequence, a normal GCSC injector was designed and fabricated as the first model.

2.3 Simple Solid Cone (SSC) Injector

A commercially available spray nozzle with an orifice diameter of 0.79 mm and cone angle of 60° was used to inject liquid fuel from the injector flange. Four holes at right angles were drilled in the injector plate at a pitch circle diameter (PCD) of 22 mm for GOX injection. In the previous study [1-3], the exit plane of the solid cone nozzle was located at a distance of 21 mm from the injector flange. Hence almost 25% of the chamber length did not encounter any breakup or atomization. This region was referred to as non-breakup length in subsequent section. This was the preliminary design which has been used in the initial test runs. The current test runs with gasoline as fuel with an area ratio of 4, chamber pressure of 8.46 bar, an oxygen mass flow rate of 8.73 g/s, and a fuel mass flow rate of 3.83 g/s provides an I_{sp} efficiency of 80.39 % and a c^* efficiency of 94.95 % with this injector.

2.4 Impinging Solid Cone (ISC) Injector

With the aim of improving upon the current atomization using the available commercial nozzle, a revised design was explored. The exit plane of the nozzle was set to coincide with the injector flange with the aim of reducing the non-breakup region inside the chamber. GOX was injected at 60° to the injector plane as opposed to the 90° in the previous design. This caused the liquid and oxidizer streams to impinge, resembling a pendant impinging injector. This design was expected to reduce breakup lengths and provide better atomization. A major drawback expected from this injector was the symmetry of spray. Since it was similar to the pendant impinging injector, the atomization in the two impinging planes was expected to be coarser and with a higher droplet density than any other plane. The mixing efficiency in such an injector suffers a setback because of the asymmetry in the flow.

3 Experimental Setup

The sprays generated by the four types of injectors were diagnosed by shadowgraphy and photography. A La Vision® spray diagnostic facility was used for the shadowgraphy experiments. It uses an Nd: YAG double cavity Evergreen 145 laser. The pulse energy of the laser was 2x145 mJ at a wavelength of 532 nm. It generates a 50ns pulse and captures images at frequency up to 15 Hz. A CCD camera, Imager ProSX 5M, was used in the setup. The full resolution of the camera was 2448 x 2050 pixels. The sensor used in the camera was SONY ICX625 with a pixel size of 3.45 X 3.45 µm. Raw images from the camera were processed in the setup using an inbuilt software Da Vis 8.1.5 for shadowgraphy.

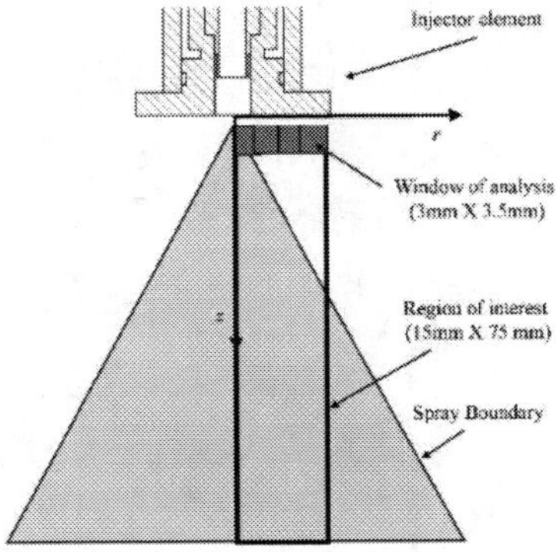

Figure 1: Region of analysis of spray using shadowgraphy

The window size or zone of interrogation at each point for shadowgraphy experiments was 3.5 x 3 mm. The region of interest was marked from the exit plane of the injector to an axial distance of 75 mm (the length of the combustion chamber) and a radial distance of 15 mm (the radius of combustion chamber). Surrogates were used to simulate the behaviour of fuel and oxidisers at atmospheric conditions. Water was used as the surrogate for fuel and nitrogen as the surrogate for GOX. The mass flow rates of the surrogates were kept at 9 g/s for nitrogen and 3 g/s for water at atmospheric conditions. The injector was mounted on an XY translational stage to gather data in radial and axial coordinates from the exit plane of the injector. The camera was then focussed on the central plane of the injector. A burst of 100 images were taken for a particular location and analysed using the in-built software. The software processes the boundaries of droplets based on the intensity difference and calculates the SMD subsequently. The breakup length is also determined using shadowgraphy. The length of liquid column before the start of secondary atomisation was recorded as the breakup length. The quantification of breakup length based on shadowgraphy remains apocryphal because of the highly dense sprays hampering visibility. However, a comparative study can lead to substantial results. The spray angles for the coaxial injectors are calculated by analysing the laser sheet illuminated images in ImageJ software.

4 Results and Discussions

The nature of spray from each injector was understood by inspecting the pertinent parameters of atomisation and mixing. The SMD and number density plots for each location were used for comparing the nature of sprays. However, the number density plots could not be shown due to lack of space.

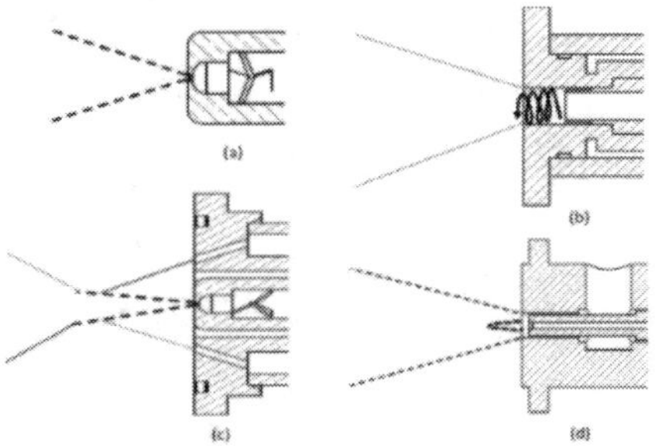

Figure 2: Injector schematic for (a) SSC injector, (b) GCSC injector, (c) ISC injector, (d) SC injector

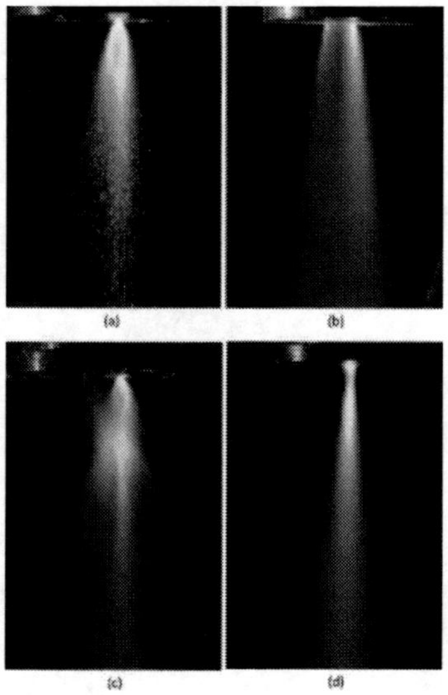

Figure 3: Laser-sheet illuminated spray image for (a) SSC injector, (b) GCSC injector, (c) ISC injector, (d) SC injector

Property/ Injector type	SSC	GCSC	ISC	SC
Spray angle	18°	7.3°	24.1°	6.2°

Table 1: Spray angles for different injectors

As expected, the solid cone spray nozzle was found to possess average atomization and mixing efficiency. The breakup length for this nozzle was considerably large, further reducing the effective area filled with liquid droplets within the chamber. Because of this, data was acquired without the assisted oxygen ports. This was found to generate bigger droplets and did not exhibit proper mixing or uniformity of flow as seen in Fig. 4.

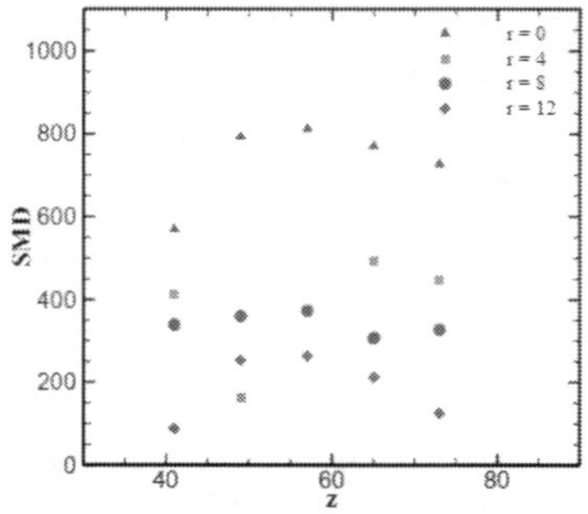

Figure 4: SMD data for SSC injector

As seen in Fig. 2(c) and Fig. 3(c), with the solid cone nozzle and oxygen injection at 60°, the existing nozzle provided spray characteristics that were convincingly better than the SSC injector setup. The non-breakup region was removed and the breakup length was considerably shortened by allowing the oxygen streams to impinge at 60° to the spray. However, because of the impinging nature, the spray is concentrated in two planes, rendering the mixing efficiency of this system low. The particle density in the two planes was observed to be very high, but reduced drastically in the non-impinging planes. Only the sheared away droplets were seen in this region and were thus, very small in diameter. The variation of number density in the axial direction was not found to follow a particular trend and was expected to be the result of the impinging streams. In the central plane, the number density is seen to increase up to the distance where the streams impinge and decrease thereafter. In the radially outward locations, the number density was noted to increase first and then decreased. The SMD variation was observed to decrease initially, while attaining a relatively constant value in the axially extreme locations as seen in Fig. 5.

As seen in Fig. 2(d) and Fig. 3(d), the primary breakup region of the SC injector possessed a liquid core close to the injector exit plane. This was expected since the instabilities instigating the primary breakup need a region to grow and break the liquid stream into ligaments. Owing to the stripping of this core with the high speed adjacent gas flow, a large droplet number density was expected. Since the core contained large ligaments accompanied by smaller stripped droplets, the SMDs for such a case were found to be higher. As seen in Fig. 5, with the increasing axial coordinate, the ligaments further breakup and form smaller and smaller droplets due to the effect of aerodynamic drag. This results in the breakup and spreading of larger ligaments into smaller droplets, thus making the prediction of number density difficult. Although the disintegration of droplets increases the number of droplets in a region, the effect of spreading is counterproductive and makes the assessment of number density inscrutable. However, it was seen in the experimental trends that the number density increases giving an idea about the overshadowing effect of droplet disintegration. The radial variation of number density was more predictable than the axial variation. The radial coordinates were expected to see a drop in the number density since the spray disintegrated from a central liquid jet.

For a GCSC injector, the swirling liquid sheet from the annular region tends to move outwards because of the centrifugal force and because of continuity, decreases in thickness as it moves down. This simple breakup mechanism assisted strongly when a high velocity gas stream was imposed on the flow. The central gas stream not only sheared the cross liquid flow, it also took a copious amount of these droplets to the central part of the flow, thus increasing the mixing efficiency of the injector. A typical spray pattern for GCSC injector is shown in Fig. 2(b).

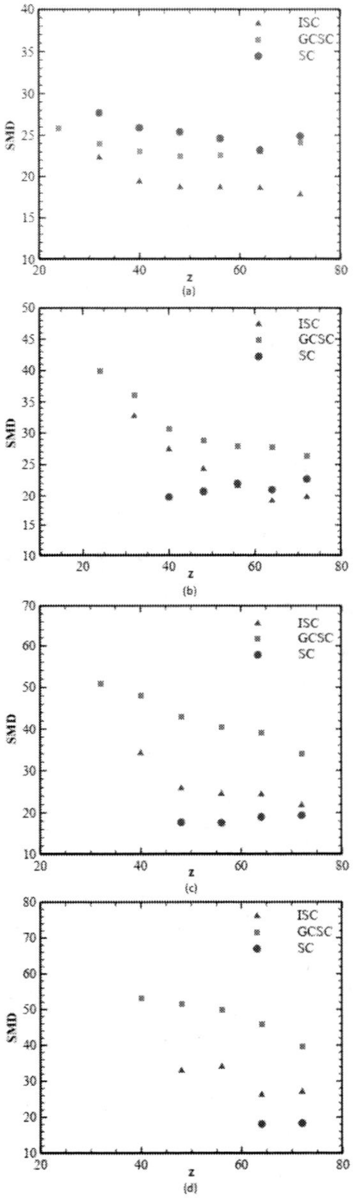

Figure 5: SMD data for injectors at (a) r = 2 mm (b) r = 6 mm (c) r = 10 mm (d) r = 14 mm

The number density in the central region near the injector was found to be low because of the lack of a central liquid flow. This number density was, however, expected to rise, as verified by the experiments because of the spreading and disintegration of droplets. The radial variation in number density may be directly correlated with the fact that the cone produced by a GCSC injector imposes a region between the boundary and central part where the droplet density is the highest. This is due to the conical structure of the swirling liquid flow. It is also worthwhile to note that the introduction of a large recess contributed to an important attribute of this injector. Since a recess prevents the swirling liquid flow from expanding and increases its interaction with the central gas core, the breakup length was seen to be really short in this case. The trends of variation of SMD were also noted to decrease with the axial distance owing to secondary atomisation as seen in Fig. 5. The variation was expected to increase and then decrease radially outwards owing to the conical nature of the swirling flow. However, in the current experiments, the SMD was seen to increase within the region of interest.

5 Conclusions

Four different injectors were compared for selecting the best design to improve the efficiencies of a GLRE setup. Their designs were based on an extensive literature review and examples from previously characterized injectors. The existing injector was redesigned to develop an impinging gas injector. An SC injector and a GCSC injector were also fabricated. All the injectors performed better than the existing solid cone injector. The ISC injector achieved good atomization but with the drawback of a substantial breakup length and non-uniform flow. The SC injector also had comparable atomization but with a smaller cone angle and a comparatively larger breakup length. The GCSC injector had the lowest breakup length and generated the most uniform flow. However, the values of SMD for this injector were a bit higher than the other two injectors in certain regions. Since the combustion chamber is 75 mm in length, a breakup length of 20 to 25 mm can largely affect the combustion and thus the efficiency of the engine. Hence the GCSC injector was chosen for further tests, as it provided the lowest breakup length, despite slightly higher SMDs.

References

1. T. Shaikh, L. Patidar, A. Chowdhury, Applied Thermal Engineering, 2017. 110: p. 1554-1567.
2. L. Patidar, T. Shaikh, A. Chowdhury, Numerical Simulation of Combustion in an n-Heptane and Gaseous Oxygen Rocket, 10th ASPACC Beijing, China, July 2015.
3. T. Shaikh, L. Patidar, A. Chowdhury, Experimental Characterization of a Gasoline-GOX Rocket, 10th ASPACC Beijing, China, July 2015.
4. A. Falk, R. Burick, Injector design guidelines for gas/liquid propellant systems. NASA CR-120968, 1973.
5. P. Strakey, R. Cohn, D. Talley. 5th international symposium on liquid space propulsion. Reno, Nevada, Jan. 2003.
6. M. Lightfoot, S. Danczyk, D. Talley, Atomization in Gas-Centered Swirl-Coaxial Injectors, Air Force Research Lab Edwards AFB CA Propulsion Directorate, 2006.
7. Jeon, J., et al., Journal of Fluids Engineering, 2011. 133(12): p. 121303.
8. M. Lightfoot, S. Danczyk, D. Talley. Atomization rate of gas-centered swirl-coaxial injectors. in 21st Annual Conference on Liquid Atomization and Spray Systems, Orlando, FL. 2008.
9. J. Muss, et al., Swirl Coaxial Injector Development. Part I: Test Results. 2002, Air Force Research Lab Edwards AFB CA Propulsion Directorate West.
10. W. Mayer, Coaxial atomization of a round liquid jet in a high speed gas stream: a phenomenological study. Experiments in Fluids, 1994. 16(6): p. 401-410.
11. M. Hong, et al. A Study of the Internal Two-Phase Flow in Gas-Centered Swirl Coaxial Injectors. in 8th International Conf. on Multiphase Flow, ICMF 2013, Jeju, Korea, May 2013.

11th Asia-Pacific Conference on Combustion,
The University of Sydney, NSW Australia
10th -14th December 2017.

On the Effects of High-Pressure on Turbulent Inhomogeneous Flames Computed using PDF-RANS Methods

S. Al-Dawsari [1], M.J. Cleary[1], A.R. Masri[1]

[1]School of Aerospace, Mechanical and Mechatronic Engineering,

The University of Sydney, NSW 2006, Australia

Abstract

This paper uses the PDF-RANS approach to compute the structure of turbulent inhomogeneous flames of methane at pressures up to 10 bars. The Monte Carlo particle method is employed to solve for the composition PDF where the chemistry is represented by DRM19 which consists of 19 species and 84 reactions. The calculations attempt to represent the Sydney piloted inhomogeneous burner which has been studied extensively at atmospheric pressure. A key feature of the burner is the ability to stabilize flames with different levels of composition inhomogeneities at the jet exit plane. This is introduced by varying the recession of a fuel tube within an annulus of air. At high pressure, mixing inside the pipe upstream of the jet exit plane so that the recess distance needed to bring about the same level of inhomogeneity is about 50mm longer at 10bars. The structure of inhomogeneous flames and the occurrence of local extinction as they approach global blow is similar at all pressures. However, the region where extinction occurs move downstream at higher pressure due to the different mixing patterns

1. Introduction

Accounting for high-pressure effects in combustion is essential given its relevance to most applications from engines to gas turbines[1]. Compositional inhomogeneities which exist in such combustors either due to intentional stratification or imposed by instabilities makes this task even harder. This is because the models must then be able to account for mixed-mode combustion at high pressure. Such capabilities are gradually being developed aided by experimental platforms which are providing data for model validation.

The Sydney piloted inhomogeneous burner is providing such a platform with its ability to stabilize turbulent flames that could transition from fully homogeneous to inhomogeneous[2]. While extensive data already exists for a range of flames where mixed-mode prevails, such measurements are all performed at atmospheric pressure. More recently, a duplicate of this burner was installed at the high-pressure facility at King Abdulla University of Science and Technology (KAUST) in Saudi Arabia but detailed measurements are yet to be made. The objective of this paper is to perform preliminary calculations that guide such measurements and provide an incentive for future developments.

Corresponding author.
E-mail address: sald0396@uni.sydney.edu.au

Numerically, two broad modelling approaches are employed to account for mixed-mode combustion, namely flamelet modelling[3] and probability density function methods PDF[4]. With the flamelet approach, mixture fraction and a reaction progress variable are employed to account for non-premixed and premixed modes of burning. A key limitation, however, lies in the need to develop a flame indicator that determines the extent of prevalence of each mode[5, 6].

PDF methods have the ability to calculate the complexities of nonlinear interactions between turbulent fluctuations and chemical reactions. These have been proven in many applications. Also, PDF methods have several advantages that the reaction terms and turbulent mixing terms are treated exactly. However, PDF methods need closure models for these terms [7]. Early and simple mixing models are Modified Curl (MC) [8] model and the Interaction by Exchange with the Mean (IEM) [9]. Both models do not achieve localness which is achieved in mixture fraction space by Euclidean Minimum Spanning Trees (EMST) model [10]. These models have been investigated with different flames and in general, the EMST model shows better results compared to the IEM and MC. The model constant C_φ has been investigated by different values (standard value $C_\varphi = 2$) and generally, it is found that the mean temperature value have proportional relation with the increasing value of C_φ. However, the values of mixture fraction rms have an inverse relationship with C_φ [7, 11].

This paper uses the Reynolds-averaged Navier Stokes, RANS-PDF approach to compute the structure of turbulent inhomogeneous flames at high pressure. RANS-PDF methods have achieved a good accuracy with low computational cost compared to other CFD approach[12]. The standard k-ε model has shown to results in better agreement with experiment when employed with a modified $C_{\varepsilon 1}$=1.6 [13]. Details are provided about mixing patterns and the structure of the flames at different pressures and varying departures from global blow-off.

2. Numerical approach

The Composition PDF transport equations is utilized to account species composition and RANS equations to solve the mean momentum, conservation mass and energy as well as turbulence kinetic energy and turbulence dissipation rate. Lagrangian particle-based Monte Carlo method is used to solve C-PDF transport equations. The unclosed terms in C-PDF transport equation is solved by closure models with the following assumption: 1) Gradient diffusion assumption based on K-ε with modified $C_{\varepsilon 1}$=1.6 is used for turbulent diffusion term; and 2) EMST model with constant $C\varphi$=2.0 is assumed for molecular mixing term and this value is used at all pressures studied here for the lack of better guidance. The number of particles per cell is 30 PPC. The reduced chemical mechanism DRM19, using 84-steps reactions with 19 species has been applied in the calculation utilizing the In-Situ-Adaptive Tabulation (ISAT) algorithm with an error tolerance of 1e-4 [14] which is deemed adequate here given that the cases studied here have modest departures from finite rate chemistry. We have optimized the grid resolution for 10 bars and used it for all other pressure cases. With respect to the number of particles per cell PPC, it was found that 20 PPC are needed at 10 bars compared to 30 PPC at 1 bar. Also, the following modeling parameters sensitivity have been studied: values of mixing constant C_φ and turbulence models, etc. All these methods and assumptions are applied by using Fluent 17.1.

A key feature of the burner used here is the ability to stabilize the entire range of flames from homogeneous to fully inhomogeneous. This is achieved by the injecting the fuel and air separately in concentric tubes. The inner tube, carrying fuel, can be recessed to different distances upstream of the jet exit plane to vary the mixture fraction profile. The computational domain contains the recessed inner tube (D=4mm) to feed the methane, while air is fed through an annulus (D=7.5mm) encircling the inner tube and the volumetric air/fuel ratio is V_a/V_f=2.0. The pilot utilized (stoichiometric mixture of acetylene, hydrogen and air-same C/H ratio as methane) and the diameter of tube surrounding the pilot is 18 mm. The remaining distance in radial (r) direction is the Co-flow. The computational solution domain is rectangular, symmetric and consists of 42,414 quadrilateral cells. It extends 50 jet diameter in the axial (z) direction and 10 jet diameter in the radial (r) direction. The jet (pipe) flow calculations are started 600mm upstream of the jet exit plane.

3. Results and discussion.

Two separate calculations were performed over a range of pressures from 1 to 10 bars, one for mixing inside the concentric tubes upstream of the jet exit plane and another for actual turbulent flames at fixed recess distances and hence fixed levels of inhomogeneity. Further details are given here:

3.1 Mixing inside the pipe:

Calculations for non-reacting flows inside two concentric pipes are performed here to resolve the extent of mixing at various recess distances. Fuel and air injected separately are allowed to mix at different pressues and the resulting mixture fraction profile is then monitored at different distances between the tips of the inner and outer tubes (referred to as recess distance, Lr. There are two aspects of immediate relevance here: (i) the recess distance required to produce a homogeneous mixture at high pressure, and (ii) the recess distance which leads to the established of a reactive mixture at the out edge of the annulus. Both of these issues are discussed next.

Figure 1 shows the computed local equivalent ratio at the outer edge of the annulus (r/D=0.5), as a function of recession distance, Lr. Three curves are related to 1, 5, and 10 bars pressure values. Calculations are performed for a bulk velocity and volumetric ratio are kept constant, as 82 m/s and 2.0 respectively. This correspond to 88.65 m/s in fuel stream and 78.89 m/s in the air stream.

The horizontal dotted line on Figure 1 corresponds to an equivalence ratio φ=1.1 which is taken as the most reactive mixture[15]. It is evident from the profiles of Figure 1 that the optimal recess distance which results in a mean equivalence ratio of 1.1 at the outer rim of the annulus (and hence in contact with the pilot) shifts from Lr=100mm at 1bar to 150mm at 10 bars.

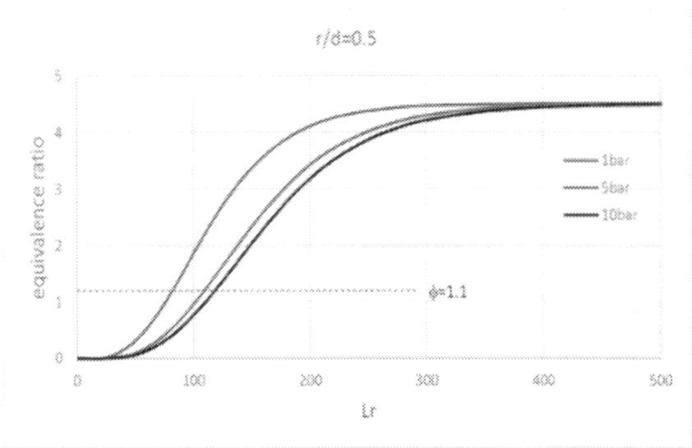

Figure 1: Computed local equivalent ratio at r/d=0.5 as a function of recession distance, Lr for different pressure.

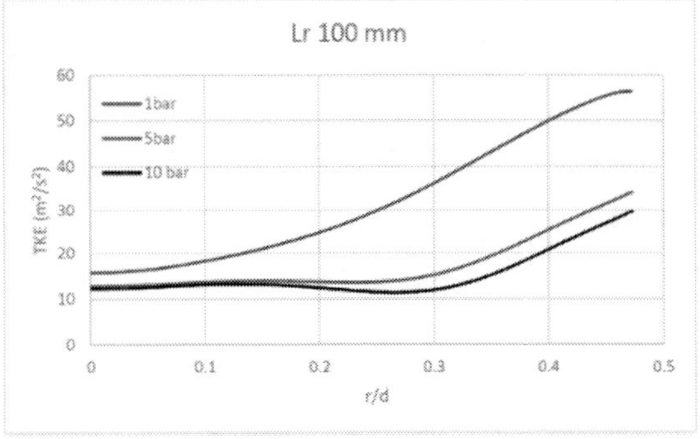

Figure 2: Computed radial profiles of the turbulent kinetic energy at Lr = 100mm for different pressure.

Additionally, Figure 1 shows that between air and fuel at 1 bar reach homogeneity at Lr =300mm. However, at 5 and 10 bars the mixing becomes complete at Lr = 400mm. At 1 bar, the level of mixture inhomogeneity at Lr =100mm has stability of flame higher than flame with homogeneous inlet by 50% at the same pressure [2, 16]. However, at 5 and 10 bars the same level of mixture inhomogeneity is reached at Lr=150mm and this implies that mixing is slower at these higher pressures. To further elucidate this, Figure 2 shows computed radial profiles of turbulent kinetic energy at Lr=100mm and for the three pressures in question, namely 1, 5 and 10bars. It is evident from these profiles that the turbulent kinetic energy is higher at 1 bar hence contributing to the faster mixing at that pressure compared to 5 and 10 bars. This is explain the need for longer values of Lr needed at higher pressures to achieve the same level of inhomogeneity.

3.2 Flame Structure

In the absence of experimental blow-off limits for the flames at high pressure, we assume this to be the same at 1 bar and we proceed with calculations for the 82 m/s jet velocity for all pressures of 1-10 bars. This corresponds to a nominal departure from blow-off of about 50%. The recess distance is taller as Lr=100 for 1 bar and Lr=150 for 5 and 10 bars.

Figures 3, 4 and 5 show the computed scatter plots for temperature versus mixture fraction at different axial locations in three flames 1bar, 5bars, and 10bars. The three flames have the same bulk jet velocity 82 m/s and the same inhomogeneous inlet conditions however, the recess distances are different, Lr= 100mm for first flame and Lr=150mm for the other flames. The mixture fraction axis of the plots has been trimmed to span $\xi = 0 - 0.3$ to highlight the prominent features of the plots.

The first flame shows mixed-mode combustion at x/d<5 while the other continue to show mixed-mode combustion to x/d <10. In Figures 3-5, it is apparent that the premixed/stratified combustion behavior is found at upstream locations and the behavior switches to diffusion mode further downstream. It is also found that, at higher pressures, the transition from premixed to diffusion happens latter compared with the 1 bar case.

Figure 3 shows that the local extinction starts to occur at x/d=10 and re- ignition at x/d=30. However, Figures (4-5) show that at x/d=20 the local extinction starts to occur and re- ignition at x/d=30.

Both of turbulence length scales and flame thickness have an inverse relationship with pressure value as a result the local extinction has been vanished at x/d=10 which means that the reaction and mixing times are faster than turbulence time. Also, when pressure value increase, the turbulence effects become less on the flame positions zones as shows in Figure 5. It is worth noting that when the pressure increases the amount of locally extinguished samples at a given axial location in the flames generally decreases, implying more stable flames. This aspect requires further study and will be ascertained by the measurements as these become available.

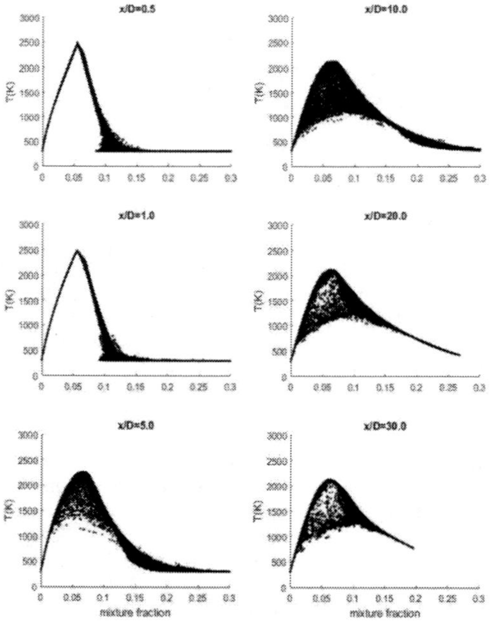

Figure 3: Computed scatter plots for temperature at different axial locations at 1bar

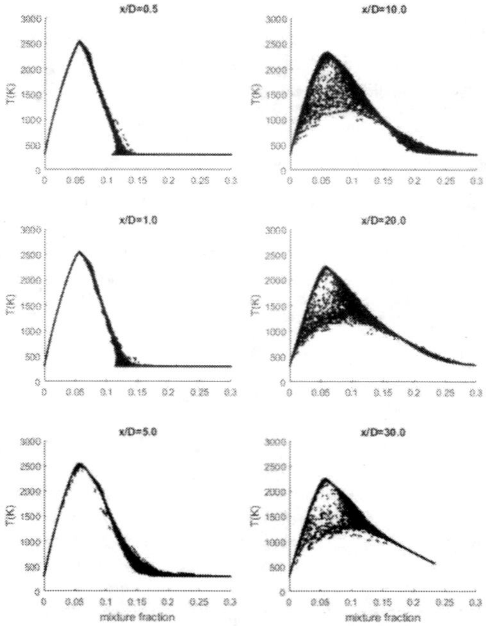

Figure 4: Computed scatter plots for temperature at different axial locations at 5bars

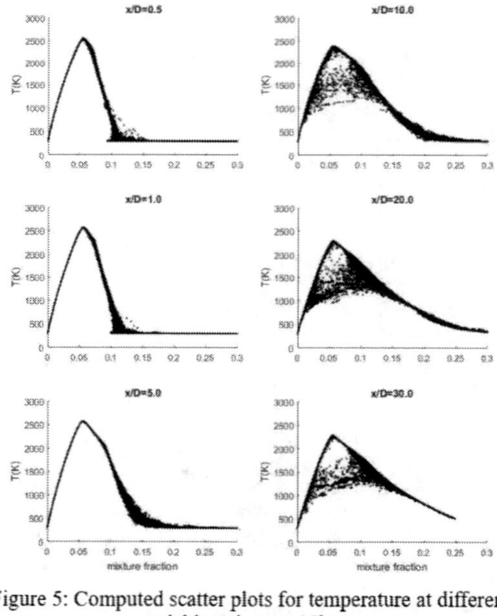

Figure 5: Computed scatter plots for temperature at different axial locations at 10bars

4. Conclusions

The joint RANS-PDF approach, combined with closure models, K-ε and EMST, and ISAT method, is utilized to compute the pressure effects on mixing, flames stability and local extinction. Between 1 bar and 5 bars the effects of pressure on mixing gradually increases while from 5 bars to 10 bars the effects of pressure on mixing slightly increases. At 5 bars and 10 bars the mixing become complete at Lr=400 which is further delayed compared to the 1bar case. At high pressures, local extinction decreases at the same bulk jet velocity considered here hence implying more stable conditions.

5. Acknowledgment

This research is supported by the Australian Research Council. The authors acknowledge the Sydney Informatics Hub and the University of Sydney's high performance computing cluster Artemis for providing the high performance computing resources that have contributed to the research results reported within this paper.

References

1. Rosenberg, D.A., P.M. Allison, and J.F. Driscoll. Method to measure flame index in a partially-premixed gas turbine combustor. in 51st AIAA Aerospace Sciences Meeting, Grapevine, TX. 2013.
2. Meares, S. and A.R. Masri, A modified piloted burner for stabilizing turbulent flames of inhomogeneous mixtures. Combustion and Flame, 2014. 161(2): p. 484-495.
3. Bray, K. and N. Peters, Laminar flamelets in turbulent flames. Turbulent reacting flows, 1994: p. 63-113.
4. Pope, S.B., PDF methods for turbulent reactive flows. Progress in Energy and Combustion Science, 1985. 11(2): p. 119-192.
5. Perry, B.A., M.E. Mueller, and A.R. Masri, A two mixture fraction flamelet model for large eddy simulation of turbulent flames with inhomogeneous inlets. Proceedings of the Combustion Institute, 2017. 36(2): p. 1767-1775.
6. Kleinheinz, K., et al., Computational study of flame characteristics of a turbulent piloted jet burner with inhomogeneous inlets. Proceedings of the Combustion Institute, 2017. 36(2): p. 1747-1757.
7. Celis, C. and L.F.F. da Silva, Lagrangian mixing models for turbulent combustion: Review and prospects. Flow, Turbulence and combustion, 2015. 3(94): p. 643-689.
8. Curl, R.L., Dispersed phase mixing: I. Theory and effects in simple reactors. AIChE journal, 1963. 9(2): p. 175-181.
9. Aubry, C. and J. Villermaux, Représentation du mélange imparfait de deux courants de réactifs dans un réacteur agité continu. Chemical Engineering Science, 1975. 30(5-6): p. 457-464.
10. Subramaniam, S. and S. Pope, A mixing model for turbulent reactive flows based on Euclidean minimum spanning trees. Combustion and Flame, 1998. 115(4): p. 487-514.
11. Cao, R.R., H. Wang, and S.B. Pope, The effect of mixing models in PDF calculations of piloted jet flames. Proceedings of the combustion institute, 2007. 31(1): p. 1543-1550.
12. Pope, S.B., Turbulent flows. 2001, IOP Publishing.
13. Pope, S., An Explanation of the Turbulent. Journal, Vol, 1968: p. 1549-1555.
14. Liu, B. and S. Pope, The performance of in situ adaptive tabulation in computations of turbulent flames. Combustion Theory and Modelling, 2005. 9(4): p. 549-568.
15. Guiberti, T., et al., Fuel effects on the stability of turbulent flames with compositionally inhomogeneous inlets. Proceedings of the Combustion Institute, 2017. 36(2): p. 1777-1784.
16. Meares, S., et al., Stabilization of piloted turbulent flames with inhomogeneous inlets. Proceedings of the Combustion Institute, 2015. 35(2): p. 1477-1484.

11th Asia-Pacific Conference on Combustion,
The University of Sydney, NSW Australia
10th -14th December 2017.

Effects of fuel Lewis number on flame spread over solids in microgravity environment

B Praveen Kumar[1], Amit Kumar[2]

[1]Indian Institute of Technology Madras
Chennai, Tamil Nadu, India
[2]Indian Institute of Technology Madras
Chennai, Tamil Nadu, India

Abstract

A detailed three-dimensional numerical model is used, to study the effect of fuel Lewis number ($Le_F = \alpha/D_F$) and mass transfer on flame spread over thin solids. The focus of this work is to investigate the sensitivity of opposed flame spread rate to fuel Lewis number in microgravity. By using computational model as a tool, the effects of fuel mass diffusion in three directions are isolated by means of fuel Lewis number, Le_F. In general, $Le_{Fx} = Le_{Fy} = Le_{Fz} = Le_F$ however in this study Lewis numbers in x, y and z directions are varied independently. Sensitivity analysis is performed to quantify the influence of Le_{Fx}, Le_{Fy} and Le_{Fz} on flame spread rate. Numerical simulations are performed to analyze the behavior of spread rate with Le_F for 2D limit and 2cm wide fuel.

1. Introduction

In combustion of solid fuels, the solid goes through a pyrolysis process. The pyrolyzed fuel vapor next to the solid surface can have a different molecular structure from that of the solid, and as the fuel vapor diffuses toward the reaction zone it undergoes further decomposition. The fuel vapor in gas phase can have a complex chemical composition that may vary with location. Hence fuel diffusivity or fuel Lewis number is not known. After pyrolysis, the transport and the mixing of fuel vapor with oxygen depends on the mass transfer rate, which is a function of the fuel diffusivity or the fuel Lewis number. The fuel Lewis number appears in the solid phase boundary condition and the fuel species equation. Splitting of overall Le_F into components was done by Tolejko et al. [1] for 2D opposed flame spread to explain the dependence of flame spread rate on fuel Lewis number. Similar study is carried out here for 3D / finite width fuels to obtain more physical insight on three-dimensional aspect of fuel mass transfer on flame spread. Self-spreading flame under limiting oxygen conditions exhibits unsteady oscillatory behavior in microgravity over wide fuels [2]. Several researchers have identified thermo-diffusive instability to be the cause for flame oscillations. As this instability is related to fuel Lewis number [3], numerical simulations are running to study the effect of fuel mass diffusion on flame oscillations (not included in the present work).

Corresponding author. Fax: +91-44-2257-4002
E-mail address: praveen1bharadwaj@gmail.com

2. The geometry and the numerical model

2.1 Geometry

Figure 1 shows the schematic of self-spreading flame in quiescent microgravity environment. The figure shows a thin fuel strip of length l = 42 cm (along x) and width w (along z) bounded by 6 mm wide metallic strips on either side. The fuel has thickness of 0.0076 cm and density 0.263 g/cm3. The flame propagates along x with rate V_f. In the flame fixed frame of reference the flame is stabilized at location close to x = 0. The longitudinal and cross-sectional view of the tunnel showing the computational domain (shaded region), boundary types and mesh are also shown in Fig. 1. The governing equations are non-dimensionalized based on scaling laws of convection-diffusion [5, 6]. The grids points are clustered near the fuel surface and walls to resolve sharp gradients in the computed variables. The smallest grid size is 0.05 thermal lengths in the flame anchoring in the vicinity of the fuel surface.

2.2 Numerical Model

The details of the basic flame spread model and the numerical approach can be found in [5], here a short description is presented. The gas-phase consists of 3D governing equations of full Navier-Stokes equations for laminar flow along with the conservation equations of mass, species (fuel vapor, oxygen, carbon dioxide, and water vapor) and energy. A single-step, second-order finite rate Arrhenius reaction between fuel vapor and oxygen is considered. The set of coupled elliptic partial differential equations for the reactive flow in the gas phase is solved numerically by PISO algorithm. The equation of thin solid fuel model comprises that of mass and energy in 2D along with a solid fuel pyrolysis law of zeroth order. The solid fuel considered here is an aerodynamically and thermally thin cellulosic material. The solid is assumed to burn ideally to form fuel vapors without melting or charring. The solid-phase equations are solved by the finite-difference technique. The fuel properties are identical to one used in [4] with fuel half thickness τ =3.8x10-3cm. The 3D Radiative Transfer Equation models the radiation transport in the gas and is solved using the Discrete Ordinates Method. The non-dimensional governing equations are solved in flame fixed coordinates. The transient

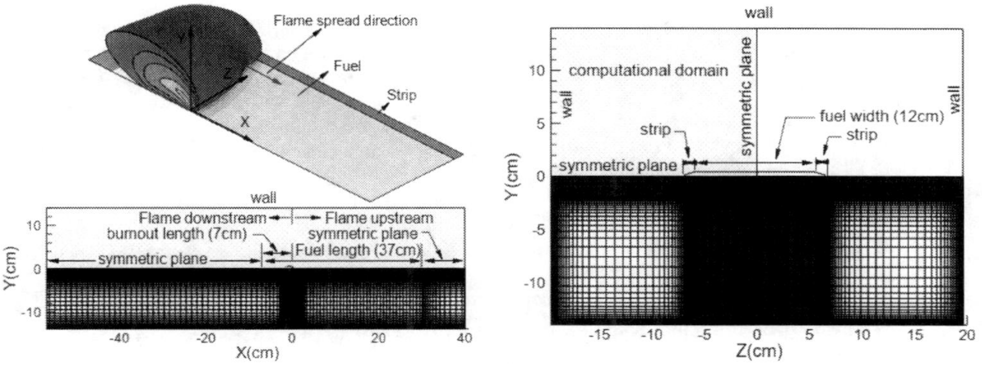

Figure 1: Schematic diagram of computational domain and mesh

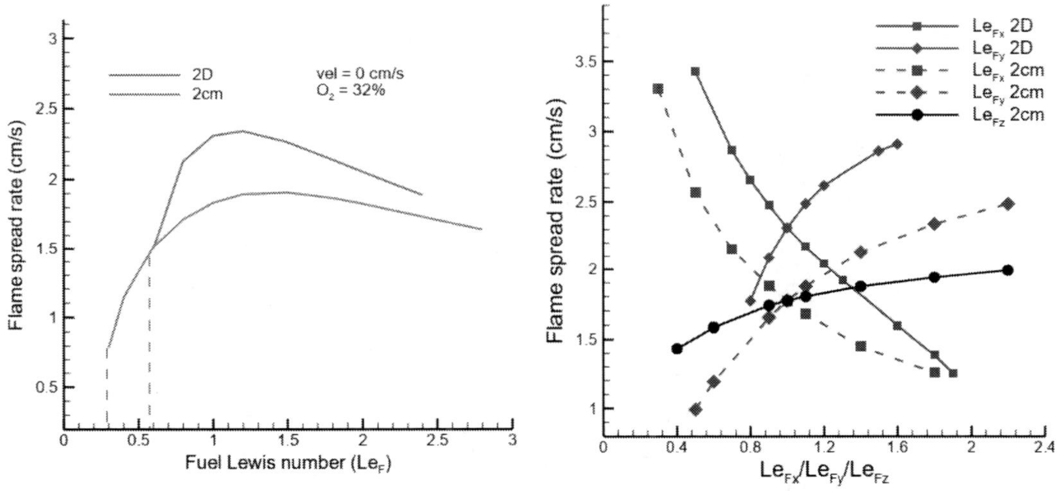

Figure 2: Variation of flame spread rate for 2D and 2cm width fuel with (a) Le_F (b) Le_{Fx}/ Le_{Fy}/ Le_{Fz}

simulations are second order accurate in space and time. The model was tested at various stages [4, 5] and was successfully used to simulate transient flame spread phenomena such sudden transition from 1ge to 0ge and effect of g-jitter [6]. Flame-tracking algorithm is developed to capture the fastest spreading section of the flame. The grid size and time steps were chosen based on refinement study to ensure consistent qualitative trend and reasonably time accurate solution over the period of simulation. The time step for simulation is 0.01s and for each time step, 80 iterations were carried out to ensure convergence.

2.3 Equations

The primary effect of the fuel Lewis number comes from its influence over the fuel species equation:

$$\frac{\partial(\rho Y_F)}{\partial t} + \rho u(Y_F)_x + \rho v(Y_F)_y + \rho w(Y_F)_z = \frac{1}{Le_{Fx}}(\rho D_F(Y_F)_x)_x + \frac{1}{Le_{Fy}}(\rho D_F(Y_F)_y)_y + \frac{1}{Le_{Fz}}(\rho D_F(Y_F)_z)_z + \dot{\omega}_F \quad (1)$$

and the boundary condition over the surface of the fuel:

$$\dot{m}Y_{F,w} = \dot{m} + \frac{\rho D_F}{Le_F}\left(\frac{\partial Y_F}{\partial y}\right)_w \quad (2)$$

3. Results and Discussion

The case of zero-g self-propagation in a quiescent atmosphere is chosen to examine the details of three-dimensional opposed flame spread. From figure 2(a), it can be observed that the flame spread curve is non-monotonic with Le_F with a maximum value occurring at an intermediate value of $Le_F = 1.2$ for 2D and $Le_F = 1.5$ for 2cm wide fuel. To explain this behavior, Lewis numbers in x, y and z directions are varied independently. When the x-direction fuel Lewis number (Le_{Fx}) is varied, Le_{Fy} and Le_{Fz} are held at unity. Similar approach is adopted for varying Le_{Fy} and Le_{Fz}. Figure 2(b) shows that the spread rate curves of Le_{Fy} and Le_{Fz} have same trend, while Le_{Fx} has opposite trend. From figure 3, it can be seen that Le_{Fx} and Le_{Fy} have significant effect on spread rate compared to Le_{Fz}. For 2cm wide fuel case because of third dimension, sensitivity of Le_{Fx} and Le_{Fy} is reduced compared to 2D. The curve of varying Le_F ($Le_{Fx} = Le_{Fy} = Le_{Fz}$) lies in between Le_{Fx} and Le_{Fy} curves, thus the individual curves (Le_{Fx} , Le_{Fy} and Le_{Fz}) are critical to the understanding of the over-all Le_F curve.

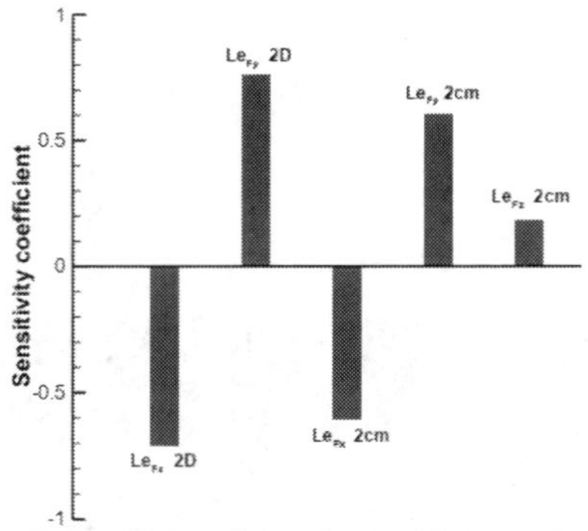

Figure 3: Sensitivity of spread rate with Le_{Fx}/ Le_{Fy}/Le_{Fz} for 2D and 2cm wide fuels at $Le_F = 1$.

As Le_{Fx} decreases (D_{Fx}, increases), more fuel diffuses ahead of the flame and the flame front moves upstream relative to the pyrolysis front (x = 0) as shown in fig 4(a). Therefore, the flame front advances and increases the heat transfer in the preheat zone and flame spread rate increases. There is no appreciable change in the movement of flame front in y and z directions. From figure 2(a), if Le_{Fx} is large, spread rate is lower. Lower spread rate implies, less fuel is burnt and length of the flame decreases (see the contour of Le_{Fx} = 1.8 in figure 4(a)). When flame length becomes too short, the ratio of heat loss to heat generated increases leading to extinction of flame.

Decreasing Le_{Fy} (D_{Fy} increases), increases the flame standoff distance from the solid surface, (see close up of fig 4(b) xy plane). The conductive heat flux to the preheat region decreases,

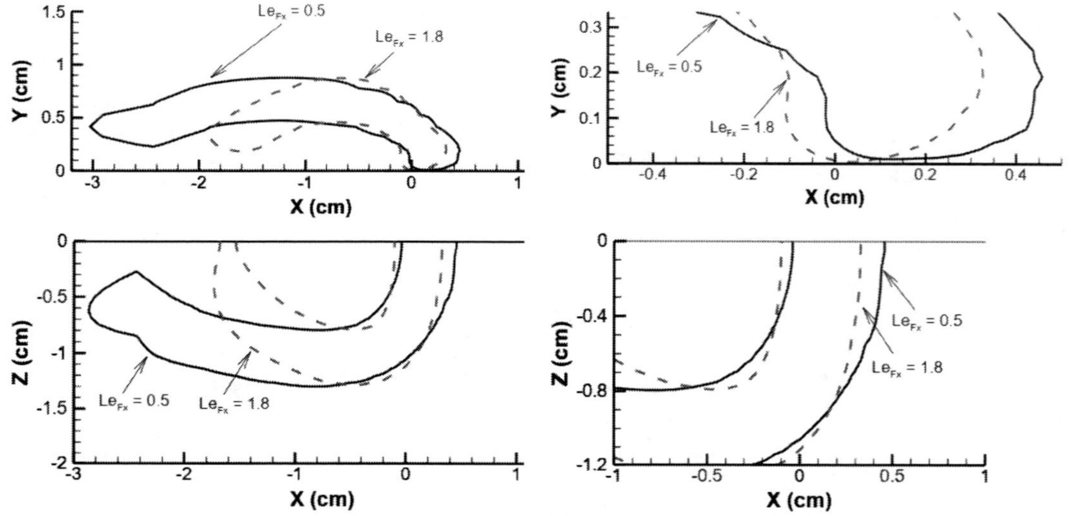

(a)

Figure 4: Flame shape and size (defined by reaction rate contours of 0.0001 g/cm³/s) for flame spread over 2 cm wide fuel at 32% oxygen for (a) Le_{Fx} = 0.5 and 1.8 (b) Le_{Fy} = 0.5 and 1.8 (c) Le_{Fz} = 0.5 and 1.8. Flame section shown at Z = 0 cm (top part of figures) and at Y = 0.2 cm (bottom part of figures). Figures on the right column are close up views (zoom in) of the flame base of the corresponding left column figures. Note: figures 4(b) and 4(c) are below.

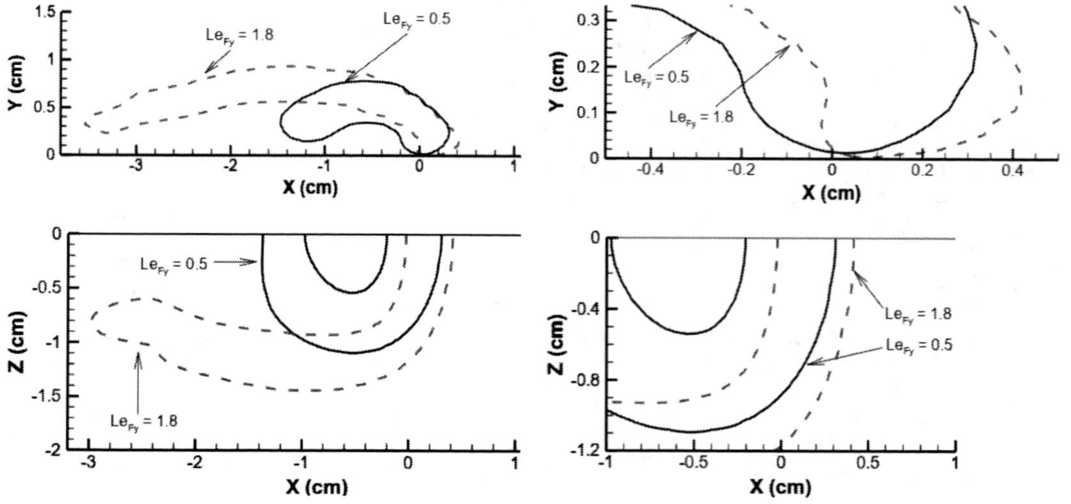

(b)

which leads to a decrease in the flame spread rate. Although only Le_{Fy} is varied, y and z-direction diffusion are also present. It can also be observed that as D_{Fy} increases, it hinders the fuel diffusion to the upstream and flame front retracts in x direction. This decreases the heat feedback further and the spread rate decreases furthermore.

effects in x and y-direction are dominant compared to z. Mass transfer affects the location and shape of the flame front, and as a result the spread rate changes. Conventionally, heat transfer has been considered as the main driving mechanism for flame spread over condensed fuels. The present work indicates that mass transfer can also significantly affect flame spread and extinction.

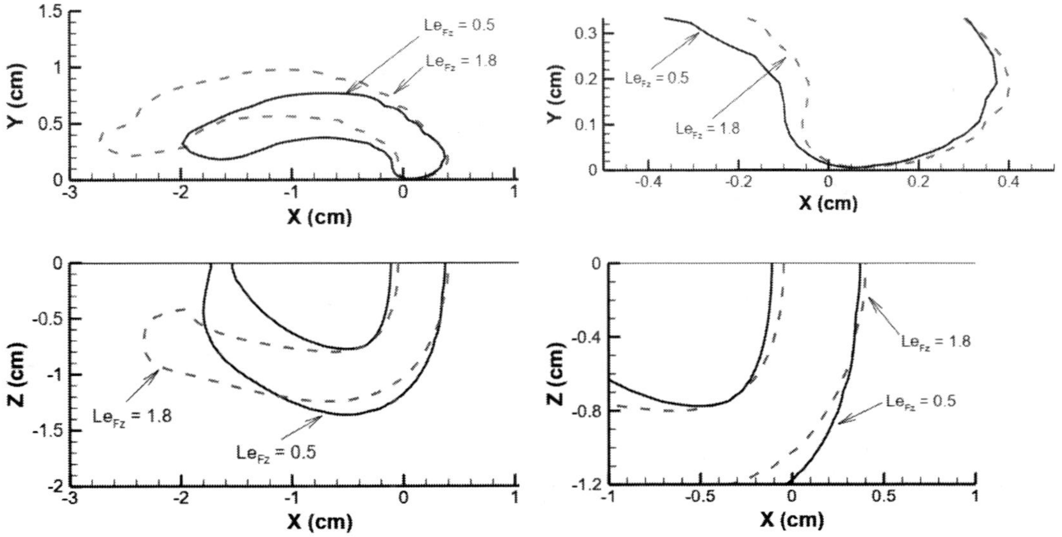

(c)

As Le_{Fz} decreases (D_{Fz}, increases), fuel diffusion in x and y-direction is reduced. As a result the flame front moves downstream in x direction, reducing the heat transfer to the preheat zone and flame spread rate decreases. There is no noticeable change in flame base position in y direction. Figure 2(a) shows if Le_{Fy} or Le_{Fz} is small, spread rates are low. Therefore, flame length shrinks (see the contour of $Le_{Fy} = 0.5$ in figure 4(b)) and flame quenches due to similar argument discussed for Le_{Fx}.

Conclusions

An investigation is made to study the effect of fuel Lewis number (Le_F) and mass transfer on flame spread over thin solids using a detailed three-dimensional numerical model. The curve of flame spread rate with Le_F is non-monotonic. As Le_F is decreased the flame spread rate first increases, reaches a peak, and then decreases. In the high fuel Lewis number range, the spread rate curve follows the trend of the x-direction fuel diffusion (Le_{Fx}). As Le_F decreases, more fuel vapor diffuses upstream of the pyrolysis front, and hence the flame front is anchored further upstream, thereby increasing the heat flux to the preheat zone and spread rate increases. In the low fuel Lewis number range, the spread rate curve follows the trend of the y-direction fuel diffusion (Le_{Fy}). As Le_F decreases, the flame is pushed further away from the solid surface and the flame heat feedback is reduced leading to decrease in flame spread rate. While Le_{Fx} and Le_{Fy} have direct effect on spread rate, Le_{Fz} doesn't influence the spread rate directly. Increasing the fuel diffusivity in z direction, hinders the fuel diffusion in x and y. As a result flame front retracts in x, reducing the heat transferred to the preheat region and hence flame spread rate decreases. Sensitivity analysis demonstrated that mass transport

References

1. Kevin Tolejko, Ioan I. Feier, James S. T'ien, Proc. Combust. Inst., 30(2), (2005) 2263-2270.
2. Kiran Kumar M N and Amit Kumar, Proc. Combust. Inst., 36(2), (2015) 3081-3087.
3. Kirkby, L. L and R. A Schmitz, Combust. Flame., 10(3), (1966) 205-220
4. Chenthil Kumar and Amit Kumar, Combust. Theory and Modelling, 16(3), (2012) 537-569.
5. Vinayak Malhotra, Chenthil Kumar and Amit Kumar, Combust. Theory and Modelling, 17(5), (2013) 837-857.
6. Chenthil Kumar, Amit Kumar, Proc. Combust. Inst., 34(2), (2013) 2675-2682

Response to reviewer's comments:

1. As the work presented in the paper is hypothetical, there are no experiments results available. Numerical results in 2D are available in literature. Splitting of overall Le_F into components was done by Tolejko et al. [1] for 2D opposed flame spread to explain the dependence of flame spread rate on fuel Lewis number. Similar study is carried out here for 3D / finite width fuels to obtain more physical insight on three-dimensional aspect of fuel mass transfer on flame spread. There is no significant effect due to third dimension (Z direction) and the trend of flame spread rate with Le_F didn't change (remained non-monotonic) because Le_{Fx} and Le_{Fy} have direct effect on spread rate whereas Le_{Fz} doesn't influence the spread rate directly. As the spread rate curve of Le_{Fz}, follows the trend of Le_{Fy}, the fuel Lewis number at which maximum spread rate occurs is shifted further away from unity compared to 2D. Note:

that the 2D and 3D results presented in this work (Figure 2) are at O_2 = 32% and the 2D results in literature (Tolejko et al. [1]) are at O_2 = 21%.

2. Heat transfer and mass transfer are inherently coupled and in the present work they are not decoupled. In this paper, the fuel mass diffusion in three directions are isolated by means of fuel Lewis number.

11th Asia-Pacific Conference on Combustion,
The University of Sydney, NSW Australia
10th -14th December 2017.

Numerical investigation of thermal and chemical effects of nanosecond repetitively pulsed discharges on a laminar premixed counterflow flame

Sylvain Heitz[1,2*], Jonas P. Moeck[1,3], Anne Bourdon[4], Deanna A. Lacoste[2]

[1]Institut für Strömungsmechanik und Technische Akustik, Technische Universität Berlin
Müller-Breslau Straße 8, 10623 Berlin, Germany

[2]Clean Combustion Research Center, King Abdullah University of Science and Technology
Thuwal 23955-6900, Saudi Arabia

[3]Department of Energy and Process Engineering, Norwegian University of Science and Technology
N-7491 Trondheim, Norway

[4]LPP, CNRS, École polytechnique, UPMC Univ Paris 06, Univ. Paris-Sud, Observatoire de Paris, Université Paris-Saclay, Sorbonne Universités, PSL Research University, 91128 Palaiseau, France

Abstract

This paper focuses on modelling the influence of nanosecond repetitively pulsed (NRP) corona discharges on a lean premixed methane-air flame at atmospheric pressure and ambient temperature. The discharges are produced by high voltage pulses of 3.5 kV amplitude and 10 ns duration, with a repetition frequency of 30 kHz. Taking into account the geometry of the electrodes, at breakdown, the reduced electric field is then comprised between 40 and 500 Td. The flame has an equivalence ratio of 0.7. A numerical model is developed to determine the effect of the discharges on the unburnt gas. In a first scenario, all the plasma energy goes to chemical reactions, eventually resulting in ozone production. A second scenario considers that all the plasma energy is thermal, thus leading to a temperature increase of the fresh gases. The third scenario considers that half the energy goes to ozone production and the other half to fresh gases heating. The ozone concentrations or gas heating resulting from each of these scenarios are then implemented into the simulation of a freely propagating flame, allowing the comparison of thermal and chemical effect of plasma discharges, as well as their combined effect, on the laminar flame speed and adiabatic flame temperature. The high ozone concentrations (up to 10,000 ppm) resulting from the NRP corona discharges enhance the laminar flame speed up to 40 % of its initial value, while the temperature increase (up to 10 K) results into a maximum flame speed enhancement of 25 %. The effect on the adiabatic flame temperatures follow a similar trend.

1 Introduction

During the last decade, non-equilibrium plasma discharges have drawn increasing attention for their potential in influencing combustion in gas turbines and aeronautical engines. Studies have demonstrated the effect of these discharges on various combustion-related phenomena such as mitigation of thermoacoustic instabilities [1,2], extension of a flame's stability domains [3] and modification of the pollutant emissions, in particular the nitric oxides [4,5].

Non-equilibrium plasma discharges are known to affect the flame in two different ways. In the first one, the discharge locally heats the inter-electrode gas and can be considered as a heating source. In the second one, plasma-produced species such as excited or ionised species and radicals modify the reaction pathways of the flame.

In Lacoste *et al.* [6], the effect of NRP glow discharges on laminar premixed flames in terms of the step response has been investigated. An upstream movement of the flame was observed once the discharges were activated. This was attributed to an increase in the burning velocity resulting from higher reactivity of the gas treated by the plasma. These NRP glow discharges are known to have a negligible thermal effect, so that an increase in laminar flame speed could not be attributed to heating of the unburnt gas alone, but rather to the generation of active species by the discharges.

The experimental work performed prior to the present study is described in Heitz *et al.* [7]. The experimental setup is displayed in Fig. 1(a): a laminar, methane and air flow, set at an equivalence ratio of 0.7, was confined at the quartz nozzle of a burner of 10-mm diameter. A cooled stagnation plate was located 4.5 mm above the top of this quartz tube, allowing for stabilisation of an approximately flat flame. The flame rim was attached to the quartz tube and its centre was curved due to the presence of the central electrode upstream of the flame. NRP plasma discharges in the corona regime were generated between two tungsten pin electrodes. The upstream high-voltage electrode had a diameter of 100 μm and the downstream grounded electrode had a diameter of 1 mm. The pulses had peak amplitudes, U, of up to 3.5 kV, a duration of 10 ns and were applied at a repetition frequency of 30 kHz. A photograph of the flame and the NRP corona discharges is shown in Fig. 1(b) with all the significant dimensions (in mm).

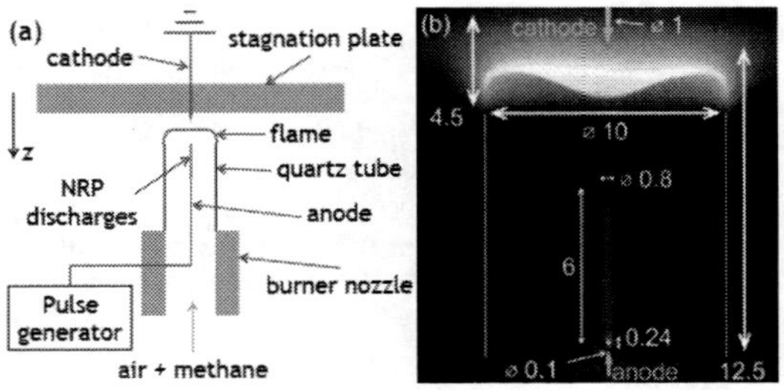

Figure 1: (a) Experimental setup. (b) Photograph of the flame

* Corresponding author.
E-mail address: sylvain.heitz@kaust.edu.sa

(blue) and the NRP corona discharges (violet), dimensions are given in millimetres.

The displacement of the flame during application of NRP corona discharges has been quantified for two different applied voltages 2.5 and 3.5 kV. Figure 2 presents the evolution of the flame's displacement, Δz, of the flame's centre as a function of the number of applied plasma discharges, n. For both voltage amplitudes considered, a fast flame displacement over the first 2,000 pulses is observed, and the maximum flame displacement is reached after approximately 10,000 pulses; beyond, the flame displacement remains within the range of uncertainty of the experiments. The upstream displacement at a given number of applied pulses increases with the voltage amplitude. This action has been attributed to an increase in laminar flame speed induced by the NRP corona discharges. The influence of a laminar flame speed enhancement on the evolution of flame displacement can be calculated and fitted to the experimental data of Fig. 2 (see dashed red line) in [7]. The flame speed enhancements are 5 and 9 % for 2.5 and 3.5 kV applied voltage, respectively.

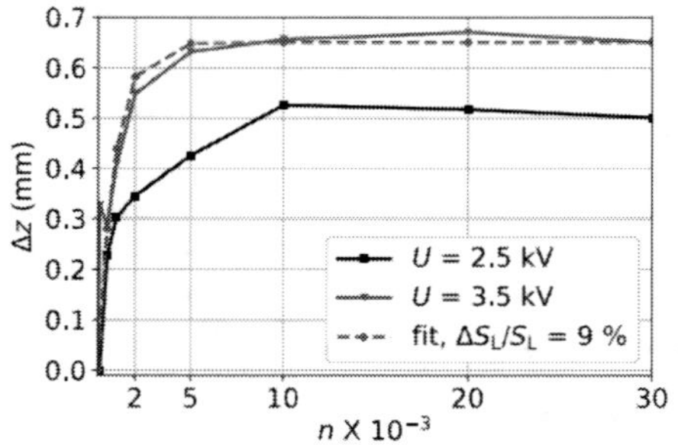

Figure 2: Evolution of flame front displacement, Δz, as a function of the number of applied pulses, n. The laminar flame speed enhancement is estimated by fitting the experimental curves (red line) with a model for the flame displacement [7].

The effect of NRP plasma discharges on a flame has thus been assessed experimentally and the laminar flame speed modification established as a possible cause for the observed phenomenon of flame displacement. This influence has been shown previously but there is a lack of information in the literature as to the thermal and chemical effects one can expect from NRP corona discharges. In this study, a numerical model is developed in order to predict the thermal and chemical effects of NRP corona discharges in the configuration used in [7], on laminar flame speed and adiabatic flame temperature.

2 Numerical model

A numerical model is developed to evaluate the thermal and chemical effects of NRP corona discharges and their potential influence on the burning velocity and the adiabatic temperature of a freely propagating flame. In the first scenario considered (named chemical scenario), all the energy of the NRP corona discharge is used to excite electrons, thus inducing subsequent reactions eventually leading to ozone production. This unique species has been chosen, in order to simplify the model and because, in the experiments of [7], as the end of the plasma plume is far from the flame, only long duration species can reach the combustion front. In addition, as no strong heating of the gas can be expected, O_3 is likely to be the dominant species. Another model (named thermal scenario) is then developed to evaluate the effect of the NRP discharges on the flame, if all the electrical energy of the discharges were to be converted to thermal energy and thus increased the temperature of the unburnt gases. A third, intermediate scenario (named mix scenario) considers that one half of the plasma energy goes to electron excitation and the other half to unburnt gas heating. In this approach, no losses either of heat or of active species, are considered.

2.1 Chemical scenario

In this first scenario, all the plasma energy serves to excite the electrons. The focus of this scenario is on the ozone concentration resulting from the reactions in the plasma discharge. Ozone is a species which has been proved to be one of the main products of non-equilibrium plasma discharges [8] and is also known to have an influence on a flame's burning velocity [9]. Due to the 10-ms convection time of the particles from the high-voltage electrode to the flame front, only the long-lived species can affect the combustion. In this study we only focus on the effect of ozone. Other long-lived species such as NOx will be considered in a future extended version of the chemical scenario. Excited molecular and atomic oxygen could also present some interest towards modifying the considered flame's properties. In a first step, two reactions of electron-impact dissociation of oxygen and excitation of nitrogen, happening during the 10 ns of the pulse, are considered:

$$N_2 + e^- \rightarrow N_2^* + e^-, \qquad (1)$$

$$O_2 + e^- \rightarrow 2\,O + e^-. \qquad (2)$$

A path flux analysis in [10] shows that more than 70 % of plasma-produced atomic oxygen originates from reaction (2). A third reaction of nitrogen quenching is added, providing additional atomic oxygen:

$$N_2^* + O_2 \rightarrow N_2 + 2\,O. \qquad (3)$$

The authors of [11] have shown that it takes 25 ns for the excited nitrogen to be totally quenched, so this duration is considered in the model for reaction (3). Finally, ozone production is considered with the three-body reaction during the 10-ms convection of the gases from the plasma discharge to the flame:

$$O + O_2 + M \rightarrow O_3 + M, \qquad (4)$$

where M is either nitrogen or oxygen. Several studies [12,13] mention this reaction as the most significant in terms of ozone production. The electron number density is supposed to be constant for the calculations of reactions (1) and (2) which take place during 10 ns of the high voltage pulse.

The reaction rate constants from (1) and (2) as a function of the applied reduced electric field were determined by solving the Boltzmann equation with the BOLSIG+ code [12]. A set of cross sections for electron collisions with nitrogen and oxygen molecules were taken from the IST-Lisbon database on the BOLSIG+ website [13]. The reaction rates for (3) and (4) are taken respectively from [11] and [14]. The BOLSIG+ code also provides electron temperatures as a function of the reduced electric field. For electric fields estimated between 40 and 500 Td, the electron temperature, T_{e^-}, is comprised between 1.2 and 10 eV. The number density of electrons, n_{e^-}, is taken as 1.2×10^{15} cm^{-3}, an average value estimated from [15] for NRP glow discharges with an electric field similar to the range considered in this study. The volume considered for the discharge energy deposition, V_1, is defined as the region in Fig. 2 where the light emission is maximum. This volume is assimilated to a cylinder of 400 µm radius and 240 µm height, so that its value is 0.12 mm^3. Multiplying the electron number density by the discharge volume and the electron temperature allows an estimation of the discharges' deposited energy, E_{pl}, as comprised between 0.03 and 0.25 μJ per pulse. It has not been possible to measure the deposited energy in [7] in a quantitative manner, but the electrical energy of NRP corona discharges is known to be of the order of a few microjoules per pulse [16]. Considering the dimensions of the volume V_1 and the 30-kHz PRF of the discharge, a volume of gas flowing into the discharge region is subjected to 8 high-voltage pulses before exiting it.

2.2 Thermal scenario

In this second scenario, the energy is considered to be deposited integrally in the form of thermal energy within the discharge volume, V_1, defined in Part 2.1. Therefore, the unburnt gases flowing through the discharge are heated with the same energy range as in Part 2.1. The temperature increase of the gas is then comprised between 1 and 11 K. After this temperature increase, the gas at temperature T_1 exits from the discharge volume and expands to a volume V_2, decreasing the gas temperature in the process. Assuming an adiabatic expansion of the gas, the new volume, V_2, is determined with:

$$p_1 \times {V_1}^{\gamma_1} = p_2 \times {V_2}^{\gamma_2}, \quad (7)$$

where p_2 is the atmospheric pressure and γ_1 and γ_2 are the heat capacity ratios, before and after expansion of the gas, respectively. The new gas temperature after expansion, T_2, is then deduced from the ideal gas law.

2.3 Mix scenario

This scenario considers an intermediate situation where 50 % of the energy goes into electron excitation, resulting in approximately half the final ozone concentration compared to the chemical scenario. The other half of the discharge energy goes into unburnt gas heating, leading to approximately half of the temperature increase from the thermal scenario. This case is considered in order to observe the possible existence of a synergetic effect between thermal and chemical effects on the flame characteristics.

2.4 Effects of the different scenarios on the flame

Once the ozone concentrations and temperature increases are estimated over the entire energy range considered, they are implemented in a Cantera [17] script allowing to determine laminar flame speed and adiabatic flame temperature of a freely propagating flame. The GRI-Mech 3.0 mechanism [18] is modified to include reactions between ozone and air molecules from [14] and validated with experimental data from [9]. In the simulations, the pressure is one atmosphere, the initial fresh gas temperature is 283 K and the equivalence ratio of the methane-air mixture is 0.7. The initial conditions of the gas are modified according to the results from Sections 2.1, 2.2 and 2.3.

3 Results and discussion

3.1 Chemical scenario

As the electric field—and so the energy deposited by the discharges—is increased, so is the ozone production. Figure 3 displays the final concentration of ozone, $[O_3]$, as a function of the energy per pulse, E_{pl}, also expressed in terms of reduced electric field, E/n. The ozone concentration comes entirely from the recombination reaction of atomic oxygen with molecular oxygen (4). For this reason, two pathways of ozone formation are distinguished: the first one where atomic oxygen is produced by direct electron-impact dissociation, as expressed in (2), and the other one coming from the excitation then quenching of nitrogen, as in (1) and (3). Ozone generated by the first pathway is represented by the dotted blue line and ozone originating from the second pathway by the dashed green line.

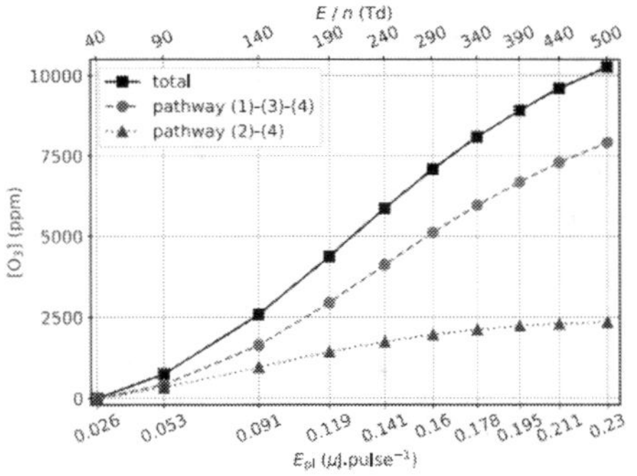

Figure 3: Ozone production, $[O_3]$, as a function of deposited energy per pulse, E_{pl}, and reduced electric field, E/n.

Ozone production is strongly dependent on the input energy of the plasma discharges. The total concentration of ozone reaches 10,000 ppm at the highest energy considered. The pathways decomposition shows that ozone originating from direct electron-impact dissociation of oxygen (blue curve) provides for up to 45 % of the total ozone concentration at low energies and that its share is reduced to no more than 25 % of the total ozone concentration for higher reduced electric fields.

3.2 Thermal scenario

The linear increase in the unburnt gases temperature, ΔT, as a function of the deposited energy, E_{pl}, is displayed in Fig. 4. For the same range of deposited energies as in Section 3.1, the resulting temperature increase reaches a maximum of about 10 K.

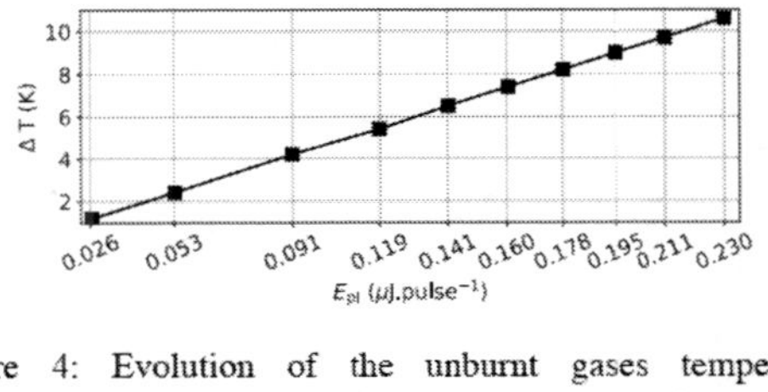

Figure 4: Evolution of the unburnt gases temperature increase, ΔT, as a function of the plasma-deposited energy per pulse, E_{pl}.

3.3 Comparison of thermal and chemical scenarios

Figures 5 and 6 show the evolution of the laminar flame speed enhancement and the adiabatic flame temperature increase as a function of the deposited energy per pulse.

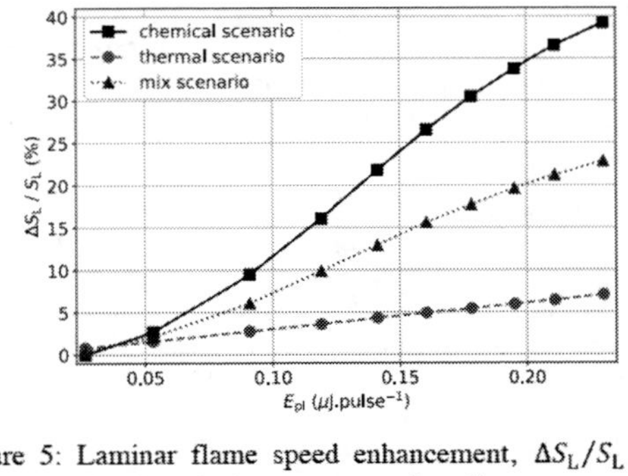

Figure 5: Laminar flame speed enhancement, $\Delta S_L/S_L$, as a function of the deposited energy per pulse, E_{pl}.

The flame speed enhancement provided by ozone (solid black line) reaches 40 %, a value far higher than the thermally induced flame speed increase (dashed green line) which reaches 25 % for a similar deposited energy. The enhancement at low energy deposition is similar for all three scenarios, but increases far more when all the energy goes to ozone production, than when it goes to the heating of the unburnt gases. The enhancement provided by the mix scenario (dotted blue line) corresponds to the intermediary value between the 100 % thermal and 100 % chemical cases. The model in its current state does not display the existence of a synergetic effect between plasma-induced unburnt gas heating and plasma-produced ozone. The flame speed enhancement is higher than the experimental enhancement of 4 to 9 % found in [7], a difference attributed to the radial diffusion of species and heat, which are not considered in the model in its present state, and to the reduced electric field, whose value features some uncertainty in the experimental study.

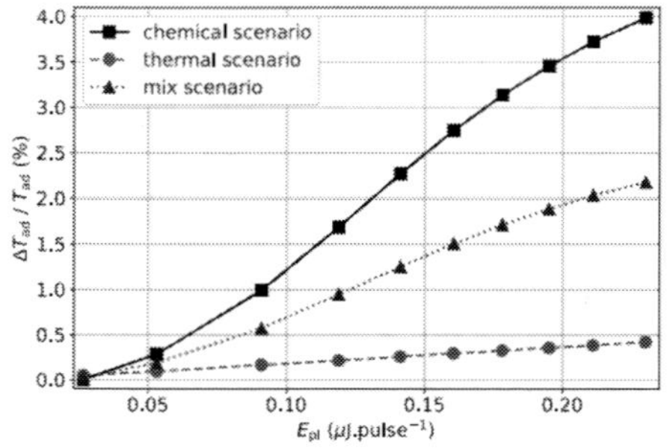

Figure 6: Adiabatic temperature enhancement, $\Delta T_{\mathrm{ad}}/T_{\mathrm{ad}}$, as a function of the deposited energy per pulse, E_{pl}.

The trend of adiabatic flame temperature evolution is similar to the one of the laminar flame speed, but smaller by a factor of ten. The ozone-induced flame temperature increase is 8 times higher than the one induced by unburnt gases heating. Due to the linear evolution of both the laminar flame speed and the adiabatic flame temperature with the ozone concentration in the fresh gases and with the fresh gas temperature, they follow similar trends.

4 Conclusion

The effect of NRP corona discharges on a laminar premixed methane-air flame has been modelled by considering, on the one hand, a purely thermal effect of the discharges, and on the other hand, a purely chemical effect, through the production of ozone. The simplified model shows that a greater effect of the chemical scenario is to be expected, thus emphasizing the interest of using the essentially non-thermal corona and glow discharges. Even at the low deposited energies considered, an increase of up to 40 % in the laminar flame speed was obtained with 10,000 ppm ozone addition. The thermal enhancement for a similar energy is 25 %. The adiabatic temperature increase of the flame follows a trend similar to the flame speed enhancement but is less significant.

Future numerical work will focus on refining the model by including losses of heat and species, which would result in a smaller and more realistic enhancement, compared to what was obtained in this study. The inclusion of other combustion-enhancing species in the chemical scenario, in particular the nitric oxides, will also be considered. The simulations will also need to be implemented on a stagnation flame instead of a freely propagating flame, to consider a configuration closer to the previous experimental study. Experimental work will be performed in order to obtain quantitative data of the energy deposited by the plasma, as well as the ozone concentration and the temperature increase resulting from the plasma discharges. These quantitative data will be used for comparison with the results yielded by the model.

5 Acknowledgments

This work was supported by the Agence Nationale de la Recherche (ANR) and the German Research Foundation (DFG) through the DRACO project (grant numbers ANR-13-IS09-0004, MO2551/1) as well as by the King Abdullah University of Science and Technology. The authors would also like to thank the École Normale Supérieure (ENS) de Cachan and the Université Franco-Allemande (UFA) for the stipend of Sylvain Heitz.

References

[1] J. P. Moeck, D. A. Lacoste, C. O. Laux, C. O. Paschereit, AIAA paper 2013-0565 (2013).

[2] D. A. Lacoste, J. P. Moeck, D. Durox, C. O. Laux, T. Schuller, J. Eng. Gas. Turb. Power 135 (2013) 101501 (7 pp).

[3] S. Barbosa, G. Pilla, D. A. Lacoste, P. Scouflaire, S. Ducruix, C. O. Laux, D. Veynante, Phil. Trans. R. Soc. A 373 (2015) 20140335 1-11.

[4] W. Kim, H. Do, M. G. Mungal, M. A. Cappelli, Proc. Comb. Inst. 31 (2007) 3319-3326.

[5] D. A. Lacoste, J. P. Moeck, C. O. Paschereit, C. O. Laux, J. Propul. Power 29 (2013) 748-751.

[6] D. A. Lacoste, J. P. Moeck, W. L. Roberts, S. H. Chung, M. S. Cha, Proc. Combust. Inst. 36 (2017) 4145-4153.

[7] S. Heitz, J. P. Moeck, D. A. Lacoste, Proc. 8th Europ. Combust. Meeting, Dubrovnik, Croatia (2017) 1378-1383.

[8] J. Chen, J. H. Davidson, Plasma Chem. Plasma Process. 22 (2002) (4) 495-522.

[9] T. Ombrello, S. H. Won, Y. Ju, S. Williams, Combust. Flame 157 (2010) 1906-1915.

[10] S. Nagaraja, T. Li, J. A. Sutton, I. V. Adamovich, V. Yang, Proc. Combust. Inst. 35 (2015) 3471-3478.

[11] D. L. Rusterholtz, D. A. Lacoste, G. D. Stancu, D. Z. Pai, C. O. Laux, J. Phys. D: Appl. Phys. 46 (2013) 464010 (21pp).

[12] G. J. M. Hagelaar, L. C. Pitchford, Plasma Sources Sci. Technol. 14 (2005) 722-733.

[13] L. L. Alves, J. Phys. Conf. Series (2014) 565, 1, www.lxcat.net.

[14] L. B. Ibraguimova, G. D. Smekhov, O. P. Shatalov, http://chemphys.edu.ru/media/published/Ибрагимова-2003-01-20-001.pdf, 2003.

[15] F. Tholin, A. Bourdon, J. Phys. D: Appl. Phys. 44 (2011) 385203 (17pp).

[16] D. Z. Pai, G. D. Stancu, D. A. Lacoste, C. O. Laux, Plasma Sources Sci. Technol. 18 (2009) 045030 (7pp).

[17] D. G. Goodwin, H. K. Moffat, R. L. Speth, http://www.cantera.org, 2017. Version 2.3.0.

[18] G. P. Smith, D. M. Golden, M. Frenklach, N. W. Moriarty, B. Eiteneer, M. Goldenberg, C. T. Bowman, R. K. Hanson, S. Song, W. C. Gardiner, Jr., V. V. Lissianski, Z. Qin, http://www.me.berkeley.edu/gri_mech/.

11th Asia-Pacific Conference on Combustion,
The University of Sydney, NSW Australia
10th -14th December 2017.

Derivation of accurate rate constants for char CO_2, H_2O gasification using a pressurized drop tube furnace

Geun Sohn[1], Insoo Ye[1], Changkook Ryu[1], Howon Ra[2], Sungmin Yoon[2]

[1]School of Mechanical Engineering, Sungkyunkwan University

2066 Seobu-ro, Suwon-si, Gyeonggi-do, 164-19, Republic of Korea

[2]Clean Fuel Department, Korea Institute of Energy Research

140 Yuseong-daero, Daejeon, 341-29, Republic of Korea

Abstract

Drop tube furnace (DTF) test is a routinely applied technique to analyze the reactivity of solid fuels such as coal or biomass under controlled atmospheres. For char conversion, the rate constants for a reaction model are derived from the reaction conditions varying the temperature and the partial pressure of reactants. The rate constants can be used in computational fluid dynamic (CFD) simulations for combustion or gasification systems utilizing the fuel. However, the reaction conditions of particles inside a DTF are often assumed to be identical to the controlled reactor conditions. This study assesses the validity of such approach and proposes a new method to derive accurate rate constants. For this purpose, coal char was gasified by CO_2 and H_2O in a pressurized DTF, and the rate constants of a char conversion model was derived for the controlled reactor conditions. CFD simulations using the rate constants significantly underestimated the char conversion. Three reasons were identified for the discrepancies: i) shorter residence time of char particles concentrated more on the high velocity region, ii) lower particle temperatures by the endothermic reactions, and iii) lower partial pressure of reactants by its consumption. A new method was proposed to estimate the actual reaction conditions. The CFD results adopting the corrected rate constants based on the method well matched with the measured data.

1 Introduction

Solid fuels such as coal and biomass undergo pyrolysis and char conversion as two major reaction processes during combustion or gasification [1]. Pyrolysis or devolatilization is the first step that releases various volatile species including tar and light gases by thermal decomposition of the chemical structure gradually progressing at temperatures typically above 250°C. Char is the carbonaceous solid residue including inorganic matter remained after pyrolysis. It is slowly converted into gases by heterogeneous reactions with O_2, CO_2, H_2O, and H2, expressed by the following global reactions.

[R1] $C(s) + O_2 \rightarrow CO_2$ Δhr= -393.5 kJ/mol

[R1'] $C(s) + O_2 \rightarrow CO$ Δhr= -110.5 kJ/mol

[R2] $C(s) + H_2O \rightarrow CO + H_2$ Δhr= 131.4 kJ/mol

[R3] $C(s) + CO_2 \rightarrow 2CO$ Δhr= 172.5 kJ/mol

[R4] $C(s) + 2H_2 \rightarrow CH_4$ Δhr= -87.4 kJ/mol

Under the typical combustion condition at high temperatures, [R1'] is dominant whereas two endothermic reactions of [R2] and [R3] becomes important under gasification conditions. The actual char conversion of a particle involves complicated physical and chemical mechanisms, including the gas diffusion and structural changes (diameter, density, and pore development) of a particle as well as the char reactivity. It is described by various global or intrinsic models such as apparent rate model, shrinking core model (SCM), random pore model (RPM), char burnout kinetics (CBK) model, etc. [2-5].

Understanding the reaction characteristics of target fuel is essential in designing and operating combustors or gasifiers. However, the reaction process and kinetics of solid fuel conversion differ between samples and are difficult to predict in details. This is because the chemical structures vary widely and complicated physical/chemical phenomena are involved including heat transfer, molecular diffusion, influence of inorganic matter, and structural changes of the solid particle. Because these factors are influenced by reaction conditions (temperature, pressure, heating rate, particle size, gas concentrations etc.), it is better to evaluate the reaction characteristics under the conditions close to the actual reactor as far as possible.

This study assessed the validity of the simplified method in determining the rate constants of char conversion from DTF tests based on the controlled reactor conditions. For this purpose, coal char was gasified by CO_2 and H_2O in a pressurized DTF for a set of partial pressures and temperatures. The flow conditions in the DTF were also examined in separate tests. The rate constants for char gasification were derived using the conventional method for SCM as an indicative model. CFD simulations were performed to evaluate the rate constants and to find the reasons for the discrepancies in char conversion. Then, an analytical method was proposed to determine the actual reaction conditions of char particles. The rate constants based on the corrected reaction conditions were validated by repeating the CFD simulations.

Corresponding author. Fax: +82-31-290-5889
E-mail address: cryu@skku.edu

2 Experimental method

2.1 Pressurized drop tube furnace

Fig. 1 shows the pressurized DTF used in this study. It consists of a vertical reactor (ID 40.9 mm x H 1000 mm) installed in electric heaters, coal feeding and gas supply system at the top, and particle trap at the bottom. Char particles are fed by a screw feeder inside a pressurized chamber and transported by N_2 to the water-cooled inlet tube (ID 14 mm) at the center of the reactor. The blend of gases with specific composition is supplied through a mixer and preheater to the annulus of char supply tube. At the exhaust of the heating section, the char particles are collected using a water-cooled receiver.

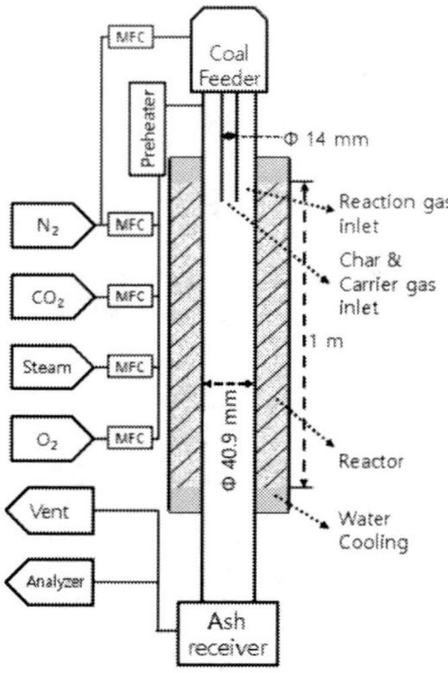

Figure 1: Schematic of the pressurized drop tube furnace.

2.2 Coal and char properties

Table 1 shows the properties of a coal used in this study. It was air-dried, pulverized, and sieved to have a size range of 45-75μm. The char was then produced from the prepared coal sample in another DTF having the similar configuration with that in Fig. 1, but operating at atmospheric pressure and 1200°C under N_2 atmosphere (1 l/min). The coal throughput was 3 g/min.

Parameter		Coal
Proximate analysis [wt.%$_{wet}$]	Moisture	2.37
	Volatile	27.11
	Fixed Carbon	58.23
	Ash	12.29
Ultimate analysis [wt.%$_{daf}$]	C	82.38
	H	4.37
	O	11.67
	N	1.17
	S	0.40

Table 1: Properties of coal and char tested in the PDTF.

2.3 Pressurized char gasification tests

Table 2 summarizes the test conditions to determine the reactivity of char gasification. Under the operating pressure of 4 bar$_{abs}$, 1 g/min of coal was transported by 1 l/min N_2. The reactant containing CO_2 or H_2O was supplied at a flow rate of 5 l/min at room temperature. Two sets of tests were performed by varying reactor temperatures (T_R) and partial pressures (P_i), respectively, for each char gasification reaction by CO_2 and by H_2O. In the first set of tests, T_R was varied between 1000 and 1300°C while P_i was fixed in order to derive the pre-exponential factor (A) and activation energy (E). The range of T_R was determined to have a sufficient reactivity while the reaction rate remains below the diffusion limit of the reactant gas. In the second set, P_i was varied between P_{CO2} of 0.4 and 2.4 bar or P_{H2O} of 0.2 and 1.6 bar while T_R was fixed at 1200°C. The results can be used to determine the reaction order (n).

Parameter	Set #1	Set #2	Set #3	Set #4
Gas	CO_2	CO_2	H_2O	H_2O
T [°C]	1000-1300	1200	1000-1300	1200
Partial Pressure [bar]	0.1	0.05-0.3	0.05	0.025-0.1
Average residence time [sec]	2.68-3.31	2.86	2.68-3.31	2.86

Table 2: Test conditions of char gasification by CO_2 and H_2O in PDTF.

2.4 Char conversion model

As an indicative char conversion model, SCM was selected in this study. This model assumes that the char size gradually shrinks by carbon conversion on its surface while the density remains constant [6]. Therefore, the surface area decreases by the power of 2/3 of the char conversion ratio (x). The reaction rate can be expressed as follows.

$$\frac{dx}{dt} = A\exp(-\frac{E}{RT})P_i^n(1-x)^{2/3} \quad (1)$$

Using time (t) and char converson (x) acquired from the sets of DTF tests, the three rate constants (A, E, and n) can be calculated from an integrated form of Eq.(1).

3 Numerical Method

The validity of the rate constants for SCM was tested using CFD simulations. The mesh was constructed for a quarter of the cylindrical geometry of DTF using 80,000 hexahedral cells with periodic conditions. The CFD code used was ANSYS Fluent version 16 [7]. The analysis of flow and combustion takes into consideration mass, momentum and energy conservation as well as particle behavior, turbulence, radiation, and gas reaction. The particle behavior was tracked using the Lagrangian method for a total of 240 individual particles and the gasification reaction was calculated by coding the shrinking core model with a user defined function (UDF). Realizable k-ε model was used for the gas flow [8], and radiation was considered with the gas absorption of H_2O and CO_2 [9]. The emissivity of the particles was set to 0.9. The wall temperature in the DTF tests was applied as the boundary condition of the CFD analysis.

4 Results and discussion

For correction of reaction conditions for the char particles, it is required to evaluate the particle residence time during active char conversion (after the entry region for heating), partial pressure, and temperature. A method to determine these parameters was established only using the test data prior to CFD.

Because the gas and particle supply at the inlet has a low temperature, it is required first to estimate the effective reaction length after they attain the temperature close to the target value. This is not straightforward because i) the gas is gradually heated by convection from the wall to develop a temperature profile, ii) the particles are directly heated by radiation from the wall, and iii) the particles and gas exchanges heat by convection. Assuming that the convection between the gas and particles is sufficiently fast to have the same temperature, a simplified energy conservation can be established for their temperature (T) in the heating zone.

$$(\dot{m}_p C_p + \dot{m}_g C_g)dT = [h_{conv}(T_w - T) + \varepsilon\sigma A_{sp}(T_w^4 - T^4)]\pi D dy \tag{2}$$

Eq (2) can be expressed by

$$\ln\left(\frac{T(y) - T_w}{T_{in} - T_w}\right) = -Ky \tag{3}$$

The convection coefficient (h_{conv}) was estimated from the mixed condition of natural and forced convection [10-13], as follows.

$$h_{conv} = Nu_{mixed} k_g / D \tag{4}$$

$$Nu^3_{Mixed} = Nu^3_{Natural} - Nu^3_{forced,entry} \tag{5}$$

$$Nu_{forced,entry} = 3.66 + \frac{0.104\, Gz}{1 + 0.016 Gz^{0.8}} \tag{6}$$

$$Nu_{natural} = 4.36\left(1 + \frac{Gr^{0.468}}{750 + 0.24\mathrm{Re}}\right) \tag{7}$$

Where Gz is Graetz number and Gr is Grashof number.

Finally, the reactor length required for the gas and particle to reach 5% of the dimensionless temperature, $\theta=(T_y-T_w)/(T_{in}-T_w)$ becomes

$$y_{\theta=0.95} = -\ln 0.95 / K \tag{8}$$

Note that the choice of θ=5% is arbitrary, but was found appropriate for the correction. From Eq. (8), the effective reaction length for actual char conversion to take place can be calculated.

$$L_r = L_R - y_{\theta=0.95} \tag{9}$$

Lr is later used to determine the particle residence time during actual char conversion under different reaction conditions. Particle residence time is deterimined by the average velocity of the laminar flow.

$$\tau_{p,avg} = L_r / v_{p,avg} \tag{10}$$

The particle temperature was applied lower than the wall temperature considering the reaction based on the energy equation. And the similar to the temperature change, the partial pressures of CO_2 and H_2O around the char particles gradually decrease by their consumption and products formation. In this study, partial pressures were calculated using the char conversion of the tests

The corrected rate constants for char gasification by CO_2 and H_2O are listed in Table 3, together with those before correction. The pre-exponential factors are significantly increased by the decreases in $\tau_{p,avg}$. Corresponding to the changes in $P_{i,avg}$ and $T_{p,avg}$, the reaction orders (n) are slightly decreased whereas the activation energies (E) are increased. Fig. 2 compares the measured and predicted carbon conversion before correction (a) and after correction (b). Note here that the comparison of CFD results should be made to those of curve-fit data rather than measured data, because the rate constants were derived by the curve-fit. For all char-CO_2, H_2O reaction conditions, the discrepancies are less than 4.3%p which can be considered very satisfactory.

Rate constants	Before correction	After correction
Gasifying agent	CO_2	
A (s^{-1})	2.205×10^5	9.432×10^5
E (kJ/mol)	158.51	163.83
n	0.716	0.774
Gasifying agent	H_2O	
A (s^{-1})	6.138×10^7	2.334×10^9
E (kJ/mol)	216.57	249.13
n	0.858	0.846

Table 3: Rate constants for char gasification.

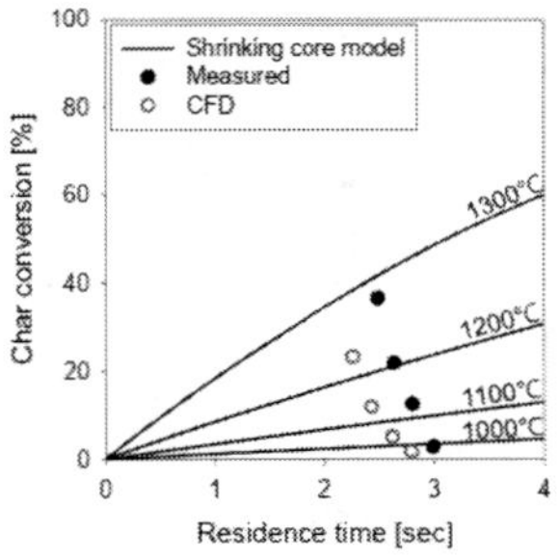

(a) Before correction

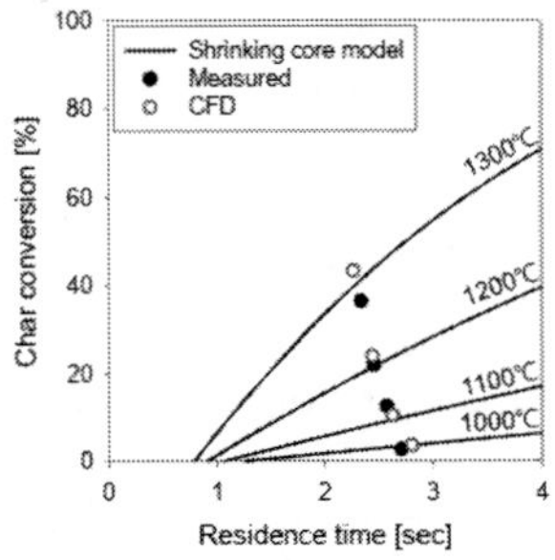

(b) After correction

Figure 2: Comparison of measured and predicted char conversion for gasification by CO_2 at different temperature

The char conversion by char gasification CO_2 and H_2O at each residence time are listed in Table 4. The char conversion rate of measured value and CFD value was quite improved than before correction. Therefore, it is essential to calibrate the reaction constant derivation through the DTF reactor.

CO_2	By Temperature (P_i = 0.8 bar)			
Parameter [°C]	1000	1100	1200	1300
Measured [%]	2.7	12.5	22.0	36.5
Curve-fit [%]	3.4	9.0	21.5	41.2
CFD [%]	3.6	10.4	24.0	43.4
CO_2	By Partial Pressure (T_w = 1200 ℃)			
Parameter [bar]	0.4	0.8	1.6	2.4
Measured [%]	12.9	22.0	31.5	42.6
Curve-fit [%]	13.1	21.3	33.3	41.4
CFD [%]	15.0	24.0	35.9	41.8
H_2O	By Temperature (P_i = 0.4 bar)			
Parameter [°C]	1000	1100	1200	1300
Measured [%]	1.4	7.1	19.7	48.1
Curve-fit [%]	1.4	6.7	22.5	45.1
CFD [%]	1.2	6.4	22.3	49.4
H_2O	By Partial Pressure (T_w = 1200 ℃)			
Parameter [bar]	0.2	0.4	0.8	1.6
Measured [%]	10.2	19.7	35.9	50.1
Curve-fit [%]	10.9	19.0	32.4	53.8
CFD [%]	13.5	22.3	35.3	52.4

Table 4: Comparison of Char conversion .

5 Conclusions

This study evaluated the flow and reaction conditions for char gasification in a DTF by experiments, analytical methods, and CFD in order to derive accurate rate constants. The char conversion ratios acquired under different temperatures and partial pressures of CO_2 and H_2O were used to determine the rate constants for shrinking core model as an exemplary char conversion model. Overall, it is essential to derive the rate constants from DTF tests by careful evaluation of reaction time, partial pressure, and temperature for actual particle reactions. The key findings are as follows.

-The rate constants derived from the conventional method, assuming the average residence time, target reactor temperature, and input partial pressure of reagent, significantly underestimated the char conversion predicted by CFD for the DTF tests.

-There were significant discrepancies between the controlled reactor conditions and actual reaction conditions of the particles. Without addressing the discrepancies, the reaction conditions contribute to the overestimation of the rate constants for char gasification, which has been often neglected in previous studies in the literature.

-In determining the particle residence time, the length for the actual reaction should exclude the initial region near the inlet where the input gas and char particles are heated. Also, the velocity profile on the cross-section should be considered in which char particles are concentrated more on the high velocity region at the center.

-The particle temperature and partial pressure of reagent surrounding the char particles were lower than the target or input values by gasification reactions.

-When the particle residence time, temperature and partial pressure were corrected by the proposed method, the rate constants for the shrinking core model provided good agreement between the measured and predicted values of char conversion ratio.

6 Acknowledgment

This work was supported by the Establishment and Utilization of Clean Fuel Test Bed project funded by the Korea Institute of Energy Technology Evaluation and Planning (KETEP) affiliated with the Ministry of Knowledge Economy of the Korean Government (No. 20152010201970).

References

[1] K.L. Smith, L.D. Smoot, T.H. Fletcher, R.J. Pugmire, Ther structure and reaction processes of coal, 1994.

[2] S. Kajitani, N. Suzuki, M. Ashizawa, S. Hara, CO2 gasification rate analysis of coal char in entrained flow coal gasifier, Fuel 85 (2) (2006) 163-169.

[3] R. Hurt, J.K. Sun, M. Lunden, A kinetic model of carbon burnout in pulverized coal combustion, Combust Flame 113 (1) (1998) 181-97.

[4] R.C. Shurtz, *Effects of pressure on the properties of coal char under gasification conditions at high initial heating rates. Dissertation*, Brigham Young University, USA, 2011.

[5] A. Molina, F. Mondragon, Reactivity of coal gasification with steam and CO2, Fuel 77 (15) (1998) 1831-39.

[6] H.Y. Sohn, J. Szekely, The effect of intragrain diffusion on the reaction between a porous solid and a gas, Chem Eng Sci 29 (2) (1974) 630-34.

[7] ANSYS FLUENT 12.1 User Guide, Canonsburg, Unit state, 2009.

[8] T.H. Shih, W.W. Liou, A. Shabbir, Z. Yang, J. Zhu, A new k-ε eddy-viscosity model for high Reynolds number turbulent flows, Comput Fluids 24 (3) (1995) 227-38.

[9] E.H. Chui, G.D. Raithby, Computation of radiant heat transfer on a non-orthogonal mesh using the finite-volume method, Numer Heat Transfer 23 (3) (1993) 269-88.

[10] B. Metais, E.R.G. Eckert, Forced, mixed, and free convection regimes, J Heat Transfer 86 (2) (1964) 295-96.

[11] Behzadmehr, A. M. I. N., N. Galanis, and A. Laneville, Low Reynolds number mixed convection in vertical tubes with uniform wall heat flux, International journal of heat and mass transfer 46(25) (2003) 4823-4833

[12] Kakac, Sadik, Yaman Yener, and Anchasa Pramuanjaroenkij. Convective heat transfer CRC press 2013.

[13] E.L.I. Ruckenstein, R.A.J. Rajagopalan, A simple algebraic method for obtaining the heat or mass transfer coefficients under mixed convection, Chem Eng Commun 4 (1-3) (1980) 15-39.

11th Asia-Pacific Conference on Combustion,
The University of Sydney, NSW Australia
10th -14th December 2017.

PIV Investigation of Effects of Circular DBD Plasma Actuator on Turbulent Swirling Premixed Flame

Sujoy Chakraborty, Masayasu Shimura, Mamoru Tanahashi

Dept. of Mechanical and Aerospace Engineering, Tokyo Institute of Technology

2-12-1 Ookayama, Meguro-ku, Tokyo, 152-8550, Japan

Abstract

In lean premixed combustion, combustion instability is a crucial issue. And control method of this phenomena is important for materialize high efficiency low-emission combustors. There are many approaches for solving this problem. Many of them are based on changing the configuration of fuel injection in some way, this includes multiple point injection or zero axial velocity injection. These techniques primarily aim to phase-shift control the flow by decoupling the instabilities. Secondary fuel injection, or in general unsteady/variable fuel injection is also similar approach. Other technique includes excitation of pressure using speaker and magnetic flap actuation. This study focuses on controlling the oscillatory combustion using a circular dielectric barrier discharge plasma actuator (DBDPA). The DBDPA was first tested for non-combustion system then implemented to a combustion field. The effects were investigated using pressure measurement, time-resolved chemiluminescence (CH) and stereoscopic particle image velocimetry (SPIV). It is found that the DBDPA was able to lower the peak frequency of pressure oscillation while also reducing the energy at the peak frequency. The results suggest that the DBDPA can work as a control device for this problem.

1 Introduction

The importance of protecting the environment is getting higher and higher. To achieve cleaner and environment friendly heat engine, lean premixed combustion is getting more attention recently. However, the issue associated with the lean premixed combustion is that as the flow rate increases it induces thermo-acoustic instabilities like oscillatory combustion which increases noise and can eventually destroy the combustor. Passive techniques like optimize design of combustor chamber were originally considered but such technique was very limited in its robustness making each design taking very long time to be completed. Thus, to achieve more robust and adaptable technique which can be easily applied to various combustor Active Combustion Control (ACC) is now considered. Previous research showed that the fluctuation in recirculation zone, that is the generation of large-scale vortex has strong correlation with the existence of oscillatory combustion [1]. This indicates that controlling the fluctuation in recirculation zone is effective in suppressing the oscillatory combustion, and it has been shown that secondary fuel injection is an effective way to achieve this goal [2]. Though introducing secondary fuel injection cause the combustion to become partially non-premixed which may have negative impact on HC generation. But also, its response time may not adequate for active control.

To develop a more effective control system dielectric barrier discharge plasma actuator (DBDPA) [3,4] is considered in this study. The basic principle of DBDPA is it has two electrodes divided by a dielectric, by applying high frequency high voltage input signal, it generates plasma which generates body force. The strength of DBDPA is that it does not have any mechanically moving part and has short response time. However, there is not much researches on its application to turbulent combustion thus development of the device which can sustain the high temperature of combustion field has been required.

In this study, a circular DBDPA for control of swirling premixed flames is proposed, and effects of DBDPA on characteristics of cold flow and combustion fields are investigated by time resolved particle image velocimetry (PIV) for overall flow field and time-resolved high spatial resolution stereoscopic PIV with pressure and CH chemiluminescence measurements.

2 Experimental setup

2.1 Swirl burner setup

The schematic of swirl burner and the placement of the DBDPA is shown in Fig.1. The methane and air are introduced to the mixing chamber and then the mixed gas forms the premixed flame on the burner inside the duct. The duct has cross-section of 120 mm by 120 mm and has length of 575 mm which is water cooled. The outer wall of lower part of the duct has window made of fused quartz to give optical access to the flame. The burner has the capability to perform secondary fuel injection but it is not used in this experiment. In this experiment swirl angle was fixed at 45 degrees. The detail description on the swirl burner can be found in the literature [2].

2.2 DBDPA setup

The geometry of DBDPA is also shown in Fig.1. The DBDPA is configured to induce the surrounding gas inward. Because the DBDPA directly contacts with the flame it had to be designed to withstand the heat. Thus, the electrode is made of 70 μm cupper plate and the dielectric is made of 500 μm alumina. Also, to cover the entire periphery of the swirl the DBDPA was designed circular. To place the DBDPA inside the burner ceramic stand is designed. On the top of the stand we placed the DBDPA setup which consists one dielectric plate sandwiched with two electrodes, the input signal is provided by cupper strip placed in between the stand and the dielectric plate, the dielectric has a cut off to connect the top electrode with the signal.

Corresponding author. Fax: +81-3-5734-3183
E-mail address: Chakraborty.s.ab@m.titech.ac.jp

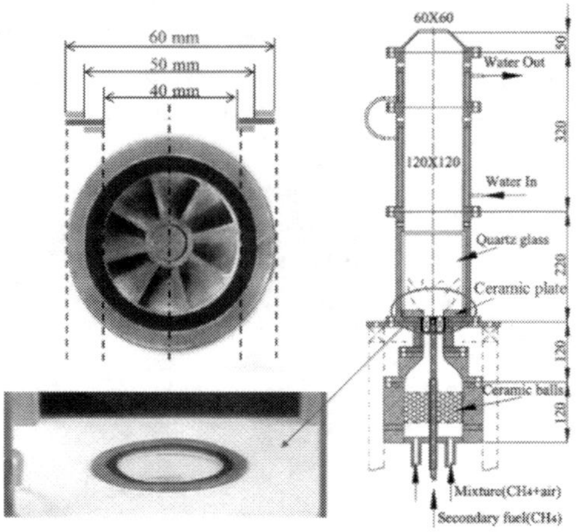

Figure 1: The schematic of swirl burner and DBDPA setup. The DBDPA is placed at the bottom of the duct where flame sits.

To drive the DBDPA, input signal is generated by a function generator then amplified with a power amp and finally the voltage is raised with a transformer to desired kilovolt order. This system is capable of generating 20 kHz, 10 kV peak to peak sin wave. The input used in this experiment was 10 kHz, 10 kV peak to peak. The output signal is clean enough to be used as a DBDPA input.

2.3 Operation condition and measurement technique

For time-resolved PIV, high repetition rate Nd:YAG laser (Lee Laser, LDP-100MQG, 532 nm, 6 mJ/pulse) and high speed CMOS cameras (Photoron SA-X2, 1024×1024 pixel) were used with stereoscopic configuration. And two Nd:YAG lasers (LOTIS TII LS-2131M, 532 nm, 125 mJ/pulse) with two CCD Camera (Roper Scientific MegaPlus II ES4020, 2048×2048 pixel) were used for high spatial resolution SPIV. For combustion case, a bandpass filter (Semrock, FF01-525/20-25, centre wavelength 532 nm, half width 20 nm) is installed to the cameras for PIV. The tracer particle used for PIV is 2 μm SiO_2. For chemiluminescence measurements, the light gathered by a lens (Nikon, AF Micro Nikkor105 mm, F#4.5) with a bandpass filter (Semrock, FF01-434/17-50.8-D, centre wavelength 434, half width 17 nm) is amplified by an image intensifier (Hamamatsu Photonics, C10880-03F), and is detected by a CMOS camera (Photron, FASTCAM SA-X2). For both cold flow and combustion cases, the pressure measurement was performed using a pressure sensor (KISTLER, 7061C) and an amplifier (KISTLER, 5018).

Prior to this study the effect of DBDPA on cold flow was investigated. The flow rate of methane-air mixture was fixed to 200 L/min throughout the experiments. Characteristics of flow fields can be found in the literature [5].

The coordinate system of every test is taken as the axial direction being x, the parallel to laser sheet direction being y and perpendicular to leaser sheet direction being z. The origin is taken at the centre of swirl burner. And velocity vector is taken as u, v, w respectively.

For the time-resolved PIV the size of measurement region is 100mm*100 mm and the centre is at (52.0,0.0,12.0). The interrogation region is set to 16 pixel by 16 pixel yielding the optical resolution of 1.5 mm and the repetition rate was set 10 kHz for both case. For SPIV the dimension of the region is 30 mm*30 mm and the centre is at (17.0,7.5,12.0) for the inner region and (17.0,30.0,12.0) for the outer region. These regions correspond to the inner and outer recirculation zones. The interrogation region is set to 24 pixel by 24 pixel yielding optical resolution of 350 μm and repetition rate was set to 4Hz for non-combustion case. And 420 μm and 10 kHz for combustion case due to the measurement region reduced to 18 mm* 18mm and centre set to (11.0,9.0,0.0) and (11.0.25.0,0.0) respectively. The equivalence ratio for combustion case was set to 0.6.

3 Results and discussion

3.1 Chemiluminescence and Pressure

The distribution of time averaged and RMS value of chemiluminescence intensity measurement is shown in Fig.2. All images are normalized with the max value which for time averaged occurred when DBDPA is on and for RMS it is DBDPA off. There is no clear difference in overall shape for two conditions. But the time averaged intensity increased while RMS value decreased when DBDPA is on.

The power spectrum of pressure fluctuation is shown in Fig.3. The red line represents DBDPA off and blue line is on. It is clear that existence of DBDPA reduces the peak frequency from 207 Hz to 179 Hz and also reduces its energy at peak. These results indicate that DBDPA has the potential to control oscillatory combustion.

3.2 Effects of DBDPA in combustion flow field

Next discussion is on flow characteristic of recirculation zone. In past experiment on cold flow showed that DBDPA enlarges

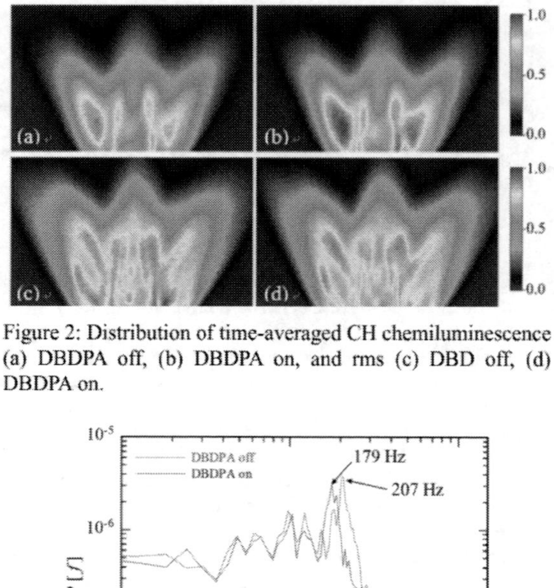

Figure 2: Distribution of time-averaged CH chemiluminescence (a) DBDPA off, (b) DBDPA on, and rms (c) DBD off, (d) DBDPA on.

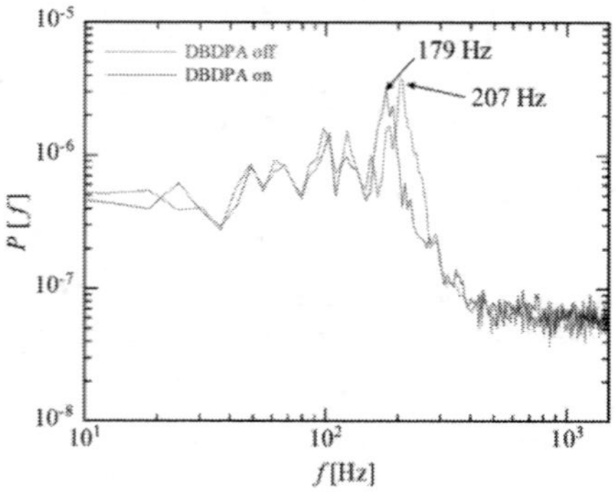

Figure 3: Power spectrum of pressure fluctuation inside burner. The general trend is similar but at the peak two lines shows clear difference.

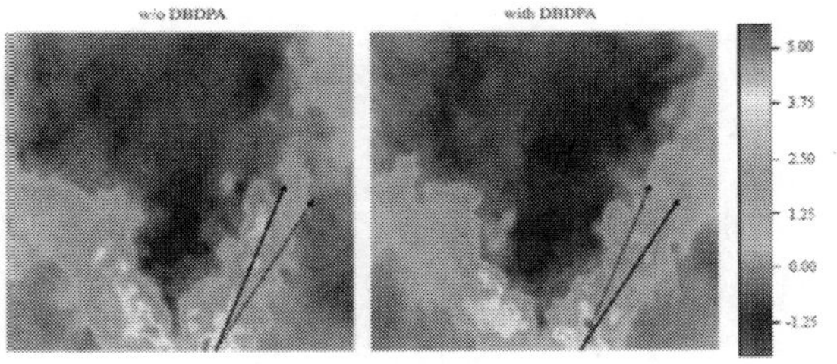

Figure 4: Instantaneous u field obtained from PIV. The solid line represents the expansion angle of that condition, while dashed line represents the expansion angle of counter condition.

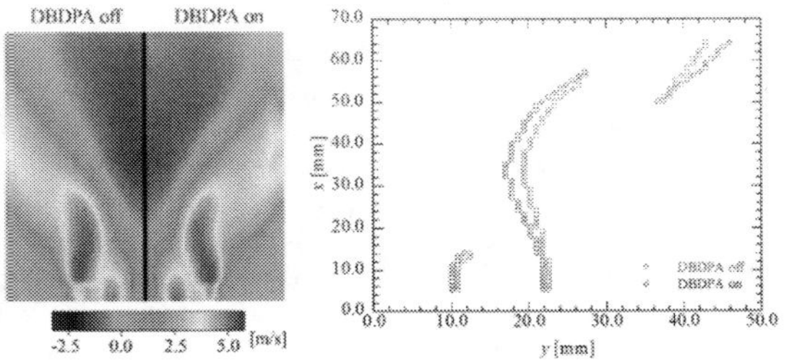

Figure 5: Distribution of maximum value of u in x-y plane. DBDPA on condition left side image is flipped to right side.

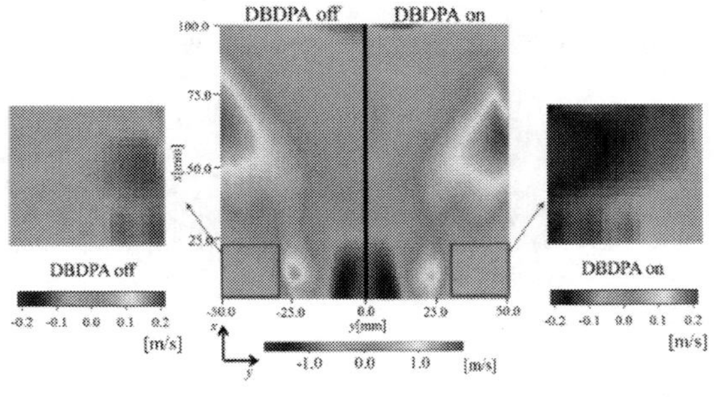

Figure 6: v directional time averaged mean velocity distribution. Again DBDPA on is flipped to right side.

the expansion angle of the main flow. This is shown in Fig.4. This can be considered as DBDPA needs to suck gas from top due to it is constrain inside the chamber, this suction coincides with the outer recirculation zone making it more effective, which lower the swirling main flow causing the increase in expansion angle. In addition, we took the spatially averaged power spectrum of velocity fluctuation at the outer sear layer near the exhaust. That result showed reduction of peak frequency from 175 Hz to 117 Hz, this can mean the change of vortex generation at shear layer is also the cause of above phenomena.

Moving to combustion field the maximum value and second peak of u on x-y plane obtained from PIV measurement is shown in Fig.5. It can be observed that when DBDPA is on, upstream u max shifts inner while at downstream it shifts outer. This can be think as at the upstream the main flow is affected by the jet generated from DBDPA which is flowing inward, while at downstream the flow is affected by the suction due to the entrainment of the DBDPA.

This also means that expansion angle of the main flow become larger which conforms the finding in cold flow test.

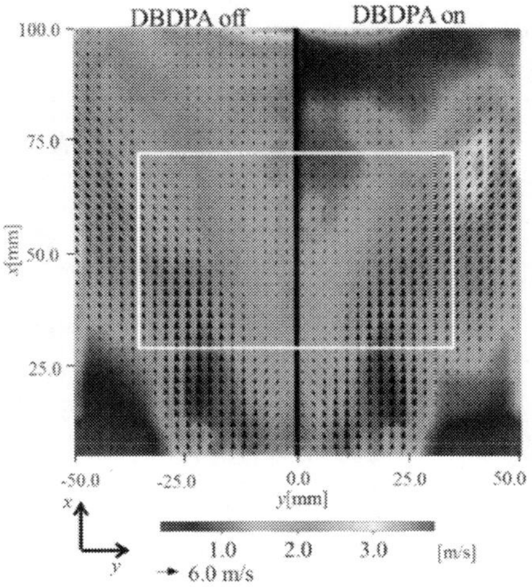

Figure 7: Turbulence intensity in wide region based on two-components of velocity field. DBDPA on condition is flipped to right side and clear decrees in turbulence intensity can be observed.

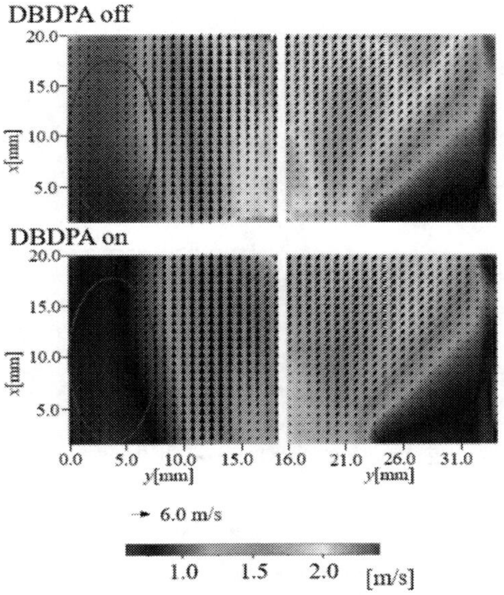

Figure 8: Distributions of in-plane velocity vectors and turbulence intensity based on three-components of velocity fields. Red ellipse shows the main flow region.

The PIV result of v is presented in Fig.6. It shows that at the outer recirculation zone inward flow velocity is increased by DBDPA, meaning that DBDPA successfully strengthen the flow in outer recirculation zone.

Figure.7 shows the turbulence intensity computed from u, v. DBDPA on condition is mirrored to right side for comparison. The vector represents the time averaged velocity distribution. And background colour represents the turbulence intensity. The white box indicates the region where chemiluminescence is taken (Fig.2). The result indicates that the turbulence intensity at inner recirculation zone has decreased by applying DBDPA. This is think to be caused by the inner shifting of upstream flow

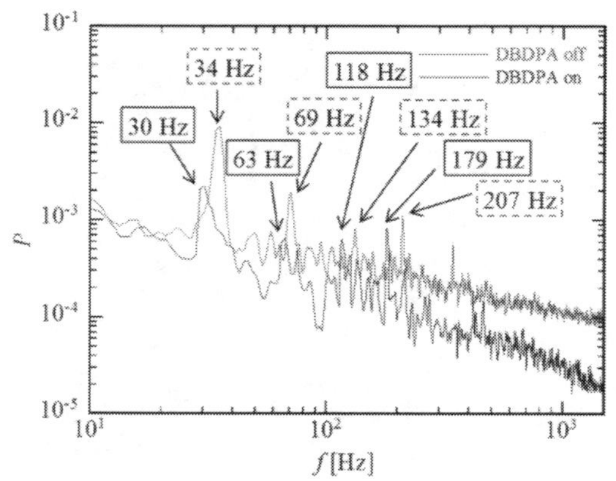

Figure 9: Power spectrum of velocity fluctuation (inner recirculation zones). Clear decrease in fluctuation can be observed when DBDPA is on.

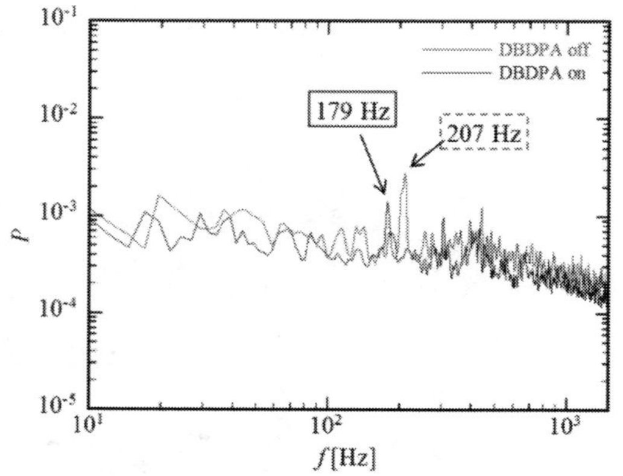

Figure 10: Power spectrum of velocity fluctuation (outer recirculation zones). Blue line is consistently lower than red.

discussed in Fig.5 which suppresses the fluctuation.

Finally moving to the result of SPIV. Figure.8 shows distributions of averaged in-plane velocity vectors and of turbulence intensity computed from three components of velocity, which is illustrated with background colour. The upper two corresponds to DBDPA off and lower two is on. Inner zones are paced left side and outer zones are on right side. The two identical red ellipses are to visualize the inner shifting due to DBDPA. This image shows that the turbulence intensity is decreased by half at inner recirculation zone, and it is also reduced in large portion of main flow region. Indicating that DBDPA can be an effective control device to suppress velocity fluctuation.

Figure.9 and Fig.10 shows the power spectrum of velocity fluctuation in both inner and outer recirculation zones. This is computed form the spatially averaged velocity distribution inside the measurement region. It matches with the result from pressure fluctuation, having peak at 207 Hz which reduces to 179 Hz when DBDPA is on. This reduction in frequency is same trend as in cold flow experiment. Also, there are several other local peaks at inner zone all of which decreases in both frequency and power when DBDPA is on. Because oscillatory combustion occurs at a certain frequency determined by geometry of burner or other characteristics, been able to shift the frequency of velocity fluctuation has crucial importance in controlling the combustion.

4 Conclusions

The effects of circular DBD plasma actuator on turbulent swirling flame were investigated. The non-combustion testing showed that DBDPA affects the expansion angle of swirling flow. And combustion testing backed this finding. From the SPIV measurement under combustion field showed that DBDPA not only affects the inner recirculation zone and main flow but also influences the outer recirculation zone. It was also observed that the DBDPA reduces the turbulence intensity at the inner recirculation zone. The resulting reduction in fluctuation in inner recirculation zone is think to be the cause for chemiluminescence fluctuation reduction and pressure fluctuation reduction. All of this is a strong indication that DBDPA can be used as a control device. But on the other hand, our DBDPA design suffered from poor heat resistivity and designing such DBDPA which can sustain inside burner is crucial for further testing. It was also found that because the resistivity and other characteristics changes as the temperature become higher, the DBDPA was unable to work as efficiently as in room temperature condition when placed inside burner. Another important point is that at this experiment condition oscillatory combustion does not occurs, thus it is crucial to investigate the effect of DBDPA at such condition. It is also important to separate the influence of plasma flame chemical reaction against the effect of induced flow by DBDPA. Finally, further experiment with more powerful DBDPA is necessary for applying this technique to more general burner.

5 Acknowledgment

This work was supported by Japan Society for the Promotion of Science(JSPS) through Grant-in-Aid for Young Scientists (A). Under JSPS KAKENHI Grant Number 16H06069.

References

[1] M. Shimura, M. Tanahashi, G.-M. Choi, T. Miyauti, J. Therm, Sci. Technol. 4(2009), pp.494-506.

[2] G.-M. Choi, M. Tanahashi, T. Miyauchi, Proc. Combust. Inst. 30 (2005), pp1807-1814.

[3] J. Reece Roth, Daniel M. Sherman, Stephen P. Wikison, AIAA paper 98(1998), 0328.

[4] Thomas C. Corke, Eric J. Jumper, Martiqua L. Post, Dmitriy Orlow, Thomas E. McLaughlin, AIAA paper 2002, 0350

[5] K. Komatsu, M. Shimura, M. Tanahashi, Annual Meeting of Japan society of Fluid Mechanics 2016

11th Asia-Pacific Conference on Combustion,
The University of Sydney, NSW Australia
10th -14th December 2017.

Suppression of Flame Propagation in a Long Duct by an Inert Gas Plug

Yu-Jhen Lin, Sheng-Hsun Wang, Chien-Ho Liu, Hsiao-Yun Tsai, Jenq-Renn Chen*

Department of Safety, Health and Environmental Engineering
National Kaohsiung First University of Science and Technology
1 University Rd., Yenchau, Kaohsiung, 824, Taiwan

Abstract

Ignition and explosion from accidental leaks of large amount of flammables into a long, underground trench may result in extensive damage and fatalities. Mitigation of such leak and explosion is difficult although not impossible. Typical explosion suppression measures such as water spray and inert powder cannot be readily applied owing to underground location. In this work, we proposed a concept of suppression the flame propagation rather than the explosion overpressure by injecting inert gas into the underground duct. Experimental studies were carried for the suppression by an inert gas of flame propagation in one section of a tube to another section with flammable mixtures. The critical length of the inert gas section required for successful suppression is determined to be less than 0.3 m for a 3-m ignition section containing propylene/air mixture near stoichiometric concentration. Application of the results in responding to large-scale leak into duct is discussed.

1 Introduction

Large scale leak of flammables into a duct such as underground trench and its subsequent ignition and explosion may cause significant damage. Although such type of explosion was rare compared with unconfined vapor cloud explosion, the damages were far more severe as shown by the incidents in Guadalajara, Mexico in 1992[1], Qingdao, China in 2013 [2], and more recently in Kaohsiung, Taiwan in 2016 [3]. For the later case [3], a total of 4.5 km of road surface was blow out causing extensive damage in lives and properties. The above three cases have a common source of leak: a corroded underground pipeline transporting flammable fluids. Another striking common feature of the above three cases is that the delay time between leak and ignition is considerably longer than unconfined vapor cloud explosion. Part of the reason is attributed to the fact that there were usually more ignition sources in an open space than the confined duct. Although the long delay time of ignition was part of the causes of extensive damage, it also opens up an opportunity to mitigate or to prevent the explosion, provided that proper action can be taken.

Extensive work has been done on explosion suppression which focuses mainly on reduction of explosion intensity upon the ignition and explosion was initiated. For long duct filled with flammable gases or vapors, such suppression is possible but usually special devices are needed for injecting the suppressant. For underground trench lasted for several kilometers, the number of suppression device required can rendered the suppression impractical.

In this work, we explore the possibility of mitigating the explosion by suppressing the flame propagation with inert gas plugs. The idea is simple: by injecting nonflammable gases such as nitrogen or carbon dioxide into the duct at different section *before* the ignition, the flammable gases or vapors in the duct are segmented. Should ignition occurred in any section of the duct, flame will only propagate in that section but not other sections owing to the isolation by the inert gas plug. The explosion is thus limited to a few meters rather than a few kilometers. Similar ideas have been carried out Du et al. [4] and Zhang et al. [5] in a duct for the suppressions of the gasoline/air mixture explosion by non-premixed nitrogen. However, both sides of the duct were closed which prevent the outflow of the gases and gave an excessively long suppression section for a successful suppression. It is reported that failure of suppression was caused by insufficient dilution from the inert gas. For an open duct such as an underground trench, flame propagation resulted in overpressure, volume expansion and gas outflow. It is expected that the suppression mechanism will be different from those of closed ends.

In this work, the idea is explored by using a pipe comprising an ignition section, suppression section, flammable section with ignition at the closed end and flammable section at the open end. Critical length is determined for a fixed length of 3-m ignition section filled with propylene/air mixture. Implication of the results for responding to large scale leaks are discussed.

2 Experimental Setup

A stainless-steel pipe with internal diameter of 4.26 cm is used as the duct. The total pipe length is adjustable through assembly of different length of pipe by clamping flanges. Flame sensors and/or pressure sensors were installed on each section of pipe. The pressure sensors used were Kistler Type 211B quartz sensors with response time of 1 μs. The flame sensors used were high-speed silicon photodiodes from OSI Optoelectronics Type PIN-HR008 with spectral range from 350 nm to 1100 nm and response time of 0.6 ns. A total of 14 flame sensors and 2 pressure sensors were installed. All sensor data were acquired by a Yokogawa DL850E data acquisition recorder at a rate of 5,000 Hz.

The whole pipe comprised of four sections depending on test configuration. The first section consists of a 3-m long ignition section with one end sealed with a blind flange and an inserted Nichrome wire igniter. The igniter provided a glowing ignition source with energy around 10 J which is sufficient to ignite flammable mixture but did not disturb the mixture. The first

Corresponding author. Fax: +886-7-601-1236
E-mail address: jrc@nkfust.edu.tw

section was filled with propylene/air mixture. The second section was filled with an inert gas such as nitrogen or carbon dioxide and the length was varied between 0.1 m to 6 m. The third section consisted of another 3-m long flammable section also filled with propylene/air mixture. The fourth section was an empty pipe open to ambient air with a length of 18 m which acted as a buffer section to avoid the outflow of flammable gas in the third section. All sections were separated by a thin polyethylene film.

Propylene with a purity of 99.5% was used as the flammable gas. The flammable sections comprised of air from an air compressor and then propylene fed by pressure to a concentration around 4.5±0.15%, which was closed to the stoichiometric concentration. The flammable mixture was mixed by a diaphragm suction pump for at least 20 min. Exact concentration was determined from a sampling and analysis by GC-TCD. After the sampling, the mixture was left quiescent for at least 1 min before ignition. The third section also filled with propylene/air mixture but at slightly higher concentrations of 5.3±0.1% to prevent possible dilution from the inert section.

3 Results

3.1 General Observation of Flame Propagation

The flame propagation in a duct with ignition at the closed end and open at the other end has been studied extensively in classical experiments by Guénoche [6], and more recently by Jones and Thomas [7] and Kerampran et al. [8]. The flame propagation behavior was quite complex and subjected to by affected by several factors such as flammable concentration, duct length, roughness, etc. The most notable feature is the oscillating flame propagation which was resulted from tube acoustics initially set in motion by the expansion due to ignition at the closed end [9].

To validate the current setup, several blank tests with the third section of flammable mixture replaced by an inert gas. The results were shown in Figure 1 in which it is clear that the flame from the 3-m flammable section may propagated deep into the inert gas section. The propagation distance for the flame from first section is however not affected by the length of the inert section as shown in Table 1 in which all flame propagation terminated at downstream distance of 4.8~6 m. Clearly, the flame propagation outside the flammable section is mainly caused by expansion and outflow from the burnt gases. In addition, the flame propagation is also oscillatory as seen in Figure 1 where the flame at both 4.95 m and 5.45 m were both diminished at 400 ms but grow again at 4.95 m. Such complicated flame propagation behavior is in consistent with existing literatures [6-8] except that all previous literatures deal with a duct filled with flammable gas mixture for the whole length of duct.

3.2 Results of Flame Propagation Suppression

The major aim of this work was to find the required length of inert section that can prevent the flame propagation in the ignition section to ignite the third section with flammable mixture. An initial trial started with 6-m of inert section and tests repeated with reduced length of inert section until the third section with flammable was also ignited. The full results of maximum flame sensor signal at different location of the pipe were summarized in Table 1.

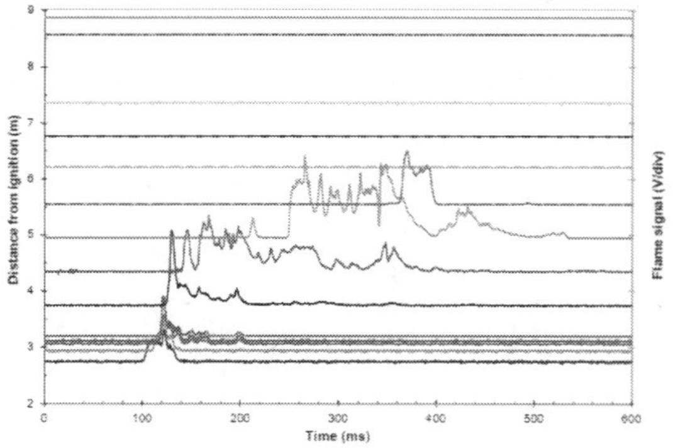

Figure 1: Typical results of flame propagation for a blank test, Test 10 in Table 1. The inert section is a 0.1 m nitrogen plug.

All tests with 6-m of inert section showed no ignition and flame in the third section. Subsequent tests with 0.6-m and 0.3-m of inert section also showed similar results. It is also interested to note that these tests with the third section filled with flammable mixture showed roughly the same flame propagation distance compared with the corresponding blank tests. Clearly, the flame propagated well into the third section with flammable mixture yet the flammable mixture was not ignited but instead pushed downstream.

Further tests, Test 12 and 13, with 0.1-m of inert section filled with nitrogen and carbon dioxide, respectively, failed to suppress the flame propagation and the flammable section was ignited in both cases. Figure 2 shows the results of Test 12. It is clear that the flame propagation prior to 5.53 m was roughly the same as the corresponding blank test in Figure 1. However, the flame sustained and created stronger flame further downstream. This is also evident from the comparison of overpressure measurement for Tests 10 and 12 in Figure 3. Overpressure measurement for both tests was roughly the same before 330 ms from ignition. After 330 ms, Test 12 showed slightly larger overpressure and the flame propagated to further downstream of 6.2 m and eventually all the flammable section in the downstream was ignited producing significant overpressure at 800 ms from ignition. Comparison of the flame signal and overpressure measurement for Tests 10 and 12 confirmed that the flammable section in Test 12 was ignited and the flame propagation suppression by 0.1 m inert section was failed.

Finally, Test 18 with 0.2-m of inert section filled with nitrogen showed clear flame in all downstream flame sensors as shown in Figure 4. However, two repeated tests, Test 19 and 20, with a 0.2-m inert section filled with carbon dioxide showed flame propagation up to less than 6 m as shown in Figure 5. Figure 6 shows the comparison of overpressure for the two tests. Again, the overpressure for Test 19 diminished gradually while Test 18 showed sustained overpressure until when the flammable section was ignited and generated significant overpressure. Therefore, the 0.2 m inert section is near the boundary of failed/successful suppression of flame propagation. The 0.3 m inert section is considered as the critical length for a successful suppression of flame propagation from a 3-m propylene ignition section.

Test no.	Ignition section		Inert section		Flammable section		Buffer section	Total length	Maximum flame sensor signal													
	L (m)	Fuel (%)	L (m)	Inert gas	L (m)	Fuel (%)	L (m)	(m)	f1	f2	f3	f4	f5	f6	f7	f8	f9	f10	f11	f12	f13	f14
Flame sensor location from ignition									2.74	2.93	5.34	5.93	6.51	7.10	7.69	8.09	8.88	9.05	9.60	10.20	11.10	14.41
1	3	4.32	6	N_2	0	-	18	27	0.04	0.05	0	0	0	0	0	0	0	0	0	0	0	0
2	3	4.64	6	CO_2	0	-	18	27	0.07	0.05	0.06	0	0	0	0	0	0	0	0	0	0	0
3	3	4.67	6	N_2	6	5.17	12	27	0.05	0.05	0.17	0.09	0	0	0	0	0	0	0	0	0	0
4	3	4.55	6	CO_2	6	5.38	12	27	0.04	0.04	0	0	0	0	0	0	0	0	0	0	0	0
Flame sensor location from ignition									2.74	2.93	3.14	3.33	3.52	3.69	4.23	4.83	5.43	6.03	6.69	7.25	7.85	9.05
5	3	4.78	0.6	N_2	0	-	23.4	27	0.04	0.04	0.09	0.22	0.25	0.21	0.16	0.13	0.14	0	0	0	0	0
6	3	4.45	0.6	CO_2	0	-	23.4	27	0.03	0.04	0.06	0.09	0.13	0.11	0.11	0.11	0	0	0	0	0	0
7	3	4.45	0.6	N_2	6	5.38	17.4	27	0.05	0.04	0.07	0.10	0.16	0.11	0.09	0.13	0.16	0.11	0	0	0	0
8	3	5.04	0.6	CO_2	6	5.12	17.4	27	0.03	0.03	0.04	0.05	0.05	0.04	0.16	0.15	0.02	0	0	0	0	0
9	3	4.47	0.6	CO_2	6	5.27	17.4	27	0.04	0.05	0.10	0.15	0.17	0.12	0.25	0.11	0.09	0	0	0	0	0
Flame sensor location from ignition									2.74	2.93	3.06	3.11	3.20	3.74	4.34	4.95	5.54	6.20	6.76	7.36	8.56	8.86
10	3	4.54	0.1	N_2	0	-	24	27.1	0.06	0.06	0.09	0.08	0.04	0.13	0.10	0.15	0.10	0	0	0	0	0
11	3	4.65	0.1	CO_2	0	-	24	27.1	0.06	0.06	0.11	0.10	0.04	0.08	0.11	0.17	0.11	0	0	0	0	0
12	3	4.50	0.1	N_2	6	5.22	18	27.1	0.07	0.07	0.11	0.11	0.05	0.13	0.29	0.17	0.12	0.12	0.16	0.29	0.27	0.23
13	3	4.38	0.1	CO_2	6	5.38	18	27.1	0.06	0.05	0.08	0.09	0.04	0.09	0.09	0.08	0.07	0.12	0.08	0.11	0.24	0.09
Flame sensor location from ignition									2.74	2.93	3.13	3.24	3.38	3.93	4.53	5.13	5.73	6.40	6.95	7.55	8.75	9.05
14	3	4.68	0.3	N_2	0	-	24	27.3	0.10	0.15	0.24	0.26	0.10	0.18	0.25	0.27	0.17	0	0	0	0	0
15	3	4.40	0.3	CO_2	0	-	24	27.3	0.05	0.05	0.05	0.12	0.04	0.07	0.08	0.07	0	0	0	0	0	0
16	3	4.35	0.3	N_2	6	5.44	18	27.3	0.04	0.04	0.08	0.11	0.03	0.05	0.05	0.06	0	0	0	0	0	0
17	3	4.47	0.3	CO_2	6	5.19	18	27.3	0.05	0.03	0.04	0.06	0.02	0.04	0.05	0.11	0	0	0	0	0	0
Flame sensor location from ignition									2.74	2.93	3.07	3.16	3.30	3.85	4.45	5.05	5.65	6.31	6.87	7.46	8.67	8.97
18	3	4.27	0.2	N_2	6	5.13	18	27.2	0.04	0.05	0.07	0.08	0.02	0.03	0.06	0.06	0.07	0.07	0.07	0.10	0.07	0.08
19	3	4.47	0.2	CO_2	6	5.26	18	27.2	0.05	0.05	0.08	0.10	0.04	0.09	0.11	0.13	0.07	0	0	0	0	0
20	3	4.40	0.2	CO_2	6	5.33	18	27.2	0.04	0.04	0.07	0.06	0.03	0.06	0.08	0.10	0	0	0	0	0	0
Flame sensor location from ignition									2.74	2.93	3.17	3.28	3.48	4.03	4.63	5.23	5.83	6.49	7.04	7.64	8.84	9.14
21	3	4.39	0.4	N_2	6	5.30	18	27.4	0.05	0.05	0.10	0.15	0.07	0.07	0.04	0.01	0	0	0	0	0	0
22	3	4.58	0.4	CO2	6	5.26	18	27.4	0.07	0.05	0.06	0.06	0.04	0.05	0.06	0.07	0	0	0	0	0	0
Flame sensor location from ignition									2.64	2.83	3.00	3.09	3.93	4.12	4.94	5.13	5.95	6.14	8.98	9.17	11.00	11.19
23	3	4.67	0.3	Air	9	4.51	-	12.3	0.32	0.29	0.39	0.39	0.35	0.33	0.27	0.22	0.23	0.28	0	0	0	0

Table 1: Summary of test results. The flame sensor locations were given based on the ignition end.

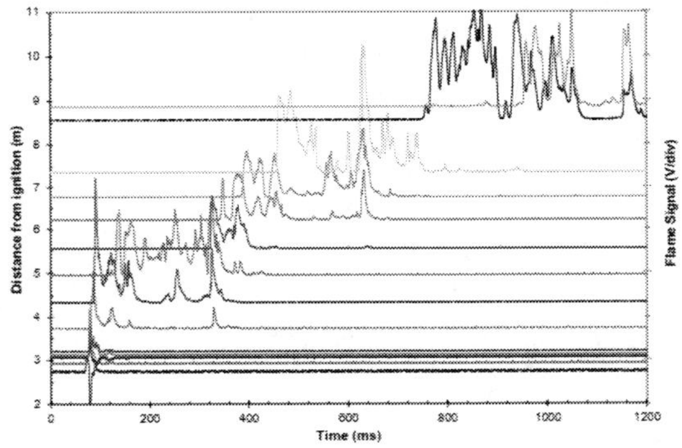

Figure 2: Results of flame propagation for a failed suppression test, Test 12 in Table 1. The inert section is a 0.1 m nitrogen plug.

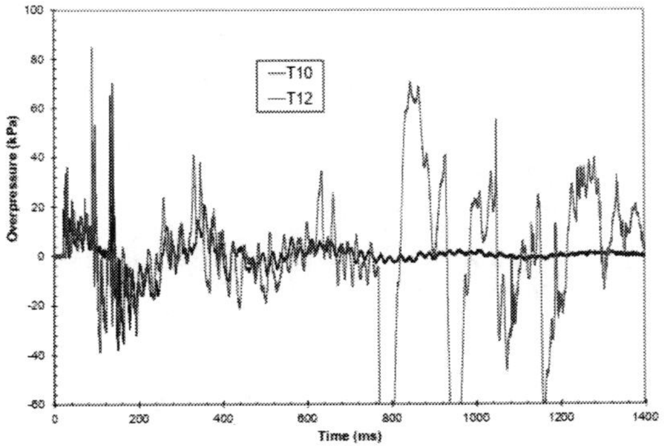

Figure 3: Results of overpressure for a failed suppression test, Test 12, and a successful suppression test, Test 10. The overpressure was measured at 8.75 m from ignition.

The current, short suppression section is far smaller than the studies done by Du et al. [4] which used a sealed pipe at both ends. The critical length of inert section is only 1/10 of the ignition section in the current work while the ratio was 1~4, depending on the fuel concentration in the ignition section, in the closed tube tests of Du et al. [4]. Zhang et al. [5] performed similar tests but with a section for visualization of flame propagation. Analysis of photos of the flame front by Zhang et

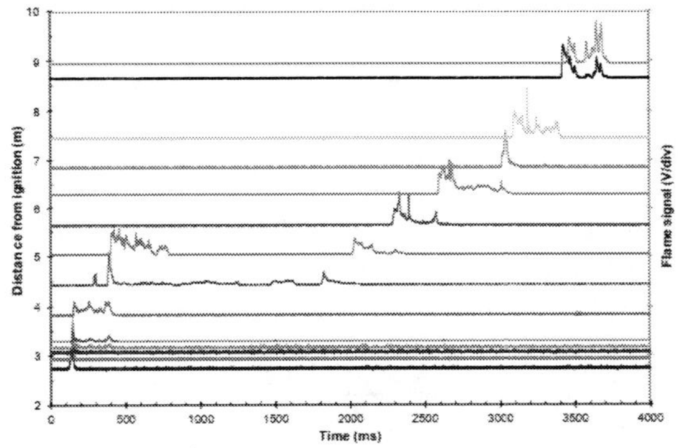

Figure 4: Results of Test 18 with failed suppression of flame propagation. The inert section is a 0.2 m N_2 plug.

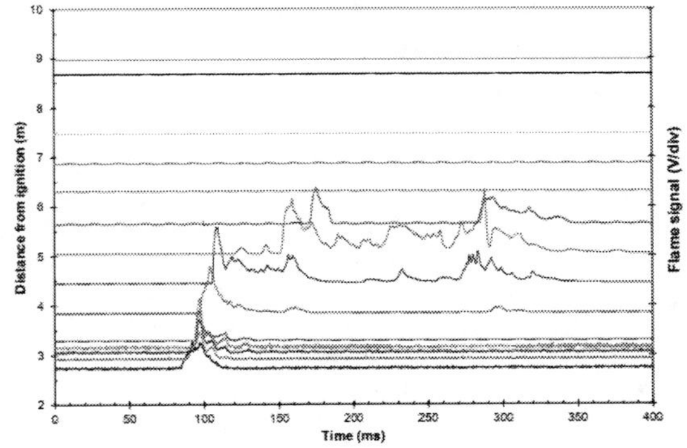

Figure 5: Results of Test 19 with a successful suppression of flame propagation. The inert section is a 0.2 m CO_2 plug.

al. [5] showed that the whole suppression process can be divided into three periods, inertia maintenance stage, suppression attenuation stage and diffusion extinguishment stage. They suggested that nitrogen molecule took away the energy of the high-energy radicals of the combustion reactions during the suppression process, resulting in the termination of the main reaction chains [5]. Another factor that contributes to the discrepancy between the current results and those of Du et al. [4]

is the criterion of successful suppression. Du et al. [4] required that flame sensors behind the suppression section did not detect any flame signal as the condition for successful suppression. This criterion will be ambiguous in the present cases with an open duct and gas outflow. In fact, the stringent requirement of no flame signal after inert section is only achievable with sufficient long inert section and sealed pipe.

Clearly, the mechanism of flame propagation suppression in the open duct is completely different from those of a sealed duct. For a sealed duct, the suppression is achieved through complete flame quench by the inert gas. For an open duct, the suppression is achieved through isolation during out flow and is thus called *inertia isolation*. Different inert gas did show slightly different results as 0.2 m section of CO_2 successfully suppressed the flame while 0.2 m section of N_2 failed. Such results are also in consistent with the fact that minimum oxygen concentration in CO_2 is larger than that of N_2 [10]. To further validate the proposed mechanism, a separate test, Test 23 in Table 1, was performed with the inert gas replaced by air. The results showed almost identical and successful result of suppressing the flame propagation from the ignition section to the flammable section. The only notable difference between Test 23 and Tests 14~17 is that flame propagation distance with air is larger than those with inert gas, i.e. 6.14 m vs. 5.13 m. The inert gas did contribute to the flame quench and thus giving smaller flame propagation distance.

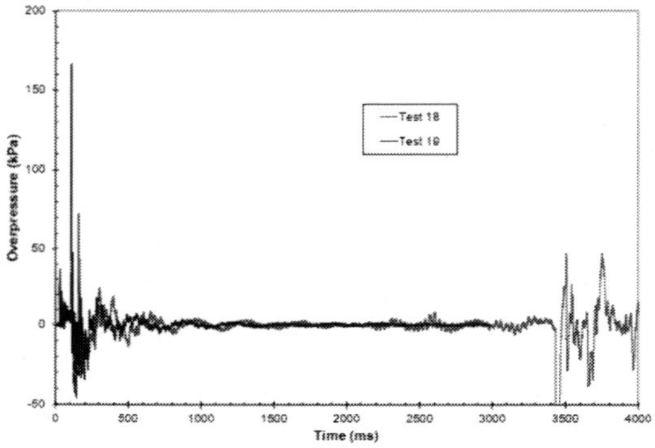

Figure 6: Comparison of overpressure for Tests 18 and 19. The overpressure was measured at 8.67 m from ignition.

3.3 Implications for Responding Large Scale Leaks

The current study is limited to a fixed fuel concentration and fixed length of ignition section. Certainly, the critical length for successful suppression in an open pipe may varied depending on the type of fuel, fuel concentration, and the length of ignition section. Continued work with different fuel, fuel concentration and ignition length is underway. Nevertheless, the short critical length required for suppression offer excellent opportunity for field application of suppression in large-scale flammable leak into a long duct.

In large-scale leaks of flammables into an underground duct such as the incidents described in [1-3], the long trench was never sealed but was similar to the open duct as the current study. It is impossible to inert all the length of duct as the required inert gas can be excessive. To suppress the flame propagation, one simply need to inject inert gas into the duct at fixed intervals. The best point of injection of inert gas will be the manhole to the underground trench. Volume of the injected inert gas should be at least 1/10 volume of the duct interval section. Clearly, field test to verify the effectiveness of suppression will be extremely difficult. Tests are planned by using a large duct with an internal diameter of 0.5 m to provide a near-field-scale validation for the current results. Injection of inert gas is however far easier than other explosion suppression methods such as water spray, water mist, dry power etc. All these suppression methods require continuing injection in order to maintain the effectiveness. Inert gas injection remained effective as long as the diffusion across the inert/fuel gas interface is insignificant.

4 Conclusions

Leak of flammables from a pipeline into a confined duct such as an underground trench has been a major safety issue across different countries. Upon ignition, flame and explosion propagated along the duct will create a significant impact to community and environment. We proposed the concept of suppressing flame propagation in a long duct by injecting an inert gas into the duct at multiple locations to act as inert gas sections which segregate the flammable mixture. The segregation limits the flame propagation to only one section without reaching nearby sections and thus prevents the successive ignition and flame propagation throughout the duct. The concept is demonstrated in a small duct with a 3-m ignition section, a 0.1~6 m inert section, a 3-m flammable section, and a 18-m air buffer section. The critical length required for successful suppression of flame propagation is found to be 0.3 m, which is 1/10 of the ignition section and is also far smaller than the case of suppression in a sealed duct. Further work is underway to cover different parameters that may affect the flame suppression.

5 Acknowledgment

This work is supported by Ministry of Science and Technology, Taiwan, through grant no. MOST-104-2221-E-327-002-MY2 and MOST-105-2221-E-327-001-MY3. All supports are gratefully acknowledged.

References

[1] N. Andersson, A. Morales, Lancet 339 (8801) (1992) 1103.

[2] Y. Zhu, X.M. Qian, Z.Y. Liu, P. Huang, M.Q. Yuan, J. Loss Prev. Process Ind. 33 (2015) 289-303.

[3] H.N. Yang, J.H. Chen, H.J. Chiu, T.J. Kao, H.Y. Tsai, J.R. Chen, J. Loss Prev. Process Ind. 41 (2016) 107-120.

[4] Y. Du, P. Zhang, Y. Zhou, S. Wu, J. Xu, G. Li, J. Loss Prev. Process Ind. 31 (2014) 113-120.

[5] P. Zhang, Y. Du, S. Wu, Y. Zhou, J. Zhou, J. Xu, J. Therm. Anal. Calorimetry 118 (2014) 1773–1780

[6] H. Guénoche, in: G.H. Markstein (ed.), Non-steady flame propagation, Pergamon, Oxford, 1964, p. 107.

[7] S.A.S. Jones, G.O. Thomas, Combust. Flame 87 (1) (1991) 21–32.

[8] S. Kerampran, D. Desbordes & B. Veyssière, Combust. Sci. Tech. 158 (1) (2000) 71-91.

[9] C.R.L. Bauwens, L. Bauwens, I. Wierzba, Proc. Combust. Inst. 31 (2007) 2381–2388.

[10] D.A. Crowl, J.L. Louvar, Chemical Process Safety, 3rd. Ed., Prentice-Hall, Upper Saddle River, NJ, 2011.

11th Asia-Pacific Conference on Combustion,
The University of Sydney, NSW Australia
10th -14th December 2017.

Sparse-Lagrangian MMC-LES Modelling of Reacting Acetone Spray

M.N. Khan and M.J. Cleary

The University of Sydney, School of Aerospace, Mechanical and Mechatronic Engineering

New South Wales, Australia

Abstract

Modelling the combustion of liquid sprays is a challenging task due to a wide range of complexities in both the liquid and gas phases and their interactions. For the liquid phase, the physical processes include turbulent dispersion, evaporation and exchange of heat and mass with the gas phase. On the other hand, there is turbulence, mixing and chemical reactions in the gas phase. The model that is validated in this study is comprised of an Eulerian large eddy simulation (LES) in combination with a sparse-Lagrangian stochastic particle scheme with a multiple mapping conditioning (MMC) mixing closure for the filtered density function (FDF) of the gaseous composition field and a Lagrangian fuel particle (LFP) scheme for the dispersion of droplets. A mixture-fraction-based subgrid model is introduced for the unresolved heat and mass transfer between the gaseous and liquid phases. The heat and mass coupling between LES, MMC and LFP are achieved elegantly following the OpenFOAM templated class structure, producing a robust numerical tool. The model and implementation are validated for an experimental reacting acetone spray case of Gounder *et al.* [1]. Here we demonstrate the consistency of the interphase heat and mass coupling schemes by examining the LES and MMC results for mixture fraction and compare predictions of mean and RMS of droplet velocity and gas phase temperature to the experimental data.

1. Introduction

Impressive progress has been made in recent decades in the field of turbulent combustion modelling [2] made possible by advances in computational fluid dynamics. LES, which resolves the energy containing turbulent motions while introducing subgrid models for the universal turbulent motions in the inertial range, is now widely used. However, LES of reacting sprays is less mature and its modelling is complicated by the many competing processes, not least the subgrid scale modelling of the heat and mass transfer between the carrier gas and the discrete liquid phase. In a turbulent flow, closure of the non-linear interactions between the phases is not trivial and is compounded by the fact that the droplets are affected by inertial forces and have a different trajectory to the carrier, necessitating separate numerical implementations with complex two-way coupling between them. Added to this, liquid fuels typically have greater molecular mass than the simple gaseous fuels such as methane and hydrogen, and as a result the detailed chemical kinetics schemes of liquid fuels are more complex and expensive to integrate. Models that inherently address the twin demands of having accurate turbulent heat and mass transfer coupling and efficient simulation of the turbulent gas phase chemistry are highly sought after. In the present paper we combine a mixture-fraction-based Spalding transfer model for the subgrid interphase heat and mass transfer and a sparse-Lagrangian MMC-LES turbulent combustion model [3]. It is applied to the piloted acetone spray flame (AcF1) of Gounder *et al.* [1]. The use of the mixture fraction, which is a conserved scalar, allows for a consistent treatment of the interphase coupling irrespective of the presence of a flame in the diffusive layer around the droplets. We previously presented it for a non-reacting evaporating acetone case [4] and in the present work it is extended to the reacting case for the first time. Sparse MMC-LES has evolved as a highly accurate and efficient turbulent combustion model. It is a FDF method which uses a stochastic particle Monte Carlo implementation and MMC plays the part of a mixing model whereby a low computational-cost, sparse distribution of particles may be used by virtue of the enforced localness in a reference mixture fraction space.

Corresponding author. Fax: +612-9351-7060

E-mail address: mkha1785@uni.sydney.edu.au

2. The MMC-LES Spray Combustion Model

The governing equations for two-phase flow may be derived through the separated flow model first introduced by Kataoka [5] and later by Mortensen and Bilger to formulate the spray conditional moment closure (CMC) model [6]. When cast in the LES context, the phase-weighted, Favre-filtered continuity, momentum and composition equations are

$$\frac{\partial\bar{\theta}\bar{\rho}}{\partial t}+\frac{\partial\bar{\theta}\bar{\rho}\tilde{u}_i}{\partial x_i}=\widetilde{\dot{S}_\rho} \quad (1)$$

$$\frac{\partial\bar{\theta}\bar{\rho}\tilde{u}_i}{\partial t}+\frac{\partial\bar{\theta}\tilde{u}_j\tilde{u}_j}{\partial x_j}=-\bar{\theta}\frac{\partial\bar{P}}{\partial x_i}+\frac{\partial\bar{\theta}\tilde{\tau}_{ij}}{\partial x_j}-\frac{\partial\bar{\theta}\tau_{ij}^{sgs}}{\partial x_j}+\tilde{\dot{S}}_{u,i} \quad (2)$$

$$\frac{\partial\bar{\theta}\bar{\rho}\tilde{\phi}_\alpha}{\partial t}+\frac{\partial\bar{\theta}\bar{\rho}\tilde{u}_i\tilde{\phi}_\alpha}{\partial x_i}=-\frac{\partial\bar{\theta}\tilde{J}_{\alpha,i}}{\partial x_i}-\frac{\partial\bar{\theta}J_{\alpha,i}^{sgs}}{\partial x_i}+\tilde{\dot{S}}_\alpha \quad (3)$$

where the overbar and tilde represent Reynolds and Favre filtered quantities, respectively, ρ is density, u_i is the velocity in coordinate x_i , P is pressure and $\phi=(\boldsymbol{Y},h)$ is the composition tensor where $\boldsymbol{Y}$ denotes the mass fraction of chemical species and h is the standardised enthalpy. The resolved viscous shear stress and mass or heat diffusion are denoted by $\tilde{\tau}_{ij}$ and $\tilde{J}_{\alpha,i}$ and the related subgrid quanties are given the *sgs* supersript. θ is the volume fraction of the gaseous mixture and in the dilute spray investigated here its filtered value is assumed to be unity. Finally, the filtered source terms are

$$\widetilde{\dot{S}_\rho}=\bar{\rho}\tilde{\pi} \quad (4)$$

$$\widetilde{\dot{S}_{u,i}}=\bar{\theta}\bar{\rho}\tilde{a}_{D,i}+\bar{\rho}\widetilde{u_{D,i}\pi} \quad (5)$$

$$\widetilde{\dot{S}_\alpha}=\bar{\theta}\bar{\rho}\tilde{W}_\alpha+\bar{\rho}\widetilde{\phi_{\alpha,D}^{*}\pi} \quad (6)$$

where $a_{D,i}$ and $u_{D,i}$ are the acceleration due to the drag imparted by the droplets and droplet velocity, respectively, W_α

is the source term due to chemical reaction or gas-phase radiation, π is the evaporative mass flow rate per unit gaseous volume, and $\phi^*_{\alpha,D} = \phi_{\alpha,D}$ for $\phi_\alpha \in \boldsymbol{Y}$ and $\phi^*_{\alpha,D} = \phi_{\alpha,D} - q_D + q_R$ for $\phi_\alpha = h$ where q_D and q_R are the rates of droplet internal heating and radiation per unit evaporative mass. As the momentum interactions, chemical reactions and gas production due to evaporation occur at the subgrid scales, closure models are required for the filtered source terms in Eq. (4) to (6). These closures are obtained in the context of an Eulerian LES scheme, a stochastic sparse-Lagrangian MMC scheme, a Lagrangian fuel particle scheme and two-way coupling schemes. In the following, the superscripts *eu*, *sp* and *fp* denote values at Eulerian grid cells, stochastic particles and fuel particles, respectively. Subscripts D, G and S denote values within the dispersed liquid, in the gaseous state at the outer edge of a droplet diffusion layer and on the gaseous side of a droplet surface, respectively.

Eulerian LES Scheme: The filtered mass and momentum source terms in Eq.(1) and (2) are evaluated with information from the $N_{fp,LES}$ fuel particles in each LES cell [7]:

$$\tilde{\bar{S}}^{eu}_{\rho} = \bar{\rho}^{eu}\tilde{\pi}^{eu} \simeq \frac{1}{V^{eu}}\sum_{fp=1}^{N_{fp,LES}} \dot{m}^{fp}_{D} \tag{7}$$

$$\begin{aligned}\tilde{\bar{S}}^{eu}_{u,i} &= \bar{\theta}^{eu}\bar{\rho}^{eu}\tilde{a}^{eu}_{D,i} + \bar{\rho}^{eu}\tilde{u}^{eu}_{D,i}\pi^{eu} \\ &= \frac{1}{V^{eu}}\sum_{fp=1}^{N_{fp,LES}}\left(-\dot{m}^{fp}_{D}a^{fp}_{D,i} + \dot{m}^{fp}_{D}u^{fp}_{D,i}\right).\end{aligned} \tag{8}$$

Here, V^{eu} is the filter volume and $\dot{m}^{fp}_D$, $a^{fp}_{D,i}$ and $u^{fp}_{D,i}$ are obtained from the fuel particle scheme. The composition field is simulated in the stochastic MMC scheme and, hence, we do not solve Eq.(3) for chemical species mass fractions and enthalpy, but we do solve a similar equation for the Eulerian filtered mixture fraction, $\tilde{f}$, which becomes the reference variable in the stochastic MMC scheme. Its filtered source term is

$$\tilde{\bar{S}}^{eu}_{f} = \bar{\rho}^{eu}\widetilde{f_D\pi}^{eu} = \bar{\rho}^{eu}\tilde{\pi}^{eu} \simeq \frac{1}{V^{eu}}\sum_{fp=1}^{N_{fp,LES}} \dot{m}^{fp}_{D} \tag{9}$$

where we have used the fact that the mixture fraction of the fuel droplet is $f_D = 1$. The remaining LES closures are conventional, including the Smagorinsky model for the subgrid viscosity.

Stochastic MMC Scheme: For the turbulent composition field, the sparse-Lagrangian MMC model is introduced through the stochastic differential equations for increments in stochastic particle position, composition and mass (due to interphase transfer):

$$dx^{sp}_i = \left[\tilde{u}^{sp}_i + \frac{1}{\bar{\rho}^{sp}}\left(\frac{\partial}{\partial x_i}\left(\bar{\rho}(\mathcal{D}+\mathcal{D}_t)\right)\right)^{sp}\right]dt + \left[\sqrt{2(\mathcal{D}+\mathcal{D}_t)^{sp}}\right]dw_i \tag{10}$$

$$d\phi^{sp}_\alpha = \left[W^{sp}_\alpha + M^{sp}_\alpha + (\phi^{*,sp}_{\alpha,D} - \phi^{sp}_\alpha)\pi^{sp}\right]dt \tag{11}$$

$$dm^{sp} = \frac{m^{sp}\pi^{sp}dt}{1-\pi^{sp}dt} \tag{12}$$

Here, M^{sp}_α is the MMC micromixing operation as described in detail in Ref. [3]. $\pi^{sp} = \dot{m}^{sp}/(\rho V)^{sp}_G$ is the conditional transfer rate from the LFP to MMC. Note that the diffusion is assumed to be Fickian where $\mathcal{D}$ and $\mathcal{D}_t$ are the molecular and subgrid diffusivities, respectively, and the unity Lewis number assumption is applied.

Lagrangian Fuel Particle Scheme: The dispersion of fuel particles is modelled by the approach developed by Dukowicz [8] whereby parcels represent ensembles of identical fuel particles or droplets. The governing equations for the location, velocity and temperature of each parcel are given by

$$\frac{dx^{fp}_i}{dt} = u^{fp}_{D,i} \tag{13}$$

$$du^{fp}_{D,i} = (a^{fp}_{D,i} + g_i)dt + b^{fp}_{D,i}dw_i \tag{14}$$

$$m^{fp}C_{p,D}\frac{dT^{fp}_D}{dt} = \dot{m}^{fp}\; q^{fp}_D. \tag{15}$$

In the above $a^{fp}_{D,i}$ and g_i are the accelerations of due to drag forces imparted by the gas phase and gravity, and $b^{fp}_{D,i}dw_i$ is a stochastic term to account for subgrid interactions. The latter can be neglected as a first approximation [9]. Following the transfer approach of Spalding [10], the evaporative mass flux is given by

$$\dot{m}^{fp} = (\rho\mathcal{D}\; Sh)^{fp}_G \frac{1}{d^{fp}} ln(1 + B^{fp}_M)A^{fp} \tag{16}$$

where the Sherwood number, density and diffusivity are obtained for the gas phase at a prescribed film state (here the 1/3 law is used), d is the droplet diameter and A is its surface area. B_M is the Spalding transfer number that is defined strictly for a conserved scalar such as the mixture fraction, viz.

$$B^{fp}_M = \frac{f^{fp}_G - f^{fp}_S}{f^{fp}_S - f^{fp}_D} \tag{17}$$

Finally, the droplet internal heating is given by

$$q^{fp}_D = \frac{h^{fp}_G - h^{fp}_S}{B^{fp}_M} - (h^{fp}_S - h^{fp}_D) + q^{fp}_R \tag{18}$$

Coupling schemes: There is two-way coupling of mass and enthalpy between the three schemes listed above. The Eulerian LES passes velocity, reference mixture fraction (to localise the micromixing operation) and turbulence quantities to the stochastic MMC and density are coupled back to the LES by way of a conditionally averaged equivalent enthalpy method [3, 11]. The dispersion in the LFP scheme uses the cell-based filtered fields from the LES and in return the LES source terms are obtained from the LFP as described by Eq.(7) – (9). Heat and mass is exchanged between the LFP and MMC schemes due to evaporation. Each fuel parcel is paired with a single stochastic MMC particle based on proximity in physical and reference mixture fraction spaces. We set $f^{fp}_G = z^{sp}$, where z is the stochastic mixture fraction solved according to Eq.(10) – (12), and f^{fp}_S is obtained by assuming a Clausius-Clapeyron equilibrium state exists at the droplet surface. The equilibrium will vary with the state of reaction in the droplet diffusive layer where an envelope flame may or may not exist. For simplicity we assume that in such conditions that $f^{fp}_S > f_{stoic} > f^{fp}_G$ that an envelope flame is present.

3 Numerical Implementation

The boundary conditions for the reacting acetone case, AcF1, are shown in Table 1. The numerical simulation is performed in mmcFoam, a branch of OpenFOAM developed at Sydney

Reacting acetone case	AcF1
Bulk jet Velocity, Ujet (m/s)	24
Carrier mass flow rate (g/min)	150
Liquid fuel injection rate (g/min)	75
Measured liquid flow rate at exit (g/min)	18
Measured temperature at jet exit plane (0C)	293 K
Co-flow bulk velocity (m/s) of air	4.5
Bulk velocity in pilot stream (m/s) of air	11.9
Vapour fuel flow rate at jet exit (g/min)	57.0
Mixture fraction at exit plane	0.275
Pilot mixture fraction	0.0858

Table 1: Boundary conditions for the AcF1 case [1].

University. Finite rate chemistry is applied in the gas phase using the mechanism developed by Chong and Hochgreb [12]. The numerical domain is taken as a rectangular box with a dimension of *14Dx14Dx35D*, where, *D* is the diameter of the main jet. The grid resolution includes approximately 1.3M LES cells with 320 cells in the axial direction and 4161 cells in the xy-plane. The simulations employ a sparse distribution of stochastic particles with approximately 1 particle per 10 LES cells (1L/10E). Sensitivity to both the LES and stochastic particle resolution have been investigated with minimal variation to the present results for resolutions greater than those stated above.

4 Results and Analysis

Evaporation is a source of gas phase mixture fraction. In order to assess the consistency of the coupling schemes, an instantaneous image of the LES mixture fraction, *f*, (at the stochastic particle locations) and the stochastic particle mixture fraction, *z*, are shown in Fig. 1 and the radial profiles of their mean and RMS are shown in Fig. 2 at three axial stations. There is clearly a strong correlation between the *f* and *z* fields giving a good indication that evaporative mass coupling has been implemented consistently.

While experimental data is not available for the mixture fraction in these droplets laden flows the accuracy of the simulations is assessed through comparison of droplet velocities and gaseous temperature. The predicted mean axial centerline velocity of droplets in two size bins, 0-10μm and 30-40μm, is compared to the experimental data in Fig. 3. For comparison the predicted gaseous mean velocity is also shown and found to be very similar to the small droplet velocity on account of their low Stokes numbers. The predictions are in good agreement with the data except for slight underprediction downstream. Radial profiles of mean and RMS velocity of droplets in 10 – 20 μm range are shown in Fig. 4 at *z/D =10, 20* and *30*. The trends are accurately captured but quantitatively there is some underprediction of the mean while the RMS shows good agreement at all locations. While the results of mean velocity profiles are similar to the study of Ukai *et al.* [9] the RMS velocity has improved.

Radial profiles of mean and RMS of gaseous temperature are shown in Fig. 5 at *z/D=10, 20* and *30*. There is an excellent agreement of mean temperature at the first two stations while the predictions indicate somewhat greater jet spreading relative to the data at *z/d = 30*. The mean peak temperature in the shear layer is well predicted at all axial locations. The predicted RMS profiles of temperature at *z/D=10* and *20* show peak values on the inner side of the shear layer while at *z/D=30* the RMS is more

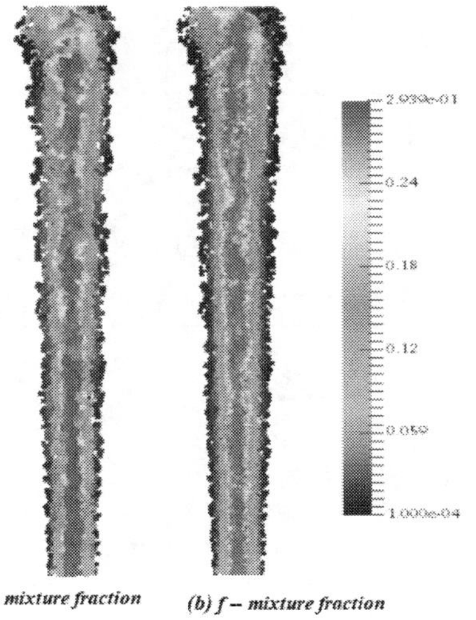

Figure 1: Instantaneous contours of *f* and *z* mixture fraction fields at the stochastic particle locations.

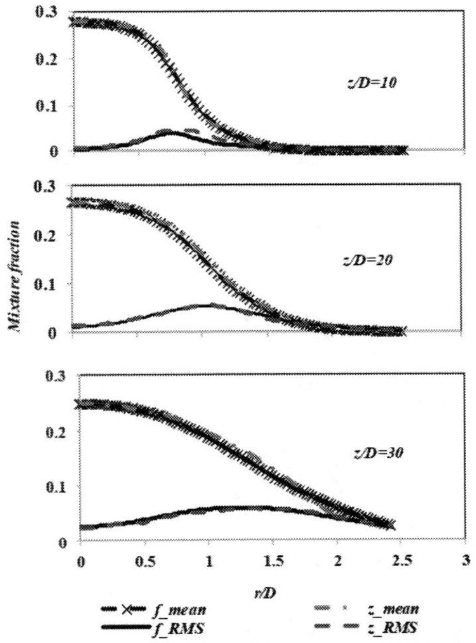

Figure 2: The radial profiles of the mean and RMS mixture fraction at different locations.

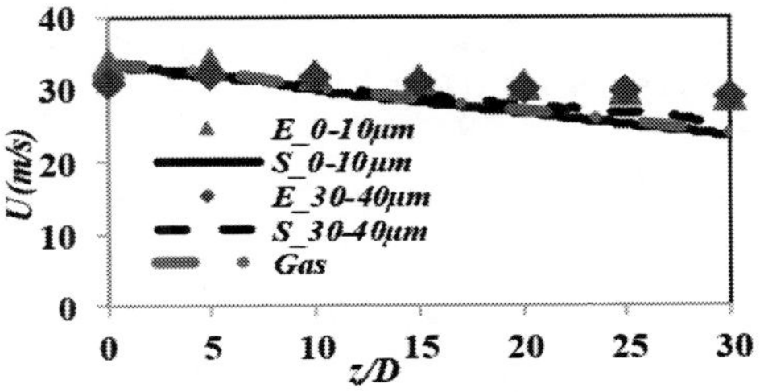

Figure 3: Axial centerline profiles of droplets and gas velocity. The letter *E* and *S* represent for the experimental and simulation results.

even distributed across the radial coordinate with a smaller peak on the outer edge the shear layer. The mean temperature profiles are also similar to the study of Ukai *et al.* [9].

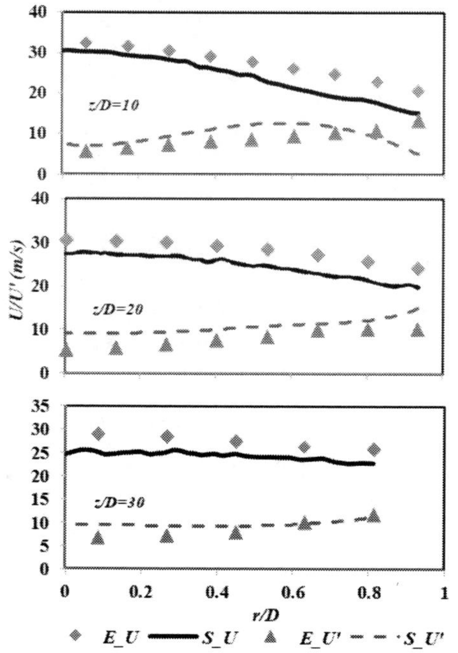

Figure 4: Radial profiles of the mean axial droplet velocity (U) and RMS (U') in the 10-20 μm size range. RMS values have been multiplied by four for clarity.

5 Conclusions

The simulation of reacting acetone case of Grounder *et al.* [1] is performed with a novel sparse-Lagrangian MMC-LES model incorporating a mixture-fraction-based subgrid heat and mass transfer model between the gaseous and dispersed phases. A comparison of the instantaneous and averaged mixture fraction fields demonstrates good consistency between the Eulerian LES and stochastic MMC schemes indicating the robustness of coupling algorithms. The results of mean and RMS velocity of droplets and the mean gas temperature are shown to be in good agreement with the experimental data [1].

6 Acknowledgment

The work is supported by Australian Research Council under Grant Number DP130100763. We also acknowledge the facilities and technical assistance of the Sydney Informatics Hub at the University of Sydney and, in particular, access to the high-performance computing facility Artemis.

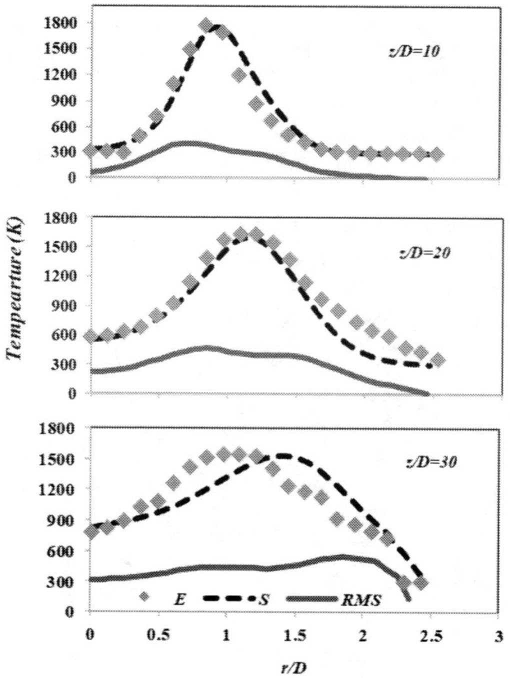

Figure 5: Radial profiles of mean (up) and RMS (down) of gaseous temperature.

References

1. J.D. Gounder, A. Kourmatzis, and A.R. Masri, Combustion and Flame 159(11) (2012) 3372-3397.

2. T. Echekki and E. Mastorakos, Turbulent combustion modeling - Advances, New Trends and Perspectives, Springer Dordrecht Heidelberg London New York, 2010, p. 19.

3. M. Cleary and A. Klimenko, Physics of Fluids 23(11) (2011) 115102.

4. M.N. Khan and M.J. Cleary, Large Eddy Simulation of Evaporating Acetone in a Turbulent Multiphase Flow, The University of Melbourne, Year, P.304-307.

5. I. Kataoka, M. Ishii, and A. Serizawa, International Journal of Multiphase Flow 12(4) (1986) 505-529.

6. M. Mortensen and R.W. Bilger, Combustion and Flame 156(1) (2009) 62-72.

7. M. Sharma and D. Stock, Journal of Fluids Engineering (1977) 325.

8. J.K. Dukowicz, Journal of Computational Physics 35(2) (1980) 229-253.

9. S. Ukai, A. Kronenburg, and O.T. Stein, Flow, Turbulence and Combustion 93(3) (2014) 405-423.

10. S.D. B., Some Fundamentals of Combustion, Buttersworth Scientific Publications, London, 1955, p.

11. M. Muradoglu, S.B. Pope, and D.A. Caughey, Journal of Computational Physics 172(2) (2001) 841-878.

12. C.T. Chong and S. Hochgreb, Combustion and Flame 158(3) (2011) 490-500.

11th Asia-Pacific Conference on Combustion,
The University of Sydney, NSW Australia
10th -14th December 2017.

Characteristics of Liftoff and NOx Emission in Microwave Enhanced Methane Micro-Jet Flames

Young Hoon Jeon[1], Eui Ju Lee[1]

[1]Department of Safety Engineering, Pukyong National University

45, Youngso-Ro, Nam-Gu, Busan, 48513, Republic of Korea

Abstract

The use of electromagnetic energy is one of methods to enhance the combustion stability and a microwave as electromagnetic wave is receiving increased attention recently because of its high performance and low-cost system. The microwave effect on the stability and NOx concentration was experimentally investigated in the methane micro jet diffusion flames. The results show that micro jet flames had three different modes with increasing oxidizer velocity; attached yellow flame, lifted flame, and lifted partially premixed flame. The overall flame stability and blowout limit were extended with the higher microwave power. Especially the interaction between a flame and a microwave was shown clearly in the partially premixed flame, in which the lift-off height decreased and NOx emission measured in post flame region increased with increasing microwave power. It might be attributed to increase of reactivity due to the abundance of radical pool and the enhanced absorption of thermal energy.

1. Introduction

High efficient and environment friendly combustion technologies are used to be operated an extreme condition, which results in unintended flame instability such as extinction and oscillation. The use of electromagnetic energy and non-equilibrium plasma for enhancing ignition and combustion stability are receiving increased attention recently. The conventional technologies have adapted the electrical devices to make the electromagnetic field, which resulted in various safety issues such as high-maintenance, additional high-cost system, electric shock and explosion. Therefore, an electrodeless microwave technology has an advantage for economics and reliability compared with conventional one because of no oxidation.

The electromagnetic wave such as microwave has been widely applied to various combustion systems like plasma torch [1,2]. The insufficient power of microwave cannot reach the reactants into plasma states, but it still effects on the extension of flammability and the enhancement of ignition [3]. Even though the previous study provided the mechanism of chemical reaction under microwave as an electromagnetic wave [4], the characteristics of various combustion system by microwave has not been investigated sufficiently. Ogawa et al. [5] measured the gas temperature, luminosity and electron temperature of methane/air laminar flames in the cavity resonator excited at 2.45 GHz, and Takita et al. [6] showed the increase of burning velocity due to high temperature radical pool. Hemawan et al. [7] observed plasma jet flames according to the microwave power in the cavity shaped microwave burner. Also, a number of study was conducted for burning velocity and flame stability in the view of interaction between microwave and hydrocarbon flames [8,9]. Recently, an experimental study was performed with jet diffusion flames induced by microwaves to clarify the effect of microwave on the combustion stability and pollutant emissions such as soot formation and NOx [10].

However, the application of microwave has been still limited because of lack of interaction mechanism between flame and microwave. In this study, an experiment was performed with jet diffusion flames induced by microwaves to clarify the effect of microwave on the combustion stability and pollutant emissions such as NOx emission. Micro jet was introduced to simulate the high velocity of industrial combustor.

2. Experiments

Figure 1 shows the schematic of experimental set up. The burner used in this experiment consisted of coaxial tubes, where 1.0 mm of inner diameter for fuel and 26 mm of outer diameter for oxidizer. The fuel nozzle was made by quartz to minimize the antenna effect of microwave. Methane (99.9% purity) was used as a fuel and compressed air was used as an oxidizer, of which flow rates were controlled by using MFC (Mass Flow Controller). MFCs were calibrated with a flow meter (PFM, Definer 220H) and the accuracy of the flow rate was ±1%. Flame images were recorded and analyzed with the 60fps camcorder (SONY, HXR-NX30N).

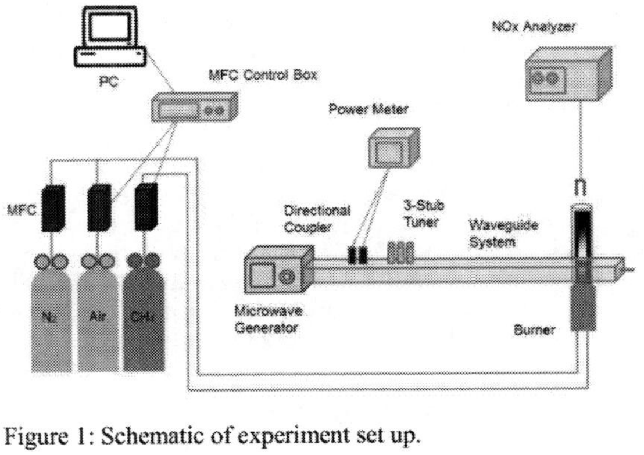

Figure 1: Schematic of experiment set up.

Corresponding author. Fax: +82-51-629-6463
E-mail address: ejlee@pknu.ac.kr

Schematic waveguide system is shown in Fig. 2. It was designed to provide for more efficient coupling of the microwave power into the flame. As increasing the electric field strength and minimizing the ion losses by the wall, it ensures that practically all of the power delivered from the source passes into the flame. WR-430 (10.92cm x 5.46.cm) rectangular waveguide was used for the experiment, which is a standard size for high power microwave transmission at 2.45 GHz. Microwaves produces a standing wave within the waveguide, and the flame experiences a sinusoidal time-varying electric field at a frequency of 2.45 GHz. The sliding short was used to control the location of the standing wave such that a field maximum is coincident with the location of the laminar flame burner. The field maximum is stationary in a position but switches field direction at a frequency of 2.45 GHz. The 3-stub tuner also served as a means to match the phase of the incident wave with the standing wave. Directional coupler measures intensities of incident and reflected waves, which provides the net power absorbed by an objects in waveguide. Circulator prevents the propagation of reflected waves to magnetron and water dummy load absorbs reflected waves because magnetron can be broken by the reflected waves.

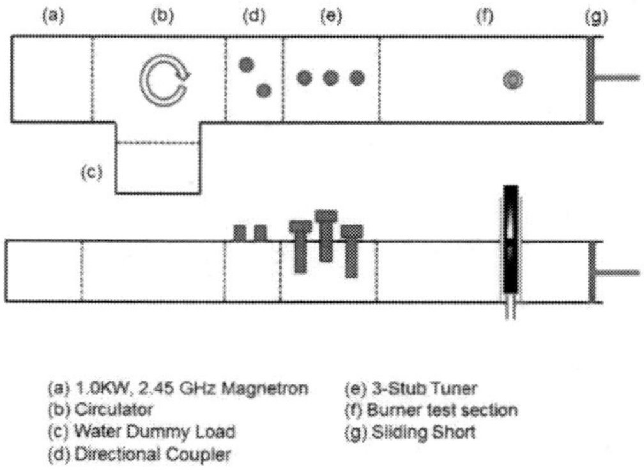

Figure 2: Configuration and components of waveguide system.

3. Results and Discussion

Figure 3 shows the direct photographs of methane/air micro jet diffusion flames without microwave flames as oxidizer velocity increases. As the oxidizer velocity increased, three flame modes were observed. At low oxidizer velocity, the base of the diffusion flame was attached on the burner tip, and the typical yellow diffusion flame was observed as shown in Fig. 3(a). Beyond a certain value of the oxidizer velocity the jet flame lifted from the nozzle tip and the yellow luminosity was reduced as shown in Fig. 3(b). The liftoff height just before transition to next flame mode was relatively small around 5 mm and slightly increased with increasing the oxidizer velocity. As shown in Fig. 3(c), a turbulent premixed flame was observed at a large oxidizer velocity. The flame length was very short and the blue colored flame was vibrating itself. The flame lifted before blowout, and the liftoff height was about 3cm, which is larger than that of a lifted yellow flames. The higher oxidizer velocity induced the flames to blowout finally. Although it was not presented in this paper, the flame modes of a microwave induced flames were similar to Fig. 3, but the microwave induced flames showed more stable and bright color.

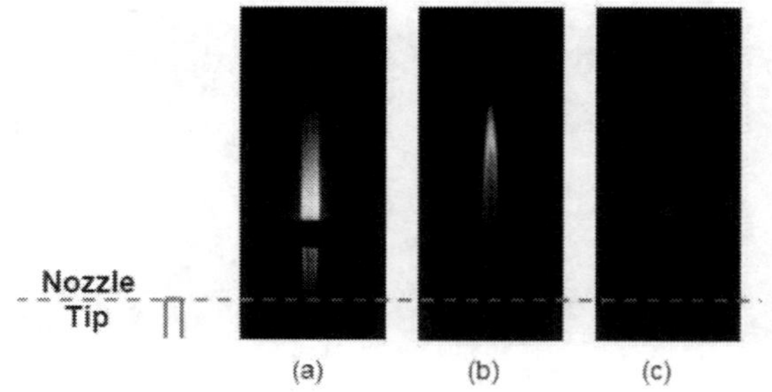

Figure 3: Direct photos of CH_4/air micro jet flames as increasing oxidizer velocity at V_f=1274 cm/s. (a) Attached yellow flame, V_{oxi}=21 cm/s (b) Lifted yellow flame, V_{oxi}=78 cm/s (c) Partially premixed lifted flame, V_{oxi}=79 cm/s.

Figure 4 shows the flame modes with respect to the fuel and oxidizer velocities. The attached yellow flame (Regime I) was observed at low oxidizer and fuel velocities. The higher velocity made the flames lift from the nozzle tip and change the color from yellow to blue (Regime II). The boundaries between the lift diffusion flames (Regime II-1) and vibrating partially premixed flames (Regime II-2) could not distinguished clearly so that the regime was depicted together in Fig. 4. At the higher fuel and oxidizer velocities, the lifted flames blow out eventually (Regime III). When the microwave induced flame set to high oxidizer velocity and low fuel velocity, the lifted flame regime became narrow and disappeared. The flame modes under 1 kW microwave was also presented in Fig. 4. The fuel and oxidizer velocities at the whole boundary of flame mode moved to the larger velocity, and hence it might be concluded that the microwave induced flame became stable and the flame extinction such as blow off could be attenuated especially at the higher fuel velocity.

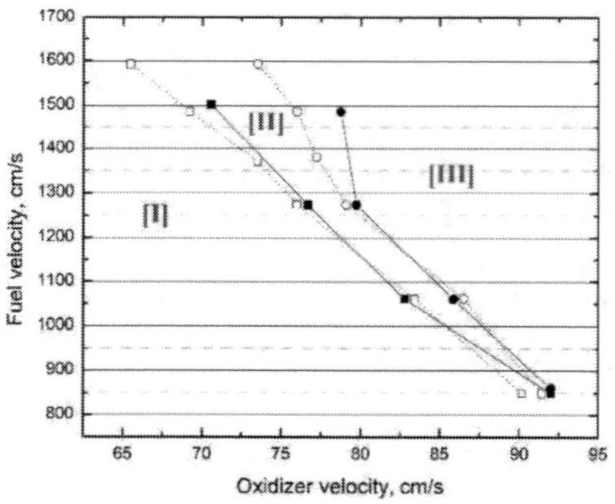

Figure 4: Flame map of micro jet diffusion flames. Solid symbols indicate the microwave induced flames at 1 kW and open symbols indicate the flame without microwave.

Figure 5 shows the NOx concentration and liftoff height according to the microwave intensity. The higher microwave intensity results to increase the NOx emission and reduce the liftoff height. The reduction of liftoff height can be attributed to the increase of burning velocity. The detailed previous experiment of burning velocity by microwave [6] reported that 35% enhancement of burning velocity resulted from induction of 1.2 kW microwave in the methane/air premixed flames. Therefore, partially premixed flames of Fig. 5 are similar to the premixed flames and hence the reduction of liftoff height might be understood mainly due to increase of the burning velocity. Increase of NOx emission may be analyzed the variation of reactivity by microwave. This can be supported by the fact that the post flame temperature increased with microwave intensity.

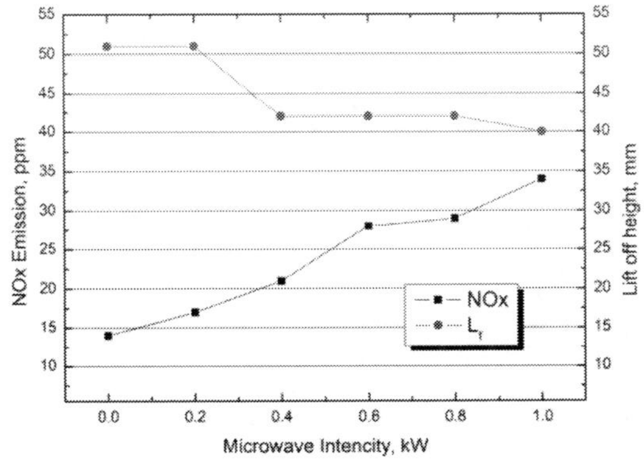

Figure 5: Variation of NOx emission and liftoff heights with respect to the microwave intensity at V_f=1486 cm/s and V_{oxi}=74.1 cm/s.

The effect of microwave location against flames on the NOx emission and liftoff heights was investigated as shown in Fig. 6. In the experiment, the induced microwave location was changed by using adjusting burner height. When the microwave directed the partially premixed flames at the center of waveguide cross section, NOx emission has maximum value, which means that microwave induction is most efficient to irradiate to the primary reaction zone instead of upstream or far down stream.

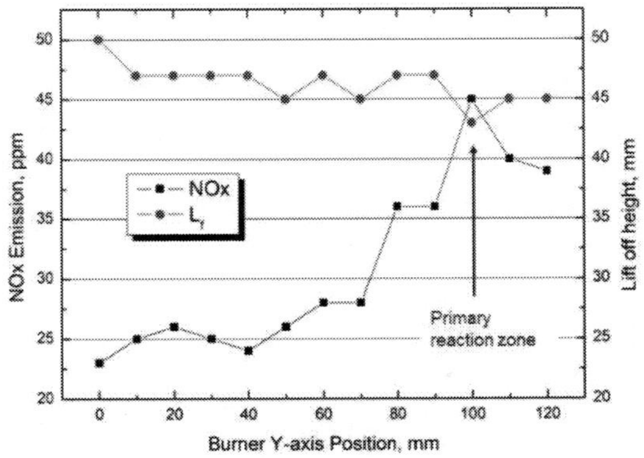

Figure 6: Variation of NOx emission and liftoff heights with respect to the microwave induction location at V_f=1486 cm/s and V_{oxi}=74.1 cm/s.

It has been reported that microwave induced flames results in the increase of reactive species and high temperature simultaneously [5,6,9]. These chemical and thermal effects of microwave are investigated on the NOx emission and liftoff height of micro jet flames. The partially premixed flame among various flame modes is selected for this experiment because the entire primary reaction zone of the flame mode can be interacted with microwave in the waveguide. To try to isolate the thermal effect of microwave, nitrogen was diluted in the oxidizer stream, in which the downstream temperature and the total flow rate kept same as those of micro jet flames without microwave at each microwave intensity. As shown in Fig. 7 and Fig. 8, NOx emission by thermal and chemical effect increases with microwave intensity, but the concentrations of diluted flames are saturated at a lower concentration regardless of microwave intensity. The liftoff height shows the opposite behavior to the NOx emission, and all effects are obvious at high microwave intensity.

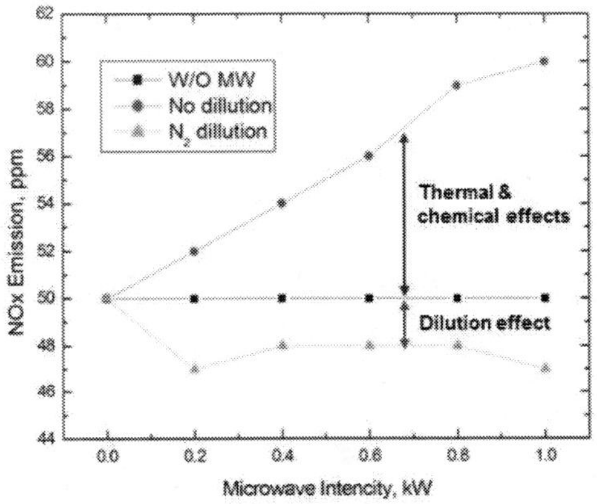

Figure 7: Effect of microwave on NOx emissions of nitrogen diluted micro jet flames.

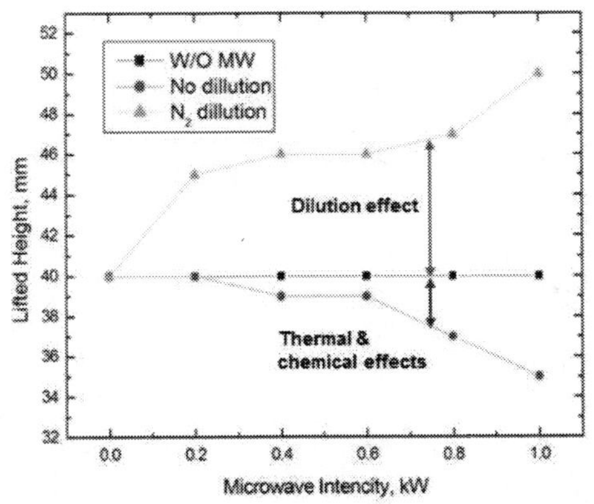

Figure 8: Effect of microwave on liftoff heights of nitrogen diluted micro jet flames.

4. Conclusions

The microwave effect on the stability and NOx concentration was experimentally investigated in the methane micro jet diffusion flames. Waveguide system was designed to provide for more efficient coupling of the microwave power into the flame and to adjust microwave intensity up to 1kW at a frequency of

2.45 GHz. The experimental and numerical findings were as follows:

1) For different fuel and oxidizer velocity, micro jet diffusion flames can be classified into three modes: Attached flames, Lifted flames (yellow flames and partially premixed flames), Double flames.

2) NOx emissions of the lifted flames increased with increasing microwave intensity, but lifted height decreased. These are representative evidence of enhanced flame stability due to the induction of microwave.

3) As the microwave position irradiated on the flame was changed, local maximum of NOx emission and the minimum of liftoff height corresponded to the irradiation position on the primary reaction zone exactly.

4) Dilution of nitrogen provides the decoupling between thermo-chemical effect and dilution effect, of which results shows the thermo-chemical effect responds on the microwave intensity. Therefore, NOx emission of microwave induced flames is mainly affected by thermal mechanism.

5. Acknowledgment

This research was supported by the Fire Fighting Safety & 119 Rescue Technology Research and Development Program funded by the Ministry of Public Safety and Security (MPSS-2015-66).

References

[1] F.J. Weinberg, K. Hom, A.K. Oppenheim, K. Teichman, Nature 272(5651) (1978) 341-343.

[2] P. Fauchais, A. Vardelle, IEEE Trans. Plasma Sci. 25(6) (1997) 1258-1280.

[3] S.M. Starikovskaia, J. Phy. D: Appl. Phys. 39 (2006) 265-299.

[4] X. Raoa, K. Hemawanb, I. Wichmana, C. Carterc, T. Grotjohnb, J. Asmussenb, T. Leea, Proc. Combust. Inst. 33 (2011) 3233-3240.

[5] S. Ogawa, Y. Sakai, K. Sato, S. Sega, Jpn. J. Appl. Phys. 37(1) (1998) 179-185.

[6] K. Takita, G. Masuya, T. Sato, Y. Ju, AIAA J. 39(4) (2001) 742-744.

[7] K.W. Hemawan, I.S. Wichman, T. Lee, T.A. Grotjohn, J. Asmussen, Rev. Sci. Instrum., 80(5) (2009) 503-507.

[8] Y.C. Hong, S.C. Cho, C.U. Bang, D.H. Shin, J.H. Kim, H.S. Uhm, W.J. Yi, Appl. Phys. Lett. 88(20) (2006) 201502-201504.

[9] E.S. Stockmana, S.H. Zaidia, R.B. Milesa, C.D. Carterb, M.D. Ryanc, Combust. Flame, 156(7) (2009) 453-1461.

[10] Y.H. Jeon, E.J. Lee, J. Korean Soc. Safety. 29(4) (2014) 2 3-27.

11th Asia-Pacific Conference on Combustion,
The University of Sydney, NSW Australia
10th -14th December 2017.

Experimental study on the effects of equivalence ratio & reactor length on flame speed in micro scale reactors

Maryam Yeganeh[1], Sadegh Tabejamaat[1], Amin Aramesh[1], Mohammadreza Baigmohammadi[2]

[1]Department of aerospace engineering,

Amirkabir University of technology (Tehran Polytechnic), Hafez Ave., Tehran 15875-4413, Iran

[2]Combustion Chemistry Centre, School of Chemistry,

National University of Ireland, Galway, Ireland

Abstract

The main purpose of this study is investigating methane-oxygen premixed flame characteristics in micro scale quartz reactors (combustion chamber). Experimental study is carried out on reactors with the diameter of 1 mm and lengths of 5, 10, 15, 20cm. Also, total flow rate is fixed at 0.15 SLM and equivalence ratios (φ) are considered between 0.5 and 2. The result is consist of stating different flame regimes, the effects of equivalence ratio on flame stability, the effects of reactor length on flame speed and stable flame locations . The results show that equivalence ratio has a considerable effect on flame propagation speed in the reactor. Also, by increasing reactor length, flame propagation speed decreases as a result of heat losses and other instabilities.

1 Introduction

Nowadays, the requirements of reducing power systems dimensions and weight has encouraged researchers to produce micro power systems which results in optimizing energy consumption. Focusing on such concerns about energy saving has led to create a variety types of micro power systems such as micro gas turbines, micro fuel cells, micro thermo electric systems, micro thrusters, micro thermo photovoltaic systems, micro electromagnetic systems, micro electro mechanical systems (MEMS), etc. Many of these systems are based on micro combustion which is formation of flame in micro scale reactors and has a wide application area such as in chemical process industry (such as gas reforming), chemical propulsion (such as micro air vehicles), humanoid robots, lab-on-chip, communication devices (such as mobile phones), armor equipment, etc.

The power suppliers of these systems are different kinds of batteries with low specific energy (energy density (W.hr/kg)) or fuel cells with low power density (W/kg). Some disadvantages of these systems is high weight and limited performance time such as Ni-Cd, Ni-MH, Lithium-Ion, Lithium-poly, Zinc-Air, etc. High energy density (W.hr/kg) together with high power density (W/kg) is the main factor of producing high power with long performance time. According to the above features, an internal combustion engine with a hydrocarbon fuel seems to be an appropriate choice as shown in Fig. 1 and Fig. 2.

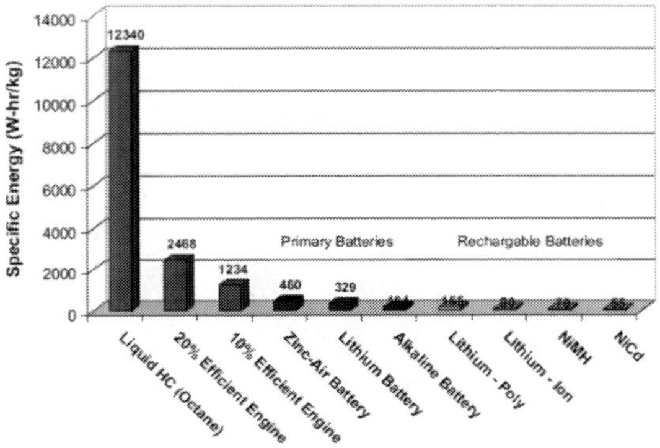

Figure 1: specific energy of different power suppliers [1&2]

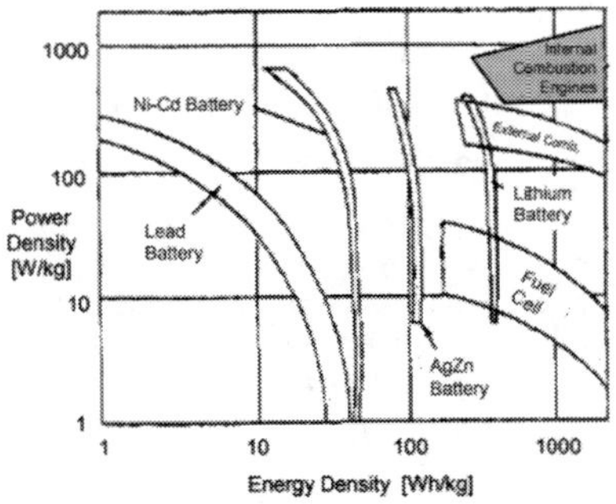

Figure 2: power density of different power suppliers [3]

Such investigations and other advantages of micro combustion such as high area to volume ratio, suitable equipment removal, optimization of different systems performance, etc. were the most important motivations of this study.

First of all, the definition of micro scale must be explained. There are several definitions of micro scale based on different references but in this study we rely on the definition of micro scale as the characteristic diameter of reactor less than 5mm [4&5].

There are various studies on micro combustion characteristics such as investigating the effects of different fuel and oxidizers, post flame formation, heat transfer between flame and reactor wall, etc.. A brief review on several similar studies on methane-

Corresponding author. Fax: +98-21-6454-3212
E-mail address: sadegh@aut.ac.ir

oxygen premixed flames is given here.

On one hand, Melish et al [6], studied on cylindrical quartz, alumina and stainless steel reactors with the diameter of 0.5mm.On the other hand, Miesse et al. and Prakash et al [6], both studied on square gapped stainless steel, alumina and quartz reactors with the length of 0.75mm. These studies were about reactor geometry, heat transfer between reactor and surrounding, equivalence ratio of fuel and oxidizer mixture, flow speed and flame characteristics in the reactor.

Another research on effect of equivalence ratio on reactor heat loss, dependence of heat transfer to flow rate, flame location in reactor and the effect of radiation on heat loss was investigated by Li and Zhong on three types of micro scale reactors [7].

Evans et al also studied on methane-oxygen fuel reached mixtures, especially on flame fluctuations frequency modes as a function of equivalence ratio and flame speed [8].

In this study, after different flame regimes investigation, we place emphasis on effect of equivalence ratio and reactor length on flame speed and finally present flame location graphs that can be useful for designing any kinds of micro power system.

2 Test Stand

This investigation carried out the influence of equivalence ratio (φ=0.5-2) and different reactor lengths (5, 10, 15, 20cm) on methane-oxygen premixed flame characteristics specially on flame speed in a reactor with diameter of 1mm and total flow rate of 0.15SLM.

The test stand is shown in Fig. 3

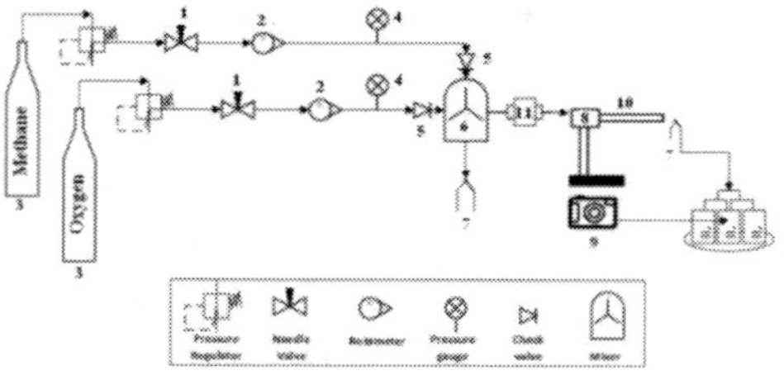

Figure 3: schematic of test stand

Main components of this test stand are as below:

1- Fuel and oxidizer tanks with appropriate regulators to check both tank and line pressure during each test
2- Needle valves to open/close fuel and oxidizer lines
3- Rota meters for setting fuel and oxidizer flow rate according to calculated amounts
4- Dial gauge pressures to check fuel and oxidizer line pressure
5- Mixer for premixing fuel and oxidizer completely
6- Thermocouples for checking temperature
7- Reactors entrance manifold
8- Flame arrestor to prevent flash-back
9- Check valve to prevent flow returning
10- High speed camera

According to the test stand capacity, by choosing a total flow rate (0.15 SLM in this test), we can calculate equivalence ratio, fuel and oxidizer flow rate, average flow speed and Reynolds number of the flow. So, each test can begin by starting test stand, setting the flow rate of fuel and oxidizer, switching the high speed camera on and making an ignition at the open end of the reactor.

About the test pressure, since the reactor is exposed to the air the gage pressure is zero (atmospheric), fuel line pressure is fixed at 3.5bar and oxidizer line pressure is fixed at 2bar during all tests. These fixed numbers are because of different valves, flow meters, etc. between the fuel and oxidizer tanks and the reactor.

For expressing the results of experimental data, we need to process the recorded films or images for investigating the flame behavior and calculating the flame speed.

After image processing we can calculate the flame speed from (1).

$$v = \frac{L}{n\Delta t} \qquad (1)$$

In this equation, L is the length of flame propagation distance in the reactor, n is the number of frames that has been recorded and Δt is flame propagation time.

3 Results

Five flame regimes were observed during the test:

1- RERI (Repetitive Extinction Re-Ignition): when the local reactor wall temperature reaches to auto-ignition temperature and flame propagation speed is larger than incoming flow speed, the flame propagates to the upstream but as a result of reactor wall heat loss, temperature falls gradually and the flame extinguishes. If downstream temperature is high enough, the process can begin again. This flame regime is like the FREI (Flame repetitive Extinction and ignition) regime in [9] but without the limitations of FREI definition such as the temperature gradient.
2- Stationary RERI: this flame regime happens when flame propagation speed is near to incoming flow speed but the flame cannot be stable.
It is obvious that the calculation of flame speed is for RERI flame regimes.
3- Stationary: when the flame propagation speed is equal to incoming flow speed despite all heat losses, we have a stable flame at a special location in the reactor.
4- Marginal: a stable flame that is exactly located at the open end of the reactor
5- Blow out: this flame regime happens when incoming flow speed is much larger than the flame propagation speed that the flame cannot even enters the reactor.

These five flame regimes are mapped in Fig. 4 for mixture equivalence ratio from 0.5 to 2 and four lengths of reactor (5, 10, 15, 20cm).

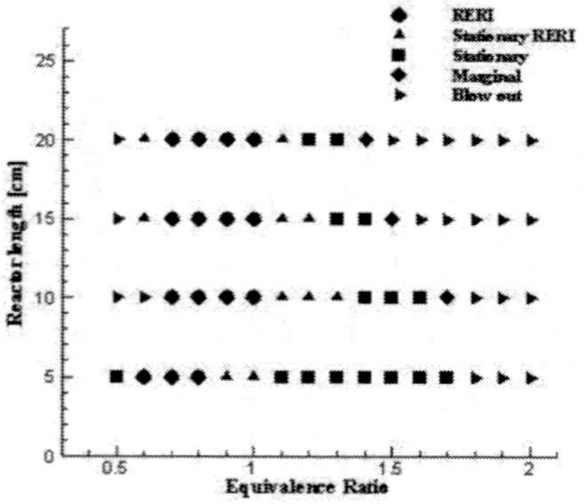

Figure 4: flame map for a reactor with the diameter of 1mm and total flow rate of 0.15 SLM

As shown in Fig. 4 the flame stability was affected by increasing both equivalence ratio and reactor length. Stable flames was formed at fuel reached mixtures (φ>1) that it may be as a result of thermal support of post flame which is formed at the open end of the reactor or intensified thermal coupling between the flame and the reactor wall. This flame map also stated that increasing the reactor length limits the flame stability that the reason may be enough time for heat loss and other instabilities in longer reactors.

In addition, observations stated that flame propagation in reactor can be divided to three main phases: at the first phase flame penetrates to the reactor that it takes time because of the hydraulic resistance of the reacting flow. Equivalence ratio and incoming flow speed have a high influence on this phase, too. The second phase is flame acceleration in the reactor that is affected by incoming flow peed and the thermal coupling between the flame and the reactor wall. Finally, the third phase is flame deceleration in the reactor that it can be as a result of equivalence ratio, thermal coupling between the flame and the reactor wall and incoming flow speed.

The effect of equivalence ratio for a constant reactor length (5cm) on these phases is shown in Fig. 5 and 6.

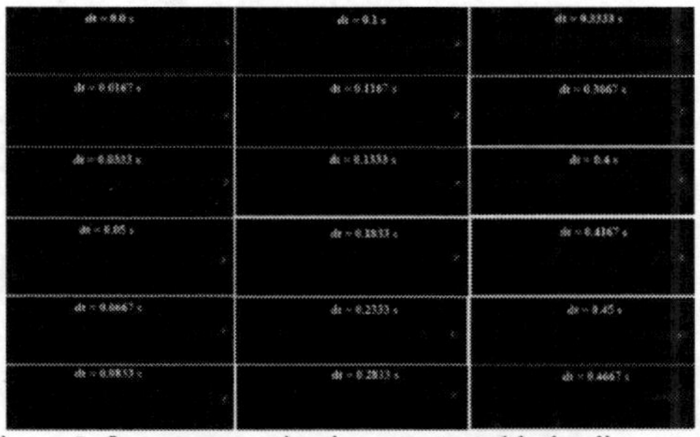

Figure 5: fame propagation in a reactor with the diameter of 1mm, total flow rate of 0.15 SLM and φ=0.8

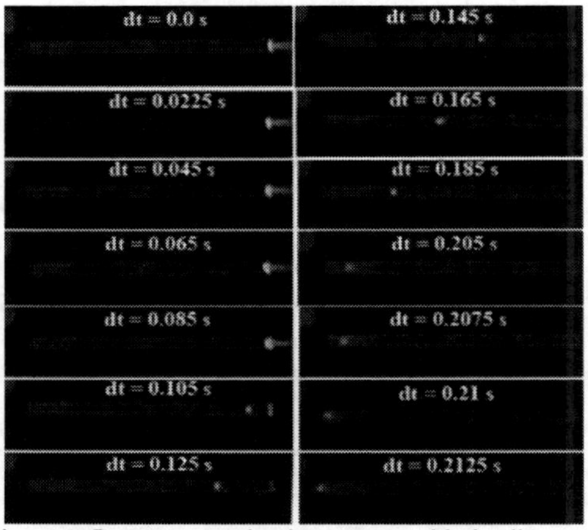

Figure 6: flame propagation in a reactor with the diameter of 1mm, total flow rate of 0.15 SLM and φ=1

As shown in Fig. 5 and Fig. 6, increasing the equivalence ratio has decreased the flame penetration time in the reactor and also made the flame shape symmetric that shows the flame tendency for being stable in the reactor. Thermal coupling between the flame and the reactor wall is also affected by increasing equivalence ratio that we can observe more chemical light reflection in Fig. 6.

Investigating the effect of reactor length on flame characteristics is shown in Fig. 7

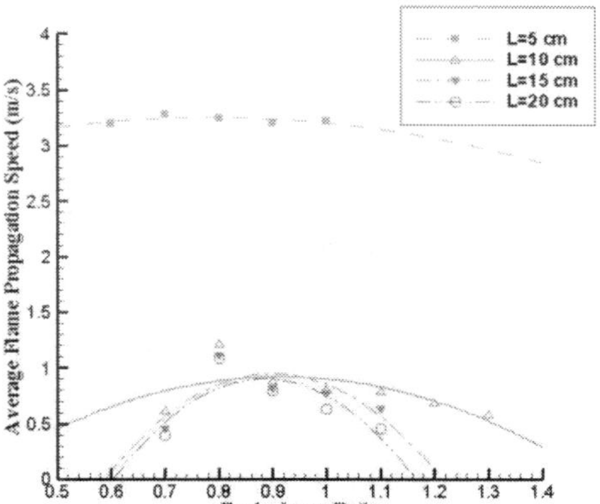

Figure 7: flame speed graphs for a reactor with the diameter of 1mm and total flow rate of 0.15SLM

As shown in Fig. 7 the flame speed was decreased by increasing the reactor length as a result of both hydrodynamic and thermal-diffusive instabilities. The flame has enough time for heat transfer with the reactor wall in longer reactors. So, thermal-diffusive instabilities developed in this situation and simultaneously, the differences between flame propagation speed in the reactor and incoming flow speed led to hydrodynamics instability development. Therefore the average flame speed is larger in shorter reactor, but without a regular and predictable behavior. For example, we can observe a huge average flame speed difference between a 5cm reactor and a 10 cm reactor but a slight average flame speed difference between a 15cm reactor and a 20 cm reactor.

As final part of results we can also observe stable flame locations in Fig. 8 to Fig. 11

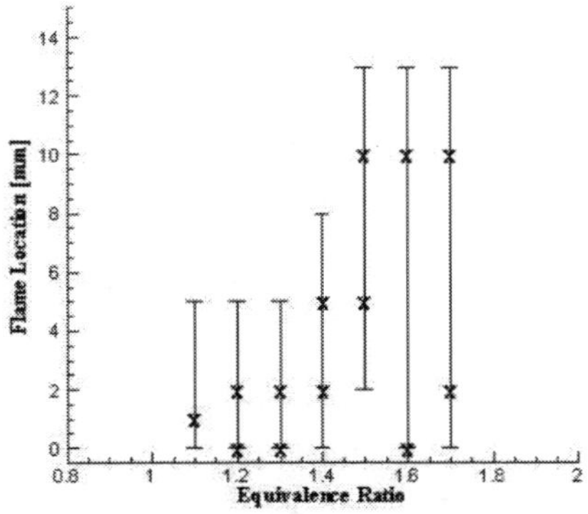

Figure8: stable flame locations in a reactor with the diameter of 1mm, length of 5cm and total flow rate of 0.15 SLM

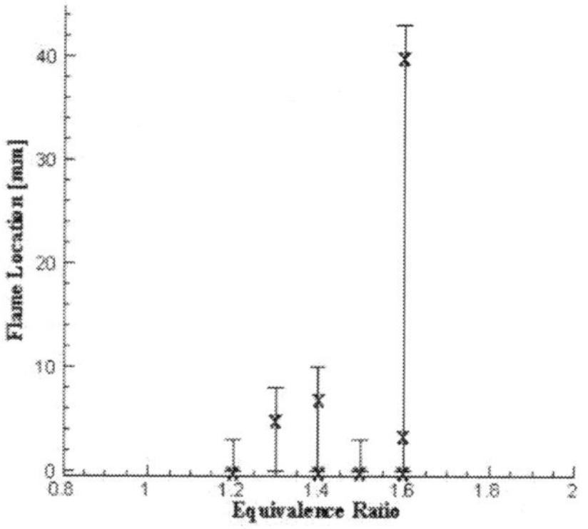

Figure9: stable flame locations in a reactor with the diameter of 1mm, length of 10cm and total flow rate of 0.15 SLM

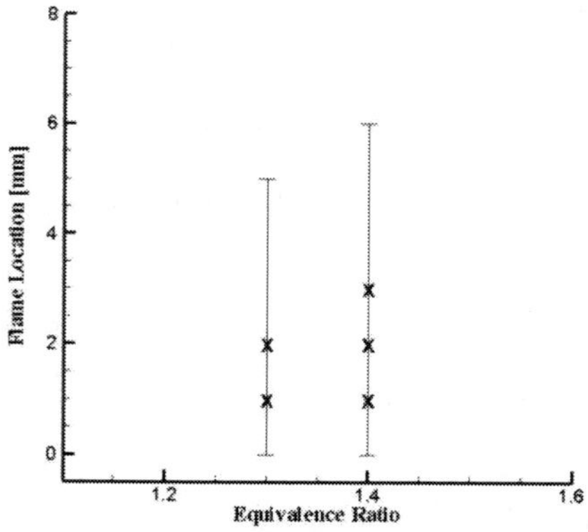

Figure10: stable flame locations in a reactor with the diameter of 1mm, length of 15cm and total flow rate of 0.15 SLM

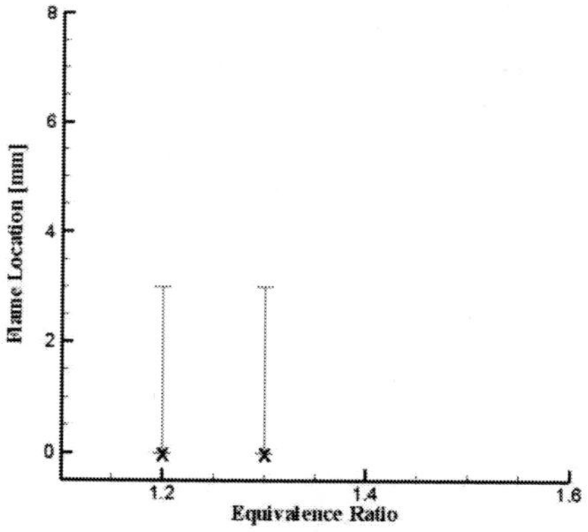

Figure11: stable flame locations in a reactor with the diameter of 1mm, length of 20cm and total flow rate of 0.15 SLM

From these flame location figures, we can investigate the domain of stable flame locations. For example as shown in Fig. 10 domain of flame location in a reactor with the diameter of 1mm, length of 15cm, φ=1.4 and total flow rate of 0.15 SLM is from 0 to 6mm of reactor length that flame was exactly observed at 1,2&3mm from where the incoming flow enters the reactor. According to stable flame location graph and type of application, we can select the best choice for designing a micro power system.

4 Conclusions

Observations of this experimental study generally stated that there are five flame regimes in a quartz reactor with a diameter of 1mm, length of 5,10,15,20cm, equivalence ratios from 0.5 to 2 and total flow rate of 0.15 SLM. These flame regimes are RERI, stationary RERI, stationary, marginal and blowout.

Also the flame characteristic is affected by equivalence ratio and reactor length. Increasing equivalence ratio makes the flame shape symmetric and increases chemical light reflection of the flame and the flame tendency for being stable. By increasing the reactor length which results in thermal-diffusive and hydrodynamic instabilities, the average flame speed decreases without a regular behavior.

5 Acknowledgments

Sincere thanks to Mr.mohammadzadeh for his help on data analysis.

References

[1] Chia, L. C., Feng, B., "Review: The development of a micro power (micro-thermo photovoltaic) device", *Journal of Power Sources*, 165, p. p. 455-480, 2007.

[2] Walther, D. C., Ahn, J., "Advances and challenges in the development of power generation systems at small scales", *Progress in Energy and Combustion Science*, 37, p. p. 583-610, 2011.

[3] Isomura, K., Tanaka, S., Togo, S. I., Kanebako, H., "Development of Microachine Gas Turbine at Tohoku University", *NATO/OTAN Organization*, RTO-EN-AVT-131, 10-1-10-34, 2005.

[4] Wang, Y., Zhou, Z., Yang, W., Zhou, J., Liu, J., Wang, Z., Cen, K.,"Instability of flame in micro-combustor under different external thermal Environment", *Experimental Thermal and Fluid Science*, 35, p. p. 1451–1457, 2011.

[5] Taywade, U. W., Deshpande, A. A., Kumar, S.," Thermal performance of a micro combustor with heat recirculation", *Fuel Processing Technology*, 109, p. p. 179–188, May 2013.

[6] Chou, S. K., Yang, W. M., Chua, K. J., Li, J., Zhang, K. L. ,"Development of micro power generators-A review", *Applied Energy*, 88, p. p. 1-16, 2011.

[7] Li, J., Zhong, B.," Experimental investigation on heat loss and combustion in methane/Oxygen micro-tube combustor", *Applied Thermal Engineering*, 28(7), p. p. 707-716, 2008.

[8] Evans, C. J., Kyritsis, D. C., "Operational regimes of rich methane and propane/oxygen flames in mesoscale non-adiabatic ducts", *Proceedings of the Combustion Institute*, 32, p. p. 3107–3114, 2009.

[9] Minaev, S., Maruta, K., Fursenko, R. ,"Nonlinear dynamics of flame in a narrow channel with a temperature gradient", *Combustion Theory and Modelling*, 11, p. p. 187–203, 2007.

11th Asia-Pacific Conference on Combustion,
The University of Sydney, NSW Australia
10th -14th December 2017.

Flame Stretch Statistics in Premixed Jet Flames at High Reynolds Number

Stefano Luca[1], Antonio Attili[2], Fabrizio Bisetti[3]

[1]King Abdullah University of Science and Technology (KAUST)

Thuwal, 23955-6900, Saudi Arabia

[2]RWTH Aachen University

Aachen, 52056, Germany

[3]University of Texas at Austin

Austin, Texas, 78712, USA

Abstract

A set of Direct Numerical Simulations (DNS) of turbulent jet flames is performed to investigate the effect of Reynolds number on flame characteristics. The simulations feature finite rate chemistry with 16 species and up to 22 billion grid points. The jet consists of a methane/air mixture with equivalence ratio $\varphi = 0.7$ and unburnt temperature of 800 K. The temperature and species concentrations in the coflow correspond to the equilibrium state of the burnt mixture. All the simulations are performed at 4 atm. The flame length, normalized by the jet width, decreases significantly as the Reynolds number increases. This is consistent with an increase of the turbulent flame speed due to the increased integral scale of turbulence. This behavior is typical of flames in the thin-reaction zone regime, which are affected by turbulent transport in the preheat layer. Statistics of stretch at the flame surface are investigated and the dependence of these quantities on the Reynolds number is assessed.

1 Introduction

The goal of the present investigation is to understand the behavior and the dynamics of turbulent flames at high Reynolds number. This is motivated by the evidence that technical devices, ranging from internal combustion engines for automotive applications to stationary gas turbines for electricity production, operate at high Reynolds numbers. In order to formulate reliable models and to assess their accuracy, detailed DNS data at high Reynolds number are extremely important. On the other hand, it should be recognized that well controlled laboratory experiments and DNS will unlikely be able to achieve the Reynolds number of typical devices of industrial relevance in the near future. This observation motivates the investigation of the trends of key flame properties with respect to variations of the Reynolds number of the flow. One of these key quantities, the turbulent burning velocity, is affected by stretch induced by both the tangential strain rate and the propagation terms. Stretch affects the burning velocity directly through the production/destruction of Σ and the effect on I_0 in the equation discussed by Bray and Cant [1]:

$$\frac{S_T}{S_L} = I_0 \int_{-\infty}^{\infty} \Sigma \, d\eta \qquad (1)$$

where Σ is the flame surface density, I_0 is the stretch factor and the coordinate η is normal to the flame brush. The behavior of the flame stretch is a topic that has received considerable interest in the premixed combustion community. Few DNS studies, mostly of statistically 1D planar premixed flames in the thin reaction zone regime [2,3], have shown that the tangential strain rate tends to be a production term, while the propagative term tends to act as a sink. Wang et al. [4] observed a similar behavior in a turbulent jet flame. The objective of this study is to examine the key statistics in the DNS, in particular the effect of Reynolds number on the flame stretch.

2 DNS Database

	R1-K1	R2-K1	R3-K1	R4-K1
Jet Reynolds number, Re_{jet}	2800	5600	11200	22400
Jet bulk velocity, U_b	100 m/s	100 m/s	100 m/s	100 m/s
Slot width, H	0.6 mm	1.2 mm	2.4 mm	4.8 mm
L_x	24 H	24 H	24 H	24 H
L_y	16 H	16 H	16 H	16 H
L_z	8.52 H	4.26 H	4.26 H	4.26 H
N_x	720	1440	2880	5760
N_y	480	960	1920	3840
N_z	256	256	512	1024
Total number of points	88 M	350 M	2.8 B	22 B
Karlovitz number, Ka	40-50	40-50	40-50	40-50

Table 1: Parameters of the simulations. L_x, L_y, and L_z indicate the domain size and N_x, N_y, and N_z are the number of discretization points in the three directions. The Karlovitz number is computed in the fully developed region of the flame, in points where the mean temperature is 1800 K, the temperature of maximum heat release rate in the 1D planar unstretched flame.

The flame configuration studied in the present work is a slot jet surrounded by a coflow of burnt gases [5]. This arrangement is similar to piloted flames used in experiments. The simulations feature finite rate chemistry with 16 species and 73 reactions [6]. The gas phase hydrodynamics are modeled with the reactive, unsteady Navier-Stokes equations in the low Mach number limit [7]. The species obey the ideal gas equation of state and all transport properties are computed with a mixture-average approach. The jet consists of a methane/air mixture with equivalence ratio $\varphi = 0.7$ and temperature $T = 800$ K. All simulations were performed at 4 atm. Relevant parameters are summarized in Table 1.

The bulk jet velocity is $U_b = 100$ m/s, while the coflow has a uniform velocity of $U_c = 15$ m/s. The jet Reynolds number based on the slot width and the jet bulk velocity $Re_{jet} = U_bH/\nu$ varies between 2800 and 22400. The flow is periodic in the spanwise (z) direction, open boundary conditions are prescribed at the outlet in the streamwise (x) direction and no-slip conditions are imposed at the boundaries in the crosswise (y) direction. The jet and the coflow are separated at the inlet by walls with thickness $H/10$ and length $2H/3$.

The domain size expressed in terms of the slot width H is $24H$ in the x direction, $16H$ in the y direction and $4.3H$ in the z direction ($8.5H$ for R1-K1). The grid is uniform in all three directions with a resolution of 20 µm resulting in a total number of points that goes from 88 Millions to 22 Billions, and in a laminar flame thickness to cell size ratio of $\delta_L/\Delta x \approx 6$ and in a cell size to Kolmogorov scale ratio of $\Delta x/\eta_{min} < 2$. The three smaller simulations were performed at double resolution with a grid spacing of 10 µm and the same results were observed, thereby verifying grid convergence.

The variation of the Reynolds number is obtained varying the jet slot width H between 0.6 and 4.8 mm, while the jet exit velocity U_b is maintained constant at 100 m/s. Since the root mean square of velocity fluctuations is expected to be a fraction of U_b, (≈10%) independently of the Reynolds number, the simulations will be characterized by an approximately equal turbulence intensity u'.

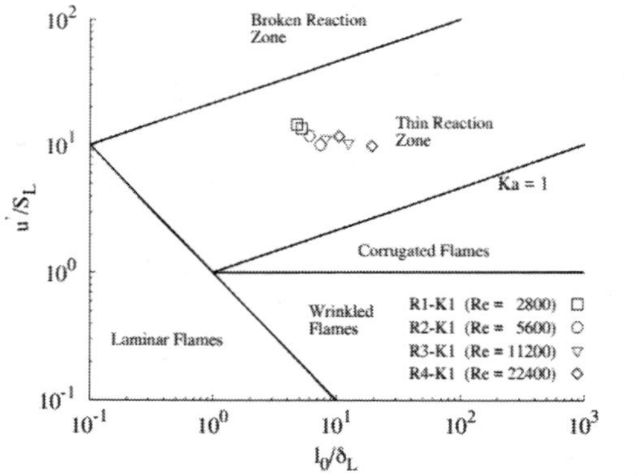

Figure 1: Borghi-Peters diagram for the four flames simulated.

All the four flames considered are in the Thin Reaction Zone (TRZ) regime as shown in the Borghi-Peters diagram in Fig. 1. The turbulent intensity u' is approximately 10 times the laminar flame speed and the integral scale varies between 5 and 20 times the thickness of a planar 1D freely-propagating unstretched laminar flame with the same fresh-gas conditions of the DNS.

Due to the rather small variation of the Kolmogorov scale among the cases all the simulations have a Karlovitz number ($Ka = (\delta_L/\eta_k)^2$) between 40 and 50.

Figure 2 shows the mass fraction of oxygen in the four flames at different Reynolds numbers. The range of scales characterizing the fields increases with the Reynolds number. In addition, the figure shows that the length of the flames, measured in terms of the slot width H, decreases as the Reynolds number increases. This is consistent with an increase of the turbulent flame speed due to the increased flame surface area density caused by the larger Reynolds number. The comparison of the 4 fields reveals that the flame-tip dynamics change for the different cases as more and more flame-flame interactions are observed as the Reynolds number increases.

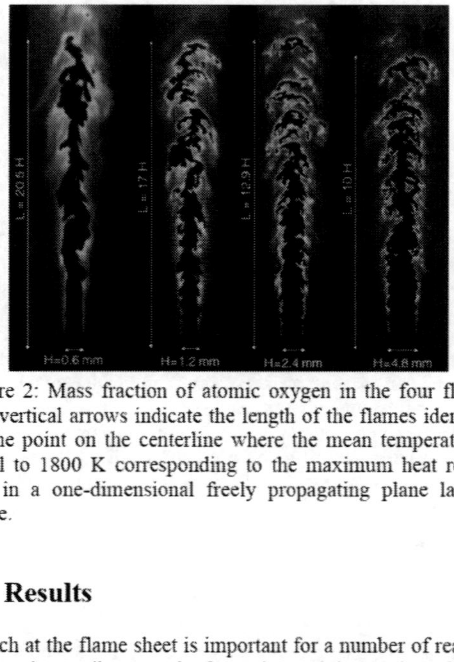

Figure 2: Mass fraction of atomic oxygen in the four flames. The vertical arrows indicate the length of the flames identified by the point on the centerline where the mean temperature is equal to 1800 K corresponding to the maximum heat release rate in a one-dimensional freely propagating plane laminar flame.

3 Results

Stretch at the flame sheet is important for a number of reasons: (i) stretch contributes to the formation and destruction of flame surface, appearing in the surface density balance equation; (ii) in the theory of laminar flames, stretch is known to affect the local displacement speed, so that it has been suggested that turbulence affects the local flame propagation through stretch. Candel and Poinsot [8] obtained an expression for the flame stretch and can be computed as:

$$K = -\boldsymbol{n}\nabla\boldsymbol{u}\boldsymbol{n} + \nabla\cdot\boldsymbol{u} - 2K_M S_D = A_T + S_d\nabla\cdot\boldsymbol{n} \qquad (2)$$

where $\boldsymbol{n}$ is the normal to the flame surface defined positive pointing toward the reactants, K_M is the mean curvature and S_D is the propagation speed of the surface.

Stretch features three contributions: first the local velocity gradient tensor, projected on the normal to the flame and then further dotted with the normal ($-\boldsymbol{n}\nabla\boldsymbol{u}\boldsymbol{n}$) called the "normal strain rate"; second the divergence of the velocity (non-zero in variable density flows) ($\nabla\cdot\boldsymbol{u}$); and third the propagation of the surface in the presence of curvature ($S_D\nabla\cdot\boldsymbol{n}$). The sum of the two terms

$$A_T = -\boldsymbol{n}\nabla\boldsymbol{u}\boldsymbol{n} + \nabla\cdot\boldsymbol{u} \qquad (3)$$

is interpreted as the "tangential strain rate", A_T, i.e. stretch in the plane tangential to the flame surface.

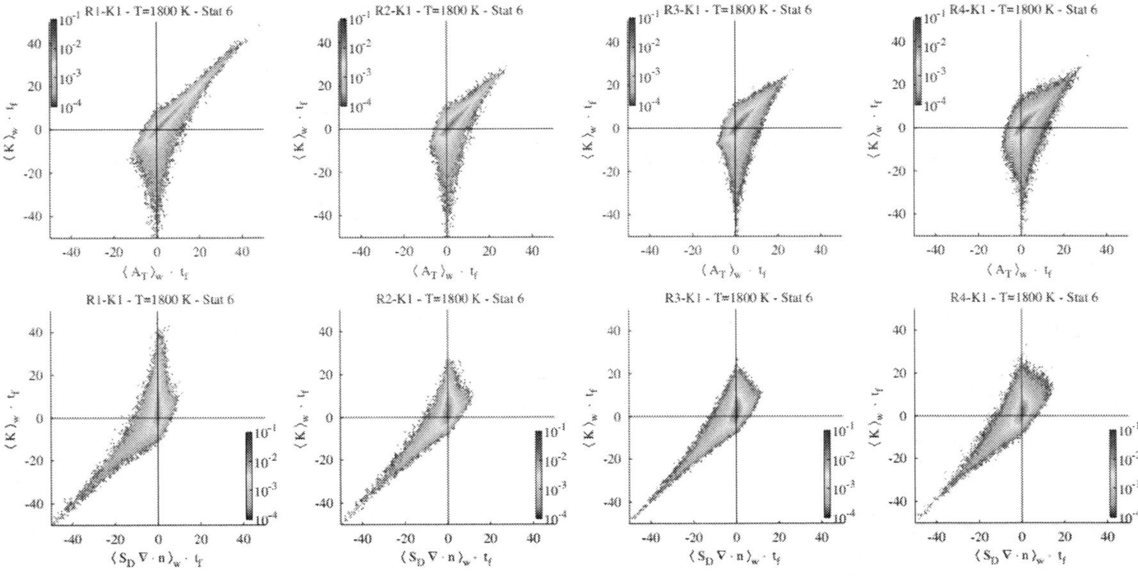

Figure 3: Conditional joint probability density functions of stretch K and tangential strain rate term A_T (top) and stretch K and propagative term $S_D \nabla \cdot n$ (bottom) for the four flames. The quantities are area weighted $(\langle \cdot \rangle_w)$ and the statistics are conditioned on axial location (x/L=0.6) and flame temperature (1800 K).

Statistics of the area weighted stretch, tangential strain rate term and propagative term are presented in the form of conditional joint probability density functions. The statistics are accumulated in time once the simulations reach the statistically steady state.

Figure 3 shows the joint PDF of the stretch with the

contribution of the tangential strain rate term and the stretch with the propagative term for the four flames considered.

Due to the Bunsen configuration the turbulence statistics change moving downstream from the nozzle exit. So the statistics are presented only at one axial location defined as x/L=0.6 and called station 6. This axial location is far enough from the inlet and the tip of the flames and is representative of a wider region of the flame where the statistics are seen to be similar. The statistics are also conditioned on the temperature representative of the flame location 1800 K, that is the temperature at which the heat release rate peaks in the one-dimensional freely-propagating unstretched laminar flame.

Figure 3 allow us to make observation on the correlation between stretch and the two components but also on the effect of Reynolds number on this quantity.

Few main features can be observed in the joint PDF that are seen in each flame: large *positive* values of the stretch occur when the tangential strain rate term is large and positive, while the propagative term is very close to zero; on the other hand large *negative* values of stretch occur when the propagative term assumes large negative values and the tangential strain rate is close to zero. Furthermore, in the neighborhood of the most probable value of stretch, which is positive, the tangential strain rate term is highly (positively) correlated with the stretch, indicating that for modest values of positive stretch, tangential strain rate term is responsible.

So the picture that emerges is that the most probable region of the total stretch is close to zero, that low and high positive values are due to the tangential strain rate term, while large negative values are associated with the propagative term; and overall large values, positive and negative, are infrequent, since the value of the joint PDF is three orders of magnitude smaller.

Considering now the joint PDF for different flames, i.e. the effect of Reynolds number, we can see that the correlation between stretch and the single terms decreases. High, positive and negative, values of stretch becomes less probable moving from R1-K1 to R4-K1 (increasing Reynolds number), and the region of the most probable values becomes wider and closer to zero. Overall the behavior of the flames is similar: this is expected due to how these flames are designed, i.e. increasing Reynolds number but keeping Karlovitz number almost constant.

Stretch affects the surface area and appears as a source term in the flame surface density equation. The sign of the mean stretch is an indication of the global creation or destruction of surface due to the velocity gradient. At this location of the flame, stretch is, on average, positive, meaning that a global production of surface is occurring.

4 Conclusions

A set of Direct Numerical Simulations of three-dimensional spatially developing turbulent slot Bunsen burner methane/air flames is performed. The flames differ for the slot width, i.e. the jet Reynolds number.

The flames are found to belong to the thin reaction zone of the regime diagram. The length of the flames, measured in terms of the slot width H, is found to decrease as the Reynolds number increases. This is consistent with an increase of the turbulent flame speed due to the increased flame surface area density caused by the larger Reynolds number.

Stretch statistics are presented, and the picture that emerges is that the most probable region of the total stretch is close to zero. Moreover, contribution of the single terms is taken in consideration: positive values of stretch are found to be due to the tangential strain rate term, while large negative values are associated with the propagative term. The joint PDF suggest that the effect of Reynolds number is to decrease the correlation between stretch and the single terms, increasing Reynolds number.

5 Acknowledgment

We acknowledge valuable support from KAUST Supercomputing Laboratory (KSL) in the form of assistance with code development and computational time on the Cray XC40 Shaheen.

References

[1] K.N.C. Bray, R.S. Cant, "Some Applications of Kolmogorov Turbulence Research in the Field of Combustion", Proceedings: Mathematical and Physical Science, (1991): 217-240

[2] N. Chakraborty, S. Cant, "Effects of Strain Rate and Curvature on Surface Density Function Transport in Turbulent Premixed Flames in the Thin Reaction Zones Regime", Physics Fluids 19.6, (2005)

[3] N. Chakraborty, N. Swaminathan, "Influence of the Damkohler Number on Turbulence Scalar Interaction in Premixed Flames. I. Physical insight", Physics of Fluids 19.4, (2007)

[4] H. Wang, E.R. Hawkes, J.H. Chen, B. Zhou, Z. Li, M. Alden, "Direct Numerical Simulation of a High Karlovitz Number Laboratory Premixed Jet Flame – an Analysis of Flame Stretch and Flame Thickening", Journal of Fluid Mechanics 815, (2017): 511-536

[5] S. Luca, A. Attili, F. Bisetti, "Direct Numerical Simulation of Turbulent Lean Methane-Air Bunsen Flames with Mixture Inhomogeneities", 54th AIAA Aerospace Sciences Meeting, AIAA SciTech Forum, (AIAA 2016-0189), 2016.

[6] S. Luca, A. Al-Khateeb, A. Attili, F. Bisetti, "Comprehensive Validation of a Skeletal Mechanism for Turbulent Premixed Methane-Air Flame Simulations", Journal of Propulsion and Power, (2017), accessed September 6, 2017. doi: http://arc.aiaa.org/doi/abs/10.2514/1.B36528

[7] A. Attili, F. Bisetti, M.E. Mueller, H. Pitsch, "Effects of non-unity Lewis number of gas-phase species in turbulent non-premixed sooting flames" Combust. Flame 166, (2016).

[8] S.M. Candel, T.J. Poinsot, "Flame Stretch and the Balance Equation for the Flame Area", Combustion Science and Technology 70(1-3), (1990): 1-15

11th Asia-Pacific Conference on Combustion,
The University of Sydney, NSW Australia
10th -14th December 2017.

Laminar Burning Characteristics of 2-Methyltetrohydrofuran Compared with 2-Methylfuran and Isooctane

Xiangshan Fan[1], Kangkang Yang[1], Ximing Jiao[1], Jingshan Wang[1], Zhiyi Guo[1], Xibin Wang[1, *], Xiaomin Wu[1]

[1] Xi'an Jiaotong University

Xianning West Road, Xi'an, Shannxi, 710049, China

Abstract

Based on laminar flame propagation theory and high-speed schlieren photography, the laminar burning velocities and Markstein lengths of 2-methyltetrahydrofuran (MTHF), 2-methylfuran (MF) and isooctane were compared at the temperature of 373K, the pressure of 1bar and equivalence ratio of 0.7-1.5. MF has the highest laminar burning velocity and adiabatic flame temperature, followed by MTHF and then isooctane. Adiabatic temperature and chemical kinetics contribute to the discrepancies of the laminar burning velocities among the three fuels. The Markstein lengths of MF are generally lower than that of MTHF and isooctane.

1 Introduction

Bio-derived oxygenates attracted rising attention as alternative transportation fuels for their advantages in releasing the energy crisis and emissions. 2-methyltetrahydrofuran (MTHF) and 2-methylfuran (MF) were considered as promising second-generation oxygenated biofuels for their attractive physical properties and wild range of raw materials such as straw, sugarcane bagasse and other agricultural waste.

Many researches were carried out for MTHF on the combustion characteristics and the performances on engines. Janssen et al. [1] realized nearly soot-free diesel combustion over a wide load range by combining homogeneous low-temperature combustion process with mixtures of MTHF and dibutylether. Leppard et al. [2] Thewes et al. [3] and Sudholt et al. [4] investigated the anti-knock and ignition character of MTHF by using motored-engine technique, a homogeneous engine combustion system and an ignition quality tester respectively. Moshamme et al. [5] developed a detailed high-temperature oxidation model for MTHF and verified this model by measuring the intermediate species involved in the MTHF flame using mass spectrometry technology. Brassat et al.[6], Wang et al. [7] and Tripathi et al. [8] explored the high-temperature ignition delay characteristics of MTHF using rapid compression machine and shock tube. Tripathi et al. [8] developed a detailed chemical kinetic model describing both low-and high-temperature chemistry of MTHF oxidation. A detailed chemical mechanism for combustion and pyrolysis of MTHF was proposed by Bruycker et al. [9] and has been verified by using a plug flow reactor and a premixed burner.

MF is also a promising gasoline alternative produced from furfural and warrants an extensive discussion. Despite of similar molecular structure to MTHF, MF is of higher octane number and could be used as octane booster. Many studies in SI engine [6, 10-14] reported reduced HC emission and raised NO emission by blending gasoline with MF. Tran et al. [15], Ma et al. [16]and Wei et al. [17] studied the laminar flame structure and characteristics of MF. Somers et al. [18] researched the ignition characteristics of MF in high-pressure shock tube. Somers et al. [19] and Tran et al. [20] developed their detailed MF chemical mechanism independently. Jouzdani et al. [21] and Sudholt et al. [4] studied the ignition characteristics of MF and MTHF by using shock tube and ignition quality tester, respectively. Despite the additional two C=C double bonds on the ring of MF molecular, the molecular structure of MF is similar to that of MTHF.

Although the above studies have significantly improved the understanding into the combustion chemistry of MTHF and MF, there are rarely comparative studies between the two fuels on laminar flame characteristics. In this research, the spherically propagating flames of MTHF and isooctane in a constant volume chamber were captured using high-speed photography technique. The experiment data of MTHF were compared with that of MF to reveal the effect of molecule structure on laminar flame characteristics.

Table 1 Properties of the MTHF, MF, isooctane and gasoline

	2-MTHF	2-Methylfuran	Iso-octane	RON95
Chemical formula				-
Density, kg/m^3	849	910	691.9	737
Viscosity, mPa*s	0.48	4	0.48	~0.42
Surface tension, mN/m	24.56	25.51	18.35	-
Heat of vaporization@25℃, kJ/kg	375.3	389	309	~397
Stoichiometric air requirement	11.21	10.08	15.13	14.14
Lower heating value, MJ/kg	34.12	30.37	44.3	42.13
Research octane number	86	100.7	100	96.3

Xibin Wang. Fax: 029-82668789
E-mail address: xbwang@xjtu.edu.cn

2 Experiment Setup and Procedures

The experimental apparatus consists of a cylinder chamber of $0.0055m^3$ in volume, a gas supply system, a schlieren optical system, an ignition system and a control system. Heating-tapes surrounding the chamber were utilized for elevated initial temperature. Pressure transducer with the deviation of 1% and thermocouples with the accuracy of ±2K were installed inside the chamber to monitor the initial temperature and pressure. Two circular quartz windows of 80 mm in diameter on each side of the chamber supplied the optical path for the schlieren system. The fuels prepared for this experiment were MTHF and isooctane with the purity of 99%. N2 and O2 of ultra-high purity (99.999%) were induced in the ratio of 3.76 to mimic the composition of dry air. Each component of the mixture to be measured was induced into the chamber individually with the amount controlled by Dalton's law. Before every ignition, the procedure suspended for 5 minutes to ensure the complete blending of the mixture. A high speed camera (Phantom V611) was utilized to capture the ignition and flame propagating process at a sample rate of 10 kHz and a resolution of 800 × 800 pixels. After each ignition, fresh air was induced to scavenge the residual gas for at least 3 times. The experiment was conducted at the initial temperature of 373K, initial pressure of 1bar and equivalence ratio of 0.7-1.5. For each operating condition, the experiment repeated at least 3 times to reduce the random uncertainty.

2.1 Data processing

The flame propagations were captured by the high speed camera, so the flame radius (r_u) at every time (t) could be acquired by image processing. Therefore, the stretched flame propagation speed (Sb) could be determined by differential:

$$S_b = dr_u / dt$$

The flame front is stretched as the spherical flame propagating. Considering A_f as the surface of the flame front, the stretch rate (α) could be connected with Sb:

$$\alpha = dA_f / (A_f \cdot dt) = 2S_b / r_u$$

To eliminate the effect of stretch rate, the unstretched flame propagation speed (S_b^0) could be determined from the unlinear relationship of S_b with r_u.

$$\ln(S_b) = \ln(S_b^0) - S_b^0 L_b \cdot 2/(r_u S_b)$$

Where the L_b is the Markstein length representing the stability of laminar burning velocity.

From mass conservation, following relationship could be carried out:

$$S_u = S_b^0 \cdot \rho_u / \rho_b$$

Where Su is the laminar burning velocity and ρb and ρu are the densities of the burned and unburned gas, respectively.

2.2 Simulation method

The kinetic models used for MTHF, MF and isooctane combustion were from Burycker et al. [9], Somers et al. [19] and Chaos et al.[22], respectively. Flame simulations were performed using PREMIX code [23]. Grid independent solutions were found with GRAD and CURV set as 0.05. Soret effect was taken into account and mixture-averaged transport was adopted.

3 Results and Discussion

3.1 Experiment validation

High-quality measurement of laminar flame speed is affected by various factors including mixture preparation, ignition, buoyancy, instability, confinement, radiation, nonlinear stretch behavior, and extrapolation [24]. To minimize the effects of ignition, chamber confinement and the flame radius range for data collection, the flame radius range of 8–24 mm was adopted for data collection under all the conditions in this experiment. Unlinear method was adopted in determination of unstretched flame propagation speed (S_b^0) to reduce the uncertainty from nonlinear stretch behavior. Some factors like radiation, accuracy of transducers, extrapolation method and random error which could induce certain amount of uncertainties have been discussed in previous works [25, 26], and the total uncertainty of the laminar burning velocity measured by this system was evaluated to be 1–4.5 cm/s.

In order to verify the accuracy of this study, the laminar burning velocities of isooctane in 373K,1bar were compared with previous researches under similar operating condition. The data in this work agreed well with previous works as shown in figure 1.

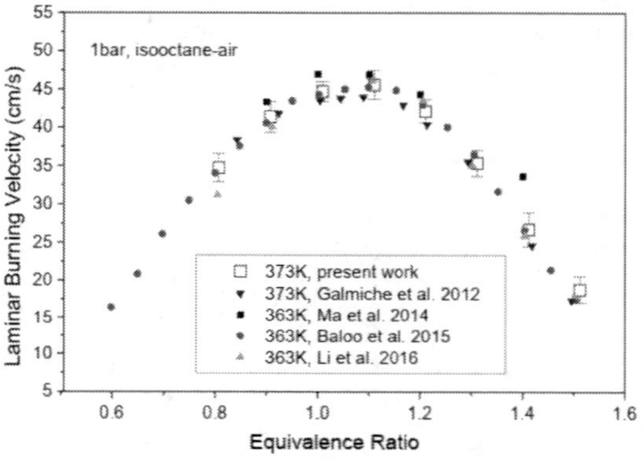

Figure 1. Laminar burning velocity of isooctane in this and previous works [26-29] at 1bar.

3.2 Laminar burning velocity

Laminar burning velocity is one of the most important parameters representing laminar combustion characteristic of fuels and influenced by initial temperature, pressure, equivalence ratio and the type of fuels. As shown in figure 2, the laminar burning velocities of the three kind of fuels varies with equivalence ratio and reached their peak at equivalence ratio near 1.1. Under all the operating conditions, the laminar burning velocities of MF are the fastest and that of isooctane is the slowest. The maximum laminar burning velocity of MF, MTHF and isooctane is 64.45, 52.32 and 45.56 cm/s, respectively. It can be inferred that the burning velocity of gasoline could be enhanced by blending with MF or MTHF.

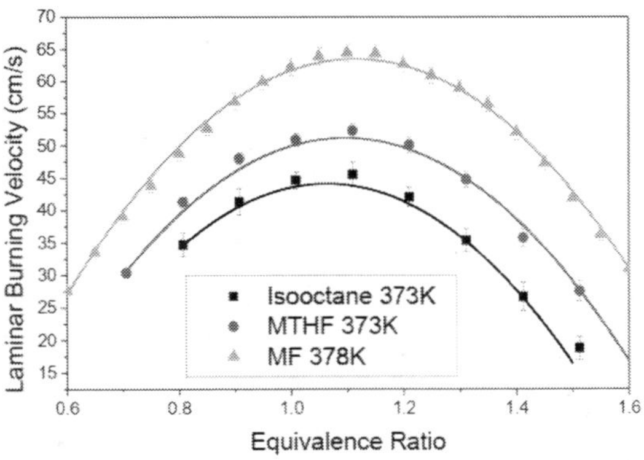

Figure 2. Laminar burning velocities of MTHF, MF [18] and isooctane at ≈373K, 1bar.

3.3 Influencing factors on laminar burning velocity

Adiabatic flame temperature is considered having a strong function with laminar burning velocity. The adiabatic flame temperatures of the three fuels were depicted in figure 3. The trend of adiabatic flame temperatures versus equivalence ratio is consistent with that of the laminar burning velocities. The adiabatic flame temperatures of MTHF and MF reach their peaks at the equivalence ratio of 1.1. The maximum adiabatic flame temperature of isooctane is observed when closer to stoichiometric ratio. The adiabatic flame temperature of MF is higher than that of MTHF, which should be part of the factor leading to higher laminar burning velocity of MF than that of MTHF.

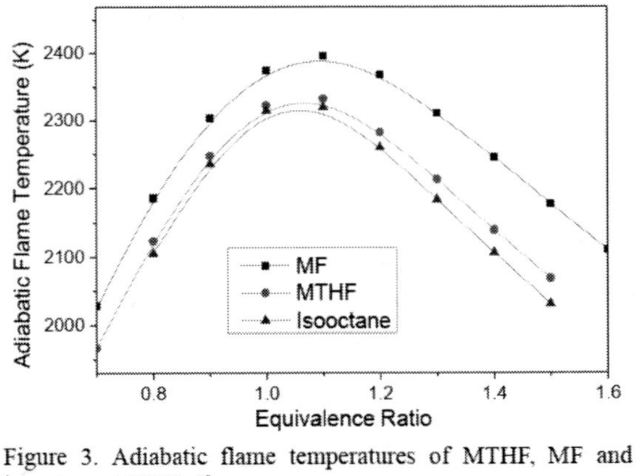

Figure 3. Adiabatic flame temperatures of MTHF, MF and isooctane at 373K, 1bar.

Reactive radicals play an important role in initial reactions and chain branching reactions, thus could promote the overall reaction rate and combustion intensity. The concentrations of reactive radicals can reflect the effect of chemical kinetic on laminar burning velocity. The concentration profiles of H and OH in MF, MTHF and isooctane flames were obtained based on the MF model, MTHF model and the isooctane model from Somers et al. [19], Bruycker et al.[9] and Chaos et al.[22], respectively. Figure 4 (a) shows the sum of the maximum mole fractions of H and OH at different equivalence ratios in MF, MTHF and isooctane flames. The fractions of reactive radicals MTHF produced is slightly lower than that of MF, but higher than that of isooctane, which should be the reason of higher laminar burning velocity of MTHF than that of isooctane.

Figure 4 (b) shows the maximum mole fraction of CH3 at different equivalence ratios in MF, MTHF and isooctane flames. The mole fraction of CH3 of the three fuels are in reverse order of laminar burning velocities. The concentrations of CH3 of MF are the lowest among the three fuels while that of isooctane are the highest at most conditions. The sensitivity of the chain termination reaction, CH3+ H (+ M) = CH4 (+ M), is negative to laminar burning velocity for the consumption of reactive H and the high reaction barrier for decomposition of stable CH4. As a consequence, the high CH3 concentration of isooctane contributes to the slowdown of reaction intensity and the lowest laminar burning velocity among the three fuels at most conditions.

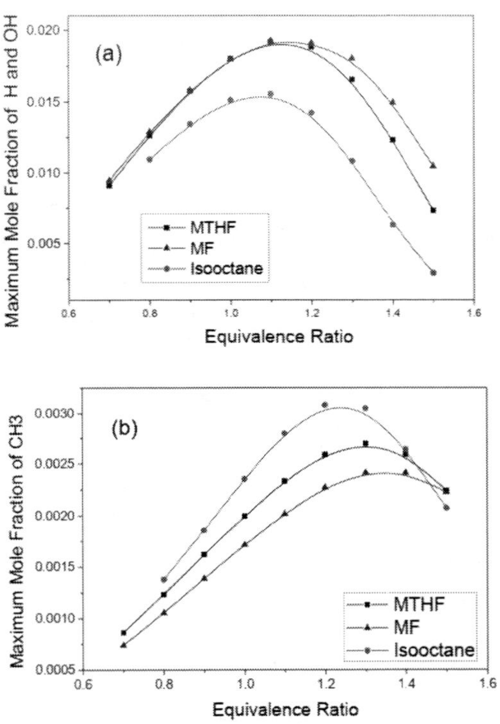

Figure 4. Maximum concentrations of H+OH (a) and CH3 (b) of MTHF, MF and isooctane at 373K, 1bar.

3.4 Markstein length

Markstein length is used to represent the thermal diffusion instability of the flame and the sensitivity of laminar flame speed on stretch rate. In this study, the unstretched flame speeds were obtained by linear extrapolation of the stretched flame speeds to the stretching rate of 0 and the negative value of the slope was taken as the Markstein length. For heavy hydrocarbons, the Markstein lengths reduce with the increase of equivalence ratio while that would be opposite for light hydrocarbons. As shown in figure 5, the Markstein lengths of the three fuels in this work decrease with the increase of equivalence ratio. For lean mixture, the Markstein lengths of MTHF and isooctane are similar and that of MF is the lowest. At equivalent ratio higher than 1.2, the Markstein lengths of the three fuels show minor differences.

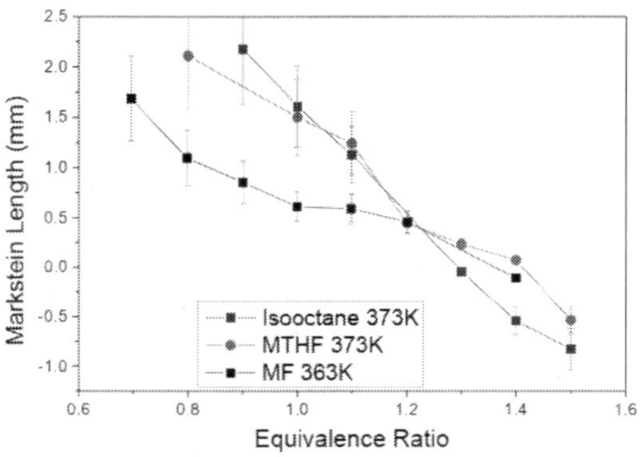

Figure 5. Markstein lengths of MTHF, MF [26] and isooctane with 25% error bar at ≈373K, 1bar.

4 Conclusions

In this research, the propagating flames of MTHF and isooctane in a constant volume chamber were recorded by high speed photography technology and the laminar flame characteristics of MTHF, MF and isooctane were compared based on laminar burning velocity, Markstein length and other parameters.

In this experimental condition, the maximum laminar burning velocities of the MTHF, MF and isooctane occur at the equivalence ratio near 1.1. The laminar burning velocities of the three fuels are in the order of MF>MTHF>isooctane. The adiabatic flame temperature and chemical kinetics are the factors leading to this phenomenon.

The Markstein lengths of the three fuels decrease with the increase of equivalence ratio. For lean mixtures, the Markstein lengths of MF are lower than that of MTHF and isooctane. While at equivalence ratio higher than 1.2, the Markstein lengths of the three fuels are similar.

5 Acknowledgment

The authors acknowledge the financial support from National Natural Science Foundation of China (NSFC) (No. 61235003).

References

[1]. Janssen AJ, Kremer FW, Baron JH, Muether M, Pischinger S, Klankermayer J, Energ Fuel 10 (25) (2011), 4734-4744.

[2]. Leppard WR, The Autoignition Chemistries of Octane-Enhancing Ethers and Cyclic Ethers: A Motored Engine Study, SAE Technical Paper, 1991.

[3]. Thewes M, Muther M, Brassat A, Pischinger S, Sehr A, SAE Int. J. Fuels Lubr. 1 (5) (2011), 274-288.

[4]. Sudholt A, Lee C, Klankermayer J, Fernandes RX, Pitsch H, Combust Flame (171) (2016), 133-136.

[5]. Moshammer K, Vranckx S, Chakravarty HK, Parab P, Fernandes RX, Kohse-Höinghaus K, Combust Flame 12 (160) (2013), 2729-2743.

[6]. Brassat A, Thewes M, Muther M, Pischinger S, Lee C, et al,'Presented at' SAE 2012 World Congress and Exhibition, April 24, 2012 - April 26, 2012,Detroit, MI, United states2012.

[7]. Wang J, Wang X, Fan X, Yang K, Zhang Y, FUEL (186) (2016), 758-769.

[8]. Tripathi R, Lee C, Fernandes RX, Olivier H, Curran HJ, et al, P Combust Inst (2016). 1–9

[9]. Bruycker RD, Tran LS, Carstensen HH, Glaude PA, Monge F, et al, Combust Flame (176) (2017), 409-428.

[10]. Hoppe F, Burke U, Thewes M, Heufer A, Kremer F, Pischinger S, FUEL (167) (2016), 106-117.

[11]. Ma X, Xu H, Jiang C, Shuai S, Appl Energ (122) (2014), 247-260.

[12]. Wei H, Feng D, Shu G, Pan M, Guo Y, et al, Appl Energ (132) (2014), 317-324.

[13]. Wang C, Xu H, Daniel R, Ghafourian A, Herreros JM, et al, FUEL (103) (2013), 200-211.

[14]. Thewes M, Muether M, Pischinger S, Budde M, Brunn A, et al, Energ Fuel 12 (25) (2011), 5549-5561.

[15]. Tran LS, Sirjean B, Glaude P, Fournet R, Battin-Leclerc F, ENERGY 1 (43) (2012), 4-18.

[16]. Ma X, Jiang C, Xu H, Shuai S, Ding H, Energ Fuel 10 (27) (2013), 6212-6221.

[17]. Wei L, Tang C, Man X, Huang Z, Energ Fuel 12 (27) (2013), 7809-7816.

[18]. Somers KP, Simmie JM, Gillespie F, Burke U, Connolly J, et al, P Combust Inst 1 (34) (2013), 225-232.

[19]. Somers KP, Simmie JM, Metcalfe WK, Curran HJ, Phys Chem Chem Phys 11 (16) (2014), 5349.

[20]. Tran L, Togbé C, Liu D, Felsmann D, Oßwald P, et al, Combust Flame 3 (161) (2014), 766-779.

[21]. Jouzdani S, Eldeeb MA, Zhang L, Akih Kumgeh B, Int J Chem Kinet 48 (9) (2016). 491-503.

[22]. Chaos M, Kazakov A, Zhao Z, Dryer FL, Int J Chem Kinet 7 (39) (2007), 399-414.

[23]. Kee RJ, Grcar JF, Smooke M, Miller JA, Meeks E, Sandia Report SAND85-8240, Sandia National Labs 5 (143) (1985), 65.

[24]. Chen Z, Combust Flame 6 (162) (2015), 2442-2453.

[25]. Meng X, Hu E, Li X, Huang Z, FUEL (185) (2016), 916-924.

[26]. Li Q, Zhang W, Jin W, Xie Y, Huang Z, FUEL (184) (2016), 836-845.

[27]. Baloo M, Dariani BM, Akhlaghi M, Chitsaz I, FUEL (144) (2015), 264-273.

[28]. Ma X, Jiang C, Xu H, Ding H, Shuai S, FUEL (116) (2014), 281-291.

[29]. Galmiche B, Halter F, Foucher F, Combust Flame 11 (159) (2012), 3286-3299.

11th Asia-Pacific Conference on Combustion,
The University of Sydney, NSW Australia
10th -14th December 2017.

Electro-hydrodynamic instability of premixed flames under manipulations of DC electric fields

Yihua Ren[1], Wei Cui[1], Shuiqing Li[1,*]

[1]Key Laboratory for Thermal Science and Power Engineering of Ministry of Education, Department of Thermal Engineering, Tsinghua University, Beijing 100084, China

Abstract

A novel electro-hydrodynamic instability of flames is investigated by measuring electric and hydrodynamic responses of a premixed stagnation flame to a sub-breakdown DC electric field. When the applied voltage increases beyond 1.5 kV, the flame front suddenly transits from a substrate-stabilized near-flat mode to an electrode-stabilized conical mode, accompanied by a current jump from a space-charge regime to a saturation one. At the transition moment, the flow velocity decreases to zero and the flame spontaneously flashbacks towards the nozzle, evincing a balanced relation between the fluidic inertia and the electric body force. The electric body force maximizes at the upstream of the flame wrinkle, which can be explained by equipotentiality of the wrinkled flame front. The non-uniform electric body force gradually pulls the flame upstream, while the inclined flame front in turn enhances the local electric body force at the same time. This self-looping between electric and hydrodynamic responses finally develops into a positive feedback process and triggers the instability when both growth rates and magnitudes of electric pressure difference exceeds that of fluidic inertia.

1 Introduction

Manipulating flames by electric fields has been investigated for decades in various ways [1-18]. Generally speaking, electric fields have three main effects on flames [19]: (i) an electro-hydrodynamic effect caused by electric body forces [20,21], (ii) a chemical kinetic effect due to electric reactions [22,23], and (iii) a thermal heating effect [24]. Despite of many debates, it is believed that the electro-hydrodynamic effect is predominantly responsible for the flame changes under weak sub-breakdown electric fields. In detail, ions and electrons (dominated by CHO^+, H_3O^+, electrons, O_2^-, with a characteristic density of $\sim 10\times10^{-11}$ cm^{-3} [25]) generated in flames collide with neutral molecules, which, in a macroscopic view, exert a net body force on the fluids.

Previous studies on the electro-hydrodynamic effects mainly focused on the hydrodynamic response of flames under the manipulations of electric fields. Steady or slowly varied electric fields can significantly induce gas convection, i.e., the so-called "ionic winds" [1,2,4,21,26]. The convection flows can substitute gravity [2], crush down flame shape [4], and extend the flame stabilization zone [6]. Analyses in these studies concentrated on the one-way action of electric fields on flames. However, the reverse impact of the flame front on its electric responses, though recognized in several experiments and simulations, is still not well addressed. In our recent work [27], it is found that non-uniform electric body forces localize at the corner position of an envelope flame front, which then significantly perturb the flame. Similar phenomenon of non-uniform current density has also been discovered in the simulation of a jet diffusion flame by Belhi et al. [28]. Therefore, the retroaction of flame fronts to their electric responses cannot be ignored, as the curved plasma surface can guide current densities and generate non-uniform body forces. More importantly, the non-uniform electric body forces can in turn affect the flow field and manipulate the flame front. As a result, the two-way interactions between the hydrodynamic and electric responses of flames may form a positive feedback process, namely, an electro-hydrodynamic instability of flame fronts.

The object of this work is to demonstrate the novel electro-hydrodynamic instability of flames under external sub-breakdown electric fields. To achieve this, we utilize a premixed stagnation flame and observe its behavior under a uniform DC electric field. Flame chemiluminescence, electric characteristics of the flame, and the flow patterns are examined at different external voltages. Dynamic characteristics of the electro-hydrodynamic instability phenomenon are further recorded by a high-speed camera. Based on the static and dynamic features of the flame, two necessary conditions for the instability are identified as well as ways to avoid it.

2 Experiment

The experimental assembly of the stagnation flame and the static electric field is shown in Fig. 1. Premixed gases CH4, O_2, and N_2 (molar composition 0.13:0.18:0.69) flowed through a contoured nozzle with a velocity of 1 m/s. A special-designed honeycomb, placed at the outlet of the nozzle, generated a non-uniform axial velocity distribution at the inlet of the flame field. The axis velocity is lowest at the center, inducing the wrinkled flame front. The flow field was measured by a Particle Image Velocimetry (PIV) system. In the PIV measurement, 2-4 μm alumina particles were seeded into the stagnation flow field. The trace particles are enlightened by a 2W 450nm CW laser sheet and imaged by a Phantom V331 high-speed camera with a frame rate of 2000 fps and an exposure time of 489 μs.

Luminescence of chemically excited CH radicals (A2Δ) at the wavelength of 431 nm was used to demonstrate the flame front structure and the chemi-ionization layer, as the reaction $CH^* + O \rightarrow CHO^+ + e^-$ dominates the charge generation in methane-air flames [25]. Hence, the flame was first filtered by a bandpass filter (centered at 430 nm with a FWHM of 10 nm) before imaged by a Nikon D300s camera with an AF-S NIKKOR 100mm f/3.2 lens. Two typical images of flame

Corresponding author. Shuiqing Li. 8610-62788506
E-mail address: lishuiqing@tsinghua.edu.cn

chemiluminescence are illustrated in Fig. 1. Dynamic behaviors of the flame front were recorded by a Phantom V311 high-speed camera with a frame rate of 1000 fps and exposure time of 990 μs.

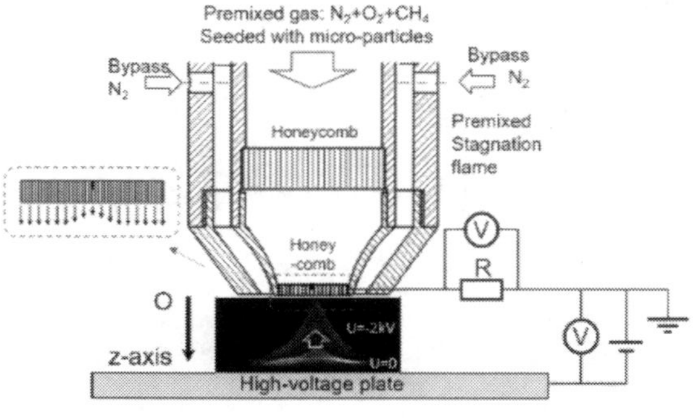

Figure 1. Schematic of the stagnation flame setup, the electric circuit system, and cameras (a Nikon camera for static images and a high-speed camera for dynamic images). The peuso-color image in the schematic indicates the chemiluminescence of CH*.

A high-voltage power, controlled by a signal generator (RIGOL DG-4162), applied a sinusoidal electric field between the honeycomb at the nozzle outlet and the substrate. The sinusoidal voltage has a period ~ 102 s, which is long enough to be regarded as a DC voltage. According to an electrostatic simulation, the electric field in absence of the flame is uniform in space especially at the center zone of the flow field. A resistor of 100 kΩ was added in the electric circuit to prevent breakdown and protect the power. The applied voltages U(t) and the electric currents I(t) that flowed through the circuit were monitored

3 Result and Discussion

3.1 Steady features

As previously mentioned, adding electric fields on flames can drive the movement of positive charges (dominated by CHO+ and H3O+) and negative charge (dominated by electrons) in flames. The movement causes electric and hydrodynamic responses of flames simultaneously. Steady features of these responses are demonstrated in this section.

Fig. 2 shows the current responses of the stagnation flame together with the flame chemiluminescence images at different applied voltages. When the applied voltages are smaller than 0.3 kV, the current is approximately proportional to the applied voltage, as shown in the double logarithmic plot of the I-U curve at the inset of Fig. 2(a). This proportionality holds as the net charge density is too small to influence the external electric field. In the linear regime, recombination of charges rivals the ion generation. When the voltages are larger than 0.3 kV, the linear relation breaks down and the I-U curve approaches a quadratic function, indicating a space-charge regime. In this regime, the net charge density is approximately proportional to the applied field strength, i.e., n_c~E, and the current density then scales as j~encEμ~E^2 (where μ is the mobility). For the above two regimes, the current responses are well consistent with previous theoretical and simulation studies [21,29], as the flame surfaces keep almost unchanged (as illustrated in the flame images A-C of Fig. 2(b)), which complies with basic assumptions in these studies

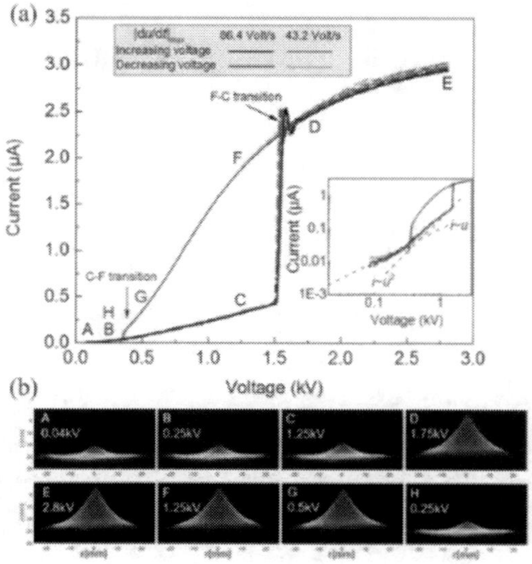

Figure 2. (a) Electric current as a function of the applied voltage. Solid, dashed, and dotted lines stand for the 1st, 2nd and 3rd cycles of changing voltages, respectively. The black and blue lines indicate the increasing and decreasing lines when the periodic time of applied voltages is 200s, i.e., the maximum |dU/dt| is 86.4 V/s; the red and green lines indicate the increasing and decreasing lines when the periodic time is 100s, i.e., the maximum |dU/dt| is 43.2 V/s. The inlet figure shows the double logarithm curves of the same I-U curve; the purple dashed line indicates the fitting proportional and quadratic line. (b) Flame chemiluminescence at different voltages corresponding to the points A to H at the upper I-U curves. The pseudo-color images indicate the intensities of the flame chemiluminescence.

However, when the voltage further increases, the classical theory does not hold any longer. The current jumps at a critical voltage about 1.5 kV, accompanied by a flame transition from a near-flat structure to a conical one. The flame images (D-G) in Fig. 2(b) indicate that the conic flame is stable and does not depend on the voltages. The current responses of the flame almost lie in another I-U curve completely different from the previous one. In this curve, the flame current plateaus with further increasing voltages, indicating the approaching of a saturation regime [21]. When the voltage decreases to another transition point, the flame transits back from the conical to the near-flat structure when the current falls back to the original one. It is noteworthy that the flat-to-conical (F-C) and conical-to-flat (C-F) transition points do not coincide with each other. This bifurcation phenomenon implies a nonlinear effect of the electric field on the flame.

To better explain the F-C transition, flow fields of the flame at different voltages were measured by PIV, as shown in Fig. 3(a-c). In absence of the electric field, the velocity jumps across the flame front due to the thermal expansion effect, and then approaches an axisymmetric divergence flow pattern before reaching the substrate. Velocities at U = 1.2 kV are similar with that at U = 0, as both flow fields have low-velocity zones in the central region at upstream of the flame front. Under the electric field, the low-velocity zone moves slightly to the nozzle accompanied with the flame front. Moreover, it should be noted that the velocity at the low-velocity zone normal to the flame front keeps almost constant under different voltages, which indicates a constant flame speed at U = 1.2 kV. Therefore, the sub-breakdown electric field manipulates the flame front by altering the upstream flow field other than directly changing the flame speed, consistent with previous studies [30,31].

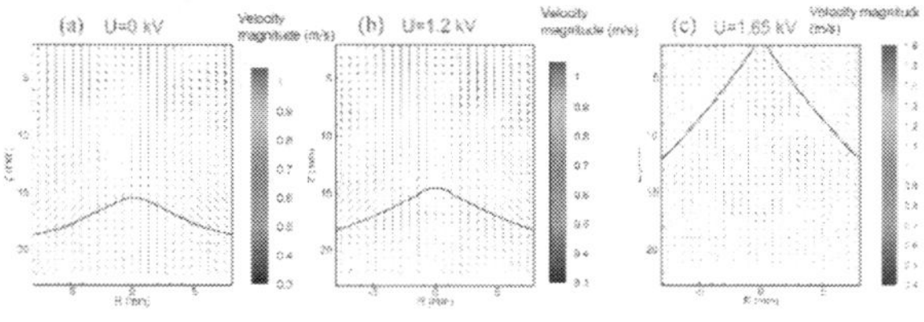

Figure 3. PIV measured flow field at U = 0 (a), 1.2 kV (b), and 1.65 kV (c).

After the F-C transition, the flow field significantly changes due to the flame front transition. According to the classical combustion theory, the velocity component perpendicular to the flame front increases to while the velocity component tangential to the front keeps constant. The heavily inclined flame front gives rise to large tangential velocities which finally results in the large velocity magnitudes at downstream of the flame front, as shown in Fig. 3(c). Similar to the electric responses of the near-flat flame front, the electric current flows along the inclined flame front. Therefore, effects of electric fields on flames strongly depend on the original flame structures, which can well explain the bifurcated flame stabilization modes for increasing and decreasing voltages.

3.2 Dynamic features

Dynamic behaviors of the flame transition between the two stabilization modes are further investigated by tracking the flame front transition. Fig. 4 illustrates variations of the flame front and the corresponding current responses within 0.5 s in one typical F-C transition. The flame stabilizes in the near-flat mode from 0 ms to ~220 ms and in the conical mode after ~260 ms, while the currents keep almost constant in each mode. As shown in the flame snapshots of c1-c3, the flame wrinkle inclines and develops into the conical structure during the transition within 40 ms, while the current rapidly increase simultaneously. The snapshots s1-s12 demonstrate that the flame front is fiercely perturbed before reaching the conical stabilization mode. The current strongly oscillates accompanied by the unsteady flame front, as seen in the inset of Fig. 4(a). The synchronization between currents and flame fronts validates the strong dependency of the electric responses on the flame front.

We noted that the position of the flame front in the center transits from 6.527 mm (c1) to 1.436 mm at (c3) within 20 ms, which corresponds to a transition speed of 0.255 m/s. The flame speed is calculated to be 0.259 m/s by CHEMKIN, and measured to be 0.3±0.07 m/s by the PIV result (the unburned gas velocity normal to the flame front). Consistence between the flame speed and the transition speed indicates that the F-C transition is a flame flashback process when the upstream unburned gas velocity is reduced to zero by the electric body force.

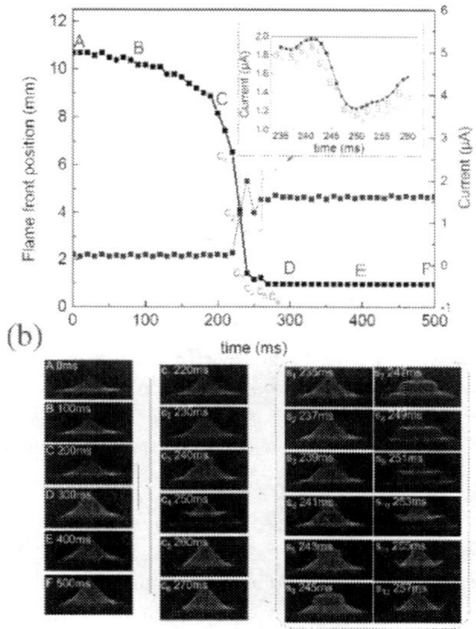

Figure 4. Transition process of flame structure. (a) Position of the flame front in the center (black squares and line) and the current response (blue squares and line) during a typical F-C transition. The current responses from 235 ms to 257 ms are zoomed-in at the inset figure. (b) flame front structure imaged by a high-speed camera.

More importantly, since the applied voltage increases less than 1.6 V within 40 ms from c1 to c3, the transition can be regarded as a spontaneous process, and must be triggered by a positive feedback mechanism. When the flame front gradually inclines and moves upstream, both the upstream fluidic inertia and the electric body force increase. As the two factors drive the flame front movement in reverse directions, the critical point for triggering the spontaneous process is that growth rates of the electric body force exceed that of the fluidic inertia. If this condition can be satisfied, the electric body force is able to reduce the local flow velocity and to move the flame upstream. Then, the more inclined flame fronts can be induced larger electric body forces by the same field, which forms a positive feedback loop

3.3 Electro-hydrodynamic instability

The positive feedback loop originates from the non-uniform distributions of electric body forces. This non-uniformity is caused by the electrostatic screening effect of the flame front. Because the Debye length of the flame plasma is much smaller than the flame thickness, the flame front can be considered as a quasi-neutral equipotential plane, as demonstrated in Fig. 3(d-e). Under external electric fields, the charges tend to redistribute in the equipotential plane to minimize their electric potential energy until there is no potential difference among various parts of the plane. Outside the flame front, electric field strengths gradually increase, together with the generations of net charges in this region. Drews et al. [1] demonstrated this effect and proposed an expression for a characteristic electric pressure difference as $\Delta p_E = 1/2 \varepsilon_0 E_0^2$ based on the one-dimensional reaction layer assumption. However, when the surface is not strictly perpendicular with the external electric field, the strongest electric field is generated between the electrode and the flame wrinkles at the smallest distance. If the flame wrinkle has a length scale of δ in the electric field directions, the local electric field can be approximately expressed by $E_z \approx E_0 \cdot z_0/(z_0-\delta)$. The electric pressure difference can be estimated by

$$\Delta p_E = \frac{1}{2}\varepsilon_0 E_0^2 \frac{z_0^2}{(z_0-\delta)^2}, \tag{1}$$

where E_0 is the original electric field without flame wrinkle and z_0 is the distance between the unwrinkled flame front and the electrode. As a result, the flame wrinkle closer to the electrode bears a larger electric body force, which leads to the non-uniform hydrodynamic response of the flame.

The non-uniform responses can further trigger the electro-hydrodynamic flame instability after satisfying two necessary conditions. Firstly, as mentioned above, the growth rates of electric pressure difference need to be equal or larger than that of fluidic inertias. After the 1st condition is satisfied, the electric pressure difference grows faster and finally exceeds the fluidic inertia at the flame flashback position, which is the 2nd necessary condition.

Finally, the electro-hydrodynamic instability of the flame occurs between two different stabilization modes. The near-flat stagnation stabilization mode is generated by the jet-impinging flow in absence of the electric field, while the conical stabilization mode is created by the electric body force near the nozzle electrode. Under the electric field, the original stagnation mode cannot stabilize the flame any longer after satisfying the above two conditions. Based on the above analyses, two solutions are proposed here to depress the instability and broaden the stabilization limit in other electric field/plasma assisted combustion systems. The first method is to increase distances between flames and electrodes (z0 in our case). In this way, electric pressure differences Δp_E are less dependent on the non-uniformity of flame shapes (the flame wrinkle size δ in our case) according to Eq. (1). The second method, from the opposite perspective, is to directly stabilize flames by electrodes (in absence of electric fields) which can impede the flame transition to other potential modes after adding the electric field.

4 Conclusions

In summary, we report the electro-hydrodynamic instability of the premixed stagnation flame under the sub-breakdown DC electric field. When the applied voltage increases beyond 1.5 kV, the flame front suddenly transits from the near-flat mode to the conical mode, accompanied by the current jump from the space-charge regime to the saturation regime. The electric body force is non-uniformly distributed at the upstream of the flame wrinkle, which expands the low-velocity zone and pulls the flame front upstream. A theoretical analysis of the Debye length of the flame plasma reveals that the non-uniformity of the electric body force originates from equipotentiality of the wrinkled flame front. Consequently, both the local electric field and the space charge density are largest at the flame wrinkle closest to the electrode, which can be described by a simplified model.

The electric responses of the flame, coupled with the hydrodynamic responses, further trigger the electro-hydrodynamic instability of the flame front. High-speed camera measurements show that at the transition moment, the flow velocity decreases to zero and the flame front spontaneously flashbacks to the nozzle, indicating a balanced relation between the electric pressure difference and the fluidic inertia. This flame transition phenomenon is driven by a positive feedback loop between electric and hydrodynamic responses of the flame front: The increased electric body force pulls the flame wrinkle upstream which in turn enhance the local electric body force. Two necessary conditions are required to trigger the loop: (1) The growth rates of the electric pressure difference exceed than that of the fluidic inertia; (2) The electric pressure difference can balance or become larger than the fluidic inertia to reduce unburned gas velocities to zero. Finally, two solutions are proposed to prevent the instability: (i) placing flames far from the electrode to weaken the dependence of the electric response on the shapes of flame fronts, and (ii) stabilizing flame by electrodes to isolate the flame in a single stabilization mode.

5 Acknowledgment

This work is mainly funded by the National Natural Science Funds of China (Nos. 91641204).

References

[1] A.M. Drews, L. Cademartiri, M.L. Chemama, M.P. Brenner, G.M. Whitesides, and K.J.M. Bishop, Phys. Rev. E 86, 36314 (2012).

[2] F. Carleton and F. Weinberg, Nature (1987).

[3] S. Karnani, D. Dunn-Rankin, F. Takahashi, Z.-G. Yuan, and D. Stocker, Combust. Sci. Technol. 184, 1891 (2012).

[4] J. Schmidt and B. Ganguly, Combust. Flame 160, 2820 (2013).

[5] M.K. Kim, S.K. Ryu, S.H. Won, and S.H. Chung, Combust. Flame 157, 17 (2010).

[6] M.K. Kim, S.H. Chung, and H.H. Kim, Proc. Combust. Inst. 33, 1137 (2011).

[7] D. a. Lacoste, J. Eng. Gas Turbines Power 135, 101501 (2013).

[8] C.L. Maupin and H.H. Harris, Combust. Flame 97, 435 (1994).

[9] J. Hu, B. Rivin, and E. Sher, Exp. Therm. Fluid Sci. 21, 124 (2000).

[10] E. Sher, A. Pokryvailo, E. Jacobson, and M. Mond, Combust. Sci. Technol. 87, 59 (1993).

[11] H. Zhao, X. Liu, and S.D. Tse, J. Nanoparticle Res. 10, 907 (2008).

[12] D.R. Hardesty and F.J. Weinberg, Symp. Combust. 14, 907 (1973).

[13] Y. Wang, G. Sharma, C. Koh, V. Kumar, R. Chakrabarty, and P. Biswas, Aerosol Sci. Technol. 51, 1 (2017).

[14] S. Li, Y. Ren, P. Biswas, and S.D. Tse, Prog. Energy Combust. Sci. 55, 1 (2016).

[15] Y. Wang, G.J. Nathan, Z.T. Alwahabi, K.D. King, K. Ho, and Q. Yao, Combust. Flame 157, 1308 (2010).

[16] M. Saito, T. Arai, and M. Arai, Combust. Flame 119, 356 (1999).

[17] E. V. Vega, S.S. Shin, and K.Y. Lee, Fuel 86, 512 (2007).

[18] A. Sakhrieh, G. Lins, F. Dinkelacker, T. Hammer, A. Leipertz, and D.W. Branston, Combust. Flame 143, 313 (2005).

[19] Y. Ju and W. Sun, Prog. Energy Combust. Sci. 48, 21 (2015).

[20] H. a. Wilson, Rev. Mod. Phys. 3, 156 (1931).

[21] J. Lawton and F.J. Weinberg, Proc. R. Soc. London A Math. Phys. Eng. Sci. 277, 468 (1964).

[22] H.C. Jaggers and A. von Engel, Combust. Flame 16, 275 (1971).

[23] B.N. Ganguly, Plasma Phys. Control. Fusion 49, B239 (2007).

[24] S. Marcum and B. Ganguly, Combust. Flame 143, 27 (2005).

[25] A. Fialkov, Prog. Energy Combust. Sci. 23, 399 (1997).

[26] D.A. Lacoste, Y. Xiong, J.P. Moeck, S.H. Chung, W.L. Roberts, and M.S. Cha, Proc. Combust. Inst. 0, 1 (2016).

[27] Y. Ren, S. Li, W. Cui, Y. Zhang, and L. Ma, Combust. Flame 176, 479 (2017).

[28] M. Belhi, P. Domingo, and P. Vervisch, Combust. Flame 157, 2286 (2010).

[29] Y. Xiong, D.G. Park, B.J. Lee, S.H. Chung, and M.S. Cha, Combust. Flame 163, 317 (2016).

[30] J. Kuhl, T. Seeger, L. Zigan, S. Will, and A. Leipertz, Combust. Flame 176, 391 (2017).

[31] E.N. Volkov, V.N. Kornilov, and L.P.H. De Goey, Proc. Combust. Inst. 34, 955 (2013).

Replies to the 1st reviewers' comments

1. *In this paper, there are only the simulation results, can you describe the simulation process and conditions?*

Sorry for the careless mistakes. Limited by the length of paper, we have to enlarge the figure and delete the simulation results. Thus we are unable to present the simulation part in this conference paper.

2. *As shown in Fig.2, the I-U curve with decreasing voltage is different from the increasing voltage, and it doesn't approach quadratic function even not at plateaus region. How to explain that result?*

This is very good comment. Actually, as seen in the inlet of the Fig. 2, the I-U curve with decreasing voltage has the same slope with that with increasing voltage in the double-log diagram. Therefore, the flame with different structures can approach the quadratic function but with different coefficients, indicating different electric responses.

3. In this study, the PIV technique is used. Are the particles used in PIV affected by the electric field? I mean that charges in the flame may transfer to the trace particles. Thus the results are influenced. Is it possible?

This is a good point. Actually, PIV is the only method that we can use to demonstrate the flow field. We have considered this particle charging effect and choose the micro-sized particles with the migration velocity of ~1-5 cm/s at the transition zone between the attachment and field charging mechanisms.

4. The transition time discussed in section 3.2 is 40ms, and the flame response time under the electric field is 10-16ms according to previous researchs, Could it influence the transition time if the voltage change faster?

Yes. In that case, the transition is caused by the ionic wind directly but not by the electro-hydrodynamic flame instability.

5. Some suggestions:The first paragraph in the experiment section should be delected, I think that you forgot it. And it is better to change figure 2's title. The format of index should be altered , eg : m3→m3. And the larger figures may be better for some details can not be observed.

Thanks for the suggestions. We have corrected these careless mistakes accordingly.

Replies to the 2nd reviewers' comments

The size of the figures should be increased, it is hard to see any information in them, let alone read the microscopic text.

We have corrected the figure accordingly.

It is argued that the flame's transition from flat to conical comes from an alteration in the flow field and not from a change in the flame speed, but the causality is not very well explained and remains a little unclear. Maybe you could try to re-write this paragraph to increase the causality you want to display.

We have revised this part according to the suggestion.

A part of the template remains in the beginning of Part 2.

We have deleted it according to the suggestion.

Various typos in the text should be corrected.

Thanks for the reviewer's suggestions. We have corrected these careless mistakes accordingly.

11th Asia-Pacific Conference on Combustion,
The University of Sydney, NSW Australia
10th -14th December 2017.

Experimental Study on the Improvement of Fire Retardant Performance for Boron-Phosphorous Chemicals Treated Wood-based Materials

Hyun Jeong Seo[1], Jeong Min Jo[1], Wuk Hwang[1], Min Chul Lee[1,2,*]

[1] Department of Safety Engineering, College of Engineering, Incheon National University
[2] Fire Disaster Prevention Research Center, Incheon National University

119 Academy-ro, Incheon, 22012, Republic of Korea

Abstract

This study presents the experiemental results of the fire retardant performance of boron-phosphorous chemicals treated wood based materials according to the KOEFIS 1001 standard and the test procedure of ISO/TR 16312-2 for incombustibility of interior finishing materials and the elements of buildings. Environmental friendliness and detective whiteness on the surface of wood as well as the composition of the boron-phosphorus chemical compounds were carefully adjusted when FRCs were manufactured. As a result of flame retardant performance test, the flame spread characteristics of flame retardant materials were improved in both carbonization area and length. Furthemore, the experimental results of gas hazardousness showed that FRCs are not highly toxic to human body after burning. The purpose of this study is to develop appropriate mixing ratio of fire-retardant solution for wood and wood-based material of building applications which can sufficiently suppress flame spread and reduce the emission of harmful gases.

1 Introduction

Wood is natural resource and it has been widely used for various field including flooring, celling, wall covering, in building construction [1-3]. These wood and wood-based materials have advantages of environment-friendly and renewable materials, but also have disadvantages of being easily corroded or vulnerable to fire. Therefore, it is necessary for long-term use of wood to improve both preservation and flame retardant performance of wood utilizing some chemicals, such as boric acid, di-ammonium phosphate, borax, phosphoric acid, that sort of thing. Many previous studies have been conducted to solve these problem and develop fire-retardant performance of wood and wood-based materials [4-8]. The flame-retardant chemical composites(FRCs) are widely used to enhance the fire-resistance and they can be classified into bromine-based, phosphorus-based, ammonium-based and carbon-based chemicals. Among these flame retardants, brominated flame retardants was prohibited in Europe since it generats a large amount of toxic gas such as HBr, dioxins, etc. when combusted [9-11].

The fire was occurred from the building more than 60% of the number of fires in Korea, so in this study examined the previous studies for the flame retardant and manufacturing mixture ratio of them. Also, it was investigated multifunctional flame retardants with eco-friendly performance, of the ability to release less harmful gases when they burned [12, 13]. This is because the hazardous gases are as dangerous as the heat generated during combustion to the occupants in the building [13, 14]. Therefore, in this study, boron and phosphorous chemicals were prepared in the state of aqueous solution and applied to the surface of wood-based materials. The objective of this study is to measure toxic gases contents and toxicity index of flame retardant treated wood and untreated wood to provide an initial data base for wood-based building materials.

Corresponding author. Fax: +82-32-835-0779
E-mail address: LMC@inu.ac.kr

2 Experiment

In this study conducted flame retardant performance evaluation and gas hazard evaluation test. Thus, the specimens were produced in two sizes.

2.1 Materials

The 5-plywoods and wall covering woods were used as the test specimens. The plywood was manufactured with a hot press machine that utilized red pine (*Pinus densiflora*). When that specimen was produced, phenolic resin was applied between each of the veneer woods. Also, wall covering woods were consisted of various kinds of soft wood including Japanese larch (*Larix kaempferi*), Korean pine (*Pinus koraiensis*), and red cedar (*Cryptomeria japonica*). They are prepared by 290 mm × 190 mm according to KOFEIS 1001 (2013) to conduct measurement of flame retardant performance. The specimens, for gas hazard assessment, were 220 mm square and not more than 15 mm thick according to ISO/TR 16312-2 (2007). Also, the thickness of the manufactured plywood is 0.8 cm, those of wall covering woods were 1 cm.

The flame retardant chemical composites (FRCs) were prepared using a stirrer at a speed of 1000 RPM with water as a solvent. Table 1 shows the FRCs proportion and formulas. Also, each of the chemical composites listed above are applied to the wood as a single mixed species. As a method of applying FRCs to wood, a method of applying the FRCs to the surface by spraying was used. FRCs were applied on the surface of the plywood to estimate the throughput of the prepared FRCs at 100, 200, and 300 g/m^2, respectively. In addition, these specimens were dried by air-dry oven for 2 weeks at a temperature of 25 °C, until the moisture content of the surface fell below 15%.

Table 1. Chemical species constituting FRCs.

Chemicals	Formula	Composition Rate (wt.%)
Boric acid	H_3BO_3	5
Di-ammonium phosphate	$(NH_4)H_2PO_4$	1
Ammonium borate octahydrate	$(NH_4)_2O{\cdot}5B_2O_3{\cdot}8H_2O$	3
Borax decahydrate	$Na_2B_4O_7{\cdot}10H_2O$	6
Ammonium carbonate	$(NH_4)_2CO_3$	5
Phosphoric acid	H_3PO_4	30
Water	H_2O	Used as a solvent

2.2 Methods

Fire retardant performance test methods by measuring the flame spread from the surface of the materials have been used in the United States, Britain, China, France, Japan, and Korea. These parameters can predict the possibility of flame propagation and spread on the surface of materials as stated by KOFEIS 1001 (2013) [15], which was made according to JIS A 1322 (1982). The specimens were clamped with steel holders and exposed to the flame at an angle of 45 degree. The liquefied petroleum gas flame was used as the heat source. According to the standard, the flame length generated by a Mecker burner should be adjusted to a height of 6.5 cm was applied for 2 min, and the times for the specimens to be charred and ignited were recorded. The results of carbonization area and carbonization length are calculated and exposed after three repeated experiments.

To evaluate the hazardous combustion gases, the specimens were exposed in moderately vitiated air to convective and radiative heating. The exhaust gases were introduced an animal –exposure chamber in which there were 8 rotary cages, each containing a mouse. Also, the animals are experimental rats, they are female and weigh between 18 and 20 grams. The test is in progress to measure the times to incapacitation of the mice. In this study, according to the standard, the test is run for 15 minutes [16].

3 Results and Discussion

3.1 The Spread of Flame

Among the various parameters that indicated fire risk, flame spread is one of the important factors when assessing the degree of fire spread in buildings. Especially because wood and wood-based materials are flammable, and they can serve as a flame propagation medium and interrupt the passage of time. According to KOFEIS 1001(2013), three specimens are randomly selected from the test specimens produced, and the result is obtained by repeating the flame spread test three times. In this experiment, the test is carried out three times for each test specimen, and three times the results are averaged to graph them. Both the carbonized area and the carbonized length of the test specimens were calculated using this method. Figure 1 shows the results of the 45 degree flammability test. The results of carbonization area of the plywood show that there is a flame diffusion suppression capability of FRCs.

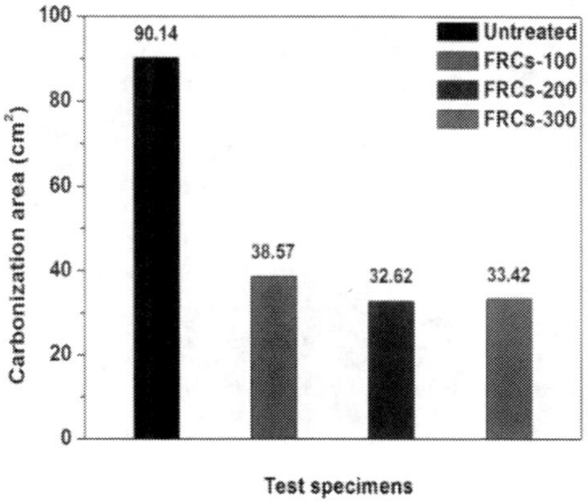

(a) Plywood

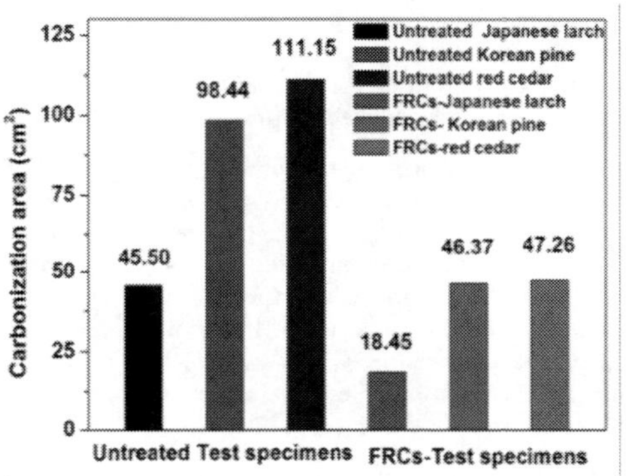

(b) Wall covering woods

Figure 1: Carbonization area after 45 degree flammability test

And, according to the results in figure 1 (a), it is determined that its throughput is about 200 g/m^2. In the case of FRCs treated wall covering specimens, the carbonization area of the larch was measured to be 1/3 on comparison to that of untreated specimen. And compared to untreated specimens, those of pine and cedar were confirmed to reduce one-half the size, respectively.

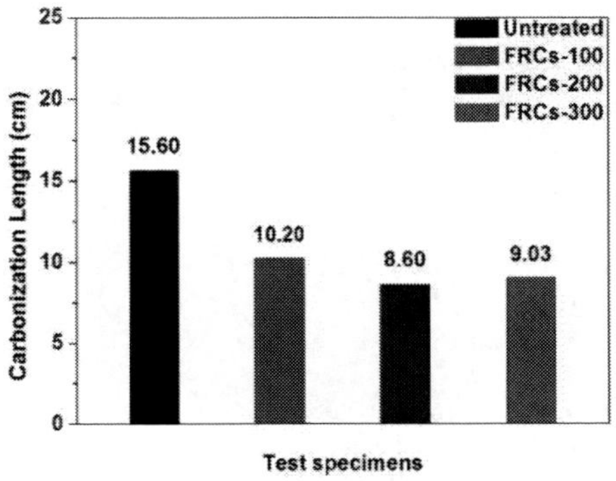

(a) Plywood

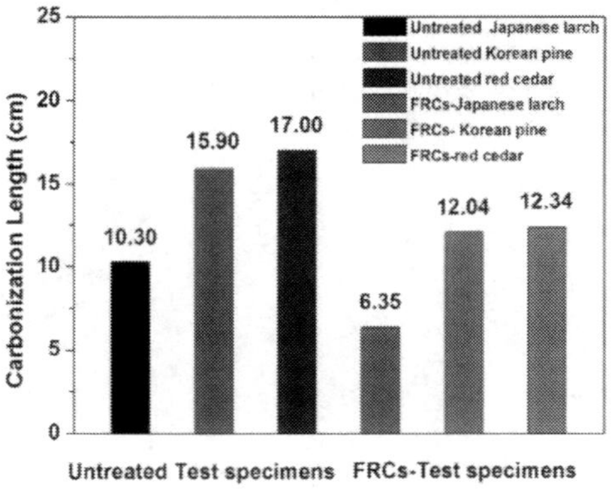

(b) Wall covering woods

Figure 2: Carbonization length after 45 degree flammability test

This reduction in flame spread is determined that it is attributable to the phosphorus chemicals among the various chemicals in FRCs. These results are also due to the HPO_3, HPO_2 and PO, which is formed by pyrolysis of the phosphate series of chemical species. These pyrolysis products positively affect the formation of char on the surface of the materials, and eventually prevent or reduce the spread of the flame.

As a result of measuring the carbonization length, FRCs have been proven for its flame spread suppression performance. In this study, flame retardant aqueous solution was developed attributed to the di-ammonium phosphate, ammonium carbonate and phosphoric acid in FRCs. According to the report of Hsiue et al. (2000) and Lu and Hamerton (2002), in the case of the phosphorous-based chemicals, they decomposed chemical reactions involved in decomposition in favor of reactions to form the char layers as pyrolysis [17, 18].

In the case of ammonium-based chemicals, also, they can combine with phosphorus based chemicals to form char at the initial combustion [3]. The results of this experiment were confirmed to be similar to those of the previous studies.

3.2 Evaluation of Gas Hazard

The standard procedure, according to ISO/TR 16312-2, is to measure the times to incapacitation of the mice (ICR, female). Currently, Korea's Gas Hazard Assessment Standard applies mice test, which is evaluating the gas harmfulness through the average downtime of experimental mice among the tests specified in ISO/TR 16312-2. The gas hazard test was conducted twice for each test specimen, and the test was conducted on the FRCs treated specimens and the untreated specimens, respectively. Equation (1) shows the formula for calculating the average activity stopping time of experimental rats, which is the standard for evaluating the hazard of combustion gases.

$$x = \bar{X} - \sigma \qquad (1)$$

where x is the average mice stopping time (min), $\bar{X}$ is the mean value of the downtime (min), and σ is the standard deviation of the downtime (min).

The assessment criteria for the combustion gases in the test method are 7 minutes, which is based on the Laurent wood species, and a determination is made when the average mouse stopping time of the mice exceeds 7 minutes. Table 2 shows the results of acting stopping time data for experimental specimens.

From the results of two experiments in total, the average stopping time of the experimental rats was confirmed to be greater than 7 minutes for both specimens. Among the chemical species including the FRCs, ammonium-based chemicals can be main cause of NO_X emission during the combustion.

Table 2. Results of mice activity stopping time.

Distribution	Test inspection results			
	1		2	
	Average time (m:s)	Standard Deviation (m:s)	Average time (m:s)	Standard Deviation (m:s)
Untreated Specimen	8:16	1:36	12:44	0:44
FRCs treated specimen	8:17	0:29	8:22	0:45

However, they are mixed with phosphorous-based fire retardant chemicals, they affect the occurrence of faster char formation [3, 19, 20]. If a char layer is formed on the surface of the burned materials quickly, less hazardous gases can be generated because pyrolysis does not affect the inside of the materials. The results of this study are also caused by that fact.

4 Conclusions

In this study, fire retardant solution was developed that can be treated on wood and wood-based materials. FRCs treated amount is considered that the throughput of 200 g/m^3 is enough based on the results of this study. In addition, it showed improved flame retardant performance through the flame spread test. FRCs, also, has lower human hazard based on the results of mice activity experiment. However, additional analysis is required for this experiment, and further analysis will be conducted in future studies about quantitative and qualitative analysis of harmful components of gas e.g. NOx.

5 Acknowledgment

This work was supported by the National Research Foundation of Korea (NRF) grant funded by the Korea government (MSIP) (No. NRF-2016R1C1B1006636). This research was supported by Korea Electric Power Corporation through Korea Electrical Engineering & Science Research Institute (grant number: R15XA03-13).

References

[1] J.H. Kim, J.H. Lee, S. Kim. Journal of Thermal Analysis and Calorimetry, 110 (2) (2012) 677-683.

[2] A.P. Pereyra, C.A. Giudice. Fire Safety Journal, 44 (4) (2009) 497-503.

[3] L.A. Lowden, T.R. Hull. Fire Science Reviews, 2 (4) (2013) 1-19.

[4] J. Kawahara, Y. Yanagisawa, Journal of Society of Indoor Environment, 6 (2003) 180-181.

[5] Z.A. Nagieb, M.A. Nassar, M.G. El-Meligy. International Journal of Carbohydrate Chemistry, 2011 (2011) 1-6.

[6] J. Jiang, J. Li, J. Hu, D. Fan. Construction and Building Materials, 24 (12) (2010) 2633-2637.

[7] M. N. Islam, K. Ando, H. Yamauchi, D. Kamikawa, T. Harada, N. Hattori, Bioresources, 8 (1) (2013) 395-404.

[8] H.J. Seo, W. Hwang, G. Park, M.C. Lee. KOSCO SYPOSIUM, 53 (2016) 41-44.

[9] E. Mfiso, M.E. Mngomezulu, J.J. Maya, S.L. Adriaan. Carbohydrate Polymers, 111 (2014) 149-182.

[10] A. Dassari, Z.Z. Yu, G.P. Cai, Y.W. Mai. Progress in Polymer Science, 38 (2013) 1357-1387.

[11] S. Liang, N.M. Neisius, S. Gaan. Progress in Organic Coatings, 76 (11) (2013) 1642-1665.

[12] H.H. Liang, M.C. Ho. Construction and Building Materials, 21 (6) (2007) 1254-1261.

[13] H.J. Seo, D.W. Son, Fire Sci. Eng., 29 (4) (2015) 49-56.

[14] H.J. Park. Journal of the Korean Society of Safety, 25 (1) (2010) 38-43.

[15] KOFEIS 1001. Korean Fire Equipment Inspection Corporation, South Korea, 2013.

[16] Guidance for assessing the validity of physical fire models for obtaining fire effluent toxicity data for fire hazard and risk assessment - Part 2: Evaluation of individual physical fire models, 2007. ISO/TR 16312-2.

[17] G.-H. Hsiue, Y.-L. Liu, J. Tsiao, J. Appli. Polym. Sci., 78 (13) (2000) 1-7.

[18] S.-Y. Lu, I. Hamerton. Prog. Polym. Sci., 27 (2002) 1661-1712.

[19] H.J. Seo, W. Hwang, M.C. Lee. Bioresources, 12 (3) (2017) 5417-5427.

[20] S. Liodakis, V. Agiovlasitis, D. Vorisis. Thermochimica Acta, 568 (20) (2013), 156-160.

11th Asia-Pacific Conference on Combustion,
The University of Sydney, NSW Australia
10th -14th December 2017.

Numerical simulation on heat release rate in triple point of triple flame

Naoki Hayashi[1]

[1]Kanagawa Institute of Technology

1030, Shimo-ohino, Atsugi, Kanagawa, 243-0292, Japan

Abstract

Flame structures in triple point of triple flame are discussed using a numerical simulation. Methane, air and their mixture were injected with uniform velocity in inlet boundary of numerical model, and the inlet velocity was changed. First, the flame structure is discussed using mixedness, flame index and so on. Next, the maximum heat release rate in triple flame was discussed. As a result, the maximum heat release rate is smaller than that in stoichiometric premixed flame. Correlation between maximum heat release rate and flame stretch rate is also investigated. The influence of flame stretch is small, and thus, the influence of the distribution of radicals is important.

1 Introduction

Diffusion flames and premixed flames, which have mixture gradients in front of flame front, have a complicated flame structure by local distributions of fuel and oxygen concentration. A lifted flame in diffusion flame so unstable that it is easily influenced by disturbance. Thus, it is important to elucidate its characteristics. The flame-edge usually has a triple flame structure, which consists of lean and rich premixed flame branches, and a diffusion flame branch. The structure is interesting phenomena from a viewpoint of flame modelling.

Many researchers have studied about lifted flame or triple flame structure [1-11]. For example, Plessing et al. [1] have investigated the triple flame structure in a lifted flame using various experimental method and detailed numerical simulation.

For many triple flame studies, it is usually considered that the minimal velocity in front of triple flame is local flame speed [3-5]. This is based that it is close the laminar burning velocity [3]. However, the velocity profile in counterflow premixed flame also has minimal before flame front as well as triple flames, and this minimal velocity is used to estimate laminar burning velocity by extrapolation to zero stretch [12], and some researches considered as local burning velocity for the counterflow flame[13]. However, it has been reported that the minimal velocity have increased with the increase in stretch for various Lewis number conditions for which the flame is weakened by Lewis number effect [14].

In this study, the maximum heat release rate is used for evaluation of flame intensity. In authors previous study, the maximum heat release rate in triple flame was smaller than that in stoichiometric premixed flame. Therefore, in the present paper, the reason of these phenomena is investigated. The Coloration between flame stretch rate and maximum heat release rate is discussed.

Corresponding author. Fax: +81-46-281-3080
E-mail address: hayashi@me.kanagawa-it.ac.jp

2 Numerical model and Method

Figure 1 shows a schematic diagram of a triple flame in two-dimensional planar coordinates (x, y). The size of the computational domain is 18.5 x 14.3 mm. The number of grid points is 200 x 200. In the x direction, the equally-spaced girds are used from 1 to 170 grid points, and minimum grid size is 0.05 mm. In the y direction, the equally-spaced girds are also used from 31 to 170 grid points. The other grid points is unequaly-spaced grid, and the common ratio is 1.05.

In the left boundary, pure air is spouted in upper side of domain (2 mm to 7.15 mm), and pure methane is spouted from bottom side (-2 mm to 7.15 mm). The mixing layer, l, is set up between two layers (l = 4.0 mm). The mass fraction of Fuel is 1 inearly changed from 1 to 0 in the layer. Here, the flame is attached at inlet boundary for u_0 = 0.05 m/s. It corresponds to flash back. All calculations are carried out at atmospheric pressure and 300 K.

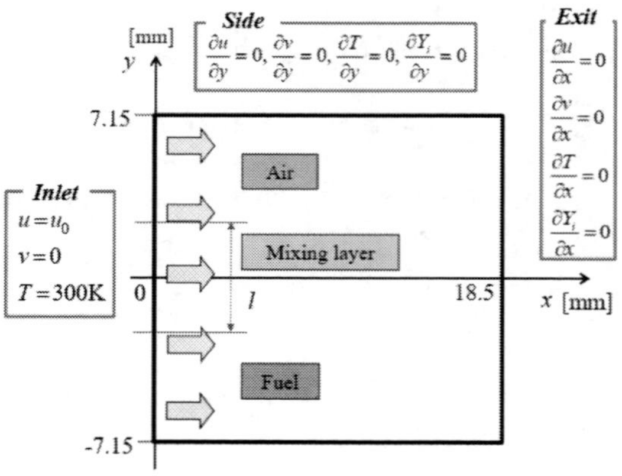

Figure 1: Analytical model of a triple flame

The thermodynamic properties for the species are obtained from the CHEMKIN database[15, 16]. The transport properties are calculated according to a Smooke's simplified transport model [17]. The chemical kinetics model used in the present study is the Smooke's skeletal methane-air reaction mechanism[17]. This mechanism contains of 16 species and 25 elementary reaction steps. The numbers of species and reactions in the mechanism are fewer than those in detailed mechanisms. The authors have confirmed the validity of the kinetic model in Ref. [18].

3 Results and Discussion

3.1 Flame structure

The distributions of heat release rate, Q, temperature, T, Mixedness, Z_{FO}, Flame index, G_{FO}, are shown in Fig. 2 to 5. In Fig. 3, the velocity vector is also shown. The Mixedness and Flame index [12] are defined as follows;

$$\begin{cases} Z_{FO} = Y_O / j & (Y_O / j \le Y_F) \\ Z_{FO} = -Y_F & (Y_O / j \ge Y_F) \end{cases} \tag{1}$$

$$G_{FO} = \text{grad}\, Y_F \cdot \text{grad}\, Y_O \qquad (Q > 0.01 Q_{max}) \tag{2}$$

Here, Y_F, and Y_O are mass fraction of fuel and oxidizer, and j is stoichiometric coefficient. The flame index is calculated only for the region of 0.01 or more to the maximum heat release rate, Q_{max}. The negative value of Z_{FO} means fuel lean mixture and the positive value means rich mixture. For flame index, the positive value is diffusion flames, and negative value is premixed flame, and the large absolute value corresponds to large reaction rate.

The heat release rate distributions (Fig.2) show the flame forms on the air side. The flame moves to downstream with the increase in the inlet velocity. At this time, the flame curvature becomes small, and large structure is formed. The triple flame structure, which consists of rich, lean premixed flame and diffusion flame, cannot be distinguished clearly from heat release rate. However,

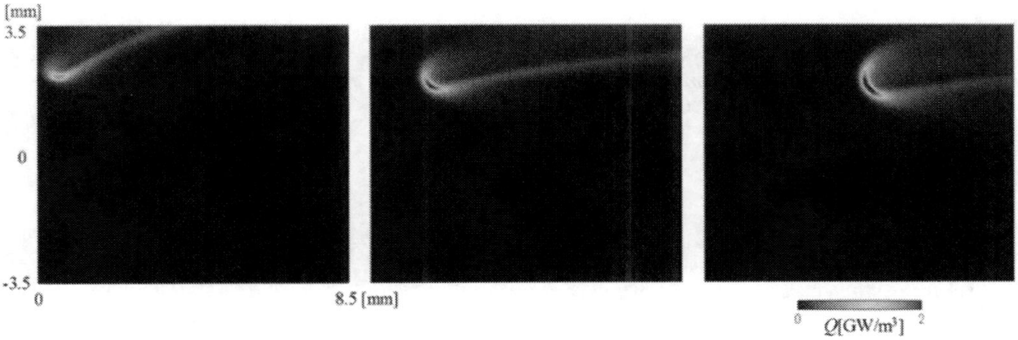

u_0 = 0.1 m/s 0.3m/s 0/5 m/s

Figure 2: Distributions of heat release rate

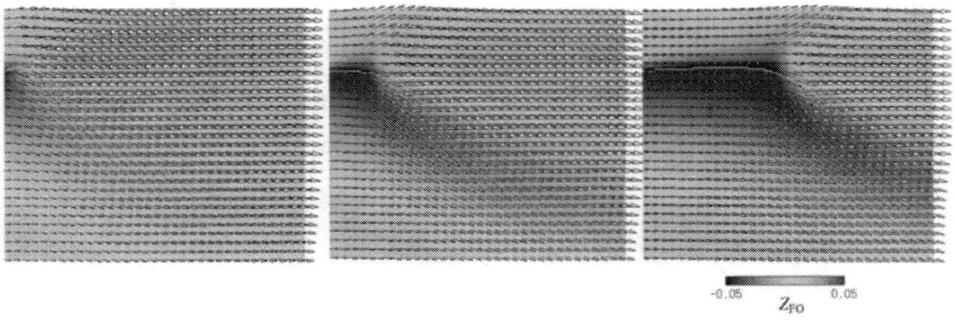

Figure 3: Distributions of mixedness

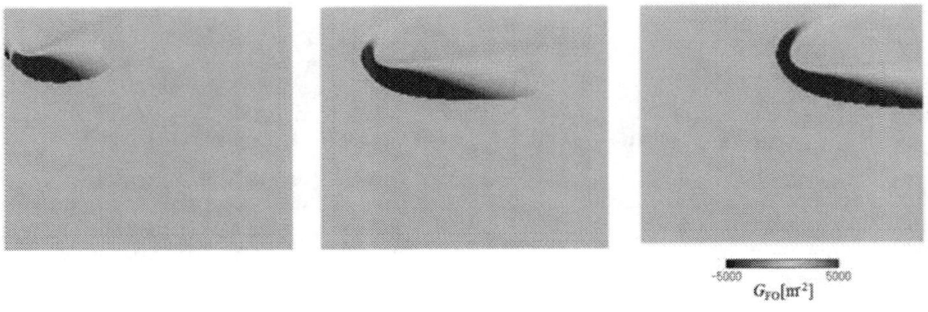

Figure 4: Distributions of flame index

Figure 5: Distributions of temperature

from the mixedness and the flame index, we can see the triple flame structure. From the flame index shown in Fig.4, it can be seen that the diffusion flame branches, which is indicated by the blue region, and rich premixed flame branches (the bottom branch) are almost integrated. It is confirmed that the positions of maximum heat release rate are almost corresponds to the triple point.

In Fig. 2, the flame front in the triple point is parallel to the x-axis when the small inlet velocity, whereas it becomes parallel to the y-axis with the increase in the velocity. In addition, mixing between fuel and oxidizer, the diffusion flame branch becomes weak.

The temperature increases in the downstream, and thus there is no clear correlation between the inlet velocity and maximum value of the temperature.

3.2 Correlation between heat release rate and flame stretch

Figure 6 shows variation of maximum heat release rate, Q_{max}, with the increase in the inlet velocity. In this figure, the maximum heat release rate for the laminar premixed flame (equivalence ratio, ϕ, is unity) is also shown. The Q_{max} increases with the increase in u_0, and the value is much smaller than that in premixed flames.

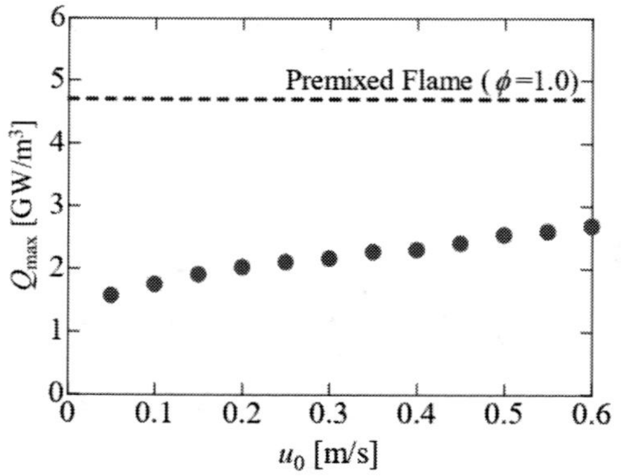

Figure 6: Correlation between Maximum heat release rate and inlet velocity

We attempt to evaluate using flame stretch rate. The flame stretch rate, κ, is defined as follows;

$$\kappa = \frac{1}{\delta A}\frac{d\delta A}{dt}$$

$$= -(\boldsymbol{n}\cdot\nabla)\boldsymbol{v}_{\mathrm{flame}}\cdot\boldsymbol{n} + \nabla\cdot\boldsymbol{v}_{\mathrm{flame}} \quad (3)$$

Here, A is flame area, and $\boldsymbol{n}$ is unit vector in normal direction for flame front, $\boldsymbol{v}_{\mathrm{flame}}$ is flame propagation rate vector. These vectors are positive toward unburned region.

By decomposing Eq. (3) in the direction perpendicular to the flame front and tangential direction,

$$\kappa = \nabla_t \cdot \boldsymbol{v}_{flame,t} + \boldsymbol{v}_{flame,n}(\nabla_t \cdot \boldsymbol{n}) \quad (4)$$

t and n show the tangential and normal direction, respectively. Assuming that the flame front tangential component of the flame propagation velocity is equal to the tangential direction component of the fluid velocity at the flame surface,

$$\kappa = \nabla \cdot \boldsymbol{v}_{fluid,t} + \boldsymbol{v}_{flame,n}(\nabla \cdot \boldsymbol{n}) \quad (5)$$

Because the flame in this study is steady, the second term of equation (5) becomes zero and the flame stretch rate becomes the divergence of the tangential component on the flame front of the fluid velocity in the flame. Therefore, the author can calculate the flame stretch rate near the triple point, where the heat release rate becomes the maximum, and can examine the relation with the flame structure.

Figure 7 shows the correlation between flame stretch rate and inlet velocity. Since the inlet velocity distribution is uniform for y direction in this study, the flame stretch rate stems from the change in velocity distribution due to the flame front having curvature. From the figure, the flame stretch rate is extremely small under the condition of low inlet velocity, $u_0 < 0.15$ m/s. This is because the flame surface exists parallel to the flow. On the other hand, at $u_0 > 0.15$ m/s, the flame stretch rate tends to increase as u_0 increases.

Figure 8 shows the correlation between maximum heat release rate and flame stretch rate. The maximum heat release rate increases with the increase in flame stretch rate. On the other hand, the maximum heat release rate in triple flame is smaller than that in premixed flame with zero stretch rate. Therefore, the reason of the weakened flame is not only flame stretch, but also the distributions of radicals need to be considered.

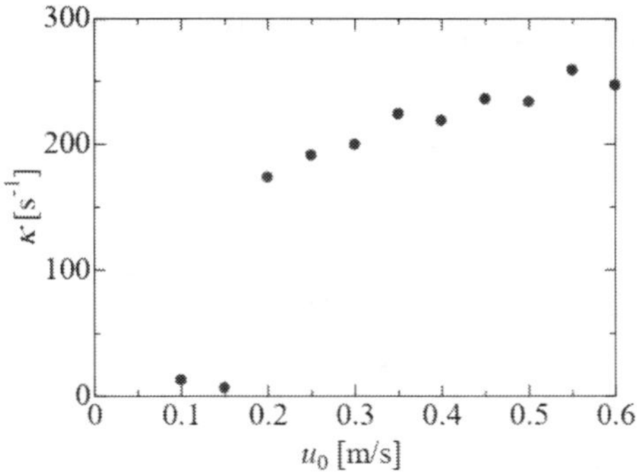

Figure 7: Correlation between flame stretch rate and inlet velocity

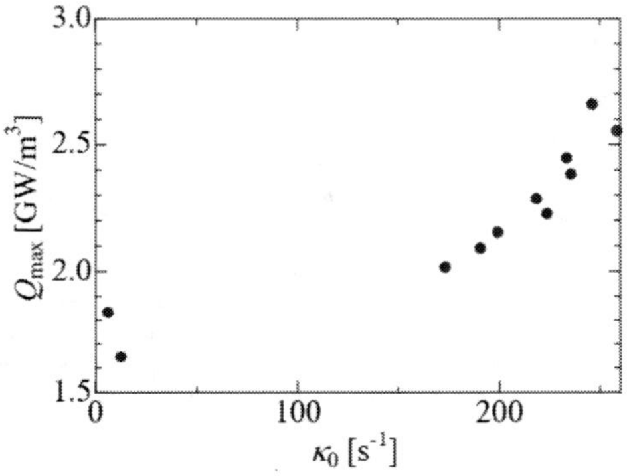

Figure 8: Correlation between flame stretch rate and maximum heat release rate

4 Conclusions

The numerical simulation of methane triple flame is conducted with chemical kinetics model. The following conclusions were derived:

(1) The triple flame structure is observed, but the tree branch cannot be distinguished clearly in this condition.

(2) The maximum heat release rate is smaller than that in the stoichiometric laminar premixed flame. On the other hand, it increases with the increase in flame stretch rate.

5 Acknowledgment

This work was supported by JSPS KAKENHI Grant Number 15H02977.

References

[1] T. Plessing, P. Terhoeven, N. Peters, M.S. Mansour, Combust. Flame, 115(1998), 335-353.

[2] N.I. Kim, J.I. Seo, K.C. Oh, H.D. Shin, Proc. Combust. Inst. 30 (2005) 367-374.

[3] G.R. Ruetsch, L. Vervisch, A. Linan, Phys. Fluids 7 (6) (1995) 1447-1454.

[4] M. Hirota, T. Yokomori, K. Yasuda, Y. Nagai, M. Mizomoto, G. Masuya, Proc. Combust. Inst. 31 (2007) 893-899.

[5] K. Yamamoto, S. Kato, Y. Isobe, N. Hayashi, H. Yamashita, Proc. Combust. Inst. 33 (2011) 1195-1201.

[6] P.N. Kinoni, K.N.C. Bray, D.A. Greenhalth, B. Rogg, Combust. Flame 116 (1999) 192-206.

[7] H. Yamashita, M. Shimada, T. Takeno, Proc. Combust. Inst. 26(1996) 27-34.

[8] Y. Mizobuchi, J. Shinio, S. Ogawa, T. Takeno, Proc. Combust. Inst. 30 (2005) 611-619.

[9] S. Kostra Jr., W.F. Carnell Jr., M.W. Renfro, Combust. Flame, 154 (2008) 82-95.

[10] W.M. Cha, S.H. Chung, Proc. Combust. Inst. 26 (1996) 121-128.

[11] F. Bisetti, S.M. Sarathy, M. Toma, S.H. Chung, Combust. Inst. 35 (2015) 1023-1032.

[12] K. Prasad, E.W. Price, Combust. Flame 90 (1992) 155-173.

[13] T. Echekki, J.H. Chen, Combust. Flame 114(1998) 231-245.

[14] M.D. Smooke, Y. Xu, R.M. Zurn, P. Lin, J.H. Frank, M.B. Long, Proc. Combust. Inst. 24 (1992) 813-821.

[15] R.J. Kee, et al., Sandia Report, SAND 86-8246 (1986).

[16] R.J. Kee, et al., Sandia Report, SAND 89-8009 (1989).

[17] M.D. Smooke, Reduced Kinetics Mechanisms and Asymptotic Approximations for Methane-Air Flames, Springer-Verlag, (1991) 1-28.

[18] N. Hayashi, C. Miyata, H. Yamashita, K. Yamamoto, Progress in Computational Fluid Dynamics, 10 (2010) 225-231.

11th Asia-Pacific Conference on Combustion,
The University of Sydney, NSW Australia
10th -14th December 2017.

Combustion and Reformulation Characteristics of Biogas in Plasma Assisted Combustion Using Microwave Induced Non-Equilibrium Plasma

T. Yamamoto[1], S. Uchiyama[1], H. Matsune[1], M. Kishida[1]

[1]Department of Chemical Engineering, Kyushu University

744 Motooka, Nishi-ku, Fukuoka, 819-0395, Japan

Abstract

To utilize biogas validly, we have applied plasma assisted combustion using microwave to the combustion and reformulation of biogas. In this study, the characteristics of combustion and reformulation of biogas have been investigated by comparing of large and small nozzle. In case of small nozzle, the combustion and reformulation of biogas stably occurs over a wider equivalence ratio area than that in case of large nozzle, because the damping of microwave is inhibited. N_2 and OH radical decrease, and H_α, CO, CH and C_2 radicals increase as equivalence ratio. H_2 and CO concentrations increase, and CO_2 concentration decreases with increasing equivalence ratio. In this method, since the reaction of $CO_2 + e \rightarrow CO + O$ occurs, the amount of CO_2 of produced gas is less than the amount of supplied CO_2 in high equivalence ratio region. Cold gas efficiency increases with increasing H_2 and CO concentrations. When CH_4 is completely reformed into H_2 and CO, cold gas efficiency is about 95 %. However, the maximum value of cold gas efficiency is approximately 110 % at small nozzle, because CO_2, which does not have calorific value, is decomposed into CO, which has calorific value.

1 Introduction

Recently, a demand for the beneficial use of low grade fuels and biomass has increased through a perspective of alternative fuel sources and greenhouse gas reduction. There are a lot of efforts to expand the use of low grade fuel and biomass, such as tree prunings, waste plastic, plant waste, landfills, agricultural residues, and non-recycle paper, all over the world. Although there is a huge amount of low grade fuels and biomass, the poor stability of ignition and combustion, and low calorific value of low grade fuels and biomass make it difficult to apply to a conventional energy utility system. In order to utilize low grade fuel and biomass effectively, it is necessary to develop the promotion technology of the ignition and combustion of low grade fuel and biomass.

Plasma assisted combustion using non-equilibrium plasma [1-7] has considerably been studied by researchers all over the world. A pulse discharge is commonly applied in the area of non-equilibrium plasma assisted combustion [1, 2]. A pulsed corona discharge technology was used to remove tar from biomass derived fuel gas [3] and a glow discharge was used to investigate the plasma chemistry of the excited species [4]. The plasma assisted combustion by microwave plasma under atmospheric pressure was studied to reform various hydrocarbon fuels [5, 6] and investigate the combustion chemistry [7]. Electron is activated and excited by energy of electrical discharge, and it is deactivated by the collision with neutral particle during its drift in non-equilibrium plasma. The lifetime of non-equilibrium plasma under low pressure is longer than that under atmospheric pressure, because the probability of collision between electrons and neutral particles under low pressure is lower than that under atmospheric pressure. Furthermore, non-equilibrium plasma using microwave under reduced pressure has high plasma density and high uniformity, because microwave plasma is created by the electromagnetic wave of 2.45 GHz. When non-equilibrium plasma using microwave is overlapped with flame at reduced pressure, greater enhancement of ignition and combustion is expected to be brought by non-equilibrium plasma assisted combustion method using microwave.

There have been little studies that applied the plasma assisted combustion using microwave plasma to the low grade fuel. Therefore, we have applied non-equilibrium plasma using microwave under reduced pressure to the biogas to improve the combustion and reformulation states [8, 9]. In the present work, an experimental study has been conducted to investigate the combustion and reformulation characteristics of the biogas.

2 Experimental

Figure 1 shows the schematic diagram of experimental apparatus. Table 1 and 2 show the equipment specifications and the experimental conditions. The experimental apparatus is composed of a microwave generator (NISSHIN, MPS-15D), a diaphragm pressure gauge (ULVAC, GP-1000), a rotary vacuum pump (ULVAC, VD401), a double pipe condenser, and a quartz reaction tube. The quartz reaction tube is 50 mm in diameter and 1000 mm in length. The pressure in the quartz reaction tube, which is monitored by a diaphragm pressure gauge, is reduced by a rotary vacuum pump and the quartz reaction tube is irradiated with a microwave. The microwave induced non-equilibrium plasma is generated in the quartz reaction tube. Air is fed into microwave induced non-equilibrium plasma and air non-equilibrium plasma is created. Biogas (CH_4:60%, CO_2:40%) is used as fuel in this study. After biogas is mixed with air in mixer, mixed gas of fuel and air is supplied from the combustion nozzle. The ignition of biogas automatically occurs due to air non-equilibrium plasma and the combustion of biogas initiates [8, 9]. Two cylindrical nozzles, which are 1 and 3/4 inches in diameters, are used in this study. To maintain equal flow rate at nozzle exit, the total flow rates

Corresponding author. Fax: +81-92-802-2752
E-mail address: yamamoto@chem-eng.kyushu-u.ac.jp

of fuel and air at large and small nozzle are constant rates of 2.88 and 1.62 L/min, respectively. The equivalence ratio varies from 0.8 to 8.8 with a change in the ratio between fuel and air. The pressure in the quartz reaction tube is at a constant pressure 4.0 kPa. The microwave input power is 700 W.

The emission spectra of flame are analyzed by a polychromator (BUNKO-KEIKI Co. MK-300) at 3 observation windows, which are from 10 mm to 83.8 mm downstream of the nozzle at intervals of 36.9 mm. When the spectra of the entire area of visible light are measured, the emission spectra are analyzed between 130 and 740 nm with a focus on 435.8 nm by the diffraction grating with 150 lines/mm. The exposure times are 0.01 seconds at intervals of 0.029 seconds. The emission spectra are measured 100 times per condition and are averaged out 100 measurements.

The produced gas volume is measured at the exit of rotary vacuum pump by a gas meter (SHINAGAWA Co. DC-1). The compositions of produced gas are detected at the exit of rotary vacuum pump by gas chromatograph (SHIMADZU GC-2014AT TCD; Molecular Sieve 13X, Porapak N). The detected compositions of produced gas are H_2, N_2, O_2, CO, CO_2, CH_4, C_2H_2, C_2H_4 and C_2H_6.

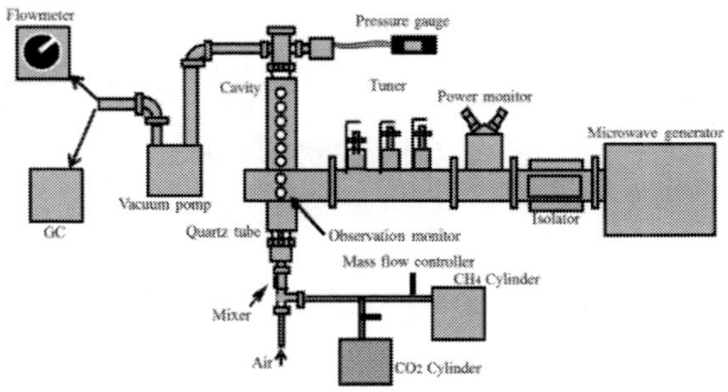

Figure 1 Schematic diagram of experimental apparatus.

Table 1 Equipment specifications.

Microwave generator	
Input power	4.5 kVA
Oscillation frequency	2.455 GHz ± 20 MHz
Output range	200 – 1500 W
Rotary vacuum pump	
Pumping speed	800 L/min
Heat exchanger	
Double-pipe type temperature	283 – 293 K

Table 2 Experimental conditions.

Microwave input power	700 W
Pressure	4.0 kPa
Fuel	Biogas (CH_4:60%, CO_2:40%)
Nozzle size	1, 3/4 inches
Total flow rate	2.88, 1.62 L/min
Equivalence ratio	0.8 – 8.8

3 Results and discussion

Figure 2 shows the emission spectra between 200 and 700 nm of the plasma assisted combustion of biogas using microwave under equivalence ratios 1.6, 4.0 and 8.0 at small nozzle. The emission intensities of various radicals, such as N_2, H_α, CO, CH, OH and C_2 radicals, are detected. In equivalence ration 1.6, N_2 and OH radical are mainly detected as shown in Fig. 2 (a). The emission intensities of N_2 and OH radicals have tendency to decrease with increasing equivalence ratio, because the amount of supplied air, which includes N_2 and O_2, decreases. The emission intensities of H_α, CO, CH and C_2 radicals have tendency to increase as equivalence ratio, because the amount of supplied biogas, which includes CH_4 and CO_2, increases.

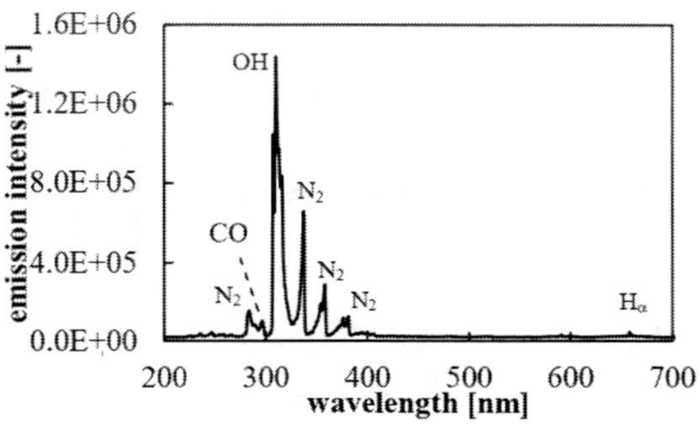

(a) equivalence ratio 1.6

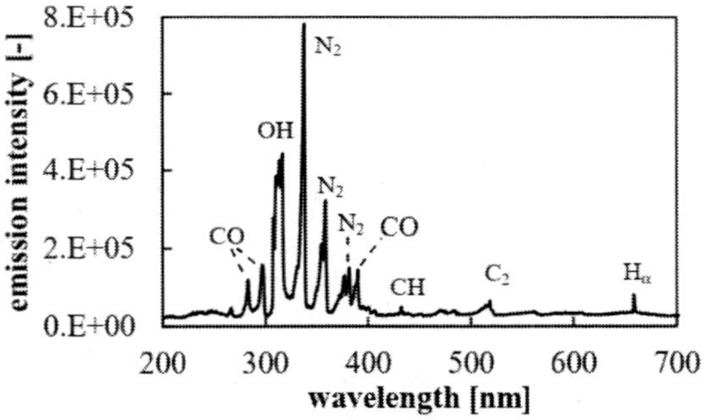

(b) equivalence ratio 4.0

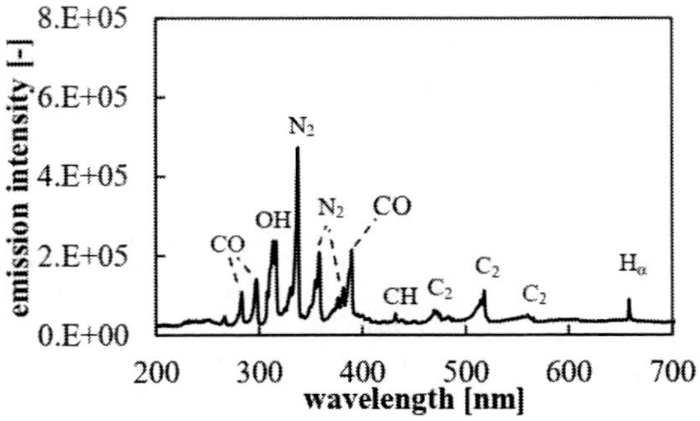

(c) equivalence ratio 8.0

Figure 2 Emission spectra of the plasma assisted combustion of biogas using microwave.

Figure 3 shows the effect of equivalence ratio for mole fractions of H_2, CO, CO_2 and cold gas efficiency under large and small cylindrical nozzle. Here, the cold gas efficiency is the ratio between low calorific value of combustion exhaust gas and low calorific value of fuel. The results in Figs. 3 are plotted at equivalence ratio ranges of 0.8 – 8.8. Because the experiments were stopped to confirm that the combustion state becomes worse at high equivalence ratio condition. The mole fractions of H_2 and CO increase, and the mole fraction of CO_2 decreases with increasing the equivalence ratio in both nozzle. The concentration of CO_2 at large nozzle and small nozzle are lowest values at around equivalence ratios 4.0 and 6.0, respectively. After lowest value, CO_2 concentration increases with increasing equivalence ratio at large nozzle. On the other hand, CO_2 concentration maintains until equivalence ratio 8.0 and increases at equivalence ratio 8.8 at small nozzle. Therefore, the experiments were stopped at equivalence ratio 8.8. The cold gas efficiency increases with increasing the equivalence ratio in all cases, because the production amount of combustible gas, such as H_2 and CO, increase with increasing the equivalence ratio. In case of large nozzle, cold gas efficiency is the maximum value, which is approximately 93 %, at equivalence ratio 4.4 and then decreases with increasing equivalence ratio. In case of small nozzle, cold gas efficiencies are more than 105 % at equivalence ratio more than 5.6. The maximum value of cold gas effieicncy is approximately 110 % at equivalence ratio 8.0. When CH_4 is completely reformed into H_2 and CO by the reaction (R1), cold gas efficiency is approximately 95 %. Cold gas efficiency at small nozzle is larger than that at large nozzle, and its value is more than 95 %. Because CO_2, which does not have calorific value, is decomposed into CO, which has the calorific value, to occur the electron impact reaction (R2) in this method. In particular, CO_2 is decomposed nearly 90 % under high cold gas efficiency conditions. Furthermore, O radical produced by reaction (R2) is used the oxidation of CH_4.

$$CH_4 + 1/2O_2 \rightarrow CO + 2H_2 \quad \text{(R1)}$$

$$CO_2 + e \rightarrow CO + O \quad \text{(R2)}$$

In this study, the plasma assisted combustion using microwave is conducted by irradiating the flame with the microwave from one side. Microwave power is damped with moving forward in the flame. Plasma density and electron temperature decrease with moving forward in the flame, and the ununiformities of plasma density and electron temperature are yielded in radial direction. In case of large nozzle, the flame thickness increases to the direction of forward movement of microwave, and the damping of microwave is larger than that in case of small nozzle. As a consequence, the effect of combustion enhancement becomes large with decreasing the nozzle size to inhibit the damping of microwave.

4 Conclusions

In this study, to utilize low grade fuel validly, plasma assisted combustion using microwave has been applied to the combustion and reformulation of biogas. The combustion and reformulation of biogas maintains good condition over a wide range of equivalence ratio. N_2 and OH radical are mainly detected in equivalence ratio 1.6 and have tendency to decrease as equivalence ratio. On the other hand, H_α, CO, CH and C_2 radicals have tendency to increase as equivalence ratio. The concentrations of H_2 and CO increase with increasing the equivalence ratio. The concentration of CO_2 decreases with increasing the equivalence ratio. The amount of CO_2 of

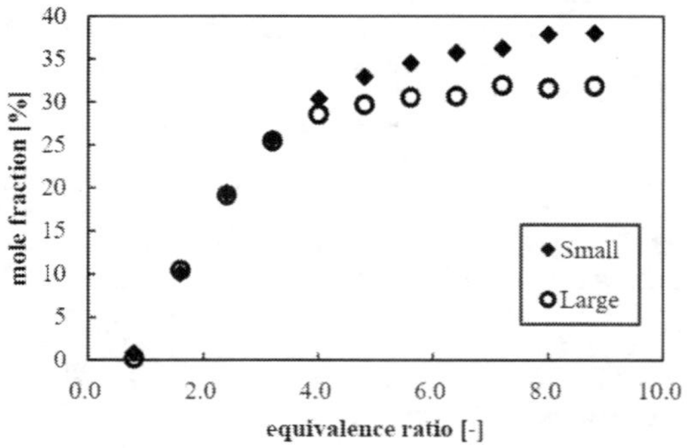

(a) H_2

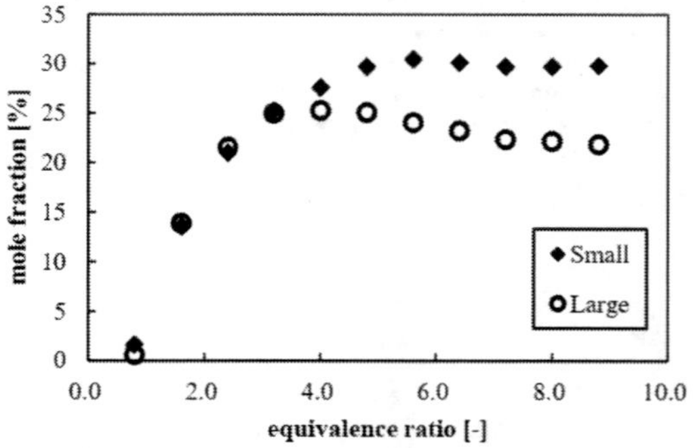

(b) CO

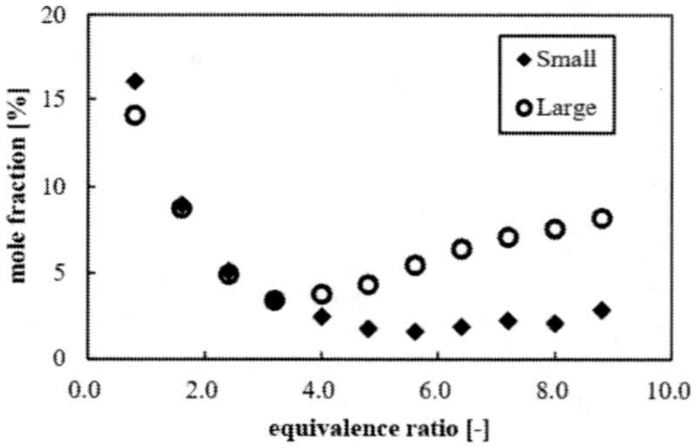

(c) CO_2

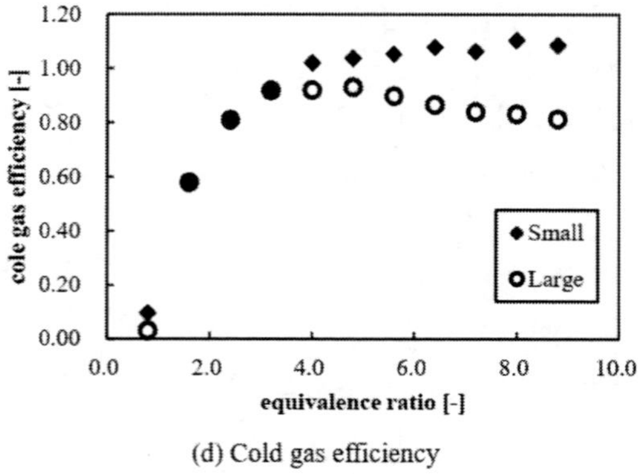

(d) Cold gas efficiency

Figure 3 Effect of equivalence ratio for mole fractions of H_2, CO, CO_2 and cold gas efficiency at large and small nozzle.

produced gas is less than the amount of supplied CO_2 in high equivalence ratio, because CO_2 is decomposed into CO by microwave plasma. Cold gas efficiency increases with increasing the combustible gas, such as H_2 and CO. In case of small nozzle, biogas is stably burned over a wider equivalence ratio than that in case of large nozzle. This is why the damping of microwave is inhibited. When CH_4 is completely converted into H_2 and CO, cold gas efficiency is approximately 95 %. However, in this method, since CO_2, which does not have calorific value, is decomposed into CO, which has calorific value, the maximum value of cold gas efficiency is approximately 110 %.

References

[1] A.Y. Starikovskii, Proc. Combust. Inst. 30 (2005) 2405-2417.

[2] Y. Zuzeek, I. Choi, M. Uddi, I.V. Adamovich, W.R. Lempert, J. Phys. D: Appl. Phys. 43 (2010) 124001.

[3] S.A. Nair, K. Yan, A.J.M. Pemem, van E.J.M. Heesch, K.J. Ptasinski, A.A.H. Drinkenburg, Ind. Emg. Chem. Res. 43 (2004) 1649-1658.

[4] D.Z. Pai, G.D. Stancu, D.A. Lacoste, C.O. Laux, Plasma sources Sci. Technol. 18 (2009) 045030.

[5] T.S. Kim, S. Song, K.M. Chun, S.H. Lee, Energy 35 (2010) 2734-2743.

[6] Y.F. Wang, C.H. Tsai, W.Y. Chang, Y.M. Kuo, Int. J. Hydrogen Energy 35 (2010) 135-140.

[7] X. Rao, K. Hemawan, I. Wichman, C. Carter, T. Grotjohn, J. Asmussen, T. Lee, Proc. Combust. Inst. 33 (2011) 3233-3240.

[8] T. Yamamoto, R. Imazato, T. Yamamoto, R. Tanaka, Energy Fuels 27 (2013) 2283-2289.

[9] T. Yamamoto, T. Tsuboi, Y. Iwama, R. Tanaka, Energy Fuels 30 (2016) 3495-3501.

11th Asia-Pacific Conference on Combustion,
The University of Sydney, NSW Australia
10th -14th December 2017.

Dual-Angle Micro-Particle Tracking Velocimetry in the Primary Atomization Zone of Electrostatically Charged Diesel Sprays

Phuong X. Pham; Agisilaos Kourmatzis; Shehzaib Y. Khan; and Assaad R. Masri

School of Aerospace, Mechanical and Mechatronic Engineering,

The University of Sydney, NSW 2006, Australia

Abstract

Electrically insulating (dielectric) charged sprays have been experimentally investigated for commercial Diesel fuel using a charge injection atomizer with a copper electrode. Application of the dual-angle imaging technique has allowed, for the first time, a detailed characterization of the primary atomization zone of electrified sprays. The paper first presents the global structure including spray cone angle and the onset of spray dispersion, showing agreement with previous work. This is followed by new quantitative information on (i) liquid jet tip breakup location, (ii) initial size of non-spherical droplets generated from the jet tip and (iii) the velocity of produced droplets relative to that of liquid jet tip. Global images of spray structure are collected using a CCD camera while microscopic information is quantified using a dual-angle particle tracking velocimetry technique developed previously by the authors [1,2]. In agreement with previous work, the charge creates electrostatic repulsion between the surface and the tip of the liquid jet and this significantly affects the cone angle and breakup length. Findings on the relative velocity of atomized droplets to that of the liquid jet tip demonstrate an effect of the initial electrostatic charge present on the liquid jet surface on the subsequent velocity of generated droplets. The applied voltage significantly affects the relative position of break-up to dispersion onset. However, for the charge levels examined here, a minimal effect of charge on droplet size is noted.

1 Introduction

Electrostatic charging of liquid fuels may have a significant role to play in the context of developing combustors with improved control of dispersion characteristics, break-up location, and droplet size. This is particularly true in the context of micro to meso-scale combustion, given that the power requirements of electrostatic atomization can be as low as just a few milliWatts [3]. The presence of electric charge on a liquid jet has been known for some time to improve the primary atomization and dispersion process [4-6], having been used both to enhance spraying performance of electrically conducting liquids [7,8] and electrically insulating liquids [6,9]. The former type of spray is most commonly referred to as an 'electrospray' and in the context of atomizing commercial fuels such as Diesels, kerosene or newer synthetic fuels, electrospraying is only feasible with the use of conductive additives, such as STADIS [10].

On the other hand, Electrohydrodynamics (EHD, also may be referred to as a charge injection atomization) operates without the requirement of high fuel electrical conductivity. A very high electric field (MV/m) applies potentials of up to 10-20kV across two electrodes typically separated by a space as small as 100μm-250μm, known as the 'inter-electrode gap'. In such an atomizer, the charging and grounded electrodes are fully immersed in the working fluid. With a flow imposed to the charged fuel, electric charge is advected outside of the atomizer, through the inter-electrode gap, prior to dielectric breakdown, which would occur where the charge is stagnant between two electrodes. This forced advection is made possible by drilling an orifice into the grounded electrode, resulting in typical orifice mean velocities of up to 10m/s. Recent work has extended EHD sprays to air assisted-EHD spraying at higher velocities [12]. A schematic of a charge injection atomizer is shown in Fig. 1. This is a 'third generation' atomizer [12] following on from decades of research in charge injection atomizer development by many groups [9, 11-14].

To date, work on charge injection atomizer development has focused on measurement and analysis of break-up length and droplet size for both continuous flow or 'steady flow' atomizers [11]. This has used a variety of electrode configurations and orifice hole patterns [15] including pulsed flow atomizers [3,16]. Recent work also demonstrated a link between the electrohydrodynamic flow in the inter-electrode gap and the primary atomization zone [17]. However, it still remains unclear under which conditions electrohydrodynamic instabilities can be advected to the primary atomization zone and whether this process would influence droplet size.

With the recent development of a dual-angle particle tracking velocimetry technique by the current authors [1,2], detailed measurements of the primary atomization zone in electrically charged sprays are now possible. Quantitative information on ligament and droplet dynamics, as well as measurement of the primary instability wavelengths and their relation to subsequent droplet formation can be measured accurately for a variety of conditions. This contribution presents an initial effort towards obtaining improved fundamental insights on the primary atomization zone of electrified fuel sprays through measurement of both local velocity and volume of liquid structures as well as more accurate measurements of break-up length. Such results are also necessary for the development of computational models used in simulation of the atomization of electrified fuels [18].

2 Experimental Description

A new charge injection atomizer, herein referred to as an electrohydrodynamic (EHD) atomizer is schematically shown in Fig. 1 and has been designed and manufactured at The University of Sydney.

Corresponding author.
E-mail address: x.pham@sydney.edu.au

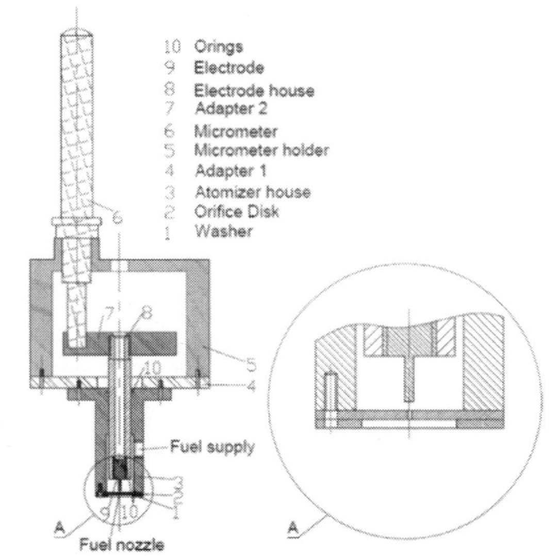

Fig. 1. Assembly of a charge injection atomizer with a plane-plane electrode layout used in this study. Inset-A shows an enlarged view of the tip.

This includes a negative electrode (#9), a grounded electrode or a disk with a central nozzle (#2) and other components shown in Fig. 1. The electrode housing (#8) was made of Perspex in order to assist future visualization of the charging process occurring between the electrode and the disk. The most important controlling parameters used in the EHD atomizer here include the electrode material, the gap between negatively charged electrode tip to the disk (grounded electrode) surface (L), the applied voltage, and the fuel physical and electrical properties. In the initial study reported here, both electrodes were made of copper, the nozzle diameter (D) is 250μm, and the normalized electrode gap (L/D) is 2.35, with the liquid tested being commercial Diesel. Macro spray structure information to be presented includes spray cone angle and the location of onset of dispersion. Micro atomization characteristics include the jet breakup location, the fragment velocity and the fragment volume. Macroscopic images are taken using a CCD camera while a dual angle particle tracking velocitmetry (PTV) reported in [1,2] was adopted to measure the liquid jet and droplet characteristics.

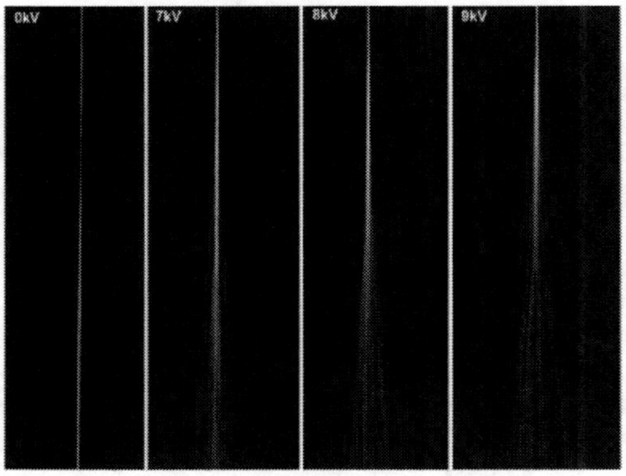

Fig. 2. Sprays of Diesel. Dj=250μm; U_L=3.4m/s; L/D=2.5; left to right: applied voltage = 0, 7, 8 and 9kV, respectively.

Samples of the spray images captured by the CCD camera are shown in Fig. 2. From left to right, the applied voltages are 0, 7, 8 and 9kV. A Matlab code has been developed to binarize the images taken by the CCD camera and quantify the locations where the spray starts to disperse (referred to hereon as: onset of dispersion) and the spray cone angle. This is a standard technique which has also been employed earlier to quantify breakup length of EHD sprays [12,21]. However, the breakup length obtained using this technique is limited to a qualitative interpretation, due to the low resolution of the images. In this paper, the break-up length marked by the first liquid jet breakup location is measured using the dual-angle PTV technique developed in [1,2]. The macroscopic CCD images are used only to measure the onset of dispersion and the cone angle.

The dual-angle particle tracking velocimetry technique includes two lasers, two lenses and two cameras both operated in a particle imaging velocimetry (PIV) mode and synchronized to the lasers using a LaVISION high-speed controller [1,2]. Both cameras have the same fields of view with a spatial resolution of 4.1μm/pixel and a field of view of 3.15mm. Figure 3 shows an example of one realization captured by both cameras. Each camera, under PIV mode, collects two images at two different times t_1 and t_2, respectively, and this leads to a set of four 2D-images per realization captured by both cameras. The resulting four 2D-images will be referred to as an `imaging set'. In each imaging set, two images captured at the same time form an imaging pair (eg. images in column 1 of Fig. 3) representing the front and side views of the liquid fragments. Measurement of the locations of a particular fragment at two different times (t_1 or t_2, respectively) allows computation of fragment velocity from the known delay time, $dt=t_2-t_1$. The image processing script, tracking locations of fragments, makes this a particle tracking velocimetry (PTV) method. Further details on the PTV technique can be found elsewhere [1,2].

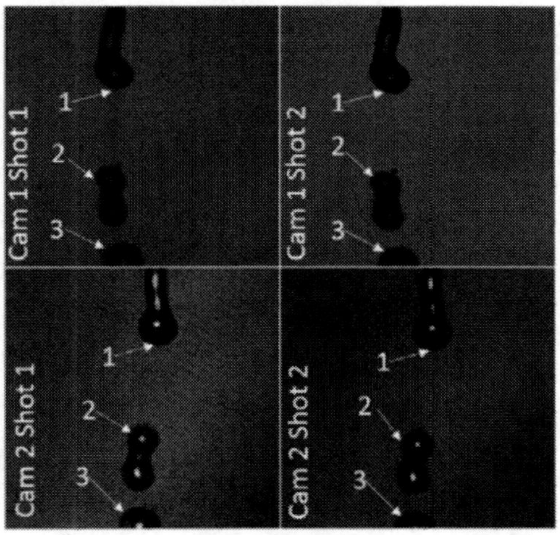

Fig. 3. Example of an imaging set taken using the particle tracking velocimetry setup described elsewhere [1,2] (#1 represents the liquid jet tip; #2 & #3 are fragments generated from an EHD atomizer used in this study; D_J=250μm; U_L=3.4m/s; L/D=2.5; applied voltage=10kV).

In this contribution, the PTV technique will distinguish and track the liquid jet (#1 shown in Fig. 3) and fragments generated from the jet (#2 and 3 shown in Fig. 3). Thousands of "imaging sets" have been collected at the spray centre line and different downstream locations. The first frame upstream was carefully located so that most the liquid jet tip appears in that frame. It is noted that "liquid jet tip" and "'jet tip location" are used throughout this work as synonyms to represent the lowest vertical position of the boundary of liquid jet tip (#1 shown in Fig. 3). This location is tracked to quantify the tip velocity. Other frames downstream are equally spaced 3mm apart from the first frame and as such the frames cover most of the atomization events in the near field of the sprays. Only the first frame is analysed in this current work with the aim of developing a link between fragments and the jet tip, however further analysis for frames downstream will give supplementary information and this will be subject to a further study.

3 Results and discussion

3.1 Spray breakup length, onset of spray dispersion and spray cone angle.

In this and the following sections, macro information such as onset of dispersion and spray cone angle measured using the CCD camera and micro information obtained by the PTV technique (droplets size, jet tip location, and break up length) will be presented. As mentioned earlier, the jet tip location is identified as the lowest vertical point in the boundary of the liquid jet (#1 shown in Fig. 3). The location is also used to quantify the tip velocity (U_T). Fig. 4 shows results of the probability distribution (PD) of jet tip location under four selected different applied voltages (0, 7, 8 and 10 kV). As can be seen, the jet tip locations span a wide range of axial positions with the distinguished peak of those PDs being an appropriate measure of the mean jet breakup length. This technique also enables quantification of the standard deviation of the break-up length. It is also evident that the peak location in the distribution of axial position of the tip is significantly affected by the level of applied voltage. Under these testing conditions, the peak location shifts closer to the nozzle exit plane with increasing applied voltage, which is to be expected. However, it is of particular interest to note that the PD is consistently wider for the charged cases compared to the case without charge. This is likely an effect of electrostatic repulsion in the primary breakup zone creating instability on the liquid jet surface. This repulsion would occur due to an interplay between electrostatic forces and the surface tension force, the latter of which tries to `recede' the liquid jet as has been discussed in [19,20]. Further analysis of the variation of this break-up location is clearly merited given the significant impact that it would have on flame stabilization and height, particularly if such an atomizer were to be applied in a micro-combustion environment.

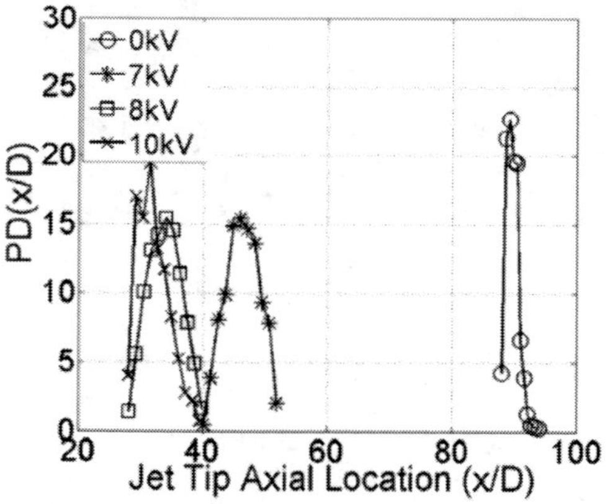

Fig. 4. Probability density (PD) of liquid jet tip's axial location (x/D). (D=250μm; U_L=3.4m/s; L/D=2.5)

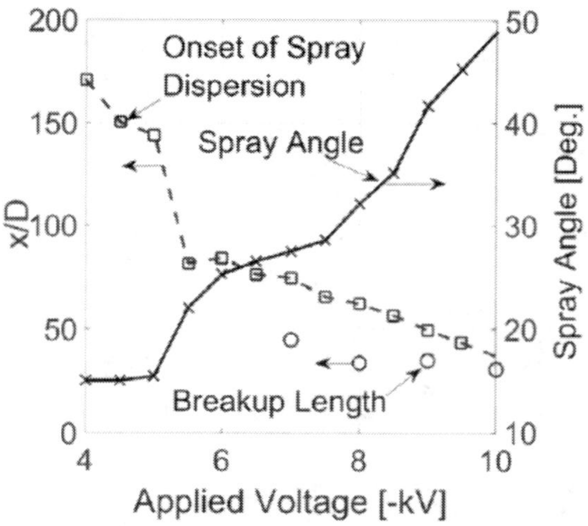

Fig. 5. Onset of spray dispersion (start of dispersion), spray angle and breakup length versus applied voltage. (Dj=250μm; U_L=3.4m/s; L/D=2.35)

The peak locations shown in Fig. 4 for PD of jet tip location are used as markers for the jet breakup length summarized in Fig. 5 for the various conditions studied, along with the spray cone angle and locations of dispersion onset measured using the CCD camera's images for all of the sprays. It is clear that the applied electric charge significantly affects the dispersion location, the breakup location, as well as the spray cone angle. These are all in agreement with previous findings [11, 21]. The cone angle is significantly increased with applied voltage due to the mutual electrostatic repulsion occurring between the surface and the tip of liquid jet. An example of the repulsion is shown in Fig. 3 where the jet tip in the top two frames is bent in contrast to any image these authors examined from an uncharged case where the liquid jet was consistently vertical. Detailed analysis of this primary atomization zone will be investigated in the future.

A particularly interesting result from Fig. 5 is that the variation in the location of the onset of dispersion with respect to the initial breakup location. This difference decreases with applied voltage, which can most likely be attributed to electrostatic repulsion, as discussed above. A higher applied voltage leads to stronger repulsion and, as such, the spray will start to disperse at a location closer to the breakup point.

3.2 Size of droplets

The PTV technique is capable of measuring fragment sizes and velocity [1,2] and Fig. 6 shows the probability density (PD) of drop size (arithmetic mean size, d10) of the EHD sprays tested here versus applied voltage. It is notable here that the drop size information shown in Fig. 6 is measured at the location where the liquid jet tip most frequently appears. Fig. 6 shows that the droplets formed by the EHD sprays tested here are nearly identical in size to that of an uncharged mono-dispersed spray despite the significantly improved spray dispersion and the jet break-up length as will be shown in the following section. The negligible effect on droplet size is mostly due to the low specific charge levels, which can result in droplet populations that are of the same order as the orifice size [11] but this must be investigated further for a wider range of charge conditions. The small sibling droplets, shown in the first peak of the PDs shown in Fig. 6, are formed during the Rayleigh breakup process.

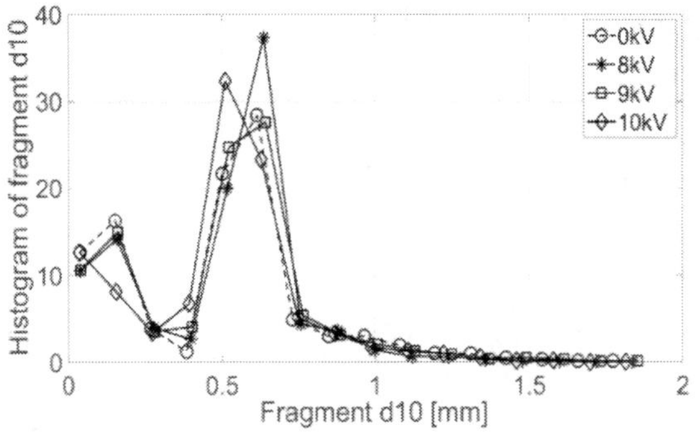

Fig. 6. Histogram of fragment diameter (d10). (D=250μm; U_L=3.4m/s; L/D=2.5)

3.3 Fragment velocity

Figure 7 shows the normalized velocity of droplets (U_{Frg}) to that of the jet tip (U_T). Although the charge range tested here is not wide enough to cover the range from 0 to 10kV, the relative velocity seems to increase with electrostatic charge under these conditions. Under dripping mode break-up (which would be relevant when no charge is applied, 0kV), the conventional spray shows a relative velocity of almost unity and this is in direct agreement with Shinjo et al. [19]. When a charge is applied, however, the fragments show some acceleration with respect to the liquid jet tip, which is evidently due to the mutual repulsion of electrostatic charge between the parent jet and the generated residual, resulting in a local Lorentz force at the primary tip break-up location. At higher charge levels, the relative velocity seems to decrease again and this could likely be attributed to the added sinusoidal instability which is imparted on to the liquid jet at higher charge levels, thereby increasing the radial fragment velocity with respect to the axial velocity. Again, this needs to be further investigated with a wider range of applied voltages, and through closer analysis of relative velocity magnitudes and both radial and axial electrostatic repulsion as opposed to axial repulsion only.

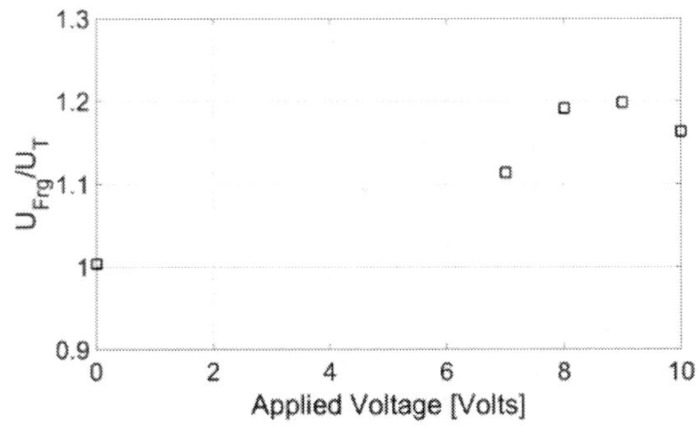

Fig. 7. Normalized velocity of fragments and liquid jet tip versus applied voltage. (Dj=250μm; U_L=3.4m/s; L/D=2.35)

4 Conclusions

Macro and micro characteristics of EHD Diesel sprays have been investigated successfully in this study. Results from this particular study demonstrate that in addition to the electric charge significantly affecting cone angle and break-up length, which is known, a significant influence on the relative velocity generated of droplets to that of jet tip was also noted. The position of dispersion onset relative to the break-up length also varies with applied voltage. Future work will include analysis of a wider range of electric charge levels, measurement of the spray and injected current as well as extension of this work to biodiesel component fuels (fatty acid methyl esters), ultimately aiming towards development of stable combustion of electrically charged insulating fuels.

References

1. P.X. Pham. A. Kourmatzis and A.R. Masri. 20th Australasian Fluid Mechanics Conference, Perth, Australia Dec. 2016

2. P.X. Pham. A. Kourmatzis and A.R. Masri. Accepted (DOI: 0.1088/1361-6501/aa8893), Measurement Science and Technology, 2017

3. A. Kourmatzis and J.S. Shrimpton,, Atomization and Sprays, 22(4) (2012) 351-370

4. Lefebvre, A.H. "Atomization and Sprays", Taylor and Francis, 1989

5. A.J. Kelly, Journal of the Institute of Energy, 57(431) (1984) 312-320

6. K. Kim and R. Turnbull, Journal of Applied Physics, 47 1964(1976)

7. A. Gomez and K. Tang, Physics of Fluids, 6 404(1994)

8. D.C. Kyritsis, B. Coriton, F. Faure, S. Roychoudhury, and A. Gomez, Combustion and Flame, 139(1-2) (2004) 77-89

9. J.S. Shrimpton, Atomization and Sprays, 11(4) (1995) 365-396

10. E.K. Anderson, J.A. Koch, and D.C. Kiritsis, Combustion and Flame, 154(3) (2008) 624-629

11. J.S. Shrimpton and A.J. Yule, Experiments in Fluids, 26(5) (1999) 460-469

12. A. Kourmatzis, E.L. Ergene, J.S. Shrimpton, D.C. Kyritsis, F. Mashayek and M. Huo, Experiments in Fluids, 53(1) (2012) 221-235

13. G. Malkawi, A.L. Yarin, F. Mashayek, Journal of Applied Physics, 108, 064910 (2010)

14. G. Al-Ahmad, J.S. Shrimpton, E.L. Ergene and F. Mashayek, Atomization and Sprays, 19(6) (2009) 547-566

15. A Kourmatzis, J Allen, JS Shrimpton, Atomization and Sprays, 20(4) (2010) 269-280

16. A. Kourmatzis and J.S. Shrimpton, Journal of Electrostatics, 69(3) (2011) 157-167

17. A. Kourmatzis and J.S. Shrimpton, Experiments in Fluids, 55:1688 (2014)

18. B.P.V. Poppel, O. Desjardins, J.W. Daily Journal of Computational Physics, 229(20) (2010) 7977-7996

19. J. Shinjo and A. Umemura, International Journal of Multiphase Flow 36(7) (2010) 513-532

20. A. Umemura, Journal of Fluid Mechanics, 757 (2014) 665-700

21. A.R.H. Rigit, J.S. Shrimpton, Atomization and Sprays, 16(4) (2006) 421-442

11th Asia-Pacific Conference on Combustion,
The University of Sydney, NSW Australia
10th -14th December 2017.

Joint Volume-Velocity Analysis of Liquid Jet and Fragment Dynamics in Air-Assisted Sprays

Phuong X. Pham; Agisilaos Kourmatzis; and Assaad R. Masri

School of Aerospace, Mechanical and Mechatronic Engineering,

The University of Sydney, NSW 2006, Australia

Abstract

A dual-angle particle tracking velocimetry (PTV) method developed by the authors has been extended in this work to study the dynamics of the liquid jet tip and the fragments generated from an air-blast atomizer. The PTV tool enables tracking of fragment locations, and quantification of their volume (V_{Frg}) and velocity (U_{Frg}) simultaneously regardless of their shapes. Results reported here include probability distributions (PDs) of jet tip locations; fragment velocity with respect to the tip velocity (U_T); the arithmetic mean fragment size ($d10=\sqrt[3]{\frac{6V_{Frg}}{\pi}}$) with respect to liquid nozzle diameter (D_j); and local Weber number of fragments. The PDs of the tip locations reported here represent a new approach to quantify the jet breakup length. It is evident that the ratios U_{Frg}/U_T and $d10/D_j$ are dependent on jet breakup modes (such as tip breakup, base breakup and surface breakup) with further investigations being required to fully understand those breakup regimes. Taking the bulk mean gas phase velocity to represent the local gas phase velocity around the fragments, a linear correlation is observed for the local Weber number of fragments and the jet global Weber number. These new quantities provide a mean for quantifying the local characteristics of liquid fragments and droplets in the primary and secondary atomization zones of sprays, leading to improved models to predict droplet size, formation rate and concentration.

1 Introduction

A key challenge in the analysis of atomizing sprays is to accurately characterize the breakup of ligaments, which are generated from primary breakup of the liquid jet. Measurement challenges are present due to the three-dimensionality of the ligaments [1] as well as the nature of the atomization process which is usually transient, stochastic [2] and can occur over micro-second time-scales. A common approach is to use information on the breakup mechanism of the liquid jet itself to define the subsequent characteristics of ligaments [3,4]. However, information on the (i) jet tip location and velocity, (ii) the relative velocity between ligaments and the primary liquid jet tip, (iii) the relative diameter between ligaments and the liquid nozzle; and (iv) the local Weber number of ligaments, is largely absent in the literature due to limitations with imaging technologies. Recent work done by the authors aims to quantify these parameters using a new dual-angle particle tracking velocimetry (PTV) technique hence enabling simultaneous volume and velocity measurement of arbitrary shapes [1].

Previous work by Salam et al. [5] focused on experimentally examining a turbulent jet spray and identified two mechanisms of ligament breakup, namely tip and base breakup. A very high probability of the tip breakup mode (90% approximately) was observed where the fragmentation occurs at the ligament tip, similarly to a liquid jet broken up under a Rayleigh breakup mode [4]. The latter mode, base breakup, occurs due to the change of flow direction at the ligament base before the tip breakup mode is triggered hence leading to the formation of large ligaments and/or droplets.

Umemura [6, 7] investigated the breakup length and time of a laminar liquid jet. They observed the recession velocity in which the jet transition from jetting-to-dripping mode with a stepwise decrease in jet issue speed and they noted this as the first Plateau-Rayleigh instability [6]. The liquid Weber number in this study was defined as $We=\rho*U_0^2*a/\sigma$, where U_0 is the jet exit velocity and "a" is the jet radius, ρ and σ are liquid density and surface tension, respectively. The jet tip recession velocity was defined in that study as $U_t = [\sigma/(\rho* a)]^{0.5}$ so that $We^{0.5} = U_t/U_o$. It is interesting to note that the occurrence of jetting-to-dripping transition was noted by Umemura at $We^{0.5}=0.8$ [6, 7].

Using pulsed shadowgraphy, Sallam et al. [8] examined the jet surface breakup mode of a round liquid jet in standard temperature and pressure air environment. They have measured droplet velocity after the turbulent primary breakup of the jet. The droplet velocity quantified in that work was the local streamwise drop surface velocity. The ratio of the surface velocity to the bulk liquid jet velocity was found to be 0.88. This is comparable to the $We^{0.5}$ value at the transition regime of a laminar liquid jet observed by Umemura [7]. Moreover, in the direct numerical simulations (DNS) of Shinjo et al. [9], the velocity ratio or square root of local Weber number, We_{local}, of fragments matched the experimental data described in Umemura [7] and Sallam et al. [8].

While it is clear that both DNS and simple experiments have addressed the need to identify local conditions in the spray, measurements of local conditions in more complex atomizing sprays, dominated by high speed non-spherical fragments are rare. This is largely due to limitations in the techniques used for characterization of atomization. Simultaneous high-speed shadowgraph and micro-PIV of in-nozzle cavitation has also been applied to investigate diesel spray nozzles [17-19]. Droplet size and distribution in more dispersed sprays can be measured using structured laser illumination planar imaging (SLIPI) [2]. However, information from single-viewing images is insufficient to characterize 3D-structures of irregular shapes like ligaments. Non-spherical fragments such as ligaments which are normally observed in atomizing zones [10] make traditional single-angle imaging techniques insufficient to characterize their 3D-structure. For example, an out-of plane ligament may appear as a drop from its top viewing angle with multi-angle imaging techniques being required to re-construct that type of non-spherical fragment. Sipperley and Bachalo [11] introduced multi-angle 3D volumetric imaging using an array of CMOS sensors. Liu et al. [12] used 4D X-ray imaging to image and reconstruct a dense fuel spray structure, however this did not entail simultaneous imaging from multiple angles.

Kourmatzis et al. [1, 10, 13] have developed an imaging tool to observe atomizing fragments from both front and side views. A particle tracking technique (PTV) capable of simultaneously measuring volume and velocity of irregular shapes has also recently been developed [14]. A two-angle holography technique has also recently been introduced by Sojka and co-workers [15]. In the contribution presented here, the PTV technique is extended to track the jet tip and fragments generated from the jet in order to examine dynamic characteristics of liquid jet tip and fragments broken up from the jet. This information can lead to significantly improved models of droplet size and formation rate given that both spatially and temporally resolved measurements of the local conditions of liquid fragments are being provided.

The overall objective of this work is to extend the capability of the PTV technique to enable joint measurements of volume and velocity of liquid fragments shedding from an atomizing jet. The specific aims of this work are to: (i) to track the jet tip location and link this to breakup length; (ii) study dynamic characteristics of the liquid jet tip and fragments generated from the jet and (iii) to quantify the local conditions of the fragments under different atomizing conditions of sprays generated using an air-blast atomizer.

2 Experiment Description

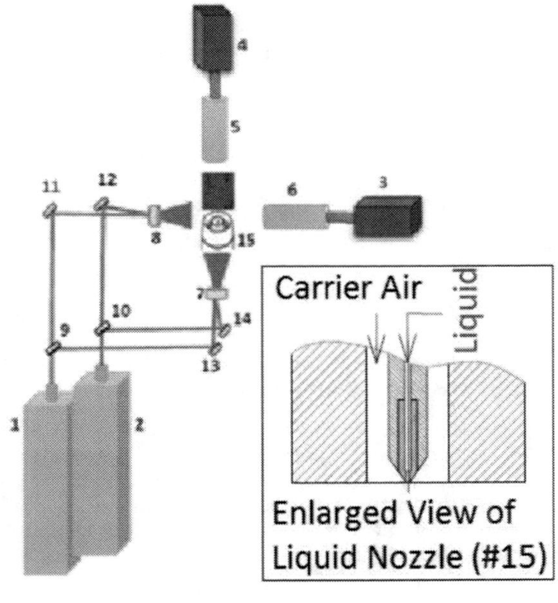

Fig. 1. Schematic of experimental setup for joint volume and velocity measurement of liquid fragments and the enlarged view of airblast atomizer used in this study. (1&2: 532nm-15W-5kHz Edgewave lasers; 3&4: High speed cameras (HSS6); 5&6: QM100 long-distance microscopes; 7&8: Opal glass-diffusing optics; 9&10: 523nm 50:50 beam-splitters; 11-14: 532nm mirrors; and 15: Atomizer and probe volume)

The experimental setup is schematically shown in Fig. 1. This layout is an extension of a one-dimensional backlight illumination system developed in [10, 13] to include two lasers, two lenses and two cameras both operated in a PIV mode and synchronized to the lasers using a LaVISION high-speed controller. Two x 5kHz Edgewave laser heads (items #1 & #2 shown in Fig. 1) are used in order to provide two 532nm-beams. Each beam is split into two using a 50-50 beam-splitter (item #9 or #10 shown in Fig. 1). The four split beams are guided into two diffusing optics (items #7 or #8), using laser-line mirrors (items from #11 to #14) and finally into the probe volume (#15). The two laser heads are typically time shifted by 5μs (for high Weber number) to 40μs (for dripping mode break-up). On the collection side of each viewing angle, there is a long-distance microscope (QM100, item #5 or #6) coupled to a LaVISION HSS6 camera (item #3 or #4) operating in PIV mode.

The sprays tested in this study are generated using an airblast atomizer and its nozzle exit's enlarged view is schematically shown in the right bottom corner of Fig. 1. Two sizes of the liquid exit nozzle diameter (Dj=250μm and 500μm, respectively) were tested in this work. The diameter of carrier air nozzle, D_a, is fixed at 12mm. Water is used in this study and sprays under different atomizing conditions tested here are summarized in Table 1. Seven different carrier air velocities (0; 20; 30; 49; 53; 64; and 76m/s respectively) are investigated for both Dj=250μm and 500μm. Three liquid loadings (1.3; 6; 10.3m/s) are used for the smaller liquid nozzle while four conditions (1.1; 3.6; 5.6; 8.3m/s respectively) were tested for the bigger counterpart. Those conditions were selected to cover atomization regimes from Rayleigh to multimode breakup.

Table 1. Sprays conditions investigated in this study

U_L (m/s)	Dj=250μm	1.3; 6; 10.3
	Dj=500μm	1.1; 3.6; 5.6; 8.3
U_G (m/s)	Both Dj=250μm & 500μm	0; 20; 30; 49; 53; 64; and 76

The two cameras have the same fields of view with a spatial resolution for both cameras at 4.1μm/pixel with a field of view of 3.15mm. Fig. 2 shows an example of one realization captured by both cameras. One camera, under PIV mode, captures two images at two different times t_1 and t_2, respectively, and this leads to a set of four 2D-images per realization captured by both cameras. These four 2D-images will be referred to as an `imaging set'. In each imaging set, two images captured at the same time form an imaging pair representing the front and side views of the fragments. Measurement of the locations of a particular fragment at two different times (t_1 or t_2, respectively) allows computation of fragment velocity from the known delay time, $dt=t_2-t_1$. The image processing script, tracking locations of fragments, makes this a particle tracking velocimetry (PTV) method.

This PTV method has been proven as a useful tool to quantify 3D-volume [1] and joint volume-velocity [14] of fragments in the near-field of atomizing sprays. One challenge of the PTV technique is matching fragments among 4 frames in each imaging set [14] and another extension has recently been done to improve this by sub-ranging the fragments according to their characteristic size [16]. The "larger" fragments (e.g. fragments with their arithmetic mean diameter, d10, >150μm) normally appear in the central core of the sprays and they are easily tracked by a 3D PTV script. Most of the smaller fragments (e.g. d10<<150μm) obtained from one viewing angle cannot be easily identified in the other imaging view which is due to occlusion [16]. Fortunately, the smaller fragments are normally spherical and they are found to be better processed with a 2D processor,

Corresponding author.
E-mail address: x.pham@sydney.edu.au (Phuong Pham)

which separately processes the PIV images taken from each camera [16].

In this contribution, the tracking function will be extended to distinguish and track the liquid jet (e.g. #1 shown in Fig. 2) and fragments generated from the jet (e.g. #2 and 3 shown in Fig. 2). A number of probing locations have been chosen at the spray centre line and different downstream locations. The first location upstream was carefully located so that most the liquid jet tip appears in that frame. Other locations down-stream are equally spaced 3mm apart from the first frame and as such the frames cover most of the atomization events in the near field of the sprays. Only images taken at the first location are analysed in this current work with the aim of developing a link between fragments and the jet tip, however further analysis for frames downstream will give supplementary information and this will be subject to a further study. The jet is identified using the highest vertical point of its boundary which is exactly the same amongst four frames in one imaging set as it is located in the top edge of those frames. The jet tip, on the other hand, is defined as the lowest vertical locations of the jet boundary and the lowest locations are also used to quantify the tip velocity.

3 Results and discussion

3.1 Jet breakup locations

Fig. 3 shows selected results of the probability distribution (PD) of jet tip locations under the same liquid jet velocity (U_L=10.3m/s) but different gas velocities (U_G=20m/s, 53m/s and 76m/s). For each spray condition, the PDs shown in Fig. 3 is output by processing 2,000 imaging sets and an approximate standard deviation of 0.7 is obtained for the x/D tip locations. As can be seen, the jet tip locations span a wide range of axial locations with the distinguished peak of those PDs being an appropriate measure of the mean jet breakup length. This technique also enables quantification of the standard deviation of the break-up length. It is also evident that the distribution in the axial location of the tip narrows at higher U_G (higher jet bulk Weber number) and this suggests that the breakup location is more consistent with higher gas velocity however further work is merited to confirm this over a wider range of conditions.

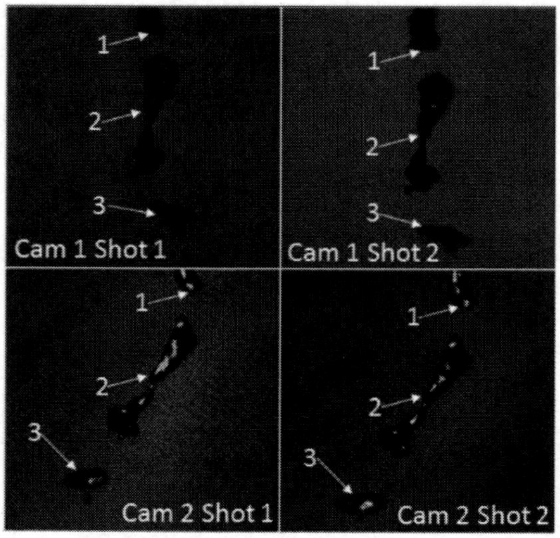

Fig. 2. Example of an imaging set taken using the particle tracking velocimetry setup shown in Fig. 1 (#1 represents the liquid jet; #2 & #3 are fragments generated from an air-assisted atomizer used in this study; D_j=250µm; U_G=49m/s, U_L=10.3m/s).

This distribution in axial tip location may be driven by changes related to the liquid jet recession velocity discussed at length by Shinzo and co-workers [6,7,9]. Unlike the case with U_G=53m/s, the PD of jet tip locations with U_G=53m/s shown in Fig. 3 does not approach zero at x/D=12. This means that some tip locations can be seen at x/D>12 confirming that the PD in the lower gas velocity case shown in Fig. 3 is wider compared to those with higher gas velocities.

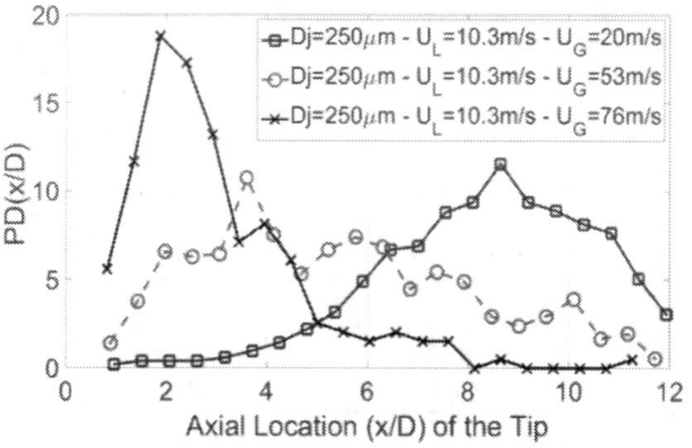

Fig. 3. Probability density (PD) of jet tip's axial location (x/D)

3.2. Relative fragment velocity and size

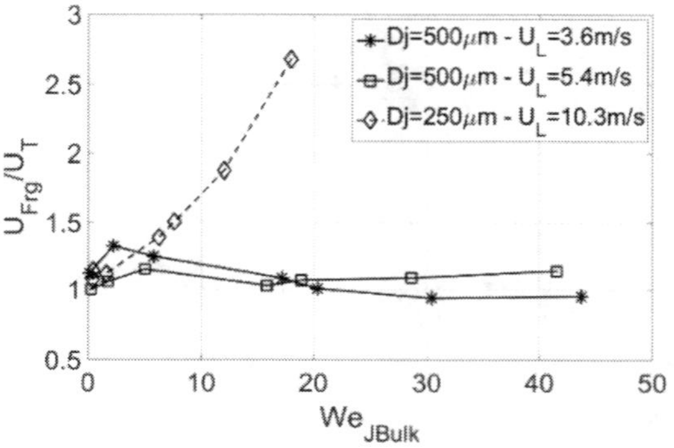

Fig. 4. Normalized velocity of fragments by jet tip velocity. (Only fragments with d10>150µm are used in this analysis)

Information quantifying the local velocity of ligaments with respect to the parent liquid jet is absent in the literature, despite the importance of liquid ligament velocity on subsequent droplet formation, local Weber numbers and the mass peeling rate, all factors which dictate the rate of vapor formation in the near-field, and therefore the onset of heat release in a reacting flow. Fig. 4 shows the average fragment velocity normalized by the tip velocity at different liquid jet velocities and bulk jet Weber numbers. It is quite clear at U_G=0m/s the fragment velocity is similar to the tip velocity and this agrees well with previous work [6-8]. With a larger diameter nozzle (D_j=500µm), the relative velocity fluctuates around 1 and even decreases slightly to about 0.9 at We_{JBulk}=45. Two scenarios are possible here: (i) with low gas velocity cases, the tip breakup mode is dominant so that the drops formed at the tip will retain the velocity of the liquid jet tip due to minimal entrainment from the surroundings (ii) at high gas velocity the liquid rapidly atomizes, generating fragments such that U_{Frg}/U_T may represent the ratio of fragment velocity to that of jet surface velocity which was reported earlier to be approximately 0.9 for a turbulent liquid jet [8]. With D_j=250µm, the experiments done are only for one liquid case, U_L=10.3m/s.

With the smaller nozzle, the base breakup is dominant, with fragments being more prone to follow the gas phase due to more entrainment compared to the case of D_j=500μm.

Fig. 5 shows the averaged fragment diameter (d10) normalized by the nozzle diameter. For further information on how the corrected dimension, therefore volume, of a fragment is determined using the dual angle technique, the reader is directed elsewhere [1, 16]. The larger fragments (d10>150μm) consist of a mean size of approximately 2 nozzle diameters in the case of D_j=500μm and this may suggest that those fragments are mainly formed by jet tip breakup. With the smaller nozzle, a bigger relative diameter is shown, possibly confirming the earlier statement that the base breakup of the jet is dominant in this case. However, this must be subject to further investigation for a wider range of cases and more importantly, for smaller fragments (d10<150μm) and for the full population of the atomized spray.

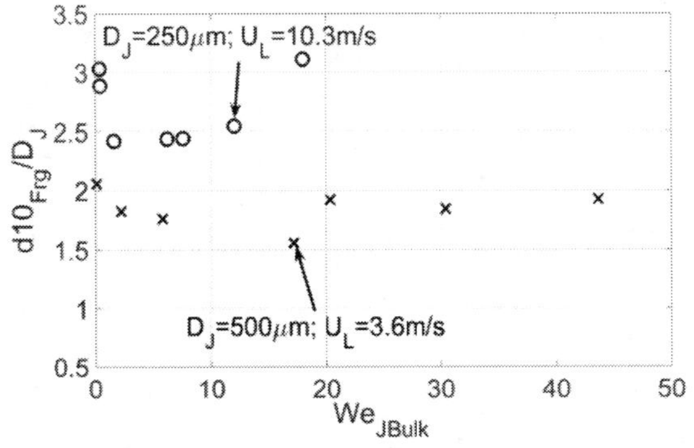

Fig. 5. Normalized diameter of fragments ($d10_{Frg}$) by nozzle diameter (D_j=500μm with U_L=3.6m/s and D_j=250μm with U_L=10.3m/s). (Only fragments with d10>150μm are used in this analysis).

3.4. Local Weber number of fragments

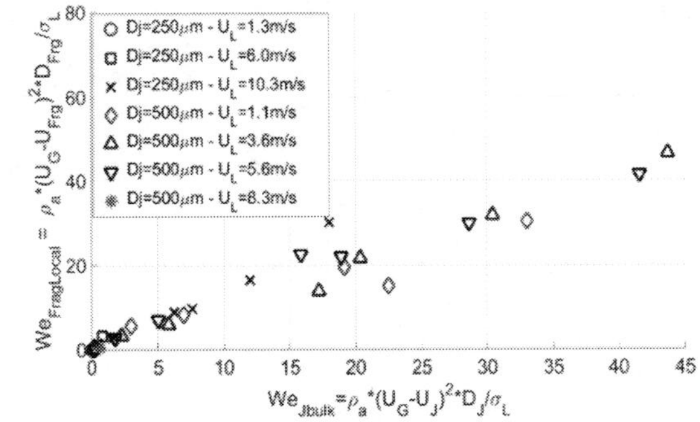

Fig. 6. Average local fragments velocity versus We_{JBulk}. (Only fragments with d10>150μm are used in this analysis)

It was reported earlier [16] that there are inconsistencies in describing the local Weber number for fragments and this is due to difficulties in quantifying the local gas phase velocity and therefore the local slip velocity for fragments. In this work, we use the bulk mean gas velocity to describe the slip velocity. This is not ideal. However, in the near-field of the liquid jet, it is a reasonable approximation given the minimum decay in the axial velocity of the air jet over the atomization length-scale. It is clear from Fig. 6 that the local Weber number of fragments follows a linear relation with the jet bulk Weber number. This is potentially a very useful correlation to ascertain local conditions, which are challenging to measure, using only the global parameters of the spray, which are known a priori. Further work will focus on obtaining real-time local measurements of the gas phase simultaneously with the dispersed phase.

4 Conclusions

The PTV technique developed earlier by these authors [1,16] is extended further to track the jet tip and fragments generated from a liquid jet exposed to a co-flowing air stream. Key achievements from this study can be summarized as follows:

(i) Using the probability distribution of the tip location, accurate statistics on the liquid jet breakup length can be achieved, demonstrating how the variation in the location of the liquid jet tip relates to the global Weber number.

(ii) Three dominant breakup modes are observed including: tip breakup, base breakup and surface breakup. We have reported relationships between the size and velocity of fragments with respect to the liquid jet tip under these different break-up modes. However, further investigations are required to understand these jet breakup modes in detail.

(iii) A linear correlation is observed here for local Weber number of fragments and jet global Weber number. However, it is noted here that the bulk mean gas velocity has been used as the relevant velocity surrounding all fragments in the atomization zone.

References

1. A. Kourmatzis, P.X. Pham, and A.R. Masri, Meas. Sci. Technol, 28(3) (2017) 035302
2. T. D. Fansler and S. E. Parrish. Spray measurement technology: a review. Meas. Sci. Technol., 26(012002) (2015).
3. J. Shinjo and A. Umemura, International Journal of Multiphase Flow 36(7) (2010) 513-532
4. A. Umemura, Journal of Fluid Mechanics, 757 (2014) 665-700
5. K.A. Sallam and G.M. Faeth, AIAA Journal 41(8) (2003) 1514-1524
6. A. Umemura, J. Osaka, J. Fluid Mech. 752 (2014) 184-218
7. A. Umemura, Physical Review E, 83 046307(2011)
8. K. A. Sallam, Z. Dai, G.M. Faeth, Intl J. of Multiphase Flow 28(2002) 427-449
9. J. Shinjo and A. Umemura, Intl. J. of Multiphase Flow 36 (2010) 513–532
10. A. Kourmatzis, P. X. Pham, and A. R. Masri, Fuel, 108 (2013) 758-770
11. C.M. Sipperley and W.D. Bachalo. 17th International Symposium on Applications of Laser Techniques to Fluid Mechanics, Lisbon, Portugal, 07-10 July, 2014.
12. X. Liu, K.S. Im, J. Wang, M.W. Tate, A. Ercan, D.R. Schuette, and S.M. Gruner, Applied Physics Letters, 94 084101(2009)
13. A. Kourmatzis, P.X. Pham, A.R. Masri, Combustion and Flame 162(4) (2015) 978-996
14. P.X. Pham. A. Kourmatzis and A.R. Masri, Accepted (DOI: 0.1088/1361-6501/aa8893), Measurement Science and Technology, 201715. D. R. Guildenbecher, J. Gao, J. Chen and P. E. Sojka, Intl J. of Multiphase Flow 94 (2017) 107-122
16. P.X. Pham. A. Kourmatzis and A.R. Masri, 10th Mediterranean Combustion Symposium, Naples, Italy, September 17-21, 2017
17. T. Hayashi, M. Suzuki, and M. Ikemoto. International J. of Engine Research, 14(6) (2013) 646-654
18. Y. Hardalupas, S. Sahu, A. M. K. P. Taylor, and K. Zarogoulidis. Experiments in Fluids, 49 (2010) 417-434
19. S. Sahu, Y. Hardalupas, and A. M. K. P. Taylor, Experiments in Fluids, 55 (2014) 1673

11th Asia-Pacific Conference on Combustion,
The University of Sydney, NSW Australia
10th -14th December 2017.

Multiple Optical Diagnostics on both Partially Premixed Combustion (PPC) and Reactivity Controlled Compression Ignition (RCCI)

Haifeng Liu, Xingwang Ran, Qinglong Tang, Chao Geng, Zhi Yang, Zunqing Zheng

State Key Laboratory of Engines, Tianjin University

Weijin Road, Nankai District, Tianjin, 300072, China

Abstract

The mixing and combustion process of partially premixed combustion (PPC) and reactivity controlled compression ignition (RCCI) was investigated on a light-duty optical engine. In PPC mode, PRF70, a mixture of 70% iso-octane and 30% n-heptane by volume, was directly injected into the cylinder. In RCCI mode, 70% iso-octane was supplied by the port fuel injector and 30% n-heptane was directly injected. The overall fuel reactivity is the same to PPC and RCCI and the overall fuel/air equivalence ratio was kept at 0.77 for both combustion modes. The direct-injection timing was set to -25°CA ATDC. The results found that the local fuel-rich regions were formed before high temperature heat release (HTHR) and the maximum fuel/air equivalence ratio was lower than 1.2 for PPC and was lower than 1.6 for RCCI. The PRF number covered 35-80 before HTHR for RCCI, while the PRF number was always 70 for PPC. Distinct flame front propagation for PPC was seen during the early stages of the flame development process by the high-speed images, formaldehyde and OH planar laser-induced fluorescence (PLIF) imaging. However, for RCCI, there was no distinct flame front propagation based on high-speed images, formaldehyde and OH PLIF, and RCCI was a staged multi-sites auto-ignition combustion process.

1 Introduction

HCCI combustion has been drawing the considerable attention due to high efficiency and lower nitrogen oxide (NOx) and particulate matter (PM) emissions. However, there are still tough challenges in the successful operation of HCCI engines, such as controlling the combustion phasing, extending the operating range. It has been realized that advanced control strategies of fuel/air mixture are more important than simple homogeneous charge in the process of the controlling of HCCI combustion processes. The stratification strategy has the potential to control the time of auto-ignition and extend the HCCI operation range to higher loads [1, 2]. Based on the idea of stratification, some new combustion modes have been proposed. Kalghatgi et al. [3] proposed a gasoline partially premixed compression ignition (PPC) strategy in which gasoline was directly injected in the cylinder of a diesel engine forming stratification and ignited by compression. It was indicated by Manente et al. [4] that PPC can be operated at loads as high as 25 bar gross IMEP with low emissions of soot and NOx. In addition, a gross indicated engine efficiency of more than 50% could be achieved with acceptable maximum pressure rise rate by moderate EGR. Meantime, fuels of different reactivity are required under different operating conditions, i.e., a high cetane fuel at low load and a low cetane fuel at high load. To address this issue, Inagaki et al. [5] employed a dual-fuel (premixed iso-octane and direct injected diesel) compression ignition strategy, which was able to operate the engine up to 12 bar indicated mean effective pressure (IMEP) with ultra-low NO_x and soot emissions, using a reduced amount of EGR. This dual-fuel strategy was named as reactivity controlled compression ignition (RCCI) by Kokjohn et al. [6] and RCCI could also achieve a higher engine loads with lower emissions like PPC.

However, although PPC and RCCI have been intensely studied, most of the works reported in the literature only concern the engine performance and emissions. The study of the in-cylinder combustion processes of PPC and RCCI are still very limited. Tanov et al. [7] explored the impact of split fuel injections on the combustion stratification by OH chemiluminescence imaging using PRF55 and found that combustion with a triple injection was more homogenous compared to single and double injections. Lu et al. [8] compared PRF70 and diesel split injection strategies by high speed imaging technique and found that the evaporation process of PPC was faster than that of diesel operation. Kokjohn et al. [9] studied the mixture distribution in RCCI by planar fuel-tracer laser-induced fluorescence (PLIF) techniques and the combustion process by the high-speed imaging. They found that the ignition sites firstly appeared in the downstream portion of the injected fuel jets, where the local fuel reactivity was high, and the rate of heat release could be controlled by fuel stratification. However, for both PPC and RCCI, the high-speed imaging is a line-of-sight technique, which presents the two-dimensional projection of the natural flame luminosity from the three dimensional reaction zones. Meanwhile it is unable to distinguish the low-temperature heat release (LTHR) phase and high-temperature heat release (HTHR) phase, respectively, which are very important features of the combustion.

Therefore, in the current work, detailed optical diagnostics about fuel injection, mixture formation and combustion characteristics have been conducted, which will help to promote better understanding of the combustion phenomena in PPC and RCCI and provide experimental support for further simulation under both PPC and RCCI conditions.

2 Experimental setup

2.1 Optical engine setup

The experiment was conducted in a naturally aspirated, single-cylinder optical engine (Bore×Stroke: 92×100 mm). A Bowditch-piston was used, and a cylindrical combustion chamber (diameter: 66 mm, height: 9 mm) was designed with a

Corresponding author: Haifeng Liu. Fax: +86-22-2738-3362
E-mail address: haifengliu@tju.edu.cn

flat quartz window at the bottom, allowing optical access to the combustion chamber through the piston. To permit the entrance of a laser sheet into the combustion chamber around the crank angle near TDC, a cut-out of 40 mm was set on the right of combustion chamber (see Fig. 1.). A quartz ring window replaced a part of the top cylinder liner, making the upper part of cylinder passable for the laser sheet. The combustion chamber modifications reduced the compression ratio to about 11. Due to the low compression ratio, the intake air was preheated to 75 °C, producing a motored temperature of about 820 K at TDC at RCCI mode. However, the preheated temperature was not enough to keep stable auto-ignition for PPC, so the intake temperature was set to 125 °C and the motored temperature of about 870 K at TDC. The overall fuel/air equivalence ratio was kept at 0.77 for both PPC and RCCI. The engine was motored by a dynamometer at 1200 rpm and no EGR was introduced. To reduce the risk of window failure caused by thermal and mechanical stresses, and to extend the experimental period before window fouling, the engine was operated in a skip-fire mode where only one fuel injection event was performed every 20 cycles. For RCCI mode, iso-octane (70% by volume) was supplied by the port fuel injector at -360 °CA ATDC with 3 bar injection pressure, and 30% n-heptane was directly injected at -25 °CA ATDC with 600 bar injection pressure. For PPC mode, PRF70, a mixture of 70% iso-octane and 30% n-heptane by volume was directly injected at -25 °CA ATDC with 600 bar injection pressure.

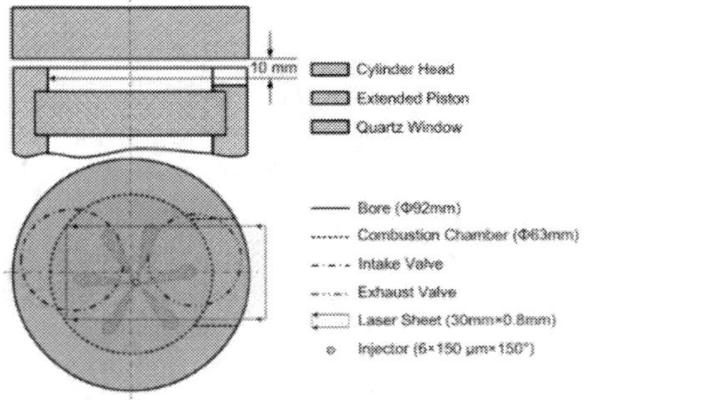

Figure 1. Geometric structure of the combustion chamber and the field of view for the optical diagnostics.

2.2 Optical diagnostic techniques

Three laser beams at 266 nm, 355 nm and 282.95 nm were used for three different PLIF imaging techniques of fuel-tracer, formaldehyde and OH, separately. Each beam was formed into a 30 mm wide and less than 1 mm thick horizontal laser sheet by a group of cylindrical lens, and was directed horizontally into the cylinder though the quartz ring window. The relative position of cylinder head, extended piston and laser diagnostics region is shown in Figure 1. The horizontal laser sheet was maintained at a plane 10 mm below the firedeck. The distance between the firedeck and laser sheet was carefully chosen so that the laser sheet just passed through the main combustion regions. The PLIF signals were reflected by a 45° UV-mirror to an ICCD camera coupled with different fluorescence filters. Meantime, a high-speed camera (Photron SA1) was used to record the natural flame luminosity of PPC and RCCI combustion. A Nikon lens of 50 mm focal length was used. The frame rate and exposure time were set to 10000 frames per second and 100 μs, respectively. Thus, the temporal resolution between adjacent two acquired images was 0.72° CA at the engine speed of 1200 rpm. Detailed introduction on optical diagnostic techniques can be found in Refs. 9-10.

3 Results and discussion

3.1 Fuel stratification for PPC and RCCI

Figure 2 presents the distribution of fuel concentration (left) and reactivity (right) quantified by fuel-tracer PLIF for both PPC and RCCI. Ten selected single-shot fuel-tracer PLIF images at the specific crank angles were averaged and processed to get the fuel/air equivalence ratio and PRF number. It can be seen that the local fuel-rich region is reduced from 0° to 5° for PPC and from -10° to -5° for RCCI. For PPC, it has longer ignition delay and higher intake temperature, so the local maximum equivalence ratio is lower than 1.2 before HTHR at 5°, while the shorter ignition delay and lower intake temperature lead that the local maximum equivalence ratio reach 1.6 before HTHR at -5° for RCCI. In PPC mode, all fuel is injected directly, so the local equivalence ratio is quite low where fuel can not reach. But the lowest equivalence ratio is 0.55 for RCCI due to its port-injection of iso-octane. For the stratification of fuel reactivity, the PRF number covers 35-80 before HTHR for RCCI, while the PRF number is always 70 for PPC. In RCCI, the PRF number is quite low in these fuel-rich regions due to the direct-injection of n-heptane.

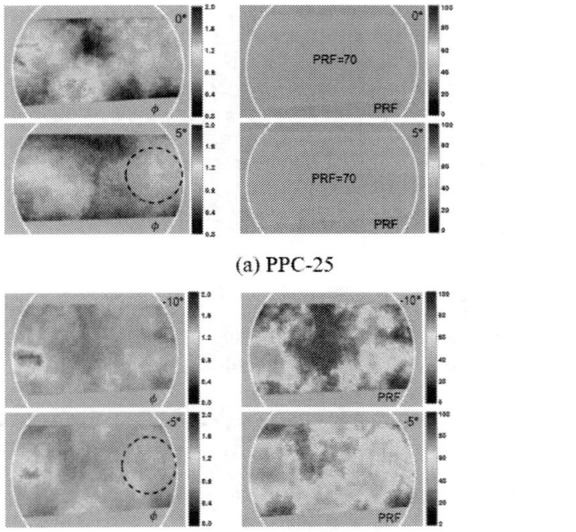

(a) PPC-25

(b) RCCI-25

Figure 2. The distribution of fuel concentration (left) and reactivity (right) quantified by fuel-tracer (1% toluene addition) PLIF.

Figure 3 presents the in-cylinder pressure and heat release rate for both PPC and RCCI. It shows that the ignition delay of PPC is quite longer than that of RCCI, even if the PPC mode has higher intake temperature. The heat release rate of PPC is smooth than that of RCCI due to the retarded ignition. The main reason is attributed to the reactivity stratification of RCCI.

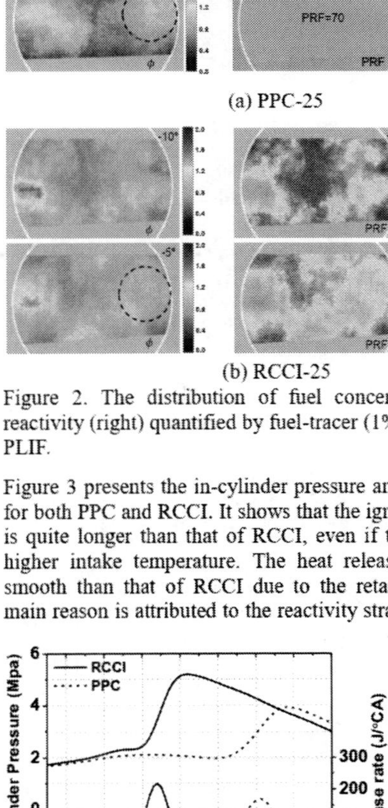

Figure 3. In-cylinder pressure and heat release rate

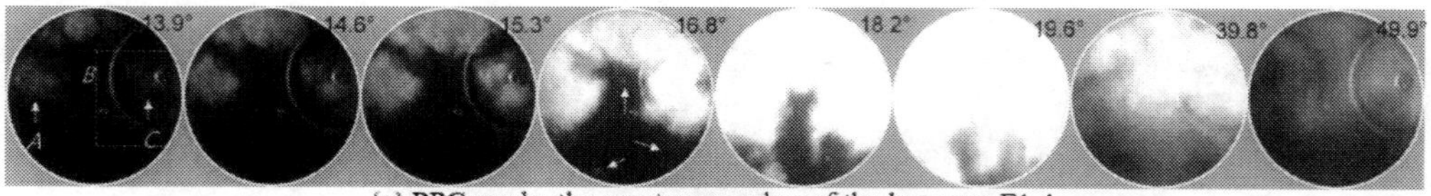

(a) PPC mode, the aperture number of the lens was F1.4

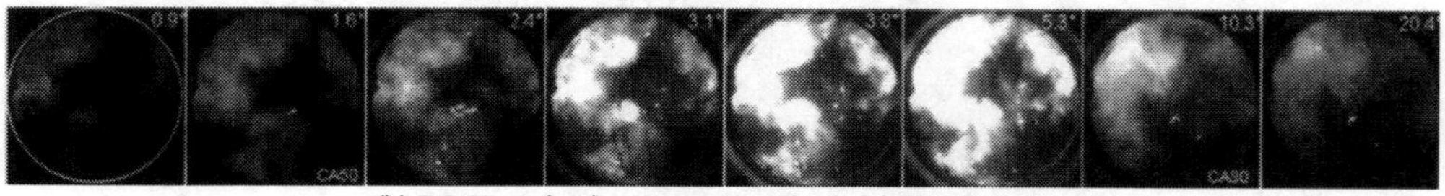

(b) RCCI mode, the aperture number of the lens was F2

Figure 4. Time-resolved single-shot true-color images of PPC and RCCI combustion from a typical fired cycle.

Figure 4 presents the time-resolved single-shot images of PPC and RCCI combustion from a typical fired cycle. It shows that the local high reactivity region with low PRF number forms the clear blue flame very early at 0.9° in RCCI. Then, in the next images at 1.6° and 2.4°, the flame occupies the whole combustion chamber very fast besides the center region where the PRF number is higher and fuel reactivity is lower. After, the blue flame can be seen in the center region at 3.1°. However, for PPC mode, it can also seen that multiple auto-ignition kernels emerge in the combustion chamber at 13.9°. These ignition kernels are located in the regions marked with "A", "B" and "C". The ignition regions "A" and "C" correspond well with the two local fuel-rich zones as shown in Figure 2(a) at 5°. Zone "B" is out of the laser-sheet region, but it can be inferred that there is a fuel-rich zone. After the auto-ignition, in the next two images at 14.6° and 15.3°, the flame front of these three regions gradually propagates to locations with lower fuel/air equivalence ratio. This distinct flame front propagation can not be seen in RCCI mode. Therefore, it can be found that the stratification of fuel reactivity in RCCI induces more simultaneously auto-ignition in the early stage of the combustion, resulting in faster combustion reaction rates. However, for PPC, there is no reactivity stratification and the auto-ignition only depends on the local equivalence ratio. The local fuel-rich region induces the initial auto-ignition, then the local equivalence ratio support the subsequent flame propagation. But new auto-ignition sites are hard to be formed in the other fuel-lean regions until the combustion temperature is high enough at 16.8°, where additional auto-ignition kernels, pointed to by arrows, appear in the fuel lean region before the flame front arrives. These new auto-ignition sites grow and merge with the previous flame zones at 18.2° and the whole combustion chamber is occupied by combustion flames at 19.6°. In the late stage of combustion, the bright flame luminosity can be seen for both PPC and RCCI, indicating the emergence of soot in the fuel-rich regions. On one hand, the bright luminosity can not be seen at homogeneous HCCI combustion. On the other hand, the emergence of soot has been proved by the results of the emission spectra in previous study [10].

Figure 5 presents the flame boundary development extracted from the combustion images for PPC and RCCI. In the early stage of PPC (from 13.9° to 16.0°), a clear flame propogation can be seen, while it is not obtained in RCCI mode. The RCCI presents a staged combustion process that auto-ignition first happens in the region of high reactivity around the combustion chamber and then another auto-ignition process takes place in the region of low reactivity in the central part of the combustion chamber. From auto-ignition to the flame occupying the whole combustion chamber, PPC needs 5.7°CA but RCCI only needs 2.9°CA.

Figure 6 and 7 present the typical non-simultaneous, single-shot formaldehyde and OH PLIF images in false-color for PPC

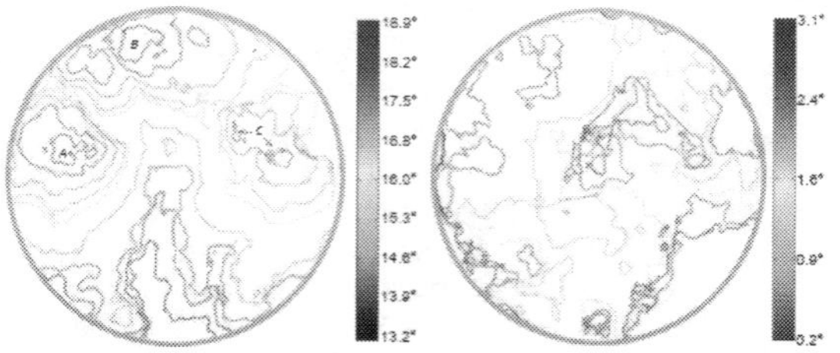

(a) PPC (b) RCCI

Figure 5. Flame boundary development extracted from the combustion images for PPC and RCCI

and RCCI. Formaldehyde and OH are often used to be the indicator of LTHR and HTHR stage, respectively, in HCCI and LTC. Furthermore, the consumption of formaldehyde or the formation of OH can also be used to determine the flame front development in spark-ignition engines. In Figure 6(a), the weak formaldehyde signal is shown at -5° and -9° when the LTHR is under progress for both PPC and RCCI. Then the PLIF intensity of formaldehyde gets stronger later when the LTHR has basically finished. The distribution of formaldehyde PLIF region corresponds well with the fuel distribution and the regions with higher fuel/air equivalence ratios present higher formaldehyde PLIF intensity. Sequentially at 14°, the formaldehyde PLIF vanishes in a small region with a wrinkled boundary and the decomposition of formaldehyde in this region is regarded as the emergence of ignition kernel. Accordingly, the ignition kernel represented by OH PLIF appears at 14° as shown in Figure 7(a). Then, more formaldehyde in the field of view is consumed and the ignition kernel expands to a larger region as time goes by from 15° to 17° as shown in Figure 6(a) and 7(a). At 17°, only a small residue of formaldehyde exists at the lower part and the OH PLIF signal occupies the most diagnostic region. For both PPC and RCCI, the OH-LIF can be detected in the whole diagnostic regions and persists more than 30°CA. The OH radical is an important radical for soot oxidation. The wide distribution and late persistence of OH can explain why PPC and RCCI have low soot emissions compared to the traditional mixing-controlled diesel combustion and LTC. The difference between PPC and RCCI is that the clear boundary of formaldehyde consumption and OH formation can not be seen in RCCI mode, which again validates that there is no flame propogation in RCCI mode and it is multi-sites staged auto-ignition in the current conditions. However, our recent study found that the flame propagation can also be seen as the port premixed equivalence ratio is increased to 1.0 or 1.25. Therefore, it can be concluded that the flame development of PPC and RCCI always changes from pure auto-ignition such as HCCI to pure flame propagation such as SI engines. Both PPC and RCCI combine auto-ignition and flame propagation, which mainly depends on the local distribution of equivalence ratios.

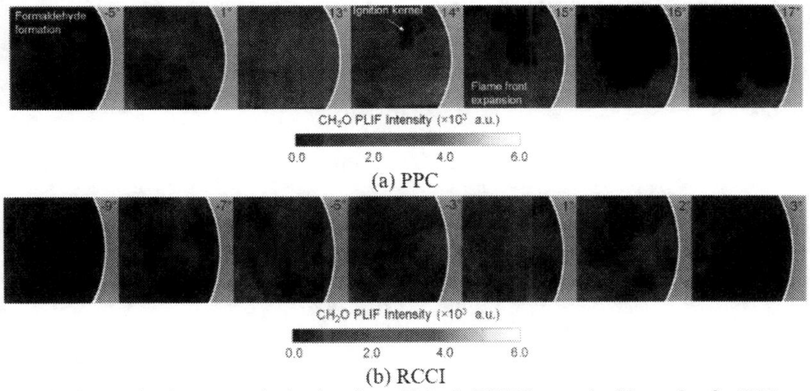

Figure 6. Typical non-simultaneous, single-shot formaldehyde PLIF images in false-color for PPC and RCCI

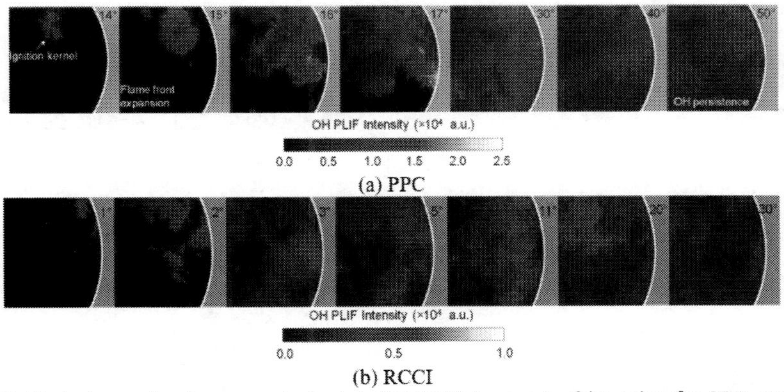

Figure 7. Typical non-simultaneous, single-shot OH PLIF images in false-color for PPC and RCCI

4 Conclusions

The mixing and combustion process of partially premixed combustion (PPC) and reactivity controlled compression ignition (RCCI) was investigated on a light-duty optical engine. In PPC, PRF70, a mixture of 70% iso-octane and 30% n-heptane by volume, was directly injected into the cylinder. In RCCI, 70% iso-octane was supplied by the port fuel injector and 30% n-heptane was directly injected. The overall fuel reactivity is the same and the overall fuel/air equivalence ratio was kept at 0.77 for both PPC and RCCI. The direct-injection timing was set to -25°CA ATDC. The main conclusions are that the local fuel-rich regions are formed before high temperature heat release (HTHR) and the maximum fuel/air equivalence ratio is lower than 1.2 for PPC and is lower than 1.6 for RCCI. The PRF number covers 35-80 before HTHR for RCCI, while the PRF number is always 70 for PPC. Distinct flame front propagation for PPC is seen during the early stages of the flame development process by the high-speed images, formaldehyde and OH planar laser-induced fluorescence (PLIF) imaging. However, for RCCI, there is no distinct flame front propagation based on high-speed images, formaldehyde and OH PLIF, and RCCI is a staged multi-sites auto-ignition combustion process. PPC presents a larger stratified images of formaldehyde and OH PLIF compared with that of RCCI and the wide distribution and late persistence of OH can be seen for both PPC and RCCI. The different combustion process dominated by auto-ignition or flame propagation are mainly depends on the local distribution of equivalence ratios.

5 Acknowledgment

This work was supported by National Natural Science Found of China (NSFC) through the Projects of 91541111 and 51576138.

References

[1] M. Richter, J. Engström, A. Franke, SAE 2000-01-2868 (2000).

[2] H.F. Liu, Z.L. Zheng, M.F. Yao, P. Zhang, Z.Q. Zheng, B.Q. He, et al, App. Therm. Eng. 33-34 (2012) 135-143.

[3] G.T. Kalghatigi, P. Risberg, H. Angstrom, SAE 2006-04-3385 (2006).

[4] V. Manente, B. Johansson, W. Cannella, Int. J Engine Res. 12 (2011) 194-208.

[5] K. Inagaki, T. Fuyuto, K. Nishikawa, K. Nakakita, I. Sakata, SAE 2006-01-0028 (2006).

[6] S.L. Kokjohn, R.M. Hanson, D.A. Splitter, R.D. Reitz, Int. J Engine Res. 12 (2011) 209-226.

[7] S. Tanov, R. Collin, B. Johansson, M. Tuner, SAE Int. J Engines 7 (2014) 1911-1919.

[8] P. Lu, H. Zhao, M. Herfatmanesh, SAE 2015-01-1819, (2015).

[9] S.L. Kokjohn, M.P.B. Musculus, R.D. Reitz, Combust. Flame 162 (2015) 2729-2742.

[10] Q.L. Tang, H.F. Liu, M.K. Li, M.F. Yao, Z.S. Li, Combust. Flame 177 (2017) 98-108.

11th Asia-Pacific Conference on Combustion,
The University of Sydney, NSW Australia
10th -14th December 2017.

The Mechanism of Electrophilic Addition of Singlet Oxygen to Unsubstituted Pyrrole

Nassim Zeinali, Jomana Al-Nu'airat, Zhe Zeng, Mohammednoor Altarawneh*, Dan Li, Jakub Skut,

Bogdan Z. Dlugogorski[1]

School of Engineering and Information Technology

Murdoch University, 90 South Street, Murdoch, WA 6150, Australia

Abstract

The essential role of pyrrole in comprehending the chemistry of living organisms, coal surrogates and novel drugs is indisputable. However, literature reports a few studies on its reactivity toward prominent oxidizing agents. Herein we present a comprehensive mechanistic study of oxidation of unsubstituted pyrrole with singlet oxygen (O_2 $^1\Delta_g$) by deploying a quantum chemical framework leading to the production of succinimide, as the major products, through a Diels-Alder addition of O_2 $^1\Delta_g$ to the aromatic ring. Other products, comprising maleimide, hydroperoxide, formamide or epoxide adducts, appear to form via rather minor channels. Additionally, we determined that, the primary Diels-Alder channel encompasses a barrier of 41 kJ/mol with a fitted rate constant of $k(T)$=1.87×10^{-13} exp(-48 000/RT) cm^3 molecule^{-1} s^{-1}.

1 Introduction

Pyrrole (Fig. 1) possess a pentagonal heterocyclic structure featuring a planar aromatic ring which principally constitutes the core skeleton of many naturally and synthetically occurring cyclic-nitrogen compounds such as coal tar, and various natural pigments (i.e. heme and chlorophyll). Therefore, structural entities featuring pyrrole assumes a fundamental importance in many systems encompassing nitrogen chemistry in biomass and nitrogen cycle in many biological systems [1, 2].

In general, nitrogen speciation in coal mainly exists as heterocyclic compounds such as pyrrole and pyridine structural entities; rather than in the form of saturated amines prevailing in biomass. Aromatic pyrrolic nitrogen is the most abundant nitrogen-bearing fragments in coal [3] and the content of pyrrole type nitrogen increases with the coal rank [4].

Literature provides very little experimental and theoretical investigations on the reactivity of pyrrolic systems towards singlet oxygen, for application pertinent to spontaneous ignition of heterocyclic N-containing fuels and biochemistry of pyrrole-like compounds in living systems. The electron-rich structure of pyrrole remarkably facilitates the electrophilic attack of singlet oxygen to its aromatic core. Obviously, the very reactive nature of pyrrole, [5] and the excessively transient nature of singlet oxygen [6] renders experimental investigation into this title reaction very challenging.

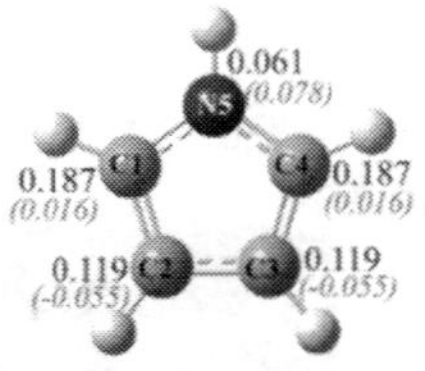

Figure 1: Pyrrole chemical structure and its acquired Fukui (f^{-1}) indices (black) and Hirshfeld charges (red between brackets).

Few studies have attempted to identify major products of pyrrole oxidation induced by singlet oxygen [7-9]. Quistad and Lightner [10] reported potential formation of methoxy-lactam as the major product from photoxidation of unsubstituted pyrrole in methanol along with a small amount of maleimide, in addition to the formation of hydroxy-lactam in the aqueous medium (Fig. 2).

Figure 2: Possible pyrrole photoxidation products.

Moreover, the formation of free radicals has been observed through ESR spectroscopy [11] during photosentisized oxidation of pyrrole in ethanol or pyridine, which is generally attributed to the commonly discussed role of biradicals in pyrrole photo-induced oxidation.

The objective of the present study is to provide comprehensive theoretical insights into the photo-induced activation of pyrrole. In doing so, we have employed quantum chemical frameworks to ascertain all feasible reaction routes and their corresponding products. Results presented herein will assist in attaining a better understanding for reactions operating during self-ignition of nitrogen-contain aromatic compounds; i.e., major N-carriers in biomass.

2 Methodology

The Gaussin09 [12] code carries out all structural optimization and energy calculations. The commonly deployed DFT functional of B3LYP with the polarized 6-311+G(d,p) basis set [13] is utilized to investigate all geometry optimizations. The nature of minimum energy points have been confirmed via estimation of vibrational frequencies in which transition structures contain one and only one negative eigenvalue along the designated reaction pathway.

We corrected the final energies of the biradical systems formed by addition of O_2 $^1\Delta_g$ to pyrrole, based on the Yamaguchi's

Corresponding author: Mohammednoor Altarawneh
Phone: 9360 7507
E-mail address: m.altarawneh@murdoch.edu.au

criteria [14, 15] via the application of a simple approximate spin-projection (AP) scheme, derived from the energies of the broken-symmetry (BS) and pure high-spin (HS) states. We have also validated the identity of each transition structure employing intrinsic reaction coordinate (IRC) calculations. The ChemRate [16] code facilitates the calculation of the reaction rate constants and Arrhenius parameters, in the narrow temperature range of 300 K to 600 K.

The effect of solvation on reaction energies for the main pathway is investigated by deploying a polarisable continuum model (PCM) [17] at the B3LYP/6-311g+(d,p) level of theory. According to this approach, the solvent effect is incorporated by generating multiple overlapping spheres inside of a dielectric continuum, around the atoms [18, 19] To obtain governing electronic descriptors, we computed Fukui [20] indices for electrophilic attack (f^{-1}) and Hirschfeld partial charges by the DMol3 code. [21] The MESMER 3.0 code delivers the RRKM-ME kinetic analysis for the leading reaction pathway. [22]

3 Results and discussion

Literature values for the bond lengths for pyrrole reported by Nygaard et al. [23] and Bak et al. [24] in Table 1 are compatible with the values obtained by our adapted computational approach supporting the reliability of the method in optimization of the involved chemical structures.

Table 1: The atomic bond lengths (Å) in pyrrole molecular structure (refer to Fig. 1 for atomic numberings).

Atomic bond	Ref. [16]	Ref. [17]	Modeled
N-C_2	1.370	1.383	1.375
C_2-C_3	1.382	1.371	1.377
C_3-C_4	1.417	1.429	1.425
N-H	0.996	0.993	1.006
C_2-H	1.076	1.075	1.078
C_3-H	1.077	1.075	1.079

The Fukui indices and Hirshfeld charges on pyrrole atoms are displayed in Fig. 1, as indications of tendency of a compound to sustain electrophilic attack by singlet oxygen. The Fukui indices (f^{-1}), predicts the most electron rich sites in a chemical species. These sites are more susceptible for undergoing an electrophilic addition. Likewise, the Hirshfeld charge is a descriptor to measure the charge's distribution in the molecular structure and stems from generated spin densities. In general, the flat five-membered ring of pyrrole facilitates the occurrence of 1,2-cycloaddition, Diels-Alder reaction and the so called ene reaction. Figure 3 portrays the mechanism of the reactions of pyrrole with singlet oxygen. Higher f^{-1} values on C_1 and C_4 atoms verifies that O_2 $^1\Delta_g$ inclines to attack pyrrole ring via these sites to confront a lower activation barrier (in reference to other opening routes), yielding the corresponding endoperoxide structure (P_5) via TS_5 route (Fig. 3). The ene-type addition of singlet oxygen to pyrrole proceeds via activation barrier of 104 kJ/mol characterized by the transition structure TS_1. It is generally viewed that such ene-type addition dominantly favors a stepwise mechanism resulting in the formation of biradical intermediates [25]. Formation of the biradical P_1 is an endothermic process by 66 kJ/mol and passes through an enthalpic barrier of 104 kJ/mol. An intramolecular H movement from the NH site to the outer oxygen atom in the proxy group in P_1 forms the allylic hydroperoxide, P_2.

In a parallel channel, O_2 $^1\Delta_g$ attacks the double bond through 1,2-cycloaddition mode (TS_3) leading to the dioxetane structure of P_3. A ring-opening reaction in the former affords the ketoamide P_4 intermediate in a highly facile and exothermic reaction. Production of P_4-like structures has been previously observed via photoxidation of tetr-butyl-pyrroles as well as 3,4-diethyl-2,5-dimethylpyrrolle in acetone with about 10% of yield. [26]

P_4 could further undergo a 1,2-cycloaddition by singlet oxygen forming the P_{17} dioxetane structure (Fig. 4) through a relatively sizable enthalpic barrier of 99 kJ/mol (TS_{13}). Simultaneous C-C/O-O bond fissions in P_{17} produces P_{18} and P_{19} adducts. Reaction of water (i.e., hydrolysis) with the former generates the formamide structure of P_{21} through a substantial barrier of 204 kJ/mol (TS_{15}), as depicted in Fig. 4. Formation of similar carboxiamides has also been reported by Lightner and Pak [26] during photoxidation of substituted pyrroles. However, the yield of formation of such product is observed to be negligible (~3%). This observation could be rationalised in view of the relatively large barriers of the opening transition structures i.e., 99 kJ/mol and 204 kJ/mol (the uppermost part in Figure 4).

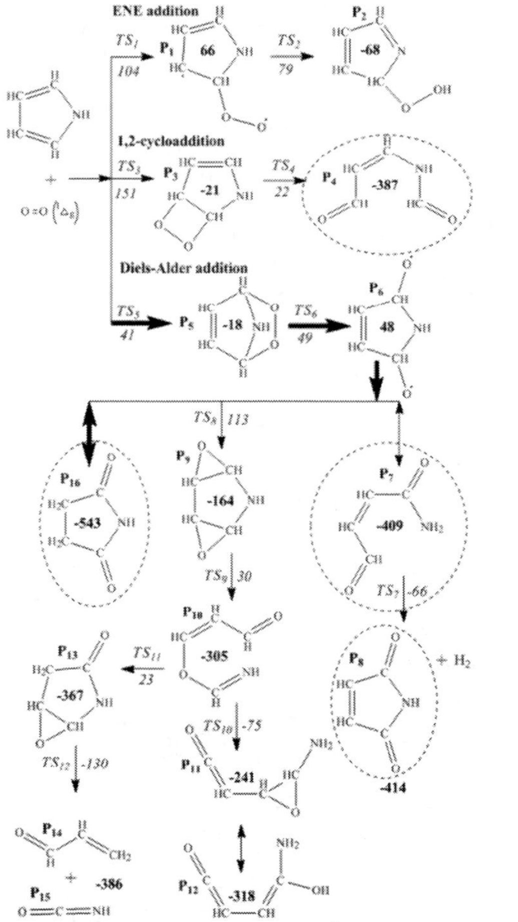

Figure 3: Initial channels of pyrrole + O_2 $^1\Delta_g$ reaction obtained by B3LYP/6-311+G(d,p) level of theory. Values in bold/italic denote reaction/activation enthalpies [kJ/mol] in reference to the initial separated reactants.

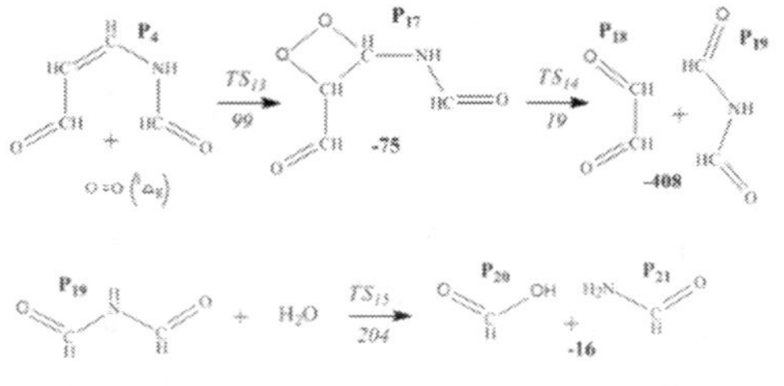

Figure 4: Mechanism of further oxidation of side product (P_4) and hydrolysis of P_{18}, leading to production of foramide (P_{20}).

According to the kinetic parameters enlisted in Table 2, the pyrrole + O_2 $^1\Delta_g$ reaction eminently initiates by the Diels-Alder channel and elusively proceeds via the reaction sequence $P_5 \rightarrow P_6 \rightarrow P_{16}$. Thus, our results suggest the formation of succinimide (P_{16}) as the principle product through instant conversion of highly unstable biradical via the $P_6 \rightarrow P_{16}$ channel that is immensely exothermic by 588 kJ/mol. Succinimide is also identified as the product of pyrrole autoxidation in aqueous phase by Ciamician and Silber. [27]

Unimolecular emission of a hydrogen molecule from P_7 produces maleimide (P_8) that is reported to be one of the major products of pyrrole photoxidation. [10]. However, a tremendous activation enthalpy of TS_7 (343 kJ/mol) indicates that formation of P_8 from P_7 most likely proceeds via alternatives routes including successive abstraction of two hydrogen atoms by the singlet oxygen. Nonetheless, TS_7 resides 66 kJ/mol below the entrance channel. The P_7 moiety is in a great well-depth (409 kJ/mol) with respect to the initial reactants; and thus it signifies an energised intermediate which could facilitate the unimolecular evolution of a hydrogen molecule.

Along the same line of enquiry, It was suggested that formation of maleimide could directly be originated from endoperoxide (P_5) in a process that characterizes intramolecular H movements [28]. However, we were unable to locate a relevant transition state. Instead, our foremost efforts resulted in the production of the biradical adduct (P_6) through TS_6, which subsequently converts to P_7. This finding concurs with the fact that H-transfer in the $^1\pi\pi^*$ state of such systems is a barrierles process. [29]

Table 2: Arrhenius parameters fitted for temperature range of 300-600K.

Reaction	A [s^{-1} or cm^3 $molecule^{-1}$ s^{-1}]	E_a [kJ/mol]
Pyrrole + O_2 $^1\Delta_g \rightarrow P_1$	5.40×10^{-13}	111
$P_1 \rightarrow P_2$	1.52×10^{12}	15
Pyrrole + O_2 $^1\Delta_g \rightarrow P_3$	2.48×10^{-13}	158
$P_3 \rightarrow P_4$	2.13×10^{14}	47
Pyrrole + O_2 $^1\Delta_g \rightarrow P_5$	1.87×10^{-13}	48
$P_5 \rightarrow P_6$	6.32×10^{13}	71
$P_7 \rightarrow P_8 + H_2$	3.35×10^{10}	345
$P_6 \rightarrow P_9$	1.11×10^{14}	68
$P_9 \rightarrow P_{10}$	1.58×10^{14}	198
$P_{10} \rightarrow P_{11}$	3.51×10^{11}	233
$P_{10} \rightarrow P_{13}$	3.37×10^{11}	331
$P_{13} \rightarrow P_{14} + P_{15}$	3.79×10^{14}	241

An additional channel, that is of probably a less significance concerns the unimolecular cyclisation of P_6 leading to the epoxide-type adduct (P_9) and subsequent unimolecular rearrangement to form intermediate product of isoimides (P_{10}). Insight into such an analogous transformation comes from a study by Wasserman and Floyd [30] who identify isoimides from the reaction of substituted oxazoles with singlet oxygen. As portrayed in Fig. 3, the isoimide (P_{10}) preferentially proceeds through TS_{10} which leads to production of epoxide intermediate P_{11}. Formation of a an analogous epoxide with relatively high yields(70%) was observed during a selective photosensitized oxidation of N-phenylpyrrole[7]. Additionally, P_{12} is generated via a cascade hydrogen transfer/cyclization in P_{10}, leading to the formation of P_{12} and P_{13} via exothermic fission of relevant C-C, C-N and C-O bonds.

In Fig. 5, we have contrasted the acquired energies of the main reaction channel in gas and liquid mediums. An energy reduction in the range of 10 kJ/mol (P_5) – 20 kJ/mol (TS_5) is predicted for transition states, intermediates and final products by PCM model when methanol was chosen as a solvent. In general, solvent effect reduces energy levels, while the product distribution and the reaction kinetic values for most steps are not considerably affected in reference to the gas phase system.

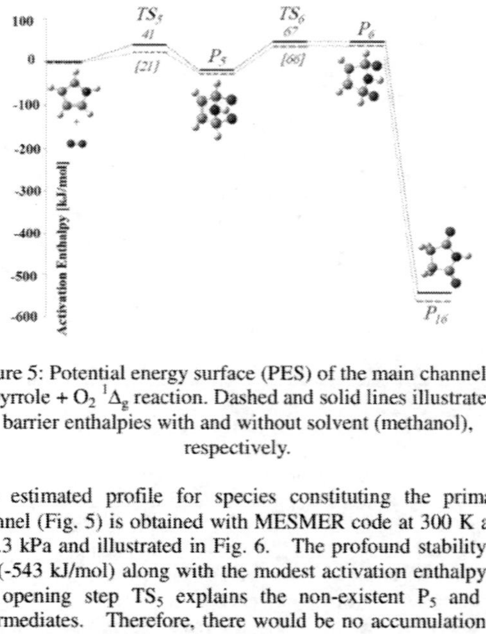

Figure 5: Potential energy surface (PES) of the main channel of pyrrole + O_2 $^1\Delta_g$ reaction. Dashed and solid lines illustrate barrier enthalpies with and without solvent (methanol), respectively.

The estimated profile for species constituting the primary channel (Fig. 5) is obtained with MESMER code at 300 K and 101.3 kPa and illustrated in Fig. 6. The profound stability of P_{16} (-543 kJ/mol) along with the modest activation enthalpy of the opening step TS_5 explains the non-existent P_5 and P_6 intermediates. Therefore, there would be no accumulation of these intermediates in the system. Conversation of the reactants to P_{16} occurs enduring an overall rate constant at 8.3×10^{-4} cm^3 mol^{-1} s^{-1}. Clearly, the modest barriers of TS_5 and TS_6 will be engulfed by the highly exothermic conversion of $P_6 \rightarrow P_{16}$ (-591 kJ/mol) forming a highly stable succinimide product.

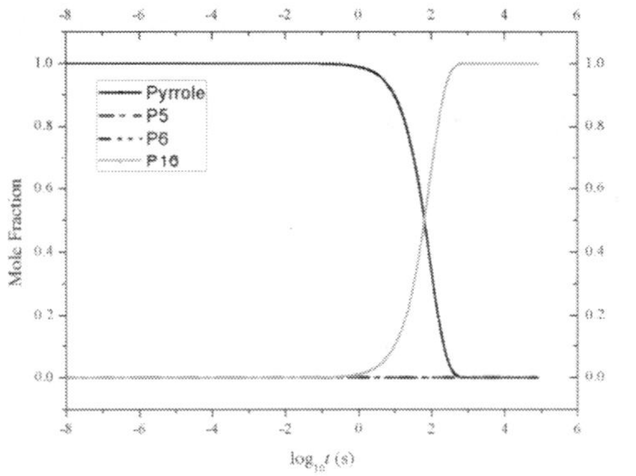

Figure 6: Time-dependent profile for species produced from pyrrole + O_2 $^1\Delta_g$ reaction at 300 K and 101.3 kPa.

4 Conclusions

The suggested mechanism of singlet oxygen induced oxidation of unsubstituted pyrrole discerns the dominance of endoperoxides formation through singlet oxygen electrophilic attack to the aromatic ring at C1 and C4 positions. Evidently, the major foreseeable products (i.e. succinimide and maleimide) will be derived from the endoperoxide (P_5) as the most attainable intermediate originated from singlet oxygen reaction with the dienic system of pyrrole and its derivatives. A simplified kinetic model indicates conversion of P5 – via the biradical P_6 - into the succinimide structure (P_{16}) at relatively long reaction time. Production of maleimide (P_8), as another principle product is anticipated to be hindered by very substantial energy barrier required for unimolecular elimination of H_2 molecule.

5 Acknowledgment

This study has been supported by grants of computing time from the National Computational Infrastructure (NCI) and from the Pawsey Supercomputing Centre, Australia as well as funds from the Australian Research Council (ARC). N.Z., J.N. and Zh.Z. thank Murdoch University for the award of postgraduate scholarships.

References

[1] K. Wang, M. Xu, Z. Gu, P. Ahrenkiel, J. Lee, W. Gibbons, J. Croat, Q. Fan, Int. J. Hydrogen Energ. 41(30) (2016) 13109-13115.

[2] J. Reindel, R. Roth, Am. J. Path. 138(3) (1991) 707.

[3] S.M. Kirtley, O.C. Mullins, J. van Elp and S.P. Cramer, Fuel, 1993. 72(1): p. 133-135.

[4] J.G. Speight, CRC Press 2012.

[5] R.A. Jones, G.P. Bean, Academic Press 2013.

[6] E.L. Clennan, A. Pace, Tetrahedron 61(28) (2005) 6665-6691.

[7] H.H. Wasserman, A.H. Miller, J. Chem. Soc. D: Chemical Communications (5) (1969) 199-200.

[8] H.H. Wasserman, R.W. DeSimone, D.L. Boger, C.M. Baldino, J. Am. Chem. Soc. 115(18) (1993) 8457-8458.

[9] D.L. Boger, C.M. Baldino, J. Org. Chem. 56(24) (1991) 6942-6944.

[10] G.B. Quistad, D.A. Lightner, J. Chem. Soc. D: Chemical Communications (18) (1971) 1099-1100.

[11] Ber GC., Acta chem. scand; (1962) 16(2).

[12] M. Frisch, G. Trucks, H.B. Schlegel, G. Scuseria, M. Robb, J. Cheeseman, G. Scalmani, V. Barone, B. Mennucci, G. Petersson, Gaussian 09, revision a. 02, gaussian, Inc., Wallingford, CT 200 (2009).

[13] J. Montgomery Jr, J. Ochterski, G. Petersson, J. Chem. Phys. 101(7) (1994) 5900-5909.

[14] B.D. Beaver, J.V. Cooney, J.M. Watkins, J. Heterocyclic Chem.. 23(4) (1986) 1095-1097.

[15] S. Yamanaka, T. Kawakami, H. Nagao, K. Yamaguchi, Chemical physics letters 231(1) (1994) 25-33.

[16] V. Mokrushin, V. Bedanov, W. Tsang, M. Zachariah, V. Knyazev, ChemRate, version 1.19, NIST: Gaithersburg, MD (2002).

[17] V. Barone, M. Cossi, J. Tomasi, J. Chem. Phys. 107(8) (1997) 3210-3221.

[18] N. Zeinali, M. Altarawneh, D. Li, J. Al-Nu'airat, B.Z. Dlugogorski, ACS Omega 1(2) (2016) 220-225.

[19] J. Al-Nu'airat, M. Altarawneh, X. Gao, P.R. Westmoreland, B.Z. Dlugogorski, J. Phys. Chem. A 121(17) (2017) 3199-3206.

[20] K. Fukui, Science (New York, NY) 218(4574) (1982) 747-754.

[21] B. Delley, J. Chem. Phys. 113(18) (2000) 7756-7764.

[22] D.R. Glowacki, C.-H. Liang, C. Morley, M.J. Pilling, S.H. Robertson, J. Phys. Chem. A 116(38) (2012) 9545-9560.

[23] U. Nygaard, J.T. Nielsen, J. Kirchheiner, G. Maltesen, J. Rastrup-Andersen, G.O. Sørensen, J. Mol. Struct. 3(6) (1969) 491-506.

[24] B.r. Bak, D. Christensen, L. Hansen, J. Rastrup-Andersen, J. Chem. Phys. 24(4) (1956) 720-725.

[25] A.G. Leach, K. Houk, Chem. Commun. (12) (2002) 1243-1255.

[26] D.A. Lightner, C.-S. Pak, J. Org. Chem. 40(19) (1975) 2724-2728.

[27] G. Ciamician, P. Silber, Chemische Lichtwirkungen. XXIV., Eur. J. Inorg. Chem. 45(2) (1912) 1842-1845.

[28] J.K. Howard, K.J. Rihak, A.C. Bissember, J.A. Smith, Chem. Asian J. 11(2) (2016) 155-167.

[29] T. Elsässer, H. Van den Akker, Springer Science & Business Media2013.

[30] H. Wasserman, H. and M. Floyd, Tetrahedron, 1966. 22: p. 441-448.

11th Asia-Pacific Conference on Combustion,
The University of Sydney, NSW Australia
10th -14th December 2017.

Effect of CO_2 Dilution on End-gas Auto-ignition in a Rapid Compression Machine

Yunliang Qi[1], Yingdi Wang[1], Yanfei Li[1], Hui Liu[1], Zhi Wang[1, 2]

[1] State Key Laboratory of Automotive Safety and Energy, Tsinghua University

Beijing, 100084, China

[2] Center for Combustion Energy, Tsinghua University

Beijing, 100084, China

Abstract

The effect of carbon monoxide (CO_2) dilution on end-gas auto-ignition of stoichiometric iso-octane/oxygen/nitrogen mixture was studied using a rapid compression machine. Both the combustion process and pressure traces were recorded synchronously. Nine cases with the same energy density but with different compression ratios and CO_2 dilution ratios were tested. The results showed that CO_2 dilution could reduce the knock intensity but prolong the combustion duration. The reason for the knock intensity reduction was attributed to the suppression of detonation caused by the CO_2 dilution. Additionally, the dimensionless analysis was carried out using two parameters (ε and ξ) proposed by Bradley [1]. The results showed that ε decreased with the increasing CO_2 dilution ratio, and under high CO_2 dilution ratio ξ decreased with increasing compression ratio.

1 Introduction

Super-knock has become one of the main factors restricting the R&D on high efficiency gasoline engines. Pre-ignition is believed to be the cause of super-knock. Once the pre-ignition occurs, the probability of end-gas auto-ignition increases. Under the high energy density conditions especially under those in highly boosted engines, the end-gas auto-ignition can cause detonation [2, 3]. Thus, two ways can be taken to suppress the super-knock. One is to prevent pre-ignition and the other one is to prevent the detonation formation. With regard to the suppression of pre-ignition, the adopted measures included improving the oil and fuel components, optimizing the injection strategy and using exhaust gas recirculation (EGR) and so on [4]. The researches related to detonation suppression are relatively rare. Zaccardi and Escudié [5] increased the EGR ratio to 10% at the constant engine loads and found the pre-ignition frequency was reduced by more than 50%. Amman et al [6] used 10% EGR and increased the engine indicated mean effective pressure (IMEP) from 1.24 MPa to 1.45 MPa. The results showed that the pre-ignition frequency was reduced by 80% and the knock intensity was lowered by 30%. Thus, EGR can not only suppress the pre-ignition, but also can reduce the tendency of end-gas auto-ignition to develop into detonation.

However, more fundamental studies are needed to understand the effect of EGR on the suppression of super-knock. In this study, CO_2 was used as the surrogate for the EGR components and dilute the stoichiometric iso-octane mixture. The effects of CO_2 dilution ratio and compression ratios (CR) on the end-gas auto-ignition were studied in a rapid compression machine (RCM). Throughout the tests, the mixture energy densities were kept constant.

Corresponding author: Zhi Wang
E-mail address: wangzhi@tsinghua.edu.cn

2 Experimental Setup

The experimental setup, as shown in Figure 1, consists of a RCM, a high speed camera (Photron SA-X2) and a data acquisition system.

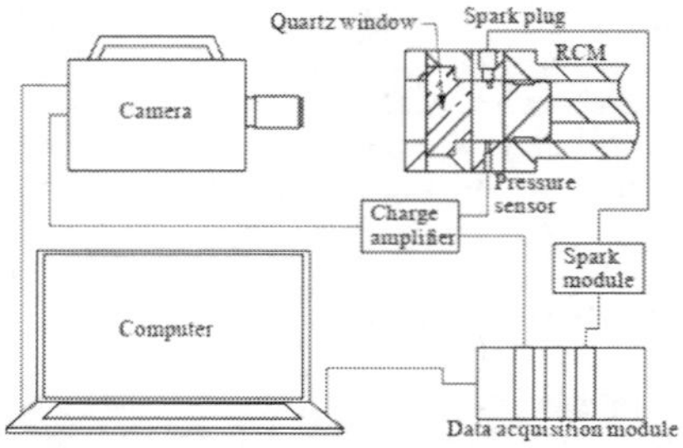

Figure 1: Schematic of experimental setup

2.1 Rapid Compression Machine

The RCM has a stroke of 500 mm and a combustion chamber bore of 50.8 mm. A quartz window was mounted in the end cap of the combustion chamber, which has an effective observation diameter of 50.8 mm and can provide the optical access to the whole combustion chamber. A spark plug (Denso K20R) was mounted in the circular wall of the combustion chamber to ignite the mixture after the end of the compression (EOC). A pressure sensor (Kistler 6125C) was also mounted in the circular wall of the combustion chamber, where the location is opposite to the spark plug. The adjustment of CR was achieved by changing the height of the combustion chamber.

2.2 Data Acquisition

Pressure traces and spark ignition signals were recorded synchronously using a NI data acquisition system with a sample rate of 500 kHz. The spark ignition was triggered by the pressure signal. By setting an appropriate pressure threshold, the spark timing could be controlled within 1 ms after the EOC.

The high speed camera, coupled with a 50 mm lens (Nikon

Micro Nikkor 1:2.8G), was triggered by the falling edge of the TTL spark signal, then the combustion images was recorded at a frame rate of 288 kHz.

2.3 Test Conditions and Mixture Preparation

Nine test conditions including three CRs (10, 11.4 and 14.3) and three CO_2 dilution ratios (0%, 10% and 20%) as listed in Table 1 were examined. The mixture preparation was as follows. First, a stoichiometric mixture was prepared by the partial pressures of iso-octane, oxygen and nitrogen. Then the desired content of CO_2 was added into the stoichiometric mixture. The CO_2 dilution ratio denotes the ratio of the CO_2 partial pressure to the pre-prepared stoichiometric mixture

Case No.	CR	CO_2 content	p_{EOC} (MPa)	T_{EOC} (K)	Camera Settings FR (fps)	Shutter (μs)
1	10	0%	2.54	641	288 k	0.29
2		10%	2.72	634	288 k	0.29
3		20%	3.02	633	36 k	26.7
4	11.4	0%	2.35	681	288 k	0.29
5		10%	2.58	677	288 k	2.45
6		20%	2.82	675	36 k	26.7
7	14.3	0%	2.28	729	288 k	0.29
8		10%	2.43	724	288 k	2.45
9		20%	2.7	723	36 k	26.7

Table 1: Experimental conditions of nine test cases

pressure. To prevent the quartz window from being cracked by the end-gas auto-ignition induced detonation, a pressure upper limit at the EOC timing was set. For the stoichiometric mixture without CO_2 dilution, the limit was 2.5 MPa.

This paper intends to study the effect of CO_2 dilution on the end-gas auto-ignition under the same energy density condition at the spark timing, i.e. the EOC timing. Therefore, to maintain the same energy density for different CRs and CO_2 ratios, the target pressure at EOC timing was decreased as CR and increased as the CO_2 ratio.

The camera shutter and frame rate were also adjusted based on the certain conditions to avoid overexposure (image saturation) or underexposure.

3 Results and Discussion

3.1 End-gas Auto-ignition Characteristics

Figure 2 shows the pressure traces of the nine test cases and the corresponding images of the end-gas auto-ignition are shown in Figure 3. The main knock intensity parameters are summarized in Table 2. As seen in Figure 2 and Table 2, for the cases with the same CO_2 ratio the knock onset timing is advanced with increasing CR. This can be attribute to two aspects. The first one is the increased spark-ignited flame speed due to the increased CR and the resultant temperature increase at EOC; the other one is the shortening of the end-gas ignition delay due to the temperature increase. Another obvious trend can also be found in Figure 2 and Table 2. For the cases with the same CR, the knock onset timing is retarded as the CO_2 content increases. In addition, the knock intensity (both peak pressure and the maximum pressure amplitude) is also decreased, except Case 2.

Figure 2 and Table 2 show that Case 1 has the same peak pressure (21.3 MPa) as Case 2. Besides, the three cases without

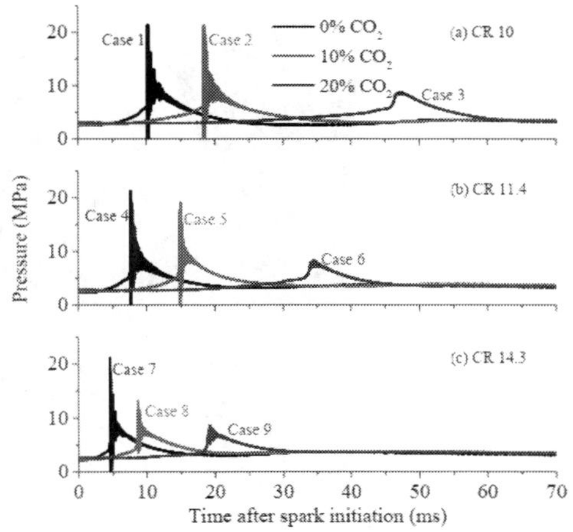

Figure 2: Pressure traces of the nine test cases with different CO_2 contents and CRs

Case No.	p_{max} (MPa)	t_{pmax} (ms)	p_{ko} (MPa)	t_{ko} (ms)	p_{amp} (MPa)
1	21.29	10.17	6.58	10.17	14.47
2	21.29	18.38	11.35	18.30	12.59
3	8.79	47.10	8.23	46.55	0.16
4	21.27	7.60	7.09	7.59	12.92
5	19.17	14.93	9.23	14.87	10.16
6	8.45	34.56	7.28	33.97	0.70
7	21.25	4.58	5.92	4.56	13.75
8	13.19	8.71	7.85	8.63	5.18
9	8.71	19.19	6.69	18.9	1.56

Table 2: Knock intensity parameters

CO_2 dilution (Cases 1, 4 and 7) also have almost the same peak pressure (21.3 MPa). It should be noted that they are not the real peak pressures due to the signal cutoff. The occurrence of the detonations could generate the waves with extremely high pressures, which could exceed the measurement range (up to 20 MPa) of the data acquisition system. Figure 3 clearly shows the detonation initiation processes of Cases 1, 2, 4 and 7, where the highly bright areas near the circular wall are the detonation waves. By extracting the detonation wave front using the in-house image processing code, the detonation waves are estimated to propagate at the speeds over 1800 m/s, which is very close to the C-J detonation speed.

For the three cases without CO_2 dilution, it is impossible to compare the peak pressures due to the pressure signal cutoff. For the cases with 10% CO_2 content (Cases 2, 5 and 8), the peak pressure decreases as the CR increases. This can be attributed to the change of end-gas auto-ignition combustion mode at high CRs. For Case 2, very clear detonation waves can be observed in the end-gas area in Figure 3. However, for Cases 5 and 8, the end-gas auto-ignition generated flames have much less clear front in contrast to that in Case 2. Moreover, less time is used for the flame to occupy the combustion chamber than the detonation wave of Case 2 does (17.36 μs for Cases 5 and 8, and 38.19 μs for Case2). In other words, the speed of the end-gas auto-ignition generated flames for Cases 5 and 8 is higher than a detonation wave. Bradley et al [1] and Gu et al [7] identified a reaction wave propagation mode, supersonic autoignitive deflagration, of which the combustion wave propagated much faster but had lower peak pressure than a detonation wave. In consideration of the speed and pressure

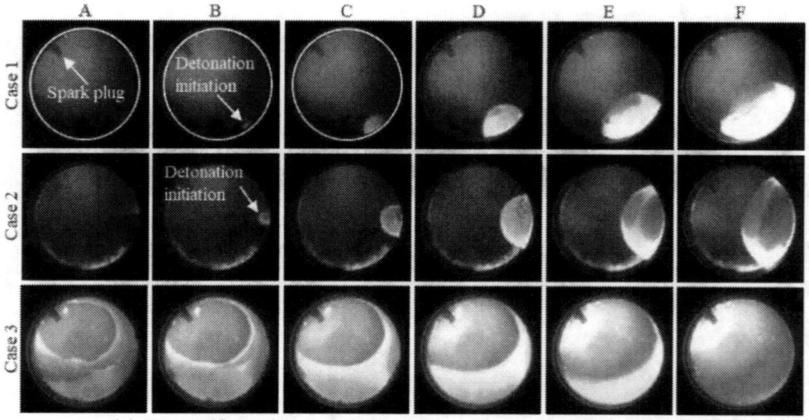

(a) CR=10 (frame interval: 3.47 μs (Case 1, Case 2), 27.78 μs (Case 3))

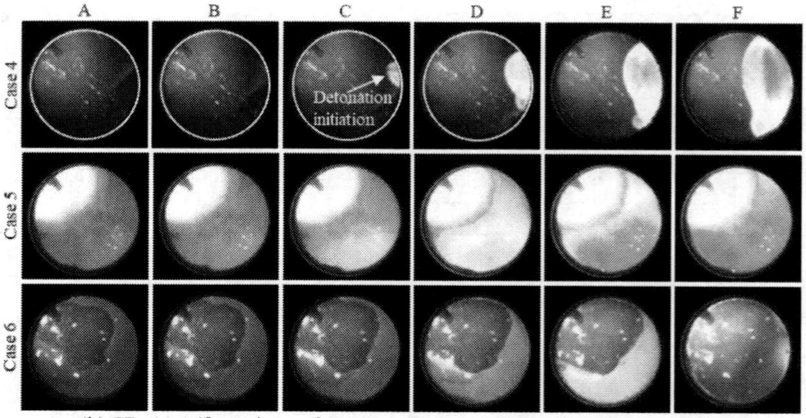

(b) CR=11.4 (frame interval: 3.47 μs (Case 4, Case 5), 27.78 μs (Case 6))

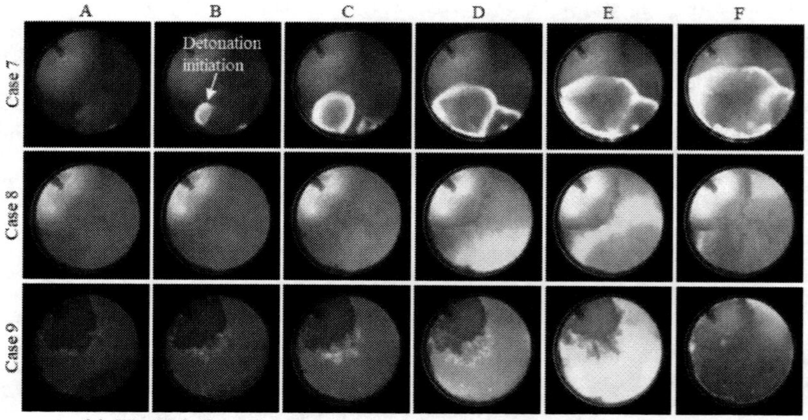

(c) CR=14.3 (frame interval: 3.47 μs (Case 7, Case 8), 27.78 μs (Case 9))

Figure 3: Combustion images during end-gas auto-ignition of the nine test cases

characteristics of Cases 5 and 8, the end-gas auto-ignition generated flames of the two cases can be regard as supersonic autoignitive deflagration.

For the cases with 20% CO_2 content, the peak pressure slightly changes with the increasing CR, but the pressure amplitude increases as the CR increases, which is contrast to the cases with 10% CO_2 content. Although the camera settings of longer shutter duration and lower frame rate have increased the motion blur for Cases 3, 6 and 9, no clear wave front is spreading out from the end- gas area, as shown in Figure 3, indicating that no detonation occured.

The above results show that the CO_2 dilution can suppress the knock intensity and reduce the detonation probability under the same energy density conditions. However, it should be noted that the CO_2 dilution also prolongs the combustion duration, which can deteriorate the engine thermal efficiency. When CO_2 dilution is used in an engine, other measures should be taken to accelerate the combustion.

3.2 Analysis on the End-gas Auto-ignition Mode

Kalghatgi and Bradley [8] used two parameters, ε and ξ, based on hot-spot auto-ignition theory to characterize the relationship between knock intensity and end-gas combustion mode. The two parameters are calculated using the following equations:

$$\varepsilon = \frac{r_0/a}{\tau_e} \quad (1)$$

$$\xi = a\frac{\partial \tau_i}{\partial T}\frac{\partial T}{\partial r} \quad (2)$$

where r_0 is hot spot radius; a is sound speed; τ_e is excitation time; τ_i is ignition delay; T is temperature; r is the radial direction of the hot spot.

In this study, τ_e and τ_i are calculated using Wisconsin PRF mechanism [9]; and r_0 and $\partial T/\partial r$ are assumed to be 5 mm and 2 K/mm, respectively, the same to those adopted by Kalghatgi and Bradley [8]. Table 3 summarizes the characteristic parameters for the nine tested cases at knock onset timing, including the calculated ε and ξ. Figure 4 shows the locations of ε and ξ in the detonation peninsula [8] for the nine test cases. Obviously, Cases 2, 4 and 5 are located in detonation area while Cases 3, 6, 8 and 9 in deflagration area, which is in accordance with the analysis in Section 3.1. Although the locations of Cases 1 and 7 are outside the detonation area, they are still very close to the detonation-deflagration boundary. Considering that the boundary between detonation and deflagration in this peninsula was not a strictly distinct curve [1], there is a chance that the detonation points are located outside the boundary.

Case No.	p_{ko} (MPa)	T_{ko} (K)	a (m/s)	τ_e (μs)	$d\tau_i/dT$ (μs/K)	ε	ξ
1	6.58	793	529	1.21	-8.49	7.83	8.97
2	11.35	851	534	1.70	-2.83	5.52	3.03
3	8.23	764	497	4.53	-27.02	2.22	26.85
4	7.09	853	547	1.11	-2.19	8.25	2.40
5	9.23	862	536	1.93	-3.05	4.83	3.27
6	7.28	787	504	4.71	-14.64	2.11	14.76
7	5.92	885	556	1.20	-8.28	7.49	9.22
8	7.85	891	545	2.08	-7.15	4.42	7.79
9	6.69	828	516	4.73	-6.17	2.05	6.37

Table 3: Parameters of the nine test cases at knock onset timing

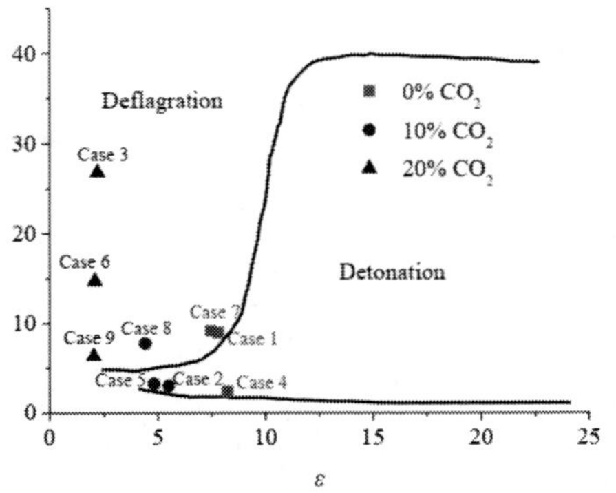

Figure 4: Location of ε-ξ in detonation peninsula for the nine test cases at knock onset timing

Figure 4 also clearly shows that ε decrease with the increase of CO_2. This is mainly due to the increase in τ_e. According to Equation (1), ε is proportional to r_0/a and inversely proportional to τ_e. Table 3 shows that τ_e continuously increases with the increase of CO_2, although a increases first then decreases.

With regard to ξ, Figure 4 shows that under 20% CO_2 dilution (Cases 3, 6 and 9), it decreases as the CR increases, indicating the increasing probability in the occurrence of detonation. This is in accordance with the change of pressure amplitude with the increasing CR. However, for the other two CO_2 dilution ratios, it is difficult to see a clear trend over CR in Figure 4.

4 Conclusions

1. Under the same energy density conditions, CO_2 dilution can reduce the knock intensity but prolong the combustion duration. The knock intensity decreases as the compression ratio increases under low CO_2 dilution ratio but increases as the compression ratio increases under high CO_2 dilution ratio.
2. The reason for the knock intensity reduction is that the CO_2 dilution can suppress the end-gas auto-ignition developing into detonation. Under the same compression ratio, the suppressing effect becomes more prominent under higher CO_2 ratios.
3. ε decreases with the increasing CO_2 dilution ratio. ξ decreases with increasing compression ratio under high CO_2 dilution condition.

5 Acknowledgment

This work is supported by National Science Foundation of China (Grant No.: 91541206).

References

[1] D. Bradley; C. Morley; X. J. Gu, et al., SAE Technical Paper 2002-01-2868 (2002)

[2] Z. Wang; Y. L. Qi; X. He, et al., Fuel 144 (0) (2015) 222-227.

[3] Y. Qi; Z. Wang; J. Wang, et al., Combustion and Flame 162 (11) (2015) 4119-4128.

[4] Z. Wang; H. Liu; R. D. Reitz, Prog Energy Combust Sci. 61 (2017) 78-112.

[5] J.-M. Zaccardi; D. Escudié, Int J Engine Res 16 (2) (2014) 152-165.

[6] M. Amann; T. Alger; D. Mehta, SAE Int. J. Engines 4 (1) (2011) 235-245.

[7] X. J. Gu; D. R. Emerson; D. Bradley, Combustion and Flame 133 (1-2) (2003) 63-74.

[8] G. T. Kalghatgi; D. Bradley, Int J Engine Res 13 (4) (2012) 399-414.

[9] Y. Ra; R. D. Reitz, Combustion and Flame 155 (4) (2008) 713-738.

11th Asia-Pacific Conference on Combustion,
The University of Sydney, NSW Australia
10th -14th December 2017.

Combustion of Methanol in Diesel Engine Using Diethyl Ether as Ignition Enhancer

R.Vallinayagam, S.Vedharaj, Mohammed Jaasim, Hong.G. Im, S. M. Sarathy, R. W. Dibble

King Abdullah University of Science and Technology, Clean Combustion Research Centre,

Thuwal, 23900, Saudi Arabia.

Abstract

In this study, combustion of methanol in diesel engine is investigated using diethyl ether (DEE) as an ignition enhancer. Both experimental and numerical studies were performed and the results are compared with baseline diesel fuel. DEE has higher derived cetane number (DCN) of 139 when compared to methanol (DCN=8). Two blends DM75 (DEE 75% v/v + Methanol 25% v/v) and DM50 (DEE 50% v/v+ Methanol 50% v/v) were tested in diesel engine. For the same fuel injection timing of -20 CAD (aTDC), the intake air pressure is boosted to 1.75 bar to support auto ignition of DM mixtures. The peak cylinder pressures were comparable for diesel and DM75, while DM50 exhibited lower pressure with delayed start of combustion (SOC). Full cycle numerical simulations were performed using 3-D computational fluid dynamics (CFD) code (CONVERGE). DM75 and diesel showed similar flame liftoff and combustion phasing, and the numerical simulation results matched with the experimental results. Diesel ignited at a rich equivalence ratio range of 2-3, whereas DM75 and DM50 ignited at an equivalence ratio range of 1.0-1.5 and 0.3-0.6, respectively. The DM50 blend was found to exhibit stratified combustion with low in-cylinder and exhaust temperature when compared to DM75 and diesel.

1 Introduction

Stringent emission regulations and impact of global warming have paved way for alternative biofuel research [1]. Multiple research areas (low temperature combustion, after-treatment, and lean combustion) are proposed to curb the emissions from diesel fueled compression ignition (CI) engines. CI engines operating on diesel is widely used due to its higher thermal efficiency, but NO_X and particulate emissions are highly undesirable [2]. This has lead to alternative fuel research that can substitute diesel with similar efficiency, but yields lower emissions than diesel fuels. Soot is reported as a carcinogen [3] and advanced combustion concepts such as partially premixed combustion (PPC) and homogeneous charge compression ignition (HCCI) enables simultaneously reduce soot and NO_x emissions [4, 5]. Fuel stratification studies indicate that stratification reduces emission and operation at LTC regime is effective in abating emissions [6]. Fuels with longer mixing time are also an area of interest, since it results in a lean mixture, which facilitates reduced emissions [7]. However, such fuel that creates longer ignition delay has a disadvantage of misfire, which increases unburned hydrocarbons (UHC) emissions [8]. Operation of low reactive fuels in diesel engine requires strategies for improving the auto ignition of fuel [9]. While engine modification techniques to support ignition are complex, blending ignition promotes is a simple strategy.

Various types of biofuels are widely studied but very few biofuels are the suitable alternative to diesel [10]. Alcohols such as ethanol and methanol are such type of biofuels that can be used efficiently after proper design and optimization of the engine. The advantages of using alcohol fuels in engines are well reported in the literature [11] . A very high laminar speed of alcohol fuels facilitates faster combustion and increases the thermal efficiency [12]. Lower viscosity and boiling point of alcohol fuels improves fuel atomization, evaporation and air/fuel mixing process . Evaporative cooling of alcohols due to its higher latent heat of vaporization can help in increasing the volumetric efficiency of the engine that reduces the work required during the compression stroke [13].

Demonstration of methanol in SI engines has been done, which showed that methanol could be introduced in gasoline fuel market [14]. Similarly, methanol can be utilized in compression ignition engines. Methanol's lower cetane number inhibits its use directly in CI engines. To improve the cetane number, it is blended with diesel and used in CI engines [15]. In most of the studies, it is reported that oxygenated methanol fuel increases the brake thermal efficiency (BTE). Methanol also has the tendency to lower engine out emissions when compared to diesel due to high air/fuel stoichiometric ratio, high H/C ratio and lower sulfur content [16]. Some commercially available ignition promotes such as nitrates and peroxides can be blended with methanol to support auto ignition. Ethers such as diethyl ether (DEE) and dimethyl ether (DME) are recognized as ignition promoters [17, 18] and can be used to promote ignition of pure methanol. DEE is in liquid form and can be easily handled and stored. DEE can be derived from ethanol easily and used as an ignition promoter.

In this study, methanol was selected as base fuel with DEE as an ignition enhancer to promote ignition. Suitable blends of methanol and DEE are prepared and used in CI engine. An attempt to understand the auto-ignition and combustion characteristics of methanol-DEE blends in CI engine is performed and reported. In the present work, DEE was blended with methanol at 75% and 50% by volumes, DM75 (DEE 75%+ Methanol 25%) and DM50 (DEE 50% + Methanol 50%) and their combustion characteristics were compared. The experimental conditions were simulated using 3D CFD code to for gaining a better understanding of the combustion and emission characteristics. A detailed chemical kinetic model that was developed in-house for DEE and methanol blends was reduced and used in the CFD code. Analysis of in-cylinder temperature and equivalence ratio distribution complements the experimental combustion results.

Corresponding author. Phone: +966-544-701-225
E-mail address: vallinayagam.raman@kaust.edu.sa

2 Methodology

2.1 Fuel properties

Table 1 shows the comparison physical and thermal properties between DEE, methanol and US NO.2 diesel. The boiling point of DEE and methanol is closer to that of gasoline. Thus, fuel evaporation and the subsequent air/fuel-mixing process might be improved. Compression ignition of 100% methanol is not possible in CI engines as the cetane number of it is too low. In order to overcome the poor ignition characteristics of methanol, DEE is added into methanol. These two fuels, when blended together, are expected to have improved ignition characteristics due to DEE acting as an ignition promoter. The derived cetane number (DCN) of these mixtures was measured from ignition quality tester based on ASTM D6890. The measured DCN are DM75 = 58, DM50 = 21 and DM25 = 9. Furthermore, the latent heat of vaporization of methanol is higher, which could produce a charge cooling effect during fuel injection. DEE is blended with methanol in following proportions: DM75 (75% DEE + 25% methanol), DM50 (50% DEE+ 50 % methanol) and DM 25 (25% DEE + 75% methanol).

Property	US NO. 2 Diesel	Methanol	Diethyl ether
Density (kg/m^3)	843	792	720
Viscosity (*10^{-6}m^2/s)	2.3	0.44	0.2
Boiling point (°C)	180-300	64.7	35
Flash point (°C)	69	11	-45
Enthalpy of combustion (MJ/kg)	45.6	22.5	33.9
H/C	1.75	4	2.5
O/C	0	1.3	0.25
Derived cetane number	48	8	139
Latent heat of vaporization (kJ/kg)	250	922	376
Stoichiometric air – fuel ratio	14.7	6.66	11

Table 1: Fuel properties of diesel, methanol and diethyl ether

2.2 Experimental Methodology

A single cylinder AVL research engine with a compression ratio (CR) of 17 is used for investigating DM mixtures. Other specifications of the engine can be found in the previous study by the author [18]. The engine is operated at a speed of 1200 rpm at an ambient intake air temperature of 35°C. The intake air pressure is boosted to 1.75 bar to support the auto ignition of DM mixtures. Fuel is injected at a pressure of 300 bar and fuel injection timing is maintained at -20 CAD (aTDC). A piezo electric pressure transducer measures the in-cylinder pressure with a sensitivity of 19.35 pC/bar. From the pressure data recorded for 300 consecutive cycles, heat release is calculated using AVL indicom software. The operation of biofuel blends has limitations since the CI engines are optimized for diesel operation. The poor lubrication properties of DM mixtures could damage the fuel injection equipment's. To overcome this, 500 ppm of lubricating additive (Lubrizol 539M) was added to the DM mixtures.

2.3 Computational Modeling

Figure 1 (a) shows the CAD geometry of the engine and Figure 1 (b) shows the mesh generated in-situ for the simulations. Full cycle CI engine simulations were performed using CFD code CONVERGE accounting for mass, momentum and energy conservation. Lagrangian spray droplets are used to model the spray with Kelvin- Helmholtz Rayleigh-Taylor (KH-RT) spray break up to represent the spray breakup during injection. A base mesh size of 4 mm is used with AMR of level 4 for the simulations which refined the mesh locally based on the temperature difference of 5K between the neighboring cells (New cell size = Base mesh/2$^{(level)}$). Embedding of grid near nozzle was employed during spray injection event to capture the near nozzle behavior of the spray. Flow inside the nozzle is not considered in the study. The minimum mesh size obtained was 0.25 mm in the simulations, which ensured grid converged result (not shown). The gas phase chemical kinetic reaction rate calculation was performed using the multi-zone combustion model, which assumes each cell as well stirred reactor and groups them into zones based on bin range of 0.05 ϕ and 5 K increments. Heat loss from the walls is accounted using wall heat loss model and the in-cylinder turbulence is modeled using renormalized group k-epsilon model. A skeletal mechanism with 88 species and 470 reactions was used in the engine simulations to account for gas phase chemical kinetics of DM mixtures [19]. The n-heptane mechanism that consists of 40 species and 165 reactions was used to represent the gas phase combustion chemistry of diesel fuel. More details about the numerical models and chemical kinetic models used can be found in the previous study by authors [20].

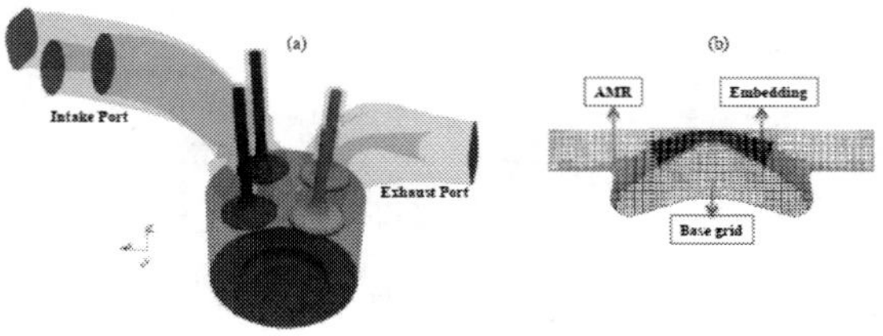

Figure 1: (a) CAD geometry of the AVL CI engine, (b) Mesh generated for AVL CI engine from top-view.

3 Results and Discussion

3.1 Experimental Study of methanol combustion

Diesel and DM75 have higher cetane number and required no boosted pressure or increased intake temperature for auto ignition. DM50 and DM25 have lower cetane number due to increase in the proportion of methanol that suppresses the auto ignition. The intake pressure was boosted from 1 bar to 2 bar at increments of 0.25 bar with the constant intake air temperature of 25 °C. The auto ignition was achieved by increasing the intake air pressure. No successful ignition of DM25 was achieved for all intake pressures (P_{in}) and temperature conditions. Auto-ignition of DM 50 was not possible at P_{in} = 1 bar, but the mixture auto ignited at P_{in} = 1.75 bar. For comparison of results, all tests for diesel, DM 75 and DM 50 were conducted at P_{in} = 1.75 bar and T_{in} = 25 °C.

The in-cylinder pressure and heat release rate for diesel and DM mixtures at indicated mean effective pressure (IMEP) = 4 bar are shown in Figures 2 and 3. DM50 was able to achieve auto-ignition only at advanced injection timing of -20 CAD aTDC due to its lower DCN. Therefore, diesel and DM75 were injected at same injection timing to make a fair comparison. Note that diesel and DM75 were able to achieve combustion at all ranges of fuel injection timing. When diesel and DM75 were operated, ignition delay was noted to be 6 CAD and 8 CAD, respectively. Given the DCN of DM75 is comparable to diesel, the ignition delay of it is shorter and comparable to diesel. DM 50 showed a longer ignition delay of 22 CAD due to its lower DCN. Diesel and DM75 showed similar pressure rise rate whereas DM50 exhibited 30% reduced in-cylinder pressure rise rate when compared to diesel and DM75.

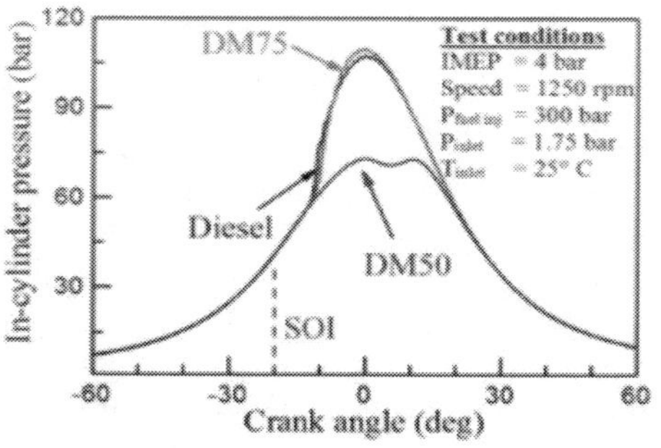

Figure 2: In-cylinder pressures for DM mixtures

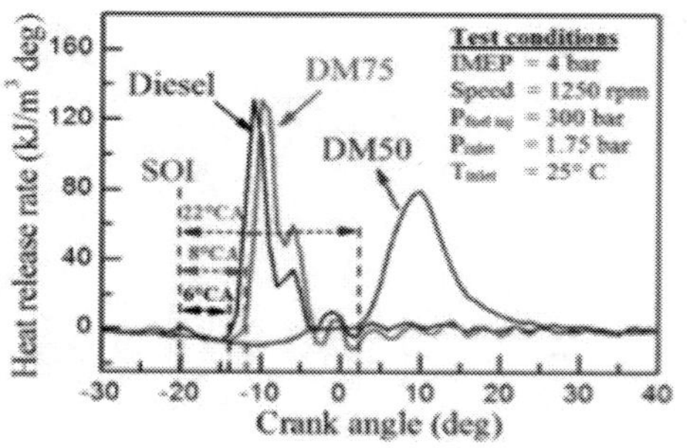

Figure 3: Heat release rate for DM mixtures

DM mixtures require a higher quantity of fuel to produce the same IMEP due to lower enthalpies of combustion of DEE and methanol when compared to diesel. The maximum heat release rate is lower for DM50 due to longer ignition delay and lower enthalpy of combustion. The DCN of DEE is higher than that of methanol, but methanol suppresses the auto-ignition of DEE by radical scavenging. The higher enthalpy of vaporization of methanol increases the auto-ignition time of methanol-enriched mixtures. The lower viscosity and bulk modulus of the DM mixtures delayed the start of injection of fuel, which results in 2 CAD delayed ignition delay than diesel in case of DM75. Note that cetane number of DM75 is higher than diesel. The combustion duration of DM75 is also smaller than diesel for the same injection timing.

3.2 Computational Study of methanol combustion

Simulations were performed for improved understanding on the combustion characteristics of DM blends in comparison with diesel fuel. Figure 4 shows the in-cylinder pressure comparison of diesel and DM mixtures between the experiments and simulations. The numerical simulations were able to reproduce the experimental results satisfactorily for SOC, peak pressure and ignition delay. DEE has a carbon that is loosely bonded to oxygen atom, which leads to effective H-atom abstraction. The intra-molecular hydrogen migration and bimolecular reactions lead to formation of highly reactive peroxides as a result of chain branching. However, as methanol proportion increases the radical formation is suppressed by the scavenging effect of methanol, which results in increasing ignition delay. This is reproduced in the simulations, supporting the chemical model and the simulation approach. In the computations, DM25 was not able to achieve ignition, which is in line with the experimental findings (not shown).

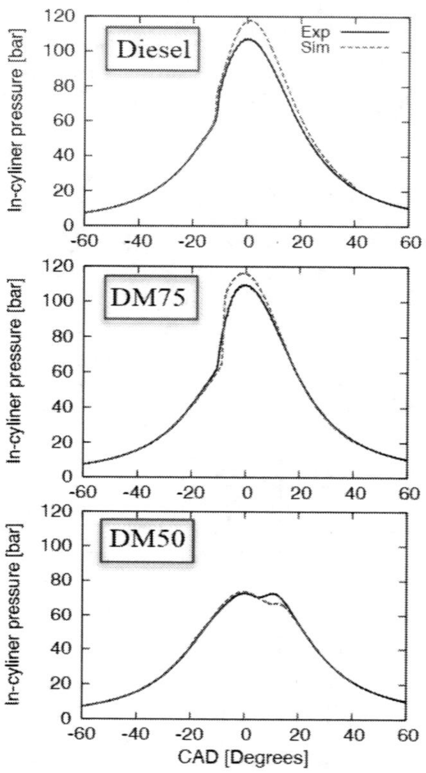

Figure 4: In-cylinder pressures for diesel and DM mixtures

The in-cylinder temperature and equivalence ratio distribution are investigated at CA10 (crank angle at 10% of fuel burnt), CA50 (crank angle at 50% fuel burnt) and CA90 (crank angle at 90% fuel burnt) and shown in Figure 5. Diesel and DM75 exhibits flame liftoff structure whereas no flame liftoff is observed for the DM50. Diffusion flame is clearly observed for diesel and DM75 whereas lean combustion is noted for DM50. Thus, spray dominant combustion is noted for diesel and DM75 with similar combustion characteristics. With DM50, the ignition delay is longer and this increased the air/fuel mixing process. The increased premixing effect reduces the in-cylinder temperature, as observed from the temperature contour. The contours show that the combustion clouds are scattered and the mixture is lean. DM50 exhibits a lean premixed charge at SOI of -20 CAD aTDC, which is compared using a histogram of volume ratio to equivalence ratio. This plot was drawn with a bin size of 0.1, as shown in Figure 6, for diesel and DM75 at SOC, to obtain a more quantitative comparison of the mixture distribution. Diesel exhibits lean to rich mixture distribution of φ between 0.1 to 3 when compared to DM75, which has a distribution (φ) range of 0.1 to 2.2. DM50 has a very lean distribution range of 0.1-0.8.

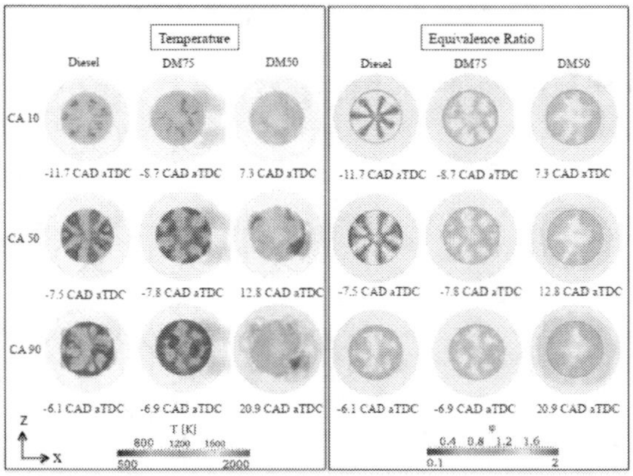

Figure 5: Volume contour of temperature and equivalence ratio distribution for diesel and DM mixtures at CA 10, CA 50, CA 90 (bottom view).

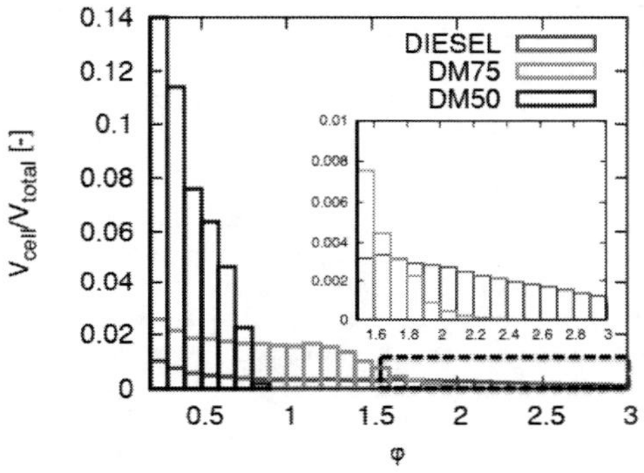

Figure 6: Volume ratio and φ histogram of bin width of 0.1

4 Conclusions

Full cycle numerical simulations were performed using CONVERGE for DM mixtures and diesel. The numerical results were compared with experiments conducted in CI engine at SOI of -20 CAD aTDC. DE75 exhibits similar combustion characteristics like diesel. DM50 was able to achieve ignition at an intake pressure of 1.75 bar. DM25 was not able to auto-ignite at any of the intake air pressure and temperature conditions. The SOC for diesel and DM75 blend was comparable whereas the DM50 had delayed SOC. The peak in-cylinder pressure of diesel and DM75 was comparable whereas peak pressure decreased by 30% when compared to diesel. The addition of DEE to methanol improves the cetane number whereas methanol suppresses the auto-ignition by radical scavenging effect. Numerical simulations by CFD were able to reproduce the experimental results in terms of combustion. Diesel and DM75 showed flame liftoff whereas combustion is premixed and lean for DM50. Diesel ignites at rich equivalence ratio range at φ = 2-3 and DM75 ignites near the stoichiometric range of φ = 1-1.5. DM50 blend ignited at lean region of φ=0.1-0.6 and the longer ignition delay period increased the mixing time leading to lower in-cylinder temperature. DM50 blend is promising due to its low temperature combustion regime operation, which could simultaneously reduce the NO_X and soot emissions.

5 Acknowledgments

The work was supported by competitive research funding from King Abdullah University of Science and Technology (KASUT) under Clean Combustion Research Center's Future Fuel's program. The simulations were performed using clusters in KAUST IT research computing facility. We thank our research technician Mr. Adrian Ichim, for his support in carrying out the engine experiments at KAUST engine lab.

References

[1] A.K. Agarwal, Progress in energy and combustion science 33 (3) (2007) (233-271).

[2] K. Okude, K. Mori, S. Shiino, T. Moriya, 2004-01-1907, SAE Technical Paper (2004)

[3] P. Gerde, B. Muggenburg, M. Lundborg, A. Dahl, Carcinogenesis 22 (5) (2001) (741-749).

[4] V. Manente, B. Johansson, P. Tunestal, Journal of Engineering for Gas Turbines and Power 132 (8) (2010) (082802).

[5] M. Yao, Z. Zheng, H. Liu, Progress in Energy and Combustion Science 35 (5) (2009) (398-437).

[6] A. Sarangi, *Diesel low temperature combustion: an experimental study*, PhD thesis, University of Loughborough, Leicestershire, United Kingdom, 2012.

[7] G.T. Kalghatgi, 2005-01-0239, SAE Technical Paper (2005)

[8] J.E. Dec, Proceedings of the combustion institute 32 (2) (2009) (2727-2742).

[9] L. Xing-cai, Y. Jian-Guang, Z. Wu-Gao, H. Zhen, Fuel 83 (14) (2004) (2013-2020).

[10] R. Vallinayagam, S. Vedharaj, W. Yang, W.L. Roberts, R. Dibble, Renewable and Sustainable Energy Reviews 51 (2015) (1166-1190).

[11] A.C. Hansen, Q. Zhang, P.W. Lyne, Bioresource technology 96 (3) (2005) (277-285).

[12] S. Liu, E.R.C. Clemente, T. Hu, Y. Wei, Applied Thermal Engineering 27 (11) (2007) (1904-1910).

[13] C. Yao, C. Cheung, C. Cheng, Y. Wang, T. Chan, S. Lee, Energy conversion and management 49 (6) (2008) (1696-1704).

[14] R. Vallinayagam, S. Vedharaj, N. Naser, W.L. Roberts, R.W. Dibble, S.M. Sarathy, Fuel 187 (2017) (9-15).

[15] A. Kowalewicz, Proceedings of the Institution of Mechanical Engineers, Part D: Journal of Automobile Engineering 207 (1) (1993) (43-52).

[16] C. Sayin, Fuel 89 (11) (2010) (3410-3415).

[17] C. Cheng, C. Cheung, T. Chan, S. Lee, C. Yao, Science of the total environment 389 (1) (2008) (115-124).

[18] S.C. Sorenson, Transactions-american society of mechanical engineers journal of engineering for gas turbines and power 123 (3) (2001) (652-658).

[19] R. Vallinayagam, S. Vedharaj, S.M. Sarathy, R.W. Dibble, (2016) (V001T002A004-V001T002A004).

[20] F.H. Perez, R. Vallinayagam, S. Vedharaj, B. Johansson, H. Im, SAE Technical Paper 2017-01-0573 (2017)

11th Asia-Pacific Conference on Combustion,
The University of Sydney, NSW Australia
10th -14th December 2017.

Characteristics of oxygen-enriched laminar ethylene diffusion sooting flames

Zhiwei Sun[1,2], Bassam Dally[1,2], Zeyad Alwahabi[1,3], Graham Nathan[1,2]

[1]Centre for Energy Technology (CET), [2]School of Mechanical Engineering, and [3]School of Chemical Engineering,

The University of Adelaide, Adelaide, S.A. 5005, Australia

Abstract

This paper reports on an experimental study of the effects of oxygen enrichment on laminar steady ethylene diffusion flames using optical methods. The volumetric concentration of oxygen the oxidizing stream was varied from 21%, i.e. ambient air, to 25.0%, 28.9% and 36.8%, while the volume flow rates of the fuel and the oxidizing stream were held constant with a velocity ratio of 1:6. Soot luminosity in the near infrared, excited hydroxyl radicals (OH^*) chemiluminescence, radial distribution of the ground-state OH, soot volume fraction and flame temperature (*gaseous*) were measured using optical techniques with high spatial resolutions. The experimental methods include direct imaging combined with inversion re-construction, line-wise laser-induced fluorescence, planar laser-induced incandescence and planar two-line atomic laser-induced fluorescence using atomic indium as tracer. Results show that oxygen-enrichment can significantly influence flame structure, soot formation and flame temperature. The datasets, together with those not presented here (including soot particle temperature, primary particle diameter and results in other six flames with oxygen concentration of less than 21%) will help shed light on the role of O_2 in soot evolution in flames. The data may also serve as benchmark for developing soot models due to the well-defined boundary conditions, high spatial resolution, systematic nature and combination of measured scalars.

1 Introduction

Oxygen-enriched combustion (OEC) offers the potential advantages of fuel savings, increased thermal efficiency, improved flame characteristics and efficient carbon capture [1]. OEC is particularly attractive in those industries where high temperatures and high-intensity heat are required, such as for the manufacture of steel, glass and cement. While the combustion performance is greatest with nearly pure oxygen (i.e. > 90% by volume), termed oxy-fuel combustion, the high cost of cryogenic air separation has prevented its penetration into practical industries. However, the recent emergence of membrane-based air separation techniques (e.g. [2-4]) can provide oxygen-enriched air (~25%-50%) at low cost, raising interest in OEC, particularly given recent results suggesting that most of the benefits are achieved for oxygen concentrations below 35% when burning natural gas [5]. Other results showed that, in turbulent non-premixed ethylene flames, the thermal radiation only increases by 14% with an increase in oxygen in the co-flow from 50% to 100% [6]. Lin *et al.* showed that, for a flue gas temperature of 1922 K, OEC of natural gas can save energy by 35% and make an economic benefit of 29%, compared with natural gas burned in pure air [7].

Corresponding author. Phone: +61-8-8313 2310
E-mail address: zhiwei.sun@adelaide.edu.au

The advantages of OEC have been attributed not only to the increase in flame temperature, but also with the increase in soot formation in flames, particularly for non-premixed flames. This is because the soot loading controls the flame emissivity, to enhance the radiation heat transfer between the flame and the load. This has prompted recent investigations of oxygen-enriched sooting flames, both laminar and turbulent. For example, measurements have been performed of soot loading [8-10], flames structure [10-12], particle morphology [6, 9, 10, 13] and (spectrally resolved) thermal radiation [6, 8, 14]. Nevertheless, significant gaps remain in the understanding of soot evolution under OEC, which the present investigation aims to address.

While it is well established that the oxygen concentration influences the two competing effects of soot formation and oxidation, the two effects have not been studied together. That is, it is well known that an increase in flame temperature on the rich side of the reaction zone enhances fuel pyrolysis and soot formation (i.e. inception and surface growth), while on the lean side, an increase in both oxygen concentration and flame temperature increase the rate of soot oxidation [1, 11]. However, no previous investigation has reported the influence of oxygen enrichment on both soot formation and its oxidation. Instead, most previous experimental investigations report only the formation of soot and are also limited mostly to the investigation of soot volume fraction.

There is also a lack of data of flame temperature measurements in two dimensions in OEC flames, despite the importance of temperature in soot flames [11]. Temperature not only affects the viscosities and diffusion rates of flame species, but also is a controlling factor of most chemical reactions. However, while flame is known to increase with oxygen concentration, little (if any) quantitative data of this effect is available. This may be due to the challenges of measuring temperature in the highly luminous environment of flames that are laden with soot particles. Indeed, even measurements of soot particle temperature in oxygen-enriched steady flames are lacking, even though they can be relatively easily measured using two-color pyrometry through soot luminosity.

A further gap in previous investigations of OEC is lack of data of multiple key parameters in the same flames. This is desirable because of the non-linear, tightly coupled nature of the complex processes of combustion, particularly for sooting flames. Planar measurement with high spatial resolution is also desirable, because it can provide more information than point- or line- wise measurements, improving the capacity to validate models. In particular, for the small steady diffusion flames investigated here, soot formation and oxidation are both strongly influenced by gaseous diffusion and residence time of particles (associated with flame length). Thus, good spatial resolution in both two dimensions of the flames is necessary to accurately interpret results and assess the two competing effects of oxygen enrichment on sooting flames.

Finally, systematic data in OEC flames is also desirable. Du *et al.* performed a systematic study of oxygen concentration in oxidizing streams on soot formation in ethylene and propane diffusion flames, but only over the range of oxygen concentration between 15% - 21% (by volume). They concluded that the influence on soot inception limit is almost entirely thermal, with the chemical effect being negligible [15]. Glassman *et al.* showed that increasing oxygen concentration above that of air can enhance the inception rate and growth rate of soot particles, resulting in higher soot loading, although oxidation also becomes more pronounced. So that an maximum total soot loading appears at an oxygen concentration which is generally much less than pure oxygen and depends on both the fuel type and the turbulence strength [8].

In light of the above background, the present investigation aims to provide new insight into sooting flames by detailed measurement of four co-flowing oxygen-enriched ethylene (C_2H_4) diffusion flames. We aim to find the influence of oxygen concentration in the oxidizer on soot luminosity in the near infrared, on the distribution of the excited and ground-state hydroxyl (OH) radicals, on soot volume fraction and flame temperature. It is worth noting that the present investigation is closely related to our previous studies on effects of blending hydrogen and nitrogen into the fuel steam of ethylene/air diffusion flames [16].

2 Experiments

2.1 Burner and flames

Four laminar C_2H_4 diffusion flames were investigated. The oxidizing stream comprises a mixture of O_2 and N_2 whose ratio was varied by replacing ambient air with pure O_2 to achieve oxygen concentrations of 21% (i.e. ambient air), 25.0%, 28.9% and 36.8%. The variation was performed at a constant co-flow velocity of 24 cm/s, which is 6 times that of the fuel, being 3.98 cm/s. The burner is a Santoro-type concentric-tube burner. The central fuel tube has an inner diameter (ID) of 10.5 mm, while the oxidizing flow was fed through the outer tube with ID = 97.7 mm. More details of the burner have been reported previously work [16].

2.2 Experimental methods

The LII+TLAF setup used for simultaneous measurement of soot volume fraction and temperature is the same as that used previously [16]. Both are non-intrusive, laser-based diagnostic techniques. Soot volume fraction was measured through the incandescence intensity of soot particles heated by a pulsed laser, while in TLAF gas temperature was derived from the fluorescence ratio of two LIF processes of atomic indium that was seeded into the flames. More details can also be found elsewhere [17-19]. Measurement uncertainties are estimated to be within 100 K for temperature and ±10% for soot volume fractions, for which detailed discussion can also be found in the reference [16].

The soot luminosity was measured in the spectral range 680-720 nm using a band-pass filter (700±20 nm) and an intensified CCD camera (ICCD). The OH* chemiluminescence was recorded with another ICCD at 310 nm using an interference filter (IF) centered at 310 nm.

The OH laser-induced fluorescence (OH-LIF) was measured line-wise, with excitation performed with a Nd:YAG pumped dye laser system to generate a beam at 283.553 nm and excite the temperature-insensitive $Q_1(8)$ line of OH. The laser beam was focused and directed into the flame using a spherical lens (f = 500 mm). A beam was chosen over a sheet to increase the laser fluence and reduce the influence of quenching on the LIF signal. Nevertheless, the results are still qualitative. The OH-LIF signal was collected using an ICCD camera with the IF filter.

3 Results and discussion

3.1 Soot luminosity

Figure 1 presents the soot luminosity recorded from the four flames in the spectral range from 680 nm to 720 nm. These images were used to obtain the flame height (length) and the flame diameter (measured at the height with maximum soot loading). It can be seen that both the flame height and diameter decrease with an increase in oxygen concentration, particularly the flame height. These trends are consistent with Rope's model [20] and with previous measurements [16, 21, 22] that are, as the stoichiometric mole ratio (oxidizer/fuel) decrease, both the flame height and the diameter increase. Glassman *et al.* also attributed the decrease of flame height to the increased oxygen concentration gradient [11]. Also noted in Fig. 1 is the change of the 'wing' structure on the flame tip. Increasing oxygen makes the flame tip (the central part) be lower than the wing layers beside, forming a *W-like* structure. This interesting phenomenon is more clearly revealed in the following images of soot volume fractions.

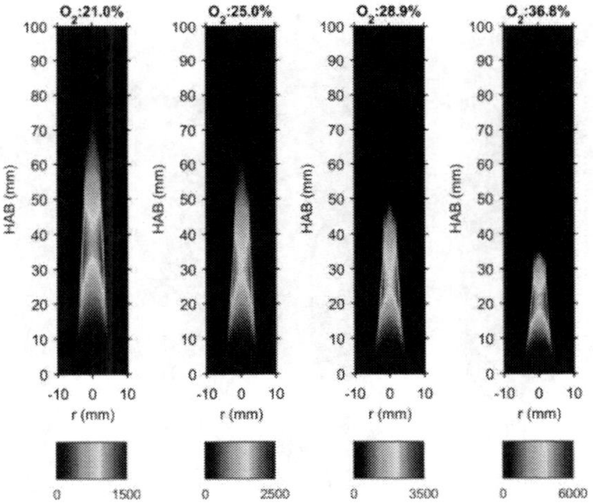

Figure 1. Soot luminosity recorded in the spectral range from 680 to 720 nm. Color scale is different from each images.

Figure 1 also presents the apparent enhancement of maximum soot luminosity by increasing oxygen. Locally, soot luminosity increases approximately by 4 times from the oxygen-21% to 36.8% flames. The total soot luminosity that is integrated over the flame volume also presents an increase with oxygen increased. Figure 2 presents the total soot luminosities for the four flames shown in Fig. 1, together with those of another three similar flames with less oxygen (< 21%). It can be seen that even taking the significant decrease of flame size into account, soot luminosity is still enhanced by increasing oxygen concentration. From 21% to 36.8% oxygen concentration, which is a 1.75 times increase, the total soot luminosity increases approximately by 1.62 times. The enhancement effect is non-linearly dependent on oxygen concentration, which is weaker at high oxygen concentration than at low concentration.

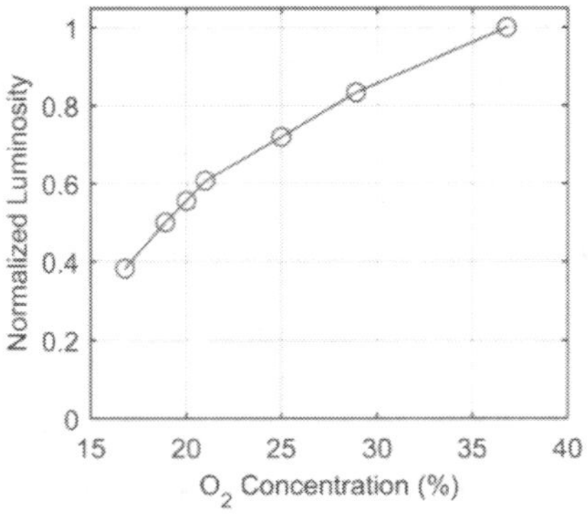

Figure 2. Normalized total soot luminosity that is integrated over the whole flame luminous region as a function of the oxygen concentration in the oxidizing stream. Results from another three flames with less oxygen (i.e. < 21%) are also shown.

Figure 3 presents the images of soot volume fractions measured using planer LII. Compared with soot luminosity in Fig. 1, these planar results reveal soot distribution more clearly with high spatial resolution and without the influences of flame temperature that varies significantly as shown later. Besides of the consistent findings with previous works, such as the increased rates of inception, growth and oxidation of soot, as well as the shortened sooting region, Fig. 3 also shows the significant change of the *W-like* structure on the flame tip and the narrowed soot layers on the flame wings with increasing oxygen. Moreover, these quantitative results should be beneficial for soot modelling validations.

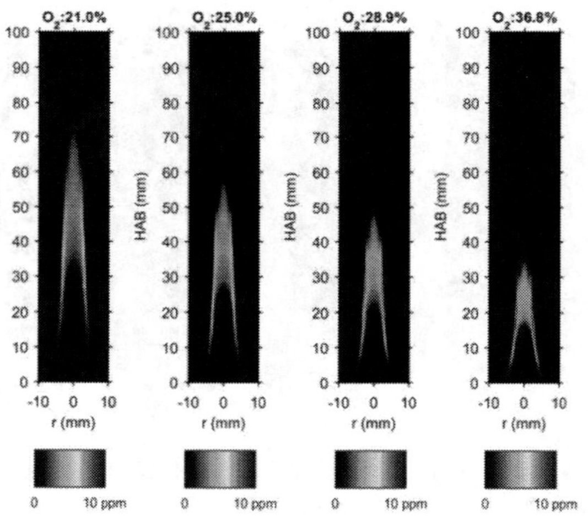

Figure 3. Soot volume fractions measured in the four oxygen-enriched flames.

Figure 4 presents the results of flame temperature. Temperature significantly rises when increasing oxygen, and it approaches to 2400 K in the oxygen-36.8% flame. Meanwhile, the hot region with maximum temperature becomes broaden and encloses at the flame tips, which is clearly shown in the oxygen-36.8% flame. Similar phenomena was also found in the laminar ethylene diffusion flames, where nitrogen was diluted into the fuel flow [16, 23], which was interpreted as the shrink of the stoichiometric surface as a result of that fuel is diluted or oxidizer is enriched.

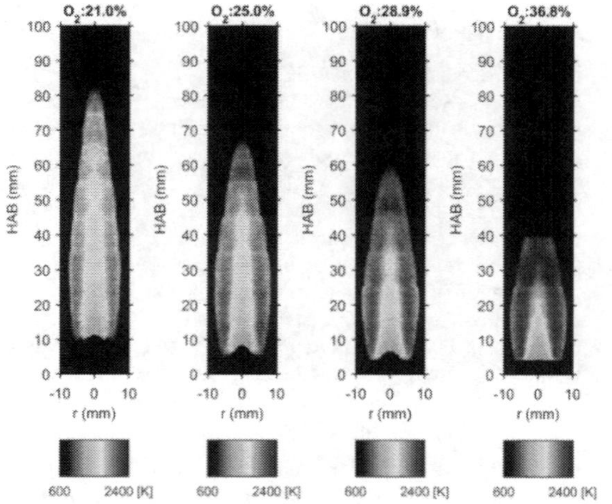

Figure 4. Flame temperature measured in the four oxygen-enriched flames.

OH radials, O_2 and atomic O are regarded as the principle oxidizers in soot oxidation, see [24, 25] and references therein. OH* chemiluminescence is a good marker of hot oxidation zone, but its distribution in flames may be different from that of ground state OH radicals which are relatively more stable and distributes more broadly. Figure 5 presents OH* chemiluminescence reconstructed from the direct images of line-of-sight signals. OH* appears immediately above the burner, i.e. around the flame feet, and mainly distributes on the upstream of the flames (or say the first half of flame length) and away from the sooting region. Increasing oxygen shortens the distribution of OH* chemiluminescence but significantly enhances its intensity. OH* region clearly separates from the soot region. This is different with the distribution of ground-state OH (see the following discussion). Between the two dominant OH* wings, an apparent structure appears in the three oxygen-enriched flames, and becomes significant in the oxygen-36.8% flame. This signal might be the OH* chemiluminescence generated in the 'second' oxidation zone, for which further assessment is needed.

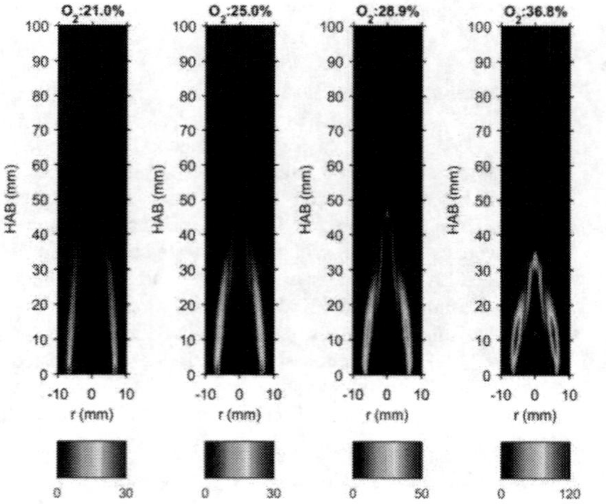

Figure 5. Planar OH* chemiluminescence in the four flames. The images are shown in different color bar scales.

Figure 6 presents the radial distributions of ground-state OH radicals at several representative heights in two of the oxygen-enriched flames. Due to the strong laser scattering at 283 n from soot particles, some OH-LIF signals are interfered with soot scattering. However, these two type signals can be easily identified based on the planar results of soot volume fractions as shown in Fig. 3. Beneficially, the scattering and LIF co-existing curves clearly shows the overlap regions of soot and ground-state OH, where soot is oxidized by OH radicals. Above the sooting region, e.g. at the height of 41 mm in the oxygen-36.8% flam, OH radicals are still broadly distributed with high intensity. This is different from the OH* distribution. It is also interesting that with increasing oxygen concentration from 25.0% to 36.8% i.e. by 1.47 times, the peak of OH-LIF signal increases accordingly with similar magnitude, but OH* chemiluminescence greatly increases by approximately 5 times as shown in Fig. 5.

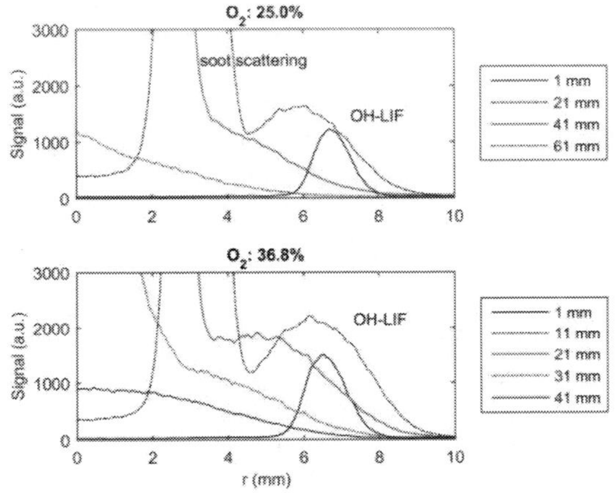

Figure 6. Radial profiles of OH at different heights in the oxygen-25.0% and 36.8% flames. The signals at 21 and 41 mm in the oxygen-25.0% flame and those at 11, 21 and 31 mm in the oxygen-36.8% flame at 310 nm.

4 Conclusions

Besides that the quantitative results of soot volume fraction and flame temperature can serve as benchmark for soot model development, the following key points can be drawn from the present investigation. With the increase of oxygen concentration in the oxidizing stream,

(1) The ethylene diffusion flames are significantly influenced. While the shape of soot luminosity changes and the flame length becomes short, both local and total soot luminosities increase.

(2) Soot volume fraction increases and the sooting layers with maximum soot loading becomes narrow.

(3) Flame temperature increase significantly and the high temperature region encloses above the tip of sooting region in the oxygen-28.9% and 36.8% flames.

(4) OH* chemiluminescence becomes broadly and mainly distributes in the first half of the flames. They have clear sharp structures to the sooting side. However, ground-state OH apparently overlaps with soot in all flames. OH* chemiluminescence might reveal a second oxidation zone on the tip of the oxygen-enriched flames, particularly in the oxygen-36.8% flame. The intensity of OH* chemiluminescence has a large increase, while that of OH-LIF only increases moderately.

5 Acknowledgment

The Australia Research Council is gratefully acknowledged for their funding support through DP130100198. The authors also thank Mr. KaeKen Foo for his assistance in the experiment.

References

[1]. C.E. Baukal Jr., *Oxygen-Enhanced Combustion*, Second, CRC Press, (2013).
[2]. M. Eddaoudi, H. Li, O.M. Yaghi, Journal of the American Chemical Society 122 (7) (2000) 1391-1397.
[3]. M.V. Parkes, D.F. Sava Gallis, J.A. Greathouse, T.M. Nenoff, The Journal of Physical Chemistry C 119 (12) (2015) 6556-6567.
[4]. D.F. Sava Gallis, M.V. Parkes, J.A. Greathouse, X. Zhang, T.M. Nenoff, Chemistry of Materials 27 (6) (2015) 2018-2025.
[5]. S.G. Kimura, W.R. Browall, Journal of Membrane Science 29 (1) (1986) 69-77.
[6]. C.R. Shaddix, T.C. Williams, Proceedings of the Combustion Institute 36 (3) (2017) 4051-4059.
[7]. H. Lin, M. Zhou, J. Ly, J. Vu, J.G. Wijmans, T.C. Merkel, J. Jin, A. Haldeman, E.H. Wagener, D. Rue, Industrial & Engineering Chemistry Research 52 (31) (2013) 10820-10834.
[8]. L. Wang, N.E. Endrud, S.R. Turns, M.D. D'Agostini, A.G. Slavejkov, Combustion Science and Technology 174 (8) (2002) 45-72.
[9]. K.C. Oh, H.D. Shin, Fuel 85 (5–6) (2006) 615-624.
[10]. K.-O. Lee, C.M. Megaridis, S. Zelepouga, A.V. Saveliev, L.A. Kennedy, O. Charon, F. Ammouri, Combustion and Flame 121 (1–2) (2000) 323-333.
[11]. Glassman, P. Yaccarino, Combustion Science and Technology 24 (3-4) (1980) 107-114.
[12]. C.E. Baukal, B. Gebhart, International Journal of Heat and Mass Transfer 40 (11) (1997) 2539-2547.
[13]. M. Si, Q. Cheng, J. Song, Y. Liu, M. Tao, C. Lou, Combustion and Flame 183 (2017) 261-270.
[14]. S.S. Hwang, J.P. Gore, Proceedings of the Institution of Mechanical Engineers Part a-Journal of Power and Energy 216 (A5) (2002) 379-386.
[15]. D.X. Du, R.L. Axelbaum, C.K. Law, Symposium (International) on Combustion 23 (1) (1991) 1501-1507.
[16]. Z. Sun, B. Dally, G. Nathan, Z. Alwahabi, Combustion and Flame 175 (2017) 270-282.
[17]. Z.W. Sun, Z.T. Alwahabi, B.B. Dally, G.J. Nathan, *Joint statistics of soot volume fraction, primary particle diameter and temperature in flames. Part I: experimental setup*, in Mediterranean Combustion Conference Naples, Italy, (2017)
[18]. D. Gu, Z. Sun, G.J. Nathan, P.R. Medwell, Z.T. Alwahabi, B.B. Dally, Combustion and Flame 167 (2016) 481-493.
[19]. Z.W. Sun, D.H. Gu, G.J. Nathan, Z.T. Alwahabi, B.B. Dally, Proceedings of the Combustion Institute 35 (3) (2015) 3673-3680.
[20]. F.G. Roper, Combustion and Flame 29 (1977) 219-226.
[21]. Ö.L. Gülder, D.R. Snelling, Combustion and Flame 92 (1–2) (1993) 115-124.
[22]. J.H. Kent, H.G. Wagner, Symposium (International) on Combustion 20 (1) (1985) 1007-1015.
[23]. M.D. Smooke, M.B. Long, B.C. Connelly, M.B. Colket, R.J. Hall, Combustion and Flame 143 (4) (2005) 613-628.
[24]. H. Ghiassi, D. Lignell, J.S. Lighty, Energy & Fuels 31 (3) (2017) 2236-2245.
[25]. A. Khosousi, S.B. Dworkin, Proceedings of the Combustion Institute 35 (2) (2015) 1903-1910.

11th Asia-Pacific Conference on Combustion,
The University of Sydney, NSW Australia
10th -14th December 2017.

Complete catalytic oxidation of propene over thin film catalyst

Achraf El Kasmi[1,a], Guan-Fu Pan[3,a], Zhen-Yu Tian[1,2,*]

[1]Institute of Engineering Thermophysics, Chinese Academy of Sciences, Beijing 100190, China

[2]University of Chinese Academy of Sciences, Beijing 100190, China

[3]Energy Conservation Corporation Ltd. of China Coal Research Institute, Beijing 100013, China

Abstract

Complete catalytic oxidation of propene (C_3H_6) has been studied over thin film CuO catalysts that synthesized in one-step at low temperature using pulsed-spray evaporation chemical vapor deposition (PSE-CVD). To better understand the interplay between properties and catalytic performance, the prepared thin film catalysts have been comprehensively characterized in terms of structure and morphology. The XRD analysis showed that the obtained catalysts are well indexed in face centered cubic (fcc) structure with nano-crystallite size. The catalytic performance was evaluated through the total oxidation of C_3H_6 with varying the oxygen concentration in the gas flow by on-line GC analysis. The results indicated that the CuO catalyst was very active for the total oxidation of C_3H_6 at low temperature, with good reusability and reproducibility. It is of interest to note that no CO was observed during the oxidation processes, and the catalytic performance exhibited a good consistence with the apparent activation energies (E_a). The findings revealed that CuO could be a promised catalyst for complete oxidation of other volatile organic compounds (VOCs) at low temperature.

1 Introduction

Volatile organic compounds (VOCs) have been widely recognized as major source of air pollution due to their high potential toxicity, as well as their roles in the formation of photochemical smog and destruction of ozone layer [1,2]. Due to the progressive increase of their emissions, more strict legislations have been established and led to the development of various treatment technologies allowing higher efficiency and lower cost [3,4]. Catalytic oxidation is known as one of the most promising technologies for the abatement of VOCs [5–7]. Accordingly, low-temperature catalytic oxidation of VOCs is highly desirable to proceed due to the consideration of energy savings, low cost, safety, and environmental friendliness [8]. Currently, precious metals were widely tested as efficient catalysts for VOCs oxidation due to their high activity at low-temperature [8–10]. However, because of the high price, limited resources and tendency to poisoning of such noble metals, considerable attention has been paid to develop catalysts based on transition metal oxides (TMOs) for deep oxidation of VOCs, regarding their competitive activity, easy manufacturing, high thermal stability and low cost [11]. Furthermore, the facile and high reproducibility to synthesize highly active catalysts makes single oxides much attractive catalysts [12]. Among the available single oxides based-transition metal oxides, CuO is abundantly available, non-toxic and widely exploited for a broad field of applications [13]. However, the application of CuO as catalyst for the abatement of VOCs is not much known [14].

Several methods have been used to synthesize CuO including thermal oxidation [15], solvothermal synthesis [16], solution-phase [17], sol-gel [18], magnetron sputtering [19], microplasma [20], precipitation [13], wet impregnation [21], hydrothermal [22], electrospinning [23] and etc. In most of these studies, a mixture of Cu, CuO and Cu_2O was generally obtained and the synthesis of pure CuO is of interest regarding its applications. Also, CuO films could be formed just at relatively higher temperature [24], while at lower temperature the CuO could convert to the Cu_2O phase and/or mixture of CuO-Cu_2O [14,24]. Thus, the synthesis of pure CuO with these methods is not easy to be achieved, which in turn renders the direct synthesis of pure CuO.

Compared to the above-mentioned techniques, pulsed-spray evaporation chemical vapor deposition (PSE-CVD) is a facile, direct and inexpensive technique for the preparation of continuous thin film oxides on a variety of substrates without any after-treatment. It offers potential for producing films with high uniformity of thickness, high purity, conformal step coverage, minimal substrate damage, the possibility of growth in selected area and the flexibility to use different precursors in the feedstock simultaneously. Recently, PSE-CVD has been successfully utilized to synthesize a series of TMOs, such as Co_3O_4 [25], Mn_3O_4 [26], Fe_2O_3 [27] and etc. Thus, it shows great potential to prepare pure CuO films.

The present study aims to synthesize thin CuO films at low temperature using one-step PSE-CVD synthesis method. Furthermore, the attention was paid to investigate the catalytic performance of the synthesized film catalysts towards complete oxidation of C_3H_6 using an empirical catalytic screening based on light-off curves. The good catalytic performance was discussed by linking with the material properties to reveal the key role played by the catalyst in the oxidation process.

2 Experimental

The synthesis of CuO thin film catalysts was carried out in a cold-wall stagnation point-flow CVD reactor employing a PSE unit for the delivery of liquid precursor feedstock. Details of the experimental procedure can be found elsewhere [25], and only a brief description is given here. In the deposition process, commercially available $Cu(acac)_2$ was used as precursor and dissolved in ethanol. The PSE delivery of the feedstock was achieved using a four-pinhole injector (Bosch) with a frequency of 4 Hz and a valve opening time of 2.5 ms. The feedstock was injected as a fine spray into a 30 cm long evaporation chamber kept at 200 °C. The resulting vapor was transported to the

Corresponding author. Fax: +86-10-8254-3184
E-mail address: tianzhenyu@iet.cn

deposition chamber kept at 200 °C, with N_2/O_2 flow rates of 0.5/1.0 standard liter per minute (slm) under a total pressure of 3.0 kPa in the reactor. During the deposition process, planar glass, and stainless steel plates and mesh grid were used as substrates and heated to 325 °C using a flat resistive heater. The thickness of the deposited films was estimated gravimetrically by measuring the weight change of the substrates before and after deposition with a microbalance (Sartorius). To verify the crystalline structure of the deposited films, XRD analysis was performed using Bruker D8 Focus with Cu Kα radiation, scan angle of 5-90° and scanning step of 0.02°. The surface morphology of the deposited catalyst film on a mesh of stainless steel was analyzed using SEM.

The catalytic performance of the CuO films were investigated by a fixed bed quartz reactor for C_3H_6 oxidation. A 60 cm long alundum tube (8.0 mm inner diameter) was fixed in a digital electrical furnace. CuO catalyst (12.8 mg) is supported on grid mesh of stainless steel was located at the isothermal area. A K-thermocouple was used to measure the temperature of CuO film. A gas mixture consisted of 1 % C_3H_6, different concentrations of O_2 (from 5 to 40 %) in flow of Ar was introduced into the alundum tube with a total flow rate of 15 ml/min, corresponding to the gaseous hourly space velocity (GHSV) of 70312 ml g^{-1}_{cat} h^{-1}. The temperature of the furnace was rised with a ramp of 2 °C/min. Finally, the exhaust gas was measured by a Gas Chromatograph (Agilent, GC 7890B) equipped with TCD and FID detector, and capillary columns of Al_2O_3-KCl and HP-INNOWax, for qualitative and quantitative analysis.

3 Results and Discussion

3.1 Structure

XRD analysis was performed to investigate the crystalline structure of the thin films deposited at 300 °C. The XRD patterns of the thin films coated on glass obtained are presented in Fig. 1. The diffraction peaks of the as-deposited films are observed at 2θ of 32.74°, 35.57°, 38.75°, 49.01°, 61.64° and 66.05° and 68.26°, corresponding to (110), (11-1), (111), (202), (11-3), (31-1) and (220) orientations of CuO structure. All the observed peaks fit well with the cubic structure of CuO (JCPDS No. 48-1548), indicating that a pure CuO was readily synthesized. In addition, the XRD patterns show no peaks of impurity phases, which reveals a high purity of the all deposited films.

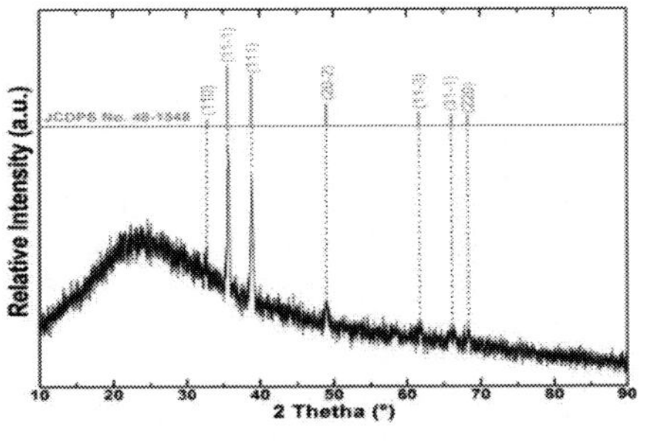

Figure 1: XRD pattern of the CuO thin film deposited on glass at 300 °C.

The crystalline size of the deposited film is determined by applying the Debye-Scherrer equation, $D = 0.9\lambda/\beta\cos\theta$, to two intense diffraction peaks, where λ = 0.154056 nm, θ is the Bragg angle and β is the full width at half maximum of the peaks. The average crystalline size was determined to be about 51 nm. It was reported that the catalytic activity is generally higher with smaller particle size [28]. Furthermore, the small grain size is also known to have a beneficial effect on active sites dispersion and it would improve the efficiency of the catalytic performance [29]. In this regard, the small grain size obtained here could further benefit for the catalytic reaction.

3.2 Morphology

The representative images of the obtained films are shown in Fig. 2. It can be seen that the deposited film is morphologically homogeneous, uniform and continuous. Moreover, the deposited film presents an open and porous texture. The film seems to be consisted of nano-sized cubic embedded grains in matrix. Therefore, the observed morphology with small grain size and open porosity are expected to offer high specific surface area, which in turn can engender much accessible active catalytic sites as well as providing predisposed space holding great amount of oxygen, thus will be beneficial to the catalytic performance of VOCs.

Figure 2: SEM images of deposited CuO thin film.

3.3 Catalytic performance

The catalytic performance of the as-deposited CuO thin films on mesh of stainless steel was performed for complete oxidation of C_3H_6 at atmospheric pressure and the formed products during the catalytic tests were monitored by on-line GC analysis. Considering the effect of non-coated mesh, the tests of complete oxidation of C_3H_6 were carried out over non-coated mesh as reference. In attempts to investigate the effect of oxygen concentration on the catalytic oxidation of C_3H_6, we have performed different concentrations of oxygen as 5, 10, 20, 30 and 40% in gas mixture. Figure 3 shows the light-off curves corresponding to the temperature-dependent conversion of C_3H_6 to CO_2 as final product over CuO film catalysts. The results showed that CuO coated mesh permits to shift total oxidation temperatures towards lower values than that on non-coated mesh, with the onset of C_3H_6 conversion almost at 200 °C and the complete conversion takes place at 430 °C. The reproduciblity of the obtained results was verified by reusing the CuO catalyst for several times and some light-off curves are shown in Fig. 4.

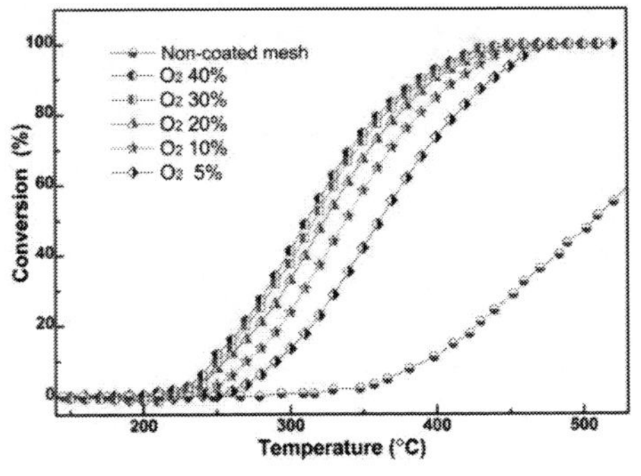

Figure 3: Light-off curves of C_3H_6 deep oxidation over CuO-coated mesh of stainless steel and non-coated mesh.

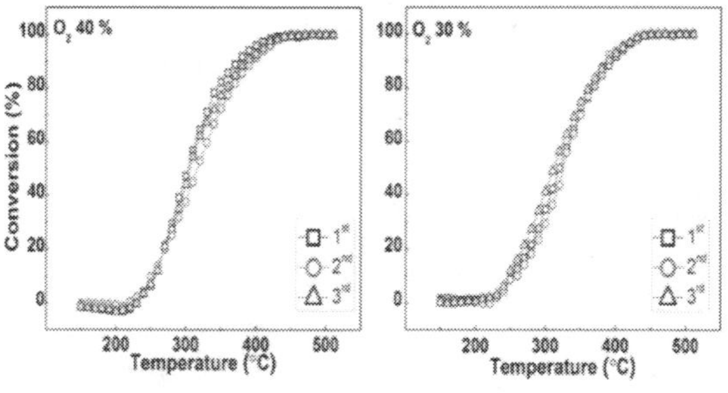

Figure 4: Reproducibility of C_3H_6 conversion over CuO-coated mesh of stainless steel.

In addition, it should be noted that when using CuO as catalyst, no trace of CO was detected during the oxidation process, which reflects that CuO is a very active catalyst for C_3H_6 oxidation; while with non-coated mesh, CO can be detected.

Moreover, it can be seen from Fig. 3 that conversion temperature values of T_{10}, T_{50} and T_{90} (correspond respectively to 10, 50 and 90% C_3H_6 conversion) decrease gradually with the increase of oxygen concentration, which means the oxygen promotes the reactivity and reduces the consumption energy during reaction. The latter was also reflected by the increase in reaction rate values from 13.64 to 31.22 μmol/(m^2·h) for 5 and 30%, respectively; while, above 30% oxygen concentration, the conversion temperature and conversion rate values for 30% and 40% oxygen concentrations have been remained quite the same, which further confirms that oxygen concentration has been close to oxygen demand for complete conversion of C_3H_6 over CuO catalyst.

Furthermore, the difference in light-off curves behavior may also be linked to the difference observed in apparent activation energies (E_a), that were calculated by applying Arrhenius equation within the range data up to 15% conversion, as shown in Fig. 5. The lower E_a, the lower C_3H_6 conversion temperature [14,30]. Thus, compared to other catalysts that present higher activation energy E_a, the as-deposited CuO catalyst is more efficient in terms of activity.

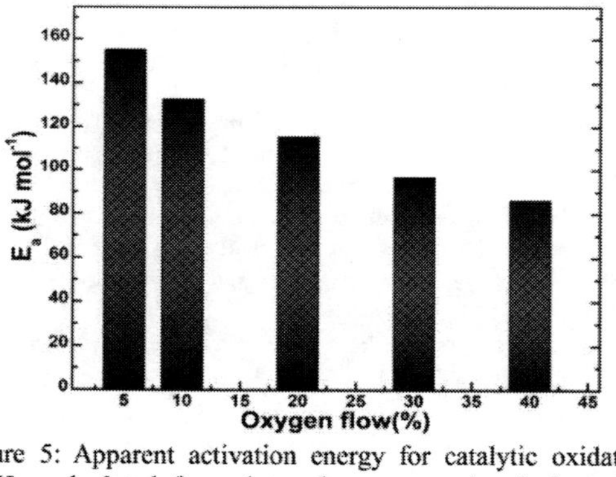

Figure 5: Apparent activation energy for catalytic oxidation C_3H_6, calculated from data where conversion is less than 15%.

In comparison to the catalytic C_3H_6 oxidation behavior for a number of interesting oxides as noble [31,32] and/or non-noble [26,33–35] oxides, the CuO catalyst prepared in the present work exhibits quite promising catalytic performance.

It has been commonly accepted that the catalytic oxidation of hydrocarbons employing transition metal oxide catalysts can follow the Mars-van Krevelen mechanism [36], in which the C-H bond activation is related to the rate of oxidation for hydrocarbons. According to the SEM results, the open porosity can engender much accessible active catalytic sites as well as provide pre-disposed space holding great amount of oxygen, and the catalytic oxidation of C_3H_6 over CuO is assumed here involving alternate reduction and oxidation of the oxides with formation of surface oxygen vacancies and their replenishment by the adsorption and dissociation of gas phase oxygen [37]. Hence, the previous explanations demonstrate that the open porosity texture of CuO deposit films can play an important role in the C_3H_6 oxidation process, which is in accordance to the literature [14]. Furthermore, the simple synthesized CuO catalyst with these promising properties could be a good catalyst for other catalytic reactions.

4 Conclusions

Thin film catalysts based CuO have been successfully synthesized at low deposition temperature using PSE-CVD method as a facile synthesis approach. The deposited films exhibited pure crystalline structure of CuO with cubic nano-sized crystallite of 51 nm. The catalytic tests indicate that the as-deposited CuO is a very active catalyst for the complete oxidation of propene, with a performance comparable to that of supported noble metals. The saturation commencement of reaction rate above 30% oxygen can confirm that the oxygen concentration is close to oxygen demand for complete conversion of propene over CuO catalyst. Moreover, the good catalytic activity of CuO could be attributed to the low activation energy of CuO.

5 Acknowledgment

The authors thank the financial support from the Ministry of Science and Technology of China (2017YFA0402804) and Recruitment Program of Global Youth Experts. Achraf El Kasmi would kindly thank the support of Chinese Academy of Sciences for senior international scientists.

[a] Both authors have contributed equally to this work.

References

[1] M.E. Jenkin, R.G. Derwent, T.J. Wallington, Photochemical ozone creation potentials for volatile organic compounds: Rationalization and estimation, Atmos. Environ. 163 (2017) 128–137.

[2] W.P.L. Carter, Development of ozone reactivity scales for volatile organic compounds, Air Waste. 44 (1994) 881–899.

[3] K. Everaert, J. Baeyens, Catalytic combustion of volatile organic compounds, J. Hazard. Mater. 109 (2004) 113–139.

[4] C.S. Heneghan, G.J. Hutchings, S.H. Taylor, The destruction of volatile organic compounds by heterogeneous catalytic oxidation, Catalysis. 17 (2004) 105–151.

[5] Z. Guo, B. Liu, Q. Zhang, W. Deng, Y. Wang, Y. Yang, Recent advances in heterogeneous selective oxidation catalysis for sustainable chemistry, Chem. Soc. Rev. 43 (2014) 3480–3524.

[6] L.F. Liotta, Catalytic oxidation of volatile organic compounds on supported noble metals, Appl. Catal. B Environ. 100 (2010) 403–412.

[7] M.S. Kamal, S.A. Razzak, M.M. Hossain, Catalytic oxidation of volatile organic compounds (VOCs)–A review,

Atmos. Environ. 140 (2016) 117–134.
[8] H. Huang, Y. Xu, Q. Feng, D.Y.C. Leung, Low temperature catalytic oxidation of volatile organic compounds: a review, Catal. Sci. Technol. 5 (2015) 2649–2669.
[9] A. Wittstock, V. Zielasek, J. Biener, C.M. Friend, M. Bäumer, Nanoporous gold catalysts for selective gas-phase oxidative coupling of methanol at low temperature, Science. 327 (2010) 319–322.
[10] Z. Zhang, Z. Jiang, W. Shangguan, Low-temperature catalysis for VOCs removal in technology and application: a state-of-the-art review, Catal. Today. 264 (2016) 270–278.
[11] W.B. Li, J.X. Wang, H. Gong, Catalytic combustion of VOCs on non-noble metal catalysts, Catal. Today. 148 (2009) 81–87.
[12] X. Xie, Y. Li, Z.-Q. Liu, M. Haruta, W. Shen, Low-temperature oxidation of CO catalysed by Co_3O_4 nanorods, Nature. 458 (2009) 746–749.
[13] L. Wang, H. Gong, C. Wang, D. Wang, K. Tang, Y. Qian, Facile synthesis of novel tunable highly porous CuO nanorods for high rate lithium battery anodes with realized long cycle life and high reversible capacity, Nanoscale. 4 (2012) 6850–6855.
[14] Z.Y. Tian, H.J. Herrenbrück, P.M. Kouotou, H. Vieker, A. Beyer, A. Gölzhäuser, K. Kohse-Höinghaus, Facile synthesis of catalytically active copper oxide from pulsed-spray evaporation CVD, Surf. Coat. Technol. 230 (2013) 33–38.
[15] J. Liu, D. Xue, Thermal Oxidation Strategy towards Porous Metal Oxide Hollow Architectures, Adv. Mater. 20 (2008) 2622–2627.
[16] X.-Y. Yu, R.-X. Xu, C. Gao, T. Luo, Y. Jia, J.-H. Liu, X.-J. Huang, Novel 3D Hierarchical Cotton-Candy-Like CuO: Surfactant-free solvothermal synthesis and application in As(III) removal, ACS Appl. Mater. Interfaces. 4 (2012) 1954–1962.
[17] X. Wang, G. Xi, S. Xiong, Y. Liu, B. Xi, W. Yu, Y. Qian, Solution-phase synthesis of single-crystal CuO nanoribbons and nanorings, Cryst. Growth Des. 7 (2007) 930–934.
[18] Y.-F. Lim, C.S. Chua, C.J.J. Lee, D. Chi, Sol–gel deposited Cu_2O and CuO thin films for photocatalytic water splitting, Phys. Chem. Chem. Phys. 16 (2014) 25928–25934.
[19] J.K. Feng, H. Xia, M.O. Lai, L. Lu, Electrochemical performance of CuO nanocrystal film fabricated by room temperature sputtering, Mater. Res. Bull. 46 (2011) 424–427.
[20] T. Koh, E. O'Hara, M.J. Gordon, Growth of nanostructured CuO thin films via microplasma-assisted, reactive chemical vapor deposition at high pressures, J. Cryst. Growth. 363 (2013) 69–75.
[21] H. Zhu, D. Han, Z. Meng, D. Wu, C. Zhang, Preparation and thermal conductivity of CuO nanofluid via a wet chemical method, Nanoscale Res. Lett. 6 (2011) 181.
[22] T. Jiang, Y. Wang, D. Meng, X. Wu, J. Wang, J. Chen, Controllable fabrication of CuO nanostructure by hydrothermal method and its properties, Appl. Surf. Sci. 311 (2014) 602–608.
[23] R. Sahay, P. Suresh Kumar, V. Aravindan, J. Sundaramurthy, W. Chui Ling, S.G. Mhaisalkar, S. Ramakrishna, S. Madhavi, High Aspect Ratio Electrospun CuO Nanofibers as Anode Material for Lithium-Ion Batteries with Superior Cycleability, J. Phys. Chem. C. 116 (2012) 18087–18092.
[24] V. Figueiredo, E. Elangovan, G. Gonçalves, N. Franco, E. Alves, S. h. k. Park, R. Martins, E. Fortunato, Electrical, structural and optical characterization of copper oxide thin films as a function of post annealing temperature, Phys. Status Solidi A. 206 (2009) 2143–2148.
[25] N. Bahlawane, P.H.T. Ngamou, V. Vannier, T. Kottke, J. Heberle, K. Kohse-Höinghaus, Tailoring the properties and the reactivity of the spinel cobalt oxide, Phys. Chem. Chem. Phys. 11 (2009) 9224–9232.
[26] Z.Y. Tian, P. Mountapmbeme Kouotou, N. Bahlawane, P.H. Tchoua Ngamou, Synthesis of the catalytically active Mn_3O_4 spinel and its thermal properties, J. Phys. Chem. C. 117 (2013) 6218–6224.
[27] P. Mountapmbeme Kouotou, Z.Y. Tian, H. Vieker, A. Beyer, A. Gölzhäuser, K. Kohse-Höinghaus, Selective synthesis of α-Fe_2O_3 thin films and effect of the deposition temperature and lattice oxygen on the catalytic combustion of propene, J. Mater. Chem. A. 1 (2013) 10495-10504.
[28] A. El Kasmi, Z.Y. Tian, H. Vieker, A. Beyer, T. Chafik, Innovative CVD synthesis of Cu_2O catalysts for CO oxidation, Appl. Catal. B Environ. 186 (2016) 10–18.
[29] K. Qadir, S.H. Joo, B.S. Mun, D.R. Butcher, J.R. Renzas, F. Aksoy, Z. Liu, G.A. Somorjai, J.Y. Park, Intrinsic Relation between Catalytic Activity of CO Oxidation on Ru Nanoparticles and Ru Oxides Uncovered with Ambient Pressure XPS, Nano Lett. 12 (2012) 5761–5768.
[30] G.F. Pan, S.-B. Fan, J. Liang, Y.X. Liu, Z.Y. Tian, CVD synthesis of Cu_2O films for catalytic application, RSC Adv. 5 (2015) 42477–42481.
[31] S. Ivanova, C. Petit, V. Pitchon, Application of heterogeneous gold catalysis with increased durability: Oxidation of CO and hydrocarbons at low temperature, Gold Bull. 39 (2006) 3–8.
[32] A.C. Gluhoi, N. Bogdanchikova, B.E. Nieuwenhuys, The effect of different types of additives on the catalytic activity of Au/Al_2O_3 in propene total oxidation: transition metal oxides and ceria, J. Catal. 229 (2005) 154–162.
[33] Z.Y. Tian, P.H.T. Ngamou, V. Vannier, K. Kohse-Höinghaus, N. Bahlawane, Catalytic oxidation of VOCs over mixed Co–Mn oxides, Appl. Catal. B Environ. 117 (2012) 125–134.
[34] Z.Y. Tian, P. Mountapmbeme Kouotou, A. El Kasmi, P.H. Tchoua Ngamou, K. Kohse-Höinghaus, H. Vieker, A. Beyer, A. Gölzhäuser, Low-temperature deep oxidation of olefins and DME over cobalt ferrite, Proc. Combust. Inst. 35 (2015) 2207–2214.
[35] S.B. Fan, G.F. Pan, J. Liang, Z.Y. Tian, Tailored synthesis of CoO_X thin films for catalytic application, RSC Adv. 5 (2015) 97272–97278.
[36] C. Doornkamp, V. Ponec, The universal character of the Mars and Van Krevelen mechanism, J. Mol. Catal. Chem. 162 (2000) 19–32.
[37] H. Bao, X. Chen, J. Fang, Z. Jiang, W. Huang, Structure-activity relation of Fe_2O_3–CeO_2 composite catalysts in CO oxidation, Catal. Lett. 125 (2008) 160–167.

11th Asia-Pacific Conference on Combustion,
The University of Sydney, NSW Australia
10th -14th December 2017.

Experimental Investigation of Explosion Characteristics of Methane–Air Mixtures in a Vented Vessel Connected to a Curved Duct

Sazal K. Kundu, Jafar Zanganeh*, Daniel Eschebach, Behdad Moghtaderi

Priority Research Centre for Frontier Energy Technologies & Utilisation, Discipline of Chemical Engineering, School of Engineering, Faculty of Engineering & Built Environment, University of Newcastle, Callaghan, NSW 2308, Australia

Abstract

The explosion characteristics of methane-air mixtures were explored in a 20 litre vented vessel connected to a curved duct (one form of ducted vessel). With the aim of relating the experimental outcome with the real world explosion scenario, a wafer check valve was employed. The valve was used to keep the methane-air mixture initially confined and open at the time of explosion. The experiments were carried out in turbulent methane-air mixtures with strong igniters. The turbulent mixture of methane-air, strong ignition energies and curved flow pathways impacted on explosion parameters. The explosion pressure rise was found to be higher in the investigated system when compared to quiescent methane-air mixtures, weak ignition energies and straight duct.

1 Introduction

Methane-air mixtures bear explosive characteristics within a range of methane concentrations. The lower explosive limit of methane is 4.6 ± 0.3 % while the upper explosive limit of this flammable gas is 15.8 ± 0.4 % [1]. In presence of an ignition source, methane-air mixtures within this range can produce a massive explosion. In presence of another flammable substance such as coal dust particles, explosion can occur at much lower concentration of methane.

Several methane explosion accidents and their associated damages are documented in literature and in the media. In 2016, a natural gas explosion in Stafford Township in New Jersey injured 15 people and levelled the house where the accident occurred (the main constituent of natural gas is methane) [2]. Explosions in coal and gold mines killed a massive number of people and destroyed infrastructures [3-5]. In 1902, a coal mine explosion in Mount Kembla Mine of Australia took lives of 96 people and it is the deadliest coal mine explosion in Australian history [3]. The highest number of people killed in human history due to a coal mine explosion was 1549 at the Benxihu Colliery, China [4].

In process industries, various units are connected with pipes and ducts [6]. If explosion occurs in a process unit, it can spread to the connected ducts or pipelines. This type of phenomenon, where explosion initiates in a process unit and transmits to the connected duct or pipeline, is known as ducted explosion [7]. Previous studies suggest that the pressure rise in ducted vessels is higher than in vented vessel [8]. For ducted vessels, the severity of explosion transmits from the vessel to the duct. In the case of longer duct, a transition from deflagration to detonation can occur in a ducted vessel [9].

In the ducted explosion, the combustion products and the burning materials release from the vessel to the duct. Therefore, the pressure rise in ducted vessels is lower than in confined vessels. This phenomenon leads to the definition of the pressure rise in the vessel component of the ducted vessel, called 'reduced explosion pressure (P_{red})' [10]. In the duct component of the ducted vessel, there is no association of pressure reduction phenomenon. Thus the pressure rise in the duct component is simply defined as explosion pressure rise.

Tube bends are commonly employed in process industries. Previous studies suggest that if a bend is present in the flame propagation path, the explosion characteristics are impacted [11]. In this article, we are reporting the explosion studies of methane-air mixtures in a vented vessel connected with a curved duct. The effect of ignition energy and concentration of methane were investigated. Explosion pressure rise, rate of pressure rise and flame speed were explored and discussed in this article.

2 Experimental

Figure 1 shows an image of the experimental setup. In the experimental investigations, a 20 litre explosion sphere apparatus (ANKO Trading Limited) was employed. This apparatus is a modified setup of our previous works [8, 12] and the modification includes the integration of a bend segment in the duct.

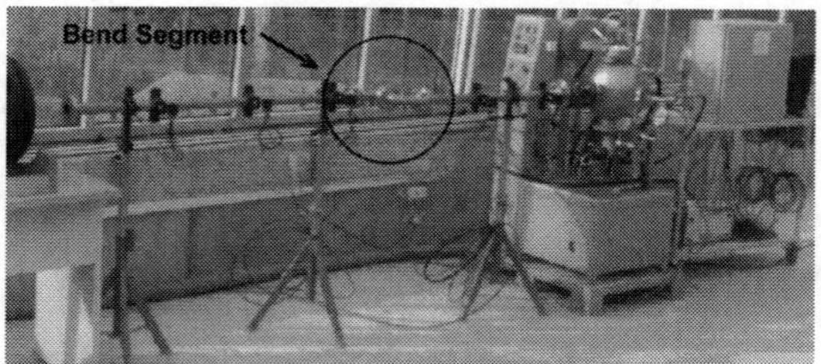

Figure 1: Ducted vessel apparatus employed in the experimental investigations.

The vessel was connected to the duct with a connector and a valve. The duct was comprised of two straight tubes and a bend segment. The bend segment includes two 90 degree bends and it was employed to connect the straight portions of the duct as shown in the figure. The total duct length was 3063 mm. This

*Corresponding author. Fax: +61-2-4033-9095
E-mail address: Jafar.Zanganeh@newcastle.edu.au

includes; connector – 70 mm, valve – 203 mm, first straight portion of the duct – 1000 mm, bend segment – 250 mm and second straight portion of the duct – 1540 mm. The internal diameter of the valve was 50 mm. The connector and the three portions of the duct had an identical internal diameter of 42.72 mm. The schematic of the experimental apparatus is presented in Fig.2.

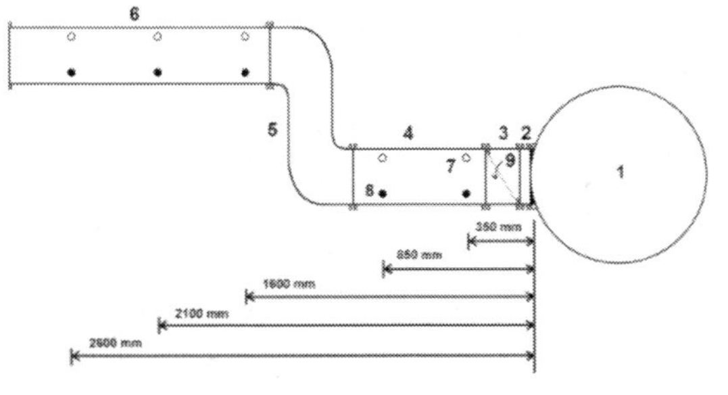

Figure 2: Experimental setup employed in the investigations: 1) 20 litre sphere, 2) connector, 3) valve, 4) first portion of the duct, 5) bend segment, 6) second portion of the duct, 7) photodiodes (open circles), 8) pressure transducers (closed markers), 9) flexible disc that closes when the vessel is in vacuum and opens as soon as the vessel's pressure increased from vacuum to atmospheric pressure.

The valve, integrated in the test apparatus, was a wafer check valve with rubber-coated flexible disc acting in its closure mechanism. The check valve was mounted upside down to allow gravity assisted opening. In the vessel, two piezoelectric pressure transducers (Kistler 701A) were mounted. Both of these pressure transducers were employed for the purpose of measuring pressure rise in the vessel. ANKO Explosion Plotter (version 1.0.0.0) software was employed to record pressure rise in the vessel.

The pressure rises along the length of the duct were measured by five piezoresistive pressure transducers (model – ATM.ECO/Ex, manufacturer – STS Sensor Technik Sirnach AG, 0-10 bar). Five photodiode sensors (model – DET10A, manufacturer – Thorlabs, 200-1100 nm) were integrated with the duct to measure the light intensity at various locations of the duct. A data acquisition system, capable to collecting data at a rate of 10 MS s^{-1}, was employed to record data of various sensors along the length of the duct. ANKO Dvmsoft (version 2.0.1.0) software was employed in this data acquisition.

The methane gas was sourced from Coregas and the gas was of 99.95 % purity. Pyrotechnic igniters of 1, 5 and 10 kJ (Simex Control) were used in the investigations. The pressure rise values due to igniters were deducted in the reported explosion pressure data.

The details of the operation method can be found in our earlier publication [8]. Briefly, methane was introduced to the system via partial pressure method. To initiate methane injection, the sphere was initially evacuated to a desired pressure level. For instance, if the experiment was for 8% methane concentration, the sphere was initially evacuated to – 68.1 kPag. Once the methane injection was completed, the pressure of the vessel reached – 60 kPag. The 20 litre sphere apparatus was connected to a dust chamber. Before the initiation of ignition, this dust chamber was pressurised to 20 barg. In the next stage, the pressurised air in the dust chamber was released into the vessel by opening a fast acting valve. After 60 ms of pressurised air discharge, the vessel's pressure equilibrated to atmospheric pressure (1 atm) and the flexible disc of the check valve opened just prior to the time the igniter was initiated. After ignition, the explosion event started and the combustion wave propagated from the vessel to the duct. Pressure rise and light intensity data with respect to time were recorded and analysed.

3 Results and Discussion

3.1 Pressure Profile of the Vessel

Figure 3 illustrates the pressure profile of the vessel. This pressure profile was developed by deducting the explosion pressure profile of the igniter from the experimentally obtained pressure profile for methane-air mixtures from the ignition point (87 ms in the figure). In the experimental investigations, the vessel to duct opening was open at the time of explosion. This approach assisted to examine the explosion parameters at zero static activation overpressure. In the explosion studies of ducted vessels, membrane or disc is often used for the purpose of the initial confinement of the flammable gas. Static activation overpressure is defined as the pressure to crack that membrane or disc [13].

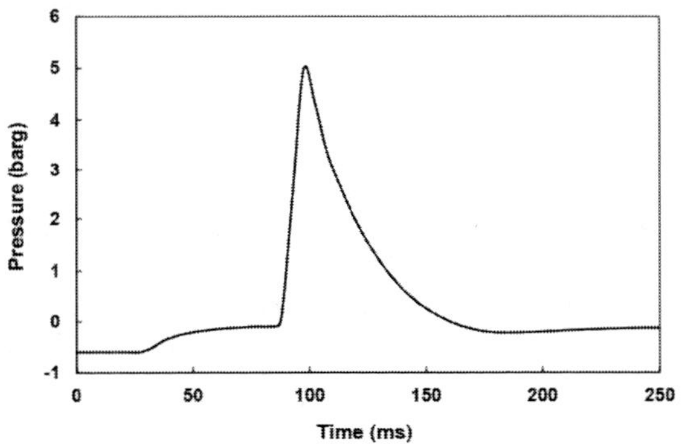

Figure 3: Pressure profile of the vessel as a function of time for a methane concentration of 8% with an ignition energy of 10 kJ.

The developed pressure profile of the vessel can be described as the reduced explosion pressure profile as the pressure rise of the vessel component of a ducted vessel is termed as reduced explosion pressure. Investigating explosion parameters at zero static activation overpressure is beneficial, as the experimental data can be correlated with a real world explosion scenario. The time to detect the reduced peak explosion pressure since the ignition is known as explosion time [14]. The reduced pressure profile of the vessel shows that explosion time for methane-air mixtures is very low. For the 8% methane experiment, explosion time was only 7.6 ms (Fig. 3).

The evolution of the reduced explosion pressure profile depends on a number of factors including the vessel's volume length, geometry of the duct, concentration of the flammable substance and the reactivity of the flammable material. In the present investigation, the release of pressurised gas to the vessel introduced turbulence to the gaseous medium. In addition, pyrotechnic igniters were used in the experimental investigations and these igniters are considered as strong igniters. Turbulence and strong ignition energies can be present in the real world explosion scenario. Both turbulence to the gaseous medium and the application of strong igniters impacted the development of the reduced explosion pressure profile.

3.2 Pressure and Flame Histories

The pressure rise along the length of the duct was mapped by five pressure transducers (Fig. 4). Steep pressure rises were observed in all pressure transducers. This suggests that the pressure rises at any point of the duct occurs sharply and a high level of turbulence is produced along the length of the duct. The peak explosion pressure recorded at the first pressure transducer was 5.69 barg for stoichiometric methane-air mixtures (9.5% methane in air) when an ignition energy of 10 kJ was applied. At the same ignition energy, the fifth pressure transducer detected a peak explosion pressure of 3.35 barg for stoichiometric methane-air mixture. This suggests that a significant explosion pressure hazard remains up to the end of the duct.

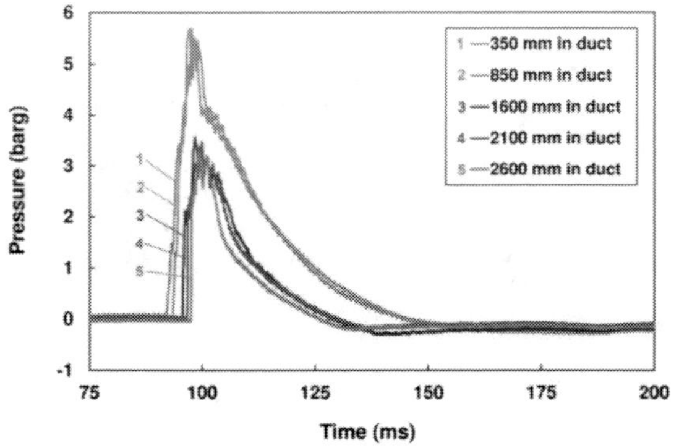

Figure 4: Explosion pressure profiles at various locations on the duct as a function of time as recorded by the pressure transducers for stoichiometric methane–air mixtures applying an ignition energy of 10 kJ.

The light intensity profiles recorded at the various photodiode sensors is presented in Fig. 5. The burning characteristics of methane can be understood from these profiles. The spans of explosion light signal reduced from the first sensor to the fifth. The recorded flame durations from the first to the fifth sensors were 75, 54, 36, 14 and 10 ms, respectively. The amount of flammable materials in the flow stream gradually reduced from the first to the fifth sensors. Thus the flame durations reduced from the first to the fifth sensor. The third photodiode sensor (1600 mm in the duct) experienced the highest light intensity compared to other photodiode sensors. This finding suggests that the maximum burning rate of methane-air mixture was in the vicinity of the third sensor [8].

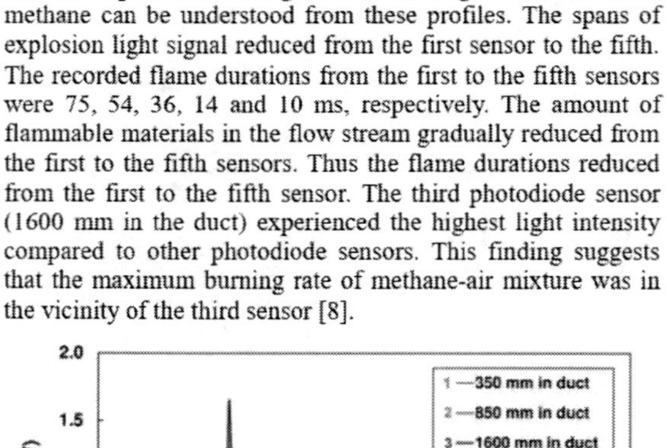

Figure 5: Light intensity profiles at various locations of the duct as a function of time as recorded by photodiode sensors for stoichiometric methane–air mixtures applying an ignition energy of 10 kJ.

3.3 The Effects of Ignition Energies on Pressure Rise and Flame Speed

Pyrotechnic igniters impact on explosion pressure in two ways. Firstly, they produce gas and this formation of gas increases explosion pressure rise. Figure 6 was developed deducting the igniters' pressure rise. Secondly, the dispersal of explosive powders from these igniters corresponds to the flame area they produce. The rating of these igniters is linked to the amount of their explosive powder. A pyrotechnic igniter of high rating (e.g., 10 kJ) contains a high amount of explosive powder and produces high flame area. The overall burning rate is proportional to the flame area produced [15] and thus the 10 kJ igniter produced the highest reduced peak explosion pressure as illustrated in Fig. 6. The experimentally obtained data was employed to develop a correlation between reduced peak explosion pressure and ignition energy for stoichiometric methane-air mixture. The correlation is presented below:

$$P_{red,peak} = 0.098\ln(I_g) + 5.91 \qquad (1)$$

where $P_{red,peak}$ is the reduced peak explosion pressure in barg and I_g is the ignition energy in kJ.

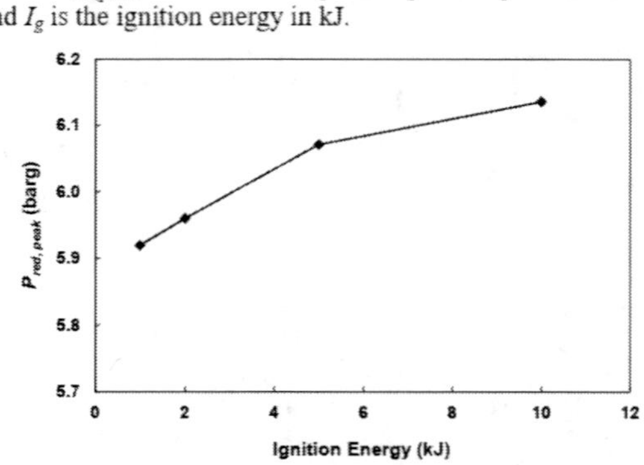

Figure 6: Effect of pyrotechnic ignitor's ignition energy on $P_{red,peak}$ for stoichiometric methane-air mixtures.

Average flame speed along the length of the duct was estimated based on the signals of two successive photodiode sensors (Fig. 7). The lowest flame speed was found by a 1 kJ igniter with a value of 204 m s^{-1}. This flame speed corresponds to a high level of Reynolds number and therefore indicates the presence of a high level of turbulence along the length of the duct. With the increase of ignition energy, flame speed increases. The highest flame speed estimated was 468 m s^{-1} at 10 kJ. This value of flame speed corresponds to the deflagration state of an explosion. When the flame speed is lower than the speed of sound, the state of explosion is known as deflagration. In this case, the speed of sound is estimated at temperature of combustion products [1].

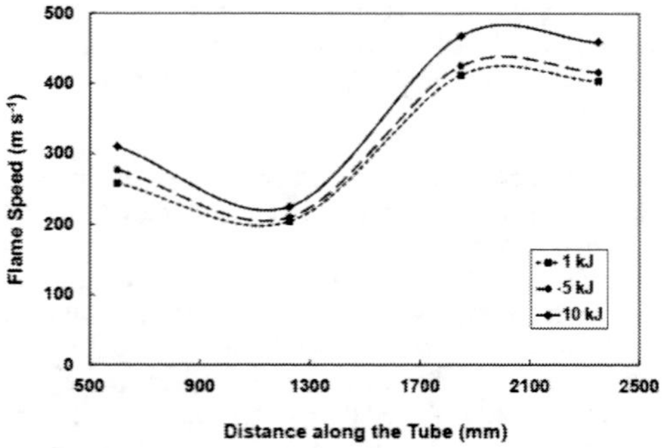

Figure 7: Flame speed along the length of the duct for stoichiometric methane-air mixtures.

3.4 Rates of pressure rise in the vessel

The geometry of the explosion apparatus plays an important role in development of the rate of pressure rise profile. The rate of pressure rise mechanism in ducted vessels is quite different from that in confined vessels. The flow velocity decays to zero at the wall in a confined vessel. In contrast, turbulence is generated in the ducted vessel due to the process of emitting the burning materials from the vessel to the duct.

Figure 8 illustrates the rate of pressure rise profile for the 8% methane experiment at 10 kJ ignition energy. This rate of pressure rise profile was developed by deducting the rate of pressure rise of the igniter from the experimentally obtained rate of pressure rise for methane-air mixture. The evolution of this rate of pressure rise profile was not only impacted by the geometry of the explosion apparatus, but also the operation method. The inclusion of turbulence in the gaseous medium and the use of strong igniters in the course of experimental operation significantly impacted the rate of pressure rise. This resulted in a high value of the maximum rates of pressure rise or $(dP/dt)_{max}$ as illustrated by Fig. 8. The integration of bend with the straight duct impacts the rate of pressure rise in the vessel and this phenomenon is reflected in Fig. 8.

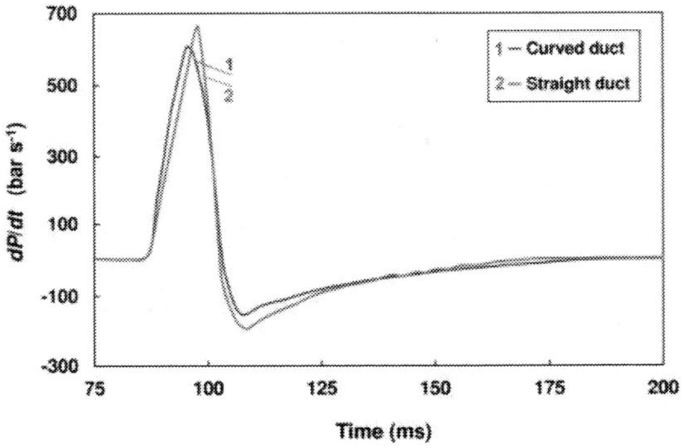

Figure 8: Rates of pressure rise (dP/dt) in the vessel as a function of time for a methane concentration of 8% with an ignition energy of 10 kJ.

3.5 The Effects of Pipe Bends on Explosion Parameters

The presence of a duct with a vented vessel introduces resistance to the flow stream [16]. The integration of a bend segment increases the flow resistance further. This was reflected in the obtained results. In our previous study with a straight duct (L=2813 mm), the reduced peak explosion pressure was 4.72 barg at 10 kJ for the stoichiometric methane-air experiment [8]. In contrast, the reduced peak explosion pressure was 6.14 barg at 10 kJ with the curved duct for stoichiometric methane-air experiment.

While the present study was carried out in a turbulent state, the presence of bend segment reduced the turbulence slightly. Thus the rate of pressure rise decreased slightly compared to our previous study with a straight duct [8]. For 8% methane experiment with 10 kJ igniter, the values of $(dP/dt)_{max}$ were 660 bar s^{-1} with the straight duct of previous study and 610 bar s^{-1} in the present study.

The presence of the bend segment also impacted on the flame speed values. The flame speed value between the third and the fourth sensors was 468 m s^{-1} for stoichiometric methane-air mixtures in the present study. The velocity at the same location in the previous study with straight duct was 327 m s^{-1} [8]. The bend segment increased the pressure up to the bend portion. Once the flow stream passed the restriction, a higher flame speed was yielded.

4 Conclusion

A spherical vented vessel connected to a curved duct was employed in understanding the explosion characteristics of methane-air mixtures in the system. The experiments are aligned with real world scenario of the presence of turbulent mixtures of methane-air, strong ignition energies and curved flow pathways. The turbulent state of gaseous medium and ignition energies increased the explosion pressure when compared to quiescent methane-air mixture and weak ignition energies. The bend segment increased the explosion pressure further. However, the bend reduced the turbulence slightly and the rate of pressure rise data was found to be slight lower when compared to the rate of pressure rise data with straight duct.

Acknowledgement

This work was supported by the Department of Resources Energy and Tourism, Australia (grant number – G1201029) and ACA Low Emissions Technologies Ltd., Australia (grant number – G1400523).

References

[1] S. Kundu, J. Zanganeh, B. Moghtaderi, J. Loss Prev. Process Ind., 40 (2016) 507-523.

[2] J. Einiger, http://abc7ny.com/news/stafford-township-house-explosion-blamed-on-underground-gas-leak/532311/, Accessed: 19 Oct 2016.

[3] J. Radford, http://www.illawarramercury.com.au/story/2441786/what-happened-tragic-day-of-mt-kembla-mine-disaster/, Accessed: 14 August 2015.

[4] B. Dhillon, Mine Safety: A Modern Approach, Springer, 2010, p. 59.

[5] J. du Plessis, D. van Greuning, J S Afr I Min Metall, 111 (2011) 887-894.

[6] S.K. Kundu, Process Saf. Environ. Protect., 88 (2010) 91-96.

[7] R. Kasmani, G. Andrews, H. Phylaktou, Process Saf. Environ. Protect., 91 (2013) 245-252.

[8] S. Kundu, J. Zanganeh, B. Moghtaderi, Process Saf. Environ. Protect., 111 (2017) 85-93.

[9] B. Ponizy, N. Henneton, A. Claverie, B. Veyssiere, Combust. Flame, 161 (2014) 1348-1364.

[10] J. Barton, Dust Explosion Prevention and Protection: A Practical Guide, Institution of Chemical Engineers (IChemE), Warwickshire, UK, 2002.

[11] R. Blanchard, D. Arndt, R. Grätz, M. Poli, S. Scheider, J. Loss Prev. Process Ind., 23 (2010) 253-259.

[12] S. Kundu, J. Zanganeh, D. Eschebach, B. Moghtaderi, in: Proceedings of The Australian Mine Ventilation Conference, 2017.

[13] W. Bartknecht, Explosions: course, prevention, protection, Springer-Verlag, New York, USA, 1981.

[14] D. Razus, C. Movileanu, V. Brinzea, D. Oancea, J. Hazard. Mater., 135 (2006) 58-65.

[15] J.H. Lee, C.M. Guirao, Plant/Operations Prog., 1 (1982) 75-85.

[16] G. Ferrara, A. Di Benedetto, E. Salzano, G. Russo, J. Hazard. Mater., 137 (2006) 654-665.

11th Asia-Pacific Conference on Combustion,
The University of Sydney, NSW Australia
10th -14th December 2017.

Methane-Air Flame Propagation Characteristics in Presence of Restriction in a Duct

Mohammadreza Shirzaei, Jafar Zanganeh*, Sazal Kundu, Behdad Moghtaderi

Priority Research Centre for Frontier Energy Technologies & Utilisation, Discipline of Chemical Engineering

School of Engineering, Faculty of Engineering & Built Environment, University of Newcastle

Callaghan, NSW 2308, Australia

Abstract

Many process industries including coal and gold mines deal with life-threatening methane-air mixtures. In human history, methane-air mixtures caused numerous explosions leading to property damage and loss of lives. The severity of fire and explosion is impacted by some variables such as fuel concentration and turbulence. This work aims to obtain an enhanced understanding of methane-air explosion phenomena in a straight duct. Orifice plate obstacles of various blockage ratios (30%, 50%, and 70%) were employed in the duct and the effect of those obstacles on flame propagation characteristics were investigated. The experimental glass tube has a volume of 23 L and obstacles were placed at various locations of the test tube. It was observed that the increase of blockage ratio increased the explosion pressure and flame speed. The number of obstacles also influenced the explosion parameters and it was found that the explosion pressure and flame speed increased with the number of obstacles. In particular, with the test of 10% methane in air, the maximum overpressure increased from 181 to 216 kPa and flame speed increased from 149 to 170 m/s when a single orifice plate was placed in the flame propagation path increasing the blockage ratio from 30% to 70%. The maximum overpressure and flame speed increased further when the number of orifice plates of 70% blockage ratio was increased from one to two. In this case, the values of maximum overpressure and flame speed were 240 kPa and 175 m/s respectively.

1 Introduction

Methane-air mixtures are of great safety concern in the process industries including coal and gold mines. A large number of accidents have been documented in literature and media due to explosions caused by methane-air mixtures [1-4]. Those accidents not only damaged infrastructure but also injured and killed a massive number of people. Often the death and injury toll exceeded one thousand and massive destruction occurred in the accident zone [5-7].

The severity of methane-air explosion has been recognised by the research community. Since the last century, researchers are involved in understanding the explosion characteristics of methane-air mixtures as well as finding pathways to eliminate or reduce this fatal problem by experimental and theoretical studies [5, 8-15]. Valuable information has been gained through those works. For instance, it was found that there is a methane concentration range in which explosion occurs and the concentration range is 4.6±0.3 to 15.8±0.8 percent methane in air[5]. In addition, explosion pressure and flame speed are a maximum at the stoichiometric conditions [5].

The effect of obstacles on explosion parameters was explored and it was found that explosion pressure increases in presence of obstacles. In the case of flame propagation in a duct, the flame speed increases if there is any obstacle in the path of flame propagation [5]. The scale of experimental apparatus further impacts on these explosion parameters [16, 17] and study shows that the burning velocity can be increased significantly with increased diameter of the reactor [18]. The number of obstacles in the path of flame propagation has significant impact on explosion parameters. Flame speed and explosion pressure increases with the increase of obstacle number [19]. Similarly, blockage ratio of an obstacle impacts on explosion pressure and flame speed. These explosion parameters, in general, increases with increasing blockage ratio [20, 21].

In process industries, materials are transported by pipes and often these pipes include obstacles [16, 22]. Tunnels and roadways are present in coal and gold mines. Therefore, understanding flame propagation characteristics in presence and in the absence of obstacles can provide vital information and assist in developing safety strategies to these industries. Flame propagation time in a real world scenario is very short and previous experiments show that flame can propagate long distance in the order of milliseconds [7]. Therefore, sophisticated instruments need to be deployed to understand flame propagation phenomena in a duct. With this aim, piezo-resistive pressure transducers, a high speed data acquisition system and a high-speed camera were employed in this experimental investigation. A series of methane-air flame propagation experiments were conducted in a straight tube in both the presence and absence of obstacles. Data obtained from the study was analysed and presented with detailed discussion.

2 Experimental setup

The flame propagation apparatus employed for this study is illustrated in Fig. 1. This apparatus was comprised of four clear borosilicate glass modules and the modules were connected to each other by stainless steel flanges maintaining an internal diameter of 75 mm and constituting a total length of approximately 5 m. The glass tubes were sourced from Schott Duran. The glass tubes were of 5 mm wall thickness with a temperature tolerance of 800°C and a pressure tolerance of more

*Corresponding author. Fax: +61-2-4033-9095
E-mail address: Jafar.Zanganeh@newcastle.edu.au

than 8 bar. For each module, one glass tube was employed with a length of 1 m.

A Bronkhorst F-201CV mass flow controller was used to control the feed rate methane gas into the system. The concentration of methane in the system was confirmed by an inline methane monitor. The pressure rise from the explosion tests were measured by three pressure transducers (manufacturer – STS Sensor Technik Sirnach, model – ATM.ECO/EX with pressure range of 0 to 5 bar). These pressure transducers were installed on the connection flanges as depicted in Fig. 1. The distances of these pressure transducers from the ignition source were 1.2, 2.4 and 3.6 m. Data from these pressure transducers were recorded by a high speed data acquisition system. Flame propagation through the apparatus was filmed by a sophisticated high speed colour video camera (Phantom V1211, 12,000 frames-per-second at full resolution).

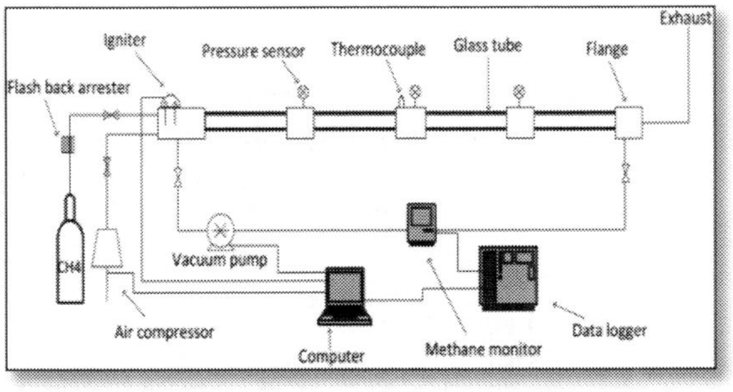

Figure 1: Schematic diagram of the flame propagation apparatus employed in the experimental investigation.

With the aim of ensuring that the apparatus was free from any contaminant of earlier experiments, the duct was purged five times prior to commencing any experiment. Compressed air was used in this purging purposes. Before introducing methane gas to the system, the open end of the tube was closed by a thin plastic wrap. Once the methane gas injection was completed, a homogeneous methane-air mixture was prepared by circulating the gaseous medium for 5 minutes employing an intrinsically safe (EX rated) vacuum air pump (manufacturer – Air Dimensions Incorporated, model – Dia-Vac Diaphragm R252-FP-GB2). In this investigation, orifice plates were introduced at the flanges to understand their impact on explosion characteristics. The blockage ratios developed by those orifice plates were 30, 50 and 70% of the glass tube circular opening.

Figure 2: Image of flame propagation captured from the experimental setup.

The methane gas was sourced from Coregas, Australia with 99.95% purity. A 50 mJ energy pyrotechnic ignitor, connected to a 24 voltage power, was employed to ignite the mixture. Error bars, presented in this article, were calculated from three data points and they represent errors from average to maximum/minimum value. Figure 2 shows a sample image of flame propagation through the duct during an experiment.

3 Result and discussion

3.1 Effect of obstacle on the flame propagation time and the maximum overpressure

Figures 3 and 4 assist in understanding the observed flame speed in the presence of obstacles. It is clear that the flame with faster speed passes through the tube in shorter time. Both of these figures were constructed by eight sequential images of flame propagation through the experimental duct in the absence of an obstacle and in the presence of an obstacle of 30% blockage ratio, respectively, employing 10% methane-air mixtures. The measured elapsed time to pass the flame from one end to the opposite was approximately 40 ms when there was no obstacle (Fig. 3). The elapse time for the flame to pass the experimental duct reduced to 35 ms in the presence of an orifice plate of 30% blockage ratio (Fig. 4). The difference in time is related to the application of obstacles in the duct because of increased interaction between the flame and obstacle as discussed previously.

Figure 3: Flame propagation for 10% methane concentration in the absence of obstacle.

Figure 4: Flame propagation for 10% methane concentration in presence of obstacle of 30% blockage ratio positioned at 1.2 m distance from ignition source.

Figure 5 illustrates the maximum overpressure as a function of methane concentration. In the absence of orifice plates, the maximum overpressure was found to be 129 kPa for the 7.5% methane concentration. The maximum overpressure increased to 175 kPa for 10% methane and dropped to 111 kPa for 15% methane concentration. This pressure profile is consistent with the literature [5]. The maximum overpressure for methane-air mixture is expected to be observed near or at stoichiometric methane-air mixture (9.5%) due to complete combustion reaction. For a lean methane-air mixture, the lower percentage of methane produces less maximum overpressure when compared to stoichiometric methane-air mixtures. For a rich methane-air mixture, partial combustion of methane reduces the pressure rise in the system when compared to stoichiometric methane-air reactions.

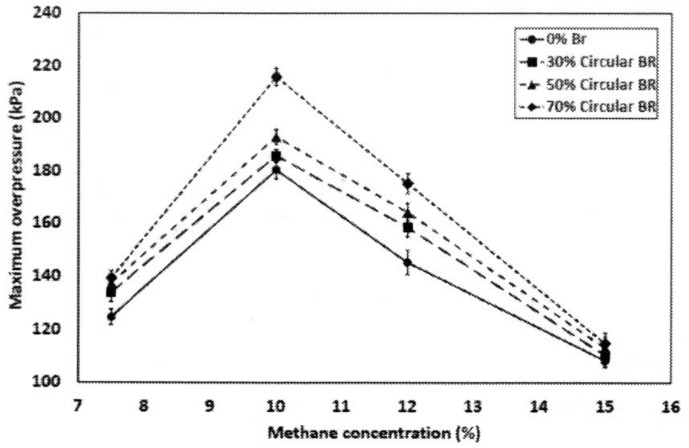

Figure 5: Maximum overpressure as a function of methane concentration. Orifice plates, if used, were placed at 1.2 m distance from ignition source.

The presence of obstacles magnifies the severity of explosion [5]. As illustrated in Fig. 5, the maximum pressure increased with increasing blockage ratio, in agreement with previous work [20]. The impact of the position of a single orifice plate along the length of the duct was investigated. It was observed that the maximum overpressure at 10% methane increased when the orifice plate was moved from 1.2 m to 2.4 m distance from ignition source. The number of orifice plates also impacted the maximum overpressure. When two orifice plates of 70% blockage ratio were placed at 1.2 m and 2.4 m respectively, a pressure rise value of 240 kPa was recorded. For one orifice plate, the pressure rise was lower than the pressure rise for two orifice plates.

The position and number of obstacles impact on flame propagation characteristics in many ways. The movement of obstacles outward from the ignition source increased the confinement due to the trapping of more unburned gas. The increase of confinement due to the presence of obstacle increases the explosion pressure. In presence of an obstacle, the flame surface area increases as it passes through the obstacle, this leads to increased burning rate. The level of turbulence increased due to the interaction between the obstacles and the flame. Therefore, increased number of obstacles increased the explosion pressure, as supported by literature [19, 23].

Figure 6 shows maximum overpressure along the length of the duct for 10% methane experiment. The highest variation of maximum overpressure from the first sensor to the third sensor was related to the largest blockage ratio which was 70%. For 0%, 30% and 50% blockage ratios, the variations of the maximum overpressure were nearly the same although the maximum overpressure increased with the increase of blockage ratio.

In the absence of obstacle, the pressure rise at the first sensor was 180 kPa. The pressure rise increased to 216 kPa when an orifice plate of 70% blockage ratio was placed at 1.2 m from the ignition source.

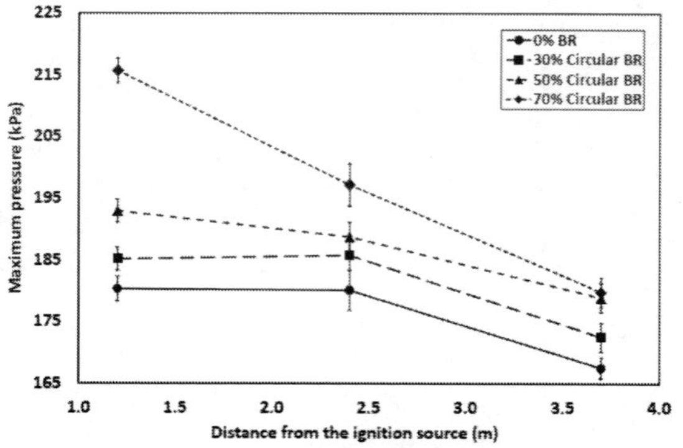

Figure 6: Maximum overpressure along the length of the flame propagation duct for 10% methane in air. Orifice plate, if employed, was placed at 1.2 m distance from ignition source.

3.2 Effect of obstacle on the flame speed

In explosion studies, multiple approaches can be found in literature to estimate flame speed to meet specific purposes. In the present investigation, we estimated flame speed employing the ignition source as the reference point. This approach is important from the safety point of view as it can provide an estimation on the time requirement for the flame to travel from the point of origin of explosion to any point in a tunnel or pipeline.

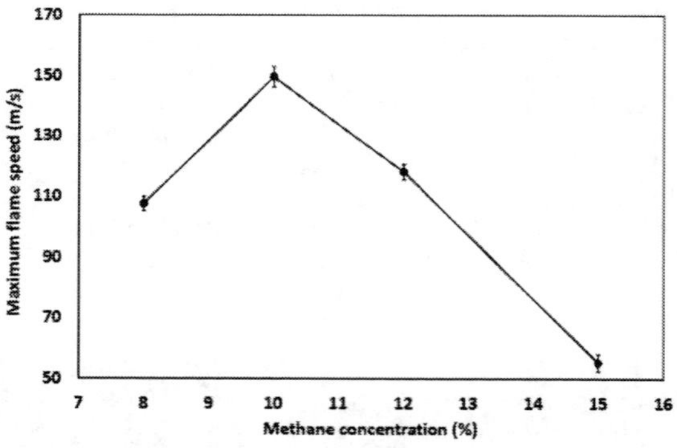

Figure 7: Variation of maximum flame speed with the change of methane concentration.

Figure 7 shows the effect of the concentration of methane on flame speed. The highest flame speed was observed for 10% methane experiments (145 m/s). This experiment was a near stoichiometric methane-air experiment [5]. Stoichiometric or near stoichiometric methane-air mixture produces the highest flame speed due to complete combustion and reaction kinetics.

The impact of obstacles on flame speed is illustrated in Fig. 8. In general, flame speed increases in the presence of obstacles in a straight duct when compared to flame speed without any obstacles, which is an increase from 170 m/s for a single blockage at 1.2 m. While Fig. 8 was constructed applying a single orifice plate, experiments were also carried out with two orifice plates as described earlier. In the presence of two orifice plates, placed at 1.2 m and 2.4 m from the ignition source, a flame speed value of 175 m/s was obtained when the blockage ratio of both plates were 70%.

In presence of an obstacle, the flame front is fragmented. The consequential vortices and turbulence lead to an increased flame surface area that enhances the burning rate [20]. This is how the flame speed increases in presence of obstacles.

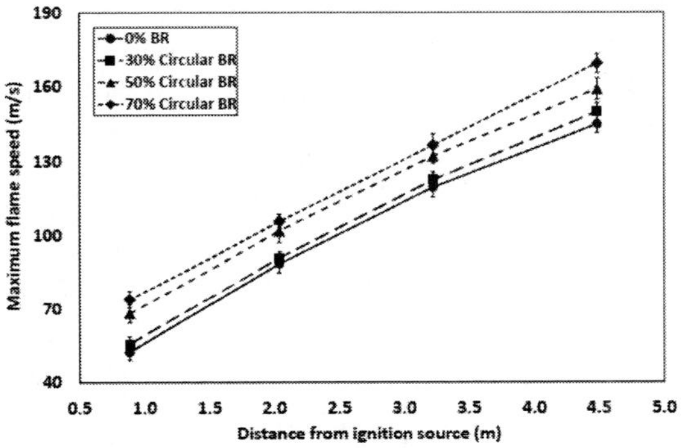

Figure 8: Maximum flame speed along the length of the flame propagation duct for 10% methane in air. Orifice plate, if employed, was placed at 1.2 m distance from ignition source.

4 Conclusion

Explosion parameters such as explosion pressure and flame speed were investigated in a clear glass tube in presence and in the absence of obstacles. Maximum overpressure and flame speed increased in the presence of obstacles when compared those parameters in the absence of obstacles.

The effect of the number of obstacles of identical blockage ratio was investigated. It was observed that the increase of obstacles increased maximum overpressure and flame speed. However, the relative increase in explosion pressure rise and flame speed values for second obstacle was less than those for a single obstacle. It is therefore expected that there is a limit of the number of obstacles above which the explosion pressure rise and flame speed do not change significantly.

5 Acknowledgment

This work was supported by the Department of Resources Energy and Tourism, Australia (grant number – G1201029) and ACA Low Emissions Technologies Ltd., Australia (grant number – G1400523).

References

[1] B. Dhillon, Mine Safety: A Modern Approach, Springer, (2010) 59-71.

[2] J. du Plessis, D. van Greuning, J S Afr I Min Metall, 111 (2011) 887-894.

[3] J. Radford, http://www.illawarramercury.com.au/story/2441786/what-happened-tragic-day-of-mt-kembla-mine-disaster/, Accessed: 14 August 2015.

[4] J. Einiger, http://abc7ny.com/news/stafford-township-house-explosion-blamed-on-underground-gas-leak/532311/, Accessed: 19 Oct 2016.

[5] S. Kundu, J. Zanganeh, B. Moghtaderi, J. Loss Prev. Process Ind., 40 (2016) 507-523.

[6] M.J. Ajrash, J. Zanganeh, B. Moghtaderi, J. Loss Prev. Process Ind., 40 (2016) 317-328.

[7] M. Shirzaei, J. Zanganeh, S. Kundu, B. Pitkin , B. Moghtaderi, in: Interflam, 2016, pp. 1-6.

[8] M.J. Ajrash, J. Zanganeh, B. Moghtaderi, Fuel, 194 (2017) 491–502.

[9] W. Bartknecht, Explosionsschutz, Grundlagen und Anwendung, Springer Verlag, Berlin, Heidelberg, New York, 1993.

[10] S. Dorofeev, V. Sidorov, M. Kuznetsov, I. Matsukov, V. Alekseev, Shock Waves, 10 (2000) 137-149.

[11] R. Eckhoff, Explosion Hazards in the Process Industries, Gulf Publishing Company, Houston, Texas, 2005.

[12] J. Lee, R. Knystautas, C. Chan, in: Symposium (International) on Combustion, Elsevier, 1985, pp. 1663-1672.

[13] I. Moen, M. Donato, R. Knystautas, J. Lee, Combust. Flame, 39 (1980) 21-32.

[14] E. Oran, V. Gamezo, R. Zipf Jr, Combust. Sci. Technol., 187 (2015) 324-341.

[15] B. Vanderstraeten, D. Tuerlinckx, J. Berghmans, S. Vliegen, E. Van't Oost, B. Smit, J. Hazard. Mater., 56 (1997) 237-246.

[16] G. Li, Y. Du, S. Qi, Y. Li, S. Wang, B. Wang, J. Loss Prev. Process Ind., 43 (2016) 529-536.

[17] Q. Zhang, L. Pang, H.M. Liang, J. Loss Prev. Process Ind., 24 (2011) 43-48.

[18] M.K. Fig, G.E. Bogin, J.F. Brune, J.W. Grubb, J. Loss Prev. Process Ind., 41 (2016) 241-251.

[19] C. Wang, F. Huang, E.K. Addai, X. Dong, J. Loss Prev. Process Ind., 43 (2016) 302-310.

[20] A.R. Masri, S.S. Ibrahim, N. Nehzat, A.R. Green, Exp. Thermal Fluid Sci., 21 (2000) 109-116.

[21] M. Yu, K. Zheng, T. Chu, J. Nat. Gas Sci. Eng., 30 (2016) 221-227.

[22] S.K. Kundu, Process Saf. Environ. Protect., 88 (2010) 91-96.

[23] X. Wen, M. Yu, Z. Liu, G. Li, W. Ji, M. Xie, J. Loss Prev. Process Ind., 26 (2013) 1335-1340.

11th Asia-Pacific Conference on Combustion,
The University of Sydney, NSW Australia
10th -14th December 2017.

Effect of the presence of nanosilver on nanoaluminium powder vented explosion

Khairiah Mohd Mokhtar[1], Rafiziana Md Kasmani[2], Che Rosmani Che Hassan[1], Mahar Diana Hamid[1], Mohamad Iskandr Mohamad Nor[1]

[1]Department of Chemical Engineering, Faculty of Engineering, University of Malaya
Kuala Lumpur, 50603, Malaysia

[2]Department of Energy Engineering, Faculty of Chemical and Energy Engineering, Universiti Teknologi Malaysia
Johor, 81310, Malaysia

Abstract

Studies have found that nanomaterials – in this case dusts and powders with nanosized particles – exhibit higher explosion severity due to very low ignition energies and temperatures. Due to their susceptibility for explosion hazard, prevention of accidental nano-dust explosions should be assessed, as these mixtures of two or more nanoparticle metals are common in industrial practices. To date, the effects of combining two different metals on explosion parameters are still ambiguous. Thus, this paper aims to quantify the effect of addition of nanosilver on aluminium nanoparticle explosion. Experimental works were carried out on modified 1.2L Hartmann test vessel, by varying the mixing ratio and dust concentration for vented explosion. However, this work only considered a filter paper as the vent cover. It was found that there was a significant decrease in reduced explosion pressure (P_{red}) ~ 55% of aluminium explosion with the addition of silver nanopowder. This particular condition occurred when the mixtures ratio was 40:60 of nanoaluminium and nanosilver, respectively, as compared to pure aluminium nanoparticles.

1 Introduction

Nanosized metal is getting more attention in various industrial applications as it poses improved properties. For instance, aluminium and magnesium are used as metal-based fuels in propellants [1-3], while silver is widely used in the pharmaceutical industry for its effective antibacterial properties [4, 5]. The small sizes of the particles increase the reactive surface area, subsequently modifying their physical and chemical properties [6]. Therefore, it is important to seriously consider accidental nano-dust explosions to assess their safety hazard in the industrial process. However, in ensuring the safety requirement is met, the existing prevention and control techniques are still insufficient for nano-dust production, transportation, and storage. Studies have shown significant effects on explosion likelihood and severity if particles' sizes are changed from micron to nanoscale [2, 7-9]. For instance, Wu et al.[9] demonstrated an increase of maximum explosion pressure (P_{max}) and rate of pressure rise (dP/dt) from 5.9 bar and 282 bar/s to 12.5 bar and 1090 bar/s, respectively in nanosized particles as compared to micron particles. Meanwhile, minimum ignition energy (MIE) gives a decrease value in nanoparticles, as low as 1 mJ for aluminium and magnesium, indicating that nanoparticles pose higher ignition risk compared to microparticles as suggested by Wu et al.[9] and Mittal et al.[1].

Extensive studies have been conducted on explosion of metal nanoparticles, but they are only considered based on a single nanometal explosion. If compared with the understanding of micron-dust explosions, the current level of physical understanding of the nano-dust explosion severity if there are two different nanometal particles involved in the process is still in a rudimentary state. The literature is insufficient for us to fully understand the explosion risk and severity magnitude if two nanoparticle metals are combined. This is because it is a norm to have two or more different metal nanoparticles in the industrial application, particularly in processing and storage. Studies on the effect of combining metal powder with another type of powder showed reduction of P_{max} [10] and minimum ignition temperature (MIT) [11] as compared to single type powder explosion. In addition, Kuai et al. [10] showed that by adding calcium carbonate and potassium chloride to micron size of magnesium particles, approximately 83% of P_{red} was reduced. Meanwhile, another study conducted on MIT by Yuan et al.[11] presented about 13% increment of MIT value with addition of 60% of titanium oxide into pure microtitanium powder. It implies that the presence of other metal powder would give significant changes on the explosibility and ignition sensitivity of pure metal powder. Knowing the effect of agglomeration exited in nano-dust [12], one big question arises; what would be the flame propagation behaviour of two different nano-dust metals cloud should be, and worst, how are the physics and dynamics explosibility mechanism if the two metals are in different groups? In this study, nanoaluminium dust vented explosion in the presence of nanosilver particles was carried out to characterise their combustion behaviours and explosion severity magnitude, considering their different mixing ratios, concentrations and particle sizes.

2 Method

2.1 Experimental apparatus and materials

The experimental setup is shown schematically in Figure 1. It consists of a 1.2L stainless steel test vessel, dispersion system, an ignition system, time controller system, pressure transducer, data logger and high-speed video camera. A gas nozzle is mounted in the vessel and connected to air compressor for dust

Corresponding author. Fax: +60-75581463
E-mail address: rafiziana@utm.my

dispersion. The ignition system consists of voltage transformer and a pair of tungsten wire electrodes. An ignition spark is initiated with 60ms delay time after dispersion of powder. Pressure time history of the powder is measured by pressure transducer mounted in the vessel wall and data generated is recorded by a data acquisition system for each run, whereas the high speed camera is used to record flame propagation process and vented flame behaviours of the particles.

Three different sizes of aluminium powder (40 nm, 70 nm and 100 nm) and silver powder (20 nm, 100 nm and 10 μm) used in this experiment were purchased from manufacturer. The properties of both metals powder were tabulated in Table 1. On the other hand, filter paper with venting pressure, P_v 0.1bar is mounted on the top of the vessel as venting membrane.

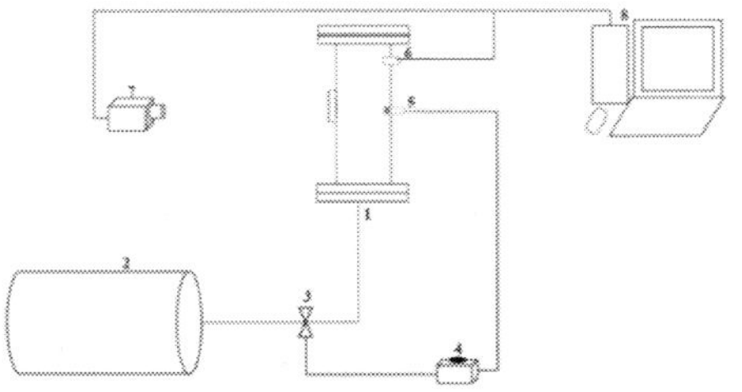

Figure 1: Experimental apparatus : 1. Gas nozzle, 2. Air compressor, 3. Solenoid valve, 4. Time controller, 5. Ignitor electrode, 6. Pressure transducer, 7. High speed camera, 8. Data recorder

2.2 Experimental procedures

The venting membrane was assembled on top of the vessel prior to insertion of metal powder mixtures. Then, aluminium nanopowder was weighed based on five concentrations (100 g/m^3, 200 g/m^3, 300 g/m^3, 400 g/m^3 and 500 g/m^3). The test vessel was tightly screwed after the powder has been inserted in the test vessel. At 6 bar, the powder was dispersed and ignited after 60ms delay of powder dispersion (i.e ignition delay time, t_d). The same procedures were repeated for mixture of aluminium and silver with mixing ratio of 70:30, 50:50 and 40:60. Table 1 shows the physical and chemical properties of the studied metal dusts.

	Aluminium	Silver
Chemical Formula	Al	Ag
Molar mass	26.98 g/mol	107.87 g/mol
Melting point	660.3°C	961.8°C
Boiling point	2, 470°C	2,162°C
Enthalpy of vaporization	293.72 kJ/mol	255.1 kJ/mol

Table 1: Physical and chemical characteristics of aluminium and silver.

3 Results and discussion

3.1 Effect of particle size on P_{red}

An increase of P_{red} value of aluminium nanopowder could be observed in Table 2 as the particle size decreased from 100 nm to 70 nm but no similar observation on Al at 40nm. Further decrease of the particle size to 40 nm resulted in smaller P_{red} value than Al 70 nm. Rate of pressure rise of aluminium particles was found to decrease when the particle size changed from 70 nm to 40 nm. Previous studies have demonstrated that when the particle size is reduced, the severity of explosion is increased due to influence of the increase of specific areas for oxidation reaction [1, 9]. However, in the present study, it contradicted with the previous studies. It can be depicted that when the aluminium particle changed from 70 nm to 40 nm, or when the particles were getting smaller, it can be suggested that agglomeration process became rapid, and consequently, gave a smaller surface area, limiting the mass burning rate and thus, affecting the overall explosion pressure and rate of pressure rise.

According to Eckhoff (2012), agglomeration effect of the particles is associated with Van der Waals' forces between particles. As shown in equation (1), where A was a constant, d was the diameter of the particles and a was the smallest distance between the surface of particles [12]. Van der Waals' forces would be directly proportional to diameter of the particles. Hence, when aluminium particles decreased from 70nm to 40nm, the Van der Waals' forces was increased and led to an increase of the effective particle size. Consequently, specific surface area of particle was reduced and caused overpressure to be declined [13].

$$Fw = \frac{Ad}{2a^2} \tag{1}$$

Metal	Size of particle	P_{red} (mbarg)	Rate of pressure rise (barg/s)
Aluminium (Al)	40 nm	133.682	4.020
	70 nm	145.888	4.437
	100 nm	109.213	3.332
Silver (Ag)	20 nm	64.217	0.255
	100 nm	82.023	0.269
	10 μm	88.984	0.321

Table 2: Reduced explosion pressure (P_{red}) and rate of pressure rise of different sizes of aluminium and silver particles.

In contrast, the P_{red} value of silver particles was found to be decreased as the particles' sizes reduced. When 100 nm of silver particle was used, P_{red} value recorded was 82.023 mbarg, 7.8% smaller than P_{red} value of Ag of 10 μm. The P_{red} value was further decreased to 64.217 mbarg as the size of the silver particle decreased to 20 nm. Overall, it can be observed that the P_{red} value of silver particle was approximately one third to half of the P_{red} value of aluminium particle. Meanwhile, the rate of pressure rise value of silver particles recorded about 90% lower than the aluminium particles. It can be said that this condition was due to the molar mass of silver particles which is approximately four times heavier than aluminium particles (as tabulated in Table 1) and silver is less reactive metals. Moreover, the heavier molar mass of silver particles might affect the dispersion process prior to ignition. The poor dispersion of the particles may cause the effective particle size to increase thus, decreasing the reactivity of the particles and explosion severity [14]. It can also be postulated that the poor dispersion of silver particles, which gives suggestion that agglomeration is prone on silver nanoparticles is the cause of the significant difference on pressure-time profile of vented explosion of silver particles compared of that aluminium explosion, as shown in Figure 2. For brevity, only graphs shown the pressure-time histories of Al 70 nm and Ag 20nm are presented as they represent the general trend of pressure time profiles for all particles' sizes of aluminium and silver.

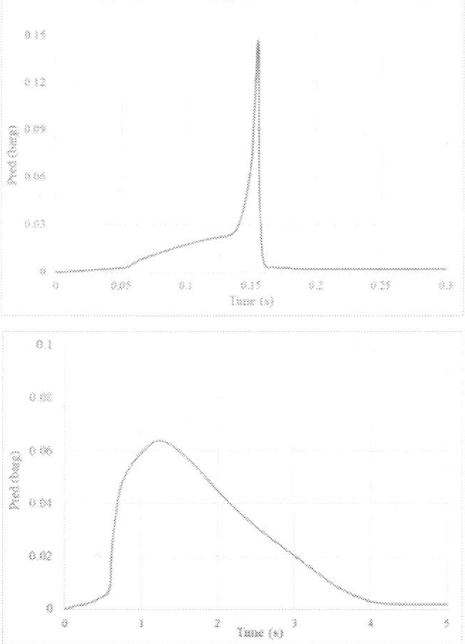

Figure 2 : Pressure-time profile of vented explosion of Al 70 nm particles (above) and Ag 20 nm particles (below)

The aforementioned statement leads to an assumption that by mixing silver particle to aluminium particle at certain ratios, explosion severity of aluminium particle, both P_{red} and rate of pressure rise, might reduce. Since Al at 70 nm demonstrated the highest P_{red} value compared to the other two sizes of aluminium particles, Al at 70 nm was chosen to be further investigated with different sizes of silver particles at different mixing ratios.

3.2 Effect of the presence of nanosilver to P_{red} of nanoaluminium vented explosion

Al at 70 nm is mixed with 10 µm, 100 nm and 20 nm of silver particles at three mixing ratios; 70:30, 50:50 and 40:60. The result of the investigations is presented in Figure 3-5. Generally, gradual increase of P_{red} value can be observed at concentration of 300 g/m^3 to 500 g/m^3. This trend is demonstrated at 70:30 and 50:50 mixing ratio of aluminium and silver nanoparticles, respectively. The maximum P_{red} value is recorded when concentration reached 500 g/m^3. It is essential to note that there is no significant effect on P_{red} value of the mixture at 40:60 mixing ratio despite of changes in the aluminium concentration.

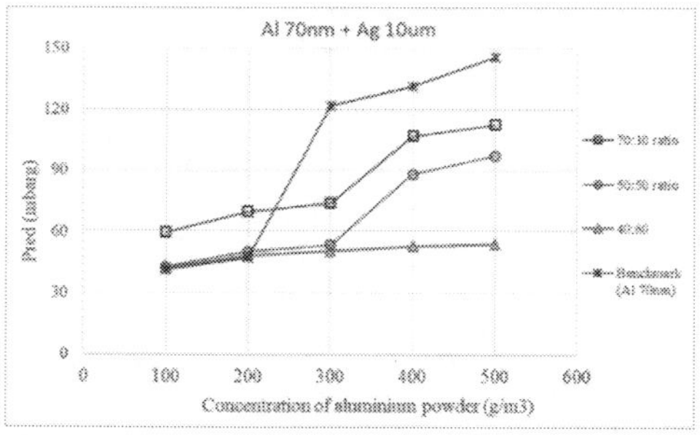

Figure 3: P_{red} value of mixture of Al 70nm and Ag 10µm at three different mixing ratios

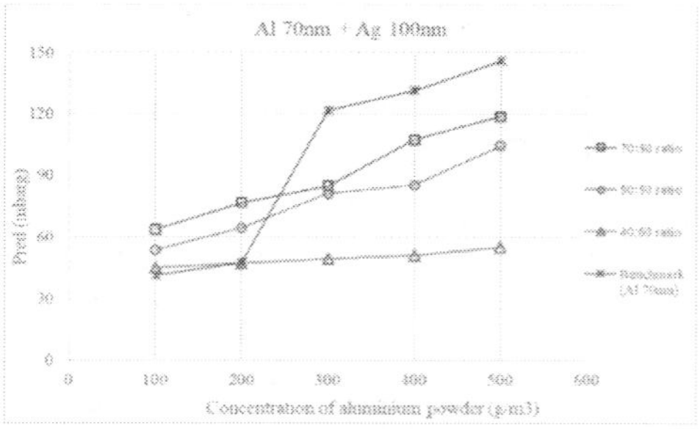

Figure 4: P_{red} value of mixture of Al 70nm and Ag 100nm at three different mixing ratios

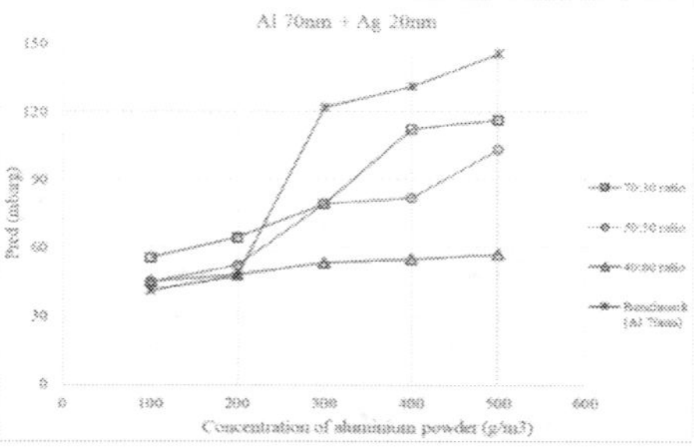

Figure 5: P_{red} value of mixture of Al 70nm and Ag 20nm at three different mixing ratios

From Figure 3, P_{red} value of aluminium nanoparticles at 500g/m^3 decreased from 145 mbarg to 113 mbarg at 70:30 mixing ratio. Approximately 50% of the P_{red} value was further reduced when 40:60 mixing ratio was applied compared to 50:50 mixing ratio. Similar trend is illustrated in Figure 4-5, in which different sizes of silver particle were added with Al at 70 nm. Overall, the P_{red} values of 40:60 mixing ratio of mixture at 300 g/m^3 to 500 g/m^3 concentration were reduced by almost two third of the single aluminium particle explosion (as the benchmark). This condition shows that the heavier silver particles contributed to less effective surface area, and interestingly, when combined with Al nanoparticles, rapid agglomeration and accumulation of silver particles around aluminium particles became more apparent as shown in Figure 6. The particles' morphology obviously shows that heavier molar mass of silver particles somehow gives a 'suppressing effect' to the combustion reaction of the silver-aluminium nanoparticles and hence, causing lower P_{red} value when nanosilver is added to the vented aluminium explosion mixtures. It can be suggested that the chemical reactivity of silver-aluminium particles plays an important role on the overall thermodynamic and chemical kinetics of the combined mixtures. In consequence, maximum combustion of highly reactive aluminium is hindered by the presence of less reactive silver particles. However, the analysis of thermodynamic and chemical kinetics is not the scope in the present paper.

Figure 6 : Accumulation of silver particles (whitish particles) around aluminium particles (greyish particles) at a) 70:30 ratio, b) 50:50 ratio, c) 40:60 ratio of mixtures from Field Emission Scanning Electron Microscope (FESEM) analysis

4 Conclusions

The presence of silver particles gave a decrease to the reduced explosion pressure (P_{red}) of nanoaluminium vented explosion. It was found that P_{red} reduced its value corresponding with the mixing ratio, in which 40:60 mixing ratio of mixture was the most notable on P_{red} reduction as compared to others~ 55% reduction. Agglomeration and rapid accumulation of heavier molar mass of silver particles around the aluminium were postulated to give a 'suppressing effect', which consequently resulted in lower P_{red} values of the mixtures. It can be said that the chemical reactivity of silver particles affects the overall combustion mechanism of aluminium and silver nanoparticle vented explosion by changing their thermodynamic properties and chemical kinetic of the metal mixtures. However, the details for the association have yet to be studied in the future.

5 Acknowledgment

The authors would like to express the gratefulness and appreciation to FRGS Grant (Project No: FP013-2014B) and PPP Grant (Project No: PG270-2015B) from University of Malaya (UM).

References

[1] M. Mittal, Explosion characteristics of micron-and nano-size magnesium powders, Journal of Loss Prevention in the Process Industries, 27 (2014) 55-64.

[2] B. Jiang, B. Lin, S. Shi, C. Zhu, W. Li, Explosive characteristics of nanometer and micrometer aluminum-powder, Mining Science and Technology (China), 21 (2011) 661-666.

[3] K. Jayaraman, S. Chakravarthy, R. Sarathi, Accumulation of nano-aluminium during combustion of composite solid propellant mixtures, Combustion, Explosion, and Shock Waves, 46 (2010) 21-29.

[4] J.S. Kim, E. Kuk, K.N. Yu, J.-H. Kim, S.J. Park, H.J. Lee, S.H. Kim, Y.K. Park, Y.H. Park, C.-Y. Hwang, Y.-K. Kim, Y.-S. Lee, D.H. Jeong, M.-H. Cho, Antimicrobial effects of silver nanoparticles, Nanomedicine: Nanotechnology, Biology and Medicine, 3 (2007) 95-101.

[5] S. Shrivastava, T. Bera, A. Roy, G. Singh, P. Ramachandrarao, D. Dash, Characterization of enhanced antibacterial effects of novel silver nanoparticles, Nanotechnology, 18 (2007) 225103.

[6] S.P. Boilard, P.R. Amyotte, F.I. Khan, A.G. Dastidar, R.K. Eckhoff, Explosibility of micron-and nano-size titanium powders, Journal of Loss Prevention in the Process Industries, 26 (2013) 1646-1654.

[7] Q. Li, B. Lin, W. Li, C. Zhai, C. Zhu, Explosion characteristics of nano-aluminum powder–air mixtures in 20 L spherical vessels, Powder Technology, 212 (2011) 303-309.

[8] J. Bouillard, A. Vignes, O. Dufaud, L. Perrin, D. Thomas, Explosion risks from nanomaterials, in: Journal of Physics: Conference Series, IOP Publishing, 2009, pp. 012032.

[9] H.-C. Wu, H.-J. Ou, H.-C. Hsiao, T.-S. Shih, Explosion characteristics of aluminum nanopowders, Aerosol and Air Quality Research, 10 (2010) 38-42.

[10] N. Kuai, J. Li, Z. Chen, W. Huang, J. Yuan, W. Xu, Experiment-based investigations of magnesium dust explosion characteristics, Journal of Loss Prevention in the Process Industries, 24 (2011) 302-313.

[11] C. Yuan, P.R. Amyotte, M.N. Hossain, C. Li, Minimum ignition temperature of nano and micro Ti powder clouds in the presence of inert nano TiO2 powder, Journal of Hazardous Materials, 275 (2014) 1-9.

[12] R.K. Eckhoff, Does the dust explosion risk increase when moving from µm-particle powders to powders of nm-particles?, Journal of Loss Prevention in the Process Industries, 25 (2012) 448-459.

[13] X. Zhang, J. Yu, X. Yan, Q. Xie, W. Gao, Flame propagation behaviors of nano- and micro-scale PMMA dust explosions, Journal of Loss Prevention in the Process Industries, 40 (2016) 101-111.

[14] R. Eckhoff, Dust explosions in the process industries: identification, assessment and control of dust hazards, Gulf professional publishing, 2003.

11th Asia-Pacific Conference on Combustion,
The University of Sydney, NSW Australia
10th -14th December 2017.

The sintering characteristics of ash during co-firing wheat straw and coal

Nijie Jing[1]*, Hongmei Zhu[1], Heping Li[2]

[1] School of Automation, Hangzhou Dianzi University, Hangzhou, Zhejiang, China

[2] School of Science, Hangzhou Dianzi University, Hangzhou, Zhejiang, China

Abstract

This contribution presents an investigation into the sintering behaviour of ash from coal and wheat straw blends with different ash preparation methods. Blends of Xiaolongtan coal (XLT), and a wheat straw were subjected to two different ash preparation procedures, namely, combustion in a muffle furnace at 815 °C and in a drop tube furnace at 1500 °C. The resulting ash samples were then conducted to the sintering temperature measurement using a pressure-drop sintering device and morphological characterisation using scanning electron microscope (SEM) fitted with X-ray energy dispersive spectroscopy (EDS). For the same XLT coal and wheat straw blends, the sintering temperatures in the muffle furnace were always the lower than those in the drop tube furnace. The addition of the wheat straw to the blends lowered the sintering temperature of corresponding ash. The SEM imaging showed that the texture of ash samples from the muffle furnace was irregular and more fibrous, but the drop tube furnace ash particles were mostly in spherical-shape. The EDS analysis confirmed that the ash samples with wheat straw addition contained more K-bearing minerals. K-bearing minerals reacted with other minerals in the ash to form low temperature eutectics, leading to the reduction of sintering temperatures of the blends.

1 Introduction

Co-firing biomass with coal is a short term option to reduce greenhouse-gas emissions from existing coal-fired boilers used in power generation. Combusting biomass, a renewable and sustainable energy source with coal in coal-fired boilers, is a practical approach to partial replacement of fossil fuel for power generation [1]. Although co-firing of biomass and coal represents a cheaper and low risk sustainable energy option, technical issues associated with co-firing, namely difficulty in fuel handling and storage, decrease in overall combustion efficiency, pollutant emissions and carbon burnout still need to be resolved. In addition, the focal problem during biomass combustion is the high alkali content of its ash. The high alkali content can significantly worsen ash deposition, reducing the heat transfer in the furnace and convective surfaces of a boiler [2–5].

The ash-deposition phenomena are influenced by coal type (ash composition, melting temperature, and distribution of mineral matter), reaction atmosphere, particle temperature, flow dynamics, and so forth. Sintering, a very important ash characteristic related with ash-deposition, defines a thermal treatment for bonding particles into a solid structure via mass transport, occurring on a solid, liquid, or gaseous phase resulting in shrinkage. Sintering is associated with fusion and melting of the ash particles during combustion, occurring when particle temperatures are high enough to provoke partial melting [6]. Although recent research efforts have resulted in new knowledge, because of the large heterogeneity of high-rank and low-rank coals, the knowledge is still fragmentary.

Several studies on the ash agglomeration and sintering characteristics of lignite and biomass have been conducted during co-firing [7–10]. Selvakumaran [7] conducted the research on the combustion, agglomeration and clogging/ blockage behavior of lignite. The sintering tendency of the lignite coal ashes below 1000°C is due to the amount and type of clay minerals present in the coal ashes and the extent of dehydroxylation. Duan et al., [11] investigate bed agglomeration characteristics of pelletized rice straw conducted in a bench-scale vortexing fluidized bed. The results revealed that eutectic composition with low melting point materials promote defluidization at high temperatures. Zhang et al. [12] investigated the melting behavior of ashes from German brown coal and biomass (wheat straw) and the results showed that biomass addition has significantly affected the ash melting behavior. Wang et al., [13] investigated the ash characteristics of rye straw and effects of additives. Severe sintering of the rye straw ash was attributed to the formation and fusion of low temperature K-silicates and K-phosphates with high K/Ca ratios.

Therefore, the scope of this paper is to look at the effect of ash preparation method on the sintering behaviour of the ashes from XLT and wheat straw blends. With further improvement beyond the scope of this paper, the theory can be used to provide technical advice to organizations that implement biomass feed-stock in coal fired boilers.

2 Experimental

2.1 Preparation of samples

Xiaolongtan (XLT) coal, a lignite from Yunnan Province, China and a wheat straw, were chosen for this study. The preparation of samples, the experimental setup and most of the data used were as presented in Reference 14. The proximate and ultimate analysis data of the coal and two biomass samples as well as their ash chemistry are presented in Table 1.

The ash samples were prepared using two different methods, namely, combustion in a Muffle furnace and in a drop tube furnace. The preparation of ashes of the individual coal, biomass and coal/biomass blends by combustion in a muffle furnace followed the Ref. [14]. The ashes thus prepared were denoted as MFA.

High temperature ashes were also prepared using a drop tube

Corresponding author. Fax: +86-571-86919133;

E-mail: njjing@hdu.edu.cn

furnace. The drop tube furnace facility was reported in reference [14]. Briefly, the drop tube furnace was first preheated to 1500 °C and then the sample was fed from a hopper into the furnace through the vibration screw feeder at a feeding rate of 0.1gs^{-1} in a typical combustion experiment. Combustion air was supplied from an air cylinder at a flow rate of 15 Lmin^{-1} and preheated to 200 °C before entering the furnace. Preliminary experiments confirmed that complete combustion was achieved under such conditions. The ash collected in the ash collector was analysed for unburnt carbon content and the LOI (loss on ignition) was always less than 0.5%. The ashes thus prepared were denoted as DTFA.

Table 1 Proximate and ultimate analysis and ash compositions of XLT coal and wheat straw

	Proximate analysis / w_{ad}/%				Ultimate analysis / w_{ad}/%						
	M	V	A	FC	C	H	O	N	S		
XLT	2.05	25.07	39.95	32.93	45.59	3.32	18.09	1.06	1.65		
straw	9.89	68.39	5.22	16.5	45.55	5.7	46.53	1.32	0.15		
	Ash Composition / w/%										
	SiO_2	Al_2O_3	Fe_2O_3	CaO	MgO	Na_2O	K_2O	SO_3	TiO_2	P_2O_5	Cl
XLT	41	13.2	10.2	14.8	1.41	0.44	1.18	16.1	0.95	0.43	-
straw	30.09	1.67	1.01	4.8	5.09	13.24	32.6	3.45	0.15	7.72	1.97

2.2 Measurement of sintering temperature

A pressure-drop sintering temperature measurement technique [15-19] was used to determine the sintering temperatures of the various ash samples prepared. Briefly, a compacted ash pellet was pressed into the centre of a mullite tube. A K-type thermocouple was inserted into the ash pellet to monitor the temperature of the ash pellet during the heating process. The mullite tube was then placed in an electrically heated furnace and heated from ambient temperature at 8 °C min^{-1} in air. The air flow rate was maintained constant at 4 cm^3 min^{-1} controlled by a mass flow meter. The upstream pressure in the mullite tube prior to the ash pellet was continuously recorded using a pressure transducer. The ash sintering temperature was defined as the temperature at which an abrupt change in the rate of pressure drop occurred versus was occurred. The sintering temperatures thus determined have been shown to have a repeatability of ± 10 °C.

2.3 SEM-EDS analysis

The texture, morphological characteristics and spot composition of the ashes were analysed using a FEI-SIRION-100 field emission Scanning Electronic Microscopy (SEM) analyser aided with an EDAX GENESIS4000 X-ray energy dispersive spectroscopy (EDS).

3 Results and discussion

3.1 Sintering temperatures of ash samples from different ash preparation methods

Figure 1 compares the sintering temperatures of the ashes of XLT coal/wheat straw blends prepared using the two different ashing methods. As can be seen, the sintering temperature of XLT coal ash was much higher than those of the blends and biomass ashes. For a given sample, the sintering temperatures of the DTFAs were higher than the MFAs. This is thought to be due to some interactions of the minerals, such as formation of anorthite and mullite, may occur in the drop tube furnace ashing processes due to the high combustion temperatures.

It is evident that adding either of the biomass to the coal lowered the sintering temperature of the corresponding ash, regardless the ash preparation method employed. The higher the proportion of biomass addition, the lower the ash sintering temperature was. This is because the ashes of the biomass contained higher amounts of K, Na and lower Si and Al, as shown in Table 1, resulting in the ash of the blended fuel to form low melting point eutectics during the ashing process [20], decreasing the sintering temperature of the ash. In order to understand the effect of the ashing method and biomass addition on the sintering temperature of ashes, the detailed morphology and minerology of ashes will be discussed in detail in the following sections.

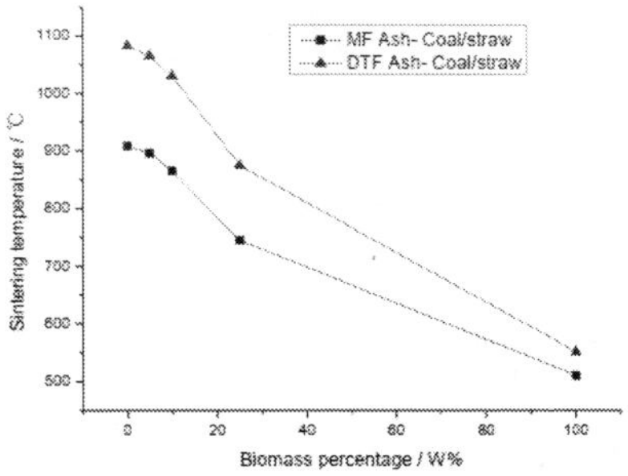

Figure 1 The sintering temperatures of ash of coal and biomass blends from different ash preparation methods: MF: muffle furnace; DTF: drop tube furnace

3.2 Ash morphology by SEM-EDS analysis

Figure 2 shows typical SEM images of XLT ash from the two ash preparation methods. The MFA of the XLT coal showed loose irregular and some agglomerated nature. The DTFA of XLT coal showed more big blocky-shaped particles than the MFA. This indicates that a greater degree of inorganic reactions and mineral transformation occurred in the high temperature ashing process in the drop tube furnace, leading to the formation of more high-melting minerals and thus the high sintering temperature of the corresponding ash as shown in Figure 1. In addition, EDS results show that the DTFA contained higher amounts of Na, K, Fe and Si and lower Ca and S than the MFA.

Figure 2 SEM-EDS micrographs of MFA and DTFA of XLT coal

Figure 3 shows typical SEM images of the ashes of XLT coal and wheat straw blends prepared in the muffle furnace. The particles were more irregular and loose particles with 5% wheat straw addition of MFA. With the addition to 10%, there were more agglomerated particles and more agglomerated and bigger particles were detected in the ash of the XLT coal and biomass blends with 25% wheat straw addition. This indicated that the addition of wheat straw accelerated the agglomeration and melting of ash. More Na and K and less Ca and Fe were detected in the EDS results with 25% wheat straw addition. It is therefore deduced that some fluxing minerals had formed in the process of ashing in the muffle furnace, such as K and Na-bearing minerals, which can produce the low temperature co-fused mineral phases [20], reducing the sintering temperature of the ash samples.

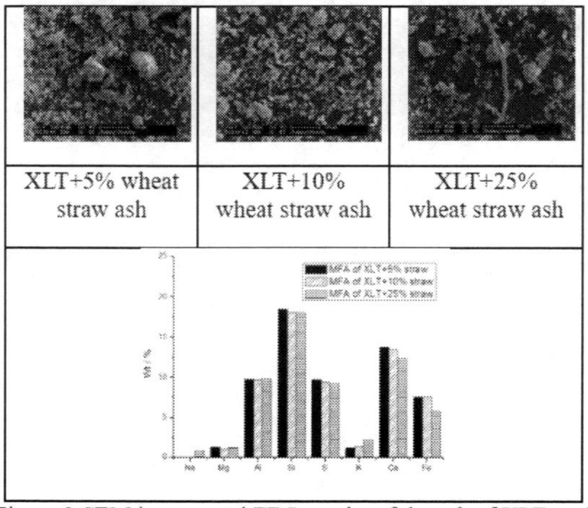

Figure 3 SEM images and EDS results of the ash of XLT coal and wheat straw prepared in the muffle furnace

Figure 4 shows typical SEM images of the ashes of XLT coal and wheat straw blends prepared in the drop tube furnace. Compared to the MFAs, the particles of the DTFAs were significantly larger in size and denser and more agglomerated. This indicates that sintering and agglomeration had occurred during the ash preparation in the drop tube furnace due to the high combustion temperature. The EDS analysis revealed that the DTFA with 25% wheat straw was higher in Si, K and Na and lower in Fe and Ca. It is therefore deduced that some minerals had formed in the process of ashing in the drop tube furnace, such as Si, K and Na-bearing minerals, which can produce the low temperature eutectics [21], reducing the sintering temperature of the ash samples.

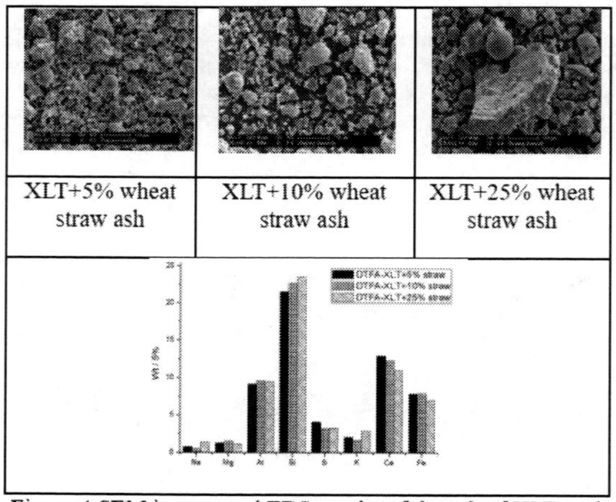

Figure 4 SEM images and EDS results of the ash of XLT coal and wheat straw prepared in the drop tube furnace

The DTFAs of the blends of XLT with the wheat straw were more spherical than the corresponding MFAs, indicating that the ash fusion had occurred as soon as the ash was formed during the combustion. The particles of the DTFA with the addition of wheat straw contained relatively high contents of Si, Na, K, Ca and Fe, suggesting that this can lead to the severe sintering and deposition at the combustion temperature studied.

4 Conclusions

The addition of wheat straw to the coal lowered the sintering temperature of the corresponding ash; the higher the proportion of wheat straw addition, the lower the ash sintering temperature. For a given blend of XLT coal and wheat straw but different ash preparation methods, the sintering temperatures of the ash from the muffle furnace were always the lower than the ash from the drop tube furnace.

SEM imaging analysis showed that the texture of the MFA samples was irregular and more fibrous but the DTFA ash particles were in spherical-shape, indicating ash melting had occurred in the drop tube furnace. The EDS analysis confirmed that the ash samples with the wheat straw addition contained more K-bearing minerals.

5 Acknowledgement

This work was financially supported by Public Projects of Zhejiang Province (No. LGG18E060004) and Zhejiang Province Natural Science Foundation Project of China (No. LQ14E060004).

References

[1] H.B. Vuthaluru, V. Doshi, R. Korbee, J.H.A. Kiel, K.Shah, B. Moghtager. Co-firing of coal and biomass: Development of a conceptual model for ash formation prediction. Fuel 139 (2015) 594-605.

[2] Zheng Y J, Jensen P A, Jensen A D, Sander B, Junker H. Ash transformation during co-firing coal and straw. Fuel (2007) 86: 1008-1020.

[3] Bryers R. Fireside slagging, fouling and high temperature corrosion of heat transfer surface due to impurities in steam-raising fuels. Prog Energy Combust Sci (1996) 22(120):29.

[4] Jenkins B, Baxter L, Miles T, Miles TJ. Combustion Properties of Biomass. Fuel Process. Technol. (1998)54:17-46.

[5] Robinson A, Junker H, Baxter L. Pilot-scale investigation of the influence of coal-biomass co firing on ash deposition. Energy Fuels (2002)16:343–55.

[6] Pronobis M. Evaluation of the influence of biomass co-combustion on boiler furnace slagging by means of fusibility correlations. Biomass Bioenergy (2005)28:375–383.

[7] P. Selvakumaran, A.K. Bakthavatsalam. Effect of chemical composition of ash on sintering of lignites in circulating fluid bed combustion and successful operation of large CFBC boilers. Applied Thermal Engineering. (2015) 85: 135-147

[8] W. Lin, K. Dam-Johansen, F. Frandsen, Agglomeration in bio-fuel fired fluidized bed combustors, Chemical Engineering Journal (2003) 96: 171– 185.

[9] F. Ishoma, T. Harada, T. Aoyagi, K. Sakanishi, Y. Korai, I. Mochida, Problems in PFBC boiler (2): characterization of bed materials found in a commercial PFBC boiler at different load levels, Fuel 83 (2004)1019-1029.

[10] E.J. Anthony, L. Jia, Agglomeration and strength development of deposits in CFBC boilers firing high-sulfur fuels, Fuel (2000) 79: 1933– 1942.

[11] Feng Duan, Chien-Song Chyang, Li-hui Zhang, Siang-Fong Yin. Bed agglomeration characteristics of rice straw combustion in a vortexing fluidized-bed combustor. Bioresource Technology (2015) 183: 195–202.

[12] Guanjun Zhang, Markus Reinmöller, Mathias Klinger, Bernd Meyer. Ash melting behavior and slag infiltration into alumina refractory simulating co-gasification of coal and biomass. Fuel (2015) 139:457-465.

[13] Liang Wang, Øyvind Skreiberg, Michael Becidan, Hailong Li. Investigation of rye straw ash sintering

characteristics and the effect of additives. Applied Energy (2016) 162 :1195-1204.
[14] Nijie Jing, Mingming Zhu, Guoqing Shen, Qinhui Wang, Dongke Zhang. Effect of ash preparation method on the sintering characteristics of ashes from combustion of coal and biomass blends. Fuel (2016) 186: 830-837.
[15] Al-Otoom A Y, Elliott L K, Wall T F, Moghtaderi B. Measurement of the Sintering Kinetics of Coal Ash. Energy & Fuels (2000)14: 994-1001.
[16] Luan C, You C F, Zhang D K. Composition and sintering characteristics of ashes from co-firing of coal and biomass in a laboratory-scale drop tube furnace. Energy (2014) 69: 562-570.
[17] Jing N J, Wang Q H, Luo Z, Cen K F. Effect of different reaction atmospheres on the sintering temperature of Jincheng coal ash under pressurized conditions. Fuel (2011) 90:2645-2651.
[18] Li J B, Zhu M M, Zhang Z Z, Zhang K, Shen G Q, Zhang D K. The mineralogy, morphology and sintering characteristics of ash deposits on a probe at different temperatures during combustion of blends of Zhundong lignite and a bituminous coal in a drop tube furnace. Fuel Processing Technology (2016)149:176-186.
[19] Li J B, Zhu M M, Zhang Z Z, Zhang, K., Shen, G.Q., Zhang D K.Characterisation of ash deposits on a probe at different temperatures during combustion of a Zhundong lignite in a drop tube furnace. Fuel Processing Technology. (2016) 144:155-163.
[20] Raask E. Mineral impurities in coal combustion; behaviour, problems and remedial measures. Hemisphere Publishing.Bristol, PA: 1985.
[21] Teixeira P, Lopes H, Gulyurtlu I, Lapa N, Abelha P. Evaluation of slagging and fouling tendency during biomass co-firing with coal in a fluidized bed. Biomass and Bioenergy (012)39: 192-203.

11th Asia-Pacific Conference on Combustion,
The University of Sydney, NSW Australia
10th - 14th December 2017.

Examine the Minimum Ignition Energy Required to Ignite the Coal Dust Cloud in Ventilation Air Methane

Mohammed Jabbar Ajrash[1], Jafar Zanganeh[1]* and Behdad Moghtaderi[1]

[1]The Frontier Energy Technologies Centre, Chemical Engineering,

School of Engineering, Faculty of Engineering & Built Environment,

The University of Newcastle, Callaghan, NSW 2308, Australia

Abstract

This work aims to examine the minimum ignition energy (MIE) required to ignite a coal dust cloud in a ventilation air methane capture duct. For this purpose, a Hartmann apparatus was employed to find the MIE of a coal dust collected from a NSW, Australian mine. This study also predicts the deposition of coal dust inside a Ventilation Air Methane (VAM) capture duct by assuming the particles are spherical and VAM flow rate is 150 $m^3 s^{-1}$. The deposition prediction indicates that coal dust particles (50-212 μm) could be deposited in the VAM capture duct within 55 m to 217 m depending on the particles size, particle velocity. MIE tests for three size ranges of coal dust particles were undertaken. Results showed that the MIE of coal dust particles in the 0-74 μm size range, ignited in a range of 100-300 mJ. However, for the coal dust particles in the size range from 74-125 μm and 125-212 μm the MIE measured was in the range of 300-1000 mJ. The ignition coal dust concentration was varied between 150 $g.m^{-3}$ to 1500 $g.m^{-3}$ for the tests.

1 Introduction

Considering the location of a VAM capture duct between the mine fans and thermal oxidiser units, great attention has to be given to the probability of fire and explosion in the VAM capture duct. Coal dust ignition represents one threat that could trigger fires and explosions in the system. Previous literature of fire and explosion in VAM capture ducts have investigated the possibility of auto ignition of coal dust in layer and cloud form, where it has been found that the concentration of coal dust particles play an important role on the minimum ignition temperature of coal dust. It has been found that a coal dust cloud could be ignited even at low concentrations (i.e. 15 $g.m^{-3}$) [1]. The deposited coal dust increases the likelihood of VAM capture duct fire and explosion, as the coal particles may cause a second explosion, in addition to their threat as an ignition source [2]. Previous literature show that the coal dust layers consisting of particles less than 74 μm diameter could be self-ignited when the contacted surface temperature reached 235 °C[3]. Dust cloud ignition by electrical sparks was first investigated by Boyle and Llewellyn[4]. The energy of the sparks was controlled via a capacitor. A wide range of dust clouds were examined, such as aluminum, magnesium, ferromanganese, zinc, silicon and sulphur. Furthermore, a number of experimental studies were conducted to investigate the behavior and the properties of dust explosions caused by electrical sparks. They concluded that even low energies i.e. 1 mJ could ignite the dust cloud. The system utilised the Hartmann Glass tube. The researchers also indicated that the system could be used for measuring the minimum explosion concentration in addition to the minimum ignition energy of duct clouds [5]. Other groups of researchers investigated the influence of ignition energy on the coal dust explosion characteristics, and they found that the minimum explosion concentration could be reduced from 90 $g.m^{-3}$ to about 50 $g.m^{-3}$ as the initial ignition energy increased from 1 kJ to 10 kJ[5,6].

This work delivers more information in terms of coal particle deposition and minimum ignition energy of coal dust. Firstly, coal dust deposition along the VAM capture duct for coal dust particles size up to 212 μm are predicted. Also, the minimum ignition energy of three size ranges of coal dust particles i.e. <74 μm, 74-124 μm and 125-212 μm is experimentally measured.

2 Experimental Setup

The experimental procedure of this work follows ASTM E2019[7]. The minimum ignition energy of coal dust was measured by employing a Hartmann glass tube as seen in Figure 1. The coal dust was collected from the supplier in a sealed pack and kept in cool conditions to reduce the weathering and oxidation effects during the work progress. The coal dust was sieved to three sizes fractions, i.e. <74 μm, 74-125 μm and 125-212 μm; by using sieves and a sieve shaker. A Malvern particle sizer was used to find the particle size distribution of the coal dust groups (shown in Figure 2). The proximate analysis of the coal dust, as shown in Table 1, was determined by a third party consultant (ALS Global). Instrument air at a pressure of 5 bar was used to disperse the dust to the Hartmann tube.

Figure 1: Hartmann Glass tube apparatus, University of Newcastle, Australia

Corresponding author. Phone: (+61) 2 4033 9335
Email: jafar.zanganeh@newcastle.edu.au

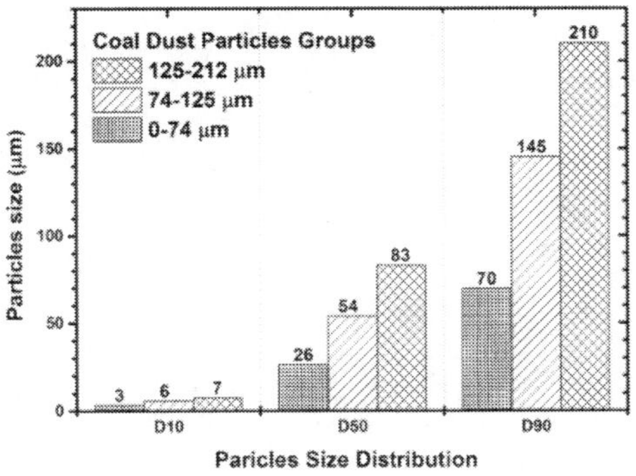

Figure 2: Coal Dust Particles Size Distribution

The energy of ignition was controlled via a series of capacitors able to deliver 1, 3, 10, 30, 100, 300 and 1000 mJ. Each test point was repeated 10 times starting at high energy to confirm if there was ignition or not. In the event of the occurrence of ignition, the test series was terminated and further tests were conducted at lower ignition energies.

Table 1: Proximate analysis of coal dust

Moisture%	Ash%	Volatile Matter%	Fixed Carbon
3.6	15	29.8	51.5

It is worth mentioning that the methane percentage in VAM is about 1%, however, this study was conducted under methane free conditions,

3 Results and Discussion

3.1 Coal Dust Settling

Coal dust particles carried by ventilation air methane pass through the VAM capture duct at a certain velocity, which is equal to the velocity of the ventilation air methane gas in the VAM capture ducts. As these particles have varied particle size, the deposition of particles occurs depending on the drag and settling velocity. The drag velocity could be correlated to the particle size and shape, particle velocity and the height. The particle deposition was estimated according to the drag force coefficient equation developed by Cheng et al [8].

$$C_D = \frac{24}{\text{Re}}(1+0.27\,\text{Re})^{0.43} + 0.47[1-\exp(-0.04\,\text{Re}^{0.38})] \quad (1)$$

Where C_D is the drag coefficient and Re Reynold number.

Particle size deposition calculations based on Equation (1) were carried out with the following assumptions: the particles are spheres, particles enter at a height of 0.2 m from the bottom surface and the sizes varied between 50 microns and 212 microns. The average velocity was 75 m.s^{-1}. Equation (1) comprises of two terms to predict the drag coefficient of particles over the entire subcritical region; the first term is applicable to Re>100 and the second term is applicable for higher Re up to Re<1x10^{6}. Coal dust particle deposition along the VAM capture duct is shown in Figure 3.

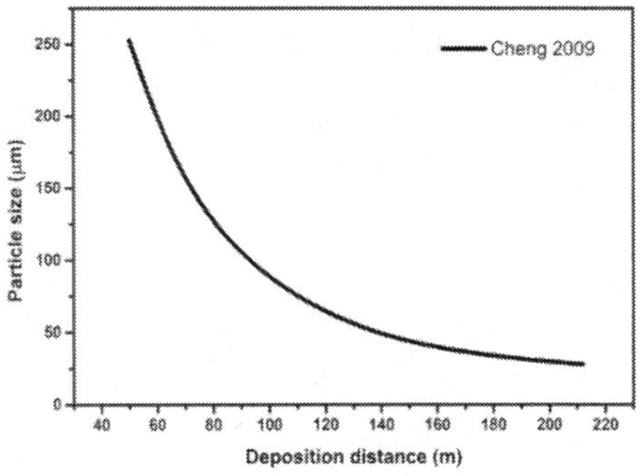

Figure 3: Coal Dust Particle Deposition Along The VAM Capture Duct

Figure 3 shows that the smallest particle sizes are deposited at a distance of 217 m, while the largest particle sizes, i.e. 212 μm, deposit at a shorter distance of 57 m.

3.2 Minimum Ignition Energy

Gaining more information about the minimum ignition energy of coal dust and its relationship to the particle size distribution and concentration could assist in reducing the likelihood of fires and explosions in a VAM capture duct. In this work the coal dust concentrations were varied from 25 g.m^{-3} to 4200 g.m^{-3}.

The minimum ignition energy of fine coal dust particles (<74 μm) at varying concentrations is shown in Figure 4. The results indicate that the minimum ignition energy of coal dust particles less than 74 μm were between 100-300 mJ. The minimum concentration for ignition was 150 g.m^{-3}, however, the optimum coal dust concentration was 300 g.m^{-3}. Also, for coal dust concentrations of 300 g.m^{-3} to 750 g.m^{-3}, the ignition energy was in the range of 1000 mJ to 300 mJ, while no ignition was observed at 100 g.m^{-3}.

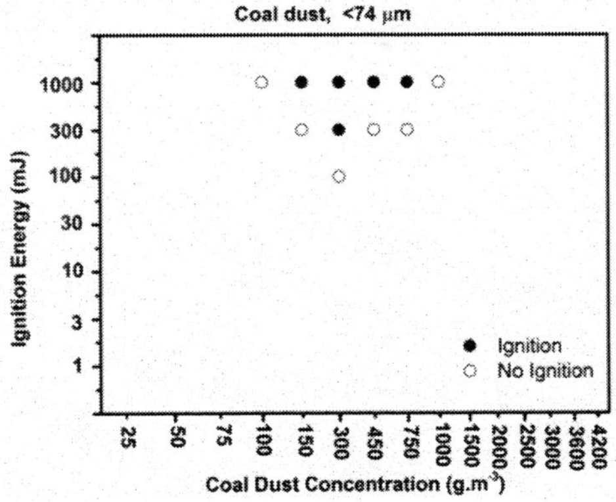

Figure 4: Minimum Ignition Energy of Coal Dust Particles (<74 μm)

The minimum ignition energy of coal dust particles sized between 74 -125 μm with varying coal dust concentrations is shown in Figure 5. The results show that energy in the range of 300 - 1000 mJ was required to ignite the particular coal dust. No ignitions were observed for concentrations below 150 g.m^{-3} or higher than 750 g.m^{-3}. The literature review shows that the range of ignitable coal dust concentrations in the Hartmann glass tube range is wider from 50 g.m^{-3} to 1800 g.m^{-3}, despite using similar ignition energies and a similar configuration as this work. The difference in ignitable concentration range between this work and literature could be attributed to two main facts, the coal dust D_{50} size used in literature was lower (21 µm) as compared to 26.2 µm (see Figure 2). The coal dust in the literature had higher volatile matter contents, 37% as compared to only 29.3% for this work. Coal dust particle size D_{50} and volatile matter play a vital role on the extent of explosions and ignitions [9,10].

The minimum ignition energy of coal dust sized between 125 - 212 µm for varying coal dust concentrations is shown in Figure 6. A similar minimum ignition energy behavior for the coal dust range 74-125 µm is observed. The minimum ignition energy was in the range of 300 - 1000 mJ. However, the range of ignitable concentrations is different. The lower ignitable concentration was 450 g.m^{-3} and higher ignitable concentration 1500 g.m^{-3}.

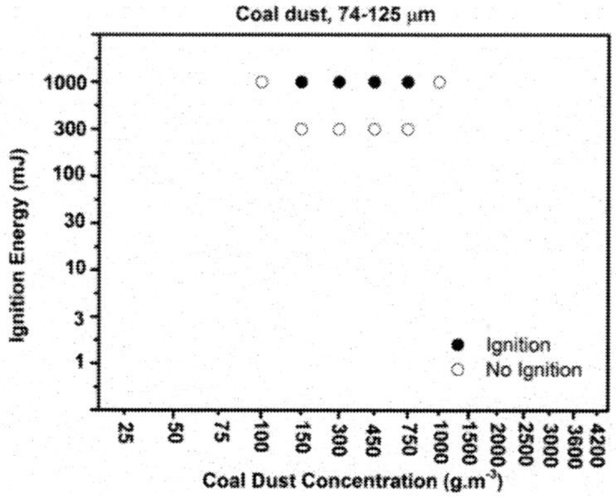

Figure 5: Minimum Ignition Energy of Coal Dust Particles (74-125 µm)

Overall, the coal dust ignition energy was in the range of 100-300 mJ for coal dust particles below 74 µm, and in the range of 300-1000 mJ for coal particles in the range of 74-125 µm and 125-212 µm. Flame propagation footage has been recorded by a high speed camera at a rate of 2000 frame/s. Average upward flame propagation for coal particle size of 74-125 µm was determined as approximately 1.85 m.s^{-1}. See Figure 8.

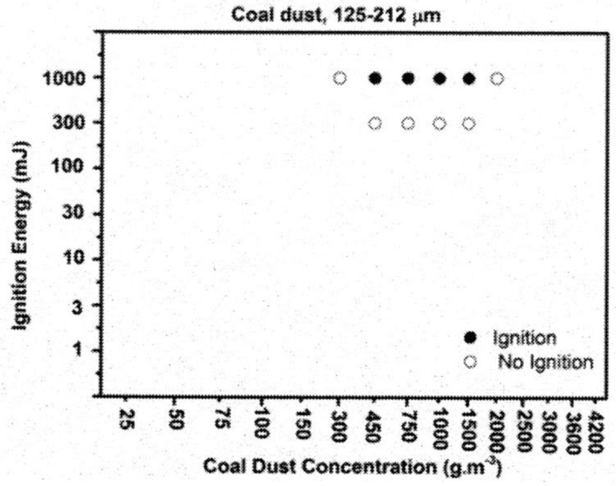

Figure 6: Minimum Ignition Energy of Coal Dust Particles (125-212 µm)

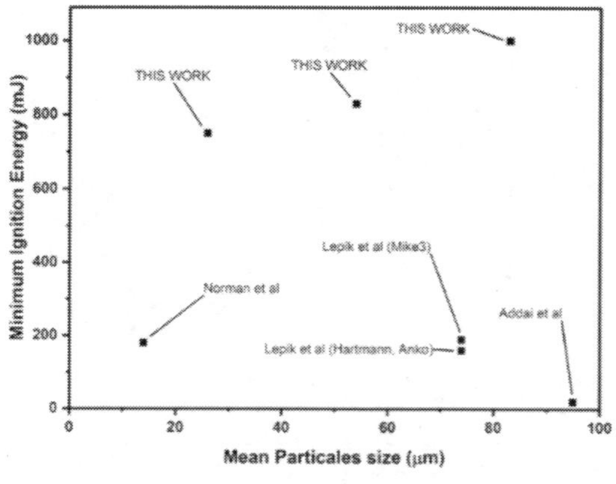

Figure 7: Relevant data available for coal dust MIE

The current MIE work indicates a quite good agreement in terms of MIE measurement in Hartmann glass tubes from different suppliers[11]. The lowest MIE value in Figure 7 has been reported by Addai et al[12], this could be attributed to the high volatile content (89.4%) of the coal dust used indicating that volatile matter content plays an important role in coal dust ignition[13]. The relatively low volatile content of coal dust used in this study has a major impact on the MIE, where the lowest MIE achieved was about 750 mJ (as shown in Figure 7) as compared to MIE range of 200-20mJ in the literature.

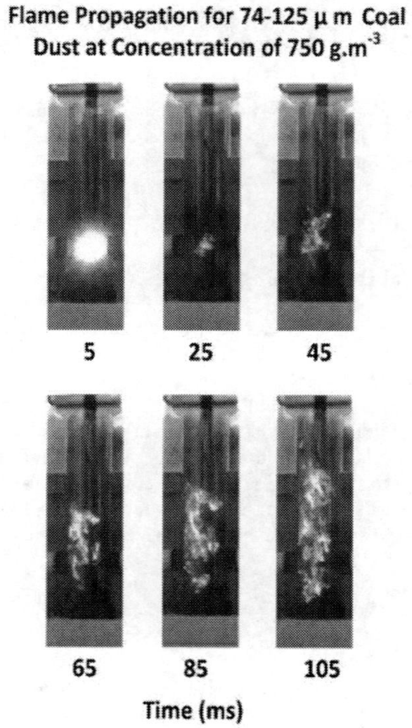

Figure 8: Upward Flame Propagation in Hartmann Glass Tube

4 Conclusions

The theoretical and experimental work of coal dust deposition and ignition energy in a VAM capture duct has been carried out at the University of Newcastle, Australia. The outcome of this work could be summarized as follows.

- Fine coal dust particles (50 µm) deposit at about 217m distance from the point of entry to a VAM capture duct when introduced at 0.2 m height. Larger particles (212 µm) deposit faster, within 55 m distance from the point of entering the VAM capture duct.
- The ignition energy required to ignite fine coal dust particles (<74µm) is between 300-1000 mJ, however, energies between 300-100 mJ is required for the ignition of coal dust particles of size range 74-212 µm
- The flame propagated upward in the Hartmann tube at an average velocity of 1.85 $m.s^{-1}$.

5 Acknowledgment

The authors wish to acknowledge the financial support provided to them by Australian Coal Association and Low Emission Technology (ACALET), Australian Department of Industry, and the University of Newcastle (Australia). In addition, special gratitude is given to the Higher Committee for Education Development (HCED) and Midland Refineries Company (MRC) from the Iraqi government for sponsoring a postgraduate candidate working in this project.

References

[1] M.J. Ajrash, J. Zanganeh, B. Moghtaderi, The effects of coal dust concentrations and particle sizes on the minimum auto-ignition temperature of a coal dust cloud, Fire Mater. 1 (2017).

[2] M.J. Ajrash, J. Zanganeh, B. Moghtaderi, Experimental investigation of the minimum auto-ignition temperature (MAIT) of the coal dust layer in a hot and humid environment, Fire Saf. J. 82 (2016) 12–22.

[3] Miron, Lazzara, Hot-surface Ignition Temperatures of Dust Layers, Fire Mater. 12 (1988) 115–126.

[4] A.R. Boyle, F.J. Llewellyn, The electrostatic ignitibility of dust clouds and powders, J. Chem. Technol. Biotechnol. 69 (1950) 173–181.

[5] K.L. Cashdollar, Overview of dust explosibility characteristics, J. Loss Prev. Process Ind. 13 (2000) 183–199.

[6] M.J. Ajrash, J. Zanganeh, B. Moghtaderi, Effects of ignition energy on fire and explosion characteristics of dilute hybrid fuel in ventilation air methane, J. Loss Prev. Process Ind. 40 (2016) 207–216.

[7] ASTM-E2019, Standard Test Method for Minimum Ignition Energy of a Dust Cloud in Air, 3 (2013).

[8] N. Cheng, Comparison of formulas for drag coef fi cient and settling velocity of spherical particles, Powder Technol. 189 (2009) 395–398.

[9] K.L. Cashdollar, I.A. Zlochower, G.M. Green, R.A. Thomas, M. Hertzberg, Flammability of methane , propane , and hydrogen gases, 13 (2000) 327–340.

[10] M.J. Ajrash, J. Zanganeh, B. Moghtaderi, Influences of the Initial Ignition Energy on Methane Explosion in a Flame Deflagration Tube, Energy & Fuels. (2017).

[11] P. Lepik, D. Gabel, W. Adamus, L. Mokos, M. Mynarz, J. Serafin, Determination of the minimum ignition energy on different devices, X (2015) 8–14.

[12] E.K. Addai, D. Gabel, U. Krause, Experimental investigations of the minimum ignition energy and the minimum ignition temperature of inert and combustible dust cloud mixtures, J. Hazard. Mater. 307 (2016) 302–311.

[13] M.J. Ajrash, J. Zanganeh, B. Moghtaderi, The flame deflagration of hybrid methane coal dusts in a large-scale detonation tube (LSDT), Fuel. 194 (2017) 491–502.

11th Asia-Pacific Conference on Combustion,
The University of Sydney, NSW Australia
10th -14th December 2017.

Laminar burning velocity of n-pentanol air mixtures using meso-scale channels

Amit Katoch[1], Ayush Chauhan[2], Rohit Kumar[1], Sudarshan Kumar[1]

[1]Combustion Research Laboratory, Department of Aerospace Engineering, Indian Institute of Technology, Bombay
Mumbai, Maharashtra, 400076, India

[2]Department of Mechanical Engineering, National Institute of Technology Hamirpur, Himachal Pradesh, 177005, India

Abstract

In the present work, the meso-scale diverging channel method is used to measure the laminar burning velocity of n-pentanol+air mixtures. The experiments were carried out at atmospheric pressure and for unburnt mixture temperatures up to 480K. PREMIX was employed to compare the experimental results with the kinetic mechanisms given by Liu et al. , Togbe et al. and Heufer et al. . The burning velocity data was fitted with power law correlations to study the effect of temperature with varying equivalence ratios. The results show a minimum value of temperature exponent for slightly rich mixtures.

1 Introduction

The ever-increasing energy consumption rate and the alarming anthropogenic pollution levels require stringent emission control and determined scientific advancement towards efficient fuel consumption and targeted research. A favorable transition for the environmental consideration is, from coal to hydrocarbon fuels such as biofuels, which will cause a reduction in greenhouse gases, mainly CH_4 and CO_2. This transition will happen only if hydrocarbon fuel prices are low in the future [1]. But the prices of crude oil are bound to increase as its reserves are forecasted to last for only another 27 years [2]. Hence it is imperative, now more than ever to reduce the dependency on fossil fuels, by opting for and improving the alternate and preferably renewable sources of energy. Presently, a global initiative to switch from crude oil and coal to blends of biofuels in the transport sector is being promoted by the International Energy Agency for energy sustainability, to increase the present worldwide biodiesel consumption to more than 10 times, by 2050 [3].

Up till now, primary focus for biofuels was on the lower carbon chain alcohols, mainly ethanol (C_2H_5OH), as it is easy and inexpensive to produce using bacterial fermentation of sugars [4]. The oxygen content in the biofuel blend is found to control desirable fuel properties such as knock resistance and flame speed [5]. When compared against ethanol, pentanol has more desirable thermal and chemical properties such as higher energy density, higher boiling point, lesser hygroscopicity, and corrosiveness [6,7]. Now, with the proposed possible advancement in production of pentanol in microorganisms, by increasing the substrate range [8], it is possibly the upcoming representative of alcohol based biofuels, termed as Second-generation biofuel, since, a higher blend of pentanol with diesel of up to 45% is expected to be achievable for CI engines. Pentanol has been found to have lesser particulate emission and soot formation than the $C_1 - C_4$ alcohols [9,10,11]. Togbe et al. in JSR experiment showed pentanol producing lesser amount of possible pollutants than ethanol [12].

Knowledge of the combustion characteristics is a prerequisite for improving the performance of the biodiesel blends. One such characteristic parameter is the laminar burning velocity, which depicts important information for chemically reacting flows such as diffusivity, reactivity and energy release rates. Experimental data from laminar flame speed measurement provides information about the thermal boundary layer regarding flame quenching, thermal losses at the wall and the range of flammability [13]. If a kinetic model is not able to reproduce the experimental parameters, then its comprehensiveness and applicability needs further examination. Therefore, accurate burning velocity data helps in the validation of kinetic mechanisms [14].

$$S_u = S_{u,o}\left(\frac{T_u}{T_{u,o}}\right)^{\propto} \qquad (1)$$

Equation (1) describes the dependence of laminar burning velocity S_u on initial mixture temperature T_u [15], where $\propto$ is the temperature exponent and its value depends on pressure and mixture equivalence ratio ϕ. $S_{u,o}$ is the laminar burning velocity at reference temperature and $T_{u,o}$ is the reference temperature, taken as 300 K. The ideal conditions for measurement of laminar flame speeds are: a) planar flame b) Zero flame stretch c) Adiabatic conditions.

To the authors' knowledge only four studies [6,7,12,16] have been reported for burning velocity measurements of n-pentanol-air mixtures at atmospheric pressure and only one study [7] assessed the temperature exponent variation with equivalence ratio. Thus, data for this biofuel is scarce. Besides, all these studies were carried out with the spherical bomb method which suffers from shortcomings induced by extrapolation to zero stretch rate, which causes erroneous predictions.

The meso-scale diverging channel employs stretch-free planar and nearly adiabatic flames to measure the burning velocities. It has been successfully used for elevated temperature measurements of gaseous as well as liquid fuels [17-21]. The present study reports new data for n-pentanol up to 480K.

2. Experimental methodology

In the present work quartz diverging channels with a depth of

Corresponding author. Fax: +91-22-2572-2602
E-mail address: amitkatoch15@gmail.com

0.002 m were used. The divergence angles of channels used were 10° and 15° with an inlet dimension of 0.025m×0.002m and aspect ratio 12.5. An infrared heater is used to heat the channel with varying combinations of heating rates along with variations in vertical and horizontal overlap to create different temperature gradient profiles. Quartz material is chosen due to its high heat capacity, low thermal expansion and low thermal conductivity which facilitates the creation of steady temperature gradient across the length of the channel. N-pentanol from HiMedia with a purity of 99% is used.

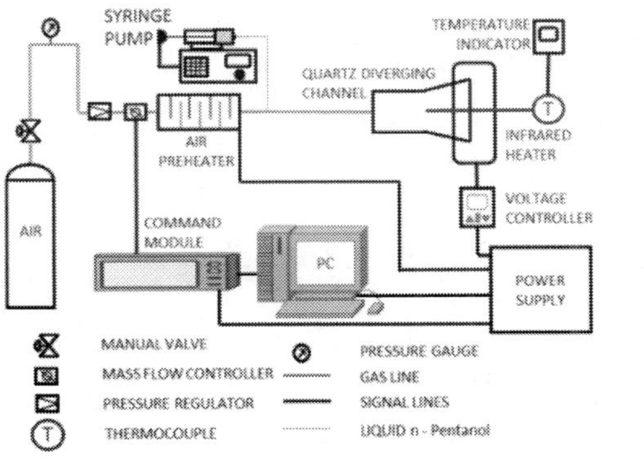

Figure 1: Schematic diagram of the experimental setup

Since the fuel is initially in the liquid form it requires vaporization and mixing before entering the channel. The upstream air is heated using plate type heaters to a temperature lower than the boiling point of the fuel such that the required mixture can be formed based on saturation vapor pressure at that temperature. The fuel is fed using an infusion pump to ensure accurate flow control since the fuel flow rate is of the order of few milliliters per hour. The hot jet of air is then directed to a 2mm internal diameter tube. The high temperature and velocity of the incoming air jet impinges onto the liquid fuel and causes instant vaporization and mixing. The mixture is then fed to the channel. The entire passage of the hot air and subsequent mixture is heated by tape heaters and insulated to prevent any condensation. The fuel-air mixture is ignited at the exit of the channel using a nichrome ignitor. The temperature profile in the channel is measured using a 0.25 mm K-type thermocouple (accuracy ± 5 K). The starting of the diverging section is referred to as inlet. The Peclet number (Pe) for the present flow configuration is small since the mixture flow rates are very low. Therefore, the unburnt mixture temperature is assumed to be equal to the channel surface temperature at the flame stabilization location. The laminar burning velocity is calculated using mass conservation equation as:

$$S_u = U_{inlet}\left(\frac{A_{inlet}}{A_{flame}}\right)\left(\frac{T_{flame}}{T_{inlet}}\right) \qquad (2)$$

3. Numerical prediction of burning velocity:

Laminar burning velocities were computed using a FORTRAN based PREMIX code [22] with detailed kinetic mechanism-Liu et al. [11], Togbe et al. [12] and Heufer et al. [23]. The computational results were compared with experimentally measured values. Multicomponent diffusion and thermal diffusion options were used. The adaptive mesh parameters were grid independent for GRAD = 0.01 and CURV= 0.03.

Laminar burning velocities were measured for equivalence ratios – 0.8,1.0 and 1.2 for an unburnt mixture temperature range of 350-480 K.

4. Results and Discussion

4.1 Variation of burning velocity with unburnt mixture temperature

The effect of initial mixture temperature on laminar burning velocity for n-pentanol+air mixture at ambient pressure is obtained experimentally. Figure 2,3 & 4 show the variation of LBV with temperature ratio for different equivalence ratios. The experimental values are fitted with a power law correlation and reported with an uncertainty band of ±5%. The details of the uncertainty analysis have been presented by the authors in a previous study [21]. The burning velocity was found to increase with increasing temperature across all equivalence ratios.

For Φ=0.8 (as shown in Fig. 2) the fitted correlation predicts higher values than all mechanisms used for comparison. However, it is interesting to note that the present data is in good agreement with the experimental data reported by other researchers. The data and mechanism by Togbe et al. [12] both show significant underprediction.

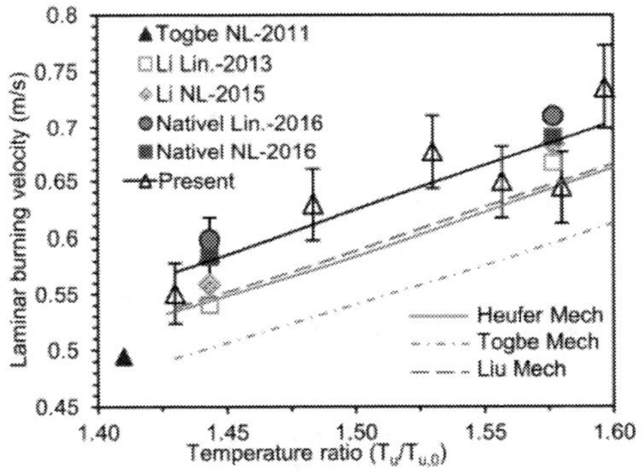

Figure 2: Laminar burning velocity of lean (φ=0.8) n-pentanol+air mixture

For a stoichiometric mixture (as shown in Fig. 3) the fitted correlation shows an excellent match with the Liu et al. [11] mechanism and other experimental data. The mechanism by Heufer et al. also shows a close match although slightly lower than present values. The only exception is the experimental data by Togbe et al. [12] along with their proposed mechanism, both of which show significant underprediction.

For Φ=1.2 (as shown in Fig. 4) the fitted power law correlation shows a very good match with the predictions from Liu et al. mechanism [11] and Heufer et al. mechanism whereas the underprediction by Togbe mechanism [12] and the experiments by the same group are significant. The data by Nativel et al. [7] using both linear and non-linear extrapolations for spherical flames shows a close match with the correlation fitted to the present data. The measurements by Li et al. [11] using both linear and non-linear extrapolations show a close match with the present values.

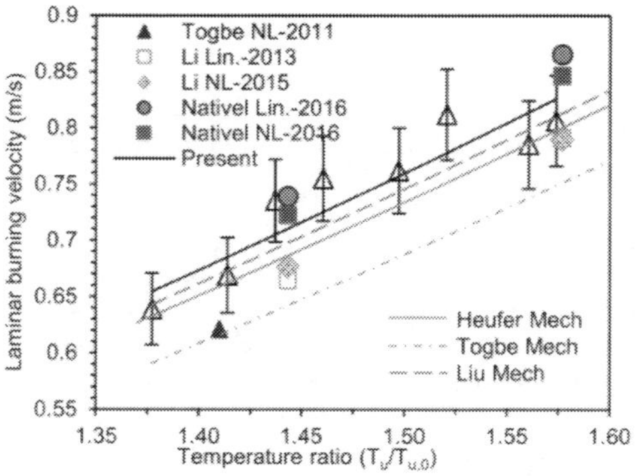

Figure 3: Laminar burning velocity of stoichiometric (φ=1.0) n-pentanol+air mixture

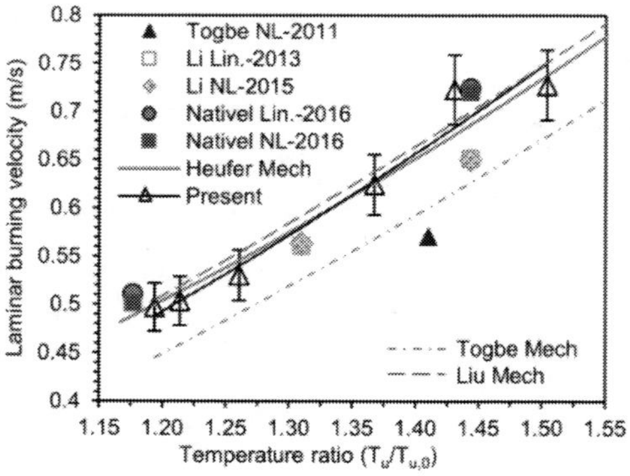

Figure 4: Laminar burning velocity of rich (φ=1.2) n-pentanol+air mixture

4.2 Effect of equivalence ratio on temperature exponent

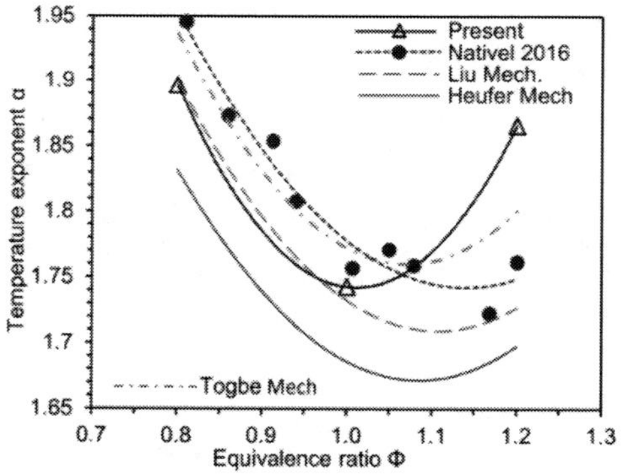

Figure 5: Temperature exponent variation of n-pentanol+air mixture

Temperature exponents were evaluated from the power law correlations for each equivalence ratio. As shown in Fig. 5 the temperature exponent variation with equivalence ratio for the present data was fitted as:

$$\alpha = 3.4588\phi^2 - 6.9948\phi + 5.2786$$

The fitted data shows an inverted parabolic variation with minima at Φ≈ 1.01 (slightly rich mixture). A similar trend can be observed for data by Nativel et al. [7] as well both the mechanisms. For the lean and stoichiometric case, the present α values in close agreement with Liu et al. mechanism and slightly lower than Togbe mechanism [11,12] predications and direct values reported by Nativel et al. [7]. However, for the richer case (Φ=1.2) the present value was higher by ≈ 0.15 as compared to the lowest value derived from Liu et al. mechanism. The mechanism by Heufer et al. shows the lowest values for temperature exponent across all equivalence ratios.

4. Conclusions

Laminar burning velocities of n-pentanol+air mixtures at atmospheric pressure were measured experimentally from planar flames established in meso-scale channels up to unburnt mixture temperature of 480 K. The variation of burning velocity with temperature was analyzed using power-law correlations. The present data shows a good agreement with the predictions from Liu mechanism [11] whereas the Togbe mechanism [12] shows a consistent underprediction across all equivalence ratios. Heufer mechanism [23] shows close match on stoichiometric and rich sides with lowest temperature exponents across all mixture ratios. The burning velocity was found to increase with increasing temperature ratio. The temperature exponent exhibits an inverted parabolic variation with minima at 1.01.

Nomenclature

S_u= Laminar burning velocity (m/s)
A_{inlet}= Cross-section area at the channel inlet (m^2)
U_{inlet}= Flow velocity at flame stabilization location (m/s)
A_{flame}= Cross-sectional area of flame (m^2)
T_{flame}= Mixture temperature at flame stabilization location (K)
T_{inlet}= Mixture temperature at channel inlet (K)

References

[1] B. van Ruijven, D. P. van Vuuren, Energy Policy 37 (2009) 4797–4808.
[2] S. Shafiee, E. Topal, Energy Policy 37 (2009) 181–189.
[3] International Energy Agency (IEA) (2011) Technology Roadmap: Biofuels for Transport. OECD/IEA. http://www.iea.org/publications/freepublications/publication/technology-roadmap-biofuels-for-transport.html/
[4] J. C. Serrano-Ruiz, R. M. West, J. A. Dumesic, Annu. Rev. Chem. Biomol. Eng. 2010;1:79–100
[5] M. Gautam, D.W. Martin, Proc. Instn. Mech. Engrs. Part A 214 (2000) 497–511
[6] Q. Li, E. Hu, X. Zhang, Y. Cheng, Z. Huang, Energy Fuels 2013, 27, 1141−1150.
[7] D. Nativel, M. Pelucchi, A. Frassoldati, A. Comandini, A. Cuoci, E. Ranzi, N. Chaumeix, T. Faravelli, Combust. Flame 166 (2016) 1–18.
[8] K. Zhanga, M. R. Sawayab, D. S. Eisenbergb, and J. C. Liaoa, Proc, Natl, Acad, Sci, U S A 105:20653–20658.
[9] Y M. R. Huang, S. Huang, Y. Zhang, S. Xu, Z. Wang, Energy Conversion and Management 148 (2017) 440–455.
[10] V. Babu M, M. Murthy K, A. P. Rao G, Renewable and Sustainable Energy Reviews, 78 (2017) 1068–1088.
[11] X. Liu, H. Wang, Z. Zheng, J. Liu, R. D. Reitz, M. Yao, Energy 114 (2016) 542-558.
[12] C. Togbe, F. Halter, F. Foucher, C. Mounaim-Rousselle, P.

Dagaut, Proc. Combust. Inst. 33 (2011) 367–374.
[13] C. K. Westbrook, Y. Mizobuchi, T. J. Poinsot, P. J. Smith, J. Warnatz. Proc. Combust. Inst. 30 (2005) 125–157.
[14] F.N. Egolfopoulos, N. Hansen,Y.Ju, K. Kohse-Höinghaus, C.K. Law, F.Qi, Prog. Energy Combust. Sci. 43 (2014) 36-67.
[15] S.P. Sharma, D.D. Agrawal, and C.P. Gupta, Symposium (International) on Combust., 18 (1981), 493 - 501,
[16] Q. Li, C. Tang, Y. Cheng, L. Guan, and Z. Huang, Energy Fuels, 29 (2015) 5334–5348
[17] M. Akram, P. Saxena, S. Kumar, Energy Fuels, 27 (2013), 3460-3466
[18] M. Akram, S. Kumar, Energy Fuels, 26 (2012), 3267-3274
[19] M. Akram, Kumar S., Combustion & Flame 158 (2011) 915-924
[20] Akram M., Minaev S., Kumar, Combust. Sci. Technol. 2013; 185:645-60
[21] A. Katoch, M. Asad, S. Minaev, S. Kumar, Fuel, Volume 182, (2016) 57-63
[22] R.J. Kee, J.F. Grear, M.D. Smooke, J.A. Miller, A FORTRAN Program for Modeling Steady Laminar One-Dimensional Premixed Flames, Report No. SAND85-8240, Sandia National Laboratories, 1993.
[23] Heufer, K. A., Sarathy, S. M., Curran, H. J., Davis, A. C., Westbrook, C. K. Pitz, W. J. Energy Fuels, 26 (11), (2012), 6678-6685

11th Asia-Pacific Conference on Combustion,
The University of Sydney, NSW Australia
10th -14th December 2017.

Surface mechanism for the ammonia oxidation over Pt(111)

Juan D. Gonzalez, B. S Haynes, and Alejandro Montoya

The University of Sydney, School of Chemical and Biomolecular Engineering, NSW, 2006, Australia.

Abstract

The mechanism of ammonia oxidation catalysed by Pt(111) has been revised using recent computational methods to locate transition state structures, thereby improving predictions of activation energy of elementary reactions. We found good agreement with literature reports on the site selectivity for species binding and surface reactions. However, the new calculated energy barriers indicate that the OH^* species has a stronger promoting effect on the NH_3 oxidation than previously predicted. The energy barriers of products formation are nevertheless consistently higher than experimental observations.

Keywords: Density Functional Theory, Ammonia oxidation, Mechanism, Microkinetics, Modelling.

1. Introduction

The combustion of NH_3 results in a product mixture having a high concentration of NOx. The NH_3 oxidation to NO catalysed by a Pt based catalyst is a key step in the industrial production of nitric acid [1]. Despite the vast amount of research devoted to this process, the numerical accuracy of some rate constants of surface reactions needs further refinement. Molecular modelling has emerged as a powerful tool and a complement to the experimental techniques to study elementary reactions of surface catalysed processes. Most of the molecular studies of NH_3 oxidation have focused on Pt(111) because this surface is the most thermodynamically stable at high temperatures[2, 3]. The mechanism for NH_3 dissociation in the absence of oxygen on the Pt(111) surface has been investigated by Michaelides and Hu [4] and by Falsig *et al.* [5]. There is only one comprehensive study of the NH_3 oxidation by surface O^* and OH^* species over Pt(111) carried out by Offermans *et al.* [6]. In previous studies, the Nudged Elastic Band (NEB) method [7] was used to locate transition states (TS) along with minimum energy pathways (MEPs). It has been shown in the literature that using the NEB method instead of the more accurate Climbing Image NEB method (CI-NEB) can lead to a systematic error in activation energy (ΔEa) of around 30 kJ mol^{-1} [8]. Considering the uncertainty of DFT methods in the estimation of ΔEa values (~30 kJ mol^{-1}), the impact of these additive uncertainties on the calculation of reaction rates can be significant. Kinetic parameters for the NH_3 dissociation on the bare surface and for the oxidation assisted by surface O* and OH* species is revised by using the CI-NEB method. The elementary N_2, NO and N_2O formation pathways are also described. Finally, the impact of the reaction parameters on the rate of NO production is analysed. The reaction kinetic parameters provided in this contribution are the most accurate set of data for the construction of a micro-kinetic model of NH_3 combustion catalyzed by Pt to predict conversion and product yields under a broad range of pressure and temperature conditions.

Corresponding author. Fax: +61 2 9351 2854
E-mail address: alejandro.montoya@sydney.edu.au

2. Methodology

The PBE functional within the periodic supercell approach as implemented in the Vienna Ab-initio Simulation Package VASP (Version 5.3) was used [9-11]. The Pt(111) surface was modelled using a p(2x2) unit cell consisting of five layers with a space of 15 Å normal to the surface; created from a fully relaxed Pt bulk structure. The relaxed DFT bulk lattice constant of 3.97 Å is consistent with the experimental value of 3.92 Å[12]. The adsorbate together with the two top-most Pt layers was relaxed while the bottom three layers were kept at the DFT-bulk positions. A Methfessel-Paxton method with a smearing of 0.2 eV was used, with a (8×8×1) k-point mesh and an energy cut-off of 400 eV. The MEP for each elementary step was obtained using the CI-NEB method [7, 8] with at least five images to locate transition states. The climbing image approach was turned on when the forces of the NEB computation reached 0.5 eV/ Å. The convergence of the CI-NEB was achieved when the force of the band was lower than 0.05 eV/Å. The validity of the transition states was confirmed with a single imaginary frequency in the direction of the reaction coordinate. All the energies presented here are corrected by vibrational energy at zero Kelvin.

3. Results

3.1 NH_3 decomposition in the absence of oxygen

Figure 1 shows molecular representations of the minimum energy pathways of NH_3^*, NH_2^* and NH^* dehydrogenation over the bare Pt(111). While NH_3^* binds strongly to top site, NH_2^* and NH^* bind strongly on a bridge site and a three hollow (fcc) site respectively. The dehydrogenation is assisted by adjacent Pt, leaving the nitrogen containing products occupying the bridge and the fcc sites, and the surface hydrogen binding on an adjacent position. The site selectivity for binding energy and the active sites of reaction are consistent with previous predictions [6]. Although experimental activation energies for single elementary steps of the NH_3 oxidation over Pt(111) are scarce in the literature, Löffler and Schmidt [13] have reported an experimental activation energy for NH_3 dissociation on the bare Pt(111) of 88 kJ mol^{-1}, which is only 15 kJ mol^{-1} lower than the value predicted in this study. In addition, the energy barriers and reaction energies are relatively consistent with molecular modelling literature values [4-6]. For instance, the energy barrier of NH_3^*, NH_2^* and NH^* dehydrogenation are 10, 12 and 11 kJ mol^{-1} more activated than those reported previously [6]. Similarly, the reaction energies for of NH_3^*, NH_2^* and NH^* dehydrogenation (27, -5 and 21 kJ mol^{-1}) are consistent with the values previously reported values (35, -8, 23 kJ mol^{-1}) [6].

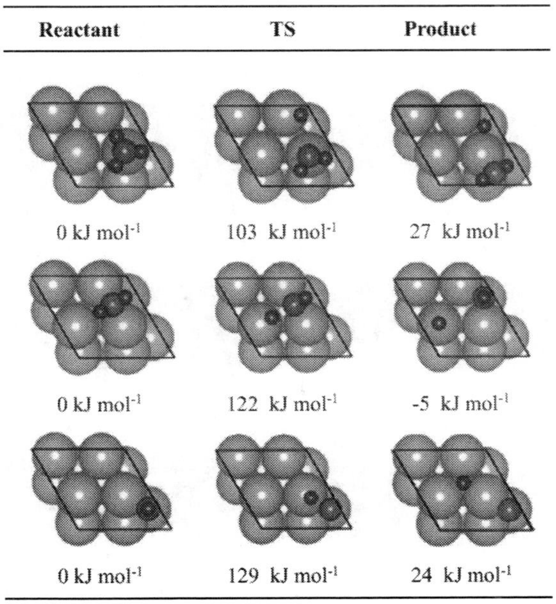

Figure 1. Schematic representation of the NH_3^*, NH_2^* and NH^* dehydrogenation on the bare Pt(111) surface. Blue: nitrogen atom, green: hydrogen atom and grey: platinum atom.

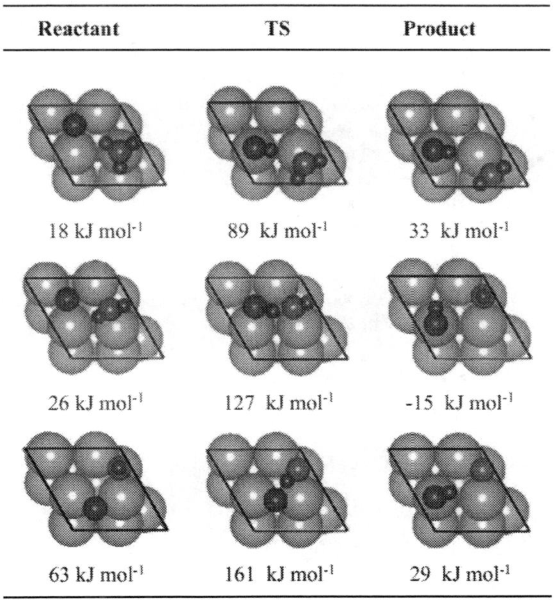

Figure 2. Schematic representation of the minimum energy pathways of NH_3^*, NH_2^* and NH^* oxidation by O^* on Pt(111). The energies are relative to the energy of the NH_x species and O^* in separate unit cells.

3.2 NH_3 oxidation assisted by surface oxygen species

Figure 2 shows the MEPs of NH_3^*, NH_2^* and NH^* dehydrogenation assisted by adjacent surface O^* species. The reported energies are relative to the energy of NH_x^* and O^* species in separate cells. Nitrogen containing products and surface hydroxyl, OH^*, are formed after the dehydrogenation process. The products can further react in subsequent stages. In all cases, the lateral interaction of NH_x with O^* is repulsive. Despite this, the energy barrier of NH_3^* decomposition in the presence of O^* (Fig. 2) is 14 kJ mol^{-1} lower than the energy barrier predicted for NH_3^* dehydrogenation assisted by Pt (Fig. 1). The kinetic parameters indicate that the rate constant for the first NH_3 dehydrogenation on the oxidised Pt surface is 16 times faster at 300 K than over the clean surface. The NH_2^* and NH^* dehydrogenation assisted by O^* are 5 and 32 kJ mol^{-1} more activated than the dissociation on the clean surface, mainly due to the strong repulsive interaction between NH_2^* and NH^* with O^*. Moderate agreement is found with previously reported values [6]. The ΔEa values for NH_3^*, NH_2^* and NH^* dehydrogenation assisted by O^* are 63, 18 and 30 kJ mol^{-1} respectively more activated than previously reported. These differences will have a strong impact on the calculated rate constant and therefore on the reaction rate. This is especially true for the NH_3^* and NH^* oxidation reactions where the lower ΔE_a previously reported lead to an overestimation of the rate constant of the order of 10^{10} and 10^5 s^{-1} at 300 K respectively.

3.2 NH_3 oxidation by surface hydroxyl species

Figure 3 shows the molecular representations of the dehydrogenation processes assisted by OH* species.

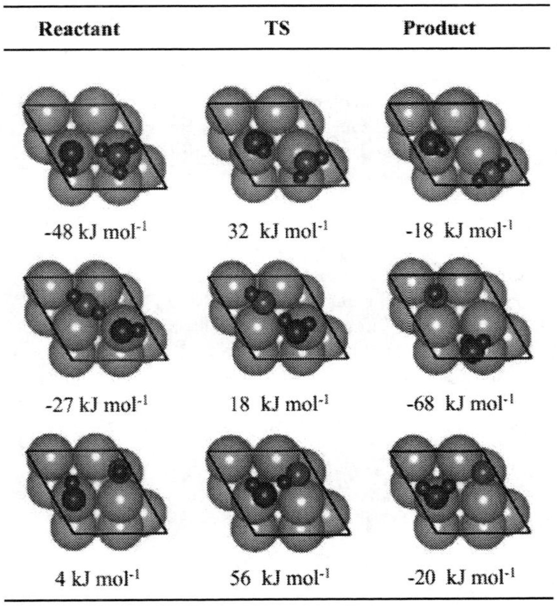

Figure 3. Minimum energy pathways of NH_3^*, NH_2^* and NH^* oxidation by OH^* on Pt(111). The energies are relative to the energy of the NH_x species and OH* in separate unit cells.

The energy barriers for NH_3^*, NH_2^* and NH^* dehydrogenation in the presence of OH^* are decreased by 71, 104 and 73 kJ mol^{-1} respectively compared to the dissociation reaction on the bare Pt(111). Similarly, the dehydrogenations in the presence of OH^* are 57, 109 and 105 kJ mol^{-1} respectively less activated compared to the dehydrogenation by O^*. Differences of 12-22 kJ mol^{-1} were found between the ΔEa values calculated here and those reported elsewhere [6].

The kinetic parameters for the ammonia dissociation on the bare Pt(111) and oxidation assisted by O^* and OH^* species are presented in table 1.

Table 1. Selected kinetic parameters of elementary reactions of NH_3 dehydrogenation assisted by Pt, O^* and OH^* surface species.

Reaction	A	ΔE_a
$NH_3^* + Pt \rightleftharpoons NH_2^* + H^*$	1.60×10^{12}	103
$NH_2^* + Pt \rightleftharpoons NH^* + H^*$	6.74×10^{13}	122
$NH^* + Pt \rightleftharpoons N^* + H^*$	1.78×10^{14}	129
$NH_3^* + O^* \rightleftharpoons NH_2^* + OH^*$	9.52×10^{10}	89
$NH_2^* + O^* \rightleftharpoons NH^* + OH^*$	3.84×10^{12}	127
$NH^* + O^* \rightleftharpoons N^* + OH^* + Pt$	2.82×10^{12}	161
$NH_3^* + OH^* \rightleftharpoons NH_2^* + H_2O^*$	1.33×10^{10}	32
$NH_2^* + OH^* \rightleftharpoons NH^* + H_2O^*$	4.15×10^{11}	18
$NH^* + OH^* \rightleftharpoons N^* + H_2O^*$	5.36×10^{11}	56

A= pre-exponential factor in s^{-1}. ΔE_a = activation energy in kJ mol^{-1}, from reactants without lateral interactions and zero-point energy (ZPE) corrected.

Even though the interaction of NH_3^* and NH_2^* with OH^* in the reactant state is attractive, and the interaction of NH^* with OH^* is nearly thermo-neutral, the kinetic parameters point to a strong promoting effect of the OH^* towards decomposition of NH_3. The promoting effect can be seen more clearly in Figure 4, where the energy of NH_3 decomposition in the presence of OH^* (green lines) are compared with the decomposition in the presence of O^* (red line) and Pt alone (blue line).

Atomic nitrogen (N^*) is the precursor of $N_{2(g)}$, $NO_{(g)}$ and $N_2O_{(g)}$ during the NH_3 oxidation. NO is the primary product at high temperatures while N_2 and trace amounts of N_2O dominate at low temperatures. Figure 5 shows the MEPs of N_2, NO and N_2O formation. The formation of N_2 is the most activated of all product species, followed by NO and N_2O. The activation energies for N_2 and NO are at most 4 kJ mol^{-1} higher than literature values [6]. In addition, the N_2O formation energy barrier is in excellent agreement with a recent study [14].

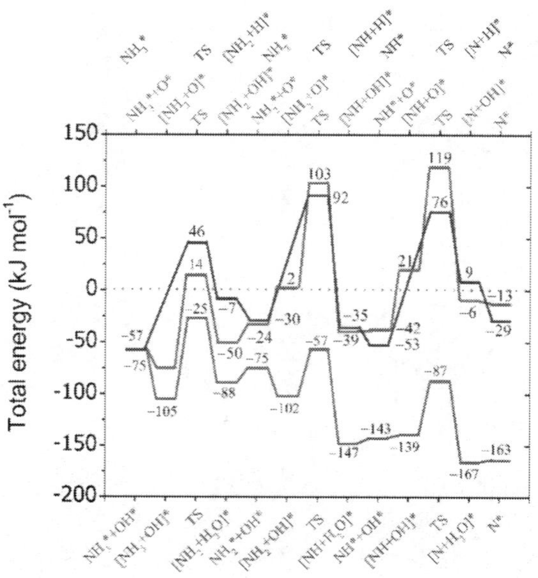

Figure 4. Combined reaction energy diagram for the ammonia decomposition on Pt(111), O^*-Pt(111) and OH^*-Pt(111). Total energies are referenced to the energy of $NH_{3(g)}$, Pt(111), $3O^*$, $3OH^*$ and $3H^*$. The square brackets represent coadsorbed states with lateral interactions. All energies are zero-point energy (ZPE) corrected.

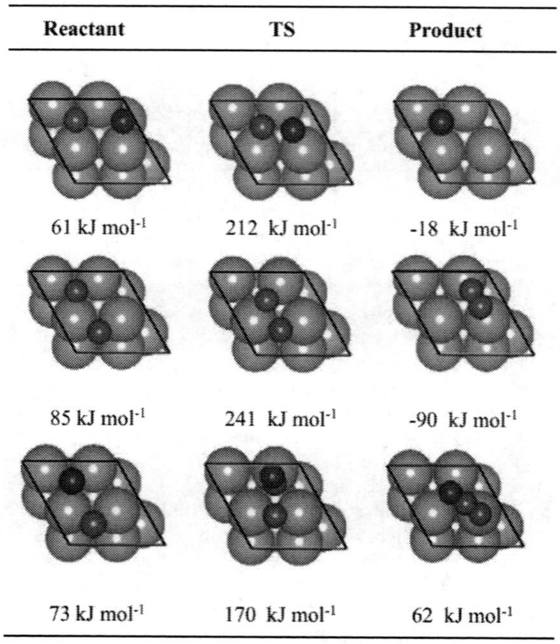

Figure 5. Schematic representation of the relaxed images of the reactant (left), transition state TS (middle) and product (right) NO^*, $N_{2(g)}$ and $N_2O_{(g)}$ formation on Pt(111). Blue: nitrogen atom, green: hydrogen atom, grey: platinum atom and red: oxygen atom.

We have pointed out that previous information on the DFT-based rate of NO formation over the Pt(111) surface via $N^* + O^* \rightleftharpoons NO_{(g)}$ is inconsistent with experimental derivations [15], and that $HNO^* \rightleftharpoons NO^* + H^*$ and $NOH^* \rightleftharpoons NO_{(g)} + H^*$ reactions do not have a primary role in the overall rate of NO formation [16]. A comprehensive kinetic mechanism for the ammonia oxidation using our DFT-calculated kinetic parameters was implemented in the Chemkin-Pro software package [17] using a Shear Flow Reactor (SFR). The formation of NO is negligible under a NH_3/O_2 molar ratio of 2 at 1000K and 1atm, inconsistent with experimental observations. The nitrogen flow is directed toward $N_{2(g)}$ formation. Although the activation energy of NO formation is lower than the N_2 formation (Figure 5), the NO formation rate is slower than the N_2 formation rate due to the high surface coverage of nitrogen species. The mismatch between the DFT-calculated and the experimentally derived values may, therefore, be related to surface coverage effects or morphological effects such as the presence of stepped surfaces. The coverage effect on the activation energy of NO formation and incorporation of these effects in a microkinetic model could provide further evidence of the likely activity of the Pt(111) surface under industrial conditions of ammonia oxidation.

4. Conclusions

The surface mechanism for the ammonia oxidation over the Pt(111) surface has been revised by updating and complementing rate parameters of relevant elementary steps. Differences ranging from 10-30 kJ mol^{-1} in the energy barriers were found n comparison with previous studies. Adsorbed O^* and OH^* promote the NH_3^* dehydrogenation reactions. The NH_3^*, NH_2^* and NH^* dehydrogenation steps were found to be less activated in the presence of OH^* than in the presence of O^* species. The desorption of products N_2 and NO are highly endothermic, limiting the ammonia oxidation process.

5. Acknowledgment

This research was supported by funding from the Faculty of Engineering & Information Technologies, The University of Sydney, under the Faculty Research Cluster Program. The authors acknowledge the University of Sydney HPC resources and the National Computational Infrastructure Facility (NCI) supported by the Australian Government for providing resources that have contributed to the results reported in this paper. Juan D. Gonzalez would like to thank the Australia Awards Scholarship and Australian Leadership Awards programs for the PhD scholarship.

6. References

[1] S.T. Hatscher, T. Fetzer, E. Wagner, H.-J. Kneuper, Ammonia Oxidation, in: Handbook of Heterogeneous Catalysis, Wiley-VCH Verlag GmbH & Co. KGaA, 2008.

[2] D.W. Blakely, G.A. Somorjai, Surface Science, 65 (1977) 419-442.

[3] M. Flytzani-Stephanopoulos, S. Wong, L.D. Schmidt, Journal of Catalysis, 49 (1977) 51-82.

[4] A. Michaelides, P. Hu, Journal of the American Chemical Society, 122 (2000) 9866-9867.

[5] S. Wang, V. Petzold, V. Tripkovic, J. Kleis, J.G. Howalt, E. Skulason, E.M. Fernandez, B. Hvolbaek, G. Jones, A. Toftelund, H. Falsig, M. Bjorketun, F. Studt, F. Abild-Pedersen, J. Rossmeisl, J.K. Norskov, T. Bligaard, Physical Chemistry Chemical Physics, 13 (2011) 20760-20765.

[6] W.K. Offermans, A.P.J. Jansen, R.A. van Santen, Surface Science, 600 (2006) 1714-1734.

[7] G. Henkelman, H. Jónsson, The Journal of Chemical Physics, 113 (2000) 9978-9985.

[8] G. Henkelman, B.P. Uberuaga, H. Jónsson, The Journal of Chemical Physics, 113 (2000) 9901-9904.

[9] J. Hafner, Journal of Computational Chemistry, 29 (2008) 2044-2078.

[10] G. Kresse, J. Furthmüller, Physical Review B, 54 (1996) 11169-11186.

[11] G. Kresse, D. Joubert, Physical Review B, 59 (1999) 1758-1775.

[12] W.M. Haynes, CRC Handbook of Chemistry and Physics, 95th Edition, CRC Press 2015.

[13] D.G. Löffler, L.D. Schmidt, Surface Science, 59 (1976) 195-204.

[14] C.A. Farberow, J.A. Dumesic, M. Mavrikakis, ACS Catalysis, 4 (2014) 3307-3319.

[15] J.D. Gonzalez, M. Warner, B.S. Haynes, A. Montoya, Proceedings of the Combustion Institute, 36 (2017) 637-644.

[16] M.W. J.D. Gonzalez, B.S. Haynes, A. Montoya, in: Proceedings of the Australian Combustion Symposium, The University of Melbourne, 2015, pp. 132-135.

[17] Reaction Workbench 15131, Reaction Design, San Diego, 2013.

A Comparative Study of Obstacle-induced Detonation Initiation and Flame Jet Detonation Initiation in Chambers at the Same Scale

Yeqing Chi, Wei Fan*, Jiannan He, Jiawei Zheng

School of power and energy, Northwestern Polytechnical University

127 West Youyi Road, Xi'an, 710072, P. R. China

Abstract

A series of experiments were carried out to investigate the characteristics of detonation initiation with pre-detonators and obstacles. Chambers with a 20mm×70mm rectangular cross-section and smooth tubes with 6 mm inner diameters were used as main detonation chambers and pre-detonators, respectively. The gas mixture used in the study is oxygen/ethylene/nitrogen mixture and the nitrogen diluted ratio β is set as a variable. Flame acceleration process was captured by a high-speed camera. Three initiation modes were observed in the flame jet detonation, and only one form of detonation initiation was found in detonations triggered by obstacles. The DDT distance and flame velocity all indicate that when $\beta \leq 0.7$, hot jet can initiate detonation in a short time, while obstacles perform better when $\beta \geq 1$.

1 Introduction

Flame jet detonation initiation and deflagration to detonation transition (DDT) are considered to be two most popular detonation initiation methods which are widely used in the field of detonation application, such as pulse detonation engines (PDE) and rotating detonation engines (RDE). The flame jet, or pre-detonator, can trigger detonation in short distance but with higher ignition energy, while the DDT can initiate detonation with low initial energy and longer DDT distance. Both detonation initiation methods require extra facilities, which are pre-detonation tube and obstacles, respectively. Those initiation methods would lead to thrust loss, extra weight and other disadvantages. Meanwhile, the mechanisms of these two methods are still unclear, which also limit their application. Therefore, it is essential to study their characteristics and control laws in order to promote the development of detonation engines.

The research on these two detonation initiation methods has a long history. The study of flame jet detonation initiation can date back to 1970s [1].However, previous researchers mostly focus on the diffraction of detonation wave and critical expansion ratio of direct detonation initiation. It is well known that critical tube diameter d_c is 9 to 13 times the detonation cell size λ to ensure detonation re-initiation after the expansion section [2]. In a research presented by Gavrilenko et al. [3], the phenomenon of C-J detonation wave's transmission from a circular tube to a chamber with rectangular cross-section was researched and they established the relationship between cellular size and overdriven ratio of detonation. In 1998, the pre-detonator using ethylene/oxygen mixture was installed on the PDE made by United States Naval Research Laboratory and the PDE worked successfully. [4] An investigation carried out by Lieberman et al. [5] shows that the hot jet detonation was sensitive to dilution ratio instead of the initial pressure inside pre-detonator. It was also found that, comparing to obstacles, the hot jet can initiate detonation with little energy loss in the study. Some other experimental and numerical study shows that the position and structure of pre-detonator do not obviously influence the detonation initiation. [6, 7] In a simulation research carried out by Y. Lv [8], it was found that comparing with using Argon, using Nitrogen as diluent can prolong the ignition time via reducing the temperature behind the shock.

The concept of obstacle-induced detonation initiation is proposed even earlier. In 1940s, the researchers had paid attention to the DDT process in rough tubes. [9] But it was since 1970s that the obstacle-induced detonation initiation had been widely investigated.

According to the research of Matsukov et al. [10], in the rough tube, obstacles can curve the flame and obtain more eddies, which can accelerate the flame and initiate detonation. With a large number of experiments, Lee et al. [11] found that the DDT distance was the shortest in their experiments when block ratio is 0.43, and the flame speed of $0.5V_{CJ}$ is essential to trigger detonation, in which V_{CJ} refers to the speed of C-J detonation wave. The relationship of obstacles spacing and DDT distance were also investigated in different ways [12, 13], and it was found that the long spacing helps the formation of Mach stem while the short spacing can strengthen the turbulence. As a result, an ideal obstacles spacing exists when those two effects achieve a balance. .

It can be seen that lots of results have been obtained in previous researches, which are helpful in designing our experiment. However, most researches just focus on obstacle-induced detonation initiation or flame jet detonation initiation respectively. Little research compares these two detonation initiation methods, and provides a guideline on detonation engines design. Meanwhile, the phenomenon of detonation initiation is so complicated that it is very hard to be fully understood. As a result, it is essential to carry out a research on the two initiation method in the same scale to explore the DDT process and acquire a better understanding of detonation initiation.

2 Experimental setup

The schematic diagram of experimental apparatus is shown in Fig. 1. The experimental system has been described in detail in [14]. Stoichiometric ethylene and oxygen mixture was applied in this work, into which nitrogen was added as diluted gas. The

Corresponding author. Fax: +86 29 88431116
E-mail address: weifan419@nwpu.edu.cn

dilution ratio β is defined as $\beta=V_{N2}/V_{O2}$, in which V_{N2} and V_{O2}represent volume of N_2 and O_2, respectively. In present work, β=0, 0.4, 0.7, 1 and 1.375, while the pressure and temperature of the mixture is 1atm and 300K, respectively.

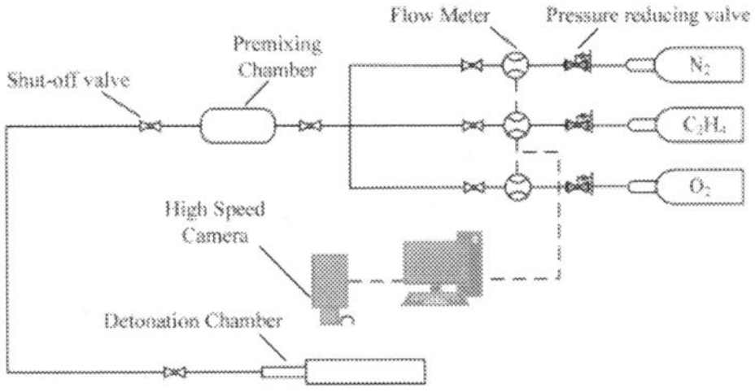

Figure 1: Schematic of the experimental apparatus

The detonation chambers used in present research are shown in Fig. 2. Both detonation chambers have a rectangular cross-section (height = 70mm and width = 20mm). The length of the chamber in which detonation was initiated with flame jet is 350 mm, while obstacle-included detonation chamber is 1000mm long. For the chamber in which detonation ignited by flame jet, a pre-detonator with 6mm inner diameter is connected to the head. As a result, the expansion ratio σ for this detonator-chamber system is σ=11.67, which is the ratio of the height of main chamber and the inner diameter of pre-detonator. The lengths of pre-detonator L' is 700mm and 1000 mm, which can affect the ignition energy. An ordinary spark igniter is utilized to fire up mixture at the head of pre-detonator. For the detonation chamber with obstacle, eight pairs of flat plate obstacles are mounted on the top and the bottom wall. The blockage ratios are 0 and 0.4, while the obstacle spacing is 40mm. An ordinary spark igniter was applied to initiate the combustion.

In this investigation, all chambers and pre-detonators are made of acrylics. The flame acceleration and detonation are captured by a high-speed camera (Phantom V7.2), whose sample rate in this work is 40000~90000 fps.

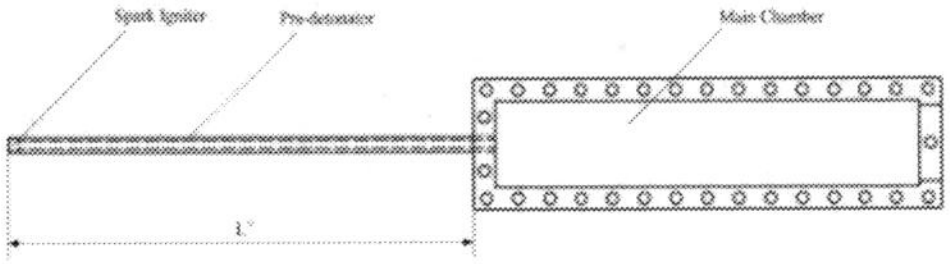

(a) Detonation chamber with a pre-detonator tube

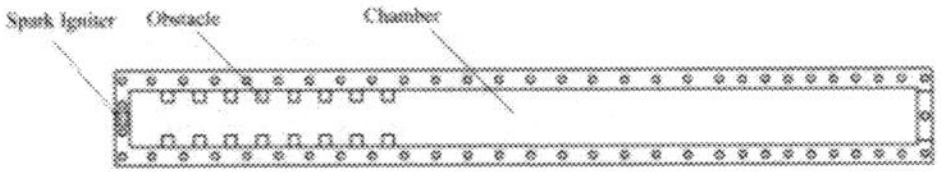

(b) Obstacle-included detonation chamber

Figure 2: Detonation chambers

3 Results and discussion

For each of detonation initiation methods described above, a series of dilution ratios were investigated. Their flame acceleration process and detonation trigger characterizations are recorded by a high-speed camera. The local flame velocity over chambers and DDT distance is measured, analyzed and compared below.

3.1 Characterizations of detonation initiation with two initiation methods

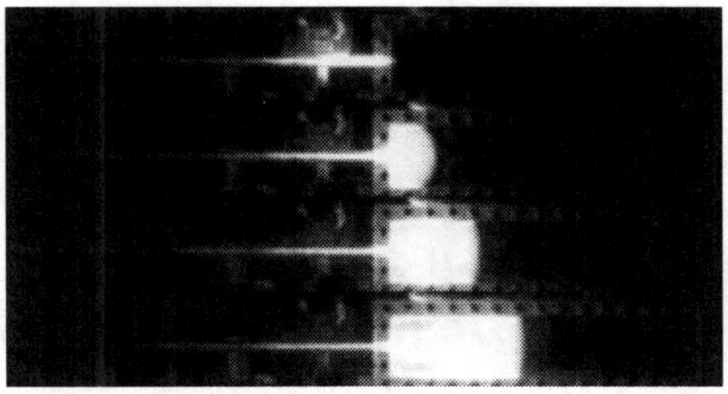

(a) β=0.7, L'=700 mm

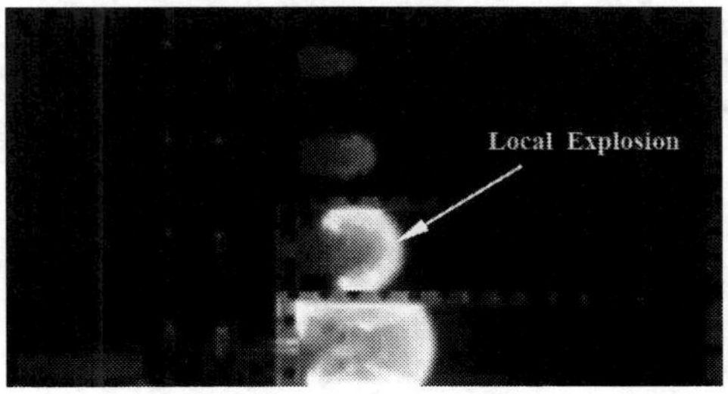

(b)β=0.4, L'=400 mm

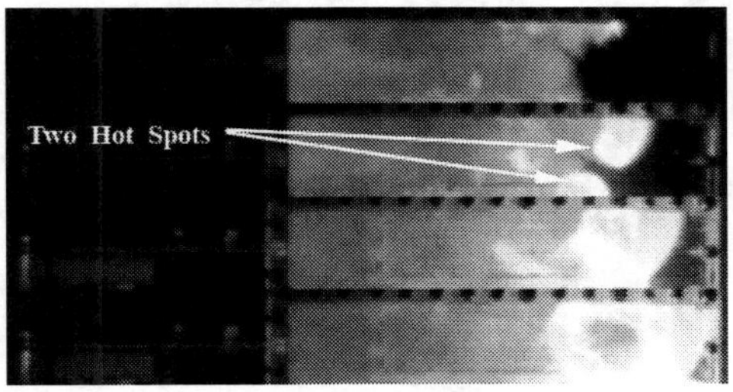

(c) β=0.4, L'=400 mm

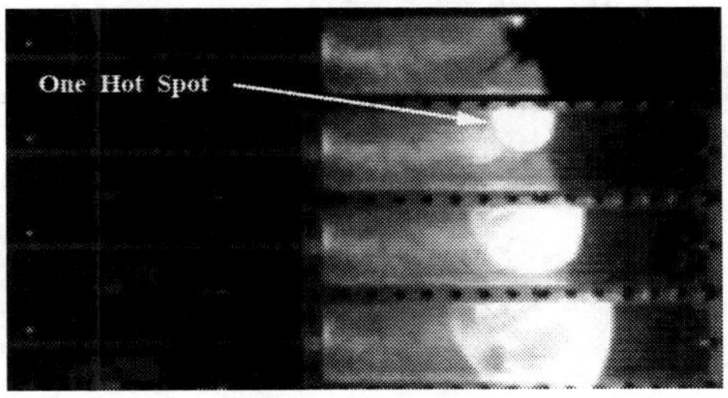

(d) β=0.4, L'=700 mm

Figure 3: Characterizations of the detonation initiation with flame jet

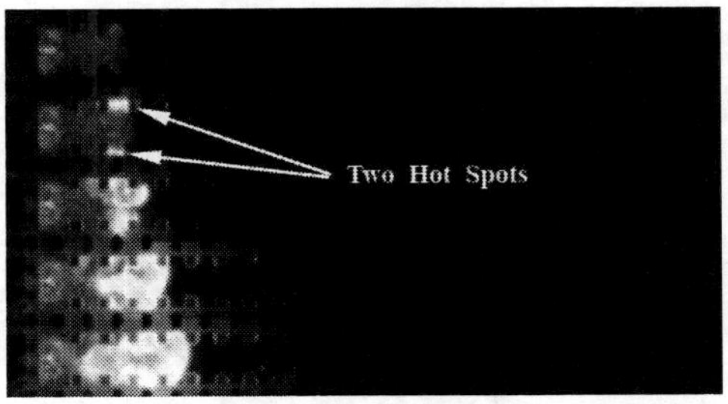

Figure 4: Characterizations of the obstacle-induced detonation initiation

Figure 3 and Figure 4 illustrate the different detonation trigger modes with flame jet and obstacle, respectively. As can be seen in Fig. 3, three kinds of detonation initiation mode are obtained during multiple shots. One is direct detonation initiation, as shown in Fig. 3 (a). The ejected flame is very bright and nearly spherical, with no obvious decoupling captured. This initiation mode occurs when the β is high and the DDT distance in the pre-detonator is longer. As a result of long DDT distance, the overdriven detonation appears very close to the joint section and ejects into the main chamber before reduction, which leads to higher ignition energy. Another is detonation initiation caused by local explosion. As shown in Fig. 3(b), after the flame propagates in the chamber for a short distance, a local explosion happens and initiates detonation immediately. The local explosion results from the re-coupling of shock wave and reaction front. It is discovered that local explosion appears when the flame speed V_j at the joint section satisfies the condition of $M_1< V_j<V_{CJ}$, in which M_1 stands for the local acoustic velocity and V_{CJ} is the speed of C-J detonation wave. The last mode is DDT-induced detonation initiation, showed in Fig.3 (c) and 3(d). When the flame velocity at the joint section is V_{CJ}, the C-J detonation would also decouple, and after a long DDT distance, one or two hot spots arise at the top or bottom wall, triggered the detonation. It is noteworthy that the number and position of hot spots are random.

In Fig. 4, it is observed that, in obstacle-included detonation chamber, only DDT-induced detonation initiation is found. Similar to Fig. 3 (c), detonation was triggered via two hot spots. However, the difference is that hot spots always appear symmetrically between two obstacles, and the DDT distance is shorter due to the obstacles. That is because the energy of detonation wave in Fig. 3 (c) is much lower than that in Fig. 3 (a), though higher than that in Fig. 4. The detonation initiation in those condition all experience an flame acceleration process, while the flame acceleration is promoted by high ignition energy and turbulence generated by the obstacles, respectively. The DDT distance shows that obstacle-induced detonation initiation can be better when the pre-detonator cannot produce a direct detonation initiation.

3.2 Comparison of flame velocities

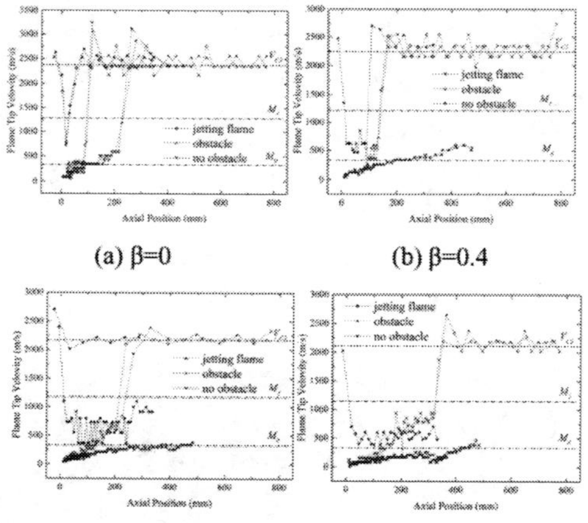

(a) β=0 (b) β=0.4

(c) β=0.7 (d) β=1

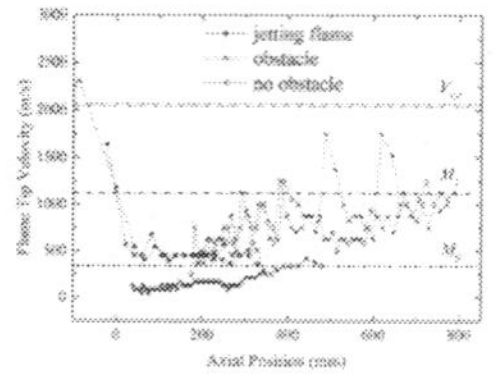

(e) β=1.375

Figure 5: The flame velocity in different initiation method and dilution ratio β

Evolutions of flame velocities in the detonation chambers at different operation conditions are shown in Figure 5. In the diagrams, M_1 and M_0 are the sound speed of the products and reactants, respectively.

The flame velocity evolution of flame jet detonation initiation in Fig. 5(a), (b), (c) illustrates local explosion, DDT process and direct detonation initiation, respectively, which supports the result in section 3.1.The steep fall of velocity shown in Fig.5(b), (c), (d) correspond the decoupling process, which is also consistent with the phenomenon in Fig. 3.Comparing the flame velocity in chamber with pre-detonator and obstacles, shown in Figs. 5(a) ~ (e), it is found that the speed of flame ignited by flame jet was higher before the first 100 mm of the main chamber, because of the high energy provided by the pre-detonator. As for the chamber without pre-detonator or obstacles, this difference is more apparent.

In Figs. 5(a) and (c), where β=0 and 0.7, the overdriven ratio of jetting flame, which is the ratio of the max velocity and V_{CJ}, is lower than that of the flame in the chamber with obstacles. However, it is interesting to notice that the overdriven ratio of jetting flame is highest when β=0.4, as shown in Fig. 5(b). This phenomenon illustrates that the overdriven ratio may not only rely on the ignition method and dilution ratio, but also influenced by the detonation trigger modes.

3.3 Comparison of DDT distances

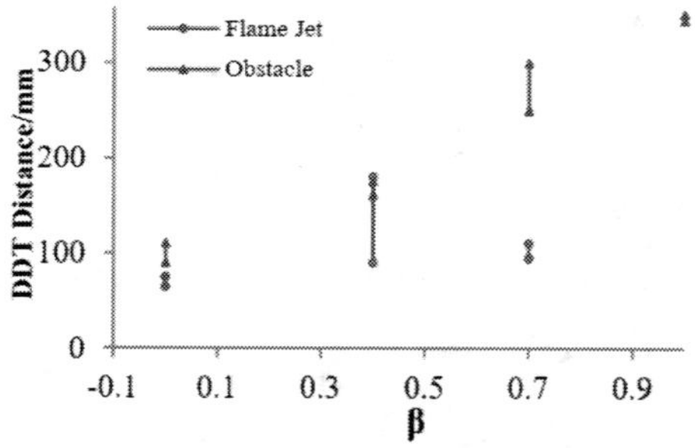

Figure 6: The DDT distances under different initiation methods and dilution ratios β

Figure 6 reveals the DDT distance in the main chambers under different dilution ratios β. Because of the failure of detonation initiation when β=1.375, only the data of β=0~1 is shown in the diagram. The length of the pre-detonator is 700mm.

As shown in Fig. 6, when β ≤ 0.7, detonation in chambers with two different detonation initiation methods can always be initiated successfully. In most cases, the flame jet detonation initiation results in a shorter DDT distance comparing to the obstacle-induced detonation initiation. The reason is that the pre-detonator can provide higher ignition energy as well as a shock wave, and the reaction zone can couple with the shock in a short distance. Meanwhile, comparing the flame velocity in cases that both methods cannot trigger detonation, it is found that the flame ignited by pre-detonator is obviously faster than that in obstacle-included chamber. As a result, flame jet detonation initiation has better detonation performance than that of obstacle-induced detonation initiation, when $\beta \leq 0.7$.

Generally speaking, the DDT distance would increase with the increase of dilution ratio, because of the decrease of the chemical activity. The DDT distance of obstacle-induced detonations perfectly matches this rule. However, when β changes from 0.4 to 0.7, the DDT distance of flame jet ignition detonation decreases in average. It is because in pre-detonator, the DDT distance would also increase with the raise of β. when β=0.7, the overdriven detonation does not reduce to C-J detonation at the joint section. As a result, the ignition energy would be higher at the cases of β=0.7, than that of β=0.4. High ignition energy results in a shorter and more stable detonation initiation.

Whilst β=1, detonation cannot be initiated in the main chamber by flame jet, yet the obstacle-induced detonation is successfully triggered at 350mm. For the case of β=1.375, though both ignition methods cannot initiate detonation at 350mm, data in Fig. 5(e) reveals that the flame velocity in obstacle-included detonation chamber is still higher than that ignited by flame jet at the same point. Accordingly, when $\beta \geq 1$, well-designed obstacles can trigger detonation faster than flame jet.

4 Conclusions

Two detonation initiation methods were used in this research to investigate the initiation characterizations, flame velocities and DDT distances experimentally. The dilution ratio is changed to reveal the benefits and defects of initiation methods in different dilution ratios.

There are four main conclusions of present research:

(1) Via photographic observation, it is discovered that for flame jet detonation, it can be triggered in three different ways, including direct detonation initiation, local explosion and DDT-induced detonation initiation, but for obstacle-induced detonation, only DDT-induced detonation initiation was observed.

(2) This research demonstrates that pre-detonator performs better in initiating detonation when $\beta \leq 0.7$, whereas well-designed obstacles can trigger detonation easier when $\beta \geq 1$.

(3) The obstacles can initiate detonation more stably, and even when $\beta \leq 0.7$, the obstacles can trigger detonation with a shorter DDT distance in some cases.

(4) The overdriven ratio is influenced by ignition method, dilution ratio and detonation initiation method.

5 Acknowledgment

The authors wish to thank National Natural Science Foundation of China (91441201, 91641101, and 51376151) for financial support of this work.

References

[1]S.P. Medvedev, S.V. Khomik, H. Olivier et al., Shock Waves 14(3) (2005) 193-203.

[2] C. Conrad, S. Saretto, S. Lee, R. Santoro, 40th AIAA/ASME/SAE/ASEE Joint Propulsion Conference and Exhibit, Fort Lauderdale, Florida, 2004, p. 2004-3397.

[3] T.P. Gavrilenko, 8th ICOGER, Minsk, USSR, 1981, p. 23-36.

[4] C. Brophy, 34th AIAA/ASME/SAE/ASEE Joint Propulsion Conference and Exhibit, Joint Propulsion Conferences, Cleveland, OH, 1998, p. 1998-4003.

[5] D.H. Lieberma, K. Parkiny, J.E. Shepherd, 38th AIAA/ASME/SAE/ASEE Joint Propulsion Conference, Indianapolis, Indiana, 2002, p. 2002-3909.

[6] K. Ishii, T. Akiyoshi, M. Gonda, et al., Shock Waves (2009) 239-244.

[7] X. Wang. ACTA Aerodynamica Sinica 27 (03) (2009) 340-345.

[8] Y. Lv, M. Ihme, Computational analysis of re-ignition and re-initiation mechanisms of quenched detonation waves behind a backward facing step, Proceedings of the Combustion Institute, 2015, p. 1963-1972.

[9] K.I. Shelkin, J. Exp. Theor. Phys. 10 (1940) 823-829

[10] I. Matsukov, M. Kuznetsov, V. Alekseev, et al., the 17th International Colloquium on the Dynamics of Explosions and Reactive Systems, 1999, p. 143.

[11] S.Y. Lee, J. Watts, S. Saretto, Journal of Propulsion and Power, 20(6) (2004) 1026-1036.

[12] R.S. Chue, J. H. Lee, T. Scarinci, et al., Dynamic Aspects of Detonations (1993) 270-282.

[13] V.N. Gamezo, T. Ogawa, E.S. Oran, Combustion and flame 155 (1-2) (2008) 302-315.

[14] J. He, W. Fan, Y. Chi, et al., 21st AIAA International Space Planes and Hypersonics Technologies Conference, Xiamen, 2017, p. 2017-2149.

11th Asia-Pacific Conference on Combustion,
The University of Sydney, NSW Australia
10th -14th December 2017.

Rheological properties and ageing of bioslurry fuels prepared from crude glycerol/methanol/bio-oil blend and biochar for stationary combustion applications

Wenran Gao, Mingming Zhang, Hongwei Wu*

Department of Chemical Engineering, Curtin University

Kent Street, Perth, Western Australia, 6102, Australia

Abstract

This study investigates the rheological properties and ageing of bioslurry fuels prepared from crude glycerol/methanol/bio-oil (FCGMB) blend and biochar, benchmarking against the glycerol/methanol/bio-oil/biochar (GMBB) bioslurry for stationary combustion applications. Both FCGMB/biochar (FCGMBB) bioslurry and GMBB bioslurry fuels exhibit non-Newtonian and thixotropic behavior. The viscosities of FCGMBB bioslurry fuels are 222–268 mP.s which are lower than that of GMBB bioslurry due to the presence of water in formulated crude glycerol (FCG). Ageing leads to decreases in the total acid number (TAN) and viscosity but an increase in water content of both FCGMBB and GMBB bioslurry fuels due to various reactions. The changes of the water content (4.4–4.8%) and viscosity (6.0–11.7%) of FCGMBB, and the TAN (2.6% and 4.3%) of FCG1MBB/FCG3MBB bioslurry fuels before and after ageing are lower than that of GMBB bioslurry. This is due to the presence of soap and NaOH in FCGs. Overall, FCGMBB bioslurry fuels have better rheological property than GMBB bioslurry, and the presence of crude glycerol in the FCGMBB slurry fuels slows down the ageing.

1 Introduction

Crude glycerol is a major by-product of biodiesel industry and its utilization is of great importance to both economic and environmental performance of a biodiesel plant [1-5]. Direct utilization of crude glycerol is highly desired because of the large amount of glycerol surplus and the high cost associated with crude glycerol purification [4-6]. Recently, Wu et al. [7] developed a plausible way for crude glycerol direct utilization via mixing formulated crude glycerol (FCG) with fast pyrolysis bio-oil (with the addition of methanol) to produce FCG/methanol/bio-oil (i.e. FCGMB) blend and also mixing the resultant FCGMB blend with biochar to prepare FCGMB/biochar (FCGMBB) bioslurry fuels. Those blend and bioslurry fuels meet the specifications for stationary combustion applications [7]. Such strategies for producing FCGMBB bioslurry fuels not only help the effective utilization of byproducts (e.g. biochar from fast pyrolysis bio-oil production and crude glycerol from biodiesel production) but also address issues associated with those by-products (e.g. biochar being dusty and prone to spontaneous combustion and crude glycerol being of high viscosity) [8, 9]. However, little is reported on the rheological properties and ageing of FCGMBB bioslurry fuels in the open literature. These are the objectives of this study, i.e. to determine the rheological properties and ageing of FCGMBB bioslurry fuels.

Corresponding author. Fax: +61-8-92662681
E-mail address: h.wu@curtin.edu.au

2 Experimental Section

2.1 Samples preparation

Bio-oil was produced from pine wood biomass fast pyrolysis at 500 °C and supplied by a commercial supplier. Before experiments, the bio-oil sample was stored at ~4 °C. Biochar was also prepared from pine wood fast pyrolysis at 500 °C using a reactor outlined elsewhere [9, 10]. The biochar sample was ground (using a ball mill, Retsch MM400) and sieved to below 75 μm. Both Glycerol (G5516) and sodium oleate (26125) were supplied by Sigma-Aldrich. Analytical grade methanol, sodium chloride (NaCl) and sodium hydroxide (NaOH) were all purchased from Chem-Supply. Three formulated crude glycerol (FCG) samples (referred to as FCG1–3) were prepared as listed in Table 1, according to the compositions of crude glycerols [1]. FCGMB blends were prepared with composition of 4.6% FCG, 7% methanol and 88.4% bio-oil. FCGMBB bioslurry fuels were prepared by further adding 10% biochar to the FCGMB blends. Glycerol/methanol/bio-oil/biochar (GMBB) bioslurry with same compositions were also prepared as blanks. The fuel blends and bioslurries were mixed via magnetic stirring in a plastic vial at ambient temperature. The fuel properties of biochar, bio-oil, GMBB and FCGMBB bioslurry fuels (studied previously [7]) are listed in Table 2.

2.2 Sample characterization

The rheological properties of bioslurry fuels were determined via a rheometer (model Haake Mars II) with a Z20 cylinder sensor system and a Thermocontroller (Haake TC501). The viscosity at 100 s^{-1} are reported for bioslurry fuel samples. The flow and thixotropic behavior of bioslurry fuel samples were also studied according to a method outlined elsewhere [8]. A Karl Fisher titrator (Metter V30) and an acid-base titrator (MEP Oil Titrino plus 848) were used to determine the water content and total acid number (TAN) of the fuel samples, respectively, according to the methods detailed previously [10]. The accelerated aging experiment of bioslurry fuels were conducted at 80 °C for 24 h which is equal to the long-term stability via changes in the water content [10].

FCG	Glycerol	Soap	Water	NaOH	NaCl
FCG1	83.3	10.4	4.2	2.1	
FCG2	81.2		14.6		4.2
FCG3	66.7	10.4	20.8	2.1	

Table 1. Compositions of different types of formulated crude glycerol (FCG) samples in this study (wt %).

3 Results and Discussion

3.1 Rheologies of FCGMBB bioslurry fuels

Table 2 shows that FCGMBB bioslurry fuels have similar fuel properties with GMBB bioslurry except water content and viscosity. Figure 1a–b present the viscosity and shear stress as a function of shear rate for FCGMBB bioslurry fuels, benchmarking against GMBB bioslurry fuels. Both FCGMBB and GMBB bioslurry fuels exhibit non-Newtonian behavior. Compared to GMBB bioslurry, FCGMBB bioslurry samples have lower viscosities. The variances in the viscosity of bioslurry fuels are mainly attributed to the differences in water contents (see Table 2). FCGMBB bioslurry fuels also show thixotropic behavior which is characterized by the hysteresis loop (see Figure 1c). The pseudo-plastic behavior is attributed to the breakdown of agglomerated char particles along the upward ramp [11]. Such findings are consistent with those on bio-oil/biochar (BB) and GMBB bioslurry fuels [8, 11, 12]. It is worth noting that the area of hysteresis loop of FCGMBB bioslurry fuels are smaller than that of GMBB bioslurry, also attributed to high water contents in FCGs. Overall, the FCGMBB bioslurry fuels have better rheological properties than the GMBB bioslurry fuel.

3.2 Ageing of FCGMBB bioslurry fuels

The ageing of FCGMBB bioslurry fuels was investigated via changes in water content, TAN and viscosity, benchmarking against the ageing of GMBB bioslurry fuel. As is shown in Figure 2, after ageing, the water contents of FCGMBB bioslurry samples increase while the TAN and viscosity decrease. This can be attributed to the previously-reported reactions taken place during ageing, such as esterification, acetalization and/or polycondensation reactions [10]. Such a trend is also consistent with those of GMBB and BB bioslurry fuels reported earlier [8, 10]. It is also worth noting that the changes of the water content (4.4–4.8%) and viscosity (6.0–11.7%) of FCGMBB bioslurry fuels before and after ageing are reduced than that of GMBB bioslurry fuel (5.4% and 24.0%, respectively). This is likely attributed to the presence of soap and NaOH in FCGs which affect the acid-base reactions during ageing or even before ageing [1]. The reductions in the TAN for the FCG1MBB and FCG3MBB bioslurry fuels before and after ageing are 2.6% and 4.3% respectively. These are considerably lower than that of GMBB bioslurry (8.8%) due to sodium salts (e.g. soap and NaOH) addition. Such observations are in consistent with previous findings on GMB and

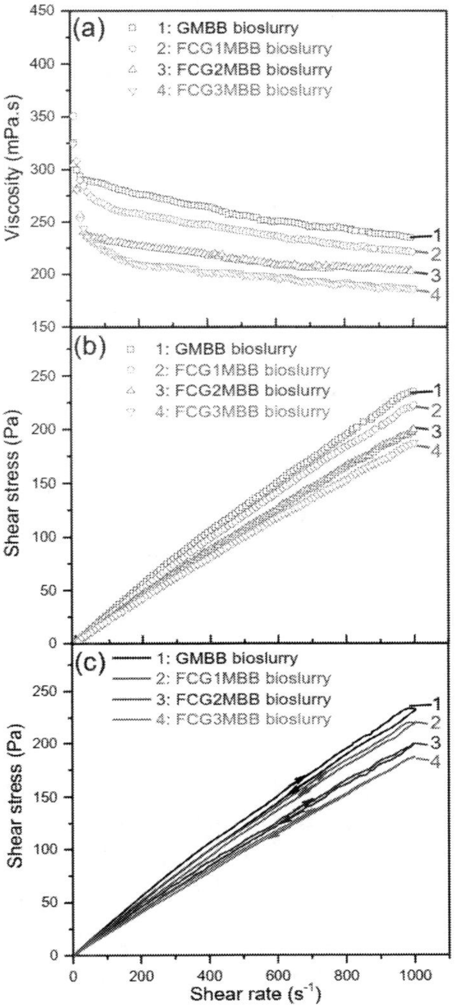

Figure 1. Viscosity (a), shear stress (b) and thixotropic behavior (c) of FCGMBB bioslurry fuels benchmarking against GMBB bioslurry as a function of shear rate (at 25 °C). GMBB bioslurry stands for glycerol/methanol/bio-oil/biochar bioslurry; FCG1–3MBB bioslurry stands for FCG1–3/methanol/bio-oil/biochar bioslurry; FCG stands for formulated crude glycerol.

	Elemental analysis (wt%)[a]				HHV[c]	Water content	TAN[d]	Viscosity	Surface Tension	Density
	C	H	N	O[b]	MJ/Kg	wt%	mgNaOH/g	mP.s,@25°C	mN/m,@25°C	g/cm³,@25°C
Biochar	82.14	3.17	0.11	14.28	27.2	2.3	nd[e]	nd[e]	nd[e]	nd[e]
Bio-oil	42.50	7.10	0.06	50.34	18.5	23.6	49.4	174.2	36.5	1.20
GMBB [f]	46.00	7.14	0.06	46.79	19.6	19.4	36.3	288.1	35.7	1.19
FCG1MBB [g]	46.04	7.11	0.06	46.66	19.5	19.8	35.9	268.2	33.5	1.19
FCG2MBB [g]	45.67	7.14	0.06	46.94	19.5	20.0	36.1	233.7	35.5	1.19
FCG3MBB [g]	45.75	7.17	0.06	46.94	19.5	20.3	35.9	222.1	33.6	1.19

[a] measured value for biochar and bio-oil, calculated value for others; [b] by difference; [c] calculated high heating value, HHV stands for higher heating value; [d] total acid number; [e] not determined; [f] GMBB stands for glycerol/methanol/bio-oil/biochar bioslurry; [g] FCG1–3MBB stands for FCG1–3/methanol/bio-oil/biochar bioslurry; FCG stands for formulated crude glycerol.

Table 2. Fuel Properties of FCGMBB bioslurry fuels benchmarking against the biochar, bio-oil and GMBB bioslurry, cited from Ref.[7].

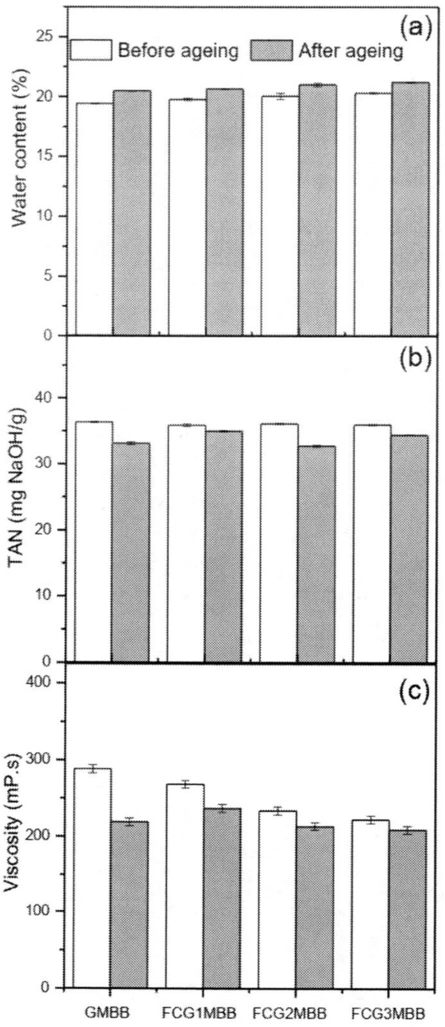

Figure 2. Water content (a), TAN (b) and viscosity (c) of FCGMBB bioslurry samples benchmarking against GMBB bioslurry before and after ageing. GMBB bioslurry stands for glycerol/methanol/bio-oil/biochar bioslurry; FCG1–3MBB bioslurry stands for FCG1–3/methanol/bio-oil/biochar bioslurry; FCG stands for formulated crude glycerol.

FCGMB fuel blends.[1] Overall, the presence of crude glycerol in FCGMBB bioslurry fuels slow down the ageing.

4 Conclusions

This study reports the rheological properties and ageing of FCGMBB bioslurry fuels prepared from FCGMB blends and biochar for stationary combustion applications. The results show that FCGMBB bioslurry samples have lower viscosities than the GMBB bioslurry and exhibit non-Newtonian and thixotropic behavior. After ageing, the water contents of FCGMBB bioslurry samples increase while the TAN and viscosity decrease. In comparison to the GMBB bioslurry fuel, the ageing of FCGMBB bioslurry fuels slows down due to the presence of soap and NaOH in crude glycerol.

5 Acknowledgment

The authors are grateful to the partial support from the Australian Research Council via its Discovery Projects scheme. W.G. also acknowledge the Curtin Strategic International Research Scholarship (CSIRS) for supporting her PhD studies. The authors would also like to thank the bio-oil provider (who has chosen to remain anonymous) for supplying the fast pyrolysis bio-oil samples and WA&J King Pty Ltd (Western Australia) for supplying the pine wood sample.

References

[1] M. Zhang, H. Wu, Fuel 159 (2015) 118-127.

[2] M. Zhang, H. Wu, Energy & Fuels 30 (2016) 8419-8424.

[3] F. Yang, M.A. Hanna, R. Sun, Biotechnol Biofuels 5 (13) (2012) 1–10.

[4] Z. Gholami, A.Z. Abdullah, K.-T. Lee, Renewable and Sustainable Energy Reviews 39 (2014) 327-341.

[5] A.B. Leoneti, V. Aragão-Leoneti, S.V.W.B. de Oliveira, Renewable Energy 45 (2012) 138-145.

[6] P.U. Okoye, B.H. Hameed, Renewable and Sustainable Energy Reviews 53 (2016) 558-574.

[7] W. Gao, M. Zhang, H. Wu, Fuel 207 (2017) 240-243.

[8] W. Gao, M. Zhang, H. Wu, Fuel 176 (2016) 72-77.

[9] H. Wu, Y. Yu, K. Yip, Energy & Fuels 24 (10) (2010) 5652-5659.

[10] M. Zhang, S.B. Liaw, H. Wu, Energy & Fuels 27 (12) (2013) 7560-7568.

[11] H. Abdullah, D. Mourant, C.-Z. Li, H. Wu, Energy & Fuels 24 (10) (2010) 5669-5676.

[12] H. Abdullah, H. Wu, Energy & fuels 25 (4) (2011) 1759-1771.

11th Asia-Pacific Conference on Combustion,
The University of Sydney, NSW Australia
10th -14th December 2017.

Research on Explosive Characteristics of Combustible Gas within the Restricted Space in Wells

Lyu Shuran, Ma Shuqi

School of Safety and Environmental Engineering, Capital University of Economics and Business

121 Zhangjialukou, Huaxiang Fengtai District, Beijing 100070, China

Abstract

The explosion of the combustible gas within the restricted space in the city wells will endanger the people's life and fortune. The syncretic gas detector of MX4 and MX6 are used in this study to detect the harmful gas in the well. The result shows that the main harmful gases are mainly CH_4, H_2S and CO_2, and their content range is given. Then the explosion test of the combustible gas is done in the spherical explosion vessel of 20L in volume in the lab, which studies the effect of the content of the methane and CO_2 on the explosion maximum pressure and its increase rate. Finally, FLACS code is adopted to study the effect of the inert gas CO_2 in the well on effectively restricting the methane combustion and explosive reaction. It is found that CO_2 has the restrict effect on the methane explosion.

1 Introduction

In recent years, our city scale is growing larger and larger with the rapid development of the urban construction. Meanwhile the underground pipeline system of the city is becoming complex and enormous, which is called the city lifeline. However lots of combustible gas even hazardous gas is unceasingly accumulated in the city wells at the same time, which will produce very destructive explosive reaction when it encounters the ignition source. Therefore, the operational state of the city well system not only reflects the feature of the whole city but also closely relates to the safety of the citizens and their property.

Because lots of slush and household garbage are prolonged kept in the city wells, a series of combustible and hazardous gases such as CH_4, H_2S, CO_2, CO and SO_2 are produced because they are decomposed by the microbe. Lots of these gases are accumulated in the wells, and the significant impact on the safety of the citizens and their property will occur when it encounters the ignition source. According to the statistic results, the blasting incidents in the well often happen in recent years. In seriousness, tens of accidents happen which causes the direct and indirect economic loss even to several hundred million RMB[1]. So, it is no time to delay to study the explosion characteristic of the combustible gas within the restricted space in wells. Through the intensive study on the explosion characteristic, parameter change law and its specific effect factor, the effective countermeasures are obtained which can prevents the explosion of the gas within the restricted space in the well to happen. Meanwhile the corresponding theory is put forward to ensure the safety of the citizens and their property, which will fundamentally provide the safeguard for the urbanization construction.

Many researchers studied this issue with theory, experiment and numerical simulation. In order to obtain the more detailed temperature information from the image shot from the high speed photography, Baisheng Nie[2] proposed a method to calculate the two dimensional temperature field of the pre-mixed gas explosion flame. Russel[3] proposed a thermal isolation compound calculation model of the gas explosion in the restricted space from the theoretical viewpoint which can effectively predict fuel minimum mass producing the necessary explosive pressure. Jin Guo[4] made an experiment in a small cylindrical vessel to study the effect of the explosion pressure on the explosion of the mixed gas of methane and air. R.M. Kasmani[5] made an experiment in a cylindrical vessel to study the effect of the explosion pressure and ignition position on the maximum explosion overpressure. Salzan[6] and Michele[7] adopted AutoReagas code to simulate the explosion process of the combustible gas in the small vessel pipeline. Fairweather[8] studied the flame propagation law during the gas explosion process in the pipeline with the numerical simulation. Qiuju Ma[9] did a numerical simulation to study the explosion characteristic of the mixed gas of methane and hydrogen in a vessel with 5L in volume and a blasting chamber with $64m^3$ in volume.

Therefore on basis of the previous research, the ingredient detection of the gas in the well is firstly investigated to determine the main ingredients of the gas in the well, and then the explosion effect under different conditions in the well is simulated with FLACS code to investigate the effect of many factors on the explosion characteristic of the combustible gas within the restricted space in the well.

2 Detection of the Gas in the Well and its Ingredients

Take one college of Beijing as an example, and the gas ingredient detection is done in some representative regions including the complex area, dining area, education area and apartment area. The high accurate and advanced gas detector of iBridMX4 and iBridMX6 produced in America is selected in this study. The detection results in different areas are shown in Tab.1.

Corresponding author. Fax: +86-013098507822
E-mail address: shuqi_10@163.com

item / area	H_2S/ppm	CH_4/LEL	CO_2/VOL
Complex area	0.2~34.3	0~22	0~1.81
Dining area	1.6~52.5	0~24	0~2.71
Education area	1.2~1.8	0~27	0~2.50
Apartment area	2.9~61.6	0~32	0~2.31

Table 1: Detection result of the gas concentration

3 The Explosion Test of the Combustible Gas in the Spherical Explosion Vessel

3.1 Experiment Equipment

The experiment equipment consists of the spherical explosion vessel of 20L in volume and the controlling and data collection system. Through the inspection window, the flame effect in the explosion chamber can be inspected. The piezoelectric pressure sensors are fixed at the internal face of the vessel, which can measure the dynamic pressure produced in the explosion.

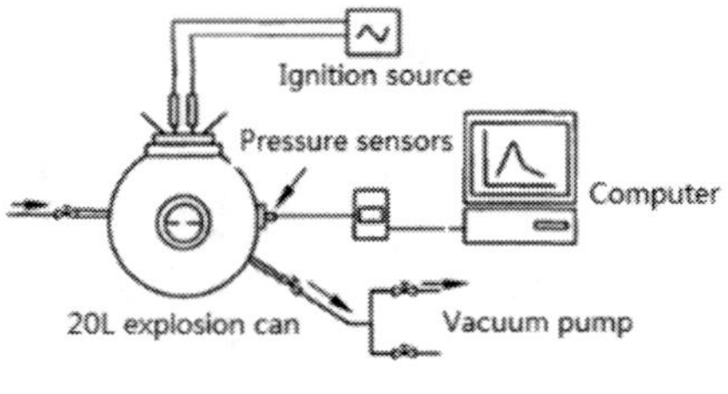

Figure 1: 20L spherical experimental device

3.2 Experimental Procedure

(1) Gas preparation

According to the Dalton's laws of partial pressure, the absolute pressure of the mixed gas equals to the sum of that of each one. The pressure balance method is adopted to make the mixed gas in this experiment, and the calculation formulae are shown as (1) and (2).

$$P = \sum P_i \quad (1)$$

$$P_i = P\varnothing_i \quad (2)$$

where, P is the absolute pressure of the mixed gas, Pa; P_i is the absolute pressure of the ith gas of the mixed gas, Pa; $\varnothing_i$ is the volume percentage of the ith gas of the mixed gas, %.

From (1) and (2), the partial pressure of each gas can be calculated, and according to the result, each gas is injected and stirred for a few minutes, finally the mixed gas is done.

(2) Detection of the ignition source

After the explosion experiment, the ignition source needs to change and it is necessary to detect the performance of the ignition source. Detecting the pressure produced by the ignition source itself after pumping the explosion vessel into vacuum, and it value range is about 0.11 ± 0.01MPa.

(3) Gas tightness detection of the apparatus

After detecting the performance of the ignition source, the gas tightness detection of the experiment apparatus must be done in order to ensure the good gas tightness of the apparatus. It can prevent the bad results because of the unsuitable gas tightness detection of the apparatus.

(4) Gas replacement

Firstly turn on the inflow and drain taps, and then incessantly pour the pure air into the reaction system to clean the explosion chamber. Then blowing the explosion vessel with the mixed gas through the inflow tap.

(5) Pouring the mixed gas

After pumping the explosion vessel into vacuum, pouring the mixed gas into the explosion chamber through the inflow tap. During it, monitoring the pressure of the explosion chamber till it reaches the required value, and turning off the inflow tap.

(6) Ignition

After turning off the inflow tap, we should fully and evenly make the gas mixed for 50ms. After that, ignition is made. The pressure sensors start to record the pressure and the test result is done by the data-collecting system.

3.3 Experiment Results and Analysis

Since the H2s content is very small in the test， it is negligible at the experimental dose, so H_2S is not used in this experiment. The gases used in this experiment are mainly CH_4 and CO_2.

(1) Lower explosion limit test of CH_4 and CO_2 mixed gas

Four groups of experiments are done in the spherical explosion vessel of 20L in volume in the lab to test the lower explosion limit test of CH_4 and CO_2 mixed gas when the content of CO_2 are 3%, 6%, 9% and 12%. The setting ratio of the mixed gas adopted in the experiment is slightly different from that in reality. Recording every gas explosion case and obtaining the following results. When 3%, 6%, 9% and 12% of CO_2 is pouring into the blast vessel, the lower explosion limit test of CH_4 are 4.3%, 4.8%, 5.2% and 5.9%.

(2) The effect of the methane content on the explosion characteristic of the combustible gas

The explosion experiment is done on CH_4 and air mixed gas with content of 5.5%, 7.5%, 9.5%, 11.5% and 13.5% respectively at the normal pressure and temperature. After the explosion, the explosion pressure increased rapidly, and then the pressure of explosion began to decrease because of the loss of heat in practice. The pressure and its climbing speed both firstly increase and then decrease.

The maximum explosion pressure P_{max} and the maximum climbing speed of the pressure $(dp/dt)_{max}$ are obtained shown in Fig.2 and Fig.3. It can be found that the explosion is weak when the content of methane in the mixed gas reaches the lower explosion limit. The waveform produced by the explosion shows that there is a long precursor pressure wave region during the explosion, and the pressure in the positive and negative pressure region is little.

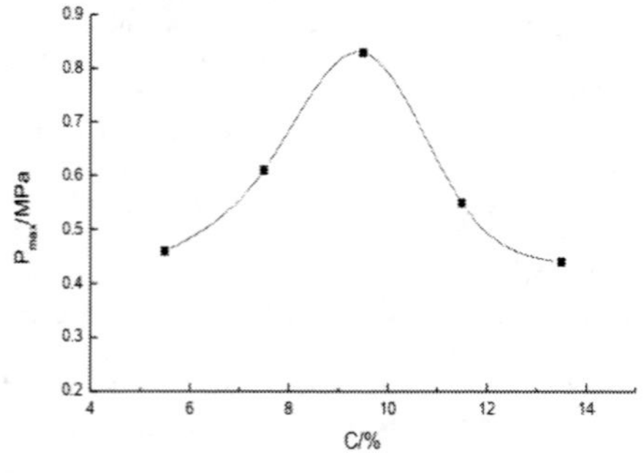

Figure 2: The maximum explosive pressure of CH_4

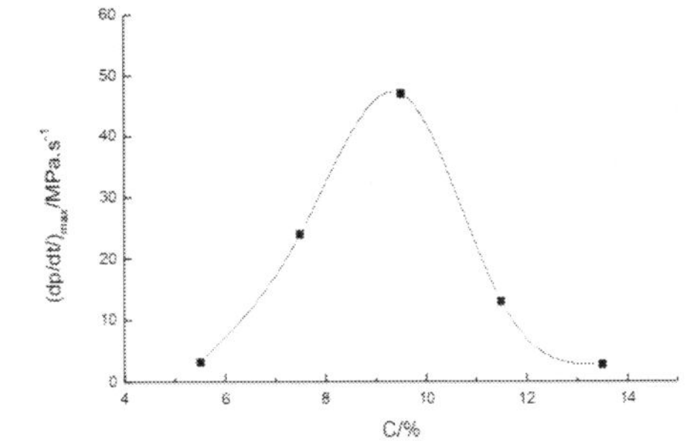

Figure 3: The maximum climbing speed of the pressure of CH_4

When the content of CH_4 is 5.5% and 13.5%, P_{max} is 0.46MPa and 0.44MPa, and $(dp/dt)_{max}$ is 3.2MPa·s^{-1} and 2.8 MPa·s^{-1} respectively. The explosion is relatively strong when the content of methane in the mixed gas is in the middle of the lower explosion limit.

(3) The effect of CO_2 on the explosion characteristic

From the detection result of the gas in the well, it can be known that the maximum content inert gas ingredient in the well is CO_2. Therefore CO_2 is selected to do the explosion test to study the effect of CO_2 on the maximum explosion pressure P_{max} and the maximum climbing speed of the pressure $(dp/dt)_{max}$. Five mixed gas with different CO_2 content of 0%, 5%, 10%, 15% and 20%, and different CH_4 content of 5%, 6%, 8%, 10% and 12% are used respectively in the experiment. The experiment results are shown in Tab.1 and Tab.2. The following conclusions can be obtained.

(1) Fixing the content of CH_4 in the mixed gas and increasing the content of CO_2 from 0 in every group of experiments, we can find that the maximum explosion pressure of the mixed gas of methane and air constantly increases. It can be known that CO_2 can make the explosion pressure of the mixed gas of methane and air decrease, which has certain explosion suppression. Meanwhile it can be seen from Tab.1 that the explosion pressure is evidently affected by CO_2 and the decreasing extent of the pressure is rather large when the content of CH_4 is 10% and 12% respectively.

(2) Fixing the content of CO_2 in the mixed gas and increasing the content of CH_4 from 5% in every group of experiments, we can find that the maximum explosion pressure of the mixed gas of methane and air all increase firstly and then decrease. The change trend is in accordance with that without CO_2. Similarly, accordingly the maximum explosion pressure of CH_4 and air mixed gas decrease with increasing the content of CO_2.

(3) When there is CO_2 pouring into the system, the maximum explosion pressure of the mixed gas of methane and air reaches 0.75MPa when the content of CH_4 increases to 10%. At this moment, pouring CO_2 into the system from its content being 5%, it can be found that the maximum explosion pressure of the mixed gas of methane and air occurs when the content of methane is 8%. The decrease from 10% to 8% indicates that CO_2 has the function of explosion suppression.

(4) Fixing the content of CH_4 in the mixed gas and increasing the CO_2 content from 0 in every group of experiments, we can find that the maximum climbing speed of the explosion pressure of the mixed gas of methane and air constantly decreases. It can be known that CO_2 can make the maximum climbing speed of explosion pressure of the mixed gas and air decrease, which has certain explosion suppression. Meanwhile it can be seen from Fig.2 that the explosion pressure is evidently affected by CO_2 and the decreasing extent of the climbing speed of explosion pressure is rather evident when the content of CH_4 is 10% and 12% respectively.

(5) Fixing the content of CO_2 in the mixed gas and increasing the content of CH_4 from 5% in every group of experiments, we can find that the maximum climbing speed of the explosion pressure of the mixed gas of methane and air all increase firstly and then decrease. The change trend is in accordance with that without CO_2. Similarly, accordingly the maximum climbing speed of the explosion pressure of CH_4 and air mixed gas decrease with increasing the content of CO_2.

(6) When there is CO_2 constantly pouring into the system from 0% to 10%, the maximum climbing speed of the explosion pressure of the mixed gas of methane and air reaches 39.7MPa/s when the content of CH_4 increases to 10%. At this moment, continuously pouring CO_2 into the system to 15% and 20% respectively, it can be found that the maximum climbing speed of the explosion pressure of the mixed gas of methane and air occurs when the content of methane is 8%. The decrease from 10% to 8% indicates that CO_2 has the function of explosion suppression.

P_{max}/ MPa	CO_2/ 0%	CO_2/ 5%	CO_2/ 10%	CO_2/ 15%	CO_2/ 20%
5%CH_4	0.31				
6%CH_4	0.50	0.32	0.26	0.23	
8%CH_4	0.61	0.42	0.32	0.28	0.27
10%CH_4	0.75	0.41	0.31		
12%CH_4	0.52	0.33			

Table 2: Experimental results of the maximum explosion pressure of the mixed gas of CH_4 and air

$\dot{p}_{max}$ / MPa/s	CO_2/ 0%	CO_2/ 5%	CO_2/ 10%	CO_2/ 15%	CO_2/ 20%
5%CH_4	2.3				
6%CH_4	10.9	8.7	6.5	5.4	
8%CH_4	25.3	18.1	16.1	10.4	3.7
10%CH_4	39.7	25.2	19.1		
12%CH_4	33.1	13.4			

Table 3: Experimental results of the maximum explosion pressure increase rate of the mixed gas of CH_4 and air

4 Numerical Simulation of the Explosion of the Combustible Gas in the Well

4.1 Numerical Simulation model

According to the design scheme of the underground drainage pipeline and the well system and the measurement in field, the partial three dimensional numerical model of the well is established with FLACS code. In the process of modeling, uniform grids are used in the whole calculation area.

The dimension of the calculation model is as Fig. 4. The diameters of the inspection well, cleaning well and drainage pipeline are 0.6m, 0.6m and 0.8m respectively. The diameter of the two air holes on the well lid is 0.03m, and the distances from their position to the edge of the well lid are both 0.1m. The depth of the well is 2.5m, and it extends to 2m left and 10m right from the bottom the well. The digestion tank with the cross-section of 4m×2m is set on the right side of the inspection well. On the top of the digestion tank is the cleaning well with 0.6m in diameter and 2m in depth. The cleaning well is in the middle of the digestion tank top. The dirt deposit well with 0.8n in diameter is set on the left side of 2m away. Because this calculation model is the partial one of the well, the link gradient among the wells is not considered. Moreover, in order to reduce the calculation time, the length of each region is not established according to the actual measurement distance.

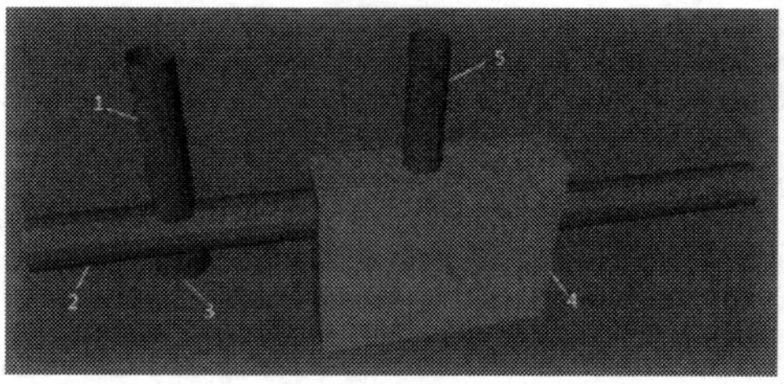

Figure 4: Model of the well system. 1—inspection well; 2—drainage pipeline; 3—dirt deposit well; 4—digestion tank; 5—cleaning well

4.2 Simulation Cases

The simulation of gas explosion by pouring CO_2 into the mixed gas of methane and air is done. The content of CH_4 is 9.5%, and CO_2 is selected as the depressor. Six groups of simulation cases of CH_4 explosion simulation experiment with 0%, 6%, 9%, 12%, 15% and 18% CO_2 are done. The simulation calculation shows that explosion all occurs in these six conditions.

4.3 Results

According to the recorded data by the pressure detector at the entrance of the well, the explosion pressure curves produced by these six groups of mixed gas with different content matching are shown in Fig.5. It can be seen that: (1) The content of the explosion outcome is only 26% when only methane explodes. With increasing CO_2, the content of the explosion outcome increases. With the content of CO_2 is 18%, the content of the explosion outcome increases to 46%. The proportion of CO_2 is much in the outcome produced by the explosion of the mixed gas of methane and air. So pouring CO_2 can effectively prevent the forward proceeding of the explosion and lead to the increase of the content of outcome. (2) CO_2 has some effect on the temperature during the methane explosion. With increasing CO_2, the temperature during the explosion decreases. The peak temperature decreases from 2260℃ to 1700℃ when the content of CO_2 increases from 0 to 18%.

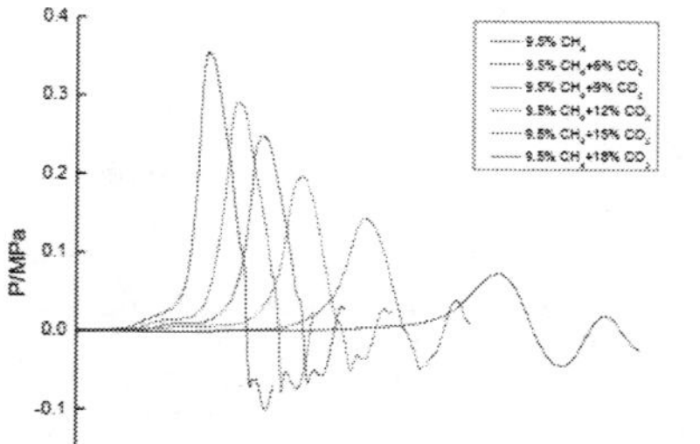

Figure 5: Effect of CO_2 on the maximum explosive pressure of the mixed gas of CH_4 and air

Then we study the effect of the initial pressure on the explosion pressure in the well with numerical simulation method. Assuming the other conditions are the same, the explosion process is studied only when the initial pressure increases from one standard atmospheric pressure to five standard atmospheric pressures. The calculation pressure and velocity of every measurement points are show in Fig.6. It can be seen that the maximum explosion pressure increases linearly with the increase in the initial pressure which indicates that the initial pressure has much effect on the maximum explosive pressure of the mixed gas.

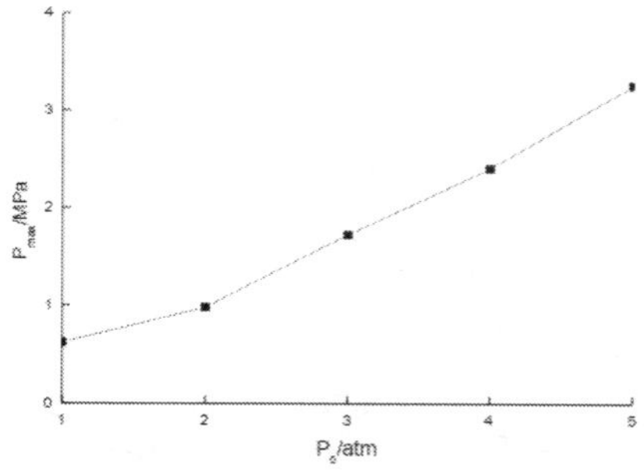

Figure 6: Effect of the initial pressure on the maximum explosive pressure

5 Conclusions

(1) From the detection of the gas in the city well, it can be known that the main harmful gas is CH_4, CO_2 and a small amount H_2S.

(2) CO_2 has some effect on the methane explosion limit. It has a certain ability to suppress explosion, and with the increase of CO_2 content in the mixed gas, the explosion pressure decreases.

(3) CO_2 has the suppression action on the methane explosion. The delay time of the methane explosion ignition is longer, the effect on the flame temperature is larger. The effect of the explosion initial pressure is in almost linear relationship with the maximum explosion pressure.

6 Acknowledgment

This study is supported by "National Natural Science Funds (51474151)".

References

[1] K. Yang, Safety Environ. 22 (2) (2015) 108-111

[2] B.S. Nie, Int. J. Therm. Sci. 2 (78) (2014) 132-144

[3] A.O. Russell, Process Saf Prog. 18 (3) (1999) 170-177

[4] J. Guo, C.J. Wang, Q. Li, et al, J. Loss Prevent. Proc. 5(40) (2016) 82-88

[5] R.M. Kasmani, G.E. Andrews, Process Saf. Environ. Protect. 2(91) (2013) 245-252

[6] E. Salzano, F.S. Marra, G. Russo, et al, J. Hazard. Mat. 95(3) (2002) 233-247.

[7] M. Maremonti, G. Russo, Eesto, et al, J. Loss Prevent. Proc. 12(3) (1999) 189-194.

[8] M. Fairweather, C.K. Hargrave, S.S. Brahim, et al, Combust. Flame 116 (4) (1999) 504-518

[9] Q.J. Ma, Q. Zhang, L. Pang, et al, J. Loss Prevent. Proc. 5(27) (2014) 65-73

11th Asia-Pacific Conference on Combustion,
The University of Sydney, NSW Australia
10th -14th December 2017.

Characteristics of Hydrogen produced by Methanol Reformation in Compact Whirling Orbital Plate Fluidized Bed Reactor

Prashant Nehe[1], Sudarshan Kumar[2], V. Mahendra Reddy[3]

[1]Department of Mechanical Engineering, Gokhale Education Society's, R. H. Sapat College of Engineering, Management Studies & Research, Nashik, Maharashtra 422005, India

[2]Department of Aerospace Engineering, Indian Institute of Technology Bombay, Mumbai 400076, India

[3]School of Energy Science and Engineering, Indian Institute of Technology Kharagpur, Kharagpur 721302, India

Abstract

The characteristics and conversion of methanol to hydrogen in a compact Fluidized Bed Reactor (FBR) is investigated in this study. This type of reactor is much more efficient in heat transfer, resulting in more uniform temperature distribution and larger surface volume ratio than packed bed type of reactor. The steam reforming reaction was carried out over $CuO/ZnO/Al_2O_3$ catalyst in a novel FBR system with whirling orbital plate type of design. The effect of various operating parameters including catalyst density and distribution on overall methanol conversion rate and hydrogen generation are studied. The orbital plate type of design assist in proper mixing of gases and longer residence time for the methanol reformation compared to mesh type distributor plate in conventional FBR system. The suitable minimum fluidization velocity for the catalyst particles in the reactor for different particle sizes and masses were studied to estimate critical inlet flow rate of methanol and water mixture. The experimental study indicates that the methanol conversion is 90 % and hydrogen production rate is 870 sccm for operating reaction temperature of 260 ^{0}C, water-methanol molar ratio of 1.1 and feed flow rate of 350 ml/hr.

1 Introduction

A significant attention is currently being paid towards the development of newer technologies for renewable energies and for increasing the efficiency of overall energy usage. Hydrogen can be produced from the reformation of hydrocarbons such as methanol, natural gas or oil by means of partial oxidation, auto-thermal and steam reforming [1]. Methanol is a highly suitable liquid fuel for onboard production of hydrogen, offering a high hydrogen-carbon ratio (H/C = 4), exists as liquid at room temperature, biodegradable, free from sulphur, absence of carbon-carbon bonds and its reactivity allowing reformation at relatively lower temperatures (200-350 ^{0}C) [2].

The design of fuel processors or reactors can be attributed to small volume, small weight, low cost, long lifetime, low byproduct (carbon monoxide), ease of manufacture, low flow resistance, low thermal resistance and short start-up time [3]. The conventional packed-bed reactor design consists of a packed bed filled with catalyst particles. The advantages of using conventional catalysts include moderate cost, better catalyst availability and reproducibility and greater understanding of catalyst performance. For portable power supply requirements, many researchers have recently focused on minimizing the size and weight of the reactor [4,5].

Karim et al. [6] used a commercial $CuO/ZnO/Al_2O_3$ catalyst for experiments on a packed bed reactor to investigate the kinetics of methanol-steam reforming reaction on a catalyst bed. Micro-reactors prepared by packing catalyst particles into micro channels have the advantage of easy catalyst loading. However, high flow resistance resulting from the packed catalyst particles may limit the design and performance of micro-reactors [6]. Since methanol-steam reforming (MSR) is an endothermic reaction, external heat must be supplied to the reactor to activate the methanol conversion. Methanol conversion can be improved by enhancing the heat transfer characteristics within the catalyst bed [7-9]. Many engineering applications find importance of convective heat transfer in domains partially filled with porous material [10-12]. Steam reforming of methanol is accomplished by a chemical reaction in the catalyst packed bed and the required heat is supplied from an external source. The gaseous mixture of methanol vapor and steam is normally heated from a packed catalyst bed wall which results in lower temperatures within the bed and non-uniform temperature distribution within the reactor. They suffer other disadvantages like cold spot, diffusion resistance, catalyst sintering and replenishment [13].

Fluidized Bed Reactor (FBR) is a type of reactor device that can be used to carry out a variety of multiphase chemical reactions. In this type of reactor, a fluid (gas or liquid) is passed through a granular solid material (usually a catalyst possibly shaped as tiny spheres) at high enough velocities to suspend the solid and cause it to behave as though it were a fluid. It is much more efficient in heat transfer and exhibits much more uniform temperature distribution in the reactor. Higher reaction rate in fluidized bed reactor than fixed bed reactor was shown by Liu et al. [14], representing transport controlled reaction in the fixed bed. Ye et al. [15] and Chen et al. [16] studied steam methane reforming for hydrogen production in a fluidized-bed reactor for various operating parameters. A review of literature shows a very few theoretical and experimental studies on hydrogen production by methanol steam reforming in compact fluidized bed reactor.

The characteristics and conversion of methanol to hydrogen in compact fluidized bed reactor is investigated in this study. The steam reforming reaction was carryout out over

Corresponding author. Tel: +91-09371667753
E-mail address: prashantnehe@gmail.com (Prashant Nehe)

$CuO/ZnO/Al_2O_3$ catalyst in a novel FBR system with whirling orbital plate type of design. The minimum fluidization velocity required for the catalyst particles in the reactor for various particle sizes and masses was estimated experimentally. The effect of various operating parameters including catalyst density and distribution on overall methanol conversion rate and hydrogen generation are studied. An externally heated reactor has issues of adverse lower temperature in the core of the reactor and significant heat loss to the environment thus impacting its performance [17]. Therefore, internal heating arrangement for reformer section of the whirling orbital plate FBR has been selected in this study for improved performance in terms of higher methanol conversion and reduced heat losses to surroundings.

2 Experimental system

Methanol reforming refers to the chemical reaction between methanol (CH_3OH) and water vapour (H_2O) for the production of hydrogen (H_2) gas. This process is typically carried out in the presence of metal oxide catalyst at temperatures ranging from 195 ^{0}C to 260 ^{0} [18]. The chemical reactions taking place during reforming process are outlined below:

$$CH_3OH + H_2O \xleftrightarrow{k_R} CO_2 + 3H_2 \qquad \Delta H^0_{298} = +49.4 \text{ KJ mol}^{-1} \quad (1)$$

$$CH_3OH \xleftrightarrow{k_D} CO + 2H_2 \qquad \Delta H^0_{298} = +90 \text{ KJ mol}^{-1} \quad (2)$$

$$CO + H_2O \xleftrightarrow{k_{WGS}} CO_2 + H_2 \qquad \Delta H^0_{298} = -41.1 \text{ KJ mol}^{-1} \quad (3)$$

k_R, k_D and k_{WGS} are rate constants for reforming, decomposition and water-gas shift reactions respectively while ΔH is the enthalpy of reaction. Reaction (1) is the main reforming reaction which gives the stoichiometric conversion of methanol to hydrogen. It can be regarded as the effect of overall reactions of methanol decomposition (2) and water-gas shift reaction (3). The relative kinetics of these reactions determines the amount of intermediate product carbon monoxide (CO) formed in the process. Proper control of methanol to water feed ratios and reactor operating temperature and pressure conditions can be used to minimize the amount of CO formed in the process.

The minimum fluidization velocity at which the entire bed of catalyst particles is suspended and behaves as if its weight is counterbalanced by the force of buoyancy is obtained before carrying out reforming reaction in the reactor. For the same bed height, the pressure drop increases with flow velocity for fixed bed. With a further increase in gas velocity, the packed bed suddenly unlocks (at the onset of minimum fluidization condition), resulting in a decrease in pressure drop. With gas velocities beyond minimum fluidization, the bed expands and gas bubbles starts to rise, resulting in non-homogeneity in the bed. With the increase in gas flow, the pressure drop should remain unchanged, signifying the start point if fluidization bed.

2.1 Catalyst preparation

Cu/Zn based metal oxide catalysts are generally used for steam reforming of methanol. The role of Zn is to improve the Cu dispersion and reducibility which are important attributes of highly active catalysts for these reactions. Typically, Al_2O_3 is added to these catalysts to hinder the deactivation due to sintering. The proportion of the $CuO/ZnO/Al_2O_3$ catalyst for use in the reactor is 40:40:20 [17]. The procedure of preparing catalyst powder in laboratory is as follows. 40 gm of copper is completely dissolved in 60 mL of 70-72 % concentrated nitric acid (HNO_3). The prepared solution is diluted and then 25% ammonia solution is slowly added till the complete precipitation occurs. After allowing the setting time of half an hour, the precipitate is filtered and separated from the solution. This precipitate is dried in an oven at 110 ^{0}C for about 4-5 h. The dried product is heated in the furnace at 750-800 ^{0}C for 6-7 h. Finally, the cooling of the dried product produces 50 gm of CuO. This procedure is used for producing oxides of Zn and Al. 40 gm of Zn is dissolved in 60 mL of 35% concentrated hydrochloric acid which provides 48 g of ZnO. Finally, 20 gm of Al is dissolved in concentrated hydrochloric acid which provides 36 gm of Al_2O_3. The three types of catalyst powders obtained are mixed with a binder to form catalyst particles of diameter 1-2 mm.

2.2 Reactor configuration and experimental set up

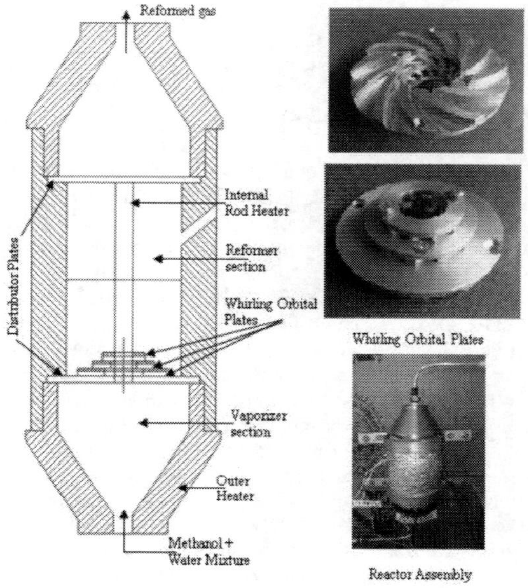

Figure 1: Whirling Orbital Plate Fluidized Bed Reactor

The Fluidized Bed Reactor (Fig. 1) has an i.d. of 6 cm and a total length of 26.6 cm including 10 cm of catalyst bed. The lower part of the reactor serves as a preheating/vaporizer zone for methanol-water mixture while the middle part of the reactor contains the $CuO/ZnO/Al_2O_3$ catalyst bed and serves as the reaction zone. External Mica heater and internal cartridge type rod heater is used for supplying heat for endothermic steam reforming reaction. To monitor independently the temperatures in the vaporization and reaction zones, Type K Inconel sheathed 1/16" o.d. thermocouples (Omega) (error ±0.5 %) were inserted in the reactor. A fine mesh perforation/distributor plate, fixed approximately 8 cm above the reactor inlet, supports the catalyst bed. Another removable distributor plate is located near the exit of the reactor. The conventional system of perforation plate type design which has laminar flow inside the reactor is replaced by orbital plate system for increasing efficiency of mixing by creating whirling flow in the fluidized bed reactor.

The experimental system (Fig. 2) was developed to study the characteristics of hydrogen produced by methanol reformation in compact whirling orbital plate fluidized bed reactor. The operating parameters controlling the methanol conversion in the reactor include reaction temperature, feed flow rate and the catalyst mass. Reforming of methanol-steam mixture was carried out in the temperature range of 200-240 ^0C using heater and a precise temperature controller (accuracy of 0.1 % of displayed value). A peristaltic pump with flow range of 100-700 ml/hr (accuracy of ±1 %) supplied the mixture of methanol and water to the inlet of the reactor. The vaporized mixture reacts inside the fluidized $CuO/ZnO/Al_2O_3$ catalyst bed section where the methanol-steam reforming reaction takes place and gets converted to a hydrogen rich stream of reformed gas. The liquid components in the produced stream were separated using a cold trap. The flow rate of reformed gas was measured by a flow meter (error ±1 %). The composition of the dry reformed gas was analyzed by a gas chromatograph (Shimadzu GC-2025). The temperature of the reactor unit was varied to measure the effect of temperature change on the composition of the reformed gas. The flow rate of the inlet feed mixture was varied to investigate optimum reforming condition. The steam to methanol molar ratio was fixed at 1.1 to obtain higher methanol conversion rate [19].

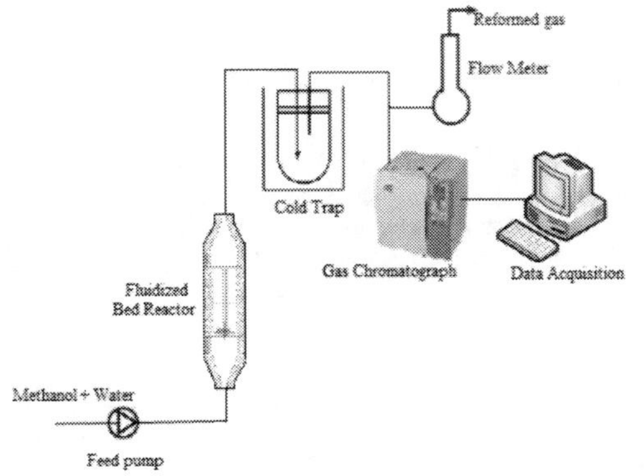

Figure 2: Schematic of experimental set up

3 Results and discussion

Initial experiments were performed with air as a fluidizing gas to find out minimum fluidization velocity in the reactor for different catalyst particle sizes of 1-2 mm and catalyst masses of 60 gm and 80 gm. Figure 3 shows the variation of pressure drop for different gas velocities at the outlet of the reactor.

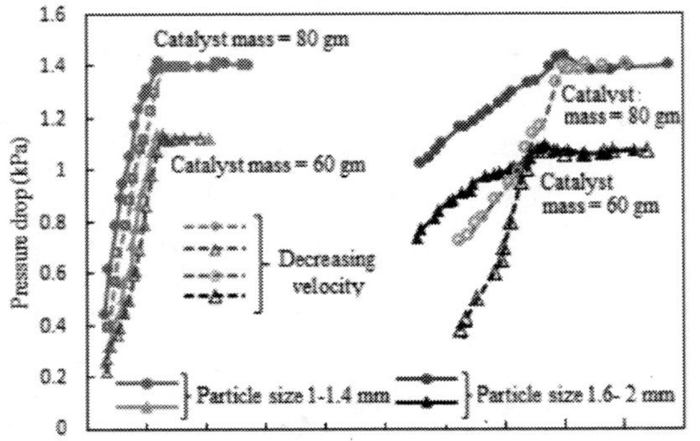

Figure 3: Pressure drop variation for different catalyst particle sizes and masses

The resistance to flow of gas increases with increase in gas velocity initially and after certain value of gas velocity the pressure drop stabilizes. For decreasing gas velocity, pressure drop decreases and the bed resistance for this case is lower than that for increasing gas velocity. Alternatively, the minimum fluidization velocity slightly increases with increase of catalyst mass. The minimum fluidization velocity for larger catalyst size particles is much higher than that of smaller size particles. Furthermore, the minimum fluidization velocity increases with increase of catalyst size but catalyst mass has no effect on the minimum fluidization velocity in the same catalyst particle size. This minimum fluidization velocity is further used to estimate critical inlet flow rate of methanol and water mixture in the reactor. For the critical inlet feed flow rate of 450 ml/hr, the catalyst with particle size 1-1.4 mm is in the fluidized state and catalyst with particle size 1.6-2 mm is nearly in static state.

There is possibility of deactivation of the $CuO/ZnO/Al_2O_3$ based catalyst in the fluidized bed reactor, particularly the degradation caused by sintering [20]. Therefore, in order to test the repeatability of the results, the repeatability test for the catalyst developed is carried out at 260 ^0C and inlet flow rate of 300 ml/hr for methanol-water mixture and molar ratio of 1.1. Figure 4 shows the methanol conversion (%) for this test. The extent of reduction rate of methanol conversion which is also the degradation rate of catalyst activity, declines steadily with respect to test time for both catalyst particle sizes (1 mm and 2 mm). After half an hour of reforming, the conversion was reduced from 82 to 76 %. All other tests in this work were done within half hour and each time the reactor is loaded with catalyst. It is also expected that the rate of reduction of the catalyst activity increases with increasing reactor temperature. Therefore, results shown in this work were obtained from the experiments performed at temperatures not higher than 260 ^{0} C.

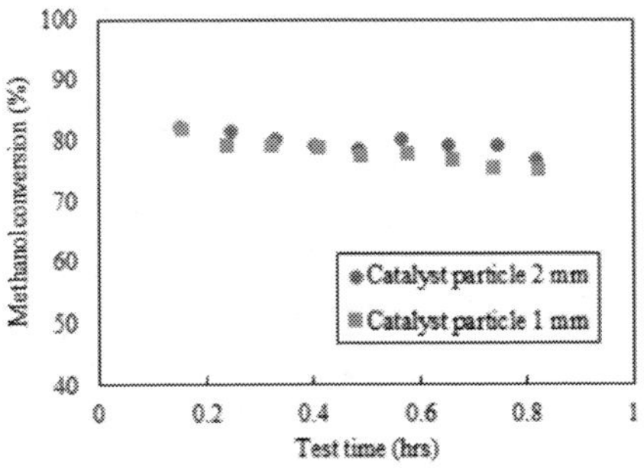

Figure 4: Results of repeatability test for the reactor, testing at 260 ^0C, inlet flow rate of 300 ml/hr, molar ratio of methanol-water 1.1, catalyst loading = 80 gm

To determine the conditions for maximum hydrogen output from the developed fluidized bed reactor, the effect of inlet feed flow rate on methanol conversion was investigated. Figure 5 shows that the methanol conversion decreases with a decrease in the reactor temperature. Therefore, a higher temperature is required to achieve higher methanol conversion rates. Nevertheless, the forward reaction of (3) is exothermic and this reaction is more favorable at lower temperatures. Therefore, considering the effect of reaction temperature, there is a trade-off relationship between methanol conversion and H_2 production. The H_2 production rate increases with an increase in the inlet feed flow rate. If the feed flow rate is lowered to values less than 350 ml/hr methanol conversion is expected to increase due to higher residence time of the methanol-steam mixture, however, hydrogen production rate is expected to decrease further as evident from the trend of the curves shown in Fig. 5. Hence for a reforming temperature 260 ^0C, the feed flow rate may be limited to 450-500 ml/hr range to facilitate

~82 % methanol conversion and improved hydrogen concentration in the reformed gas. For these operating conditions, the hydrogen production rate was around 700 sccm.

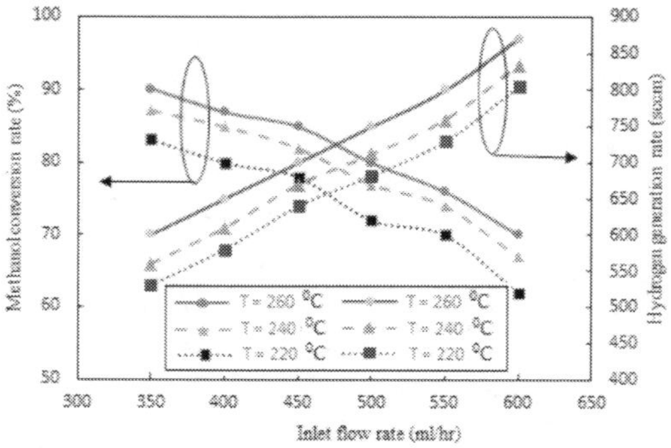

Figure 5: Methanol conversion and hydrogen generation rate

The comparison between whirling orbital plate FBR and packed bed reformer with the same operating conditions was carried out. Figure 6 shows the methanol conversion rate and hydrogen generation rate in the orbital plate reactor at fluidized bed state is much higher than in packed bed state. For reforming temperature of 260 ^{0}C, the methanol conversion rate in fluidized bed reactor is 25 % higher than the packed bed reactor. Compared to fluidized bed reactor, packed bed reactors suffer the drawback of uneven temperature distribution and sintering; therefore, it is essential that the particle size of the catalyst should not be too small to obstruct the passage. Furthermore, the catalyst has larger contact area in fluidized bed reactor due to much smaller catalyst particle size and more even temperature distribution in the reactor resulting in better methanol conversion rate.

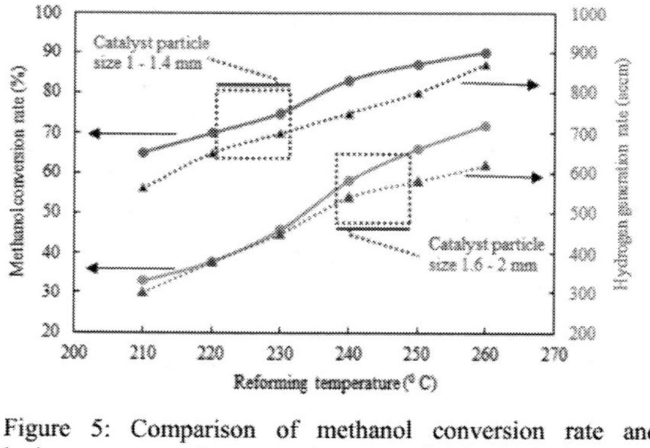

Figure 5: Comparison of methanol conversion rate and hydrogen generation rate between fluidized bed reactor (catalyst particle size 1-1.4 mm) and packed bed reactor (catalyst particle size 1.6–2 mm) (Feed flow rate of 450 ml/hr, water-methanol molar ratio 1.1 and catalyst mass 60 gm)

4 Conclusions

The experimental investigation to study characteristics and conversion of methanol to hydrogen in a compact whirling orbital plate Fluidized Bed Reactor (FBR) is carried out in this study. The steam reforming reaction was carried out over $CuO/ZnO/Al_2O_3$ catalyst prepared in the laboratory. The following conclusions are obtained.

- The minimum fluidization velocity of the catalyst in the reactor for different particle sizes and mass was studied. This can be further used to obtain the critical inlet flow rate of methanol-water mixture.
- For reforming temperature of 260 ^{0}C, the methanol conversion rate in fluidized bed reactor is 25 % higher than the packed bed reactor.
- The orbital plate type of design assist in proper mixing of gases and longer residence time for the methanol reformation compared to mesh type distributor plate in conventional FBR system.
- The experimental study indicates that the methanol conversion is 90 % and hydrogen production rate is 870 sccm for operating reaction temperature of 260 ^{0}C, water-methanol molar ratio of 1.1 and feed flow rate of 350 ml/hr.

5 Acknowledgment

The partial financial support from the Board of College and University Development (BCUD), Savitribai Phule Pune University, Pune, India is gratefully acknowledged.

References

[1] Y.T. Cheng, C.K. Yang, H.T. Su, J. Energy 25 (1995) 158–180.

[2] J.S. Han, I.S. Kim, K.S. Choi, J. Power Sources 86 (2000) 223–227.

[3] J. B. Hansen. Handbook of fuel cells - fundamentals, technology and applications, John Wiley & Sons, Ltd., New York, 2003, p. 141.

[4] W. Ehrfeld, V. Hessel, H. Lowe. Microreactors, Wiley-VCH, Weinheim, Germany, 2000.

[5] K. Shah, R. Besser, J. Power Sources 166 (2007) 177-193.

[6] A. Karim, J. Bravo, A. Datye, Appl Catal A -Gen 282 (2005) 101-109.

[7] P. A. Erickson, C. Liao, ASME J Heat Transf 129 (2007) 995-1003.

[8] C. Liao, P. A. Erickson, Int J Hydrogen Energy 33 (2008) 1652-1260.

[9] R. Chein, L. Chen, Y. Chen, J. N. Chung JN, Int J Hydrogen Energy 13 (2009) 5398-5408.

[10] S. Chikh, A. Boumedien, K. Bouhadef, G. Lauriat, Int J Heat Mass Transf 38 (1995) 1543-1551.

[11] K. Vafai , S. J. Kim, Int J Heat Mass Transf 11 (1990) 254-256.

[12] M. Kaviany, Principle of heat transfer in porous media, Springer, New York, 1995.

[13] J. S. Suh, M. T. Lee, R. Greif, C. P. Grigoropoulos, Int J Hydrogen Energy 34 (2009) 314-322.

[14] S. W. Lui, Y. D. Li, J Chem Ind Eng 58 (2007) 102-107.

[15] G. Y. Ye, D. L. Xie, W. Y. Qiao, J. R. Grace, L. C. Jim, Int J Hydrogen Energy 34 (11) (2009) 4755-4762.

[16] Z. X. Chen, J. R. Grace, L. C. Jim, L. Anwu, Int J Hydrogen Energy 32 (13) (2007) 2359-2366.

[17] P. Nehe, V. M. Reddy, S. Kumar, Int J Hydrogen Energy 40 (2015) 5715-5725.

[18] J. C. Amphlett, M. J. Evans, R. A. Jones, R. F. Mann, R. D. Weir, Can J Chem Eng 59 (1981) 720-727.

[19] P. Nehe, S. Kumar, Int J Hydrogen Energy 38 (30) (2013) 13216-13229.

[20] D.G. Loffler, S.D. McDermott, C.N. Renn, J. Power Sources 114 (2003)15–20.

11th Asia-Pacific Conference on Combustion,
The University of Sydney, NSW Australia
10th -14th December 2017.

Research on the effects of diluents and dilution ratios on laminar burning velocities and flame instabilities of hydrogen flame

Xin Lu, Erjiang Hu*, Xiaotian Li, Zuohua Huang

State Key Laboratory of Multiphase Flow in Power Engineering, Xi'an Jiaotong University,
Xi'an, People's Republic of China

Abstract

Laminar burning velocities and flame instabilities of H_2/O_2/diluents mixture were studied at initial pressure of 0.3 MPa, initial temperature of 373 K and equivalence ratio of 1.0. The He diluted mixtures have the largest laminar burning velocities and the N_2 diluted mixtures have the smallest, while He diluted mixtures have the largest mole fraction of H and OH radical and N_2 diluted mixtures have the smallest. And laminar burning velocities fall when the dilution ratio rises, due to the mole fractions of H and OH radical fall with increasing dilution ratio. The cellular structure appeared on both N_2 and Ar diluted mixture, while only wrinkles grew on He diluted mixtures. When dilution ratio rose, the appearance of cellular structure deferred. Thus, He diluted mixtures are more stable than N_2 and Ar diluted mixtures, and higher dilution ratios restrain the flame instabilities.

1 Introduction

H_2/O_2 reaction system is perhaps the most important block in the hierarchy of hydrocarbon oxidation. Since H_2 and its intermediate oxidation species are also intermediate species in the oxidation of all hydrocarbon and oxygenated fuels, the H_2/O_2 mechanism forms an essential subset of any hydrocarbon or oxygenate oxidation mechanism.[1] Therefore, research on H_2 laminar burning velocities is the basic of research on ICE alternative fuels, such as gases(methane[2], propane[3]and so on) mixed with H_2 and syngas[4].

Laminar burning velocity provides supplements of basic diffusion and chemical reaction information of combustible mixtures, and verification information for chemical reaction kinetics.[5] With the initial conditions of premixed mixture and laminar flame characteristics, the combustion of fuel at different conditions could be predicted, guiding the researches on how fuel burns in ICE. Laminar flame builds the basics of turbulence combustion researches, and many turbulence combustion researches were proposed based on the further laminar flame research.

Since flame instabilities grow on H_2 flame surface at elevated pressures, it's not easy to measure the laminar burning velocities. Thus, different diluents and diluent concentrations are applied to study flame characters of H_2/O_2 flame in this work. The objective of this work is to study the effects of diluents and the dilution ratios on (1) laminar burning velocities (2) and the flame instabilities. The experiments were conducted at T=373 K, p=0.3 MPa and ϕ=1.0.

Corresponding author. Fax:86-29-82668789
E-mail address:hujiang@mail.xjtu.edu.cn

2 Experimental and simulation methods

2.1 Experimental approach

The objective of the experiment was to record the flame images, and thus to obtain the laminar burning velocities and instability parameters such as the flame thickness and the critical Peclet numbers. The experiments were conducted on a constant chamber with quartz glasses on both ends. The chamber was heated to the designed temperature by the heat tape wrapping outside the chamber. Mixtures of hydrogen, oxygen, nitrogen, helium and argon were prepared with partial pressure method, and purities of all gases are above 99.99%. After intensive mixing of 5 minutes, the distribution of components and the temperature reached relative uniformity. A pair of electrodes were placed in the central to ignite the well-mixed gases. A Phantom V611 high-speed camera was utilized to record the flame images. The accuracy of experiment apparatus has already been validate, and the details of apparatus could been accessible in[2, 6].

Since different diluents and dilution concentration were utilized in this work. The dilution ratio is defined as:

$$D = V_{\text{diluent}}/V_{O_2} \tag{1}$$

Where V_{diluent} and V_{O2} is the volume of diluent and O_2. For example, in H_2/air mixture, D is 3.762.

2.2 Laminar burning velocities

The streched burned burning velocity could be obtained from:

$$S_{\text{n}} = dr_{\text{f}}/dt \tag{2}$$

Where r_f is the flame radius obtained from flame pictures, and t is the time after the spark appeared. The unstreched burned burning velocity S_b was extrapolated using nonlinear equation of Frankel et.al[7]and Chen et.al[8]:

$$S_{\text{n}} = S_{\text{b}} - 2 \cdot S_{\text{b}} L_{\text{b}}/r_{\text{f}} \tag{3}$$

Where L_b is the Markstein length. Thus the laminar burning velocities S_u could be obtained by dividing S_b by the density ratio caculated from GasEq[9]:

$$S_{\text{u}} = S_{\text{b}} \cdot \rho_{\text{b}}/\rho_{\text{u}} \tag{4}$$

Where ρ_u and ρ_b represents the density of unburned and burned mixtures.

The uncertainty methods of laminar burning velocites were proposed by Moffat[10] and Yu[11]. The radiation, systematic and random uncertainties were all taken into accout, and the details could be accesible in their work.

2.3 Instabilities of flame front

The instabilities of flame front could be caused by viscous diffusion, thermal diffusion, mass diffusion and hydrodynamic factors. Gravity also affects the flame instabilities, and usually occurs near the inflammability limits. The effects of gravity is neglected in this paper since the experiments were conducted avoid from inflammability limits.

The expansion coefficient (σ) can be expressed as:

$$\sigma = \rho_b/\rho_u \quad (5)$$

And the flame thickness (l_0) can be defined as:

$$l_0 = \lambda/(\rho_u \cdot S_u \cdot C_p) \quad (6)$$

Where λ is the thermal conductivities of unburned mixtures, and C_p is the heat capacities of unburned mixtures at constant pressure. And the mass and thermal diffusion can be described by the Lewis number *(Le)*, which is expressed as:

$$Le = \lambda/(\rho_u \cdot Dif \cdot C_p) = \alpha/Dif \quad (7)$$

Where *Dif* is the binary diffusion coefficient of insufficient reactant and the diluent, and α is the rate of thermal diffusion. When ϕ=1, both hydrogen and oxygen are completely consumed and there is no insufficient reactant. Thus *Dif* is obtained by averaging the the binary diffusion coefficients of H_2-diluent and O_2-diluent.

$$Dif = (Dif_{H2\text{-}diluent} + Dif_{O2\text{-}diluent})/2 \quad (8)$$

The critical Peclet number (Pe_{cr}) is a non-dimensional number represents the time when the flame started to be instable, and is defined as:

$$Pe_{cr} = r_{cr}/l_0 \quad (9)$$

Where r_{cr} is the critical radius of flame front about to be cellular.

2.4 Numerical methods

Laminar burning velocity predictions were calculated using Premix code[12]and CHEMKIN II[13]. The Burke model[14] was used for simulation in this study. Soret diffusion and Multi-component transport were taken into consideration. The final gird numbers were above 900 to ensure the accuracy of simulation. And the thermal conductivities, densities and heat capacities at constant pressure were all calculated using GasEq[9].

3 Laminar burning velocities

3.1 Experimental and calculated results

The laminar burning velocities of mixtures with different diluents and dilution ratios were shown in Fig. 1. When D=4 and the diluents were Ar and N_2, the laminar burning velocities couldn't be extrapolation since cellular structure developed on the flame surface. Laminar burning velocities fall when the dilution ratio increase. And He diluted mixtures have the fastest laminar burning velocities while N_2 diluted mixtures have the slowest. The experiment results match the calculated results well thus the mole fractions of radicals were calculated using Burke model.

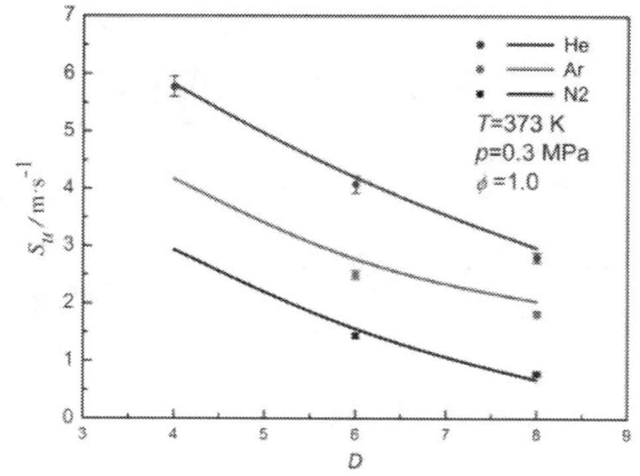

Figure 1: Laminar burning velocities of mixtures with different diluent and dilution ratio. The symbols represent the experiment results and the solid lines represent the calculated results with Burke model. The error bars represent the uncertainty of experimental laminar burning velocities.

3.2 Flame structure analysis

The mole fraction of H and OH radicals can imply overall reaction activity, which strongly affects the laminar burning velocities. The larger the mole fractions are, the stronger the reaction activity is. The mole fractions of mixtures of D=6 and different diluents are shown in Fig. 2. The He diluted mixtures have the highest mole fractions of both H and OH radicals while the N_2 diluted mixtures have the smallest mole fractions. Thus the He diluted mixtures have the largest laminar burning velocities and the N_2 diluted have the slowest.

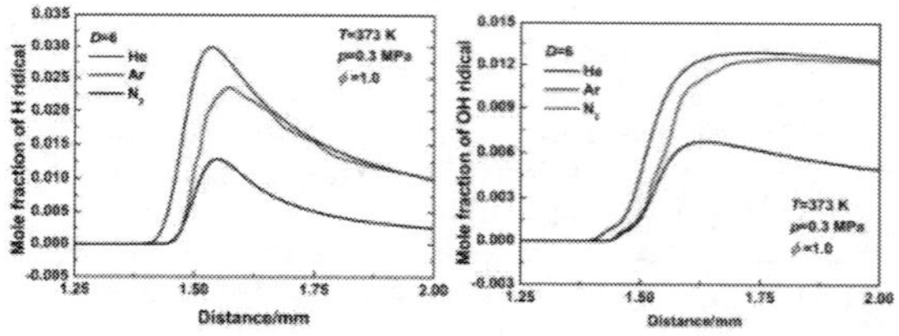

(a) (b)

Figure 2: Mole fractions of H(a) and OH(b) radicals of mixtures of D=6 and different diluents.

The mole fractions of H and OH radicals of He diluted mixtures and different dilution ratios are shown in Fig. 3. The mixtures of D=4 have the largest mole fractions of both H and OH radicals, while mixtures of D=8 have the smallest. Thus the laminar burning velocities falls with increasing dilution ratios.

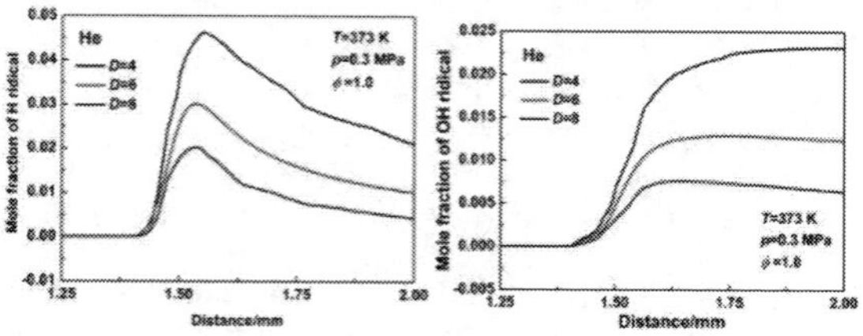

(a) (b)

Figure 3: Mole fractions of H(a) and OH(b) radicals of He diluted mixtures and different dilution ratios

4 Flame instabilities

The Lewis number represents the thermal diffusion and the mass diffusion of the mixtures. Figure 4 gives the Lewis number. Since *Dif* is independent of *D*, α of He diluted mixtures increases when *D* rises. On the contrary, α of N_2 and Ar diluted mixtures decreases when *D* rises. Lewis number of He diluted mixtures rises when dilution ratio increases, while those of N_2 and Ar diluted mixtures fall. Thus the thermal diffusion instability of He diluted mixtures trend to be stable. And that of N_2 and Ar diluted mixtures trend to be unstable. However, Le of all mixtures are larger than 1, indicating that the thermal diffusion of all mixtures trend to be stable.

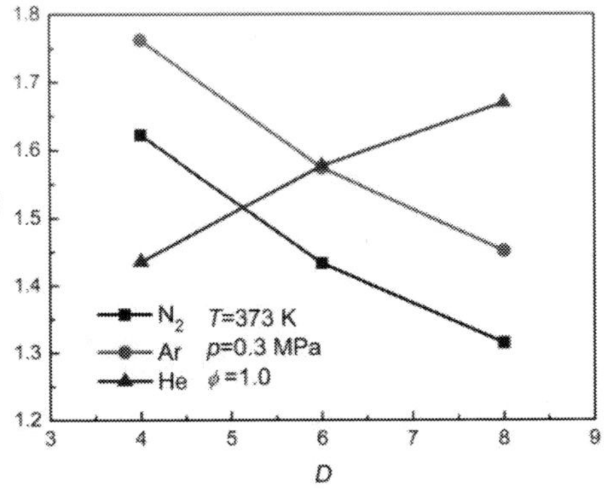

Figure 4: Lewis number of mixtures of different diluents and dilution ratios.

The expansion coefficient and flame thickness are shown in Fig. 5. The hydrodynamic instabilities are enhanced when σ grows and l_0 decreases. When *D* rises, σ falls and l_0 increases, thus the hydrodynamic instabilities were restrained. Specially, σ of the He and Ar diluted mixtures reached superposition due to the similar molecular structure of He and Ar. And N_2 diluted mixtures have the lowest σ. In addition, it starts form Ar and goes to He in ascending order of the flame thickness. In conclusion, hydrodynamic instabilities of Ar diluted mixture is stronger than those of N_2 diluted mixtures. The He diluted mixtures have large σ and l_0, mainly because of the large S_u and small density. However, the flame instabilities are relative to the coupling of thermal diffusion instabilities and hydrodynamic instabilities, which could be represented by Pe_{cr}.

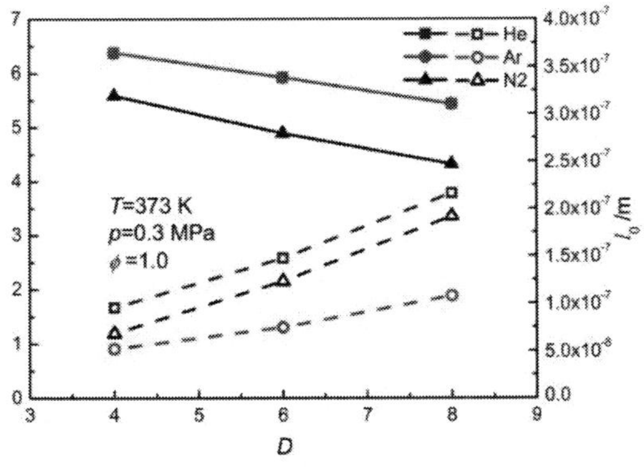

Figure 5: The expansions ratios and flame thickness of mixtures of different diluents and dilution ratios. The solid symbols represent the expansion ratios and the open symbols represent the flame thickness.

The Peclet number is obtained with the critical radius and represents the appearance of cellular structure. The images of flame propagation and the Peclet number are shown in Fig.6 and Fig.7. In Fig.6, the cellular structure appeared on both N_2 and Ar diluted mixtures. And the Pe_{cr} of Ar diluted flame is larger than that of N_2 diluted flame, which implied the cellular structure appeared earlier on N_2 diluted flame than on Ar diluted flame. However, the He diluted flame remained stable but with some cracks on the surface.

Figure 7 shows the flame images of N_2 diluted flame of different dilution ratios. Cellular structure developed on all mixtures. The Pe_{cr} increases from *D*=4 to *D*=8, indicating that the cellular structure developed earlier at smaller dilution ratio.

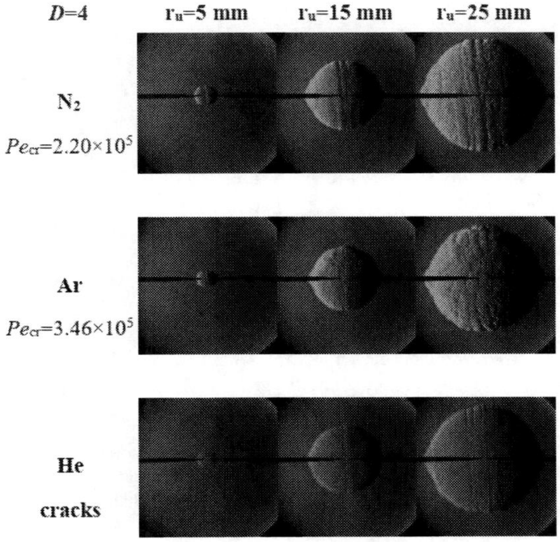

Figure 6 The critical Peclet numbers and flame images of mixtures of different diluents and *D*=4.

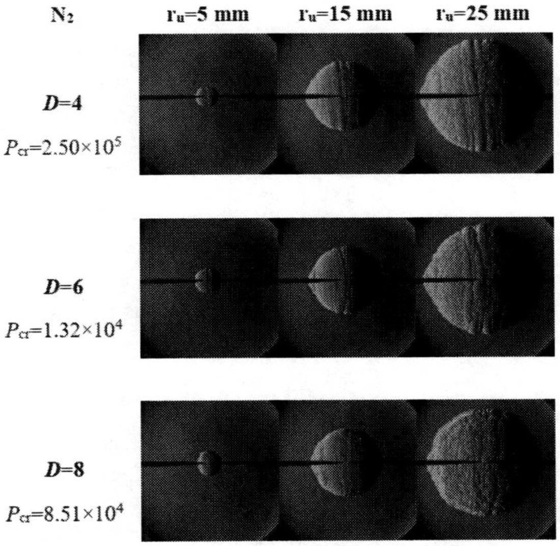

Figure 7: The critical Peclet numbers and flame images of N_2 diluted mixtures of different dilution ratios.

5 Conclusions

Laminar burning velocities of hydrogen/oxygen/diluent mixtures are measured with a constant volume chamber and a high-speed camera. The calculated results match the experimental results well. The laminar burning velocity falls with increasing dilution ratios. And He diluted mixtures have the fastest laminar burning velocities while N_2 diluted mixtures have the slowest.

The flame instabilities were also discussed with Lewis number, expansion ratio, flame thickness and critical Peclet number. Lewis numbers of all mixtures were larger than 1, which implied that the thermal instabilities trend to be stable. The cellular structure appeared on both N_2 and Ar diluted flames, while only few cracks appeared on He diluted mixtures. And the cellular structure grew earlier on N_2 diluted mixture than on Ar diluted mixtures. When dilution ratio rose, the expansion ratio decreased, the flame thickness increased, and the critical Peclet number increased thus the appearance of cellular structure delayed.

6 Acknowledgment

This study is supported by the National Natural Science Foundation of China (91441118, and 91441203). Authors appreciate the funding support from the Fundamental Research Funds for the Central Universities.

References

[1] Westbrook CK, Dryer FL. Chemical kinetic modeling of hydrocarbon combustion. Progress in Energy & Combustion Science. 1984;10:1-57.

[2] Hu E, Huang Z, He J, Jin C, Zheng J. Experimental and numerical study on laminar burning characteristics of premixed methane–hydrogen–air flames. International Journal of Hydrogen Energy. 2009;34:4876-88.

[3] Tang C, He J, Huang Z, Jin C, Wang J, Wang X, et al. Measurements of laminar burning velocities and Markstein lengths of propane–hydrogen–air mixtures at elevated pressures and temperatures. International Journal of Hydrogen Energy. 2008;33:7274-85.

[4] Xie Y, Wang J, Nan X, Yu S, Huang Z. Comparative study on the effect of CO_2 and H_2O dilution on laminar burning characteristics of CO/H_2/air mixtures. International Journal of Hydrogen Energy. 2014;39:3450–8.

[5] Aung KT, Hassan MI, Faeth GM. Flame stretch interactions of laminar premixed hydrogen/air flames at normal temperature and pressure . Combustion & Flame. 1997;109:1-24.

[6] Tang C, Huang Z, Jin C, He J, Wang J, Wang X, et al. Laminar burning velocities and combustion characteristics of propane–hydrogen–air premixed flames. International Journal of Hydrogen Energy. 2008;33:4906-14.

[7] Frankel ML, Sivashinsky GI. On Quenching of Curved Flames. Combustion Science & Technology. 1984;40:257-68.

[8] Chen Z. On the extraction of laminar flame speed and Markstein length from outwardly propagating spherical flames. Combustion and Flame. 2011;158:291-300.

[9] Morley C. Gaseq,Version 0.76. http://wwwgaseqcouk/. 2005.

[10] Moffat RJ. Describing the uncertainties in experimental results. Experimental thermal and fluid science. 1988;1:3-17.

[11] Yu H, Han W, Santner J, Gou X, Sohn CH, Ju Y, et al. Radiation-induced uncertainty in laminar flame speed measured from propagating spherical flames. Combustion and Flame. 2014;161:2815-24.

[12] Kee RJ, Grcar JF, Smooke MD, Miller JA, Meeks E. PREMIX. A Fortran Program for Modeling Steady Laminar One-dimensional Premixed Flames. 1985;143:65.

[13] Rupley FM, Kee RJ, Miller JA. CHEMKIN-2: A FORTRAN Chemical Kinetics Package for the Analysis of Gas-phase Chemical Kinetics. 1989.

[14] Burke MP, Chaos M, Ju Y, Dryer FL, Klippenstein SJ. Comprehensive H_2/O_2 kinetic model for high - pressure combustion. International Journal of Chemical Kinetics. 2011;44:444-74.

11th Asia-Pacific Conference on Combustion,
The University of Sydney, NSW Australia
10th -14th December 2017.

Flame front evolution and self-acceleration of hydrogen spherically expending flames

Xiao Cai[1], Jinhua Wang[1,*], Haoran Zhao[1], Zuohua Huang[1,*]

[1] State Key Laboratory of Multiphase Flow in Power Engineering, Xi'an Jiaotong University,
NO.28, Xianning West Road, Xi'an 710049, China

Abstract

An experimental study on flame front evolution and self-acceleration of hydrogen spherically expending flames was conducted on a constant volume combustion chamber over a wide range of Lewis number at elevated pressures. Flame morphology, onset of flame acceleration and flame self-similar acceleration were analyzed. The results shows that the critical radius for onset of flame acceleration increases with the increase in effective Lewis number and the decrease in initial pressure. And the critical Peclet number increases nonlinearly with the increasing Markstein length. Thus, the onset of flame acceleration is controlled by the couple effects of hydrodynamic and diffusional-thermal instabilities. Furthermore, the self-similar acceleration exponent increases with the decrease in the effective Lewis number (or Markstein number) and almost keeps constant with initial pressure. It is concluded that the self-similar acceleration exponent is mainly affected by diffusional-thermal instability.

1 Introduction

The acceleration motion of spherically expending flames due to flame intrinsic instabilities is often occurred within a high pressure environment, especially for large-scale facilities [1]. The flame intrinsic instabilities include hydrodynamic, diffusional-thermal, and buoyancy instabilities, which are characterized by density jump through the flame front, non-equidiffusive of heat and mass, and density gradient between flame and environment, respectively [2]. The hydrodynamic and diffusional-thermal instabilities will induce cellular structure over the flame front. The flame cells over the flame front increases the total area of flame surface [3, 4]. Consequently, the global flame propagation speed increases with the total area of flame surface, and the local front propagation speed remaining largely at the constant laminar flame speed [5]. Since the flame cells continuously evolve and distribute over the entire flame front as the flame expends, the flame propagation speed increases continuously, then the flame will self-accelerate. In the initial stage of flame acceleration, the flame radius is small, the flame evolution is expected to be controlled mainly by small scale diffusional-thermal instability. As the flame expends, the hydrodynamic instability gradually controls the flame self-acceleration motion [6].

The self-acceleration of flame propagation has been investigated extensively. Gosintsev et al. [7] found the self-similarity in flame self-acceleration propagation, and the self-acceleration was quantified through $r_f = At^{\alpha}$, where r_f is flame radius, A is a constant value, and α is acceleration exponent. The results showed the acceleration exponent is 1.5, and the flame has undergone self-turbulization. Wu et al. [6] carefully extracted the acceleration exponents at elevated pressures. They found the acceleration ranges from 1.2 to 1.4, and the flame propagation is self-similar but does not reach self-turbulent. Recently, Kim et al. [1] experimentally investigated the self-similar propagation with a large-scale combustion chamber. The self-similar propagation was confirmed. A transition flame acceleration stage was found between onset flame acceleration and self-similar acceleration. They suggested that flame acceleration is affected by flame intrinsic instabilities. However, the separate effects of hydrodynamic and diffusional-thermal instability on flame self-similar propagation is still unknown.

In this work, the evolutions of hydrogen cellular expanding flames were obtained using a constant volume combustion chamber over a wide range of Lewis number under elevated pressures. The experimental conditions were carefully controlled to study the separate effects of hydrodynamic and diffusional-thermal instabilities. The flame morphology, onset of flame self-acceleration, and flame self-similar acceleration were analyzed.

2 Experimental and calculation Methods

2.1 Experimental setup

The experiments were conducted on a constant volume bomb. The experimental setup was described in detail in [8, 9]. In brief, this experimental setup is consisted of a cylinder combustion chamber, ignition system, heating system, high-speed schlieren camera, and data acquisition system. The mixtures are quantified with a pressure transducer based on the partial pressure law. The temperature is controlled using a K type thermocouple and a heating tape wrapped on the cylinder chamber. The ignition spark is provided by two tungsten electrodes in the center of the chamber. Two opposite quartz windows with diameters of 80 mm provide optical access. The flame evolution images are recorded with a high-speed Phantom V611 camera operating at 10,000 franes/s. The instantaneous flame radius is defined as $r_f = \sqrt{A/\pi}$, where A is the flame area of projection of schlieren flame. Then the flame propagation speed can be calculated as $S_b = \mathrm{d}r_f/\mathrm{d}t$.

In this experiment, initial temperature is set as 353 K, and initial pressures are 0.1, 0.2, 0.4 and 0.6 MPa. The adiabatic flame temperature is kept constant as 1800 K. The equivalence ratios of $H_2/O_2/N_2$ mixtures ranges from 0.4 to 2.5. The experimental conditions are shown in Table 1 in detail. At least three times of experiments were measured under each condition.

2.2 Calculation method

The flame basic parameters, such as density ratio σ, flame thickness l_f, and effective Lewis number Le_{eff}, characterize the

Corresponding author. Fax: +86 29 82668789
E-mail address: jinhuawang@mail.xjtu.edu.cn

flame intrinsic instabilities. The density ratio $\sigma = \rho_u/\rho_b$ and $l_f = (T_{ad} - T_u)/(\mathrm{d}T/\mathrm{d}x)_{max}$ are calculated using the PREMIX code [10] in the CHEMKIN II package [11] with Burke Mech [12]. Effective Lewis number Le_{eff}, is the weighted average of the Lewis number of the excess reactant, Le_E and of the deficient reactant, Le_D [13].

ϕ	P (MPa)	$X(H_2)$	$X(O_2)$	$X(N_2)$
0.4	0.1-0.6	0.18957	0.23697	0.57346
0.6	0.1-0.6	0.18868	0.15723	0.65409
0.8	0.1-0.6	0.18824	0.11765	0.69412
1.0	0.1-0.6	0.18797	0.09398	0.71805
1.5	0.1-0.6	0.27985	0.09328	0.62687
2.5	0.1-0.6	0.46211	0.09242	0.44547

Table 1: Experimental conditions of $H_2/O_2/N_2$ flames: Initial temperature is 353 K, adiabatic flame temperature is 1800 K.

3 Results and discussions

3.1 Flame morphology

In spherically expanding flames, the smooth flame surface will develop into cellular type due to the formation of flame intrinsic instabilities. The flame surface is increased, thus the global flame propagation speed is accelerated. Fig. 1 shows the effects of the initial pressure and Lewis number on the flame front evolution. In order to study the flame evolution affected by hydrodynamic instability, the $H_2/O_2/N_2$ mixtures with near-unity Lewis number is presented in Fig. 1(a). It is observed that the flame front is smooth at large flame radius and 0.1 MPa. As the initial pressure increases, the cracks and branches develop gradually on the flame front, then the flame cells distribute over the entire flame front. Because the density ratio is kept constant, the hydrodynamic instability is controlled by flame thickness. The flame thickness decreases dramatically as the increase in the initial pressure. Thus, the hydrodynamic instability is promoted, and the cellular surface is appeared at smaller flame radius. To investigate the effects of diffusional-thermal instability on flame front evolution, the initial pressure and density ratio are kept constant. It is shown in Fig. 1(b) that flame front becomes cellular at large flame radius with Le_{eff}=1.12. For mixtures with Le_{eff}=1.79, the flame front is smooth, only a small amount of large cracks are observed on the flame surface due to the effect of spark ignition. While the cellular surface is developed at small flame radius with Le_{eff}=0.55. It reveals that the diffusional-thermal effects will suppress the hydrodynamic instability for mixtures with large Le, while destabilize the hydrodynamic instability for mixtures with small Le. The flame stretch also affects the formation of cellular surface. As the flame propagates outwardly, the flame stretch decreases and stretch stabilized effect decreases. Furthermore the local flame thickness decreases gradually as fame stretch decreases, thus the hydrodynamic instability is promoted.

3.2 Onset of flame self-acceleration

The evolution of flame front will affected by intrinsic instabilities and may lead to the cellular surface especially at high pressure, small Lewis number, and large flame radius. In the initiation stage of flame front evolution, the flame radius is small and local flame stretch is large. The strong flame stretch effect can stabilize the flame with smooth flame surface. As the flame expands, the flame stretch stabilized effect decreases and the local flame thickness decreases. Thus, the intrinsic instabilities play roles, the large cracks on the surface start to branch, and then flame cells uniformly distribute over the flame surface. Beyond a certain flame radius, the fame surface area increases, thus the flame starts to accelerate. Critical flame radius, r_{cr} for onset of flame self-acceleration is defined as the flame speed increases significantly. Fig. 2 shows the effects of the initial pressure and Lewis number on the critical radius for onset of self-acceleration. The r_{cr} decreases with the increase in pressure, especially under relatively lower pressures. While the r_{cr} increases approximately linearly with the Le. The cellular surface is more prone to occur early with small fame thickness and Le. Thus, the critical radius is always used to characterize the strength of flame intrinsic instabilities. However, critical radius can not be measured under some special conditions, see Fig. 2(b) with Le_{eff}=0.55 and P=0.6 MPa.

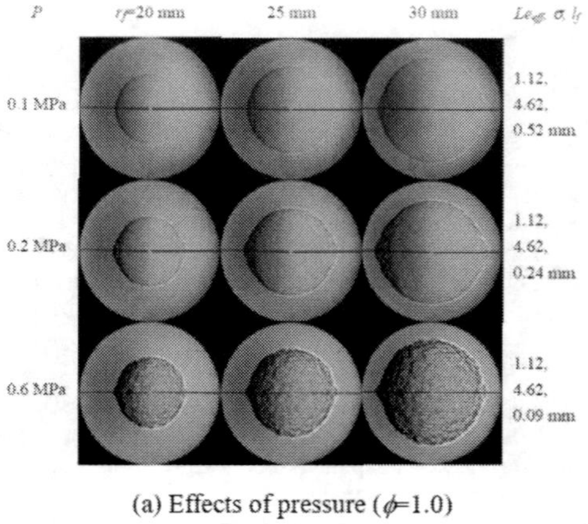

(a) Effects of pressure (ϕ=1.0)

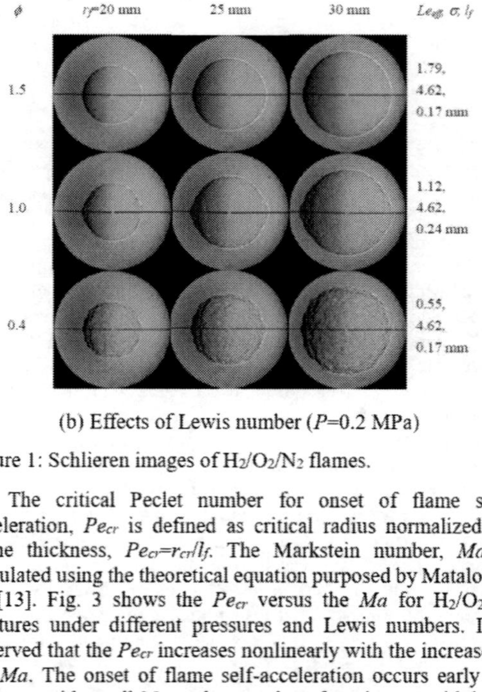

(b) Effects of Lewis number (P=0.2 MPa)

Figure 1: Schlieren images of $H_2/O_2/N_2$ flames.

The critical Peclet number for onset of flame self-acceleration, Pe_{cr} is defined as critical radius normalized by flame thickness, $Pe_{cr}=r_{cr}/l_f$. The Markstein number, Ma is calculated using the theoretical equation purposed by Matalon et al. [13]. Fig. 3 shows the Pe_{cr} versus the Ma for $H_2/O_2/N_2$ mixtures under different pressures and Lewis numbers. It is observed that the Pe_{cr} increases nonlinearly with the increase in the Ma. The onset of flame self-acceleration occurs early for mixtures with small Ma, and occurs later for mixtures with large Ma. It is concluded that the flame self-acceleration depends on the couple effects of hydrodynamic and diffusional-thermal instabilities. It also reveals that the flame front will become cellular as the flame radius large enough even under normal pressure.

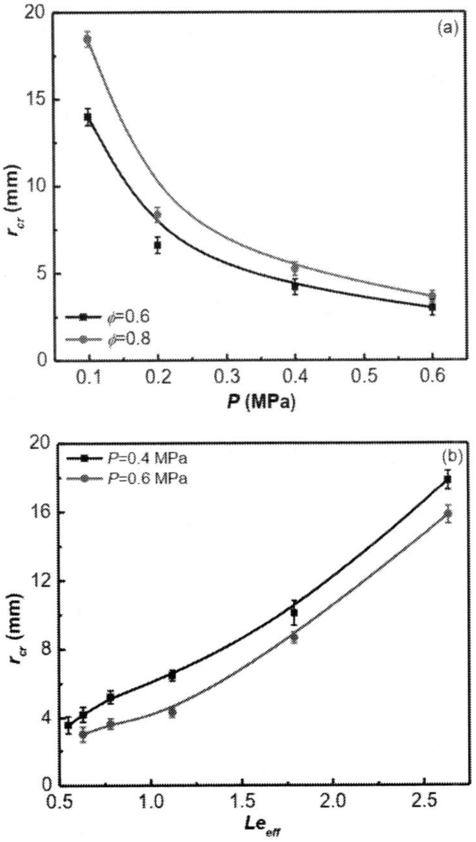

Figure 2: Critical flame radii for onset of flame self-acceleration.

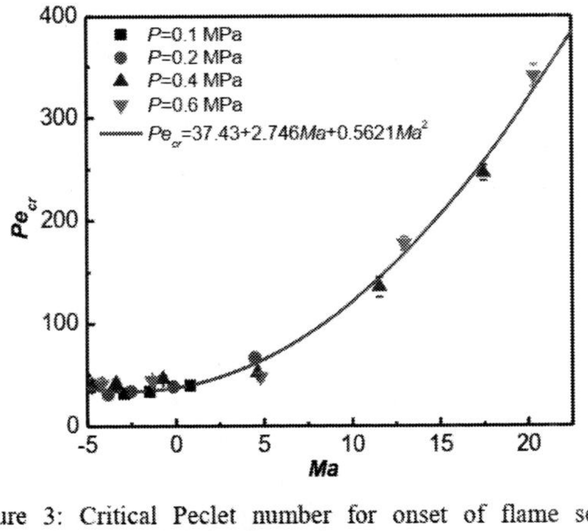

Figure 3: Critical Peclet number for onset of flame self-acceleration versus Markstein number.

3.3 Flame self-acceleration exponents

As the flame propagates outwardly, the flame intrinsic instabilities set in, and gradually dominate the self-acceleration process. The intensity of flame self-acceleration can be characterized by the flame acceleration exponent α. The acceleration exponent can be extracted from a power-law expression, $r_f = At^{\alpha}$, where A is a constant. However, the extraction result depends largely on the starting point. Thus, in this study, we extract the acceleration exponent using the power-law expression of the flame propagation speed and flame radius, $S_b = \alpha A^{1/\alpha} r_f^{\,d}$, where d=1-1/α is the fractal excess. The acceleration exponent changes with the flame radius before reach a constant value. And a transition period exists between the onset flame acceleration radius and self-similar acceleration regime. In order to extract accurate self-similar acceleration exponent, the flame radii beyond the transition period are processed.

Figure 4(a) shows the flame self-acceleration exponent of self-similar regime changes with effective Lewis numbers and initial pressures. The self-similar exponent decreases with the increase in the effective Le. While the self-similar exponent almost keeps constant as the initial pressure increases from 0.4 MPa to 0.6 MPa. This results imply that the diffusional-thermal instability have a major influence on flame self-similar acceleration exponent. Furthermore, the self-similar acceleration exponent decreases with the increase in Markstein number, as shown in Fig. 4(b). This result reconfirms that the diffusional-thermal instability has significant influence on the flame self-similar propagation. Moreover, the self-similar acceleration exponent is always larger than 1 and smaller than 1.5. Thus, the self-acceleration of cellular expanding flame does not reach the self-turbulizztion condition in present investigation.

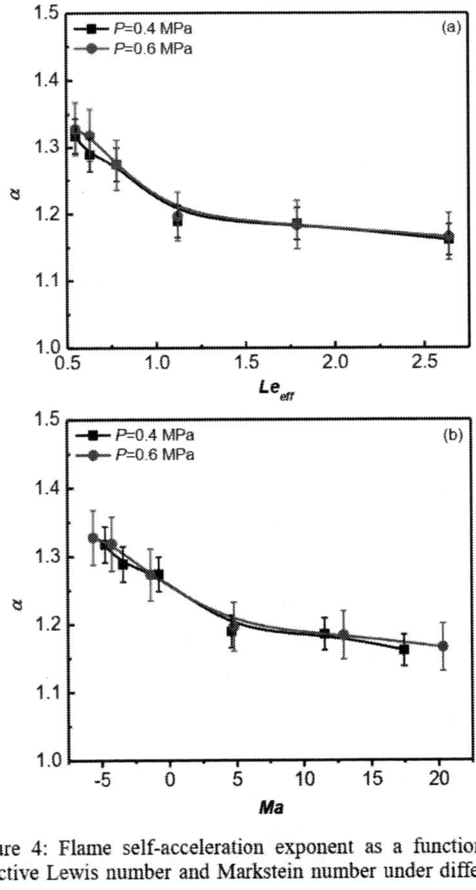

Figure 4: Flame self-acceleration exponent as a function of effective Lewis number and Markstein number under different pressures.

4 Conclusions

An experimental study on flame front evolution and self-acceleration of hydrogen spherically expanding flame was conducted in a constant volume combustion chamber. The main conclusions are summarized as follows:

(1) The critical radius for onset of flame self-acceleration increases with the increase in the effective Lewis number, and decreases with the increase in the initial pressure. The critical Peclet number increases nonlinearly with the Markstein number. Thus, the onset of flame self-acceleration is controlled by the couple effects of diffusional-thermal and hydrodynamic instabilities.

(2) The self-similar acceleration exponent decreases with the increase in effective Lewis number (or Markstein number) and changes little with the initial pressure. It is concluded that the self-similar acceleration intensity is mainly affected by diffusional-thermal instability for hydrogen spherically expanding flames.

5 Acknowledgment

This study is supported by National Natural Science Foundation of China (No. 51376004, 91441203). The support from The Fundamental Research Funds for the Central Universities is also appreciated.

References

[1] W.K. Kim, T. Mogi, K. Kuwana, R. Dobashi, Proc. Combust. Inst. 35 (2015) 2051-2058.

[2] C. Law, C. Sung, Prog. Energy Combust. Sci. 26 (2000) 459-505.

[3] J. Wang, Y. Xie, X. Cai, Y. Nie, C. Peng, Z. Huang, Combusti. Sci. Technol. 188 (2016) 1054-1072.

[4] Y. Xie, J. Wang, X. Cai, Z. Huang, Int. J. Hydrogen Energy 41 (2016)18250-18258.

[5] Y.X. Xin, C.S. Yoo, J.H. Chen, C.K. Law, Proc. Combust. Inst. 35 (2015)753-760.

[6] F. Wu, G. Jomaas, C.K. Law, Proc. Combust. Inst. 34 (2013) 937-945.

[7] Y.A. Gostintsev, A.G. Istratov, Y.V. Shulenin, Combust. Expl. Shock Waves 24 (1988) 563-569.

[8] X. Cai, J. Wang, W. Zhang, Y. Xie, M. Zhang, Z. Huang, Fuel 184 (2016) 466-473.

[9] Y. Xie, J. Wang, N. Xu, S. Yu, M. Zhang, Z. Huang, Energy Fuels 28 (2014)3391-3398.

[10] R.J. Kee, J.F. Grcar, M. Smooke, J. Miller, E. Meeks, Sandia National Laboratories Report 1985.

[11] R.J. Kee, F.M. Rupley, J.A. Miller, Sandia National Labs., Livermore, CA (USA), 1989.

[12] M.P. Burke, M. Chaos, Y. Ju, F.L. Dryer, Int. J. Chem. Kinet. 44 (2012) 444-474.

[13] J. Bechtold, M. Matalon, Combusti. Flame 127 (2001) 1906-1913.

11th Asia-Pacific Conference on Combustion,
The University of Sydney, NSW Australia
10th -14th December 2017.

Shock-Tube Measurement and Kinetic Study of NO_2 promoting-effect on ethane ignition

Fuquan Deng, Xue Jiang, Yousun Pan, Feiyu Yang, Yingjia Zhang*, Zuohua Huang[a]

Xianning West Road, Xi'an, Shanxi, 710049, China

Abstract

Ignition delay times have been measured in a high-pressure shock tube for $NO_2/C_2H_6/O_2/Ar$ mixtures at p of 1.2 – 20 atm- and T of 1050 – 1600 K. NO_2 is added to the stoichiometric baseline mixtures (neat ethane mixtures) as 30% and 100% of the ethane concentration by volume. The results indicate that NO_2 significantly promotes the ethane ignition 1) with higher NO_2 addition and 2) at higher pressure and lower temperature. Nevertheless, the NO_2 addition shows ignorable effect on the ethane ignition at lower pressure and higher temperature. A detail kinetic mode is developed against the data measured in this study and is used to perform sensitivity analysis and flux analysis. The kinetic analysis indicates that the interaction between NO_2 and ethane is very strong over the whole conditions investigated in this study. And the promoting-principle of NO_2 on the ethane ignition can be mainly attributed to two aspects: the first one is that reaction ($NO + HO_2 <=> OH + NO_2$) converts the stable radicals HO_2 into the chain initiating OH radicals; the second one is that the presence of NO_2 accelerates the consumption of these stable fuels (C_2H_6, CH_4, CH_2O, CH_3) in the system.

1 Introduction

Nitrogen oxides (NO_x) is well known as main gas emissions in engine combustion, and it can be recirculated into the fresh cycle particularly at internal combustion engines with exhaust gas recirculation. Our previous study [1] demonstrably indicates that even a very small amounts of nitrogen oxides can cause a dramatic impact on combustion characteristics of hydrocarbons. Although the significant importance of the interaction chemistry between nitrogen oxides and ethane, there are only a few studies regarding the C_2H_6/NO_x system. T. Faravelli* et al. [2] performed study on the combustion characteristics of ethane/NO mixture in a PFR reactor. They also proposed a low-temperatures detailed NO_x kinetic model to investigate the influence of NO on the ethane combustion. Sivaramakrishnan et al.[3] studied the mutual sensitization between NO and natural gas in a high pressure shock tube and jet stirred reactor (JSR). They assembled a detailed kinetic model to describe the influence of NO on the oxidation of CH_4/C_2H_6 mixtures. They reported that the addition of NO significantly promotes the oxidation of natural gas by the formation of abundant OH radical via reaction HO_2 + NO = OH +NO_2. J. Herzler and C. Naumann [4] investigated the NO_2 promoting effect on the $CH_4/C_2H_6/O_2/Ar$ mixtures ignition in a shock tube. They noted that even addition with a little NO_2 tremendously drive the ethane ignition, and the promoting effect is closely related with equivalence ratio. S. Gersen et al.[5] conducted study on the promoting effect of nitrogen oxide on the $CH_4/C_2H_6/O_2/N_2/Ar$ mixtures and $C_2H_6/O_2/N_2/Ar$ mixtures in a rapid compression machine. For the $CH_4/C_2H_6/O_2/N_2/Ar$ mixtures, a great promoting effect of NO_2 was observed, which dramatically increased with decreasing temperature. While, their results indicated that the NO_2 only impacted a modest promoting effect on the $C_2H_6/O_2/N_2/Ar$ mixture. As discussed above, the NO_2 promoting effect on the methane/ethane mixture is differ with that on the pure ethane mixture. So the first objective of this study is to obtain the ignition delay times of $NO_2/C_2H_6/O_2/Ar$ mixtures at p of 1.2 -20 atm. The second target is to evaluate the performances of four recently reported NO_x-models by validating against the data measured in this study. Finally, a detailed kinetic model will be developed to perform the kinetic analysis to interpret the kinetic principle of the interaction between NO_2 and C_2H_6. Table 1 lists the detailed components of the tested mixtures.

Table 1: The detail constitutes of these mixtures.

Mix	Mole blending ratio	C_2H_6	NO_2	O_2	Ar
$N_{0/1.0}$	0% NO_2/100% C_2H_6	1.0	0.0	3.5	95.5
$N_{30/10}$	0% NO_2/100% C_2H_6	1.0	0.3	3.5	95.2
$N_{100/0}$	0% NO_2/100% C_2H_6	1.0	1.0	3.5	94.5

2 Experimental

All the ignition delay times were measured in a double-diaphragm high pressure shock tube. Specifically, its driver section and driven section are 4.0 and 4.8 m, respectively. Four fast-response pressure transducers (PCB 113B26) are equally distributed (30.0 cm) at the last 1.3 m of the driven section, which were used to detect the shock velocity of the incident shock wave together with three time-interval counters (FLUKE PM 6690). The typical incident-shock attenuation ratio is within 4%/m. The driver and driven sections are evacuated to below 0.025 kPa before each run. And the typical leak rate is 1 Pa/min. A piezoelectric pressure transducer (PCB 113B03) is located at the end-wall to monitor the reflected shock wave. The pressure rise is typical within 4%/ms. The OH* signal during ignition is recorded by a photomultiplier (HAMAMASSU CR131) equipped with a 307 ± 10 nm narrowband filter. The starting time of the occurrence main ignition is defined as the time by extrapolating the maximum slope of OH* emission to based line. As shown in Fig.1, the measured ignition delay times in this study were defined as the time interval between the time of

Yingjia Zhang. Fax: +00-00-000-000
E-mail address: yjzhang_xjtu@xjtu.edu.cn

incident shock wave arrive at the end-wall, indicating a jump of the pressure recorded by the end-wall piezoelectric pressure transducer, and the onset of ignition mentioned above.

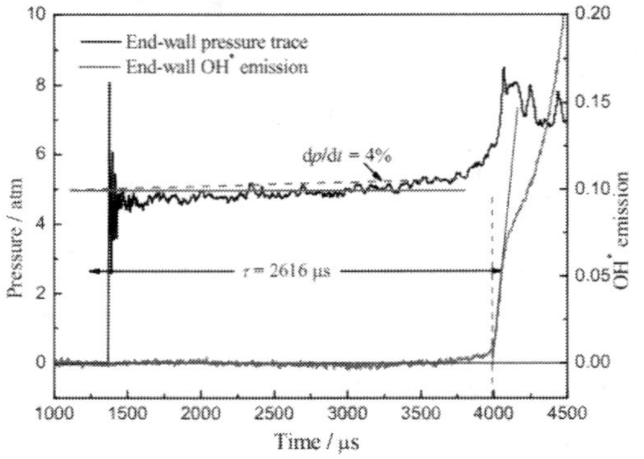

Figure 1: Typical end-wall profiles of pressure and OH* emission for N30/1.0 at 5.0 atm and 1044 K.

3 Modeling

The simulations were performed using CHEMKIN program with SENKIN code. In this study, the NUIG Aramco Mech 2.0 model [6] is chosen as the based model for ethane chemistry. Then, four recent literature NO_x kinetic models (Sivaramakrishnan et al. [3], J. Giménez-López et al. [7], Mathieu et al [8] and Faravelli et al. [2] model) are added into the NUIG Aramco Mech 2.0 to eliminate the influence of hydrocarbon mechanism. The four assembled models are called Aramco-S, Aramco-G, Aramco-O and Aramco-F, respectively. A model is also developed in this study, assembled by using the NUIG Aramco Mech 2.0 as the based model, the $NH_3/H_2/O_2/NO_x$ chemistry from Zhang et al. [9], and C_2/NO_x interaction reactions from J. Giménez-López et al [7]; and the $C_2H_5NO_2$ sub-mechanism taken from Zhang et al [10-11].

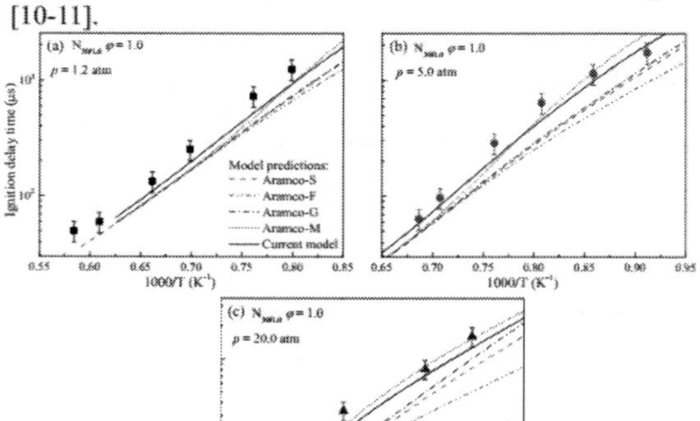

Figure 2: Comparisons between the experimental data and the model predictions for the $N_{30/1.0}$ mixtures at different pressures.

For the $N_{30/1.0}$ mixture, Fig. 2, the new model and Aramco-O give very close predictions and reproduce very well the measured data at any pressure and temperature studied. The Aramco-F model terribly under-predicts the measured ignition delay times especially at the low temperatures and high pressures condition. In more detail, the τ_{ing} simulated by Aramco-F almost has a three-fold difference with the measured data at 20 atm and 1050 K. The ignition delay times predicted by Aramco-S and Aramco-G are very close but obviously under-predict (around 56%) the measured data at 20 atm and T > around 1100 K. Fig. 3 shows the comparison between the five models and the ignition delay times of $N_{100/1.0}$ at p of 1.2- 20 atm. It can been seen that the new model, Aramco-S, Aramco-G and Aramco-O can well capture the data in consideration of measurement errors. While the simulation results of the new model, Aramco-S and Aramco-G are very similar and are smaller (typically within around 30%) than that predicted by the Aramco-O. However, the discrepancies between the simulations by Aramco-F and the measured data are very big, which increase with increasing pressure or decreasing temperature.

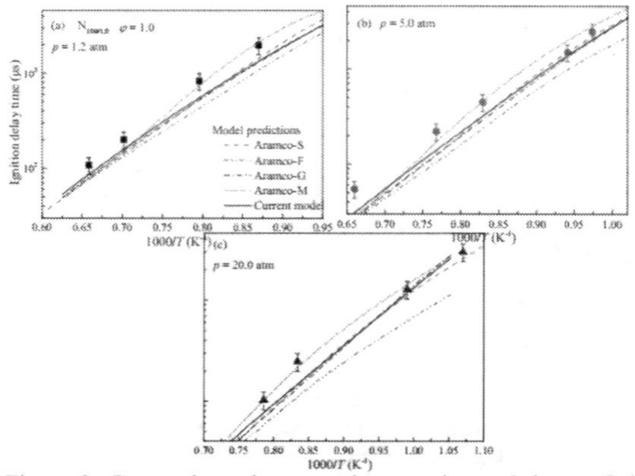

Figure 3: Comparisons between the experimental data and the model predictions for the $N_{100/1.0}$ mixtures at different pressures.

4 Results and Discussion

Fig. 4 depicts the NO_2 effect on ethane ignition with 30%-100% NO_2 addition at p of 1.2 - 20 atm. The NO_2 shows greater promoting effects at higher pressures and lower temperatures. On the whole, the NO_2 imposes very weak effect on the ethane ignition at 1.2 atm over the temperatures measured in this study. At 5 atm, the 30% NO_2 addition gives ignorable effect at T > 1160 K, while obviously reduces the ethane ignition delay times at T < 1160 K.

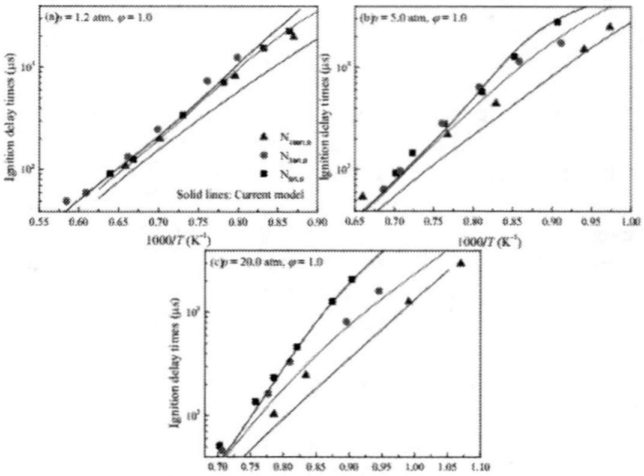

Figure 4: Effects of NO2 mixing ratio on the ethane ignition at p of 1.2 to 20.0 atm. (a) 1.2 atm; (b) 5.0 atm; (c) 20 atm.

With 100% NO_2 addition, the τ_{ing} further reduces with more NO_2 addition. The critical temperature indicating that the NO_2 starts to exert apparent effect on the ethane ignition increases from about 1160 K for $N_{30/1.0}$ to around 1250 K for $N_{100/1.0}$. At 20 atm, the promoting effect of NO_2 largely increases, see in Fig. 4. In more detail, the reductions in the ignition delay times are 54.1% and 75.9% at about 1100 K and 20 atm for $N_{30/1.0}$ and $N_{100/1.0}$, respectively.

5 Kinetic analysis

In an attempt to interpret the different promoting-effect of NO_2 on the ethane ignition at various conditions, the brute-force sensitivity analysis and flux analysis are conducted for the C_2H_6 mixtures with and without NO_2. The kinetic analysis is performed at the high pressure and lower temperature where the NO_2 greatly drives the ethane ignition, and at low pressure and higher temperature where the NO_2 only gives ignorable impact on the ethane ignition. The flux analysis are studied at the time of 10% consumption of ethane. As highlighted in Fig.5 (a), the chain-branching reaction R5 ($O_2 + H = O + OH$) with no surprise dominates the pure ethane ignition. And the importance of the HO_2 chemistry is identified by more than half of the most-sensitive reactions containing HO_2 radical. The reaction between ethylene and OH radical to form vinyl ranks the second big promoting reaction, and its product vinyl will react with molecular oxygen to form vinoxy radicals and oxygen atoms. The reactions between methyl and HO_2 radical are also very significant for the ethane ignition, with the chain termination reaction producing methane and molecular oxygen being the most-inhibiting reaction.

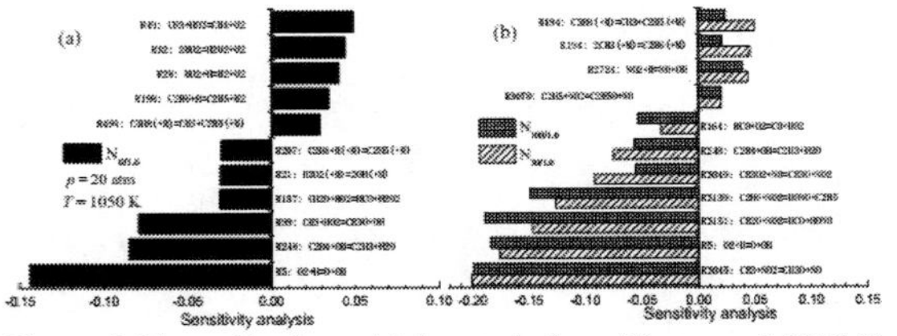

Figure 5: Normalized sensitivity analysis at 20 atm and 1050 K using current model.

For the mixtures with NO_2 addition (see Fig. 5 (b)), the differences between the top eleven most-sensitive reactions of the pure ethane and that of NO_2/C_2H_6 mixtures are very great. NO_2 addition impacts a great perturbation on the ethane ignition, indicating by four in five of the top five most-promoting reaction involving NO_x. As discussed above, the HO_2 chemistry plays an important role on the ethane ignition at the high pressure and lower temperature condition. While the presence of NO_2 significantly reduces the importance of HO_2 chemistry for a two-fold effect. Firstly, the nitrogen dioxide is more reactive than molecular oxygen, which causes that the fuel will more readily reacts NO_2 rather than O_2 resulting in much less HO_2 radical will be produced during the ignition of NO_2/C_2H_6 mixture. Secondly, almost all the HO_2 radical formed in the system will immediately react with NO to produce the chain-initiating radical OH and NO_2, which is a primary promoting-principle of NO_x on the hydrocarbon ignition. In addition, the reaction between methyl and nitrogen dioxide (R3045: $CH_3 + NO_2 <=> CH_3O + NO$) becomes the top promoting reaction. And its products CH_3O will decompose to form H radical, which feeds R5 and greatly drives the development of the radical pool of system. CH_3 radicals has very low reactivity and its production and consumption are very vital to control the reactivity of system. So R3045 will significantly promotes the ethane ignition for it not only drives the consumption of low reactive CH_3 radical but also promotes the development of racial pool through the formation of H radical *via* R174 $CH_3O (+ M) = CH_2O + H (+ M)$. The reactions between NO_2 and formaldehyde or ethane give comparable sensitivity coefficients with R5 because that they significantly drive the consumption of the stable fuels (ethane and formaldehyde) for the high reactive of NO_2 and its product HONO will quickly decompose to form the chain initiating radical OH and NO.

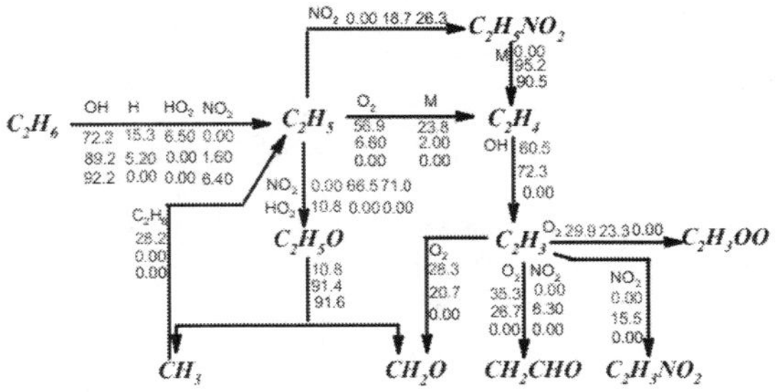

Figure 6 Flux analysis for C_2H_6/NO_2 mixtures at 20 atm and 1050 K. N0/1.0: red, N30/1.0: blue and N100/1.0: black.

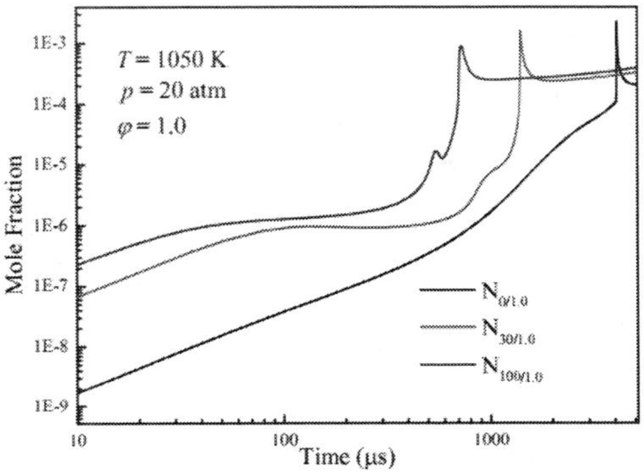

Figure 7 Effects of NO_2 on the mole fraction of the radical pool (H, O and OH) at 20 atm and 1050 K.

For the $N_{0/1.0}$ mixture at 1050 K and 20 atm (see in Fig. 6), the C_2H_6 was mainly consumed through H-atom abstraction with OH, H and HO_2 radicals to produce C_2H_5 radicals. Subsequently, the C_2H_5 radical largely undergoes three pathways: first one is up to 56% C_2H_5 reacts with O_2 to ethylene and HO_2 radical; the second one is about 24% C_2H_5 directly decompose into ethylene and H radical, leading to a great increase in the reactivity of system; the third one is around 11% C_2H_5 react with HO_2 radical to form C_2H_5O radical. With NO_2 addition, the presence of NO_2 induces noticeable shifts in the main flux of NO_2/C_2H_6 mixtures. Firstly, the branching ratio of H-atom abstraction between fuels and OH radical largely increases, while the branching ratio between fuels and HO_2 radical becomes negligible. This can be attributed to two main reasons: the first one is that almost all the HO_2 produced in the system readily reacts with NO via R2722 ($NO + HO_2 = NO_2 + OH$) to form the chain initiating OH radical; the second one is that the NO_2 reacts with fuels for its high reactivity, which drives the consumption of fuels and its products HONO directly decompose to form OH radical and NO. In addition, almost all the C_2H_5 reacts with NO_2 even with 30% NO_2 addition to produce C_2H_5O or $C_2H_5NO_2$ at the presence of NO_2, which indicates the interaction between nitrogen oxide and ethane is very strong.

The mole fractions of radical pool (H, O, OH, HO_2) are hence illustrated at 20.0 atm and 1050 K for three stoichiometric C_2H_6 mixtures, Fig. 7. It is observed that the NO_2 addition increases the mole fractions of the radical pool by around two orders of magnitude during the initiating ignition induced time. This indicates that the NO_2 addition dramatically promotes the ethane ignition at the high pressure and lower temperature condition.

At the Low pressure and high temperature condition, for $N_{0/1.0}$

(see Fig.8 (a)), R5 absolutely controls the ethane ignition and the other ten reactions only have very small sensitivity coefficients. For the mixtures with NO_2 addition (see in Fig.6 (b)), the results demonstrate that the differences between the eleven most-sensitive reactions for the neat ethane and mixtures doped with NO_2 are very small, which indicates that the presence of NO_2 has very limited impact on the control kinetic of mixtures at the low pressure and higher temperature. To further study the small effect, the mole fractions of radical pool are displayed in Fig. 9 at 1.2 atm and 1500 K for the three stoichiometric NO_2/C_2H_6 mixtures. A little increase in the radical pool can been seen at the presence of NO_2 even doped with very high NO_2 concentrations. So only ignorable promoting of NO_2 on the ethane ignition is observed at the low pressure and higher temperature.

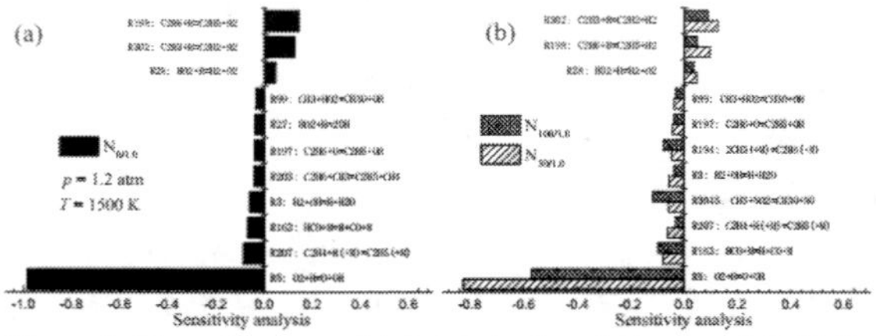

Figure 8: Normalized sensitivity analysis at 1.2 atm and 1500 K using the current model. (a) $N_{0/1.0}$; (b) $N_{30/1.0}$ and $N_{100/1.0}$.

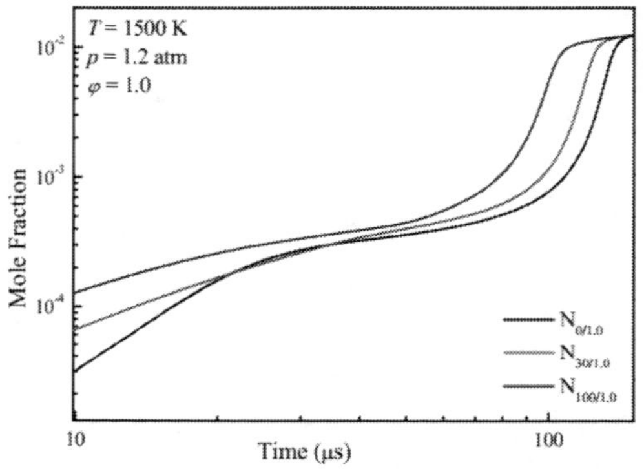

Figure 9: Effects of NO_2 on the mole fraction of the radical pool at 1.2 atm and 1500 K.

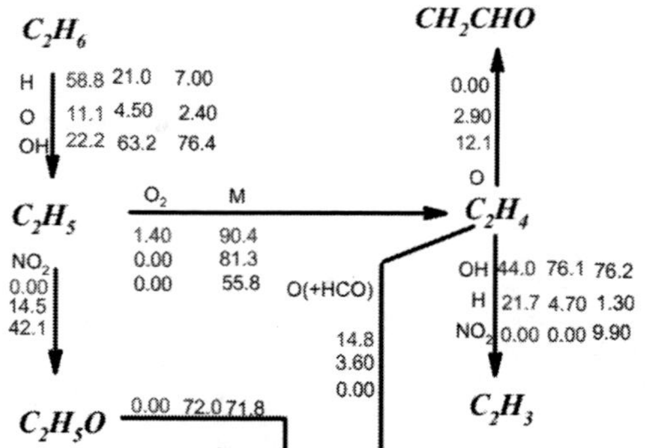

Figure 10: Flux analysis for stoichiometric C_2H_6/NO_2 mixtures at 1.2 atm and 1500 K using current model. $N_{0/1.0}$: red, $N_{30/1.0}$: blue and $N_{100/1.0}$: black.

Figure 10 shows the flux analysis of stoichiometric C_2H_6/NO_2 mixtures at 1.2 atm and 1500 K during main ignition stage. For $N_{0/1.0}$, fuel mainly undergoes H-atom abstractions by H atoms (58.8%), OH radicals (22.2%) and O atoms (11.1%) forming C_2H_5 radicals followed by decomposition to form two active species, C_2H_4 and Ḣ atom, via R207. However, when adding NO_2 to the reaction system, the branching ratio of C_2H_5 consumption is changed notably. The direct decomposition of C_2H_5 reduces from 90.4% for $N_{0/1.0}$ to 55.8% for $N_{100/1.0}$ while the reaction of C_2H_5 and NO_2 increases from 0% to 42.1%. As a result, more stable species including CH_2O, CH_3 and CH_3CHO can be formed via C-C beta scission and C-H beta scission in the C_2H_5O generated by the reaction of C_2H_5 with NO_2. In contrast, the formation of C_2H_4 and Ḣ atoms is inhibited by the NO_2 addition.

6 Conclusions

In this study, ignition delay times of $NO_2/C_2H_6/O_2/Ar$ mixtures with mole blending ratio of NO_2:C_2H_6 = 0.3/1.0 and 1.0/1.0 were measured behind reflected shock waves at p of 1.2 – 20 atm and T of 950 – 1700 K. The results indicate that NO_2 addition generally promotes the reactivity of ethane at high pressures and lower temperatures, but plays a limited impact at lower pressures and higher temperatures. Secondly, sensitivity and flux analyses exhibit that NO_2 changes the dominant reactions and critical branching ratios of key reactions during ethane oxidation. And the promoting principles of NO_2 at different conditions have been revealed.

6 Acknowledgment

The authors would like to thank the support from the National Natural Science Foundation of China (Nos. 91541115).

References

[1] F. Deng; F. Yang; P. Zhang; Y. Pan; J. Bugler; H. J. Curran; Y. Zhang; Z. Huang, Fuel 181 (2016) 188-198.

[2] T. Faravelli; A. Frassoldati; E. Ranzi, Combust. Flame 132 (2003) 188-207.

[3] R. Sivaramakrishnan; K. Brezinsky; G. Dayma; P. Dagaut, Phys. Chem. Chem. Phys. 9 (2007) 4230-4244.

[4] J. Herzler; C. Naumann, Combust. Sci. Technol. 184 (2012) 1635-1650.

[5] S. Gersen; A. Mokhov; J. Darmeveil; H. Levinsky; P. Glarborg, Proc. Combust. Inst. 33 (2011) 433-440.

[6] NUI Galway, Home combustion chemistry centre.http://c3.nuigalway.ie/.

[7] J. Giménez-López; M. Alzueta; C. Rasmussen; P. Marshall; P. Glarborg, Proc. Combust. Inst. 33 (2011) 449-457.

[8] O. Mathieu; J. M. Pemelton; G. Bourque; E. L. Petersen, Combust. Flame 162 (2015) 3053-3070.

[9] Y. Zhang, O. Mathieu, E. L. Petersen, G. Bourque, H. J. Curran, Combust. Flame182 (2017) 122-141.

[10] K. Zhang; L. Zhang; M. Xie; L. Ye; F. Zhang; P. Glarborg; F. Qi, Proc. Combust. Inst. 34 (2013) 617-624.

[11] K. Zhang; Y. Li; T. Yuan; J. Cai; P. Glarborg; F. Qi, Proc. Combust. Inst. 33 (2011) 407-414.

11th Asia-Pacific Conference on Combustion,
The University of Sydney, NSW Australia
10th -14th December 2017.

Study on the torrefaction characteristics of raw and pelletized biomass

Yangtian Ye, Kai Liu , Yuxuan Wang, Ping Lu*

Nanjing Normal University

Bancang Street 78, Nanjing, Jiangsu, 210042, China

Abstract

Torrefaction is a thermo-chemical conversion which conducted in the absence of oxygen, at temperature of 200 to 300 °C, and heating rate less than 50 °C/min. Torrefaction of six biomass materials (poplar branch, willow branch, rice husk, corn stalk, saw dust briquette, rice husk briquette) was carried out in a fixed-bed reactor in nitrogen atmosphere. Two main torrefaction parameters of temperature and residence time were varied in the range of 200-300°C and 20-40 min, respectively. The results obtained indicate that the volatile matters, O/C and H/C ratios of biochar decrease with increasing temperature and residence time. The fixed carbon and heating value of biochars increase with temperature and residence time. It is noticed that relative energy density of pelletized biomass is only 1.10 and raw biomass is 1.20-1.40 at the same torrefied conditions. This indicates pelletized one is hard to torrefy, and woody biomass could be a potential high-quality fuel after torrefaction.

1 Introduction

The carbon dioxide emissions from the consumption of fossil fuels around the world grow at an average rate of 2.1% per year, and the rate continues to increase [1-2]. Among renewable resources, biomass behaving low nitrogen and sulphur content, high volatiles, low ash content, high char activity, has been considered as a feasible option due to its carbon neutrality, sustainable and vastly available [3]. However, biomass has its limitations to be utilized particularly as a direct feedstock for power generation. These include low combustion efficiency attributable to its high moisture content, low energy density, high hydrophilic behavior, high oxygen content which makes it susceptible to biological attack and biodegradation [4-6].

Torrefaction is a thermo-chemical conversion method where biomass is subjected to thermal heating in the absence of air, typically in the temperature range of 200 to 300 °C, at atmospheric pressure and heating rate less than 50 °C/min, which improves the thermochemical properties of biomass, making it more suitable for energy generation[7]. Torrefaction differs from pyrolysis process where the former's purpose is to retain solid mass yield while enabling its energy content to be conserved and with incomplete removal of volatile matters. This process releases both moisture and carbon dioxide, both of which remove oxygen from the biomass, resulting in a fuel with a lower O/C ratio [8]. Torrefaction also partially decomposes the hemicellulose in the biomass fiber. The torrefaction process may also be called mild pyrolysis, which releases a certain of producing compounds and forms a solid product (call biochar, or char coal). A variety of biomass, such as wood [9-11], crop straw [8, 12-13] and grass [14-15], has been studied, and indicated that biomass torrefaction is strongly dependent on pyrolysis temperature, residence time and biomass species, and the biochar were characterized by a more brownish color, the decrease of moisture and volatile matter, the increase of ash, fixed carbon, and energy density [2, 16-17]. However, less attention has been paid to torrefaction of the pelletized biomass.

In this study, six biomass materials contain four biomass and two pelletized ones were torrefied in a fixed bed reactor to investigate their potential utilization as biofuel. Results obtained from this study may provide the influence of torrefaction conditions on physico-chemical properties of raw and pelletized biomass.

2 Experimental

Six biomass materials including four biomass (poplar branch (PB), willow branch (WB), rice husk (RH), corn stalk (CS)) and two pelletized biomass (saw dust briquette (SDB), rice husk briquette (RHB)) were selected as test samples. The proximate analysis, ultimate analysis and low heating value of test biomass on the air dry basis were shown in Table 1 and Table 2, respectively. Before torrefaction, all biomass samples were air-dried and cut to obtain the sample particle size less than 10 mm.

Sample	M_{ad}	V_{ad}	A_{ad}	FC_{ad}	LHV /MJ•kg^{-1}
PB	6.33	78.56	1.64	13.47	16.85
WB	7.26	73.25	2.43	17.06	19.53
CS	10.89	69.77	9.04	10.30	16.21
RH	11.58	68.99	7.51	11.92	16.93
SDB	10.60	75.18	2.75	11.47	17.89
RHB	8.74	75.81	2.48	12.97	16.52

Table 1: Proximate analysis and LHV of test biomass (Wt %)

Sample	C_{ad}	H_{ad}	O_{ad}	N_{ad}	S_{ad}
PB	45.31	6.19	47.32	0.55	0.63
WB	44.74	6.42	47.55	0.61	0.68
CS	39.50	5.75	53.73	0.91	0.11
RH	44.20	5.92	49.19	0.57	0.12
SDB	43.82	6.21	49.57	0.36	0.04
RHB	44.67	6.19	48.55	0.53	0.07

Table 2: Ultimate analysis test biomass (Wt %)

A vertical fixed-bed reactor (SZGL-1200C, Shanghai, China) with a heating chamber of 50 mm (ID) ×300 mm (in length) was used to torrefy the biomass sample. For each experiment, about 50 g biomass sample was placed into the reactor. The carried gas of nitrogen (99.999%) with the flow rate of 2 L/min was fed into the reactor to guarantee the oxygen-free atmosphere. While oxygen content at the outlet of reactor is less than 0.5%, the furnace was heated at the heating rate of 10 °C/min to the desired torrefaction temperatures (T) of 200, 250, and 300 °C and kept the temperature at the desired residence time (τ) of 20 or 40 min. The torrefied sample was cooled to room temperature at continue nitrogen feeding. The different torrefied samples are named as BB-XXX-YY, in which, BB is biomass species, XXX means

Corresponding author. Tel/Fax: +86-25-8548-1124
E-mail address: luping@njnu.edu.cn

torrefaction temperature, YY means residence time. For example, PB-200-20 represents torrefied poplar branch at T=200 °C and τ= 40 min.

The mass yield (η_m), energy yield (η_e) and energy density (E_D) of torrefied samples are calculated by:

$$\eta_m = (m_2 / m_1)_{daf} \times 100\% \quad (1)$$

$$\eta_e = \eta_m (LHV_2 / LHV_1)_{daf} \times 100\% \quad (2)$$

$$E_D = \eta_m / \eta_e \quad (3)$$

where m_1 is the mass of raw biomass on the day ash free (daf), g; m_2 is the mass of torrefied biomass, LHV_1 and LHV_2 are lower heating values of raw and torrefied biomass , respectively, MJ/kg.

3 Results and discussion

3.1 Effect of temperature

Figure 1 shows the effect of torrefaction temperature on volatile matter (on the dry ash free basis) of different biomass. It is clear that volatile matter (VM) contents remarkably decrease with increasing torrefaction temperature at the same residence time. As well known, the VM decrease is due to the release of moisture and light volatiles, such as CO, CO_2 and CH_4, which are significantly affected by temperature. Meanwhile, there is a big difference in VM decrease degree for different biomass species. For example, at the residence time of 20 min, woody biomass (PB and WB) gets the high VM decrease of 39.88-43.64 percentage points, the agriculture ones (CS and RH) come secondary VM decrease of 21.81-27.6 percentage points, and the pelletized ones (SDB and RHB) get the lowest VM decrease of 4.48-5.58. The differences of VM decrease of four raw biomass may be due to different chemical structure, and the VM release of two pelletized one may be attribute to heat transfer of particles.

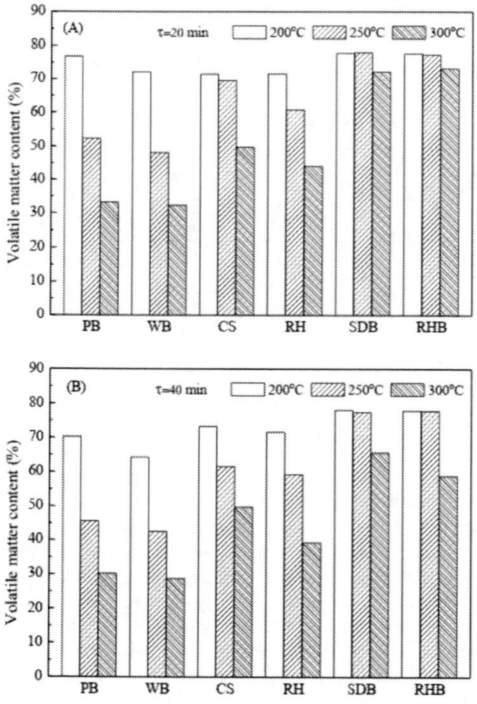

Figure 1: Influence of temperature on VM content of biochar

Figure 2 describes the influence of temperature on the fixed carbon (FC) content. It shows that FC content increase with increasing temperature, and the order of FC content increase is woody > agriculture > pelletized, which is consistent with the results of volatiles release. Normally, the higher FC content gets, the lower VM content is. As shown in Table 2, biomass contains plenty of oxygen, which could be released in form of radicals of OH and CO, and resulting in carbon concentration effect.

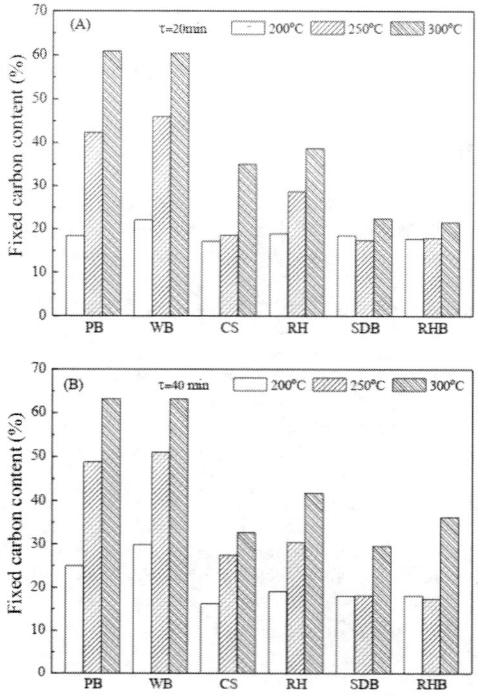

Figure 2: Influence of temperature on fixed carbon of biochar

Figure 3 depicts the relationship of H/C and O/C of all biochars in a van Krevelen diagram. It shows that H/C and O/C decrease significantly with increasing temperature, and values of H/C and O/C at lowest point are 1.01 and 0.30, respectively, which closes to the values of lignite. This indicates the releases of H and O atoms are faster than C atom with increasing temperature during torrefaction process. This appearance can attribute to release in hydroxyl (OH) groups during torrefaction [8]. From the combustion point of view, the decrease in O/C and H/C ratios are favorable since less smoke and water vapor are formed, which results in improving the performance of the torrefied biomass.

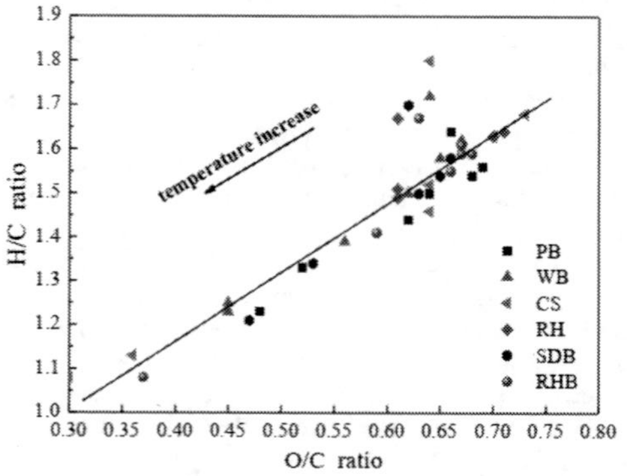

Figure 3: Influence of temperature on O/C and H/C ratios

Figure 4 shows the influence of temperature on LHV of biomass and torrefied ones. It shows that increasing temperature is generally helpful to increase the LHV of torrefied biomass, typically, the LHV of woody biomass (PB and WB) increase significantly, and the LHVs of PB and WB torrefied at 300 °C increase about 30% compared with the ones at the 200 °C. However, the LHV increase a little for pelletized biomass, whose increase locates in the range of 1.47%-7.62%. The above results indicate that torrefaction has little influence on elemental components and lower heating value of pelletized biomass.

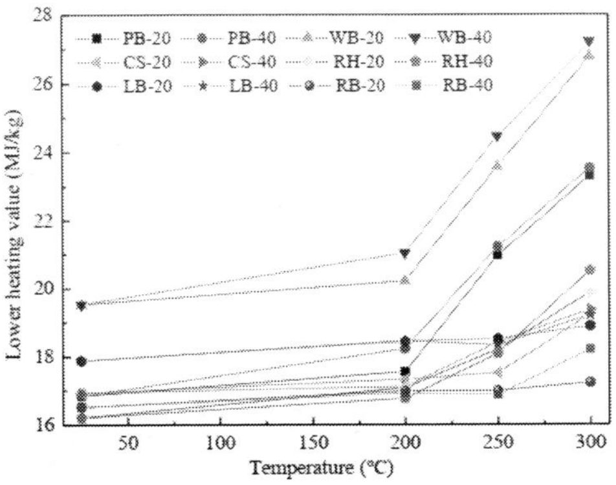

Figure 4: Influence of temperature on LHV of biochar

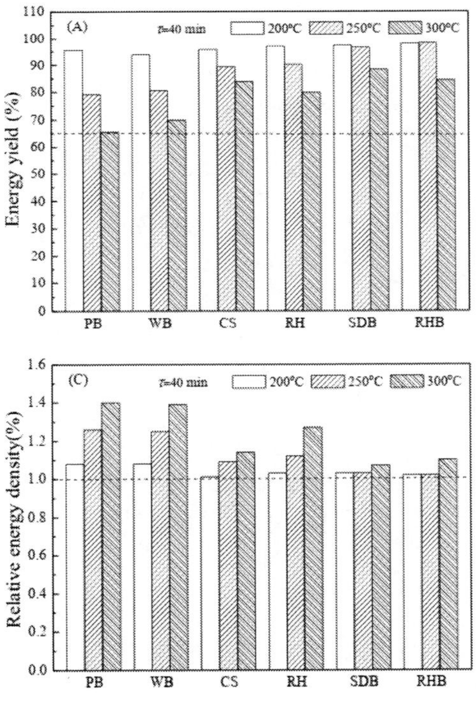

Figure 5: Influence of temperature on energy yield and relative energy density of biochar

Figure 5 shows energy yield and relative energy density of biochars at temperatures of 200-300 °C and residence time of 40 min. As can be seen from Figure 5(A), it's clear that energy yield decrease with increasing temperature, and all biochars get more than 65% of energy yield. Energy yield can be regarded as a significant indicator to the amount of energy retained after torrefaction [11]. In general, the higher torrefaction temperature is, the less the energy yield gets. This suggests torrefaction at low temperature will keep more energy in biochar, and it is benefit for energy utilization of biochar. As can be seen from Fig. 5(B), the relative energy density increases with increasing temperature, and the relative energy density is larger than 1 for all biochars. However, there is a big difference for three kind of biomass. The woody biomass gets the highest relative energy density, and the pelletized get the lowest one. This indicates that, on the one hand, torrefaction can increase the energy in unit bulk volume, on the other hand, biomass pretreatment has a significant effect on biomass torrefaction except biomass species.

3.2 Effect of residence time

Residence time is also a very important influence factor for biomass torrefaction. As can be seen from Figure 1, volatile matter contents of all biomass torrefied at the same temperature increase with the increase of residence time. For example, at 300°C, VM of PB biochar decreases 3 percentage points as the residence time increases from 20 to 40 min. However, it can be observed from Figure 2 that the fixed carbon content of torrefied biomass changes very little when the residence time increases from 20 to 40min. this indicates that torrefaction temperature has more significant influence on torrefaction than residence time. As can be seen from Figure 4, the LHV of torrefied biomass increases between 0.7% and 5.1% while the increase in the residence time from 20 to 40 min.

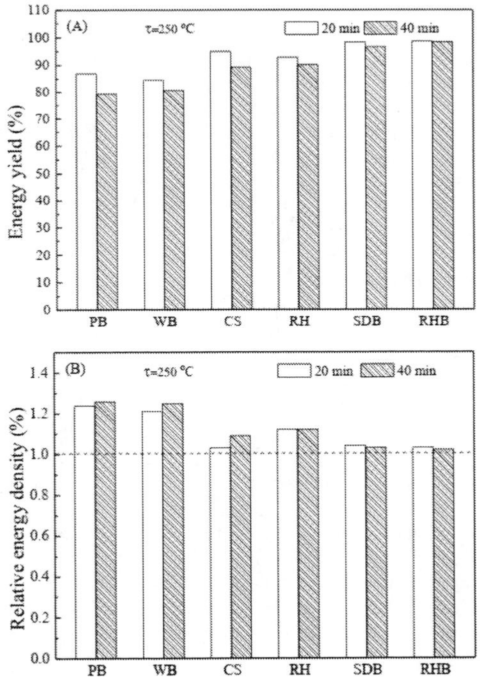

Figure 6: Influence of residence time on energy yield and relative energy density of biochar

Figure 6 depicts the influence of the residence time on energy yield and relative energy density of torrefied biomass at temperatures of 250 °C. It can be seen that when the residence time increase from 20 min to 40 min, energy yields of different biochar decrease, but the relative energy density get a little bit increase. This indicates long torrefaction time will make more energy release from the biomass, but the energy loss is less than mass loss, therefore, it is helpful to get high relative energy density. In general, the influence of residence time is less than torrefaction temperature based on the above results.

3.3 Effect of biomass species

Biomass species also has very significant on theirs torrefaction characteristics. As can be seen from Figure 1, the volatile matter present in torrefied woody biomass (PB and WB) and agricultural biomass (CS and RH)) has noticeable reduction with the temperature increases from 200 to 250°C, while pelletized ones (SDB and RHB) are almost unchanged. As well known, hemicellulose is the most reactive part of biomass and is subjected to limited devolatilization and carbonization below 250°C. Pelletized biomass is detrimental to devolatilize because its compactness resulting in bad heating and mass transfer. As can be seen from Figure 5 and Figure 6, the relative energy density of woody biomass is the highest, and the pelletized ones are the lowest. Particularly, at 300°C and 40 min, the relative energy densities of PB and WB are 1.40 and 1.39, however, the relative energy density of SDB and RHB are only 1.07 and 1.10, respectively. This big difference between woody biomass and pelletized ones indicates woody biomass may be a potential high-quality fuel after torrefaction. However, torrefaction has no significant energy improvement on pelletized biomass.

4 Conclusions

Torrefaction of six biomass materials was carried out in a fixed-bed reactor. The proximate and ultimate analyses as well as heating value was measured. The obtained results show that torrefaction has a significantly improvement on the thermo-physical of the biomass. The mass and energy yields of biomass decrease with the increase in the torrefaction temperature and residence time. However, the temperature had a stronger impact than residence time on the increase in the relative energy density of the torrefied biomass. The LHV of the torrefied biomass increases much faster with the increase in the torrefaction temperature than that the increase in the residence time. Pelletized biomass gets a little change than woody and agricultural biomass under the same torrefaction conditions. Highest relative energy density of pelletized one was 1.10 at the torrefied temperature of 300 °C and residence time of 40 min. however, woody biomass obtains the highest of relative energy density of 1.40 at the same torrefied conditions. The results indicates woody biomass could be potential high-quality fuel after torrefaction, but the pelletized one is hard to torrefy, and the higher temperature and the higher residence time are necessary.

5 Acknowledgment

The financial supports of the National Natural Science Foundation of China (51476079), Jiangsu Province Prospective Cooperation Research Project of China (BY2015001-01), and Jiangsu Province Major Project of University Natural Science Research of China (15KJA610002) are gratefully acknowledged.

References

[1] A. Pimchuai, A. Dutta, P. Basu, Torrefaction of agriculture residue to enhance combustible properties. Energy Fuels 24 (9) (2010) 4638-4645.

[2] S. Matali, N. A. Rahman, S. S. Idris, et al., Lignocellulosic biomass solid fuel properties enhancement via torrefaction. Procedia Engineering 148 (2016) 671-678.

[3] N.S. Harding, R.A. Bradley. Biomass as a reburning fuel: a specialized cofiring application. Biomass Bioenergy 19 (2000) 429-445.

[4] G. Toscano, A. Pizzi, E. Foppa Pedretti, et al., Torrefaction of tomato industry residues. Fuel 143 (143) (2015) 89-97.

[5] W.H. Chen, J. Peng, X.T. Bi, A state-of-the-art review of biomass torrefaction densification and applications. Renewable & Sustainable Energy Reviews 44 (2015) 847-866.

[6] Y. Uemura, W.N. Omar, T. Tsutsui, et al., Torrefaction of oil palm wastes. Fuel 90 (8) (2011) 2585-2591.

[7] J.S. Tumuluru, Comparison of chemical composition and energy property of torrefied switchgrass and corn stover. Frontiers in Energy Research 3 (2015) 1-11.

[8] F.F. Felfli, C.A. Luengo, J.A. Suárez, et al, Wood briquette torrefaction. Wood briquette torrefaction. Energy for Sustainable Development 9 (3) (2005) 19-22.

[9] B. Arias, C. Pevida, J. Fermoso, et al., Influence of torrefaction on the grindability and reactivity of woody biomass. Fuel Process Techno. 89 (2) 2008 169-175.

[10] G. Almeida, J.O. Brito, P. Perré, Alterations in energy properties of eucalyptus wood and bark subjected to torrefaction: the potential of mass loss as a synthetic indicator. Bioresource. Techno. 101 (24) (2010) 9778-9784.

[11] J. Wannapeera, B. Fungtammasan, N. Worasuwannarak, Effects of temperature and holding time during torrefaction on the pyrolysis behaviors of woody biomass. J Anal. Applied Pyrolysis 92 (1) (2011) 99-105.

[12] J. Poudel, S. Oh, Effect of Torrefaction on the properties of corn stalk to enhance solid fuel qualities. Energies 7 (9) (2014) 5586-5600.

[13] D. Medic, M. Darr, A. Shah, et al, Effects of torrefaction process parameters on biomass feedstock upgrading. Fuel 91 (1) (2012) 147-154.

[14] H. Nam, S. Capareda, Experimental investigation of torrefaction of two agricultural wastes of different composition using RSM (response surface methodology). Energy 91 (2) (2015) 507-516.

[15] T.G. Bridgeman, J.M. Jones, I. Shield, et al., Torrefaction of reed canary grass, wheat straw and willow to enhance solid fuel qualities and combustion properties. Fuel 87 (6) (2008) 844-856.

[16] P.C.A. Bergman, J.H.A. Kiel, Torrefaction for Biomass Upgrading [R]. Energy research Centre of the Netherlands (ECN), Unit ECN Biomass, ECN Report: ECN-RX-05-180.

[17] M. Phanphanich, S. Mani, Impact of torrefaction on the grindability and fuel characteristics of forest biomass. Bioresource Technology 102 (2) (2011) 1246-1253.

11th Asia-Pacific Conference on Combustion,
The University of Sydney, NSW Australia
10th -14th December 2017.

Study on Cellular Structure Characteristics of Expanding Flame under Dynamic Pressure in Limited space

Yan-Huan Jiang, Guo-Xiu Li*, Hong-Meng Li, Fu-Sheng Li

Beijing Jiaotong University

No.3,Shangyuancun, Haidian District, Beijing, 100044, China

Abstract

To investigate the cellular structure evolution characteristics of the expanding flame under the dynamic pressure, experiments of methane/hydrogen/air laminar premixed flame at various equivalence ratios and hydrogen fractions in a spherical combustion chamber were conducted. The mean area of cells and evenness degree were defined to quantitatively study the cellular structure. The results show that with an increase of the equivalence ratio or hydrogen fraction, the maximum pressure increase, and the moment of the maximum pressure decrease. When the flame develop to a certain extent, the mean area of cells and the evenness degree tend to remain a constant. With the evolution of the flame, the increase in pressure has little influence on the evolution of the cellular structure. In the early stage after the flame past the quartz window, the mean area of cells more sensitive to the hydrogen addition.

1 Introduction

The automobile industry is a pillar industry of the national economy and has made great contributions to the development of human society. But with a growing shortage of fossil fuels and the increasing pollution of the environment, the search for clean and renewable alternative fuels is imminent [1,2]. Natural gas reserves are abundant, it is a promising vehicle alternative fuel, and has been widely used in the automotive industry. However, for the low propagation speed, leads to a series of problems, such as lower thermal efficiency, higher cyclic variation, etc. Hydrogen addition is a good solution, and the relevant experimental study is a good proof of this [3,4].

For the expanding flame of methane and hydrogen mixture, most previous works focused on the laminar flame speed [5,6] and the intrinsic instabilities [7,8]. In order to avoid the influence of pressure and flame intrinsic instabilities on the flame propagation velocity in the limited space, all the research on the flame propagation velocity is mainly concentrated in the constant pressure section. For the quantitative study of flame intrinsic instabilities, the critical radius is mainly used. It can be seen that the existing research results mainly focus on the early stage of the flame evolution, which pressure in the limited space is near to constant. For the expanding flame in a limited space, the initial disturbance is applied to the flame surface due to the flame inherent instabilities in the early stage of flame development. As the flame is further developed, the stretch in the flame decrease, the disturbances in the flame front increase rapidly and cellular structures form eventually [9]. When the combustion develops to a certain extent, the pressure in the confined space rises rapidly. The pressure wave will affect the structural characteristics of the flame. However, the relevant research has not yet been found.

In order to study the evolution of flame cellular structure and the correlation between pressure and cellular structure under dynamic pressure, a series of laminar premixed flame experiments are carried out on a mixture of methane/hydrogen/air under an initial pressure of 0.1 MPa and an initial temperature of 300K in a spherical combustion chamber. The effects of hydrogen fractions and equivalence ratios have been considered.

2 Experimental setup and procedures

All the experiments were conducted in a constant combustion chamber system. The inner diameter of the constant volume chamber is 380 mm (Fig. 1). Two pair of quartz windows with an inner diameter of 120mm are installed on the chamber horizontally to ensure the light pass through. Kistler 6052c type high transient pressure sensor was used to measure the pressure of the mixture in the combustion chamber. A type 5018 charge amplifier and a DPO 3054 digital phosphor oscilloscope were combined with the pressure sensor to display and store pressure data. FASTCAME SA-X2 high-speed digital camera was used to record the schlieren photos of the expanding flame. The volumetric percentage of methane, hydrogen and dry air controlled by the pressure gauge was supplied into the chamber through an inlet valve according to the given equivalence ratio. The mixture was ignited by the electrodes located in the center of the chamber after the mixture had been kept for 5 minutes. The combustion products were removed by a vacuum pump after complete combustion, and a nitrogen flow was used to purge the constant combustion chamber in order to eliminate the impact of residual gas on the experiments.

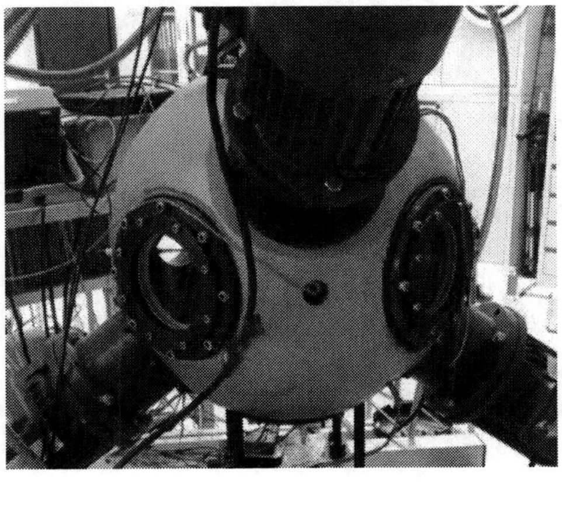

Corresponding author. Fax: +86-010-51682047
E-mail address: Li_guoxiu@yahoo.com

Figure 1 Combustion chamber used in the experiment

The burning velocity of hydrogen is higher than that of methane, which makes the hydrogen percentage a prominent factor influencing the explosion characteristics of the methane/hydrogen mixture. The hydrogen fraction of the methane/hydrogen mixture is defined as

$$X_{H2}=V_{H2}/(V_{H2}+V_{CH4}) \quad (1)$$

To deepen the comprehensive understanding of the inherent instabilities of flames, extraction of the flame front has a great significance. Due to the huge amounts of flame images, MATLAB was selected to deal with the flame. Figure 2 shows the comparison between the original flame and the extraction edge. The results show that the flame extracted by using the program has a high degree of accuracy, which laid a foundation for the following study.

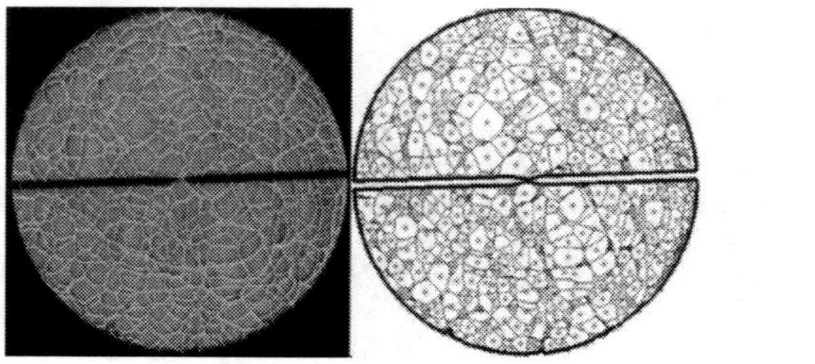

Figure 2 Comparison between the original flame and extraction edge

At the beginning of the flame development, the pressure in the constant combustion bomb remains a constant. When the flame develops to a certain extent the flame front gradually appears cracks under the influence of the flame inherent instabilities, the cracks further develop into cellular structure. The flame surface cellular structure is relatively simple in the early stage. The number and uniformity of the cells in the flame front are increase with the pressure to further increase. With the consumption of the fuel and the radiation in the combustion bomb, the pressure gradually decreases after the pressure reaches the maximum pressure (P_{max}), and the cellular structure remains develop (Fig. 3). The combustion duration (t_d) is set as the time elapsed from the ignition signal to the instant when maximum pressure is reached, which involves both the constant pressure stage and the pressure rise stage.

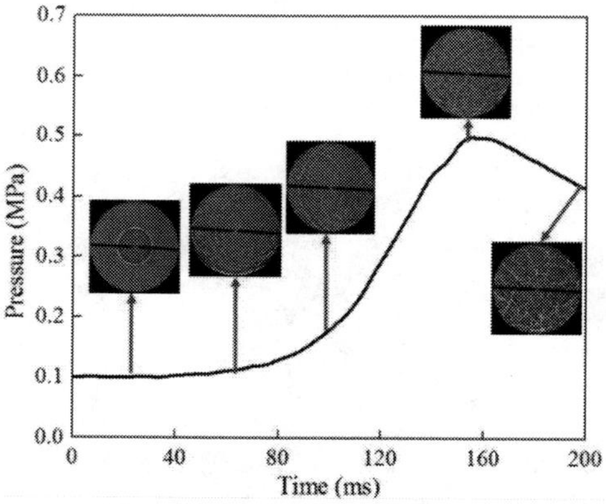

Figure 3 Pressure and flame evolution of 50% H_2 and 50% CH_4 mixture under equivalence ratio of 0.6

In order to further study the change of the structural characteristics of the flame under dynamic pressure, the cellular structure of the flame surface needs to be quantified. To quantitative study the cellular structure on the expanding flame, the mean area of cells and evenness degree are defined by Eqs (1) and (2), respectively, as follows.

$$\bar{A} = \sum_{i=0}^{n} A_i/n \quad (2)$$

$$Var = \sqrt{\sum_{i=0}^{n}(A_i - \bar{A})^2/n} \quad (3)$$

where A_i represents the area of i th cell on the flame surface, and n is the number of cells on the flame front. Considering that the pressure remains substantially constant at the beginning of the flame propagation in the constant combustion chamber, all the flame photos which past the quartz window are chosen.

3 Results and discussion

The evolution of pressure in limited spaces is the basis for studying the correlation between dynamic pressure and the cellular structure of the flame surface, which is of great significance for studying the combustion pressure of methane / hydrogen mixture under different conditions. Fig.4 shows that in the initial period of flame propagation, the pressure remains essentially unchanged, and when the combustion developed to a certain extent, the pressure in the constant combustion chamber increases rapidly. After the pressure reaches its maximum value, the pressure gradually reduces due to the fuel consumption and heat dissipation. With the increase of the equivalent ratio, the maximum pressure increases and the moment of maximum pressure decreases. This is mainly due to the fact that the initial radius of the flame kernel is small and the combustion consumes less fuel. As the combustion progresses, the amount of fuel consumed per unit of time increases non-linearly, which causes the pressure to grow slowly at the beginning of the flame propagation, and when the flame develops to a certain extent, the pressure in the limited space increases rapidly. In addition, with the increase of the equivalent ratio, the laminar combustion rate of the mixture gradually increased [10].

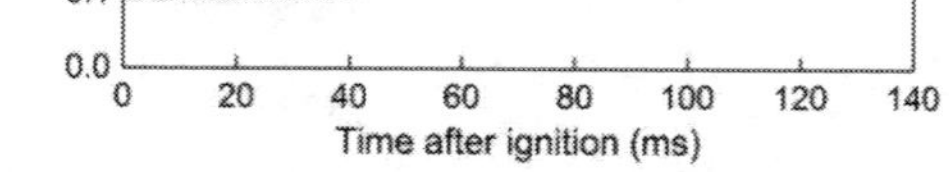

Figure 4 Effects of the equivalence ratio on the explosion pressure under 70% hydrogen fractions

The effects of the equivalence ratio on P_{max} and t_d under different hydrogen fraction values are illustrated in Figure 5. P_{max} increases monotonically with the increase of the equivalence ratio, and the growth rate of the maximum P_{max} decreases. The highest value of P_{max} is found at an equivalence ratio of 1.0. t_d decreases monotonically with the increase of the equivalence ratio, and its absolute growth rate value decreases. This is mainly because hydrogen fraction influences the concentration of H and OH produced during the combustion of the mixture. With the increase of the hydrogen fraction, the concentration of H and OH increase, which accelerates the combustion process, leading the laminar speed increased significantly [11].

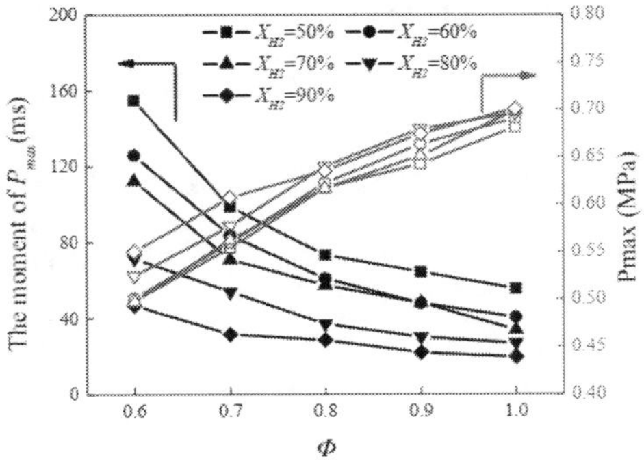

Figure 5 Effects of equivalence ratio on the Pmax and td under different hydrogen fractions

Cellular structure of the flame is a result of the evolution of initial disturbance applied to the flame front. The quantitative study of cellular structure is of great significance to deepen the effect of dynamic pressure on the cellular structure of flame. Fig.6 shows that the mean area of cells gradually decreases with the development of the flame under the lower equivalence ratios and the value tends to a constant when the flame develop to a certain extent. When the equivalent ratios range from 0.8 to 1.0, the mean area of cells has a tendency to increase first and then decrease, and the value tends to a constant when the flame develop to a certain extent. With the increase of the equivalent ratio, the maximum value of the mean area of cells increased rapidly. This is mainly due to with the decrease of the equivalence ratio, the thermal diffusion instability increases, and the disturbances on the flame front increase which leads to a more complexed cellular structure of the flame front. Due to the difference in the flame propagate speed, the moment of the maximum value of mean area of cells have a significance difference.

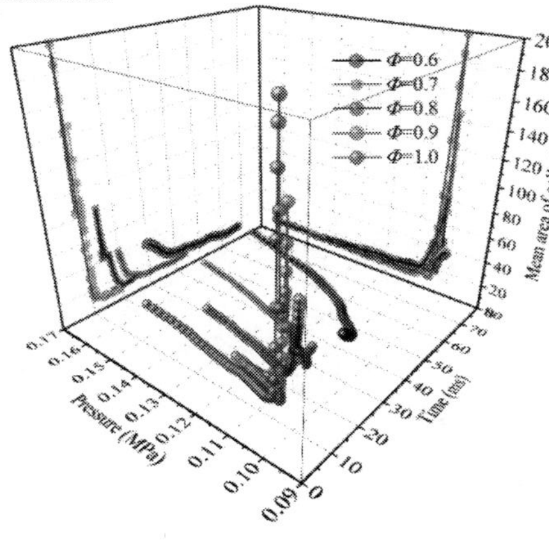

Figure 6 Relation between pressure and mean area and of cells in flame front with 70% hydrogen fraction under different equivalence ratios

The mean area of cells tends to reach a constant value, which gradually decreases with an increase in the equivalence ratio. It indicates that the influence of pressure on the evolution of the cellular structure becomes less significant with the evolution of the flame. In the early stage after the flame past the quartz window the change of pressure has a significant effect on the cellular structure of the flame front. With the development of the flame, the effect of pressure on the cellular structure of the flame fronts is gradually reduced (Fig.6).

There are complex disturbances on the flame surface during the evolution of the expanding flame. The complex interactions between the different disturbances, such as merging, offset, leading to a complex cellular structure of the flame surface. There are significant differences in the number and shape of the cells in the flame surface under different initial conditions. The mean area of cells can only give the evolution of the cellular structure as a whole, the difference between cells does not reflect.

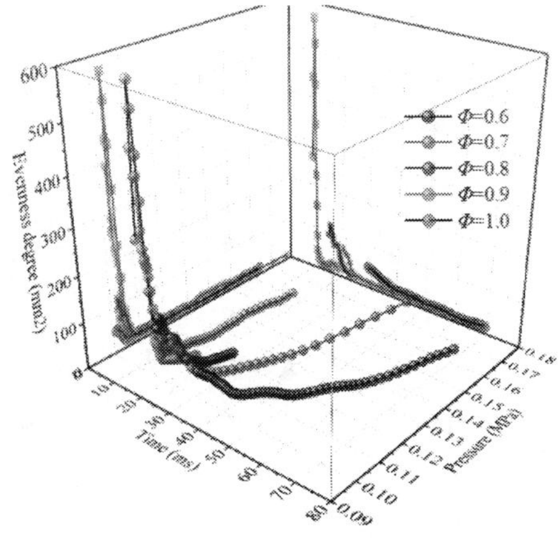

Figure 7 Relation between pressure and evenness degree in the flame front of 70% hydrogen addition under different equivalence ratios

The evenness degree can reflect the microscopic information of the cellular structure. Fig.7 shows that the curve shape of the evenness degree-time is similar to the ones of the mean area of cells. When the flame develops to a certain extent, the evenness degree of the cellular structure tends to a constant, indicating that the difference in the area among the cells reduces. With the increase of the equivalent ratio, the evenness degree of the cellular structure tends to decrease. The difference in the evenness degree and the moment of the maximum value is caused by the competition between the flame propagation velocity and the growth rate of the cracks.

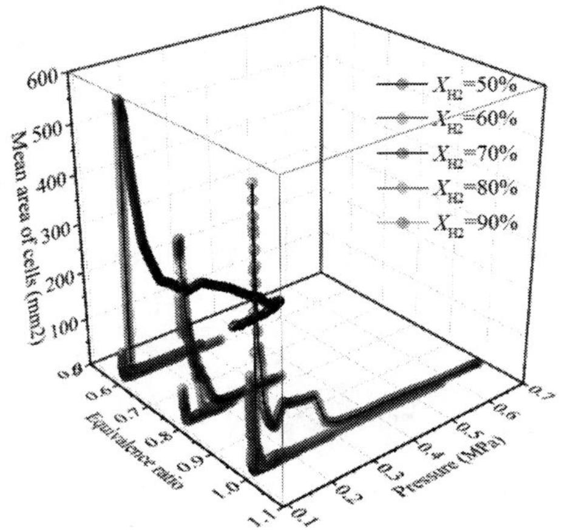

Figure 8 Relation between pressure and mean area of cells under 70% hydrogen fractions

Fig.8 shows that the influence of pressure on the evolution of the cellular structure of the flame front becomes gradually weaker when the flame develop to a certain extent. In the early stage after the flame past the quartz window, the mean area of cells more sensitive to the hydrogen addition. Under the lower hydrogen fraction (50%) and lower equivalence ratio (0.6), the mean area of cells increases sharply after the flame past the quartz window, and then follows a decrease. With the consumption of the mixture and the heat dissipation, the pressure in the constant combustion chamber reduces. However, the mean area of cells continuously reduces and tend to a constant. Under the higher equivalence (1.0) and lower hydrogen fraction (50%), the curve shape of the pressure - mean area of cells presents bimodal pattern. This is mainly due to the weak instability in the flame front.

4 Conclusions

To study the structural characteristics of the flame under dynamic pressure, a series of experiments under different equivalence ratio and hydrogen fraction were conducted at ambient temperature and pressure. The main conclusions from this work can be summarized as follows:

1. With the increase of the equivalent ratio or hydrogen fraction, the maximum pressure increases and the moment of maximum pressure decreases. Duo to the consumption of the mixture and the heat dissipation, the curve shape pf the pressure – time presents a single peak form.
2. The mean area of cells tends to a constant when the flame develop to a certain extent, when the difference of the cells is the least. In the early stage after the flame past the quartz window, the mean area of cells of different condition and the evenness degree have significant difference.
3. The influence of pressure on the evolution of the cellular structure of the flame front becomes gradually weaker when the flame develop to a certain extent. In the early stage after the flame past the quartz window, the mean area of cells more sensitive to the hydrogen addition.

5 Acknowledgment

This work is supported by the Fundamental Research Funds for the Central Universities (No. 2017JBZ102) and the Fundamental Research Funds for the Central Universities (No. 2017YJS161).

References

[1] C.W. Ji, X.L. Liu, S.F. Wang, et al. Fuel 133 (2014):139-142.

[2] Z.Y. Sun, F.S. Liu, B.G. Sun, et al. Int J Hydrogen Energy 37 (2012): 664-681.

[3] B. Korb, S. Kawauchi, G. Wachtmeister. Fuel 164 (2016): 410-418

[4] B.W. Fan, J.F. Pan, W.M. Yang, et al. Int J Hydrogen Energy 41 (2016) :4039-4053.

[5] E.J. Hu, Z.H. Huang, J.J. He, et al. Int J Hydrogen Energy 34 (2009): 8741-8755

[6] H.Y. Miao, Q. Jiao, Z.H. Huang, et al. Int J Hydrogen Energy 34 (2009):507-518.

[7] Z.Y. Sun, F.S. Liu, X.C. Bao, et al. Int J Hydrogen Energy 37 (2012):7889-7899.

[8] E.J. Hu, Z.H. Huang, J.J. He, et al. Int J Hydrogen Energy 34 (2009):5574-5584.

[9] F.S. Li, G.X. Li, Y.H. Jiang, et al. Energies 2017 (10):678-693.

[10] E.J. Hu, Z.H. Huang, J.J. He, et al. Int J Hydrogen Energy 34 (2009):4876-4888.

[11] H.M. Li, G.X. Li, Z.Y. Sun, et al. Energy 2016 (112):146-152.

[12] H.Y. Miao, Q. Jiao, Z.H. Huang, et al. Int J Hydrogen Energy 33 (2008):3876-3885.

11th Asia-Pacific Conference on Combustion,
The University of Sydney, NSW Australia
10th -14th December 2017.

Study On the Statistical Analysis Methodology of the Swirl Flame Dynamics

Zheng Yu[1], Zhang Chi[1], Huang Min[2]

[1]School of Energy and Power Engineering, Beihang University, 37 Xueyuan Road, Beijing, China

[2]School of Reliability and System Engineering, Beihang University, 37 Xueyuan Road, Beijing, China

Abstract

Combustion instability is now being widely studied. Combustion instability includes static and dynamic instability. On dynamic instability, the traditional research methods are to measure the dynamic pressure and heat release of the swirl flame through experiments, and then the experimental results are obtained by Fast Fourier Transform, and the influence of the parameters on the pressure fluctuation and the thermal acoustic oscillation is obtained. However, this method has some defects, such as the large error, the dynamic characteristics of the flame is not intuitive and so on. Therefore, a statistical image processing method can be developed based on large sample size to describe the flow characteristics of swirl flame. By using two methods of proper orthogonal decomposition and dynamic mode decomposition, the dynamic characteristics of the swirl flame are described. Meanwhile, due to the large sample size, the error of processing experimental data is small.

1 Introduction

Because of the low-emission requirements for aero engines, low-emission combustion chambers are widely developed. The mechanism of combustion instability is shown in Figure 1. And the instability of the combustion seriously affected the performance and life of the gas turbine [1]. Now in the burning field, the burning instability problem is a hot spot.

The burning instability is divided into dynamic instability and static instability. The research mainly concentrates on the dynamic instability of the swirl flame. In order to further understand the dynamic instability of the swirl flame, the pulsating characteristics of the swirling flame are studied by using the digital image processing technique.

As the digital image of the flame is taken by the high-frequency camera, the number of samples is very large, and the conventional statistical data processing method is difficult to analyse the experimental data efficiently. In this field, Proper Orthogonal Decomposition (POD) and Dynamic Modal Decomposition (DMD) method are two commonly used decomposition methods, which have been widely used in the analysis of flow and flow field, but there are few applications in the combustion field.

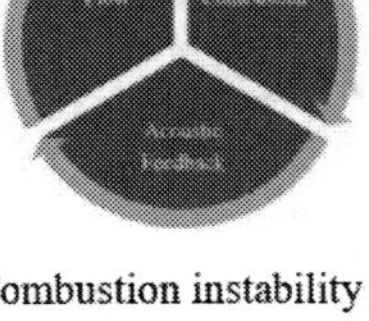

Figure 1: Combustion instability mechanism

2 POD and DMD methods

2.1 POD method

Proper Orthogonal Decomposition is a statistical method, which can take the means of orthogonal decomposition on a large number of sample data and extract the most part of the contribution. The concrete process is as follows: First, assume that a matrix A records the variation of the signal x over time t, and it can be written as an $m \times n$ matrix, where m represents the data at each time and n represents the time coordinates, so that $m \times n$ represents the number of all the data in that period of time. The intensity distribution of the high-frequency flame images is stored in the data matrix, expressed as $A_{m\times n}$, x is the sample size, and t corresponds to the time of characteristic image. After decomposition, it is also a function $\varphi(x)$ that can describe the characteristics of the flame energy ripple. At the same time, the time coefficient $\alpha_i(x)$ can be achieved, so the matrix can be decomposed as follows:

$$A_{m\times n} = \alpha_0(x)\varphi_0(x) + \sum_{i=1}^{m} \alpha_i(x)\varphi_i(x) \qquad (1)$$

In order to study the dynamics of swirl flame, Sirovich[2] snapshot method is taken. The purpose is to linearly represent the data matrix with all snapshots and to reduce the dimension of the matrices which need the eigenvalues and eigenvectors. In order to facilitate the solution, further decompositions can be done on (1):

$$A_{m\times n} = U_{m\times m} S_{m\times n} V_{n\times n} \qquad (2)$$

$U_{m\times m}$ represents the various modes of the signal, reflecting the periodic structure of the signal. $S_{m\times n}$ represents the singular value of the signal, describing the energy distribution of each mode. $V_{n\times n}$ represents the time mode, which characterizes the pattern of mode changes over time. $U_{m\times m}$ and $V_{n\times n}$ are orthogonal matrices and their inverse matrices are equal to the transposed matrices. So the three matrices of $U_{m\times m}$, $S_{m\times n}$ and $V_{n\times n}$ can be approached according to their orthogonality. The solution is as follows:

We calculate the product of the matrix and its transpose matrix so that we can get the following two equations:

$$AA^T = US^2U^T \rightarrow AA^TU = US^2 \qquad (3)$$

$$A^TA = VS^2V^T \rightarrow A^TAV = VS^2 \qquad (4)$$

As can be seen from the above equation, S represents the square root of the eigenvalues of the two matrices, we make it equal to $\sqrt{\lambda}$, and the matrix U and V are the eigenvectors of the two matrices. So the two equations can be written in the form shown below:

$$AA^TU = U\lambda \quad (5)$$

$$A^TAV = V\lambda \quad (6)$$

Therefore, we can do eigenvalue decomposition on matrices AA^T and A^TA. Then we can get U, S and V these three matrices. According to the physical meaning of the three matrices obtained, the matrix U represents the spatial feature of the mode, the matrix V represents the temporal feature of the mode, and the eigenvalues of the matrix S can approximate the eigenvalues of the matrix A. From the above conclusion, the formula for each modal can be obtained as follows:

$$\varphi_i = U_i \quad (7)$$

And then we find out the eigenvalues of all the mode matrices and arrange them in descending order:

$$\lambda_1 > \lambda_2 > \cdots > \lambda_n \quad (8)$$

The eigenvalues of mode matrices represent the contribution of the mode to the overall pulsating energy, which we define as follows:

$$E_i = \frac{\lambda_i}{\sum_{j=1}^{n} \lambda_j} \quad (9)$$

By sorting the contribution rate of each mode to the overall pulsating energy, we can extract the main modes and decompose and analyse them.

2.2 DMD method

Dynamic Mode Decomposition method was proposed by Schmid[3] to estimate the flow evolution and to perform stability analysis through dynamic system eigenvalues. Meanwhile, it is a low-dimensional system decomposition method developed on the basis of Koopman's overall stability analysis[4]. Because the flow field is decomposed based on the dynamic characteristics of the flow field, the substructures that are orthogonal to each other can be obtained. Therefore, Dynamic Mode Decomposition method can describe the dynamic characteristics of the flow field.

The experimental data acquisition process of Dynamic Mode Decomposition method is also using a high-frequency camera to take snapshots of the flame, and then the light intensity distribution of high-frequency flame images is stored in a matrix. The high-frequency flame images of the former n-1 moments and the after n-1 moments obtained by the test can be written as the image sequence matrix X and Y and the time interval between any two images is Δt:

$$X = [v_1, v_2, v_3, \cdots, v_{n-1}] \quad (10)$$

$$Y = [v_2, v_3, v_4, \cdots, v_n] \quad (11)$$

We assume that the high-frequency image matrices X and Y can be represented by linear mapping A:

$$v_{i+1} = Av_i \quad (12)$$

If the dynamic system itself is a nonlinear system, then (12) is a linear estimation process, which is through a linear assumption to achieve a nonlinear estimate. From the assumption of (12), we can know:

$$Y = AX \quad (13)$$

Equation (13) indicates that the matrix A can move the velocity field in the time dimension. At the same time, matrix A can only be calculated by dynamic analysis. So in DMD method, the dimension of matrix A need to be approximately reduced:

$$Y = AX = XS + re_{n-1}^T \approx XS \quad (14)$$

$$S = \begin{bmatrix} 0 & 0 & \cdots & 0 & 0 & a_1 \\ 1 & 0 & \cdots & 0 & 0 & a_2 \\ 0 & 1 & \cdots & 0 & 0 & a_3 \\ \vdots & \vdots & \ddots & \vdots & \vdots & \vdots \\ 0 & 0 & \cdots & 1 & 0 & a_{n-2} \\ 0 & 0 & \cdots & 0 & 1 & a_{n-1} \end{bmatrix} \quad (15)$$

In the matrix S, a_1, a_2,$\cdots$, a_{n-1} is unknown and needs to be calculated. re_{n-1}^T is the remainder of dimension reduction for A. The original velocity matrix X is projected onto the eigenvector y_j of the matrix S, and the modes with high contribution to the combustion field are selected and the spatial mode φ_j of the combustion field can be calculated:

$$\varphi_j = Xy_j \quad (16)$$

We assume that the rank of matrix X is r. In order to simplify the calculation, Dynamic Mode Decomposition method requires finding the low-dimensional matrix S instead of the high-dimensional matrix A, and the eigenvalue λ of S is a complex number, which can be expressed as:

$$\lambda = g_j + i \cdot \omega_j \quad (17)$$

g_j represents the system rate of change and ω_j represents the system frequency [5]. The following will explain the way to replace the high-dimensional matrix with low-dimensional matrices.We perform Q-R decomposition on matrix X:

$$X = QR \quad (18)$$

We assume that X is an m × n matrix, and after Q-R decomposition, Q is an m×m unitary matrix, and R is an m×n upper triangular matrix. Matrix Q contains the proper orthogonal features of X. According to the method to calculate eigenvalues in the Krylov subspace, you can construct an upper Hessenberg matrix, that is:

$$H = RSR^{-1} \quad (19)$$

The eigenvalues of the matrix A can be approximated by the eigenvalues of the matrix H, which can be represented as:

$$AQ \approx QR \quad (20)$$

Then the low dimensional approximation matrix S of the high-dimensional matrix A can be represented as:

$$S = R^{-1}Q^HY \quad (21)$$

The eigenvalues of the matrix S are part of the eigenvalues of the matrix A, which contains most of the information of the matrix A. Matrix S is an approximately alternative matrix of A. So the next step is to find the eigenvalues μ_j and eigenvectors y_j of the matrix S. Then according to equation (16), each mode can be calculated.

In order to express the contribution of each mode to the

combustion field, we define the modal amplitude as:

$$a = R^{-1}Q^H v_n \tag{21}$$

The vector a is the last column of matrix S. According to the mode amplitude of the modes with different frequencies, we can pick out the mode with high contribution to the combustion field. The frequency of each modal can be calculated from the eigenvalues of the matrix S, and we can calculate the rate of growth and frequency from the logarithmic form of the eigenvalues:

$$g_j = \frac{Re\{log(\mu_j)\}}{\Delta t} \tag{23}$$

$$\omega_j = \frac{Im\{log(\mu_j)\}}{\Delta t} \tag{24}$$

In fact, the reduction of A to the Hessenberg matrix by Arnoldi method is not only done by a simple Q-R decomposition, but also by a series of projections to continuous Krylov subspaces [4]. This algorithm has a relatively good stability. So in the following program preparation, we also chose this way.

3 Analysis of swirl flame characteristics

3.1 Analysis of light intensity

According to the analysis of the combustion characteristics of flame, the characteristics of a flame image can be represented by the following variables: average grey level, maximum grey level, variance, flame abundance and energy. These characteristic parameters comprehensively reflect the basic characteristics of flame. Among them, the average grey level reflects the average intensity of radiation of fire. The variance reflects the uneven intensity distribution of the flame intensity.

The average value of all the image grey values can be calculated, and the flame intensity is replaced by the grey level of each image. Each picture then corresponds to a transient light intensity of swirl flame at one moment. The range of grey level of the image pixel is 0-255, and the stronger the flame reaction is, the lower the grey value is. The flame height, fuel reactivity and combustion chamber radiation are the main factors affecting the flame intensity. The mean value of flame intensity is expressed as:

$$\bar{x} = \frac{1}{n}\sum_{i=1}^{n} x_i \tag{25}$$

The time averaged figure represents the mean intensity of the swirling flame, and the swirl flame in figure 2 is axisymmetric, indicating that the intensity distribution is also axisymmetric.

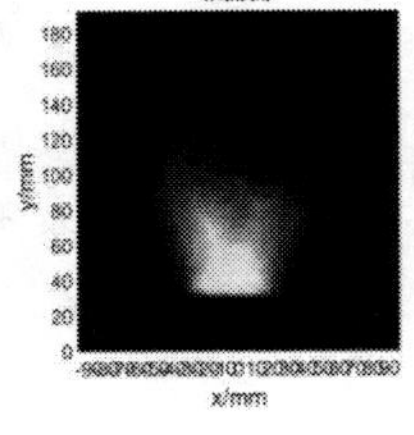

Figure 2: Time averaged figure

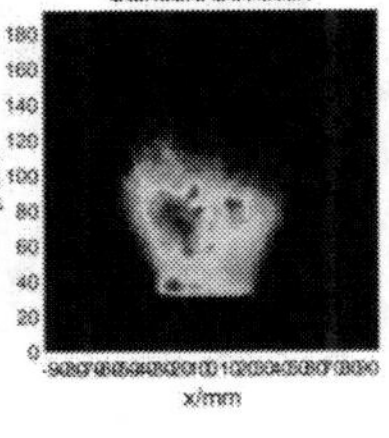

Figure 3: Standard deviation figure

In order to sort the POD mode, we need to calculate the energy variance of each spatial position. Here intensity of light can be used to replace energy. The larger the variance is, the larger the energy pulse amplitude is, so the higher the energy contribution to the system is. Variance can be expressed as:

$$\sigma^2 = \frac{1}{n}\sum_{i=1}^{n}(x_i - \bar{x})^2 \tag{25}$$

The standard deviation is the square root of the variance. The redder part of the standard deviation figure means the larger energy fluctuation of flame and the more unstable flame energy. From the standard deviation figure, we can observe which parts of the flame are not stable enough and the instability can be improved objectively.

3.2 Analysis by POD method

Proper Orthogonal Decomposition can extract modes of the highest energy contribution from the sample data. We first sort the energy contributed by each mode, and then accumulate the energy contribution of each mode in turn. When the cumulative total energy reaches 90% -99%, we believe that the selected modes have been fully embodied the swirl flame energy pulse characteristics. As shown in figure 4, we selected the swirl flame modes with the top-ten contribution of energy. Although it is not possible to know exactly the specific energy values of the modes, the ten modes' cumulative energy contribution have reached 90% of the total energy. When we analyse the influence of a mode on the swirling flame, this energy distribution figure can intuitively show the mode which has the greatest influence on the system, which is convenient for our subsequent analysis and screening of modes.

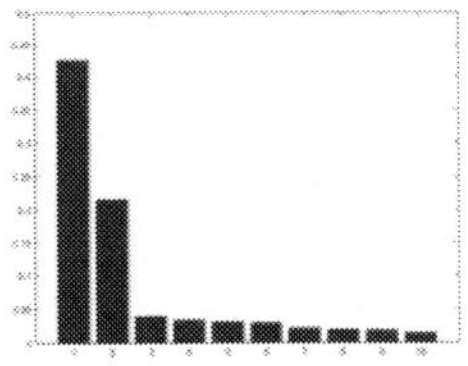

Figure 4: Mode energy distribution

The mode figures are reflections of the intensity of reaction of each mode at different positions of the figure. Blue and red indicate the intensity of the reaction. The intensity is quantified to dimensionless number, and the range is [-1, 1]. Red is positive and blue is negative. We can obtain the corresponding characteristics of swirl flame through each mode figure. At the same time, we can know that it is necessary to calculate the average intensity of each flame image. Because the average value is significant for describing the pulsation of the flame energy, we can subtract the corresponding average in each data matrix to obtain the energy fluctuation value of the flame in the spatial position. The energy fluctuation of swirl flame can be described in a more complete way. We can also understand the change of the energy fluctuation of the swirling flame in time series and space position by each mode figure, which is helpful for us to study dynamics of swirl flame.

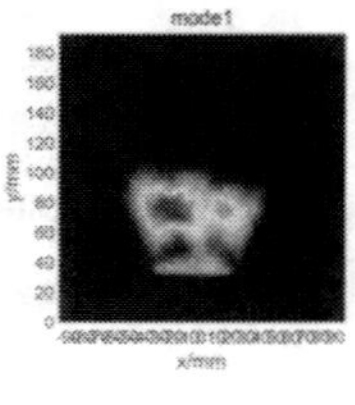

Figure 5: Mode 1

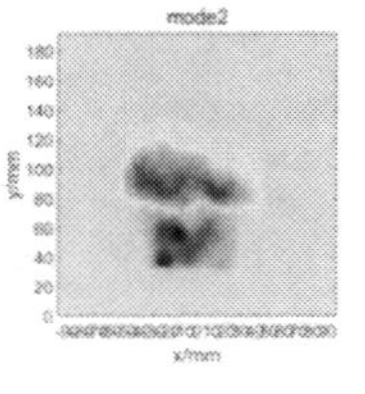

Figure 6: Mode 2

As shown in Figure 5 to Figure 6, the first two modes correspond to the top six energy contribution. The first two modes can represent the heat release of the swirl flame, and the last modes mainly represent the details of the swirl flame changes.

3.3 Analysis by DMD method

Dynamic Mode Decomposition can separate modes at different frequencies. At the same time, the energy contribution of the modes at each frequency can be sorted, and the order is based on the amplitude of each mode. Therefore, it is possible to know the frequency of the mode which has the greatest impact on the swirl flame. As shown in Figure 11, the eigenvalue distribution of the processed flame images is approximately a unit circle. This shows that the swirl flame system is mainly in the critical steady state under simple harmonic motion, which is consistent with the actual physical phenomena. A small number of points deviate from the unit circle because of the low numerical accuracy and the system error in the experiment. The growth rate and frequency of each mode can be calculated by eigenvalues. By means of Dynamic Mode Decomposition method, the sound pressure oscillation of swirl flame pulsation can be described more accurately and more intuitively.

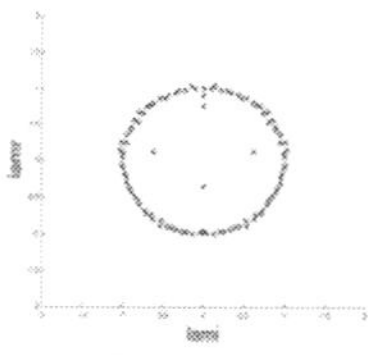

Figure 7: Scatter diagram of all eigenvalues

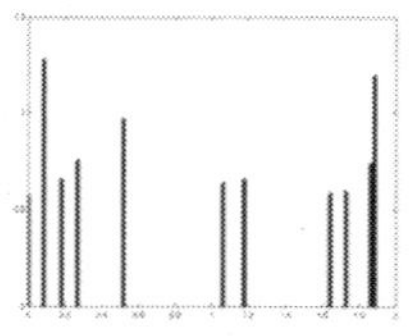

Figure 8: The relation between mode amplitude and frequency

The relation between frequency and amplitude cannot be represented in the scatter diagram, so the histogram is used to represent the relation between frequency and amplitude. As shown in Figure 8, eleven modes with the highest energy contribution to the swirl flame pulse are extracted by sorting each mode according to the mode amplitude. We can find that the energy contribution of low frequency fluctuation to swirl flame is higher than that of high frequency fluctuation. The main frequency of swirl flame can also be obtained by extracting the modes according to amplitude sorting. But for the observation of swirl flame, the mode figures are more intuitive.

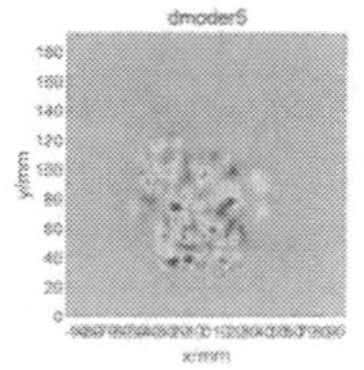

Figure 9. Static mode

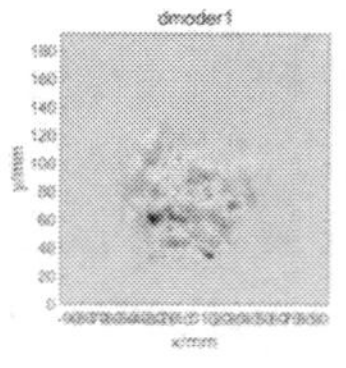

Figure 10. Mode 2-3

Among the three modes extracted, there is a static mode and a pair of dynamic modes. The positive and negative effects are more uniform in the static mode than that in the dynamic modes. The static mode is the one in which the combustion is the most stable. The pair of modes are sorted according to the mode amplitude. Since the eigenvalues of matrix S contain complex numbers, each pair of conjugate complex numbers corresponds to a pair of dynamic modes, the growth rates of which are equal and the frequencies are opposite numbers.

4 Conclusions

This paper mainly discusses the application and necessity of the two statistical methods to the study of dynamics of swirling flame. The mathematical principle of the two methods, the program realization idea and the result of image analysis are briefly introduced and analysed.

In the decomposition and reconstruction of some flow fields, POD method is widely used because it is easy to understand, easy to implement and has the advantages of small overall error. But POD method is less accurate than DMD method for the variation of flow field with time. Therefore, DMD method is more suitable to describe the dynamics of flow field, and POD method is more suitable to describe the energy distribution. In the description of dynamics of swirling flame, DMD method is more suitable to describe the acoustic pressure oscillation of swirl flame, and POD method is more suitable to describe the heat release oscillation of swirling flame.

References

[1]Lieuwen. T. C and Vigor. Y, Combustion instabilities in gas turbine engines: operational experience, fundamental mechanisms, and modelling. *US: AIAA*. (2005) 1-3.

[2]Mode. P. O. D, OH* chemiluminescence in a multipoint combustion system: Steady state and limit cycle behavior.

[3]Schmid. P. J, Dynamic mode decomposition of numerical and experimental data. Journal of Fluid Mechanics. 656(1)(2010)5-128.

[4]Jovanovicm. R, Schmid. P. J and Nichols. J. W, Sparsity-promoting dynamic mode decomposition. Physics of Fluids. 26(2) (2014) 024103-1-22.

[5]Schmid. P. J, Li. L, Juniper. M. P and Pust. O, Applications of the dynamic mode decomposition. Theor Comput Fluid Dyn, 25(2011) 249–259.

11th Asia-Pacific Conference on Combustion,
The University of Sydney, NSW Australia
10th -14th December 2017.

Simulation Study on Influence of Natural Gas Pipeline Pressure on Jet Fire

Shuran Lyu, Shuqi Ma, Youbo Huang

School of Safety and Environmental Engineering, Capital University of Economics and Business

121 Zhangjiakou Street, Huaxiang Fengtai District, Beijing, 100070, China

Abstract

In order to study the influences of natural gas pipeline pressure on jet fire with FDS(fire dynamics simulation) software, based on principle of fluid mechanics and Thornton flame model, characteristics and hazard scope of the jet fire were comparatively studied for different pipe pressure. The results show that characteristics and hazard scope of the jet fire are largely affected by different pressure. When the pipe pressure is 0.2MPa and 0.4MPa, the size of both leak holes are same, during a same period, the latter flow is greater than the former. When the pipe pressure is greater than 0.8MPa, the leakage hole size is 100mm, and the maximum temperature, diameter and the high temperature affected area of the flame are increased.

1 Introduction

Pipeline transportation is widely used in oil and gas conveying system with the advantages of high efficiency and strong continuity[1], but there are many hidden dangers. When the pipeline failure, the leaked inflammable and explosive gas can easily cause fire and explosion accidents[2], including jet fire, fireball fire, and vapor cloud explosion. The jet flame refers to a fire that is formed at the leak hole when the gas is leaking[3].

Researchers have conducted the research of natural gas fires for many years, and some developed countries have already developed the theoretical model of natural gas fire and the experimental study of natural gas jet fire[4]. For example, Pfenning performed a large scale jet fire experiment to obtain the temperature field and radiation field [5]. B.J. Lowesmith, G. and Hankinson have conducted large scale jet fire experiments on different methane concentrations in natural gas pipelines[6]. Based on the theory of Becker, Kalghatgi does small-scale jet fire experiments in wind tunnels, and studies the influence of different injection angles and wind speed on the characteristics of the jet fire[7].Based on the Kalghatgi jet fire experimental data and Shell Research mathematical model, Chamberlain developed the formula of geometric characteristics of jet fire [8].In addition, some foreign researchers have analyzed the hazards of natural gas injection, such as Tom Bajcar, who has established a risk assessment system for natural gas metering stations[9].Y.Z Hao et al. established the calculation model of injection fire, analyzed the distribution rule of thermal radiation, and compared the numerical simulation results with the experimental results[10].

Much research has been done to improve the model derived from previous studies, mainly about the characteristics and the harm of jet fire. For example, Wang Wenhe studied the possibility of fireball combustion and vapor cloud explosion after a gas leak in an open space pipeline[11]. The numerical model of combustion and jet fire in open space is studied by Niu Kun[12]. Using the Thornton model to study the fire of a real instance, Wang Zhaoqin and other researchers analyzed the damage radius of the steam cloud explosion, the injection fire and the fireball fire in gas pipeline[13-14].

The existing researches on natural gas jet fire are mainly focused on the experimental study of the concentration of substance, the angle of injection and the size of voids in natural gas, and the study of fire hazard in open space. It is necessary to perform the research on the leakage of different pressure pipelines, especially the researches about this area are rare.

2 Mathematical Model

The jet velocity and air injection of natural gas injection can be calculated based on the pressure in the pipeline, as the initial parameters of FDS simulation. In FDS simulation, large eddy simulation is used as the numerical model of turbulent flow, and the field simulation theory was followed in the operation, namely, the three law of conservation of natural.

2.1 Jet Velocity Model

The characteristic parameters of natural gas jet fire are related to gas leaking velocity, and the gas leakage rate is related to gas velocity and medium pressure in the pipeline. There are different formulas for calculating the gas leakage rate at different flow rates.

When the pressure difference between the pipe inside and outside meets $\frac{P_0}{P} \le \left(\frac{2}{k+1}\right)^{\frac{k}{k-1}}$, the gas in the tube is sonic flow, and the flow velocity calculation formula is:

$$Q_0 = C_d A P \sqrt{\left(\frac{Mk}{RT}\right)\left(\frac{2}{k+1}\right)^{\frac{k+1}{k-1}}} \tag{1}$$

When the pressure difference between the pipe inside and outside meets $\frac{P_0}{P} > \left(\frac{2}{k+1}\right)^{\frac{k}{k-1}}$, the gas in the tube is subsonic flow, and the flow velocity calculation formula is:

$$Q_0 = C_d A P \sqrt{\frac{2k}{k-1}\frac{Mk}{RT}\left[\left(\frac{P_0}{P}\right)^{\frac{2}{k}} - \left(\frac{P_0}{P}\right)^{\frac{k+1}{k}}\right]} \tag{2}$$

Where P_0 is local atmospheric pressure, P is gas pressure in pipe, k is adiabatic index, Q_0 is leakage mass rate, C_d is leakage coefficient, M is gas molar mass, T is gas initial temperature, A is leak area, R is gas constant.

Corresponding author. Fax: +86-013098507822
E-mail address: shuqi_10@163.com

The gas leakage velocity is related to the Maher number, which is related to the static pressure at the leak hole. The static pressure at the leakage hole is determined by the initial pressure of the pipeline. The formula is as follows.

$$P_c = P \times \left[2/(k+1)\right]^{\frac{k}{k-1}} \tag{3}$$

$$M_j = \left\{\left[(k+1)\times(P_c/P_0)^{\frac{k-1}{k}} - 2\right]/(k-1)\right\}^{\frac{1}{2}} \tag{4}$$

$$T_j = T\times(P_0/P)^{\frac{k-1}{k}} \tag{5}$$

$$u_j = M_j \times (k\times 8.314\times T_j/M)^{\frac{1}{2}} \tag{6}$$

Where P_c is static pressure of leakage hole, M_j is Maher number, T_j is leakage hole pre expansion temperature, u_j is gas flow at the leak hole.

2.2 Governing Equation

FDS fire calculation follows the conservation equation of mass, momentum conservation equation, energy conservation equation and chemical reaction law.

1) Mass Conservation Equation

$$\frac{\partial \rho}{\partial t} + \nabla \bullet \rho u = 0 \tag{7}$$

Where ρ is leakage gas density, t is the simulation time, ∇ is Laplace operator, u is leakage of gas velocity vector.

2) Momentum Conservation Equation

$$\frac{\partial}{\partial t}(\rho u) + \nabla \bullet \rho uu + \nabla p = \rho f + \nabla \bullet \tau_{ij} \tag{8}$$

Where P is pressure, f is the external force vector, τ_{ij} is Viscous vector.

3) Energy Conservation Equation

$$\frac{\partial}{\partial t}(\rho h) + \nabla \bullet \rho hu = \frac{\partial p}{\partial t} + u \bullet \nabla p + q - \nabla \bullet q + \theta \tag{9}$$

Where h is gas component enthalpy, q is heat release rate of gas unit volume, q is radiant heat flux, Ø is dissipation rate.

3 Jet Fire Simulation of Different Pressure

3.1 Conditions set

We conducted multiple simulations of natural gas leakage and jet fire under 6 working conditions. When the pipe pressure value is 4.0Mpa, 2.5Mpa, 1.6Mpa and 0.8Mpa, the diameter of leak hole is 100mm, and when the pressure value of the pipe is 0.4Mpa and 0.2Mpa, the diameter of leak hole is 40mm. Set the wind speed at 2m/s and the pipeline is steel pipe. Since the gas injection tends to be stable in 5s, the simulation time 50s is sufficient to show the basic characteristics of the fire injection. Two hypotheses are made for the injection of pipeline gas leakage, one is that gas leaks and burns, the injection hole at the top of the pipeline and ejection of gas without any barriers, the other one is that the gas leakage is continuous leakage.

3.2 Flame Geometry

The simulation results show that the jet fire propagation distance is long at the beginning of leak, and then the distance gradually attenuates to an unstopped pulsation range. The larger the pipe pressure is, and the wider the flame contour is, and the more obvious the flame pulsation is. When the pipe pressure is greater than 0.4Mpa, the leakage hole becomes larger, and the leakage rate of gas increases exponentially. The combustion consumes a large amount of air around a wide range of the leakage. The flame contours of X direction and Y direction of 40mm leakage at 0.4Mpa and 100mm leakage at 0.8Mpa shows as Fig. 1.

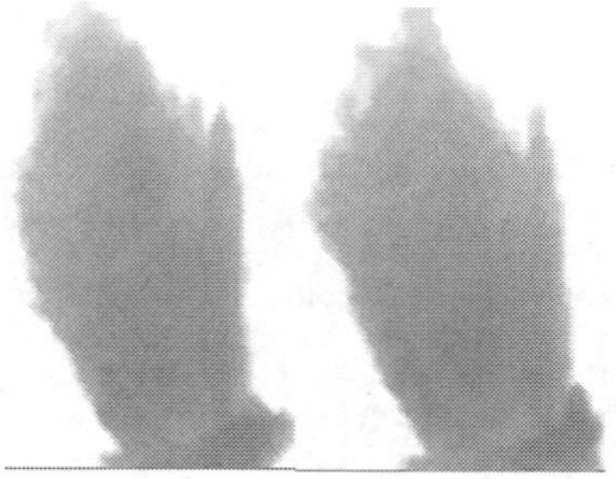

(a) Flame contour at 0.4Mpa

(b) Flame contour at 0.8Mpa

Figure 1: Flame contour at 2 different pressure

3.3 Temperature Analysis

Through temperature analysis, there is a small increase about temperature of the gas jet center near the leakage hole, and no obvious combustion phenomenon. The reason is that injection velocity of the gas at the hole is too high, the ejected gas is not sufficiently mixed with the air in an instant, and the internal oxygen concentration of the injection stream does not reach the range of combustion concentration.

The speed flow field is obviously at the jet center near the leakage. Gas velocity decreases gradually in flow field, and because of the Buoyancy and air resistance, the velocity of the gas will gradually decrease to 0m/s after a certain distance. When the pipeline pressure become larger, the more obvious velocity flow field of the gas leak, the longer time needed that velocity decreases to 0m/s, and the larger flow field. The temperature and velocity distribution of the x-z axis of the leakage center at 0.8Mpa are shown as Fig. 2.

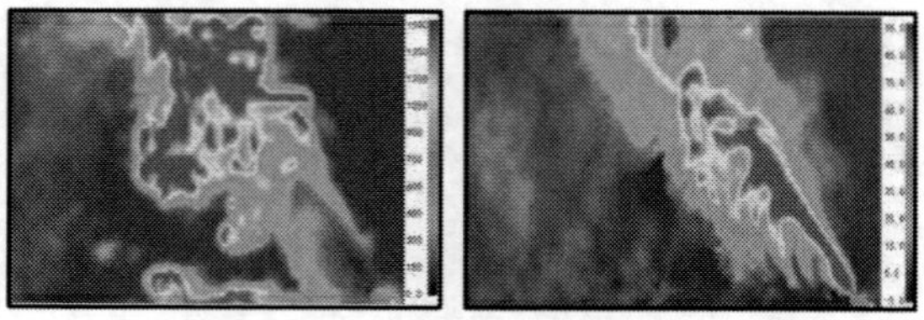

(a) Temperature monitoring (b) Velocity monitoring

Figure 2: Temperature and velocity distribution at 0.8Mpa

3.4 Thermal Radiation Analysis

Through monitoring the thermal radiation of fire by 50 seconds, we found that heat flux in the simulation first reached a peak, and then gradually decreased and fluctuated within a certain range. As the pressure of pipeline increases, the fluctuation of thermal flux become obvious, and thermal flux fluctuates in a large range. The heat flux monitoring results of the pipeline under 0.2Mpa and 0.4Mpa at the distance of 12m of the leakage hole is shown as Fig. 3.

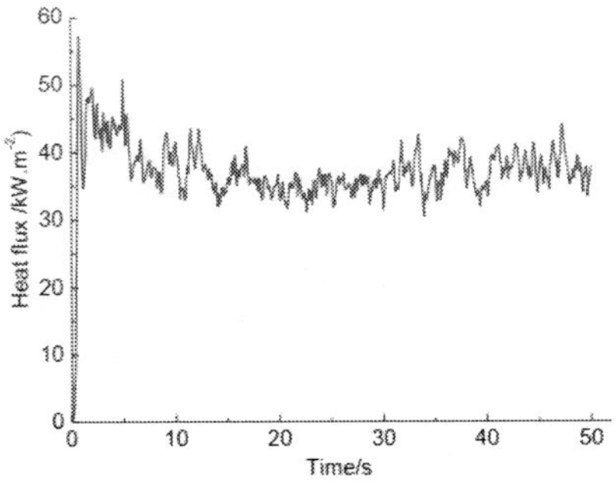

(a) The heat flux at 0.2Mpa

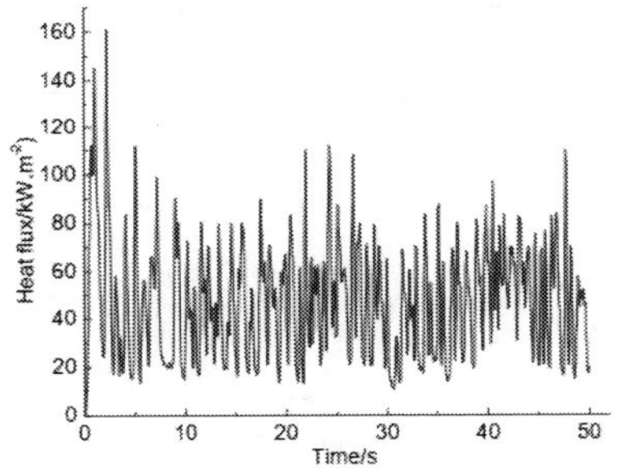

(b) The heat flux at 0.4Mpa

Figure 3: Heat flux monitoring results

The leakage of combustible gas is greatly influenced by the diameter of leakage hole. The thermal flux value of two different leakage hole dimensions is analyzed as follows.

1) Diameter of leakage hole is 40mm

The pipeline pressure is 0.2Mpa and 0.4Mpa, and the leakage hole is 40mm. The thermal flux values of the same distance of the leakage hole are measured respectively. The statistical results are shown in Tab. 1.

Distance / Pressure	7m	8m	9m	12m	16m	18m	28m
0.2Mpa	300	220	151	37.1	13.2	9.35	2.19
0.4Mpa	420	300	210	54.4	26.4	19.6	8.72

Table 1: Calorimeter value under 0.2mpa and 0.4mpa /kW.m^{-2}

As the distance between the monitoring point and the leakage hole increases, the heat flux gradually decreases. The relation curve of the jet heat flux and leakage hole distance is shown as Fig. 4. It can be seen that the distance between the thermal flux value and the leakage hole is exponential. When the distance increases, the value of the heat flux decreases rapidly.

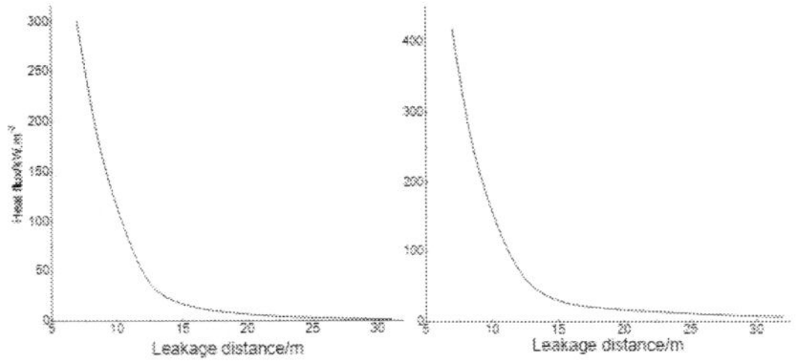

(a) Pressure of 0.2Mpa (b) Pressure of 0.4Mpa

Figure 4: Relation between heat flux and hole spacing

Under the same leakage hole condition, the thermal radiation value increases along with the increase of pipe pressure in the same distance of the leakage hole. With the increase of leakage hole distance, the heat flux difference between the two pressure conditions increases, the curve is shown as Fig. 5. It can be seen from the figure that the distance between the heat flux value and the leakage hole has a hyperbolic exponential relation. When the distance from the leakage hole is small, the heat flux value is large, and as the distance increases, the heat flux decreases obviously, and the decrease margin become small and stable.

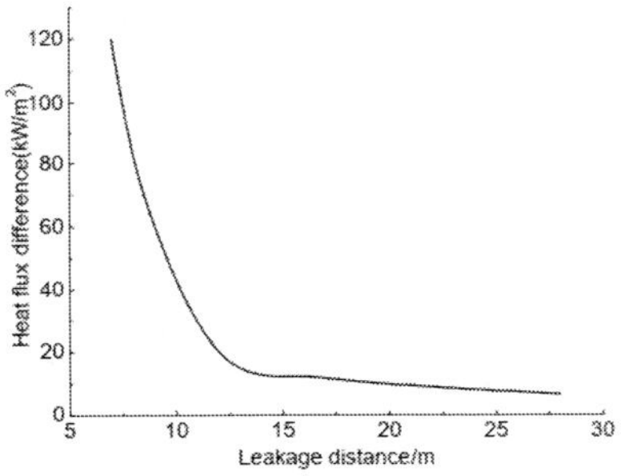

Figure 5: Variation of heat flux difference

2) Diameter of leakage hole is 100mm

When the pipeline pressure is equal or greater than 0.8Mpa, diameter of leakage hole is 100 mm and the pipeline pressure is 0.8mpa, 1.6Mpa, 2.5Mpa and 4Mpa. The jet fire heat fluxes are measured at the same discharge position. The result is shown as Fig. 6. The relation between heat fluxes values and leak spacing is exponential. With the increasing of distance, heat flux values first decrease greatly, and then decrease rate becomes small, eventually, the heat flux gradually close to a stable value. The variation trend of heat flux is consistent with the increase of pipeline pressure from 0.2mpa to 0.4Mpa.

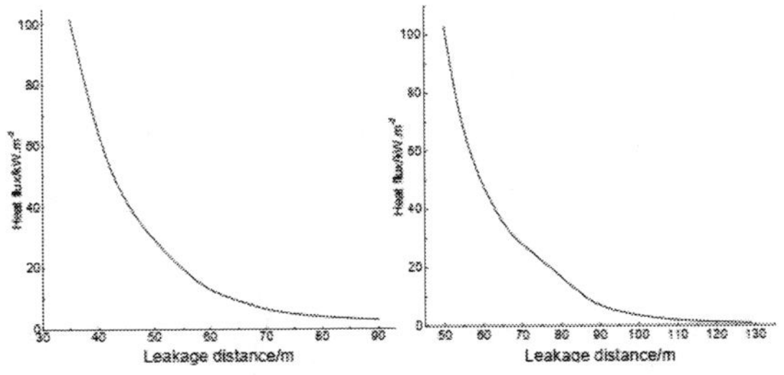

(a) Pressure of 0.8Mpa (b) Pressure of 1.6Mpa

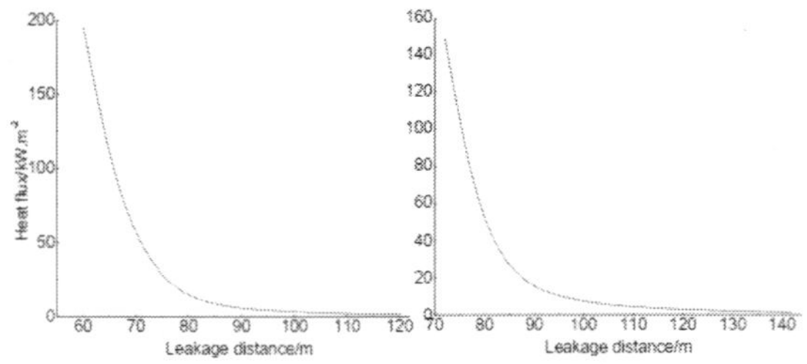

(c) Pressure of 2.5Mpa (d) Pressure of 4.0Mpa

Figure 6: Relation between heat flux and hole spacing

3.5 Hazard Analysis

The heat flux value is used to determine the standard value of heat radiation to personnel injury, by comparing and analyzing the position of the critical value, we found that the variation of thermal radiation simulation value is consistent with the trend of the damage radius. The figure 7 shows that near 0.8Mpa, as the pressure increase, the pipeline damage radius increase sharply, mainly due to increase size of leakage hole, leading to larger thermal radiation intensity, radiation scope and the damage radius of the combustion. When the pipeline pressure is greater than 0.8Mpa, leakage hole size maintains consistent, leakage rate of quality for the quality of pipeline pressure control rate, jet heat radiation damage radius increases with the increase of pressure pipeline is basically linear relation.

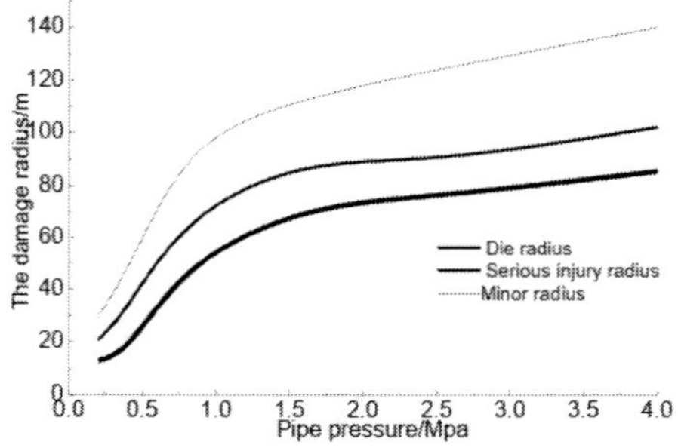

Figure 7: Variation trend of personnel injury radius

4 Conclusions

By using FDS software to simulate jet fire under the different pipeline pressure, we analyzed the influence law of pipeline pressure on characteristics of jet fire and personnel injury radius, including analysis of jet fire flame pulsation, temperature and radiation characteristics. The main conclusions are as follows:

1) Jet fire flame geometry size affected by the pipeline pressure. When diameter of leakage hole is fixed, the larger the pipeline pressure is, the larger the flame propagation distance of the jet flame is, the more obvious the flame pulsation is, and the more obvious and formation of the contour.

2) When the pipeline pressure is different, the flame temperature distribution is different. With the increase of the pipeline pressure, the flame high temperature influence area expands, and the flame pulsation phenomenon is more obvious.

3) The leakage rate of combustible gas is affected by the diameter of leakage hole, and the heat flux value is exponential in relation to the spacing of the leakage hole. The variation trend of thermal radiation value and damage range is consistent under different pipeline pressure. According to the results of hazard analysis, as the pressure of the pipeline increases, the radiation of the jet hot radiation increases the hazard range of the personnel.

5 Acknowledgment

We gratefully acknowledge the National Nature Science Foundation of China under Grant 51474151 for financial support.

References

[1] R. He, CH. XIU, The testing and assessment of oil and gas pipelines, China Petrochemical Press, Beijing, 2010, p.1-5.

[2] X. Chu, Site reconnaissance and trace characteristic of gas fire and explosion. Fire Science and Technology 33(4) (2014)466-469.

[3] Liu Jun-e, Consequence analysis of jet fire accident on urban gas pipeline. Journal of Hebei University of Engineering: Natural Science Edition 24(4) (2007)53-56.

[4] V. S. Ito A, Saito K, Flame Base Structure of Small-Scale Pool Fires. Combustion Science and Technology 26(1) (1996)1437-1443.

[5] P. D. B, Final report for blowout fire simulation tests Report No.NBS-GCR-85-484. Catlersburg, 1985

[6] B.J. Lowesmith, G. Hankinson, Large scale experiments to study fires following the rupture of high pressure pipelines conveying natural gas and natural gas-hydrogen mixture. Process Safety and Environmental Protection 9(1)(2013) 101-111.

[7] L. T. Cowley, A. D. Johnson. Oil and gas fires characteristics and impact. OTI92596, London: HSE. 1992.

[8] C. G. Soares, A.P. Teixeira. Probabilistic modelling of offshore fires. Fire Safety Journal 34(1) (2000) 25-45

[9] T. Bajcar, F. Cimerman, B. Širok, Model for quantitative risk assessment on naturally ventilated metering-regulation stations for natural gas. Safety Science 64(3) (2014) 50–59.

[10] Y. Zhao, J. Lai, L. Lu, Numerical modeling of fires on gas pipelines. Applied Thermal Engineering 31(6) (2011) 1347-1351.

[11] W. Wang, ZH. Xu, Consequences evaluation of fireball accidents made by leakage of natural gas pipelines. Journal of Safety Science and Technology 08(1) (2012) 18-21.

[12] N. Kun, *Study on dispersion and Deflagration of Flammable Gas*, phD thesis, Hebei University of Technology, Tianjin, China, 2007.

[13] W. Qihu, ZH. Li, Simulation of the Blowout Jet Fire Temperature Field and Radiation Field Based on CFD, Fire Safety Science, 18(1) (2009) 34-38.

[14] SH. Yang, Q. Fang Q, Consequence Quantitative Analysis of Jet Fire Caused by Gas Leakage from Underground Gas Storages in Salt Rock, Chines Journal of Rock Mechanics and Engineering,31(9)(2012) 1846-1853.

11th Asia-Pacific Conference on Combustion,
The University of Sydney, NSW Australia
10th -14th December 2017.

Application of spatially resolved emission spectroscopy to study low-pressure premixed ethylene/air sooting flames

S. Algoraini [1, 3], **S. Zhiwei** [2, 3], **Z. T. Alwahabi** [1, 3]

[1]School of Chemical Engineering,

[2]School of Mechanical Engineering,

[3]Centre for Energy Technology

The University of Adelaide, Adelaide, S.A. 5005, Australia

Abstract

Spatially resolved emission spectroscopy was applied to investigate C_2H_4/air flames under two low-pressures, i.e. 48 kPa and 32 kPa. Flame emission spectra were recorded at several heights above the burner surface (HABs). CH* and C_2* chemiluminescence around 430 nm and 560 nm and broadband soot luminosity were clearly identified in the spectra. It was found that CH* peaks at HABs of 3 mm and 4 mm for the pressure of 48 kPa and 32 kPa, respectively, while C_2* appears 4 mm and 5 mm with 1 mm higher than CH*. CH* and C_2* shows similar thickness (the width at half peak value) in the two flames. Soot particle temperature was evaluated from the spectrally broadband soot luminosity. At the inception region, the fitting of the broadband soot luminosity points out to soot particles temperature of 2432.8 K and 2385.7 K for pressures 48 kPa and 32 kPa, respectively. At the soot growth region, the fitting yields a soot particles temperature of 1600 K.

1 Introduction

Soot generated in combustion is harmful to human health [1], therefore it is desirable to mitigate soot emission from combustion. Meanwhile, soot generated during combustion process is one of dominant sources to enhance thermal radiation from flames to boilers. Therefore, it is also desirable to enhance soot formation in such applications. Soot formation and oxidation in flames are very complex, in which physical and chemical processes are non-linearly coupled together. To understand the complex processes of soot inception, nucleation, growth, and oxidation, as well as the integrations of these processes with flame turbulence, sooting flames have been extensively studied both experimentally and computationally, particularly focusing on atmospheric and elevated pressure flames, whose conditions are closely related to practical applications. As per fundamental studies of the chemistry of soot formation/oxidation, low-pressure sooting flames are also suitable because flames become extend to be thick at low pressure than atmospheric pressure condition. This benefits the measurement of the process taking place within a short distance. However, less low-pressure sooting flames have been reported compared with atmospheric and elevated pressure flames. In addition, it is already known that soot formation in flame is strongly dependent on pressure [2-5], investigation extending into low pressure flames can also help to understand the pressure dependence on soot formation.

Flame emission spectroscopy is a relatively simple optical method for combustion diagnostics. It can be easily applied to characterise flames, for example, flame structure [6, 7] and equivalence ratio [8] without the need of extra laser system. It is particularly useful in flames that have well characterized structures, for example one-dimensional flames. Chemiluminescence of OH*, CH* and C_2* in flames are often used to characterize flame structures. While OH* distributes in above combustion production zone, both CH* and C_2* are regarded as the distributes at primary combustion zone of hydrocarbon flames [9]. Thus, the use of Chemiluminescence in flame measurement present advantages to obtained the information from flame emission spectra properties for instance temperature and relative species concentrations.

This work utilizes spatially resolved emission spectroscopy (SRES) to investigate CH* and C_2* spatially resolved spectra and soot temperature profiles in well-controlled, low-pressure C_2H_4/air laminar premixed flames.

2 Experiments

Two ethylene/air premixed flames were stabilized on a water-cooled McKenna burner installed in a low-pressure chamber. The mixture of ethylene and air were supplied through the central part of the burner, i.e., a porous plate with a diameter of 60 mm. A steel plate was employed at a height of 25 mm to stabilize the flames. Fuel mixing was achieved by three mass-flow controllers (Bronkhorst). The air and fuel were mixed in a 2-m long tube before being supplied to the water-cooled McKenna burner. The flame equivalence ratio (Φ) was kept constantly as 2.1. To control the pressure, the burner was housed in a stainless-steel vacuum chamber of a 39.8 Liters volume with four optical accesses. The chamber was connected to a vacuum pump (Edwards, EDM12) to generate low pressure conditions. Pressure was measured using a Baratron (MKS, 122AA-0100AB), the pressure was kept with an accuracy of ± 0. 2 kPa. To achieve this accuracy, a pressure controller was connected to the vacuum line to regulate the pressure at desired values. These criteria were followed to determine the flow rates and the ranges of pressure, which were dependent on achieving excellent flame stability.

(Figure 1 presents spatially resolved emission spectroscopy set up.) The radiation emitted from the flame was collected with a lens (L1; f = 300 mm), then was reflected by a mirror and focused by a second lens (L2; f = 250 mm) to focus onto the end of a circular to linear fibre bundle (Thorlabs, BFL200HS02). The fibre was positioned on the axis of the equipment to allow for vertical translation. In the test, the identified flames were imaged

E-mail address: safa.algoraini@adelaide.edu.au

precisely at focus to focus separations running 23 mm high from the light fibre. The fibre conveyed light from the McKenna burner source to the spectrometer (Andor, Shamrock 500 lines/mm diffraction grating) equipped with an ICCD camera (Andor, iStar).

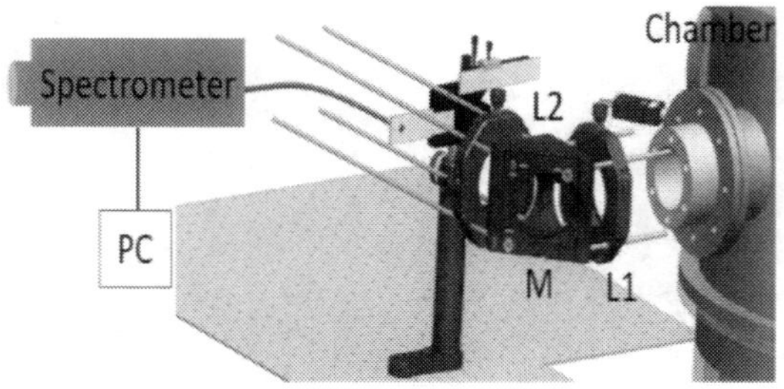

Figure 1: Experimental setup. L and M present spherical lens and mirrors, respectively.

3 Results and Discussion

3.1 Emission Spectra of CH* and C_2* radicals

Typical spectra are shown in Figure 2. At heights from 0 to 23 mm above the burner surface with an increasing step of 1 mm, at each height 200 spectra were collected and averaged. We did a calculation based on the configuration of the optical collection system and found that only the measurements below 3.2 mm was minorly influenced by the blocking edge of the burner about 12%. Also, as shown in flames images (Figure. 2) collected using a digital camera, no clear blue emission was found below 3 mm. CH* chemiluminescence from 418 to 430 nm, where the C_2* is from 500 to 525 nm.

Also shown in Figure 2, at pressure of 48 kPa, the CH* radicals started to form at a height of 2 mm HAB, and then the C_2* radicals appeared at 3 mm HAB. The peaks of CH* and C_2* increased from 3 mm HAB to 5 mm HAB. While at 4 mm HAB, the soot emission profile was shown beside the CH* and C_2*. The C_2* emission profile disappeared at 6 mm HAB; while the CH* emission was still recorded, The CH* emission decreased until it disappeared at 9 mm HAB. At 4 mm HAB, the emission of soot increased in intensity until it reached 10 mm HAB from where it started to decrease until the signal disappeared at 19 mm.

At 32 kPa, the CH* radicals began at a height of 2 mm HAB, and the C_2* started to form at 3 mm HAB. They emitted a clear signal at 4 mm, and the peak increased until a height of 6 mm HAB. At a height of 6 mm HAB, a soot emission was shown beside CH* and C_2* radicals. The C_2* vanished at 8 mm HAB, while only CH* and the soot emission were recorded at this height. The peak of CH* diminished once a height of 11 mm was reached, where at this height we could see only the emission of the soot profile. The soot emission started to decrease from 12 mm and disappeared at 17 mm.

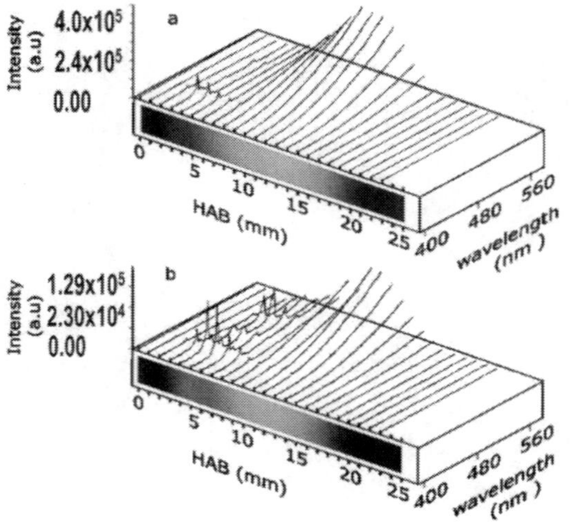

Figure 2: Flame photos and spatially resolved emission spectra recorded at $\Phi = 2.1$ for two different pressures, (a), 48 kPa and (b), 32 kPa.

Figure 3 presents the intensity of CH* chemiluminescence as a function of the height above the burner, the thickness of the CH* layer was found to be 6 mm and 7 mm at the pressure of 48 and 32 kPa, respectively. From Figure 3 it clear that the intensity of the CH* spectra decreased with increasing pressure. At 32 kPa, the highest intensity of the spectra was noted. A sharp spike in intensity was noted for all pressure points from 2 mm, peaking at 3 mm with pressure at 48 kPa. At the pressure 32 kPa, it reached its peaking at 4 mm (fast reaction HAB) in the 2.1 equivalence ratio. After 5 mm, there was a gradual decline in the intensity for all pressure points. The intensity became constant from around 14 mm HAB. Increasing pressure from 32 to 48 kPa had a huge impact on the distribution of the ethylene fuel components. This is because of a reduction in the quenching efficiency of the compacted fuel components. The effect of increasing the pressure could therefore be seen in the reduction of the flame temperature, which could be an indicator of the flame intensity. To satisfactorily quench the CH*, the side reactions ran in a similar manner, with the higher-pressure points determining the rate at which the combustion took place, if the fuel content was constant.

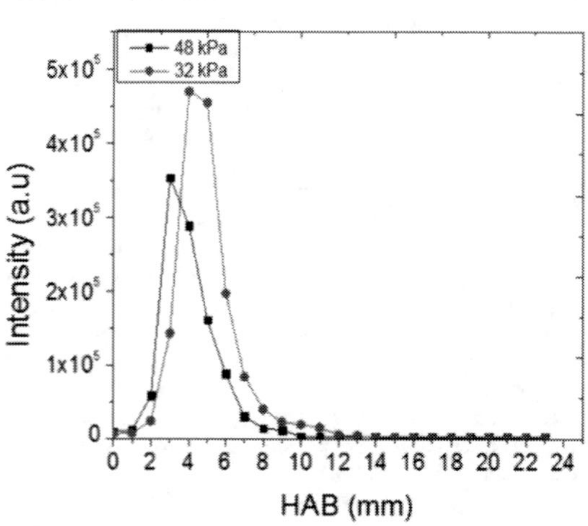

Figure 3: The intensity of the CH* at different pressure as function of the height above the surface burner.

Figure 4 presents the intensity of C_2* chemiluminescence as a function of the height above the burner, the thickness of the C_2* layer was found to be 2 mm and 4 mm at pressures 48 and 32 kPa, respectively. C_2* inception shifted at 1 mm at a comparably from CH* inception for both pressures. A spike in intensity was noted for all pressure points from 3 mm HAB. At pressure 48 and 32 kPa, C_2* was peaking at 4 mm and 5 mm HAB, respectively. The levels of intensity were considerably different at higher distances from the burner. Increasing the pressure resulted in the compaction of the fuel components, which then led to a reduced flame intensity regardless of the height above burner. Since the formation of the C_2* was part of the chain reactions that characterized the combustion process, it was noted that an increase in pressure had a concurrent effect on the distribution of the fuel components. For the 5 mm HAB, increasing pressure had caused a much more significant effect to the flame intensity compared to the 10 mm.

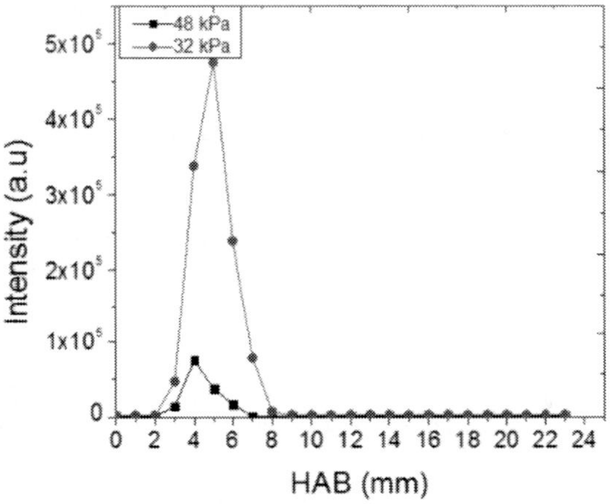

Figure 4: The intensity of the C_2* at different pressure as function of the height above the surface burner.

The results of the laminar premixed ethylene/air flame carried out under a well-controlled pressure indicated a distinct presence of CH* and C_2* excited hydrocarbon radicals. Figure 5 shows a strong presence of CH* at 430 nm wavelength. This observation is consistent for both the 5 mm and 10 mm HBA. CH* was formed as a product of a side reaction of ethylene and not a direct breakdown as expected in the combustion process and can be used as a marker for the flame front [10]. Biomolecular reactions in the fast reaction zone could also yield CH* radicals, which were evidenced by a blue flame color.

Figure 5 also shows C_2* bands observed in the 436 nm to 560 nm wavelength for the 5 mm HAB. These bands, known as the Swan bands, were formed as a result of the transition between triplet levels, or between singlet levels [10]. It is important to note that there was no inter-combination of any kind between the singlet and triplet levels. The wavelength of the C_2* Swan band made the flame turn into an intense greenish-blue color. The Swan band system is a core part of the triplet side system, and typically exists in the range of 435 nm and 686 nm. This, therefore, means that the results obtained in the laminar premixed ethylene/air flame were consistent with the previous experiments carried out. The spectral characteristics of the Swan band have been similar for a wide array of experimental sources [11]. The excitation mechanism was therefore expected to be the same, and the resultant flame color confirmed this assumption. The C_2* radicals were formed in an experimental environment where oxygen levels were below the specified stoichiometric proportions. The energy transfer process accounted for the reduction in the flame intensity at high pressure points. There was no apparent difference in the wavelength of the CH* excited radicals, hence a consistent blue flame was observed at 5 mm and 10 mm HAB. The C_2* Swan band was only observed in the 5 mm HAB, while the 10 mm HAB was characterized by soot formation.

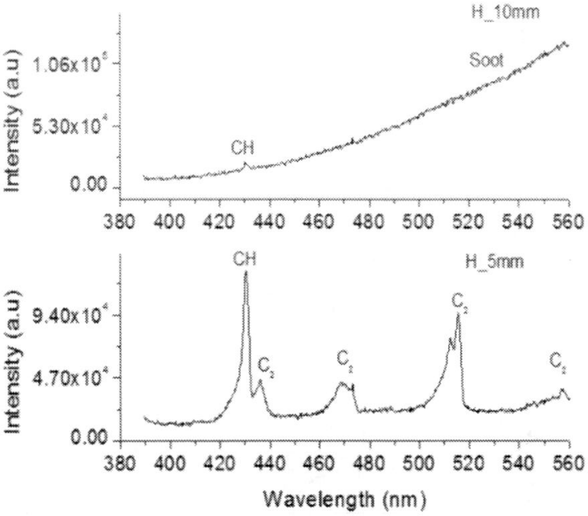

Figure 5: Flame emission at pressure 32 kPa with different height above the surface burner, 5 mm and 10 mm.

3.2 Soot Temperature

Soot luminosity at difference wavelength can be calculated using the Planck relation shown in equation (1), where M (λ) is radiative emission, T is temperature, λ is wavelength, h is Planck's constant, c is speed of light, k is the Boltzmann constant , ε_λ is the emissivity [12].

$$M(\lambda) = \varepsilon_\lambda \frac{2\pi^2 hc^2}{\lambda^5 e^{\frac{hc}{\lambda kt}} - 1} \quad (1)$$

Therefore, soot particle temperature can be derived by fitting the measured soot emission spectra after the correction of spectral response of the detection system. In the evaluation of particle temperature based on luminous spectra, a temperature change of ±10 K can cause clear larger fitting residuals than does the best fitting value. Also, the calibration broadband source has an uncertainty of ±15 K on this specific colour temperature, according to the specification provided by the manufacture. Therefore, the uncertainty of the particle temperature measured in this work should be within ± 25 K. Figure 6 presents a typical fitting curve with the corresponding recorded spectrum, showing a fitting quality.

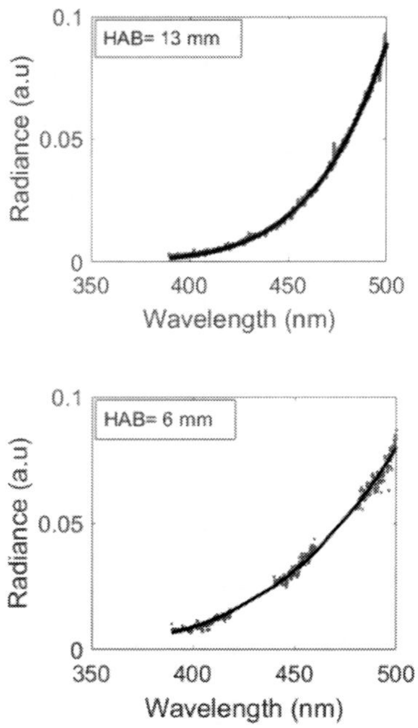

Figure 6: Soot emission spectra, red, at a height above the burner of 13 mm and 6 mm and pressure 32 kPa, together with the fitting curve, Black. The fitting points out to soot's particle temperature of 1607.9 ± 15 K and 2385.7, for the soot-growth and soot-inception regions, respectively.

Figure 7 presents the soot particle temperature derived using the above fitting process. An identical trend of temperature against HAB can be found in the two flames. The particle temperature decreases with the increase of HAB. Interestingly, the soot particle temperature is quite high around 5 mm, i.e. in the soot inception regions, which is higher than gas temperature (measured in another work). This may be due to the uncertainty of the emissivity of these nascent small soot particles. In the soot growth region, the soot particle temperature decreases to ~ 1600 K, which appears not to be affected by the pressure.

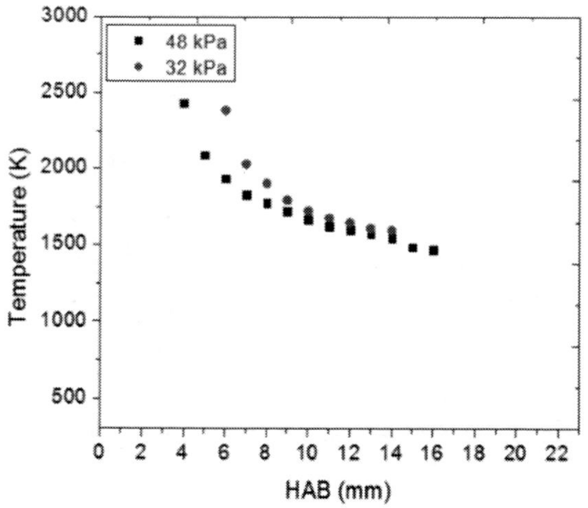

Figure 7: Soot particle temperature at different HABs for pressure of 48 kPa and 32 kPa.

4 Conclusions

Spatially resolved emission spectroscopy was applied to investigate C_2H_4/air flames under low-pressure condition. The location and the thickness of the of the CH* and C_2* chemiluminescence was recorded prior to soots inception. The CH* appears approximately 1 mm before C_2*. Soot particle temperature was evaluated from the fitting of luminous spectra. In the soot inception, the obtained temperature appears to be high, which might be due to the low luminous signals and the uncertainty of the emissivity of nascent small soot particles in this region. The soot particle temperature decreases to ~ 1600 K in the soot growth region.

5 References

[1] Barfknecht, T.R., Toxicology of soot. Progress in energy and combustion science, 1983. 9(3): p. 199-237.

[2] Joo, H.I. and Ö.L. Gülder, Soot formation and temperature field structure in co-flow laminar methane–air diffusion flames at pressures from 10 to 60 atm. Proceedings of the Combustion Institute, 2009. 32(1): p. 769-775.

[3] Steinmetz, S.A., T. Fang, and W.L. Roberts, Soot particle size measurements in ethylene diffusion flames at elevated pressures. Combustion and Flame, 2016. 169: p. 85-93.

[4] Karataş, A.E. and Ö.L. Gülder, Dependence of sooting characteristics and temperature field of co-flow laminar pure and nitrogen-diluted ethylene–air diffusion flames on pressure. Combustion and Flame, 2015. 162(4): p. 1566-1574.

[5] Desgroux, P., et al., Soot volume fraction measurement in low-pressure methane flames by combining laser-induced incandescence and cavity ring-down spectroscopy: Effect of pressure on soot formation. Combustion and Flame, 2008. 155(1–2): p. 289-301.

[6] Zhang, T., et al., Distribution Characteristics of OH*, CH*, and C2* Luminescence in CH4/O2 Co-Flow Diffusion Flames. Energy & Fuels, 2012. 26(9): p. 5503-5508.

[7] Kojima, J., Y. Ikeda, and T. Nakajima, Spatially resolved measurement of OH*, CH*, and C2* chemiluminescence in the reaction zone of laminar methane/air premixed flames. Proceedings of the Combustion Institute, 2000. 28(2): p. 1757-1764.

[8] Hardalupas, Y., et al., Chemiluminescence sensor for local equivalence ratio of reacting mixtures of fuel and air (FLAMESEEK). Applied Thermal Engineering, 2004. 24(11–12): p. 1619-1632.

[9] Zizak, G., Flame emission spectroscopy: fundamentals and applications. 2000.

[10] Gaydon, A., The spectroscopy of flames. 2012: Springer Science & Business Media.

[11] Appel, J., H. Bockhorn, and M. Frenklach, Kinetic modeling of soot formation with detailed chemistry and physics: laminar premixed flames of C 2 hydrocarbons. Combustion and Flame, 2000. 121(1): p. 122-136.

[12] Goulay, F., et al., A data set for validation of models of laser-induced incandescence from soot: temporal profiles of LII signal and particle temperature. Applied Physics B, 2013. 112(3): p. 287-306.

11th Asia-Pacific Conference on Combustion,
The University of Sydney, NSW Australia
10th -14th December 2017.

MILD Combustion of Grape Marc as a Source of Renewable Energy

Manabendra Saha*, Giovanni Gitto and Bassam Dally

Centre for Energy Technology, School of Mechanical Engineering,

The University of Adelaide, SA 5005, Australia

Abstract

The present study focuses on the experimental investigation of Moderate or Intense Low oxygen Dilution (MILD) combustion of grape marc as a potential biomass source of renewable energy. A vertical furnace with a cross section of 260 x 260 mm^2 and 1200 mm long was used in this study. The fine particles of grape marc, 150-250 μm in size, are fed through a central jet, using CO_2 as the carrier gas, and burned in a hot and vitiated co-flow. The furnace walls, as well as co-flow temperature and local oxygen concentrations, are controlled by a secondary swirling burner using non-premixed natural gas combustion. The co-flow stream temperature was fixed at 1320K with an excess O_2 mole fraction of 6%. The jet exit velocity was varied from 5.75 ms^{-1} (~Re_{jet} = 13,040) to 7.63 ms^{-1} (~Re_{jet} = 17,300) to investigate the impact of turbulent jet velocity, as well as the particle residence time, on the burning characteristics of grape marc under MILD combustion conditions. Detailed measurements of in-furnace temperatures and in-furnace CO and NO_x concentrations are presented and discussed, together with visual observations at the bottom, middle and top region of the furnace. It is found that, for both cases, a stable MILD combustion is established with a semi-uniform temperature profile. The measured NO_x emission of the higher jet velocity case is found to be 38% higher than that of lower jet velocity case. While the CO emission for the higher jet velocity case is considerably, around 51%, lower than the lower jet velocity case.

1 Introduction

Grape marc, includes grape skin, seeds, pulp, and stems, is the solid waste that remains after the juice is collected from the pressing of grapes for wine production. The estimated global production of grape marc is 13 million tonnes each year, while nationally, it is estimated that several hundred thousand tonnes are generated annually, and it is generally disposed of at a cost to the winery [1]. In recent years, grape marc has been proposed as a raw material for bio-energy production [2]. However, only a few studies, in the literature, have dealt with the combustion of grape marc [3-5] with very limited information available on energy conversion efficiency and the environmental impact of grape marc combustion.

Moderate or Intense Low oxygen Dilution (MILD) combustion is an advanced technique that offers ultra-low pollutant emissions, improved thermal efficiency, enhanced combustion stability, thermal field uniformity, and broad fuel flexibility. MILD combustion differs from conventional combustion because of the absence of any visible or audible flame at optimised conditions by creating a distributed volumetric reaction leading to an increase in the net radiant heat flux [6]. There is a progressive base of knowledge of the MILD combustion regime for fossil fuels [7-9] and the MILD combustion technology has been successfully implemented in various commercial sectors for gaseous, liquid, and solid fossil fuels [10]. However, the application of this combustion mode to biomass energy has received much less attention than that of conventional fuels. Abuelnuor et al. [11] reviewed the latest developments on the MILD combustion of biomass as a mean to reduce pollutant emissions and to decrease the rate of fossil fuel consumption. They found that this relatively new technology has many unresolved issues that require further attention and consideration. Some of those issues relate to the impact of turbulence on the devolatilisation as well as volatiles' release and reaction under MILD combustion conditions.

To the best of authors' knowledge, no study has been conducted to probe the MILD combustion characteristics of grape marc as a potential fuel. The objective of the current work is to investigate the influence of jet inlet velocities of CO_2 carried grape marc (at room temperature) on the flame stability and pollutants emissions under MILD combustion conditions. The turbulent bulk jet Reynolds number was varied from Re_{jet} = 13,040 to Re_{jet} = 17,300 to investigate the impact of turbulence on the burning characteristics of grape marc. For a fixed fuel flow rate, detailed measurements of in-furnace species concentration and temperatures along with exhaust emissions are analysed and discussed.

2 Experimental Setup

A schematic diagram of the MILD combustion furnace used in this project is shown in Figure 1.

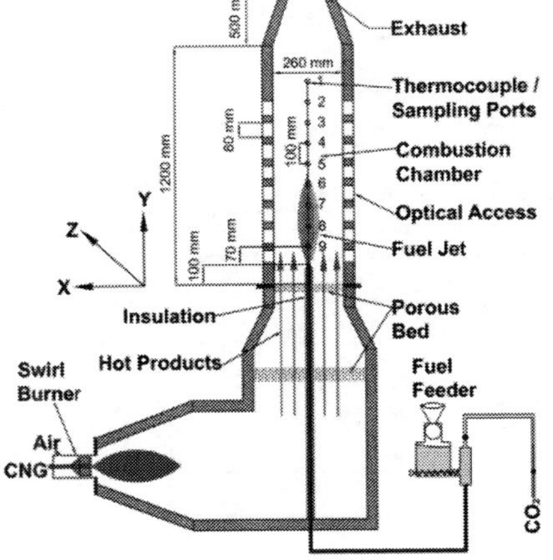

Figure 1: Schematic diagram of the MILD combustion furnace.

*Corresponding author. Fax: +61 (0)8 8313 4367
E-mail address: manabendra.saha@adelaide.edu.au

The furnace has an inner cross-section of 260 × 260 mm^2 and a height of 1200 mm. The furnace wall is well insulated, with 100-mm-thick high-temperature ceramic fibre-board refractory. A secondary swirling burner operating on the non-premixed combustion mode supplies hot combustion products under different stoichiometry. The water cooled and well insulated central jet with an internal diameter (D_j) of 19 mm is used to inject grape marc into the furnace. A summary of the operating conditions and measured gas concentrations in the co-flow is shown in Table 1. The in-furnace temperature was measured using 6-mm-diameter sheathed R-Type thermocouples at heights of 70, 170, 270, 370, 470, 570, 670, 770, and 870 mm above the fuel jet exit. All temperature measurements are logged using a PC and USB-TC data logger. A water cooled sampling probe was introduced through the thermocouple ports to measure in-furnace O_2, CO, CO_2 and NO_x mole fraction using a TESTO Model 350XL gas analyser. The absolute errors of these measurements, according to the manufacturer's specifications, are ±0.8% (by volume) for O_2, ±10 ppmv for CO, and ±5 ppmv for NO of the measured value. The sampling probe was traversed in the horizontal directions (Z-axis) at three positions (Z = 0, Z = 50, and Z = 100 mm) across eight different planes of various vertical locations (Y-axis) of the furnace to provide a total of 24 points of measurement.

2.1 Characteristics of Grape Marc

The grape marc as a biomass fuel used in this work was collected in April 2015 from the Riverland region of South Australia. Tarac Technologies Pty Ltd. (Tarac) collects grape marc from wineries in the main wine producing states of South Australia, Victoria and New South Wales. Tarac process grape marc for the recovery of grape alcohol and tartaric acid. The bulk of the experiments were performed with the Post-Tarac red marc (processed by Tarac before collection). The Post-Tarac red grape marc was milled and sieved into the 150 µm <d ≤250 µm size range. The grape marc was dried in an oven at 105 °C for 6 hours to remove all moisture and then stored in a sealed container under ambient conditions. Proximate and ultimate analyses of the grape marc are presented in Table 2. The gross dry calorific value of grape marc is measured to be 22.2 MJ/kg. A twin-screw volumetric particles feeder was specially built for this project to decouple the carrier gas flow rate from the solid particle supply rate. The feeder is operated using an AC motor with a speed controller to regulate the amount of grape marc particles fed into the furnace. The grape marc feeding rate was constant and set to 1.64 kg/ hr corresponding to 10.1 kW heat input from the biomass particulates. The additional heat input from the hot co-flow, supplied by the secondary burner was 40 kW.

Parameters	**Case-1**	**Case-2**
Fuel jet velocity (m/s)	5.75	7.63
Bulk jet Reynolds number	13040	17300
Particle residence time (ms)	209	157
Global equivalence ratio, ϕ	0.85 ± 0.01	0.90 ± 0.01
Fuel jet inlet temperature (K)	325	325
Co-flow velocity (m/s)	1.58	1.58
Co-flow temperature (K)	1323	1320
Co-flow O_2 (v/v%)	6.01	6.04
Co-flow NO (ppmv)	116	106

Table 1: Furnace operating parameters.

Ultimate analysis *(wt% dry ash free)*	
C	51.49
H	6.52
O (by difference)	37.71
N	2.74
S	1.54
Proximate analysis *(wt% dry ash free)*	
Fixed carbon	24.98
Volatile matter	65.66
Ash	9.36

Table 2: Grape marc analyses.

3 Results and Discussion

Shown in Figure 2 are photographs taken from a side window close to the bottom (Y = 80 mm), middle (Y = 500 mm), and top (Y = 995 mm) of the furnace for both jet Reynolds number (Re_{jet}) cases. The photographs were taken with the same camera settings (F/4.6–16 ms–ISO 250). It is clear that there is no visible flame apparent inside the furnace for either case, which can be described as stable flameless or MILD combustion. Although visual inspection through the windows of the furnace could not detect any significant differences in terms of particles' ignition, the combustion of the higher Re_{jet} case is far more luminous in comparison with the combustion of lower Re_{jet} case. This difference indicates a higher reaction rate induced by the enhanced mixing, which will have direct impact on devolatilisation rate, ignition and burnout.

Shown in Figure 3 is the measured temperature contours for the two Re_{jet} cases. It is clear that temperature inside the furnace is semi-uniform and the temperature gradient along the furnace is very low with a maximum difference of less than 100K. The contour plots also depict the trend of decreasing temperature with increasing fuel injection velocities, leading to a large temperature difference at the top part of the furnace. The difference is attributed to the carrier gas stream, which adds 32% more CO_2 to the system for the Re_{jet} = 17,300 case, as compared with Re_{jet} = 13,040 case. The thermal mass of the additional CO_2 accounts for the heat sink and most of the resultant reduction in temperature.

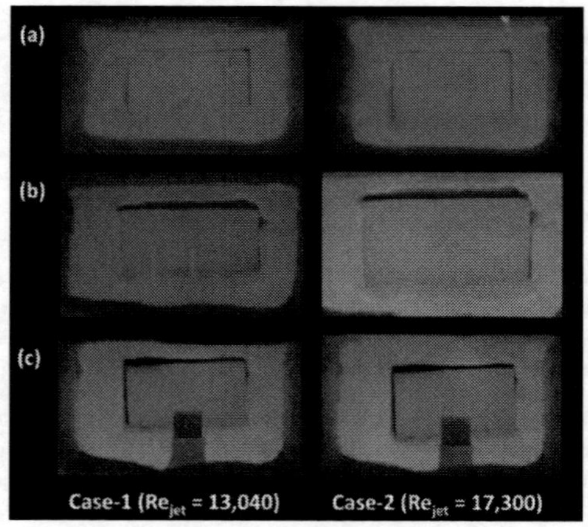

Figure 2: Photographs of the inside of the furnace; (a) top, (b) middle, and (c) bottom part of the furnace.

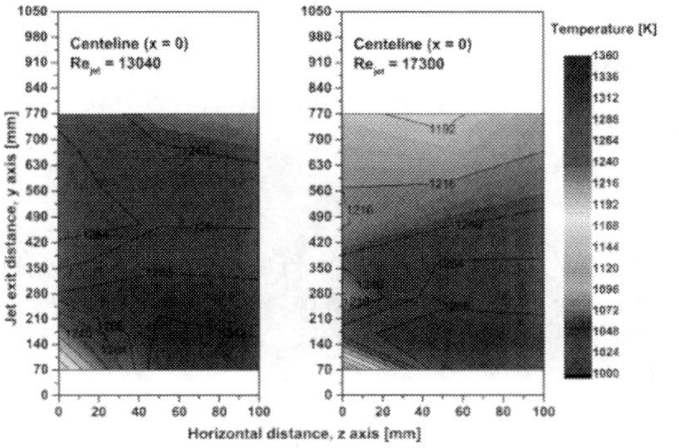

Figure 3: Measured mean temperature distribution for one-half in the middle of the furnace

In addition, the temperature contours relate to volatiles' release and reaction which is extended further downstream with increasing jet velocity and moreover, the high-velocity cold jet penetrates further downstream for the largest Reynolds number case. The increased penetration of the jet extends the volatiles' release zone, enhances mixing with the surrounding flow and delays ignition and start of the reaction.

Figure 4 represents a comparison of in-furnace CO dry mole fraction between two different jet velocities along the centre line of the furnace. The measured in-furnace CO level for the Re_{jet} = 13,040 case is generally higher in these experiments than the Re_{jet} = 17,300 case. CO is predominantly formed by the oxidation reactions of the volatile contents and char. The lower jet Reynolds number leads to a higher residence time of grape marc and consequently provides sufficient time to volatile combustion and char burnout. In the top part of the furnace, CO levels are found at their lowest concentration, suggesting that the gaseous volatiles are almost burned by then.

Figure 5 compares the measured in-furnace NO dry molar fraction between the two jet Reynolds number cases and for all measurement locations.

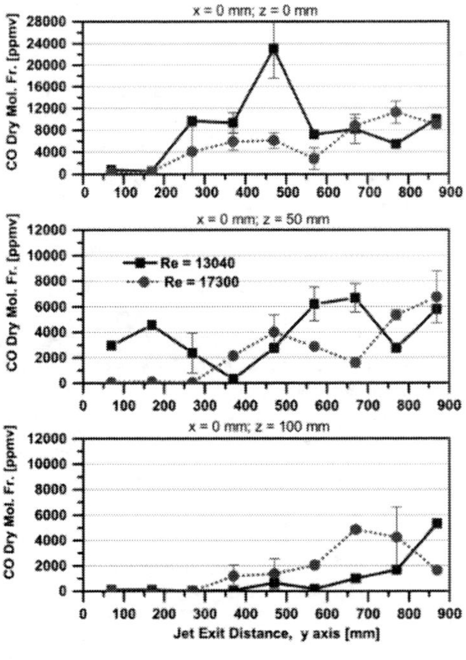

Figure 4: Measured profiles of in-furnace CO molar fraction.

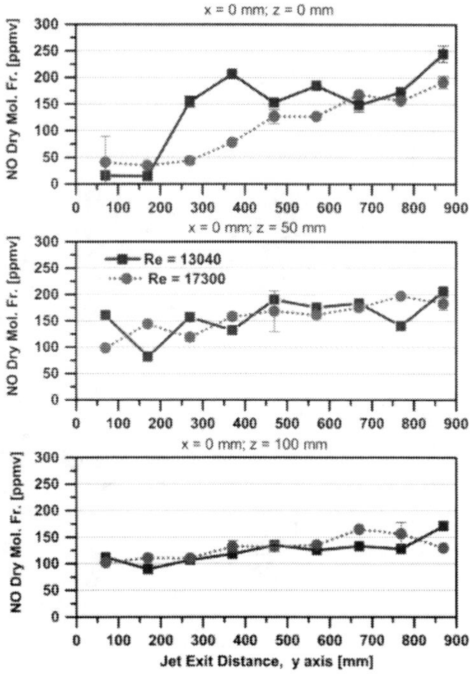

Figure 5: Measured profiles of in-furnace NO molar fraction.

An increasing trend of NO measured concentration is found along the furnace height. It is interesting to note that the first NO concentration peak for the Re_{jet} = 13,040 case is found to be ≈213 ppm at 370 mm above the jet exit along the furnace centreline (X = 0 mm, Z= 0 mm). Whilst, for the case of Re_{jet} = 17,300, the first NO concentration peak is measured to be ≈133 ppm at 470 mm above the jet exit which can be explained by the early devolatilisation of grape marc particles for the lowest Reynolds number case. Considering the NO profiles at 50 mm (X = 0 mm, Z= 50 mm) and 100 mm (X = 0 mm, Z= 100 mm) away from the furnace centreline, minor differences between two cases can be observed, although the overall NO formation for the Re_{jet} = 13,040 case is dominating over the Re_{jet} = 17,300 case. This can be explained by the difference of residence time of grape marc particles between two cases where the particle residence time is around 32% higher for the lowest jet Reynolds number case. The long particle residence time allows the production of higher NO. It is worth noting that almost all NO is produced through the fuel-NO_x mechanism for both cases. Because, thermal-NO_x formation is insignificant for this experiments as the local temperatures are below 1800 K. Indeed, the highest mean temperature of the present experiments is measured to be lower than 1350 K and consequently the contribution of thermal NO_x is negligible. It is speculated that the incorporated nitrogen (N) in the grape marc is entirely discharged throughout the devolatilisation process and diffused into the volatile matters and the char. Thus, the fuel-NO_x is produced by two evolution paths such as oxidation reaction of volatile-N and char-N.

Figure 6 illustrates the comparison of measured concentrations of exhaust gases including; NO (measured value of NO_2 was negligible in this experimental campaign), CO, O_2, and CO_2 for both cases. The largest NO emission was measured for the Re_{jet} = 17,300 case which is ≤ 507 ppm, whilst for the case of Re_{jet} = 13,040, the value is ≤ 368 ppm. The overall NO formation is impacted by the mixing of fuel streams with the surrounding hot combustion gases. Whilst, co-flow contributes additional

115 ppm NO to the system. The higher jet Reynolds number attributed to the strong mixing of hot co-flow combustion products with the volatile matter of the grape marc and extended the char burnout zone close to the exhaust port. As a consequence, the NO concentrations at the exhaust is found to be significantly higher than in-furnace measurements.

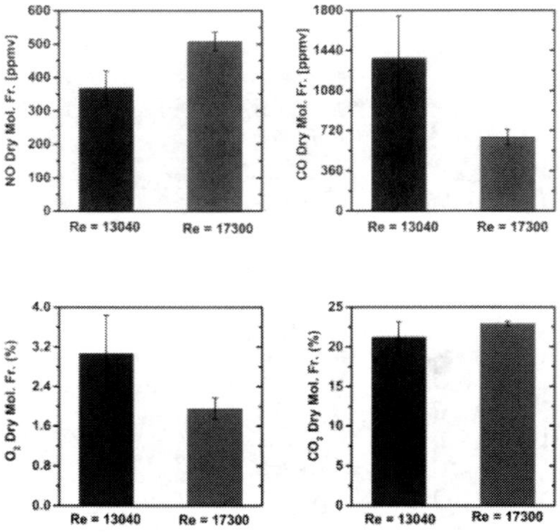

Figure 6: Exhaust gas concentrations at the furnace exit.

The exhaust CO level was measured to be 1250~1650 ppmv for the Re_{jet} = 13,040 case while for the highest Reynolds number case the CO level was 610~690 ppmv. It is noticeable that the exhaust CO emission for the lowest Reynolds number case is considerably higher than that of largest Reynolds number case. The reason of higher CO emission for the lowest Reynolds number case is most likely related to the more eminent char burnout by the heterogeneous reactions between char ($C_{(s)}$) and gaseous products ($C_{(s)} + 1/2O_2 \rightarrow CO$, $C_{(s)} + CO_2 \rightarrow 2CO$, $C_{(s)} + H_2O \rightarrow CO + H_2$) of the combustion reaction. A lower value of exhaust O_2 was measured for the largest Reynolds number case. For the same boundary conditions, the lower value of exhaust O_2 concentration indicates that more intense combustion of grape marc for the higher jet Reynolds number case owing to the strong mixing of jet with surrounding hot combustion products. The concentration of CO_2 emission was measured to be higher for the highest Reynolds number case which is related to the largest volume of incoming CO_2 concentration as a carrier gas.

4 Conclusions

This study aim is to explore the effect of turbulence on the burning characteristics of grape marc as a biomass fuel in a vertical furnace operating under MILD combustion conditions. The interaction of the volatile matters of grape marc with the vitiated co-flow and its impact on the formation and emission of NO and other pollutants have been investigated. A stable MILD combustion of grape marc is achieved for both cases. No visible flame is found in the furnace for any case. Higher luminosity and interior furnace radiation is observed for the higher Reynolds number case than the lower. A semi-uniform temperature distribution is measured inside the furnace with <100 K difference from the bottom to the top of the furnace for both cases. The NO emission is substantially, around 38%, higher for higher Reynolds number case than lower Reynolds number case. The CO emission for the larger Reynolds number case is considerably, around 51%, lower than the lower Reynolds number case. This indicates char burnout rate is significantly lower for the higher Reynolds number case in comparison with the lower Reynolds number case. Detailed modelling studies are currently underway to better understand the dependence of turbulence on emissions and carbon burnout of grape marc under MILD combustion conditions.

5 Acknowledgment

The authors would like to thank Mr Marc Simpson, Manager, Thebarton Research Laboratory, University of Adelaide for his help throughout the experiments.

References

[1] K. R. Corbin, Y. S. Hsieh, N. S. Betts, C. S. Byrt, M. Henderson, J. Stork, *et al.*, "Grape marc as a source of carbohydrates for bioethanol: chemical composition, pre-treatment and saccharification," *Bioresource Technology,* vol. 193, pp. 76-83, 2015.

[2] G. Toscano, G. Riva, D. Duca, E. F. Pedretti, F. Corinaldesi, and G. Rossini, "Analysis of the characteristics of the residues of the wine production chain finalized to their industrial and energy recovery," *Biomass and Bioenergy,* vol. 55, pp. 260-267, 2013.

[3] A. Celma, S. Rojas, and F. López-Rodríguez, "Waste-to-energy possibilities for industrial olive and grape by-products in Extremadura," *Biomass and Bioenergy,* vol. 31, pp. 522-534, 2007.

[4] M. Valente, A. Brillard, C. Schönnenbeck, and J.-F. Brilhac, "Investigation of grape marc combustion using thermogravimetric analysis. Kinetic modeling using an extended independent parallel reaction (EIPR)," *Fuel Processing Technology,* vol. 131, pp. 297-303, 2015.

[5] L. Fiori and L. Florio, "Gasification and combustion of grape marc: comparison among different scenarios," *Waste and Biomass Valorization,* vol. 1, pp. 191-200, 2010.

[6] A. Cavaliere and M. de Joannon, "Mild combustion," *Progress in Energy and Combustion Science,* vol. 30, pp. 329-366, 2004.

[7] R. Weber, J. P. Smart, and W. Kamp, "On the (MILD) combustion of gaseous, liquid, and solid fuels in high temperature preheated air," *Proceedings of the Combustion Institute,* vol. 30, pp. 2623-2629, 2005.

[8] M. Saha, B. B. Dally, P. R. Medwell, and A. Chinnici, "Effect of particle size on the MILD combustion characteristics of pulverised brown coal," *Fuel Processing Technology,* vol. 155, pp. 74–87, 2017.

[9] M. Saha, B. B. Dally, P. R. Medwell, and E. M. Cleary, "Moderate or Intense Low Oxygen Dilution (MILD) Combustion Characteristics of Pulverised Coal in a Self-Recuperative Furnace," *Energy & Fuels,* vol. 28, pp. 6046–6057, 2014.

[10] P. F. Li, J. C. Mi, B. Dally, F. F. Wang, L. Wang, Z. H. Liu, *et al.*, "Progress and recent trend in MILD combustion," *Science China Technological Sciences,* vol. 54, pp. 255-269, 2011.

[11] A. Abuelnuor, M. Wahid, S. E. Hosseini, A. Saat, K. M. Saqr, H. H. Sait, *et al.*, "Characteristics of biomass in flameless combustion: A review," *Renewable and Sustainable Energy Reviews,* vol. 33, pp. 363-370, 2014.

11th Asia-Pacific Conference on Combustion,
The University of Sydney, NSW Australia
10th-14th December 2017.

Transient Simulation of a Heavy Duty Turbocharged Multi Cylinder Diesel Engine

Sandeep V[1], Sunil Kumar Pandey[2], Devanandh V[2], Brijesh P[3]

[1]ARAI Student

Kothrud, Pune, Maharashtra, 411038, India

[2]Ashok Leyland Technical Centre

Vellivoyalchavadi, Chennai, Tamil Nadu, 600103, India

[3]ARAI Faculty

Kothrud, Pune, Maharashtra, 411038, India

Abstract

Diesel engines are presently facing the challenge of controlling oxides of nitrogen (NO_x) and soot emissions on transient cycles, to meet stricter emission norms and to control emissions during field operations. Development of a simulation tool for prediction of NO_x and soot emissions on transient operating cycles has become the most important objective, which can significantly reduce the experimentation time and cost required for tuning these emissions. Hence, in this work, 0D comprehensive predictive model has been formulated with selection and coupling of appropriate combustion and emissions models to engine cycle models. The numerical simulation was performed with the standard specification of a diesel engine by using MICROSOFT EXCEL software with least time. Selected combustion and emissions models were further modified to improve their prediction accuracy in the engine full operating regime. Responses of the combustion and emissions models have been validated for load and start of injection changes. Predicted transient fuel consumption, NO_x and soot emissions with model were found in good agreement with measured data on the turbocharged common rail diesel engine for the European transient cycle. Results of the study indicated that 0D models can be used for prediction of transient emissions coming out from diesel engines.

1 Introduction

Diesel engines have distinct advantages of better fuel economy, higher torque back-up and high reliability over the other prime movers. For off-road and commercial vehicles, mostly diesel engine is used as the prime mover and it may maintain its dominance for many years to come. However, these engines require very complex after-treatment devices to meet stringent emission targets for NO_x and soot, on transient legislative cycles as well as on different transient cycles representing field operations. There are differences in NO_x and soot emissions between operation under transient condition with speed or load increase and their corresponding steady state speed-load points [1]. Hence, to meet transient cycle emissions with low cost after-treatment devices as well as to achieve lower emissions in real operations, engines are required to be optimized. For prediction of transient emissions, interactions and responses of different combustion related systems are required to be considered. Quasi-stationary steady state based models with correction for transient conditions fail to predict transient emissions [2]. Hence, transient NO_x and soot prediction model will be of great help to reduce experimentation time and cost. Study of emissions during transient operation with development of suitable models is the most important objective for the future and has received a little attention so far [3, 4].

Corresponding author. Brijesh P; Fax: +91-20-3023-1104
E-mail address: patel.pga@araiindia.com

While selecting the simulation approach, tuning cost and time become extremely important when a large number of cycles for complete engine are to be evaluated. 3D models demand large amount of computing power and time; also their benefit lowers due to requirement of empirical correlations [5]. Hence, 0D multi-zone models are used for these analyses; however the majority of available 0D transient models have not included emissions prediction [1].Generally, diagnostic models prepared by statistical processing on experimental data are used for analyses of transient emissions. It is required to change in statistical function parameters of diagnostic combustion model to predict apparent heat release rate (AHRR) with varying load operating conditions [6, 7]. NO_x and soot diagnostic simulation models prepared from regression of experimental data [7] or by neural network trained thermodynamics mimicked model [4] are required for compensations with transient operations. Additionally, such diagnostic models have prediction applicability only for the engine on which that particular empirical model is prepared. Empirical model also requires large amount of engine test data and regression efforts. It is also experienced that the commercially available software for transient emission analysis requires extensive efforts for tuning constants to match with the experimental data, throughout the operating regime. On the contrary, predictive models do not require test data for model formulation and do not require compensations with transient operation or for any particular operation zone. Hence, such models can be used as a design tool; however appropriateness of 0D models as a tool for prediction of transient emissions of modern engines is required to be assessed. Hence, in this work, a 0D comprehensive predictive model has been formulated with selection and coupling of appropriate combustion and emissions models to engine cycle models. Selected combustion and emissions models are further evaluated for their prediction accuracies in the engine full operating regime. European transient cycle (ETC) has been selected for model validation.

2 Simulation Methodology

The formulated comprehensive model consists of torque equilibrium, friction, combustion and emissions models. Coupling of these models and sub-models used in each of these models are shown in Fig. 1. Fuel consumption is predicted using torque equilibrium and modified friction sub-models. Single-zone combustion model and two-zone emissions models are used for predicting combustion parameters and emissions respectively.

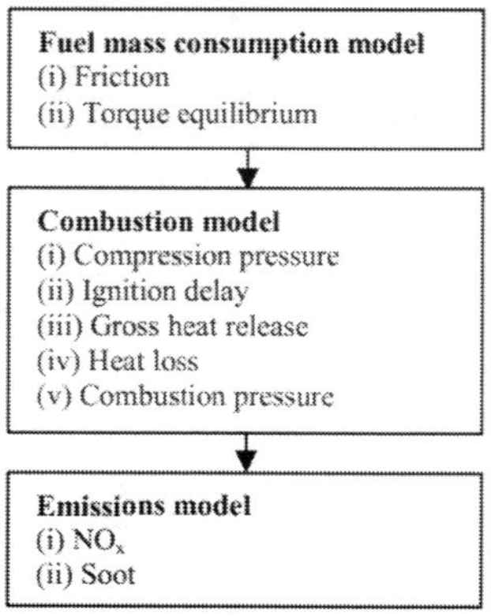

Figure 1: Comprehensive model structure

2.1 Fuel Mass Consumption Model- In this model, fuel mass consumption is predicted with the help of total torque required to be produced by the engine and its rpm. This total torque is addition of operating cycle torque demand(τ_{cycle}), frictional torque (τ_{fr}) and torque required for engine acceleration, as given in Eq. 1 [8]. τ_{fr} is estimated using modified friction mean effective pressure (FMEP) model as described in Eq. 2. FMEP model uses brake mean effective pressure (BMEP) as one of the model input (refer Eq. 3).

$$\dot{m}_{fuel} = \frac{C_{IP}\pi N\left(\tau_{cycle}+\tau_{fr}+G_E\left(\frac{dw}{dt}\right)\right)}{30000} \tag{1}$$

$$\tau_{fr} = \frac{FMEP_m A_{pist} rn_{cyl}}{2\pi} \tag{2}$$

$$FMEP = \alpha + \beta_{1m} e^{\beta_{1f} BMEP} + \gamma U_{pist} \tag{3}$$

The followings are the fuel consumption model constants derived by experimental investigations which area modified version of [9] and the values may get changed with respect to engines.

$C_{IP} = 6.1358\times10^{-3}$ $\beta_{1m} = 5.00\times10^{3}$ $\gamma = 2.84\times10^{4}$

$\alpha = 4.5\times10^{4}$ $\beta_{1f} = 2.64\times10^{-6}$

2.2 Combustion Model- Combustion and emission parameters are calculated at the interval of 0.1° crank angle (CA) from intake valve closing (IVC) to exhaust valve opening (EVO). Combustion model starts from prediction of compression pressure, considering a polytropic compression process from IVC to start of combustion (SOC). Hardenburg and Hase correlation as referred by Lakshminarayan and Aghav [10] as well as correlation proposed by Jung and Assanis [11] have been considered for prediction of the polytropic index (n) of the compression process. After investigations with experimental results, it has been observed that predictions by Jung correlation are closer to the measured data. However, the predicted compression pressure matches the experimental data only at one speed. Hence, to improve the model prediction accuracy at other operating speeds, Jung correlation model is modified by introducing engine speed (N) dependent correction, as given in Eq.4.

$$n = 1.4 - \frac{60}{N} - 6\times10^{-5}T_{cyl} + 1\times10^{-8}T_{cyl}^2 \tag{4}$$

For predicting ignition delay, the empirical formula used in the present model is provided in Eq.5 which was given by Hardenberg [12, 13]. It gives the value of ignition delay in terms of values of CA using temperatures (T) in Kelvin and pressure (P) in bar.

$$\tau_{id} = \left(0.36 + 0.22u_{pis}\right)\exp\left[E_A\left(\frac{1}{RT_{cyl}} - \frac{1}{17190}\right)\left(\frac{21.2}{P_{cyl}-12.4}\right)^{0.63}\right] \tag{5}$$

The rate of heat release from the cylinder walls has been calculated using the Eq. 7 [14].

$$\frac{dQ_h}{d\theta} = hA_\theta(T_\theta - T_w) \tag{6}$$

where h, the heat transfer coefficient is found using Woshni model, as described in Eq. 8[15].

$$h = 129.9B^{-0.2}p^{0.8}T_g^{-0.53}\left(C_1V_p + \frac{C_2V_sT_r(P-P_r)}{P_rV_r}\right)^{0.8} \tag{7}$$

In this expression, P and T_g correspond to instantaneous in-cylinder conditions and P_r and T_r are known working values corresponding to a volume V_r of a reference state, such as inlet valve closure or beginning of combustion. T_w is the wall temperature which have assumed as 450K.

The engine combustion characteristics are mainly understood through the AHRR that is determined from the measured cylinder pressure data as a function of CA (θ). The rate of heat release has been estimated from Weibe's heat release model provided in 'Eqs.9 to 11.

$$AHHR = (LHV)M_f\frac{df}{d\theta} \tag{8}$$

$$\frac{df}{d\theta} = \frac{a(m+1)}{\Delta\theta}\left(\frac{\theta-\theta_0}{\Delta\theta}\right)^m e^{-a\left(\frac{\theta-\theta_0}{\Delta\theta}\right)^{m+1}} \tag{9}$$

$$AHHR = (LHV)M_f\frac{a(m+1)}{\Delta\theta}\left(\frac{\theta-\theta_0}{\Delta\theta}\right)^m e^{-a\left(\frac{\theta-\theta_0}{\Delta\theta}\right)^{m+1}} \tag{10}$$

Gas properties during the compression and expansion have been calculated using isentropic relations given in Eqs. 11 and 12 [16-17].

$$P_{\theta+1} = P_\theta\left[\frac{V_\theta}{V_{\theta+1}}\right]^n \tag{11}$$

$$T_{\theta+1} = T_\theta\left[\frac{V_{\theta+1}}{V_\theta}\right]^n \tag{12}$$

Pressure is predicted by using first law (refer Eq. 14) with heat release rate modelled with Weibe function and compared with experimental pressure data at different loads. Pressure curves for various heat release pattern can be calculated by solving ordinary differential equation using forth order Runge-Kutta method.

$$\frac{dp}{d\theta} = \frac{\gamma-1}{V}\left(\frac{dQ_{gr}}{d\theta} - \frac{dQ_W}{d\theta}\right) + \gamma\frac{P}{V}\frac{dV}{d\theta} \tag{13}$$

where, Q_{gr} is gross heat release and Q_W is heat loss through walls.

Once it is solved for pressure, temperature can be calculated using the Eq. 14 obtained from the equation of state.

$$T_{cal} = P_{cal}V_{cal}\left(\frac{T_{ref}}{P_{ref}V_{ref}}\right) \tag{14}$$

2.3 Emissions Model- NO formation for each thermodynamic cycle is modelled using the Zeldovich mechanism. After establishing the temperature and pressure data at each crank angle using zero dimensional model the equilibrium concentrations of the nitrogen $[N_2]_e$ and oxygen$[O_2]_e$in moles can be calculated (refer Eqs. 19 and 20) to predict the NO formation. The nitric oxide emissions are computed using Eq. 15 [13,18].

$$\frac{d[NO]}{dt} = 2k_1^+[O]_e[N_2]_e \tag{15}$$

$$[O]_e = \frac{K_{p(o)}[O_2]_e^{\frac{1}{2}}}{(RT)^{\frac{1}{2}}} \tag{16}$$

$$k_1^+ = 1.82\times10^{14}e^{\left[-\frac{38370}{T(K)}\right]} \tag{17}$$

$$K_{p(o)} = e^{-\frac{\Delta G^0 T}{R_u T}} \tag{18}$$

$$[O_2]_e = \frac{0.21\times P}{RT} \tag{19}$$

$$[N_2]_e = \frac{0.79\times P}{RT} \tag{20}$$

The engine out smoke has been measured in steady state condition at the full throttle position in experimental setup using AVL Smokemeter 415S. The sampled smoke in filter smoke number (FSN) has been converted to elementary carbon concentration (C) [mg/m^3] using the following correlation proposed by Christian et al. [19]. For the calculation of the net soot that is formed inside the cylinder, the modified model of Hiroyasu et al. [20] by Lipkea and DeJoodecan be used. The net soot formation rate is expressed as the difference between the formation and oxidation rates, which can be calculated using Eqs. 21 to 24.

$$\frac{dm_{sn}}{dt} = \frac{dm_{sf}}{dt} - \frac{dm_{so}}{dt} \quad (21)$$

$$\frac{dm_{sf}}{dt} = A_{sf} dm_f^{\,0.8} P_g^{\,0.5} \exp\left(-\frac{E_{sf}}{R_{mol} T_g}\right) \quad (22)$$

$$\frac{dm_{so}}{dt} = A_{so} m_{sn} \left(\frac{P_{O_2}}{P_g}\right) P_g^{\,n} \exp\left(-\frac{E_{so}}{R_{mol} T_g}\right) \quad (23)$$

$$A_{sf} = 18.8e^{1.94\times10^{-4}N} \quad (24)$$

Activation energies E_{sf} and E_{so} are taken as 82000J/mol and 120000J/mol respectively, where A_{so} is taken as 1000, pressures are in bar and masses in kg, dm_f is the fuel vapour mass to be burned in the current time step, and P_{O_2} is the partial pressure of oxygen in the zone. Also, constants A_{sf}, A_{so}, activation energies E_{sf}, E_{so} and the exponent n are evaluated after experimental matching at steady-state conditions.

3 Model Validation and Discussions

3.1 Fuel Consumption Model- Figure 2 shows comparison of predicted fuel mass injected with the same output acquired from engine electronic control unit (ECU). It can be seen that transient fuel consumption predicted by the model is closely matching with the ECU output by which it can be concluded that the formulated fuel mass flow model predicts fuel consumption under transient operations with very good accuracy.

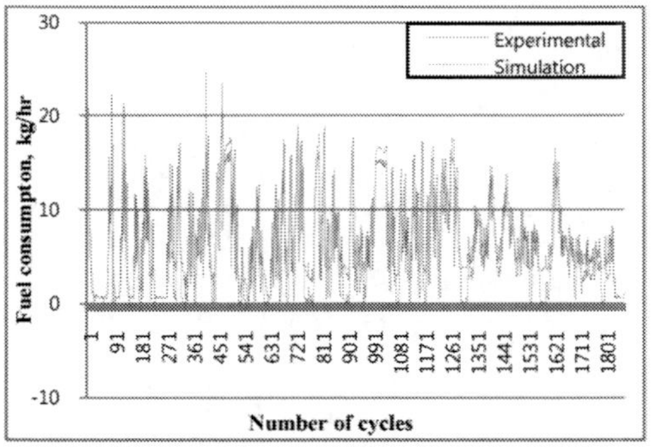

Figure 2: Transient fuel consumption for 1800 cycles

3.2 Combustion Model- For emissions prediction on transient cycle, the emission model should have capability to predict correct emission trends over load and start of injection. Hence, before validation of the comprehensive model on ETC, combustion and emission models were checked for their responses to load and start of injection (SOI) changes in steady state cycle and later validated for the ETC cycle. But, due to space limit, the validated plots could not be plotted and explained in this manuscript. Experimental data for 2400 RPM and 1700 RPM have been used to fit the simulated heat release rate curve and coefficients are adjusted. Resulting model is then validated with the experimental in-cylinder pressure data for other rpm. Figures 3 and 4 shows the in-cylinder pressure traces with crank angle for Engine 1 and Engine 2 respectively. Very good matching in simulated and experimental in-cylinder pressure traces can be observed with both the engines.

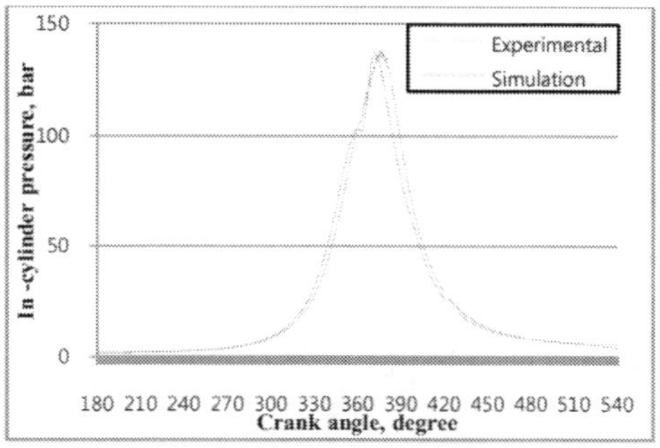

Figure 3: Experimental vs simulated in-cylinder pressure for Engine 1

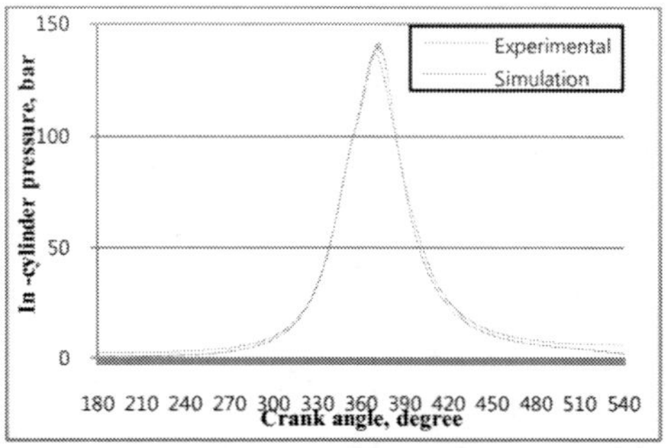

Figure 4: Experimental vs simulated in-cylinder pressure for Engine 2

Predicted temperature distribution inside the engine cylinder at different crank angle position for Engine 1 and Engine 2 is also found close matching with that of experimental traces, as shown in Figs. 5 and 6 respectively.

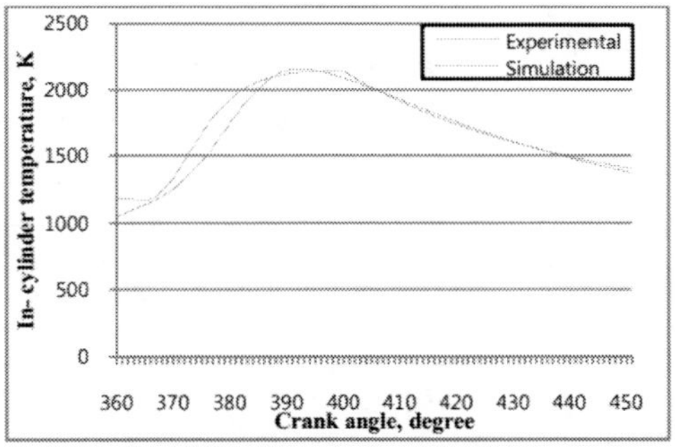

Figure 5: Experimental vs simulated in-cylinder temperature for Engine 1 of 2400 rpm

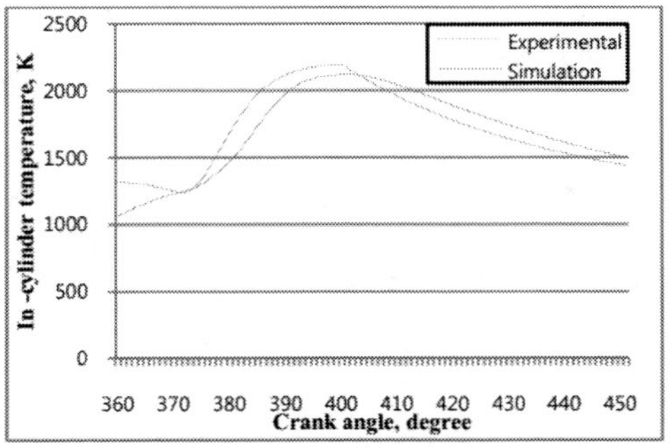

Figure 6: Experimental vs simulated in-cylinder temperature for Engine 2 of 2400 rpm

3.3 Emission Model- Figure 7 shows comparison of predicted NO_x with measured values. Trend of the predicted NO_x emissions is matching with the measured trends; however there are differences in the amplitudes of NO_x peaks, which could be due to peaks in the predicted ignition delay model.

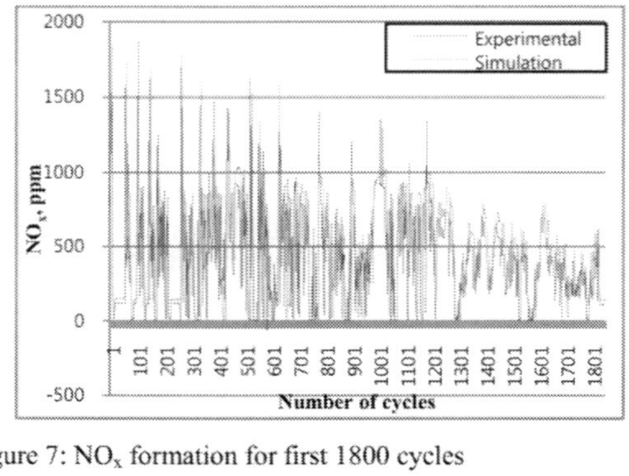

Figure 7: NO_x formation for first 1800 cycles

The soot emissions with FTP values are listed in Table 1. The values were matching much with its experimental at lower RPM. But at higher RPM, values were mismatching and it is needed to modify the model in order to get correlation for different speeds.

Sl. No.	Speed (RPM)	Load (Nm)	Exp-soot (mg/sec)	Sim-soot (mg/sec)	% Deviati on
1	1900	658.3	0.000287	0.00034	18.5
2	1800	657.0	0.000261	0.000307	22.28
3	1700	653.7	0.000197	0.00025	22.9
4	1600	651	0.000193	0.000236	22.4
5	1500	650.4	0.000643	0.000714	11.04
6	1400	657.7	0.00036	0.00042	16.66

Table 1: Experimental vs simulated soot values

4 Summary and Conclusions

A comprehensive zero-dimensional predictive model has been formulated for soot and NO_x emissions prediction under transient operating cycle. Responses of combustion and emissions models have been validated for load and SOI changes and prediction results of the comprehensive model have been validated using a turbocharged common rail engine on ETC with an overall correlation of about 81%. The formulated model has predicted transient fuel consumption, in-cylinder pressure and temperature traces, NO_x and soot emissions with good accuracies. Results of the study showed that 0D models can be used for prediction of transient emissions on modern diesel engines, to reduce time and cost of experimentations. Though the model has been validated on a diesel engine for ETC, it can be extended for other applications and fuels. With modifications in sub-models, more prediction parameters can be added to the model to make it an effective design tool.

5 Acknowledgements

The authors thank "Ashok Leyland Technical Centre" for allowing publishing the work. The authors also thank the laboratory staff of Engine Development Centre, Ashok Leyland Technical Centre, for their help during the work.

References

1. C.D. Rakopoulos, A.M. Dimaratos, E.G. Giakoumis, D.C. Rakopoulos, Exhaust emissions estimation during transient turbocharged diesel engine operation using a two-zone combustion model, 49(3) (2009) 125–149.
2. C. Ericson, B. Westerberg, R. Egnell, Transient emission predictions with quasi stationary models, SAE technical paper, 2005-01-3852.
3. C.D. Rakopoulos, E.G. Giakoumis, Review of thermodynamic diesel engine simulations under transient operating conditions, SAE technical paper, 2006-01-0884.
4. K. Nikzadfar, A.H. Shamekhi, An extended mean value model (EMVM) for control-oriented modeling of diesel engines transient performance and emissions, 154 (2009) 275–292.
5. K. Ebrahimi, M. Bazargan, S.A. Jazayeri, A new phenomenological model for combustion and performance studies of direct injection diesel engines, SAE technical paper, 2007-01-1904.
6. J. R. Serrano, F. J. Arnau, V. Dolz, Methodology for characterization and simulation of turbocharged diesel engines combustion during transient operation. Part 1, Applied Thermal Engineering, 29(1) (2009) 142–149.
7. C. Ericson, M. Andersson, R. Egnell, Modelling diesel engine combustion and NO_x formation for model based control and simulation of engine and exhaust after-treatment systems, SAE technical paper 2006-01-0687.
8. C.D. Rakopoulos, E.G. Giakoumis, Diesel Engine Transient Operation, Springer, London, 2009
9. N.H Walke, NO_x, soot, and fuel consumption predictions under transient operating cycle for common rail high power density diesel engines, Hindawi publications, (2016) 2-3.
10. P.A. Lakshminarayanan, Y.V. Aghav, Modelling Diesel Combustion, Springer, Amsterdam, 2010.
11. D. Jung, D. N. Assanis, Multi-zone DI diesel spray combustion model for cycle simulation studies of engine performance and emissions, SAE technical paper, 2001-011246.
12. J.B. Heywood, Internal Combustion Engine Fundamentals, McGraw-Hill, 1988.
13. H. Hardenberg, F. Hase, An empirical formula for computing the pressure rise delay of a fuel from its cetane number and from the relevant parameters of direct-injection diesel engines, SAE technical paper, 1979-790493.
14. W.J.D. Annand, Heat transfer in the cylinders of reciprocating internal combustion engines, 177 (1963) 973-990.
15. G. Eichelberg, Some investigations on old combustion-engine problems engineering 148(1) (1939) 463-466.
16. F.P. Incropera, D.P. DeWitt, Introduction to heat transfer, John Wiley & Sons Inc., 1990.
17. C.D. Rakopoulos, K.A. Antonopoulos, D.C. Rakopoulos, E.G. Giakoumis, Study of combustion in a divided chamber turbocharged diesel engine by experimental heat release analysis in its chambers, Applied Thermal Engineering 150(26) (2006) 1611–1620.
18. C. Ericson, NO_x modelling of a complete diesel engine /SCR system [Licentiate thesis], Lund University, Sweden, 2007.
19. V. R Christian, F. Knopf, A. Jasche, and W. Schindler, 1993, Eine neue Messmethodik der Bosch-Zahl miterhohter Empfindlichkeit, Motortech. 54 (1993) 16–22
20. H. Hiroyasu, T. Kadota, M. Arai, Development and use of a spray combustion modeling to predict diesel engine efficiency and pollutant emissions,130(26) (1983)569–570.

11th Asia-Pacific Conference on Combustion,
The University of Sydney, NSW Australia
10th -14th December 2017.

Large Eddy Simulation of MILD Combustion of Syngas

Santanu Pramanik, R. V. Ravikrishna

Department of Mechanical Engineering, Indian Institute of Science

Bangalore, Karnataka, 560012, India

Abstract

The present work numerically investigates the reaction zone structure of syngas combustion under MILD conditions using the wall-modelled LES (WMLES) and the flamelet generated manifold (FGM) combustion model. The fuel contains 20% CO, 20% H_2 and 60% N_2, while the O_2 percentage in the coflow has been reduced to 7.6% by the addition of N_2. The reactants are injected into the chamber in the non-premixed mode where the fuel and coflow are heated to 544 and 768 K, respectively. The LES model reproduces Kolmogorov's (-5/3) decay of the axial turbulent kinetic energy in the shear layer. The temperature in the chamber is nearly uniform due to the intense recirculation (recirculation ratio = 2.4). The comparison of the OH contours shows that the flame ignites early, although the global characteristics of the reaction zone such as the peak intensity and OH distribution are predicted reasonably well. The probable causes for the differences in the experiments and the predictions have been proposed and require further investigation.

1 Introduction

The stringent emission standards for NO_x and other pollutants have led to the development of new combustion technologies for low emissions while maintaining combustion stability. One such technology is moderate or intense low oxygen dilution (MILD) combustion [1] that reduces the formation of thermal NO_x by the recirculation of exhaust gases. Technologies that use a similar principle of internal and external exhaust gas recirculation have variously been called in the literature as flameless oxidation [2], colorless distributed combustion [3] or high-temperature air combustion [4]. In this mode of combustion, the reactants are preheated and injected at high velocities such that there is sufficient mixing between the incoming fresh reactants and hot recirculating exhaust before the onset of combustion. The combustion is sustained as the temperature is maintained higher than the autoignition temperature either due to preheating or mixing with the exhaust gases. This leads to a reduction in the maximum temperature and consequent decrease in the emission of thermal NO_x.

MILD combustion is said to occur in the low Damköhler number regime ($Da \sim 1$) due to the highly vitiated and diluted environment [5][6] such that both the turbulent mixing and kinetic time scales control the combustion processes. This poses a challenge from a modeling perspective since the inclusion of finite-rate kinetic models for practical combustion systems can be expensive. Also, the necessity to resolve the mixing by the use of scale-resolved models such as LES can increase the computational costs significantly. Several studies have investigated the ability of the turbulence and combustion models to capture the characteristics of MILD combustion. Coelho and Peters [7] used the Eulerian Particle Flamelet model with RANS to numerically investigate a MILD combustion furnace. The EDC and the mixture fraction/PDF models have been used by Christo et al. [8] for a jet in hot coflow (JHC) burner in which the EDC model was observed to predict the experimental results better. Later, the JHC burner was simulated by Ihme et al. [9] using LES and an extended flamelet/progress variable model (FPV) where an additional conserved scalar is introduced to account for the three streams. A diluted homogeneous reactors (DHR) model for flameless combustion has also been proposed by Locci et al. [10] within an LES framework and validated for the Sandia D flame and a small scale flameless combustor. LES studies of a reverse-flow combustion chamber using the steady flamelet (SF), eddy breakup (EBU) and linear eddy mixing (LEM) models have been performed by Undapalli et al. [11]. They concluded that the LEM model is superior in comparison to the SF and EBU models in capturing the combustion. While combustion models such as EDC, composition PDF transport (CPDF) and LEM are superior to the flamelet based models in capturing finite-rate kinetic effects, they are significantly more expensive in terms of computation cost. An improvement over the steady flamelet models is the flamelet generated manifold (FGM) model in which the chemistry is parameterized in terms of mixture fraction and the progress variable [9][12]. The model can be applied to the stirred reactor limit or even to capture ignition and extinction. While the FGM model has been used with LES for the JHC burner [13], further investigations are necessary to understand the applicability of this model under MILD conditions with different fuel and combustor configurations, which is the focus of the present work.

The present study numerically investigates the flow field and reaction-zone structure of a laboratory-scale 3.26 kW reverse flow MILD combustion chamber operating on syngas using LES and the FGM model. The results have been compared with OH-PLIF measurements performed under the same conditions.

2 Problem setup

2.1 Physical model

The experimental test rig modeled consists of a cylindrical reverse-flow combustion chamber. The reverse-flow configuration is generic for MILD combustors [6] as it induces entrainment of the exhaust gases into the incoming fresh reactants due to shear. The chamber has a length of 265 mm and a diameter of 80 mm. The injector assembly has a central fuel port (D_f = 3 mm) surrounded by a 1 mm wide annular port for air injection. The air nozzle is again surrounded by an annular exhaust manifold of width 5 mm. The air is mixed with

Corresponding author.
E-mail address: ravikris@mecheng.iisc.ernet.in

additional nitrogen (N_2) to vary the level of dilution within the chamber and simulates exhaust gas recirculation. The fuel contains 20% H_2, 20% CO and 60% N_2 by volume and is heated by the pre-heated coflow and the returning exhaust gases. The fuel flow rate is maintained constant at 42 slpm (Table 1) and results in a fuel thermal input of 3.26 kW. The average velocity of the fuel jet is ~ 175 m/s while that of the coflow varies from ~ 170 m/s. This results in a Re_f ~ 10000 for the fuel jet and Re_{ox} ~ 2000 for the coflow jet. Further details of the experimental setup can be found in an accompanying publication in ASPACC 2017 (Title: Reaction Zone Structure of Syngas Combustion under MILD and Conventional Conditions).

Case		A45N80
Fuel (slpm)		42
Coflow	Air (slpm)	45
	N_2 (slpm)	80
O_2% in coflow (by vol.)		7.6
$T_{in, fuel}$		544
$T_{in, ox}$		768

Table 1: Experimental conditions simulated.

2.2 Numerical model

The three-dimensional computational domain was divided into 1.45 million hexahedral cells with suitable biasing near the inlet, walls, and the centreline to resolve the turbulent structures. The cells near the injector are nearly isotropic with Δx = 0.125 mm. The computational domain and grid are shown in Fig. 1. The turbulence was generated at the inlet using the spectral synthesizer method. The power spectrum of the axial turbulent kinetic energy in the shear layer follows Kolmogorov's -5/3 profile for at least one decade and is considered reasonable for LES (Fig. 2). The temperatures measured at the inner wall of the combustion chamber (6 thermocouples) have been fitted into polynomial profiles and provided as boundary conditions for the lateral wall. The walls at the bottom and the top were considered adiabatic as the chamber was surrounded by a 2" thick layer of glass fiber insulation. The heat transfer between the reactant streams was accounted for by performing precursor RANS simulations with the full injector and inlet temperatures measured at the upstream. These walls were subsequently removed in the LES simulations and the temperature obtained from the RANS simulations were provided as boundary conditions.

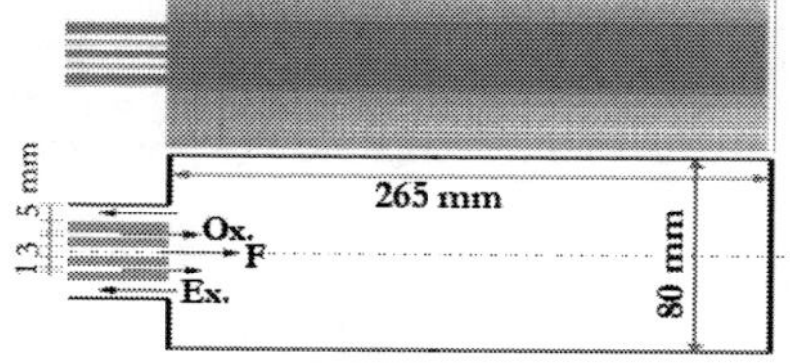

Figure 1: Schematic diagram of the computational domain and the grid.

The simulations were conducted using the finite-volume CFD code, Ansys Fluent. The momentum equation was discretized using the central difference scheme, while the bounded central difference/second order upwind schemes were used for the other scalars. The PISO scheme was used for the pressure-velocity coupling. The time discretization was performed using the second-order implicit scheme. While the implicit scheme provides stability even at higher Courant numbers (CFL), a time step size $\Delta t = 10^{-6}$ seconds was chosen such that CFL < 1 throughout the domain. This led to a run-time of over a month on an AMD FX-8350 octa-core processor with a clock-speed of 4 GHz.

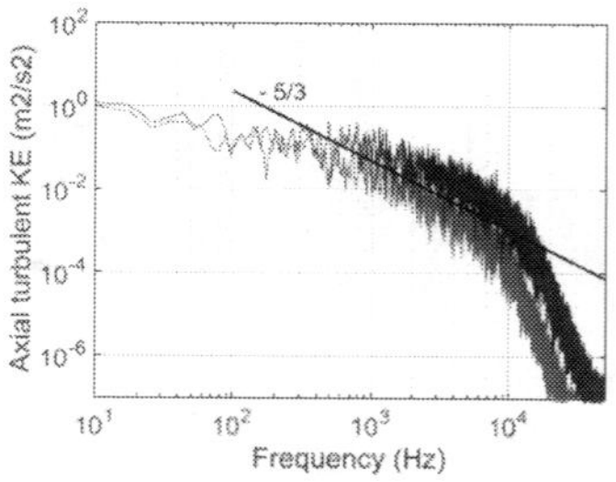

Figure 2: Power spectrum of the axial turbulent kinetic energy at two different locations in the outer shear layer, (r, z) = (7, 64) mm and (7, 144) mm from the injector. The spectrum follows Kolmogorov's –5/3 decay at least for one decade.

2.2 Governing equations

The unsteady, three-dimensional governing equations of continuity, momentum, energy, mixture fraction, and progress variable are solved on the computational domain. The subgrid-scale stresses are modeled using the wall-modelled LES (WMLES) model. The near-wall region is treated with a mixing length model while the regions away from the wall are modeled using a modified Smagorinsky model. The eddy viscosity is calculated with the use of a hybrid length scale

$$\nu_t = min\left[(Kd_w)^2, (C_{Smag}\Delta)^2\right].S.\left\{1 - exp\left[-\left({y^+}/{25}\right)^3\right]\right\} \quad (1)$$

where d_w is the wall distance, S is the strain rate, $K = 0.41$, $C_{Smag} = 0.2$ and y^+ is the normal to the wall inner scaling. The FGM model has earlier been used for the simulation of non-premixed laminar co-flow flames [14]. The FGM model uses a low-dimensional manifold created by solving the flamelet equations and is parameterized by the mixture fraction (f) and reaction progress (c). We generate the flamelet table by solving the flamelet equations for a counterflow diffusion flame due to the non-premixed configuration of the current setup. The reaction progress is determined by solving an equation for the un-normalized progress variable, Y_c

$$\bar{\rho}\frac{\partial \tilde{Y}_c}{\partial t} + \frac{\partial}{\partial x_i}\left(\bar{\rho}\tilde{u}_i\tilde{Y}_c\right) = \frac{\partial}{\partial x_i}\left(\bar{\rho}D_{eff}\frac{\partial \tilde{Y}_c}{\partial x_i}\right) + \bar{S}_c \quad (2)$$

The mean source term $\bar{S}_c$ is modelled as

$$\bar{S}_c = \bar{\rho}\iint S_{FR}(c,f)P(c,f)dc\,df = \bar{S}_{FR} \quad (3)$$

where S_{FR} is the finite-rate flamelet source term from the flamelet library, $c = Y_c/Y_c^{eq}$ is the normalized reaction progress variable, f is the mixture fraction, and P is the joint PDF of mixture fraction (f) and the reaction progress variable (c). The mixture fraction (f) and reaction progress (c) are assumed to be independent beta (β) functions and the joint PDF is expressed as the product of these two β-PDFs. The source term $\bar{S}_c$ is also modelled using a turbulent flame speed model as follows:

$$\bar{S}_c = \rho_u S_t|\nabla\tilde{c}| = \bar{S}_{TFS} \quad (4)$$

Here, ρ_u is the unburnt density and S_t is the turbulent flame speed which is modelled by the correlation by Zimont [15]. Separate laminar flame speed simulations were conducted for syngas using GRI 3.0 [16] at various mixture fractions and provided as input for calculation of the turbulent flame speed. The minimum of the source terms $\bar{S}_{FR}$and $\bar{S}_{TFS}$ is used as the mean source term $\bar{S}_c$. The variance of the un-normalized progress variable Y_c is required for the β-PDF and is calculated using an algebraic equation:

$$\widetilde{Y_c''^2} = C_{var} \frac{l_{turb}^2}{Sc_t}\left(\nabla\tilde{Y_c}^2\right) \quad (5)$$

where l_{turb} is the turbulent length scale and C_{var} is a constant with a value of 0.1.

3 Results and Discussions

3.1 Velocity field and mixing

The instantaneous velocity and mixture fraction fields are shown in Fig. 3. The fuel and coflow enter the domain co-axially in the non-premixed configuration. The velocity at the inlet of the domain can be as high as ~200 m/s. The high relative velocity of the jets with respect to the returning flow increases the shear and results in the formation of turbulent vortices in the outer shear layer. This leads to entrainment of the returning exhaust gases into the incoming fresh reactants. The maximum recirculation ratio of 2.4 $(R = \dot{m}_{net} - \dot{m}_{in}/\dot{m}_{in})$ was obtained at around 150-mm from the injector inlet. The returning flow causes the coflow jet to be pushed towards the central fuel jet. This also enhances mixing between the two reactant streams. The jet velocity decreases with axial distance due to the presence of the stagnation plane downstream.

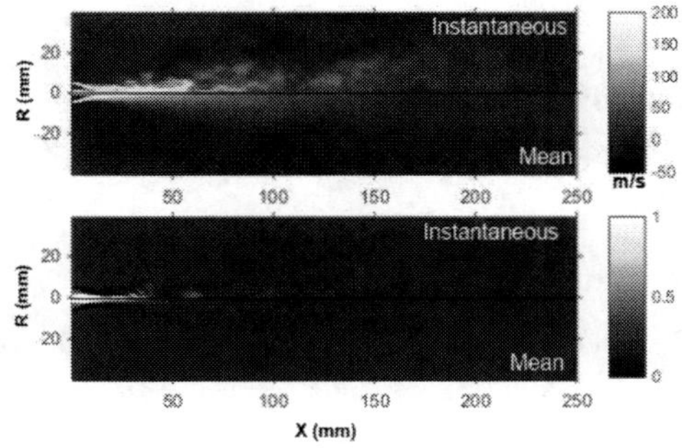

Figure 3: Contours of the axial velocity (top) and the mixture fraction (bottom).

The mean and instantaneous mixture fraction are shown in Fig. 3 (bottom). It is observed that the fuel and coflow mix within the first 50-mm from the nozzle inlet to form a partially premixed mixture which then auto-ignites to a form a lifted turbulent flame. The observation is in accordance with other studies on MILD combustion in which the reactants are injected at high velocity to enhance mixing while preventing the flame from stabilizing at the nozzle.

3.2 Temperature field

The mean and instantaneous temperature fields are shown in Fig. 4. The peak temperature is around 1300 K. There is some heat loss to the walls where the temperature varies in the range of 700-900 K. Overall, the temperature within the chamber is nearly uniform due to the internal recirculation of the gases. The maximum fluctuation in temperature is observed near the injector (~ 40 mm, 130 K), which progressively decreases along the axial distance. The fluctuations are below 20 K after 100-mm of the combustion chamber which is also indicative of the MILD combustion regime [17]. The overall low temperature in the combustion chamber results in near-zero thermal NO_x production, as also observed in the experiments.

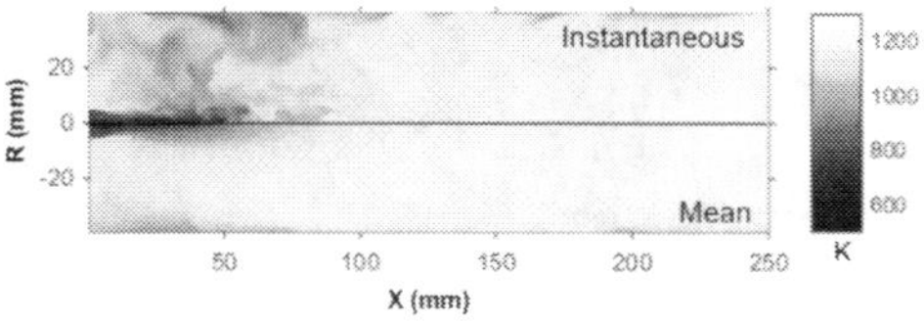

Figure 4: Contours of instantaneous and mean temperature.

3.3 Comparison with OH-Planar Laser Induced Fluorescence

The instantaneous OH-PLIF image obtained from the experiments is shown in Fig. 5 (top). The images have been acquired at 5 locations, covering nearly 210-mm length of the chamber. The individual images have been combined to obtain the instantaneous image. The OH signal represents the primary location of the reaction zone and heat release. The numerically predicted OH distribution is also shown in Fig. 5 (bottom). Overall, there is a good match between the experimentally obtained and numerically predicted OH profiles. The results show that there exists a central core near the nozzle (upto ~ 60 mm) before the onset of reactions which is devoid of OH signal. Downstream of this region, the reactions are initiated due to mixing between the fuel, the coflow, and the heated exhaust gases. The OH concentration is maximum in this region (~ 100 mm). After this peak OH region, the OH concentration decreases and becomes diffuse, signifying a more distributed reaction zone. An OH concentration of around 250-ppm can also be observed to be present throughout the chamber in the numerical simulations. This also signifies a distributed heat release within the chamber. The instantaneous images show that the current models are able to capture the essential features of MILD combustion under the present conditions.

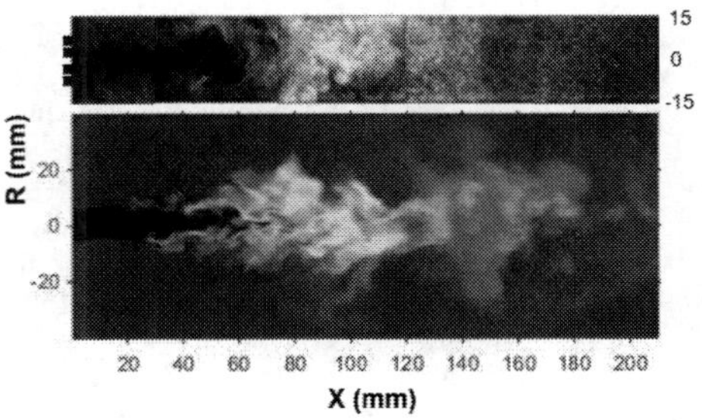

Figure 5: Instantaneous OH-PLIF and comparison with the predicted distribution.

We have compared the variation of the normalized mean OH signal from the experiments in the axial and radial direction with the numerical predictions (Fig. 6). Around 200 experimental images acquired at 10 Hz have been averaged to generate the mean profiles. The numerical simulations have been averaged for 10-ms which correspond to around one flow-through time. As evident from the profiles, more time is required to obtain smoother mean profiles.

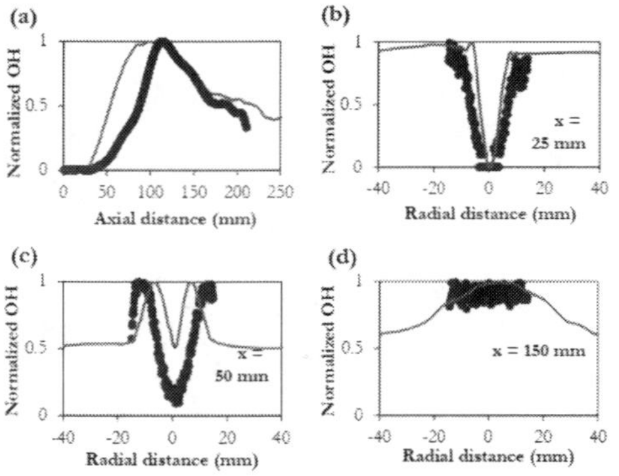

Figure 6: Comparison of normalized mean OH profiles obtained from experiments (blue markers) and simulations (red line) at various axial and radial locations.

The simulations are able to predict the location of the peak OH intensity within 10-mm of the experimentally-observed peak (Fig. 6a). However, the rise in the OH signal is early in the simulations, thereby giving rise to a larger reaction zone. This early rise in the OH can be both due to the combustion model and improper prediction of the mixing. From the experimental images, it appears that the central jet core propagates further into the chamber before the onset of reaction. While the central fuel jet is fully turbulent (Re ~ 10000), the coflow is in the transition regime (Re ~ 2000). The inability of the turbulence models to capture this transition regime may lead to overprediction of the turbulence intensity in the shear layer and the subsequent mixing. Preferential diffusion in the combustion of H_2 containing fuels has also been cited as a reason for the early onset of ignition in JHC burners [13] which is a limitation of the present model. Further studies using different turbulence and kinetic models along with other experimental data sets such as major species and temperature will be required to identify the reasons for early ignition. The decay of the OH signal matches well with the simulations and is gradual compared to the rise in the OH.

The radial variation of the OH signal very near to the nozzle (25-mm) is captured well by the model (Fig. 6b). This also corroborates well with the instantaneous contour images in Fig. 5 where the core central jet propagates into the chamber without narrowing. However, after around 30-mm, the ignition in the shear layer leads to a progressive narrowing of this central core, unlike the experiments. This leads to underprediction of the radial penetration of the OH signal at 50-mm (Fig. 6c). The early onset of ignition as observed in Fig. 6a is also related to this observed mismatch in the radial profiles after 50 mm. Further downstream at x = 150 mm, the radial profiles are nearly uniform (Fig. 6d) and a similar trend has been captured in the numerical predictions. Predictions using the EDC combustion model (not presented for the sake of brevity) show a better match with the OH-PLIF signal near the nozzle compared to the FGM model.

4 Conclusions

The present work numerically investigates the reaction zone structure of syngas combustion under MILD conditions in a reverse-flow combustion chamber. The fuel (syngas: 20% CO, 20% H_2 and 60% N_2) and the coflow (7.6% O_2, 92.4% N_2) are injected into the chamber in the non-premixed configuration co-axially. Both the streams are preheated (544 and 768 K, respectively) and are injected at high velocities (~ 170 m/s). The WMLES and FGM models have been used to capture the turbulent mixing and the combustion in the chamber. The results have been compared with the experimentally-obtained OH-PLIF images. The velocity and mixture fraction contours show good mixing between the fuel, the coflow and the exhaust gases due to shear. The temperature in the chamber is nearly uniform at around 1200 K due to the intense recirculation caused by the reverse flow configuration. The OH predictions are able to capture the essential features of the reaction zone when compared with the OH-PLIF images. The peak OH signal is predicted within 10-mm of the experimentally observed values, although the early onset of ignition results in a larger flame volume and lower penetration of the unreacted central core in the radial direction. This can be attributed both to the turbulence and the combustion model and requires further investigation. Finally, the study shows that the models used are able to capture the essential features of syngas combustion under MILD conditions and can be applied to realistic MILD combustion chambers and furnaces.

References

[1] A. Cavaliere, M. De Joannon, Prog. Energy Combust. Sci. 30 (2004) 329–366.

[2] J. Wünning, Prog. Energy Combust. Sci. 23 (1997) 81–94.

[3] V.K. Arghode, A.K. Gupta, K.M. Bryden, Appl. Energy. 92 (2012) 822–830.

[4] M. Katsuki, T. Hasegawa, Proc. Combust. Inst. 27 (1998) 3135–3146.

[5] C. Galletti, A. Parente, L. Tognotti, Combust. Flame. 151 (2007) 649–664.

[6] G.G. Szegö, B.B. Dally, G.J. Nathan, Combust. Flame. 154 (2008) 281–295.

[7] P.J. Coelho, N. Peters, Combust. Flame. 124 (2001) 503–518.

[8] F.C. Christo, B.B. Dally, Combust. Flame. 142 (2005) 117–129.

[9] M. Ihme, J. Zhang, G. He, B. Dally, Flow, Turbul. Combust. 89 (2012) 449–464.

[10] C. Locci, O. Colin, J.B. Michel, Flow, Turbul. Combust. 93 (2014) 305–347.

[11] S. Undapalli, S. Srinivasan, S. Menon, Proc. Combust. Inst. 32 I (2009) 1537–1544.

[12] E. Abtahizadeh, P. De Goey, J. Van Oijen, Combust. Flame. 162 (2015) 4358–4369.

[13] E. Abtahizadeh, P. de Goey, J. van Oijen, Fuel. 191 (2017) 36–45.

[14] L.M. Verhoeven, W.J.S. Ramaekers, J.A. van Oijen, L.P.H. De Goey, Combust. Flame. 159 (2012) 230–241.

[15] V. Zimont, W. Polifke, M. Bettelini, W. Weisenstein, J. Eng. Gas Turbines Power. 120 (1998) 526.

[16] G.P. Smith, D.M. Golden, M. Frenklach, N.W. Moriarty, B. Eiteneer, GRI-Mech 3.0, (2000). http://combustion.berkeley.edu/gri-mech/version30/text30.html.

[17] S. Kumar, P.J. Paul, H.S. Mukunda, Proc. Combust. Inst. 29 (2002) 1131–1137.

11th Asia-Pacific Conference on Combustion,
The University of Sydney, NSW Australia
10th -14th December 2017.

Reaction Zone Structure of Syngas Combustion under MILD and Conventional Conditions

Santanu Pramanik, R. V. Ravikrishna

Department of Mechanical Engineering, Indian Institute of Science

Bangalore, Karnataka, 560012, India

Abstract

The present work experimentally investigates the reaction zone structure of syngas combustion under near-stoichiometric (Φ_{global} = 0.8), ultra-lean (Φ_{global} = 0.32) and diluted (MILD) conditions. The fuel contains 20% CO, 20% H_2 and 60% N_2, while the O_2 percentage in the coflow has been varied from 21% (air) to 7.6% by the addition of N_2. The reactants are injected into the chamber in the non-premixed mode where the coflow is heated in the range of ~600-800 K. We report OH* chemiluminescence and OH-planar laser-induced fluorescence (PLIF), which have been shown to correlate well with the heat release under the given conditions. The average OH* and OH-PLIF signal intensity moves downstream and decreases as we transition to MILD conditions. The instantaneous images, however, show thin reaction zones with sharp gradients, although these regions progressively become less intense and thicker. Another distinct region can be observed after the primary thin reaction zones where the OH distribution and intensity increase with dilution. Plug-flow reactor simulations indicate an increase in the lifetime of the OH radical that results in a distributed reaction zone. This also indicates that the convection of OH radical can be a possible mechanism for the distributed reactions.

1 Introduction

The stringent emission standards for NO_x and other pollutants have led to the development of new combustion technologies for low emissions while maintaining combustion stability. One such technology is moderate or intense low oxygen dilution (MILD) combustion [1] that reduces the formation of thermal NO_x by the recirculation of exhaust gases. Technologies that use a similar principle of internal and external exhaust gas recirculation have variously been called in the literature as flameless oxidation [2], colourless distributed combustion [3] or high-temperature air combustion [4]. In this mode of combustion, the reactants are preheated and injected at high velocities such that there is sufficient mixing between the incoming fresh reactants and hot recirculating exhaust before the onset of combustion. The combustion is sustained as the temperature is maintained higher than the autoignition temperature either due to preheating or mixing with the exhaust gases. This leads to a reduction in the maximum temperature and consequent decrease in the emission of thermal NO_x.

While most studies on MILD combustion have focussed on methane [3][5], higher hydrocarbons and liquid fuels [6], MILD combustion of low calorific value fuels such as syngas is of particular interest and has not been widely explored. Syngas represents a class of fuels containing a mixture of $CO/H_2/N_2/CH_4/CO_2$ that is used in blast furnaces, power plants, and IC engines. The low calorific value and large specific flow rate of syngas influences the flame stability, ignition delay and the laminar flame speed depending on the particular composition. A few studies have focussed on the MILD combustion of syngas from the level of system performance [7][8]. Several studies [9][10] have also focussed on the effect of H_2 addition in the fuel stream in the jet in hot coflow configuration that mimics the recirculation of exhaust gases in MILD and FLOX combustion systems. However, in-situ investigation of the reaction zone during MILD combustion of syngas in practical configurations has not been reported in the literature, and is the focus of the present study. The reaction zone in MILD combustion is said to be distributed and volumetric with very low luminosity such that it may appear transparent, colourless or flameless. The reason is attributed to the slowing down of the kinetics in relation to the turbulence due to the recirculation of exhaust gases. The presence of different reactants in syngas, particularly H_2 is expected to produce kinetic interactions different from that of other fuels. Also in light of recent experimental [5] and DNS studies [11] that suggest MILD combustion occurs in thin reaction zones similar to conventional flames, this investigation is significant.

The present study experimentally investigates the reaction zone structure of syngas combustion under MILD and conventional conditions using OH* chemiluminescence and OH PLIF techniques in an optically-accessible laboratory scale combustor. We investigate a wide range of combustion conditions from near-stoichiometric, to ultra-lean and ultra-diluted combustion to identify the differences between the reaction zone structures. We also correlate the heat release zone to the presence of the flame markers OH and OH* and try to identify if the heat release is indeed distributed. Finally, we provide explanations regarding the differences in the reaction zone using insights from kinetics and idealized reactors.

2 Experimental and Numerical Details

The experimental test rig consists of a cylindrical reverse-flow combustion chamber (Fig. 1) with optical access. The reverse-flow configuration is generic for MILD combustors [12][13] as it induces entrainment of the exhaust gases into the incoming fresh reactants due to shear. The chamber has a length of 265-mm and a diameter of 80-mm. The inside of the chamber is viewed through a 30-mm wide quartz window. Two additional 10-mm quartz windows provide access to the laser sheet in the perpendicular plane along the chamber diameter. The injector assembly has a central fuel port (D_f = 3-mm) surrounded by a 1-mm wide annular port for air injection. The air nozzle is

Corresponding author.
E-mail address: ravikris@mecheng.iisc.ernet.in

again surrounded by an annular exhaust manifold of width 5-mm. The air is mixed with additional nitrogen to vary the level of dilution within the chamber and simulates varying degree of exhaust gas recirculation. The chamber is insulated on all sides using a 2-inch thick layer of ceramic wool.

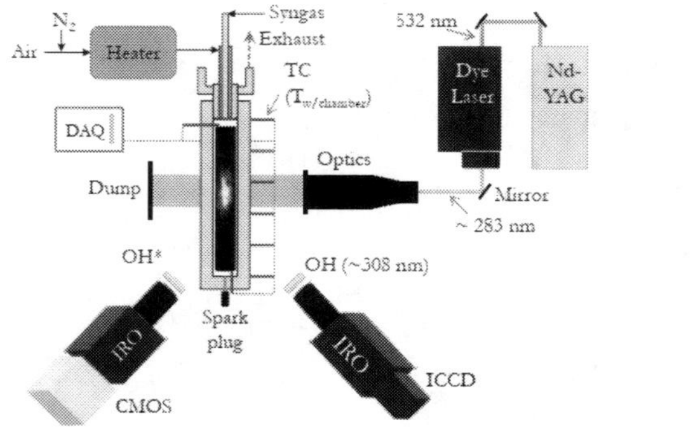

Figure 1: Schematic diagram of the experimental setup. (Acronyms: TC – thermocouple, DAQ – data acquisition system, IRO – intensified relay optics)

The fuel containing 20% H_2, 20% CO and 60% N_2 by volume enters the chamber at ambient temperature. The fuel flow rate is maintained constant at 42 SLPM (Table 1) and results in a fuel thermal input of 3.26 kW. A maximum thermal input of 1.23 kW enters with the coflow due to preheating (Case A125). The average velocity of the fuel jet is ~ 99 m/s, while that of the coflow varies from ~ 47 – 177 m/s. This results in a Re_f ~ 15600 for the fuel jet and Re_{ox} ~ 1880 – 4590 for the coflow jet. The strain rate at the inlet for the fuel and coflow streams varies in the range of ~ 9000 - 15000 for these cases. The global equivalence ratio within the chamber varies from ~ 0.3 - 0.8. The minimum percentage of O_2 (by vol.) in the coflow is 7.6% and simulates the conditions pertaining to MILD combustion. The oxygen concentration has been varied in Cases A45N80 – A125 while maintaining the flow rates of fuel and coflow constant. This allows us to maintain similar velocity and temperature fields so that the effect of dilution and its role in the transition to MILD combustion can be clearly identified.

Cases		A50	A125	A85N40	A45N80
Fuel (slpm)		42	42	42	42
Coflow (slpm)	Air	50	125	85	45
	N_2	0	0	40	80
O_2% in coflow		21	21	14.3	7.6
Φ_{global}		0.8	0.32	-	-
$T_{in,ox}$(K)		605	789	749	784

Table 1: Experimental conditions.

Counterflow diffusion flame simulations were conducted at atmospheric pressure using CHEMKIN using the kinetic mechanism GRI 3.0 with the OH* sub-mechanism from Panoutsos et al. [14]. As Fig. 2 shows, the peak of the heat release rate (HRR) is closely correlated with the OH and OH* profiles which motivated the use of OH* chemiluminescence and OH-PLIF. This is because the primary reaction for the conversion of H_2 to H_2O is $OH + H_2 \rightarrow H + H_2O$, while that for the conversion of CO to CO_2 is $CO + OH \rightarrow H + CO_2$. Both these reactions are exothermic and major sources of heat release.

The natural flame luminosity images have been acquired using a Canon EOS 60D DSLR camera with an exposure time of 0.5 seconds. The OH* chemiluminescence images were captured at 5 kHz with a high-speed CMOS camera (Photron Fastcam SA-5, 1024 x 1024 pixels upto 7000 fps) that was connected to a high-speed intensified relay optics system (make LaVision) for signal amplification. The exposure time was 76-µs. A bandpass filter at 308-nm was used to collect the chemiluminescence. The resolution of the images was around 80 µm/pixel. The setup for OH-PLIF consisted of a pulsed Quanta-Ray Nd:YAG laser that pumps a 532-nm beam to a dye laser (Sirah Lasertechnik, GmbH). The pump beam had energy of ~350 mJ/pulse and was operated at 10 Hz. The output from the dye laser was ~ 25 mJ/ pulse at a wavelength of 283.553 nm that excited the $Q_1(8)$ line belonging to the $X^2\Pi - A^2\Sigma^+(0,1)$ band. The linewidth of the beam is 0.001 nm while the laser fluence is approximately 0.25 J/cm^2. The beam was expanded into a collimated sheet of height 50-mm and thickness 0.5-mm using a collimator and sheet optics. The fluorescence signal was in the linear regime. The intensifier gate time was 50-ns and the signal was captured using an ICCD camera. A total of 150 images were captured for each case with a resolution of around 100 µm/pixel. The acquired images were corrected for sheet non-uniformities and laser energy.

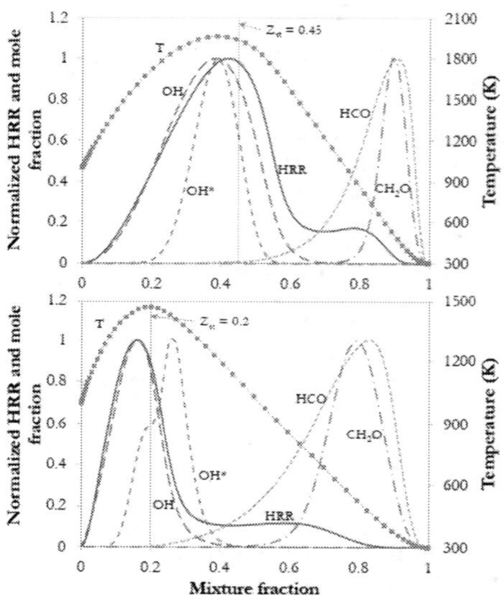

Figure 2: Species, temperature, and heat release rate profiles in a laminar counterflow diffusion flame. The fuel is syngas with 20% H_2, 20% CO and 60% N_2. The oxidizer is air (top) and air/N_2 with 6% O_2 (bottom) at 1000 K. The strain rate is maintained at 230s^{-1}. The heat release rate profiles closely follow that of OH and OH* profiles.

3 Results and Discussions

3.1 Flame Visualization

The reactants are ignited in the non-premixed configuration and the initial flame is observed to be attached to the nozzle. As we increase the flow rates, the reaction zone moves downstream although a small diffusion flame remains attached to the nozzle. This is both due to the high velocities and the shear induced by the recirculating gases as can be observed for Case A50 in Fig. 3. The diffusion flame attached to the nozzle also extinguishes as we reduce the O_2 concentration by the progressive dilution with N_2. The flame is now completely lifted (A45N80) and burns with very low luminosity as compared to the other cases. The flame also becomes transparent such that the wall behind the flame can be visualized. This is indicative of the MILD combustion regime [6]. Finally, we reduce the flow rate of N_2 and replace it with air to transition to the Cases A85N40 and A125. The flame progressively moves closer to the nozzle, although it remains lifted.

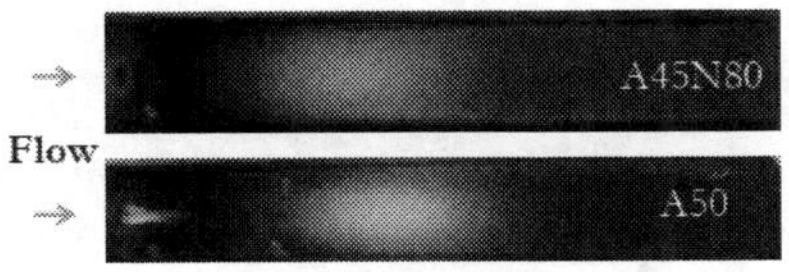

Figure 3: Images of the flame with pure air (Case A50) and diluted air (Case A45N80) with 7.5% O_2. The camera exposure time was 0.5 seconds.

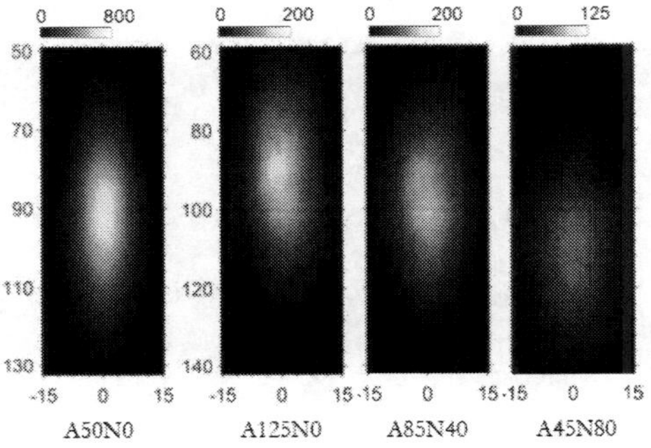

Figure 4: Average OH* chemiluminescence signal for the four cases in Table 1. The vertical dimensions are from the jet exit plane.

3.2 Mean OH* chemiluminescence

The peak of the OH* chemiluminescence signal correlates well with the OH signal and the HRR when air is used as the oxidizer (Fig. 2) although the peak is away from the stoichiometric surface towards the oxidizer side. When the oxidizer is diluted, the peak of the OH* signal is on the fuel side away from the stoichiometric surface. For the A50N0 case, a strong OH* signal can be observed which peaks at around 90-mm from the injector. The OH* intensity decreases with a decrease in the equivalence ratio (A125). The OH* intensity also peaks at around 90-mm for the A125 case, although the intensity has reduced by a factor of 4. The intensity further decreases with increase in dilution (A85N40 and A45N80). The OH* signal also becomes wider in the lateral direction. This observation is different from that of Sidey et al. [5] who found that the OH* intensity is not very sensitive to the O_2 concentration in the combustion of methane with heated and diluted cross flow. This may be attributed to the varying temperature in their flames. The distance between the peak of OH* and HRR increases with dilution (Fig. 2). This suggests that the heat release zone is further downstream for the A45N80 case. The difference in the OH* signal despite the presence of similar temperature and flow conditions for the A125–A45N80 cases suggests that kinetics plays a key role in the reaction zone distribution. The investigation of the instantaneous images suggested the presence of ignition kernels for all the cases (not presented here for the sake of brevity).

3.3 OH-Planar Laser Induced Fluorescence

The instantaneous OH-PLIF images have been acquired at 5 locations, covering nearly 210-mm of the chamber (Fig. 5). The primary observation from the OH-PLIF images is that the peak OH intensity decreases as we move from the A50 case to the lean (A125) and diluted cases (A85N40 and A45N80). The peak intensity is similar from A125 to A85N40, but thereafter reduces significantly in A45N80 case. A similar observation was made in the chemiluminescence images. In all the OH-PLIF images, the flame structure looks remarkably similar to that of premixed combustion of methane in a stagnation point reverse flow combustor as reported by Bobba et al. [15]. For the A50 case, the OH contour is characterized by sharp gradients characterizing distinct flame fronts (Fig. 5). A clear diffusion flame can be observed in the inner shear layer attached to the wall between the fuel and air nozzles. This can also be observed in the flame luminosity images in Fig. 3. The unburnt fuel and coflow further mix with the recirculating exhaust and form an intensely burning lifted flame downstream. The average OH distribution, however, suggests that a weak OH signal connects the diffusion flame at the injector tip and the lifted flame downstream. On closer inspection of the instantaneous images, it was observed that the several ignition kernels were formed between the diffusion flame and the downstream flame that gave the impression of continuity between the two on the average. This also suggests that the auto-ignition of the partially-premixed fuel-air-exhaust mixture led to the stabilization of the downstream flame.

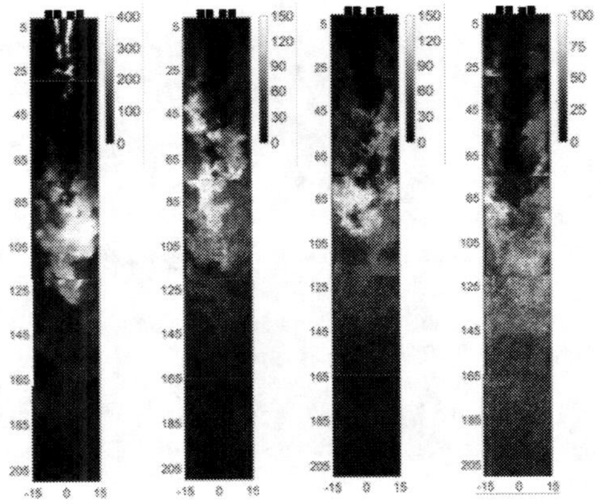

Figure 5: Instantaneous OH-PLIF for the four cases in Table 1. The images from left to right are for the cases A50, A125, A85N40, and A45N80, respectively.

For the A125 case, the diffusion flame at the tip of the nozzle is extinguished which is also corroborated by the absence of OH radicals there. The flame appears completely lifted although thin reaction zones appear within 15-mm from the injector near the outer shear layer between the coflow and the exhaust. The peak intensity of the OH signal also decreases in comparison to the A50 case. The central region near the injector without OH signal represents the incoming fresh reactants. Similar to the A50 case, the flame is characterized by sharp gradients signifying localized reaction zones. A weak OH signal can be detected near the exhaust manifold in the returning flow. Another interesting observation is the presence of weak OH signal near the bottom of the chamber. For the A50 case, the signal at this region is still localized and seems to represent turbulent structures. For the A125 case, however, the OH signal appears more diffuse and fills the width of the window (> 30 mm). This seems to indicate a distributed presence of OH in the chamber away from the primary reaction zones near the injector.

For the transition case A85N40, the flame structure is similar to that of the A125 case. Thin reaction zones can also be observed in this condition in the vicinity of the fresh reactants with occasional intensely burning regions. The O_2% in the coflow has now been reduced to 14.3%, which is in between that of the A125 case and the A45N80 case. The peak OH count is similar to that of the A125 case (Fig. 6), although the penetration of the central reactant jet is slightly higher. The distributed layer of OH further downstream has higher counts compared to the A125 case. This indicates that the heat release is more distributed in A85N40 case in comparison to the A125 case (Fig. 2). For the maximum diluted case A45N80, the fresh reactants penetrate further into the chamber (Fig. 6). The O_2% in the coflow has been further reduced to 7.6% although the mixture is still lean with around 0.7% of excess O_2 in the exhaust. The flame also appears completely lifted from the injector. The peak counts of OH have further decreased, signifying a weaker reaction zone and reduced heat release rate. However, even with such high levels of dilution, we observe thin reaction zones and distinct flame fronts instead of a distributed reaction zone. A similar observation of thin reaction zones was made by Sidey et al. [5] in the jet in heated and diluted (3% O_2) cross flow experiments. The distribution of OH radical is however evident in the downstream section of the chamber where the intensity of OH signal is higher compared to the A125 and A85N40 cases (Fig. 6).

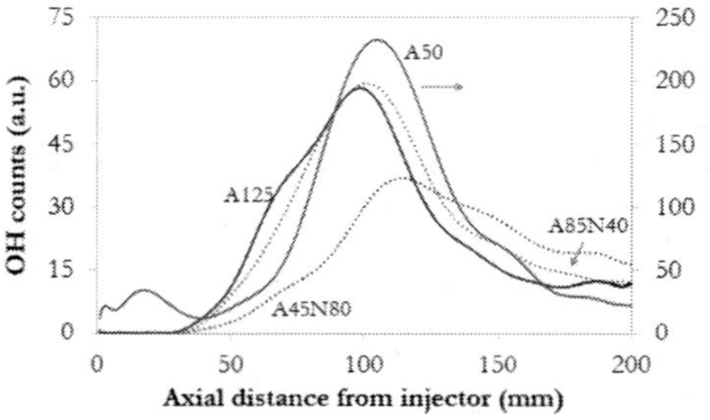

Figure 6: Mean OH profiles along the chamber centreline for the four cases in Table 1.

We performed isothermal (1370 K) plug-flow reactor simulations to investigate the distribution of OH radicals with dilution. The O_2 percentage was 3% and 21% in the oxidizer while the mixture entered the reactor at 8 m/s. The primary observation was that the OH lifetime increased with a decrease in the O_2 percentage. The OH concentration was higher over a longer distance of the reactor in the 3%-O_2 case as compared to the 21%-O_2 case. This also corroborates the standard description of MILD combustion in literature with a typical volumetric reaction zone. From the net production curves, we observe that the region with positive OH production is localized, although it widens for the 3% O_2 case. However, the consumption of OH is gradual and spreads over a longer distance. The slow consumption leads to a passive convection of OH along the channel length that is also responsible for the distributed OH and heat release. The ratio of the peak rates of consumption and production also reduces with increase in dilution, strengthening the explanation. This can also be observed in the OH-PLIF images (Fig. 5), where a distinction can be made between the thin reaction zones near the inlet (net production) and the diffuse region further downstream (net consumption). The near-nozzle regions become weaker and wider with dilution but still maintain their distinct structure. The OH distribution downstream can be attributed to the reduction in the rate of consumption which leads to passive convection of OH in the chamber. Hence, convection appears to be a possible mechanism for the distribution of the reaction zone. This also highlights the importance of finite-rate chemistry under MILD conditions.

4 Conclusions

The present work experimentally investigates the reaction zone structure of syngas combustion under MILD and conventional conditions using OH* chemiluminescence and OH-PLIF which have been shown to correlate well with the heat release. The OH* signal moves downstream and decreases with an increase in the dilution that suggests a weak and more distributed reaction zone. The average OH-PLIF signals also show a similar trend with a more uniform profile as a result of dilution. However, the instantaneous images show thin reaction zones with sharp gradients, although these regions become less intense and wider with dilution. Another distributed region can be observed after the thin reaction zones where the OH signal increases with dilution. Plug-flow reactor simulations indicate a higher lifetime of OH with dilution owing to a greater reduction in the consumption rates. This suggests that convection of the OH radicals can lead to a more distributed reaction zone with heat release due to the coupled effect of slower kinetics and higher velocities.

References

[1] A. Cavaliere, M. De Joannon, Prog. Energy Combust. Sci. 30 (2004) 329–366.

[2] J. Wünning, Prog. Energy Combust. Sci. 23 (1997) 81–94.

[3] V.K. Arghode, A.K. Gupta, K.M. Bryden, Appl. Energy. 92 (2012) 822–830.

[4] M. Katsuki, T. Hasegawa, Proc. Combust. Inst. 27 (1998) 3135–3146.

[5] J. Sidey, E. Mastorakos, Proc. Combust. Inst. 35 (2015) 3537–3545.

[6] V. Mahendra Reddy, D. Sawant, D. Trivedi, S. Kumar, Proc. Combust. Inst. 34 (2013) 3319–3326.

[7] T. Zornek, T. Monz, M. Aigner, Appl. Energy. 159 (2015) 276–284.

[8] K. Kwiatkowski, E. Mastorakos, Energy and Fuels. 30 (2016) 4386–4397.

[9] P.R. Medwell, P.A.M. Kalt, B.B. Dally, Combust. Flame. 152 (2008) 100–113.

[10] L.D. Arteaga Mendez, M.J. Tummers, E.H. Van Veen, D.J.E.M. Roekaerts, Proc. Combust. Inst. 35 (2015) 3557–3564.

[11] Y. Minamoto, N. Swaminathan, Combust. Flame. 161 (2014) 1063–1075.

[12] G.G. Szegö, B.B. Dally, G.J. Nathan, Combust. Flame. 154 (2008) 281–295.

[13] M. Castela, A. S. Veríssimo, A. M. A. Rocha, M. Costa, Combust. Sci. Technol. 184 (2012) 243–258.

[14] C.S. Panoutsos, Y. Hardalupas, A.M.K.P. Taylor, Combust. Flame. 156 (2009) 273–291.

[15] Thesis: M.K. Bobba, *Flame stabilization and mixing characteristics in a stagnation point reverse flow combustor*, Georgia Institute of Technology, 2007.

11th Asia-Pacific Conference on Combustion,
The University of Sydney, NSW Australia
10th -14th December 2017.

Study on evaporation and combustion of n-heptane droplet in a heated tube

Mang Feng*, Junwei Li, Ningfei Wang, Xinjian Chen, Rong Yao

School of Aerospace Engineering, Beijing Institute of Technology

Beijing 100081, PR China

Abstract

When a suspended droplet of n-heptane burns in a heated tube, oscillatory flame is formed at the tube exit due to periodic droplet dripping and colliding with hot wall, along with heat and mass transfer processes among solid, gaseous and liquid. In order to understand the mechanism of oscillatory flames, an evaporation model of suspended droplet and deposited liquid film were established, with careful consideration of both evaporation and combustion processes. Experiments and calculations were conducted to examine the effects of vaporization on heat and mass transfer processes for droplet and liquid films. Results indicate that oscillatory flames result from both suspension and dripping processes. Increasing fuel flow rate can increase flame temperature and benefit flame stability. While increasing air velocity shortens oscillating period and impedes flame stability.

1 Introduction

Spray combustion of liquid fuels is of interest in various engineering fields related to liquid rocket motor and gas turbine, etc. The evaporation and combustion of a single droplet in hot gas flows are the most fundamental phenomena in a fuel spay system[1].

During the process of droplet oscillatory combustion, dynamic behavior of the impinging droplets and heat transfer between the liquid droplets and the hot surfaces are important factors, which affect the mass transfer associated with liquid-vapor phase change. Wachters and Westerling[2] were among the first to investigate the impact of a saturated water droplet of about 2mm in diameter impinging on a polished gold surface heated to 400°C. Xiong and Yuen[3] measured time history of an n-heptane droplet impinging on a stainless-steel surface heated to temperatures between 63°C and 605°C. Ko and Chumg[4] investigated experimentally the effect of wall temperature on break-up process of n-decane fuel, in the Leidenfrost temperature range of 220-330°C.

In this paper, we studied the process that fuel droplet deposited on hot surface and formed liquid film, and then the film evaporated and formed stable oscillatory flames by using liquid n-heptane as fuel. According to experimental studies conducted before[5], we carried out more experiments about flame oscillation and built an evaporation model for periodic droplet deposition. Influences of fuel flow rate and air velocity are also investigated through calculation.

2 Experimental work

A series of experiments were conducted to investigate the evaporation and combustion characteristics of liquid n-heptane droplet in a heated tube. The schematic diagram of our experimental setup is shown in Fig.1. Compressed air enters into the quartz tube from the left side, and mass flow rate was monitored by a controller with ±2% accuracy. A digital syringe pump and a 20ml plastic syringe filled with liquid n-heptane are used to supply fuel droplets. The syringe is connected with a stainless steel capillary tube, which is inserted into the quartz tube concentrically. The syringe pump pushes plunger of the syringe at a desired speed to provide fuel at a desired volumetric flow rate. Temperatures of wall and exhaust gas are measured using 5 shielded K-type thermocouples (TC1, TC3-TC6).

Shapes and positions of droplet and flames are recorded by a digital microscope and a high-definition camcorder Canon HF200. An electric stove is used to provide external heating for the quartz tube. Data acquisition system developed by Labview software is employed to measure fuel flow rate, air velocity and temperature profiles.

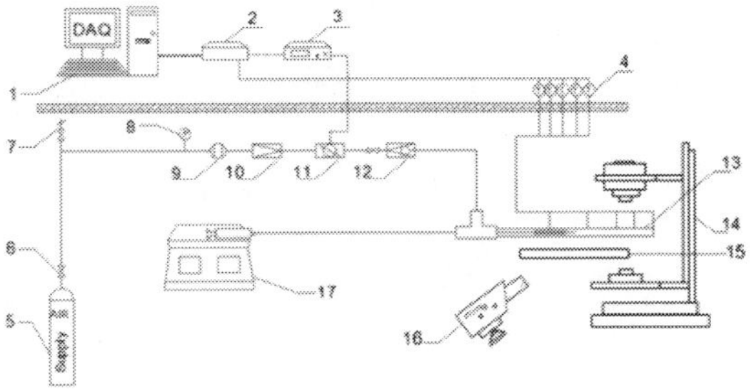

Figure 1: Schematic of experimental system

1. Computer 2.DAQ card 3.mass flow rate controller 4. thermocouples 5. Air tank 6. Hand valve 7. Relief valve 8.Pressure gauge 9.filter 10.Regulator 11. Flow control valve 12.Check valve 13.Micro-tube 14.Digital microscope 15. Electric stove 16.Camcorder 17.Syringe pump

Here, theoretical equivalence ratio is calculated from mass flow rate of liquid n-heptane and air flow rate, as expressed in (1)

$$er_{theo} = \frac{\left(m_{fuel}/m_{air}\right)_{real}}{\left(m_{fuel}/m_{air}\right)_{stio}} \quad (1)$$

Under conditions with external heating, evaporation and combustion characteristics of liquid n-heptane are plotted in Fig.1. According to the picture, there're six types of combustion phenomena, (1) no droplet and stable flame in tube (NDSF), (2) no droplet and unstable flame within tube (NDUF), (3) droplet suspension and unstable flame (DSUF), (4) droplet suspension and stable flame at tube exit (DSSFE), (5) droplet dripping and stable flame at tube exit (DDSFE), and (6) flame extinction when dripping (FED). There are two conditions where stable flame could exist. NDSF or DSSFE is stable flame inside the tube with or without droplets, and DDSFE is oscillatory combustion with periodic droplet dripping (green

Corresponding author. Fax: +00-00-000-000
E-mail address: abc@abcde.com

diamond scatters in Fig.1).

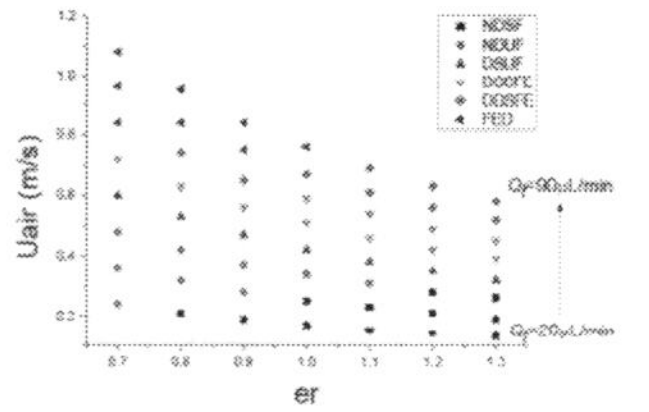

Figure 2: Evaporation and combustion characteristics with equivalence ratio with external heating

Our previous work[5] mainly focused on the first type of combustion, i.e. stable flame inside the tube. The second one is discussed in this paper, i.e. the periodic oscillatory flame when fuel flow rate increases to a certain value and droplet drips continuously. As shown in Fig.3, we chose a case where $er_{theo}=1.3$ and fuel flow rate was 1.13mg/s. With heating outside, shapes of flame and droplets are listed as Fig.3(a) and Fig.3(b).

When fuel flow rate increases to a certain value, droplet dripping and flame oscillation are inevitable[5]. It's easy to understand why oscillatory flame occurs. When fuel flow rate is limited, fuel supply is much smaller than fuel consumption caused by evaporation and combustion. Therefore, no droplet drips and flame stays steady in tube. However, with the increment of fuel flow rate, flame moves downstream till the tube exit and droplet starts to drip. At the moment of droplet colliding with the tube wall, a liquid film is formed and vaporizes quickly due to high wall temperature. The instant evaporation of liquid film causes a sudden increment of mass concentration of n-heptane vapor, which makes the flame zone become larger and brighter in a sudden, and flame temperature rises sharply at the same time That's how flame oscillation happens[5]. When liquid fuel is injected into the capillary continuously, droplets keep dripping and periodic oscillatory flame is formed.

Fig.3(a) and 3(b) shows flame and droplet shapes at different moments during one oscillation cycle. Fig.4 shows temperature variations with time. It can be seen that temperature profiles of TC1~TC5 are relatively smooth, and the values are under 650°C, indicating that oscillatory flame is located at the tube exit. Temperature at the exit is measured by TC6. Variations of temperature and flame shape in Fig.4 and Fig.3(a) can be well explained by variations of droplet shape and size in Fig.3(b). From $t=0s$ to $t=4.8s$, droplet size keeps increasing and so does the droplet surface area, which leads to the enlargement of evaporation area. Mass concentration of gaseous n-heptane in unburnt gas increases. Flame zone becomes larger and brighter during this period and temperature profile of TC6 is rising. After the droplet size increases to a certain value, it drips at the moment of $t=6.4s$. The droplet collides with tube wall and form a liquid film. Because wall temperature is higher than the boiling point of n-heptane, the liquid film evaporates quickly and mass concentration of gaseous n-heptane experiences a sudden rise. At the moment of $t=6.4s$, flame temperature reaches its maximum value and combustion becomes very fierce, in corresponding to the peak value of temperature profile measured by TC6. Size of liquid film decreases quickly as shown in moment $t=8.0s$. After gaseous n-heptane generated by instant evaporation of liquid film was burnt out, droplet size starts increasing from zero again and flame temperature begins to fall, which corresponds to the declining parts of temperature profile measured by TC6, and the oscillation starts a new cycle.

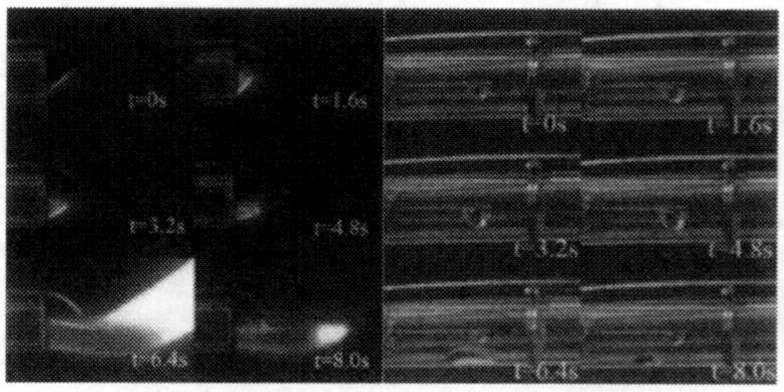

(a) Flame shapes (b)Droplet shapes

Figure 3: $er_{theo}=1.3$, fuel flow rate fixed at 1.13mg/s, shape variations in one cycle

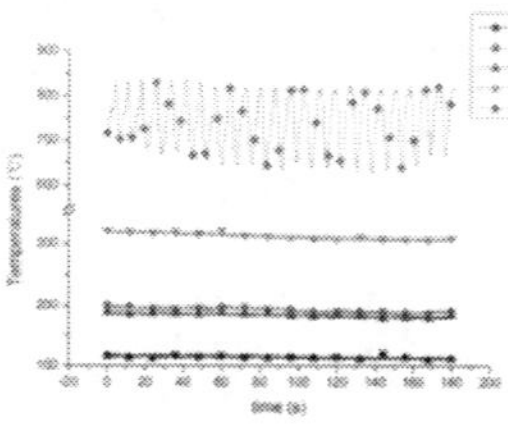

Figure 4: $er_{theo}=1.3$, fuel flow rate fixed at 1.13mg/s, temperature profiles measured by different thermocouples

3 Evaporation models

3.1 Droplet suspension model

In order to further explore characteristics and mechanism of oscillatory flame of n-heptane droplet, an evaporation model of periodic droplet is developed according to the experimental setup. A simplified model is shown in Fig.5. Droplet shape is assumed as spherical, and droplet diameter is assumed as equal to droplet height. Because the droplet is suspended on a capillary tip, droplet size is affected both by fuel flow rate $\dot{m}_f$ and evaporation rate $\dot{m}_f'$. Injection of liquid fuel increases droplet size, while evaporation at the droplet surface decreases droplet size. When fuel flow rate was larger than evaporation rate, droplet size increases continuously. When gravity of the droplet is larger than the adhesive force of capillary, droplet drips and collides with tube wall.

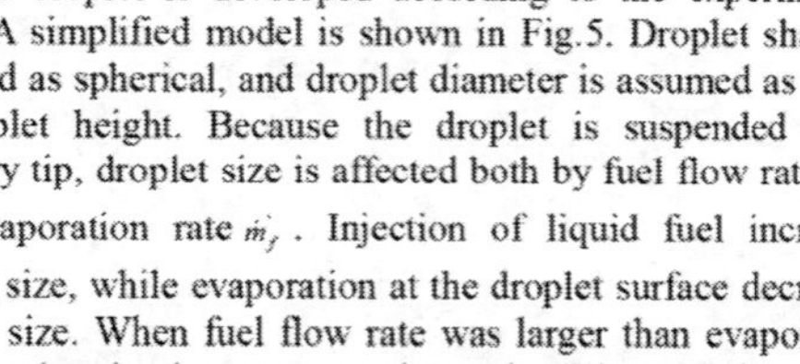

Figure 5: Droplet evaporation model

In every certain case, mass flow rate of liquid n-heptane $\dot{m}_f$ is fixed, indicating that the mass increment rate of liquid droplet is constant. The evaporation rate of liquid droplet can be expressed as following equation according to Ref[6]

$$\dot{m}_f' = \frac{\pi d \lambda_{mix}}{c_{p,mix}} Nu \ln[1 + B_T] \tag{2}$$

Where λ_{mix} and $c_{p,mix}$ are thermal conductivity and constant-pressure specific heat of gases around the droplet, respectively. They are calculated according to Ref[7]. Spalding heat transfer number B_T and Nusselt number Nu are taken from Ref[8]. Thus the mass increasing rate of droplet can be expressed as (3)

$$\dot{m} = \dot{m}_f - \dot{m}_f' \tag{3}$$

At the moment of t_i, droplet diameter is taken as d_i and droplet weight is m_i. Evaporation rate is $\dot{m}_{f,i}^{'}$ and mass increasing rate is $\dot{m}_i$. Thus Eq.(3) can be rewritten as $\dot{m}_i = \dot{m}_f - \dot{m}_{f,i}^{'}$. Then the weight of droplet at next moment t_{i+1} ($t_{i+1} = t_i + \Delta t$) can be expressed as $m_{i+1} = m_i + \dot{m}_i \cdot \Delta t$. Droplet volume and diameter at moment of t_{i+1} can be calculated by equations $V_{i+1} = m_{i+1}/\rho_l$ and $d_{i+1} = (6V_{i+1}/\pi)^{1/3}$.

At the beginning of injection $t_0 = 0$, droplet weight m_0=0 and diameter d_0=0. After several time steps of Δt ($\Delta t = 0.001s$) droplet weight m_i and diameter d_i can be calculated by iteration of $\dot{m}_i$ and V_i. Obviously, droplet weight m_i increases with time. When gravity of droplet exceeds the adhesive force of capillary, the droplet will drip down. Then droplet weight and droplet size return to zero, entering next cycle for new iteration.

The value of surface tension (σ) for n-heptane at room temperature is 20.3mN/m. The characteristic length (l) can be taken as droplet diameter d approximately. Characteristic length is the contact length between droplet and capillary, as shown in Fig.5. Then adhesive force of the droplet attached on capillary can be expressed as $F = \sigma l$.

A case of equivalence ratio $er_{theo} = 1.3$ and fuel flow rate $\dot{m}_f = 1.13mg/s$ is chosen to calculate above data. Mathematical software Matlab is used to program above equations. Predicted periodic variations of droplet diameter with time are plotted by black line in Fig.6, while measured droplet diameters within one cycle from experimental photos are plotted by scatters. There is a slight deviation between predicted and measured data due to our spherical assumption, but it is acceptable.

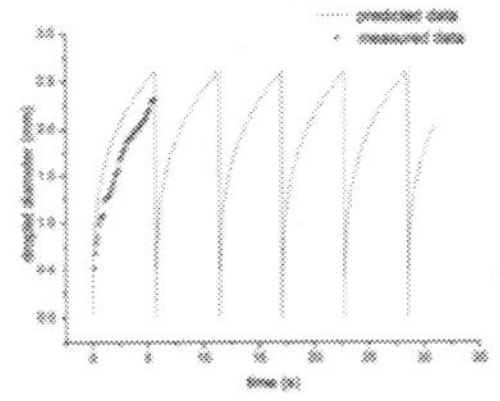

Figure 6: Periodic variation of droplet diameter with time

3.2 Deposited film model

After the droplet drips and collides with tube wall, a liquid film is formed on the wall and vaporizes rapidly due to high wall temperature. Wall temperature measured in experiments is higher than the boiling point of n-heptane. It is well-known that when a drop of liquid is deposited on a hot solid, if the solid temperature is much higher than the boiling point, the drop is not anymore in contact with the solid, but levitates above its own vapor, as shown in Fig.7. Because the heat transfer through vapor is less than the heat transfer between hot solid and liquid film, evaporation rate of the film is reduced considerably, so that the drop does not boil but evaporates quietly. Such floating drops are called Leidenfrost drops[9]. A vapor film appears when wall temperature exceeds 160°C, and the optimum temperature for Leidenfrost drop of n-heptane is 210°C[10].

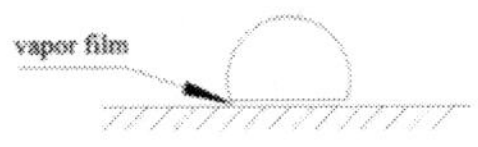

Figure 7: Schematic of vapor film

On the basis of droplet suspension model in section 3.1, an evaporation model for liquid film is developed as shown in Fig.8. Volume of the liquid film at dripping moment equals to the volume of droplet before dripping. Shape of the drop deposited on tube wall is taken as spherical approximately, and its size is characterized by drop diameter d.

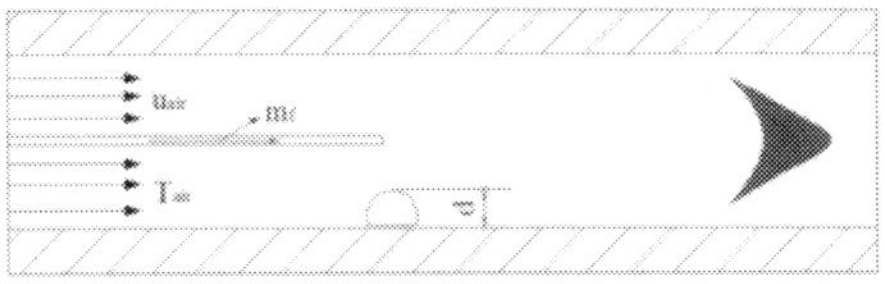

Figure 8: Evaporation model for liquid film

A case with equivalence ratio er_{theo}=1.3 and fuel flow rate being 1.13mg/s is chosen to build the model. The average temperature of wall in contact with droplet (T_w) is 191°C measured by TC1. Since T_w is in the temperature range of a Leidenfrost drop, we need to calculate evaporation time of the drop as levitating on its own vapor. According to calculation model of evaporation time for spherical drop proposed by AL Biance et.al[10], the evaporation time τ can be expressed as (4)

$$\tau = \mathrm{A} \cdot \frac{\rho q_e}{\lambda \cdot \Delta T} R_0^2 \quad (4)$$

Based on our experimental results, modified equation of evaporation time can be expressed as (5)

$$\tau = \mathrm{A} \cdot \frac{\rho q_e}{\lambda \cdot \Delta T} R_0^2 + B \cdot R_0 \quad (5)$$

Where A and B represent fitting coefficients. ρ and q_e are density and latent heat of liquid n-heptane, and λ denotes thermal conductivity of gaseous n-heptane. ΔT is temperature difference between wall temperature and drop temperature. It is assumed that there's no internal temperature gradient inside the film, and temperatures inside the drop equals to the surface temperature T_s. Then ΔT can be expressed as $\Delta T = T_w - T_s$. R_0 is radius of the spherical drop at dripping moment and $R_0 = d_0/2$. Since the liquid film vaporizes quickly and evaporation time is short, it is assumed that evaporation rate is constant when liquid drop is deposited on tube wall. Therefore, the mean evaporation rate $\dot{m}_{film}$ of liquid film can be expressed as $\dot{m}_{film} = \rho V_0 / \tau$. Substitute the expression of τ, the detailed expression of $\dot{m}_{film}$ can be expressed as (6)

$$\dot{m}_{film} = \frac{\rho V_0}{\mathrm{A} \cdot \frac{\rho q_e}{\lambda \cdot \Delta T} R_0^2 + B \cdot R_0} \quad (6)$$

Where $V_0 = 4\pi d_0^3/3$ represents the volume of droplet at dripping moment. Combining (2) and (6), the changing process of evaporation rate during the whole period can be figured out. Still the case of er_{theo}=1.3 and fuel flow rate $\dot{m}_f = 1.13$ is chosen. Periodic variation of evaporation rate can be obtained in Fig.9.

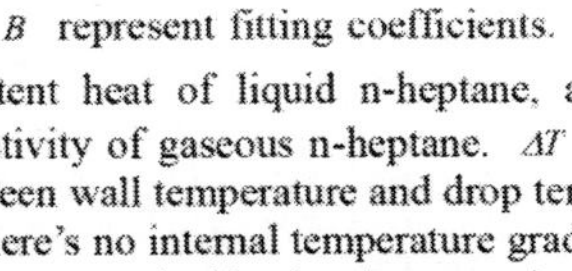

Figure 9. $er_{theo} = 1.3$, fuel flow rate fixed at 1.13mg/s , periodic variation of evaporation rate for droplet and film

4 Results and discussion

Evaporation rate means mass increment of gaseous n-heptane converted from liquid fuel in unit time. If air velocity is fixed, real equivalence ratio of chemical reaction will change with evaporation rate of n-heptane. The periodic variation of real equivalence ratio with time can be obtained from evaporation rate profile in Fig.9. Theoretical equivalence ratio from Eq.(1) serves as a comparison. Calculation method for real equivalence ratio can be expressed as (7)

$$er_{real} = \frac{\left(m_g/m_{air}\right)_{real}}{\left(m_g/m_{air}\right)_{stio}} \tag{7}$$

4.1 Influence of fuel flow rate

To explore the effects of equivalence ratio on oscillatory combustion characteristics of n-heptane droplet, cases with fixed air velocity ($u = 1.06m/s$) and varied theoretical equivalence ratio are tested. Variations of real equivalence ratio over time are plotted in Fig.10(a), and temperature oscillation profiles are shown in Fig.10(b).

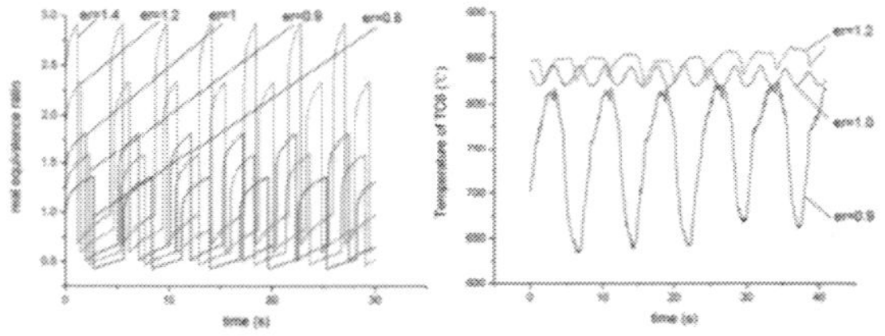

(a) Real equivalence ratios (b) Temperature profiles

Figure 10: Effects of theoretical equivalence ratio on oscillatory flames

It is obvious in Fig.10(a) that when theoretical equivalence ratio increases from 0.8 to 1.4, dripping period of droplet decreases gradually. Real equivalence ratio also increases with theoretical equivalence ratio, which are 1.4, 1.6, 1.8, 2.3, 2.9 (based on the maximum calculation value). According to Fig.10(b), it can be seen that when theoretical equivalence ratio varies from er=0.9 to er=1.0, average flame temperature rises and oscillation amplitude decreases. This indicates that when air velocity is fixed, increasing of theoretical equivalence ratio can raise flame temperature and benefit flame stability. Moreover, when theoretical equivalence ratio changes from er=1.0 to er=1.2, mean flame temperature decreases and the oscillation amplitude is only 21°C, demonstrating that flame stability is improved.

4.2 Influence of air velocity

In this section fuel flow rate was fixed at 1.13mg/s and air velocity varied gradually. Variations of real equivalence ratio with time are plotted in Fig.11(a), and temperature oscillation profiles are shown in Fig.11(b).

It can be observed in Fig.11(a) that with the increasing of air velocity, dripping period of droplet becomes shorter, and a complete evaporation cycle of droplet consists of two parts. Lower part of the curve represents vaporization of suspended droplet and higher part denotes liquid film. Evaporation time of suspended droplet decreases with air velocity, while that of liquid film increases instead. Moreover, real equivalence ratio also decreases with the increasing of air velocity. Fig.11(b) shows that temperature oscillation becomes fiercer with increasing of air velocity. The amplitude of temperature oscillation rises from 83.8°Cto 101.8°C, and then to 121.1°C. Oscillation period exhibits the same tendency with dripping period, decreasing with air velocity.

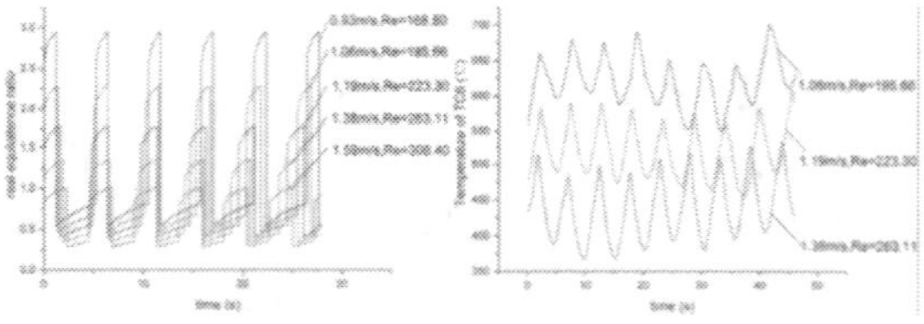

(a) Real equivalence ratios (b) Temperature profiles

Figure 11 Effects of air velocity on oscillatory flames

5 Conclusions

When a liquid droplet of n-heptane burns inside a micro-tube, a liquid film is formed after it drips and collides with the hot wall. The film will vaporize rapidly and cause flame oscillation. Based on this process, evaporation models of droplet and liquid film were built.

The main cause for oscillatory flame is the instant increase of real equivalence ratio during combustion. After a droplet drips, a liquid film comes into being and vaporizes quickly. Thus the mass concentration of gaseous n-heptane increases and a more intense combustion appears.

With increasing of fuel flow rate, dripping period for liquid n-heptane decreases, but the real equivalence ratio increases, leading to fiercer combustion and more heat generation.

Increasing of air velocity accelerates the blown-off process of droplets and oscillation frequency of temperature increases. However, higher air velocity enhances the disturbing effect of air, and magnifies temperature oscillation amplitude. The average flame temperature also decreases with air velocity.

References

[1] S. Hyunkyu, L. Changsik, A review on atomization and exhaust emissions of a biodiesel-fueled compression ignition engine, Renewable & Sustainable Energy Reviews, 58 (2016) 1601-1620.

[2] L.H.J. Wachters, N.A.J. Westerling, The heat transfer from a hot wall to impinging water drops in the spheroidal state, Chemical Engineering Science, 21 (1966) 1047-1056.

[3] T.Y. Xiong, M.C. Yuen, Evaporation of a liquid droplet on a hot plate, International Journal of Heat & Mass Transfer, 34 (1991) 1881-1894.

[4] Y.S. Ko, S.H. Chung, An experiment on the breakup of impinging droplets on a hot surface, Rowman & Littlefield.

[5] J. Li, J. Huang, M. Yan, D. Zhao, J. Zhao, Z. Wei, N. Wang, Experimental study of n -heptane/air combustion in meso-scale burners with porous media, Experimental Thermal & Fluid Science, 52 (2014) 47-58.

[6] S.S. Sazhin, Advanced models of fuel droplet heating and evaporation, Progress in Energy & Combustion Science, 32 (2006) 162-214.

[7] D.B. Spalding, The combustion of liquid fuels, Symposium on Combustion, 4 (1953) 847-864.

[8] A.H. Lefebvre, Atomization and sprays, Atomization & Sprays, 16 (1989) 127-145.

[9] A.L. Biance, C. Clanet, D. Quéré, Leidenfrost drops, Physics of Fluids, 15 (2003) 1632-1637.

[10] S.L. Manzello, J.C. Yang, An experimental study of high Weber number impact of methoxy-nonafluorobutane C4F9OCH3 (HFE-7100) and n-heptane droplets on a heated solid surface* 1, International Journal of Heat & Mass Transfer, 45 (2002) 3961–3971.

11th Asia-Pacific Conference on Combustion,
The University of Sydney, NSW Australia
10th -14th December 2017.

Simultaneous removal of soot and NOx with a $La_{0.7}Ag_{0.3}MnO_3$ perovskite-like catalyst: Effect of synthesis method.

Laura Urán[1], Jaime Gallego[1], Alexander Santamaria[1]

[1] Química de Recursos Energéticos y Medio Ambiente, Instituto de Química, Facultad de Ciencias Exactas y Naturales, Universidad de Antioquia UdeA

Calle 70 No.52-21, Medellín, Antioquia, 050010, Colombia

Abstract

$La_{0.7}Ag_{0.3}MnO_3$ perovskite-like catalyst was prepared using two synthetic routes, microwaved-assisted hydrothermal (HMW) and solid state (SS) with the aim of studying catalyst synthesis method effect for the simultaneous removal of soot and NOx under loose contact condition. The materials were characterized using various analysis techniques, such as X-ray diffraction (XRD), X-ray photoelectron spectroscopy (XPS), temperature-programmed reduction under hydrogen atmosphere (H_2-TPR) and oxygen temperature-programmed desorption (O_2-TPD). The results indicated that the main effect of the catalyst synthesis methodology was the degree of incorporation of silver in the perovskite structure. In this sense, the larger the silver incorporation into the perovskite structure the higher the oxygen vacancies. The presence of oxygen vacancies on catalyst surface was the determining factor for the catalytic activity during the simultaneous removal of soot and NOx. The HMW synthesized catalyst that had a α-oxygen species desorption as high as 702 μmol/g, and an Oads/Otot ratio of 0.49 was the material with the best catalytic activity performance: temperature of 50% of formation of CO_2 (T50) at 359°C, 100% soot conversion to CO_2 and 41% conversion efficiency from NOx to N_2.

1 Introduction

Due to the environmental problems associated with the use of diesel engines, and the strengthening of the regulations concerning pollutant emissions in the transport field. The study of the catalytic simultaneous removal of soot and NOx at relatively low temperatures (within the range of diesel engine exhaust: 150-450°C) has taken a paramount importance in pollution control area.

Mixed oxides coming from perovskite-like materials seem to be a promising catalyst for such purpose due to their ability to provided active oxygen species at a low temperature range (below 500°C). The performance of these materials depends of surface properties, size and stoichiometry, which are determined by the synthesis method [1].

In this way, the present work tested the $La_{0.7}Ag_{0.3}MnO_3$ perovskite-like material, prepared for different synthetic routes, in order to evaluate the response of the synthesized materials on the catalytic activity for the simultaneous removal of soot and NOx. Two different synthetic routes are compared: microwave-assisted hydrothermal (HMW) and solid state (SS).

2 Experimental

$La_{0.7}Ag_{0.3}MnO_3$ perovskite-like catalyst was prepared using two synthetic routes, microwaved-assisted hydrothermal (HMW) and solid state (SS) using as starting materials $La(NO_3)_3{\cdot}6H_2O$, $AgNO_3$ and $Mn(NO_3)_2{\cdot}4H_2O$. The resulting material of each synthetic route was calcinated at 600°C for 10h.

The catalytic activity of the prepared catalysts was tested in a temperature-controlled furnace inside which a soot-catalyst-sand fixed bed was placed. The mixture of 5 mg of carbonaceous material and 45 mg catalysts (1:9 weight ratio, in accordance with other authors [2], [3]) was carried out with spatula (i.e. loose contact) by mixing for 5 min. Afterwards, the 50 mg of the soot-catalyst mixture was mixed with spatula with 150 mg of sand for another 5 minutes. The sand (SiO_2), which is an inert material, was added in order to reduce the specific pressure drop across the reactor and to reduce thermal runaways [3].

The soot model used in this study consisted of a carbon black, FW200, from Orion Engineering. The soot-catalyst-sand mixture was exposed at a constant heating rate of 5°C/min up to 600 °C, under a 2000 ppm NO/ 10% vol.O_2/ He atmosphere at ambient pressure. Even though the temperature range of interest is the same of engine exhaust temperature, i.e. up to 450 °C, the experiments were carried out up to 600 °C to monitor completely the catalysts performance during soot oxidation process without exceeding their calcination temperature. The inlet gas composition was chosen in accordance with other catalytic soot oxidation studies that suggest similar gas compositions to simulate engine exhaust conditions [1], [4], [5].

The evolution of CO_2, CO, NO, NOx, N_2, and O_2 gases with temperature were followed by mass spectrometry and a KIGAZ 310 gas analyzer. The catalyst materials were characterized by X-ray diffraction (XRD) to determine whether the desired perovskite-like structure was formed, X-ray photoelectron spectroscopy (XPS) was used to characterize the surface composition of the catalysts, temperature-programmed reduction under hydrogen atmosphere (H_2-TPR) and oxygen temperature-programmed desorption (O_2-TPD) to obtain an

Corresponding author. Fax: +574-2196611
E-mail address: alexander.santamaria@udea.edu.co

Table 1: Characterization of $La_{0.7}Ag_{0.3}MnO_3$ catalyst synthesized by HMW and SS methodologies and their catalytic activity in the simultaneous removal of soot and NOx

Catalyst	Tig (°C)	T50 (°C)	T90 (°C)	S_{CO2} (%)	Max NOx conv (%)	TMax NOx conv (°C)	H_2-TPR (μmol/g)	α-O_2 TPD (μmol/g)	Mn^{4+}/Mn_{tot}	O_{add}/O_{tot}	Ag^0 (segregated)/Ag_{tot})	Ag^+/Ag_{tot}	Ag^{2+}/Ag_{tot})
HMW	200	359	406	100	41	371	674	702	0.34	0.43	UD[a]	1	UD[a]
SS	200	404	452	100	38	394	736	345	0.37	0.13	0.37	0.44	0.19
Non-catalytic	250	578	639	40	D/A[b]	--	--	--	--	--	--	--	--

a. undetected
b. does not apply

insight into the catalysts reducible sites and oxygen active species respectively. Equation 2 was used to calculate NOx conversion efficiency.

$$NOx\ conversion\ (\%) = (1 - [NO_x]_{out}/[NO_x]_{in}) \times 100 \quad (2)$$

Where $[NO_x]_{in}$ and $[NO_x]_{out}$ are the NOx concentration in parts per million (ppm) at the inlet and the outlet of the experimental setup respectively.

3 Results and discussion

3.1 X-ray diffraction (XRD)

Figure 1 presents the X-ray diffraction patterns of the synthesized materials. The catalysts present the diffraction peaks corresponding to the perovskite-like structure. The XRD patterns present signals at 38.3° and 64.6°, corresponding to Ag_2O; and a peak at 44.5° that corresponds to metallic silver (Ag^0), indicating that not all the silver was incorporated into the perovskite-like structure. Furthermore there are weak signals at 29° corresponding to lanthanum oxide (La_2O_3) and 37° and 39° indicating manganese oxide (MnO_2).

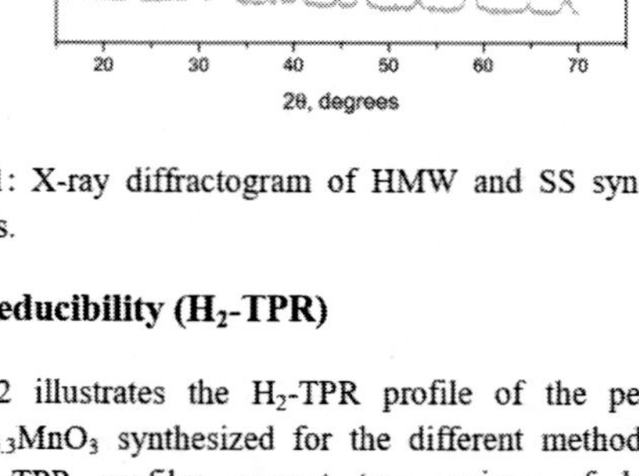

Figure 1: X-ray diffractogram of HMW and SS synthesized materials.

3.2 Reducibility (H_2-TPR)

Figure 2 illustrates the H_2-TPR profile of the perovskite $La_{0.7}Ag_{0.3}MnO_3$ synthesized for the different methodologies. The H_2-TPR profiles present two regions of hydrogen consumption, below and above 335 °C. In the low temperature region; the HMW synthesized sample presents a predominant hydrogen consumption peak at 216 °C, meanwhile for the SS synthesized material the main peak appears at 207°C. Above 335 °C, both samples presents a weak hydrogen consumption peak around 410 °C. Considering the TPR profiles of the undoped perovskite $LaMnO_3$ presented for other authors [5], [6], where the main reduction peak is located about 380 °C, the weak peak located around 410 °C of the samples synthesized in this study suggests that most of the Mn species were reduced at lower temperatures.

Table 1 lists the total hydrogen consumption of the synthesized catalyst. The SS sample had a hydrogen consumption of 736 μmol/g while the HMW consumed 634 μmol/g. The higher hydrogen consumption of the SS material could be due to the higher concentration of metallic silver (Ag^0), as was evidenced by XPS analysis; the higher proportion of Ag_2O and AgO segregated, which means a less silver incorporation into the perovskite structure. Furthermore, the higher Mn^{4+}/Mn_{tot} ratio (see Table 1), consequence of the less silver incorporation, that means the presence of nonstoichiometric excess oxygen that also reacts with hydrogen during the H_2-TPR analysis. Actually, it was found that when the ratio of Mn^{4+}/Mn_{tot} increased from 0.34 to 0.37 (i.e. an increase of 9.2%), in the same percentage was the increment in the hydrogen consumption, i.e. from 674 to 736 μmol/g between the SS and the HMW synthesized samples.

3.3 Oxygen desorption (O_2-TPD)

The literature indicates [7]–[9] that the desorbed oxygen species during a O_2-TPD experiment can be classified according to their desorption temperature, as follows: The ones that desorb below 550 °C are attributed to physical and chemical weakly adsorbed oxygen species on surface catalyst (O^-), named as α-oxygen. These oxygen species have an important interest in catalyst area because it has been reported [2], [7]–[10] to enhance the catalytic activity, and also are an indicative of surface oxygen vacancies. On the other hand, the oxygen species desorbed above 550 °C correspond to bulk oxygen species (O^{2-}), also known as β-oxygen. β-oxygen species are of minor interest in catalyst studies due to their higher desorption temperature. As the oxygen species of interest for this study are those that are available at a low temperature, the O_2-TPD experiment was done up to 600 °C, besides, with the idea of not exceeding samples calcination temperature.

Figure 3 presents the oxygen desorption profiles (O_2-TPD) of the HMW and SS synthesized samples. These oxygen-desorption-profiles present mainly two regions, below and above 500 °C. For the region of interest, i.e., below 500 °C, the samples present a predominant desorption peak at 121 °C, and 128 °C for HMW, and SS synthesized samples respectively. The desorption profile located above 500 °C, which corresponds to β-oxygen species, is not completely defined because the test was performed only up to 600 °C. Table 1 presents the total amount of oxygen species desorbed up to 500 °C. HMW sample has a α-oxygen desorption as high as 702 μmol/g, meanwhile SS has desorption amount of 345 μmol/g. It is important to highlight that the sample with the highest α-oxygen content is the one with the best soot oxidation catalytic activity, measured as T50, and the best conversion efficiency of NOx to N_2, as can be evidenced in Table 1.

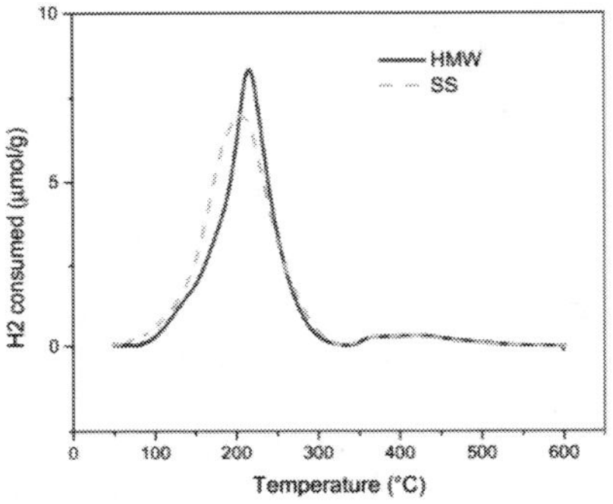

Figure 2: Hydrogen consumption (H_2-TPR) profile

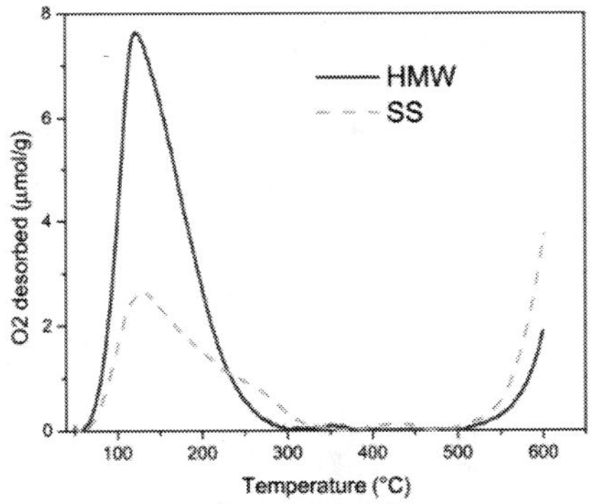

Figure 3: α-oxygen desorption (O_2-TPD) profile

O_2-TPD and H_2-TPR tests only include either oxidizing or reductive specie, while soot oxidation experiment includes oxidizing (O_2) and reductive (soot) species at the same time. This could be a reason why O_2-TPD and H_2-TPR peak profiles appear in a range of temperature (100 °C–300 °C) lower than the peaks during soot oxidation process (200 °C-450 °C). In this sense, the O_2-TPD and H_2-TPR profile peaks are an indicative of the oxygen vacancies and reducible sites availability that promotes α-oxygen formation in a more dynamic process as soot oxidation; rather than the total α-oxygen species availability for the whole oxidation process.

3.4 Surface elemental composition and oxidation states (XPS)

Table 1 presents Mn^{4+} atomic ratio, along with the adsorbed oxygen and silver species (Ag°, Ag^{+}, Ag^{2+}) content on catalyst surface. According to Fino et al. 2003 [2], when the A ion of a perovskite is partially substituted with an ion of different oxidation state, as for the $La_{0.7}Ag_{0.3}MnO_3$ catalyst where La^{3+} is partially substituted by Ag^{+}, a charge compensation is requiered to achieve electroneutrality. This can be either by formation of oxygen vacancies or by a shift of B metal toward higher valences (e.g., Mn^{3+} to Mn^{4+}). For the catalysts synthesized in this study, the charge compensation mechanism due to incorporation of silver (Ag) into the perovskite structure results in the cases mentioned before: Mn^{4+} represented the 34% and 37% of the total manganese content; and the adsorbed oxygen species, that suggest the presence of oxygen vacancies, were the 49% and 13% of the total surface oxygen amount for the HMW and SS synthesized materials respectively.

There is an important difference in the content of adsorbed oxygen species (from 13% to 49%), suggesting more oxygen vacancies for the HMW sample compared to SS synthesized material, this result agrees well with the O_2-TPD analysis, where HMW sample has the highest oxygen desorption content, 702 μmol/g, also shown in Table 1.

The increase in the content of adsorbed oxygen species in the HMW sample could be explained by the large incorporation of silver into the lattice structure of the perovskite, contrary to that seen for the SS sample where the combined segregation of metallic silver and AgO over the catalyst surface was as high as 56% (see Table 1). The lower incorporation of silver into the SS perovskite structure results in A site cation deficiency that is compensated with the shift to a higher oxidation states of the B site cation, in this case, from Mn^{3+} to Mn^{4+} providing more redox properties and less oxygen vacancies.

3.5 Simultaneous soot and NOx removal

Figure 4 and Table 1 present the temperature at which soot ignition begins (T_{ig}), and the temperatures at which the conversion of soot to CO_2 is 50% and 90%, named as T50 and T90, respectively. When the catalytic activity results are compared with the non-catalytic one; it was found a reduction in the ignition temperature and T50. Highlighting HMW synthesized material that was able to decrease in about 220°C the T50 of the non-catalytic test, meanwhile SS prepared catalyst decreased this temperature in 174°C, i.e., a difference of 45°C between the catalytic systems. According to this, we found the following decreasing order for T50: non-catalytic > SS > HMW. Another important result regarding catalytic soot oxidation, it is that no CO emission is found in the gas stream, indicating 100% selectivity toward CO_2. In contrast with the non-catalytic one test where the selectivity to CO_2 is only 40%. To assure that the soot charged to the furnace has been eliminated completely during the catalytic test, the remained solid mixtures at the end of the experiments were recovered and analyzed in a thermogravimetric instrument (TGA) under air at a heating rate of 10°C/min up to 700°C. The recovered solid mixtures presented a weight loss of only 0.1%, indicating that just contain catalyst and sand and that all the soot charged to the catalytic experiments was oxidized. In addition, considering that no CO was detected during the catalytic tests, then soot oxidation was only toward CO_2.

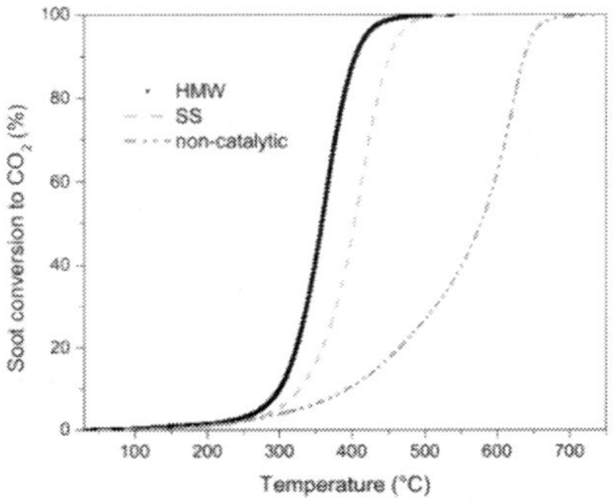

Figure 4: soot conversion to CO_2 (%) as function of temperature for the catalytic and non-catalytic tests.

Regarding NOx removal, Figure 5 shows the NOx evolution with temperature, for the catalytic experiments carried out with $La_{0.7}Ag_{0.3}MnO_3$ synthesized by the different methodologies. Figure 5 shows that during soot oxidation, the catalysts tested present some NOx adsorption/desorption events followed by a NOx reduction toward N_2 around the soot oxidation peak. As Table 1 indicates, the maximum NOx reduction temperatures were around 371°C and 395°C, meanwhile the maximum NOx reduction to N_2 were 41% and 38%, for HMW and SS materials respectively.

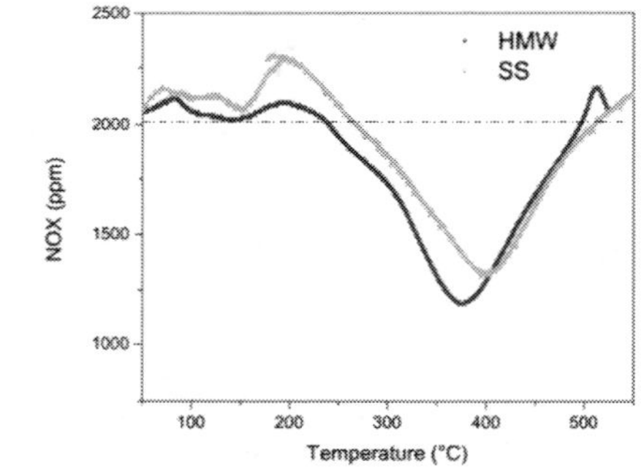

Figure 5: NOx concentration profile with temperature during the catalytic test

Figure 6 shows the CO_2, O_2, NOx and N_2 profiles as function of temperature when the HMW-synthesized catalyst was used. In this figure it is possible to appreciate that the temperature for the maximum NOx reduction and N_2 peak is closed to the temperature at which the evolution of CO_2 is maximum, at the values mentioned before and presented in Table 1.

This result shows the ability for the simultaneous removal of soot and NOx of the $La_{0.7}Ag_{0.3}MnO_3$ perovskite-like material synthetized by the HMW method.

The good results of the HMW material could be related with the presence of more oxygen vacancies on catalyst surface, demonstrated by the desorption of more α-oxygen species during O_2-TPD test and the higher O_{ads}/O_{tot} ratio as was evidenced by XPS analysis.

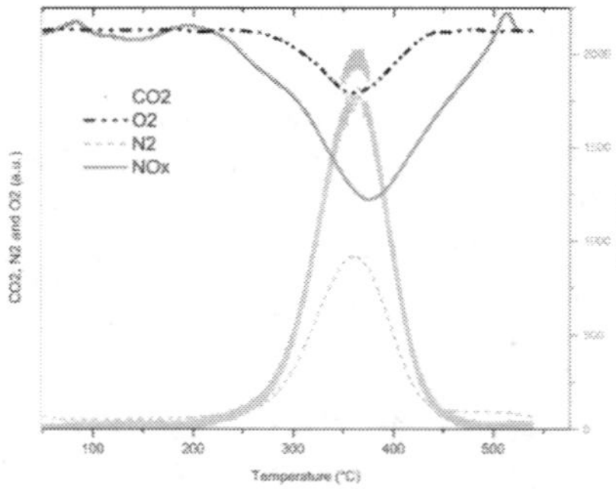

Figure 6: CO_2, O_2, NOx and N_2 profiles with temperature for the HMW-synthesized catalyst

4 Conclusions

The main effect of the catalyst synthesis methodology on the perovskite-like material, $La_{0.7}Ag_{0.3}MnO_3$, was the degree of incorporation of silver into the perovskite structure. When silver incorporation was larger, the material results with higher oxygen vacancies; whereas when silver incorporation was lower, the material results in A site cation deficiency that is compensate with the shift to a higher oxidation states of the B site cation, in this case from Mn^{3+} to Mn^{4+} providing more redox properties.

The presence of oxygen vacancies on catalyst surface, evidenced by the desorption of α-oxygen species during O_2-TPD test and the O_{ads}/O_{tot} ratio obtained by XPS analysis, was the determining factor for catalytic activity in the simultaneous removal of soot and NOx. The HMW synthesized catalyst that had a α-oxygen species desorption as high as 702 μmol/g, and a O_{ads}/O_{tot} ratio of 0.49 was the material with the best catalytic activity. It was able to reduce in 220 °C the temperature at which 50% of soot is converted to CO_2 (T50), when it was compared with the non-catalytic test. Besides, HMW material presents 100% soot conversion to CO_2 since there was no evidence of CO formation; and it exhibits 41% conversion efficiency from NOx to N_2.

5 Acknowledgment

The authors thank to University of Antioquia for the financial support received through the CODI project No 2015-7828. Laura U. wishes to thank the Colombian Administrative Department of Science, Technology and Innovation (COLCIENCIAS) for her Ph.D. Scholarship.

References

[1] V. Albaladejo-Fuentes, F. E. López-Suárez, M. S. Sánchez-Adsuar, and M. J. Illán-Gómez, Appl. Catal. A Gen 519 (2016) 7–15.

[2] D. Fino, N. Russo, G. Saracco, and V. Specchia, J. Catal. 217 (2) (2003) 367–375.

[3] N. Russo, S. Furfori, D. Fino, G. Saracco, and V. Specchia, Appl. Catal. B Environ 83 (1–2) (2008) 85–95.

[4] R. Matarrese, S. Morandi, L. Castoldi, P. Villa, and L. Lietti, Appl. Catal. B Environ 201 (2017) 318–330.
[5] K. Wang, L. Qian, L. Zhang, H. Liu, and Z. Yan, Catal. Today 158 (3–4) (2010) 423–426.
[6] J. Chen, M. Shen, X. Wang, G. Qi, J. Wang, and W. Li, Appl. Catal. B Environ 134–135 (2013) 251–257.
[7] J. A. Onrubia and J. R. González-velasco, Appl. Catal. B, Environ 213 (2) (2017) 198–210.
[8] R. Dinamarca, X. Garcia, R. Jimenez, J. L. G. Fierro, and G. Pecchi, Mater. Res. Bull 81 (2016) 134–141.
[9] W. Y. Hernández, M. N. Tsampas, C. Zhao, A. Boreave, F. Bosselet, and P. Vernoux, Catal. Today 258 (2015) 525–534.
[10] J. M. López, A. L. Gilbank, T. García, B. Solsona, S. Agouram, and L. Torrente-Murciano, Appl. Catal. B Environ 174–175 (2015) 403–412.

11th Asia-Pacific Conference on Combustion,
The University of Sydney, NSW Australia
10th -14th December 2017.

Study of combustion of brown coal using the ReaxFF reactive force field

Longzhen Guo, Liping Chen, Chongna Wo, Zhijun Zhou*

State Key Laboratory of Clean Energy Utilization, Zhejiang University,

Hangzhou 310027, PR China

Abstract

Based on the Wender coal model, the processes of brown coal combustion were studied by molecular dynamics simulations with a reactive force field. To examine the combustion process and the initiation mechanism of brown coal, some large-scale reactive systems containing different numbers of brown coal and oxygen were built in this work. A relatively high simulation temperature, which was proven reasonable in other studies, was used to control the simulation within an acceptable period of time. The gas production was analyzed. The related initial reaction mechanism and factors were discussed. It was observed that temperature enhanced the reactions in the combustion process. There were more carbon monoxide rather than carbon dioxide generated at higher temperatures. It was also found that higher excess air coefficient could promote carbon monoxide and carbon dioxide generation and oxygen consumption. Some of the important results agreed with other experimental data from the literature.

1 Introduction

Brown coal is a kind of low-rank coal, with a low condensation degree of aromatic rings and numerous bridged bonds and functional groups, it behaves very differently from many other solid fuels. Brown coal generally contains high volatiles and high moisture content, which makes it difficult to achieve long-distance transportation. Therefore, the most prominent utilization process of brown coal worldwide is combustion for power generation in mine-mouth power plants. Hence an in-depth understanding of brown coal combustion mechanism is in demand. So far, researchers have done a lot of work on it. Paul A. Morgan and his co-authors studied the coal oxidation by thermogravimetric analysis, they found that some features of the profiles were clearly related to coal properties especially maceral composition[1]. R.J. Hamor investigated the kinetics of combustion of a pulverized brown coal char between 630 and 2200 K[2].

Because of the complex structures of brown coal and the numerous influential factors on the combustion reaction, traditional experimental methods cannot provide a deep comprehension of the mechanisms of brown coal combustion. Therefore, researchers have been looking for a suitable way to describe the possible reaction pathways and products. Many quantum chemistry (QC) methods as well as molecular dynamics (MD) methods have been applied to coal combustion simulations. However, the two kind of methods have their own limitation, QC methods can be extremely time consuming and expensive, they are better at predicting the geometries, energies, and vibrational energies quite accurately for small molecules and systems. Tradition molecular force fields, for example DREIDING[3], can hardly any describe the chemical reactivity. Benner potential[4], as the only exception, does not take van der Waals and Coulomb interactions into consideration. In recent years, researchers pay more attention to ReaxFF reactive. This was developed to simulate the molecular dynamics of large scale reactive chemical systems (more than 1000 molecules) practically. It was proposed by Van Duin in 2001[5] and was first developed for hydrocarbons and sooner generalized to other kind of systems. The reaction force field does not need to fix the atom connectivity within molecules, and chemical bonds between atoms are free to fracture and generate. The ReaxFF reactive force field is based on bond order, and all the interactions are taken into account. Therefore, it is a suitable way to describe the reaction pathways and intermediate products of the reactions. Bhoi studied the oxygen coal combustion products and the effects of temperature and equivalence ratio using ReaxFF force field, but there was lack of analysis on reaction pathways and intermediates in his work[6]. So far, such kind of work is in great shortage, more work is needed to reveal the coal combustion mechanism.

In this paper, the processes of brown coal combustion were studied by molecular dynamics simulations with a reactive force field. The effect of temperature and excess air coefficient on reaction pathways and products of brown coal combustion was discussed; the forming mechanism of some main products was also shown. Our work is expected to be beneficial for the clean utilization of coal.

2 Computational details

In our work, a brown coal model based on Wender model[7] was constructed and optimized. Several reaction systems containing different number of brown coal and oxygen were also established and equilibrated and were set as the initial model for simulations.

2.1 Construction of reaction systems

The structure of coal is always complex and the construction of coal model is generally based on the characteristic information of fragments and the way of molecular bonding. Wender brown coal model is a generic and simplified coal model instead of a specific structure of a certain coal. There is only one aromatic ring and numerous side chains and bridge in the basic unit of coal structure. Some characteristic structure such as aromatic methoxyl and some C_3 - aliphatic side chains are also described.

Original Wender model is a plain structural model, so the model was saturated with hydrogen truncation and then optimized using QC and MD methods. The software package are Gaussian and materials studio respectively. The structure was firstly

Corresponding author. Fax: +86-0571-87952889
E-mail address: zhouzj@zju.edu.cn

subjected to DFT calculations with the method of B3LYP and the basis set of 6-31G(d) for basic geometric optimization. However, preliminary structural optimization for brown coal could only obtain a local minimum point due to its complex potential energy surface. Therefore, a further annealing processing was needed. The temperature of annealing process was from 300K to 600K and the simulation lasted for 10ps, 10 times. Then, the minimum energy configuration after the above optimization was chosen as the final geometric configurations. Which is shown in Fig. 1.

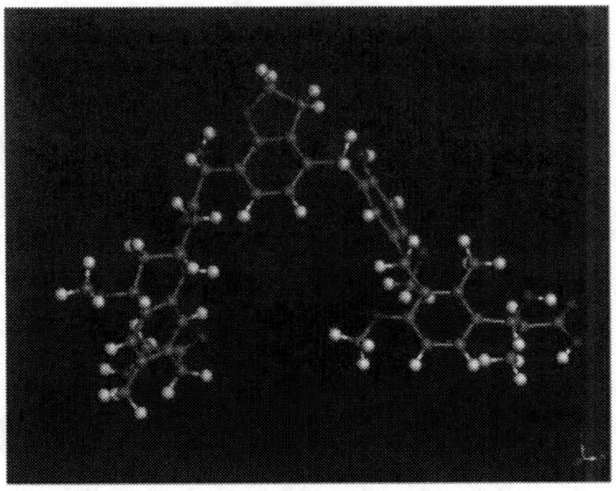

Figure 1: The final optimal 3-D geometric structure of brown coal (different colors represent different atoms, for instance, gray-C, white-H, red-O)

In order to understand the effect of some important factors on coal combustion, several reaction systems were constructed. Some important parameters are shown in the table 1, and S1-5 represent systems of different excess air coefficient ranging from 0.25 to 2 .

systems	coal	O_2	Geometries
S1	16	216	69.41×69.41×69.41 Å
S2	16	432	76.58×76.58×76.58 Å
S3	16	864	87.87×87.87×87.87 Å
S4	16	1296	96.84×96.84×96.84 Å
S5	16	1728	104.39×104.39×104.39Å

Table 1: Some parameters of systems built.

All the systems were equilibrated through an annealing cycle as described above and then set as the initial models. The ReaxFF reaction force field simulations can be very time-consuming, fortunately, researchers found that rising temperature can accelerate reaction rate without effecting the accuracy of computation, which was named " temperature accelerated reaction dynamics principle"[8]. Therefore, in order to control the simulation time in an acceptable period, the simulations were performed at relatively high temperatures range between 2000K to 4000K. Every 500K a simulation would be performed so we could obtain the temperature influence. Using Berendsen thermostat, temperature was controlled with a damping constant of 100 fs. A 0.25 fs time step along with a total time of 225 ps was used to study the coal combustion process[9].

2.2 An introduction of ReaxFF force field

ReaxFF is a kind of reactive force field based on bond order. All types of bond energy were related to bond lengths, valence angles and torsion angle. Similarly, the potential energy of the system was divided into the sum of each several energies as shown in (1).

$$E_{system} = E_{bond} + E_{over} + E_{under} + E_{val} + E_{pen} + E_{tors} + E_{conj} + E_{vaWaals} + E_{Coulomb} \quad (1)$$

In which: E_{bond} is the chemical bond energy; E_{over} and E_{under} are the over- and under- conditioned energies; E_{val} is the valance angle energy; E_{pen} is the penalty energy; E_{tors} is the torsion energy; E_{conj} is the conjugate effect of the molecular energy; $E_{vdWaals}$ is the van der Waals interaction energy; $E_{Coulomb}$ is the shielded Coulomb potential energy.

All the energies are based on bond order, so the expression of bond order is of the most importance, which is given in (2).

$$BO'_{ij} = exp\left[p_{bo,1} \cdot \left(\frac{r_{ij}}{r_o}\right)^{p_{bo,2}}\right] + exp\left[p_{bo,3} \cdot \left(\frac{r_{ij}^{\pi}}{r_o}\right)^{p_{bo,4}}\right] + exp\left[p_{bo,5} \cdot \left(\frac{r_{ij}^{\pi\pi}}{r_o}\right)^{p_{bo,6}}\right] \quad (2)$$

In which: $p_{bo,1-6}$ are constant coefficients; r_{ij} is the distant between atoms; r_o is the average of the radius of neighboring atoms.

3 Results and discussions

3.1 Products of brown coal combustion at 3000K

In order to obtain the generating way of products and intermediates during brown coal combustion, a certain condition (3000K, α=1) was picked as a typical example. The frequency corresponds to the percentage of the total number of molecules and the number of inorganic compounds at a certain moment. Due to the enormous amount of oxygen, it was not taken into consideration during statistical process.

As shown in Fig.2, along with simulation time, the number of macromolecular structures reduced while that of micro molecular structures increased. At the end of simulation, inorganic compounds take the biggest part of the products, and compound of more than six carbon molecules had been almost out of sight. As for inorganic compounds, the primary products were CO, CO_2 and H_2O. In the initial phase, the number of total inorganic compounds was relatively few while the reactions were relatively intense, so the frequency of products was in hight-variability. However, the proportion of all kind of products tended to stabilize as the reaction progress. The number of CO was huge, it conformed to the theory that more CO rather than CO_2 generated at high temperature[10]. Water was another important product, it can be seen from the chart that OH- free radical translated into H_2O.

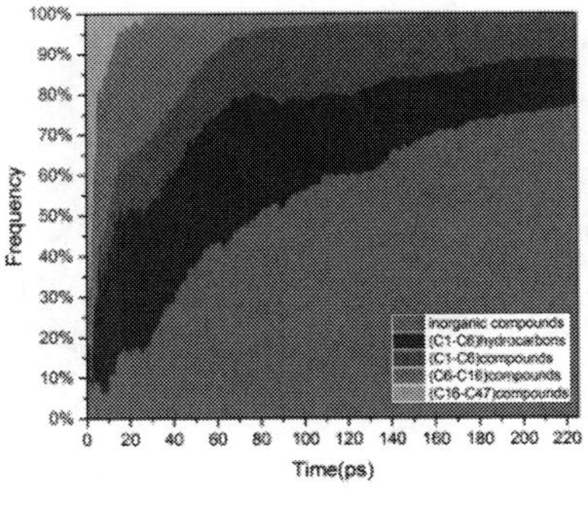

(a)

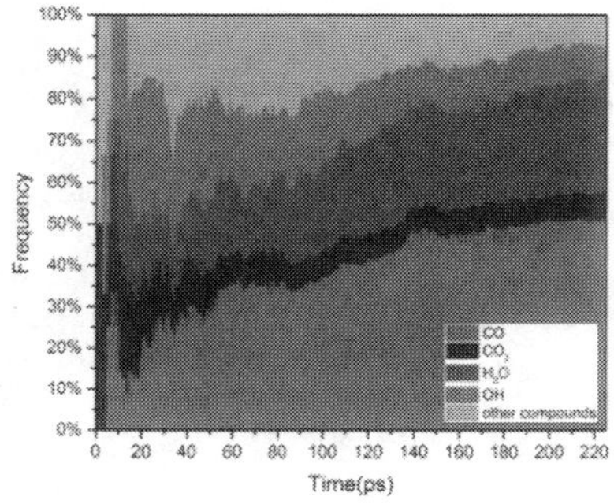

(b)

Figure 2. The frequency of (a) compounds and (b) inorganic compounds obtained from the ReaxFF MD simulation during combustion of brown coal at 3000K with time.

3.1 Effect of temperature on brown coal combustion

Temperature is a very important influencing factor of combustion, and an in-depth study of temperature is necessary. In our study, five typical temperatures range from 2000K-4000K were chosen to study the change of products.

In our study, it was found that the number of CO molecules were much more than the number of CO_2 molecules. Dav indicated that the radio of the number might have something to do with the type of carbon, and that CO could have advantage at high temperature however[11].

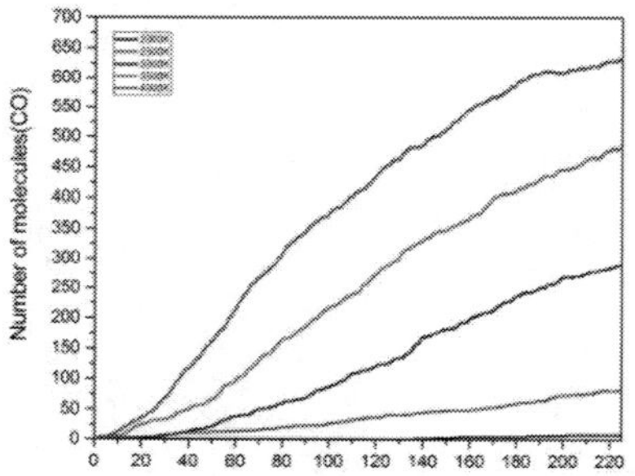

(a)

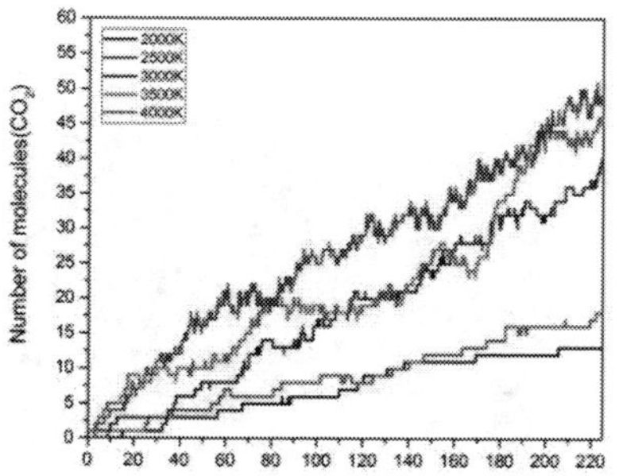

(b)

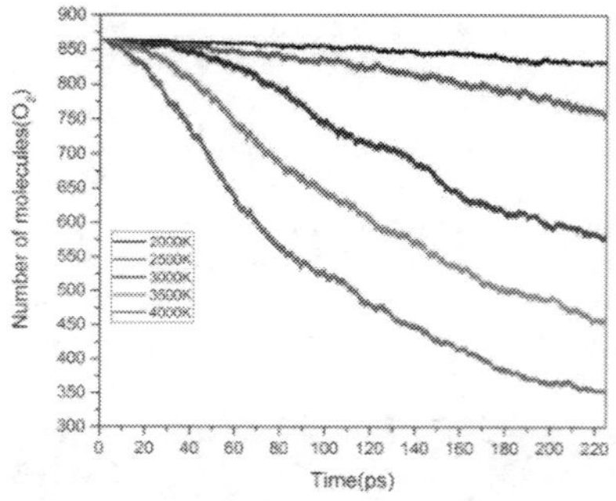

(c)

Figure 3. Time evolution of (a) CO, (b) CO_2 and (c) O_2 molecules during the combustion of brown coal from ReaxFF MD simulations at different temperatures.

As show in Fig.3, the yield of products showed an increasing trend with temperature. At 2000K, the yield of products was much smaller than that of high temperature. It was because that at lower temperature, the reaction rate was low as well, the reaction was not going deep enough. On the contrary, temperature could accelerate reaction rate, more small fragment generated and thus generating more products at higher temperature. The rate of O_2 decreased more rapidly at higher temperature, which showed that more O_2 involved in the reactions.

3.2 Effect of excess air coefficient on brown coal

Excess air coefficient is another important parameter in brown coal combustion. In our study, excess air coefficient reflected in the number of oxygen. Theoretically, 16 brown coal molecules need 864 oxygen molecules to burn out. Therefore, reaction systems that contained 216, 432, 864, 1296 and 1728 oxygen represented the excess air coefficient range from 0.25-2 respectively.

As shown in fig.4, the productions of CO and CO_2 increased with excess air coefficient, and the number of CO was much more than CO_2. On the basis of previous research, oxygen reacted with carbon-based molecules and generate CO firstly at high temperature, then oxidized into carbon dioxide in the gas phase. In our work, the bulk density was set as $0.1g/cm^3$ in order to prevent cross stack of aromatic structure. Therefore, the CO released from the carbon surface was hard to react with oxygen at such a relatively low density. Moreover, CO_2 could be reduced to CO by carbon atom at high temperature. Therefore, the results of the simulations was reasonable whatever.

With more oxygen molecules added into reactive systems, more CO and CO generated. High concentration of oxygen mad it easy to have carbon oxygen reactions and generate more gas products[12]. More oxygen were consumed at high excess air coefficient, which can be seen in Fig.4(c), it was because more oxygen reacted with brown coal. However, the effect of excess air coefficient was not that significant compared with temperature. It might because that temperature can accelerate the process of reaction process, thus, reactions transferred into deeper processing.

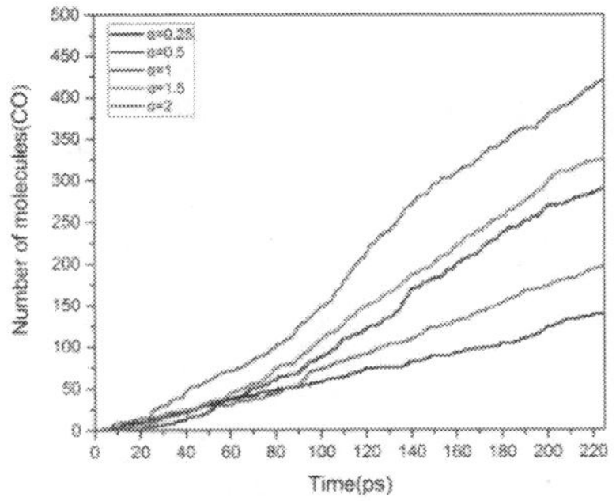

(a)

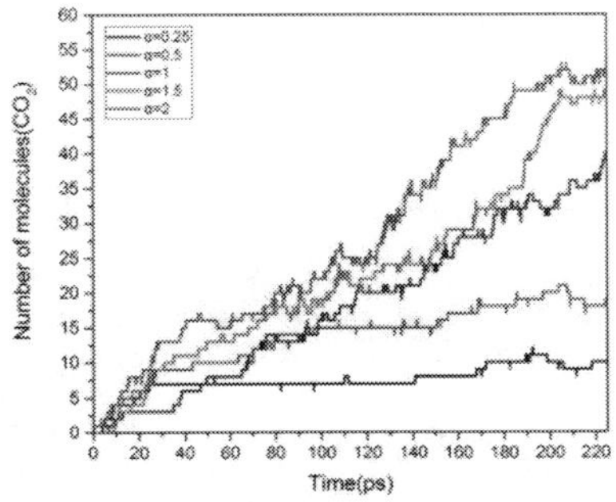

(b)

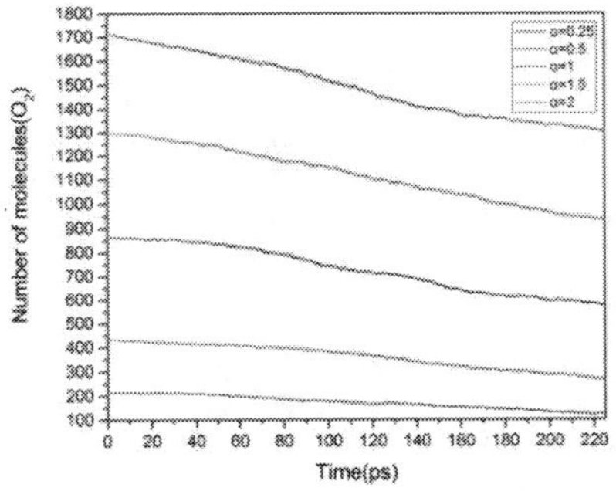

(c)

Figure 4. Time evolution of (a) CO, (b) CO_2 and (d) O_2 molecules during the combustion of brown coal from ReaxFF MD simulations at different excess air coefficient.

4 Conclusions

In this work, the ReaxFF force field was used to study the combustion process of brown coal. The brown coal combustion was conducted at the temperature range of 2000K-4000K. Both temperature and excess air coefficient were took into consideration. Some conclusions and processes observed are as follows:

1. The number of macromolecular structures reduced while that of micro molecular structures increased with time, CO, CO_2 and H_2O were main productions in the combustion process. More CO rather than CO_2 generated at high temperature.

2. The yield of main products such as CO and CO_2 increased with temperature, and the consumption of O_2 also increased at higher temperature.

3. Excess air coefficient was an important parameter in brown coal combustion. With more oxygen molecules put into the reaction system, the yield of CO and CO_2 increased. More oxygen molecules took part in reactions at higher excess air coefficient.

4. Compared with excess air coefficient, the effect of temperature was much more significant

5 Acknowledgment

The authors are grateful for financial support from the National Science-Technology Support Plan Projects in China (Project Number: 2015BAA04B03).

References

[1] Morgan P A, Robertson S D, Unsworth J F. Fuel, 1986, 65(11):1546-1551.

[2] Hamor R J, Smith I W, Tyler R J. Combustion & Flame, 1973, 21(2):153-162.

[3] Mayo S L, Olafson B D, Goddard W A. Journal of Physical Chemistry, 1990, 94(26):8897-8909.

[4] Brenner D W. Physical Review B Condensed Matter, 1990, 42(42):9458-9471.

[5] van Duin, A.C.T., et al., The Journal of Physical Chemistry A, 2001. 105(41): p. 9396-9409.

[6] Bhoi, S., T. Banerjee and K. Mohanty. Fuel, 2014. 136: p. 326-333.

[7] Mathews J P, Chaffee A L. Fuel, 2012, 96(7):1-14.

[8] So/Rensen M R, Voter A F. Journal of Chemical Physics, 2000, 112(21):9599-9606.

[9] Salmon E, Duin A C T V, Lorant F, et al. Organic Geochemistry, 2009, 40(12):1195-1209.

[10] Jiang D E, Duin A C T V, Goddard W A, et al. ournal of Physical Chemistry A, 2009, 113(25):6891-6894.

[11] R.J.Dav. Ph.D.Thesis. The Pennsylvania State University, University Park, Pa. 1949

[12] Liu G S, Niksa S. Progress in Energy & Combustion Science, 2004, 30(6):679-717.

11th Asia-Pacific Conference on Combustion,
The University of Sydney, NSW Australia
10th -14th December 2017.

Emission Spectroscopy of the C_2 Swan Bands to Estimate Temperature of the Near-Extinction Flamelets of Turbulent Premixed Flames

Yuzo Kawasoe[1], Hideki Hashimoto[2], Osamu Moriue[1], Eiichi Murase[1], Junichi Furukawa[3]

[1]Kyushu University

744 Motooka, Nishi-Ku, Fukuoka, 819-0395, Japan

[2]Kobe City College of Technology

8-3 Gakuen-Higashimachi, Nishi-Ku, Kobe, Hyogo, 651-2194, Japan

[3]Tokyo Metropolitan College of Industrial Technology

1-10-40 Higashi-Ohi, Shinagawa-Ku, Tokyo, 140-0011, Japan

Abstract

In order to establish a novel technique that estimates the temperature of the near-extinction flamelet of turbulent premixed flames by measuring the emission of the C_2 Swan bands, the relation between the temperature and the emission intensity-ratio of the 515.5 nm/470.5 nm bands of C_2 of strained premixed flames has been examined by using a counter-flow burner. A unique relation has been found to exist between the flame temperature and the emission intensity-ratio of the 515.5 nm/470.5 nm bands of C_2 for the Propane-air strained flame. Therefore, it can be concluded that the temperature of the near-extinction flamelets of Propane-air turbulent premixed flames can be estimated by measuring the emission intensity-ratio of the 515.5 nm/470.5 nm bands of C_2.

1 Introduction

It is well known that band spectrum of intermediates such as OH, CH and C_2 are observed in hydrocarbon/air flames [1]. In order to explore the flame structure or combustion conditions, a number of studies have been conducted to measure the emission of OH, CH and C_2 [2, 3]. However, the optical systems used in these studies are composed of a single lens, the spatial resolution of the measurement does not satisfy that required to clarify the structure of the local reaction zone of turbulent premixed flames. Adopting the Cassegrain optics, which consists of a pair of convex and concave mirrors, the spatial resolution of the measurement is extremely improved to the order of 10^{-1} mm [4]. The distribution of emission intensities of OH, CH and C_2 across the local reaction zone can be measured by using the spectroscopy system adopting the Cassegrain optics.

It has been shown by assuming thermal equilibrium that a unique relation exists between the emission intensity-ratio of the 515.5 nm/470.5 nm bands of C_2 and the temperature of hydrocarbon-air premixed flames [5], implying that the temperature of the flamelet of hydrocarbon-air turbulent premixed flames can be estimated by measuring the emission intensity-ratio of the 515.5 nm/470.5 nm bands of C_2.

In order to establish a novel technique that estimates the temperature of the flamelets of hydrocarbon-air turbulent premixed flames, the relation between the temperature and the emission intensity-ratio of the 515.5 nm/470.5 nm bands of C_2 of the unstrained Methane-air and Propane-air premixed flames has been examined by using the rectangular nozzle burner in our previous study [6]. It has been found that a unique relation exists between the temperature and the emission intensity-ratio of the 515.0 nm/470.5 nm bands of C_2 of unstrained premixed flames [6]. However, the temperature of the strained flames, such as the flamelets of turbulent premixed flames, cannot be estimated by using the relation obtained in our previous study.

Therefore, in the present study, an attempt has been made to examine the dependence of both the flame temperature and the emission intensity-ratio of the 515.0 nm/470.5 nm bands of C_2 on the strain rate by using the counter-flow burner.

2 Experimental

The schematic illustration of the spectroscopy system, which is composed of the optics, the spectrometer and the signal recorder, used in the present study is shown in Fig. 1. The Cassegrain optics is adopted in order to conduct the measurement of the distribution of the emission intensity at a high spatial resolution of the order of 10^{-1} mm. The spectroscopy system used in the present study is identical to that used in our previous study [6]. Signals detected by the spectroscopy system are amplified 10^5 to 10^6 times by current-voltage preamplifier, and are monitored by a separate parallel 12 bit analogue-digital (A/D) converter. Thus signals are recorded in a personal computer. The counter-flow burner, which is composed of two identical co-axial nozzle burners [7], is utilized to establish stable strained premixed flames. The mixture is supplied to the inner nozzle, whose diameter d is 10 mm, of the upper and lower burners. Thus a strained twin premixed flame is stabilized near a stagnation plane between two burners. The strain rate ε can be operated by changing the axial velocity gradient between two burners because the velocity gradient that is perpendicular to the flame front in the counter-flow field is the axial velocity gradient between two burners. The axial velocity gradient between two

Corresponding author. Fax: +81-92-802-0001
E-mail address: yuzo.kawasoe.577@s.kyushu-u.ac.jp

burners can be varied by changing the flow velocity of inner nozzle of two burners and/or the separation distance L between two inner nozzles. When the axial distribution of the axial component of the gas velocity between two burners satisfies the condition of the potential flow, the strain rate ε can be obtained by Eq. (1), where V_u, V_l and L denote the flow velocities of the upper and lower inner nozzles and the separation distance between two inner nozzles, respectively.

$$\varepsilon = \frac{V_u + V_l}{L} \quad (1)$$

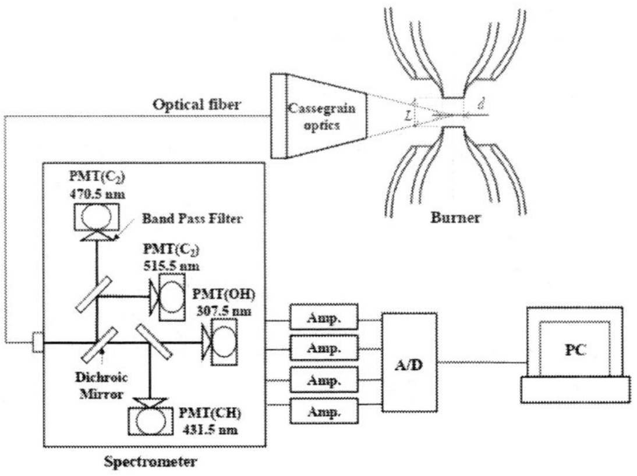

Figure 1: Schematic illustration of the experimental setup of the emission spectroscopy of the counter-flow flame

Uniform Methane-air and Propane-air mixture, whose equivalence ratio ϕ is 0.8 to 1.2, is supplied at velocity of 0.5 to 5.0 m/s to the inner nozzle, while the separation distance L is changed from 10 to 4.0 mm. Thus the stain rate is varied from 125 $[1/s]$ to 2500 $[1/s]$. Nitrogen is supplied to the outer nozzle whose diameter is 28 mm in order to prevent diffusion of surrounding oxygen and to erase the tail flame, which is due to the combustion of unburnt mixture. The distribution of the emission intensities of the 470.5 nm and 515.5 nm bands of C_2 across the stable strained flame is obtained simultaneously, as the measurement volume of the Cassegrain optics is moved perpendicular to the strained flame at the velocity of 30 mm/s. The A/D converter is operated with the sampling frequency of 10 kHz, which gives the spatial resolution of the measurement of the distribution of the emission intensity across the strained flame is of the order of 3×10^{-3} mm. Temperature measurement is conducted by using an uncoated Pt-Rh40%/Pt-Rh20% thermocouple of 0.1mm in diameter. Correction for the radiation loss of temperature measurement is not conducted in the present study.

3 Results and Discussions

As the strain rate ε increases, the flame temperature decreases, which causes the burning velocity to decrease. The twin flame moves toward the downstream position, where the reduced burning velocity balances to the local gas velocity. Consequently, the separation distance between the twin flame decreases. When the separation distance between the twin flame is small, the emission from one flame becomes to prevent the correct measurement of the emission of the target flame.

Therefore, the extinction behaviors of the twin and single flames due to the strain rate ε are examined and compared in Fig. 2. The single flame is established near a stagnation plane by supplying Nitrogen from the upper inner-nozzle. It is clearly seen that the strain rate ε at the extinction limit of the twin flame is much larger than that of the single flame because the extinction of the former is due only to the strain while that of the latter is due both to the strain and the heat loss to the downstream. However, the flame temperature at the extinction limit of both

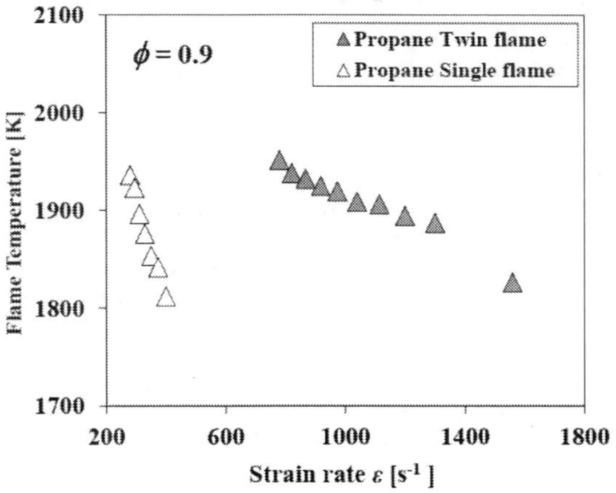

Figure 2: Dependence of the temperature of the Propane-air single and twin flames on the strain rate ε

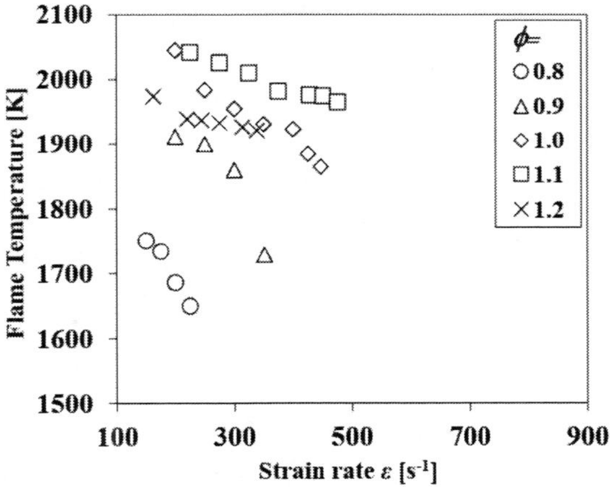

(a) Methane-air single flame

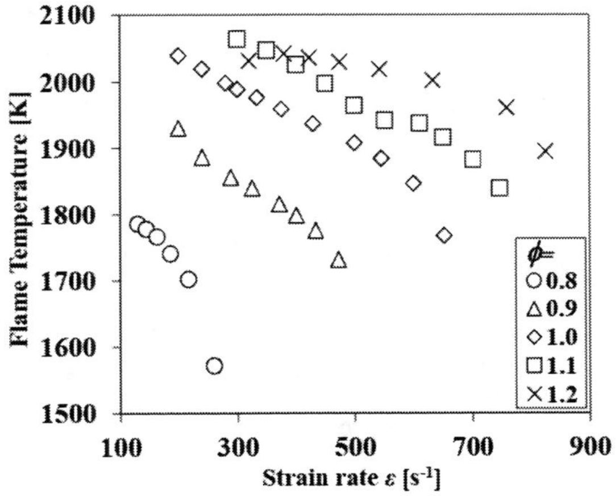

(b) Propane-air single flame

Figure 3: Dependence of the temperature of (a) Methane-air and (b) Propane-air single flames on the strain rate ε

the single and twin flames are almost the same. It seems that no matter how much the heat loss to the downstream, the flame extinction occurs at the same temperature. The influence of the strain rate ε on the flame temperature and the emission intensity-ratio of the 515.5nm/470.5nm bands of C_2 is examined for the single flame throughout the present study.

The dependence of the temperature of the Methane-air and the Propane-air single flames on the strain rate ε is shown in Fig. 3. The temperature of the unstrained Methane-air flame is maximum around the equivalence ratio of 1.0 and decreases to both the leaner and richer sides [6]. The temperature of the strained Methane-air flame is maximum around the equivalence ratio of 1.1 and decreases to both the leaner and richer sides. The temperature of the unstrained Propane-air flame is maximum around the equivalence ratio of 1.1 and decreases to both the leaner and richer sides [6]. When the strain rate ε is low, the temperature of the strained Propane-air flame is maximum around the equivalence ratio of 1.1, and decreases to both the leaner and richer sides. On the other hand, when the strain rate ε is high, the temperature of the strained Propane-air flame is maximum around the equivalence ratio of 1.2 and decreases to both the leaner and richer sides. As the strain rate ε increases, the temperature of the Methane-air and Propane-air flames decreases monotonically, then extinction occurs. The temperature of the lean flames at the extinction limit is lower than that of the rich flames. The strain rate ε at the extinction limit of the Propane flame is higher than that of the Methane flame.

The dependence of the emission intensities of the 470.5 nm and 515.5 nm bands of C_2 on the strain rate ε are shown in Figs. 4 and 5, respectively. The emission intensities obtained from the strained flames are normalized by the emission intensities obtained from the unstrained flames where the flame temperature is maximum, i.e. the equivalence ratio is 1.0 for the Methane flame and the equivalence ratio is 1.1 for the Propane flame [6]. Due to the heat loss to the downstream region, the temperature of the strained flame is lower than that of the unstrained flame. The emission intensity of the strained flame is lower than that of the unstrained flame. The emission intensities of the 307 nm band of OH and the 431 nm band of CH obtained from unstrained flame are maximum where the flame temperature is maximum, on the other hand, the emission intensities of the 470.5 nm and 515.5 nm bands of C_2 are maximum when the equivalence ratio is 1.3 and decrease to the leaner side [6]. The emission intensities of the 470.5 nm and 515.5 nm bands of C_2 of lean flames are extremely feeble [6]. Thus, in the present study, the emission intensities of the 470.5 nm and 515.5 nm bands of C_2 are maximum when the equivalence ratio is 1.2. The emission intensities of the 470.5 nm and 515.5 nm bands of C_2

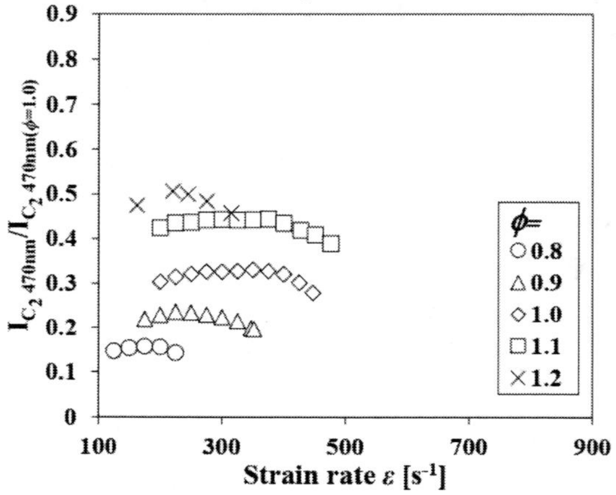

(a) Methane-air single flame

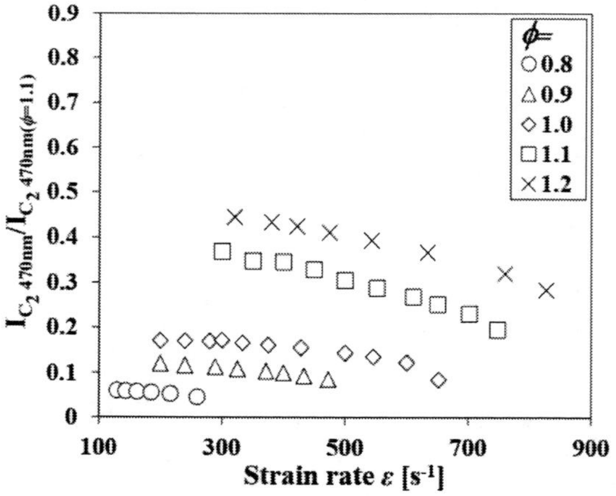

(b) Propane-air single flame

Figure 4: Dependence of the emission intensity of the 470.5 nm band of C_2 obtained from (a) Methane-air and (b) Propane-air single flames on the strain rate ε

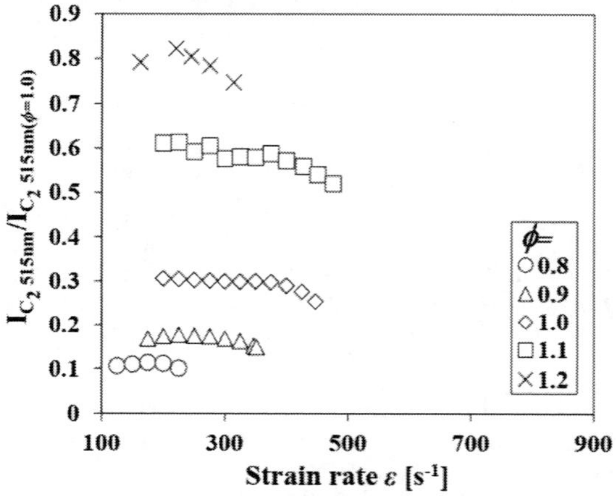

(a) Methane-air single flame

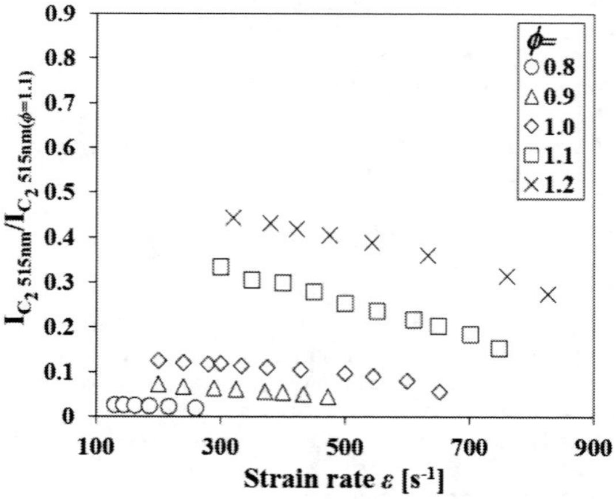

(b) Propane-air single flame

Figure 5: Dependence of the emission intensity of the 515.5 nm band of C_2 obtained from (a) Methane-air and (b) Propane-air single flames on the strain rate ε

obtained from the Propane-air strained flames decrease monotonically, then extinction occurs, as the strain rate ε increases. On the other hand, the emission intensities of the 470.5 nm and 5145.5 nm bands of C_2 obtained from the Methane-air strained flames slightly increase, are maximum and decrease, as the strain rate ε increases. Temperature change with the strain rate ε of the Methane flames has been re-examined carefully, however, the flame temperature does not increase in the low strain rate slightly but decreases monotonically, as the strain rate ε increases. The emission intensity of the 470.5 nm and 515.5 nm bands of C_2 obtained from the Propane-air strained flame decrease more rapidly than those obtained from the Methane-air strained flames. It has been first shown experimentally by the present study that the temperature and the emission intensities of the 470.5 nm and 515.5 nm bands of C_2 of the Propane-air strained flame decrease monotonically, as the strain rate ε increases.

The relation between the flame temperature and the emission intensity-ratio of the 515.5 nm/470.5 nm bands of C_2 is shown in Fig. 6. It is clearly seen that, for each equivalence ratio of the Propane-air strained flame, a unique relation exists between the flame temperature and the emission intensity-ratio of the 515.5 nm/470.5 nm bands of C_2. As the strain rate ε increases, both the temperature of the Propane-air flames and the emission intensity

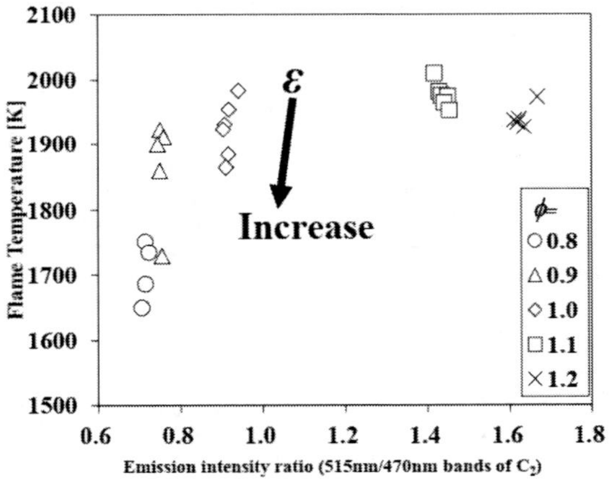

(a) Methane-air single flame

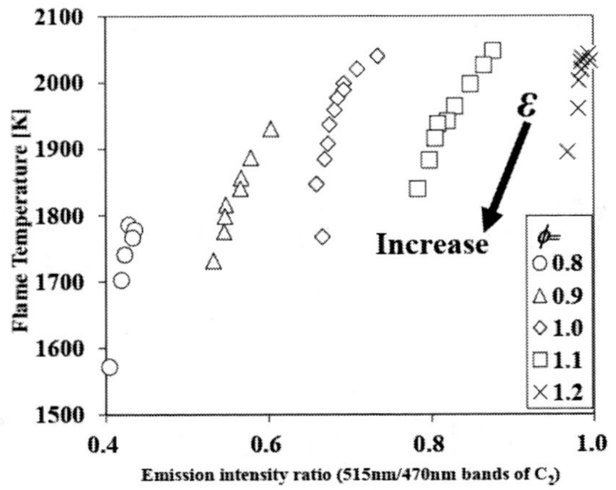

(b) Propane-air single flame

Figure 6: Relation between the temperature and the emission intensity-ratio of the 515.5 nm/470.5 nm bands of C_2 for (a) Methane-air and (b) Propane-air single flames

-ratio of the 515.5 nm/470.5 nm bands of C_2 decrease. We conclude that the temperature of the strained Propane-air flames can be estimated by measuring the emission intensities of the C_2 Swan bands, if the equivalence ratio of the mixture is known. On the other hand, for the Methane-air flames, a unique relation can not be seen between the flame temperature and the emission intensity-ratio of the 515.5 nm/470.5 nm bands of C_2. As the strain rate ε increases, the emission intensity-ratio of the 515.5 nm/470.5 nm bands of C_2 remains almost constant in lean mixtures or slightly increases especially for the equivalence ratio of 1.2. Therefore, we conclude that the temperature of the strained Methane-air flames can not be estimated by measuring the emission intensities of the C_2 Swan bands.

4 Conclusions

In order to accumulate the fundamental knowledge for estimating the temperature of strained flames, such as the flamelet of hydrocarbon-air turbulent premixed flames by measuring the emission intensities of the C_2 Swan bands, the influence of the strain rate ε on both the flame temperature and the emission intensity-ratio of the 515.0 nm/470.5 nm bands of C_2 has been examined for the strained Methane-air and Propane-air flames in detail. The temperature and the emission intensities of the 470.5nm and 515.5nm bands C_2 obtained from the strained Propane-air flames decrease monotonically, as the strain rate ε increases. A unique relation exists for each equivalence ratio of the strained Propane-air flames between the flame temperature and the emission intensity-ratio of the 515.5 nm/470.5 nm bands of C_2. It can be concluded that the temperature of the near-extinction flamelets of Propane-air turbulent premixed flames can be estimated by measuring the emission intensity of the C_2 Swan bands.

5 Acknowledgments

This work is supported by the Grant-in-Aid for Scientific Research Grant Number 15K17988 and the contribution of the fifth author is supported by the Grant-in-Aid for Scientific Research Grant Number 16K06136 from the Japan Society for the Promotion of Science.

References

1] A. G. Gaydon, The Spectroscopy of Flames, Chapman & Hall, London, U.K., 1974, p. 144.

2] H. Itoh, Y. Hommo, K. Noda, T. Gomi, Transactions of Society of Automotive Engineers of Japan **29** (1984) 10-17.

3] K. Itoh, H. Ihara, S. Tatsuta, O. Fujita, Transactions of the Japan Society of Mechanical Engineers, **56**-536B (1990) 3508-3513.

4] T. Wakabayashi, F. Akamatsu, M. Katsuki, Y. Mizutani, Y. Ikeda, N. Kawahara, T. Nakajima, Transactions of the Japan Society of Mechanical Engineers, **64**-619B (1998) 925-930.

5] T. Ishiguro, S. Tsuge, T. Furuhata, K. Kitagawa, N. Arai, T. Hasegawa, R. Tanaka, A. K. Guputa, *Proc. Combust. Inst.* **27** (1998) 3205–3213.

6] H. Hashimoto, T. Shinagawa, J. Furukawa, Transactions of the Japan Society of Mechanical Engineers, **76**-769B (2010) 1420-1425.

7] H. Kobayashi, M. Kitano, E. Mori, Transactions of the Japan Society of Mechanical Engineer, **57**-535B (1991) 1141-1146.

11th Asia-Pacific Conference on Combustion,
The University of Sydney, NSW Australia
10th -14th December 2017.

On the Effects of Shear and Compositional Inhomogeneities in Turbulent, Lean, Premixed Flames

S.A. Steinmetz[1], M. Juddoo[1], W. Jin[1,2], A.R. Masri[1]

[1]School of Aerospace, Mechanical, and Mechatronic Engineering

The University of Sydney, NSW 2006 Australia

[2]State Key Laboratory of Multiphase Flow in Power Engineering,

Xi'an Jiaotong University, Xian 710049,China

Abstract

This paper presents some initial results on the separate effects of shear and compositional inhomogeneities on the global structure of turbulent lean piloted premixed flames. Shear is introduced by injecting the mixture in two streams through two concentric tubes at different velocities. Compositional inhomogeneity, for the same shear rate, is obtained by injecting air in the annulus and fuel or rich fuel-air mixtures through the inner jet which may be recessed upstream of the jet exit plane. It is found that increasing the shear inhomogeneity can lead to a shorter flame where the tip may transition from "open" to "closed". The latter refers to a "Bunsen-like" flame while an "open" flame has reacting edges that do not close at the tip. Compositional inhomogeneity seems to have little effects on the global structure despite the presence of mixture fraction gradients a noted from Rayleigh scattering performed the jet exit plane. Measurements of velocity fields and imaging of reaction zones will be performed to enable a better understanding of the structure of these flames and to locate them on the premixed flames regime diagram.

1 Introduction

Practical combustion devices operate under conditions of shear and compositional inhomogeneity which have significant impact not only on combustion efficiency but also on combustor stability and the formation of pollutants. Lean combustors also experience such effects because of intentionally introduced stratification (as in stratified engines) or because of combustor instabilities that lead to incomplete mixing [1]. Resolving the effects of shear and compositional inhomogeneities on the structure of lean flames is important to advance modelling capabilities of such flows. In premixed flames, higher turbulence levels impacts the burning rate and changes the structure of reaction zones, as shown on relevant regime diagrams [6]. A novelty of the current configuration is the ability to independently increase shear and compositional inhomogeneity within the fuel-air mixture. This is made possible with the Sydney piloted inhomogeneous burner where the reacting mixture is introduced through two annular tubes such that the inner tubes can slide upstream of the jet exit plane. This burner has been studied extensively for globally rich mixture and the focus here is on lean counterparts. Both shear and compositional inhomogeneity can be introduced independently by varying the composition, the velocity ratio and the recess distances of the inner tube with respect to the annulus.

Earlier research in compositionally inhomogeneous but overall rich flames have shown that stability is enhanced by the introduction of mixture fraction gradients that places stoichiometric fluid samples in the proximity of the pilot at the jet exit plane. The present work seeks to understand any possible enhancement of lean premixed flame combustion through the introduction of shear or compositional inhomogeneity in lean mixtures. Shear inhomogeneity is ensured by injecting the same equivalence ratio through the inner and outer streams abut at different jet velocities. Compositional inhomogeneity requires that air is introduced in the annulus and a rich mixture in the recessed central jet such that the overall equivalence ratio in unchanged.

Essentially lean and premixed, the flame is Bunsen-like and does not lift-off since it is anchored to the burner by the pilot. As the inhomogeneity increases, the visible length of the flame changes along with the possibility of the tip change from "closed" where the edges of the reaction zones meet into a cone to "open" where the edges remain apart and the flame looks somewhat cylindrical in shape. The paper explores the qualitative changes of the flames with varying shear and compositional inhomogeneities. Rayleigh measurements are also reported to show the mixture fraction distribution at the jet exit plane.

2 Methods

The modified Sydney piloted burner [2-5] is used for these experiments. The burner consists of concentric "jet" and "annulus" tubes, with outside diameters (OD) of 4.5 mm and 8.0 mm, respectively. The inside diameter (ID) of the annular tube is 7.5 mm, while the ID of the jet tube is 2.5 mm. The tubes are concentric to an 18 mm diameter pilot section. The entire burner is enclosed in a 150 mm x 150 mm wind tunnel, which provides a constant 15 m/s velocity. The exit plan of the jet tube is recessed from the exit plane of the annulus at a distance of L_r. Further details of the burner can be found in [2-5]. Compressed natural gas (CNG) is employed as fuel. The composition of the CNG is 88% CH_4, 7.8% C_2H_4, 1.9% CO_2, 1.2% N_2, and 1.1% H_2, water, and other hydrocarbons [5]. The pilot is a stoichiometric mixture of acetylene, hydrogen, air, nitrogen and carbon dioxide with a bulk velocity of 3m/s for the non-reacting mixture. The pilot provides a heat release rate of Hr = 2.2kW,

Corresponding author.
E-mail address: scott.steinmetz@sydney.edu.au

and is designed to have the same C/H ratio and adiabatic equilibrium temperature as a stoichiometric methane/air mixture [2]. Flame height, H_f, and stability is evaluated from flame images which are taken with a Canon EOS 70D CMOS camera fixed with an 85 mm lens. The exposure time is set to 0.8 s, while f number is 1/6.3.

Mixture fraction is measured by Rayleigh scattering performed only at the jet exit place. The 532 nm output of a frequency doubled Nd:YAG laser is formed into a sheet perpendicular to the flow, and directed over the burner, 5 mm above the outlet. Two-dimensional images of Rayleigh scattering are collected with a CCD camera behind a line filter. Scattering is calibrated from the known cross-sections of helium, nitrogen, and CNG. When performed at the jet exit plane where the mixture is non-reacting and the temperature is uniform, Rayleigh scattering is proportional to the Rayleigh cross section which, in turn, is proportional to the mixture fraction. The fact that methane has a higher cross section than air, leads to the formation of a measurable gradient which then provides a measure of mixture fraction.

For blow off limit experiments, bulk equivalence ratio and bulk velocity are varied. For shear and compositional inhomogeneity experiments, the global equivalence ratio is fixed at 0.65. Flames are investigated with fixed percentages of the total flow in either the jet or annulus. Jet flowrate percentages (Q_j) of 0, 6, 30, and 45% are investigated, where 6% corresponds to the fuel air ratio at the chosen bulk equivalence ratio. Each Q_j results in a different velocity ratio, U_r, between jet and annulus. For "premixed" flames, both the jet and annulus streams are at the bulk equivalence ratio, such that there is no compositional inhomogeneity. For "inhomogeneous" flames, all fuel flows through the jet, and air flows through both jet and annulus in order to produce the desired Q_j. As such, there is a varying equivalence ratio (>1) in the jet.

Q_j	φ_{jet} inhomogeneous	φ_{jet} premixed	U_r
0	0	0	-
6	∞	0.65	0.4
30	2.6	0.65	2.5
45	1.6	0.65	4.7

Table 1: Experimental conditions for inhomogeneous and premixed flames.

3 Results

3.1 Lean flame blow-off

The lean flames studied here look more like Bunsen types and do not blow-off like the rich flames studied earlier which extinguish just downstream of the pilot while still burning on either sides. Blow-off in these lean flames is defined as the condition when the flame which is otherwise closed at the tip, open up and look cylindrical. Example images of a lean flame approaching blow-off are shown in Fig. 1. The flame is initially closed at the tip. As bulk velocity increases, turbulence intensity within reactants increases. This would increase turbulent burning velocity, resulting in a shorter flame. As velocity is further increased, the flame opens, allowing some unreacted fuel to pass through. The open flame will continue to shorten at higher velocities, until only the pilot remains. This behavior is different from that of rich premixed flames, where there is a distinct extinguishing of the flame above the blow-off velocity [2-5]. In these lean flames, blow-off is more subjective and harder to determine because of the difficulty of visually detecting whether the flame's tip is open or partially open.

Figure 1. Piloted premixed CNG flames with φ = 0.65, and bulk velocity of (l to r) 50, 60, 70, 80, 90, and 100 m/s. The top of each image corresponds to a length of 300 mm. As the flame approaches blow-off, it continuously shortens. Eventually, the flame loses its conical tip, and begins to open. This results in two distinct "legs" of the flame, with some unburned reactants escaping through the center.

3.2 Compositional inhomogeneity

Previous studies have demonstrated an enhancement in blow-off limits in rich flames through inhomogeneity [2-5]. This inhomogeneity is controlled through recession of the central fuel jet, which changes the profile of mixture fraction at the burner outlet. The optimal recession, L_r, is a function of fuel, and is typically between 75 and 150 mm. Similarly, inhomogeneity alters the structure of lean flames, as shown in Fig. 3. While setting recession to L_r ~75 mm closes the open flame shown in Fig. 1, it is a more subtle change than what was observed earlier in rich flames.

Figure 2. Inhomogeneous flames with φ = 0.65, Q_j = 6%, bulk velocity of 90 m/s, and Lr = (l to r) 300, 250, 200, 150, 125, 100, 75, 50, 25, and 0 mm. As the central fuel jet is recessed, the structure of the flame changes. The flame is open at the near homogeneous limit (L_r = 300 mm), but begins to close at L_r ~ 75 mm.

Enhanced stability in rich flames has been attributed to an increased probability of stoichiometric samples appearing near the pilot [3]. The mixture fraction near the outlet plane of the burner for lean flames is measured by Raleigh scattering. PDFs of the mixture fraction for different L_r are shown in Fig. 3.

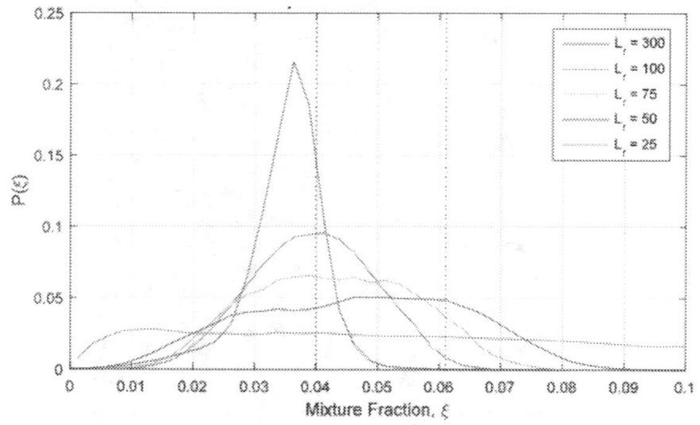

Figure 3. PDFs of mixture fraction across the burner outlet for bulk φ = 0.65, Q_j = 6%, and velocity of 90 m/s. The vertical dotted lines represent the global mixture fraction (l), and stoichiometric mixture fraction (r).

Intuition tells us that inhomogeneity introduced through decreased L_r will not increase the observed mixture fraction near the pilot (which is also confirmed by Rayleigh). However, the overall probability of stoichiometric samples in the mixture does increase through reduced recession because the mixture is no longer homogeneous. However, as these are located near jet centerline, they would likely affect the flame downstream, in agreement with the observed change in the flame structure.

The significant limitations to understanding what can be gleaned from visual inspection of the flame is apparent. Further investigation will require measurement of more precise flame markers, such as radical species.

3.2 Shear and inhomogeneity

The effect of shear for premixed flames is shown in Fig. 4. Consistent with previous discussion, flame height decreases with increasing bulk velocity for a given configuration. At some velocity the flame tip opens, and can be considered blown off. The top plot of Fig. 4 shows that blowoff occurs at lower velocities as Q_j is increased. As there is no compositional difference with varied L_r or Q_j, the observed change can be attributed to varied turbulence and this needs to be confirmed later with measurements of velocity and turbulence fields. The top plot of Fig. 4 shows how the dissipation of a fixed amount of injected turbulence affects blowoff. An ideal recession length, between 50 and 75 mm, is observed, where blowoff velocity is highest. This suggests an optimal profile of turbulence, which enhances turbulent burning velocity without excessively straining the flame. The bottom plot of Fig. 4 shows how different levels of turbulence injection affect blowoff velocity. The highest turbulence injections resulted in highest blowoff velocities. One way to interpret the data is as follows: For optimal blowoff, sufficient shear and hence turbulence must be injected, and it must be distributed across the exit plane of the burner. If high levels of turbulence are injected to close to the burner outlet, blowoff limits are not enhanced optimally.

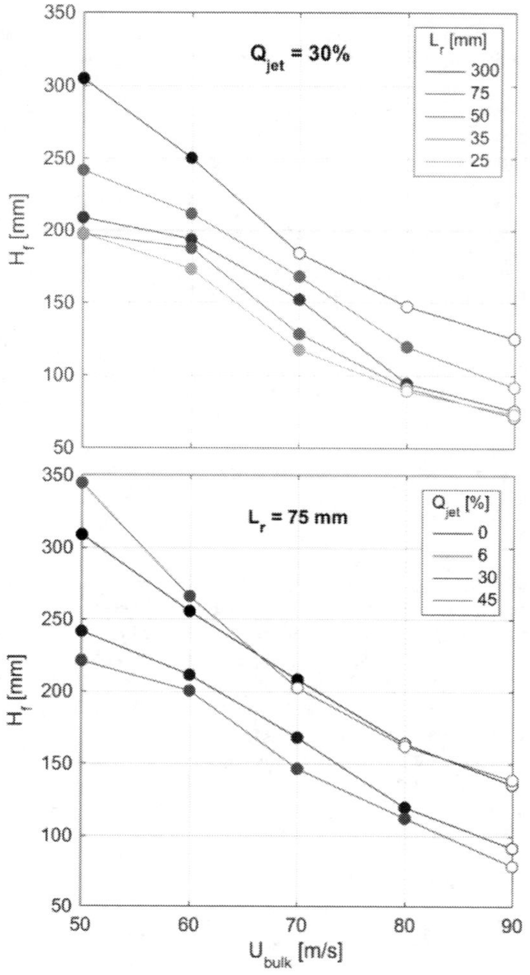

Figure 4. H_f of premixed flames as a function of U_{bulk}. Top) H_f for different Q_j. Bottom) H_f for different L_r. Closed symbols – closed flames, open symbols – open flames.

Similar data is presented for inhomogeneous flames in Fig. 5. The only difference in the configuration of these flames is that all fuel issues from the central jet, resulting in compositional inhomogeneity. As with the premixed cases, there is an optimal L_r for blowoff enhancement in inhomogeneous flames. However, blowoff is much more sensitive to L_r due to the addition of compositional inhomogeneity. These can be seen in the reduced blowoff velocities at L_r 25-50 mm, compared with the premixed cases. Likewise, these flames are much more sensitive to the levels of turbulence injected, as seen in the blowoff velocities at 6 and 45% Q_j. While compositionally inhomogeneous flames can enhance blowoff compared to fully premixed flames, this enhancement seems to be entirely due to changes in turbulence profiles, and sheared premixed flames provide better enhancement.

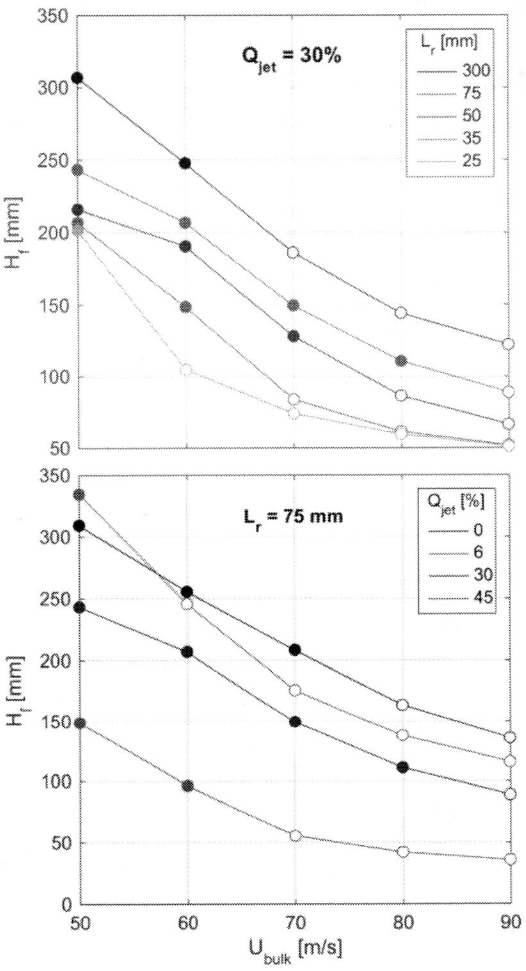

Figure 5. H_f of inhomogeneous flames as a function of U_{bulk}. Top) H_f for different Q_j. Bottom) H_f for different L_r. Closed symbols – closed flames, open symbols – open flames.

4 Conclusions and Future Work

Lean, turbulent, piloted flames are imaged, and reactant mixture fraction profiles are measured. Compositional and velocity field inhomogeneities are introduced through recession of the central fuel jet.

Blowoff is characterized by a change in the flame structure from a closed tip to an open tip. The reason for the difference in flame behavior compared to rich flames warrants further investigation.

As with rich flames previously investigated, enhancement of lean flame stability may be possible through the introduction of compositional inhomogeneity. However, this enhancement is less distinct, and is suggested to be caused by the accompanying enhanced turbulence. Further investigation is necessary to understand these observations.

Further studies will include Laser Doppler Velocimetry to characterize turbulence intensities and characteristic length scales for each condition. In order to improve our understanding of the stabilization mechanism in lean flames, simultaneous measurements of OH and CH radicals are currently being performed. Such measurements will reveal essential features about the pre-heat and reaction zones which will complement the flow-field measurements and provide a more comprehensive picture about the structure of these flames.

5 Acknowledgment

This work is supported by the Australian Research Council.

References

[1] A.R. Masri, Proc. Combust. Inst. 35 (2015) 1115-1136.

[2] S. Meares, A.R. Masri, Combust. Flame 161 (2014) 484-495.

[3] S. Meares, V.N. Prasad, G. Magnotti, R.S. Barlow, A.R. Masri, Proc. Combust. Inst 35 (2015) 1477-1484

[4] H.C. Cutcher, R.S. Barlow, G. Magnotti, A.R. Masri, Proc. Combust. Inst. 36 (2017) 1737-1745.

[5] T.F. Guiberti, M. Juddoo, D.A. Lacoste, M.J. Dunn, W.L. Roberts, A.R. Masri, Proc. Combust. Inst. 36 (2017) 1777-1784.

[6] N. Peters, J. Fluid. Mech.384 (1999) 107-132.

11th Asia-Pacific Conference on Combustion,
The University of Sydney, NSW Australia
10th -14th December 2017.

An Experimental Study on Flame Behavior with Porosity of Center Plate in a Low-swirl Combustor

Minsoek Han[1], Chul-Ho, Kim[2], Keeman Lee[3*]

[1] Department of Aerospace Engineering, Sunchon National University

255, Jungang-ro, Suncheon-si, Jeollanam-do, 57938, , Korea Rep.

[2] Dept. of Mechanical & Automotive Engineering, Seoul National University of Science and Technology

232, Gongneung-ro,Nowon-gu, Seoul, 01811, Korea Rep.

[3*]School of Aerospace and mechanical Engineering, Sunchon National University

255, Jungang-ro, Suncheon-si, Jeollanam-do, 57938, , Korea Rep.

Abstract

This study investigated the effect of porosity of the center plate on the combustion characteristics in a low-swirl combustor. The swirl number was 0.42 while the swirler angle was fixed to 37degree. For the turndown ratio, it was confirmed that the flame as a laboratory scale is stable up to 21 kW and it's also stable even over higher thermal power. As a result, it was found that the flame height increased and the flame shape was changed from V shape to flat shape with increasing porosity of center plate. Although the flash back caused by increased porosity was suppressed, the lean flammability limit in stability map was decreased. Because the low swirl combustion has no internal recirculation zone, the middle flow pattern of the swirler had relatively uniform flow with a constant velocity across the center plate. So, as the porosity of center plate increases the heat loss of the combustor wall decrease with a reduction of the external recirculation zone which made the inner temperature conditions of combustor to increase. Consequently, the emission of NOx and CO increased with porosity increasing.

1 Introduction

In recent years, lean premix technology has been applied to many gas turbine combustors for the low pollution combustion. Lean premixed combustion is one of the ways to reduce thermal NOx by lowering the residence time in high temperature region and flame temperature. [1] It has been mostly used in high-swirl combustion with strong internal recirculation zones. However, as the regulation of NOx is gradually strengthened, the advanced research group have suggested low-swirl combustion that is able to perform lean premixed combustion more effectively and exhibits NOx emission performance similar to or better than that of the high swirl combustion. [2-4] The low swirl combustion has a turbulent combustion that is newly regenerated by the divergent flow. There is seldom flash back due to the characteristics of the low swirl combustor. It was the flame itself that has self-similarity to find its own position.[5] Low swirl burning has advantages over high swirl combustion, in that it needs low swirling strength which leads to drop pressure loss. As a result, low_swirl burning can significantly reduce the power necessary for the oxidant supply of the combustion system. Also, the combustion load ratio of the low swirl combustor is 1:20, which has more advantageous than high swirl combustor, 1:5.[6] And also, because the flame is lifted off from the nozzle it is considered positively that it protect the nozzle tip from overheating caused by high temperatures and high-pressure of gas turbine engines. The low swirler has a tendency to generate a divergent flow through a burner and makes a turbulence that prevents the formation of internal recirculating flow paths. Assuming the axial velocity distribution is uniform, the swirl number of the low swirl combustor can be defined as follows.

$$S = \int_0^{R_b} UWrdr/R\left(\int_{R_c}^{R_b} U^2\,rdr + \int_0^{R_b} U^2 rdr\right)$$

$$= \frac{2/3\ tan\alpha(1-R^3)}{(1-R^2+U_c^2R^2\ /\ U_a^2)}$$

$$= \frac{2}{3}\,tan\alpha\frac{1-R^3}{1-R^3+\left[m^2\left(\frac{1}{R^2-1}\right)^2\right]R^2} \quad (1)$$

$$U = U_c/U_a \qquad m = \dot{m}_c/\dot{m}_a$$

In equation (1), U is the average speed in the axial direction, and W is the average speed in the radial direction. R_c is the radius of the center plate in the middle of the swirler, and R_b is the radius of the exit of the burner nozzle. U_c is the average axial velocity of the fluid from the center plate, and U_a is the average axial velocity through the vane. m is the mass flow rate, $\dot{m}_c$ is the flow rate through the center plate, and $\dot{m}_a$ is the mass flow rate through the vane in the swirler. The swirl strength of the swirler used in this study could be calculated from the above equation.

2 Experimental approach

*Corresponding author. Fax: +82-61-750-3820
E-mail address: kmlee@scnu.ac.kr

2.1 Low swirl combustor and experimental conditions

Figure 1 shows the swirl part and the exit part of the low swirl combustor. A plate with porosity is set at the center of the swirler. On the outer side of this plate, there is a vane for swirling flow, resulting in weak low-swirl motion. With this characteristic of the low swirl combustor, flames meet the propagation characteristics of the spread to the combustion chamber and the propagation characteristics of premixed flames. The velocity characteristics of this lifting flame were confirmed by the flow field measurement which will be described later. The resources of the low swirl combustor in this study are as follows. CH4 is used as a fuel, and compressed air is for the oxydant. The nozzle exit diameter of the combustor is 28 mm, the diameter of the combustion chamber is 74 mm, and the vane angle of the swirler is 37°[7] The swirl strength (S) of the burner is approximately 0.42 by calculating with equation (1). The exit length (L) of the combustor is 33.5 mm. The combustion chamber used a quartz tube to visualize the flame. The thermal power (kW), the equivalence ratio (Φ) and the porosity of center plate were experimentally examined as the main variables affecting the shape and behavior of the flame. In this study, the chemiluminescence images were acquired by the ICCD cameras to identify the characteristics of flame heat release. The CH* chemiluminescence was measured by attaching a CH* optical filter (434/17 nm, Semrock Inc.) in front of an ICCD camera (PI-MAX Princeton Instrument). Direct flame images were taken under the constant exposure condition through a DSLR camera, and a combustion exhaust gas analyzer (330-2LL, Testo Inc.) was used to examine the emission performance such as CO and NOx. The concentrations of CO and NOx in this work are converted into dry-basis of O_2 15%, and NOx is the sum of NO and NO_2. In this study under various porosities which considered as important parameters were compared with 0%, 10%, 15%, 20%, 25%, 30%. The shape of center plate is summarized in table 1.

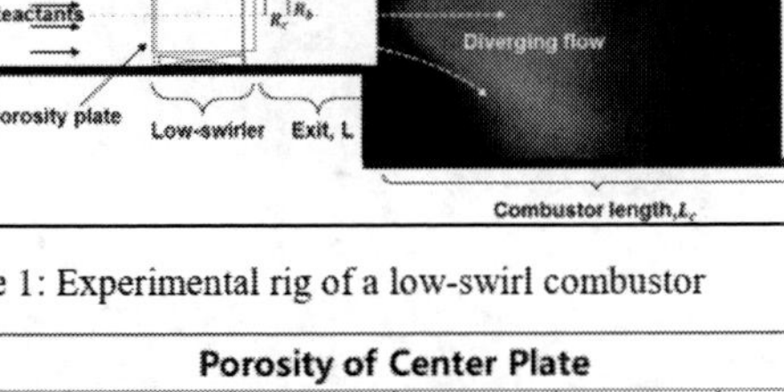

Figure 1: Experimental rig of a low-swirl combustor

Porosity of Center Plate					
10%	15%	20%	25%	30%	35%

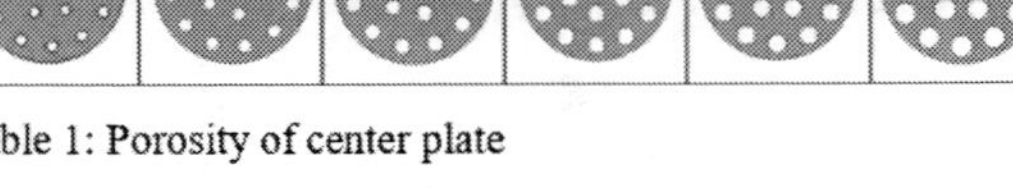

Table 1: Porosity of center plate

3 Result and Discussions

3.1 Pressure drop in the swirler

One of the advantages of low swirl combustor is relatively low pressure drop. The pressure drop, ΔP, which is a measure of the input power required to generate the enhanced turbulence, is measured for the different turbulence generators. This is done by a precision differential pressure meter that measures the static pressure at 10 mm upstream and 10 mm downstream of the swirler respectively. Fig 2 shows a pressure drop as function of the porosity for individual volumetric flow rate. As shown in figure, the pressure drop was decreased with porosity increment in all volumetric flow rates. And, flow rate was higher than small flow rate in pressure drop. When the flame was lifted, the pressure drop of the non-reaction field and the reaction flied were the same, but the pressure drop of the reaction flied was increased in the attached flame.

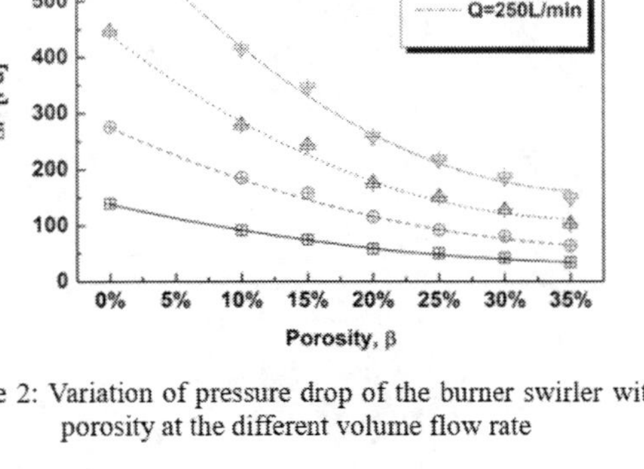

Figure 2: Variation of pressure drop of the burner swirler with porosity at the different volume flow rate

3.2 Flame stability

Figure 3 shows the flame stability map with porosity. The flash back in this study is defined as flame attached to the nozzle. If the porosity is more than 30% it is confirmed that the flashback is suppressed even at high thermal power. It can be shown that the flame region increases with porosity increasing. However, at 35% porosity, the lean flammability limit is reduced. In this study, 25% and 30% porosity were checked that have the widest flammability limit. The factors cause flash back in the lifted off flame of low swirl combustor are the non-uniformity of the flame propagation speed and the velocity field. The flame propagation speed in low swirl combustion is very important factor because the flame height is determined by the propagation characteristics of the premixed flame. Since the flame propagation speed is proportional to the turbulence intensity, assuming that the velocity fields are the same, the turbulence intensity is considered to be the smallest because the flow passing through the swirler is decreased due to the increase of the flow rate through the plate when the 35% of porosity.

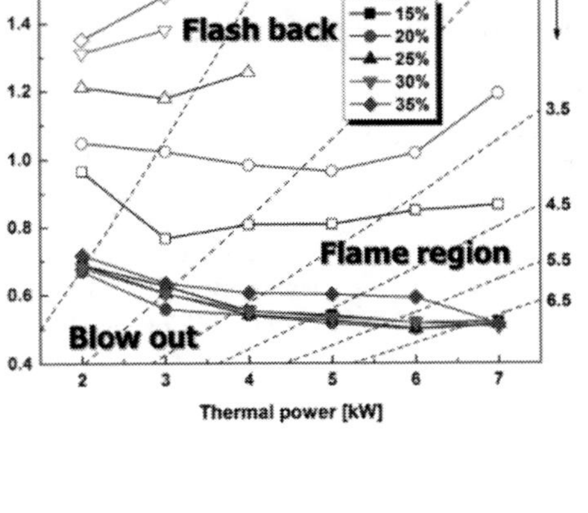

Figure 3: Stability map of the combustion with the porosity rate

3.3 Flame behavior and CH* radical intensity

Figure 4 shows flame behavior and CH* with porosity. The combustion chamber was simulated using a quartz tube to visualize the flame. Flashback occurred as $\Phi > 0.8$ at 15% of porosity. When the porosity is over 20%, V shape appears. As the porosity gradually increased, the flow of the center plate increased and the flame gradually changed in a flat shape. When the flame is in V shape, it is found that the CH* is strong in the brush part of the flame, and the CH* is concentrated in the flat shape flame. As the porosity increases, the lift-off height increases from the nozzle as shown in Fig 5.

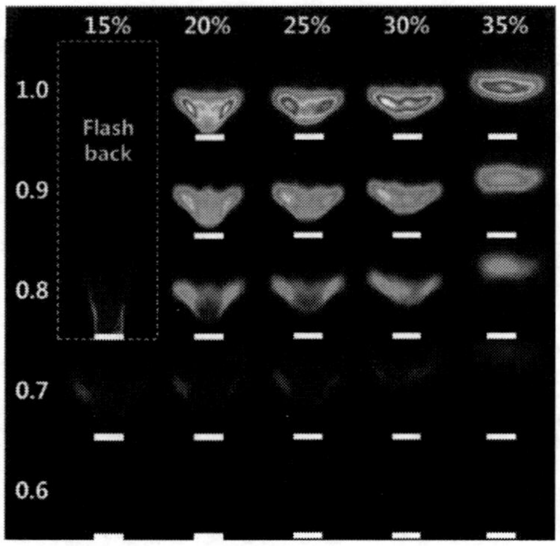

Figure 4: CH * chemiluminescence images.

3.4 Flame lift off height & velocity profiles

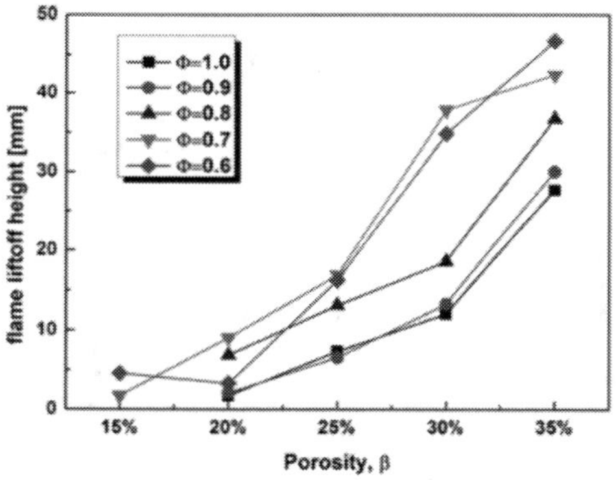

Figure 5: Variation of the lift-off height of the flame with porosity rate at each equivalence ratio

Figure 5 shows that the left-off height increases with porosity increment. Also, at the same porosity, it was found that the lift-off height was increased as the equivalence ratio was lowered. The velocity distribution was measured at the exit of a low swirl combustor which affects to the lift-off height and is shown in Fig 6. The measurement of the velocity distribution was carried out in the axial direction with respect to the radial direction by hot wire anemometer. The center plate of the low swirl combustor causes uniformity of velocity in the central part of the swirler. It can be seen that the internal recirculation zone which usually appears in the high-swirl flow field is not formed. In addition, it can be seen that as the porosity increases, the center part of the swirler and the velocity of the swirling flow field are similar. This factor results in an increase in flow rate and a change in velocity due to an increase in the porosity of the plate. Thereby there are reducing in the swirl strength and swirl number (S.N.). The decrease in S.N means a decrease in the external recirculation zone. The decrease in external recirculation area makes the flame or exhaust gas temperature rise due to the decrease of heat loss at the inner wall of the combustion chamber. The temperature measurement is being tested.

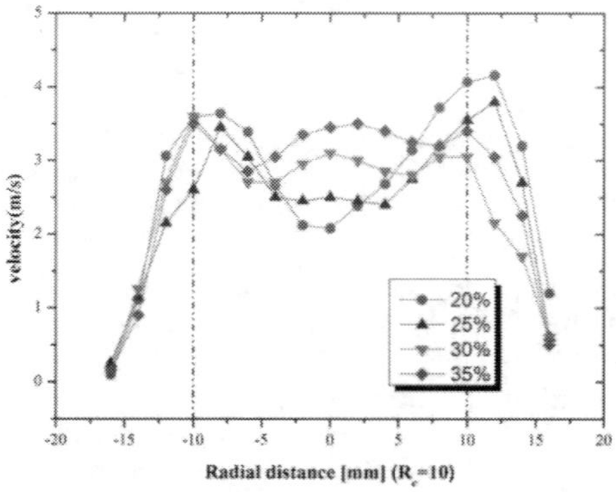

Figure 6: Variation of the velocity distribution with the porosity rate

3.5 Flow visualization with PIV system

Figure 7 shows that the images were acquired by PIV system with porosity at of 3m/s in non-reaction condition. At 0% porosity, the intensity of swirler was stronger with there was no flow of porosity. However, it appears to be vortex due to lack of flow over the plate. As the porosity increases, the width of the flow field decreases with the decrease of the flow rate of the swirler. Also, it can be seen that the flow of the plate increases with increasing porosity in the velocity profile, and the vector value gradually becomes similar to the vector value of the swirler. (A) Mie scattering images, (B) is velocity vector images of PIV.

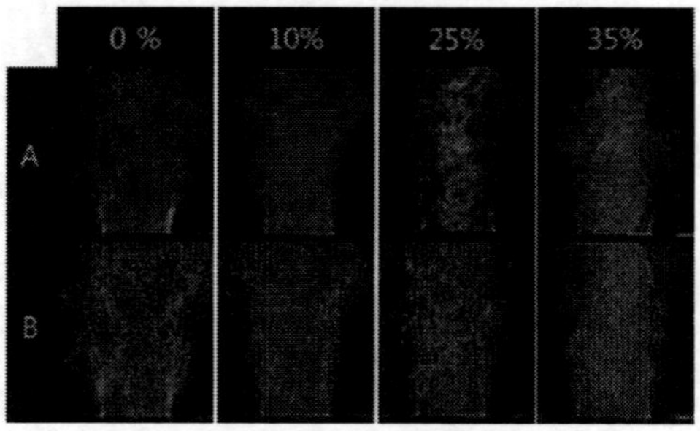

Figure 7: PIV images of the velocity distribution.

3.6 CO and NOx emissions

The recirculation zone effects to the temperature of the flame, resulting in changes in CO and NOx emission, as shown in Fig 8 and Fig 9. First, it was confirmed that CO emission decreased as the equivalence ratio decreased, but it increased with the unstable flame from blow off and re-ignition by recirculation in the flame near the blow out when an equivalence ratio was less than 0.6. It could be seen once again that the lean flammable limit was reduced through the rapid appearance of these flames, which are repeated in extinction and ignition. As the porosity increases, the

CO emission increases because of the decrease of the swirl intensity and the mixing rate of reactants. Fig. 9 shows NOx emission with the porosity as a function of the equivalence ratio. As the equivalence ratio decreases, the NOx emission decreases due to the decrease of flame temperature. As the porosity increases, the NOx emission slightly increases. It is considered that the influence of outer recirculation is reduced and the flame temperature is increased when the generation of NOx is increased.

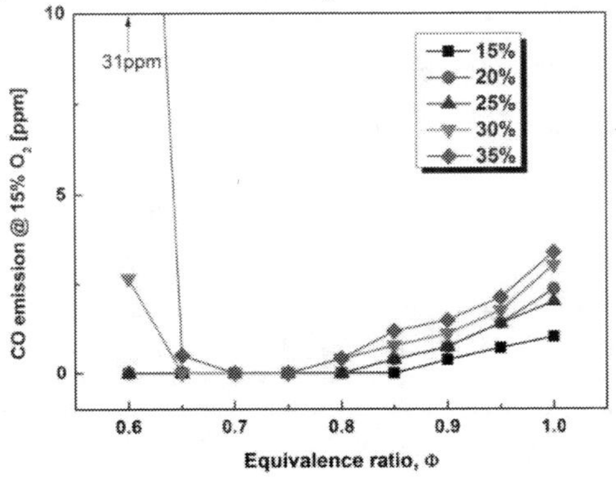

Figure 8: Variation of CO emission concentration with the equivalence ratio at each porosity rate

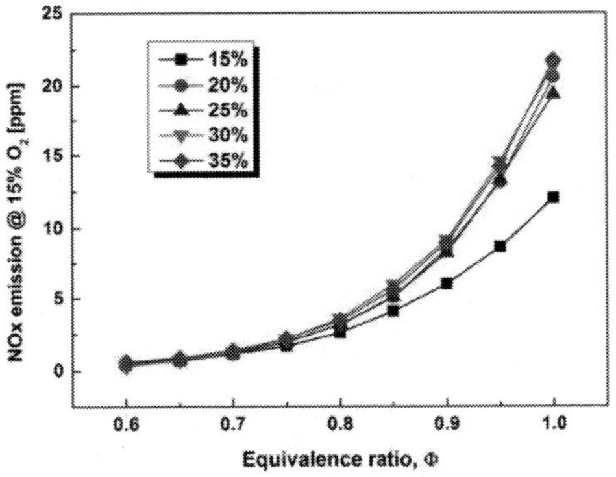

Figure 9: Variation of NOx emission concentration with the equivalence ratio at each porosity rate

4 Conclusions

The aim of this work is to investigate the effect of porosity of center plate in low swirling burner. To do, the experimental study with ICCD camera, hot-wire and PIV system was employed. The conclusions are as follows

1. When the porosity increased the flash back is suppressed while the lean flammability limit decreased and the flow rate through the center plate increased with the increase of flame height.

2. As the porosity increased, the flame changed from V shape to flat shape. Also, as the equivalence ratio was lower, the helical flame formed in the whole flame field of the combustion chamber, with the increase of CO emission.

3. It is confirmed through hot wire velocity fields that the uniform velocity of the center plate reduced the internal recirculation zone.

4. As the porosity increased in the PIV images, the center flow in the low swirler showed almost uniform flow. With the flow rate of the center increases, the flow in the swirler vane get smaller and the swirl strength get weaker.

5 Acknowledgment

This research was supported by Basic Science Research Program through the National Research Foundation of Korea (NRF) funded by the Ministry of Education (No. 2015R1D1A1A01058610)

References

[1] C.H. Hwang, C.E. Lee, Journal of The Korean Society for Aeronautical & Space science, Vol.34, No. 7, (2006)79-88

[2] Cheng, R.K., Yegian, D.T., Miyasato, M.M., Samuelsen, G.S., Benson, C.E., Pellizzari, R. and Loftus, P.,Proceedings of the Combustion Institute, Vol. 28, (2000)1305-1313(2000).

[3] Cheng, R.K., Fable, S.A., Schmidt, D.,Arellano, L. and Smith, K.O., 2001 International Joint Power Conference, New Orleans, L.A., U.S.A., IJPGC2001/FACT-19055, June 2001.

[4] Yegian, D.T. and Cheng, R.K., Energy and Environment Division Lawrence Berkeley National Laboratory berkeley, LBNL-39354, 1996.

[5] Day, M., Tachibana, S., Bell, J., Lijewski, M., Becjner, V. and Cheng,R.K.,Combustion & Flame,Vol.159, Issue 1,(2012) 275-290,.

[6] Rankin, D.D., Lean Combustion : Technologyand Control, 1th ed., Elsevier Inc., Irvine,C.A., U.S.A., 2007

[7] H.H. Jeong, K,M, Lee, Journal of the Korean society of propulsion Engineers, Vol. 20, No. 4, (2016) 40-49

11th Asia-Pacific Conference on Combustion,
The University of Sydney, NSW Australia
10th -14th December 2017.

Liquid Jet in Swirling Cross Flow – An Experimental Study

Surya Prakash R.[1a], Prasad Boggavarapu.[2], Raghunandan B. N.[1], Gaurav Tomar[2], Ravikrishna R. V.[2],

[1]Department of Aerospace Engineering, Indian Institute of Science

Bangalore-560054, Karnataka, India.

[2]Department of Mechanical Engineering, Indian Institute of Science

Bangalore-560054, Karnataka, India.

Abstract

The current study focusses on the experimental investigation of liquid jet injected into swirling flow – a configuration similar to the ones used in recent combustors. Swirling flow is generated in the confined annular space with the aid of axial swirlers employing contoured vanes. The swirl number is varied between 0 and 1, covering the range considered as low-intensity swirl thus avoiding the issues associated with recirculation zones that occur at high-intensity swirl flows. Water is used as the working liquid and is injected radially outwards with a turbulent mean velocity profile, from the surface of the inner tube into the annular space. The velocities of both liquid and air flow are varied to obtain different liquid-to-gas momentum ratios. Important spray characteristics such as spray trajectory and drop size distribution are studied with the aid of laser diagnostics. Correlations are derived for the spray to represent its radial penetration as well as angular deflection. Drop size data shows evidence of coalescence of droplets. These experiments provide insight on the effect of swirl flow on essential spray characteristics.

1 Introduction

Liquid jet injected into gaseous crossflow finds applications in various fields ranging from aerospace engineering to agriculture, where spraying of pesticides in an effective manner is important. In the field of aerospace engineering, one may find this configuration in gas turbine engines, afterburners, ramjets and scramjets. Liquid jet in crossflow (JICF) has been extensively studied by various experimental groups and not so extensively by the computational and theoretical groups due to the associated difficulties and complexities. While experimental studies have mostly been conducted at ambient conditions [1, 2] there have been few investigations at elevated pressures [3] and temperatures [4, 5]. The underlying objective of the present experimental investigation is to characterize the resultant spray in terms of its trajectory, penetration and drop size distribution. The trajectory of a liquid jet in crossflow is found to be primarily dependent on the liquid-to-air momentum ratio by Wu et al. [1, 2]. Several other studies have also corroborated this observation, albeit with modifications and fine-tuning of the constants [5, 6].

Though there are many groups that have studied the behaviour of the liquid jet in cross flow, most of them have been done so in the rectangular domain, with the liquid injected from one of the faces, and with no swirl. In the current investigation, an annular domain is utilized and the liquid is radially injected outwards from the surface of the inner tube. This is an ideal configuration to generate swirling air flow and then explore the effects of such a flow on the liquid jet breakup processes. While a similar annular domain was also used by Tambe et. al. [7] the outer confinement consisted of flat walls to aid optical access. There have been other studies [8, 9] which have investigated the twin swirler arrangement, however, not particularly focused on spray characterization of jet in swirling crossflow. The present study involves the analyses of spray trajectory and drop size distribution of a jet in swirling crossflow in a realistic cylindrical geometry. The following sections describe the experimental setup and a discussion of the results obtained.

2 Experimental Setup

The experimental setup broadly consists of the flowlines for the operating fluids - air and water, the test section, swirl-generators and the required diagnostics for various measurements. The flow line for air is basically a wind tunnel capable of handling high mass flow rates. A schematic of the experimental setup is shown in Fig. 1. All the design principles for obtaining a uniform laminar flow are followed from the established rules available in the literature. The wind tunnel comprises of the essential components such as the wire meshes for breaking down the eddies and honeycomb mesh for straightening the flow. All the wind-tunnel components are of cylindrical cross section. The flow at the exit of the contraction section is found to be uniform and near-laminar (with turbulence intensity < 3%).

Water is the operating liquid used in the current experimental investigation. Compressed air storage is used to pressurize the liquid tank. This liquid is then routed through a flow meter to the injector. Since the study involves swirling flows the test section is designed to be of cylindrical geometry. The swirl flow itself is generated in an annular space. The inner cylinder is 60 mm in diameter and the outer section has 150 mm diameter. This provides an aspect ratio of 2:5 for inner diameter to outer diameter. The outer section is made of Acrylic material which is transparent and hence provides us with the required optical access. The liquid is injected through an injector, which is essentially a long pre-fabricated stainless-steel tube (make: Supelco). This tube injects water radially outwards into the annular space, from the surface of the inner cylinder. The

Corresponding author. Fax: +00-00-000-000
E-mail address: surya@aero.iisc.ernet.in

diameter of this liquid injector is 1-mm and the L/D ratio is ~150. This long L/D is decided based on the study presented in the previous chapter to ensure a turbulent jet at the exit of the injector. The Reynolds number is greater than 2500 for all cases, which is a general criterion for a laminar fluid flow in a pipe to undergo transition to turbulent flow.

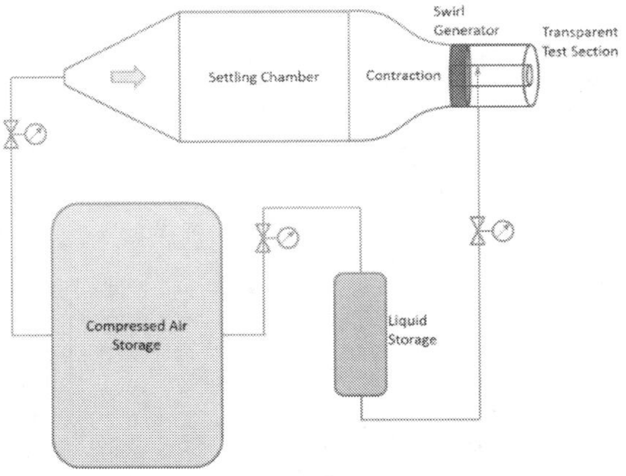

Figure 1: Schematic of the experimental setup used to study the behaviour of liquid jet breakup in swirling flows.

The vital part of these experiments lies in the generation of swirling flows. Swirl generators henceforth referred to as the swirlers are employed to do this job. The current wind tunnel setup permits use of an axial flow swirler which is assembled in-line with all the components. Swirlers are designed to obtain a velocity profile matching that of a forced vortex (solid-body rotation velocity profile). This also resembles a helical path when the flow moves along the axis. The basic principles of fluid flow are kept in consideration while designing the contoured vane profile of swirlers. It is to be noted here that the vanes have a 3-D contour which is a function of both the axial and radial directions. Any swirling flow is characterized by its angle of swirl and consequently by the swirl number. The swirlers are assigned an equivalent angle which is the integral average of the varying flow across the annular space. The present study, consequently, utilizes three swirlers – 0, 30 and 45 degrees – corresponding to swirl numbers of 0, 0.42 and 0.84. The swirl number is defined as the ratio of axial flux of angular momentum to axial flux of the axial momentum.

3 Results and Discussion

The most important characteristics of the LJICF configuration are: trajectory/penetration and drop size distribution, each of which are detailed here.

3.1 Trajectory

Since the spray path can no longer be captured in a two-dimensional plane, the conventional method of imaging will not help in obtaining its complete information. Figure 2 shows the schematic of the diagnostic setup arranged to get this information. A pulsed Nd:YAG laser beam is converted into a thin sheet of 0.5-mm thickness and passed through the spray cross sections (y-z plane) at different axial locations (x). The scattered signal is imaged using a DSLR camera in each of the measurement planes, which are located 5-mm apart and covered till a distance of 75-mm from the jet injection location. Multiple long-exposure images are acquired to capture the scattered signal at each of the cross-sectional planes. These images are then processed to locate the centroid of the spray cross-section and the spray edge in the radial direction, providing information on the spray trajectory. It is to be noted that the laser sheet and the camera are traversed together as a single unit to maintain the focal distance ensuring consistency in measurements.

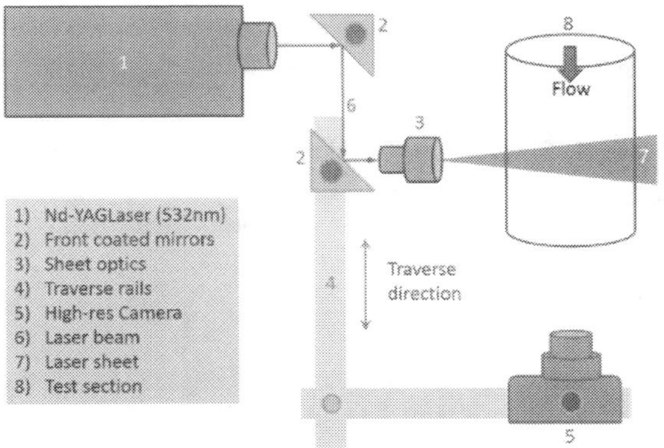

Figure 2: Schematic of the experimental arrangement used to measure the trajectory of the resultant spray.

The windward edge of the spray is conventionally considered to define the spray trajectory and correlations are then derived to express the radial penetration in terms of axial distance. Most of these expressions have been with reference to the rectangular geometry, where the penetration may be shown on a two-dimensional plane and as a function of stream wise distance (x) alone. However, in the annular configuration the spray traverses a three-dimensional path, which implies that this single expression does not suffice to show the complete picture of the trajectory. Therefore, the trajectory is hereby expressed in terms of both the radial penetration - r and the angular deflection θ. A schematic representing this convention is shown in Figure 3.

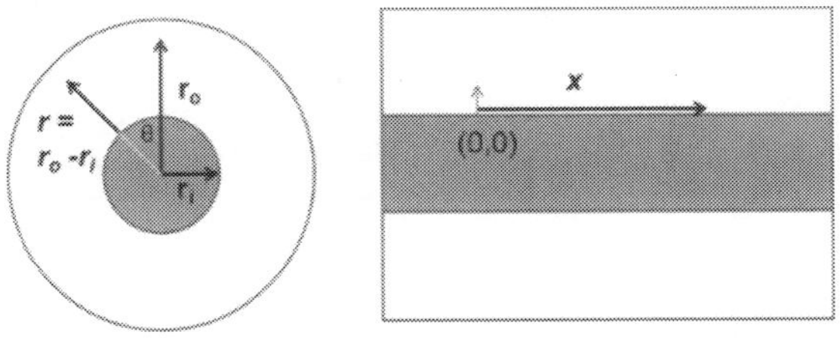

Figure 3 Axes-convention used to express trajectory in the current study.

Thus, the expressions used to capture the trajectory are:

$r = a\,Q^{0.5}x^{c}$ and $\theta = f\,(SN, x)$

The exponent of momentum ratio, Q is assumed to be 0.5 taking cue from the already established expressions available in the literature, though not involving swirling flows. The expression for angular deflection is to be derived based on the observations.

Figures 4, 5 and 6 show trajectories for momentum ratio 20, for all the three swirl numbers. The experimental data obtained is also fitted with the correlation representing the above expression. It may be observed that the experimental data and the corresponding correlations match satisfactorily. The swirl cases are also analyzed for the angular deflection of the spray and a correlation for its variation is derived. Figure 7 shows the variation of spray angular deflection with axial distance. It may be observed that the angle of deflection varies linearly with axial distance and their slopes are proportional to the corresponding swirl numbers. The values of b, c and k for the best fits are calculated and the resultant expressions for radial penetration and angular deflection are given as

$$r = 1.4\, Q^{0.3} x^{c} \qquad \text{and} \qquad \theta = SN\, x$$

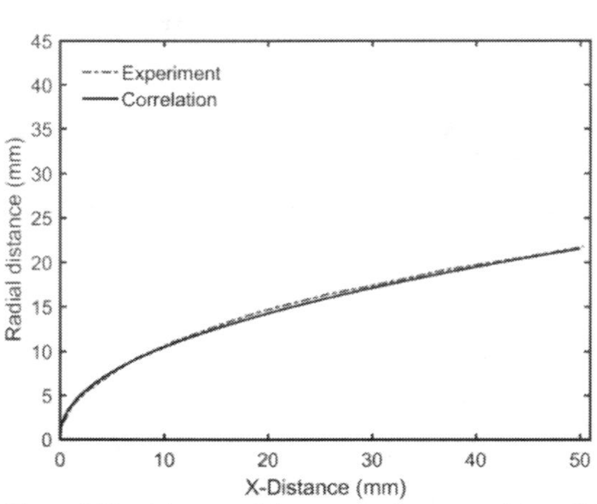

Figure 4 The trajectory of the spray under zero-swirl condition for Q = 20.

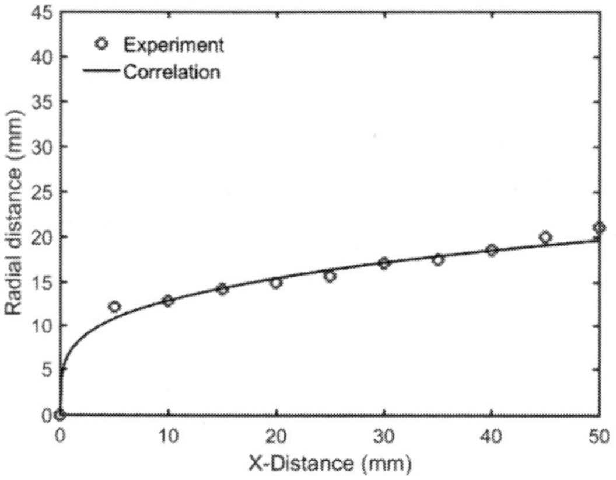

Figure 5 The trajectory of the spray under 30^0 swirl flows for Q = 20.

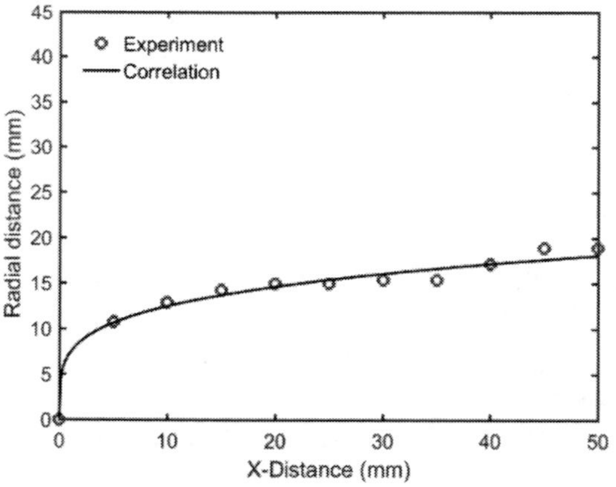

Figure 5 The trajectory of the spray under 45^0 swirl flows for Q = 20.

3.2 Drop size distribution

Drop sizing constitutes the other important parameter of spray characterization. Figure 8 shows the schematic of the experimental setup used to carry out drop size measurements [10]. Shadowgraphy is employed to obtain the information on drop sizing. Nd:YAG laser pulse (532-nm) is used as a source of backlighting after diffusing it through a fluorescent plate. The effective exposure time is around 20-ns. This provides us with drop images frozen in time which are captured with a 4 MP CCD camera. These images are post-processed to calculate the droplet sizes.

Drop sizes are measured at three different downstream locations – 35-mm. 55-mm and 75-mm from the location of jet injection. The information of the spread of the spray obtained in the process of analyzing the trajectory aids in appropriately deciding upon the measurement locations within these cross sections. There are 5 measurement windows used in each of the axial location including a central position and four positions along the y-z axes corresponding to the spray edge, but remaining well within the confines of the presence of spray droplets.

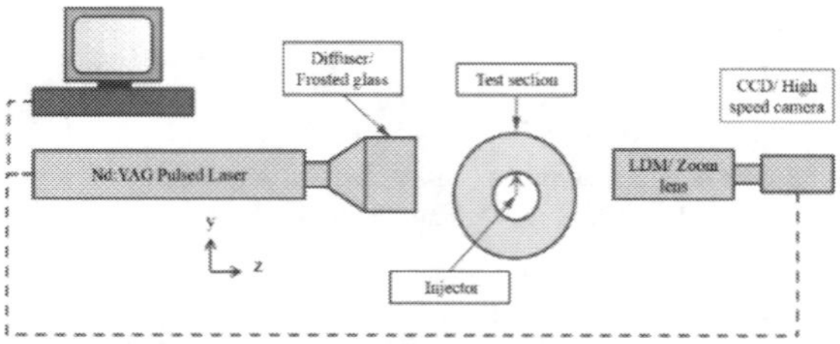

Figure 8 Schematic of the shadowgraphy setup used to make drop size measurements and also carry out high-speed imaging.

Figures 9-11 show the variation of drop size distribution along the axial distance for all three swirl numbers. Firstly, considering the zero-swirl case, we may observe that the spray drop distribution is closely adhering to the log-normal fit. This is found to be true even for SN = 0.42 and 0.84 conditions. Further, it is observed that the Sauter Mean Diameter (SMD) follows an increasing trend as we move downstream along the spray. This behavior is attributed to droplet coalescence. This is clearly exhibited in the SN = 0 and 0.42 cases, while there is some variation in behaviour for the SN=0.84 condition. This may be attributed to the different air flow profiles that are encountered during the jet breakup. While it is a uniform axial velocity profile with the zero swirl condition, there is a linear variation in axial and angular velocities along the radius for both the swirl flows. Thus, there is a recognizable difference in the air velocity profiles that cause the breakup of the liquid jet. It appears that a small change in the profile from uniform to a low swirl number (SN=0.42) is causing the drop size to increase, whereas, a larger variation towards high swirl number (SN=0.84) is resulting in the opposite effect of smaller drop sizes. This is a consequence of trajectories of the large droplets, especially under the higher swirl number conditions, where several droplets impinge on the outer wall of the domain and cause the observed trend.

4 Conclusions

Though there have been many studies in the past on liquid jet in cross flow, there have been very few which have investigated liquid jet breakup transversely injected into a swirling flow. The current study throws more light on this aspect making use of an annular domain – the best possible configuration to work with swirling flows. Two different swirlers are used – 30° and 45° - along with a base case of no swirl condition. Liquid is injected radially outwards from the surface of the inner tube. The important characteristics of the spray including trajectory, drop-sizing and transient behavior, are investigated. Due to the three-dimensional nature of the spray path, its trajectory is expressed in terms radial penetration and angular deflection, thus giving complete information of the spray location. Correlations are successfully derived for both the radial penetration and angular deflection. It is observed that the radial penetration may be expressed on similar lines as that of the existing correlations for rectangular domain. There is however a lag in angular deflection of the spray with respect to the swirling air flow, indicating partial transfer of momentum to the spray drops. However, the angular deflection of the spray is also found to vary linearly proportional with the corresponding swirl numbers. Drop size measurements showed that their distribution mostly follows the log-normal curve. The experiments also revealed an interesting phenomenon of coalescence of drops. This is manifested in the form of increasing magnitude of SMD that is measured as the spray travels downstream along its path.

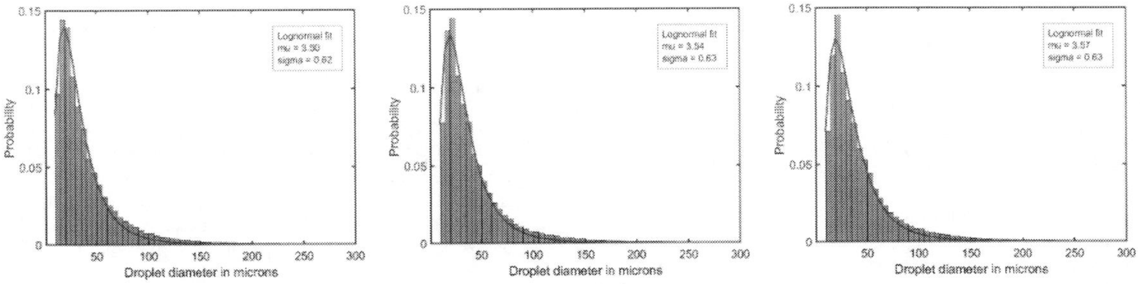

Figure 9 Drop size distributions for Q = 20 and SN = 0 at z = 35mm, 55mm and 75mm

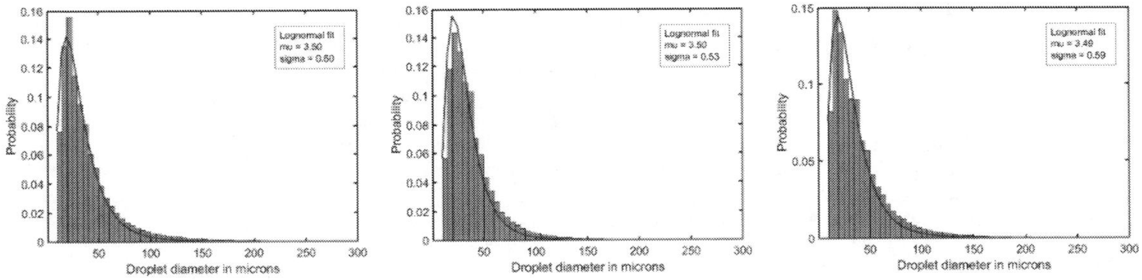

Fig. 10 Drop size distributions for Q = 20 and SN = 0.84 at z = 35mm, 55mm and 75mm

References

[1] P. K. Wu, K. A. Kirkendall, R. P. Fuller and A. S. Nejad, Breakup Processes of Liquid Jets in Subsonic Crossflows, Journal of Propulsion and Power, 13(1) (1997) 64-72.

[2] P. K. Wu, K. A. Kirkendall, R. P. Fuller and A. S. Nejad, Spray Structures of Liquid Jets Atomized in Subsonic Crossflows, Journal of Propulsion and Power, 14(2), (1998) 173-181.

[3] J. Becker and C. Hassa, Breakup and Atomization of a Kerosene Jet in Crossflow at Elevated Pressure, Atomization and Sprays, 12(1-3) (2002) 49-67.

[4] Y. Gopala.,E. Lubarsky , O. Bibik and B.T. Zinn, Measurements of Spray Characteristics in Preheated Crossflowing Air, AIAA-2007-1179, 45th AIAA Aerospace Sciences Meeting and Exhibit, Reno, Nevada, 2007.

[5] Eslamian, M., A. Amighi, and N. Ashgriz., Atomization of Liquid Jet in High-Pressure and High-Temperature Subsonic Crossflow, AIAA Journal (52.7) (2014): 1374-1385.

[6] Lee, K., Aalburg, C., Diez, F. J., Faeth, G. M., & Sallam, K. A., Primary breakup of turbulent round liquid jets in uniform crossflows, AIAA journal, (45-8) (2007).

[7] S. B. Tambe and S. M. Jeng, Three-Dimensional Penetration and Velocity Distribution of Liquid Jets Injected Transversely into a Swirling Crossflow, ILASS Americas, 22nd Annual Conference on Liquid Atomization and Spray Systems, Cincinnati, OH, May 2010.

[8] J. Becker and C. Hassa, Liquid Fuel Placement and Mixing of Generic Aeroengine Premix Module at Different Operating Conditions, Journal of Engineering for Gas Turbines and Power, (124-4) (2003) 901-908.

[9] S. K. Dhanuka, J. E. Temme, J. F. Driscoll, H. C. Mongia, Vortex-shedding and mixing layer effects on periodic flashback in a lean premixed prevaporized gas turbine combustor, Proceedings of the Combustion Institute (32) (2009) 2901-2908.

[10] M. Jain, R. Surya Prakash, G. Tomar, and R. V. Ravikrishna, Secondary Breakup of a Drop at moderate Weber numbers, Proceedings of the Royal Society A. (2015). 471:2177.

11th Asia-Pacific Conference on Combustion,
The University of Sydney, NSW Australia
10th -14th December 2017.

Investigation of Ignition Process in a Premixed Swirl Configuration

Masoud EidiAttarZade[1], Sadegh Tabejamaat[1], Mahmoud Mani[1], Mohammad Farshchi[2]

[1] Amirkabir University of Technology

424 Hafez Ave., Tehran, 15875-4413, Iran

[2]Sharif University of Technology

Azadi Ave., Tehran, 11365-11155, Iran

Abstract

In this paper, we study the ignition process in a premixed methane-air swirl configuration. A large eddy simulation method with a developed thickened flame combustion model are used. At first non-reacting mean and RMS axial and tangential velocity profiles are validated against experimental results as well. It is shown that 4 zones exist: inner recirculation zone, inner shear layer, outer shear layer and corner recirculation zone. Ignition sequence with sparking in these four zones are investigated with the focus on flame propagation. It is shown that sparking in IRZ has the fastest flame growth and minimum time for flame stabilizing. Flame propagating in all cases are like edge flames. Moreover, there isn't any hysteresis for flame position.

1 Introduction

Ignition is an important process of startup phase in gas turbine and thrusters. In aircraft gas turbine, re-ignition in high-altitude is an important requirement that very complicated with modern low emission lean premixed combustion chambers [1]. Also, modern laser igniting devices take the advance of desire energy deposition to reactants. Fast and safe start are the main issues in the ignition process. There are four major steps to a successful ignition [2]: (1) energy transfer to reactants, (2) formation of a flame kernel, (3) flame growth and propagation, and (4) flame stabilization.

A considerable number of experimental and numerical studies have been performed on ignition process in swirl jet flow [3, 4] and gas turbine combustors [5, 6]. Various mechanisms that drive ignition failure or success are local mixture fraction, flow strain rate, flow velocity.

In liquid fueled systems, a range of interlinked phenomena, including droplet dispersion and evaporation, the subsequent mixing of kerosene vapor with air are added to ignition process [1]. Jones and Tyliszczak [1] studied spark ignition in the liquid fuel combustor. They used LES-pdf methodology to for both dispersed (liquid) and gas phases. The results suggest that the spark size is a critical parameter for liquid fuel combustion chamber ignition. Eyssartier et al. [4] developed a model to predict igniting of real configuration base on cold flow conditions for two phase flows. Linassier et al. [3] developed an ignition model for liquid non-premixed swirl configuration. Their results show that if sparking position is in the downstream region of the combustor, the flame will be blown out.

Corresponding author. Fax: +98-21-6454-3212
E-mail address: sadegh@aut.ac.ir

Esclapez et al. [7] investigated ignition process in an academic lean partial premixed swirl burner numerically. They found that the ignition is successful if the flame kernel have minimum size and can enter the inner recirculation zone. Boileau et al. [5] investigated the ignition in a full annular chamber with LES. They found that the flame propagation from burner to burner is higher than turbulent flame speed. Also the aerodynamics have a major effect on flame leading point.

In the present work the ignition process of premixed methane-air swirl configuration is studied using LES based numerical calculation. The physical model was studied experimentally by Roux et al. [8]. The setup of Roux et al. [8] consists of a high swirl configuration with an overall $\varphi = 0.7$. Its combustion chamber is rectangular with 86x86x110 mm dimensions. The main goal of the current study is investigation of flame growth and propagation after successful flame kernel initiation. The effect of sparking position on propagation history will be studied.

2 Numerical Approach

To model the ignition process, a 3-dimensional pressure-based compressible flow solver is used. The filtered governing equations are as follows [9]:

Continuity Equation:

$$\frac{\partial \bar{\rho}}{\partial t} + \frac{\partial}{\partial x_i}(\bar{\rho}\tilde{u}_i) = 0 \tag{1}$$

Momentum equation:

$$\frac{\partial \bar{\rho}\tilde{u}_j}{\partial t} + \frac{\partial}{\partial x_i}\left(\bar{\rho}\tilde{u}_i\tilde{u}_j\right) + \frac{\partial \bar{p}}{\partial x_j} = \frac{\partial}{\partial x_i}\left[\bar{\tau}_{ij} + 2\nu_t \bar{S}_{ij}\right] \tag{2}$$

Species equation:

$$\frac{\partial\left(\bar{\rho}\tilde{Y}_k\right)}{\partial t} + \frac{\partial}{\partial x_i}\left(\bar{\rho}\tilde{u}_i\tilde{Y}_k\right) = \frac{\partial}{\partial x_i}\left[\frac{\nu}{Sc}\frac{\partial \tilde{Y}_k}{\partial x_i} + \frac{\nu_t}{Sc}\frac{\partial \tilde{Y}_k}{\partial x_i}\right] + \bar{\dot{w}}_k \quad k = 1, N \tag{3}$$

Sensible enthalpy equation:

$$\begin{aligned}\frac{\partial\left(\bar{\rho}\tilde{h}_s\right)}{\partial t} + \frac{\partial}{\partial x_i}\left(\bar{\rho}\tilde{u}_i\tilde{h}_s\right) &= \frac{\partial \bar{p}}{\partial t} + \tilde{u}_i(\partial\bar{p}/\partial x_i) \\ &+ \frac{\partial}{\partial x_i}\left[\bar{\lambda}\frac{\partial \bar{T}}{\partial x_i} + \frac{\nu_t}{\mathrm{Pr}}\frac{\partial \tilde{h}_s}{\partial x_i}\right] + \overline{\tau_{ij}\frac{\partial u_i}{\partial x_j}} + \bar{\dot{w}}_T + \dot{Q}\end{aligned} \tag{4}$$

Where Sc and Pr denote Schmidt and Prandtl numbers are assumed to be unity. $\dot{Q}$ is the energy addition during the sparking time and $\dot{\omega}_T$ is the heat release due to combustion. ν_t is subgird scale viscosity modeled by Smagorinsky [10, 11].

A 2-steps and 6-species chemical mechanism [12] is used to

calculate chemical source term. Turbulence combustion interaction is modelled by thickened flame approach [13, 14]. In this model, the flame is thickened such that several grid points fall in the flame front so the front is resolved explicitly. If thermal diffusivity, D_{th}, is multiplied by a factor F and pre-exponential constant, A, is divided by F, the flame speed remains constant while the flame thickness is increased by F. The efficiency function, E, was defined by Colin et al. [13] which takes the sub-grid scale wrinkling into account. Accordingly, D_{th} is modified to EFD_{th} which can be decomposed: $EFD_{th} = D_{th} + (EF - 1)\ D_{th}$. $(EF-1)D_{th}$ can be considered as sub-grid scale diffusivity.

OpenFOAM package [15] is used to solve the above equations. A 3-D Cartesian compressible Navier-Stokes with multi-species formulation is applied. The thickened flame model is developed and embedded in the code. In order to model the energy addition during sparking, a simple model is used. A source term, $\dot{Q}$, is added to sensible enthalpy equation. This term is activated only in sparking duration in a desired spherical region in the flow field. During sparking, in each time step, the sensible enthalpy of cells in the spherical region increases uniformly in space. A temperature limit is defined equal to 3000 K.

No slip velocity boundary conditions are considered for the all wall and zero gradient for the outlet. To reduce the computational cost, the swirler blades are simulated separately and the results of mean velocity in x = -24 mm plane are extracted and are used as inlet velocity. A synthetic turbulence generation method has been used for inlet velocity boundary condition by 8% turbulence intensity. For all species mass fractions, Dirichlet type boundary condition for inlet and Neumann type for other boundaries are used. Zero gradient pressure boundary conditions are used for all boundaries. For the outlet, the infinity pressure is set to 1e5 Pa. To damp the acoustic waves from the outlet, a wave transmissive is used for the outlet. Premixed methane-air mixture with $\varphi = 0.7$ is set for inlet with 12 gr/s air mass-flow rate and $T = 298$ K.

3 Result and Discussion

In this section, at first, non-reacting simulations are presented and validated and then the ignition process is discussed in ignition subsection. Two coarse and fine grids have been used for simulations. The minimum cell size of swirler exit is 0.9 and 0.4 mm for coarse and fine grids, respectively. The Kolmogorov length scale is about 0.1 mm in this configuration.

3.1 Non-reacting flow

Two coarse and fine grids have been set up for non-reacting simulation. For these simulations, at first the domain is solved for 100 ms that equals to about 3 flow through time. Then averaging is turned on for another 100 ms. Because of the highly swirled flow, there exist an adverse pressure gradient and consequently an inner recirculation zone (IRZ) forms as it can be seen in Fig. 1. The IRZ is bounded by swirler jet (SWJ) and a corner recirculation zone (CRZ) exist between walls and SWJ. Because of high velocity difference between SWJ and recirculation zones, two strong shear layer forms. These shear layers are called outer shear layer (OSL) and inner shear layer (ISL).

Mean and RMS velocity profiles at various sections of combustion chambers are validated by experimental results of Roux et al. [8] for both fine and coarse grids. Fig. 2 shows mean axial velocity. The magnitude of velocity is well captured by both grids, but the fine grid can predict it better. Also the RMS of axial velocity predicted as well (Fig. 3). At x = 3.5 mm, the maximum RMS is in the middle while in the other sections, the maximums are in SWJ zone. LES and experimental [8] mean tangential velocity profiles have been compared in Fig. 4. Both grids predicted the velocity as well, but at x = 25 & 35 mm, the coarse one has some errors.

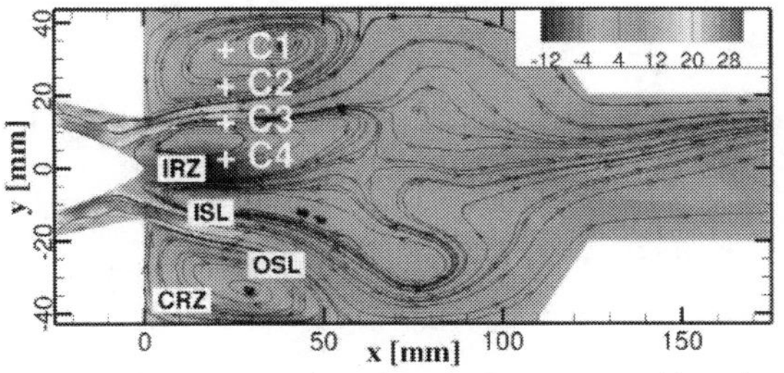

Fig. 1: Non-reacting mean axial velocity contour with various flow zones and sparking positions

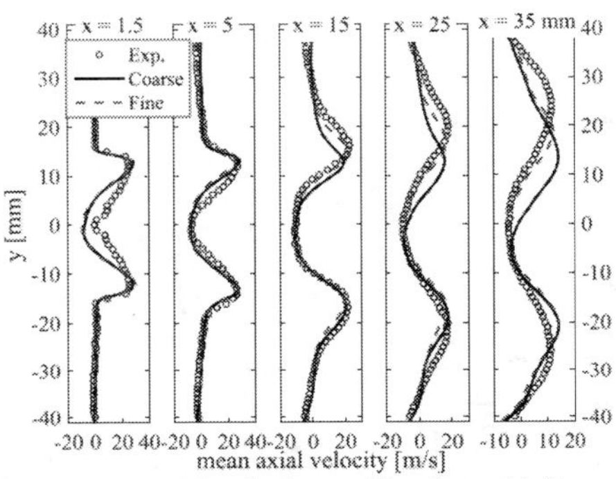

Fig. 2 Mean axial velocity profiles for cold flow at different distance from swirler

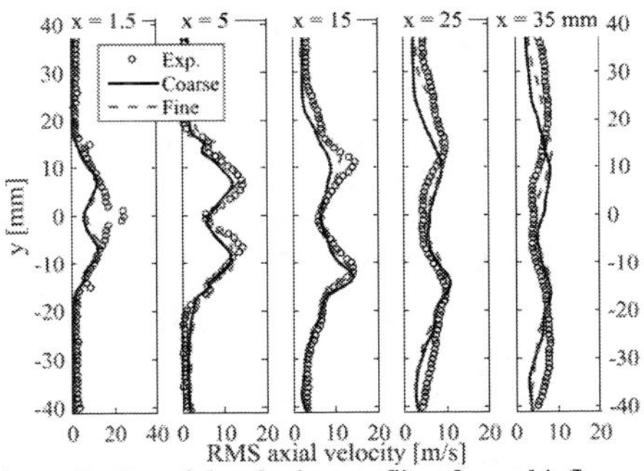

Fig. 3 RMS axial velocity profiles for cold flow at different distance from swirler

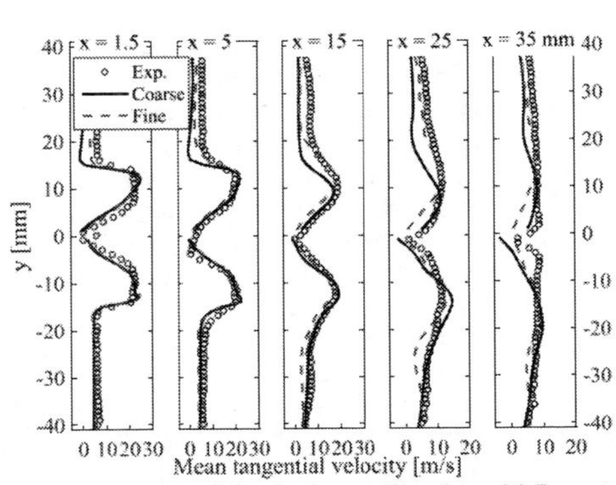

Fig. 4 Mean tangential velocity profiles for cold flow at different distance from swirler

Fig. 5 shows power spectra for both coarse and fine grids. As it represents, both grids' spectra resolve the -5/3 law for more than 2 orders of magnitude of length scales. So it can be concluded that both grids can capture the turbulent flow structure correctly, but the fine one have better precision. So for ignition simulations, we use the fine grid.

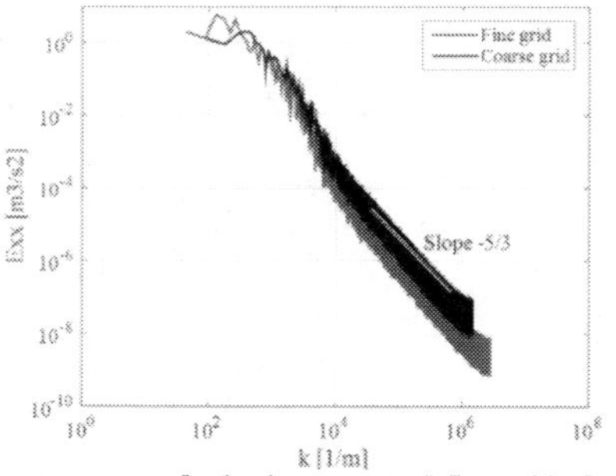

Fig. 5 power spectra for both coarse and fine grids for instant axial velocity at (10, 0, 0) mm position

3.2 Ignition

Four ignition positions have been chosen to study the ignition process in swirl flow. The positions are chosen such that each case has different zones but have the same axial distance to swirler. Table 1 presents details about sparking position. All cases have 2 mm spark diameter and 3 ms spark duration. The energy transferred to reactants is about 40 mJ.

Name	Position [mm]	zone	U mean [m/s]	U instant [m/s]
Case1	(20,30,0)	CRZ	(0.5, -0.5, 4.5)	(3, 9, 9)
Case2	(20,20,0)	OSL	(12, 1.4, 7.5)	(8, 20, 11)
Case3	(20,10,0)	ISL	(6, -2, 15)	(3, 8, 15)
Case4	(20,0,0)	IRZ	(-13,-0.5,1)	(-5, -0.5, -3)
Case5	(20,20,0)	OSL	Same as Case2	(10, 2, 9)

Table 1 specification of ignition simulation cases

Snapshots of ignition in IRZ and CRZ are presented in Fig. 6a and b, respectively. As seen, the flame kernel growth is very fast for spark in IRZ. At t = 2 ms, the flame of IRZ case reaches swirler cone, but the CRZ one is still in the kernel growth phase. In IRZ, the flame kernel grows in IRZ until stabilizes at swirler cone and then from its downstream edge, propagates to the CRZ zone (t = 4 ms) and then ignite the whole of domain. But in the CRZ case, flame kernel divides into two sections (as seen in t = 4 ms) then both IRZ and CRZ zones begin to burn (t = 7 ms). Another interesting is that, at all cases, the flame propagation is not spherical but it is like an edge flam shape however the domain is fully premixed.

Fig. 7a & b show upstream and downstream positions of flame edge during time, respectively. In failure case5, the flame position is disappeared after sparking turned off, but in successful cases, flames propagate upstream until they reach the swirler cone and stabilize. For IRZ and ISL sparking cases, the flame stabilizes very fast about 2 ms but for CRZ and OSL cases, it takes more than 7 ms. Because in ISL and IRZ, the flame kernel forms in reverse direction flow and also their growth rate is higher as seen in Fig. 9. Since in case IRZ, the IRZ zone burned up rapidly, the flame at the end of this zone propagates toward the downstream so the IRZ case is the first case that flame reaches the end of the domain (Fig. 7b). At t = 15 ms, all cases have the same position in upstream and downstream so there isn't any hysteresis.

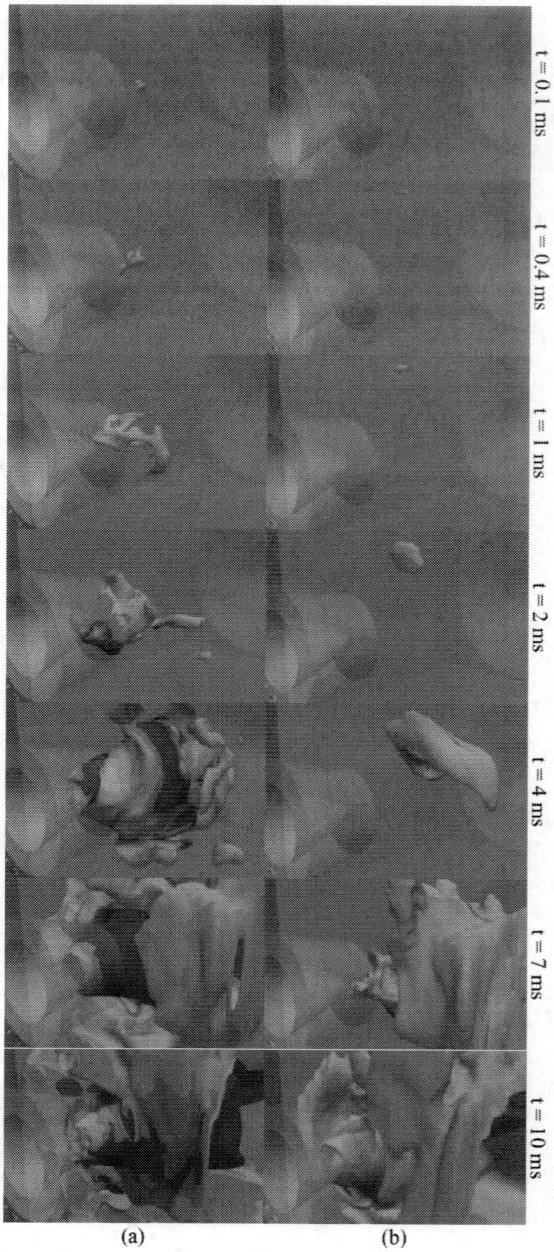

Fig. 6: Snapshots of iso-surface T = 1200 K colored by axial velocity, (a) Case4, IRZ and (b) Case1, CRZ.

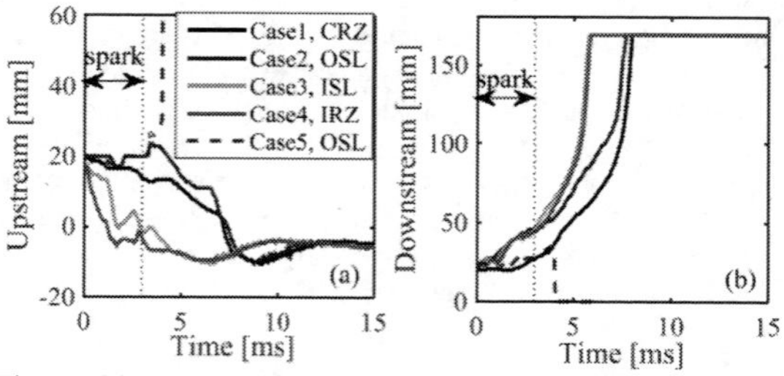

Fig. 7: (a) Upstream and (b) downstream positions of flame during time

Heat release is maximum at flame propagation edge [14] so it can be used as an indicator of flame front. Fig. 8 presents the maximum position of heat release at each time step. As seen in Fig. 8a, in all cases, the flame propagates downstream, but in the Case4 (IRZ) the flame move upstream. Also, in all cases, the flame rotates clockwise in the same direction of tangential velocity Fig. 8b. Therefore, it can be concluded that flame propagation obey the direction of flow because the velocity magnitude is too large.

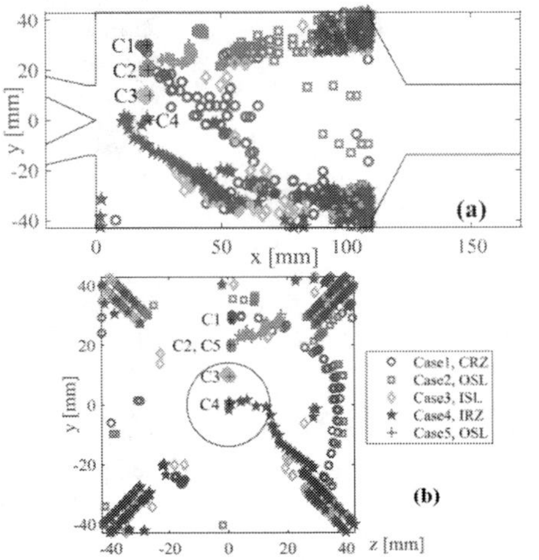

Fig. 8 Position of maximum heat release point at each time step for various ignition cases (a) x-y plane, (b) y-z plane

Flame volume growth during the time is presented in Fig. 9. The flame volume is defined by T = 1200 K. As seen, two IRZ and ISL cases have the fastest flame growth and CRZ has the most latency to grow. It seems in IRZ, the turbulence intensity is in a middle range that increase the turbulent flame speed, but no too much to quench the flame. In all cases, flame volume is not fluctuating because it is an integral parameter. Except in case5, the flame volume is fluctuating and don't have enough growth to sustain a flame kernel.

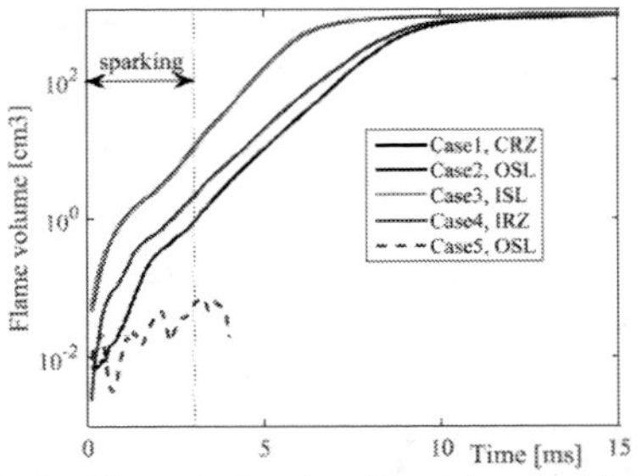

Fig. 9 Flame volume during time for various ignition cases

Integrated heat release is a global parameter that determines the total heat release in the domain. Heat release intensity is calculated from integrated heat release divided by flame volume. Despite the fast flame growth for IRZ and ISL, all successful cases have the same order integrated heat releases (Fig. 10a). Therefore, the combustion intensity of IRZ and ISL cases are less than CRZ and OSL as seen in Fig. 10b. In the first millisecond of sparking, IRZ and ISL cases have sharp rise, but CRZ and OSL have a little oscillating. This is the reason of latency for CRZ and OSL cases in flame growth. All successful cases have a peak value for integrated heat release that is the same time as the flame reaches the end of the combustor so nearly all of the reactants are consumed and only fresh reactants from inlet remain for burning. Fig. 10a shows that flame becomes statistically steady state about t = 20 ms that is nearly two times when the flame reaches the end of the combustor.

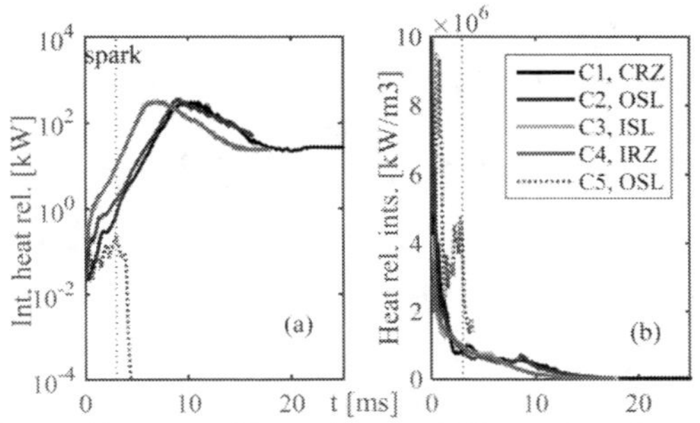

Fig. 10: (a) Integrated heat release and (b) eat release intensity in whole of domain for different ignition cases

4 Conclusions

A LES method with thickened flame combustion approach have been used to model ignition process in a premixed methane-air swirl configurations. Non-reacting mean and RMS velocity profiles are in good agreement with experimental results. Also power spectra shows that grids are fine enough to capture turbulent energy. Ignition in four zones of the flow field is investigated. Results show that sparking in IRZ has the fastest flame growth and minimum residence time to flame stabilization so it can be concluded that energy release in IRZ by devices such as laser igniter is the best case for fast and reliable ignition. Maximum heat release point during time presented the trajectory of flame propagation, shows the flame propagation path follows flow direction.

References

[1] W. P. Jones, A. Tyliszczak, Flow Turbulence Combustion, 85 (2010) 711–734.

[2] E. Mastorakos, Progress in Energy and Combustion Science, 35 (1) (2009) 57–97.

[3] G. Linassier, A. Bruyat, P. Villedieu, N. Bertier, C. Laurent, O. Rouzaud, R. Lecourt, H. Verdier, G. Laergne, Comptes Rendus Mecanique, 34 (1-2) (2013) 201-210.

[4] A. Eyssartier, B. Cuenot, L. Y. Gicquel, T. Poinsot, Combust. Flame, 160 (7) (2013) 1191-1207.

[5] M. Boileau, G. Staffelbach, B. Cuenot, T. Poinsot, C. Bérat, Combust. Flame, 154 (1-2) (2008) 2–22.

[6] J. F. Bourgouin, D. Durox, T. Schuller, J. Beaunier, S. Candela, Combust. Flame, 160 (8) (2013) 1398–1413.

[7] L. Esclapez, E. Riber, B. Cuenot, Proc. Combust. Inst. 35 (2015) 3133-3141.

[8] S. Roux, G. Lartiguea, T. Poinsot, U. Meier, C. Bérat, Combust. Flame, 141 (1-2) (2005) 40-54.

[9] T. Poinsot, D. Veynante, Theoretical and Numerical Combustion, Edwards, Philadelphia, USA, 2005.

[10] S. B. Pope, Turbulent Flows, Cambridge University Press, Cambridge, United Kingdom, 2000.

[11] J. Smagorinsky, Monthly Weather Review, 91 (1963) 99-164.

[12] M. Shahsavari, M. Farshchi, M. H. Arabnejad, Flow, Turbulence and Combustion, 98 (3) (2017) 817–840.

[13] O. Colin, F. Ducros, D. Veynante, T. Poinsot, Physics of Fluids, 12 (7) (2000) 1843.

[14] M. Attarzade, S. Tabejamaat, M. Mani, M. Farshchi, Flow, Turbulence and Combustion, (2017) 1-26.

[15] H. G. Weller, G. Tabor, H. Jasak, C. Fureby, Computers in Physics, 12 (6) (1998) 620-631.

11th Asia-Pacific Conference on Combustion,
The University of Sydney, NSW Australia
10th -14th December 2017.

Thermogravimetric Analysis of Sludge Pyrolysis Oil Mixed with Heavy Fuel Oil

Samuel Chatelier[1], Yong Hao Kuan[1], Guan-Bang Chen[2], Hsien-Tsung Lin[3], Ta-Hui Lin[1,2]

[1]Department of Mechanical Engineering, National Cheng Kung University

[2]Research Center for Energy Technology and Strategy (RCETS), National Cheng Kung University

[3]Department of Aeronautics and Astronautics, National Cheng Kung University

No. 1, University Road, East District, Tainan City, 70101, Taiwan, R.O.C.

Abstract

Pyrolysis of sewage sludge may be a viable solution to the different problems related to its management and handling. The aim of this study is to investigate the combustion characteristics of sludge pyrolysis oil (SPO), heavy fuel oil (HFO), and the blends of SPO and HFO. Different combustion characteristic parameters such as ignition temperature (T_i), burnout temperature (T_e), flammability index (C), and combustion characteristics index (S) are evaluated by thermogravimetric analysis. The results show that the combustion of SPO and the blends of SPO and HFO can be divided into three stages: dewaterization, lightweight components (LWC) burning and heavyweight components (HWC) combustion. In the LWC burning stage, the fuel blends with a higher SPO content exhibits greater values for C and S due to the release and oxidation of the LWC. However, their C and S were quite similar for the HWC combustion stage. In general, adding any quantity of SPO to the SPO/HFO fuel blend enhances the combustion efficiency.

1 Introduction

Sewage sludge is the residual matter, semi-solid slurry that is produced as by-product from the wastewaters treatment processes. The production of sewage sludge is increasing, and all indications suggest that it will be the trend as the global population density keeps rising. Sewage sludge is already causing disposal problems in many industrial countries [1, 2, 3] as it represents a major threat to the environment and human health if managed inadequately.

The most common methods to handle sewage sludge disposal are incineration, agricultural use and landfill. However, none of these methods totally solve the issue as they may cause subsequent problems and require complicated pre-/post-treatment processes [4].

To address these aforementioned problems, pyrolysis is regarded as having particular potential for sewage sludge disposal since it does not only reduce the volume of the sewage sludge by up to 50% [5], but also stabilizes the organic matter, concentrates the heavy metals into the char obtained, and produces valuable byproducts such as pyrolysis oil, chemical materials, and the char itself [6].

Due to limited reserves of fossil fuels, environmental and climatic regulations, and recent awareness in promoting circular economy model, there has been an increase interest in the utilization of sewage sludge, as alternative energy, as a means of both protecting human health and developing a greener and more sustainable energy supply framework.

In this study, combustion characteristics of sludge pyrolysis oil (SPO), heavy fuel oil (HFO) and the mixture of SPO and HFO are investigated. The investigation considers various thermal characteristics of the fuels and their blends, including the ignition temperature (T_i), the burnout temperature (T_e), the flammability index (C), and the combustion characteristics index (S). For each fuel, these characteristics are determined using the results data of thermogravimetric analysis (TGA). The ignition temperature is obtained using the intersection method (IM), while the burnout temperature is determined using the conversion method (CM).

2 Materials and Methods

2.1 Materials and Properties

The HFO used in the present experiments was obtained from the Chinese Petroleum Corporation (CPC), which was the same fuel used in the previous experiments of Li [7], and one can directly refer to Ref. [7] for its properties.

Dewatered sewage sludge was obtained from a wastewater treatment plant in Tainan City (Taiwan). In the as-received condition, the sewage sludge had a moisture content of more than 80 wt.%. However, prior to the experiments, the sewage sludge was sun-dried for about 10 days. As a result, the final moisture content of the sludge was reduced to 7 wt.%.

Then, the final sun-dried sewage sludge was pyrolyzed. The pyrolysis parameters and process were set in accordance with the results obtained by Li [7] for the optimization of the SPO yield. The liquid fraction of the pyrolysis products include water, some hydrosoluble organic compounds, and an immiscible organic fraction [5]. The immiscible organic fraction was separated from the other fractions using a funnel and is referred as sludge pyrolysis oil (SPO) in this study. The average yield of SPO was 10.19 wt.%.

The properties of SPO are shown in Table 1 and are compared to those found in Ref. [8] and HFO found in Ref. [7]. As shown, all these results are similar with those reported in Ref. [8].

Corresponding author. Fax: +886-6-2389940
E-mail address: gbchen26@gmail.com

Sample	Elemental Analysis (wt. %)					Density (g/cc, @15 °C)	Viscosity (cSt, @40 °C)	Pour Point (°C)	Flash Point (°C)	Water Content (vol. %)	pH Value	HHV (MJ/kg)
	N	C	H	O	S							
SPO	5.92	58.44	9.08	18.98	1.25	1.04	64.86~ 66.81	6	>80	11.6	8.22	30.71
SPO [8]	5.08	69.91	8.97	14.66	1.39	1.207[a]	6.83	-	35	0.377	4.03[b]	31.08
HFO [7]	0~ 0.46	86.7~ 86.91	11.8~ 11.56	0.26~ 0.73	0.28~ 0.46	0.93	321.1	6	160	0.309	-	43.99

[a] unit in g/cc, @20 °C [b] acid value, unit in mg KOH/g

Table 1: Properties of SPO and HFO.

Besides, the high heating value (HHV) of HFO (43.99 MJ/kg) is higher than that of SPO (30.71 MJ/kg) because HFO has higher contents of carbon and hydrogen, and lower contents of nitrogen, oxygen and sulfur.

Furthermore, SPO has a similar density to that of HFO. However, the viscosity of SPO at 40 °C is lower than that of HFO and their pour point are the same. By contrast, the flash point of SPO is greater than 80°C and the pH value is 8.22. The water content of SPO is much higher than HFO. Compared to the properties of SPO from Ref. [8], some differences can be seen. These discrepancies are due most probably to the differences in the source of the original sewage sludge, its treatment process, and the pyrolysis process parameters.

2.2 Thermogravimetric Analysis (TGA)

The thermal analyzer (PerkinElmer STA 8000) was used to conduct the TGA and differential scanning calorimetry (DSC) processes under an oxidation environment (air, as the carrier gas). The sample weight was around 35.5~39 mg, and was heated at a rate of 10 °C/min. The air flow rate was set to 50 ml/min and the temperature range was between 30 to 800 °C.

For this study, each case was repeated for three times to ensure consistency. Although the TGA curves were not identical, they all showed a similar tendency. Therefore, the results were divided into three stages for explanation purpose.

The first stage (stage 1) is the dewaterization stage, in which the water content or moisture in the fuel is released. The second stage (stage 2) is the volatile decomposition and burning stage, in which the volatiles or other lower-boiling-point components (lightweight components (LWC)) are decomposed, vaporized and released, and then burned. The third stage (stage 3) is the combustion stage. This stage is associated with the release and burnout of the heavyweight components (HWC) that are in the fuel and a significant heat release.

2.3 Combustion Characteristics Parameters

The intersection method was used in this study to determine the ignition temperature (T_i) of the fuels. For further readings regarding this method, one can refer to the studies of Lu et al. [9]. While the burnout temperature (T_e) was estimated by the conversion method [9, 10].

Knowing both T_i and T_e, two combustion characteristic parameters, namely the combustion characteristics index (S) [11] and the flammability index (C) [12], can be found. Typically, a fuel with a larger value of S and C has an improved combustion performance and combustion stability, respectively.

3 Results and Discussion

3.1 TGA Result of SPO

Figure 1 (a) shows the TGA result of SPO. Stage 1, the dewaterization stage, starts at 30°C and ends at 134°C. In this stage, the DSC curve indicates an endothermic reaction and the derivative thermogravimetric (DTG) curve has two peaks at 105°C and 115°C, respectively. The weight loss (10.799 wt.%) in this stage is mainly the result of moisture evaporation.

Stage 2, the lightweight components (LWC) burning stage, has a temperature range from 134°C to 485°C. In this stage, the DSC curve shows a slight exothermic reaction between 300~450°C, with a peak intensity at a temperature of 395°C and an oscillating heat release period leading up to this peak. For the DTG curve, a concave feature can be seen and the peak appears at 280°C. Thereafter, the DTG curve undulates until the end of the stage; with significant peaks at 386°C and 425°C, respectively. Comparing the DSC and DTG curves, the undulating nature of the DTG curve is related to the oscillating heat release characteristic shown in the DSC curve, and is the result most likely of the vaporization and burning of the relative lightweight components in the SPO. In this stage, the ignition and burnout temperatures are 200°C and 466°C, respectively; and the total weight loss is 77.188 wt.%.

The third stage, which is the heavyweight components (HWC) combustion stage, ranges from 485°C to 800°C. In this stage, the remaining vapor fuel that was released in stage 2 is mixed with the decomposed gaseous fuel released in this stage, then this mixed gaseous fuel surrounds the SPO. The DSC curve shows a burning of the mixed gaseous fuel and an intense combustion heat release. For the DTG curve, a significant peak occurs at 624°C due to a further decomposition of the relative HWC, which subsequently react with the oxygen under the effects of the combustion heat energy and the chain effect of the present stage. In this stage, the TG, DTG and DSC curves all remain relatively stable at temperatures higher than 700°C, and the ignition and burnout temperatures are 563°C and 706°C, respectively; and the total weight loss is 11.9 wt.%.

3.2 TGA Result of HFO

In contrast to the TGA result of SPO, the one of HFO, as shown in Figure 1 (b), can only be divided into two main stages as the dewaterization stage is indistinct since the moisture content of HFO is very low.

Stage 2, the volatile decomposition and burning stage, ranges from 30°C to 530°C. In this stage, the DSC curve first presents a very small concave characteristic, and then (at temperatures higher than 100°C) remains approximately constant. This straight line behavior terminates at a temperature of around 287°C. An obvious exothermic peak then occurs at 359°C. Finally, primary endothermic reactions persist until the end of the stage. For the DTG curves, a very small concave feature corresponding to an endothermic reaction initially, but then have a concave characteristic exhibit from 61~381°C and has a peak located at 354°C. At higher temperatures, two further concave features occur. The first feature has a peak at 396°C, while the latter corresponds to the primary endothermic reaction and has two peaks located at 467°C and 485°C, respectively. Observing the DSC and DTG curves, it can be seen that the peak at 396°C in the DTG curve can be attributed to the oxidation and heat released at 359°C resulting in the continuous vaporization and burning of the relative lightweight component in the HFO over the temperature range of 61~381°C. In this stage, the ignition and burnout temperatures are 250°C and 505°C, respectively; and the total weight loss is 88.605 wt.%.

Stage 3 starts at 530°C and ends at 800°C. In this stage, the remaining vapor fuel that released in stage 2 is mixed with the decomposed gaseous fuel released in this stage, then this mixed gaseous fuel surrounds the HFO. The DSC curve shows a burning of the mixed gaseous fuel and an intense combustion heat release. The DTG curve displays a significant peak at 671°C as a result of the further decomposition of the relative heavyweight components, which then react with the oxygen under the effects of the combustion heat energy and the chain effect of the current stage. In this stage, the TG, DTG and DSC curves all remain relatively stable at temperatures higher than 675°C, and the ignition and burnout temperatures are 581°C and 669°C, respectively; and the total weight loss is 10.21 wt.%.

a.

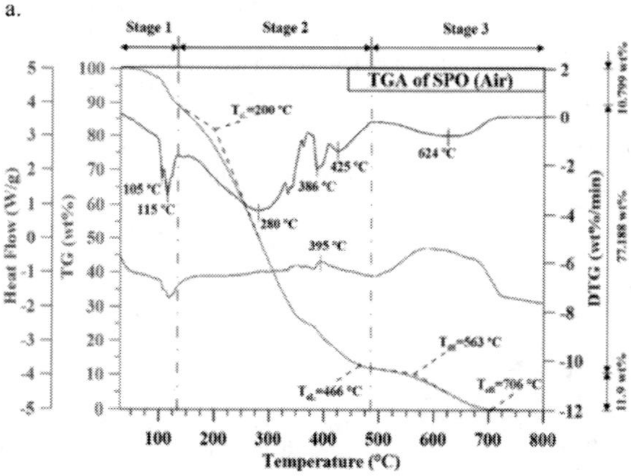

b.

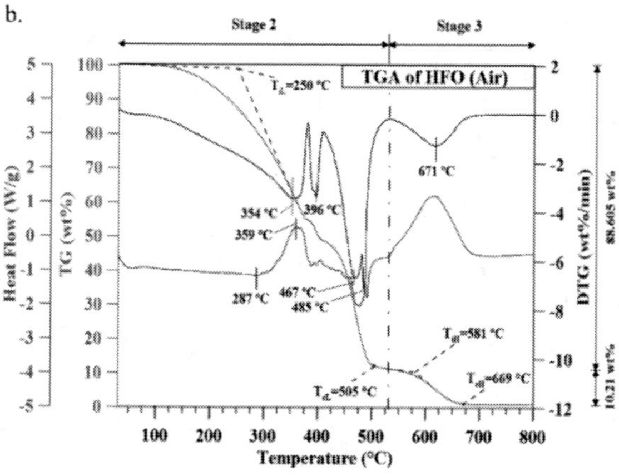

c.

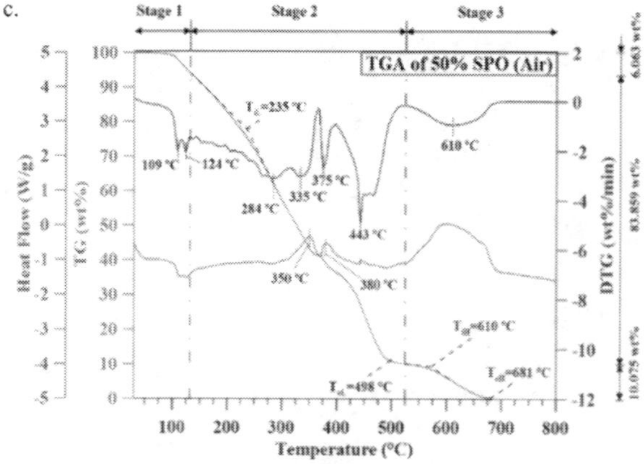

Figure 1: TGA result for SPO (a), HFO (b) and 50% SPO fuel blend (c).

3.3 TGA Result of 50% SPO Fuel Blend

Figures 1 (c) shows the TGA result for the fuel blend consisting of 50% SPO and 50% HFO. As shown, it can be divided into three main stages.

Stage 1 ranges from 30°C to 132°C. In this stage, the DSC curve shows the existence of endothermic reactions, the DTG curve has two peaks at 109°C and 124°C. The weight loss (6.063 wt.%) in this stage is mainly the result of moisture evaporation.

Stage 2 starts at 132°C and ends at 524°C. In this stage, the DSC curve shows a straight-line characteristic at lower temperatures. However, two obvious exothermic peaks are observed at 350°C and 380°C. At temperatures higher than 400°C, the DSC curve presents two concave features. For the initial part of the DTG curve, a concave characteristic is shown, which has two peaks located at 284°C and 335°C, respectively, and the curve shows an oscillating decline. Then, the DTG curve has two further concave features with peaks located at approximately 375°C and 443°C, respectively. Comparing the DSC and DTG curves, it can be seen that the concave feature with a peak at 375°C is the result of the heat released at 350°C and 380°C caused by the vaporization and burning of the relative lightweight components in the fuel blend. In this stage, the ignition and burnout temperatures are 235°C and 498°C, respectively, and the total weight loss is 83.859 wt.%.

Stage 3, the HWC combustion stage, has a temperature range from 524°C to 800°C. In this stage, the remaining vapor fuel that released in stage 2 is mixed with the decomposed gaseous fuel released in this stage, then this mixed gaseous fuel surrounds the fuel blend. For the DTG curve, a significant peak can be seen at 610°C. It is due to a further decomposition of the relative heavyweight components and a subsequent reaction with the oxygen under the effects of the combustion heat energy and the chain effect of the current stage. In the stage, all three curves remain relatively stable at temperatures higher than 675°C. The ignition and burnout temperatures are 610°C and 681°C, respectively; and the total weight loss is 10.075 wt.%.

3.4 Combustion Characteristic Parameter

Figure 2 shows the ignition and burnout temperature profile of the fuel blends with different SPO:HFO ratios (0:100, 20:80, 40:60, 50:50, 60:40, 80:20, 100:0). As shown, the ignition and burnout temperatures of the LWC in the SPO (T_{iL}=200°C and T_{eL}=466°C) are lower than those in the HFO (T_{iL}=250°C and T_{eL}=505°C). As a result, the values of C and S of the LWC in the

SPO (C_L=9.603×10^{-7}, S_L=4.545×10^{-5}) are higher than those in the HFO (C_L=5.512×10^{-7}, S_L=2.057×10^{-5}); as shown in Figure 3.

In the experiments of Li [7], it is shown that SPO contains certain amounts of 3-Methyl Indole, toluene, and many other components having boiling point between 80°C to 265°C. Such low boiling point components might constitute the LWC released in the LWC vaporization period of the present SPO/HFO fuel blends, and thus reduce their ignition temperature.

Figure 2 also shows that the ignition temperature of the HWC in the SPO (T_{iH}=563°C) is lower than those in the HFO (T_{iH}=581°C). However, the burnout temperature of the HWC in the SPO (T_{eH}=706°C) is higher than that in the HFO (T_{eH}=669°C). As a result, the values of C and S for the SPO (C_H=0.26×10^{-7}, S_H=0.016×10^{-5}) are lower than those for the HFO (C_H=0.38×10^{-7}, S_H=0.021×10^{-5}) for the HWC.

Besides, comparing the results of the Figure 2 and Figure 3, it can be found that a higher SPO ratio results in a lower ignition temperature (for both the LWC and the HWC), a lower burnout temperature of the LWC, and a higher burnout temperature of the HWC. The flammability index (C) and combustion characteristics index (S) both increase with an increasing SPO addition in the second stage (LWC); particularly for SPO ratios greater than 50 wt.%. However, in the third stage (HWC), both indexes are essentially insensitive to the SPO content.

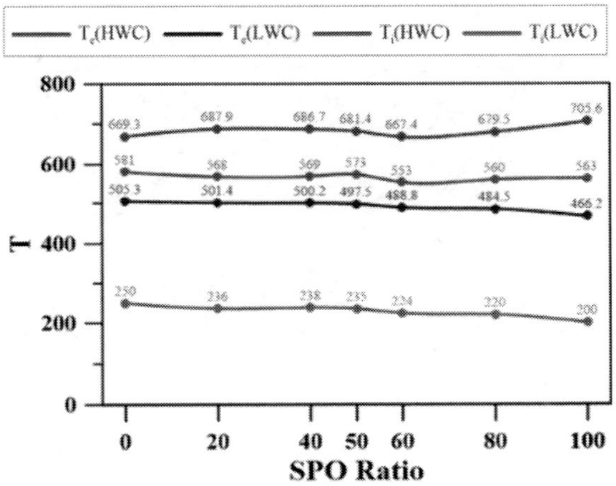

Figure 2: Ignition and burnout temperature profile of blended fuels.

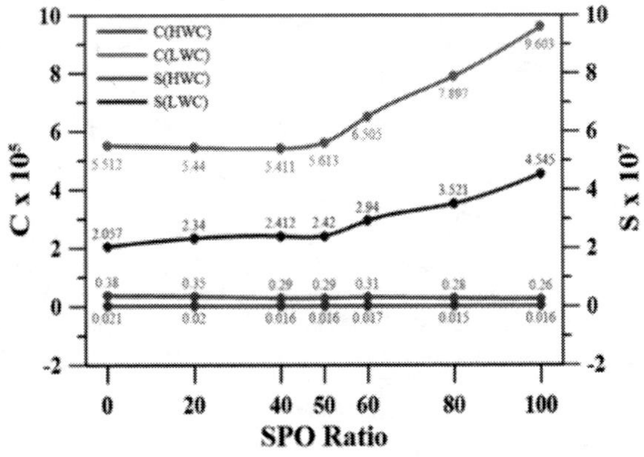

Figure 3: Combustion characteristic parameters profile of blended fuels.

4 Conclusions

The TGA results show that the weight loss of the fuel comprises three stages. A dewaterization stage that is mainly characterizes by moisture evaporation, a volatile decomposition and burning stage where lightweight components in the fuel are released and burned, and a combustion stage associated with the release and burnout of the heavyweight components (HWC) that are in the fuel and a significant heat release.

Fuel blends with a greater SPO content have a higher flammability index and combustion characteristics index in the volatile burning stage as low boiling point components of the SPO are released, and consequently reduce the ignition temperature; particularly when the SPO ratio exceeds 50%. However, in the combustion stage, all of the fuel blends have a similar flammability index and combustion characteristics index irrespective of the SPO content. Generally speaking, the addition of SPO to HFO enhances the combustion characteristics of fuel blends.

5 Acknowledgment

This work was supported by the Ministry of Science and Technology, Taiwan, R.O.C., under contract MOST 105-ET-E-006-003-ET.

References

1. M. Lundin, M. Olofsson, G. J. Pettersson, H. Zetterlund, Resources, Conservation & Recycling 41 (4) (2004) 255-278.
2. N. Kress, B. Herut, B.S. Galil, Mar. Environ. Res. 57 (3) (2004) 213–233.
3. M.F. Gómez-Rico, R. Font, A. Fullana, I. Martín-Gullón, J. Anal. App. Pyrolysis 74 (1) (2005) 421-428.
4. P. Thipkhunthod, V. Meeyoo, P. Rangsunvigit, B. Kitiyanan, K. Siemanond, T. Rirksomboon, Chemosphere 64 (6) (2006) 955-962.
5. M. Inguanzo, A. Domınguez, J.A. Menéndez, C.G. Blanco, J.J. Pis, J. Anal. App. Pyrolysis 63 (1) (2002) 209-222.
6. I. Fonts, G. Gea, M. Azuara, J. Ábrego, J. Arauzo, Renew. Sustainable Energy Rev. 16 (5) (2012) 2781-2805.
7. J. W. Li, *A Study of Thermal Process for Sewage Sludge Pyrolytic Oil and its Combustion Characteristic*, Master's thesis, Department of Aeronautics and Astronautics, National Cheng Kung University, Taiwan, R.O.C., 2016.
8. Y. Wang, G. Chen, Y. Li, B. Yan, D. Pan, Waste Management 33 (11) (2013) 2408-2415.
9. J.J. Lu, W.H. Chen, Appl. Energy 160 (2015) 49-57.
10. G.B. Chen, Y.H. Li, C.H. Lan, H.T. Lin, Y.C. Chao, Appl. Therm. Eng. 114 (2017) 1053-1063.
11. J. Cheng, X. Sun, Power Engineering 5 (1987) 002.
12. Y. Jin, Y. Li, F. Liu, Front. Environ. Sci. Eng. 10 (1) (2016) 201–210.

11th Asia-Pacific Conference on Combustion,
The University of Sydney, NSW Australia
10th -14th December 2017.

A pilot-scale study on the oxidation of ventilation air methane (VAM) using ilmenite

Priscilla Tremain[1], Andrew Maddocks[1], Behdad Moghtaderi[1]

[1]Priority Research Centre for Frontier Energy Technologies and Utilisation, The University of Newcastle
University Drive, Callaghan, NSW, 2038, Australia

Abstract

Ventilation air methane (VAM) is a dilute stream of methane expelled from underground coal mines in large ventilation shaft air flows (100-600 m^3/s). Fugitive VAM emissions pose a significant carbon liability to underground coal mines. To overcome the shortcomings of current VAM abatement technologies, the VAM chemical looping oxidizer (VAMCO) was developed at The University of Newcastle, Australia. To demonstrate the concept at scale, a pilot-scale 1 m^3/s VAMCO unit was constructed. In this study, ilmenite, a naturally occurring iron and titanium rich mineral sand was used as the bed material for VAM abatement in the 1 m^3/s VAMCO unit. Two sets of experimental trials were completed. Firstly, commissioning of the VAMCO unit was completed with ilmenite at bed temperatures of 630-740°C, for which methane conversions of 93-100% were achieved. Secondly, extended run time experiments were performed to examine ilmenite deactivation. No change in ilmenite reactivity was observed over the 500 hours of operation, and 100% methane conversion was achieved in all extended run time experiments.

1 Introduction

Methane is the second highest greenhouse gas emitted by anthropogenic activities after carbon dioxide. It is a potent, short lived greenhouse gas with a global warming potential 25 times greater than carbon dioxide. In the coal mining industry, methane emissions are an economic and environmental liability. In particular, fugitive emissions of ventilation air methane (VAM) from underground coal mines are a major carbon liability.

VAM is the dilute stream of methane expelled from underground coal mines in large ventilation shaft air flows (100-600 m^3/s). The large volumes and dilute nature of VAM pose challenges for its mitigation. A number of technologies are currently available for VAM mitigation such as:

- Thermal and catalytic flow reversal reactors [1, 2];
- Catalytic monolith reactors [3, 4];
- Catalytic lean-burn gas turbines [4].

However a number of shortcomings exist with these technologies such as high operating temperatures, sensitivity to fluctuations in VAM concentration and contaminants in the VAM stream [5, 6]. To overcome the shortcomings of current VAM abatement technologies, a novel VAM chemical looping oxidizer (VAMCO) for mitigation and/or utilization of methane was developed. VAMCO was designed with the view of developing an innovative solution that is inherently safe, can be used over a wider range of VAM concentrations, is insensitive to fluctuations in the VAM concentration, and reduces the energy footprint and capital cost of VAM mitigation.

At the laboratory scale, the VAMCO process primarily consists of a fluidized bed of metal oxide particles. Upon the introduction of VAM as the fluidizing gas, the metal oxide particles undergo simultaneous redox reactions oxidizing VAM to form carbon dioxide. During extensive laboratory scale investigations, iron oxide (Fe_2O_3) as the active metal supported on alumina (Al_2O_3) at 20% loading was found to be the most promising for VAM oxidation.

Following successful completion of laboratory scale trials, a pilot-scale 1 m^3/s VAMCO unit was designed and constructed at The University of Newcastle, pictured in Fig. 1.

Figure 1: Pilot-scale 1 m^3/s VAMCO unit at the University of Newcastle.

A final consideration for the operation of the pilot plant was the choice of the metal oxide particles to be used in the fluidized bed. At the laboratory scale, 20% Fe_2O_3/Al_2O_3 metal oxides were synthesized via wet impregnation methods to produce up to 50 g of material. However at pilot-scale, 1-2 tonnes of metal oxide particles were required, for which wet impregnation methods incur a significant cost. As such, an alternative metal oxide material was sought.

Ilmenite is a naturally occurring iron and titanium rich mineral sand. Titanium oxide is the major component of ilmenite, typically 55-65%, while iron oxide is typically between 25-40%, which is comparable to the iron content of synthesized Fe_2O_3/Al_2O_3 particles. The remaining ilmenite components consist of silica and oxides of other metals such as manganese, aluminum, magnesium and chromium. Ilmenite is available at a relatively low cost ~$1000/tonne. Hence, ilmenite was chosen as the bed material for the pilot-scale VAMCO unit, primarily due to its low cost compared with the synthesized iron oxide material. Additionally, ilmenite (or more specifically iron oxide within ilmenite) has been shown to be a suitable oxygen carrier for chemical looping reactions [7-9].

Corresponding author.
E-mail address: behdad.moghtaderi@newcastle.edu.au

In this study, the results of commissioning of the pilot-scale 1 m^3/s VAMCO unit with ilmenite are presented. Extended run time experiments with ilmenite were also conducted. For all experiments VAM conversion efficiency is presented.

2 Experimental Methods

A simplified process flow diagram of the VAMCO pilot-scale unit is presented in Fig. 2. Air from a blower is passed through the heat exchanger prior to entering the combustion chamber. The maximum achievable air flowrate was 1 Nm^3/s. A natural gas burner provides additional heat to the incoming air within the combustion chamber. Natural gas was injected into the hot gas stream immediately before the plenum to simulate the VAM stream. For safety purposes an interlock in the PLC limited methane injection to 2 vol.%, which is well below the lower explosive limit of methane of 5 vol.%. From the plenum, the VAM (methane-air) mixture flowed upward through the distributor plate and into the fluid bed.

Above the fluid bed a series of lamella plates were installed to allow for greater flowrate flexibility and retention of particles in the bed. The gas exiting the fluid bed and lamella sections then passes through a cyclone to return any entrained particles to the fluid bed via a loop seal (LS). The gas from the cyclone passes through the heat exchanger to transfer heat to the incoming air before being vented via a baghouse.

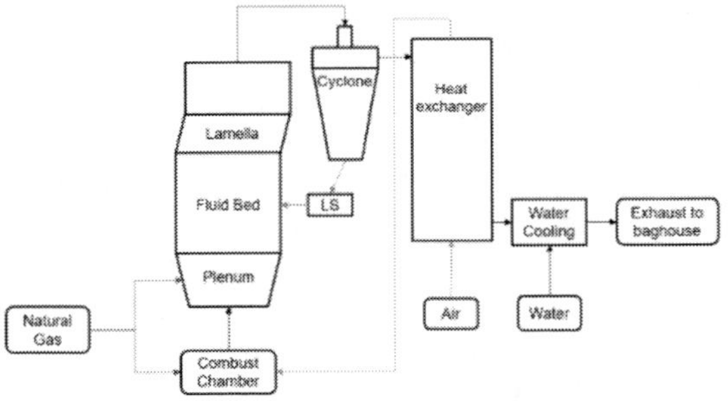

Figure 2: Simplified process flow diagram of the VAMCO pilot-scale unit.

2.1 Bed material

Ilmenite with a mean particle size of 161 μm was used as the bed material for commissioning and extended run time experiments. X-ray fluorescence (XRF) of the raw ilmenite sample is presented in Table 1.

Table 1: XRF analysis of raw ilmenite sample

Element	Raw Ilmenite
Al_2O_3	1.15
Fe	30.47
TiO_2	63.86
MgO	1.91
Ba	0.05
CaO	0.08
Cr_2O_3	0.67
Mn	0.79
Na_2O	0.11
P	0.05
S	0.02
SiO_2	0.58
V	0.15
Zn	0.03
Zr	0.08

2.2 Commissioning experiments

Commissioning of the 1 m^3/s VAMCO unit with ilmenite was performed over a period of 10 days. The first-time operation of the pilot plant was successful, with some trial-and-error operating undertaken during the early period of operation to fully understand the operational requirements of the plant.

For the commissioning experiments an ilmenite inventory of 2 t was loaded into the reactor prior to start up. Once the particles were loaded and the reactor was sealed, a series of pre-start up checks were conducted including; check of air blower oil level, check natural gas compressor oil level, confirm water supply, confirm power supply and confirm compressed air supply for instruments.

Start-up consisted of heating the bed to the desired temperature before the VAM stream was introduced and experimentation commenced. The start-up procedure was conducted over a period of three days as the bed temperature was slowly heated from ambient temperature to 720°C. This process is illustrated in Fig. 3. Once the bed temperature reached approximately 600°C, there was difficulty in continuing to heat up the bed, as evidenced in Fig. 3. A small amount of methane was introduced to the plenum in an attempt to speed up the heating process. Consequently the temperature in the plenum (located below the fluid bed and distributor plate) increased above the combustion chamber temperature up to approximately 850°C. This significant increase in plenum temperature allowed the bed to heat faster to the desired bed temperature of 720°C. Once the desired bed temperature was reached, the methane supply to the VAM stream was shut off and the input from the burner reduced to allow the plenum to cool below the auto ignition temperature of methane (560°C).

Once the desired bed temperature was reached the VAM stream was then reintroduced in order to maintain the bed at the desired operating temperature. Gas sampling points were located in the plenum, bed and outlet of the reactor. Methane and carbon dioxide concentrations were monitored periodically over the course of each experiment. A matrix of the bed temperatures and flowrates investigated during the commissioning experiments are presented in Table 2. Each experiment was completed for a minimum of 8 h of continuous operation.

Table 2: Experimental conditions investigated using the pilot-scale VAMCO unit.

Zone (see Fig. 3)[a]	Bed temperature (°C)	Air flow (Nm^3/s)	Air flow @ temperature (m^3/s)
A	740	0.53	1.8
B	735	0.56	1.9
C	720	0.56	1.9
D	720	0.54	1.8
E	645	0.56	1.7
F	665	0.60	1.9
G	630	0.61	1.8
H	735	0.59	2.0

a. At least 8 hours of continuous operation completed for each zone.

2.3 Extended runtime experiments

Following the successful completion of commissioning of the 1 m^3/s VAMCO unit with ilmenite, a series of extended runtime experiments were completed to investigate ilmenite deactivation over time. For the extended run time experiments, an inventory size of 1 t was utilized. A summary of the experimental conditions examined is presented in Table 3.

Table 3: Extended run time experimental conditions for 1 t ilmenite inventory size using the pilot-scale VAMCO unit.

Exp.	Bed temperature (°C)	Air flow (Nm³/s)	Air flow @ temperature (m³/s)	Run time (h)
1	700	0.33	1.1	306
2	660	0.33	1.0	143
3	660	0.50	1.6	43

3 Results and Discussion

3.1 Commissioning experiments

For the commissioning experiments, the combustion chamber temperature, plenum temperature, bed temperature and air flowrate (at temperature) over the course of the 10 day period are presented in Fig. 3. The different experimental zones are also labelled in Fig. 3. The start- up period was conducted over three days with the bed temperature reaching 600°C relatively quickly (one day) and a further two days being required to heat the bed to 720°C.

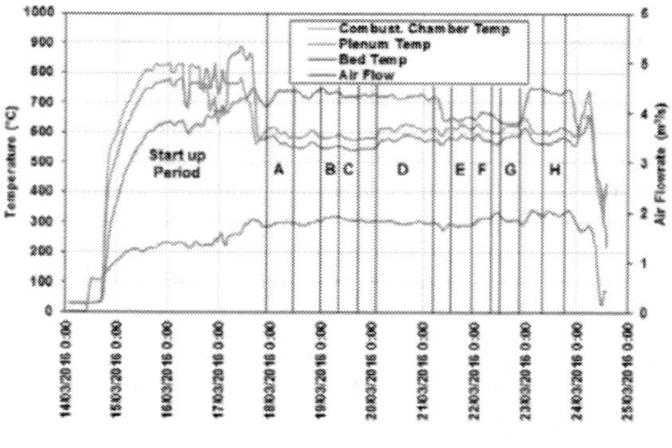

Figure 3: Reactor temperatures and air flowrate profile over 10 day commissioning period for an ilmenite inventory of 2 t.

For the duration of experiments, the methane and carbon dioxide concentrations at the inlet and outlet of the VAMCO reactor were monitored. This allowed for the determination of methane conversion. Table 4 presents the methane conversion achieved for each of the experiments performed during commissioning. Initially a conversion of 93% was achieved at a bed temperature of 740°C. However, contrary to what was expected, the methane conversion increased for the subsequent experiments at lower temperatures. For all subsequent experiments, methane conversion of >99.5% was achieved.

Table 4: Methane conversion for each of the experimental conditions examined for 2 t ilmenite inventory.

Zone*	Bed temp (°C)	Air flow (m3/s)	Conversion (%)
A	740	1.8	93.1
B	735	1.9	99.5
C	720	1.9	99.5
D	720	1.8	99.5
E	645	1.7	100
F	665	1.9	100
G	630	1.8	100
H	735	2.0	100

The high rate of methane conversion achieved for the latter experiments was attributed to the activation of ilmenite. It has been reported that activation of ilmenite in air at temperatures above 700°C change the active phase on the surface of the ilmenite particles from $FeTiO_3$ to Fe_2O_3 and TiO_2 [7]. Hence presence of Fe_2O_3 allowed for greater oxidation of methane at lower temperatures.

3.2 Extended run time experiments

Following the successful commissioning experiments, a series of extended run time experiments were performed to investigate changes to ilmenite reactivity over time. Three operating conditions were investigated and results of these experiments are presented in Table 5.

Table 5: Summary of results for ilmenite trials.

Bed Temp (°C)	Air flow (Nm³/s)	Air flow @ temp (m³/s)	Run time (h)	Conv. in bed (%)	Conv. at outlet (%)
700	0.33	1.1	306	85%	>99%
660	0.33	1.0	143	77%	>99%
660	0.5	1.6	43	65%	>99%

For the first extended run experiment, a total of 306 h of stable operation was completed at 700°C and 0.33 Nm³/s with an average outlet methane conversion of greater than 99% achieved. During operations it was observed that the temperature of the bed material was lower than the temperature of the region between the bed material the inclined lamella plates. This phenomenon was also observed during the commissioning with ilmenite. The temperature difference is a result of oxidation of methane that bypasses the bed. The oxidation mechanism is not clearly understood and could be either due to:

- Gas-solid contact with entrained particles (i.e. particles ejected from the surface of the fluidized bed) returning to the bed from the inclined plates, or
- Gas phase reactions between methane and oxygen, or
- A combination of the above two mechanisms.

Regardless of the oxidation mechanism, methane conversion based on outlet gas analyzer measurements was greater than 99%. The quantity of methane required to be oxidized to increase from the bed temperature to the above bed temperature was calculated and used to determine the methane conversion in the bed, by subtracting the methane oxidized above the bed from the incoming methane quantity. At a bed temperature of 700°C and flowrate of 0.33 Nm3/s, the in bed methane conversion was approximately 85 % (Fig. 4).

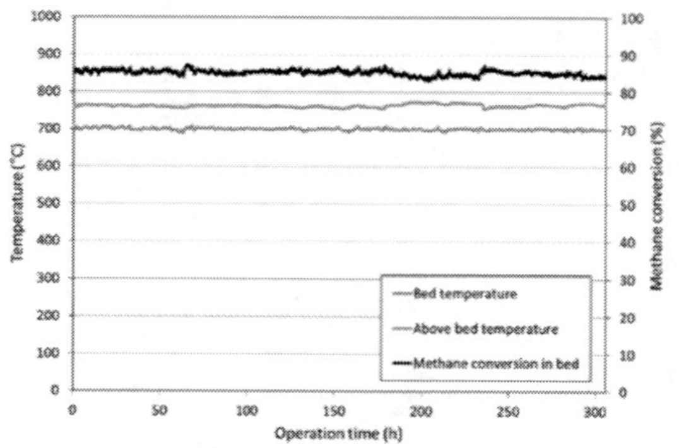

Figure 4: Reactor temperatures and in bed methane conversion as a function of operational time for a bed temperature of 700°C and flowrate of 0.33 Nm³/s.

Following successful continuous operation for 300 h at 700°C and 0.33 Nm^3/s, the bed temperature was decreased to 660°C and the flowrate maintained at 0.33 Nm^3/s. As shown in Fig. 5, an increase in the temperature difference between the bed and above bed region was observed and an in-bed conversion of approximately 77 % was achieved. The lower in-bed conversion (and consequential increase in temperature difference) was most likely due to a reduction in the rate of oxidation at lower temperatures. Although the in bed conversion decreased with a decrease in in-bed temperature, full methane conversion (>99 %) at the reactor outlet was maintained for the entire 143 h of stable operation at 660°C and 0.33 Nm^3/s.

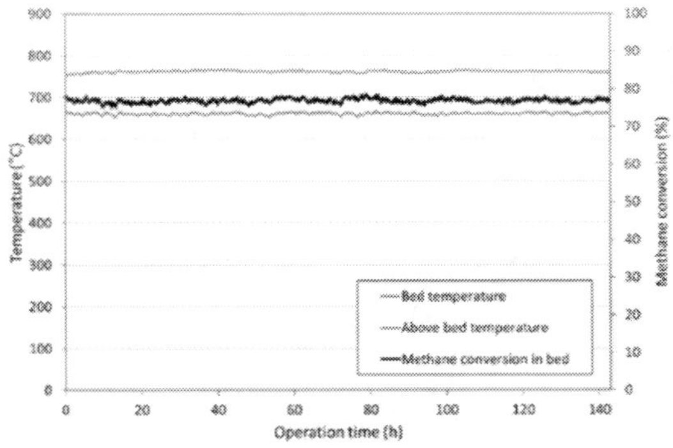

Figure 5: Reactor temperatures and in bed methane conversion as a function of operational time for a bed temperature of 660°C and flowrate of 0.33 Nm^3/s.

A final experiment was conducted at a bed temperature of 660°C and 0.5 Nm^3/s. As can be seen in Fig. 6, the increased flowrate resulted in a temperature differential of approximately 160°C between the bed and above bed region and an in-bed methane conversion of 65 %. Increasing the flowrate, increases the quantity of gas that bypasses the bed. However similar to the other conditions tested, full methane conversion (>99 %) was achieved at the reactor outlet for the entire 43 h of stable operation at 660°C and 0.5 Nm^3/s.

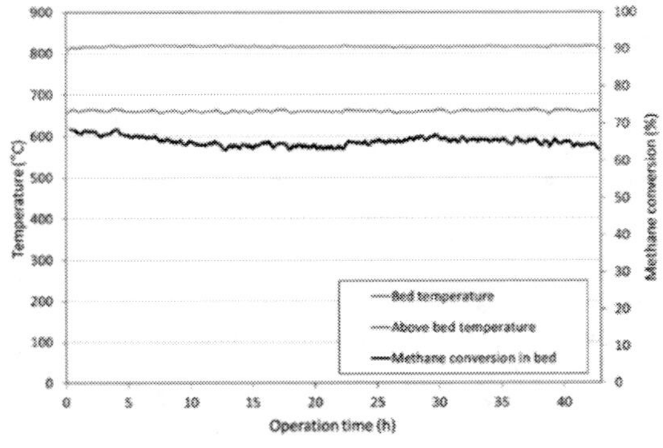

Figure 6: Reactor temperatures and in bed methane conversion as a function of operational time for a bed temperature of 660°C and flowrate of 0.5 Nm^3/s.

4 Conclusions

A pilot-scale 1 m^3/s VAMCO unit was constructed at the University of Newcastle for VAM abatement. Commissioning of the plant with ilmenite revealed full conversion of methane can be achieved in the VAMCO unit after an initial activation phase. As a result of the extended run time trials, in which 492 h of stable operation was achieved, it was concluded that ilmenite did not undergo deactivation and that full conversion of methane can be maintained in the VAMCO unit. On this basis, it was determined that ilmenite, which is a low cost natural mineral, could be used as the bed material for scale up of the VAMCO technology to demonstration scale, which will be conducted as future work to this study.

5 Acknowledgments

The authors wish to acknowledge the financial support provided by ACALET, DIIS and The University of Newcastle, Australia for the work presented in this paper.

References

[1] Somers, J.M. and H.L. Schultz, Coal mine methane ventilation air emissions: new mitigation technologies, in Ninth International Mine Ventilation Congress. 2009: New Dehli, India.

[2] Somers, J.M. and H.L. Schultz, Thermal oxidation of coal mine ventilation air methane, in 12th United States/ North American Mine Ventilation Symposium. 2008.

[3] Su, S., et al., Characteristics of coal mine ventilation air flows. Journal of Environmental Management, 2008. 86(1): p. 44-62.

[4] Su, S. and J. Agnew, Catalytic combustion of coal mine ventilation air methane. Fuel, 2006. 85(9): p. 1201-1210.

[5] Zhongqing, Y., Z. Li, and T. Qiang, Research progress in the utilization of ventilation air methane as an energy source. Natural Gas Industry, 2010. 2: p. 33.

[6] Karakurt, I., G. Aydin, and K. Aydiner, Mine ventilation air methane as a sustainable energy source. Renewable and Sustainable Energy Reviews, 2011. 15(2): p. 1042-1049.

[7] Leion, H., et al., The use of ilmenite as an oxygen carrier in chemical-looping combustion. Chemical Engineering Research and Design, 2008. 86(9): p. 1017-1026.

[8] Cuadrat, A., et al., Behavior of ilmenite as oxygen carrier in chemical-looping combustion. Fuel Processing Technology, 2012. 94(1): p. 101-112.

[9] Abad, A., et al., Kinetics of redox reactions of ilmenite for chemical-looping combustion. Chemical Engineering Science, 2011. 66(4): p. 689-702.

11th Asia-Pacific Conference on Combustion,
The University of Sydney, NSW Australia
10th -14th December 2017.

The Study of Fire Resistance of Building Integrated Photovoltaics

Ming-Tsung Yang[1], Yu-Liang Liou[1], Chun-Kuei Chen[2], Hsun-Ku Lee[2], Ta-Hui Lin[1,2], Cho-Fang Tsai[3], Ming-Yuan Lei[3], Tien-Chih Wang[3]

[1]Department of Mechanical Engineering, National Cheng Kung University

No.1, University Road, Tainan City 701, Taiwan (R.O.C.)

[2]Research Center for Energy Technology and Strategy, National Cheng Kung University

No.1, University Road, Tainan City 701, Taiwan (R.O.C.)

[3]Architecture and Building Research Institute, Ministry of the Interior, Taiwan

13F., No.200, Sec. 3, Beisin Rd., Sindian District, New Taipei City 23143, Taiwan (R.O.C.)

Abstract

In this study, the fire and heat performance of PV modules is evaluated using the CNS 12514-1 testing method. In the test of PV module with insulation materials, there are 3 kinds of fire-resistant methods of polysilicon PV modules. Polysilicon PV modules with a calcium silicate board, a phenova insulation board, a gypsum board as exposed surfaces, respectively, were tested. There were wire holes on the back board of the PV modules in the three designs. The holes degraded fire resistance. With the aluminum and gypsum back board, the PV modules caught fire and their integrity failed. To enhance the fire and heating performance of PV modules, it is necessary to further improve the designs. The PV module with the calcium silicate board had no flames spill out during the test, but the surface temperature was higher than that of the CNS-12514-1 specification.

1 Introduction

In building energy generation, solar photovoltaic (PV) systems are the main equipment. Building-integrated photovoltaics (BIPV) have become increasingly popular. There has been many research on the heat dissipation, integral designs, and so on for BIPV. However, past research studies on the BIPV's fire-stop performance in certain walls and are very limited. [1-3]

According to Building Technical Regulations Article 70 and 79 [4], if general commercial solar modules are directly applied to buildings, there will be fire safety problems. Solar PV systems have caused fires due to short circuits, arcs, and wire fires. Therefore, it is worthy to further study the fire safety of BIPV. The fire resistance of BIPV can be increased to more than 30 min using exterior wall or roof designs. Expanding the application of PV panels for buildings is an important issue. This research assesses BIPV module fire resistance for buildings and develops fire-resistant PV modules to reduce the gap between the safety rules of PV modules and construction codes, and expand the applications of PV modules for buildings. Yang et al. [5] improved the use of the light reflection high principle, for PV modules. The insulation film has reflectivity behind the light transmission module, so that the power generation module generates secondary power generation, enhancing power generation efficiency and providing insulation.

Corresponding author. Fax: +886-6-2352973
E-mail address: thlin@mail.ncku.edu.tw

To avoid the thermal storage problems of PV modules, Yang et al. increased the electrical storage capacity of the system. Then, they conducted a fire resistance test for both sides of the PV module in a small-scale fire resistance furnace to simulate indoor fire spread and outdoor fire spread. The results show that the PV module has a certain fire resistance, compared with general glass, and that the damage time from 30 seconds to 20 minutes up and down.

Murata et al. [6] improved a commercial PV module whose size was 2 m× 4 m. An intermediate air layer was used between the metal separator and the PV module and the module was surrounded by fire-resistant silicone. The results show that when the maximum temperature of the metal shield and other shelters was 840 °C and the loading pressure reached 637 Pa, the PV module maintained its complete shape without damage, indicating that it is a general fire-resistant building material with fire and heat resistance if not directly exposed to fire.

2 Experimental Equipment and Method

The full-scale heating furnace used in this study is mainly for general fire-resistant doors, walls, and fire-resistant glass, in accordance with the standard temperature curve of CNS 12514-1 testing method. The inner dimensions of the heating furnace are 4.3 m (W) × 4.5 m (H) × 1 m (L).

All three polysilicon PV modules have dimensions of 164.5 cm (L) × 98.5 cm (W) × 4.5 cm (H). The PV modules are combined with a fire-resistant material and a fire-resistant backplane (phenova insulation, gypsum, and calcium silicate board). The fire-resistant design of PV module as shown in Figure 1 and Figure 2.

Phenova insulation board is a very versatile material suitable for a large range of applications where thermal performance, moisture resistance, and fire performance is key performance criteria. It has low smoke toxicity, low thermal conductivity (0.02~0.03 W/m-K), good heat insulation, difficult-flammability, and high chemical resistant. It is a green building material. The limit operating temperature of phenova insulation board is 180 °C. There will be a fire resistance test for polysilicon PV module with a 30 mm-thick phenova insulation board and a 1 mm-thick aluminum board (Its thermal transmittance is 0.511 W/m^2-K.)

Gypsum board and calcium silicate board are both commonly used green building material. These materials were selected because of its good heat insulation, low thermal conductivity, and durability. There will be a fire resistance test for polysilicon PV module with a 15 mm-thick gypsum board and a 1 mm-thick aluminum board (thermal transmittance is 1.730 W/m2-K.); and another polysilicon PV module with a 9 mm-thick calcium silicate board and a 1 mm-thick aluminum board (thermal transmittance is 1.066 W/m2-K.).

The PV module fire resistance tests conformed to the CNS 12514-1 standard [7] to test with the full-scale heating furnace. Figure 3 shows the positions of the temperature measurements for Design 1-3. The 5 red points measured the temperature of the watch glass unexposed surface of Design 1 using infrared thermography. Since the test for Design 2 and Design 3 were done simultaneously, only two points were measured by the thermography for each design; both positions were shown as the black points. The test time was 2 h. The test was stopped earlier if the specimens broke or there were some risks. Photographs were taken during and after the test. Finally, the data were analyzed and the specimen fire integrity and insulation were evaluated.

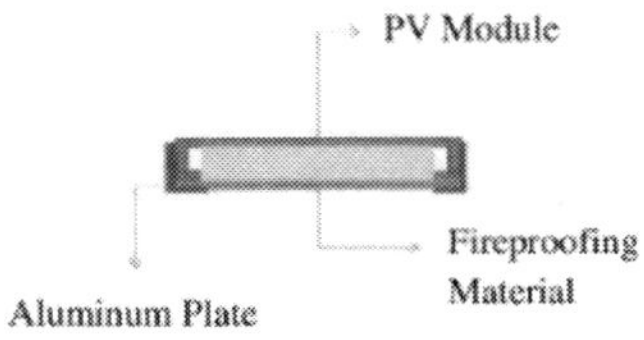

Figure 1: Fire-resistant Design of PV Module.

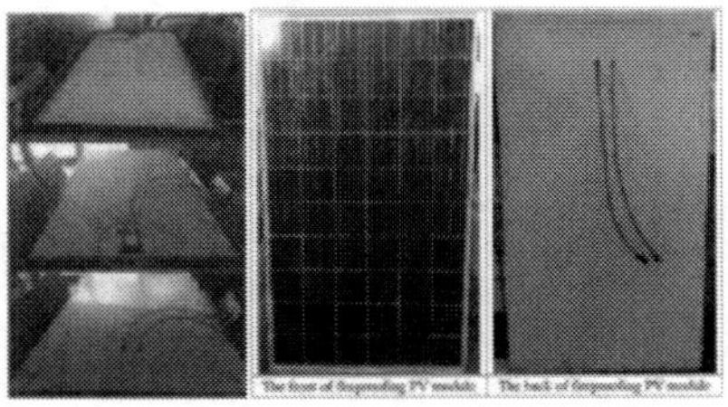

Figure 2: PV Module Image for three designs.

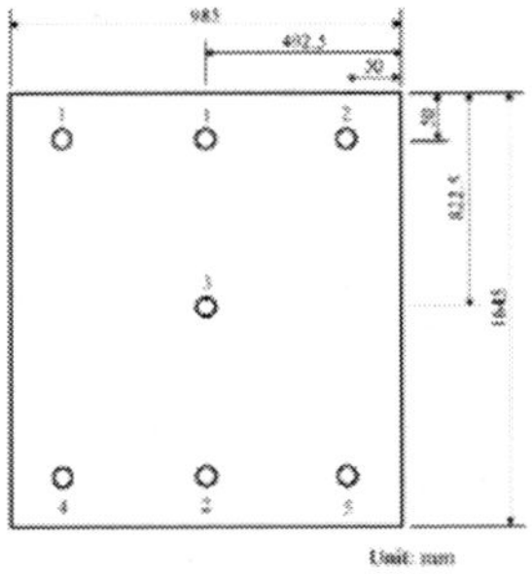

Figure 3: Measurement points of Infrared Thermograph on Unexposed Surface

3 Results and Discussion

3.1 Fire Resistance Test of PV Module with Phenova Insulation Board (Design 1)

The overall test time was 1 h 3 min. At 14 min, flames spit out for 10 s at the top of the frame, indicating that the fire integrity of the PV module failed. Severe deformation on the front glass of that PV module occurred. The overall process and events of the PV module test are shown in Figure 4. At 3 min 37 s, smoke appeared above the PV module and the surface of the glass at point 2 deformed and the amount of smoke became large. Figure 5 shows that there is a high-temperature region at the top of the PV module. The back of this area is the hole of the wire penetrating the phenova insulation board. The hole degraded fire resistance. At 14 min, flames sprung up for 10 s on the top of the PV module. From 16 min, the failure of the fire-resistant material exposed the polycrystalline silicon wafers of the PV module to heat, so they ruptured. The fire-resistant material then began to significantly burn and peel, and the polycrystalline silicon wafers continued to break down. At 42 min 20 s, the watch glass of the PV module began to deform due to heat. At 1 h 3 min, there was excessive deformation of the watch glass and thus the test ended.

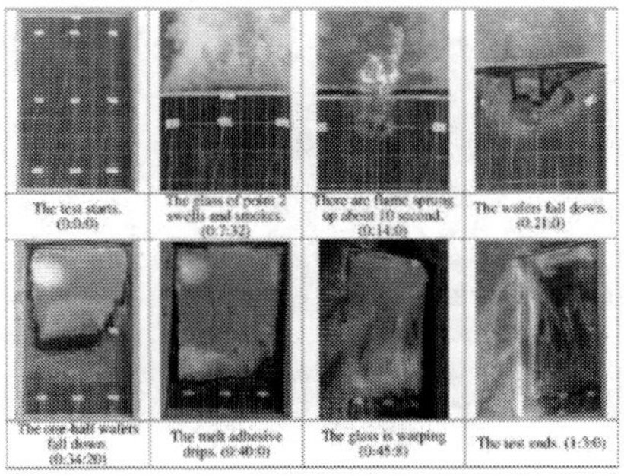

Figure 4: Image of PV Module During Fire Resistance Test (Phenova Insulation Board)

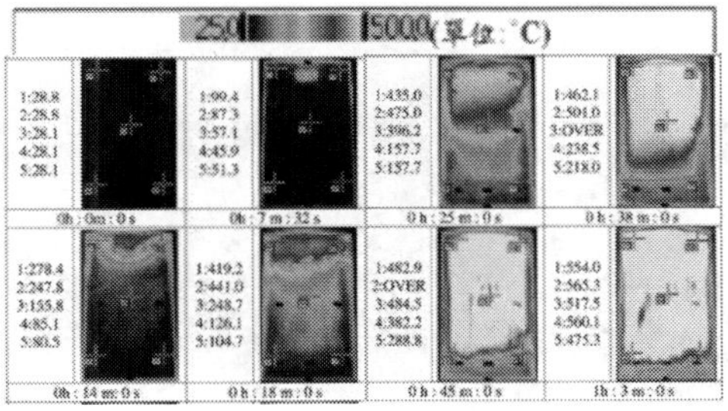

Figure 5: Process of the PV Module During Fire Resistance Test Imaged using Infrared Thermograph (Phenova Insulation Board)

The limit operating temperature of phenova insulation board was 180 °C. See Figure 6, the exposed surface of phenova insulation board impact over 350 °C at which phenova insulation board would be pyrolyzed. Because the pyrolysis, the heat resistance of pyrolyzed phenova insulation board was worse than that of perfect phenova insulation board. So, this is a reason the fire resistance of this design failure.

Figure 6 shows the temperature changes obtained from infrared thermography. As the reason for the wire hole behind point 2, it is found that the hole would be a defect for fire in the course of the test, measured the surface temperature of this area begins to rise. The temperature of the upper part of PV module thus rises faster than that of the lower part. Figure 6 also shows that the temperature changes at all points are similar in the first 5 min. However, after 15 min, the temperature changes for each point diverge. The rates of temperature rise at points 1, 2, and 3 are faster than that at point 4. The final surface temperature of the PV module is about 500 °C.

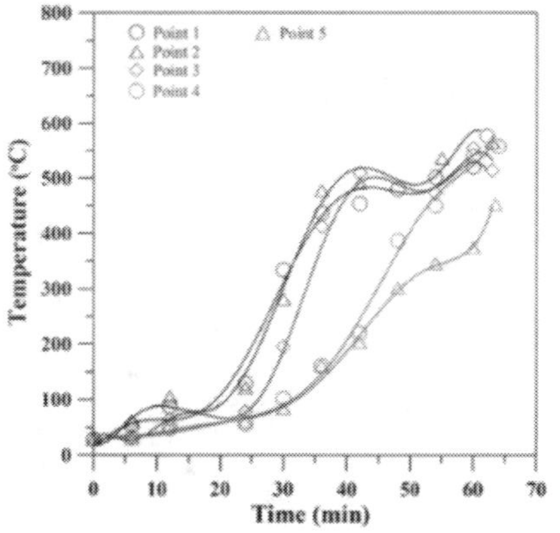

Figure 6: Temperature Profile of Each Point on Unexposed Surface

The CNS 12514-1 test standard [7] was applied to determine the fire and heat resistance of the PV module. At 14 min, flames sprung up for more than 10 s, so the fire integrity of the module failed. At the end of the test, the surface temperature of the PV module was about 520 °C. For the CNS 12514-1 standard [7], the temperatures of unexposed surface exceed 209 °C and the average temperature of unexposed surface also exceeds 179 °C; therefore, the insulation failed. So, the fire resistance of the design failed.

3.2 Fire Resistance Test of PV Module with Gypsum Board (Design 2)

The total time of the test was 1 h 18 min. At 21 min 15 s, the upper side of the PV module had flames, which continued for 1 min 40 s. The fire integrity of the PV module thus failed. The test ended at 1 h 18 min because the watch glass of the PV module ruptured and flames sprung up. The process and events are shown in Figure . At 10 min 14 s, there was a lot of white smoke above the PV module. At 19 min 5 s, the upper glass of the PV module deformed. The temperature was the highest at the upper part of the PV module, as shown in Figure 8. This result is the same as that for design 1. The wire hole degraded fire resistance. At 21 min 15 s, the top of the PV module had flame spring up for more than 10 s, and the time is about 1 min 40 s. At 30 min 12 s, the upper glass began to deform due to heat. The rate of the temperature rise of the upper part eventually became higher than that of the lower part. At 1 h 18 min, the watch glass of the PV module ruptured, making flames spring up, and thus the test ended.

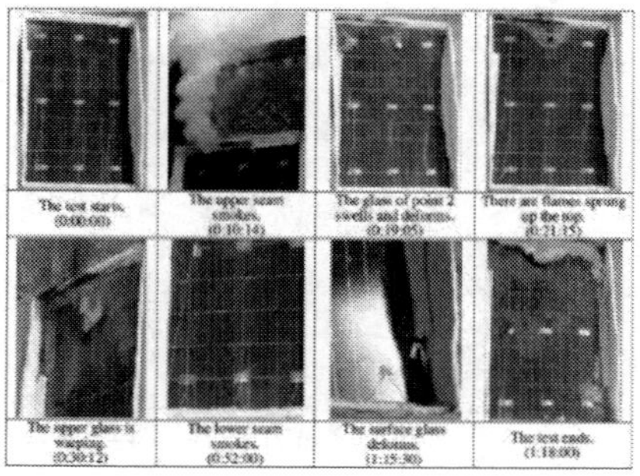

Figure 7: Image of the PV Module During Fire Resistance Test (Gypsum Board)

See Figure 8, the temperature rise of the exposed surface was lower than that in the design 1. Because, in high heat all the water had been released from its core, gypsum board continued to serve as a heat-insulating barrier on the exposed surface. But the strength of gypsum board was not enough to resist high heat for a long time, gypsum board still broke in the test.

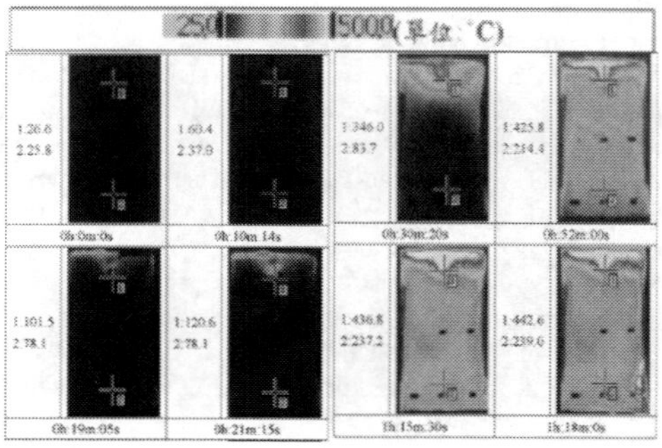

Figure 8: Process of PV Module During Fire Resistance Test Imaged using Infrared Thermograph (Gypsum Board)

Figure 9 shows the temperature change recorded using infrared thermography. The temperature rise rate of the upper part is faster than that of the lower part after 15 min. After 45 min, the temperature of the upper part remains at about 420 °C; that of the lower part remains at about 200 °C after 55 min.

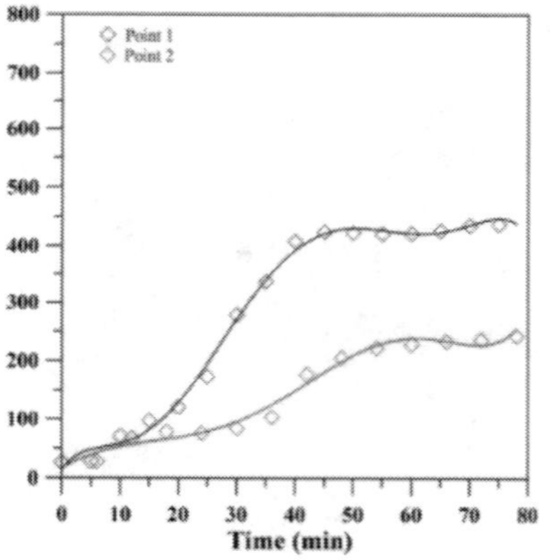

Figure 9: Temperature Profile of Each Point on Unexposed Surface (Gypsum Board)

CNS 12514-1 [7] was used to determine the performance of the design. Fire sprung up on the unexposed surface for 10 s at 21 min 15 s, indicating fire integrity failure. At 26 min, the temperature at point 1 increased to about 220 °C. At the end of the test, the temperature at point 1 was about 420 °C and the average temperature of unexposed surface also exceeds 179 °C, indicating insulation failure. So, the fire resistance of the design failed.

3.3 Fire Resistance Test of PV Module with Calcium Silicate Board (Design 3)

The test procedure and events are shown in Figure 10 and Figure 11. At 17 min 40 s, the upper surface of the PV module was deformed by heat and the polysilicon wafers of this part peeled off. At 30 min 55 s, the watch glass became discolored. At 51 min 10 s, the underside of the PV module began to smoke. The PV module did not catch fire in the test, with only discoloration of the watch glass and some deformation.

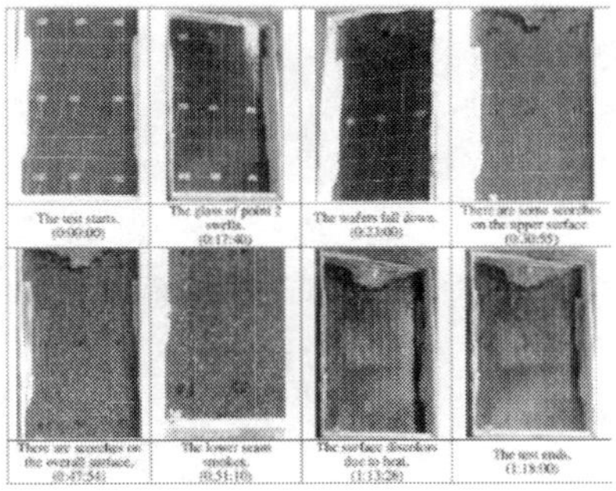

Figure 10: Image of PV Module During Fire Resistance Test (Calcium Silicate Board)

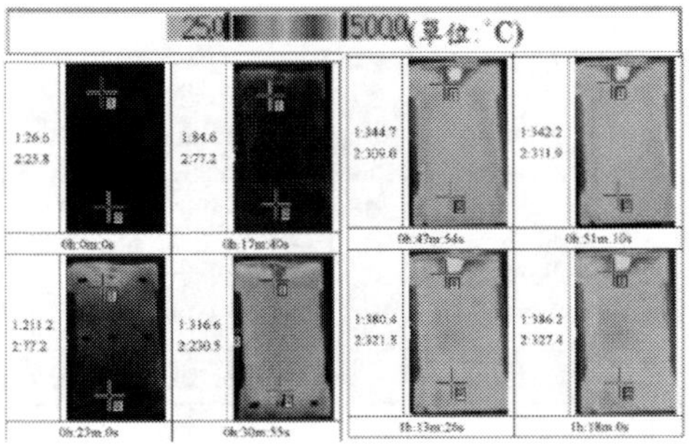

Figure 11: Process of PV Module During Fire Resistance Test Imaged using Infrared Thermograph (Calcium Silicate Board)

Figure 11 shows the temperature changes at each point measured using infrared thermography. The temperature rise rate of the upper part is higher than that of the lower part after 10 min. After 40 min, the temperature in the upper and lower parts remained at about 320 °C and 300 °C. The surface temperature change at each point for design 3 has a smaller difference. The strength of calcium silicate board was better than that of gypsum board, and it did not break in the test. But, the temperature rise in the design is higher than that in design 2.

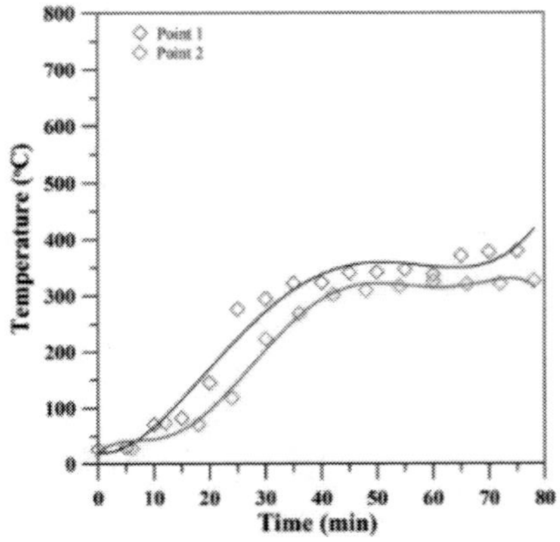

Figure 12: Temperature Profile of Each Point on Unexposed Surface (Calcium Silicate Board)

According to CNS 12514-1 [7], no fire sprung up for the PV module of design 3 in the test. At the end of the test, the surface temperature rose to about 310 °C, exceeding the standard (209 °C) and the average temperature of unexposed surface also exceeds 179 °C. Therefore, the PV module of design 3 has more than one hour of fire resistance, but it does not have one hour of insulation. So, the fire resistance period of the design failed.

4 Conclusions

There were wire holes on the back board of the PV modules in the three designs. The holes degraded fire resistance. With back boards of phenova insulation and gypsum, the PV modules caught fire and their fire integrity failed. To enhance fire and heat performance of PV modules, it is necessary to further improve the designs, especially at the sealing of wire holes and the assembly of PV modules. The PV module with a calcium silicate board had no flames spill out during the test, but the surface temperature was higher than that in the CNS-12514-1 specification. The insulation thus failed. So, the fire resistance of the design failed as well.

5 Acknowledgment

This work was supported by the Architecture and Building Research Institute, Ministry of the Interior, Taiwan, under contract. 10415b0009

References

[1] B. Norton, Solar Energy 85 (2011) 1629-1664.

[2] J. H. Kim, J. T. Kim, Energy procedia, 30 (2012), 1016-1024.

[3] D. Iencinella, E. Centurioni, M. G. Busana, Sol. Energ. Mat. Sol. C. 93 (2009), 206-210.

[4] Building Technical Regulations, Construction and Planning Agency Ministry of the Interior, R.O.C., 2015. [Text in Chinese]

[5] C. H. Young, Y. L. Chen, P. C. Chen, Energ. Buildings 78 (2014) 66-78

[6] K. Murata, T.Yagiura, , K. Takeda, M. Tanaka, S. Kiyama, Sol. Energ. Mat. Sol. C. 75 (2003) 647-653.

[7] CNS 12514-1, "Fire-resistance tests – elements of building construction – part 1: general requirements." National Standards of the Republic of China, 2014. [Text in Chinese].

11th Asia-Pacific Conference on Combustion,
The University of Sydney, NSW Australia
10th -14th December 2017.

The effect of EGR on ignition characteristics of gasoline under lean burn conditions

Masaki Naruke[1], Shoichi Yoshida[1], Yuusuke Wachi[1], Kotaro Tanaka[1], Mitsuru Konno[1]

[1]Ibaraki University

4-12-1 Nakanarusawa, Hitachi, Ibaraki, 316-8511, Japan

Abstract

Exhaust gas recirculation (EGR) and lean burn strategies are used to achieve high thermal efficiency in internal combustion engines. To investigate the effect of EGR on ignition characteristics of gasoline under lean burn conditions, ignition delay times of gasoline and gasoline surrogate mixtures consisting of n-heptane, isooctane, toluene, diisobutylene, and methylcyclohexane were measured using a rapid compression machine under EGR and lean burn conditions at a pressure of 4 MPa, a temperature of 720 K, and an equivalence ratio of 0.5. In addition, the factor that contributed to the change of ignition delay times under the EGR conditions was experimentally investigated. Ignition delay times were also calculated using a detailed chemical kinetic model of gasoline surrogate mixtures. Ignition delay times of gasoline and gasoline surrogate mixtures under lean burn condition with EGR were longer than those without EGR. The most influential factor under EGR conditions was found to be the reduction of fuel concentration in the gas mixtures. When NO, which was included in the EGR gas in the practical engines, was added to the EGR gas mixtures, the ignition delay times of gasoline and gasoline surrogate mixtures decreased compared to those without NO. Based on the chemical kinetic analysis, the reaction of NO with HO_2 to form OH contributed to the decrease of ignition delay times under EGR with NO conditions.

1 Introduction

In spark ignition (SI) engines, exhaust gas recirculation (EGR) and lean burn strategies are used to improve the thermal efficiency. When the EGR level increases, the thermal efficiency increases owing to the reduction in pumping and cooling losses. In lean burn SI engines, the thermal efficiency is also improved owing to an increase in the specific heat ratio of mixtures and reduction of cooling losses.

The thermal efficiency of a SI engine is limited by knock. To realize high thermal efficiency of SI engines using EGR and lean burn strategies, it is necessary to predict the knock phenomenon under EGR and lean burn conditions. Therefore, understanding ignition characteristics of practical fuel under EGR and lean burn conditions is required. Gauthier et al. [1] showed that there appears to be no discernible difference between experiments with and without EGR for gasoline and gasoline surrogate mixtures under pressures of 15 - 25 atm and temperatures of 1031 - 1227 K at an equivalence ratio of 0.5. However, there has been no study conducted on the effects of EGR on the ignition characteristics of practical fuel under lean burn conditions of high pressure and low temperature.

In this study, we focus on the effect of EGR on the ignition characteristics of gasoline under lean burn conditions, at a pressure of 4 MPa, which is equivalent to that in a lean burn boosted engine [2]. In addition, the effects of CO and NO on the ignition characteristics of the gasoline were also investigated.

Kotaro Tanaka. Fax: +81-294-38-5080
E-mail address: kotaro.tanaka.comb@vc.ibaraki.ac.jp

2 Experimental method

2.1 Fuel characterization

The compositions of gasoline and gasoline surrogate mixtures are shown in Table 1. Gasoline surrogate mixtures consist of n-heptane, isooctane, toluene, diisobutylene (DIB), and methylcyclohexane (MCH). The octane numbers of gasoline are RON 99.8 and MON 87.5. The octane numbers of gasoline surrogate mixtures are RON 100.2 and MON 88.8.

Gasoline		Gasoline surrogate mixtures	
Component	Vol%	Component	Vol%
n,i-parafin	43.8	n-heptane	10.0
		isooctane	31.0
		MCH	5.0
olefin	19.6	DIB	14.0
aromatic	36.6	toluene	40.0

Table 1: Compositions of gasoline and gasoline surrogate mixtures.

2.2 Rapid compression machine (RCM)

The RCM used in this study employs a pneumatically driven piston. The bore and stroke are 100 mm and 130 mm, respectively, the displacement is 1021 cc, and the compression ratio is 10.7:1. The compression is a single stroke event with a compression time of 32.9 ± 1.3 ms and the t50 is 7.0 ± 0.6 ms. The pressure in the reaction cylinder is measured using a pressure transducer (Kistler 6061BU20) with a time resolution of 0.05 ms. To reduce the noise level in the pressure signals, the 13 signals were averaged. Homogeneous mixtures are prepared separately in a mixing chamber.

2.3 Experimental conditions

The ignition delay times of gasoline and gasoline surrogate mixtures were measured at an equivalence ratio of 0.5 and a pressure of 4 MPa, over a temperature range 660 - 720 K. The ignition delay times of gasoline and gasoline surrogate mixtures under EGR conditions were also measured at an equivalence ratio of 0.5, pressure of 4 MPa, temperature of 720 K, and EGR ratio of 20 %. In addition, the ignition delay times of gasoline and gasoline surrogate mixtures with CO and NO, respectively, under EGR conditions were measured at an equivalence ratio of 0.5, pressure of 4 MPa, temperature of 720 K, and EGR ratio of 20%.

The compositions of gas mixtures are shown in Table 2 and Table 3. The composition of mixtures used in the EGR conditions was determined by theoretical calculation, assuming that the gasoline surrogate mixtures were completely oxidized to form CO_2 and H_2O. Since it is difficult to include H_2O in the gas mixtures, the mole fraction of CO_2 was adjusted to achieve the same ratio of specific heat as those of the mixtures that included H_2O. In the gas mixtures under EGR conditions, the fuel and O_2 concentrations decrease and the CO_2 concentration increases. To investigate the factor of the EGR effect on the ignition characteristics of gasoline and gasoline surrogate mixtures, their ignition delay times were measured using the following gas mixtures, O_2 reduction mixture, fuel reduction mixture, and CO_2 addition mixture, as shown in Table 2 and Table 3. Moreover, to investigate the effects of CO and NO on the ignition characteristics of the gasoline, CO (160 ppm) and NO (20ppm) were added to the mixtures used in EGR conditions, respectively. The concentrations of CO and NO were determined from the results of the lean burn engine experiments [2].

The compressed temperature (T_c) was calculated using the adiabatic core hypothesis [3],

$$\ln\left(\frac{P_c}{P_0}\right) = \int_{T_0}^{T_c} \frac{\gamma}{\gamma - 1} \cdot \frac{dT}{T} \tag{1}$$

where P_0 is the initial pressure, P_c is the compressed pressure, T_0 is the initial temperature, γ is the specific heat ratio of mixtures.

	Base	EGR	O_2 reduction	Fuel reduction	CO_2 addition
Fuel	1.15	1.03	1.15	1.03	1.15
O_2	20.76	18.61	18.60	20.76	20.76
N_2	78.09	78.51	80.24	78.21	76.24
CO_2	0	1.85	0	0	1.85

Table 2: Compositions of mixtures using gasoline.

	Base	EGR	O_2 reduction	Fuel reduction	CO_2 addition
Fuel	0.99	0.89	0.99	0.89	0.99
O_2	20.79	18.63	18.63	20.79	20.79
N_2	78.22	78.63	80.38	78.32	76.38
CO_2	0	1.84	0	0	1.84

Table 3: Compositions of mixtures using gasoline surrogate mixtures.

2.4 Definition of ignition delay time

The experimental pressure and derivative of pressure profiles are shown in Fig. 1. The end of compression is defined as the point of maximum pressure during compression. In this study, two-stage ignition was observed. The first ignition delay time (τ_1) is defined as the time difference between the end of compression and the first maximum point of the pressure derivative. The total ignition delay time (τ_2) is defined as the time difference between the end of compression and the second maximum point of the pressure derivative.

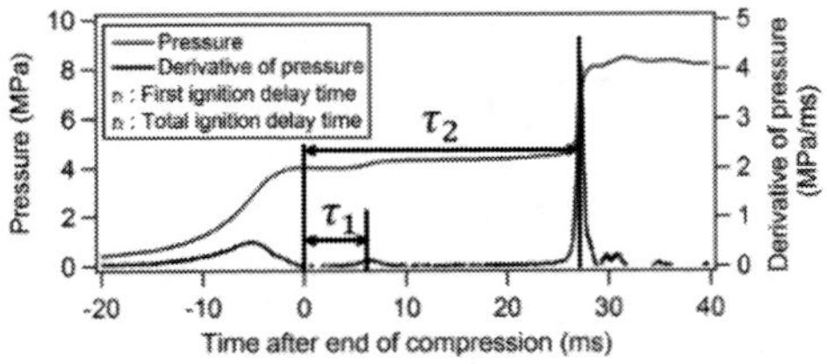

Figure 1: Definition of first and total ignition delay time (φ = 0.5, T_c = 720 K, P_c = 4 MPa, EGR conditions).

3 Simulations

The RCM simulations were performed using the gasoline surrogate model constructed by Miyoshi et al. [4] that includes n-heptane, isooctane, toluene, DIB, and MCH reactions and consists of 1761 species and 5802 reactions. In the case of the simulation with added NO, since there were no reactions involving NO in this model, the TRF model constructed by Anderlohr et al. [5] that includes TRF and NO reactions and consists of 538 species and 2987 reactions was used. The RCM simulations were conducted with an imposed effective volume profile, which was calculated from an experimental pressure profile to consider the compression stroke and heat loss after compression, as described by Tanaka et al. [6]. In the RCM simulations, ignition delay times were determined based on the same definition in the experiment. The CHEMKIN-PRO package [7] was used to conduct simulations.

4 Results and discussion

4.1 Ignition delay times of gasoline at 4 MPa

The experimental and simulated ignition delay times of the gasoline and gasoline surrogate mixtures are shown in Fig. 2. The experimental ignition delay times of both fuels decreased with increasing temperature. The experimental total ignition delay times of gasoline surrogate mixtures were approximately the same as those of gasoline, while the experimental first ignition delay times of gasoline surrogate mixtures were shorter than those of gasoline. Although the simulated total and first ignition delay times were longer than those of the experiments for all experimental conditions, the simulated ignition delay times had the same tendency as that of the experiments.

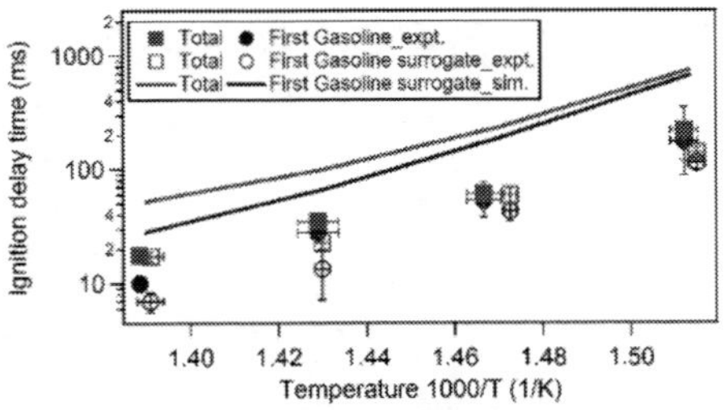

Figure 2: Experimental and simulated ignition delay times of gasoline and gasoline surrogate mixtures (φ = 0.5, T_c = 660 - 720 K, P_c = 4 MPa).

4.2 Effect of EGR on ignition characteristics of gasoline

The experimental pressure profiles of gasoline for each of the conditions are shown in Fig. 3(a). The ignition delay time for the EGR condition was longer than that for the base condition. The fuel reduction condition resulted in the longest ignition delay time of all the conditions tested to investigate the factors of EGR effects. The experimental pressure profiles of gasoline surrogate mixtures for each of the conditions [8] are also shown in Fig. 3(b). The results of gasoline surrogate mixtures had the same tendencies as those of gasoline. The simulated pressure profiles of gasoline surrogate mixtures for each of the conditions are shown in Fig. 3(c). The results of the simulations also had the same tendencies as those of the experiments. The reduction of fuel concentration in the initial gas mixtures was found to have the strongest influence on the increase in ignition delay times of gas mixtures containing EGR gas out of several factors under EGR conditions.

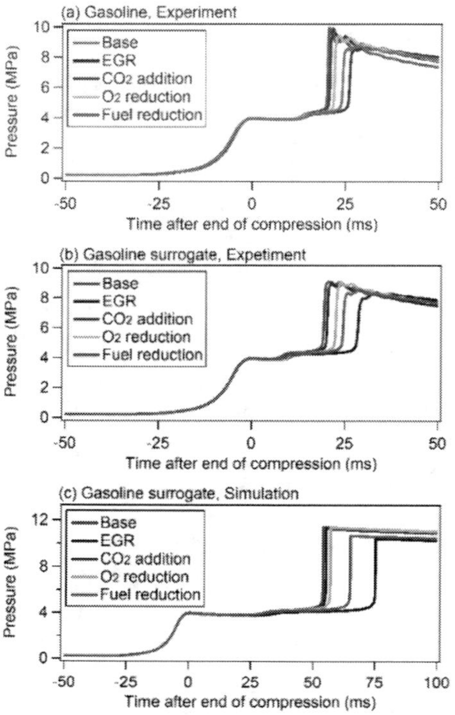

Figure 3: Experimental and simulated pressure profiles of gasoline and gasoline surrogate mixtures under EGR conditions ($\varphi = 0.5$, $T_c = 720$ K, $P_c = 4$ MPa). (a) Experimental results of gasoline, (b) experimental results of gasoline surrogates, and (c) simulated results.

4.3 Effects of CO and NO on ignition characteristics of gasoline

The experimental pressure profiles of gasoline under EGR with CO and NO conditions are shown in Fig. 4(a). The ignition delay time of gasoline under EGR with CO conditions was the same as that under EGR conditions without CO, while the ignition delay time under EGR with NO conditions was shorter than that under EGR conditions without NO. The experimental pressure profiles of gasoline surrogate mixtures under EGR with CO, and with NO conditions, respectively, are shown in Fig. 4(b) [8]. The tendencies of the results of the gasoline surrogate mixtures were similar to those of gasoline. The simulated pressure profiles of gasoline surrogate mixtures, which were calculated using the TRF model constructed by Anderlohr et al. [5], are also shown in Fig. 4(c). The results of the simulation showed the same tendencies as those of experiments, while the simulated ignition delay times were much longer than those of the experiments. These results indicate that NO leads to a decrease in the ignition delay time under EGR conditions, while CO has little influence on the ignition delay time under EGR conditions.

Since the chemical kinetic model of TRF with NO had the same tendencies as those of the experiments, a chemical kinetic analysis for the decrease of ignition delay time under the conditions with NO was performed. The simulated OH and HO_2 concentration profiles are shown in Fig. 5. The profiles of OH and HO_2 concentration under base and EGR conditions showed no significant differences. On the other hand, the OH concentration under EGR with NO conditions increased dramatically during the compression stroke, even more than under the base and EGR conditions without NO. The simulated profiles of H_2O_2 and HCHO concentrations are also shown in Fig. 6. The concentrations of H_2O_2 and HCHO under EGR with NO conditions were exactly the same as those under EGR conditions, which indicates that there was no difference between the reaction rates of the end of low temperature oxidation and the start of high temperature oxidation. Thus, the decrease in ignition delay times under the conditions where NO

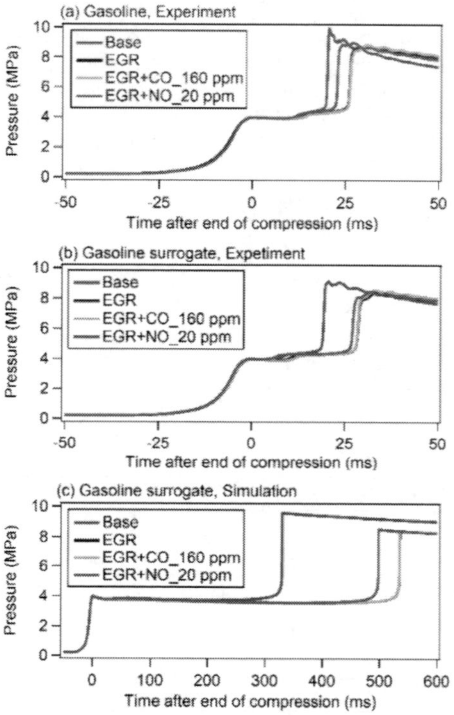

Figure 4: Experimental and simulated pressure profiles of gasoline and gasoline surrogate mixtures under EGR conditions with CO, and NO, respectively ($\varphi = 0.5$, $T_c = 720$ K, $P_c = 4$ MPa). (a) Experimental results of gasoline, (b) experimental results of gasoline surrogates, and (c) simulated results.

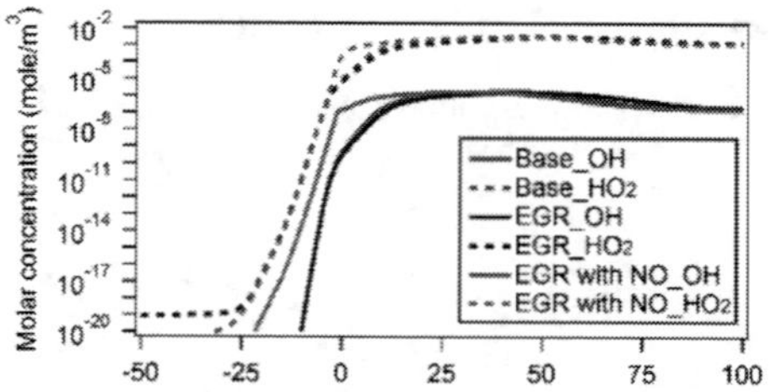

Figure 5: Simulated profiles of OH and HO_2 concentrations using the TRF model under base, EGR, and EGR with NO conditions ($\varphi = 0.5$, $T_c = 720$ K, $P_c = 4$ MPa).

was added to the gas mixture was derived from the increase of OH concentration during the compression stroke, which promotes the low temperature oxidation of gasoline.

A sensitivity analysis was performed to elucidate the reaction contributing to the formation of OH under EGR with NO conditions, 33 ms after the start of the calculation (at the end of compression) corresponding to 0.18% fuel consumption. Positive coefficients indicate a reaction that contributes to the formation of OH, and negative coefficients indicate a reaction that contributes to the consumption of OH. The results of sensitivity analysis are shown in Fig. 7. It was found that the reaction with the highest positive sensitivity coefficient was the reaction of NO with HO_2 radicals to form OH, which is considered to be the key reaction for the decrease of ignition delay time under EGR with NO conditions.

Finally, the temperature effects on the ignition delay times under EGR with NO conditions was investigated. The simulated ignition delay times under EGR with and without NO conditions at 4 MPa in adiabatic constant volume reactor are shown in Fig.8. The effect of NO addition on the ignition delay times were markedly observed in the temperature range 850 - 1100 K that is the NTC region.

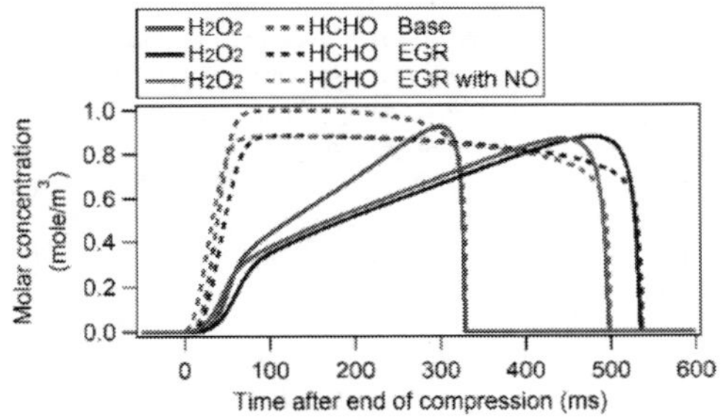

Figure 6: Simulated profiles of H_2O_2 and HCHO concentrations using the TRF model under base, EGR, and EGR with NO conditions ($\varphi = 0.5$, $T_c = 720$ K, $P_c = 4$ MPa).

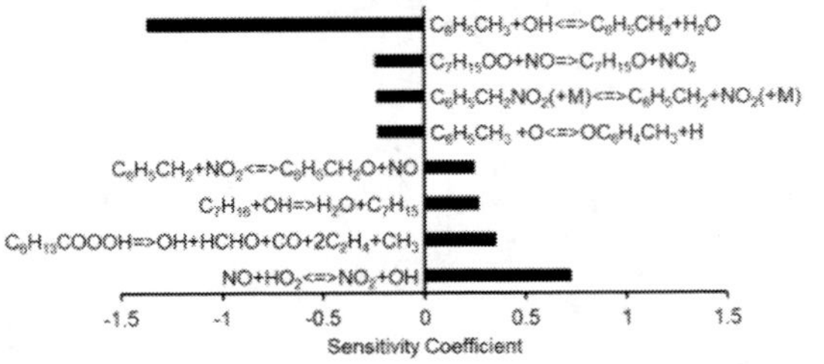

Figure 7: Sensitivity analysis of OH formation calculated using the TRF model under EGR with NO condition, 33 ms after the start of the calculation (at the end of compression) corresponding to 0.18% fuel consumption ($\varphi = 0.5$, $T_c = 720$ K, $P_c = 4$ MPa).

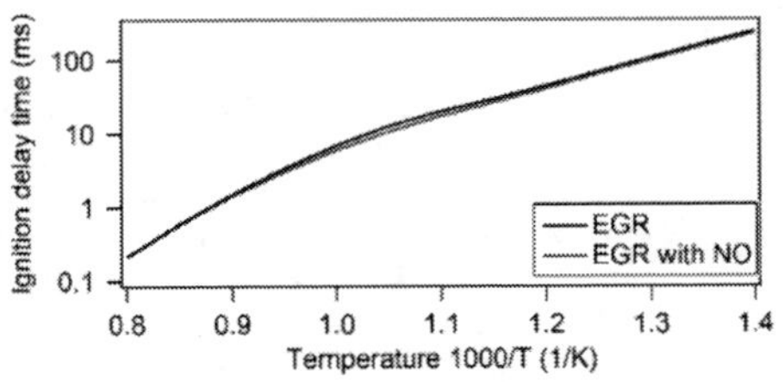

Figure 8: The simulated ignition delay times under EGR with and without NO conditions in adiabatic constant volume reactor ($\varphi = 0.5$, $P = 4$ MPa).

5 Conclusions

The effect of EGR on the ignition characteristics of gasoline at 4 MPa under lean burn conditions was investigated. In addition, the effects of CO and NO on the ignition characteristics of gasoline were investigated.

The experimental total ignition delay times of gasoline surrogate mixtures were approximately the same as those of gasoline, while the experimental first ignition delay times of gasoline surrogate mixtures were shorter than those of gasoline. Although the simulated total and first ignition delay times were longer than those of experiments for all experimental conditions, the simulated ignition delay times had the same tendencies as those of the experiments.

The ignition delay time under EGR conditions was longer than that under base conditions. Of the conditions used to investigate the factor of the EGR effect, the fuel reduction resulted in the longest ignition delay time, which indicates that the reduction of fuel concentration in the initial gas mixtures had the strongest influence on the increase of ignition delay time of gas mixtures containing EGR gas among several factors under the EGR conditions.

The ignition delay time under EGR with NO conditions was shorter than that under EGR conditions. The OH concentration under EGR with NO conditions increased dramatically in the initial stage of the fuel oxidation. The results of the sensitivity analysis showed that the reaction of NO with HO_2 contributed to the formation of OH, which led to the decrease of ignition delay time under EGR with NO conditions.

6 Acknowledgment

This study was supported by the council for Science, Technology and Innovation (CSTI), Cross-ministerial Strategic Innovation Promotion Program (SIP), "Innovative Combustion Technology" (Funding agency: JST).

References

[1] B.M. Gauthier, D.F. Davidson, R.K. Hanson, Combustion and flame 139 (2004) 300-411.

[2] N.Iida, The 9th international conference on modeling and diagnostics for advanced engine systems (2017) PL-1.

[3] G. Mittal, J. Sung, Combustion and Science and Technology 179 (2007) 497-530.

[4] SIP Gasoline Combustion Team, available at <http://sip.st.keio.ac.jp/topics/sipg1/>.

[5] J.M. Anderlohr, R. Bounaceur, A. Pires Da Cruz, F. Buttin-Leclerc, Combustion and Flame 156 (2009) 505-521.

[6] S. Tanaka, F. Ayala, J.C. Keck, Combustion and flame 133 (2003) p.467-481.

[7] ANSIS Chemkin-Pro, available at <http://www.ansys.com/de-DE/products/fluids/ansys-chemkin-pro>.

[8] M. Naruke, S. Yoshida, Y. Wachi, K. Tanaka, M. Konno, 2017 JSAE Annual Congress (Spring) (2017) 191.

11th Asia-Pacific Conference on Combustion,
The University of Sydney, NSW Australia
10th -14th December 2017.

Turbulence Structures in Turbulent Premixed Bunsen Flames

Nobuhiro Makita[1], Hiroshi Saito[1], Hideki Hashimoto[2], Junichi Furukawa[1]

[1]Tokyo Metropolitan College of Industrial Technology

1-10-40 Higashi-Ohi, Shinagawa-Ku, Tokyo, 140-0011, Japan

[2] Kobe City College of Technology

8-3 Gakuen-Higashimachi, Nishi-Ku, Kobe, Hyogo, 651-2194, Japan

Abstract

Turbulence structures in turbulent premixed Bunsen flames in the corrugated-flamelet regime were investigated by use of a three-color six beam laser-Doppler velocimetry. Two different mixtures with identical laminar burning velocities were selected to facilitate comparisons between lean and rich propane-air flames. A bimodal distribution was observed in the radial component of gas velocity in the turbulent flame brush. The low-velocity mode was identified as velocity fluctuations of the unburnt mixture and the high-velocity mode as fluctuations of the burnt gas radial velocity. The conditioned average and conditioned RMS (root-mean-square) fluctuation of the axial, tangential and radial components of the gas velocity and the shear component of the Reynolds stress were calculated form these bi-modal distributions. Average velocities are larger in the reacting flow especially in the turbulent flame brush, indicating that the acceleration due to thermal expansion to the downstream is significant. The turbulence intensities and the shear component of the Reynolds stress of the non-reacting flow are much larger than those conditioned on the unburnt mixture and burnt gas.

1 Introduction

Structure of turbulent flames are affected by the characteristics of turbulence [1]. For this reason, flame-flow interaction constitute an essential problem for turbulent premixed flames. To understand the flame-flow interaction, accumulation of reliable information of the turbulence structure is indispensable. An attempt has been made to examine the turbulence structure of lean and rich propane-air turbulent premixed Bunsen flames of identical laminar burning velocity in the corrugated-flamelet regime by use of a three-color six beam laser-Doppler velocimetry.

2 Experimental

The burner used in the present study is the same cylindrical burner, 26 mm in diameter, that was employed in our previous investigations [2-7]. Propane is selected as the fuel used in the present study. The oxygen and inert have approximately the same diffusivity, except for propane, whose diffusivity is smaller than that of oxygen, enabling preferential diffusion to occur in the propane flames. Values of the equivalence ratio employed for propane-air flames were 0.80 and 1.40. These values correspond to identical laminar burning velocities S_L of about 0.30 m/s. A uniform propane-air mixture was supplied to the burner at an average non-reacting flow exit velocity of 4.0 m/s, corresponding to fully developed turbulent pipe flow at a Reynolds number of about 7,000. The characteristics of the turbulence were examined by a hot-wire anemometer in non-reacting flow at the centerline position. The average non-reacting flow velocity U on the centerline was 4.5 m/s, the turbulence intensity (root-mean-square velocity fluctuations) u' was 0.22 m/s, the integral scale L_x was 10.9 mm, the Taylor scale λ_x was 3.5 mm, the Kolmogorov scale η was 260 μm, and the Reynolds number R_L based on the integral scale was 154 [2]. The laminar-flame thickness and Karlovitz number [1] are 53 μm and 0.04 respectively, placing the system well into the flamelet regime.

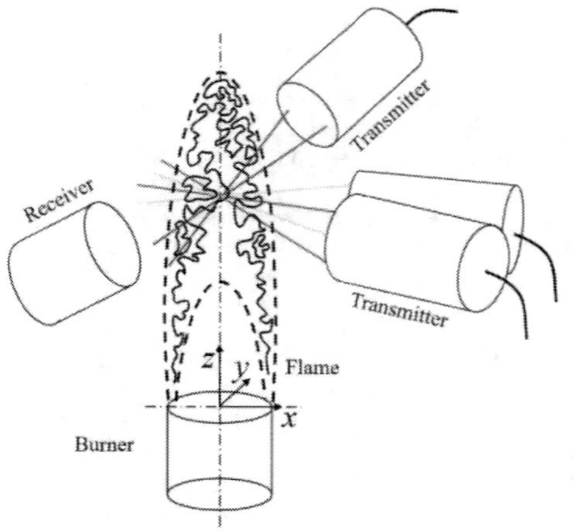

Fig. 1 Schematic illustration of the experimental setup

An LDV system (Artium LDV-300SC), composed of three semiconductor lasers operating at wavelengths 491 nm (100 mW), 532 nm (150 mW) and 561 nm (100 mW), along with a three-color, six-beam, forward-scattering optical system, and three Doppler signal analyzers provided the three-dimensional measurement of the local gas velocity. The measuring volume of this 3-D LDV system is approximately a sphere 0.2 mm in diameter, determined by the intersection of two football-like measuring volumes of the 2-D and 1-D transmitters, whose length and diameter are 2.2 mm and 0.2 mm, respectively. The micron-size scattering particles, seeded in the unburnt mixture stream, were obtained from SiO_2 powder. Coincident measurement of all three components of the gas velocity were conduct-

Corresponding author. Fax: +81-3-3471-6338
E-mail address: s17029@g.metro-cit.ac.jp

ed at z = 50 mm above the burner exit. The axial, tangential and radial components of the average velocity, RMS (root-mean-square) fluctuations and the shear component of the Reynolds stress conditioned on the unburnt mixture and the burnt gas can be calculated from the coincident data of the gas velocity.

3 Results and Discussions

Representative probability distributions of the radial component of the gas velocity obtained in the present study is shown in Fig. 2. In the figure, N and N_s represent the number of samples in a bin of interval 0.05 m/s and the total number of samples, respectively. Mono-modal distributions (not shown) are observed in the axial and tangential components of the gas velocity. Contrary to the axial and tangential components, the radial component of the gas velocity is characterized by a bi-modal distribution in the interior of the flame brush [7]. When the radial position x is less than 5 mm, a mono-modal distribution, which is composed of the velocity fluctuations of the unburnt mixture, is observed in the radial component. When the radial position x is between 5 mm and 11 mm, a bimodal distribution, composed of a low-velocity mode and a high-velocity mode, is observed in the radial component. The zone in which the bimodal distribution is observed can be interpreted as the turbulent flame brush, with the low-velocity data coming primarily from measurements made when the fresh mixture is present and the high-velocity data corresponding mainly to the presence of the burnt gas. When the radial position x is larger than 11 mm, a mono-modal distribution again develops in this radial component, which now is dominated by the velocity fluctuations of the burnt gas.

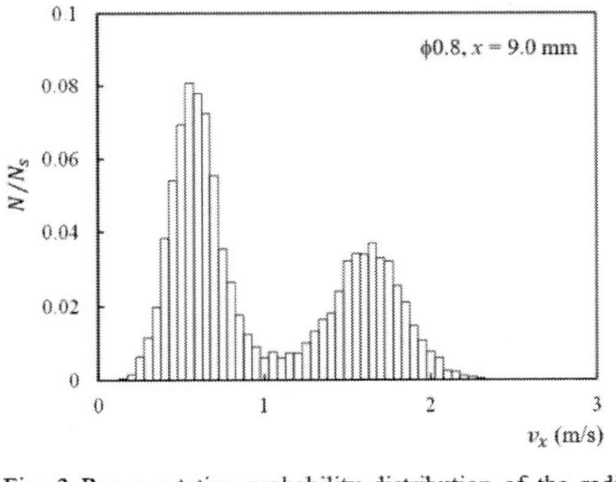

Fig. 2 Representative probability distribution of the radial component of the gas velocity.

It may be difficult to understand why a bi-modal distribution appears only in the radial velocity component. Since the tangential component of velocity is distributed symmetrically about 0 m/s, the bi-modal distribution, composed of the low-velocity mode of the unburnt mixture and the high-velocity mode of the burnt gas, can appear only in the vertical or radial velocity-component distributions. The gas velocity increases in a direction perpendicular to the flamelet because of thermal expansion, leading to anisotropic enhancement of gas flow across the flamelet [3]. The direction in which the gas velocity increases most depends on the flamelet orientation. When the polar angle of the flamelet velocity vector in the spherical coordinate system is less than 45 degrees, the gas velocity increases more in the vertical direction. On the other hand, when the polar angle of the flamelet velocity vector is larger than 45 degrees, the gas velocity increases more in the horizontal direction. We found in a previous study [4] that the polar angle of the flamelet velocity vector peaks around 53 to 55 degrees, indicating that the gas velocity increases more in the horizontal direction and less in the vertical direction. The bimodality in the radial component (rather than the vertical component) thus arises from the fact that that the gas flow is bent outward more than upward across the flamelet, that is, the surfaces of the flamelets are more nearly vertical sheets than horizontal sheets. This is also the source of the overall flow divergence in the turbulent flame brush.

The LDV measurements revealed that our data can provide velocity-component separately conditioned on the unburnt mixture and the burnt gas [7], under the assumption that the measurements are essentially only in the unburnt mixture or burnt gas. Since the coincident measurement of all three components of the gas velocity were conducted, i.e., all three components of the gas velocity were detected simultaneously, the axial and tangential components of the velocity fluctuations can be retrieved to be conditioned on the unburnt mixture or the burnt gas by observing the radial component.

Radial distributions of the axial $\overline{v_z}$ and radial $\overline{v_x}$ components of the average velocity conditioned on the unburnt mixture and the burnt gas are shown in Fig. 3. Values measured at z = 50 mm above the burner exit in the non-reacting flow are shown by the dotted line. The tangential $\overline{v_y}$ component is not shown here because it is symmetry about 0 m/s.

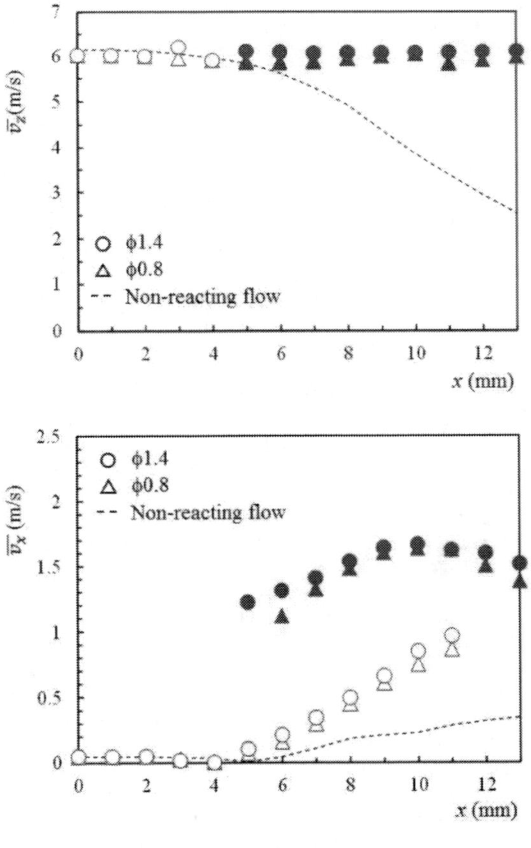

Fig. 3 Radial distribution of the axial (upper) and radial (lower) components of the average velocity conditioned on the unburnt mixture and the burnt gas; open and solid symbols represent the unburnt mixture and the burnt gas, respectively. Values measured in the non-reacting flow are shown by dotted line.

The axial $\overline{v_z}$ component of the average velocity of the non-reacting flow is maximum on the centerline and remains almost constant when the radial position x is less than 5 mm. It decreases toward the outer side when the radial position x is larger than 5 mm. The axial $\overline{v_z}$ component of the average velocity conditioned on the unburnt mixture agrees well with that of the non-reacting flow in the unburnt gas, where the radial position x is less than 5 mm. The axial $\overline{v_z}$ components of the average velocity conditioned on both the unburnt mixture and the burnt gas remain constant in the turbulent flame brush, where the radial position x is between 5 mm to 11 mm, indicating that the acceleration due to thermal expansion to the downstream is significant in the turbulent flame brush. The axial $\overline{v_z}$ component of the average velocity conditioned on the burnt gas remains constant in the burnt gas, where the radial position x is larger than 11 mm. The radial $\overline{v_x}$ component of the average velocity of the non-reacting flow is zero when the radial position x is less than 5 mm and it increases slightly toward outer side. The radial $\overline{v_x}$ component of the average velocity conditioned on the unburnt mixture agrees well with that of the non-reacting flow when the radial position x is less than 5 mm. It increases toward outer side much more than that of the non-reacting flow in the turbulent flame brush. The radial $\overline{v_x}$ component of the average velocity conditioned on the burnt gas increases toward outer side in the turbulent flame brush. The radial $\overline{v_x}$ component of the average velocity conditioned on the burnt gas is larger than that conditioned on the unburnt mixture, indicating that the acceleration due to thermal expansion is much larger in the radial direction. The radial $\overline{v_x}$ component of the average velocity conditioned on the burnt gas decreases in the burnt gas. The radial component $\overline{v_x}$ of the average velocity conditioned on the burnt gas of the rich flame is about 4% larger than that of the lean flame.

Radial distributions of the axial, tangential and radial components of the RMS (root-mean-square) fluctuations conditioned on the unburnt mixture and the burnt gas are shown in Fig. 4. Values measured at z = 50 mm above the burner exit in the non-reacting flow are shown by the dotted line. The axial, tangential and radial components of the RMS fluctuations of the non-reacting flow is 0.22 m/s on the centerline and increase toward the outer side and are maximum around the radial position x is 10 to 11 mm. The axial component of the RMS fluctuations conditioned on the unburnt mixture agrees well with that of the non-reacting flow when the radial position is less than 4 mm. It remains almost constant when the radial position is between 5 mm and 11 mm. The axial component of the RMS fluctuations conditioned on the burnt gas increases slightly toward outer side in the turbulent flame brush and is maximum when the radial position x is 8 mm, the center of the turbulent flame brush, where the heat release is maximum. The tangential component of the RMS fluctuations conditioned on the unburnt mixture agrees well with that of the non-reacting flow when the radial position x is less than 4 mm. It remains almost constant in the turbulent flame brush. The tangential component of the RMS fluctuations conditioned on the burnt gas increases toward outer side in the turbulent flame brush and is maximum when the radial position x is 8 mm, the center of the turbulent flame brush. The radial component of the RMS fluctuations conditioned on the unburnt mixture agrees well with that of the non-reacting flow when the radial position x is less than 4 mm. It remains almost constant in the turbulent flame brush. The radial component of the RMS fluctuations conditioned on the burnt gas increases toward outer side in the turbulent flame brush and is maxim-mum when the radial position x is 8 mm, the center of the turbulent flame brush. The axial, tangential and radial components of the RMS fluctuations is maximum at the center of the turbulent flame brush. The RMS fluctuations conditioned on the burnt gas is larger than that on the unburnt mixture. The RMS fluctuation conditioned on the burnt gas of the rich flame are about 10% larger than those on the lean flame. Turbulence is enhanced across the flamelet due to thermal expansion [2, 7]. The RMS fluctuations of the non-reacting flow is much larger than those conditioned on the burnt gas.

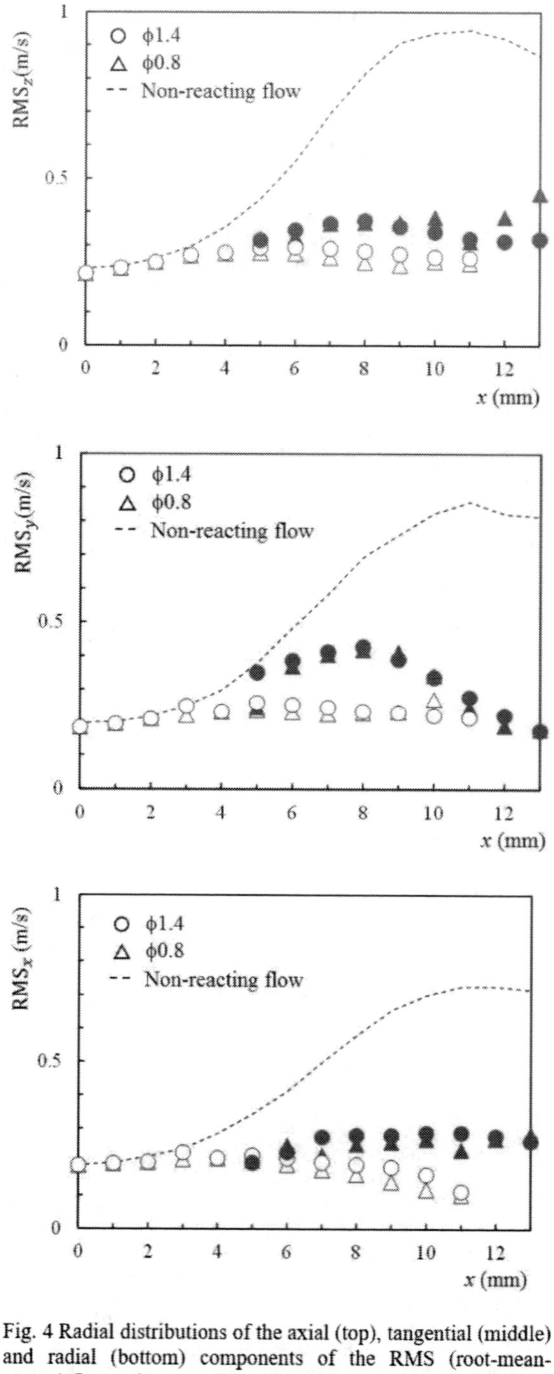

Fig. 4 Radial distributions of the axial (top), tangential (middle) and radial (bottom) components of the RMS (root-mean-square) fluctuations conditioned on the unburnt mixture and the burnt gas; open and solid symbols represent the unburnt mixture and the burnt gas, respectively. Values measured in the non-reacting flow are shown by dotted line.

The i-th component v_i of the gas velocity can be written as;

$$v_i = \bar{v}_i + v_i'$$

where $\bar{v}_i$ and v_i' represent the time average and fluctuation, respectively. The shear component $-\overline{v_x' v_z'}$ of the Reynolds stress can be calculated from the conditioned quantities of the axial and radial components of the fluctuations of gas velocity. Radial distribution of the shear component $-\overline{v_x' v_z'}$ of the Reynolds stress conditioned on the unburnt mixture and the burnt gas are shown in Fig. 5. Values measured at z = 50 mm above the burner exit in the non-reacting flow are shown by the dotted line.

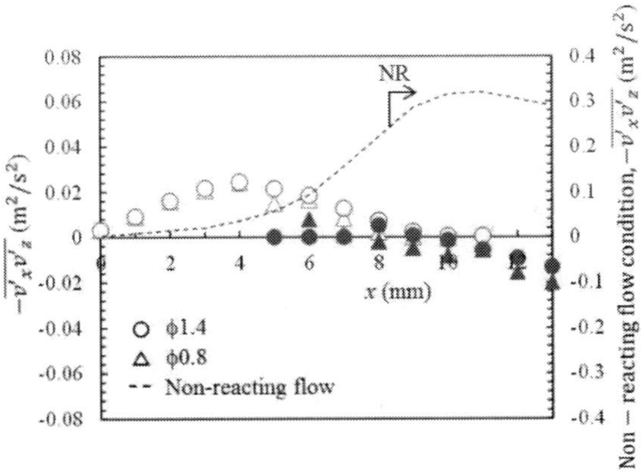

Fig. 5 Radial distribution of the shear component $-\overline{v_x' v_z'}$ of the Reynolds stress conditioned on the unburnt mixture and the burnt gas; open and solid symbols represent the unburnt mixture and the burnt gas, respectively. Values measured in the non-reacting flow are shown by dotted line.

The shear component $-\overline{v_x' v_z'}$ of the Reynolds stress of the non-reacting flow is zero on the centerline. It increases towards the outer side of the burner and reaches maximum when the radial position x is around 10 to 11 mm. The shear component $-\overline{v_x' v_z'}$ of the Reynolds stress of the non-reacting flow is much larger than those conditioned on the unburnt mixture and the burnt gas. The shear component $-\overline{v_x' v_z'}$ of the Reynolds stress conditioned on the unburnt mixture is zero on the centerline. It increases towards the outer side and reaches maximum when the radial position x is 4 mm. The shear component $-\overline{v_x' v_z'}$ of the Reynolds stress conditioned on the unburnt mixture is larger than that conditioned on the burnt gas. The shear component $-\overline{v_x' v_z'}$ of the Reynolds stress conditioned on the burnt gas changes from positive to negative toward outer side. It is about zero when the radial position x is 8 mm, where the heat release is maximum in the turbulent flame brush. The difference of the shear components $-\overline{v_x' v_z'}$ of the Reynolds stress conditioned on the burnt gas of the rich flame and that of the lean flame is about 0.007 m^2/s^2 when the radial position x is 8 mm.

It has been shown in the present study that the average velocities conditioned on the burnt gas are much larger than those of the non-reacting flow. On the other hand, the turbulence intensities and the shear component of the Reynolds stress conditioned on the burnt gas are smaller than those of the non-reacting flow. The shear component of the Reynolds stress can be written as;

$$-\rho\overline{v_x' v_z'} = \mu_T \frac{\partial \bar{v}_z}{\partial x}$$

where μ_T denotes the turbulence viscosity. In the turbulent flame brush, as shown in Fig. 3, distributions of the axial component $\bar{v}_z$ of the average velocity conditioned on the unburnt mixture and the burnt gas remain almost flat, therefore the velocity gradient $\partial\bar{v}_z/\partial x$ become smaller than that of the reacting flow, consequently the shear component $-\overline{v_x' v_z'}$ of the Reynolds stress become small. The velocity gradient $\partial\bar{v}_z/\partial x$ of the axial component $\bar{v}_z$ of the average velocity conditioned on the unburnt mixture is maximum when the radial position x is 4 mm, therefore, the shear component $-\overline{v_x' v_z'}$ of the Reynolds stress conditioned on the unburnt mixture is maximum when the radial position x is 4 mm.

4 Conclusions

The present study has been performed to accumulate reliable information of the turbulence structure in turbulent premixed Bunsen flames in the corrugated-flamelet regime. The axial, tangential and radial components of the average velocity, the root-mean-square fluctuations and the shear component of the Reynolds stress conditioned on the unburnt mixture and the burnt gas were calculated from the conditioned quantities in the axial and tangential components of the velocity fluctuations extracted from the bi-modal distributions of the radial component of the velocity fluctuations. The following conclusions are obtained in the present study;

1) The average velocities conditioned on the burnt gas are much larger than those of the non-reacting flow.

2) The turbulence intensities and the shear component of the Reynolds stress of the non-reacting flow are much larger than those conditioned on the burnt gas.

5 Acknowledgment

This work is supported by the Grant-in-Aid for Scientific Research Grant Number 16K06136 from the Japan Society for the Promotion of Science in Japan.

References

1] Peters N. (2000). Turbulent Combustion. Cambridge University Press (ISBN 0-521-66082-3)

2] Furukawa J., Noguchi Y., Hirano T., Williams F. A., *Journal of Fluid Mechanics*, **462** (2002), 209-243

3] Furukawa J., Williams F. A., *Combustion Science and Technology,* **175** (2003), 1835-1858

4] Furukawa J., Hashimoto H., and Williams F. A., *Combustion Science and Technology,* **182** (2010), 1-38

5] Furukawa J., Yoshida Y., Amin V., Williams F. A., *Combustion Science and Technology,* **185** (2013), 200-211

6] Furukawa J., Yoshida Y., Williams F. A., *Combustion Science and Technology,* **185** (2013), 661-671

7] Furukawa J., Yoshida Y., Williams F. A., *Combustion Science and Technology,* **188** (2016), 1538-1564

11th Asia-Pacific Conference on Combustion,
The University of Sydney, NSW Australia
10th -14th December 2017.

Numerical Simulation of Yttrium Oxide Nanoparticle Growth in Gas-Phase Flame Synthesis Using a Conditional Quadrature Method of Moments (CQMOM)

K. Takigawa[1], H. Matsushita[2] and T. Yokomori[3]

[1,2,3]Department of Mechanical Engineering Keio University

3-14-1 Hiyoshi, Kohoku-ku, Yokohama, 223-8522, Japan

Abstract

Gas-phase flame synthesis is a useful method for manufacturing metal oxide fine nanoparticles. In order to investigate the particle growth mechanism in detail, we conducted a numerical analysis of laminar flame and particle growth in gas-phase flame synthesis. The analysis focused on the synthesis of Y_2O_3 nanoparticles from $Y(DPM)_3$. Particle growth was described using the conditional quadrature method of moment (CQMOM) approach, which can represent the local distribution of particle volume and surface area. In terms of the particle growth process, we considered nucleation, collision-based agglomeration, and sintering. Additionally, to describe them precisely, we applied Gauss-Radau quadrature to the CQMOM. The simulation results showed that there are both spatial and local dispersion of particle size. The simulations also revealed the detailed mechanism of collision that particles undergo, and collisions between large particles and small ones were proved to have a substantial effect to particle growth. Increasing precursor partial pressure increased not only particle concentration, but also flame height and spatial dispersion of the primary particle diameter.

1 Introduction

Nanoparticles have numerous applications to medical treatment, electronics, etc., due to their differing characteristics from bulk materials. For example, Yttrium Oxide (Y_2O_3) nanoparticles have commonly been used as the host crystal of nanophosphors. The functionality of nanophosphors strongly depends on its crystallinity, size and morphology. Several methods have been used to synthesize them, including spray pyrolysis, the sol-gel method, and other wet processes. However, those methods generally require long hours of calcination in order to improve the crystallinity and photoluminescence intensity, which creates large and more aggregated particles.

Gas-phase flame synthesis, which produces nanoparticles with the vapor of the precursor, is expected to prevent this problem. This is a one-step operation of continuous synthesis that doesn't require drying or annealing. In addition, it has the potential to synthesize quasi-stable state crystalline or multicomponent nanoparticles. Nevertheless, the particle growth mechanism in flame synthesis is not well understood, and it is difficult to control particle size. Therefore, the mechanism is being elucidated with several methods of numerical simulation. The conditional quadrature method of moments (CQMOM) [1] is a recently developed approach. It has been applied to the particle growth of, for example TiO_2 [2]. The results showed that CQMOM enabled acquisition of accurate results at lower numerical cost. The quadrature method of moments with Gauss-Radau quadrature, which fixes a node to a constant value, was used for soot formation in the flame, and its accuracy was validated by comparing with those by the Monte Carlo method [3].

The scope of this research is to analyze the particle growth process in detail by numerical simulation using CQMOM with Gauss-Radau quadrature and to investigate the effect of precursor partial pressure on particle growth.

2 Numerical model

2.1 Gas-phase combustion

In this study, thermal fluid analysis, including decomposition of the precursor and combustion, was performed with the laminarSMOKE [4], which was developed as a laminar reacting flow solver in the open-source CFD software package OpenFOAM. We employed the comprehensive mechanism with C4 species [5] for methane combustion. In addition, we considered the decomposition of $Y(DPM)_3$ and the oxidation of Y as irreversible first-order reactions shown below in reference to [6, 7]:

$$Y(DPM)_3 \rightarrow Y + 3CO + 3HCCO + 6C(CH_3)_3 \quad (1)$$

$$4Y + 3O_2 \rightarrow 2Y_2O_3 \quad (2)$$

2.2 CQMOM

In this study, we assumed that the particle motion did not affect the flow field for the particle size, and conducted simulation of particle growth using the results of gas-phase combustion, (i.e., temperature, velocity, nucleation rate, etc.)

We used CQMOM with Gauss-Radau quadrature in order to represent the properties of particle, volume and surface area, and to reduce the computational cost. Particle growth is represented as the evolution of the bivariate number density function (NDF) $n(v,a;x,t)$, with particle volume v and surface area a as the variables. It is governed by population balance equation (PBE):

$$\frac{\partial n}{\partial t} + \nabla\cdot(\mathbf{u}n) - \nabla\cdot(\Gamma\nabla n) = \dot{n}_{nuc} + \dot{n}_{agg} + \dot{n}_{sint} \quad (3)$$

where $\mathbf{u}$ is the velocity, Γ is the diffusivity, and $\dot{n}_{nuc}$, $\dot{n}_{agg}$ and $\dot{n}_{sint}$ are the changes of particle number by nucleation,

Corresponding author. Fax: +81-45-566-1640
E-mail address: takigawa-kotaro@keio.jp

agglomeration and sintering, respectively.

Since the NDF is the unknown function, in CQMOM, it is represented by

$$n(v,a)=\sum_{i=1}^{N_v}\sum_{j=1}^{N_a}w_i w_{ij}\delta(v-v_i)\delta(a-a_{ij}) \tag{4}$$

and the bivariate moments are defined as

$$m_{kl}=\int_0^\infty\int_0^\infty v^k a^l n(v,a)dvda \tag{5}$$

Here N_v and N_a are conditions of volume and surface area, w_i and v_i are the volume weights and nodes, and w_{ij} and a_{ij} are the surface ones. With a Gauss-Radau quadrature, the lowest nodes (v_1 and a_{11}) are fixed at the monomer's ones.

Equation (3) is transformed to the transport equation for the moments using (4) and (5).

$$\frac{\partial m_{kl}}{\partial t}+\nabla\cdot(\boldsymbol{u}m_{kl})-\nabla\cdot(\Gamma\nabla m_{kl}) = \dot{m}_{kl,\mathrm{nuc}}+\dot{m}_{kl,\mathrm{agg}}+\dot{m}_{kl,\mathrm{sint}} \tag{6}$$

In this study, we chose N_v = 4 and N_a = 1 and therefore the second subscript of the surface nodes are omitted below.

2.3 Particle kinetics

In (6), the source terms by nucleation and agglomeration are written as functions of nodes and weights:

$$\dot{m}_{kl,\mathrm{nuc}}=\dot{\omega}_{Y_2O_3}N_a v_0^k a_0^l=k[\mathrm{Y}][\mathrm{O_2}]N_a v_0^k a_0^l \tag{7}$$

$$\dot{m}_{kl,\mathrm{agg}}=\sum_i^{N_v}\sum_j^{N_v}w_i w_j\beta_{ij}\begin{bmatrix}(v_i+v_j)^k(a_i+a_j)^l\\ -v_i^k a_i^l-v_j^k a_j^l\end{bmatrix} \tag{8}$$

where k is the reaction rate of (2), [Y] and [O_2] are the molar concentration of each species, N_a is the Avogadro's number, and β_{ij} is the collision frequency between particle i and j found in [9].

In this study, sintering isn't treated as a change of moments, but of surface area nodes [10]:

$$\frac{\partial a_i}{\partial t}=-\frac{a_i-a_{s,i}}{\tau_i} \tag{9}$$

where $a_{s,i}$ is the surface area when the i-th particle is fully coalesced, and τ_i is the characteristic sintering time of the i-th particle shown below:

$$\tau_i=8.347\times10^{12}d_{p,i}^{\,4}T\exp\left(\frac{49300}{T}\left(1-\left(\frac{3.4\times10^{-9}}{d_{p,i}}-\frac{T}{4100}\right)^{3.76}\right)\right) \tag{10}$$

where $d_{p,i}$ is the primary particle diameter of the i-th particle, and T is the Temperature. Equation (10) is formulated in [11] and the dependency on particle size is taken into account [12].

3 Experimental and simulation setup

The numerical model was applied to simulate the growth of nanoparticles in the laminar axisymmetric diffusion flame made by the co-flow burner [14]. Figure 1 shows the experimental setup and computational domain. The inlet fuel was a mixture of methane and vaporized Y(DPM)3. The oxidizer was pure oxygen. Each of the gases had a uniform exit velocity of 0.3 m/s but a different temperature, 573 K in the case of former and 293 K in that of the latter. The ambient pressure as the atmospheric pressure. The concentration of the vapor Y(DPM)3 was varied from 62.5 to 500 mg/L, corresponding to partial pressure p_p = 0.44 to 3.54 kPa.

The transport equations were discretized with the second-order upwind total variation-diminishing (TVD) scheme, using the van Leer method for the convection term and the central difference for the diffusion term. The time difference schemes for the transport equations were the forward Euler scheme for the simulation of gas-phase combustion, and the backward Euler scheme for particle growth.

4 Results and discussion

4.1 Particle growth

In this section, we show the results of p_p = 0.44 and examine the process of particle growth. Those results are the values along the streamline generated from (r, z) = (0.0002, -0.0149). Figure 2 represents particle number densities for each node. Particles were massively generated around z = 0.002 m for each node, with a sharp decrease just after. In particular, w_1 decreased more than other nodes and became less than w_2 and w_3. This implied that monomer nucleation occurred in a narrow region.

The results of temperature and primary particle diameter of each node along the streamline are shown in Fig. 3. Here, the primary particle diameter $d_{p,i}$ of the i-th node and mean primary particle diameter d_p are defined as

$$d_{p,i}=\frac{6v_i}{a_i} \tag{11}$$

$$d_p=\frac{6m_{10}}{a_{01}} \tag{12}$$

They drastically increased around z = 0.004 m. This is because the particles experienced the high temperatures, and sintering occurred at that position. This is supported by the results shown below. Figure 4 shows the temperature and total particle surface area along the streamline.

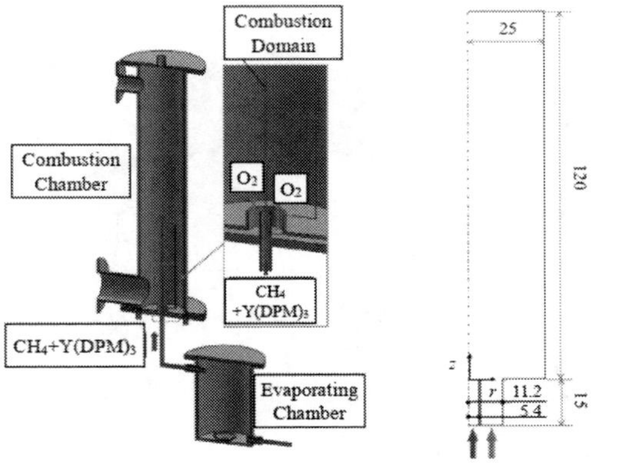

(a) (b)

Figure 1: Experimental and numerical setup (a) Co-flow burner; (b) Computational domain

The figure shows that sintering occurred around z = 0.004 m and finished at z = 0.01 m. This is because the sintering depends on temperature and doesn't progress much under low temperature. From Fig. 4, it may be seen that each primary particle diameter was the same before sintering. In this region, agglomerates of each node consisted of primary particles, which had the same size but different numbers. Thus, they became bigger primary particles according to their volume when fully sintered. The difference of primary particle diameter between each node was a result of this.

Figure 5 shows the ratio of collision frequency between each pair for the sum of all collision frequencies along the streamline. In this figure, the collision frequency ratio of (i-j) corresponds to the one between the i-th node to the j-th node. In addition, the pairs of smaller ratio are uncontained. From the figure, it may be seen that there are 3 steps in agglomeration process. First, monomers agglomerate with each other: the (1-1) process. Second, monomers collide with the agglomerates: (1-2) and (1-3) processes. Finally, the agglomerates collide with each other and grow: (2-2) and (2-3) processes.

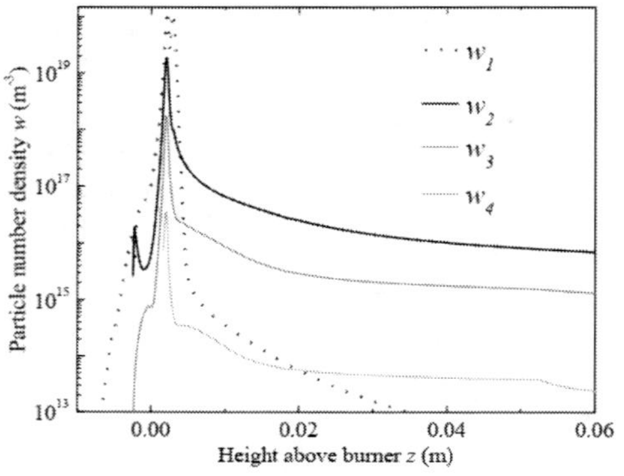

Figure 2: Particle number density of each nodes along the streamline

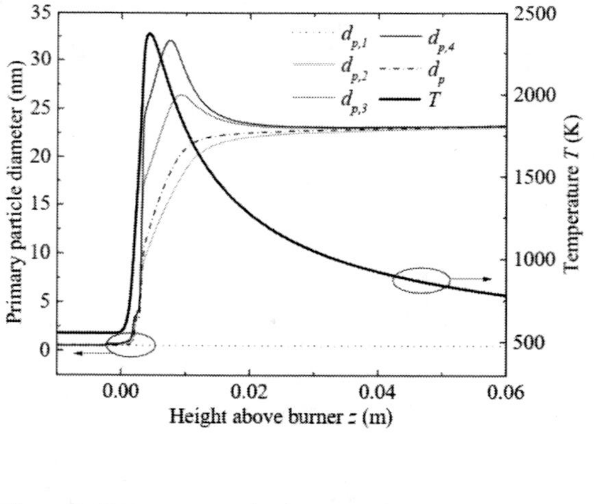

Figure 3: Temperature and primary particle diameter of each nodes along the streamline

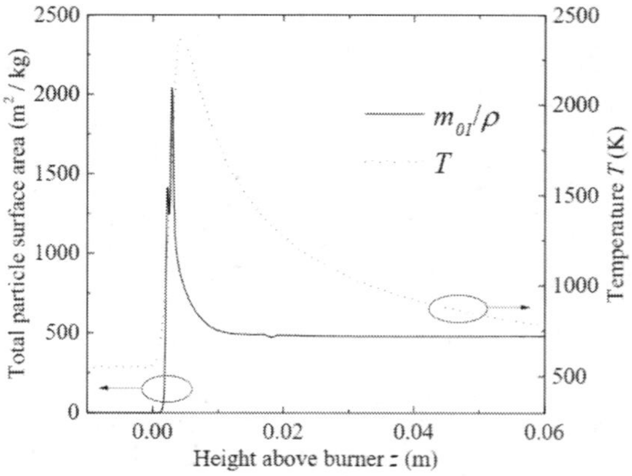

Figure 4: Temperature and total particle surface area along the streamline

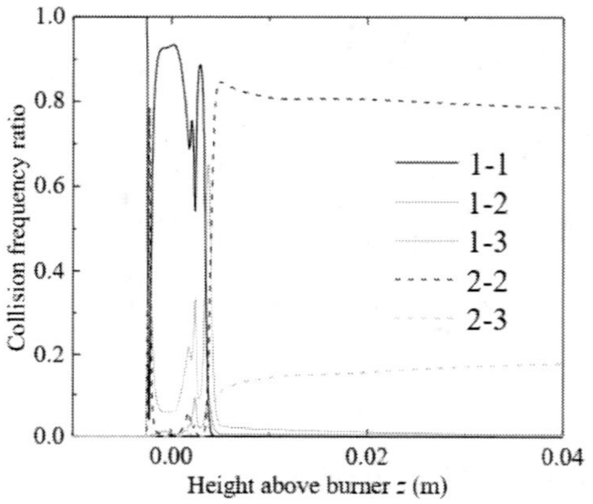

Figure 5: Collision frequency ratio of each pair along the streamline

4.2 Partial pressure of precursor

In this section, we show the results of each p_p and consider how the variation of p_p influences the flame properties and particle growth.

The temperature distributions for each p_p are shown in Fig. 6. The images focus on the exit of the burner and the flame. From Fig. 6, it may be seen that flame height and the high-temperature zone of the flame became larger as p_p increased. This is because the precursor is an organic complex, and therefore the by-products react as fuels. As p_p increased, so too did the amount of by-products and heat emission by their reaction.

Figure 7 shows the distributions of particle generation rate for each p_p. The value near the inner wall is larger than the one at the center and becomes bigger as p_p increases. In this study, we assumed nucleation occurs immediately when yttrium and oxygen atoms collide. Thus, more particles are generated in regions rich with both types of atoms. Therefore, the particle generation rate became larger near the inner tube as p_P increased.

Figure 8 shows the distribution of mean primary particle diameter for each p_P. In the figure, mean primary particle diameter become larger as p_P increases. This is because of larger particle-generation rate and broader high-temperature zone. The former facilitates agglomeration, and the latter promotes the sintering of agglomerates to make larger primary particles. Incidentally, there were larger primary particles around r = 2.7 mm (above the inner tube). As we mentioned earlier, more particles are generated near the inner tube than at the other regions, and that high concentration of particles facilitates agglomeration. For this reason, there is a spatial dispersion of primary particle diameters in gas-phase flame synthesis.

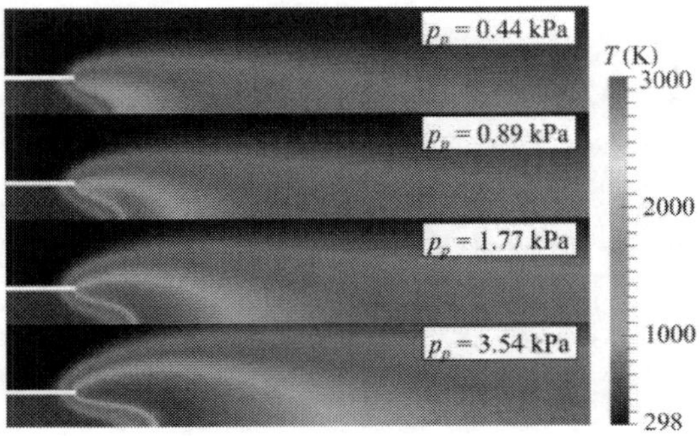

Figure 6: Temperature distribution in various precursor partial pressures

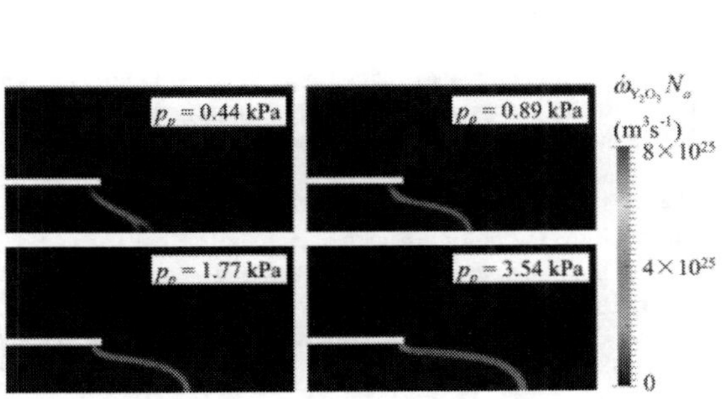

Figure 7: Particle-generation rate in various precursor partial pressures

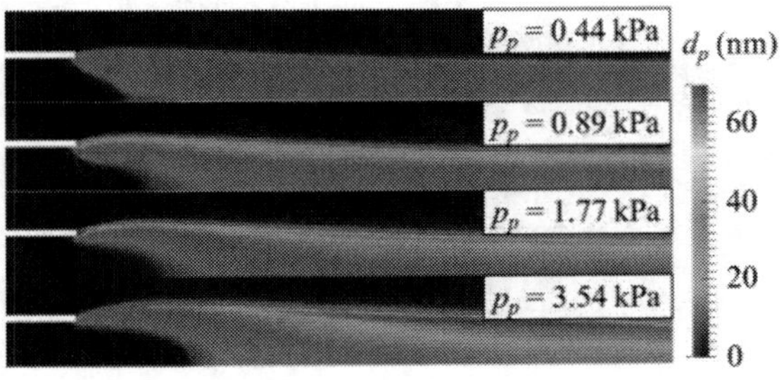

Figure 8: Mean primary particle diameter distribution in various precursor partial pressures

5 Conclusions

We conducted numerical simulations of laminar diffusion flame and particle growth for gas-phase flame synthesis. In the simulations, we used the CQMOM with Gauss-Radau quadrature to represent the bivariate NDF and its evolution. Through particle growth, agglomeration transits from collisions between monomers, to collisions between monomers and agglomerates, to collisions between agglomerates. In addition, it was implied that agglomeration between different size particles occurs frequently, and strongly influences particle growth.

The partial pressure of the precursor also had an enormous effect on particle growth. It was found that particle concentration and temperature increased together with the partial pressure. In addition, the particle nucleation occurred actively in the oxygen-rich region, and leading to increased spatial dispersion of the primary particle diameter.

References

[1] C. Yuan., R. O. Fox., J. Comput. Phys., 230 (2011), 8216-8246.

[2] Y. Sung., V. Raman., Koo, H., M. Mehta., R. O. Fox., AIChE J., 60 (2014), 459-472.

[3] S. Salenbauch., A. Cuoci., A. Frassoldati., C. Saggese., T. Faravelli., C. Hasse., Combust. Flame, 162 (2015), 2529-2543.

[4] A. Cuoci., A. Frassoldati., T. Faravelli., E. Ranzi., Combust. Flame, 160 (2013), 870-886.

[5] P. Zhao., W. Yuan., H. Sun., Y. Li., A. P. Kelley., X. Zheng., C. K. Law., Proc. Combust. Inst., 35 (2015), 309-316.

[6] D. K. Russell., A. Yee., New J. Chem., 29 (2005), 485-492.

[7] A. F. Bykov., P. P. Semyannikov., I. K. Igumenov., J. Therm. Anal., 38 (1992), 1477-1486.

[8] B. Buesser., A. J. Grohn., Chem. Eng. Technol., 35 (2012), 1133-1143.

[9] Fuchs, N. A., The Mechanics of Aerosols. New York: Pergamon Press (1964).

[10] W. Koch., S. K. Friedlander., J. Colloid Interface Sci., 140 (1990), 419-427.

[11] X. H. Wang., P. L. Chen., I. W. Chen., J. Am. Ceram. Soc., 89 (2006), 431-437.

[12] B. Buesser., A. J. Grohn., S. E. Pratsinis., J. Phys. Chem. C, 115 (2011), 11030-11035.

[13] V. Vikas., Z. J. Wang., R. O. Fox., J. Comput. Phys., 249 (2013), 162–179.

[14] Yokomori, T., Tajima, J., Iwako, Y., Mizomoto, M., Ueda, T., Flame Synthesis of Y2O3: Yb, Er Nanophosphors with Vapor Precursors, Proceedings of 7th Asia-Pacific Conference on Combustion, Chinese Taipei Section of the Combustion Institute, 2009, ID: 1018

11th Asia-Pacific Conference on Combustion,
The University of Sydney, NSW Australia
10th-14th December 2017.

Mixing Model Evaluation for Transported PDF Modelling of Sooting Flames

Joshua C.K. Tang[1], Haiou Wang[1], Michele Bolla[2], Evatt R. Hawkes[1], David O. Lignell[3]

[1] School of Mechanical and Manufacturing Engineering

University of New South Wales NSW 2052 Australia

[2]Aerothermochemistry and Combustion Systems Laboratory

ETH Zurich Sonneggstrasse 38092 Zurich

[3]Chemical Engineering Department

Brigham Young University UT 84602 United States

Abstract

Transported probability density function (TPDF) methods are well suited to modelling turbulent, reacting, variable density flows. In particular, local processes such as soot formation and radiation emission can be modelled without approximations resulting from the effect of unresolved turbulent fluctuations on the source terms. One of the main challenges to the successful deployment of TPDF methods is accurately modelling the unclosed molecular mixing term. These models have received considerable attention for gaseous flames, and while not a solved problem, significant progress has been made. The situation is much less clear for soot, as soot is not very diffusive, having a very high Schmidt number, and therefore it is not clear that models used for gaseous species will also work well for soot. This study examines three of the most widely used mixing models: the Interaction by Exchange with the Mean (IEM), Modified Curl (MC) and Euclidean Minimum Spanning Tree (EMST) models in predicting soot formation in non-premixed flames. A Direct Numerical Simulation (DNS) data-set was used to provide both initial conditions and inputs needed over the course of the runs, including the mean flow velocities, mixing frequency, and the turbulent diffusion coefficient. The same chemical mechanism and thermodynamic properties were used, allowing the study to focus on the mixing model. The simulation scenario was a statistically one-dimensional, sooting, non-premixed, turbulent jet flame burning an ethylene fuel stream. The models were tested with the mixing frequency defined from the dissipation rate and variance of mixture fraction. In all simulations, the soot moments were treated with zero mixing frequency, therefore they were not mixed. This definition, this study finds that the TPDF method is successful at predicting the temperature and soot mass fraction profiles when using the EMST mixing model. The IEM and MC over predicted the amount of extinction compared to the DNS and therefore could not predict the soot production accurately.

1 Introduction

Soot formation in turbulent flames is a process of practical importance. Soot is a known health hazard, and it is believed to contribute to climate change. Soot formation in turbulent flames is a complex process starting from precursor chemistry, through inception, surface growth, agglomeration, and oxidation. The timescales of these processes is typically long compared with other combustion processes, which creates both a scale-separation issue and may invalidate the use of fast chemistry models of combustion for soot prediction [1].

Transported probability density function (TPDF) methods [2,3] are an attractive method to simulate sooting flames. The principle behind TPDF modelling is that the single point statistics and thermochemical state variables of the flow can be evaluated by solving transport equations for the joint probability density function of these quantities. These transport equations are applied to a large ensemble of notional stochastic particles representing one-point, one-time samples of the fluid.

The advantage of this approach is that processes which require local information about the fluid composition, such as the chemical source term, appear in closed form [2]. Since soot formation is governed solely by the local composition and temperature, TPDF methods appear to be ideally suited to modelling sooting flames. Another key advantage is that the TPDF method can in principle treat both fast and slow processes simultaneously, which may be important considering the relatively slow soot processes. A major obstacle when using TPDF methods is accounting for the effect of molecular mixing, which is unclosed in the TPDF formulation. Models used to close this term are collectively referred to as "mixing models" [4]. Soot in particular presents a problem in that it has a very high Schmidt number ($Sc > 100$). As a result, even in simple homogeneous mixing, it is expected to have a different small-scale behaviour than typical gaseous species, for instance exhibiting a viscous-convective subrange in its energy spectrum [5]. At sufficiently high Reynolds numbers (Re), such small-scale dissipative differences would not be expected to require corrections to mixing models in the case of non-reacting mixing, since the small scales adjust to large-scales regardless of Sc. However, it is not clear that Re in typical combusting flows is sufficient to observe this limit, and in addition soot is coupled to small-scale chemical processes, which might not result in a vanishing differential diffusion effect even at very high Re. As such, the existing models should be tested to determine whether they need improvement.

This study will compare three of the most widely used mixing models against a Direct Numerical Simulation (DNS) data set [6]. The three models tested were: the Interaction by Exchange with the mean (IEM) [7], Modified Curl's (MC) [8], and Euclidean Minimum Spanning Tree (EMST) [9] models.

These models were implemented within a Composition TPDF (C-TPDF) method [10] which solves a transport equation for the joint PDF of scalar variables (species mass fractions and enthalpy) within the flow. This method requires inputs of the mean velocity field, turbulent diffusion coefficient, and turbulent mixing frequency. Typically, a two equation turbulence model such as the k-ϵ model is used to provide this information, however in this study they will be extracted directly from the DNS [6], thereby eliminating the turbulence modelling as a potential source of error. Chemical kinetic and thermodynamic models for the TPDF method are kept identical to the DNS, which eliminates yet more sources of error when compared to a test against experimental data. This work constitutes an extension with respect to previous studies of this type [11] as it considers soot formation.

2 Simulation Scenario

The DNS simulated a temporally evolving, turbulent plane-jet flame. The full details of the simulations are given in Ref. [6], and only a brief description will be given here to help orient the reader. The DNS was initialised with a three-dimensional planar slab of ethylene fuel in counter flow with oxidiser streams on either side. A small turbulent velocity fluctuation was imposed at the initial time which generated the instabilities in the shear layer between the fuel and oxidant streams.

The Damköhler number (Da) is the ratio of the chemical to mixing timescales and is defined as, $Da = \chi_q \tau_{jet}$, where χ_q is the extinction scalar dissipation rate and τ_{jet} is the jet time, which is defined as $\tau_{jet} = \frac{H}{U}$, where H is the jet height and U is the characteristic jet velocity. The DNS had a Damköhler number of 0.056. The parameters were chosen to maximise the residence time of the fluid and to minimise the extinction within the flame.

This configuration is statistically one dimensional, and allows for ensemble averaging in both the streamwise and spanwise directions, leaving a statistical dependence on the transverse direction, $\hat{\mathbf{y}}$, and the time, **t**. The ensemble average is performed on a mass (Favre) basis.

3 Solution Methods

3.1 Modelled Transport Equations

The TPDF model is based on a Reynolds-Averaged Navier-Stokes (RANS) formulation and the turbulence closure is provided directly from the Favre-averaged DNS data. This method has been used in previous mixing model studies by Krisman *et al.* [11] and Kuron *et al.* [12]. Only a brief presentation of the methodology is given here, for more details on the implementation and sensitivities, the reader is referred to Ref. [10]. Stochastic differential equations, (1) and (2), are solved for each notional particle.

$$dx(t) = [\tilde{V} + \nabla\tilde{\Gamma}_T]dt + \sqrt{2\tilde{\Gamma}_T W dt} \quad (1)$$

$$d\phi(t) = [\boldsymbol{M}]dt + [\boldsymbol{R}]dt \quad (2)$$

Equation (1) represents the transport in physical space and is composed of deterministic and stochastic components. The deterministic component, $[\tilde{V} + \nabla\tilde{\Gamma}_T]dt$, accounts for the transport due to the mean velocity field and gradients of the turbulent diffusion coefficient. The stochastic component, $\sqrt{2\tilde{\Gamma}_T W dt}$, is the increment of a Weiner process that models transport by turbulent fluctuations [8]. Equation (2) represents the change in composition space by molecular mixing, [**M**], and chemical reaction, [**R**]. The mixing term, [**M**], is the focus of this study. The chemical reaction term, [**R**], is solved using a 6-stage, 4[th]order Runge-Kutta numerical method [13]. The equations were supplemented with the thermodynamic equation of state and the ideal gas law (assuming constant pressure).

Equations (1) and (2) represent three coupled processes: advection, molecular mixing and chemical reaction. A symmetric Strang's splitting scheme [14] is used to solve equations (1) and (2) by subdividing each time step into partial time steps in which each process (transport, mixing and reaction) can be evaluated individually. This scheme is denoted as **TMRMT** [8] where each letter represents a sub-step. **T** represents transport in physical space, **M**, molecular mixing and **R**, chemical reaction. Transport and mixing are evaluated twice for each iteration and reaction once in a symmetrical manner.

In order to solve the transport sub-step, **T**, the turbulent diffusivity coefficient, Γ_T, and mean velocity field were required. To reduce potential modelling errors, these three inputs were extracted directly from the DNS. The turbulent diffusivity is defined as $\Gamma_t = \frac{\overline{\rho v Z} - \bar{\rho}\tilde{v}\tilde{Z}}{\bar{\rho}\nabla\tilde{Z}}$, where v is the velocity, Z the mixture fraction and ρ is the density.

The mixing sub-step, **M**, required the input of the mixing frequency, Ω_ϕ. The mixing frequency, Ω_ϕ, is taken from the DNS as $\Omega_\phi = \widetilde{\chi_\phi}/Z''^2$ where $\widetilde{Z''^2}$ is the Favre-averaged variance of the mixture fraction and $\widetilde{\chi_\phi}$ is the Favre-averaged scalar dissipation, which is defined as, $\widetilde{\chi_\phi} = 2\widetilde{D|\nabla Z|^2}$, where D is the mixture averaged thermal diffusivity.

The reaction step (**R**) is solved for each particle using the same chemical kinetic mechanism as the DNS. This mechanism is a 19 species reduced ethylene mechanism [6]. The soot mechanism was originally developed by Leung and Lindstedt [15] and is also identical to the DNS. The soot Particle Size Distribution (PSD) is modelled using the method of moments, where the first three mass moments of the PSD are transported and closure of fractional moments is performed by assuming a lognormal size distribution [16].

3.2 Numerical Parameters

The C-TPDF method was implemented using a hybrid particle-mesh approach. An Eulerian mesh was populated with Lagrangian particles, which evolved according to Equations (1) and (2).

The key numerical parameters in this study were the number of cells (N_{cell}), the number of particles per cell (N_{pc}) and the timestep per iteration, t_{step}. All results presented in this study are based on simulations using 2000 particles per cell, 384 cells and 1.5×10^{-8} $[s]$ timestep. A sensitivity analysis was undertaken and these values were deemed to provide statistical convergence.

In this work, the soot moments are considered as additional variables transported on the notional particles. However, these moments are assumed to have zero diffusivity and thus soot moment information is not transferred between particles during the mixing process. This approach has been used in Ref. [17].

Joshua Tang.
E-mail address: joshua.tang@unsw.edu.au

3.3 Initial and Boundary Conditions

As there are no initial composition fluctuations, the TPDF simulation is initialised with Favre-averaged thermo-chemical, turbulence, and mean flow information from the DNS [6].

Boundary conditions are only required in the cross-stream ($\hat{y}$) direction due to the statistical one-dimensionality of the flame. The left boundary is a symmetry boundary corresponding to centre of the fuel stream. The right boundary is an outflow to the system and allows particles to leave the domain.

4 Results and Discussion

Figure 1 presents the spatial profiles of mean (top) and Root Mean Square (RMS) (bottom) mixture fraction at 30 and 50 t_j. These times are chosen as they occur well after the turbulence has fully developed. Figure 1 (top) shows that all mixing models correctly predict the mean mixture fraction profile. Since the mean mixture fraction is only affected by particle transport, the results presented provide confidence that the gradient diffusion hypothesis is valid and the data has been correctly extracted from the DNS.

Figure 1 (bottom) presents the spatial profile of mixture fraction RMS. All mixing models perform similarly for both times. At 30 t_j the mixing models slightly over predict the peak RMS, however at 50 t_j, the mixture fraction profile becomes similar to the DNS throughout the domain. While there are some slight differences between the profiles, good agreement between the PDF and DNS profiles is found, which shows that the mixing frequency has been correctly extracted from the DNS. The minor differences observed may be due to the fact that turbulence Schmidt number required for transport of the variance is slightly different for that of the mean.

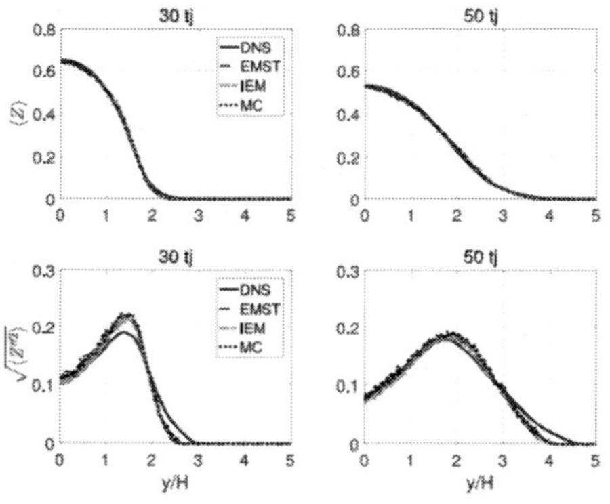

Figure 1: Mean (top) and RMS (bottom) mixture fraction profile at 30 and 50 t_j.

Figure 2 compares the spatial and temporal evolution of mean temperature from different model predictions to that from the DNS. The EMST model predicts the mean temperature very accurately, although the mean temperature in the flame appears to be slightly under predicted between 10 and 20 t_j. The IEM and MC model both over predict the amount of extinction, producing low temperatures compared to the DNS, from 20 t_j until the end of the simulation.

In the study by Krisman *et al.* [11] the IEM and MC models were able to capture extinction and re-ignition in syngas flames but over predicted the amount of extinction in the ethylene flame series. The difference was thought to be due to the broad reaction zone in the syngas flames compared to the relatively thin reaction zones in the ethylene flames. Since the IEM and MC models allow mixing between particles of different composition, burning and non-burning particles can mix. This causes these particles to move away from the burning solution, to an intermediate state. In the syngas cases, these intermediate states may still be sufficiently reactive to allow reignition, moving these partially burning particles back towards the burning solution. In the ethylene case, the burning and extinct states are more clearly separated due to the fuel's higher effective Zeldovich number [11]. Therefore, as the mixing process creates intermediate states, the particles are no longer reactive and quickly travel towards the extinguished, mixing only solution. This results in global extinction in the models, despite the DNS showing little to no extinction.

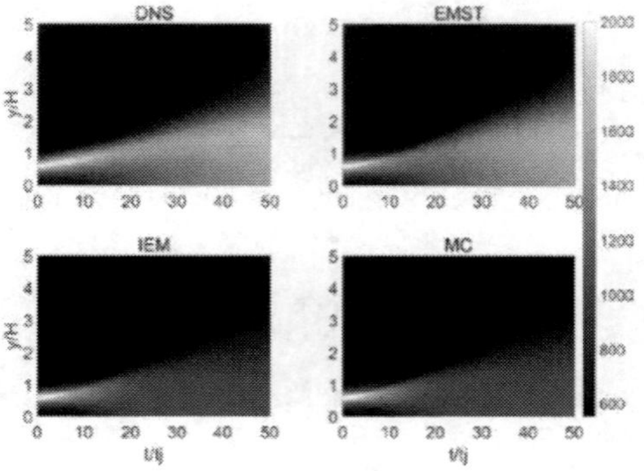

Figure 2: Temporal evolution of the mean temperature (K).

Since soot formation strongly depends on the temperature, the soot formation predictions vary strongly depending on the mixing model used. The IEM and MC models cannot accurately predict the mean temperature and, as a result, they are not able to produce the correct soot formation rates. Therefore, the results of the IEM and MC models will not be presented in the context of soot formation.

Figure 3 presents the spatial and temporal evolution of the mean soot mass fraction, Y_{soot}. The EMST model produces qualitatively similar results to the DNS, with the shape of the profile varying only slightly compared to the DNS.

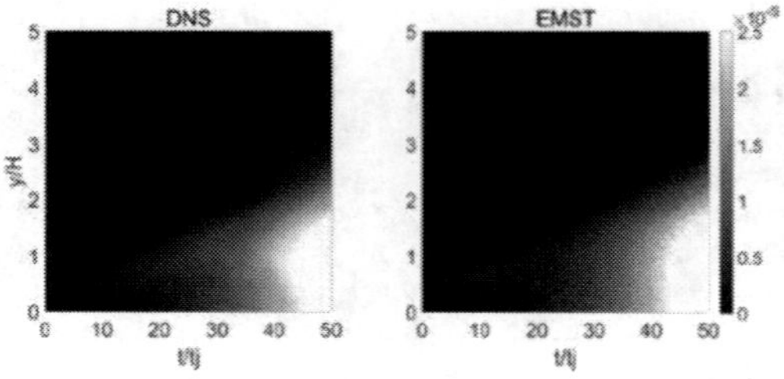

Figure 3: Temporal evolution of the mean soot mass fraction.

Figure 4 presents the spatial profiles of mean (top) and RMS (bottom) soot mass fraction at 30 and 50 t_j. The EMST results agree very well with the DNS results and the soot mass fraction is accurately predicted throughout the domain at both times. The soot mass fraction statistics shows a greater amount of statistical noise compared to the other quantities presented, such as mean temperature or mixture fraction, which is a result of the high intermittency associated with soot formation and oxidation. The soot mass fraction RMS is predicted well by the EMST model at 30 t_j, however at 50 t_j the EMST model significantly over predicts the soot mass fraction RMS within the jet core.

The unconditional statistics show that the EMST model can produce good results for both temperature and soot formation. Statistics conditioned on mixture fraction can provide additional information on the model performance. Figure 5 presents the PDF of temperature (top) and soot mass fraction (bottom) conditioned on mixture fraction at 50 t_j. The conditional mean of the temperature is predicted well by the EMST model, however the conditional fluctuations are significantly under predicted as demonstrated by the extremely narrow PDF. This behaviour is typical of this model and has been seen in previous mixing model studies [11]. Interestingly, the EMST model is unable to realise very rich mixtures (Z > 0.6) at t = 50 t_j stopping abruptly at Z = 0.6.

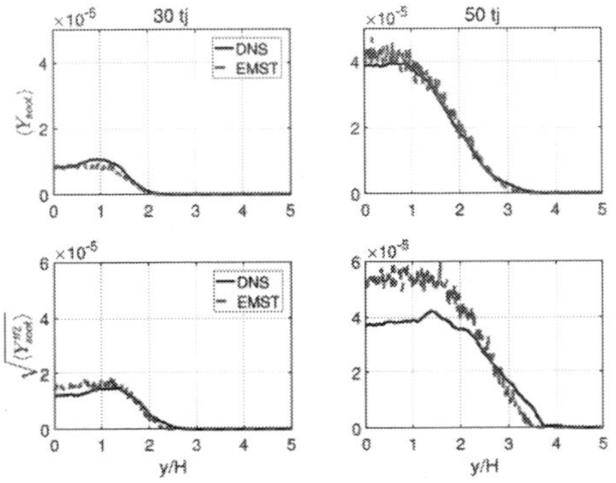

Figure 4: Mean (top) and RMS (bottom) soot mass fraction profile at 30 and 50 t_j.

Figure 5 (bottom) shows that the EMST model generally predicts the conditional mean of soot mass fraction well, slightly over predicting the conditional mean in the region Z > 0.4. The DNS shows that the soot mass fraction exhibits very large conditional fluctuations as result of the high intermittency of soot formation. This large fluctuation is captured by the EMST model, which is to be expected as the soot properties are not mixed in the results presented.

THE GRAPHIC ON THIS PAGE WAS NOT AVAILABLE ON THE E-MEDIA VERSION OF THIS PAPER.

Figure 5: PDF of temperature (K) (top) and soot mass fraction (bottom) conditioned on mixture fraction at 50 t_j. Red solid line represents the conditional mean taken from the DNS, green dashed line represents the model conditional mean.

4 Conclusions

An evaluation of mixing models was performed in the context of a RANS-based C-TPDF method. The C-TPDF model was compared with DNS of a sooting, turbulent non-premixed jet flames with ethylene fuel. In the present work the mixing frequency, mean velocity, and turbulent diffusion coefficient were taken directly from a DNS database, enabling the elimination of several possible sources of modelling error.

The TPDF method was found to be an attractive method for soot simulations, however the choice of mixing model was found to greatly impact the results. In the present study, soot mixing was set to zero, while gaseous phase mixing was considered with three mixing models, EMST, IEM and MC. The EMST mixing model performs significantly better than the other models. The EMST model can correctly predict the unconditional means of temperature and soot mass fraction. The conditional means of temperature and soot mass fraction are also well predicted by the EMST model.

Both the IEM and MC models predict large scale extinction, not present in the DNS. This results in lower temperatures and correspondingly low soot mass fractions. This highlights the importance of the mixing model for gaseous species in sooting flame simulations as only the correct prediction of the thermochemical state can produce correct soot predictions.

Future work will consider variations on the model for soot mixing, which here has been taken as zero.

5 Acknowledgment

This work was supported by the Australian Research Council. The research benefited from computational resources provided through the National Computational Merit Allocation Scheme, supported by the Australian Government. The computational facilities supporting this project included the Australian NCI National Facility, the partner share of the NCI facility provided by Intersect Australia Pty Ltd., Pawsey Supercomputing Centre (with funding from the Australian Government and the Government of Western Australia), and the UNSW Faculty of Engineering.

References

[1] K. J. Young, J. B. Moss, Combust. Sci. Tech. 105 (1995) 33–53
[2] S. Pope, Prog. Energy Combust. Sci. 11 (1985) 119-192.
[3] D. Haworth, Prog. Energy Combust. Sci. 36 (2009) 168-259.
[4] R.R. Cao, H. Wang, S.B. Pope, Proc. Combust. Inst. 31 (2006) 1543-1550.
[5] G. K. Batchelor, Phys. Fluids. 12 (1969) 233-239.
[6] D. O. Lignell, J. H. Chen, P. J. Smith, Combust. Flame 155 (2008) 316-333.
[7] Villermaux, J. Devillion, Proc. Int. Symp. Chem. React. Eng. (ISCRE), 1972, 1-13.
[8] J. Janicka, W. Kolbe, W. Kollman, J. Non-Equilib. Thermodyn. 4 (1) (1979). 47-66.
[9] S. Subramaniam, S. Pope, Proc. Combust. Inst. 115 (1998) 487-512.
[10] S. Viswanathan, H. Wang, S.B. Pope, J. of Comput. Phys. 230 (17) (2011) 6916-6957.
[11] A. Krisman, J. C. Tang, E. R. Hawkes, D. O. Lignell, J. H. Chen, Combust. Flame 161 (2014) 2085-2106.
[12] M. Kuron, E. R. Hawkes, Z. Ren, J. C. Tang, H. Zhou, J. H. Chen, T. Lu, Proc. of the Combust. Inst. 36 (2017) 1987-1995.
[13] C.A. Kennedy, M.H. Carpenter, R.M. Lewis, Appl. Numer. Math. 35 (3) 2000 177-219
[14] G. Strang, SIAM J. Numer. Anal. 5 (3) (1968) 506-517

[15] R. Lindstedt, S. Louloudi, Proc. Combust. Insti. 30 (2005) 775-783.
[16] S. E. Pratsinis, Journal of Colloid and Interface Science 124 (1988) 416-427.
[17] R.S. Mehta, D.C. Haworth, M.F. Modest, Combust. Flame 157 (2010) 982 - 994.

11th Asia-Pacific Conference on Combustion,
The University of Sydney, NSW Australia
10th -14th December 2017.

The Dynamic Characteristics of Jet-diffusion Flame and Premixed Flame

Myunggeun Ahn[1], Taesung Kim[1], Deahong Lim[1], Youngbin Yoon[1,2]

[1]Seoul National University Department of Mechanical and Aerospace Engineering

1, Gwanak-ro, Gwanak-gu, Seoul, 08826, Republic of Korea

[2] Seoul National University Department of Mechanical and

Aerospace Engineering and Institute of Advanced Aerospace Technology

1, Gwanak-ro, Gwanak-gu, Seoul, 08826, Republic of Korea

Abstract

This study investigated the dynamic characteristics of jet-diffusion flame and premixed flame. The heat release rate was measured using PMT (Photomultiplier Tubes). Because of the external excitation, we used a loudspeaker to observe flame dynamic characteristics. Experimental conditions include various amplitudes defined in this paper is the magnitude of velocity perturbation over the mean velocity ($u'/\bar{u}$). Experimental results show that the jet-diffusion flame has no effect when the amplitude is greater than 0.15. The premixed flame had no effect on the gain value despite the change in amplitude.

1 Introduction

Early in development of combustion system, preceding studies had focused more on mechanical operations than its completeness. Most combustion systems use diffusion flames which have the side effect of producing soot generated by combustion. Recently, the importance of energy conservation is emphasized worldwide, the type of combustion system was changed from diffusion flame to premixed flame. However, nitrogen oxide (NOx) emission phenomenon was increased near a theoretical air-fuel ratio of the premixed flame. As problems of environmental pollution issues are emerging recently, exhaust gas regulations have been strengthened, and several combustion systems have been revised as countermeasures. In lean-burn combustion system, the maximum temperature was not relatively high. Therefore, the lean-burn combustion system with less formation of nitrogen oxides had a drawback which unstable phenomenon appeared at a lean burn.

Environmental problems as well as combustion instabilities during combustion system operation are also being studied in many countries recently. The research about combustion instability was conducted at rocket engine system in the beginning. The combustion instability research is divided into a prediction and a control of combustion instability. It is necessary to study the factors causing combustion instability in order to predict combustion instability.

Factors causing combustion instability include acoustic perturbation, heat release perturbation, and velocity perturbation. When any one of these three factors arises, it would affect other factors, and intensity of factors will be amplified through the positive feedback process which would come back first. Because periodic oscillation occurs, the combustion system is not done, known that stability develops. These problems are deemed to have a fatal influence on the operation of combustion system which ultimately might cause damage.

Many researches on jet diffusion flame and premixed flame have been conducted as a basic study of combustion instability. You et al. [1] conducted an experimental study on the effect of single vortex interaction on the diffusion flame. Lakshminarasimhan et al. [2] experimentally investigated the effect of flame length, luminescence, and strong amplitude on the mixing of fuel and oxidizer in a flame spreading flame with strong amplitude. Joo et al. [4] conducted an experimental study of flame length, EINOx and ANN (Artificial Neural Network) in a partially premixed gas turbine combustor using H_2 / CO / CH_4 mixed gas. Yoon et al. [5] conducted an experimental study on the mode transition phenomenon of combustion instability in a partial premixed combustor using H_2 / CH_4 mixed gas. Hwang et al. [6,7] conducted an experimental study on flame stabilization and flame length of a turbulent jet diffusion flame using H_2 / CO as fuel. Kim et al. [8] confirmed that numerical and experimental studies show a similar tendency through comparison of numerical and experimental studies of the Burke-Schumann flame in diffusion flame combustors. Cho et al. [9] conducted a numerical study on the flame response characteristics according to the equivalence ratio perturbation in premixed flames. Wang et al. [10] numerically studied the flame response characteristics of premixed flames with speed perturbations. Durox et al. [11] experimentally confirmed the characteristics of the premixed Flame Transfer Function (FTF) with various shapes. Schuller et al. [12] experimentally studied the flame response characteristics of premixed flames by nonlinear processing based on excited-state G-equations. Cuquel et al. [13-14] experimentally studied the flame response characteristics of premixed flames by varying the diameter of the quartz tube that blocks air inflow.

There are not many examples studies of flame response characteristics for diffusion flame in the previous works, and there are not many experimental cases of composition with flame response characteristics of premixed flame and jet-diffusion flame. Therefore, in this study, flame characteristics

Corresponding author. Fax: +82-2-872-8032
E-mail address: ybyoon@snu.ac.kr

of premixed flame and jet-diffusion flame were compared. The flame transfer function was calculated by measuring the heat release rate using PMT (Photomultiplier Tubes). Flame response characteristics of various velocity perturbations were studied by using mixed fuel with CH_4 / H_2 of 50/50%.

2 Experimental apparatus and method

2.1 Experimental model combustors

Figure 1 is a cross-sectional schematic combustor used in the experiment. Figure 1 (a) is a premixed flame combustor. Fuel was supplied to the down part, and oxidizer was supplied from both sides. Fuel and oxidizer were immediately mixed in the mixing space, and the mixed fuel and oxidizer ignite premixed flame on the tip of the nozzle. Figure 1 (b) is jet-diffusion flame combustor. Fuel was supplied from the fuel supply line and oxidant was supplied from both sides. The supplied fuel and oxidizer ignite a jet-diffusion flame from the tip of the nozzle. The outer diameters of oxidant supply line arranged coaxially with the fuel supply line were 6.35 mm and 12.7 mm respectively. A circular quartz tube (diameter: 35 mm) was used to remove the impact of any eternal airflows which maintained a flame attached to the nozzle of the combustor. To investigate flame perturbations phenomenon, both sides of combustor were equipped with two loudspeakers.

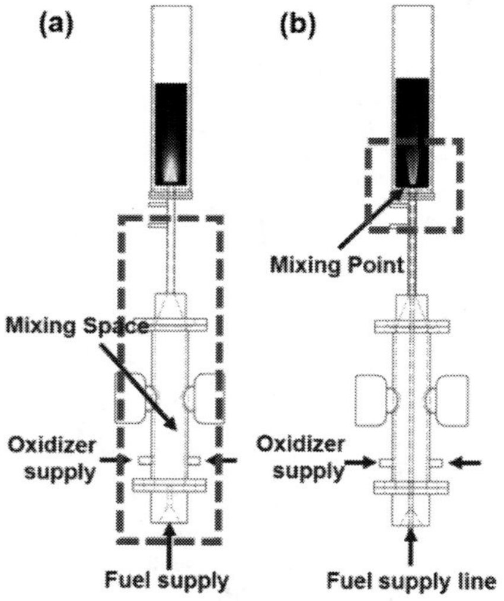

Figure 1: Schematic of a cross-section of a combustor.
a) premixed flame (b) jet-diffusion flame.

Since the design of two combustors were different, forcing frequency directions were also different. In the case of a premixed flame combustor, traveling frequency of supplied forcing frequency flows in a single nozzle line. However, in the case of a jet-diffusion flame combustor, since fuel and oxidant are supplied separately, forcing frequency must be supplied to the fuel or oxidant supply pipe. While the supply flow rate of the fuel is 0.98 slpm, the supply flow rate of the oxidizer is relatively high which is at 11.40 slpm. Therefore, in this study, forcing frequency was supplied to the oxidizer supply line with a relatively larger flow rate than fuel supply line to observe flame perturbation. Because supplying a forcing frequency to the oxidizer supply line, which is supplied at a large flow rate, allows better observation of flame perturbation than a fuel supply line supplied at low flow rates. A flow rate measuring device for speed measurement was placed in the possible location of the nozzle where the flame was formed, and it was easily configured for speed measurement and speaker periodicity.

2.2 Experimental Conditions

Table 1 shows experimental conditions. The fuel used in the experiment was CH_4 / H_2, and the oxidizer was air. The fuel and oxidizer were supplied via Mass Flow Controller (MFC), and the fuel mixture composition ratio was 50/50% of CH_4 / H_2 respectively. The flow rate of fuel was supplied at CH_4 / H_2 of 0.98 slpm, and the air was supplied at 11.40 slpm. The fuel jet was set to an equivalence ratio of 1 for the premixed flame. In order to quantitatively compare the premixed flame and the jet-diffusion flame, the equivalent ratio of the premixed flame was set to 1. Therefore, the fuel and air supply flow rate of premixed flame were supplied same as to the diffusion flame. The forcing frequency for flame perturbation was gradually increased per 20 Hz at between 40 - 300 Hz when supplied. In order to observe the magnitudes of flame perturbations variously, the speaker voltage was increased than multiple amplitudes were generated. The amplitude defined in this paper is the magnitude of velocity perturbation over the mean velocity ($u'/\bar{u}$). The amplitude of premixed flame was supplied at 0.15, 0.30, 0.45, and the amplitude of jet-diffusion flame was supplied at 0.12, 0.15, 0.25, 0.35.

Parameter	Change
Fuel & Oxidizer	CH_4 / H_2 & Air
Fuel compositions	CH_4 / H_2 (50/50%)
Supply rate of fuels	0.98 slpm (respectively)
Supply rate of air	11.40 slpm
Forcing frequency	40 – 300 Hz (20 Hz steps)
Magnitude of amplitude	
Premixed flame	0.15, 0.30, 0.45
Jet-diffusion flame	0.12, 0.15, 0.25, 0.35

Slpm : standard liter per minute (L/min)

Table 1: This is an example of a table and table title. For tables, the recommended size is 75mm wide.

3 Result and Discussion

3.1 Premixed Flame and Jet-Diffusion Flame Shape on Nozzle End

Figure 2 shows the result of visualizing the premixed flame and jet-diffusion flame applied in this study through DSLR camera and OH-chemiluminescence measurement. Figure 2 (a) and (c) show the images obtained by setting the exposure time of the DSLR camera to 10 seconds. (B) and (d) are visualized images obtained through OH-chemiluminescence measurement.

Figure 2 (a) and (b) show photographs of the premixed flame visualized. The excitation frequency for the flame perturbation is supplied through a single nozzle. The fuel and air supplied via the MFC control are mixed immediately in the mixing space of the combustor thereby ignite the premixed flame at the end of the nozzle. The premixed flame in this study is a conical flame shape, which is the shape of a general premixed flame

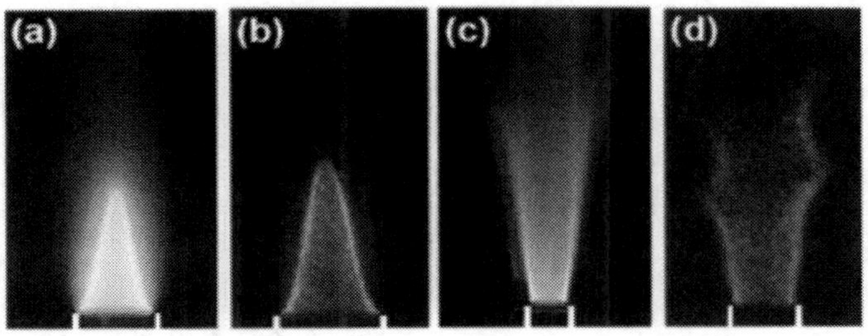

Figure 2: Premixed flame & jet-diffusion flame visualization results; captured by (a), (c) DSLR camera, (b), (d) OH-chemiluminescence in 120 Hz, Amplitude of 0.15.

supplied after mixing fuel and oxidant.

Figure 2 (c) and (d) show a visualization of jet-diffusion flame. The jet-diffusion flame combustor used in this study is fueled in the fuel supply line and air is supplied from a coaxially aligned air supply line to form jet-diffusion flame at the nozzle end. To better observe the perturbation of the flame over the frequency, the frequency supplied to the air supply line which has a higher flow rate than the fuel supply line. The jet-diffusion flame shape of this study is observed in the form of a jet flame, but it is not the shape of a general jet flame with the flame end closed like a candle. The phenomenon that the end of the flame does not close can be explained for two reasons. The first reason is the experimental condition. In this study, the fuel and air flow rate were tested on the basis of the fuel and air supply flow rate of the premixed flame equivalence ratio 1. For the purpose of quantitative comparison of two flames, the experiment was applied to the diffusion flame based on the fuel and air supply flow rate, which is equal to 1 for the premixed flame, because there is a concept of equivalence ratio in the premixed flame, but not in the jet-diffusion flame. Experimental conditions of the premixed flame (equivalence ratio 1) are applied to jet-diffusion flame, and when the experiment is carried out, a jet flame with an open end is formed because a sufficient amount of air is not supplied to close the end of the jet flame. If enough air is supplied, the end of the flame closes and a flame length elongated jet flame is formed. The second reason is due to the quartz tube that blocks external air inflow during combustion. When the quartz tube is not installed, the shape of the common diffusion flame is formed, in which the end of the flame is closed. However, when the quartz tube is attached, the outside air flow for closing the end of the flame is blocked and the coaxially arranged air supply line Fig. 2 (c) and (d).

3.2 Premixed flame response and phase difference

In this study, the transfer function of control engineering is applied to the flame to understand the relation between flow perturbation and heat perturbation. This can be expressed as a Flame Transfer Function (FTF) that can be expressed as a function of the perturbation frequency. The flame transfer function is expressed as

$$F(\omega) = \frac{q'/\bar{Q}}{u'/\bar{U}} = G(\omega)e^{i\emptyset(\omega)}$$

In this equation, $G(\omega)$ is the gain of the flame transfer function, $\emptyset(\omega)$ is the phase difference, the real part is the natural frequency of the system, the imaginary part is the growth rate of the initial instability factor, and the instability is amplified when the imaginary part is negative.

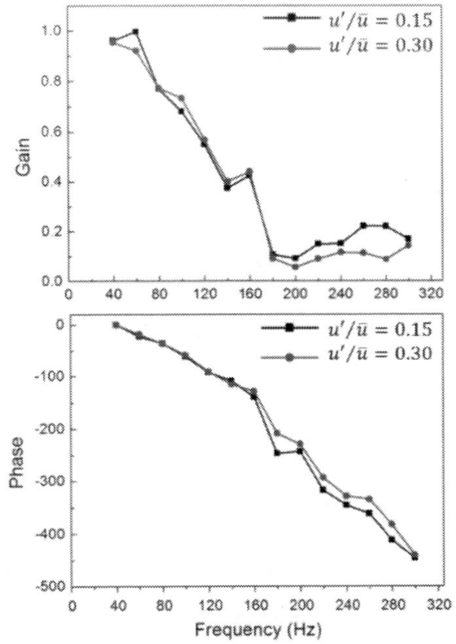

Figure 3: The result of premixed flame response characteristics with gain and phase.

Figure 3 shows the response and phase difference of the flame transfer function with the excitation frequency of the premixed flame. The response value of the premixed flame can be observed that the response value decreased before 180 Hz and the response value does not change much after 180 Hz. In addition, when the frequency having both of the amplitudes of 0.15 and 0.30 increases, the response value tends to decrease and there is almost no difference in the response value between the two amplitudes. Results with the same value even if the amplitudes are different also appear in the phase difference. As a result of the phase difference, it can be observed that the two amplitudes have the same value as the response value. It can be concluded that the premixed flame satisfies the assumption of the flame transfer function in the sense that the response value and the phase difference value are the same in spite of the difference in the magnitude of amplitude because the heat release rate to the velocity perturbation is proportional to the assumption of the flame transfer function. It is experimentally confirmed that in the previous study [14], the explanation that the change of the velocity perturbation value does not have a great influence on the heat release for the amplitude of 0.20 or more is valid even when the amplitude is 0.30.

3.3 Jet-diffusion flame response and phase difference

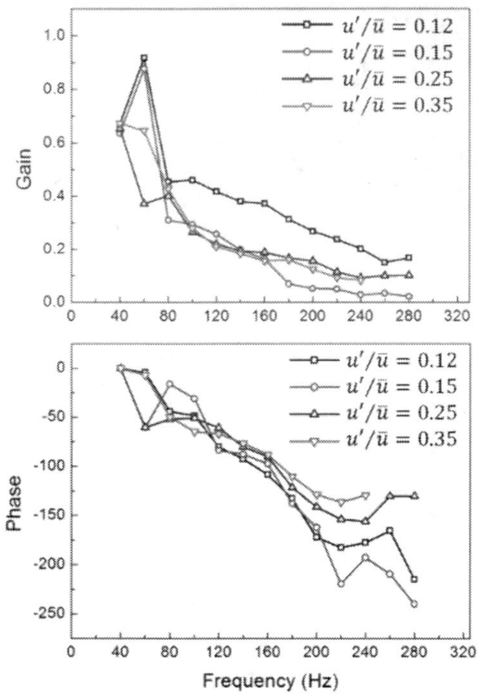

Figure 4: The result of jet-diffusion flame response characteristics with gain and phase.

Figure 4 shows the result of the flame transfer function of the response and the phase difference according to the excitation frequency of the diffusion flame. The response value has the same value for all amplitudes at 40 Hz. At 60 Hz, 0.12 and 0.15 increase with similar tendency, but 0.25 and 0.35 show completely different trends. After 80 Hz, the tendency of the response value becomes clear. It can be seen that the three amplitudes of 0.15, 0.25, and 0.35 tend to decrease to similar response values, but the amplitude of 0.12 has a larger response value than the other amplitudes and appears to decrease. The reason why the amplitude of 0.12 differs between the three amplitudes and the response value is because the magnitude of the heat release rate to a given velocity perturbation is different. That is, it can be judged that the response value of the diffusion flame is a nonlinear behavior. As a result of the phase difference, 0.12 and 0.15 have similar values and 0.25 and 0.35 have similar values. Based on the result of the response value, it can be observed that the flame response characteristic can be closely observed by using the Flame Describing Function having the velocity perturbation as a variable rather than the Flame Transfer Function which is proportional to the velocity perturbation and the heat release perturbation.

4 Conclusions

The dynamic characteristics of the premixed flame and the jet-diffusion flame were investigated by heat release rate measurement. The premixed flame confirmed the linear behavior of the gain value regardless of the amplitude. The jet-diffusion flame has nonlinear behavior when the gain is less than 0.20 and linear behavior when the gain is greater than 0.25.

5 Acknowledgment

This work was supported by Advanced Research Center Program (NRF-2013R1A5A1073861) through the National Research Foundation of Korea(NRF) grant funded by the Korea government(MSIP) contracted through Advanced Space Propulsion Research Center at Seoul National University. This work was supported by the National Research Foundation of Korea(NRF) grant funded by the Korea government(MSIP) (No. 2015R1A2A2A010043). This work was supported by NRF(National Research Foundation of Korea) Grant funded by the Korean Government(NRF-2017-Fostering Core Leaders of the Future Basic Science Program/Global Ph.D. Fellowship Program).

References

•Y.H. You, D.K. Lee, H.D. Shin, Combust. Sci. and Tech., 139 (1) (1998) 365-393.

•K. Lakshminarasimhan, N. T. Clemens, O. A. Ezekoye, Experiments in fluids, 41 (4) (2006) 523-542.

•K. Lakshminarasimhan, M.D. Ryan, N.T. Clemens, O.A. Ezekoye, Proc. Combust. Inst., 31 (1) (2007) 1617-1624.

•S. Joo, J. Yoon, J. Kim, M.C. Lee, Y. Yoon, Appl. Therm. Eng., 80 (2015) 436-444.

•J. Yoon, S. Joo, J. Kim, M.C. Lee, J. Lee, Y. Yoon, Proc. Combust. Inst., 36 (3) (2017) 3753-3761.

•J. Hwang, N. Bouvet, K. Sohn, Y. Yoon, 17 (1) (2012) 22-29.

•J. Hwang, K. Sohn, N. Bouvet, Y. Yoon, J. Korean Soc. Combust., 17 (4) (2012) 30-37.

•T. Kim, M. Ahn, J. Hwang, S. Kim, Y. Yoon, Proc. Combust. Inst., 36 (1) (2017) 1629-1636.

•J.H. Cho, T. Lieuwen, Combust. Flame, 140 (1) (2005) 116-129.

•H.Y. Wang, C.K. Law, T. Lieuwen, Combust. Flame, 156 (4) (2009) 889-895.

•D. Durox, T. Schuller, N. Noiray, S. Candel, Proc, Combust. Inst., 32 (1) (2009) 1391-1398.

•T. Schuller, S. Ducruix, D. Durox, S. Candel, Proc, Combust, Inst., 29 (1) (2002) 107-113.

•A. Cuquel, D. Durox, T. Schuller, Comptes Rendus Mécanique, 314 (1-2) (2013) 171-180.

•A. Cuquel, D. Durox, T. Schuller, Proc. Combust. Inst., 34 (1) (2013) 1007-1014.

11th Asia-Pacific Conference on Combustion,
The University of Sydney, NSW Australia
10th-14th December 2017.

The effects of positive electric field on the spherical flame propagation at elevated pressures

Yiming Li[1], Jinhua Wang[1], Xuxing Wei[1], Xiaomin Wu[1], Zuohua Huang[1], Haibo Mu[2], Guanjun Zhang[2]

[1]School of Energy and Power Engineering, Xian Jiaotong University,
Xi'an, 710049, China

[2]State KeyLaboratory of Electrical Insulation and Power Equipment, Xian Jiaotong University,
Xi'an, 710049, China

Abstract

The effects of positive DC electric field, which the electric field vector points to the burned region, on the spherical flame propagation at elevated pressures have been investigated. The results show that positive electric field shows similar effects on the enhancement on flame propagation with negative electric field, and the effects of electric field are also evident at high pressures. Cracks on the flame front increase with voltage at high pressures, and it is pronounced at higher pressures and leaner mixtures. The smaller physical scale cracks could be observed at 0.5MPa with 10kV electric field, which is quite different with previous study [8] at 0.2MPa pressure and AC electric field. The flame initiation time, which is analyzed from pressure data, decreases with voltage increases. And it also shows more obvious at higher pressures and leaner mixtures.

1 Introduction

Flames are sometimes described as weakly ionized plasmas, so the sub-breakdown electric field, which can be used to manipulate their properties, has become of interest in the combustion field. Electric field assisted combustion has been reported in terms of increased premixed flame speeds [1], reduction in the emission of nitrogen oxides, carbon oxides and soot formation [2][3], and modifications in the flame stability [4].

For DC or low frequency AC electric field, the so-called ionic wind effects, a result of a body force induced by electric field acting on charged particles, is recognized as an explanation for the enhancements on flame under electric field. The ionic wind effect is a bulk flow which induced by oriented migration and collisions of charged particles under the electric field. The charged particles in the flame front migrate towards the electrodes and there cause the momentum transfer during the multi-collision processes between the charged ions and the neutral molecules. The intensity of ionic wind is associated with ion mobility and electric field intensity. For the electrons, which are main elements of negative charged particles, have negligible mass compared with positive ions. Thus there exist cathode-directed convective flows for DC electric field. Many previous studies [5][6] usually put the cathode on the unburned region compared with the cathode on the burned region. But the latter configuration also shows the similar enhancement on the flame [7].

Corresponding author: Jinhua Wang.
E-mail address: jinhuawang@xjtu.edu.cn

Most previous researches are conducted at atmospheric pressure, however many industrial combustors are not restricted to atmospheric pressure. Pressures variation would change the flame properties (e.g. flame temperature, flame thickness, flame speed, ion mole fraction…), and the variation of flame properties would change the effect of electric field, which is proved by the different effects with various excess air ratios [1][7]. Thus it would be a meaningful work to ascertain whether electric fields are effective at elevated pressures.

2 Experimental Setup

The experimental system, including a fuel supply, a constant-volume combustion chamber, an ignition control circle, an optical Schlieren system, a high-speed camera, a pressure acquisition and a high-voltage supply system, is shown in Fig.1.

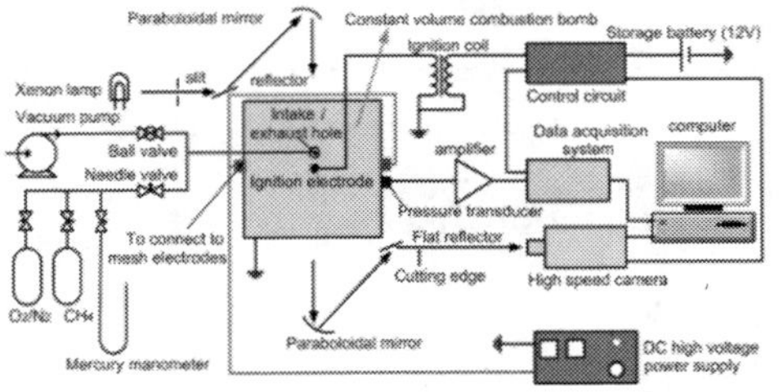

Figure 1: Schematic of the experimental setup

The constant-volume combustion chamber, which is shown in Fig.2, is a cylinder with a diameter of 130 mm and a length of 130 mm. The chamber is covered with an insulating layer which is made of polytetrafluoroethylene (PTFE). Two ignition electrodes covered with PTFE are installed at the vertical center. A pair of quartz glasses with a diameter of 80 mm is installed on both sides of the combustion chamber to provide an optical access.

To visualize flame fronts during flame propagation, a high-speed digital camera (HG-100K) is adopted. The frequency of the high-speed digital camera is set to be 5000 frames per second. A piezoelectric-type pressure sensor (Kistler 4075A10) is used to record the pressure history during combustion in the chamber, which has an error of less than ± 0.3% and the collection frequency of 20 kHz.

Two mesh electrodes with the outer diameter of 60 mm are installed horizontally and symmetrically placed 35 mm away from the ignition electrodes, which are connected to the high-voltage DC power supply (Wisman DEL15P15) with the voltage range from 0 kV to 15 kV, so that the positive DC electric field can be generated between the ignition electrodes and the mesh electrodes. The electric field is from mesh electrode oriented to ignition electrode.

Methane is used in this study, excess air ratio and applied electrical potential are varied. Each experiment is repeated at least three times for the same condition. Table 1 summarized the experimental conditions of this study.

Mass burning rate is calculated from pressure history with Elbe's equation. And flame initiation time, which is defined as the time between the moment of ignition and 10% mass burning rate, is used to evaluate the effect of electric field on the initial flame propagation process.

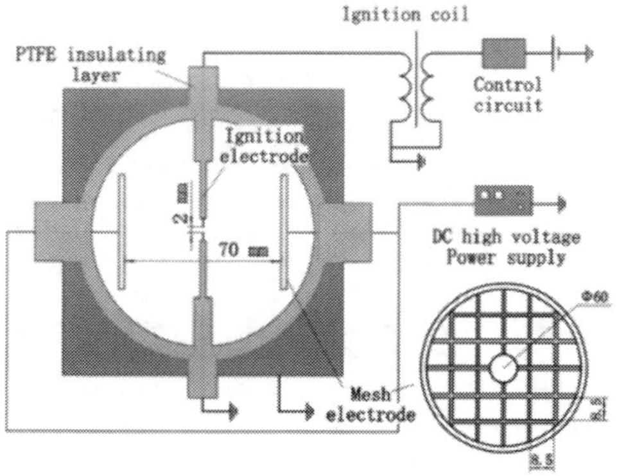

Figure 2: Schematic of the experimental setup

Parameters	Specifications
Fuel	methane
Excess air ratio	1.0,1.2,1.4
Electrical potential	0,5,10kV
Initial pressure	0.1,0.3,0.5MPa
Initial temperature	300K

Table 1: Experimental conditions

3 Results and Discussions

3.1 Flame Propagation Characteristics

Spherical flame propagation images at elevated pressures for the excess air ratio of 1.2 are shown in Fig.3. For the 0kV condition, which corresponds to the images in the first row, the flame front spreads nearly spherical, as expected for laminar methane/air mixtures. Something different is that the horizontal propagation is suppressed when the flame front near the mesh electrodes, it is due to the heat loss effect. For U=5 kV, the shape of flame front shows slight deformation in the horizontal direction and it indicates that flame propagation speed increases. When the voltage increases to 10 kV, obvious changes in the shape of the flame front as well as its enhancement of propagation speed could be observed. Remarkable enhancement in flame propagation can be found in the horizontal direction where the mesh electrodes are installed, and it is due to the body force induced by the motion of charged particles along the electric field direction. And it shows that the enhancement of the flame propagation induced by positive DC electric field has similar effects with the negative DC electric field[7]. Having considered that the intensity of ion wind induced by negative ions is much weaker due to the negative ions concentration, the effect of electrons should be noted for the similar effect between positive and negative electric field.

the effect of electric field on the flame is similar at elevated pressures. However, there are still some differences at high pressures. Flame front cellular crack was occurred when the initial pressure is up to 0.3 MPa even without electric field. Since Lewis numbers of methane and oxygen in methane/air mixtures are close to unity, the effect of thermodiffusive instability can be neglected. It may be caused by hydrodynamic instability. The flame thickness decreases with pressure and the thermal expansion increases with pressure, and they promote hydrodynamic instability. When the electric field is applied, the cracks in the flame surface would increase, and the time corresponding to crack initiation will be in advance. And this effect is more obvious for 10 kV voltage. And the cracks are much more intensive when electric field is applied at 0.5 MPa. What's more, the physical scale of cracks is much smaller than that without electric field condition at 0.5 MPa. It is inconsistent from the Cha's [8] study, in which they found that the physical dimensions of the electrically induced cracked structure are much larger than those found in the mixture without electric field. In consideration of the different condition, it may be related to the higher pressures and different electric field in our experiment.

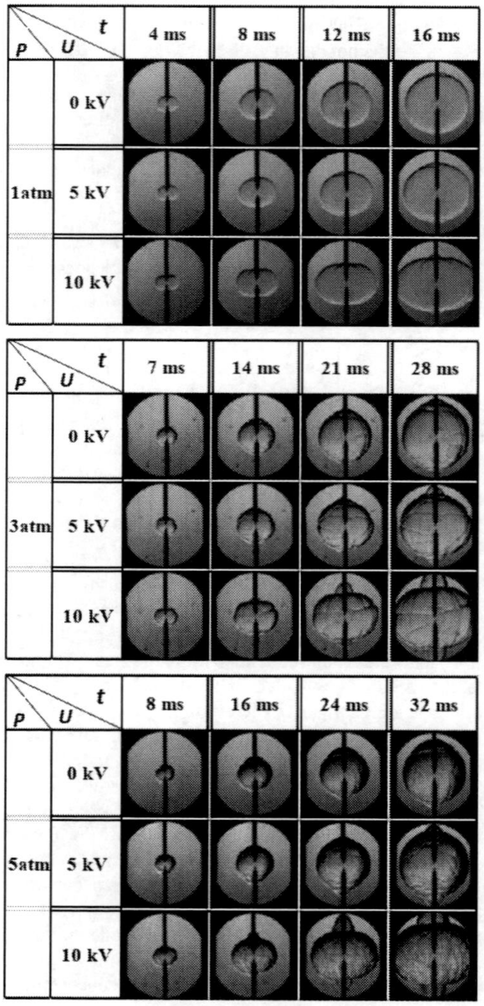

Figure 3: Images of flame propagation: $\lambda = 1.2$

There are several hypotheses to account for the phenomenon. Firstly, the electric field triggered flame instability could be attributed to hydrodynamic instability which is induced by the ionic wind effect, and it is similar with the connection between flame hydrodynamic instability and turbulent flow. The flame front is affected by the bulk flow and results in the cracks promotion with electric field. Though the number of negative ions is less than the positive ions, the Park's study[9] showed that the negative ions can also induce the ionic wind. Secondly, the electric field would change the flame thickness and thermal expansion due to the heat/mass transfer from the electrons/ions collisions. These would influence the flame hydrodynamic instability. But it need further study. For Lewis numbers of methane and oxygen in methane/air mixtures are close to unity, and it shows no obvious change in cracks at atmosphere pressure with electric field. The effect of electric field on thermodiffusive instability proposed by Ganguly[10] may not be a key factor in this study, but it should also make a difference.

Some unique phenomenon is found in this study. As shown in Fig.3, obvious bulging deformation in the vertical direction, and it is more obvious with the initial pressure increases. This phenomenon is not observed in the previous research which put the cathode on the unburned region. The explanation for this phenomenon is not clear and needs further study.

The different results with several excess air ratios at 0.3MPa are shown in Fig.4. It seems that the tendency with excess air ratio is independent of pressure. And the effect of electric field assisted combustion shows more obvious on leaner flame. And the positive electric field shows the same tendency with excess air ratio compared with negative electric field. There may be competitive relation of decrease in flame speed, increase in flame thickness and decrease in ion mole fraction with excess air ratio, which are related to the effect of electric field. And it seems that the former two factors are more important.

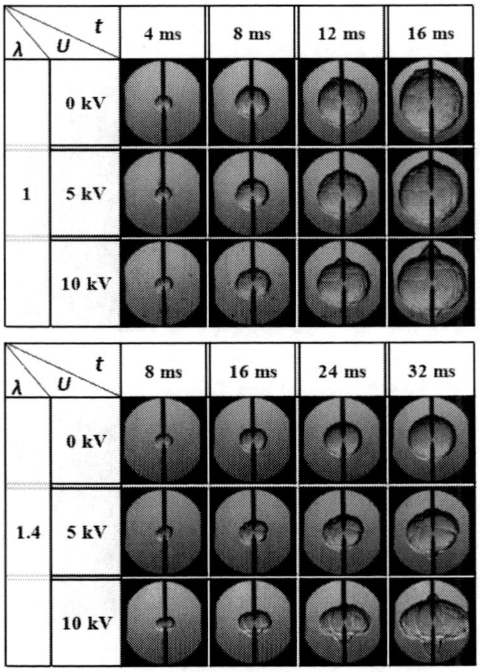

Figure 4: Images of flame propagation: P=0.3MPa

3.2Pressure Variation Inside the Chamber

The pressure history during the combustion period is shown in Fig.5. Subfigures are listed by initial pressure, and tested conditions under the electric field with V=0, 5, and 10kV, and the excess air ratio is 1.2. The tendency of the pressure with electric field is similar with the base condition. After ignition, the pressure is nearly constant at initiation time, then sharp increase in pressure can be observed, and it slows down again before reaching its maximum value. Finally, pressure decreases due to heat loss after complete burn out. When the electric field is applied, the peak pressure value shows slightly increase, which may be due to the decrease in the heat lose to mesh electrode. And the time which pressure is up to the maximum is in advance, which shows the advance is related to voltages. As shown in Fig.5, the change in pressure is more evident when the initial pressure increases.

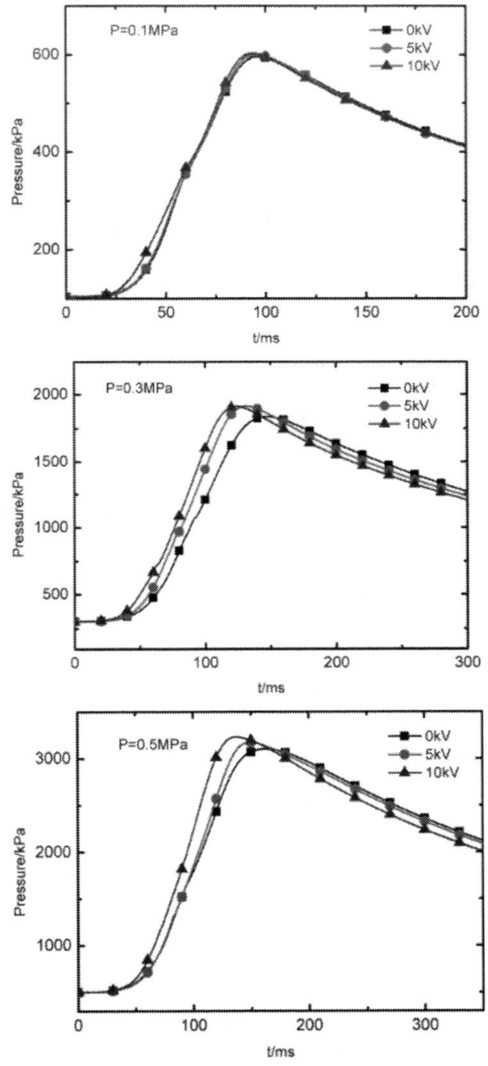

Figure 5: Pressure data with different initial pressures

The pressure variation with several excess air ratios at 0.3MPa is shown in Fig.6. The tendency of change in pressure is similar with the alteration of flame propagation characteristics shown in Fig.4. The effects of electric field are pronounced in the fuel lean mixture.

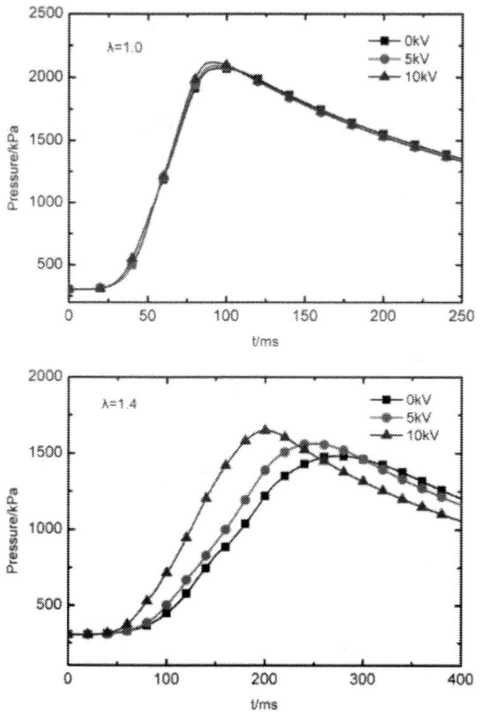

Figure 6: Pressure data with different excess air ratios

It should be noted that the main effect of the pressure variation occurred at the initial stage during the combustion period. It could be concluded that the electric field significantly affects the flame at initial stage during the combustion and that it governs overall combustion characteristics. The results are consistent with the previous study[8]. The flame initiation time, as defined in the experimental section, is obtained from the mass burning rate, which is gained from the pressure data. The result is shown in Fig.7, and the excess air ratio is 1.2 with several initial pressures.

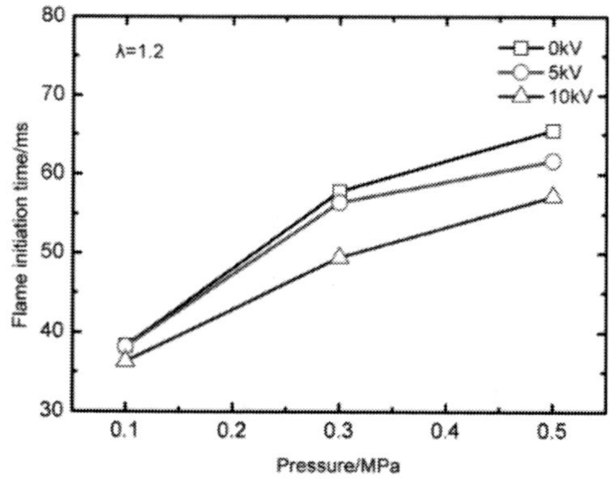

Figure 7: Flame initiation time with different initial pressures

The initiation time decreases with voltage increases, and the effect is pronounced at the high pressures compared with the slightly variation at atmosphere pressure. Considered the flame structure change as shown in Fig.3, it may be due to the intrinsic instability induced by electric field at the high pressures, which has influence on the combustion characteristics. Otherwise, the effects are also pronounced in the fuel lean mixture.

4 Conclusions

The effects of positive electric field, which the electric field vector points to burned region, on the spherical flame propagation and combustion have been investigated.

It was found that the positive electric field shows similar enhancement on flame propagation though the negative ions are much less, which indicates that electrons may make some differences on the effect of electric field on the flame.

The effects of electric field are also found at high pressure. It is sensitive to excess air ratios, and it is pronounced in the fuel lean mixture. The cracks, which are related to instability, are observed at high pressures, and it is pronounced at higher pressures. The physical scale of cracks at 0.5MPa with electric field is contrary to previous study[8] due to the different pressures.

The flame initiation time, which is obtained from pressure data, decreases with voltage increases. And it shows more obvious at higher pressures and leaner mixture.

The detailed explanation for pressure on the effect of electric field on flame needs further investigation.

5 Acknowledgment

The authors gratefully acknowledge the National Natural Science Foundation of China (Grants No: 51376004, 51476126)

References

[1] Hao D, Wu X, Sun T, Fuel 158 (2015)807-815.

[2] Sakhrieh A, Lins G, Dinkelacker F, Combustion. Flame143(3) (2005) 313-322.

[3] Wang Y, Nathan G J, Alwahabi Z T, Combustion. Flame 157(7) (2010) 1308-1315.

[4] Kribs J D, Shah P V, Hutchins A R,Journal of Electrostatics 84 (2016) 1-9.

[5] Luo Y, Gan Y, Jiang X,Fuel188 (2017)621-627.

[6] Kuhl J, Seeger T, Zigan L, Combustion. Flame 176 (2017) 391-399.

[7] Li C, Wu X, Li Y, Fuel193 (2017)22-30.

[8] Min S C, Lee Y, IEEE Transactions on Plasma Science40(12) (2012) 3131-3138.

[9] Park D G, Chung S H, Min S C, Combustion. Flame 168(2016) 138-146.

[10] Wismana D L, Marcuma S D, Combustion. Flame 151(4) (2007)639-648.

11th Asia-Pacific Conference on Combustion,
The University of Sydney, NSW Australia
10th -14th December 2017.

Experimental and kinetic modeling investigation on premixed tetralin flames

Yuyang Li[1,2,*], Wenhao Yuan[2], Chuangchuang Cao[3], Yan Zhang[3], Jiabiao Zou[2], Yizun Wang[2]

[1] Collaborative Innovation Center for Advanced Ship and Deep-Sea Exploration (CISSE), Shanghai 200240, PR China

[2] Key Laboratory for Power Machinery and Engineering of MOE, Shanghai Jiao Tong University, Shanghai 200240, P. R. China

[3] National Synchrotron Radiation Laboratory, University of Science and Technology of China, Hefei, Anhui 230029, P. R. China

Abstract

This work reports the investigation on premixed flames of tetralin at 30 Torr and equivalence ratios of 0.7 and 1.7. Measurements of the flame chemical structure, i.e. identification and mole fraction measurements of radicals, isomers and polycyclic aromatic hydrocarbons (PAHs), were performed using synchrotron vacuum ultraviolet photoionization mass spectrometry (SVUV-PIMS). A kinetic model with 296 species and 1571 reactions was developed and validated by the flame chemical structure data in this work. Modeling analysis demonstrates the key reaction pathways in tetralin decomposition and aromatics growth. It is concluded that H-atom abstraction reactions dominate the consumption of tetralin under premixed flame conditions, while oxidation reactions play more important roles in the lean flame. Indene and naphthalene are abundantly produced in the tetralin flames since they are key decomposition products of tetralin. Consequently their radicals play important roles in the aromatics growth process and stimulate the production of many typical large PAHs.

1 Introduction

Tetralin is a typical bicyclic aromatic hydrocarbon and has wide applications in both combustion and energy conversion. It is a typical hydronaphthalenic component in petroluem or synthetic transportation fuels and their surrogate mixtures [1-6]. It was found to effectively reduce the formation of soot in the pyrolysis of actual jet fuels [7]. On the other hand, tetralin is widely recognized as an effective hydrogen donor [8, 9] and serves as a representative of hydrogen donors in the coal liquefaction processes [10, 11]. It is also used as a model compound for hydroaromatic structures in coal [12]. Furthermore, tetralin combustion can abundantly produce naphthalene which makes it an ideal system to investigate naphthalene chemistry [9, 13, 14].

Previous experimental studies on tetralin combustion have been reviewed in detail by several groups [12, 14, 15]. In general, the investigations on tetralin combustion before 2013 are mainly focused on the pyrolysis. After 2013, there are several oxidation studies for tetralin. Dagaut et al. investigated the oxidation of tetralin in a jet-stirred reactor at 1 and 10 atm, 700-1400 K and equivalence ratios of 0.5-1.5 [16]. Gas chromatography (GC) was used in their work to measure stable products in tetralin oxidation. Wang et al. measured the ignition delay times for tetralin/air mixtures in a shock tube at equivalence ratios of 0.5 and 1.0, pressures of 13 and 37-39 bar and temperatures from 978 to 1277 K [17]. Zhu et al. conducted shock tube ignition delay measurements of decalin/air mixtures over temperatures of 769-1202 K, pressures of 11.7-51.2 atm, and equivalence ratios of 0.5, 1.0 and 2.0 [18]. Kukkadapu et al. investigated the ignition characteristics of tetralin at pressures of 15-50 bar, temperatures of 762-950 K and equivalence ratios of 0.5 and 1.0 using a rapid compression machine (RCM) [19]. The previous modeling efforts are also focused on the pyrolysis models of tetralin [9, 13-15, 20]. Up to now, only Dagaut et al. reported a kinetic model for tetralin oxidation [16]. It is recognized that there is no investigation on tetralin flames, thus the validation of tetralin combustion models under high temperature conditions is also limited.

Regarding above issues, an experimental and kinetic modeling investigation on tetralin combustion was carried out in this work. The first target of this work is to investigate the chemical structures of lean and rich premixed tetralin/O_2/Ar flames. Synchrotron vacuum ultraviolet photoionization mass spectrometry (SVUV-PIMS) was adopted for the identification and mole fraction measurements of flame intermediates such as radicals, isomers and polycyclic aromatic hydrocarbons (PAHs). The second target is to develop a kinetic model for tetralin combustion with good prediction abilities for tetralin pyrolysis, oxidation and flames under intermediate-to-high temperature region. Key reaction pathways in both tetralin decomposition and aromatics growth processes were analyzed using modeling tools.

2 Experimental method

In this work, chemical structure measurements of tetralin flames were carried out at the Combustion and Flame endstation of National Synchrotron Radiation Laboratory. Detailed descriptions of the synchrotron facility and flame apparatus can be found elsewhere [21, 22]. In brief, the flame apparatus consisted of three major parts, i.e. the flame chamber with a McKenna burner, the differentially pumped chamber with a molecular sampling system and the photoionization chamber with a home-made reflectron time-of-flight mass spectrometer (RTOF-MS). Tetralin with a purity of 99% was provided by Aladdin Reagent Co., Ltd. Ar and O_2 were provided by Nanjing Special Gas Factory Co., Ltd. The experimental conditions are listed in Table 1. During

* Corresponding author. Fax: +86-21-34205503
E-mail address: yuygli@sjtu.edu.cn

experiments, premixed tetralin/O_2/Ar flames were stabilized on the McKenna burner and the flame species were sampled by a quartz nozzle with a 40° cone angle and a 500 μm tip orifice. The molecular beam formed downstream passed through a nickle skimmer and entered the photoionization chamber where the flame species were ionized by the synchrotron light and detected by the RTOF-MS. Flame temperature profiles were measured with a Pt-6%Rh/Pt-30%Rh thermocouple coated with Y_2O_3-BeO anti-catalytic ceramic to inhibit the catalytic effects [23]. The temperature profiles were corrected for radiative heat loss [24] and cooling effects of the sampling nozzle [25]. The uncertainties of maximum flame temperatures were estimated to be within ±100 K. Intermediate identification and mole fraction evaluation methods were reported in our previous work [21, 26]. The mole fraction uncertainties were estimated to be within ±10% for major species, ±25% for intermediates with known photoionization cross sections (PICS) and a factor of 2 for those with estimated PICS. The PICSs of flame species are available in the online database [27].

Table 1: Experimental conditions of the investigated flames.

ϕ	C/O	P/Torr	X_{Fuel}	X_{O2}	X_{Ar}	V/cm·s^{-1}
0.70	0.27	30	2.55%	47.45%	50.00%	35.00
1.70	0.65	30	5.78%	44.22%	50.00%	35.00

Note: X_i is the percentage of species i in the inlet mixture; V is the flow velocity of inlet mixture at 300 K.

3 Kinetic model

The present model of tetralin (H4A2, nomenclature in the present model) combustion was developed from our previously published tetralin pyrolysis model [14], ethylbenzene combustion model [28] and *o*-xylene combustion model [29]. The development of the tetralin oxidation sub-mechanism will be introduced herein. The rate constants for the H-atom abstraction reactions of tetralin by OH, O, HO_2, O_2 and HCO forming 1-tetralyl (H4A2-1) radical (R1) were referred to the similar reactions of ethylbenzene [28], while those forming 2-tetralyl (H4A2-2) radical (R2) were referred to the similar reactions of cyclohexane [30]. Similarly, the rate constants for the H-atom abstraction reactions of dihydronaphthalenes (H2A2 and H2A223) were estimated from the analogous reactions of ethylbenzene [28], styrene [31] and cyclohexene [30]. The rate constants for the H-atom abstraction reactions of *o*-allyltoluene ($CH_3A1C_3H_5$) were referred to the similar reactions of ethylbenzene [28] and allylbenzene [32]. The rate constants for the oxidation reactions of 1-tetralyl and 2-tetralyl radicals, e.g. the reactions with O and HO_2 forming tetrahydronaphthalenoxyl radicals (H4A2-O1 and H4A2-O2) and the concerted elimination reactions, were estimated from the analogous reactions of benzyl and 2-phenylethyl radicals [28], respectively. The rate constants for the decomposition reactions of tetrahydronaphthalenoxyl radicals were referred to the similar reactions of phenylethoxyl radicals [28]. The oxidation reactions of *o*-allyltoluene were referred to the similar reactions of *o*-xylene [29] and allylbenzene [32]. Furthermore, thermodynamic data of species in the tetralin oxidation sub-mechanism were mainly estimated using the THERGAS software [33], while the transport data of species in the tetralin oxidation sub-mechanism were taken from literature [34] or estimated using the method propos[35]ed by Sarathy et al. [36]. The final model consists of 296 species and 1571 reactions. Premixed Laminar Burner-Stabilized Flame module in Chemkin-Pro software [37] was used for the flame simulation.

H4A2 + X = H4A2-1 + HX   R1

H4A2 + X = H4A2-2 + HX   R2

4 Results and discussion

Dozens of flame species were detected in this work, including a series of radicals, isomers and PAHs. Their mole fraction profiles along the burner centerline were also measured which constitute the chemical structure of the flames. In general, the model has good performance in simulating the present flame data. Rate of production (ROP) analysis was conducted for both flames in order to reveal the key pathways in the fuel decomposition and aromatics growth processes.

4.1. Fuel decomposition

Figure 1 shows the experimental and simulated mole fraction profiles of major flame species in this work, along with the measured flame temperature. The simulation results capture the depletion positions of fuel and O_2 and the inflection or peak positions of Ar and major products, i.e. H_2, H_2O, CO and CO_2. The equilibrium mole fractions of major flame species are also well predicted by the present model. As expected, the lean flame produces more abundant complete combustion products (i.e. H_2O and CO_2), while the rich flame generates incomplete combustion products (e.g. H_2 and CO) with much higher concentration levels.

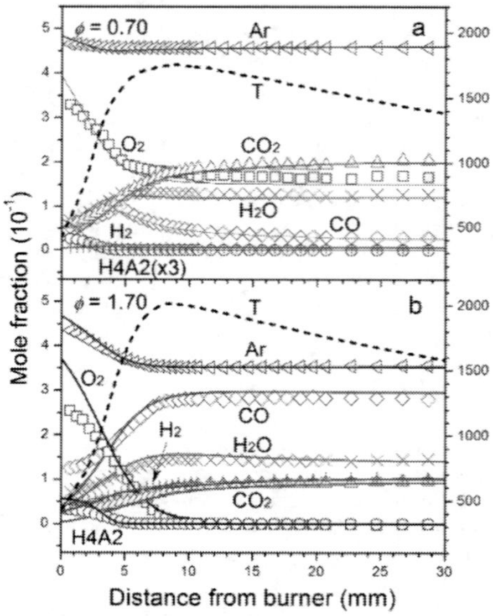

Figure 1: Experimental (symbols) and simulated (solid lines) mole fraction profiles of major species in the lean and rich tetralin flames, along with the measured temperature profiles (dash lines).

Figure 2 shows the experimental and simulated mole fraction profiles of C_{10}-C_8 intermediates which are all decomposition products of tetralin. In both flames, the primary decomposition of tetralin are dominated by the H-atom abstraction reactions (R1 and R2) producing 1-tetralyl (H4A2-1) and 2-tetralyl (H4A2-2) radicals. Among the H-atom abstraction reactions of tetralin, those by H and OH are the most important ones in the lean flame, contributing over 50% and 30% to the consumption of tetralin, respectively. Furthermore, the H-atom abstraction reactions by O also consume 16% of fuel in the lean flame. Due to the stronger pyrolytic circumstance, the primary decomposition of tetralin in the rich flame is controlled by the

H-atom abstraction reactions by H with a total contribution of 70%, while those by OH and O only contributes 17% together. The unimolecular decomposition reactions, mainly the dehydrogenation reaction (R3), also consumes 13% of tetralin. However the C-H bond scission reactions of tetralin producing 1- and 2-tetralyl radicals (R4 and R5) are not important for the consumption of tetralin. In fact, the two reactions produce tetralin through the reverse direction in flames due to the abundant production of tetralyl radicals via R1 and R2. This is quite different with the pyrolysis of tetralin [14] where R4 is the driven force of radical production and fuel decomposition.

$H4A2 = H2A2 + H_2$ R3

H4A2 = H4A2-1 + H R4

H4A2 = H4A2-2 + H R5

The tetralyl radicals have different key consumption pathways as the reaction circumstance changes. In the rich flame, tetralyl radicals mainly suffers β-C-H scission reactions producing 1,2-dihydronaphthalene (R6, R7) or 1,4-dihydronaphthalene (R8) and β-C-C scission reaction producing 2-(2'-vinylphenyl)ethyl ($C_2H_4A1C_2H_3$) radical (R9) or *o*-allylbenzyl ($CH_2A1C_3H_5$) radical (R10). This is the same as that in the pyrolysis of tetralin [14]. Besides the unimolecular decomposition reactions, tetralyl radicals can also be oxidized by O to produce tetrahydronaphthalenoxyl radicals (R11 and R12). In particular, R11 is the dominant consumption reaction of 1-tetralyl radical in the lean flame.

H4A2-1 = H2A2 + H R6

H4A2-2 = H2A2 + H R7

H4A2-2 = H2A223 + H R8

$H4A2\text{-}1 = C_2H_4A1C_2H_3 + H$ R9

$H4A2\text{-}2 = CH_2A1C_3H_5 + H$ R10

H4A2-1 + O = H2A2-O1 + H R11

H4A2-2 + O = H2A2-O2 + H R12

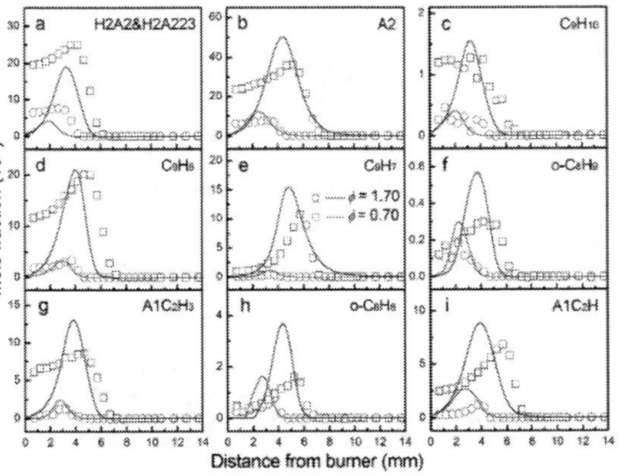

Figure 2: Experimental (symbols) and simulated (solid lines) mole fraction profiles of C_{10}-C_8 intermediates in the lean and rich tetralin flames.

The fate of dihydronaphthalene isomers in the flames is the same as that in the pyrolysis of tetralin [14], that is, both of them suffer stepwise H-loss reactions (H2A2→H2A2-1/H2A2-2→A2 and H2A2→H2A2-2→A2) and eventually produce naphthalene (A2). These pathways are also the dominant formation pathways of naphthalene in both flames. The fate of 2-(2'-vinylphenyl)ethyl radical is also the same as that in the pyrolysis of tetralin [14] which dominates the formation of styrene ($A1C_2H_3$) and eventually phenylacetylene ($A1C_2H$). As for *o*-allylbenzyl radical, its further decomposition in the lean flame mainly produces *o*-xylylene (o-C_8H_8) and 2-methylindene ($C_9H_8CH_3$) via R13 and R14. The former is the most important formation pathway of *o*-xylylene in both lean and rich flames. In contrast, the unimolecular decomposition of tetralin (R15) producing *o*-xylylene and ethylene (C_2H_4) which is exclusively important for *o*-xylylene formation in pyrolysis [14] has negligible contribution in the lean flame and only minor contribution in the rich flame. The latter mainly undergoes the unimolecular decomposition reaction to produce indene (R16) which is the dominant formation pathway of indene in both flames. Furthermore, the results of C_7-C_1 intermediates are shown in Figs. 3 and 4. The formation and consumption pathways of these intermediates are similar to those in the flames of smaller aromatic hydrocarbons [28, 31, 35] and will not be introduced in detail herein.

$CH_2A1C_3H_5 = o\text{-}C_8H_8 + C_2H_3$ R13

$CH_2A1C_3H_5 = C_9H_8CH_3$ R14

$H4A2 = o\text{-}C_8H_8 + C_2H_4$ R15

$C_9H_8CH_3 = C_9H_8 + CH_3$ R16

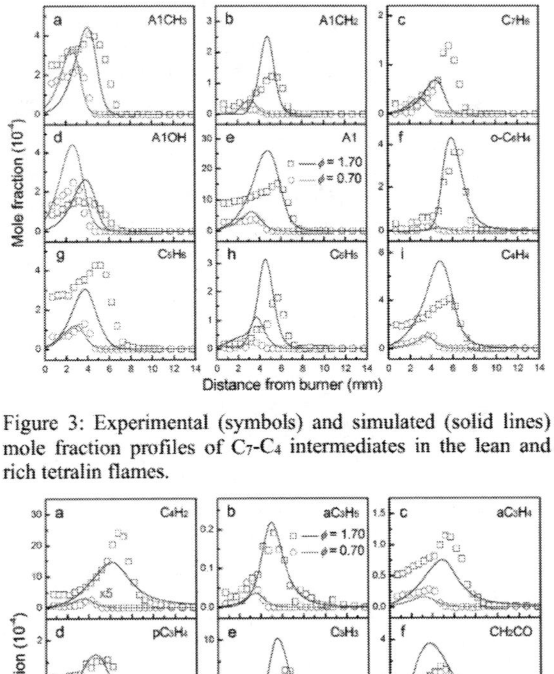

Figure 3: Experimental (symbols) and simulated (solid lines) mole fraction profiles of C_7-C_4 intermediates in the lean and rich tetralin flames.

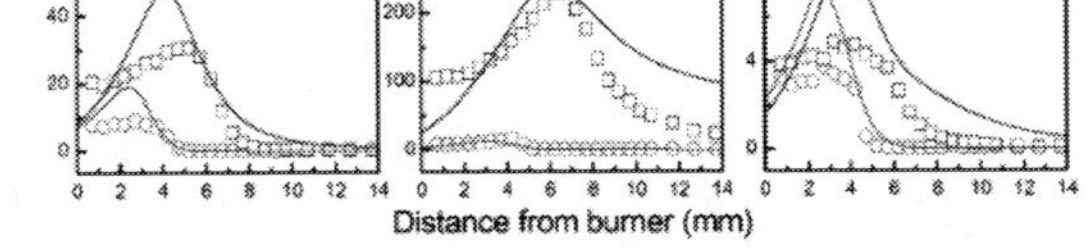

Figure 4: Experimental (symbols) and simulated (solid lines) mole fraction profiles of C_4-C_1 intermediates in the lean and rich tetralin flames.

4.2. Aromatics growth

There are many typical PAHs observed in this work. As discussed above, the two most important bicyclic PAHs, i.e. indene and naphthalene, can be largely produced from the decomposition of tetralin and their maximum mole fractions both reach more than 2×10^{-3} in the rich flame. Besides, a series of PAHs with molecular weights greater than the fuel were also detected. Figure 5 shows the experimental and simulated results of ethynylindene ($C_9H_7C_2H$), methylnaphthalene ($A2CH_3$), acenaphthylene (A2R5), biphenylene (BIPHEN), fluorene ($C_{13}H_{10}$) and phenanthrene (A3) in the rich flame. It is noted that methylnaphthalene is the simplification of two $C_{11}H_{10}$ isomers (1- and 2-methylnaphthalene) in this work since the two isomers have very similar features. The present model can reasonably simulate the formation of these PAHs. ROP analysis were also performed to explore the important reaction pathways in aromatics growth process.

The ROP analysis shows that indene and naphthalene are mainly consumed by H-atom abstraction reactions to produce indenyl (C_9H_7) and naphthyl (A2-1 and A2-2) radicals. Besides indene and naphthalene, methylnaphthalene and acenaphthylene are also abundantly produced in the rich flame with maximum mole fractions both exceeding 1×10^{-4}. According to the ROP analysis, methylnaphthalene are mainly produced from the CH_3 addition on naphthyl radicals directly or via naphthylmethyl ($A2CH_2$) radical, while acenaphthylene is mainly produced from the addition-cyclization reaction between indenyl radical and propargylene (C_3H_2) radical. Therefore it is concluded that the abundant formation of methylnaphthalene and acenaphthylene is benefited from the easy formation of their precursors, i.e. naphthyl and indenyl radicals, respectively. Similarly, the relatively high maximum mole fraction of ethynylindene (almost 1×10^{-4}) also results from the easy formation of naphtyl radical since it is the other major decomposition product of naphthylmethyl radical.

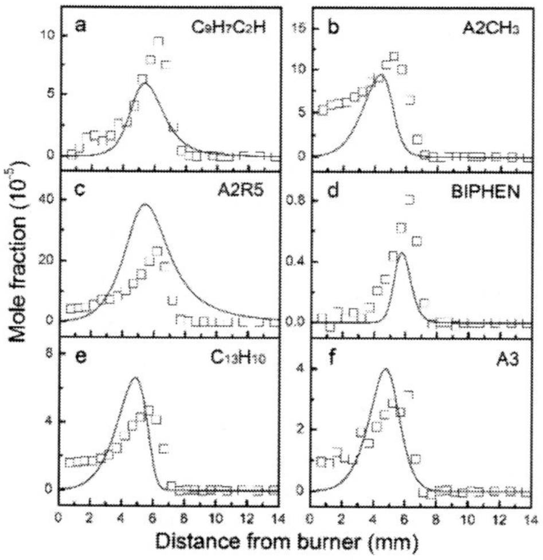

Figure 5: Experimental (symbols) and simulated (solid lines) mole fraction profiles of large aromatics in the rich tetralin flame.

Biphenylene and fluorene are two PAHs whose formation are highly related to phenyl (A1-) radical. The ROP analysis shows that most of biphenylene is produced from the self-combination of benzyne which is a major decomposition product of phenyl radical. The reaction between biphenylene and methyl is the most significant formation pathway for the fluorene, while the reaction between benzyne and benzyl ($A1CH_2$) radical is the second important pathway. Phenanthrene is the most important tricyclic PAH formed in hydrocarbon flames. There are a variety of formation pathways for phenanthrene, e.g. the reaction between phenyl radical and phenylacetylene, the reaction between indenyl and cyclopentadienyl radicals, the self-combination of fulvenallenyl (C_7H_5) radical and so on. Among them, the reaction between phenyl radical and phenylacetylene is the most important one since both precursors are abundantly produced, while the reaction between indenyl and cyclopentadienyl radicals is the second important one due to the easy formation of indenyl radical.

5 Conclusions

Two premixed flames of tetralin with equivalence ratios of 0.7 and 1.7 were investigated at 30 Torr in this work. Flame species were identified with SVUV-PIMS and their mole fraction profiles were measured. A kinetic model of tetralin combustion including 296 species and 1571 reactions was developed from our previous kinetic models of aromatic fuels. The model can reasonably simulate the experimental data in this work. Based on the ROP analysis, it is concluded that the H-atom abstraction reactions of tetralin are the dominant consumption pathways of the fuel in flames. In particular, oxidation reactions of fuel and its decomposition products play important roles in the lean flame. Due to the fused ring structure of tetralin, indene and naphthalene are abundantly produced in its flames through direct decomposition reaction sequences from tetralin. As a result, their radicals play important roles in the aromatics growth process, leading to high concentration levels of many typcial large PAHs.

5 Acknowledgment

This work was supported by National Natural Science Foundation of China (91541201, 51622605, 51476155), National Postdoctoral Program for Innovative Talents (BX201600100) and China Postdoctoral Science Foundation (2016M600312). The authors are also grateful to Dr. Jiuzhong Yang and Dr. Zhanjun Cheng for technical assistance.

References

[1] Toxicological profile for fuel oils-Jet Fuels, U.S. Department of Health and Human Services, Agency for Toxic Substances and Disease Registry (ATSDR). 1995.

[2] A. Violi, S. Yan, E.G. Eddings, A.F. Sarofim, S. Granata, T. Faravelli, E. Ranzi, Experimental formulation and kinetic model for JP-8 surrogate mixtures, Combustion Science and Technology 174 (2002) 399-417.

[3] W.J. Pitz, C.J. Mueller, Recent progress in the development of diesel surrogate fuels, Prog. Energ. Combust. Sci. 37 (2011) 330-350.

[4] P. Dagaut, G. Dayma, F. Karsenty, Z. Serinyel, Asme, The combustion of synthetic jet fuels (gas to liquid and coal to liquid) and multi-component surrogates: Experimental and modeling study, Asme Turbo Expo: Turbine Technical Conference And Exposition, New York, 2015, Paper No. GT2015-42004.

[5] D.J. Luning Prak, P.J. Luning Prak, P.C. Trulove, J.S. Cowart, Formulation of Surrogate Fuel Mixtures Based on Physical and Chemical Analysis of Hydrodepolymerized Cellulosic Diesel Fuel, Energy Fuels 30 (2016) 7331-7341.

[6] P. Dagaut, P. Dievart, Asme, Experimental and modeling study of the combustion of synthetic jet fuels: Naphtenic cut and blend with a GTL jet fuel, Proceedings of the Asme Turbo Expo: Turbine Technical Conference and Exposition, New York, 2016, Paper No. GT2016-56086.

[7] E.M. Yoon, L. Selvaraj, C.S. Song, J.B. Stallman, M.M. Coleman, High-temperature stabilizers for jet fuels and similar hydrocarbon mixtures .1. Comparative studies of hydrogen donors, Energy

Fuels 10 (1996) 806-811.
[8] D.F. McMillen, R. Malhotra, S.J. Chang, W.C. Ogier, S.E. Nigenda, R.H. Fleming, Mechanisms of hydrogen transfer and bond scission of strongly bonded coal structures in donor solvent systems, Fuel 66 (1987) 1611-1620.
[9] R. Bounaceur, G. Scacchi, P.-M. Marquaire, F. Dominé, Mechanistic Modeling of the Thermal Cracking of Tetralin, Industrial & Engineering Chemistry Research 39 (2000) 4152-4165.
[10] G.P. Curran, R.T. Struck, E. Gorin, Mechanism of hydrogen-transfer process to coal and coal extract, Industrial & Engineering Chemistry Process Design and Development 6 (1967) 166-173.
[11] J.J. Devlieger, A.P.G. Kieboom, H. Vanbekkum, Behavior of tetralin in coal-liquefaction - Examination in long-run batch-autoclave experiments, Fuel 63 (1984) 334-340.
[12] M.L. Poutsma, Free-radical thermolysis and hydrogenolysis of model hydrocarbons relevant to processing of coal, Energy Fuels 4 (1990) 113-131.
[13] V. Burklé-Vitzthum, R. Michels, R. Bounaceur, P.-M. Marquaire, G. Scacchi, Experimental Study and Modeling of the Role of Hydronaphthalenics on the Thermal Stability of Hydrocarbons under Laboratory and Geological Conditions, Industrial & Engineering Chemistry Research 44 (2005) 8972-8987.
[14] Y.Y. Li, L.D. Zhang, Z.D. Wang, L.L. Ye, J.H. Cai, Z.J. Cheng, F. Qi, Experimental and kinetic modeling study of tetralin pyrolysis at low pressure, Proc. Combust. Inst. 34 (2013) 1739-1748.
[15] M.L. Poutsma, Progress toward the Mechanistic Description and Simulation of the Pyrolysis of Tetralin, Energy Fuels 16 (2002) 964-996.
[16] P. Dagaut, A. Ristori, A. Frassoldati, T. Faravelli, G. Dayma, E. Ranzi, Experimental Study of Tetralin Oxidation and Kinetic Modeling of Its Pyrolysis and Oxidation, Energy Fuels 27 (2013) 1576-1585.
[17] H.W. Wang, W.J. Gerken, W.J. Wang, M.A. Oehlschlaeger, Experimental Study of the High-Temperature Autoignition of Tetralin, Energy Fuels 27 (2013) 5483-5487.
[18] Y. Zhu, D.F. Davidson, R.K. Hanson, Pyrolysis and oxidation of decalin at elevated pressures: A shock-tube study, Combust. Flame 161 (2014) 371-383.
[19] G. Kukkadapu, B.W. Weber, C.J. Sung, Autoignition study of tetralin in a rapid compression machine at elevated pressures and low-to-intermediate temperatures, Fuel 159 (2015) 436-445.
[20] R. Bounaceur, G. Scacchi, P.-M. Marquaire, F. Dominé, O. Brévart, D. Dessort, B. Pradier, Inhibiting Effect of Tetralin on the Pyrolytic Decomposition of Hexadecane. Comparison with Toluene, Industrial & Engineering Chemistry Research 41 (2002) 4689-4701.
[21] Y.Y. Li, F. Qi, Recent Applications of Synchrotron VUV Photoionization Mass Spectrometry: Insight into Combustion Chemistry, Acc. Chem. Res. 43 (2010) 68-78.
[22] F. Qi, Combustion chemistry probed by synchrotron VUV photoionization mass spectrometry, Proc. Combust. Inst. 34 (2013) 33-63.
[23] R.A. Shandross, J.P. Longwell, J.B. Howard, Noncatalytic thermocouple coating for low-pressure flames, Combust. Flame 85 (1991) 282-284.
[24] R.M. Fristrom, Flame Structure and Processes, Oxford University Press, New York, 1995.
[25] A.T. Hartlieb, B. Atakan, K. Kohse-Höinghaus, Effects of a sampling quartz nozzle on the flame structure of a fuel-rich low-pressure propene flame, Combust. Flame 121 (2000) 610-624.
[26] Y.Y. Li, L.D. Zhang, Z.Y. Tian, T. Yuan, J. Wang, B. Yang, F. Qi, Experimental Study of a Fuel-Rich Premixed Toluene Flame at Low Pressure, Energy Fuels 23 (2009) 1473-1485.
[27] Photonionization Cross Section Database (Version 1.0), National Synchrotron Radiation Laboratory, Hefei, China, 2011, http://flame.nsrl.ustc.edu.cn/en/database.htm.
[28] W.H. Yuan, Y.Y. Li, G. Pengloan, C. Togbé, P. Dagaut, F. Qi, A comprehensive experimental and kinetic modeling study of ethylbenzene combustion, Combust. Flame 166 (2016) 255-265.
[29] L. Zhao, Z.J. Cheng, L.L. Ye, F. Zhang, L.D. Zhang, F. Qi, Y.Y. Li, Experimental and kinetic modeling study of premixed o-xylene flames, Proc. Combust. Inst. 35 (2015) 1745-1752.
[30] Z.D. Wang, L.L. Ye, W.H. Yuan, L.D. Zhang, Y.Z. Wang, Z.J. Cheng, F. Zhang, F. Qi, Experimental and kinetic modeling study on methylcyclohexane pyrolysis and combustion, Combust. Flame 161 (2014) 84-100.
[31] W.H. Yuan, Y.Y. Li, P. Dagaut, J.Z. Yang, F. Qi, Experimental and kinetic modeling study of styrene combustion, Combust. Flame 162 (2015) 1868-1883.
[32] Z.D. Wang, Y.Y. Li, F. Zhang, L.D. Zhang, W.H. Yuan, Y.Z. Wang, F. Qi, An experimental and kinetic modeling investigation on a rich premixed n-propylbenzene flame at low pressure, Proc. Combust. Inst. 34 (2013) 1785-1793.
[33] C. Muller, V. Michel, G. Scacchi, G.M. Côme, THERGAS: a computer program for the evaluation of thermochemical data of molecules and free radicals in the gas phase, J. Chim. Phys. 92 (1995) 1154-1178.
[34] E. Ranzi, A. Frassoldati, R. Grana, A. Cuoci, T. Faravelli, A.P. Kelley, C.K. Law, Hierarchical and comparative kinetic modeling of laminar flame speeds of hydrocarbon and oxygenated fuels, Prog. Energ. Combust. Sci. 38 (2012) 468-501.
[35] W. Yuan, Y. Li, P. Dagaut, J. Yang, F. Qi, Investigation on the pyrolysis and oxidation of toluene over a wide range conditions. I. Flow reactor pyrolysis and jet stirred reactor oxidation, Combust. Flame 162 (2015) 3-21.
[36] S.M. Sarathy, S. Vranckx, K. Yasunaga, M. Mehl, P. Oßwald, W.K. Metcalfe, C.K. Westbrook, W.J. Pitz, K. Kohse-Höinghaus, R.X. Fernandes, H.J. Curran, A comprehensive chemical kinetic combustion model for the four butanol isomers, Combust. Flame 159 (2012) 2028-2055.
[37] CHEMKIN-PRO 15092, Reaction Design: San Diego, 2009.

11th Asia-Pacific Conference on Combustion,
The University of Sydney, NSW Australia
10th -14th December 2017.

The impact of biochars on the immobilisation of Zn(II) in contaminated soil

Hong Wang, Hui Zhao, Yuwei Bu, Ping Lu*

School of Energy & Mechanical Engineering, Nanjing Normal University, Nanjing 210042, China

Abstract

Two kinds of biochars (called water hyacinth biochar(WHC) and corn straw biochar(CSC)) were prepared in a fixed-bed pyrolyzer at different pyrolysis temperature of 300~700 ℃. The vegetable soil (collected from lead-zinc-silver mining area, Nanjing, China) was incubated with these biochars, then the modified BCR three-step sequential extraction procedure(SEP) was used to investigate the effect of biochars on chemical fractions of Zn(II) in soil under different incubation time. The obtained results indicate that when the soil is incubated for one week, the ability of WHC to reduce the content of available Zn(II) in soil is better than that of CSC, when the soil is incubated for two weeks, the ability of WHC to reduce the content of available Zn(II) in soil is no different from that of CSC and when the soil is incubated for one month, the ability of CSC to reduce the content of available Zn(II) in soil is better than that of WHC. WHC prepared under low-temperature and CSC prepared under high-temperature have better improvement effect on Zn (II) contaminated soil when the incubation time is one week, while WHC and CSC prepared under high-temperature have better improvement effect on Zn (II) contaminated soil when the incubation time is two weeks and one month. The CSC700 may be a good choice to be used improvement Zn(II) contaminated soil.

1 Introduction

Heavy metals (HMs) in soils may come from various human activities, such as industrial and energy production, construction, vehicle exhaust, waste disposal, as well as coal and fuel-oil combustion. HMs are persistent and difficult to remove or decompose when introduced into soils and high concentrations of HMs in soils may cause long term risks to ecosystems and humans [1, 2]. Zeolite, lime and red-mud, and chicken manure compost reported as adsorbents of HMs, can have large surface areas, many functional groups or high pH values and thus can effectively fix HMs of contaminated soils [3]. Biochar is a solid material produced by biomass thermochemical process under the condition of lack of oxygen (gasification) or oxygen free (pyrolysis), which possesses good pore structure, large specific surface area and a variety of surface oxygen-containing functional groups. This excellent physico-chemical characteristics of biochar is helpful for adsorption and immobilization of HMs in soil. Therefore, biochar has the potential to be an ideal modifier [4]. HMs in soils may exist in different chemical forms or ways of binding. In unpolluted soils HMs are mainly bound to silicates and primary minerals which form relatively immobile species, whereas in polluted ones HMs are generally more mobile and bound to other soil phases. The determination of total concentration of the metals gives no information on their various forms. Therefore, the determination of the different ways of binding gives more information on HMs mobility, as well as on their availability or toxicity, in comparison with the total element content [5]. The modified BCR (the Community Bureau of Reference) three-step SEP has proved a useful prediction tool to monitor chemical forms of metals in order to evaluate the proper extraction efficiency of contaminants [6]. Recently, some studies have reported the applications of biochar in soil remediation. Rees et al. [7] observed a reduction of extractable Cd(II), Pb(II), Cu(II), Ni(II) and Zn(II) in soils after the addition of 80% coniferous and 20% hardwood biochar in one week. While Lucchini et al. [8] did not observe any significant changes of total metal concentrations in soil or plant three years after wood biochar treatment on a lightly contaminated agriculture land. Therefore, the immobilisation of HMs in contaminated soil treated by biochar remains unclear and needs further investigation.

In this study, the vegetable soil (collected from lead-zinc-silver mining area, Nanjing, China) was incubated with WHC and CSC biochars, then the modified BCR three-step SEP was used to investigate the effect of biochars on chemical fractions of Zn(II) in soil under different incubation time.

2 Materials and methods

2.1 Soil and biochar

Surface samples of soil (0~20 cm) was collected from vegetable garden at lead-zinc-silver mining area (32°02'50"~32°14'41"N, 118°45'42"~119°14'50"E), Nanjing, China. The soil sample was air-dried and ground to pass a 100-mesh sieve. Soil pH was measured in 1:2.5 ratio soil solutions (with de-ionized water) with a pH meter [3] and the value is 6.78. The concentrations of Zn(II) in soil sample was analyzed by Atomic absorption spectrophotometer(AAS, TAS-986, Beijing Purkinje General Instrument Co., Ltd., China). The total concentration of Zn(II) in soil samples was determined after a series of solutions of $HCl/HNO_3/HF/HClO_4$ digestion and is 852.545 $\mu g \cdot g^{-1}$, which is 17 times higher than that of the second level criterion regulated by the China standard of Environment Quality Standard for Soil (GB 15618-2008)[9] .

The water hyacinth (WH) and corn straw (CS) were collected from Wenzhou and Lianyungang in China, respectively. The biomass samples were air-dried at room temperature and ground to pass a 4-mesh sieve. The biomass samples pyrolyzed under oxygen-limited conditions in a fixed-bed pyrolyzer (SLGL-1200, Shanghai crystal Precision Instrument Manufacturing Co., Ltd., China). The pyrolysis temperature was raised to 300℃, 400℃, 500℃, 600 ℃ and 700℃ at approximately 20℃/min and held constant for 30min, Then the biochar samples were allowed to cool to room temperature and ground to pass a 100-mesh sieve. The biochars prepared from different biomass and pyrolysis temperature were called as WHC300, WHC400, CSC500, CSC600

Corresponding author. Fax: +86-25-8548-1124
E-mail address: luping@njnu.edu.cn

etc., in which, WHC and CSC represent WH biochar and CS biochar, respectively. The biochars pH were measured in 1:100 ratio soil solutions (with de-ionized water) with a pH meter [10].

2.2 Incubation experiments

Air-dried soil samples of 50 g were placed in beakers and biochar was added at 5%. The soil and biochar were mixed thoroughly and then wetted with deionized water to 70% of field water holding capacity of the soil. All beakers did not covered to allow gas exchange but to minimize moisture loss, and then incubated at a constant 25 °C. The beakers were weighed every day, with water added to maintain constant moisture content throughout the incubation period. After one week, two weeks and one month incubation, the soil samples were removed from the beakers. Then samples of 5g were air-dried and ground to pass a 100-mesh sieve for the modified BCR three-step SEP to determine the chemical fractions of Zn(II) present in the soils. The Zn(II) were partitioned into four fractions, Exchangeable fraction (EF), Fe/Mn oxide fraction (FMOF), Organic/Sulfides fraction (OSF) and Residue fraction (RF).

3 Results and discussion

3.1 Characterization of the samples

sample	Proximate analysis Wt/%				$Q_{net.ad}$
	M_{ad}	V_{ad}	FC_{ad}	A_{ad}	MJ/Kg
WH	13.93	60.79	12.84	12.45	10.52
WHC300	2.63	44.35	32.25	20.76	19.20
WHC400	2.69	28.87	41.44	27.00	19.71
WHC500	3.87	18.20	46.62	31.31	18.95
WHC600	3.84	13.14	49.17	33.85	19.21
WHC700	4.97	8.76	52.09	34.17	19.35
CS	13.14	64.68	15.82	6.36	14.11
CSC300	3.12	41.34	42.21	13.33	22.76
CSC400	1.71	22.69	58.76	16.84	20.58
CSC500	1.84	12.85	65.01	20.30	22.84
CSC600	3.19	8.14	68.12	20.55	25.37
CSC700	3.52	6.74	67.86	21.88	24.73

Table 1: Proximate analysis of the biomass and biochars

sample	ultimate analysis Wt/%				
	C_{ad}	H_{ad}	O_{ad}^{a}	N_{ad}	S_{ad}
WH	36.94	5.63	29.56	1.33	0.16
WHC300	50.15	4.34	19.78	2.22	0.11
WHC400	52.31	3.72	12.02	2.19	0.07
WHC500	51.89	2.79	7.98	2.05	0.11
WHC600	53.24	1.98	5.16	1.92	0.01
WHC700	55.57	1.66	2.02	1.51	0.10
CS	40.71	5.64	33.59	0.50	0.06
CSC300	58.24	4.70	20.05	0.49	0.06
CSC400	64.89	4.01	12.06	0.48	0.01
CSC500	67.77	3.06	6.56	0.47	0.01
CSC600	68.96	2.56	4.27	0.46	0.01
CSC700	69.20	1.84	3.06	0.43	0.06

a. indicates the results obtained by subtraction.

Table 2: Ultimate analysis of the biomass and biochars

The proximate analysis and ultimate analysis on air dried basis of the biomass and biochars are listed in Table 1 and Table 2. It shows that two biomass have a higher volatile matters. WH possesses higher ash content and lower fixed carbon content than CS. After pyrolysis, the water contents of biomass decrease obviously, the volatile contents decrease greatly, and the fixed carbon contents and ash contents increase. WHC biochars possess higher ash contents and lower fixed carbon contents than CSC biochars at the same pyrolysis temperature. The contents of N and S in the two kinds of biomass are pretty low. After pyrolysis, the contents of C increase obviously, the contents of H and O decrease, and the contents of N and S change very little. The lower heating value (LHV) of WH is lower than that of CS, and the LHV of two biomass increase after pyrolysis.

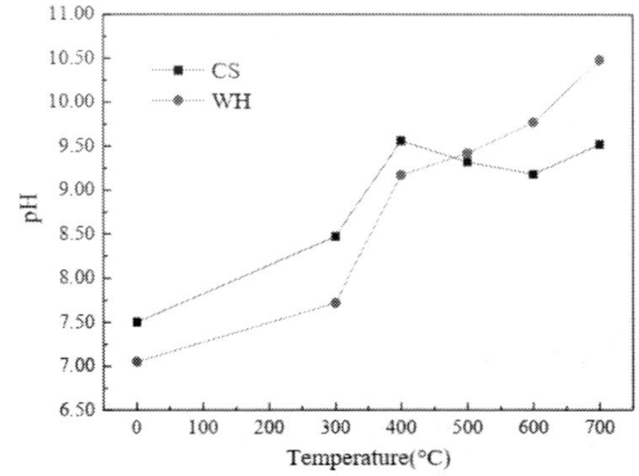

Figure 1: The pH value of biochars under different pyrolysis temperature

The pH value of biochars under different pyrolysis temperature are show in Fig. 1. It can be seen from the Fig. 1, the pH value of biomass increase after pyrolysis. With the increase of pyrolysis temperature (300~400 °C), the pH value of CS increases from 7.50 to 9.56, and when the pyrolysis temperature continues to rise, the pH value decreases slightly, but the change is very little. The pH value of WH increases with increasing pyrolysis temperature and reaches the maximum value of 10.48 at 700 °C.

Fig. 2 shows the XRD curves of two kinds of biomass raw materials and biochars pyrolyzed at 300°C, 400°C, 500°C, 600°C and 700°C. It can be seen from Fig. 2(a)-(b) that biochars contain KCl、$CaCO_3$、$MgCl_2$、SiO_2 and other substances, indicate that biochars contain base ion K, Ca, Mg and Si, etc. Therefore, it can decrease the content of effective state heavy metal ions in soil by ion exchange. It can be seen from Fig.2 that WHC biochars contain the more base ions than CSC biochars. So compare with CSC biochar, WHC biochar can adsorb more HMS in soil through ion exchange.

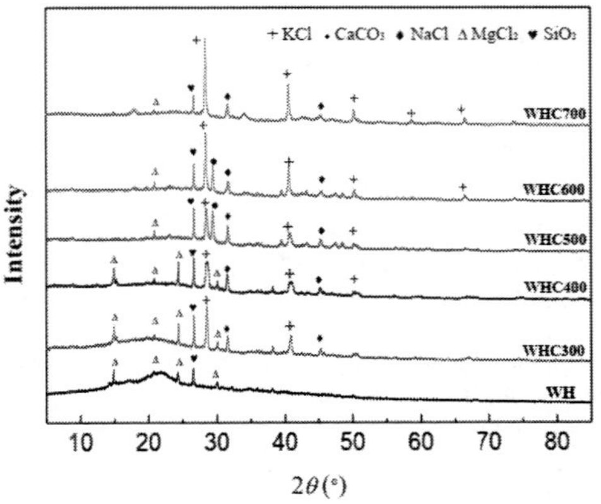

(a)

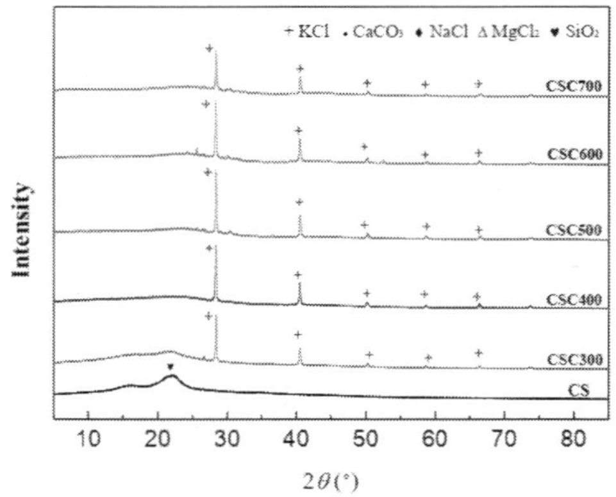

(b)

Figure 2: XRD curves of the biochars

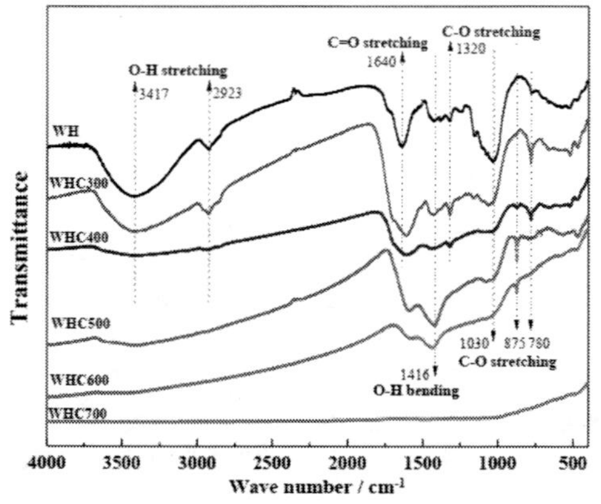

(a)

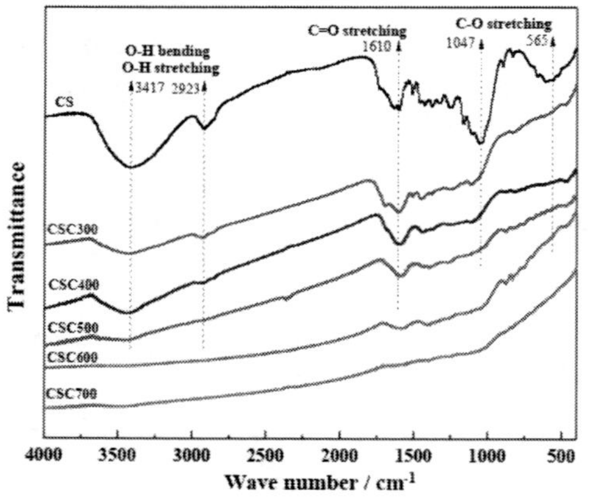

(b)

Figure 3: FTIR curves of the biochars

Fig. 3 shows the FTIR curves of two kinds of biomass raw materials and biochars pyrolyzed at 300°C, 400°C, 500°C, 600°C and 700°C. It can be seen from Fig. 3(a)-(b) that biochars contain the carboxyl group, phenolic hydroxyl, lactone and carbonyl group, etc. So biochars can adsorb heavy metal ions through complexing reaction of surface oxygen containing functional groups and heavy metal ions. Biochars under the five kinds of pyrolysis temperature have obviously different absorption spectra, in which, the intensity and number of biochars pyrolyzed at 700 °C are significantly weaker than that of other biochars pyrolyzed at 300 °C~600 °C. This indicates that too high pyrolysis temperature is not helpful for forming of oxygen-containing functional groups and biochars pyrolyzed at low temperature can adsorb more HMs through complexing reaction.

3.2 The effect of biochars on fractions of Zn(II) in the soil

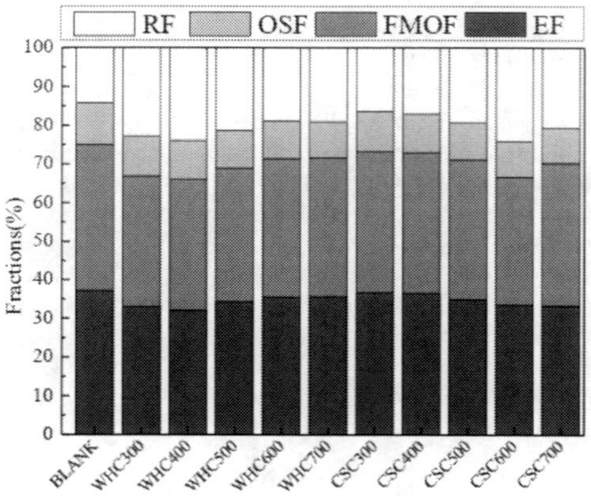

(a) One week

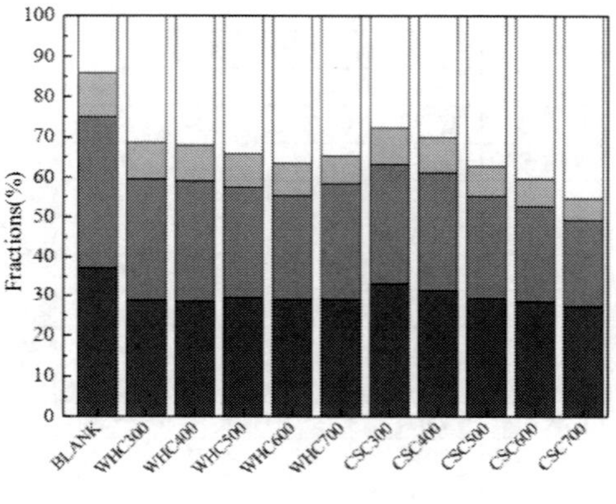

(b) Two weeks

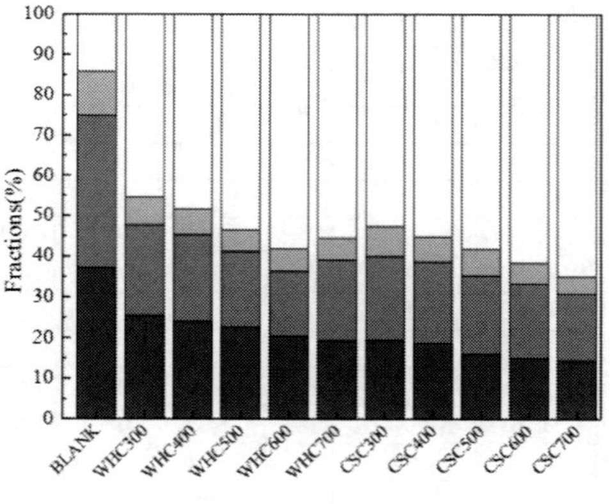

(c) One month

Figure 4: Fractions of Zn in soil under different incubation time: (a) one week, (b) two weeks, (c) one month

The results of chemical fractions of Zn(II) in the soil after different incubation time with biochars investigated by the modified BCR three-step SEP are shown in Fig. 4. The distribution pattern in each fraction of blank soil sample(CK) follows the order as FMOF (37.82%) > EF (37.20%) > RF(14.14%) > OSF(10.84%). Only 14.14% of the total Zn(II) in soil is in the residual fraction while most of Zn(II) (85.86%) is found in the non-residual fraction, which indicates Zn(II) in the sample soil has greater bioavailability. It can be seen from Fig.4 that biochars can significantly reduce the available Zn(II) content in soil. As shown in Fig.4 (a) , the WHC prepared at low-temperature has better improvement effect on Zn(II) contaminated soil, and WHC300 and WHC400 decrease the content of Zn(II) in EF about 11.19%~13.62% and increase the content of Zn(II) in RF about 61.24%~ 69.34% compared with CK. While the CSC prepared at high-temperature has better improvement effect on Zn(II) contaminated soil and CSC700 decrease the content of Zn(II) in EF about 9.75% and increase the content of Zn(II) in RF about 70.52% compared with CK. WHC reduce the content of Zn(II) in EF about 4.25%~13.62% and increase the content of Zn(II) in RF about 33.33%~69.34%. While CSC reduce the content of Zn(II) in EF about 1.22%~10.29% and increase the content of Zn(II) in RF about 15.34%~70.52%. It can be seen that when the soil is incubated for one week, the ability of WHC to reduce the content of available Zn(II) in soil is better than that of CSC. Therefore, combined with the results of Fig. 2-3, the good adsorption of WHC on Zn(II) may be attribute to the co-effects of surface oxygen-containing functional groups and exchangeable metal ions of biochar.

As shown in Fig. 4(b), the WHC prepared at high-temperature has better improve-ment effect on Zn (II) contaminated soil and WHC600 decrease the content of Zn(II) in EF about 21.82% and increase the content of Zn(II) in RF about 158.521% compared with CK. The CSC prepared at high-temperature has better improve-ment effect on Zn(II) contaminated soil and CSC700 decrease the content of Zn(II) in EF about 26.75% and increase the content of Zn(II) in RF about 220.58% compared with CK. WHC reduce the content of Zn(II) in EF about 20.76%~22.39% and increase the content of Zn(II) in RF about 122.30% ~158.52%. While CSC reduce the content of Zn(II) in EF about 10.83%~26.75% and increase the content of Zn(II) in RF about 96.10%~220.58%. It can be seen that when the soil is incubated for two weeks, the ability of WHC to reduce the content of available Zn(II) in soil is no different from that of CSC.

As shown in Fig.4 (c), the WHC prepared at high-temperature has better improve-ment effect on Zn(II) contaminated soil and WHC600 decrease the content of Zn(II) in EF about 44.90% and increase the content of Zn(II) in RF about 310.66% compared with CK. The CSC prepared at high-temperature has better improvement effect on Zn(II) contaminated soil and CSC700 decrease the content of Zn(II) in EF about 61.18% and increase the content of Zn(II) in RF about 358.74% compared with CK. WHC reduced the content of Zn(II) in EF about 31.34%~47.87% and increase the content of Zn(II) in RF about 221.57%~310.66%. While CSC reduce the content of Zn(II) in EF about 47.67%~61.18% and increase the content of Zn(II) in RF about 272.37%~358.74%. It can be seen that when the soil is incubated for one month, the ability of CSC to reduce the content of available Zn(II) in soil is better than that of WHC. As a whole, for a long incubation time, the improvement effect of biochars prepared under high-temperature on Zn(II) contaminated soil are better than that of biochars prepared at low-temperature. It can be seen from the Fig.1, the pH value of biochars prepared under high-temperature higher than that of biochar prepared at low temperature, which is consistent with the previous experimental results, and it can be concluded that the pH value of biochar has an influence on the improvement effect of biochars on Zn(II) contaminated soil in long incubation time. WHC600 shows the best improvement effect in all WHC biochars, which indicates the 600°C may be a good pyrolysis temperature for WH improvement Zn(II) contaminated soil. Similarly, the 700°C may be a good pyrolysis temperature for CS improvement Zn(II) contaminated soil.

4 Conclusions

Biochars is believed to play a key role in reducing the available Zn(II) content in soil and this was confirmed by the modified BCR three-step SEP. When the soil is incubated for one month, the ability of WHC to reduce the content of available Zn(II) in soil is better than that of WHC. WHC biochars decrease the content of Zn (II) in EF by an average of 8.24% and increase the content of Zn (II) in RF by an average of 49.82%. When the soil is incubated for two weeks, the ability of WHC to reduce the content of available Zn(II) in soil is different from that of CSC. When the soil is incubated for one month, the ability of WHC to reduce the content of available Zn(II) in soil is weaker than that of WHC. CSC biochars decrease the content of Zn (II) in EF by an average of 54.96% and increase the content of Zn (II) in RF by an average of 313.045%. WHC prepared under low-temperature and CSC prepared under high-temperature have better improvement effect on Zn(II) contaminated soil when the incubation time is one week, while WHC and CSC prepared under high-temperature have better improvement effect on Zn(II) contaminated soil when the incubation time is two weeks and one month. The pH value of biochar may has an influence on the improvement effect of biochar on Zn(II) contaminated soil in long incubation time. The CSC700 may be a good choice to be used improvement Zn(II) contaminated soil.

5 Acknowledgment

The financial supports of the National Natural Science Foundation of China (51476079), Jiangsu Province Prospective Cooperation Research Project of China (BY2015001-01), and Jiangsu Province Major Project of University Natural Science Research of China (15KJA610002) are gratefully acknowledged.

References

[1] S. Tokalıoğlu, V. Yılmaz, S. Kartal, CLEAN - Soil, Air, Water 38 (8) (2015) 713-718.

[2] L.C.A. Melo, A.P. Puga, A.R. Coscione, L. Beesley, C.A. Abreuand, O.A. Camargo, Journal of Soils & Sediments 16 (1) (2016) 226-234.

[3] T.Y. Jiang, J. Jiang, R.K. Xu, Chemosphere 89 (3) (2012) 249-256.

[4] R.K. Xu, A.Z. Zhao, Environmental Science & Pollution Research International 20 (12) (2013) 8491-8501.

[5] Y. Zong, Q. Xiao, S. Lu, Environmental Science & Pollution Research 23 (23) (2016) 1-8.

[6] M. Pueyo, J. Mateu, A. Rigol, Environmental Pollution 152 (2) (2008) 330-341.

[7] E. Rees, M.O. Simonnot, J.L. Morel, European Journal of Soil Science 65 (1) (2014) 149-161.

[8] P. Lucchini, R.S. Quilliam, T.H. Deluca, Agriculture Ecosystems & Environment 184 (1) (2014) 149-157.

[9] China MEP. GB15618-2008: Environmental quality standards for soils, Beijing, China (2008)

[10] A. Masulili, W.H. Utomo, M.S. Syechfani, Journal of Agricultural Science 2 (1) (2010) 39-47

11th Asia-Pacific Conference on Combustion,
The University of Sydney, NSW Australia
10th -14th December 2017.

Effect of Spark Gap on Turbulent Facilitated Ignition through Differential Diffusion

S.S. Shy[1*], M.T. Nguyen[1], S.Y. Huang[1], C.C. Liu[2]

[1]Department of Mechanical Engineering, National Central University

300 Jhong-da Road, Jhong-li District, Tao-yuan City 32001, Taiwan

[2]Max Planck Institute for Dynamics and Self-Organization

Am Fassberg 17, D-37077 Goettingen, Germany

Abstract

This note is based on a recent brief communication published in Combustion and Flame (Shy et al. 2107) to discuss an unexpected result discovered by Wu et al. (2014). Contrary to traditional views, Wu et al. found that turbulence can facilitate ignition through differential diffusion when the effective Lewis number (*Le*) of mixtures is sufficiently greater than unity, where ignition experiments were conducted in a near-isotropic turbulence generated by a fan-stirred burner by applying small spark electrode gaps ($d_{gap} \leq 0.8$ mm). This implied that the associated minimum ignition energy (MIE) in intense turbulence could be smaller than that in quiescence. In this work, the effect of d_{gap} on the turbulent facilitated ignition (TFI) for $Le > 1$ is measured and discussed. We apply the same hydrogen mixtures at the equivalence ratio $\phi = 5.1$ ($Le \approx 2.3$) and $\phi = 0.18$ ($Le \approx 0.3$) as Wu et al. in our large fan-stirred cruciform bomb capable of generating near-isotropic turbulence to measure values of MIE as a function of d_{gap} at both quiescence and intense turbulence with the rms turbulent fluctuating velocity $u' = 5.4$ m/s. When d_{gap} is increased from 0.58 mm to about 6 mm, a drastic fall of values of laminar and turbulent MIE (MIE_L and MIE_T) is found. TFI only occurs for $Le > 1$ ($\phi = 5.1$) in support of Wu et al.'s finding, but it is restricted at smaller $d_{gap} = 0.58$ mm. We discover that the MIE_L and MIE_T curves *versus* d_{gap} can cross each other at larger d_{gap}, at which no TFI for $Le > 1$ at $d_{gap} \geq 2$ mm.

1 Introduction

A recent brief communication published in Combustion and Flame [1] addresses a question: Is turbulent facilitated ignition through differential diffusion independent of spark gap? To further disseminate the message in [1] in our combustion community, this note expands the results in [1] for more details.

Law and co-workers [2] discovered that turbulence can facilitate ignition through differential diffusion when the effective Lewis number (*Le*) of mixtures is sufficiently greater than unity. They applied a pair of cantilever electrodes of 0.25-mm in diameter with small electrode gaps ($d_{gap} \leq 0.8$ mm) in near-isotropic turbulence generated by a fan-stirred burner. They found that a fixed spark discharge voltage at $d_{gap} = 0.58$ mm which is assured of being unable to ignite the hydrogen/air mixture at the equivalence ratio $\phi = 5.1$ ($Le \approx 2.3 > 1$) in both quiescence and weak turbulence can nevertheless ignite the same mixture in intense turbulence up to the rms turbulent fluctuating velocity $u' = 5.4$ m/s. This suggested that the ignition energy (E_{ig}) required for successful ignition in intense turbulence can be smaller than that in quiescence, which is opposite to the classic conclusion of the turbulent effect on E_{ig} [e.g., 3-11]. For the hydrogen/air mixture at $\phi = 0.18$ with $Le \approx 0.3 < 1$, no turbulent facilitated ignition (TFI) was found with $d_{gap} \leq 0.8$ mm [2]. A question arises. Is the aforesaid TFI for $Le > 1$ independent of d_{gap}? Larger d_{gap} is expected to play a different role in turbulent ignition. This is because Shy and co-workers [7-9,11] reported that turbulence renders ignition more difficult and thus leads to an ignition transition regardless of *Le*, by using 2-mm electrodes with a larger $d_{gap} = 2$ mm in near-isotropic turbulence generated in a large fan-stirred cruciform bomb (see Fig. 1 of [12]). Across the transition, the increase of $E_{ig,T}$ with increasing u' varies from linearly to exponentially, where the subscript T represents the turbulent property. Using the same rich and lean H_2/air mixtures and electrodes as in [2] in our cruciform bomb together with the well-established ignition system [11] allows measuring E_{ig} here of high accuracy and/or minimum ignition energy (MIE) as a function of d_{gap} at both quiescence ($u' = 0$) and intense turbulence ($u' = 5.4$ m/s) conditions, which reveals the subtle detail of spark ignition phenomena.

It is known that laminar MIE (MIE_L) data increase drastically when $d_{gap} < d_q$, where d_q is a critical d_{gap} called the quenching distance that may be related to the critical radius of the developing flame kernel for successful flame initiation in the classic thermal-diffusion theory [3,13]. When $d_{gap} > d_q$, MIE_L is roughly the same over a considerable range of d_{gap} (Figs. 163,165 of [3]). Note that for the H_2/air mixture used in [2] at $\phi = 0.18$ ($Le \approx 0.3$) and 5.1 ($Le \approx 2.3$), $d_q \approx 2.8$ mm and 4 mm, respectively (Table 9 in [3] and Fig. 12 in [14]). Since Wu et al. [2] applied a small $d_{gap} = 0.58$ mm at $Le \approx 2.3$, it may be reasonable to postulate that MIE_L could be possibly higher than turbulent MIE (MIE_T) when $d_{gap} << d_q$. However, the factor of $d_{gap} << d_q$ alone fails to explain the *Le* dependence of TFI [2]. To account for the *Le* effect, we further consider the critical flame radius (R_c), below which flame propagation may not be possible [15,16]. As noted by Ju and co-workers, R_c increases with *Le* [16]. For very rich and very lean H_2/air mixtures, it appears that $R_c >> d_q$ when $Le >> 1$ and conversely $R_c << d_q$ when $Le << 1$ [16]. As such, TFI could occur at $Le >> 1$ when $d_{gap} << d_q$, but it is not likely to occur at $Le << 1$ even with $d_{gap} << d_q$. Here it is interesting to know if TFI remains existent at $Le \approx 2.3$ for larger $d_{gap} > 0.58$ mm and to explore the sensitivity of distance between electrodes on TFI.

Corresponding author. Fax: +886-3-427-6157
E-mail address: sshy@ncu.edu.tw

2 Experimental Consideration

To substantiate the aforesaid discussion on whether TFI depends on d_{gap}, a series of well-controlled turbulent ignition experiments are conducted using the H_2/air mixture at $\phi = 5.1$ at $T = 300$ K in near-isotropic turbulence over a wide range of u'/S_L. We aim to obtain the impact of turbulence on MIE of the rich H_2/air mixture with Le $\approx$ 2.3 >> 1 by applying two different kinds of the electrodes. The first kind is a pair of cantilevered electrodes of 0.25-mm diameter with flat ends having a suspended length of about 32 mm long for each electrode same as that used in [2]. The second kind is a pair of much thicker electrodes of 2-mm diameter with sharp ends same as that used in [7-9,11]. We measure MIE as a function of d_{gap} varying from 0.58 mm to 6.5 mm under both quiescence (u' = 0) and intense near-isotropic turbulence (u' = 5.4 m/s).

The large dual-chamber, constant-temperature/pressure, fan-stirred 3D cruciform burner capable of generating near-isotropic turbulence has been described and the reader is directed to [9,11-12] and references therein for detail treatment on the facility and associated turbulence properties. Thus, they are not elaborated upon here.

Two points deserve to comment. First, many valuable data are available in literature, not only for MIE itself but also for the effects of many parameters on MIE (e.g., [4-11,17,18] among others). These parameters are: (1) electrical breakdown characteristics i.e. type of discharge, discharged voltage/current, pulse duration time; (2) electrode characteristics i.e. material, geometry, size, gap; (3) flow characteristics i.e. type of flow, turbulent velocity/length scales, pressure, temperature; (4) mixture characteristics i.e. equivalence ratio, phase of fuel. In order to accurate reproduction of a spark ignition experiment, these aforesaid parameters as well as the discharged E_{ig} are needed. Each value of the discharged E_{ig} should be directly measured across the gap between electrodes by integrating the product of discharged current and voltage waveforms (best in square waveforms with little fluctuations) within the pulse duration time (Δt_p) [7-9,11]. An example is given in Figure 1.

Since measured values of E_{ig} can vary from shot to shot due to the breakdown perturbations, it is necessary to measure E_{ig} in situ directly from the discharged electrodes via the integration of voltage and current waveforms across the two electrodes for every single experiment when studying spark ignition and measuring MIE. Figure 1 presents typical near-square voltage and current waveforms directly measured across the two electrodes with sharp ends (right bottom image), where the ignition circuit (right top) is connected to a precision high-voltage pulse generator with a maximum breakdown voltage of 25 kV and a loading resistance R_Ω. The loading resistors in series are connected between the negative electrode and the ground (Fig. 1) to vary and control the discharged value of E_{ig}. The larger the value of R_Ω is, the smaller the value of E_{ig} at a given condition. In Fig. 1, the breakdown voltage used is 15 kV and the pulse duration time Δt_p = 100 μs. Accurate E_{ig} can be obtained by directly integrating the product of the discharged current $I(t)$ and the voltage difference $[V_1(t) - V_2(t)]$ across the gap between electrodes within $\Delta t_p = t_2 - t_1$ (Fig. 1). $I(t)$ is measured by the Pearson current monitor and V_1(t) and V_2(t) are measured by two Tektronix high-voltage probes. Also plotted in Fig. 1 is the integration value of E_{ig} as a function of time (the light blue line), where an accurate E_{ig} = 17.5 mJ can be obtained.

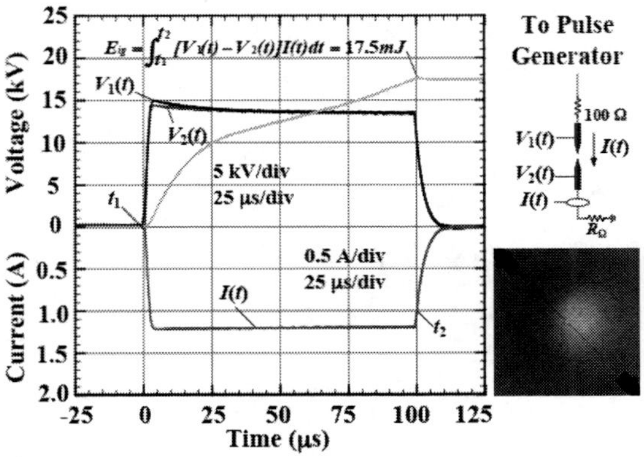

Figure 1: The determination of E_{ig} via direct measurements of the discharged current waveform I(t) and the voltage difference between positive and negative electrodes [V_1(t) - V_2(t)]. R_Ω in the ignition circuit is the controlled loading resistance.

Second, MIE is a statistical quantity in nature owing to inherent perturbations in the electrical breakdown characteristics (e.g., [3,5,11,18]). These perturbations can result in either ignition or non-ignition even applying the "same discharged E_{ig}" for a given condition. Thus, MIE measurements should be approached as a statistical rather than a threshold phenomenon, because MIE is a probabilistic variable, not a threshold value. Repeated ignition experiments for a given condition with a range of E_{ig} are required to identify MIE at 50% ignitability (e.g., [8,17,18] among others). This work determines MIE ≡ $E_{ig(50\%)}$ using the logistic regression method [9,11,19]. Also, this work applies the same spark ignition circuit with one-shot mode as in our previous studies [9,11], of which the range of E_{ig} lies between the smallest E_{ig} = 0.2 mJ (breakdown voltage = 15 kV, R_Ω = 2 MΩ, Δt_p = 10 μs) and the largest E_{ig} = 300 mJ (15 kV, R_Ω = 5 kΩ, Δt_p = 500 μs) where R_Ω is the loading resistance.

3 Results and Discussion

Using the logistic regression method, Fig. 2 presents a typical example of spark ignition probability versus different energy

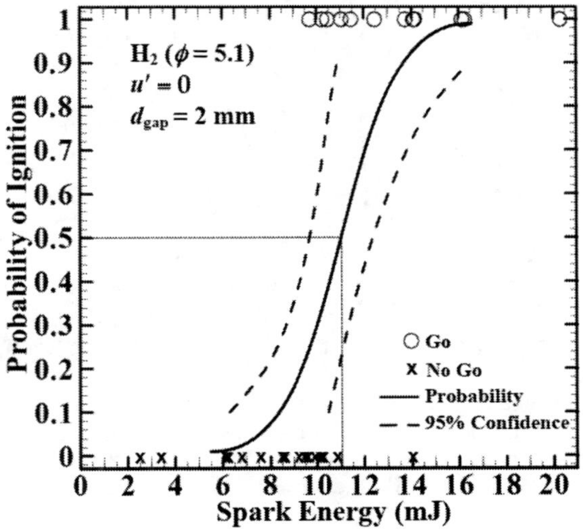

Figure 2: A typical case for the statistical determination of MIE ≡ $E_{ig(50\%)}$ using the logistic regression method, where the hydrogen/air mixture at ϕ = 5.1 with $Le \approx 2.3$ is applied in quiescence.

level at d_{gap} = 2 mm in quiescence using rich hydrogen as a fuel at ϕ = 5.1 with $Le \approx 2.3$. Each MIE data point is determined by 18-40 repeated ignition experiments using a range of E_{ig} with either "Go" indicated by an open circle symbol or "No Go" representing by a cross symbol, as shown in Fig. 2. Note that sufficient samples within the overlapping energy band are required to obtain an accurate cumulative probability, distribution (the solid curve in Fig. 2) versus stimulus level (E_{ig}). When the cumulative probability distribution is obtained, percentiles and confidence intervals can be then calculated to determine MIE at 50% ignitability (MIE $\equiv E_{ig(50\%)}$). In this work, 95% confidence intervals are also plotted in Fig. 2. It can be seen from Fig. 2 that both "Go" (ignition) and "No Go" (non-ignition) coexist within the overlapping energy band.

Figure 3 shows time sequential schlieren images of rich hydrogen spark ignition dynamics under (a) laminar (u' = 0) and (b) intense turbulent (u' = 5.4 m/s) conditions using 0.25-mm thin electrodes at dgap ≈ 2 mm, where the hydrogen/air mixture at ϕ = 5.1 with $Le \approx 2.3$ is applied. By comparing these "Go" and "No Go" images at t = 0.1 ms, it can be seen that the radii of these "Go" and "No Go" flame kernels in either Fig. 3a or Fig. 3b are about the same, but their subsequent flame developments are completely different. One goes with flame propagation (the upper row of Figs. 3a and 3b), while the other leads to flame global quench (the lower row of Figs. 3a and 3b). These results show that spark ignition is statistical in nature, where "Go" and "No Go" coexist even at the same E_{ig} within the overlapping energy band. Specifically, for laminar cases (Fig. 3a), both "Go" and "No Go" have the same $E_{ig} \approx 10.2$ mJ. Similarly, the same $E_{ig} \approx 9.3$ mJ is applied for both "Go" and "No Go" intense turbulent cases in Fig. 3b. Hence, as shown in Figs. 2 and 3, MIE is a probabilistic variable, not a threshold value. Hence, the determination of the value of MIE should be approached as a statistical rather than a threshold phenomenon.

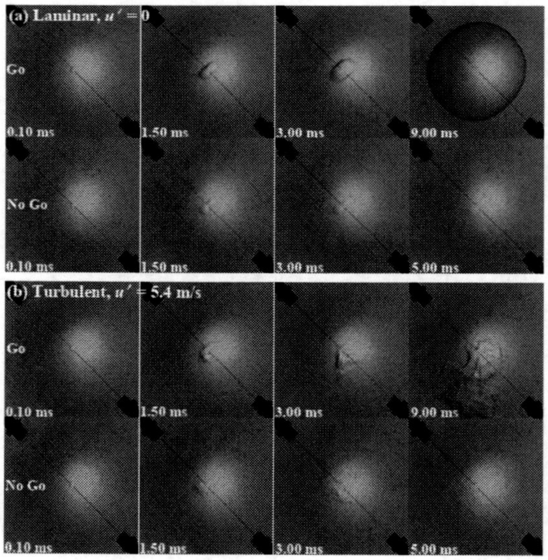

Figure 3: Sequential schlieren images of (a) laminar and (b) intense turbulent centrally-ignited hydrogen/air mixtures at ϕ = 5.1 using 0.25-mm electrodes at $d_{gap} \approx 2$ mm. Each case includes both "Go" example: spark → kernel → flame propagation and "No Go" example: spark → kernel → complete quench, both examples applying the same E_{ig}. Field of view: 62 x 62 mm^2.

Figure 4 shows the effect of d_{gap} on both MIE_L and MIE_T at u' = 5.4 m/s of the hydrogen/air mixture at ϕ = 5.1 with $Le \approx 2.3$ using the thin 0.25-mm tungsten electrodes, same as that used in [2]. A quick fall of MIE_L with increasing d_{gap} is found when $d_{gap} < d_q \approx 4$ mm, where MIE_L = 61.5 mJ at d_{gap} = 0.58 mm that is 77 folds larger than MIE_L = 0.8 mJ at d_{gap} = 3 mm. Further, when d_{gap} = 4 mm, 5.2 mm and 6.5 mm, such rich hydrogen mixture is found to be extremely easy to ignite having 100% ignitability to our lowest E_{ig} = 0.2 mJ (see the marked square symbols in Fig. 4), suggesting MIE_L is much less than 0.2 mJ within 4 mm ≤ d_{gap} ≤ 6.5 mm. Note that the above 100% ignitability is based on more than 20 repeated runs at fixed d_{gap} and conditions. The monotonic decrease in MIE_L at $Le \approx 2.3$ with increasing d_{gap} in Fig. 4 is consistent with the effect of R_c [15,16] or d_q [3,13].

As to the intense turbulence case (u' = 5.4 m/s), when d_{gap} = 0.58 mm, MIE_T = 26.0 mJ << MIE_L = 61.5 mJ. This quantitative MIE result supports the previous qualitative result of [2] that for $Le > 1$ mixtures, turbulence can facilitate ignition. But we discover that such TFI phenomenon only occurs at small d_{gap}. Interestingly, the MIE_L and MIE_T curves in Fig. 4 cross each other when $d_{gap} \geq 2$ mm. At d_{gap} = 2 mm, MIE_T = 9.20 mJ is only slightly smaller than MIE_L = 11 mJ, of which both values of MIE_T are within their overlapping ignition energy band. Moreover, when d_{gap} = 3 mm, MIE_T = 6.3 mJ which is much larger than MIE_L = 0.8 mJ. Similarly, when d_{gap} = 5.2 mm, MIE_T = 3.05 mJ >> MIE_L, where MIE_L is much less than 0.2 mJ. In summary, at $Le \approx 2.3$, the effect of $d_{gap} << d_q \approx R_c$ (4-5 mm estimated from Fig. 1(b) in [2]) severely increases MIE_L, owing possibly to the huge heat loss to spark electrodes. It might then be easier to, as argued in [2], survive for a flame kernel that moves with the turbulent wind away from the spark gap, leading to TFI. However, TFI seems to weaken and then to disappear with the continual enlargement of d_{gap}, as shown in Fig. 4.

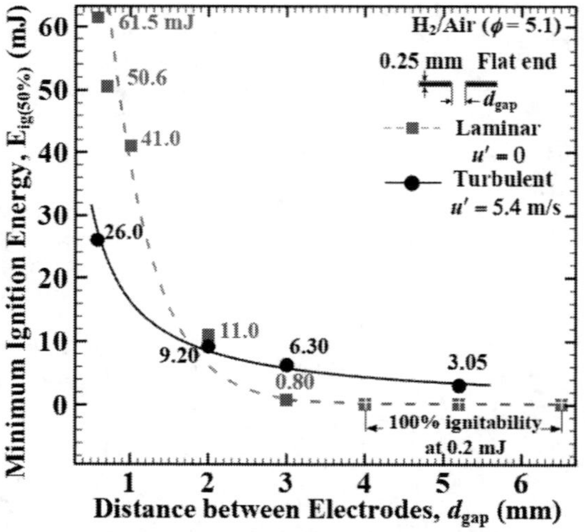

Figure 4: MIE comparison between laminar and intense turbulence cases over a wide range of d_{gap} for the H_2/air mixture at ϕ = 5.1 with $Le \approx 2.3$ using the 0.25-mm electrodes same as [2].

Here we present the results with a pair of thick tungsten electrodes (2-mm in diameter) to test the thickness effect of spark electrodes on TFI and to compare the former result ($Le > 1$) with a very lean case ($Le < 1$), which are shown in Fig. 5. At ϕ = 0.18, when d_{gap} = 0.3 mm, MIE_L = 1.11 mJ at u' = 0 << MIE_T = 14 mJ at u' = 5.4 m/s, while MIE_L = 1.06 mJ at u' = 0 < MIE_T = 1.42 mJ at u' = 5.4 m/s when d_{gap} = 0.58 mm. Although there is no TFI for $Le < 1$ mixture, here, MIE_T seems to approach MIE_L toward larger d_{gap}. Note that when $Le \approx 0.3 << 1$ (H_2/air mixture at $\phi = 0.18$), due to $R_c << d_q$, $MIE_L \sim R_c^3$ is less sensitive to the effect of $d_{gap} << d_q$ [16]. Therefore, it is not surprising that at $Le \approx 0.3$ TFI does not occur in Fig. 5 even when $d_{gap} << d_q$, consistent with the finding in [2]. As shown in Fig. 5 for $Le > 1$ mixture, TFI is still restricted at smaller d_{gap} = 0.58 mm, where MIE_L = 255.5 mJ >> MIE_T = 36.8 mJ at u' = 5.4 m/s having the same trend as Fig. 4. We find no TFI at d_{gap} = 2 mm (Fig. 5), because MIE_L is only 0.52 mJ which is much less than MIE_T = 17.3 mJ at u' = 5.4 m/s. The possible reason for ignition facilitated in intense turbulence for $Le > 1$ flame may, as argued in [2], be that the flame segments intensified by the negative stretch move away from incipient extinction and serve as local ignition sources. Since the state of incipient extinction may be a direct result of $d_{gap} << R_c$, the present finding that TFI may not hold for arbitrary d_{gap} thus confirms this point. In short, TFI depends on d_{gap}, which deserves to be disseminated in our combustion community.

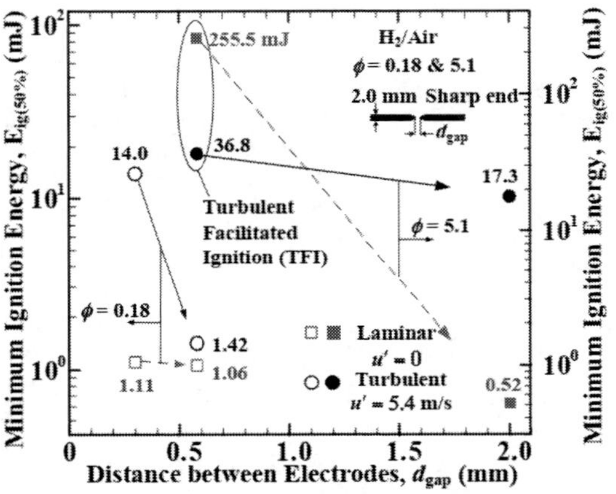

Figure 5: Effect of ϕ and u' on MIE at three selected d_{gap} for both lean and rich H_2/air mixtures at $\phi = 0.18$ and 5.1 with $Le \approx 0.3 < 1$ and $Le \approx 2.3 > 1$.

4 Concluding Remarks

Turbulence can facilitate ignition through differential diffusion for $Le >> 1$ mixture, but such peculiar phenomenon is restricted at small d_{gap}. When d_{gap} > 2 mm (0.25-mm tungsten electrodes) and d_{gap} = 2 mm (2-mm tungsten electrodes), there is no TFI and turbulence always renders ignition more difficult even for the rich hydrogen/air mixture at $\phi = 5.1$ with $Le \approx 2.3 >> 1$. For $Le << 1$ mixture, $MIE_T > MIE_L$ but MIE_T seems to approach MIE_L toward larger d_{gap}. These interesting results deserve to be disseminated in our combustion community.

5 Acknowledgment

The authors would like to thank Prof. C.K. Law for his valuable comments. The financial support from the Ministry of Science and Technology, Taiwan, under grants (MOST 101-2221-E-008-049-MY3; 103-2221-E-008-099-MY3; 104-3113-E-008-001; 106-3113-E-002-013) is greatly appreciated.

References

[1] S.S. Shy, M.T. Nguyen, S.Y. Huang, C.C. Liu, Combust. Flame 185 (2017) 1-3.

[2] F. Wu, A. Saha, S. Chaudhuri, C.K. Law, Phys. Rev. Lett. 113 (2014) 024503(1-5).

[3] B. Lewis, G. von Elbe, Combustion, Flame and Explosions of Gases, Academic Press, New York, 1987 (1951,1961), pp. 333-361 (see Figs. 163 and 165 on p. 336 and p. 339; Table 9 on pp. 356-360).

[4] D.R. Ballal, A.H. Lefebvre, Proc. R. Soc. A 357 (1977) 163-181.

[5] M. Kono, K. Hatori, K. Iinuma, Proc. Combust. Inst. 20 (1984) 133-140.

[6] D. Bradley, F.K.K. Lung, Combust. Flame 69 (1987) 71-93.

[7] C.C. Huang, S.S. Shy, C.C. Liu, Y.Y. Yan, Proc. Combust. Inst. 31 (2007) 1401-1409.

[8] S.S. Shy, C.C. Liu, W.T. Shih, Combust. Flame 157 (2010) 341-350.

[9] M.W. Peng, S.S. Shy, Y.W. Shiu, C.C. Liu, Combust. Flame 160 (2013) 1755-1766.

[10] C. Cardin, B. Renou, G. Cabot, A.M. Boukhalfa, Combust. Flame 160 (2013) 1414-1427.

[11] S.S. Shy, Y.W. Shiu, L.J. Jiang, C.C. Liu, S. Minaev, Proc. Combust. Inst. 36 (2016) 1785-1791.

[12] L.J. Jiang, S.S. Shy, W.Y. Li, H.M. Huang, M.T. Nguyen, Combust. Flame 172 (2016) 173-182.

[13] J. Han, H. Yamashita, N. Hayashi, Combust. Flame 157 (2010) 1414-1421.

[14] I.L. Drell, F.E. Belles, Survey of hydrogen combustion properties, NACA Research Memorandum E57D24, Report No. 1383, 1957.

[15] A.P. Kelley, G. Jomaas, C.K. Law, Critical radius for sustained propagation of spark-ignited spherical flames, Combust. Flame 156 (2009) 1006–1013.

[16] Z. Chen, M.P. Burke, Y. Ju, On the critical flame radius and minimum ignition energy for spherical flame initiation, Proc. Combust. Inst. 33 (2011) 1219–1226.

[17] E. Mastorakos, Prog. Energy Combust. Sci. 35 (1) (2009) 57-97.

[18] S.P.M. Bane, J.L. Ziegler, P.A. Boettcher, S.A. Coronel, J.E, J. Loss Prevent. Process Ind. 26 (2013) 290-294.

[19] S.P.M. Bane, Spark ignition: Experimental and numerical investigation with application to aviation safety, PhD dissertation, California Institute of Technology, 2010. http://thesis.library.caltech.edu/5868/1/thesis SBane.pdf.

11th Asia-Pacific Conference on Combustion,
The University of Sydney, NSW Australia
10th -14th December 2017.

The Model Development for a Bottom-blowing Converter

Zhi-Hong Yang[1], Xin-Yi Lin[1], Sheng-Yen Hsu[1*], Jyun-Sheng Wang[1], Yung-Chang Liu[2], Chien-Hsiung Tsai[3]

[1]Department of Mechanical & Electro-Mechanical Engineering, National Sun Yat-sen University, Kaohsiung, Taiwan
No.70, Leinhai Rd., Gushan Dist., Kaohsiung 804, Taiwan (R.O.C.)

[2]Iron & Steel Research & Development Department, China Steel Corporation, Kaohsiung, Taiwan
No.88, Chenggong 2nd Rd., Qianzhen Dist., Kaohsiung 806, Taiwan (R.O.C.)

[3]Department of Vehicle Engineering, National Pingtung University of Science and Technology, Pingtung, Taiwan
No.1, Xuefu Rd., Neipu Township, Pingtung 912, Taiwan (R.O.C.)

Abstract

Using the scrap can not only reduce the total amount of scrap, but also save the energy costs in the steelmaking process. The bottom-blowing converter is a future trend. The carbon particles and carrier gas (nitrogen) are injected from the bottom into the converter to increase the carbon content of the liquid iron and improve carbon distribution. In the same time, oxygen and protective gas (methane) are injected to accelerate the melting of the scrap and protect the converter bottom tuyeres from damage induced by a high temperature.

The steelmaking process usually accompanies high temperature and high pressure, which makes it impossible to directly observe the condition inside the converter. Therefore, in this study a 3-D single-phase model is developed. The injected carbon and gas were assumed as particles using discrete phase model (DPM). The carbon particles could melt into the liquid iron. The gas bubbles can also change their states thermodynamically due to different temperature and pressure. The bottom-blowing model developed in this study can give further understanding of the converter and improve the operation in the steelmaking process.

1 Introduction

With the development of human societies, the amount of scrap steel/iron becomes more and more, so to increase the use of scrap in the steelmaking processes is the tendency of world situation. Currently, the scrap is usually recycled by using the Electric-Arc Furnace (EAF) and the Basic Oxygen Furnace (BOF). In the EAF process, the scrap is the primary raw material (>80%) and melt by using the electric power. On the other hand, the scrap is added as the supplementary (<20%) in the BOF process. The melting heat is provided by the oxidation of carbon, silicon, etc. contained in the liquid metal as oxygen is blown into the converter furnace. Although the EAF can use whole scrap in the steelmaking process, the energy cost, process time and the quality of liquid iron are not beneficial as compared to those in the BOF process. Therefore, the whole-scrap steelmaking technology based on the BOF process have been developed. The additional carbonaceous sources and oxygen are proposed to be injected into liquid iron for melting scrap efficiently [1, 2] and required to meet the desired process time in the production line.

The carburization process not only provide the melting heat but also lower the melting temperature [3]. In addition, for the high scrap ratio of steelmaking processes, several important indicators need to be considered to evaluate the converter performance, such as thermal efficiency, carburization rate, and ability of mixing [4-6]. The carbonaceous materials, injected into liquid iron as small particles with carrier gas, can maximize the area of carbonaceous solids and agitate the flow of liquid iron to increase the dissolution rate. On the other hand, the bottom-blown oxygen is injected via a tuyere made by two concentric tubes and shrouded by the hydrocarbon gas. Oxygen gas can quickly react with the dissolving carbon atoms in liquid iron and effectively heat up the liquid iron. Therefore, to optimize the process operation, it is required to know the distributions of carbon concentration and temperature in the liquid.

Due to the high temperature condition in the steelmaking converter, it is difficult to observe and measure the internal flow field of liquid iron. Therefore, in this study a numerical model including the computational fluid dynamics (CFD), chemical kinetics and thermodynamics is used to further understand the complex flow characteristics with heat and mass transfer in the process. In addition, the process with the bottom-blowing techniques includes the complicated multiphase-flow phenomenon, complex homogeneous and heterogeneous reactions, etc., so it is also difficult to establish a detailed model to simulate the process. Therefore, in this study a simplified 3-D model is developed for further understanding the flow characteristics of liquid iron in the bottom-blowing converter.

2 Numerical model

The bottom-blowing converter contains liquid iron, gas phase and tuyeres, but both models in this study are simplified gas phase and only simulate the liquid iron region. This study divides into two parts. The first model only simulate the control volume near the tuyere. The control volume is cylindrical and both of the top and the bottom boundary conditions are wall. The side of the converter is the pressure outlet, and this study lets the pressure be modified with depth. The height of cylinder is 1.5 meter and the diameter is 1.2 meter. The boundary condition of second model is similar to previous model, but in the side of the converter, the pressure outlet is changed to wall, the height is 1.5 meter and the diameter is 5.8 meter. The tuyeres located at the

Corresponding author. Fax: +886-7-525-4299
E-mail address:syhsu@mail.nsysu.edu.tw

bottom of the converter were used to inject gas and particles. To simplify the model, the interface between gas and liquid iron is assumed to be flat during the process and only the liquid iron is simulated, as shown in Fig. 1. There are two models used in this study. One is to demonstrate the bottom-blowing effects for different gases and/or particles. The other is to simulate the heat and mass transfer in the stirred flow field of liquid iron when the gases and carbonaceous particles are injected from the tuyeres at the bottom.

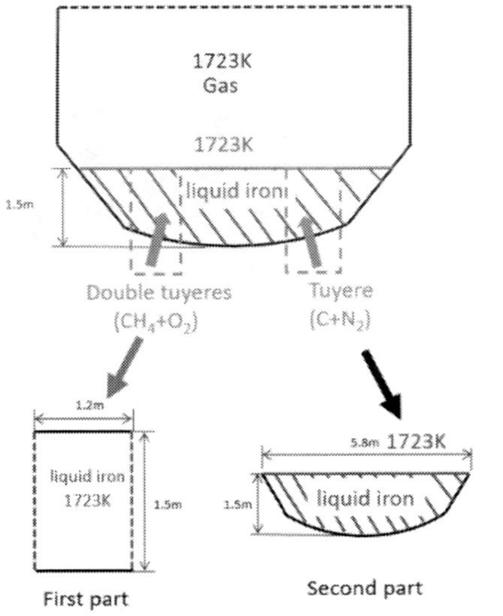

Figure 1: A schematic diagram of bottom-blowing converter

This study uses Gambit 2.4.6 to build the model and create the grids, and then use the commercial CFD software FLUENT to perform the numerical calculations.

The governing equations are the following:

Continuity equation:

$$\frac{\partial \rho}{\partial t}+\nabla\cdot(\rho\vec{v})=S_m \tag{1}$$

Momentum equation:

$$\frac{\partial}{\partial t}(\rho\vec{v})+\nabla\cdot(\rho\vec{v}\vec{v})=-\nabla p+\nabla\cdot\left(\overline{\overline{\tau}}\right)+\rho\vec{g}+\vec{F} \tag{2}$$

Energy equation:

$$\frac{\partial}{\partial t}(\rho E)+\nabla\cdot\left(\vec{v}(\rho E+p)\right)= \nabla\cdot\left(k_{eff}\nabla T-\sum_j h_j\vec{J}_j+\left(\overline{\overline{\tau_{eff}}}\cdot\vec{v}\right)\right)+S_h \tag{3}$$

Species transport equation:

$$\frac{\partial}{\partial t}(\rho Y_i)+\nabla\cdot(\rho\vec{v}Y_i)=-\nabla\cdot\vec{J}_i+R_i+S_i \tag{4}$$

In this study, the flow field is simulated by the detached eddy simulation (DES) turbulence model [7], and the bottom-blown gases and solid particles are simulated by using the discrete phase model (DPM) model. The particle force balance equation is:

$$\frac{d\vec{u}_p}{dt}=\frac{\vec{u}-\vec{u}_p}{\tau_r}+\frac{\vec{g}(\rho_p-\rho)}{\rho_p}+\vec{F} \tag{5}$$

Oxygen and methane are injected at the same location and the carbon particles are injected by using nitrogen as the carrier gas. In addition, the carbon particles can dissolve into the liquid iron. Oxygen can react with carbon atom in liquid iron to form CO exothermically and methane can be decomposed endothermically into carbon atom and hydrogen at a proper temperature. Here are the chemical reactions:

$$C\rightarrow C_{(Fe)} \tag{6}$$

$$C_{(Fe)}+0.5O_2\rightarrow CO \tag{7}$$

$$CH_4\rightarrow C+2H_2 \tag{8}$$

For the gas bubbles, the ideal gas law is applied:

$$\rho=\frac{P_{op}}{\frac{R}{M_w}T} \tag{9}$$

The above chemical reactions and the volume of bubble gases in the simulation are applied in the FLUENT calculation by using the user-defined function (UDF).

3 Results and discussion

In this study, the converter contains liquid iron and gas phase, and both temperature is 1723K. All the particles temperature is 300K initially. The injected bubble diameters varies from 0.001m to 0.01m. The mass flow rate of methane, oxygen and nitrogen are respectively 0.028 kg/s, 0.2 kg/s and 0.042kg/s. The initially velocity and diameter of carbon particle are 300 m/s 0.0002m and the mass flow rate is 0.4 kg/s. In the following, the simulations near the tuyeres and the effects of the bottom-blowing tuyeres will be discussed separately.

3.1 Simulation near the tuyeres

In this study, two types of tuyeres are simulated. One is for blowing the carbon particles by using nitrogen as the carrier gas and the other is for the injection of oxygen shrouded by propane. Fig. 2 shows the diameter distributions of carbon particle and nitrogen bubbles. The carbon particles can dissolve into liquid iron (Eq. 6), and therefore, the carbon particle size decreases after injected, as shown in Fig. 2 (a). Fig. 2 (b) shows the diameter distribution of the nitrogen in the liquid iron. The nitrogen bubbles expands as they rise in liquid iron due to the pressure decrease and the temperature increase. The oxygen and methane bubbles have the same expansion effects to nitrogen bubbles. Besides, the oxygen bubbles can react with carbon in liquid metal (Eq. 7) and methane bubbles can decompose into hydrogen and carbon (Eq. 8). The volume changes of bubbles due to the variation of the molecular number are also considered.

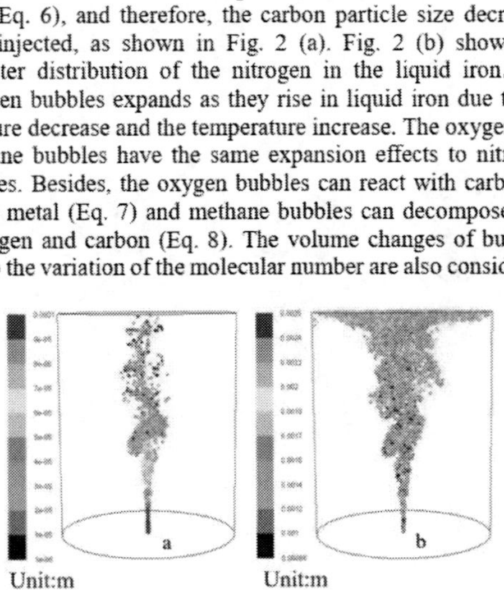

Figure 2: The Diameter (m) of the particles.
(a)Carbon. (b)Nitrogen.

The initial temperature of the injected carbon or gas bubbles are 300K and Fig. 3 shows the temperature distribution near the tuyeres after they are injected. One can see that the tuyere blowing carbon by using nitrogen as the carrier gas significantly cools down the liquid iron because of the carbon particles, as shown in Fig. 3(a). On the other hand, the tuyere injecting oxygen gas shrouded by methane gas heats up the liquid metal because of the exothermic reaction (Eq.7). In this study, the injected carbon particles and methane bubbles are the carbon sources in liquid metal. Fig. 4(a) shows that the carbon particle dissolve into liquid metal and increase the mass fraction of carbon. However, oxygen react with carbon in liquid metal to form carbon monoxide, decreasing the carbon in liquid metal near the oxygen tuyere, as shown in Fig. 4(b). In the steelmaking process, carbon particle can increase the mass fraction of liquid. In addition to lower the melting temperature, it is also the heat source to melting the scrap. Bottom blowing oxygen causes the exothermic reaction occurs in liquid metal, heating it up effectively.

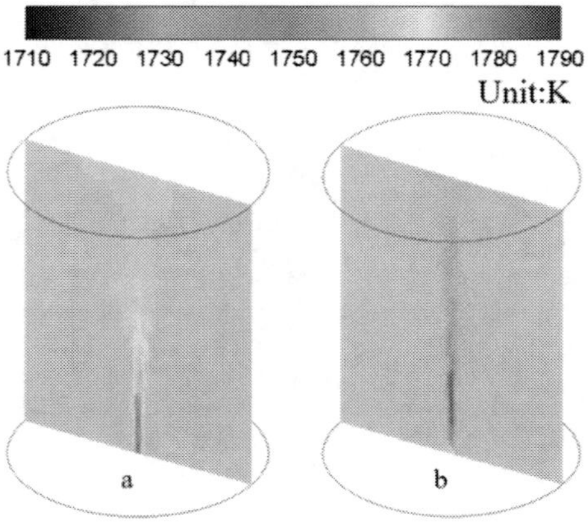

Figure 3: Temperature (K) distribution. (a)Injection of carbon by using nitrogen gas. (b)Blowing oxygen gas shrouded by methane gas.

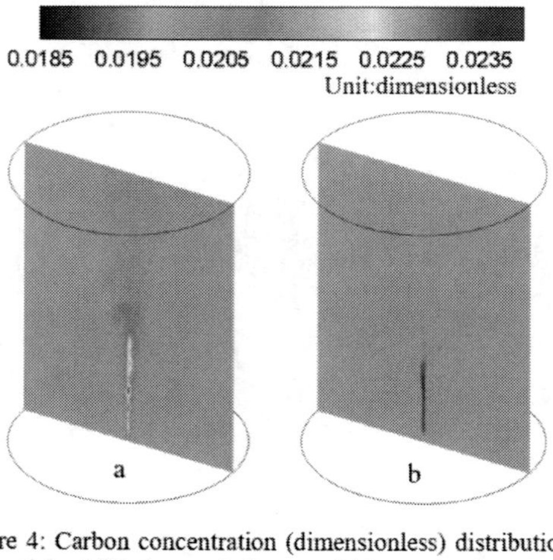

Figure 4: Carbon concentration (dimensionless) distribution of the particles.(a)Carbon and nitrogen. (b)Oxygen and methane.

3.2 A converter with ten tuyeres

In this section, a converter with ten bottom blowing-tuyeres are simulated. The distribution of the tuyeres are shown in Fig. 5. Four tuyeres inject the carbon particles with nitrogen as the carriser gas and the others blow oxygen gas shrouded by methane gas.

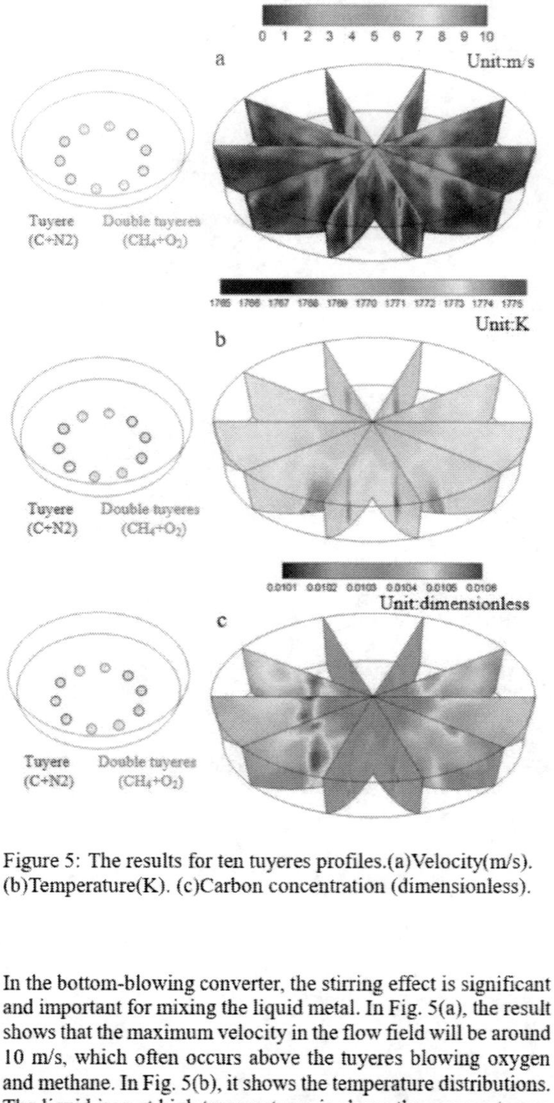

Figure 5: The results for ten tuyeres profiles.(a)Velocity(m/s). (b)Temperature(K). (c)Carbon concentration (dimensionless).

In the bottom-blowing converter, the stirring effect is significant and important for mixing the liquid metal. In Fig. 5(a), the result shows that the maximum velocity in the flow field will be around 10 m/s, which often occurs above the tuyeres blowing oxygen and methane. In Fig. 5(b), it shows the temperature distributions. The liquid iron at high temperatures is above the oxygen tuyere because of the exothermic reaction. According to the mass fraction of carbon in liquid iron in Fig. 5(c), due to the carbon dissolutions the mass fraction of carbon near the blowing tuyere is higher. In addition, because the carbon particles do not complete dissolve into liquid metal during the rise from the bottom to the liquid surface, and then spread to the sides of the converter. Therefore, one can see that the mass fraction of carbon near the sides of the converter is higher than that in the middle. As mentioned in the introduction, the mass fraction of carbon in liquid metal can alter the melting temperature, so it may important in the steelmaking to know where the scrap should be located. In addition, it may also be considered where the liquid metal of high temperature is to optimize the scrap melting. Fig. 6 shows the distribution of gas bubbles. The injected oxygen reacts with carbon to form carbon monoxide (Fig. 6(a)) and the injected methane decomposes into hydrogen gas(Fig. 6(b)). The results shows that the gas bubbles may move with the flow field and stay longer in the upper part of liquid metal.

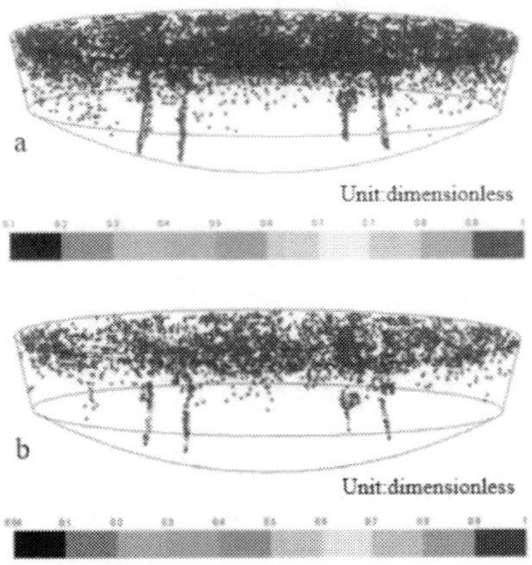

Figure 6: The mass fraction (dimensionless) in the bubbles. (a)Carbon monoxide. (b)Hydrogen.

4 Conclusions

In this study, a numerical model for simulating the bottom-blowing tuyeres in a converter by using the Discrete Phase Model (DPM) for the carbon particles and gas bubbles is preliminarily developed. In addition, the carbon dissolution and the bobble reactions are also considered, so the numerical results can give the distributions of temperature and mass fraction of carbon in liquid, which is useful for the steelmaking design or operation.

At current stage, further development on the reliability of the model should be modified and verified. In addition, in authors' opinion, the model can be further added scrap to investigate the melting efficiency.

5 Acknowledgment

This research was supported by the Ministry of Science and Technology under Project No.: MOST 105-2622-8-006-001, Taiwan.

References

[1] K. Ohnuki, K. Umezawa, T. Hiraoka, N. Matsumoto and T. Inoue, Development of steel scrap melting process, Nippon Steel Technical Report No. 61, 1994.

[2] R. J. Fruehan, The Making, Shaping and Treatment of Steel – Vol. 2: Steelmaking and Refining Volume (11th edition), AISE Foundation, 1998.

[3] H. Sun, Analysis of reaction rate between solid carbon and molten Iron by mathematical models, ISIJ International, Vol. 45 (2005), No. 10, pp. 1482-1488.

[4] E.G. Schempp, Increased Scrap Melting Capacity for Oxygen Steelmaking--the New OBM-S Process, Iron Steel Eng., 56.12 (1979) 43-47.

[5] H. Klein, J.F. Liesch, H. Iso, K. Nakamura, Scrap ratio increase by coal blowing in the BOF, Steelmaking Proceedings. (1985) 129-136.

[6] Nozaki, Bottom-Blowing Converter method introduction · Stirring effect · Top-end compound blowing, Metallurgical Industry Press, China, 2008

[7] ANSYS FLUENT 16, Theory Guide.

11th Asia-Pacific Conference on Combustion,
The University of Sydney, NSW Australia
10th -14th December 2017.

The Role of Co-injected Hydrocarbon Gas with Oxygen in a Furnace

Le-Kuan Lin[1], Cheng-Hao Hou[1], Sheng-Yen Hsu[1], Jyun-Sheng Wang[1], Yung-Chang Liu[2], Chien-Hsiung Tsai[3]

[1] Department of Mechanical & Electro-Mechanical Engineering, National Sun Yat-sen University

No.70, Lianhai Rd., Kaohsiung City, Gushan Dist, 804, Taiwan (R.O.C.)

[2] Iron and Steel R&D Department, China Steel Corporation

No.1, Zhonggang Rd., Kaohsiung City, Xiaogang Dist, 812, Taiwan (R.O.C.)

[3] Department of Vehicle Engineering, National Pingtung University of Science and Technology

No.1, Xuefu Rd., Pingtung County, Neipu Township, 912, Taiwan (R.O.C.)

Abstract

In this study, the role of con-injected hydrocarbon gas with oxygen in a furnace is investigated based on the shape of ice in the water experiments [5, 6] by using a two-dimensional axisymmetric model with the detailed propane-oxygen chemical kinetics

The simulation results represent that in the beginning of the formation of the solid iron, the temperature near the tip of tuyere is high and the hydrocarbon gas in this region does not mixed with oxygen. Consequently, the endothermic decomposition can cause important cooling effect on chilling the local liquid metal to solidify, which provide the heat absorption about 30% higher than that by using inert gas. However, as the size of solid iron grows, the temperature at the base of mushroom decreases, and therefore, the decomposition moves downstream and becomes weaker. Eventually, the decomposition of hydrocarbon gas is insignificant in a sizable solid iron because of the mixing with oxygen in the downstream. It suggests that the solid iron is repeatedly formed at the tip of tuyere and blown away during the steelmaking process and the importance of hydrocarbon gas is to increase the existence time of solid iron on the tip of the tuyere.

However, in authors' opinion, the structure of solid iron is different from that of ice in the water experiments [5, 6]. The further investigation on the structure of solid iron is needed.

1. Introduction

With the development of human societies, the amount of scrap steel/iron becomes noticeable in many developed and developing countries. To save energy and reduce emissions, steelmaking companies have demonstrated new steelmaking processes using scrap iron/steel as the primary raw material, decreasing the use of fresh liquid iron supplied by the blast furnace. Compared to the pig iron from blast furnace, recycling scrap can reduce 50%~60% of the energy cost when making the same amount of liquid metal. Therefore, as expected that the scrap will continuously increase in the future, to develop a sophisticated whole-scrap steelmaking process is needed. Although the electric arc furnace is a matured and relatively simple steelmaking process for scrap iron, it uses expensive energy source and the quality of liquid iron depends on the used scrap greatly. Therefore, the other process based on the basic oxygen furnace has been developed [1].

The bottom-blown oxygen (OBM-S) method can effectively stir the liquid metal and burn the carbon inside the liquid metal, consequently increasing the rates of carbon dissolution and scrap melt. However, bottom oxygen injection is easy to burn the tuyere and the nearby refractory. Therefore, the bottom-blown oxygen tuyere made by two concentric tubes, have been successfully developed and applied [2]. The hydrocarbon gas is injected via the annular section between the two concentric tubes. Ideally, the endothermic decomposition of hydrocarbon gas can chill the liquid metal and help with the formation of solid iron at the tip of the tuyere [3]. The existence of solid iron is important to reduce the burn back rate of the tuyere and the wear of the surrounding refractory [4]. However, the protection mechanism in the steelmaking process is still not well understood. One would think that if the hydrocarbon gas is well mixed with oxygen before decomposition, the exothermic oxidation reaction will burn the tuyere away more easily. Therefore, it is encouraged to clearly understand the effects of hydrocarbon gas when co-injected with oxygen in the steelmaking process.

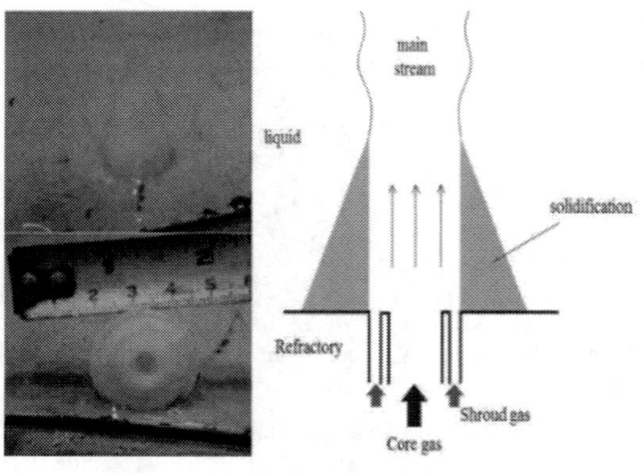

Figure 1: The configuration of the solidification.

Nevertheless, the high temperature of liquid iron causes the difficulties in designing experiments to investigate the protection mechanism directly. Therefore, Huang [5] and Huang et al. [6] designed water experiments by bottom-blowing gases of extremely low temperature into a water tank. The blowing rates

Corresponding author. Fax: +886-7-525-4299
E-mail address: syhsu@mail.nsysu.edu.tw

were designed according the similarity of the endothermic heat of hydrocarbon gases in the steelmaking process. In their studies [5, 6], the cone shape of ice with a gas channel in the middle was observed at the tip of tuyere, as shown in Fig. 1. Therefore, in this study, the tuyere and the solid iron according to the ice structure are established in the model. The purpose of this study is to investigate the possibility of the solid-iron structure similar the ice structure (Fig. 1) by using the developed model.

2. Numerical model

In this study, a 2-D axisymmetric model is developed to simulate the flow field and heat transfer around the tip of the tuyere. The model includes the tuyere, nearby refractory [7], solid iron [8], and the main gas channel in the middle (according the water experiments [5, 6]), as shown in Fig. 2. The solidification of liquid iron starts from the tip of the tuyere, so the reactions and heat transfer there are of importance. In the model, the shape of the solid iron is simplified and the thickness of solid iron can be varied to investigate the formation of solid iron at different stage.

The tuyere injects propane via the outer tube and oxygen through the inner tube. In order to catch the endothermic and exothermic reactions, the detailed chemical kinetics is applied for the propane-oxygen mixture. The refractory boundary is assumed to be adiabatic. The solid iron boundary is set the same as the temperature of liquid iron (1800K). Therefore, as the gases flow close to the tip of the tuyere, they will be heated up and react according to the mixture composition. To accurately simulate the chemical reactions of the gases/mixture at different conditions/positions, the detailed propane combustion mechanisms with 247 step reactions is used [9]. The catalytic effect of solid walls and radiative heat transfer are neglected.

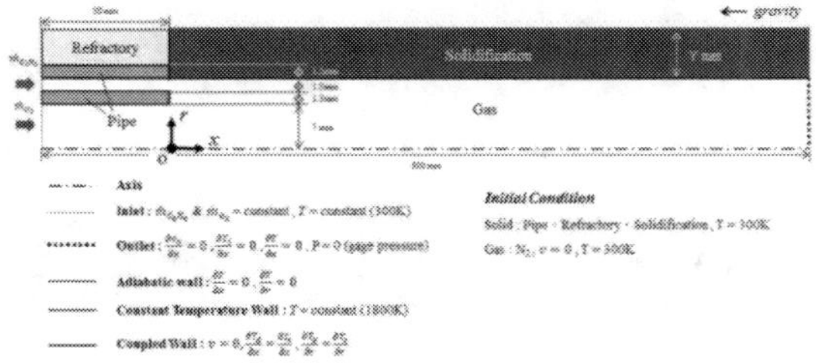

Figure 2: The model configuration for the investigating the effect of hydrocarbon on the formation of solid iron at the tip of tuyere.

In this study, the governing equations for the two-dimensional axisymmetric model are the following.

The continuity equation:

$$\frac{\partial}{\partial x}(\rho v_x) + \frac{\partial}{\partial r}(\rho v_r) + \frac{\rho v_r}{r} = 0 \tag{1}$$

The momentum equation:

$$\nabla \cdot (\rho \vec{v} \vec{v}) = -\nabla P + \mu_{eff} \nabla^2 \vec{v} + \rho \vec{g} \tag{2}$$

The energy equation:

$$\nabla \cdot (\vec{v}(\rho E + P)) = \nabla \cdot \left(K_{eff} \nabla T - \sum_i h_i \vec{J}_i + \left(\overline{\overline{\tau}}_{eff} \cdot \vec{v} \right) \right) + S_h \tag{3}$$

The species equation:

$$\nabla \cdot (\rho \vec{v} Y_i) = -\nabla \cdot \vec{J}_i + R_i \tag{4}$$

The equation of ideal gas:

$$\rho = \frac{P_{op} + P}{\frac{R}{M_W} T} \tag{5}$$

R_i is the rate of production of species i and $\vec{J}_i$ is the diffusion flux of species i.

$$\vec{J}_i = -\left(\rho D_{i,m} + \frac{\mu_t}{Sc_t} \right) \nabla Y_i - D_{T,i} \frac{\nabla T}{T} \tag{6}$$

$$R_i = M_{w,i} \sum_{R=1}^{N_R} R_{i,r} \tag{7}$$

$M_{w,i}$ is the molecular weight of species i ; $R_{i,r}$ is the rate at species i are generated or consumed by reaction.

$D_{i,m}$ is the mass diffusion coefficient of the species in the mixture; $D_{T,i}$ is the thermal diffusivity of the species; Sc_t is turbulent Schmidt number.

$$Sc_t = \frac{\mu_t}{\rho D_t} \tag{8}$$

D_t is turbulence diffusion coefficient. The realized k-ε model is used for the turbulence flow [10].

In the study, the commercial software ANSYS FLUENT 16 [10] is used to help with the model development and carry out the numerical results.

3. Result and Discussion

In this study, the flow rate of propane via the outer channel is 0.01 kg/s and oxygen through the inner channel is 0.2 kg/s. The thickness of inner tube is 1.5 mm, which separates the two gases, and the operating pressure is 2 atm. In addition, the solid iron thickness (Y) is set 1.5mm, which can represent the early stage of iron solidification. To clearly understand the endothermic effect of propane on the formation of solid iron on the tip of tuyere, two cases, without and with considering the reactions of propane-oxygen mixture, are simulated and compared.

Fig. 3 shows the numerical results without the chemical reactions. One can see that propane and oxygen mix quickly after flowing out the exits, as shown in Fig. 3(a) and (b). Fig. 3(c) shows that propane from the outer tube significantly is heated up by the solid iron near the tip of the tuyere and still receives heat after mixing with oxygen. The distribution of density (Fig. 3(d)), velocity (Fig. 3(e)) and pressure (Fig. 3(f)) show that the flow rate of oxygen in the inner tube is fast enough to reach sonic speed and a supersonic jet is formed after the outlet.

Fig. 4 shows the numerical results with considering the chemical reactions. The mass fractions of propane and oxygen are shown in Fig. 4 (a) and (b) respectively. It is found that the mass fraction of propane near the solid iron decreases due to propane decomposition before mixing with oxygen (Fig. 4(a)). The temperature distribution in Fig. 4(c) also shows the endothermic effect at the wall of solid iron. However, after mixing with oxygen, propane reacts with oxygen and releases heat (Fig 4(d)) and the gas temperature reaches ~3400K. The exothermic reactions dominate after a short distance (~ 0.7cm) above the tuyere tip. Including the chemical reaction, one can see the flow characteristics, shown in Fig. 4(e)(f)(g) are drastically different from those in Fig. 3(d)(e)(f). No shack waves above the tuyere is observed when considering the chemical reactions.

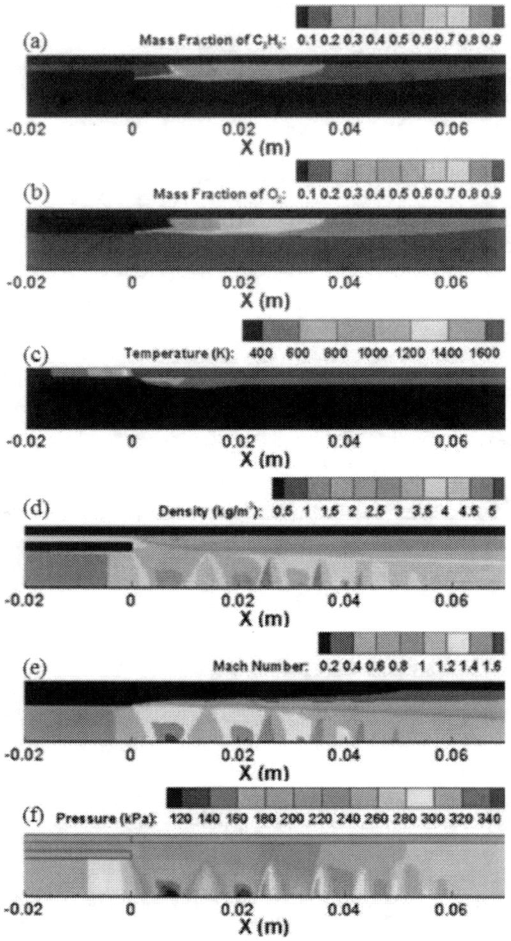

Figure 3: The distributions of the results without considering chemical reactions. (a) Mass fraction of propane. (b) Mass fraction of oxygen. (c) Temperature. (d) Density. (e) Velocity. (f) Pressure.

From Fig. 4(a)(b)(c), the endothermic reaction seems only near the wall of solid iron and the region is very small. Fig. 5(a) shows the distribution of temperature and heat of reaction on the wall of tuyere and solid iron. It is found that the endothermic reactions occur as propane is heated up above 1300K, where is 0.2 cm before the tuyere exit. Significant endothermic reactions continue to 0.7 cm above the tip of tuyere. After that the exothermic combustion reactions dominate because of the mixing with oxygen. The distribution of mass fraction of propane on the wall of solid iron is shown in Fig. 5(b). It is found that the region of the endothermic reactions on the walls is less than 1 cm. If the thickness of solid iron increases, the heat transfer from iron liquid decreases and the endothermic region will move downstream. However, the mixing of oxygen and propane after tuyere exit will lead to the weaker decomposition and significant combustion. Therefore, if the structure of solid iron is similar to that of ice in Fig. 1, the endothermic decomposition seems to be ineffective as the solid iron growing up to a certain size.

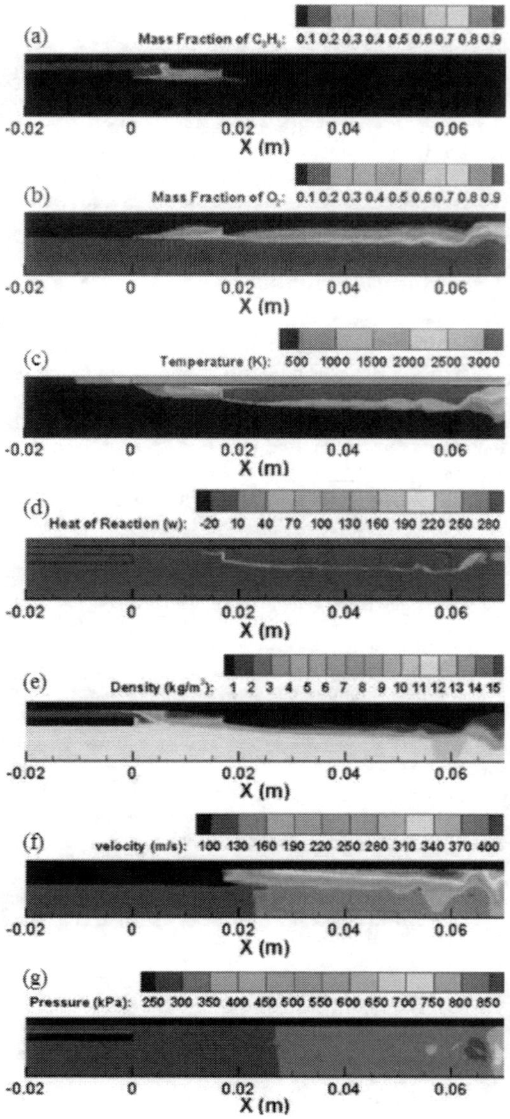

Figure 4: The distributions of the results with considering the detailed chemical kinetics. (a) Mass fraction of propane. (b) Mass fraction of oxygen. (c) Temperature. (d) Heat of reaction. (e) Density. (f) Velocity. (g)Pressure.

Fig. 6 shows the comparison of heat flux at the walls of tuyere and solid iron between the two cases (with and without the chemical reactions). The positive value of heat flux represents the heat transfer from solid to gas. In the small decomposition region (~0.9 cm), the heat fluxes on the wall with and without chemical reactions are respectively 20.8 kW and 15.7 kW. The decomposition of propane helps with the additional 32% of heat absorbed from solid iron, as compared to that by using inert gas (no chemical reaction). However, one can find that the combustion heat of propane will lead to high gas temperature and melt the solid iron after the decomposition zone, while using inner gas remains to cool the solid iron. From the comparison, it is hardly to accept that the co-injected hydrocarbon gas with oxygen can help with the formation of solid iron due to the endothermic decomposition reactions in the similar structure of ice (Fig. 1).

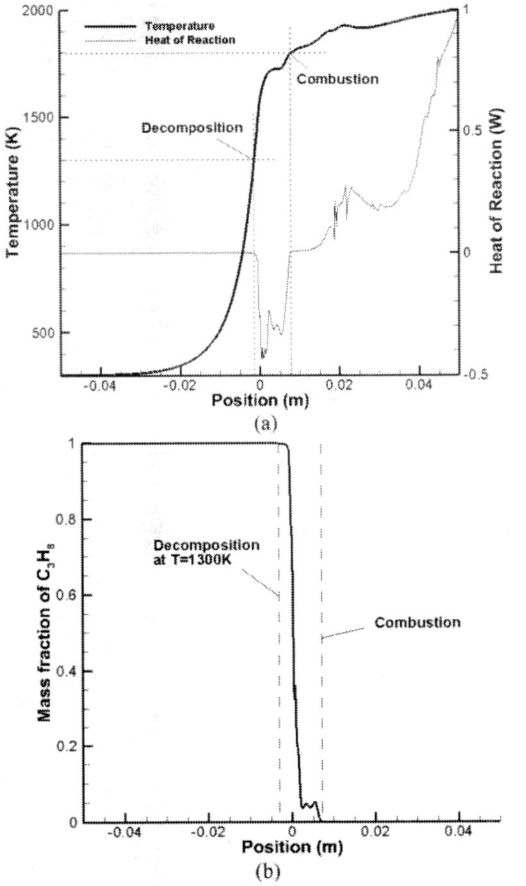

Figure 5: The distribution of the numerical results on the wall of tuyere and solid iron with considering the chemical reactions. (a) Temperature and heat of reaction. (b) Mass fraction of propane.

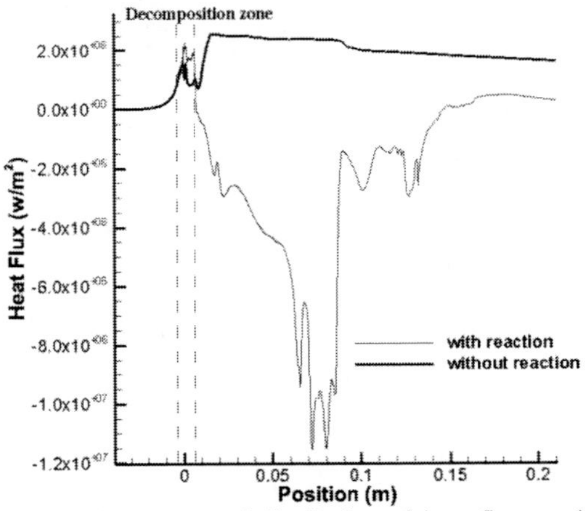

Figure 6: The contour of distribution of heat flux on the solidification wall.

4. Conclusions

In this study, a two-dimensional axisymmetric model is developed according to the ice structure in the water model experiments [5, 6]. The cases, with and without chemical reactions, are simulated to investigate the solid iron structure and the effect of co-injected hydrocarbon fuel on the formation of solid iron at the tip of the tuyere base on the ice structure in Fig.1.

According to the model results, the conclusions are the following:

1. The endothermic decomposition of propane provides the heat absorption about 32% higher than that by using an inert gas.
2. The mixing length of propane and oxygen after injected by a tuyere is approximately 0.7 cm. After the mixing length, the propane-oxygen mixture will have exothermic combustion reactions.
3. As the size of solid iron grows, the temperature at the base of solid iron decreases, and therefore, the decomposition moves downstream and becomes weaker. So, the decomposition of hydrocarbon gas is insignificant in a sizable solid iron because of the mixing with oxygen.
4. According to the experiment data and model results, the authors thinks that the structure of liquid iron may not similar to that of ice in Fig. 1. The structure of solid iron should be further investigated.
5. However, if the structure of solid iron is correct and the co-injected hydrocarbon gas is important, the growing mechanism of solid iron and its existence period in the steelmaking process need to be further studied.

Acknowledgment

This research was supported by the Ministry of Science and Technology under Project No.: MOST 105- 2622-8-006-001.

References

[1] D.H. Wakelin, The Making, Shaping and Treating of Steel : Steelmaking and Refining Volume. 1999: AIST.

[2] H.I. Iso, et al., Development of Bottom-blowing Nozzle for Combined Blowing Converter. Transactions ISIJ, 1988

[3] T. Ohta, et al., Improvement of Higher Productivity Using Bottom Blown Converter. Tetsu-to-Hagane, 1981.

[4] T. Aoki, The Mechanism of the Back-attack Phenomenon on a Bottom Blowing Tuyere Investigated in Model Experiments. 鉄と鋼, 1990. 11.

[5] Y.L. Huang, Water Modeling of Mushroom Formations in Gas Bottom-Blowing BOF Steelmaking Process. China Steel Technical Report, 2014. P27.

[6] Y.P. Huang, et al., Development and Verification of Accretion Similarity Conversion Method for Gas Bottom-Blowing Process inside Pyrometallurgical Vessels. Materials Transactions, 2007. P48.

[7] Y. Saito, K. Kanematsu, and T. Matsui, Measurement of Thermal Conductivity of Magnesia Brick with Straight Brick Specimens by Hot Wire Method. Materials Transactions, 2009. 50

[8] D.H. Wakelin, The Making, Shaping and Treating of Steel : Ironmaking Volume. 1999: AIST.

[9]UCSD.http://web.eng.ucsd.edu/mae/groups/combustion/mechanism.html. The San Diego Mechanism Version 20141004. 2014.

[10] ANSYS FLUENT 16, User's Guide.

11th Asia-Pacific Conference on Combustion,
The University of Sydney, NSW Australia,
$10^{th} - 14^{th}$ December 2017

Metallic mesh and quartz wafer as emitter-filter for a thermophotovoltaic system

J.R. Llobet,* [1], X. Kang,[2] and A. Veeraragavan[1]

[1]The University of Queensland, Brisbane, Queensland, Australia.
[2]Wuhan University of Technology, Wuhan, China.

Abstract

The use of metallic meshes as radiative material in a mesoscale combustor and thermophotovoltaic (TPV) system for power generation is investigated. A previous study showed promising results combining a silicon carbide (SiC) foam porous material as a radiator, and quartz wafers as optical filters. High temperature metallic meshes have a similar emissivity in the usable wavelength region for the TPV system. In addition, these present a clear advantage over SiC foams in terms of availability and cost. The previous study focusing in the SiC foam has been extended using two different thread count Kanthal meshes, also featuring different surface treatments. The results show higher achievable power with similar values of efficiency to the SiC cases. Moreover, the spatial discontinuities of the metallic mesh allow to stabilize the flame at different combustor heights independently of fuel-air mixture velocity. The effect of the flame location on power and efficiency has been assessed. The ideal flame location and importance of TPV placement with respect to the flame is described.

1 Introduction

Portable and micro propulsion systems requiring low power supply are traditionally powered by an electrochemical battery [1]. However, batteries have low energy densities compared to micro/mesoscale combustion devices [2, 3]. The advantages of micro/mesoscale combustion have lead to multiple studies of several different systems, such as micro-gas turbines, micro-rotary engines, micro-free-piston engines, micro-internal swing engines, and combustors paired with a TPV system [1, 4]. All these approaches except the combustor-TPV system use moving elements (rotors, pistons). Frictional losses caused by these moving parts become more important as the scale is reduced. For this reason, the combustor-TPV system presents an advantage for very small power requirements.

Combustor-TPV systems work by heating the combustor walls to a high temperature in order to produce radiative emissions that are harvested in the TPV to produce power. The TPV system useful band gap is limited. Therefore, high temperatures in the radiative element are required to improve the radiation output and radiation spectrum. Higher temperatures improve the radiation spectrum producing a higher portion of the total emission within the TPV's useful band gap [4].

Porous media has been used previously to increase TPV power generation, by increasing wall temperatures [5], or by improving the radiation spectrum [4, 6].

The use of porous media for spectral-control can be enhanced even further using a spectral filter, which transmits the radiation in the useful band for the TPV while reflecting or absorbing the radiation outside this band, keeping this energy in the combustor to increase the flame and wall temperature. Quartz wafers have been shown to be very suitable for micro/mesoscale combustor-TPV systems thanks to its thermal resistance and its optical transmission properties [4, 7].

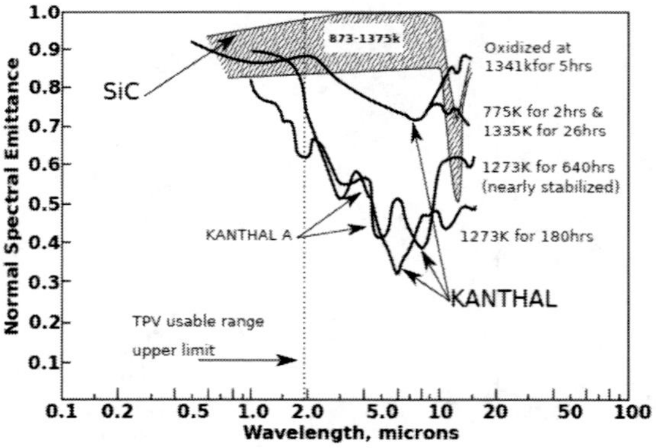

Figure 1: Normal spectral emittance of Kanthal and Silicon carbide. Adapted from [8]

The work by Xin and Veeraragavan [4] is extended in this paper studying metallic meshes as radiation source instead of SiC foam. The selected material is a high temperature alloy commercially known as Kanthal (0Cr25Al5). The spectral emittance curves in In Fig. 1 show the normal emittance of Kanthal and SiC. The emittance within the region of wavelengths use-

*Corresponding author. Phone: +61 434 998 447.
E-mail address: j.r.llobet@uq.edu.au

ful for the TPV system is similar for both materials. Additionally, metallic meshes are more easily sourced than SiC foam, and their cost is generally two orders of magnitude lower.

2 Experimental Set Up

The experimental set up is the same used in previous works by Xin and Veeraragavan [1, 4]. The combustion region is bounded by two quartz wafers and two lateral ceramic mounts (Fig. 2).

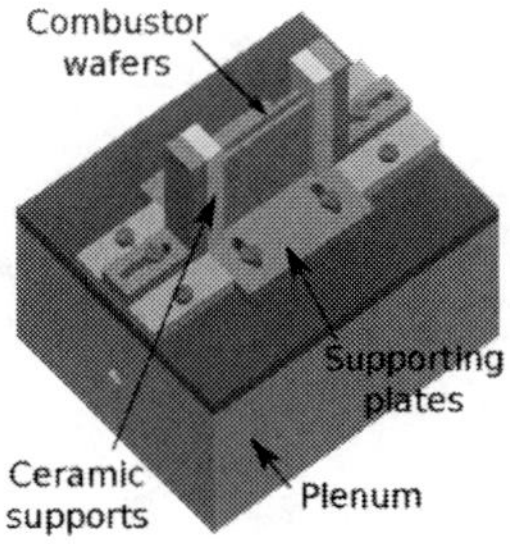

Figure 2: Experimental set up sketch. Adapted from [4]

The wafers, purchased from Knight Optical Ltd, are sized 80 mm × 60 mm × 1 mm. The gap between the wafers is set to 3.5 mm and 3.3 mm for the SiC foam and metallic mesh cases respectively. The SiC foam used in [4] was purchased from ERG Aerospace Inc. Two pieces of 50 mm × 25 mm × 1.7 mm are placed side by side to cover an area of 50 mm × 50 mm. The characteristics of the metallic meshes are summarized in Table 1. The foam and metallic mesh pieces are inserted into the combustor using a metallic groove attached to one of the wafers.

The CH_4-air mixture is set using two mass flow controllers. The two gases traverse a gas mixer to ensure complete mixing, and reach a plenum where mass flow instabilities are damped. A metallic stainless steel screen below the entrance of the combustion region flattens the velocity profile and serves as a flashback arrestor [4]. The mixture equivalence ratio was 1.0 for all cases.

Infrared images are acquired to obtain the temperature distribution on the quartz wafer. A high-performance FLIR Systems A655 infrared camera with a resolution of 640 × 480 and accuracy of 2% was used. The range of radiation detected is 7.5 − 13 µm. The emissivity of the quartz wafer was set to 0.81 [4].

The TPV cells are fabricated with gallium antimonide (GaSb) by JX Crystals Inc. The useful bandgap of these cells cuts off above ∼ 1.8 µm.The TPV is placed approximately 15 mm from the wafers. The output of the TPV system was measured using a FLUKE 77 multimeter with a DC voltage accuracy of ±0.3 %, and a DC current accuracy of ±1.5 %.

3 Results

3.1 TPV performance

The power and efficiency obtained using the two metallic meshes with quartz wafers are compared to the results obtained with SiC foam, and quartz and silicon wafers by Xin and Veeraragavan [4] in Fig. 3.

The power output at the maximum power point (mpp) of the TPV, is calculated using the open circuit voltage (V_{OC}) and short circuit current (I_{SC}) measurements, and using a manufacturer provided fill factor ($FF = P_{mpp}/(V_{OC} \cdot I_{SC})$) of 0.7[9]. The power generated with quartz + Mesh 2 is slightly lower than the quartz + SiC combination. The power generated with quartz + Mesh 1 seems to closely match the trend of the quartz + SiC case. Nonetheless, the gap between wafers in the metallic mesh cases is 0.2 mm smaller than in the SiC foam cases. Therefore, the mixture mass flow rate is lower for the same inflow velocity. This provides an advantage in terms of efficiency. The efficiency in Fig 3b is calculated considering the radiation is available from both sides of the combustor. The methane Low Heat Value and density at standard ambient pressure and temperature are $LHV_{CH_4} = 50\,\mathrm{MJ\,kg^{-1}}$, and $\rho_{CH_4} = 0.656\,\mathrm{kg\,m^{-3}}$. The maximum efficiency for the quartz+Mesh 1 combination is very similar to the maximum efficiency of the quartz+SiC case while producing 10 % higher power.

The thermal conductivity of Kanthal is very high compared to SiC, however the large gaps between the mesh threads highly reduce the global conductivity of the mesh. The large spacial discontinuities induce the flame to stabilize just downstream of a horizontal mesh thread. The mixture is heated continuously by the wafer and vertical threads as it moves dowstream. However, the heating produced by the crossing of each horizontal thread is discontinuous, and induces the stabilization of the flame just downstream of one of the horizontal threads. The flame location and high temperature regions on the mesh are depicted in the sketch in Fig. 4.

When comparing the two meshes, the effect of the surface finish was not relevant, as both surfaces presented a layer of oxide after brief periods of use. Therefore, the mesh count is the main source of performance difference. The low velocity limit (LVL), the inflow velocity at which the flame moves upstream up to the combustor inlet, is smaller in Mesh 2. This is due to the larger distance between threads, which reduces the global thermal conductivity. The lower number of vertical threads also reduce the effective radiative surface, reducing its performance .

Table 1: Mesh characteristics

	Material.	Surf. Treat.	Wire ⌀	Mesh Count $[\text{in}^{-1}]$	Size
Mesh 1	Kanthal	Oxidized	0.71 mm	8, Plain weaving	$55 \times 90 \times 1.42$ mm
Mesh 2	Kanthal	Hydrogen annealing	1.04 mm	5, Plain weaving	$55 \times 90 \times 2.08$ mm

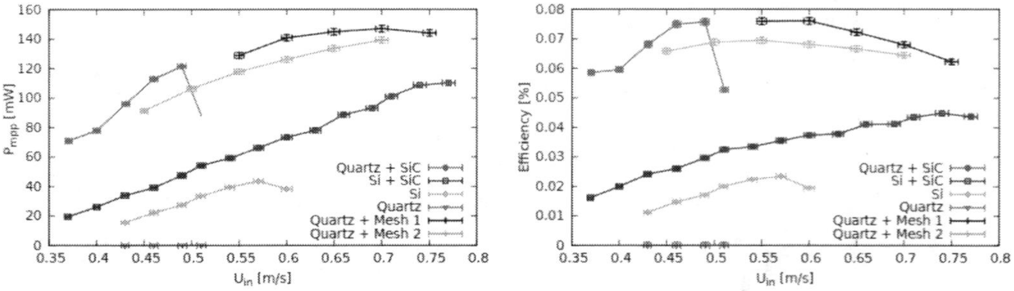

(a) Power.

(b) Efficiency.

Figure 3: TPV performance as a function of mixture velocity. SiC data from [4]

3.2 Effect of flame location on TPV performance

The nearly continuous nature of the SiC foam couples the velocity inlet and the location of the flame on the SiC plate. Higher inflow velocities move the flame away from the inlet, while the opposite happens for lower velocities. In the case of the metallic meshes, the large spacial discontinuities allow to 'trap' the flame in any of the spaces between horizontal threads. This allows a new degree of freedom affecting the performance of the combustor-TPV system.

In order to assess the effect of the flame location on the TPV performance, a number of IR temperature images, and voltage and current measurements were obtained for different flame locations. Fig. 4 shows the spatial temperature distribution on the wafer surface for a single case obtained with the IR camera.

The temperature on the wafer is not equal to the metallic mesh temperature. However, the temperature distribution along the wafer is a good indication of the temperature distribution on the metallic mesh. The average temperature distribution in the streamwise direction is obtained for different flame locations. Fig. 5 shows the curves obtained for Mesh 1.

The temperature distribution can be used to calculate the potential thermal radiation power of each flame location. For this, the radiation of a black body at the temperature on the wafer surface is considered ($P = \sigma T^4$, with sigma the Stefan-Boltzman constant). This value is integrated on the wafer surface to obtain the reference potential radiation power for each flame location. The normalized values are plotted in Fig. 6a, labeled as 'Wafer'. These curves show that the potential radiation power increases as the flame moves closer to the inlet.

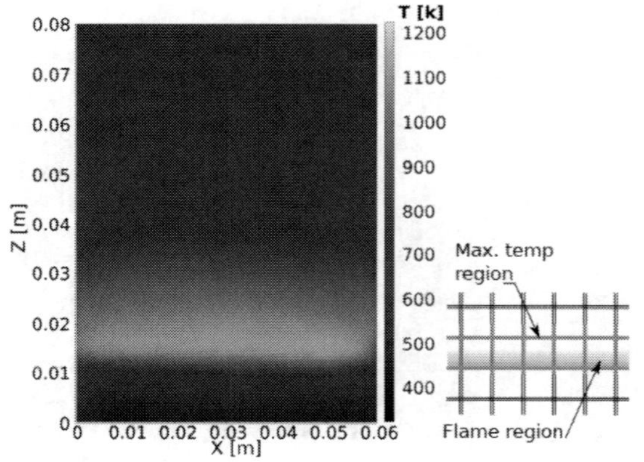

Figure 4: Temperature distribution of the quartz wafer. Mesh 1, $U_{in} = 0.60\,\mathrm{m\,s^{-1}}$, flame 16 mm from the inlet; and flame and mesh sketch.

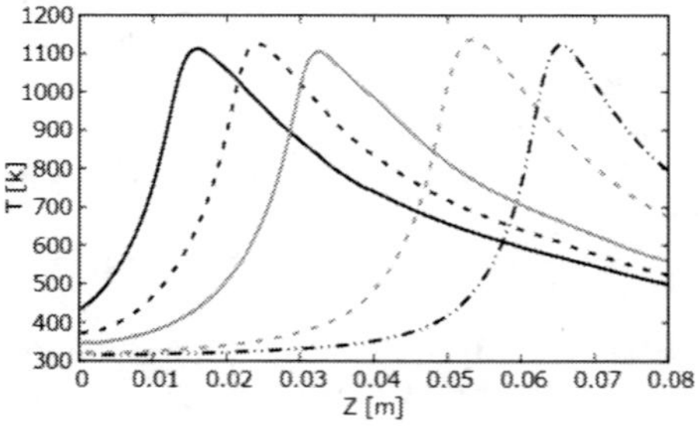

Figure 5: Average temperature distribution in the streamwise direction. Mesh 1, $U_{in} = 0.60\,\mathrm{m\,s^{-1}}$.

However, the measured TPV power output does not show the same trend. Fig. 6b shows the normalized measured power. This curves present a maximum value

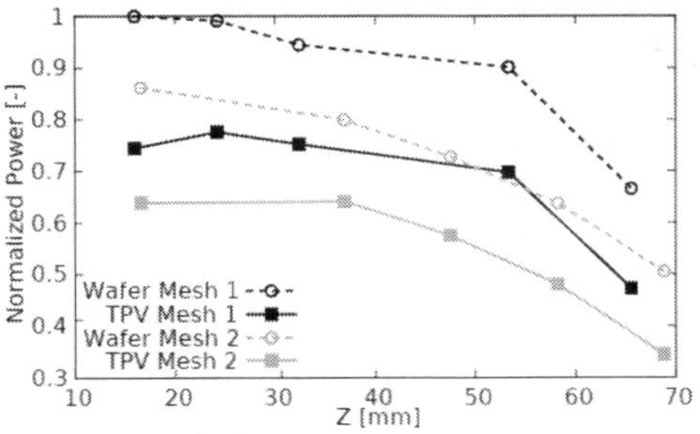

(a) Calculated from temperature distribution. Normalized by the maximum power calculated for Mesh 1

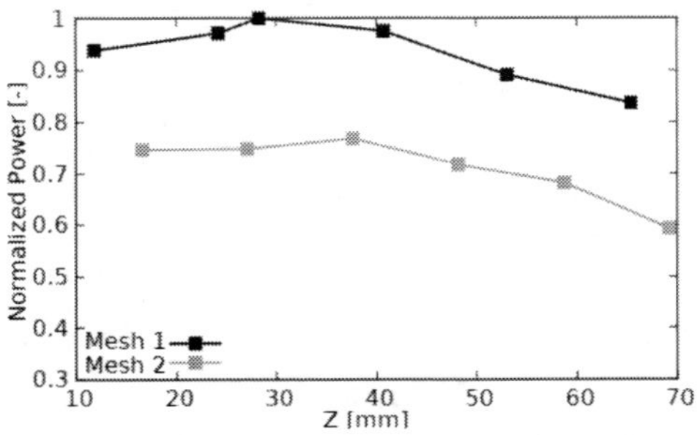

(b) Measured from temperature distribution. Normalized by the maximum power measured for Mesh 1

Figure 6: Normalized power for Mesh 1 at $U_{in} = 0.60\,\mathrm{m\,s^{-1}}$ and Mesh 2 at $U_{in} = 0.45\,\mathrm{m\,s^{-1}}$

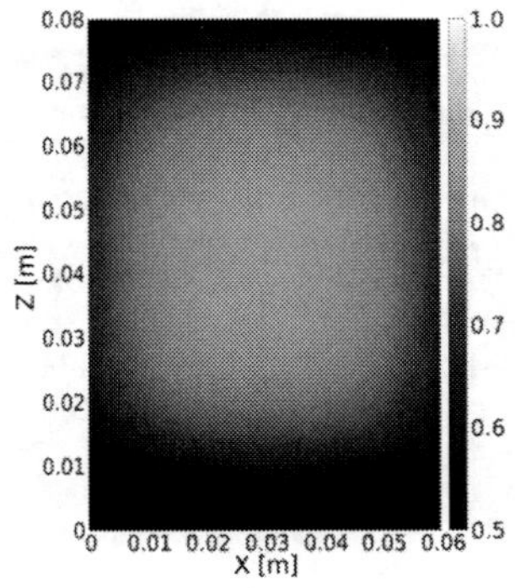

Figure 7: View factor map.

of power for a flame location of approximately $Z = 30\,\mathrm{mm}$. This discrepancy is caused by the view factor between the wafer and the TPV cells. The view factor between two given points in the combustor and the TPV can be calculated as $F_{1-2} = \frac{\cos(\theta_1)\cos(\theta_1)}{\pi s^2} dA_2$, where θ is the angle between the surface normal and a ray between the two considered differential areas. The view factor map for the wafer surface is presented in Fig. 7.

Applying the view factor to the potential radiation, the curves in Fig. 6a labeled as 'TPV' are obtained. These present a similar trend to the measured power, showing the loss in performance as the flame reaches the edges of the TPV. This result shows the ability to improve the TPV performance for metallic meshes by selecting the most appropriate flame location. Moreover, it highlights the possibility of improving performance by displacing the TPV to locate the region of maximum view factor in front of the region of maximum radiation.

4 Conclusions

The performance of a mesoscale combustor and TPV system using metallic meshes as radiative source coupled with quartz wafers has been investigated. The results have been compared to the values obtained using SiC as radiative source, showing higher achievable power with similar efficiency. This, together with the low cost of metallic mesh, makes it a very promising material for future applications. Additionally, using metallic meshes in the combustor allows controlling the location of the flame. The influence of flame location has been assessed, showing the location of maximum performance. Moreover, the relevance of the positioning of the TPV with respect to the flame has been identified.

References

[1] Kang, X. & Veeraragavan, A., *Experimental investigation of flame stability limits of a mesoscale combustor with thermally orthotropic walls.*, Appl Therm Eng, 85, (2015), pp.234-242.

[2] Kaisare, N.S. & Vlachos, D.G., *A review on microcombustion: fundamentals, devices and applications.*, Prog Energy Combust Sci, 38, 3, (2012), pp.321-359.

[3] Veeraragavan, A. & Cadou, C.P., *Flame speed predictions in planar micro/mesoscale combustors with conjugate heat transfer.*, Combust Flame, 158, 11, (2011), pp.2178-2187.

[4] Kang, X. & Veeraragavan, A., *Experimental demonstration of a novel approach to increase power conversion potential of a hydrocarbon fuelled, portable, thermophotovoltaic system.*, Energy Convers Manage, 133, (2017), pp.127-137.

[5] Li, J., Wang, Y., Shi, J. & Liu, X., *Dynamic behaviors of premixed hydrogen-air flames in a planar microcombustor filled with porous medium.*, Fuel, 145, (2015), pp.70-78.

[6] Bermel, P., Ghebregrhan, M., Chan, W., Yeng, Y.X., Araghchini, M., Hamam, R., et al. *Design and global optimization of high-efficiency thermophotovoltaic systems.*, Optics Express, 18, (2010), pp.A314-A334.

[7] Li, Y.H., Wu, C.Y., Lien, Y.S. & Chao, Y.C., *Development of a high-flame-luminosity thermophotovoltaic power system.*, Chem Eng J, 162, 1, (2010), pp.307-313.

[8] Touloukian, Y.S. & DeWitt, D.P., *Thermophysical Properties of Matter - The TPRC Data Series. Volume 7 & 8.*, IFI/PLENUM, New York-Washington, 1972.

[9] JX Crystals Inc. GaSb cells and circuits. http://jxcrystals.com/drupal/products/ <[accessed 05.09.17]>.

11th Asia-Pacific Conference on Combustion,
The University of Sydney, NSW Australia
10th -14th December 2017.

Predication of sooting tendency of gasoline surrogate fuels

Muhammad Kashif[1,2], **Guillaume Legros**[2], **Jérôme Bonnety**[2]

[1] Department of Mechanical Engineering Faculty of Engineering, University of Central Punjab, Lahore, Pakistan

1-Khayabane e Jinnah, Lahore, Punjab, 54000, Pakistan

[2] Sorbonne Univ, UPMC Univ Paris 06, UMR 7190, Inst Jean Le Rond d'Alembert,

Paris, F-75005, France

Abstract

This study presents the sooting tendencies exhibited by binary mixtures of synthetic fuels i.e. n-heptane/toluene and isooctane/toluene, in terms of Yield Sooting Indices (YSIs) measured in nitrogen diluted diffusion flames of methane and seeded with the vapors of the aforementioned blends, probed using light extinction method using a Santoro's burner. These values are then extrapolated to reveal the soot production tendencies of tri-component surrogate gasoline fuels through an interpolation technique. Soot volume fraction maps in these flames are then inferred from line-of-sight attenuation measurements and are converted into apparatus-independent YSIs which affords the possibility to compare the relative influence of fuel formulation on soot production. The trends in soot production capacity as a function of toluene mole fraction ($\mathbf{X_{tol.}}$) are first analyzed by curve fitting YSI data and a non-linear evolution of the sooting tendency as a function of $\mathbf{X_{tol.}}$ in the binary mixtures reveals especially for isooctane/toluene blends at lower $\mathbf{X_{tol.}}$. Two Quadratic equations adequately model the evolution of YSI as a function of toluene mole fraction in n-heptane/toluene and isooctane/toluene blends. A synergistic effect is only visible when certain types of hydrocarbons are present in the fuel mixture (branched-chain alkanes and alkyl-benzenes in our case). These non-linear trends are adequately translated into predicted sooting tendencies of tri-component blends and can be very helpful while formulating fuel mixing strategies in view of the future engine and environmental exigencies.

1 Introduction

Internal combustion engines are a major source of locomotion and power generation but the combustion process in both gasoline and diesel engine leads to the formation of number of pollutants e.g. Carbon monoxide (CO), nitric oxide (NO) and nitrogen oxide (NO_2) collectively called as NOx, unburnt hydrocarbon (UHC) and Particulate matter. These pollutants are ultimately emitted into the atmosphere where they pose a serious health hazard to human health. Many recent studies have hinted toward a strong link between high mortality rates to increased concentration of fine and ultra-fine particles in city atmosphere [1, 2].

Complex physical and chemical phenomena lead to the formation of soot and play important role in the inception, growth and subsequent oxidation of soot particles from the species formed during the pyrolysis of a complex blend of hydrocarbons in fuel [3]. The recent development in computational power now affords the possibility to develop comprehensive methodologies to refine the understanding of soot formation pathways by developing numerical models and testing them against a set of experimental measurements performed on laboratory reference flames [4]. A strategy based upon sophisticated computational and experimental methodology is required to provide accurate and meaningful data at reduced cost and time. All commercial fuels contain hundreds of hydrocarbons at atmospheric conditions [5]. Every chemical compound present in fuel mixture has a variable and coupled effect on combustion chemistry, engine performance and emissions, especially soot production [6]. Many works now employ surrogate fuels to minimize the computational cost and time needed to develop a molecular-level understanding of fuel combustion process. Dedicated efforts are being made to develop optimized surrogate fuels for gasoline, diesel and jet engines based on available reaction kinetics data and experimental validation results [5]. The cumulative effort has resulted in the development of blends which are mixtures of at least n-heptane (n-C_7H_{16}, isooctane (i-C_8H_{18}) and toluene (C_7H_8) and have shown to be capable to reproduce quite adequately the various aspects of combustion behavior of gasoline.

Smoke point height, threshold sooting index (TSI) and Yield sooting Index (YSI) measured in laboratory flames have been used in order to measure and compare sooting propensity of a large number of liquid compounds and their mixtures [7-11]. Among these methodologies, YSI methodology has shown to be more reliable with uncertainties which are lower as compared to the ones normally associated with TSI methodology. Therefore, an experimental procedure to reproduce the methodology used by McEnally and Pfefferle is implemented to measure sooting tendencies of blends of n-heptane/isooctane with toluene. The influence of the blending is then assessed for these aforementioned binary mixtures.. The measured values of YSIs for the aforementioned blends are then used to estimate the YSIs for blends containing n-heptane, isooctane and toluene in the form of ternary diagram.

2 MATERIAL AND METHOD

The methodology detailed by McEnally and Pfefferle [9] and extended and assessed for consistency at higher doping concentrations by [10] is reproduced to measure soot production capacity of binary mixtures. The experimental setup consisting of axis-symmetric coflow burner, fuel metering, fuel evaporation and mixing, and optical diagnostics has been carefully explained in [10]. A schematic of fuel metering, evaporation and mixing diagram is shown in Fig. 1. A Santoro type burner is used in this

Corresponding author. Fax: +92-42-35954892
E-mail address: muhammad.kashif@ucp.edu.pk

with a vertical axis of symmetry located at the center of inner tube of 11 mm diameter. The height above the burner (HAB) is measured from the top of the burner fuel tube along the fuel tube. The distance from the axis of symmetry along the horizontal axis is the cross-stream coordinate r. The reactants are $\mathbf{CH_4}$ from gas cylinder; (99.9 % stated purities), n-heptane and isooctane and toluene from regent-grade bottles (99.5 % purity), and filtered air from laboratory compressor bottles.

The YSI methodology suggested by McEnally and Pfefferle provide an apparatus-independent value for a fuel, based upon the maximum measured value of soot volume fraction ($f_{v,max}$ = $\mathbf{max\{fv(r,z_{meas.})|r \in [0, R]\}}$) in the doped methane flame at a fixed height above burner (HAB) termed as z_{meas} using Eq.(1):

$$YSI = Cf_{v,max} + D \quad (1)$$

Here, C and D are constants which are apparatus-specific parameters and can be calculated using two reference fuels with arbitrarily assigned YSI values [11].

2.2 Experimental and post-processing procedure

As depicted in the Fig. 1, the liquid fuel being investigated for its sooting tendency is stored in fuel reservoir and is pushed by an inert gas, i.e. Argon, and flows through a Coriolis mass flow controller. The fuel is then vaporized and mixed with a carrier gas, i.e. methane, using Bronkhorst controlled evaporation and mixing (CEM) system. The gaseous mixture flows via a heated line to the central fuel tube of the burner. This liquid fuel doped flame is then probed by light extinction method (LEM) to get soot volume fraction; necessary to calculate YSI of the liquid fuel blend being investigated. Table 1 lists the constant experimental parameters for the whole experimental campaign. In brief, two reference mass flow rates of benzene are used for the present investigation resulting in two different dopant concentrations in the flame, i.e. **0.42mg/sec** (1.5g/hr) and **0.56mg/sec** (2.0g/hr), at room temperature. The corresponding vapor mole fractions of benzene vapors are $\mathbf{X_{vap,1} = 1.86 \times 10^{-2}}$ and $\mathbf{X_{vap,2} = 2.47 \times 10^{-2}}$, respectively. The dopant vapor mole fraction $\mathbf{X_{vap,\,i}}$ **(i = 1,2)** is then kept constant for all liquid fuel blends to be investigated and the corresponding mass flow rate for any blend is calculated from compiled liquid phase densities and molecular weights [12] with a relative uncertainty of $\pm$0.1%.

The details of the methods to calculate the YSI of a given fuel using Eq.(1), suggesting different alternatives to the evaluation of $\mathbf{f_{v,max}}$ are reproduced here and have been explained in detail in [10]. The first method -an extension of the original definition of McEnally and Pfefferle [9] is referred as $\mathbf{YSI_{Max.Ax.fv}}$ and is based on the maximum axial soot volume fraction of the flame (but not at a fixed location). The second method is identified by $\mathbf{YSI_{GlobalMax.fv}}$ based on maximum soot volume fraction $\mathbf{f_{v,max}}$ in the whole soot field. The third method is referred as $\mathbf{YSI_{Max.Int.fv}}$. Here, the integrated soot volume fraction $\mathbf{F_v(z)}$ was first computed as follows:

$$F_v(z) = 1/(\pi R^2) \int_0^R 2\pi f_v(r,z)dr \quad (2)$$

Then the maximum value of $F_v(z)$ was used to evaluate YSI. $\mathbf{F_v(z)}$ is an evaluation of the actual soot load at a given HAB into the flame. For every flame investigated, 400 consecutive fields of raw integrated extinction coefficient are first processed from a 20 sec long LEM recording. Then a deconvolution procedure based on the onion peeling method stabilized using Tikhonov regularization [13] is applied to every field, providing 400 maps of local extinction coefficient $\mathbf{\kappa_\lambda^{ext}(r,z)}$ and is converted to soot concentration using Eq.(3).

$$f_v(r,z) = \frac{\lambda\kappa_\lambda^{ext}(r,z)}{6\pi(1+\alpha_{sa})E(m)} \quad (3)$$

Where **E(m)** is a function of the complex refractive index **m** of soot. $\boldsymbol{\alpha_{sa}}$ is the relative contribution of scattering to extinction, therefore is directly related to soot morphology [10].

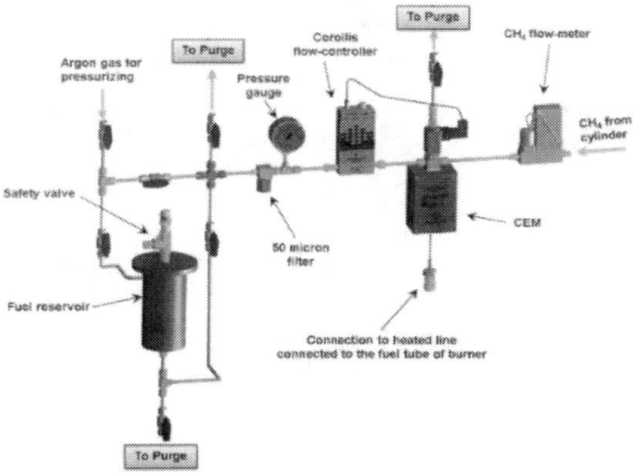

Fig. 1: Various components of liquid metering and doping system for the measurement of soot production capacity of different liquid fuels in doped methane diffusion flames.

Three different maximum measured soot volume fractions, axial, global and integrated are extracted from the soot maps in doped flame for all fuel blends. The value of $\boldsymbol{\alpha_{sa}}$ were adjusted following the procedure outlined in [10]. However, the difference in the YSIs calculated -with and without taking account the relative contribution of $\boldsymbol{\alpha_{sa}}$- are of the order of experimental uncertainty.

Parameter	**Present study**	**(McEnally, et al., 2011)**
Carrier gas Composition	CH_4 (50 %), N_2 (50 %)	CH_4 (55 %), N_2 (45 %)
Carrier gas flow rate (cm^3/min)	400$\pm$ 2.4	605
Coflowing air flow rate (cm^3/min)	60000 $\pm$ 360	30000
Evaporator temperature (°C)	150 $\pm$1	N.A.
Heated line temperature (°C)	150 $\pm$2	145
Reference YSIs	YSI(Benzene)=30 YSI(toluene)=43.5	YSI(Benzene)=30 YSI(DHN)=100

Table 1: Values of the constant experimental parameters for the evaluation of YSI.

3. RESULTS AND DISCUSSION

In summary, the study reveals the presence of synergistic effect of fuel blending on the evolution of soot production capacity (measured here as YSI). This trend is more pronounced for binary mixtures of isooctane and toluene. This trend is in agreement with the available literature [14]. However the blends of n-heptane and toluene show linear evolution of YSI with increasing toluene mole fraction. This observation hints at the absence of synergistic effect of blending on soot formation. The same linear trends have been reported for n-heptane and isooctane in an early investigation [10]. This clearly shows that the effect of chemical structure of a compound on the sooting tendency are different for a liquid fuel blend especially in the presence of aromatic hydrocarbons.

3.1. Effect of blending on sooting propensity

The soot production capacity of a fuel blend, as measured in a laboratory flame using LEM technique, increases with increasing X_{tol} as shown in Fig. 2. The soot maps in flames doped with benzene are shown in the extreme right column for reference. These fields show the global sooting behaviors of n-heptane, isooctane, toluene, benzene, and binary mixtures of n-heptane or isooctane and toluene in coflowing diffusion flames established over modified Santoro type burner. There is a monotonic increase in soot production and visible flame height as the mole fraction of toluene in the dopant liquid mixture increase, both for heptane or isooctane, which is qualitatively in accordance with the available data [14].

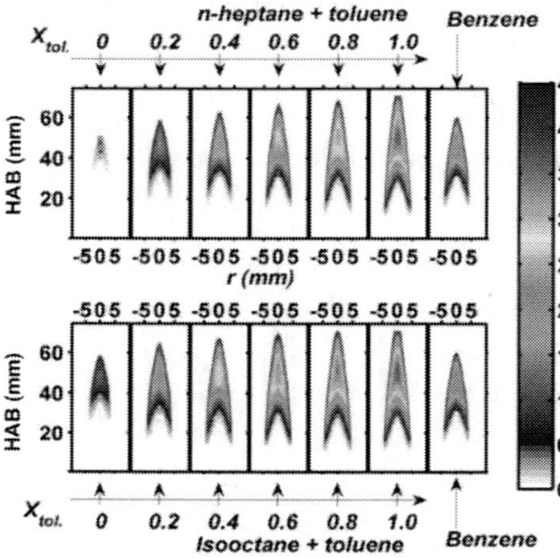

Fig. 2: Soot maps extracted using measurements of LEM in doped methane diffusion flames of different binary blends of n-heptane or isooctane with toluene. $X_{tol.}$, here represents the mole fraction of toluene in a binary fuel blend.($X_{vap,2} = 2.47 \times 10^{-2}$).

Some important chemical reactions leading to formation of poly-aromatic hydrocarbons (PAHs) have been resumed by Choi, et al. [14]. These PAHs are the precursors to the soot particles and hence, any increase in the rate of PAH formation increases soot production capacity. The leading chemical pathwas leading to the production of benzene (A1) and naphthalene (A2) are: **$C_6H_5CH_3$ + H → A1(C_6H_6) + CH_3; A1 -CH_2 + C_3H_3→ A2($C_{10}H_8$) + H_2; and A1 - C_6H_5 + C_4H_4→ A2($C_{10}H_8$) + H; and A1 - C_6H_5 + C_4H_4→ A2($C_{10}H_8$) + H**. The formation reactions of naphthalene (A2) and larger PAHs are accelerated especially in case of isooctane/toluene blends due to the higher concentrations of propargyl, **C_3H_3** and vinylacetylene (**C_4H_4**). This phenomenon reduces with the increase in toluene mole fraction and also is more dominant in case of isooctane/toluene blends because of the pyrolysis of heptane forms species which have little effect on PAH formation pathways. This synergistic behavior on soot production then becomes important in the light of the fact that most of the surrogate fuels simulation and recommended contain toluene fraction in this range. Hence, a more prudent approach is necessary while formulating a surrogate fuel matching the sooting tendencies of commercial gasoline.

3.2. Curve fit models for YSI ($X_{tol.}$) for binary blends of n-heptane or isooctane with toluene

Figure 3 show the YSIs exhibited by the different blends investigated at both vapor mole fractions. Equation (4) is a second-order polynomial used to model the trends in the evolution of YSIs with $X_{tol.}$ and is expected to decently model the non-linearity observed in Fig.3:

$$YSI = a(X_{tol.})^2 + b(X_{tol.}) + c \qquad (4)$$

Table 2 gives the values of coefficients ***a*** and ***b*** for fuel blends n-heptane or isooctane with toluene. The detail explanation of the curve-fitting procedure and its implementation in MatLab environment can be found in [15, 16].

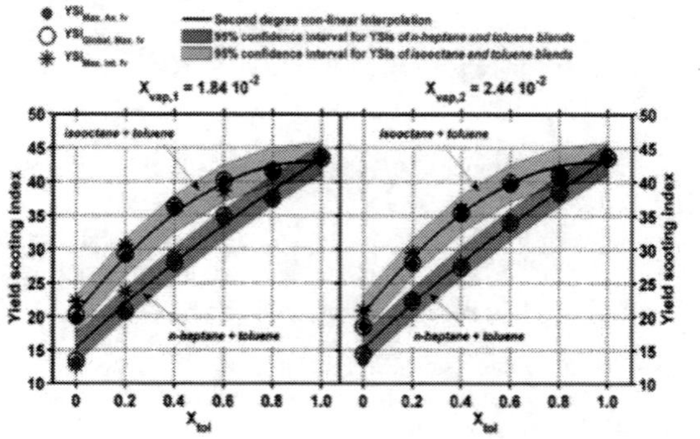

Fig. 3: Evolution of YSIs as function of X_{tol} in blends of n-heptane or isooctane and toluene. The red and green patches around experimental measurements (symbols) and trend lines represent 95% interval.

Consequently, Fig.3 exhibits trends in soot production in terms of YSIs as a function of X_{tol} at both $X_{vap.}$ for investigated binary blend of n-heptane or isooctane with toluene. The 95% confidence intervals (***CI***) for the blends is calculated using Eq.(5):

$$CI(X_{tol.}) = \overline{YSI}(X_{tol.}) \pm 1.96 \times \sigma_{max} \qquad (5)$$

where $\overline{YSI}$ represents mean value of YSIs at a given set of conditions (**$X_{vap,i}$, $X_{tol.}$**) and standard deviation is calculated at every set of YSI values and σ_{max} is the maximum among all the standard deviations. The 95% confidence interval is calculated by **$\pm 1.96 \times \sigma_{max}$** and is equal to 95% of the area under the standard distribution curve [16]. The magnitude and sign of the ratio of coefficients ***a*** and ***b*** demonstrate quantitatively the non-linear trends of YSI as a function of toluene mole fraction.

Coefficients	heptane/toluene	Isooctane/toluene
a	-6.07	-24.30
b	34.05	46.82
c	15.34	20.40

Table 2: Calculated values of the coefficients for the polynomial equation fitted to the measured YSIs (see Eq.(4)).

The ratio of the coefficients of the quadratic equation modeling YSI as a function of X_{tol} .shows that while YSIs for the binary mixtures of n-heptane/toluene evolve almost linearly as function of toluene mole fraction (**$(a / b)_{heptane}$**=-0.178), this trend is far from linear for isooctane/toluene blends (**$(a / b)_{isooctane}$**= -

0.51).This clearly shows that the parabolic term in the YSI model (Eq. 4) in case of isooctane/toluene blends is important, hinting at a non-linear evolution which dampens YSI at higher toluene mole fraction

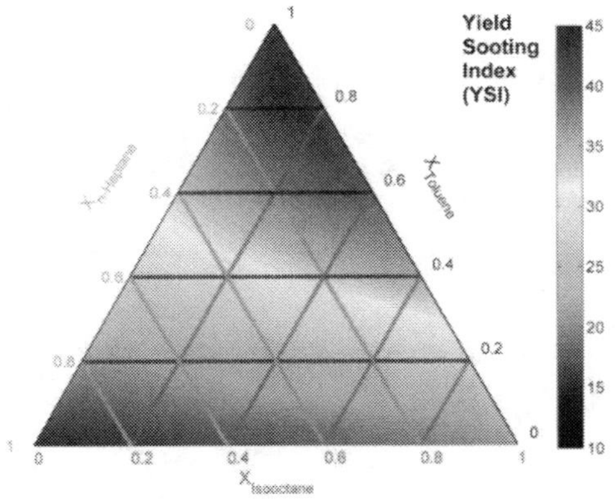

Fig. 4: Tri-component surface plot representing YSI values of n-heptane, isooctane and toluene (at the vertices) and binary blends (along the edges), and interpolated YSI values from tri-component blends (area enclosed inside the triangle).

Figure 4 shows the triangular surface plot of the interpolated YSI values for tri-component blends of n-heptane, isooctane and toluene. The predicted YSI values of tri-component blends are found by interpolating the YSI values of binary blends reported above. The mole fraction of one of three component is 0 along any of three axis e.g. the mole fraction of toluene along isooctane is zero, hence the $\mathbf{X_{toL}}$= 0 for any point lying on this axis. Thus the horizontal axis represent YSI values for blends of n-heptane and isooctane and so on for other two faces of the triangle. To draw the tri-surface map shown in fig. 4, the values along the all three axis are required. The values for blends of n-heptane and isooctane are calculated by taking volume-weighted average of YSI values of pure n-heptane and isooctane since the sooting tendency for the blend of n-heptane/isooctane evolve linearly with the increasing mole fraction of isooctane in the blend.

The sooting propensity of any blend can be readily found on the surface plot by locating the point corresponding to the mole fractions of n-heptane, isooctane and toluene following the iso-concentration lines which are at 45° to the respective axis (the iso-concentration lines and axis label for a compound are of the same color i.e. black for toluene, green for n-heptane and magenta for isooctane). The YSI value at the point of intersection of three iso-concentration lines is the predicted YSI value for the tri-component synthetic fuel. Hence, a quick comparison of the soot production capacity can be readily made by projecting the points onto the surface plot in Fig. 4.

Another interesting aspect of the graph is the bands of iso-YSI values evident by same colored zones on the surface plots. This clearly shows the existence of a number of different tri-component blends having the same soot production capacity. Therefore, other criteria of selection like auto-ignition behavior, ignition delay, octane rating, laminar burning velocity and strain extinction rate etc. need to be included in the selection criteria for a surrogate or alternative fuels.

4. CONCLUSIONS

A methodology of predicting soot production capacity of tri-component fuel blend from the measured YSIs binary mixtures of n-heptane/ isooctane with toluene in doped methane diffusion flames has been presented. A synergistic effect on soot production tendencies is quantified in terms of measured YSI and an equation modeling YSI as function of toluene mole fraction for these binary mixtures is fitted to the experimental values. A quadratic equation has been modeled by fitting a parabolic curve to the measured YSIs for binary fuel blends. The modeled YSI equations provide YSI values of a binary mixture of n-heptane or isooctane with toluene with an uncertainty of ±2.5 YSI units. The results presented highlights the non-linear effect on soot production capacity by fuels containing branched-chain and aromatic hydrocarbons. The presence of strong synergistic effect on sooting propensity of binary mixtures necessitates a detail study of fuel formulation policies in view of the evolving emission regulations. A ternary surface plot showing interpolated YSI values of tri-component blends of n-heptane, isooctane and toluene is also presented which can be used to get an early prediction of sooting propensities. It is clearly evident from Fig. 4 that non-linear trends are quite reasonably translated into the ternary surface graph showing their potential for the prediction of gasoline surrogate fuels.

REFERENCES:

1. Patricia M., Sierra-Vargas and L. M. Teran, (2012) Respirology, vol. 17, no. 7, 1031–1038.

2. OŚhaughnessy, P.T., (2012), Environ. Sci.: Processes Impacts, vol. 15, no. 1, 49-62.

3. D'Anna, A. (2009) Proc. Combust.Inst., vol. 32, no. 1, 593–613.

4. Liu, F., H. Guo, G. J. Smallwood, and L. Gülder, (2001), Combust. Flame, vol. 125, no. 1-2, 778–787.

5. Pitz, W. J., N. P. Cernansky, F. L. Dryer, F. N. Egolfopoulos, J. T. Farrell, D. G. Friend, and H. Pitsch, (2007, SAE Paper No. 2007-01, vol. 175.

6. Ußner M. and F. Müller-Langer, (2009), Accred. Qual. Assur., vol. 14, no. 12, 685–691.

7. Dotson, K., P. Sunderland, Z.G. Yuan, and D. Urban, (2011), Fire Safety J., vol. 46, no. 8, 550–555.

8. Calcote H. F. and D. M. Manos, (1983), Combust. Flame, vol. 49, no. 1-3, 289–304.

9. McEnally, C. S. and L. D. Pfefferle, (2007), Combust. Flame, vol. 148, no. 4, 210–222.

10. Kashif, M., P. Guibert, J. Bonnety, and G. Legros, (2015), Combust. Flame, vol. 162, no. 5, 1840–1847.

11. McEnally, C. S., L. D. Pfefferle, (2011), Environ. Sci. Technol., vol. 45, no. 6, 2498–2503.

12. CRC Handbook of Chemistry and Physics, 94th Edition, (2004) CRC Press, Cleveland, OH, 85th edition.

13. Daun, K. J., K. A. Thomson, F. Liu, and G. J. Smallwood, (2006), Appl. Opt., vol. 45, no. 19, 4638–4646,

14. Choi, B. C., S. K. Choi, and S. H. Chung, (2011), Proc. Combust. Inst., vol. 33, no. 1, 609–616.

15. Coleman T. and Y. Li, (1996), SIAM J. Optim., vol. 6, no. 2, 418–445.

16. Kenney J. F. and E. S. Keeping, (1962), Mathematics of Statistics, part 1, 3rd ed. Princeton, NJ, USA: D. Von Nostrand, 167–169.

11th Asia-Pacific Conference on Combustion,
The University of Sydney, NSW Australia
10th -14th December 2017.

Effect of floating perlites on the occurrence of thin-layer boilover

Tzu-Yan TSENG, Kuang-Chung TSAI

National Kaohsiung First University of Science and Technology, Kaohsiung, Taiwan

Abstract

Boilover is the most dangerous phenomenon in oil tank fires because it expands the burning field and may hurt fire fighters by flame or severe radiant exposure. Because of the low density and cost, floating perlites were used on the surface of burning oil to study whether the floating perlites can delay the occurrence of thin-layer boilover and consequently prevent the flame enlargement. Two round stainless steel pans with a diameter of 0.1 or 0.3 m and depth of 0.1 m were used. 0.02 m thick diesel was used as fuel, below which 0.05 m water existed. None, 0.5, 1, 2 or 4 layers of perlites were put above the fuel to investigate their effects. The results showed that the temperature under the fuel surface with perlites was lower than that without perlites. The use of perlites can produce a favorable thermal insulation. Additionally, the boilover phenomenon did not occur after using the perlites while it occurred in the none-perlites test. The use of the perlites can decrease the mass loss rate and the flame enlargement. Therefore, the use of the perlites can enhance the fire safety of oil tanks.

Keywords: thin-layer boilover, perlites, oil tank fire

1 Introduction

It is well known that boilover is the most dangerous scenario in oil tank fires. The phenomenon of boilover is caused by the vaporization of water from the bottom of oil tanks when the water is heated by the heat transfer from the oil tank fire. The temperature of water is increased quicker and quicker after a period of time, because the thickness of fuel is gradually decreased with time. Once the temperature of water reaches the boiling temperature of water, water will vaporize immediately and the burnt oil will be ejected or spilled out of the oil tank. The spilled fuel with fire may cause fire fighters injured, ignite other fuels, form fireballs, and consequently enlarge the burning field. Figure 1 shows the mechanism of boilover.

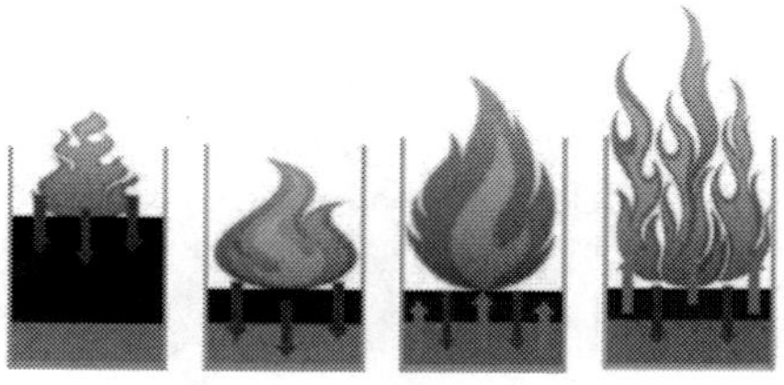

Figure 1: The mechanism of boilover occurrence

In 1925, Hall [1] proposed that the occurrence of boilover had to follow three conditions: the presented water in the oil, heat wave moved down into the oil well in advance of the burning surface, and the viscosity of oil in nature. Additionally, boilover can be divided in two types, i.e. hot zone boilover and thin-layer boilover. The major difference between the hot zone boilover and thin-layer boilover is the thickness of the fuel layer above the water sub-layer. Only thin-layer boilover was investigated in this study.

Some previous studies [2-5] have conducted experiments to understand the mechanism of thin-layer boilover. Broeckmann [2] summarized some conditions for thin-layer boilover:

- upper boiling range of the first phase (i.e. fuel) exceeding 250 ℃
- small tank diameter enhancing the influence of the tank wall
- long period of burning and thus increased heating of the first phase via the tank wall
- low oil level and intense contact of flame and tank wall
- fires in the surroundings giving an additional heat input
- the second phase (i.e. water or methanol) of higher density and a lower boiling point than the first phase
- the second phase with the ability of superheat
- the second phase not covering the whole tank bottom so that intense burning can go on after starting the water vaporization

Garo and Vantelon [3] conducted an experiment in order to study the effect of the fuel boiling point on the occurrence of thin-layer boilover. Their result depicted that boilover was due to the heterogeneous nucleation of the water at the fuel-water interface and the fuel boiling point must be at least higher than the saturation temperature of water. Moreover, the increasing boiling point of fuel resulted in a decreasing burning rate which may cause the thicker of fuel at the onset of boilover and more disruptive ejection of the fuel. Consequently, the intensity of boilover increases as the boiling point of fuel increases. Garo et al. [4] also proposed that there are some major parameters which can affect the thin-layer boilover, i.e. the initial fuel thickness, the pool diameter and the boiling point of fuel. Laboureur et al. [5] considered physical understanding of the thin-layer boilover by visualization. The results of visualization depicted that flame enlargement is caused by the piston effect of the homogenous vaporization of water layer. In their experiment, the bubble of the water vaporization generated at the fuel-water interface and can be clearly observed that bubbles raised up to the fuel surface and the size of bubble increased as time increased.

For the aforementioned studies focused on the physical understanding and the major effective parameters of the thin-layer boilover, the research of preventing or delaying the thin-layer boilover is scarce. Nowadays, foam system is usually used to extinguish pool fires. In fact, the foam system does not work very well in the oil tank fire and it does not have a good effect on preventing boilover. Furthermore, the agents of foam system may not be cleaned up easily after extinguishing the oil tank fire

Corresponding author. Fax: (07)-6011708
E-mail address: tsaikc@nkfust.edu.tw

and consequently cause water pollution of the entire environment. Thus, Tsai et al. [6] used a new design, vapor suppression spheres, for delaying and preventing thin-layer boilover. Their results showed that the vapor suppression spheres are useful to prevent thin-layer boilover, but the cost of vapor suppression spheres is high. This study investigates the effect of using floating perlites for delaying the occurrence of thin-layer boilover and preventing the flame enlargement. The effect of layer number of perlites on the characteristic of thin-layer boilover was particularly investigated.

2 Experimental

2.1 Perlite

Perlite is an amorphous, porous, light and white particle which is produced by the volcanic eruption and rapid cooling. Perlite has steady physical and chemical performance, and does not react with other chemicals. Moreover, perlite is non-toxic and has well thermal insulation performance (thermal conductivity is 0.045~0.047 W/(m • k)) and fire performance (the refractoriness is 1300~1380 °C). Therefore, it is a commonly used insulation building material. The effective diameter of perlite used in this study is 3.85±1.4 mm and the density of perlite is 0.17~0.25 g/cm^3. Figure 2 shows the picture of perlites.

Figure 2: The picture of perlites.

2.2 Experimental design

The experiment was conducted at the Fire Safety Laboratory in the National Kaohsiung First University of Science and Technology, Kaohsiung, Taiwan. All fire test facilities in this lab were installed indoor, and the whole experiment in this study can be conducted without external unexpected conditions.

Figure 3 and 4 show the experimental setups. Two round stainless steel pans with a diameter of 0.1 and 0.3 m and a depth of 0.1 m were used. The depths of fuel layer and water layer were 0.02 and 0.05 m, respectively. Diesel oil was used as fuel, and its nominal close-cup flashpoint is 70 ℃. Diesel does not cause hot zone boilover during its burning in oil tank but cause thin layer boilover [7]. Garo et al. [8] have shown fuel splash and increased burning rate (mass loss rate) and radiation after long burning when diesel thin-layer boilover occurs. Gasoline of 0.020 and 0.060 g were added onto the diesel to help ignite diesel for the 0.1 m- and 0.3 m-fuel-pan experiments, respectively. To know effects of perlites, none, 0.5, 1, 2 or 4 layers of perlites were put above the fuel. The weight of one-layer perlites for the fuel pans with a diameter of 0.1 and 0.3 m is 2.78 and 20.6 g, respectively. The 0.5 layer of perlites was produced by the perlites with weight of a half of that of 1 layer of perlites. Figure 4 shows the schematic of the facilities for measuring external radiation, mass burning loss rate, heat release rate, fuel temperature and water temperature in this experiment. Horizontal distance between pan rim and heat flux meter was set to be L/D = 3, where D is pan diameter and L is the distance from pan rim to the heat flux meter. Temperature was measured by a K-type thermocouple tree, which is set along the center of fuel pan with an interval of 0.1 m.

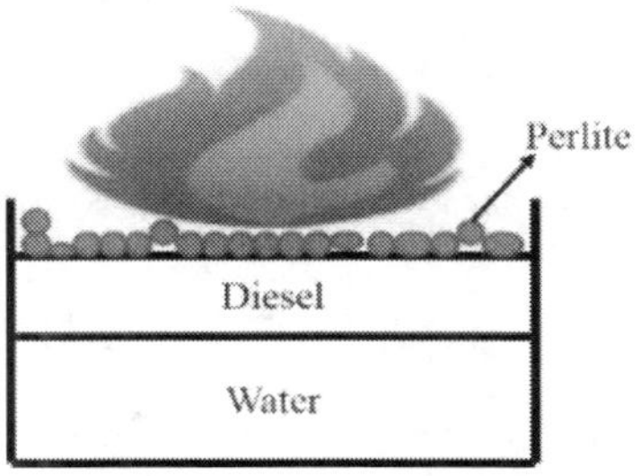

Figure 3: Schematic of the setups of the thin-layer boilover test

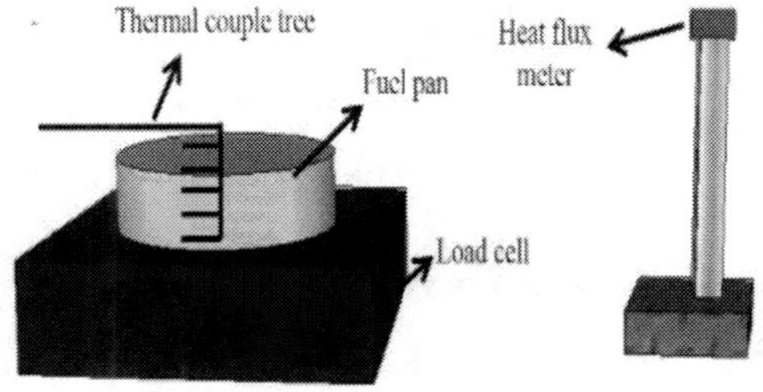

Figure 4: Schematic of the experimental design

3 Experimental results and discussion

3.1 Experimental observations

Figure 5 and 6 show the observations of explosive burning for the 10 cm and 30 cm fuel pan tests with 0, 0.5 and1 layer perlites, respectively. The explosive burning with the 0 and 0.5 layer perlites can be grouped as thin-layer boilover because the burnt oil was ejected and spilled out of the fuel pan. The time to thin-layer boilover for the 10 cm fuel pan tests with 0 and 0.5 layer of perlites was 1338 and 3582 s, respectively. For the 30 cm fuel pan tests with 0 and 0.5 layer of perlites was 746 and 811 s, respectively. The tests with 1, 2 and 4 layers of perlites did not perform thin-layer boilover. This observation might be caused by the decreased radiation from flame to fuel and water because of the existence of perlite layer. Further, the decreased radiation lowered the rate of temperature rise of water, and consequent time to reach the boiling point of water. Once the water started to vaporize, the amount of vaporizing water was low, and the thin layer boilover was not severe. Additionally, 0.5 layer of perlites still delayed the time to thin-layer boilover. Additionally, the time to thin-layer boilover with large fuel pans occurred early.

3.2 Temperature history of fuel without perlites

Figure 7 shows the fuel temperature history for the 30 cm fuel pan test without perlites. The distance was measured from the bottom of the fuel pan. The fuel surface located at 7 cm from the bottom at the beginning of the tests. The temperatures at 8, 9 and 10 cm actually measured the temperature of flames, and the temperature of the flame reached 700 ℃. The temperature at a specific level higher than 7 cm decreased because the fuel surface descended. Additionally, the temperature at 5 cm level displayed the temperature of the fuel-water interface.

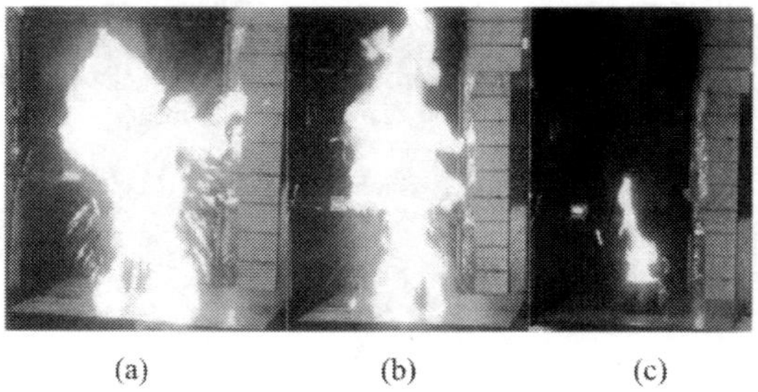

(a) (b) (c)

Figure 5 (a) - (c): The observation of thin-layer boilover for 10 cm fuel pan tests with 0, 0.5 and1 layer perlites, respectively

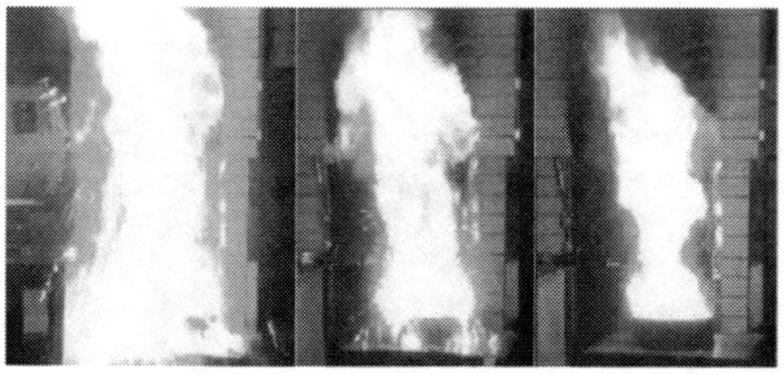

(a) **(b)** **(c)**

Figure 6 (a) - (c): The observation of thin-layer boilover for 30 cm fuel pan tests with 0, 0.5 and 1 layer perlites, respectively

Thin layer boilover occurred at 746 s in this 30 cm fuel pan test. Dramatic flame fluctuation was observed. The temperature at levels above the fuel-water interface subsequently started to decrease sharply when thin-layer boilover occurred and then increased. This oscillation may be caused by the ejection of water vapors. The thermocouples may expose to flames at some times and to water vapors at other times. Additionally, the temperature of fuel-water interface, which located at 5 cm from the bottom, reached approximately 100 ~ 120 ℃ when thin-layer boilover occurred. Consistent observation was seen in this experiment of Garo et al. [9], reporting that the temperature at the fuel-water interface would reach a value of approximately 120 ℃ when thin-layer boilover occurs. Additionally, the experimental observations in the 10 cm fuel pan test were consistent with the 30 cm fuel pan test.

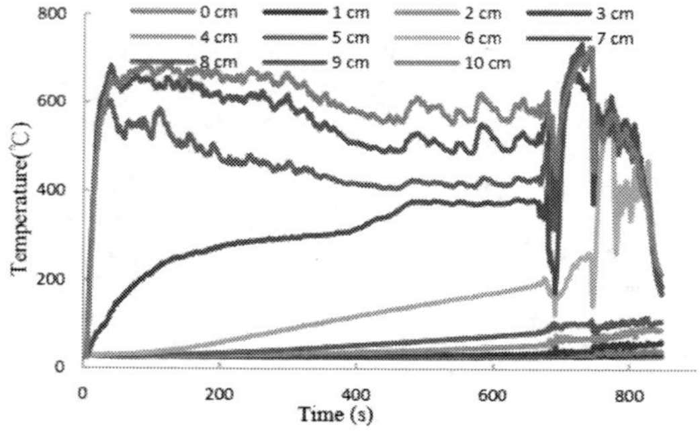

Figure 7: The fuel temperature history for the 30 cm fuel pan tests without perlites

3.3 Temperature history of fuel with perlites

Figure 8 and 9 show the temperature histories of fuel-water interface for the 10 and 30 cm fuel pan tests with perlites, respectively. Clearly, adding perlites delayed or prevented the time to reach 120 ℃. The performance of delaying or preventing the temperature rise to 120 ℃ was more significant with more layers of perlites. The thicker the perlite layer was produced, the better radiation blockage was formed. The radiation blockage can decrease the heat feedback from the flame to the fuel.

Figure 10 shows the temperature history at the level of 1 cm below the fuel surface for the 30 cm fuel pan tests. It can be clearly seen that the temperature decreased as the layers of perlites increased. It depicted that perlites performed significant thermal isolation during burning.

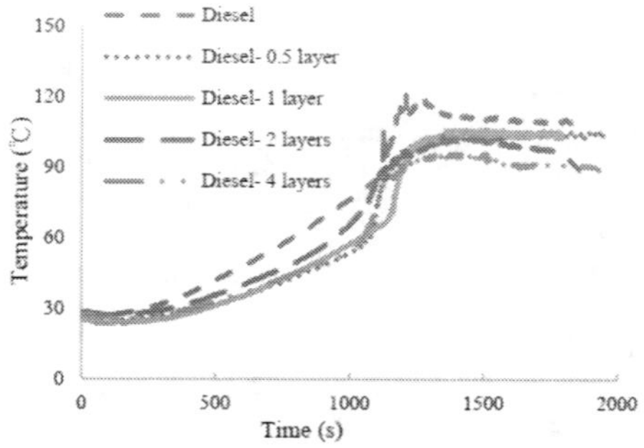

Figure 8: The effect of layer number of perlites on the temperature history of fuel-water interface for the 10 cm fuel pan test.

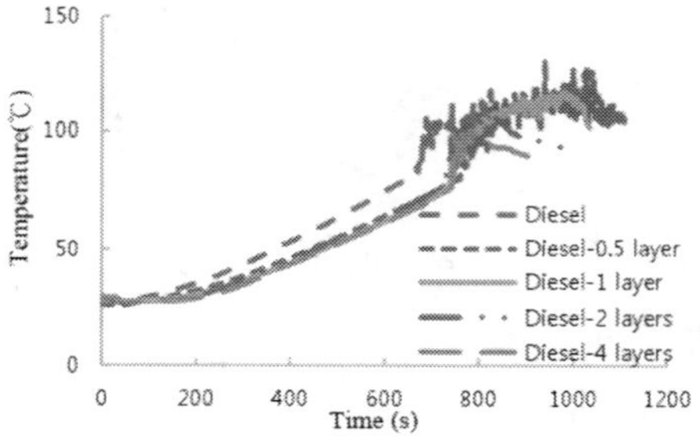

Figure 9: The effect of layer number of perlites on the temperature history of fuel-water interface for the 30 cm fuel pan test.

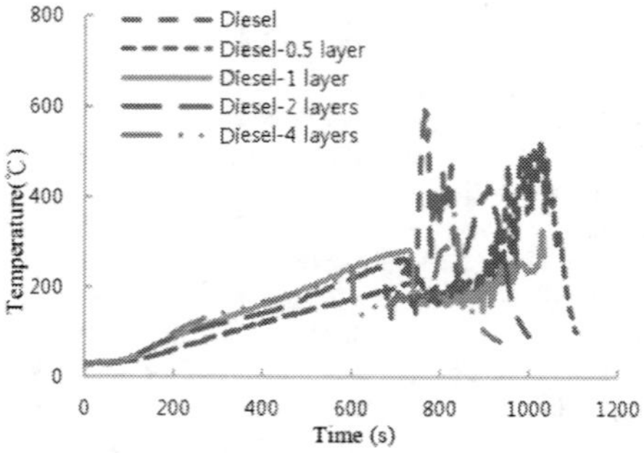

Figure 10: The effect of layer number of perlites on temperature history at the level of 1 cm below the fuel surface for the 30 cm fuel pan tests.

Figure 11 shows the effect of layer number of perlites on the heat release rate during the 30 cm fuel pan test. Figure 12 shows the effect of layer number of perlites on the heat flux during the 30 cm fuel pan tests. There are some peaks in Fig. 11-12. These peaks could response to the occurrence of thin-layer boilover or an explosive burning. Although the values of peaks were different, the occurrence time of peaks were nearly identical in these tests. The results depict that the effect of perlite layer to the time to thin-layer boilover was not obvious. It might be caused by the effect of the heat conduction through the tank wall. Nakakuki [10] has shown that the heat conduction through the tank wall influenced the heat transfer and the occurrence of boilover. Therefore, added perlites can decrease the radiation heat but water can be heated to boiling point by the conduction heat of the tank wall. The perlite layer has little effect on the time to thin-layer boilover, but it has effect on the severity of boilover.

Figure 13 shows the effect of layer number of perlites on the mass loss rate during the 30 cm test. The mass loss rates for all tests gave close values during steady state burning, so the effect of perlites addition was not significant and the mass loss rate without perlites is much higher than those with perlites during the occurrence of thin-layer boilover.

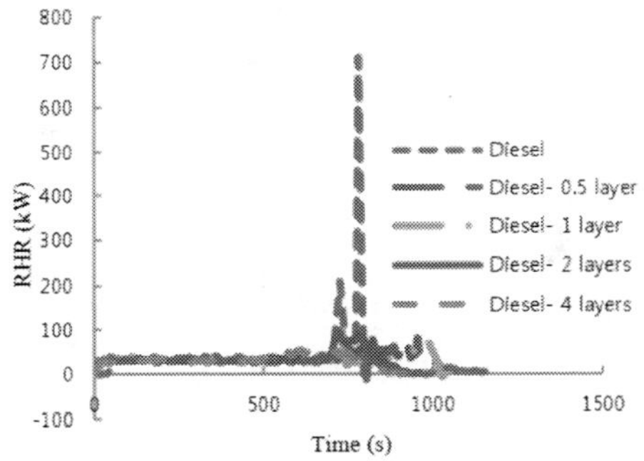

Figure 11: The effect of layer number of perlites on the heat release rate during the 30 cm test

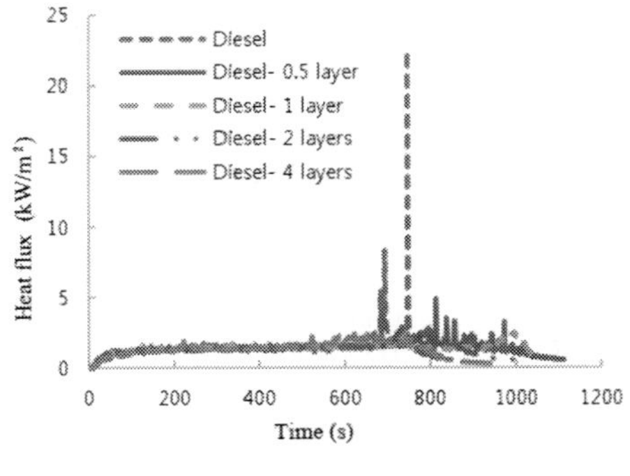

Figure 12: The effect of layer number of perlites on the heat flux during the 30 cm test

4 Conclusions

This study investigated the effect of floating perlites on preventing the occurrence of thin-layer boilover by conducting small-scale tests. Our data showed that thin-layer boilover occurred with none and 1/2 layer perlites, while it did not occur when 1, 2 and 4 layers were added. The use of floating perlites decreased the heat feedback to the fuel, and the speed of temperature rise in the fuel. Further research is needed to conduct full-scaled tests to confirm the performance of perlites in preventing the thin-layer boilover in realistic tank fires.

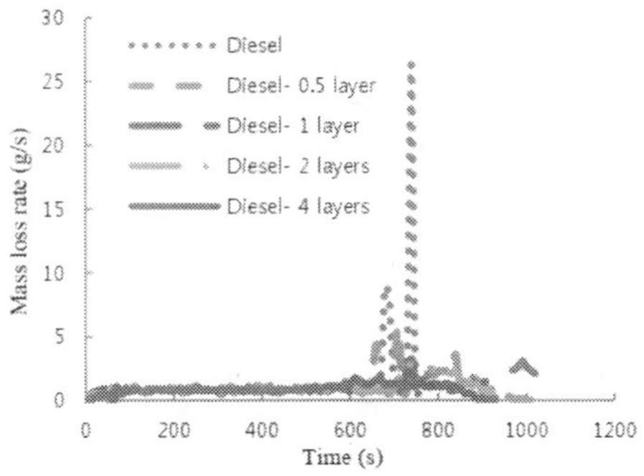

Figure 13: The effect of layer number of perlites on the mass loss rate during the 30 cm fuel pan test

5 Acknowledgment

The author wishes to thank Mr. Bo-Cheng Huang for assistance in carrying out experiments.

References

•1. Hall, H.H. ,"Oil-Tank-Fire Boilover", Mech. Engineering 47(7) , 1925, pp.540 – p.544.

•2. Broeckmann, B., Schecker, H.G., "Heat Transfer Mechanism and Boil over in Burning Oil-Water Systems", Journal of Loss Prevention and Process Industries, vol. 8, 1995, No. 3, pp.137-147.

•3. Garo J.P., Vantelon J. P., Proc. Combust. Inst. 26 (1996) 1461-1467.

•4. Garo, J.P., Gillard, P., Vantelon, J.P. and Fernandez-Pello, A.C., "On The Thin Layer Boilover". Fire Safety Science 6, 2006, pp.579-590.

•5. Laboureur D., Aprin L., Osmont A., Buchlin J.M., Rambaud P., "Small scale thin-layer boilover experiments: Physical understanding and modeling of the water sub-layer boiling and the flame enlargement", Journal of Loss Prevention and Process Industries, 26(6), 2013, pp.1380-1389.

•6. Tsai K.C., Koseki H., Tseng T.Y., Kelly R., "Effects of floating beads on the flash/fire temperatures and occurrence of boilover", Annual International Symposium, Texas, 2016, pp.279-289.

•7. Hasegawa K., "Experimental Study on the Mechanism of Hot Zone Formation in Open-Tank Fires", Fire Safety Science 2, 1991, pp.221-230

•8. Garo J.P., Vantelon J.P., Koseki H., "Thin-layer boilover: Prediction of its onset and intensity", Combustion Science and Technology, 178, 2006, pp.1217-1235.

•9. Garo J.P., Koseki H., Vantelon J.P., Fernandez-Pello C., "Combustion of liquid fuels floating on water", Thermal Science, 11(2), 2007, pp.119-140.

•10. Nakakuki A., "Heat Transfer in Hot-Zone-Forming Pool Fire", COMBUSTION AND FLAME, 1997, pp.353-369.

11th Asia-Pacific Conference on Combustion,
The University of Sydney, NSW Australia
10th -14th December 2017.

Thermal stratification in corridor of typical building structure under wind effect

Bei Cao[1], Xiaodong Zhou[1], Yubiao Huang[1], Yuan Zheng[1], Kun Zhao[1], Fei Peng[1], Kai Ye[1], Lizhong Yang[1]

[1]University of Science and Technology of China

Huang Shan Road, Hefei, Anhui, 230027, P.R. China

Abstract

The knowledge of thermal stratification is essential for building fire detection, smoke control; human evacuation and fire hazard evaluation. The present research aims at exploring thermal stratification in corridor of typical building structure under wind effect. CFD simulation data and Froude number are examined. The results indicates that in the simulated cases, there existed two wind speed ranges. In the range of relatively low wind velocity, there is a certain region in the vicinity of the spilled-out smoke in the corridor that presents a weak stratification phenomenon in terms of the values of $\frac{\Delta T_{cf}}{\Delta T_{avg}}$. In the range of relatively large wind velocity, there exists a region dominated by the wind-driven force where the wind pressure is strong and has a significant interaction with the spilled-out smoke. And with the increasing of the heat release rate of the fire source, the region affected by the strong ambient wind will shrink. $\frac{\Delta T_{cf}}{\Delta T_{avg}}$ and Fr show a linear relationship except in above two regions. Under same conditions, the stratification is more significant in the higher heat release rate cases.

1 Introduction

The basic law of smoke behavior, which underlies the smoke control strategies, attracts lots of researchers worldwide. Some provide basic knowledge of smoke movement in local space of buildings including staircase [1-4], elevator shaft [5], and atria [6] and in full-scale structure such as multi-storey buildings [7] and rooms [8]. However, because smoke movement in building space can be influenced by many factors of building structure and environmental conditions, etc., some efforts still need to be made. As for the thermal stratification characteristics under wind conditions, Jeffrey S. Newman has quantified the stratification in a ventilated mine passages by a Froude number and some dimensionless parameters of temperature. The correlations between $\frac{\Delta T_{cf}}{\Delta T_{avg}}$, $Fr = \frac{V_{avg}}{[\frac{\Delta T_{cf}}{T_{avg}} gH]^{1/2}}$ (the meaning of the symbols can be seen in the original research work and in the following section 3: Stratification characteristics analysis in which the symbols mentioned have the similar meaning with the original work but are directed against the research object in this study) are established to investigate the relative importance of the fire-induced buoyancy and wind-driven force in combustion products stratification process [9]. Hans Nyman and Haukur Ingason have investigated the applicability of Jeffrey S. Newman's established correlations with tunnel fire scenarios [10]. It seems that the two dimensionless parameters of $\frac{\Delta T_{cf}}{\Delta T_{avg}}$, $\frac{V_{avg}}{[\frac{\Delta T_{cf}}{T_{avg}} gH]^{1/2}}$ can describe the thermal stratification phenomenon under wind conditions well. After reviewing substantial research articles [11-12] and careful field investigations, among all the building styles, compartment-corridor structure form can be extracted from most of the commonly used buildings. In this study, the stratification characteristics in corridor of typical building structure under wind effect is detailed by dimensionless parameters of $\frac{\Delta T_{cf}}{\Delta T_{avg}}$, $\frac{V_{avg}}{[\frac{\Delta T_{cf}}{T_{avg}} gH]^{1/2}}$.

2 Fire scenario description

Taking Yang's experimental platform as a prototype [12] and using Fire Dynamics Simulator (FDS, version 6.1.1) as a computational tool, the almost full-scale physical model of a typical building structure of compartment-corridor is built with some details changing to explore thermal stratification in corridor under wind conditions. The compartment, with dimensions of 3200 *3200* 3200 mm, is situated at the windward side of the corridor, forming an L-shape compartment-corridor structure. Two rectangular windows of size of 1500mm * 1000mm is opened in the center of the windward and leeward wall of the corridor, respectively. The detailed information of dimensions of the model and the arrangement of the measured temperature points are shown in Fig. 2 and Fig. 3.

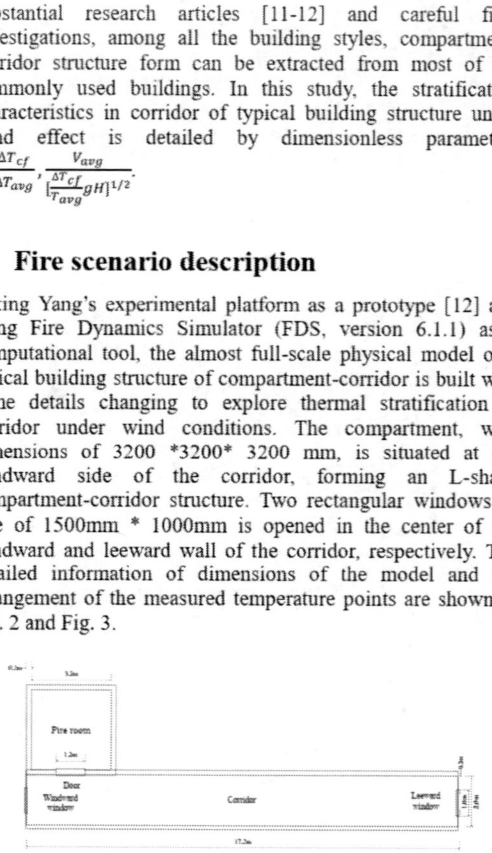

Figure 1: Top view of the physical model

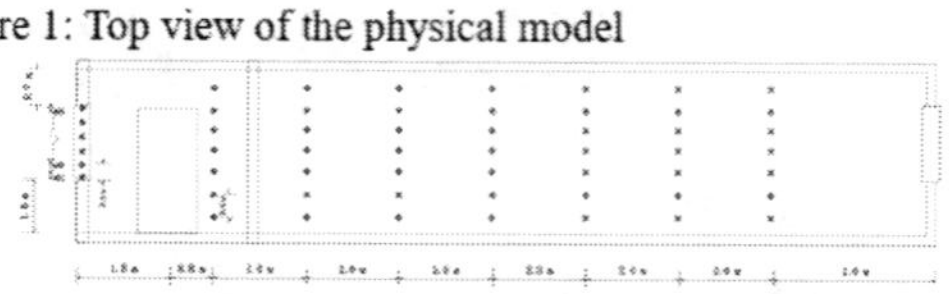

Figure 2: Arrangement of the measured points in the physical model

As for the initial and boundary conditions of the computational model, the ambient temperature is set at 20℃. And all the walls and floors and roofs of the compartment and corridor are made of concrete materials with thickness of 200mm; the thermal-

Corresponding author. Fax: +86-551-6360-1669
E-mail address: yanglz@ustc.edu.cn (L Yang)

physical properties of the concrete refer to the material libraries of FDS. Representing approximately the heat release rate of the typical combustible materials of sofa and bookshelf, the fire source of 500kW, 700kW are set respectively in the middle center of the compartment [13]. The leeward side of the corridor is set to be naturally opened without initial velocity. A non-dimensional expression of $D^*/\delta x$ is used to define the mesh resolution of the computational domain, where $D^* = \left(\frac{Q}{\rho_\infty c_p T_\infty \sqrt{g}}\right)^{\frac{2}{5}}$, is a characteristic fire diameter, δx is the nominal size of the mesh cell, Q is the total heat release rate of the fire and $\rho_\infty, c_p, T_\infty$ is the density, specific heat, temperature of the ambient air, respectively. In high-rise buildings, the fire source is usually in one floor or one compartment, and the height of the floor or compartment is relatively small compared with the total height of the building. Therefore, the variations of the wind velocity along the vertical direction of the building can be neglected. The default wind profile in FDS model is constant and thus this setting is adopted in the present research. When the ambient wind blows to the building, because of the blockage of the buildings, the wind conditions may be changed to some extent. In this paper, only the wind direction, which is perpendicular to the windward surface, is considered. To limit the computational domain, an inlet velocity boundary is set at the windward side of the corridor domain, which is inspired by previous researches [14] concerning wind effect on fire smoke movement in similar structure. The wind velocity is set at between 0-3m/s to heat release rate of 500kW, 0-4m/s to heat release rate of 700kW.

3 Stratification characteristics analysis

Varying degrees of stratification of fire smoke can be described by a dimensionless parameter of $\frac{\Delta T_{cf}}{\Delta T_{avg}}$, where ΔT_{cf} is the difference between smoke temperature near the ceiling of the corridor at height of 2.8m and smoke temperature near the floor at height 0.4m, ΔT_{avg} is the average smoke temperature rise relative to ambient environment. The value of $\frac{\Delta T_{cf}}{\Delta T_{avg}}$ reflects the magnitude of fire-induced buoyancy. To exam the wind-driven force, a non-dimension parameter of $\mathrm{Fr} = \frac{V_{avg}}{[\frac{\Delta T_{cf}}{T_{avg}} gH]^{1/2}}$ is used. Where , T_{avg} is the average gas temperature, $V_{avg}$is the average "hot" velocity across the corridor, and $V_{avg} = T_{avg} V_0 / T_0$, in which V_0 is the initial wind velocity, T_0 is the ambient temperature and g is the acceleration due to gravity. The stratification characteristics in corridor of typical building structure under wind effect is supposed to be explored by these two dimensionless parameters of $\frac{\Delta T_{cf}}{\Delta T_{avg}}, \frac{V_{avg}}{[\frac{\Delta T_{cf}}{T_{avg}} gH]^{1/2}}$. And further, the correlation between $\frac{\Delta T_{cf}}{\Delta T_{avg}}$ and $\frac{V_{avg}}{[\frac{\Delta T_{cf}}{T_{avg}} gH]^{1/2}}$ is also studied.

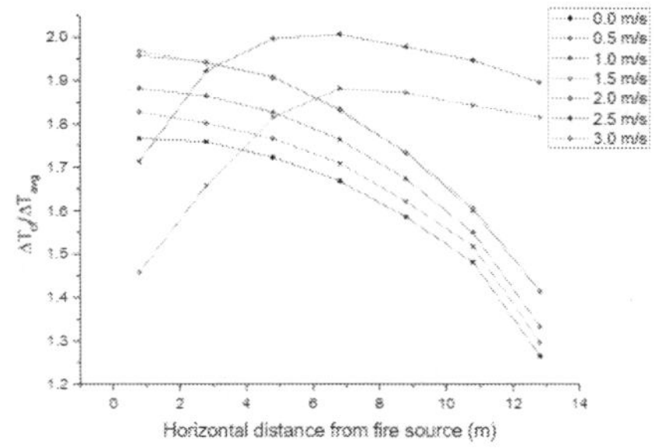

Figure 3: Calculated $\frac{\Delta T_{cf}}{\Delta T_{avg}}$ value along the corridor with different wind velocity (HRR=500kW)

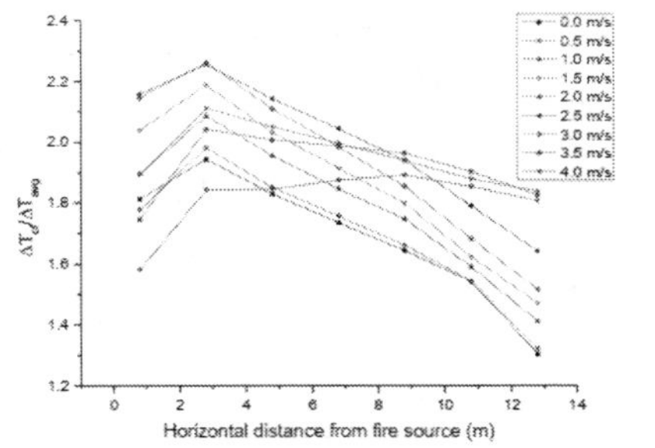

Figure 4: Calculated $\frac{\Delta T_{cf}}{\Delta T_{avg}}$ value along the corridor with different wind velocity (HRR=700kW)

As can be seen from Fig. 3 and Fig. 4.

(1) The values of $\frac{\Delta T_{cf}}{\Delta T_{avg}}$ follow almost the same trend in the range of relatively low wind velocity. In cases with heat release rate of 500kW, this range includes from 0 to 2m/s. As for cases with heat release rate of 700kW, this range extends to 2.5 m/s. In cases with heat release rate of 500kW, the farther the horizontal distance from the fire source, the less evident thermal stratification will be observed. This is mainly because there are heat exchange between the cold air and convective heat transfer with the walls and roofs as the fire smoke travels along the corridor. The buoyancy force is decrease and smoke descends along the corridor. But there is a slight difference in cases of heat release of 700kW, the curves first ascend and then decline. This can be attributed to the relatively large heat release rate. In cases of heat release rate of 700kW, smoke spilled out of the compartment holds larger upper space of the connected door, and certainly occupies larger upper space and spreads farther in the corridor compared with the cases of 500kW. There is a certain region dominated by the fire-induced buoyancy in the vicinity of the spilled-out smoke in the corridor that presents a weak stratification phenomenon in terms of the value of $\frac{\Delta T_{cf}}{\Delta T_{avg}}$ and the first calculated point is just within the region in cases of heat release rate of 700kW.

(2) There are relatively large wind velocity of 2.5-3.0m/s in cases of heat release rate of 500kW, of 3.0-4.0m/s in cases of heat release rate of 700kW, in which cases, the values of $\frac{\Delta T_{cf}}{\Delta T_{avg}}$ first increase then decrease. The main reason is that in the relatively large wind velocity conditions, there

exists a region dominated by the wind-driven force where the wind pressure is strong and has a significant interaction with the spilled-out smoke. The larger the heat release rate of the fire source, the higher temperature the spilled-out smoke will get, which means more blockage the wind-driven force will confront and the region affected by the strong wind velocity will shrink. This can be used to explain why in cases of heat release rate of 500kW the curves begin to decline at the fourth calculated point while in cases of 700kW the curve began to decline at second calculated point with the exception of case of wind velocity of 4 m/s, which presents a stronger wind-driven force. Meanwhile, out of the region, the reason of the drop of the curves in the two sets of cases is the same as those mentioned in the analysis of cases of relatively low wind velocity.

(3) It is obvious in the two sets of cases, in the relatively low wind velocity conditions, the stratification of fire smoke becomes more evident when the wind velocity increases. The reason is that smoke transport in the vertical direction in the corridor is caused by gradient differences and held by diffusivity, the relatively large wind blows the smoke in the lower layer away which can strengthen the stratification. While in the relatively large wind velocity cases, the effect of ambient wind of cooling the smoke by heat removal and dilution becomes more obvious with the wind velocity increases, which determines the trend of the curves.

(4) In the two sets of cases, under same conditions, the stratification is more significant in the higher heat release rate cases.

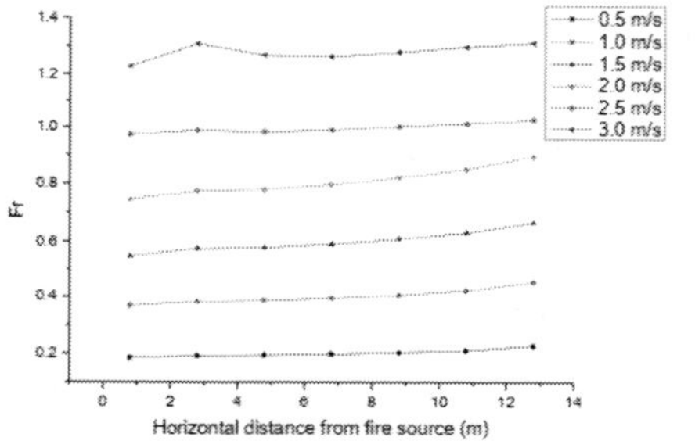

Figure 5: Calculated Fr value along the corridor with different wind velocity (HRR=500kW)

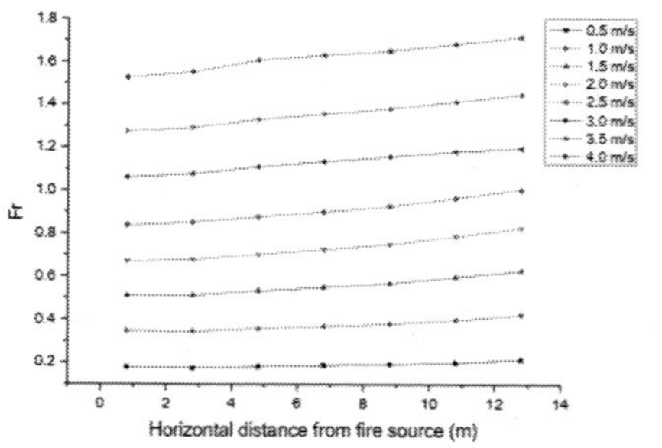

Figure 6: Calculated Fr value along the corridor with different wind velocity (HRR=700kW)

Figure 5 and Fig. 6 show that with the wind velocity increases, the Fr becomes larger. Along the corridor downstream the fire source, the Fr increases slightly. And Fr is not very sensitive to the changes of values of $\frac{\Delta T_{cf}}{\Delta T_{avg}}$.

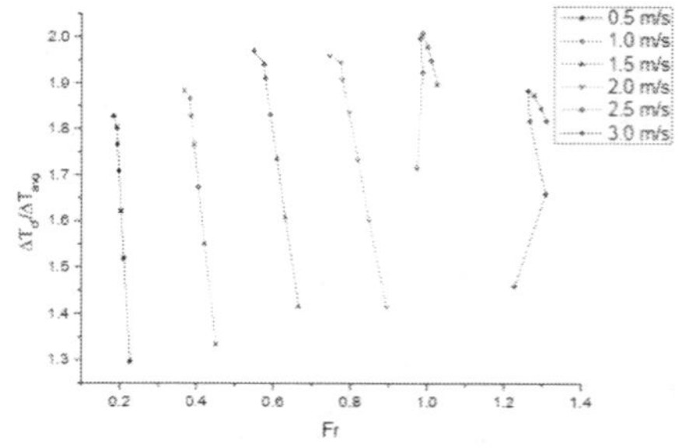

Fig7. Temperature stratification versus Froude number (HRR=500kW)

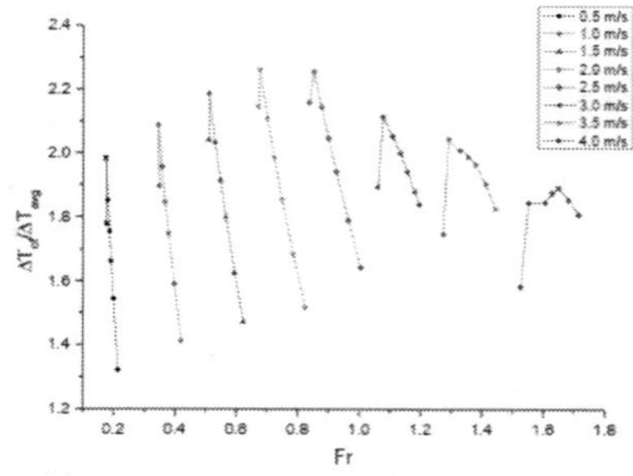

Fig8. Temperature stratification versus Froude number (HRR=700kW)

Figure 7 and Fig. 8 indicates that along the corridor downstream the fire source, the values of $\frac{\Delta T_{cf}}{\Delta T_{avg}}$ and Fr show a linear relationship except in the two regions: the first region is controlled by the fire-induced buoyancy in the vicinity of the spilled-out smoke in the corridor that presents a weak stratification phenomenon; the other is dominated by the wind-driven force where the wind-driven force is strong and has a significant interaction with the fire-induced buoyancy.

The above computational results revealed some effects of ambient wind on thermal stratification in corridor. As it is known, when there is no wind or the wind speed is relatively small, there is bidirectional flow across the windward window of corridor, that is, the hot smoke gas escapes out from upper part of windward window, and the cold air blows in from lower part of windward window. However, in the cases under large wind velocity conditions, the high wind pressure will suppress the outflow from the windward window so that there is only air inflow across the windward window. This phenomenon, which is believed to influence the smoke motion greatly. It seems that, for a specific heat release rate, there is a certain critical wind velocity which is just large enough to prevent the smoke outflow from the windward vent. When the set wind speed exceeds this value, thermal stratification in the corridor will be deeply influenced by smoke motion change. Some theoretical considerations on the prediction of the critical wind velocity will be studied in the future work.

4 Conclusions

In this study, computational data from CFD simulations and Froude number are examined to explore the thermal stratification characteristics in corridor of typical building structure under wind effect. The main conclusions can be summarized as follows:

(1) In the range of relatively low wind velocity, the farther the horizontal distance from the fire source, the less evident thermal stratification will be observed. An exception of the calculated point shows that there is a certain region in the vicinity of the spilled-out smoke in the corridor that presents a weak stratification phenomenon in terms of the value of $\frac{\Delta T_{cf}}{\Delta T_{avg}}$. Meanwhile, the stratification of the fire smoke becomes more evident when the wind velocity increases.

(2) In the range of relatively large wind velocity, there exists a region dominated by the wind-driven force where the wind pressure is strong and has a significant interaction with the spilled-out smoke. The larger the heat release rate of the fire source, the higher temperature the spilled out smoke will get, which means more blockage the wind-driven force will confront and the region affected by the strong wind velocity will shrink in the corridor.

(3) Under same conditions, the stratification is more significant in the higher heat release rate cases. Along the corridor downstream the fire source, the Fr increases slightly. And Fr is not very sensitive to the change of $\frac{\Delta T_{cf}}{\Delta T_{avg}}$. $\frac{\Delta T_{cf}}{\Delta T_{avg}}$ and Fr show a linear relationship except in the regions affected by the spilled-out smoke and strong wind-driven force.

(4) Further analysis on some other parameters which may indicate the thermal stratification in corridor under wind conditions and additional experimental work still need to be made for the validation of these results.

5 Acknowledgment

This research was supported by National Key R&D Program of China under Grant No.2016YFC0800603、National Natural Science Foundation of China under Grant No.51323010 and the Fundamental Research Funds for the Central Universities under Grant No. WK2320000033. The authors deeply appreciate the supports.

References

[1] Y.Q. Chen, X.D. Zhou, T.L. Zhang, Y.Q. Hu, L.Z. Yang, International Journal of Numerical Methods for Heat & Fluid Flow 25(3) (2015) 534-549.

[2] J. Ji, H.X. Wan, Y.F. Li, K.Y. Li, J.H. Sun, International Journal of Thermal Sciences 89 (2015) 23-33.

[3] R.S. Miller, D. Beasley, Building and Environment 44(6) (2009) 1306-1317.

[4] X.Q. Sun, L.H. Hu, Y.Z. Li, R. Huo, W.K. Chow, N.K. Fong, G.C.H. Lui, K.Y. Li, Applied Thermal Engineering 29(13) (2009) 2757-2765.

[5] W.Z. Black, Fire Safety Journal 44(2) (2009) 168-182.

[6] L.Z. Wang, G.C. Zhao, Fire Safety Journal 61 (2013) 45-53.

[7] Y. Wang, G. Hadjisophocleous, E. Zalok, Safety Science 52 (2013) 13-27.

[8] C.M. Lai, C.J. Chen, M.J. Tsai, M.H. Tsai, T.H. Lin, Fire Safety Journal 58 (2013) 1-14.

[9] J.S. Newman, Combustion and flame (1984) 33-39.

[10] H. Nyman, H. Ingason, Fire Safety Journal 48 (2012) 30-37.

[11] T.Y. Fang, R. Huang, W.X. Feng, G.L. Zhai, W.C. Fan, Journal of Fire Sciences 21 (2002) 305-18.

[12] L.Z. Yang, W.X. Feng, J.Q. Ye, Journal of Fire Sciences 26(1) (2008) 45-62.

[13] T. Beji, S. Verstockt, R.V. de Walle, B. Merci, Fire Safety Journal 76 (2015) 9-18.

[14] C.G. Fan, J. Ji, W. Wang, J.H. Sun, International Journal of Heat and Mass Transfer 73 (2014) 158-169.

11th Asia-Pacific Conference on Combustion,
The University of Sydney, NSW Australia
10th -14th December 2017.

A Study of Natural Gas Mixing Percentage on Combustion and Emission Characteristics of a CNG/Diesel Dual-Fuel Engine

Ocktaeck Lim[1], Shubhra Kanti Das [2], Kyeonghun Jwa [2]

[1]School of Mechanical Engineering, University of Ulsan

Daehak-ro, Nam-gu, Ulsan, 44610, South Korea

[2]Graduate School of Mechanical Engineering, University of Ulsan

Daehak-ro, Nam-gu, Ulsan, 44610, South Korea

Abstract

A compression ignition engine was modified into a dual-fuel engine that allows small pilot injection of diesel fuel followed by combustion of compressed natural gas as a primary fuel with a pressurized common rail direct injection system. The modified engine sprayed a small amount of pilot diesel fuel as an ignition source, while CNG was injected through the intake manifold under different operating conditions. During the dual fuel operating mode, the effects of different amounts of natural gas mixing percentages were varied with a maximum CNG substitution of 90%. The effects of different CNG mixing contents (CNG 65%, CNG 80%, and CNG 90%) on dual-fuel performance and emission were carried and compared with a base engine at increasing engine speed of 1000-3500 rpm. The in-cylinder pressure, combustion duration, heat release rate and ignition delay were analyzed and compared to both engines.

1 Introduction

An increase in oil prices, the rapid extinction of fossil fuel and the negative impact of crude oil based fuels on environment compel automotive industry to drive on various alternative fuels. As a result, diesel engine-based advancements have undergone a noticeable shift in attempts to achieve future emission legislation and the outstanding fuel economy by automotive manufacturers. When improving NOx and particulate emissions of CI-based diesel engines, it is difficult to increase the thermal efficiency simultaneously. Therefore, the present situation demands an extraordinary technique that can use alternative fuel in existing diesel engines without requiring any major modifications. Among the many established alternative fuels, CNG is highly reliable due to the abundance of its reserves. For achieving high thermal efficiency, CNG based dual-fuel engine is one of the prominent choices; it has been explored over the last couple of decades by engine researchers [1]. When implementing natural gas as an ignition and combustion source in diesel engines, three primary methods can be used: dual-fuel [2,3] high-pressure direct injection (HPDI) [4,5] and hot surface-assisted compression ignition [6,7]. The most common technique is the dual-fuel mode, which allows natural gas to be injected directly into the intake manifold and mixed homogeneously with air where it is then ignited by a directly-injected fuel with high CN [8]. This dualfuel phenomenon is also applicable in a diesel engine without any modification of the engine cylinder because the pilot diesel fuel works like a liquid spark plug at the start of ignition [9]. Significant experimental and numerical efforts have been carried out to understand the impact of different engine operating parameters of a dual-fuel engine, such as injection timing [10] intake air heating [11], pilot fuel quantity [12–14] and exhaust gas recirculation [15–18].

Corresponding author. Fax: +00-00-000-000
E-mail address: otlim@ulsan.ac.kr

2 Experimental system and test conditions

2.1 Test engine system

The base system for the experiment was a single–cylinder, turbocharged engine with a displacement volume of 2.5 L. The engine was remodeled to operate in the CNG/diesel dual-fuel approach by adding a CNG-MPI to the existing diesel CRDI fuel injection control system. The engine shaft is coupled to a Froude Hofmann AG150 eddy current dynamometer, capable of delivering 150 kW and has a maximum torque of 500 Nm with a maximum rotating speed of 8000 rpm. A DYTEK-3000K dynamometer controller is used to control the dynamometer. The key engine specifications are, In-cylinder data is measured via a KISTLER-type 6056A-41 high-temperature pressure sensor and a KISTLER charge amplifier (Type 5015). Fig. 1 represents a scheme of the experimental arrangement of the modified CNG/diesel dualfuel engine. The rate of intake air flow is measured using a Meriam 50MC2-4 LFM, which ranges from 0-314 kg/h with a standard accuracy of ± 0.57%. The supplied natural gas flow rate is measured by an Oval Ultra mass MKII CNG flow meter with an accuracy of ±0.10%; this uses Karman's vortex flow detection method with its piezoelectric element sensor. The diesel fuel consumption is measured with Oheus precision plus equipment provided by Hwanwoong Mechatronics Co. Ltd (South Korea). To inject the pilot fuel during the compression stroke and to supply natural gas as a primary fuel, a dedicated CNG injector is installed at the intake port. For supplying CNG to the intake manifold as the main fuel, a CNG fuel tank, tank pressure regulator, and emergency shut-off valve with solenoid valves are used. When injecting CNG, the supply pressure is maintained at 8 bar using two regulators for decompression. The exhaust emission is measured using an exhaust gas analyzer (HORIBA MEXA-7100DEGR) The exhaust gas analyzer can measure CO, CO2, NO /NOx, THC, and O2 simultaneously. An automotive opacity smoke meter

(Qrotech, OPA-102) is connected to the exhaust pipe to measure the PM concentration. All data acquisitions are conducted by using a National Instrument DAQ board (NIC DAQ-9172) and a National Instrument LabVIEW software.

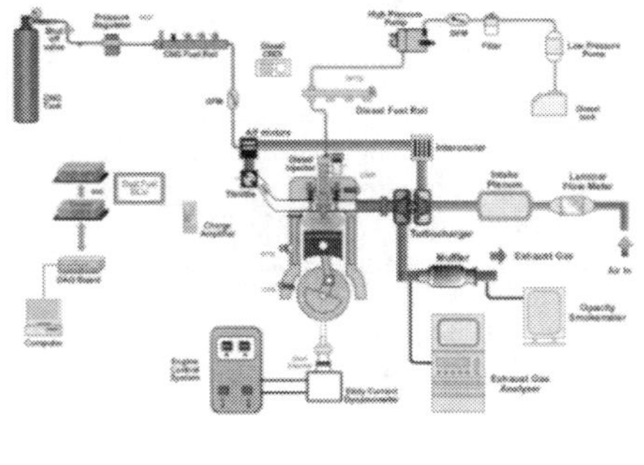

Figure 1: Schematic diagram of the experimental test engine.

2.2 Test fuel

The diesel fuel and natural gas used in this experiment are obtained from the commercial distribution network and industrial sources in South Korea. All the fuel properties are measured and certified by the Korea Institute of Petroleum Authority. The evaluated properties for diesel fuel and natural gas are mentioned in Tab. 1 and 2.

Fuel	Diesel	Natural gas
Boiling point(℃)	180-361.1	-162
Kinematic viscosity at 40℃(mm^2)	57.9	-
Lower heating value(MJ/kg)	40	50.9
Octane Index	42.8	130
Auto ignition temperature(℃)	30	650
Density at 15℃(kg/m^2)	316	0.7-0.9
Stoichimetric air-fuel ratio (kg/kg)	14.5	16.82
Lubricity(μm)	238	-

Table 1: Properties of the test fuel.

Constituent	Volume Percentage(%)
Methane	89.57
Ethane	5.89
Propane	1.85
i-Butane	0.56
n-Butane	0.42
Nitrogen	1.08
Carbon dioxide	0.63

Table 2: Natural gas constituent of the test system.

2.3 Test method and conditions

Experiments are performed in the single-fuel diesel mode and then in the CNG/diesel dual-fuel mode. All experiments are conducted while increasing the engine speed from 1000 to 3500 rpm to measure the torque, power, and exhaust emissions (THC, CO, THC and NOx) for different CNG mixing ratios. The in-cylinder combustion pressures are measured for a diesel engine under a full load with a SOI 15°CA BTDC. Also, the combustion pressures are measured for 25%, 50%, 75%, and full load conditions for the dual-fuel CNG/diesel mode using a pilot diesel flow of 0.5 L/h at 11°CA BTDC diesel pilot timing for a fixed engine speed with an increasing load. The pilot injection timing is altered from 11 11°CA BTDC to 15°CA BTDC in the dual-fuel CNG/diesel mode to compare the cylinder pressure at injection timing. Cycle analysis is carried out for 150 cycles for the in-cylinder pressure data obtained via data acquisition, and the combustion duration is estimated. The HRR is determined using the cylinder pressure data by applying the first law of thermodynamics. Tests are performed over the entire operating range (1000 to 3500 rpm) to compare the power output and emission characteristics for both single fuel and dual-fuel condition. The testing interval is 500 rpm, and the CNG substitution is calculated using (1). The diesel reduction rate for each CNG/diesel dual–fuel mode is calculated using (2), which compares the diesel reduction with the power required.

Equation (1) & (2) is

$$CNG\% = \frac{\dot{m}_{ng}}{\dot{m}_{ng}+\dot{m}_{diesel}} \quad (1)$$

$$D_{Red} = \left(\frac{P_{dual-fuel}}{P_{dual-fuel}}\right) \times D_{diesel} + D_{dual-fuel} \quad (2)$$

Here, $\dot{m}_{ng}$ and $\dot{m}_{diesel}$ are the mass flow rates of CNG and diesel. D_{Red} is the diesel reduction rate. $D_{dual-fuel}$ & D_{diesel} are the flow rates of the base diesel engine and the modified dual fuel engine. $P_{dual-fuel}$ & P_{diesel} are the power outputs.

3 Results and discussions

3.1 Performance characteristics

Fig. 2 and 3 compare the torque and power output between the diesel and CNG/diesel dual-fuel tests. According to the comparison, the output of the diesel test shows a maximum torque of 289 N·m at 2500 rpm and a maximum power of 128 PS at 3500 rpm. The output of the dual-fuel test shows a maximum torque of 300 N·m at 3000 rpm and a maximum power of 134 PS at 3500 rpm. The performance variation of the modified dual-fuel engine is 5% higher compared to base engine. In Fig. 2, for engine speeds ranging from 1000 to 2500 rpm, the trends are similar (except CNG 90%). The torque data gathered for different mixing show similar trends with conventional diesel operation. However, as the engine speed is increased, the torque of the CNG 90% condition appears to increase more sharply compared to the other ratios (Fig. 3). At 2500 rpm, CNG 90%, CNG 65%, and conventional diesel all lie at the same point; however, when the engine speed is increased from 3000 to 3500 rpm, the conventional diesel power is lower than CNG 65% and CNG 90%; this can be interpreted as representing friction loss with higher engine speeds. Fig. 4 illustrates the diesel reduction rate of the dual-fuel engine for CNG substitutions of 65%, 80%, and 90%. The maximum diesel reduction rate is obtained at 90% in all experimental conditions.

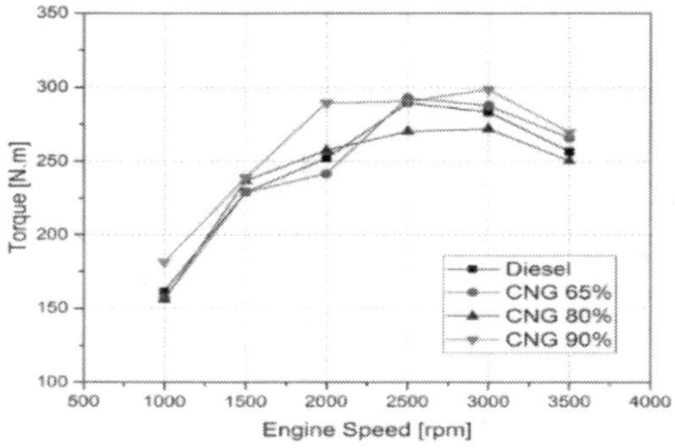

Figure 2: Brake torque output of diesel and different percentage natural gas mixed dual-fuel mode as increasing engines speeds.

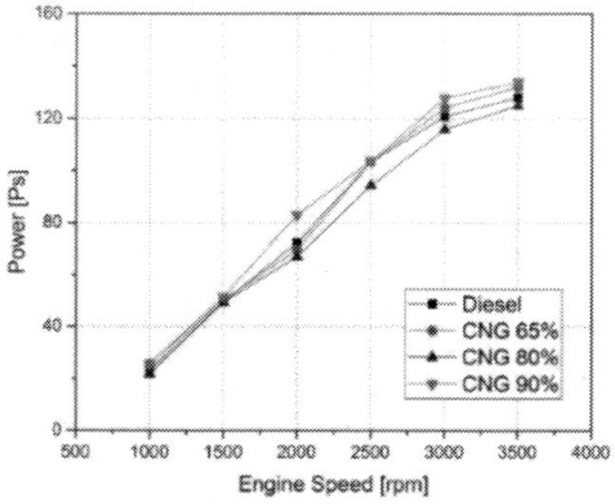

Figure 3: Brake power output of diesel and different percentage natural gas mixed dual-fuel mode as increasing engines speeds.

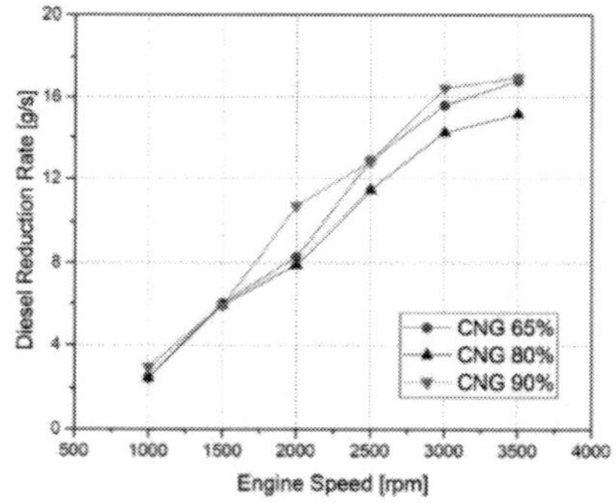

Figure 4: Diesel reduction rate consumed under different percentage natural gas mixed dual-fuel operations.

3.2 Combustion characteristics

The experimental dual-fuel combustion behavior is compared to the single-fuel diesel combustion mode in Fig. 5. The results obtained for the single-fuel diesel mode are under full load conditions at 2000 rpm, while the dual-fuel engine load is varied from 25%, 50%, 75%, and full load conditions at 2000 rpm. Single-mode diesel fuel injection timing is maintained at 15°CA BTDC, and the dual-fuel combustion mode pilot fuel timing is applied for 11°CA BTDC. The pilot fuel flow rate is observed to be around 0.5 L/h. The amount of pilot fuel and the CNG percentage are not increased, and the SOI is held constant at increased loads in the dual-fuel mode. By comparing the results, it is evident that the combustion pressure is much lower for the dual-fuel combustion mode, which implies a much lower peak pressure. In the dual-fuel combustion regime, the initial premixed combustion phase sometimes experiences a weak burning rate, which influences the rate at which the pressure rises, resulting in a lower combustion pressure compared to the single fuel conventional diesel engine mode. Fig. 6 represents the combustion pressures and HRRs for different CNG mixing ratios; this experiment was performed at 1500 rpm (7 bar BMEP). The data is plotted and then generalized for both modes. Injection timing is carried out in the same way for both the diesel fuel and pilot fuel at 4°CA BTDC. The in-cylinder pressure peak is plotted in Fig. 6, which shows a small peak pressure rise for 65% CNG substitution compared to the diesel mode. The pressure is decreased with increasing CNG%, but both CNG 65% and CNG 80% achieved higher peak pressures than the single-fuel diesel mode. Alternatively, the single-fuel diesel mode has a much higher HRR than any of the CNG substitution conditions. Increasing CNG, the HRR decreases gradually.

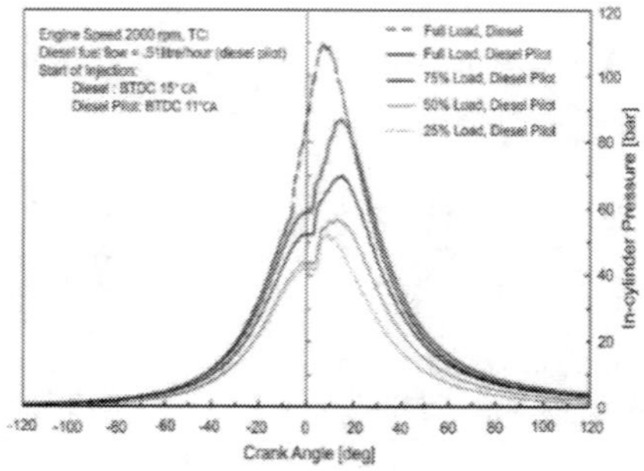

Figure 5: In-cylinder pressure comparison for diesel and dual-fuel combustion mode at increasing engine load.

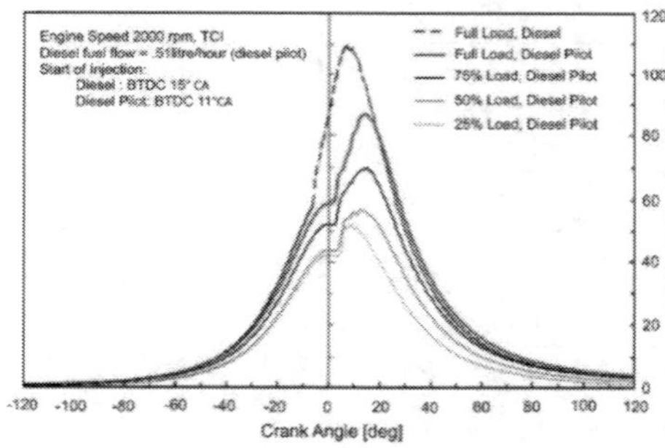

Figure 6: Comparison of in-cylinder pressure trace and heat release rate for diesel and different percentage natural gas mixed dual-fuel mode.

Fig. 7 shows the combustion duration and ignition delay for different CNG mixing operations with the diesel-fuel mode. These data indicate that the combustion phase is delayed with an increase in CNG mass ratio. The ignition delay is comparatively long, but the combustion duration is shorter for CNG 65%. The difference between CNG 65% and the diesel mode is 1.61 degrees. As the mixing percentage is further increased, the ignition delay is only marginally increased. The combustion duration is prolonged for both CNG 80% and CNG 90%. The reason behind this deterioration is likely caused by the smaller amount of pilot diesel fuel, which is insufficient for the combustion of CNG and the intake air mixture.

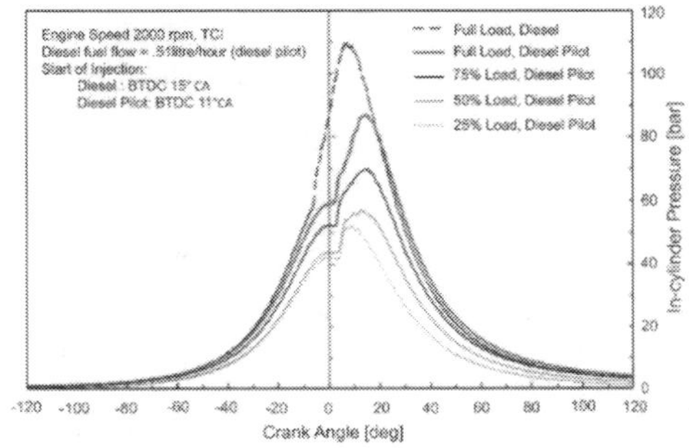

Figure 7: Ignition delay combustion duration for diesel and different percentage natural gas mixed dual-fuel mode.

3.3 Emission characteristics

Fig. 8 shows the comparison between the NOx emissions of both base and modified engine and the comparison between the NOx emissions of both base and modified engine. The NOx emission of the diesel test is less than 4 g/KWh in the experimental condition. However, in the CNG/diesel dual-fuel engine, the NOx emission is higher than the diesel test for the entire operating range, except at 1000 rpm. The auto ignition temperature of the CNG/diesel engine was 650 °C which is much higher compare to diesel engine (250 °C), resulting in a longer ignition delay. The ignition delay is determined from the measured injector needle lift, and the start of combustion is considered the minimum point of the cumulative heat release where the gross heat release rate becomes positive. Also, due to the presence of the homogeneous CNG premixture in the cylinder, combustion rapidly increases after ignition. Therefore, due to the rapid combustion which leads a rapid increase in pressure and temperature resulting higher NOx emissions in the dual-fuel engine. Previously, significant improvements in THC emissions, a slight reduction in CO emissions, and a considerable increase in NOx emissions were observed for the proposed engine. Fig. 9 shows the THC emissions of the diesel and dual-fuel engines in our proposed system. The THC emission of the diesel engine is less than 500 ppm throughout the operating range. However, more THC emissions are released by the modified dual-fuel engine because of the long ignition delay. With increases in engine speed and combustion temperature, THC emission decreases automatically.

Fig. 10 shows the CO emissions of the diesel and dual-fuel engines. The CO emissions appear to increase with increasing engine speed for both conditions. However, up to 2500 rpm, the CO emissions of the dual-fuel and conventional diesel engines do not fluctuate. Alternatively, for the remainder of the operating range, the CO value for CNG 90% is increased dramatically.

Fig. 11 shows the PM emissions of the diesel and dual-fuel engines. The PM emissions obtained from of the dual-fuel engine were remarkably lower than those obtained from the diesel engine because only diesel is used for ignition. Also, a local high temperature/rich region was not formed due to the absence of CNG premixture. The potential for PM-specific emission reduction in dual-fuel mode is corroborated in transient engine operation. Throughout the operating range, the PM concentration obtained with the dual-fuel modified system was 93% less than that of the base diesel engine. With increasing engine speeds, the data vary somewhat but still contain much smaller variations compared to the diesel mode.

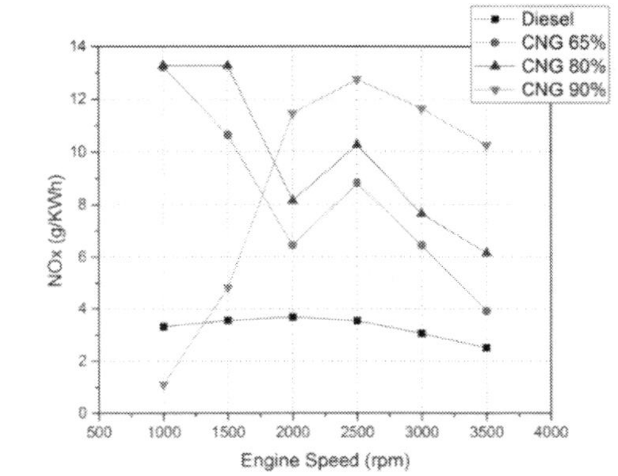

Figure 8: Comparison of the NOx emissions for diesel and different percentage natural gas mixed dual-fuel mode at increasing engine speeds.

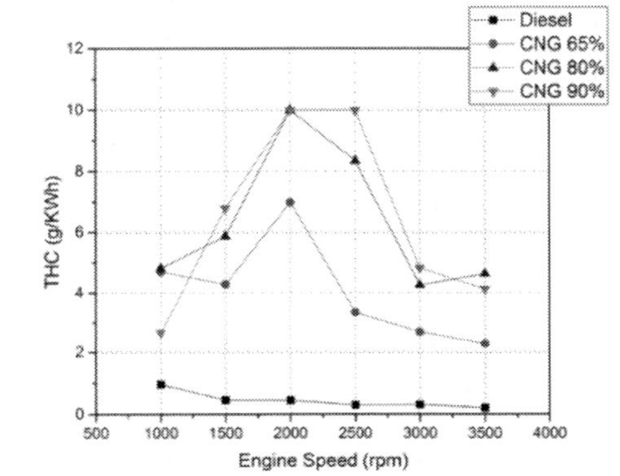

Figure 9: Comparison of the THC emissions for diesel and different percentage natural gas mixed dual-fuel mode at increasing engine speeds.

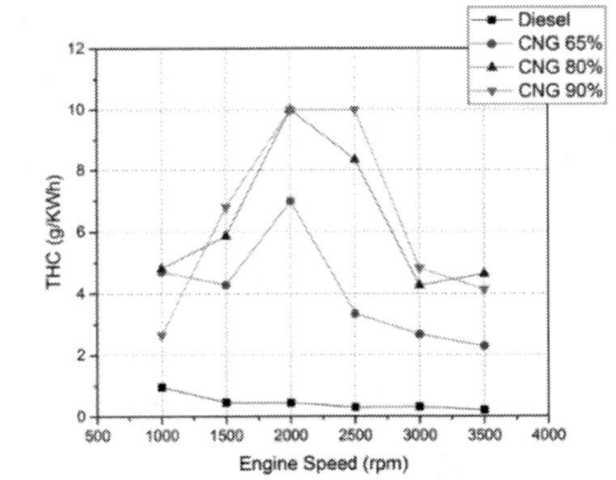

Figure 10: Comparison of the CO emissions for diesel and different percentage natural gas mixed dual-fuel mode at increasing engine speeds.

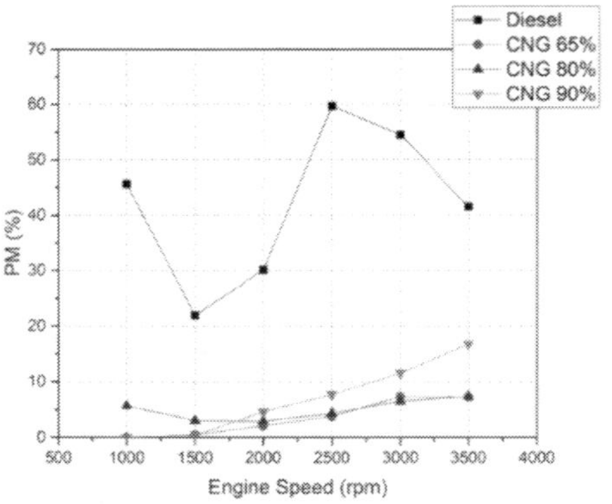

Figure 11: Comparison of the PM emissions for diesel and different percentage natural gas mixed dual-fuel mode at increasing engine speeds.

4 Conclusions

The results revealed that the torque and power output of the modified engine were significantly higher than the conventional diesel mode. Also, the diesel reduction rate was greater than 89% throughout the operating conditions of the dual-fuel engine. Particulate matter emissions of the CNG/diesel modified engine were 94% lower, but the NOx emissions were comparatively higher compared to only diesel fuel mode.

5 Acknowledgment

This research was financially supported by the Ministry of Education (MOE) and the National Research Foundation of Korea (NRF) through the Human Resource Training Project for Regional Innovation. This work was supported by the Industrial Strategic technology development program (10053151, Development of the 800 kPa Fuel System of a High Pressure Precision Control for NGV) funded By the Ministry of Trade, industry & Energy (MI, Korea). This research was financially supported by the CEFV(Center for Environmentally Friendly Vehicle) as Global-Top Project of KMOE(Ministry of Environment, Korea).

References

1. L. Wei and P. Geng, A review on natural gas/diesel dual fuel combustion, emissions and performance, Fuel Process. Technol. 142 (2016) 264–278.

2. R.G. Papagiannakis et al., Theoretical study of the effects of engine parameters on performance and emissions of a pilot ignited natural gas diesel engine, Energy 35 (2) (2010) 1129–1138.

3. R.G. Papagiannakis and D.T. Hountalas, Combustion and exhaust emission characteristics of a dual fuel compression ignition engine operated with pilot Diesel fuel and natural gas, Energy Convers. Manag. 45 (18–19) (2004) 2971–2987.

4. B. Douville et al., Performance and Emissions of a Two-Stroke Engine Fueled Using High-Pressure Direct Injection of Natural Gas, (1998).

5. G.P. McTaggart-Cowan et al., The Effects of High-Pressure Injection on a Compression Ignition, Direct Injection of Natural Gas Engine, J. Eng. Gas Turbines Power 129 (2) (2006) 579–588.

6. V. Aesoy and H. Valland, Hot Surface Assisted Compression Ignition of Natural Gas in a Direct Injection Diesel Engine, (1996).

7. V. Aesoy and H. Valland, The Influence of Natural Gas Composition on Ignition in a Direct Injection Gas Engine Using Hot Surface Assisted Compression Ignition, SAE Tech. Pap. Ser. 961934 (1996).

8. G.A. Karim, A review of combustion processes in the dual fuel engine—The gas diesel engine, Prog. Energy Combust. Sci. 6 (3) (1980) 277–285.

9. T. Lennox, A. Post, and M. Wickstone, Method and apparatus for delivering two fuels to a direct injection internal combustion engine, (2008).

10. K. Ryu, Effects of pilot injection timing on the combustion and emissions characteristics in a diesel engine using biodiesel–CNG dual fuel, Appl. Energy 111 (2013) 721–730.

11. K.K. Srinivasan, S.R. Krishnan, and K.C. Midkiff, Improving low load combustion, stability, and emissions in pilot-ignited natural gas engines, Proc. Inst. Mech. Eng. Part D J. Automob. Eng. 220 (2) (2006) 229–239.

12. G.H. Abd Alla et al., Effect of pilot fuel quantity on the performance of a dual fuel engine, Energy Convers. Manag. 41 (6) (2000) 559–572.

13. C. Sayin and M. Canakci, Effects of injection timing on the engine performance and exhaust emissions of a dual-fuel diesel engine, Energy Convers. Manag. 50 (1) (2009) 203–213.

14. R.G. Papagiannakis, D.T. Hountalas, and C.D. Rakopoulos, Theoretical study of the effects of pilot fuel quantity and its injection timing on the performance and emissions of a dual fuel diesel engine, Energy Convers. Manag. 48 (11) (2007) 2951–2961.

15. M.M. Abdelaal and A.H. Hegab, Combustion and emission characteristics of a natural gasfueled diesel engine with EGR, Energy Convers. Manag. 64 (2012) 301–312.

16. V. Pirouzpanah and R.K. Sarai, Reduction of emissions in an automotive direct injection diesel engine dual-fuelled with natural gas by using variable exhaust gas recirculation, Proc. Inst. Mech. Eng. Part D J. Automob. Eng. 217(8) (2003) 719–725.

17. V. Pirouzpanah et al., Comparison of thermal and radical effects of EGR gases on combustion process in dual fuel engines at part loads, Energy Convers. Manag. 48 (7) (2007) 1909–1918.

18. R.G. Papagiannakis, Study of air inlet preheating and EGR impacts for improving the operation of compression ignition engine running under dual fuel mode, Energy Convers. Manag. 68 (2013) 40–53.

11th Asia-Pacific Conference on Combustion,
The University of Sydney, NSW Australia
10th -14th December 2017.

Understanding the effect of high environmental pressure on supercritical jet

Liang Li[1], Maozhao Xie[1], Ming Jia[1], Wu Wei[1]

[1] School of Energy and Power Engineering, Dalian University of Technology

Ling Gong Road on the 2nd, Dalian City, Liaoning Province, 116024, China

Abstract

Understanding and predicting fuel spray characteristics under trans/supercritical conditions is crucial to the design of rocket and diesel engines. In this paper, the effects of different supercritical environmental pressure on the thermodynamics and flow characteristics of a cryogenic liquid jet are numerically simulated and analyzed. In particular, we discussed the development and differences of the mixing layer on the liquid jet surface. Results demonstrate that, under a ultrahigh pressure, the pseudo-boiling behavior of the jet is weakened, and the intensity of thermal diffusion increase, also the length of the cold core is shorter. However, in the downstream of the cold core, the thickness of the mixing layer is increased due to the reduction of the damping introduced by density gradient. An enthalpy analysis of the mixing layer indicates that under ultrahigh pressure conditions the cryogenic liquid absorbs more energy from the surrounding high temperature environment, and more energy is used for the temperature rise, while the energy used to volume expansion is reduced. All of the above features make the high pressure jets enter the self-similar state earlier. Therefore, it is expected that a high chamber pressure contributes to improving the utilization of the fuel for supercritical combustion.

1 Introduction

For a long time, due to the requirements of hypersonic velocity, the study of transcritical and supercritical jets is mainly concentrated in the field of liquid rocket engines[1]. And, in the aspect of numerical simulation, Zong[2] simulated the mixing process of liquid nitrogen injected into a supercritical chamber by means of LES. The results showed that there was a large density gradient at the surface of the jet, which plays a role similar to a "solid wall", enhancing the axial oscillation of the jet and weakening the radial oscillation. Also, as the supercritical pressure increases, the length of the liquid core decreases and the instability of the mixing layer increases. Dahms et al[3] confirmed that there is also a continuous gas-liquid mixed layer under high-pressure diesel engine conditions. Moreover, the non-equilibrium of the gas-liquid interface is also described. The analysis reveals that the development of the mixing diffusion layer was not necessarily because of the weakening of the surface tension, but because of the combination of a reduction in mean free molecular path and broadening interfaces, and is dominated by the transport processes. Kima[4]and Park[5],using RANS models and based on the Mayer experiments[6], compared the effects of different equations of state and turbulence models. The authors pointed out that the suitable adoption of real-fluid equation of state is more significant than the selection of turbulence model.

Although there are many studies on supercritical jets, most of them analyzed the fluid physical properties and structure of supercritical jet. However, detailed investigations on the effects of different supercritical pressures on the jet structure, particularly on the mixing layer and its thickness are still rare. Here, in this paper we applied a real fluid equation of state and the related theories of computational fluid dynamics and thermodynamics to address these problems in the framework of OpenFOAM.

Maozhao Xie . Fax: +00-86-0411-84708460
E-mail address: xmz@dlut.edu.cn

2 Theoretical formulation

Based on the OpenFOAM software package, a CFD solver is created based on the theory of real fluid equation of state, thermodynamics and computational fluid dynamics.

As pointed out by the Kima[4]and Park[5], the selection of turbulence model is not as significant as the equation of state. Also in order to reduce the cost of calculation, in the code, the RANS k- ε model is used to close the turbulence. And the classical PISO algorithm is used to couple the iterative calculation of pressure and velocity.

2.1 Governing equations

$$\frac{\partial\rho}{\partial t}+\nabla\cdot(\rho U)=0 \quad (1)$$

$$\frac{\partial(\rho U)}{\partial t}+\nabla\cdot(\rho u\otimes U)=-\nabla P+\nabla\cdot(\mu_{eff}[(\nabla U+\nabla U^T)-\frac{2}{3}\mu_{eff}(\nabla\cdot U)I]) \quad (2)$$

$$\frac{\partial(\rho h)}{\partial t}+\nabla\cdot(\rho hU)=\nabla\cdot(\alpha_{eff}\nabla h)+\frac{Dp}{Dt} \quad (3)$$

where ρ , u_i , t , P, h , μ_{eff} , α_{eff}, C_p represent the density, velocity component in direction i, time, pressure, enthalpy, effective thermal diffusivity, effective dynamic viscosity and constant-pressure specific heat respectively. The superscript T represents the transpose of the matrix. I is the unit tensor.

2.2 Compressible pressure correction equation

$$\frac{\partial\rho_{n-1}}{\partial t}+\varphi\frac{\partial(P_n-P_{n-1})}{\partial t}+\nabla\cdot(\frac{\rho}{A_p}(H(U))-\nabla\cdot(\frac{\rho}{A_p}\nabla P)=0 \quad (4)$$

The subscript n-1 represents the previous time step, and n represents the current time step; A_P is the current mesh coefficient matrix; A_N is the adjacent grid coefficient matrix; φ is a compressibility coefficient, the detailed expression is as follows.

$$\varphi=\frac{M_w(M_w-b\rho)^2(M_w+b\rho)^2}{M_w^2R_uT(M_w+b\rho)^2-a\alpha\rho(2M_w+b\rho)(M_w-b\rho)^2} \quad (5)$$

where M_w, R_u, T, represent the molar mass of substance, general gas constant and temperature. a, α, b represent the coefficient functions in the SRK equation[7]. Here, the density ρ is given by an iterative calculation of the SRK equation.

2.3 Real-fluid thermodynamic properties

All thermodynamic properties are calculated by the equation of SRK. For closure of the system of the conservation equations, the constant-pressure specific heat Cp has to be provided[7].

$$C_p(T,\rho) = C_{p0}(T) - \int_{\rho_0}^{\rho}\left[\frac{T}{\rho^2}\left(\frac{\partial p}{\partial T}\right)_\rho\right]_T d\rho + \frac{T}{\rho^2}\frac{(\frac{\partial P}{\partial T})_\rho^2}{(\frac{\partial p}{\partial \rho})_T} \qquad (6)$$

$C_{p0}(T)$ is the ideal reference value and can be solved by the TRC (Thermodynamics Research Center) polynomial equation(http://treweb.tamu.edu).

Thermal conductivity and dynamic viscosity have been modeled by the empirical formula of Zébergmikkelsen[8] and Vasserman et al[9].

2.4 Geometric model and boundary conditions

The geometric model of concern follows the experiments described in references[6]. Cryogenic nitrogen fluid is injected into supercritical gaseous nitrogen through a circular duct with an inner diameter of 2.2 mm. In this study, the mixing chamber is an axisymmetric cylinder with a diameter of 122 mm and a length of 500 mm to ensure a sufficient distance for the far field. Due to the enormous computational cost for calculating the entire three-dimensional flow field, in the present simulation, only an axisymmetric computational domain is treated.

Based on the grid-independent research of Park[5],the nozzle is divided into 20 grids along the diameter, and the minimum size of the grid is 0.1mm. The chamber is divided into 120 grids in the radial direction and 200 in the axial direction. And details of the initial and boundary conditions are shown in Table 1 and Reference[5].

Case	V_{inj} [m/s]	T_{inj} [k]	P_∞ [MPa]	T_∞[k]
1	5.72	126.9	4	298
2	5.47	128.7	6	298
3	5.72	126.9	9	298

Table 1: Initial and boundary conditions.

3 Results

3.1 Supercritical jet analysis

Figure 1 presents the axial distributions of the normalized density, temperature, and constant-pressure specific heat along the centerline for three different pressures. Obviously, there is a cold core near the nozzle by observing the normalized density and temperature curves, and the length of the cold core decreases gradually as the chamber supercritical pressure increases. The cold core length is about 7D, 6D and 5D in cases 1, 2 and 3, respectively. That is the higher the chamber pressure, the earlier the starting position of the density drop. Moreover, with the increase in the chamber supercritical pressure, the temperature rise along the centerline is relatively faster.

The constant-pressure specific heat curve shows a rapid increase at the end of the cold core, and the lower the chamber pressure, its increase is more intensive, this is induced by the pseudo-boiling behavior(129.8K at 4 MPa; 139.1K at 6 MPa; 148.3 k at 9 MPa). Moreover, it can be clearly seen that the peak and non-linear amplitudes of the constant-pressure specific heat curve decrease as the chamber supercritical pressure increases, resulting in a decrease in the degree of pseudo-boiling, that is the thermal conductivity increases, eventually lead to the rising trend of the normalized temperature along the centerline.

For further verification and comparison, the figure also gives the simulation results of Park[5] under the supercritical pressure 6MPa. In the cold core region, the current simulation and Park simulation show a similar result. However, the Tae simulation shows a faster density drop and temperature rise in the downstream of the cold core.

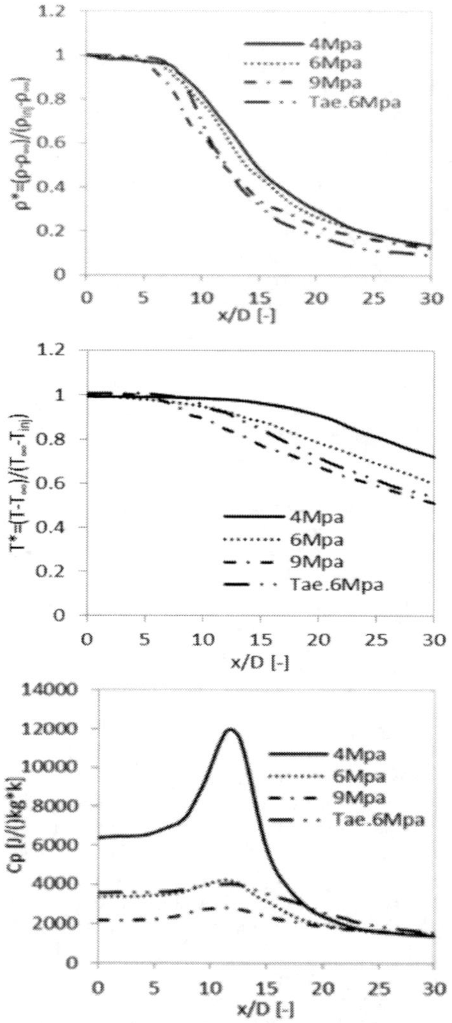

Fig. 1 Thermodynamic profiles at the centerline under different pressures

3.2 Mixing layer analysis

According to Oefelein[10] and Yang et.al[2], the density gradient has a critical effect on the radial diffusion of the fluid in the spray and jet. It can be seen from Figure 2, no matter at low or high supercritical pressures, that the density gradient in the mixing layer has a peak, and as the jet continues downstream in the axial direction, the peak drops gradually and moves to the center line. And the curve gradually developed from a steep shape to a horizontal one. This shows that the density difference between the fluids inside and outside the mixing layer has been very limited at the far downstream of the jet, the fluid properties in the area near the centerline and those in the surrounding far-field tend to be the same, suggesting that the jet enters the self-similar state.

At the position x=5D, the three cases are all in the cold core region, and each has a higher density gradient peak. This high density gradient produces a "solid wall effect" that limits the radial turbulent diffusion and the development of the mixing layer. Therefore, it can be considered that there is still a widened gas-liquid interface on the surface of the cold core near the nozzle, and the peak point of the density gradient can be assumed to be the interface point. Also find that the higher the chamber pressure, the higher the density gradient in the mixing layer of the cold core. Thus, the "solid wall effect" is stronger under the higher supercritical pressure, and then the thickness of the mixing layer is lower.

However, in the downstream of the cold core, the higher the chamber pressure, the faster the density gradient drops, the lower the peak, so that the increase of the mixing layer thickness is faster.

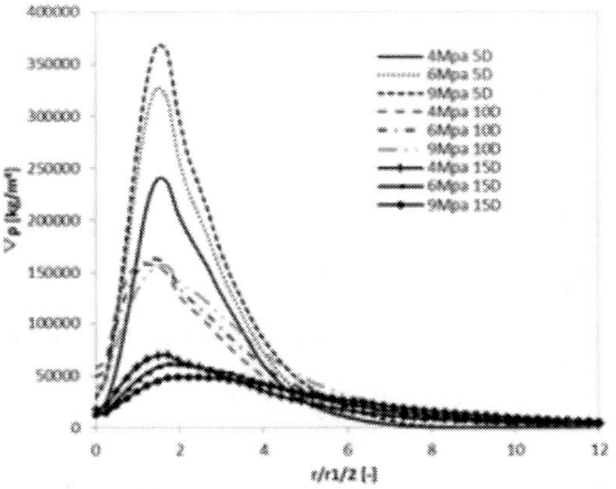

Fig.2 Radial profiles of density gradient at different axial positions

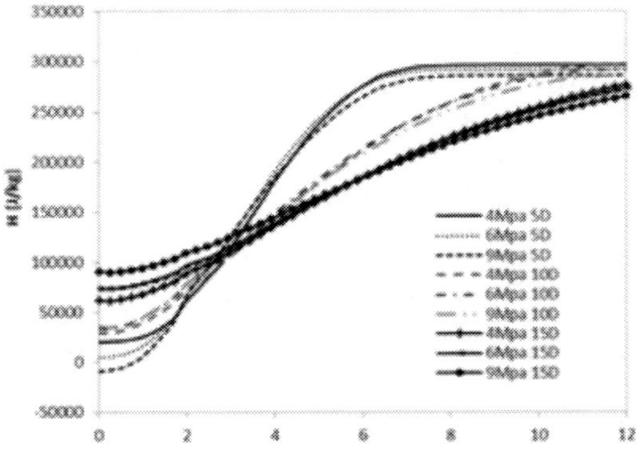

Fig.3 Radial profiles of enthalpy at different axial positions

Figure 3 compares the enthalpy distribution over the mixing layer for three different pressures. According to Banuti [11], for the supercritical jet, the thermal disintegration mechanism in the mixing layer is directly related to the initial energy level of the jet. For all three cases, the initial enthalpy is lower than the enthalpy corresponding to the pseudo-boiling point(34865 J/kg), then upon entering the chamber, the jet experiences firstly a temperature rising and then transits across the pseudo-boiling point, producing the large density gradient. So the mixing layer of the jets is affected by both the thermal disintegration mechanism and the mechanical disintegration mechanism.

Comparing the enthalpy curves for the three different pressures at the same locations, we can find that the slop of the case 3 at the position 5D is larger, while at the far downstream of the jet its slop is smaller. In other words, the mixing layer of high pressure jet has a greater energy gradient as well as greater density gradient in the cold core region; hence, it is easier to produce a fluid stratification, which results in a thin thickness of the mixing layer. However, at the downstream of the cold core it is opposite.

Also, observing the increase of enthalpy at the center-line from 5D to 15D, we can find that the increased value of the enthalpy for case 3 (9MPa) is about 2.5 times of that in case 1 (4MPa). This indicates that the mixing layer of high pressure supercritical jet is more likely to absorb energy from the surrounding warm gas. These absorbed energies are used to the development of mixing layer, resulting in a wider thickness.

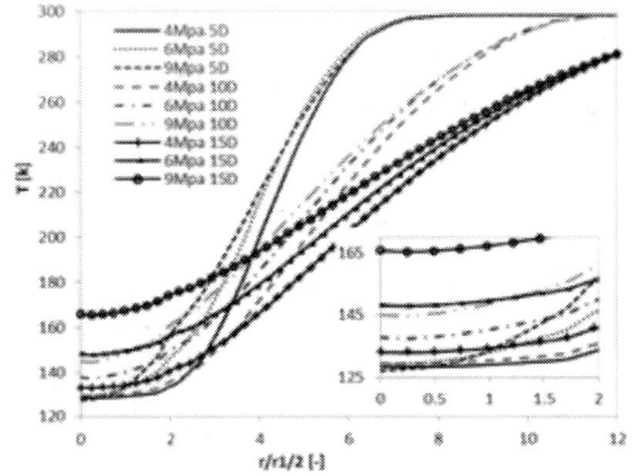

Fig.4 Radial profiles of temperature at different axial positions

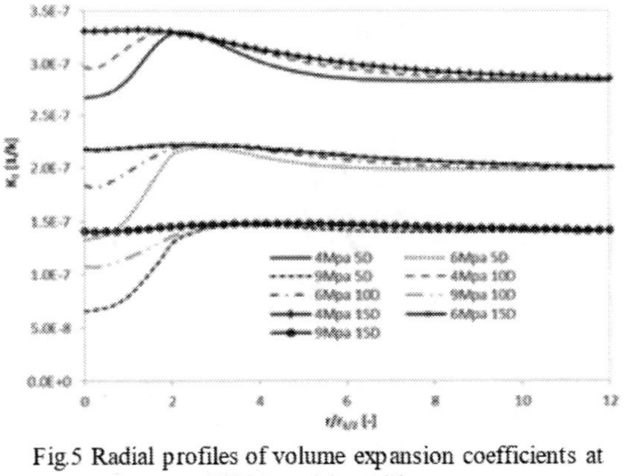

Fig.5 Radial profiles of volume expansion coefficients at different axial positions

It can be seen from figure 2 that the energy the cold liquid absorbs from the surrounding warm gas is mainly used to rise the temperature and expand volume at the pseudo boiling. In order to explore the effect of chamber pressure on temperature rise and volume expansion, figures 4 and 5 shows the radial distributions of the temperature and volume expansion coefficients at different axial positions, respectively. The volume expansion coefficient is defined as follows : $k_T = -\frac{1}{v}(\frac{\partial v}{\partial P})_T$

For the cold core area within 5D, the temperature rise in the mixing layer must cross the pseudo-boiling point. As mentioned above, the higher the chamber pressure, the weaker the pseudo-boiling behavior. Thus, under the same energy absorption, more energy is used to raise the temperature under the high supercritical chamber pressure, and the proportion of the energy in the isothermal expansion decreases. Moreover, from figure 3, the cryogenic liquid core has a lower energy value under high pressure, and therefore, the proportion of energy used to temperature rise is greater. This is also shown in figure 4 and figure 5, at the position of 5D, with the increase of the chamber pressure, the temperature rapidly rising stage starts earlier in the radial direction, and the rise rate gradually increased. However, the volume expansion coefficient is gradually reduced. Therefore, it can be considered that the change of the density gradient in the mixing layer of the cold core surface under high chamber pressure is greatly influenced by the temperature rise.

In the downstream of the cold core, since the pseudo-boiling in the mixing layer disappears, the absorbed energy by the cold liquid from the surroundings is rapidly increased under high supercritical chamber pressure; so there still has a faster temperature rise. Also, as is shown in figure 5 that the volume expansion coefficient of case 3 (9MPa) is substantially horizontal at the position 15D, which further indicates that the high pressure supercritical jet enters the self-similar state earlier in the axial direction.

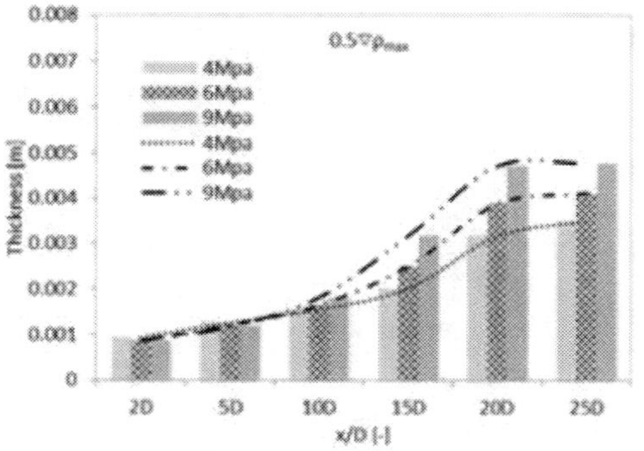

Fig.6 Thickness of the mixing layer determined by $0.5\nabla\rho_{max}$

In order to intuitively display the relationship between the thickness of the mixing layer and the chamber pressure, a cutoff value, which is 0.5 times of the maximum absolute value of the density gradient in the radial mixing layer, is used to determine the thickness of the mixing layer. Form the figure 6, for a single pressure condition, it can be observed that the thickness of the radial mixing layer is increasing along the axial direction of the jet.

However, there are significant differences for different supercritical pressure conditions. Firstly, in the cold core region(x/D<5D), the calculated thickness of the mixing layer shows a slight decrease with the increased chamber pressure, when the two cutoff values are applied. This corresponds to the larger density gradient of the cold core surface under the higher pressure conditions, which is shown in figure 2. That is, the greater the density gradient, the stronger the "solid wall effect", the less the mixing layer. Secondly, in the downstream of the cold core, the thickness of the mixed layer increases sharply under the high supercritical pressure. Also, it can be seen that the higher of the supercritical chamber pressure, the earlier of the position appears where the thickness of the mixing layer increase nonlinearly. This is closely related to the lower density gradient value and the higher energy absorption capacity of the mixing layer downstream of the cold core under the high supercritical pressure.

4 Conclusions

The main conclusions are the following.

1. As the supercritical pressure increases, the pseudo-boiling behavior weakens and the intensity of turbulent mixing and thermal diffusion increases, resulting in the jet produces a faster temperature rise and the length of the cold core decrease.

2. The thickness of the mixing layer under the high supercritical pressure is slightly reduced in the cold core region, which has a higher density gradient. While in the downstream of the cold core, the density gradient within the mixing layer has a faster rate of descent, and the damping of the solid wall effect is reduced. That causes a higher thickness of the mixing layer.

3. From the view of energy, a cryogenic liquid "absorbs" energy faster, which is caused by turbulent mixing, from the surrounding high temperature environment under higher supercritical pressure conditions. Coupled with the reduction of pseudo-boiling behavior under higher pressure conditions, so more energy is used for temperature rise, while the energy proportion of volume expansion is reduced.

5 Acknowledgment

This work is supported by National Natural Science Foundation of China (grant number: 51376029).

References

[1] V. Yang, Proc. Combust. Inst. 28(1) (2008) 925-942.

[2] N. Zong, H. Meng, S.Y. Hsieh, V. Yang, Physics of Fluids 16(12) (2004) 4248-4261.

[3] R.N. Dahms, J.C. Oefelein, Proceedings of the Combustion Institute 35(4) (2015) 1587–1594.

[4] T. Kima, Y. Kima, S.K. Kimb, Journal of Supercritical Fluids 56(2) (2011) 152-163.

[5] T.S. Park, Journal of Supercritical Fluids 72(12) (2012) 232-247.

[6] W. Mayer, J. Telaar, R. Branam, Heat and Mass Transfer 39 (8-9) (2003) 709–719.

[7] S.K. Kim, H.S. Choi, Y. Kim, Combustion & Flame 159(3) (2012) 1351-1365.

[8] C.K. Zébergmikkelsen, S.E. QuinonesCisneros, E.H. Stenby, Industrial & Engineering Chemistry Research 40(17) (2001) 3848-3854.

[9] A.A. Vasserman, V.I. Nedostup, Journal of Engineering Physics and Thermophysics 20(1) (1971) 89-92.

[10] J.C. Oefelein, R.N. Dahms, G. Lacaze, SAE International Journal of Engines 5 (3) (2012) 1410—1419.

[11] D.T. Banuti, K. Hannemann, PHYSICS OF FLUIDS, 28(3) (2016) 101-353.

11th Asia-Pacific Conference on Combustion,
The University of Sydney, NSW Australia
10th -14th December 2017.

Outwardly Propagating Spherical Flame with Cellular Instability and Laminar Burning Velocities in Methane/ethylene/air Premixed Flames

K. H. Van[1], H. J. Kim[1], J. Park[2†], Oh Boog Kwon[1], Dae Keun Lee[2], Seung Gon Kim[2], Young Tea Guahk[2], Dong-Soon Noh[2], S. H. Chung[3]

[1]Dept. of Mechanical Engineering, Pukyong National University, Busn, Korea.

[2]Advanced Combustion Lab, Korea Institute of Energy Research, Daejeon, Korea.

[3]Clean Combustion Research Center, King Abdullah University of Science and Technology, Saudi Arabia

Abstract

Experiments in expanding spherical flame were carried out to investigate unstretched laminar burning velocities by adding ethylene mole fractions of 0.1, 0.2, and 0.3 to methane-air mixtures at room temperature and elevated pressures up to 4 atm. Our experimental methodology was validated through comparison of a variety of flame radius ranges from previous experimental studies and three kinds of extrapolation model with numerical ones in methane-air premixed flames. Based on them, also the wrinkling of methane/air with increase of ethylene addition is examined experimentally and theoretically under increased pressure at various equivalence ratios.

1 Introduction

Lean-burn which refers to the burning of lean fuel with excess of air must be bright technology to solve the problems on serious depletion fossil fuel and environmental regulation of air pollutant emissions (low nitric oxide emissions) in the current situation. Natural gas with methane being its major component, has been widely used to solve global-warning problem. However, there still remain some assignment for solution such as fragmentary heat losses to reduce laminar burning velocities and thermal efficiency as well as poor lean-burn capability [1]. In this situation, the comprehensive study has been conducted to solve these problems via using blended fuels on methane or hydrogen [2]. In interaction between lean premixed methane-air and hydrogen-air flames, lean flammable limits were extended significantly while the extinction behaviors were influenced complicatedly by the existence of hydrogen, e.g. preferential diffusion of H and H_2 as well as complicated chemical effects of sharing the radicals such as H, O, and OH on extinction behavior [2]. In CH_4/H_2-air laminar premixed flames, lean flammable limit was extended greatly with enrichment of H_2, resulting in significant reduction of NO formation. The effects of additional reformate gas reduced NO formation greatly [3].

Ethylene is the primary olefin feedstock of the modern organic chemical industry and is derived massively in the United States from the advanced extraction technology with steam cracking [4]. Also, ethylene as a key intermediate in the oxidation of higher alkanes has some merits: ignition temperature is lower than that of methane. In this situation, some research efforts have been devoted to grasp fundamental combustion characteristics in ethylene flames, e.g. ignition characteristics [5] and unstretched laminar flame speed [6]. However, those in CH_4/H_2-air flames, have been seldom found in the literature. Considering the feasibility of using methane/ethylene mixtures as a fuel to secure the fuel flexibility particularly in gas turbine combustor, it may be required to extensively study fundamental combustion characteristics in CH_4/H_2-air flames. In the present study, experimental study was conducted to investigate unstretched laminar burning velocities, and cellular instabilities by varying fuel composition at room temperature and elevated pressure (up to 4 atm) in outwardly propagating spherical CH_4/C_2H_4-air premixed flames.

2 Experimental Facility

In the present study, the experiments were conducted in stainless, cylindrical constant-volume chamber with 200 and 220 mm in an inside diameter and a length as shown schematically in Fig. 1. Two quartz windows with 100 mm in diameter and 40 mm in thickness were mounted oppositely on both flat sides of the chamber for visual access. Two movable, opposed tungsten electrodes diameter were connected to a high

Corresponding author. Fax: +00-00-000-000
E-mail address: abc@abcde.com

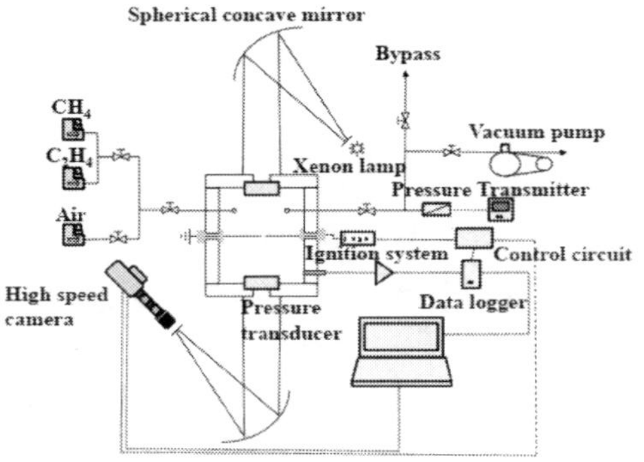

Figure 1. Schemaric diagram of the experimental setup.

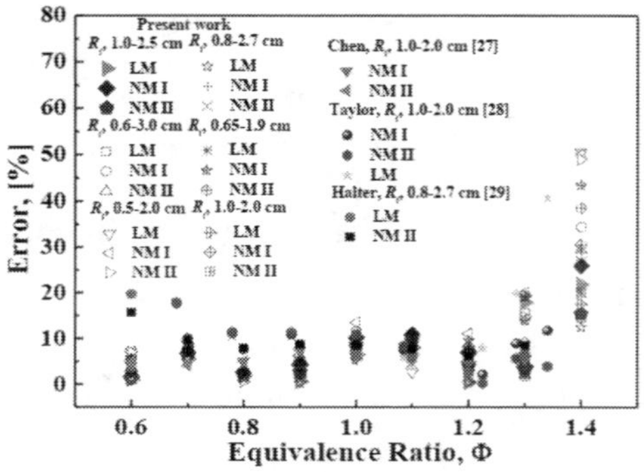

Figure 2. Error with experimental results as a function of equivalence ratio in CH4-air at Pu = 1 atm.

voltage source to ignite the mixture at the center of the chamber. Purity of methane and ethylene is 99.99 % and air with that of 99.95 %, respectively. Fuels and air were supplied to the inside chamber such that the partial pressure of each reactant was adjusted by a pressure transmitter to achieve the desired initial pressure. To ensure complete mixing of supplied reactants to the chamber and their quiescence, the experiments was conducted after 15 minutes elapsed. After the flame propagation was completed, the chamber was ventilated to the laboratory exhaust system and purged with an air compressor to get rid of condensed water for the next experiment. The outwardly propagating spherical flame was visualized using a Schlieren photography with a xenon light source and a pair of concave mirrors, and taken with framing rate of 10,000 fps by a high–speed camera. The edge of the flame front was determined by a Matlab-based code and then converted to flame radius. Operating the high-speed camera and measuring the instantaneous chamber pressure were triggered by the signal of the signal of the spark ignition source.

3 Results and Discussion

The error in S_u^0 due to discrepancies in equivalence ratio, mixture preparation, buoyancy, ignition transition, radiation-induced reduction and effect of confinement [7, 8]. Resultantly, to reduce the discrepancy of uncertainty on flame speed, it may be required to determine the range of flame radius for extrapolation model. Three extrapolation linear and nonlinear models (LM, NM I, and NM II) in the followings were investigated:

$$LM:\ S_b^0 - S_b = L_b K \tag{1}$$

$$NM\ I: S_b = S_b^0 - S_b^0 L_b \times \left(\frac{2}{R_f}\right) \tag{2}$$

$$NM\ II: ln(S_b) = \ln(S_b^0) - S_b^0 L_b \times \left(\frac{2}{R_f S_b}\right) \tag{3}$$

The LM can be used for the spherical flame thickness assumed to be infinitesimally thin, weakly stretched and The NM I can be also valid for outwardly spherical flame propagating along with thermal expansion at large flame radius [9, 10]. Then the L_b and S_b^0 can be determined by the linear relationship of S_b and 2/Rf, and such a nonlinear stretch effect in Su0 was analyzed by Frank and Sivashinsky [11]. The NM II was derived for quasisteady, adiabatic flame condition using an asymptotic method [12, 13], and its capability in predicting Su0 was investigated by the previous studies [14]. In Fig. 2, the error was also defined the difference of Su0 measured by varying Rf for the three extrapolation models from that calculated with GRI v 3.0. The error with LM in increase of ϕ increased in fuel-rich conditions. Resultantly, the results indicated that the error in Su0 measured with NM II was reduced significantly even at fuel-rich conditions. However, at fuel-rich conditions, further investigation could be required to clarify the sensitivity of flame radius to Su0. The current study also investigated them with several flame radii and the three extrapolation models (LM and NMs I and II). Similarly to the above results [15, 16], the error from all ranges of taken differently is not sensitive so much at 0.6 , 1.0. Resultantly, Su0 is best-fitted when the NM II is used at 10-25 mm. In the present study, the NM II and 10-25 mm were adopted for further study of CH_4/C_2H_4/air premixed flames. In the foregoing section, the extrapolation method and the range of flame radius in measuring S_u^0 were evaluated in methane-air premixed flame because of lack of experimental data on Su0 in CH_4/C_2H_4/air premixed flames. Using the NM II and 10-25 mm taken in the previous section, Su0 was measured for various ambient pressures and equivalence ratios in CH_4/C_2H_4/air premixed flames. Three Detailed chemistries (GRI v-3.0 [17], USC Mech II [18] and Sung Mech [19]) were compared to attain the priority in CH_4/C_2H_4-air flames. Figure 3 (a) compares experimental and numerical unstretched laminar at various equivalence ratios in CH_4/C_2H_4/air premixed flames. Numerical results with the three detailed chemisties were calculated with PREMIX code [20]. The USC mech II in Fig. 6 is evenly distributed near linear line regardless of equivalence ratio and additional ethylene volume fraction to mixture, whereas in cases of GRI v 3.0, the discrepancy between experimental and numerical burning velocity increase with increasing additional ethylene volume fraction. Sung mechanism exhibits stark difference particularly at 1 atm. The average error in simulation value with USC mech II, in comparison with experimental data, is measured to 4.94 %. Sung mech is rather more matched to experimental flame speeds with increasing initial pressure. The error with GRI v–3.0 also increases with increasing ethylene volume fraction regardless of initial pressures. Resultantly, USC mech II describes 1-D expanding spherical flame speeds well in methane/ethylene-air premixed flames, in that the average error of simulation flame speeds with USC mech II is less than 6.63 %. The cellular instability in premixed flames results in local flame acceleration due to buoyancy-induced, hydrodynamic, and diffusive-thermal effects. The hydrodynamic instability is well represented by the baroclinic torque term in vorticity equation. It is related to the density jump between burned and unburned mixture which is represend as $\sigma = (\rho_u / \rho_b)$ and characteristic

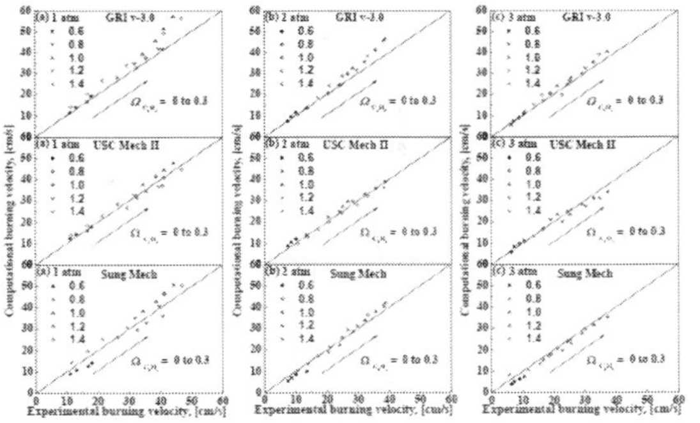

Figure 3. Comparison of Laminar burning velocities between eaperiment and calculation with cariation of detiled mehanims at 1 (a), 2 (b), and 3 (c) atm.

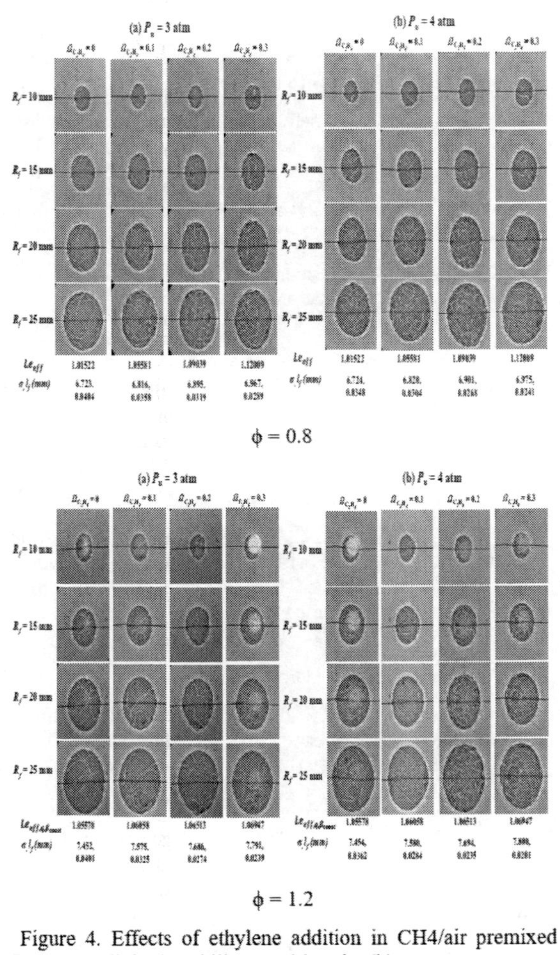

Figure 4. Effects of ethylene addition in CH4/air premixed flame on cellular instability at 3 (a) and 4 (b) atm.

flame thickness is defined as lf = (λ/C_p)/ρuSu0.Another factor is the diffusive-thermal instability attributed to unbalance between mass and thermal diffusivity. Figure 4 shows sequential images of expanding spherical flames at 2, 3, and 4 atm at various ethylene volume fractions. Effective Lewis number, thermal expansion ratio, and flame thickness are indicated at the bottom of each photos. At 3 and 4 atm, the flame surface remains smooth at early stages, and then some cracks occur and branch until cells

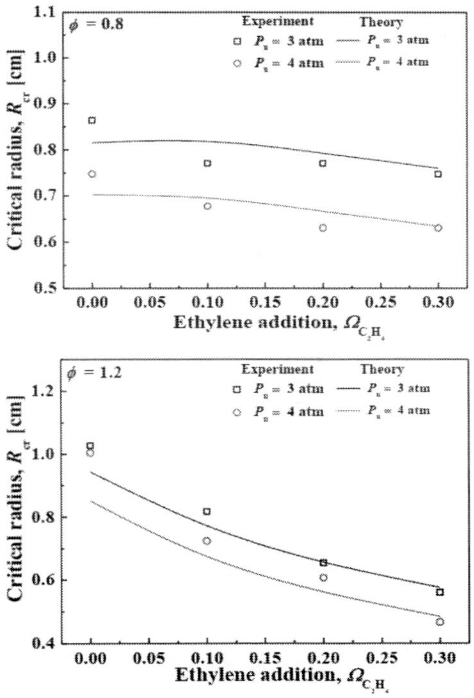

Figure 5. Mesured and theoretical critical radius with various mxing ratio and elevated pressure.

become uniform over the entire flame surface, even if the effective Lewis numbers is larger than unity. Therefore, propensity to destabilize the flame was not affected by the diffusive-thermal instability but by the hydrodynamic instability which relates to the thermal expansion ratio, and the flame thickness. This instability would be promoted because of the significant decrease in flame thickness and increase in thermal expansion ratio with increasing initial pressure and ethylene volume fraction. However, the effect of additional ethylene volume fraction is less influential in cells formation at the same pressure Thus, further onset of cellular instabilities was investigated through evaluating the critical Peclet number. And also the instability due to the substantial decrease of the flame thickness and increase of the thermal expansion ratio. Theory applied here accounts for both hydrodynamic and diffusive-thermal instabilities which are amalgamating the dependency of temperature transport coefficient. A theoretical analysis of the onset of instability in a spherically expanding flame was represented in [21, 22], where a hydrodynamic instability with Markstein correction was considered. The disturbed flame front can be written in the form r = R_f(t)[1 + A(t)S_n(θ, ϕ)] [32], where A(t) is the amplitude of the disturbance and Sn is the spherical surface harmonics with n as an integer. As the flame is expanded, the rate of growth of the relative amplitude of a disturbance of wavenumber n is given as follows in eq (4). The coefficient depends on the thermal expansion ratio, the Prandtl number *Pr*, and the wavenumber n. Definitions for ω_{DL} and Ω are given in [21, 22]. Figure 5 compares experimental and theoretical critical flame radii against ethylene volume. The experimental critical radius is defined at the moment when the burning velocity increased appreciably and lost the linear relationship between Sb and K, and theoretical one

$$\frac{1}{A}\frac{dA}{dt} = \frac{\dot{R}_f}{R_f}\left\{\omega_{DL} - \frac{l_f}{R_f}\Omega\right\} \qquad (4)$$

is shown in eq (4). As shown in Fig. 5, in both lean and rich conditions, experimental and theoretical critical radii decrease with increase in additional ethylene volume fraction and initial pressure. This means that the flame instabilities were promoted due to increase in thermal expansion ratio and decrease in flame thickness. However, in case of fuel-lean conditions, both experimental and theoretical critical radii slightly decreases and then have nearly constant with additional ethylene volume fraction regardless of decrease in flame thickness and increase in thermal expansion ratio. It was noted increasing effective Lewis number results in slight decrease in flame radius, despite reduction of flame thickness and increase in thermal expansion ratio. Consequently, measured and theoretical critical flame radii shows excellent agreement.

4 Conclusions

To experimentally validate effects of ethylene addition on methane-air premixed laminar burning velocities and cellular instability of expanding spherical flames, the most reliable detailed reaction mechanism to CH_4/C_2H_4-air premixed flame is USC Mech II after determination to the range of flame radius (10-25 mm) and extrapolation model (NM II) via comparison of between measured and predicted laminar burning velocities. The Markstein length has always positive value, because of effective Lewis number is larger than unit in all cases. Based on them, When corrugated flames represented by cellular instabilities was occurred in high initial pressures, the hydrodynamic instability was enhanced due to the increases in thermal expansion ratio and the decreases in flame thickness. Cellular instability was enhanced in increase of chamber pressure and ethylene volume fraction, i.e., the critical flame radius for onset of cellular instability could be reduced. Matalon's theory described measured critical flame radii for onset of cellular instability qualitatively well. However, the theoretical critical flame radii overpredicted roughly 1.5 times experimental ones.

5 Acknowledgment

This research was supported by SGER Program through the National Research Foundation of Korea (NRF) founded by the Ministry of Education.

References

[1] Shy SS, Chen YC, Liu CC, Huang CM (2008). Experimental and numerical investigation of the effect of H2 enrichment on laminar methane-air flame thickness. Combust Flame. 153: 510.

[2] Kim JS, Park J, Kwon OB, Lee EJ, Yun JH, Keel SI (2008). Preferential diffusion effects in opposed-flow diffusion flame with blended fuels of CH4 and H2. Int J Hydrogen Energy 33:842.

[3] Coppens FHV, Ruyck DJ, Konnov AA. The effects of composition on burning velocity and nitric oxide formation in laminar premixed flames of CH4 + H2 + O2 + N2. Combust Flame (2007) 149-409.

[4]Scott J. http://www.americanchemistry.com/Media/PressReleasesTranscripts/ACC-news-releases/US -Chemical-Invest ment-Linked-to-Shale-Gas-Reaches-100-Billion.html. American Chemical Council February (2014).

[5] Fotache CG, Kreutz TG, Law CK. Ignition of counterflowing methane versus heated air under reduced and elevated pressures. Combust Flame. (1997) 108-442.

[6] Davis SG, Law CK. Determination of and fuel structure effects on laminar flame speeds of C1 to C8 hydrocarbons. Combust Sci Technol (1998) 140-427.

[7] Chen Z, Qin X, Ju Y, Zhao Z, Chaos M, Dryer FL. High temperature ignition and combustion enhancement by dimethyl ether addition to methane-air mixtures. Proc Combust Flame. (2007) 31-1215.

[8] Chen Z. On the accuracy of laminar flame speeds measured from outwardly propagating spherical flames: Methane/air at normal temperature and pressures. Combust Flame. (2015) 162-2442.

[9] Burke MP, Chen Z, Ju Y, Dryer FL. Combust. Effect of cylindrical confinement on the determination of laminar flame speeds using outwardly propagating flames. Combust Flame. (2009) 156-771.

[10] Ai Y, Zhou Z, chen Z, Kong W. Laminar flame speed and Markstein length of syngas at normal and elevated pressures and temperatures. Fuel (2014) 137-339.

[11] Markstein GH. Experimental and theoretical studies of flame-front stability. J Aeronaut Sci (1951) 18-199.

[12] Chen Z, On the extraction of laminar flame speed and Markstein length from outwardly propagating spherical flames. Combust Flame (2011) 291-300.

[13] Ronney PD, Sivashinsky GI. A theoretical study of propagation and extinction of nonsteady spherical flame fronts. SIAM J Appl Math (1989) 49-1029.

[14] Kelley AP, Law CK. Nonlinear effects in the extraction of laminar flame speeds from expanding spherical flames. Combust Flame (2009).156-1844.

[15] Rozenchan G, Zhu DL, Law CK, Tse SD. Outward propagation, burning velocities, and chemical effects of methane flames up to 60 ATM. Proc Combust Inst. (2003) 29-1461.

[16] Halter F, Chauveau C, Djebaili-Chaumeix N, Gokalp I. Characterization of the effects of pressure and hydrogen concentration on laminar burning velocities of methane-hydrogen-air mixtures. Proc Combust Inst (2005) 30-201.

[17] Smith GP et al, GRI-MECH 3.0 <http://www.me.berkeley.edu/gri_mech/>.

[18] Wang H, You X, Joshi AV, Davis SG, Salkin A, Egolfopoukos F, Law CK, USC MECH II <http://ignis.use.edu/USC_II.htm>.

[19] Sung CJ, Li B, Law CK, Wang H. Wang MECH < https://www.princeton.edu/~cklaw/kinetics/slw001/index.html>.

[20] Kee RJ, Grcar JF, Smooke MD, Miller JA. A Program for modeling steady, laminar, one-dimensional premixed flames. Sandia National Laboratory Report SAND85-8240 (1985).

[21] Matalon M. Intrinsic flame instabilities in premixed and nonpremixed combustion. Annu Rev Fluid Mech. (2007) 39-163.

[22] Addabbo R, Bechtold JK, Matalon M. Wrinkling of spherical expanding flames. Proc Combust Int. (2002) 29-1527.

11th Asia-Pacific Conference on Combustion,
The University of Sydney, NSW Australia
10th -14th December 2017.

Mixture-Fraction Based Hybrid Binomial-Langevin–MMC Modelling Applied to Auto-ignition in Vitiated Flow

Andrew P. Wandel[1], R. Peter Lindstedt[2]

[1]Computational Engineering and Science Research Centre, School of Mechanical and Electrical Engineering,
University of Southern Queensland
West St, Toowoomba, Queensland, 4350, Australia
[2]Department of Mechanical Engineering, Imperial College London
Exhibition St, London, SW7 2AZ, United Kingdom

Abstract

A new hybrid binomial Langevin–MMC (Multiple Mapping Conditioning) modelling approach is proposed. The mixture fraction derived from the binomial Langevin model is used to specify the reference variable for MMC. The modified Curl's model is used to close the stochastic MMC mixing term. The new model is applied to a jet burner with a vitiated co-flow (the "Cabra burner") with methane as the fuel. The first- and second-order statistics show good agreement with experimental data.

1 Introduction

With the pressure to reduce emissions becoming a key factor in the design of modern combustion systems, engineers are moving closer to the combustion limits. This results in finite-rate chemistry effects (e.g. extinction and reignition) becoming important and such phenomena are increasingly examined through computational methods [1,2]. Because simple models generally cannot completely describe these effects, transport probability density function (PDF) models are often required [3]. Extinction and reignition processes amplify the sensitivity of the results to different closure elements, including molecular mixing [4], so developments continue to be made in this area. A hybrid model [5,6] was recently proposed that used the Multiple Mapping Conditioning (MMC) [7] and binomial Langevin (BL) [8] models as its basis. The proposal was to use the binomial Langevin model to solve joint velocity-scalar statistics with one scalar (a mixture fraction), while the MMC model was used to solve for all scalars (including the mixture fraction). The goal was to overcome the implementation difficulties inherent in solving bounded scalars in the BL model, while simultaneously overcoming difficulties with specifying certain coefficients in the MMC model. This was achieved by forcing the mixture fraction in the MMC component of the model to approach the BL value, while using the dominant velocity component from the BL solution as a basis for the MMC reference variable. The current approach is consistent with this methodology. However, it is simplified with the mixture fraction derived from the BL solution used directly to define the reference quantity rather than via an explicit transport equation [9].

Corresponding author. Fax: +61-7-4631-2526
E-mail address: andrew.wandel@usq.edu.au

2 Theory

The hybrid binomial Langevin–MMC model is described below. First the BL model is outlined, then the MMC model, both of which are common to previous work [5,6]. Finally, the new link between the BL and MMC models is described.

2.1 Binomial Langevin Model

A generalised form of the binomial Langevin model for the joint-PDF (Probability Density Function) of velocity and multiple scalars was developed by Hůlek and Lindstedt [10]. The stochastic differential equation for velocity component u_i is:

$$du_i^p = \left(\alpha_1\delta_{ij} + \alpha_2\beta_{ij}\right)\left(u_j^p - \langle u_j\rangle\right)\frac{dt}{\tau_u} + \left(C_0\langle\varepsilon\rangle\right)^{1/2} dw_i \quad (1)$$

where p denotes the pth particle, $\tau_u = \langle k\rangle/\langle\varepsilon\rangle$ is the turbulent timescale, k the turbulent kinetic energy, ε the turbulent dissipation rate, $\langle\varepsilon\rangle$ its average, dw_i a Wiener process, and β_{ij} the Reynolds stress anisotropy tensor:

$$\beta_{ij} = \langle u_i'u_j'\rangle / \langle u_k'u_k'\rangle - \tfrac{1}{3}\delta_{ij} \quad (2)$$

The remaining coefficients are $\alpha_1 = -(\frac{1}{2} + \frac{3}{4}C_0) - \alpha_2(\beta_{II})^2$, $\alpha_2 = 3.7$ and $C_0 = 2.1$. The corresponding stochastic differential equation for any scalar η is:

$$d\eta^p = G_\eta\left(\eta^p - \langle\eta\rangle\right)\frac{dt}{\tau_\eta} + \left(B_\eta\langle\varepsilon_\eta\rangle\right)^{1/2} dw_{\text{bin}}. \quad (3)$$

The mean scalar dissipation rate is $\langle\varepsilon\eta\rangle = \langle\eta'^2\rangle/\tau_\eta$, the scalar timescale was modelled as $\tau_\eta = \tau_u/C_\phi$ ($C_\phi = 2.3$ was chosen [3]) and dw_{bin} is a binomial Wiener process [7]. The drift coefficient G_η is

$$G_\eta = -\left\{K_\eta\left[1 - \left\langle\left(\frac{\eta'^p}{\eta'_*}\right)^2\right\rangle\right] + 1\right\} \quad (4)$$

while the diffusion coefficient B_η is

$$B_\eta = K_\eta\left[1 - \left(\frac{\eta'^p}{\eta'_*}\right)^2\right] \quad (5)$$

where, with $K_0 = 2.1$ and $C_K = 0.76$. The other quantities are:

$$\eta'^p = \eta^p - \langle \eta \rangle^p$$

$$\eta'_* = \begin{cases} \eta'^p_{\max}, & \eta'^p > 0 \\ \eta'^p_{\min}, & \eta'^p < 0 \end{cases}$$

$$\langle \eta \rangle^p = \eta_{\min}\big|_{c=c^p} + \left(\langle \eta \rangle - \eta_{\min}\big|_{c=c^p}\right) \frac{\eta_{\max}\big|_{c=c^p} - \eta_{\min}\big|_{c=c^p}}{\eta_{\max}\big|_{c=\langle c \rangle} - \eta_{\min}\big|_{c=\langle c \rangle}}$$

where c is a basis scalar; usually the mixture fraction or a reaction progress variable.

The above approach reproduces many physical processes well [10]. However, a practical difficulty arises with the definition of the max and min values used to define η'_* and $\langle \eta \rangle_p$ for reactive scalars. The total mass fraction is unity and compositions are further constrained by the mass of each element (e.g. C, O and H). Accordingly, the permissible range for a particular scalar depends on the values of all other scalars. By contrast, for the hybrid model [5,6], only the mixture fraction is required and the problem is avoided.

2.2 MMC model

The MMC concept is that all scalars Z_I can be transported in a mathematical space (i.e. the reference space) thereby making the transport simpler because the reference space can be defined to take any properties. The simplest reference space ξ is one-dimensional and is conventionally related to the mixture fraction. For this one-dimensional reference space, the deterministic form of the conditional MMC transport equation is [7]:

$$\frac{\partial \overline{Z}_I}{\partial t} + \mathbf{U} \cdot \nabla \overline{Z}_I + A \frac{\partial \overline{Z}_I}{\partial \xi} - B \frac{\partial^2 \overline{Z}_I}{\partial \xi^2} = W_I\left(\overline{Z}\right) \qquad (6)$$

where Z_I represents each scalar I, $\overline{Z}_I \equiv \langle Z_I \,|\, \xi \rangle$ is the conditional average of Z_I given the value of the reference variable ξ, A and B are the drift and diffusion coefficients respectively and W_I is the chemical source term for species I and is a function of all scalars. Because ξ and the velocity U are both taken to have Gaussian distributions, they are modelled to be linearly related to each other (Z_0 is the mixture fraction):

$$\mathbf{U} = \langle \mathbf{u} \rangle + \left(\langle \mathbf{u}' Z'_0 \rangle / \langle \xi' Z'_0 \rangle \right) \xi \qquad (7)$$

In the current implementation, the scalars are transported stochastically:

$$dZ_I^p = \left(S + W_I\right) dt \qquad (8)$$

where S represents the mixing process, for which the Modified Curl's model [11, 12] was used. To enforce locality in the mixing process, particle pairs p and q were chosen so that the following was satisfied:

$$\left|\xi^p - \xi^q\right| \le \left(B \Delta t\right)^{1/2} \qquad (9)$$

This process mimics the diffusive term of a stochastic differential equation (SDE). Note that (9) is not a minimisation: the specification is that particles p and q are close, not the closest possible pairing. In practice, the inequality may be violated by outliers, but this does not pose any numerical difficulty.

2.3 Mixture-fraction–based hybrid model

All the details up to this point are identical to the previous, velocity-based hybrid model [5,6]. The key difference with the current approach is that the reference variable ξ is defined to be the BL mixture fraction η. Following the approach of Wandel [13], mixing proceeds via the Modified Curl's model with the exchange between selected paired particles specified as 50% of full mixing. The fraction of particles to be mixed is treated as a parameter that can be specified independently. Work is underway to determine a relationship between the fraction of particles to be mixed and scalar statistics to enable *a priori* determination.

3 Results

The Cabra burner in Fig. 1 and with methane as the primary fuel [14] was used as the test case with base conditions specified. A parabolic code [3] was used for a single realisation with 1865 streamwise locations and 80 cross-stream cells (the width of the domain expanded with the entrainment width); and 400 particles/cell. The fraction of particles mixed each time step was 6%; this is at least 10 times as many particles as Modified Curl's model would mix for the same mixing parameters.

The mean mixture fraction is presented is quantitatively similar to results [15] using the same code but with the Modified Curl's mixing model. The similarity is essentially due to the fluid mechanics of the solver. The mixture fraction rms (Fig. 2) has its peak in the correct location, but tends to be significantly under-predicted for most of the domain.

The mean temperature results (Fig. 3) are arguably excellent and again similar to previous results [15]. The temperature rms results (Fig. 4) appear outstanding. The authors could not find any other RANS-based approach that correctly predicts the locations of both the rise at $z/d = 40$ and the decrease at $z/d = 60$. Of those LES simulations which report the temperature rms, some cannot simultaneously predict both these locations [16,17]; the remainder do not predict a local maxima in the radial profile of mean temperature at $z/d = 40$ [18,19]. The current results appear unique.

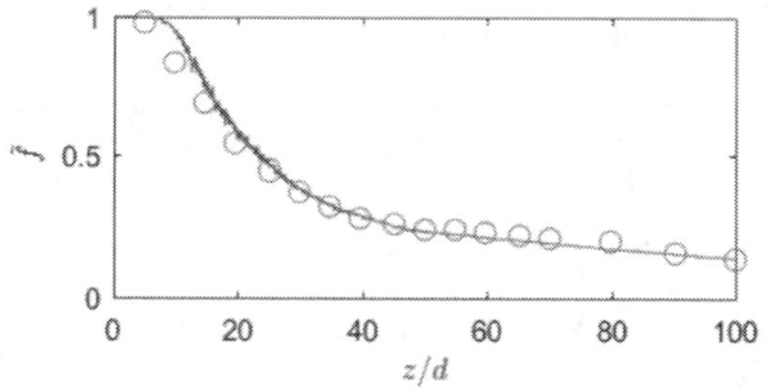

Figure 1: Centreline mean mixture fraction profile. MMC, line; experiment, symbol [14].

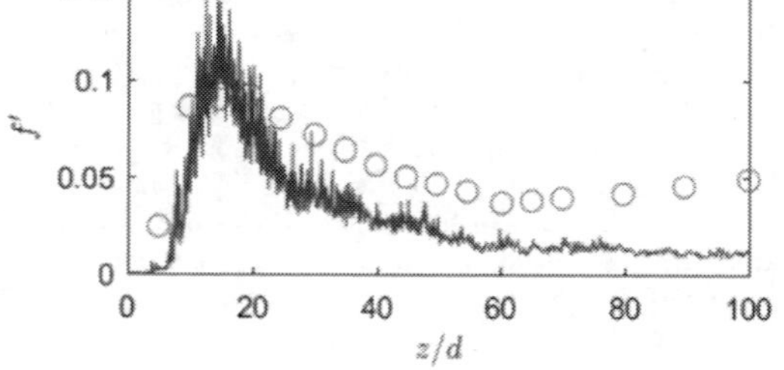

Figure 2: Centreline mixture fraction rms profile. As per Fig. 1.

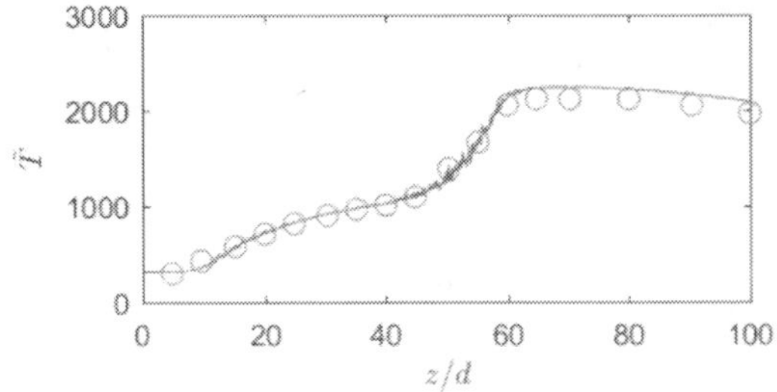

Figure 3: Centreline mean temperature profile. Symbols as per Fig. 1.

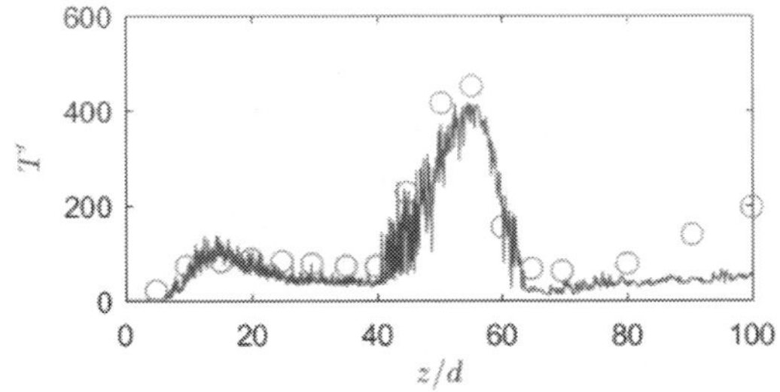

Figure 4: Centreline temperature rms profile. Symbols as per Fig. 1.

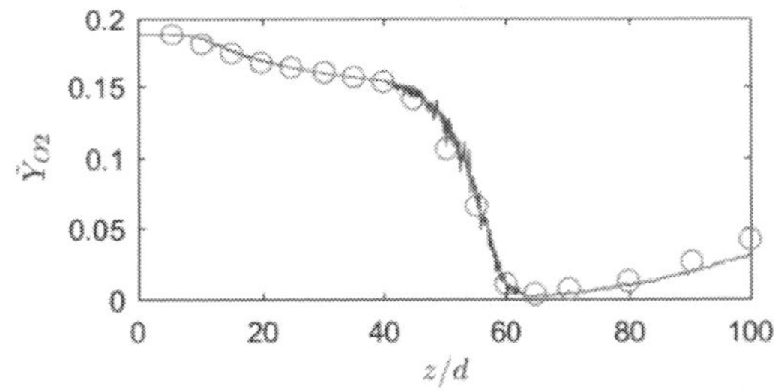

Figure 5: Centreline O_2 mean mass fraction profile. Symbols as per Fig. 1.

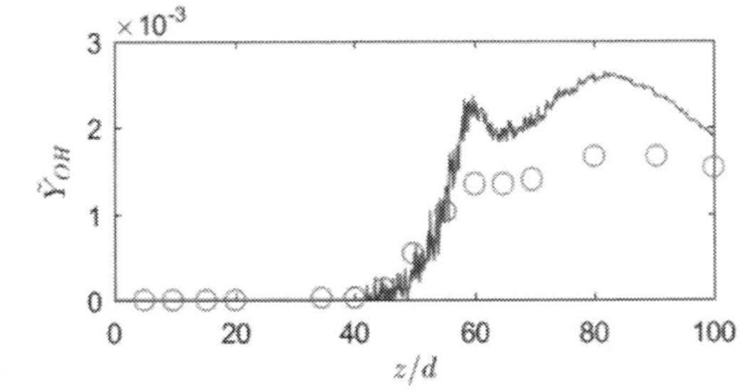

Figure 6: Centreline OH mean mass fraction profile. Symbols as per Fig. 1.

The prediction of the O_2 profile (Fig. 5) is very satisfactory and quantiatively similar to previous results [15]. The OH profile (Fig. 6) shows accurate predictions of the trend, but with a clear over-prediction for $z/d > 60$.

The radial profiles of mixture fraction (Fig. 7) show good agreement with experimental data with the trends matches the centreline results in Figs. 1 and 2. Similarly, the radial profiles of temperature (Fig. 8) show good agreement with experimental data. The results are essentially similar to the Modified Curl's results [15]. However, the rms temperature profile at $z/d = 70$ is substantially over-predicted by the Modified Curl's model for $r/d < 3$, while the current results are substantially improved.

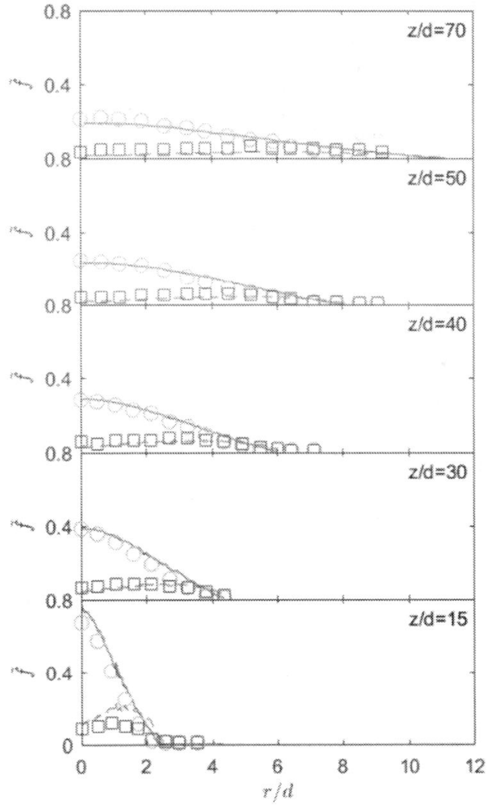

Figure 7: Radial mixture fraction profiles. Mean: MMC, line; expt, circles [14]. rms: MMC, dashed line; expt, squares [14].

The procedure to determine the liftoff height follows the methodology of [14]: it is the average of the axial locations where C_2H_4 is 100 ppm and C_2H_2 is 2 ppm. For the current simulation, this was found to be 42 diameters, as opposed to the experimental value of 35 [14]. Other measurements of this configuration [20] indicate that the value could be closer to 50. Because this variable is highly sensitive to the boundary conditions of velocity and temperature, a parametric study will be performed in the future.

4 Conclusions

A new hybrid binomial Langevin–MMC model is proposed. The linkage between the models is by virtue of the BL mixture fraction being used as the MMC reference variable. The model is closed by using the Modified Curl's model with the amount of mixing half the original Curl's model [13]. To close the model, the fraction of particles mixed each time step is an input parameter; ongoing efforts are investigating a relationship which can be used to govern this value.

A jet burner with vitiated co-flow [14] was simulated to test the model. The first-order statistics of mixture fraction, temperature and species mass fraction were matched very well, which is fundamentally due to the fluid mechanics, since similar results were obtained using the Modified Curl's model [15]. The location of the peak in centreline mixture fraction rms was predicted well, but the value far downstream was severely under-predicted. However, the centreline temperature rms was predicted extraordinarily well, which is largely due to the performance of the mixing model, since the directly comparable Modified Curl's model did not predict the location of the significant decrease at $z/d = 60$. The authors have not found another RANS simulation which is able to predict the shape of this quantity. Some LES models are also incapable of predicting this shape [16,17]; those which are able to predict the shape of this quantity fail to predict the radial profiles correctly by not producing the local maxima in mean temperature at $z/d = 40$ [18,19]. The liftoff height was predicted within accepted uncertainties associated with the burner exit boundary conditions. The behaviour of the flame is known to be highly sensitive to the inflow conditions; future work will assess the model for this parametric study.

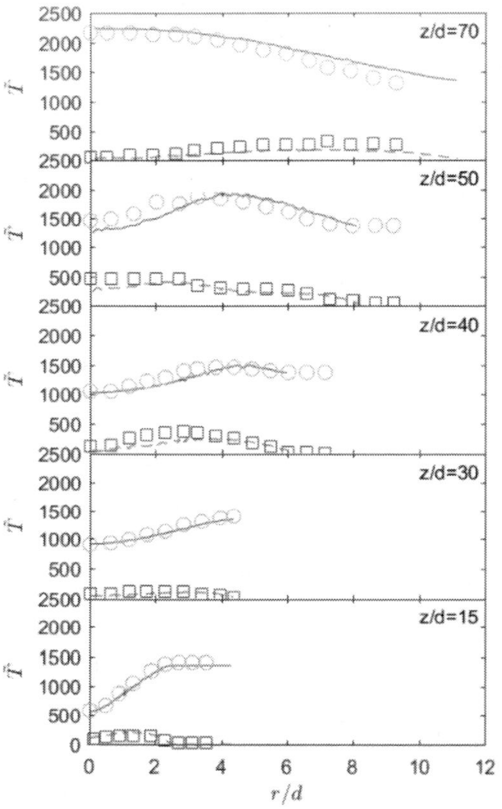

Figure 8: Radial temperature profiles. Symbols as per Fig. 7.

5 Acknowledgment

The current work was in part performed at Imperial College while Dr Wandel was on sabbatical. The authors are grateful to Mr. P. Simatos for technical support in conducting the simulations.

References

[1] S.T. Smith and R.O. Fox, Phys. Fluids 19 (2007) p. 085102.
[2] B. Merci, B. Naud and D. Roekaerts, Flow, Turbul. Combust. 79 (2007) 41–53.
[3] R.P. Lindstedt, S.A. Louloudi, J.J. Driscoll and V. Sick, Flow, Turbul. Combust. 72 (2004), 407–426.
[4] Z. Ren and S.B. Pope, Combust. Flame 136 (2004), 208–216.
[5] A.P. Wandel and R.P. Lindstedt, Phys. Fluids 21 (2009) p. 015103.
[6] A.P. Wandel and R.P. Lindstedt, Proc. Combust. Inst. 34 (2013) 1365–1372.
[7] A.Y. Klimenko and S.B. Pope, Phys. Fluids 15 (2003) 1907–1925.
[8] L. Valiño and C. Dopazo, Phys. Fluids A 3 (1991) 3034–3037.
[9] M.J. Cleary and A.Y. Klimenko, Flow, Turbul. Combust. 82 (2009) 477–491.
[10] T. Hůlek, and R.P. Lindstedt, Comb. Sci. Tech. 136 (1998) 303–331.
[11] C. Dopazo, Phys. Fluids 22 (1979) 20–30.
[12] J. Janicka, W. Kolbe and W. Kollmann, J. Non-Equilib. Thermodyn. 4 (1979) 47–66.
[13] A.P. Wandel, Combust. Theory Model. 17 (2013) 707–748.
[14] R. Cabra, J.-Y. Chen, R.W. Dibble, A.N. Karpetis and R.S. Barlow, Combust. Flame 143 (2005) 491–506.
[15] K. Gkagkas and R.P. Lindstedt, Proc. Combust. Inst. 31 (2007) 1559–1566.
[16] M. Ihme and Y.C. See, Combust. Flame 157 (2010) 1850–1862.
[17] O. Schulz, T. Jaravel, T. Poinsot, B. Cuenot and N. Noiray, Proc. Combust. Inst. 36 (2017) 1637–1644.
[18] P. Domingo, L. Vervisch and D. Veynante, Combust. Flame 152 (2008) 415–432.
[19] N. Enjalbert, P. Domingo and L. Vervisch, Combust. Flame 159 (2012) 336–352.
[20] R.L. Gordon, S.H. Stårner, A.R. Masri and R.W. Bilger, 5th Asia-Pacific Conference on Combustion, The University of Adelaide, 2005.

11th Asia-Pacific Conference on Combustion,
The University of Sydney, NSW Australia
10th -14th December 2017.

Flow and Flame Dynamics of Confined Buoyant Inverse Diffusion Flames

Xuren Zhu[1#], Xi Xia[1#], Peng Zhang[1*]

[1] Department of Mechanical Engineering, the Hong Kong Polytechnic University, Hong Kong

(# Both authors equally contributed to the paper)

Abstract

PIV experiments and simulation were performed to investigate non-reacting and reacting buoyant inverse diffusion flames, with emphasis on flow and flame dynamics. The results show that in non-reacting inverse diffusion flames the initial shear flow and the buoyancy effect induce opposite-direction vorticities, which interact with each other and cause flow instability. This instability can be quantified by the Richardson number, *Ri*, which measures the gravitational effect relative to the initial shear flow. So the instability is strongest at $Ri = 1$ when the two mechanisms of vorticity are comparable to each other. In the reacting inverse diffusion flame, the density gradient is reversed due to chemical heat release and so is the buoyancy-induced vorticity that it shares the same direction with the vorticity of the initial shear flow. As a result, the near-field flow instability is suppressed while the initial shear flow would continue growing under the gravitational effect until instability develops in the far field.

1 Introduction

In a wide range of accidental fire and industrial flames, it is common to find buoyant diffusion flames. Near field characteristics have important influence on the flame dynamics [1]. The different density between fuel and air and the concentrated heat release are the two main contributors to the buoyancy effect, which strongly influences the dynamics of jet diffusion flames with low Reynolds number [2].

By analyzing the vorticity transport equation for buoyant flows, Jiang and Luo [1] concluded that in buoyant normal diffusion flames (NDFs) enhanced tendency of flow transition to turbulence is due to combustion induced buoyancy.

Back to 1984, Wu and Essenhigh [3] found that with very low fuel inlet velocity (< 1cm/s) the IDFs started to oscillate with a frequency of 1-2 Hz. Katta et al. [4] investigated the dynamics of an IDF, focused on the polycyclic-aromatic-hydrocarbon and soot formation. However, no detailed analysis on flow or flame dynamics of the IDF was presented.

In the present study, both experimental and numerical simulation will be used to investigate the fundamental physics responsible for the unsteady flow in buoyant IDFs.

2 Experimental Specifications and Numerical Methodology

As shown in Fig.1, the burner consists of two concentric stainless-steel tubes. Air flow was injected through a 10-mm-diameter and 200-mm-length central tube. The outer tube has an inner diameter of 40 mm. Fuel flow was injected through an annulus between the outer tube and the central tube. A quartz glass tube was installed above the burner. Detailed dimensions of the burner are presented in Fig. 1. The operation conditions for all the cases are 293 K and 101 kPa.

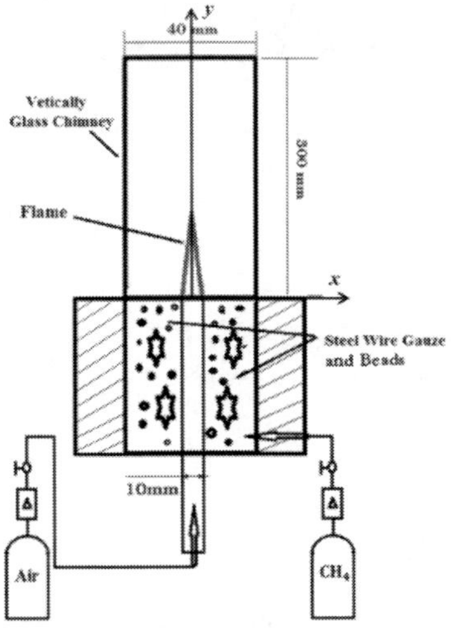

Figure 1: Schematic of the IDF burner

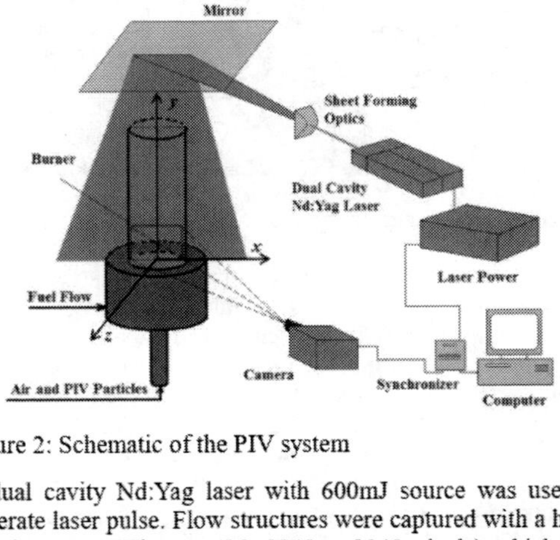

Figure 2: Schematic of the PIV system

A dual cavity Nd:Yag laser with 600mJ source was used to generate laser pulse. Flow structures were captured with a high-speed camera (Hisense 4M, 2048 × 2048 pixels) which was synchronized using a Berkeley Nucleonics Corporation 575 pulse/delay generator with a time delay of 200μs between two frames. The Cartesian coordinates for the burner have been defined in Fig. 1 and 2.

LES was performed with the open source code OpenFOAM (version 2.3.0) [5]. PISO method [6] was used to solve the time

Corresponding author.
E-mail address: pengzhang.zhang@polyu.edu.hk

dependent Navier-Stokes equations. The discretization of unsteady terms was performed by a second order backward scheme [7]. NVD scheme [8] and the second order central difference scheme were used to discretize the convection and diffusion terms. Second-order central difference linear interpolation was applied for mass flux calculation. Infinite fast and irreversible chemical reaction was applied and the mesh with 0.5million grid nodes is shown in Fig. 3. Boundary conditions for a reacting IDF are listed in Tab. 1.

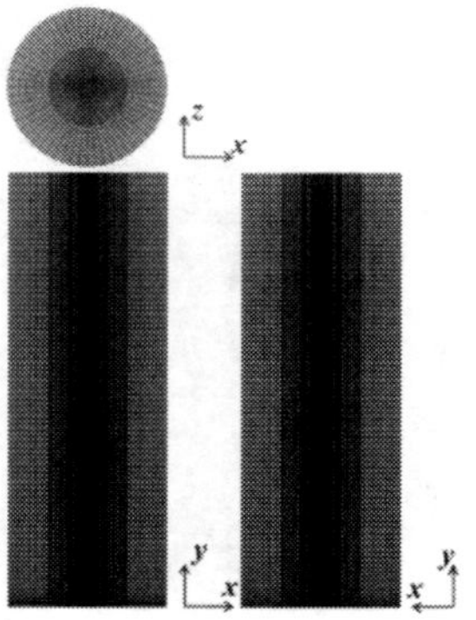

Figure 3: Mesh for the LES calculation.

	V (m/s)	P (Pa)	T (K)	Yi (-)
Air inlet	0.3	0 Grad.	300	air
Fuel inlet	0.02	0 Grad.	300	Methane
Outlet	0 Grad.	101325	0 Grad.	0 Grad.
Chimney	0	0 Grad.	450	0 Grad.

Table 1: Boundary conditions for a reacting IDF (0 Grad. means zero gradient).

3 Results and Discussion

3.1 Flow and Flame Dynamics in IDFs

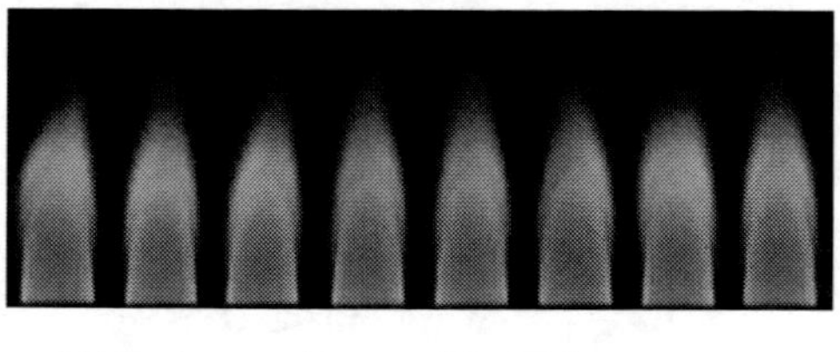

Figure 4: Instantaneous images of the IDF

To examine the flame oscillation found in [3], an experimental case was performed with the air inlet velocity v_1=30cm/s and the fuel inlet velocity v_2=2cm/s (or $Re_1 = 192$ and $Re_2 = 46$), respectively. It is observed in Fig. 4 that flame oscillation only occurs in the downstream of the flame (yellow cap), where the direction of the flame tip swings and the height varies; however, the blue region at the bottom seems to remain stable. In Fig.4, the instantaneous cross-streamwise vorticity (Ω_z) results for a non-reacting IDF ($Re_1 = 192$ and $Re_2 = 46$) are plotted in Fig.5, where $\tilde{x} = x/r$ and $\tilde{y} = y/r$, r is the radius of the center air tube. It is seen that the flow is also highly unsteady as the shear layer between fuel and air wiggles and deforms dramatically. Qualitatively, the flow in this case does seem to be more unstable than the reacting flow as the bottom half of the shear layer also tends to be unsteady.

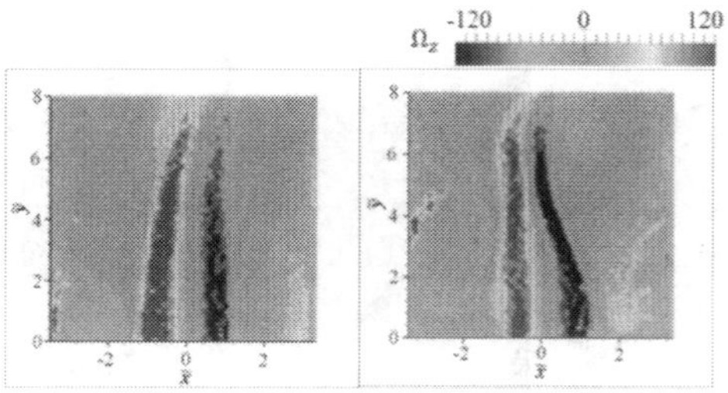

Figure 5: Instantaneous cross-streamwise vorticity (Ω_z)

3.2 Validation of Numerical Simulation

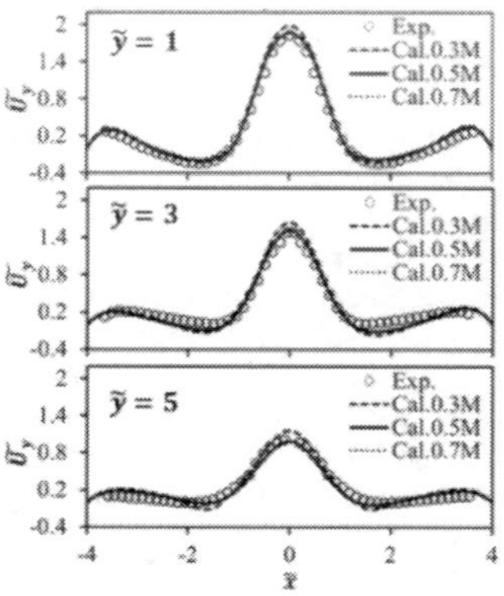

Figure 6: Validation of averaged streamwise velocity

For a benchmark case considered, $Re_1 = 192$ and $Re_2 = 46$, respectively, its normalized streamwise velocity as a function of x at different y locations has been validated in Fig. 6. Here, $\widetilde{U_y} = U_y/v_1$ with U_y is the streamwise velocity and $v_1 = 30$cm/s is the air inlet velocity. Cal. 0.3M, Cal.0.5M, and Cal.0.7M represent the simulation results with 0.3million, 0.5 million, and 0.7 million grid nodes, respectively. It is seen that the simulation results are in good agreement with experiment. The convergence of the simulations of different meshes confirms that the grid resolution for the current study is sufficient, thus, the mesh level with 0.5 million grid nodes will be applied in the following investigations.

Since non-premixed flame is diffusion controlled, the flame height of a non-premixed flame presents one of the most important flame characteristics. According to Mikofski et al. [9], the edge of the blue flame approximates the reaction zone of methane IDFs, where massive heat is released to cause a local peak of temperature. In this respect, the outer boundary of the blue flame on the left side of Fig. 7 approximately matches with the ridge of the temperature contour on the right, verifying the good agreement between simulation and experiment.

3.3 Non-reacting Flows of IDF

In buoyant flow, the flow characteristics are directly related to Richardson number (*Ri*) or Froude number (*Fr*). Following the derivation of vorticity transport relation in [1], the *Ri* is defined as $Ri = \frac{(\rho_1-\rho_2)gd}{\overline{\rho}(v_1{}^2-v_2{}^2)}$, where g is the gravitational acceleration and d is the diameter of the center air tube. $\overline{\rho} = \frac{(\rho_1+\rho_2)}{2}$ where ρ_1, ρ_2 denote the air and fuel densities, respectively. v_1, v_2 are the inlet air and fuel velocities, respectively. Thus, the Richardson number gives the ratio of buoyancy term to the initial shear term.

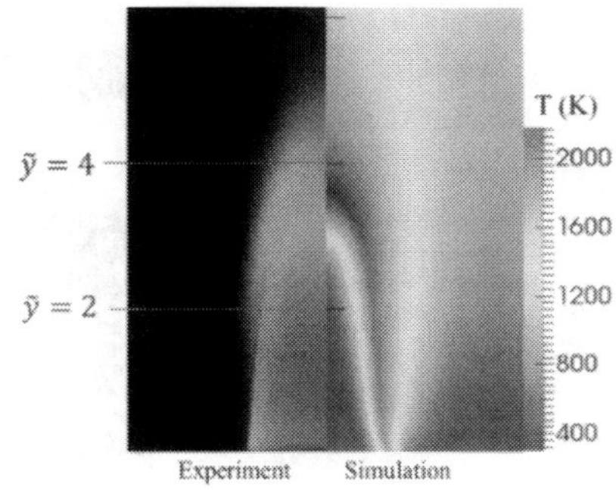

Figure 7: Validation of flame height for the benchmark case

In Fig. 8, five cases were simulated with varying buoyancy term by diluting air into fuel flow while fixing initial shear term. With 25%-100% air diluted into fuel flow, the density ratio changes from 1.8 to 1.0 hence the *Ri*. In Fig. 8, the contours of the cross-streamwise vorticity (Ω_z) in xy plane (bottom) and y=30mm plane (top) are plotted. It is observed that flow at xy plane becomes more unstable as *Ri* increases from 0 to 0.81. Specifically, the initial shear layer between fuel and air tends to be more affected by an external shear layer in the near field as the buoyancy effect becomes more profound.

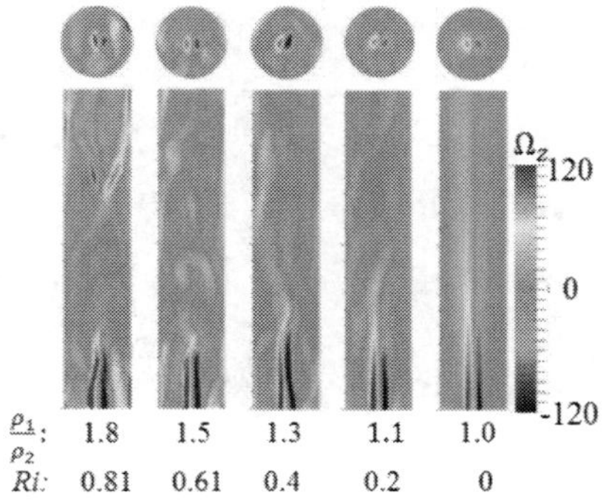

Figure 8: Instantaneous cross-streamwise vorticity (Ω_z) contours in the xy plane (bottom) and the y=30mm plane (top).

Alternatively, by fixing other inlet flow conditions, so that the buoyancy term is unchanged, and increasing v_2 to decrease the initial shear term, *Ri* can be adjusted from 0.81 to infinity. Again, we simulated the non-reacting flow in this case and presented the vorticity contours in Fig. 9. It is seen that the shear flows in the near field for *Ri*= 0.81, 1.02 and 1.59 all display significant unsteady feature. As the Richardson number further increases to *Ri*=6.38, flow in the near filed tends to become relatively stable. When *Ri* further increases to infinity, the shear layers in the near field grow into a symmetric pattern, indicating the establishment of a stable flow. Apparently, in the situation of Fig. 9 larger *Ri* would result in a more stable flow in the near field, in contrast to the conclusion made about Fig. 8. This non-monotonic influence of the *Ri* on the stability of the non-reacting IDF flow is quantitatively supported by Figs. 10 (a) and (b), where the y-direction turbulent kinetic energy, $\frac{1}{2}U_y'^2$, is plotted as a function of time at point (0,0.03,0) for different *Ri*. We indeed observed that the unsteadiness of the flow is strongest as *Ri* approaches 1, for which detailed explanation will be provided below.

According to the vorticity transport equation in [1], the term related to buoyancy effect is $\nabla\rho \times \mathbf{g}$, which is dictated by the density gradient as gravity is assumed to be a constant. For non-reacting flow, the density gradient is purely caused by the different densities between fuel and air. On the other hand, for the term associated with the initial shear flow, its vorticity is given by $\Omega_z = \left(\frac{\partial U_y}{\partial x} - \frac{\partial U_x}{\partial y}\right)\vec{k} \approx (\frac{\partial U_y}{\partial x})\vec{k}$. Considering this specific problem, we have the relation $\rho_1 > \rho_2$ and $v_1 > v_2$, which gives that the two terms, $(\frac{\partial U_y}{\partial x})\vec{k}$ and $\nabla\rho \times \mathbf{g}$, always have opposite directions. To this end, the buoyancy effect and the initial shear flow are two competing mechanisms in the near field of the present flow. It is further noted that the dominance of any single effect (*Ri* being very small or very large) would promote flow stabilization in the near field, although instability might still develop in the far field as the nature of the flow transits from laminar to turbulent. However, if the two effects are comparable to each other (*Ri* approaches 1), instability would develop from the near field because of the strong interaction between vortices of different signs. This explains why the instability of the near field is most significant near *Ri* = 1.

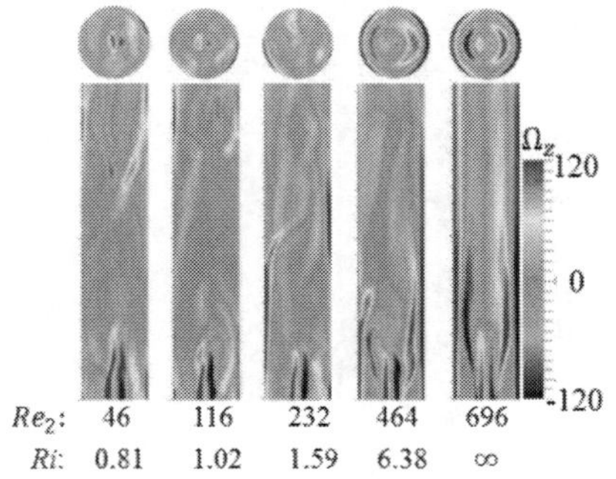

Figure 9: Instantaneous cross-streamwise vorticity (Ω_z) contours in the xy plane (bottom) and the y=30mm plane (top).

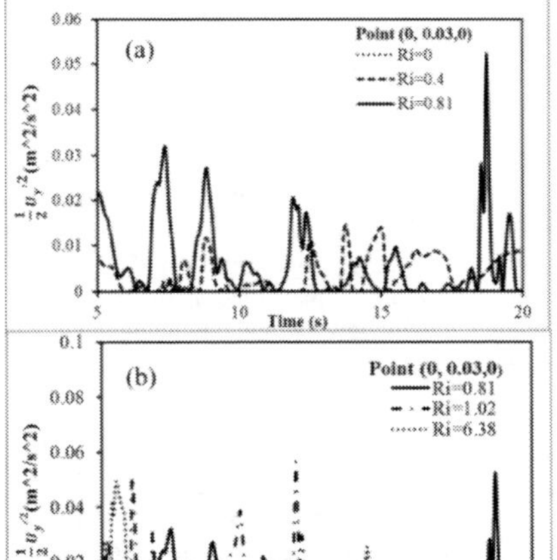

Figure 10: Time trace of the turbulent kinetic energy, $\frac{1}{2}U_y'^2$, for different *Ri*.

To justify this hypothesis, a non-reacting NDF was simulated by injecting fuel flow through the center tube with a velocity of 30cm/s while air flow enters the annular with a velocity of 2cm/s. Note that the vorticities of the initial shear flows, induced $(\frac{\partial U_y}{\partial x})\vec{k}$, are the same between the IDF and the NDF, while the density gradients are opposite so that the directions of the $\nabla\rho \times \mathbf{g}$ term for the IDF and the NDF are different. The outcome is that the two competing mechanisms for an IDF are of assist to each other in the corresponding NDF. Thus, it can be implied that the near field of an NDF is stable. This is indeed observed in Fig. 11 that the reversed buoyancy effect in the NDF suppresses the flow instability occurring to the IDF.

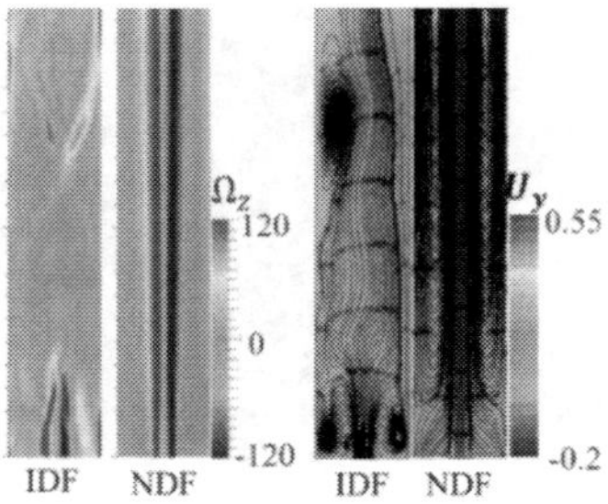

Figure 11: Instantaneous Ω_z (left) contour and stream line (right) for an IDF and a NDF.

3.4 Reacting IDFs

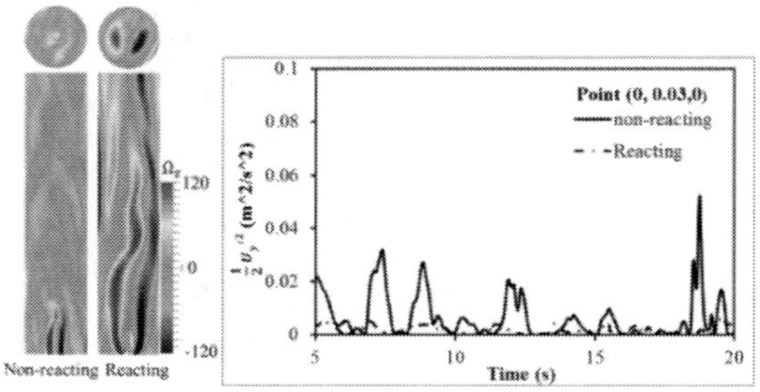

Figure 12: Instantaneous contour of vorticity, Ω_z (left), and time trace of the turbulent kinetic energy, $\frac{1}{2}U_y{'}^2$ (right), for non-reacting and reacting IDFs.

The two mechanisms for the non-reacting flow will be applied to a reacting IDF to understand the effect of the flame on flow stability. Fig. 12 illustrates the comparison between a non-reacting IDF flow and its corresponding reacting IDF with the same inlet flow conditions. It is observed that in the reacting IDF the initial shear flow is notably enhanced due to the increased flow velocity by chemical heat release. However, the density gradient is reversed so that $\nabla\rho \times \mathbf{g}$ is opposite of that for the non-reacting flow. This happens as the chemical reaction of the IDF heats up the inner flow and, consequently, decreases its density significantly so that $\nabla\rho \times \mathbf{g}$ and the vorticity $(\frac{\partial U_y}{\partial x})\vec{k}$ of the initial shear flow share the same direction. This shear layer configuration resembles the non-reacting NDF flow, discussed in section 3.3, in the sense that the gravitational effect in both cases would enhance the initial shear flow. Based on previous analysis, we conjecture that the gravitational effect for a reacting IDF also help to stabilize the initial shear flow in the near field. This is qualitatively supported by the left plot of Fig. 12, where the shear layer of the reacting flow displays good symmetric feature in the near field compared with the non-reacting flow. This is further verified by the right plot of Fig. 12 which implies that the unsteadiness of the reacting flow is less than the non-reacting flow. It is concluded that the combustion-induced buoyancy effect suppresses flow instability in the near field of IDFs. In Fig. 13, the oscillation frequency for the reacting case is about 0.5 Hz which approximates to the observation in [3] and the present experiment.

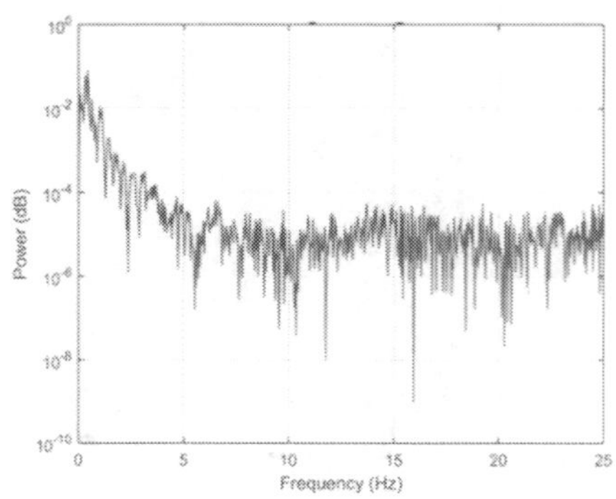

Figure 13: Energy spectra of the streamwise velocity at (0, 30, 0) for the reacting case.

4 Conclusions

Inverse diffusion flames (IDFs) were experimentally and numerically investigated to understand the flow and flame dynamics in the near field. In buoyant non-reacting IDFs the initial shear flow and the buoyancy effect induce opposite-direction vorticities and the interaction between them causes flow instability in the near field; the flow instability (turbulent intensity) increases as Ri approaches unity, at which the buoyancy term equals the initial shear term. Chemical reaction would result in an enhancement of the initial shear layer in the near field and reversed density gradient, thereby suppressing the development of the near-field flow instability for an IDF.

5 Acknowledgment

This work was supported by the Hong Kong RGC/GRF (operating under contract numbers PolyU 152217/14E and 152651/16E) and partly by the Hong Kong Polytechnic University (G-UA2M and G-YBGA).

References

[1] X. Jiang, K. Luo, Combust. Flame 133(1) (2003) 29-45

[2] G. Cox, Combustion Fundamentals of Fire. Academic Press, London, U.K, 1995.

[3] K.T. Wu, R.H. Essenhigh. Proc. Combust. Inst. 20 (1985) 1925-1932.

[4] V.R. Katta, L.G. Blevins, W.M. Roquemore, Combust. Flame.142(1) (2005) 33-51.

[5] H.G. Weller, G. Tabor, H. Jasak, C. Fureby. Comp. in Phys., 12 (6) (1998) 620-631

[6] R.I. Issa, J. Comput. Phys., 62(1)(1986) 40-65.

[7] H. Jasak, Error analysis and estimation in the Finite Volume method with applications to fluid flow, 1996.

[8] B. Leonard, Int. J. Num. Meth. Fluids 8(10)(1988) 1291-1318.

[9] M.A. Mikofski, T.C. Williams, C.R. Shaddix, A.C. F. Pello, L.G. Blevins. Combust. Flame, 149 (2007), 463-478

11th Asia-Pacific Conference on Combustion,
The University of Sydney, NSW Australia
10th -14th December 2017.

Numerical simulation research on fire risk in a super long river-crossing subway tunnel

Xinxin Guo[1], Xuhai Pan[1,2,3], Zhilei Wang[1], Juan Yang[1], Min Hua*[1], Juncheng Jiang[1,2,3]

[1]College of Safety Science and Engineering, Nanjing Tech University
Zhongshan North Road, Nanjing, Jiangsu, 210009, China
[2]Institute of Fire Science and Engineering, Nanjing Tech University
Zhongshan North Road, Nanjing, Jiangsu, 210009, China
[3]Jiangsu Key Laboratory of Urban and Industrial Safety
Zhongshan North Road, Nanjing, Jiangsu, 210009, China

Abstract

In recent years, with the gradually increasing number of river-crossing tunnels, the risk of tunnel fire continues to rise, so it is necessary to research the control effect of fire smoke for different kinds of tunnel structure and the ventilation and smoke exhaustion system. The FDS software was used to simulate the scene of fire at different slope segment of Nanjing Metro Line 10 in the case of single-way and two-way structure, and the variation of the temperature, concentration of CO, visibility and the height of smoke layer were analyzed [1]. Research results show that: in terms of the temperature, concentration of CO and visibility in the -7.5‰ tunnel, single-way tunnel structure has a certain advantage, however, for the +8.9‰ tunnel, the two-way tunnel is lower than single-way tunnel in connection with the risk parameter values. And in general, fire risk in the positive (+8.9‰) slope segment is less than that in the negative (-7.5‰) slope segment. By combining the structure of the two tunnels, in the actual design and application process, the choice of two-way scheme is better.

Key words: river-crossing tunnel; longitudinal ventilation; numerical simulation; slope segment; risk

1 Introduction

Nowadays, the traditional traffic mode has been very difficult to withstand the rapid increase of traffic pressure, so more and more tunnels have become an important part of the urban public transport system. However, the small space and low visibility result in extremely serious social impact and economic loss. In view of the different tunnel structures, the impact of the traditional ventilation mode on the tunnel temperature was studied by Vladimir Maslak [2]. The effect of the exhaust duct size and the position of the fire source in tunnels was studied by R. Harish [3]. This paper mainly uses FDS software to carry out a series of research on ventilation and smoke exhaust system of Nanjing Metro Line 10, and through the comparison of different subway tunnels, finally determine the suitable design structure. Nanjing Metro Line 10 tunnel across the river is from the Riverside Road station to Jiangxinzhou station, full length of 3850m, is a super long interval, divided into 1130m, 740m, 1030m, 950m four gradient tunnel, corresponding slope is -28‰, -7.5‰, +8.9‰, +28‰ [4], marking A, B, C, D. This article considers -7.5‰ and +8.9‰ two high-risk tunnels, schematic diagram is shown in Fig. 1. According to existing subway tunnel structure, this paper's main structure of tunnel was divided into two types: single-hole double-line (two-way) and single-hole single-track (single-way) structure.

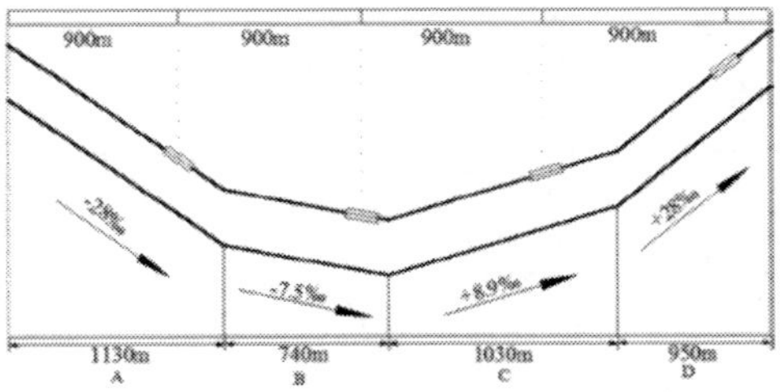

Figure 1: Schematic diagram of tunnel section

2 Different tunnel design schemes

2.1 Tunnel structure

Corresponding author:Min Hua Fax: +86-025-83239949
E-mail address: mhminhua@163.com

The main body of single-way tunnel was two single-line circular shield tunnel. There was a contact channel between the two single-way tunnel, and the inner diameter of the tunnel was 6.2m, because there was no condition to set up a dedicated exhaust duct, it was only suitable for longitudinal ventilation [5]. The two-way tunnel was a single-hole two-line circular shield tunnel, it's inner diameter was 9.8m, and the exhaust duct was installed at the top of the tunnel. Once a fire, the smoke outlet of the downstream of the fire source was opened and the other exhaust vents were closed. The interior partition was designed as a firewall, so that the tunnel separated into two independent part of upper and lower row. Relying on the partition to set up a lateral evacuation platform, the height of platform was consistent with the compartment floor for evacuation.

2.2 Simulation process

2.2.1 Grid size

$$D^* = \left(\frac{Q}{\rho_\infty C_p T_\infty \sqrt{g}} \right)^{2/5} \tag{1}$$

Among them, Q represent the Heat Release Rate (HRR), ρ_∞ is the ambient air density (20℃), c_p is the specific heat of air, T_∞ is the ambient temperature, g is the acceleration of gravity. Generally, the mesh size ranges from $0.1D^*$ to $0.2D^*$ can obtain better simulation results. In this paper, the D^* is equal to 1.95m by calculating the value of the parameter with the Q, so the appropriate mesh size is between 0.195m and 0.39m. Finally, the calculated grid dimension of 0.25m is within the precision range. The partition of the specific grid system is shown in table 1.

Grid System	Single-way precision	Total cells	Two-way precision	Total cells
-7.5‰	0.25m×0.25 m×0.25m	32834 56	0.25m×0.25 m×0.25m	33914 88
+8.9‰	0.25m×0.25 m×0.25m	42588 16	0.25m×0.25 m×0.25m	47370 24

Table 1: The division of the grid system

2.2.2 Initial and boundary conditions

According to the types of combustible materials of urban rail transit vehicles and the professional data of other cities in China, the power of fire source is selected to be 7.5MW, the time required for the corresponding heat release rate of the ignition to reach the peak can be calculated by formula (2). Referring to NFPA classification, the fire development grade is super fast fire, the corresponding α value is 0.1876, and the t is 199.95s when flame burning peak is reached.

$$Q_c = \alpha t^2 \tag{2}$$

Among them, Q_c is the Heat Release Rate (KW), α is the fire development rate (KW/s^2), t is the fire development time (s). In this paper, the size of the exhaust duct for the two-way tunnel is 3m×5m, at the same time, in order to ensure discharging effect and prolong effective evacuation time, the spacing of exhaust duct is set to 900m. Heselden and Danziger [6] proposed the semi-empirical formula based on Froude number and experimental data for critical wind speed [7]. Finally, longitudinal wind speed was determined to be 2.3m/s. In order to effectively organize ventilation and exhaust smoke, prevent smoke from passing through the exhaust duct and continuing to swim down, the smoke exhaust wind speed was set to 6m/s. Combined with "SFPE Fire Engineering Manual" [8], the critical safe temperature was set to 60℃, and the critical safe CO concentration was set to 600ppm. The fire source was located in the middle of the two adjacent exhaust ducts.

3 Results and discussion

3.1 B-section risk parameter

3.1.1 Top temperature

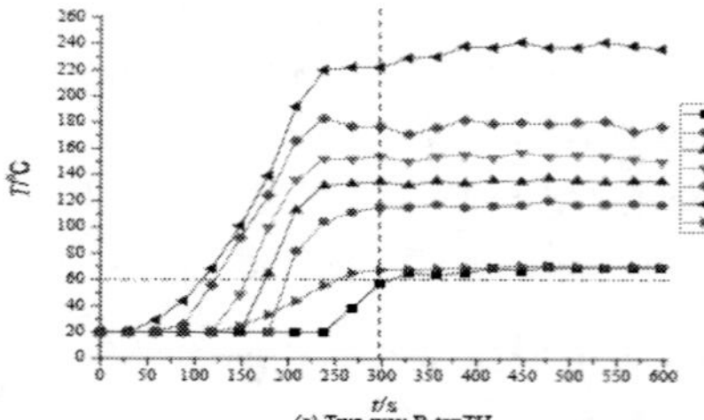

(a) Two-way-B-topTH

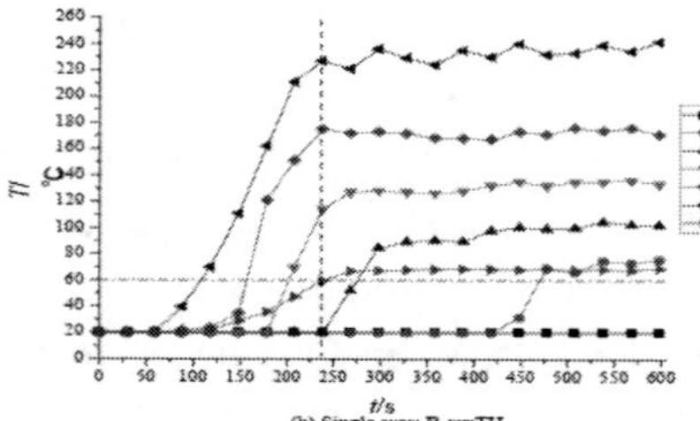

(b) Single-way-B-topTH

Figure 2: Top temperature of B

Compared with the two tunnels, in the single-way mode, it is found that the average temperature of B-section roof is relatively low, and the distance between the measuring point and fire source is more obvious to the temperature.

3.1.2 CO concentration

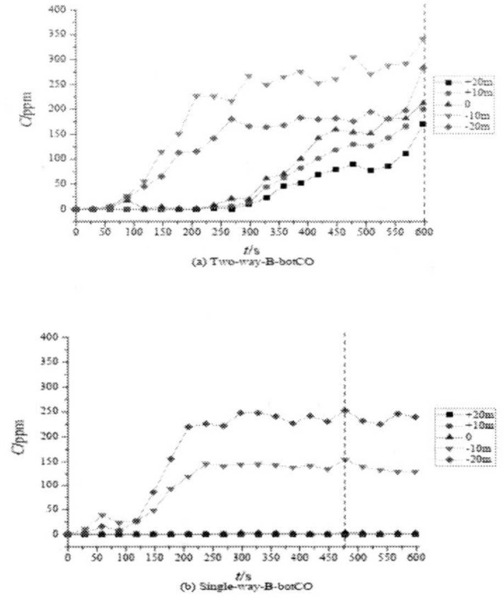

Figure 3: CO concentration from the ground 2m height of B

In comparison with the two figures, the CO concentration in single-way tunnel is lower, but the concentration in the two tunnels is considered to be within acceptable range.

3.1.3 Smoke layer

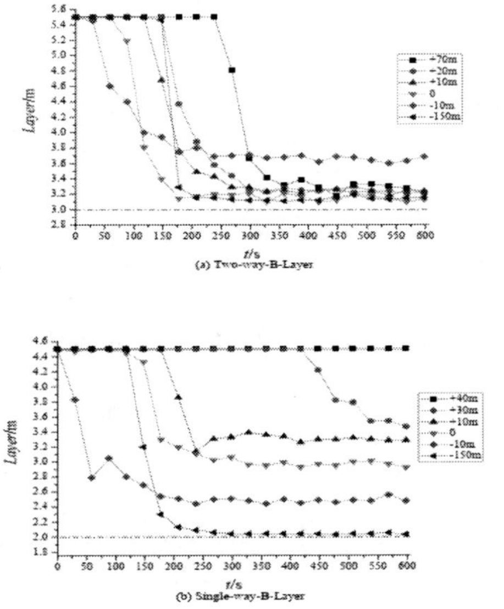

Figure 4: Smoke layer sedimentation height of B

Based on different structures for the two tunnels, it can be concluded that the smoke layer after a fire of the two-way tunnel was higher than that of the single-way tunnel.

3.2 C-section risk parameter

3.2.1 Top temperature

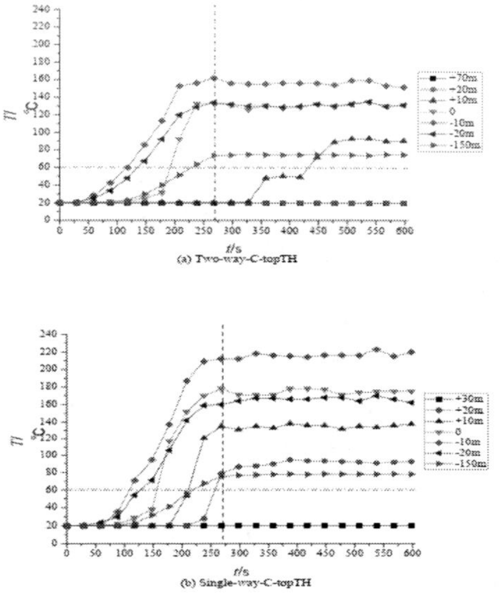

Figure 5: Top temperature of C

In the two-way mode, it is found that the average temperature of C-section roof is relatively low.

3.2.2 CO concentration

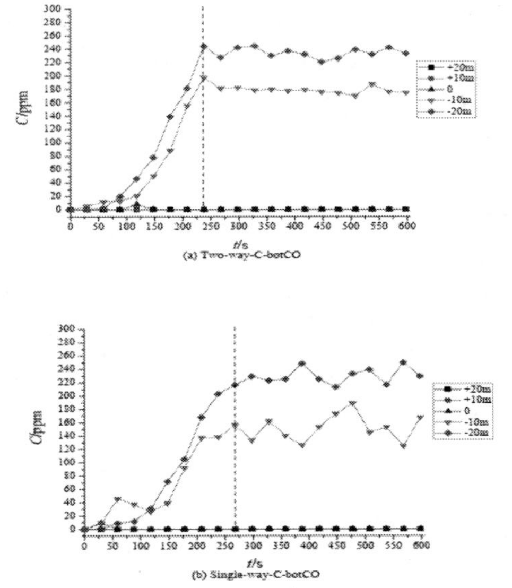

Figure 6: CO concentration from the ground 2m height of C

The CO concentration is much lower than 600ppm in the two groups of data, and the evacuation is not affected by the CO concentration at the 2m height of the two tunnels.

3.2.3 Smoke layer

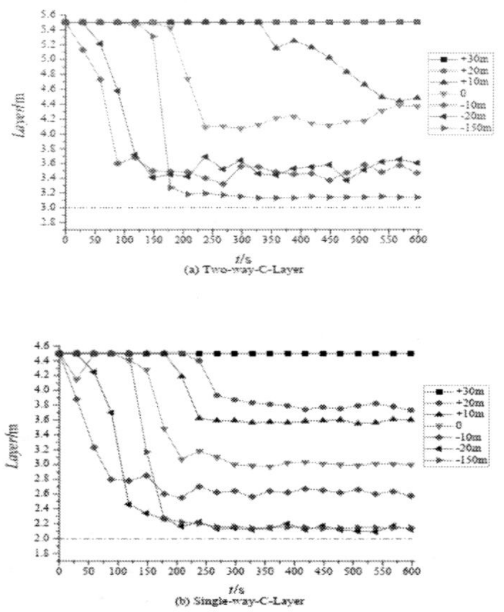

Figure 7: Smoke layer sedimentation height of C

In combination with the above diagram, the spread of the distance for the upstream of the fire source in the two-way tunnel is less than that of the single-way tunnel. And in general, the smoke layer after a fire of the two-way tunnel was higher than that of the single-way tunnel.

3.3 Discussion

From the B-slope section, the two-way tunnel structure of the internal risk parameters are higher than single-way, however, in terms of the C-slope section, the result is the opposite. According to the different position of the fire, it is known that the different tunnel structures have their own advantages in the positive and negative slopes. But because of the large diameter of the two-way tunnel, the height of the smoke layer after a fire is higher, and it's interior is equipped with exhaust duct, evacuation platform and fireproof door, so it has more advantages in the evacuation of personnel. Finally, the design scheme of two-way tunnel is considered better in practical engineering application.

4 Conclusions

(1) In view of the -7.5‰ slope tunnel, after a fire, in single-way mode, the roof average temperature is relatively low, and the distance between the measuring point and the fire source has more obvious influence on the temperature.

(2) In terms of the +8.9‰ slope tunnel, once a fire, in two-way mode, the roof average temperature is relatively low, and the upstream temperature is generally lower than the single-way roof temperature in the same position.

(3) The results show that the risk of the positive slope segment is lower than that in the negative slope segment after a fire, and the harm to personnel is relatively low.

(4) Because of the structural characteristics of the two-way tunnel, including the smoke exhaust duct, evacuation platform and fireproof door, etc., in the event of fire, it is better to choose two-way design plan from the angle of ensuring the safety evacuation of the personnel.

5 Acknowledgment

This work was supported by the National Natural Science Foundation of China (No.51476075).

References

• Gao R, Li A, Hao X, et al. Fire-induced smoke control via hybrid ventilation in a huge transit terminal subway station[J]. Energy & Buildings, 2012, 45(2):280-289.

• Maslak V, Boytsov D, Danilov A, et al. Innovative Engineering Solutions for Improving Operational Safety and Efficiency of Subways with Two-way Tunnels ☆[J].Procedia Engineering, 2016, 165: 214-223.

• Harish R, Venkatasubbaiah K. Effects of buoyancy induced roof ventilation systems for smoke removal in tunnel fires[J]. Tunnelling & Underground Space Technology, 2014, 42(5): 195-205.

• Yi L, Xu Q, Xu Z, et al. An experimental study on critical velocity in sloping tunnel with longitudinal ventilation under fire[J]. Tunnelling & Underground Space Technology Incorporating Trenchless Technology Research, 2014, 43(1):198-203.

• Luo N, Li A, Gao R, et al. Smoke confinement utilizing the USME ventilation mode for subway station fire[J]. Safety Science, 2014, 70:202-210.

• Danziger N H. Fire Life Safety[M]// Tunnel Engineering Handbook. Springer US, 1996:369-383.

• Xi Y H, Mao J, Bai G, et al. Safe velocity of on-fire train running in the tunnel[J]. Tunnelling & Underground Space Technology Incorporating Trenchless Technology Research, 2016, 60:210-223.

• Hurley M J, Gottuk D T, et al. SFPE Handbook of Fire Protection Engineering[M]. Springer New York,2016.

11th Asia-Pacific Conference on Combustion,
The University of Sydney, NSW Australia
10th -14th December 2017.

The Influence of Swirl Flow on Soot Formation Processes inside the Piston Bowl of a Small-bore Diesel Engine

Y. Zhang[1], D. Kim[1], L. Rao[1], S. Kook[1,*], K.S. Kim[2], and C.-B. Kweon[2]

[1]School of Mechanical and Manufacturing Engineering, The University of New South Wales
[2]US Army Research Laboratory

Abstract

Although swirling flow is known to enhance fuel-air mixing in diesel engines and thereby reducing the engine-out soot emissions, its influence on in-cylinder soot formation is poorly understood. The present study provides a direct comparison of soot particle structures between the wall-jet heads developing on up- and down-swirl sides via the thermophoresis-based soot particles sampling inside the piston bowl of a small-bore diesel engine. The sampled soot particles are imaged using a transmission electron microscope and post-processed to obtain the size of soot primary particles and aggregates as well as the fractal dimension. The TEM images show more significant reduction in soot concentration for the wall-jet head developing from the flame-wall impingement region in the up-swirl direction. The statistical analysis also shows that the increased size of soot aggregates and primary particles as well as the decreased fractal dimension are more apparent on the up-swirl side as the wall-jet travels against the swirl flow.

1. Introduction

The adverse impacts of diesel soot emissions have called for strict regulations in various countries. Although diesel-powered vehicles are mostly equipped with aftertreatment filters to reduce tailpipe soot emissions, the sacrifice of engine performance due to back pressure and the extra cost have turned intensive interests to soot reduction inside the engine cylinder. The fuel-air mixing enhancement has been found to play a key role to mitigate in-cylinder soot yield [1]. While the high pressure fuel injection using a common-rail system [2] and small nozzle injectors [3] is a primary cause for the improved air entrainment and reduced soot yield, the swirl flow induced by the advanced intake port designs has also been effective in enhancing the mixing and reducing soot [4].

Previous studies reported that the fuel jet developing on the up-swirl side shows enhanced pre-combustion mixing and thereby lower soot formation than those on the down-swirl side [2, 5]. The laser induced incandescence images presented in our previous study showed a consistent trend that the soot formation is lower on the up-swirl side due to locally leaner mixtures [6]. Although it is apparent that the swirl flow helps increase fuel-air mixing, one study reported the complex vortex structure on the up-swirl side of the penetrating wall jet could cause limited oxygen and thus higher soot concentration [7]. The mixed conclusions in the previous studies therefore present the need for a close investigation of soot formation process on each side of a reacting jet under the influence of swirl flow. The present study aims to bridge this gap by sampling soot particles directly from the reacting diesel flame in a running engine.

In a swirl-supported engine, the present study performed soot particle sampling within the piston bowl, in which particles are anticipated to incept and grow while the flame develops along the bowl wall on both up-swirl and down-swirl sides. The soot particles were sampled from the region of flame-wall impingement and the developing wall-jet heads in the directions along and against swirl flow before they are imaged using a transmission electron microscope (TEM). The images are post-processed to obtain soot primary particle diameter, aggregate radius of gyration, and fractal dimension for soot particle aggregates sampled at each location.

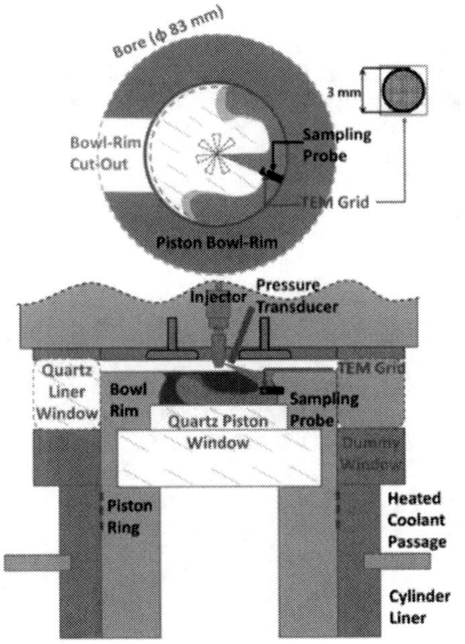

Figure 1. Illustration of the optical diesel engine and soot sampling system in the bottom-view (top) and side-view (bottom) access.

A/Prof S. Kook
E-mail address: s.kook@unsw.edu.au

2. Methodology

The in-bowl soot sampling experiments were conducted in a single-cylinder automotive-size diesel engine with optical access, as illustrated in Fig. 1. The specifications and operating conditions of the engine are summarised in Table 1. The engine has a displacement volume of 498 cm^3 with 83-mm bore and 92-mm stroke. The piston bowl was designed with a curved-lip wall onto which a sampling probe was installed to store a TEM grid for the sampling of soot particles directly from the flame. The combustion chamber is bounded by the piston bowl, flat cylinder head and cylindrical liner leading to a geometric compression ratio of 15.5. The engine was naturally aspired with no throttling and the estimated swirl ratio was 1.4. The coolant water was heated to 90°C and then circulated through the engine head, engine block and cylinder liner to simulate warmed up engine operating conditions. The intake air temperature was measured at 30°C throughout the experiments. The engine speed was held constant at 1200 revolutions per minute (rpm) using an electric motor.

Methyl decanoate ($C_{11}H_{22}O_2$) was selected as a diagnostics fuel for the soot sampling experiments in the present study. A common-rail system (Bosch CP3) and a solenoid-type injector was used for the direct fuel injection with the timing and duration of injection being controlled by a universal signal controller (Zenobalti 9013-P). Centrally-mounted on the cylinder head was the injector that originally had 7 holes with a nominal hole-diameter of 134 μm and an included angle of 150°. The nozzle holes were laser welded to have only one of them open for fuel injection, which isolates the soot formation process of a single jet flame from complex jet-jet interactions. Also, this approach allowed for long fuel injection duration of 2.34 ms (actual) with no concern on the safety limit of optical windows. The injection timing was fixed at 9 crank angle degrees before top dead centre (°CA bTDC). The piston top can be easily removed from the piston body when the cylinder liner is dropped down, which enables the rotation of the sampling probe to collect soot particles from the wall-impinged flame and the wall-jet heads 60 angular degrees downstream from the jet axis in both up- and down-swirl directions, as illustrated in Fig. 2. The sampling locations are carefully selected based on our previous planar laser-induced incandescence (PLII) images [6] in which the initial soot formation region was found near the flame-wall impingement point and soot pocket in wall-jet head was noticed at 60° downstream of flame-wall impingement location on both up- and down-swirl sides. Therefore, the comparison of soot formation process can be made between the soot particles on the up-swirl and down-swirl side of the wall-jet. The present study abbreviates the sampling location of flame-wall impingement as FW, while the other two 60° downstream sampling locations in up- and down-swirl directions are labeled using US and DS, respectively.

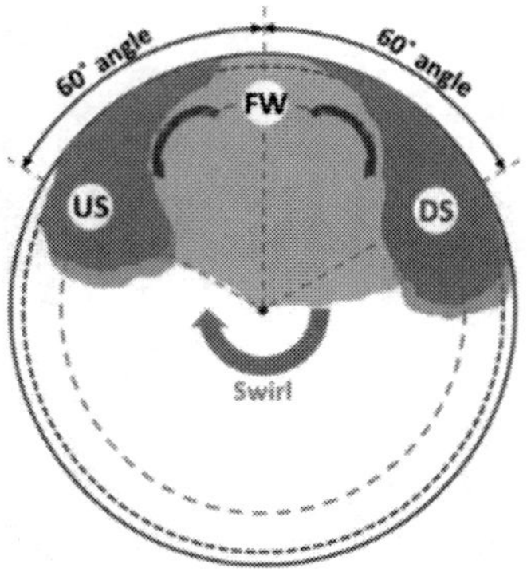

Figure 2. Illustration of the sooting flame development and the three sampling locations of the present study: Flame-wall impingement location (FW), up-swirl location (US) and down-swirl location (DS).

Table 1. Engine specifications and operating conditions

Displacement volume	498 cm^3 (single cylinder)
Bore/stroke	83 mm/92 mm
Compression ratio	15.5
Swirl ratio	1.4
Coolant temperature	90°C
Intake air temperature	30°C
Engine speed	1200 rpm
Fuel	Methyl decanoate
Cetane number	48-51.7
Injection system	Common-rail injector (Bosch CP3)
Nozzle hole diameter	134 μm (nominal)
Included angle	150°
Injection pressure	100 MPa
Injection duration	2.34 ms
Injection timing	9°CA bTDC

Similar to our previous study [8], the sampling probe tip has a shallow circular pocket in which a TEM grid (Emgrid CF400-Cu) can be placed. The cover plate is used to hold the grid in place while a 2-mm opening on the probe body provides a small passage for the direct exposure of the grid to sooting flames. This restricted passage successfully protects the TEM grid (3 mm in diameter) from excessive heat and maintains the low grid temperature needed for the temperature gradient – i.e. thermophoresis. The engine was motored for about 90 seconds to create thermally stable in-cylinder conditions prior to fuel injection. The fuel injection was executed every 10th cycle (i.e. 9 motoring and 1 injection cycles) for up to 20 injections (i.e. 20 injections onto the TEM grid) for each sampling experiment. This skip-firing approach was used to expel the residual gases from the previous firing cycle and the selected number of injections was to ensure sufficient soot particles deposited on TEM grids for statistical data analysis while preventing an overloading issue.

The sampled soot particles onto the standard TEM grids were observed using a TEM (JEOL 1400) with a point resolution of 0.38 nm and an accelerating voltage of 120 kV under magnification at 100,000 times, which were recorded in a CCD camera with a pixel resolution of 1350 by 1040. Details about the post-processing of standard TEM images and structural analysis of soot aggregates and primary particles are found in our previous study [9]. The same in-house-developed Matlab code was used in which the soot aggregates are selected from the raw images and then the edge of the soot primary particles is highlighted via unsharp filtering, intensity inversion, and self-subtraction. The primary particles are automatically detected using the Canny Edge Detection (CED) and Circular Hough Transform (CHT) with the fine tuning of the six image processing parameters [9]. The obtained size of the soot primary particles and aggregates was used to calculate the fractal dimension of the soot aggregates. For each standard TEM grid sample, a total of 30 images were obtained from various on-grid locations.

3. Results and Discussion

The in-cylinder gas conditions during the in-bowl soot sampling experiments are shown in Fig. 3 using the ensemble-averaged in-cylinder pressure. The blue lines correspond to the three sampling experiments performed for the three in-bowl locations. For each sampling experiment, the ensemble average of 20 firing cycles was compared to the motoring pressure trace (black line) and other ensemble-averaged in-cylinder pressure traces. To show the overall in-cylinder phenomenon of the selected combustion conditions for the sampling experiments, the average of all the in-cylinder pressure traces is plotted using a red line in Fig. 3. Also shown at the bottom of Fig. 3 are the corresponding apparent heat release rates (aHRR). It is shown that upon the start of fuel injection at 9 °CA bTDC, the firing cycle in-cylinder pressure becomes lower than the motoring pressure while the aHHR becomes negative due to the evaporative cooling. The start of combustion occurs at around 6 °CA aTDC resulting in a significant increase in the aHRR and in turn in-cylinder pressure. The injection ended at 8 °CA aTDC, slightly after the aHRR reaches its maximum value of about 25 J/°CA, before the end of combustion led to the decreased aHRR and in-cylinder pressure. Its peak in-cylinder pressure exceeding the motoring peak pressure and its timing observed after the TDC show a typical in-cylinder condition of diesel engine combustion. From Fig. 3, it is evident that the soot sampling experiments, despite being performed from different engine cycles/runs, were conducted at the same in-cylinder conditions regardless of the sampling location.

Figure 4 (middle) shows the selected TEM images for the soot particles sampled at location FW. The standard TEM images

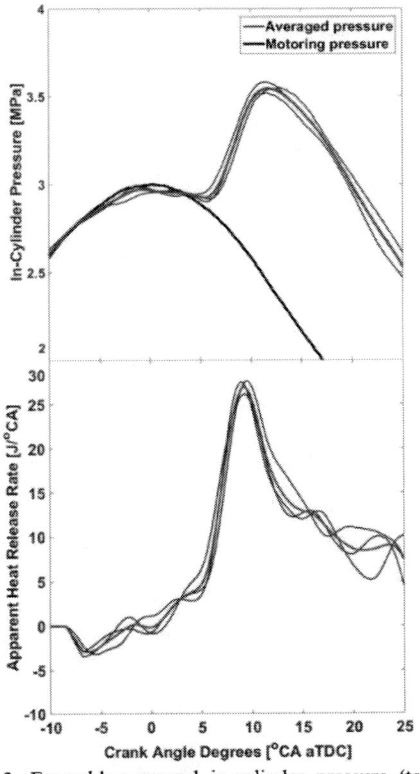

Figure 3. Ensemble-averaged in-cylinder pressure (top) and apparent heat release rate (aHRR, bottom) traces (blue) for each of the 3 different sampling engine runs. The average of the measured in-cylinder pressure and aHRR are displayed using a red line.

show many small particles with most of them being single-primary particles or soot aggregates comprising only two to three primary particles. There are a few larger soot aggregates with fractal-like structures, which however are outnumbered by very small particles appearing as black dots for the used TEM magnification setting. Noticeably, the sampled particles exhibit low contrast signals with no clear boundaries. This faint appearance was reported as soot precursors and soot particles in the early formation stage [10]. The TEM images of soot particles sampled at location US are shown in the left columns of Fig. 4. Compared to location FW, location US show greatly reduced soot concentration. It is noticeable that the single and double primary particles are rare in location US but larger soot aggregates with chain-like structures comprising of many soot primary particles that are larger in size with clear boundaries, suggesting a combination of aggregates and surface growth as well as the carbonisation of primary particles [11]. Shown on the right column of Fig.4 are the TEM images of soot particles sampled at location DS. The soot amount in location DS also decreased apparently from location FW while it is still much larger than that in location US. The higher soot concentration in location DS than location US is primarily because of the presence of many small soot particles. Compared to location FW, the growth in the size of soot aggregates and primary particles and the clearly defined boundaries of the primaries in location DS also suggest the aggregate combination and the primary particle surface growth and carbonisation.

The size of soot aggregates and soot primary particles are reduced to the probability density functions (pdfs) of soot primary particle diameter (d_p) and aggregate radius of gyration (R_g) in the left and middle of Fig. 5, respectively, for the three sampling locations. The mean values with the error range of 95% confidence are annotated next to each pdf plots. The pdfs and the mean values show increased d_p and R_g from location FW to the two downstream locations, which confirms the visual inspection that the size of soot aggregate and primary particles increased. However, the soot particles in wall-jet heads developing in US and DS directions experienced different levels of increase in d_p and R_g. For instance, the mean d_p increased for only 1 nm from location FW to DS, whereas increase of 4 nm is noticed in d_p for soot particles developing in US direction. Also, the growth in R_g of 33 nm from location FW to US is much more than 24 nm for soot travelling in DS direction. The overall smaller d_p and R_g of the soot particles in the wall-jet head traveling in DS direction is likely due to the existence of many small aggregates and primary particles that can be hardly found in the sample of US location. The absence of small soot particles could be the results of enhanced mixing that lowered the soot formation and promoted oxidation to remove those small particles on the US side of the jet.

The results of the soot particles fractal demission (D_f) for the three sampling locations are shown in the left of Fig. 5. The D_f was calculated using the measured R_g and d_p to descript the structural compactness of soot aggregates. For instance, more compacted soot aggregate has higher fractal dimension. The D_f is lower for the two downstream locations than location FW, suggesting the aggregation of small soot particles from the early formation stage to form larger, stretched aggregates. The soot particles sampled at location US has the lowest D_f likely due to the weak formation and the oxidative elimination of those small cluster-like aggregates that are very common in the other two locations (see Fig. 4).

4. Conclusion

The soot sampling was conducted inside the piston bowl of a small-bore diesel engine to compare the soot formation

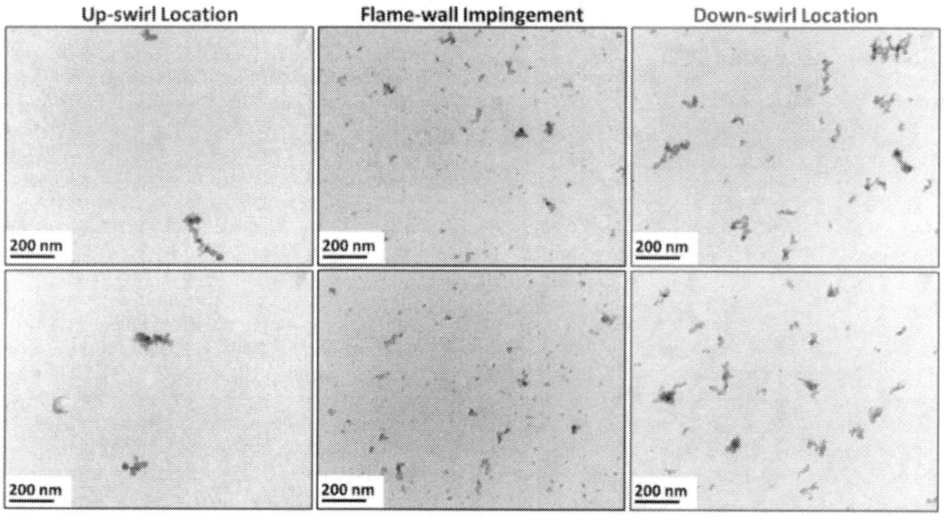

Figure 4. The TEM images for soot particles sampled at up-swirl location (left), flame-wall impingement location (middle), and down-swirl location (right).

processes between the wall-jet heads developing in up- and down-swirl directions. Soot samples were imaged using a transmission electron microscope and then post-processed for the statistical analysis of soot particle size and fractal dimension. The TEM images show a decrease in soot concentration from flame-wall impingement region to downstream locations, and the lowest soot amount of up-swirl location. Soot aggregate and primary particle size increased in the wall-jet heads on both up- and down-swirl sides, while the growth is more apparent on the up-swirl side mainly due to the less formation and more oxidation to eliminate small soot particles. The fractal dimension decreased for proceeding soot formation in the downstream locations, which is more obvious on the up-swirl side.

Acknowledgement

Experiments were performed at the UNSW Engine Research Laboratory, Sydney, Australia. Support for this research was provided by the US Army RDECOM ITC-PAC and in part by the Australian Research Council.

References

[1] Pickett, L.M., Manin, J., Genzale, C.L., Siebers, D.L., Musculus, M.P.B., Idicheria, C.A., SAE Int. J. Engines 4 (2011) 764–799.

[2] Rusly, A., Le, M.K., Kook, S., SAE Int. J. Fuels Lubr. 6 (2013) 615–626.

[3] Pickett, L.M., Siebers, D.L., Proc. Combust. Inst. 29 (2002) 655–662.

[4] Mccracken, M.E., Abraham, J., SAE Paper 2001-01-0996.

[5] Musculus, M.P.B., SAE Paper 2003-01-0074.

[6] Le, M.K., Zhang, R., Rao, L., Kook, S., Hawkes, E.R., Fuel 166 (2015) 320–332.

[7] Bergin, M.J., Reitz, R.D., Miles, P.C., Hildingsson, L., Hultqvist, A., SAE Paper 2007-01-0912.

[8] Kook, S., Zhang, R., Szeto, K., Pickett, L.M., Aizawa, T.,SAE Int. J. Fuels Lubr. 6 (2013)80–97.

[9] Zhang, Y., Zhang, R., Rao, L., Kim, D., Kook, S., Fuel 194 (2017) 423–435.

[10] Kook, S., Pickett, L.M., SAE Int. J. Fuels Lubr. 5 (2012) 647–664.

[11] Abid, A.D., Heinz, N., Tolmachoff, E.D., Phares, D.J., Campbell, C.S., Wang, H., Combust. Flame 154 (2008) 775–788.

[12] Echavarria, C.A., Jaramillo, I.C., Sarofim, A.F., Lighty, J.S., Combust. Flame 159 (2012) 2441–2448.

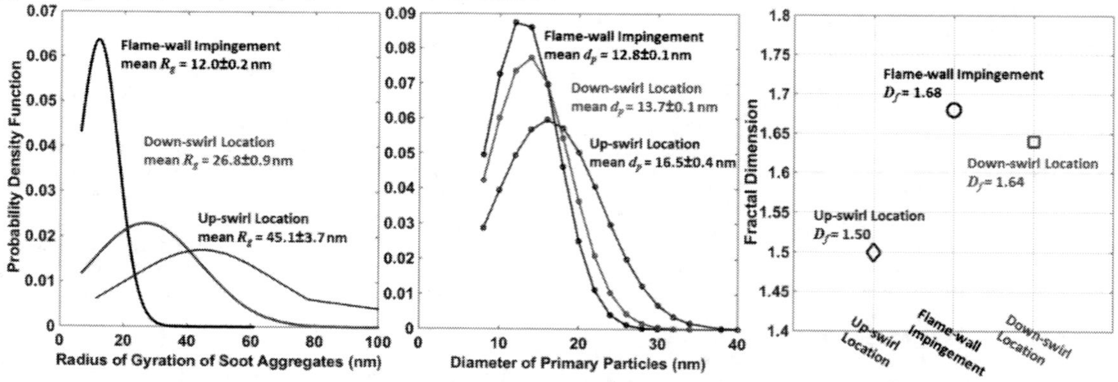

Figure 5. The TEM images for soot particles sampled at up-swirl location (left), flame-wall impingement location (middle), and down-swirl location (right).

11th Asia-Pacific Conference on Combustion,
The University of Sydney, NSW Australia
10th -14th December 2017.

Investigation of Methane and Coal Dust Concentration Impact on Explosion Pressure Rise in a Confined Spherical Space

Yusuf Badat[1], Mohammed Jabbar Ajrash[1], Jafar Zanganeh [1*], Behdad Moghtaderi[1]

[1] Priority Research Centre for Frontier Energy Technologies & Utilisation, Chemical Engineering, School of Engineering, Faculty of Engineering & Built Environment

The University of Newcastle, Callaghan , NSW, 2308, Australia

Abstract

Preventing explosions from happening requires a good understanding of the characteristics of a gas explosion.

The effect of methane concentration, coal dust concentration and igniters energy on explosion characteristics were investigated using a 1m^3 spherical vessel.

The pressure rise increased from 6.8 bar for a 6.5% methane mixture to 8.1 and 8.3 bar for 9.5% and 11.0% methane mixtures respectively. Additionally; the rates of pressure rise increased from 222 to 517 bar.s^{-1} for the range of methane mixtures exploded.

The severity of the explosions increased with increasing igniter energy. The maximum pressure rise rate increased from 83 bar.s^{-1} for 1 kJ igniter to 90 bar.s^{-1} for 2 kJ igniter, to 155 bar.s^{-1} for 5 kJ igniter and to 222 bar.s^{-1} for 10 kJ igniter.

The severity of the explosions increased with increased coal dust concentration. At 2.5% methane-air mixture, the pressure rise rate increased from 4 bar.s^{-1} to 114 bar.s^{-1} when the coal dust concentration increased from 50 g.m^{-3} to 100 g.m^{-3}, respectively.

1. Introduction

An understanding of the explosions that occur in underground coal mines is critical to the successful mitigation of such explosions and to the prevention of resultant loss of life. These explosions may result from the ignition of the methane contained within the coal seams when concentrations reach the flammability range of methane. It has been shown that coal dust plays a significant role in reducing the minimum energy required for methane explosions [1-3, 5 and 7].

Ajrash, et al [2] investigated the effects of ignition energy on fire and explosion characteristics of dilute hybrid fuel in ventilation air methane in a closed 20L spherical vessel. In this study, the Minimum Explosion Concentration (MEC), Pressure Rise (PR), deflagration index for gas and dust hybrid mixtures (K_{st}) and explosive region of hybrid fuel mixtures present in Ventilation Air Methane (VAM) were investigated. Results obtained from this study have shown that the presence of methane significantly affects explosion characteristics of coal dust clouds. Dilute concentrations of methane, 0.75-1.25%, resulted in coal dust clouds PR increasing from 0.3 bar to 2.2 bar and boosting the K_{st} value from 10 bar m s^{-1} to 25 bar m s^{-1}.

The explosion characteristics were also affected by the ignitors' energy; for instance, for a coal dust cloud concentration of 50 g m^{-3} the PR recorded was 0.09 bar when a 1 kJ chemical ignitor was used, while, 0.75 bar (PR) was recorded when a 10 kJ chemical ignitor was used. The explosive concentration range significantly increased as the energy of ignitors increased. Dilute concentrations of methane were shown to be capable of reducing the MEC of coal dust clouds and the MEC decreased with increasing energy of the ignition source.

Ajrash, et al [3] studied the influences of the initial ignition energy on methane explosions in a 30-m long flame deflagration tube. The initial ignition energy was delivered by three alternative chemical ignitors' energies, which were 1, 5, and 10 kJ. The results of the study revealed the notable influences of the initial ignition energies on the flame deflagrations, over pressure rises, and pressure wave velocities along the detonation tube. When the initial ignition energy was increased from 1 kJ to 10 kJ, the maximum pressure rise increased by 45% and 56%, respectively, for the 7.5% and 9.5% methane concentrations. For a 9.5% methane concentration, the increased ignition energy enhanced the pressure wave velocity from 130 m·s^{-1} to 359 m·s^{-1} and enhanced the flame deflagration velocity from 105 m·s^{-1} to 179 m·s^{-1}.

Hertzberg, et al [7] studied flammability limit measurements for dusts and gases in relation to ignition energy requirements and pressure dependencies in 20 L and 120 L chambers. Pressure rise measurements were made with methane in air using igniters having different source strengths (energies of 0.5, 1.0, 2.5 and 5.0 kJ). The pressure rise increased from 0.5 bar at 0.5 kJ to 4.2 bar at 5.0 kJ. Additionally the pressure rise rate in bar m s^{-1} increased for varying methane concentrations.

Amyotte, et al [1] studied the ignitability of two coals by examining the influence of ignition energy on the lean flammability and the effects of methane mixtures on the ignitability characteristics of each coal. The apparent lean limit at a given ignition energy was found to decrease in the presence of methane. Also, the ignitability of the coal dust-air mixtures was enhanced by the presence of methane, meaning that ignition of a given coal dust concentration was possible with lower energies.

Li, et al [9] examined the explosion characteristics of methane/air/coal dust mixtures in a 20L spherical vessel. They found that explosion parameters such as pressure rise rate increases with increasing coal dust concentration.

A 1 m^3 closed spherical vessel recently acquired by the University of Newcastle was used to investigate the explosions resulting from the ignition of methane-air mixtures. The aim of this study is to examine the impact of methane concentration, coal dust concentration and ignition energy on the maximum

Corresponding author. Phone: +61-2-4033-9335
E-mail address: jafar.zanganeh@newcastle.edu.au

explosion pressure rise and the deflagration index (K_{st}).

2. Experimental Setup and Instrumentation

Figures 1 and 2 show the 1 m^3 dust explosion apparatus used in this study. A methane-air gas mixture is ignited inside the vessel designed to a maximum pressure of 35 bar. A data acquisition system linked to the vessel records the maximum pressure rise, maximum pressure rise rate and K_{st} for each test. Coal dust at 21 bar is injected via the dust chamber located on the side of the vessel. The vessel, although designed for dust explosions, can be used for gas explosions as well. For gas explosions, the dust injection system is used to inject high pressure gas to create turbulence and enhance mixing. The vessel includes pressure transducers, thermocouples, vacuum and pressurizing circuits, dust dispersion system, dust or gas ignition system, water cooling system, gas supply circuit and operating system linked to a computer.

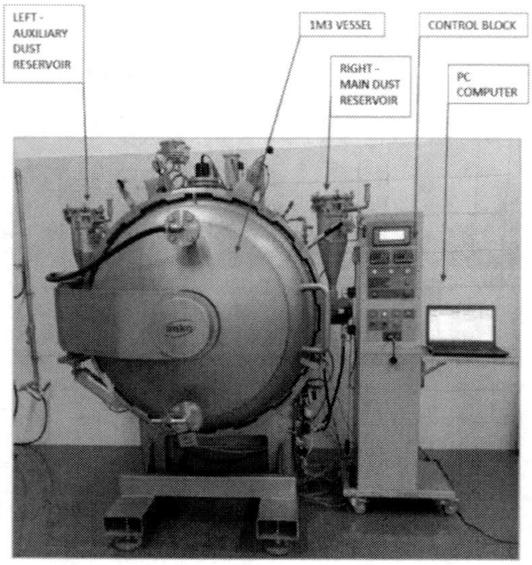

Figure 1: Spherical 1 m^3 test apparatus – external

Igniters of ignition energies 1, 2, 5 and 10 kJ were used to ignite methane-air mixtures of varying concentration within the vessel.

Methane-air gas compositions were controlled by measuring and controlling the partial pressure of methane added to an initially evacuated system.

For each test, a chemical igniter is installed at the end of an ignition stick. A short electric pulse (1s) initiates the ignition of the detonator inside the vessel containing the required concentration of methane gas. Rapid introduction of air at 21 bar via the dust chamber provides turbulence to ensure uniform gas concentration within the vessel. Combustion products are removed from the vessel after the test. The vessel is then purged with air and prepared for the next test.

Figure 2: Spherical 1 m^3 test apparatus – internal

Coal dust, sourced from a NSW mine in Australia, used for this study was of size less than 74 μm. The proximate analysis is shown in Table 1.

Table 1: Coal dust analysis

Proximate analysis (%)			
Ash	Moisture	Volatile matter	Carbon
21.4	3.1	28	47.5

3. Results and Discussion

3.1 Influence of Methane Concentration on Pressure profile

Figure 3 shows the pressure-time history data obtained when a 10kJ igniter was used as the methane concentration was varied at 6.5, 9.5 and 11.0 % methane.

An ignition delay time of 600 ms was set for the tests to allow for the injection of high pressure (21 bar) methane-air mixture in order to create turbulence and good mixing before ignition. This resulted in a slight increase in the pressure (0.2 bar) before ignition during each test.

The igniter is energized in about 40 ms during which the pressure rises a further 0.2-0.3 bar.

A pressure rise of 6.8 bar was recorded when a 6.5% methane mixture was exploded. The pressure rise increased to 8.1 and 8.3 bar for 9.5% and 11.0% methane mixtures respectively. Additionally; the rates of pressure rise increased from 222 to 473 and 517 $bar.s^{-1}$ for the range of methane mixtures exploded.

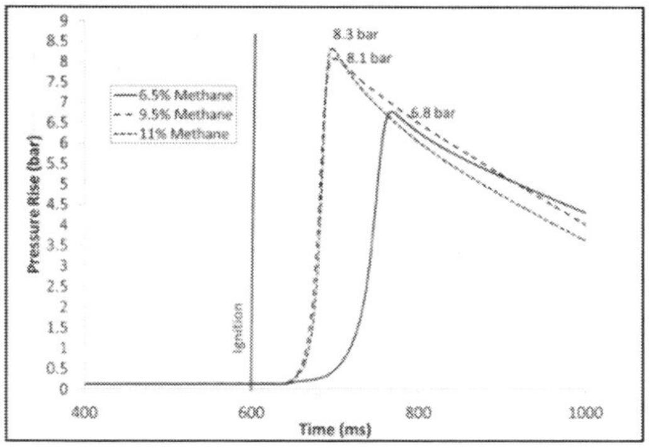

Figure 3: Pressure-Time history for 10 kJ igniter

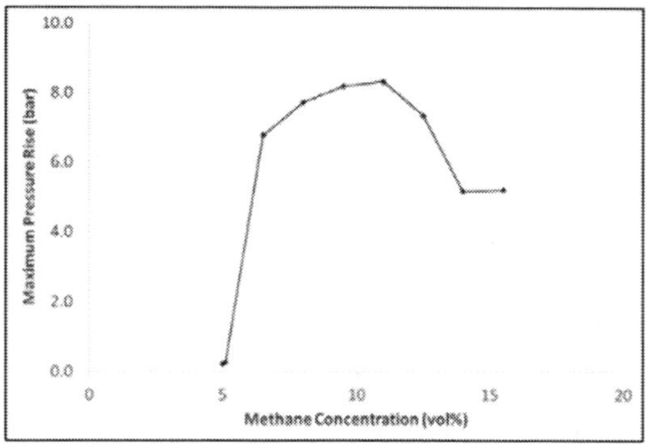

Figure 4: Pressure rise vs concentration - 10 kJ igniter

The deflagration index (K_{st}) is an important characteristic in the evaluation of dust explosions. It is dependent on the value of the maximum pressure of the explosion and the time at which this occurs for a fixed volume within the explosion chamber.

Deflagration index (K_{st}), bar.m.s^{-1} is the normalized pressure rise to 1 m^3, and is calculated from the following equation [2]:

$$K_{st} = \left(\frac{dp}{dt}\right)_{\max} .V^{\left(\frac{1}{3}\right)} \quad (1)$$

where $\left(\frac{dp}{dt}\right)_{max}$ is the maximum rate of pressure rise of a single explosion, bar.s^{-1}; and V is the volume of the explosion chamber, m^3.

For this series of experiments; since the tests are carried out in a 1 m^3 spherical vessel, the deflagration index of the explosions becomes equal to the maximum pressure rise rate measured.

The K_{st} value can be used as an indicator of explosion severity. Nifuku, et al [10] classified the severity of explosions based on K_{st} values. This is shown in Table 1.

Table 2: Explosion severity

Explosion Class	K_{st} (bar.m.s^{-1})	Explosion Severity
St 0	0	non explosive
St 1	1-200	weak
St 2	201-300	strong
St 3	> 300	extreme

For the explosions resulting from the ignition of 6.5%, 9.5% and 11% methane-air mixtures using a 10 kJ igniter; the maximum rate of pressure rise (and subsequently the K_{st}) was measured to be 222, 473 and 517 bar.s^{-1} respectively. These explosions can therefore be classified as strong to extreme according to Table 2.

3.2 Influence of Methane Concentration on Pressure Rise

The effect of increasing concentration of methane in the methane-air mixture on the maximum pressure rise and pressure rise rate using a 10 kJ igniter were investigated. The results are shown in Figures 4 and 5.

As expected, the maximum pressure rise increased with increasing methane concentrations up to about 11% percent, then decreased for further increase in concentrations. The maximum pressure and pressure rise rates measured are highest near the stoichiometric ratio of methane-air mixture. A maximum value of 8.3 bar with a corresponding rate of pressure rise of 517 bar.s^{-1} was measured at a methane concentration of 11%. These results are in good agreement with Cashdollar and Herzberg (1985) [7]. As the fuel reaches the stoichiometric concentration, the flame burning rate reaches a maximum value and the fuel is converted to combustion products in a shorter time. When fuel concentrations higher or lower than the stoichiometric ratio are used; the flame burning rate decelerates resulting in lower pressure rise rates.

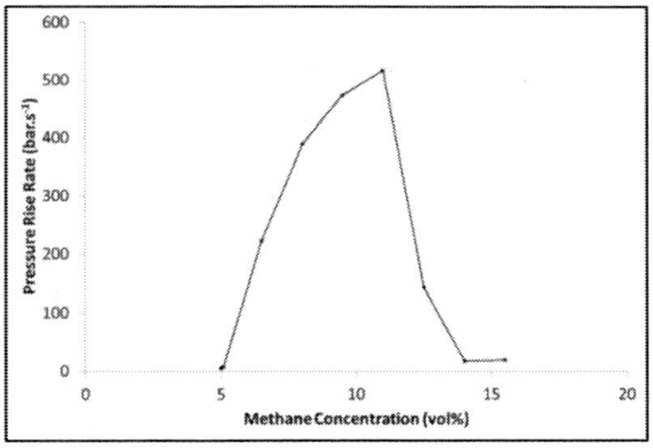

Figure 5: Pressure rise rate vs concentration - 10kJ igniter

3.3 Influence of Igniter Energy on Pressure Rise

The effect of igniter energy for a given concentration of methane-air mixture (6.5%) is shown in Figure 5. The maximum pressure rise rate (bar.s^{-1}) increased from 83 bar.s^{-1} for 1 kJ igniter to 90 bar.s^{-1} for 2 kJ igniter, 155 bar/s for 5 kJ igniter to 222 bar.s^{-1} for 10 kJ igniter.

These results support the fact that increasing initial ignition energy, especially for chemical ignitors, enhances the turbulence of the fluid resulting in high flame deflagration.

Amyotte, et al (1993) [1] postulated that the ignition of chemical igniters are volumetric in action. The radius of influence of the higher energy igniters is greater i.e. the ignition volume as well as the energy is increased. Both factors can lead to a decrease in the measured lean limit and subsequently higher flame deflagration.

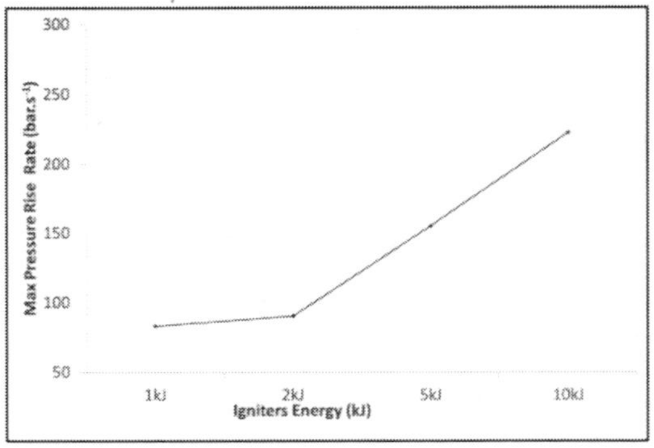

Figure 6: Pressure rise rate vs igniter energy at 6.5% methane

3. Influence of Coal Dust Concentration on Pressure Rise

The effect of increasing concentrations of coal dust in varying methane-air mixtures on the pressure rise rate using a 10 kJ igniter was investigated. The results are shown in Figure 7.

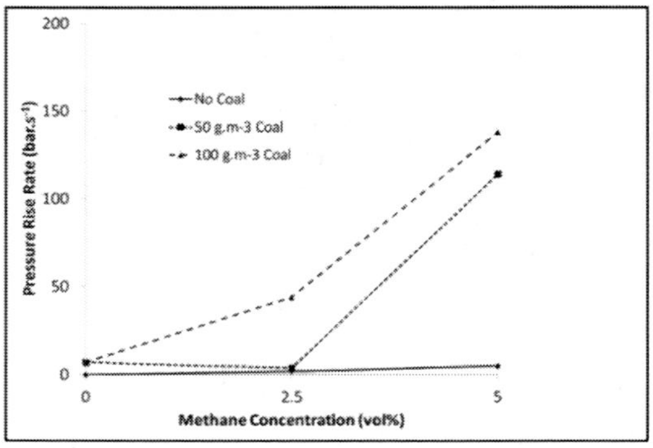

Figure 7: Effect of coal dust concentration on pressure rise rate

At 2.5 and 5 % methane concentrations no explosions occurred without coal dust injection. At 2.5% methane, the pressure rise rate increased to 4 $bar.s^{-1}$ when 50 $g.m^{-3}$ of coal dust was injected into mixture; the pressure rise rate increased further to 114 $bar.s^{-1}$ when 100 $g.m^{-3}$ of coal dust was used. For a 5% methane-air mixture, the pressure rise rate increased to 44 $bar.s^{-1}$ and 138 $bar.s^{-1}$ when 50 $g.m^{-3}$ and 100 $g.m^{-3}$ of coal dust was injected respectively. These results show good agreement with Ajrash, et al [2].

The combustion of the coal particles can be summarized in three stages: devolatilisation and volatile matter combustion, active site combustion and finally fixed carbon combustion. In the presence of methane at low concentrations; even though below the lower flammability limit; the first stage of coal combustion is enhanced. High heat is liberated which contributes to the acceleration of the subsequent stages of coal combustion.

Additionally, once devolatilization of sufficient coal particles occur, the flame characteristics are dominated by the volatiles with combustion taking place mainly in the gas phase, Amyotte et al (1993) [1]. Smaller sized coal particles enhance the devolatilisation process and impacts on the ignitability of the mixture.

All the explosions, although considered weak, increased in severity with increasing coal dust concentration. (see Table 2).

4. Conclusions

The effect of methane concentration, coal dust concentration and igniters energy on explosion characteristics were investigated.

- The explosion severity class moved from weak to strong as the methane concentration increased above 6%, when a 10 kJ ignitor was used. The maximum deflagration index achieved was 520 $bar.m.s^{-1}$ at 10% methane.
- Increased igniters energy for a given methane-air fuel mixture causes increased pressure rise and deflagration pressure rise rate. When the ignition energy for a 6.5% methane-air mixture was increased from 5 kJ to 10 kJ, a subsequent increase in explosion severity class from weak to strong is observed.
- Introducing coal dust concentrations of 50 $g.m^{-3}$ and 100 $g.m^{-3}$ to the ignition of 5% methane, significantly enhances the pressure rise rate from 5 $bar.s^{-1}$ to 114 $bar.s^{-1}$ and 138 $bar.s^{-1}$, respectively.

5. Acknowledgment

The researchers would like to acknowledge the financial support provided to this project by Australian Coal Association Low Emission Technologies Ltd (ACALET), the Australian Department of Industry, and the University of Newcastle, Australia.

References

[1] Amyotte, P.R., Mintz, K.J., Pegg, M.J., Sun, Y.H., The ignitability of coal dust-air and methane-coal dust-air mixtures. Fuel 72 (1993) 671-679

[2] Ajrash, M.A., Zanganeh, J., Moghtaderi B., Effects of ignition energy on fire and explosion characteristics of dilute hybrid fuel in ventilation air methane. J. Loss Prev. Process Ind. 40 (2016) 207-216

[3] Ajrash, M.A., Zanganeh, J., Moghtaderi B., Influences of the Initial Ignition Energy on Methane Explosion in a Flame Deflagration Tube. Energy & Fuels

[4] Cashdollar, K.L., Overview of dust explosibility characteristics. J. Loss Prev. Process Ind. 13 (2000) 183-199.

[5] Cashdollar, K.L., Coal dust explosibility. J. Loss Prev. Process Ind. 9, (1996) 65-76.

[6] Cashdollar, K.L., Hertzberg, M., 20-L Explosibility test chamber for dusts and gases. (1985) Rev. Sci. Instrum. 56 596.

[7] Hertzberg, M., Cashdollar, K.L., Zlochower, I., Flammability limit measurements for dusts and gases: Ignition energy requirements and pressure dependences. Symp. Combust. 21 (1998) 303-313.

[8] Kundu, S.; Zanganeh, J.; Moghtaderi, B. A Review on Understanding Explosions from Methane-Air Mixture. J. Loss Prev. Process Ind. 40 (2016) 507–523.

[9] Li, Q., Lin, B., Dai, H., Zhao, S., Explosion characteristics of H_2/CH_4/air and CH_4/coal dust/air mixtures. Powder Technology 229 (2012) 222-228

[10] Nifuku, M., Matsuda, T., Enomoto, H., Recent development of standardization of testing methods for dust explosion in Japan. J. Loss Prev. Process Ind. 13 (2000) 243-251.

11th Asia-Pacific Conference on Combustion,
The University of Sydney, NSW Australia
10th -14th December 2017.

Using Sparse-Lagrangian MMC-LES to Simulate Swirling Jets and Flames

Z. Huo, F. Salehi, M.J. Cleary, A.R. Masri

School of Aerospace, Mechanical and Mechatronic Engineering, The University of Sydney, NSW, 2006 Australia

Abstract

The sparse-Lagrangian MMC-LES approach is applied here, and for the first time, to swirling jets and flames stabilized on the Sydney swirl burner. Non-reacting case N16S159 is firstly simulated to demonstrate the ability to compute the recirculating flow fields. Reacting methane-air case SMA2 is then simulated using the sparse-Lagrangian MMC-LES model with detailed chemistry involving 19 species. Results show good predictions of the flow field including vortex breakdown and the extent of the recirculation zone. The results are sensitive to the specification of turbulence on the boundary. The mixture fraction and temperature fields are in excellent agreement with experimental mean and rms hence validating the applicability of sparse-Lagrangian MMC mixing model to flames with complex flow.

1 Introduction

Swirl and bluff-body stabilizers are employed to generate recirculation zones in the combustor hence assisting in anchoring the flame to the burner and increasing its resistance to blow-off [1], which is a major concern in the design of gas turbines. Swirling jets and flames are, therefore, highly relevant to practical flows and are of interest to numerical and experimental researchers. The Sydney swirling jet series are among the most investigated swirling laboratory cases for both isothermal and combusting cases [2-6], providing a detailed platform for validating computational models. Large eddy simulation (LES) has been widely accepted as a powerful numerical tool to investigate such complex flows due to its ability to capture unsteady behavior. Stein and Kempf [7, 8] employed LES and the flamelet approach with a presumed probability density function (PDF) closure for the sub-grid distribution of mixture fraction for one non-reacting case (N29S054) and two reacting cases (SM1 and SMH1). Good agreement with the experimental flow field data and reasonable prediction for the temperature and composition fields are achieved, except that it fails to predict the location of the vortex breakdown for case SMH1. El-Asrag and Menon [9] employed the linear eddy mixing LES, for two reacting cases (SM1 and SMA2), demonstrating a reasonable agreement with the experiment. Hu et al. [10] tested various subgrid-scale stress (SGS) models (Smagorinsky-Lilly and dynamic kinetic energy model) and combustion models (second-order moment and presumed PDF method), however, limited sensitivity was observed. Zhang and Mastorakos [11] implemented conditional moment closure (CMC) with LES demonstrating good agreement to the data for the flow and major species fields, although the prediction of hydroxyl mass fraction is underpredicted. A thorough review [12] on this topic is available for the interested reader.

Corresponding author. Fax: +00-00-000-000
E-mail address: zhuo3164@uni.sydney.edu.au

When applying LES to reacting flow cases it is necessary to choose a subgrid combustion model and a number of such models are under active investigation, each with its own distinct advantages. The flamelet approach tabulates the filtered chemical reaction rate by assuming locally balanced reaction and mixing at the stoichiometric contour, leading to low computational cost but less universality than some other methods [13]. The PDF method is associated with the widest generality, however, it also has a high computational cost [14]. In the context of LES, the filtered density function (FDF) method may be used [15]. CMC, which introduces conditional averaging on a conserved scalar to close the filtered reaction rates, may be associated with medium cost and medium universality [16]. In this work we use a generalized multiple mapping conditioning (MMC) model which is a PDF/FDF-type model that also utilizes aspects of CMC. A stochastic particle scheme is used and, while the reaction rate is evaluated without simplification, conditioning on a conserved scalar (namely the mixture fraction) achieves an accurate and local closure of the micromixing model allowing for a low cost, sparse particle implementation. The sparse MMC-LES model is applied here to two Sydney swirl cases; a non-reacting case (N16S159) and a reacting case (SMA2). The capability of sparse methods to simulate flames with complex swirling and recirculating flow structures has not been investigated previously and that is the objective of the present work. It is also one of the first simulations of a laboratory flame in which the newly derived anisotropic mixing time scale model [17] is used.

2 Model and Numerical Implementation

This study employs a hybrid technique, MMC-LES, that combines the finite volume Eulerian LES and Monte-Carlo Lagrangian FDF with mixing closed by sparse MMC for simulations. The important formulations are presented here. More details can be found in Ref.[18].

2.1 LES

The Favre filtered equations solved for continuity, momentum and reference mixture fraction can be written as:

$$\frac{\partial \bar{\rho}}{\partial t}+\frac{\partial \bar{\rho}\tilde{u}_i}{\partial x_i}=0 \tag{1}$$

$$\frac{\partial \bar{\rho}\tilde{u}_j}{\partial t}+\frac{\partial \bar{\rho}\tilde{u}_i\tilde{u}_j}{\partial x_i}=-\frac{\partial \bar{P}}{\partial x_j}+\frac{\partial}{\partial x_i}\left(\tilde{\tau}_{ij}-\tau_{ij}^{sf}\right) \tag{2}$$

$$\frac{\partial \bar{\rho}\tilde{f}}{\partial t}+\frac{\partial \bar{\rho}\tilde{u}_i\tilde{f}}{\partial x_i}=-\frac{\partial}{\partial x_i}\left(\tilde{J}_{f,i}-J_{f,i}^{sf}\right) \tag{3}$$

where $\tilde{\tau}_{ij}$ and $\tilde{J}_{f,i}$ are the resolved scale viscous stress tensor and diffusive flux. $\tilde{\tau}_{ij}$ can be rewritten as

$$\tilde{\tau}_{ij}=\bar{\rho}\nu\left(2\tilde{S}_{ij}-\frac{2}{3}\tilde{\tau}_{kk}\delta_{ij}\right) \tag{4}$$

where ν represents the kinematic viscosity; $\tilde{S}_{ij}$ represents the Favre-averaged strain rate. $\bar{J}_{f,i}$ can be found using

$$\bar{J}_{f,i} = -\bar{\rho}\mathfrak{D}\frac{\partial \tilde{f}}{\partial x_i} \tag{5}$$

where $\mathfrak{D}$ represents the molecular diffusivity defined as $\mathfrak{D} = \nu/\sigma$ where $\sigma = 0.7$ is the Schmidt number. The eddy viscosity model is applied to close the subfilter stress term:

$$\tau_{ij}^{sf} - \frac{\delta_{ij}}{3}\tau_{kk}^{sf} = -\bar{\rho}\nu_t\left(2\tilde{S}_{ij} - \frac{2}{3}\tilde{\tau}_{kk}\delta_{ij}\right) \tag{6}$$

where the closure for ν_t is achieved by applying the standard Smagorinsky model [19]

$$\nu_t = C_s\Delta_E^2\sqrt{2\tilde{S}_{ij}\tilde{S}_{ij}} \tag{7}$$

where C_s is the Smagorinsky constant and Δ_E is the filter width. A corresponding eddy diffusivity model is used to close the subfilter mass fluxes

$$J_{f,i}^{sf} = -\bar{\rho}\mathfrak{D}_t\frac{\partial \tilde{f}}{\partial x_i} \tag{8}$$

where $\mathfrak{D}_t = \nu_t/\sigma_t$, is the turbulent diffusivity and $\sigma_t = 0.4$ is the turbulent Schmidt number.

2.2 Stochastic Lagrangian FDF

A Monte Carlo implementation is applied to solve the FDF. The deterministic FDF equations are replaced by stochastic differential equations that govern the evolution of the notional particles:

$$dx_i^p = A_i^p dt + b_{ij}^p d\omega_j \tag{9}$$

$$d\phi_\alpha^p = (W_\alpha^p + S_\alpha^p)dt. \tag{10}$$

Equation (9) governs the stochastic evolution of the locations of particles where $d\omega_i$ is the independent Wiener process which introduces randomness to the equations. The coefficients A_i and b_{ij} are

$$A_i = \tilde{u}_i + \frac{1}{\bar{\rho}}\frac{\partial}{\partial x_i}(\bar{\rho}(\mathfrak{D} + \mathfrak{D}_t)) \tag{11}$$

$$b_{ij} = \delta_{ij}\sqrt{2(\mathfrak{D} + \mathfrak{D}_t)} \tag{12}$$

The composition of the particles evolves due to conditional dissipation of scalar variance at the subgrid scale, denoted by S_α, and chemical reactions which proceed at the rate W_α. Importantly W_α appears in closed form, however the conditional dissipation term is unclosed and a mixing model is introduced for that purpose. Here we use a generalized form of the MMC mixing model. It involves an exchange of composition between two particles, p and q, so that over a time step the composition of each particle varies as

$$\phi_\alpha^p(t + \Delta t) = \phi_\alpha^p(t) + \mu\left(\bar{\phi}_\alpha^{p,q}(t) - \phi_\alpha^p(t)\right) \tag{13}$$

$$\phi_\alpha^q(t + \Delta t) = \phi_\alpha^q(t) + \mu\left(\bar{\phi}_\alpha^{p,q}(t) - \phi_\alpha^q(t)\right) \tag{14}$$

where $\bar{\phi}_\alpha^{p,q}$ is the two-particle mean and $\mu = -exp\,(-\Delta t/\tau_L^{p,q})$ is the extent of mixing. This introduces the mixing time scale, $\tau_L^{p,q}$, for which a newly-derived, anisotropic model [17] is used:

$$\tau_L = \frac{C_f d_x^2}{2(\mathfrak{D} + \mathfrak{D}_{t,L})}. \tag{15}$$

Here $\mathfrak{D}_{t,L} = \frac{d_x}{\Delta_E}\mathfrak{D}_t$. An important feature of the MMC mixing model is that it enforces localness of mixing in both physical space and a reference variable space. For the non-premixed flames investigated here the LES filtered mixture fraction is chosen as the reference variable. Localness of mixing is achieved by selecting particle pairs, p and q, such that the effective square distance defined by

$$\hat{d}_{p,q}^2 = \sum_{i=1}^{3}\left(\frac{d_{x_i}^{p,q}}{r_i}\right)^2 + \left(\frac{d_{\tilde{f}}^{p,q}}{f_m}\right)^2 \tag{16}$$

is minimized. Here d_{x_i} is the distance between p and q in physical space; $d_{\tilde{f}}$ is the corresponding distance in the reference mixture fraction space and r_i and f_m are characteristic global parameters which are related by [18]

$$r_i = \frac{C_m}{\sqrt{3}}\left(\frac{d\tilde{f}}{dn}\frac{\Delta_L^3}{r_c^{2-D_f}}\frac{1}{f_m}\right)^{1/D_f} \tag{17}$$

where $C_m = 0.5$ is a constant, n is the direction of the steepest mixture fraction gradient, Δ_L is the nominal distance between adjacent particles in physical space, r_c is the inner cut off scale (taken to be the LES grid resolution) and $D_f = 2.36$ is a fractal dimension. In the present work we set $f_m = 0.03$ [17].

2.3 Numerical implementation

The sparse Lagrangian MMC-LES solver, called *mmcFoam*, is coupled with a low-Mach compressible LES solver available in the object orientated C++ open source software, OpenFOAM. The flow field and reference mixture fraction field are solved by LES, while the composition is solved in a Lagrangian sense. There is two-way coupling between the Eulerian and Lagrangian schemes. The filtered flow field data are interpolated from the Eulerian grid to the particles and filtered density is then fed back to the Eulerian scheme to complete the loop. The particle data is stochastic and passing particle-obtained filtered density directly to the flow solver may lead to numerical instability. In sparse methods this issue is acute as many LES cells do not contain any stochastic particles. To overcome this the density coupling is achieved through an adaptation of the equivalent enthalpy method [18]. In contrast to previous MMC-LES simulations, where the so-called target enthalpy values are calculated using flamelet profiles in what could be called a flamelet parametric regression, the target enthalpy is computed in the present work as conditional means conditioned on mixture fraction using a kernel estimation that draws ideas from smoothed particle hydrodynamics.

3 Case setup

Figure 1 shows a schematic structure of the swirling flame burner. The fuel jet is ejected from a 3.6 mm-diameter nozzle located at the center of the cylindrical bluff-body with diameter of 50 mm, while the swirling air stream is ejected from the 5mm wide annulus surrounding the bluff-body. The burner is housed in a wind tunnel with square cross section of $130\times130mm^2$[2].

A non-reacting case N16S159 and a reacting case SMA2 are selected as the benchmark of the MMC-LES simulations. These cases have the same jet velocity of 66 m/s and swirl number of 1.59. Case N16S159 has pure air jet while case SMA2 has methane-air jet (1:1 by volume). The Reynolds number based on the central jet is 15400 and the flame is sufficiently far from global blow-off which occurs at 216 m/s. This selection tests the ability of the approach to compute both reacting and non-reacting swirling flows without the additional complication of extensive local extinction.

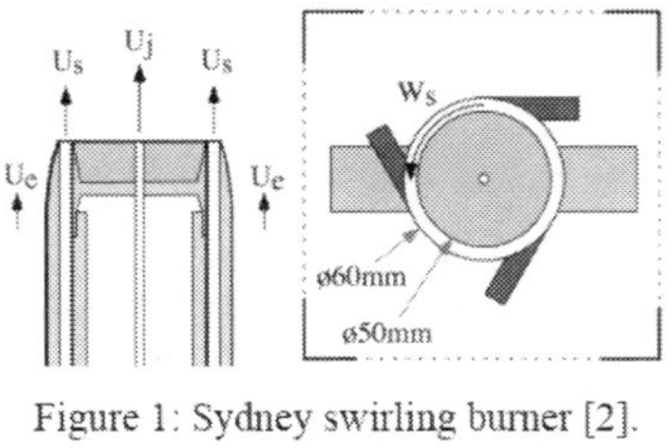

Figure 1: Sydney swirling burner [2].

The computational domain is defined as a cylindrical volume with a diameter of 180 mm and a length of 200 mm. As an ordinary cylindrical mesh would inevitably have either coarse resolution or large aspect ratio cells at the locations of the swirling stream, we instead use an unstructured sweep mesh, as shown in Fig. 2. The chosen mesh has 4.91 million cells and our tests have confirmed that there is low sensitivity to further refinement. On the particle side, we use one stochastic particle per 32 LES cells (1L/32E) on average, with as dense as 1L/2E for region close to the centerline where has finer LES mesh. Time-varying realistic turbulent velocity boundary conditions are applied for the fuel jet based on a turbulence intensity of 0.18 and an integral length-to-orifice diameter ratio of 0.125. A reduced kinetic scheme DRM19 is used to simulate the chemistry, which introduces 19 species + nitrogen and argon with 84 reactions.

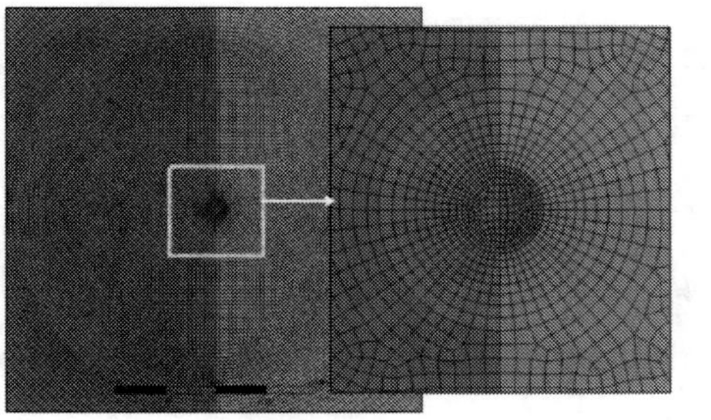

Figure 2: Cross section view of the LES mesh.

4 Results and discussion

The non-reacting case, N16S159, is tested first to validate the mesh and LES implementation. The flow through time for this case is estimated from the mean velocity along the central line (Fig. 3) as about 0.07 s, therefore the physical time simulated for transient development is set to be 0.1 s, after which another 0.1s is used to sample the flow field statistics. The centreline profile of the mean axial velocity is shown in Fig. 3. The axial coordinate is normalized by the bluff-body diameter. The jet breakup point and subsequent decay are captured quite well. At axial locations between $x/D_b = 0.5$ and $x/D_b = 2$, the predicted mean velocity tracks slightly below the experimentally obtained values. This profile is sensitive to the jet turbulent inflow conditions. We have used synthetic turbulent inflow data. While the present results are quite satisfactory it may be possible to produce more accurate results if pipe flow simulations were conducted instead. The radial profiles of axial velocity are shown in Fig. 4, and the swirl velocity profiles are shown in Fig. 5. Overall, the LES results are in very good agreement with the experimental data although, inline with the comments above, the axial velocity at $x/D_b = 0.8$ close to centerline is underpredicted. The accuracy of the swirl velocity is excellent and, based on our comparison to simulations with a coarser grid, may be attributed to the fine resolution in the tangential direction around the swirl stream.

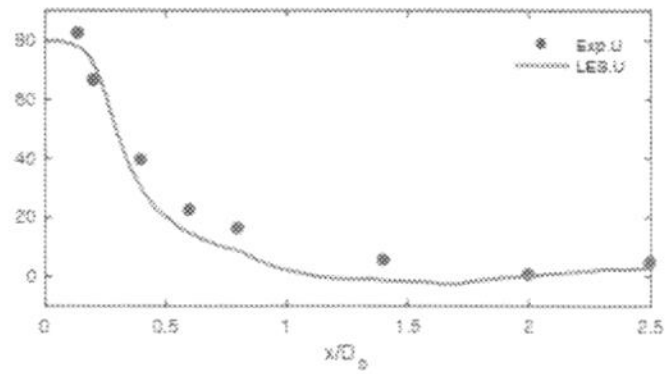

Figure 3: (N16S159) Mean axial velocity (m/s) along centerline.

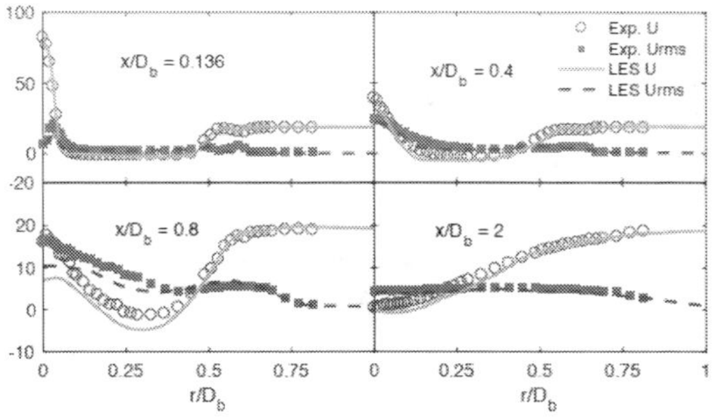

Figure 4: (N16S159) Radial profiles of axial velocity (m/s).

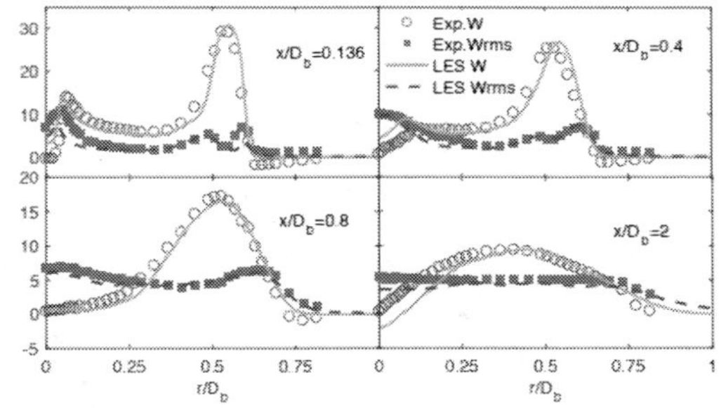

Figure 5: (N16S159) Radial profiles of swirl velocity (m/s).

The reacting case, SMA2, has exactly the same flow setup as the N16S159, therefore it can be expected that the good agreement to experimental flow field data is inherited. The centreline profile of the mean axial velocity is shown in Fig. 6. Once again we see the trend, especially the rate of decay, is accurately predicted although there is a slight underprediction throughout. The corresponding radial profiles of both axial and swirl velocity are shown in Fig. 7 and Fig. 8. The predictions match the experimental data well, especially for the swirl velocity. The predictions of mixture fraction radial profiles are shown in Fig. 9. There is generally very good agreement with the experimental data for both the mean and rms, indicating the effectiveness of the sparse MMC-LES mixing model to capture the correct scalar evolution in this complex flow configuration. It is noted that there is slight underprediction of the mean near the centreline further downstream and this is attributed to the underprediction of convective transport due to the slightly too low mean velocity in that region. Finally, the temperature profiles are shown in Fig. 10. There is excellent qualitative agreement to the data throughout the flame. Quantitative agreement is also very good although the mean is slightly underpredicted in the shear layer at $x/D_b = 2$.

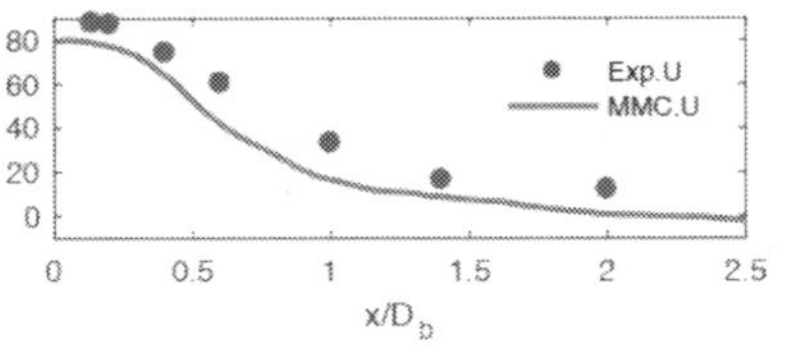

Figure 6: (SMA2) Mean axial velocity (m/s) along centerline

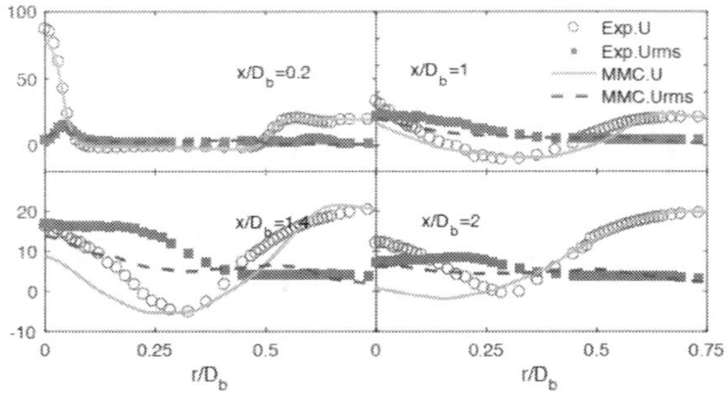

Figure 7: (SMA2) Radial profiles of axial velocity (m/s).

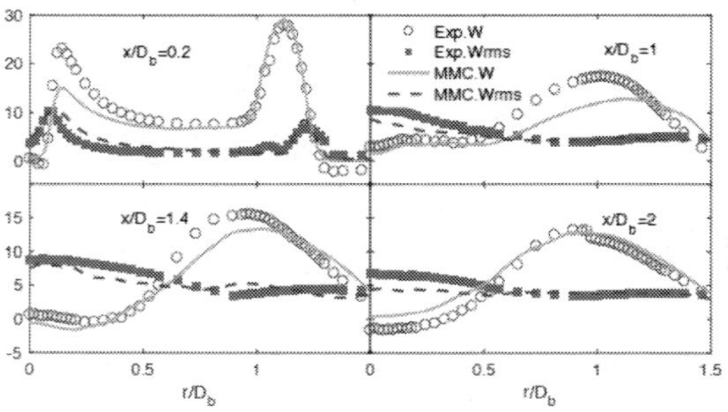

Figure 8: (SMA2) Radial profiles of swirl velocity (m/s).

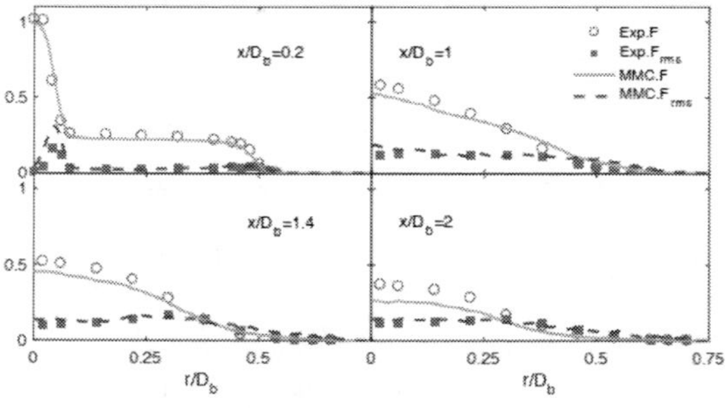

Figure 9: (SMA2) Radial profiles of mixture fraction.

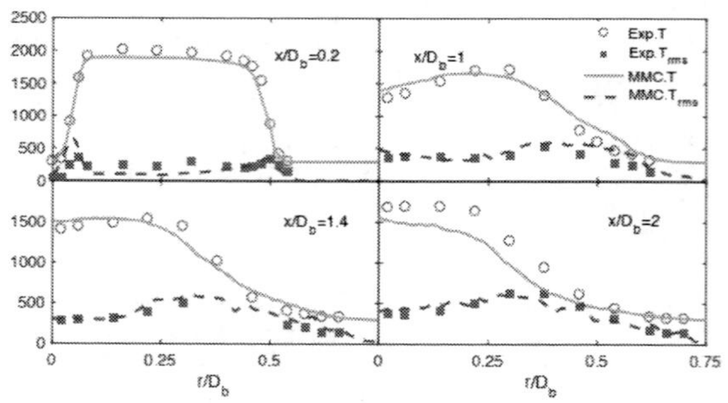

Figure 10: (SMA2) Radial profiles of temperature (K).

5 Conclusions

Sparse MMC-LES is applied to complex swirling flows for the first time. Simulation results of a non-reacting case validate the LES setup including the unstructured mesh. A fine resolution of the grid in the tangential direction is needed to capture the swirl velocity accurately. The predictions of the axial and swirl velocity profiles in the reacting case are also very good, although there is some underprediction of the axial component along the centreline of the flame. Results for both the mean and rms of mixture fraction and temperature are overall excellent and demonstrate the ability of sparse MMC-LES to accurately capture scalar profiles in complex swirling flames.

6 Acknowledgments

The research is supported by the Australian Research Council. The support of the Sydney Informatics Hub at the University of Sydney and, in particular, access to the high-performance computing facility Artemis is gratefully acknowledged.

References

[1] H. Edmondson, M. Heap, Combust. and Flame, 14 (2) (1970) 191-194.

[2] P.A. Kalt, Y.M. Al-Abdeli, A.R. Masri, R.S. Barlow, Proc. Combust. Inst., 29 (2) (2002) 1913-1919.

[3] Y.M. Al-Abdeli, A.R. Masri, Exp. therm. Fluid Sci., 27 (5) (2003) 655-665.

[4] Y.M. Al-Abdeli, A.R. Masri, Combust. Theory Model., 7 (4) (2003) 731-766.

[5] Y.M. Al-Abdeli, A.R. Masri, Combust. Sci. Tech., 176 (5-6) (2004) 645-665.

[6] A.R. Masri, P.A. Kalt, R.S. Barlow, Combust. and Flame, 137 (1) (2004) 1-37.

[7] O. Stein, A. Kempf, J. Janicka, Combust. Sci. Tech., 179 (1-2) (2007) 173-189.

[8] O. Stein, A. Kempf, Proc. Combust. Inst., 31 (2) (2007) 1755-1763.

[9] H. El-Asrag, S. Menon, Proc. Combust. Inst., 31 (2) (2007) 1747-1754.

[10] L. Hu, L. Zhou, Y. Luo, Num. Heat Transf. Part B: Fundamen., 53 (1) (2008) 39-58.

[11] H. Zhang, E. Mastorakos, Proc. Combust. Inst., 36 (2) (2017) 1669-1676.

[12] Y.M. Al-Abdeli, A.R. Masri, Experimental Thermal and Fluid Science, 69 (2015) 178-196.

[13] N. Peters, Progress in Energy and Combustion Science, 10 (3) (1984) 319-339.

[14] S.B. Pope, Progress in Energy and Combustion Science, 11 (2) (1985) 119-192.

[15] P.J. Colucci, F.A. Jaberi, P. Givi, S.B. Pope, Phys. Fluids, 10 (2) (1998) 499-515.

[16] A.Y. Klimenko, R.W. Bilger, Progress in energy and combustion science, 25 (6) (1999) 595-687.

[17] S. Vo, O. Stein, A. Kronenburg, M.J. Cleary, Combust. and Flame, 179 (2017) 280-299.

[18] M.J. Cleary, A.Y. Klimenko, Phys. Fluids, 23 (11) (2011) 115102.

[19] J. Smagorinsky, Monthly Weath. Rev., 91 (3) (1963) 99-164.

11th Asia-Pacific Conference on Combustion,
The University of Sydney, NSW Australia
10th -14th December 2017.

A numerical study of initial temperature distribution inside the cylindrical constant volume bomb and its effect on H_2-air outwardly propagating spherical flames

Jian Tan, Gesheng Li, Junjie Liang, Peng Cheng, Zunhua Zhang*

School of Energy and Power Engineering, Wuhan University of Technology,

Wuhan, Hubei 430063, PR China.

Abstract

Based on CFD, the temperature distribution of fluid inside the cylindrical constant volume bomb was revealed and the numerical results were validated. In addition the 2-D H_2-air outwardly propagating spherical flame at the non-uniform initial temperature field was investigated by using CFD. The temperature distribution of fluid inside the chamber exhibits significant gradient along the gravity direction, and it first decreases to the minimum near the bomb center and then rises along gravity direction. With the increase of the overall temperature inside the chamber, the non-uniformity of the temperature field gradually increases as well. When the identified initial temperature inside the chamber is 373, 413 and 473 K, the maximum temperature differences along the gravity direction reach 14, 23 and 35 K, respectively. The shape of flame is influenced by the non-uniformity of fluid temperature distribution, and the flame front is not a standard circle. With the rise of overall temperature inside the chamber, this trend is more obvious. When the identified initial temperatures inside the chamber are 413 and 473 K, the relative changes in S_u^0 for H_2-air at P_0=0.1 MPa and Φ=1.0 between uniform and non-uniform initial temperature fields are around 2% and 4% , respectively.

1 Introduction

The laminar burning velocity, S_u^0, is one of the most important characteristic parameters of premixed combustible mixture[1], and the initial temperature of the fuel has an important influence on the laminar burning velocity [2-3]. The outwardly propagating spherical flame method is popularly used to measure S_u^0. In this method, the combustible mixture in a closed chamber is ignited by the ignition kernel. The history of outwardly propagating spherical flame, $R_f = R_f(t)$, is recorded by using high-speed shadowgraphy.

The cylindrical constant volume bomb is always heated by band heaters when the objective initial temperature of experiment is higher than normal temperature. Due to natural convection in closed constant volume bomb, the temperature distribution of the combustible gases is not absolutely uniform. Usually, the non-uniformity of the temperature distribution is not considered in the determination of the laminar burning velocity. The initial temperature inside the chamber is always controlled by several thermocouples to ensure the uniformity of the temperature distribution [4-5] .Therefore, the first purpose of this paper is to reveal the temperature distribution of the fluid medium inside the chamber.

In recent years, the influence of the initial temperature disturbance on the measurement of the laminar burning velocity through the outwardly propagating spherical flame method has also attracted the attention of relevant scholars [6]. However, there is almost no research on the uncertainty induced by the non-uniformity of the initial temperature distribution in the closed chamber. So the second purpose of this paper is to analyze the effect of non-uniformity of the initial temperature field on outwardly propagating spherical flame and evaluate the uncertainty of the laminar burning velocity due to the non-uniformity of the initial temperature field.

Based on the above purposes, CFD model was established to reveal the temperature distribution inside a cylindrical constant volume bomb when the heater voltage reaches 90, 120 and 150 V, respectively, and the relevant measurement was carried out to validate the simulation. Furthermore, two-dimensional H_2-air propagating spherical flame process at the uniform and non-uniform initial temperature field corresponding to different heater voltages was calculated by CFD to investigate the effect of non-uniformity of the initial temperature field on the determination of the laminar burning velocity.

2 Numerical model

2.1 Simulation of the temperature distribution in the constant volume bomb

The modeled cylindrical constant volume bomb is made of carbon steel and the two sides of this chamber are fitted with optical quartz glass. The heating device is a band resistance heater with a length of 3.1 m and a width of 25 mm, and its rated power is 1440 W. The band heater is attached to the outside surface of the chamber, and the chamber is covered by thermal insulation materials.

The initial condition inside the chamber is same with that outside the chamber. The initial temperature is 298 K, the initial pressure is 0.1 MPa. The thermo-physical parameters of both the fluid and the thermal insulation materials are defined in the form of polynomial functions due to their variation with temperature. But the thermo-physical properties of quartz glass and carbon steel are considered as constant.

The convective heat transfer coefficient between the surface outside chamber, optical quartz glass windows and air is calculated by the experimental correlation: $Nu_m = C(Gr \cdot Pr)_m^n$, in which Nu_m is the average surface *Nusselt* number, Gr is *Grashof* number, Pr is *Prandtl* number, and the coefficient C and the exponent n are determined by experiments. The temperatures on outside surfaces of different parts are determined by K-type thermocouples. The band heater is considered as an energy source.

Corresponding author. Fax: +86-27-8652-3305
E-mail address: zunhuazhang@whut.edu.cn

Fluent 15.0 is used to perform the numerical simulation, the laminar flow and Discrete Ordinates radiation model are employed. The conservation equations of mass, momentum and energy in this model are solved implicitly with a 3-D steady-state segregated solver and the numerical method is based on the SIMPLE algorithm. The governing equations are discretized with second order upwind scheme. The conservation equations are checked for convergence. Convergence is determined from the residuals of all conservation equations.

In order to verify the accuracy of this model, a confirmatory temperature measurement is conducted. One K-type thermocouple is placed between the external surface of the chamber and thermal insulation materials to measure the temperature outside the chamber, the other K-type thermocouple temperature probe is fixed into the chamber to measure the temperature inside the chamber. In the process of heating, the chamber is filled with air. The temperature measuring device is arranged as Fig. 1. Moreover, a K-type thermocouple compensating conductor is penetrated into the chamber through the exhaust valve to measure temperatures of the four monitor points (a, b, c and d).

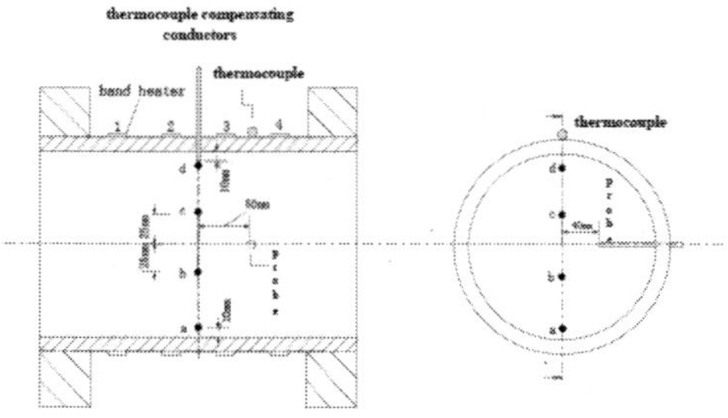

Figure 1: Experimental apparatus.

When the heater voltage is 90, 120 and 150 V, the measuring value of the temperature probe inside the chamber is 373, 413 and 473 K. That is, the identified initial temperature inside the chamber is 373, 413 and 473 K. The maximum absolute error between numerical and experimental values is no more than 6 K, and the relative error is less than 5%. As is shown in Fig. 2, the simulation values agree well with experimental ones.

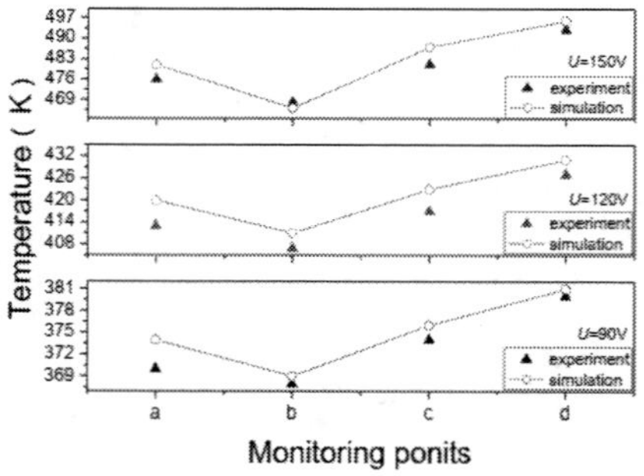

Figure 2: The comparison between the experiment and simulation for temperature at monitoring points.

2.2 Simulation of outwardly propagating spherical flame

The 2-D outwardly propagating spherical flame process of premixed H_2–air at adiabatic condition is calculated by Fluent 15.0. The computational domain is a circle with a diameter of 176 mm. The transient-state 2-D continuity, momentum, energy and species equations in the fluid phase are discretized, respectively, using a finite-volume method. The combustion inside the chamber is simulated by using EDC (Eddy-Dissipation-Concept) and second-order upwind discretization model. A time increment of 1 μs is used. The detailed chemical kinetic mechanism which consists of 19 elementary reactions [7] and 9 species was used to describe the chemical reaction process. The 2-D non-uniform initial temperature field at different voltages of band heater is defined by UDF (User Defined Functions) according to the numerical results of 3-D temperature distribution inside the chamber. The ignition kernel is assumed to be a circular region with a radius of 1mm in 2000 K.

In outwardly propagating spherical flame, the stretched flame speed is defined as: $S_b = dr_u/dt$, where r_u is the flame radius. The unstretched laminar flame speed S_b^0 can be extracted from the linear extrapolation based on the following correlation: $S_b = S_b^0 - L_b\alpha$, where L_b is Markstein length and $\alpha = \frac{dlnA}{dt} = \frac{2}{r_u}S_b$ is the stretch rate. The unstretched laminar burning velocity, S_u^0, is deduced from S_b^0 using $S_u^0 = \sigma S_b$, where $\sigma = \rho_b/\rho_u$ is the density ratio between the burned gas (at adiabatic equilibrium condition) and unburned gas. As shown in Fig. 3, the variation of the stretched flame propagation speed S_b with the stretch rate α obtained from the numerical results of 2-D H_2-air propagating spherical flame at NTP (normal temperature and pressure) and equivalence ratio Φ=1.0 is compared with the experimental result of literature [8]. The numerical results show a good agreement with the experimental results, indicating this model of 2-D outwardly propagating spherical flame can be considered to be reasonable and credible.

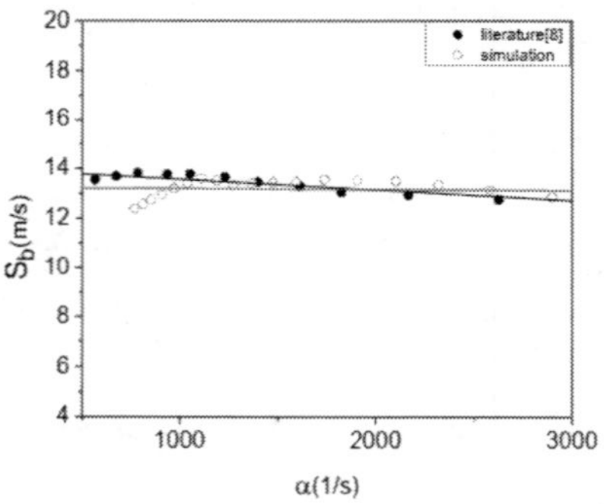

Figure 3: S_b obtained from simulation as a function of α along with experimental results in literature [8], for H_2-air at Φ=1.0 and NTP.

3 Results and discussions

3.1 The temperature distribution inside the constant volume bomb

In order to reflect the distribution of the temperature field inside the constant volume bomb more intuitively, a cross-sectional perpendicular to the X direction is created where the Y-direction is the opposite direction of gravity.

Figure. 4 shows predicted temperature contours of the constant volume bomb on the cross-sectional at different heater voltages. In Fig. 4, since the quartz glass is installed at both ends of the constant volume bomb and insulation measures cannot be taken to the glass, the heat loss at both ends is more serious. Therefore, the temperature field exhibits higher temperature in the middle of the bomb wall and symmetrically presents the lower temperature at the glass and the walls near the ends. The temperature of fluid medium in the constant volume bomb exhibits significant gradient along the gravity direction, and it first decreases to the minimum near the bomb center and then rises along gravity direction. With the increase of the heater voltage, the overall temperature inside the constant volume bomb rises, and the non-uniformity of the temperature field gradually increases as well.

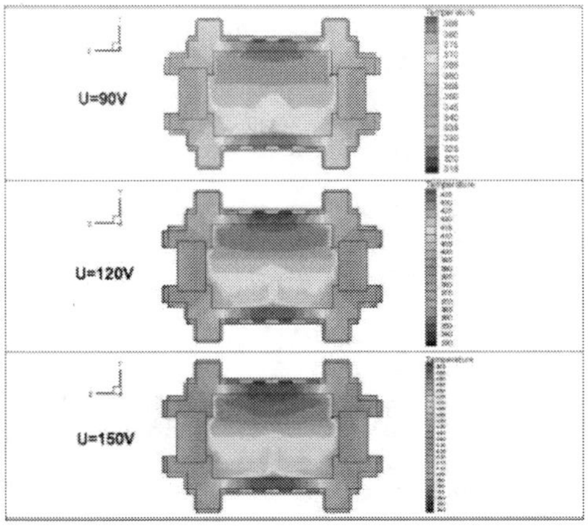

Figure 4: Predicted temperature (K) contours on the cross-sectional at different heater voltages.

In order to reflect the variation trend of the fluid temperature along the gravity direction more clearly, this paper creates a monitoring line in the fluid region of the constant volume bomb model which cross the constant volume bomb center in the vertical direction. The temperature variation along the monitoring line is shown in Fig. 5, where the x-axis corresponds to the coordinates of different heights on the monitoring line, the x-axis coordinates 0 point is the height of the model center, and the x-axis 88 mm and -88 mm correspond to the upper and lower wall inside surfaces of the chamber respectively. As seen in Fig. 5, the temperature of the fluid medium is higher in the area near the lower wall, and as the height rises, the temperature drops rapidly. The temperature drops to a minimum in the vicinity of the horizontal center, and as the height continues to rise, the temperature of the fluid medium in the projectile increases gradually until it coincides with the temperature of surface of upper wall. It can be seen from Fig. 5 that the fluid gets its maximum temperature when x=±88 mm and minimum temperature when x= -10 mm. When the heater voltage is 90, 120 and 150 V, the maximum temperature difference reaches 14, 23 and 35 K, respectively. Considering the correction for confinement inducing pressure rise and flow, flame data is only used for flame radius less than 25% of the chamber radius to calculate laminar burning velocity [9], and thus the temperature of upper region is general higher than that of the lower region in laminar flame propagation space available for analysis according to Fig. 5.

Due to the existence of certain non-uniformity in the temperature field of the fluid medium inside the constant volume bomb, if the temperature of the measuring probe inside chamber is considered as the initial temperature of the fuel mixture when conducting the flame experiments at high temperature condition, the location of the temperature probe will have a considerable influence on the identified value of initial temperature in experiment.

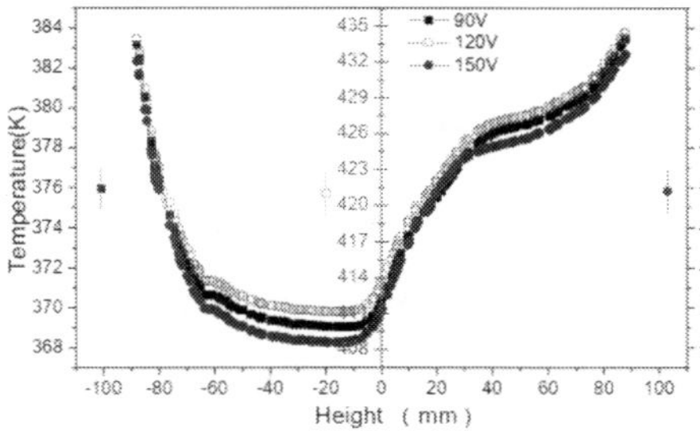

Figure 5: The temperature variation of the fluid medium on the monitoring line along the gravity direction.

3.2 Effect of non-uniformity of the initial temperature field on outwardly propagating spherical flame

The H_2-air propagating spherical flame process at non-uniform and uniform initial temperature field corresponding to different heater voltage is simulated by the above model, and the initial pressure and equivalence ratio are fixed at P_0=0.1 MPa and Φ=1.0, respectively. The initial temperature in simulation is determined by the value of the K-type thermocouple temperature probe inside the chamber at different the heater voltage. The predicted density contours of propagating spherical flame at non-uniform initial temperature field when the heater voltage is 150 V are shown in Fig. 6. It suggests that the shape of the flame front does not tend to be rounded at non-uniform temperature field conditions. Thus the distance between the edge pixel of the flame front and the center of flame at different flame radii is calculated as shown in Fig. 7, and the number of pixel points is rotated clockwise a circle from point A. As shown in Fig. 7, when the heater voltage is 90 V, the variability of the distance from flame front to the center of flame is generally low. When the heater voltage is 120 and 150 V, the variability is more obvious over the time of flame propagation, and the distance from upper edges of flame front to the center of flame is longer than that of lower edges. The shape of the flame tends to be an ellipse rather than a standard circle. Therefore, the analysis method based on the shape of flame will bring some uncertainty at this condition. The direct cause of this phenomenon is that the temperature of upper region is higher than that of lower region in laminar flame propagation space available for analysis according to Fig. 5, and therefore the flame front spreads slightly faster at upper region than at lower region. Accordingly, the laminar burning velocity calculated from the history of outwardly propagating spherical flame is bound to be influenced by non-uniform initial temperature field.

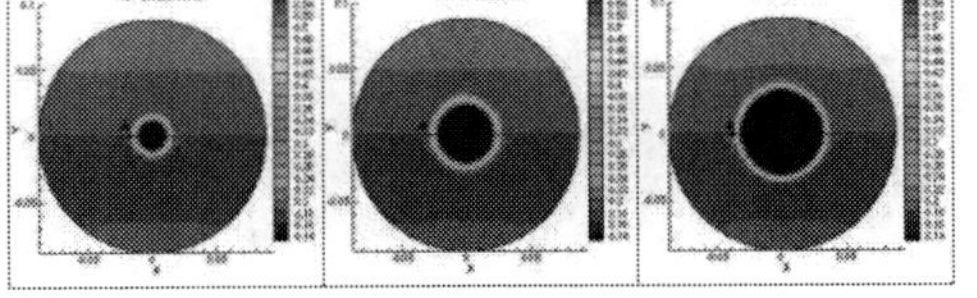

Figure 6: Predicted density (kg/m³) contours of propagating spherical flame for H_2-air at Φ=1.0, P_0=0.1 MPa and non-uniform initial temperature field when the heater voltage is 150 V.

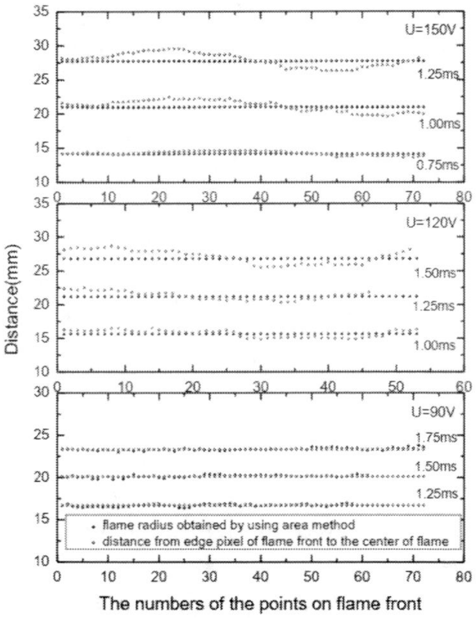

Figure 7: The distance between the edge pixel of the flame front and the center of flame of predicted density contours at different flame radii.

Figure. 8 shows the comparison of predicted laminar burning velocities S_u^0 for the H_2-air at non-uniform and uniform initial temperature fields at P_0=0.1 MPa and Φ=1.0. It is emphasized that area method is used in this study to obtain the flame radii, and thus the influence of the non-uniformity of initial temperature field on the shape of flame can be partly eliminated. Though, it is observed that the predicted laminar burning velocities at non-uniform temperature field are slightly higher than those at uniform temperature fields. As the heater voltage increases, the difference in laminar burning velocity between uniform and non-uniform temperature fields becomes bigger accordingly. When the heater voltage is 90 V, the relative change in S_u^0 is below 1%. When the heater voltage is 120 and 150 V, (i.e., the identified initial temperature inside chamber is 413 and 473 K) the relative changes in S_u^0 are around 2% and 4%, respectively.

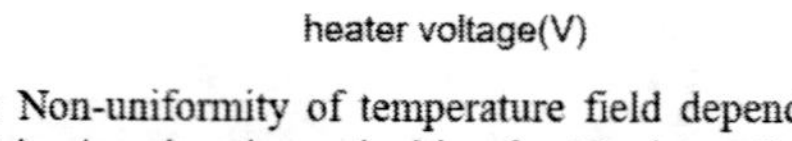

Figure 8: Non-uniformity of temperature field dependence of predicted laminar burning velocities for H_2-air at Φ=1.0 and P_0=0.1 MPa, and the band heater is 90, 120 and 150 V, respectively.

4 Conclusions

The temperature distribution inside a cylindrical heated constant volume bomb at different heater voltages is revealed by using CFD, and the outcome of simulation is in good agreement with the experimental values. The temperature distribution of fluid inside the constant volume bomb exhibits significant gradient along the gravity direction. With the increase of the heater voltage, the overall temperature inside constant volume bomb rises, and the non-uniformity of the temperature field gradually increases as well. When the heater voltages are 90, 120 and 150 V, the maximum temperature differences along the gravity direction reach 14, 23 and 35 K respectively. Due to the existence of certain non-uniformity in the temperature field of the fluid medium, the location of the temperature probe will have a certain influence on the identified value of initial temperature in experiment.

The propagating process of the H_2-air spherical flame at non-uniform initial temperature field is investigated by using CFD. The shape of flame is influenced by the non-uniformity of fluid temperature distribution, and the flame front is not a standard circle. The flame front spreads slightly faster at upper region than at lower region. When the identified initial temperature inside the chamber are 413 and 473 K, the relative changes in S_u^0 induced by the non-uniformity of temperature distribution for H_2-air at P_0=0.1 MPa and Φ=1.0 are around 2% and 4% , respectively.

5 Acknowledgment

This work was supported by National Natural Science Foundation of China (Nos. 51479149 and 51509198).

References

[1] Pugh D, Crayford A P, Bowen P J, et al, Appl. Energ. 136 (2014) 1026-1034.

[2] Konnov A A, Meuwissen R J, De Goey L P H, Proc. Combust. Inst. 33(1) (2011) 1011-1019.

[3] Konnov A A, Fuel 89(9) (2010) 2211-2216.

[4] Galmiche B, Halter F, Foucher F, Combust. Flame 159(11) (2010) 3286-3299.

[5] Baloo M, Dariani B M, Akhlaghi M, et al, Fuel 144 (2015) 264-273.

[6] Chen Z, Combust. Flame 162(6) (2015) 2442-2453.

[7] Tien J H, Stalker R J, Combust. Flame 131(3) (2002) 329-348.

[8] Hu E, Huang Z, He J, He J, et al, Int. J. Hydrogen Energ. 34(20) (2009) 8741-8755.

[9] Chen Z, Burke M P, Ju Y, Combust. Theory Model. 13(2) (2009) 343-364.

AUTHOR INDEX

AUTHOR INDEX

AUTHOR INDEX

AUTHOR INDEX

AUTHOR INDEX

AUTHOR INDEX

AUTHOR INDEX

AUTHOR INDEX

AUTHOR INDEX

Combustion Institute China Section
c/o Section Chair: Fei Qi, Shanghai Jiao Tong University
fqi@sjtu.edu.cn

ISBN 978-1-5108-5646-2

9 781510 856462

CURRAN ASSOCIATES INC.
proceedings.com